Perry's Chemical Engineers' Handbook

OTHER McGRAW-HILL CHEMICAL ENGINEERING BOOKS OF INTEREST

PERRY'S CHEMICAL ENGINEERS' HANDBOOK

SEVENTH EDITION

McGraw-Hill
New York
San Francisco
Washington, D.C.
Auckland
Bogotá
Caracas
Lisbon
London
Madrid
Mexico City
Milan
Montreal
New Delhi
San Juan
Singapore
Sydney
Tokyo
Toronto

**Prepared by a staff of specialists
under the editorial direction of**

Late Editor
Robert H. Perry

Editor
Don W. Green
Deane E. Ackers Professor of Chemical
and Petroleum Engineering,
University of Kansas

Associate Editor
James O. Maloney
Professor Emeritus of Chemical Engineering,
University of Kansas

Library of Congress Cataloging-in-Publication Data

Perry's chemical engineers' handbook. — 7th ed. / prepared by a staff
of specialists under the editorial direction of late editor Robert H.
Perry : editor, Don W. Green : associate editor, James O'Hara
Maloney.
    p.   cm.
    Includes index.
    ISBN 0-07-049841-5 (alk. paper)
    1. Chemical engineering—Handbooks, manuals, etc. I. Perry,
Robert H., date. II. Green, Don W. III. Maloney, James O.
TP151.P45   1997
660—dc21   96-51648
                                      CIP

McGraw-Hill

A Division of The McGraw·Hill Companies

ISBN 0-07-049841-5

INTERNATIONAL EDITION

Copyright © 1997. Exclusive rights by The McGraw-Hill Companies, Inc.,
for manufacture and export. This book cannot be re-exported from the
country to which it is consigned by McGraw-Hill. The International Edition is not available in North America.

When ordering this title, use ISBN 0-07-115448-5.

*The sponsoring editors for this book were Zoe Foundotos and Robert
Esposito, the editing supervisor was Marc Campbell, and the production
supervisor was Pamela A. Pelton. It was set in Caledonia by North Market
Street Graphics.*

Printed and bound by R. R. Donnelley & Sons Company.

This book was printed on acid-free paper.

*Dedicated to
Robert H. Perry*

ABOUT THE EDITORS

The late **Robert H. Perry** served as chairman of the Department of Chemical Engineering at the University of Oklahoma and program director for graduate research facilities at the National Science Research Foundation. He was a consultant to various United Nations and other international organizations. From 1973 until his death in 1978 Dr. Perry devoted his time to a study of the cross impact of technologies within the next half century. The subjects under his investigation on a global basis were energy, minerals and metals, transportation and communications, medicine, food production, and the environment.

Don W. Green is Chair and the Deane E. Ackers distinguished professor of chemical and petroleum engineering and codirector of the Tertiary Oil Recovery Project at the University of Kansas in Lawrence, Kansas, where he has taught since 1964. He received his doctorate in chemical engineering in 1963 from the University of Oklahoma, where he was Dr. Perry's first doctoral student. Dr. Green has won several teaching awards at the University of Kansas, and he is a fellow of the AIChE and a distinguished member of the Society of Petroleum Engineers. He is the author of numerous articles in technical journals.

James O. Maloney is Professor Emeritus of the Department of Chemical and Petroleum Engineering, University of Kansas. He holds a Ph.D. degree in chemical engineering from Pennsylvania State University. In 1941 he began his professional career at the DuPont de Nemours Company, before joining the University of Kansas in 1945, where he taught for 40 years. He served as department chairman for nineteen years. He is a fellow of the AIChE.

List of Contributors

Michael M. Abbott, Ph.D., Howard P. Isermann Department of Chemical Engineering, Rensselaer Polytechnic Institute; Member, American Institute of Chemical Engineers (Section 4, Thermodynamics)

Terry Allen, Ph.D., Senior Research Associate (retired), DuPont Central Research and Development (Section 20, Size Reduction and Size Enlargement)

John D. Bacha, Ph.D., Consulting Scientist, Chevron Products Company; Member, ASTM (American Society for Testing and Materials), Committee D02 on Petroleum Products and Lubricants; American Chemical Society; International Association for Stability and Handling of Liquid Fuels, Steering Committee (Section 27, Energy Resources, Conversion, and Utilization)

Glenn W. Baldwin, M.S., P.E., Staff Engineer, Union Carbide Corporation; Member, American Institute of Chemical Engineers (Section 12, Psychrometry, Evaporative Cooling, and Solids Drying)

Scott D. Barnicki, Ph.D., Senior Research Chemical Engineer, Eastman Chemical Company (Section 13, Distillation)

Kenneth J. Bell, Ph.D., P.E., Regents Professor Emeritus, School of Chemical Engineering, Oklahoma State University; Member, American Institute of Chemical Engineers (Section 11, Heat-Transfer Equipment)

Richard C. Bennett, B.S., Ch.E., Registered Professional Engineer, Illinois; Member, American Institute of Chemical Engineers (AIChE); President of Crystallization Technology, Inc.; Former President of Swenson Process Equipment, Inc. (Section 18, Liquid-Solid Operations and Equipment)

Charles E. Benson, M.Eng., M.E., Director, Combustion Technology, Arthur D. Little, Inc.; Member, American Society of Mechanical Engineers, Combustion Institute (Section 27, Energy Resources, Conversion, and Utilization)

Patrick M. Bernhagen, P.E., B.S.M.E., Senior Mechanical Engineer, Foster Wheeler USA Corporation, American Society of Mechanical Engineers (Section 11, Heat-Transfer Equipment)

Heinz P. Bloch, P.E., B.S.M.E., M.S.M.E., Consulting Engineer, Process Machinery Consulting; American Society of Mechanical Engineers, Vibration Institute; Registered Professional Engineer (New Jersey, Texas) (Section 29, Process Machinery Drives)

Frank T. Bodurtha, Sc.D., E.I. DuPont de Nemours and Co., Inc. (retired), Wilmington, Delaware (retired); Consultant, Frank T. Bodurtha, Inc. (Section 26, Process Safety)

Meherwan P. Boyce, P.E., Ph.D., President, Boyce Engineering International; ASME Fellow; Registered Professional Engineer (Texas, Oklahoma) (Section 10, Transport and Storage of Fluids; Section 29, Process Machinery Drives)

Laurence G. Britton, Ph.D., Research Scientist, Union Carbide Corporation (Section 26, Process Safety)

Evan Buck, M.S.Ch.E., Manager, Thermophysical Property Skill Center, Central Technology, Union Carbide Corporation (Section 2, Physical and Chemical Data)

Henry R. Bungay, P.E., Ph.D., Professor of Chemical and Environmental Engineering, Rensselaer Polytechnic Institute; Member, American Institute of Chemical Engineers, American Chemical Society, American Society for Microbiology, American Society for Engineering Education, Society for General Microbiology (Section 24, Biochemical Engineering)

Anthony J. Buonicore, M.Ch.E., P.E., Diplomate AAEE, CEO, Environmental Data Resources, Inc.; Member, American Institute of Chemical Engineers, Air and Waste Management Association (Section 25, Waste Management)

Michael M. Calistrat, B.S.M.E., M.S.M.E., Owner, Michael Calistrat and Associates; Member, American Society of Mechanical Engineers (Section 29, Process Machinery Drives)

Giorgio Carta, Ph.D., Professor, Department of Chemical Engineering, University of Virginia; Member, American Institute of Chemical Engineers, American Chemical Society, International Adsorption Society (Section 16, Adsorption and Ion Exchange)

Vincent Conrad, Ph.D., Group Leader, Technical Services Development Laboratory, CONSOL, Inc.; Member, Spectroscopy Society of Pittsburgh, Society for Analytical Chemistry of Pittsburgh, Society for Applied Spectroscopy (Section 27, Energy Resources, Conversion, and Utilization)

Harrison Cooper, Ph.D., Harrison R. Cooper Systems, Inc., Salt Lake City, Utah (Section 19, Solid-Solid Operations and Equipment)

B. B. Crocker, S.M., P.E., Consulting Chemical Engineer; Fellow, American Institute of Chemical Engineers; Member, Air Pollution Control Association (Section 14, Gas Absorption and Gas-Liquid System Design)

Daniel A. Crowl, Ph.D., Professor of Chemical Engineering, Chemical Engineering Department, Michigan Technological University; Member, American Institute of Chemical Engineers, American Chemical Society (Section 26, Process Safety)

Roger W. Cusack, Vice President, Glitsch Process Systems, Inc.; Member, American Institute of Chemical Engineers (Section 15, Liquid-Liquid Extraction Operations and Equipment)

Donald A. Dahlstrom, Ph.D., Research Professor, Chemical and Fuels Engineering Department and Metallurgical Engineering Department, University of Utah; Member, National Academy of Engineering, American Institute of Chemical Engineers (AIChE), American Chemical Society (ACS), Society of Mining, Metallurgic Exploration (SME) of the American Institute of Mining, Metallurgical and Petroleum Engineers (AIME), American Society of Engineering Education (Section 18, Liquid-Solid Operations and Equipment)

Thomas E. Daubert, Ph.D., Professor, Department of Chemical Engineering, The Pennsylvania State University (Section 2, Physical and Chemical Data)

R. H. Daugherty, Ph.D., Consulting Engineer, Research Center, Reliance Electric Company; Member, Institute of Electrical and Electronics Engineers (Section 29, Process Machinery Drives)

James F. Davis, Ph.D., Professor of Chemical Engineering, Ohio State University (Section 3, Mathematics)

James B. Dunson, B.S., Principal Consultant, E. I. duPont de Nemours & Co.; Member American Institute of Chemical Engineers; Registered Professional Engineer (Delaware) (Section 17, Gas-Solid Operation and Equipment)

Thomas F. Edgar, Ph.D., Professor of Chemical Engineering, University of Texas, Austin, Texas (Section 8, Process Control)

Robert C. Emmet, Jr., B.S., Ch.E., Senior Process Consultant, EIMCO Process Equipment Co.; Member, American Institute of Chemical Engineers (AIChE), American Institute of Mining, Metallurgical and Petroleum Engineers (AIME), Society of Mining, Metallurgical and Exploration Engineers (SME) (Section 18, Liquid-Solid Operations and Equipment)

Stanley M. Englund, M.S., Ch.E., Fellow, American Institute of Chemical Engineers; Process Consultant, The Dow Chemical Company (retired) (Section 26, Process Safety)

Bryan J. Ennis, Ph.D., President, E&G Associates, and Adjunct Professor of Chemical Engineering, Vanderbilt University; Member and Chair of Powder Technology Programming Group of the Particle Technology Forum, American Institute of Chemical Engineers (Section 20, Size Reduction and Size Enlargement)

William Eykamp, Ph.D., Adjunct Professor of Chemical Engineering, Tufts University; Formerly President, Koch Membrane Systems; Member, American Institute of Chemical Engineers, American Chemical Society, American Association for the Advancement of Science, North American Membrane Society, European Society of Membrane Science and Technology (Section 22, Alternative Separation Processes)

James R. Fair, Ph.D., P.E., Professor of Chemical Engineering, University of Texas; National Academy of Engineering; Fellow, American Institute of Chemical Engineers; Member, American Chemical Society, American Society for Engineering Education, National Society of Professional Engineers (Section 14, Gas Absorption and Gas-Liquid System Design)

Bruce A. Finlayson, Ph.D., Rehnberg Professor and Chair, Department of Chemical Engineering, University of Washington; Member, National Academy of Engineering (Section 3, Mathematics)

Thomas M. Flynn, Ph.D., P.E., Cryogenic Engineer, President CRYOCO, Louisville, Colorado; Member, American Institute of Chemical Engineers (Section 11, Heat-Transfer Equipment)

Anthony G. Fonseca, Ph.D., Director, Coal Utilization, CONSOL, Inc.; Member, American Chemical Society, Society for Mining, Metallurgy, and Extraction (Section 27, Energy Resources, Conversion, and Utilization)

D. G. Friend, National Institutes of Standards and Technology, Boulder, Colorado (Section 2, Physical and Chemical Data)

George W. Gassman, B.S.M.E., Senior Research Specialist, Final Control Systems, Fisher Controls International, Inc., Marshalltown, Iowa (Section 8, Process Control)

Fred K. Geitner, P.Eng., B.S.M.E., M.S.M.E., Consulting Engineer, Registered Professional Engineer (Ontario, Canada) (Section 29, Process Machinery Drives)

Victor M. Goldschmidt, Ph.D., P.E., Professor of Mechanical Engineering, Purdue University, West Lafayette, Indiana (Section 11, Heat-Transfer Equipment)

Stanley Grossel, President, Process Safety & Design, Inc.; Fellow, American Institute of Chemical Engineers; Member, American Chemical Society; Member, The Combustion Institute; Member, Explosion Protection Systems Committee of NFPA (Section 26, Process Safety)

Peter Harriott, Ph.D., Professor, School of Chemical Engineering, Cornell University; Member, American Institute of Chemical Engineering, American Chemical Society (ACS) (Section 18, Liquid-Solid Operations and Equipment)

T. Alan Hatton, Ph.D., Ralph Landau Professor and Director of the David H. Koch School of Chemical Engineering Practice, Massachusetts Institute of Technology; Founding Fellow, American Institute of Medical and Biological Engineering; Member, American Institute of Chemical Engineers, American Chemical Society, International Association of Colloid and Interface Scientists, American Association for the Advancement of Science, Neutron Scattering Society of America (Section 22, Alternative Separation Processes)

Joseph D. Henry, Jr., Ph.D., P.E., Senior Fellow, Department of Engineering and Public Policy, Carnegie Mellon University; Member, American Institute of Chemical Engineers, American Society for Engineering Education (Section 22, Alternative Separation Processes)

W. G. High, C.Eng., B.Sc., F.I.Mech.E., Burgoyne Consultants Ltd., W. Yorks, England (Section 26, Process Safety)

Richard Hogg, Ph.D., Professor, Department of Mineral Engineering, The Pennsylvania State University, University Park, PA (Section 19, Solid-Solid Operations and Equipment)

F. A. Holland, D.Sc., Ph.D., Consultant in Heat Energy Recycling; Research Professor, University of Salford, England; Fellow, Institution of Chemical Engineers, London (Section 9, Process Economics)

Hoyt C. Hottel, S.M., Professor Emeritus of Chemical Engineering, Massachusetts Institute of Technology; Member, National Academy of Sciences, American Academy of Arts and Sciences, American Institute of Chemical Engineers, American Chemical Society, Combustion Institute (Section 5, Heat and Mass Transfer)

Colin S. Howat, Ph.D., P.E., John E. & Winfred E. Sharp Professor, Department of Chemical and Petroleum Engineering, University of Kansas; Member, American Institute of Chemical Engineers; Member, American Society of Engineering Education (Section 30, Analysis of Plant Performance)

Predrag S. Hrnjak, Ph.D., V.Res., Assistant Professor, University of Illinois at Urbana Champaign and Principal Investigator—U. of I. Air Conditioning and Refrigeration Center, Assistant Professor, University of Belgrade; Member, International Institute of Refrigeration, American Society of Heating, Refrigeration and Air Conditioning (Section 11, Heat-Transfer Equipment)

Arthur E. Humphrey, Ph.D., Retired, Professor of Chemical Engineering, Pennsylvania State University; Member, U.S. National Academy of Engineering, American Institute of Chemical Engineers, American Chemical Society, American Society for Microbiology (Section 24, Biochemical Engineering)

Eric Jenett, M.S.Ch.E., Manager, Process Engineering, Brown & Root, Inc.; Associate Member, AIChE, Project Management Institute; Registered Professional Engineer (Texas) (Section 29, Process Machinery Drives)

John S. Jeris, Sc.D., P.E., Professor of Environmental Engineering, Manhattan College; Environmental Consultant; Member, American Water Works Association, Water Environment Federation Section Director (Section 25, Waste Management)

T. L. P. Jespen, M.S., Min. Proc., Metallurgical Engineer, Basic, Inc., Gabbs, Nevada (Section 19, Solid-Solid Operations and Equipment)

Keith P. Johnston, Ph.D., P.E., Professor of Chemical Engineering, University of Texas (Austin); Member, American Institute of Chemical Engineers, American Chemical Society, University of Texas Separations Research Program (Section 22, Alternative Separation Processes)

Trevor A. Kletz, D.Sc., Senior Visiting Research Fellow, Department of Chemical Engineering, Loughborough University, U.K.; Fellow, American Institute of Chemical Engineers, Royal Academy of Engineers (U.K.), Institution of Chemical Engineers (U.K.), and Royal Society of Chemistry (U.K.) (Section 26, Process Safety)

Edgar B. Klunder, Ph.D., Project Manager, Energy Technology Center (Pittsburgh), U.S. Department of Energy (Section 27, Energy Resources, Conversion, and Utilization)

Kent S. Knaebel, Ph.D., President, Adsorption Research, Inc.; Member, American Institute of Chemical Engineers, American Chemical Society, International Adsorption Society. Professional Engineer (Ohio) (Section 5, Heat and Mass Transfer)

Frank Knoll, M.S., Min. Proc., President, Carpco, Inc., Jacksonville, Florida (Section 19, Solid-Solid Operations and Equipment)

James G. Knudsen, Ph.D., Professor Emeritus of Chemical Engineering, Oregon State University; Member, American Institute of Chemical Engineers, American Chemical Society; Registered Professional Engineer (Oregon) (Section 5, Heat and Mass Transfer)

Michael Krumpelt, Ph.D., Manager, Fuel Cell Technology, Argonne National Laboratory; Member, American Institute of Chemical Engineers, American Chemical Society, Electrochemical Society (Section 27, Energy Resources, Conversion, and Utilization)

Irwin J. Kugelman, Sc.D., Professor of Civil Engineering, Lehigh University; Member, American Society of Civil Engineering, Water Environmental Federation (Section 25, Waste Management)

Tim Laros, M.S. Mineral Processing, Senior Process Consultant, EIMCO Process Equipment Co.; Member, Society of Mining, Metallurgy and Exploration (SME of AIME) (Section 18, Liquid-Solid Operations and Equipment)

Richard M. Lemert, Ph.D., P.E., Assistant Professor of Chemical Engineering, University of Toledo; Member, American Institute of Chemical Engineers, American Chemical Society, Society of Mining Engineers, American Society for Engineering Education (Section 22, Alternative Separation Processes)

Robert Lemlich, Ph.D., P.E., Professor of Chemical Engineering Emeritus, University of Cincinnati; Fellow, American Institute of Chemical Engineers; Member, American Chemical Society, American Society for Engineering Education, American Chemical Society (Section 22, Alternative Separation Processes)

Wallace Leung, Sc.D., Director, Process Technology, Bird Machine Company; Member, American Filtration and Separation Society (Director) (Section 18, Liquid-Solid Operations and Equipment)

M. Douglas LeVan, Ph.D., Professor, Department of Chemical Engineering, University of Virginia; Member, American Institute of Chemical Engineers, American Chemical Society, International Adsorption Society (Section 16, Adsorption and Ion Exchange)

Peter E. Liley, Ph.D., D.I.C., School of Mechanical Engineering, Purdue University (Section 2, Physical and Chemical Data)

James D. Litster, Ph.D., Associate Professor, Department of Chemical Engineering, University of Queensland; Member, Institute of Chemical Engineers—Australia (Section 20, Size Reduction and Size Enlargement)

Peter J. Loftus, D. Phil., Arthur D. Little, Inc.; Member, American Society of Mechanical Engineers (Section 27, Energy Resources, Conversion, and Utilization)

Hsue-peng Loh, Ph.D., P.E., Federal Energy Technology Center (Morgantown), U.S. Department of Energy; Member, American Institute of Chemical Engineers, American Society of Information Sciences (Section 27, Energy Resources, Conversion, and Utilization)

Douglas E. Lowenhaupt, M.S., Group Leader, Coke Laboratory, CONSOL, Inc.; Member, American Society for Testing and Materials, Iron and Steel Making Society, International Committee for Coal Petrology (Section 27, Energy Resources, Conversion, and Utilization)

James O. Maloney, Ph.D., P.E., Emeritus Professor of Chemical Engineering, University of Kansas; Fellow, American Institute of Chemical Engineering; Fellow, American Association for the Advancement of Science; Member, American Chemical Society, American Society for Engineering Education (Section 1, Conversion Factors and Mathematical Symbols)

Thomas J. McAvoy, Ph.D., Professor of Chemical Engineering, University of Maryland, College Park, Maryland (Section 8, Process Control)

Chad McCleary, EIMCO Process Equipment Company, Process Consultant (Section 18, Liquid-Solid Operations and Equipment)

Thomas F. McGowan, P.E., Senior Consultant, RMT/Four Nines; Member, American Institute of Chemical Engineers, American Society of Mechanical Engineers, Air and Waste Management Association (Section 25, Waste Management)

Howard G. McIlvried, III, Ph.D., Senior Engineer, Burns and Roe Services Corporation, Federal Energy Technology Center (Pittsburgh), Member, American Chemical Society, American Institute of Chemical Engineers (Section 27, Energy Resources, Conversion, and Utilization)

John D. McKenna, Ph.D., President and Chairman, ETS International, Inc., Member, American Institute of Chemical Engineers, Air and Waste Management Association (Section 25, Waste Management)

Shelby A. Miller, Ph.D., P.E., Resident Retired Senior Engineer, Argonne National Laboratory; American Association for the Advancement of Science (Fellow), American Chemical Society, American Institute of Chemical Engineers (Fellow), American Institutes of Chemists (Fellow), Filtration Society, New York Academy of Sciences, Society of Chemical Industry (Section 18, Liquid-Solid Operations and Equipment; Section 27, Energy Resources, Conversion, and Utilization)

Booker Morey, Ph.D., Senior Consultant, SRI International; Member, Society of Mining, Metallurgy and Exploration (SME of AIME), The Filtration Society, Air and Waste Management Association; Registered Professional Engineer (California and Massachusetts) (Section 18, Liquid-Solid Operations and Equipment)

Charles G. Moyers, Ph.D., P.E., Principal Engineer, Union Carbide Corporation; Fellow, American Institute of Chemical Engineers (Section 12, Psychrometry, Evaporative Cooling, and Solids Drying; Section 22, Alternative Separation Processes)

John Newman, Ph.D., Professor of Chemical Engineering, University of California, Berkeley; Principle Investigator, Inorganic Materials Research Division, Lawrence Berkeley Laboratory (Section 22, Alternative Separation Processes)

James Y. Oldshue, Ph.D., President, Oldshue Technologies International, Inc.; Member, National Academy of Engineering; Adjunct Professor of Chemical Engineering at Beijing Institute of Chemical Technology, Beijing, China; Member, American Chemical Society (ACE), American Institute of Chemical Engineering (AIChE), Traveler Century Club, Executive Committee on the Transfer of Appropriate Technology for the World Federation of Engineering Organizations (Section 18, Liquid-Solid Operations and Equipment)

Robert W. Ormsby, M.S., Ch.E. P.E., Manager of Safety, Chemical Group, Air Products and Chemicals, Inc.; Air Products Corp.; Fellow, American Institute of Chemical Engineers (Section 26, Process Safety)

John E. Owens, B.E.E., Electrostatic Consultant, Condux, Inc.; Member, Institute of Electrical and Electronics Engineers, Electrostatics Society of America (Section 26, Process Safety)

Bhupendra Parekh, Ph.D., Associate Director, Center for Applied Energy Research, University of Kentucky, Lexington, Kentucky (Section 19, Solid-Solid Operations and Equipment)

Mel Pell, Ph.D., Senior Consultant, E. I. duPont de Nemours & Co.; Fellow, American Institute of Chemical Engineers; Registered Professional Engineer (Delaware) (Section 17, Gas-Solid Operations and Equipment)

W. R. Penney, Ph.D., P.E., Professor of Chemical Engineering, University of Arkansas; Member, American Institute of Chemical Engineers (Section 14, Gas Absorption and Gas-Liquid System Design)

Walter F. Podolski, Ph.D., Chemical Engineer, Electrochemical Technology Program, Argonne National Laboratory; Member, American Institute of Chemical Engineers (Section 27, Energy Resources, Conversion, and Utilization)

Herbert A. Pohl, Ph.D. (deceased), Professor of Physics, Oklahoma State University (Section 22, Alternative Separation Processes)

Kent Pollock, Ph.D., Member of Technical Staff, Group 91, Space Surveillance Techniques, MIT Lincoln Laboratory (Section 22, Alternative Separation Processes)

George Priday, B.S., Ch.E., EIMCO Process Equipment Company; Member, American Institute of Chemical Engineering (AIChE), Instrument Society of America (ISA) (Section 18, Liquid-Solid Operations and Equipment)

Michael E. Prudich, Ph.D., Professor and Chair of Chemical Engineering, Ohio University; Member, American Institute of Chemical Engineers, American Chemical Society, Society of Mining Engineers, American Society for Engineering Education (Section 22, Alternative Separation Processes)

Raj K. Rajamani, Ph.D., Professor, Department of Metallurgy and Metallurgical Engineering, University of Utah, Salt Lake City, Utah (Section 19, Solid-Solid Operations and Equipment)

Lawrence K. Rath, B.S., P.E., Federal Energy Technology Center (Morgantown), U.S. Department of Energy; Member, American Institute of Chemical Engineers (Section 27, Energy Resources, Conversion, and Utilization)

Grantges J. Raymus, M.E., M.S., President, Raymus Associates, Incorporated, Packaging Consultants; Adjunct Professor and Program Coordinator, Center for Packaging Science and Engineering, College of Engineering, Rutgers, The State University of New Jersey; formerly Manager of Packaging Engineering, Union Carbide Corporation; Registered Professional Engineer, California; Member, Institute of Packaging Professionals, ASME (Section 21, Handling of Bulk Solids and Packaging of Solids and Liquids)

Lanny A. Robbins, Ph.D., Research Fellow, Dow Chemical Company; Member, American Institute of Chemical Engineers (Section 15, Liquid-Liquid Extraction Operations and Equipment)

Joseph J. Santoleri, P.E., Senior Consultant, RMT/Four Nines; Member, American Institute of Chemical Engineers, American Society of Mechanical Engineers, Air and Waste Management Association (Section 25, Waste Management)

Adel F. Sarofim, Sc.D., Lammot DuPont Professor of Chemical Engineering and Assistant Director, Fuels Research Laboratory, Massachusetts Institute of Technology; Member, American Institute of Chemical Engineers, American Chemical Society, Combustion Institute (Section 5, Heat and Mass Transfer)

Kalanadh V. S. Sastry, Ph.D., Professor, Department of Materials Science and Mineral Engineering, University of California, Berkeley, CA; Member, American Institute of Chemical Engineers, Society for Mining, Metallurgy and Exploration (Section 19, Solid-Solid Operations and Equipment)

Paul J. Schafbuch, Ph.D., Senior Research Specialist, Final Control Systems, Fisher Controls International, Inc., Marshalltown, Iowa (Section 8, Process Control)

Carl A. Schiappa, B.S., Ch.E., Process Engineering Associate, Michigan Division Engineering, The Dow Chemical Company; Member, AIChE and CCPS (Section 26, Process Safety)

David K. Schmalzer, Ph.D., P.E., Fossil Energy Program Manager, Argonne National Laboratory; Member, American Chemical Society, American Institute of Chemical Engineers (Section 27, Energy Resources, Conversion, and Utilization)

J. D. Seader, Ph.D., Professor of Chemical Engineering, University of Utah, Salt Lake City, Utah; Fellow, American Institute of Chemical Engineers; Member, American Chemical Society; Member, American Society for Engineering Education (Section 13, Distillation)

Dale E. Seborg, Ph.D., Professor of Chemical Engineering, University of California, Santa Barbara, California (Section 8, Process Control)

Richard L. Shilling, P.E., B.S.M., B.E.M.E., Manager of Engineering Development, Brown Fintube Company—a Koch Engineering Company; Member, American Society of Mechanical Engineers (Section 11, Heat-Transfer Equipment)

F. Greg Shinskey, B.S.Ch.E., Consultant (retired from Foxboro Co.), North Sandwich, New Hampshire (Section 8, Process Control)

Oliver W. Siebert, P.E., B.S.M.E., Washington University, Graduate Metallurgical Engineering, Sever Institute of Technology; Professor, Department of Chemical Engineering, Washington University, St. Louis, Missouri; President, Siebert Materials Engineering, Inc., St. Louis, Missouri; Senior Engineering Fellow (retired), Monsanto Co.; Mechanical Designer, Sverdrup Corp.; Metallurgist, Carondelet Foundry; United Nations Consultant to the People's Republic of China; Fellow, American Institute of Chemical Engineers; Life Fellow, American Society of Mechanical Engineers; Past Elected Director and Fellow, National Association of Corrosion Engineers, Int'l; American Society for Metals, Int'l; American Welding Society; Pi Tau Sigma, Sigma Xi, and Tau Beta Pi (Section 28, Materials of Construction)

Jeffrey J. Siirola, Ph.D., Research Fellow, Eastman Chemical Company; Member, National Academy of Engineering; Fellow, American Institute of Chemical Engineers, American Chemical Society, American Association for Artificial Intelligence, American Society for Engineering Education (Section 13, Distillation)

Charles E. Silverblatt, M.S., Ch.E., Peregrine International Associates, Inc.; Consultant to WesTech Engineering, Inc., American Institute of Mining, Metallurgical and Petroleum Engines (AIME) (Section 18, Liquid-Solid Operations and Equipment)

Richard Siwek, M.S., Explosion Protection Manager, Corporate Unit Safety and Environment, Ciba-Geigy Ltd., Basel, Switzerland (Section 26, Process Safety)

J. Stephen Slottee, M.S., Ch.E., Manager, Technology and Development, EIMCO Process Equipment Co.; Member, American Institute of Chemical Engineers (AIChE) (Section 18, Liquid-Solid Operations and Equipment)

Cecil L. Smith, Ph.D., Principal, Cecil L. Smith Inc., Baton Rouge, Louisiana (Section 8, Process Control)

Julian C. Smith, B. Chem., Ch.E., Professor Emeritus Chemical Engineering, Cornell University; Member, American Chemical Society (ACS), American Institute of Chemical Engineers (AIChE) (Section 18, Liquid-Solid Operations and Equipment)

Richard H. Snow, Ph.D., Engineering Advisor, IIT Research Institute; Member, American Chemical Society, Sigma Xi; Fellow, American Institute of Chemical Engineers (Section 20, Size Reduction and Size Enlargement)

Thomas Sorenson, M.B.A., Min. Eng., President, Galigher Ash (Canada) Ltd. (Section 19, Solid-Solid Operations and Equipment)

Rameshwar D. Srivastava, Ph.D., Fuels Group Manager, Burns and Roe Services Corporation, Federal Energy Technology Center (Pittsburgh) (Section 27, Energy Resources, Conversion, and Utilization)

F. C. Standiford, M.S., P.E., Member, American Institute of Chemical Engineers, American Chemical Society (Section 11, Heat-Transfer Equipment)

D. E. Steinmeyer, M.A., M.S., P.E., Distinguished Fellow, Monsanto Company; Fellow, American Institute of Chemical Engineers; Member, American Chemical Society (Section 14, Gas Absorption and Gas-Liquid System Design)

Gary J. Stiegel, M.S., P.E., Program Coordinator, Federal Energy Technology Center (Pittsburgh), U.S. Department of Energy (Section 27, Energy Resources, Conversion, and Utilization)

John G. Stoecker II, B.S.M.E., University of Missouri School of Mines and Metallurgy; Principal Consultant, Stoecker & Associates, St. Louis, Missouri; Principal Materials Engineering Specialist (retired), Monsanto Co.; High-Temperature Design/Application Engineer, Abex Corporation; Member, NACE International, ASM International (Section 27, Energy Resources, Conversion, and Utilization)

Judson S. Swearingen, Ph.D., Retired President, Rotoflow Corporation (Section 29, Process Machinery Drives)

Louis Theodore, Sc.D., Professor of Chemical Engineering, Manhattan College; Member, Air and Waste Management Association (Section 25, Waste Management)

Michael P. Thien, Sc.D., Senior Research Fellow, Merck & Co., Inc.; Member, American Institute of Chemical Engineers, American Chemical Society, International Society for Pharmaceutical Engineers (Section 22, Alternative Separation Processes)

George H. Thomson, AIChE Design Institute for Physical Property Data (Section 2, Physical and Chemical Data)

James N. Tilton, Ph.D., P.E., Senior Consultant, Process Engineering, E. I. duPont de Nemours & Co.; Member, American Institute of Chemical Engineers; Registered Professional Engineer (Delaware) (Section 6, Fluid and Particle Dynamics)

Klaus D. Timmerhaus, Ph.D., P.E., Professor and President's Teaching Scholar, University of Colorado, Boulder, Colorado; Fellow, American Institute of Chemical Engineers, American Society for Engineering Education, American Association for the Advancement of Science; Member, American Astronautical Society, National Academy of Engineering, Austrian Academy of Science, International Institute of Refrigeration, American Society of Heating, Refrigerating and Air Conditioning Engineers, American Society of Environmental Engineers, Engineering Society for Advancing Mobility on Land, Sea, Air, and Space, Sigma Xi, The Research Society (Section 11, Heat-Transfer Equipment)

David B. Todd, Ph.D., President, Todd Engineering; Member, American Association for the Advancement of Science (AAAS), American Chemical Society (ACS), American Institute of Chemical Engineering (AIChE), American Oil Chemists Society (AOCS), Society of Plastics Engineers (SPE), and Society of the Plastics Industry (SPI); Registered Professional Engineer, Michigan (Section 18, Liquid-Solid Operations and Equipment)

George T. Tsao, Ph.D., Director, Laboratory for Renewable Resource Engineering, Purdue University; Member, American Institute of Chemical Engineers, American Chemical Society, American Society for Microbiology (Section 24, Biochemical Engineering)

Hendrick C. Van Ness, D.Eng., Howard P. Isermann Department of Chemical Engineering, Rensselaer Polytechnic Institute; Fellow, American Institute of Chemical Engineers; Member, American Chemical Society (Section 4, Thermodynamics)

Stanley M. Walas, Ph.D., Professor Emeritus, Department of Chemical and Petroleum Engineering, University of Kansas; Fellow, American Institute of Chemical Engineers (Section 7, Reaction Kinetics; Section 23, Chemical Reactors)

Phillip C. Wankat, Ph.D., Professor of Chemical Engineering, Purdue University; Member, American Institute of Chemical Engineers, American Chemical Society, International Adsorption Society (Section 5, Heat and Mass Transfer)

Ionel Wechsler, M.S., Min. and Met., Vice President, Sala Magnetics, Inc., Cambridge, Massachusetts (Section 19, Solid-Solid Operations and Equipment)

Arthur W. Westerberg, Ph.D., Swearingen University Professor of Chemical Engineering, Carnegie Mellon University; Member, National Academy of Engineering (Section 3, Mathematics)

John M. Wheeldon, Ph.D., Electric Power Research Institute (Section 27, Energy Resources, Conversion, and Utilization)

Robert E. White, Ph.D., Principal Engineer, Chemistry and Chemical Engineering Division, Southwest Research Institute (Section 26, Process Safety)

J. K. Wilkinson, M.Sc., Consultant Chemical Engineer; Fellow, Institution of Chemical Engineers, London (Section 9, Process Economics)

David Winegarder, Ph.D., Engineering Associate, Michigan Division Engineering, The Dow Chemical Company; Member AIChE and CCPS (Section 26, Process Safety)

John L. Woodward, Ph.D., Principal, DNV Technica, Inc. (Section 26, Process Safety)

Yoshiyuki Yamashita, Ph.D., Associate Professor of Chemical Engineering, Tohoku University, Sendai, Japan (Section 3, Mathematics)

Carmen M. Yon, M.S., Development Associate, UOP, Des Plaines, Illinois; Member, American Institute of Chemical Engineers (Section 16, Adsorption and Ion Exchange)

Preface to the Seventh Edition

Perry's has been an important source for chemical engineering information since 1934. The significant contributions of the editors who have guided preparation of the previous editions is acknowledged. These include John H. Perry (first to third editions), Robert H. Perry (fourth to sixth editions), Cecil H. Chilton (fourth and fifth editions), and Sidney D. Kirkpatrick (fourth edition). Ray Genereaux (DuPont) contributed to each of the first six editions, and Shelby Miller (Argonne National Lab) worked on the second through the seventh. The current editors directed both the sixth and seventh editions. Advances in the technology of chemical engineering have continued as we have moved toward the twenty-first century, and this edition will carry us into that century.

The *Handbook* has been reorganized. The first group of sections focuses on chemical and physical property data and the fundamentals of chemical engineering. The second and largest group of sections deals with processes, generally divided as heat transfer operations, distillation, kinetics, liquid-liquid, liquid-solid, and so on. The last group treats auxiliary information such as materials of construction, process machinery drives, waste management, and process safety. All sections have been revised and updated, and several sections are entirely new or have been extensively revised. Examples of these sections are mathematics, mass transfer, reaction kinetics, process control, transport and storage of fluids, alternative separation processes, heat-transfer equipment, chemical reactions, liquid-solid operations and equipment, process safety, and analysis of plant performance. Significant new information has also been included in the physical and chemical data sections.

Several section editors and contributors worked on this seventh edition, and these persons and their affiliations are listed as a part of the front material. Approximately one-half of the section editors are fellows of the AIChE. In addition, the following chemical engineering students at the University of Kansas assisted in the preparation of the index: Jason Canter, Pau Ying Chong, Mei Ling Chuah, Li Phoon Hor, Siew Pouy Ng, Francis J. Orzulak, Scott C. Renze, Page B. Surbaugh, and Stephen F. Weller. Shari L. Gladman and Sarah Smith provided extensive secretarial assistance.

Much of Bob Perry's work carries over into this edition and his influence is both recognized and remembered.

DON W. GREEN
JAMES O. MALONEY
University of Kansas
April, 1997

Perry's
Chemical
Engineers'
Handbook

Conversion Factors and Mathematical Symbols*

James O. Maloney, Ph.D., P.E., *Emeritus Professor of Chemical Engineering, University of Kansas; Fellow, American Institute of Chemical Engineering; Fellow, American Association for the Advancement of Science; Member, American Chemical Society, American Society for Engineering Education*

* Much of the material was taken from Sec. 1. of the fifth edition. The contribution of Cecil H. Chilton in developing that material is acknowledged.

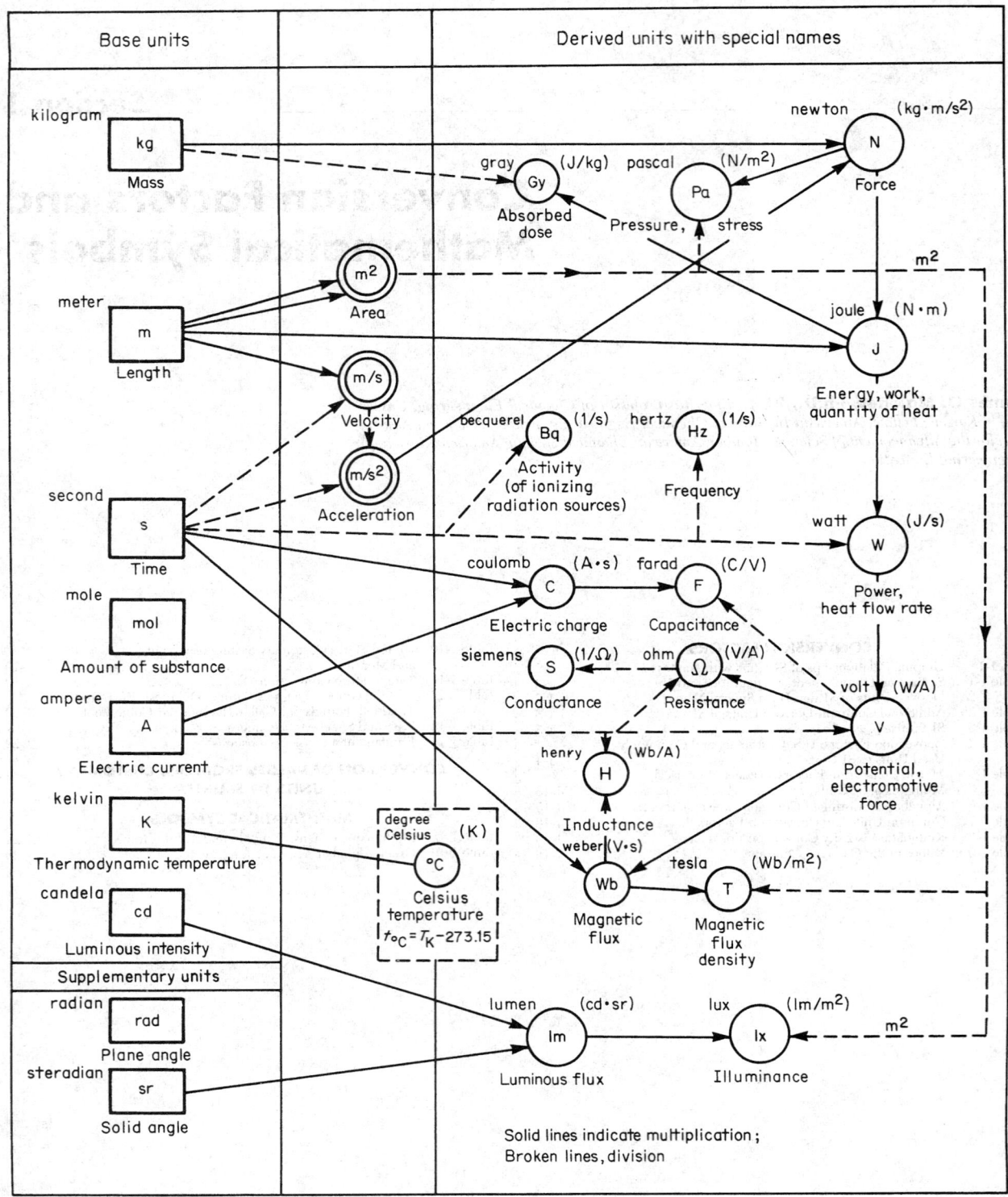

FIG. 1-1 Graphic relationships of SI units with names (*U.S. National Bureau of Standards, LC 1078, December 1976.*)

TABLE 1-1 SI Base and Supplementary Quantities and Units

Quantity or "dimension"	SI unit	SI unit symbol ("abbreviation"); Use roman (upright) type
Base quantity or "dimension"		
length	meter	m
mass	kilogram	kg
time	second	s
electric current	ampere	A
thermodynamic temperature	kelvin	K
amount of substance	mole°	mol
luminous intensity	candela	cd
Supplementary quantity or "dimension"		
plane angle	radian	rad
solid angle	steradian	sr

° When the mole is used, the elementary entities must be specified; they may be atoms, molecules, ions, electrons, other particles, or specified groups of such particles.

TABLE 1-2a Derived Units of SI that Have Special Names

Quantity	Unit	Symbol	Formula
frequency (of a periodic phenomenon)	hertz	Hz	1/s
force	newton	N	$(kg \cdot m)/s^2$
pressure, stress	pascal	Pa	N/m^2
energy, work, quantity of heat	joule	J	$N \cdot m$
power, radiant flux	watt	W	J/s
quantity of electricity, electric charge	coulomb	C	$A \cdot s$
electric potential, potential difference, electromotive force	volt	V	W/A
capacitance	farad	F	C/V
electric resistance	ohm	Ω	V/A
conductance	siemens	S	A/V
magnetic flux	weber	Wb	$V \cdot s$
magnetic-flux density	tesla	T	Wb/m^2
inductance	henry	H	Wb/A
luminous flux	lumen	lm	$cd \cdot sr$
illuminance	lux	lx	lm/m^2
activity (of radionuclides)	becquerel	Bq	1/s
absorbed dose	gray	Gy	J/kg

TABLE 1-2b Additional Common Derived Units of SI

Quantity	Unit	Symbol
acceleration	meter per second squared	m/s^2
angular acceleration	radian per second squared	rad/s^2
angular velocity	radian per second	rad/s
area	square meter	m^2
concentration (of amount of substance)	mole per cubic meter	mol/m^3
current density	ampere per square meter	A/m^2
density, mass	kilogram per cubic meter	kg/m^3
electric-charge density	coulomb per cubic meter	C/m^3
electric-field strength	volt per meter	V/m
electric-flux density	coulomb per square meter	C/m^2
energy density	joule per cubic meter	J/m^3
entropy	joule per kelvin	J/K
heat capacity	joule per kelvin	J/K
heat-flux density, irradiance	watt per square meter	W/m^2
luminance	candela per square meter	cd/m^2
magnetic-field strength	ampere per meter	A/m
molar energy	joule per mole	J/mol
molar entropy	joule per mole-kelvin	$J/(mol \cdot K)$
molar-heat capacity	joule per mole-kelvin	$J/(mol \cdot K)$
moment of force	newton-meter	$N \cdot m$
permeability	henry per meter	H/m
permittivity	farad per meter	F/m
radiance	watt per square-meter-steradian	$W/(m^2 \cdot sr)$
radiant intensity	watt per steradian	W/sr
specific-heat capacity	joule per kilogram-kelvin	$J/(kg \cdot K)$
specific energy	joule per kilogram	J/kg
specific entropy	joule per kilogram-kelvin	$J/(kg \cdot K)$
specific volume	cubic meter per kilogram	m^3/kg
surface tension	newton per meter	N/m
thermal conductivity	watt per meter-kelvin	$W/(m \cdot K)$
velocity	meter per second	m/s
viscosity, dynamic	pascal-second	$Pa \cdot s$
viscosity, kinematic	square meter per second	m^2/s
volume	cubic meter	m^3
wave number	1 per meter	1/m

TABLE 1-3 SI Prefixes

Multiplication factor	Prefix	Symbol
1 000 000 000 000 000 000 = 10^{18}	exa	E
1 000 000 000 000 000 = 10^{15}	peta	P
1 000 000 000 000 = 10^{12}	tera	T
1 000 000 000 = 10^9	giga	G
1 000 000 = 10^6	mega	M
1 000 = 10^3	kilo	k
100 = 10^2	hecto°	h
10 = 10^1	deka°	da
0.1 = 10^{-1}	deci°	d
0.01 = 10^{-2}	centi	c
0.001 = 10^{-3}	milli	m
0.000 001 = 10^{-6}	micro	μ
0.000 000 001 = 10^{-9}	nano	n
0.000 000 000 001 = 10^{-12}	pico	p
0.000 000 000 000 001 = 10^{-15}	femto	f
0.000 000 000 000 000 001 = 10^{-18}	atto	a

° Generally to be avoided.

TABLE 1-4 Conversion Factors: U.S. Customary and Commonly Used Units to SI Units

Quantity	Customary or commonly used unit	SI unit	Alternate SI unit	Conversion factor; multiply customary unit by factor to obtain SI unit	
Space,† time					
Length	naut mi	km		1.852°	E + 00
	mi	km		1.609 344°	E + 00
	chain	m		2.011 68°	E + 01
	link	m		2.011 68°	E − 01
	fathom	m		1.828 8°	E + 00
	yd	m		9.144°	E − 01
	ft	m		3.048°	E − 01
		cm		3.048°	E + 01
	in	mm		2.54°	E + 01
	in	cm		2.54	E + 00
	mil	μm		2.54°	E + 01
Length/length	ft/mi	m/km		1.893 939	E − 01
Length/volume	ft/U.S. gal	m/m³		8.051 964	E + 01
	ft/ft³	m/m³		1.076 391	E + 01
	ft/bbl	m/m³		1.917 134	E + 00
Area	mi²	km²		2.589 988	E + 00
	section	ha		2.589 988	E + 02
	acre	ha		4.046 856	E − 01
	ha	m²		1.000 000°	E + 04
	yd²	m²		8.361 274	E − 01
	ft²	m²		9.290 304°	E − 02
	in²	mm²		6.451 6°	E + 02
		cm²		6.451 6°	E + 00
Area/volume	ft²/in³	m²/cm³		5.699 291	E − 03
	ft²/ft³	m²/m³		3.280 840	E + 00
Volume	cubem	km³		4.168 182	E + 00
	acre·ft	m³		1.233 482	E + 03
		ha·m		1.233 482	E − 01
	yd³	m³		7.645 549	E − 01
	bbl (42 U.S. gal)	m³		1.589 873	E − 01
	ft³	m³		2.831 685	E − 02
		dm³	L	2.831 685	E + 01
	U.K. gal	m³		4.546 092	E − 03
		dm³	L	4.546 092	E + 00
	U.S. gal	m³		3.785 412	E − 03
		dm³	L	3.785 412	E + 00
	U.K. qt	dm³	L	1.136 523	E + 00
	U.S. qt	dm³	L	9.463 529	E − 01
	U.S. pt	dm³	L	4.731 765	E − 01
	U.K. fl oz	cm³		2.841 307	E + 01
	U.S. fl oz	cm³		2.957 353	E + 01
	in³	cm³		1.638 706	E + 01
Volume/length (linear displacement)	bbl/in	m³m		6.259 342	E + 00
	bbl/ft	m³/m		5.216 119	E − 01
	ft³/ft	m³/m		9.290 304°	E − 02
	U.S. gal/ft	m³/m		1.241 933	E − 02
		L/m		1.241 933	E + 01
Plane angle	rad	rad		1	
	deg (°)	rad		1.745 329	E − 02
	min (′)	rad		2.908 882	E − 04
	sec (″)	rad		4.848 137	E − 06
Solid angle	sr	sr		1	
Time	year	a		1	
	week	d		7.0°	E + 00
	h	s		3.6°	E + 03
		min		6.0°	E + 01
	min	s		6.0°	E + 01
		h		1.666 667	E − 02
	mμs	ns		1	
Mass, amount of substance					
Mass	U.K. ton	Mg	t	1.016 047	E + 00
	U.S. ton	Mg	t	9.071 847	E − 01
	U.K. cwt	kg		5.080 234	E + 01
	U.S. cwt	kg		4.535 924	E + 01
	lbm	kg		4.535 924	E − 01
	oz (troy)	g		3.110 348	E + 01
	oz (av)	g		2.834 952	E + 01
	gr	mg		6.479 891	E + 01

TABLE 1-4 Conversion Factors: U.S. Customary and Commonly Used Units to SI Units (Continued)

Quantity	Customary or commonly used unit	SI unit	Alternate SI unit	Conversion factor; multiply customary unit by factor to obtain SI unit
Amount of substance	lbm·mol	kmol		4.535 924 E − 01
	std m³(0°C, 1 atm)	kmol		4.461 58 E − 02
	std ft³ (60°F, 1 atm)	kmol		1.195 30 E − 03
Enthalpy, calorific value, heat, entropy, heat capacity				
Calorific value, enthalpy (mass basis)	Btu/lbm	MJ/kg		2.326 000 E − 03
		kJ/kg	J/g	2.326 000 E + 00
		kWh/kg		6.461 112 E − 04
	cal/g	kJ/kg	J/g	4.184° E + 00
	cal/lbm	J/kg		9.224 141 E + 00
Caloric value, enthalpy (mole basis)	kcal/(g·mol)	kJ/kmol		4.184° E + 03
	Btu/(lb·mol)	kJ/kmol		2.326 000 E + 00
Calorific value (volume basis—solids and liquids)	Btu/U.S. gal	MJ/m³	kJ/dm³	2.787 163 E − 01
		kJ/m³		2.787 163 E + 02
		kWh/m³		7.742 119 E − 02
	Btu/U.K. gal	MJ/m³	kJ/dm³	2.320 800 E − 01
		kJ/m³		2.320 800 E + 02
	Btu/ft³	kWh/m³		6.446 667 E − 02
		MJ/m³	kJ/dm³	3.725 895 E − 02
		kJ/m³		3.725 895 E + 01
		kWh/m³		1.034 971 E − 02
	cal/mL	MJ/m³		4.184° E + 00
	(ft·lbf)/U.S. gal	kJ/m³		3.581 692 E − 01
Calorific value (volume basis—gases)	cal/mL	kJ/m³	J/dm³	4.184° E + 03
	kcal/m³	kJ/m³	J/dm³	4.184° E + 00
	Btu/ft³	kJ/m³	J/dm³	3.725 895 E + 01
		kWh/m³		1.034 971 E − 02
Specific entropy	Btu/(lbm·°R)	kJ/(kg·K)	J/(g·K)	4.186 8° E + 00
	cal/(g·K)	kJ/(kg·K)	J/(g·K)	4.184° E + 00
	kcal/(kg·°C)	kJ/(kg·K)	J/(g·K)	4.184° E + 00
Specific-heat capacity (mass basis)	kWh/(kg·°C)	kJ/(kg·K)	J/(g·K)	3.6° E + 03
	Btu/(lbm·°F)	kJ/(kg·K)	J/(g·K)	4.186 8° E + 00
	kcal/(kg·°C)	kJ/(kg·K)	J/(g·K)	4.184° E + 00
Specific-heat capacity (mole basis)	Btu/(lb·mol·°F)	kJ/(kmol·K)		4.186 8° E + 00
	cal/(g·mol·°C)	kJ/(kmol·K)		4.184° E + 00
Temperature, pressure, vacuum				
Temperature (absolute)	°R	K		5/9
	K	K		1
Temperature (traditional)	°F	°C		5/9(°F − 32)
Temperature (difference)	°F	K, °C		5/9
Pressure	atm (760 mmHg at 0°C or 14,696 psi)	MPa		1.013 250° E − 01
		kPa		1.013 250° E + 02
		bar		1.013 250° E + 00
	bar	MPa		1.0° E − 01
		kPa		1.0° E + 02
	mmHg (0°C) = torr	MPa		6.894 757 E − 03
		kPa		6.894 757 E + 00
		bar		6.894 757 E − 02
	μmHg (0°C)	kPa		3.376 85 E + 00
	μ bar	kPa		2.488 4 E − 01
	mmHg = torr (0°C)	kPa		1.333 224 E − 01
	cmH₂O (4°C)	kPa		9.806 38 E − 02
	lbf/ft² (psf)	kPa		4.788 026 E − 02
	mHg (0°C)	Pa		1.333 224 E − 01
	bar	Pa		1.0° E + 05
	dyn/cm²	Pa		1.0° E + 00
Vacuum, draft	inHg (60°F)	kPa		3.376 85 E + 00
	inH₂O (39.2°F)	kPa		2.490 82 E − 01
	inH₂O (60°F)	kPa		2.488 4 E − 01
	mmHg (0°C) = torr	kPa		1.333 224 E − 01
	cmH₂O (4°C)	kPa		9.806 38 E − 02
Liquid head	ft	m		3.048° E − 01
	in	mm		2.54° E + 01
		cm		2.54° E + 00
Pressure drop/length	psi/ft	kPa/m		2.262 059 E + 01

TABLE 1-4 Conversion Factors: U.S. Customary and Commonly Used Units to SI Units (*Continued*)

Quantity	Customary or commonly used unit	SI unit	Alternate SI unit	Conversion factor; multiply customary unit by factor to obtain SI unit
Density, specific volume, concentration, dosage				
Density	lbm/ft^3	kg/m^3		1.601 846 E + 01
		g/m^3		1.601 846 E + 04
	lbm/U.S. gal	kg/m^3		1.198 264 E + 02
		g/cm^3		1.198 264 E − 01
	lbm/U.K. gal	kg/m^3		9.977 633 E + 01
	lbm/ft^3	kg/m^3		1.601 846 E + 01
		g/cm^3		1.601 846 E − 02
	g/cm^3	kg/m^3		1.0° E + 03
	lbm/ft^3	kg/m^3		1.601 846 E + 01
Specific volume	ft^3/lbm	m^3/kg		6.242 796 E − 02
		m^3/g		6.242 796 E − 05
	ft^3/lbm	dm^3/kg		6.242 796 E + 01
	U.K. gal/lbm	dm^3/kg	cm^3/g	1.002 242 E + 01
	U.S. gal/lbm	dm^3/kg	cm^3/g	8.345 404 E + 00
Specific volume (mole basis)	L/(g·mol)	m^3/kmol		1
	ft^3/(lb·mol)	m^3/kmol		6.242 796 E − 02
Specific volume	bbl/U.S. ton	m^3/t		1.752 535 E − 01
	bbl/U.K. ton	m^3/t		1.564 763 E − 01
Yield	bbl/U.S. ton	dm^3/t	L/t	1.752 535 E + 02
	bbl/U.K. ton	dm^3/t	L/t	1.564 763 E + 02
	U.S. gal/U.S. ton	dm^3/t	L/t	4.172 702 E + 00
	U.S. gal/U.K. ton	dm^3/t	L/t	3.725 627 E + 00
Concentration (mass/mass)	wt %	kg/kg		1.0° E − 02
		g/kg		1.0° E + 01
	wt ppm	mg/kg		1
Concentration (mass/volume)	lbm/bbl	kg/m^3	g/dm^3	2.853 010 E + 00
	g/U.S. gal	kg/m^3		2.641 720 E − 01
	g/U.K. gal	kg/m^3	g/L	2.199 692 E − 01
	lbm/1000 U.S. gal	g/m^3	mg/dm^3	1.198 264 E + 02
	lbm/1000 U.K. gal	g/m^3	mg/dm^3	9.977 633 E + 01
	gr/U.S. gal	g/m^3	mg/dm^3	1.711 806 E + 01
	gr/ft^3	mg/m^3		2.288 351 E + 03
	lbm/1000 bbl	g/m^3	mg/dm^3	2.853 010 E + 00
	mg/U.S. gal	g/m^3	mg/dm^3	2.641 720 E − 01
	gr/100 ft^3	mg/m^3		2.288 351 E + 01
Concentration (volume/volume)	ft^3/ft^3	m^3/m^3		1
	bbl/(acre·ft)	m^3/m^3		1.288 931 E − 04
	vol%	m^3/m^3		1.0° E − 02
	U.K. gal/ft^3	dm^3/m^3	L/m^3	1.605 437 E + 02
	U.S. gal/ft^3	dm^3/m^3	L/m^3	1.336 806 E + 02
	mL/U.S. gal	dm^3/m^3	L/m^3	2.641 720 E − 01
	mL/U.K. gal	dm^3/m^3	L/m^3	2.199 692 E − 01
	vol ppm	cm^3/m^3		1
		dm^3/m^3	L/m^3	1.0° E − 03
	U.K. gal/1000 bbl	cm^3/m^3		2.859 403 E + 01
	U.S. gal/1000 bbl	cm^3/m^3		2.380 952 E + 01
	U.K. pt/1000 bbl	cm^3/m^3		3.574 253 E + 00
Concentration (mole/volume)	(lb·mol)/U.S. gal	kmol/m^3		1.198 264 E + 02
	(lb·mol)/U.K. gal	kmol/m^3		9.977 644 E + 01
	(lb·mol)/ft^3	kmol/m^3		1.601 846 E + 01
	std ft^3 (60°F, 1 atm)/bbl	kmol/m^3		7.518 21 E − 03
Concentration (volume/mole)	U.S. gal/1000 std ft^3 (60°F/60°F)	dm^3/kmol	L/kmol	3.166 91 E + 00
	bbl/million std ft^3 (60°F/60°F)	dm^3/kmol	L/kmol	1.330 10 E − 01
Facility throughput, capacity				
Throughput (mass basis)	U.K. ton/year	t/a		1.016 047 E + 00
	U.S. ton/year	t/a		9.071 847 E − 01
	U.K. ton/day	t/d		1.016 047 E + 00
		t/h		4.233 529 E − 02
	U.S. ton/day	t/d		9.071 847 E − 01
		t/h		3.779 936 E − 02
	U.K. ton/h	t/h		1.016 047 E + 00
	U.S. ton/h	t/h		9.071 847 E − 01
	lbm/h	kg/h		4.535 924 E − 01

TABLE 1-4 Conversion Factors: U.S. Customary and Commonly Used Units to SI Units (*Continued*)

Quantity	Customary or commonly used unit	SI unit	Alternate SI unit	Conversion factor; multiply customary unit by factor to obtain SI unit
Throughput (volume basis)	bbl/day	t/a		5.803 036 E + 01
		m³/d		1.589 873 E − 01
	ft³/day	m³/h		1.179 869 E − 03
	bbl/h	m³/h		1.589 873 E − 01
	ft³/h	m³/h		2.831 685 E − 02
	U.K. gal/h	m³/h		4.546 092 E − 03
		L/s		1.252 803 E − 03
	U.S. gal/h	m³/h		3.785 412 E − 03
		L/s		1.051 503 E − 03
	U.K. gal/min	m³/h		2.727 655 E − 01
		L/s		7.576 819 E − 02
	U.S. gal/min	m³/h		2.271 247 E − 01
		L/s		6.309 020 E − 02
Throughput (mole basis)	(lbm·mol)/h	kmol/h		4.535 924 E − 01
		kmol/s		1.259 979 E − 04

<table>
<tr><td colspan="5" align="center">Flow rate</td></tr>
</table>

Quantity	Customary or commonly used unit	SI unit	Alternate SI unit	Conversion factor; multiply customary unit by factor to obtain SI unit
Flow rate (mass basis)	U.K. ton/min	kg/s		1.693 412 E + 01
	U.S. ton/min	kg/s		1.511 974 E + 01
	U.K. ton/h	kg/s		2.822 353 E − 01
	U.S. ton/h	kg/s		2.519 958 E − 01
	U.K. ton/day	kg/s		1.175 980 E − 02
	U.S. ton/day	kg/s		1.049 982 E − 02
	million lbm/year	kg/s		5.249 912 E + 00
	U.K. ton/year	kg/s		3.221 864 E − 05
	U.S. ton/year	kg/s		2.876 664 E − 05
	lbm/s	kg/s		4.535 924 E − 01
	lbm/min	kg/s		7.559 873 E − 03
	lbm/h	kg/s		1.259 979 E − 04
Flow rate (volume basis)	bbl/day	m³/d		1.589 873 E − 01
		L/s		1.840 131 E − 03
	ft³/day	m³/d		2.831 685 E − 02
		L/s		3.277 413 E − 04
	bbl/h	m³/s		4.416 314 E − 05
		L/s		4.416 314 E − 02
	ft³/h	m³/s		7.865 791 E − 06
		L/s		7.865 791 E − 03
	U.K. gal/h	dm³/s	L/s	1.262 803 E − 03
	U.S. gal/h	dm³/s	L/s	1.051 503 E − 03
	U.K. gal/min	dm³/s	L/s	7.576 820 E − 02
	U.S. gal/min	dm³/s	L/s	6.309 020 E − 02
	ft³/min	dm³/s	L/s	4.719 474 E − 01
	ft³/s	dm³/s	L/s	2.831 685 E + 01
Flow rate (mole basis)	(lb·mol)/s	kmol/s		4.535 924 E − 01
	(lb·mol)/h	kmol/s		1.259 979 E − 04
	million scf/D	kmol/s		1.383 45 E − 02
Flow rate/length (mass basis)	lbm/(s·ft)	kg/(s·m)		1.488 164 E + 00
	lbm/(h·ft)	kg/(s·m)		4.133 789 E − 04
Flow rate/length (volume basis)	U.K. gal/(min·ft)	m²/s	m³/(s·m)	2.485 833 E − 04
	U.S. gal/(min·ft)	m²/s	m³/(s·m)	2.069 888 E − 04
	U.K. gal/(h·in)	m²/s	m³/(s·m)	4.971 667 E − 05
	U.S. gal/(h·in)	m²/s	m³/(s·m)	4.139 776 E − 05
	U.K. gal/(h·ft)	m²/s	m³/(s·m)	4.143 055 E − 06
	U.S. gal/(h·ft)	m²/s	m³/(s·m)	3.449 814 E − 06
Flow rate/area (mass basis)	lbm/(s·ft²)	kg/(s·m²)		4.882 428 E + 00
	lbm/(h·ft²)	kg/(s·m²)		1.356 230 E − 03
Flow rate/area (volume basis)	ft³/(s·ft²)	m/s	m³/(s·m²)	3.048° E − 01
	ft³/(min·ft²)	m/s	m³/(s·m²)	5.08° E − 03
	U.K. gal/(h·in²)	m/s	m³/(s·m²)	1.957 349 E − 03
	U.S. gal/(h·in²)	m/s	m³/(s·m²)	1.629 833 E − 03
	U.K. gal/(min·ft²)	m/s	m³/(s·m²)	8.155 621 E − 04
	U.S. gal/(min·ft²)	m/s	m³/(s·m²)	6.790 972 E − 04
	U.K. gal/(h·ft²)	m/s	m³/(s·m²)	1.359 270 E − 05
	U.S. gal/(h·ft²)	m/s	m³/(s·m²)	1.131 829 E − 05

TABLE 1-4 Conversion Factors: U.S. Customary and Commonly Used Units to SI Units (*Continued*)

Quantity	Customary or commonly used unit	SI unit	Alternate SI unit	Conversion factor; multiply customary unit by factor to obtain SI unit
		Energy, work, power		
Energy, work	therm	MJ		1.055 056 E + 02
		kJ		1.055 056 E + 05
		kWh		2.930 711 E + 01
	U.S. tonf·mi	MJ		1.431 744 E + 01
	hp·h	MJ		2.684 520 E + 00
		kJ		2.684 520 E + 03
		kWh		7.456 999 E − 01
	ch·h or CV·h	MJ		2.647 780 E + 00
		kJ		2.647 780 E + 03
		kWh		7.354 999 E − 01
	kWh	MJ		3.6° E + 00
		kJ		3.6° E + 03
	Chu	kJ		1.899 101 E + 00
		kWh		5.275 280 E − 04
	Btu	kJ		1.055 056 E + 00
		kWh		2.930 711 E − 04
	kcal	kJ		4.184° E + 00
	cal	kJ		4.184° E − 03
	ft·lbf	kJ		1.355 818 E − 03
	lbf·ft	kJ		1.355 818 E − 03
	J	kJ		1.0° E − 03
	(lbf·ft^2)/s^2	kJ		4.214 011 E − 05
	erg	J		1.0° E − 07
Impact energy	kgf·m	J		9.806 650° E + 00
	lbf·ft	J		1.355 818 E + 00
Surface energy	erg/cm^2	mJ/m^2		1.0° E + 00
Specific-impact energy	(kgf·m)/cm^2	J/cm^2		9.806 650° E − 02
	(lbf·ft)/in^2	J/cm^2		2.101 522 E − 03
Power	million Btu/h	MW		2.930 711 E − 01
	ton of refrigeration	kW		3.516 853 E + 00
	Btu/s	kW		1.055 056 E + 00
	kW	kW		1
	hydraulic horsepower—hhp	kW		7.460 43 E − 01
	hp (electric)	kW		7.46° E − 01
	hp [(550 ft·lbf/s]	kW		7.456 999 E − 01
	ch or CV	kW		7.354 999 E − 01
	Btu/min	kW		1.758 427 E − 02
	(ft·lbf)/s	kW		1.355 818 E − 03
	kcal/h	W		1.162 222 E + 00
	Btu/h	W		2.930 711 E − 01
	(ft·lbf)/min	W		2.259 697 E − 02
Power/area	Btu/(s·ft^2)	kW/m^2		1.135 653 E + 01
	cal/(h·cm^2)	kW/m^2		1.162 222 E − 02
	Btu/(h·ft^2)	kW/m^2		3.154 591 E − 03
Heat-release rate, mixing power	hp/ft^3	kW/m^3		2.633 414 E + 01
	cal/(h·cm^3)	kW/m^3		1.162 222 E + 00
	Btu/(s·ft^3)	kW/m^3		3.725 895 E + 01
	Btu/(h·ft^3)	kW/m^3		1.034 971 E − 02
Cooling duty (machinery)	Btu/(bhp·h)	W/kW		3.930 148 E − 01
Specific fuel consumption (mass basis)	lbm/(hp·h)	mg/J	kg/MJ	1.689 659 E − 01
		kg/kWh		6.082 774 E − 01
Specific fuel consumption (volume basis)	m^3/kWh	dm^3/MJ	mm^3/J	2.777 778 E + 02
	U.S. gal/(hp·h)	dm^3/MJ	mm^3/J	1.410 089 E + 00
	U.K. pt/(hp·h)	dm^3/MJ	mm^3/J	2.116 806 E − 01
Fuel consumption	U.K. gal/mi	dm^3/100 km	L/100 km	2.824 807 E + 02
	U.S. gal/mi	dm^3/100 km	L/100 km	2.352 146 E + 02
	mi/U.S. gal	km/dm^3	km/L	4.251 437 E − 01
	mi/U.K. gal	km/dm^3	km/L	3.540 064 E − 01

Quantity	Customary or commonly used unit	SI unit	Alternate SI unit	Conversion factor; multiply customary unit by factor to obtain SI unit
Velocity (linear), speed	knot	km/h		1.852° E + 00
	mi/h	km/h		1.609 344° E + 00
	ft/s	m/s		3.048° E − 01
		cm/s		3.048° E + 01
	ft/min	m/s		5.08° E − 03
	ft/h	mm/s		8.466 667 E − 02
	ft/day	mm/s		3.527 778 E − 03
		m/d		3.048° E − 01
	in/s	mm/s		2.54° E + 01
	in/min	mm/s		4.233 333 E − 01
Corrosion rate	in/year (ipy)	mm/a		2.54° E + 01
	mil/year	mm/a		2.54° E − 02
Rotational frequency	r/min	r/s		1.666 667 E − 02
		rad/s		1.047 198 E − 01
Acceleration (linear)	ft/s²	m/s²		3.048° E − 01
		cm/s²		3.048° E + 01
Acceleration (rotational)	rpm/s	rad/s²		1.047 198 E − 01
Momentum	(lbm·ft)/s	(kg·m)/s		1.382 550 E − 01
Force	U.K. tonf	kN		9.964 016 E + 00
	U.S. tonf	kN		8.896 443 E + 00
	kgf (kp)	N		9.806 650° E + 00
	lbf	N		4.448 222 E + 00
	dyn	mN		1.0 E − 02
Bending moment, torque	U.S. tonf·ft	kN·m		2.711 636 E + 00
	kgf·m	N·m		9.806 650° E + 00
	lbf·ft	N·m		1.355 818 E + 00
	lbf·in	N·m		1.129 848 E − 01
Bending moment/length	(lbf·ft)/in	(N·m)/m		5.337 866 E + 01
	(lbf·in)/in	(N·m)/m		4.448 222 E + 00
Moment of inertia	lbm·ft²	kg·m²		4.214 011 E − 02
Stress	U.S. tonf/in²	MPa	N/mm²	1.378 951 E + 01
	kgf/mm²	MPa	N/mm²	9.806 650° E + 00
	U.S. tonf/ft²	MPa	N/mm²	9.576 052 E − 02
	lbf/in² (psi)	MPa	N/mm²	6.894 757 E − 03
	lbf/ft² (psf)	kPa		4.788 026 E − 02
	dyn/cm²	Pa		1.0° E − 01
Mass/length	lbm/ft	kg/m		1.488 164 E + 00
Mass/area structural loading, bearing capacity (mass basis)	U.S. ton/ft²	Mg/m²		9.764 855 E + 00
	lbm/ft²	kg/m²		4.882 428 E + 00
	Miscellaneous transport properties			
Diffusivity	ft²/s	m²/s		9.290 304° E − 02
	m²/s	mm²/s		1.0° E + 06
	ft²/h	m²/s		2.580 64° E − 05
Thermal resistance	(°C·m²·h)/kcal	(K·m²)/kW		8.604 208 E + 02
	(°F·ft²·h)/Btu	(K·m²)/kW		1.761 102 E + 02
Heat flux	Btu/(h·ft²)	kW/m²		3.154 591 E − 03
Thermal conductivity	(cal·cm)/(s·cm²·°C)	W/(m·K)		4.184° E + 02
	(Btu·ft)/(h·ft²·°F)	W/(m·K)		1.730 735 E + 00
		(kJ·m)/(h·m²·K)		6.230 646 E + 00
	(kcal·m)/(h·m²·°C)	W/(m·K)		1.162 222 E + 00
	(Btu·in)/(h·ft²·°F)	W/(m·K)		1.442 279 E − 01
	(cal·cm)/(h·cm²·°C)	W/(m·K)		1.162 222 E − 01
Heat-transfer coefficient	cal/(s·cm²·°C)	kW/(m²·K)		4.184° E + 01
	Btu/(s·ft²·°F)	kW/(m²·K)		2.044 175 E + 01
	cal/(h·cm²·°C)	kW/(m²·K)		1.162 222 E − 02
	Btu/(h·ft²·°F)	kW/(m²·K)		5.678 263 E − 03
		kJ/(h·m²·K)		2.044 175 E + 01
	Btu/(h·ft²·°R)	kW/(m²·K)		5.678 263 E − 03
	kcal/(h·m²·°C)	kW/(m²·K)		1.162 222 E − 03

Quantity	Customary or commonly used unit	SI unit	Alternate SI unit	Conversion factor; multiply customary unit by factor to obtain SI unit
Volumetric heat-transfer coefficient	Btu/(s·ft³·°F)	kW/(m³·K)		6.706 611 E + 01
	Btu/(h·ft³·°F)	kW/(m³·K)		1.862 947 E − 02
Surface tension	dyn/cm	mN/m		1
Viscosity (dynamic)	(lbf·s)/in²	Pa·s	(N·s)/m²	6.894 757 E + 03
	(lbf·s)/ft²	Pa·s	(N·s)/m²	4.788 026 E + 01
	(kgf·s)/m²	Pa·s	(N·s)/m²	9.806 650° E + 00
	lbm/(ft·s)	Pa·s	(N·s)/m²	1.488 164 E + 00
	(dyn·s)/cm²	Pa·s	(N·s)/m²	1.0° E − 01
	cP	Pa·s	(N·s)/m²	1.0° E − 03
	lbm/(ft·h)	Pa·s	(N·s)/m²	4.133 789 E − 04
Viscosity (kinematic)	ft²/s	m²/s		9.290 304° E − 02
	in²/s	mm²/s		6.451 6° E + 02
	m²/h	mm²/s		2.777 778 E + 02
	ft²/h	m²/s		2.580 64° E − 05
	cSt	mm²/s		1
Permeability	darcy	μm²		9.869 233 E − 01
	millidarcy	μm²		9.869 233 E − 04
Thermal flux	Btu/(h·ft²)	W/m²		3.152 E + 00
	Btu/(s·ft²)	W/m²		1.135 E + 04
	cal/(s·cm²)	W/m²		4.184 E + 04
Mass-transfer coefficient	(lb·mol)/[h·ft²(lb·mol/ft³)]	m/s		8.467 E − 05
	(g·mol)/[s·m²(g·mol/L)]	m/s		1.0 E + 01
Electricity, magnetism				
Admittance	S	S		1
Capacitance	μF	μF		1
Charge density	C/mm³	C/mm³		1
Conductance	S	S		1
	℧ (mho)	S		1
Conductivity	S/m	S/m		1
	℧/m	S/m		1
	m℧/m	mS/m		1
Current density	A/mm²	A/mm²		1
Displacement	C/cm²	C/cm²		1
Electric charge	C	C		1
Electric current	A	A		1
Electric-dipole moment	C·m	C·m		1
Electric-field strength	V/m	V/m		1
Electric flux	C	C		1
Electric polarization	C/cm²	C/cm²		1
Electric potential	V	V		1
	mV	mV		1
Electromagnetic moment	A·m²	A·m²		1
Electromotive force	V	V		1
Flux of displacement	C	C		1
Frequency	cycles/s	Hz		1
Impedance	Ω	Ω		1
Linear-current density	A/mm	A/mm		1
Magnetic-dipole moment	Wb·m	Wb·m		1
Magnetic-field strength	A/mm	A/mm		1
	Oe	A/m		7.957 747 E + 01
	gamma	A/m		7.957 747 E − 04
Magnetic flux	mWb	mWb		1

TABLE 1-4   Conversion Factors: U.S. Customary and Commonly Used Units to SI Units (*Continued*)

Quantity	Customary or commonly used unit	SI unit	Alternate SI unit	Conversion factor; multiply customary unit by factor to obtain SI unit	
Magnetic-flux density	mT	mT		1	
	G	T		1.0°	E − 04
	gamma	nT		1	
Magnetic induction	mT	mT		1	
Magnetic moment	A·m²	A·m²		1	
Magnetic polarization	mT	mT		1	
Magnetic potential difference	A	A		1	
Magnetic-vector potential	Wb/mm	Wb/mm		1	
Magnetization	A/mm	A/mm		1	
Modulus of admittance	S	S		1	
Modulus of impedance	Ω	Ω		1	
Mutual inductance	H	H		1	
Permeability	μH/m	μH/m		1	
Permeance	H	H		1	
Permittivity	μF/m	μF/m		1	
Potential difference	V	V		1	
Quantity of electricity	C	C		1	
Reactance	Ω	Ω		1	
Reluctance	H⁻¹	H⁻¹		1	
Resistance	Ω	Ω		1	
Resistivity	Ω·cm	Ω·cm		1	
	Ω·m	Ω·m		1	
Self-inductance	mH	mH		1	
Surface density of change	mC/m²	mC/m²		1	
Susceptance	S	S		1	
Volume density of charge	C/mm³	C/mm³		1	
Acoustics, light, radiation					
Absorbed dose	rad	Gy		1.0°	E − 02
Acoustical energy	J	J		1	
Acoustical intensity	W/cm²	W/m²		1.0°	E + 04
Acoustical power	W	W		1	
Sound pressure	N/m²	N/m²		1.0°	
Illuminance	fc	lx		1.076 391	E + 01
Illumination	fc	lx		1.076 391	E + 01
Irradiance	W/m²	W/m²		1	
Light exposure	fc·s	lx·s		1.076 391	E + 01
Luminance	cd/m²	cd/m²		1	
Luminous efficacy	lm/W	lm/W		1	
Luminous exitance	lm/m²	lm/m²		1	
Luminous flux	lm	lm		1	
Luminous intensity	cd	cd		1	
Radiance	W/m²·sr	W/m²·sr		1	
Radiant energy	J	J		1	
Radiant flux	W	W		1	
Radiant intensity	W/sr	W/sr		1	
Radiant power	W	W		1	

TABLE 1-4 Conversion Factors: U.S. Customary and Commonly Used Units to SI Units *(Concluded)*

Quantity	Customary or commonly used unit	SI unit	Alternate SI unit	Conversion factor; multiply customary unit by factor to obtain SI unit	
Wavelength	Å	nm		1.0°	E − 01
Capture unit	10^{-3} cm^{-1}	m^{-1}		1.0°	E + 01
			10^{-3} cm^{-1}	1	
	m^{-1}	m^{-1}		1	
Radioactivity	Ci	Bq		3.7°	E + 10

° An asterisk indicates that the conversion factor is exact.

† Conversion factors for length, area, and volume are based on the international foot. The international foot is longer by 2 parts in 1 million than the U.S. Survey foot (land-measurement use).

NOTE: The following unit symbols are used in the table:

Unit symbol	Name	Unit symbol	Name
A	ampere	lm	lumen
a	annum (year)	lx	lux
Bq	becquerel	m	meter
C	coulomb	min	minute
cd	candela	′	minute
Ci	curie	N	newton
d	day	naut mi	U.S. nautical mile
°C	degree Celsius	Oe	oersted
°	degree	Ω	ohm
dyn	dyne	Pa	pascal
F	farad	rad	radian
fc	footcandle	r	revolution
G	gauss	S	siemens
g	gram	s	second
gr	grain	″	second
Gy	gray	sr	steradian
H	henry	St	stokes
h	hour	T	tesla
ha	hectare	t	tonne
Hz	hertz	V	volt
J	joule	W	watt
K	kelvin	Wb	weber
L, ℓ, l	liter		

TABLE 1-5 Metric Conversion Factors as Exact Numerical Multiples of SI Units

The first two digits of each numerical entry represent a power of 10. For example, the entry "−02 2.54" expresses the fact that 1 in = 2.54 × 10⁻² m.

To convert from	To	Multiply by	To convert from	To	Multiply by
abampere	ampere	+01 1.00	fluid ounce (U.S.)	meter³	−05 2.957 352
abcoulomb	coulomb	+01 1.00	foot	meter	−01 3.048
abfarad	farad	+09 1.00	foot (U.S. survey)	meter	−01 3.048 006
abhenry	henry	−09 1.00	foot of water (39.2°F)	newton/meter²	+03 2.988 98
abmho	mho	+09 1.00	footcandle	lumen/meter²	+01 1.076 391
abohm	ohm	−09 1.00	footlambert	candela/meter²	+00 3.426 259
abvolt	volt	−08 1.00	furlong	meter	+02 2.011 68
acre	meter²	+03 4.046 856	gal (galileo)	meter/second²	−02 1.00
ampere (international of 1948)	ampere	−01 9.998 35	gallon (U.K. liquid)	meter³	−03 4.546 087
			gallon (U.S. dry)	meter³	−03 4.404 883
angstrom	meter	−10 1.00	gallon (U.S. liquid)	meter³	−03 3.785 411
are	meter²	+02 1.00	gamma	tesla	−09 1.00
astronomical unit	meter	+11 1.495 978	gauss	tesla	−04 1.00
atmosphere	newton/meter²	+05 1.013 25	gilbert	ampere turn	−01 7.957 747
bar	newton/meter²	+05 1.00	gill (U.K.)	meter³	−04 1.420 652
barn	meter²	−28 1.00	gill (U.S.)	meter³	−04 1.182 941
barrel (petroleum 42 gal)	meter³	−01 1.589 873	grad	degree (angular)	−01 9.00
barye	newton/meter²	−01 1.00	grad	radian	−02 1.570 796
British thermal unit (ISO/TC 12)	joule	+03 1.055 06	grain	kilogram	−05 6.479 891
			gram	kilogram	−03 1.00
British thermal unit (International Steam Table)	joule	+03 1.055 04	hand	meter	−01 1.016
			hectare	meter²	+04 1.00
British thermal unit (mean)	joule	+03 1.055 87	henry (international of 1948)	henry	+00 1.000 495
British thermal unit (thermochemical)	joule	+03 1.054 350	hogshead (U.S.)	meter³	−01 2.384 809
British thermal unit (39°F)	joule	+03 1.059 67	horsepower (550 ft lbf/s)	watt	+02 7.456 998
British thermal unit (60°F)	joule	+03 1.054 68	horsepower (boiler)	watt	+03 9.809 50
bushel (U.S.)	meter³	−02 3.523 907	horsepower (electric)	watt	+02 7.46
cable	meter	+02 2.194 56	horsepower (metric)	watt	+02 7.354 99
caliber	meter	−04 2.54	horsepower (U.K.)	watt	+02 7.457
calorie (International Steam Table)	joule	+00 4.1868	horsepower (water)	watt	+02 7.460 43
			hour (mean solar)	second (mean solar)	+03 3.60
calorie (mean)	joule	+00 4.190 02	hour (sidereal)	second (mean solar)	+03 3.590 170
calorie (thermochemical)	joule	+00 4.184	hundredweight (long)	kilogram	+01 5.080 234
calorie (15°C)	joule	+00 4.185 80	hundredweight (short)	kilogram	+01 4.535 923
calorie (20°C)	joule	+00 4.181 90	inch	meter	−02 2.54
calorie (kilogram, International Steam Table)	joule	+03 4.186 8	inch of mercury (32°F)	newton/meter²	+03 3.386 389
			inch of mercury (60°F)	newton/meter²	+03 3.376 85
calorie (kilogram, mean)	joule	+03 4.190 02	inch of water (39.2°F)	newton/meter²	+02 2.490 82
calorie (kilogram, thermochemical)	joule	+03 4.184	inch of water (60°F)	newton/meter²	+02 2.4884
			joule (international of 1948)	joule	+00 1.000 165
carat (metric)	kilogram	−04 2.00	kayser	1/meter	+02 1.00
Celsius (temperature)	kelvin	$t_K = t_c + 273.15$	kilocalorie (International Steam Table)	joule	+03 4.186 74
centimeter of mercury (0°C)	newton/meter²	+03 1.333 22	kilocalorie (mean)	joule	+03 4.190 02
centimeter of water (4°C)	newton/meter²	+01 9.806 38	kilocalorie (thermochemical)	joule	+03 4.184
chain (engineer's)	meter	+01 3.048	kilogram mass	kilogram	+00 1.00
chain (surveyor's or Gunter's)	meter	+01 2.011 68	kilogram-force (kgf)	newton	+00 9.806 65
			kilopond-force	newton	+00 9.806 65
circular mil	meter²	−10 5.067 074	kip	newton	+03 4.448 221
cord	meter³	+00 3.624 556	knot (international)	meter/second	−01 5.144 444
coulomb (international of 1948)	coulomb	−01 9.998 35	lambert	candela/meter²	+04 1/π
			lambert	candela/meter²	+03 3.183 098
cubit	meter	−01 4.572	langley	joule/meter²	+04 4.184
cup	meter³	−04 2.365 882	lbf (pound-force, avoirdupois)	newton	+00 4.448 221
curie	disintegration/second	+10 3.70			
day (mean solar)	second (mean solar)	+04 8.64	lbm (pound-mass, avoirdupois)	kilogram	−01 4.535 923
day (sidereal)	second (mean solar)	+04 8.616 409	league (British nautical)	meter	+03 5.559 552
degree (angle)	radian	−02 1.745 329	league (international nautical)	meter	+03 5.556
denier (international)	kilogram/meter	−07 1.111 111			
dram (avoirdupois)	kilogram	−03 1.771 845	league (statute)	meter	+03 4.828 032
dram (troy or apothecary)	kilogram	−03 3.887 934	light-year	meter	+15 9.460 55
dram (U.S. fluid)	meter³	−06 3.696 691	link (engineer's)	meter	−01 3.048
dyne	newton	−05 1.00	link (surveyor's or Gunter's)	meter	−01 2.011 68
electron volt	joule	−19 1.602 10	liter	meter³	−03 1.00
erg	joule	−07 1.00	lux	lumen/meter²	+00 1.00
Fahrenheit (temperature)	kelvin	$t_K = (5/9)(t_F + 459.67)$	maxwell	weber	−08 1.00
			meter	wavelengths Kr 86	+06 1.650 763
Fahrenheit (temperature)	Celsius	$t_c = (5/9)(t_F - 32)$	micrometer	meter	−06 1.00
farad (international of 1948)	farad	−01 9.995 05	mil	meter	−05 2.54
faraday (based on carbon 12)	coulomb	+04 9.648 70	mile (U.S. statute)	meter	+03 1.609 344
			mile (U.K. nautical)	meter	+03 1.853 184
faraday (chemical)	coulomb	+04 9.649 57	mile (international nautical)	meter	+03 1.852
faraday (physical)	coulomb	+04 9.652 19	mile (U.S. nautical)	meter	+03 1.852
fathom	meter	+00 1.828 8	millibar	newton/meter²	+02 1.00
fermi (femtometer)	meter	−15 1.00	millimeter of mercury (0°C)	newton/meter²	+02 1.333 224

To convert from	To	Multiply by	To convert from	To	Multiply by
minute (angle)	radian	−04 2.908 882	second (ephemeris)	second	+00 1.000 000
minute (mean solar)	second (mean solar)	+01 6.00	second (mean solar)	second (ephemeris)	Consult
minute (sidereal)	second (mean solar)	+01 5.983 617			American
month (mean calendar)	second (mean solar)	+06 2.628			Ephemeris
nautical mile (international)	meter	+03 1.852			and Nautical
nautical mile (U.S.)	meter	+03 1.852			Almanac
nautical mile (U.K.)	meter	+03 1.853 184	second (sidereal)	second (mean solar)	−01 9.972 695
oersted	ampere/meter	+01 7.957 747	section	meter²	+06 2.589 988
ohm (international of 1948)	ohm	+00 1.000 495	scruple (apothecary)	kilogram	−03 1.295 978
ounce-force (avoirdupois)	newton	−01 2.780 138	shake	second	−08 1.00
ounce-mass (avoirdupois)	kilogram	−02 2.834 952	skein	meter	+02 1.097 28
ounce-mass (troy or apothecary)	kilogram	−02 3.110 347	slug	kilogram	+01 1.459 390
ounce (U.S. fluid)	meter³	−05 2.957 352	span	meter	−01 2.286
pace	meter	−01 7.62	statampere	ampere	−10 3.335 640
parsec	meter	+16 3.083 74	statcoulomb	coulomb	−10 3.335 640
pascal	newton/meter²	+00 1.00	statfarad	farad	−12 1.112 650
peck (U.S.)	meter³	−03 8.809 767	stathenry	henry	+11 8.987 554
pennyweight	kilogram	−03 1.555 173	statmho	mho	−12 1.112 650
perch	meter	+00 5.0292	statohm	ohm	+11 8.987 554
phot	lumen/meter²	+04 1.00	statute mile (U.S.)	meter	+03 1.609 344
pica (printer's)	meter	−03 4.217 517	statvolt	volt	+02 2.997 925
pint (U.S. dry)	meter³	−04 5.506 104	stere	meter³	+00 1.00
pint (U.S. liquid)	meter³	−04 4.731 764	stilb	candela/meter²	+04 1.00
point (printer's)	meter	−04 3.514 598	stoke	meter²/second	−04 1.00
poise	(newton-second)/meter²	−01 1.00	tablespoon	meter³	−05 1.478 676
pole	meter	+00 5.0292	teaspoon	meter³	−06 4.928 921
pound-force (lbf avoirdupois)	newton	+00 4.448 221	ton (assay)	kilogram	−02 2.916 666
			ton (long)	kilogram	+03 1.016 046
pound-mass (lbm avoirdupois)	kilogram	−01 4.535 923	ton (metric)	kilogram	+03 1.00
			ton (nuclear equivalent of TNT)	joule	+09 4.20
pound-mass (troy or apothecary)	kilogram	−01 3.732 417	ton (register)	meter³	+00 2.831 684
			ton (short, 2000 lb)	kilogram	+02 9.071 847
poundal	newton	−01 1.382 549	tonne	kilogram	+03 1.00
quart (U.S. dry)	meter³	−03 1.101 220	torr (0°C)	newton/meter²	+02 1.333 22
quart (U.S. liquid)	meter³	−04 9.463 529	township	meter²	+07 9.323 957
rad (radiation dose absorbed)	joule/kilogram	−02 1.00	unit pole	weber	−07 1.256 637
			volt (international of 1948)	volt	+00 1.000 330
Rankine (temperature)	kelvin	$t_K = (5/9)t_R$	watt (international of 1948)	watt	+00 1.000 165
rayleigh (rate of photon emission)	1/second-meter²	+10 1.00	yard	meter	−01 9.144
			year (calendar)	second (mean solar)	+07 3.1536
rhe	meter²/(newton-second)	+01 1.00	year (sidereal)	second (mean solar)	+07 3.155 815
			year (tropical)	second (mean solar)	+07 3.155 692
rod	meter	+00 5.0292	year 1900, tropical, Jan., day 0, hour 12	second (ephemeris)	+07 3.155 692
roentgen	coulomb/kilogram	−04 2.579 76			
rutherford	disintegration/second	+06 1.00	year 1900, tropical, Jan., day 0, hour 12	second	+07 3.155 692
second (angle)	radian	−06 4.848 136			

TABLE 1-6 Alphabetical Listing of Common Conversions

To convert from	To	Multiply by
Acres	Square feet	43,560
Acres	Square meters	4074
Acres	Square miles	0.001563
Acre-feet	Cubic meters	1233
Ampere-hours (absolute)	Coulombs (absolute)	3600
Angstrom units	Inches	3.937×10^{-9}
Angstrom units	Meters	1×10^{-10}
Angstrom units	Microns	1×10^{-4}
Atmospheres	Millimeters of mercury at 32°F	760
Atmospheres	Dynes per square centimeter	1.0133×10^{6}
Atmospheres	Newtons per square meter	101,325
Atmospheres	Feet of water at 39.1°F	33.90
Atmospheres	Grams per square centimeter	1033.3
Atmospheres	Inches of mercury at 32°F	29.921
Atmospheres	Pounds per square foot	2116.3
Atmospheres	Pounds per square inch	14.696
Bags (cement)	Pounds (cement)	94
Barrels (cement)	Pounds (cement)	376
Barrels (oil)	Cubic meters	0.15899
Barrels (oil)	Gallons	42
Barrels (U.S. liquid)	Cubic meters	0.11924
Barrels (U.S. liquid)	Gallons	31.5
Barrels per day	Gallons per minute	0.02917
Bars	Atmospheres	0.9869
Bars	Newtons per square meter	1×10^{5}
Bars	Pounds per square inch	14.504
Board feet	Cubic feet	1/12
Boiler horsepower	B.t.u. per hour	33,480
Boiler horsepower	Kilowatts	9.803
B.t.u.	Calories (gram)	252
B.t.u.	Centigrade heat units (c.h.u. or p.c.u.)	0.55556
B.t.u.	Foot-pounds	777.9
B.t.u.	Horsepower-hours	3.929×10^{-4}
B.t.u.	Joules	1055.1
B.t.u.	Liter-atmospheres	10.41
B.t.u.	Pounds carbon to CO_2	6.88×10^{-5}
B.t.u.	Pounds water evaporated from and at 212°F	0.001036
B.t.u.	Cubic foot-atmospheres	0.3676
B.t.u.	Kilowatt-hours	2.930×10^{-4}
B.t.u. per cubic foot	Joules per cubic meter	37,260
B.t.u. per hour	Watts	0.29307
B.t.u. per minute	Horsepower	0.02357
B.t.u. per pound	Joules per kilogram	2326
B.t.u. per pound per degree Fahrenheit	Calories per gram per degree centigrade	1
B.t.u. per pound per degree Fahrenheit	Joules per kilogram per degree Kelvin	4186.8
B.t.u. per second	Watts	1054.4
B.t.u. per square foot per hour	Joules per square meter per second	3.1546
B.t.u. per square foot per minute	Kilowatts per square foot	0.1758
B.t.u. per square foot per second for a temperature gradient of 1°F. per inch	Calories, gram (15°C.), per square centimeter per second for a temperature gradient of 1°C. per centimeter	1.2405
B.t.u. (60°F.) per degree Fahrenheit	Calories per degree centigrade	453.6
Bushels (U.S. dry)	Cubic feet	1.2444
Bushels (U.S. dry)	Cubic meters	0.03524
Calories, gram	B.t.u.	3.968×10^{-3}
Calories, gram	Foot-pounds	3.087
Calories, gram	Joules	4.1868
Calories, gram	Liter-atmospheres	4.130×10^{-2}
Calories, gram	Horsepower-hours	1.5591×10^{-6}
Calories, gram, per gram per degree C.	Joules per kilogram per degree Kelvin	4186.8
Calories, kilogram	Kilowatt-hours	0.0011626
Calories, kilogram per second	Kilowatts	4.185
Candle power (spherical)	Lumens	12.556
Carats (metric)	Grams	0.2
Centigrade heat units	B.t.u.	1.8
Centimeters	Angstrom units	1×10^{8}
Centimeters	Feet	0.03281
Centimeters	Inches	0.3937
Centimeters	Meters	0.01
Centimeters	Microns	10,000
Centimeters of mercury at 0°C.	Atmospheres	0.013158
Centimeters of mercury at 0°C.	Feet of water at 39.1°F.	0.4460
Centimeters of mercury at 0°C.	Newtons per square meter	1333.2
Centimeters of mercury at 0°C.	Pounds per square foot	27.845
Centimeters of mercury at 0°C.	Pounds per square inch	0.19337
Centimeters per minute	Feet per minute	1.9685
Centimeters of water at 4°C.	Newtons per square meter	98.064
Centistokes	Square meters per second	1×10^{-6}
Circular mils	Square centimeters	5.067×10^{-6}
Circular mils	Square inches	7.854×10^{-7}
Circular mils	Square mils	0.7854
Cords	Cubic feet	128
Cubic centimeters	Cubic feet	3.532×10^{-5}
Cubic centimeters	Gallons	2.6417×10^{-4}
Cubic centimeters	Ounces (U.S. fluid)	0.03381
Cubic centimeters	Quarts (U.S. fluid)	0.0010567
Cubic feet	Bushels (U.S.)	0.8036
Cubic feet	Cubic centimeters	28,317
Cubic feet	Cubic meters	0.028317
Cubic feet	Cubic yards	0.03704
Cubic feet	Gallons	7.481
Cubic feet	Liters	28.316
Cubic foot-atmospheres	Foot-pounds	2116.3
Cubic foot-atmospheres	Liter-atmospheres	28.316
Cubic feet of water (60°F.)	Pounds	62.37
Cubic feet per minute	Cubic centimeters per second	472.0
Cubic feet per minute	Gallons per second	0.1247
Cubic feet per second	Gallons per minute	448.8
Cubic feet per second	Million gallons per day	0.64632
Cubic inches	Cubic meters	1.6387×10^{-5}
Cubic yards	Cubic meters	0.76456
Curies	Disintegrations per minute	2.2×10^{12}
Curies	Coulombs per minute	1.1×10^{12}
Degrees	Radians	0.017453
Drams (apothecaries' or troy)	Grams	3.888

TABLE 1-6 Alphabetical Listing of Common Conversions (*Concluded*)

To convert from	To	Multiply by
Drams (avoirdupois)	Grams	1.7719
Dynes	Newtons	1×10^{-5}
Ergs	Joules	1×10^{-7}
Faradays	Coulombs (abs.)	96,500
Fathoms	Feet	6
Feet	Meters	0.3048
Feet per minute	Centimeters per second	0.5080
Feet per minute	Miles per hour	0.011364
Feet per (second)^2	Meters per (second)^2	0.3048
Feet of water at 39.2°F.	Newtons per square meter	2989
Foot-poundals	B.t.u.	3.995×10^{-5}
Foot-poundals	Joules	0.04214
Foot-poundals	Liter-atmospheres	4.159×10^{-4}
Foot-pounds	B.t.u.	0.0012856
Foot-pounds	Calories, gram	0.3239
Foot-pounds	Foot-poundals	32.174
Foot-pounds	Horsepower-hours	5.051×10^{-7}
Foot-pounds	Kilowatt-hours	3.766×10^{-7}
Foot-pounds	Liter-atmospheres	0.013381
Foot-pounds force	Joules	1.3558
Foot-pounds per second	Horsepower	0.0018182
Foot-pounds per second	Kilowatts	0.0013558
Furlongs	Miles	0.125
Gallons (U.S. liquid)	Barrels (U.S. liquid)	0.03175
Gallons	Cubic meters	0.003785
Gallons	Cubic feet	0.13368
Gallons	Gallons (Imperial)	0.8327
Gallons	Liters	3.785
Gallons	Ounces (U.S. fluid)	128
Gallons per minute	Cubic feet per hour	8.021
Gallons per minute	Cubic feet per second	0.002228
Grains	Grams	0.06480
Grains	Pounds	1/7000
Grains per cubic foot	Grams per cubic meter	2.2884
Grains per gallon	Parts per million	17.118
Grams	Drams (avoirdupois)	0.5644
Grams	Drams (troy)	0.2572
Grams	Grains	15.432
Grams	Kilograms	0.001
Grams	Pounds (avoirdupois)	0.0022046
Grams	Pounds (troy)	0.002679
Grams per cubic centimeter	Pounds per cubic foot	62.43

To convert from	To	Multiply by
Horsepower (British)	Pounds water evaporated per hour at 212°F	2.64
Horsepower (metric)	Foot-pounds per second	542.47
Horsepower (metric)	Kilogram-meters per second	75.0
Hours (mean solar)	Seconds	3600
Inches	Meters	0.0254
Inches of mercury at 60°F	Newtons per square meter	3376.9
Inches of water at 60°F	Newtons per square meter	248.84
Joules (absolute)	B.t.u. (mean)	9.480×10^{-4}
Joules (absolute)	Calories, gram (mean)	0.2389
Joules (absolute)	Cubic foot-atmospheres	0.3485
Joules (absolute)	Foot-pounds	0.7376
Joules (absolute)	Kilowatt-hours	2.7778×10^{-7}
Joules (absolute)	Liter-atmospheres	0.009869
Kilocalories	Joules	4186.8
Kilograms	Pounds (avoirdupois)	2.2046
Kilograms force	Newtons	9.807
Kilograms per square centimeter	Pounds per square inch	14.223
Kilometers	Miles	0.6214
Kilowatt-hours	B.t.u.	3414
Kilowatt-hours	Foot-pounds	2.6552×10^6
Kilowatts	Horsepower	1.3410
Knots (international)	Meters per second	0.5144
Knots (nautical miles per hour)	Miles per hour	1.1516
Lamberts	Candles per square inch	2.054
Liter-atmospheres	Cubic foot-atmospheres	0.03532
Liter-atmospheres	Foot-pounds	74.74
Liters	Cubic feet	0.03532
Liters	Cubic meters	0.001
Liters	Gallons	0.26418
Lumens	Watts	0.001496
Micromicrons	Microns	1×10^{-6}
Microns	Angstrom units	1×10^4
Miles (nautical)	Feet	6080
Miles (nautical)	Miles (U.S. statute)	1.1516
Miles	Feet	5280
Miles	Meters	1609.3
Miles per hour	Feet per second	1.4667
Miles per hour	Meters per second	0.4470
Milliliters	Cubic centimeters	1
Millimeters	Meters	0.001

To convert	Into	Multiply by
Grams per cubic centimeter	Pounds per gallon	8.345
Grams per liter	Grams per gallon	58.42
Grams per liter	Pounds per cubic foot	0.0624
Grams per square centimeter	Pounds per square foot	2.0482
Grams per square centimeter	Pounds per square inch	0.014223
Hectares	Acres	2.471
Hectares	Square meters	10,000
Horsepower (British)	B.t.u. per minute	42.42
Horsepower (British)	B.t.u. per hour	2545
Horsepower (British)	Foot-pounds per minute	33,000
Horsepower (British)	Foot-pounds per second	550
Horsepower (British)	Watts	745.7
Horsepower (British)	Horsepower (metric)	1.0139
Horsepower (British)	Pounds carbon to CO_2 per hour	0.175
Pounds (avoirdupois)	Grains	7000
Pounds (avoirdupois)	Kilograms	0.45359
Pounds (avoirdupois)	Pounds (troy)	1.2153
Pounds per cubic foot	Grams per cubic centimeter	0.016018
Pounds per cubic foot	Kilograms per cubic meter	16.018
Pounds per square foot	Atmospheres	4.725×10^{-4}
Pounds per square foot	Kilograms per square meter	4.882
Pounds per square inch	Atmospheres	0.06805
Pounds per square inch	Kilograms per square centimeter	0.07031
Pounds per square inch	Newtons per square meter	6894.8
Pounds force	Newtons	4.4482
Pounds force per square foot	Newtons per square meter	47.88
Pounds water evaporated from and at 212°F.	Horsepower-hours	0.379
Pound-centigrade units (p.c.u.)	B.t.u.	1.8
Quarts (U.S. liquid)	Cubic meters	9.464×10^{-4}
Radians	Degrees	57.30
Revolutions per minute	Radians per second	0.10472
Seconds (angle)	Radians	4.848×10^{-6}
Slugs	Gee pounds	1
Slugs	Kilograms	14.594
Slugs	Pounds	32.17
Square centimeters	Square feet	0.0010764
Millimeters of mercury at 0°C.	Newtons per square meter	133.32
Millimicrons	Microns	0.001
Mils	Inches	0.001
Mils	Meters	2.54×10^{-5}
Minims (U.S.)	Cubic centimeters	0.06161
Minutes (angle)	Radians	2.909×10^{-4}
Minutes (mean solar)	Seconds	60
Newtons	Kilograms	0.10197
Ounces (avoirdupois)	Kilograms	0.02835
Ounces (avoirdupois)	Ounces (troy)	0.9115
Ounces (U.S. fluid)	Cubic meters	2.957×10^{-5}
Ounces (troy)	Ounces (apothecaries')	1.000
Pints (U.S. liquid)	Cubic meters	4.732×10^{-4}
Poundals	Newtons	0.13826
Square feet	Square meters	0.0929
Square feet per hour	Square meters per second	2.581×10^{-5}
Square inches	Square centimeters	6.452
Square inches	Square meters	6.452×10^{-4}
Square yards	Square meters	0.8361
Stokes	Square meters per second	1×10^{-4}
Tons (long)	Kilograms	1016
Tons (long)	Pounds	2240
Tons (metric)	Kilograms	1000
Tons (metric)	Pounds	2204.6
Tons (metric)	Tons (short)	1.1023
Tons (short)	Kilograms	907.18
Tons (short)	Pounds	2000
Tons (refrigeration)	B.t.u. per hour	12,000
Tons (British shipping)	Cubic feet	42.00
Tons (U.S. shipping)	Cubic feet	40.00
Torr (mm. mercury, 0°C.)	Newtons per square meter	133.32
Watts	B.t.u. per hour	3.413
Watts	Joules per second	1
Watt-hours	Kilogram-meters per second	0.10197
Watt-hours	Joules	3600
Yards	Meters	0.9144

TABLE 1-7 Common Units and Conversion Factors*

Mass (M)

1 pound mass	= 453.5924 grams
	= 0.45359 kilograms
	= 7000 grains
1 slug	= 32.174 pounds mass
1 ton (short)	= 2000 pounds mass
1 ton (long)	= 2240 pounds mass
1 ton (metric)	= 1000 kilograms
	= 2204.62 pounds mass
1 pound mole	= 453.59 gram moles

Length (L)

1 foot	= 30.480 centimeters
	= 0.3048 meters
1 inch	= 2.54 centimeters
	= 0.0254 meters
1 mile (U.S.)	= 1.60935 kilometers
1 yard	= 0.9144 meters

Area (L²)

1 square foot	= 929.0304 square centimeters
	= 0.09290304 square meters
1 square inch	= 6.4516 square centimeters
1 square yard	= 0.836127 square meters

Volume (L³)

1 cubic foot	= 28,316.85 cubic centimeters
	= 0.02831685 cubic meters
	= 28.31685 liters
	= 7.481 gallons (U.S.)
1 gallon	= 3.7853 liters
	= 231 cubic inches

Time (θ)

1 hour	= 60 minutes
	= 3600 seconds

Temperature (T)

1 centigrade or Celsius degree	= 1.8 Fahrenheit degree
Temperature, Kelvin	$= T°C + 273.15$
Temperature, Rankine	$= T°F + 459.7$
Temperature, Fahrenheit	$= 9/5\ T°C + 32$
Temperature, centigrade or Celsius	$= 5/9\ (T°F - 32)$
Temperature, Rankine	$= 1.8\ T\ K$

Force (F)

1 pound force	= 444,822.2 dynes
	= 4.448222 Newtons
	= 32.174 poundals

Pressure (F/L²)
Normal atmospheric pressure

1 atm	= 760 millimeters of mercury at 0°C (density 13.5951 g/cm³)
	= 29.921 inches of mercury at 32°F
	= 14.696 pounds force/square inch
	= 33.899 feet of water at 39.1°F
	$= 1.01325 \times 10^6$ dynes/square centimeter
	$= 1.01325 \times 10^5$ Newtons/square meter

Density (M/L³)

1 pound mass/cubic foot	= 0.01601846 grams/cubic centimeter
	= 16.01846 kilogram/cubic meter

Energy (H or FL)

1 British thermal unit	= 251.98 calories
	= 1054.4 joules
	= 777.97 foot-pounds force
	= 10.409 liter-atmospheres
	= 0.2930 watt-hour

Diffusivity (L²/θ)

1 square foot/hour	= 0.258 cm²/s
	$= 2.58 \times 10^{-5}$ m²/s

Viscosity (M/Lθ)

1 pound mass/foot hour	= 0.00413 g/cm s
	0.000413 kg/m s
1 centipoise	= 0.01 poise
	= 0.01 g/cm s
	= 0.001 kg/m s
	= 0.000672 lbm/ft s
	= 0.0000209 lb_f s/ft²

Thermal conductivity [H/θL²(T/L)]

1 Btu/hr ft² (°F/ft)	= 0.00413 cal/s cm² (°C/cm)
	= 1.728 J/s m² (°C/m)

Heat transfer coefficient

1 Btu/hr ft² °F	= 5.678 J/s m² °C

Heat capacity (H/MT)

1 Btu/lbm °F	= 1 cal/g °C
	= 4184 J/kg °C

Gas constant

1.987 Btu/lbm mole °R	= 1.987 cal/mol K
	= 82.057 atm cm³/mol K
	= 0.7302 atm ft³/lb mole °F
	= 10.73 (lb_f/in.²) (ft³)/lb mole °R
	= 1545 (lb_f/ft²) (ft³)/lb mole °R
	= 8.314 (N/m²) (m³)/mol K

Gravitational acceleration

g	= 9.8066 m/s²
	= 32.174 ft/s²

NOTE: U.S. customary units; or British units, on left and SI units on right.
*Adapted from Faust et al., *Principles of Unit Operations*, John Wiley and Sons, 1980.

TABLE 1-8 Kinematic-Viscosity Conversion Formulas

Viscosity scale	Range of t, sec	Kinematic viscosity, stokes
Saybolt Universal	$32 < t < 100$	$0.00226t - 1.95/t$
	$t > 100$	$0.00220t - 1.35/t$
Saybolt Furol	$25 < t < 40$	$0.0224t - 1.84/t$
	$t > 40$	$0.0216t - 0.60/t$
Redwood No. 1	$34 < t < 100$	$0.00260t - 1.79/t$
	$t > 100$	$0.00247t - 0.50/t$
Redwood Admiralty		$0.027t - 20/t$
Engler		$0.00147t - 3.74/t$

TABLE 1-9 Values of the Gas-Law Constant

Temp. scale	Press. units	Vol. units	Wt. units	Energy units	R
Kelvin			g-moles	calories	1.9872
			g-moles	joules (abs)	8.3144
			g-moles	joules (int)	8.3130
	atm.	cm³	g-moles	atm cm³	82.057
	atm.	liters	g-moles	atm liters	0.08205
	mm. Hg	liters	g-moles	mm Hg-liters	62.361
	bar	liters	g-moles	bar-liters	0.08314
	kg/cm²	liters	g-moles	kg/(cm²)(liters)	0.08478
	atm	ft³	lb-moles	atm-ft³	1.314
	mm Hg	ft³	lb-moles	mm Hg-ft³	998.9
			lb-moles	chu or pcu	1.9872
Rankine			lb-moles	Btu	1.9872
			lb-moles	hp-hr	0.0007805
			lb-moles	kw-hr	0.0005819
	atm	ft³	lb-moles	atm-ft³	0.7302
	in Hg	ft³	lb-moles	in Hg-ft³	21.85
	mm Hg	ft³	lb-moles	mm Hg-ft³	555.0
	lb/in²abs	ft³	lb-moles	(lb)(ft³)/in²	10.73
	lb/ft²abs	ft³	lb-moles	ft-lb	1545.0

TABLE 1-10 United States Customary System of Weights and Measures

Linear Measure
12 inches (in) or (″) = 1 foot (ft) or (′)
3 feet = 1 yard (yd)

$\left.\begin{array}{l}16.5 \text{ feet} \\ 5.5 \text{ yards}\end{array}\right\} = 1 \text{ rod (rd)}$

$\left.\begin{array}{l}5280 \text{ feet} \\ 320 \text{ rods}\end{array}\right\} = 1 \text{ mile (mi)}$

1 mil = 0.001 inch

Nautical:

6080.2 feet = 1 nautical mile
6 feet = 1 fathom
120 fathoms = 1 cable length
1 knot = 1 nautical mile per hour
60 nautical miles = 1° of latitude

Square Measure
144 sq. inches (sq. in) or (in²) or (□″) = 1 sq. foot (ft²) or (□′)
9 sq. feet (ft²) (□′) = 1 sq. yard (yd²)
30.25 sq. yards = 1 sq. rod, pole, or perch

$160 \text{ sq. rods} = \left\{\begin{array}{l}10 \text{ sq. chains} \\ 43{,}560 \text{ sq. ft}\end{array}\right\} = 1 \text{ acre}$

640 acres = 1 sq. mile = 1 section
1 circular inch (area of
circle of 1 inch diameter) = 0.7854 sq. inch
1 sq. inch = 1.2732 circular inch
1 circular mil = area of circle of 0.001
inch diameter
1,000,000 circular mils = 1 circular inch

Circular Measure
60 seconds (″) (sec) = 1 minute (min) or (′)
60 minutes (′) = 1 degree (°)
90 degrees (°) = 1 quadrant
360 degrees (°) = 1 circumference

$57.29578 \text{ degrees} \left\{\begin{array}{l} = 1 \text{ radian (rad.)} \\ = 57° \, 17' \, 44.81''\end{array}\right.$

Volume Measure

Solid:
1728 cubic in (cu. in) (in³) = 1 cubic foot (cu. ft)(ft³)
27 cu. ft = 1 cubic yard (cu. yd)

Dry Measure:
2 pints = 1 quart
8 quarts = 1 peck
4 pecks = 1 bushel
1 United States
Winchester bushel = 2150.42 cubic inches

Liquid:
4 gills = 1 pint (pt)
2 pints = 1 quart (qt)
4 quarts = 1 gallon (gal)
7.4805 gallons = 1 cubic foot

Apothecaries' Liquid:
60 minims (min. or ♏) = 1 fluid dram or drachm
8 drams (℥) = 1 fluid ounce
16 ounces (oz. ℥) = 1 pint

Avoirdupois Weight
16 drams = 437.5 grains = 1 ounce (oz)
16 ounces = 7000 grains = 1 pound (lb)
100 pounds = 1 hundredweight (cwt)
2000 pounds = 1 short ton: 2240 pounds = 1 long ton

Troy Weight
24 grains = 1 pennyweight (dwt)
20 pennyweights = 1 ounce (oz)
12 ounces = 1 pound (lb)

Apothecaries' Weight
20 grains (gr) = 1 scruple (℈)
3 scruples = 1 dram (℈)
8 drams = 1 ounce (℥)
12 ounces = 1 pound (lb)

TABLE 1-11 Temperature Conversion Formulas

$°F = (°C \times 5/9) + 32$
$°C = (°F - 32) \times 5/9$
$°R = °F + 459.69$
$°K = °C + 273.15$
$°K = °R \times 5/9$

Temperature difference, ΔT
$°F = °C \times 9/5$

NOTE: An extensive table of temperature conversions may be found in the sixth edition of the *Handbook* (Table 1-12).

TABLE 1-12 Specific Gravity, Degrees Baumé, Degrees API, Degrees Twaddell, Pounds per Gallon, Pounds per Cubic Foot*

$$°Bé = 145 - \frac{145}{sp\ gr} \text{ (heavier than } H_2O\text{)};\quad °Bé = \frac{140}{sp\ gr} - 130 \text{ (lighter than } H_2O\text{)};\quad °Tw = \frac{sp\ gr\ 60°/60°F - 1}{0.005}\quad °API = \frac{141.5}{sp\ gr} - 131.5$$

Sp gr 60°/60°	°Bé	°API	Lb per gal at 60°F wt in air	Lb per ft³ at 60°F wt in air	Sp gr 60°/60°	°Bé	°API	Lb per gal at 60°F wt in air	Lb per ft³ at 60°F wt in air	Sp gr 60°/60°	°Bé	°API	Lb per gal at 60°F wt in air	Lb per ft³ at 60°F wt in air	Sp gr 60°/60°	°Bé	°API	Lb per gal at 60°F wt in air	Lb per ft³ at 60°F. wt. in air
0.600	103.33	104.33	4.9929	37.350	0.700	70.00	70.64	5.8268	43.587	0.800	45.00	45.38	6.6606	49.825	0.900	25.56	25.72	7.4944	56.062
.605	101.40	102.38	5.0346	37.662	.705	68.58	69.21	5.8685	43.899	.805	43.91	44.28	6.7023	50.137	.905	24.70	24.85	7.5361	56.374
.610	99.51	100.47	5.0763	37.973	.710	67.18	67.80	5.9101	44.211	.810	42.84	43.19	6.7440	50.448	.910	23.85	23.99	7.5777	56.685
.615	97.64	98.58	5.1180	38.285	.715	65.80	66.40	5.9518	44.523	.815	41.78	42.12	6.7857	50.760	.915	23.01	23.14	7.6194	56.997
.620	95.81	96.73	5.1597	38.597	.720	64.44	65.03	5.9935	44.834	.820	40.73	41.06	6.8274	51.072	.920	22.17	22.30	7.6612	57.310
.625	94.00	94.90	5.2014	39.910	.725	63.10	63.67	6.0352	45.146	.825	39.70	40.02	6.8691	51.384	.925	21.35	21.47	7.7029	57.622
.630	92.22	93.10	5.2431	39.222	.730	61.78	62.34	6.0769	45.458	.830	38.67	38.98	6.9108	51.696	.930	20.54	20.65	7.7446	57.934
.635	90.47	91.33	5.2848	39.534	.735	60.48	61.02	6.1186	45.770	.835	37.66	37.96	6.9525	52.008	.935	19.73	19.84	7.7863	58.246
.640	88.75	89.59	5.3265	39.845	.740	59.19	59.72	6.1603	46.082	.840	36.67	36.95	6.9941	52.320	.940	18.94	19.03	7.8280	58.557
.645	87.05	87.88	5.3682	40.157	.745	57.92	58.43	6.2020	46.394	.845	35.68	35.96	7.0358	52.632	.945	18.15	18.24	7.8697	58.869
.650	85.38	86.19	5.4098	40.468	.750	56.67	57.17	6.2437	46.706	.850	34.71	34.97	7.0775	52.943	.950	17.37	17.45	7.9114	59.181
.655	83.74	84.53	5.4515	40.780	.755	55.43	55.92	6.2854	47.018	.855	33.74	34.00	7.1192	53.255	.955	16.60	16.67	7.9531	59.493
.660	82.12	82.89	5.4932	41.092	.760	54.21	54.68	6.3271	47.330	.860	32.79	33.03	7.1609	53.567	.960	15.83	15.90	7.9947	59.805
.665	80.53	81.28	5.5349	41.404	.765	53.01	53.47	6.3688	47.642	.865	31.85	32.08	7.2026	53.879	.965	15.08	15.13	8.0364	60.117
.670	78.96	79.69	5.5766	41.716	.770	51.82	52.27	6.4104	47.953	.870	30.92	31.14	7.2443	54.191	.970	14.33	14.38	8.0780	60.428
.675	77.41	78.13	5.6183	42.028	.775	50.65	51.08	6.4521	47.265	.875	30.00	30.21	7.2860	54.503	.975	13.59	13.63	8.1197	60.740
.680	75.88	76.59	5.6600	42.340	.780	49.49	49.91	6.4938	48.577	.880	29.09	29.30	7.3277	54.815	.980	12.86	12.89	8.1615	61.052
.685	74.38	75.07	5.7017	42.652	.785	48.34	48.75	6.5355	48.889	.885	28.19	28.39	7.3694	55.127	.985	12.13	12.15	8.2032	61.364
.690	72.90	73.57	5.7434	42.963	.790	47.22	47.61	6.5772	49.201	.890	27.30	27.49	7.4111	55.438	.990	11.41	11.43	8.2449	61.676
.695	71.44	72.10	5.7851	43.275	.795	46.10	46.49	6.6189	49.513	.895	26.42	26.60	7.4528	55.750	.995	10.70	10.71	8.2866	61.988
															1.000	10.00	10.00	8.3283	62.300

Sp gr 60°/60°	°Bé	°Tw	Lb per gal at 60°F wt in air	Lb per ft³ at 60°F wt in air	Sp gr 60°/60°	°Bé	°Tw	Lb per gal at 60°F wt in air	Lb per ft³ at 60°F wt in air	Sp gr 60°/60°	°Bé	°Tw	Lb per gal at 60°F wt in air	Lb per ft³ at 60°F wt in air	Sp gr 60°/60°	°Bé	°Tw	Lb per gal at 60°F wt in air	Lb per ft³ at 60°F. wt. in air
1.005	0.72	1	8.3700	62.612	1.255	29.46	51	10.4546	78.206	1.505	48.65	101	12.5392	93.800	1.755	62.38	151	14.6238	109.394
1.010	1.44	2	8.4117	62.924	1.260	29.92	52	10.4963	78.518	1.510	48.97	102	12.5809	94.112	1.760	62.61	152	14.6655	109.705
1.015	2.14	3	8.4534	63.236	1.265	30.38	53	10.5380	78.830	1.515	49.29	103	12.6226	94.424	1.765	62.85	153	14.7072	110.017
1.020	2.84	4	8.4950	63.547	1.270	30.83	54	10.5797	79.141	1.520	49.61	104	12.6643	94.735	1.770	63.08	154	14.7489	110.329
1.025	3.54	5	8.5367	63.859	1.275	31.27	55	10.6214	79.453	1.525	49.92	105	12.7060	95.047	1.775	63.31	155	14.7906	110.641
1.030	4.22	6	8.5784	64.171	1.280	31.72	56	10.6630	79.765	1.530	50.23	106	12.7477	95.359	1.780	63.54	156	14.8323	110.953
1.035	4.90	7	8.6201	64.483	1.285	32.16	57	10.7047	80.077	1.535	50.54	107	12.7894	95.671	1.785	63.77	157	14.8740	111.265
1.040	5.58	8	8.6618	64.795	1.290	32.60	58	10.7464	80.389	1.540	50.84	108	12.8310	95.983	1.790	63.99	158	14.9157	111.577
1.045	6.24	9	8.7035	65.107	1.295	33.03	59	10.7881	80.701	1.545	51.15	109	12.8727	96.295	1.795	64.22	159	14.9574	111.889
1.050	6.91	10	8.7452	65.419	1.300	33.46	60	10.8298	81.013	1.550	51.45	110	12.9144	96.606	1.800	64.44	160	14.9990	112.200
1.055	7.56	11	8.7869	65.731	1.305	33.89	61	10.8715	81.325	1.555	51.75	111	12.9561	96.918	1.805	64.67	161	15.0407	112.512
1.060	8.21	12	8.8286	66.042	1.310	34.31	62	10.9132	81.636	1.560	52.05	112	12.9978	97.230	1.810	64.89	162	15.0824	112.824
1.065	8.85	13	8.8703	66.354	1.315	34.73	63	10.9549	81.948	1.565	52.35	113	13.0395	97.542	1.815	65.11	163	15.1241	113.136
1.070	9.49	14	8.9120	66.666	1.320	35.15	64	10.9966	82.260	1.570	52.64	114	13.0812	97.854	1.820	65.33	164	15.1658	113.448
1.075	10.12	15	8.9537	66.978	1.325	35.57	65	11.0383	82.572	1.575	52.94	115	13.1229	98.166	1.825	65.55	165	15.2075	113.760
1.080	10.74	16	8.9954	67.290	1.330	35.98	66	11.0800	82.884	1.580	53.23	116	13.1646	98.478	1.830	65.77	166	15.2492	114.072
1.085	11.36	17	9.0371	67.602	1.335	36.39	67	11.1217	83.196	1.585	53.52	117	13.2063	98.790	1.835	65.98	167	15.2909	114.384
1.090	11.97	18	9.0787	67.914	1.340	36.79	68	11.1634	83.508	1.590	53.81	118	13.2480	99.102	1.840	66.20	168	15.3326	114.696
1.095	12.58	19	9.1204	68.226	1.345	37.19	69	11.2051	83.820	1.595	54.09	119	13.2897	99.414	1.845	66.41	169	15.3743	115.007
1.100	13.18	20	9.1621	68.537	1.350	37.59	70	11.2467	84.131	1.600	54.38	120	13.3313	99.725	1.850	66.62	170	15.4160	115.318
1.105	13.78	21	9.2038	68.849	1.355	37.99	71	11.2884	84.443	1.605	54.66	121	13.3730	100.037	1.855	66.83	171	15.4577	115.630
1.110	14.37	22	9.2455	69.161	1.360	38.38	72	11.3301	84.755	1.610	54.94	122	13.4147	100.349	1.860	67.04	172	15.4993	115.943
1.115	14.96	23	9.2872	69.473	1.365	38.77	73	11.3718	85.067	1.615	55.22	123	13.4564	100.661	1.865	67.25	173	15.5410	116.255
1.120	15.54	24	9.3289	69.785	1.370	39.16	74	11.4135	85.379	1.620	55.49	124	13.4981	100.973	1.870	67.46	174	15.5827	116.567
1.125	16.11	25	9.3706	70.097	1.375	39.55	75	11.4552	85.691	1.625	55.77	125	13.5398	101.285	1.875	67.67	175	15.6244	116.879
1.130	16.68	26	9.4123	70.409	1.380	39.93	76	11.4969	86.003	1.630	56.04	126	13.5815	101.597	1.880	67.87	176	15.6661	117.191
1.135	17.25	27	9.4540	70.721	1.385	40.31	77	11.5386	86.315	1.635	56.32	127	13.6232	101.909	1.885	68.08	177	15.7078	117.503
1.140	17.81	28	9.4957	71.032	1.390	40.68	78	11.5803	86.626	1.640	56.59	128	13.6649	102.220	1.890	68.28	178	15.7495	117.814
1.145	18.36	29	9.5374	71.344	1.395	41.06	79	11.6220	86.938	1.645	56.85	129	13.7066	102.532	1.895	68.48	179	15.7912	118.126
1.150	18.91	30	9.5790	71.656	1.400	41.43	80	11.6637	87.250	1.650	57.12	130	13.7483	102.844	1.900	68.68	180	15.8329	118.438
1.155	19.46	31	9.6207	71.968	1.405	41.80	81	11.7054	87.562	1.655	57.39	131	13.7900	103.156	1.905	68.88	181	15.8746	118.740
1.160	20.00	32	9.6624	72.280	1.410	42.16	82	11.7471	87.874	1.660	57.65	132	13.8317	103.468	1.910	69.08	182	15.9163	119.062
1.165	20.54	33	9.7041	72.592	1.415	42.53	83	11.7888	88.186	1.665	57.91	133	13.8734	103.780	1.915	69.28	183	15.9580	119.374
1.170	21.07	34	9.7458	72.904	1.420	42.89	84	11.8304	88.498	1.670	58.17	134	13.9150	104.092	1.920	69.48	184	15.9996	119.686
1.175	21.60	35	9.7875	73.216	1.425	43.25	85	11.8721	88.810	1.675	58.43	135	13.9567	104.404	1.925	69.68	185	16.0413	119.998
1.180	22.12	36	9.8292	73.528	1.430	43.60	86	11.9138	89.121	1.680	58.69	136	13.9984	104.715	1.930	69.87	186	16.0830	120.309
1.185	22.64	37	9.8709	73.840	1.435	43.95	87	11.9555	89.433	1.685	58.95	137	14.0401	105.027	1.935	70.06	187	16.1247	120.621
1.190	23.15	38	9.9126	74.151	1.440	44.31	88	11.9972	89.745	1.690	59.20	138	14.0818	105.339	1.940	70.26	188	16.1664	120.933
1.195	23.66	39	9.9543	74.463	1.445	44.65	89	12.0389	90.057	1.695	59.45	139	14.1235	105.651	1.945	70.45	189	16.2081	121.245
1.200	24.17	40	9.9960	74.775	1.450	45.00	90	12.0806	90.369	1.700	59.71	140	14.1652	105.963	1.950	70.64	190	16.2498	121.557
1.205	24.67	41	10.0377	75.087	1.455	45.34	91	12.1223	90.681	1.705	59.96	141	14.2069	106.275	1.955	70.83	191	16.2915	121.869
1.210	25.17	42	10.0793	75.399	1.460	45.68	92	12.1640	90.993	1.710	60.20	142	14.2486	106.587	1.960	71.02	192	16.3332	122.181
1.215	25.66	43	10.1210	75.711	1.465	46.02	93	12.2057	91.305	1.715	60.45	143	14.2903	106.899	1.965	71.21	193	16.3749	122.493
1.220	26.15	44	10.1627	76.022	1.470	46.36	94	12.2473	91.616	1.720	60.70	144	14.3320	107.210	1.970	71.40	194	16.4166	122.804
1.225	26.63	45	10.2044	76.334	1.475	46.69	95	12.2890	91.928	1.725	60.94	145	14.3737	107.522	1.975	71.58	195	16.4583	123.116
1.230	27.11	46	10.2461	76.646	1.480	47.03	96	12.3307	92.240	1.730	61.18	146	14.4153	107.834	1.980	71.77	196	16.5000	123.428
1.235	27.59	47	10.2878	76.958	1.485	47.36	97	12.3724	92.552	1.735	61.34	147	14.4570	108.146	1.985	71.95	197	16.5417	123.740
1.240	28.06	48	10.3295	77.270	1.490	47.68	98	12.4141	92.864	1.740	61.67	148	14.4987	108.458	1.990	72.14	198	16.5833	124.052
1.245	28.53	49	10.3712	77.582	1.495	48.01	99	12.4558	93.176	1.745	61.91	149	14.5404	108.770	1.995	72.32	199	16.6250	124.364
1.250	29.00	50	10.4129	77.894	1.500	48.33	100	12.4975	93.488	1.750	62.14	150	14.5821	109.082	2.000	72.50	200	16.6667	124.676

*Prepared by Lewis V. Judson, Ph.D., Chief of Length Section of National Bureau of Standards with the advice and assistance of E. L. Peffer, B.S., A.M., late Chief of Capacity and Density Section, National Bureau of Standards.

TABLE 1-13 Wire and Sheet-Metal Gauges*

Values in approximate decimals of an inch

As a number of gauges are in use for various shapes and metals, it is advisable to state the thickness in thousandths when specifying gauge number.

Gauge number	American (AWG) or Brown & Sharpe (B & S) (for nonferrous wire and sheet)†	U.S. Steel Wire (Stl WG) or Washburn & Moen or Roebling or Am. Steel & Wire Co. [A. (steel) WG] (for steel wire)	Birmingham (BWG) (for steel wire) or Stubs Iron Wire (for iron or brass wire)‡	U.S. Standard (for sheet metal, plate metal, wrought iron)	Standard Birmingham (BG) (for sheet and hoop metal)	Imperial Standard and Wire Gauge (SWG) (British legal standard)	Gauge number
0000000	—	.4900	—	—	.6666	0.500	0000000
000000	—	.4615	—	—	.6250	.464	000000
00000	—	.4305	—	—	.5883	.432	00000
0000	0.460	.3938	0.454	—	.5416	.400	0000
000	.410	.3625	.425	—	.5000	.372	000
00	.365	.3310	.380	—	.4452	.348	00
0	.325	.3065	.340	—	.3964	.324	0
1	.289	.2830	.300	—	.3532	.300	1
2	.258	.2625	.284	—	.3147	.276	2
3	.229	.2437	.259	.239	.2804	.252	3
4	.204	.2253	.238	.224	.2500	.232	4
5	.182	.2070	.220	.209	.2225	.212	5
6	.162	.1920	.203	.194	.1981	.192	6
7	.144	.1770	.180	.179	.1764	.176	7
8	.128	.1620	.165	.164	.1570	.160	8
9	.114	.1483	.148	.150	.1398	.144	9
10	.102	.1350	.134	.135	.1250	.128	10
11	.091	.1205	.120	.120	.1113	.116	11
12	.081	.1055	.109	.105	.0991	.104	12
13	.072	.0915	.095	.090	.0882	.092	13
14	.064	.0800	.083	.075	.0785	.080	14
15	.057	.0720	.072	.067	.0699	.072	15
16	.051	.0625	.065	.060	.0625	.064	16
17	.045	.0540	.058	.054	.0556	.056	17
18	.040	.0475	.049	.0478	.0495	.048	18
19	.036	.0410	.042	.0418	.0440	.040	19
20	.032	.0348	.035	.0359	.0392	.036	20
21	.0285	.0317	.032	.0329	.0349	.032	21
22	.0253	.0286	.028	.0299	.0313	.028	22
23	.0226	.0258	.025	.0269	.0278	.024	23
24	.0201	.0230	.022	.0239	.0248	.022	24
25	.0179	.0204	.020	.0209	.0220	.020	25
26	.0159	.0181	.018	.0188	.0196	.018	26
27	.0142	.0173	.016	.0172	.0175	.0164	27
28	.0126	.0162	.014	.0156	.0156	.0148	28
29	.0113	.0150	.013	.0141	.0139	.0136	29
30	.0100	.0140	.012	.0125	.0123	.0124	30
31	.0089	.0132	.010	.0109	.0110	.0116	31
32	.0080	.0128	.009	.0102	.0098	.0108	32
33	.0071	.0118	.008	.0094	.0087	.0100	33
34	.0063	.0104	.007	.0086	.0077	.0092	34
35	.0056	.0095	.005	.0078	.0069	.0084	35
36	.0050	.0090	.004	.0070	.0061	.0076	36
37	.0045	.0085	—	.0066	.0054	.0068	37
38	.0040	.0080	—	.0062	.0048	.0060	38
39	.0035	.0075	—	—	.0043	.0052	39
40	.0031	.0070	—	—	.0039	.0048	40
41	—	.0066	—	—	.0034	.0044	41
42	—	.0062	—	—	.0031	.0040	42
43	—	.0060	—	—	.0027	.0036	43
44	—	.0058	—	—	.0024	.0032	44
45	—	.0055	—	—	.0022	.0028	45
46	—	.0052	—	—	.0019	.0024	46
47	—	.0050	—	—	.0017	.0020	47
48	—	.0048	—	—	.0015	.0016	48
49	—	.0046	—	—	.0014	.0012	49
50	—	.0044	—	—	.0012	.0010	50

Metric wire gauge is 10 times the diameter in millimeters.
° Courtesy of Dr. Lewis V. Judson with I. H. Fullmer, National Bureau of Standards.
† Sometimes used for iron wire.
‡ Sometimes used for copper plate and for steel plate 12 gauge and heavier and for steel tubes.

TABLE 1-14 Fundamental Physical Constants

1 sec = 1.00273791 sidereal seconds

g_0 = 9.80665 m/sec^2

1 liter = 0.001 cu. m

1 atm = 101,325 newtons/sq m

1 mm Hg (pressure) = ($1/760$) atm
$\qquad\qquad\qquad$ = 133.3224 newtons/sq m

1 int ohm = 1.000495 ± 0.000015 abs ohm

1 int amp = 0.999835 ± 0.000025 abs amp

1 int coul = 0.999835 ± 0.000025 abs coul

1 int volt = 1.000330 ± 0.000029 abs volt

1 int watt = 1.000165 ± 0.000052 abs watt

1 int joule = 1.000165 ± 0.000052 abs joule

$T_{0°C}$ = 273.150 ± 0.010°K

$(PV)_{0°C}{}^{P=0}$ = $(RT)_{0°C}$ = 2271.16 ± 0.04 abs joule/mole
$\qquad\qquad\qquad$ = 22,414.6 ± 0.4 cu. cm atm/mole
$\qquad\qquad\qquad$ = 22.4146 ± 0.0004 liter atm/mole

R = 8.31439 ± 0.00034 abs joule/deg mole
$\quad$ = 1.98719 ± 0.00013 cal/deg mole
$\quad$ = 82.0567 ± 0.0034 cu. cm atm/deg mole
$\quad$ = 0.0820567 ± 0.0000034 liter atm/deg mole

ln 10 = 2.302585

R ln 10 = 19.14460 ± 0.00078 abs joule/deg mole
$\qquad\quad$ = 4.57567 ± 0.00030 cal/deg mole

N = (6.02283 ± 0.0022) × 10^{23}/mole

h = (6.6242 ± 0.0044) × 10^{-34} joule sec

c = (2.99776 ± 0.00008) × 10^8 m/sec

$(h^2/8\pi^2 k)$ = (4.0258 ± 0.0037) × 10^{-39} g sq cm deg

$(h/8\pi^2 c)$ = (2.7986 ± 0.0018) × 10^{-39} g cm

Z = Nhc = 11.9600 ± 0.0036 abs joule cm/mole
$\qquad\quad$ = 2.85851 ± 0.0009 cal cm/mole

(Z/R) = (hc/k) = c_2 = 1.43847 ± 0.00045 cm deg

$\mathscr{F}$ = 96,501.2 ± 10.0 int coul/g-equiv or int joule/int volt g-equiv
$\quad$ = 96,485.3 ± 10.0 abs coul/g-equiv or abs joule/abs volt g-equiv
$\quad$ = 23,068.1 ± 2.4 cal/int volt g-equiv
$\quad$ = 23,060.5 ± 2.4 cal/abs volt g-equiv

e = (1.60199 ± 0.00060) × 10^{-19} abs coul
$\quad$ = (1.60199 ± 0.00060) × 10^{-20} abs emu
$\quad$ = (4.80239 ± 0.00180) × 10^{-10} abs esu

1 int electron-volt/molecule = 96,501.2 ± 10 int joule/mole
$\qquad\qquad\qquad\qquad\qquad$ = 23,068.1 ± 2.4 cal/mole

1 abs electron-volt/molecule = 96,485.3 ± 10. abs joule/mole
$\qquad\qquad\qquad\qquad\qquad$ = 23,060.5 ± 2.4 cal/mole

1 int electron-volt = (1.60252 ± 0.00060) × 10^{-12} erg

1 abs electron-volt = (1.60199 ± 0.00060) × 10^{-12} erg

hc = (1.23916 ± 0.00032) × 10^{-4} int electron-volt cm
$\quad$ = (1.23957 ± 0.00032) × 10^{-4} abs electron-volt cm

k = (8.61442 ± 0.00100) × 10^{-5} int electron-volt/deg
$\quad$ = (8.61727 ± 0.00100) × 10^{-5} abs electron-volt/deg
$\quad$ = (R/N) = (1.38048 ± 0.00050) × 10^{-23} joule/deg

1 IT cal = ($1/860$) = 0.00116279 int watt-hr
$\qquad\quad$ = 4.18605 int joule
$\qquad\quad$ = 4.18674 abs joule
$\qquad\quad$ = 1.000654 cal

1 cal = 4.1840 abs joule
$\qquad$ = 4.1833 int joule
$\qquad$ = 41.2929 ± 0.0020 cu. cm atm
$\qquad$ = 0.0412929 ± 0.0000020 liter atm

1 IT cal/g = 1.8 Btu/lb

1 Btu = 251.996 IT cal
$\qquad$ = 0.293018 int watt-hr
$\qquad$ = 1054.866 int joule
$\qquad$ = 1055.040 abs joule
$\qquad$ = 252.161 cal

1 horsepower = 550 ft-lb (wt)/sec
$\qquad\qquad\quad$ = 745.578 int watt
$\qquad\qquad\quad$ = 745.70 abs watt

1 in = (1/0.3937) = 2.54 cm

1 ft = 0.304800610 m

1 lb = 453.5924277 g

1 gal = 231 cu. in
$\qquad$ = 0.133680555 cu. ft
$\qquad$ = 3.785412 × 10^{-3} cu. m
$\qquad$ = 3.785412 liter

sec = mean solar second
Definition: g_0 = standard gravity

Definition: atm = standard atmosphere
mm Hg (pressure) = standard millimeter mercury

int = international; abs = absolute
amp = ampere
coul = coulomb

Absolute temperature of the ice point, 0°C

PV product for ideal gas at 0°C

R = gas constant per mole

ln = natural logarithm (base e)

N = Avogadro number
h = Planck constant
c = velocity of light
Constant in rotational partition function of gases
Constant relating wave number and moment of inertia
Z = constant relating wave number and energy per mole

c_2 = second radiation constant
$\mathscr{F}$ = Faraday constant

e = electronic charge

Constant relating wave number and energy per molecule

k = Boltzmann constant

Definition of IT cal: IT = International steam tables

cal = thermochemical calorie
Definition: cal = thermochemical calorie

Definition of Btu: Btu = IT British Thermal Unit

cal = thermochemical calorie
Definition of horsepower (mechanical): lb (wt) = weight of 1 lb at standard gravity
Definition of in: in = U.S. inch
ft = U.S. foot (1 ft = 12 in)

Definition; lb = avoirdupois pound
Definition; gal = U.S. gallon

CONVERSION OF VALUES FROM U.S. CUSTOMARY UNITS TO SI UNITS

American engineers are probably more familiar with the magnitude of physical entities in U.S. customary units than in SI units. Consequently, errors made in the conversion from one set of units to the other may go undetected. The following six examples will show how to convert the elements in six dimensionless groups. Proper conversions will result in the same numerical value for the dimensionless number. The dimensionless numbers used as examples are the Reynolds, Prandtl, Nusselt, Grashof, Schmidt, and Archimedes numbers.

Table 1-7 provides a number of useful conversion factors. To make a conversion of an element in U.S. customary units to SI units, one multiplies the value of the U.S. customary unit, found on the left side in the table, by the equivalent value on the right side. For example, to convert 10 British thermal units to joules, one multiplies 10 by 1054.4 to obtain 10544 joules.

In each example, the initial values of the factors are expressed in U.S. customary units, and the dimensionless value is calculated. Then the factors are converted to SI units, and the dimensionless value is recalculated. The two dimensionless values will be approximately the same. (Small variations occur due to the number of significant figures carried in the solution.)

Example 1. Calculation of a Reynolds Number

$$N_{Re} = \frac{DV\rho}{\mu}$$

U.S. customary units
$D = 3$ in. $= \frac{3}{12}$ ft
$V = 6$ ft/s
$\rho = 0.08$ lbm/ft³
$\mu = 0.015$ cp $= (0.015)(0.000672)$ lbm/ft·s

$$N_{Re} = \frac{(3/12)(6)(0.08)}{(0.015)(0.000672)} = 11,904$$

SI units
$D = (3)(0.0254)$ m
$V = (6)(0.3048)$ m/s
$\rho = (0.08)(16.018)$ kg/m³
$\mu = (0.015)(0.001)$ kg/m·s

$$N_{Re} = \frac{(3 \times 0.0254)(6 \times 0.3048)(0.08 \times 16.018)}{(0.015)(0.001)} = 11,904$$

Example 2. Calculation of a Prandtl Number

$$N_{Pr} = \frac{C_p \mu}{k}$$

U.S. customary units
$\gamma_p = 0.47$ Btu/lbm °F
$\mu = 15$ centipoise $= (15)(0.000672)(3600)$ lbm/ft·hr
$k = 0.065$ Btu/hr·ft² (°F/ft)

$$N_{Pr} = \frac{(0.47)(15 \times 0.000672 \times 3600)}{0.065} = 262.4$$

SI units
$\gamma = (0.47)(4184)$ J/kg °C
$\mu = (15)(0.001)$ kg/m·s
$k = (0.065)(1.728)$ J/s·m² (°C/m)

$$N_{Pr} = \frac{(0.47)(4184)(15)(0.001)}{(0.065)(1.728)} = 262.6$$

(Difference due to rounding)

Example 3. Calculation of a Nusselt Number

$$N_{Nu} = \frac{hD}{k}$$

U.S. customary units
$h = 200$ Btu/hr·ft²·°F
$D = 1.5$ in. $= 1.5/12$ ft
$k = 0.07$ Btu/hr·ft² (°F/ft)

$$N_{Nu} = \frac{(200)(1.5/12)}{0.07} = 357.1$$

SI units
$h = (200)(5.678)$ J/(s·m²·°C)
$D = (1.5)(0.0254)$ m
$k = (0.07)(1.728)$ J/s·m² (°C/m)

$$N_{Nu} = \frac{(200)(5.678)(1.5)(0.0254)}{(0.07)(1.728)} = 357.7$$

(Difference due to rounding)

Example 4. Calculation of a Grashof Number

$$N_{Gr} = L^3 \rho^2 g \beta (\Delta T) / \mu^2$$

U.S. Customary units
$L = 3$ ft
$\rho = 0.0725$ lbm/ft³
$g = 32.174$ ft/s²
$\beta = 0.00168$/°R
$\Delta T = 99$ °R
$\mu = 0.019$ centipoise $= 0.019 \times 0.000672$ lbm/ft·s
$\quad = 1.277 \times 10^{-5}$ lbm/ft·s

$$N_{Gr} = \frac{(3^3)(0.0725)^2(32.174)(0.00168)(99)}{(1.277 \times 10^{-5})^2} = 4.66 \times 10^9$$

SI units
$L = (3)(0.3048) = 0.9144$ m
$\rho = (0.0725)(16.018) = 1.1613$ kg/m³
$g = 9.807$ m/s²
$\beta = (0.00168)/(1.8) = 0.000933$/°K
$\Delta T = (99)(1.8) = 178.2$ °K
$\mu = (0.019)(0.001) = 1.9 \times 10^{-5}$ kg/m·s

$$N_{Gr} = \frac{(0.9144)^3(1.1613)^2(9.807)(0.000933)(178.2)}{(1.9 \times 10^{-5})^2} = 4.66 \times 10^9$$

Example 5. Calculation of a Schmidt Number

$$N_{Sc} = \frac{\mu}{\rho D}$$

U.S. customary units
$\mu = 0.02$ centipoise $= (0.02)(2.42)$ lbm/ft·hr
$\rho = 0.08$ lbm/ft³
$D = 1.0$ ft²/hr (diffusivity)

$$N_{Sc} = \frac{(0.02)(2.42)}{(0.08)(1.0)} = 0.605$$

SI units
$\mu = (0.02)(0.001)$ kg/m·s
$\rho = (0.08)(16.02)$ kg/m²
$D = (1.0)(2.58 \times 10^{-5})$ m²/s

$$N_{Sc} = \frac{(0.02)(0.001)}{(0.08)(16.02)(1.0)(2.58 \times 10^{-5})} = 0.605$$

Example 6. Calculation of an Archimedes Number

$$N_{Ar} = \frac{d^3 \rho_f (\rho_p - \rho_f) g}{\mu^2}$$

U.S. customary units
$d = 2$ mm $= 2/[(1000)(0.3048)] = 0.00656$ ft
$\rho_f = 0.0175$ lbm/ft³
$\rho_p = 168.5$ lbm/ft³
$g = 32.174$ ft/s²
$\mu = 0.04$ centipoise $= 0.04 \times 0.000672 = 2.688^{-5}$ lbm/ft·s

$$N_{Ar} = \frac{(0.00656)^3(0.0175)(168.5 - 0.017)(32.174)}{(2.688 \times 10^{-5})^2} = 37,064$$

SI units
$d = 2/1000$ m
$\rho_p = 168.5 \times 16.02 = 2699.37$ kg/m³
$\rho_f = 0.0175 \times 16.02 = 0.2804$ g/m³
$g = 9.807$ m/s²
$\mu = 0.04 \times 0.001 = 4 \times 10^{-5}$ kg/m·s

$$N_{Ar} = \frac{(2/1000)^3(0.2804)(2699.37 - 0.28)(9.807)}{(4 \times 10^{-5})^2} = 37,118$$

(Difference due to rounding)

MATHEMATICAL SYMBOLS

TABLE 1-15 Mathematical Signs, Symbols, and Abbreviations

$\pm$ ($\mp$)	plus or minus (minus or plus)
:	divided by, ratio sign
::	proportional sign
$<$	less than
$\not<$	not less than
$>$	greater than
$\not>$	not greater than
$\cong$	approximately equals, congruent
$\sim$	similar to
$\Leftrightarrow$	equivalent to
$\neq$	not equal to
$\doteq$	approaches, is approximately equal to
$\propto$	varies as
∞	infinity
$\therefore$	therefore
$\sqrt{}$	square root
$\sqrt[3]{}$	cube root
$\sqrt[n]{}$	nth root
$\angle$	angle
$\perp$	perpendicular to
$\parallel$	parallel to
$\|x\|$	numerical value of x
log or log$_{10}$	common logarithm or Briggsian logarithm
log$_e$ or ln	natural logarithm or hyperbolic logarithm or Naperian logarithm
e	base (2.178) of natural system of logarithms
$a°$	an angle a degrees
a' a	prime, an angle a minutes
a'' a	double prime, an angle a seconds, a second
sin	sine
cos	cosine
tan	tangent
ctn or cot	cotangent
sec	secant
csc	cosecant
vers	versed sine
covers	coversed sine
exsec	exsecant
sin^{-1}	anti sine or angle whose sine is
sinh	hyperbolic sine
cosh	hyperbolic cosine
tanh	hyperbolic tangent
sinh^{-1}	anti hyperbolic sine or angle whose hyperbolic sine is
$f(x)$ or $\phi(x)$	function of x
Δx	increment of x
$\sum$	summation of
dx	differential of x
dy/dx or y'	derivative of y with respect to x
d^2y/dx^2 or y''	second derivative of y with respect to x
d^ny/dx^n	nth derivative of y with respect to x
$\partial y/\partial x$	partial derivative of y with respect to x
$\partial^n y/\partial x^n$	nth partial derivative of y with respect to x
$\dfrac{\partial^n y}{\partial x \partial y}$	nth partial derivative with respect to x and y
$\int$	integral of
$\int_a^b$	integral between the limits a and b
$\dot{y}$	first derivative of y with respect to time
$\ddot{y}$	second derivative of y with respect to time
Δ or ∇^2	the "Laplacian"
	$\left(\dfrac{\partial^2}{\partial x^2} + \dfrac{\partial^2}{\partial y^2} + \dfrac{\partial^2}{\partial z^2}\right)$
δ	sign of a variation
$\oint$	sign for integration around a closed path

TABLE 1-16 Greek Alphabet

Alpha	= A, α = A, a	Nu	= N, ν = N, n
Beta	= B, β = B, b	Xi	= Ξ, ξ = X, x
Gamma	= Γ, γ = G, g	Omicron	= O, o = O, o
Delta	= Δ, δ = D, d	Pi	= Π, π = P, p
Epsilon	= E, ε = E, e	Rho	= P, ρ = R, r
Zeta	= Z, ζ = Z, z	Sigma	= Σ, σ = S, s
Eta	= H, η = E, e	Tau	= T, τ = T, t
Theta	= Θ, θ = Th, th	Upsilon	= Y, υ = U, u
Iota	= I, ι = I, i	Phi	= Φ, ϕ = Ph, ph
Kappa	= K, κ = K, k	Chi	= X, χ = Ch, ch
Lambda	= Λ, λ = L, l	Psi	= Ψ, ψ = Ps, ps
Mu	= M, μ = M, m	Omega	= Ω, ω = O, o

Physical and Chemical Data*

Peter E. Liley, Ph.D., D.I.C., *School of Mechanical Engineering, Purdue University.* (physical and chemical data)

George H. Thomson, *AIChE Design Institute for Physical Property Data. (Tables 2-6, 2-30, 2-164, 2-193, 2-196, 2-198, 2-221)*

D.G. Friend, *National Institutes of Standards and Technology, Boulder, CO. (Tables 2-333, 2-334, Figs. 2-25, 2-26)*

Thomas E. Daubert, Ph.D., *Professor, Department of Chemical Engineering, The Pennsylvania State University. (Prediction and Correlation of Physical Properties)*

Evan Buck, M.S.Ch.E., *Manager, Thermophysical Property Skill Center, Central Technology, Union Carbide Corporation. (Prediction and Correlation of Physical Properties)*

* The contributions of J.K. Fink, Argonne National Laboratory; U. Grigull, Tech. Universität, Munich, Germany; and H. Sato, Keio University, Japan, are acknowledged.

GENERAL REFERENCES

Considerations of reader interest, space availability, the system or systems of units employed, copyright considerations, etc., have all influenced the revision of material in previous editions for the present edition. Reference is made at numerous places to various specialized works and also, when appropriate, to more general works. A listing of general works may be useful to readers in need of further information.

ASHRAE Handbook—Fundamentals, IP and SI editions, ASHRAE, Atlanta, various dates; Beaton, C.F. and G.F. Hewitt, *Physical Property Data for the Design Engineer,* Hemisphere, New York, 1989 (394 pp.); Benedek, P. and F. Olti, *Computer-Aided Chemical Thermodynamics of Gases and Liquids,* Wiley, New York, 1985 (731 pp.); Daubert, T.E. and R.P. Danner, Physical and Thermodynamic Properties of Pure Chemicals, 4 vols., Hemisphere, New York, 1989 (2030 pp.); suppl. 1, 1991 (456 pp.); suppl. 2, 1992 (736 pp.); Gmehling, J., *Azeotropic Data,* 2 vols., VCH Weinheim, Germany, 1994 (1900 pp.); Kaye, S.M., *Encyclopedia of Explosives and Related Items,* U.S. Army R&D command, Dover, NJ, 1980; King, M.B., *Phase Equilibrium in Mixtures,* Pergamon, Oxford, 1969; Lyman, W.J., W.F. Reehl et al., *Handbook of Chemical Property Estimation Methods,* McGraw-Hill, N.Y., 1982 (929 pp.); Ohse, R.W., *Handbook of Thermodynamic and Transport Properties of Alkali Metals,* Blackwell Sci. Pubs., Oxford, England, 1985 (1020 pp.); Reid, R.C., J.M. Prausnitz et al., The Properties of Gases and Liquids, McGraw-Hill, New York, 1987 (742 pp.); Sterbacek, Z., B. Biskup et al., *Calculation of Properties Using Corresponding States Methods,* Elsevier, Amsterdam, 1979.

Compilations of critical data include Ambrose, D., "Vapor-Liquid Critical Properties," N.P.L. Teddington, Middx. rept. Chem 107, 1980 (62 pp.); Brule, M.R., L.L Lee et. al., Chem. Eng., **86,** 25 (Nov. 19, 1979) 155–164; Kudchaker, A.P., G.H. Alani et al., *Chem. Revs.,* **68** (1968) 659–735; Matthews, J.F., *Chem. Revs.,* **72** (1972) 71–100; Reid, R.C., J.M. Prausnitz et al., *The Properties of Gases and Liquids,* 4th ed., McGraw-Hill, New York, 1987 (741 pp.); Ohse, R.W. and H. von Tippelskirch, *High Temp.—High Press.,* **9** (1977) 367–385; Young, D.A., "Phase Diagrams of the Elements," UCRL Rept. 51902, 1975 (64 pp.); republished in expanded form by the University of California Press, 1991. Rothman, D. et al., Max Planck Inst. f. Stromungsforschung, Ber 6, 1978 (77 pp.).

PUBLICATIONS ON THERMOCHEMISTRY

Pedley, J.B., Thermochemical Data and Structures of Organic Compounds, 1, Thermodyn. Res. Ctr., Texas A&M Univ., 1994 (976 pp., 3000 cpds.); Frenkel, M., G.J. Kabo et al., *Thermodynamics of Organic Compounds in the Gas State,* 2 vols., Thermodyn. Res. Ctr., Texas A&M Univ., 1994 (1825 pp., 2000 cpds.); Barin, I., *Thermochemical Data of Pure Substances,* 2 vols., 2d ed., VCH Weinheim, Germany 1993 (1834 pp., 2400 substances); and Gurvich, L.V., I.V. Veyts et al., *Thermodynamic Properties of Individual Substances,* 3 vols., 4th ed., Hemisphere, New York, 1989, 1990, and 1993 (2520 pp.). See also Lide, D.R. and G.W.A. Milne, *Handbook of Data on Organic Compounds,* 7 vols., 3d ed., Chemical Rubber, Miami, 1993 (7000 pp.); Daubert, T.E., R.P. Danner et al., *Physical and Thermodynamic Properties of Pure Chemicals: Data Compilation,* extant 1995, Taylor & Francis, Bristol, PA, 1995; Database 11, N.I.S.T. Gaithersburg, MD. U.S. Bureau of Mines publications include Bulletins 584, 1960 (232 pp.); 592, 1961 (149 pp.); 595, 1961 (68 pp.); 654, 1970 (26 pp.) Chase, M.W., C.A. Davies et al., *JANAF Thermochemical Tables,* 3d ed., *J. Phys. Chem. Ref. Data* 14 suppl 1., 1986 (1896 pp.).

PHYSICAL PROPERTIES OF PURE SUBSTANCES

TABLE 2-1 Physical Properties of the Elements and Inorganic Compounds*

Abbreviations Used in the Table

a, acid
A, specific gravity with reference to air = 1
abs, absolute
ac, acetic acid
act, acetone
al, 95 percent ethyl alcohol
alk, alkali (*i.e.*, aq. NaOH or KOH)
ann, amyl (C_5H_{11})
amor, amorphous
anh, anhydrous
aq, aqueous or water
aq. reg., aqua regia

atm, atmosphere or 760 mm. of mercury pressure
bk, black
bm, brown
bz, benzene
c., cold
cb, cubic
cc, cubic centimeter
chl, chloroform
col, colorless or white
conc, concentrated
cr, crystals or crystalline
d., decomposes
D, specific gravity with reference to water at 4°C, or hydrogen (D) = 1.

d. 50, decomposes at 50°C; 50 d., melts at 50°C with decomposition
delq, deliquescent
dil, dilute
dk, dark
eff, effloresces or efflorescent
et, ethyl ether
expl, explodes
gel, gelatinous
gly, glycerol (glycerin)
gn, green
h., hot
hex, hexagonal

hyg, hygroscopic
i., insoluble
ign, ignites
lq, liquid
lt, light
m. al, methyl alcohol
mn, monoclinic
nd, needles
NH_3, liquid ammonia
NH_4OH, ammonium hydroxide solution
oct, octahedral
or, orange
pd, powder

pl, plates
pr, prisms or prismatic
rhb, rhombic (orthorhombic)
s, soluble
satd, saturated
sl, slightly
soln, solution
subl, sublimes
sulf, sulfides
tart. a., tartaric acid
tet, tetragonal
tr, transition
tri, triclinic

trig, trigonal
v, very
vac, in vacuo
vl, violet
volt, volatile or volatilizes
wh, white
yel, yellow
∞, soluble in all proportions
$<$, less than
$>$, greater than
$42\pm$, about or near 42
$-3H_2O$, 100, loses 3 moles of water per formula weight at 100°C

Formula weights are based upon the International Atomic Weights of 1941 and are computed to the nearest hundredth.

Refractive index, where given for a uniaxial crystal, is for the ordinary (ω) ray; where given for a biaxial crystal, the index given is for the median (β) value. Unless otherwise specified, the index is given for the sodium D-line ($\lambda = 589.3$ mμ).

Specific gravity values are given at room temperatures (15° to 20°C) unless otherwise indicated by the small figures which follow the value; thus, "$5.6\ ^{18°}/_4$" indicates a specific gravity of 5.6 for the substance at 18°C referred to water at 4°C. In this table the values for the specific gravity of gases are given with reference to air (A) = 1, or hydrogen (D) = 1.

Melting point is recorded in a certain case as "82 d." and in some other case as "d. 82," the distinction being made in this manner to indicate that the former is a melting point with decomposition at 82°C, while in the latter decomposition only occurs at 82°C. Where a value such as "$-2H_2O, 82$" is given it indicates loss of 2 moles of water per formula weight of the compound at a temperature of 82°C.

Boiling point is given at atmospheric pressure (760 mm. of mercury) unless otherwise indicated; thus, "$82^{15mm.}$" indicates the boiling point is 82°C when the pressure is 15 mm.

Solubility is given in parts by weight (of the formula shown at the extreme left) per 100 parts by weight of the solvent; the small superscript indicates the temperature. In the case of gases the solubility is often expressed in some manner as "$5^{10°cc}$" which indicates that at 10°C, 5 cc. of the gas are soluble in 100 g. of the solvent. The symbols of the common mineral acids: H_2SO_4, HNO_3, HCl, etc., represent dilute aqueous solutions of these acids. See also special tables on Solubility.

REFERENCES: The information given in this table has been collected mainly from the following sources: Mellor, *A Comprehensive Treatise on Inorganic and Theoretical Chemistry*, Longmans, New York, 1922. Abegg, *Handbuch der anorganischen Chemie*, S. Hirzel, Leipzig, 1905. Gmelin-Kraut, *Handbuch der anorganischen Chemie*, 7th ed., Carl Winter, Heidelberg, Berlin, 1924. Friend, *Textbook of Inorganic Chemistry*, Griffin, London, 1914. Winchell, *Microscopic Character of Artificial Inorganic Solid Substances or Artificial Minerals*, Wiley, New York, 1931. *International Critical Tables*, McGraw-Hill, New York, 1926. *Tables annuelles internationales de constants et donnes numeriques*, McGraw-Hill, New York. *Annual Tables of Physical Constants and Numerical Data*, National Research Council, Princeton, N.J. 1943. Comey and Hahn, *A Dictionary of Chemical Solubilities*, Macmillan, New York, 1921. Seidell, *Solubilities of Inorganic and Metal Organic Compounds*, Van Nostrand, New York, 1940.

Name	Formula	Formula weight	Color, crystalline form and refractive index	Specific gravity	Melting point, °C	Boiling point, °C	Solubility in 100 parts		
							Cold water	Hot water	Other reagents
Aluminum	Al	26.97	silv., cb.	$2.70^{20°}$	660	2056	i.	i.	s. HCl, H_2SO_4, alk.
acetate, normal	$Al(C_2H_3O_2)_3$	204.10	wh. pd.		d. 200		s.	d.	
acetate, basic	$Al(OH)(C_2H_3O_2)_2$	162.07	wh., amor.		d.		i.		s.a.; i. NH_4 salts
bromide	$AlBr_3$	266.72	trig.	$3.01^{25°}/_4$	97.5	268	s.		s. al., act., CS_2
bromide	$AlBr_3·6H_2O$	374.82	col., delq. cr.		d. 100		s.	s.	s. al., CS_2
carbide	Al_4C_3	143.91	yel., hex., 2.70	$2.95^{25°}/_4$	d. >2200		d. to CH_4		s. a.; i. act.
chloride	$AlCl_3$	133.34	wh., delq., hex.	$2.44^{25°}/_4$	$194^{2.5atm.}$	$182.7^{753mm.}$, subl. 178	$69.87^{15°}$	s. d.	s. et., chl., CCl_4, i. bz.
chloride	$AlCl_3·6H_2O$	241.44	col., delq., trig., 1.560				400	v. s.	50 al., s. et.
fluoride (fluellite)	$AlF_3·H_2O$	101.99	col., rhb., 1.490	2.17			sl. s.	sl. s.	
fluoride	$Al_2F_6·7H_2O$	294.05	wh., cr. pd.		$-4H_2O$, 120	$-6H_2O$, 250	$0.000104^{18°}$	i.	s. a., alk.; i. a.
hydroxide	$Al(OH)_3$	77.99	wh., mn.	2.42	$-2H_2O$, 300		i.		s. al., CS_2
nitrate	$Al(NO_3)_3·9H_2O$	375.14	rhb., delq.		73	d. 134	v. s.	v. s. d.	s. alk. d.
nitride	Al_2N_2	81.96	yel., hex.	$3.05^{25°}/_4$	$2150^{4atm.}$	d. >1400	d. slowly		v. sl. s. a., alk.
oxide	Al_2O_3	101.94	col., hex., 1.67–8	3.99	1999 to 2032	2210	i.	i.	v. sl. s. a., alk.
oxide (corundum)	Al_2O_3	101.94	wh., trig., 1.768	4.00	1999 to 2032		i.	i.	v. sl. s. a., alk.
phosphate	$AlPO_4$	121.95	col., hex.	2.59			i.	i.	s. a., alk.; i. ac.

*By N. A. Lange, Ph.D., Handbook Publishers, Inc., Sandusky, Ohio. Abridged from table of Physical Constants of Inorganic Compounds in Lange, "Handbook of Chemistry."

Name	Formula	Formula weight	Color, crystalline form and refractive index	Specific gravity	Melting point, °C	Boiling point, °C	Solubility in 100 parts		
							Cold water	Hot water	Other reagents
Aluminum (Cont.)									
potassium silicate (muscovite)	$3Al_2O_3 \cdot K_2O \cdot 6SiO_2 \cdot 2H_2O$	796.40	mm, 1.590	2.9	d.		i.		
potassium silicate (orthoclase)	$Al_2O_3 \cdot K_2O \cdot 6SiO_2$	556.49	col., mn., 1.524	2.56	1450 (1150)			s.	
Aluminum potassium tartrate	$AlK(C_4H_4O_6)_2$	362.21	col.		1000		s.		i. HCl
sodium fluoride (cryolite)	$AlF_3 \cdot 3NaF$	209.96	wh., mn., 1.3389	2.90	1000		sl. s.	i.	d. a.
sodium silicate	$Al_2O_3 \cdot Na_2O \cdot 6SiO_2$	524.29	col., tri., 1.529	2.61	1100		i.	i.	i. al.
sulfate	$Al_2(SO_4)_3$	342.12	wh. cr.	2.71	d. 770		31.3^{0}	$89^{100°}$	i. al.
Alum, ammonium (tschermigite)	$Al_2(SO_4)_3 \cdot (NH_4)_2SO_4 \cdot 24H_2O$	906.64	col., oct., 1.4594	1.64^{20}_{4}	93.5	$-20H_2O$, 120; $-24H_2O$, 200	3.9^{0}	$\infty 100°$	i. al.
ammonium chrome	$Cr_2(SO_4)_3 \cdot (NH_4)_2SO_4 \cdot 24H_2O$	956.72	gn. or vl., oct., 1.4842	1.72	100 d.		$21.2^{25°}$		s. al.
ammonium iron	$Fe_2(SO_4)_3 \cdot (NH_4)_2SO_4 \cdot 24H_2O$	964.40	vl., oct., 1.485	1.71	40		$124^{25°}$		i. al.
potassium (kalinite)	$Al_2(SO_4)_3 \cdot K_2SO_4 \cdot 24H_2O$	948.76	col., mn., 1.4564	1.76^{26}_{4}	92	$-18H_2O$, 64.5	5.7^{00}	$\infty 93°$	i. al.
potassium chrome	$Cr_2(SO_4)_3 \cdot K_2SO_4 \cdot 24H_2O$	998.84	red or gn., cb., 1.4814	1.83	89		20	50	i. al.
sodium	$Al_2(SO_4)_3 \cdot Na_2SO_4 \cdot 24H_2O$	916.56	col., oct., 1.4388	1.675^{20}_{4}	61		$106.4^{0°}$	$121.7^{45°}$	i. al.
Ammonia†	NH_3	17.03	col. gas, 1.325 (lq.)	$0.817^{-79°}$ 0.5971 (A)	-77.7	-33.4	$89.9^{0°}$	$7.4^{96°}$	$14.8^{20°}$ al.; s. et.
Ammonium acetate	$NH_3C_2H_3O_2$	77.08	wh., hyg. cr.	1.073	114		$148^{4°}$	v. s.	s. al.; sl. s. act.
auricyanide	$NH_4CN \cdot Au(CN)_3 \cdot H_2O$	337.33	pl.		d. 200	d.			i. al.
bicarbonate	NH_4HCO_3	79.06	mn. or rhb., 1.5358	1.573	d. 35–60		$11.9^{0°}$	$27^{30°}$	i. al.
bromide	NH_4Br	97.96	col., cb., 1.7108	2.327^{15}_{4}	subl. 542		$68^{10°}$	$145.6^{100°}$	s. al., et., act.
carbonate, carbamate	$(NH_4)_2CO_3 \cdot H_2O$	114.11	col. pl.		d. 58		$100^{15°}$		i. al, CS_2, NH_3
	$NH_4CO_2NH_4$‡	157.11	wh. cr.		subl.		$25^{15°}$	$67^{65°}$	
carbonate, sesqui-	$(NH_4)_2CO_3 \cdot 2NH_4HCO_3 \cdot H_2O$	272.22	wh.		d.		$20^{15°}$	$50^{49°}$	s. al.
chloride (salammoniac)	NH_4Cl	53.50	wh., cb., 1.639, 1.6426	1.53^{17}	d. 350	subl. 520	$29.4^{0°}$	$77.3^{100°}$	s. NH_3; sl. s. al., m. al.
chloroplatinate	$(NH_4)_2PtCl_6$	444.05	yel., cb.	3.065	d.		$0.7^{15°}$	$1.25^{100°}$	0.005 al.
chloroplatinite	$(NH_4)_2PtCl_4$	373.14	tet.		d.		s.	v. s.	
chlorostannate	$(NH_4)_2SnCl_6$	367.52	pink-, cb.	2.4	d.		$33.3^{15°}$		sl. s. act., NH_3; i. al.
chromate	$(NH_4)_2CrO_4$	152.09	yel., mn.	$1.917^{12°}$	d. 180		$40.5^{30°}$	d.	s. al.
cyanide	NH_4CN	44.06	col., cb.	$0.79^{100°}$ (A)	36		s.	v. s.	s. al.; i. act.
dichromate	$(NH_4)_2Cr_2O_7$	252.10	or., mn.	2.15	d. 185		$47.2^{30°}$	v. s.	i. al.
ferrocyanide	$(NH_4)_4Fe(CN)_6 \cdot 6H_2O$	392.21	mn.		d.		v. s.	d.	s. al.; i. NH_3
fluoride	NH_4F	37.04	wh., hex.		d.		v. s.	d.	i. al.
fluoride, acid	$NH_4F \cdot HF$	57.05	wh., rhb., 1.390	2.21^{12}_{2}			v. s.		s. al.
formate	HCO_2NH_4	63.06	col., mn., delq.	1.266	114–116	d. 180; subl. in vac.	$102^{0°}$	$531^{80°}$	s. al.
hydrosulfide	NH_4HS	51.11	col., rhb.		d.	subl. 120	v. s.		s. al.
hydroxide	NH_4OH	35.05	in soln. only				s.		i. al, NH_3
molybdate	$(NH_4)_2MoO_4$	196.03	mn.	2.27	d.		d.	d.	i. al.
molybdate, hepta-	$(NH_4)_6Mo_7O_{24} \cdot 4H_2O$‡	1235.95	col., mn.		d.		$44^{25°}$		i. al.
nitrate (α), stable -16° to 32°	NH_4NO_3	80.05	col., tet., 1.611	$1.66^{25°}_{4}$	169.6	d. 210	$118.3^{0°}$	$241.8^{30°}$	$3.8^{20°}$ al., $17.1^{20°}$ m. al.; v. s. NH_3
nitrate (β), stable 32° to 84°	NH_4NO_3	80.05	col., rhb. or mn.	1.725^{25}_{4}		d. 210	$365.8^{35°}$	$580^{80°}$	s. al.
nitrite	NH_4NO_2	64.05	wh. nd.	1.69	expl.		s.	d.	
osmochloride	$(NH_4)_2OsCl_6$	439.02	cb.	2.93^{20}_{4}			s.		sl. s. al.; i. NH_3
oxalate	$(NH_4)_2C_2O_4 \cdot H_2O$	142.12	col., rhb.	1.501	d.		$2.5^{0°}$	$11.8^{50°}$	sl. s. al.; i. NH_3
oxalate, acid	$NH_4HC_2O_4 \cdot H_2O$	125.08	col., trimetric	1.556	d.		s.		
perchlorate	NH_4ClO_4	117.50	col., rhb., 1.4833	1.95	d.		$10.9^{0°}$	$46.9^{100°}$	$2^{20°}$ al.; s. act.; i. et.
persulfate	$(NH_4)_2S_2O_8$	228.20	wh., mn., 1.5016	1.98	d. 120		$58.2^{0°}$	d.	i. ac.
phosphate, monobasic	$NH_4H_2PO_4$	115.04	col., tet., 1.5246	1.803^{19}_{4}	d.		$22.7^{0°}$	$173.2^{100°}$	

Name	Formula	Mol. wt.	Color, crystalline form, n	Density	M.p.	B.p.	Sol. cold water	Sol. hot water	Sol. other solvents
phosphate, dibasic	$(NH_4)_2HPO_4$	132.07	col., mn., 1.53	1.619	d.		131^{15}	i.	i. act.
phosphate, meta-	$(NH_4)_4P_4O_{12}$	388.08	col., mn.	2.21			s.	55.5	s. alk.; i. al., HNO_3
Ammonium phosphomolybdate	$(NH_4)_3PO_4 \cdot 12MoO_3 \cdot 3H_2O$ (?)	1930.55	yel.				0.03^{15}		s. al.; i. act.
silicofluoride	$(NH_4)_2SiF_6$	178.14	cb, 1.3696	2.01		subl.	$18.5^{17.5}$		s. al.; i. act.
sulfamate	$NH_4 \cdot SO_3NH_2$	114.12	col. pl.		132	d. 160	$134°$	357^{50}	i. al., act., CS_2
sulfate (mascagnite)	$(NH_4)_2SO_4$	132.14	col., rhb., 1.5230	1.769^{20}_{4}	235 d.		$70.6°$	103.3^{100}	v. sl. s. al.; i. act. 120^{25} NH_3
sulfate, acid	NH_4HSO_4	115.11	col., rhb., 1.480	1.78	146.9	490	100		
sulfide	$(NH_4)_2S$	68.14	yel.-wh.		d.		v. s.		i. al., act.
sulfide, penta-	$(NH_4)_2S_5$	196.38	or-red pr.		d.		s.		
sulfite	$(NH_4)_2SO_3 \cdot H_2O$	134.16	col., mn.	1.41	d.		100^{12}	87^{60}	sl. s. al.
sulfite, acid	NH_4HSO_3	99.11	rhb.	2.03^{12}_{4}			$45°$	170^{20}	s. al., act., NH_3, SO_2
tartrate	$(NH_4)_2C_4H_4O_6$	184.15	col., mn.	1.60			$120°$	3.05^{70}	i. al., NH_4Cl
thiocyanate	NH_4CNS	76.12	col., mn., 1.685±	1.305	149.6	d. 170		i.	
vanadate, meta-	NH_4VO_3	116.99	col. cr.	2.326	d.		0.44^{18}		
Antimony	Sb	121.76	tin wh., trig.	6.684^{25}_{4}	630.5	1380	i.		s. aq. reg., h. conc. H_2SO_4
chloride, tri- (butter of antimony)°	$SbCl_3$	228.13	col., rhb., delq.	3.14^{20}_{4}	73.4	220.2	$601.6°$	$∞^{72}$	s. al., HCl, HBr, $H_2C_4H_4O_6$
oxide, tri- (valentinite)	Sb_2O_3	291.52	rhb., 2.35	5.67	656	1570	v. sl. s.	sl. s.	s. HCl, KOH, $H_2C_4H_4O_6$
oxide, tri- (senarmontite)	Sb_2O_3	291.52	cb., 2.087	5.2	652		0.00017^{18}	d.	s. HCl; alk. NH_4HS, K_2S, NH_4HS
sulfide, tri- (stibnite)	Sb_2S_3	339.70	bk., rhb., 4.046	4.64	550		i.	i.	s. HCl, alk., NH_4HS
sulfide, penta-	Sb_2S_5	403.82	golden	$4.120°$	−2S, 135				
telluride, tri-	Sb_2Te_3	626.35	gray		629				
Antimonyl potassium tartrate (tartar emetic)	$(SbO)KC_4H_4O_6 \cdot ½H_2O$	333.94	wh., rhb.	2.60	−½H_2O, 100		$5.26^{8.7}$	35.7^{100}	s. gly.; i. al.
sulfate, normal	$(SbO)_2SO_4$	371.58	wh., pd.	4.89			d.	d.	
sulfate, basic	$(SbO)_2SO_4 \cdot Sb_2(OH)_4$	683.13	wh., pd.				i.	d.	
Argon	A	39.94	col. gas	$1.65^{-298°}$; $1.402^{-185.7°}$; 1.38 (A)	−189.2	−185.7	$5.6°$ cc	2.23^{50} cc	5.15^{15} gly.; 24^{25} cc al.
Arsenic (crystalline) (α)	As_4	299.64	met., hex.	5.727^{14}	814^{36atm}	subl. 615	i.	i.	s. HNO_3
Arsenic (black) (β)	As_4	299.64	bk., amor.	$4.7°$			i.	i.	s. HNO_3, aq. reg., aq. Cl_2, h. alk.
Arsenic (yellow)(γ)	As_4	299.64	yel., cb.	$2.0^{3°}$	d. 358				s. alk.
acid, ortho-	$H_3AsO_4 \cdot ½H_2O$	150.94	col., hyg.	2.0–2.5	35.5	−H_2O, 160	16.7	50 H_3AsO_4	
acid, meta-	$HAsO_3$	123.92	wh., hyg.		d. 206		d. to form	H_3AsO_4	
acid, pyro-	$H_4As_2O_7$	265.85	col.				d. to form		
pentoxide	As_2O_5	229.82	wh., amor.	4.086		d. 565	$59.5°$	76.7^{100}	s. alk., al.
sulfide, di- (realgar)	As_2S_2	213.94	red, mn., 2.68	$(α)3.506^{19°}$; $(β)3.254^{19°}$	(α tr. 267°; (β)307		i.		s. K_2S, $NaHCO_3$
sulfide, penta-	As_2S_5	310.12	yel.			d. 500	$0.000136°$		s. HNO_3, alk.
Arsenious chloride (butter of arsenic)	$AsCl_3$	181.28	oily lq.	lq. 2.163	−18	130	d.	d.	s. HCl, HBr, PCl_3
hydride (arsine)	AsH_3	77.93	col. gas	2.695 (A)	−113.5	−55; d. 230	20 cc	sl. s.	sl. s. alk.
oxide (arsenolite)	As_2O_3	197.82	col., cb., fibrous, 1.755	3.865^{25}_{4}	subl.		sl. s.	sl. s.	i. al., et.
oxide (claudetite)	As_2O_3	197.82	col., mn., 1.92	3.85	subl.		sl. s.	sl. s.	i. al., et.
oxide	As_2O_3	197.82	amor. or vitreous	3.738	315		$1.21°$	2.93^{40}	s. HCl, alk., Na_2CO_3; i. al., et
Auric chloride	$AuCl_3 \cdot 2H_2O$	339.60	or. cr.		d. 50	d. 290	v. s.		s. HCl, al., et.; sl. s. NH_3
cyanide	$Au(CN)_3 \cdot 6H_2O$	383.35					v. s.		s. al.
Aurous chloride	$AuCl$	232.66	yel. cr.	7.4	$AuCl_3$, 170		d.	d.	s. HCl, HBr; d. al.
cyanide	$AuCN$	223.22	yel. cr.		d.		i.	i.	s. KCN; i. al., et.
Cf. also under *Gold*									
Barium	Ba	137.36	silv. met.	3.5	850	1140	d.	d.	s. a.; d. al.
acetate	$Ba(C_2H_3O_2)_2$	255.45	col.	2.468			$58.8°$	75.0^{100}	i. al.
acetate	$Ba(C_2H_3O_2)_2 \cdot H_2O$	273.46	wh., tri, pr., 1.517	2.19	−H_2O, 41		$75°$(anh.)	79^{40}(anh.)	
bromide	$BaBr_2$	297.19	col.	4.781^{24}_{4}	847	d.	$98°$	149^{100}	v. s. m. al.; v. sl. s. act.

*Usually the solution.
†See special tables.
‡Usual commercial form.

TABLE 2-1 Physical Properties of the Elements and Inorganic Compounds (*Continued*)

Name	Formula	Formula weight	Color, crystalline form and refractive index	Specific gravity	Melting point, °C	Boiling point, °C	Solubility in 100 parts		
							Cold water	Hot water	Other reagents
Barium (*Cont.*)									
bromide	$BaBr_2 \cdot 2H_2O$	333.22	col., mn., 1.7266	3.69	$-2H_2O$, 100	d. 1450	v. s.	v. s.	s. al.
carbonate (witherite)	$BaCO_3$	197.37	wh., rhb., 1.676	4.29	tr. 811 to α	d. 1450	0.0022^{18}	0.0065^{100}	s. a.; i. al.
carbonate (α)	$BaCO_3$	197.37	wh., hex.		tr. 982 to β		0.0022^{18}	0.0065^{100}	s. a.; i. al.
carbonate (β)	$BaCO_3$	197.37	wh.		$1740^{90\,atm}$				
Barium chlorate	$Ba(ClO_3)_2 \cdot H_2O$*	304.27	col.	3.179	d. 120		20.35^{18}	84.8^{80}	sl. s. al., act.
chlorate		322.29	col., mn., 1.577		414		s.	s.	
chloride	$BaCl_2$	208.27	col., mn., 1.7361	3.856^{24}_{4}	tr. 925	1560	31^{0}	59^{100}	sl. s. HCl, HNO_3; i. al.
chloride	$BaCl_2$	208.27	col., cb.		962	1560			sl. s. HCl, HNO_3; i. al.
chloride	$BaCl_2 \cdot 2H_2O$†	244.31	col., mn., 1.646	3.097^{24}_{4}	$-2H_2O$, 100		39.3^{0}	76.8^{100}	sl. s. HCl, HNO_3; i. al.
hydroxide	$Ba(OH)_2$	171.38	col., mn.	4.495	77.9		1.67^{0}	101.4^{80}	v. sl. s. al.; i. et.
hydroxide	$Ba(OH)_2 \cdot 8H_2O$	315.50	col., mn., 1.5017	2.188^{16}		$-8H_2O$, 550	5.6^{15}	34.2^{100}	s. a.; i. al.
nitrate (nitrobarite)	$Ba(NO_3)_2$	261.38	col., cb., 1.572	3.244^{28}	592	d	5.0^{0}	90.8^{80}	s. a. NH_4Cl; i. al.
oxalate	BaC_2O_4	225.38	wh. cr.	2.658			0.0016^{8}	0.0024^{24}	s. HCl, HNO_3, abs. al.; i. NH_3, act.
oxide	BaO	153.36	col., cb., 1.98	5.72	1923	2000±	1.5^{0}		
peroxide	BaO_2*	169.36	gray or wh. pd.	4.958	$-O$, 800		v. sl. s.	d.	s. dil. a.; i. act.
peroxide	$BaO_2 \cdot 8H_2O$	313.49	pearly sc.	2.9[a]	$-8H_2O$, 100		0.168	d.	s. dil. a.; i. al., et., act.
phosphate, monobasic	$BaH_4(PO_4)_2$	331.35	tri.				d.	d.	s. a.
phosphate, dibasic	$BaHPO_4$	233.35	wh., rhb. nd., 1.635	4.165^{15}			0.015		s. a. NH_4 salts
phosphate, tribasic	$Ba_3(PO_4)_2$	602.04	wh., cb.	4.1^{16}			i.		s. a.
phosphate, pyro-	$Ba_2P_2O_7$	448.68	wh., rhb.	3.9^{30}			0.01		s. a. NH_4 salts
silicofluoride	$BaSiF_6$	279.42	pr.	4.279^{15}			0.026^{17}	0.09^{100}	s. s. HCl, NH_4Cl; i. al.
sulfate (barite, barytes)	$BaSO_4$	233.42	col., rhb., 1.636	4.499^{15}	1580 d.		0.000115^{0}	0.000285^{30}	s. conc. H_2SO_4, 0.006, 3% HCl
sulfide, mono-	BaS	169.42	col., cb., 2.155	4.25^{15}			d.	d.	d. HCl; i. al.
sulfide, tri-	BaS_3	233.54	yel.-gn.		d. 400			s.	i. al., CS_2
sulfide, tetra-	$BaS_4 \cdot 2H_2O$	301.63	red, rhb.	2.988^{20}	d. 200		41^{15}	v. s.	s. dil. a., alk.
Beryllium (glucinum)	Be (Gl)	9.02	gray, met., hex.	1.816	1284	2767	i.	sl. s. d	s. dil. a., alk.
Bismuth	Bi	209.00	silv. wh. or reddish, hex.	9.80^{20}	271	1450	i.	i.	s. aq. reg., conc. H_2SO_4, HNO_3
carbonate, sub-	$Bi_2O_3 \cdot CO_2 \cdot H_2O$	528.03	wh. pd.	6.86	d.		i.	i.	s. a.
chloride, di-	$BiCl_2(?)$	279.91	bk. nd.	4.86	163	d. 300	d.	d.	s. dil. a., i. al., et., act.
chloride, tri-	$BiCl_3$*	315.37	wh. cr.	4.75	230	447	d.	d.	s. a.; i. al., et., act.
nitrate	$Bi(NO_3)_3 \cdot 5H_2O$	485.10	col., tri.	2.82	d. 30	$-5H_2O$, 80	d.		42^{19} act.; s. a.; i. al.
nitrate, sub-	$BiONO_3 \cdot H_2O$	305.02	hex. pl.	4.928^{15}	d. 260		i.	i.	s. a.
oxide, tri-	Bi_2O_3	466.00	yel., rhb.	8.9	820	1900±	i.	i.	s. a.
oxide, tri-	Bi_2O_3	466.00	yel., tet.	8.55	860		i.	i.	s. a.
oxide, tri-	Bi_2O_3	466.00	yel., cb.	8.20	tr. 704		i.	i.	s. a.
oxychloride	$BiOCl$	260.46	wh., amor.	7.72^{15}			i.	sl. s.	s. a.; i. act, NH_3, $H_2C_4H_4O_6$
Boric acid	H_3BO_3	61.84	wh., tri.	1.435^{15}	185 d.		2.66^{20}	40.2^{100}	s. al., et., alk., NH_3, $H_2C_4H_4O_6$ 22.2^{20}[f] gly., 0.24^{25} et.; s. al.
Boron	B	10.82	gray or bk., amor. or mn.	2.32	2300	2550	i.	i.	i. a.
carbide	B_4C	55.29	bk. cr.	2.54	2450	>3500	i.	i.	i. a.
oxide	B_2O_3	69.64	col. glass, 1.459	1.85	577	>1500	1.1^{0}	15.7^{100}	s. a., al., gly.
oxide (sassolite)	$B_2O_3 \cdot 3H_2O$	123.69	tri., 1.456	1.49	d.		sl. s.	s.	
Bromic acid	$HBrO_3$	128.92	col.; in soln. only		d. 100		v. s.		s. HNO_3; i. al.
Bromine	Br_2	159.83	rhb., or red lq.	3.119^{30}[f]; 5.87 (A)	-7.2	58.78	4.22^{0}	3.13^{30}	s. al., et., alk., CS_2
hydrate	$Br_2 \cdot 10H_2O$	339.99	red, oct.		d. 6.8		s.		
Cadmium	Cd	112.41	silv. met., hex.	8.65^{30}	320.9	767	i.	i.	s. a. NH_4NO_3
acetate	$Cd(C_2H_3O_2)_2$	230.50	col.	2.341	256	d.	v. s.		s. m. al.
acetate	$Cd(C_2H_3O_2)_2 \cdot 2H_2O$*	266.53	col., mn.	2.01	$-H_2O$, 130		v. s.		s. al.
carbonate	$CdCO_3$	172.42	wh., trig.	4.258^{4}	d. <500		i.	i.	s. a. KCN, NH_4 salts; i. NH_3
chloride	$CdCl_2$	183.32	wh., cb.	4.047^{25}_{4}	568	960	90^{0}	147^{100}	1.52^{15} al.; i. et., act.

Name	Formula	Mol. wt.	Color, crystalline form, index of refraction	Density	M.p., °C	B.p., °C	Solubility, cold water	Solubility, hot water	Solubility in other solvents
chloride	CdCl₂·2½H₂O	228.36	col., mn., 1.6513	3.327	tr. 34		168$^{20°}$	180$^{100°}$	2.05^{15} m. al.; s. a.; NH₄OH, KCN
cyanide	Cd(CN)₂	164.45	wh., trig.		d. >200		0.0247$^{18°}$		s. a., NH₄ salts; i. alk.
hydroxide	Cd(OH)₂	146.43	col.	4.79$^{15/4}$		d. 300	0.00026$^{25°}$		v. s. a.
nitrate	Cd(NO₃)₂	236.43	col. nd.		350		109.7°	326$^{59.5°}$	s. al., NH₄ salts; i. HNO₃
nitrate	Cd(NO₃)₂·4H₂O°	308.49	brn., cb.	2.455$^{17/4}$	59.4	132	215°	v. s.	s. a., NH₄ salts; i. alk.
oxide	CdO	128.41	brn., cb.	8.15	d. 900–1000		i.	i.	s. a., NH₄ salts; i. alk.
oxide	CdO	128.41	brn., amor, 2.49	6.95			i.	i.	d. a., alk.
oxide, sub-	Cd₂O	240.82	gn., amor.	8.192$^{18/4}$					i.act., NH₃
Cadmium sulfate	CdSO₄	208.47	rhb.	4.691$^{18/4}$	1000		76.5°	60.8$^{100°}$	i. al.
sulfate	CdSO₄·H₂O	226.49	col., mn., 1.565	3.786$^{20°}$	tr. 108		s.	s.	i. al.
sulfate	3CdSO₄·8H₂O°	769.54	col.	3.09			114.2°	127.6$^{60°}$	i. al.
sulfate	CdSO₄·8H₂O°	280.53	mn.	3.05	tr. 41.5		s.	s.	s. a.; v. s. NH₄OH
sulfate	CdSO₄·7H₂O	334.58	mn.	2.48$^{20/4}$			s.		sl. s. al.
sulfide (greenockite)	CdS	144.47	yel.-or., hex., 2.506	4.58	1750^{100atm}	subl. in N₂, 980	0.000001	Colloidal	s. HCl
Calcium	Ca	40.08	silv. met., cb.	1.55$^{20°}$	810	1200 ± 30	d.	d.	s. dil. a.
acetate	Ca(C₂H₃O₂)₂·H₂O	176.18	wh. nd.		d.		52°	45.5$^{80°}$	s. al., act.; sl. s. NH₃
aluminate	Ca(AlO₂)₂	158.02	col., rhb. or mn.	3.67$^{20°}$	1600	1810	d.	d.	s. a., NH₄Cl
aluminum silicate (anorthite)	CaO.Al₂O₃.2SiO₂	278.14	tri, 1.5832	2.765	1551		i.	i.	s. a., NH₄Cl
arsenate	Ca₃(AsO₄)₂	398.06	wh. pd.				0.013$^{25°}$	i.	s. al.
bromide	CaBr₂	199.91	delq. nd.	3.353$^{25/4}$	760	>1600	125°	312$^{105°}$	s. al.
carbonate (aragonite)	CaCO₃	100.09	col., rhb., 1.6809	2.93	d. 825		0.0012$^{20°}$†	0.002$^{100°}$	0.0065$^{18°}$ al.
carbonate (calcite)	CaCO₃	100.09	col., hex., 1.550	2.711$^{25/4}$	1339^{100atm}		0.0014$^{25°}$	0.002$^{100°}$	i. al.
chloride (hydrophilite)	CaCl₂	110.99	wh., delq., cb, 1.52	2.152$^{15/4}$	772	>1600	59.5°	347$^{260°}$	s. al.
chloride	CaCl₂·H₂O	129.01	col., delq.	1.68^{17}			v. s.	s.	s. al.
citrate	Ca₃(C₆H₅O₇)₂·4H₂O	570.50	col., trig, 1.417	1.7	−4H₂O, 185		0.085$^{18°}$	v. s.	sl. s. a.
cyanamide	CaCN₂	80.11	col., rhombohedral		1330		s. d.	d.	i. al., et.
ferrocyanide	Ca₂Fe(CN)₆·12H₂O	508.31	yel., tri, 1.5818		d.		s.	150$^{0°}$	d. a.; i. bz.
fluoride (fluorite)	CaF₂	78.08	wh., cb., 1.4339	3.180$^{20°}$			0.0016$^{18°}$	0.0017$^{26°}$	s. NH₄Cl
formate	Ca(HCO₂)₂	130.12	col., rhb.	2.015	d.		16.1⁰	18.4^{100}	s. HCl, H₃PO₆
hydride	CaH₂	42.10	wh. cr. or pd.	1.7	d. 675		d.	d.	∞ h. al.; i. et.
hydroxide	Ca(OH)₂	74.10	col., hex., 1.574	2.2	−H₂O, 580		0.185⁰	0.077$^{100°}$	i. al.
hypochlorite	Ca(ClO)₂·4H₂O	215.06	wh., feathery cr.		−2H₂O, 130		delq.; d.	d.	
hypophosphate	Ca₂P₂O₆·2H₂O	274.15	granular	2.23^{4}	−2H₂O, 200		i.		
lactate	Ca(C₃H₅O₃)₂·5H₂O	308.30	col., eff.	2.2^{4}	−3H₂O, 100		10.5		
magnesium carbonate (dolomite)	CaO.MgO.2CO₂	184.42	trig, 1.68174	2.872	d. 730–760		0.032$^{18°}$		14$^{15°}$ al.; s. amyl al., NH₃
magnesium silicate (diopside)	CaO.MgO.2SiO₂	216.52	wh., mn.	3.3	1391		i.		s. dil. a.; i. al., ac.
nitrate (nitrocalcite)	Ca(NO₃)₂	164.10	col., cb.	2.36	561		102°	376$^{151°}$	90% al.
nitride	Ca₃N₂	148.26	brn., cr.	2.63^{17}	900		d.	v. s.	s. a.; i. ac.
nitrite	Ca(NO₂)₂·H₂O	150.11	delq., hex.	1.82	42.7		266°	417$^{90°}$	s. a.; i ac
oxalate	CaC₂O₄	128.10	col., cb.				0.00067$^{13°}$	0.0014$^{95°}$	s. a.; i. al.
oxalate	CaC₂O₄·H₂O	146.12	col., mn., 1.498	2.2	−H₂O, 200		i.	i.	s. a.; i. al.
oxide	CaO	56.08	col., cb, 1.837	3.32	2570	2850	Forms Ca(OH)₂		
peroxide	CaO₂·8H₂O	216.21	pearly, tet.		expl. 275 / d. 200	−8H₂O, 100 / −H₂O, 100	sl. s.	d.	s. a. d.; i. al., et.
phosphate, monobasic	CaH₄(PO₄)₂·H₂O	252.09	wh., tri.	2.220$^{16/4}$			i.		s. a.; i. al., ac.
phosphate, dibasic	CaHPO₄·2H₂O	172.10	wh., mn. pl.	2.306$^{16/4}$			0.02$^{24.5°}$	0.075$^{100°}$	s. a.
phosphate, tribasic	Ca₃(PO₄)₂	310.20	wh., amor.	3.14	1670		0.0025	d.	s. a.; i. NH₄Cl
phosphate, meta-	Ca(PO₃)₂	198.04	wh., tet, 1.588	2.82	975		i.		
phosphate, pyro-	Ca₂P₂O₇	254.12	col., biaxial, 1.60	3.09	1230		i.		
phosphate, pyro- (brushite)	Ca₂P₂O₇·5H₂O	344.20	wh., mn.				sl. s.	i.	s. dil. a.; i. al., et.
phosphide	Ca₃P₂	182.20	red cr.	2.51^{15}	>1600		d.	d.	s. HCl
silicate (α) (pseudowollastonite)	CaSiO₃	116.14	col., pseudo hex., 1.6150 or mn.(?)	2.905	1540		i.		
silicate (β) (wollastonite)	CaSiO₃	116.14	col., mn., 1.610	2.915	tr. 1190 to α		0.0095$^{17°}$	0.075$^{100°}$	s. a., Na₂S₂O₃, NH₄ salts
sulfate (anhydrite)	CaSO₄	136.14	col., rhb, 1.576, or mn, 1.50	2.96	1450(mn.)	tr. 1193 to rhb.	0.298$^{20°}$	0.1619$^{100°}$	s. HCl

° Usual commercial form.

† The solubility of CaCO₃ in H₂O is greatly increased by increasing the amount of CO₂ in the H₂O.

TABLE 2-1 Physical Properties of the Elements and Inorganic Compounds (*Continued*)

Name	Formula	Formula weight	Color, crystalline form and refractive index	Specific gravity	Melting point, °C	Boiling point, °C	Cold water	Hot water	Other reagents
Calcium (*Cont.*)									
sulfate (gypsum)	$CaSO_4 \cdot 2H_2O$	172.17	col., mn., 1.5226	2.32	$-1\frac{1}{2}H_2O$, 128	$-2H_2O$, 163	$0.223^{0°}$	$0.257^{50°}$	s. a., gly., $Na_2S_2O_3$, NH_4 salts
sulfhydrate	$Ca(SH)_2 \cdot 6H_2O$	214.31	col. pr.				v. s.	v. s.	s. al.
sulfide (oldhamite)	CaS	72.14	col., cb.	$2.8^{15°}$	d. 15	d. 650	d.	d.	s. a.
sulfite	$CaSO_3 \cdot 2H_2O$	156.17	wh., cr., 1.595		$-2H_2O$, 100		$0.0043^{18°}$	$0.0027^{90°}$	s. H_2SO_3
tartrate	$CaC_4H_4O_6 \cdot 4H_2O$	260.22	col., rhb.		d.		$0.037^{0°}$	$0.22^{85°}$	sl. s. al.
thiocyanate	$Ca(CNS)_2 \cdot 3H_2O$	210.28	wh., delq. cr.	$1.873^{16°}$			s.	v. s.	v. s. al.
thiosulfate	$CaS_2O_3 \cdot 6H_2O$	260.30	col., tri., 1.56	6.06	d.		$71.2^{9°}$	d.	i. al.
tungstate (scheelite)	$CaWO_4$	288.00	wh., tet., 1.9200				0.2		s. NH_4Cl; i. a.
Carbon, *cf.* table of organic compounds									
Carbon, amorphous	C	12.01	bk., amor.	1.8–2.1	>3500	4200	i.	i.	i. a., alk.
Carbon, diamond	C	12.01	col., cb., 2.4195	$3.51^{20°}$	>3500	4200	i.	i.	i. a., alk.
Carbon, graphite	C	12.01	bk., hex.	$2.26^{20°}$	>3500	4200	i.	i.	i. a., alk.
dioxide	CO_2	44.01	col. gas	lq. $1.101^{-87°}$; solid $1.56^{-79°}$	$-56.6^{5.2atm}$	subl. -78.5	$179.7^{0°}$ cc	$90.1^{20°}$ cc	s. a., alk.
disulfide	CS_2	76.13	col. lq.	lq. $1.261\frac{22°}{20°}$; 2.63 (A)	-108.6	46.3	$0.2^{0°}$	$0.014^{50°}$	s. al.; et.
monoxide	CO	28.01	col., poisonous, odorless gas	lq. $0.814\frac{-195°}{4}$; 0.968	-207	-192	$0.0044^{0°}$; $3.5^{0°}$ cc	$0.0018^{50°}$; $2.32^{20°}$ cc	s. al., Cu_2Cl_2
oxychloride (phosgene)	$COCl_2$	98.92	poisonous gas	$1.392\frac{19°}{4}$	-104	8.2^{756mm}	v. s. sl. d.	d.	s. ac., CCl_4, bs.; d.a.
oxysulfide	COS	60.07	gas	lq. $1.24^{-87°}$; 2.10 (A)	-138.2	-50.2^{760mm}	$133^{0°}$ cc	$40.3^{0°}$ cc	v. s. alk., al.
suboxide	C_3O_2	68.03	gas	lq. $1.114^{0°}$	-107	7^{761mm}	d.		s. et.
thionyl chloride	$CSCl_2$	114.98	yel.–red lq.	$1.509^{15°}$		73.5			
Ceric hydroxide	$2CeO_2 \cdot 3H_2O$	398.31	yel. gelatinous						s. a.; sl. s. alk. carb.; i. alk
hydroxynitrate	$Ce(OH)(NO_3)_3 \cdot 3H_2O$	397.21	red, mn.				d.		s. H_2SO_4, HCl
oxide	CeO_2	172.13	wh. or pa. yel., cb.	7.3	1950		i.	i.	s. dil. H_2SO_4
sulfate	$Ce(SO_4)_2 \cdot 4H_2O$	404.31	yel., rhb.	3.91	645		s. d.		s. dil. H_2SO_4
Cerium	Ce	140.13	steel gray, cb. or hex.	$6.9^{20°}$ cb; 6.7 hex.		1400	i.	Slowly oxidized	s. a.; i. al.
Cerous sulfate	$Ce_2(SO_4)_3$	568.44	wh., mn. or rhb.	3.91	$-8H_2O$, 630		$18.98^{0°}$	$0.4^{100°}$	i.
sulfate	$Ce_2(SO_4)_3 \cdot 8H_2O$	712.57	tri.	$2.886^{17°}$	28.5		$25^{0°}$	$7.6^{40°}$	
Cesium	Cs	132.91	silv. met., hex.	$1.90^{20°}$	<-20	670	d.		s. a., al., NH_3
Chloric acid	$HClO_3 \cdot 7H_2O$	210.58	lq.	$1.282^{14.2°}$	-101.6	d. 40	v. s.		
Chlorine	Cl_2	70.91	rhb., or gn.–yel. gas	lq. $1.56^{-33.6°}$; 2.49$^{0°}$ (A)	-101.6	-34.6	$1.46^{0°}$; $310^{10°}$ cc	$0.57^{30°}$; $177^{30°}$ cc	s. alk.
hydrate	$Cl_2 \cdot 8H_2O$	215.04	rhb.	1.23	d. 9.6		s.		s. alk.
Chloroplatinic acid	$H_2PtCl_6 \cdot 6H_2O$	518.08	red-brn., delq.	2.431	60		v. s.	v. s.	s. al., et.
Chlorostannic acid	$H_2SnCl_6 \cdot 6H_2O$	441.55	delq.	$1.971^{28°}$	19.2		s.		d. al.; i. CS_2
Chlorosulfonic acid	$HO \cdot SO_2Cl$	116.52	col. lq.	$1.787^{23°}$	-80	151.5^{765mm}	d.		$4.76^{15°}$ m. al.
Chromic acetate	$Cr_2(C_2H_3O_2)_6 \cdot 2H_2O$	494.32	gn.				s.		i. a., act., CS_2
chloride	$CrCl_3$	158.38	pink, trig.	$2.757^{15°}$	subl. 83	1200–1500 d.	i.§	sl. s.	s. al.; i. et.
chloride	$CrCl_3 \cdot 6H_2O$	266.48	vl. or gn., hex. pl.	$1.835\frac{95°}{4}$	>1000	d.	v. s. d.		sl. s. a.; i. al., NH_3
fluoride	CrF_3	109.01	gn., rhb.	3.8			i.		i. a.
hydroxide	$Cr(OH)_3$	103.03	gn. or blue, gelatinous				i.		s. a., alk.; sl. s. NH_3
hydroxide	$Cr(OH)_3 \cdot 2H_2O$	139.07	gn.		$-2H_2O$, 100		i.	i.	s. a., alk.
nitrate	$Cr(NO_3)_3 \cdot 9H_2O$	400.18	purple pr.		36.5	d. 100	s.	s.	s. a., alk., al., act.
nitrate	$Cr(NO_3)_3 \cdot 7\frac{1}{2}H_2O$	373.15	purple, mn.		100	d.	s.	s.	sl. s. a.
oxide	Cr_2O_3	152.02	dark gn., hex.	5.21	1900		i.	i.	i. a.
sulfate	$Cr_2(SO_4)_3$	392.20	rose pd.	3.012			s.		s. al., H_2SO_4
sulfate	$Cr_2(SO_4)_3 \cdot 15H_2O$	482.28	vl.	$1.867^{17°}$	100	$-10H_2O$, 100	s.	d. $67°$	sl. s. al.
sulfate	$Cr_2(SO_4)_3 \cdot 18H_2O$	716.49	vl. cb. 1.564	$1.7\frac{25°}{}$		$-12H_2O$, 100	$120^{20°}$		s.

Name	Formula	Mol. wt.	Color, crystalline form, index of refraction	M.p., °C	B.p., °C	Density	Solubility in cold water	Solubility in hot water	Solubility in other solvents
sulfide	Cr_2S_3	200.02	brn.-bk. pd., cb.	−S. 1350		$3.77^{19°}$	i.	d.	s. h. HNO_3;
Chromium	Cr	52.01	gray, met., rhb.	1615	2200	7.1	i.	i.	s. HCl, dil. H_2SO_4; i. HNO_3
trioxide (chromic acid)	CrO_3	100.01	red, rhb.	197 d.	d.	2.70	$164.9°$	$206.7^{100°}$	s. H_2SO_4, al., et.
Chromous chloride	$CrCl_2$	122.92	wh., delq.			2.75	v. s.	v. s.	sl. s. al.; i. et.
hydroxide	$Cr(OH)_2$	86.03	yel.-brn.				i.		s. conc. a.
oxide	CrO	68.01	bk. pd.	d.			d.	i.	i. dil. HNO_3
sulfate	$CrSO_4·7H_2O$	274.18	blue				$12.35°$	i.	sl. s. al.
sulfide (daubrelite)	CrS	84.07	bk. pd.	1550		3.97	i.	i.	v. s. a.
Chromyl chloride	CrO_2Cl_2	154.92	dark red lq.	−96.5	117.6	1.92	d.	d.	s. et.
Cobalt	Co	58.94	silv. met., cb.	1480	2900	$8.9^{20°}$	d.	d.	s. a.
carbonyl	$Co(CO)_4$	170.98	or. cr.	51	d. 52	$1.73^{18°}$	i.	i.	s. al., et., CS_2
sulfide, di-	CoS_2	123.06	bk., cb.			4.269	i.	d.	s. HNO_3; aq. reg.
Cobaltic chloride	$CoCl_3$	165.31	red cr.	subl.		2.94	s.	s.	
Cobaltic chloride, dichro	$Co(NH_3)_4Cl_3·H_2O$	234.42							s. a.; al.
chloride, luteo	$Co(NH_3)_6Cl_3$	267.50	or., mn.			$1.7016^{29°}$	$4.26^{0°}$	$12.74^{46.5°}$	i. al. NH_4OH
chloride, praseo	$Co(NH_3)_4Cl_3·H_2O$	251.46	gn., rhb.			1.847	v. s.	i.	s. a.; i. al.
Cobaltic chloride, purpureo	$Co(NH_3)_5Cl_3$	250.47	rhb.			1.819^{25}_{25}	$0.232^{0°}$	$1.031^{46.5°}$	i. al.
chloride, roseo	$Co(NH_3)_5Cl_3·H_2O$	268.49	brick red				$16.12^{0°}$	$24.87^{16°}$	sl. s. HCl
hydroxide	$Co(OH)_3$	109.96	bk.	d. 100		5.18	i.	i.	s. a.; i. al.
oxide	Co_2O_3	165.88	bk.				i.	i.	s. a.
sulfate	$Co_2(SO_4)_3$	406.06	bk. cr.	−1¼H_2O, 100			d.	d.	s. H_2SO_4
sulfide	Co_2S_3	214.06	bk. cb.			4.8	i.	i.	d. a.
Cobalto-cobaltic oxide	Co_3O_4	240.82	bk.-bk.	d. 900		6.07	i.	i.	s. H_2SO_4; i. HCl, HNO_3
Cobaltous acetate	$Co(C_2H_3O_2)_2·4H_2O$	249.09	red-vl., mn., 1.542	−4H_2O, 140		$1.7053^{18.7°}$	s.	s.	s. a., al.
chloride	$CoCl_2$	129.85	blue cr.	subl.	1049	3.356	$45^{0°}$	$105^{96°}$	31 al.; 8.6 act.
chloride	$CoCl_2·6H_2O°$	237.95	red, mn.	86	−6H_2O, 110	1.924^{25}_{25}	$116.5^{0°}$	$177^{80°}$	v. s. et., act.
nitrate	$Co(NO_3)_2·6H_2O$	291.05	red, mn., 1.4	<100	d.	1.883^{25}_{25}	$84.03^{0°}$ (anh.)	$334.9^{90°}$ (anh.)	$100^{12.5°}$ al.; s. act.; sl. s. NH_3
oxide	CoO	74.94	brn., cb.	d. 1800		5.68	i.	i.	s. a., NH_4OH; i. al.
sulfate	$CoSO_4·H_2O$	155.00	red pd.	d. 880		$3.710^{25°}$	$25.6^{0°}$	$83^{100°}$	$1.04^{18°}$ m. al.; i. NH_3
sulfate	$CoSO_4$	173.02	red pd., mn.(?), 1.639	d.		3.13	s.	s.	s.
sulfate (bieberite)	$CoSO_4·7H_2O°$	281.11	red, mn., 1.483	96.8	−7H_2O, 420	1.948^{25}_{25}	$33^{80°}$	s.	$2.5°$ al.
sulfide (syeporite)	CoS	91.00	brn. nd.	>1100		$5.45^{18°}$	$0.00035^{18°}$	i.	s. a., aq. reg.
Copper	Cu	63.57	yel.-red met., cb.	1083	2300	$8.92^{20°}$	i.		s. HNO_3, h. H_2SO_4
Cupric acetate	$Cu(C_2H_3O_2)_2$	181.66	dark gn., mn.	115		1.930^{20}_{4}	s.	s.	7 al.; s. et.; gly.
acetate	$Cu(C_2H_3O_2)_2·H_2O$	199.67	gn.		240 d.	1.882	7.2	20	s. a., NH_4OH
aceto-arsenite (Paris green)	$(CuO·As_2O_3)_3·Cu(C_2H_3O_2)_2°$	1013.83					i.		s. a.
ammonium chloride	$CuCl_2·2NH_4Cl·2H_2O$	277.51	blue, tet., 1.670, 1.744	d. 110		1.98	$33.8^{80°}$	$99.3^{80°}$	i. al.
ammonium sulfate	$CuSO_4·4NH_3·H_2O$	245.77	blue, rhb.	d. 150		1.81	$18.05^{21.5°}$		s. NH_4OH, h. aq. $NaHCO_3$
carbonate, basic (azurite)	$2CuCO_3·Cu(OH)_2$	344.75	blue, mn., 1.758	d. 220		3.88	i.		s. KCN; 0.03 aq. CO
carbonate, basic (malachite)	$CuCO_3·Cu(OH)_2$	221.17	dark gn., mn., 1.875	d.		3.9	d.	d.	$53^{15°}$ al.; $68^{15°}$ m. al.
chloride (eriochalcite)	$CuCl_2$	134.48	brn.-yel. pd.	498	Forms Cu_2Cl_2 993	3.054	$70.7^{0°}$	$107.9^{100°}$	s. al.; et., NH_4Cl
chloride	$CuCl_2·2H_2O$	170.52	gn., rhb., 1.684	−2H_2O, 110		$2.39^{22.4°}$	$110.4^{0°}$	$192.4^{100°}$	s. HNO_3, NH_4OH
chromate, basic	$CuCrO_4·2CuO·2H_2O$	374.75	yel.-gn.	−2H_2O, 260			i.		s. KCN, C_6H_5N
cyanide	$Cu(CN)_2$	115.61	bk., tri	d.		$2.286^{18°}$	sl. s.	d.	s. a.; NH_4OH
dichromate	$CuCr_2O_7·2H_2O$	315.62	yel.-gn.	−2H_2O, 100			i.		s. NH_4OH; i. HCl
ferricyanide	$Cu_3[Fe(CN)_6]_2$	614.63	red-brn.				i.	i.	s. NH_4OH; i. a., NH_3
ferrocyanide	$Cu_2Fe(CN)_6·7H_2O$	465.21							
formate	$Cu(HCO_2)_2$	153.61	blue, mn.			1.831	12.5		0.25 al.
hydroxide	$Cu(OH)_2$	97.59	blue, gelatinous	−H_2O		3.368		d.	s. a., NH_4OH, KCN, al.
lactate	$Cu(C_3H_5O_3)_2·2H_2O$	277.74	dark blue, mn.				16.7	$45^{100°}$	sl. s. al.
nitrate	$Cu(NO_3)_2·3H_2O°$	241.63	blue, rhb.	114.5	−HNO_3, 170	$2.047^{3.9°}$	$381^{40°}$	$666^{89°}$	$100^{12.5°}$ al.
nitrate	$Cu(NO_3)_2·6H_2O$	295.68	blue, rhb.	−3H_2O, 26.4		2.074	$243.7°$	∞	s. al.

° Usual commercial form.
† Also a soluble modification.

TABLE 2-1 Physical Properties of the Elements and Inorganic Compounds (*Continued*)

Name	Formula	Formula weight	Color, crystalline form and refractive index	Specific gravity	Melting point, °C	Boiling point, °C	Solubility in 100 parts		
							Cold water	Hot water	Other reagents
Cupric acetate (*Cont.*)									
oxide (paramelaconite)	CuO	79.57	bk., cb.	6.40	d. 1026		i	i	s. a.; KCN, NH$_4$Cl
oxide (tenorite)	CuO	79.57	bk., tri., 2.63	6.45	d. 1026		i	i	s. a., KCN, NH$_4$Cl
oxychloride	$CuCl_2\cdot2CuO\cdot4H_2O$	365.69	blue-gn.		$-3H_2O$, 140		i		s. a.
phosphide	Cu_3P_2	252.67	bk.	6.35					s. HNO$_3$; i. HCl
sulfate (hydrocyanite)	$CuSO_4$	159.63	gn.-wh., rhb., 1.733	3.606^{15}	d. >600	Forms CuO, 650	$14.3^{0°}$	$75.4^{100°}$	i. al.
sulfate (blue vitriol or chalcanthite)	$CuSO_4\cdot5H_2O$°	249.71	blue, tri., 1.5368	$2.286\,^{15.6°}_{4}$	$-4H_2O$, 110	$-5H_2O$, 250	$24.3^{0°}$	$205^{100°}$	1.1^{18} al.
sulfide (covellite)	CuS	95.63	blue, hex. or mn., 1.45	4.6	tr. 103	d. 220	$0.000033^{18°}$		s. HNO$_3$, KCN
tartate	$CuC_4H_4O_6\cdot3H_2O$	265.69	1 gn. pd.		d.		$0.02^{15°}$	$0.14^{85°}$	s. a.; KOH
Cuprous ammonium iodide	$CuI\cdot NH_4I\cdot H_2O$	353.47	rhb. pl.		d.		d.		s. NH$_4$I
carbonate	Cu_2CO_3	187.15	yel.	4.4	d.		i.	i.	s. a., NH$_4$OH
chloride (nantokite)	Cu_2Cl_2	198.05	wh., cb., 1.973	3.53	422	1366	$1.52^{25°}$		s. HCl, NH$_4$OH, al.
cyanide	$Cu_2(CN)_2$	179.16	wh., mn.	2.9	474.5	d.	i		s. KCN, HCl, NH$_4$OH; sl. s. NH$_3$
ferricyanide	$Cu_3Fe(CN)_6$	402.67	brm-red				i		s. NH$_4$OH; i. HCl
ferrocyanide	$Cu_4Fe(CN)_6$	466.24	brm-red				i	i	s. NH$_4$OH; i. NH$_4$Cl
fluoride	CuF_2	165.14	red cr.	3.4	908	subl. 1100	i		s. HF, HCl, HNO$_3$; i. al.
hydroxide	$CuOH$	80.58	yel.		$-\frac{1}{2}H_2O$, 360		i		s. a., NH$_4$OH
oxide (cuprite)	Cu_2O	143.14	red, cb., 2.705	6.0	1235	$-O$, 1800	$0.0005^{18°}$		s. HCl, NH$_4$Cl, NH$_4$OH
Cuprous phosphide	Cu_4P_2	443.38	gray-bk.	6.4 to 6.8					s. HNO$_3$; i. HCl
sulfide (chalcocite)	Cu_2S	159.20	bk., rhb.	5.6	1100		$0.0005^{18°}$		s. HNO$_3$, NH$_4$OH; i. act.
sulfide	Cu_2S	159.20	bk., cb.	5.80	1130				s. HNO$_3$, NH$_4$OH; i. act.
Cyanogen	C_2N_2	52.02	poisonous gas	lq. $0.866^{-17.8°}$; 1.806 (A)	-34.4	-20.5	$450^{20°}$ cc		$2300^{20°}$ cc al.; $500^{18°}$ cc et.
Cyanogen compounds, *cf.* table of organic compounds									
Ferric acetate, basic	$Fe(OH)(C_2H_3O_2)_2$	190.95	brn., amor.				i		s. a.; al.
ammonium sulfate, *cf.* Alum									
chloride (molysite)	$FeCl_3$	162.22	bk-brm., hex. delq.	$2.804^{11°}$	282	315	$74.4^{0°}$	∞	v. s. al. et. +HCl
chloride	$FeCl_3\cdot6H_2O$°	270.32	red-yel., delq.		37	280	$246^{3°}$	$535.8^{100°}$	s. al., act., gly.
ferrocyanide (Prussian blue)	$Fe_4[Fe(CN)_6]_3$	859.27	dark blue		d.		i	d	s. HCl, conc. H$_2$SO$_4$; i. al., et.
hydroxide	$Fe(OH)_3$	106.87	brm-brn.	3.4 to 3.9	$-1\frac{1}{2}H_2O$, 500		i		s. a.; i. al., et.
lactate	$Fe(C_3H_5O_3)_3$	323.06	brm., amor., delq.	$1.684^{20°}$			v. s.	v. s.	i. et.
nitrate	$Fe(NO_3)_3\cdot6H_2O$	349.97	rhb., delq.		35	d.	$150^{0°}$	∞	s. al., act.
oxide (hematite)	Fe_2O_3	159.70	red or bk., trig., 3.042	5.12	1560 d.		i		s. HCl
sulfate	$Fe_2(SO_4)_3$	399.88	rhb., 1.814	$3.097^{18°}$	d. 480		sl. s.	d	i. H$_2$SO$_4$, NH$_3$
sulfate (coquimbite)	$Fe_2(SO_4)_3\cdot9H_2O$	562.02	yel., trig.	2.1	d. 50		440	d	s. abs. al.
Ferroso-ferric chloride	$FeCl_2\cdot2FeCl_3\cdot18H_2O$	775.49	yel., delq.		d. 180		s.	s.	s. d. h. HCl
ferricyanide (Prussian green)	$Fe_4'''Fe_3''[Fe(CN)_6]_6$	1662.70	gn.				i	i	i. al.
oxide (magnetite; magnetic iron oxide)	Fe_3O_4	231.55	bk., cb., 2.42	5.2	1538 d.		i	i	s. a.
oxide, hydrated	$Fe_3O_4\cdot4H_2O$	303.61	bk.		d.		i		i. al.
Ferrous ammonium sulfate	$FeSO_4\cdot(NH_4)_2SO_4\cdot6H_2O$	392.15	blue-gn., mn., 1.4915	1.864	d.		$18^{0°}$	$100^{75°}$	
chloride (lawrencite)	$FeCl_2$	126.76	gn.-yel., hex., 1.567	2.7		delq.	$64.4^{10°}$	$105.7^{100°}$	100 al.; s. act.; i. et.
chloroplatinate	$FePtCl_6\cdot6H_2O$	571.92	yel., hex.	2.714	d.		v. s.	v. s.	i. dil. a., al.
ferricyanide (Turnbull's blue)	$Fe_3[Fe(CN)_6]_2$	591.47	dark blue				i.		
ferrocyanide	$Fe_2Fe(CN)_6$	323.66	blue-wh., amor.				sl. s.		
formate	$Fe(HCO_2)_2\cdot2H_2O$	181.92		3.4	d.		sl. s.		
hydroxide	$Fe(OH)_2$	89.87	lt. gn.				0.00067		s. a., NH$_4$Cl
nitrate	$Fe(NO_3)_2\cdot6H_2O$	287.96	cr.		60.5		$200^{0°}$	$300^{25°}$	
oxide	FeO	71.85	bk.	5.7	1420		i		s. a.; i. alk.

Name	Formula	Mol. wt.	Color, crystalline form, refractive index, etc.	Density	Melting point, °C	Boiling point, °C	Solubility in cold water	Solubility in hot water	Solubility in other solvents
phosphate (vivianite)	$Fe_3(PO_4)_2 \cdot 8H_2O$	501.64	blue, mn., 1.592, 1.603	2.58			i.	i.	s. a.; i. ac.
silicate	$FeSiO_3$	131.91	mn.	3.5	1550		s.		i. al.
sulfate (siderotile)	$FeSO_4 \cdot 5H_2O$	241.99	gn., tri., 1.536	2.2	64	$-5H_2O$, 300			i. al.
sulfate (copperas)	$FeSO_4 \cdot 7H_2O$°	278.02	blue-gn., mn.	$1.899^{14.5°}$		$-7H_2O$, 300	$32.8^{0°}$	$149^{50°}$	s. a.; i. NH_3
sulfide	FeS	87.91	bk., hex.	4.84	1193	d.	$0.000616^{18°}$	d.	
cf. also under iron									
Fluoboric acid	HBF_4	87.83	col. lq.			130 d.	∞	∞	s. al.
Fluorine	F_2	38.00	gn.-yel. gas	lq. $1.51^{-187°}$; $1.31^{15°}$ (A)	-223	-187	d.	s.	
Fluosilicic acid	H_2SiF_6	144.08					s.	s.	
Gadolinium	Gd	156.9					s.		
Gallium bromide	$GaBr_3$	309.47	delq. cr.				s.	s.	
Glucinum cf. Beryllium									
Gold	Au	197.20	yel. met., cb.	$19.3^{20°}$	1063	2600	i.	i.	s. aq. reg, KCN; i. a.
Gold, colloidal	Au	197.20	blue to vl.				s.		s. aq. reg, KCN; i. a.
Gold salts cf. under Auric and Aurous									
Hafnium	Hf	178.6	hex.	12.1	>1700	>3200(?)	i.		Absorbed by Pt
Helium	He	4.00	col. gas	0.1368 (A)	<-272.2	-268.9	$0.97^{0°}$ cc	$1.08^{50°}$ cc	s. al.
Hydrazine	N_2H_4	32.05	col. lq.	$1.011^{15°/4}$	1.4	113.5	∞	∞	∞ al.; i. et.
formate	$N_2H_4 \cdot 2HCO_2H$	124.10	cb.		128		s.		sl. s. al.
hydrate	$N_2H_4 \cdot H_2O$	50.06	col.	$1.03^{21°}$	-40	$118.5^{739.5mm}$	∞		s. al.
hydrochloride	$N_2H_4 \cdot HCl$	68.51	yel. lq.		198		v.s.		i. al.
hydrochloride, di-	$N_2H_4 \cdot 2HCl$	104.98	wh., cb.			subl. 140	v.s.		v. sl. s. abs. al.
nitrate	$N_2H_4 \cdot HNO_3$	95.06	cr.	1.42	70.7		$174.9^{10°}$		∞ al.
nitrate, di-	$N_2H_4 \cdot 2HNO_3$	158.08	nd.		104		v.s.		∞ al.
sulfate	$N_2H_4 \cdot \frac{1}{2}H_2SO_4$	81.09	delq. pl.		85		v.s.		s. al.
sulfate	$N_2H_4 \cdot H_2SO_4$	130.12	rhb.	1.378	254	d.	$3.055^{22°}$	$27.65^{60°}$	s. al.
Hydrazoic acid (azoimide)	HN_3	43.03	col. gas		-80	37	∞		s. al.
Hydriodic acid	HI	127.93	col. lq.	$4.4^{0°}$ (A)	-50.8	-35.5	$42500^{10°}$ cc	v.s.	s. al.
Hydriodic acid	$HI \cdot H_2O$	145.94	col. lq.	$1.7^{15°}$	-43	127^{774mm}	∞		s. al.
Hydriodic acid	$HI \cdot 2H_2O$	163.96	col. lq.		-48		∞		s. al.
Hydriodic acid	$HI \cdot 3H_2O$	181.98	col. lq.		-36.5		∞		s. al.
Hydriodic acid	$HI \cdot 4H_2O$	199.99	col. lq.						s. al.
Hydrobromic acid	HBr	80.92	col. gas; 1.325 (lq.)	$2.71^{0°}$ (A)	-86	-67	$221^{0°}$	$130^{100°}$	Stable at $-15.5°$ and 1 atm., and at $-11.3°$ and 2.5 atm.
Hydrobromic acid	$HBr \cdot H_2O$	98.94	col. lq.	1.78	-11		∞		s. al.
Hydrobromic acid	HBr (47.8% in H_2O)	80.92	col. lq.	1.486		126	s.		s. a., et.
Hydrobromic acid	$HBr \cdot 2H_2O$	116.96	wh. cr.	$2.11^{-15°}$					s. al.
Hydrochloric acid	HCl‡	36.47	col. gas: 1.256 (lq.)	$1.268^{0°}$ (A)	-111	-85	$82.3^{0°}$	$56.1^{60°}$	s. al.
Hydrochloric acid	HCl (45.2% in H_2O)	36.47	col. lq.	1.48			∞		s. al.
Hydrochloric acid	$HCl \cdot 2H_2O$	72.50	col. lq.	$1.46^{-18.3°/4}$	-15.35	d.	∞		s. al.
Hydrochloric acid	$HCl \cdot 3H_2O$	90.51	col. lq.		-24.4	d.	∞		s. al.
Hydrocyanic acid (prussic acid)	HCN	27.03	poisonous gas or col. lq., 1.254	$0.697^{18°}$	-14	26	∞		∞ al., et.
Hydrofluoric acid	HF	20.01	gas or col. lq.	$0.988^{13.6°}$	-83	19.4	∞ 0° to 19.4°		s. al.
Hydrofluoric acid	HF (35.35% in H_2O)	20.01	col. lq.	1.15	-35	120	v.s.	v.s.	
Hydrogen	H_2	2.016	col. gas or cb.	$0.0709^{-252.7°}$/0.06948 (A)	-259.1	-252.7	$2.1^{0°}$ cc	$0.85^{80°}$ cc	sl. s. Fe, Pd, Pt
peroxide	H_2O_2‡	34.02	col. lq., 1.333	$1.438^{9°/4}$	-0.89	151.4^{760mm}	∞		s. a., et.; i. petr. et s. CS_2, $COCl_2$
selenide	H_2Se	81.22	col. gas	$2.12^{-42°}$	-64	-42	$377^{4°}$ cc	$270^{22.5°}$ cc	$9.54^{15°}$ cc al.; s. CS_2
sulfide	H_2S	34.08	col. gas	1.1895 (A)	-82.9	-59.6	$437^{40°}$ cc	$186^{40°}$ cc	s. a., al.
Hydroxylamine	NH_2OH	33.03	rhb., delq.	$1.35^{18°}$	34	56.5^{22mm}	s.	v.s.	s. al.; i. et.
hydrochloride	$NH_2OH \cdot HCl$	69.50	col., mn.	$1.67^{17°}$	151	d.	$83.3^{17°}$	d.	v. s. al.; abs. al.
nitrate	$NH_2OH \cdot HNO_3$	96.05	col. cr.		48	d.<100	v.s.		
sulfate	$NH_2OH \cdot \frac{1}{2}H_2SO_4$	82.07	col., mn.		170 d.		$32.9^{0°}$	$68.5^{90°}$	v. sl. s. al.; i. et, abs. al.

°Usual commercial form.
†Usual commercial form about 31 percent.
‡Usual commercial forms 3 or 30 percent.

TABLE 2-1 Physical Properties of the Elements and Inorganic Compounds (*Continued*)

The last three columns (Cold water, Hot water, Other reagents) fall under the heading **Solubility in 100 parts**.

Name	Formula	Formula weight	Color, crystalline form and refractive index	Specific gravity	Melting point, °C	Boiling point, °C	Cold water	Hot water	Other reagents
Hypobromous acid	$HBrO$	96.92	yel.			40^{90mm}	s.	d.	s. a.
Illinium	Il	146(?)							
Indium	In	114.76	soft, tet. met.	$7.3^{20°}$	155	1450	i.	i.	s. a.
Iodic acid	HIO_3	175.93	col., rhb.	$4.629^{0°}$	110 d		$286^{0°}$	576^{101}	v. s. 87% al.; i. abs. al. et., chl.
Iodine	I_2	253.84	blue-bk., rhb.	$4.93^{20°}$	113.5	184.35	$0.0162^{0°}$	$0.09566^{50°}$	s. al., KI, et.
oxide, penta-	I_2O_5	333.84	wh., trimetric	$4.799^{25°}_{4}$	d. 300		$187.4^{12°}$	i.	i. abs. al., et., chl.
Iodoplatinic acid	$H_2PtI_6\cdot 9H_2O$	1120.91	brn., delq. mn.				s. d.		
Iridium	Ir	193.10	gray	$22.4^{20°}$	2350	>4800	i.	i.	sl. s. aq. reg.; aq. Cl_2
Iron, cast†	Fe	55.85	gray	7.03	1275		i.	i.	s. a.; i. alk.
pure	Fe	55.85	silv. met., cb.	$7.86^{20°}$	1535	3000	i.	i.	s. a.; i. alk.
steel	Fe	55.85	silv. gray	7.6 to 7.8	1375		i.	i.	s. a.; i. alk.
white pig	Fe	55.85	gray	7.6 to 7.8	1075		i.	i.	s. a.; i. alk.
wrought	Fe	55.85	gray	7.86	1505		i.	i.	s. a.; i. alk.
carbide (cementite)	Fe_3C	179.56	pseudo hex.	7.4	1837		i.	i.	s. a.
carbonyl	$Fe(CO)_5$	195.90	pa. yel. lq.	$1.457^{21°}$	-21	102.5^{760mm}	d.		s. al., H_2SO_4, alk.
nitride	Fe_2N	125.71	gray	6.35	d. >560		d.		s. HCl, H_2SO_4
Cf. also under ferric and ferrous									
silicide	FeSi	83.91	yel.-gray, oct.	$6.1^{20°}_{4}$	tr. 450		0.00049	i.	i. aq. reg.
sulfide, di- (marcasite)	FeS_2	119.97	yel., rhb.	4.87	1171		0.0005		i. dil. a.
sulfide, di- (pyrite)	FeS_2	119.97	yel., cb.	5.0	d. >700		i.		i. dil. a.
sulfide (pyrrhotite)	Fe_7S_8	647.43	hex.	$4.6^{20°}_{4}$					
Krypton	Kr	83.70	col. gas	2.818 (A)	-169	-151.8	$11.05^{0°}$ cc	$3.57^{60°}$ cc	sl. s. al., bz.
Lanthanum	La	138.92	lead gray	$6.15^{20°}$	826	1800	d.	i.	s. a.
Lead	Pb	207.21	silv. met., cb.	$11.337^{20°}_{25°}$	327.5	1620	i.	i.	s. HNO_3; i. c. HCl, H_2SO_4
acetate	$Pb(C_2H_3O_2)_2$	325.30	wh. cr.	$3.251^{20°}_{4}$	280		$19.7^{0°}$	$221^{50°}$	s. gly.; v. sl. s. al.
acetate (sugar of lead)	$Pb(C_2H_3O_2)_2\cdot 3H_2O$†	379.35	wh. mn.	2.55	$-3H_2O$, 75		$45.64^{15°}$	$200^{100°}$	s. gly.; sl. s. al.
acetate	$Pb(C_2H_3O_2)_2\cdot 10H_2O$	505.46	wh., rhb.	1.689	22		s.	s.	
acetate, basic	$Pb_2(C_2H_3O_2)_3OH$	608.56	wh.				v. s.		sl. s. al.
acetate, basic	$Pb(C_2H_3O_2)_2\cdot Pb(OH)_2\cdot H_2O$	584.54	wh. nd.				v. s.		
acetate, basic	$Pb(C_2H_3O_2)_2\cdot 2Pb(OH)_2$	807.75	wh. nd.				5.55	18.2	s. al.
arsenate, monobasic	$PbH_4(AsO_4)_2$	489.06	tri. 1.82	$4.46^{15°}$	d. 140		d.	sl. s.	s. HNO_3
arsenate, dibasic (schultenite)	$PbHAsO_4$	347.13	wh., mn., 1.9097	5.94	d. >200	$-H_2O$, 280	i.	i.	s. HNO_3, NaOH
arsenate, meta-	$Pb(AsO_3)_2$	453.03	hex.	$6.42^{15°}$	802		d.	d.	s. HNO_3
pyro-	$Pb_2As_2O_7$	676.24	rhb., 2.03	$6.85^{15°}_{15°}$			i.	d.	s. HCl, HNO_3; i. sc.
Lead azide	PbN_6	291.26	col. nd.		expl. 350		i.	$0.05^{100°}$	v. s. ac.; NH_4OH
bromide	$PbBr_2$	367.05	col., rhb.	6.66	373	918	$0.4554^{0°}$	$4.75^{100°}$	s. a., KBr.; sl. s. NH_3; i. al.
carbonate (cerussite)	$PbCO_3$	267.22	wh., rhb., 2.0763	6.6	d. 315		$0.00011^{20°}$	d.	s. a., alk.; i. NH_3, al.
carbonate, basic (hydrocerussite; white lead)	$2PbCO_3\cdot Pb(OH)_2$†	775.67	wh., hex.	6.14	d. 400		i.	$3.34^{100°}$	s. ac.; sl. s. aq. CO_2
chloride (cotunnite)	$PbCl_2$	278.12	wh., rhb., 2.2172	5.80	501	954^{760mm}	$0.673^{0°}$	i.	sl. s. dil. HCl, NH_3, i. al.
chromate (crocoite)	$PbCrO_4$	323.22	yel., mn., 2.42	6.12	844		$0.000007^{20°}$	i.	s. a., alk.; i. NH_3, ac.
chromate, basic	$PbCrO_4\cdot PbO$	546.43	or.-yel. nd.				i.		s. a., alk.
formate	$Pb(HCO_2)_2$	297.25	wh., rhb.	4.56	d. 190		$1.6^{16°}$	$18^{100°}$ d.	i. al.
hydroxide	$3PbO\cdot H_2O$	687.65	cb.	7.592	$-H_2O$, 130		0.014		s. a., alk.
nitrate	$Pb(NO_3)_2$	331.23	col., cb. or mn., 1.7815	4.53	d. 470		$38.8^{0°}$	$138.8^{100°}$	$8.8^{22°}$ al.
oxide, sub-	Pb_2O	430.42	bk., amor.	8.34	d. red heat		i.	i.	s. a., alk
oxide, mono- (litharge)	PbO	223.21	yel., tet.	9.53	888		$0.0068^{18°}$	i.	s. alk., PbAc, NH_4Cl, $CaCl_2$
oxide, mono (massicotite)	PbO	223.21	yel., rhb., 2.61	8.0			i.		

Name	Formula	Mol. wt.	Color, crystalline form, index of refraction	Density	M.P., °C	B.P., °C	Solubility, cold water	Solubility, hot water	Solubility, other solvents
oxide, mono-	PbO	223.21	amor.	9.2 to 9.5	d. 500		i.	i.	s. alk, PbAc, NH₄Cl, CaCl₂
oxide, red (minium)	Pb₃O₄	685.63	red, amor.	9.1	d. 360		i.	i.	s. ac., h. HCl
oxide, sesqui-	Pb₂O₃	462.42	red-yel, amor.	9.375	d. 290		i.	i.	s. a., alk.
oxide, di- (plattnerite)	PbO₂	239.21	brn, tet., 2.229	9.375	d. 290		i.	i.	s. ac., h. alk.; i. al.
silicate	PbSiO₃	283.27	col, mn, 1.961	6.49	766		i.		s. a.
sulfate (anglesite)	PbSO₄	303.27	wh, mn or rhb., 1.8823	6.2	1170		0.0028^{18}	0.0056^{40}	s. conc. a., NH₄ salts; i. al.
sulfate, acid	Pb(HSO₄)₂·H₂O	419.36	cr.		d.		0.0001^{18}		sl. s. H₂SO₄
sulfate, basic (lanarkite)	PbSO₄·PbO	526.48	col, mn.	6.92	977		0.0044^{18}	i.	sl. s. H₂SO₄
sulfide (galena)	PbS	239.27	lead gray, cb., 3.912	7.5	1120		0.00009^{18}	s.	s. KCNS, HNO₃
thiocyanate	Pb(CNS)₂	323.37	col, mn.	3.82	d. 190		0.05^{20}	d.	s. a., NH₃
Lithium	Li	6.94	silv. met. cb.	0.53^{20}	186	1336 ± 5	d.	d.	7.7^{25}, 10^{78} al.
benzoate	LiC₇H₅O₂	128.05	wh. leaflets		33^{35}		40^{100}		s. al., act.
bromide	LiBr	86.86	wh, delq, cb., 1.784	$3.464 \tfrac{25}{4}$	547	1265	143^{0} (2H₂O)	266^{100} (1H₂O)	s. al.
bromide	LiBr·2H₂O	122.89	wh. pr.		44		246^{20}	d.	s. dil. a.; i. al., act., NH₃
carbonate	Li₂CO₃	73.89	wh, mn, 1.567	$2.11^{0}\tfrac{25}{4}$	618		1.54^{0}	0.72^{100}	2.48^{15} al.; s. et.
chloride	LiCl	42.40	wh, delq, cb., 1.662	$2.068\tfrac{25}{4}$	614	1360	67^{0}	127.5^{100}	sl. s. al., et.
citrate	Li₃C₆H₅O₇·4H₂O	281.98	wh. cr.		d.		61.2^{15}	66.7^{100}	s. HF; i. act.
fluoride	LiF	25.94	wh, cb, 1.3915	$2.295^{21.5°}$	870	1670	0.27^{18}	$0.135^{35°}$	sl. s. a., et.
formate	LiHCO₂·H₂O	69.97	col, rhb.	1.46	−H₂O, 94		49.2^{0}	$346.6^{04°}$	i. et.
hydride	LiH	7.95	wh, cb.	0.820	680	925±	d.		sl. s. al.
hydroxide	LiOH	23.95	wh. cr.	2.54	445		12.7^{0}	17.5^{100}	sl. s. al.
hydroxide	LiOH·H₂O	41.96	col, mn.	1.83	261		$22.3^{10°}$	$26.8^{80°}$	s. al., NH₃
nitrate	LiNO₃	68.95	col, trig., 1.735	2.38	261		53.4^{0}	$194^{70°}$	s. al., NH₃
nitrate	LiNO₃·3H₂O	123.00	col.		29.88		v. s.	∞	
oxide	Li₂O	29.88	col, 1.644	$2.013\tfrac{25}{4}$		subl. <1000	forms LiOH		
phosphate, monobasic	LiH₂PO₄	103.94	col.	2.461	>100		0.034^{18}		s. a., NH₄Cl; i. act.
phosphate, tribasic	Li₃PO₄	115.80	wh, rhb.	$2.537^{7.5°}$	837		v. sl. s.	v. sl. s.	
phosphate, tribasic	Li₃PO₄·12H₂O	331.99	wh, trig.	1.645	100		v. sl. s.		
salicylate	LiC₇H₅O₃	144.05	col.		d.		$128^{25°}$		v. s. al.
sulfate	Li₂SO₄	109.94	col, mn, 1.465	2.22	860		35.34^{0}	29.9^{100}	i. act, 80% al.
sulfate	Li₂SO₄·H₂O	127.96	col, mn, 1.477	2.06	−H₂O, 130		43.6^{0}	35^{100}	i. 80% al.
sulfate, acid	LiHSO₄	104.01	pr.	$2.123^{13°}$	170.5		d.		
Lutecium	Lu	174.99							
Magnesium	Mg	24.32	silv. met., hex.	$1.74^{20°}$	651	1110	i.		s. a., NH₄ salts; i. al.
acetate	Mg(C₂H₃O₂)₂	142.41	wh.	1.42	323		v. s.		5.25^{15} m. al.
acetate	Mg(C₂H₃O₂)₂·4H₂O†	214.47	wh, mn, pr, 1.491	1.454	80		v. s.		v. s. al.
aluminate (spinel)	MgO·Al₂O₃	142.26	col cb, 1.718–23	3.6	2135		i.		v. sl. s. dil. HCl; i. dil. HNO₃
ammonium chloride	MgCl₂·NH₄Cl·6H₂O	256.83	wh, rhb, delq.	1.456	−4H₂O, 195		16.7	s.	
ammonium phosphate (struvite)	MgNH₄PO₄·6H₂O	245.44	col, rhb, 1.496	1.715	d. 100		0.0231^{0}	0.0195^{90}	s. a.; i. al.
ammonium sulfate (boussingaultite)	MgSO₄·(NH₄)₂SO₄·6H₂O	360.62	col, mn.	1.72	>120		16.86^{0}	130^{100}	
benzoate	Mg(C₇H₅O₂)₂·3H₂O	320.59	wh, pd.		−3H₂O, 110		$4.5^{25°}$ (anh.)	s.	s. act.
carbonate (magnesite)	MgCO₃	83.43	wh, trig, 1.700	3.037	d. 350		0.0106	d.	s. a., aq, CO₂; i. act., NH₃
carbonate (nesquehonite)	MgCO₃·3H₂O	138.38	wh, rhb, 1.501	1.852	−H₂O, 100		$0.1518^{19°}$	d.	s. a., aq, CO₂
carbonate, basic (hydromagnesite)	3MgCO₃·Mg(OH)₂·3H₂O	365.37	wh, rhb, 1.530	2.16	d.		0.04	0.011	s. a., NH₄ salts; i. al.
Magnesium chloride (chloromagnesite)	MgCl₂	95.23	col, hex, 1.675	$2.325^{25°}$	712	1412	52.8^{0}	73^{100}	50 al.
chloride (bischofite)	MgCl₂·6H₂O†	203.33	wh, delq, mn, 1.507	1.56	118 d.		281^{0}	918^{100}	50 al.
hydroxide (brucite)	Mg(OH)₂	58.34	wh, trig, 1.5617	2.4	d.		0.0009^{18}	d.	s. NH₄ salts, dil. a.
nitride	Mg₃N₂	100.98	gn-yel, amor.		d.		i.		s. a.; i. al.
oxide (magnesia; periclase)	MgO	40.32	col, cb, 1.7364	3.65	2800	3600	0.00062	d.	s. a., NH₄ salts; i. al.
perchlorate	Mg(ClO₄)₂†	223.23	wh, delq.	$2.60^{25°}$	d.		$99.6^{5\,25°}$	v. s.	24^{25} al., 51.8^{25} m. al.; 0.29 et.

*See also a table of alloys.
†Usual commercial form.

2-17

TABLE 2-1 Physical Properties of the Elements and Inorganic Compounds (Continued)

Name	Formula	Formula weight	Color, crystalline form and refractive index	Specific gravity	Melting point, °C	Boiling point, °C	Solubility in 100 parts		
							Cold water	Hot water	Other reagents
Magnesium chloride (*Cont.*)									
peroxide	MgO_2	56.32	wh. pd.	2.598^{22}	expl. 275		i.	i.	s. a.
phosphate, pyro-	$Mg_2P_2O_7$	222.60	col., mn., 1.604	2.56	1383		i.	i.	s. a.; i. alk.
phosphate, pyro-	$Mg_2P_2O_7 \cdot 3H_2O$	276.65	wh., amor.		$-3H_2O$, 100		sl. s.		s. a.; i. al.
potassium chloride (carnallite)	$MgCl_2 \cdot KCl \cdot 6H_2O$	277.88	delq., rhb., 1.475	$1.60^{19.4°/4}$	265		64.5^{19} d.	d.	d. al.
potassium sulfate (picromerite)	$MgSO_4 \cdot K_2SO_4 \cdot 6H_2O$	402.73	mn., 1.4629	2.15	d. 72		19.26° d.	81.75°	d. HF
silicofluoride	$MgSiF_6 \cdot 6H_2O$	274.48	col., trig., 1.3439	$1.788^{17.5°/4}$	d.		$64.8^{17.5°}$	s.	
sodium chloride	$MgCl_2 \cdot NaCl \cdot H_2O$	171.70	col.				s.	s.	s. al.
sulfate	$MgSO_4$	120.38	col.	2.66	1185		26.9°	68.3^{100}	s. al.
sulfate (epsom salt; epsomite)	$MgSO_4 \cdot 7H_2O°$	246.49	col., rhb., 1.4554	1.68	70 d.		72.4°	178^{40}	s. dil. a.
Manganese	Mn	54.93	gray-pink met.	$7.2^{20°}?$	1260	1900	d.	s.	
acetate	$Mn(C_2H_3O_2)_2$	173.02	pa. pink. mn.	$1.74^{20°/4}$			s.		
acetate	$Mn(C_2H_3O_2)_2 \cdot 4H_2O$	245.08	rose, trig., 1.817	1.589	d.			64.5^{50}	
carbonate (rhodochrosite)	$MnCO_3$	114.94	rose red, delq., mn., 1.575	3.125			0.0065^{25}		s. al, CO_2, dil. a.; l. NH_3, al.
chloride (scacchite)	$MnCl_2$	125.84	rose, delq., cb.	$2.977^{25°/4}$	650	1190	63.4°	123.8^{100}	s. al.; i. et., NH_3
chloride	$MnCl_2 \cdot 4H_2O°$	197.91	rose red, mn.	2.01	58.0	$-H_2O$, 106; $-4H_2O$, 200	151^{8}	∞	s. al.; i. et.
chloride, per-	$MnCl_4$	196.76	gn.		d.		s.	s.	s. al., et.
hydroxide (ous) (pyrochroite)	$Mn(OH)_2$	88.95	wh., trig., 2.24	$3.258^{18°}$			0.002^{20}	i.	s. a., NH_4 salts; i. alk.
hydroxide (ic) (manganite)	$Mn_2O_3 \cdot H_2O$	175.88	brm., rhb., 2.24	4.3					s. h. H_2SO_4
nitrate	$Mn(NO_3)_2 \cdot 6H_2O$	287.04	rose red, mn.	$1.82^{21°}$	25.8	129.5	426°	∞	v. s. al.
oxide (ous) (manganosite)	MnO	70.93	gray-gn., cb., 2.16	5.18	1650		i.	i.	s. a., NH_4Cl
oxide (ic)	Mn_2O_3	157.86	brn-bk., cb.	4.81	−O, 1080		i.	i.	s. a.; i. act.
oxide, di- (pyrolusite; polianite)	$MnO_2°$	86.93	bk., rhb.	5.026	−O, >230		i.	i.	s. HCl; i. HNO_3, act.
sulfate (ous)	$MnSO_4$	150.99	red-wh.	3.235	700	d. 850	53°	73^{50}	s. al.; i. et.
sulfate (ous) (szmikite)	$MnSO_4 \cdot H_2O$	169.01	pa. pink, mn., 1.595	2.87	Stable 57 to 117		98.47^{48}	79.77^{100}	
sulfate (ous)	$MnSO_4 \cdot 2H_2O$	187.02	pink, rhb. or mn., 1.518	$2.526^{15°}$	Stable 40 to 57		85.27^{35}	106.8^{55}	
sulfate (ous)	$MnSO_4 \cdot 3H_2O$	205.04	pink, tri., 1.508	$2.356^{15°}$	Stable 30 to 40		74.22^{5}	99.31^{57}	
sulfate (ous)	$MnSO_4 \cdot 4H_2O°$	223.05	pink, rhb. or mn.	2.107	Stable 18 to 30	$-4H_2O$, 450	136^{16}	169^{90}	i. al.
sulfate (ous)	$MnSO_4 \cdot 5H_2O$	241.07		$2.103^{15°}$	Stable 8 to 18		142^{5}	200^{85}	
sulfate (ous)	$MnSO_4 \cdot 6H_2O$	259.09			Stable −5 to +8		204^{0}	247^{8}	
sulfate (ous)	$MnSO_4 \cdot 7H_2O$	277.10	pink. mn. or rhb.	2.092	Stable −10 to −5; 19 d.	$-7H_2O$, 280	176^{0}	251^{14}	
sulfate (ic)	$Mn_2(SO_4)_3$	398.04	gn., delq. cr.	3.24	d. 160		v. s.	d.	i. al.
Masurium	Ma	98–99.5		11.5	2300 (?)				s. HCl, dil. H_2SO_4; l. conc. H_2SO_4, HNO_3
Mercuric acetate	$Hg(C_2H_3O_2)_2$	318.70	wh., pl.	3.270	d.		25^{10}	100^{100}	s. al. sl. d.
bromide	$HgBr_2$	360.44	wh., rhb.	6.053	237	322	0.5^{20}	25^{100}	25.2° al.; v. sl. s. et.
carbonate, basic	$HgCO_3 \cdot 2HgO$	693.84	brm-red	5.44			i.	i.	s. aq. CO_2, NH_4Cl
chloride (corrosive sublimate)	$HgCl_2$	271.52	wh., rhb., 1.859	5.44	277	304	3.6°	61.3^{100}	33^{25} 99% al., 33 et.
fulminate	$Hg(CNO)_2$	284.65	cb.	4.42	expl.		sl. s.	i.	s. NH_4OH, al.
hydroxide	$Hg(OH)_2$	234.63			$-H_2O$, 175		i.		s. a.
oxide (montroydite)	HgO	216.61	yel. or red, rhb., 2.5	11.14	d. 100		0.0052^{25}	0.041^{100}	s. a.; i. al.
oxychloride (kleinite)	$HgCl_2 \cdot 3HgO$	921.35	yel., hex.	7.93	d. 260		i.	d.	s. HCl
silicofluoride, basic	$HgSiF_6 \cdot HgO \cdot 3H_2O$	613.33	yel. nd.		d.		d.		s. a.
sulfate	$HgSO_4$	296.67	wh., rhb.	6.47	d.				s. a.; i. al., act, NH_3
sulfate, basic (turpeth)	$HgSO_4 \cdot 2HgO$	729.89	yel., tet.	6.44			0.005	0.167^{100}	s. a.; i. al.
Mercurous acetate	$HgC_2H_3O_2$	259.65	yel., sc.		subl. 345		0.75^{13}	i.	s. H_2SO_4, HNO_3; i. al.
bromide	$HgBr$	280.53	wh., tet.	7.307	d. 130		7×10^{-9}	d.	s. a.; i. al., act,
carbonate	Hg_2CO_3	461.23	yel. pd.				i.		s. NH_4Cl

Name	Formula	Mol. wt.	Color, crystalline form, index of refraction	Density	M.P., °C	B.P., °C	Solubility cold water	Solubility hot water	Solubility in other solvents
chloride (calomel)	HgCl	236.07	wh, tet, 1.9733	7.150	302	383.7	0.0014^{0}	0.0007^{43}	s. aq. reg, Hg(NO₃)₂; sl. s. HNO₃, HCl; i. al., etc.
iodide	HgI	327.53	yel, tet.	7.70	290 d.	subl. 140; 310d.	2×10^{-8}	v. sl. s.	s. KI; i. al.
nitrate	HgNO₃·H₂O	280.63	wh. mn.	$4.785^{3.9°}$	70	expl.	v. s.	d.	s. HNO₃; i. al., et.
Mercurous oxide	Hg₂O	417.22	bk.	9.8	d. 100		i.	0.0007	s. h. ac.; i. alk, dil. HCl; NH₃
sulfate	Hg₂SO₄	497.28	wh., mn.	7.56	d.		$0.055^{16.5}$	$0.092^{100°}$	s. H₂SO₄, HNO₃
Mercury†	Hg	200.61	silv. lq. or hex.(?)	$13.546^{20°}_{4}$	-38.87	356.9	i.	i.	s. HNO₃; i. HCl
Molybdenum	Mo	95.95	gray, cb.	10.2	2620 ± 10	3700	i.	i.	s. h. conc. H₂SO₄; i. HCl, HF, NH₃, dil. H₂SO₄, Hg
chloride, di-	MoCl₂	166.85	yel, amor.		d.		i.	i.	s. HCl, H₂SO₄, NH₄OH, al., et.
chloride, tri-	MoCl₃	202.32	dark red pd.	$3.714^{25°}_{4}$	d.		i.	d.	s. HNO₃, H₂SO₄; v. sl. s. al., et.
chloride, tetra-	MoCl₄	237.78	brn., delq.	$3.578^{25°}_{4}$	volt.	d.	s.	s.	s. HNO₃, H₂SO₄, sl. s. al., et.
chloride, penta-	MoCl₅	273.24	bk. cr.		194	268	s.	d.	s. HNO₃, H₂SO₄; i. abs. al., et.
oxide, tri- (molybdite)	MoO₃	143.95	col., rhb.	$4.50^{19.5}$	795	subl.	0.107^{18}	2.106^{70}	s. a., NH₄OH
sulfide, di- (molybdenite)	MoS₂	160.07	bk., hex., 4.7	4.80^{114}	1185		i.	i.	s. H₂SO₄ aq. reg.
sulfide, tri-	MoS₃	192.13	red-brn.		d.		sl. s.	s.	s. alk. sulfides
sulfide, tetra-	MoS₄	224.19	brn. pd.		d. 115		v. sl. s.	i.	s. alk. sulfides; i. NH₃
Molybdic acid	H₂MoO₄	161.97	yel-wh., hex.	3.124^{15}	-H₂O, 70	-2H₂O, 200	0.133^{18}	sl. s.	s. NH₄OH, H₂SO₄; i. NH₃
Molybdic acid	H₂MoO₄·H₂O	179.98	yel., mn.		840		d.	2.13^{70}	s. a., NH₄OH, NH₄, salts
Neodymium	Nd	144.27	yellowish	$6.9^{3°}$	840			d.	s. a.
Neon	Ne	20.18	col. gas	lq. $1.204^{-245.9}$ (A), 0.674 (A)	-248.67	-245.9	2.6^{0} cc	1.1^{45} cc	s. lq. O₂, al., act., bz.
Neptunium	Np²³⁹	239			Produced by Neutron bombardment of U²³⁸		i.	i.	
Nickel	Ni	58.69	silv. met., cb.	8.90^{20}	1452	2900	i.	i.	s. dil. HNO₃; sl. s. H₂SO₄, HCl; i. al.
acetate	Ni(C₂H₃O₂)₂	176.78	gn. pr.	1.798	d.		16.6	v. s.	i. NH₃
ammonium chloride	NiCl₂·NH₄Cl·6H₂O	291.20	gn., delq., mn.	1.645			$150^{25°}$	v. s.	v. sl. s. (NH₄)₂SO₄
ammonium sulfate	NiSO₄·(NH₄)₂SO₄·6H₂O	394.99	blue-gn., mn., 1.5007	1.923	d.		$2.5^{3.5°}$	$39.2^{88°}$	s. NH₄OH
bromate	Ni(BrO₃)₂·6H₂O	422.62	gn., cb.	2.575			28	s.	s. al., et., NH₄OH
bromide	NiBr₂	218.52	yel., delq.	$4.64^{25°}_{4}$	d.		$112.8^{9°}$	$156^{100°}$	s. al., et., NH₄OH
bromide	NiBr₂·3H₂O	272.57	gn., delq.	1.837	-3H₂O, 200		$199^{9°}$	$316^{100°}$	i. c. NH₄OH
bromide, ammonia	NiBr₂·6NH₃	320.71	vl. pd.	3.715	d.		v. s.	d.	s. a.
bromoplatinate	NiPtBr₆·6H₂O	841.51	trig.				v. s.		s. a., NH₄ salts
carbonate	NiCO₃	118.70	lt. gn., rhb.		d.		i.	i.	s. a.
carbonate, basic	2NiCO₃·3Ni(OH)₂·4H₂O	587.58	lt. gn.		d.		$0.0093^{35°}$	d.	s. a., NH₄ salts
carbonyl	Ni(CO)₄	170.73	lq.	1.31^{17}	-25	43^{751mm}	i.	d.	s. aq. reg, HNO₃, al., et.
chloride	NiCl₂	129.60	yel., delq.	3.544	subl.	973	53.8^{0}	$87.6^{100°}$	s. NH₄OH, al.; i. NH₃
chloride	NiCl₂·6H₂O°	237.70	gn., delq., mn., 1.57±				180	v. s.	v. s.
chloride, ammonia	NiCl₂·6NH₃	231.80	gn. pl.		d.		s.	d.	s. NH₄OH; i. al.
cyanide	Ni(CN)₂·4H₂O	182.79	scarlet red cr.		-4H₂O, 200		i.	i.	s. KCN; i. dil. KCl
dimethylglyoxime	NiC₈H₁₄O₄N₄	288.91			subl. 250		i.	i.	s. abs. al., a.; i. ac., NH₄OH
formate	Ni(HCO₂)₂·2H₂O	184.76	gn. cr.	2.154	d.		s.	s.	s. a., NH₄OH, NH₄Cl
hydroxide (ic)	Ni(OH)₃	109.71	bk.	4.36	d.		i.	i.	s. a., NH₄OH; i. alk.
hydroxide (ous)	Ni(OH)₂·¼H₂O	97.21	lt. gn.		d.		v. sl. s.	i.	s. a., NH₄OH; i. abs. al.
nitrate	Ni(NO₃)₂·6H₂O	290.80	gn., mn.	2.05	56.7	136.7	243.0^{0}	v. s.	i. al.
nitrate, ammonia	Ni(NO₃)₂·4NH₃·2H₂O	286.87	red yel, mn.		d.		v. s.	i.	s. a., NH₄OH
oxide, mono- (bunsenite)	NiO	74.69	gn-bk., cb., 2.37	7.45	Forms Ni₂O₃ at 400		i.	i.	i. al.
potassium cyanide	Ni(CN)₂·2KCN·H₂O	258.97		$1.875^{11°}$	-H₂O, 100		v. sl. s. $\infty^{56.7}$	i.	s. a., NH₄OH
sulfate	NiSO₄	154.75	yel., cb.	3.68	-SO₃, 840		27.2^{0}	$76.7^{100°}$	i. al., et., act.

°Usual commercial form.
†See also Tables 2-28 and 2-280.

TABLE 2-1 Physical Properties of the Elements and Inorganic Compounds (Continued)

Name	Formula	Formula weight	Color, crystalline form and refractive index	Specific gravity	Melting point, °C	Boiling point, °C	Cold water	Hot water	Other reagents
								Solubility in 100 parts	
Nickel (*Cont.*)									
sulfate	$NiSO_4·6H_2O°$	262.85	gn. mn. or blue, tet., 1.5109	2.07	tr. 53.3	$-6H_2O$, 280	$131^{30°}$	$280^{100°}$	v. s. NH_4OH, al.
sulfate (morenosite)	$NiSO_4·7H_2O$	280.86	gn. rhb., 1.4893	1.948	98–100	$-6H_2O$, 103	$63.5^{0°}$	$117.8^{30°}$	s. al.
Nitric acid	HNO_3	63.02	col. lq.	1.502	−42	86	∞	∞	expl. with al.
Nitric acid	$HNO_3·H_2O$	81.03	col. lq.		−38		∞	∞	d. al.
Nitric acid	$HNO_3·3H_2O$	117.06	col. lq.		−18.5				d. al.
Nitro acid sulfite	NO_2HSO_3	127.08	col., rhb.		73 d.		d.	$1.55^{20°}$ cc	s. H_2SO_4
Nitrogen	N_2	28.02	col. gas or cb. cr.	$1.026^{-252.5°}$ $0.808^{-195.8°}$ $12.5°$ (D)	−209.86	−195.8	$2.35^{0°}$ cc	$1.55^{20°}$ cc	sl. s. al.
Nitrogen oxide, mono- (ous)	N_2O	44.02	col. gas	lq. $1.226^{-89°}$ 1.530 (A)	−102.3	−90.7	$130.52^{0°}$ cc	$60.82^{24°}$ cc	s. H_2SO_4, al.
oxide, di- (ic)	NO or $(NO)_2$	30.01 (60.02)	col. gas	lq. $1.269^{-150.2°}$ 1.0367 (A)	−161	−151	$7.34^{0°}$ cc	$0.00^{0°}$	26.6 cc al.: 3.5 cc H_2SO_4; s. aq. $FeSO_4$
oxide, tri-	N_2O_3	76.02	red-brn. gas or blue lq. or solid	$1.447°$	−102	3.5	s.		s. a., et.
oxide, tetra- (per- or di-)	NO_2 or $(NO_2)_2$	46.01 (92.02)	yel. lq., col. solid, red-brn. gas	$1.448^{20°}$	−9.3	21.3	d.	d.	s. HNO_3, H_2SO_4, chl., CS_2
oxide, penta-	N_2O_5	108.02	wh., rhb,	1.63^{18}	30	47	s.	Forms HNO_3	s.
oxybromide	$NOBr$	109.92	brn. lq.	>1.0	−55.5	−2	d.	d.	s. fuming H_2SO_4
oxychloride	$NOCl$	65.47	red-yel. lq. or gas	$1.417^{-12°}$ 2.31 (A)	−64.5	−5.5	d.	d.	
Nitroxyl chloride	NO_2Cl	81.47	yel-brn. gas	lq. $1.32^{14°}$	<−30	5	d.		sl. s. aq. reg., HNO_3; i. NH_3
Osmium	Os	190.2	blue, hex.	$22.48^{20°}$	2700	>5300	i.	i.	s. $NaCl$, al., et.
chloride, di-	$OsCl_2$	261.11	gn., delq.		d. 560–600		s. d.		s. a. alk., al.; sl. s. et.
chloride, tri-	$OsCl_3$	296.57	brn., cb.				sl. s.		s. HCl, d.
chloride, tetra-	$OsCl_4$	332.03	red-yel. nd.				s. d.		sl. s. al., s. fused Ag
Oxygen	O_2	32.00	col. gas or hex. solid	$1.14^{-188°}$ $1.426^{-252.5°}$ 1.1053 (A)	−218.4	−183	$4.89^{0°}$ cc	$2.6^{30°}$ cc $1.7^{100°}$ cc	
Ozone	O_3	48.00	col. gas	$1.71^{-183°}$ $3.03^{-80°}$ 1.658 (A)	−251	−112	$0.494^{0°}$ cc	$0^{60°}$ cc	s. oil turp., oil cinn.
Palladium	Pd	106.70	silv. met., cb.	$12.0^{20°}$ $11^{1550°}$	1555	2200	i.	i.	s. aq. reg., h. H_2SO_4; i. NH_3
bromide (ous)	$PdBr_2$	266.53	brn.				i.	i.	s. HBr
chloride	$PdCl_2$	177.61	brn., cb.		500 d.		s.	s.	s. HCl, act., al.
chloride	$PdCl_2·2H_2O$	213.65	brn. pr.				s.	s.	s. HCl, act., al.
cyanide	$Pd(CN)_2$	158.74	yel.		d.		i.	i.	s. HCN, KCN, NH_4OH; i. dil. a.
hydride	Pd_2H	214.41	met.	11.06	d.				
Palladous dichlorodiammine	$Pd(NH_3)_2Cl_2$	211.68	red or yel. tet.	2.5			s.		s. a., NH_4OH
Perchloric acid	$HClO_4$	100.46	unstable, col. lq	$1.768^{\frac{22°}{4}}$	−112	16^{18mm}	s.	v. s.	
Perchloric acid	$HClO_4·H_2O$	118.48	fairly stable nd.	1.88	50	d.	s.	d.	
Perchloric acid	$HClO_4·2H_2O°$ 73.6% anh.	136.50	stable lq., col.	$1.71^{\frac{25°}{4}}$	−17.8	200	v. s.	v. s.	s. al.
Periodic acid	HIO_4	191.93	wh. cr.		d. 138	subl. 110	s.	s.	sl. s. al., et.
Periodic acid	$HIO_4·2H_2O$	227.96	delq., mn.		d. 110		v. s.	v. s.	d. al.
Permanganic acid	$HMnO_4$	119.94	exists only in solution				v. s.	v. s.	
Permolybdic acid	$HMoO_4·2H_2O$	196.99	wh. cr.		<60		v. s.	v. s.	
Persulfuric acid	$H_2S_2O_8$	194.14	hyg. cr.		d.		v. s.	v. s. d.	
Phosphamic acid	$PONH_2'(OH)_2$	97.02	cb.		78	$-25H_2O$, 140	v. s. d.		i. al.
Phosphatomolybdic acid	$H_7P(Mo_2O_7)_6·28H_2O$	2365.88	yel. cb.		d.	−85	$26^{17°}$ cc	$i.^{100°}$	s. HNO_3
Phosphine	PH_3	34.00	col. gas	lq. $0.746^{-90°}$ 1.146 (A)	−132.5	subl.	d.		s. Cu_2Cl_2, al., et.
Phosphonium chloride	PH_4Cl	70.47	wh., cb.		28^{64atm}	subl.	d.		

Name	Formula	Mol. wt.	Color, crystalline form, refractive index	Density	M.p., °C	B.p., °C	Solubility in cold water	Solubility in hot water	Solubility in other solvents
Phosphoric acid, hypo-	$H_4P_2O_6$	161.99	cr.	2.2–2.5	55	d. 70	s.	$450^{62°}$	i. lq. CO_2
Phosphoric acid, meta-	HPO_3	79.99	vitreous, delq		subl.	$-½H_2O$, 213	s.	Forms H_3PO_4	s. al.
Phosphoric acid, ortho-	H_3PO_4†	98.00	col., rhb.	$1.834^{18.2°}$	42.35		$2340^{20°}$	v. s.	v. s. al. et.
Phosphoric acid, pyro-	$H_4P_2O_7$†	177.99	wh. nd.		61	d.	$800^{28°}$	Forms H_3PO_4	
Phosphorous acid, hypo-	H_3PO_2	66.00	syrupy	$1.493^{18.8°}$	26.5	d. 200	∞	$730^{40°}$	
Phosphorous acid, ortho-	H_3PO_3	82.00	col.	$1.651^{21.2°}$	74	d. 130	$307.3^{0°}$		
Phosphorous acid, pyro-	$H_4P_2O_5$	145.99	nd.		38		d.		
Phosphorus, black	P_4	123.92	rhombohedral	2.69	590^{43atm}	ign. in air, 400	i.	i.	i. CS_2
Phosphorus, red	P_4	123.92	red, cb.	$2.20^{20°}$		ign. in air, 725	i.	i.	
Phosphorus, yellow	P_4	123.92	yel., hex., 2.1168	$1.82^{20°}$; lq. $1.745^{44.5°}$	44.1; ign. 34	280	0.0003	sl. s.	s. alk.; i. CS_2, NH_3, et. 0.4 al.; $1000^{10°}$ CS_2; $1.5^{0°}$, 10^{87} bs.; s. NH_3
chloride, tri-	PCl_3	137.35	col., fuming lq.	$1.574^{20.8°}/4$	-111.8	75.95^{760mm}	d.	v. s.	s. et., chl., CS_2
chloride, penta-	PCl_5	208.27	delq., tet.	solid 1.6; $3.60^{295°}$(A)	148 under pressure	subl. 160	d.	Forms H_3PO_4	s. CS_2, C_6H_5COCl
oxide, penta-	P_2O_5	141.96	wh., delq., amor.	2.387	subl. 250		Forms H_3PO_4	v. s.	s. H_2SO_4; i. NH_3, act.
oxychloride	$POCl_3$	153.35	col., fuming lq.	1.675	2	107.2^{760mm}	d.	i.	d. al.
Phosphotungstic acid	$P_2O_5·2WO_3·42H_2O$	3681.67	yel.-gn. cr.				s.	v. s.	s. al., et.
Platinum	Pt	195.23	silv. met., cb.	$21.45^{20°}$ lq. $19^{1755°}$	1755	4300	i.	i.	s. aq. reg., fused alk.
chloride (ic)	$PtCl_4$	337.06	brn.		d. 370		$140^{25°}$		s. al., act.; sl. s. NH_3; i. et.
chloride (ous)	$PtCl_2$	266.14	brn.	$5.87^{11°}$	d. 581		i.		s. HCl, NH_4OH; sl. s. NH_3; i. al., et.
chloride (ic)	$PtCl_4·8H_2O$	481.19	red, mn.	2.43	$-4H_2O$, 100		v. s.		s. al., et.
cyanide (ous)	$Pt(CN)_2$	247.27	yel.-brm.				i.		i. alk.
Plutonium	Pu	238	Produced by deuteron bombardment on U^{238}						
Plutonium	Pu	239	Produced by neutron bombardment on U^{238}						
Potassium	K	39.10	silv. met., cb.	$0.86^{20°}$ lq. $0.83^{42°}$	62.3	760	d.	Forms KOH	s. a., al., Hg
acetate	$KC_2H_3O_2$	98.14	wh. pd.	1.8	292		$217^{0°}$	$396^{90°}$	33 al.; i. et.
acetate, acid	$KH(C_2H_3O_2)_2$	158.19	delq. nd. or pl.		148	d. 200	d.	d.	s. ac.
aluminate	$K_3(AlO_2)_2·3H_2O$	250.18	cr.				s.	v. s.	s. alk.; i al.
amide	KNH_2	55.12	yel.-grn.		338	subl. 400			d. al.; $3.6^{25°}$ NH_3
arsenate (monobasic)	KH_2AsO_4	180.02	col., tet., 1.5674	2.867	288		$18.87^{0°}$	v. s.	i. al.
auricyanide	$KAu(CN)_4·1.5H_2O$	367.39	pl.		d. 200		14.3	$200^{100°}$	sl. s. al.; i. et.
aurocyanide	$KAu(CN)_2$	288.33	rhb.					$60^{60°}$	i. satd. K_2CO_3, al.
bicarbonate	$KHCO_3$	100.11	mn., 1.482	2.17	d. 100–200		$22.4^{0°}$	$60^{60°}$	d. al.
bisulfate	$KHSO_4$	136.16	rhb. or mn., 1.480	2.35	210		$36.3^{0°}$	$121.6^{100°}$	sl. s. al.; i. act.
bromate	$KBrO_3$	167.01	trig.	$3.27^{17.5°}$	370 d.		$3.11^{0°}$	$49.75^{100°}$	sl. s. al., et.
bromide	KBr	119.01	col., cb., 1.5594	$2.75^{25°}$	730	1380	$53.5^{0°}$	$104^{100°}$	i. al.
carbonate	K_2CO_3	138.20	wh., delq. pd., 1.531	2.29	891	d.	$105.5^{0°}$	$156^{100°}$	0.83 al.; s. alk.
carbonate	$K_2CO_3·2H_2O$	174.23	rhb.	2.043			$183^{0°}$	$331^{100°}$	s. al., alk.
carbonate	$2K_2CO_3·3H_2O$	330.45	mn.	2.13			$129.4^{0°}$	$268^{100°}$	i. al., et.
chlorate	$KClO_3$	122.56	col., mn., 1.5167	2.32	368	d. 400	$3.3^{0°}$	$57^{100°}$	i. al.
chloride (sylvite)	KCl	74.56	col., cb., 1.4904	1.988	790	1500	$27.6^{0°}$	$56.7^{100°}$	v. sl. s. al.
chloroplatinate	K_2PtCl_6	486.16	yel., cb., 1.825±	3.499	d. 250		$0.74^{0°}$	$5.2^{100°}$	s. gly.; $0.9^{19.5°}$ al.; 1.3 h. al.
chromate (tarapacaite)	K_2CrO_4	194.20	yel., rhb., 1.7261	$2.732^{18°}$	975		$58.0^{0°}$	$75.6^{100°}$	i. al.
cyanate	$KCNO$	81.11	wh., tet.	2.048				d.	s. act.; sl. s. al.; i. NH_3
cyanide	KCN	65.11	wh., cb., delq., 1.410	$1.52^{16°}$	634.5		s.	$122.2^{108.8°}$	s. act.; i. NH_3, al., et.
dichromate	$K_2Cr_2O_7$	294.21	red, tri.	2.676	398		$4.9^{0°}$	$80^{100°}$	sl. s. al.; i. et.
ferricyanide	$K_3Fe(CN)_6$	329.25	red, mn. pr., 1.5689	1.84	d.		$33.4^{0°}$	$77.5^{100°}$	i. et., bz., CS_2
ferrocyanide	$K_4Fe(CN)_6·3H_2O$	422.39	yel., mn., 1.5772	$1.853^{17°}$	$-3H_2O$, 70		$27.8^{12.2°}$	$90.6^{96.8°}$	s. al.
formate	$KHCO_2$	84.11	col., rhb.	1.91	167.5		$331^{18°}$	$657^{90°}$	s. al., et.; i. NH_3
hydride	KH	40.10	cb., 1.453	0.80	d.		d.		s. KI; i. al., NH_3
hydrosulfide	KHS	72.16	wh., delq., rhb.	2.0	455		s.	s. d.	
hydroxide	KOH	56.10	wh., delq., rhb.	2.044	380	1320	$97^{0°}$	$178^{100°}$	
iodate	KIO_3	214.02	col., mn.	3.89	560		$4.73^{0°}$	$32.2^{100°}$	
iodide	KI	166.02	wh., cb., 1.6670	3.13	723	1330	$127.5^{0°}$	$208^{100°}$	$4^{20°}$ al.; s. NH_3; sl. s. et.

°One commercial form 70 to 72 per cent.
†Common commercial form 85 per cent H_3PO_4 in aqueous solution.

Name	Formula	Formula weight	Color, crystalline form and refractive index	Specific gravity	Melting point, °C	Boiling point, °C	Solubility in 100 parts — Cold water	Solubility in 100 parts — Hot water	Solubility in 100 parts — Other reagents
Potassium (*Cont.*)									
tri-iodide	KI_3	419.86	dark blue, delq., mn.	3.498	45	d. 225	v. s.	v. s.	s. KI, al.
iodoplatinate	K_2PtI_6	1034.94	cb.	5.18			s.		s. KOH
manganate	K_2MnO_4	197.12	gn., rhb.		d. 190		d.	$120^{94°}$	sl. s. al.; i. et.
metabisulfite	$K_2S_2O_5$	222.31	mn., pl.		d. 150	d. 400	$25°$	$246^{100°}$	$0.13^{0°}$ al.; i. et.
nitrate (saltpeter)	KNO_3	101.10	col., rhb., 1.5038	$2.11^{10.6°}$	tr. 129, 333	d. 400	$13.3°$	$413^{100°}$	v. s. NH_3; sl. s. al.
nitrite	KNO_2	85.10	pr.	1.915	297	d. 350	$281°$	$83.2^{100°}$	
oxalate	$K_2C_2O_4·H_2O$	184.23	wh., mn.	2.13	d.		$28.7°$	$48.1^{100°}$	s. al. et.
oxalate, acid	KHC_2O_4	128.12	mn., 1.545	2.0	d.		$14.3^{0°}$	$51.5^{100°}$	$0.105^{20°}$ m. al.; i. et.
oxalate, acid	$KHC_2O_4·½H_2O$	137.13	trimetric		d.		$2.2°$	v. s.	s. H_2SO_4; d. al.
oxide	K_2O	94.19	wh., cb.	$2.32^{20°/4°}$	d. 400		Forms KOH	$21.8^{100°}$	i. al
perchlorate	$KClO_4$	138.55	col., rhb., 1.4737	$2.524^{11°/4°}$			$0.75°$	$32.35^{75°}$	i. al
permanganate	$KMnO_4$	158.03	purple, rhb.	2.703	d. <240		$2.83°$	$10^{40°}$	
persulfate	$K_2S_2O_8$	270.31	wh., tri., 1.4669		d. <100		$1.77°$	$83.5^{50°}$	
phosphate, monobasic	KH_2PO_4	136.09	col., delq., tet., 1.5095	2.338	256		$14.8°$		
phosphate, dibasic	K_2HPO_4	174.18	wh., delq.	$2.564^{17°}$	d.		$33^{25°}$	v. s.	sl. s. al.
phosphate, tribasic	K_3PO_4	212.27	wh., rhb.	$2.258^{14.5°}$	1340	1320	$193.1^{25°}$	v. s.	i. al.
phosphate, meta-	KPO_3	118.08	wh. pd.	$2.264^{14.5°}$	tr. 450; 798	d.	s.	s.	
phosphate, meta-	$K_4P_4O_{12}·2H_2O$	508.34	amor.	2.33	$-2H_2O$, 100		s.		s. a.
phosphate, pyro-	$K_4P_2O_7·3H_2O$	384.39	delq.	1.63	$-2H_2O$, 180	$-3H_2O$, 300	s.	v. s.	i. al
phthalate, acid	$KHC_8H_4O_4$	204.22	wh. cr.	$2.45^{16°}$	d.		$10.2^{25°}$	36	s. al., et.
platinocyanide	$K_2Pt(CN)_4·3H_2O$*	431.54	yel., rhb., 1.62±				sl. s.		i. al.
silicate	K_2SiO_3	154.25	hyg. 1.521±	2.417	976		s.	$24.1^{100°}$	i. al., act, CS_2
silicate, tetra-	$K_2Si_4O_9·H_2O$	352.45	rhb., 1.530	2.662	d. 400	$-3H_2O$, 150	$7.35°$	d.	
sulfate (arcanite)	K_2SO_4	174.25	col., rhb., 1.4947	2.277	tr. 558		s.		s. al., gly.; i. et.
Potassium sulfate, pyro-	$K_2S_2O_7$	254.21	col.		300		s.		sl. s. al.; i. NH_3
sulfide, mono-	$K_2S·5H_2O$	200.33	rhb., delq.		60		100	>100	i. abs. al.
sulfite	$K_2SO_3·2H_2O$	194.28	wh., rhb.	1.98	d.		$45.1^{15°}$	$91.5^{75°}$	sl. s. al.
sulfite, acid	$KHSO_3$	120.16	wh., mn.		d. 190		$12.5^{7.5°}$		
tartrate	$K_2C_4H_4O_6·½H_2O$	235.27	col., mn., 1.526	1.956		d.		$278^{100°}$	s. a., alk.; i. al., ac.
tartrate, acid	$KHC_4H_4O_6$*	188.18	col., rhb.	1.886			$0.37°$	$6.1^{100°}$	$20.8^{22°}$ act; s. al.
thiocyanate	$KCNS$	97.17	col., delq., mn., 1.660±		172.3	d. 500	$177°$	$217^{20°}$	
thiosulfate	$K_2S_2O_3$	190.31	col., cb.	2.23	d. 400	d.	$96.1°$	$311.2^{90°}$	i. al.
thiosulfate	$3K_2S_2O_3·H_2O$	558.95	delq., mn.		$-H_2O$, 180		d.		
Praseodymium	Pr	140.92	yel.	$6.5^{20°}$	940		d. $+H_2$		d. a.
Radium	Ra	226.05	wh., met.	5?	960	1140			s. al.
bromide	$RaBr_2$	385.88	wh., mn.	5.79	728	subl. 900	$70^{20°}$		
Radon (Niton)	Rn	222.0	gas	lq. 5.5; 111 (D)	-71	-62	$51°$ cc	$8.5^{60°}$ cc	
Rhenium	Re	186.31	hex.		3440	>2500	i.		i. HF, HCl; s. H_2SO_4; HNO_3
Rhodium	Rh	102.91	gray-wh., cb.	12.5	1955	>2500	i.	i.	sl. s. aq. reg., a.
chloride	$RhCl_3$	209.28	red		d. 450	subl. 800±	i.		v. sl. s. alk; i. aq. reg., a.
chloride	$RhCl_3·4H_2O$	281.35	dark red				v. s.		s. HCl, al.; i. et.
Rubidium	Rb	85.48	silv. wh.	lq. $1.475^{88.5}$; $1.53^{20°}$ (D)	38.5	700	d.		s. a., al
Ruthenium	Ru	101.70	bk., porous	8.6	>1950		i.	i.	
Ruthenium	Ru	101.70	gray, hex.	$12.2^{20°}$	2450	>2700	i.	i.	sl. s. aq. reg., a.
Samarium (also Sa)	Sm (also Sa)	150.43		7.7	>1300	2400			
Scandium	Sc	45.10		2.5?	1200				
Selenic acid	H_2SeO_4	144.98	hex. pr.	$2.950^{15°/4°}$	58	260	$1300^{30°}$	$\infty^{60°}$	s. H_2SO_4; d. al.; i. NH_3
Selenic acid	$H_2SeO_4·H_2O$	162.99	nd.	$2.627^{15°/4°}$	26	205	v. s.		
Selenium	Se_8	631.68	red pd., amor, 2.92	$4.26^{25°}$	50	688	i.	i.	s. CS_2, H_2SO_4, CHJ_2
Selenium	Se_8	631.68	gray, trig., 3.00; red, hex.	4.80; 4.50	220	688	i.	i.	s. CS_2, H_2SO_4

Name	Formula	Mol. wt.	Color, crystalline form, index of refraction	Density	M.P. °C	B.P. °C	Solubility in cold water	Solubility in hot water	Solubility in other solvents
Selenium	Se_8	631.68	steel gray; hex.	4.8^{25}	217	688	i.	i.	i. CS_2; s. H_2SO_4; v. s. al.; i. NH_3
Selenous acid	H_2SeO_3	128.98		3.004^{15}_{4}	d.		90^{0}	400^{90}	s. alk.; i. NH_4Cl
Silicic acid, meta-	H_2SiO_3	78.08	amor., 1.41	2.1–2.3			i.	i.	s. alk.; i. NH_4Cl
Silicic acid, ortho-	H_4SiO_4	96.09		1.576^{17}			sl. s.	sl. s.	s. $HNO_3 + HF$; Ag; sl. s. Pb, Zn; i. HF
Silicon, crystalline	Si	28.06	gray, cb., 3.736	2.4^{20}	1420	2600	i.	i.	s. $HNO_3 + HF$, fused alk.; i. HF
Silicon, graphitic	Si	28.06	cr.				i.		s. HF, KOH
Silicon, amorphous	Si	28.06	brn., amor.	2		2600	i.	i.	s. fused alk.; i. a.
carbide	SiC	40.07	blue-bk., trig., 2.654	3.17	>2700	subl. 2200	i.	i.	d. alk.
chloride, tri-	Si_2Cl_6	268.86	lf. or lq.	1.58^{0}	−1	144^{760mm}	d.	d.	d. conc. H_2SO_4, al.
chloride, tetra-	$SiCl_4$	169.89	col. fuming lq., 1.412	1.50	−70	57.6	d.	d.	s. HNO_3, al., et.
fluoride	SiF_4	104.06	gas	3.57 (A)	−95.7	-65^{810mm}	v. s. d.		i. al., et.; d. KOH
hydride (silane)	SiH_4	32.09	col. gas	lq. 0.68^{-185}	−185	-112^{760mm}	i.		s. HF; h. alk., fused $CaCl_2$
oxide, di- (opal)	$SiO_2 \cdot xH_2O$		iridescent, amor.	2.2	1600–1750	subl. 1750	i.	i.	s. HF; i. alk.
oxide, di- (cristobalite)	SiO_2	60.06	col., cb. or tet., 1.487	2.32	1710	2230	i.	i.	s. HF; i. alk.
oxide, di- (lechatelierite)	SiO_2	60.06	hex, 1.5442	2.20	tr. <1425	2230	i.		s. HF; i. alk.
oxide, di- (quartz)	SiO_2	60.06	trig., rhb., 1.469	2.650^{30}_{4}	tr. 1670	2230	i.	i.	s. HF; i. alk.
oxide, di- (tridymite)	SiO_2	60.06	silv. met., cb., 2.252	2.26		1950	i.	i.	s. HF; i. alk.
Silver	Ag	107.88	pa. yel., cb., 2.252	10.5^{30}_{4}	960.5	d. 700	i.	i.	s. HNO_3, h. H_2SO_4; i. alk. 0.51^{18} NH_4OH; s. KCN, $Na_2S_2O_3$
bromide (bromyrite)	AgBr	187.80	yel. pd.	6.473^{25}_{4}	434	d. 700	0.00002^{20}	0.00037^{100}	s. NH_4OH, $Na_2S_2O_3$; i. al.
carbonate	Ag_2CO_3	275.77	col., rhb., 1.744	6.077	218 d.		0.003^{27}	0.05^{100}	s. NH_4OH, $Na_2S_2O_3$; i. al.
chloride (cerargyrite)	AgCl	143.34	silv. met, cb.	5.56	455	1550	0.000089^{10}	0.00217^{100}	s. NH_4OH, KCN; sl. s. HCl
cyanide	AgCN	133.90	wh., 1.685±	3.95	−(CN)₂, 320	d., forms NaOH	0.000022^{20}		s. NH_4OH, KCN, HNO_3
nitrate (lunar caustic)	$AgNO_3$	169.89	col., rhb., 1.744	4.352^{19}_{4}	212	444 d.	122^{0}	952^{100}	s. gly.; v. sl. s. al.; i. bz.; d. al.
Sodium	Na	22.997	silv. met, cb.	0.97^{20}_{4}	97.5	880	d., forms NaOH		s. gly.; i. al., et.
acetate	$NaC_2H_3O_2$	82.04	wh., mn., 1.464	1.528	324		46.5^{20}	170^{100}	2.1^{18} al.
acetate	$NaC_2H_3O_2 \cdot 3H_2O$	136.09	wh., mn.	1.45	58	−3H₂O, 120	v. s.	v. s.	7.8^{25} abs. al.
aluminate	$NaAlO_2$	81.97	amor.		1650		s.	v. s.	i. al.
amide	$NaNH_2$	39.02	olive gn.		210	400	d.		d. al.
Sodium ammonium phosphate	$NaNH_4HPO_4 \cdot 4H_2O$	209.09	col., mn.	1.574	79 d.		16.7	100	sl. s. al., NH_3 salts; i. ac.
antimonate, meta-	$2NaSbO_3 \cdot 7H_2O$	511.63	cb.				$0.031^{12.8}$		1.67 al., 50^{15} gly.
arsenate	$Na_3AsO_4 \cdot 12H_2O$	424.09	hex., 1.4589	1.759	86.3		26.7^{17}		sl. s. al.
arsenate, acid (monobasic)	$NaH_2AsO_4 \cdot H_2O$*	181.94	rhb., 1.5535	2.535	d. 100		s.		sl. s. al.
arsenate, acid (dibasic)	$Na_2HAsO_4 \cdot 7H_2O$	312.02	col., mn., 1.4658	1.871	125	−7H₂O, 100	61^{15}	v. s.	sl. s. al.
arsenate, acid (dibasic)	$Na_2HAsO_4 \cdot 12H_2O$	402.10	mn., 1.4496	1.72	28	−12H₂O, 100	$5.59^{0.1}$	140.7^{90}	
arsenite, acid	Na_2HAsO_3	169.91	col.	1.87			v. s.		
benzoate	$NaC_7H_5O_2$	144.11	col. cr.				62.5^{25}	76.9^{100}	2.3^{25}, 8.3^{78} al.
bicarbonate	$NaHCO_3$	84.01	wh., mn., 1.500	2.20	−CO₂, 270		6.9^{0}	16.4^{60}	i. al.
bifluoride	$NaHF_2$	62.00	col. cr.		d.		3.7^{20}	s.	
bisulfate	$NaHSO_4$	120.06	col., tri.	2.742	>315	d., −H₂O	50^{0}	100^{100}	d. al.; i. NH_3
bisulfite	$NaHSO_3$	104.06	col., mn., 1.526	1.48	d.		s.		i. al., act.
borate, tetra-	$Na_2B_4O_7$	201.27		2.367	741		1.3^{0}	8.79^{40}	i. al.
borate, tetra	$Na_2B_4O_7 \cdot 5H_2O$	291.35	col., rhb., 1.461	1.815			22^{62}	52.3^{100} (anh.)	
borate, tetra- (borax)	$Na_2B_4O_7 \cdot 10H_2O$*	381.43	wh., mn., 1.4694	1.73	75	−10H₂O, 200	$1.3^{0.5}$	20.3^{80} (anh.)	s. gly.; i. abs. al.
bromate	$NaBrO_3$	150.91	col., cb.	$3.339^{17.5}$	381		27.5^{0}	90.9^{100}	i. al.
bromide	NaBr	102.91	col., cb., 1.6412	$3.205^{17.5}$	755	1390	90^{20}	121^{100}	sl. s. al.
bromide	$NaBr \cdot 2H_2O$	138.95	col., mn.	2.176	50.7		79.5^{0}	118.3^{80} (anh.)	sl. s. al.
carbonate (soda ash)	Na_2CO_3	106.00	wh. pd., 1.535	2.533	851		7.1^{0}	48.5^{104}	i. al., et.
carbonate	$Na_2CO_3 \cdot H_2O$	124.02	wh., rhb., 1.506–1.509	1.55	−H₂O, 100		s.	s.	s. gly.; i. al., et.
carbonate	$Na_2CO_3 \cdot 7H_2O$	232.12	rhb. or trig.	1.51	d. 35.1		s.	s.	i. al.
carbonate (sal soda)	$Na_2CO_3 \cdot 10H_2O$	286.16	wh., mn., 1.425	1.46			21.5^{0}	238^{30}	i. al.

*Usual commercial form.

TABLE 2-1 Physical Properties of the Elements and Inorganic Compounds (Continued)

Name	Formula	Formula weight	Color, crystalline form and refractive index	Specific gravity	Melting point, °C	Boiling point, °C	Solubility in 100 parts		
							Cold water	Hot water	Other reagents
Sodium ammonium phosphate (Cont.)									
carbonate, sesqui- (trona)	$Na_3H(CO_3)_2 \cdot 2H_2O$	226.05	wh., mn., 1.5073	2.112	d.		$13°$	$42^{100°}$	s. al.
chlorate	$NaClO_3$	106.45	wh., cb., or trig., 1.5151	$2.490^{15°}$	248	d.	$79°$	$230^{100°}$	
chloride	$NaCl$	58.45	col., cb., 1.5443	2.163	800.4	1413	$35.7°$	$39.8^{100°}$	sl. s. al.; i. conc. HCl
chromate	Na_2CrO_4	162.00	yel., rhb.	2.723	392		$32°$	$126^{100°}$	sl. s. al.
chromate	$Na_2CrO_4 \cdot 10H_2O$	342.16	yel., delq., mn.	1.483	19.9	d.	v.s.	∞	i. al.
citrate	$2Na_3C_6H_5O_7 \cdot 11H_2O$	714.36	wh., rhb.	$1.857^{23.5°}_{4}$	$-11H_2O$, 150	d.	$91^{25°}$	$250^{100°}$	i. al.
cyanide	$NaCN$	49.02	wh., cb., 1.452	$2.52^{18°}$	563.7	1496	$48^{10°}$	$82^{35°}$	s. NH_3; sl. s. al.
dichromate	$Na_2Cr_2O_7 \cdot 2H_2O$	298.05	red, mn., 1.6994		$-2H_2O$, 84.6; 356 (anh.)	d. 400	$238°$	$508^{80°}$	s. NH_3; sl. s. al.
ferricyanide	$Na_3Fe(CN)_6 \cdot H_2O$	298.97	red, delq.		d.		$18.9°$	$67^{100°}$	i. al.
ferrocyanide	$Na_4Fe(CN)_6 \cdot 10H_2O$	484.11	yel., mn.	1.458			$17.9^{37°}$ (anh.)	$63^{18.5°}$ (anh.)	i. al.
fluoride (villiaumite)	NaF	42.00	tet., 1.3258	2.79	992		$4°$	$5^{100°}$	i. al.
formate	$NaHCO_2$	68.01	wh., mn.	1.919	253		$44°$	$160^{100°}$	sl. s. al.; i. et.
hydride	NaH	24.005	silv. nd., 1.470	0.92	d. 800		d.		i. bz., CS_2, CCl_4, NH_3; s. molten metal
hydrosulfide	$NaSH \cdot 2H_2O$	92.10	col., delq., nd.		d.		s.	s.	s. al.; d. a.
hydrosulfide	$NaSH \cdot 3H_2O$	110.15	rhb.		22	d.	s.	s.	s. al.; d. a.
hydrosulfite	$Na_2S_2O_4 \cdot 2H_2O$	210.15	col. cr.		d.		$22^{20°}$	s.	i. al.
hydroxide	$NaOH$	40.00	wh., delq.	2.130	318.4	1390	$42°$	$347^{100°}$	v. s. al., et., gly.; i. act.
hydroxide	$NaOH \cdot 3½H_2O$	103.06	mn.		15.5		v.s.	v.s.	
hypochlorite	$NaOCl$	74.45	pa. yel., in soln. only		d.		$26°$	$158^{50°}$	v. s. al., act.
iodide	NaI	149.92	col., cb., 1.7745	$3.667°$	651	1300	$158.7°$	$302^{100°}$	v. s. NH_3
iodide	$NaI \cdot 2H_2O$	185.95	col., amor.	2.448	d.		v.s.	v.s.	v. s. NH_3
lactate	$NaC_3H_5O_3$	112.07	col., amor.				v.s.	v.s.	s. al.; i. et.
nitrate (soda niter)	$NaNO_3$	85.01	col., trig., rhb.	2.257	308	d. 380	$73°$	$180^{100°}$	s. NH_3; sl. s. gly., al.
nitrite	$NaNO_2$	69.01	pa. yel., rhb.	$2.168°$	271	d. 320	$72.1°$	$163.2^{100°}$	$0.3^{20°}$ et.; 0.3 abs. al.; $4.4^{20°}$ m. al.; v. s. NH_3
oxide	Na_2O	61.99	wh., delq.	2.27	subl.		Forms NaOH	d.	d. al.
perborate	$NaBO_3 \cdot H_2O$	99.83	wh. pd.		d. 40		sl. s.	d.	s. gly., alk.
perchlorate	$NaClO_4$	122.45	rhb., 1.4617	2.02	482 d.		$170°$	$320^{100°}$	s. al.; 51 m. al.; 52 act.; i. et.
perchlorate	$NaClO_4 \cdot H_2O$	140.47	hex.	2.805	d. 130		$209^{15°}$	$284^{50°}$	s. al.
peroxide	Na_2O_2	77.99	yel.-wh. pd.		d. 30		s. d.	d.	s. dil. a.
peroxide	$Na_2O_2 \cdot 8H_2O$	222.12	wh., hex.	2.040	$-H_2O$, 100	d. 200	s. d.	d.	
phosphate, monobasic	$NaH_2PO_4 \cdot H_2O$	138.01	col., rhb., 1.4852	1.91	60		$71°$	$390^{83°}$	i. al.
phosphate, monobasic	$NaH_2PO_4 \cdot 2H_2O$	156.03	col., rhb., 1.4629	1.679			$91.1°$	$308^{40°}$	
phosphate, dibasic	$Na_2HPO_4 \cdot 7H_2O$	268.09	col., mn., 1.4424	1.52		$-12H_2O$, 180	$185^{40°}$	$2000^{100°}$	i. al.
phosphate, dibasic	$Na_2HPO_4 \cdot 12H_2O$	358.17	col., mn., 1.4361	1.52	34.6		$4.3°$	$76.7^{30°}$	
phosphate, tribasic	Na_3PO_4	163.97	wh.	$2.537^{17.5°}$	1340		$4.5°$	$77^{100°}$	i. al.
phosphate, tribasic	$Na_3PO_4 \cdot 12H_2O$	380.16	wh., trig., 1.4458	1.62	73.4	$-11H_2O$, 100	$28.3^{15°}$	∞	
phosphate, meta-	$Na_4P_4O_{12}$	407.91	col.	2.476	616 d.		$2.26°$	$45^{46°}$	i. CS_2
phosphate, pyro-	$Na_4P_2O_7$	265.95	wh.	2.45	988		$5.4°$	$93^{100°}$	s. a., alk.
phosphate, pyro-	$Na_4P_2O_7 \cdot 10H_2O$	446.11	mn., 1.4525	1.82	d. 220		$4.5°$	$21^{49°}$	d. a.
phosphate (pyrodisodium)	$Na_2H_2P_2O_7$	221.97	col., mn., 1.510	1.862			$6.9°$	$36^{60°}$	i. al., NH_3
phosphate (pyrodisodium)	$Na_2H_2P_2O_7 \cdot 6H_2O$	330.07	col., mn., 1.4645	1.848	70 to 80	$-4H_2O$, 215	$26°$	$66^{28°}$	
potassium tartrate	$NaKC_4H_4O_6 \cdot 4H_2O$	282.23	rhb., 1.493	1.790		$-6H_2O$, 100	s. d.	s. d.	sl. s. al.
silicate, meta-	Na_2SiO_3	122.05	col., rhb., 1.520		1088		v. s.	v. s.	i. Na or K salts, al. $29^{18°}$ ½N NaOH
Sodium silicate, meta-	$Na_2SiO_3 \cdot 9H_2O$	284.20	rhb.		47		s.	s.	
silicate, ortho-	Na_4SiO_4	184.05	col., hex., 1.530		1018		s.	s.	i. al.
silicofluoride	Na_2SiF_6	188.05	wh., hex., 1.312	2.679	d. 140		$0.44°$	$2.45^{100°}$	i. al., act.
stannate	$Na_2SnO_3 \cdot 3H_2O$	266.74	wh., hex. tablets		tr. 100 to mn.		$50°$	$67^{50°}$	i. al., alk.
sulfate (thenardite)	Na_2SO_4	142.05	col., rhb., 1.477	2.698	tr. 500 to hex.		$5°$	$42^{100°}$	i. al.
sulfate	Na_2SO_4	142.05	col., mn.				$48.8^{40°}$	$42.5^{100°}$	d. HI; s. H_2SO_4

Physical Constants of Inorganic Compounds (continued)

Name	Formula	Mol. wt.	Color, crystalline form, refractive index	Density	M.p., °C	B.p., °C	Solubility, cold water	hot water	Solubility in other reagents
sulfate	Na_2SO_4	142.05	col. hex.		884		19.4^{20}	45.3^{60}	i. al.
sulfate	$Na_2SO_4,7H_2O$	268.17	tet.				44.9^{0}	202.6^{25}	sl. s. al.; i. et.
sulfate (Glauber's salt)	$Na_2SO_4,10H_2O$	322.21	col., mm, 1.396	1.464	32.4	−10H_2O, 100	36^{15}	412^{34}	s. al.
sulfide, mono-	Na_2S	78.05	pink or wh., amor.	1.856			15.4^{0}	57.3^{0}	s. al.
sulfide, tetra-	Na_2S_4	174.23	yel., cb.		275		s.	s.	i. al., NH
sulfide, penta-	Na_2S_5	206.29	yel.		251.8		s.	s.	i. al.
sulfite	Na_2SO_3	126.05	hex. pr., 1.565	$2.633^{15/4}$	d.		13.9^{0}	28.3^{84}	i. al.
sulfite	$Na_2SO_3,7H_2O$	252.17	mn.	1.561	−7H_2O, 150		34.7^{2}	67.8^{15}	
tartrate	$Na_2C_4H_4O_6,2H_2O$	230.10	rhb.	1.818			29^{0}	66^{15}	v. s. al.
thiocyanate	$NaCNS$	81.08	delq., rhb., 1.625±		287		110^{10}	225^{80}	s. NH_3; v. sl. s. al.
thiosulfate	$Na_2S_2O_3$	158.11	mn.	1.667			50^{0}	231^{80}	sl. s. NH_3; i. a., al.
thiosulfate (hypo)	$Na_2S_2O_3,5H_2O°$	248.19	mn. pr., 1.5079	1.685	d. 48.0		74.7^{0}	301.8^{60}	s. NH_3; i. a., al.
tungstate	Na_2WO_4	293.91	wh., rhb.	4.179	692		57.58^{0}	97^{100}	s. alk. carb., dil. a.
tungstate	$Na_2WO_4,2H_2O°$	329.95	wh., rhb.	3.245	−2H_2O, 100		88^{0}	123.5^{100}	i. al.
tungstate, para-	$Na_6W_7O_{24},16H_2O$	2097.68	wh., tri.	$3.987^{14/4}$	−16H_2O, 300		8	d.	i. al.
uranate	Na_2UO_4	348.06	yel.		866 (anh.)		i.	i.	s. abs. al., act., NH_3; s. ∞ CS_2
vanadate	$Na_3VO_4,16H_2O$	472.20	col. nd.				v. s.	d.	s. conc. H_2SO_4; i. alk.; NH_4OH, NH_3
vanadate, pyro-	$Na_4V_2O_7$	305.89	hex.		654		s.	d.	s. dil. H_2SO_4, HCl; d. abs. al.
Stannic chloride	$SnCl_4$	260.53	col. fuming lq.	2.226	−30.2	114.1		d.	s. C_6H_5N
oxide (cassiterite)	SnO_2	150.70	wh., tet., 1.9968	7.0	1127		i.	i.	s. alk., abs. al., et.
sulfate	$Sn(SO_4)_2,2H_2O°$	346.85	col., delq, hex.				v. s.	d.	s. tart. a., alk., al
Stannous bromide	$SnBr_2$	278.53	yel., rhb.	5.12^{17}	215.5	620	s.	d.	s. H_2SO_4
chloride	$SnCl_2$	189.61	wh., rhb.		246.8	623	83.9^{0}	269.8^{15}	s. al., a.
chloride (tin salt)	$SnCl_2,2H_2O°$	225.65	wh., tri.	$2.71^{15.5}$	37.7	d.	118.7^{0}	∞	0.26^{15} m. al.
sulfate	$SnSO_4$	214.76	wh. cr.		−SO_2, 360		19^{19}	18^{100}	s. a., NH_4 salts, aq, CO_2
Strontium	Sr	87.63	silv. met.	2.6	800	1150	d.	Forms $Sr(OH)_2$	v. sl. s. act., abs. al.; i. NH_3
acetate	$Sr(C_2H_3O_2)_2$	205.72	wh. cr.	2.099	d.		36.9^{0}	36.4^{97}	s. NH_4Cl
carbonate (strontianite)	$SrCO_3$	147.64	wh., rhb., 1.664	3.70	1497^{60atm}	−CO_2, 1350	0.0011^{18}	0.065^{100}	s. NH_4Cl; i. act.
chloride	$SrCl_2$	158.54	wh., cb., 1.6499	3.052	873		43.5^{0}	100.8^{100}	s. NH_3; 0.012 abs. al.
chloride	$SrCl_2,6H_2O°$	266.64	wh., rhb., 1.5364	1.933^{17}	−4H_2O, 61	−6H_2O, 100	104^{0}	198^{40}	i. HNO_3
hydroxide	$Sr(OH)_2$	121.65	wh., delq	3.625	375		0.41^{0}	21.83^{100}	sl. s. al.; i. et.
hydroxide	$Sr(OH)_2,8H_2O°$	265.77	col., tet., 1.499	1.90	−7H_2O in dry air		0.90^{0}	47.7^{100}	s. al., NH_4Cl; i. act.
nitrate	$Sr(NO_3)_2°$	211.65	col., cb., 1.5878	2.986	570		40^{0}	100^{89}	s. al.; i. NH_4OH
nitrate	$Sr(NO_3)_2,4H_2O$	283.71	wh., mn.	2.2			62.2^{0}	124^{20}	s. al. s. a.; i. dil. H_2SO_4, al. 14^{70} H_2SO_4
oxide (strontia)	SrO	103.63	col., cb., 1.870	4.7	2430		Forms $Sr(OH)_2$		sl. s. al., act.; i. et.
peroxide	SrO_2	119.63	wh. pd.		d.		0.008^{20}	d.	
peroxide	$SrO_2,8H_2O$	263.76	wh. cr.		−8H_2O, 100		0.018^{20}	d.	
sulfate (celestite)	$SrSO_4$	183.69	col., rhb., 1.6237	3.96	1580 d.		0.0113^{0}	0.0114^{32}	
sulfite, acid	$Sr(HSO_4)_2$	281.77	col., granular		d.		d.		
Sulfamic acid	NH_2SO_4H	97.09	wh., rhb.	$2.03^{12/4}$	205 d.		20^{0}	40^{70}	
Sulfur, amorphous	S	32.06	pa. yel. pd., 2.0–2.9	2.046	120	444.6	i.	i.	s. al. s. CS_2
Sulfur, monoclinic	S_8	256.48	pa. yel., mn.	1.96	119.0	444.6	i.	i.	s. CS_2, al
Sulfur, rhombic	S_8	256.48	pa. yel., rhb.	2.07	112.8	444.6	i.	i.	24^{0}, 181^{55} CS_2
Sulfur bromide, mono-	S_2Br_2	223.95	red-yel. lq.	2.635	−46	$54^{0.18mm}$	d.		s. CS_2, et., bz.
chloride, mono-	S_2Cl_2	135.03	red-yel. lq.	1.687	−80	138	d.		d. al.
chloride, di-	SCl_2	102.97	dark red fuming lq.	$1.621^{15/15}$	−78	59	d.		
chloride, tetra-	SCl_4	173.89	yel.-brn. lq.		−30	d. >−20	d.		
oxide, di-	SO_2	64.06	col. gas	lq., 1.434^{0}; 2.264 (A)	−75.5	−10.0	22.8^{0}	4.5^{50}	s. H_2SO_4; al., ac.
oxide, tri-(α)	SO_3	80.06	col. pr.	lq., 1.923; 2.75 (A)	16.83	44.6	d.	d.	s. H_2SO_4
oxide, tri-(β)	$(SO_3)_2°$	160.12	col., silky, nd.	1.97^{20}	50		Forms H_2SO_4		s. H_2SO_4
Sulfuric acid	$H_2SO_4°$	98.08	col., viscous lq.	$1.834^{18/4}$	10.49	d. 340	∞	∞	d. al.
Sulfuric acid	H_2SO_4,H_2O	116.09	pr. or lq.	$1.842^{15/4}$	8.62	290	∞	∞	d. al.

° Usual commercial form.

TABLE 2-1 Physical Properties of the Elements and Inorganic Compounds (*Concluded*)

Name	Formula	Formula weight	Color, crystalline form and refractive index	Specific gravity	Melting point, °C	Boiling point, °C	Solubility in 100 parts		
							Cold water	Hot water	Other reagents
Sulfuric acid	H2SO4.2H2O	134.11	col. lq.	$1.650^{0°}_{4}$	−38.9	167	∞	∞	d. al.
Sulfuric acid, pyro-	H2S2O7	178.14	cr.	1.9^{38p}	35	d.	d.	d.	d. al.
Sulfuric oxychloride	SO2Cl2	134.97	col. lq.	$1.667^{20°}_{4}$	−54.1	69.1^{760mm}	d.	d.	s. bz., CS2, CCl4; d. act.
Sulfurous oxybromide	SOBr2	207.89	or.-yel. lq.	$2.68^{18°}_{4}$	−50	68^{40mm}	d.	d.	s. bz., chl.
oxychloride	SOCl2	118.97	col. lq.	1.638	−104.5	78.8	d.	d.	
Tantalum	Ta	180.88	bk.-gray, cb.	16.6	2850	>4100	i.	i.	s. fused alk., HF; i. HCl, HNO3, H2SO4
Tellurium	Te	127.61	met., hex.	(α) 6.24; (β) 6.00	452	1390	i.	i.	s. H2SO4, HNO3, KCN, KOH, aq. reg.; i. CS2
Terbium	Tb	159.20							
Thallium	Tl	204.39	blue-wh., tet.	11.85	303.5	1650	i.	i.	s. HNO3, H2SO4; i. NH3
acetate	TlC2H3O2	263.43	silky nd.	3.68	110		v. s.	i.	v. s. al.
chloride, mono-	TlCl	239.85	wh., cb.	7.00	430	806	$0.21^{0°}$	$1.8^{100°}$	sl. s. HCl; i. al, NH4OH
chloride, sesqui-	Tl2Cl3	515.15	yel., hex.	5.9	400–500	d.	$0.26^{15°}$	$1.9^{100°}$	
chloride, tri-	TlCl3	310.76	hex. pl.		25		v. s.	v. s.	s. al., et.
chloride, tri-	TlCl3.4H2O	382.83	nd.		−6H2O, 200	−4H2O, 100	$86.2^{17°}$	d.	s. al., et.
sulfate (ic)	Tl2(SO4)3.7H2O	823.07	lf.	6.77	632	d.	d.	d.	s. dil. H2SO4
sulfate (ous)	Tl2SO4	504.84	col., rhb., 1.8671		115 d.	d.	$2.70^{0°}$	$18.45^{100°}$	v. sl. s. dil. H2SO4
sulfate, acid	TlHSO4	301.46	trimorphous						
Thio, *cf.* sulfo or sulfur									
Thorium	Th	232.12	cb.	11.2	1845	>3000	i.	i.	s. HCl, H2SO4; sl. s. HNO3; i. HF, alk.
oxide, di- (thorianite)	ThO2	264.12	wh., cb.	9.69	>2800	4400	i.	i.	s. h. H2SO4; i. alk.
sulfate	Th(SO4)2	424.24	mn. pr.	$4.225^{17°}$			$0.74^{0°}$	$5.22^{50°}$	
sulfate	Th(SO4)2.9H2O	586.38		2.77	−9H2O, 400		sl. s.	sl. s.	s. HCl, H2SO4, dil. HNO3
Thulium	Tm	169.40	silv. met., tet.	7.31		2260	i.	i.	s. a., h. alk. solns.
Tin	Sn	118.70	gray, cb.	5.750	231.85	2260	i.	i.	s. alk.; v. sl. s. dil. a.; i. al.
Tin salts, *cf.* stannic and stannous									
Titanic acid	H2TiO3	97.92	wh. pd.		1800	>3000	i.	i.	s. a.
Titanium	Ti	47.90	dark gray, cb.	$4.50^{17.5°}$	Unstable in air	>3000	i.	d.	i. CS2, et, chl.
chloride, di-	TiCl2	118.81	bk., delq.		d. 440		d.		
chloride, tri-	TiCl3	154.27	vl., delq.		−30		s.	s.	s. dil. HCl
chloride, tetra-	TiCl4*	189.73	col. lq.	lq, 1.726		136.4	s.	d.	sl. s. alk.
oxide, di- (anatase)	TiO2	79.90	brn. or bk., tet., 2.534–2.564	3.84			i.	i.	
oxide, di- (brookite)	TiO2	79.90	brn. or bk., rhb., 2.586	4.17			i.	i.	
oxide, di- (rutile)	TiO2	79.90	col. if pure, tet., 2.615	4.26	1640 d.	<3000	i.	i.	s. H2SO4, alk.
Tungsten	W	183.92	gray-bk., cb.	19.3	3370	5900	i.	i.	s. h. conc. KOH; sl. s. NH3, HNO3, aq. reg.
carbide	WC	195.93	gray pd., cb.	$15.7^{18°}$	2777	6000	i.	i.	s. F2; i. a.
carbide	W2C	379.85	iron gray	$16.06^{18°}$	2877	6000	i.	i.	s. h. HNO3; sl. s. HCl, H2SO4
oxide, tri-	WO3	231.92	yel., rhb.	7.16	>2130		i.	i.	s. alk.; i. a.
Tungstic acid (tungstite)	H2WO4	249.94	yel., rhb. 2.24	5.5	−½H2O, 100; 1473		sl. s.	i.	s. HF, alk., NH3
Uranic acid	H2UO4	304.09	yel. pd.	$5.926^{15°}$	−H2O, 250 to 300		i.	sl. s.	s. a., alk. carb.; i. alk.
Uranium	U	238.07	wh. cr.	$18.485^{13°}_{4}$	1133	3500	i.	i.	s. a.; i. alk.
carbide	U2C3	512.14	cr.	11.28	2400		d.	d.	d. a.
oxide, di- (uraninite)	UO2	270.07	bk., rhb.	10.9	2176		i.	i.	s. HNO3, conc. H2SO4

Name	Formula	Mol. wt.	Color, crystalline form, index of refraction	Density	Melting point, °C	Boiling point, °C	Solubility cold water	Solubility hot water	Solubility in other solvents
oxide (pitchblende)	U₃O₈	842.21	olive gn.	8.30					s. HNO₃, H₂SO₄
sulfate (ous)	U(SO₄)₂·4H₂O	502.25	gn., rhb.	2.89^{15}			23^{15}	9.63^{0}	s. dil. a.
Uranyl acetate	UO₂(C₂H₃O₂)₂·2H₂O°	424.19	yel., rhb.	5.6	−4H₂O, 300; −2H₂O, 110		9.21^{17}	d.	s. al., act.
carbonate (rutherfordine)	UO₂CO₃	330.08	tet.						
nitrate	UO₂(NO₃)₂·6H₂O	502.18	yel., rhb., 1.4967	2.807	60.2	118	170.3^{0}	∞^{60}	v. s. ac., al., et.; i. dil., alk.
sulfate	UO₂SO₄·3H₂O	420.18	yel. cr.	$3.28^{16.5}$	d. 100		$18.9^{13.2}$	230^{25}	4 al.; s. a.
Vanadic acid, meta-	HVO₃	99.96	yel. scales						s. a., alk.; i. NH₃
Vanadic acid, pyro-	H₄V₂O₇	217.93	pa. yel. amor.						s. a., alk., NH₄OH
Vanadium	V	50.95	lt. gray, cb.	5.96	1710	3000	i.		s. HNO₃, H₂SO₄; i. aq., alk.
chloride, di-	VCl₂	121.86	gn., hex., delq.	3.23^{18}					s. al., et.
chloride, tri-	VCl₃	157.23	pink, tabular, delq.	3.00^{18}	d.		s.	i.	s. abs. al., et.
chloride, tetra-	VCl₄	192.78	red lq.	1.816^{30}	−109	148.5^{755mm}	s. d.	d.	s. abs. al., et., chl., ac.
oxide, di-	V₂O₂	133.90	lt. gray cr.	3.64	ign.				s. a.
oxide, tri-	V₂O₃	149.90	bk. cr.	4.87^{18}_{4}	1970		i.	d.	s. HNO₃, HF, alk.
oxide, tetra-	V₂O₄	165.90	blue cr.	4.399	1967		sl. s.	s.	s. a., alk.
oxide, penta-	V₂O₅	181.90	red-yel., rhb.	3.357^{18}_{4}	800	d. 1750	0.8^{20}	i.	s. a., alk.; i. abs. al.
oxychloride, mono-	VOCl	102.41	brn. pd.	2.824	d. in air		i.		v. s. HNO₃
Vanadyl chloride	(VO)₂Cl	169.36	yel. cr.	3.64					s. HNO₃
chloride, di-	VOCl₂	137.86	gn., delq.	2.88^{13}			d.	d.	s. abs. al., dil. HNO₃
chloride, tri-	VOCl₃	173.32	yel. lq.	1.829	< −15	127.19	s. d.		s. al., et., ∞ Br₂
Water†	H₂O	18.016	col. lq., 1.33300²⁰ᵈ; 0.915⁰ᵈ (ice)	1.00^{0} (lq.); 0.915^{0} (ice)	0	100	∞	∞	∞ al.; sl. s. et.
Water, heavy	D₂O	20.029	col. lq., 1.32844²⁰ᵈ	1.107^{20}	3.82	101.42	∞	∞	∞ al.
Xenon	Xe	131.30	col. gas	lq., $3.06^{-109.1}$; 2.7^{-149}; 4.53 (A)	−140	−109.1	24.2^{0} cc	7.3^{0} cc	
Ytterbium	Yb	173.04	dark gray, hex.	5.51					
Yttrium	Y	88.92	silv. met., hex.		1490	2500	d.	∞	
Zinc	Zn	65.38	mn.	7.140	419.4	907	i.	d.	v. s. dil. a., h. KOH
acetate	Zn(C₂H₃O₂)₂	183.47	wh., mn., 1.494	1.840	242	subl. in vac.	30^{25}	44.6^{100}	s. a., ac., alk
acetate	Zn(C₂H₃O₂)₂·2H₂O°	219.50	wh., mn.	1.735	237	−2H₂O, 100	40^{25}	66.6^{100}	2.8^{25}; 166^{79} al.
bromide	ZnBr₂	225.21	wh., trig., 1.818	4.219^{18}	394	650	390^{100}	670^{100}	v. s. al.
carbonate	ZnCO₃	125.39		4.42	−CO₂, 300		0.001^{15}	i.	v. s. NH₄OH, al., et.
chloride	ZnCl₂	136.29	wh., delq., 1.687, uniaxial	2.91^{25}_{4}	283	732	432^{25}	615^{100}	s. a., alk., NH₄ salts; i. act., NH₃
cyanide	Zn(CN)₂	117.42	col., rhb.		d. 80		0.0005^{18}	sl. s.	$100^{12.5}$ al.; v. s. et.; i. NH₃
hydroxide	Zn(OH)₂	99.40	col., rhb.	3.053	d. 125		0.00052^{18}	i.	s. KCN, NH₃, alk.; i. al.
iodide	ZnI₂	319.22	cb.	4.666^{14}_{2}	446	624	430^{0}	510^{100}	s. a., alk. NH₄OH
nitrate	Zn(NO₃)₂·6H₂O	297.49	col., tet.	2.065^{14}_{4}	36.4	−6H₂O, 105	324.5	$\infty^{36.4}$	s. a., al., NH₃, aq. (NH₄)₂CO₃
oxide (zincite)	ZnO	81.38	wh., hex., 2.004	5.606	>1800		0.00042^{18}		v. s. al.
oxide	ZnO°	81.38	wh., amor.	5.47	>1800		0.00042^{18}		s. a., alk., NH₄Cl; i. NH₃
peroxide	ZnO₂	97.38	yel.	1.571	expl. 212		0.0022		i. NH₄OH; d. a.
phosphide	Zn₃P₂	258.10	steel gray, cb.	4.55^{13}_{4}	>420	1100	i.		s. dil. a.
silicate	ZnSiO₃	141.44	hex. or rhb.; glass, 1.650	3.52	1437		i.		
sulfate (zincosite)	ZnSO₄	161.44	wh., rhb., 1.669	3.74^{15}_{4}	d. 740		42^{0}	61^{100}	sl. s. al.; s. gly.
sulfate	ZnSO₄·H₂O	179.46	col.	3.28^{15}_{4}	d. 238		s.	89.5^{100}	sl. s. al.; i. act.; NH₃
sulfate	ZnSO₄·6H₂O	269.54	mn.	2.072^{15}_{4}	−5H₂O, 70		s.		sl. s. al.; i. act.; NH₃
sulfate (goslarite)	ZnSO₄·7H₂O°	287.55	rhb., 1.4801	$1.966^{16.5}$	tr. 39	−7H₂O, 280	115.2^{0}	653.6^{100}	v. s. a.; i. ac.
sulfide (α) (wurtzite)	ZnS°	97.44	wh., hex., 2.356	4.087	1850^{150atm}	subl. 1185	0.00069^{18}	i.	v. s. a.; i. ac.
sulfide (β) (sphalerite)	ZnS	97.44	wh., cb.; glass (P), 2.18–2.25	4.102^{25}_{4}	tr. 1020		i.	i.	s. a.
sulfide (blende)	ZnS	97.44		4.04			i.	i.	v. s. a.; i. ac.
sulfite	ZnSO₃·2½H₂O	190.48	wh., granular		−2½H₂O, 100	d. 200	0.16	d.	s. H₂SO₃, NH₄OH; i. al.
Zirconium	Zr	91.22	mn.	6.4	1700	>2900	i.	i.	s. HF, aq. reg.; sl. s. a.
oxide, di- (baddeleyite)	ZrO₂	123.22	gy. or brn., mn., 2.19	5.49	2700		i.	i.	s. H₂SO₄, HF
oxide, di- (free from Hf)	ZrO₂	123.22	wh., mn.	5.73		4300	i.	i.	s. H₂SO₄, HF

° Usual commercial form.
† Cf. special tables on water and steam, Tables 2-3, 2-4, 2-5, 2-185, 2-186 and 2-351 through 2-357.
NOTE: °F = 9/5 °C + 32.

TABLE 2-2 Physical Properties of Organic Compounds*

Abbreviations Used in the Table

(A), density referred to air	cr., crystalline	nd., needles	i-, iso-, containing the group (CH₃)₂CH-	s, sec-, secondary
al., ethyl alcohol	d., decomposes	o-, ortho	i., insoluble	silv., silvery
amor., amorphous	d-, dextrorotatory	or., orange	ign., ignites	sl., slightly
aq., aqua, water	dl-, dextro-laevorotatory	p-, para	l-, laevorotatory	subl., sublimes
brn., brown	et., ethyl ether	pd., powder	lf., leaflets	sym., symmetrical
bz., benzene	expl., explodes	pet., petroleum ether	lq., liquid	t-, tertiary
c., cubic	gn., green	pl., plates	m-, meta	tet., tetragonal
cc., cubic centimeter	h., hot	pr., prisms	mn., monoclinic	tri., triclinic
chl., chloroform	hex., hexagonal	rhb., rhombic	n-, normal	uns., unsymmetrical
col., colorless		s., soluble		v., very

v. s., very soluble v. sl. s., very slightly soluble wh., white yel., yellow (+), right rotation >, greater than <, less than ∞, infinitely

This table of the physical properties includes the organic compounds of most general interest. For the properties of other organic compounds, reference must be made to larger tables in Lange's *Handbook of Chemistry* (Handbook Publishers), *Handbook of Chemistry and Physics* (Chemical Rubber Publishing Co.), Van Nostrand's *Chemical Annual*, *International Critical Tables* (McGraw-Hill), and similar works.

The **molecular weights** are based on the 1941 atomic weight values. The **densities** are given for the temperature indicated and are usually referred to water at 4°C, e.g., $1.028^{0.854}$ a density of 1.028 at 95°C referred to water at 4°C, the 4 being omitted when it is not clear whether the reference is to water at 4°C or at the temperature indicated by the upper figure. The melting and boiling points given have been selected from available data as probably the most accurate. The **solubility** is given in grams of the substance in 100 g. of the solvent. In the case of gases, the solubility is often expressed in some manner as "5^{10} cc." which indicates that, at 10°C, 5 cc. of the gas are soluble in 100 g. of the solvent.

Name	Synonym	Formula	Formula weight	Form and color	Specific gravity	Melting point, °C	Boiling point, °C	Water	Alcohol	Ether
Abietic acid	sylvic acid, abietinic acid	$C_{20}H_{30}O_2$	302.44	lf.	$1.069^{95_{95}}$	182	278–9	i.	v. s.	v. s.
Acenaphthene	naphthylene ethylene	$C_{12}H_8(CH_2)_2$	154.20	rhb./al.		95	102.2	6^{25}	s. h.	s. chl.
Acetal	acetaldehyde diethylacetal	$CH_3CH(OC_2H_5)_2$	118.17	lq.	0.821^{224}	–123.5	20.2	∞	∞	∞
Acet-aldehyde	ethanal	CH_3CHO	44.05	col. lq.	0.783^{384}	10.5–12	124.4^{762}	12^{13}	v. s.	sl. s.
-aldehyde, par-	paraldehyde	$(C_2H_4O)_3$	132.16	col. cr.	0.994^{204}	97	100–10 d.	s.	v. s.	v. sl. s.
-aldehyde ammonia		$CH_3CHOHNH_2$	61.08	col. cr.		81(69.4)	222	s.	s.	7^{25}
-amide	ethanamide	CH_3CONH_2	59.07	rhb./al.	1.159	113–4	305	0.5^6	21^{20}	s.
-anilide	antifebrin	$C_6H_5NHCOCH_3$	135.16	lf./al.	1.21^4	79	>250	i.	21^{20}	s.
-phenetidine (o-)	o-ethoxyacetanilide	$CH_3CONHC_6H_4OC_2H_5$	179.21	rhb.		96–7	296	sl. s.	s.	s.
(m-)	acetyl-m-phenetidine	$CH_3CONHC_6H_4OC_2H_5$	179.21	rhb. or mn.		110	306–7	0.86^{10}	10^{25}	s.
-toluidide (o-)	N-tolylacetamide	$CH_3C_6H_4NHCOCH_3$	149.19	col. lq.	1.168^{15}	153	118.1	0.09^{22}	∞	∞
(p-)	N-tolylacetamide	$CH_3C_6H_4NHCOCH_3$	149.19	col. lq.	1.212^{15}	16.7	139.6	12 c.	∞	∞
Acetic acid	ethanoic acid, vinegar acid	CH_3CO_2H	60.05	col. lq.	1.049^{204}	–73	81.6–2.0	∞	∞	∞
anhydride	acetyl oxide, acetic oxide	$(CH_3CO)_2O$	102.09	col. lq.	1.082^{204}	–41	56.5	s.	s.	s.
nitrile	methyl cyanide	CH_3CN	41.05	col. lq.	0.783^{204}	–94.6	subl.	∞	∞	v. s.
Acetone	propanone, dimethyl ketone	CH_3COCH_3	58.08	tri./al.	0.792^{204}	175	202.3^{49}	∞	∞	∞
Acetonyl urea	dimethyl hydantoin	$<NHCONHCOC>(CH_3)_2$	128.13	lf.		20.5	51–2	s.	s.	s.
Acetophenone benzoyl hydride	methyl-phenyl ketone	$CH_3COC_6H_5$	120.14	col. lq.	1.033^{1515}	–112.0	-84^{760}	i.	∞	∞
Acetyl-chloride	ethanoyl chloride	CH_3COCl	78.50	nd./aq.	1.105^{204}	162	60.3	s. h.	d.	v. sl. s.
-phenylenediamine (-p)	amino-acetanilide (p)	$C_2H_5ONHC_6H_4NH_2$	150.18	col. gas	(A) 0.906	-81.5^{981}	48.4	$100\ cc.^{18}$	$600\ cc.^{18}$	s.
Acetylene	ethyne; ethine	$HC{:}CH$	26.04	col. lq.	1.291^{154}	–80.5	346	0.35^{20}	sl. s.	0.6^{15}
dichloride (cis)	1,2-dichloroethene	$CHCl{:}CHCl$	96.95	col. lq.	1.265^{154}	–50	52.5	0.63^{20}	s.	v. sl. s.
(trans)	diform	$CHCl{:}CHCl$	96.95	cr./aq.		192 d.	141–2	33^{15}	v. s.	i.
Aconitic acid	equisetic acid; citric acid	$C_3H_3(CO_2H)_3$	174.11	rhb./aq. al.		110–1	78–9	sl. s. h.	v. sl. s.	∞
Acridine		$C_6H_4{<}(CH)(N){>}C_4H_4$	179.21	col. lq.	0.841^{204}	–87.7	265^{10}	40	v. sl. s.	s.
Acrolein ethylene aldehyde	acrylic aldehyde, propenal	$CH_2{:}CH{\cdot}CHO$	56.06	col. lq.	1.062^{164}	12–13	295	s.	v. s.	∞
Acrylic acid	propenoic acid	$CH_2{:}CH{\cdot}CO_2H$	72.06	col. lq.	0.811^{20}	–82	subl. >200	1.4^{15}	0.03³⁰	∞
nitrile	vinyl cyanide	$CH_2{:}CH{\cdot}CN$	53.06	mn. pr.	1.360^{254}	151–3	83^{20}	0.4^{12}	v. s.	v. sl. s.
Adipic acid	hexandioic acid, adipinic acid	$(CH_2{\cdot}CH_2{\cdot}CO_2H)_2$	146.14	cr. pd.		226–7	430	v. sl. s.	∞	i.
amide		$(CH_2{\cdot}CONH_2)_2$	144.17	col. oil	0.951^{1919}	1	96.6	0.03^{20}	s.	∞
nitrile	tetramethylene	$(CH_2{\cdot}CN)_2$	108.14	col. pd.		d. 207–11	$70-1^{753}$	22^{17}	s. h.	v. sl. s.
Adrenaline (1-) (3,4,1)	1-suprarenine	$C_6H_3(OH)_2(CHOHCH_2NHCH_3)$	183.20	nd./aq.		295 d.	44.6	0.03^{100}	s.	i.
Alanine (α) (dl-)		$CH_3CH(NH_2)CO_2H$	89.09	col. lq.	1.103^{204}	289–90	152	∞	i.	i.
Aldol acetaldol	2-hydroxybutyraldehyde	$CH_3CH(OH)CH_2CO_2H$	88.10	red rhb.		–129	subl.	i.	s. h.	s.
Alizarin	Anthraquinoic acid	$C_6H_4(CO)_2C_6H_2(OH)_2$	240.20	col. lq.	0.854^{204}	–119.4	subl.	<0.1	s.	∞
Allyl alcohol	propen-1-ol-3, propenyl alcohol	$CH_2{:}CH{\cdot}CH_2OH$	58.08	lq.	1.398^{204}	–136.4	$200-5^{10}$	0.2	s.	∞
bromide	3-bromo-propene-1	$CH_2{:}CH{\cdot}CH_2Br$	120.99	col. lq.	0.938^{204}	–80	subl.	3^0	v. sl. s.	v. sl. s.
chloride	3-chloro-propene-1	$CH_2{:}CH{\cdot}CH_2Cl$	76.53	col. oil	1.013^{204}	77–8	subl.	d.	i.	v. sl. s.
thiocyanate (i)	mustard oil	$CH_2{:}CH{\cdot}CH_2NCS$	99.15	col. pr.	1.219^{2020}	150–60	225^{120}	i.	s. h.	i.
thiourea	thiosinamine	$CH_2{:}CH{\cdot}CH_2NHCSNH_2$	116.18	pd.	1.142^{200}	256		sl. s. h.	s. h.	s.
Aluminum ethoxide		$Al(OCH_2CH_3)_3$	164.15	red nd.		302		v. sl. s.	s. h.	1.8^6
Amino-anthraquinone (α)		$C_6H_4(CO)_2C_6H_3{\cdot}NH_2$	223.22	red nd.		126–7		sl. s. s.	2^{10}	8.2^6
(β)		$C_6H_4(CO)_2C_6H_3{\cdot}NH_2$	223.22	yel. mn.		173–4		v. sl. s.	11^{10}	
-azobenzene		$C_6H_5{\cdot}N{:}N{\cdot}C_6H_4NH_2$	197.23	nd./aq.		187–8		0.3^{13}		
-benzoic acid (m-)	aminodracylic acid	$H_2N{\cdot}C_6H_4CO_2H$	137.13	mn. pr.	$1.511^{4°}$					
(p-)		$H_2N{\cdot}C_6H_4CO_2H$	137.13							

Name	Formula	M.W.	Form	Sp. gr.	M.P.	B.P.	Sol. H₂O	Sol. alc.	Sol. eth.	Sol. bz.
Amino-diphenylamine (p-)	$H_2N \cdot C_6H_4 \cdot NH \cdot C_6H_5$	184.23	nd./aq.al.		67	354	sl. s.		s.	
-G-acid (2-)(6-,8-), Na₂ salt	$C_{10}H_5(NH_2)(SO_3Na)_2$	347.28					v. s.	sl. s.		i. bz.
-mono-potassium salt	$C_{10}H_5(NH_2)S_2O_6HK$	341.39								
-J-acid (2-)(5-,7-)	$C_{10}H_5(NH_2)(SO_3H)_2$	325.29								
-sodium salt	$C_{10}H_5(NH_2)(SO_3H)_2$	303.30					12.8[20]			
-mono-potassium salt	$C_{10}H_5(NH_2)S_2O_6HK$	341.39					10.0[20]			
-naphthol sulfonic (1-,2-,4-)(α-)	$C_{10}H_5OHNH_2SO_3H, \frac{1}{2}H_2O$	248.25					3.4[18]			
(1-,4-,2-)	$NH_2(OH)C_{10}H_5SO_3H \cdot H_2O$	239.24					v. s.			
(1-,4-,3-)										
(1-,8-,4-)										
-phenol (o-) — 2-aminophenol	$H_3N \cdot C_6H_4 \cdot OH$	109.12	col. nd.		173	subl.	1.7[70]	v. sl. s.		
(m-) — 3-aminophenol	$H_3N \cdot C_6H_4 \cdot OH$	109.12	pr.		122–3	subl.	2.6[0]	v. sl. s.		
(p-) — p-hydroxyaniline	$H_3N \cdot C_6H_4 \cdot OH$	109.12	lf.		184–6 d.		1.1[0]	v. sl. s.		
-toluene sulfonic acid (1-,2-,3-)	$C_6H_4(CH_3)(NH_2)SO_3H$	187.21	nd.		d.		0.97[11]	sl. s.		
(1-,4-,2-)	$C_6H_4(CH_3)(NH_2)SO_3H \cdot H_2O$	205.23	mm.		–H₂O, 120		0.5[20]	sl. s.		
(1-,4-,3-)		205.23	nd.				0.47			
(1-,2-,5-)	$C_6H_4(CH_3)(NH_2)SO_3H \cdot H_2O$	196.22	tri./aq.				3[11]			
Amyl acetate (n-)	$CH_3CO_2CH_2(CH_2)_3CH_3$	130.18	col. lq.	0.879[20/20]	–70.8	148.4[737]	0.3[15]	∞	∞	
(i-)	$CH_3CO_2CH_2CH_2CH(CH_3)_2$	130.18	col. lq.	0.876[15⁴]		142[757]	v. sl. s.	∞	∞	
common amyl acetate	$CH_3CO_2CH(CH_3)C_2H_5$	130.18	col. lq.	0.880[13]		141–2	sl. s.	∞	∞	
(s-)	$CH_3CO_2CH(CH_3)_2$	130.18	col. lq.	0.922[0]		133.5	v. sl. s.	∞	∞	
α-Me-Bu-acetate	$CH_3CO_2C(CH_3)_2C_2H_5$	130.18	col. lq.	0.871[20/0]		133		∞	∞	
(s-)		130.18	col. lq.			124.5[749]		∞	∞	
di Et-carbinol acetate	$CH_3CO \cdot CH(CH_2)_2CH_3$	130.18	col. lq.	0.874[19]		137.9		∞	∞	
alcohol (n-) fuel oil — pentanol-1	$CH_3(CH_2)_3CH_2OH$	88.15	col. lq.	0.817[20/20]	–78.5	119.5	2.7[22]	∞	∞	
(s-,n-) methyl-propyl carbinol — pentanol-2	$C_2H_5CH_2CH(OH)CH_3$	88.15	col. lq.	0.810[20/20]	–117.2	132.0	4[20]	∞	∞	
(prim.-i-) isobutyl carbinol — 2-methyl-butanol-4	$(CH_3)_2CHCH_2CH_2OH$	88.15	col. lq.	0.813[15⁴]		115.6	2[14]	∞	∞	
(s-i-) — 2-methyl-butanol-3	$(CH_3)_2CHCH(OH)CH_3$	88.15	col. lq.	0.815[25/4]	–11.9	113–4	5.5[30]	∞	∞	
(t-) — 2-methyl-butanol-2	$(CH_3)_2C(OH)C_2H_5$	88.15	col. lq.	0.819[19]	52–3	102	2.8[30]	∞	∞	
active amyl alcohol	$(CH_3)_2CHCH_2 \cdot CH_2OH$	88.15	col. lq.	0.809[20⁴]		113–4	sl. s.	∞	∞	
(d-)	$C_2H_5CH(CH_3)CH_2OH$	88.15	cr.	0.816[20⁴]		128	3.6[30]	∞	∞	
-amine (n-) — 1-NH₂-2-Me-butane	$CH_3(CH_2)_4NH_2$	87.16	col. lq.	0.766[19]	–55	103–4	∞	∞	∞	
(s-,n-) — 3-amino pentane	$(C_2H_5)_2CHNH_2$	87.16	col. lq.	0.749[20⁴]		91–2		∞	∞	
(i-)	$(CH_3)_2CH(CH_2)_2NH_2$	87.16	col. lq.	0.751[18⁴]	–105	95		∞	∞	
(t-) — 3-NH₂-2-Me-butane	$(C_2H_5)CHNH_2$	87.16	col. lq.	0.731[25⁴]		77–8		∞	∞	
	$C_2H_5CH(CH_3)CH_2NH_2$	87.16	col. lq.	0.755[18]		95–6		∞	∞	
	$(C_2H_5)CHNH_2$	87.16	col. lq.	0.749[20⁴]		90–1		∞	∞	
	$(CH_3)_2CHCH(CH_3)NH_2$	87.16	col. lq.	0.757[18]		83–4		∞	∞	
aniline (i-)	$C_6H_5NHC_5H_{11}$	163.25	lq.	0.928[15⁴]		254.5	i.	∞	∞	
benzoate (i-)	$C_6H_5CO_2C_5H_{11}$	192.25	col. lq.	0.992[14/14]		261[746]	i.	∞	∞	
bromide (n-) — 1-bromopentane	$CH_3(CH_2)_3CH_2Br$	151.05	col. lq.	1.218[20⁴]	–95	129.7	i.	∞	∞	
(i-) — 4-Br-2-Me-butane	$(CH_3)_2CH \cdot CH_2Br \cdot C_2H_5$	151.05	col. lq.	1.220[17/15]		120[745]	i.	∞	∞	
(s-) — 2-Br-2-Me-butane	$(CH_3)_2C \cdot Br \cdot C_2H_5$	151.05	col. lq.	1.216[19/0]		108[765]	i.			
n-butyrate (n-)	$C_3H_7CO_2CH_2(CH_2)_3CH_3$	158.23	col. lq.	0.871[15⁴]	–73.2	186.4	0.02[16]	∞	∞	
(i-)	$C_2H_5CH_2CO_2 \cdot C_5H_{11}$	158.23	col. lq.	0.866[19/15]		178.6	0.05[50]	∞	∞	
(s-)	$C_2H_5CO_2C(CH_3)_2C_2H_5$	158.23	col. lq.	0.865[15/0]		164	i.	∞	∞	
(s-,i-)	$(CH_3)_2CHCO_2 \cdot C_5H_{11}$	158.23	lq.	0.876[0⁴]		168.8	sl. s.	∞	∞	
(t-)	$(CH_3)_3CHCO_2C_5H_{11}$	158.23	lq.	0.878[20⁴]		108.4	i.			
i-butyrate (i-)										
chloride (n-) — 1-chloropentane	$CH_3(CH_2)_3CH_2Cl$	106.60	col. lq.	0.870[20⁴]	–99	108.4	i.	∞	∞	
(s-) — 2-chloropentane	$(C_2H_5)_2CHCl$	106.60	col. lq.	0.595[21]		96.7	i.	∞	∞	
(s-) — 3-chloropentane	$(CH_2)_3CHCl$	106.60	col. lq.	0.893[20⁴]		97.3	i.	∞	∞	
(s-i-) — 4-Cl-2-Me-butane	$(CH_3)_2CHCH_2 \cdot Cl$	106.60	col. lq.	0.883[0]		99.7[758]	i.	∞	∞	
(s-i-) — 3-Cl-2-Me-butane	$(CH_3)_2CHCHClCH_3$	106.60	lq.	0.871[20⁴]		91[753]	i.	∞	∞	
(t-) — 1-Cl-2-Me-butane	$(CH_3)_2CClC_2H_5$	106.60	lq.	0.881[17/5]	–72.9	85.7	i.			
i-cyanide iso-nitrile — iso-caproic iso-nitrile	$(CH_3)_2CH(CH_2)_2NC$	97.16	lq.	0.902[0]		98–9	i.	∞	∞	
formate (i-)	$HCO_2CH_2(CH_2)_3CH_3$	116.16	lq.	0.882[20⁴]	–73.5	137–9	i.	∞	∞	
(i-)	$HCO_2CH_2CH_2CH(CH_3)_2$	116.16	col. lq.		–93.5	132	i.			
iodide (n-) — 1-iodopentane	$CH_3(CH_2)_3CH_2I$	198.06	col. lq.	1.510[20⁴]	–86	123.5	i.	∞	∞	
(i-) — 4-I-2-Me-butane	$(CH_3)_2CHCH_2CH_2I$	198.06	col. lq.	1.515[18⁴]		157.0	i.	∞	∞	
(s-,n-) — 2-iodopentane	$C_2H_5CH_2CHICH_3$	198.06	col. lq.	1.507[17/4]		147[765]	i.	∞	∞	
(t-) — 2-I-2-Me-butane	$(CH_3)_2CIC_2H_5$	198.06	col. lq.	1.471[19/15]		144–5	i.			
mercaptan (n-) — pentanthiol-1	$CH_3(CH_2)_3CH_2SH$	104.21	col. lq.	1.524[20⁴]		127[765]	i.			
(n-) — pentanthiol-3	$(C_2H_5)_2CHSH$	104.21	col. lq.	0.857[20]		148	i.			
(n-) — 2-Me-butanthiol-4	$(CH_3)_2CHCH_2 \cdot SH$	104.21				126[767]				
phenol (t-,t-)/(p-) — pentaphen	$C_5H_{11} \cdot C_6H_4 \cdot OH$	164.24	cr.	0.835[20⁴]	93	105	v. sl. s.			
(i-)	$C_2H_5CO_2(CH_2)_4CH_3$	144.21	col. lq.	0.876[15⁴]	–73.1	120	0.3[22]			
(act.)	$C_2H_5CO_2 \cdot C_5H_{11}$	144.21	col. lq.	0.870[20⁴]		265–7	i.			
propionate (t-,t-)/(p-)	$C_2H_5CO_2(CH_2)_2CH(CH_3)_2$	144.21	lq.	0.866[20⁴]		168.7	i.			
salicylate (n-)	$HOC_6H_4CO_2C_5H_{11}$	208.25	lq.	1.065[15]		160.2	0.125	v. sl. s.		
Amyl i-valerate (i)	$C_4H_9CO_2C_5H_{11}$	172.26	col. lq.	0.858[20/15]		58[16]	i.	v. sl. s.		
	$C_4H_9CO_2C_5H_{11}$	172.26	col. lq.	0.861[14⁰]		265	sl. s.			
						194				
						173–4				

*By N. A. Lange, Ph.D., Handbook Publishers, Inc., Sandusky, Ohio. Abridged from table of Physical Constants of Organic Compounds in Lange's "Handbook of Chemistry."

TABLE 2-2 Physical Properties of Organic Compounds *(Continued)*

Name	Synonym	Formula	Formula weight	Form and color	Specific gravity	Melting point, °C	Boiling point, °C	Solubility in 100 parts: Water	Alcohol	Ether
Amylene (*n*-)(α-)	pentene-1	$C_2H_5CH_2CH:CH_2$	70.13	lq.	0.644^{20}		30–1	i.	∞	∞
(*i*-)	2-methyl-butene-3	$(CH_3)_2CHCH:CH_2$	70.13	col. lq.	0.632^{15}	−135	20.5^{771}	i.	∞	∞
(α-)	2-methyl-butene-1	$(C_2H_5)(CH_3)C:CH_2$	70.13	col. lq.	$0.667^{0/0}$		$31–2^{758}$	i.	∞	∞
(-*n*)(β-)	pentene-2	$C_2H_5CH:CHCH_3$	70.13	col. lq.	$0.650^{20/4}$	−139	36.4	v. sl. s.	∞	∞
(*i*-)(β-)	2-methyl-butene-2	$(CH_3)_2C:CHCH_3$	70.13	col. lq.	$0.663^{19/4}$	−124	37–8	i.	s.	∞
Anethole (*p*-)	*p*-propenyl anisole	$CH_3CH:CH{\cdot}C_6H_4OCH_3$	148.20	lf./al.	$0.991^{20/20}$	22.5	235.3	v. sl. s.	s.	∞
Anhydroformald-aniline	methylene aniline	$(CH_2NC_6H_5)_3$	315.40	pr./al.		143	185	i.	sl. s.	s.
Aniline	amino benzene, phenyl amine, cyanol	$C_6H_5NH_2$	93.12	col. oil	$1.022^{20/4}$	−6.2	184.4	3.6^{18}	∞	∞
hydrochloride	aniline salt, aniline chloride	$C_6H_5NH_2{\cdot}HCl$	129.59	cr.	1.222^{4}	198	245	18^{15}	s.	i.
nitrate		$C_6H_5NH_2{\cdot}HNO_3$	156.14	rhb.	1.356^{4}	d. 190		s.	s.	sl. s.
sulfate		$(C_6H_5NH_2)_2{\cdot}H_2SO_4$	284.32	lf./al.	1.377^{4}	d.		5^{14}	sl. s.	i.
Anisal-acetone (*p*-)	MeO-benzalacetone	$CH_3OC_6H_4CH:CHCOCH_3$	176.22	lf./et.		73–4		i.	v. s.	v. s.
Anisic acid (*p*-)		$CH_3OC_6H_4CO_2H$	152.14	mn./aq.	1.385^{4}	184.2	275–80	0.03^{19}	v. s.	v. s.
aldehyde (*p*-)		$CH_3OC_6H_4CHO$	136.14	col. oil	$1.123^{20/4}$	2.5	247–8	v. sl. s.	∞	∞
Anisidine (*o*-)	2-amino-anisole	$CH_3OC_6H_4NH_2$	123.15	col. lq.	$1.098^{15/15}$	5.2	225	v. sl. s.	∞	∞
(*m*-)	MeO-aniline(*m*)	$CH_3OC_6H_4NH_2$	123.15	oil	$1.096^{20/4}$	<−12	251	v. sl. s.	s.	s.
(*p*-)	4-amino anisole	$CH_3OC_6H_4NH_2$	123.15	pl./aq.	$1.089^{55/55}$	57.2	243	s. h.	s.	s.
Anisole	methyl phenyl ether	$CH_3OC_6H_5$	108.13	col. lq.	$0.990^{22/4}$	−37.3	154–5	i.	s.	s.
Anthracene	paranaphthalene, anthracin green oil	$C_6H_4{:}(CH)_2{:}C_6H_4$	178.22	col. mn.	$1.25^{27/4}$	217–8	340–2	i.	1.5^{20}	
Anthramine (α)	α-amino-anthracene	$C_6H_4{:}(CH)_2{:}C_6H_3NH_2$	193.24	yel./al.		130±		i.	s.	
(β)	β-amino-anthracene	$C_6H_4{:}(CH)_2{:}C_6H_3NH_2$	193.24	yel./al.		238	subl.	i.	sl. s.	sl. s.
Anthranil		$C_6H_4{:}(NH)CO$	119.12	col. oil	$1.187^{15/4}$	<−18	d. >215	sl. s. h.	s.	s.
Anthranilic acid (*o*-)		$H_2NC_6H_4CO_2H$	137.13	col. rhb.		144–5	subl.	0.35^{14}	11^{10}	16^{7}
Anthrapurpurin (1-,2-,7-)		$C_{14}H_5O_2(OH)_3$	256.20	or. nd./al.		369	462	sl. s. h.	v. s. h.	sl. s.
Anthraquinone	diphenyleneketone, dihydrodiketoanthracene	$C_6H_4{:}(CO)_2{:}C_6H_4$	208.20	yel. rhb.	$1.438^{20/4}$	286	379–81	i.	0.05^{18}	v. sl. s.
disulfonate Na_2 (1-,5-)	ρ-anthraquinone disulfonate	$C_{14}H_6O_2(SO_3Na)_2{\cdot}5H_2O$	502.38	yel. lf.				v. s.	i.	i.
(1-,8-)	*x*-anthraquinone disulfonate	$C_{14}H_6O_2(SO_3Na)_2{\cdot}4H_2O$	484.37	yel. pr.				sl. s.		
(2-,6-)		$C_{14}H_6O_2(SO_3Na)_2{\cdot}7H_2O$	538.41	col. cr.				3.9^{20}		
(2-,7-)		$C_{14}H_6O_2(SO_3Na)_2{\cdot}4H_2O$	484.37	cr.				30.5^{20}	v. sl. s.	i.
sulfonate Na (1-)		$C_{14}H_7O_2SO_3Na$	310.25	yel. lf.				0.53^{20}	i.	i.
(2-)		$C_{14}H_7O_2SO_3Na$	310.25	silv. lf.				0.84^{25}	i.	i.
Anthrarufin (1-,5-)		$C_{14}H_6O_3(OH)_2$	240.20	yel. lf.		280	subl.	i.	sl. s.	s.
Antipyrene	1-ph-2,3-diMepyrazolone-5	$C_{11}H_{12}ON_2$	188.22	mn./aq.	$1.088^{113/4}$	113(109)	319^{174}	100^{25}	100	sl. s.
Apiole	1-allyl-2, 5-diMeO-3,4 methylenedioxybenzene	$C_{12}H_{14}O_2$	222.23	col. nd.	$1.02^{20/4}$	30	294	i	s.	s.
Arabinose (α)(*d*- or *l*-)		$CH_2OH(CHOH)_3CHO$	150.13	rhb. pr.	$1.585^{20/4}$	159.5		46^{0}	$0.5^{9°}$	i.
(*dl*-)		$CH_2OH(CHOH)_3CHO$	150.13			164.5		16.9^{10}		
Arachidic acid	eicosanoic acid	$CH_3(CH_2)_{18}CO_2H$	312.52	col. lf.		77	328	i.	s. h.	v. s.
Arsanilic acid (*p*-)		$H_2N{\cdot}C_6H_4.AsO_3H_2$	217.04	nd./aq.		232		v. s. h.	v. s. h.	i.
Asparagine (*l*-)		$HO_2C{\cdot}C_2H_3(NH_2){\cdot}CONH_2$	132.12	rhb.	$1.543^{15/4}$	227–35	d. 235	3.1^{28}	i. c.	
Aspirin (*o*-)		$CH_3CO_2{\cdot}C_6H_4{\cdot}OH$	180.15	nd./aq.		135–6		1^{37}	s.	5^{20}
Atropic acid	α-phenyl acrylic acid	$C_6H_5C({:}CH_2){\cdot}CO_2H$	148.15	nd./aq.		106–7	267 d.	0.1 c.	s.	s.
Auramine	4,4′-dimethylaminobenzophenomide	$[(CH_3)_2NC_6H_4]_2C:NH$	267.36	col./al.		136		i.	7^{20}	2.3^{20}
Aurine, coralline (4-,4′-)		$(HOC_6H_4)_2C:C_6H_4:O$	290.30	red		310 d.		i.	s.	s.
Azo-anisole (2-,2′-)	diMeO-azobenzene	$(CH_3O{\cdot}C_6H_4N:)_2$	242.27	or. pr.		153		i.	s.	s.
benzene	diphenyldiimide	$C_6H_5N:N{\cdot}C_6H_5$	182.22	or. mn.	$1.203^{20/4}$	68	297	i.	4.2^{20}	
Azoxybenzene		$(C_6H_5)_2N_2O$	198.22	yel. rhb.	$1.248^{20/20}$	36	d.	i.	11.4^{15}	
Barbituric acid	malonyl urea	$CO{:}(NHCO)_2{:}CH_2{\cdot}2H_2O$	164.12	col./aq.		d. 245		s. h.	sl. s.	s.
Benzal acetone	Me-cinnamyl ketone	$C_6H_5CH:CHCOCH_3$	146.18	pl.	$1.035^{20/20}$	41–2	260–2	i.	s.	s.
Benzaldehyde	artificial almond oil	C_6H_5CHO	106.12	col. lq.	$1.046^{20/4}$	−26	179	0.3	∞	∞
Benzamide		$C_6H_5CONH_2$	121.13	col. pr.	1.341	130	290	1.35^{25}	17^{25}	sl. s.
Benzanilide		$C_6H_5CONHC_6H_5$	197.23	lf./al.	1.31^{4}	163	$117–9^{10}$	i.	4^{30}	sl. s.
Benzene	benzol, phenyl hydride, cyclohexatriene	C_6H_6	78.11	col. lq.	$0.879^{20/4}$	5.5	80.1	0.07^{22}	s.	∞
sulfinic acid		$C_6H_5SO_2H$	142.17	pr./aq.		83–4	d. > 100	v. s. h.	v. s.	v. s.
sulfonic acid		$C_6H_5SO_3H$	158.17	col. nd.		65–6	d.	v. s.	v. s.	i.
sulfonic amide	benzene sulfonamide	$C_6H_5SO_2NH_2$	157.18	mn./aq.		156		0.43^{16}	v. s.	v. s.
sulfonic chloride	benzene sulfonyl chloride	$C_6H_5SO_2Cl$	176.62	cr.	$1.384^{15/15}$	14.5	251.5	i.	v. s.	s.
Benzidine (4-,4′-)		$NH_2{\cdot}C_6H_4{\cdot}C_6H_4{\cdot}NH_2$	184.23	cr./aq.		128–9	400^{740}	1 h.	1 h.	2
disulfonic acid (2-,2′-)		$({\cdot}C_6H_3(NH_2)SO_3H)_2{\cdot}3H_2O$	398.40	pr./aq.		d. >175		0.09^{25}	i.	i.
(3-,3′-)		$({\cdot}C_6H_3(NH_2)SO_3H)_2$	344.35					v. sl. s.		
Benzil	dibenzoyl	$C_6H_5CO{\cdot}COC_6H_5$	210.22	pr.	1.23^{15}	95	348 d.	i.	v. s.	v. s.
Benzoic acid		$C_6H_5CO_2H$	122.12	mn. pr.	$1.266^{15/4}$	121.7	249.2	0.2^{17}	46^{15}	66^{15}
anhydride		$(C_6H_5CO)_2O$	226.22	rhb./et.	$1.199^{15/4}$	42	360	i.	s.	s.
nitrile	phenyl cyanide	C_6H_5CN	103.12	col. lq.	$1.001^{25/6}$	−12.9	190.7	1^{100}	∞	∞

Name	Synonym	Formula	M.W.	Form	Density	M.P.	B.P.	Sol. (water)	Sol.	Sol.	Sol.
Benzoin (dl-)		$C_6H_5CO \cdot CHOHC_6H_5$	212.24	mn. rhb.		133–7	344^{768}	sl. s. 15^{13}	s. h. 6.5^{15}	v. sl. s.	v. sl. s.
Benzophenone	diphenyl ketone	$C_6H_5 \cdot CO \cdot C_6H_5$	182.21	col. rhb.	1.083	48.5	305.4	i.	s.	i.	∞
Benzotrichloride	phenyl chloroform	$C_6H_5 \cdot CCl_3$	195.48	col. lq.	1.380^{14}	−4.75	220.7	sl. s.	d. h.	sl. s.	∞
Benzoyl-benzoic acid (o-)		$C_6H_5COC_6H_4CO_2H \cdot H_2O$	244.24	tri./aq.		93(128)		d.	s. h.	d.	s.
-chloride		C_6H_5COCl	140.57	col. lq.	$1.212^{20/4}$	−0.5	197.2	i.	∞	i.	∞
-peroxide		$(C_6H_5CO)_2O_2$	242.22	rhb./et.		108 d.	expl.	4^{17}	∞	4^{17}	s.
Benzyl acetate		$CH_3CO_2CH_2C_6H_5$	150.17	col. lq.	1.057^{17}	−51.5	213.5	∞	∞	∞	∞
alcohol	phenyl carbinol	$C_6H_5CH_2OH$	108.13	col. lq.	$1.043^{20/4}$	−15.3	204.7	4^{17}	∞	∞	∞
amine	o-amino toluene	$C_6H_5CH_2NH_2$	107.15	col. lq.	$0.982^{20/4}$	37–8	184.5	s.	∞	s.	∞
aniline	phenyl-benzylamine	$C_6H_5CH_2NHC_6H_5$	183.24	mn. pr.	$1.065^{25/25}$	21	306^{780}	i.	∞	i.	∞
benzoate		$C_6H_5CO_2CH_2C_6H_5$	212.24	col. lq.	$1.12^{20/4}$	238–40	323–4	i.	v. s.	i.	∞
butyrate		$C_3H_7CO_2CH_2C_6H_5$	178.22	col. lq.	$1.016^{16/16}$			i.	∞	i.	∞
chloride	o-chlorotoluene	$C_6H_5CH_2Cl$	126.58	col. lq.	$1.100^{20/20}$	−39	179.4	i.	∞	i.	∞
ether	dibenzyl ether	$(C_6H_5CH_2)_2O$	198.25	col. lq.	1.036^{16}	3.6	295–8	i.	s. h.	i.	∞
formate		$HCO_2CH_2C_6H_5$	136.14	col. lq.	1.081^{23}		$202-3^{747}$	i.	∞	i.	∞
propionate		$C_2H_5CO_2CH_2C_6H_5$	164.20	col. lq.	$1.036^{16/17}$		220–2	s.	∞	s.	∞
Berberonic acid (2-,4-,5-)		$C_5H_2N(CO_2H)_3 \cdot 2H_2O$	247.16	tri.		243	subl.	v. sl. s.	sl. s. h.	v. sl. s.	i.
Biuret	allophanamide	$NH(CONH_2)_2$	103.08	ndl./al.		192–3 d.		1.3^0	v. s.	1.3^0	i.
Borneol (dl-)		$C_{10}H_{17} \cdot OH$	154.24	col. cr.	$1.011^{20/4}$	210.5	212–3	v. sl. s.	v. s.	v. sl. s.	v. s.
(d- or l-)		$C_{10}H_{17} \cdot OH$	154.24	col. cr.	$1.011^{20/4}$	208–9	212–3	v. sl. s.	v. s.	v. sl. s.	v. s.
(iso-)		$C_{10}H_{17} \cdot OH$	154.24	col. cr.		212		i. c.	v. s.	i.	v. s.
Bornyl acetate (d-)		$CH_3CO_2C_{10}H_{17}$	196.28	rhb.	0.991^{15}	29	226–7	i.	∞	i.	∞
Bromo-aniline (p-)		$BrC_6H_4NH_2$	172.03	col. lq.	1.8^{20}	63–4		i.	∞	i.	∞
-benzene		C_6H_5Br	157.02	col. lq.	$1.495^{20/4}$	−30.6	156.2	i.	∞	i.	∞
-camphor (3-)(d-)		$BrC_{10}H_{15}O$	231.14	cr.	$1.449^{20/4}$	77–8	274	i.	20^{26}	i.	s.
-diphenyl (p-)		$BrC_6H_4 \cdot C_6H_5$	233.11	cr./al.		90–1	310	i.	v. s.	i.	∞
-naphthalene (α-)		$C_{10}H_7 \cdot Br$	207.07	col. oil	$1.482^{20/4}$	5–6	281.1	i.	v. s.	i.	∞
-naphthalene (β-)		$C_{10}H_7 \cdot Br$	207.07	lf./al.		59	281–2	i.	6^{20}	i.	s.
-phenol (o-)		BrC_6H_4OH	173.02	col. lq.	1.605^0	5.6	194–5	s.	v. s.	s.	∞
(m-)		BrC_6H_4OH	173.02	cr.	1.553^{80}	32–3	236–7	1.4^{15}	v. s.	i.	v. s.
(p-)		BrC_6H_4OH	173.02	tet. cr.	1.588^{80}	63.5	238	i.	v. s.	i.	s.
-styrene (ω)(1)		$C_6H_5CH \cdot CHBr$	183.05	lq.	$1.422^{20/4}$	7	221	i.	s.	i.	∞
(2)		$C_6H_5CH \cdot CHBr$	183.05	lq.	$1.427^{20/4}$	−7.5	108^{26}	i.	s.	i.	∞
-toluene (o-)		$CH_3 \cdot C_6H_4Br$	171.04	col. lq.	$1.422^{20/4}$	−28	181.8	0.1 c.	s.	i.	∞
(m-)		$CH_3 \cdot C_6H_4Br$	171.04	col. lq.	$1.410^{20/4}$	−39.8	183.7	i.	s.	i.	∞
(p-)		$CH_3 \cdot C_6H_4Br$	171.04	cr./al.	$1.390^{20/4}$	28.5	184–5	i.	s.	i.	∞
Bromoform		$CHBr_3$	252.77	col. lq.	$2.890^{20/4}$	8–9	150.5	0.1 c.	∞	i.	∞
Butadiene (1-2-)	methyl-allene	$CH_3CH:C:CH_2$	54.09	col. gas	$0.621^{20/4}$	−108.9	18–9	i.	s.	i.	∞
(1-3-)	erythrene	$CH_2:CHCH:CH_2$	54.09	col. gas	$0.773^{20/4}$	−4.41	−4.41	i.	s.	i.	∞
Butadienyl acetylene		$CH_2:CHCH:CH \cdot C:CH$	78.11	col. gas	0.60^0		83–6	i.	s.	i.	s.
Butane (i-)	trimethyl-methane	$(CH_3)_3CH$	58.12	col. gas	0.60^0	−145	−10	i.	∞	i.	∞
(n-)	diethyl	$CH_3CH_2CH_2CH_3$	58.12	col. gas	0.60^0	−135	−0.6	i.	∞	i.	∞
Butyl acetate (n-)		$CH_3CO_2(CH_2)_3CH_3$	116.16	col. lq.	0.882^{20}	−76.3	125^{740}	0.7	∞	0.7	∞
(s-)		$CH_3CO_2CH(CH_3)C_2H_5$	116.16	col. lq.	$0.865^{25/4}$		112^{744}	0.6^{25}	∞	i.	∞
(i-)		$CH_3CO_2CH_2CH(CH_3)_2$	116.16	col. lq.	$0.871^{20/4}$		118	0.6^{25}	∞	i.	∞
(tert-)		$CH_3CO_2C(CH_3)_3$	116.16	col. lq.	$0.866^{20/4}$		$95-6^{760}$	i.	∞	i.	∞
alcohol (n-)	butanol-1	$C_3H_7CH_2OH$	74.12	col. lq.	$0.810^{20/4}$	−79.9	117	9^{15}	∞	9^{15}	∞
(s-)	butanol-2	$C_2H_5CH(OH)CH_3$	74.12	col. lq.	$0.808^{20/4}$	−114.7	99.5	12.5^{30}	∞	12.5^{30}	∞
(i-)	2-methyl-propanol-1	$(CH_3)_2CHCH_2OH$	74.12	col. lq.	$0.805^{17.5}$	−108	107–8	10^{15}	∞	10^{15}	∞
(tert-)	2-methyl-propanol-2	$(CH_3)_3COH$	74.12	col. lq.	0.779^{26}	25.5	82.9	∞	∞	∞	∞
amine (n-)		$C_3H_7CH_2NH_2$	73.14	col. lq.	$0.739^{25/4}$	−50	77.8	∞	∞	∞	∞
(s-)		$C_2H_5CH(NH_2)CH_3$	73.14	col. lq.	$0.724^{20/4}$	−104	66^{772}	∞	∞	∞	∞
(i-)		$(CH_3)_2CHCH_2NH_2$	73.14	col. lq.	$0.732^{20/20}$	−85	68–9	s.	∞	s.	∞
(t-)		$(CH_3)_3CNH_2$	73.14	col. lq.	$0.698^{20/4}$	−67.5	45.2	∞	∞	∞	∞
p-aminophenol (N)(n)		$C_4H_9NH \cdot C_6H_4 \cdot OH$	165.23	lq.		71		i.	s.	i.	s.
(N)(i-)		$C_4H_9NH \cdot C_6H_4 \cdot OH$	165.23	oil		79		i.	s.	i.	s.
aniline (n-)		$C_4H_9NHC_6H_5$	149.23	col. lf.	$0.940^{20/4}$		235^{720}	i.	∞	i.	∞
(i-)		$C_4H_9NHC_6H_5$	149.23	col. oil			231–2	i.	∞	i.	∞
arsonic acid (n-)		$C_4H_9AsO(OH)_2$	182.04	col. oil		158–9		0.01^{15}	v. s.	0.01^{15}	s.
benzoate (n-)		$C_6H_5CO_2C_4H_9$	178.22	col. lq.	$1.005^{25/25}$	−22	249–50	i.	∞	i.	∞
(i-)		$C_6H_5 \cdot CO_2 \cdot C_4H_9$	178.22	col. lq.	$0.997^{25/25}$		241.5	i.	∞	i.	∞
bromide (n-)	1-bromo-butane	$C_2H_5CH_2CH_2Br$	137.03	lq.	$1.277^{20/4}$	−112.4	101.6	0.06^{16}	∞	0.06^{16}	∞
(s-)	2-bromo-butane	$C_2H_5CH(Br)CH_3$	137.03	lq.	$1.251^{25/4}$	−112	91.3	i.	∞	i.	∞
(i-)	1-Br-2-Me-propane	$(CH_3)_2CHCH_2Br$	137.03	lq.	$1.258^{25/4}$	−118.5	91.5	0.06^{18}	∞	0.06^{18}	∞
(i-)	2-Br-2-Me-propane	$(CH_3)_3CBr$	137.03	lq.	$1.211^{20/4}$	−16.2	73.3	i.	∞	i.	∞
butyrate (n-)(n-)		$C_2H_5CH_2CO_2CH_2CH_2CH_2CH_3$	144.21	col. lq.	$0.872^{20/20}$		165.7^{736}	i.	∞	i.	∞
(n-)(i-)		$C_3H_7CO_2CH_2CH_2CH(CH_3)_2$	144.21	col. lq.	$0.863^{20/4}$		156.9	i.	∞	i.	∞
(i-)(i-)		$CH(CH_3)_2 \cdot CO_2CH(CH_3)_2$	144.21	col. lq.	0.875^0	−80.7	148–9	i.	∞	i.	∞
caproate (i-)		$CH(CH_3)_2CO_2CH(CH_3)_2C_4H_9$	172.26	col. lq.	0.882^{20}		204.3	i.	s.	i.	∞
carbamate (i-)		$NH_2CO_2CH_2CH(CH_3)_2$	117.15	col. lf.	$0.956^{?/64}$	65	206–7	i.	s.	i.	∞
cellosolve (n-)	2-BuO-ethanol-1	$C_4H_9OCH_2CH_2OH$	118.17	col. lf.	$0.903^{20/4}$		171.2	∞	∞	∞	∞

TABLE 2-2 Physical Properties of Organic Compounds (*Continued*)

Name	Synonym	Formula	Formula weight	Form and color	Specific gravity	Melting point, °C	Boiling point, °C	Solubility in 100 parts		
								Water	Alcohol	Ether
chloride (*n-*)	1-chloro-butane	$C_2H_5CH_2CH_2Cl$	92.57	col. lq.	0.887^{20}	-123.1	77.9^{763}	0.07^{18}	∞	∞
(*s-*)	2-chloro-butane	$C_2H_5CHClCH_3$	92.57	col. lq.	0.871^{204}	-131	67.8^{787}	i.	∞	∞
(*t-*)	1-Cl₂-2-Me-propane	$(CH_3)_2CHCH_2Cl$	92.57	col. lq.	0.884^{15}	-131.2	68.9	i.	∞	∞
(*t-*)	2-Cl₂-2-Me-propane	$(CH_3)_3CCl$	92.57	col. lq.	0.847^{15}	-26.5	51-2	i.	∞	∞
dimethylbenzene (*t-*)(1,3,5-)		$(CH_3)_3C{\cdot}C_6H_3{\cdot}(CH_3)_2$	162.26	lq.			$200\text{-}2^{147}$	i.	∞	∞
formate (*n-*)		$HCO_2CH_2CH_2C_2H_5$	102.13	lq.	0.911^{0}		106.9		∞	∞
(*i-*)		$HCO_2CH(CH_3)C_2H_5$	102.13	lq.	0.882^{204}		97	sl. s.	∞	∞
(*t-*)		$HCO_2CH_2CH(CH_3)_2$	102.13	lq.	0.885^{204}	-95.3	98.2	1.1^{22}	∞	∞
furoate (*n-*)		$OC_4H_3CO_2C_4H_9$	168.19	col. lq	1.056^{204}		$118\text{-}20^{25}$	i.	∞	∞
iodide (*n-*)	1-iodo-butane	$C_2H_5CH_2CH_2I$	184.03	lq.	1.617^{204}	-103.5	129.9	i.	∞	∞
(*s-*)	2-iodo-butane	$C_2H_5CHICH_3$	184.03	lq.	1.595^{20}	-104	118-9	i.	∞	∞
(*i-*)	1-iodo-2-Me-propane	$(CH_3)_2CHCH_2I$	184.03	lq.	1.606^{204}	-90.7	120	i.	∞	∞
(*t-*)	2-iodo-2-Me-propane	$(CH_3)_3CI$	184.03	lq.	$1.370^{19/15}$	-34	99	i.	∞	∞
lactate (*n-*)		$CH_3CH(OH)CO_2C_4H_9$	146.18	col. lq.	0.968		$75\text{-}6^{6}$	sl. s.	∞	∞
mercaptan (*n-*)	butanthiol-1	$C_2H_5CH_2CH_2SH$	90.18	col. lq.	0.837^{254}	-116	97-8	sl. s.	v. s.	v. s.
(*i-*)	2-Me-propanthiol-1	$(CH_3)_2CHCH_2SH$	90.18	lq.	0.836^{204}	<-79	88	v. sl. s.	s.	s.
(*t-*)		$(CH_3)_3CSH$	90.18	lq.			65-7			s.
methacrylate (*n-*)		$CH_2{:}C(CH_3)CO_2C_4H_9$	142.19	lq.	$0.889^{15.6}$		155	i.	s.	s.
(*i-*)		$CH_2{:}C(CH_3)CO_2C_4H_9$	142.19	lq.	$0.889^{15.6}$		155	i.	∞	∞
phenol (*p-*)(*t-*)		$(CH_3)_3C{\cdot}C_6H_4{\cdot}OH$	150.21	nd./aq.	$0.908^{11/24}$	99	236-8	sl. s.	s.	s.
propionate (*n-*)		$C_2H_5CO_2C_4H_9$	130.18	col. lq.	0.883^{15}	-89.55	146	i.	∞	∞
(*i-*)		$C_2H_5CO_2C_4H_9$	130.18	col. lq.	0.866^{204}	-71.4	132.5	i.	∞	∞
stearate (*n-*)		$CH_3(CH_2)_{16}CO_2C_4H_9$	340.57	col. lq.	0.888^{64}	27.5	136.8	i.	s.	s.
(*i-*)		$CH_3(CH_2)_{16}CO_2C_4H_9$	340.57	wax	0.555^{5525}	25	$220\text{-}5^{25}$	0.3^{25}	s.	s.
iso-thiocyanate (*n-*)	butyl mustard oil	$C_2H_5CH_2CH_2{\cdot}N{:}CS$	115.19	lq.	0.956^{11}		165^{734}	i.	∞	∞
(*s-*)(*d-*)	iso-Bu mustard oil	$(CH_3)_2CHCH_2{\cdot}N{:}CS$	115.19	lq.	0.964^{144}		162	i.	∞	∞
(*i-*)		$C_3H_7N{:}CS$	115.19	lq.	0.943^{204}		159-63	i.	∞	∞
(*t-*)		$(CH_3)_3C{\cdot}N{:}CS$	115.19	lq.	0.919^{10}	10.5	140^{770}	i.	s.	s.
valerate (*n-*)(*n-*)		$CH_3(CH_2)_3CO_2(CH_2)_3CH_3$	158.23	lq.	0.870^{154}	-93	186	v. sl. s.	s.	s.
(*i-*)(*n-*)		$(CH_3)_2CHCH_2CO_2(CH_2)_3CH_3$	158.23	lq.	0.862^{254}		168.8	i.	∞	∞
(*i-*)(*s-*)		$(CH_3)_2CHCH_2CO_2CH(CH_3)C_2H_5$	158.23	col. lq.	0.848^{204}		$163\text{-}4^{702}$	i.	∞	∞
(*t-*)(*i-*)		$C_2H_5CO_2C_4H_9$	158.23	col. lq.	0.874^{04}		168.7	i.	s.	s.
Butylene (α-)	butene-1	$C_2H_5CH{:}CH_2$	56.10	col. gas	0.6^{9}	-130	-5^{736}	i.	v. s.	v. s.
(β-)	butene-2	$CH_3CH{:}CHCH_3$	56.10	col. gas		-127	3^{746}	i.		
Butyraldehyde (*n-*)		$CH_3CH_2CH_2CHO$	72.10	col. lq.	0.817^{204}	-99	75.7	4	∞	∞
(*i-*)		$(CH_3)_2CHCHO$	72.10	col. lq.	0.794^{204}	-65.9	64^{757}	11^{20}	∞	∞
Butyric acid (*n-*)	butanoic acid	$CH_3CH_2CH_2CO_2H$	88.10	col. lq.	0.964^{204}	-4.7	163.5^{757}	∞	∞	∞
(*i-*)	2-Me-propanoic acid	$(CH_3)_2CHCO_2H$	88.10	col. lq.	0.949^{204}	-47	154.5	20^{20}	∞	∞
amide (*n-*)	n-butyramide	$C_2H_5CH_2CONH_2$	87.12	rhb.	1.032	115-6	216	16.3^{15}	s.	sl. s.
(*i-*)	iso-butyramide	$(CH_3)_2CHCONH_2$	87.12	mn. pl.	1.013	129-30	216-20	v. s.	s.	sl. s.
anhydride (*n-*)		$(C_2H_5CH_2CO)_2O$	158.19	col. lq.	0.968^{2020}	-75	199.5	d	d	∞
(*i-*)		$[(CH_3)_2CHCO]_2O$	158.19	col. lq.	0.950^{254}	-53.5	181.5^{734}	d	d	∞
anilide (*n-*)	n-butyranilide	$C_3H_7CONHC_6H_5$	163.21	mn. pr.	1.134	92	189^{15}	i.	∞	∞
Caffeic acid (3-,4-)		$(HO)_2C_6H_3CH_2CH_2CO_2H$	180.15		1.23^{19}	195-213	d.	s. h.	v. s.	sl. s.
Caffeine		$C_8H_{10}O_2N_4{\cdot}H_2O$	212.21	ndl./al.		237	subl.	2	2	0.3
Camphene (*dl-*)		$C_{10}H_{16}$	136.23	cr.	0.845^{204}	50	160	i.	s.	v. s.
Camphor (*d-* or *l-*)		$C_{10}H_{16}O$	152.23	trig.	0.999^{99}	178-9	209.1^{739}	0.1	∞	∞
Camphoric acid (*d-*)		$C_8H_{14}(CO_2H)_2$	200.23	mn.	1.186	187	159.6	0.6^{12}	120^{12}	v. s.
Cantharidine		$C_{10}H_{12}O_4$	196.20	cr.		212	subl.	0.003	s.	s.
Capric acid	decanoic acid	$CH_3(CH_2)_8CO_2H$	172.26	col. nd.	0.889^{87}	31.5	268-70	0.003	s.	s.
Caproic acid (*n-*)	hexanoic acid	$CH_3(CH_2)_4CO_2H$	116.16	oily lq.	0.922^{204}	-1.5	202^{761}	1.1^{20}	∞	∞
	2-Me-pentanoic-5 acid	$(CH_3)_2CH(CH_2)_2CO_2H$	116.16	col. oil	0.925^{204}	-35	207.7	v. sl. s.	s.	s.
Caprylic acid (*n-*)	octanoic acid	$CH_3(CH_2)_6CO_2H$	144.21	col. lf.	0.910^{204}	16	237.5	0.07^{15}	∞	∞
Carbazole	diphenylenelimine, dibenzopyrrole	$(C_6H_4)_2NH$	167.20	lf.		244.8	354.8	d	d	sl. s.
Carbitol	diethylene glycol mono-Et ether	$C_2H_5O(CH_2)_2O(CH_2)_2OH$	134.17	col. lq.	0.990^{2020}	-108.6	201.9	∞	∞	∞
Carbon disulfide		CS_2	76.13	col. lq.	1.263^{204}	-108.6	46.3	0.2^{0}	∞	∞
monoxide		CO	28.01	col. gas	$0.81^{-185.4}$	-207	-192	3.5^{50} cc.	0.92^{14}	
suboxide		$OC{:}C{:}CO$	68.03	gas	1.114^{0}	-107	7^{01}	d.		
tetrabromide		CBr_4	331.67	col. mn.	3.42	90.1(48)	189.5	0.02^{30}	s.	s.
tetrachloride		CCl_4	153.84	col. lq	1.595^{204}	-22.6	76.8	0.08^{20}	s.	∞
tetrafluoride		CF_4	88.01	gas			-128	sl. s.		
Carbonyl sulfide		COS	60.07	col. gas	1.24^{-87}	-138.2	-50.2^{760}	80^{14} cc.	s.	s.
Carminic acid		$C_{22}H_{20}O_{13}$	492.40	red pd.		d.136			s.	v. sl. s.
Carvacrol (1-,2-,4-)		$CH_3{\cdot}C_6H_3(OH)CH(CH_3)_2$	150.21	col. lq.	0.977^{204}	0.5	238	v. sl. s.	∞	∞

Name	Formula	M.W.	Form	Sp. gr.	M.P.	B.P.	Cold H₂O	Hot H₂O	Alcohol	Ether
Carvacrylamine (2-,1-,4-)	$H_2NC_6H_3(CH_3)C_3H_7$	149.23	oil	0.994^{20}	−16	241	v. sl. s.	s.	s.	s.
Carvone (d-)	$C_{10}H_{14}O$	150.21	col. lq.	0.961^{204}	−70	230^{766}	i.	s.	∞	∞
Cellosolve	$C_2H_4O(CH_3)OH$	90.12	col. lq.	0.931^{204}		135.1	∞	∞	∞	∞
acetate	$CH_3CO_2CH_2CH_2OC_2H_5$	132.16	col. lq.	0.975^{204}		156.3	22		∞	∞
Cellulose	$(C_6H_{10}O_5)x$	162.14	amor.	1.3–1.4			i.	i.	i.	i.
Cetyl acetate	$CH_3CO_2(CH_2)_{15}CH_3$	284.47	nd.	0.858^{30}	22–3	200^{15}	i.	v. sl. s. c.	s.	s.
alcohol	$CH_3(CH_2)_{15}CH_2OH$	242.43	lf.	0.818^{504}	49–50	189.5^{15}	i.	i.	s.	∞
Chloral	$CCl_3\cdot CHO$	147.40	col. lq.	1.505^{254}	−57	97.6^{708}	v. s.	v. s.	∞	∞
hydrate	$CCl_3\cdot CH(OH)_2$	165.42	mn. pr.	1.619^{504}	51.7	d. 98	474^{17}	i. c.	v. s.	i. c.
Chloralil	$OC_2\cdot CCl\cdot CCl_3\cdot CO$	245.89	yel./bz.		290		0.8 c.	111		111
Chloretone	$Cl_3C\cdot C(OH)(CH_3)_2$	177.47	col. cr.	1.385^{22}	97	167	sl. s.	s.	s.	s.
Chloro-acetanilide (p-)	$CH_3CO_2NHC_6H_4Cl$	169.61	rhb.	1.58^{2020}	175–6	189.5	v. s.	v. s.	∞	∞
-acetic acid	$ClCH_2CO_2H$	94.50	col. cr.	1.162^{16}	61.2	189.5	v. s.	v. s.	∞	∞
-acetone	CH_3COCH_2Cl	92.53	col. lq.	1.324^{15}	−44.5	121	d.	d.	v. s.	v. s.
-acetophenone (ω-)	$C_6H_5COCH_2Cl$	154.59	rhb.	1.498^{2020}	58–9	245–7	0.11	0.11	v. s.	v. s.
-acetyl chloride	$ClCH_2COCl$	112.95	col. lq.			105	d.	d.		
-aniline (o-)	$ClC_6H_4NH_2$	127.57	lq.	1.213^{204}	0	210.5	i.	i.	s.	s.
(m-)	$ClC_6H_4NH_2$	127.57	lq.	1.216^{204}	−10.4	230^{767}	s. h.	s. h.	s.	s.
(p-)	$ClC_6H_4NH_2$	127.57	rhb.	1.427^{19}	70–1	230–1	sl. s. h.		s.	s.
-anthraquinone (1-)	$C_6H_4(CO)_2C_6H_3Cl$	242.65	yel. nd.		162	subl.				
(2-)	$C_6H_4(CO)_2C_6H_3Cl$	242.65	nd./al.		208–9					
-benzaldehyde (o-)	ClC_6H_4CHO	140.57	nd.	1.29^{8}	11	208^{748}	v. sl. s.	v. s.	v. s.	v. s.
(m-)	ClC_6H_4CHO	140.57	pr.	1.250^{15}	17–8	213–4	v. sl. s.	v. s.	v. s.	v. s.
(p-)	ClC_6H_4CHO	140.57	pr.	1.196^{61}	47.8	213^{748}	s. h.	v. s.	v. s.	v. s.
-benzene	C_6H_5Cl	112.56	col. lq.	1.107^{204}	−45.2	132.1	0.049^{20}		∞	∞
-benzoic acid (o-)	$ClC_6H_4CO_2H$	156.57	mn./aq.	1.544^{254}	141–2	subl.	0.208^{25}	v. s.	s.	s.
(m-)	$ClC_6H_4CO_2H$	156.57	col. lq.	1.496^{254}	158	59.4	0.041^{25}	v. s.	∞	∞
(p-)	$ClC_6H_4CO_2H$	156.57	pr.	1.541^{24}	242–3	69	0.008^{25}	v. s.	s.	s.
-buta-1,3-diene (2-)	$CH_2:CCl\cdot CH:CH_2$	88.54	tri.	0.958^{2020}		88	v. sl. s.		v. s.	v. s.
-buta-1,2-diene (1-)	$CH_2:CH\cdot CH:CHCl$	88.54	col. lq.	0.965^{2020}			v. sl. s.		v. s.	s.
-dimethylhydantoin	$-C(CH_3)_2N(Cl)CON(Cl)CO-$	88.54	col. lq.	0.991^{2020}			d.		v. s.	v. s.
-dinitrobenzene (α)(1-2-)(4-)	$ClC_6H_3(NO_2)_2$	197.03	col. lq.	1.5^{2020}	130	315 d.	0.21^{25}	d.	v. s. h.	v. s.
(α)(1-,3-)(4-)	$ClC_6H_3(NO_2)_2$	202.56	cr./et.	1.697^{22}	39(36)	315 d.	i.		s. h.	s. h.
-diphenyl (o-)	$C_6H_5\cdot C_6H_4Cl$	202.56	rhb./et.		53(43)	267–8	i.		v. s.	v. s.
(m-)	$C_6H_5\cdot C_6H_4Cl$	188.65	cr.		34	284–5	i.		v. s.	v. s.
(p-)	$C_6H_5\cdot C_6H_4Cl$	188.65	cr.		89	282	i.		v. s.	v. s. h.
-hydroquinone	$ClC_6H_3(OH)_2$	144.56	lf.		77.5	263 sl. d.	v. s.	v. s.	v. s.	v. s.
-naphthalene (α-)	$C_{10}H_7Cl$	162.61	mn.	1.194^{204}	106	259.3	i.		s.	s.
(β-)	$C_{10}H_7Cl$	162.61	col. lq.	1.266^{16}	−20	264^{751}	i.		v. s. h.	s.
-nitrobenzene (o-)	$ClC_6H_4NO_2$	157.56	lf./al.	1.305^{504}	56–7	245.5^{753}	i.		v. s. h.	∞
(m-)	$ClC_6H_4NO_2$	157.56	mn. nd.	1.343^{504}	32.5	235.6	i.		v. s. h.	∞
(p-)	$ClC_6H_4NO_2$	157.56	yel./al.	1.298^{91}	44.4(24)	242^{761}	i.		v. s. h.	∞
-nitrotoluene (2-,4-)	$CH_3C_6H_3(NO_2)(Cl)$	171.56	mn. pr.	1.256^{80}	83–4	240^{718}	i.		s.	s.
(2-,6-)	$CH_3C_6H_3(NO_2)(Cl)$	171.56	cr.	1.241^{1815}	38.2	238	i.		v. s.	v. s.
-phenol (o-)	ClC_6H_4OH	128.56	col. lq.	1.268^{25}	37.5	175–6	2.85^{20}	s.	∞	∞
(m-)	ClC_6H_4OH	128.56	nd.	1.306^{204}	7(0)	214	2.60^{20}	s.	∞	∞
(p-)	ClC_6H_4OH	128.56	nd.	1.306^{9}	32–3	217	2.71^{20}	v. s.	∞	∞
-propionic acid (α)/(dl-)	$CH_3\cdot CHCl\cdot CO_2H$	108.53	col. lq.	1.082^{204}	41–3	186	∞	∞	∞	∞
-toluene (o-)	$CH_3\cdot C_6H_4Cl$	126.58	col. lq.	1.072^{204}	<−20	159.5	i.		∞	∞
(m-)	$CH_3\cdot C_6H_4Cl$	126.58	col. lq.	1.070^{204}	−34	161.6	i.		∞	∞
(p-)	$CH_3\cdot C_6H_4Cl$	126.58	col. lq.	1.489^{20}	−47.8	162.2	i.		∞	∞
Chloroform	$CHCl_3$	119.39	col. lq.		7.5	61.2	0.82^{20}		∞	∞
Chlorophyll (α-)	$C_{55}H_{72}O_5N_4\cdot Mg$	893.48			−63.5		i.		s.	s.
Chloropicrin	Cl_3CNO_2	164.39	lq.	1.651^{234}	d.	112.3^{766}	0.17^{18}	1.1^{17}	v. s.	18
Cholesterol	$C_{27}H_{45}OH\cdot H_2O$	404.65	rhb./al.	1.067	−64	subl.	0.26^{20}	0.1^{16}	sl. s.	v. sl. s.
Chrysene	$C_{18}H_{12}$	228.28	col. rhb.		149–51	448	i.		sl. s. h.	sl. s.
Chrysoidine (2-,4-)	$C_{12}H_5\cdot N:N\cdot C_6H_3(NH_2)_2$	212.25	col. lq.		253–4	subl.	sl. s. h.		v. sl. s.	i.
Chrysophanic acid	$C_{15}H_5(OH)_2(CH_3)O_2$	254.23	yel. cr.		117.5	subl. d.	i. c.		v. sl. s.	sl. s.
Cinchomeronic acid (3-,4-)	$C_5H_3N(CO_2H)_2$	167.12	yel./al.	0.927^{20}	195	176–7	v. sl. s.		v. sl. s.	∞
Cineole, eucalyptole	$C_{10}H_{18}O$	154.24	col. oil	1.284^{4}	258–9 d.	125^{19}	1.9^{15}	24^{20}	∞	∞
Cinnamic acid (cis-)	$C_6H_5\cdot CH:CHCO_2H$	148.15	mn. pr.	1.245	1.5	300				
(trans-)	$C_6H_5\cdot CH:CHCO_2H$	148.15	mn. pr.	1.110^{2079}	68	252 sl. d.	0.04^{18}	s.	s.	v. s.
aldehyde	$C_6H_5\cdot CH:CHCHO$	132.15	lq.	1.040^{3535}	133	257.5	v. sl. s.	v. sl. s.	∞	∞
Cinnamyl alcohol	$C_6H_5\cdot CH:CHCH_2OH$	134.17	nd. or pr.	1.085^{165}	−7.5		sl. s.	s.	∞	∞
cinnamate	$C_6H_5\cdot CO\cdot C_9H_9$	264.31	nd.	1.617	33	229	i.	4 c.	v. s.	v. s.
Citraconic acid (cis-)	$CH_3C(CO_2H):CHCO_2H$	130.10	col. oil	0.890^{174}	44	d.	4 c.		s.	s.
Citral (α)	$C_9H_{15}CHO$	152.23	cr.	1.542^{204}	92–3	204–8	360^{25}	76^{15}	∞	∞
Citric acid	$C_3H_4(OH)(CO_2H)_3$	192.12	col. oil	0.855^{175}	153	224–5	207.7^{25}	2^{15}	∞	2^{15}
Citronellal (d-)	$C_9H_{18}O$	154.24	col. oil	0.848^{204}		166–7	v. sl. s.	v. sl. s.	∞	∞
Citronellol (d-)	$C_{10}H_{20}O$	156.26	col. lq.	0.847^{17}	−2		v. sl. s.	v. sl. s.	∞	∞
Coniine (d-)/(2-)	$C_8H_7\cdot C_5H_{10}N$	127.22	col. lq.				1.1	v. s.	∞	v. s.

TABLE 2-2 Physical Properties of Organic Compounds (Continued)

Name	Formula	Formula weight	Form and color	Specific gravity	Melting point, °C	Boiling point, °C	Solubility in 100 parts — Water	Alcohol	Ether
Coumaric acid (o-)	$HOC_6H_4CH{:}CHCO_2H$	164.15	nd/aq.		207–8	subl.	sl. s. c.	s.	v. sl. s.
(p-)	$HOC_6H_4CH{:}CHCO_2H$	164.15	cr./aq.		206–7 d.		s. h.	v. s. h.	v. s.
Coumarin	$C_9H_6O_2$	146.14	rhb./et.	0.935^{20}_{4}	70	290–1	0.3 c.	s.	s.
Coumarone	C_8H_6O	118.13	oil	$1.078^{15/15}$	<−18	173–4			s.
Creatine	$C_4H_9N_3O_2{\cdot}H_2O$	149.15	mn.		295		1.4^{18}	0.01^{17}	i.
Creatinine	$C_4H_7N_3O$	113.12			260 d.		8.7^{16}	1^{16}	
Creosol (3-,1-,4-)	$CH_3O{\cdot}C_6H_3(CH_3)OH$	138.16	pr.	$1.092^{20/20}$	5.5	$221{-}2^{765}$	v. sl. s.	∞	∞
Cresidine (1-,2-,4-)	$CH_3(NH_2)C_6H_3{\cdot}OCH_3$	137.18	nd./pet.		93–4	235	v. sl. s.	∞	
Cresol (o-)	$CH_3C_6H_4OH$	108.13	cr.	1.048^{20}_{4}	30.8	190.8	2.5	$∞^{30}$	$∞^{30}$
(m-)	$CH_3C_6H_4OH$	108.13	lq.	1.034^{20}_{4}	10.9	202.8	0.5		
(p-)	$CH_3C_6H_4OH$	108.13	pr.	1.035^{20}_{4}	35–6	202	1.8	$∞^{36}$	$∞^{36}$
Cresyl benzoate (o-)	$C_6H_5CO_2C_6H_4CH_3$	212.24	lq.			308	i.		
(m-)	$C_6H_5CO_2C_6H_4CH_3$	212.24	lq.		55	314	i.		
(p-)	$C_6H_5CO_2C_6H_4CH_3$	212.24	cr.		71.5	316	i.	s.	s.
Crotonic acid (α-)	$CH_3CH{:}CHCO_2H$	86.09	col. mn.	$0.964^{79.7}$	72	189	8.3^{15}	∞	∞
acid (β-)(cis-)	$CH_3CH{:}CHCO_2H$	86.09	nd.	1.031^{15}_{4}	15.5	170–1 d.	$s.^{20}$	s.	
aldehyde (α)	$CH_3CH{:}CHCHO$	70.09	col. lq.	$0.853^{20/20}$	−69	102.2	18	∞	∞
Cumene	$C_6H_5CH(CH_3)_2$	120.19	col. lq.	0.862^{20}_{4}	−96.9	152.5	i.	∞	∞
Cumic acid (p-)	$(CH_3)_2CH{\cdot}C_6H_4{\cdot}CO_2H$	164.20	tri.	1.162^{4}	116–7	subl	0.02^{25}	v. s.	v. s.
Cumidine (p-)	$(CH_3)_2CH{\cdot}C_6H_4{\cdot}NH_2$	135.20	lq.	0.953	<−20	225^{761}	i.	s.	
Cyanamide	$H_2N{\cdot}CN$	42.04	col. nd.	$1.073^{58/4}$	44–5	140^{19}	v. s.	v. s.	v. s.
Cyanic acid	$HOCN$ or $HNCO$	43.03	gas	1.140^{0}	−80	-64^{0}	sl. s.	v. s.	v. s.
Cyanoacetic acid	$CH_2(CN)CO_2H$	85.06	col. lq.	0.866^{17}	65–6	108^{92}	v. s.	v. s.	v. s.
Cyanogen	$(CN)_2$	52.04	col. gas	2.015^{20}_{4}	−34.4	−21	450^{20} cc.	2300^{20} cc.	500^{20} cc.
bromide	$BrCN$	105.93	nd.	1.222^{0}	52	61.3^{750}	s.	s.	s.
chloride	$ClCN$	61.48	col. gas	1.768^{194}	−6.5	12.5–13	$250^{0.9\,20}$ cc.	v. s.	5000^{20} cc.
Cyanuric acid	$C_3H_3O_3N_3{\cdot}2H_2O$	165.11			>360	d.	0.27^{17}	0.1^{22}	i.
Cyclo-butane	$CH_2{<}(CH_2)_2{>}CH_2$	56.10	mn./aq.	0.703^{94}	−50	$11{-}12^{726}$	i.	v. s.	v. s.
-heptane	$CH_2{<}(CH_2{\cdot}CH_2)_2{>}CH_2$	98.18	col. gas	0.810^{20}_{4}	−12	118–20	i.	∞	∞
-hexane	$CH_2{<}(CH_2{\cdot}CH_2)_2{>}CH_2$	84.16	col. lq.	0.779^{20}_{4}	6.5	80–1	i.	∞	∞
-hexanol	$CH_2{<}(CH_2{\cdot}CH_2)_2{>}CHOH$	100.16	col. nd.	0.962^{20}_{4}	23.9	160–1	3.6^{20}	v. s.	v. s.
-hexanone	$CH_2{<}(CH_2{\cdot}CH_2)_2{>}CO$	98.14	col. oil	0.947^{18}_{4}	−45	155–6	i.	∞	∞
-hexene	$(-CH_2{\cdot}CH_2{\cdot}CH{:})_2$	82.14	lq.	0.810^{20}_{4}	−103.7	83.3	i.	v. s.	v. s.
-hexyl acetate	$CH_3{\cdot}CO_2{\cdot}C_6H_{11}$	142.19	oil	0.985^{0}_{4}		174^{750}	v. sl. s.	v. s.	v. s.
amine	$CH_2{<}(CH_2{\cdot}CH_2)_2{>}CHNH_2$	99.17	col. lq.	0.865^{200}		134	i.	s.	s.
bromide	$CH_2{<}(CH_2{\cdot}CH_2)_2{>}CHBr$	163.06	col. lq.	$1.324^{20/20}$		165^{714}	i.	∞	∞
chloride	$CH_2{<}(CH_2{\cdot}CH_2)_2{>}CHCl$	118.61	col. lq.	0.977^{18}_{4}	−43.9	142	0.01^{19}	s.	s.
-pentadiene (1-,3-)	$CH_2{<}(CH{:}CH)_2{>}$	66.10	col. oil	0.805^{194}	−85	41–2	2^{12}	s.	s.
-pentane	$<(CH_2{\cdot}CH_2)_2{>}CH_2$	70.13	col. lq.	0.745^{20}_{4}	−93.3	49–50	i.	s.	s.
-pentanone	$<(CH_2{\cdot}CH_2)_2{>}CO$	84.11	col. lq.	0.948^{89}	−58.2	129–30	i.	∞	∞
-propane	$<CH_2{\cdot}CH_2{\cdot}CH_2>$	42.08	col. gas	0.720^{-79}	−126.6	-34^{749}	i.	s.	s.
Cymene (o-)	$CH_3{\cdot}C_6H_4{\cdot}CH(CH_3)_2$	134.21	col. lq.	0.875^{20}_{4}	<−25	177	i.	s.	s.
(m-)	$CH_3{\cdot}C_6H_4{\cdot}CH(CH_3)_2$	134.21	col. lq.	0.862^{20}	−73.5	175–6	i.	s.	s.
(p-)	$CH_3{\cdot}C_6H_4{\cdot}CH(CH_3)_2$	134.21	col. lq.	0.857^{20}_{4}		176–7	i.	∞	∞
Cystine (l-)	$[{\cdot}SCH_2{\cdot}CH(NH_2)CO_2H]_2$	240.29	pl.		d.258–61		sl. s. h.	i.	i.
Dambose	$C_6H_6(OH)_6$	180.16	mn./aq.	1.752	253	319^{25}	s.	∞	∞
Decahydronaphthalene (cis-)	$C_{10}H_{18}$	138.24	lq.	0.895^{184}	−43	193.3	i.	s.	s.
(trans-)	$C_{10}H_{18}$	138.24	lq.	0.872^{20}_{4}	−32	185.3	i.	s.	s.
Decane (n-)	$CH_3(CH_2)_8CH_3$	142.28	col. lq.	0.730^{2}	−29.7	174.0	i.	s.	s.
Decyl alcohol	$CH_3(CH_2)_8CH_2OH$	158.28	col. lq.	0.830^{20}_{4}	7	232.9	i.	∞	∞
Dextrin	$(C_6H_{10}O_5)x$	162.14	amor.	1.038			s.	i.	i.
Diacetone alcohol	$(CH_3)_2C(OH){\cdot}CH_2COCH_3$	116.16	lq.	0.931^{25}	−47	167.9	∞	∞	∞
Diamino-benzophenone (4-,4'-)	$H_2NC_6H_4COC_6H_4NH_2$	212.24	yel. nd.		237–9	d.	i.	sl. s.	sl. s.
-diphenylamine (4-,4'-)	$H_2NC_6H_4NHC_6H_4NH_2$	199.25	lf./aq.		158		sl. s.	s.	s.
-diphenylmethane (4-,4'-)	$H_2NC_6H_4CH_2C_6H_4NH_2$	198.26	nd./aq.		93–4	$249{-}53^{35}$	sl. s.	s.	v. s.
-diphenylurea (4-,4'-)	$(H_2NC_6H_4NH)_2CO$	242.28	cr.		subl. 310		i.	s.	
Diamyl-amine (n-)	$[(CH_3)_2CHCH_2CH_2]_2NH$	157.29	col. lq.	0.767^{21}_{4}	−44	188–90	sl. s. c.	s.	s.
ether (n-)	$[(CH_3)_2CH(CH_2)_2]_2O$	158.28	col. lq.	0.774^{20}_{4}	−69	190	i.	∞	∞
(i-)	$[(CH_3)_2CHCH_2CH_2]_2O$	158.28	col. lq.	0.777^{20}_{4}		173.4	i.	∞	∞
Diamyl ketone (i-)	$[(CH_3)_2CHCH_2CH_2]_2CO$	170.29	yel. oil	0.821^{25}_{4}	14.6	228	i.	s.	s.
phthalate (i-)	$C_6H_4(CO_2C_5H_{11})_2$	306.39	col. lq.			$204{-}6^{11}$	i.		
Dianisidine (o-)(4-,3-)₂	$[NH_2(OCH_3)C_6H_3{\cdot}]_2$	290.35	col. lf.		131.5	225^{40}	sl. s.	s.	s.
tartrate (i-)	$C_4H_6{\cdot}{\cdot}{\cdot}$	244.28		1.03		195^{16}	0.05	s. h.	v. s.
Diazo-aminobenzene (2-,2'-)	$C_6H_5N{:}N{\cdot}NHC_6H_5$	197.23	yel. lf.	1.063^{15}_{4}	96–8	expl.	i.	s.	s.
-aminotoluene (2-,2'-)	$C_6H_5N{:}N{\cdot}NHC_7H_7$	225.28	vit. or. cr.		51		i.		
-methane	$CH_2{:}N_2$	42.04	gas		−145	−23	d.		

Name	Formula	Mol. wt.	Form, color	Sp. gr.	M.P. °C	B.P. °C	Sol. water	Sol. alcohol	Sol. ether
Dibenzothiazyl-disulfide (2-2'-)	(C₆H₄NSC)₂S₂ → $(C_6H_4NSC)_2S_2$	232.46	cr.	1.50	180	d.	i.		s.
Dibensoyl methane	$(C_6H_5CO)_2CH_2$	224.25	rhb./al.	$1.028^{25/25}$	78	$219\text{–}21^{18}$	i.	4.4^{20}	s.
Dibensyl-amine	$(C_6H_5CH_2)_2NH$	197.27	col. oil		−26	$268\text{–}71^{250}$	i.	s.	s.
-aniline	$C_6H_5N(CH_2C_6H_5)_2$	273.36	pr./al.		70–1	>300	i.	v. s. h.	s.
ketone	$(C_6H_5CH_2)_2CO$	210.26	cr.		34–5	330.6	v. sl. s.	s.	s.
phthalate	$C_6H_4(CO)_2(CH_2C_6H_5)_2$	346.36	lf./al.		42–3	274^{12}	i.	s.	s.
succinate	$(\text{-}CH_2CO_2CH_2C_6H_5)_2$	298.32	col. lq.		45–6	238^{4}	i.	s.	s.
Dibromo-benzene (o-)	$C_6H_4Br_2$	235.92	col. lq.	$1.956^{20/4}$	1.8	221–2	i.	s.	71^{25}
(m-)	$C_6H_4Br_2$	235.92	pl./al.	$1.952^{20/4}$	−6.9	219^{755}	i.	1.6	
(p-)	$C_6H_4Br_2$	235.92	mn. pr.	2.261^{18}	87–8	218.6^{755}	i.	v. sl. s. h.	∞
-diphenyl (4-4'-)	$BrC_6H_4{\cdot}C_6H_4Br$	312.02	col. lq.	1.897	164–5	355–60	i.	∞	∞
Dibutyl-adipate (n-)	$(\text{-}CH_2CH_2CO_2C_4H_9)_2$	258.35	col. lq.	0.965^{25}	−38	183^{14}	v. sl. s.	∞	∞
(i-)	$(\text{-}CH_2CH_2CO_2C_4H_9)_2$	258.35	col. lq.	0.950^{25}	−20	278–80	i.	s.	∞
-amine (n-)	$(C_2H_5CH_2CH_2)_2NH$	129.24	col. lq.	$0.768^{20/20}$		159^{761}	i.	s.	∞
(i-)	$[(CH_3)_2CHCH_2]_2NH$	129.24	lq.	$0.741^{25/4}$	−70	139–40	i.	∞	∞
-p-aminophenol (s-)	$(C_4H_9)_2N{\cdot}C_6H_4OH$	221.33	col. lq.			170^{10}	∞	s.	s.
-aniline (n-)	$C_6H_5N(C_4H_9)_2$	205.33	col. lq.	$0.924^{20/4}$		262.8	i.	∞	∞
carbonate (n-)	$CO(OC_4H_9)_2$	174.23	col. lq.	0.919^{15}		207^{740}	i.	s.	s.
(i-)	$CO(OC_4H_9)_2$	174.23	col. lq.			190	i.	∞	∞
(s-)	$CO(OC_4H_9)_2$	174.23	lq.		−98	178–80			
ether (n-)	$(C_3H_5CH_2CH_2)_2O$	130.22	lq.	$0.769^{20/20}$		142.4			
(i-)	$[(CH_3)_2CHCH_2]_2O$	130.22	lq.	0.762^{15}		122.5			
(s-)	$(C_2H_5CH_2CH_2)_2O$	130.22	lq.	0.756^{21}	−5.9	121			
ketone (n-)	$(C_3H_5CH_2CH_2)_2CO$	142.23	lq.	$0.827^{18/4}$		187.7			
(i-)	$[(CH_3)_2CHCH_2]_2CO$	142.23	lq.	$0.805^{21/4}$	−29.6	168.1	<0.05	s.	v. s.
malate (l-)(n-)	$C_2H_4O(CO_2C_4H_9)_2$	246.30	lq.	$1.038^{20/4}$		$170\text{–}1^{18}$	i.	s.	s.
oxalate (n-)	$(\text{-}CO_2C_4H_9)_2$	202.24	col. lq.	$0.986^{20/4}$		245.5	<0.06	∞	∞
phthalate (n-)	$C_6H_4(CO_2C_4H_9)_2$	278.34	pr.	1.045^{21}		340	v. sl. s.	v. s.	s.
tartrate (d-)(n-)	$(CHOHCO_2C_4H_9)_2$	262.30	lq.	1.098^{15}	22–2.5	$200\text{–}3^{18}$	0.04^{25}	v. s.	∞
(d-)(i-)	$(CHOHCO_2C_4H_9)_2$	262.30	lq.	$1.031^{75/4}$	73–4	323–5	i.		
Dichloro-acetic acid	$Cl_2CH\text{-}CO_2H$	128.95	lq.	$1.560^{25/25}$	9.7(−4)	194.4	v. sl. s.	∞	∞
-acetone (αα-)	$Cl_2CHCOCH_3$	126.98	nd.	1.234^{15}	50	120	v. sl. s.		
-aniline (2,5-)	$Cl_2C_6H_3NH_2$	162.02	yel. nd.		208–9	251			
-anthraquinone (1,3-)	$C_6H_2(CO)_2C_6H_2Cl_2$	277.10	yel. nd.		187.5				
(1,4-)	$C_6H_4{:}(CO)_2{:}C_6H_2Cl_2$	277.10	yel. nd.		251				
(1,5-)	$C_6H_4Cl{:}(CO)_2{:}C_6H_4Cl$	277.10	yel. nd.		203–4				
(1,6-)	$C_6H_4Cl{:}(CO)_2{:}C_6H_4Cl$	277.10	yel. nd.		202–3				
(1,8-)	$C_6H_4Cl{:}(CO)_2{:}C_6H_4Cl$	277.10	yel. nd.		268–70				
(2,3-)	$C_6H_4{:}(CO)_2{:}C_6H_3Cl$	277.10	yel. nd.		282				
(2,6-)	$C_6H_4{:}(CO)_2{:}C_6H_3Cl$	277.10	yel. nd.		210–11				
(2,7-)	$C_6H_4Cl{:}(CO)_2{:}C_6H_4Cl$	277.10	col. lq.	$1.305^{20/4}$	−17.6	179	i.	∞	s.
-benzene (o-)	$C_6H_4Cl_2$	147.01	col. lq.	$1.288^{20/4}$	−24.8	172^{766}	i.	∞	∞
(m-)	$C_6H_4Cl_2$	147.01	col. mn.	1.458^{21}	53	174^{764}	i.	∞	∞
(p-)	$C_6H_4Cl_2$	147.01	lq.		−38.7	161–3	i.	∞	s.
-butane (n-)(1,4-)	$ClCH(CH_2)_2CH_2Cl$	127.02	pr.	$1.442^{0/4}$	148	315–9	0.9^{0}	v. s.	v. s.
-diphenyl (4,4'-)	$ClC_6H_4{\cdot}C_6H_4Cl$	223.10	col. lq.	$1.256^{20/20}$	−35.3	83.7	i.	v. sl. s.	4
-ethane (1,2-)	$ClCH_2{\cdot}CH_2Cl$	98.97	nd./al.	$1.300^{70/4}$	67–8	$286\text{–}7^{740}$	i.	s.	∞
-naphthalene (β-)(1,4-)	$C_{10}H_6Cl_2$	197.06	lf./al.	1.669^{22}	107	subl.	i.	s. h.	s.
(γ-)(1,5-)	$C_{10}H_6Cl_2$	197.06	tri/al.	$1.094^{25/4}$	54.6	266	i.	s.	∞
-nitrobenzene (2,5-)	$Cl_2C_6H_3NO_2$	192.01	nd.	$1.383^{60/25}$	45	180–1	0.45^{20}	s.	∞
-pentane (1,5-)	$ClCH_2(CH_2)_3CH_2Cl$	141.04	col. lq.	1.40^{14}	83	209–10	sl. s.	v. s.	
-phenol (2,4-)	$Cl_2C_6H_3OH$	163.01	cr.	$1.097^{20/4}$	207–8	d.	∞	∞	∞
Dichloramine T (p-)	$CH_3C_6H_4SO_2NCl_2$	240.11	mn. pl.	$1.009^{20/4}$	28	270^{748}	sl. s.	∞	s.
Dicyandiamide	$H_2N{\cdot}C({:}NH){\cdot}NH{\cdot}CN$	84.08	pr.	$0.712^{15/15}$	−21	$239\text{–}41^{761}$	s.	s.	∞
Diethanolamine	$HN(C_2H_4{\cdot}OH)_2$	105.14	syrup		−38.9	55.5^{709}	∞	∞	s.
Diethyl adipate	$(\text{-}CH_2CH_2CO_2C_2H_5)_2$	202.24	col. lq.	$0.934^{20/4}$	78	276–80	i.	v. s.	∞
-amine	$(C_2H_5)_2NH$	73.14	col. lq.	$0.975^{20/4}$	−34.4	216	0.9	s.	∞
-aminophenol (m-)	$(C_2H_5)_2N{\cdot}C_6H_4{\cdot}OH$	165.23	rhb.	$0.985^{20/4}$	270 d.	126^{709}	i.	v. s.	∞
-aniline	$(C_2H_5)_2NC_6H_5$	149.23	oil	$0.994^{25/25}$	−43	230	i.	∞	∞
sulfonic acid (m-)	$(C_2H_5)_2NC_6H_4SO_3H$	229.29	cr.	1.025^{11}		196.7			
carbonate	$OC(OC_2H_5)_2$	118.13	col. lq.	$0.816^{19/4}$	125	237	0.88^{20}	∞	∞
diethyl malonate	$(C_2H_5)_2C(CO_2C_2H_5)_2$	216.27	col. lq.	$1.055^{20/4}$		101.7	4.7^{20}	∞	∞
Diethyl dimethyl malonate	$(CH_3)_2C(CO_2C_2H_5)_2$	188.22	col. lq.		−24	198.9	i.	v. s.	v. s.
glutarate	$CH_2(CH_2CO_2C_2H_5)_2$	188.22	col. lq.	1.005	−42	d. 170–80	2.08^{20}	v. s.	v. s.
ketone	$(C_2H_5)_2CO$	86.13	col. lq.	1.026	−49.8	285–90	65^{16}	∞	∞
malonate	$CH_2(CO_2C_2H_5)_2$	160.17	col. lq.	$1.079^{20/4}$		318	i.	∞	∞
-malonic acid	$(C_2H_5)_2C(CO_2H)_2$	160.17	pr./aq.	$1.121^{25/25}$		186	i.	v. s.	∞
-naphthylamine (α-)	$C_{10}H_7{\cdot}N(C_2H_5)_2$	199.28	col. oil	$1.172^{25/4}$	−40.6	298–9	i.	v. s.	∞
(β-)	$C_{10}H_7{\cdot}N(C_2H_5)_2$	199.28	col. lq.				v. sl. s.	v. s.	∞
oxalate	$(\text{-}CO_2C_2H_5)_2$	146.14	col. lq.			210	i.	∞	∞
phthalate (o-)	$C_6H_4(CO_2C_2H_5)_2$	222.23	col. lq.		−25		i.	∞	∞
sulfate	$O_2S(OC_2H_5)_2$	154.18	col. lq.	$0.837^{20/4}$	−99.5	$92\text{–}3^{754}$	0.31^{20}	∞	∞
sulfide	$(C_2H_5)_2S$	90.18	col. lq.						

TABLE 2-2 Physical Properties of Organic Compounds (Continued)

Name	Formula	Formula weight	Form and color	Specific gravity	Melting point, °C	Boiling point, °C	Solubility in 100 parts — Water	Alcohol	Ether
tartrate (d-)	$(CHOH \cdot CO_2C_2H_5)_2$	206.19	lq.	1.204^{204}	17	280	sl. s.	∞	∞
-toluidine (o-)	$CH_3 \cdot C_6H_4 \cdot N(C_2H_5)_2$	163.25	lq.			$208\text{-}9^{755}$	i.	s.	s.
(m-)	$CH_3 \cdot C_6H_4 \cdot N(C_2H_5)_2$	163.25	lq.	$0.924^{15.5}$		231-2	i.	s.	s.
(p-)	$CH_3 \cdot C_6H_4 \cdot N(C_2H_5)_2$	163.25	lq.			228-9	i.		
Diethyleneglycol dinitrate	$O(CH_2CH_2ONO_2)_2$	196.12	lq.	1.377^{254}			s.	s.	s.
Difluorodichloromethane	F_2CCl_2	120.92	gas		-155	-29.2	5.7 cc.^{28}		i.
Diglycerol	$((HO)_2C_3H_5)_2O$	166.17		1.486^{-30}		$220\text{-}30^{10}$	s. h.		
Dihydroxy-dinaphthyl (α- -2,-2',-1,-1')	$(HO \cdot C_{10}H_6 \cdot)_2O$	286.31	pl./al.		300	subl.	i.	v. s.	v. s.
-diphenyl (4,-4')	$(HO \cdot C_6H_4 \cdot)_2$	186.20	nd/al.	1.25	270-2	subl.	i.	v. s.	v. s.
-ethyl formal (β-)	$CH_2(OCH_2CH_2OH)_2$	136.15	rhb./al.	1.154^{25}	-5.3	264	∞	v. s.	v. s.
-naphthalene (1,-5-)	$C_{10}H_6(OH)_2$	160.16	lq.		258-60	d.	i.		
(1,-8-)	$C_{10}H_6(OH)_2$	160.16	pr./aq.		140		sl. s. h.	v. s.	v. s.
Dimethoxy-benzene (p-)	$(CH_3O)_2C_6H_4$	138.16	lf.	1.053^{5555}	56	212.6	v. sl. s.	v. s.	v. s.
-diphenylamine (4,-4'-)	$HN(C_6H_4OCH_3)_2$	229.26	cr.		103		i.		
Dimethyl adipate	$(CH_2)_4(CO_2CH_3)_2$	262.30	lq.	$1.075^{15.6}$	10-1	115^{18}	i.	sl. s.	sl. s.
-amine	$(CH_3)_2NH$	174.19	col. lq.	1.063^{204}	-96	7.4	v. s.		
-aminoazobenzene (p-)	$C_6H_5N:N \cdot C_6H_4N(CH_3)_2$	45.08	col. lq.	0.680^{04}	116-7	d.	i.		
-aminoethanol	$(CH_3)_2NCH_2CH_2OH$	225.28	yel./al.	0.887^{204}	85	135^{756}	sl. s. h.	s.	s.
-aminophenol (m-)	$(CH_3)_2NC_6H_4OH$	89.14	nd.			265-8	i.		
-aniline	$(CH_3)_2NC_6H_5$	137.18	yel. lq.	0.956^{204}	2.5	193	i.		
sulfonic acid (m-)	$(CH_3)_2NC_6H_4SO_3H$	121.18	cr.		d. 266		s. h.		
(p-)	$(CH_3)_2NC_6H_4SO_3H \cdot H_2O$	201.24	pr.		257		i.		
carbonate	$OC(OCH_3)_2$	219.25	col. lq.	1.070^{204}	0.5	89-90	i.	∞	∞
ether	CH_3OCH_3	90.08	gas		-138.5	-23.7	3700 cc.^{18}	∞	∞
-formamide	$HCON(CH_3)_2$	46.07	col. lq.	0.945^{25}	-58.3	152.8	v. s.	v. s.	s.
fumarate	$(\cdot CHCO_2CH_3)_2$	73.09	col. tri.		102	192	sl. s.	sl. s.	sl. s.
glutarate	$(CH_2)_3(CO_2CH_3)_2$	144.12	lq.	$1.089^{15.6}$	-37	130^{50}	i.		
glyoxime	$(CH_3 \cdot C:NOH)_2$	160.17	col. cr.		240-6	d.	v. s.		v. s.
-naphthalene (1,-4-)	$C_{10}H_6(CH_3)_2$	116.12	lq.	1.016^{204}	<-18	264-6	i.		
(2,-3-)	$C_{10}H_6(CH_3)_2$	156.22	lf./al.		104	265^{767}	i.	sl. s.	s.
-naphthylamine (α-)	$C_{10}H_7 \cdot N(CH_3)_2$	156.22	col. oil	1.042^{20}		274.5^{711}	i.	sl. s.	s.
(β-)	$C_{10}H_7 \cdot N(CH_3)_2$	171.23	col. cr.	1.039^{7070}	46	304-5	i.	s.	s.
oxalate	$(CO_2CH_3)_2$	171.23	col. cr.	1.148^{54}	54	163.3	6	s.	s.
phthalate (o-)	$C_6H_4(CO_2CH_3)_2$	118.09	col. lq.	1.189^{2525}	-26.8	280^{734}	0.43	s.	∞
sulfate	$(CH_3O)_2SO_2$	194.18	col. oil	1.352^{204}	-83.2	188.3	v. sl. s.		
sulfide	$(CH_3)_2S$	126.13	oil	0.846^{214}	61.5	37.3	i.	∞	∞
tartrate (d-)	$(CHOH \cdot CO_2CH_3)_2$	62.13	cr.	1.328^{204}	160	280	6^{20}	s.	
-vinyl-ethenyl carbinol	$(CHOH \cdot COH \cdot C:CH \cdot CH_2$	178.14	lq.	0.887^{204}	109	150	i.	200^{15}	v. s.
Dinaphthyl (αα'-methane)	$(C_{10}H_7)_2CH_2$	110.15	lf./al.		92	$240\text{-}4^{12}$	i.	s. h.	s.
(ββ'-)	$(C_{10}H_7)_2CH_2$	254.31	nd./al.		94-5	>360	i.	0.8 c.	v. s.
Dinitro-anisole (1-/2,-4-)	$CH_3OC_6H_3(NO_2)_2$	268.34	col. mm.	1.341^{20}	117-8	319^{74}	sl. s. h.	1.5^{20}	v. s.
-benzene (o-)	$C_6H_4(NO_2)_2$	268.34	col. mm.	1.59^{18}	89.8	300-2	0.01 c.	1.9^{21}	v. s. h.
(m-)	$C_6H_4(NO_2)_2$	198.13	col. rhb.	1.575^{204}	173-4	299^{77}	0.3^{99}	3^{20}	v. s. h.
(p-)	$C_6H_4(NO_2)_2$	168.11	col. mm.	1.625^{18}	106-8		0.18^{100}	0.18^{21}	s.
sulfonic acid (2,-4-/1-)	$(NO_2)_2C_6H_3SO_3H \cdot 3H_2O$	168.11	pr.		179-80	subl.	s.	s.	v. sl. s.
-benzoic acid (2,-4-)	$(NO_2)_2C_6H_3CO_2H$	168.11	cr./aq.		204-5		1.85^{25}	s.	sl. s.
(3,-5-)	$(NO_2)_2C_6H_3CO_2H$	302.22	mn. pr.		189		s. h.		
-benzophenone (4,-4'-)	$(NO_2)_2C_6H_3CO$	212.12	col. nd.		233		i.		
-diphenyl (4,-4'-)	$(NO_2)_2C_6H_3 \cdot$	212.12	nd./al.		93.5		i.	v. s.	
(2,-4'-)	$(NO_2)_2C_6H_3 \cdot$	272.21	mn.	1.445	216		i.	v. s.	
-naphthalene (1,-5-)	$C_{10}H_6(NO_2)_2$	244.20	nd.	1.474	170-2	subl.	i.	1.5^{20}	
(1,-8-)	$C_{10}H_6(NO_2)_2$	244.20	rhb.		144-5		i.	0.2 c.	v. s. h.
-phenol (2,-3-)	$(NO_2)_2C_6H_3OH$	218.16	yel. mm.	1.681^{20}	114-5	d.	sl. s.	v. s. h.	v. s. h.
(2,-4-)	$(NO_2)_2C_6H_3OH$	218.16	yel. rhb.	1.683^{34}	63-4	subl.	0.5 c.	4^{20}	v. s.
(2,-6-)	$(NO_2)_2C_6H_3OH$	184.11	yel. rhb.		173 d.		s. c.	v. s.	v. s.
-salicylic acid (3,-5-)	$(NO_2)_2C_6H_2(OH)(CO_2H) \cdot H_2O$	184.11	pl./aq.		210-6		i.	v. s.	v. sl. s.
-stilbene (4,-4'-)		184.11	yel. lf.		70		s.	1.2^{15}	9^{16}
-toluene (2,-4-)	$(NO_2)_2C_6H_3CH_3$	246.13	nd.	1.321^{71}	60-1	300	i.	s. h.	s.
(3,-4-)	$(NO_2)_2C_6H_3CH_3$	270.24	nd.	1.259^{111}	92-3	subl.	i.		
(3,-5-)	$(NO_2)_2C_6H_3CH_3$	182.13	mn. pr.	1.277^{111}	9.5-10.5	subl.	i.	s. h.	s.
Dioxane	$O<(CH_2 \cdot CH_2)_2>O$	182.13	col. lq.	1.033^{204}		101.1	∞		
Dipentene	$C_{10}H_{16}$	136.23	col. lq.	0.865^{18}		178	i.		s.

Physical properties of organic compounds (continued). Columns: name · formula · mol. wt. · crystalline form/color · density · melting point (°C) · boiling point (°C) · solubility in water · solubility in alcohol · solubility in ether.

Name	Formula	Mol. wt.	Form	Density	M.p., °C	B.p., °C	Water	Alcohol	Ether
Diphenyl	$C_6H_5 \cdot C_6H_5$	154.20	col. mn.	$0.992^{7.34}$	69–70	254.9	i.	10^{20}	6.6^{20}
-amine	$C_6H_5NHC_6H_5$	169.22	col. mn.	$1.160^{20/20}$	52.9	302	0.03^{25}	$56^{19.5}$	s.
carbonate	$O(COC_6H_5)_2$	214.21	nd./al.	1.272^{14}	80	302–6	i.	v. s.	s.
-chlorarsine	$(C_6H_5)_2AsCl$	264.57	rhb.	1.583^{40}	43–4	d. 327	i.	20	v. s.
-ethane	$C_6H_5CH_2 \cdot 2$	182.25	col. pr.	$0.978^{50/50}$	52–3	284	v. sl. s.	∞	∞
ether	$C_6H_5OC_6H_5$	170.20	col. rhb.	1.073^{20}	27	259	v. sl. s.	9^{20}	sl. s.
guanidine	$(C_6H_5 \cdot NH)_2C{:}NH$	211.26	mn./al.		147–8	d. > 170	i.	v. s.	v. s.
-methane	$(C_6H_5)_2CH_2$	168.23	col. pr.	$1.001^{26/4}$	26–7	265	i.		
phenylenediamine (p-)	$C_6H_4(NH_2)_2C_6H_4$	260.32	cr.		152		i.		
succinate	$(\cdot CH_2CO_2C_6H_5)_2$	270.27	lf./al.	$1.119^{15/15}$	122–3	330	i.	s. h.	s.
sulfide	$(C_6H_5)_2S$	186.26	col. lq.	1.248^{254}	<-40	296–7	sl. s.	s. h.	∞
sulfone	$(C_6H_5)_2SO_2$	218.26	nd./aq.	1.276	128–9	379	v. sl. s.		
urea (uns.)	$(C_6H_5)_2NCONH_2$	212.24	rhb.		189		i.	s. h.	s.
Diphenylene oxide	$<(C_6H_4)_2O$	168.18	col. lq.	$0.979^{20/4}$	86–7	287–8	s.	s.	v. s.
Dipropyl adipate (n-)	$(\cdot CH_2CH_2CO_2C_3H_7)_2$	230.30	col. lq.	$0.739^{20/4}$	-20.3	$143{-}5^{10}$	i.	∞	∞
-amine (n-)	$(C_3H_7)_2NH$	101.19	col. lq.	0.722^{22}	-39.6	110–1	sl. s.	∞	∞
(i-)	$[(CH_3)_2CH]_2NH$	101.19	col. lq.	0.910^{20}	-61	83.5^{743}	sl. s.	∞	∞
aniline (n-)	$C_6H_5N(C_3H_7)_2$	177.28	yel. oil	0.968^{22}		245.4	i.		
carbonate (n-)	$O(COC_3H_7 \cdot C_3H_7)_2$	146.18	col. lq.	$0.744^{21.0}$		168.2	i.		
(i-)	$(C_3H_7CH_2)_2O$	102.17	col. lq.	$0.725^{21.0}$			i.		
ether (n-)	$[(CH_3)_2CH]_2O$	102.17	col. lq.	$0.822^{20/4}$	-122	91	0.2	∞	∞
(i-)	$(C_2H_5CH_2)_2CO$	114.18	col. lq.	$0.806^{20/4}$	-60	69	0.43	∞	∞
ketone (n-)	$[(CH_3)_2CH]_2CO$	114.18	col. lq.	$1.038^{0/0}$	-32.6	144.2	v. sl. s.	s.	s.
(i-)	$(CO_2C_3H_7)_2$	174.19	col. lq.			123.7	v. sl. s.	s.	s.
oxalate (n-)	$[CO_2CH(CH_3)_2]_2$	174.19	col. lq.		-51.7	213.5	d. h.	∞	∞
(i-)		268.30	cr.			190			
Disalicylal ethylenediamine	$[HOC_6H_4CH{\cdot}NCH_2 \cdot]_2$	239.31	cr.	1.34	125–6		i.	s.	s.
Ditolyl guanidine (o-)	$(C_7H_7 \cdot NH)_2C{:}NH$	78.11	lq.	$1.10^{39/4}$	178–9		v. sl. s.	v. s.	v. s.
Divinyl acetylene	$(H_2C{:}CH \cdot C{:})_2$	310.59	lq.	$0.776^{44/4}$		85	i.	v. sl. s.	i.
Docosane (n-)	$CH_3(CH_2)_{20}CH_3$	170.33	lq.	$0.778^{44/4}$	44.5	224.5^{15}	i.	v. s.	v. s.
Dodecane (n-)	$CH_3(CH_2)_{10}CH_3$	182.17	mn.	$0.751^{20/4}$	-9.6	214.5	i.		
Dulcitol	$CH_2OH(CHOH)_4CH_2OH$	134.21	lf./al.	1.466^{15}	189	$290{-}5^{1}$	3.2^{15}	s.	i.
Durene (1-,2-,4-,5-)	$(CH_3)_4C_6H_2$	282.45	col. cr.	$0.838^{81/4}$	79–80	193–5	i.	i.	i.
Elaidic acid	$C_7H_{14} \cdot CH{:}CH(CH_2)_7CO_2H$	647.93	cr./et.	$0.851^{79/4}$	51–2	288^{100}	i.	s.	s.
Eosine	$C_{20}H_8O_5Br_4$	165.23	lq.				s.	500	∞
Ephedrine (l-)	$C_6H_5CHOHCH(CH_3)NHCH_3$	92.53	col. lq.	$1.183^{25/25}$	40	255	5	sl. s. c.	i.
Epichlorhydrin (α-)	$CH_2 \cdot O \cdot CH_2Cl$	110.98	tet. pr.	1.204^{25}	-25.6	117^{756}	<5	s.	1
Epidichlorohydrin (α-)	$CH_2 \cdot CCl \cdot CH_2Cl$	122.12	lf./al.	$1.451^{20/4}$		94	i.	150 cc.	
Erythritol (dl-)	$CH_2OH(CHOH)_2CH_2OH$	302.12	col. gas		126	329–31	60	∞	
tetranitrate	$C_4H_6(ONO_2)_4$	30.07	col. oil	0.546^{-88}	61	expl.	i. c.		
Ethane	CH_3CH_3	61.08	lq.	1.022^{20}	-172	-88.6	4.7 cc.^{20}	∞	∞
Ethanol-amine	$HOCH_2CH_2NH_2$	89.09	col. lq.	1.169^{25}	10.5	171^{757}	∞	∞	i.
formamide	$HCONHCH_2CH_2OH$	74.12	col. lq.	$0.708^{25/4}$	<-40	d.	∞	∞	i.
Ether	$(CH_3CH_2)_2O$	330.49	col. lq.	$1.020^{20/20}$	-116.3	34.6	7.5^{20}	∞	∞
Ethyl abietate	$C_{19}H_{29}CO_2C_2H_5$	88.10	col. lq.	$0.901^{20/4}$		200^{4}	i.		
acetate	$CH_3CO_2C_2H_5$	130.14	col. lq.	$1.025^{20/4}$	-82.4	77.1	8.5^{15}	∞	∞
acetoacetate	$CH_3COCH_2CO_2C_2H_5$	46.07	col. lq.	$0.789^{20/4}$	-45	180^{755}	13^{17}	∞	∞
alcohol	CH_3CH_2OH	45.08	col. lq.	$0.689^{15/15}$	-112	78.4	∞	∞	∞
-amine	$C_2H_5NH_2$	81.55	mn.	1.216	-80.6	16.6	∞	∞	∞
hydrochloride	$C_2H_5NH_2 \cdot HCl$	121.18	nd./aq.	$0.963^{20/4}$			240^{17}	v. s.	s.
aniline	$C_6H_5NHC_2H_5$	201.24	lq.		-63.5	204	2.15^{15}	∞	∞
sulfonic acid (m-)	$C_6H_5NHC_6H_4SO_3H$	180.20	cr.		d. 294		v. sl. s.	s.	s.
anisate (p-)	$CH_3OC_6H_4CO_2C_2H_5$	165.19	col. lq.	$1.103^{25/25}$	7–8	269–70	0.01^{15}	s.	∞
anthranilate (o-)	$NH_2C_6H_4CO_2C_2H_5$	106.16	col. lq.	$1.117^{20/4}$	13	266–8	i.	∞	∞
benzoate	$C_6H_5 \cdot C_6H_5$	211.29	col. lq.	$1.052^{15/15}$	-34.6	211–2	0.08^{20}	∞	∞
-benzyl-aniline	$C_6H_4N(C_2H_5)CH_2C_6H_5$	108.98	yel. oil	$1.034^{18.5}$		285^{10}	i.	18	i.
bromide	C_2H_5Br	116.16	lf.	$1.431^{20/4}$	-117.8	38.4	1.06^{0}	∞	∞
butyrate (n-)	$C_2H_5CH_2CO_2C_2H_5$	116.16	col. lq.	$0.879^{20/4}$	-93.3	120–1	0.68^{25}	∞	∞
(i-)	$(CH_3)_2CHCO_2C_2H_5$	200.31	col. lq.	$0.871^{20/4}$	-88.2	110–1	sl. s.	∞	∞
caprate (n-)	$CH_3(CH_2)_8CO_2C_2H_5$	144.21	lq.	0.859^{25}	-20	244.6^{736}	0.002^{20}	∞	∞
Ethyl caproate (n-)	$CH_3(CH_2)_4CO_2C_2H_5$	172.26	col. lq.	$0.873^{20/20}$	-67.5	$165{-}6^{736}$	i.	∞	∞
caprylate (n-)	$CH_3(CH_2)_6CO_2C_2H_5$	64.52	col. lq.	0.878^{17}	-45	$207{-}8^{753}$	0.45^{0}	∞	∞
chloride	CH_3CH_2Cl	122.55	col. lq.	$0.917^{66/5}$	-139	13	d.	∞	∞
chloroacetate	$ClCH_2CO_2C_2H_5$	108.53	col. lq.	$1.159^{20/4}$	-26	144	i.	∞	∞
chlorocarbonate	$ClCO_2C_2H_5$	176.21	col. lq.	$1.138^{20/4}$	-80.6	94–5	i.	∞	∞
cinnamate (trans-)	$C_6H_5CH{:}CHCO_2C_2H_5$	113.11	col. lq.	$1.049^{20/4}$	12	271	i.	∞	∞
cyanoacetate	$C_2H_5(CN)CO_2C_2H_5$	74.08	col. lq.	$1.062^{20/4}$	-22.5	208^{753}	2^{25}	∞	∞
formate	$HCO_2C_2H_5$	140.13	col. lq.	$0.923^{20/4}$	-79	54^{760}	11^{18}	∞	∞
furoate (α)	$OC_4H_3CO_2C_2H_5$	158.23	lf.	$1.117^{21/4}$	34	195^{766}	i.	∞	∞
heptoate	$CH_3(CH_2)_5CO_2C_2H_5$	80.52	col. lq.	$0.872^{20/20}$	-66.1	187–8	0.029^{20}	∞	∞
hypochlorite	$ClOCH_2CH_3$	155.98	yel. lq.	$1.013^{4/64}$	expl.	36^{762}	i.	∞	∞
iodide	CH_3CH_2I	118.13	col. lq.	$1.933^{20/4}$	-105	72.4	0.4^{20}	∞	∞
lactate	$CH_3CH(OH)CO_2C_2H_5$		oil	$1.030^{25/4}$		155	∞	∞	∞

TABLE 2-2 Physical Properties of Organic Compounds (Continued)

Name	Formula	Formula weight	Form and color	Specific gravity	Melting point, °C	Boiling point, °C	Solubility in 100 parts		
							Water	Alcohol	Ether
laurate	$CH_3(CH_2)_{10}CO_2C_2H_5$	228.36	oil	0.868^{134}	−10.7	269	i.	s.	∞
mercaptan	CH_3CH_2SH	62.13	lq.	0.839^{204}	−121	36–7	1.5	s.	s.
methacrylate	$CH_2{:}C(CH_3)CO_2C_2H_5$	114.14	col. lq.	0.913^{156}		118	i.	s.	s.
naphthylamine (α-)	$C_{10}H_7{\cdot}NHC_2H_5$	171.23	oil	1.060^{204}		303^{733}	i.	s.	s.
naphthyl ether (α-)	$C_{10}H_7{\cdot}OC_2H_5$	172.22	oil	1.061^{2020}	5.5	276.4	i.	s.	∞
nitrate	$C_2H_5ONO_2$	91.07	cr.	1.100^{254}	−102	87–8	1.3^{35}	∞	∞
nitrite	C_2H_5ONO	75.07	col. lq.	0.900^{155}		17	v. sl. s.	s.	s.
oleate	$C_{17}H_{33}CO_2C_2H_5$	310.50	lq.	0.867^{25}	<−15	$216{-}8^{15}$	i.	∞	∞
palmitate	$CH_3(CH_2)_{14}CO_2C_2H_5$	284.47	oil	0.858^{254}	24–5	191^{10}	i.	s.	∞
pelargonate	$CH_3(CH_2)_7CO_2C_2H_5$	186.29	col. nd.	0.866^{175}	−44.5	$227{-}8^{757}$	i.	∞	s.
propionate	$CH_3CH_2CO_2C_2H_5$	102.13	col. lq.	0.891^{204}	−72.6	99.1	i.	∞	∞
salicylate (o-)	$HOC_6H_4CO_2C_2H_5$	166.17	col. lq.	1.136^{154}	1.3	233–4	2.4^{20}	s.	∞
stearate	$CH_3(CH_2)_{16}CO_2C_2H_5$	312.52	col. cr.	0.848^{363}	33.4(31)	201^{10}	i.	∞	s.
toluate (o-)	$CH_3{\cdot}C_6H_4CO_2C_2H_5$	164.20	col. lq.	1.032^{2525}	<−10	227	i.	∞	∞
(m-)	$CH_3{\cdot}C_6H_4CO_2C_2H_5$	164.20	lq.	1.030^{2020}		231^{790}	i.	∞	∞
toluene sulfonate (p-)	$CH_3{\cdot}C_6H_4SO_3C_2H_5$	200.25	pr./al.	1.166^{454}	33–4	221.3	i.	s.	s.
toluidine (o-)	$CH_3{\cdot}C_6H_4NHC_2H_5$	135.20	lq.	0.948^{254}	<−15	215–6	i.	∞	∞
(p-)	$CH_3{\cdot}C_6H_4NHC_2H_5$	135.20	lq.	0.942^{254}		217	i.	s.	∞
urea	$C_2H_5NH{\cdot}CO{\cdot}NH_2$	88.11	nd.	1.213^{18}	92		v. s.	80	i.
valerate (n-)	$CH_3(CH_2)_3CO_2C_2H_5$	130.18	col. lq.	0.877^{20}	−91.2	145.5	0.24^{25}	∞	∞
(i-)	$(CH_3)_2CHCH_2CO_2C_2H_5$	130.18	lq.	0.867^{204}	−99.3	135	0.17^{20}	∞	∞
Ethylal	$CH_2(OC_2H_5)_2$	104.15	lq.	0.824^{254}	−66.5	89	9^{18}	∞	∞
Ethylene	$H_2C{:}CH_2$	28.05	col. gas	0.57^{-1024}	−169	−103.9	26 cc.^{0}	360 cc.	s.
bromide	$BrCH_2{\cdot}CH_2Br$	187.88	col. lq.	2.180^{204}	10	131.5	0.43^{80}	s.	∞
bromohydrin	$BrCH_2{\cdot}CH_2OH$	124.98	col. lq.	1.772^{204}		150.3	0.69^{80}	∞	∞
chlorobromide	$ClCH_2{\cdot}CH_2Br$	143.43	col. lq.	1.689^{19}	−16.6	106.7	i.	∞	∞
chlorohydrin	$ClCH_2{\cdot}CH_2OH$	80.52	col. lq.	1.213^{204}	−69	128.8	∞	∞	s.
diamine	$H_2NCH_2{\cdot}CH_2NH_2$	60.10	col. lq.	0.900^{2020}	8.5	117.2	∞	∞	∞
oxide	$<(CH_2)_2>O$	44.05	col. lq.	0.887^{774}	−111.3	13.5^{347}	∞	∞	0.3
Ethylidene diacetate	$CH_3CH(O_2CCH_3)_2$	146.14	col. lq.	1.061^{13}	18.85	168^{740}	sl. s.	∞	v. s.
Eugenol (1-4-3-)	$C_3H_5{\cdot}C_6H_3(OH)OCH_3$	164.20	oil	1.070^{1515}	10.3	253.5	v. sl. s.	∞	∞
(+-)(1-3-4-)	$C_3H_5{\cdot}C_6H_3(OCH_3)OH$	164.20	oil	1.091^{1515}	−10	267.5	v. sl. s.	s.	∞
Fenchyl alcohol (dl-)	$C_{10}H_{17}OH$	154.24	col. cr.	0.935^{40}	35	201	sl. s.		s.
(d-)(α-)	$C_{10}H_{17}OH$	154.24	col. pr.	0.964^{204}	45–7	201–2	i.		s.
(+-)(l-)	$C_{10}H_{17}OH$	154.24	col. cr.	0.961	61–2	201–2	v. sl. s.		∞
Ferric dimethyl-dithiocarbamate	$Fe[SSCN(CH_3)_2]_3$	416.47	cr.		d. 100–30	ign. >150	i.		i.
Fluorene	$(C_6H_4)_2{>}CH_2$	166.21	cr./al.	1.203^{04}	115–6	293–5	i.	s. h.	s.
Fluorescein	$C_{20}H_{12}O_5$	332.30	yel. red		d. > 290		v. sl. s.	s. h.	i.
Fluoro-dichloromethane	$FCHCl_2$	102.93	gas	1.426^0	−127	14.5	i.	v. s.	v. s.
-trichloromethane	Cl_3CF	137.38	col. lq.	$1.494^{17.2}$	−92	24.9	i.	∞	∞
Formaldehyde	HCHO	30.03	gas	0.815^{-20}		−21	∞	∞	s.
(m-)	$(CH_2O)_3$	90.08	wh.	1.17^{65}	64	114.5^{709}	21^{25}	v. s.	v. s.
(p-)	$(CH_2O)_x{\cdot}xH_2O$	(30.03)	amor.		150–60	subl.	$20{-}30^{18}$	∞	i.
Formamide	$HCONH_2$	45.04	lq.	1.139^{204}	2	193	∞	∞	v. s.
Formanilide	$HCONHC_6H_5$	121.13	nn.	1.147^{1515}	47	216^{120}	sl. s.	∞	∞
Formic acid	HCO_2H	46.03	col. lq.	1.220^{204}	8.6	100.8	∞	∞	∞
Fructose	$CH_2OH(CHOH)_3COCH_2OH$	180.16	nd./aq.	1.669^{175}	95–105		v. s.	8^{18}	i.
Fuchsin	$C_{20}H_{19}N_3HCl$	337.84	red		d. >200		0.3	∞	i.
Fulminic acid	$C{:}NOH$	43.03		1.22					
Fumaric acid ($trans$-)	$HO_2CCH{:}CHCO_2H$	116.07	col. pr.	1.635^{204}	286–7	290	0.7^{17}	5.8^{30}	0.7^{25}
Furfural	$C_4H_3O{\cdot}CHO$	96.08	lq.	1.159^{204}	−38.7	161.7^{760}	9.1^{13}	∞	∞
Furfuran	C_4H_4O	68.07	col. lq.	0.937^{204}		$31{-}2^{796}$	i.	∞	∞
Furfuryl acetate	$CH_3CO_2CH_2C_4H_3O$	140.13	col. oil	1.118^{204}		175–7	i.	s.	s.
alcohol	$C_4H_3O{\cdot}CH_2OH$	98.10	oil	1.129^{254}		169.5^{752}	v. sl. s.	∞	s.
butyrate	$C_4H_7CO_2CH_2{\cdot}C_4H_3O$	168.19	col. lq.	1.053^{204}		212–3	v. sl. s.	s.	s.
propionate	$C_2H_5CO_2CH_2{\cdot}C_4H_3O$	154.16	col. lq.	1.109^{204}		195–6	i.	s.	s.
Furoic acid	$C_4H_3O{\cdot}CO_2H$	112.08	mn. pr.		133–4	230–2	3.6^{15}	s.	i.
G-acid, K salt (2-)(6-8-)	$HOC_{10}H_5(SO_3K)_2$	380.46	cr.				8^{25}	sl. s.	
Na salt (2-)(6-8-)	$HOC_{10}H_5(SO_3Na)_2$	348.26	pr.				34^{20}	v. sl. s.	
Galactose (d-)(α-)	$C_5H_{11}O_5{\cdot}CHO$	180.16	cr.		165.5		10.3^0	0.6^{40}	i.
Gallic acid (3-4-5-)	$(HO)_3C_6H_2CO_2H{\cdot}H_2O$	188.13	col. lq.	1.694^{44}	d. 220		1^{13}	28^{15}	2.5^{15}
Gamma acid (2-8-6-)	$C_{10}H_5(NH_2)(OH)SO_3H$	239.24	cr.				i.	sl. s.	i.
Geraniol	$C_8H_{15}CH_2OH$	154.24	col. lq.	0.883^{15}	<−15	230	i.	∞	∞
Glucose (d-)(α-)	$C_6H_{12}O_6{\cdot}H_2O$	180.16	rhb.	1.544^{25}	146		82^{175}	sl. s.	i.
(d-)(β-)	$C_6H_{12}O_6$	198.17	cr.	1.562^{154}	150		154^{145}		
Glucuronic acid	$CHO(CHOH)_4CO_2H$	194.14	cr.	1.460	154	d.	v. s.	v. sl. s.	
Glutam(in)ic acid (dl-)	$[{\cdot}CHNH_2(CH_2)_2{\cdot}](CO_2H)_2$	147.13	cr./aq.		199 d.		1.5^{20}	v. sl. s.	v. sl. s.

Name	Formula	M. wt.	Form	Density	m.p.	b.p.	Solubility: Water (cold)	Water (hot)	Alcohol	Ether
Glutaric acid	$CH_2(CH_2CO_2H)_2$	132.11	col. cr.	1.429^{15}	97.5	200^{30}	63.9^{30}	v. s.	v. s.	v. s.
Glycerol	$CH_2OH\cdot CHOH\cdot CH_2OH$	92.09	col. oil	1.260^{504}	17.9	290	∞	∞	v. s.	i.
acetate (mono-)	$C_5H_{10}O_4$	134.13	col. oil	1.20^{204}	40	158^{165}	v. s.	v. s.	v. s.	sl. s.
(di-)	$(CH_2CO_2)_2C_3H_5OH$	176.17	col. lq.	1.178^{1515}	58–9	175–6^{40}	s.	v. s.	v. s.	s.
nitrate (mono-) (α-)	$CH_2OH\cdot CHOH\cdot CH_2NO_3$	137.09	col. pr.	1.40^{15}	54	155–60	v. s.	v. s.	v. s.	v. sl. s.
(β-)	$CH_2OH\cdot CHNO_3\cdot CH_2OH$	137.09	lf.	1.47^{15}	<–30	155–60	v. s.	v. s.	v. s.	s.
dinitrate (1-3-)	$CHOH(CH_2ONO_2)_2$	182.09	oil		–78	146–8^{15}		∞	∞	∞
Glyceryl triacetate	$(CH_3CO_2)_3C_3H_5$	218.20	col. lq.	1.161^{174}	75–6	258–9	7.17^{15}	s. h.	s. h.	s.
tribenzoate	$(C_6H_5CO_2)_3C_3H_5$	404.40	nd.	1.228^{12}	<–75	d.	i.	s. h.	s. h.	v. s.
tributyrate	$[C_3H_7CO_2]_3C_3H_5$	302.36	col. cr.	1.032^{204}	31(25)	305–9	i.	s.	s. s. c.	s.
tricaprate	$[CH_3(CH_2)_8CO_2]_3C_3H_5$	554.83	col. lq.	0.921^{404}	–25		i.	s.	s.	s.
tricaproate	$[CH_3(CH_2)_4CO_2]_3C_3H_5$	386.51	col. lq.	0.987^{204}	8.3(–21)		i.	s.	s.	s.
tricaprylate	$[CH_3(CH_2)_6CO_2]_3C_3H_5$	470.67	col. lq.	0.954^{204}	45–6		i.	s.	s.	s.
trilaurate	$[CH_3(CH_2)_{10}CO_2]_3C_3H_5$	638.98	col. lq.	0.894^{604}	56.5		i.	s.	i.	∞
trimyristate	$[CH_3(CH_2)_{12}CO_2]_3C_3H_5$	723.14	col. nd.	0.885^{606}	13.3(2)		i.	s.	sl. s.	v. s.
trinitrate	$CH_2NO_3\cdot CHNO_3\cdot CH_2NO_3$	227.09	lf.	1.601^{15}	–4	160^{15}	0.18^{20}	50^{39}	v. s.	v. s.
trinitrite	$CH_2NO_2\cdot CHNO_2\cdot CH_2NO_2$	179.09	yel. oil	1.291^{1016}	65.1	150 sl. d.	d.	d.	s. h.	s. h.
trioleate	$(C_{17}H_{33}CO_2)_3C_3H_5$	885.40	yel. lq.	0.915^{15}	70.8(55)	240^{15}	i.	sl. s.	sl. s.	∞
tripalmitate	$[CH_3(CH_2)_{14}CO_2]_3C_3H_5$	807.29	col. nd.	0.866^{504}	22	310–$20^{.1}$	i.	0.004^{21}	i.	∞
tristearate	$[CH_3(CH_2)_{16}CO_2]_3C_3H_5$	891.45	col. pr.	0.862^{504}		166 sl. d.	i.	s. h.	i.	i.
Glycide	$C_2H_3O\cdot CH_2OH$	74.08	col. lf.	1.114^{1616}		197.4	23 c.	0.1 c.	∞	
Glycine, Glycocoll	$NH_2\cdot CH_2\cdot CO_2H$	75.07	mn.	1.161	232–6 d.	190.5	∞	v. s.	i.	1.0
Glycol	$CH_2OH\cdot CH_2OH$	62.07	col. lq.	1.113^{194}	–15.6	>360	∞	∞	v. s.	∞
diacetate	$(C_2H_4CO_2CH_2\cdot)_2$	146.14	col. lq.	1.109^{144}	–31	240	14.3²²	s. d.	v. s.	v. s.
dibenzoate	$(C_6H_5CO_2CH_2\cdot)_2$	270.27	rhb./et.		73–4		i.	∞	s.	∞
dibutyrate	$(C_3H_5CO_2CH_2\cdot)_2$	202.24	lq.	1.024^{0}		174	i.	∞	s.	v. s.
dicaprylate	$(HCO_2CH_2\cdot)_2$	314.45	lq.			188^{20}	i.			s.
diformate	$(C_{11}H_{23}CO_2CH_2\cdot)_2$	118.09	amor.	1.482^{212}	52–4	expl. 114	v. sl. s.			i.
dilaurate	$(O_2NO\cdot CH_2\cdot)_2$	426.66	yel. lq.	1.216^{0}	–20	96–8	i.			∞
dinitrate	$(ONO\cdot CH_2\cdot)_2$	152.07	nd.		<–15	$260^{.1}$	sl. s.			∞
dinitrite	$(C_{15}H_{31}CO_2CH_2\cdot)_2$	120.07	lq.	1.045^{25}	71–2	211–2	i.			s.
dipalmitate	$(C_2H_4CO_2CH_2\cdot)_2$	538.87	lq.	1.118^{3020}		244.8	∞	∞	∞	∞
dipropionate	$(HO\cdot CH_2CH_2O)_2$	174.19	lq.	1.060^{204}	–10.5	75–6	∞	∞	∞	∞
ether	$<O\cdot CH_2CH_2OCH_2>$	106.12	lq.	1.199^{154}		180	i.		s.	s.
formate (mono-)	$HCO\cdot CH_2CH_2OH$	74.08	nd./aq.		79(63)	d.	∞	∞	∞	∞
Glycolic acid	$HOCH_2CO_2H$	90.08	pr.	1.140^{1515}	28.3	205	1.7^{15}	90^{25}	s.	s.
Guaiacol (o-)	$CH_3O\cdot C_6H_4OH$	76.05	col. cr.		50		v. s.	v. s.	∞	∞
Guanidine	$NH\cdot C(NH_2)_2$	124.13	cr.			270¹⁵	0.17^{20}	s.	s.	s.
H-acid, Na salt (1-8-3-6-)	$C_{10}H_4O\cdot NS_2Na\cdot 1\frac{1}{2}H_2O$	59.07	col. lq.	0.780^{604}	59.5	98.4^{760}	v. s.	v. s.		
Heptacosane (n-)	$CH_3(CH_2)_{25}CH_3$	368.31	col. lq.	0.684^{204}	–90.6	90.0	0.005^{15}	sl. s.	i.	s.
Heptane (n-)	$CH_3(CH_2)_5CH_3$	380.72	col. lq.	0.679^{204}	–118.2	91.8	i.	s.	i.	∞
(i-)	$(CH_3)_2CH(CH_2)_3CH_3$	100.20	col. lq.	0.687^{204}	–119.4	79.1	i.	s.	i.	∞
Heptoic acid	$C_3H_7\cdot CH(CH_3)\cdot C_2H_5$	100.20	col. lq.	0.674^{204}	–125	80.8	i.	s.	i.	∞
aldehyde	$(CH_3)_3C\cdot CH_2\cdot C_2H_5$	130.18	col. lq.	0.675^{204}	–119.4	86.0	i.	s.	i.	∞
Heptyl acetate (n-)	$[(CH_3)_2CH]_2CH_2$	114.18	col. lq.	0.693^{204}	–135.0	93.5	i.	s.	i.	∞
alcohol (n-)	$(C_2H_5)_3CH$	158.24	col. lq.	0.698^{204}	–118.7	80.8	i.	s.	i.	∞
mercaptan	$(CH_3)_3C\cdot CH(CH_3)_2$	116.20	col. lq.	0.690^{204}	–25	221–2	0.25^{15}	s.	i.	∞
Hexachloro-benzene	$CH_3\cdot CH_2\cdot CO_2H$	116.20	col. lq.	0.918^{20}	–10	155	0.02^{20}	s.	i.	s.
-ethane	$CH_3(CH_2)_5CHO$	132.26	lq.	0.850^{207}	–42	191.5^{759}	0.18^{25}	s.	i.	s.
Hexacosane (n-)	$CH_3CO_2CH_2(CH_2)_5CH_3$	284.80	lq.	0.874^{1616}		175^{756}	v. sl. s.	s.	v. sl. s.	s.
Hexadecane (n-)	$CH_3(CH_2)_5CH_2OH$	236.76	lq.	0.824^{204}	34.6	140	i.	s.	i.	s.
Hexaethylbenzene	$[(CH_2)_5CH]CH_2OH$	366.69	lq.	0.829^{204}		156	i.	s.	i.	s.
Hexamethylbenzene	$CH_3CH(SH)\cdot C_5H_{11}$	226.43	lq.	0.820^{204}	–37	174–5^{765}	v. sl. s. h.	s.	v. s.	s. h.
Hexamethylene-diamine	C_6Cl_6	246.42	mn.	0.835^{20}	228–31	186^{777}	i.	v. s.	v. s.	v. s.
-diisocyanate	$CCl_3\cdot CCl_3$	162.26	rhb.	2.044^{24}	186–7	262^{15}	i.	v. s.	v. sl. s.	∞
-glycol	$CH_3(CH_2)_{24}CH_3$	116.20	cr.	2.091^{204}	56.6	287.5	i.	∞	∞	∞
tetramine	$CH_3(CH_2)_{14}CH_3$	118.17	lf.	0.779^{574}	18.5	298.3	i.	0.75^{25}	8^{25}	i.
Hexane (n-)	$C_6(C_2H_5)_6$	140.19	pr./al.	0.774^{304}	130	265	i.	0.2^{0}	v. s.	s.
	$C_6(CH_3)_6$	86.17	pl./al.	0.831^{1304}	166	204–5	v. s.	s.	sl. s. h.	s.
	$NH_2(CH_2)_6NH_2$	86.17	lf.		42	143–4^{20}	d.	d.	v. sl. s.	sl. s.
-diisocyanate	$OCN(CH_2)_6NCO$	168.19	lq./aq.	1.04^{28}	subl.	250	81^{12}	s.	∞	∞
-glycol	$HO(CH_2)_6OH$	118.17	col. rhb.		–94	69	0.014^{15}	3	s.	s.
tetramine	$(CH_2)_6N_4$	140.19	col. lq.	0.659^{204}	–153.7	60.2	i.	50^{33}	v. s.	s.
Hexane (n-)	$CH_3(CH_2)_4CH_3$	86.17	lq.	0.654^{204}	–98.2	49.7	i.		∞	s.
(i-)	$(CH_3)_3C\cdot C_2H_5$	86.17	lq.	0.649^{2020}	–129.8	58.0^{760}	i.		∞	s.
(neo-)	$(C_2H_5)_2CH\cdot CH(CH_3)_2$	86.17	lq.	0.662^{204}	–118	63.2	i.		v. s.	s.

TABLE 2-2 Physical Properties of Organic Compounds (Continued)

Name	Formula	Formula weight	Form and color	Specific gravity	Melting point, °C	Boiling point, °C	Water	Alcohol	Ether
Hexyl acetate (n-)	$CH_3CO_2(CH_2)_5CH_3$	144.21	col. liq.	0.890^{20}	-51.6	169.2	i.	v. s.	v. s.
alcohol (n-)	$CH_3(CH_2)_4CH_2OH$	102.17	col. liq.	$0.820^{20/20}$	-14	157.2	0.6^{20}	∞	∞
formate (n-)	$HCO_2(CH_2)_5CH_3$	130.18	liq.	0.821^{200}	-107	120-1	v. sl. s.	∞	∞
resorcinol (2,4-)	$C_6H_{13}C_6H_3(OH)_2$	194.26	liq.	0.809^{204}	68-70	179^{7}	v. sl. s.	∞	∞
Hippuric acid	$C_6H_5CONHCH_2CO_2H$	179.17	col. nd.	1.371^{204}	187-8	d.	0.05	v. s.	s.
Histidine (l-)	$C_6H_9O_2N_3$	155.16	rhb.		d. 287	d.	0.4^{20}	s. h.	0.25^{18}
Homophthalic acid (o-)	$HO_2C\cdot C_6H_4\cdot CH_2CO_2H$	180.15	lf./aq.		175-80	d.	s. h.	v. s.	i.
Hydracrylic acid	$HOCH_2CH_2CO_2H$	90.08	syrup		-12	d.	s. h.		sl. s.
Hydro-cyanic acid	HCN	27.03	col. liq.	0.697^{18}	-14	25-6	∞	∞	
-quinone (p-)	$C_6H_4(OH)_2$	110.11	cr.	1.332^{15}	170.3	285^{730}	6^{15}	v. s.	v. s.
Hydroxy-benzaldehyde (p-)	$HO\cdot C_6H_4\cdot CHO$	122.12	nd./aq.	1.129^{130}	116-7	subl.	1.38^{31}		
-benzanilide (o-)	$HO\cdot C_6H_4\cdot CONHC_6H_5$	213.23	pr./al.		135	d.	v. sl. s. h.	s.	s.
-quinoline (2-)(α-)	$C_9H_6N\cdot OH$	145.15	pr./al.	1.35	199-200	subl.	v. sl. s. h.	i.	v. s.
(8-)(o-)	$C_9H_6N\cdot OH$	145.15	pr.		75-6	266.6^{762}	v. sl. s. c.	i.	v. s.
Indigo	$[C_6H_4(CO)(NH)C]_2$	262.26	gray		390-2	subl.	i.		i.
White	$C_{16}H_{12}O_2N_2$	264.27					s. h.		
Indole	C_8H_7N	117.14	yel. pr.		52	253-4	0.034^{20}	s. h.	s.
Indoxyl	C_8H_6NOH	133.14	col. liq.		85	110	sl. s.	s.	s.
Iodo-benzene	IC_6H_5	204.02	nd./aq.	1.824^{254}	-28.5	188.6	0.01^{25}	i.	s.
-phenol (p-)	IC_6H_4OH	220.02	yel. hex.	1.857^{112}	93-4	d.	sl. s.	s.	s.
Iodoform	CHI_3	393.78	yel. hex.	4.008^{17}	119	subl.	sl. s.	i.	13.6^{25}
Ionone (α-)	$C_{10}H_{16}{:}CHCOCH_3$	192.29	col. oil	0.930^{20}		136.1^{17}	s. h.	s. h.	v. s.
Irone (β-)	$C_{10}H_{16}{:}CHCOCH_3$	206.32	col. oil	0.939^{20}		144^{16}	d.	1.5^{17}	
Isatin	$C_6H_4{<}(CO)(N){>}COH$	147.13	yel. red		200-1	subl.	7.2^{30}		v. s.
Isoprene	$CH_2{:}C(CH_3)CH{:}CH_2$	68.11	col. liq.	0.681^{204}	-120	34	v. sl. s.	v. s.	sl. s.
Ketene	$H_2C{:}CO$	42.04	col. gas		-151	-56	d.	v. s. h.	∞
Koch acid (1-)(3-,6-,8-)	$C_{10}H_4(NH_2)S_2O_6HNa_2$	427.34	cr.				17^{10}	d.	
Lactic acid (dl-)	$CH_3CH(OH)CO_2H$	90.08	hyg.	1.249^{154}	16.8	122^{14}	∞		
Lactide (dl-)	$C_6H_8O_4$	144.12	yel. oil		124.5	d. 250	v. sl. s.		s.
Lactose	$C_{12}H_{22}O_{11}\cdot H_2O$	360.31	tri./al.		202		17^{10}	s.	∞
Lauric acid	$CH_3(CH_2)_{10}CO_2H$	200.31	col. nd.	0.869^{504}	48(44)	225^{100}	i.	i.	∞
Laurone	$[CH_3(CH_2)_{10}]_2CO$	338.60	pl.	0.809^{694}	69-70	255^{757}	i.	i. c.	s. h.
Lauryl alcohol	$CH_3(CH_2)_{10}CH_2OH$	186.33	lf.	0.831^{244}	24	255-9	i.	i.	s.
Lead tetraethyl	$Pb(C_2H_5)_4$	323.45	col. liq.	1.659^{184}	-136	152^{281}	i.	sl. s.	
tetramethyl	$Pb(CH_3)_4$	267.35	col. liq.	1.995^{204}	-27.5	110^{780}	i.		s. h.
Lecithin (protagon)	$C_{42}H_{84}O_9NP$	778.08	wax		150-200 d.		sl. s.	s. h.	∞
Lepidine (py-4)	$C_9H_6N\cdot CH_3$	143.18	liq.	1.086^{20}	9-10	261-3		∞	∞
Leucine (l-)	$(CH_3)_2CHCH_2CH(NH_2)CO_2H$	131.17	cr.	1.293^{15}	295	subl.	2.2^{18}	v. s.	i.
Levulinic acid	$CH_3CO(CH_2)_2CO_2H$	116.11	lf.	$1.140^{20/20}$	33.5	245-6	v. s.	i.	
Limonene (d- or l-)	$C_{10}H_{16}$	136.23	liq.	0.842^{204}	-96.9	177	v. sl. s.		∞
Linalool (d- or l-)	$C_{10}H_{17}\cdot OH$	154.24	col. oil	0.868^{20}	-9.5	198-200	i.	s.	∞
Linalyl acetate	$CH_3CO_2C_{10}H_{17}$	196.28	col. liq.	0.895^{20}		220^{762} d.		v. s.	∞
Linoleic acid	$C_{17}H_{31}CO_2H$	280.44	yel. oil	0.903^{184}	-9.5	$229\text{-}30^{16}$	i.	∞	
Maleic acid	$HO_2C\cdot CH{:}CH\cdot CO_2H$	116.07	mn.	1.609	130.5	135 d.	79^{25}	70^{30}	8^{25}
anhydride	${<}(CHCO)_2{>}O$	98.06	cr.	1.5	57-60	202	16.3^{90}	v. s.	v. s.
Malic acid (dl-)	$HO_2CCH_2CH(OH)CO_2H$	134.09	col. cr.	1.601^{204}	128-9	150 d.	144^{25}	v. sl. s. c.	8.4^{15}
(d- or l-)	$HO_2CCH_2CH(OH)CO_2H$	134.09	col. cr.	1.595^{204}	99-100	140 d.	v. s.	s. h.	8^{15}
Malonic acid	$H_2C(CO_2H)_2$	104.06	col. tri.	1.631^{15}	130-5 d.		138^{16}	∞	s.
Maltose	$C_{12}H_{22}O_{11}\cdot H_2O$	360.31	col. nd.	1.540^{17}	d.		108^{25}	v. s.	i.
Mandelic acid (dl-)	$C_6H_5CH(OH)CO_2H$	152.14	rhb./aq.	1.300^{204}	118.1	d.	16^{20}	v. s.	i.
Mannitol (d-)	$CH_2OH(CHOH)_4CH_2OH$	182.17	col. rhb.	1.489^{204}	166	$290\text{-}5^{3}$	13^{14}	0.01^{14}	v. s.
Mannose (d-)	$CH_2OH(CHOH)_4CHO$	180.16	rhb.	1.539^{204}	132		248^{17}	v. s.	
Margaric acid	$CH_3(CH_2)_{15}CO_2H$	270.44	col. pl.	0.853^{60}	60-1	227^{100}	i.	32^{25}	v. s.
Mellitic acid	$C_6(CO_2H)_6$	342.17	nd./al.		286-8			v. s.	sl. s.
Menthol (l-)(α-)	$C_{10}H_{19}OH$	156.26	col. cr.	$0.890^{15/15}$	42-3	212	0.04 c.	v. s.	
Mercapto-benzothiazole (2-)	${<}C_6H_4N{:}C(SH)S{>}$	167.24	nd.	1.42^{204}	179	d.		s.	
-thiazoline (2-)	${<}CH_2N\cdot C(SH)SCH_2{>}$	119.20	cr.	1.50	106		1.6^{60}		
Mercuric cyanide	$Hg(CN)_2$	252.65	cr./aq.	4.003^{22}	d. 320		12.5^{15}		
fulminate	$Hg(ONC)_2\cdot\tfrac12 H_2O$	293.65	cr.	4.4	expl.		0.07^{12}		
Mesityl oxide	$(CH_3)_2C{:}CHCOCH_3$	98.14	liq.	0.858^{204}	-59	130^{750}	3^{20}	s.	
Mesitylene (1-,3-,5-)	$C_6H_3(CH_3)_3$	120.19	col. liq.	0.865^{204}	-45(-52)	164.8	2^{15}	v. sl. s.	
Metanilic acid (m-)	$H_2NC_6H_4SO_3H$	173.18	col. nd.		d.			s.	
Methane	CH_4	16.04	gas	0.415^{-164}	-182.6	-161.4	0.4^{20} cc.	47^{20} cc.	104^{10} cc.

2-40

Name	Formula	M. wt.	Form	Density	m.p., °C	b.p., °C	Cold water	Hot water	Alcohol	Ether
Methoxy-methoxyethanol	$CH_3(OCH_2)_2CH_2OH$	106.12	lq.	1.038^{25}	<-70	167.5	∞	∞	∞	∞
Methyl acetate	$CH_3CO_2CH_3$	74.08	col. lq.	0.924^{20}	-98.7	57.1	33^{22}	∞	∞	∞
— acrylic acid (α-)	$CH_2C(CH_3)CO_2H$	86.09	pr.	1.015^{20}	15–16	161–3	s. h.	v. s.	∞	∞
— alcohol	CH_3OH	32.04	col. lq.	0.792^{20}	-97.8	64.7	∞	∞	∞	∞
— -amine	CH_3NH_2	31.06	col. gas	0.699^{-11}	-92.5	-6.7^{758}	v. s.	v. s.	v. s.	i.
— -amine hydrochloride	$CH_3NH_2 \cdot HCl$	67.52	lf./al.	1.23	226–8	230^{15}	v. s.	23 h.	s.	∞
— aniline	$C_6H_5NHCH_3$	107.15	col. lf.	0.989^{20}	-57	195.5	0.01^{25}	s.	s.	s.
— anthracene (α-)	$C_6H_4(CH_3)_2C_6H_3CH_3$	192.25	col. lf.	1.047^{29}	86	135.5^{15}	i.		s.	∞
— (β-)	$C_6H_4(CH_3)_2C_6H_3CH_3$	192.25	col. nd.	1.181^{20}	207	subl.	i.		s.	s.
— anthranilate (o-)	$NH_2C_6H_4CO_2CH_3$	151.16	col. lq.	1.168^{19}	24	198–9	sl. s.		v. sl. s.	∞
— anthraquinone (2-)	$C_6H_4(CO)_2C_6H_3CH_3$	222.23	lq.	1.087^{25}	176–7	305–6	0.02^{30}		1.7	s.
— benzoate	$C_6H_5CO_2CH_3$	136.14	col. lq.	1.732^{00}	-12.5	4.5^{738}	i.	∞	∞	∞
— benzylaniline	$C_6H_5N(CH_3)CH_2C_6H_5$	197.27	lq.	0.898^{20}	9.2	92.6	v. sl. s.	s.	s.	s.
— bromide	CH_3Br	94.95	col. lq.	0.891^{20}	-93	102.3	v. sl. s.	v. s.	v. s.	v. s.
— butyrate (n-)	$CH_3(CH_2)_2CO_2CH_3$	102.13	col. lq.	0.904^{00}	<-95	223–4	i.	s.	∞	∞
— (i-)	$(CH_3)_2CHCO_2CH_3$	102.13	col. lq.	0.887^{18}	-84.7	149.5	i.	s.	∞	∞
— caprate	$CH_3(CH_2)_8CO_2CH_3$	186.29	col. lq.	0.965^{20}	-18	192–4	i.		∞	∞
— caproate (n-)	$CH_3(CH_2)_4CO_2CH_3$	130.18	col. lq.	0.904^{00}	-40	124–5	i.		∞	∞
— caprylate	$CH_3(CH_2)_6CO_2CH_3$	158.23	col. lq.	0.887^{18}		-24	i.		∞	∞
— cellosolve	$CH_3OCH_2CH_2OH$	76.09	col. lq.	0.965^{20}		130^{740}	∞	∞	∞	v. s.
— chloride	CH_3Cl	50.49	gas	0.952^{0}	-97.7	71–2	280^{16} cc.	s.	s.	s.
— chloroacetate	$ClCH_2CO_2CH_3$	108.53	col. lq.	1.236^{20}	-32.7	263	d.		v. s.	v. s.
— chloroformate	$ClCO_2CH_3$	94.50	col. lq.	1.236^{15}	33.4	101	v. sl. s.		s.	v. s.
— cinnamate	$C_6H_5CH{:}CHCO_2CH_3$	162.18	cr.	1.042^{300}	-126.3	109.2	i.		s.	s.
— cyclohexane	$CH_3{<}(CH_2CH_2)_2{>}CHCH_3$	98.18	lq.	0.769^{20}	-14.5	79.6	i.		v. s.	v. s.
— ethyl carbonate	$CH_3O{\cdot}CO{\cdot}OC_2H_5$	72.10	col. lq.	1.002^{27}	-85.9	173.7	i.		∞	∞
— ethyl ketone	$CH_3COC_2H_5$	132.11	col. lq.	0.805^{20}		32	35^{10}		∞	∞
— ethyl oxalate	$CH_3OCO{\cdot}CO_2C_2H_5$	60.05	lq.	1.156^{00}	-99.8	181.3	i.		∞	s.
— formate	HCO_2CH_3	126.11	col. lq.	0.974^{20}			30^{20}		v. s.	v. s.
— furoate	$C_4H_3O{\cdot}CO_2CH_3$	195.21	lq.	1.179^{214}		151.2	i.		s.	s.
— glucamine	$CH_3NH(CHOH)_4CH_2OH$	90.08	lq.	1.168^{15}		172–3				
— glycolate	$HOCH_2CO_2CH_3$	144.21	lq.	0.881^{154}		12^{726}				
— heptoate	$CH_3(CH_2)_5CO_2CH_3$	66.49	gas			42.4	i.		∞	∞
— hypochlorite	$ClOCH_3$	141.95	lq.	2.279^{20}	-64.4	144.8	i.	s.	s.	v. s.
— iodide	CH_3I	214.34	lq.	1.090^{19}		148^{18}	1.8^{15}	s.	s.	s.
— laurate	$CH_3(CH_2)_{10}CO_2CH_3$	48.10	lq.		5	5.8^{752}	i.		∞	∞
— mercaptan	CH_3SH	100.11	gas	0.896^{0}	-121	100.3	∞	s.	∞	∞
— methacrylate	$CH_2{:}C(CH_3)CO_2CH_3$	242.39	lq.	0.950^{156}	-48	295^{715}	i.		∞	∞
— myristate	$CH_3(CH_2)_{12}CO_2CH_3$	142.19	cr./al.	1.025^{144}	18–9	244.6	i.		∞	∞
— naphthalene (α-)	$C_{10}H_7{\cdot}CH_3$	142.19	oil	0.994^{404}	-19	241–2	i.		∞	∞
— (β-)	$C_{10}H_7{\cdot}CH_3$	77.04	mn.	1.203^{25}	35–6	65	i.		∞	∞
— nitrate	CH_3ONO_2	61.04	lq.	0.991^{15}	expl.	-12	sl. s.		∞	∞
— nitrite	CH_3ONO	170.29	gas	0.828^{2020}	13.5	228	i.		∞	∞
— nonyl ketone (n-)	$CH_3(CH_2)_8COCH_3$	296.48	col. oil	0.879^{18}		$190{-}1^{10}$	i.		s.	∞
— oleate	$C_{17}H_{33}CO_2CH_3$	327.33	oil		30–1	196^{15}	i.		∞	∞
— orange	$(CH_3)_2NC_6H_4N{:}NC_6H_4SO_3Na$	270.44	red pd.			-14^{759}	0.2 c.	s.	v. sl. s.	i.
— palmitate	$CH_3(CH_2)_{14}CO_2CH_3$	48.03	col. cr.		-87.5	79.7	i.		∞	∞
— phosphine	CH_3PH_2	88.10	gas	0.915^{20}	-77.8	102	0.5^{20}		s.	s.
— propionate	$CH_3CH_2CO_2CH_3$	86.13	col. lq.	0.812^{1515}	-8.3	222.2	v. sl. s.		∞	∞
— propyl ketone (n-)	$CH_3COCH_2CH_2CH_3$	152.14	col. lq.	1.182^{25}	38–9	215^{15}	0.07^{30}		∞	∞
— salicylate (o-)	$HO{\cdot}C_6H_4CO_2CH_3$	298.49	col. cr.	1.073^{15}	<-50	213	i.		∞	∞
— stearate	$CH_3(CH_2)_{16}CO_2CH_3$	150.17	col. cr.	1.066^{15}	33–4	215	i.		s.	∞
— toluate (o-)	$CH_3{\cdot}C_6H_4CO_2CH_3$	150.17	col. lq.			217	i.		∞	∞
— (m-)	$CH_3{\cdot}C_6H_4CO_2CH_3$	150.17	col. lq.			206–7	i.		∞	∞
— (p-)	$CH_3{\cdot}C_6H_4CO_2CH_3$	121.18	cr.	0.973^{15}		206–7	i.		∞	∞
Methyl toluidine (o-)	$CH_3{\cdot}C_6H_4{\cdot}NHCH_3$	121.18	lq.	0.935^{554}		211^{761}	i.		∞	∞
— (m-)	$CH_3{\cdot}C_6H_4{\cdot}NHCH_3$	121.18	lq.	0.895^{154}	-91	127.3	i.	v. s.	∞	∞
— (p-)	$CH_3{\cdot}C_6H_4{\cdot}NHCH_3$	116.16	lq.	0.881^{20}		116.7^{764}	v. sl. s.	s.	∞	∞
— valerate (n-)	$CH_3(CH_2)_3CO_2CH_3$	116.16	col. lq.	0.836^{20}		81	v. sl. s.	sl. s.	s.	s.
— (i-)	$(CH_3)_2CHCH_2CO_2CH_3$	70.09	col. lq.	0.866^{154}	-104.8	42–3	>85		∞	∞
— vinyl ketone	$CH_3COCH{:}CH_2$	76.09	lq.	1.222^{30}		$210{-}2^{13}$	33		∞	∞
Methylal	$HCH(OCH_3)_2$	250.25	lq.		-52.8	98.5^{136}	d.	d.	v. s.	∞
Methylene-bis-(phenyl-4-isocyanate)	$(OCN{\cdot}C_6H_4)_2CH_2$	173.86	col. lq.	2.495^{20}	-96.7	40–1	1.17^{70}		s.	∞
— bromide	CH_2Br_2	84.94	col. lq.	1.336^{20}	65	208–9 d.	2^{20}		v. s.	∞
— chloride	CH_2Cl_2	198.26	cr.		5.7	180 d.	i.		s.	s.
— dianiline	$(C_6H_5NH)_2CH_2$	267.87			96–7		1.4^{20}		v. sl. s.	i.
— iodide	CH_2I_2	270.36	gn. lf./al.	3.325^{20}	174	>360 d.	i.		sl. s.	s.
Michler's hydrol (p-,p'-)	$[(CH_3)_2NC_6H_4]_2CHOH$	268.35	pr./al.				i.		sl. s.	i.
— ketone	$[(CH_3)_2NC_6H_4]_2CO$	303.35	pd.		254 d.		0.02^{20}		s.	i.
Morphine	$C_{17}H_{19}O_3N{\cdot}H_2O$	210.14	pr./al.	1.317	206–14		0.33^{14}		i.	i.
Mucic acid	$(CHOHCHOHCO_2H)_2$		pd.							

TABLE 2-2 Physical Properties of Organic Compounds (*Continued*)

Name	Formula	Formula weight	Form and color	Specific gravity	Melting point, °C	Boiling point, °C	Solubility in 100 parts		
							Water	Alcohol	Ether
Mustard gas	$(ClCH_2CH_2)_2S$	159.08	oil	1.275^{20}_{4}	13–4	217	0.07^{25}	s.	s.
Myricyl alcohol	$C_{30}H_{61}OH(?)$	452.82	cr.	0.777^{95}	88		i.	v. s.	v. s.
Myristic acid	$CH_3(CH_2)_{12}CO_2H$	228.36	col. lf.	0.853^{70}_{4}	57–8	250.5^{100}	i.	v. s.	v. s.
Myristyl alcohol	$CH_3(CH_2)_{12}CH_2OH$	214.38	cr.	0.824^{38}_{4}	38	167^{15}	<0.02	sl. s.	v. s.
Naphthalene	$C_{10}H_8$	128.16	pl./al.	1.145^{30}_{4}	80.2	217.9	0.003^{25}	9.5^{20}	i.
disulfonic acid (1-,5-)	$C_{10}H_6(SO_3H)_2$	288.28	lf.		d.		102^{20}	s.	i.
(1-,6-)	$C_{10}H_6(SO_3H)_2$	288.28	cr.		d. 125		164^{20}	v. s.	sl. s.
sulfonic acid (α-)	$C_{10}H_7·SO_3H·2H_2O$	244.26	cr.		90		v. s.	v. s.	s.
(β-)	$C_{10}H_7·SO_3H·H_2O$	226.24	nd.		125		77^{30}	sl. s.	
Naphthasultam (1-,8-)	$C_{10}H_7·O·NS$	205.22	lf.		177–8		v. s.		s.
disulfonate Na (1-,8-)	$C_{10}H_5O_3NS_2Na_2·2H_2O$	445.35	nd.				v. s.	sl. s.	
(2-,4-)	$C_{10}H_4O_3NS_2Na_3·8½H_2O$	554.45	nd.				v. sl. s. h.		
Naphthoic acid (α-)	$C_{10}H_7·CO_2H$	172.17	mn.	1.077^{100}_{4}	160–1	300	0.007^{25}	sl. s. h.	v. s.
(β-)	$C_{10}H_7·CO_2H$	172.17	mn.	1.224^{4}	184	>300	0.074^{25}	v. s.	v. s.
Naphthol (α-)	$C_{10}H_7·OH$	144.16	lf.	1.217^{4}	96	278–80	v. s. h.	v. s.	v. s.
(β-)	$C_{10}H_7·OH$	144.16	pl./aq.		122–3	285–6	sl. s. h.	v. s.	i.
sulfonic acid (α-)/(1-,2-)	$HO·C_{10}H_6SO_3H$	224.22	nd./al.		>250		v. s.	v. s.	
(β-)/(2-,6-)	$HO·C_{10}H_6SO_3H$	224.22	nd./al.		125		s. h.	s.	
Naphthyl acetate (α-)	$CH_3CO·C_{10}H_7·$	186.20	rhb.	1.123^{25}_{25}	46–9		i.	v. s.	v. s.
(β-)	$CH_3CO·C_{10}H_7·$	186.20	lf./aq.	1.061^{96}_{4}	69–70		i.	s.	v. s.
amine (α-)	$C_{10}H_7·NH_2$	143.18	nd.		50	300.8	0.17 c.	v. s.	s.
(β-)	$C_{10}H_7·NH_2$	143.18	lf.		111–2	306.1	v. s. h.	v. s.	s.
amine hydrochloride (α-)	$C_{10}H_7·NH_2·HCl$	179.65	nd.		d.	subl.	v. s. h.	s.	i.
(β-)	$C_{10}H_7·NH_2·HCl$	179.65	lf.				3.8^{20}	i.	
amine sulfonic acid (1-,4-)	$NH_2·C_{10}H_6·SO_3H·H_2O$	223.24	cr.				0.2^{100}	i.	
(1-,5-)	$NH_2·C_{10}H_6·SO_3H·H_2O$	241.26	cr.				sl. s.		
(1-,7-)	$NH_2·C_{10}H_6·SO_3H·H_2O$	241.26	cr.				0.46^{25}		
(1-,8-)	$NH_2·C_{10}H_6·SO_3H·H_2O$	241.26	cr.				0.42^{100}		
(2-,5-)	$NH_2·C_{10}H_6·SO_3H$	223.24	cr.				0.08		
(2-,6-)	$NH_2·C_{10}H_6·SO_3H·H_2O$	241.26	cr.				0.38^{100}		
(2-,7-)	$NH_2·C_{10}H_6·SO_3H·H_2O$	241.26	cr.				0.28^{100}		
isocyanate (α-)	$C_{10}H_7·N·CO$	169.17	col. lq.	1.18		269–70	d.	s. h.	s.
Nicotine	$C_{10}H_{14}N_2$	162.23	oil	1.009^{20}_{4}	<–80	246^{30}	∞	∞	∞
Nicotinic acid (3-)	$C_5H_4NCO_2H$	123.11	nd./al.		235.2	subl.	s. h.	s. h.	v. sl. s.
(i-)(4-)	$C_5H_4NCO_2H$	123.11	rhb.		317	d.	s. h.	sl. s. h.	v. sl. s.
Nitro-acetanilide (p-)	$CH_3CONHC_6H_4NO_2$	180.16	nd./aq.		215–6		s. h.	v. s.	s.
-acetophenone (m-)	$CH_3COC_6H_4NO_2$	165.14	red nd.		80–1	202	i.	s.	
-aminoanisole (4-,1-,2-)	$NO_2·C_6H_3(OCH_3)·NH_2$	168.15	yel. nd.	1.207^{156}	139–40		sl. s.	v. s.	v. s.
(5-,1-,2-)	$NO_2·C_6H_3(OCH_3)·NH_2$	168.15	red	1.211^{156}	123		sl. s. c.	v. s.	v. s.
(3-,1-,4-)	$NO_2·C_6H_3(OCH_3)·NH_2$	168.15	or. pr.		142–3		s. h.	s.	7.9^{20}
-aminophenol (4-,2-,1-)	$NO_2·C_6H_3(NH_2)·OH$	154.12	yel. rhb.				0.11^{20}	v. s.	6.1^{10}
-aniline (o-)	$NO_2·C_6H_4·NH_2$	138.12	yel. rhb.	1.442^{15}	71.5	284.1	0.08^{10}	v. s.	∞
(m-)	$NO_2·C_6H_4·NH_2$	138.12	yel. mm.	1.43	114	306.4	0.17^{30}	7.1^{20}	v. sl. s.
(p-)	$NO_2·C_6H_4·NH_2$	138.12	col. cr.	1.437^{14}	146–7	331.7	0.06^{30}	5.8^{20}	v. s.
-anisole (o-)	$CH_3O·C_6H_4·NO_2$	153.13	pr./al.	1.254^{20}_{4}	9.4	272–3	i.	v. s.	v. s.
(p-)	$CH_3O·C_6H_4·NO_2$	153.13	nd.	1.233^{20}	54	274	s.	sl. s.	s.
-anthraquinone (α-)	$C_6H_4(CO)_2C_6H_3·NO_2$	253.20	yel. cr.		230	270^{7}	i.	i.	v. s.
-anthraquinone sulfonic acid (1-,5-)	$NO_2·C_{14}H_6O_2·SO_3H$	333.26	mn.				s.	v. s. h.	v. s.
-benzal chloride (m-)	$NO_2·C_6H_3·CHCl_2$	206.03	nd./aq.		65			v. s. h.	v. s.
-benzaldehyde (m-)	$NO_2·C_6H_4·CHO$	151.12	yel. lq.	1.205^{18}_{4}	58	164^{23}	1.95^{112}	v. s.	i.
Nitro-benzene	$C_6H_5NO_2$	123.11	red nd.		5.7	210.9	0.19^{20}	∞	∞
benzidine (2-)	$NH_2C_6H_4·C_6H_4(NH_2)NO_2$	229.23	tri/aq.		143		sl. s. h.	sl. s. h.	v. s.
-benzoic acid (o-)	$NO_2·C_6H_4·CO_2H$	167.12	mm.	1.575^{44}	147.5		0.65^{20}	28^{11}	22^{11}
(m-)	$NO_2·C_6H_4·CO_2H$	167.12	yel. mm.	1.494^{44}	140–1		$0.24^{46.5}$	31^{12}	25^{10}
(p-)	$NO_2·C_6H_4·CO_2H$	167.12	cr.	1.550^{22}_{4}	240–2		0.02^{15}	0.9^{10}	2.2^{18}
-benzyl alcohol (m-)	$NO_2·C_6H_4·CH_2OH$	153.13	nd./al.		27	175–80³	i.	2^{10}	v. s.
-benzyl bromide (p-)	$NO_2·C_6H_4·CH_2Br$	216.04	yel.		99–100		v. sl. s.	v. s.	v. s.
-chlorotoluene (1-,2-,6-)	$CH_3·C_6H_3(NO_2)·Cl$	171.58	oil	1.240^{96}_{4}	37.5	238	i.	v. s.	s.
-cresol (1-,3-,4-)	$CH_3·C_6H_3(NO_2)·OH$	153.13	yel.	1.067^{20}_{4}	32	125^{22}	v. sl. s.	s. h.	v. s.
-cymene (1-,2-,4-)	$CH_3·C_6H_3(NO_2)·CH(CH_3)_2$	179.21	yel. oil	1.179^{20}_{4}		152^{15}	i.	s.	v. s.
-dimethylaniline (o-)	$NO_2·C_6H_4·NHCH_3$	166.18	red mm.	1.313^{17}	60–1	151–3⁸⁰	v. sl. s.	v. s. h.	∞
(p-)	$NO_2·C_6H_4·NHCH_3$	166.18	yel. nd.		163–4	280–5	i.	v. s. h.	v. s.
-diphenyl (o-)	$C_6H_4·C_6H_4·NO_2$	199.20	rhb.		37	320	i.	v. s.	v. s.
(p-)	$C_6H_4·C_6H_4·NO_2$	199.20	nd./al.	1.44	113–4	340	i.	s. h.	s.
-diphenylamine (o-)	$C_6H_5·NH·C_6H_4·NO_2$	214.22	or. cr.		75–6		i.	sl. s. c.	v. s.
-guanidine	$H_2NC(NH)NHNO_2$	104.07	nd./aq.		246–7		9^{100}	sl. s.	v. sl. s.

Name	Formula	Mol. wt.	Form	Sp. gr.	M.P., °C	B.P., °C	Water (cold)	Water (hot)	Alcohol	Ether
-naphthalene (α-)	$C_{10}H_7NO_2$	173.16	yel./al.	1.223^{62}	59–60	304	i.	s.	s.	s.
(β-)	$C_{10}H_7NO_2$	173.16	col./al.	1.295^{45}	79	165^{15}	1.08^{100}	v.s.	v.s.	v.s.
-phenol (o-)	$NO_2C_6H_4OH$	139.11	col.mn.	1.485^{20}	44–5	214.5	1.35^{20}	v.s.	v.s.	v.s.
(m-)	$NO_2C_6H_4OH$	139.11	yel.pr.	1.485^{20}	96–7	194^{70}	1.6^{25}	v.s.	v.s.	v.s.
(p-)	$NO_2C_6H_4OH$	139.11	nd.	1.479^{20}	113–4	subl.		v.s.h.	sl.s.	sl.s.
-phenol sulfonic acid (1-4-,2-)	$HO·C_6H_3(NO_2)SO_3H·3H_2O$	273.22	nd./aq.		d. 110			v.s.h.	sl.s.	sl.s.
(1-2-,4-)	$HO·C_6H_3(NO_2)SO_3H·3H_2O$	273.22	yel./aq.		51.5			v.s.h.	sl.s.	sl.s.
-phthalic acid (3-)	$NO_2C_6H_3(CO_2H)_2$	211.13	yel.cr.		222	222.3	2.05^{25}	v.s.	∞	s.
(4-)	$NO_2C_6H_3(CO_2H)_2$	211.13	yel.lf.		164–5	230–1	0.07^{90}	8.6^{15}	∞	∞
-toluene (o-)	$CH_3·C_6H_4NO_2$	137.13	lq.	$1.163^{20/4}$	–4.1	237.7	0.05^{90}	v.s.	80.8^{15}	∞
(m-)	$CH_3·C_6H_4NO_2$	137.13	lq.	$1.160^{18/4}$	15–16		0.04^{80}	v.s.	v.s.	v.s.
(p-)	$CH_3·C_6H_4NO_2$	137.13	rhb.	$1.139^{35/55}$	51.9		47.7^{28}	v.s.	v.s.	v.s.
-toluene sulfonic acid (1-4-,2-)	$CH_3·C_6H_3(NO_2)SO_3H·2H_2O$	253.23	pl./aq.	1.365^{15}	130		v.sl.s.	s.h.	v.sl.s.	sl.s.h.
-toluidine (4-,1-,2-)	$NO_2·C_6H_3(CH_3)NH_2$	152.15	yel.mn.	1.312^{17}	105–7	189–90 d.	sl.s.h.	s.h.	v.s.	s.
(3-,1-,2-)	$NO_2·C_6H_3(CH_3)NH_2$	152.15	red mn.		116–7		i.	s.	s.	s.
Nitron	$C_{20}H_{16}N_4$	312.36	gn.tri.		189–90 d.	330	i.	2.4^{18}	s.	s.
Nitroso-dimethylaniline (p-)	$ON·C_6H_4N(CH_3)_2$	150.18	brn.pr.		86–7	150.5^{759}	0.1^{20}	sl.s.	s.	s.
-naphthol (β-)/(1-)	$ON·C_{10}H_6OH$	173.16	cr.	$0.777^{20/4}$	109.5	317	i.	sl.s.	s.	s.
Nonadecane (n-)	$CH_3(CH_2)_{17}CH_3$	268.51	col.lq.	$0.718^{20/4}$	32	125.7	i.	sl.s.	s.	∞
Nonane (n-)	$CH_3(CH_2)_7CH_3$	128.25	cr.	$0.775^{20/4}$	–53.7	99.3^{780}	i.	∞	∞	∞
Octadecane (n-)	$CH_3(CH_2)_{16}CH_3$	254.48	col.lq.	$0.703^{20/4}$	28	210	i.	v.s.	∞	∞
Octane (n-)	$(CH_3)_3CCH_2CH(CH_3)_2$	114.22	col.lq.	$0.692^{20/4}$	–56.5	195	i.	∞	∞	∞
(iso-)	$(CH_3)_3CCH_2CH(CH_3)_2$	114.22	col.lq.	$0.885^{0/4}$	–107.4	194–5	i.	∞	∞	∞
Octyl acetate (n-)	$CH_3CO_2CH_2(CH_2)_6CH_3$	172.26	col.lq.	$0.863^{14/4}$	–38.5	179–80	0.002^{16}	v.sl.s.	∞	∞
alcohol (n-)	$CH_3CO_2CH(CH_3)C_6H_{13}$	172.26	col.lq.	$0.827^{20/4}$	–16	126	i.	∞	∞	i.
(sec-)	$CH_3(CH_2)_6CH_2OH$	130.22	col.lq.	$0.822^{20/4}$	–38.6	$285–6^{100}$	i.	v.s.	∞	∞
Octylene (n-)	$CH_3(CH_2)_5CH(OH)CH_3$	130.22	col.lq.	$0.721^{18/4}$		287–90	i.	v.s.	∞	∞
(sec-)	$CH_3(CH_2)_5CH·CH_2$	112.21	lq.	$0.854^{70/4}$		subl.		sl.s.	∞	∞
Oleic acid	$C_7H_{15}CH·CH(CH_2)_7·CO_2H$	282.45	col.nd.	1.290^4	14	271.5^{100}	0.054^{25}	sl.s.	∞	∞
Orcinol (1-,3-,5-)	$(OH)_2C_6H_3·CH_3$	124.13	pr./bz.	$1.653^{19/4}$	107–8	253–4	0.096^{25}	v.s.	∞	i.
Oxalic acid	$HO_2C·CO_2H·2H_2O$	126.07	col.mn.	$0.849^{70/4}$	101.5	162	i.	s.	v.s.	∞
Palmitic acid	$CH_3(CH_2)_{14}CO_2H$	256.42	col.pl.	$0.906^{20/4}$	63–4	270.5	i.	9^{20}	∞	∞
Pelargonic acid	$CH_3(CH_2)_7CO_2H$	158.23	col.oil	$1.671^{25/4}$	12.5	276^{90}	v.sl.s.	v.sl.s.	∞	∞
Penta-chloroethane	$CHCl_2·CCl_3$	202.31	col.lq.	$0.770^{20/4}$	–22	239.4	0.05^{20}	v.s.	v.sl.s.	i.
-decane (n-)	$C_5H_{11}(CH_2)_5CH_3$	212.41	col.lq.		10	36.3	5.6^{15}	v.sl.s.	i.	∞
-erythritol	$C(CH_2OH)_4$	136.15	cr.	$0.994^{20/4}$	262	27.95	∞	∞	∞	∞
Pentanediol	$HOCH_2(CH_2)_3CH_2OH$	104.15	lq.	$0.630^{18/4}$	–129.7	9.5	0.036^{16}	∞	∞	1.6^{25}
Pentane (n-)	$CH_3(CH_2)_3CH_3$	72.15	col.lq.	0.621^{19}	–160.0	d.	i.	v.s.	v.s.	v.s.
(i-)	$(CH_3)_2CHCH_2CH_3$	72.15	col.lq.	$0.613^{20/4}$	–20	340	i.	9^{20}	s.	s.
(neo-)	$(CH_3)_4C(CH_3)_2$	72.15	col.mn.		134–5	228–9	0.7^{20}	v.sl.s.	∞	∞
Phenacetin	$C_2H_5OC_6H_4NHCOCH_3$	179.21	pl./al.	1.179^{25}	99–100	254–5	i.	40 h.	∞	∞
Phenanthrene	$<(C_6H_4CH)_2>$	178.22	oil	1.061^{15}	<–21	172	i.	10 h.	s.	i.
Phenetidine (o-)	$C_2H_5O·C_6H_4·NH_2$	137.18	lq.	$0.967^{20/4}$	3–4	181.4	i.	8.2^{15}	∞	5.9 c.
(p-)	$C_2H_5O·C_6H_4·NH_2$	137.18	lf.	$0.930^{20/4}$	–30.2	193–4	8.2^{15}	v.s.	∞	∞
Phenetole	$C_2H_5O·C_6H_5$	122.16	cr.	$1.299^{25/4}$	42–3	265.5	0.2^{20}	v.s.	∞	v.s.
Phenol	C_6H_5OH	94.11	lf.		261–2	142–3	v.s.	v.s.	∞	∞
-phthalein (o-)	$C_{20}H_{14}O_4$	318.31	col.oil	1.025^{20}	50 d.	299^{760}	v.sl.s.	v.s.	sl.s.	sl.s.
-sulfonic acid (o-)	$HO·C_6H_4·SO_3H·½H_2O$	187.68	cr.	$1.081^{90/4}$	76–7	302	1.66^{20}	v.s.	d.	v.s.h.
Phenyl acetaldehyde	$C_6H_5CH_2CHO$	120.14	yel.oil	$0.930^{20/4}$	–43	$219–21^{750}$	i.	v.s.	v.s.h.	∞
acetic acid	$C_6H_5CH_2CO_2H$	136.14	cr./al.		45–6	243.5	v.sl.s.	d.	∞	∞
aniline (o-)	$C_6H_5·C_6H_4·NH_2$	169.22	lq.	$1.023^{18/4}$	50–2	166^{769}	s.h.	v.s.	s.	sl.s.
(p-)	$C_6H_5·C_6H_4·NH_2$	169.22	col.lq.		127	191^{17}	1.6^{20}	sl.s.	v.sl.s.	s.h.
Phenyl-ethyl alcohol	$C_6H_5CH_2CH_2OH$	122.16	col.mn.	$1.097^{22/4}$	19.6	219–20	sl.s.h.	d.	v.s.	v.sl.s.
-glycine	$C_6H_5NHCH_2CO_2H$	151.16	col.rhb.		286	336–7	0.6^{12}	v.s.h.	i.	v.s.
-hydrazine	$C_6H_5NH·NH_2$	108.14	cr.	$1.096^{20/4}$	128	335^{535}	d.	d.	i.	v.s.h.
-hydrazine sulfonic acid (p-)	$H_2NNHC_6H_4SO_3H$	188.20	lq.		–21	399.5	1^{20}		i.	∞
isocyanate	$C_6H_5N·CO$	119.12	lf.		45	275	i.	s.	sl.s.	s.
-methylpyrazolone (3-)/(N-)	$C_6H_5N_2C_3H_2O$	174.20	cr.	$1.138^{15/15}$	102.5	305–8	i.	v.s.	sl.s.	s.h.
-mustard oil	$C_6H_5N_2CS$	135.18	waxy		62	235–7	i.	d.	sl.s.	v.s.
naphthalene (α-)	$C_{10}H_7·C_6H_5$	204.26	lf./al.	1.17	107–8	363	0.08^{60}	v.s.h.	sl.s.	v.s.h.
(β-)	$C_{10}H_7·C_6H_5$	204.26	pr./al.	1.18	56–7	283^{187}	0.4^{60}	∞	v.s.	∞
naphthylamine (α-)	$C_{10}H_7NHC_6H_5$	219.27	rhb.		<–18	$172–3^{312}$	i.	s.h.	s.h.	s.
(β-)	$C_{10}H_7NHC_6H_5$	219.27	nd.		86	267^{15}	i.	v.s.	s.	s.
phenol (o-)	$C_6H_5·C_6H_4OH$	170.20	nd.	$1.008^{20/4}$	42–3	237–8	sl.s.	sl.s.	∞	s.
(p-)	$C_6H_5·C_6H_4OH$	170.20	oil		52		sl.s.	s.h.	s.	s.
propyl alcohol (γ-)	$C_6H_5·C_3H_6OH$	136.19	nd.		52–3		sl.s.	v.s.	s.	s.
quinoline (2-)/(α-)	$C_6H_5·C_9H_6N$	205.25	lq.	$1.250^{20/4}$			0.015^{25}	v.s.	sl.s.	s.
(8-)/(0-)	$C_6H_5·C_9H_6N$	205.25	rhb./al.							
salicylate, salol	$HO·C_6H_4·CO·C_6H_5$	214.21	cr.	$1.106^{20/4}$			i.c.	s.	v.s.	s.
stearate	$CH_3(CH_2)_{16}CO_2C_6H_5$	360.56	pl./al.							
urethane	$C_6H_5NHCO_2C_2H_5$	165.19								

TABLE 2-2 Physical Properties of Organic Compounds (Continued)

Name	Formula	Formula weight	Form and color	Specific gravity	Melting point, °C	Boiling point, °C	Solubility in 100 parts — Water	Solubility in 100 parts — Alcohol	Solubility in 100 parts — Ether
Phenylene-diamine (o-)	$C_6H_4(NH_2)_2$	108.14	lf./aq. rhb.	$1.139^{15/15}$	103–4	256–8	733^{81}	v.s.	v.s.
(m-)	$C_6H_4(NH_2)_2$	108.14	rhb.		62.8	284–7	35.1^{25}	v.s.	s.
(p-)	$C_6H_4(NH_2)_2$	108.14	mm.		140	267	669^{107}	s.	v.s.
Phloroglucinol (1,3,5-)	$C_6H_3(OH)_3 \cdot 2H_2O$	162.14	rhb.		117	subl.	1.13^{25}	v.s.	v.s.
Phorone	$[(CH_3)_2C{:}CH]_2CO$	138.20	yel. pr.	$0.885^{20/4}$	28	197.2^{743}	0.1^{80}	v.s.	s.
Phosgene	$OCCl_2$	98.92	gas	$1.392^{19/4}$	−104	8.2^{796}	d.	s.	0.68^{15}
Phthalic acid (o-)	$C_6H_4(CO_2H)_2$	166.13	mm./aq.	$1.593^{20/4}$	208	d.	0.70^{25}	12^{18}	sl.s.
anhydride (o-)	$C_6H_4{<}(CO)_2{>}O$	148.11	nd./aq. rhb.	1.527^{4}	130.8	284.5	0.2^{100}	s.	s.h.
nitrile (o-)	$C_6H_4(CN)_2$	128.13	cr.		141	subl.	sl.s.c.	s.	sl.s.
Phthalide	$C_8H_6O_2$	134.13	cr./et.	$1.164^{99/4}$	73(65)	290	0.04^{25}	s.	1^{13}
Phthalimide (o-)	$C_6H_4(CO)_2NH$	147.13	nd./aq.		238	subl.	v.s.	5	7^{17}
Picoline (α-)	$C_5H_4N{\cdot}CH_3$	93.12	col. lq.	$0.950^{15/4}$	−70	128.8	v.s.	∞	v.s.
(β-)	$C_5H_4N{\cdot}CH_3$	93.12	col. lq.	$0.961^{15/4}$		143.5	∞	∞	∞
(γ-)	$C_5H_4N{\cdot}CH_3$	93.12	lq.	$0.957^{15/4}$		143.1	v.s.	s.	∞
Picramic acid (1-2-,4-,6-)	$HO{\cdot}C_6H_2(NH_2)(NO_2)_2$	199.12	red nd.		169	expl.	0.14^{22}	6^{20}	s.
Picric acid (2-,4-,6-)	$HO{\cdot}C_6H_2(NO_2)_3$	229.11	yel. rhb.	$1.763^{20/4}$	121.8	d.	1.23^{20}	4.8^{17}	s.
Picryl chloride (2,4-,6-)	$ClC_6H_2(NO_2)_3$	247.56	yel. mm.	1.797^{20}	83	$171–2^{789}$	0.018^{15}	v.s.	v.s.
Pinacol	$(CH_3)_2C{\cdot}OH]_2$	118.17	col. nd.	0.967^{15}	43(38)	$171–2^{789}$	v.s.	s.	s.
Pinacoline	$CH_3COC(CH_3)_3$	100.16	col. lq.	0.800^{18}	−52.5	106.2	2.5^{15}	s.	s.
Pinene (α-)(dl-)	$C_{10}H_{16}$	136.23	col. lq.	$0.878^{20/4}$	−55	154–6	i.	33	∞
Pinol (dl-)	$C_{10}H_{16}O$	152.23	lf.	$0.953^{20/20}$	131–2	207–8	i.	s.	∞
hydrochloride	$C_{10}H_{16}{\cdot}HCl$	172.69	lq.	$0.860^{20/4}$	−9	183–4	i.	∞	∞
Piperidine	$CH_2{<}(CH_2CH_2)_2{>}NH$	85.15	lq.		264	106	∞	∞	∞
carboxylic acid (α-)(dl-)	$HO_2C{\cdot}CH{<}(CH_2CH_2)_2{>}NH$	129.16	cr.	1.13	175	s.	6^{28}	s.	s.
Piperidinium pentamethylene dithiocarbamate	$(CH_2)_5N{\cdot}CS_2H{\cdot}HN(CH_2)_5$	232.41	cr.				i.		
Propane	$CH_3CH_2CH_3$	44.09	gas	$0.585^{-45/4}$	−187.1	−42.2	6.5^{18} cc.	∞	v.s.
Propionic acid	$CH_3CH_2CO_2H$	74.08	col. lq.	$0.992^{20/4}$	−22	141.1	∞	∞	∞
aldehyde	CH_3CH_2CHO	58.08	col. lq.	$0.807^{20/4}$	−81	49.5^{740}	20^{20}	∞	∞
anhydride	$(CH_3CH_2CO)_2O$	130.14	col. lq.	$1.012^{20/4}$	−45	168.8^{780}	d.	∞	∞
Propyl acetate (n-)	$CH_3CO_2CH_2CH_2CH_3$	102.13	col. lq.	$0.886^{20/4}$	−92.5	101.6	1.6^{16}	∞	∞
(i-)	$CH_3CO_2CH(CH_3)_2$	102.13	col. lq.	$0.874^{20/20}$	−73.4	88.4	3^{20}	∞	∞
alcohol (n-)	$CH_3CH_2CH_2OH$	60.09	col. lq.	$0.804^{20/4}$	−127	97.8	∞	∞	∞
(i-)	$(CH_3)_2CHOH$	60.09	col. lq.	$0.789^{20/4}$	−85.8	82.5	∞	∞	∞
amine (n-)	$CH_3CH_2CH_2NH_2$	59.11	col. lq.	$0.718^{20/20}$	−83	$49–50^{761}$	∞	∞	∞
(i-)	$(CH_3)_2CHNH_2$	59.11	col. lq.	$0.694^{15/4}$	−101	33–4	∞	∞	∞
aniline (n-)	$C_6H_5NHCH_2CH_2CH_3$	135.20	lq.	0.949^{15}		222	i.	∞	∞
benzoate (n-)	$C_6H_5CO_2CH_2CH_2CH_3$	164.20	col. lq.	$1.021^{25/25}$	−51.6	231	i.	∞	∞
(i-)	$C_6H_5CO_2CH(CH_3)_2$	164.20	col. lq.	$1.010^{20/25}$		218.5	i.	∞	∞
bromide (n-)	$CH_3CH_2CH_2Br$	123.00	col. lq.	$1.353^{20/4}$	−109.9	70.8	0.25^{20}	∞	∞
(i-)	$(CH_3)_2CHBr$	123.00	col. lq.	$1.310^{20/4}$	−89	60	0.32^{20}	∞	∞
n-butyrate (n-)	$C_2H_5CH_2CO_2CH_2CH_2CH_3$	130.18	col. lq.	0.879^{15}	−95.2	142.7	0.17^{17}	∞	∞
i-butyrate (n-)	$(CH_3)_2CHCO_2CH_2CH_2CH_3$	130.18	col. lq.	$0.884^{20/4}$		134–5	v.sl.s.	∞	∞
n-butyrate (i-)	$C_2H_5CH_2CO_2CH(CH_3)_2$	130.18	col. lq.	0.865^{18}		128	v.sl.s.	∞	∞
i-butyrate (i-)	$(CH_3)_2CHCO_2CH(CH_3)_2$	130.18	col. lq.	$0.869^{20/4}$		120.8	v.sl.s.	∞	∞
chloride (n-)	$CH_3CH_2CH_2Cl$	78.54	col. lq.	$0.890^{20/4}$	−122.8	46.4	0.27^{20}	∞	∞
(i-)	CH_3CH_2CHCl	78.54	col. lq.	0.859^{20}	−117	36.5	0.31^{20}	∞	∞
Propyl formate (n-)	$HCO_2CH_2CH_2CH_3$	88.10	col. lq.	$0.901^{20/4}$	−92.9	81.3	12.2^{12}	∞	∞
(i-)	$HCO_2CH(CH_3)_2$	88.10	col. lq.	$0.873^{20/4}$		$68–71^{751}$	2.1^{12}	∞	∞
furoate (n-)	$C_4H_3O{\cdot}CO_2C_3H_7$	154.16	col. lq.	$1.075^{20/4}$		211	s.	s.	v.s.
lactate (n-)	$CH_3CH(OH)CO_2CH_2C_2H_5$	132.16	col. lq.			$122–3^{150}$	v.sl.s.	∞	s.
(i-)	$CH_3CH(OH)CO_2CH(CH_3)_2$	132.16	col. oil			167.5	i.	∞	s.
mercaptan (n-)	$CH_3CH_2CH_2SH$	76.15	lq.	$0.836^{25/4}$	−112	67–8	0.25^{20}	∞	∞
(i-)	$(CH_3)_2CHSH$	76.15	lq.	$0.809^{25/4}$	−130.7	58–60	0.56^{25}	∞	∞
propionate (n-)	$C_2H_5CO_2CH_2CH_2CH_3$	116.16	col. lq.	$0.883^{20/4}$	−76	122–3	0.6^{25}	∞	∞
(i-)	$C_2H_5CO_2CH(CH_3)_2$	116.16	col. lq.	0.893^{0}		$109–11^{750}$	i.	∞	∞
thiocyanate (i-)	$(CH_3)_2CH{\cdot}CNS$	101.16	lq.	0.963^{20}		$152–3^{754}$	i.	∞	∞
n-valerate (n-)	$CH_3(CH_2)_3CO_2C_3H_7$	144.21	lq.	0.874^{15}	−70.7	67.5	i.	∞	∞
i-valerate (n-)	$(CH_3)_2CHCH_2CO_2C_3H_7$	144.21	col. lq.	$0.863^{20/4}$		155.9	i.	∞	∞
i-valerate (i-)	$(CH_3)_2CHCH_2CO_2C_3H_7$	144.21	col. lq.	0.854^{17}		142^{756}	i.	∞	∞
Propylene	$CH_3CH{:}CH_2$	42.08	gas	$0.609^{-47/4}$	−185	$−48^{749}$	44.6 cc.	∞	∞
bromide	$CH_3CHBrCH_2Br$	201.91	col. lq.	$1.933^{20/4}$	−55.5	141.6	0.25^{20}	s.	v.s.
chlorohydrin	$CH_3CHClCH_2OH$	94.54	col. lq.	1.103^{20}		133–4	s.	s.	s.
chloride	$CH_3CHClCH_2Cl$	112.99	col. lq.	$1.159^{20/20}$	<−70	96.8	0.27^{20}	v.s.	v.s.
glycol	$CH_3CH(OH)CH_2OH$	76.09	col. oil	$1.040^{19/4}$		188–9	∞	∞	8
oxide	CH_3CHCH_2O	58.08	col. lq.	$0.831^{20/20}$		35	33^{20}	∞	∞
Protocatechuic acid (3-,4-)	$(HO)_2C_6H_3CO_2H{\cdot}H_2O$	172.13	nd./aq.	$1.542^{4/4}$	199 d.		1.82^{14}	v.s.	s.

Name	Formula	Mol. wt.	Form/color	Density	m.p., °C	b.p., °C	Sol. cold H₂O	Sol. hot H₂O	alcohol	ether	other
Pulegol (iso-)(d-)	$C_{10}H_{17}\cdot OH$	154.24	col. liq.	$0.911^{20/4}$		$86\text{-}9^{10}$			∞	∞	v. sl. s.
Pulegone	$C_{10}H_{16}O$	152.23	col. liq.	$0.932^{20/20}$		224^{734}			s.	s.	s.
Pyrazole	$-NH\cdot N{:}CH\cdot CH{:}CH-$	68.08	nd./et.		70	186-8	s.	s.	∞	s.	sl. s.
Pyrazoline	$-NH\cdot N{:}CH\cdot CH_2CH_2-$	70.09	nd.			144	s.	v. s.	∞	v. s.	v. sl. s.
Pyrazolone	$-NH\cdot CO\cdot CH_2CH{:}N-$	84.08	yel. pr.		165	subl. d.	i.	3 h.	v. s.	3 h.	v. s.
Pyrene	$C_{16}H_{10}$	202.24	col. liq.		149-50	>360			v. sl. s.		s.
Pyridazine	$N_2{<}(CHCH)_2{>}N$	80.09	yel. pr.	$1.277^{20/4}$	-8	208			∞	∞	s.
Pyridine	$CH{<}(CHCH)_2{>}N$	79.10	nd./aq.	$1.107^{20/4}$	-42	115-6	∞	∞	∞	∞	v. s.
Pyrocatechol (o-)	$C_6H_4(OH)_2$	110.11	nd.	$0.982^{20/4}$	104-5	240-5	45.1^{20}	v. s.	v. s.	v. s.	v. s.
Pyrogallol (1-,2-,3-)	$C_6H_3(OH)_3$	126.11	cr.	1.344^{4}	133-4	309	40^{13}	s.	v. s.	v. s.	v. s.
Pyrone	$CO{<}(CHCH)_2{>}O$	96.08	lq.	1.453^{4}	32.5	215-7			v. s.	v. s.	v. s.
Pyrrole	${<}(CH\cdot CH)_2{>}NH$	67.09	lq.	$1.190^{40/3}$		131	v. sl. s.	v. s.	∞	∞	∞
Pyrrolidine	${<}(CH_2\cdot CH_2)_2{>}NH$	71.12	lq.	$0.948^{20/4}$		87-8	∞	∞	∞	∞	∞
Pyrroline	${<}(CH\cdot CH_2)_2{>}NH$	69.10	lq.	$0.852^{22/5}$		90-1	∞	∞	∞	∞	sl. s.
Pyruvic acid	CH_3COCO_2H	88.06	col. liq.	$0.910^{20/4}$	13.6	165	∞	∞	∞	∞	s. s.
Quercitrin	$C_{21}H_{20}O_{11}\cdot 2H_2O$	484.40	yel. nd.	$1.267^{20/4}$	182-5		0.04^{20}		v. sl. s.		s.
Quinaldine (py-2)	$CH_3\cdot C_9H_6N$	143.18	lq.	$1.059^{20/4}$	-1	244.5^{750}	v. sl. s.	v. s.	∞	∞	∞
Quinoline	C_9H_7N	129.15	pl.	1.095^{20}	-15	237.1^{747}	v. sl. s.	s.	∞	∞	
(iso-)	C_9H_7N	129.15	cr.	$1.099^{21/4}$	24.6	240.5^{763}	sl. s.	sl. s.	∞	s.	s.
-diol (1-,3-)	$-C_6H_4\cdot CH{:}C(OH)\cdot N{:}C(OH)-$	161.15	cr.		237						
Quinone (p-)	$CO{<}(CHCH)_2{>}CO$	108.09	yel. mn.	$1.318^{20/4}$	115.7	subl.	sl. s.	v. s. h.	v. s.	v. s. h.	v. s.
R acid Ca salt (2-,)(3-,6-,)	$HOC_{10}H_5(SO_3)_2Ca$	342.35	cr./aq.				v. s. h.	sl. s.	sl. s. h.		
K salt	$HOC_{10}H_5(SO_3K)_2$	380.40	col. rhb.				v. s.	v. s. h.	v. s. h.		s.
Na salt	$HOC_{10}H_5(SO_3Na)_2$	348.26	lf./al.				v. s. h.	sl. s.	sl. s. h.		
Raffinose	$C_{18}H_{32}O_{16}\cdot 5H_2O$	594.52	col. mm.	1.465^{0}	119	d. 130	14.3^{20}	0.1^{20}	30.6^{25}	i.	
Resorcinol (m-)	$C_6H_4(OH)_2$	110.11	red lf.	1.272^{15}	110.7	276.5	147^{12}	v. s.	147^{12}	v. s.	v. s. h.
Retene	$C_{18}H_{18}$	234.32	mn.	1.13^{16}	98-9	390-4	i.				i.
Rhamnose (β-)	$CH_3(CHOH)_4CHO\cdot H_2O$	182.17	col. mm.	$1.471^{20/4}$	126		60.8^{21}				∞
Ricinoleic acid	$C_{17}H_{32}(OH)CO_2H$	298.45	col. lq.	0.954^{16}	4-5	$226\text{-}8^{10}$	i.	i.	v. s.	v. s.	s.
Rosaniline	$C_{20}H_{20}ON_3$	319.39	mn.		186 d.		v. sl. s.	v. sl. s.	sl. s.		1.05 c.
Rosolic acid	$C_{20}H_{16}O_3$	304.33	col. oil		308-10 d.		0.12^{25}	v. s.	v. s. h.	v. s.	
Saccharin	$C_6H_4(CO)(SO_2){>}NH$	183.18	rhb./aq.	$1.100^{20/4}$	225-8	subl.	0.4^{25}	3.1 c.	30.6^{25}	0.4^{25}	
Safrole (1-,3-,4-)	$CH_2{:}CHCH_2\cdot C_6H_3\cdot O_2\cdot CH_2$	162.18	cr.	$1.100^{20/4}$	11.2	233-4	i.	i.	i.	i.	∞
(iso-)(1-,3-,4-)	$CH_3\cdot CH{:}CH\cdot C_6H_3\cdot O_2\cdot CH_2$	162.18	col. mm.	$1.122^{20/4}$	6-7	252-3	i.	s.	i.	s.	∞
Salicylic acid (o-)	$HO\cdot C_6H_4\cdot CO_2H$	138.12	col. mm.	$1.443^{20/4}$	159	211^{20}	0.2^{23}	∞	i.	∞	51^{15}
aldehyde (o-)	$HO\cdot C_6H_4\cdot CHO$	122.12	col. oil	$1.153^{20/4}$	-7	196.5	1.7^{96}	49^{15}	0.2^{23}		
Saligenin	$HO\cdot C_6H_4\cdot CH_2OH$	124.13	rhb./aq.	1.161^{25}	86-7	subl.	6.6^{15}	6.6^{15}	1.7^{96}	v. s.	
Schaeffer's salt, Ca	$(HOC_{10}H_5SO_3)_2Ca\cdot 5H_2O$	576.59	cr.				4.76^{20}	v. s.			
K	$HOC_{10}H_5SO_3K$	262.31	cr.				3.46^{25}	v. sl. s.			
Na	$HOC_{10}H_5SO_3Na$	246.21	pr./al.				6.29^{25}				
Semicarbazide hydrochloride	$NH_2\cdot CO\cdot NH\cdot NH_2\cdot HCl$	111.54	pr.		173 d.		v. s.	v. s.	v. s.	i.	i.
Skatole (3-)	$CH_3\cdot C_9H_6N$	131.17	lf.		95	$265\text{-}6^{765}$	v. s.	sl. s.	v. s.	i.	s.
Sodium methylate	CH_3ONa	54.03	pd.		d. 300	subl.	0.05 c.	d.	d.		
Sorbitol	$[CH_2OH(CHOH)_2]_2$	182.17	cr.		110-2		d.	v. s. h.	v. s.	v. s. h.	i.
Sorbose (d- or l-)	$C_6H_{12}O_6$	180.16	rhb.	1.654^{15}	165	d. 155-8	55^{17}		0.4^{15}	sl. s. h.	6^d
Starch	$(C_6H_{10}O_5)x$	162.14	amor.	1.50^{21}	d.	subl.	i.	i.	10^{15}	1^{15}	s. h.
Stearic acid	$CH_3(CH_2)_{16}CO_2H$	284.47	col. cr.	$0.847^{70/3}$	70-1	291^{110}	0.03^{25}	2^{20}	2^{20}	v. s.	∞
amide	$CH_3(CH_2)_{16}CONH_2$	283.48	col. cr.		108-9	251^{12}	i.	s. h.	s. h.	v. s.	0.8^{15}
Styrene	$C_6H_5CH{:}CH_2$	104.14	col. cr.	$0.903^{20/4}$	-31	145-6	v. sl. s.	v. sl. s.	∞	∞	0.8^{15}
Suberic acid	$HO_2C(CH_2)_6CO_2H$	174.19	col./aq.	$1.266^{25/4}$	140-4	279^{100}	0.14^{16}	6.8^{20}	6.8^{20}	9.9^{15}	1.2^{15}
Succinic acid	$HO_2C(CH_2)_2CO_2H$	118.09	col. mm.	$1.572^{25/4}$	189-90	235 d.	6.8^{20}	179^{0}	179^{0}	0.9	i.
Sucrose	$C_{12}H_{22}O_{11}$	342.30	col. cr.	1.588^{15}	170-86 d.		179^{0}	0.9	0.8^{0}		v. sl. s.
Sulfanilic acid (p-)	$H_2N\cdot C_6H_4\cdot SO_3H$	173.18	lq.		d. >280	176-7	0.8^{10}	v. sl. s.			
Sylvestrene (d-)	$C_{10}H_{16}$	136.23	cr.	$0.863^{20/4}$							
Tartaric acid (meso-)	$(CHOHCO_2H)_2$	150.09	tri.	1.737	159-60		120^{15}	120^{15}	v. s.	v. s.	i.
(racemic)	$(CHOHCO_2H)_2\cdot H_2O$	168.10	mn.	$1.697^{20/4}$	205-6		20.6^{20}	20.6^{20}	v. s.	s.	i.
(d- or l-)	$(CHOHCO_2H)_2$	150.09	pr./aq.	$1.760^{20/4}$	168-70	d.	139^{20}	139^{20}	v. s.	sl. s. h.	s.
Tartronic acid	$CH(OH)(CO_2H)_2$	120.06	rhb.	1.510	117	subl.	20.6^{20}	25^{15}	0.001 c.	10^{15}	i.
Terephthalic acid (p-)	$C_6H_4(CO_2H)_2$	166.13	cr.	0.935^{15}	38-40	$219\text{-}21$	0.001 c.	0.4^{15}	0.4^{15}	0.4^{15}	1^{15}
Terpin hydrate (cis-)	$C_{10}H_{18}(OH)_2\cdot \tfrac{1}{2}H_2O$	190.28	col. cr.	$0.935^{20/20}$	35	$218\text{-}9^{72}$	0.4^{15}	i.	i.	i.	v. s.
Terpineol (α-)(d- or l-)	$C_{10}H_{17}OH$	154.24	col. lq.	$0.966^{20/4}$	<-50	220 d.	i.	i.	v. s.	v. s.	v. s.
(dl-)	$C_{10}H_{17}OH$	154.24	col. lq.		-1.0	151^{54}	i.	v. s.	i.	v. s.	20
Terpinyl acetate (α-)(dl-)	$CH_3\cdot CO_2\cdot C_{10}H_{17}$	196.28	col. lq.		0	104^{43}	i.	i.	s.	s.	
Tetrabromo-ethane (sym)	$Br_2CH\cdot CHBr_2$	345.70	col. lq.	$2.964^{20/4}$	-36	146.3		s.	∞	∞	∞
Tetrachloro-ethane (sym)	$Cl_2CH\cdot CHCl_2$	167.86	col. lq.	$2.875^{20/4}$		129-30	0.29^{20}	0.29^{20}	∞	∞	s. s.
-ethylene	$Cl_2C{:}CCl_2$	165.85	cr.	$1.624^{15/4}$	-19	120.8	0.02^{20}	0.02^{20}	∞	∞	v. s.
Tetracosane (n-)	$CH_3(CH_2)_{22}CH_3$	338.64	col. lq.	$0.779^{53/4}$	51.1	324	i.				
Tetradecane (n-)	$CH_3(CH_2)_{12}CH_3$	198.38	cr.	$0.765^{20/4}$	5.5	252.5	i.				
Tetraethyl-thiuram disulfide	$[(C_2H_5)_2NCS]_2S_2$	296.52	cr.	1.17	70						

TABLE 2-2 Physical Properties of Organic Compounds (*Concluded*)

Name	Formula	Formula weight	Form and color	Specific gravity	Melting point, °C	Boiling point, °C	Solubility in 100 parts — Water	Alcohol	Ether
Tetrafluoro-ethylene	$F_2C{:}CF_2$	100.02	gas	1.58^{-78}	−142.5	−76.3	0.01^{30}	s.	s.
Tetrahydro-furan	—$CH_2(CH_2)_2CH_2{\cdot}O$—	72.10	col. lq.	$0.888^{21/4}$	−65	65-6	s.	s.	∞
-furfuryl alcohol	$C_4H_7O{\cdot}CH_2OH$	102.13	lq.	$1.050^{20/4}$		$177\text{-}8^{743}$	∞	∞	∞
-pyran	—$CH_2(CH_2)_3CH_2{\cdot}O$—	86.13	lq.	$0.881^{20/4}$	−31	88	i.	s.	s.
Tetralin	—$C_6H_4CH_2(CH_2)_2CH_2$—	132.20	col. lq.	$0.973^{18/4}$		206^{764}	i.	∞	∞
Tetramethyl-thiuram disulfide	$[(CH_3)_2NCS]_2S_2$	240.41	cr.	1.29	155-6	expl.	i.	s. h.	0.03 h.
Tetryl (2-4-6-)	$(NO_2)_3C_6H_2{\cdot}N(CH_3)NO_2$	287.15	yel. mn.	1.57^{19}	130.5		0.06^{15}	0.06 c.	
Theobromine	$C_7H_8O_2N_4$	180.17	rhb.		330		sl. s. h.	s.	
Thio-acetic acid	$CH_3{\cdot}CO{\cdot}SH$	76.11	yel. lq.	1.074^{10}	<−17	93			∞
-aniline (4, 4'-)	$(NH_2{\cdot}C_6H_4)_2S$	216.29	nd./aq.		108		i.	v. s.	v. s.
-carbanilide	$(C_6H_5{\cdot}NH)_2CS$	228.30	rhb./al.	1.3^{34}	154	d.	i.	v. s.	v. s.
-naphthol (β-)	$C_{10}H_7{\cdot}SH$	160.22	cr/al.	$1.074^{23/4}$	81	286-8	v. sl. s.	v. s.	v. s.
-phenol	$C_6H_5{\cdot}SH$	110.17	col. lq.			168-9	v. sl. s.	s.	v. s.
-salicylic acid (o-)	$HS{\cdot}C_6H_4{\cdot}CO_2H$	154.18	yel. nd.	$1.405^{20/4}$	164	subl.	sl. s. h.	s.	sl. s.
-urea	$NH_2{\cdot}CS{\cdot}NH_2$	76.12	rhb./al.	$1.070^{15/4}$	180-2	d.	9.2^{13}	s.	s.
Thiophene	$<(CH{:}CH)_2> S$	84.13	col. lq.		−30	84			
Thymol (5.-2.-1.-)	$(CH_3)(C_3H_7)C_6H_3{\cdot}OH$	150.21	cr.		51.5	232^{752}	0.09^{19}	s.	v. s.
-Tolidine (o-)(3.-3'.-4.-4'.-)	$[CH_3(NH_2)C_6H_3]_2$	212.28	lf.		128-9		v. sl. s.	v. s.	s.
Toluene	$C_6H_5{\cdot}CH_3$	92.13	col. lq.	$0.866^{20/4}$	−95	110.8	0.05^{16}	∞	∞
sulfonic acid (o-)	$CH_3{\cdot}C_6H_4SO_3H{\cdot}2H_2O$	208.23	cr.		d.	128.8^0	v. s.	s.	
(p-)	$CH_3{\cdot}C_6H_4{\cdot}SO_3H{\cdot}H_2O$	190.21	mn.		104-5	$146\text{-}7^0$	v. s.	7.4^5	
sulfonic amide (p-)	$CH_3{\cdot}C_6H_4{\cdot}SO_2{\cdot}NH_2$	171.21	mn.		137		0.2^9		
sulfonic chloride (p-)	$CH_3{\cdot}C_6H_4{\cdot}SO_2Cl$	190.64	tri.		69	134.5^{10}	i.		s.
Toluic acid (o-)	$CH_3{\cdot}C_6H_4{\cdot}CO_2H$	136.14	cr/aq.	$1.062^{115/4}$	104-5	259^{751}	2.17^{100}	v. s.	v. s.
(m-)	$CH_3{\cdot}C_6H_4{\cdot}CO_2H$	136.14	pr/aq.	$1.054^{112/4}$	110-1	263	1.6^{100}	v. s.	v. s.
(p-)	$CH_3{\cdot}C_6H_4{\cdot}CO_2H$	136.14	cr/aq.		179-80	274-5	1.3^{100}	v. s.	v. s.
Toluidine (o-)	$CH_3{\cdot}C_6H_4{\cdot}NH_2$	107.15	col. lq.	$0.999^{20/4}$	−16.3	199.7	1.5^{25}	∞	∞
(m-)	$CH_3{\cdot}C_6H_4{\cdot}NH_2$	107.15	col. lq.	$0.989^{20/4}$	−31.5	203.3	0.74^{21}	v. s.	v. s.
(p-)	$CH_3{\cdot}C_6H_4{\cdot}NH_2$	107.15	cr.	$1.046^{20/4}$	44-5	200.3	s.	sl. s.	v. s.
hydrochloride (o-)	$CH_3{\cdot}C_6H_4{\cdot}NH_3Cl$	143.62	mn. pr.		218-20	242	s.		
sulfonic acid (1.-2.-3.-)	$CH_3(NH_2{\cdot}C_6H_3)SO_3H$	187.21	cr.				s. h.		
Toluylenediamine (1.-2.-4.-)	$CH_3{\cdot}C_6H_3(NH_2)_2$	122.17	lq.	1.23^{25}	99	283-5	d.	d.	s.
Tolylene diisocyanate (1.-2.-4.-)	$CH_3{\cdot}C_6H_3(NCO)_2$	174.15	rhb./al.		97	134.5^{30}	s. h.	s. h.	i.
Trehalose	$C_{12}H_{22}O_{11}{\cdot}2H_2O$	378.33					d.	d.	
Triamylamine (n-)	$[CH_3(CH_2)_3CH_2]_3N$	227.42	col. lq.	$0.786^{20/4}$		240-5	i.		s.
(i-)	$[(CH_3)_2CH(CH_2)_2]_3N$	227.42	col. lq.	$0.778^{20/20}$		235	i.		i.
Tributyl-amine (n-)	$[CH_3(CH_2)_2CH_2]_3N$	185.34	lq.			216.5^{761}	i.	s.	s.
phosphite	$[CH_3(CH_2)_3O]_3P$	250.32	lq.	$0.925^{20/4}$		$122\text{-}3^{12}$	i.		s.
Trichloro-acetic acid	$Cl_3C{\cdot}CO_2H$	163.40	cr.	$1.617^{46/15}$	58	195.5^{734}	120^{25}	v. s.	s.
-benzene (s-)(1.-3.-5.-)	$C_6H_3Cl_3$	181.46	nd.	$1.325^{26/4}$	63.5	208.5^{784}	i.	sl. s.	sl. s.
-ethane (1.-1.-1.-)	$Cl_3C{\cdot}CH_3$	133.42	lq.	$1.466^{20/20}$	−73	74.1	i.	s.	s.
-ethylene	$Cl_2C{:}CHCl$	131.40	lq.	$1.490^{75/4}$		87.2	0.1^{25}	∞	∞
-phenol	$Cl_3C_6H_2{\cdot}OH$	197.46	nd.		68-9	246	0.09^{25}	v. s.	v. s.
Tricosane (n-)	$CH_3(CH_2)_{21}CH_3$	324.61	lf.	$0.779^{40/4}$	47.7	234^{45}	i.		
Tricresyl phosphate (o-)	$OP(OC_6H_4CH_3)_3$	368.36	lq.	$1.137^{20/4}$	−6.2		i.	s.	s.
Tridecane (n-)	$CH_3(CH_2)_{11}CH_3$	184.35	col. lq.	$0.757^{20/4}$	20-1	234	i.		
Triethanol amine	$(HOCH_2{\cdot}CH_2)_3N$	149.19	col. oil	$1.126^{20/20}$	−114.8	$277\text{-}9^{150}$	∞ > 19°	v. s.	v. s.
Triethyl-amine	$(CH_3CH_2)_3N$	101.19	col. oil	$0.729^{20/20}$		89.4	∞ > 19°	∞	∞
-benzene (1.-3.-5.-)	$(C_2H_5)_3C_6H_3$	162.26	lq.	$0.861^{20/4}$	−73	215	i.	∞	∞
(1.-2.-4.-)	$(C_2H_5)_3C_6H_3$	162.26	lq.	$0.882^{17/4}$	68-9	217-8	i.	∞	v. s.
borate	$B(OCH_2CH_3)_3$	146.00	oil	$0.864^{20/20}$		120	d.		
citrate	$HOC_3H_4(CO_2C_2H_5)_3$	276.28	col. lq.	$1.137^{20/4}$		294	i.	∞	∞
Triethylene glycol	$({\cdot}OCH_2CH_2OH)_2$	150.17	col. lq.	$1.125^{20/20}$	−5	290	∞	∞	s.
Trifluoro-chloromethane	CF_3Cl	104.47	gas	1.726^{-130}	−182	−80	i.		
chloroethylene	$F_2C{:}CFCl$	116.48	gas		−157.5	−27.9	d.		∞
-trichloroethane	$Cl_2CF{\cdot}CClF_2$	187.39	lq.	$1.576^{20/4}$	−35	47.6	i.	∞	∞
Trimethoxybutane (1.-3.-3.-)	$CH_3(OCH_3)_2CH_2C(OCH_3)_2CH_3$	148.20	lq.	0.932		$63\text{-}5^{25}$	i.		
Trimethylamine	$(CH_3)_3N$	59.11	gas	0.662^{-5}	−124	3.5	41^{19}	v. s.	s.
Trimethylene bromide	$BrCH_2CH_2CH_2Br$	201.91	lq.	$1.987^{15/4}$	−34.4	167.5	0.17^{30}	s.	s.
chloride	$ClCH_2CH_2CH_2Cl$	112.99	lq.	1.201^{15}		123-5	0.27^{25}	s.	s.
glycol	$HOCH_2CH_2CH_2OH$	76.09	lq.	$1.060^{20/4}$		214	∞	∞	∞
Trinitro-benzene (1.-3.-5.-)	$C_6H_3(NO_2)_3$	213.11	col. rhb.	$1.688^{20/4}$	121	d.	0.03^{15}	sl. s.	v. sl. s.
-benzoic acid (2.-4.-6.-)	$(NO_2)_3C_6H_2{\cdot}CO_2H$	257.12	rhb./aq.		210-20 d.		2.05^{24}		
-tert-butylxylene	$(NO_2)_3C_6(CH_3)_2C_4H_9$	297.26	nd./al.		110		i.	sl. s.	v. sl. s.
-naphthalene (α-)(1.-3.-5.-)	$C_{10}H_5(NO_2)_3$	263.16	rhb.		122-3		i.	1.9^{18}	1.5^{18}
(β-)(1.-3.-8.-)	$C_{10}H_5(NO_2)_3$	263.16	cr./al.		218-9		0.02^{100}	0.05^{23}	0.13^{15}
(γ-)(1.-4.-5.-)	$C_{10}H_5(NO_2)_3$	263.16	yel. cr.		148-9		i.	0.11^{19}	0.4^{19}

Name	Formula	Mol. wt.	Form	Density	m.p. °C	b.p. °C	Water	Alcohol	Ether
-phenol (2-,3-,6-)	(NO₂)₃C₆H₂OH	229.11	nd.	1.620[20/4]	117-8	expl.	s. h.	v. s.	v. s.
-toluene (β-)(2-,3-,4-)	CH₃C₆H₂(NO₂)₃	227.13	cr.	1.620[20/4]	112	expl.	i.	sl. s. c.	s.
(γ-)(2-,4-,5-)	CH₃C₆H₂(NO₂)₃	227.13	yel. pl.	1.654	104	expl.	0.01[20]	s. h.	v. s.
(α-)(2-,4-,6-)	CH₃C₆H₂(NO₂)₃	227.13	cr./al.	1.199[85/4]	80.8	expl.	0.3[15]	1.5[22]	5[33]
Trional	C₂H₅(CH₃)C(SO₂C₂H₅)₂	242.34	pl./al.	1.306	76	d.	i.	5[0]	6.6[15]
Triphenyl-arsine	(C₆H₅)₃As	306.21	cr.	1.188[20/4]	59-60	>360	i.	s.	v. s.
carbinol	(C₆H₅)₃COH	260.32	rhb./al.	1.13	162.5	>360	i.	v. s.	v. s.
guanidine (α-)	C₆H₅N:C(NHC₆H₅)₂	287.35	cr.	1.014[99/4]	144-5	d.	i.	4[0]	v. s.
methane	(C₆H₅)₃CH	244.32	col. cr.		93.4	359[75/4]		sl. s. h.	∞
methyl	(C₆H₅)₃C·	243.31	pr./al.		145-7	d.		155[25]	sl. s.
phosphate	OP(OC₆H₅)₃	326.28	col. lq.	1.206[56/4]	49-50	245[11]		∞	i.
Tripropylamine (n-)	(CH₃CH₂CH₂)₃N	143.27	col. lq.	0.757[20/4]	-93.5	156.5	v. sl. s.	∞	∞
Undecane (n-)	CH₃(CH₂)₉CH₃	156.30	col. lq.	0.741[20/4]	-25.6	194.5		∞	∞
Urea	H₂N·CO·NH₂	60.06	col. pr.	1.335[20/4]	132.7	d.	100[17]	20[20]	s.
nitrate	CO(NH₂)₂·HNO₃	123.07	col. mn.		152 d.			s.	s.
Uric acid	C₅H₄O₃N₄	168.11	cr.	1.893[20]	d.		0.06 h.	i.	v. s.
Valeric acid (n-)	C₂H₅CH₂CH₂CO₂H	102.13	col. lq.	0.939[20/4]	-34.5	187	3.3[16]	∞	v. s.
(i-)	(CH₃)₂CHCH₂CO₂H	102.13	lq.	0.931[20/20]	-37.6	176	4.2[20]	∞	v. s.
aldehyde (n-)	C₂H₅CH₂CH₂CHO	86.13	col. lq.	0.819[11]	-92	103.4	sl. s.	s.	v. s.
(i-)	(CH₃)₂CHCH₂CHO	86.13	mn. pl.	0.803[17]	-51	92.5	v. s.	v. s.	s.
amide (n-)	C₂H₅CH₂CH₂CONH₂	101.15	mn.	1.023	106	232	s.	s.	∞
(i-)	(CH₃)₂CHCH₂CONH₂	101.15	nd./aq.	0.965[20/4]	135-7	subl.	0.12[14]	v. s. h.	∞
Vanillic acid (3-,4-,1-)	CH₃O(OH)C₆H₃·CO₂H	168.14	mn./aq.		207	d.	v. s. h.	v. s.	∞
alcohol (3-,4-,1-)	CH₃O(OH)C₆H₃·CH₂OH	154.16	cr.		115		i.	v. s.	v. s.
hyl-thiuram disulfide	[(C₂H₅)₂NCS]₂S₂	296.52	mn.	1.17	70		1[14]	v. s.	∞
Vanillin (3-,4-,1-)	CH₃O(OH)C₆H₃·CHO	152.14	col. lq.	1.056	81-2	285	2[20]	∞	∞
Veratrole (o-)	C₆H₄(OCH₃)₂	138.16	col. lq.	1.091[15/15]	22.5	207.1	i.	∞	v. s.
Vinyl acetate	CH₃CO₂CH:CH₂	86.09	col. lq.	0.932[20/4]	<-60	72-3	0.67[0.6]	s.	s.
(poly-)	(CH₃CO₂CH:CH₂)x	(86.09)		1.19[20]	100-25			s.	i.
acetic acid	CH₂:CHCH₂CO₂H	86.09	lq.	1.013[15/15]	-39	163	s.	s.	v. sl. s.
acetylene	CH₂:CH·C:CH	52.07	gas	0.705[15]		5.5	sl. s.	v. s.	
alcohol	CH₂:CHOH	44.06					v. sl. s.		
(poly-)	(CH₂:CHOH)x	(44.06)		1.3[20]	d. >200				
chloride	CH₂:CHCl	62.50	lq.	0.908[25/25]	-160	-12			
(poly-)	(CH₂:CHCl)x								
propionate	C₂H₅CO₂CH:CH₂	100.11	col. lq.			93-5			
Xylene (o-)	C₆H₄(CH₃)₂	106.16	col. lq.	0.881[20/4]	-25	144	i.		
(m-)	C₆H₄(CH₃)₂	106.16	col. lq.	0.867[17/4]	-47.4	139.3	i.		
(p-)	C₆H₄(CH₃)₂	106.16	col. lf.	0.861[20/4]	13.2	138.5			
sulfonic acid (1-,4-,2-)	(CH₃)₂C₆H₃SO₃H·2H₂O	222.25	lq.		86	149[0.1]	s.		
Xylidine (1:2)(3-)	(CH₃)₂C₆H₃NH₂	121.18	pr.	0.991[15]	<-15	223	v. sl. s.		
(1:2)(4-)	(CH₃)₂C₆H₃NH₂	121.18	lq.	1.076[17/5]	49-50	224-6	v. sl. s.		
(1:3)(2-)	(CH₃)₂C₆H₃NH₂	121.18	lq.	0.980[15]		216-7	v. sl. s.		
(1:3)(4-)	(CH₃)₂C₆H₃NH₂	121.18	oil	0.978[20/4]		213-4	v. sl. s.		
(1:3)(5-)	(CH₃)₂C₆H₃NH₂	121.18	oil	0.973[20/4]		221-2	v. sl. s.		
(1:4)(2-)	(CH₃)₂C₆H₃NH₂	121.18		0.979[20/4]	15.5	215[789]	v. sl. s.		
Xylose (l-)(+)	CH₂OH(CHOH)₃CHO	150.13	nd.	1.535[0]	153-4		117[20]	s.	
Xylylene dichloride (p-)	C₆H₄(CH₂Cl)₂	175.06	mn.	1.417[0]	100.5	240-5 d.	i.		
Zinc diethyl	Zn(C₂H₅)₂	123.50	col. lq.	1.182[18]	-28	118	d.	v. sl. s.	v. sl. s.
dimethyl	Zn(CH₃)₂	95.45	col. lq.	1.386[11]	-40	46	d.	d.	s.
dimethyl-dithiocarbamate	Zn[S₂CN(CH₃)₂]₂	305.79		2.00[49/4]	248-50		i.		

NOTE: °F = ⅘ °C + 32.

VAPOR PRESSURES OF PURE SUBSTANCES

UNITS CONVERSIONS

For this subsection, the following units conversions are applicable:

$$°F = \frac{9}{5} °C + 32.$$

To convert millimeters of mercury to pounds-force per square inch, multiply by 0.01934.

ADDITIONAL REFERENCES

Additional compilations of vapor-pressure data include Boublik, Fried, and Hala, *The Vapor Pressures of Pure Substances*, Elsevier, Amsterdam, 1984. See also Hirata, Ohe, and Nagahama, *Computer Aided Data Book of Vapor-Liquid Equilibria*, Kodansha/Elsevier, Tokyo, 1975; Weishaupt, *Landolt-Börnstein New Series Group IV*, vol. 3; *Thermodynamic Equilibria of Boiling Mixtures*, Springer-Verlag, Berlin, 1975; Wichterle, Linek, and Hala, *Vapor-Liquid Equilibrium Data Bibliography*, Elsevier, Amsterdam, 1973; suppl. 1, 1976; suppl. 2, 1982.

TABLE 2-3   Vapor Pressure of Water Ice from −15 to 0°C*

mmHg

t, °C	0.0	0.1	0.2	0.3	0.4	0.5	0.6	0.7	0.8	0.9
−14	1.361	1.348	1.336	1.324	1.312	1.300	1.288	1.276	1.264	1.253
−13	1.490	1.477	1.464	1.450	1.437	1.424	1.411	1.399	1.386	1.373
−12	1.632	1.617	1.602	1.588	1.574	1.559	1.546	1.532	1.518	1.504
−11	1.785	1.769	1.753	1.737	1.722	1.707	1.691	1.676	1.661	1.646
−10	1.950	1.934	1.916	1.899	1.883	1.866	1.849	1.833	1.817	1.800
−9	2.131	2.112	2.093	2.075	2.057	2.039	2.021	2.003	1.985	1.968
−8	2.326	2.306	2.285	2.266	2.246	2.226	2.207	2.187	2.168	2.149
−7	2.537	2.515	2.493	2.472	2.450	2.429	2.408	2.387	2.367	2.346
−6	2.765	2.742	2.718	2.695	2.672	2.649	2.626	2.603	2.581	2.559
−5	3.013	2.987	2.962	2.937	2.912	2.887	2.862	2.838	2.813	2.790
−4	3.280	3.252	3.225	3.198	3.171	3.144	3.117	3.091	3.065	3.039
−3	3.568	3.539	3.509	3.480	3.451	3.422	3.393	3.364	3.336	3.308
−2	3.880	3.848	3.816	3.785	3.753	3.722	3.691	3.660	3.630	3.599
−1	4.217	4.182	4.147	4.113	4.079	4.045	4.012	3.979	3.946	3.913
−0	4.579	4.542	4.504	4.467	4.431	4.395	4.359	4.323	4.287	4.252

*For data at 0(0.2)–30(2)–98°C see p. 2324, *Handbook of Chemistry and Physics*, 40th ed., Chemical Rubber Publishing Co.

TABLE 2-4   Vapor Pressure of Liquid Water from −16 to 0°C*

mmHg

t, °C	0.0	0.1	0.2	0.3	0.4	0.5	0.6	0.7	0.8	0.9
−15	1.436	1.425	1.414	1.402	1.390	1.379	1.368	1.356	1.345	1.334
−14	1.560	1.547	1.534	1.522	1.511	1.497	1.485	1.472	1.460	1.449
−13	1.691	1.678	1.665	1.651	1.637	1.624	1.611	1.599	1.585	1.572
−12	1.834	1.819	1.804	1.790	1.776	1.761	1.748	1.734	1.720	1.705
−11	1.987	1.971	1.955	1.939	1.924	1.909	1.893	1.878	1.863	1.848
−10	2.149	2.134	2.116	2.099	2.084	2.067	2.050	2.034	2.018	2.001
−9	2.326	2.307	2.289	2.271	2.254	2.236	2.219	2.201	2.184	2.167
−8	2.514	2.495	2.475	2.456	2.437	2.418	2.399	2.380	2.362	2.343
−7	2.715	2.695	2.674	2.654	2.633	2.613	2.593	2.572	2.553	2.533
−6	2.931	2.909	2.887	2.866	2.843	2.822	2.800	2.778	2.757	2.736
−5	3.163	3.139	3.115	3.092	3.069	3.046	3.022	3.000	2.976	2.955
−4	3.410	3.384	3.359	3.334	3.309	3.284	3.259	3.235	3.211	3.187
−3	3.673	3.647	3.620	3.593	3.567	3.540	3.514	3.487	3.461	3.436
−2	3.956	3.927	3.898	3.871	3.841	3.813	3.785	3.757	3.730	3.702
−1	4.258	4.227	4.196	4.165	4.135	4.105	4.075	4.045	4.016	3.986
−0	4.579	4.546	4.513	4.480	4.448	4.416	4.385	4.353	4.320	4.289

*Computed from the above table with the aid of the thermodynamic equation

$$\log_{10} \frac{p_w}{p_i} = \frac{-1.1489t}{273.1 + t} - 1.330 \times 10^{-5}t^2 + 9.084 \times 10^{-8}t^3$$

TABLE 2-5 Vapor Pressure of Liquid Water from 0 to 100°C*

mmHg

t, °C	0.0	0.1	0.2	0.3	0.4	0.5	0.6	0.7	0.8	0.9
0	4.579	4.613	4.647	4.681	4.715	4.750	4.785	4.820	4.855	4.890
1	4.926	4.962	4.998	5.034	5.070	5.107	5.144	5.181	5.219	5.256
2	5.294	5.332	5.370	5.408	5.447	5.486	5.525	5.565	5.605	5.645
3	5.685	5.725	5.766	5.807	5.848	5.889	5.931	5.973	6.015	6.058
4	6.101	6.144	6.187	6.230	6.274	6.318	6.363	6.408	6.453	6.498
5	6.543	6.589	6.635	6.681	6.728	6.775	6.822	6.869	6.917	6.965
6	7.013	7.062	7.111	7.160	7.209	7.259	7.309	7.360	7.411	7.462
7	7.513	7.565	7.617	7.669	7.722	7.775	7.828	7.882	7.936	7.990
8	8.045	8.100	8.155	8.211	8.267	8.323	8.380	8.437	8.494	8.551
9	8.609	8.668	8.727	8.786	8.845	8.905	8.965	9.025	9.086	9.147
10	9.209	9.271	9.333	9.395	9.458	9.521	9.585	9.649	9.714	9.779
11	9.844	9.910	9.976	10.042	10.109	10.176	10.244	10.312	10.380	10.449
12	10.518	10.588	10.658	10.728	10.799	10.870	10.941	11.013	11.085	11.158
13	11.231	11.305	11.379	11.453	11.528	11.604	11.680	11.756	11.833	11.910
14	11.987	12.065	12.144	12.223	12.302	12.382	12.462	12.543	12.624	12.706
15	12.788	12.870	12.953	13.037	13.121	13.205	13.290	13.375	13.461	13.547
16	13.634	13.721	13.809	13.898	13.987	14.076	14.166	14.256	14.347	14.438
17	14.530	14.622	14.715	14.809	14.903	14.997	15.092	15.188	15.284	15.380
18	15.477	15.575	15.673	15.772	15.871	15.971	16.071	16.171	16.272	16.374
19	16.477	16.581	16.685	16.789	16.894	16.999	17.105	17.212	17.319	17.427
20	17.535	17.644	17.753	17.863	17.974	18.085	18.197	18.309	18.422	18.536
21	18.650	18.765	18.880	18.996	19.113	19.231	19.349	19.468	19.587	19.707
22	19.827	19.948	20.070	20.193	20.316	20.440	20.565	20.690	20.815	20.941
23	21.068	21.196	21.324	21.453	21.583	21.714	21.845	21.977	22.110	22.243
24	22.377	22.512	22.648	22.785	22.922	23.060	23.198	23.337	23.476	23.616
25	23.756	23.897	24.039	24.182	24.326	24.471	24.617	24.764	24.912	25.060
26	25.209	25.359	25.509	25.660	25.812	25.964	26.117	26.271	26.426	26.582
27	26.739	26.897	27.055	27.214	27.374	27.535	27.696	27.858	28.021	28.185
28	28.349	28.514	28.680	28.847	29.015	29.184	29.354	29.525	29.697	29.870
29	30.043	30.217	30.392	30.568	30.745	30.923	31.102	31.281	31.461	31.642
30	31.824	32.007	32.191	32.376	32.561	32.747	32.934	33.122	33.312	33.503
31	33.695	33.888	34.082	34.276	34.471	34.667	34.864	35.062	35.261	35.462
32	35.663	35.865	36.068	36.272	36.477	36.683	36.891	37.099	37.308	37.518
33	37.729	37.942	38.155	38.369	38.584	38.801	39.018	39.237	39.457	39.677
34	39.898	40.121	40.344	40.569	40.796	41.023	41.251	41.480	41.710	41.942
35	42.175	42.409	42.644	42.880	43.117	43.355	43.595	43.836	44.078	44.320
36	44.563	44.808	45.054	45.301	45.549	45.799	46.050	46.302	46.556	46.811
37	47.067	47.324	47.582	47.841	48.101	48.364	48.627	48.891	49.157	49.424
38	49.692	49.961	50.231	50.502	50.774	51.048	51.323	51.600	51.879	52.160
39	52.442	52.725	53.009	53.294	53.580	53.867	54.156	54.446	54.737	55.030
40	55.324	55.61	55.91	56.21	56.51	56.81	57.11	57.41	57.72	58.03
41	58.34	58.65	58.96	59.27	59.58	59.90	60.22	60.54	60.86	61.18
42	61.50	61.82	62.14	62.47	62.80	63.13	63.46	63.79	64.12	64.46
43	64.80	65.14	65.48	65.82	66.16	66.51	66.86	67.21	67.56	67.91
44	68.26	68.61	68.97	69.33	69.69	70.05	70.41	70.77	71.14	71.51
45	71.88	72.25	72.62	72.99	73.36	73.74	74.12	74.50	74.88	75.26
46	75.65	76.04	76.43	76.82	77.21	77.60	78.00	78.40	78.80	79.20
47	79.60	80.00	80.41	80.82	81.23	81.64	82.05	82.46	82.87	83.29
48	83.71	84.13	84.56	84.99	85.42	85.85	86.28	86.71	87.14	87.58
49	88.02	88.46	88.90	89.34	89.79	90.24	90.69	91.14	91.59	92.05

t, °C	0	1	2	3	4	5	6	7	8	9
50	92.51	97.20	102.09	107.20	112.51	118.04	123.80	129.82	136.08	142.60
60	149.38	156.43	163.77	171.38	179.31	187.54	196.09	204.96	214.17	223.73
70	233.7	243.9	254.6	265.7	277.2	289.1	301.4	314.1	327.3	341.0
80	355.1	369.7	384.9	400.6	416.8	433.6	450.9	468.7	487.1	506.1
90	525.76	527.76	529.77	531.78	533.80	535.82	537.86	539.90	541.95	544.00
91	546.05	548.11	550.18	552.26	554.35	556.44	558.53	560.64	562.75	564.87
92	566.99	569.12	571.26	573.40	575.55	577.71	579.87	582.04	584.22	586.41
93	588.60	590.80	593.00	595.21	597.43	599.66	601.89	604.13	606.38	608.64
94	610.90	613.17	615.44	617.72	620.01	622.31	624.61	626.92	629.24	631.57
95	633.90	636.24	638.59	640.94	643.30	645.67	648.05	650.43	652.82	655.22
96	657.62	660.03	662.45	664.88	667.31	669.75	672.20	674.66	677.12	679.69
97	682.07	684.55	687.04	689.54	692.05	694.57	697.10	699.63	702.17	704.71
98	707.27	709.83	712.40	714.98	717.56	720.15	722.75	725.36	727.98	730.61
99	733.24	735.88	738.53	741.18	743.85	746.52	749.20	751.89	754.58	757.29
100	760.00	762.72	765.45	768.19	770.93	773.68	776.44	779.22	782.00	784.78
101	787.57	790.37	793.18	796.00	796.82	801.66	801.66	804.50	807.35	810.21	813.06

*From the Physikalisch-technische Reichsanstalt, Holborn, Scheel, and Henning, *Wärmetabellen*, Friedrich Vieweg & Sohn, Brunswick, 1909. By permission. For data at 50(0.2)101.8°C, see *Handbook of Chemistry and Physics*, 40th ed., p. 2326, Chemical Rubber Publishing Co. For a tabulation of temperature for pressures 700(1)779 mm Hg, see Atack, *Handbook of Chemical Data*, p. 117, Reinhold, New York, 1957. For a tabulation of pressure for 105(5)200(10)370°C, see Atack, p. 134, and for 100(1)374°C, see *Handbook of Chemistry and Physics*, 40th ed., pp. 2328–2330, Chemical Rubber Publishing Co.

2-49

TABLE 2-6 Vapor Pressure of Inorganic and Organic Liquids

Cmpd. no.	Name	Formula	CAS no.	C1	C2	C3	C4	C5	T_{min}, K	P_s at T_{min}	T_{max}, K	P_s at T_{max}
1	Methane	CH_4	74828	39.205	−1324.4	−3.4366	3.1019E-05	2	90.69	1.1667E+04	190.56	4.5897E+06
2	Ethane	C_2H_6	74840	51.857	−2598.7	−5.1283	1.4913E-05	2	90.35	1.1273E+00	305.32	4.8522E+06
3	Propane	C_3H_8	74986	59.078	−3492.6	−6.0669	1.0919E-05	2	85.47	1.6788E-04	369.83	4.2135E+06
4	n-Butane	C_4H_{10}	106978	66.343	−4363.2	−7.046	9.4509E-06	2	134.86	6.7441E-01	425.12	3.7699E+06
5	n-Pentane	C_5H_{12}	109660	78.741	−5420.3	−8.8253	9.6171E-06	2	143.42	6.8642E-02	469.7	3.3642E+06
6	n-Hexane	C_6H_{14}	110543	104.65	−6995.5	−12.702	1.2381E-05	2	177.83	9.0169E-01	507.6	3.0449E+06
7	n-Heptane	C_7H_{16}	142825	87.829	−6996.4	−9.8802	7.2099E-06	2	182.57	1.8269E-01	540.2	2.7192E+06
8	n-Octane	C_8H_{18}	111659	96.084	−7900.2	−11.003	7.1802E-06	2	216.38	2.1083E+00	568.7	2.4673E+06
9	n-Nonane	C_9H_{20}	111842	109.35	−9030.4	−12.882	7.8544E-06	2	219.66	4.3058E-01	594.6	2.3054E+06
10	n-Decane	$C_{10}H_{22}$	124185	112.73	−9749.6	−13.245	7.1266E-06	2	243.51	1.3930E+00	617.7	2.0908E+06
11	n-Undecane	$C_{11}H_{24}$	1120214	131	−11143	−15.855	8.1871E-06	2	247.57	4.0836E-01	639	1.9493E+06
12	n-Dodecane	$C_{12}H_{26}$	112403	137.47	−11976	−16.698	8.0906E-06	2	263.57	6.1534E-01	658	1.8223E+06
13	n-Tridecane	$C_{13}H_{28}$	628505	137.45	−12549	−16.543	7.1275E-06	2	267.76	2.5096E-01	675	1.6786E+06
14	n-Tetradecane	$C_{14}H_{30}$	629594	140.47	−13231	−16.859	6.5877E-06	2	279.01	2.5268E-01	693	1.5693E+06
15	n-Pentadecane	$C_{15}H_{32}$	629629	135.57	−13478	−16.022	5.6136E-06	2	283.07	1.2984E-01	708	1.4743E+06
16	n-Hexadecane	$C_{16}H_{34}$	544763	156.06	−15015	−18.941	6.8172E-06	2	291.31	9.2265E-02	723	1.4106E+06
17	n-Heptadecane	$C_{17}H_{36}$	629787	156.95	−15557	−18.966	6.4559E-06	2	295.13	4.6534E-02	736	1.3438E+06
18	n-Octadecane	$C_{18}H_{38}$	593453	157.68	−16093	−18.954	5.9272E-06	2	301.31	3.3909E-02	747	1.2555E+06
19	n-Nonadecane	$C_{19}H_{40}$	629925	182.54	−17897	−22.498	7.4008E-06	2	305.04	1.5909E-02	758	1.2078E+06
20	n-Eicosane	$C_{20}H_{42}$	112958	203.66	−19441	−25.525	8.8382E-06	2	309.58	9.2574E-03	768	1.1746E+06
21	2-Methylpropane	C_4H_{10}	75285	100.18	−4841.9	−13.541	2.0063E-02	1	113.54	1.4051E-02	408.14	3.6199E+06
22	2-Methylbutane	C_5H_{12}	78784	72.35	−5010.9	−7.883	8.9795E-06	2	113.25	1.1569E-04	460.43	3.3709E+06
23	2,3-Dimethylbutane	C_6H_{14}	79298	77.235	−5695.9	−8.5109	8.0163E-06	2	145.19	1.5081E-02	499.98	3.1255E+06
24	2-Methylpentane	C_6H_{14}	107835	77.36	−5791.7	−8.4912	7.9399E-06	2	119.55	9.2204E-06	497.5	3.0192E+06
25	2,3-Dimethylpentane	C_7H_{16}	565593	78.282	−6347	−8.502	6.4169E-06	2	160	1.2631E-02	537.35	2.8823E+06
26	2,3,3-Trimethylpentane	C_8H_{18}	560214	83.105	−6903.7	−9.1858	6.4703E-06	2	172.22	1.6820E-02	573.5	2.8116E+06
27	2,2,4-Trimethylpentane	C_8H_{18}	540841	87.868	−6831.7	−9.9783	7.7729E-06	2	165.78	1.6187E-02	543.96	2.5630E+06
28	Ethylene	C_2H_4	74851	74.242	−2707.2	−9.8462	2.2457E-02	1	104	1.2361E+02	282.34	5.0296E+06
29	Propylene	C_3H_6	115071	57.263	−3382.4	−5.7707	1.0431E-05	2	87.89	9.3867E-04	365.57	4.6346E+06
30	1-Butene	C_4H_8	106989	68.49	−4350.2	−7.4124	1.0503E-05	2	87.8	7.1809E-07	419.95	4.0391E+06
31	cis-2-Butene	C_4H_8	590181	102.62	−5260.3	−13.764	1.9183E-02	2	134.26	2.4051E-01	435.58	4.2388E+06
32	trans-2-Butene	C_4H_8	624646	70.589	−4530.4	−7.7229	1.0928E-05	2	167.62	7.4729E+01	428.63	4.0811E+06
33	1-Pentene	C_5H_{10}	109671	120.15	−6192.4	−16.597	2.1922E-02	2	107.93	3.5210E-06	464.78	3.5557E+06
34	1-Hexene	C_6H_{12}	592416	85.3	−6171.7	−9.702	8.9604E-06	2	133.39	2.5272E-04	504.03	3.1397E+06
35	1-Heptene	C_7H_{14}	592767	92.68	−7055.2	−10.679	8.4459E-06	2	154.27	1.2810E-03	537.29	2.8225E+06
36	1-Octene	C_8H_{16}	111660	97.57	−7836	−11.272	7.7267E-06	2	171.45	2.7570E-03	566.65	2.5735E+06
37	1-Nonene	C_9H_{18}	124118	144.45	−9676.2	−19.446	1.8031E-02	2	191.78	8.5514E-03	593.25	2.3308E+06
38	1-Decene	$C_{10}H_{20}$	872059	78.808	−8367.9	−7.9553	8.7442E-18	6	206.89	1.7308E-02	616.4	2.2092E+06
39	2-Methylpropene	C_4H_8	115117	102.5	−5021.8	−13.88	2.0296E-02	1	132.81	6.2213E-01	417.9	3.9760E+06
40	2-Methyl-1-butene	C_5H_{10}	563462	97.33	−5631.8	−12.589	1.5395E-02	1	135.58	1.9687E-02	465	3.4544E+06
41	2-Methyl-2-butene	C_5H_{10}	513359	82.605	−5606.6	−9.4236	1.0512E-05	2	139.39	1.9447E-02	471	3.3789E+06
42	1,2-Butadiene	C_4H_6	590192	39.714	−3769.9	−2.6407	6.9379E-18	6	136.95	4.4720E-01	452	4.3613E+06
43	1,3-Butadiene	C_4H_6	106990	73.522	−4564.3	−8.1958	1.1580E-05	2	164.25	6.9110E+01	425.17	4.3041E+06
44	2-Methyl-1,3-butadiene	C_5H_8	78795	79.656	−5239.6	−9.4314	9.5550E-03	1	127.27	2.4766E-03	484	3.8509E+06
45	Acetylene	C_2H_2	74862	172.06	−5318.5	−27.223	5.4109E-02	1	192.4	1.2603E+05	308.32	6.1467E+06
46	Methylacetylene	C_3H_4	74997	119.42	−5364.5	−16.81	2.5523E-02	2	170.45	3.7264E+02	402.39	5.6206E+06
47	Dimethylacetylene	C_4H_6	503173	66.592	−4999.8	−6.8387	6.6793E-06	2	240.91	6.1128E+03	473.2	4.8699E+06
48	3-Methyl-1-butyne	C_5H_8	598232	69.459	−5250	−7.1125	7.9289E-17	6	183.45	4.3351E+01	463.2	4.1986E+06
49	1-Pentyne	C_5H_8	627190	82.805	−5683.8	−9.4301	1.0767E-05	2	167.45	2.3990E+00	481.2	4.1701E+06
50	2-Pentyne	C_5H_8	627214	137.29	−7447.1	−19.01	2.1415E-02	1	163.83	2.0462E-01	519	4.0198E+06
51	1-Hexyne	C_6H_{10}	693027	133.2	−7492.9	−18.405	2.2062E-02	1	141.25	3.9157E-04	516.2	3.6352E+06
52	2-Hexyne	C_6H_{10}	764352	123.71	−7639	−16.451	1.6495E-02	1	183.65	5.4026E-01	549	3.5301E+06
53	3-Hexyne	C_6H_{10}	928494	47.091	−5104	−3.6451	5.1621E-04	1	170.05	2.1950E-01	544	3.5397E+06
54	1-Heptyne	C_7H_{12}	628717	66.447	−6395.6	−6.3848	1.1250E-17	6	192.22	6.7026E-01	559	3.1343E+06
55	1-Octyne	C_8H_{14}	629050	82.353	−7240.6	−9.1843	5.8038E-03	6	193.55	1.0092E+00	585	2.8202E+06
56	Vinylacetylene[1]	C_4H_4	689974	55.682	−4439.3	−5.0136	1.9650E-17	6	173.15	6.6899E-01	454	4.8874E+06

No.	Name	Formula									
57	Cyclopentane	C_5H_{10}	287923	51.434	-4770.6	-4.3515	1.9605E-17	6	179.28	9.4420E+00	511.76	4.5028E+06
58	Methylcyclopentane	C_6H_{12}	96377	79.673	-6086.6	-8.7933	7.4046E-06	2	130.73	6.7059E-05	532.79	3.7808E+06
59	Ethylcyclopentane	C_7H_{14}	1640897	88.622	-7011	-10.038	7.4481E-06	2	134.71	3.7061E-06	569.52	3.3970E+06
60	Cyclohexane	C_6H_{12}	110827	116.51	-7103.3	-15.49	1.6959E-02	1	279.69	5.3802E+03	553.58	4.0958E+06
61	Methylcyclohexane	C_7H_{14}	108872	92.611	-7077.8	-10.684	8.1239E-06	2	146.58	1.5256E-04	572.19	3.4828E+06
62	1,1-Dimethylcyclohexane	C_8H_{16}	590669	81.184	-6927	-8.8498	5.4580E-06	2	239.66	6.0584E+01	591.15	2.9387E+06
63	Ethylcyclohexane	C_8H_{16}	1678917	80.208	-7203.2	-8.6023	4.5901E-06	2	161.84	3.5747E-04	609.15	3.0411E+06
64	Cyclopentene	C_5H_8	142290	49.88	-4649.7	-4.1191	1.9564E-17	6	138.13	1.6884E-02	507	4.8062E+06
65	1-Methylcyclopentene	C_6H_{10}	693890	52.732	-5286.9	-4.4509	1.0883E-17	6	146.62	3.9787E-03	542	4.1303E+06
66	Cyclohexene	C_6H_{10}	110838	88.184	-6624.9	-10.059	8.2566E-06	2	169.67	1.0377E-01	560.4	4.3922E+06
67	Benzene	C_6H_6	71432	83.918	-6517.7	-9.3453	7.1182E-06	2	278.68	4.7620E+03	562.16	4.8819E+06
68	Toluene	C_7H_8	108883	80.877	-6902.4	-8.7761	5.8034E-06	2	178.18	4.2348E-02	591.8	4.1012E+06
69	o-Xylene	C_8H_{10}	95476	90.356	-7948.7	-10.081	5.9756E-06	2	247.98	2.1968E+01	630.33	3.7424E+06
70	m-Xylene	C_8H_{10}	108383	84.782	-7598.7	-9.2612	5.5445E-06	2	225.3	3.2099E+00	617.05	3.5286E+06
71	p-Xylene	C_8H_{10}	106423	85.475	-7595.8	-9.378	5.6875E-06	2	296.41	5.8144E+02	616.23	3.4984E+06
72	Ethylbenzene	C_8H_{10}	100414	88.09	-7688.3	-9.7708	5.8844E-06	2	178.15	4.0140E-03	617.2	3.5968E+06
73	Propylbenzene	C_9H_{12}	103651	136.83	-9544.8	-18.190	1.6590E-02	1	324.18	2.0014E+03	638.32	3.2001E+06
74	1,2,4-Trimethylbenzene	C_9H_{12}	95636	60.658	-7260.4	-5.3772	4.5816E-18	6	229.33	7.9735E-01	649.13	3.2533E+06
75	Isopropylbenzene	C_9H_{12}	98828	143.62	-9687.7	-19.305	1.7703E-02	1	177.14	3.8034E 04	631.1	3.1837E+06
76	1,3,5-Trimethylbenzene	C_9H_{12}	108678	48.603	-6545.2	-3.6412	1.9307E-18	6	228.42	1.1889E+00	637.36	3.1116E+06
77	p-Isopropyltoluene	$C_{10}H_{14}$	99876	107.71	-9402.7	-12.545	6.6661E-06	2	205.25	9.9261E-03	653.15	2.7957E+06
78	Naphthalene	$C_{10}H_8$	91203	62.447	-8109	-5.5571	2.0800E-18	6	353.43	9.9229E+02	748.35	3.9941E+06
79	Biphenyl	$C_{12}H_{10}$	92524	76.811	-9878.5	-7.4384	2.0436E-18	6	342.2	9.3752E+01	789.26	3.8615E+06
80	Styrene	C_8H_8	100425	105.93	-8685.9	-12.42	7.5583E-06	2	242.54	1.0613E+01	636	3.8234E+06
81	m-Terphenyl	$C_{18}H_{14}$	92068	88.044	-13367	-8.6482	8.7874E-19	6	360	1.0112E+00	924.85	3.3297E+06
82	Methanol	CH_4O	67561	81.768	-6876	-8.7078	7.1926E-06	2	175.47	1.1147E-01	512.64	8.1402E+06
83	Ethanol	C_2H_6O	64175	74.475	-7164.3	-7.327	3.1340E-06	2	159.05	4.8459E-04	513.92	6.1171E+06
84	1-Propanol	C_3H_8O	71238	88.134	-8498.6	-9.0766	8.3303E-18	6	146.95	3.0828E-07	536.78	5.1214E+06
85	1-Butanol	$C_4H_{10}O$	71363	93.173	-9185.9	-9.7464	4.7796E-18	6	184.51	5.7220E-04	563.05	4.3392E+06
86	2-Butanol	$C_4H_{10}O$	78922	152.54	-11111	-19.025	1.0426E-05	2	158.45	1.1323E-06	536.05	4.2014E+06
87	2-Propanol	C_3H_8O	67630	76.964	-7623.8	-7.4924	5.9436E-18	6	185.28	3.6606E-02	508.3	4.7905E+06
88	2-Methyl-2-propanol	$C_4H_{10}O$	75650	172.31	-11590	-22.118	1.3709E-05	2	298.97	5.9356E+03	506.21	3.9910E+06
89	1-Pentanol	$C_5H_{12}O$	71410	168.96	-12659	-21.366	1.1591E-05	2	195.56	3.1816E-04	586.15	3.8657E+06
90	2-Methyl-1-butanol	$C_5H_{12}O$	137326	410.44	-20262	-62.366	6.3353E-02	1	203	3.7992E-04	565	3.5749E+06
91	3-Methyl-1-butanol	$C_5H_{12}O$	123513	107.02	-10237	-11.695	6.8003E-18	6	155.95	2.1036E-08	577.2	3.9013E+06
92	1-Hexanol	$C_6H_{14}O$	111273	117.31	-11239	-13.149	9.3676E-18	6	228.55	3.7401E-02	611.35	3.4557E+06
93	1-Heptanol	$C_7H_{16}O$	111706	160.08	-14095	-19.211	1.7043E-17	6	239.15	1.6990E-02	631.9	3.1810E+06
94	Cyclohexanol	$C_6H_{12}O$	108930	135.01	-12238	-15.702	1.0349E-17	6	296.6	7.9382E+01	650	4.2456E+06
95	Ethylene glycol	$C_2H_6O_2$	107211	79.276	-10105	-7.521	7.3408E-19	6	260.15	2.4834E-01	719.7	7.7100E+06
96	1,2-Propylene glycol	$C_3H_8O_2$	57556	212.8	-15420	-28.109	2.1564E-05	2	213.15	9.2894E-05	626	6.0413E+06
97	Phenol	C_6H_6O	108952	95.444	-10113	-10.09	6.7603E-18	6	314.06	1.8798E+02	694.25	6.0585E+06
98	o-Cresol	C_7H_8O	95487	210.88	-13928	-29.483	2.5182E-02	1	304.19	6.5326E+01	697.55	5.0583E+06
99	m-Cresol	C_7H_8O	106394	95.403	-10581	-10.004	4.3032E-18	6	285.39	5.8624E+00	705.85	4.5221E+06
100	p-Cresol	C_7H_8O	106445	118.53	-11957	-13.293	8.6988E-18	6	307.93	3.4466E+01	704.65	5.1507E+06
101	Dimethyl ether	C_2H_6O	115106	44.704	-3525.6	-3.4444	5.4574E-17	6	131.65	3.0496E+00	400.1	5.2735E+06
102	Methyl ethyl ether	C_3H_8O	540670	205.79	-9834.5	-28.739	3.5317E-05	2	160	3.5423E-01	437.8	4.4658E+06
103	Methyl n-propyl ether	$C_4H_{10}O$	557175	50.83	-4781.7	-4.1773	9.4076E-18	6	133.97	4.8875E-03	476.3	3.7721E+06
104	Methyl isopropyl ether	$C_4H_{10}O$	598538	55.096	-4793.2	-4.8689	2.9518E-17	6	127.93	2.4971E-03	464.5	3.8892E+06
105	Methyl n-butyl ether	$C_5H_{12}O$	628284	102.04	-6954.9	-12.278	1.2131E-05	2	157.48	1.9430E-02	510	3.3089E+06
106	Methyl isobutyl ether	$C_5H_{12}O$	625445	58.165	-5362.1	-5.2568	2.0194E-17	6	150	1.9801E-02	497	3.4130E+06
107	Methyl tert-butyl ether	$C_5H_{12}O$	1634044	55.875	-5131.6	-4.9604	1.9123E-17	6	164.55	5.3566E-01	497.1	3.4106E+06
108	Diethyl ether	$C_4H_{10}O$	60297	136.9	-6954.3	-19.254	2.4508E-02	1	156.85	3.9545E-01	466.7	3.6412E+06
109	Ethyl propyl ether	$C_5H_{12}O$	628320	143.11	-8353.7	-18.751	2.0620E-05	2	145.65	7.3931E-04	500.23	3.3729E+06
110	Ethyl isopropyl ether	$C_5H_{12}O$	625447	57.723	-5236.9	-5.2136	2.2998E-17	6	140	4.3092E-03	489	3.4145E+06
111	Methyl phenyl ether	C_7H_8O	100663	128.06	-9307.7	-16.693	1.4919E-02	1	235.65	2.4466E+00	645.6	4.2731E+06
112	Diphenyl ether	$C_{12}H_{10}O$	101848	59.969	-8585.5	-5.1538	1.9983E-18	6	300.03	7.0874E+00	766.8	3.0971E+06

TABLE 2-6 Vapor Pressure of Inorganic and Organic Liquids (*Continued*)

Cmpd. no.	Name	Formula	CAS no.	C1	C2	C3	C4	C5	T_{min}, K	P_s at T_{min}	T_{max}, K	P_s at T_{max}
113	Formaldehyde	CH_2O	50000	101.51	−4917.2	−13.765	2.2031E−02	1	181.15	8.8700E+02	408	6.5935E+06
114	Acetaldehyde	C_2H_4O	75070	193.69	−8036.7	−29.502	4.3678E−02	1	150.15	3.2320E−01	466	5.5652E+06
115	1-Propanal	C_3H_6O	123386	80.581	−5896.1	−8.9301	8.2236E−06	2	170	1.3133E+00	504.4	4.9189E+06
116	1-Butanal	C_4H_8O	123728	99.33	−7083.6	−11.733	1.0027E−05	2	176.75	3.1699E−01	537.2	4.3232E+06
117	1-Pentanal	$C_5H_{10}O$	110623	149.58	−8890	−20.697	2.2101E−02	1	182	5.2282E−02	566.1	3.9685E+06
118	1-Hexanal	$C_6H_{12}O$	66251	81.507	−7776.8	−8.4516	1.5143E−17	6	217.15	1.2473E+00	591	3.4607E+06
119	1-Heptanal	$C_7H_{14}O$	111717	107.17	−9070.3	−12.503	7.4446E−06	2	229.8	1.1177E+00	617	3.1829E+06
120	1-Octanal	$C_8H_{16}O$	124130	250.25	−16162	−33.927	2.2349E−05	2	246	4.1640E−01	638.1	2.9704E+06
121	1-Nonanal	$C_9H_{18}O$	124196	337.71	−18506	−50.224	4.7345E−02	1	255.15	3.4172E−01	658	2.7430E+06
122	1-Decanal	$C_{10}H_{20}O$	112312	201.64	−15133	−26.264	1.4625E−05	2	267.15	4.8648E−01	674.2	2.5959E+06
123	Acetone	C_3H_6O	67641	69.006	−5599.6	−7.0985	6.2237E−06	2	178.45	2.7851E+00	508.2	4.7091E+06
124	Methyl ethyl ketone	C_4H_8O	78933	72.698	−6143.6	−7.5779	5.6476E−06	2	186.48	1.3904E+00	535.5	4.1201E+06
125	2-Pentanone	$C_5H_{10}O$	107879	84.635	−7078.4	−9.3	6.2702E−06	2	196.29	7.5235E−01	561.08	3.7062E+06
126	Methyl isopropyl ketone	$C_5H_{10}O$	563804	308.74	−13693	−47.557	5.7002E−02	1	181.15	2.2648E−02	553	3.8413E+06
127	2-Hexanone	$C_6H_{12}O$	591786	65.841	−7042	−6.1376	7.2196E−18	6	217.35	1.5111E+00	587.05	3.3120E+06
128	Methyl isobutyl ketone	$C_6H_{12}O$	108101	153.23	−10055	−19.848	1.6426E−05	2	189.15	3.3536E−02	571.4	3.2659E+06
129	3-Methyl-2-pentanone	$C_6H_{12}O$	565617	64.641	−6457.4	−6.218	3.4543E−06	2	167.15	3.2662E−03	573	3.3213E+06
130	3-Pentanone	$C_5H_{10}O$	96220	44.286	−5415.1	−3.0913	1.8580E−18	6	234.18	7.3422E+01	560.95	3.6993E+06
131	Ethyl isopropyl ketone	$C_6H_{12}O$	565695	206.77	−12537	−27.894	2.2462E−05	2	200	6.0339E−02	567	3.3424E+06
132	Diisopropyl ketone	$C_7H_{14}O$	565800	96.919	−8014.2	−11.093	7.3452E−06	2	204.81	3.9036E−01	576	3.0606E+06
133	Cyclohexanone	$C_6H_{10}O$	108941	95.118	−8300.4	−10.796	6.5037E−06	2	242	6.9667E+00	653	4.0126E+06
134	Methyl phenyl ketone	C_8H_8O	98862	62.688	−8088.8	−5.5434	2.0074E−18	6	292.81	3.5599E−01	709.5	3.8451E+06
135	Formic acid	CH_2O_2	64186	50.323	−5378.2	−4.203	3.4697E−06	2	281.45	2.4024E+03	588	5.8074E+06
136	Acetic acid	$C_2H_4O_2$	64197	53.27	−6304.5	−4.2985	8.8865E−18	6	289.81	1.2769E+03	591.95	5.7390E+06
137	Propionic acid	$C_3H_6O_2$	79094	54.552	−7149.4	−4.2769	1.1843E−18	6	252.45	1.3142E+01	600.81	4.6080E+06
138	n-Butyric acid	$C_4H_8O_2$	107926	93.815	−9942.2	−9.8019	9.3124E−18	6	267.95	6.7754E+00	615.7	4.0705E+06
139	Isobutyric acid	$C_4H_8O_2$	79312	110.38	−10540	−12.262	1.4310E−17	6	227.15	7.8244E−02	605	3.6834E+06
140	Benzoic acid	$C_7H_6O_2$	65850	88.513	−11829	−8.6826	2.3248E−19	6	395.45	7.9550E+02	751	4.4691E+06
141	Acetic anhydride	$C_4H_6O_3$	108247	100.95	−8873.2	−11.451	6.1316E−06	2	200.15	2.1999E−02	606	3.9702E+06
142	Methyl formate	$C_2H_4O_2$	107313	77.184	−5606.1	−8.392	7.8468E−06	2	174.15	6.8808E+00	487.2	5.9929E+06
143	Methyl acetate	$C_3H_6O_2$	79209	61.267	−5618.6	−5.6473	2.1080E−17	6	175.15	1.0170E+00	506.55	4.6948E+06
144	Methyl propionate	$C_4H_8O_2$	554121	70.717	−6439.7	−6.9845	2.0129E−17	6	185.65	6.3409E+00	530.6	4.0278E+06
145	Methyl n-butyrate	$C_5H_{10}O_2$	623427	71.87	−6885.7	−7.0944	1.4903E−17	6	187.35	1.3435E−01	554.5	3.4797E+06
146	Ethyl formate	$C_3H_6O_2$	109944	73.833	−5817	−7.809	6.3200E−06	2	193.55	1.8119E+01	508.4	4.7080E+06
147	Ethyl acetate	$C_4H_8O_2$	141786	66.824	−6227.6	−6.41	1.7914E−17	6	189.6	1.4318E+00	523.3	3.8502E+06
148	Ethyl propionate	$C_5H_{10}O_2$	105373	105.64	−8007	−12.477	9.0000E−06	2	199.25	7.7988E−01	546	3.3365E+06
149	Ethyl n-butyrate	$C_6H_{12}O_2$	105544	57.661	−6346.5	−5.032	8.2534E−18	6	175.15	1.0390E−02	571	2.9352E+06
150	n-Propyl formate	$C_4H_8O_2$	110747	104.08	−7535.9	−12.348	9.6020E−06	2	180.25	2.1101E−01	538	4.0310E+06
151	n-Propyl acetate	$C_5H_{10}O_2$	109604	115.16	−8433.9	−13.934	1.0346E−05	2	178.15	1.7113E−02	549.73	3.3657E+06
152	n-Butyl acetate	$C_6H_{12}O_2$	123864	71.34	−7285.8	−6.9459	9.9895E−18	6	199.65	1.4347E−01	579.15	3.1097E+06
153	Methyl benzoate	$C_8H_8O_2$	93583	82.976	−9226.1	−8.4427	5.9115E−18	6	260.75	1.8653E+00	693	3.5896E+06
154	Ethyl benzoate	$C_9H_{10}O_2$	93890	53.024	−7676.8	−4.1593	1.6850E−18	6	238.45	1.4385E−01	698	3.2190E+06
155	Vinyl acetate	$C_4H_6O_2$	108054	57.406	−5702.8	−5.0307	1.1042E−17	6	180.35	7.0586E−01	519.13	3.9298E+06
156	Methylamine	CH_5N	74895	75.206	−5082.8	−8.0919	8.1130E−06	2	179.69	1.7671E+02	430.05	7.4139E+06
157	Dimethylamine	C_2H_7N	124403	71.738	−5302	−7.3324	6.4200E−17	6	180.96	7.5575E+01	437.2	5.2585E+06
158	Trimethylamine	C_3H_9N	75503	134.68	−6055.8	−19.415	2.8619E−02	1	156.08	9.9206E+00	433.25	4.1020E+06
159	Ethylamine	C_2H_7N	75047	81.56	−5596.9	−9.0779	8.7920E−06	2	192.15	1.5183E−02	456.15	5.5937E+06
160	Diethylamine	$C_4H_{11}N$	109897	49.314	−4949	−3.9256	9.1978E−18	6	223.35	3.7411E+02	496.6	3.6744E+06
161	Triethylamine	$C_6H_{15}N$	121448	56.55	−5681.9	−4.9815	1.2363E−17	6	158.45	1.0646E−02	535.15	3.0373E+06
162	n-Propylamine	C_3H_9N	107108	58.398	−5312.7	−5.2876	1.9913E−06	2	188.36	1.3004E+01	496.95	7.4381E+06
163	di-n-Propylamine	$C_6H_{15}N$	142847	54	−6018.5	−4.4981	9.9684E−18	6	210.15	3.6942E+00	550	3.1113E+06
164	Isopropylamine	C_3H_9N	75310	136.66	−7201.5	−18.934	2.2255E−02	1	177.95	7.7251E+00	471.85	4.5404E+06
165	Diisopropylamine	$C_6H_{15}N$	108189	462.84	−18227	−73.734	9.2794E−02	1	176.85	4.4724E−03	523.1	3.1987E+06
166	Aniline	C_6H_7N	62533	66.287	−8207.1	−6.0132	2.8414E−18	6	267.13	7.1322E+00	699	5.5314E+06

No.	Name	Formula	CAS									
167	N-Methylaniline	C_7H_9N	100618	70.843	-8517.5	-6.7007	5.6411E-18	6	216.15	1.0207E-02	701.55	5.1935E+06
168	N,N-Dimethylaniline	$C_8H_{11}N$	121697	51.352	-7160	-4.0127	8.1481E-07	2	275.6	1.7940E+01	687.15	3.3262E+06
169	Ethylene oxide	C_2H_4O	75218	91.944	-5293.4	-11.682	1.4902E-02	1	160.65	7.7879E+00	469.15	7.2553E+06
170	Furan	C_4H_4O	110009	74.738	-5417	-8.0636	7.4700E-06	2	187.55	5.0026E+01	490.15	5.5497E+06
171	Thiophene	C_4H_4S	110021	89.171	-6860.3	-10.104	7.4769E-06	2	234.94	1.8538E+02	579.35	5.7145E+06
172	Pyridine	C_5H_5N	110861	82.154	-7211.3	-8.8646	5.2528E-06	2	231.51	2.0535E+01	619.95	5.6356E+06
173	Formamide	CH_3NO	75127	100.3	-10763	-10.946	3.8503E-06	2	275.6	1.0350E+00	771	7.7514E+06
174	N,N-Dimethylformamide	C_3H_7NO	68122	82.762	-7955.5	-8.8038	4.2431E-06	2	212.72	1.953E-01	649.6	4.3653E+06
175	Acetamide	C_2H_5NO	60355	125.81	-12376	-14.589	5.0824E-06	2	353.33	3.3637E+02	761	6.5688E+06
176	N-Methylacetamide	C_3H_7NO	79163	79.128	-9523.9	-7.7355	3.1616E-16	6	301.15	2.8618E+02	718	4.9973E+06
177	Acetonitrile	C_2H_3N	75058	58.302	-5385.6	-5.4954	5.3634E-06	2	229.32	1.8694E+02	545.5	4.8517E+06
178	Propionitrile	C_3H_5N	107120	82.699	-6703.5	-9.1506	7.5424E-06	2	180.26	1.6936E-01	564.4	4.1906E+06
179	n-Butyronitrile	C_4H_7N	109740	66.32	-6714.9	-6.3087	1.3516E-17	6	161.25	6.1777E-04	582.25	3.7870E+06
180	Benzonitrile	C_7H_5N	100470	55.463	-7430.8	-4.548	1.7501E-17	6	260.4	5.1063E+00	699.35	4.2075E+06
181	Methyl mercaptan	CH_4S	74931	54.15	-4337.7	-4.8127	4.5000E-17	6	150.18	3.1479E+00	469.95	7.2309E+06
182	Ethyl mercaptan	C_2H_6S	75081	65.551	-5027.4	-6.6853	6.3208E-06	2	125.26	1.1384E-03	499.15	5.4918E+06
183	n-Propyl mercaptan	C_3H_8S	107039	62.165	-5624	-5.8595	2.0597E-17	6	159.95	6.5102E-02	536.6	4.6272E+06
184	n-Butyl mercaptan	$C_4H_{10}S$	109795	65.382	-6282.4	-0.2585	1.4943E-17	6	157.46	2.3532E-03	570.1	3.9730E+06
185	Isobutyl mercaptan	$C_4H_{10}S$	513440	61.736	-5909.2	-5.7554	1.5119E-17	6	128.31	4.7502E-06	559	4.0603E+06
186	sec-Butyl mercaptan	$C_4H_{10}S$	513531	60.649	-5785.9	-5.6113	1.5577E-17	6	133.02	3.3990E-05	554	4.0598E+06
187	Dimethyl sulfide	C_2H_6S	75183	83.485	-5711.7	-9.4999	9.8449E-06	2	174.88	7.9009E+00	503.04	5.5324E+06
188	Methyl ethyl sulfide	C_3H_8S	624895	79.07	-6114.1	-8.631	6.5333E-08	2	167.23	2.2456E-01	533	4.2610E+06
189	Diethyl sulfide	$C_4H_{10}S$	352932	60.867	-5969.6	-5.5979	1.4530E-17	6	169.2	4.3401E-02	557.15	3.9629E+06
190	Fluoromethane	CH_3F	593533	59.123	-3043.7	-6.1845	1.6637E-05	2	131.35	4.3287E+02	317.42	5.8754E+06
191	Chloromethane	CH_3Cl	74873	64.697	-4048.1	-6.8066	1.0371E-05	2	175.43	8.7091E+02	416.25	6.6905E+06
192	Trichloromethane	$CHCl_3$	67663	146.43	-7792.3	-20.614	2.4578E-02	1	207.15	5.2512E+01	536.4	5.5543E+06
193	Tetrachloromethane	CCl_4	56235	78.441	-6128.1	-8.5766	6.8465E-06	2	250.33	1.1225E+03	556.35	4.5436E+06
194	Bromomethane	CH_3Br	74839	72.586	-4698.6	-7.9966	1.1553E-05	2	179.47	1.9544E+02	467	7.9972E+06
195	Fluoroethane	C_2H_5F	353366	56.639	-3576.5	-5.5501	9.8969E-06	2	129.95	8.3714E+00	375.31	5.0060E+06
196	Chloroethane	C_2H_5Cl	75003	70.159	-4786.7	-7.5387	9.3370E-06	2	134.8	1.1165E-01	460.35	5.4578E+06
197	Bromoethane	C_2H_5Br	74964	62.217	-5113.3	-5.9761	4.7174E-17	6	154.55	3.7155E-01	503.8	6.2903E+06
198	1-Chloropropane	C_3H_7Cl	540545	79.24	-5718.8	-8.789	8.4486E-06	2	150.35	6.9630E-02	503.15	4.5512E+06
199	2-Chloropropane	C_3H_7Cl	75296	46.854	-4445.5	-3.6533	1.3260E-17	6	155.97	9.0844E-01	489	4.5097E+06
200	1,1-Dichloropropane	$C_3H_6Cl_2$	78999	83.495	-6661.4	-9.2386	6.7652E-06	2	200	4.5248E+00	560	4.2394E+06
201	1,2-Dichloropropane	$C_3H_6Cl_2$	78875	65.955	-6015.6	-6.5509	4.3172E-06	2	172.71	8.2532E-02	572	4.2319E+06
202	Vinyl chloride	C_2H_3Cl	75014	91.432	-5141.7	-10.981	1.4318E-05	2	119.36	1.9178E-02	432	5.7495E+06
203	Fluorobenzene	C_6H_5F	462066	51.915	-5439	-4.2896	8.7527E-18	6	230.94	1.5142E+02	560.09	4.5437E+06
204	Chlorobenzene	C_6H_5Cl	108907	54.144	-6244.4	-4.5343	4.7030E-18	6	227.95	8.4456E+00	632.35	4.5293E+06
205	Bromobenzene	C_6H_5Br	108861	63.749	-7130.2	-5.879	5.2136E-18	6	242.43	7.8364E+00	670.15	4.5196E+06
206	Air[3]		132259100	21.662	-692.39	-0.39208	4.7574E-03	1	59.15	5.6421E+03	132.45	3.7934E+06
207	Hydrogen	H_2	1333740	12.69	-94.896	1.1125	3.2915E-04	2	13.95	7.2116E+03	33.19	1.3154E+06
208	Helium-4[4]	He	7440597	11.533	-8.99	0.6724	2.7430E-01	1	1.76	1.4625E+03	5.2	2.2845E+05
209	Neon	Ne	7440019	29.755	-271.06	-2.6081	5.2700E-04	2	24.56	4.3800E+04	44.4	2.6652E+06
210	Argon	Ar	7440371	42.127	-1093.1	-4.1425	5.7254E-05	2	83.78	6.8721E+04	150.86	4.8963E+06
211	Fluorine	F_2	7782414	42.393	-1103.3	-4.1203	5.7815E-05	2	53.48	2.5272E+02	144.12	5.1674E+06
212	Chlorine	Cl_2	7782505	71.334	-3855	-8.5171	1.2378E-02	1	172.12	1.3660E+03	417.15	7.7930E+06
213	Bromine	Br_2	7726956	108.26	-6592	-14.16	1.6043E-02	1	265.85	5.8534E+03	584.15	1.0276E+07
214	Oxygen	O_2	7782447	51.245	-1200.2	-6.4361	2.8405E-02	1	54.36	1.4754E+02	154.58	5.0206E+06
215	Nitrogen	N_2	7727379	58.282	-1084.1	-8.3144	4.1277E-02	1	63.15	1.2508E+04	126.2	3.3906E+06
216	Ammonia	NH_3	7664417	90.483	-4669.7	-11.607	1.7194E-02	1	195.41	6.1111E+03	405.65	1.1301E+07
217	Hydrazine	N_2H_4	302012	76.858	-7245.2	-8.22	6.1557E-03	1	274.69	4.0847E+02	653.15	1.4731E+07
218	Nitrous oxide	N_2O	10024972	96.512	-4045	-12.277	2.8860E-05	2	182.3	8.6908E+04	309.57	7.2782E+06

TABLE 2-6 Vapor Pressure of Inorganic and Organic Liquids (*Concluded*)

Cmpd. no.	Name	Formula	CAS no.	C1	C2	C3	C4	C5	T_{min}, K	P_s at T_{min}	T_{max}, K	P_s at T_{max}
219	Nitric oxide	NO	10102439	72.974	−2650	−8.261	9.7000E−15	6	109.5	2.1956E+04	180.15	6.5156E+06
220	Cyanogen	C_2N_2	460195	88.589	−5059.9	−10.483	1.5403E−05	2	245.25	7.3385E+04	400.15	5.9438E+06
221	Carbon monoxide	CO	630080	45.698	−1076.6	−4.8814	7.5673E−05	2	68.15	1.5430E+04	132.92	3.4940E+06
222	Carbon dioxide	CO_2	124389	140.54	−4735	−21.268	4.0909E−02	1	216.58	5.1867E+05	304.21	7.3896E+06
223	Carbon disulfide	CS_2	75150	67.114	−4820.4	−7.5303	9.1695E−03	1	161.11	1.4944E+00	552	8.0408E+06
224	Hydrogen fluoride	HF	7664393	59.544	−4143.8	−6.1764	1.4161E−05	2	189.79	3.3683E+02	461.15	6.4872E+06
225	Hydrogen chloride	HCl	7647010	104.27	−3731.2	−15.047	3.1340E−02	1	158.97	1.3522E+04	324.65	8.3564E+06
226	Hydrogen bromide[2]	HBr	1035106	29.315	−2424.5	−1.1354	2.3306E−18	6	185.15	2.9501E+04	363.15	8.4627E+06
227	Hydrogen cyanide	HCN	74908	36.75	−3927.1	−2.1245	3.8948E−17	6	259.83	1.8687E+04	456.65	5.3527E+06
228	Hydrogen sulfide	H_2S	7783064	85.584	−3839.9	−11.199	1.8848E−02	1	187.68	2.2873E+04	373.53	8.9988E+06
229	Sulfur dioxide	SO_2	7446095	47.365	−4084.5	−3.6469	1.7990E−17	6	197.67	1.6743E+03	430.75	7.8596E+06
230	Sulfur trioxide	SO_3	7446119	180.99	−12060	−22.839	7.2350E−17	6	289.95	2.0934E+04	490.85	8.1919E+06
231	Water	H_2O	7732185	73.649	−7258.2	−7.3037	4.1653E−06	2	273.16	6.1056E+02	647.13	2.1940E+07

All substances are listed in alphabetical order in Table 2-6a.
Compiled from Daubert, T. E., R. P. Danner, H. M. Sibul, and C. C. Stebbins, DIPPR Data Compilation of Pure Compound Properties, Project 801 Sponsor Release, July, 1993, Design Institute for Physical Property Data, AIChE, New York, NY; and from Thermodynamics Research Center, "Selected Values of Properties of Hydrocarbons and Related Compounds," Thermodynamics Research Center Hydrocarbon Project, Texas A&M University, College Station, Texas (extant 1994).

Temperatures are in K; vapor pressures are in Pa.

$P_a \times 9.869233E-06 = $ atm; $P_a \times 1.450377E-04 = $ psia; vapor pressure = $\exp[C1 + (C2/T) + C3 \times \ln(T) + C4 \times T^{C5}]$.

[1] Decomposes violently on heating. Forms explosive peroxides with air or oxygen. Polymerizes under pressure and heat.

[2] Coefficients are hypothetical above the decomposition temperature.

[3] At the bubble point.

[4] Exhibits superfluid properties below 2.2 K.

TABLE 2-6a Alphabetical Index to Substances in Tables 2-6, 2-30, 2-164, 2-193, 2-196, 2-198, and 2-221

Name	Synonym	Cmpd. no.	Formula	Name	Synonym	Cmpd. no.	Formula
Acetaldehyde	Ethanal	114	C_2H_4O	1-Hexene		34	C_6H_{12}
Acetamide		175	C_2H_5NO	1-Hexyne		51	C_6H_{10}
Acetic acid	Ethanoic acid	136	$C_2H_4O_2$	2-Hexyne		52	C_6H_{10}
Acetic anhydride		141	$C_4H_6O_3$	3-Hexyne		53	C_6H_{10}
Acetone	2-Propanone	123	C_3H_6O	Hydrazine		217	N_2H_4
Acetonitrile	Methyl cyanide	177	C_2H_3N	Hydrogen		207	H_2
Acetophenone	Methyl phenyl ketone	134	C_8H_8O	Hydrogen bromide		226	HBr
Acetylene		45	C_2H_2	Hydrogen chloride		225	HCl
Air		206		Hydrogen cyanide		227	HCN
Ammonia		216	NH_3	Hydrogen fluoride		224	HF
Aniline		166	C_6H_7N	Hydrogen sulfide		228	H_2S
Anisole	Methyl phenyl ether	111	C_7H_8O				
Argon		210	Ar	Isobutyl mercaptan	2-Methyl-1-propanethiol	185	$C_4H_{10}S$
				Isobutyric acid	2-Methylpropanoic acid	139	$C_4H_8O_2$
Benzene		67	C_6H_6	Isooctane	2,2,4-Trimethylpentane	27	C_8H_{18}
Benzoic acid		140	$C_7H_6O_2$	Isoprene	2-Methyl-1,3-butadiene	44	C_5H_8
Benzonitrile	Phenyl cyanide	180	C_7H_5N	Isopropylamine		164	C_3H_9N
Biphenyl	1,1'-Biphenyl	79	$C_{12}H_{10}$				
Bromine		213	Br_2	Mesitylene	1,3,5-Trimethylbenzene	76	C_9H_{12}
Bromobenzene		205	C_6H_5Br	Methane		1	CH_4
Bromoethane	Ethyl bromide	197	C_2H_5Br	Methanol	Methyl alcohol	82	CH_4O
Bromomethane	Methyl bromide	194	CH_3Br	N-Methylacetamide		176	C_3H_7NO
1,2-Butadiene		42	C_4H_6	Methyl acetate		143	$C_3H_6O_2$
1,3-Butadiene		43	C_4H_6	Methylacetylene		46	C_3H_4
n-Butane		4	C_4H_{10}	Methylamine		156	CH_5N
1-Butanol		85	$C_4H_{10}O$	N-Methylaniline		167	C_7H_9N
2-Butanol	sec-Butyl alcohol	86	$C_4H_{10}O$	Methyl benzoate		153	$C_8H_8O_2$
1-Butene		30	C_4H_8	2-Methylbutane	Isopentane	22	C_5H_{12}
cis-2-Butene	Z-2-Butene	31	C_4H_8	2-Methyl-1-butanol		90	$C_5H_{12}O$
trans-2-Butene	E-2-Butene	32	C_4H_8	3-Methyl-1-butanol	Isoamyl alcohol	91	$C_5H_{12}O$
n-Butyl acetate		152	$C_6H_{12}O_2$	2-Methyl-1-butene		40	C_5H_{10}
n-Butyl mercaptan	1-Butanethiol	184	$C_4H_{10}S$	2-Methyl-2-butene	Amylene	41	C_5H_{10}
sec-Butyl mercaptan	2-Butanethiol	186	$C_4H_{10}S$	Methyl butyl ether		105	$C_5H_{12}O$
Butyraldehyde	Butanal	116	C_4H_8O	3-Methyl-1-butyne		48	C_5H_8
n-Butyric acid		138	$C_4H_8O_2$	Methyl butyrate		145	$C_5H_{10}O_2$
n-Butyronitrile	Propyl cyanide	179	C_4H_7N	Methylcyclohexane		61	C_7H_{14}
				Methylcyclopentane		58	C_6H_{12}
Carbon dioxide		222	CO_2	1-Methylcyclopentene		65	C_6H_{10}
Carbon disulfide		223	CS_2	Methyl ethyl ether		102	C_3H_8O
Carbon monoxide		221	CO	Methyl ethyl ketone	2-Butanone	124	C_4H_8O
Carbon tetrachloride	Tetrachloromethane	193	CCl_4	Methyl ethyl sulfide	2-Thiabutane	188	C_3H_8S
Chlorine		212	Cl_2	Methyl formate		142	$C_2H_4O_2$
Chlorobenzene		204	C_6H_5Cl	Methyl isobutyl ether		106	$C_5H_{12}O$
Chloroethane	Ethyl chloride	196	C_2H_5Cl	Methyl isobutyl ketone		128	$C_6H_{12}O$
Chloroform	Trichloromethane	192	$CHCl_3$	Methyl isopropyl ether		104	$C_4H_{10}O$
Chloromethane	Methyl chloride	191	CH_3Cl	Methyl isopropyl ketone		126	$C_5H_{10}O$
1-Chloropropane	Propyl chloride	198	C_3H_7Cl	Methyl mercaptan	Methanethiol	181	CH_4S
2-Chloropropane	Isopropyl chloride	199	C_3H_7Cl	2-Methylpentane	Isohexane	24	C_6H_{14}
o-Cresol	2-Methylphenol	98	C_7H_8O	3-Methyl-2-pentanone	Methyl sec-butyl ketone	129	$C_6H_{12}O$
m-Cresol	3-Methylphenol	99	C_7H_8O	2-Methylpropane	Isobutane	21	C_4H_{10}
p-Cresol	4-Methylphenol	100	C_7H_8O	2-Methyl-2-propanol	tert-Butyl alcohol	88	$C_4H_{10}O$
Cumene	Isopropylbenzene	75	C_9H_{12}	2-Methylpropene	Isobutene	39	C_4H_8
Cyanogen		220	C_2N_2	Methyl propionate		144	$C_4H_8O_2$
Cyclohexane		60	C_6H_{12}	Methyl propyl ether		103	$C_4H_{10}O$
Cyclohexanol	Cyclohexyl alcohol	94	$C_6H_{12}O$	Methyl tert-butyl ether		107	$C_5H_{12}O$
Cyclohexanone	Cyclohexyl ketone	133	$C_6H_{10}O$				
Cyclohexene		66	C_6H_{10}	Naphthalene		78	$C_{10}H_8$
Cyclopentane		57	C_5H_{10}	Neon		209	Ne
Cyclopentene		64	C_5H_8	Nitric oxide		219	NO
p-Cymene	p-Isopropyltoluene	77	$C_{10}H_{14}$	Nitrogen		215	N_2
				Nitrous oxide		218	N_2O
1-Decanal		122	$C_{10}H_{20}O$	n-Nonadecane		19	$C_{19}H_{40}$
n-Decane		10	$C_{10}H_{22}$	1-Nonanal	n-Nonaldehyde	121	$C_9H_{18}O$
1-Decene		38	$C_{10}H_{20}$	n-Nonane		9	C_9H_{20}
1,1-Dichloropropane		200	$C_3H_6Cl_2$	1-Nonene		37	C_9H_{18}
1,2-Dichloropropane		201	$C_3H_6Cl_2$				
Diethylamine		160	$C_4H_{11}N$	n-Octadecane		18	$C_{18}H_{38}$
Diethyl ether	Ethyl ether	108	$C_4H_{10}O$	1-Octanal	n-Octaldehyde	120	$C_8H_{16}O$
Diethyl sulfide	Ethyl sulfide	189	$C_4H_{10}S$	n-Octane		8	C_8H_{18}
Diisopropylamine		165	$C_6H_{15}N$	1-Octene		36	C_8H_{16}
Diisopropyl ketone	2,4-Dimethyl-3-pentanone	132	$C_7H_{14}O$	1-Octyne		55	C_8H_{14}
Dimethylacetylene	2-Butyne	47	C_4H_6	Oxygen		214	O_2
Dimethylamine		157	C_2H_7N				
N,N-Dimethylaniline	N,N-Dimethylbenzamine	168	$C_8H_{11}N$	n-Pentadecane		15	$C_{15}H_{32}$
2,3-Dimethylbutane	Diisopropyl	23	C_6H_{14}	1-Pentanal	Valeraldehyde	117	$C_5H_{10}O$
1,1-Dimethylcyclohexane		62	C_8H_{16}	n-Pentane		5	C_5H_{12}
Dimethyl ether	Methyl ether	101	C_2H_6O	1-Pentanol	n-Amyl alcohol	89	$C_5H_{12}O$

TABLE 2-6a   Alphabetical Index to Substances in Tables 2-6, 2-30, 2-164, 2-193, 2-196, 2-198, and 2-221 (*Concluded*)

Name	Synonym	Cmpd. no.	Formula	Name	Synonym	Cmpd. no.	Formula
N,N-Dimethylformamide		174	C_3H_7NO	2-Pentanone	Methyl n-propyl ketone	125	$C_5H_{10}O$
2,3-Dimethylpentane		25	C_7H_{16}	3-Pentanone	Diethyl ketone	130	$C_5H_{10}O$
Dimethyl sulfide	Methyl sulfide	187	C_2H_6S	1-Pentene		33	C_5H_{10}
Diphenyl ether		112	$C_{12}H_{10}O$	1-Pentyne		49	C_5H_8
n-Dodecane		12	$C_{12}H_{26}$	2-Pentyne		50	C_5H_8
				Phenol		97	C_6H_6O
n-Eicosane		20	$C_{20}H_{42}$	1-Propanal	Propionaldehyde	115	C_3H_6O
Ethane		2	C_2H_6	n-Propane		3	C_3H_8
Ethanol	Ethyl alcohol	83	C_2H_6O	1-Propanol	n-Propyl alcohol	84	C_3H_8O
Ethyl acetate		147	$C_4H_8O_2$	2-Propanol	Isopropyl alcohol	87	C_3H_8O
Ethylamine		159	C_2H_7N	n-Propionic acid		137	$C_3H_6O_2$
Ethylbenzene	Phenylethane	72	C_8H_{10}	n-Propionitrile	Ethyl cyanide	178	C_3H_5N
Ethyl benzoate		154	$C_9H_{10}O_2$	n-Propyl acetate		151	$C_5H_{10}O_2$
Ethyl butyrate		149	$C_6H_{12}O_2$	n-Propylamine		162	C_3H_9N
Ethylcyclohexane		63	C_8H_{16}	di-n-Propylamine		163	$C_6H_{15}N$
Ethylcyclopentane		59	C_7H_{14}	n-Propylbenzene		73	C_9H_{12}
Ethylene		28	C_2H_4	Propylene		29	C_3H_6
Ethylene glycol	1,2-Ethanediol	95	$C_2H_6O_2$	1,2-Propylene glycol	1,2-Propanediol	96	$C_3H_8O_2$
Ethylene oxide	1,2-Epoxyethane	169	C_2H_4O	n-Propyl formate		150	$C_4H_8O_2$
Ethyl formate		146	$C_3H_6O_2$	n-Propyl mercaptan	Propanethiol	183	C_3H_8S
Ethyl isopropyl ether		110	$C_5H_{12}O$	Pyridine		172	C_5H_5N
Ethyl isopropyl ketone		131	$C_6H_{12}O$				
Ethyl mercaptan	Ethanethiol	182	C_2H_6S	Styrene		80	C_8H_8
Ethyl propionate		148	$C_5H_{10}O_2$	Sulfur dioxide		229	SO_2
Ethyl propyl ether		109	$C_5H_{12}O$	Sulfur trioxide		230	SO_3
Fluorine		211	F_2	m-Terphenyl		81	$C_{18}H_{14}$
Fluorobenzene		203	C_6H_5F	n-Tetradecane		14	$C_{14}H_{30}$
Fluoroethane	Ethyl fluoride	195	C_2H_5F	Thiophene		171	C_4H_4S
Fluoromethane	Methyl fluoride	190	CH_3F	Toluene		68	C_7H_8
Formaldehyde	Methanal	113	CH_2O	n-Tridecane		13	$C_{13}H_{28}$
Formamide		173	CH_3NO	Triethylamine		161	$C_6H_{15}N$
Formic acid	Methanoic acid	135	CH_2O_2	Trimethylamine		158	C_3H_9N
Furan		170	C_4H_4O	1,2,4-Trimethylbenzene		74	C_9H_{12}
				2,3,3-Trimethylpentane		26	C_8H_{18}
Helium-4		208	He				
n-Heptadecane		17	$C_{17}H_{36}$	n-Undecane		11	$C_{11}H_{24}$
1-Heptanal	n-Heptaldehyde	119	$C_7H_{14}O$				
n-Heptane		7	C_7H_{16}	Vinyl acetate		155	$C_4H_6O_2$
1-Heptanol	n-Heptyl alcohol	93	$C_7H_{16}O$	Vinylacetylene		56	C_4H_4
1-Heptene		35	C_7H_{14}	Vinyl chloride		202	C_2H_3Cl
1-Heptyne		54	C_7H_{12}				
n-Hexadecane		16	$C_{16}H_{34}$	Water		231	H_2O
1-Hexanal	Caproaldehyde	118	$C_6H_{12}O$				
n-Hexane		6	C_6H_{14}	o-Xylene	1,2-Dimethylbenzene	69	C_8H_{10}
1-Hexanol	n-Hexyl alcohol	92	$C_6H_{14}O$	m-Xylene	1,3-Dimethylbenzene	70	C_8H_{10}
2-Hexanone	Methyl n-butyl ketone	127	$C_6H_{12}O$	p-Xylene	1,4-Dimethylbenzene	71	C_8H_{10}

TABLE 2-7 Vapor Pressures of Inorganic Compounds, up to 1 atm*

Compound		Pressure, mm Hg										Melting point, °C
		1	5	10	20	40	60	100	200	400	760	
Name	Formula	Temperature, °C										
Aluminum	Al	1284	1421	1487	1555	1635	1684	1749	1844	1947	2056	660
borohydride	Al(BH₄)₃		−52.2	−42.9	−32.5	−20.9	−13.4	−3.9	+11.2	28.1	45.9	−64.
bromide	AlBr₃	81.3	103.8	118.0	134.0	150.6	161.7	176.1	199.8	227.0	256.3	97.
chloride	Al₂Cl₆	100.0	116.4	123.8	131.8	139.9	145.4	152.0	161.8	171.6	180.2	192.4
fluoride	AlF₃	1238	1298	1324	1350	1378	1398	1422	1457	1496	1537	1040
iodide	AlI₃	178.0	207.7	225.8	244.2	265.0	277.8	294.5	322.0	354.0	385.5	
oxide	Al₂O₃	2148	2306	2385	2465	2549	2599	2665	2766	2874	2977	2050
Ammonia	NH₃	−109.1	−97.5	−91.9	−85.8	−79.2	−74.3	−68.4	−57.0	−45.4	−33.6	−77.7
heavy	ND₃						−74.0	−67.4	−57.0	−45.4	−33.4	−74.0
Ammonium bromide	NH₄Br	198.3	234.5	252.0	270.6	290.0	303.8	320.0	345.3	370.8	396.0	
carbamate	N₂H₆CO₂	−26.1	−10.4	−2.9	+5.3	14.0	19.6	26.7	37.2	48.0	58.3	
chloride	NH₄Cl	160.4	193.8	209.8	226.1	245.0	256.2	271.5	293.2	316.5	337.8	520
cyanide	NH₄CN	−50.6	−35.7	−28.6	−20.9	−12.6	−7.4	−0.5	+9.6	20.5	31.7	36
hydrogen sulfide	NH₄HS	−51.1	−36.0	−28.7	−20.8	−12.3	−7.0	0.0	+10.5	21.8	33.3	
iodide	NH₄I	210.9	247.0	263.5	282.8	302.8	316.0	331.8	355.8	381.0	404.9	
Antimony	Sb	886	984	1033	1084	1141	1176	1223	1288	1364	1440	630.5
tribromide	SbBr₃	93.9	126.0	142.7	158.3	177.4	188.1	203.5	225.7	250.2	275.0	96.6
trichloride	SbCl₃	49.2	71.4	85.2	100.6	117.8	128.3	143.3	165.9	192.2	219.0	73.4
pentachloride	SbCl₅	22.7	48.6	61.8	75.8	91.0	101.0	114.1				2.8
triiodide	SbI₃	163.6	203.8	223.5	244.8	267.8	282.5	303.5	333.8	368.5	401.0	167
trioxide	Sb₄O₆	574	626	666	729	812	873	957	1085	1242	1425	656
Argon	A	−218.2	−213.9	−210.9	−207.9	−204.9	−202.9	−200.5	−195.6	−190.6	−185.6	−189.2
Arsenic	As	372	416	437	459	483	498	518	548	579	610	814
Arsenic tribromide	AsBr₃	41.8	70.6	85.2	101.3	118.7	130.0	145.2	167.7	193.6	220.0	
trichloride	AsCl₃	−11.4	+11.7	+23.5	36.0	50.0	58.7	70.9	89.2	109.7	130.4	−18
trifluoride	AsF₃					−2.5	+4.2	13.2	26.7	41.4	56.3	−5.9
pentafluoride	AsF₅	−117.9	−108.0	−103.1	−98.0	−92.4	−88.5	−84.3	−75.5	−64.0	−52.8	−79.8
trioxide	As₂O₃	212.5	242.6	259.7	279.2	299.2	310.3	332.5	370.0	412.2	457.2	312.8
Arsine	AsH₃	−142.6	−130.8	−124.7	−117.7	−110.2	−104.8	−98.0	−87.2	−75.2	−62.1	−116.3
Barium	Ba		984	1049	1120	1195	1240	1301	1403	1518	1638	850
Beryllium borohydride	Be(BH₄)₂	+1.0	19.8	28.1	36.8	46.2	51.7	58.6	69.0	79.7	90.0	123
bromide	BeBr₂	289	325	342	361	379	390	405	427	451	474	490
chloride	BeCl₂	291	328	346	365	384	395	411	435	461	487	405
iodide	BeI₂	283	322	341	361	382	394	411	435	461	487	488
Bismuth	Bi	1021	1099	1136	1177	1217	1240	1271	1319	1370	1420	271
tribromide	BiBr₃		261	282	305	327	340	360	392	425	461	218
trichloride	BiCl₃		242	264	287	311	324	343	372	405	441	230
Diborane hydrobromide	B₂H₅Br	−93.3	−75.3	−66.3	−56.4	−45.4	−38.2	−29.0	−15.4	0.0	+16.3	−104.2
Borine carbonyl	BH₃CO	−139.2	−127.3	−121.1	−114.1	−106.6	−101.9	−95.3	−85.5	−74.8	−64.0	−137.0
triamine	B₃N₃H₆	−63.0	−45.0	−35.3	−25.0	−13.2	−5.8	+4.0	18.5	34.3	50.6	−58.2
Boron hydrides												
dihydrodecaborane	B₁₀H₁₄	60.0	80.8	90.2	100.0	117.4	127.8	142.3	163.8			99.6
dihydrodiborane	B₂H₆	−159.7	−149.5	−144.3	−138.5	−131.6	−127.2	−120.9	−111.2	−99.6	−86.5	−169
dihydropentaborane	B₅H₉		−40.4	−30.7	−20.0	−8.0	−0.4	+9.6	24.6	40.8	58.1	−47.0
tetrahydropentaborane	B₅H₁₁	−50.2	−29.9	−19.9	−9.2	+2.7	10.2	20.1	34.8	51.2	67.0	
tetrahydrotetraborane	B₄H₁₀	−90.9	−73.1	−64.3	−54.8	−44.3	−37.4	−28.1	−14.0	+0.8	16.1	−119.9
Boron tribromide	BBr₃	−41.4	−20.4	−10.1	+1.5	14.0	22.1	33.5	50.3	70.0	91.7	−45
trichloride	BCl₃	−91.5	−75.2	−66.9	−57.9	−47.8	−41.2	−32.4	−18.9	−3.6	+12.7	−107
trifluoride	BF₃	−154.6	−145.4	−141.3	−136.4	−131.0	−127.6	−123.0	−115.9	−108.3	−100.7	−126.8
Bromine	Br₂	−48.7	−32.8	−25.0	−16.8	−8.0	−0.6	+9.3	24.3	41.0	58.2	−7.3
pentafluoride	BrF₅	−69.3	−51.0	−41.9	−32.0	−21.0	−14.0	−4.5	+9.9	25.7	40.4	−61.4
Cadmium	Cd	394	455	484	516	553	578	611	658	711	765	320.9
chloride	CdCl₂		618	656	695	736	762	797	847	908	967	568
fluoride	CdF₂	1112	1231	1286	1344	1400	1436	1486	1561	1651	1751	520
iodide	CdI₂	416	481	512	546	584	608	640	688	742	796	385
oxide	CdO	1000	1100	1149	1200	1257	1295	1341	1409	1484	1559	
Calcium	Ca		926	983	1046	1111	1152	1207	1288	1388	1487	851
Carbon (graphite)	C	3586	3828	3946	4069	4196	4273	4373	4516	4660	4827	
dioxide	CO₂	−134.3	−124.4	−119.5	−108.6	−108.6	−104.8	−100.2	−93.0	−85.7	−78.2	−57.5
disulfide	CS₂	−73.8	−54.3	−44.7	−34.3	−22.5	−15.3	−5.1	+10.4	28.0	46.5	−110.8
monoxide	CO	−222.0	−217.2	−215.0	−212.8	−210.0	−208.1	−205.7	−201.3	−196.3	−191.3	−205.0
oxyselenide	COSe	−117.1	−102.3	−95.0	−86.3	−76.4	−70.2	−61.7	−49.8	−35.6	−21.9	
oxysulfide	COS	−132.4	−119.8	−113.3	−106.0	−98.3	−93.0	−85.9	−75.0	−62.7	−49.9	−138.8
selenosulfide	CSeS	−47.3	−26.5	−16.0	−4.4	+8.6	17.0	28.3	45.7	65.2	85.6	−75.2
subsulfide	C₃S₂	14.0	41.2	54.9	69.3	85.6	96.0	109.9	130.8			+0.4
tetrabromide	CBr₄					96.3	106.3	119.7	139.7	163.5	189.5	90.1
tetrachloride	CCl₄	−50.0	−30.0	−19.6	−8.2	+4.3	12.3	23.0	38.3	57.8	76.7	−22.6
tetrafluoride	CF₄	−184.6	−174.1	−169.3	−164.3	−158.8	−155.4	−150.7	−143.6	−135.5	−127.7	−183.7
Cesium	Cs	279	341	375	409	449	474	509	561	624	690	28.5
bromide	CsBr	748	838	887	938	993	1026	1072	1140	1221	1300	636
chloride	CsCl	744	837	884	934	989	1023	1069	1139	1217	1300	646
fluoride	CsF	712	798	844	893	947	980	1025	1092	1170	1251	683
iodide	CsI	738	828	873	923	976	1009	1055	1124	1200	1280	621

*Compiled from the extended tables published by D. R. Stull in *Ind. Eng. Chem.,* **39,** 517 (1947).

TABLE 2-7 Vapor Pressures of Inorganic Compounds, up to 1 atm (*Continued*)

Compound		Pressure, mm Hg										Melting point, °C
		1	5	10	20	40	60	100	200	400	760	
Name	Formula	Temperature, °C										
Chlorine	Cl_2	−118.0	−106.7	−101.6	−93.3	−84.5	−79.0	−71.7	−60.2	−47.3	−33.8	−100.7
fluoride	ClF		−143.4	−139.0	−134.3	−128.8	−125.3	−120.8	−114.4	−107.0	−100.5	−145
trifluoride	ClF_3		−80.4	−71.8	−62.3	−51.3	−44.1	−34.7	−20.7	−4.9	+11.5	−83
monoxide	Cl_2O	−98.5	−81.6	−73.1	−64.3	−54.3	−48.0	−39.4	−26.5	−12.5	+2.2	−116
dioxide	ClO_2			−59.0	−51.2	−42.8	−37.2	−29.4	−17.8	−4.0	+11.1	−59
heptoxide	Cl_2O_7	−45.3	−23.8	−13.2	−2.1	+10.3	+18.2	29.1	44.6	62.2	78.8	−91
Chlorosulfonic acid	HSO_3Cl	32.0	53.5	64.0	75.3	87.6	95.2	105.3	120.0	136.1	151.0	−80
Chromium	Cr	1616	1768	1845	1928	2013	2067	2139	2243	2361	2482	1615
carbonyl	$Cr(CO)_6$	36.0	58.0	68.3	79.5	91.2	98.3	108.0	121.8	137.2	151.0	
oxychloride	CrO_2Cl_2	−18.4	+3.2	13.8	25.7	38.5	46.7	58.0	75.2	95.2	117.1	
Cobalt chloride	$CoCl_2$					770	801	843	904	974	1050	735
nitrosyl tricarbonyl	$Co(CO)_3NO$				−1.3	+11.0	18.5	29.0	44.4	62.0	80.0	−11
Columbium fluoride	CbF_5			86.3	103.0	121.5	133.2	148.5	172.2	198.0	225.0	75.5
Copper	Cu	1628	1795	1879	1970	2067	2127	2207	2325	2465	2595	1083
Cuprous bromide	Cu_2Br_2	572	666	718	777	844	887	951	1052	1189	1355	504
chloride	Cu_2Cl_2	546	645	702	766	838	886	960	1077	1249	1490	422
iodide	Cu_2I_2		610	656	716	786	836	907	1018	1158	1336	605
Cyanogen	C_2N_2	−95.8	−83.2	−76.8	−70.1	−62.7	−57.9	−51.8	−42.6	−33.0	−21.0	−34.4
bromide	CNBr	−35.7	−18.3	−10.0	−1.0	+8.6	14.7	22.6	33.8	46.0	61.5	58
chloride	CNCl	−76.7	−61.4	−53.8	−46.1	−37.5	−32.1	−24.9	−14.1	−2.3	+13.1	−6.5
fluoride	CNF	−134.4	−123.8	−118.5	−112.8	−106.4	−102.3	−97.0	−89.2	−80.5	−72.6	
Deuterium cyanide	DCN	−68.9	−54.0	−46.7	−38.8	−30.1	−24.7	−17.5	−5.4	+10.0	26.2	−12
Fluorine	F_2	−223.0	−216.9	−214.1	−211.0	−207.7	−205.6	−202.7	−198.3	−193.2	−187.9	−223
oxide	F_2O	−196.1	−186.6	−182.3	−177.8	−173.0	−170.0	−165.8	−159.0	−151.9	−144.6	−223.9
Germanium bromide	$GeBr_4$		43.3	56.8	71.8	88.1	98.8	113.2	135.4	161.6	189.0	26.1
chloride	$GeCl_4$	−45.0	−24.9	−15.0	−4.1	+8.0	16.2	27.5	44.4	63.8	84.0	−49.5
hydride	GeH_4	−163.0	−151.0	−145.3	−139.2	−131.6	−126.7	−120.3	−111.2	−100.2	−88.9	−165
Trichlorogermane	$GeHCl_3$	−41.3	−22.3	−13.0	−3.0	+8.8	16.2	26.5	41.6	58.3	75.0	−71.1
Tetramethylgermane	$Ge(CH_3)_4$	−73.2	−54.6	−45.2	−35.0	−23.4	−16.2	−6.3	+8.8	26.0	44.0	−88
Digermane	Ge_2H_6	−88.7	−69.8	−60.1	−49.9	−38.2	−30.7	−20.3	−4.7	+13.3	31.5	−109
Trigermane	Ge_3H_8	−36.9	−12.8	−0.9	+11.8	26.3	35.5	47.9	67.0	88.6	110.8	−105.6
Gold	Au	1869	2059	2154	2256	2363	2431	2521	2657	2807	2966	1063
Helium	He	−271.7	−271.5	−271.3	−271.1	−270.7	−270.6	−270.3	−269.8	−269.3	−268.6	
para-Hydrogen	H_2	−263.3	−261.9	−261.3	−260.4	−259.6	−258.9	−257.9	−256.3	−254.5	−252.5	−259.1
Hydrogen bromide	HBr	−138.8	−127.4	−121.8	−115.4	−108.3	−103.8	−97.7	−88.1	−78.0	−66.5	−87.0
chloride	HCl	−150.8	−140.7	−135.6	−130.0	−123.8	−119.6	−114.0	−105.2	−95.3	−84.8	−114.3
cyanide	HCN	−71.0	−55.3	−47.7	−39.7	−30.9	−25.1	−17.8	−5.3	+10.2	25.9	−13.2
fluoride	H_2F_2		−74.7	−65.8	−56.0	−45.0	−37.9	−28.2	−13.2	+2.5	19.7	−83.7
iodide	HI	−123.3	−109.6	−102.3	−94.5	−85.6	−79.8	−72.1	−60.3	−48.3	−35.1	−50.9
oxide (water)	H_2O	−17.3	+1.2	11.2	22.1	34.0	41.5	51.6	66.5	83.0	100.0	0.0
sulfide	H_2S	−134.3	−122.4	−116.3	−109.7	−102.3	−97.9	−91.6	−82.3	−71.8	−60.4	−85.5
disulfide	HSSH	−43.2	−24.4	−15.2	−5.1	+6.0	12.8	22.0	35.3	49.6	64.0	−89.7
selenide	H_2Se	−115.3	−103.4	−97.9	−91.8	−84.7	−80.2	−74.2	−65.2	−53.6	−41.1	−64
telluride	H_2Te	−96.4	−82.4	−75.4	−67.8	−59.1	−53.7	−45.7	−32.4	−17.2	−2.0	−49.0
Iodine	I_2	38.7	62.2	73.2	84.7	97.5	105.4	116.5	137.3	159.8	183.0	112.9
heptafluoride	IF_7	−87.0	−70.7	−63.0	−54.5	−45.3	−39.4	−31.9	−20.7	−8.3	+4.0	5.5
Iron	Fe	1787	1957	2039	2128	2224	2283	2360	2475	2605	2735	1535
pentacarbonyl	$Fe(CO)_5$		−6.5	+4.6	16.7	30.3	39.1	50.3	68.0	86.1	105.0	−21
Ferric chloride	Fe_2Cl_6	194.0	221.8	235.5	246.0	256.8	263.7	272.5	285.0	298.0	319.0	304
Ferrous chloride	$FeCl_2$			700	737	779	805	842	897	961	1026	
Krypton	Kr	−199.3	−191.3	−187.2	−182.9	−178.4	−175.7	−171.8	−165.9	−159.0	−152.0	−156.7
Lead	Pb	973	1099	1162	1234	1309	1358	1421	1519	1630	1744	327.5
bromide	$PbBr_2$	513	578	610	646	686	711	745	796	856	914	373
chloride	$PbCl_2$	547	615	648	684	725	750	784	833	893	954	501
fluoride	PbF_2		861	904	950	1003	1036	1080	1144	1219	1293	855
iodide	PbI_2	479	540	571	605	644	668	701	750	807	872	402
oxide	PbO	943	1039	1085	1134	1189	1222	1265	1330	1402	1472	890
sulfide	PbS	852	928	975	1005	1048	1074	1108	1160	1221	1281	1114
Lithium	Li	723	838	881	940	1003	1042	1097	1178	1273	1372	186
bromide	LiBr	748	840	888	939	994	1028	1076	1147	1226	1310	547
chloride	LiCl	783	880	932	987	1045	1081	1129	1203	1290	1382	614
fluoride	LiF	1047	1156	1211	1270	1333	1372	1425	1503	1591	1681	870
iodide	LiI	723	802	841	883	927	955	993	1049	1110	1171	446
Magnesium	Mg	621	702	743	789	838	868	909	967	1034	1107	651
chloride	$MgCl_2$	778	877	930	988	1050	1088	1142	1223	1316	1418	712
Manganese	Mn	1292	1434	1505	1583	1666	1720	1792	1900	2029	2151	1260
chloride	$MnCl_2$		736	778	825	879	913	960	1028	1108	1190	650
Mercury	Hg	126.2	164.8	184.0	204.6	228.8	242.0	261.7	290.7	323.0	357.0	−38.9
Mercuric bromide	$HgBr_2$	136.5	165.3	179.8	194.3	211.5	221.0	237.8	262.7	290.0	319.0	237
chloride	$HgCl_2$	136.2	166.0	180.2	195.8	212.5	222.2	237.0	256.5	275.5	304.0	277
iodide	HgI_2	157.5	189.2	204.5	220.0	238.2	249.0	261.8	291.0	324.2	354.0	259
Molybdenum	Mo	3102	3393	3535	3690	3859	3964	4109	4322	4553	4804	2622
hexafluoride	MoF_6	−65.5	−49.0	−40.8	−32.0	−22.1	−16.2	−8.0	+4.1	17.2	36.0	17
oxide	MoO_3	734	785	814	851	892	917	955	1014	1082	1151	795

TABLE 2-7   Vapor Pressures of Inorganic Compounds, up to 1 atm (Continued)

Name	Formula	1	5	10	20	40	60	100	200	400	760	Melting point, °C
Neon	Ne	−257.3	−255.5	−254.6	−253.7	−252.6	−251.9	−251.0	−249.7	−248.1	−246.0	−248.7
Nickel	Ni	1810	1979	2057	2143	2234	2289	2364	2473	2603	2732	1452
carbonyl	Ni(CO)$_4$					−23.0	−15.9	−6.0	+8.8	25.8	42.5	−25
chloride	NiCl$_2$	671	731	759	789	821	840	866	904	945	987	1001
Nitrogen	N$_2$	−226.1	−221.3	−219.1	−216.8	−214.0	−212.3	−209.7	−205.6	−200.9	−195.8	−210.0
Nitric oxide	NO	−184.5	−180.6	−178.2	−175.3	−171.7	−168.9	−166.0	−162.3	−156.8	−151.7	−161
Nitrogen dioxide	NO$_2$	−55.6	−42.7	−36.7	−30.4	−23.9	−19.9	−14.7	−5.0	+8.0	21.0	−9.3
Nitrogen pentoxide	N$_2$O$_5$	−36.8	−23.0	−16.7	−10.0	−2.9	+1.8	7.4	15.6	24.4	32.4	30
Nitrous oxide	N$_2$O	−143.4	−133.4	−128.7	−124.0	−118.3	−114.9	−110.3	−103.6	−96.2	−85.5	−90.9
Nitrosyl chloride	NOCl					−60.2	−54.2	−46.3	−34.0	−20.3	−6.4	−64.5
fluoride	NOF	−132.0	−120.3	−114.3	−107.8	−100.3	−95.7	−88.8	−79.2	−68.2	−56.0	−134
Osmium tetroxide (yellow)	OsO$_4$	3.2	22.0	31.3	41.0	51.7	59.4	71.5	89.5	109.3	130.0	56
(white)	OsO$_4$	−5.6	+15.6	26.0	37.4	50.5	59.4	71.5	89.5	109.3	130.0	42
Oxygen	O$_2$	−219.1	−213.4	−210.6	−207.5	−204.1	−201.9	−198.8	−194.0	−188.8	−183.1	−218.7
Ozone	O$_3$	−180.4	−168.6	−163.2	−157.2	−150.7	−146.7	−141.0	−132.6	−122.5	−111.1	−251
Phosgene	COCl$_2$	−92.9	−77.0	−69.3	−60.3	−50.3	−44.0	−35.6	−22.3	−7.6	+8.3	−104
Phosphorus (yellow)	P	76.6	111.2	128.0	146.2	166.7	179.5	197.3	222.7	251.0	280.0	44.1
(violet)	P	237	271	287	306	323	334	349	370	391	417	590
tribromide	PBr$_3$	7.8	34.4	47.8	62.4	79.0	89.8	103.6	125.2	149.7	175.3	−40
trichloride	PCl$_3$	−51.6	−31.5	−21.3	−10.2	+2.3	10.2	21.0	37.6	56.9	74.2	−111.8
pentachloride	PCl$_5$	55.5	74.0	83.2	92.5	102.5	108.3	117.0	131.3	147.2	162.0	
Phosphine	PH$_3$					−129.4	−125.0	−118.8	−109.4	−98.3	−87.5	−132.5
Phosphonium bromide	PH$_4$Br	−43.7	−28.5	−21.2	−13.3	−5.0	+0.3	7.4	17.6	28.0	38.3	
chloride	PH$_4$Cl	−91.0	−79.6	−74.0	−68.0	−61.5	−57.3	−52.0	−44.0	−35.4	−27.0	−28.5
iodide	PH$_4$I	−25.2	−9.0	−1.1	+7.3	16.1	21.9	29.3	39.9	51.6	62.3	
Phosphorus trioxide	P$_4$O$_6$			39.7	53.0	67.8	84.0	94.2	108.3	129.0	150.3	22.5
pentoxide	P$_4$O$_{10}$	384	424	442	462	481	493	510	532	556	591	569
oxychloride	POCl$_3$			2.0	13.6	27.3	35.8	47.4	65.0	84.3	105.1	2
thiobromide	PSBr$_3$	50.0	72.4	83.6	95.5	108.0	116.0	126.3	141.8	157.8	175.0	38
thiochloride	PSCl$_3$	−18.3	+4.6	16.1	29.0	42.7	51.8	63.8	82.0	102.3	124.0	−36.2
Platinum	Pt	2730	3007	3146	3302	3469	3574	3714	3923	4169	4407	1755
Potassium	K	341	408	443	483	524	550	586	643	708	774	62.3
bromide	KBr	795	892	940	994	1050	1087	1137	1212	1297	1383	730
chloride	KCl	821	919	968	1020	1078	1115	1164	1239	1322	1407	790
fluoride	KF	885	988	1039	1096	1156	1193	1245	1323	1411	1502	880
hydroxide	KOH	719	814	863	918	976	1013	1064	1142	1233	1327	380
iodide	KI	745	840	887	938	995	1030	1080	1152	1238	1324	723
Radon	Rn	−144.2	−132.4	−126.3	−119.2	−111.3	−106.2	−99.0	−87.7	−75.0	−61.8	−71
Rhenium heptoxide	Re$_2$O$_7$	212.5	237.5	248.0	261.0	272.0	280.0	289.0	307.0	336.0	362.4	296
Rubidium	Rb	297	358	389	422	459	482	514	563	620	679	38.5
bromide	RbBr	781	876	923	975	1031	1066	1114	1186	1267	1352	682
chloride	RbCl	792	887	937	990	1047	1084	1133	1207	1294	1381	715
fluoride	RbF	921	982	1016	1052	1096	1123	1168	1239	1322	1408	760
iodide	RbI	748	839	884	935	991	1026	1072	1141	1223	1304	642
Selenium	Se	356	413	442	473	506	527	554	594	637	680	217
dioxide	SeO$_2$	157.0	187.7	202.5	217.5	234.1	244.6	258.0	277.0	297.7	317.0	340
hexafluoride	SeF$_6$	−118.6	−105.2	−98.9	−92.3	−84.7	−80.0	−73.9	−64.8	−55.2	−45.8	−34.7
oxychloride	SeOCl$_2$	34.8	59.8	71.9	84.2	98.0	106.5	118.0	134.6	151.7	168.0	8.5
tetrachloride	SeCl$_4$	74.0	96.3	107.4	118.1	130.1	137.8	147.5	161.0	176.4	191.5	
Silicon	Si	1724	1835	1888	1942	2000	2036	2083	2151	2220	2287	1420
dioxide	SiO$_2$			1732	1798	1867	1911	1969	2053	2141	2227	1710
tetrachloride	SiCl$_4$	−63.4	−44.1	−34.4	−24.0	−12.1	−4.8	+5.4	21.0	38.4	56.8	−68.8
tetrafluoride	SiF$_4$	−144.0	−134.8	−130.4	−125.9	−120.8	−117.5	−113.3	−170.2	−100.7	−94.8	−90
Trichlorofluorosilane	SiFCl$_3$	−92.6	−76.4	−68.3	−59.0	−48.8	−42.2	−33.2	−19.3	−4.0	+12.2	−120.8
Iodosilane	SiH$_3$I		−53.0	−47.7	−33.4	−21.8	−14.3	−4.4	+10.7	27.9	45.4	−57.0
Diiodosilane	SiH$_2$I$_2$			3.8	18.0	34.1	52.6	64.0	79.4	101.8	125.5	149.5	−1.0
Disiloxan	(SiH$_3$)$_2$O	−112.5	−95.8	−88.2	−79.8	−70.4	−64.2	−55.9	−43.5	−29.3	−15.4	−144.2
Trisilane	Si$_3$H$_8$	−68.9	−49.7	−40.0	−29.0	−16.9	−9.0	+1.6	17.8	35.5	53.1	−117.2
Trisilazane	(SiH$_3$)$_3$N	−68.7	−49.9	−40.4	−30.0	−18.5	−11.0	−1.1	+14.0	31.0	48.7	−105.7
Tetrasilane	Si$_4$H$_{10}$	−27.7	−6.2	+4.3	15.8	28.4	36.6	47.4	63.6	81.7	100.0	−93.6
Octachlorotrisilane	Si$_3$Cl$_3$	46.3	74.7	89.3	104.2	121.5	132.0	146.0	166.2	189.5	211.4	
Hexachlorodisiloxane	(SiCl$_3$)$_2$O	−5.0	17.8	29.4	41.5	55.2	63.8	75.4	92.5	113.6	135.6	−33.2
Hexachlorodisilane	Si$_2$Cl$_6$	+4.0	27.4	38.8	51.5	65.3	73.9	85.4	102.2	120.6	139.0	−1.2
Tribromosilane	SiHBr$_3$	−30.5	−8.0	+3.4	16.0	30.0	39.2	51.6	70.2	90.2	111.8	−73.5
Trichlorosilane	SiHCl$_3$	−80.7	−62.6	−53.4	−43.8	−32.9	−25.8	−16.4	−1.8	+14.5	31.8	−126.6
Trifluorosilane	SiHF$_3$	−152.0	−142.7	−138.2	−132.9	−127.3	−123.7	−118.7	−111.3	−102.8	−95.0	−131.4
Dibromosilane	SiH$_2$Br$_2$	−60.9	−40.0	−29.4	−18.0	−5.2	+3.2	14.1	31.6	50.7	70.5	−70.2
Difluorosilane	SiH$_2$F$_2$	−146.7	−136.0	−130.4	−124.3	−117.6	−113.3	−107.3	−98.3	−87.6	−77.8	
Monobromosilane	SiH$_3$Br		−85.7	−77.3	−68.3	−57.8	−51.1	−42.3	−28.6	−13.3	+2.4	−93.9
Monochlorosilane	SiH$_3$Cl	−117.8	−104.3	−97.7	−90.1	−81.8	−76.0	−68.5	−57.0	−44.5	−30.4	
Monofluorosilane	SiH$_3$F	−153.0	−145.5	−141.2	−136.3	−130.8	−127.2	−122.4	−115.2	−106.8	−98.0	
Tribromofluorosilane	SiFBr$_3$	−46.1	−25.4	−15.1	−3.7	+9.2	17.4	28.6	45.7	64.6	83.8	−82.5
Dichlorodifluorosilane	SiF$_2$Cl$_2$	−124.7	−110.5	−102.9	−94.5	−85.0	−78.6	−70.3	−58.0	−45.0	−31.8	−139.7
Trifluorobromosilane	SiF$_3$Br								−69.8	−55.9	−41.7	−70.5

TABLE 2-7   Vapor Pressures of Inorganic Compounds, up to 1 atm (*Concluded*)

Compound		Pressure, mm Hg										Melting point, °C
		1	5	10	20	40	60	100	200	400	760	
Name	Formula	Temperature, °C										
Trifluorochlorosilane	SiF_3Cl	−144.0	−133.0	−127.0	−120.5	−112.8	−108.2	−101.7	−91.7	−81.0	−70.0	−142
Hexafluorodisilane	Si_2F_6	−81.0	−68.8	−63.1	−57.0	−50.6	−46.7	−41.7	−34.2	−26.4	−18.9	−18.6
Dichlorofluorobromosilane	$SiFCl_2Br$	−86.5	−68.4	−59.0	−48.8	−37.0	−29.0	−19.5	−3.2	+15.4	35.4	−112.3
Dibromochlorofluorosilane	$SiFClBr_2$	−65.2	−45.5	−35.6	−24.5	−12.0	−4.7	+6.3	23.0	43.0	59.5	−99.3
Silane	SiH_4	−179.3	−168.6	−163.0	−156.9	−150.3	−146.3	−140.5	−131.6	−122.0	−111.5	−185
Disilane	Si_2H_6	−114.8	−99.3	−91.4	−82.7	−72.8	−66.4	−57.5	−44.6	−29.0	−14.3	−132.6
Silver	Ag	1357	1500	1575	1658	1743	1795	1865	1971	2090	2212	960.5
chloride	AgCl	912	1019	1074	1134	1200	1242	1297	1379	1467	1564	455
iodide	AgI	820	927	983	1045	1111	1152	1210	1297	1400	1506	552
Sodium	Na	439	511	549	589	633	662	701	758	823	892	97.5
bromide	NaBr	806	903	952	1005	1063	1099	1148	1220	1304	1392	755
chloride	NaCl	865	967	1017	1072	1131	1169	1220	1296	1379	1465	800
cyanide	NaCN	817	928	983	1046	1115	1156	1214	1302	1401	1497	564
fluoride	NaF	1077	1186	1240	1300	1363	1403	1455	1531	1617	1704	992
hydroxide	NaOH	739	843	897	953	1017	1057	1111	1192	1286	1378	318
iodide	NaI	767	857	903	952	1005	1039	1083	1150	1225	1304	651
Strontium	Sr		847	898	953	1018	1057	1111	1192	1285	1384	800
Strontium oxide	SrO	2068	2198	2262	2333	2410						2430
Sulfur	S	183.8	223.0	243.8	264.7	288.3	305.5	327.2	359.7	399.6	444.6	112.8
monochloride	S_2Cl_2	−7.4	+15.7	27.5	40.0	54.1	63.2	75.3	93.5	115.4	138.0	−80
hexafluoride	SF_6	−132.7	−120.6	−114.7	−108.4	−101.5	−96.8	−90.9	−82.3	−72.6	−63.5	−50.2
Sulfuryl chloride	SO_2Cl_2		−35.1	−24.8	−13.4	−1.0	+7.2	17.8	33.7	51.3	69.2	−54.1
Sulfur dioxide	SO_2	−95.5	−83.0	−76.8	−69.7	−60.5	−54.6	−46.9	−35.4	−23.0	−10.0	−73.2
trioxide (α)	SO_3	−39.0	−23.7	−16.5	−9.1	−1.0	+4.0	10.5	20.5	32.6	44.8	16.8
trioxide (β)	SO_3	−34.0	−19.2	−12.3	−4.9	+3.2	8.0	14.3	23.7	32.6	44.8	32.3
trioxide (γ)	SO_3	−15.3	−2.0	+4.3	11.1	17.9	21.4	28.0	35.8	44.0	51.6	62.1
Tellurium	Te	520	605	650	697	753	789	838	910	997	1087	452
chloride	$TeCl_4$		233	253	273	287	304	330	360	392		224
fluoride	TeF_6	−111.3	−98.8	−92.4	−86.0	−78.4	−73.8	−67.9	−57.3	−48.2	−38.6	−37.8
Thallium	Tl	825	931	983	1040	1103	1143	1196	1274	1364	1457	3035
Thallous bromide	TlBr		490	522	559	598	621	653	703	759	819	460
chloride	TlCl		487	517	550	589	612	645	694	748	807	430
iodide	TlI	440	502	531	567	607	631	663	712	763	823	440
Thionyl bromide	$SOBr_2$	−6.7	+18.4	31.0	44.1	58.8	68.3	80.6	99.0	119.2	139.5	−52.2
Thionyl chloride	$SOCl_2$	−52.9	−32.4	−21.9	−10.5	+2.2	10.4	21.4	37.9	56.5	75.4	−104.5
Tin	Sn	1492	1634	1703	1777	1855	1903	1968	2063	2169	2270	231.9
Stannic bromide	$SnBr_4$		58.3	72.7	88.1	105.5	116.2	131.0	152.8	177.7	204.7	31.0
Stannous chloride	$SnCl_2$	316	366	391	420	450	467	493	533	577	623	246.8
Stannic chloride	$SnCl_4$	−22.7	−1.0	+10.0	22.0	35.2	43.5	54.7	72.0	92.1	113.0	−30.2
iodide	SnI_4		156.0	175.8	196.2	218.8	234.2	254.2	283.5	315.5	348.0	144.5
hydride	SnH_4	−140.0	−125.8	−118.5	−111.2	−102.3	−96.6	−89.2	−78.0	−65.2	−52.3	−149.9
Tin tetramethyl	$Sn(CH_3)_4$	−51.3	−31.0	−20.6	−9.3	+3.5	11.7	22.8	39.8	58.5	78.0	
trimethyl-ethyl	$Sn(CH_3)_3 \cdot C_2H_5$	−30.0	−7.6	+3.8	16.1	30.0	38.4	50.0	67.3	87.6	108.8	
trimethyl-propyl	$Sn(CH_3)_3 \cdot C_3H_7$	−12.0	+10.7	21.8	34.0	48.5	57.5	69.8	88.0	109.6	131.7	
Titanium chloride	$TiCl_4$	−13.9	+9.4	21.3	34.2	48.0	58.0	71.0	90.5	112.7	136.0	−30
Tungsten	W	3990	4337	4507	4690	4886	5007	5168	5403	5666	5927	3370
Tungsten hexafluoride	WF_6	−71.4	−56.5	−49.2	−41.5	−33.0	−27.5	−20.3	−10.0	+1.2	17.3	−0.5
Uranium hexafluoride	UF_6	−38.8	−22.0	−13.8	−5.2	+4.4	10.4	18.2	30.0	42.7	55.7	69.2
Vanadyl trichloride	$VOCl_3$	−23.2	+0.2	12.2	26.6	40.0	49.8	62.5	82.0	103.5	127.2	
Xenon	Xe	−168.5	−158.2	−152.8	−147.1	−141.2	−137.7	−132.8	−125.4	−117.1	−108.0	−111.6
Zinc	Zn	487	558	593	632	673	700	736	788	844	907	419.4
chloride	$ZnCl_2$	428	481	508	536	566	584	610	648	689	732	365
fluoride	ZnF_2	970	1055	1086	1129	1175	1207	1254	1329	1417	1497	872
diethyl	$Zn(C_2H_5)_2$	−22.4	0.0	+11.7	24.2	38.0	47.2	59.1	77.0	97.3	118.0	−28
Zirconium bromide	$ZrBr_4$	207	237	250	266	281	289	301	318	337	357	450
chloride	$ZrCl_4$	190	217	230	243	259	268	279	295	312	331	437
iodide	ZrI_4	264	297	311	329	344	355	369	389	409	431	499

TABLE 2-8 Vapor Pressures of Organic Compounds, up to 1 atm*

| Compound | | Pressure, mm Hg | | | | | | | | | | Melting point, °C |
Name	Formula	1	5	10	20	40	60	100	200	400	760	
Acenaphthalene	$C_{12}H_{10}$		114.8	131.2	148.7	168.2	181.2	197.5	222.1	250.0	277.5	95
Acetal	$C_6H_{14}O_2$	−23.0	−2.3	+8.0	19.6	31.9	39.8	50.1	66.3	84.0	102.2	
Acetaldehyde	C_2H_4O	−81.5	−65.1	−56.8	−47.8	−37.8	−31.4	−22.6	−10.0	+4.9	20.2	−123.5
Acetamide	C_2H_5NO	65.0	92.0	105.0	120.0	135.8	145.8	158.0	178.3	200.0	222.0	81
Acetanilide	C_8H_9NO	114.0	146.6	162.0	180.0	199.6	211.8	227.2	250.5	277.0	303.8	113.5
Acetic acid	$C_2H_4O_2$	−17.2	+6.3	17.5	29.9	43.0	51.7	63.0	80.0	99.0	118.1	16.7
anhydride	$C_4H_6O_3$	1.7	24.8	36.0	48.4	62.1	70.8	82.2	100.0	119.8	139.6	−73
Acetone	C_3H_6O	−59.4	−40.5	−31.1	−20.8	−9.4	−2.0	+7.7	22.7	39.5	56.5	−94.6
Acetonitrile	C_2H_3N	−47.0	−26.6	−16.3	−5.0	+7.7	15.9	27.0	43.7	62.5	81.8	−41
Acetophenone	C_8H_8O	37.1	64.0	78.0	92.4	109.4	119.8	133.6	154.2	178.0	202.4	20.5
Acetyl chloride	C_2H_3OCl	−50.0	−35.0	−27.6	−19.6	−10.4	−4.5	+3.2	16.1	32.0	50.8	−112.0
Acetylene	C_2H_2	−142.9	−133.0	−128.2	−122.8	−116.7	−112.8	−107.9	−100.3	−92.0	−84.0	−81.5
Acridine	$C_{13}H_9N$	129.4	165.8	184.0	203.5	224.2	238.7	256.0	284.0	314.3	346.0	110.5
Acrolein (2-propenal)	C_3H_4C	−64.5	−46.0	−36.7	−26.3	−15.0	−7.5	+2.5	17.5	34.5	52.5	−87.7
Acrylic acid	$C_3H_4O_2$	+3.5	27.3	39.0	52.0	66.2	75.0	86.1	103.3	122.0	141.0	14
Adipic acid	$C_6H_{10}O_4$	159.5	191.0	205.5	222.0	240.5	251.0	265.0	287.8	312.5	337.5	152
Allene (propadiene)	C_3H_4	−120.6	−108.0	−101.0	−93.4	−85.2	−78.8	−72.5	−61.3	−48.5	−35.0	−136
Allyl alcohol (propen-1-ol-3)	C_3H_6O	−20.0	+0.2	10.5	21.7	33.4	40.3	50.0	64.5	80.2	96.6	−129
chloride (3-chloropropene)	C_3H_5Cl	−70.0	−52.0	−42.9	−32.8	−21.2	−14.1	−4.5	10.4	27.5	44.6	−136.4
isopropyl ether	$C_6H_{12}O$	−43.7	−23.1	−12.9	−1.8	+10.9	18.7	29.0	44.3	61.7	79.5	
isothiocyanate	C_4H_5NS	−2.0	+25.3	38.3	52.1	67.4	76.2	89.5	108.0	129.5	150.7	−80
n-propyl ether	$C_6H_{12}O$	−39.0	−18.2	−7.9	+3.7	16.4	25.0	35.8	52.6	71.4	90.5	
4-Allylveratrole	$C_{11}H_{14}O_2$	85.0	113.9	127.0	142.8	158.3	169.6	183.7	204.0	226.2	248.0	
iso-Amyl acetate	$C_7H_{14}O_2$	0.0	+23.7	35.2	47.8	62.1	71.0	83.2	101.3	121.5	142.0	
n-Amyl alcohol	$C_5H_{12}O$	+13.6	34.7	44.9	55.8	68.0	75.5	85.8	102.0	119.8	137.8	
iso-Amyl alcohol	$C_5H_{12}O$	+10.0	30.9	40.8	51.7	63.4	71.0	80.7	95.8	113.7	130.6	−117.2
sec-Amyl alcohol (2-pentanol)	$C_5H_{12}O$	+1.5	22.1	32.2	42.6	54.1	61.5	70.7	85.7	102.3	119.7	
tert-Amyl alcohol	$C_5H_{12}O$	−12.9	+7.2	17.2	27.9	38.8	46.0	55.3	69.7	85.7	101.7	−11.9
sec-Amylbenzene	$C_{11}H_{16}$	29.0	55.8	69.2	83.8	100.0	110.4	124.1	145.2	168.0	193.0	
iso-Amyl benzoate	$C_{12}H_{16}O_2$	72.0	104.5	121.6	139.7	158.3	171.4	186.8	210.2	235.8	262.0	
bromide (1-bromo-3-methylbutane)	$C_5H_{11}Br$	−20.4	+2.1	13.6	26.1	39.8	48.7	60.4	78.7	99.4	120.4	
n-butyrate	$C_9H_{18}O_2$	21.2	47.1	59.9	74.0	90.0	99.8	113.1	133.2	155.3	178.6	
formate	$C_6H_{12}O_2$	−17.5	+5.4	17.1	30.0	44.0	53.3	65.4	83.2	102.7	123.3	
iodide (1-iodo-3-methylbutane)	$C_5H_{11}I$	−2.5	+21.9	34.1	47.6	62.3	71.9	84.4	103.8	125.8	148.2	
isobutyrate	$C_9H_{18}O_2$	14.8	40.1	52.8	66.6	81.8	91.7	104.4	124.2	146.0	168.8	
Amyl isopropionate	$C_8H_{16}O_2$	+8.5	33.7	46.3	60.0	75.5	85.2	97.6	117.3	138.4	160.2	
iso-Amyl isovalerate	$C_{10}H_{20}O_2$	27.0	54.4	68.6	83.8	100.6	110.3	125.1	146.1	169.5	194.0	
n-Amyl levulinate	$C_{10}H_{18}O_3$	81.3	110.0	124.0	139.7	155.8	165.2	180.5	203.1	227.4	253.2	
iso-Amyl levulinate	$C_{10}H_{18}O_3$	75.6	104.0	118.8	134.4	151.7	162.6	177.0	198.1	222.7	247.9	
nitrate	$C_5H_{11}NO_3$	+5.2	28.8	40.3	53.5	67.6	76.3	88.6	106.7	126.5	147.5	
4-tert-Amylphenol	$C_{11}H_{16}O$		109.8	125.5	142.3	160.3	172.6	189.0	213.0	239.5	266.0	93
Anethole	$C_{10}H_{12}O$	62.6	91.6	106.0	121.8	139.3	149.8	164.2	186.1	210.5	235.3	22.5
Angelonitrile	C_5H_7N	−8.0	+15.0	28.0	41.0	55.8	65.2	77.5	96.3	117.7	140.0	
Aniline	C_6H_7N	34.8	57.9	69.4	82.0	96.7	106.0	119.9	140.1	161.9	184.4	−6.2
2-Anilinoethanol	$C_8H_{11}NO$	104.0	134.3	149.6	165.7	183.7	194.0	209.5	230.6	254.5	279.6	
Anisaldehyde	$C_8H_8O_2$	73.2	102.6	117.8	133.5	150.5	161.7	176.7	199.0	223.0	248.0	2.5
o-Anisidine (2-methoxyaniline)	C_7H_9NO	61.0	88.0	101.7	116.1	132.0	142.1	155.2	175.3	197.3	218.5	5.2
Anthracene	$C_{14}H_{10}$	145.0	173.5	187.2	201.9	217.5	231.8	250.0	279.0	310.2	342.0	217.5
Anthraquinone	$C_{14}H_8O_2$	190.0	219.4	234.2	248.5	264.3	273.3	285.0	314.6	346.2	379.9	286
Azelaic acid	$C_9H_{13}O_4$	178.3	210.4	225.5	242.4	260.0	271.8	286.5	309.6	332.8	356.5	106.5
Azelaldehyde	$C_9H_{13}O$	33.3	58.4	71.6	85.0	100.2	110.0	123.0	142.1	163.4	185.0	
Azobenzene	$C_{12}H_{10}N_2$	103.5	135.7	151.5	168.3	187.9	199.8	216.0	240.0	266.1	293.0	68
Benzal chloride (α,α-Dichlorotoluene)	$C_7H_6Cl_2$	35.4	64.0	78.7	94.3	112.1	123.4	138.3	160.7	187.0	214.0	−16.1
Benzaldehyde	C_7H_6O	26.2	50.1	62.0	75.0	90.1	99.6	112.5	131.7	154.1	179.0	−26
Benzanthrone	$C_{17}H_{10}O$	225.0	274.5	297.2	322.5	350.0	368.8	390.0	426.5			174
Benzene	C_6H_6	−36.7	−19.6	−11.5	−2.6	+7.6	15.4	26.1	42.2	60.6	80.1	+5.5
Benzenesulfonylchloride	$C_6H_5ClO_2S$	65.9	96.5	112.0	129.0	147.7	158.2	174.5	198.0	224.0	251.5	14.5
Benzil	$C_{14}H_{10}O_2$	128.4	165.2	183.0	202.8	224.5	238.2	255.8	283.5	314.3	347.0	95
Benzoic acid	$C_7H_6O_2$	96.0	119.5	132.1	146.7	162.6	172.8	186.2	205.8	227.0	249.2	121.7
anhydride	$C_{14}H_{10}O_3$	143.8	180.0	198.0	218.0	239.8	252.7	270.4	299.1	328.8	360.0	42
Benzoin	$C_{14}H_{12}O_2$	135.6	170.2	188.1	207.0	227.6	241.7	258.0	284.4	313.5	343.0	132
Benzonitrile	C_7H_5N	28.2	55.3	69.2	83.4	99.6	109.8	123.5	144.1	166.7	190.6	−12.9
Benzophenone	$C_{13}H_{10}O$	108.2	141.7	157.6	175.8	195.7	208.2	224.4	249.8	276.8	305.4	48.5
Benzotrichloride (α,α,α-Trichlorotoluene)	$C_7H_5Cl_3$	45.8	73.7	87.6	102.7	119.8	130.0	144.3	165.6	189.2	213.5	−21.2
Benzotrifluoride (α,α,α-Trifluorotoluene)	$C_7H_5F_3$	−32.0	−10.3	−0.4	12.2	25.7	34.0	45.3	62.5	82.0	102.2	−29.3
Benzoyl bromide	C_7H_5BrO	47.0	75.4	89.8	105.4	122.6	133.4	147.7	169.2	193.7	218.5	0
chloride	C_7H_5ClO	32.1	59.1	73.0	87.6	103.8	114.7	128.0	149.5	172.8	197.2	−0.5
nitrile	C_8H_5NO	44.5	71.7	85.5	100.2	116.6	127.0	141.0	161.3	185.0	208.0	33.5
Benzyl acetate	$C_9H_{10}O_2$	45.0	73.4	87.6	102.3	119.6	129.8	144.0	165.5	189.0	213.5	−51.5
alcohol	C_7H_3O	58.0	80.8	92.6	105.8	119.8	129.3	141.7	160.0	183.0	204.7	−15.3

*Compiled from the extended tables published by D. R. Stull in *Ind. Eng. Chem.*, **39,** 517 (1947). For information on fuels see Hibbard, N.A.C.A. Research Mem. E56I21, 1956. For methane see Johnson (ed.), WADD-TR-60-56, 1960.

TABLE 2-8 Vapor Pressures of Organic Compounds, up to 1 atm (Continued)

Compound		Pressure, mm Hg										Melting point, °C
		1	5	10	20	40	60	100	200	400	760	
Name	Formula	Temperature, °C										
Benzylamine	C₇H₉N	29.0	54.8	67.7	81.8	97.3	107.3	120.0	140.0	161.3	184.5	
Benzyl bromide (α-bromotoluene)	C₇H₇Br	32.2	59.6	73.4	88.3	104.8	115.6	129.8	150.8	175.2	198.5	−4
chloride (α-chlorotoluene)	C₇H₇Cl	22.0	47.8	60.8	75.0	90.7	100.5	114.2	134.0	155.8	179.4	−39
cinnamate	C₁₆H₁₄O₂	173.8	206.3	221.5	239.3	255.8	267.0	281.5	303.8	326.7	350.0	39
Benzyldichlorosilane	C₇H₈Cl₂Si	45.3	70.2	83.2	96.7	111.8	121.3	133.5	152.0	173.0	194.3	
Benzyl ethyl ether	C₉H₁₂O	26.0	52.0	65.0	79.6	95.4	105.5	118.9	139.6	161.5	185.0	
phenyl ether	C₁₃H₁₂O	95.4	127.7	144.0	160.7	180.1	192.6	209.2	233.2	259.8	287.0	
isothiocyanate	C₈H₇NS	79.5	107.8	121.8	137.0	153.0	163.8	177.7	198.0	220.4	243.0	
Biphenyl	C₁₂H₁₀	70.6	101.8	117.0	134.2	152.5	165.2	180.7	204.2	229.4	254.9	69.5
1-Biphenyloxy-2,3-epoxypropane	C₁₅H₁₄O₂	135.3	169.9	187.2	205.8	226.3	239.7	255.0	280.4	309.8	340.0	
d-Bornyl acetate	C₁₂H₂₀O₂	46.9	75.7	90.2	106.0	123.7	135.7	149.8	172.0	197.5	223.0	29
Bornyl n-butyrate	C₁₄H₂₄O₂	74.0	103.4	118.0	133.8	150.7	161.8	176.4	198.0	222.2	247.0	
formate	C₁₁H₁₈O₂	47.0	74.8	89.3	104.0	121.2	131.7	145.8	166.4	190.2	214.0	
isobutyrate	C₁₄H₂₄O₂	70.0	99.8	114.0	130.0	147.2	157.6	172.2	194.2	218.2	243.0	
propionate	C₁₃H₂₂O₂	64.6	93.7	108.0	123.7	140.4	151.2	165.7	187.5	211.2	235.0	
Brassidic acid	C₂₂H₄₂O₂	209.6	241.7	256.0	272.9	290.0	301.5	316.2	336.8	359.6	382.5	61.5
Bromoacetic acid	C₂H₃BrO₂	54.7	81.6	94.1	108.2	124.0	133.8	146.3	165.8	186.7	208.0	49.5
4-Bromoanisole	C₇H₇BrO	48.8	77.8	91.9	107.8	125.0	136.0	150.1	172.7	197.5	223.0	12.5
Bromobenzene	C₆H₅Br	+2.9	27.8	40.0	53.8	68.6	78.1	90.8	110.1	132.3	156.2	−30.7
4-Bromobiphenyl	C₁₂H₉Br	98.0	133.7	150.6	169.8	190.8	204.5	221.8	248.2	277.7	310.0	90.5
1-Bromo-2-butanol	C₄H₉BrO	23.7	45.4	55.8	67.2	79.5	87.0	97.6	112.1	128.3	145.0	
1-Bromo-2-butanone	C₄H₇BrO	+6.2	30.0	41.8	54.2	68.2	77.3	89.2	107.0	126.3	147.0	
cis-1-Bromo-1-butene	C₄H₇Br	−44.0	−23.2	−12.8	−1.4	+11.5	19.8	30.8	47.8	66.8	86.2	
trans-1-Bromo-1-butene	C₄H₇Br	−38.4	−17.0	−6.4	+5.4	18.4	27.2	38.1	55.7	75.0	94.7	−100.3
2-Bromo-1-butene	C₄H₇Br	−47.3	−27.0	−16.8	−5.3	+7.2	15.4	26.3	42.8	61.9	81.0	−133.4
cis-2-Bromo-2-butene	C₄H₇Br	−39.0	−17.9	−7.2	+4.6	17.7	26.2	37.5	54.5	74.0	93.9	−111.2
trans-2-Bromo-2-butene	C₄H₇Br	−45.0	−24.1	−13.8	−2.4	+10.5	18.7	29.9	46.5	66.0	85.5	−114.6
1,4-Bromochlorobenzene	C₆H₄BrCl	32.0	59.5	72.7	87.8	103.8	114.8	128.0	149.5	172.6	196.9	
1-Bromo-1-chloroethane	C₂H₄BrCl	−36.0	−18.0	−9.4	0.0	+10.4	17.0	28.0	44.7	63.4	82.7	16.6
1-Bromo-2-chloroethane	C₂H₄BrCl	−28.8	−7.0	+4.1	16.0	29.7	38.0	49.5	66.8	86.0	106.7	−16.6
2-Bromo-4,6-dichlorophenol	C₆H₃BrCl₂O	84.0	115.6	130.8	147.7	165.8	177.6	193.2	216.5	242.0	268.0	68
1-Bromo-4-ethyl benzene	C₈H₉Br	30.4	42.5	74.0	90.2	108.5	121.0	135.5	156.5	182.0	206.0	−45.0
(2-Bromoethyl)-benzene	C₈H₉Br	48.0	76.2	90.5	105.8	123.2	133.8	148.2	169.8	194.0	219.0	
2-Bromoethyl 2-chloroethyl ether	C₄H₈BrClO	36.5	63.2	76.3	90.8	106.6	116.4	129.8	150.0	172.3	195.8	
(2-Bromoethyl)-cyclohexane	C₈H₁₅Br	38.7	66.6	80.5	95.8	113.0	123.7	138.0	160.0	186.2	213.0	
1-Bromoethylene	C₂H₃Br	−95.4	−77.8	−68.8	−58.8	−48.1	−41.2	−31.9	−17.2	−1.1	+15.8	−138
Bromoform (tribromomethane)	CHBr₃		22.0	34.0	48.0	63.6	73.4	85.9	106.1	127.9	150.5	8.5
1-Bromonaphthalene	C₁₀H₇Br	84.2	117.5	133.6	150.2	170.2	183.5	198.8	224.2	252.0	281.1	5.5
2-Bromo-4-phenylphenol	C₁₂H₉BrO	100.0	135.4	152.3	171.8	193.8	207.0	224.5	251.0	280.2	311.0	95
3-Bromopyridine	C₅H₄BrN	16.8	42.0	55.2	69.1	84.1	94.1	107.8	127.7	150.0	173.4	
2-Bromotoluene	C₇H₇Br	24.4	49.7	62.3	76.0	91.0	100.0	112.0	133.6	157.3	181.8	−28
3-Bromotoluene	C₇H₇Br	14.8	50.8	64.0	78.1	93.9	104.1	117.8	138.0	160.0	183.7	39.8
4-Bromotoluene	C₇H₇Br	10.3	47.5	61.1	75.2	91.8	102.3	116.4	137.4	160.2	184.5	28.5
3-Bromo-2,4,6-trichlorophenol	C₆H₂BrCl₃O	112.4	146.2	163.2	181.8	200.5	213.0	229.3	253.0	278.0	305.8	
2-Bromo-1,4-xylene	C₈H₉Br	37.5	65.0	78.8	94.0	110.6	121.6	135.7	156.4	181.0	206.7	+9.5
1,2-Butadiene (methyl allene)	C₄H₆	−89.0	−72.7	−64.2	−54.9	−44.3	−37.5	−28.3	−14.2	+1.8	18.5	
1,3-Butadiene	C₄H₆	−102.8	−87.6	−79.7	−71.0	−61.3	−55.1	−46.8	−33.9	−19.3	−4.5	−108.9
n-Butane	C₄H₁₀	−101.5	−85.7	−77.8	−68.9	−59.1	−52.8	−44.2	−31.2	−16.3	−0.5	−135
iso-Butane (2-methylpropane)	C₄H₁₀	−109.2	−94.1	−86.4	−77.9	−68.4	−62.4	−54.1	−41.5	−27.1	−11.7	−145
1,3-Butanediol	C₄H₁₀O₂	22.2	67.5	85.3	100.0	117.4	127.5	141.2	161.0	183.8	206.5	77
1,2,3-Butanetriol	C₄H₁₀O₃	102.0	132.0	146.0	161.0	178.0	188.0	202.5	222.0	243.5	264.0	
1-Butene	C₄H₈	−104.8	−89.4	−81.6	−73.0	−63.4	−57.2	−48.9	−36.2	−21.7	−6.3	−130
cis-2-Butene	C₄H₈	−96.4	−81.1	−73.4	−64.6	−54.7	−48.4	−39.8	−26.8	−12.0	+3.7	−138.9
trans-2-Butene	C₄H₈	−99.4	−84.0	−76.3	−67.5	−57.6	−51.3	−42.7	−29.7	−14.8	+0.9	−105.4
3-Butenenitrile	C₄H₅N	−19.6	+2.9	14.1	26.6	40.0	48.8	60.2	78.0	98.0	119.0	
iso-Butyl acetate	C₆H₁₂O₂	−21.2	+1.4	12.8	25.5	39.2	48.0	59.7	77.6	97.5	118.0	−98.9
n-Butyl acrylate	C₇H₁₂O₂	−0.5	+23.5	35.5	48.6	63.4	72.6	85.1	104.0	125.2	147.4	−64.6
alcohol	C₄H₁₀O	−1.2	+20.0	30.2	41.5	53.4	60.3	70.1	84.3	100.8	117.5	−79.9
iso-Butyl alcohol	C₄H₁₀O	−9.0	+11.6	21.7	32.4	44.1	51.7	61.5	75.9	91.4	108.0	−108
sec-Butyl alcohol	C₄H₁₀O	−12.2	+7.2	16.9	27.3	38.1	45.2	54.1	67.9	83.9	99.5	−114.7
tert-Butyl alcohol	C₄H₁₀O	−20.4	−3.0	+5.5	14.3	24.5	31.0	39.8	52.7	68.0	82.9	25.3
iso-Butyl amine	C₄H₁₁N	−50.0	−31.0	−21.0	−10.3	+1.3	8.8	18.8	32.0	50.7	68.6	−85.0
n-Butylbenzene	C₁₀H₁₄	22.7	48.8	62.0	76.3	92.4	102.6	116.2	136.9	159.2	183.1	−88.0
iso-Butylbenzene	C₁₀H₁₄	14.1	40.5	53.7	67.8	83.3	93.3	107.0	127.2	149.6	172.8	−51.5
sec-Butylbenzene	C₁₀H₁₄	18.6	44.2	57.0	70.6	86.2	96.0	109.5	128.8	150.3	173.5	−75.5
tert-Butylbenzene	C₁₀H₁₄	13.0	39.0	51.7	65.6	80.8	90.6	103.8	123.7	145.8	168.5	−58
iso-Butyl benzoate	C₁₁H₁₄O₂	64.0	93.6	108.6	124.2	141.8	152.0	166.4	188.2	212.8	237.0	
n-Butyl bromide (1-bromobutane)	C₄H₉Br	−33.0	−11.2	−0.3	+11.6	24.8	33.4	44.7	62.0	81.7	101.6	−112.4
iso-Butyl n-butyrate	C₈H₁₆O₂	+4.6	30.0	42.2	56.1	71.7	81.3	94.0	113.9	135.7	156.9	
carbamate	C₅H₁₁NO₂		83.7	96.4	110.1	125.3	134.6	147.2	165.7	186.0	206.5	65
Butyl carbitol (diethylene glycol butyl ether)	C₈H₁₈O₃	70.0	95.7	107.8	120.5	135.5	146.0	159.8	181.2	205.0	231.2	
n-Butyl chloride (1-chlorobutane)	C₄H₉Cl	−49.0	−28.9	−18.6	−7.4	+5.0	13.0	24.0	40.0	58.8	77.8	−123.1
iso-Butyl chloride	C₄H₉Cl	−53.8	−34.3	−24.5	−13.8	−1.9	+5.9	16.0	32.0	50.0	68.9	−131.2

TABLE 2-8 Vapor Pressures of Organic Compounds, up to 1 atm (Continued)

| Compound | | Pressure, mm Hg | | | | | | | | | | Melting point, °C |
| | | 1 | 5 | 10 | 20 | 40 | 60 | 100 | 200 | 400 | 760 | |
Name	Formula	Temperature, °C										
sec-Butyl chloride (2-Chlorobutane)	C₄H₉Cl	−60.2	−39.8	−29.2	−17.7	−5.0	+3.4	14.2	31.5	50.0	68.0	−131.3
tert-Butyl chloride	C₄H₉Cl				−19.0	−11.4	−1.0	+14.6	32.6	51.0		−26.5
sec-Butyl chloroacetate	C₆H₁₁ClO₂	17.0	41.8	54.6	68.2	83.6	93.0	105.5	124.1	146.0	167.8	
2-tert-Butyl-4-cresol	C₁₁H₁₅O	70.0	98.0	112.0	127.2	143.9	153.7	167.0	187.8	210.0	232.6	
4-tert-Butyl-2-cresol	C₁₁H₁₅O	74.3	103.7	118.0	134.0	150.8	161.7	176.2	197.8	221.8	247.0	
iso-Butyl dichloroacetate	C₆H₁₀Cl₂O₂	28.6	54.3	67.5	81.4	96.7	106.6	119.8	139.2	160.0	183.0	
2,3-Butylene glycol (2,3-butanediol)	C₄H₁₀O₂	44.0	68.4	80.3	93.4	107.8	116.3	127.8	145.6	164.0	182.0	22.5
2-Butyl-2-ethylbutane-1,3-diol	C₁₀H₂₂O₂	94.1	122.6	136.8	151.2	167.8	178.0	191.9	212.0	233.5	255.0	
2-tert-Butyl-4-ethylphenol	C₁₂H₁₅O	76.3	106.2	121.0	137.0	154.0	165.4	179.0	200.3	223.8	247.8	
n-Butyl formate	C₅H₁₀O₂	−26.4	−4.7	+6.1	18.0	31.6	39.8	51.0	67.9	86.2	106.0	
iso-Butyl formate	C₅H₁₀O₂	−32.7	−11.4	−0.8	+11.0	24.1	32.4	43.4	60.0	79.0	98.2	−95.3
sec-Butyl formate	C₅H₁₀O₂	−34.4	−13.3	−3.1	+8.4	21.3	29.6	40.2	56.8	75.2	93.6	
sec-Butyl glycolate	C₆H₁₂O₃	28.3	53.6	66.0	79.8	94.2	104.0	116.4	135.5	155.6	177.5	
iso-Butyl iodide (1-iodo-2-methylpropane)	C₄H₉I	−17.0	+5.8	17.0	29.8	42.8	51.8	63.5	81.0	100.3	120.4	−90.7
isobutyrate	C₈H₁₆O₂	+4.1	28.0	39.9	52.4	67.2	75.9	88.0	106.3	126.3	147.5	−80.7
isovalerate	C₉H₁₈O₂	16.0	41.2	53.8	67.7	82.7	92.4	105.2	124.8	146.4	168.7	
levulinate	C₉H₁₆O₃	65.0	92.1	105.9	120.2	136.2	147.0	160.2	181.8	205.5	229.9	
naphthylketone (1-isovaleronaphthone)	C₁₅H₁₆O	136.0	167.9	184.0	201.6	219.7	231.5	246.7	269.7	294.0	320.0	
2-sec-Butylphenol	C₁₀H₁₄O	57.4	86.0	100.8	116.1	133.4	143.9	157.3	179.7	203.8	228.0	
2-tert-Butylphenol	C₁₀H₁₄O	56.6	84.2	98.1	113.0	129.2	140.0	153.5	173.8	196.3	219.5	
4-iso-Butylphenol	C₁₀H₁₄O	72.1	100.9	115.5	130.3	147.2	157.0	171.2	192.1	214.7	237.0	
4-sec-Butylphenol	C₁₀H₁₄O	71.4	100.5	114.8	130.3	147.8	157.9	172.4	194.3	217.6	242.1	
4-tert-Butylphenol	C₁₀H₁₄O	70.0	99.2	114.0	129.5	146.0	156.0	170.2	191.5	214.0	238.0	99
2-(4-tert-Butylphenoxy)ethyl acetate	C₁₄H₂₀O₃	118.0	150.0	165.8	183.3	201.5	212.8	228.0	250.3	277.6	304.4	
4-tert-Butylphenyl dichlorophosphate	C₁₀H₁₃Cl₂O₂P	96.0	129.6	146.0	164.0	184.3	197.2	214.3	240.0	268.2	299.0	
tert-Butyl phenyl ketone (pivalophenone)	C₁₁H₁₄O	57.8	85.7	99.0	114.3	130.4	140.8	154.0	175.0	197.7	220.0	
iso-Butyl propionate	C₇H₁₄O₂	−2.3	+20.9	32.3	44.8	58.5	67.6	79.5	97.0	116.4	136.8	−71
4-tert-Butyl-2,5-xylenol	C₁₂H₁₈O	88.2	119.8	135.0	151.0	169.8	180.3	195.0	217.5	241.3	265.3	
4-tert-Butyl-2,6-xylenol	C₁₂H₁₈O	74.0	103.9	119.0	135.0	152.2	163.6	176.0	196.0	217.8	239.8	
6-tert-Butyl-2,4-xylenol	C₁₂H₁₈O	70.3	100.2	115.0	131.0	148.5	158.2	172.0	192.3	214.2	236.5	
6-tert-Butyl-3,4-xylenol	C₁₂H₁₈O	83.9	113.6	127.0	143.0	159.7	170.0	184.0	204.5	226.7	249.5	
Butyric acid	C₄H₈O₂	25.5	49.8	61.5	74.0	88.0	96.5	108.0	125.5	144.5	163.5	−74
iso-Butyric acid	C₄H₈O₂	14.7	39.3	51.2	64.0	77.8	86.3	98.0	115.8	134.5	154.5	−47
Butyronitrile	C₄H₇N	−20.0	+2.1	13.4	25.7	38.4	47.3	59.0	76.7	96.8	117.5	
iso-Valerophenone	C₁₁H₁₄O	58.3	87.0	101.4	116.8	133.8	144.6	158.0	180.1	204.2	228.0	
Camphene	C₁₀H₁₆			47.2	60.4	75.7	85.0	97.9	117.5	138.7	160.5	50
Campholenic acid	C₁₀H₁₆O₂	97.6	125.7	139.8	153.9	170.0	180.0	193.7	212.7	234.0	256.0	
d-Camphor	C₁₀H₁₆O	41.5	68.6	82.3	97.5	114.0	124.0	138.0	157.9	182.0	209.2	178.5
Camphylamine	C₁₀H₁₉N	45.3	74.0	83.7	97.6	112.5	122.0	134.6	153.0	173.8	195.0	
Capraldehyde	C₁₀H₂₀O	51.9	78.8	92.0	106.3	122.2	132.0	145.3	164.8	186.3	208.5	
Capric acid	C₁₀H₂₀O₂	125.0	142.0	152.2	165.0	179.9	189.8	200.0	217.1	240.3	268.4	31.5
n-Caproic acid	C₆H₁₂O₂	71.4	89.5	99.5	111.8	125.0	133.3	144.0	160.8	181.0	202.0	−1.5
iso-Caproic acid	C₆H₁₂O₂	66.2	83.0	94.0	107.0	120.4	129.6	141.4	158.3	181.0	207.7	−35
iso-Caprolactone	C₆H₁₀O₂	38.3	66.0	80.3	95.7	112.3	123.2	137.2	157.8	182.1	207.0	
Capronitrile	C₆H₁₁N	9.2	34.6	47.5	61.7	76.9	86.8	99.8	119.7	141.0	163.7	
Capryl alcohol (2-octanol)	C₈H₁₈O	32.8	57.6	70.0	83.3	98.0	107.4	119.8	138.0	157.5	178.5	−38.6
Caprylaldehyde	C₈H₁₆O	73.4	92.0	101.2	110.2	120.0	126.0	133.9	145.4	156.5	168.5	
Caprylic acid (octanoic acid)	C₈H₁₆O₂	92.3	114.1	124.0	136.4	150.6	160.0	172.2	190.3	213.9	237.5	16
Caprylonitrile	C₈H₁₅N	43.0	67.6	80.4	94.6	110.6	121.2	134.8	155.2	179.5	204.5	
Carbazole	C₁₂H₉N							248.2	265.0	292.5	323.0	354.8	244.8
Carbon dioxide	CO₂	−134.3	−124.4	−119.5	−114.4	−108.6	−104.8	−100.2	−93.0	−85.7	−78.2	−57.5
disulfide	CS₂	−73.8	−54.3	−44.7	−34.3	−22.5	−15.3	−5.1	+10.4	28.0	46.5	−110.8
monoxide	CO	−222.0	−217.2	−215.0	−212.8	−210.0	−208.1	−205.7	−201.3	−196.3	−191.3	−205.0
oxyselenide (carbonyl selenide)	COSe	−117.1	−102.3	−95.0	−86.3	−76.4	−70.2	−61.7	−49.8	−35.6	−21.9	
oxysulfide (carbonyl sulfide)	COS	−132.4	−119.8	−113.3	−106.0	−98.3	−93.0	−85.9	−75.0	−62.7	−49.9	−138.8
tetrabromide	CBr₄					96.3	106.3	119.7	139.7	163.5	189.5	90.1
tetrachloride	CCl₄	−50.0	−30.0	−19.6	−8.2	+4.3	12.3	23.0	38.3	57.8	76.7	−22.6
tetrafluoride	CF₄	−184.6	−174.1	−169.3	−164.3	−158.8	−155.4	−150.7	−143.6	−135.5	−127.7	−183.7
Carvacrol	C₁₀H₁₄O	70.0	98.4	113.2	127.9	145.2	155.3	169.7	191.2	213.8	237.0	+0.5
Carvone	C₁₀H₁₄O	57.4	86.1	100.4	116.1	133.0	143.8	157.3	179.6	203.5	227.5	
Chavibetol	C₁₀H₁₂O₂	83.6	113.3	127.0	143.2	159.8	170.7	185.5	206.8	229.8	254.0	
Chloral (trichloroacetaldehyde)	C₂ECl₃O	−37.8	−16.0	−5.0	+7.2	20.2	29.1	40.2	57.8	77.5	97.7	−57
hydrate (trichloroacetaldehyde hydrate)	C₂H₃Cl₃O₂	−9.8	+10.0	19.5	29.2	39.7	46.2	55.0	68.0	82.1	96.2	51.7
Chloranil	C₆Cl₄O₂	70.7	89.3	97.8	106.4	116.1	122.0	129.5	140.3	151.3	162.6	290
Chloroacetic acid	C₂H₃ClO₂	43.0	68.3	81.0	94.2	109.2	118.3	130.7	149.0	169.0	189.5	61.2
anhydride	C₄H₄Cl₂O₃	67.2	94.1	108.0	122.4	138.2	148.0	159.8	177.8	197.0	217.0	46
2-Chloroaniline	C₆H₆ClN	46.3	72.3	84.8	99.2	115.6	125.7	139.5	160.0	183.7	208.8	0
3-Chloroaniline	C₆H₆ClN	63.5	89.8	102.0	116.7	133.6	144.1	158.0	179.5	203.5	228.5	−10.4
4-Chloroaniline	C₆H₆ClN	59.3	87.9	102.1	117.8	135.0	145.8	159.9	182.3	206.6	230.5	70.5
Chlorobenzene	C₆H₅Cl	−13.0	+10.6	22.2	35.3	49.7	58.3	70.7	89.4	110.0	132.2	−45.2
2-Chlorobenzotrichloride (2-α,α,α-tetrachlorotoluene)	C₇H₄Cl₄	69.0	101.8	117.9	135.8	155.0	167.8	185.0	208.0	233.0	262.1	28.7

TABLE 2-8 Vapor Pressures of Organic Compounds, up to 1 atm (*Continued*)

Compound		Pressure, mm Hg										Melting point, °C
		1	5	10	20	40	60	100	200	400	760	
Name	Formula	Temperature, °C										
2-Chlorobenzotrifluoride (2-chloro-α,α,α-trifluorotoluene)	$C_7H_4ClF_3$	0.0	24.7	37.1	50.6	65.9	75.4	88.3	108.3	130.0	152.2	−6.0
2-Chlorobiphenyl	$C_{12}H_9Cl$	89.3	109.8	134.7	151.2	169.9	182.1	197.0	219.6	243.8	267.5	34
4-Chlorobiphenyl	$C_{12}H_9Cl$	96.4	129.8	146.0	164.0	183.8	196.0	212.5	237.8	264.5	292.9	75.5
α-Chlorocrotonic acid	$C_4H_5ClO_2$	70.0	95.6	108.0	121.2	135.6	144.4	155.9	173.8	193.2	212.0	
Chlorodifluoromethane	$CHClF_2$	−122.8	−110.2	−103.7	−96.5	−88.6	−83.4	−76.4	−65.8	−53.6	−40.8	−160
Chlorodimethylphenylsilane	$C_8H_{11}ClSi$	29.8	56.7	70.0	84.7	101.2	111.5	124.7	145.5	168.6	193.5	
1-Chloro-2-ethoxybenzene	C_8H_9ClO	45.8	72.8	86.5	101.5	117.8	127.8	141.8	162.0	185.5	208.0	
2-(2-Chloroethoxy) ethanol	$C_4H_9ClO_2$	53.0	78.3	90.7	104.1	118.4	127.5	139.5	157.2	176.5	196.0	
bis-2-Chloroethyl acetacetal	$C_6H_{12}Cl_2O_2$	56.2	83.7	97.6	112.2	127.8	138.0	150.7	169.8	190.5	212.6	
1-Chloro-2-ethylbenzene	C_8H_9Cl	17.2	43.0	56.1	70.3	86.2	96.4	110.0	130.2	152.2	177.6	−80.2
1-Chloro-3-ethylbenzene	C_8H_9Cl	18.6	45.2	58.1	73.0	89.2	99.6	113.6	133.8	156.7	181.1	−53.3
1-Chloro-4-ethylbenzene	C_8H_9Cl	19.2	46.4	60.0	75.5	91.8	102.0	116.0	137.0	159.8	184.3	−62.6
2-Chloroethyl chloroacetate	$C_4H_6Cl_2O_2$	46.0	72.1	86.0	100.0	116.0	126.2	140.0	159.8	182.2	205.0	
2-Chloroethyl 2-chloroisopropyl ether	$C_5H_{10}Cl_2O$	24.7	50.1	63.0	77.2	92.4	102.2	115.8	135.7	156.5	180.0	
2-Chloroethyl 2-chloropropyl ether	$C_5H_{10}Cl_2O$	29.8	56.5	70.0	84.8	101.5	111.8	125.6	146.3	169.8	194.1	
2-Chloroethyl α-methylbenzyl ether	$C_{10}H_{13}ClO$	62.3	91.4	106.0	121.8	139.6	150.0	164.8	186.3	210.8	235.0	
Chloroform (trichloromethane)	$CHCl_3$	−58.0	−39.1	−29.7	−19.0	−7.1	+0.5	10.4	25.9	42.7	61.3	−63.5
1-Chloronaphthalene	$C_{10}H_7Cl$	80.6	104.8	118.6	134.4	153.2	165.6	180.4	204.2	230.8	259.3	−20
4-Chlorophenethyl alcohol	C_8H_9ClO	84.0	114.3	129.0	145.0	162.0	173.5	188.1	210.0	234.5	259.3	
2-Chlorophenol	C_6H_5ClO	12.1	38.2	51.2	65.9	82.0	92.0	106.0	126.4	149.8	174.5	7
3-Chlorophenol	C_6H_5ClO	44.2	72.0	86.1	101.7	118.0	129.4	143.0	164.8	188.7	214.0	32.5
4-Chlorophenol	C_6H_5ClO	49.8	78.2	92.2	108.1	125.0	136.1	150.0	172.0	196.0	220.0	42
2-Chloro-3-phenylphenol	$C_{12}H_9ClO$	118.0	152.2	169.7	186.7	207.4	219.6	237.0	261.3	289.4	317.5	+6
2-Chloro-6-phenylphenol	$C_{12}H_9ClO$	119.8	153.7	170.7	189.8	208.2	220.0	237.1	261.6	289.5	317.0	
Chloropicrin (trichloronitromethane)	CCl_3NO_2	−25.5	−3.3	+7.8	20.0	33.8	42.3	53.8	71.8	91.8	111.9	−64
1-Chloropropene	C_3H_5Cl	−81.3	−63.4	−54.1	−44.0	−32.7	−25.1	−15.1	+1.3	18.0	37.0	−99.0
2-Chloropyridine	C_5H_4ClN	13.3	38.8	51.7	65.8	81.7	91.6	104.6	125.0	147.7	170.2	
3-Chlorostyrene	C_8H_7Cl	25.3	51.3	65.2	80.0	96.5	107.2	121.2	142.2	165.7	190.0	
4-Chlorostyrene	C_8H_7Cl	28.0	54.5	67.5	82.0	98.0	108.5	122.0	143.5	166.0	191.0	−15.0
1-Chlorotetradecane	$C_{14}H_{29}Cl$	98.5	131.8	148.2	166.2	187.0	199.8	215.5	240.3	267.5	296.0	+0.9
2-Chlorotoluene	C_7H_7Cl	+5.4	30.6	43.2	56.9	72.0	81.8	94.7	115.0	137.1	159.3	
3-Chlorotoluene	C_7H_7Cl	+4.8	30.3	43.2	57.4	73.0	83.2	96.3	116.6	139.7	162.3	
4-Chlorotoluene	C_7H_7Cl	+5.5	31.0	43.8	57.8	73.5	83.3	96.6	117.1	139.8	162.3	+7.3
Chlorotriethylsilane	$C_8H_{15}ClSi$	−4.9	+19.8	32.0	45.5	60.2	69.5	82.3	101.6	123.6	146.3	
1-Chloro-1,2,2-trifluoroethylene	C_2ClF_3	−116.0	−102.5	−95.9	−88.2	−79.7	−74.1	−66.7	−55.0	−41.7	−27.9	−157.5
Chlorotrifluoromethane	$CClF_3$	−149.5	−139.2	−134.1	−128.5	−121.9	−117.3	−111.7	−102.5	−92.7	−81.2	
Chlorotrimethylsilane	C_3H_9ClSi	−62.8	−43.6	−34.0	−23.2	−11.4	−4.0	+6.0	21.9	39.4	57.9	
trans-Cinnamic acid	$C_9H_8O_2$	127.5	157.8	173.0	189.5	207.1	217.8	232.4	253.3	276.7	300.0	133
Cinnamyl alcohol	$C_9H_{10}O$	72.6	102.5	117.8	133.7	151.0	162.0	177.8	199.8	224.6	250.0	33
Cinnamylaldehyde	C_9H_8O	76.1	105.8	120.0	135.7	152.2	163.7	177.7	199.3	222.4	246.0	−7.5
Citaconic anhydride	$C_5H_4O_3$	47.1	74.8	88.9	103.8	120.3	131.3	145.4	165.8	189.8	213.5	
cis-α-Citral	$C_{10}H_{16}O$	61.7	90.0	103.9	119.4	135.9	146.3	160.0	181.8	205.0	228.0	
d-Citronellal	$C_{10}H_{18}O$	44.0	71.4	84.8	99.8	116.1	126.2	140.1	160.0	183.8	206.5	
Citronellic acid	$C_{10}H_{18}O_2$	99.5	127.3	141.4	155.6	171.9	182.1	195.4	214.5	236.6	257.0	
Citronellol	$C_{10}H_{20}O$	66.4	93.6	107.0	121.5	137.2	147.2	159.8	179.8	201.0	221.5	
Citronellyl acetate	$C_{12}H_{22}O_2$	74.7	100.2	113.0	126.0	140.5	149.7	161.0	178.8	197.8	217.0	
Coumarin	$C_9H_6O_2$	106.0	137.8	153.4	170.0	189.0	200.5	216.5	240.0	264.7	291.0	70
o-Cresol (2-cresol; 2-methylphenol)	C_7H_8O	38.2	64.0	76.7	90.5	105.8	115.5	127.4	146.7	168.4	190.8	30.8
m-Cresol (3-cresol; 3-methylphenol)	C_7H_8O	52.0	76.0	87.8	101.4	116.0	125.5	138.0	157.3	179.0	202.8	10.9
p-Cresol (4-cresol; 4-methylphenol)	C_7H_8O	53.0	76.5	88.6	102.3	117.7	127.0	140.0	157.3	179.4	201.8	35.5
cis-Crotonic acid	$C_4H_6O_2$	33.5	57.4	69.0	82.0	96.0	104.5	116.3	133.9	152.2	171.9	15.5
trans-Crotonic acid	$C_4H_6O_2$			80.0	93.0	107.8	116.7	128.0	146.0	165.5	185.0	72
cis-Crotononitrile	C_4H_5N	−29.0	−7.1	+4.0	16.4	30.0	38.5	50.1	68.0	88.0	108.0	
trans-Crotononitrile	C_4H_5N	−19.5	+3.5	15.0	27.8	41.8	50.9	62.8	81.1	101.5	122.8	
Cumene	C_9H_{12}	+2.9	26.8	38.3	51.5	66.1	75.4	88.1	107.3	129.2	152.4	−96.0
4-Cumidene	$C_9H_{13}N$	60.0	88.2	102.2	117.8	134.2	145.0	158.0	180.0	203.2	227.0	
Cuminal	$C_{10}H_{12}O$	58.0	87.3	102.0	117.9	135.2	146.0	160.0	182.8	206.7	232.0	
Cuminyl alcohol	$C_{10}H_{14}O$	74.2	103.7	118.0	133.8	150.3	161.7	176.2	197.9	221.7	246.6	
2-Cyano-2-n-butyl acetate	$C_7H_{11}NO_2$	42.0	68.7	82.0	96.2	111.8	121.5	133.8	152.2	173.4	195.2	
Cyanogen	C_2N_2	−95.8	−83.2	−76.8	−70.1	−62.7	−57.9	−51.8	−42.6	−33.0	−21.0	−34.4
bromide	$CBrN$	−35.7	−18.3	−10.0	−1.0	+8.6	14.7	22.6	33.8	46.0	61.5	58
chloride	$CClN$	−76.7	−61.4	−53.8	−46.1	−37.5	−32.1	−24.9	−14.1	−2.3	+13.1	−6.5
iodide	CIN	25.2	47.2	57.7	68.6	80.3	88.0	97.6	111.5	126.1	141.1	
Cyclobutane	C_4H_8	−92.0	−76.0	−67.9	−58.7	−48.4	−41.8	−32.8	−18.9	−3.4	+12.9	−50
Cyclobutene	C_4H_6	−99.1	−83.4	−75.4	−66.6	−56.4	−50.0	−41.2	−27.8	−12.2	+2.4	
Cyclohexane	C_6H_{12}	−45.3	−25.4	−15.9	−5.0	+6.7	14.7	25.5	42.0	60.8	80.7	+6.6
Cyclohexaneethanol	$C_8H_{16}O$	50.4	77.2	90.0	104.0	119.8	129.8	142.7	161.7	183.5	205.4	
Cyclohexanol	$C_6H_{12}O$	21.0	44.0	56.0	68.8	83.0	91.8	103.7	121.7	141.4	161.0	23.9
Cyclohexanone	$C_6H_{10}O$	+1.4	26.4	38.7	52.5	67.8	77.5	90.4	110.3	132.5	155.6	−45.0
2-Cyclohexyl-4,6-dinitrophenol	$C_{12}H_{14}N_2O_5$	132.8	161.8	175.9	191.2	206.7	216.0	229.0	248.7	269.8	291.5	
Cyclopentane	C_5H_{10}	−68.0	−49.6	−40.4	−30.1	−18.6	−11.3	−1.3	+13.8	31.0	49.3	−93.7
Cyclopropane	C_3H_6	−116.8	−104.2	−97.5	−90.3	−82.3	−77.0	−70.0	−59.1	−46.9	−33.5	−126.6
Cymene	$C_{10}H_{14}$	17.3	43.9	57.0	71.1	87.0	97.2	110.8	131.4	153.5	177.2	−68.2

TABLE 2-8 Vapor Pressures of Organic Compounds, up to 1 atm (*Continued*)

Compound		Pressure, mm Hg										Melting point, °C
		1	5	10	20	40	60	100	200	400	760	
Name	Formula	Temperature, °C										°C
cis-Decalin	$C_{10}H_{18}$	22.5	50.1	64.2	79.8	97.2	108.0	123.2	145.4	169.9	194.6	−43.3
trans-Decalin	$C_{10}H_{18}$	−0.8	+30.6	47.2	65.3	85.7	98.4	114.6	136.2	160.1	186.7	−30.7
Decane	$C_{10}H_{22}$	16.5	42.3	55.7	69.8	85.5	95.5	108.6	128.4	150.6	174.1	−29.7
Decan-2-one	$C_{10}H_{20}O$	44.2	71.9	85.8	100.7	117.1	127.8	142.0	163.2	186.7	211.0	+3.5
1-Decene	$C_{10}H_{20}$	14.7	40.3	53.7	67.8	83.3	93.5	106.5	126.7	149.2	172.0	
Decyl alcohol	$C_{10}H_{22}O$	69.5	97.3	111.3	125.8	142.1	152.0	165.8	186.2	208.8	231.0	+7
Decyltrimethylsilane	$C_{13}H_{30}Si$	67.4	96.4	111.0	126.5	144.0	154.3	169.5	191.0	215.5	240.0	
Dehydroacetic acid	$C_8H_8O_4$	91.7	122.0	137.3	153.0	171.0	181.5	197.5	219.5	244.5	269.0	
Desoxybenzoin	$C_{14}H_{12}O$	123.3	156.2	173.5	192.0	212.0	224.5	241.3	265.2	293.0	321.0	60
Diacetamide	$C_4H_7NO_2$	70.0	95.0	108.0	122.6	138.2	148.0	160.6	180.8	202.0	223.0	78.5
Diacetylene (1,3-butadiyne)	C_4H_2	−82.5	−68.0	−61.2	−53.8	−45.9	−41.0	−34.0	−20.9	−6.1	+9.7	−34.9
Diallyldichlorosilane	$C_6H_{10}Cl_2Si$	+9.5	34.8	47.4	61.3	76.4	86.3	99.7	119.4	142.0	165.3	
Diallyl sulfide	$C_6H_{10}S$	−9.5	+14.4	26.6	39.7	54.2	63.7	75.8	94.8	116.1	138.6	−83
Diisoamyl ether	$C_{10}H_{22}O$	18.6	44.3	57.0	70.7	86.3	96.0	109.6	129.0	150.3	173.4	
oxalate	$C_{12}H_{22}O_4$	85.4	116.0	131.4	147.7	165.7	177.0	192.2	215.0	240.0	265.0	
sulfide	$C_{10}H_{22}S$	43.0	73.0	87.6	102.7	120.0	130.6	145.3	166.4	191.0	216.0	
Dibenzylamine	$C_{14}H_{15}N$	118.3	149.8	165.6	182.2	200.2	212.2	227.3	249.8	274.3	300.0	−26
Dibenzyl ketone (1,3-diphenyl-2-propanone)	$C_{15}H_{14}O$	125.5	159.8	177.6	195.7	216.6	229.4	246.6	272.3	301.7	330.5	34.5
1,4-Dibromobenzene	$C_6H_4Br_2$	61.0	79.3	87.7	103.6	120.8	131.6	146.5	168.5	192.5	218.6	87.5
1,2-Dibromobutane	$C_4H_8Br_2$	7.5	33.2	46.1	60.0	76.0	86.0	99.8	120.2	143.5	166.3	−64.5
dl-2,3-Dibromobutane	$C_4H_8Br_2$	+5.0	30.0	41.6	56.4	72.0	82.0	95.3	115.7	138.0	160.5	
meso-2,3-Dibromobutane	$C_4H_8Br_2$	+1.5	26.6	39.3	53.2	68.0	78.0	91.7	111.8	134.2	157.3	−34.5
1,2-Dibromodecane	$C_{10}H_{20}Br_2$	95.7	123.6	137.3	151.0	167.4	177.5	190.2	209.6	229.8	250.4	
Di(2-bromoethyl) ether	$C_4H_8Br_2O$	47.7	75.3	88.5	103.6	119.8	130.0	144.0	165.0	188.0	212.5	
α,β-Dibromomaleic anhydride	$C_4H_2Br_2O_3$	50.0	78.0	92.0	106.7	123.5	133.8	147.7	168.0	192.0	215.0	
1,2-Dibromo-2-methylpropane	$C_4H_8Br_2$	−28.8	−3.0	+10.5	25.7	42.3	53.7	68.8	92.1	119.8	149.0	−70.3
1,3-Dibromo-2-methylpropane	$C_4H_8Br_2$	14.0	40.0	53.0	67.5	83.5	93.7	107.4	117.8	150.6	174.6	
1,2-Dibromopentane	$C_5H_{10}Br_2$	19.8	45.4	58.0	72.0	87.4	97.4	110.1	130.2	151.8	175.0	
1,2-Dibromopropane	$C_3H_6Br_2$	−7.0	+17.3	29.4	42.3	57.2	66.4	78.7	97.8	118.5	141.6	−55.5
1,3-Dibromopropane	$C_3H_6Br_2$	+9.7	35.4	48.0	62.1	77.8	87.8	101.3	121.7	144.1	167.5	−34.4
2,3-Dibromopropene	$C_3H_4Br_2$	−6.0	+17.9	30.0	43.2	57.8	67.0	79.5	98.0	119.5	141.2	
2,3-Dibromo-1-propanol	$C_3H_6Br_2O$	57.0	84.5	98.2	113.5	129.8	140.0	153.0	173.8	196.0	219.0	
Diisobutylamine	$C_8H_{19}N$	−5.1	+18.4	30.6	43.7	57.8	67.0	79.2	97.6	118.0	139.5	−70
2,6-Ditert-butyl-4-cresol	$C_{15}H_{24}O$	85.8	116.2	131.0	147.0	164.1	175.2	190.0	212.8	237.6	262.5	
4,6-Ditert-butyl-2-cresol	$C_{15}H_{24}O$	86.2	117.3	132.4	149.0	167.4	179.0	194.0	217.5	243.4	269.3	
4,6-Ditert-butyl-3-cresol	$C_{15}H_{24}O$	103.7	135.2	150.0	167.0	185.3	196.1	211.0	233.0	257.1	282.0	
2,6-Ditert-butyl-4-ethylphenol	$C_{16}H_{26}O$	89.1	121.4	137.0	154.0	172.1	183.9	198.0	220.0	244.0	268.6	
4,6-Ditert-butyl-3-ethylphenol	$C_{16}H_{26}O$	111.5	142.6	157.4	174.0	192.3	204.4	218.0	241.7	264.6	290.0	
Diisobutyl oxalate	$C_{10}H_{18}O_4$	63.2	91.2	105.3	120.3	137.5	147.8	161.8	183.5	205.8	229.5	
2,4-Ditert-butylphenol	$C_{14}H_{22}O$	84.5	115.4	130.0	146.0	164.3	175.8	190.0	212.5	237.0	260.8	
Dibutyl phthalate	$C_{16}H_{22}O_4$	148.2	182.1	198.2	216.2	235.8	247.8	263.7	287.0	313.5	340.0	
sulfide	$C_8H_{18}S$	+21.7	51.8	66.4	80.5	96.0	105.8	118.6	138.0	159.0	182.0	−79.7
Diisobutyl *d*-tartrate	$C_{12}H_{22}O_6$	117.8	151.8	169.0	188.0	208.5	221.6	239.5	264.7	294.0	324.0	73.5
Dicarvacryl-mono-(6-chloro-2-xenyl) phosphate	$C_{32}H_{34}ClO_4P$	204.2	234.5	249.3	264.5	280.5	290.7	304.9	323.8	342.0	361.0	
Dicarvacryl-2-tolyl phosphate	$C_{27}H_{33}O_4P$	180.2	209.3	221.8	237.0	251.5	260.3	272.5	290.0	309.8	330.0	
Dichloroacetic acid	$C_2H_2Cl_2O_2$	44.0	69.8	82.6	96.3	111.8	121.5	134.0	152.3	173.7	194.4	9.7
1,2-Dichlorobenzene	$C_6H_4Cl_2$	20.0	46.0	59.1	73.4	89.4	99.5	112.9	133.4	155.8	179.0	−17.6
1,3-Dichlorobenzene	$C_6H_4Cl_2$	12.1	39.0	52.0	66.2	82.0	92.2	105.0	125.9	149.0	173.0	−24.2
1,4-Dichlorobenzene	$C_6H_4Cl_2$			54.8	69.2	84.8	95.2	108.4	128.3	150.2	173.9	53.0
1,2-Dichlorobutane	$C_4H_8Cl_2$	−23.6	−0.3	+11.5	24.5	37.7	47.8	60.2	79.7	100.8	123.5	
2,3-Dichlorobutane	$C_4H_8Cl_2$	−25.2	−3.0	+8.5	21.2	35.0	43.9	56.0	74.0	94.2	116.0	−80.4
1,2-Dichloro-1,2-difluoroethylene	$C_2Cl_2F_2$	−82.0	−65.6	−57.3	−48.3	−38.2	−31.8	−23.0	−10.0	+5.0	20.9	−112
Dichlorodifluoromethane	CCl_2F_2	−118.5	−104.6	−97.8	−90.1	−81.6	−76.1	−68.6	−57.0	−43.9	−29.8	
Dichlorodiphenyl silane	$C_{12}H_{10}Cl_2Si$	109.6	142.4	158.0	176.0	195.5	207.5	223.8	248.0	275.5	304.0	
Dichlorodiisopropyl ether	$C_6H_{12}Cl_2O$	29.6	55.2	68.2	82.2	97.3	106.9	119.7	139.0	159.8	182.7	
Di(2-chloroethoxy) methane	$C_5H_{10}Cl_2O_2$	53.0	80.4	94.0	109.5	125.5	135.8	149.6	170.0	192.0	215.0	
Dichloroethoxymethylsilane	$C_3H_8Cl_2OSi$	−33.8	−12.1	−1.3	+11.3	24.4	32.6	44.1	61.0	80.3	100.6	
1,2-Dichloro-3-ethylbenzene	$C_8H_8Cl_2$	46.0	75.0	90.0	105.9	123.8	135.0	149.8	172.0	197.0	222.1	−40.8
1,2-Dichloro-4-ethylbenzene	$C_8H_8Cl_2$	47.0	77.2	92.3	109.6	127.5	139.0	153.3	176.0	201.7	226.6	−76.4
1,4-Dichloro-2-ethylbenzene	$C_8H_8Cl_2$	38.5	68.0	83.2	99.8	118.0	129.0	144.0	166.2	191.5	216.3	−61.2
cis-1,2-Dichloroethylene	$C_2H_2Cl_2$	−58.4	−39.2	−29.9	−19.4	−7.9	−0.5	+9.5	24.6	41.0	59.0	−80.5
trans-1,2-Dichloro ethylene	$C_2H_2Cl_2$	−65.4	−47.2	−38.0	−28.0	−17.0	−10.0	−0.2	+14.3	30.8	47.8	−50.0
Di(2-chloroethyl) ether	$C_4H_8Cl_2O$	23.5	49.3	62.0	76.0	91.5	101.5	114.5	134.0	156.0	178.5	
Dichlorofluoromethane	$CHCl_2F$	−91.3	−75.5	−67.5	−58.6	−48.8	−42.6	−33.9	−20.9	−6.2	+8.9	−135
1,5-Dichlorohexamethyltrisiloxane	$C_6H_{18}Cl_2O_2Si_3$	26.0	52.0	65.1	79.0	94.8	105.0	118.2	138.3	160.2	184.0	−53.0
Dichloromethylphenylsilane	$C_7H_8Cl_2Si$	35.7	63.5	77.4	92.4	109.5	120.0	134.2	155.5	180.2	205.5	
1,1-Dichloro-2-methylpropane	$C_4H_8Cl_2$	−31.0	−8.4	+2.6	14.6	28.2	37.0	48.2	65.8	85.4	106.0	
1,2-Dichloro-2-methylpropane	$C_4H_8Cl_2$	−25.8	−4.2	+6.7	18.7	32.0	40.2	51.7	68.9	87.8	108.0	
1,3-Dichloro-2-methylpropane	$C_4H_8Cl_2$	+8.0	+20.6	32.0	44.8	58.6	67.5	78.8	96.1	115.4	135.0	
2,4-Dichlorophenol	$C_6H_4Cl_2O$	53.0	80.0	92.8	107.7	123.4	133.5	146.0	165.2	187.5	210.0	45.0
2,6-Dichlorophenol	$C_6H_4Cl_2O$	59.5	87.6	101.0	115.5	131.6	141.8	154.6	175.5	197.7	220.0	

TABLE 2-8 Vapor Pressures of Organic Compounds, up to 1 atm (Continued)

Compound		Pressure, mm Hg										Melting point, °C
		1	5	10	20	40	60	100	200	400	760	
Name	Formula	Temperature, °C										
α,α-Dichlorophenylacetonitrile	$C_8H_5Cl_2N$	56.0	84.0	98.1	113.8	130.0	141.0	154.5	176.2	199.5	223.5	
Dichlorophenylarsine	$C_6H_5AsCl_2$	61.8	100.0	116.0	133.1	151.0	163.2	178.9	202.8	228.8	256.5	
1,2-Dichloropropane	$C_3H_6Cl_2$	−38.5	−17.0	−6.1	+6.0	19.4	28.0	39.4	57.0	76.0	96.8	
2,3-Dichlorostyrene	$C_8H_6Cl_2$	61.0	90.1	104.6	120.5	137.8	149.0	163.5	185.7	210.0	235.0	
2,4-Dichlorostyrene	$C_8H_6Cl_2$	53.5	82.2	97.4	111.8	129.2	140.0	153.8	176.0	200.0	225.0	
2,5-Dichlorostyrene	$C_8H_6Cl_2$	55.5	83.9	98.2	114.0	131.0	142.0	155.8	178.0	202.5	227.0	
2,6-Dichlorostyrene	$C_8H_6Cl_2$	47.8	75.7	90.0	105.5	122.4	133.3	147.6	169.0	193.5	217.0	
3,4-Dichlorostyrene	$C_8H_6Cl_2$	57.2	86.0	100.4	116.2	133.7	144.6	158.2	181.5	205.7	230.0	
3,5-Dichlorostyrene	$C_8H_6Cl_2$	53.5	82.2	97.4	111.8	129.2	140.0	153.8	176.0	200.0	225.0	
1,2-Dichlorotetraethylbenzene	$C_{14}H_{20}Cl_2$	105.6	138.7	155.0	172.5	192.2	204.8	220.7	245.6	272.8	302.0	
1,4-Dichlorotetraethylbenzene	$C_{14}H_{20}Cl_2$	91.7	126.1	143.8	162.0	183.2	195.8	212.0	238.5	265.8	296.5	
1,2-Dichloro-1,1,2,2-tetrafluoroethane	$C_2Cl_2F_4$	−95.4	−80.0	−72.3	−63.5	−53.7	−47.5	−39.1	−26.3	−12.0	+3.5	−94
Dichloro-4-tolylsilane	$C_7H_8Cl_2Si$	46.2	71.7	84.2	97.8	113.2	122.6	135.5	153.5	175.2	196.3	
3,4-Dichloro-α,α,α-trifluorotoluene	$C_7H_3Cl_2F_3$	11.0	38.3	52.2	67.3	84.0	95.0	109.2	129.0	150.5	172.8	−12.1
Dicyclopentadiene	$C_{10}H_8$		34.1	47.6	62.0	77.9	88.0	101.7	121.8	144.2	166.6	32.9
Diethoxydimethylsilane	$C_6H_{16}O_2Si$	−19.1	+2.4	13.3	25.3	38.0	46.3	57.6	74.2	93.2	113.5	
Diethoxydiphenylsilane	$C_{16}H_{20}O_2Si$	111.5	142.8	157.6	174.3	193.2	205.0	220.0	243.8	259.7	296.0	
Diethyl adipate	$C_{10}H_{18}O_4$	74.0	106.6	123.0	138.3	154.6	165.8	179.0	198.2	219.1	240.0	−21
Diethylamine	$C_4H_{11}N$			−33.0	−22.6	−11.3	−4.0	+6.0	21.0	38.0	55.5	−38.9
N-Diethylaniline	$C_{10}H_{15}N$	49.7	78.0	91.9	107.2	123.6	133.8	147.3	168.2	192.4	215.5	−34.4
Diethyl arsanilate	$C_{10}H_{16}As$ NO_3	38.0	62.6	74.8	88.0	102.6	111.8	123.8	141.9	161.0	181.0	
1,2-Diethylbenzene	$C_{10}H_{14}$	22.3	48.7	62.0	76.4	92.5	102.6	116.2	136.7	159.0	183.5	−31.4
1,3-Diethylbenzene	$C_{10}H_{14}$	20.7	46.8	59.9	74.5	90.4	100.7	114.4	134.8	156.9	181.1	−83.9
1,4-Diethylbenzene	$C_{10}H_{14}$	20.7	47.1	60.3	74.7	91.1	101.3	115.3	136.1	159.0	183.8	−43.2
Diethyl carbonate	$C_5H_{10}O_3$	−10.1	+12.3	23.8	36.0	49.5	57.9	69.7	86.5	105.8	125.8	−43
cis-Diethyl citraconate	$C_9H_{14}O_4$	59.8	88.3	103.0	118.2	135.7	146.2	160.0	182.3	206.5	230.3	
Diethyl dioxosuccinate	$C_8H_{10}O_6$	70.0	98.0	112.0	126.8	143.8	153.7	167.7	188.0	210.8	233.5	
Diethylene glycol	$C_4H_{10}O_3$	91.8	120.0	133.8	148.0	164.3	174.0	187.5	207.0	226.5	244.8	
Diethyleneglycol-bis-chloroacetate	$C_8H_{12}Cl_2O_5$	148.3	180.0	195.8	212.0	229.0	239.5	252.0	271.5	291.8	313.0	
Diethylene glycol dimethyl ether												
Di(2-methoxyethyl) ether	$C_6H_{14}O_3$	13.0	37.6	50.0	63.0	77.5	86.8	99.5	118.0	138.5	159.8	
glycol ethyl ether	$C_6H_{14}O_3$	45.3	72.0	85.8	100.3	116.7	126.8	140.3	159.0	180.3	201.9	
Diethyl ether	$C_4H_{10}O$	−74.3	−56.9	−48.1	−38.5	27.7	−21.8	−11.5	+2.2	17.9	34.6	−116.3
ethylmalonate	$C_9H_{16}O_4$	50.8	77.8	91.6	106.0	122.4	132.4	146.0	166.0	188.7	211.5	
fumarate	$C_8H_{12}O_4$	53.2	81.2	95.3	110.2	126.7	137.7	151.1	172.2	195.8	218.5	+0.6
glutarate	$C_9H_{16}O_4$	65.6	94.7	109.7	125.4	142.8	153.2	167.8	189.5	212.8	237.0	
Diethylhexadecylamine	$C_{20}H_{43}N$	139.8	175.8	194.0	213.5	235.0	248.5	265.5	292.8	324.6	355.0	
Diethyl itaconate	$C_9H_{14}O_4$	51.3	80.2	95.2	111.0	128.2	139.9	154.3	177.5	203.1	227.9	
ketone (3-pentanone)	$C_5H_{10}O$	−12.7	+7.5	17.2	27.9	39.4	46.7	56.2	70.6	86.3	102.7	−42
malate	$C_8H_{14}O_5$	80.7	110.4	125.3	141.2	157.8	169.0	183.9	205.3	229.5	253.4	
maleate	$C_8H_{12}O_4$	57.3	85.6	100.0	115.3	131.8	142.4	156.0	177.8	201.7	225.0	
malonate	$C_7H_{12}O_4$	40.0	67.5	81.3	95.9	113.3	123.0	136.2	155.5	176.8	198.9	−49.8
mesaconate	$C_9H_{14}O_4$	62.8	91.0	105.3	120.3	137.3	147.9	161.6	183.2	205.8	229.0	
oxalate	$C_6H_{10}O_4$	47.4	71.8	83.8	96.8	110.6	119.7	130.8	147.9	166.2	185.7	−40.6
phthalate	$C_{12}H_{14}O_4$	108.8	140.7	156.0	173.6	192.1	204.1	219.5	243.0	267.5	294.0	
sebacate	$C_{14}H_{26}O_4$	125.3	156.2	172.1	189.8	207.5	218.4	234.4	255.8	280.3	305.5	1.3
2,5-Diethylstyrene	$C_{12}H_{16}$	49.7	78.4	92.6	108.5	125.8	136.8	151.0	173.2	198.0	223.0	
Diethyl succinate	$C_8H_{14}O_4$	54.6	83.0	96.6	111.7	127.8	138.2	151.1	171.7	193.8	216.5	−20.8
isosuccinate	$C_8H_{14}O_4$	39.8	66.7	80.0	94.7	111.0	121.4	134.8	155.1	177.7	201.3	
sulfate	$C_4H_{10}O_4S$	47.0	74.0	87.7	102.1	118.0	128.6	142.5	162.5	185.5	209.5	−25.0
sulfide	$C_4H_{10}S$	−39.6	−18.6	−8.0	+3.5	16.1	24.2	35.0	51.3	69.7	88.0	−99.5
sulfite	$C_4H_{10}O_3S$	10.0	34.2	46.4	59.7	74.2	83.8	96.3	115.8	137.0	159.0	
d-Diethyl tartrate	$C_8H_{14}O_6$	102.0	133.0	148.0	164.2	182.3	194.0	208.5	230.4	254.8	280.0	17
dl-Diethyl tartrate	$C_8H_{14}O_6$	100.0	131.7	147.2	163.8	181.7	193.2	208.0	230.0	254.3	280.0	
3,5-Diethyltoluene	$C_{11}H_{16}$	34.0	61.5	75.3	90.2	107.0	117.7	131.7	152.4	176.5	200.7	
Diethylzinc	$C_4H_{10}Zn$	−22.4	0.0	+11.7	24.2	38.0	47.2	59.1	77.0	97.3	118.0	−28
1-Dihydrocarvone	$C_{10}H_{16}O$	46.6	75.5	90.0	106.0	123.7	134.7	149.7	171.8	197.0	223.0	
Dihydrocitronellol	$C_{10}H_{22}O$	68.0	91.7	103.0	115.0	127.6	136.7	145.9	160.2	176.8	193.5	
1,4-Dihydroxyanthraquinone	$C_{14}H_8O_4$	196.7	239.8	259.8	282.0	307.4	323.3	344.5	377.8	413.0	450.0	194
Dimethylacetylene (2-butyne)	C_4H_6	−73.0	−57.9	−50.5	−42.5	−33.9	−27.8	−18.8	−5.0	+10.6	27.2	−32.5
Dimethylamine	C_2H_7N	−87.7	−72.2	−64.6	−56.0	−46.7	−40.7	−32.6	−20.4	−7.1	+7.4	−96
N,N-Dimethylaniline	$C_8H_{11}N$	29.5	56.3	70.0	84.8	101.6	111.9	125.8	146.5	169.2	193.1	+2.5
Dimethyl arsanilate	$C_8H_{12}AsNO_3$	15.0	39.6	51.8	65.0	79.7	88.6	101.0	119.8	140.3	160.5	
Di(α-methylbenzyl) ether	$C_{16}H_{18}O$	96.7	128.3	144.0	160.3	179.6	191.5	206.8	229.7	254.8	281.0	
2,2-Dimethylbutane	C_6H_{14}	−69.3	−50.7	−41.5	−31.1	−19.5	−12.1	−2.0	+13.4	31.0	49.7	−99.8
2,3-Dimethylbutane	C_6H_{14}	−63.6	−44.5	−34.9	−24.1	−12.4	−4.9	+5.4	21.1	39.0	58.0	−128.2
Dimethyl citraconate	$C_7H_{10}O_4$	50.8	78.2	91.8	106.5	122.6	132.7	145.8	165.8	188.0	210.5	
1,1-Dimethylcyclohexane	C_8H_{16}	−24.4	−1.4	+10.3	23.0	37.3	45.7	57.9	76.2	97.2	119.5	−34
cis-1,2-Dimethylcyclohexane	C_8H_{16}	−15.9	+7.3	18.4	31.1	45.3	54.4	66.8	85.6	107.0	129.7	−50.0
trans-1,2-Dimethylcyclohexane	C_8H_{16}	−21.1	+1.7	13.0	25.6	39.7	48.7	61.0	79.6	100.9	123.4	−88.0
trans-1,3-Dimethylcyclohexane	C_8H_{16}	−19.4	+3.4	14.9	27.4	41.4	50.4	62.5	81.0	102.1	124.4	−92.0
cis-1,3-Dimethylcyclohexane	C_8H_{16}	−22.7	0.0	+11.2	23.6	37.5	46.4	58.5	76.9	97.8	120.1	−76.2
cis-1,4-Dimethylcyclohexane	C_8H_{16}	−20.0	+3.2	14.5	27.1	41.1	50.1	62.3	80.8	101.9	124.3	−87.4
trans-1,4-Dimethylcyclohexane	C_8H_{16}	−24.3	−1.7	+10.1	22.6	36.5	45.4	57.6	76.0	97.0	119.3	−36.9

TABLE 2-8   Vapor Pressures of Organic Compounds, up to 1 atm (*Continued*)

Name	Formula	1	5	10	20	40	60	100	200	400	760	Melting point, °C
Dimethyl ether	C_2H_6O	−115.7	−101.1	−93.3	−85.2	−76.2	−70.4	−62.7	−50.9	−37.8	−23.7	−138.5
2,2-Dimethylhexane	C_8H_{18}	−29.7	−7.9	+3.1	15.0	28.2	36.7	48.2	65.7	85.6	106.8	
2,3-Dimethylhexane	C_8H_{18}	−23.0	−1.1	+9.9	22.1	35.6	44.2	56.0	73.8	94.1	115.6	
2,4-Dimethylhexane	C_8H_{18}	−26.9	−5.3	+5.2	17.2	30.5	39.0	50.6	68.1	88.2	109.4	
2,5-Dimethylhexane	C_8H_{18}	−26.7	−5.5	+5.3	17.2	30.4	38.9	50.5	68.0	87.9	109.1	−90.7
3,3-Dimethylhexane	C_8H_{18}	−25.8	−4.4	+6.1	18.2	31.7	40.4	52.5	70.0	90.4	112.0	
3,4-Dimethylhexane	C_8H_{18}	−22.1	+0.2	11.3	23.5	37.1	45.8	57.7	75.6	96.0	117.7	
Dimethyl itaconate	$C_7H_{10}O_4$	69.3	94.0	106.6	119.7	133.7	142.6	153.7	171.0	189.8	208.0	38
1-Dimethyl malate	$C_6H_{10}O_5$	75.4	104.0	118.3	133.8	150.1	160.4	175.1	196.3	219.5	242.6	
Dimethyl maleate	$C_6H_8O_4$	45.7	73.0	86.4	101.3	117.2	127.1	140.4	160.0	182.2	205.0	
malonate	$C_5H_8O_4$	35.0	59.8	72.0	85.0	100.0	109.7	121.9	140.0	159.8	180.7	−62
trans-Dimethyl mesaconate	$C_7H_{10}O_4$	46.8	74.0	87.8	102.1	118.0	127.8	141.5	161.0	183.5	206.0	
2,7-Dimethyloctane	$C_{10}H_{22}$	+6.3	30.5	42.3	55.8	71.2	80.8	93.9	114.0	136.0	159.7	−52.8
Dimethyl oxalate	$C_4H_6O_4$	20.0	44.0	56.0	69.4	83.6	92.8	104.8	123.3	143.3	163.3	
2,2-Dimethylpentane	C_7H_{16}	−49.0	−28.7	−18.7	−7.5	+5.0	13.0	23.9	40.3	59.2	79.2	−123.7
2,3-Dimethylpentane	C_7H_{16}	−42.0	−20.8	−10.3	+1.1	13.9	22.1	33.3	50.1	69.4	89.8	−135
2,4-Dimethylpentane	C_7H_{16}	−48.0	−27.4	−17.1	−5.9	+6.5	14.5	25.4	41.8	60.6	80.5	−119.5
3,3-Dimethylpentane	C_7H_{16}	−45.9	−25.0	−14.4	−2.9	+9.9	18.1	29.3	46.2	65.5	86.1	−135.0
2,3-Dimethylphenol (2,3-xylenol)	$C_8H_{10}O$	56.0	83.8	97.6	112.0	129.2	139.5	152.2	173.0	196.0	218.0	75
2,4-Dimethylphenol (2,4-xylenol)	$C_8H_{10}O$	51.8	78.0	91.3	105.0	121.5	131.0	143.0	161.5	184.2	211.5	25.5
2,5-Dimethylphenol (2,5-xylenol)	$C_8H_{10}O$	51.8	78.0	91.3	105.0	121.5	131.0	143.0	161.5	184.2	211.5	74.5
3,4-Dimethylphenol (3,4-xylenol)	$C_8H_{10}O$	66.2	93.8	107.7	122.0	138.0	148.0	161.0	181.5	203.6	225.2	62.5
3,5-Dimethylphenol (3,5-xylenol)	$C_8H_{10}O$	62.0	89.2	102.4	117.0	133.3	143.5	156.0	176.2	197.8	219.5	68
Dimethylphenylsilane	$C_8H_{12}Si$	+5.3	30.3	42.6	56.2	71.4	81.3	94.2	114.2	136.4	159.3	
Dimethyl phthalate	$C_{10}H_{10}O_4$	100.3	131.8	147.6	164.0	182.8	194.0	210.0	232.7	257.8	283.7	
3,5-Dimethyl-1,2-pyrone	$C_7H_8O_2$	78.6	107.6	122.0	136.4	152.7	163.8	177.5	198.0	221.0	245.0	51.5
4,6-Dimethylresorcinol	$C_8H_{10}O_2$	49.0	76.8	90.7	105.8	122.5	133.2	147.3	167.8	192.0	215.0	
Dimethyl sebacate	$C_{12}H_{22}O_4$	104.0	139.8	156.2	175.8	196.0	208.0	222.6	245.0	269.6	293.5	38
2,4-Dimethylstyrene	$C_{10}H_{12}$	34.2	61.9	75.8	90.8	107.7	118.0	132.3	153.2	177.5	202.0	
2,5-Dimethylstyrene	$C_{10}H_{12}$	29.0	55.9	69.0	84.0	100.2	110.7	124.7	145.6	168.7	193.0	
α,α-Dimethylsuccinic anhydride	$C_6H_8O_3$	61.4	88.1	102.0	116.3	132.3	142.4	155.3	175.8	197.5	219.5	
Dimethyl sulfide	C_2H_6S	−75.6	−58.0	−49.2	−39.4	−28.4	−21.4	−12.0	+2.6	18.7	36.0	−83.2
d-Dimethyl tartrate	$C_6H_{10}O_6$	102.1	133.2	148.2	164.3	182.4	193.8	208.8	230.5	255.0	280.0	61.5
dl-Dimethyl tartrate	$C_6H_{10}O_6$	100.4	131.8	147.5	164.0	182.4	193.8	209.5	232.3	257.4	282.0	89
N,N-Dimethyl-2-toluidine	$C_9H_{13}N$	28.8	54.1	66.2	80.2	95.0	105.2	118.1	138.3	161.5	184.8	−61
N,N-Dimethyl-4-toluidine	$C_9H_{13}N$	50.1	74.3	86.7	100.0	116.3	126.4	140.3	161.6	185.4	209.5	
Di(nitrosomethyl) amine	$C_2H_5N_3O_2$	+3.2	27.8	40.0	53.7	68.2	77.7	90.3	110.0	131.3	153.0	
Diosphenol	$C_{10}H_{16}O_2$	66.7	95.4	109.0	124.0	141.2	151.3	165.6	186.2	209.5	232.0	
1,4-Dioxane	$C_4H_8O_2$	−35.8	−12.8	−1.2	+12.0	25.2	33.8	45.1	62.3	81.8	101.1	10
Dipentene	$C_{10}H_{16}$	14.0	40.4	53.8	68.2	84.3	94.6	108.3	128.2	150.5	174.6	
Diphenylamine	$C_{12}H_{11}N$	108.3	141.7	157.0	175.2	194.3	206.9	222.8	247.5	274.1	302.0	52.9
Diphenyl carbinol (benzhydrol)	$C_{13}H_{12}O$	110.0	145.0	162.0	180.9	200.0	212.0	227.5	250.0	275.6	301.0	68.5
chlorophosphate	$C_{12}H_{10}ClPO_3$	121.5	160.5	182.0	203.8	227.9	244.2	265.0	299.5	337.2	378.0	
disulfide	$C_{12}H_{10}S_2$	131.6	164.0	180.0	197.0	214.8	226.2	241.3	262.6	285.8	310.0	61
1,2-Diphenylethane (dibenzyl)	$C_{14}H_{14}$	86.8	119.8	136.0	153.7	173.7	186.0	202.8	227.8	255.0	284.0	51.5
Diphenyl ether	$C_{12}H_{10}O$	66.1	97.8	114.0	130.8	150.0	162.0	178.8	203.3	230.7	258.5	27
1,1-Diphenylethylene	$C_{14}H_{12}$	87.4	119.6	135.0	151.8	170.8	183.4	198.6	222.8	249.8	277.0	
trans-Diphenylethylene	$C_{14}H_{12}$	113.2	145.8	161.0	179.8	199.0	211.5	227.4	251.7	278.3	306.5	124
1,1-Diphenylhydrazine	$C_{12}H_{12}N_2$	126.0	159.3	176.1	194.0	213.5	225.9	242.5	267.2	294.0	322.2	44
Diphenylmethane	$C_{13}H_{12}$	76.0	107.4	122.8	139.8	157.8	170.2	186.3	210.7	237.5	264.5	26.5
Diphenyl sulfide	$C_{12}H_{10}S$	96.1	129.0	145.0	162.0	182.8	194.8	211.8	236.8	263.9	292.5	
Diphenyl-2-tolyl thiophosphate	$C_{18}H_{17}O_3PS$	159.7	179.8	201.6	215.5	230.6	240.4	252.5	270.3	290.0	310.0	
1,2-Dipropoxyethane	$C_8H_{18}O_2$	−38.8	−10.3	+5.0	22.3	42.3	55.8	74.2	103.8	140.0	180.0	
1,2-Diisopropylbenzene	$C_{12}H_{18}$	40.0	67.8	81.8	96.8	114.0	124.3	138.7	159.8	184.3	209.0	
1,3-Diisopropylbenzene	$C_{12}H_{18}$	34.7	62.3	76.0	91.2	107.9	118.2	132.3	153.7	177.6	202.0	−105
Dipropylene glycol	$C_6H_{14}O_3$	73.8	102.1	116.2	131.3	147.4	156.5	169.9	189.9	210.5	231.8	
Dipropyleneglycol monobutyl ether	$C_{10}H_{22}O_3$	64.7	92.0	106.0	120.4	136.3	146.3	159.8	180.0	203.8	227.0	
isopropyl ether	$C_9H_{20}O_3$	46.0	72.8	86.2	100.8	117.0	126.8	140.3	160.0	183.1	205.6	
Di-*n*-propyl ether	$C_6H_{14}O$	−43.3	−22.3	−11.8	0.0	+13.2	21.6	33.0	50.3	69.5	89.5	−122
Diisopropyl ether	$C_6H_{14}O$	−57.0	−37.4	−27.4	−16.7	−4.5	+3.4	13.7	30.0	48.2	67.5	−60
Di-*n*-propyl ketone (4-heptanone)	$C_7H_{14}O$	23.0	44.4	55.0	66.2	78.1	85.8	96.0	111.2	127.3	143.7	−32.6
Di-*n*-propyl oxalate	$C_8H_{14}O_4$	53.4	80.2	93.9	108.6	124.6	134.8	148.1	168.0	190.3	213.5	
Diisopropyl oxalate	$C_8H_{14}O_4$	43.2	69.0	81.9	95.6	110.5	120.0	132.6	151.2	171.8	193.5	
Di-*n*-propyl succinate	$C_{10}H_{18}O_4$	77.5	107.6	122.2	138.0	154.8	166.0	180.3	202.5	226.5	250.8	
Di-*n*-propyl *d*-tartrate	$C_{10}H_{18}O_6$	115.6	147.7	163.5	180.4	199.7	211.7	227.0	250.1	275.6	303.0	
Diisopropyl *d*-tartrate	$C_{10}H_{18}O_6$	103.7	133.7	148.2	164.0	181.8	192.6	207.3	228.2	251.8	275.0	
Divinyl acetylene (1,5-hexadiene-3-yne)	C_6H_6	−45.1	−24.4	−14.0	−2.8	+10.0	18.1	29.5	46.0	64.4	84.0	
1,3-Divinylbenzene	$C_{10}H_{10}$	32.7	60.0	73.8	88.7	105.5	116.0	130.0	151.4	175.2	199.5	−66.9
Docosane	$C_{22}H_{46}$	157.8	195.4	213.0	233.5	254.5	268.3	286.0	314.2	343.5	376.0	44.5
n-Dodecane	$C_{12}H_{26}$	47.8	75.8	90.0	104.6	121.7	132.1	146.2	167.2	191.0	216.2	−9.6
1-Dodecene	$C_{12}H_{24}$	47.2	74.0	87.8	102.4	118.6	128.5	142.3	162.2	185.5	208.0	−31.5
n-Dodecyl alcohol	$C_{12}H_{26}O$	91.0	120.2	134.7	150.0	167.2	177.8	192.0	213.0	235.7	259.0	24
Dodecylamine	$C_{12}H_{27}N$	82.8	111.8	127.8	141.6	157.4	168.0	182.1	203.0	225.0	248.0	
Dodecyltrimethylsilane	$C_{15}H_{34}Si$	91.2	122.1	137.7	153.8	172.1	184.2	199.5	222.0	248.0	273.0	
Elaidic acid	$C_{18}H_{34}O_2$	171.3	206.7	223.5	242.3	260.8	273.0	288.0	312.4	337.0	362.0	51.5

TABLE 2-8 Vapor Pressures of Organic Compounds, up to 1 atm (*Continued*)

Compound		Pressure, mm Hg										Melting point, °C
Name	Formula	1	5	10	20	40	60	100	200	400	760	Temperature, °C
Epichlorohydrin	C₃H₅ClO	−16.5	+5.6	16.6	29.0	42.0	50.6	62.0	79.3	98.0	117.9	−25.6
1,2-Epoxy-2-methylpropane	C₄H₈O	−69.0	−50.0	−40.3	−29.5	−17.3	−9.7	+1.2	17.5	36.0	55.5	
Erucic acid	C₂₂H₄₂O₂	206.7	239.7	254.5	270.6	289.1	300.2	314.4	336.5	358.8	381.5	33.5
Estragole (p-methoxy allyl benzene)	C₁₀H₁₂O	52.6	80.0	93.7	108.4	124.6	135.2	148.5	168.7	192.0	215.0	
Ethane	C₂H₆	−159.5	−148.5	−142.9	−136.7	−129.8	−125.4	−119.3	−110.2	−99.7	−88.6	−183.2
Ethoxydimethylphenylsilane	C₁₀H₁₆OSi	36.3	63.1	76.2	91.0	107.2	127.5	131.4	151.5	175.0	199.5	
Ethoxytrimethylsilane	C₅H₁₄OSi	−50.9	−31.0	−20.7	−9.8	+3.7	11.5	22.1	38.1	56.3	75.7	
Ethoxytriphenylsilane	C₂₀H₂₀OSi	167.0	198.2	213.5	230.0	247.0	258.3	273.5	295.0	319.5	344.0	
Ethyl acetate	C₄H₈O₂	−43.4	−23.5	−13.5	−3.0	+9.1	16.6	27.0	42.0	59.3	77.1	−82.4
acetoacetate	C₆H₁₀O₃	28.5	54.0	67.3	81.1	96.2	106.0	118.5	138.0	158.2	180.8	−45
Ethylacetylene (1-butyne)	C₄H₆	−92.5	−76.7	−68.7	−59.9	−50.0	−43.4	−34.9	−21.6	−6.9	+8.7	−130
Ethyl acrylate	C₅H₈O₂	−29.5	−8.7	+2.0	13.0	26.0	33.5	44.5	61.5	80.0	99.5	−71.2
α-Ethylacrylic acid	C₅H₈O₂	47.0	70.7	82.0	94.4	108.1	116.7	127.5	144.0	160.7	179.2	
α-Ethylacrylonitrile	C₅H₇N	−29.0	−6.4	+5.0	17.7	31.8	40.6	53.0	71.6	92.2	114.0	
Ethyl alcohol (ethanol)	C₂H₆O	−31.3	−12.0	−2.3	+8.0	19.0	26.0	34.9	48.4	63.5	78.4	−112
Ethylamine	C₂H₇N	−82.3	−66.4	−58.3	−48.6	−39.8	−33.4	−25.1	−12.3	+2.0	16.6	−80.6
4-Ethylaniline	C₈H₁₁N	52.0	80.0	93.8	109.0	125.7	136.0	149.8	170.6	194.2	217.4	−4
N-Ethylaniline	C₈H₁₁N	38.5	66.4	80.6	96.0	113.2	123.6	137.3	156.9	180.8	204.0	−63.5
2-Ethylanisole	C₉H₁₂O	29.7	55.9	69.0	83.1	98.8	109.0	122.3	142.1	164.2	187.1	
3-Ethylanisole	C₉H₁₂O	33.7	60.3	73.9	88.5	104.8	115.5	129.2	149.7	172.8	196.5	
4-Ethylanisole	C₉H₁₂O	33.5	60.2	73.9	88.5	104.7	115.4	128.4	149.2	172.3	196.5	
Ethylbenzene	C₈H₁₀	−9.8	+13.9	25.9	38.6	52.8	61.8	74.1	92.7	113.8	136.2	−94.9
Ethyl benzoate	C₉H₁₀O₂	44.0	72.0	86.0	101.4	118.2	129.0	143.2	164.8	188.4	213.4	−34.6
benzoylacetate	C₁₁H₁₂O₃	107.6	136.4	150.3	166.8	181.8	191.9	205.0	223.8	244.7	265.0	
bromide	C₂H₅Br	−74.3	−56.4	−47.5	−37.8	−26.7	−19.5	−10.0	+4.5	21.0	38.4	−117.8
α-bromoisobutyrate	C₆H₁₁BrO₂	10.6	35.8	48.0	61.8	77.0	86.7	99.8	119.7	141.2	163.6	
n-butyrate	C₆H₁₂O₂	−18.4	+4.0	15.3	27.8	41.5	50.1	62.0	79.8	100.0	121.0	−93.3
isobutyrate	C₆H₁₂O₂	−24.3	−2.4	+8.4	20.6	33.8	42.3	53.5	71.0	90.0	110.0	−88.2
Ethylcamphoronic anhydride	C₁₁H₁₆O₅	118.2	149.8	165.0	181.8	199.8	211.5	226.6	248.5	272.8	298.0	
Ethyl isocaproate	C₈H₁₆O₂	11.0	35.8	48.0	61.7	76.3	85.8	98.4	117.8	139.2	160.4	
carbamate	C₃H₇NO₂		65.8	77.8	91.0	105.6	114.8	126.2	144.2	164.0	184.0	49
carbanilate	C₉H₁₁NO₂	107.8	131.8	143.7	155.5	168.8	177.3	187.9	203.8	220.0	237.0	52.5
Ethylcetylamine	C₁₈H₃₉N	133.2	168.2	186.0	205.5	226.5	239.8	256.8	283.3	313.0	342.0	
Ethyl chloride	C₂H₅Cl	−89.8	−73.9	−65.8	−56.8	−47.0	−40.6	−32.0	−18.6	−3.9	+12.3	−139
chloroacetate	C₄H₇ClO₂	+1.0	25.4	37.5	50.4	65.2	74.0	86.0	103.8	123.8	144.2	−26
chloroglyoxylate	C₄H₅ClO₃	−5.1	+18.0	29.9	42.0	56.0	65.2	76.6	94.5	114.7	135.0	
α-chloropropionate	C₅H₉ClO₂	+6.6	30.2	41.9	54.3	68.2	77.3	89.3	107.2	126.2	146.5	
trans-cinnamate	C₁₁H₁₂O₂	87.6	108.5	134.0	150.3	169.2	181.2	196.0	219.3	245.0	271.0	12
3-Ethylcumene	C₁₁H₁₆	28.3	55.5	68.8	83.6	99.9	110.2	124.3	145.4	168.2	193.0	
4-Ethylcumene	C₁₁H₁₆	31.5	58.4	72.0	86.7	103.3	113.8	127.2	148.3	171.8	195.8	
Ethyl cyanoacetate	C₅H₇NO₂	67.8	93.5	106.0	119.8	133.8	142.1	152.8	169.8	187.8	206.0	
Ethylcyclohexane	C₈H₁₆	−14.5	+9.2	20.6	33.4	47.6	56.7	69.0	87.8	109.1	131.8	−111.3
Ethylcyclopentane	C₇H₁₄	−32.2	−10.8	−0.1	+11.7	25.0	33.4	45.0	62.4	82.3	103.4	−138.6
Ethyl dichloroacetate	C₄H₆Cl₂O₂	9.6	34.0	46.3	59.5	74.0	83.6	96.1	115.2	135.9	156.5	
N,N-diethyloxamate	C₈H₁₅NO₃	76.0	106.3	121.7	137.7	154.4	166.0	180.3	202.8	226.5	252.0	
N-Ethyldiphenylamine	C₁₄H₁₅N	98.3	130.2	146.0	162.8	182.0	193.7	209.8	233.0	258.8	286.0	
Ethylene	C₂H₄	−168.3	−158.3	−153.2	−147.6	−141.3	−137.3	−131.8	−123.4	−113.9	−103.7	−169
Ethylene-bis-(chloroacetate)	C₆H₈Cl₂O₄	112.0	142.4	158.0	173.5	191.0	201.8	215.0	237.3	259.5	283.5	
Ethylene chlorohydrin (2-chloroethanol)	C₂H₅ClO	−4.0	+19.0	30.3	42.5	56.0	64.1	75.0	91.8	110.0	128.8	−69
diamine (1,2-ethanediamine)	C₂H₈N₂	−11.0	+10.5	21.5	33.0	45.8	53.8	62.5	81.0	99.0	117.2	8.5
dibromide (1,2-dibromethane)	C₂H₄Br₂	−27.0	+4.7	18.6	32.7	48.0	57.9	70.4	89.8	110.1	131.5	10
dichloride (1,2-dichloroethane)	C₂H₄Cl₂	−44.5	−24.0	−13.6	−2.4	+10.0	18.1	29.4	45.7	64.0	82.4	−35.3
glycol (1,2-ethanediol)	C₂H₆O₂	53.0	79.7	92.1	105.8	120.0	129.5	141.8	158.5	178.5	197.3	−15.6
glycol diethyl ether (1,2-diethoxyethane)	C₆H₁₄O₂	−33.5	−10.2	+1.6	14.7	29.7	39.0	51.8	71.8	94.1	119.5	
glycol dimethyl ether (1,2-dimethoxyethane)	C₄H₁₀O₂	−48.0	−26.2	−15.3	−3.0	+10.7	19.7	31.8	50.0	70.8	93.0	
glycol monomethyl ether (2-methoxyethanol)	C₃H₈O₂	−13.5	+10.2	22.0	34.3	47.8	56.4	68.0	85.3	104.3	124.4	
oxide	C₂H₄O	−89.7	−73.8	−65.7	−56.6	−46.9	−40.7	−32.1	−19.5	−4.9	+10.7	−111.3
Ethyl α-ethylacetoacetate	C₈H₁₄O₃	40.5	67.3	80.2	94.6	110.3	120.6	133.8	153.2	175.6	198.0	
fluoride	C₂H₅F	−117.0	−103.8	−97.7	−90.0	−81.8	−76.4	−69.3	−58.0	−45.5	−32.0	
formate	C₃H₆O₂	−60.5	−42.2	−33.0	−22.7	−11.5	−4.3	−5.4	20.0	37.1	54.3	−79
2-furoate	C₇H₈O₃	37.6	63.8	77.1	91.5	107.5	117.5	130.4	150.1	172.5	195.0	34
glycolate	C₄H₈O₃	14.3	38.8	50.5	63.9	78.1	87.6	99.8	117.8	138.0	158.2	
3-Ethylhexane	C₈H₁₈	−20.0	+2.1	12.8	25.0	38.5	47.1	58.9	76.7	97.0	118.5	
2-Ethylhexyl acrylate	C₁₁H₂₀O₂	50.0	77.7	91.8	106.3	123.7	134.0	147.9	168.2	192.2	216.0	
Ethylidene chloride (1,1-dichloroethane)	C₂H₄Cl₂	−60.7	−41.9	−32.3	−21.9	−10.2	−2.9	+7.2	22.4	39.8	57.4	−96.7
fluoride (1,1-difluoroethane)	C₂H₄F₂	−112.5	−98.4	−91.7	−84.1	−75.8	−70.4	−63.2	−52.0	−39.5	−26.5	−117
Ethyl iodide	C₂H₅I	−54.4	−34.3	−24.3	−13.1	−0.9	+7.2	18.0	34.1	52.3	72.4	−105
Ethyl l-leucinate	C₈H₁₇NO₂	27.8	57.3	72.1	88.0	106.0	117.8	131.8	149.8	167.3	184.0	
Ethyl levulinate	C₇H₁₂O₃	47.3	74.0	87.3	101.8	117.7	127.6	141.3	160.2	183.0	206.2	
Ethyl mercaptan (ethanethiol)	C₂H₆S	−76.7	−59.1	−50.2	−40.7	−29.5	−22.4	−13.0	+1.5	17.7	35.0	−121
Ethyl methylcarbamate	C₄H₉NO₂	26.5	51.0	63.2	76.1	91.0	100.0	112.0	130.0	149.8	170.0	
Ethyl methyl ether	C₃H₈O	−91.0	−75.6	−67.8	−59.1	−49.4	−43.3	−34.8	−22.0	−7.8	+7.5	

TABLE 2-8 Vapor Pressures of Organic Compounds, up to 1 atm (Continued)

Compound		Pressure, mm Hg										Melting point, °C
		1	5	10	20	40	60	100	200	400	760	
Name	Formula	Temperature, °C										°C
1-Ethylnaphthalene	$C_{12}H_{12}$	70.0	101.4	116.8	133.8	152.0	164.1	180.0	204.6	230.8	258.1	−27
Ethyl α-naphthyl ketone (1-propionaphthone)	$C_{13}H_{12}O$	124.0	155.5	171.0	188.1	206.9	218.2	233.5	255.5	280.2	306.0	
Ethyl 3-nitrobenzoate	$C_9H_9NO_4$	108.1	140.2	155.0	173.6	192.6	205.0	220.3	244.6	270.6	298.0	47
3-Ethylpentane	C_7H_{16}	−37.8	−17.0	−6.8	+4.7	17.5	25.7	36.9	53.8	73.0	93.5	−118.6
4-Ethylphenetole	$C_{10}H_{14}O$	48.5	75.7	89.5	103.8	119.8	129.8	143.5	163.2	185.7	208.0	
2-Ethylphenol	$C_8H_{10}O$	46.2	73.4	87.0	101.5	117.9	127.9	141.8	161.6	184.5	207.5	−45
3-Ethylphenol	$C_8H_{10}O$	60.0	86.8	100.2	114.5	130.0	139.8	152.0	171.8	193.3	214.0	−4
4-Ethylphenol	$C_8H_{10}O$	59.3	86.5	100.2	115.0	131.3	141.7	154.2	175.0	197.4	219.0	46.5
Ethyl phenyl ether (phenetole)	$C_8H_{10}O$	18.1	43.7	56.4	70.3	86.6	95.4	108.4	127.9	149.8	172.0	−30.2
Ethyl propionate	$C_5H_{10}O_2$	−28.0	−7.2	+3.4	14.3	27.2	35.1	45.2	61.7	79.8	99.1	−72.6
Ethyl propyl ether	$C_5H_{12}O$	−64.3	−45.0	−35.0	−24.0	−12.0	−4.0	+6.8	23.3	41.6	61.7	
Ethyl salicylate	$C_9H_{10}O_3$	61.2	90.0	104.2	119.3	136.7	147.6	161.5	183.7	207.0	231.5	1.3
3-Ethylstyrene	$C_{10}H_{12}$	28.3	55.0	68.3	82.8	99.2	109.6	123.2	144.0	167.2	191.5	
4-Ethylstyrene	$C_{10}H_{12}$	26.0	52.7	66.3	80.8	97.3	107.6	121.5	142.0	165.0	189.0	
Ethylisothiocyanate	C_3H_5NS	13.2	+10.6	22.8	36.1	50.8	59.8	71.9	90.0	110.1	131.0	−5.9
2-Ethyltoluene	C_9H_{12}	9.4	34.8	47.6	61.2	76.4	86.0	99.0	119.0	141.4	165.1	
3-Ethyltoluene	C_9H_{12}	7.2	32.3	44.7	58.2	73.3	82.9	95.9	115.5	137.8	161.3	−95.5
4-Ethyltoluene	C_9H_{12}	7.6	32.7	44.9	58.5	73.6	83.2	96.3	116.1	136.4	162.0	
Ethyl trichloroacetate	$C_4H_5Cl_3O_2$	20.7	45.5	57.7	70.6	85.5	94.4	107.4	125.8	146.0	167.0	
Ethyltrimethylsilane	$C_5H_{14}Si$	−60.6	−41.4	−31.8	−21.0	−9.0	−1.2	+9.2	25.0	42.8	62.0	
Ethyltrimethyltin	$C_5H_{14}Sn$	−30.0	−7.6	+3.8	16.1	30.0	38.4	50.0	67.3	87.6	108.8	
Ethyl isovalerate	$C_7H_{14}O_2$	−6.1	+17.0	28.7	41.3	55.2	64.0	75.9	93.8	114.0	134.3	−99.3
2-Ethyl-1,4-xylene	$C_{10}H_{14}$	25.7	52.0	65.6	79.8	96.0	106.2	120.0	140.2	163.1	186.9	
4-Ethyl-1,3-xylene	$C_{10}H_{14}$	26.3	53.0	66.4	80.6	97.2	107.4	121.2	141.8	164.4	188.4	
5-Ethyl-1,3-xylene	$C_{10}H_{14}$	22.1	48.8	62.1	76.5	92.6	103.0	116.5	137.4	159.6	183.7	
Eugenol	$C_{10}H_{12}O_2$	78.4	108.1	123.0	138.7	155.8	167.3	182.2	204.7	228.3	253.5	
iso-Eugenol	$C_{10}H_{12}O_2$	86.3	117.0	132.4	149.0	167.0	178.2	194.0	217.2	242.3	267.5	−10
Eugenyl acetate	$C_{12}H_{14}O_3$	101.6	132.3	148.0	164.2	183.0	194.0	209.7	232.5	257.4	282.0	295
Fencholic acid	$C_{10}H_{16}O_2$	101.7	128.7	142.3	155.8	171.8	181.5	194.0	215.0	237.8	264.1	19
d-Fenchone	$C_{10}H_{16}O$	28.0	54.7	68.3	83.0	99.5	109.8	123.6	144.0	166.8	191.0	5
dl-Fenchyl alcohol	$C_{10}H_{18}O$	45.8	70.3	82.1	95.6	110.8	120.2	132.3	150.0	173.2	201.0	35
Fluorene	$C_{13}H_{10}$		129.3	146.0	164.2	185.2	197.8	214.7	240.3	268.6	295.0	113
Fluorobenzene	C_6H_5F	−43.4	−22.8	−12.4	−1.2	+11.5	19.6	30.4	47.2	65.7	84.7	−42.1
2-Fluorotoluene	C_7H_7F	−24.2	−2.2	+8.9	21.4	34.7	43.7	55.3	73.0	92.8	114.0	−80
3-Fluorotoluene	C_7H_7F	−22.4	−0.3	+11.0	23.4	37.0	45.8	57.5	75.4	95.4	116.0	−110.8
4-Fluorotoluene	C_7H_7F	−21.8	+0.3	11.8	24.0	37.8	46.5	58.1	76.0	96.1	117.0	
Formaldehyde	CH_2O			−88.0	−79.6	−70.6	−65.0	−57.3	−46.0	−33.0	−19.5	−92
Formamide	CH_3NO	70.5	96.3	109.5	122.5	137.5	147.0	157.5	175.5	193.5	210.5	
Formic acid	CH_2O_2	−20.0	−5.0	+2.1	10.3	24.0	32.4	43.8	61.4	80.3	100.6	8.2
trans-Fumaryl chloride	$C_4H_2Cl_2O_2$	+15.0	38.5	51.8	65.0	79.5	89.0	101.0	120.0	140.0	160.0	
Furfural (2-furaldehyde)	$C_5H_4O_2$	18.5	42.6	54.8	67.8	82.1	91.5	103.4	121.8	141.8	161.8	
Furfuryl alcohol	$C_5H_6O_2$	31.8	56.0	68.0	81.0	95.7	104.0	115.9	133.1	151.8	170.0	
Geraniol	$C_{10}H_{18}O$	69.2	96.8	110.0	125.6	141.8	151.5	165.3	185.6	207.8	230.0	
Geranyl acetate	$C_{12}H_{20}O_2$	73.5	102.7	117.9	133.0	150.0	160.3	175.2	196.3	219.8	243.3	
Geranyl n-butyrate	$C_{14}H_{24}O_2$	96.8	125.2	139.0	153.8	170.1	180.2	193.8	214.0	235.0	257.4	
Geranyl isobutyrate	$C_{14}H_{24}O_2$	90.9	119.6	133.0	147.9	164.0	174.0	187.7	207.6	228.5	251.0	
Geranyl formate	$C_{11}H_{18}O_2$	61.8	90.3	104.3	119.8	136.2	147.2	160.7	182.6	205.8	230.0	
Glutaric acid	$C_5H_8O_4$	155.5	183.8	196.0	210.5	226.3	235.5	247.0	265.0	283.5	303.0	97.5
Glutaric anhydride	$C_5H_6O_3$	100.8	133.3	149.5	166.0	185.5	196.2	212.5	236.5	261.0	287.0	
Glutaronitrile	$C_5H_6N_2$	91.3	123.7	140.0	156.5	176.4	189.5	205.5	230.0	257.3	286.2	
Glutaryl chloride	$C_5H_6Cl_2O_2$	56.1	84.0	97.8	112.3	128.3	139.1	151.8	172.4	195.3	217.0	
Glycerol	$C_3H_8O_3$	125.5	153.8	167.2	182.2	198.0	208.0	220.1	240.0	263.0	290.0	17.9
Glycerol dichlorohydrin (1,3-dichloro-2-propanol)	$C_3H_6Cl_2O$	28.0	52.2	64.7	78.0	93.0	102.0	114.8	133.3	153.5	174.3	
Glycol diacetate	$C_6H_{10}O_4$	38.3	64.1	77.1	90.8	106.1	115.8	128.0	147.8	168.3	190.5	−31
Glycolide (1,4-dioxane-2,6-dione)	$C_4H_4O_4$		103.0	116.6	132.0	148.6	158.2	173.2	194.0	217.0	240.0	97
Guaicol (2-methoxyphenol)	$C_7H_8O_2$	52.4	78.1	92.0	106.0	121.6	131.0	144.0	162.7	184.1	205.0	28.3
Heneicosane	$C_{21}H_{44}$	152.6	188.0	205.4	223.2	243.4	255.3	272.0	296.5	323.8	350.5	40.4
Heptacosane	$C_{27}H_{56}$	211.7	248.6	266.8	284.6	305.7	318.3	333.5	359.4	385.0	410.6	59.5
Heptadecane	$C_{17}H_{36}$	115.0	145.2	160.0	177.7	195.8	207.3	223.0	247.8	274.5	303.0	22.5
Heptaldehyde (enanthaldehyde)	$C_7H_{14}O$	12.0	32.7	43.0	54.0	66.3	74.0	84.0	102.0	125.5	155.0	−42
n-Heptane	C_7H_{16}	−34.0	−12.7	−2.1	+9.5	22.3	30.6	41.8	58.7	78.0	98.4	−90.6
Heptanoic acid (enanthic acid)	$C_7H_{14}O_2$	78.0	101.3	113.2	125.6	139.5	148.5	160.0	179.5	199.6	221.5	−10
1-Heptanol	$C_7H_{16}O$	42.4	64.3	74.7	85.8	99.8	108.0	119.5	136.6	155.6	175.8	34.6
Heptanoyl chloride (enanthyl chloride)	$C_7H_{13}ClO$	34.2	54.6	64.6	75.0	86.4	93.5	102.7	116.3	130.7	145.0	
2-Heptene	C_7H_{14}	−35.8	−14.1	−3.5	+8.3	21.5	30.0	41.3	58.6	78.1	98.5	
Heptylbenzene	$C_{13}H_{20}$	64.0	94.6	110.0	126.0	144.0	154.8	170.2	193.3	217.8	244.0	
Heptyl cyanide (enanthonitrile)	$C_7H_{13}N$	21.0	47.8	61.6	76.3	92.6	103.0	116.8	137.7	160.0	184.6	
Hexachlorobenzene	C_6Cl_6	114.4	149.3	166.4	185.7	206.0	219.0	235.5	258.5	283.5	309.4	230
Hexachloroethane	C_2Cl_6	32.7	49.8	73.5	87.6	102.3	112.0	124.2	143.1	163.8	185.6	186.6
Hexacosane	$C_{26}H_{54}$	204.0	240.0	257.4	275.8	295.2	307.8	323.2	348.4	374.6	399.8	56.6
Hexadecane	$C_{16}H_{34}$	105.3	135.2	149.8	164.7	181.3	193.2	208.5	231.7	258.3	287.5	18.5
1-Hexadecene	$C_{16}H_{32}$	101.6	131.7	146.2	162.0	178.8	190.8	205.0	226.8	250.0	274.0	4
n-Hexadecyl alcohol (cetyl alcohol)	$C_{16}H_{34}O$	122.7	158.3	177.8	197.8	219.8	234.3	251.7	280.2	312.7	344.0	49.3

TABLE 2-8 Vapor Pressures of Organic Compounds, up to 1 atm (Continued)

| Compound | | Pressure, mm Hg | | | | | | | | | | Melting point, °C |
Name	Formula	1	5	10	20	40	60	100	200	400	760	
		Temperature, °C										
n-Hexadecylamine (cetylamine)	$C_{16}H_{35}N$	123.6	157.8	176.0	195.7	215.7	228.8	245.8	272.2	300.4	330.0	
Hexaethylbenzene	$C_{18}H_{30}$		134.3	150.3	168.0	187.7	199.7	216.0	241.7	268.5	298.3	130
n-Hexane	C_6H_{14}	−53.9	−34.5	−25.0	−14.1	−2.3	+5.4	15.8	31.6	49.6	68.7	−95.3
1-Hexanol	$C_6H_{14}O$	24.4	47.2	58.2	70.3	83.7	92.0	102.8	119.6	138.0	157.0	−51.6
2-Hexanol	$C_6H_{14}O$	14.6	34.8	45.0	55.9	67.9	76.0	87.3	103.7	121.8	139.9	
3-Hexanol	$C_6H_{14}O$	+2.5	25.7	36.7	49.0	62.2	70.7	81.8	98.3	117.0	135.5	
1-Hexene	C_6H_{12}	−57.5	−38.0	−28.1	−17.2	−5.0	+2.8	13.0	29.0	46.8	66.0	−98.5
n-Hexyl levulinate	$C_{11}H_{20}O_3$	90.0	120.0	134.7	150.2	167.8	179.0	193.6	215.7	241.0	266.8	
n-Hexyl phenyl ketone (enanthophenone)	$C_{13}H_{18}O$	100.0	130.3	145.5	161.0	178.9	189.8	204.2	225.0	248.3	271.3	
Hydrocinnamic acid	$C_9H_{10}O_2$	102.2	133.5	148.7	165.0	183.3	194.0	209.0	230.8	255.0	279.8	48.5
Hydrogen cyanide (hydrocyanic acid)	CHN	−71.0	−55.3	−47.7	−39.7	−30.9	−25.1	−17.8	−5.3	+10.2	25.9	−13.2
Hydroquinone	$C_6H_6O_2$	132.4	153.3	163.5	174.6	192.0	203.0	216.5	238.0	262.5	286.2	170.3
4-Hydroxybenzaldehyde	$C_7H_6O_2$	121.2	153.2	169.7	186.8	206.0	217.5	233.5	256.8	282.6	310.0	115.5
α-Hydroxyisobutyric acid	$C_4H_8O_3$	73.5	98.5	110.5	123.8	138.0	146.4	157.7	175.2	193.8	212.0	79
α-Hydroxybutyronitrile	C_5H_9NO	41.0	65.8	77.8	90.7	104.8	113.9	125.0	142.0	159.8	178.8	
4-Hydroxy-3-methyl-2-butanone	$C_5H_{10}O_2$	44.6	69.3	81.0	94.0	108.2	117.4	129.0	146.5	165.5	185.0	
4-Hydroxy-4-methyl-2-pentanone	$C_6H_{12}O_2$	22.0	46.7	58.8	72.0	86.7	96.0	108.2	126.8	147.5	167.9	−47
3-Hydroxypropionitrile	C_3H_5NO	58.7	87.8	102.0	117.9	134.1	144.7	157.7	178.0	200.0	221.0	
Indene	C_9H_8	16.4	44.3	58.5	73.9	90.7	100.8	114.7	135.6	157.8	181.6	−2
Iodobenzene	C_6H_5I	24.1	50.6	64.0	78.3	94.4	105.0	118.3	139.8	163.9	188.6	−28.5
Iodononane	$C_9H_{19}I$	70.0	96.2	109.0	123.0	138.1	147.7	159.8	179.0	199.3	219.5	
2-Iodotoluene	C_7H_7I	37.2	65.9	79.8	95.6	112.4	123.8	138.1	160.0	185.7	211.0	
α-Ionone	$C_{13}H_{20}O$	79.5	108.8	123.0	139.0	155.6	166.3	181.2	202.5	225.2	250.0	
Isoprene	C_5H_8	−79.8	−62.3	−53.3	−43.5	−32.6	−25.4	−16.0	−1.2	+15.4	32.6	−146.7
Lauraldehyde	$C_{12}H_{24}O$	77.7	108.4	123.7	140.2	157.8	168.7	184.5	207.8	231.8	257.0	44.5
Lauric acid	$C_{12}H_{24}O_2$	121.0	150.6	166.0	183.6	201.4	212.7	227.5	249.8	273.8	299.2	48
Levulinaldehyde	$C_5H_8O_2$	28.1	54.9	68.0	82.7	98.3	108.4	121.8	142.0	164.0	187.0	
Levulinic acid	$C_5H_8O_3$	102.0	128.1	141.8	154.1	169.5	178.0	190.2	208.3	227.4	245.8	33.5
d-Limonene	$C_{10}H_{16}$	14.0	40.4	53.8	68.2	84.3	94.6	108.3	128.5	151.4	175.0	−96.9
Linalyl acetate	$C_{12}H_{20}O_2$	55.4	82.5	96.0	111.4	127.7	138.1	151.8	173.3	196.2	220.0	
Maleic anhydride	$C_4H_2O_3$	44.0	63.4	78.7	95.0	111.8	122.0	135.8	155.9	179.5	202.0	58
Menthane	$C_{10}H_{20}$	+9.7	35.7	48.3	62.7	78.3	88.6	102.1	122.7	146.0	169.5	
1-Menthol	$C_{10}H_{20}O$	56.0	83.2	96.0	110.3	126.1	136.1	149.4	168.3	190.2	212.0	42.5
Menthyl acetate	$C_{12}H_{22}O_2$	57.4	85.8	100.0	115.4	132.1	143.2	156.7	178.8	202.8	227.0	
benzoate	$C_{17}H_{24}O_2$	123.2	154.2	170.0	186.3	204.3	215.8	230.4	253.2	277.1	301.0	54.5
formate	$C_{11}H_{20}O_2$	47.3	75.8	90.0	105.8	123.0	133.8	148.0	169.8	194.2	219.0	
Mesityl oxide	$C_6H_{10}O$	−8.7	+14.1	26.0	37.9	51.7	60.4	72.1	90.0	109.8	130.0	−59
Methacrylic acid	$C_4H_6O_2$	25.5	48.5	60.0	72.7	86.4	95.3	106.6	123.9	142.5	161.0	15
Methacrylonitrile	C_4H_5N	−44.5	−23.3	−12.5	−0.6	+12.8	21.5	32.8	50.0	70.3	90.3	
Methane	CH_4	−205.9	−199.0	−195.5	−191.8	−187.7	−185.1	−181.4	−175.5	−168.8	−161.5	−182.5
Methanethiol	CH_4S	−90.7	−75.3	−67.5	−58.8	−49.2	−43.1	−34.8	−22.1	−7.9	+6.8	−121
Methoxyacetic acid	$C_3H_6O_3$	52.5	79.3	92.0	106.5	122.0	131.8	144.5	163.5	184.2	204.0	
N-Methylacetanilide	$C_9H_{11}NO$		103.8	118.8	135.1	152.2	164.2	179.8	202.3	227.4	253.0	102
Methyl acetate	$C_3H_6O_2$	−57.2	−38.6	−29.3	−19.1	−7.9	−0.5	+9.4	24.0	40.0	57.8	−98.7
acetylene (propyne)	C_3H_4	−111.0	−97.5	−90.5	−82.9	−74.3	−68.8	−61.3	−49.8	−37.2	−23.3	−102.7
acrylate	$C_4H_6O_2$	−43.7	−23.6	−13.5	−2.7	+9.2	17.3	28.0	43.9	61.8	80.2	
alcohol (methanol)	CH_4O	−44.0	−25.3	−16.2	−6.0	+5.0	12.1	21.2	34.8	49.9	64.7	−97.8
Methylamine	CH_5N	−95.8	−81.3	−73.8	−65.9	−56.9	−51.3	−43.7	−32.4	−19.7	−6.3	−93.5
N-Methylaniline	C_7H_9N	36.0	62.8	76.2	90.5	106.0	115.8	129.8	149.3	172.0	195.5	−57
Methyl anthranilate	$C_8H_9NO_2$	77.6	109.0	124.2	141.5	159.7	172.0	187.8	212.4	238.5	266.5	24
benzoate	$C_8H_8O_2$	39.0	64.4	77.3	91.8	107.8	117.4	130.8	151.4	174.7	199.5	−12.5
2-Methylbenzothiazole	C_8H_7NS	70.0	97.5	111.2	125.5	141.2	150.4	163.9	183.2	204.5	225.5	15.4
α-Methylbenzyl alcohol	$C_8H_{10}O$	49.0	75.2	88.0	102.1	117.8	127.4	140.3	159.0	180.7	204.0	
Methyl bromide	CH_3Br	−96.3	−80.6	−72.8	−64.0	−54.2	−48.0	−39.4	−26.5	−11.9	+3.6	−93
2-Methyl-1-butene	C_5H_{10}	−89.1	−72.8	−64.3	−54.8	−44.1	−37.3	−28.0	−13.8	+2.5	20.2	−135
2-Methyl-2-butene	C_5H_{10}	−75.4	−57.0	−47.9	−37.9	−26.7	−19.4	−9.9	+4.9	21.6	38.5	−133
Methyl isobutyl carbinol (2-methyl-4-pentanol)	$C_6H_{14}O$	−0.3	+22.1	33.3	45.4	58.2	67.0	78.0	94.9	113.5	131.7	
n-butyl ketone (2-hexanone)	$C_6H_{12}O$	+7.7	28.8	38.8	50.0	62.0	69.8	79.8	94.3	111.0	127.5	−56.9
isobutyl ketone (4-methyl-2-pentanone)	$C_6H_{12}O$	−1.4	+19.7	30.0	40.8	52.8	60.4	70.4	85.6	102.0	119.0	−84.7
n-butyrate	$C_5H_{10}O_2$	−26.8	−5.5	+5.0	16.7	29.6	37.4	48.0	64.3	83.1	102.3	
isobutyrate	$C_5H_{10}O_2$	−34.1	−13.0	−2.9	+8.4	21.0	28.9	39.6	55.7	73.6	92.6	−84.7
caprate	$C_{11}H_{22}O_2$	63.7	93.5	108.0	123.0	139.0	148.6	161.5	181.6	202.9	224.0	−18
caproate	$C_7H_{14}O_2$	+5.0	30.0	42.0	55.4	70.0	79.7	91.4	109.8	129.8	150	
caprylate	$C_9H_{18}O_2$	34.2	61.7	74.9	89.0	105.3	115.3	128.0	148.1	170.0	193.0	−40
chloride	CH_3Cl		−99.5	−92.4	−84.8	−76.0	−70.4	−63.0	−51.2	−38.0	−24.0	−97.7
chloroacetate	$C_3H_5ClO_2$	−2.9	19.0	30.0	41.5	54.5	63.0	73.5	90.5	109.5	130.3	−31.9
cinnamate	$C_{10}H_{10}O_2$	77.4	108.1	123.0	140.0	157.9	170.0	185.8	209.6	235.0	263.0	33.4
α-Methylcinnamic acid	$C_{10}H_{10}O_2$	125.7	155.0	169.8	185.2	201.8	212.0	224.8	245.0	266.8	288.0	
Methylcyclohexane	C_7H_{14}	−35.9	−14.0	−3.2	+8.7	22.0	30.5	42.1	59.6	79.6	100.9	−126.4
Methylcyclopentane	C_6H_{12}	−53.7	−33.8	−23.7	−12.8	−0.6	+7.2	17.9	34.0	52.3	71.8	−142.4
Methylcyclopropane	C_4H_8	−96.0	−80.6	−72.8	−64.0	−54.2	−48.0	−39.3	−26.0	−11.3	+4.5	
Methyl n-decyl ketone (n-dodecan-2-one)	$C_{12}H_{24}O$	77.1	106.0	120.4	136.0	152.4	163.8	177.5	199.0	222.5	246.5	
dichloroacetate	$C_3H_4Cl_2O_2$	3.2	26.7	38.1	50.7	64.7	73.6	85.4	103.2	122.6	143.0	
N-Methyldiphenylamine	$C_{13}H_{13}N$	103.5	134.0	149.7	165.8	184.0	195.4	210.1	232.8	257.0	282.0	−7.6

TABLE 2-8   Vapor Pressures of Organic Compounds, up to 1 atm* (Continued)

Compound		Pressure, mm Hg										Melting point, °C
		1	5	10	20	40	60	100	200	400	760	
Name	Formula	Temperature, °C										
Methyl n-dodecyl ketone (2-tetradecanone)	$C_{14}H_{28}O$	99.3	130.0	145.5	161.3	179.8	191.4	206.0	228.2	253.3	278.0	
Methylene bromide (dibromomethane)	CH_2Br_2	−35.1	−13.2	−2.4	+9.7	23.3	31.6	42.3	58.5	79.0	98.6	−52.8
chloride (dichloromethane)	CH_2Cl_2	−70.0	−52.1	−43.3	−33.4	−22.3	−15.7	−6.3	+8.0	24.1	40.7	−96.7
Methyl ethyl ketone (2-butanone)	C_4H_8O	−48.3	−28.0	−17.7	−6.5	+6.0	14.0	25.0	41.6	60.0	79.6	−85.9
2-Methyl-3-ethylpentane	C_8H_{18}	−24.0	−1.8	+9.5	21.7	35.2	43.9	55.7	73.6	94.0	115.6	−114.5
3-Methyl-3-ethylpentane	C_8H_{18}	−23.9	−1.4	+9.9	22.3	36.2	45.0	57.1	75.3	96.2	118.3	−90
Methyl fluoride	CH_3F	−147.3	−137.0	−131.6	−125.9	−119.1	−115.0	−109.0	−99.9	−89.5	−78.2	
formate	$C_2H_4O_2$	−74.2	−57.0	−48.6	−39.2	−28.7	−21.9	−12.9	+0.8	16.0	32.0	−99.8
α-Methylglutaric anhydride	$C_6H_8O_3$	93.8	125.4	141.8	157.7	177.5	189.9	205.0	229.1	255.5	282.5	
Methyl glycolate	$C_3H_6O_3$	+9.6	33.7	45.3	58.1	72.3	81.8	93.7	111.8	131.7	151.5	
2-Methylheptadecane	$C_{18}H_{38}$	119.8	152.0	168.7	186.0	204.8	216.3	231.5	254.5	279.8	306.5	
2-Methylheptane	C_8H_{18}	−21.0	+1.3	12.3	24.4	37.9	46.6	58.3	76.0	96.2	117.6	−109.5
3-Methylheptane	C_8H_{18}	−19.8	+2.6	13.3	25.4	38.9	47.6	59.4	77.1	97.4	118.9	−120.8
4-Methylheptane	C_8H_{18}	−20.4	+1.5	12.4	24.5	38.0	46.6	58.3	76.1	96.3	117.7	−121.1
2-Methyl-2-heptene	C_8H_{16}	−16.1	+6.7	17.8	30.4	44.0	52.8	64.6	82.3	102.2	122.5	
6-Methyl-3-hepten-2-ol	$C_8H_{16}O$	41.6	65.0	76.7	89.3	102.7	111.5	122.6	139.5	156.6	175.5	
6-Methyl-5-hepten-2-ol	$C_8H_{16}O$	41.9	66.0	77.8	90.4	104.0	112.8	123.8	140.0	156.6	174.3	
2-Methylhexane	C_7H_{16}	−40.4	−19.5	−9.1	+2.3	14.9	23.0	34.1	50.8	69.8	90.0	−118.2
3-Methylhexane	C_7H_{16}	−39.0	−18.1	−7.8	+3.6	16.4	24.5	35.6	52.4	71.6	91.9	
Methyl iodide	CH_3I	−55.0	−45.8	−35.6	−24.2	−16.9	−7.0	+8.0	25.3	42.4	−64.4	
laurate	$C_{13}H_{26}O_2$	87.8	117.9	133.2	149.0	166.0	176.8	190.8				5
levulinate	$C_6H_{10}O_3$	39.8	66.4	79.7	93.7	109.5	119.3	133.0	153.4	175.8	197.7	
methacrylate	$C_5H_8O_2$	−30.5	−10.0	+1.0	11.0	25.5	34.5	47.0	63.0	82.0	101.0	
myristate	$C_{15}H_{30}O_2$	115.0	145.7	160.8	177.8	195.8	207.5	222.6	245.3	269.8	295.5	18.5
α-naphthyl ketone (1-acetonaphthone)	$C_{12}H_{10}O$	115.6	146.3	161.5	178.4	196.8	208.6	223.8	246.7	270.5	295.5	
β-naphthyl ketone (2-acetonaphthone)	$C_{12}H_{10}O$	120.2	152.3	168.5	185.7	203.8	214.7	229.8	251.6	275.8	301.0	55.5
n-nonyl ketone (undecan-2-one)	$C_{11}H_{22}O$	68.2	95.5	108.9	123.1	139.0	148.6	161.0	181.2	202.3	224.0	15
palmitate	$C_{17}H_{34}O_2$	134.3	166.8	184.3	202.0							30
n-pentadecyl ketone (2-heptadecanone)	$C_{17}H_{34}O$	129.6	161.6	178.0	196.4	214.3	226.7	242.0	265.8	291.7	319.5	
2-Methylpentane	C_6H_{14}	−60.9	−41.7	−32.1	−21.4	−9.7	−1.9	+8.1	24.1	41.6	60.3	−154
3-Methylpentane	C_6H_{14}	−59.0	−39.8	−30.1	−19.4	−7.3	+0.1	10.5	26.5	44.2	63.3	−118
2-Methyl-1-pentanol	$C_6H_{14}O$	15.4	38.0	49.6	61.6	74.7	83.4	94.2	111.3	129.8	147.9	
2-Methyl-2-pentanol	$C_6H_{14}O$	−4.5	+16.8	27.6	38.8	51.3	58.8	69.2	85.0	102.6	121.2	−103
Methyl n-pentyl ketone (2-heptanone)	$C_7H_{14}O$	19.3	43.6	55.5	67.7	81.2	89.8	100.0	116.1	133.2	150.2	
phenyl ether (anisole)	C_7H_8O	+5.4	30.0	42.2	55.8	70.7	80.1	93.0	112.3	133.8	155.5	−37.3
2-Methylpropene	C_4H_8	−105.1	−96.5	−81.9	−73.4	−63.8	−57.7	−49.3	−36.7	−22.2	−6.9	−140.3
Methyl propionate	$C_4H_8O_2$	−42.0	−21.5	−11.8	−1.0	+11.0	18.7	29.0	44.2	61.8	79.8	−87.5
4-Methylpropiophenone	$C_{10}H_{12}O$	59.6	89.3	103.8	120.2	138.0	149.3	164.2	187.4	212.7	238.5	
2-Methylpropionyl bromide	C_4H_7BrO	13.5	38.4	50.6	64.1	79.4	88.8	101.6	120.5	141.7	163.0	
Methyl propyl ether	$C_4H_{10}O$	−72.2	−54.3	−45.4	−35.4	−24.3	−17.4	−8.1	+6.0	22.5	39.1	
n-propyl ketone (2-pentanone)	$C_5H_{10}O$	−12.0	+8.0	17.9	28.5	39.8	47.3	56.8	71.0	86.8	103.3	−77.8
isopropyl ketone (3-Methyl-2-butanone)	$C_5H_{10}O$	−19.9	−1.0	+8.3	18.3	29.6	36.2	45.5	59.0	73.8	88.9	−92
2-Methylquinoline	$C_{10}H_9N$	75.3	104.0	119.0	134.0	150.8	161.7	176.2	197.8	211.7	246.5	−1
Methyl salicylate	$C_8H_8O_3$	54.0	81.6	95.3	110.0	126.2	136.7	150.0	172.6	197.5	223.2	−8.3
α-Methyl styrene	C_9H_{10}	7.4	34.0	47.1	61.8	77.8	88.3	102.2	121.8	143.0	165.4	−23.2
4-Methyl styrene	C_9H_{10}	16.0	42.0	55.1	69.2	85.0	95.0	108.6	128.7	151.2	175.0	
Methyl n-tetradecyl ketone (2-hexadecanone)	$C_{16}H_{32}O$	109.8	151.5	167.3	184.6	203.7	215.0	230.5	254.4	279.8	307.0	
thiocyanate	C_2H_3NS	−14.0	+9.8	21.6	34.5	49.0	58.1	70.4	89.8	110.8	132.9	−51
isothiocyanate	C_2H_3NS	−34.7	−8.3	+5.4	20.4	38.2	47.5	59.3	77.5	97.8	119.0	35.5
undecyl ketone (2-tridecanone)	$C_{13}H_{26}O$	86.8	117.0	131.8	147.8	165.7	176.6	191.5	214.0	238.3	262.5	28.5
isovalerate	$C_6H_{12}O_2$	−19.2	+2.9	14.0	26.4	39.8	48.2	59.8	77.3	96.7	116.7	
Monovinylacetylene (butenyne)	C_4H_4	−93.2	−77.7	−70.0	−61.3	−51.7	−45.3	−37.1	−24.1	−10.1	+5.3	
Myrcene	$C_{16}H_{16}$	14.5	40.0	53.2	67.0	82.6	92.6	106.0	126.0	148.3	171.5	
Myristaldehyde	$C_{14}H_{28}O$	99.0	132.0	148.3	166.2	186.0	198.3	214.5	240.4	267.9	297.8	23.5
Myristic acid (tetradecanoic acid)	$C_{14}H_{28}O_2$	142.0	174.1	190.8	207.6	223.5	237.2	250.5	272.3	294.6	318.0	57.5
Naphthalene	$C_{10}H_8$	52.6	74.2	85.8	101.7	119.3	130.2	145.5	167.7	193.2	217.9	80.2
1-Naphthoic acid	$C_{11}H_8O_2$	156.0	184.0	196.8	211.2	225.0	234.5	245.8	263.5	281.4	300.0	160.5
2-Naphthoic acid	$C_{11}H_8O_2$	160.8	189.7	202.8	216.9	231.5	241.3	252.7	270.3	289.5	308.5	184
1-Naphthol	$C_{10}H_8O$	94.0	125.5	142.0	158.0	177.8	190.0	206.0	229.6	255.8	282.5	96
2-Naphthol	$C_{10}H_8O$		128.6	144.5	161.8	181.7	193.7	209.8	234.0	260.6	288.0	122.5
1-Naphthylamine	$C_{10}H_9N$	104.3	137.7	153.8	171.6	191.5	203.8	220.0	244.9	272.2	300.8	50
2-Naphthylamine	$C_{10}H_9N$	108.0	141.6	157.6	175.8	195.7	208.1	224.3	249.7	277.4	306.1	111.5
Nicotine	$C_{10}H_{14}N_2$	61.8	91.8	107.2	123.7	142.1	154.7	169.5	193.8	219.8	247.3	
2-Nitroaniline	$C_6H_6N_2O_2$	104.0	135.7	150.4	167.7	186.0	197.8	213.0	236.3	260.0	284.5	71.5
3-Nitroaniline	$C_6H_6N_2O_2$	119.3	151.5	167.8	185.5	204.2	216.5	232.0	255.3	280.2	305.7	114
4-Nitroaniline	$C_6H_6N_2O_2$	142.4	177.6	194.4	213.2	234.2	245.9	261.8	284.5	310.2	336.0	146.5
2-Nitrobenzaldehyde	$C_7H_5NO_3$	85.8	117.7	133.4	150.0	168.8	180.7	196.2	220.0	246.8	273.5	40.9
3-Nitrobenzaldehyde	$C_7H_5NO_3$	96.2	127.4	142.8	159.0	177.7	189.5	204.3	227.4	252.1	278.3	58
Nitrobenzene	$C_6H_5NO_2$	44.4	71.6	84.9	99.3	115.4	125.8	139.9	161.2	185.8	210.6	+5.7
Nitroethane	$C_2H_5NO_2$	−21.0	+1.5	12.5	24.8	38.0	46.5	57.8	74.8	94.0	114.0	−90
Nitroglycerin	$C_3H_5N_3O_9$	127	167	188	210	235	251					11
Nitromethane	CH_3NO_2	−29.0	−7.9	+2.8	14.1	27.5	35.5	46.6	63.5	82.0	101.2	−29
2-Nitrophenol	$C_6H_5NO_3$	49.3	76.8	90.4	105.8	122.1	132.6	146.4	167.6	191.0	214.5	45
2-Nitrophenyl acetate	$C_8H_7NO_4$	100.0	128.0	142.0	155.8	172.8	181.7	194.1	213.0	233.5	253.0	

TABLE 2-8   Vapor Pressures of Organic Compounds, up to 1 atm (Continued)

Compound		Pressure, mm Hg										Melting point, °C
		1	5	10	20	40	60	100	200	400	760	
Name	Formula	Temperature, °C										
1-Nitropropane	$C_3H_7NO_2$	−9.6	+13.5	25.3	37.9	51.8	60.5	72.3	90.2	110.6	131.6	−108
2-Nitropropane	$C_3H_7NO_2$	−18.8	+4.1	15.8	28.2	41.8	50.3	62.0	80.0	99.8	120.3	−93
2-Nitrotoluene	$C_7H_7NO_2$	50.0	79.1	93.8	109.6	126.3	137.6	151.5	173.7	197.7	222.3	−4.1
3-Nitrotoluene	$C_7H_7NO_2$	50.2	81.0	96.0	112.8	130.7	142.5	156.9	180.3	206.8	231.9	15.5
4-Nitrotoluene	$C_7H_7NO_2$	53.7	85.0	100.5	117.7	136.0	147.9	163.0	186.7	212.5	238.3	51.9
4-Nitro-1,3-xylene (4-nitro-*m*-xylene)	$C_8H_9NO_2$	65.6	95.0	109.8	125.8	143.3	153.8	168.5	191.7	217.5	244.0	+2
Nonacosane	$C_{29}H_{60}$	234.2	269.8	286.4	303.6	323.2	334.8	350.0	373.2	397.2	421.8	63.8
Nonadecane	$C_{19}H_{40}$	133.2	166.3	183.5	200.8	220.0	232.8	248.0	271.8	299.8	330.0	32
n-Nonane	C_9H_{20}	+1.4	25.8	38.0	51.2	66.0	75.5	88.1	107.5	128.2	150.8	−53.7
1-Nonanol	$C_9H_{20}O$	59.5	86.1	99.7	113.8	129.0	139.0	151.3	170.5	192.1	213.5	−5
2-Nonanone	$C_9H_{18}O$	32.1	59.0	72.3	87.2	103.4	113.8	127.4	148.2	171.2	195.0	−19
Octacosane	$C_{28}H_{58}$	226.5	260.3	277.4	295.4	314.2	326.8	341.8	364.8	388.9	412.5	61.6
Octadecane	$C_{18}H_{38}$	119.6	152.1	169.6	187.5	207.4	219.7	236.0	260.6	288.0	317.0	28
n-Octane	C_8H_{18}	−14.0	+8.3	19.2	31.5	45.1	53.8	65.7	83.6	104.0	125.6	−56.8
n-Octanol (1-octanol)	$C_8H_{18}O$	54.0	76.5	88.3	101.0	115.2	123.8	135.2	152.0	173.8	195.2	−15.4
2-Octanone	$C_8H_{16}O$	23.6	48.4	60.9	74.3	89.8	99.0	111.7	130.4	151.0	172.9	−16
n-Octyl acrylate	$C_{11}H_{20}O_2$	58.5	87.7	102.0	117.8	135.6	145.6	159.1	180.2	204.0	227.0	
iodide (1-Iodooctane)	$C_8H_{17}I$	45.8	74.8	90.0	105.9	123.8	135.4	150.0	173.3	199.3	225.5	−45.9
Oleic acid	$C_{18}H_{34}O_2$	176.5	208.5	223.0	240.0	257.2	269.8	286.0	309.8	334.7	360.0	14
Palmitaldehyde	$C_{16}H_{32}O$	121.6	154.6	171.8	190.0	210.0	222.6	239.5	264.1	292.3	321.0	34
Palmitic acid	$C_{16}H_{32}O_2$	153.6	188.1	205.8	223.8	244.4	256.0	271.5	298.7	326.0	353.8	64.0
Palmitonitrile	$C_{16}H_{31}N$	134.3	168.3	185.8	204.2	223.8	236.6	251.5	277.1	304.5	332.0	31
Pelargonic acid	$C_9H_{18}O_2$	108.2	126.0	137.4	149.8	163.7	172.3	184.4	203.1	227.5	253.5	12.5
Pentachlorobenzene	C_6HCl_5	98.6	129.7	144.3	160.0	178.5	190.1	205.5	227.0	251.6	276.0	85.5
Pentachloroethane	C_2HCl_5	+1.0	27.2	39.8	53.9	69.9	80.0	93.5	114.0	137.2	160.5	−22
Pentachloroethylbenzene	$C_8H_5Cl_5$	96.2	130.0	148.0	166.0	186.2	199.0	216.0	241.8	269.3	299.0	
Pentachlorophenol	C_6HCl_5O				192.2	211.2	223.4	239.6	261.8	285.0	309.3	188.5
Pentacosane	$C_{25}H_{52}$	194.2	230.0	248.2	266.1	285.6	298.4	314.0	339.0	365.4	390.3	53.3
Pentadecane	$C_{15}H_{32}$	91.6	121.0	135.4	150.2	167.7	178.4	194.0	216.1	242.8	270.5	10
1,3-Pentadiene	C_5H_8	−71.8	−53.8	−45.0	−34.8	−23.4	−16.5	−6.7	+8.0	24.7	42.1	
1,4-Pentadiene	C_5H_8	−83.5	−66.2	−57.1	−47.7	−37.0	−30.0	−20.6	−6.7	+8.3	26.1	
Pentaethylbenzene	$C_{16}H_{26}$	86.0	120.0	135.8	152.4	171.9	184.2	200.0	224.1	250.2	277.0	
Pentaethylchlorobenzene	$C_{16}H_{25}Cl$	90.0	123.8	140.7	158.1	178.2	191.0	208.0	230.3	257.2	285.0	
n-Pentane	C_5H_{12}	−76.6	−62.5	−50.1	−40.2	−29.2	−22.2	−12.6	+1.9	18.5	36.1	−129.7
iso-Pentane (2-methylbutane)	C_5H_{12}	−82.9	−65.8	−57.0	−47.3	−36.5	−29.6	−20.2	−5.9	+10.5	27.8	−159.7
neo-Pentane (2,2-dimethylpropane)	C_5H_{12}	−102.0	−85.4	−76.7	−67.2	−56.1	−49.0	−39.1	−23.7	−7.1	+9.5	−16.6
2,3,4-Pentanetriol	$C_5H_{12}O_3$	155.0	189.3	204.5	220.5	239.6	249.8	263.5	284.5	307.0	327.2	
1-Pentene	C_5H_{10}	−80.4	−63.3	−54.5	−46.0	−34.1	−27.1	−17.7	−3.4	+12.8	30.1	
α-Phellandrene	$C_{10}H_{16}$	20.0	45.7	58.0	72.1	87.8	97.6	110.6	130.6	152.0	175.0	
Phenanthrene	$C_{14}H_{10}$	118.2	154.3	173.0	193.7	215.8	229.9	249.0	277.1	308.0	340.2	99.5
Phenethyl alcohol (phenyl cellosolve)	$C_8H_{10}O_2$	58.2	85.9	100.0	114.8	130.5	141.2	154.0	175.0	197.5	219.5	
2-Phenetidine	$C_8H_{11}NO$	67.0	94.7	108.6	123.7	139.9	149.8	163.5	184.0	207.0	228.0	
Phenol	C_6H_6O	40.1	62.5	73.8	86.0	100.1	108.4	121.4	139.0	160.0	181.9	40.6
2-Phenoxyethanol	$C_8H_{10}O_2$	78.0	106.6	121.2	136.0	152.2	163.2	176.5	197.6	221.0	245.3	11.6
2-Phenoxyethyl acetate	$C_{10}H_{12}O_3$	82.6	113.5	128.0	144.5	162.3	174.0	189.2	211.3	235.0	259.7	−6.7
Phenyl acetate	$C_8H_8O_2$	38.2	64.8	78.0	92.3	108.1	118.1	131.6	151.2	173.5	195.9	
Phenylacetic acid	$C_8H_8O_2$	97.0	127.0	141.3	156.0	173.6	184.5	198.2	219.5	243.0	265.5	76.5
Phenylacetonitrile	C_8H_7N	60.0	89.0	103.5	119.4	136.3	147.7	161.8	184.2	208.5	233.5	−23.8
Phenylacetyl chloride	C_8H_7ClO	48.0	75.3	89.0	103.6	119.8	129.8	143.5	163.8	186.0	210.0	
Phenyl benzoate	$C_{13}H_{10}O_2$	106.8	141.5	157.8	177.0	197.6	210.8	227.8	254.0	283.5	314.0	70.5
4-Phenyl-3-buten-2-one	$C_{10}H_{10}O$	81.7	112.2	127.4	143.8	161.3	172.6	187.8	211.0	235.4	261.0	41.5
Phenyl isocyanate	C_7H_5NO	10.6	36.0	48.5	62.5	77.7	87.7	100.6	120.8	142.7	165.6	
isocyanide	C_7H_5N	12.0	37.0	49.7	63.4	78.3	88.0	101.0	120.8	142.3	165.0	
Phenylcyclohexane	$C_{12}H_{16}$	67.5	96.5	111.3	126.4	144.0	154.2	169.3	191.3	214.6	240.0	+7.5
Phenyl dichlorophosphate	$C_6H_5Cl_2O_2P$	66.7	95.9	110.0	125.9	143.4	153.6	168.0	189.8	213.0	239.5	
m-Phenylene diamine (1,3-phenylenediamine)	$C_6H_8N_2$	99.8	131.2	147.0	163.8	182.5	194.0	209.9	233.0	259.0	285.5	62.8
Phenylglyoxal	$C_8H_6O_2$		75.0	87.8	100.7	115.5	124.2	136.2	153.8	173.5	193.5	73
Phenylhydrazine	$C_6H_8N_2$	71.8	101.6	115.8	131.5	148.2	158.7	173.5	195.4	218.2	243.5	19.5
N-Phenyliminodiethanol	$C_{10}H_{15}NO_2$	145.0	179.2	195.8	213.4	233.0	245.3	260.6	284.5	311.3	337.8	
1-Phenyl-1,3-pentanedione	$C_{11}H_{12}O_2$	98.0	128.5	144.0	159.9	178.0	189.8	204.5	226.7	251.2	276.5	
2-Phenylphenol	$C_{12}H_{10}O$	100.0	131.6	146.2	163.3	180.3	192.2	205.9	227.9	251.8	275.0	56.5
4-Phenylphenol	$C_{12}H_{10}O$			176.2	193.8	213.0	225.3	240.9	263.2	285.5	308.0	164.5
3-Phenyl-1-propanol	$C_9H_{12}O$	74.7	102.4	116.0	131.2	147.4	156.8	170.3	191.2	212.8	235.0	
Phenyl isothiocyanate	C_7H_5NS	47.2	75.6	89.8	115.5	122.5	133.3	147.7	169.6	194.0	218.5	−21.0
Phorone	$C_9H_{14}O$	42.0	68.3	81.5	95.6	111.3	121.4	134.0	153.5	175.3	197.2	28
iso-Phorone	$C_9H_{14}O$	38.0	66.7	81.2	96.8	114.5	125.6	140.6	163.3	188.7	215.2	
Phosgene (carbonyl chloride)	CCl_2O	−92.9	−77.0	−69.3	−60.3	−50.3	−44.0	−35.6	−22.3	−7.6	+8.3	−104
Phthalic anhydride	$C_8H_4O_3$	96.5	121.3	134.0	151.7	172.0	185.3	202.3	228.0	256.8	284.5	130.8
Phthalide	$C_8H_6O_2$	95.5	127.7	144.0	161.3	181.0	193.5	210.0	234.5	261.8	290.0	73
Phthaloyl chloride	$C_8H_4Cl_2O_2$	86.3	118.3	134.2	151.0	170.0	182.2	197.8	222.0	248.3	275.8	88.5
2-Picoline	C_6H_7N	−11.1	+12.6	24.4	37.4	51.2	59.9	71.4	89.0	108.4	128.8	−70
Pimelic acid	$C_7H_{12}O_4$	163.4	196.2	212.0	229.3	247.0	258.2	272.0	294.5	318.5	342.1	103
α-Pinene	$C_{10}H_{16}$	−1.0	+24.6	37.3	51.4	66.8	76.3	90.1	110.2	132.3	155.0	−55
β-Pinene	$C_{10}H_{16}$	+4.2	30.0	42.3	58.1	71.5	81.2	94.0	114.1	136.1	158.3	

TABLE 2-8 Vapor Pressures of Organic Compounds, up to 1 atm (Continued)

| Compound | | Pressure, mm Hg | | | | | | | | | | Melting point, °C |
Name	Formula	1	5	10	20	40	60	100	200	400	760	
		Temperature, °C										
Piperidine	$C_5H_{11}N$		−7.0	+3.9	15.8	29.2	37.7	49.0	66.2	85.7	106.0	−9
Piperonal	$C_8H_6O_3$	87.0	117.4	132.0	148.0	165.7	177.0	191.7	214.3	238.5	263.0	37
Propane	C_3H_8	−128.9	−115.4	−108.5	−100.9	−92.4	−87.0	−79.6	−68.4	−55.6	−42.1	−187.1
Propenylbenzene	C_9H_{10}	17.5	43.8	57.0	71.5	87.7	97.8	111.7	132.0	154.7	179.0	−30.1
Propionamide	C_3H_7NO	65.0	91.0	105.0	119.0	134.8	144.3	156.0	174.2	194.0	213.0	79
Propionic acid	$C_3H_6O_2$	4.6	28.0	39.7	52.0	65.8	74.1	85.8	102.5	122.0	141.1	−22
anhydride	$C_6H_{10}O_3$	20.6	45.3	57.7	70.4	85.6	94.5	107.2	127.8	146.0	167.0	−45
Propionitrile	C_3H_5N	−35.0	−13.6	−3.0	+8.8	22.0	30.1	41.4	58.2	77.7	97.1	−91.9
Propiophenone	$C_9H_{10}O$	50.0	77.9	92.2	107.6	124.3	135.0	149.3	170.2	194.2	218.0	21
n-Propyl acetate	$C_5H_{10}O_2$	−26.7	−5.4	+5.0	16.0	28.8	37.0	47.8	64.0	82.0	101.8	−92.5
iso-Propyl acetate	$C_5H_{10}O_2$	−38.3	−17.4	−7.2	+4.2	17.0	25.1	35.7	51.7	69.8	89.0	
n-Propyl alcohol (1-propanol)	C_3H_8O	−15.0	+5.0	14.7	25.3	36.4	43.5	52.8	66.8	82.0	97.8	−127
iso-Propyl alcohol (2-propanol)	C_3H_8O	−26.1	−7.0	+2.4	12.7	23.8	30.5	39.5	53.0	67.8	82.5	−85.8
n-Propylamine	C_3H_9N	−64.4	−46.3	−37.2	−27.1	−16.0	−9.0	+0.5	15.0	31.5	48.5	−83
Propylbenzene	C_9H_{12}	6.3	31.3	43.4	56.8	71.6	81.1	94.0	113.5	135.7	159.2	−99.5
Propyl benzoate	$C_{10}H_{12}O_2$	54.6	83.8	98.0	114.3	131.8	143.3	157.4	180.1	205.2	231.0	−51.6
n-Propyl bromide (1-bromopropane)	C_3H_7Br	−53.0	−33.4	−23.3	−12.4	−0.3	+7.5	18.0	34.0	52.0	71.0	−109.9
iso-Propyl bromide (2-bromopropane)	C_3H_7Br	−61.8	−42.5	−32.8	−22.0	−10.1	−2.5	+8.0	23.8	41.5	60.0	−89.0
n-Propyl n-butyrate	$C_7H_{14}O_2$	−1.6	+22.1	34.0	47.0	61.5	70.3	82.6	101.0	121.7	142.7	−95.2
isobutyrate	$C_7H_{14}O_2$	−6.2	+16.8	28.3	40.6	54.3	63.0	73.9	91.8	112.0	133.9	
iso-Propyl isobutyrate	$C_7H_{14}O_2$	−16.3	+5.8	17.0	29.0	42.4	51.4	62.3	80.2	100.0	120.5	
Propyl carbamate	$C_4H_9NO_2$	52.4	77.6	90.0	103.2	117.7	126.5	138.3	155.8	175.8	195.0	
n-Propyl chloride (1-chloropropane)	C_3H_7Cl	−68.3	−50.0	−41.0	−31.0	−19.5	−12.1	−2.5	+12.2	29.4	46.4	−122.8
iso-Propyl chloride (2-chloropropane)	C_3H_7Cl	−78.8	−61.1	−52.0	−42.0	−31.0	−23.5	−13.7	+1.3	18.1	36.5	−117
iso-Propyl chloroacetate	$C_5H_9ClO_2$	+3.8	28.1	40.2	53.9	68.7	78.0	90.3	108.8	128.0	148.6	
Propyl chloroglyoxylate	$C_5H_7ClO_3$	9.7	32.3	43.5	55.6	68.8	77.2	88.0	104.7	123.0	150.0	
Propylene	C_3H_6	−131.9	−120.7	−112.1	−104.7	−96.5	−91.3	−84.1	−73.3	−60.9	−47.7	−185
Propylene glycol (1,2-Propanediol)	$C_3H_8O_2$	45.5	70.8	83.2	96.4	111.2	119.9	132.0	149.7	168.1	188.2	
Propylene oxide	C_3H_6O	−75.0	−57.8	−49.0	−39.3	−28.4	−21.3	−12.0	+2.1	17.8	34.5	−112.1
n-Propyl formate	$C_4H_8O_2$	−43.0	−22.7	−12.6	−1.7	+10.8	18.8	29.5	45.3	62.6	81.3	−92.9
iso-Propyl formate	$C_4H_8O_2$	−52.0	−32.7	−22.7	−12.1	−0.2	+7.5	17.8	33.6	50.5	68.3	
4,4'-iso-Propylidenebisphenol	$C_{15}H_{16}O_2$	193.0	224.2	240.8	255.5	273.0	282.9	297.0	317.5	339.0	360.5	
n-Propyl iodide (1-iodopropane)	C_3H_7I	−36.0	−13.5	−2.4	+10.0	23.6	32.1	43.8	61.8	81.8	102.5	−98.8
iso-Propyl iodide (2-iodopropane)	C_3H_7I	−43.3	−22.1	−11.7	0.0	+13.2	21.6	32.8	50.0	69.5	89.5	−90
n-Propyl levulinate	$C_8H_{14}O_3$	59.7	86.3	99.9	114.0	130.1	140.6	154.0	175.6	198.0	221.2	
iso-Propyl levulinate	$C_8H_{14}O_3$	48.0	74.5	88.0	102.4	118.1	127.8	141.8	161.6	185.2	208.2	
Propyl mercaptan (1-propanethiol)	C_3H_8S	−56.0	−36.3	−26.3	−15.4	−3.2	+4.6	15.3	31.5	49.2	67.4	−112
2-iso-Propylnaphthalene	$C_{13}H_{14}$	76.0	107.9	123.4	140.3	159.0	171.4	187.6	211.8	238.5	266.0	
iso-Propyl β-naphthyl ketone (2-isobutyronaphthone)	$C_{14}H_{14}O$	133.2	165.4	181.0	197.7	215.6	227.0	242.3	264.0	288.2	313.0	
2-iso-Propylphenol	$C_9H_{12}O$	56.6	83.8	97.0	111.7	127.5	137.7	150.3	170.1	192.6	214.5	15.5
3-iso-Propylphenol	$C_9H_{12}O$	62.0	90.3	104.1	119.8	136.2	146.6	160.2	182.0	205.0	228.0	26
4-iso-Propylphenol	$C_9H_{12}O$	67.0	94.7	108.0	123.4	139.8	149.7	163.3	184.0	206.1	228.2	61
Propyl propionate	$C_6H_{12}O_2$	−14.2	+8.0	19.4	31.6	45.0	53.8	65.2	82.7	102.0	122.4	−76
4-iso-Propylstyrene	$C_{11}H_{14}$	34.7	62.3	76.0	91.2	108.0	118.4	132.8	153.9	178.0	202.5	
Propyl isovalerate	$C_8H_{16}O_2$	+8.0	32.8	45.1	58.0	72.8	82.3	95.0	113.9	135.0	155.9	
Pulegone	$C_{10}H_{16}O$	58.3	82.5	94.0	106.8	121.7	130.2	143.1	162.5	189.8	221.0	
Pyridine	C_5H_5N	−18.9	+2.5	13.2	24.8	38.0	46.8	57.8	75.0	95.6	115.4	−42
Pyrocatechol	$C_6H_6O_2$		104.0	118.3	134.0	150.6	161.7	176.0	197.7	221.5	245.5	105
Pyrocaltechol diacetate (1,2-phenylene diacetate)	$C_{10}H_{10}O_4$	98.0	129.8	145.7	161.8	179.8	191.6	206.5	228.7	253.3	278.0	
Pyrogallol	$C_6H_6O_3$		151.7	167.7	185.3	204.2	216.3	232.0	255.3	281.5	309.0	133
Pyrotartaric anhydride	$C_5H_6O_3$	69.7	99.7	114.2	130.0	147.8	158.6	173.8	196.1	221.0	247.4	
Pyruvic acid	$C_3H_4O_3$	21.4	45.8	57.9	70.8	85.3	94.1	106.5	124.7	144.7	165.0	13.6
Quinoline	C_9H_7N	59.7	89.6	103.8	119.8	136.7	148.1	163.2	186.2	212.3	237.7	−15
iso-Quinoline	C_9H_7N	63.5	92.7	107.8	123.7	141.6	152.0	167.6	190.0	214.5	240.5	24.6
Resorcinol	$C_6H_6O_2$	108.4	138.0	152.1	168.0	185.3	195.8	209.8	230.8	253.4	276.5	110.7
Safrole	$C_{10}H_{10}O_2$	63.8	93.0	107.6	123.0	140.1	150.3	165.1	186.2	210.0	233.0	11.2
Salicylaldehyde	$C_7H_6O_2$	33.0	60.1	73.8	88.7	105.2	115.7	129.4	150.0	173.7	196.5	−7
Salicylic acid	$C_7H_6O_3$	113.7	136.0	146.2	156.8	172.2	182.0	193.4	210.0	230.5	256.0	159
Sebacic acid	$C_{10}H_{18}O_4$	183.0	215.7	232.0	250.0	268.2	279.8	294.5	313.2	332.8	352.3	134.5
Selenophene	C_4H_4Se	−39.0	−16.0	−4.0	+9.1	24.1	33.8	47.0	66.7	89.8	114.3	
Skatole	C_9H_9N	95.0	124.2	139.6	154.3	171.9	183.6	197.4	218.8	242.5	266.2	95
Stearaldehyde	$C_{18}H_{36}O$	140.0	174.6	192.1	210.6	230.8	244.2	260.0	285.0	313.8	342.5	63.5
Stearic acid	$C_{18}H_{36}O_2$	173.7	209.0	225.0	243.4	263.3	275.5	291.0	316.5	343.0	370.0	69.3
Stearyl alcohol (1-octadecanol)	$C_{18}H_{36}O$	150.3	185.6	202.0	220.0	240.4	252.7	269.4	293.5	320.3	349.5	58.5
Styrene	C_8H_8	−7.0	+18.0	30.8	44.6	59.8	69.5	82.0	101.3	122.5	145.2	−30.6
Styrene dibromide [(1,2-dibromoethyl)benzene]	$C_8H_8Br_2$	86.0	115.6	129.8	145.2	161.8	172.2	186.3	207.8	230.0	254.0	
Suberic acid	$C_8H_{14}O_4$	172.8	205.5	219.5	238.2	254.6	265.4	279.8	300.5	322.8	345.5	142
Succinic anhydride	$C_4H_4O_3$	92.0	115.0	128.2	145.3	163.0	174.0	189.0	212.0	237.0	261.0	119.6
Succinimide	$C_4H_5NO_2$	115.0	143.2	157.0	174.0	192.0	203.0	217.4	240.0	263.5	287.5	125.5
Succinyl chloride	$C_4H_4Cl_2O_2$	39.0	65.0	78.0	91.8	107.5	117.2	130.0	149.3	170.0	192.5	17
α-Terpineol	$C_{10}H_{18}O$	52.8	80.4	94.3	109.8	126.0	136.3	150.1	171.2	194.3	217.5	35
Terpenoline	$C_{10}H_{16}$	32.3	58.0	70.6	84.8	100.0	109.8	122.7	142.0	163.5	185.0	

TABLE 2-8   Vapor Pressures of Organic Compounds, up to 1 atm (Continued)

| Compound | | Pressure, mm Hg | | | | | | | | | | Melting point, °C |
Name	Formula	1	5	10	20	40	60	100	200	400	760	
1,1,1,2-Tetrabromoethane	$C_2H_2Br_4$	58.0	83.3	95.7	108.5	123.2	132.0	144.0	161.5	181.0	200.0	
1,1,2,2-Tetrabromoethane	$C_2H_2Br_4$	65.0	95.5	110.0	126.0	144.0	155.1	170.0	192.5	217.5	243.5	
Tetraisobutylene	$C_{16}H_{32}$	63.8	93.7	108.5	124.5	142.2	152.6	167.5	190.0	214.6	240.0	
Tetracosane	$C_{24}H_{50}$	183.8	219.6	237.6	255.3	276.3	288.4	305.2	330.5	358.0	386.4	51.1
1,2,3,4-Tetrachlorobenzene	$C_6H_2Cl_4$	68.5	99.6	114.7	131.2	149.2	160.0	175.7	198.0	225.5	254.0	46.5
1,2,3,5-Tetrachlorobenzene	$C_6H_2Cl_4$	58.2	89.0	104.1	121.6	140.0	152.0	168.0	193.7	220.0	246.0	54.5
1,2,4,5-Tetrachlorobenzene	$C_6H_2Cl_4$					146.0	157.7	173.5	196.0	220.5	245.0	139
1,1,2,2-Tetrachloro-1,2-difluoroethane	$C_2Cl_4F_2$	−37.5	−16.0	−5.0	+6.7	19.8	28.1	38.6	55.0	73.1	92.0	26.5
1,1,1,2-Tetrachloroethane	$C_2H_2Cl_4$	−16.3	+7.4	19.3	32.1	46.7	56.0	68.0	87.2	108.2	130.5	−68.7
1,1,2,2-Tetrachloroethane	$C_2H_2Cl_4$	−3.8	+20.7	33.0	46.2	60.8	70.0	83.2	102.2	124.0	145.9	−36
1,2,3,5-Tetrachloro-4-ethylbenzene	$C_8H_6Cl_4$	77.0	110.0	126.0	143.7	162.1	175.0	191.6	215.3	243.0	270.0	
Tetrachloroethylene	C_2Cl_4	−20.6	+2.4	13.8	26.3	40.1	49.2	61.3	79.8	100.0	120.8	−19.0
2,3,4,6-Tetrachlorophenol	$C_6H_2Cl_4O$	100.0	130.3	145.3	161.0	179.1	190.0	205.2	227.2	250.4	275.0	69.5
3,4,5,6-Tetrachloro-1,2-xylene	$C_8H_6Cl_4$	94.4	125.0	140.3	156.0	174.2	185.8	200.5	223.0	248.3	273.5	
Tetradecane	$C_{14}H_{30}$	76.4	106.0	120.7	135.6	152.7	164.0	178.5	201.8	226.8	252.5	5.5
Tetradecylamine	$C_{14}H_{31}N$	102.6	135.8	152.0	170.0	189.0	200.2	215.7	239.8	264.6	291.2	
Tetradecyltrimethylsilane	$C_{17}H_{38}Si$	120.0	150.7	166.2	183.5	201.5	213.3	227.8	250.0	275.0	300.0	
Tetraethoxysilane	$C_8H_{20}O_4Si$	16.0	40.3	52.6	65.8	81.1	90.7	103.6	123.5	146.2	168.5	
1,2,3,4-Tetraethylbenzene	$C_{14}H_{22}$	65.7	96.2	111.6	127.7	145.8	156.7	172.4	196.0	221.4	248.0	11.6
Tetraethylene glycol	$C_8H_{18}O_5$	153.9	183.7	197.1	212.3	228.0	237.8	250.0	268.4	288.0	307.8	
Tetraethylene glycol chlorohydrin	$C_8H_{17}ClO_4$	110.1	141.8	156.1	172.6	190.0	200.5	214.7	236.5	258.2	281.5	
Tetraethyllead	$C_8H_{20}Pb$	38.4	63.6	74.8	88.0	102.4	111.7	123.8	142.0	161.8	183.0	−136
Tetraethylsilane	$C_8H_{20}Si$	−1.0	+23.9	36.3	50.0	65.3	74.8	88.0	108.0	130.2	153.0	
Tetralin	$C_{10}H_{12}$	38.0	65.3	79.0	93.8	110.4	121.3	135.3	157.2	181.8	207.2	−31.0
1,2,3,4-Tetramethylbenzene	$C_{10}H_{14}$	42.6	68.7	81.8	95.8	111.5	121.8	135.7	155.7	180.0	204.4	−6.2
1,2,3,5-Tetramethylbenzene	$C_{10}H_{14}$	40.6	65.8	77.8	91.0	105.8	115.4	128.3	149.9	173.7	197.9	−24.0
1,2,4,5-Tetramethylbenzene	$C_{10}H_{14}$	45.0	65.0	74.6	88.0	104.2	114.8	128.1	149.5	172.1	195.9	79.5
2,2,3,3-Tetramethylbutane	C_8H_{18}	−17.4	+3.2	13.5	24.6	36.8	44.5	54.8	70.2	87.4	106.3	−102.2
Tetramethylene dibromide (1,4-dibromobutane)	$C_4H_8Br_2$	32.0	58.8	72.4	87.6	104.0	115.1	128.7	149.8	173.8	197.5	−20
Tetramethyllead	$C_4H_{12}Pb$	−29.0	−6.8	+4.4	16.6	30.3	39.2	50.8	68.8	89.0	110.0	−27.5
Tetramethyltin	$C_4H_{12}Sn$	−51.3	−31.0	−20.6	−9.3	+3.5	11.7	22.8	39.8	58.5	78.0	
Tetrapropylene glycol monoisopropyl ether	$C_{15}H_{32}O_5$	116.6	147.8	163.0	179.8	197.7	209.0	223.3	245.0	268.3	292.7	
Thioacetic acid (mercaptoacetic acid)	$C_2H_4O_2S$	60.0	87.7	101.5	115.8	131.8	142.0	154.0				−16.5
Thiodiglycol (2,2′-thiodiethanol)	$C_4H_{10}O_2S$	42.0	96.0	128.0	165.0	210.0	240.5	285				
Thiophene	C_4H_4S	−40.7	−20.8	−10.9	0	+12.5	20.1	30.5	46.5	64.7	84.4	−38.3
Thiophenol (benzenethiol)	C_6H_6S	18.6	43.7	56.0	69.7	84.2	93.9	106.6	125.8	146.7	168.0	
α-Thujone	$C_{10}H_{16}O$	38.3	65.7	79.3	93.7	110.0	120.2	134.0	154.2	177.8	201.0	
Thymol	$C_{10}H_{14}O$	64.3	92.8	107.4	122.6	139.8	149.8	164.1	185.5	209.2	231.8	51.5
Tiglaldehyde	C_5H_8O	−25.0	−1.6	+10.0	23.2	37.0	45.8	57.7	75.4	95.5	116.4	
Tiglic acid	$C_5H_8O_2$	52.0	77.8	90.2	103.8	119.0	127.8	140.5	158.0	179.2	198.5	64.5
Tiglonitrile	C_5H_7N	−25.5	−2.4	+9.2	22.1	36.7	46.0	58.2	77.8	99.7	122.0	
Toluene	C_7H_8	−26.7	−4.4	+6.4	18.4	31.8	40.3	51.9	69.5	89.5	110.6	−95.0
Toluene-2,4-diamine	$C_7H_{10}N_2$	106.5	137.2	151.7	167.9	185.7	196.2	211.5	232.8	256.0	280.0	99
2-Toluic nitrile (2-tolunitrile)	C_8H_7N	36.7	64.0	77.9	93.0	110.0	120.8	135.0	156.0	180.0	205.2	−13
4-Toluic nitrile (4-tolunitrile)	C_8H_7N	42.5	71.3	85.8	101.7	109.5	130.0	145.2	167.3	193.0	217.6	29.5
2-Toluidine	C_7H_9N	44.0	69.3	81.4	95.1	110.0	119.8	133.0	153.0	176.2	199.7	−16.3
3-Toluidine	C_7H_9N	41.0	68.0	82.0	96.7	113.5	123.8	136.7	157.6	180.6	203.3	−31.5
4-Toluidine	C_7H_9N	42.0	68.2	81.8	95.8	111.5	121.5	133.7	154.0	176.9	200.4	44.5
2-Tolyl isocyanide	C_8H_7N	25.2	51.0	64.0	78.2	94.0	104.0	117.7	137.8	159.9	183.5	
4-Tolylhydrazine	$C_7H_{10}N_2$	82.4	110.0	123.8	138.6	154.1	165.0	178.0	198.0	219.5	242.0	65.5
Tribromoacetaldehyde	C_2HBr_3O	18.5	45.0	58.0	72.1	87.8	97.5	110.2	130.0	151.6	174.0	
1,1,2-Tribromobutane	$C_4H_7Br_3$	45.0	73.5	87.8	103.2	120.2	131.6	146.0	167.8	192.0	216.2	
1,2,2-Tribromobutane	$C_4H_7Br_3$	41.0	69.0	83.2	98.6	116.0	127.0	141.8	163.5	188.0	213.8	
2,2,3-Tribromobutane	$C_4H_7Br_3$	38.2	66.0	79.8	94.6	111.8	122.2	136.3	157.8	182.2	206.5	
1,1,2-Tribromoethane	$C_2H_3Br_3$	32.6	58.0	70.6	84.2	100.0	110.0	123.5	143.5	165.4	188.4	−26
1,2,3-Tribromopropane	$C_3H_5Br_3$	47.5	75.8	90.0	105.8	122.8	134.0	148.0	170.0	195.0	220.0	16.5
Triisobutylamine	$C_{12}H_{27}N$	32.3	57.4	69.8	83.0	97.8	107.3	119.7	138.0	157.8	179.0	−22
Triisobutylene	$C_{12}H_{24}$	18.0	44.0	56.5	70.0	86.7	96.7	110.0	130.2	153.0	179.0	
2,4,6-Tritertbutylphenol	$C_{18}H_{30}O$	95.2	126.1	142.0	158.0	177.4	188.0	203.0	226.2	250.6	276.3	
Trichloroacetic acid	$C_2HCl_3O_2$	51.0	76.0	88.2	101.8	116.3	125.9	137.8	155.4	175.2	195.6	57
Trichloroacetic anhydride	$C_4Cl_6O_3$	56.2	85.3	99.6	114.3	131.2	141.8	155.2	176.2	199.8	223.0	
Trichloroacetyl bromide	C_2BrCl_3O	−7.4	+16.7	29.3	42.1	57.2	66.7	79.5	98.4	120.2	143.0	
2,4,6-Trichloroaniline	$C_6H_4Cl_3N$	134.0	157.8	170.0	182.6	195.8	204.5	214.6	229.8	246.4	262.0	78
1,2,3-Trichlorobenzene	$C_6H_3Cl_3$	40.0	70.0	85.6	101.8	119.8	131.5	146.0	168.2	193.5	218.5	52.5
1,2,4-Trichlorobenzene	$C_6H_3Cl_3$	38.4	67.3	81.7	97.2	114.8	125.7	140.0	162.0	187.7	213.0	17
1,3,5-Trichlorobenzene	$C_6H_3Cl_3$		63.8	78.0	93.7	110.8	121.8	136.0	157.7	183.0	208.4	63.5
1,2,3-Trichlorobutane	$C_4H_7Cl_3$	+0.5	27.2	40.0	55.0	71.5	82.0	96.2	118.0	143.0	169.0	
1,1,1-Trichloroethane	$C_2H_3Cl_3$	−52.0	−32.0	−21.9	−10.8	+1.6	9.5	20.0	36.2	54.6	74.1	−30.6
1,1,2-Trichloroethane	$C_2H_3Cl_3$	−24.0	−2.0	+8.3	21.6	35.2	44.0	55.7	73.3	93.0	113.9	−36.7
Trichloroethylene	C_2HCl_3	−43.8	−22.8	−12.4	−1.0	+11.9	20.0	31.4	48.0	67.0	86.7	−73
Trichlorofluoromethane	CCl_3F	−84.3	−67.6	−59.0	−49.7	−39.0	−32.3	−23.0	−9.1	+6.8	23.7	
2,4,5-Trichlorophenol	$C_6H_3Cl_3O$	72.0	102.1	117.3	134.0	151.5	162.5	178.0	201.5	226.5	251.8	62
2,4,6-Trichlorophenol	$C_6H_3Cl_3O$	76.5	105.9	120.2	135.8	152.2	163.5	177.8	199.0	222.5	246.0	68.5

TABLE 2-8   Vapor Pressures of Organic Compounds, up to 1 atm (Concluded)

Compound						Pressure, mm Hg						Melting point, °C
Name	Formula	1	5	10	20	40	60	100	200	400	760	
						Temperature, °C						
Tri-2-chlorophenylthiophosphate	C$_{18}$H$_{12}$Cl$_3$O$_3$PS	188.2	217.2	231.2	246.7	261.7	271.5	283.8	302.8	322.0	341.3	
1,1,1-Trichloropropane	C$_3$H$_5$Cl$_3$	−28.8	−7.0	+4.2	16.2	29.9	38.3	50.0	67.7	87.5	108.2	−77.7
1,2,3-Trichloropropane	C$_3$H$_5$Cl$_3$	+9.0	33.7	46.0	59.3	74.0	83.6	96.1	115.6	137.0	158.0	−14.7
1,1,2-Trichloro-1,2,2-trifluoroethane	C$_2$Cl$_3$F$_3$	−68.0	−49.4	−40.3	−30.0	−18.5	−11.2	−1.7	+13.5	30.2	47.6	−35
Tricosane	C$_{23}$H$_{48}$	170.0	206.3	223.0	242.0	261.3	273.8	289.8	313.5	339.8	366.5	47.7
Tridecane	C$_{13}$H$_{28}$	59.4	98.3	104.0	120.2	137.7	148.2	162.5	185.0	209.4	234.0	−6.2
Tridecanoic acid	C$_{13}$H$_{26}$O$_2$	137.8	166.3	181.0	195.8	212.4	222.0	236.0	255.2	276.5	299.0	41
Triethoxymethylsilane	C$_7$H$_{18}$O$_3$Si	−1.5	+22.8	34.6	47.2	61.7	70.4	82.7	101.0	121.8	143.5	
Triethoxyphenylsilane	C$_{12}$H$_{20}$O$_3$Si	71.0	98.8	112.6	127.2	143.5	153.2	167.5	188.0	210.5	233.5	
1,2,4-Triethylbenzene	C$_{12}$H$_{18}$	46.0	74.2	88.5	104.0	121.7	132.2	146.8	168.3	193.7	218.0	
1,3,4-Triethylbenzene	C$_{12}$H$_{18}$	47.9	76.0	90.2	105.8	122.6	133.4	147.7	168.3	193.2	217.5	
Triethylborine	C$_6$H$_{15}$B			−148.0	−140.6	−131.4	−125.2	−116.0	−101.0	−81.0	−56.2	
Triethyl camphoronate	C$_{15}$H$_{26}$O$_6$		150.2	166.0	183.6	201.8	213.5	228.6	250.8	276.0	301.0	135
citrate	C$_{12}$H$_{20}$O$_7$	107.0	138.7	144.0	171.1	190.4	202.5	217.8	242.2	267.5	294.0	
Triethyleneglycol	C$_6$H$_{14}$O$_4$	114.0	144.0	158.1	174.0	191.3	201.5	214.6	235.2	256.6	278.3	
Triethylheptylsilane	C$_{13}$H$_{30}$Si	70.0	99.8	114.6	130.3	148.0	158.2	174.0	196.0	221.0	247.0	
Triethyloctylsilane	C$_{14}$H$_{32}$Si	73.7	104.8	120.6	137.7	155.7	168.0	184.3	208.0	235.0	262.0	
Triethyl orthoformate	C$_7$H$_{16}$O$_3$	+5.5	29.2	40.5	53.4	67.5	76.0	88.0	106.0	125.7	146.0	
phosphate	C$_6$H$_{15}$O$_4$P	39.6	67.8	82.1	97.8	115.7	126.3	141.6	163.7	187.0	211.0	
Triethylthallium	C$_6$H$_{15}$Tl	+9.3	37.6	51.7	67.7	85.4	95.7	112.1	136.0	163.5	192.1	−63.0
Trifluorophenylsilane	C$_6$H$_5$F$_3$Si	−31.0	−9.7	+0.8	12.3	25.4	33.2	44.2	60.1	78.7	98.3	
Trimethallyl phosphate	C$_{12}$H$_{21}$PO$_4$	93.7	131.0	149.8	169.8	192.0	207.0	225.7	255.0	288.5	324.0	
2,3,5-Trimethylacetophenone	C$_{11}$H$_{14}$O	79.0	108.0	122.3	137.5	154.2	165.7	179.7	201.3	224.3	247.5	
Trimethylamine	C$_3$H$_9$N	−97.1	−81.7	−73.8	−65.0	−55.2	−48.8	−40.3	−27.0	−12.5	+2.9	−117.1
2,4,5-Trimethylaniline	C$_9$H$_{13}$N	68.4	95.9	109.0	123.7	139.8	149.5	162.0	182.3	203.7	234.5	67
1,2,3-Trimethylbenzene	C$_9$H$_{12}$	16.8	42.9	55.9	69.9	85.4	95.3	108.8	129.0	152.0	176.1	−25.5
1,2,4-Trimethylbenzene	C$_9$H$_{12}$	13.6	38.3	50.7	64.5	79.8	89.5	102.8	122.7	145.4	169.2	−44.1
1,3,5-Trimethylbenzene	C$_9$H$_{12}$	9.6	34.7	47.4	61.0	76.1	85.8	98.9	118.6	141.0	164.7	−44.8
2,2,3-Trimethylbutane	C$_7$H$_{16}$			−18.8	−7.5	+4.2	13.3	24.4	41.2	60.4	80.9	−25.0
Trimethyl citrate	C$_9$H$_{14}$O$_7$	106.2	146.2	160.4	177.2	194.2	205.5	219.6	241.3	264.2	287.0	78.5
Trimethyleneglycol (1,3-propanediol)	C$_3$H$_8$O$_2$	59.4	87.2	100.6	115.5	131.0	141.1	153.4	172.8	193.8	214.2	
1,2,4-Trimethyl-5-ethylbenzene	C$_{11}$H$_{16}$	43.7	71.2	84.6	99.7	106.0	126.3	140.3	160.3	184.5	208.1	
1,3,5-Trimethyl-2-ethylbenzene	C$_{11}$H$_{16}$	38.8	67.0	80.5	96.0	113.2	123.8	137.9	158.4	183.5	208.0	
2,2,3-Trimethylpentane	C$_8$H$_{18}$	−29.0	−7.1	+3.9	16.0	29.5	38.1	49.9	67.8	88.2	109.8	−112.3
2,2,4-Trimethylpentane	C$_8$H$_{18}$	−36.5	−15.0	−4.3	+7.5	20.7	29.1	40.7	58.1	78.0	99.2	−107.3
2,3,3-Trimethylpentane	C$_8$H$_{18}$	−25.8	−3.9	+6.9	19.2	33.0	41.8	53.8	72.0	92.7	114.8	−101.5
2,3,4-Trimethylpentane	C$_8$H$_{18}$	−26.3	−4.1	+7.1	19.3	32.9	41.6	53.4	71.3	91.8	113.5	−109.2
2,2,4-Trimethyl-3-pentanone	C$_8$H$_{16}$O	14.7	36.0	46.4	57.6	69.8	77.3	87.6	102.2	118.4	135.0	
Trimethyl phosphate	C$_3$H$_9$O$_4$P	26.0	53.7	67.8	83.0	100.0	110.0	124.0	145.0	167.8	192.7	
2,4,5-Trimethylstyrene	C$_{11}$H$_{14}$	48.1	77.0	91.6	107.1	124.2	135.5	149.8	171.8	196.1	221.2	
2,4,6-Trimethylstyrene	C$_{11}$H$_{14}$	37.5	65.7	79.7	94.8	111.8	122.3	136.8	157.8	182.3	207.0	
Trimethylsuccinic anhydride	C$_7$H$_{10}$O$_3$	53.5	82.6	97.4	113.8	131.0	142.2	156.5	179.8	205.5	231.0	
Triphenylmethane	C$_{19}$H$_{16}$	169.7	188.4	197.0	206.8	215.5	221.2	228.4	239.7	249.8	259.2	93.4
Triphenylphosphate	C$_{18}$H$_{15}$O$_4$P	193.5	230.4	249.8	269.7	290.3	305.2	322.5	349.8	379.2	413.5	49.4
Tripropyleneglycol	C$_9$H$_{20}$O$_4$	96.0	125.7	140.5	155.8	173.7	184.6	199.0	220.2	244.3	267.2	
Tripropyleneglycol monobutyl ether	C$_{13}$H$_{28}$O$_4$	101.5	131.6	147.0	161.8	179.8	190.2	204.4	224.4	247.0	269.5	
Tripropyleneglycol monoisopropyl ether	C$_{12}$H$_{26}$O$_4$	82.4	112.4	127.3	143.7	161.4	173.2	187.8	209.7	232.8	256.6	
Tritolyl phosphate	C$_{21}$H$_{21}$O$_4$P	154.6	184.2	198.0	213.2	229.7	239.8	252.2	271.8	292.7	313.0	
Undecane	C$_{11}$H$_{24}$	32.7	59.7	73.9	85.6	104.4	115.2	128.1	149.3	171.9	195.8	−25.6
Undecanoic acid	C$_{11}$H$_{22}$O$_2$	101.4	133.1	149.0	166.0	185.6	197.2	212.5	237.8	262.8	290.0	29.5
10-Undecenoic acid	C$_{11}$H$_{20}$O$_2$	114.0	142.8	156.3	172.0	188.7	199.5	213.5	232.8	254.0	275.0	24.5
Undecan-2-ol	C$_{11}$H$_{24}$O	71.1	99.0	112.8	127.5	143.7	153.7	167.2	187.7	209.8	232.0	
n-Valeric acid	C$_5$H$_{10}$O$_2$	42.2	67.7	79.8	93.1	107.8	116.6	128.3	146.0	165.0	184.4	−34.5
iso-Valeric acid	C$_5$H$_{10}$O$_2$	34.5	59.6	71.3	84.0	98.0	107.3	118.9	136.2	155.2	175.1	−37.6
γ-Valerolactone	C$_5$H$_8$O$_2$	37.5	65.8	79.8	95.2	101.9	122.4	136.5	157.7	182.3	207.5	
Valeronitrile	C$_5$H$_9$N	−6.0	+18.1	30.0	43.3	57.8	66.9	78.6	97.7	118.7	140.8	
Vanillin	C$_8$H$_8$O$_3$	107.0	138.4	154.0	170.5	188.7	199.8	214.5	237.3	260.0	285.0	81.5
Vinyl acetate	C$_4$H$_6$O$_2$	−48.0	−28.0	−18.0	−7.0	+5.3	13.0	23.3	38.4	55.5	72.5	
2-Vinylanisole	C$_9$H$_{10}$O	41.9	68.0	81.0	94.7	110.0	119.8	132.3	151.0	172.1	194.0	
3-Vinylanisole	C$_9$H$_{10}$O	43.4	69.9	83.0	97.2	112.5	122.3	135.3	154.0	175.8	197.5	
4-Vinylanisole	C$_9$H$_{10}$O	45.2	72.0	85.7	100.0	116.0	126.1	139.7	159.0	182.0	204.5	
Vinyl chloride (1-chloroethylene)	C$_2$H$_3$Cl	−105.6	−90.8	−83.7	−75.7	−66.8	−61.1	−53.2	−41.3	−28.0	−13.8	−153.7
cyanide (acrylonitrile)	C$_3$H$_3$N	−51.0	−30.7	−20.3	−9.0	+3.8	11.8	22.8	38.7	58.3	78.5	−82
fluoride (1-fluoroethylene)	C$_2$H$_3$F	−149.3	−138.0	−132.2	−125.4	−118.0	−113.0	−106.2	−95.4	−84.0	−72.2	−160.5
Vinylidene chloride (1,1-dichloroethene)	C$_2$H$_2$Cl$_2$	−77.2	−60.0	−51.2	−41.7	−31.1	−24.0	−15.0	−1.0	+14.8	31.7	−122.5
4-Vinylphenetole	C$_{10}$H$_{12}$O	64.0	91.7	105.6	120.3	136.3	146.4	159.8	180.0	202.8	225.0	
2-Xenyl dichlorophosphate	C$_{12}$H$_9$Cl$_2$PO	138.2	171.1	187.0	205.0	223.8	236.0	251.5	275.3	301.5	328.5	
2,4-Xyaldehyde	C$_9$H$_{10}$O	59.0	85.9	99.0	114.0	129.7	139.8	152.2	172.3	194.1	215.5	
2-Xylene (2-xylene)	C$_8$H$_{10}$	−3.8	+20.2	32.1	45.1	59.5	68.8	81.3	100.2	121.7	144.4	−25.2
3-Xylene (3-xylene)	C$_8$H$_{10}$	−6.9	+16.8	28.3	41.1	55.3	64.4	76.8	95.5	116.7	139.1	−47.9
4-Xylene (4-xylene)	C$_8$H$_{10}$	−8.1	+15.5	27.3	40.1	54.4	63.5	75.9	94.6	115.9	138.3	+13.3
2,4-Xylidine	C$_8$H$_{11}$N	52.6	79.8	93.0	107.6	123.8	133.7	146.8	166.4	188.3	211.5	
2,6-Xylidine	C$_8$H$_{11}$N	44.0	72.6	87.0	102.7	120.2	131.5	146.0	168.0	193.7	217.9	

VAPOR PRESSURES OF SOLUTIONS

UNITS CONVERSIONS

For this subsection, the following units conversions are applicable:

$$°F = \tfrac{9}{5}°C + 32.$$

To convert millimeters of mercury to pounds-force per square inch, multiply by 0.01934.

To convert cubic feet to cubic meters, multiply by 0.02832.

To convert bars to pounds-force per square inch, multiply by 14.504.

To convert bars to kilopascals, multiply by 1×10^2.

TABLE 2-9   Partial Pressures of Water over Aqueous Solutions of HCl*

$\log_{10} p\,\text{mm} = A - B/T$, $(T$ in $K)$, which, however, agrees only approximately with the table. The table is more nearly correct.

Partial pressure of H_2O, mmHg, °C

% HCl	A	B	0°	5°	10°	15°	20°	25°	30°	35°	40°	45°	50°	60°	70°	80°	90°	100°	110°
6	8.99156	2282	4.18	6.04	8.45	11.7	15.9	21.8	29.1	39.4	50.6	66.2	86.0	139	220	333	492	715	
10	8.99864	2295	3.84	5.52	7.70	10.7	14.6	20.0	26.8	35.5	47.0	61.5	80.0	130	204	310	463	677	960
14	8.97075	2300	3.39	4.91	6.95	9.65	13.1	18.0	24.1	31.9	42.1	55.3	72.0	116	185	273	425	625	892
18	8.98014	2323	2.87	4.21	5.92	8.26	11.3	15.4	20.6	27.5	36.4	47.9	62.5	102	162	248	374	550	783
20	8.97877	2334	2.62	3.83	5.40	7.50	10.3	14.1	19.0	25.1	33.3	43.6	57.0	93.5	150	230	345	510	729
22	9.02708	2363	2.33	3.40	4.82	6.75	9.30	12.6	17.1	22.8	30.2	39.8	52.0	85.6	138	211	317	467	670
24	8.96022	2356	2.05	3.04	4.31	6.03	8.30	11.4	15.4	20.4	27.1	35.7	46.7	77.0	124	194	290	426	611
26	9.01511	2390	1.76	2.60	3.71	5.21	7.21	9.95	13.5	18.0	24.0	31.7	41.5	69.0	112	173	261	387	555
28	8.97611	2395	1.50	2.24	3.21	4.54	6.32	8.75	11.8	15.8	21.1	27.9	36.5	60.7	99.0	154	234	349	499
30	9.00117	2422	1.26	1.90	2.73	3.88	5.41	7.52	10.2	13.7	18.4	24.3	32.0	53.5	87.5	136	207	310	444
32	9.03317	2453	1.04	1.57	2.27	3.25	4.55	6.37	8.70	11.7	15.7	21.0	27.7	46.5	76.5	120	184	275	396
34	9.07143	2487	0.85	1.29	1.87	2.70	3.81	5.35	7.32	9.95	13.5	18.1	24.0	40.5	66.5	104	161	243	355
36	9.11815	2526	0.68	1.03	1.50	2.19	3.10	4.41	6.08	8.33	11.4	15.4	20.4	34.8	57.0	90.0	140	212	311
38	9.20783	2579	0.53	0.81	1.20	1.75	2.51	3.60	5.03	6.92	9.52	13.0	17.4	29.6	49.1	77.5	120	182	266
40	9.33923	2647	0.41	0.63	0.94	1.37	2.00	2.88	4.09	5.68	7.85	10.7	14.5	25.0	42.1	67.3	105	158	230
42	9.44953	2709	0.31	0.48	0.72	1.06	1.56	2.30	3.28	4.60	6.45	8.90	12.1	21.2	35.8	57.2	89.2	135	195

*Accuracy, ca. 2 percent for solutions of 15 to 30 percent HCl between 0 and 100°; for solutions of > 30 percent HCl the accuracy is ca. 5 percent at the lower temperatures and ca. 15 percent at the higher temperatures. Below 15 percent HCl, the accuracy is ca. 5 percent at the lower temperatures and higher strengths to ca. 15 to 20 percent at the lower strengths and perhaps 15 to 20 percent at the higher temperatures and lower strengths.

TABLE 2-10   Partial Pressures of HCl over Aqueous Solutions of HCl*

$\log_{10} p\,\text{mm} = A - B/T$, $(T$ in $K)$, which, however, agrees only approximately with the table. The table is more nearly correct. mmHg, °C

% HCl	A	B	0°	5°	10°	15°	20°	25°	30°	35°	40°	45°	50°	60°	70°	80°	90°	100°	110°
2	11.8037	4736			0.0000117	0.000023	0.000044	0.000084	0.000151	0.000275	0.00047	0.00083	0.00140	0.00380	0.0100	0.0245	0.058	0.132	0.280
4	11.6400	4471	0.000018	0.000036	.000069	.000131	.00024	.00044	.00077	.00134	.0023	.00385	.0064	.0165	.0405	.095	.21	.46	.93
6	11.2144	4202	.000066	.000125	.000234	.000425	.00076	.00131	.00225	.0038	.0062	.0102	.0163	.040	.094	.206	.44	.92	1.78
8	11.0406	4042	.000118	.000323	.000583	.00104	.00178	.0031	.00515	.0085	.0136	.022	.0344	.081	.183	.39	.82	1.64	3.10
10	10.9311	3908	.00042	.00075	.00134	.00232	.00395	.0067	.0111	.0178	.0282	.045	.069	.157	.35	.73	1.48	2.9	5.4
12	10.7900	3765	.00099	.00175	.00305	.0052	.0088	.0145	.0234	.037	.058	.091	.136	.305	.66	1.34	2.65	5.1	9.3
14	10.6954	3636	.0024	.00415	.0071	.0118	.0196	.0316	.050	.078	.121	.185	.275	.60	1.25	2.50	4.8	9.0	16.0
16	10.6261	3516	.0056	.0095	.016	.0265	.0428	.0685	.106	.163	.247	.375	.55	1.17	2.40	4.66	8.8	16.1	28
18	10.4957	3376	.0135	.0225	.037	.060	.095	.148	.228	.345	.515	.77	1.11	2.3	4.55	8.6	15.7	28	48
20	10.3833	3245	.0316	.052	.084	.132	.205	.32	.48	.72	1.06	1.55	2.21	4.4	8.5	15.6	28.1	49	83
22	10.3172	3125	.0734	.119	.187	.294	.45	.68	1.02	1.50	2.18	3.14	4.42	8.6	16.3	29.3	52	90	146
24	10.2185	2995	.175	.277	.43	.66	1.00	1.49	2.17	3.14	4.5	6.4	8.9	16.9	31.0	54.5	94	157	253
26	10.1303	2870	.41	.64	.98	1.47	2.17	3.20	4.56	6.50	9.2	12.7	17.5	32.5	58.5	100	169	276	436
28	10.0115	2732	1.0	1.52	2.27	3.36	4.90	7.05	9.90	13.8	19.1	26.4	35.7	64	112	188	309	493	760
30	9.8763	2593	2.4	3.57	5.23	7.60	10.6	15.1	21.0	28.6	39.4	53	71	124	208	340	542	845	
32	9.7523	2457	5.7	8.3	11.8	16.8	23.5	32.5	44.5	60.0	81	107	141	238	390	623	970		
34	9.6061	2316	13.1	18.8	26.4	36.8	50.5	68.5	92	122	161	211	273	450	720				
36	9.5262	2229	29.0	41.0	56.4	78	105.5	142	188	246	322	416	535	860					
38	9.4670	2094	63.0	87.0	117	158	210	277	360	465	598	758	955						
40	9.2156	1939	130	176	233	307	399	515	627	830									
42	8.9925	1800	253	332	430	560	709	900											
44	8.8621	1681	510	655	840														
46			940																

*Accuracy, ca. 2 percent for solutions of 15 to 30 percent HCl between 0 and 100°; for solutions of > 30 percent HCl the accuracy is ca. 5 percent at the lower temperatures and ca. 15 percent at the higher temperatures. Below 15 percent HCl, the accuracy is ca. 5 percent at the lower temperatures and higher strengths to ca. 15 to 20 percent at the lower strengths and perhaps 15 to 20 percent at the higher temperatures and lower strengths.

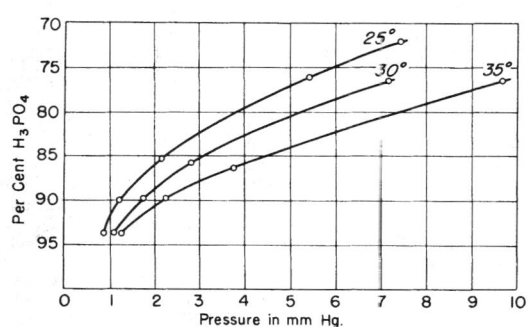

FIG. 2-1 Vapor pressures of H_3PO_4 aqueous: partial pressure of H_2O vapor. (*Courtesy of Victor Chemical Works, Stauffer Chemical Company; measurements by W. H. Woodstock.*)

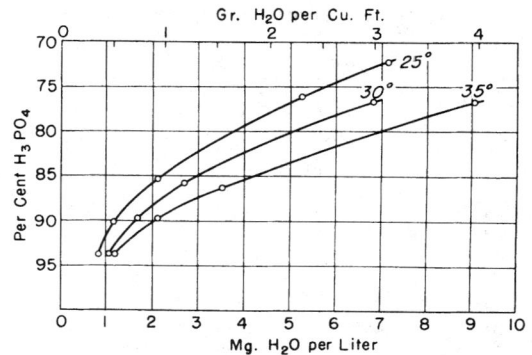

FIG. 2-2 Vapor pressures of H_3PO_4 aqueous: weight of H_2O in saturated air. (*Courtesy of Victor Chemical Works, Stauffer Chemical Company; measurements by W. H. Woodstock.*)

TABLE 2-11   Partial Pressures of H₂O and SO₂ over Aqueous Solutions of Sulfur Dioxide*

Partial pressures of H_2O and SO_2, mmHg, °C

g SO₂/ 100 g H₂O	Temperature, ∞C								
	0	10	20	30	40	50	60	90	120
0.01	0.02	0.04	0.07	0.12	0.19	0.29	0.43	1.21	2.82
0.05	0.38	0.66	1.07	1.68	2.53	3.69	5.24	12.9	27.0
0.10	1.15	1.91	3.03	4.62	6.80	9.71	13.5	31.7	63.9
0.15	2.10	3.44	5.37	8.07	11.7	16.5	22.7	52.2	104
0.20	3.17	5.13	7.93	11.8	17.0	23.8	32.6	73.7	145
0.25	4.34	6.93	10.6	15.7	22.5	31.4	42.8	95.8	186
0.30	5.57	8.84	13.5	19.8	28.2	39.2	53.3	118	229
0.40	8.17	12.8	19.4	28.3	40.1	55.3	74.7	164	316
0.50	10.9	17.0	25.6	37.1	52.3	72.0	96.8	211	404
1.00	25.8	39.5	58.4	83.7	117	159	212	454	856
2.00	58.6	88.5	129	183	253	342	453	955	
3.00	93.2	139	202	285	393	530	700		
4.00	129	192	277	389	535	720			
5.00	165	245	353	496	679				
6.00	202	299	430	602	824				
8.00	275	407	585	818					
10.00	351	517	741						
15.00	542	796							
20.00	735								

*Extracted with permission from *J. Chem Eng. Data* **8**, 1963: 333–336. Copyright 1963 American Chemical Society.

TABLE 2-12   Water Partial Pressure, bar, over Aqueous Sulfuric Acid Solutions*

Weight percent, H_2SO_4

°C	10.0	20.0	30.0	40.0	50.0	60.0	70.0	75.0	80.0	85.0
0	.582E−02	.534E−02	.448E−02	.326E−02	.193E−02	.836E−03	.207E−03	.747E−04	.197E−04	.343E−05
10	.117E−01	.107E−01	.909E−02	.670E−02	.405E−02	.180E−02	.467E−03	.175E−03	.490E−04	.952E−05
20	.223E−01	.205E−01	.174E−01	.130E−01	.802E−02	.367E−02	.995E−03	.388E−03	.115E−04	.245E−04
30	.404E−01	.373E−01	.319E−01	.241E−01	.151E−01	.710E−02	.201E−02	.811E−03	.253E−03	.589E−04
40	.703E−01	.649E−01	.558E−01	.427E−01	.272E−01	.131E−01	.387E−02	.162E−02	.531E−03	.133E−03
50	.117	.109	.939E−01	.725E−01	.470E−01	.232E−01	.715E−02	.309E−02	.106E−02	.286E−03
60	.189	.175	.152	.119	.782E−01	.395E−01	.127E−01	.565E−02	.204E−02	.584E−03
70	.296	.275	.239	.188	.126	.651E−01	.217E−01	.997E−02	.376E−02	.114E−02
80	.449	.417	.365	.290	.196	.104	.360E−01	.170E−01	.668E−02	.213E−02
90	.664	.617	.542	.434	.298	.161	.578E−01	.281E−01	.115E−01	.383E−02
100	.957	.891	.786	.634	.441	.244	.905E−01	.452E−01	.192E−01	.666E−02
110	1.349	1.258	1.113	.904	.638	.360	.138	.708E−01	.312E−01	.112E−01
120	1.863	1.740	1.544	1.264	.903	.519	.206	.108	.493E−01	.183E−01
130	2.524	2.361	2.101	1.732	1.253	.734	.301	.162	.760E−01	.291E−01
140	3.361	3.149	2.810	2.333	1.708	1.020	.481	.236	.115	.451E−01
150	4.404	4.132	3.697	3.090	2.289	1.392	.605	.339	.170	.682E−01
160	5.685	5.342	4.793	4.031	3.021	1.870	.837	.478	.246	.101
170	7.236	6.810	6.127	5.185	3.930	2.475	1.138	.662	.350	.147
180	9.093	8.571	7.731	6.584	5.045	3.233	1.525	.902	.489	.208
190	11.289	10.658	9.640	8.259	6.397	4.169	2.017	1.212	.673	.291
200	13.861	13.107	11.887	10.245	8.020	5.312	2.632	1.606	.913	.401
210	16.841	15.951	14.505	12.576	9.948	6.696	3.395	2.101	1.220	.542
220	20.264	19.225	17.529	15.287	12.217	8.354	4.331	2.714	1.609	.724
230	24.160	22.960	20.992	18.414	14.864	10.322	5.466	3.467	2.096	.952
240	28.561	27.188	24.927	21.992	17.929	12.641	6.831	4.381	2.699	1.237
250	33.494	31.939	29.364	26.056	21.452	15.351	8.458	5.480	3.435	1.587
260	38.984	37.240	34.334	30.642	25.472	18.496	10.382	6.788	4.326	2.012
270	45.055	43.116	39.865	35.784	30.030	22.121	12.640	8.333	5.395	2.525
280	51.726	49.590	45.984	41.514	35.168	26.274	15.269	10.142	6.663	3.136
290	59.015	56.681	52.715	47.865	40.926	31.003	18.311	12.242	8.155	3.857
300	66.934	64.407	60.081	54.868	47.346	36.360	21.808	14.665	9.897	4.701
310	75.495	72.781	68.100	62.553	54.470	42.395	25.804	17.438	11.912	5.680
320	84.705	81.816	76.792	70.947	62.337	49.164	30.343	20.591	14.227	6.806
330	94.567	91.518	86.172	80.077	70.988	56.721	35.473	24.153	16.867	8.093
340	105.083	101.894	96.252	89.969	80.463	65.123	41.240	28.154	19.855	9.551
350	116.251	112.946	107.043	100.646	90.802	74.426	47.692	32.622	23.217	11.193

*Vermeulen, Dong, Robinson, Nguyen, and Gmitro, AIChE meeting, Anaheim, Calif., 1982; and private communication from Prof. Theodore Vermeulen, Chemical Engineering Dept., University of California, Berkeley.

TABLE 2-12 Water Partial Pressure, bar, over Aqueous Sulfuric Acid Solutions (*Concluded*)

Weight percent, H_2SO_4

°C	90.0	92.0	94.0	96.0	97.0	98.0	98.5	99.0	99.5	100.0
0	.518E−06	.242E−06	.107E−06	.401E−07	.218E−07	.980E−08	.569E−08	.268E−08	.775E−09	.196E−09
10	.159E−05	.762E−06	.344E−06	.130E−06	.713E−07	.323E−07	.188E−07	.888E−08	.258E−08	.655E−09
20	.448E−05	.220E−05	.101E−05	.390E−06	.215E−06	.978E−07	.572E−07	.271E−07	.789E−08	.201E−08
30	.117E−04	.587E−05	.275E−05	.108E−05	.598E−06	.275E−06	.161E−06	.766E−07	.224E−07	.575E−08
40	.285E−04	.146E−04	.696E−05	.278E−05	.155E−05	.720E−06	.424E−06	.202E−06	.595E−07	.153E−07
50	.652E−04	.341E−04	.166E−04	.672E−05	.379E−05	.177E−05	.105E−05	.503E−06	.149E−06	.384E−07
60	.141E−03	.754E−04	.372E−04	.154E−04	.875E−05	.413E−05	.245E−05	.118E−05	.350E−06	.910E−07
70	.290E−03	.158E−03	.795E−04	.334E−04	.192E−04	.912E−05	.544E−05	.263E−05	.784E−06	.205E−06
80	.569E−03	.316E−03	.162E−03	.691E−04	.400E−04	.192E−04	.115E−04	.559E−05	.168E−05	.439E−06
90	.107E−02	.606E−03	.315E−03	.137E−03	.801E−04	.388E−04	.234E−04	.114E−04	.343E−05	.903E−06
100	.194E−02	.112E−02	.590E−03	.261E−03	.154E−03	.752E−04	.455E−04	.223E−04	.674E−05	.178E−05
110	.338E−02	.198E−02	.107E−02	.479E−03	.285E−03	.141E−03	.855E−04	.420E−04	.128E−04	.339E−05
120	.571E−02	.341E−02	.186E−02	.851E−03	.511E−03	.254E−03	.155E−03	.766E−04	.233E−04	.623E−05
130	.938E−02	.569E−02	.315E−02	.146E−02	.886E−03	.445E−03	.278E−03	.135E−03	.414E−04	.111E−04
140	.150E−01	.923E−02	.519E−02	.245E−02	.149E−02	.757E−03	.467E−03	.232E−03	.711E−04	.191E−04
150	.233E−01	.146E−01	.832E−02	.399E−02	.245E−02	.125E−02	.776E−03	.387E−03	.119E−03	.321E−04
160	.354E−01	.225E−01	.130E−01	.633E−02	.393E−02	.202E−02	.126E−02	.629E−03	.194E−03	.526E−04
170	.526E−01	.340E−01	.199E−01	.983E−02	.614E−02	.319E−02	.199E−02	.999E−03	.309E−03	.840E−04
180	.766E−01	.502E−01	.298E−01	.149E−01	.941E−02	.492E−02	.309E−02	.155E−02	.482E−03	.131E−03
190	.110	.729E−01	.438E−01	.222E−01	.141E−01	.744E−02	.469E−02	.236E−02	.735E−03	.201E−03
200	.154	.104	.631E−01	.325E−01	.208E−01	.110E−01	.698E−02	.352E−02	.110E−02	.300E−03
210	.213	.146	.894E−01	.467E−01	.300E−01	.161E−01	.102E−01	.516E−02	.161E−02	.442E−03
220	.290	.201	.125	.660E−01	.427E−01	.230E−01	.147E−01	.743E−02	.232E−02	.638E−03
230	.389	.273	.171	.918E−01	.598E−01	.325E−01	.208E−01	.105E−01	.329E−02	.906E−03
240	.514	.366	.232	.126	.825E−01	.451E−01	.290E−01	.147E−01	.460E−02	.127E−02
250	.673	.485	.310	.170	.112	.618E−01	.398E−01	.202E−01	.633E−02	.174E−02
260	.870	.635	.409	.227	.151	.835E−01	.540E−01	.274E−01	.858E−02	.237E−02
270	1.112	.822	.534	.300	.200	.111	.723E−01	.366E−01	.115E−01	.317E−02
280	1.407	1.052	.689	.391	.263	.147	.957E−01	.485E−01	.152E−01	.420E−02
290	1.763	1.335	.880	.505	.341	.192	.125	.634E−01	.199E−01	.548E−02
300	2.190	1.676	1.112	.646	.437	.248	.162	.820E−01	.257E−01	.708E−02
310	2.696	2.088	1.394	.817	.556	.316	.208	.105	.328E−01	.905E−02
320	3.292	2.578	1.732	1.025	.701	.400	.264	.133	.415E−01	.114E−01
330	3.990	3.159	2.133	1.274	.875	.502	.331	.167	.520E−01	.143E−01
340	4.801	3.843	2.608	1.571	1.083	.624	.413	.208	.646E−01	.178E−01
350	5.738	4.641	3.164	1.922	1.331	.770	.511	.256	.795E−01	.218E−01

TABLE 2-13 **Sulfur Trioxide Partial Pressure, bar, over Aqueous Sulfuric Acid Solutions***

Weight percent, H_2SO_4

°C	10.0	20.0	30.0	40.0	50.0	60.0	70.0	75.0	80.0	85.0
0	.644E−29	.103E−27	.205E−26	.688E−25	.368E−23	.341E−21	.784E−19	.174E−17	.531E−16	.229E−14
10	.149E−27	.223E−26	.395E−25	.113E−23	.522E−22	.415E−20	.796E−18	.158E−16	.417E−15	.141E−13
20	.278E−26	.394E−25	.626E−24	.156E−22	.621E−21	.426E−19	.685E−17	.121E−15	.280E−14	.767E−13
30	.426E−25	.577E−24	.832E−23	.181E−21	.630E−20	.376E−18	.509E−16	.808E−15	.164E−13	.371E−12
40	.549E−24	.714E−23	.941E−22	.181E−20	.555E−19	.288E−17	.331E−15	.473E−14	.851E−13	.162E−11
50	.602E−23	.757E−22	.921E−21	.158E−19	.429E−18	.195E−16	.191E−14	.246E−13	.395E−12	.643E−11
60	.573E−22	.699E−21	.789E−20	.122E−18	.294E−17	.118E−15	.985E−14	.116E−12	.165E−11	.234E−10
70	.477E−21	.567E−20	.599E−19	.843E−18	.181E−16	.643E−15	.461E−13	.492E−12	.634E−11	.791E−10
80	.352E−20	.410E−19	.408E−18	.524E−17	.101E−15	.319E−14	.197E−12	.192E−11	.223E−10	.249E−09
90	.233E−19	.266E−18	.250E−17	.296E−16	.516E−15	.145E−13	.775E−12	.693E−11	.731E−10	.734E−09
100	.139E−18	.157E−17	.140E−16	.153E−15	.242E−14	.606E−13	.283E−11	.232E−10	.223E−09	.204E−08
110	.756E−18	.844E−17	.719E−16	.730E−15	.105E−13	.236E−12	.961E−11	.729E−10	.641E−09	.538E−08
120	.377E−17	.418E−16	.340E−15	.323E−14	.424E−13	.858E−12	.307E−10	.215E−09	.174E−08	.135E−07
130	.174E−16	.191E−15	.150E−14	.133E−13	.160E−12	.293E−11	.922E−10	.601E−09	.446E−08	.324E−07
140	.743E−16	.815E−15	.615E−14	.517E−13	.569E−12	.943E−11	.262E−09	.159E−08	.109E−07	.745E−07
150	.297E−15	.325E−14	.237E−13	.188E−12	.191E−11	.287E−10	.710E−09	.403E−08	.256E−07	.165E−06
160	.111E−14	.122E−13	.862E−13	.649E−12	.608E−11	.833E−10	.183E−08	.974E−08	.575E−07	.351E−06
170	.393E−14	.430E−13	.296E−12	.212E−11	.184E−10	.231E−09	.453E−08	.226E−07	.125E−06	.725E−06
180	.131E−13	.144E−12	.967E−12	.622E−11	.532E−10	.610E−09	.107E−07	.505E−07	.260E−06	.145E−05
190	.415E−13	.458E−12	.301E−11	.197E−10	.147E−09	.155E−08	.246E−07	.109E−06	.527E−06	.282E−05
200	.125E−12	.139E−11	.893E−11	.561E−10	.391E−09	.379E−08	.542E−07	.228E−06	.103E−05	.534E−05
210	.362E−12	.404E−11	.254E−10	.154E−09	.100E−08	.894E−08	.116E−06	.462E−06	.198E−05	.986E−05
220	.100E−11	.112E−10	.695E−10	.405E−09	.246E−08	.204E−07	.240E−06	.911E−06	.368E−05	.178E−04
230	.265E−11	.301E−10	.183E−09	.103E−08	.587E−08	.450E−07	.482E−06	.175E−05	.668E−05	.314E−04
240	.678E−11	.777E−10	.465E−09	.253E−08	.135E−07	.965E−07	.944E−06	.328E−05	.119E−04	.543E−04
250	.167E−10	.193E−09	.114E−08	.602E−08	.303E−07	.201E−06	.180E−05	.600E−05	.206E−04	.923E−04
260	.399E−10	.466E−09	.272E−08	.139E−07	.660E−07	.408E−06	.336E−05	.108E−04	.352E−04	.154E−03
270	.920E−10	.109E−08	.628E−08	.312E−07	.140E−06	.807E−06	.612E−05	.189E−04	.590E−04	.253E−03
280	.206E−09	.247E−08	.141E−07	.683E−07	.288E−06	.156E−05	.109E−04	.326E−04	.973E−04	.408E−03
290	.449E−09	.545E−08	.308E−07	.145E−06	.580E−06	.295E−05	.191E−04	.553E−04	.158E−03	.649E−03
300	.953E−09	.117E−07	.657E−07	.302E−06	.114E−05	.546E−05	.329E−04	.921E−04	.253E−03	.102E−02
310	.197E−08	.245E−07	.136E−06	.614E−06	.220E−05	.990E−05	.556E−04	.151E−03	.398E−03	.158E−02
320	.397E−08	.502E−07	.277E−06	.122E−05	.414E−05	.176E−04	.923E−04	.245E−03	.621E−03	.242E−02
330	.782E−08	.100E−06	.551E−06	.237E−05	.766E−05	.308E−04	.151E−03	.391E−03	.956E−03	.367E−02
340	.151E−07	.196E−06	.107E−05	.452E−05	.139E−04	.529E−04	.243E−03	.617E−03	.145E−02	.550E−02
350	.285E−07	.376E−06	.204E−05	.846E−05	.246E−04	.893E−04	.387E−03	.963E−03	.219E−02	.815E−02

*Vermeulen, Dong, Robinson, Nguyen, and Gmitro, AIChE meeting, Anaheim, Calif., 1982; and private communication from Prof. Theodore Vermeulen, Chemical Engineering Dept., University of California, Berkeley.

TABLE 2-13 Sulfur Trioxide Partial Pressure, bar, over Aqueous Sulfuric Acid Solutions (*Concluded*)

Weight percent, H_2SO_4

°C	90.0	92.0	94.0	96.0	97.0	98.0	98.5	99.0	99.5	100.0
0	.671E–13	.216E–12	.677E–12	.240E–11	.500E–11	.124E–10	.224E–10	.502E–10	.182E–09	.755E–09
10	.345E–12	.107E–11	.326E–11	.114E–10	.234E–10	.578E–10	.104E–09	.232E–09	.839E–09	.347E–08
20	.159E–11	.475E–11	.141E–10	.482E–10	.986E–10	.241E–09	.433E–09	.961E–09	.346E–08	.142E–07
30	.664E–11	.192E–10	.557E–10	.186E–09	.376E–09	.911E–09	.163E–08	.360E–08	.129E–07	.528E–07
40	.254E–10	.709E–10	.201E–09	.655E–09	.131E–08	.315E–08	.562E–08	.123E–07	.440E–07	.179E–06
50	.897E–10	.242E–09	.669E–09	.214E–08	.424E–08	.101E–07	.179E–07	.391E–07	.139E–06	.560E–06
60	.294E–09	.771E–09	.207E–08	.647E–08	.127E–07	.299E–07	.528E–07	.115E–06	.405E–06	.163E–05
70	.904E–09	.230E–08	.602E–08	.184E–07	.357E–07	.833E–07	.146E–06	.316E–06	.111E–05	.444E–05
80	.261E–08	.643E–08	.165E–07	.492E–07	.946E–07	.218E–06	.381E–06	.820E–06	.286E–05	.114E–04
90	.712E–08	.171E–07	.426E–07	.124E–06	.237E–06	.541E–06	.940E–06	.201E–05	.698E–05	.276E–04
100	.184E–07	.430E–07	.105E–06	.300E–06	.565E–06	.127E–05	.220E–05	.470E–05	.162E–04	.638E–04
110	.456E–07	.103E–06	.247E–06	.689E–06	.128E–05	.287E–05	.494E–05	.105E–04	.359E–04	.141E–03
120	.108E–06	.238E–06	.555E–06	.152E–05	.280E–05	.619E–05	.106E–04	.224E–04	.764E–04	.298E–03
130	.244E–06	.526E–06	.120E–05	.321E–05	.586E–05	.128E–04	.219E–04	.459E–04	.156E–03	.606E–03
140	.533E–06	.112E–05	.250E–05	.656E–05	.118E–04	.257E–04	.435E–04	.910E–04	.308E–03	.119E–02
150	.112E–05	.230E–05	.504E–05	.129E–04	.231E–04	.497E–04	.837E–04	.174E–03	.588E–03	.226E–02
160	.229E–05	.459E–05	.983E–05	.247E–04	.438E–04	.932E–04	.156E–03	.324E–03	.109E–02	.416E–02
170	.453E–05	.886E–05	.186E–04	.459E–04	.806E–04	.170E–03	.283E–03	.586E–03	.196E–02	.746E–02
180	.870E–05	.166E–04	.343E–04	.829E–04	.144E–03	.301E–03	.499E–03	.103E–02	.343E–02	.130E–01
190	.163E–04	.304E–04	.615E–04	.146E–03	.252E–03	.520E–03	.859E–03	.177E–02	.587E–02	.222E–01
200	.297E–04	.543E–04	.108E–03	.251E–03	.429E–03	.878E–03	.144E–02	.296E–02	.981E–02	.370E–01
210	.528E–04	.946E–04	.185E–03	.422E–03	.714E–03	.145E–02	.237E–02	.486E–02	.161E–01	.603E–01
220	.919E–04	.161E–03	.309E–03	.694E–03	.117E–02	.235E–02	.383E–02	.781E–02	.258E–01	.965E–01
230	.157E–03	.269E–03	.508E–03	.112E–02	.187E–02	.373E–02	.605E–02	.123E–01	.405E–01	.152
240	.261E–03	.441E–03	.819E–03	.178E–02	.293E–02	.582E–02	.939E–02	.191E–01	.627E–01	.234
250	.428E–03	.708E–03	.130E–02	.276E–02	.453E–02	.891E–02	.143E–01	.291E–01	.955E–01	.356
260	.690E–03	.112E–02	.202E–02	.423E–02	.688E–02	.134E–01	.215E–01	.437E–01	.143	.532
270	.109E–02	.174E–02	.309E–02	.638E–02	.103E–01	.200E–01	.319E–01	.646E–01	.212	.786
280	.170E–02	.266E–02	.466E–02	.948E–02	.152E–01	.293E–01	.465E–01	.943E–01	.309	1.144
290	.261E–02	.401E–02	.694E–02	.139E–01	.221E–01	.423E–01	.670E–01	.136	.444	1.646
300	.395E–02	.595E–02	.102E–01	.201E–01	.318E–01	.604E–01	.953E–01	.193	.632	2.339
310	.589E–02	.873E–02	.148E–01	.287E–01	.451E–01	.852E–01	.134	.272	.889	3.289
320	.868E–02	.126E–01	.211E–01	.405E–01	.632E–01	.119	.186	.378	1.236	4.575
330	.126E–01	.181E–01	.299E–01	.565E–01	.877E–01	.164	.256	.520	1.703	6.303
340	.181E–01	.255E–01	.418E–01	.780E–01	.120	.224	.348	.708	2.323	8.603
350	.258E–01	.357E–01	.578E–01	.107	.164	.303	.470	.956	3.142	11.640

TABLE 2-14 Sulfuric Acid Partial Pressure, bar, over Aqueous Sulfuric Acid*

°C	Weight Percent, H_2SO_4									
	10.0	20.0	30.0	40.0	50.0	60.0	70.0	75.0	80.0	85.0
0	.576E−21	.843E−20	.141E−18	.344E−17	.109E−15	.438E−14	.249E−12	.200E−11	.161E−10	.121E−09
10	.634E−20	.874E−19	.131E−17	.276E−16	.769E−15	.273E−13	.135E−11	.101E−10	.743E−10	.490E−09
20	.588E−19	.769E−18	.104E−16	.193E−15	.474E−14	.149E−12	.649E−11	.447E−10	.305E−09	.179E−08
30	.468E−18	.584E−17	.721E−16	.119E−14	.259E−13	.725E−12	.278E−10	.178E−09	.113E−08	.594E−08
40	.324E−17	.389E−16	.441E−15	.649E−14	.127E−12	.317E−11	.108E−09	.643E−09	.379E−08	.181E−07
50	.197E−16	.229E−15	.241E−14	.320E−13	.562E−12	.126E−10	.380E−09	.212E−08	.117E−07	.513E−07
60	.107E−15	.121E−14	.119E−13	.144E−12	.228E−11	.462E−10	.124E−08	.646E−08	.334E−07	.135E−06
70	.526E−15	.581E−14	.535E−13	.592E−12	.851E−11	.156E−09	.373E−08	.183E−07	.888E−07	.336E−06
80	.235E−14	.254E−13	.221E−12	.225E−11	.295E−10	.492E−09	.105E−07	.485E−07	.222E−06	.786E−06
90	.960E−14	.102E−12	.844E−12	.798E−11	.956E−10	.145E−08	.279E−07	.121E−06	.522E−06	.175E−05
100	.353E−13	.381E−12	.300E−11	.264E−10	.291E−09	.402E−08	.698E−07	.287E−06	.117E−05	.371E−05
110	.127E−12	.132E−11	.997E−11	.824E−10	.835E−09	.106E−07	.166E−06	.644E−06	.249E−05	.752E−05
120	.418E−12	.432E−11	.312E−10	.243E−09	.227E−08	.264E−07	.375E−06	.138E−05	.508E−05	.147E−04
130	.129E−11	.132E−10	.924E−10	.678E−09	.589E−08	.631E−07	.814E−06	.285E−05	.995E−05	.277E−04
140	.375E−11	.385E−10	.259E−09	.181E−08	.146E−07	.144E−06	.169E−05	.565E−05	.188E−04	.503E−04
150	.103E−10	.106E−09	.694E−09	.460E−08	.346E−07	.316E−06	.340E−05	.108E−04	.343E−04	.889E−04
160	.272E−10	.279E−09	.178E−08	.112E−07	.789E−07	.670E−06	.659E−05	.200E−04	.608E−04	.152E−03
170	.682E−10	.702E−09	.436E−08	.264E−07	.174E−06	.137E−05	.124E−04	.359E−04	.104E−03	.255E−03
180	.164E−09	.170E−08	.103E−07	.599E−07	.369E−06	.271E−05	.225E−04	.627E−04	.175E−03	.416E−03
190	.378E−09	.394E−08	.234E−07	.131E−06	.760E−06	.521E−05	.400E−04	.107E−03	.286E−03	.663E−03
200	.842E−09	.883E−08	.514E−07	.278E−06	.152E−05	.975E−05	.691E−04	.177E−03	.457E−03	.104E−02
210	.181E−08	.191E−07	.109E−06	.573E−06	.295E−05	.178E−04	.117E−03	.288E−03	.715E−03	.159E−02
220	.376E−08	.401E−07	.226E−06	.115E−05	.559E−05	.316E−04	.193E−03	.459E−03	.110E−02	.239E−02
230	.758E−08	.817E−07	.455E−06	.224E−05	.103E−04	.549E−04	.311E−03	.717E−03	.166E−02	.354E−02
240	.148E−07	.162E−06	.889E−06	.427E−05	.186E−04	.935E−04	.494E−03	.110E−02	.245E−02	.515E−02
250	.283E−07	.312E−06	.170E−05	.793E−05	.329E−04	.156E−03	.770E−03	.166E−02	.358E−02	.740E−02
260	.526E−07	.588E−06	.316E−05	.144E−04	.569E−04	.255E−03	.118E−02	.247E−02	.516E−02	.105E−01
270	.954E−07	.108E−05	.577E−05	.257E−04	.965E−04	.411E−03	.178E−02	.362E−02	.733E−02	.147E−01
280	.169E−06	.194E−05	.103E−04	.450E−04	.161E−03	.650E−03	.265E−02	.524E−02	.103E−01	.203E−01
290	.294E−06	.342E−05	.180E−04	.771E−04	.263E−03	.101E−02	.389E−02	.750E−02	.143E−01	.278E−01
300	.500E−06	.591E−05	.309E−04	.130E−03	.424E−03	.156E−02	.563E−02	.106E−01	.196E−01	.376E−01
310	.834E−06	.100E−04	.522E−04	.215E−03	.672E−03	.236E−02	.805E−02	.148E−01	.266E−01	.504E−01
320	.137E−05	.167E−04	.865E−04	.352E−03	.105E−02	.352E−02	.114E−01	.205E−01	.359E−01	.670E−01
330	.220E−05	.273E−04	.141E−03	.565E−03	.162E−02	.519E−02	.159E−01	.281E−01	.480E−01	.883E−01
340	.349E−05	.440E−04	.227E−03	.895E−03	.246E−02	.757E−02	.221E−01	.382E−01	.636E−01	.116
350	.544E−05	.698E−04	.360E−03	.140E−02	.369E−02	.109E−01	.303E−01	.516E−01	.836E−01	.150

°C	Weight percent, H_2SO_4									
	90.0	92.0	94.0	96.0	97.0	98.0	98.5	99.0	99.5	100.0
0	.534E−09	.803E−09	.112E−08	.148E−08	.167E−08	.187E−08	.196E−08	.206E−08	.217E−08	.228E−08
10	.200E−08	.296E−08	.409E−08	.540E−08	.609E−08	.679E−08	.714E−08	.750E−08	.788E−08	.827E−08
20	.677E−08	.993E−08	.136E−07	.179E−07	.201E−07	.224E−07	.236E−07	.247E−07	.260E−07	.273E−07
30	.211E−07	.306E−07	.415E−07	.543E−07	.611E−07	.680E−07	.714E−07	.749E−07	.786E−07	.824E−07
40	.607E−07	.870E−07	.117E−06	.153E−06	.171E−06	.191E−06	.200E−06	.210E−06	.220E−06	.230E−06
50	.163E−06	.231E−06	.309E−06	.400E−06	.449E−06	.498E−06	.523E−06	.548E−06	.574E−06	.600E−06
60	.411E−06	.575E−06	.765E−06	.985E−06	.110E−05	.122E−05	.128E−05	.134E−05	.140E−05	.147E−05
70	.976E−06	.135E−05	.179E−05	.229E−05	.256E−05	.283E−05	.297E−05	.310E−05	.325E−05	.339E−05
80	.220E−05	.302E−05	.396E−05	.504E−05	.562E−05	.622E−05	.652E−05	.681E−05	.712E−05	.743E−05
90	.473E−05	.642E−05	.835E−05	.106E−04	.118E−04	.130E−04	.136E−04	.143E−04	.149E−04	.155E−04
100	.973E−05	.131E−04	.169E−04	.213E−04	.237E−04	.261E−04	.274E−04	.285E−04	.298E−04	.310E−04
110	.192E−04	.256E−04	.328E−04	.412E−04	.457E−04	.503E−04	.527E−04	.549E−04	.572E−04	.595E−04
120	.366E−04	.482E−04	.614E−04	.767E−04	.849E−04	.935E−04	.977E−04	.102E−03	.106E−03	.110E−03
130	.672E−04	.879E−04	.111E−03	.138E−03	.153E−03	.168E−03	.175E−03	.182E−03	.190E−03	.197E−03
140	.120E−03	.155E−03	.195E−03	.241E−03	.266E−03	.292E−03	.304E−03	.316E−03	.329E−03	.341E−03
150	.207E−03	.266E−03	.332E−03	.408E−03	.449E−03	.493E−03	.514E−03	.534E−03	.554E−03	.574E−03
160	.348E−03	.444E−03	.550E−03	.673E−03	.740E−03	.810E−03	.844E−03	.876E−03	.909E−03	.941E−03
170	.572E−03	.723E−03	.889E−03	.108E−02	.119E−02	.130E−02	.135E−02	.140E−02	.145E−02	.150E−02
180	.917E−03	.115E−02	.140E−02	.170E−02	.186E−02	.204E−02	.212E−02	.220E−02	.227E−02	.235E−02
190	.144E−02	.179E−02	.217E−02	.262E−02	.286E−02	.312E−02	.325E−02	.336E−02	.348E−02	.359E−02
200	.221E−02	.273E−02	.329E−02	.395E−02	.431E−02	.470E−02	.488E−02	.505E−02	.522E−02	.538E−02
210	.333E−02	.408E−02	.490E−02	.585E−02	.637E−02	.693E−02	.720E−02	.744E−02	.768E−02	.791E−02
220	.494E−02	.601E−02	.715E−02	.850E−02	.924E−02	.100E−01	.104E−01	.108E−01	.111E−01	.114E−01
230	.719E−02	.869E−02	.103E−01	.122E−01	.132E−01	.143E−01	.149E−01	.153E−01	.158E−01	.162E−01
240	.103E−01	.124E−01	.146E−01	.171E−01	.186E−01	.201E−01	.209E−01	.215E−01	.221E−01	.227E−01
250	.146E−01	.174E−01	.203E−01	.238E−01	.257E−01	.278E−01	.289E−01	.297E−01	.305E−01	.314E−01
260	.203E−01	.240E−01	.279E−01	.326E−01	.352E−01	.380E−01	.394E−01	.405E−01	.416E−01	.427E−01
270	.279E−01	.329E−01	.380E−01	.441E−01	.475E−01	.513E−01	.531E−01	.545E−01	.560E−01	.574E−01
280	.380E−01	.444E−01	.510E−01	.589E−01	.633E−01	.683E−01	.706E−01	.725E−01	.744E−01	.762E−01
290	.510E−01	.592E−01	.676E−01	.778E−01	.835E−01	.900E−01	.930E−01	.954E−01	.978E−01	.100
300	.678E−01	.782E−01	.888E−01	.102	.109	.117	.121	.124	.127	.130
310	.892E−01	.102	.115	.132	.141	.151	.156	.160	.164	.167
320	.116	.132	.149	.169	.180	.193	.199	.204	.209	.213
330	.150	.170	.190	.214	.228	.245	.252	.258	.263	.269
340	.192	.216	.240	.270	.287	.307	.317	.328	.330	.386
350	.243	.272	.301	.337	.358	.383	.394	.402	.410	.417

*Vermeulen, Dong, Robinson, Nguyen, and Gmitro, AIChE meeting, Anaheim, CA, 1982; and private communication from Prof. Theodore Vermeulen, Chemical Engineering Dept., University of California, Berkeley.

TABLE 2-15   Total Pressure, bar, of Aqueous Sulfuric Acid Solutions*

	Weight percent, H₂SO₄									
°C	10.0	20.0	30.0	40.0	50.0	60.0	70.0	75.0	80.0	85.0
0	.582E−02	.534E−02	.448E−02	.326E−02	.193E−02	.836E−03	.207E−03	.747E−04	.197E−04	.343E−05
10	.117E−01	.107E−01	.909E−02	.670E−02	.405E−02	.180E−02	.467E−03	.175E−03	.490E−04	.952E−05
20	.223E−01	.205E−01	.174E−01	.130E−01	.802E−02	.367E−02	.995E−03	.388E−03	.115E−03	.245E−04
30	.404E−01	.373E−01	.319E−01	.241E−01	.151E−01	.710E−02	.201E−02	.811E−03	.253E−03	.589E−04
40	.703E−01	.649E−01	.558E−01	.427E−01	.272E−01	.131E−01	.387E−02	.162E−02	.531E−03	.134E−03
50	.117	.109	.939E−01	.725E−01	.470E−01	.232E−01	.715E−02	.309E−02	.106E−02	.286E−03
60	.189	.175	.152	.119	.782E−01	.395E−01	.127E−01	.565E−02	.204E−02	.584E−03
70	.296	.275	.239	.188	.126	.651E−01	.217E−01	.997E−01	.376E−02	.114E−02
80	.449	.417	.365	.290	.196	.104	.360E−01	.170E−01	.668E−02	.213E−02
90	.664	.617	.542	.434	.298	.161	.578E−01	.281E−01	.115E−01	.383E−02
100	.957	.891	.786	.634	.441	.244	.905E−01	.452E−01	.192E−01	.666E−02
110	1.349	1.258	1.113	.904	.638	.360	.138	.708E−01	.312E−01	.112E−01
120	1.863	1.740	1.544	1.264	.903	.519	.206	.108	.493E−01	.183E−01
130	2.524	2.361	2.101	1.732	1.253	.734	.301	.162	.760E−01	.291E−01
140	3.361	3.149	2.810	2.333	1.708	1.020	.431	.236	.115	.451E−01
150	4.404	4.132	3.697	3.090	2.289	1.392	.605	.339	.170	.683E−01
160	5.685	5.342	4.793	4.031	3.021	1.870	.837	.478	.246	.101
170	7.236	6.810	6.127	5.185	3.930	2.475	1.138	.662	.350	.147
180	9.093	8.571	7.731	6.584	5.045	3.233	1.525	.902	.489	.209
190	11.289	10.658	9.640	8.259	6.397	4.169	2.017	1.212	.673	.292
200	13.861	13.107	11.887	10.245	8.020	5.312	2.633	1.606	.913	.402
210	16.841	15.951	14.505	12.576	9.948	6.696	3.396	2.101	1.221	.544
220	20.264	19.225	17.529	15.287	12.217	8.354	4.331	2.715	1.610	.726
230	24.160	22.960	20.992	18.414	14.864	10.322	5.466	3.468	2.098	.956
240	28.561	27.188	24.927	21.992	17.929	12.641	6.832	4.382	2.701	1.242
250	33.494	31.939	29.364	26.056	21.452	15.351	8.459	5.481	3.439	1.594
260	38.984	37.240	34.334	30.642	25.472	18.496	10.384	6.791	4.332	2.023
270	45.055	43.116	39.865	35.784	30.030	22.122	12.642	8.337	5.402	2.540
280	51.726	49.590	45.984	41.514	35.168	26.275	15.272	10.147	6.673	3.157
290	59.015	56.681	52.715	47.866	40.926	31.004	18.315	12.250	8.170	3.886
300	66.934	64.407	60.081	54.869	47.347	36.361	21.814	14.675	9.916	4.740
310	75.495	72.781	68.101	62.553	54.470	42.398	25.812	17.453	11.939	5.732
320	84.705	81.816	76.792	70.947	62.338	49.168	30.355	20.611	14.264	6.876
330	94.567	91.518	86.172	80.078	70.990	56.727	35.489	24.182	16.916	8.185
340	105.083	101.894	96.252	89.970	80.466	65.130	41.262	28.193	19.920	9.672
350	116.251	112.947	107.043	100.647	90.806	74.437	47.723	32.674	23.303	11.351

	Weight percent, H₂SO₄									
°C	90.0	92.0	94.0	96.0	97.0	98.0	98.5	99.0	99.5	100.0
0	.518E−06	.243E−06	.109E−06	.416E−07	.235E−07	.117E−07	.768E−08	.479E−08	.313E−08	.323E−08
10	.159E−05	.765E−06	.348E−06	.136E−06	.774E−07	.391E−07	.261E−07	.166E−07	.113E−07	.124E−07
20	.449E−05	.221E−05	.102E−05	.407E−06	.235E−06	.121E−06	.812E−07	.528E−07	.373E−07	.435E−07
30	.117E−04	.590E−05	.279E−05	.113E−05	.659E−06	.344E−06	.234E−06	.155E−06	.114E−06	.141E−06
40	.385E−04	.147E−04	.708E−05	.293E−05	.173E−05	.914E−06	.630E−06	.425E−06	.323E−06	.425E−06
50	.653E−04	.344E−04	.169E−04	.712E−05	.425E−05	.228E−05	.159E−05	.109E−05	.861E−06	.120E−05
60	.141E−03	.759E−04	.380E−04	.164E−04	.987E−05	.538E−05	.379E−05	.264E−05	.216E−05	.319E−05
70	.291E−03	.159E−03	.813E−04	.357E−04	.218E−04	.120E−04	.856E−05	.605E−05	.514E−05	.804E−05
80	.571E−03	.319E−03	.166E−03	.742E−04	.458E−04	.257E−04	.184E−04	.132E−04	.117E−04	.193E−04
90	.107E−02	.612E−03	.324E−03	.148E−03	.921E−04	.524E−04	.390E−04	.277E−04	.253E−04	.441E−04
100	.195E−02	.113E−02	.607E−03	.283E−03	.178E−03	.103E−03	.751E−04	.555E−04	.527E−04	.966E−04
110	.340E−02	.201E−02	.110E−02	.521E−03	.332E−03	.194E−03	.143E−03	.107E−03	.106E−03	.204E−03
120	.575E−02	.346E−02	.192E−02	.929E−03	.598E−03	.354E−03	.263E−03	.201E−03	.206E−03	.414E−03
130	.944E−02	.578E−02	.327E−02	.161E−02	.104E−02	.626E−03	.470E−03	.363E−03	.387E−03	.314E−03
140	.151E−01	.939E−02	.539E−02	.270E−02	.177E−02	.107E−02	.815E−03	.639E−03	.708E−03	.155E−02
150	.235E−01	.149E−01	.866E−02	.441E−02	.293E−02	.180E−02	.137E−02	.109E−02	.126E−02	.287E−02
160	.357E−01	.230E−01	.136E−01	.703E−02	.471E−02	.293E−02	.226E−02	.183E−02	.219E−02	.516E−02
170	.532E−01	.347E−01	.208E−01	.110E−01	.741E−02	.466E−02	.363E−02	.299E−02	.372E−02	.905E−02
180	.775E−01	.514E−01	.312E−01	.167E−01	.114E−01	.726E−02	.571E−02	.478E−02	.619E−02	.155E−01
190	.111	.747E−01	.460E−01	.250E−01	.172E−01	.111E−01	.880E−02	.749E−02	.101E−01	.260E−01
200	.156	.107	.665E−01	.367E−01	.255E−01	.166E−01	.133E−01	.115E−01	.161E−01	.427E−01
210	.216	.150	.944E−01	.530E−01	.371E−01	.245E−01	.198E−01	.175E−01	.253E−01	.687E−01
220	.295	.207	.132	.752E−01	.531E−01	.354E−01	.289E−01	.260E−01	.392E−01	.109
230	.396	.282	.182	.105	.749E−01	.505E−01	.417E−01	.382E−01	.596E−01	.169
240	.525	.379	.247	.145	.104	.710E−01	.592E−01	.553E−01	.895E−01	.258
250	.688	.503	.331	.197	.143	.985E−01	.830E−01	.790E−01	.132	.389
260	.881	.660	.439	.264	.193	.135	.115	.112	.193	.577
270	1.141	.856	.575	.351	.258	.153	.157	.156	.279	.846
280	1.447	1.099	.744	.460	.341	.245	.213	.215	.398	1.225
290	1.817	1.398	.954	.597	.446	.324	.285	.295	.562	1.751
300	2.261	1.761	1.211	.767	.578	.425	.379	.399	.785	2.476
310	2.791	2.199	1.524	.977	.742	.553	.498	.536	1.085	3.465
320	3.417	2.723	1.901	1.234	.944	.713	.649	.714	1.486	4.800
330	4.153	3.347	2.353	1.545	1.191	.911	.840	.944	2.018	6.586
340	5.011	4.084	2.889	1.919	1.491	1.156	1.078	1.239	2.718	8.957
350	6.006	4.949	3.523	2.366	1.852	1.456	1.374	1.614	3.631	12.079

*Vermeulen, Dong, Robinson, Nguyen, and Gmitro, AIChE meeting, Anaheim, Calif., 1982; and private communication from Prof. Theodore Vermeulen, Chemical Engineering Dept., University of California, Berkeley.

TABLE 2-16   Partial Pressures of HNO₃ and H₂O over Aqueous Solutions of HNO₃

mmHg
Percentages are weight % HNO_3 in solution.

°C	20% HNO_3	20% H_2O	25% HNO_3	25% H_2O	30% HNO_3	30% H_2O	35% HNO_3	35% H_2O	40% HNO_3	40% H_2O	45% HNO_3	45% H_2O	50% HNO_3	50% H_2O
0		4.1		3.8		3.6		3.3		3.0		2.6		2.1
5		5.7		5.4		5.0		4.6		4.2		3.6		3.0
10		8.0		7.6		7.1		6.5		5.8		5.0	0.12	4.2
15		10.9		10.3		9.7		8.9		8.0	0.10	6.9	.18	5.8
20		15.2		14.2		13.2		12.0		10.8	.15	9.4	.27	7.9
25		20.6		19.2		17.8		16.2	0.12	14.6	.23	12.7	.39	10.7
30		27.6		25.7		23.8	0.09	21.7	.17	19.5	.33	16.9	.56	14.4
35		36.5		33.8		31.1	.13	28.3	.25	25.5	.48	22.3	.80	19.0
40		47.5		44	0.11	41	.20	37.7	.36	33.5	.68	29.3	1.13	25.0
45		62	0.09	57.5	.17	53	.28	48	.52	43	.96	38.0	1.57	32.5
50		80	.13	75	.25	69	.42	63	.75	56	1.35	49.5	2.18	42.5
55	0.09	100	.18	94	.35	87	.59	79	1.04	71	1.83	62.5	2.95	54
60	.13	128	.28	121	.51	113	.85	102	1.48	90	2.54	80	4.05	70
65	.19	162	.40	151	.71	140	1.18	127	2.05	114	3.47	100	5.46	88
70	.27	200	.54	187	1.00	174	1.63	159	2.80	143	4.65	126	7.25	110
75	.38	250	.77	234	1.38	217	2.26	198	3.80	178	6.20	158	9.6	138
80	.53	307	1.05	287	1.87	267	3.07	243	5.10	218	8.15	195	12.5	170
85	.74	378	1.44	352	2.53	325	4.15	297	6.83	268	10.7	240	16.3	211
90	1.01	458	1.95	426	3.38	393	5.50	359	9.0	325	13.7	292	20.9	258
95	1.37	555	2.62	517	4.53	478	7.32	436	11.7	394	17.8	355	26.8	315
100	1.87	675	3.50	628	6.05	580	9.7	530	15.5	480	23.0	430	34.2	383
105	2.50	800	4.65	745	7.90	690	12.7	631	20.0	573	29.2	520	43.0	463
110							16.5	755	25.7	688	37.0	625	54.5	560
115									32.5	810	46	740	67	665
120													84	785

°C	55% HNO_3	55% H_2O	60% HNO_3	60% H_2O	65% HNO_3	65% H_2O	70% HNO_3	70% H_2O	80% HNO_3	80% H_2O	90% HNO_3	90% H_2O	100% HNO_3
0		1.8	0.19	1.5	0.41	1.3	0.79	1.1	2		5.5		11
5	0.14	2.5	.28	2.1	.60	1.8	1.12	1.6	3		8		15
10	.21	3.5	.41	3.0	.86	2.6	1.58	2.2	4	1.2	11		22
15	.31	4.9	.59	4.1	1.21	3.5	2.18	3.0	6	1.7	15		30
20	.45	6.7	.84	5.6	1.68	4.9	3.00	4.1	8	2.4	20		42
25	.66	9.1	1.21	7.7	2.32	6.6	4.10	5.5	10.5	3.2	27	1	57
30	.93	12.2	1.66	10.3	3.17	8.8	5.50	7.4	14	4	36	1.3	77
35	1.30	16.1	2.28	13.6	4.26	11.6	7.30	9.8	18.5	5.5	47	1.8	102
40	1.82	21.3	3.10	18.1	5.70	15.5	9.65	12.8	24.5	7	62	2.4	133
45	2.50	28.0	4.20	23.7	7.55	20.0	12.6	16.7	32	9.5	80	3	170
50	3.41	36.3	5.68	31	10.0	26.0	16.5	21.8	41	12	103	4	215
55	4.54	46	7.45	39	12.8	33.0	21.0	27.3	52	15	127	5	262
60	6.15	60	9.9	51	16.8	43.0	27.1	35.3	67	20	157	6.5	320
65	8.18	76	13.0	64	21.7	54.5	34.5	44.5	85	25	192	8	385
70	10.7	95	16.8	81	27.5	68	43.3	56	106	31	232	10	460
75	13.9	120	21.8	102	35.0	86	54.5	70	130	38	282	13	540
80	18.0	148	27.5	126	43.5	106	67.5	86	158	48	338	16	625
85	23.0	182	34.8	156	54.5	131	83	107	192	60	405	20	720
90	29.4	223	43.7	192	67.5	160	103	130	230	73	480	24	820
95	37.3	272	55.0	233	83.5	195	125	158	278	89	570	29	
100	47	331	69.5	285	103	238	152	192	330	108	675	35	
105	58.5	400	84.5	345	124	288	183	231	392	129	790	42	
110	73	485	103	417	152	345	221	278	465	155			
115	90	575	126	495	181	410	262	330	545	185			
120	110	685	156	590	218	490	312	393	640	219			
125			187	700	260	580	372	469					

TABLE 2-17   Partial Pressures of H₂O and HBr over Aqueous Solutions of HBr at 20 to 55°C

mmHg

%HBr	20°C HBr	20°C H₂O	25°C HBr	25°C H₂O	50°C HBr	50°C H₂O	55°C HBr	55°C H₂O
32			0.0016					
34			.0022					
36			.0033					
38			.0061					
40			.011					
42			.023					
44			.048					
46			.10					
48	0.09	6.2	.13	8.2	1.3	30.2	2.0	38
50	.23	4.5	.37	6.1	3.2	24.3	4.6	31
52	.71	3.3	1.1	4.5	7.2	19.3	10.2	25
54	2.2	2.4	3.2	3.3	17	16.0	23.0	21
56	6.8	1.7	9.3	2.4	40	13.3	51	18
58	21	1.3	27	1.9	91	10.4	115	14
60							260	11.4

TABLE 2-18   Partial Pressures of HI over Aqueous Solutions of HI at 25°C

mmHg

%HI	4	46	48	50	52	54	56
p_{HI}	0.00064	0.0010	0.0022	0.0050	0.013	0.035	0.10

TABLE 2-19   Vapor Pressures of the System: Water-Sulfuric Acid-Nitric Acid

For these data reference must be made to the graphs of *International Critical Tables*, vol. 3, pp. 306–308.

TABLE 2-20   Total Vapor Pressures of Aqueous Solutions of CH₃COOH

Percentages of weight % acetic acid in the solution

mmHg

°C	25%	50%	75%
20	16.3	15.7	15.3
25	22.1	21.4	20.8
30	29.6	28.8	27.8
35	39.4	38.3	36.6
40	51.7	50.2	48.1
45	67.0	65.0	62.0
50	87.2	85.0	80.1
55	110	107	102
60	141	138	130
65	178	172	162
70	223	216	203
75	277	269	251
80	342	331	310
85	419	407	376
90	510	497	458
95	618	602	550
100	743	725	666

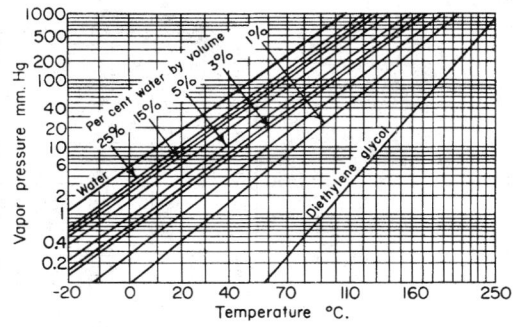

FIG. 2-3   Vapor pressure of aqueous diethylene glycol solutions. (*Courtesy of Carbide and Carbon Chemicals Corp.*)

TABLE 2-21   Partial Pressures of H₂O over Aqueous Solutions of NH₃*

Pressures are in pounds per square inch absolute

Molal concentration of ammonia in the solutions in percentages
(Weight concentration of ammonia in the solution in percentages)

t, °F	0 (0)	5 (4.74)	10 (9.50)	15 (14.29)	20 (19.10)	25 (23.94)	30 (28.81)	35 (33.71)	40 (38.64)	45 (43.59)	50 (48.57)	55 (53.58)	60 (58.62)	65 (63.69)	70 (68.79)	75 (73.91)	80 (79.07)	85 (84.26)	90 (89.47)	95 (94.72)
32	0.09	0.084	0.079	0.074	0.070	0.065	0.060	0.056	0.051	0.047	0.042	0.038	0.034	0.030	0.025	0.021	0.017	0.013	0.008	0.004
40	.12	.115	.108	.101	.095	.089	.083	.076	.070	.064	.058	.052	.046	.040	.035	.029	.023	.015	.012	.006
50	.18	.17	.16	.15	.14	.13	.12	.11	.10	.094	.085	.076	.068	.059	.051	.042	.034	.025	.017	.008
60	.26	.24	.23	.21	.20	.19	.17	.16	.15	.13	.12	.11	.097	.085	.073	.061	.049	.037	.024	.012
70	.36	.34	.32	.30	.28	.26	.25	.23	.21	.19	.17	.15	.14	.12	.10	.086	.069	.052	.034	.017
80	.51	.48	.45	.42	.40	.37	.34	.32	.29	.27	.24	.22	.19	.17	.14	.12	.096	.072	.048	.024
90	.70	.66	.63	.58	.55	.51	.47	.44	.40	.37	.33	.30	.26	.23	.20	.16	.13	.10	.066	.033
100	.95	.90	.85	.79	.74	.69	.64	.59	.55	.50	.45	.41	.36	.31	.27	.22	.18	.13	.090	.045
110	1.27	1.20	1.14	1.07	1.00	.93	.86	.80	.73	.67	.60	.54	.48	.42	.36	.30	.24	.18	.120	.061
120	1.69	1.60	1.51	1.42	1.33	1.24	1.15	1.06	.97	.89	.80	.72	.64	.56	.48	.40	.32	.24	.160	.081
130	2.22	2.10	1.98	1.86	1.74	1.62	1.51	1.39	1.28	1.17	1.05	.95	.84	.74	.63	.53	.42	.32	.210	.100
140	2.89	2.73	2.57	2.42	2.26	2.11	1.96	1.81	1.66	1.52	1.37	1.23	1.10	.96	.82	.69	.55	.41	.270	.140
150	3.72	3.51	3.31	3.11	2.91	2.72	2.52	2.33	2.14	1.95	1.76	1.59	1.41	1.24	1.06	.88	.71	.53	.350	.180
160	4.74	4.48	4.22	3.97	3.71	3.46	3.22	2.97	2.73	2.49	2.25	2.02	1.80	1.58	1.35	1.12	.90	.67	.450	.220
170	5.99	5.66	5.34	5.02	4.70	4.38	4.07	3.75	3.45	3.15	2.84	2.56	2.28	1.99	1.71	1.42	1.13	1.85	.570	.300
180	7.51	7.10	6.69	6.30	5.89	5.49	5.10	4.71	4.33	3.94	3.57	3.21	2.85	2.50	2.14	1.77	1.42	1.06		
190	9.34	8.83	8.32	7.82	7.32	6.83	6.34	5.86	5.38	4.91	4.44	3.99	3.55	3.10	2.65					
200	11.53	10.90	10.27	9.65	9.04	8.43	7.83	7.23	6.64	6.06	5.48	4.93	4.38	3.81						
210	14.12	13.35	12.58	11.82	11.07	10.32	9.59	8.86	8.13	7.42	6.71	6.04	5.34							
220	17.19	16.25	15.32	14.39	13.48	12.57	11.67	10.78	9.90	9.03	8.17	7.31								
230	20.78	19.64	18.51	17.40	16.29	15.19	14.11	13.03	11.97	10.91	9.87									
240	24.97	23.60	22.25	20.91	19.58	18.26	16.95	15.66	14.38	13.12	11.86									
250	29.83	28.20	26.58	25.00	23.39	21.82	20.25	18.71	17.18	15.67										

*Wilson, *Univ. Ill., Eng. Expt. Sta. Bull.* 146.

TABLE 2-22 Mole Percentages of H₂O over Aqueous Solutions of NH₃*

Molal concentration of ammonia in the solutions in percentages
(Weight concentration of ammonia in the solutions in percentages)

t, °F	0 (0)	5 (4.74)	10 (9.50)	15 (14.29)	20 (19.10)	25 (23.94)	30 (28.81)	35 (33.71)	40 (38.64)	45 (43.59)	50 (48.57)	55 (53.58)	60 (58.62)	65 (63.69)	70 (68.79)	75 (73.91)	80 (79.07)	85 (84.26)	90 (89.47)	95 (94.72)	100 (100.00)
32	100	24.3	13.2	7.63	4.43	2.50	1.43	0.856	0.514	0.335	0.216	0.151	0.109	0.0816	0.0585	0.0457	0.0345	0.0249	0.0146	0.00689	0.00
40	100	25.3	14.1	8.15	4.73	2.74	1.59	.943	.581	.372	.248	.172	.124	.0914	.0706	.0533	.0395	.0243	.0185	.00879	
50	100	26.6	15.2	9.09	5.24	3.03	1.78	1.060	.652	.434	.290	.202	.148	.1095	.0838	.0630	.0477	.0332	.0215	.00959	
60	100	27.9	16.2	9.50	5.69	3.42	1.97	1.210	.777	.481	.331	.238	.172	.1290	.0986	.0754	.0566	.0406	.0251	.01125	
70	100	29.1	17.4	10.30	6.14	3.65	2.27	1.390	.873	.569	.383	.266	.205	.1510	.112	.0882	.0656	.0474	.0296	.0135	
80	100	31.6	18.5	11.20	6.89	4.08	2.45	1.550	.978	.659	.444	.323	.230	.1750	.130	.103	.0772	.0528	.0351	.0167	
90	100	32.7	20.0	12.00	7.40	4.47	2.73	1.730	1.100	.742	.505	.366	.267	.2020	.157	.115	.0884	.0647	.0408	.0194	
100	100	34.4	21.0	12.90	7.92	4.85	3.00	1.890	1.250	.834	.574	.420	.307	.2290	.179	.135	.104	.0714	.0473	.0226	
110	100	35.9	22.2	13.80	8.59	5.29	3.30	2.110	1.370	.932	.644	.466	.347	.2640	.208	.157	.118	.0846	.0540	.0262	
120	100	37.5	23.4	14.70	9.22	5.75	3.63	2.320	1.520	1.044	.714	.529	.395	.3020	.233	.180	.135	.0970	.0619	.0300	
130	100	39.0	24.5	15.60	9.85	6.18	3.95	2.550	1.690	1.160	.811	.596	.444	.3430	.263	.205	.154	.1117	.0703	.0339	
140	100	40.7	25.8	16.50	10.50	6.69	4.28	2.790	1.860	1.286	.906	.663	.501	.3840	.297	.232	.175	.124	.0786	.0385	
150	100	42.3	27.1	17.50	11.20	7.19	4.63	3.080	2.040	1.410	1.004	.741	.558	.4320	.334	.257	.197	.140	.0892	.0439	
160	100	44.1	28.3	18.40	11.90	7.69	5.01	3.300	2.230	1.550	1.110	.818	.617	.4800	.372	.287	.218	.154	.1005	.0499	
170	100	45.6	29.6	19.40	12.70	8.22	5.38	3.580	2.430	1.700	1.220	.904	.689	.5300	.414	.320	.242	.174	.112	.0567	
180	100	47.3	30.9	20.40	13.40	8.76	5.78	3.870	2.640	1.850	1.340	.994	.756	.5860	.456	.352	.268	.192			
190	100	48.7	32.2	21.40	14.10	9.31	6.18	4.160	2.860	2.020	1.460	1.087	.830	.6420	.501						
200	100	50.4	33.4	22.30	14.90	9.88	6.59	4.470	3.080	2.190	1.580	1.187	.907	.7010							
210	100	52.1	34.7	23.40	15.70	10.45	7.03	4.780	3.310	2.360	1.720	1.272	.983								
220	100	53.7	36.1	24.40	16.40	11.05	7.48	5.100	3.560	2.540	1.860	1.390									
230	100	55.2	37.3	25.40	17.30	11.63	7.91	5.440	3.810	2.730	2.000										
240	100	56.8	38.6	26.50	18.00	12.24	8.36	5.780	4.060	2.920	2.150										
250	100	58.4	39.8	27.50	18.80	12.88	8.82	6.120	4.340	3.120											

*Wilson, *Univ. Ill., Eng. Expt. Sta. Bull.* 146.

TABLE 2-23 Partial Pressures of NH₃ over Aqueous Solutions of NH₃*

Pressures are in pounds per square inch absolute

t, °F	Molal concentration of ammonia in the solutions in percentages (Weight concentration of ammonia in the solutions in percentages)																		
	5 (4.74)	10 (9.50)	15 (14.29)	20 (19.10)	25 (23.94)	30 (28.81)	35 (33.71)	40 (38.64)	45 (43.59)	50 (48.57)	55 (53.58)	60 (58.62)	65 (63.69)	70 (68.79)	75 (73.91)	80 (79.07)	85 (84.26)	90 (89.47)	95 (94.72)
32	0.26	0.52	0.90	1.51	2.67	4.27	6.54	8.93	14.13	19.36	25.12	31.13	36.74	42.69	45.92	49.26	52.13	54.89	58.01
40	.33	.66	1.14	1.92	3.16	5.13	7.98	11.98	17.14	23.33	30.15	37.15	43.69	49.56	54.40	58.31	61.62	64.77	68.31
50	.47	.89	1.50	2.53	4.16	6.63	10.24	15.24	21.56	29.17	37.46	45.86	53.79	60.82	66.63	71.26	75.22	79.05	83.40
60	.62	1.19	2.00	3.21	5.36	8.48	13.06	19.15	26.92	36.14	46.12	56.22	65.81	73.99	80.90	86.44	91.04	95.67	100.65
70	.83	1.52	2.60	4.28	6.87	10.76	16.33	23.84	33.20	44.25	56.29	68.32	79.42	89.26	97.42	104.01	109.55	114.83	120.61
80	1.04	1.98	3.34	5.45	8.69	13.52	20.29	29.40	40.69	53.84	67.97	82.36	95.52	107.06	116.42	124.20	130.57	136.35	143.70
90	1.36	2.52	4.25	6.88	10.89	16.76	25.04	35.94	49.45	64.99	81.61	98.35	113.79	127.22	138.18	147.02	154.46	161.74	169.73
100	1.72	3.20	5.34	8.60	13.53	20.68	30.57	43.57	59.49	77.85	97.27	116.81	134.70	150.23	162.94	173.22	181.97	190.13	199.17
110	2.14	4.00	6.65	10.64	16.65	25.21	37.01	52.43	71.20	92.59	115.16	137.62	158.42	176.18	190.85	203.02	212.71	222.22	232.79
120	2.67	4.95	8.21	13.09	20.30	30.54	44.56	62.62	84.44	109.40	135.48	161.44	185.14	205.81	222.28	236.05	247.14	258.24	270.02
130	3.28	6.09	10.05	15.93	24.58	36.74	53.16	74.27	99.69	128.45	158.45	188.16	215.14	238.70	257.87	272.88	286.08	298.46	311.80
140	3.97	7.41	12.21	19.23	29.43	43.77	62.97	87.53	116.72	149.93	184.17	218.18	248.70	275.33	297.12	314.45	328.99	342.93	358.46
150	4.78	8.92	14.70	23.09	35.09	51.91	74.28	102.51	136.15	173.64	212.91	251.24	286.00	316.24	340.82	360.39	376.57	392.45	409.62
160	5.68	10.70	17.57	27.45	41.56	61.03	86.91	119.37	157.71	200.45	244.98	288.38	327.82	361.75	389.08	411.30	429.73	447.35	466.38
170	6.75	12.67	20.85	32.41	48.89	71.48	101.09	138.30	181.95	230.36	280.54	329.42	373.61	411.59	442.28	466.67	487.85	507.63	528.50
180	7.90	14.96	24.56	38.13	57.19	83.07	116.97	159.37	208.66	263.43	319.89	374.25	424.10	466.26	500.63	528.08	551.24		
190	9.23	17.55	28.78	44.49	66.49	96.22	134.89	182.72	238.39	299.86	363.11	424.15	479.40	526.15					
200	10.70	20.45	33.49	51.58	76.90	110.85	154.58	208.56	270.94	340.02	410.17	478.62	539.79						
210	12.26	23.68	38.76	59.65	88.48	126.83	176.24	236.97	307.08	383.99	462.36	537.56							
220	14.02	27.15	44.61	68.43	101.24	144.74	200.46	268.30	346.07	431.43	518.19								
230	15.95	31.09	51.06	78.14	115.45	164.17	226.67	302.53	389.29	483.53									
240	17.92	35.40	58.00	89.02	130.94	185.79	255.26	339.72	435.78	540.44									
250	20.12	40.09	65.74	100.69	147.66	209.37	286.89	380.42	486.73										

*Wilson, *Univ. Ill., Eng. Expt. Sta. Bull.* 146.

TABLE 2-24 Total Vapor Pressures of Aqueous Solutions of NH₃*

Pressures are in pounds per square inch absolute

Molal concentration of ammonia in the solutions in percentages
(Weight concentration of ammonia in the solutions in percentages)

t, °F	0 (0)	5 (4.74)	10 (9.50)	15 (14.29)	20 (19.10)	25 (23.94)	30 (28.81)	35 (33.71)	40 (38.64)	45 (43.59)	50 (48.57)	55 (53.58)	60 (58.62)	65 (63.69)	70 (68.79)	75 (73.91)	80 (79.07)	85 (84.26)	90 (89.47)	95 (94.72)	100 (100.00)
32	0.09	0.34	0.60	0.97	1.58	2.60	4.20	6.54	9.93	14.18	19.40	25.16	31.16	36.77	42.72	45.94	49.28	52.14	54.90	58.01	62.29
40	.12	.45	.77	1.24	2.01	3.25	5.21	8.06	12.05	17.20	23.39	30.20	37.20	43.73	49.60	54.43	58.33	61.64	64.78	68.32	73.32
50	.18	.64	1.05	1.65	2.67	4.29	6.75	10.35	15.34	21.65	29.26	37.54	45.93	53.85	60.87	66.67	71.29	75.25	79.07	83.41	89.19
60	.26	.86	1.42	2.21	3.51	5.55	8.65	13.22	19.30	27.05	36.26	46.23	56.32	65.90	74.06	80.96	86.49	91.08	95.69	100.66	107.6
70	.36	1.17	1.84	2.90	4.56	7.13	11.01	16.56	24.05	33.39	44.42	56.44	68.46	79.54	89.36	97.51	104.08	109.60	114.86	120.63	128.8
80	.51	1.52	2.43	3.76	5.85	9.06	13.86	20.61	29.69	40.96	54.08	68.19	82.55	95.69	107.20	116.54	124.30	130.64	136.40	143.72	153.0
90	.70	2.02	3.15	4.83	7.43	11.40	17.23	25.48	36.34	49.82	65.32	81.91	98.61	114.02	127.42	138.34	147.15	154.56	161.81	169.76	180.6
100	.95	2.62	4.05	6.13	9.34	14.22	21.32	31.16	44.12	59.99	78.30	97.68	117.17	135.01	150.50	163.16	173.40	182.10	190.22	199.22	211.9
110	1.27	3.34	5.14	7.72	11.64	17.58	26.07	37.81	53.16	71.87	93.19	115.7	138.10	158.84	176.54	191.15	203.26	212.89	222.34	232.85	247.0
120	1.69	4.27	6.46	9.63	14.42	21.54	31.69	45.62	63.59	85.33	110.2	136.2	162.08	185.70	206.29	222.68	236.37	247.38	258.40	270.1	286.4
130	2.22	5.38	8.07	11.91	17.67	26.20	38.25	54.55	75.55	100.86	129.5	159.0	189.00	215.88	239.33	258.40	273.3	286.4	298.67	311.9	330.3
140	2.89	6.70	9.98	14.63	21.49	31.54	45.73	64.78	89.19	118.24	151.3	185.4	219.28	249.66	276.15	297.81	315.0	329.4	343.2	358.6	379.1
150	3.72	8.29	12.23	17.81	26.00	37.81	54.43	76.61	104.65	138.1	175.4	214.5	252.65	287.24	317.3	341.7	361.1	377.1	392.8	409.8	432.2
160	4.74	10.16	14.92	21.54	31.16	45.02	64.25	89.88	122.10	160.2	202.7	247.0	290.18	329.4	363.1	390.2	412.2	430.4	447.8	466.6	492.8
170	5.99	12.41	18.01	25.87	37.11	53.27	75.55	104.84	141.75	185.1	233.2	283.1	331.7	375.6	413.3	443.7	467.8	488.7	508.2	528.8	558.4
180	7.51	15.00	21.65	30.86	44.02	62.68	88.17	121.68	163.7	212.6	267.0	323.1	377.1	426.6	468.4	502.4	529.5	552.3			
190	9.34	18.06	25.87	36.60	51.81	73.32	102.56	140.75	188.1	243.3	304.3	367.1	427.7	482.5	528.8						
200	11.53	21.60	30.72	43.14	60.62	85.33	118.68	161.81	215.2	277.0	345.5	415.1	483.0	543.6							
210	14.12	25.61	36.26	50.58	70.72	98.80	136.42	185.10	245.1	314.5	390.7	468.4	542.9								
220	17.19	30.27	42.47	59.00	81.91	113.81	156.41	211.24	278.2	355.1	439.6	525.5									
230	20.78	35.59	49.60	68.46	94.43	130.64	178.28	239.70	314.5	400.2	493.4										
240	24.97	41.52	57.65	78.91	108.60	149.20	202.74	270.92	354.1	448.9	552.3										
250	29.83	48.32	66.67	90.74	124.08	169.48	229.62	305.60	397.6	502.4											

*Wilson, *Univ. Ill., Eng. Expt. Sta. Bull.* 146.

TABLE 2-25 Partial Pressures of H₂O over Aqueous Solutions of Sodium Carbonate

mmHg

t, °C	%Na₂CO₃						
	0	5	10	15	20	25	30
0	4.5	4.5					
10	9.2	9.0	8.8				
20	17.5	17.2	16.8	16.3			
30	31.8	31.2	30.4	29.6	28.8	27.8	26.4
40	55.3	54.2	53.0	57.6	50.2	48.4	46.1
50	92.5	90.7	88.7	86.5	84.1	81.2	77.5
60	149.5	146.5	143.5	139.9	136.1	131.6	125.7
70	239.8	235	230.5	225	219	211.5	202.5
80	355.5	348	342	334	325	315	301
90	526.0	516	506	494	482	467	447
100	760.0	746	731	715	697	676	648

TABLE 2-26 Partial Pressures of H₂O and CH₃OH over Aqueous Solutions of Methyl Alcohol*

Mole fraction CH₃OH	39.9°C		Mole fraction CH₃OH	59.4°C	
	P_{H_2O}, mmHg	P_{CH_3OH}, mmHg		P_{H_2O}, mmHg	P_{CH_3OH}, mmHg
0	54.7	0	0	145.4	0
14.99	39.2	66.1	22.17	106.9	210.1
17.85	38.5	75.5	27.40	102.2	240.2
21.07	37.2	85.2	33.24	96.6	272.1
27.31	35.8	100.6	39.80	91.7	301.9
31.06	34.9	108.8	47.08	84.8	335.6
40.1	32.8	127.7	55.5	76.9	373.7
47.0	31.5	141.6	69.2	57.8	439.4
55.8	27.3	158.4	78.5	43.8	486.6
68.9	20.7	186.6	85.9	30.1	526.9
86.0	10.1	225.2	100.0	0	609.3
100.0	0	260.7			

International Critical Tables, vol. 3, McGraw-Hill, p. 290.

TABLE 2-27 Partial Pressures of H₂O over Aqueous Solutions of Sodium Hydroxide

mmHg

Conc. g NaOH/ 100 g H₂O	Temperature, °C											
	0	20	40	60	80	100	120	160	200	250	300	350
0	4.6	17.5	55.3	149.5	355.5	760.0	1,489	4,633	11,647	29,771	64,200	123,600
5	4.4	16.9	53.2	143.5	341.5	730.0	1,430	4,450	11,200	28,600	61,800	118,900
10	4.2	16.0	50.6	137.0	325.5	697.0	1,365	4,260	10,750	27,500	59,300	114,100
20	3.6	13.9	44.2	120.5	288.5	621.0	1,225	3,860	9,800	25,300	54,700	105,400
30	2.9	11.3	36.6	101.0	246.0	537.0	1,070	3,460	8,950	23,300	50,800	98,000
40	2.2	8.7	28.7	81.0	202.0	450.0	920	3,090	8,150	21,500	47,200	91,600
50		6.3	20.7	62.5	160.5	368.0	770	2,690	7,400	19,900	44,100	85,800
60		4.4	15.5	47.0	124.0	294.0	635	2,340	6,750	18,400	41,200	80,700
70		3.0	10.9	34.5	94.0	231.0	515	2,030	6,100	17,100	38,700	76,000
80		2.0	7.6	24.5	70.5	179.0	415	1,740	5,500	15,800	36,300	71,900
90		1.3	5.2	17.5	53.0	138.0	330	1,490	5,000	14,700	34,200	68,100
100		0.9	3.6	12.5	38.5	105.0	262	1,300	4,500	13,650	32,200	64,600
120			1.7	6.3	20.5	61.0	164	915	3,650	11,800	28,800	58,600
140				3.0	11.0	35.5	102	765	2,980	10,300	25,900	53,400
160				1.5	6.0	20.5	63	470	2,430	8,960	23,300	49,000
180					3.5	12.0	40	340	1,980	7,830	21,200	45,100
200					2.0	7.0	25	245	1,620	6,870	19,200	41,800
250					0.5	2.0	8	110	985	5,000	15,400	35,000
300					0.1	0.5	2.7	50	610	3,690	12,500	29,800
350							0.9	23	380	2,750	10,300	25,700
400								11	240	2,080	8,600	22,400
500									100	1,210	6,100	17,500
700										440	3,300	11,500
1000											1,470	6,800
2000											150	1,760
4000												120
8000												7

WATER-VAPOR CONTENT OF GASES

CHART FOR GASES AT HIGH PRESSURES

The accompanying figure is useful in determining the water-vapor content of air at high pressure in contact with liquid water.

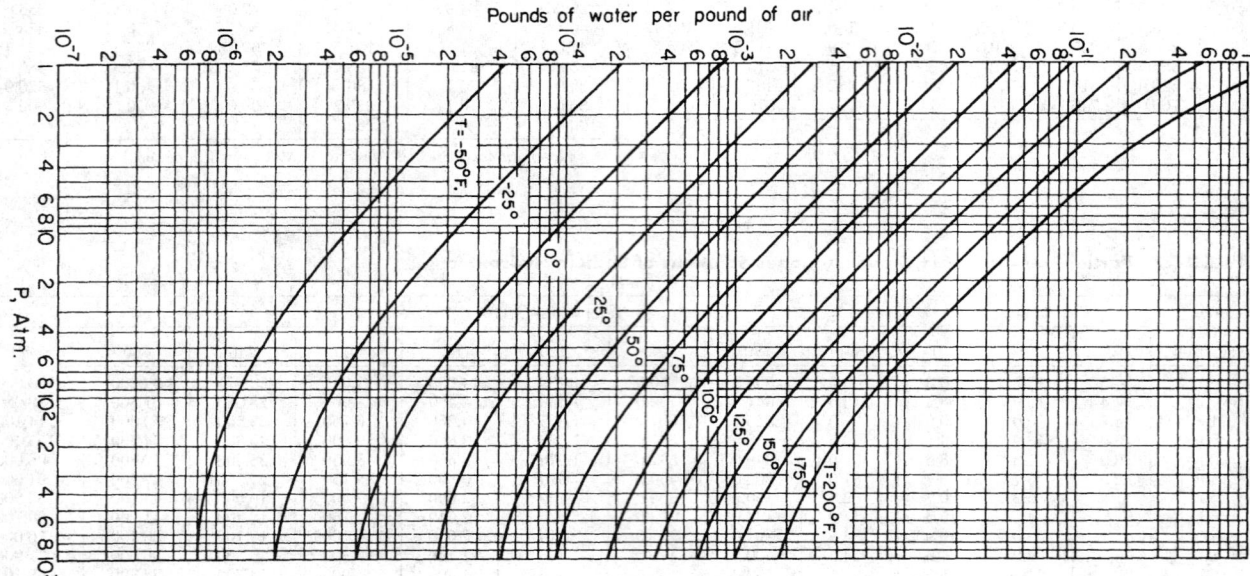

FIG. 2-4   Water content of air, °C = (°F − 32) × %. (*Landsbaum, Dadds, and Stutzman. Reprinted from vol. 47, January 1955 issue of* Ind. Eng. Chem. [*p. 192*]. *Copyright 1955 by the American Chemical Society and reproduced by permission of the copyright owner.*)

DENSITIES OF PURE SUBSTANCES

UNITS CONVERSIONS

For this subsection, the following units conversions are applicable:

$$°F = \tfrac{9}{5}\,°C + 32.$$

To convert kilograms per cubic meter to pounds per cubic foot, multiply by 0.06243.

TABLE 2-28   Density (kg/m³) of Water from 0 to 100°C*

t, °C	ρ, kg/m³									
	0.0	0.1	0.2	0.3	0.4	0.5	0.6	0.7	0.8	0.9
0	999.839	999.846	999.852	999.859	999.865	999.871	999.877	999.882	999.888	999.893
1	999.898	999.903	999.908	999.913	999.917	999.921	999.925	999.929	999.933	999.936
2	999.940	999.943	999.946	999.949	999.952	999.954	999.956	999.959	999.961	999.962
3	999.964	999.966	999.967	999.968	999.969	999.970	999.971	999.971	999.972	999.972
4	999.972	999.972	999.972	999.971	999.971	999.970	999.969	999.968	999.967	999.965
5	999.964	999.962	999.960	999.958	999.956	999.954	999.951	999.949	999.946	999.943
6	999.940	999.937	999.934	999.930	999.926	999.923	999.919	999.915	999.910	999.906
7	999.901	999.897	999.892	999.887	999.882	999.877	999.871	999.866	999.860	999.854
8	999.848	999.842	999.836	999.829	999.823	999.816	999.809	999.802	999.795	999.788
9	999.781	999.773	999.765	999.758	999.750	999.742	999.734	999.725	999.717	999.708
10	999.699	999.691	999.682	999.672	999.663	999.654	999.644	999.635	999.625	999.615
11	999.605	999.595	999.584	999.574	999.563	999.553	999.542	999.531	999.520	999.509
12	999.497	999.486	999.474	999.462	999.451	999.439	999.426	999.414	999.402	999.389
13	999.377	999.364	999.351	999.338	999.325	999.312	999.299	999.285	999.272	999.258
14	999.244	999.230	999.216	999.202	999.188	999.173	999.159	999.144	999.129	999.114
15	999.099	999.084	999.069	999.054	999.038	999.022	999.007	998.991	998.975	998.958
16	998.943	998.926	998.910	998.894	998.877	998.860	998.843	998.826	998.809	998.792
17	998.775	998.757	998.740	998.722	998.704	998.686	998.668	998.650	998.632	998.614
18	998.595	998.577	998.558	998.539	998.520	998.502	998.482	998.463	998.444	998.425
19	998.405	998.385	998.366	998.346	998.326	998.306	998.286	998.265	998.245	998.224
20	998.204	998.183	998.162	998.141	998.120	998.099	998.078	998.057	998.035	998.014
21	997.992	997.971	997.949	997.927	997.905	997.883	997.860	997.838	997.816	997.793
22	997.770	997.747	997.725	997.702	997.679	997.656	997.632	997.609	997.585	997.562
23	997.538	997.515	997.491	997.467	997.443	997.419	997.394	997.370	997.345	997.321
24	997.296	997.272	997.247	997.222	997.197	997.172	997.146	997.121	997.096	997.070
25	997.045	997.019	996.993	996.967	996.941	996.915	996.889	996.863	996.836	996.810
26	996.783	996.757	996.730	996.703	996.676	996.649	996.622	996.595	996.568	996.540
27	996.513	996.485	996.458	996.430	996.402	996.374	996.346	996.318	996.290	996.262
28	996.233	996.205	996.176	996.148	996.119	996.090	996.061	996.032	996.003	995.974
29	995.945	995.915	995.886	995.856	995.827	995.797	995.767	995.737	995.707	995.677
30	995.647	995.617	995.586	995.556	995.526	995.495	995.464	995.433	995.403	995.372
31	995.341	995.310	995.278	995.247	995.216	995.184	995.153	995.121	995.090	995.058
32	995.026	994.997	994.962	994.930	994.898	994.865	994.833	994.801	994.768	994.735
33	994.703	994.670	994.637	994.604	994.571	994.538	994.505	994.472	994.438	994.405
34	994.371	994.338	994.304	994.270	994.236	994.202	994.168	994.134	994.100	994.066
35	994.032	993.997	993.963	993.928	993.893	993.859	993.824	993.789	993.754	993.719
36	993.684	993.648	993.613	993.578	993.543	993.507	993.471	993.436	993.400	993.364
37	993.328	993.292	993.256	993.220	993.184	993.148	993.111	993.075	993.038	993.002
38	992.965	992.928	992.891	992.855	992.818	992.780	992.743	992.706	992.669	992.631
39	992.594	992.557	992.519	992.481	992.444	992.406	992.368	992.330	992.292	992.254
40	992.215	992.177	992.139	992.100	992.062	992.023	991.985	991.946	991.907	992.868
41	991.830	991.791	991.751	991.712	992.673	991.634	991.594	991.555	991.515	991.476
42	991.436	991.396	991.357	991.317	991.277	991.237	991.197	991.157	991.116	991.076
43	991.036	990.995	990.955	990.914	990.873	990.833	990.792	990.751	990.710	990.669
44	990.628	990.587	990.546	990.504	990.463	990.421	990.380	990.338	990.297	990.255
45	990.213	990.171	990.129	990.087	990.045	990.003	989.961	989.919	989.876	989.834
46	989.792	989.749	989.706	989.664	989.621	989.578	989.535	989.492	989.449	989.406
47	989.363	989.320	989.276	989.233	989.190	989.146	989.103	989.059	989.015	988.971
48	988.928	988.884	988.840	988.796	988.752	988.707	988.663	988.619	988.574	988.530
49	988.485	988.441	988.396	988.352	988.307	988.262	988.217	988.172	988.127	988.082

*From "Water: Density at Atmospheric Pressure and Temperatures from 0 to 100°C," *Tables of Standard Handbook Data*, Standartov, Moscow, 1978. To conserve space, only a few tables of density values are given. The reader is reminded that density values may be found as the reciprocal of the specific volume values tabulated in the "Thermodynamic Properties: Tables" subsection.

TABLE 2-28   Density (kg/m³) of Water from 0 to 100°C (*Concluded*)

t, °C					ρ, kg/m³					
	0.0	0.1	0.2	0.3	0.4	0.5	0.6	0.7	0.8	0.9
50	988.037	987.992	987.946	987.901	987.844	987.810	987.764	987.719	987.673	987.627
51	987.581	987.536	987.490	987.444	987.398	987.351	987.305	987.259	987.213	987.166
52	987.120	987.073	987.027	986.980	986.933	986.886	986.840	986.793	986.746	986.699
53	986.652	986.604	986.557	986.510	986.463	986.415	986.368	986.320	986.272	986.225
54	986.177	986.129	986.081	986.033	985.985	985.937	985.889	985.841	985.793	985.745
55	985.696	985.648	985.599	985.551	985.502	985.454	985.405	985.356	985.307	985.258
56	985.219	985.160	985.111	985.062	985.013	984.963	984.914	984.865	984.815	984.766
57	984.716	984.666	984.617	984.567	984.517	984.467	984.417	984.367	984.317	984.267
58	984.217	984.167	984.116	984.066	984.016	983.965	983.914	983.864	983.813	983.762
59	983.712	983.661	983.610	983.559	983.508	983.457	983.406	983.354	983.303	983.252
60	983.200	983.149	983.097	983.046	982.994	982.943	982.891	982.839	982.787	982.735
61	982.683	982.631	982.579	982.527	982.475	982.422	982.370	982.318	982.265	982.213
62	982.160	982.108	982.055	982.002	981.949	981.897	981.844	981.791	981.738	981.685
63	981.631	981.578	981.525	981.472	981.418	981.365	981.311	981.258	981.204	981.151
64	981.097	981.043	980.989	980.935	980.881	980.827	980.773	980.719	980.665	980.611
65	980.557	980.502	980.443	980.393	980.339	980.284	980.230	980.175	980.120	980.065
66	980.011	979.956	979.901	979.846	979.791	979.736	979.680	979.625	979.570	979.515
67	979.459	979.403	979.348	979.293	979.237	979.181	979.126	979.070	979.014	978.958
68	978.902	978.846	978.790	978.734	978.678	978.621	978.565	978.509	978.452	978.396
69	978.339	978.283	978.226	978.170	978.113	978.056	977.999	977.942	977.885	977.828
70	977.771	977.714	977.657	977.600	977.543	977.485	977.428	977.370	977.313	977.255
71	977.198	977.140	977.082	977.025	976.967	976.909	976.851	976.793	976.735	976.677
72	976.619	976.561	976.503	976.444	976.386	976.327	976.269	976.211	976.152	976.093
73	976.035	975.976	975.917	975.858	975.800	975.741	975.682	975.623	975.564	975.504
74	975.445	975.386	975.327	975.267	975.208	975.148	975.089	975.029	974.970	974.910
75	974.850	974.791	974.731	974.671	974.611	974.551	974.491	974.431	974.371	974.311
76	974.250	974.190	974.130	974.069	974.009	973.948	973.888	973.827	973.767	973.706
77	973.645	973.584	973.524	973.463	973.402	973.341	973.280	973.218	973.157	973.096
78	973.025	972.974	972.912	972.851	972.789	972.728	972.666	972.605	972.543	972.481
79	972.419	972.358	972.296	972.234	972.172	972.110	972.048	971.986	971.923	971.861
80	971.799	971.737	971.674	971.612	971.549	971.487	971.424	971.361	971.299	971.236
81	971.173	971.110	971.048	970.985	970.922	970.859	970.796	970.732	970.669	970.606
82	970.543	970.479	970.416	970.353	970.289	970.226	970.162	970.098	970.035	969.971
83	969.907	969.843	969.772	969.715	969.652	969.587	969.523	969.459	969.395	969.331
84	969.267	969.202	969.138	969.073	969.009	968.944	968.880	968.815	968.751	968.686
85	968.621	968.556	968.491	968.427	968.362	968.297	969.232	968.166	968.101	968.036
86	967.971	967.906	967.840	967.775	967.709	967.641	967.578	967.513	967.447	967.381
87	967.316	967.250	967.184	967.118	967.052	966.986	966.920	966.854	966.788	966.722
88	966.656	966.589	966.523	966.457	966.390	966.324	966.257	966.191	966.124	966.057
89	965.991	965.924	965.857	965.790	965.723	965.656	965.589	965.522	965.455	965.388
90	965.321	965.254	965.187	965.119	965.052	964.984	964.917	964.849	964.782	964.714
91	964.647	954.579	964.511	964.443	964.376	964.308	964.240	964.172	964.104	964.036
92	963.967	963.899	963.831	963.763	963.694	963.626	963.558	963.489	963.421	963.352
93	963.284	963.215	963.146	963.077	963.009	962.940	962.871	962.802	962.733	962.664
94	962.595	962.526	962.457	962.387	962.318	962.249	962.180	962.110	962.041	961.971
95	961.902	961.832	961.762	961.693	961.693	961.553	961.483	961.414	961.344	961.274
96	961.204	961.134	961.064	960.993	960.923	960.853	960.783	960.712	960.642	960.572
97	960.501	960.431	960.360	960.289	960.219	960.148	960.077	960.006	959.936	959.865
98	959.794	959.723	959.652	959.581	959.510	959.438	959.367	959.296	959.225	959.153
99	959.082	959.010	958.939	958.867	958.796	958.724	958.653	958.581	958.509	958.431
100	958.365									

TABLE 2-29   Density (kg/m³) of Mercury from 0 to 350°C*

t, °C	Density, kg/m³									
	0	1	2	3	4	5	6	7	8	9
0	13595.08	13592.61	13590.14	13587.68	13585.21	13582.75	13580.29	13577.82	13575.36	13572.90
10	13570.44	13567.98	13565.52	13563.06	13560.60	13558.14	13555.69	13553.23	13550.78	13548.32
20	13545.87	13543.41	13540.96	13538.51	13536.06	13533.61	13531.16	13528.71	13526.26	13523.81
30	13521.36	13518.91	13516.47	13514.02	13511.58	13509.13	13506.69	13504.25	13501.80	13499.36
40	13496.92	13494.48	13492.04	13489.60	13487.16	13484.72	13482.29	13479.85	13477.41	13474.98
50	13472.54	13470.11	13467.67	13465.24	13462.81	13460.38	13457.94	13455.51	13453.08	13450.65
60	13448.22	13445.80	13443.37	13440.94	13438.51	13436.09	13433.66	13431.23	13428.81	13426.39
70	13423.96	13421.54	13419.12	13416.69	13414.27	13411.85	13409.43	13407.01	13404.59	13402.17
80	13399.75	13397.34	13394.92	13392.50	13390.08	13387.67	13385.25	13382.84	13380.42	13378.01
90	13375.59	13373.18	13370.77	13368.36	13365.94	13363.53	13361.12	13358.71	13356.30	13353.89
100	13351.5	13349.1	13346.7	13344.3	13341.9	13339.4	13337.0	13334.6	13332.2	13329.8
110	13327.4	13325.0	13322.6	13320.2	13317.8	13315.4	13313.0	13310.6	13308.2	13305.8
120	13303.4	13301.0	13298.6	13296.2	13293.8	13291.4	13288.9	13286.6	13284.2	13281.8
130	13279.4	13277.0	13274.6	13272.2	13269.8	13267.4	13265.0	13262.6	13260.2	13257.8
140	13255.4	13253.0	13250.6	13248.2	13245.8	13243.4	13241.0	13238.7	13236.3	13233.9
150	13231.5	13229.1	13226.7	13224.3	13221.9	13219.5	13217.1	13214.7	13212.4	13210.0
160	13207.6	13205.2	13202.8	13200.4	13198.0	13195.6	13193.2	13190.8	13188.5	13186.1
170	13183.7	13181.3	13178.9	13176.5	13174.1	13171.7	13169.4	13167.0	13164.6	13162.2
180	13159.8	13157.4	13155.0	13152.6	13150.3	13147.9	13145.5	13143.1	13140.7	13138.3
190	13136.0	13133.6	13131.2	13128.3	13126.4	13124.0	13121.7	13119.3	13116.9	13114.5
200	13112.1	13109.7	13107.4	13105.0	13102.6	13100.2	13097.8	13095.4	13093.1	13090.7
210	13088.3	13085.9	13083.5	13081.1	13078.8	13076.4	13074.0	13071.6	13069.2	13066.8
220	13064.5	13062.1	13059.7	13057.3	13054.9	13052.6	13050.2	13047.8	13045.4	13043.0
230	13040.6	13038.3	13035.9	13033.5	13031.1	13028.7	13026.4	13024.0	13021.6	13019.2
240	13016.8	13014.5	13012.1	13009.7	13007.3	13004.9	13002.5	13000.2	12997.8	12995.4
250	12993.0	12990.6	12988.3	12985.9	12983.5	12981.1	12978.7	12976.3	12974.0	12971.6
260	12969.2	12966.8	12964.4	12962.0	12959.7	12957.3	12954.9	12952.5	12950.1	12947.7
270	12945.4	12943.0	12940.6	12938.2	12935.8	12933.4	12931.1	12928.7	12926.3	12923.9
280	12921.5	12919.1	12916.7	12914.4	12912.0	12909.6	12907.2	12904.8	12902.4	12900.0
290	12897.7	12895.3	12892.9	12890.5	12888.1	12885.7	12883.3	12880.9	12878.5	12876.2
300	12873.8	12871.4	12869.0	12866.6	12864.2	12861.8	12859.4	12857.0	12854.6	12852.2
310	12849.9	12847.5	12845.1	12842.7	12840.3	12837.9	12835.5	12833.1	12830.7	12828.3
320	12825.9	12823.5	12821.1	12818.7	12816.3	12813.9	12811.5	12809.1	12806.7	12804.3
330	12801.9	12799.5	12797.1	12794.7	12792.3	12789.9	12787.5	12785.1	12782.7	12780.2
340	12777.8	12775.4	12773.0	12770.6	12768.2	12765.8	12763.4	12761.0	12758.6	12756.1
350	12753.7									

*From "Mercury—Density and Thermal Expansion at Atmospheric Pressure and Temperatures from 0 to 350°C," *Tables of Standard Handbook Data,* Standartov, Moscow, 1978. The density values obtainable from those cited for the specific volume of the saturated liquid in the "Thermodynamic Properties" subsection show minor differences. No attempt was made to adjust either set.

TABLE 2-30 Densities of Inorganic and Organic Liquids

Cmpd. no.	Name	Formula	CAS no.	Mol. wt.	C1	C2	C3	C4	T_{min} K	Density at T_{min}	T_{max} K	Density at T_{max}
1	Methane	CH$_4$	74828	16.043	2.9214	0.28976	190.56	0.28881	90.69	28.18	190.56	10.082
2	Ethane	C$_2$H$_6$	74840	30.070	1.9122	0.27937	305.32	0.29187	90.35	21.64	305.32	6.845
3	Propane	C$_3$H$_8$	74986	44.097	1.3757	0.27453	369.83	0.29359	85.47	16.583	369.83	5.011
4	n-Butane	C$_4$H$_{10}$	106978	58.123	1.0677	0.27188	425.12	0.29688	134.86	12.62	425.12	3.927
5	n-Pentane	C$_5$H$_{12}$	109660	72.150	0.84947	0.26726	469.7	0.27789	143.42	10.474	469.7	3.178
6	n-Hexane	C$_6$H$_{14}$	110543	86.177	0.70824	0.26411	507.6	0.27537	177.83	8.747	507.6	2.682
7	n-Heptane	C$_7$H$_{16}$	142825	100.204	0.61259	0.26211	540.2	0.28141	182.57	7.6998	540.2	2.337
8	n-Octane	C$_8$H$_{18}$	111659	114.231	0.53731	0.26115	568.7	0.28034	216.38	6.6558	568.7	2.058
9	n-Nonane	C$_9$H$_{20}$	111842	128.258	0.48387	0.26147	594.6	0.28281	219.66	6.007	594.6	1.851
10	n-Decane	C$_{10}$H$_{22}$	124185	142.285	0.42831	0.25745	617.7	0.28912	243.51	5.3811	617.7	1.664
11	n-Undecane	C$_{11}$H$_{24}$	1120214	156.312	0.39	0.25678	639	0.2913	247.57	4.9362	639	1.519
12	n-Dodecane	C$_{12}$H$_{26}$	112403	170.338	0.35541	0.25511	658	0.29368	263.57	4.5132	658	1.393
13	n-Tridecane	C$_{13}$H$_{28}$	629505	184.365	0.3216	0.2504	675	0.3071	267.76	4.2035	675	1.284
14	n-Tetradecane	C$_{14}$H$_{30}$	629594	198.392	0.30545	0.2535	693	0.30538	279.01	3.8924	693	1.205
15	n-Pentadecane	C$_{15}$H$_{32}$	629629	212.419	0.29445	0.25269	708	0.30786	283.07	3.6471	708	1.126
16	n-Hexadecane	C$_{16}$H$_{34}$	544763	226.446	0.26807	0.25287	723	0.31143	291.31	3.4187	723	1.060
17	n-Heptadecane	C$_{17}$H$_{36}$	629787	240.473	0.2545	0.254	736	0.31072	295.13	3.2241	736	1.002
18	n-Octadecane	C$_{18}$H$_{38}$	593453	254.500	0.23864	0.25272	747	0.31104	301.31	3.0466	747	0.944
19	n-Nonadecane	C$_{19}$H$_{40}$	629925	268.527	0.22451	0.25133	758	0.3133	305.04	2.8933	758	0.893
20	n-Eicosane	C$_{20}$H$_{42}$	112958	282.553	0.21624	0.25287	768	0.31613	309.58	2.7496	768	0.855
21	2-Methylpropane	C$_4$H$_{10}$	75285	58.123	1.0463	0.27294	408.14	0.27301	113.54	12.575	408.14	3.833
22	2-Methylbutane	C$_5$H$_{12}$	78784	72.150	0.9079	0.2761	460.43	0.28673	113.25	10.776	460.43	3.288
23	2,3-Dimethylbutane	C$_6$H$_{14}$	79298	86.177	0.76929	0.27524	499.98	0.27691	145.19	9.0343	499.98	2.795
24	2-Methylpentane	C$_6$H$_{14}$	107835	86.177	0.73335	0.2687	497.5	0.28361	119.55	9.2041	497.5	2.729
25	2,3-Dimethylpentane	C$_7$H$_{16}$	565593	100.204	0.7229	0.28614	537.35	0.2713	160.00	7.8746	537.35	2.526
26	2,3,3-Trimethylpentane	C$_8$H$_{18}$	560214	114.231	0.6028	0.25287	573.5	0.2741	172.22	7.0934	573.5	2.196
27	2,2,4-Trimethylpentane	C$_8$H$_{18}$	540841	114.231	0.5586	0.27373	543.96	0.2846	165.78	6.9163	543.96	2.150
28	Ethylene	C$_2$H$_4$	74851	28.054	2.0961	0.27657	282.34	0.29147	104.00	23.326	282.34	7.579
29	Propylene	C$_3$H$_6$	115071	42.081	1.4094	0.26465	365.57	0.295	87.89	18.143	365.57	5.326
30	1-Butene	C$_4$H$_8$	106989	56.108	1.0972	0.2649	419.95	0.29043	87.80	14.326	419.95	4.142
31	cis-2-Butene	C$_4$H$_8$	590181	56.108	1.1609	0.27104	435.58	0.2816	134.26	13.895	435.58	4.283
32	trans-2-Butene	C$_4$H$_8$	624646	56.108	1.1426	0.27095	428.63	0.2854	167.62	13.1	428.63	4.217
33	1-Pentene	C$_5$H$_{10}$	109671	70.134	0.9038	0.26648	464.78	0.2905	107.93	11.543	464.78	3.392
34	1-Hexene	C$_6$H$_{12}$	592416	84.161	0.7389	0.26147	504.03	0.2902	133.39	9.6388	504.03	2.826
35	1-Heptene	C$_7$H$_{14}$	592767	98.188	0.63734	0.26319	537.29	0.27375	154.27	8.1759	537.29	2.422
36	1-Octene	C$_8$H$_{16}$	111660	112.215	0.5871	0.27005	566.65	0.27187	171.45	7.1247	566.65	2.174
37	1-Nonene	C$_9$H$_{18}$	124118	126.242	0.4945	0.26108	593.25	0.27319	191.78	6.333	593.25	1.894
38	1-Decene	C$_{10}$H$_{20}$	872059	140.269	0.44244	0.25838	616.4	0.28411	206.89	5.7131	616.4	1.712
39	2-Methylpropene	C$_4$H$_8$	115117	56.108	1.1454	0.2725	417.9	0.28186	132.81	13.506	417.9	4.203
40	2-Methyl-1-butene	C$_5$H$_{10}$	563462	70.134	0.91619	0.26752	465	0.28164	135.58	11.332	465	3.425
41	2-Methyl-2-butene	C$_5$H$_{10}$	513359	70.134	0.93322	0.27251	471	0.26031	139.39	11.218	471	3.425
42	1,2-Butadiene	C$_4$H$_6$	590192	54.092	1.187	0.2614	452	0.3065	136.95	15.123	452	4.546
43	1,3-Butadiene	C$_4$H$_6$	106990	54.092	1.2384	0.2725	425.17	0.28813	164.25	14.061	425.17	4.545
44	2-Methyl-1,3-butadiene[1]	C$_5$H$_8$	78795	68.119	0.95673	0.26488	484	0.28571	127.27	12.205	484	3.612
45	Acetylene	C$_2$H$_2$	74862	26.038	2.4091	0.27223	308.32	0.28477	192.40	23.692	308.32	8.850
46	Methylacetylene	C$_3$H$_4$	74997	40.065	1.6086	0.26448	402.39	0.279	170.45	19.027	402.39	6.082
47	Dimethylacetylene	C$_4$H$_6$	503173	54.092	1.1717	0.25895	473.2	0.27289	240.91	13.767	473.2	4.525
48	3-Methyl-1-butyne	C$_5$H$_8$	598232	68.119	0.94575	0.26008	463.2	0.30807	183.45	11.519	463.2	3.636
49	1-Pentyne	C$_5$H$_8$	627190	68.119	0.8491	0.2352	481.2	0.353	167.45	12.532	481.2	3.610
50	2-Pentyne	C$_5$H$_8$	627214	68.119	0.92099	0.25419	519	0.31077	163.83	12.24	519	3.623
51	1-Hexyne	C$_6$H$_{10}$	693027	82.145	0.84427	0.27185	516.2	0.2771	141.25	10.23	516.2	3.106
52	2-Hexyne	C$_6$H$_{10}$	764352	82.145	0.76277	0.25248	549	0.31611	183.65	10.133	549	3.021
53	3-Hexyne[1]	C$_6$H$_{10}$	928494	82.145	0.78045	0.26065	544	0.28571	170.05	10.021	544	2.994

No.	Name	Formula	CAS no.	Mol. wt.	C1	C2	C3	C4	Tmin, K	Density at Tmin	Tmax, K	Density at Tmax
54	1-Heptyne	C7H12	628717	96.172	0.67366	0.26003	559	0.29804	192.22	8.4987	559	2.591
55	1-Octyne	C8H14	629050	110.199	0.59229	0.26118	585	0.29357	193.55	7.478	585	2.268
56	Vinylacetylene[2]	C4H4	689974	52.076	1.2703	0.26041	454	0.297	173.15	15.664	454	4.878
57	Cyclopentane	C5H10	287923	70.134	1.124	0.28859	511.76	0.2506	179.28	11.883	511.76	3.895
58	Methylcyclopentane	C6H12	96377	84.161	0.84798	0.27042	532.79	0.28276	130.73	10.492	532.79	3.136
59	Ethylcyclopentane	C7H14	1640897	98.188	0.7193	0.26936	569.52	0.2777	134.71	9.018	569.52	2.670
60	Cyclohexane	C6H12	110827	84.161	0.8908	0.27396	553.58	0.2851	279.69	9.3797	553.58	3.252
61	Methylcyclohexane	C7H14	108872	98.188	0.735	0.27041	572.19	0.2927	146.58	9.018	572.19	2.718
62	1,1-Dimethyl-cyclohexane	C8H16	590669	112.215	0.55573	0.25143	591.15	0.27758	239.66	7.3417	591.15	2.222
63	Ethylcyclohexane	C8H16	1678917	112.215	0.61587	0.26477	609.15	0.28054	161.84	7.8679	609.15	2.326
64	Cyclopentene	C5H8	142290	68.119	1.1035	0.27035	507	0.29699	138.13	13.47	507	4.082
65	1-Methylcyclopentene	C6H10	693890	82.145	0.88824	0.26914	542	0.27874	146.62	10.98	542	3.300
66	Cyclohexene	C6H10	110838	82.145	0.92997	0.27056	560.4	0.25943	169.67	11.16	560.4	3.437
67	Benzene	C6H6	71432	78.114	1.0162	0.2655	562.16	0.28212	278.68	11.421	562.16	3.828
68	Toluene	C7H8	108883	92.141	0.8488	0.26655	591.8	0.2878	178.18	10.495	591.8	3.184
69	o-Xylene	C8H10	95476	106.167	0.69883	0.26113	630.33	0.27429	247.98	8.6285	630.33	2.676
70	m-Xylene	C8H10	108383	106.167	0.69555	0.26604	617.05	0.27602	225.30	8.0505	617.05	2.651
71	p-Xylene	C8H10	106423	106.167	0.6816	0.25963	616.23	0.2768	286.41	8.1616	616.23	2.625
72	Ethylbenzene	C8H10	100414	106.167	0.6952	0.26037	617.2	0.2844	178.15	9.0568	617.2	2.670
73	Propylbenzene	C9H12	103651	120.194	0.57695	0.25395	638.32	0.283	183.15	7.8942	638.32	2.272
74	1,2,4-Trimethylbenzene	C9H12	95636	120.194	0.60394	0.25955	649.13	0.27716	229.33	7.6895	649.13	2.327
75	Isopropylbenzene	C9H12	98828	120.194	0.604	0.25912	631.1	0.2914	177.14	7.9496	631.1	2.331
76	1,3,5-Trimethylbenzene	C9H12	108678	120.194	0.59579	0.25916	637.36	0.27968	228.42	7.6154	637.36	2.311
77	p-Isopropyltoluene	C10H14	99876	134.221	0.51036	0.25383	653.15	0.28816	205.25	6.8778	653.15	2.011
78	Naphthalene[6]	C10H8	91203	128.174	0.61674	0.25473	748.35	0.27355	333.15	7.7543	748.35	2.421
79	Biphenyl	C12H10	92524	154.211	0.5039	0.25273	789.26	0.281	342.20	6.4395	789.26	1.994
80	Styrene	C8H8	100425	104.152	0.7397	0.2603	636	0.3009	242.54	9.1088	636	2.842
81	m-Terphenyl	C18H14	92068	230.309	0.30826	0.23669	924.85	0.29678	360.00	4.5223	924.85	1.302
82	Methanol	CH4O	67561	32.042	2.288	0.2685	512.64	0.2453	175.47	27.912	512.64	8.521
83	Ethanol	C2H6O	64175	46.069	1.648	0.27627	513.92	0.2331	159.05	19.413	513.92	5.965
84	1-Propanol	C3H8O	71238	60.096	1.235	0.27136	536.78	0.24	146.95	15.231	536.78	4.551
85	1-Butanol	C4H10O	71363	74.123	0.965	0.2666	563.05	0.24419	184.51	12.016	563.05	3.620
86	2-Butanol	C4H10O	78922	74.123	0.966	0.26064	536.05	0.2746	158.45	12.57	536.05	3.706
87	2-Propanol	C3H8O	67630	60.096	1.24	0.27342	508.3	0.2353	185.28	14.547	508.3	4.535
88	2-Methyl-2-propanol	C4H10O	75650	74.123	0.9212	0.2544	506.21	0.276	298.97	10.555	506.21	3.621
89	1-Pentanol	C5H12O	71410	88.150	0.8164	0.2673	586.15	0.2506	195.56	10.057	586.15	3.054
90	2-Methyl-1-butanol	C5H12O	137326	88.150	0.82046	0.26829	565	0.2322	203.00	10.017	565	3.058
91	3-Methyl-1-butanol	C5H12O	123513	88.150	0.837	0.27375	577.2	0.22951	155.95	10.204	577.2	3.058
92	1-Hexanol	C6H14O	111273	102.177	0.70617	0.26901	611.35	0.2479	228.55	8.4506	611.35	2.625
93	1-Heptanol	C7H16O	111706	116.203	0.60481	0.2632	631.9	0.273	239.15	7.421	631.9	2.298
94	Cyclohexanol	C6H12O	108930	100.161	0.8243	0.26546	650	0.2848	296.60	9.4693	650	3.105
95	Ethylene glycol	C2H6O2	107211	62.068	1.3151	0.25125	719.7	0.2187	260.15	18.31	719.7	5.234
96	1,2-Propylene glycol	C3H8O2	57556	76.095	1.0923	0.26106	626	0.20459	213.15	14.363	626	4.184
97	Phenol	C6H6O	108952	94.113	1.3798	0.31598	694.25	0.32768	314.06	11.244	694.25	4.367
98	o-Cresol	C7H8O	95487	108.140	1.0861	0.30624	697.55	0.30587	304.19	9.5751	697.55	3.547
99	m-Cresol	C7H8O	108394	108.140	0.9061	0.29268	705.85	0.2707	285.39	9.6115	705.85	3.205
100	p-Cresol	C7H8O	106445	108.140	1.1503	0.31861	704.65	0.30104	307.93	9.4494	704.65	3.610
101	Dimethyl ether	C2H6O	115106	46.069	1.5693	0.2679	400.1	0.2882	131.65	18.95	400.1	5.858
102	Methyl ethyl ether	C3H8O	540670	60.096	1.2635	0.27878	437.8	0.2744	160.00	13.995	437.8	4.532
103	Methyl-n-propyl ether	C4H10O	557175	74.123	1.0124	0.27942	476.3	0.2555	133.97	11.696	476.3	3.623
104	Methyl isopropyl ether	C4H10O	598538	74.123	1.0318	0.29478	464.5	0.2444	127.93	11.568	464.5	3.623
105	Methyl-n-butyl ether[1]	C5H12O	628284	88.150	0.8281	0.27245	510	0.2827	157.48	9.8068	510	3.040
106	Methyl isobutyl ether[1]	C5H12O	625445	88.150	0.8252	0.27282	497	0.2857	150.00	9.7673	497	3.025
107	Methyl tert-butyl ether	C5H12O	1634044	88.150	0.82157	0.27032	497.1	0.2829	164.55	9.7682	497.1	3.039
108	Diethyl ether	C4H10O	60297	74.123	0.9554	0.26847	466.7	0.2814	156.85	11.487	466.7	3.559

TABLE 2-30 Densities of Inorganic and Organic Liquids (Continued)

Cmpd. no.	Name	Formula	CAS no.	Mol. wt.	C1	C2	C3	C4	T_{min}, K	Density at T_{min}	T_{max}, K	Density at T_{max}
109	Ethyl propyl ether	$C_5H_{12}O$	628320	88.150	0.7908	0.266	500.23	0.292	145.65	9.8474	500.23	2.973
110	Ethyl isopropyl ether	$C_5H_{12}O$	625547	88.150	0.82049	0.26994	489	0.30381	140.00	9.9117	489	3.040
111	Methyl phenyl ether	C_7H_8O	100663	108.140	0.77488	0.26114	645.6	0.28234	235.65	9.6675	645.6	2.967
112	Diphenyl ether	$C_{12}H_{10}O$	101848	170.211	0.52133	0.26218	766.8	0.31033	300.03	6.2648	766.8	1.988
113	Formaldehyde[3]	CH_2O	50000	30.026	1.9415	0.22309	408	0.28571	181.15	30.945	408	8.703
114	Acetaldehyde	C_2H_4O	75070	44.053	1.6994	0.26167	466	0.2913	150.15	21.499	466	6.494
115	1-Propanal	C_3H_6O	123386	58.080	1.296	0.26439	504.4	0.29471	170.00	15.929	504.4	4.902
116	1-Butanal	C_4H_8O	123728	72.107	1.0361	0.26731	537.2	0.28397	176.75	12.589	537.2	3.876
117	1-Pentanal	$C_5H_{10}O$	110623	86.134	0.83871	0.26252	566.1	0.29444	182.00	10.534	566.1	3.195
118	1-Hexanal	$C_6H_{12}O$	66251	100.161	0.71899	0.26531	591	0.27628	217.15	8.7243	591	2.710
119	1-Heptanal	$C_7H_{14}O$	111717	114.188	0.62649	0.26376	617	0.29221	229.80	7.6002	617	2.375
120	1-Octanal	$C_8H_{16}O$	124130	128.214	0.56833	0.26939	638.1	0.26975	246.00	6.6637	638.1	2.110
121	1-Nonanal	$C_9H_{18}O$	124196	142.241	0.49587	0.26135	658	0.30736	255.15	6.0165	658	1.897
122	1-Decanal	$C_{10}H_{20}O$	112312	156.268	0.46802	0.27146	674.2	0.26869	267.15	5.3834	674.2	1.724
123	Acetone	C_3H_6O	67641	58.080	1.2332	0.25886	508.2	0.2913	178.45	15.683	508.2	4.764
124	Methyl ethyl ketone	C_4H_8O	78933	72.107	0.93767	0.25035	535.5	0.29964	186.48	12.663	535.5	3.745
125	2-Pentanone	$C_5H_{10}O$	107879	86.134	0.90411	0.27207	561.08	0.30669	196.29	10.398	561.08	3.323
126	Methyl isopropyl ketone[l]	$C_5H_{10}O$	563804	86.134	0.8374	0.26204	553	0.2857	181.15	10.565	553	3.196
127	2-Hexanone	$C_6H_{12}O$	591786	100.161	0.70659	0.26073	587.05	0.2963	217.35	8.7505	587.05	2.710
128	Methyl isobutyl ketone	$C_6H_{12}O$	108101	100.161	0.71791	0.26491	571.4	0.28544	189.15	8.8579	571.4	2.710
129	3-Methyl-2-pentanone[l]	$C_6H_{12}O$	565617	100.161	0.6969	0.2587	573	0.2857	167.15	9.1722	573	2.694
130	3-Pentanone	$C_5H_{10}O$	96220	86.134	0.71811	0.24129	560.95	0.27996	234.18	10.102	560.95	2.976
131	Ethyl isopropyl ketone[l]	$C_6H_{12}O$	565695	100.161	0.66469	0.24527	567	0.34305	200.00	9.0933	567	2.710
132	Diisopropyl ketone	$C_7H_{14}O$	565800	114.188	0.56213	0.23385	576	0.2618	204.81	8.7779	576	2.404
133	Cyclohexanone	$C_6H_{10}O$	108941	98.145	0.8663	0.26941	653	0.2977	242.00	10.081	653	3.216
134	Methyl phenyl ketone	C_8H_8O	98862	120.151	0.64417	0.24863	709.5	0.28661	292.81	8.5581	709.5	2.591
135	Formic acid	CH_2O_2	64186	46.026	1.938	0.24225	588	0.24435	281.45	26.806	588	8.000
136	Acetic acid	$C_2H_4O_2$	64197	60.053	1.4486	0.25895	591.95	0.2529	289.81	17.492	591.95	5.595
137	Propionic acid	$C_3H_6O_2$	79094	74.079	1.1041	0.25659	600.81	0.26874	252.45	13.933	600.81	4.303
138	n-Butyric acid	$C_4H_8O_2$	107926	88.106	0.89213	0.25938	615.7	0.24909	267.95	11.087	615.7	3.440
139	Isobutyric acid	$C_4H_8O_2$	79312	88.106	0.88575	0.25736	605	0.26265	227.15	11.42	605	3.442
140	Benzoic acid[l]	$C_7H_6O_2$	65850	122.123	0.71587	0.24812	751	0.2857	395.45	8.8935	751	2.885
141	Acetic anhydride	$C_4H_6O_3$	108247	102.090	0.86852	0.25187	606	0.31172	200.15	11.643	606	3.448
142	Methyl formate	$C_2H_4O_2$	107313	60.053	1.525	0.2634	487.2	0.2806	174.15	18.811	487.2	5.790
143	Methyl acetate	$C_3H_6O_2$	79209	74.079	1.13	0.2593	506.55	0.2764	175.15	14.475	506.55	4.358
144	Methyl propionate	$C_4H_8O_2$	554121	88.106	0.9147	0.2594	530.6	0.2774	185.65	11.678	530.6	3.526
145	Methyl n-butyrate	$C_5H_{10}O_2$	623427	102.133	0.76983	0.26173	554.5	0.26879	187.35	9.7638	554.5	2.941
146	Ethyl formate	$C_3H_6O_2$	109944	74.079	1.1343	0.26168	508.4	0.2791	193.55	14.006	508.4	4.335
147	Ethyl acetate	$C_4H_8O_2$	141786	88.106	0.8996	0.25856	523.3	0.278	189.60	11.478	523.3	3.479
148	Ethyl propionate	$C_5H_{10}O_2$	105373	102.133	0.7405	0.25563	546	0.2795	199.25	9.6317	546	2.897
149	Ethyl n-butyrate	$C_6H_{12}O_2$	105544	116.160	0.63566	0.25613	571	0.27829	175.15	8.4912	571	2.482
150	n-Propyl formate	$C_4H_8O_2$	110747	88.106	0.915	0.26134	538	0.28	180.25	11.59	538	3.501
151	n-Propyl acetate	$C_5H_{10}O_2$	109604	102.133	0.73041	0.25456	549.73	0.27666	178.15	9.7941	549.73	2.869
152	n-Butyl acetate	$C_6H_{12}O_2$	123864	116.160	0.669	0.26028	579.15	0.309	199.65	8.3747	579.15	2.570
153	Methyl benzoate	$C_8H_8O_2$	93583	136.150	0.53944	0.23519	693	0.2676	260.75	8.2133	693	2.294
154	Ethyl benzoate	$C_9H_{10}O_2$	93890	150.177	0.4883	0.23878	698	0.28487	238.45	7.2924	698	2.045
155	Vinyl acetate	$C_4H_6O_2$	108054	86.090	0.9591	0.2593	519.13	0.27448	180.35	12.287	519.13	3.699
156	Methylamine	CH_5N	74895	31.057	1.39	0.21405	430.05	0.2275	179.69	25.378	430.05	6.494
157	Dimethylamine	C_2H_7N	124403	45.084	1.5436	0.27784	437.2	0.2572	180.96	16.964	437.2	5.556
158	Trimethylamine	C_3H_9N	75503	59.111	1.0116	0.25683	433.25	0.2696	156.08	13.144	433.25	3.939
159	Ethylamine	C_2H_7N	75047	45.084	1.1477	0.23182	456.15	0.26053	192.15	17.588	456.15	4.951
160	Diethylamine	$C_4H_{11}N$	109897	73.138	0.85379	0.25675	496.6	0.27027	223.35	10.575	496.6	3.325

No.	Name	Formula	No.								
161	Triethylamine	$C_6H_{15}N$	121448	101.192	0.7035	0.27386	535.15	0.2872	158.45	8.2943	535.15	2.569
162	n-Propylamine	C_3H_9N	107108	59.111	0.9195	0.23787	496.95	0.2461	188.36	13.764	496.95	3.851
163	di-n-Propylamine	$C_6H_{15}N$	142847	101.192	0.659	0.26428	550	0.2766	210.15	7.9929	550	2.494
164	Isopropylamine	C_3H_9N	75310	59.111	1.2801	0.2828	471.85	0.2972	177.95	13.561	471.85	4.527
165	Diisopropylamine	$C_6H_{15}N$	108189	101.192	0.6181	0.25786	523.1	0.271	176.85	8.0541	523.1	2.397
166	Aniline	C_6H_7N	62533	93.128	1.0405	0.2807	699	0.29236	267.13	11.176	699	3.707
167	N-Methylaniline	C_7H_9N	100618	107.155	0.6527	0.24324	701.55	0.25374	216.15	9.7244	701.55	2.683
168	N,N-Dimethylaniline	$C_8H_{11}N$	121697	121.182	0.4923	0.22868	687.15	0.2335	275.60	7.9705	687.15	2.153
169	Ethylene oxide	C_2H_4O	75218	44.053	1.836	0.26024	469.15	0.2696	160.65	23.477	469.15	7.055
170	Furan	C_4H_4O	110009	68.075	1.1339	0.24741	490.15	0.2612	187.55	15.702	490.15	4.583
171	Thiophene	C_4H_4S	110021	84.142	1.2875	0.28195	579.35	0.3077	234.94	13.431	579.35	4.566
172	Pyridine	C_5H_5N	110861	79.101	0.9815	0.24957	619.95	0.29295	231.51	13.193	619.95	3.933
173	Formamide[5]	CH_3NO	75127	45.041	1.2486	0.20352	771	0.25178	275.60	25.488	771	6.135
174	N,N-Dimethyl-formamide	C_3H_7NO	68122	73.095	0.89615	0.23736	649.6	0.28091	212.72	13.954	649.6	3.817
175	Acetamide	C_2H_5NO	60355	59.068	1.016	0.21845	761	0.26116	353.33	16.936	761	4.651
176	N-Methylacetamide	C_3H_7NO	79163	73.095	0.88268	0.23568	718	0.27379	301.15	13.012	718	3.745
177	Acetonitrile	C_2H_3N	75050	41.053	1.3061	0.22507	545.5	0.28678	229.32	20.628	545.5	5.781
178	Propionitrile	C_3H_5N	107120	55.079	1.0224	0.23452	564.4	0.2804	180.26	16.027	564.4	4.360
179	n-Butyronitrile	C_4H_7N	109740	69.106	0.87533	0.24331	582.25	0.28586	161.25	13.047	582.25	3.598
180	Benzonitrile	C_7H_5N	100470	103.123	0.73136	0.24793	699.35	0.2841	260.40	10.009	699.35	2.950
181	Methyl mercaptan	CH_4S	74931	48.109	1.9323	0.28018	469.95	0.28523	150.18	21.564	469.95	6.897
182	Ethyl mercaptan	C_2H_6S	75081	62.136	1.3047	0.2694	499.15	0.27866	125.26	16.242	499.15	4.843
183	n-Propyl mercaptan	C_3H_8S	107039	76.163	1.0714	0.27214	536.6	0.29481	159.95	12.716	536.6	3.937
184	n-Butyl mercaptan	$C_4H_{10}S$	109795	90.189	0.89458	0.27463	570.1	0.28512	157.46	10.585	570.1	3.257
185	Isobutyl mercaptan	$C_4H_{10}S$	513440	90.189	0.88801	0.27262	559	0.29522	128.31	10.851	559	3.257
186	sec-Butyl mercaptan	$C_4H_{10}S$	513531	90.189	0.89137	0.27365	554	0.2953	133.02	10.761	554	3.257
187	Dimethyl sulfide	C_2H_6S	75183	62.136	1.4029	0.27991	503.04	0.2741	174.88	15.556	503.04	5.012
188	Methyl ethyl sulfide	C_3H_8S	624895	76.163	1.067	0.27101	533	0.29363	167.23	12.672	533	3.937
189	Diethyl sulfide	$C_4H_{10}S$	352932	90.189	0.82413	0.26333	557.15	0.27445	169.20	10.476	557.15	3.130
190	Fluoromethane	CH_3F	593533	34.033	2.1854	0.24725	317.42	0.27558	131.35	29.526	317.42	8.839
191	Chloromethane	CH_3Cl	74873	50.488	1.817	0.25877	416.25	0.2833	175.43	22.347	416.25	7.022
192	Trichloromethane	$CHCl_3$	67663	119.377	1.0841	0.2581	536.4	0.2741	209.63	13.702	536.4	4.200
193	Tetrachloromethane	CCl_4	56235	153.822	0.99835	0.274	556.35	0.287	250.33	10.843	556.35	3.644
194	Bromomethane	CH_3Br	74839	94.939	1.6762	0.26141	467	0.28402	179.47	20.64	467	6.412
195	Fluoroethane	C_2H_5F	353366	48.060	1.6525	0.27099	375.31	0.2442	129.95	19.785	375.31	6.098
196	Chloroethane	C_2H_5Cl	75003	64.514	2.176	0.3377	460.35	0.3361	134.80	16.934	460.35	6.444
197	Bromoethane	C_2H_5Br	74964	108.966	1.1908	0.25595	503.8	0.29152	154.55	15.833	503.8	4.653
198	1-Chloropropane	C_3H_7Cl	540545	78.541	1.087	0.26832	503.15	0.28055	150.35	13.328	503.15	4.051
199	2-Chloropropane	C_3H_7Cl	75296	78.541	1.1202	0.27669	489	0.27646	155.97	12.855	489	4.049
200	1,1-Dichloropropane[1]	$C_3H_6Cl_2$	78999	112.986	0.91064	0.26561	560	0.28571	200.00	11.03	560	3.429
201	1,2-Dichloropropane	$C_3H_6Cl_2$	78875	112.986	0.89833	0.26142	572	0.2868	172.71	11.526	572	3.436
202	Vinyl chloride	C_2H_3Cl	75014	62.499	1.5115	0.2707	432	0.2716	119.36	18.481	432	5.584
203	Fluorobenzene	C_6H_5F	462066	96.104	1.0146	0.27277	560.09	0.28291	230.94	11.374	560.09	3.720
204	Chlorobenzene	C_6H_5Cl	108907	112.558	0.8711	0.26805	632.35	0.2799	227.95	10.385	632.35	3.250
205	Bromobenzene	C_6H_5Br	108861	157.010	0.8226	0.26632	670.15	0.2821	242.43	9.9087	670.15	3.089
206	Air		132259100	28.951	2.8963	0.26733	132.45	0.27341	59.15	33.279	132.45	10.834
207	Hydrogen	H_2	1333740	2.016	5.414	0.34893	33.19	0.2706	13.95	38.487	33.19	15.516
208	Helium-4[4]	He	7440597	4.003	7.2475	0.41865	5.2	0.24096	2.20	37.115	5.2	17.312
209	Neon	Ne	7440019	20.180	7.3718	0.3067	44.4	0.2786	24.56	61.796	44.4	24.036
210	Argon	Ar	7440371	39.948	3.8469	0.2881	150.86	0.29783	83.78	35.491	150.86	13.353
211	Fluorine	F_2	7782414	37.997	4.2985	0.28587	144.12	0.28776	53.48	44.888	144.12	15.005
212	Chlorine	Cl_2	7782505	70.905	2.23	0.27645	417.15	0.2926	172.12	24.242	417.15	8.067
213	Bromine	Br_2	7726956	159.808	2.1872	0.29527	584.15	0.3295	265.85	20.109	584.15	7.408
214	Oxygen	O_2	7782447	31.999	3.9143	0.2861	154.58	0.2924	54.35	40.77	154.58	13.605
215	Nitrogen	N_2	7727379	28.014	3.2091	0.28772	126.2	0.2966	63.15	31.063	126.2	11.217
216	Ammonia	NH_3	7664417	17.031	3.5383	0.25443	405.65	0.2888	195.41	43.141	405.65	13.907
217	Hydrazine	N_2H_4	302012	32.045	1.0516	0.16613	653.15	0.1898	274.69	31.934	653.15	6.330

TABLE 2-30 Densities of Inorganic and Organic Liquids (*Concluded*)

Cmpd. no.	Name	Formula	CAS no.	Mol. wt.	C1	C2	C3	C4	T_{min}, K	Density at T_{min}	T_{max}, K	Density at T_{max}
218	Nitrous oxide	N_2O	10024972	44.013	2.781	0.27244	309.57	0.2882	182.30	27.928	309.57	10.208
219	Nitric oxide	NO	10102439	30.006	5.246	0.3044	180.15	0.242	109.50	44.487	180.15	17.234
220	Cyanogen	C_2N_2	460195	52.036	1.0761	0.20984	400.15	0.20635	245.25	18.513	400.15	5.128
221	Carbon monoxide	CO	630080	28.010	2.897	0.27532	132.92	0.2813	68.15	30.18	132.92	10.522
222	Carbon dioxide	CO_2	124389	44.010	2.768	0.26212	304.21	0.2908	216.58	26.828	304.21	10.560
223	Carbon disulfide	CS_2	75150	76.143	1.7968	0.28749	552	0.3226	161.11	19.064	552	6.250
224	Hydrogen fluoride	HF	7664393	20.006	2.5635	0.1766	461.15	0.3733	189.79	60.203	461.15	14.516
225	Hydrogen chloride	HCl	7647010	36.461	3.342	0.2729	324.65	0.3217	158.97	34.854	324.65	12.246
226	Hydrogen bromide[1]	HBr	10035106	80.912	2.832	0.2832	363.15	0.28571	185.15	27.985	363.15	10.000
227	Hydrogen cyanide	HCN	74908	27.026	1.3413	0.18559	456.65	0.28206	259.83	27.202	456.65	7.216
228	Hydrogen sulfide	H_2S	7783064	34.082	2.7672	0.27369	373.53	0.29015	187.68	29.13	373.53	10.111
229	Sulfur dioxide	SO_2	7446095	64.065	2.106	0.25842	430.75	0.2895	197.67	25.298	430.75	8.150
230	Sulfur trioxide	SO_3	7446119	80.064	1.4969	0.19013	490.85	0.4359	289.95	24.241	490.85	7.873
231	Water[7]	H_2O	7732185	18.015	5.459	0.30542	647.13	0.081	273.16	55.583	333.15	54.703

All substances are listed in alphabetical order in Table 2-6a. Compiled from Daubert, T. E., R. P. Danner, H. M. Sibul, and C. C. Stebbins, DIPPR Data Compilation of Pure Compound Properties, Project 801 Sponsor Release, July, 1993, Design Institute for Physical Property Data, AIChE, New York, NY; and from Thermodynamics Research Center, "Selected Values of Properties of Hydrocarbons and Related Compounds," Thermodynamics Research Center Hydrocarbon Project, Texas A&M University, College Station, Texas (extant 1994).

Temperatures are in kelvins. Liquid densities are in $kmol/m^3$. Density formulas: $kmol/m^3 \times (mol. wt./1E+03) = g/cm^3$; $kmol/m^3 \times (mol. wt./1.601846E+01) = lb/ft^3$.

The liquid density equation is $C1/C2^{[1+(1-T/C3)^{C4}]}$ unless otherwise noted.

[1] The modified Rackett equation, density $= (P_c/RT_c)/ZRA^{[1+(1-(T/T_c))^{2/7}]}$, was used. See Spencer, C. F., and R. P. Danner, "Improved Equation for Prediction of Saturated Liquid Density," *J. Chem. Eng. Data* **17**, 236 (1972).

[2] Decomposes violently on heating. Forms explosive peroxides with air or oxygen. Polymerizes under pressure and heat.

[3] For the hypothetical pure liquid.

[4] Exhibits superfluid properties below 2.2 K.

[5] Coefficients are hypothetical above the decomposition temperature.

[6] Lower limit is for the undercooled liquid.

[7] For the temperature range 333.15 to 403.15 K, use the coefficients: $C_1 = 4.9669E+00$, $C_2 = 2.7788E-01$, $C_3 = 6.4713E+02$, $C_4 = 1.8740E-01$. For the temperature range 403.15 to 647.13 K, use $C_1 = 4.3910E+00$, $C_2 = 2.4870E-01$, $C_3 = 6.4713E+02$, $C_4 = 2.5340E-01$.

DENSITIES OF AQUEOUS INORGANIC SOLUTIONS

UNITS AND UNITS CONVERSIONS

Densities are given in grams per cubic centimeter. To convert to pounds per cubic foot, multiply by 62.43. °F = ⅗ °C + 32.

ADDITIONAL REFERENCES

For more detailed data on densities see *International Critical Tables:* tabular index, vol. 3, p. 1; abrasives, vol. 2, p. 87; air, moist, vol. 1, p. 71; building stones, vol. 2, p. 52; clays, vol. 2, p. 56; coals, vol. 2, p. 135; compounds, vol. 1. pp. 106, 176, 313, 341; elements, vol. 1, pp. 102, 340; fibers, vol. 2, p. 237; gases and vapors, vol. 3, pp. 3, 345; glass, vol. 2, p. 93; liquids and vitreous solids, vol. 3, p. 22; vol. 1, pp. 102, 340; vol. 2, pp. 456, 463; vol. 3, pp. 20, 35; liquid coolants and saturated vapors are available from WADC-TR-59-598, 1959; plastics are collected in the *Handbook of Chemistry and Physics,* Chemical Rubber Publishing Co.: solid helium, neon, argon, fluorine, and methane data are given by Johnson (ed.), WADD-TR-60-56, 1960; temperatures of maximum solubility, vol. 3, p. 107; metals, vol. 2, p. 463; oils, fats, and waxes, vol. 2, p. 201; orthobaric, vol. 3, pp. 202, 228, 237, 244; petroleums, vol. 2, pp. 137, 144; plastics, vol. 2, p. 296; porcelains, vol. 2, pp. 68, 75; refrigerating brines, vol. 2, p. 327; rubber, vol. 2, pp. 255, 259; soaps, vol. 5, p. 447; metallic solid solutions, vol. 2, p. 358; solids, vol. 3, pp. 43, 45; vol. 2, p. 456; vol. 3, p. 21; solutions and mixtures, vol. 3, pp. 17, 51, 95, 104, 107, 111, 125, 130; woods, vol. 2, p. 1. Also see the *Handbook of Chemistry and Physics,* Chemical Rubber Publishing Co., 40th ed., etc.

TABLE 2-31   Aluminum Sulfate [Al$_2$(SO$_4$)$_3$]

%	d_4^{15}	%	d_4^{15}
1	1.0093	16	1.1770
2	1.0195	20	1.2272
4	1.0404	24	1.2803
8	1.0837	26	1.3079
12	1.1293		

TABLE 2-32   Ammonia (NH$_3$)

%	−15°C	−10°C	−5°C	0°C	5°C	10°C	20°C	25°C	%	d_4^{15}
1		0.9943	0.9954	0.9959	0.9958	0.9955	0.9939	0.993	32	0.889
2		.9906	.9915	.9919	.9917	.9913	.9895	.988	36	.877
4		.9834	.9840	.9842	.9837	.9832	.9811	.980	40	.865
8	0.970	.9701	.9701	.9701	.9695	.9677	.9651	.964	45	.849
12	.958	.9576	.9571	.9561	.9548	.9534	.9501	.948	50	.832
16	.947	.9461	.9450	.9435	.9420	.9402	.9362	.934	60	.796
20		.9353	.9335	.9316	.9296	.9275	.9229		70	.755
24		.9249	.9226	.9202	.9179	.9155	.9101		80	.711
28		.9150	.9122	.9094	.9067	.9040	.8980		90	.665
30		.9101	.9070	.9040	.9012	.8983	.8920		100	.618

TABLE 2-33   Ammonium Acetate* (CH$_3$COONH$_4$)

%	d_4^{25}
1	0.9992
2	1.0013
4	1.0055
8	1.0136
12	1.0216
16	1.0294
20	1.0368
24	1.0439
28	1.0507
30	1.0540
35	1.0618
40	1.0691
45	1.0760

*For data at 16°C for 3(1)52 percent see Atack *Handbook of Chemical Data,* p. 33, Reinhold, New York, 1957.

TABLE 2-34   Ammonium Bichromate [(NH$_4$)$_2$Cr$_2$O$_7$]

%	d_4^{12}
1	1.0051
2	1.0108
4	1.0223
8	1.0463
12	1.0715
16	1.0981
20	1.1263

TABLE 2-35   Ammonium Chloride (NH$_4$Cl)

%	0°C	10°C	20°C	30°C	50°C	80°C	100°C
1	1.0033	1.0029	1.0013	0.9987	0.9910	0.9749	0.9617
2	1.0067	1.0062	1.0045	1.0018	.9940	.9780	.9651
4	1.0135	1.0126	1.0107	1.0077	.9999	.9842	.9718
8	1.0266	1.0251	1.0227	1.0195	1.0116	.9963	.9849
12	1.0391	1.0370	1.0344	1.0310	1.0231	1.0081	.9975
16	1.0510	1.0485	1.0457	1.0422	1.0343	1.0198	1.0096
20	1.0625	1.0596	1.0567	1.0532	1.0454	1.0312	1.0213
24	1.0736	1.0705	1.0674	1.0641	1.0564	1.0426	1.0327

TABLE 2-36   Ammonium Chromate [(NH$_4$)$_2$CrO$_4$]

%	°C	d_4^t
3.80	20	1.0219
10.52	13	1.0627
19.75	13.7	1.1189
28.04	19.6	1.1707

TABLE 2-37   Ammonium Nitrate (NH$_4$NO$_3$)

%	0°C	10°C	25°C	40°C	60°C	80°C
1.0	1.0043	1.0039	1.0011	0.9961	0.9870	0.9755
2.0	1.0088	1.0082	1.0051	1.0000	.9908	.9793
4.0	1.0178	1.0168	1.0132	1.0079	.9985	.9869
8.0	1.0358	1.0340	1.0297	1.0238	1.0142	1.0024
12.0	1.0539	1.0515	1.0464	1.0400	1.0301	1.0181
16.0	1.0721	1.0691	1.0633	1.0565	1.0462	1.0342
20.0	1.0905	1.0870	1.0806	1.0734	1.0627	1.0506
24.0	1.1090	1.1051	1.0982	1.0907	1.0796	1.0673
28.0	1.1277	1.1234	1.1161	1.1082	1.0968	1.0844
30.0	1.1371	1.1327	1.1252	1.1171	1.1055	1.0931
40.0	1.1862	1.1810	1.1727	1.1640	1.1515	1.1385
50.0	1.2380	1.2320	1.2229	1.2136	1.2006	1.1868

TABLE 2-38   Ammonium Sulfate [(NH$_4$)$_2$SO$_4$]

%	0°C	20°C	40°C	80°C	100°C
1	1.0061	1.0041	0.9980	0.9777	0.9644
2	1.0124	1.0101	1.0039	.9836	.9705
4	1.0248	1.0220	1.0155	.9953	.9826
8	1.0495	1.0456	1.0387	1.0187	1.0066
12	1.0740	1.0691	1.0619	1.0421	1.0303
16	1.0980	1.0924	1.0849	1.0653	1.0539
20	1.1215	1.1154	1.1077	1.0883	1.0772
24	1.1448	1.1383	1.1304	1.1111	1.1003
28	1.1677	1.1609	1.1529	1.1338	1.1232
35	1.2072	1.2800	1.1919	1.1731	1.1629
40	1.2350	1.2277	1.2196	1.2011	1.1910
50	1.2899	1.2825	1.2745	1.2568	1.2466

TABLE 2-39   Arsenic Acid (H$_3$A$_3$O$_4$)

%	d_4^{15}	%	d_4^{15}
1	1.0057	20	1.1447
2	1.0124	30	1.2331
6	1.0398	40	1.3370
10	1.0681	50	1.4602
16	1.1128	60	1.6070
		70	1.7811

TABLE 2-40 Barium Chloride (BaCl$_2$)

%	0°C	20°C	40°C	60°C	80°C	100°C
2	1.0181	1.0159	1.0096	1.0004	0.9890	0.9755
4	1.0368	1.0341	1.0275	1.0181	1.0066	.9931
8	1.0760	1.0721	1.0648	1.0551	1.0434	1.0299
12	1.1178	1.1128	1.1047	1.0948	1.0827	1.0692
16	1.1627	1.1564	1.1478	1.1373	1.1249	1.1113
20	1.2105	1.2031	1.1938	1.1828	1.1702	1.1563
24		1.2531	1.2430	1.2316	1.2186	1.2045
26		1.2793	1.2688	1.2571	1.2440	1.2298

TABLE 2-41 Cadmium Nitrate [Cd(NO$_3$)$_2$]

%	d_4^{18}	%	d_4^{18}
2	1.0154	20	1.1904
4	1.0326	25	1.2488
8	1.0683	30	1.3124
12	1.1061	40	1.4590
16	1.1468	50	1.6356

TABLE 2-42 Calcium Chloride (CaCl$_2$)

%	−5°C	0°C	20°C	30°C	40°C	60°C	80°C	100°C	120°C*	140°C
2		1.0171	1.0148	1.0120	1.0084	0.9994	0.9881	0.9748	0.9596	0.9428
4		1.0346	1.0316	1.0286	1.0249	1.0158	1.0046	.9915	.9765	.9601
8	1.0708	1.0703	1.0659	1.0626	1.0586	1.0492	1.0382	1.0257	1.0111	.9954
12	1.1083	1.1072	1.1015	1.0978	1.0937	1.0840	1.0730	1.0610	1.0466	1.0317
16	1.1471	1.1454	1.1386	1.1345	1.1301	1.1202	1.1092	1.0973	1.0835	1.0691
20	1.1874	1.1853	1.1775	1.1730	1.1684	1.1581	1.1471	1.1352	1.1219	1.1080
25			1.2376	1.2284	1.2236	1.2186	1.2079	1.1965	1.1846	
30			1.2922	1.2816	1.2764	1.2709	1.2597	1.2478	1.2359	
35				1.3373	1.3316	1.3255	1.3137	1.3013	1.2893	
40				1.3957	1.3895	1.3826	1.3700	1.3571	1.3450	

*Corrected to atmospheric pressure.

TABLE 2-43 Calcium Hydroxide [Ca(OH)$_2$]

%	d_4^{15}	d_4^{25}
0.05	0.99979	0.99773
.10	1.00044	.99838
.15	1.00110	.99904

TABLE 2-44 Calcium Hypochlorite* (CaOCl$_2$)

% total salt	d_4^{15}
2	1.0169
4	1.0345
6	1.0520
8	1.0697
10	1.0876
12	1.1060

*CaOCl$_2$ = 89.15%
CaCl$_2$ = 7.31%
Ca(ClO$_3$)$_2$ = 0.26%
Ca(OH)$_2$ = 2.92%.

TABLE 2-45 Calcium Nitrate [Ca(NO$_3$)$_2$]

%	6°C	18°C	25°C	30°C
2°	1.0157	1.0137	1.0120	1.0105
4	1.0316	1.0291	1.0272	1.0256
8	1.0641	1.0608	1.0585	1.0565
12	1.0979	1.0937	1.0911	1.0887
16	1.1330	1.1279	1.1250	1.1224
20	1.1694	1.1636	1.1602	1.1575
25	1.2168	1.2106	1.2065	1.2032
30		1.260		
35		1.311		
40		1.365		
45		1.422		
68°		1.747	1.741	1.736

*Supercooled tetrahydrate (m.p. 41.4°C).

TABLE 2-46 Chromic Acid (CrO$_3$)

%	d_4^{15}	%	d_4^{15}
1	1.006	20	1.163
2	1.014	26	1.220
6	1.045	30	1.260
10	1.076	40	1.371
16	1.127	50	1.505
		60	1.663

TABLE 2-47 Chromium Chloride (CrCl$_3$)

%	d_4^{18}		
	Violet	Green	Equilibrium mixture of violet and green
1	1.0076	1.0071	1.0075
2	1.0166	1.0157	1.0165
4	1.0349	1.0332	1.0347
8	1.0724	1.0691	1.0722
12	1.1114	1.1065	1.1111
14	1.1316		

TABLE 2-48 Copper Nitrate [Cu(NO$_3$)$_2$]

%	d_4^{20}	%	d_4^{20}
1	1.007	12	1.107
2	1.015	16	1.147
4	1.032	20	1.189
8	1.069	25	1.248

TABLE 2-49 Copper Sulfate (CuSO$_4$)

%	0°C	20°C	40°C
1	1.0104	1.0086	1.0024
4	1.0429	1.0401	1.0332
8	1.0887	1.084	1.0764
12	1.1379	1.1308	1.1222
16		1.180	
18		1.206	

TABLE 2-50 Cuprous Chloride (Cu$_2$Cl$_2$)

%	0°C	20°C	40°C
1	1.0095	1.0072	1.002
4	1.0387	1.036	1.0305
8	1.0788	1.0754	1.0682
12	1.1208	1.1165	1.107
16	1.1653	1.1595	1.151
20	1.2121	1.2052	1.1953

TABLE 2-51 Ferric Chloride (FeCl$_3$)

%	0°C	10°C	20°C	30°C
1	1.0086	1.0084	1.0068	1.0040
2	1.0174	1.0168	1.0152	1.0122
4	1.0347	1.0341	1.0324	1.0292
8	1.0703	1.0692	1.0669	1.0636
12	1.1088	1.1071	1.1040	1.1006
16	1.1475	1.1449	1.1418	1.1386
20	1.1870	1.1847	1.1820	1.1786
25	1.2400	1.2380	1.2340	1.2290
30	1.2970	1.2950	1.2910	1.2850
35	1.3605	1.3580	1.3530	1.3475
40	1.4280	1.4235	1.4175	1.4115
45		1.4920	1.4850	
50		1.5610	1.5510	

TABLE 2-52 Ferric Sulfate [Fe₂(SO₄)₃]

%	$d_4^{17.5}$
1	1.0072
2	1.0157
4	1.0327
8	1.0670
12	1.1028
16	1.1409
20	1.1811
30	1.3073
40	1.4487
50	1.6127
60	1.7983

TABLE 2-53 Ferric Nitrate [Fe(NO₃)₃]

%	d_4^{18}
1	1.0065
2	1.0144
4	1.0304
8	1.0636
12	1.0989
16	1.1359
20	1.1748
25	1.2281

TABLE 2-58 Hydrogen Fluoride (HF)

%	d_4^{20}	d_4^{0}
5	1.020	1.017
10	1.040	1.035
20	1.080	1.070
30	1.119	1.101
40	1.159	1.130
50	1.198	1.155
60	1.235	
70	1.258	
80	1.259	
90	1.178	
95	1.089	
100	1.0005	

TABLE 2-59 Hydrogen Peroxide (H₂O₂)

%	d_4^{18}	%	d_4^{18}
1	1.0022	26	1.0959
2	1.0058	28	1.1040
4	1.0131	30	1.1122
6	1.0204	35	1.1327
8	1.0277	40	1.1536
10	1.0351	45	1.1749
12	1.0425	50	1.1966
14	1.0499	55	1.2188
16	1.0574	60	1.2416
18	1.0649	70	1.2897
20	1.0725	80	1.3406
22	1.0802	90	1.3931
24	1.0880	100	1.4465

TABLE 2-54 Ferrous Sulfate (FeSO₄)

%	15°C	18°C	20°C
0.2		1.00068	1.0002
0.4		1.00275	1.0022
0.8		1.00645	1.0062
1.0	1.0090	1.0085	1.0082
4.0	1.0380	1.0375	
8.0	1.0790	1.0785	
12.0	1.1235	1.1220	
16.0	1.1690	1.1675	
20.0	1.2150	1.2135	

TABLE 2-55 Hydrogen Bromide (HBr)

%	d_4^{4}	d_4^{10}	d_4^{25}
1.0	1.0073	1.0068	1.0041
2.0	1.0146	1.0139	1.0111
4.0	1.0295	1.0285	1.0255
6.0	1.0448	1.0435	1.0402
8.0	1.0604	1.0589	1.0552
10.0	1.0764	1.0747	1.0707
12.0	1.0928	1.0910	1.0867
14.0	1.1097	1.1078	1.1032
16.0	1.1272	1.1251	1.1202
18.0	1.1453	1.1430	1.1377
20.0	1.1640	1.1615	1.1557
22.0	1.1832	1.1806	1.1743
24.0	1.2030	1.2003	1.1935
26.0	1.2235	1.2206	1.2134
28.0	1.2446	1.2415	1.2340
30.0	1.2663	1.2630	1.2552
40.0	1.3877	1.3838	1.3736
50.0	1.5305	1.5257	1.5127
60.0	1.6950	1.6892	1.6731
65.0	1.7854	1.7792	1.7613

TABLE 2-56 Hydrogen Cyanide (HCN)

%	d_4^{15}
1	0.998
2	.996
4	.993
8	.984
12	.971
16	.956
82	.752
90	.724
100	.691

TABLE 2-60 Hydrofluosilic Acid (H₂SiF₆)

%	$d_4^{17.5}$	%	$d_4^{17.5}$
1	1.0080	16	1.1373
2	1.0161	20	1.1748
4	1.0324	25	1.2235
8	1.0661	30	1.2742
12	1.1011	34	1.3162

TABLE 2-61 Magnesium Chloride (MgCl₂)

%	0°C	20°C	40°C	60°C	80°C	100°C
2	1.0168	1.0146	1.0084	0.9995	0.9883	0.9753
4	1.0338	1.0311	1.0248	1.0159	1.0050	.9923
8	1.0683	1.0646	1.0580	1.0493	1.0388	1.0269
12	1.1035	1.0989	1.0921	1.0836	1.0735	1.0622
16	1.1395	1.1342	1.1272	1.1188	1.1092	1.0984
20	1.1764	1.1706	1.1635	1.1552	1.1460	1.1359
25	1.2246	1.2184	1.2111	1.2031	1.1942	1.1847
30	1.2754	1.2688	1.2614	1.2535	1.2451	1.2360

TABLE 2-62 Magnesium Sulfate (MgSO₄)

%	0°C	20°C	30°C	40°C	50°C	60°C	80°C
2	1.0210	1.0186	1.0158	1.0123	1.0081	1.0032	0.9916
4	1.0423	1.0392	1.0362	1.0326	1.0283	1.0234	1.0118
8	1.0858	1.0816	1.0782	1.0743	1.0700	1.0650	1.0534
12	1.1309	1.1256	1.1220	1.1179	1.1135	1.1083	1.0968
16	1.1777	1.1717	1.1679	1.1637	1.1592		
20	1.2264	1.2198	1.2159	1.2117	1.2072		
26	1.3032	1.2961	1.2922	1.2879	1.2836		

TABLE 2-57 Hydrogen Chloride (HCl)

%	−5°C	0°C	10°C	20°C	40°C	60°C	80°C	100°C
1	1.0048	1.0052	1.0048	1.0032	0.9970	0.9881	0.9768	0.9636
2	1.0104	1.0106	1.0100	1.0082	1.0019	.9930	0.9819	.9688
4	1.0213	1.0213	1.0202	1.0181	1.0116	1.0026	0.9919	.9791
6	1.0321	1.0319	1.0303	1.0279	1.0211	1.0121	1.0016	.9892
8	1.0428	1.0423	1.0403	1.0376	1.0305	1.0215	1.0111	.9992
10	1.0536	1.0528	1.0504	1.0474	1.0400	1.0310	1.0206	1.0090
12	1.0645	1.0634	1.0607	1.0574	1.0497	1.0406	1.0302	1.0188
14	1.0754	1.0741	1.0711	1.0675	1.0594	1.0502	1.0398	1.0286
16	1.0864	1.0849	1.0815	1.0776	1.0692	1.0598	1.0494	1.0383
18	1.0975	1.0958	1.0920	1.0878	1.0790	1.0694	1.0590	1.0479
20	1.1087	1.1067	1.1025	1.0980	1.0888	1.0790	1.0685	1.0574
22	1.1200	1.1177	1.1131	1.1083	1.0986	1.0886	1.0780	1.0668
24	1.1314	1.1287	1.1238	1.1187	1.1085	1.0982	1.0874	1.0761
26	1.1426	1.1396	1.1344	1.1290	1.1183	1.1076	1.0967	1.0853
28	1.1537	1.1505	1.1449	1.1392	1.1280	1.1169	1.1058	1.0942
30	1.1648	1.1613	1.1553	1.1493	1.1376	1.1260	1.1149	1.1030
32				1.1593				
34				1.1691				
36				1.1789				
38				1.1885				
40				1.1980				

TABLE 2-63 Nickel Chloride (NiCl₂)

%	d_4^{18}
1	1.0082
2	1.0179
4	1.0375
8	1.0785
12	1.1217
16	1.1674
20	1.2163
30	1.353

TABLE 2-64 Nickel Nitrate [Ni(NO₃)₂]

%	d_4^{20}
1	1.0065
2	1.0150
4	1.0325
8	1.0688
12	1.1070
16	1.1480
20	1.191
30	1.311
35	1.377

TABLE 2-65 Nickel Sulfate (NiSO₄)

%	d_4^{18}
1	1.0091
2	1.0198
4	1.0415
8	1.0852
12	1.1325
16	1.1825
18	1.2090

TABLE 2-66   Nitric Acid (HNO₃)

%	0°C	5°C	10°C	15°C	20°C	25°C	30°C	40°C	50°C	60°C	80°C	100°C
1	1.0058	1.00572	1.00534	1.00464	1.00364	1.00241	1.0009	0.9973	0.9931	0.9882	0.9767	0.9632
2	1.0117	1.01149	1.01099	1.01018	1.00909	1.00778	1.0061	1.0025	.9982	.9932	.9816	.9681
3	1.0176	1.01730	1.01668	1.01576	1.01457	1.01318	1.0114	1.0077	1.0033	.9982	.9865	.9730
4	1.0236	1.02315	1.02240	1.02137	1.02008	1.01861	1.0168	1.0129	1.0084	1.0033	.9915	.9779
5	1.0296	1.02904	1.02816	1.02702	1.02563	1.02408	1.0222	1.0182	1.0136	1.0084	.9965	.9829
6	1.0357	1.03497	1.03397	1.03272	1.03122	1.02958	1.0277	1.0235	1.0188	1.0136	1.0015	.9879
7	1.0418	1.0410	1.0399	1.0385	1.0369	1.0352	1.0333	1.0289	1.0241	1.0188	1.0066	.9929
8	1.0480	1.0471	1.0458	1.0443	1.0427	1.0409	1.0389	1.0344	1.0295	1.0241	1.0117	.9980
9	1.0543	1.0532	1.0518	1.0502	1.0485	1.0466	1.0446	1.0399	1.0349	1.0294	1.0169	1.0032
10	1.0606	1.0594	1.0578	1.0561	1.0543	1.0523	1.0503	1.0455	1.0403	1.0347	1.0221	1.0083
11	1.0669	1.0656	1.0639	1.0621	1.0602	1.0581	1.0560	1.0511	1.0458	1.0401	1.0273	1.0134
12	1.0733	1.0718	1.0700	1.0681	1.0661	1.0640	1.0618	1.0567	1.0513	1.0455	1.0326	1.0186
13	1.0797	1.0781	1.0762	1.0742	1.0721	1.0699	1.0676	1.0624	1.0568	1.0509	1.0379	1.0238
14	1.0862	1.0845	1.0824	1.0803	1.0781	1.0758	1.0735	1.0681	1.0624	1.0564	1.0432	1.0289
15	1.0927	1.0909	1.0887	1.0865	1.0842	1.0818	1.0794	1.0739	1.0680	1.0619	1.0485	1.0341
16	1.0992	1.0973	1.0950	1.0927	1.0903	1.0879	1.0854	1.0797	1.0737	1.0675	1.0538	1.0393
17	1.1057	1.1038	1.1014	1.0989	1.0964	1.0940	1.0914	1.0855	1.0794	1.0731	1.0592	1.0444
18	1.1123	1.1103	1.1078	1.1052	1.1026	1.1001	1.0974	1.0913	1.0851	1.0787	1.0646	1.0496
19	1.1189	1.1168	1.1142	1.1115	1.1088	1.1062	1.1034	1.0972	1.0908	1.0843	1.0700	1.0547
20	1.1255	1.1234	1.1206	1.1178	1.1150	1.1123	1.1094	1.1031	1.0966	1.0899	1.0754	1.0598
21	1.1322	1.1300	1.1271	1.1242	1.1213	1.1185	1.1155	1.1090	1.1024	1.0956	1.0808	1.0650
22	1.1389	1.1366	1.1336	1.1306	1.1276	1.1247	1.1217	1.1150	1.1083	1.1013	1.0862	1.0701
23	1.1457	1.1433	1.1402	1.1371	1.1340	1.1310	1.1280	1.1210	1.1142	1.1070	1.0917	1.0753
24	1.1525	1.1501	1.1469	1.1437	1.1404	1.1374	1.1343	1.1271	1.1201	1.1127	1.0972	1.0805
25	1.1594	1.1569	1.1536	1.1503	1.1469	1.1438	1.1406	1.1332	1.1260	1.1185	1.1027	1.0857
26	1.1663	1.1638	1.1603	1.1569	1.1534	1.1502	1.1469	1.1394	1.1320	1.1244	1.1083	1.0910
27	1.1733	1.1707	1.1670	1.1635	1.1600	1.1566	1.1533	1.1456	1.1381	1.1303	1.1139	1.0963
28	1.1803	1.1777	1.1738	1.1702	1.1666	1.1631	1.1597	1.1519	1.1442	1.1362	1.1195	1.1016
29	1.1874	1.1847	1.1807	1.1770	1.1733	1.1697	1.1662	1.1582	1.1503	1.1422	1.1251	1.1069
30	1.1945	1.1917	1.1876	1.1838	1.1800	1.1763	1.1727	1.1645	1.1564	1.1482	1.1307	1.1122
31	1.2016	1.1988	1.1945	1.1906	1.1867	1.1829	1.1792	1.1708	1.1625	1.1542	1.1363	1.1175
32	1.2088	1.2059	1.2014	1.1974	1.1934	1.1896	1.1857	1.1772	1.1687	1.1602	1.1419	1.1228
33	1.2160	1.2131	1.2084	1.2043	1.2002	1.1963	1.1922	1.1836	1.1749	1.1662	1.1476	1.1281
34	1.2233	1.2203	1.2155	1.2113	1.2071	1.2030	1.1988	1.1901	1.1812	1.1723	1.1533	1.1335
35	1.2306	1.2275	1.2227	1.2183	1.2140	1.2098	1.2055	1.1966	1.1876	1.1784	1.1591	1.1390
36	1.2375	1.2344	1.2294	1.2249	1.2205	1.2163	1.2119	1.2028	1.1936	1.1842	1.1645	1.1440
37	1.2444	1.2412	1.2361	1.2315	1.2270	1.2227	1.2182	1.2089	1.1995	1.1899	1.1699	1.1490
38	1.2513	1.2479	1.2428	1.2381	1.2335	1.2291	1.2245	1.2150	1.2054	1.1956	1.1752	1.1540
39	1.2581	1.2546	1.2494	1.2446	1.2399	1.2354	1.2308	1.2210	1.2112	1.2013	1.1805	1.1589
40	1.2649	1.2613	1.2560	1.2511	1.2463	1.2417	1.2370	1.2270	1.2170	1.2069	1.1858	1.1638
41	1.2717	1.2680	1.2626	1.2576	1.2527	1.2480	1.2432	1.2330	1.2229	1.2126	1.1911	1.1687
42	1.2786	1.2747	1.2692	1.2641	1.2591	1.2543	1.2494	1.2390	1.2287	1.2182	1.1963	1.1735
43	1.2854	1.2814	1.2758	1.2706	1.2655	1.2606	1.2556	1.2450	1.2345	1.2238	1.2015	1.1783
44	1.2922	1.2880	1.2824	1.2771	1.2719	1.2669	1.2618	1.2510	1.2403	1.2294	1.2067	1.1831
45	1.2990	1.2947	1.2890	1.2836	1.2783	1.2732	1.2680	1.2570	1.2461	1.2350	1.2119	1.1879
46	1.3058	1.3014	1.2955	1.2901	1.2847	1.2795	1.2742	1.2630	1.2519	1.2406	1.2171	1.1927
47	1.3126	1.3080	1.3021	1.2966	1.2911	1.2858	1.2804	1.2690	1.2577	1.2462	1.2223	1.1976
48	1.3194	1.3147	1.3087	1.3031	1.2975	1.2921	1.2867	1.2750	1.2635	1.2518	1.2275	1.2024
49	1.3263	1.3214	1.3153	1.3096	1.3040	1.2984	1.2929	1.2811	1.2693	1.2575	1.2328	1.2073
50	1.3327	1.3277	1.3215	1.3157	1.3100	1.3043	1.2987	1.2867	1.2748	1.2628	1.2377	1.2118
51	1.3391	1.3339	1.3277	1.3218	1.3160	1.3102	1.3045	1.2923	1.2802	1.2680	1.2425	1.2163
52	1.3454	1.3401	1.3338	1.3278	1.3219	1.3160	1.3102	1.2978	1.2856	1.2731	1.2473	1.2208
53	1.3517	1.3462	1.3399	1.3338	1.3278	1.3218	1.3159	1.3033	1.2909	1.2782	1.2521	1.2252
54	1.3579	1.3523	1.3459	1.3397	1.3336	1.3275	1.3215	1.3087	1.2961	1.2833	1.2568	1.2296
55	1.3640	1.3583	1.3518	1.3455	1.3393	1.3331	1.3270	1.3141	1.3013	1.2883	1.2615	1.2339
56	1.3700	1.3642	1.3576	1.3512	1.3449	1.3386	1.3324	1.3194	1.3064	1.2932	1.2661	1.2382
57	1.3759	1.3700	1.3634	1.3569	1.3505	1.3441	1.3377	1.3246	1.3114	1.2981	1.2706	1.2424
58	1.3818	1.3757	1.3691	1.3625	1.3560	1.3495	1.3430	1.3298	1.3164	1.3029	1.2751	1.2466
59	1.3875	1.3813	1.3747	1.3680	1.3614	1.3548	1.3482	1.3348	1.3213	1.3077	1.2795	1.2507
60	1.3931	1.3868	1.3801	1.3734	1.3667	1.3600	1.3533	1.3398	1.3261	1.3124	1.2839	1.2547
61	1.3986	1.3922	1.3855	1.3787	1.3719	1.3651	1.3583	1.3447	1.3308	1.3169	1.2881	1.2587
62	1.4039	1.3975	1.3907	1.3838	1.3769	1.3700	1.3632	1.3494	1.3354	1.3213	1.2922	1.2625
63	1.4091	1.4027	1.3958	1.3888	1.3818	1.3748	1.3679	1.3540	1.3398	1.3255	1.2962	1.2661
64		1.4078	1.4007	1.3936	1.3866	1.3795	1.3725					

TABLE 2-66 Nitric Acid (HNO₃) (*Concluded*)

%	0°C	5°C	10°C	15°C	20°C	25°C	30°C	40°C	50°C	60°C	80°C	100°C
65		1.4128	1.4055	1.3984	1.3913	1.3841	1.3770					
66		1.4177	1.4103	1.4031	1.3959	1.3887	1.3814					
67		1.4224	1.4150	1.4077	1.4004	1.3932	1.3857					
68		1.4271	1.4196	1.4122	1.4048	1.3976	1.3900					
69		1.4317	1.4241	1.4166	1.4091	1.4019	1.3942					
70		1.4362	1.4285	1.4210	1.4134	1.4061	1.3983					
71		1.4406	1.4328	1.4252	1.4176	1.4102	1.4023					
72		1.4449	1.4371	1.4294	1.4218	1.4142	1.4063					
73		1.4491	1.4413	1.4335	1.4258	1.4182	1.4103					
74		1.4532	1.4454	1.4376	1.4298	1.4221	1.4142					
75		1.4573	1.4494	1.4415	1.4337	1.4259	1.4180					
76		1.4613	1.4533	1.4454	1.4375	1.4296	1.4217					
77		1.4652	1.4572	1.4492	1.4413	1.4333	1.4253					
78		1.4690	1.4610	1.4529	1.4450	1.4369	1.4288					
79		1.4727	1.4647	1.4565	1.4486	1.4404	1.4323					
80		1.4764	1.4683	1.4601	1.4521	1.4439	1.4357					
81		1.4800	1.4718	1.4636	1.4555	1.4473	1.4391					
82		1.4835	1.4753	1.4670	1.4589	1.4507	1.4424					
83		1.4869	1.4787	1.4704	1.4622	1.4540	1.4456					
84		1.4903	1.4820	1.4737	1.4655	1.4572	1.4487					
85		1.4936	1.4852	1.4769	1.4686	1.4603	1.4518					
86		1.4968	1.4883	1.4799	1.4716	1.4633	1.4548					
87		1.4999	1.4913	1.4829	1.4745	1.4662	1.4577					
88		1.5029	1.4942	1.4858	1.4773	1.4690	1.4605					
89		1.5058	1.4970	1.4885	1.4800	1.4716	1.4631					
90		1.5085	1.4997	1.4911	1.4826	1.4741	1.4656					
91		1.5111	1.5023	1.4936	1.4850	1.4766	1.4681					
92		1.5136	1.5048	1.4960	1.4873	1.4789	1.4704					
93		1.5156	1.5068	1.4979	1.4892	1.4807	1.4722					
94		1.5177	1.5088	1.4999	1.4912	1.4826	1.4741					
95		1.5198	1.5109	1.5019	1.4932	1.4846	1.4761					
96		1.5220	1.5130	1.5040	1.4952	1.4867	1.4781					
97		1.5244	1.5152	1.5062	1.4974	1.4889	1.4802					
98		1.5278	1.5187	1.5096	1.5008	1.4922	1.4835					
99		1.5327	1.5235	1.5144	1.5056	1.4969	1.4881					
100		1.5402	1.5310	1.5217	1.5129	1.5040	1.4952					

TABLE 2-67 Perchloric Acid (HClO₄)

%	d_4^{15}	d_4^{20}	d_4^{25}	d_4^{50}	%	d_4^{15}	d_4^{20}	d_4^{50}
1	1.0050		1.0020	0.9933	28	1.1900	1.1851	1.1645
2	1.0109		1.0070	0.9986	30	1.2067	1.2013	1.1800
4	1.0228		1.0169	0.9906	32	1.2239	1.2183	1.1960
6	1.0348		1.0270	1.0205	34	1.2418	1.2359	1.2130
8	1.0471		1.0372	1.0320	36	1.2603	1.2542	1.2310
10	1.0597		1.0475	1.0440	38	1.2794	1.2732	1.2490
12	1.0726			1.0560	40	1.2991	1.2927	1.2680
14	1.0859			1.0680	45	1.3521	1.3450	1.3180
16	1.0995			1.0810	50	1.4103	1.4018	1.3730
18	1.1135			1.0940	55	1.4733	1.4636	1.4320
20	1.1279			1.1070	60	1.5389	1.5298	1.4950
22	1.1428			1.1205	65	1.6059	1.5986	1.5620
24	1.1581			1.1345	70	1.6736	1.6680	1.6290
26	1.1738	1.1697		1.1490				

TABLE 2-68 Phosphoric Acid (H₃PO₄)

°C	2%	6%	14%	20%	26%	35%	50%	75%	100%
0	1.0113	1.0339	1.0811	1.1192					
10	1.0109	1.0330	1.0792	1.1167	1.1567	1.221	1.341		
20	1.0092	1.0309	1.0764	1.1134	1.1529	1.216	1.335	1.579	1.870
30	1.0065	1.0279	1.0728	1.1094	1.1484	1.211	1.329	1.572	1.862
40	1.0029	1.0241	1.0685	1.1048					

TABLE 2-69 Potassium Bicarbonate (KHCO₃)

°C	1%	2%	4%	6%	8%	10%
0	1.0066	1.0134	1.0270			
10	1.0064	1.0132	1.0268			
15	1.0058	1.0125	1.0260	1.0396	1.0534	1.0674
20	1.0049	1.0117	1.0252			
30	1.0024	1.0092	1.0228			
40	0.9990	1.0058	1.0195			
50	.9949	1.0017	1.0154			
60	.9901	.9969	1.0106			
80	.9786	.9855	0.9993			
100	.9653	.9722	.9860			

TABLE 2-70 Potassium Bromide (KBr)

%	d_4^{20}
1	1.0054
2	1.0127
6	1.0426
12	1.0903
20	1.1601
30	1.2593
40	1.3746

TABLE 2-71 Potassium Carbonate (K_2CO_3)

%	0°C	10°C	20°C	40°C	60°C	80°C	100°C
1	1.0094	1.0089	1.0072	1.0010	0.9919	0.9803	0.9670
2	1.0189	1.0182	1.0163	1.0098	1.0005	.9889	.9756
4	1.0381	1.0369	1.0345	1.0276	1.0180	1.0063	.9951
8	1.0768	1.0746	1.0715	1.0640	1.0538	1.0418	1.0291
12	1.1160	1.1131	1.1096	1.1013	1.0906	1.0786	1.0663
16	1.1562	1.1530	1.1490	1.1399	1.1290	1.1170	1.1049
20	1.1977	1.1941	1.1898	1.1801	1.1690	1.1570	1.1451
24	1.2405	1.2366	1.2320	1.2219	1.2106	1.1986	1.1869
28	1.2846	1.2804	1.2756	1.2652	1.2538	1.2418	1.2301
30	1.3071	1.3028	1.2979	1.2873	1.2759	1.2640	1.2522
35	1.3646	1.3600	1.3548	1.3440	1.3324	1.3206	1.3089
40	1.4244	1.4195	1.4141	1.4029	1.3913	1.3795	1.3678
45	1.4867	1.4815	1.4759	1.4644	1.4528	1.4408	1.4290
50	1.5517	1.5462	1.5404	1.5285	1.5169	1.5048	1.4928

TABLE 2-72 Potassium Chromate (K_2CrO_4)

%	d_4^{15}	d_4^{18}
1	1.0073	1.0066
2	1.0155	1.0147
4	1.0321	1.0311
8	1.0659	1.0647
12	1.1009	1.0999
16		1.1366
20		1.1748
24		1.2147
28		1.2566
30		1.2784

TABLE 2-73 Potassium Chlorate ($KClO_3$)

°C	1%	2%	3%	4%
0	1.0061	1.0124	1.0189	1.0256
10	1.0059	1.0122	1.0187	1.0254
20	1.0045	1.0109	1.0174	1.0241
30	1.0020	1.0085	1.0151	1.0218
40	0.9986	1.0051	1.0116	1.0183
60	.9895	.9959	1.0024	1.0091
80	.9781	.9845	0.9910	0.9977
100	.9646	.9709	.9774	.9840

TABLE 2-74 Potassium Chloride (KCl)

%	0°C	20°C	25°C	40°C	60°C	80°C	100°C
1.0	1.00661	1.00462	1.00342	0.99847	0.9894	0.9780	0.9646
2.0	1.01335	1.01103	1.00977	1.00471	.9956	.9842	.9708
4.0	1.02690	1.02391	1.02255	1.01727	1.0080	.9966	.9634
8.0	1.05431	1.05003	1.04847	1.04278	1.0333	1.0219	1.0888
12.0	1.08222	1.07679	1.07506	1.06897	1.0592	1.0478	1.0350
16.0	1.11068	1.10434	1.10245	1.09600	1.0861	1.0746	1.0619
20.0	1.13973	1.13280	1.13072	1.12399	1.1138	1.1024	1.0897
24.0		1.16226	1.15995	1.15299	1.1425	1.1311	1.1185
28.0				1.18304	1.1723	1.1609	1.1483

%	110°C	120°C	130°C	140°C
3.79	0.9733	0.9663	0.9583	0.9502
7.45	.9978	.9899	.9827	.9745
13.62	1.0388	1.0313	1.0238	1.0159

TABLE 2-75 Potassium Chrome Alum [$K_2Cr_2(SO_4)_4$]

%	d_4^{15}
1	1.007
2	1.016
6	1.052
10	1.089
14	1.129
20	1.193
30	1.315
40	1.456
50	1.615

TABLE 2-76 Potassium Hydroxide (KOH)

%	d_4^{15}
1.0	1.0083
2.0	1.0175
4.0	1.0359
6.0	1.0544
8.0	1.0730
10.0	1.0918
15.0	1.1396
20.0	1.1884
25.0	1.2387
30.0	1.2905
35.0	1.3440
40.0	1.3991
45.0	1.4558
50.0	1.5143
51.7	1.5355 (sat'd. soln.)

TABLE 2-77 Potassium Nitrate (KNO_3)

%	0°C	10°C	20°C	40°C	60°C	80°C	100°C
1	1.00654	1.00615	1.00447	0.99825	0.9890	0.9776	0.9641
2	1.01326	1.01262	1.01075	1.00430	.9949	.9834	.9699
4	1.02677	1.02566	1.02344	1.01652	1.0068	.9951	.9816
8	1.05419	1.05226	1.04940	1.04152	1.0313	1.0192	1.0056
12	1.08221	1.07963	1.07620	1.06740	1.0567	1.0442	1.0304
16			1.10392	1.09432	1.0831	1.0703	1.0562
20			1.13261	1.12240	1.1106	1.0974	1.0831
24			1.16233	1.15175	1.1391	1.1256	1.1110

TABLE 2-78 Potassium Dichromate ($K_2Cr_2O_7$)

%	d_4^{20}
1	1.0052
2	1.0122
4	1.0264
6	1.0408
8	1.0554
10	1.0703

TABLE 2-79 Potassium Sulfate (K_2SO_4)

%	d_4^{20}
1	1.0063
2	1.0145
4	1.0310
6	1.0477
8	1.0646
10	1.0817

TABLE 2-80 Potassium Sulfite (K_2SO_3)

%	d_4^{15}
1	1.0073
2	1.0155
4	1.0322
8	1.0667
12	1.1026
16	1.1402
20	1.1793
24	1.2197
26	1.2404

TABLE 2-81 Sodium Acetate ($NaC_2H_3O_2$)

%	d_4^{20}
1	1.0033
2	1.0084
4	1.0186
8	1.0392
12	1.0598
18	1.0807
20	1.1021
26	1.1351
28	1.1462

TABLE 2-82 Sodium Arsenate (Na_3AsO_4)

%	d_4^{17}
1	1.0097
2	1.0207
4	1.0431
8	1.0892
10	1.1130
12	1.1373

TABLE 2-83 Sodium Bichromate ($Na_2Cr_2O_7$)

%	d_4^{15}
1	1.006
2	1.013
4	1.027
8	1.056
12	1.084
16	1.112
20	1.140
24	1.166
28	1.193
30	1.207
35	1.244
40	1.279
45	1.312
50	1.342

TABLE 2-84 Sodium Bromide (NaBr)

%	d_4^{17}
1	1.0060
2	1.0139
4	1.0298
8	1.0631
10	1.0803
12	1.0981
20	1.1745
30	1.2841
40	1.4138

TABLE 2-85 Sodium Formate (HCOONa)

%	d_4^{25}
1	1.003
2	1.009
4	1.022
8	1.048
12	1.074
16	1.100
20	1.127
24	1.155
28	1.184
30	1.199
35	1.236
40	1.274

TABLE 2-86   Sodium Carbonate (Na₂CO₃)

%	0°C	10°C	20°C	30°C	40°C	60°C	80°C	100°C
1	1.0109	1.0103	1.0086	1.0058	1.0022	0.9929	0.9814	0.9683
2	1.0219	1.0210	1.0190	1.0159	1.0122	1.0027	.9910	.9782
4	1.0439	1.0423	1.0398	1.0363	1.0323	1.0223	1.0105	.9980
8	1.0878	1.0850	1.0816	1.0775	1.0732	1.0625	1.0503	1.0380
12	1.1319	1.1284	1.1244	1.1200	1.1150	1.1039	1.0914	1.0787
14	1.1543	1.1506	1.1463	1.1417	1.1365	1.1251	1.1125	1.0996
16				1.1636				
18				1.1859				
20				1.2086				
24				1.2552				
28				1.3031				
30				1.3274				

TABLE 2-87   Sodium Chlorate (NaClO₃)

%	d_4^{18}	%	d_4^{18}
1	1.0053	18	1.1288
2	1.0121	20	1.1449
4	1.0258	22	1.1614
6	1.0397	24	1.1782
8	1.0538	26	1.1953
10	1.0681	28	1.2128
12	1.0827	30	1.2307
14	1.0977	32	1.2491
16	1.1131	34	1.2680

TABLE 2-88   Sodium Chloride (NaCl)

%	0°C	10°C	25°C	40°C	60°C	80°C	100°C
1	1.00747	1.00707	1.00409	0.99908	0.9900	0.9785	0.9651
2	1.01509	1.01442	1.01112	1.00593	.9967	.9852	.9719
4	1.03038	1.02920	1.02530	1.01977	1.0103	.9988	.9855
8	1.06121	1.05907	1.05412	1.04798	1.0381	1.0264	1.0134
12	1.09244	1.08946	1.08365	1.07699	1.0667	1.0549	1.0420
16	1.12419	1.12056	1.11401	1.10688	1.0962	1.0842	1.0713
20	1.15663	1.15254	1.14533	1.13774	1.1268	1.1146	1.1017
24	1.18999	1.18557	1.17776	1.16971	1.1584	1.1463	1.1331
26	1.20709	1.20254	1.19443	1.18614	1.1747	1.1626	1.1492

TABLE 2-89   Sodium Chromate (Na₂CrO₄)

%	d_4^{18}
1	1.0074
2	1.0164
4	1.0344
8	1.0718
12	1.1110
16	1.1518
20	1.1942
24	1.2383
26	1.2611

TABLE 2-90   Sodium Hydroxide (NaOH)

%	0°C	15°C	20°C	40°C	60°C	80°C	100°C
1	1.0124	1.01065	1.0095	1.0033	0.9941	0.9824	0.9693
2	1.0244	1.02198	1.0207	1.0139	1.0045	.9929	.9797
4	1.0482	1.04441	1.0428	1.0352	1.0254	1.0139	1.0009
8	1.0943	1.08887	1.0869	1.0780	1.0676	1.0560	1.0432
12	1.1399	1.13327	1.1309	1.1210	1.1101	1.0983	1.0855
16	1.1849	1.17761	1.1751	1.1645	1.1531	1.1408	1.1277
20	1.2296	1.22183	1.2191	1.2079	1.1960	1.1833	1.1700
24	1.2741	1.26582	1.2629	1.2512	1.2388	1.2259	1.2124
28	1.3182	1.3094	1.3064	1.2942	1.2814	1.2682	1.2546
32	1.3614	1.3520	1.3490	1.3362	1.3232	1.3097	1.2960
36	1.4030	1.3933	1.3900	1.3768	1.3634	1.3498	1.3360
40	1.4435	1.4334	1.4300	1.4164	1.4027	1.3889	1.3750
44	1.4825	1.4720	1.4685	1.4545	1.4405	1.4266	1.4127
48	1.5210	1.5102	1.5065	1.4922	1.4781	1.4641	1.4503
50	1.5400	1.5290	1.5253	1.5109	1.4967	1.4827	1.4690

TABLE 2-91   Sodium Nitrate (NaNO₃)

%	0°C	20°C	40°C	60°C	80°C	100°C
1	1.0071	1.0049	0.9986	0.9894	0.9779	0.9644
2	1.0144	1.0117	1.0050	.9956	.9840	.9704
4	1.0290	1.0254	1.0180	1.0082	.9964	.9826
8	1.0587	1.0532	1.0447	1.0340	1.0218	1.0078
12	1.0891	1.0819	1.0724	1.0609	1.0481	1.0340
16	1.1203	1.1118	1.1013	1.0892	1.0757	1.0614
20	1.1526	1.1429	1.1314	1.1187	1.1048	1.0901
24	1.1860	1.1752	1.1629	1.1496	1.1351	1.1200
28	1.2204	1.2085	1.1955	1.1816	1.1667	1.1513
30	1.2380	1.2256	1.2122	1.1980	1.1830	1.1674
35	1.2834	1.2701	1.2560	1.2413	1.2258	1.2100
40	1.3316	1.3175	1.3027	1.2875	1.2715	1.2555
45		1.3683	1.3528	1.3371	1.3206	1.3044

TABLE 2-92   Sodium Nitrite (NaNO₂)

%	d_4^{15}
1	1.0058
2	1.0125
4	1.0260
8	1.0535
12	1.0816
16	1.1103
20	1.1394

TABLE 2-93   Sodium Silicates

Formula	Concentration, %												
	1	2	4	8	10	14	20	24	30	36	40	45	50
	d_4^{20}												
Na₂O/3.9SiO₂	1.006	1.014	1.030	1.063	1.080	1.116	1.172	1.211	1.275				
Na₂O/3.36SiO₂	1.006	1.014	1.030	1.065	1.083	1.120	1.179	1.222	1.290	1.365			
Na₂O/2.40SiO₂	1.007	1.016	1.034	1.071	1.090	1.130							
Na₂O/2.44SiO₂									1.309	1.387	1.445		
Na₂O/2.06SiO₂	1.007	1.016	1.035	1.073	1.093	1.134	1.200	1.247	1.321	1.397	1.450	1.520	1.594
Na₂O/1.69SiO₂	1.007	1.017	1.036	1.077	1.098	1.141	1.210	1.259	1.337	1.424			

TABLE 2-94   Sodium Sulfate (Na$_2$SO$_4$)

%	0°C	20°C	30°C	40°C	60°C	80°C	100°C
1	1.0094	1.0073	1.0046	1.0010	0.9919	0.9805	0.9671
2	1.0189	1.0164	1.0135	1.0098	1.0007	.9892	.9758
4	1.0381	1.0348	1.0315	1.0276	1.0184	1.0068	.9934
8	1.0773	1.0724	1.0682	1.0639	1.0544	1.0426	1.0292
12	1.1174	1.1109	1.1062	1.1015	1.0915	1.0795	1.0661
16	1.1585	1.1586	1.1456	1.1406	1.1299	1.1176	1.1042
20	1.2008	1.1915	1.1865	1.1813	1.1696	1.1569	
24	1.2443	1.2336	1.2292	1.2237			

TABLE 2-95   Sodium Sulfide (Na$_2$S)

%	d_4^{18}
1	1.0098
2	1.0211
4	1.0440
8	1.0907
12	1.1388
16	1.1885
18	1.2140

TABLE 2-96   Sodium Sulfite (Na$_2$SO$_3$)

%	d_4^{19}
1	1.0078
2	1.0172
4	1.0363
8	1.0751
12	1.1146
16	1.1549
18	1.1755

TABLE 2-97   Sodium Thiosulfate (Na$_2$S$_2$O$_3$)

%	d_4^{20}
1	1.0065
2	1.0148
4	1.0315
8	1.0654
12	1.1003
16	1.1365
20	1.1740
24	1.2128
28	1.2532
30	1.2739
35	1.3273
40	1.3827

TABLE 2-98 Sodium Thiosulfate Pentahydrate (Na$_2$S$_2$O$_3$·5H$_2$O)

%	d_4^{19}
1	1.0052
2	1.0105
4	1.0211
8	1.0423
12	1.0639
16	1.0863
20	1.1087
24	1.1322
28	1.1558
30	1.1676
40	1.2297
50	1.2954

TABLE 2-99   Stannic Chloride (SnCl$_4$)

%	d_4^{15}
1	1.007
2	1.015
4	1.031
8	1.064
12	1.099
16	1.135
20	1.173
24	1.212
28	1.255
30	1.278
35	1.337
40	1.403
45	1.475
50	1.555
55	1.644
60	1.742
65	1.851
70	1.971

TABLE 2-100 Stannous Chloride (SnCl$_2$)

%	d_4^{15}
1	1.0068
2	1.0146
4	1.0306
8	1.0638
12	1.0986
16	1.1353
20	1.1743
24	1.2159
28	1.2603
30	1.2837
35	1.3461
40	1.4145
45	1.4897
50	1.5729
55	1.6656
60	1.7695
65	1.8865

TABLE 2-101 Sulfuric Acid (H₂SO₄)

%	0°C	10°C	15°C	20°C	25°C	30°C	40°C	50°C	60°C	80°C	100°C
1	1.0074	1.0068	1.0060	1.0051	1.0038	1.0022	0.9986	0.9944	0.9895	0.9779	0.9645
2	1.0147	1.0138	1.0129	1.0118	1.0104	1.0087	1.0050	1.0006	.9956	.9839	.9705
3	1.0219	1.0206	1.0197	1.0184	1.0169	1.0152	1.0113	1.0067	1.0017	.9900	.9766
4	1.0291	1.0275	1.0264	1.0250	1.0234	1.0216	1.0176	1.0129	1.0078	.9961	.9827
5	1.0364	1.0344	1.0332	1.0317	1.0300	1.0281	1.0240	1.0192	1.0140	1.0022	.9888
6	1.0437	1.0414	1.0400	1.0385	1.0367	1.0347	1.0305	1.0256	1.0203	1.0084	.9950
7	1.0511	1.0485	1.0469	1.0453	1.0434	1.0414	1.0371	1.0321	1.0266	1.0146	1.0013
8	1.0585	1.0556	1.0539	1.0522	1.0502	1.0481	1.0437	1.0386	1.0330	1.0209	1.0076
9	1.0660	1.0628	1.0610	1.0591	1.0571	1.0549	1.0503	1.0451	1.0395	1.0273	1.0140
10	1.0735	1.0700	1.0681	1.0661	1.0640	1.0617	1.0570	1.0517	1.0460	1.0338	1.0204
11	1.0810	1.0773	1.0753	1.0731	1.0710	1.0686	1.0637	1.0584	1.0526	1.0403	1.0269
12	1.0886	1.0846	1.0825	1.0802	1.0780	1.0756	1.0705	1.0651	1.0593	1.0469	1.0335
13	1.0962	1.0920	1.0898	1.0874	1.0851	1.0826	1.0774	1.0719	1.0661	1.0536	1.0402
14	1.1039	1.0994	1.0971	1.0947	1.0922	1.0897	1.0844	1.0788	1.0729	1.0603	1.0469
15	1.1116	1.1069	1.1045	1.1020	1.0994	1.0968	1.0914	1.0857	1.0798	1.0671	1.0537
16	1.1194	1.1145	1.1120	1.1094	1.1067	1.1040	1.0985	1.0927	1.0868	1.0740	1.0605
17	1.1272	1.1221	1.1195	1.1168	1.1141	1.1113	1.1057	1.0998	1.0938	1.0809	1.0674
18	1.1351	1.1298	1.1271	1.1243	1.1215	1.1187	1.1129	1.1070	1.1009	1.0879	1.0744
19	1.1430	1.1375	1.1347	1.1318	1.1290	1.1261	1.1202	1.1142	1.1081	1.0950	1.0814
20	1.1510	1.1453	1.1424	1.1394	1.1365	1.1335	1.1275	1.1215	1.1153	1.1021	1.0885
21	1.1590	1.1531	1.1501	1.1471	1.1441	1.1410	1.1349	1.1288	1.1226	1.1093	1.0957
22	1.1670	1.1609	1.1579	1.1548	1.1517	1.1486	1.1424	1.1362	1.1299	1.1166	1.1029
23	1.1751	1.1688	1.1657	1.1626	1.1594	1.1563	1.1500	1.1437	1.1373	1.1239	1.1102
24	1.1832	1.1768	1.1736	1.1704	1.1672	1.1640	1.1576	1.1512	1.1448	1.1313	1.1176
25	1.1914	1.1848	1.1816	1.1783	1.1750	1.1718	1.1653	1.1588	1.1523	1.1388	1.1250
26	1.1996	1.1929	1.1896	1.1862	1.1829	1.1796	1.1730	1.1665	1.1599	1.1463	1.1325
27	1.2078	1.2010	1.1976	1.1942	1.1909	1.1875	1.1808	1.1742	1.1676	1.1539	1.1400
28	1.2160	1.2091	1.2057	1.2023	1.1989	1.1955	1.1887	1.1820	1.1753	1.1616	1.1476
29	1.2243	1.2173	1.2138	1.2104	1.2069	1.2035	1.1966	1.1898	1.1831	1.1693	1.1553
30	1.2326	1.2255	1.2220	1.2185	1.2150	1.2115	1.2046	1.1977	1.1909	1.1771	1.1630
31	1.2409	1.2338	1.2302	1.2267	1.2232	1.2196	1.2126	1.2057	1.1988	1.1849	1.1708
32	1.2493	1.2421	1.2385	1.2349	1.2314	1.2278	1.2207	1.2137	1.2068	1.1928	1.1787
33	1.2577	1.2504	1.2468	1.2432	1.2396	1.2360	1.2289	1.2218	1.2148	1.2008	1.1866
34	1.2661	1.2588	1.2552	1.2515	1.2479	1.2443	1.2371	1.2300	1.2229	1.2088	1.1946
35	1.2746	1.2672	1.2636	1.2599	1.2563	1.2526	1.2454	1.2383	1.2311	1.2169	1.2027
36	1.2831	1.2757	1.2720	1.2684	1.2647	1.2610	1.2538	1.2466	1.2394	1.2251	1.2109
37	1.2917	1.2843	1.2805	1.2769	1.2732	1.2695	1.2622	1.2550	1.2477	1.2334	1.2192
38	1.3004	1.2929	1.2891	1.2855	1.2818	1.2780	1.2707	1.2635	1.2561	1.2418	1.2276
39	1.3091	1.3016	1.2978	1.2941	1.2904	1.2866	1.2793	1.2720	1.2646	1.2503	1.2361
40	1.3179	1.3103	1.3065	1.3028	1.2991	1.2953	1.2880	1.2806	1.2732	1.2589	1.2446
41	1.3268	1.3191	1.3153	1.3116	1.3079	1.3041	1.2967	1.2893	1.2819	1.2675	1.2532
42	1.3357	1.3280	1.3242	1.3205	1.3167	1.3129	1.3055	1.2981	1.2907	1.2762	1.2619
43	1.3447	1.3370	1.3332	1.3294	1.3256	1.3218	1.3144	1.3070	1.2996	1.2850	1.2707
44	1.3538	1.3461	1.3423	1.3384	1.3346	1.3308	1.3234	1.3160	1.3086	1.2939	1.2796
45	1.3630	1.3553	1.3515	1.3476	1.3437	1.3399	1.3325	1.3251	1.3177	1.3029	1.2886
46	1.3724	1.3646	1.3608	1.3569	1.3530	1.3492	1.3417	1.3343	1.3269	1.3120	1.2976
47	1.3819	1.3740	1.3702	1.3663	1.3624	1.3586	1.3510	1.3435	1.3362	1.3212	1.3067
48	1.3915	1.3835	1.3797	1.3758	1.3719	1.3680	1.3604	1.3528	1.3455	1.3305	1.3159
49	1.4012	1.3931	1.3893	1.3854	1.3814	1.3775	1.3699	1.3623	1.3549	1.3399	1.3253
50	1.4110	1.4029	1.3990	1.3951	1.3911	1.3872	1.3795	1.3719	1.3644	1.3494	1.3348
51	1.4209	1.4128	1.4088	1.4049	1.4009	1.3970	1.3893	1.3816	1.3740	1.3590	1.3444
52	1.4310	1.4228	1.4188	1.4148	1.4109	1.4069	1.3991	1.3914	1.3837	1.3687	1.3540
53	1.4412	1.4329	1.4289	1.4248	1.4209	1.4169	1.4091	1.4013	1.3936	1.3785	1.3637
54	1.4515	1.4431	1.4391	1.4350	1.4310	1.4270	1.4191	1.4113	1.4036	1.3884	1.3735
55	1.4619	1.4535	1.4494	1.4453	1.4412	1.4372	1.4293	1.4214	1.4137	1.3984	1.3834
56	1.4724	1.4640	1.4598	1.4557	1.4516	1.4475	1.4396	1.4317	1.4239	1.4085	1.3934
57	1.4830	1.4746	1.4703	1.4662	1.4621	1.4580	1.4500	1.4420	1.4342	1.4187	1.4035
58	1.4937	1.4852	1.4809	1.4768	1.4726	1.4685	1.4604	1.4524	1.4446	1.4290	1.4137
59	1.5045	1.4959	1.4916	1.4875	1.4832	1.4791	1.4709	1.4629	1.4551	1.4393	1.4240
60	1.5154	1.5067	1.5024	1.4983	1.4940	1.4898	1.4816	1.4735	1.4656	1.4497	1.4344
61	1.5264	1.5177	1.5133	1.5091	1.5048	1.5006	1.4923	1.4842	1.4762	1.4602	1.4449
62	1.5375	1.5287	1.5243	1.5200	1.5157	1.5115	1.5031	1.4950	1.4869	1.4708	1.4554
63	1.5487	1.5398	1.5354	1.5310	1.5267	1.5225	1.5140	1.5058	1.4977	1.4815	1.4660
64	1.5600	1.5510	1.5465	1.5421	1.5378	1.5335	1.5250	1.5167	1.5086	1.4923	1.4766

TABLE 2-101 Sulfuric Acid (H₂SO₄) (Concluded)

%	0°C	10°C	15°C	20°C	25°C	30°C	40°C	50°C	60°C	80°C	100°C
65	1.5714	1.5623	1.5578	1.5533	1.5490	1.5446	1.5361	1.5277	1.5195	1.5031	1.4873
66	1.5828	1.5736	1.5691	1.5646	1.5602	1.5558	1.5472	1.5388	1.5305	1.5140	1.4981
67	1.5943	1.5850	1.5805	1.5760	1.5715	1.5671	1.5584	1.5499	1.5416	1.5249	1.5089
68	1.6059	1.5965	1.5920	1.5874	1.5829	1.5785	1.5697	1.5611	1.5528	1.5359	1.5198
69	1.6176	1.6081	1.6035	1.5989	1.5944	1.5899	1.5811	1.5724	1.5640	1.5470	1.5307
70	1.6293	1.6198	1.6151	1.6105	1.6059	1.6014	1.5925	1.5838	1.5753	1.5582	1.5417
71	1.6411	1.6315	1.6268	1.6221	1.6175	1.6130	1.6040	1.5952	1.5867	1.5694	1.5527
72	1.6529	1.6433	1.6385	1.6338	1.6292	1.6246	1.6155	1.6067	1.5981	1.5806	1.5637
73	1.6648	1.6551	1.6503	1.6456	1.6409	1.6363	1.6271	1.6182	1.6095	1.5919	1.5747
74	1.6768	1.6670	1.6622	1.6574	1.6526	1.6480	1.6387	1.6297	1.6209	1.6031	1.5857
75	1.6888	1.6789	1.6740	1.6692	1.6644	1.6597	1.6503	1.6412	1.6322	1.6142	1.5966
76	1.7008	1.6908	1.6858	1.6810	1.6761	1.6713	1.6619	1.6526	1.6435	1.6252	1.6074
77	1.7128	1.7026	1.6976	1.6927	1.6878	1.6829	1.6734	1.6640	1.6547	1.6361	1.6181
78	1.7247	1.7144	1.7093	1.7043	1.6994	1.6944	1.6847	1.6751	1.6657	1.6469	1.6286
79	1.7365	1.7261	1.7209	1.7158	1.7108	1.7058	1.6959	1.6862	1.6766	1.6575	1.6390
80	1.7482	1.7376	1.7323	1.7272	1.7221	1.7170	1.7069	1.6971	1.6873	1.6680	1.6493
81	1.7597	1.7489	1.7435	1.7383	1.7331	1.7279	1.7177	1.7077	1.6978	1.6782	1.6594
82	1.7709	1.7599	1.7544	1.7491	1.7437	1.7385	1.7281	1.7180	1.7080	1.6882	1.6692
83	1.7815	1.7704	1.7649	1.7594	1.7540	1.7487	1.7382	1.7279	1.7179	1.6979	1.6787
84	1.7916	1.7804	1.7748	1.7693	1.7639	1.7585	1.7479	1.7375	1.7274	1.7072	1.6878
85	1.8009	1.7897	1.7841	1.7786	1.7732	1.7678	1.7571	1.7466	1.7364	1.7161	1.6966
86	1.8095	1.7983	1.7927	1.7872	1.7818	1.7763	1.7657	1.7552	1.7449	1.7245	1.7050
87	1.8173	1.8061	1.8006	1.7951	1.7897	1.7842	1.7736	1.7632	1.7529	1.7324	1.7129
88	1.8243	1.8132	1.8077	1.8022	1.7968	1.7914	1.7809	1.7705	1.7602	1.7397	1.7202
89	1.8306	1.8195	1.8141	1.8087	1.8033	1.7979	1.7874	1.7770	1.7669	1.7464	1.7269
90	1.8361	1.8252	1.8198	1.8144	1.8091	1.8038	1.7933	1.7829	1.7729	1.7525	1.7331
91	1.8410	1.8302	1.8248	1.8195	1.8142	1.8090	1.7986	1.7883	1.7783	1.7581	1.7388
92	1.8453	1.8346	1.8293	1.8240	1.8188	1.8136	1.8033	1.7932	1.7832	1.7633	1.7439
93	1.8490	1.8384	1.8331	1.8279	1.8227	1.8176	1.8074	1.7974	1.7876	1.7681	1.7485
94	1.8520	1.8415	1.8363	1.8312	1.8260	1.8210	1.8109	1.8011	1.7914		
95	1.8544	1.8439	1.8388	1.8337	1.8286	1.8236	1.8137	1.8040	1.7944		
96	1.8560	1.8457	1.8406	1.8355	1.8305	1.8255	1.8157	1.8060	1.7965		
97	1.8569	1.8466	1.8414	1.8364	1.8314	1.8264	1.8166	1.8071	1.7977		
98	1.8567	1.8463	1.8411	1.8361	1.8310	1.8261	1.8163	1.8068	1.7976		
99	1.8551	1.8445	1.8393	1.8342	1.8292	1.8242	1.8145	1.8050	1.7958		
100	1.8517	1.8409	1.8357	1.8305	1.8255	1.8205	1.8107	1.8013	1.7922		

%	$d_4^{5.96}$	%	$d_4^{13.00}$	$d_4^{18.00}$
0.005	1.000 0140	0.05	0.999 810	0.999 028
.01	1.000 0576	.1	1.000 185	.999 400
.02	1.000 1434	.2	1.000 912	1.000 119
.03	1.000 2276	.3	1.001 623	1.000 820
.04	1.000 3104	.4	1.002 326	1.001 512
.05	1.000 3920	.5	1.003 023	1.002 197
.06	1.000 4726	.6	1.003 716	1.002 877
.07	1.000 5523	.8	1.005 090	1.004 227
.08	1.000 6313	1.0	1.006 452	1.005 570
.09	1.000 7098	1.2	1.007 807	1.006 909
.10	1.000 7880	1.4	1.009 159	1.008 247
.15	1.001 1732	1.6	1.010 510	1.009 583
.20	1.001 5514	1.8	1.011 860	1.010 918
.25	1.001 9254	2.0	1.013 209	1.012 252
.30	1.002 2961	2.2	1.014 557	1.013 586
.35	1.002 6639	2.4	1.015 904	1.014 919
.40	1.003 0292			
.45	1.003 3923			
.50	1.003 7534			

TABLE 2-102   Zinc Bromide (ZnBr₂)

%	0°C	20°C	40°C	60°C	80°C	100°C
2	1.0188	1.0167	1.0102	1.0008	0.9890	0.9751
4	1.0381	1.0354	1.0285	1.0187	1.0065	0.9921
8	1.0777	1.0738	1.0660	1.0554	1.0422	1.0270
12	1.1186	1.1135	1.1046	1.0932	1.0789	1.0629
16	1.1609	1.1544	1.1445	1.1320	1.1169	1.1000
20	1.2043	1.1965	1.1855	1.1720	1.1560	1.1382
30	1.3288	1.3170	1.3030	1.2868	1.2688	1.2489
40	1.477	1.462	1.445	1.427	1.406	1.385
50	1.661	1.643	1.623	1.602	1.579	1.555
60	1.891	1.869	1.845	1.822	1.797	1.771
65	2.026	2.002	1.976	1.951	1.924	1.898

TABLE 2-103   Zinc Chloride (ZnCl₂)

%	0°C	20°C	40°C	60°C	80°C	100°C
2	1.0192	1.0167	1.0099	1.0003	0.9882	0.9739
4	1.0384	1.0350	1.0274	1.0172	1.0044	.9894
8	1.0769	1.0715	1.0624	1.0508	1.0369	1.0211
12	1.1159	1.1085	1.0980	1.0853	1.0704	1.0541
16	1.1558	1.1468	1.1350	1.1212	1.1055	1.0888
20	1.1970	1.1866	1.1736	1.1590	1.1428	1.1255
30	1.3062	1.2928	1.2778	1.2614	1.2438	1.2252
40	1.4329	1.4173	1.4003	1.3824	1.3637	1.3441
50	1.5860	1.5681	1.5495	1.5300	1.5097	1.4892
60		1.749				
70		1.962				

TABLE 2-104   Zinc Nitrate [Zn(NO₃)₂]

%	18°C	%	18°C
2	1.0154	18	1.1652
4	1.0322	20	1.1865
6	1.0496	25	1.2427
8	1.0675	30	1.3029
10	1.0859	35	1.3678
12	1.1048	40	1.4378
14	1.1244	45	1.5134
16	1.1445	50	1.5944

TABLE 2-105   Zinc Sulfate (ZnSO₄)

%	20°C
2	1.019
4	1.0403
6	1.0620
8	1.0842
10	1.1071
12	1.1308
14	1.1553
16	1.1806

DENSITIES OF AQUEOUS ORGANIC SOLUTIONS*

UNITS AND UNITS CONVERSIONS

Unless otherwise noted, densities are given in grams per cubic centimeter. To convert to pounds per cubic foot, multiply by 62.43.

$$°F = ⅗ °C + 32$$

From *International Critical Tables*, vol. 3, pp. 115–129. All compositions are in weight percent in vacuo. All density values are $d_4^t = g/mL$ in vacuo.

*For gasoline and aircraft fuels see Hibbard, NACA Res. Mem. E56I21 (declassified 1958).

TABLE 2-106   Formic Acid (HCOOH)

%	0°C	15°C	20°C	30°C	%	0°C	15°C	20°C	30°C	%	0°C	15°C	20°C	30°C	%	0°C	15°C	20°C	30°C
0	0.9999	0.9991	0.9982	0.9957	25	1.0706	1.0627	1.0609	1.0540	50	1.1349	1.1225	1.1207	1.1098	75	1.1953	1.1794	1.1769	1.1636
1	1.0028	1.0019	1.0019	0.9980	26	1.0733	1.0652	1.0633	1.0564	51	1.1374	1.1248	1.1223	1.1120	76	1.1976	1.1816	1.1785	1.1656
2	1.0059	1.0045	1.0044	1.0004	27	1.0760	1.0678	1.0656	1.0587	52	1.1399	1.1271	1.1244	1.1142	77	1.1999	1.1837	1.1801	1.1676
3	1.0090	1.0072	1.0070	1.0028	28	1.0787	1.0702	1.0681	1.0609	53	1.1424	1.1294	1.1269	1.1164	78	1.2021	1.1859	1.1818	1.1697
4	1.0120	1.0100	1.0093	1.0053	29	1.0813	1.0726	1.0705	1.0632	54	1.1448	1.1318	1.1295	1.1186	79	1.2043	1.1881	1.1837	1.1717
5	1.0150	1.0124	1.0115	1.0075	30	1.0839	1.0750	1.0729	1.0654	55	1.1472	1.1341	1.1320	1.1208	80	1.2065	1.1902	1.1806	1.1737
6	1.0179	1.0151	1.0141	1.0101	31	1.0866	1.0774	1.0753	1.0676	56	1.1497	1.1365	1.1342	1.1230	81	1.2088	1.1924	1.1876	1.1758
7	1.0207	1.0177	1.0170	1.0125	32	1.0891	1.0798	1.0777	1.0699	57	1.1523	1.1388	1.1361	1.1253	82	1.2110	1.1944	1.1896	1.1778
8	1.0237	1.0204	1.0196	1.0149	33	1.0916	1.0821	1.0800	1.0721	58	1.1548	1.1411	1.1381	1.1274	83	1.2132	1.1965	1.1914	1.1798
9	1.0266	1.0230	1.0221	1.0173	34	1.0941	1.0844	1.0823	1.0743	59	1.1573	1.1434	1.1401	1.1295	84	1.2154	1.1985	1.1929	1.1817
10	1.0295	1.0256	1.0246	1.0197	35	1.0966	1.0867	1.0847	1.0766	60	1.1597	1.1458	1.1424	1.1317	85	1.2176	1.2005	1.1953	1.1837
11	1.0324	1.0281	1.0271	1.0221	36	1.0993	1.0892	1.0871	1.0788	61	1.1621	1.1481	1.1448	1.1338	86	1.2196	1.2025	1.1976	1.1856
12	1.0351	1.0306	1.0296	1.0244	37	1.1018	1.0916	1.0895	1.0810	62	1.1645	1.1504	1.1473	1.1360	87	1.2217	1.2045	1.1994	1.1875
13	1.0379	1.0330	1.0321	1.0267	38	1.1043	1.0940	1.0919	1.0832	63	1.1669	1.1526	1.1493	1.1382	88	1.2237	1.2064	1.2012	1.1893
14	1.0407	1.0355	1.0345	1.0290	39	1.1069	1.0964	1.0940	1.0854	64	1.1694	1.1549	1.1517	1.1403	89	1.2258	1.2084	1.2028	1.1910
15	1.0435	1.0380	1.0370	1.0313	40	1.1095	1.0988	1.0963	1.0876	65	1.1718	1.1572	1.1543	1.1425	90	1.2278	1.2102	1.2044	1.1927
16	1.0463	1.0405	1.0393	1.0336	41	1.1122	1.1012	1.0990	1.0898	66	1.1742	1.1595	1.1565	1.1446	91	1.2297	1.2121	1.2059	1.1945
17	1.0491	1.0430	1.0417	1.0358	42	1.1148	1.1036	1.1015	1.0920	67	1.1766	1.1618	1.1584	1.1467	92	1.2316	1.2139	1.2078	1.1961
18	1.0518	1.0455	1.0441	1.0381	43	1.1174	1.1060	1.1038	1.0943	68	1.1790	1.1640	1.1604	1.1489	93	1.2335	1.2157	1.2099	1.1978
19	1.0545	1.0480	1.0464	1.0404	44	1.1199	1.1084	1.1062	1.0965	69	1.1813	1.1663	1.1628	1.1510	94	1.2354	1.2174	1.2117	1.1994
20	1.0571	1.0505	1.0488	1.0427	45	1.1224	1.1109	1.1085	1.0987	70	1.1835	1.1685	1.1655	1.1531	95	1.2372	1.2191	1.2140	1.2008
21	1.0598	1.0532	1.0512	1.0451	46	1.1249	1.1133	1.1108	1.1009	71	1.1858	1.1707	1.1677	1.1552	96	1.2390	1.2208	1.2158	1.2022
22	1.0625	1.0556	1.0537	1.0473	47	1.1274	1.1156	1.1130	1.1031	72	1.1882	1.1729	1.1702	1.1573	97	1.2408	1.2224	1.2170	1.2036
23	1.0652	1.0580	1.0561	1.0496	48	1.1299	1.1179	1.1157	1.1053	73	1.1906	1.1751	1.1728	1.1595	98	1.2425	1.2240	1.2183	1.2048
24	1.0679	1.0604	1.0585	1.0518	49	1.1324	1.1202	1.1185	1.1076	74	1.1929	1.1773	1.1752	1.1615	99	1.2441	1.2257	1.2202	1.2061
															100	1.2456	1.2273	1.2212	1.2073

TABLE 2-107   Acetic Acid (CH₃COOH)

%	0°C	10°C	15°C	20°C	25°C	30°C	40°C	%	0°C	10°C	15°C	20°C	25°C	30°C	40°C
0	0.9999	0.9997	0.9991	0.9982	0.9971	0.9957	0.9922	50	1.0729	1.0654	1.0613	1.0575	1.0534	1.0492	1.0408
1	1.0016	1.0013	1.0006	.9996	.9987	.9971	.9934	51	1.0738	1.0663	1.0622	1.0582	1.0542	1.0499	1.0414
2	1.0033	1.0029	1.0021	1.0012	1.0000	.9984	.9946	52	1.0748	1.0671	1.0629	1.0590	1.0549	1.0506	1.0421
3	1.0051	1.0044	1.0036	1.0025	1.0013	.9997	.9958	53	1.0757	1.0679	1.0637	1.0597	1.0555	1.0512	1.0427
4	1.0070	1.0060	1.0051	1.0040	1.0027	1.0011	.9970	54	1.0765	1.0687	1.0644	1.0604	1.0562	1.0518	1.0432
5	1.0088	1.0076	1.0066	1.0055	1.0041	1.0024	.9982	55	1.0774	1.0694	1.0651	1.0611	1.0568	1.0525	1.0438
6	1.0106	1.0092	1.0081	1.0069	1.0055	1.0037	.9994	56	1.0782	1.0701	1.0658	1.0618	1.0574	1.0531	1.0443
7	1.0124	1.0108	1.0096	1.0083	1.0068	1.0050	1.0006	57	1.0790	1.0708	1.0665	1.0624	1.0580	1.0536	1.0448
8	1.0142	1.0124	1.0111	1.0097	1.0081	1.0063	1.0018	58	1.0798	1.0715	1.0672	1.0631	1.0586	1.0542	1.0453
9	1.0159	1.0140	1.0126	1.0111	1.0094	1.0076	1.0030	59	1.0805	1.0722	1.0678	1.0637	1.0592	1.0547	1.0458
10	1.0177	1.0156	1.0141	1.0125	1.0107	1.0089	1.0042	60	1.0813	1.0728	1.0684	1.0642	1.0597	1.0552	1.0462
11	1.0194	1.0171	1.0155	1.0139	1.0120	1.0102	1.0054	61	1.0820	1.0734	1.0690	1.0648	1.0602	1.0557	1.0466
12	1.0211	1.0187	1.0170	1.0154	1.0133	1.0115	1.0065	62	1.0826	1.0740	1.0696	1.0653	1.0607	1.0562	1.0470
13	1.0228	1.0202	1.0184	1.0168	1.0146	1.0127	1.0077	63	1.0833	1.0746	1.0701	1.0658	1.0612	1.0566	1.0473
14	1.0245	1.0217	1.0199	1.0182	1.0159	1.0139	1.0088	64	1.0838	1.0752	1.0706	1.0662	1.0616	1.0571	1.0477
15	1.0262	1.0232	1.0213	1.0195	1.0172	1.0151	1.0099	65	1.0844	1.0757	1.0711	1.0666	1.0621	1.0575	1.0480
16	1.0278	1.0247	1.0227	1.0209	1.0185	1.0163	1.0110	66	1.0850	1.0762	1.0716	1.0671	1.0624	1.0578	1.0483
17	1.0295	1.0262	1.0241	1.0223	1.0198	1.0175	1.0121	67	1.0856	1.0767	1.0720	1.0675	1.0628	1.0582	1.0486
18	1.0311	1.0276	1.0255	1.0236	1.0210	1.0187	1.0132	68	1.0860	1.0771	1.0725	1.0678	1.0631	1.0585	1.0489
19	1.0327	1.0291	1.0269	1.0250	1.0223	1.0198	1.0142	69	1.0865	1.0775	1.0729	1.0682	1.0634	1.0588	1.0491
20	1.0343	1.0305	1.0283	1.0263	1.0235	1.0210	1.0153	70	1.0869	1.0779	1.0732	1.0685	1.0637	1.0590	1.0493
21	1.0358	1.0319	1.0297	1.0276	1.0248	1.0222	1.0164	71	1.0874	1.0783	1.0736	1.0687	1.0640	1.0592	1.0495
22	1.0374	1.0333	1.0310	1.0288	1.0260	1.0233	1.0174	72	1.0877	1.0786	1.0738	1.0690	1.0642	1.0594	1.0496
23	1.0389	1.0347	1.0323	1.0301	1.0272	1.0244	1.0185	73	1.0881	1.0789	1.0741	1.0693	1.0644	1.0595	1.0497
24	1.0404	1.0361	1.0336	1.0313	1.0283	1.0256	1.0195	74	1.0884	1.0792	1.0743	1.0694	1.0645	1.0596	1.0498
25	1.0419	1.0375	1.0349	1.0326	1.0295	1.0267	1.0205	75	1.0887	1.0794	1.0745	1.0696	1.0647	1.0597	1.0499
26	1.0434	1.0388	1.0362	1.0338	1.0307	1.0278	1.0215	76	1.0889	1.0796	1.0746	1.0698	1.0648	1.0598	1.0499
27	1.0449	1.0401	1.0374	1.0349	1.0318	1.0289	1.0225	77	1.0891	1.0797	1.0747	1.0699	1.0648	1.0598	1.0499
28	1.0463	1.0414	1.0386	1.0361	1.0329	1.0299	1.0234	78	1.0893	1.0798	1.0747	1.0700	1.0648	1.0598	1.0498
29	1.0477	1.0427	1.0399	1.0372	1.0340	1.0310	1.0244	79	1.0894	1.0798	1.0747	1.0700	1.0648	1.0597	1.0497
30	1.0491	1.0440	1.0411	1.0384	1.0350	1.0320	1.0253	80	1.0895	1.0798	1.0747	1.0700	1.0647	1.0596	1.0495
31	1.0505	1.0453	1.0423	1.0395	1.0361	1.0330	1.0262	81	1.0895	1.0797	1.0745	1.0699	1.0646	1.0594	1.0493
32	1.0519	1.0465	1.0435	1.0406	1.0372	1.0341	1.0272	82	1.0895	1.0796	1.0743	1.0698	1.0644	1.0592	1.0490
33	1.0532	1.0477	1.0446	1.0417	1.0382	1.0351	1.0281	83	1.0895	1.0795	1.0741	1.0696	1.0642	1.0589	1.0487
34	1.0545	1.0489	1.0458	1.0428	1.0392	1.0361	1.0289	84	1.0893	1.0793	1.0738	1.0693	1.0638	1.0585	1.0483
35	1.0558	1.0501	1.0469	1.0438	1.0402	1.0371	1.0298	85	1.0891	1.0790	1.0735	1.0689	1.0635	1.0582	1.0479
36	1.0571	1.0513	1.0480	1.0449	1.0412	1.0380	1.0306	86	1.0887	1.0787	1.0731	1.0685	1.0630	1.0576	1.0473
37	1.0584	1.0524	1.0491	1.0459	1.0422	1.0390	1.0314	87	1.0883	1.0783	1.0726	1.0680	1.0626	1.0571	1.0467
38	1.0596	1.0535	1.0501	1.0469	1.0432	1.0399	1.0322	88	1.0877	1.0778	1.0721	1.0675	1.0620	1.0564	1.0460
39	1.0608	1.0546	1.0512	1.0479	1.0441	1.0408	1.0330	89	1.0872	1.0773	1.0715	1.0668	1.0613	1.0557	1.0453
40	1.0621	1.0557	1.0522	1.0488	1.0450	1.0416	1.0338	90	1.0865	1.0766	1.0708	1.0661	1.0605	1.0549	1.0445
41	1.0633	1.0568	1.0532	1.0498	1.0460	1.0425	1.0346	91	1.0857	1.0758	1.0700	1.0652	1.0597	1.0541	1.0436
42	1.0644	1.0578	1.0542	1.0507	1.0469	1.0433	1.0353	92	1.0848	1.0749	1.0690	1.0643	1.0587	1.0530	1.0426
43	1.0656	1.0588	1.0551	1.0516	1.0477	1.0441	1.0361	93	1.0838	1.0739	1.0680	1.0632	1.0577	1.0518	1.0414
44	1.0667	1.0598	1.0561	1.0525	1.0486	1.0449	1.0368	94	1.0826	1.0727	1.0667	1.0619	1.0564	1.0506	1.0401
45	1.0679	1.0608	1.0570	1.0534	1.0495	1.0456	1.0375	95	1.0813	1.0714	1.0652	1.0605	1.0551	1.0491	1.0386
46	1.0689	1.0618	1.0579	1.0542	1.0503	1.0464	1.0382	96	1.0798		1.0632	1.0588	1.0535	1.0473	1.036
47	1.0699	1.0627	1.0588	1.0551	1.0511	1.0471	1.0389	97	1.0780		1.0611	1.0570	1.0516	1.0454	1.034
48	1.0709	1.0636	1.0597	1.0559	1.0518	1.0479	1.0395	98	1.0759		1.0590	1.0549	1.0495	1.0431	1.032
49	1.0720	1.0645	1.0605	1.0567	1.0526	1.0486	1.0402	99	1.0730		1.0567	1.0524	1.0468	1.0407	1.029
								100	1.0697		1.0545	1.0498	1.0440	1.0380	1.027

TABLE 2-108 Oxalic Acid (H₂C₂O₄)

%	$d_4^{17.5}$	%	$d_4^{17.5}$
1	1.0035	8	1.0280
2	1.0070	10	1.0350
4	1.0140	12	1.0420

TABLE 2-109 Methyl Alcohol (CH₃OH)*

%	0°C	10°C	15.56°C	20°C	15°C	%	0°C	10°C	15.56°C	20°C	15°C	%	0°C	10°C	15.56°C	20°C	15°C
0	0.9999	0.9997	0.9990	0.9982	0.99913	35	0.9534	0.9484	0.9456	0.9433	0.94570	70	0.8869	0.8794	0.8748	0.8715	0.87507
1	.9981	.9980	.9973	.9965	.99727	36	.9520	.9469	.9440	.9416	.94404	71	.8847	.8770	.8726	.8690	.87271
2	.9963	.9962	.9955	.9948	.99543	37	.9505	.9453	.9422	.9398	.94237	72	.8824	.8747	.8702	.8665	.87033
3	.9946	.9945	.9938	.9931	.99370	38	.9490	.9437	.9405	.9381	.94067	73	.8801	.8724	.8678	.8641	.86792
4	.9930	.9929	.9921	.9914	.99198	39	.9475	.9420	.9387	.9363	.93894	74	.8778	.8699	.8653	.8616	.86546
5	.9914	.9912	.9904	.9896	.99029	40	.9459	.9403	.9369	.9345	.93720	75	.8754	.8676	.8629	.8592	.86300
6	.9899	.9896	.9889	.9880	.98864	41	.9443	.9387	.9351	.9327	.93543	76	.8729	.8651	.8604	.8567	.86051
7	.9884	.9881	.9872	.9863	.98701	42	.9427	.9370	.9333	.9309	.93365	77	.8705	.8626	.8579	.8542	.85801
8	.9870	.9865	.9857	.9847	.98547	43	.9411	.9352	.9315	.9290	.93185	78	.8680	.8602	.8554	.8518	.85551
9	.9856	.9849	.9841	.9831	.98394	44	.9395	.9334	.9297	.9272	.93001	79	.8657	.8577	.8529	.8494	.85300
10	.9842	.9834	.9826	.9815	.98241	45	.9377	.9316	.9279	.9252	.92815	80	.8634	.8551	.8503	.8469	.85048
11	.9829	.9820	.9811	.9799	.98093	46	.9360	.9298	.9261	.9234	.92627	81	.8610	.8527	.8478	.8446	.84794
12	.9816	.9805	.9796	.9784	.97945	47	.9342	.9279	.9242	.9214	.92436	82	.8585	.8501	.8452	.8420	.84536
13	.9804	.9791	.9781	.9768	.97802	48	.9324	.9260	.9223	.9196	.92242	83	.8560	.8475	.8426	.8394	.84274
14	.9792	.9778	.9766	.9754	.97660	49	.9306	.9240	.9204	.9176	.92048	84	.8535	.8449	.8400	.8366	.84009
15	.9780	.9764	.9752	.9740	.97518	50	.9287	.9221	.9185	.9156	.91852	85	.8510	.8422	.8374	.8340	.83742
16	.9769	.9751	.9738	.9725	.97377	51	.9269	.9202	.9166	.9135	.91653	86	.8483	.8394	.8347	.8314	.83475
17	.9758	.9739	.9723	.9710	.97237	52	.9250	.9182	.9146	.9114	.91451	87	.8456	.8367	.8320	.8286	.83207
18	.9747	.9726	.9709	.9696	.97096	53	.9230	.9162	.9126	.9094	.91248	88	.8428	.8340	.8294	.8258	.82937
19	.9736	.9713	.9695	.9681	.96955	54	.9211	.9142	.9106	.9073	.91044	89	.8400	.8314	.8267	.8230	.82667
20	.9725	.9700	.9680	.9666	.96814	55	.9191	.9122	.9086	.9052	.90839	90	.8374	.8287	.8239	.8202	.82396
21	.9714	.9687	.9666	.9651	.96673	56	.9172	.9101	.9065	.9032	.90631	91	.8347	.8261	.8212	.8174	.82124
22	.9702	.9673	.9652	.9636	.96533	57	.9151	.9080	.9045	.9010	.90421	92	.8320	.8234	.8185	.8146	.81849
23	.9690	.9660	.9638	.9622	.96392	58	.9131	.9060	.9024	.8988	.90210	93	.8293	.8208	.8157	.8118	.81568
24	.9678	.9646	.9624	.9607	.96251	59	.9111	.9039	.9002	.8968	.89996	94	.8266	.8180	.8129	.8090	.81285
25	.9666	.9632	.9609	.9592	.96108	60	.9090	.9018	.8980	.8946	.89781	95	.8240	.8152	.8101	.8062	.80999
26	.9654	.9618	.9595	.9576	.95963	61	.9068	.8998	.8958	.8924	.89563	96	.8212	.8124	.8073	.8034	.80713
27	.9642	.9604	.9580	.9562	.95817	62	.9046	.8977	.8936	.8902	.89341	97	.8186	.8096	.8045	.8005	.80428
28	.9629	.9590	.9565	.9546	.95668	63	.9024	.8955	.8913	.8879	.89117	98	.8158	.8068	.8016	.7976	.80143
29	.9616	.9575	.9550	.9531	.95518	64	.9002	.8933	.8890	.8856	.88890	99	.8130	.8040	.7987	.7948	.79859
30	.9604	.9560	.9535	.9515	.95366	65	.8980	.8911	.8867	.8834	.88662	100	.8102	.8009	.7959	.7917	.79577
31	.9590	.9546	.9521	.9499	.95213	66	.8958	.8888	.8844	.8811	.88433						
32	.9576	.9531	.9505	.9483	.95056	67	.8935	.8865	.8820	.8787	.88203						
33	.9563	.9516	.9489	.9466	.94896	68	.8913	.8842	.8797	.8763	.87971						
34	.9549	.9500	.9473	.9450	.94734	69	.8891	.8818	.8771	.8738	.87739						

*It should be noted that the values for 100 percent do not agree with some data available elsewhere, *e.g.*, *American Institute of Physics Handbook*, McGraw-Hill, New York, 1957. Also, see Atack, *Handbook of Chemical Data*, Reinhold, New York, 1957.

TABLE 2-110 Ethyl Alcohol (C₂H₅OH)*

%	10°C	15°C	20°C	25°C	30°C	35°C	40°C	%	10°C	15°C	20°C	25°C	30°C	35°C	40°C
0	0.99973	0.99913	0.99823	0.99708	0.99568	0.99406	0.99225	50	0.92126	0.91776	0.91384	0.90985	0.90580	0.90168	0.89750
1	785	725	636	520	379	217	034	51	.91943	555	160	760	353	.89940	519
2	602	542	453	336	194	031	.98846	52	723	333	.90936	534	125	710	288
3	426	365	275	157	014	.98849	663	53	502	110	711	307	.89896	479	056
4	258	195	103	.98984	.98839	672	485	54	279	.90885	485	079	667	248	.88823
5	098	032	.98938	817	670	501	311	55	055	659	258	.89850	437	016	589
6	.98946	.98877	780	656	507	335	142	56	.90831	433	031	621	206	.88784	356
7	801	729	627	500	347	172	.97975	57	607	207	.89803	392	.88975	552	122
8	660	584	478	346	189	009	808	58	381	.89980	574	162	744	319	.87888
9	524	442	331	193	031	.97846	641	59	154	752	344	.88931	512	085	653
10	393	304	187	043	.97875	685	475	60	.89927	523	113	699	278	.87851	417
11	267	171	047	.97897	723	527	312	61	698	293	.88882	446	044	615	180
12	145	041	.97910	753	573	371	150	62	468	062	650	233	.87809	379	.86943
13	026	.97914	775	611	424	216	.96989	63	237	.88830	417	.87998	574	142	705
14	.97911	790	643	472	278	063	829	64	006	597	183	763	337	.86905	466
15	800	669	514	334	133	.96911	670	65	.88774	364	.87948	527	100	667	227
16	692	552	387	199	.96990	760	512	66	541	130	713	291	.86863	429	.85987
17	583	433	259	062	844	607	352	67	308	.87895	477	054	625	190	747
18	473	313	129	.96923	697	452	189	68	074	660	241	.86817	387	.85950	407
19	363	191	.96997	782	547	294	023	69	.87839	424	004	579	148	710	266
20	252	068	864	639	395	134	.95856	70	602	187	.86766	340	.85908	470	025
21	139	.96944	729	495	242	.95973	687	71	365	.86949	527	100	667	228	.84783
22	024	818	592	348	087	809	516	72	127	710	287	.85859	426	.84986	540
23	.96907	689	453	199	.95929	643	343	73	.86888	470	047	618	184	743	297
24	787	558	312	048	769	476	168	74	648	229	.85806	376	.84941	500	053
25	665	424	168	.95895	607	306	.94991	75	408	.85988	564	134	698	257	.83809
26	539	287	020	738	442	133	810	76	168	747	322	.84891	455	013	564
27	406	144	.95867	576	272	.94955	625	77	.85927	505	079	647	211	.83768	319
28	268	.95996	710	410	098	774	438	78	685	262	.84835	403	.83966	523	074
29	125	844	548	241	.94922	590	248	79	442	018	590	158	720	277	.82827
30	.95977	686	382	067	741	403	055	80	197	.84772	344	.83911	473	029	578
31	823	524	212	.94890	557	214	.93860	81	.84950	525	096	664	224	.82780	329
32	665	357	038	709	370	021	662	82	702	277	.83848	415	.82974	530	079
33	502	186	.94860	525	180	.93825	461	83	453	028	599	164	724	279	.81828
34	334	011	679	337	.93986	626	257	84	203	.83777	348	.82913	473	027	576
35	162	.94832	494	146	790	425	051	85	.83951	525	095	660	220	.81774	322
36	.94986	650	306	.93952	591	221	.92843	86	697	271	.82840	405	.81965	519	067
37	805	464	114	756	390	016	634	87	441	014	583	148	708	262	.80811
38	620	273	.93919	556	186	.92808	422	88	181	.82754	323	.81888	448	003	552
39	431	079	720	353	.92979	597	208	89	.82919	492	062	626	186	.80742	291
40	238	.93882	518	148	770	385	.91992	90	654	227	.81797	362	.80922	478	028
41	042	682	314	.92940	558	170	774	91	386	.81959	529	094	655	211	.79761
42	.93842	478	107	729	344	.91952	554	92	114	688	257	.80823	384	.79941	491
43	639	271	.92897	516	128	733	332	93	.81839	413	.80983	549	111	669	220
44	433	062	685	301	.91910	513	108	94	561	134	705	272	.79835	393	.78947
45	226	.92852	472	085	692	291	.90884	95	278	.80852	424	.79991	555	114	670
46	017	640	257	.91868	472	069	660	96	.80991	566	138	706	271	.78831	388
47	.92806	426	041	649	250	.90845	434	97	698	274	.79846	415	.78981	542	100
48	593	211	.91823	429	028	621	207	98	399	.79975	547	117	684	247	.77806
49	379	.91995	604	208	.90805	396	.89979	99	094	670	243	.78814	382	.77946	507
								100	.79784	360	.78934	506	075	641	203

*For data from −78° to 78°C, see p. 2-142, Table 2N-5, *American Institute of Physics Handbook*, McGraw-Hill, New York, 1957.

TABLE 2-111 Densities of Mixtures of C$_2$H$_5$OH and H$_2$O at 20°C

g/mL

% alcohol by weight	0	1	2	3	4	5	6	7	8	9
			Tenths of %							
0	0.99823	804	785	766	748	729	710	692	673	655
1	636	618	599	581	562	544	525	507	489	471
2	453	435	417	399	381	363	345	327	310	292
3	275	257	240	222	205	188	171	154	137	120
4	103	087	070	053	037	020	003	°987	°971	°954
5	.98938	922	906	890	874	859	843	827	811	796
6	780	765	749	734	718	703	688	673	658	642
7	627	612	597	582	567	553	538	523	508	493
8	478	463	449	434	419	404	389	374	360	345
9	331	316	301	287	273	258	244	229	215	201
10	187	172	158	144	130	117	103	089	075	061
11	047	033	019	006	°992	°978	°964	°951	°937	°923
12	.97910	896	883	869	855	842	828	815	801	788
13	775	761	748	735	722	709	696	683	670	657
14	643	630	617	604	591	578	565	552	539	526
15	514	501	488	475	462	450	438	425	412	400
16	387	374	361	349	336	323	310	297	284	272
17	259	246	233	220	207	194	181	168	155	142
18	129	116	103	089	076	063	050	037	024	010
19	.96997	984	971	957	944	931	917	904	891	877
20	864	850	837	823	810	796	783	769	756	742
21	729	716	702	688	675	661	647	634	620	606
22	592	578	564	551	537	523	509	495	481	467
23	453	439	425	411	396	382	368	354	340	326
24	312	297	283	269	254	240	225	211	196	182
25	168	153	139	124	109	094	080	065	050	035
26	020	005	°990	°975	°959	°944	°929	°914	°898	°883
27	.95867	851	836	820	805	789	773	757	742	726
28	710	694	678	662	646	630	613	597	581	565
29	548	532	516	499	483	466	450	433	416	400
30	382	365	349	332	315	298	281	264	247	230
31	212	195	178	161	143	126	108	091	074	056
32	038	020	003	°985	°967	°950	°932	°914	°896	°878
33	.94860	842	824	806	788	770	752	734	715	697
34	679	660	642	624	605	587	568	550	531	512
35	494	475	456	438	419	400	382	363	344	325
36	306	287	268	249	230	211	192	172	153	134
37	114	095	075	056	036	017	°997	°978	°958	°939
38	.93919	899	879	859	840	820	800	780	760	740
39	720	700	680	660	640	620	599	579	559	539
40	518	498	478	458	437	417	396	376	356	335
41	314	294	273	253	232	212	191	170	149	129
42	107	086	065	044	023	002	°981	°960	°939	°918
43	.92897	876	855	834	812	791	770	749	728	707
44	685	664	642	621	600	579	557	536	515	493
45	472	450	429	408	386	365	343	322	300	279
46	257	236	214	193	171	150	128	106	085	063
47	041	019	°997	°976	°954	°932	°910	°889	°867	°845
48	.91823	801	780	758	736	714	692	670	648	626
49	604	582	560	538	516	494	472	450	428	406

% alcohol by weight	0	1	2	3	4	5	6	7	8	9
			Tenths of %							
50	0.91384	361	339	317	295	272	250	228	206	183
51	160	138	116	093	071	049	026	004	°981	°959
52	.90936	914	891	869	846	824	801	779	756	734
53	711	689	666	644	621	598	576	553	531	508
54	485	463	440	417	395	372	349	327	304	281
55	258	236	213	190	167	145	122	099	076	054
56	031	008	°985	°962	°939	°917	°894	°871	°848	°825
57	.89803	780	757	734	711	688	665	643	620	597
58	574	551	528	505	482	459	436	413	390	367
59	344	321	298	275	252	229	206	183	160	137
60	113	090	067	044	021	°998	°975	°951	°928	°905
61	.88882	859	836	812	789	766	743	720	696	673
62	650	626	603	580	557	533	510	487	463	440
63	417	393	370	347	323	300	277	253	230	206
64	183	160	136	113	089	066	042	019	°995	°972
65	.87948	925	901	878	854	831	807	784	760	737
66	713	689	666	642	619	595	572	548	524	501
67	477	454	430	406	383	359	336	312	288	265
68	241	218	194	170	147	123	099	075	052	028
69	004	°981	°957	°933	°909	°885	°862	°838	°814	°790
70	.86766	742	718	694	671	647	623	599	575	551
71	527	503	479	455	431	407	383	359	335	311
72	287	263	239	215	191	167	143	119	095	071
73	047	022	°998	°974	°950	°926	°902	°878	°854	°830
74	.85806	781	757	733	709	685	661	636	612	588
75	564	540	515	491	467	443	419	394	370	346
76	322	297	273	249	225	200	176	152	128	103
77	079	055	031	006	°982	°958	°933	°909	°884	°860
78	.84835	811	787	762	738	713	689	664	640	615
79	590	566	541	517	492	467	443	418	393	369
80	344	319	294	270	245	220	196	171	146	121
81	096	072	047	022	°997	°972	°947	°923	°898	°873
82	.83848	823	798	773	748	723	698	674	649	624
83	599	574	549	523	498	473	448	423	398	373
84	348	323	297	272	247	222	196	171	146	120
85	095	070	044	019	°994	°968	°943	°917	°892	°866
86	.82840	815	789	763	738	712	686	660	635	609
87	583	557	531	505	479	453	427	401	375	349
88	323	297	271	245	219	193	167	140	114	088
89	062	035	009	°983	°956	°930	°903	°877	°850	°824
90	.81797	770	744	717	690	664	637	610	583	556
91	529	502	475	448	421	394	366	339	312	285
92	257	230	203	175	148	120	093	066	038	010
93	.80983	955	928	900	872	844	817	789	761	733
94	705	677	649	621	593	565	537	509	480	452
95	424	395	367	338	310	281	253	224	195	166
96	138	109	080	051	022	°993	°963	°934	°905	°875
97	.79846	816	787	757	727	698	668	638	608	578
98	547	517	487	456	426	396	365	335	305	274
99	243	213	182	151	120	089	059	028	°997	°966
100	.78934									

°Indicates change in the first two decimal places.

TABLE 2-112 Specific Gravity (60°/60°F [(15.56°/15.56°C)]) of Mixtures by Volume of C_2H_5OH and H_2O

% alcohol by volume at 60°F	0	1	2	3	4	5	6	7	8	9
0	1.00000	°985	°970	°955	°940	°925	°910	°895	°880	865
1	0.99850	835	820	806	791	776	761	747	732	717
2	703	688	674	659	645	630	616	602	587	573
3	559	545	531	516	502	488	474	460	446	432
4	419	405	391	378	364	350	336	323	309	296
5	282	269	255	242	228	215	202	189	176	163
6	150	137	124	111	098	085	073	060	047	035
7	022	009	°997	°984	°972	°960	°947	°935	°923	°911
8	.98899	887	875	863	851	838	826	814	803	791
9	779	767	755	743	731	720	708	696	684	672
10	661	649	637	625	614	602	590	579	567	556
11	544	532	521	509	498	487	475	464	452	441
12	430	419	408	396	385	374	363	352	341	330
13	319	308	297	286	275	264	254	243	232	221
14	210	200	190	179	168	157	147	136	125	115
15	104	093	083	072	062	051	040	030	019	009
16	.97998	988	977	967	956	946	936	925	915	905
17	895	885	875	864	854	844	834	824	814	804
18	794	784	774	764	754	744	734	724	714	704
19	694	684	674	664	654	645	635	625	615	605
20	596	586	576	566	556	546	536	526	516	506
21	496	486	476	466	456	446	436	425	415	405
22	395	385	375	365	354	344	334	324	313	303
23	293	283	272	262	252	241	231	221	210	200
24	189	179	168	158	147	137	126	116	105	095
25	084	073	063	052	042	031	020	010	°999	°988
26	.96978	967	957	946	935	924	914	903	892	881
27	870	859	848	837	826	815	804	793	782	771
28	760	749	738	727	715	704	693	682	671	659
29	648	637	625	614	603	591	580	568	557	546
30	534	522	511	499	488	476	464	453	441	429
31	418	406	394	382	370	358	346	334	321	309
32	296	284	271	259	246	234	221	209	196	183
33	170	157	144	132	119	106	093	080	067	054
34	041	028	015	002	°988	°975	°962	°948	°935	°921
35	.95908	894	881	867	854	840	826	812	798	784
36	770	756	742	728	714	700	685	671	657	643
37	628	614	599	585	570	556	541	526	512	497
38	482	467	452	437	423	408	393	378	362	347
39	332	317	302	286	271	256	240	225	209	194
40	178	162	147	131	115	100	084	068	052	036
41	020	004	°988	°972	°956	°940	°923	°907	°891	°875
42	.94858	842	825	809	792	776	759	743	726	710
43	693	676	660	643	626	609	592	575	558	541
44	524	507	490	473	455	438	421	403	386	369
45	351	334	316	298	281	263	245	228	210	192
46	174	156	138	120	102	084	066	048	030	011
47	.93993	975	956	938	920	901	883	864	845	827
48	808	789	771	752	733	714	695	676	657	638
49	619	600	581	562	543	523	504	485	465	446

% alcohol by volume at 60°F	0	1	2	3	4	5	6	7	8	9
50	0.93426	407	387	368	348	328	309	289	270	250
51	230	210	190	171	151	131	111	091	071	051
52	031	011	°991	°971	°951	°931	°911	°890	°870	°850
53	.92830	810	789	769	749	728	708	688	667	647
54	626	605	585	564	544	523	502	482	461	440
55	419	398	377	357	336	315	294	273	252	231
56	210	189	168	147	126	105	084	062	041	020
57	.91999	978	956	935	914	892	871	849	827	806
58	784	762	741	719	697	675	653	631	610	588
59	565	543	521	499	477	455	433	410	388	366
60	344	322	299	277	255	232	210	188	165	143
61	120	097	075	052	030	007	°984	°962	°939	°916
62	.90893	870	847	825	802	779	756	733	710	687
63	664	641	618	595	572	549	526	503	480	457
64	434	411	388	365	341	318	295	272	249	225
65	202	179	155	132	108	085	061	038	014	°991
66	.89967	943	920	896	872	848	825	801	777	753
67	729	705	681	657	633	609	585	561	537	513
68	489	465	441	416	392	368	343	319	295	270
69	245	220	196	171	147	122	098	073	048	024
70	.88999	974	950	925	900	875	850	825	801	776
71	751	725	700	675	650	625	600	574	549	524
72	499	474	448	423	397	372	346	321	296	270
73	244	218	193	167	141	116	090	064	039	013
74	.87987	961	935	910	884	858	832	806	780	754
75	728	702	676	650	623	597	571	545	518	492
76	465	439	412	386	359	332	306	279	252	226
77	199	172	145	118	092	065	038	011	°984	°957
78	.86929	902	875	847	820	793	766	738	711	684
79	656	629	601	574	546	518	491	463	435	408
80	380	352	324	296	269	241	213	185	157	129
81	100	072	044	015	°987	°959	°931	°902	°874	°846
82	.85817	789	760	732	703	674	646	617	588	560
83	531	502	473	444	415	386	357	328	299	270
84	240	211	181	152	122	093	063	033	004	°974
85	.84944	914	884	854	824	794	764	734	703	673
86	642	612	581	551	520	490	459	428	398	367
87	336	305	274	243	212	181	150	119	088	056
88	025	°994	°962	°930	°899	°867	°835	°803	°771	°739
89	.83707	675	643	610	578	545	513	480	447	415
90	382	349	315	282	249	216	183	150	116	083
91	049	015	°981	°947	°913	°879	°845	°810	°776	°741
92	.82705	670	635	600	565	529	494	458	423	387
93	351	315	279	243	206	170	133	096	059	022
94	.81984	947	909	871	834	796	757	719	681	642
95	603	564	525	486	446	407	367	327	287	247
96	206	165	125	084	042	001	°960	°918	°876	°834
97	.80792	750	707	664	620	577	533	489	445	401
98	356	311	265	219	173	127	080	033	°985	°937
99	.79889	841	792	743	693	643	593	543	492	441
100	389									

°Indicates change in first two decimal places.

TABLE 2-113   n-Propyl Alcohol (C₃H₇OH)

%	0°C	15°C	30°C	%	0°C	15°C	30°C	%	0°C	15°C	30°C	%	0°C	15°C	30°C	%	0°C	15°C	30°C
0	0.9999	0.9991	0.9957	20	0.9789	0.9723	0.9643	40	0.9430	0.9331	0.9226	60	0.9033	0.8922	0.8807	80	0.8634	0.8516	0.8394
1	.9982	.9974	.9940	21	.9776	.9705	.9622	41	.9411	.9310	.9205	61	.9013	.8902	.8786	81	.8614	.8496	.8373
2	.9967	.9960	.9924	22	.9763	.9688	.9602	42	.9391	.9290	.9184	62	.8994	.8882	.8766	82	.8594	.8475	.8352
3	.9952	.9944	.9908	23	.9748	.9670	.9583	43	.9371	.9269	.9164	63	.8974	.8861	.8745	83	.8574	.8454	.8332
4	.9939	.9929	.9893	24	.9733	.9651	.9563	44	.9352	.9248	.9143	64	.8954	.8841	.8724	84	.8554	.8434	.8311
5	.9926	.9915	.9877	25	.9717	.9633	.9543	45	.9332	.9228	.9122	65	.8934	.8820	.8703	85	.8534	.8413	.8290
6	.9914	.9902	.9862	26	.9700	.9614	.9522	46	.9311	.9207	.9100	66	.8913	.8800	.8682	86	.8513	.8393	.8269
7	.9904	.9890	.9848	27	.9682	.9594	.9501	47	.9291	.9186	.9079	67	.8894	.8779	.8662	87	.8492	.8372	.8248
8	.9894	.9877	.9834	28	.9664	.9576	.9481	48	.9272	.9165	.9057	68	.8874	.8759	.8641	88	.8471	.8351	.8227
9	.9883	.9864	.9819	29	.9646	.9556	.9460	49	.9252	.9145	.9036	69	.8854	.8739	.8620	89	.8450	.8330	.8206
10	.9874	.9852	.9804	30	.9627	.9535	.9439	50	.9232	.9124	.9015	70	.8835	.8719	.8600	90	.8429	.8308	.8185
11	.9865	.9840	.9790	31	.9608	.9516	.9418	51	.9213	.9104	.8994	71	.8815	.8700	.8580	91	.8408	.8287	.8164
12	.9857	.9828	.9775	32	.9589	.9495	.9396	52	.9192	.9084	.8973	72	.8795	.8680	.8559	92	.8387	.8266	.8142
13	.9849	.9817	.9760	33	.9570	.9474	.9375	53	.9173	.9064	.8952	73	.8776	.8659	.8539	93	.8364	.8244	.8120
14	.9841	.9806	.9746	34	.9550	.9454	.9354	54	.9153	.9044	.8931	74	.8756	.8639	.8518	94	.8342	.8221	.8098
15	.9833	.9793	.9730	35	.9530	.9434	.9333	55	.9132	.9023	.8911	75	.8736	.8618	.8497	95	.8320	.8199	.8077
16	.9825	.9780	.9714	36	.9511	.9413	.9312	56	.9112	.9003	.8890	76	.8716	.8598	.8477	96	.8296	.8176	.8054
17	.9817	.9768	.9698	37	.9491	.9392	.9289	57	.9093	.8983	.8869	77	.8695	.8577	.8456	97	.8272	.8153	.8031
18	.9808	.9752	.9680	38	.9471	.9372	.9269	58	.9073	.8963	.8849	78	.8675	.8556	.8435	98	.8248	.8128	.8008
19	.9800	.9739	.9661	39	.9450	.9351	.9247	59	.9053	.8942	.8828	79	.8655	.8536	.8414	99	.8222	.8104	.7984
																100	.8194	.8077	.7958

TABLE 2-114   Isopropyl Alcohol (C₃H₇OH)

%	0°C	15°C°	15°C°	20°C	30°C	%	0°C	15°C°	15°C°	20°C	30°C	%	0°C	15°C°	15°C°	20°C	30°C
0	0.9999	0.9991	0.99913	0.9982	0.9957	35	0.9557		0.9446	0.9419	0.9338	70	0.8761	0.8639	0.86346	0.8584	0.8511
1	.9980	.9973	.9972	.9962	.9939	36	.9536		.9424	.9399	.9315	71	.8738	.8615	.8611	.8560	.8487
2	.9962	.9956	.9954	.9944	.9921	37	.9514		.9401	.9377	.9292	72	.8714	.8592	.8588	.8537	.8464
3	.9946	.9938	.9936	.9926	.9904	38	.9493		.9379	.9355	.9269	73	.8691	.8568	.8564	.8513	.8440
4	.9930	.9922	.9920	.9909	.9887	39	.9472		.9356	.9333	.9246	74	.8668	.8545	.8541	.8489	.8416
5	.9916	.9906	.9904	.9893	.9871	40	.9450		.93333	.9310	.9224	75	.8644	.8521	.8517	.8464	.8392
6	.9902	.9892	.9890	.9877	.9855	41	.9428		.9311	.9287	.9201	76	.8621	.8497	.8493	.8439	.8368
7	.9890	.9878	.9875	.9862	.9839	42	.9406		.9288	.9264	.9177	77	.8598	.8474	.8470	.8415	.8344
8	.9878	.9864	.9862	.9847	.9824	43	.9384		.9266	.9239	.9154	78	.8575	.8450	.8446	.8391	.8321
9	.9866	.9851	.9849	.9833	.9809	44	.9361		.9243	.9215	.9130	79	.8551	.8426	.8422	.8366	.8297
10	.9856	.9838	.98362	.9820	.9794	45	.9338		.9220	.9191	.9106	80	.8528	.8403	.83979	.8342	.8273
11	.9846	.9826	.9824	.9808	.9778	46	.9315		.9197	.9165	.9082	81	.8503	.8379	.8374	.8317	.8248
12	.9838	.9813	.9812	.9797	.9764	47	.9292		.9174	.9141	.9059	82	.8479	.8355	.8350	.8292	.8224
13	.9829	.9802	.9800	.9876	.9750	48	.9270		.9150	.9117	.9036	83	.8456	.8331	.8326	.8268	.8200
14	.9821	.9790	.9788	.9776	.9735	49	.9247		.9127	.9093	.9013	84	.8432	.8307	.8302	.8243	.8175
15	.9814	.9779	.9777	.9765	.9720	50	.9224		.91043	.9069	.8990	85	.8408	.8282	.8278	.8219	.8151
16	.9806	.9768	.9765	.9754	.9705	51	.9201		.9081	.9044	.8966	86	.8384	.8259	.8254	.8194	.8127
17	.9799	.9756	.9753	.9743	.9690	52	.9178		.9058	.9020	.8943	87	.8360	.8234	.8229	.8169	.8201
18	.9792	.9745	.9741	.9731	.9675	53	.9155		.9035	.8996	.8919	88	.8336	.8209	.8205	.8145	.8078
19	.9784	.9730	.9728	.9717	.9658	54	.9132		.9011	.8971	.8895	89	.8311	.8184	.8180	.8120	.8053
20	.9777	.9719	.97158	.9703	.9642	55	.9109		.8988	.8946	.8871	90	.8287	.8161	.81553	.8096	.8029
21	.9768	.9704	.9703	.9688	.9624	56	.9086		.8964	.8921	.8847	91	.8262	.8136	.8130	.8072	.8004
22	.9759	.9690	.9689	.9669	.9606	57	.9063		.8940	.8896	.8823	92	.8237	.8110	.8104	.8047	.7979
23	.9749	.9675	.9674	.9651	.9587	58	.9040		.8917	.8874	.8800	93	.8212	.8085	.8079	.8023	.7954
24	.9739	.9660	.9659	.9634	.9569	59	.9017		.8893	.8850	.8777	94	.8186	.8060	.8052	.7998	.7929
25	.9727	.9643	.9642	.9615	.9549	60	.8994		.88690	.8825	.8752	95	.8160	.8034	.8026	.7973	.7904
26	.9714	.9626	.9624	.9597	.9529	61	.8970		.8845	.8800	.8728	96	.8133	.8008	.7999	.7949	.7878
27	.9699	.9608	.9605	.9577	.9509	62	.8947	0.8829	.8821	.8776	.8704	97	.8106	.7981	.7972	.7925	.7852
28	.9684	.9590	.9586	.9558	.9488	63	.8924	.8805	.8798	.8751	.8680	98	.8078	.7954	.7945	.7901	.7826
29	.9669	.9570	.9568	.9540	.9467	64	.8901	.8781	.8775	.8727	.8656	99	.8048	.7926	.7918	.7877	.7799
30	.9652	.9551	.95493	.9520	.9446	65	.8878	.8757	.8752	.8702	.8631	100	.8016	.7896	.78913	.7854	.7770
31	.9634		.9530	.9500	.9426	66	.8854	.8733	.8728	.8679	.8607						
32	.9615		.9510	.9481	.9405	67	.8831	.8710	.8705	.8656	.8583						
33	.9596		.9489	.9460	.9383	68	.8807	.8686	.8682	.8632	.8559						
34	.9577		.9468	.9440	.9361	69	.8784	.8662	.8658	.8609	.8535						

°Two different observers; see *International Critical Tables*, vol. 3, p. 120.

TABLE 2-115 Glycerol*

Glycerol, %	Density					Glycerol, %	Density					Glycerol, %	Density				
	15°C	15.5°C	20°C	25°C	30°C		15°C	15.5°C	20°C	25°C	30°C		15°C	15.5°C	20°C	25°C	30°C
100	1.26415	1.26381	1.26108	1.15802	1.25495	65	1.17030	1.17000	1.16750	1.16475	1.16195	30	1.07455	1.07435	1.07270	1.07070	1.06855
99	1.26160	1.26125	1.25850	1.25545	1.25235	64	1.16755	1.16725	1.16475	1.16200	1.15925	29	1.07195	1.07175	1.07010	1.06815	1.06605
98	1.25900	1.25865	1.25590	1.25290	1.24975	63	1.16480	1.16445	1.16205	1.15925	1.15650	28	1.06935	1.06915	1.06755	1.06560	1.06355
97	1.25645	1.25610	1.25335	1.25030	1.24710	62	1.16200	1.16170	1.15930	1.15655	1.15375	27	1.06670	1.06655	1.06495	1.06305	1.06105
96	1.25385	1.25350	1.25080	1.24770	1.24450	61	1.15925	1.15895	1.15655	1.15380	1.15100	26	1.06410	1.06390	1.06240	1.06055	1.05855
95	1.25130	1.25095	1.24825	1.24515	1.24190	60	1.15650	1.15615	1.15380	1.15105	1.14830	25	1.06150	1.06130	1.05980	1.05800	1.05605
94	1.24865	1.24830	1.24560	1.24250	1.23930	59	1.15370	1.15340	1.15105	1.14835	1.14555	24	1.05885	1.05870	1.05720	1.05545	1.05350
93	1.24600	1.24565	1.24300	1.23985	1.23670	58	1.15095	1.15065	1.14830	1.14560	1.14285	23	1.05625	1.05610	1.05465	1.05290	1.05100
92	1.24340	1.24305	1.24035	1.23725	1.23410	57	1.14815	1.14785	1.14555	1.14285	1.14010	22	1.05365	1.05350	1.05205	1.05035	1.04850
91	1.24075	1.24040	1.23770	1.23460	1.23150	56	1.14535	1.14510	1.14280	1.14015	1.13740	21	1.05100	1.05090	1.04950	1.04780	1.04600
90	1.23810	1.23775	1.23510	1.23200	1.22890	55	1.14260	1.14230	1.14005	1.13740	1.13470	20	1.04840	1.04825	1.04690	1.04525	1.04350
89	1.23545	1.23510	1.23245	1.22935	1.22625	54	1.13980	1.13955	1.13730	1.13465	1.13195	19	1.04590	1.04575	1.04440	1.04280	1.04105
88	1.23280	1.23245	1.22975	1.22665	1.22360	53	1.13705	1.13680	1.13455	1.13195	1.12925	18	1.04335	1.04325	1.04195	1.04035	1.03860
87	1.23015	1.22980	1.22710	1.22400	1.22095	52	1.13425	1.13400	1.13180	1.12920	1.12650	17	1.04085	1.04075	1.03945	1.03790	1.03615
86	1.22750	1.22710	1.22445	1.22135	1.21830	51	1.13150	1.13125	1.12905	1.12650	1.12380	16	1.03835	1.03825	1.03695	1.03545	1.03370
85	1.22485	1.22445	1.22180	1.21870	1.21565	50	1.12870	1.12845	1.12630	1.12375	1.12110	15	1.03580	1.03570	1.03450	1.03300	1.03130
84	1.22220	1.22180	1.21915	1.21605	1.21300	49	1.12600	1.12575	1.12360	1.12110	1.11845	14	1.03330	1.03320	1.03200	1.03055	1.02885
83	1.21955	1.21915	1.21650	1.21340	1.21035	48	1.12325	1.12305	1.12090	1.11840	1.11580	13	1.03080	1.03070	1.02955	1.02805	1.02640
82	1.21690	1.21650	1.21380	1.21075	1.20770	47	1.12055	1.12030	1.11820	1.11575	1.11320	12	1.02830	1.02820	1.02705	1.02560	1.02395
81	1.21425	1.21385	1.21115	1.20810	1.20505	46	1.11780	1.11760	1.11550	1.11310	1.11055	11	1.02575	1.02565	1.02455	1.02315	1.02150
80	1.21160	1.21120	1.20850	1.20545	1.20240	45	1.11510	1.11490	1.11280	1.11040	1.10795	10	1.02325	1.02315	1.02210	1.02070	1.01905
79	1.20885	1.20845	1.20575	1.20275	1.19970	44	1.11235	1.11215	1.11010	1.10775	1.10530	9	1.02085	1.02075	1.01970	1.01835	1.01670
78	1.20610	1.20570	1.20305	1.20005	1.19705	43	1.10960	1.10945	1.10740	1.10510	1.10265	8	1.01840	1.01835	1.01730	1.01600	1.01440
77	1.20335	1.20300	1.20030	1.19735	1.19435	42	1.10690	1.10670	1.10470	1.10240	1.10005	7	1.01600	1.01590	1.01495	1.01360	1.01205
76	1.20060	1.20025	1.19760	1.19465	1.19170	41	1.10415	1.10400	1.10200	1.09975	1.09740	6	1.01360	1.01350	1.01255	1.01125	1.00970
75	1.19785	1.19750	1.19485	1.19195	1.18900	40	1.10145	1.10130	1.09930	1.09710	1.09475	5	1.01120	1.01110	1.01015	1.00890	1.00735
74	1.19510	1.19480	1.19215	1.18925	1.18635	39	1.09875	1.09860	1.09665	1.09445	1.09215	4	1.00875	1.00870	1.00780	1.00655	1.00505
73	1.19235	1.19205	1.18940	1.18650	1.18365	38	1.09605	1.09590	1.09400	1.09180	1.08955	3	1.00635	1.00630	1.00540	1.00415	1.00270
72	1.18965	1.18930	1.18670	1.18380	1.18100	37	1.09340	1.09320	1.09135	1.08915	1.08690	2	1.00395	1.00385	1.00300	1.00180	1.00035
71	1.18690	1.18655	1.18395	1.18110	1.17830	36	1.09070	1.09050	1.08865	1.08655	1.08430	1	1.00155	1.00145	1.00060	0.99945	0.99800
70	1.18415	1.18385	1.18125	1.17840	1.17565	35	1.08800	1.08780	1.08600	1.08390	1.08165	0	0.99913	0.99905	0.99823	0.99708	0.99568
69	1.18135	1.18105	1.17850	1.17565	1.17290	34	1.08530	1.08515	1.08335	1.08125	1.07905						
68	1.17860	1.17830	1.17575	1.17295	1.17020	33	1.08265	1.08245	1.08070	1.07860	1.07645						
67	1.17585	1.17555	1.17300	1.17020	1.16745	32	1.07995	1.07975	1.07800	1.07600	1.07380						
66	1.17305	1.17275	1.17025	1.16745	1.16470	31	1.07725	1.07705	1.07535	1.07335	1.07120						

*Bosart and Snoddy, *Ind. Eng. Chem.*, **20,** (1928): 1378.

TABLE 2-116 Hydrazine (N_2H_4)

%	d_4^{15}	%	d_4^{15}
1	1.0002	30	1.0305
2	1.0013	40	1.038
4	1.0034	50	1.044
8	1.0077	60	1.047
12	1.0121	70	1.046
16	1.0164	80	1.040
20	1.0207	90	1.030
24	1.0248	100	1.011
28	1.0286		

TABLE 2-117   Densities of Aqueous Solutions of Miscellaneous Organic Compounds*

d (resp., d_w, d_s) = density of the solution (resp., water; resp., the pure liquid solute) in g/mL; p_s (resp., p_w) = wt % of solute (resp., water) in the solution; range = range of applicability of the equation.

Section A $d = d_w + Ap_s + Bp_s^2 + Cp_s^3$

Name	Formula	t, °C	Range, p_s	A	B	C
Acetaldehyde	C_2H_4O	18	0– 30	$+0.0_3255$	-0.0_516	
Acetamide	C_2H_5NO	15	0– 6	$+0.0_3639$	$+0.0_4171$	
Acetone	C_3H_6O	0	0–100	-0.0_5856	-0.0_5449	-0.0_5588
		4	0–100	-0.0_77648	-0.0_41193	$+0.0_8272$
		15	0–100	-0.0_51009	-0.0_59682	-0.0_6624
		20	0–100	-0.0_31233	-0.0_53529	-0.0_75327
		25	0–100	-0.0_31171	-0.0_5904	-0.0_656
Acetonitrile	C_2H_3N	15	0– 16	-0.0_31175	-0.0_42024	
Allyl alcohol	C_3H_6O	0	0– 89	-0.0_33729	-0.0_41232	$+0.0_72984$
Benzenepentacarboxylic acid	$C_{11}H_6O_{1C}$	25	0– 0.6	$+0.0_35615$	-0.0_3117	
Butyl alcohol (n-)	$C_4H_{10}O$	20	0– 7.9	-0.0_31651	$+0.0_4285$	
Butyric acid (n-)	$C_4H_8O_2$	18	0– 10	$+0.0_3414$	$+0.0_4131$	
		25	0– 62	$+0.0_55135$	-0.0_4166	$+0.0_611$
Chloral hydrate	$C_2H_3Cl_3O_2$	0	0– 70	$+0.0_24489$	$+0.0_42802$	-0.0_71291
		15	0– 78	$+0.0_24455$	$+0.0_42198$	$+0.0_74366$
		30	0– 90	$+0.0_24401$	$+0.0_41887$	$+0.0_76549$
Chloroacetic acid	$C_2H_3ClC_2$	20	0– 32	$+0.0_23648$	$+0.0_5302$	
		25	0– 86	$+0.0_23602$	$+0.0_5552$	$+0.0_722$
Citric acid (hydrate)	$C_6H_3O_7 - H_2O$	18	0– 50	$+0.0_23824$	$+0.0_41141$	$+0.0_717$
Dichloroacetic acid	$C_2H_2Cl_2O_2$	20	0– 30	$+0.0_24427$	$+0.0_5537$	$+0.0_77534$
		25	0– 97	$+0.0_24427$	$+0.0_5537$	$+0.0_77534$
Diethylamine hydrochloride	$C_4H_{12}ClN$	21	0– 36	$+0.0_334$	$+0.0_676$	
Ethylamine hydrochloride	C_2H_8ClN	21	0– 65	$+0.0_31193$	-0.0_5307	-0.0_747
Ethylene glycol	$C_2H_6O_2$	0	0–100	$+0.0_21483$	$+0.0_52992$	-0.0_75248
		15	0– 6	$+0.0_2133$	-0.0_5108	
Ethyl ether	$C_4H_{10}O$	20	0– 5	-0.0_2221	$+0.0_448$	
		25	0– 4.5	-0.0_2221	$+0.0_435$	
tartrate	$C_8H_{14}O_6$	15	0– 95	$+0.0_22367$	$+0.0_5358$	-0.0_76005
Formaldehyde	CH_2O	15	0– 40	$+0.0_22518$	-0.0_5658	$+0.0_6542$
Formamide	CH_3NO	25	22– 96	$+0.0_21217$	$+0.0_53199$	-0.0_72529
Furfural	$C_5H_4O_2$	20	0– 8	$+0.0_21827$	$+0.0_5366$	
		25	0– 8	$+0.0_21664$	$+0.0_421$	
Isoamyl alcohol	$C_5H_{12}O$	20	0– 2.5	$+0.0_2155$	$+0.0_43$	
Isobutyl alcohol	$C_4H_{10}O$	15	0– 8	-0.0_2146	$+0.0_66$	
		20	0– 8	-0.0_2169	$+0.0_438$	
Isobutyric acid	$C_4H_8O_2$	15	0– 9	$+0.0_352$		
		18	0– 9	$+0.0_345$		
		25	0– 12	$+0.0_337$		
Isovaleric acid	$C_5H_{10}O_2$	25	0– 5	$+0.0_3253$	-0.0_4282	
Lactic acid	C_3H_6O	25	0– 9	$+0.0_2231$	$+0.0_5186$	
Maleic acid	$C_4H_4O_4$	25	0– 40	$+0.0_234$	$+0.0_575$	
Malic acid	$C_4H_6O_5$	20	0– 40	$+0.0_23933$	$+0.0_5957$	
		25	0– 40	$+0.0_23736$	$+0.0_4175$	
Malonic acid	$C_3H_4O_4$	20	0– 40	$+0.0_2389$	$+0.0_41066$	
Methyl acetate	$C_3H_6O_2$	20	0– 20	$+0.0_340$	-0.0_574	
glucoside (α-)	$C_7H_{14}O_6$	0	26– 51	$+0.0_23336$	$+0.0_5996$	$+0.01544$
		30	26– 51	$+0.0_23151$	$+0.0_5975$	$+0.0_8978$
Nicotine	$C_{10}H_{14}N_2$	20	0– 60	$+0.0_3642$	$+0.0_5454$	-0.0_7687
Nitrophenol (p-)	$C_6H_5NO_3$	15	0– 1.5	$+0.0_23216$	-0.0_455	
Oxalic acid	$C_2H_2O_4$	0	0– 4	$+0.0_55898$	-0.0_33185	$+0.0_441$
		15	0– 4	$+0.0_2494$	-0.0_58	
		17.5	0– 9	$+0.0_2494$	-0.0_58	
		20	0– 4	$+0.0_25264$	-0.0_31996	$+0.0_4254$
		25	0– 4	$+0.0_25108$	-0.0_31607	$+0.0_4208$
Phenol	C_6H_6O	15	0– 5	$+0.0_2111$	-0.0_4283	
		80	0– 65	$+0.0_3462$	-0.0_686	
Phenylglycolic acid	$C_8H_8O_3$	25	0– 11	$+0.0_2207$	$+0.0_423$	
Picoline (α-)	C_6H_7N	25	0– 70	-0.0_4386	-0.0_51405	-0.0_74167
(β-)	C_6H_7N	25	0– 60	-0.0_4683	-0.0_513	
Propionic acid	$C_3H_6O_2$	18	0– 10	$+0.0_395$	-0.0_4172	
		25	0– 40	$+0.0_39245$	-0.0_599	$+0.0_7361$
Pyridine	C_5H_5N	25	0– 60	$+0.0_3229$	-0.0_5204	-0.0_628
Resorcinol	$C_6H_6O_2$	18	0– 52	$+0.0_2201$	$+0.0_5519$	-0.0_819
Succinic acid	$C_4H_6O_4$	25	0– 5.5	$+0.0_2304$		
Tartaric acid (d, l, or dl)	$C_4H_6O_6$	15	0– 15	$+0.0_24482$	$+0.0_4185$	
		17.5	0– 50	$+0.0_24455$	$+0.0_4185$	
		20	0– 50	$+0.0_24432$	$+0.0_41837$	
		30	0– 50	$+0.0_24335$	$+0.0_4185$	
		40	0– 50	$+0.0_24265$	$+0.0_4185$	
		50	0– 50	$+0.0_24205$	$+0.0_4185$	
		60	0– 50	$+0.0_24155$	$+0.0_4185$	

*From "International Critical Tables," vol. 3, pp. 111–114.

TABLE 2-117   Densities of Aqueous Solutions of Miscellaneous Organic Compounds (*Concluded*)

Section A $d = d_w + Ap_s + Bp_s^2 + Cp_s^3$ (*Cont.*)

Name	Formula	t, °C	Range, p_s	A	B	C
Tetraethyl ammonium chloride	$C_8H_{20}ClN$	21	0– 63	$+0.0_3 1884$	$+0.0_5 6$	$+0.0_7 122$
Thiourea	CH_4N_2S	15	0– 7	$+0.0_2 2995$	$+0.0_5 374$	
Trichloroacetic acid	$C_2HCl_3O_2$	12.5	0– 61	$+0.0_2 499$	$+0.0_4 153$	
		20	10– 30	$+0.0_5 5053$	$+0.0_4 1387$	
		25	0– 94	$+0.0_5 5051$	$+0.0_5 6119$	$+0.0_6 1038$
Triethylamine hydrochloride	$C_6H_{16}ClN$	21	0– 54	$+0.0_5 6$	$+0.0_5 558$	$-0.0_6 69$
Trimethyl carbinol	$C_4H_{10}O$	20	0–100	$-0.0_2 117$	$-0.0_4 1908$	$+0.0_7 957$
		25	0–100	$-0.0_2 1286$	$-0.0_4 176$	$+0.0_7 887$
Urea	CH_4N_2O	14.8	0– 12	$+0.0_2 3213$	$-0.0_4 4802$	$+0.0_5 1216$
		18	0– 51	$+0.0_2 2718$	$+0.0_5 1552$	$+0.0_7 2573$
		20	0– 35	$+0.0_2 2702$	$+0.0_5 3712$	$-0.0_7 2285$
		25	0– 10	$+0.0_2 2728$	$-0.0_4 1817$	$+0.0_5 1379$
Urethane	$C_3H_7NO_2$	20	0– 56	$+0.0_2 1278$	$-0.0_5 245$	$-0.0_7 3437$
Valeric acid (*n-*)	$C_5H_{10}O_2$	25	0– 3	$+0.0_3 34$	$-0.0_4 27$	

Section B $d = d_s + Ap_w + Bp_w^2 + Cp_w^3$

Name	Formula	d_s	t, °C	Range, p_w	A	B	C
Butyl alcohol (*n-*)	$C_4H_{10}O$	0.8097	20	0–20	$+0.0_2 2103$	$-0.0_4 113$	
Butyric acid (*n-*)	$C_4H_8O_2$	0.9534	25	0–38	$+0.0_2 1854$	$-0.0_4 2314$	
Ethyl ether	$C_4H_{10}O$	0.7077	25	0– 1.1	$+0.0_2 34$	$+0.0_3 36$	
Isobutyl alcohol	$C_4H_{10}O$	0.8170	0	0–14	$+0.0_2 2437$	$-0.0_4 285$	
		0.8055	15	0–16	$+0.0_2 224$	$-0.0_4 129$	
Isobutyric acid	$C_4H_8O_2$	0.9425	26	0–80	$+0.0_2 1808$	$-0.0_4 2358$	$+0.0_6 1253$
Nicotine	$C_{10}H_{14}N_2$	1.0093	20	0–40	$+0.0_2 199$	$-0.0_4 331$	$+0.0_7 315$
Picoline (α-)	C_6H_7N	0.9404	25	0–30	$+0.0_2 2715$	$-0.0_4 393$	
(β-)	C_6H_7N	0.9515	25	0–40	$+0.0_2 1925$	$-0.0_4 352$	$+0.0_6 25$
Pyridine	C_5H_5N	0.9776	25	0–40	$+0.0_2 1157$	$-0.0_5 536$	$-0.0_6 2$
Trimethyl carbinol	$C_4H_{10}O$	0.7856	20	0–20	$+0.0_2 2287$	$+0.0_5 275$	

Section C $d_t = d_o + At + Bt^2$

Name	Formula	p_s	d_o	Range, °C	A	B
Allyl alcohol	C_3H_6O	76.60	0.9122	0–45	$-0.0_3 8$	$-0.0_5 27$
Butyl alcohol (*n-*)	$C_4H_{10}O$	80.95	0.8614	0–43	$-0.0_3 7292$	$-0.0_6 75$
Chloral hydrate	$C_2H_3Cl_3O_2$	2.00	1.0094	7–80	$-0.0_4 2597$	$-0.0_5 4313$
		10.00	1.0476	7–80	$-0.0_4 7955$	$-0.0_5 4253$
Ethyl tartrate	$C_7H_{14}O_6$	5.00	1.0150	15–80	$-0.0_3 2103$	$-0.0_5 2544$
		10.00	1.0270	15–80	$-0.0_3 2116$	$-0.0_6 2929$
		25.00	1.0665	15–80	$-0.0_3 401$	$-0.0_5 23$
Furfural	$C_5H_4O_2$	4.62	1.0125	22–74	$-0.0_3 232$	$-0.0_5 254$
		5.69	1.0140	22–74	$-0.0_3 221$	$-0.0_5 268$
		6.56	1.0155	22–74	$-0.0_3 211$	$-0.0_5 290$
Pyridine	C_5H_5N	9.34	1.0055	11–73	$-0.0_3 171$	$-0.0_5 3615$
		21.20	1.0115	14–73	$-0.0_3 378$	$-0.0_5 248$
		29.50	1.0145	12–72	$-0.0_3 463$	$-0.0_5 235$
		40.40	1.0182	9–74	$-0.0_3 605$	$-0.0_5 167$

DENSITIES OF MISCELLANEOUS MATERIALS

TABLE 2-118 Approximate Specific Gravities and Densities of Miscellaneous Solids and Liquids*

Water at 4°C and normal atmospheric pressure taken as unity. For more detailed data on any material, see the section dealing with the properties of that material.

Substance	Sp. gr.	Aver. weight lb/ft³	Substance	Sp. gr.	Aver. weight lb/ft³	Substance	Sp. gr.	Aver. weight lb/ft³
Metals, Alloys, Ores			**Timber, Air-dry**			**Dry Rubble Masonry**		
Aluminum, cast-hammered	2.55–2.80	165	Apple	0.66–0.74	44	Granite, syenite, gneiss	1.9–2.3	130
bronze	7.7	481	Ash, black	0.55	34	Limestone, marble	1.9–2.1	125
Brass, cast-rolled	8.4–8.7	534	white	0.64–0.71	42	Sandstone, bluestone	1.8–1.9	110
Bronze, 7.9 to 14% Sn	7.4–8.9	509	Birch, sweet, yellow	0.71–0.72	44			
phosphor	8.88	554	Cedar, white, red	0.35	22	**Brick Masonry**		
						Hard brick	1.8–2.3	128
Copper, cast-rolled	8.8–8.95	556	Cherry, wild red	0.43	27	Medium brick	1.6–2.0	112
ore, pyrites	4.1–4.3	262	Chestnut	0.48	30	Soft brick	1.4–1.9	103
German silver	8.58	536	Cypress	0.45–0.48	29	Sand-lime brick	1.4–2.2	112
Gold, cast-hammered	19.25–19.35	1205	Elm, white	0.56	35			
coin (U.S.)	17.18–17.2	1073	Fir, Douglas	0.48–0.55	32	**Concrete Masonry**		
						Cement, stone, sand	2.2–2.4	144
Iridium	21.78–22.42	1383	balsam	0.40	25	slag, etc.	1.9–2.3	130
Iron, gray cast	7.03–7.13	442	Hemlock	0.45–0.50	29	cinder, etc.	1.5–1.7	100
cast, pig	7.2	450	Hickory	0.74–0.80	48			
wrought	7.6–7.9	485	Locust	0.67–0.77	45	**Various Building Materials**		
spiegeleisen	7.5	468	Mahogany	0.56–0.85	44	Ashes, cinders	0.64–0.72	40–45
						Cement, Portland, loose	1.5	94
ferro-silicon	6.7–7.3	437	Maple, sugar	0.68	43	Lime, gypsum, loose	0.85–1.00	53–64
ore, hematite	5.2	325	white	0.53	33	Mortar, lime, set	1.4–1.9	103
ore, limonite	3.6–4.0	237	Oak, chestnut	0.74	46	Portland cement	2.08–2.25	94–135
ore, magnetite	4.9–5.2	315	live	0.87	54			
slag	2.5–3.0	172	red, black	0.64–0.71	42	Portland cement	3.1–3.2	196
						Slags, bank slag	1.1–1.2	67–72
Lead	11.34	710	white	0.77	48	bank screenings	1.5–1.9	98–117
ore, galena	7.3–7.6	465	Pine, Norway	0.55	34	machine slag	1.5	96
Manganese	7.42	475	Oregon	0.51	32	slag sand	0.8–0.9	49–55
ore, pyrolusite	3.7–4.6	259	red	0.48	30			
Mercury	13.6	849	Southern	0.61–0.67	38–42	**Earth, etc., Excavated**		
			white	0.43	27	Clay, dry	1.0	63
Monel metal, rolled	8.97	555				damp plastic	1.76	110
Nickel	8.9	537	Poplar	0.43	27	and gravel, dry	1.6	100
Platinum, cast-hammered	21.5	1330	Redwood, California	0.42	26	Earth, dry, loose	1.2	76
Silver, cast-hammered	10.4–10.6	656	Spruce, white, red	0.45	28	dry, packed	1.5	95
Steel, cold-drawn	7.83	489	Teak, African	0.99	62	moist, loose	1.3	78
machine	7.80	487	Indian	0.66–0.88	48	moist, packed	1.6	96
tool	7.70–7.73	481	Walnut, black	0.59	37	mud, flowing	1.7	108
Tin, cast-hammered	7.2–7.5	459	Willow	0.42–0.50	28	mud, packed	1.8	115
cassiterite	6.4–7.0	418				Riprap, limestone	1.3–1.4	80–85
Tungsten	19.22	1200	**Various Liquids**					
			Alcohol, ethyl (100%)	0.789	49	Riprap, sandstone	1.4	90
Zinc, cast-rolled	6.9–7.2	440	methyl (100%)	0.796	50	Riprap, shale	1.7	105
blende	3.9–4.2	253	Acid, muriatic, 40%	1.20	75	Sand, gravel, dry, loose	1.4–1.7	90–105
			nitric, 91%	1.50	94	gravel, dry, packed	1.6–1.9	100–120
Various Solids			sulfuric, 87%	1.80	112	gravel, wet	1.89–2.16	126
Cereals, oats, bulk	0.51	26						
barley, bulk	0.62	39	Chloroform	1.500	95	**Excavations in Water**		
corn, rye, bulk	0.73	45	Ether	0.736	46	Clay	1.28	80
wheat, bulk	0.77	48	Lye, soda, 66%	1.70	106	River mud	1.44	90
Cork	0.22–0.26	15	Oils, vegetable	0.91–0.94	58	Sand or gravel	0.96	60
			mineral, lubricants	0.88–0.94	57	and clay	1.00	65
Cotton, flax, hemp	1.47–1.50	93				Soil	1.12	70
Fats	0.90–0.97	58	Turpentine	0.861–0.867	54	Stone riprap	1.00	65
Flour, loose	0.40–0.50	28	Water, 4°C max. density	1.0	62.428			
pressed	0.70–0.80	47	100°C	0.9584	59.830	**Minerals**		
Glass, common	2.40–2.80	162	ice	0.88–0.92	56	Asbestos	2.1–2.8	153
			snow, fresh fallen	0.125	8	Barytes	4.50	281
plate or crown	2.45–2.72	161				Basalt	2.7–3.2	184
crystal	2.90–3.00	184	sea water	1.02–1.03	64	Bauxite	2.55	159
dint	3.2–4.7	247				Bluestone	2.5–2.6	159
Hay and straw, bales	0.32	20	**Ashlar Masonry**					
Leather	0.86–1.02	59	Bluestone	2.3–2.6	153	Borax	1.7–1.8	109
			Granite, syenite, gneiss	2.4–2.7	159	Chalk	1.8–2.8	143
Paper	0.70–1.15	58	Limestone	2.1–2.8	153	Clay, marl	1.8–2.6	137
Potatoes, piled	0.67	44	Marble	2.4–2.8	162	Dolomite	2.9	181
Rubber, caoutchouc	0.92–0.96	59	Sandstone	2.0–2.6	143	Feldspar, orthoclase	2.5–2.7	162
goods	1.0–2.0	94						
Salt, granulated, piled	0.77	48	**Rubble Masonry**			Gneiss	2.7–2.9	175
			Bluestone	2.2–2.5	147	Granite	2.6–2.7	165
Saltpeter	1.07	67	Granite, syenite, gneiss	2.3–2.6	153	Greenstone, trap	2.8–3.2	187
Starch	1.53	96	Limestone	2.0–2.7	147	Gypsum, alabaster	2.3–2.8	159
Sulfur	1.93–2.07	125	Marble	2.3–2.7	156	Hornblende	3.0	187
Wool	1.32	82	Sandstone	1.9–2.5	137	Limestone	2.1–2.86	155
						Marble	2.6–2.86	170
						Magnesite	3.0	187
						Phosphate rock, apatite	3.2	200
						Porphyry	2.6–2.9	172

*From Marks, *Mechanical Engineers' Handbook,* McGraw-Hill.

TABLE 2-118   Approximate Specific Gravities and Densities of Miscellaneous Solids and Liquids (*Concluded*)

Water at 4°C and normal atmospheric pressure taken as unity. For more detailed data on any material, see the section dealing with the properties of that material.

Substance	Sp. gr.	Aver. weight lb/ft³	Substance	Sp. gr.	Aver. weight lb/ft³	Substance	Sp. gr.	Aver. weight lb/ft³
Minerals (*Cont.*)			Bituminous Substances			Bituminous Substances (*Cont.*)		
Pumice, natural	0.37–0.90	40	Asphaltum	1.1–1.5	81	Petroleum	0.87	54
Quartz, flint	2.5–2.8	165	Coal, anthracite	1.4–1.8	97	refined (kerosene)	0.78–0.82	50
Sandstone	2.0–2.6	143	bituminous	1.2–1.5	84	benzine	0.73–0.75	46
Serpentine	2.7–2.8	171	lignite	1.1–1.4	78	gasoline	0.70–0.75	45
Shale, slate	2.6–2.9	172	peat, turf, dry	0.65–0.85	47	Pitch	1.07–1.15	69
						Tar, bituminous	1.20	75
Soapstone, talc	2.6–2.8	169	charcoal, pine	0.28–0.44	23			
Syenite	2.6–2.7	165	charcoal, oak	0.47–0.57	33	Coal and Coke, Piled		
			coke	1.0–1.4	75	Coal, anthracite	0.75–0.93	47–58
Stone, Quarried, Piled			Graphite	1.64–2.7	135	bituminous, lignite	0.64–0.87	40–54
Basalt, granite, gneiss	1.5	96	Paraffin	0.87–0.91	56	peat, turf	0.32–0.42	20–26
Greenstone, hornblende	1.7	107				charcoal	0.16–0.23	10–14
Limestone, marble, quartz	1.5	95				coke	0.37–0.51	23–32
Sandstone	1.3	82						
Shale	1.5	92						

NOTE: To convert pounds per cubic foot to kilograms per cubic meter, multiply by 16.02. °F = ⅘ °C + 32.

TABLE 2-119   Density (kg/m³) of Selected Elements as a Function of Temperature

Temperature, K°	Element symbol												
	Al	Be†	Cr	Cu	Au	Ir	Fe	Pb	Mo	Ni	Pt	Ag	Zn†
50	2736	3650	7160	9019	19,490	22,600	7910	11,570	10,260	8960	21,570	10,620	7280
100	2732	3640	7155	9009	19,460	22,580	7900	11,520	10,260	8950	21,550	10,600	7260
150	2726	3630	7150	8992	19,420	22,560	7890	11,470	10,250	8940	21,530	10,575	7230
200	2719	3620	7145	8973	19,380	22,540	7880	11,430	10,250	8930	21,500	10,550	7200
250	2710	3610	7140	8951	19,340	22,520	7870	11,380	10,250	8910	21,470	10,520	7170
300	2701	3600	7135	8930	19,300	22,500	7860	11,330	10,240	8900	21,450	10,490	7135
400	2681	3580	7120	8885	19,210	22,450	7830	11,230	10,220	8860	21,380	10,430	7070
500	2661	3555	7110	8837	19,130	22,410	7800	11,130	10,210	8820	21,330	10,360	7000
600	2639	3530	7080	8787	19,040	22,360	7760	11,010	10,190	8780	21,270	10,300	6935
800	2591		7040	8686	18,860	22,250	7690	10,430	10,160	8690	21,140	10,160	6430
1000	2365	——	7000	8568	18,660	22,140	7650	10,190	10,120	8610	21,010	10,010	6260
1200	2305		6945	8458	18,440	22,030	7620	9,940	10,080	8510	20,870	9,850	
1400	2255		6890	7920	17,230	21,920	7520		10,040	8410	20,720	9,170	
1600			6760	7750	16,950	21,790	7420		10,000	8320	20,570	8,980	
1800			6700	7600		21,660	7320		9,950	7690	20,400		
2000			7460			21,510	7030		9,900	7450	20,220		

NOTE: Above the horizontal line the condensed phase is solid; below the line, it is liquid.
°°R = ⅘ K.
†Polycrystalline form tabulated. Similar tables for an additional 45 elements appear in the *Handbook of Heat Transfer,* 2d ed., McGraw-Hill, New York, 1984.

SOLUBILITIES

UNITS CONVERSIONS

For this subsection, the following units conversions are applicable:

$$°F = ⅘ \, °C + 32.$$

To convert cubic centimeters to cubic feet, multiply by 3.532×10^{-5}.

To convert millimeters of mercury to pounds-force per square inch, multiply by 0.01934.

To convert grams per liter to pounds per cubic foot, multiply by 6.243×10^{-2}.

TABLE 2-120 Solubilities of Inorganic Compounds in Water at Various Temperatures*

This table shows the amount of anhydrous substance that is soluble in 100 g of water at the temperature in degrees Celsius as indicated; when the name is followed by †, the value is expressed in grams of substance in 100 cm³ of saturated solution. Solid phase gives the hydrated form in equilibrium with the saturated solution.

#	Substance	Formula	Solid phase	0°C	10°C	20°C	30°C	40°C	50°C	60°C	70°C	80°C	90°C	100°C	#
1	Aluminum chloride	$AlCl_3$	$6H_2O$			69.86^{15°									1
2	sulfate	$Al_2(SO_4)_3$	$18H_2O$	31.2	33.5	36.4	40.4	46.1	52.2	59.2	66.1	73.0	80.8	89.0	2
3	Ammonium aluminum sulfate	$(NH_4)_2Al_2(SO_4)_4$	$24H_2O$	2.1	4.99	7.74	10.94	14.58	20.10	26.70				109.7^{90°	3
4	bicarbonate	NH_4HCO_3		11.9	15.8	21	27								4
5	bromide	NH_4Br		60.6	68	75.5	83.2	91.1	99.2	107.8	116.8	126	135.6	145.6	5
6	chloride	NH_4Cl		29.4	33.3	37.2	41.4	45.8	50.4	55.2	60.2	65.6	71.3	77.3	6
7	chloroplatinate	$(NH_4)_2PtCl_6$			0.7									1.25	7
8	chromate	$(NH_4)_2CrO_4$					40.4								8
9	chromium sulfate	$(NH_4)_2Cr_2(SO_4)_4$	$24H_2O$			10.78^{25°									9
10	dichromate	$(NH_4)_2Cr_2O_7$					47.17								10
11	dihydrogen phosphate	$NH_4H_2PO_4$		171		$190^{4.5^\circ}$	$260^{21.0}$								11
12	hydrogen phosphate	$(NH_4)_2HPO_4$				131^{15}									12
13	iodide	NH_4I		154.2	163.2	172.3	181.4	190.5	199.6	208.9	218.7	228.8		250.3	13
14	magnesium phosphate	NH_4MgPO_4	$6H_2O$	0.023		0.052		0.030	0.000	0.010	0.016	0.019			14
15	manganese phosphate	NH_4MnPO_4	$7H_2O$			0		0		0	0.005	0.007			15
16	nitrate	NH_4NO_3		118.3		192	241.8	297.0	344.0	421.0	499.0	580.0	740.0	871.0	16
17	oxalate	$(NH_4)_2C_2O_4$	$1H_2O$	2.2	3.1	4.4	5.9	8.0	10.3						17
18	perchlorate†	NH_4ClO_4†		11.56		20.85		30.58		39.05		48.19		57.01	18
19	persulfate	$(NH_4)_2S_2O_8$		58.2											19
20	sulfate	$(NH_4)_2SO_4$		70.6	73.0	75.4	78.0	81.0		88.0		95.3		103.3	20
21	thiocyanate	NH_4CNS		119.8	144	170	207.7								21
22	vanadate (meta)	NH_4VO_3				0.48	0.84	1.32	1.78		3.05				22
23	Antimonious fluoride	SbF_3		384.7		444.7	563.6								23
24	sulfide	Sb_2S_3				0.000175^{18°									24
25	Arsenic oxide	As_2O_3		59.5	62.1	65.8	69.5	71.2		73.0		75.1		76.7	25
26	Arsenious sulfide	As_2S_3		5.17×10^{-5} at 18°											26
27	Barium acetate	$Ba(C_2H_3O_2)_2$	$3H_2O$	59	63	71	75	79	77	74	74			75	27
28	acetate	$Ba(C_2H_3O_2)_2$	$1H_2O$												28
29	carbonate	$BaCO_3$			0.0016^{18°	0.0022^{18°	0.0024 at 24.2°								29
30	chlorate	$Ba(ClO_3)_2$	$1H_2O$	20.34	26.95	33.80	41.70	49.61	43.6	66.81		84.84		104.9	30
31	chloride	$BaCl_2$	$2H_2O$	31.6	33.3	35.7	38.2	40.7	43.6	46.4	49.4	52.4		58.8	31
32	chromate	$BaCrO_4$		0.0002	0.00028	0.00037	0.00046								32
33	hydroxide	$Ba(OH)_2$	$8H_2O$	1.67	2.48	3.89	5.59	8.22	13.12	20.94		101.4			33
34	iodide	BaI_2	$6H_2O$	170.2	185.7	203.1	219.6	231.9		247.3		261.0		271.7	34
35	iodide	BaI_2	$2H_2O$												35
36	nitrate	$Ba(NO_3)_2$		5.0	7.0	9.2	11.6	14.2	17.1	20.3		27.0		34.2	36
37	nitrite	$Ba(NO_2)_2$	$1H_2O$			67.5						205.8		300	37
38	oxalate	BaC_2O_4			0.0016^{18°	0.0022^{18°	0.0024 at 24.2°								38
39	perchlorate	$Ba(ClO_4)_2$	$3H_2O$	205.8		289.1		358.7	426.3		495.2		562.3		39
40	sulfate	$BaSO_4$		1.15×10^{-4}	2.0×10^{-4}	2.4×10^{-4}	2.85×10^{-4}								40
41	Beryllium sulfate	$BeSO_4$	$6H_2O$				52		60.67		62		83	100	41
42	sulfate	$BeSO_4$	$4H_2O$				43.78	46.74				84.76	98	110	42
43	sulfate	$BeSO_4$	$2H_2O$												43
44	Boric acid	H_3BO_3		2.66	3.57	5.04	6.60	8.72	11.54	14.81	16.73	23.75	30.38	40.25	44
45	Boron oxide	B_2O_3		1.1	1.5	2.2	3.13	4.0		6.2		9.5		15.7	45
46	Bromine	Br_2		4.22	3.4	3.20									46
47	Cadmium chloride	$CdCl_2$	$4H_2O$	97.59	125.1										47
48	chloride	$CdCl_2$	$2\tfrac{1}{2}H_2O$												48
49	chloride	$CdCl_2$	$1H_2O$	90.01	135.1	134.5	132.1	135.3		136.5		140.4		147.0	49
50	cyanide	$Cd(CN)_2$				1.7^{18°									50
51	hydroxide	$Cd(OH)_2$					2.6×10^{-4} at 25°								51
52	sulfate	$CdSO_4$	$2H_2O$	76.48	76.00	76.60		78.54		83.68			63.13	60.77	52
53	Calcium acetate	$Ca(C_2H_3O_2)_2$	$2H_2O$	37.4	36.0	34.7	33.8	33.2	32.7	32.7	31.1	33.5	31.1	29.7	53
54	acetate	$Ca(C_2H_3O_2)_2$	$1H_2O$												54

* By N. A. Lange. Abridged from "Table of Solubilities of Inorganic Compounds in Water at Various Temperatures" in Lange, *Handbook of Chemistry*, 10th ed., McGraw-Hill, New York, 1961. For tables of the solubility of gases in water at various temperatures, Atack (*Handbook of Chemical Data*, Reinhold, New York, 1957) gives values at closer temperature intervals, usually 1 or 5°C, than are tabulated here. For materials marked by †, additional data are given in tables subsequent to this one. For the solubility of various hydrocarbons in water at high pressures see *J. Chem. Eng. Data*, **4**, 212 (1959).

TABLE 2-120 Solubilities of Inorganic Compounds in Water at Various Temperatures (Continued)

No.	Substance	Formula	Solid phase	0°C	10°C	20°C	30°C	40°C	50°C	60°C	70°C	80°C	90°C	100°C	No.
1	Calcium bicarbonate	Ca(HCO$_3$)$_2$		16.15		16.60		17.05		17.50		17.95		18.40	1
2	chloride	CaCl$_2$	6H$_2$O	59.5	65.0	74.5	102								2
3	chloride	CaCl$_2$	2H$_2$O						128	136.8	141.7	147.0	152.7	159	3
4	fluoride	CaF$_2$				0.0016$^{18°}$	0.0017$^{26°}$								4
5	hydroxide	Ca(OH)$_2$		0.185	0.176	0.165	0.153	0.141	0.128	0.116	0.106	0.094	0.085	0.077	5
6	nitrate	Ca(NO$_3$)$_2$	4H$_2$O	102.0	115.3	129.3	152.6	195.9							6
7	nitrate	Ca(NO$_3$)$_2$	3H$_2$O					237.5	281.5			358.7		363.6	7
8	nitrite	Ca(NO$_2$)$_2$	4H$_2$O	62.07		76.68				132.6	151.9		244.8		8
9	nitrite	Ca(NO$_2$)$_2$													9
10	nitrite	Ca(NO$_2$)$_2$	2H$_2$O												10
11	oxalate	CaC$_2$O$_4$			6.7×10^{-4} at 13°	6.8×10^{-5} at 25°	9.5×10^{-4} at 50°	14×10^{-4} at 95°							11
12	sulfate	CaSO$_4$	2H$_2$O	0.1759	0.1928		0.2090	0.2097		0.2047	0.1966			0.1619	12
13	Carbon dioxide, 760 mm ‡	CO$_2$		0.3346	0.2318	0.1688	0.1257	0.0973	0.0761	0.0576			0.0006	0	13
14	monoxide, 760 mm ‡	CO		0.0044	0.0035	0.0028	0.0024	0.0021	0.0018	0.0015	0.0013	0.0010		0	14
15	Cesium chloride	CsCl		161.4	174.7	186.5	197.3	208.0	218.5	229.7	239.5	250.0	260.1	270.5	15
16	nitrate	CsNO$_3$		9.33	14.9	23.0	33.9	47.2	64.4	83.8	107.0	134.0	163.0	197.0	16
17	sulfate	Cs$_2$SO$_4$		167.1	173.1	178.7	184.1	189.9	194.9	199.9	205.0	210.3	214.9	220.3	17
18	Chlorine, 760 mm ‡	Cl$_2$		1.46	0.980	0.716	0.562	0.451	0.386	0.324	0.274	0.219	0.125		18
19	Chromic anhydride	CrO$_3$		164.9				174.0	182.1				217.5	206.8	19
20	Cupric chloride	CuCl$_2$	2H$_2$O	70.7	73.76	77.0	80.34	83.8	87.44	91.2		99.2		107.9	20
21	nitrate	Cu(NO$_3$)$_2$	6H$_2$O	81.8	95.28	125.1		159.8		178.8		207.8			21
22	nitrate	Cu(NO$_3$)$_2$	3H$_2$O												22
23	sulfate	CuSO$_4$	5H$_2$O	14.3	17.4	20.7	25	28.5	33.3	40		55		75.4	23
24	sulfide	CuS				3.3×10^{-5} at 18°									24
25	Cuprous chloride	CuCl				1.52×10^{-3} at 25°									25
26	Ferric chloride	FeCl$_3$	4H$_2$O	74.4	81.9	91.8			315.1			525.8		535.7	26
27	chloride	FeCl$_3$		71.02											27
28	Ferrous chloride	FeCl$_2$	6H$_2$O		64.5		73.0	77.3	82.5	88.7		100	105.3	105.8	28
29	nitrate	Fe(NO$_3$)$_3$	6H$_2$O			83.8				165.6					29
30	sulfate	FeSO$_4$	7H$_2$O	15.65	20.51	26.5	32.9	40.2	48.6		50.9	43.6			30
31	sulfate	FeSO$_4$	1H$_2$O												31
32	Hydrobromic acid, 760 mm	HBr		221.2	210.3	198			171.5			115		130	32
33	Hydrochloric acid, 760 mm	HCl		82.3			67.3	63.3	59.6	56.1					33
34	Iodine	I$_2$				0.029	0.04	0.056	0.078						34
35	Lead acetate	Pb(C$_2$H$_3$O$_2$)$_2$	3H$_2$O				55.04$^{25°}$								35
36	bromide	PbBr$_2$		0.4554		0.85	1.15	1.53	1.94	2.36		3.34		4.75	36
37	carbonate	PbCO$_3$				0.00011									37
38	chloride	PbCl$_2$		0.6728		0.99	1.20	1.45	1.70	1.98		2.62		3.34	38
39	chromate	PbCrO$_4$				7×10^{-6}									39
40	fluoride	PbF$_2$			0.060	0.064	0.068								40
41	nitrate	Pb(NO$_3$)$_2$		38.8	48.3	56.5	66	75	85	95		115		38.8	41
42	sulfate	PbSO$_4$		0.0028	0.0035	0.0041	0.0049	0.0056							42
43	Magnesium bromide	MgBr$_2$	6H$_2$O	91.0	94.5	96.5	99.2	101.6	104.1	107.5		113.7		120.2	43
44	chloride	MgCl$_2$	6H$_2$O	52.8	53.5	54.5		57.5		61.0		66.0		73.0	44
45	hydroxide	Mg(OH)$_2$				0.0009$^{18°}$									45
46	nitrate	Mg(NO$_3$)$_2$	6H$_2$O	66.55				84.74					137.0		46
47	sulfate	MgSO$_4$	7H$_2$O	40.8	42.2	44.5	45.3	45.6	50.4	53.5	59.5	64.2	69.0	74.0	47
48	sulfate	MgSO$_4$	6H$_2$O			35.5						62.9		68.3	48
49	sulfate	MgSO$_4$	1H$_2$O		60.01										49
50	Manganous sulfate	MnSO$_4$	7H$_2$O	53.23		62.9	67.76	68.8							50
51	sulfate	MnSO$_4$	5H$_2$O		59.5	64.5	66.44		58.17						51
52	sulfate	MnSO$_4$	4H$_2$O												52
53	sulfate	MnSO$_4$	1H$_2$O							55.0	52.0	48.0	42.5	34.0	53
54	Mercurous chloride	HgCl		0.00014		0.0002		0.0007							54
55	Molybdic oxide	MoO$_3$	2H$_2$O			0.138	0.264	0.476	0.687	1.206	2.055	2.106			55
56	Nickel chloride	NiCl$_2$	6H$_2$O	53.9	59.5	64.2	68.9	73.3	78.3	82.2	85.2			87.6	56
57	nitrate	Ni(NO$_3$)$_2$	6H$_2$O	79.58		96.31		122.2		163.1	169.1				57
58	nitrate	Ni(NO$_3$)$_2$	3H$_2$O										235.1		58
59	sulfate	NiSO$_4$	7H$_2$O	27.22	32		42.46		50.15	54.80	59.44	63.17		76.7	59
60	sulfate	NiSO$_4$	6H$_2$O												60
61	Nitric oxide, 760 mm	NO		0.00984	0.00757	0.00618	0.00517	0.00440	0.00376	0.00324	0.00267	0.00199	0.00114	0	61
62	Nitrous oxide	N$_2$O			0.1705	0.1211									62

Dense solubility data table (continuation page). Each row lists a salt with its formula, water of crystallization, and its solubility in g (anhydrous salt) per 100 g water at successive, increasing temperatures (read left→right, low→high temperature).

No.	Salt	Formula	Water of crystn	Solubility (g per 100 g H₂O, increasing temperature →)
1	Potassium acetate	$KC_2H_3O_2$	$1\tfrac{1}{2}H_2O$	216.7, 233.9, 255.6, 283.8, 323.3, 337.3, 350, 364.8, 380.1, 396.3
2	acetate	$KC_2H_3O_2$	$\tfrac{1}{2}H_2O$	
3	alum	$K_2SO_4\!\cdot\!Al_2(SO_4)_3$	$24H_2O$	3.0, 4.0, 5.9, 8.39, 11.70, 17.00, 24.75, 40.0, 71.0, 109.0, 121.6
4	bicarbonate	$KHCO_3$		22.4, 27.7, 33.2, 39.1, 45.4, 51.4, 60.0, 67.3
5	bisulfate	$KHSO_4$		36.3, 133.1, 139.8, 147.5, 155.7
6	bitartrate	$KHC_4H_4O_6$		0.32, 0.40, 0.53, 0.90, 1.32, 1.83, 2.46, 4.6, 6.95
7	carbonate	K_2CO_3	$2H_2O$	105.5, 108, 110.5, 113.7, 116.9, 121.2, 126.8, 133.1, 139.8, 147.5, 155.7
8	chlorate	$KClO_3$		3.3, 5, 7.4, 10.5, 14, 19.3, 24.5, 38.5, 51.1, 54.0, 57
9	chloride	KCl		27.6, 31.0, 34.0, 37.0, 40.0, 42.6, 45.5, 48.3, 51.1, 54.0, 56.7
10	chromate	K_2CrO_4		58.2, 60.0, 61.7, 63.4, 65.2, 66.8, 68.6, 70.4, 72.1, 73.9, 75.6
11	dichromate	$K_2Cr_2O_7$		5, 7, 12, 20, 26, 34, 43, 52, 61, 70, 80
12	ferricyanide	$K_3Fe(CN)_6$		31, 36, 43, 50, 60, 66, 70, 80, 82.6^{104}
13	hydroxide	KOH	$2H_2O$	97, 103, 112, 126, 134, 140, 110.0, 138, 169, 202, 178
14	hydroxide	KOH	$1H_2O$	
15	nitrate	KNO_3		13.3, 20.9, 31.6, 45.8, 63.9, 85.5, 110.0, 138, 169, 202, 246
16	nitrite	KNO_2		278.8, 298.4, 334.9, 412.8
17	perchlorate	$KClO_4$		0.75, 1.05, 1.80, 2.6, 4.4, 6.5, 9, 11.8, 14.8, 18, 21.8
18	permanganate	$KMnO_4$		2.83, 4.4, 6.4, 9.0, 12.56, 16.89, 22.2
19	persulfate	$K_2S_2O_8$	†	1.62, 2.60, 4.49, 7.19, 9.89
20	sulfate	K_2SO_4		7.35, 9.22, 11.11, 12.97, 14.76, 16.50, 18.17, 19.75, 21.4, 22.8, 24.1
21	thiocyanate	$KCNS$		177.0, 217.5
22	Silver cyanide	$AgCN$		2.2×10^{-5}
23	nitrate	$AgNO_3$		122, 170, 222, 300, 376, 455, 525, 669, 952
24	sulfate	Ag_2SO_4		0.573, 0.695, 0.796, 0.888, 0.979, 1.08, 1.15, 1.22, 1.30, 1.36, 1.41
25	Sodium acetate	$NaC_2H_3O_2$	$3H_2O$	36.3, 40.8, 46.5, 54.5, 65.5, 83, 139, 146, 153, 161, 170
26	acetate	$NaC_2H_3O_2$		119, 121, 123.5, 126, 129.5, 134, 139.5
27	bicarbonate	$NaHCO_3$		6.9, 8.15, 9.6, 11.1, 12.7, 14.45, 16.4
28	carbonate	Na_2CO_3	$10H_2O$	7, 12.5, 21.5, 38.8, 48.5, 46.4, 45.5
29	carbonate	Na_2CO_3	$1H_2O$	50.5, 126, 155, 189, 230
30	chloride	$NaCl$		35.7, 35.8, 36.0, 36.3, 36.6, 37.0, 37.3, 37.8, 38.4, 39.0, 39.8
31	chloride	$NaCl$	$10H_2O$	31.70, 50.17, 88.7, 88.7
32	chromate	Na_2CrO_4	$4H_2O$	31.70, 50.17, 95.96, 104, 114.6, 123.0, 124.8, 125.9
33	chromate	Na_2CrO_4		
34	chromate	Na_2CrO_4		316.7, 376.2
35	dichromate	$Na_2Cr_2O_7$	$2H_2O$	163.0, 177.8, 244.8, 426.3
36	dichromate	$Na_2Cr_2O_7$		57.9, 69.9, 85.2, 106.5, 138.2, 158.6, 179.3, 190.3, 207.3, 225.3, 246.6
37	dihydrogen phosphate	NaH_2PO_4	$2H_2O$	57.9
38	dihydrogen phosphate	NaH_2PO_4	$1H_2O$	85
39	dihydrogen phosphate	NaH_2PO_4		
40	hydrogen arsenate	Na_2HAsO_4	$12H_2O$	7.3, 15.5, 26.5, 37, 47, 65, 85
41	hydrogen phosphate	Na_2HPO_4	$12H_2O$	1.67, 3.6, 7.7, 20.8, 51.8, 80.2, 82.9, 88.1, 92.4, 102.9, 102.2
42	hydrogen phosphate	Na_2HPO_4	$7H_2O$	
43	hydrogen phosphate	Na_2HPO_4	$2H_2O$	
44	hydrogen phosphate	Na_2HPO_4		
45	hydroxide	$NaOH$	$4H_2O$	42, 51.5, 109, 119, 129, 145, 174, 347
46	hydroxide	$NaOH$	$3\tfrac{1}{2}H_2O$	313
47	hydroxide	$NaOH$	$1H_2O$	
48	hydroxide	$NaOH$		73
49	nitrate	$NaNO_3$		73, 80, 88, 96, 104, 114, 124, 148, 180
50	nitrite	$NaNO_2$		72.1, 78.0, 84.5, 91.6, 98.4, 104.1, 124, 132.6, 163.2
51	oxalate	$Na_2C_2O_4$		3.7, 4.1, 5, 6.33
52	phosphate, tri-	Na_3PO_4	$12H_2O$	1.5, 4.1, 11, 20, 31, 43, 55, 81, 108
53	pyrophosphate	$Na_4P_2O_7$	$10H_2O$	3.16, 3.95, 6.23, 9.95, 13.50, 17.45, 21.83, 30.04, 40.26
54	sulfate	Na_2SO_4	$10H_2O$	5.0, 9.0, 19.4, 40.8, 48.8
55	sulfate	Na_2SO_4	$7H_2O$	19.5, 30, 44, 42.5
56	sulfide	Na_2S	$9H_2O$	15.42, 18.8, 22.5, 28.5, 45.3, 46.7, 45.3, 43.7, 42.5
57	sulfide	Na_2S	$5\tfrac{1}{2}H_2O$	42.69, 39.82, 45.73, 51.40, 59.23
58	sulfide	Na_2S	$6H_2O$	20, 26.9, 39.1, 36.4, 43.31, 49.14, 57.28
59	sulfide	Na_2S	$7H_2O$	
60	sulfite	Na_2SO_3	$10H_2O$	13.9, 20, 26.9, 36, 28, 28.2, 28.8, 28.3
61	sulfite	Na_2SO_3	$5H_2O$	2.7, 3.9, 10.5, 20.3
62	tetraborate	$Na_2B_4O_7$	$10H_2O$	1.3, 1.6, 2.7, 3.9, 24.4, 31.5, 41, 52.5
63	tetraborate	$Na_2B_4O_7$	$5H_2O$	
64	vanadate (meta)	$NaVO_3$	$2H_2O$	15.3^{350}, 30.2, 68.4

TABLE 2-120 Solubilities of Inorganic Compounds in Water at Various Temperatures (Concluded)

#	Substance	Formula	Solid phase	0°C	10°C	20°C	30°C	40°C	50°C	60°C	70°C	80°C	90°C	100°C
1	Sodium vanadate (meta)	$NaVO_3$				21.10^{50}		26.23		32.97	36.9	$38.8^{75°}$		
2	Stannous chloride	$SnCl_2$		83.9		$269.8^{15°}$								
3	sulfate	$SnSO_4$				19								18
4	Strontium acetate	$Sr(C_2H_3O_2)_2$	$4H_2O$	36.9	43.61	41.6								
5	acetate	$Sr(C_2H_3O_2)_2$	$\frac{1}{2}H_2O$		42.95		39.5		37.35		36.24	36.10		36.4
6	chloride	$SrCl_2$	$6H_2O$	43.5	47.7	52.9		65.3	72.4	81.8				
7	chloride	$SrCl_2$	$2H_2O$				58.7				85.9	90.5		100.8
8	nitrate	$Sr(NO_3)_2$	$1H_2O$	52.7		64.0								
9	nitrate	$Sr(NO_3)_2$	$4H_2O$	40.1		70.5	88.6	90.1		97.2			130.4	
10	nitrate	$Sr(NO_3)_2$							83.8	93.8	96	98	100	139
11	sulfate	$SrSO_4$		0.0113		0.0114	0.0114							
12	Sulfur dioxide, 760 mm †	SO_2		22.83	16.21	11.29	7.81	5.41	4.5					
13	Thallium sulfate	Tl_2SO_4		2.70	3.70	4.87	6.16		9.21	10.92	12.74	14.61	16.53	18.45
14	sulfate	$Th(SO_4)_2$	$9H_2O$	0.74	0.98	1.38	1.995	2.998	5.22					
15	sulfate	$Th(SO_4)_2$	$8H_2O$	1.0	1.25	1.62								
16	sulfate	$Th(SO_4)_2$	$6H_2O$	1.50		1.90	2.45	4.04	2.54	6.64				
17	sulfate	$Th(SO_4)_2$	$4H_2O$							1.63	1.09			
18	Zinc chlorate	$Zn(ClO_3)_2$	$6H_2O$	145.0	152.5	200.3	209.2	223.2	273.1					
19	chlorate	$Zn(ClO_3)_2$	$4H_2O$											
20	nitrate	$Zn(NO_3)_2$	$6H_2O$	94.78		118.3		206.9						
21	nitrate	$Zn(NO_3)_2$	$3H_2O$											
22	sulfate	$ZnSO_4$	$7H_2O$	41.9	47	54.4		70.1	76.8					
23	sulfate	$ZnSO_4$	$6H_2O$									86.6	83.7	
24	sulfate	$ZnSO_4$	$1H_2O$											80.8

The H in solubility tables (2-121 to 2-144) is the proportionality constant for the expression of Henry's law, $p = Hx$, where x = mole fraction of the solute in the liquid phase; p = partial pressure of the solute in the gas phase, expressed in atmospheres; and H = a proportionality constant expressed in units of atmospheres of solute pressure in the gas phase per unit concentration of the solute in the liquid phase. (The unit of concentration of the solute in the liquid phase is moles solute per mole solution.)

TABLE 2-121   Acetylene (C₂H₂)

t, °C	0	5	10	15	20	25	30
$10^{-3} \times H$°	0.72	0.84	0.96	1.08	1.21	1.33	1.46

International Critical Tables, vol. 3, p. 260, McGraw-Hill, 1928.
°H. See footnote for Table 2-122.

TABLE 2-122   Air

t, °C	0	5	10	15	20	25	30	35
$10^{-4} \times H$°	4.32	4.88	5.49	6.07	6.64	7.20	7.71	8.23

t, °C	40	45	50	60	70	80	90	100
$10^{-4} \times H$°	8.70	9.11	9.46	10.1	10.5	10.7	10.8	10.7

International Critical Tables, vol. 3, p. 257.
°H is calculated from the absorption coefficients of O_2 and N_2, taking into consideration the correction for constant argon content.

TABLE 2-123   Ammonia (NH₃)

Weight NH₃ per 100 weights H₂O	Partial pressure of NH₃, mm. Hg							
	0°C	10°C	20°C	25°C	30°C	40°C	50°C	60°C
100	947							
90	785							
80	636	987	1450			3300		
70	500	780	1170			2760		
60	380	600	945			2130		
50	275	439	686			1520		
40	190	301	470		719	1065		
30	119	190	298		454	692		
25	89.5	144	227		352	534	825	
20	64	103.5	166		260	395	596	834
15	42.7	70.1	114		179	273	405	583
10	25.1	41.8	69.6		110	167	247	361
7.5	17.7	29.9	50.0		79.7	120	179	261
5	11.2	19.1	31.7		51.0	76.5	115	165
4		16.1	24.9		40.1	60.8	91.1	129.2
3		11.3	18.2	23.5	29.6	45	67.1	94.3
2.5			15.0	19.4	24.4	(37.6)°	(55.7)	77.0
2			12.0	15.3	19.3	(30.0)	(44.5)	61.0
1.6				12.0	15.3	(24.1)	(35.5)	48.7
1.2				9.1	15.3	(18.3)	(26.7)	36.3
1.0				7.4	11.5	(15.4)	(22.2)	30.2
0.5				3.4				

°Extrapolated values.

TABLE 2-124   Ammonia (NH₃)—Low Pressures

Weight NH₃ per 100 weights H₂O	0.105	0.244	0.32	0.38	0.576	0.751	1.02
Partial pressure NH₃, mm. Hg, at 25°C	0.791	1.83	2.41	2.89	4.41	5.80	7.96
Weight NH₃ per 100 weights H₂O	1.31	1.53	1.71	1.98	2.11	2.58	2.75
Partial pressure NH₃, mm. Hg, at 25°C	10.31	11.91	13.46	15.75	16.94	20.86	22.38

"Landolt-Börnstein Physikalische-chemische Tabellen," Eg. I, p. 303, 1927. Phase-equilibrium data for the binary system NH₃-H₂O are given by Clifford and Hunter, *J. Phys. Chem.*, **37**, 101 (1933).

TABLE 2-125   Carbon Dioxide (CO₂)

Total pressure, atm	Weight of CO₂ per 100 weights of H₂O°								
	12°C	18°C	25°C	31.04°C	35°C	40°C	50°C	75°C	100°C
25		3.86		2.80	2.56	2.30	1.92	1.35	1.06
50	7.03	6.33	5.38	4.77	4.39	4.02	3.41	2.49	2.01
75	7.18	6.69	6.17	5.80	5.51	5.10	4.45	3.37	2.82
100	7.27	6.72	6.28	5.97	5.76	5.50	5.07	4.07	3.49
150	7.59	7.07		6.25	6.03	5.81	5.47	4.86	4.49
200				6.48	6.29	6.28	5.76	5.27	5.08
300	7.86	7.35					6.20	5.83	5.84
400	8.12	7.77	7.54	7.27	7.06	6.89	6.58	6.30	6.40
500				7.65	7.51	7.26			
700							7.58	7.43	7.61

°In the original, concentration is expressed in cubic centimeters of CO₂ (reduced to 0°C and 1 atm) dissolved in 1 g of water.

TABLE 2-126   Carbon Monoxide (CO)

Partial pressure of CO, mm Hg	$10^{-4} \times H$	
	17.7°C	19.0°C
900	4.77	4.88
2000	4.77	4.91
3000	4.77	4.93
4000	4.78	4.95
5000	4.80	4.97
6000	4.82	4.98
7000	4.86	5.02
8000	4.88	5.08

International Critical Tables, vol. 3, p. 260.

TABLE 2-127   Carbonyl Sulfide (COS)

t °C	0	5	10	15	20	25	30
$10^{-3} \times H$	0.92	1.17	1.48	1.82	2.19	2.59	3.04

International Critical Tables, vol. 3, p. 261.

TABLE 2-128 Chlorine (Cl$_2$)

Partial pressure of Cl$_2$, mm Hg	Solubility, g of Cl$_2$ per liter					
	0°C	10°C	20°C	30°C	40°C	50°C
5	0.488	0.451	0.438	0.424	0.412	0.398
10	.679	.603	.575	.553	.532	.512
30	1.221	1.024	.937	.873	.821	.781
50	1.717	1.354	1.210	1.106	1.025	.962
100	2.79	2.08	1.773	1.573	1.424	1.313
150	3.81	2.73	2.27	1.966	1.754	1.599
200	4.78	3.35	2.74	2.34	2.05	1.856
250	5.71	3.95	3.19	2.69	2.34	2.09
300		4.54	3.63	3.03	2.61	2.31
350		5.13	4.06	3.35	2.86	2.53
400		5.71	4.48	3.69	3.11	2.74
450		6.26	4.88	3.98	3.36	2.94
500		6.85	5.29	4.30	3.61	3.14
550		7.39	5.71	4.60	3.84	3.33
600		7.97	6.12	4.91	4.08	3.52
650		8.52	6.52	5.21	4.32	3.71
700		9.09	6.90	5.50	4.54	3.89
750		9.65	7.29	5.80	4.77	4.07
800		10.21	7.69	6.08	4.99	4.27
900			8.46	6.68	5.44	4.62
1000			9.27	7.27	5.89	4.97
1200	Cl$_2$.8H$_2$O$_2$ separates		10.84	8.42	6.81	5.67
1500			13.23	10.14	8.05	6.70
2000			17.07	13.02	10.22	8.38
2500			21.0	15.84	12.32	10.03
3000				18.73	14.47	11.70
3500				21.7	16.62	13.38
4000				24.7	18.84	15.04
4500				27.7	20.7	16.75
5000				30.8	23.3	18.46

Partial pressure of Cl$_2$, mm Hg	Solubility, g of Cl$_2$ per liter					
	60°C	70°C	80°C	90°C	100°C	110°C
5	0.383	0.369	0.351	0.339	0.326	0.316
10	.492	.470	.447	.431	.415	.402
30	.743	.704	.671	.642	.627	.598
50	.912	.863	.815	.781	.747	.722
100	1.228	1.149	1.085	1.034	.987	.950
150	1.482	1.382	1.294	1.227	1.174	1.137
200	1.706	1.580	1.479	1.396	1.333	1.276
250	1.914	1.764	1.642	1.553	1.480	1.413
300	2.10	1.932	1.793	1.700	1.610	1.542
350	2.28	2.10	1.940	1.831	1.736	1.661
400	2.47	2.25	2.08	1.965	1.854	1.773
450	2.64	2.41	2.22	2.09	1.972	1.880
500	2.80	2.55	2.35	2.21	2.08	1.986
550	2.97	2.69	2.47	2.32	2.19	2.09
600	3.13	2.83	2.59	2.43	2.29	2.19
650	3.29	2.97	2.72	2.55	2.41	2.28
700	3.44	3.10	2.84	2.66	2.50	2.37
750	3.59	3.23	2.96	2.76	2.60	2.47
800	3.75	3.37	3.08	2.87	2.69	2.56
900	4.04	3.63	3.30	3.08	2.89	2.74
1000	4.36	3.88	3.53	3.28	3.07	2.91
1200	4.92	4.37	3.95	3.67	3.43	3.25
1500	5.76	5.09	4.58	4.23	3.95	3.74
2000	7.14	6.26	5.63	5.17	4.78	4.49
2500	8.48	7.40	6.61	6.05	5.59	5.25
3000	9.83	8.52	7.54	6.92	6.38	5.97
3500	11.22	9.65	8.53	7.79	7.16	6.72
4000	12.54	10.76	9.52	8.65	7.94	7.42
4500	13.88	11.91	10.46	9.49	8.72	8.13
5000	15.26	13.01	11.42	10.35	9.48	8.84

TABLE 2-129 Chlorine Dioxide (ClO$_2$)

Vol % of ClO$_2$ in gas phase	Weight of ClO$_2$, grams per liter of solution						
	0°C	5°C	10°C	15°C	20°C	30°C	40°C
1	2.00	1.50	1.25	1.00	0.90	0.60	0.46
3	6.00	4.7	3.85	3.20	2.70	1.95	1.30
5	10.0	7.8	6.30	5.25	4.30	3.20	2.25
7	14.0	10.9	8.95	7.35	6.15	4.40	3.20
10	20.0	15.5	12.8	10.5	8.80	6.30	4.50
11		17.0	14.0	11.7	9.70	7.00	5.00
12		18.6	15.3	12.8	10.55	7.50	5.45
13		20.3	16.6	13.8	11.5	8.20	5.85
14			18.0	14.9	12.3	8.80	6.35
15			19.2	16.0	13.2	9.50	6.80
16			20.3	17.0	14.2	10.1	7.20

Ishi, *Chem. Eng. (Japan)*, **22**, 153 (1958).

TABLE 2-130 Ethane (C$_2$H$_6$)

t, °C	0	5	10	15	20	25	30	35
$10^{-4} \times H$	1.26	1.55	1.89	2.26	2.63	3.02	3.42	3.83

t, °C	40	45	50	60	70	80	90	100
$10^{-4} \times H$	4.23	4.63	5.00	5.65	6.23	6.61	6.87	6.92

International Critical Tables, vol. 3, p. 261.

TABLE 2-131 Ethylene (C$_2$H$_4$)

t, °C	0	5	10	15	20	25	30
$10^{-3} \times H$	5.52	6.53	7.68	8.95	10.2	11.4	12.7

International Critical Tables, vol. 3, p. 260.

TABLE 2-132 Helium (He)

t, °C	0	10	20	30	40	50
$10^{-4} \times H$	12.9	12.6	12.5	12.4	12.1	11.5

See also Pray, Schweickert, and Minnich, *Ind. Eng. Chem.*, **44**, 1146 (1952).

TABLE 2-133 Hydrogen (H$_2$)—Temperature

t, °C	0	5	10	15	20	25	30	35
$10^{-4} \times H$	5.79	6.08	6.36	6.61	6.83	7.07	7.29	7.42

t, °C	40	45	50	60	70	80	90	100
$10^{-4} \times H$	7.51	7.60	7.65	7.65	7.61	7.55	7.51	7.45

"International Critical Tables," vol. 3, p. 256.
See also Pray, Schweickert, and Minnich, *Ind. Eng. Chem.*, **44**, 1146 (1952).

TABLE 2-134   Hydrogen (H₂)—Pressure

Partial pressure H₂, mm Hg	$10^{-4} \times H$	
	19.5°C	23°C
900	7.42	
1100		7.75
2000	7.42	7.76
3000	7.43	7.77
4000	7.47	7.81
5000	7.56	7.89
6000	7.70	8.00
7000	7.87	8.16
8200		8.41
8250	8.17	

International Critical Tables, vol. 3, p. 256.

TABLE 2-135   Hydrogen Chloride (HCl)

Weights of HCl per 100 weights of H₂O	Partial pressure of HCl, mm Hg			
	0°C	10°C	20°C	30°C
78.6	510	840		
66.7	130	233	399	627
56.3	29.0	56.4	105.5	188
47.0	5.7	11.8	23.5	44.5
38.9	1.0	2.27	4.90	9.90
31.6	0.175	0.43	1.00	2.17
25.0	.0316	.084	0.205	0.48
19.05	.0056	.016	.0428	.106
13.64	.00099	.00305	.0088	.0234
8.70	.000118	.000583	.00178	.00515
4.17	.000018	.000069	.00024	.00077
2.04		.0000117	.000044	.000151

Weights of HCl per 100 weights of H₂O	Partial pressure of HCl, mm Hg		
	50°C	80°C	110°C
78.6			
66.7			
56.3	535		
47.0	141	623	
38.9	35.7	188	760
31.6	8.9	54.5	253
25.0	2.21	15.6	83
19.05	0.55	4.66	28
13.64	.136	1.34	9.3
8.70	.0344	0.39	3.10
4.17	.0064	.095	0.93
2.04	.00140	.0245	.280

Enthalpy and phase-equilibrium data for the binary system HCl-H₂O are given by Van Nuys, *Trans. Am. Inst. Chem. Engrs.*, **39**, 663 (1943).

TABLE 2-136   Hydrogen Sulfide (H₂S)

t, °C	0	5	10	15	20	25	30	35
$10^{-2} \times H$	2.68	3.15	3.67	4.23	4.83	5.45	6.09	6.76
t, °C	40	45	50	60	70	80	90	100
$10^{-2} \times H$	7.45	8.14	8.84	10.3	11.9	13.5	14.4	14.8

International Critical Tables, vol. 3, p. 259.

TABLE 2-137   Methane (CH₄)

t, °C	0	5	10	15	20	25	30	35
$10^{-4} \times H$	2.24	2.59	2.97	3.37	3.76	4.13	4.49	4.86
t, °C	40	45	50	60	70	80	90	100
$10^{-4} \times H$	5.20	5.51	5.77	6.26	6.66	6.82	6.92	7.01

International Critical Tables, vol. 3, p. 260.

TABLE 2-138   Nitrogen (N₂)—Temperature*

t, °C	0	5	10	15	20	25	30	35
$10^{-4} \times H$	5.29	5.97	6.68	7.38	8.04	8.65	9.24	9.85
t, °C	40	45	50	60	70	80	90	100
$10^{-4} \times H$	10.4	10.9	11.3	12.0	12.5	12.6	12.6	12.6

"*International Critical Tables*," vol. 3, p. 256. See also Pray, Schweickert, and Minnich, *Ind. Eng. Chem.*, **44**, 1146 (1952).
*Atmospheric nitrogen = 98.815 vol. % N₂ + 1.185 vol. % A.

TABLE 2-139   Nitrogen (N₂)—Pressure

Partial pressure of N₂, mm Hg	$10^{-4} \times H$	
	19.4°C	24.9°C
900	8.24	9.08
2000	8.32	9.15
3000	8.41	9.25
4000	8.49	9.38
5000	8.59	9.49
6000	8.74	9.62
7000	8.86	9.75
8100	9.04	
8200		9.91

See also Goodman and Krase [*Ind. Eng. Chem.*, **23**, 401 (1931)] for values up to 169°C and 300 atm.

TABLE 2-140   Oxygen (O₂)—Temperature

t, °C	0	5	10	15	20	25	30	35
$10^{-4} \times H$	2.55	2.91	3.27	3.64	4.01	4.38	4.75	5.07
t, °C	40	45	50	60	70	80	90	100
$10^{-4} \times H$	5.35	5.63	5.88	6.29	6.63	6.87	6.99	7.01

International Critical Tables, vol. 3, p. 257. Pray, Schweickert, and Minnich [*Ind. Eng. Chem.*, **44**, 1146 (1952)] give H = 4.46 × 10⁻⁴ at 25°C and other values up to 343°C.

TABLE 2-141   Oxygen (O₂)—Pressure

Partial pressure of O₂, mm Hg	$10^{-4} \times H$	
	23°C	25.9°C
800		4.79
900	4.58	
2000	4.59	4.80
3000	4.60	4.83
4000	4.68	4.88
5000	4.73	4.92
6000	4.80	4.98
7000	4.88	5.05
8150	4.98	
8200		5.16

International Critical Tables, vol. 3, p. 257. See also *Trans. Am. Soc. Mech. Engrs.*, **76**, 69 (1954) for solubility of O₂ for 100°F < T < 650°F, 300 < P < 2000 lb/in².

TABLE 2-142   Ozone (O₃)

t, °C	0	5	10	15	20	25	30	35	40	50
$10^{-3} \times H$	1.94	2.18	2.48	2.88	3.76	4.57	5.98	8.18	12.0	27.4

International Critical Tables, vol. 3, p. 257.

TABLE 2-143   Propylene (C₃H₆)

t, °C	2	6	10	14	18
$10^{-3} \times H$	3.04	3.84	4.46	5.06	5.69

International Critical Tables, vol. 3, p. 260.

TABLE 2-144   Partial Vapor Pressure of Sulfur Dioxide over Water, mm Hg

g SO₂/ 100 g H₂O	Temperature, °C								
	0	10	20	30	40	50	60	90	120
0.01	0.02	0.04	0.07	0.12	0.19	0.29	0.43	1.21	2.82
0.05	0.38	0.66	1.07	1.68	2.53	3.69	5.24	12.9	27.0
0.10	1.15	1.91	3.03	4.62	6.80	9.71	13.5	31.7	63.9
0.15	2.10	3.44	5.37	8.07	11.7	16.5	22.7	52.2	104
0.20	3.17	5.13	7.93	11.8	17.0	23.8	32.6	73.7	145
0.25	4.34	6.93	10.6	15.7	22.5	31.4	42.8	95.8	186
0.30	5.57	8.84	13.5	19.8	28.2	39.2	53.3	118	229
0.40	8.17	12.8	19.4	28.3	40.1	55.3	74.7	164	316
0.50	10.9	17.0	25.6	37.1	52.3	72.0	96.8	211	404
1.00	25.8	39.5	58.4	83.7	117	159	212	454	856
2.00	58.6	88.5	129	183	253	342	453	955	
3.00	93.2	139	202	285	393	530	700		
4.00	129	192	277	389	535	720			
5.00	165	245	353	496	679				
6.00	202	299	430	602	824				
8.00	275	407	585	818					
10.00	351	517	741						
15.00	542	796							
20.00	735								

Condensed from Rabe, A. E. and Harris, J. F., *J. Chem. Eng. Data*, **8** (3), 333–336, 1963. Copyright © American Chemical Society and reproduced by permission of the copyright owner.

THERMAL EXPANSION

UNITS CONVERSIONS

For this subsection, the following units conversion is applicable:

$$°F = \tfrac{9}{5} \, °C + 32.$$

ADDITIONAL REFERENCES

The tables given under this subject are reprinted by permission from the *Smithsonian Tables*. For more detailed data on thermal expansion, see *International Critical Tables*: tabular index, vol. 3, p. 1; abrasives, vol. 2, p. 87; alloys, vol. 2, p. 463; building stones, vol. 2, p. 54; carbons, vol. 2, p. 303; elements, vol. 1, p. 102; enamels, vol. 2, p. 115; glass, vol. 2, p. 93; metals, vol. 2, p. 459; petroleums, vol. 2, p. 145; porcelains, vol. 2, pp. 70, 78; refractory materials, vol. 2, p. 83; solid insulators, vol. 2, p. 310.

THERMAL EXPANSION OF GASES

No tables of the coefficients of thermal expansion of gases are given in this edition. The coefficient at constant pressure, $1/v(\partial v/\partial T)_p$, for an ideal gas is merely the reciprocal of the absolute temperature. For a real gas or liquid, both it and the coefficient at constant volume, $1/p \, (\partial p/\partial T)_v$, should be calculated either from the equation of state or from tabulated *PVT* data.

TABLE 2-145 Linear Expansion of the Solid Elements*

C is the true expansion coefficient at the given temperature; M is the mean coefficient between given temperatures; where one temperature is given, the true coefficient at that temperature is indicated; α and β are coefficients in formula $l_t = l_0(1 + \alpha t + \beta t^2)$; l_0 is length at 0°C (unless otherwise indicated, when, if x is the reference temperature, $l_t = l_x[1 + \alpha(t - t_x) + \beta(t - t_x)^2]$; l_t is length at t°C).

Element	Temp. °C	$C \times 10^4$	Temp. range, °C	$M \times 10^4$	Temp. range, °C	$\alpha \times 10^4$	$\beta \times 10^6$
Aluminum	20	0.224	100	0.235	0, 500	0.22	0.009
Aluminum	300	0.284	500	0.311			
Antimony	20	0.136‖	20	0.080⊥			
Arsenic	20	0.05					
Bismuth	20	0.014‖	20	0.103⊥			
Cadmium	0	0.54‖	−180, −140	0.59‖	20, 100	0.526‖	
Cadmium	0	0.20⊥	−180, −140	0.117⊥	20, 100	0.214⊥	
Carbon, diamond	50	0.012					
graphite	50	0.06					
Chromium			20, 100	0.068	20, 500	0.086	
Cobalt	20	0.123			6, 121	0.121	0.0064
Copper	20	0.162	100	0.166	0, 625	0.161	0.0040
Copper	200	0.170	300	0.175			
Gold	20	0.140	17, 100	0.143	0, 520	0.142	0.0022
Gold			−191, 17	0.132			
Indium	40	0.417					
Iodine			−190, 17	0.837			
Iridium	20	0.065			0, 80	0.0636	0.0032
Iridium					1070, 1720	0.0679	0.0011
Iron, soft	40	0.1210	0, 100	0.11			
cast	20	0.118			0, 750	0.1158	0.0053
wrought	20	0.119			0, 750	0.1170	0.0053
steel	20	0.114			0, 750	0.1118	0.0053
Lead (99.9)			20, 100	0.291	100, 240	0.269	0.011
	100	0.291	20, 200	0.300			
	280	0.343					
Magnesium	20	0.254	−100, +20	0.240	+20, 500	0.2480	0.0096
			20, 100	0.260			
Manganese	20	0.233	0, 100	0.228			
			−190, 0	0.159	20, 300	0.216	0.0121
Molybdenum†	20	0.053	0, 100	0.052	−142, 19	0.0515	0.0057
			25, 100	0.049	19, +305	0.0501	0.0014
			25, 500	0.055			
Nickel	20	0.126	0, 100	0.130	−190, +20	0.1308	0.0166
					+20, +300	0.1236	0.0066
					500, 1000	0.1346	0.0033
Osmium	40	0.066					
Palladium	20	0.1173			−190, +100	0.1152	0.00517
					0, 1000	0.1167	0.0022
Platinum	20	0.0887			−190, −100	0.0875	0.00314
	20	0.0893			0, +80	0.0890	0.00121
					0, 1000	0.0887	0.00132
Potassium			0, 50	0.83			
Rhodium	40	0.0850	6, 21	0.0876	−75, −112	0.0746	
Ruthenium	40	0.0963					
Selenium	0	0.439	0, 100	0.660			
Silicon	40	0.0763	−3, +18	0.0249	−75, −67	0.0182	
Silver	20	0.1846	0, 100	0.197	0, 875	0.1827	0.00479
	20	0.195			20, 500	0.1939	0.00295
Sodium			−190, −17	0.622	0, 50	0.72	
Steel, 36.4Ni			20, 260	0.031	260, 500	0.144	
			20, 340	0.055	340, 500	0.136	
Tantalum†	20	0.065	−78, 0	0.059	20, 400	0.0646	0.0009
			0, 100	0.0655			
Tellurium	20	0.016‖	20	0.272⊥			
Thallium	40	0.302					
Tin	20	0.214			8, 95	0.2033	0.0263
	20	0.305‖	20	0.154⊥			
Tungsten†	27	0.0444	0, 100	0.045	−105, +502	0.0428	0.00058
Zinc	20‡	0.643‖	−140, −100	0.656‖	+0, 400	0.354	0.010
	20‡	0.125⊥	+20, 100	0.639‖			
	20	0.358	+20, 100	0.141⊥			

*Smithsonian Tables. For more complete tabulations see Table 142, Smithsonian Physical Tables, 9th ed., 1954; Handbook of Chemistry and Physics, 40th ed., pp. 2239–2245. Chemical Rubber Publishing Co.; Goldsmith, and Waterman, WADC-TR-58-476, 1959; Johnson (ed.), WADD-TR-60-56, 1960, etc.

†Molybdenum, 300° to 2500°C; $l_t = l_{300}[1 + 5.00 \times 10^{-6}(t - 300) + 10.5 \times 10^{-10}(t - 300)^2]$

Tantalum, 300° to 2800°C; $l_t = l_{300}[1 + 6.60 \times 10^{-6}(t - 300) + 5.2 \times 10^{-10}(t - 300)^2]$

Tungsten, 300° to 2700°C; $l_t = l_{300}[1 + 4.44 \times 10^{-6}(t - 300) + 4.5 \times 10^{-10}(t - 300)^2]$

Beryllium, 20° to 100°C; 12.3×10^{-6} per °C.

Columbium, 0° to 100°C; 7.2×10^{-6} per °C.

Tantalum, 20° to 100°C; 6.6×10^{-6} per °C.

‡Two errors in the data of zinc have been corrected. These values were taken from Grüneisen and Goens, Z. Physik., **29**, 141 (1924).

TABLE 2-146 Linear Expansion of Miscellaneous Substances*

The coefficient of cubical expansion may be taken as three times the linear coefficient. In the following table, t is the temperature or range of temperature, and C, the coefficient of expansion.

Substance	$t°C$	$C \times 10^4$	Substance	$t°C$	$C \times 10^4$	Substance	$t°C$	$C \times 10^4$
Amber	0–30	0.50	Jena thermometer 59[III]	0–100	0.058	Topaz:		
	0–09	0.61	Jena thermometer 59[III]	−191 to +16	0.424	Parallel to lesser hori-		
Bakelite, bleached	20–60	0.22	Gutta percha	20	1.983	zontal axis	0–100	0.0832
Brass:			Ice	−20 to −1	0.51	Parallel to greater hori-		
Cast	0–100	0.1875	Iceland spar:			zontal axis	0–100	0.0836
Wire	0–100	0.1930	Parallel to axis	0–80	0.2631	Parallel to vertical axis	0–100	0.0472
Wire	0–100	0.1783 to 0.193	Perpendicular to axis	0–80	0.0544	Tourmaline:		
71.5 Cu + 27.7 Zn +			Lead tin (solder) 2 Pb			Parallel to longitudinal		
0.3 Sn + 0.5 Pb	40	0.1859	+ 1 Sn	0–100	0.2508	axis	0–100	0.0937
71 Cu + 29 Zn	0–100	0.1906	Limestone	25–100	0.09	Parallel to horizontal		
Bronze:			Magnalium	12–39	0.238	axis	0–100	0.0773
3 Cu + 1 Sn	16.6–100	0.1844	Manganin		0.181	Type metal	16.6–254	0.1952
3 Cu + 1 Sn	16.6–350	0.2116	Marble	15–100	0.117	Vulcanite	0–18	0.6360
3 Cu + 1 Sn	16.6–957	0.1737	Monel metal	25–100	0.14	Wedgwood ware	0–100	0.0890
86.3 Cu + 9.7 Sn + 4 Zn	40	0.1782		25–600	0.16	Wood:		
97.6 Cu + ⎰hard	0–80	0.1713	Paraffin	0–16	1.0662	Parallel to fiber:		
2.2 Sn + ⎱soft	0–80	0.1708	Paraffin	16–38	1.3030	Ash	0–100	0.0951
0.2 P			Paraffin	38–49	4.7707	Beech	2.34	0.0257
Caoutchouc		0.657 to 0.686	Platinum-iridium, 10 Pt			Chestnut	2.34	0.0649
Caoutchouc	16.7–25.3	0.770	+ 1 Ir	40	0.0884	Elm	2.34	0.0565
Celluloid	20–70	1.00	Platinum-silver, 1 Pt +			Mahogany	2.34	0.0361
Constantan	4–29	0.1523	2 Ag	0–100	0.1523	Maple	2.34	0.0638
Duralumin, 94Al	20–100	0.23	Porcelain	20–790	0.0413	Oak	2.34	0.0492
	20–300	0.25	Porcelain Bayeux	1000–1400	0.0553	Pine	2.34	0.0541
Ebonite	25.3–35.4	0.842	Quartz:			Walnut	2.34	0.0658
Fluorspar, CaF₂	0–100	0.1950	Parallel to axis	0–80	0.0797	Across the fiber:		
German silver	0–100	0.1836	Parallel to axis	−190 to + 16	0.0521	Beech	2.34	0.614
Gold-platinum, 2 Au + 1 Pt	0–100	0.1523	Perpend. to axis	0–80	0.1337	Chestnut	2.34	0.325
Gold-copper, 2 Au + 1 Cu	0–100	0.1552	Quartz glass	−190 to + 16	−0.0026	Elm	2.34	0.443
Glass:			Quartz glass	16 to 500	0.0057	Mahogany	2.34	0.404
Tube	0–100	0.0833	Quartz glass	16 to 1000	0.0058	Maple	2.34	0.484
Tube	0–100	0.0828	Rock salt	40	0.4040	Oak	2.34	0.544
Plate	0–100	0.0891	Rubber, hard	0	0.691	Pine	2.34	0.341
Crown (mean)	0–100	0.0897	Rubber, hard	−160	0.300	Walnut	2.34	0.484
Crown (mean)	50–60	0.0954	Speculum metal	0–100	0.1933	Wax white	10–26	2.300
Flint	50–60	0.0788	Steel, 0.14 C, 34.5 Ni	25–100	0.037	Wax white	26–31	3.120
Jena ther- 16[III] � mometer normal⎱	0–100	0.081		25–600	0.136	Wax white	31–43	4.860
						Wax white	43–57	15.227

*Smithsonian Tables. For a more complete tabulation see Tables 143, 144. Smithsonian Physical Tables. 9th ed., 1954, also reprinted in American Institute of Physics Handbook, McGraw-Hill, New York, 1957; Handbook of Chemistry and Physics, 40th ed., pp. 2239–2245, Chemical Rubber Publishing Co. For data on many solids prior to 1926, see Gruneisen, Handbuch der Physik, vol. 10, pp. 1–52, 1926, translation available as N.A.S.A. RE 2-18-59W, 1959. For eight plastic solids below 300 K, see Scott, Cryogenic Engineering, p. 331, Van Nostrand, Princeton, NJ, 1959. For 11 other materials to 300 K, see Scott, loc. cit., p. 333. For quartz and silica, see Cook, Brit. J. Appl. Phys., 7, 285 (1956).

TABLE 2-147 Cubical Expansion of Liquids*

If V_0 is the volume at 0°, then at t° the expansion formula is $V_t = V_0(1 + \alpha t + \beta t^2 + \gamma t^3)$. The table gives values of α, β, and γ, and of C, the true coefficient of cubical expansion at 20° for some liquids and solutions. The temperature range of the observation is Δt. Values for the coefficient of cubical expansion of liquids can be derived from the tables of specific volumes of the saturated liquid given as a function of temperature later in this section.

Liquid	Range	$\alpha \times 10^3$	$\beta \times 10^6$	$\gamma \times 10^8$	$C \times 10^8$ at 20°
Acetic acid	16–107	1.0630	0.12636	1.0876	1.071
Acetone	0–54	1.3240	3.8090	−0.87983	1.487
Alcohol:					
Amyl	−15–80	0.9001	0.6573	1.18458	0.902
Ethyl, 30% by volume	18–39	0.2928	10.790	−11.87	
Ethyl, 50% by volume	0–39	0.7450	1.85	0.730	
Ethyl, 99.3% by volume	27–46	1.012	2.20		1.12
Ethyl, 500 atm. pressure	0–40	0.866			
Ethyl, 3000 atm. pressure	0–40	0.524			
Methyl	0–61	1.1342	1.3635	0.8741	1.199
Benzene	11–81	1.17626	1.27776	0.80648	1.237
Bromine	0–59	1.06218	1.87714	−0.30854	1.132
Calcium chloride:					
5.8% solution	18–25	0.07878	4.2742		0.250
40.9% solution	17–24	0.42383	0.8571		0.458
Carbon disulfide	−34–60	1.13980	1.37065	1.91225	1.218
500 atm. pressure	0–50	0.940			
3000 atm. pressure	0–50	0.581			
Carbon tetrachloride	0–76	1.18384	0.89881	1.35135	1.236
Chloroform	0–63	1.10715	4.66473	−1.74328	1.273
Ether	−15–38	1.51324	2.35918	4.00512	1.656
Glycerin		0.4853	0.4895		0.505
Hydrochloric acid, 33.2% solution	0–33	0.4460	0.215		0.455
Mercury	0–100	0.18182	0.0078		0.18186
Olive oil		0.6821	1.1405	−0.539	0.721
Pentane	0–33	1.4646	3.09319	1.6084	1.608
Potassium chloride, 24.3% solution	16–25	0.2695	2.080		0.353
Phenol	36–157	0.8340	0.10732	0.4446	1.090
Petroleum, 0.8467 density	24–120	0.8994	1.396		0.955
Sodium chloride, 20.6% solution	0–29	0.3640	1.237		0.414
Sodium sulfate, 24% solution	11–40	0.3599	1.258		0.410
Sulfuric acid:					
10.9% solution	0–30	0.2835	2.580		0.387
100.0%	0–30	0.5758	−0.432		0.558
Turpentine	−9–106	0.9003	1.9595	−0.44998	0.973
Water	0–33	−0.06427	8.5053	−6.7900	0.207

Smithsonian Tables, Table 269. For a detailed discussion of mercury data, see Cook, *Brit. J. Appl. Phys.*, **7**, 285 (1956). For data on nitrogen and argon, see Johnson (ed.), WADD-TR-60-56, 1960.

Bromoform[1] 7.7 – 50°C.
$$V_t = 0.34204[1 + 0.00090411(t − 7.7) + 0.00000006766(t − 7.7)^2]$$
0.34204 in the specific volume of bromoform at 7.7°C.

Glycerin[2] −62 to 0°C.
$$V_t = V_0(1 + 4.83 \times 10^{-4}t − 0.49 \times 10^{-6}t^2)$$
$$0 − 80°C.$$
$$V_t = V_0(1 + 4.83 \times 10^{-4}t + 0.49 \times 10^{-6}t^2)$$

Mercury[3] 0 – 300°C.
$$V_t = V_0[1 + 10^{-8}(18153.8t + 0.7548t^2 + 0.00001533t^2 + 0.00000536t^4)]$$

[1] Sherman and Sherman, *J. Am. Chem. Soc.*, **50**, 1119 (1928). (An obvious error in their equation has been corrected.)
[2] Samsoen, *Ann. phys.*, (10) **9**, 91 (1928).
[3] Harlow, *Phil. Mag.*, (7) **7**, 674 (1929).

TABLE 2-148 Cubical Expansion of Solids*

If v_2 and v_1 are the volumes at t_2 and t_1, respectively, then $v_2 = v_1(1 + C\Delta t)$, C being the coefficient of cubical expansion and Δt the temperature interval. Where only a single temperature is stated, C represents the true coefficient of cubical expansion at that temperature.

Substance	t or Δt	$C \times 10^4$
Antimony	0–100	0.3167
Beryl	0–100	0.0105
Bismuth	0–100	0.3948
Copper†	0–100	0.4998
Diamond	40	0.0354
Emerald	40	0.0168
Galena	0–100	0.558
Glass, common tube	0–100	0.276
hard	0–100	0.214
Jena, borosilicate 59 III	20–100	0.156
pure silica	0–80	0.0129
Gold	0–100	0.4411
Ice	−20 to −1	1.1250
Iron	0–100	0.3550
Lead†	0–100	0.8399
Paraffin	20	5.88
Platinum	0–100	0.265
Porcelain, Berlin	20	0.0814
chloride	0–100	1.094
nitrate	0–100	1.967
sulfate	20	1.0754
Quartz	0–100	0.3840
Rock salt	50–60	1.2120
Rubber	20	4.87
Silver	0–100	0.5831
Sodium	20	2.13
Stearic acid	33.8–45.4	8.1
Sulfur, native	13.2–50.3	2.23
Tin	0–100	0.6889
Zinc†	0–100	0.8928

Smithsonian Tables, Table 268.
†See additional data below.

Aluminum[1] 100 – 530°C.
$$V = V_0(1 + 2.16 \times 10^{-5}t + 0.95 \times 10^{-8}t^2)$$
Cadmium[1] 130 – 270°C.
$$V = V_0(1 + 8.04 \times 10^{-5}t + 5.9 \times 10^{-8}t^2)$$
Copper[1] 110 – 300°C.
$$V = V_0(1 + 1.62 \times 10^{-5}t + 0.20 \times 10^{-8}t^2)$$
Colophony[2] 0 – 34°C.
$$V = V_0(1 + 2.21 \times 10^{-4}t + 0.31 \times 10^{-6}t^2)$$
$$34 – 150°C.$$
$$V = V_{34}[1 + 7.40 \times 10^{-4}(t − 34) + 5.91 \times 10^{-6}(t − 34)^2]$$
Lead[1] 100 – 280°C.
$$V = V_0(1 + 1.60 \times 10^{-5}t + 3.2 \times 10^{-8}t^2)$$
Shellac[2] 0 – 46°C.
$$V = V_0(1 + 2.73 \times 10^{-4}t + 0.39 \times 10^{-6}t^2)$$
$$46 – 100°C.$$
$$V = V_{46}[1 + 13.10 \times 10^{-4}(t − 46) + 0.62 \times 10^{-6}(t − 46)^2]$$
Silica (vitreous)[3] 0 – 300°C.
$$V_t = V_0[1 + 10^{-8}(93.6t + 0.7776t^2 − 0.003315t^2 + 0.000005244t^4)]$$
Sugar (cane, amorphous)[2] 0 – 67°C.
$$V_t = V_0(1 + 2.34 \times 10^{-4}t + 0.14 \times 10^{-6}t^2)$$
$$67 – 160°C.$$
$$V_t = V_{67}[1 + 5.02 \times 10^{-4}(t − 67) + 0.43 \times 10^{-6}(t − 67)^2]$$
Zinc[1] 120 – 360°C.
$$V_t = V_0(1 + 8.50 \times 10^{-5}t + 3.9 \times 10^{-8}t^2)$$

[1] Uffelmann, *Phil. Mag.*, (7) **10**, 633 (1930).
[2] Samsoen, *Ann. phys.*, (10) **9**, 83 (1928).
[3] Harlow, *Phil. Mag.*, (7) **7**, 674 (1929).

JOULE-THOMSON EFFECT

UNITS CONVERSIONS

For this subsection, the following units conversions are applicable:

To convert the Joule-Thomson coefficient, μ, in degrees Celsius per atmosphere to degrees Fahrenheit per atmosphere, multiply by 1.8.

$$°F = \%\ °C + 32; \quad °R = \%\ K$$

To convert bars to pounds-force per square inch, multiply by 14.504; to convert bars to kilopascals, multiply by 1×10^2.

TABLE 2-149 **Additional References Available for the Joule-Thomson Coefficient**

Gas	Pressure range, atm				Temp. range, °C			Unclassified
	0–10	10–50	50–200	>200	<0	0–300	>300	
Air	12, 15, 16 19, 35	12, 15, 19 35	15, 19, 35		19, 35	12, 15, 16 19, 35		3, 4, 18
Ammonia	28					28		2, 3
Argon	39	39	39		39	39		
Benzene	31	31	31			31	31	
Butane	26	26				26		
Carbon dioxide	7, 8, 28 37	7, 8, 37	7, 8, 37		7, 8, 37	7, 8, 9, 10 37		
Carbon monoxide	17	17			17	17		
Deuterium		22, 24, 25 1°	1,° 22, 24 25		1,° 22, 24, 25			
Dowtherm A	46	46				46	46	
Ethane	45	45				45		
Ethylene						9, 10		
Helium	1, 38	1, 38	38		1, 38	38		48
Hydrogen	24, 30	22, 24, 25 30	24, 30		22, 24, 25 30	24		
Methane		6	6			6		
Mixtures						9, 11		
Natural gas			33	33	33	33		
Nitrogen	13, 28, 40	13, 40	13, 40	13	13, 40	9, 10, 13 28, 40	13	19
Nitrous oxide						9, 10		
Pentane	26, 34, 44	34	34			26, 34, 44		
Propane	41	43				43		
Steam	28, 29, 42	29, 42, 47	42, 47			28, 29, 42 45	29, 42, 47	29, 47

°See also 14 (generalized chart); 18 (review, to 1919); 20–22; 23 (review, to 1948); 27 (review, to 1905); 2, 36, 41, 50.

REFERENCES: 1. Baehr. *Z. Elektrochem.*, **60**, 515 (1956). 2. Beattie, *J. Math. Phys.*, 9, 11 (1930). 3. Beattie, *Phys. Rev.*, **35**, 643 (1930). 4. Bradley and Hale, *Phys. Rev.*, **29**, 258 (1909). 5. Brown and Dean, *Bur. Stand. J. Res.*, **60**, 161 (1958). 6. Budenholzer, Sage, et al., *Ind. Eng. Chem.*, **29**, 658 (1937). 7. Burnett, *Phys. Rev.*, **22**, 590 (1923). 8. Burnett, Univ. Wisconsin Bull. 9(6), 1926. 9. Charnley, Ph.D. thesis. University of Manchester, 1952. 10. Charnley, Isles, et al., *Proc. R. Soc. (London)*, **A217**, 133 (1953). 11. Charnley, Rowlinson, et al., *Proc. R. Soc. (London)*, **A230**, 354 (1955). 12. Dalton, *Commun. Phys. Lab. Univ. Leiden*, no. 109c, 1909. 13. Deming and Deming, *Phys. Rev.*, **48**, 448 (1935). 14. Edmister, *Pet. Refiner*, **28**, 128 (1949). 15. Eucken, Clusius, et al., *Z. Tech. Phys.*, **13**, 267 (1932). 16. Eumorfopoulos and Rai, *Phil. Mag.*, **7**, 961 (1929). 17. Huang, Lin, et al., *Z. Phys.*, **100**, 594 (1936). 18. Hoxton, *Phys. Rev.*, **13**, 438 (1919). 19. Ishkin and Kaganev, *J. Tech. Phys. U.S.S.R.*, **26**, 2323 (1956). 20. Isles, Ph.D. thesis, Leeds University. 21. Jenkin and Pye, *Phil. Trans. R. Soc. (London)*, **A213**, 67 (1914); **A215**, 353 (1915). 22. Johnston, *J. Am. Chem. Soc.*, **68**, 2362 (1946). 23. Johnston, *Trans. Am. Soc. Mech. Eng.*, **70**, 651 (1948). 24. Johnston, Bezman, et al., *J. Am. Chem. Soc.*, **68**, 2367 (1946). 25. Johnston, Swanson, et al., *J. Am. Chem. Soc.*, **68**, 2373 (1946). 26. Kennedy, Sage, et al., *Ind. Eng. Chem.*, **28**, 718 (1936). 27. Kester, *Phys. Rev.*, **21**, 260 (1905). 28. Keyes and Collins, *Proc. Nat. Acad. Sci.*, **18**, 328 (1932). 29. Kleinschmidt, *Mech. Eng.*, **45**, 165 (1923); **48**, 155 (1926). 30. Koeppe, *Kältetechnik*, **8**, 275 (1956). 31. Lindsay and Brown, *Ind. Eng. Chem.*, **27**, 817 (1935). 32. Noell, dissertation, Munich, 1914, *Forschungsdienst*, 184, p. 1, 1916. 33. Palienko, *Tr. Inst. Ispol' z. Gaza, Akad. Nauk Ukr. SSR*, no. 4, p. 87, 1956. 34. Pattee and Brown, *Ind. Eng. Chem.*, **26**, 511, (1934). 35. Roebuck, *Proc. Am. Acad. Arts Sci.*, **60**, 537 (1925); **64**, 287 (1930). 36. Roebuck, see 49 below. 37. Roebuck and Murrell, *Phys. Rev.*, **55**, 240 (1939). 38. Roebuck and Osterberg, *Phys. Rev.*, **37**, 110 (1931); **43**, 60 (1933). 39. Roebuck and Osterberg, *Phys. Rev.*, **46**, 785 (1934). 40. Roebuck and Osterberg, *Phys. Rev.*, **48**, 450 (1935). 41. Roebuck, Murrell, et al., *J. Am. Chem. Soc.*, **64**, 400 (1942). 42. Sage, unpublished data, California Institute of Technology, 1959. 43. Sage and Lacy, *Ind. Eng. Chem.*, **27**, 1484 (1934). 44. Sage, Kennedy, et al., *Ind. Eng. Chem.*, **28**, 601 (1936). 45. Sage, Webster, et al., *Ind. Eng. Chem.*, **29**, 658 (1937). 46. Ullock, Gaffert, et al., *Trans. Am. Inst. Chem. Eng.*, **32**, 73 (1936). 47. Yang, *Ind. Eng. Chem.*, **45**, 786 (1953). 48. Zelmanov, *J. Phys. U.S.S.R.*, **3**, 43 (1940). 49. Roebuck, recalculated data. 50. Michels et al., van der Waals laboratory publications. Gunn, Cheuh, and Prausnitz, *Cryogenics*, **6**, 324 (1966), review equations relating the inversion temperatures and pressures. The ability of various equations of state to relate these was also discussed by Miller, *Ind. Eng. Chem. Fundam.*, **9**, 585 (1970); and Juris and Wenzel, *Am. Inst. Chem. Eng. J.*, **18**, 684 (1972). Perhaps the most detailed review is that of Hendricks, Peller, and Baron. NASA Tech. Note D 6807, 1972.

TABLE 2-150 Approximate Inversion-Curve Locus in Reduced Coordinates ($T_r = T/T_c$; $P_r = P/P_c$)*

P_r	0	0.5	1	1.5	2	2.5	3	4
T_{rL}	0.782	0.800	0.818	0.838	0.859	0.880	0.903	0.953
T_{rU}	4.984	4.916	4.847	4.777	4.706	4.633	4.550	4.401

P_r	5	6	7	8	9	10	11	11.79
T_{rL}	1.01	1.08	1.16	1.25	1.35	1.50	1.73	2.24
T_{rU}	4.23	4.06	3.88	3.68	3.45	3.18	2.86	2.24

*Calculated from the best three-constant equation recommended by Miller, *Ind. Eng. Chem. Fundam.*, **9**, 585 (1970). T_{rL} refers to the lower curve, and T_{rU}, to the upper curve.

TABLE 2-151 Joule-Thomson Data for Air*

P, atm	\multicolumn{13}{c}{t, °C}												
	−150	−100	−75	−50	−25	0	25	50	75	100	150	200	250
1		0.5895	0.4795	0.3910	0.3225	0.2745	0.2320	0.1956	0.1614	0.1355	0.0961	0.0645	0.0409
20		.5700	.4555	.3690	.3010	.2580	.2173	.1830	.1508	.1258	.0883	.0580	.0356
60	0.0450	.4820	.3835	.3195	.2610	.2200	.1852	.1571	.1293	.1062	.0732	.0453	.0254
100	.0185	.2775	.2880	.2505	.2130	.1820	.1550	.1310	.1087	.0884	.0600	.0343	.0165
140	− .0070	.1360	.1855	.1825	.1650	.1450	.1249	.1070	.0889	.0726	.0482	.0250	.0092
180	− .0255	.0655	.1136	.1270	.1240	.1100	.0959	.0829	.0707	.0580	.0376	.0174	.0027
200	− .0330	.0440	.0855	.1065	.1090	.0950							

*Free of water and CO_2. Extracted from Table 261, *Smithsonian Physical Tables*, 9th rev. ed., Washington, DC, 1954. These data are corrected from earlier publications. μ in °C/atm.

TABLE 2-152 Approximate Inversion-Curve Locus for Air

P, bar	0	25	50	75	100	125	150	175	200	225
T_L, K	(112)°	114	117	120	124	128	132	137	143	149
T_U, K	653	641	629	617	606	594	582	568	555	541

P, bar	250	275	300	325	350	375	400	425	432
T_L, K	156	164	173	184	197	212	230	265	300
T_U, K	526	509	491	470	445	417	386	345	300

°Hypothetical low-pressure limit.

TABLE 2-153 Joule-Thomson Data for Argon*

t, °C	\multicolumn{7}{c}{Pressure, atm}						
	1	20	60	100	140	180	200
−150	1.812		−0.0025	−0.0277	−0.0403	−0.0595	−0.0640
−125	1.112	1.102	.1250	.0415	.0090	− .0100	− .0165
−100	0.8605	0.8485	.6900	.2820	.1137	.0560	.0395
−75	.7100	.6895	.5910	.4225	.2480	.1537	.1215
−50	.5960	.5720	.4963	.3970	.2840	.2037	.1860
−25	.5045	.4805	.4210	.3460	.2763	.2140	.1950
0	.4307	.4080	.3600	.3010	.2505	.2050	.1883
25	.3720	.3490	.3077	.2628	.2213	.1890	.1745
50	.3220	.3015	.2650	.2297	.1947	.1700	.1580
75	.2695	.2557	.2285	.1993	.1710	.1505	.1415
100	.2413	.2277	.1975	.1715	.1490	.1320	.1255
125	.2105	.1980	.1707	.1480	.1300	.1153	.1100
150	.1845	.1720	.1485	.1285	.1123	.0998	.0945
200	.1377	.1280	.1102	.0950	.0823	.0715	.0675
250	.0980	.0910	.0785	.0665	.0555	.0485	.0468
300	.0643	.0607	.0530	.0445	.0370	.0370	.0276

*Extracted from Table 263, *Smithsonian Physical Tables*, 9th rev. ed., Washington, DC, 1954. These data are corrected from an earlier publication. μ in °C/atm.

TABLE 2-154   Approximate Inversion-Curve Locus for Argon

P, bar	0	25	50	75	100	125	150	175	200	225
T_L, K	94	97	101	105	109	113	118	123	128	134
T_U, K	765	755	744	736	726	716	705	694	683	671

P, bar	250	275	300	325	350	375	400	425	450	475
T_L, K	141	148	158	170	183	201	222	248	288	375
T_U, K	657	643	627	610	591	569	544	515	478	375

TABLE 2-155   Joule-Thomson Data for Carbon Dioxide*

t, °C	Pressure, atm							
	1	20	60	73	100	140	80	200
−75		−0.0200	−0.0200	−0.0232	−0.0228	−0.0240	−0.0250	−0.0290
−50	2.4130	−.0140	−.0150	−.0165	−.0160	−.0183	−.0228	−.0248
0	1.2900	1.4020	.0370	.0310	.0215	.0115	.0085	.0045
50	0.8950	.8950	.8800	.8225	.5570	.1720	.1025	.0930
100	.6490	.6375	.6080	.5920	.5405	.4320	.3000	.2555
125	.5600	.5450	.5160	.5068	.4750	.4130	.3230	.2915
150	.4890	.4695	.4430	.4380	.4155	.3760	.3102	.2910
200	.3770	.3575	.3400	.3325	.3150	.2890	.2600	.2455
250	.3075	.2885	.2625	.2565	.2420	.2235	.2045	.1975
300	.2650	.2425	.2080	.2002	.1872	.1700	.1540	.1505

*Extracted from Table 266, *Smithsonian Physical Tables*, 9th rev. ed., Washington, DC, 1954. These data are corrected from an earlier publication. μ in °C/atm.

TABLE 2-156   Approximate Inversion-Curve Locus for Carbon Dioxide*

P, bar	50	100	150	200	250	300	350	400	450
T_L, K	243	251	258	266	272	283	293	302	312
T_U, K	1290	1261	1233	1205	1175	1146	1111	1076	1045

P, bar	500	550	600	650	700	750	800	850	884
T_L, K	325	338	351	365	383	403	441	496	608
T_U, K	1015	983	950	914	878	840	796	739	608

*Interpolated from Vukalovich and Altunin's interpolation of data of Price, *Ind. Eng. Chem.*, **47**, 1691 (1955). T_L = lower inversion temperature, and T_U = upper inversion temperature.

TABLE 2-157   Approximate Inversion-Curve Locus for Deuterium

P, bar	0	25	50	75	100	125	150	175	94
T_L, K	(31)°	34	38	43	49	56	65	77	_08
T_U, K	216	202	189	178	168	157	146	131	_08

°Hypothetical low-pressure limit.

TABLE 2-158   Approximate Inversion-Curve Locus for Ethane

P, bar	0	25	50	75	100	125	150	175	200	225
T_L, K		249	255	262	269	275	282	290	297	306

P, bar	250	275	300	325	350	375	400	425	450	475
T_L, K	315	325	335	345	357	370	383	398	415	432

P, bar	500	525	550	575	600
T_L, K	453	477	505	545	626

TABLE 2-159   Joule-Thomson Data for Helium*

T, K	160	180	200	220	240	260	280	300
μ	−0.0574	−0.0587	−0.0594	−0.0601	−0.0608	−0.0614	−0.0619	−0.0625
T, K	320	340	360	380	400	420	440	460
μ	−0.0629	−0.0634	−0.0637	−0.0640	−0.0643	−0.0645	−0.0645	−0.0643
T, K	480	500	520	540	560	580	600	
μ	−0.0640	−0.0636	−0.0630	−0.0622	−0.0611	−0.0587	−0.0540	

*Interpolated and converted from data in Table 262, *Smithsonian Physical Tables*, 9th rev. ed., Washington, DC, 1954. These data are corrected from those in an earlier publication. μ is in °C/atm. Below about 200 atm, little change in the coefficient with pressure occurs.

TABLE 2-160   Approximate Inversion-Curve Locus for Normal Hydrogen

P, bar	0	25	50	75	100	125	150	164
T_L, K	(28)°	32	38	44	52	61	73	92
T_U, K	202	193	183	171	157	141	119	92

°Hypothetical low-pressure limit.

TABLE 2-161   Approximate Inversion-Curve Locus for Methane

P, bar	25	50	75	100	125	150	175	200	225	250	275	300
T_L, K		161	166	172	176	182	189	195	202	209	217	225
P, bar	325	350	375	400	425	450	475	500	525	534		
T_L, K	234	243	254	265	277	292	309	331	365	400		
T_U, K							505	474	437	400		

TABLE 2-162   Joule-Thomson Data for Nitrogen*

t, °C	Pressure, atm							
	1	20	33.5	60	100	140	180	200
−150	1.2659	1.1246	0.1704	0.0601	0.0202	−0.0056	−0.0211	−0.0284
−125	0.8557	0.7948	.7025	.4940	.1314	.0498	.0167	.0032
−100	.6490	.5958	.5494	.4506	.2754	.1373	.0765	.0587
−75	.5033	.4671	.4318	.3712	.2682	.1735	.1026	.0800
−50	.3968	.3734	.3467	.3059	.2332	.1676	.1120	.0906
−25	.3224	.3013	.2854	.2528	.2001	.1506	.1101	.0932
0	.2656	.2494	.2377	.2088	.1679	.1316	.1015	.0891
25	.2217	.2060	.1961	.1729	.1400	.1105	.0874	.0779
50	.1855	.1709	.1621	.1449	.1164	.0915	.0732	.0666
75	.1555	.1421	.1336	.1191	.0941	.0740	.0583	.0543
100	.1292	.1173	.1100	.0975	.0768	.0582	.0462	.0419
125	.1070	.0973	.0904	.0786	.0621	.0459	.0347	.0326
150	.0868	.0776	.0734	.0628	.0482	.0348	.0248	.0228
200	.0558	.0472	.0430	.0372	.0262	.0168	.0094	.0070
250	.0331	.0256	.0230	.0160	.0071	.0009	−.0037	−.0058
300	.0140	.0096	.0050	−.0013	−.0075	−.0129	−.0160	−.0171

*Extracted from Table 264, *Smithsonian Physical Tables*, 9th rev. ed., Washington, DC, 1954. These data are corrected from an earlier publication. μ in °C/atm.

TABLE 2-163   Approximate Inversion-Curve Locus for Propane

P, bar	0	25	50	75	100	125	150	175	200	225	250	275
T_L, K	(296)°	303	311	318	327	336	345	355	365	374	389	403
P, bar	300	325	350	375	400	425	450	475	500	525	541	
T_L, K	418	435	452	473	495	521	551	586	628	686	780	

°Hypothetical low-pressure limit.

CRITICAL CONSTANTS

ADDITIONAL REFERENCES

Other data and estimation techniques for the elements are contained in Gates and Thodos, *Am. Inst. Chem. Eng. J.*, **6** (1960):50–54; and Ohse and von Tippelskirch, *High Temperatures—High Pressures*, **9**

(1977):367–385. For inorganic substances see Mathews, *Chem. Rev.*, **72** (1972):71–100; for organics see Kudchaker, Alani, and Zwolinski, *Chem. Rev.*, **68** (1968):659–735; and for fluorocarbons see *Advances in Fluorine Chemistry*, App. B, Butterworth. Washington, 1963, pp. 173–175.

TABLE 2-164   Critical Constants and Acentric Factors of Inorganic and Organic Compounds

Cmpd. no.	Name	Formula	CAS no.	Mol. wt.	T_c, K	$P_c \times 1E\text{-}06$ Pa	V_c, m^3/Kmol	Z_c	Acentric factor
1	Methane	CH_4	74828	16.043	190.564	4.59	0.099	0.286	0.011
2	Ethane	C_2H_6	74840	30.070	305.32	4.85	0.146	0.279	0.098
3	Propane	C_3H_8	74986	44.097	369.83	4.21	0.200	0.273	0.149
4	n-Butane	C_4H_{10}	106978	58.123	425.12	3.77	0.255	0.272	0.197
5	n-Pentane	C_5H_{12}	109660	72.150	469.7	3.36	0.315	0.271	0.251
6	n-Hexane	C_6H_{14}	110543	86.177	507.6	3.04	0.373	0.269	0.304
7	n-Heptane	C_7H_{16}	142825	100.204	540.2	2.72	0.428	0.259	0.346
8	n-Octane	C_8H_{18}	111659	114.231	568.7	2.47	0.486	0.254	0.396
9	n-Nonane	C_9H_{20}	111842	128.258	594.6	2.31	0.540	0.252	0.446
10	n-Decane	$C_{10}H_{22}$	124185	142.285	617.7	2.09	0.601	0.245	0.488
11	n-Undecane	$C_{11}H_{24}$	1120214	156.312	639	1.95	0.658	0.242	0.530
12	n-Dodecane	$C_{12}H_{26}$	112403	170.338	658	1.82	0.718	0.239	0.577
13	n-Tridecane	$C_{13}H_{28}$	629505	184.365	675	1.68	0.779	0.233	0.617
14	n-Tetradecane	$C_{14}H_{30}$	629594	198.392	693	1.57	0.830	0.226	0.643
15	n-Pentadecane	$C_{15}H_{32}$	629629	212.419	708	1.47	0.888	0.222	0.685
16	n-Hexadecane	$C_{16}H_{34}$	544763	226.446	723	1.41	0.943	0.221	0.721
17	n-Heptadecane	$C_{17}H_{36}$	629787	240.473	736	1.34	0.998	0.219	0.771
18	n-Octadecane	$C_{18}H_{38}$	593453	254.500	747	1.26	1.059	0.214	0.806
19	n-Nonadecane	$C_{19}H_{40}$	629925	268.527	758	1.21	1.119	0.215	0.851
20	n-Eicosane	$C_{20}H_{42}$	112958	282.553	768	1.17	1.169	0.215	0.912
21	2-Methylpropane	C_4H_{10}	75285	58.123	408.14	3.62	0.261	0.278	0.177
22	2-Methylbutane	C_5H_{12}	78784	72.150	460.43	3.37	0.304	0.268	0.226
23	2,3-Dimethylbutane	C_6H_{14}	79298	86.177	499.98	3.13	0.358	0.269	0.246
24	2-Methylpentane	C_6H_{14}	107835	86.177	497.5	3.02	0.366	0.267	0.279
25	2,3-Dimethylpentane	C_7H_{16}	565593	100.204	537.35	2.88	0.396	0.255	0.292
26	2,3,3-Trimethylpentane	C_8H_{18}	560214	114.231	573.5	2.81	0.455	0.268	0.289
27	2,2,4-Trimethylpentane	C_8H_{18}	540841	114.231	543.96	2.56	0.465	0.264	0.301
28	Ethylene	C_2H_4	74851	28.054	282.34	5.03	0.132	0.283	0.086
29	Propylene	C_3H_6	115071	42.081	365.57	4.63	0.188	0.286	0.137
30	1-Butene	C_4H_8	106989	56.108	419.95	4.04	0.241	0.279	0.190
31	cis-2-Butene	C_4H_8	590181	56.108	435.58	4.24	0.233	0.273	0.204
32	trans-2-Butene	C_4H_8	624646	56.108	428.63	4.08	0.237	0.272	0.216
33	1-Pentene	C_5H_{10}	109671	70.134	464.78	3.56	0.295	0.271	0.236
34	1-Hexene	C_6H_{12}	592416	84.161	504.03	3.14	0.354	0.265	0.280
35	1-Heptene	C_7H_{14}	592767	98.188	537.29	2.82	0.413	0.261	0.330
36	1-Octene	C_8H_{16}	111660	112.215	566.65	2.57	0.460	0.251	0.377
37	1-Nonene	C_9H_{18}	124118	126.242	593.25	2.33	0.528	0.249	0.417
38	1-Decene	$C_{10}H_{20}$	872059	140.269	616.4	2.21	0.584	0.252	0.478
39	2-Methylpropene	C_4H_8	115117	56.108	417.9	3.98	0.238	0.272	0.192
40	2-Methyl-1-butene	C_5H_{10}	563462	70.134	465	3.45	0.292	0.261	0.237
41	2-Methyl-2-butene	C_5H_{10}	513359	70.134	471	3.38	0.292	0.252	0.272
42	1,2-Butadiene	C_4H_6	590192	54.092	452	4.36	0.220	0.255	0.166
43	1,3-Butadiene	C_4H_6	106990	54.092	425.17	4.30	0.220	0.268	0.192
44	2-Methyl-1,3-butadiene	C_5H_8	78795	68.119	484	3.85	0.277	0.265	0.158
45	Acetylene	C_2H_2	74862	26.038	308.32	6.15	0.113	0.271	0.188
46	Methylacetylene	C_3H_4	74997	40.065	402.39	5.62	0.164	0.276	0.216
47	Dimethylacetylene	C_4H_6	503173	54.092	473.2	4.87	0.221	0.274	0.239
48	3-Methyl-1-butyne	C_5H_8	598232	68.119	463.2	4.20	0.275	0.300	0.308
49	1-Pentyne	C_5H_8	627190	68.119	481.2	4.17	0.277	0.289	0.290
50	2-Pentyne	C_5H_8	627214	68.119	519	4.02	0.276	0.257	0.174
51	1-Hexyne	C_6H_{10}	693027	82.145	516.2	3.64	0.322	0.273	0.335
52	2-Hexyne	C_6H_{10}	764352	82.145	549	3.53	0.331	0.256	0.221
53	3-Hexyne	C_6H_{10}	928494	82.145	544	3.54	0.334	0.261	0.219
54	1-Heptyne	C_7H_{12}	628717	96.172	559	3.13	0.386	0.260	0.272
55	1-Octyne	C_8H_{14}	629050	110.199	585	2.82	0.441	0.256	0.323
56	Vinylacetylene	C_4H_4	689974	52.076	454	4.89	0.205	0.265	0.109

TABLE 2-164   Critical Constants and Acentric Factors of Inorganic and Organic Compounds (*Continued*)

Cmpd. no.	Name	Formula	CAS no.	Mol. wt.	T_c, K	$P_c \times$ 1E-06 Pa	V_c, m³/Kmol	Z_c	Acentric factor
57	Cyclopentane	C_5H_{10}	287923	70.134	511.76	4.50	0.257	0.272	0.196
58	Methylcyclopentane	C_6H_{12}	96377	84.161	532.79	3.78	0.319	0.272	0.230
59	Ethylcyclopentane	C_7H_{14}	1640897	98.188	569.52	3.40	0.374	0.269	0.271
60	Cyclohexane	C_6H_{12}	110827	84.161	553.58	4.10	0.308	0.274	0.212
61	Methylcyclohexane	C_7H_{14}	108872	98.188	572.19	3.48	0.368	0.269	0.236
62	1,1-Dimethylcyclohexane	C_8H_{16}	590669	112.215	591.15	2.94	0.450	0.269	0.233
63	Ethylcyclohexane	C_8H_{16}	1678917	112.215	609.15	3.04	0.430	0.258	0.246
64	Cyclopentene	C_5H_8	142290	68.119	507	4.81	0.245	0.279	0.196
65	1-Methylcyclopentene	C_6H_{10}	693890	82.145	542	4.13	0.303	0.278	0.232
66	Cyclohexene	C_6H_{10}	110838	82.145	560.4	4.39	0.291	0.274	0.216
67	Benzene	C_6H_6	71432	78.114	562.16	4.88	0.261	0.273	0.209
68	Toluene	C_7H_8	108883	92.141	591.8	4.10	0.314	0.262	0.262
69	o-Xylene	C_8H_{10}	95476	106.167	630.33	3.74	0.374	0.267	0.311
70	m-Xylene	C_8H_{10}	108383	106.167	617.05	3.53	0.377	0.259	0.325
71	p-Xylene	C_8H_{10}	106423	106.167	616.23	3.50	0.381	0.260	0.320
72	Ethylbenzene	C_8H_{10}	100414	106.167	617.2	3.60	0.375	0.263	0.301
73	Propylbenzene	C_9H_{12}	103651	120.194	638.32	3.20	0.440	0.265	0.344
74	1,2,4-Trimethylbenzene	C_9H_{12}	95636	120.194	649.13	3.25	0.430	0.259	0.380
75	Isopropylbenzene	C_9H_{12}	98828	120.194	631.1	3.18	0.429	0.260	0.322
76	1,3,5-Trimethylbenzene	C_9H_{12}	108678	120.194	637.36	3.11	0.433	0.254	0.397
77	p-Isopropyltoluene	$C_{10}H_{14}$	99876	134.221	653.15	2.80	0.497	0.256	0.366
78	Naphthalene	$C_{10}H_8$	91203	128.174	748.35	3.99	0.413	0.265	0.296
79	Biphenyl	$C_{12}H_{10}$	92524	154.211	789.26	3.86	0.502	0.295	0.367
80	Styrene	C_8H_8	100425	104.152	636	3.82	0.352	0.254	0.295
81	m-Terphenyl	$C_{18}H_{14}$	92068	230.309	924.85	3.53	0.768	0.352	0.561
82	Methanol	CH_4O	67561	32.042	512.64	8.14	0.117	0.224	0.566
83	Ethanol	C_2H_6O	64175	46.069	513.92	6.12	0.168	0.240	0.643
84	1-Propanol	C_3H_8O	71238	60.096	536.78	5.12	0.220	0.252	0.617
85	1-Butanol	$C_4H_{10}O$	71363	74.123	563.05	4.34	0.276	0.256	0.585
86	2-Butanol	$C_4H_{10}O$	78922	74.123	536.05	4.20	0.270	0.254	0.574
87	2-Propanol	C_3H_8O	67630	60.096	508.3	4.79	0.221	0.250	0.670
88	2-Methyl-2-propanol	$C_4H_{10}O$	75650	74.123	506.21	3.99	0.276	0.262	0.613
89	1-Pentanol	$C_5H_{12}O$	71410	88.150	586.15	3.87	0.327	0.260	0.592
90	2-Methyl-1-butanol	$C_5H_{12}O$	137326	88.150	565	3.87	0.327	0.270	0.678
91	3-Methyl-1-butanol	$C_5H_{12}O$	123513	88.150	577.2	3.90	0.327	0.266	0.586
92	1-Hexanol	$C_6H_{14}O$	111273	102.177	611.35	3.46	0.381	0.259	0.572
93	1-Heptanol	$C_7H_{16}O$	111706	116.203	631.9	3.18	0.435	0.263	0.592
94	Cyclohexanol	$C_6H_{12}O$	108930	100.161	650	4.25	0.322	0.253	0.371
95	Ethylene glycol	$C_2H_6O_2$	107211	62.068	719.7	7.71	0.191	0.246	0.487
96	1,2-Propylene glycol	$C_3H_8O_2$	57556	76.095	626	6.04	0.239	0.277	1.102
97	Phenol	C_6H_6O	108952	94.113	694.25	6.06	0.229	0.240	0.438
98	o-Cresol	C_7H_8O	95487	108.140	697.55	5.06	0.282	0.246	0.438
99	m-Cresol	C_7H_8O	108394	108.140	705.85	4.52	0.312	0.240	0.444
100	p-Cresol	C_7H_8O	106445	108.140	704.65	5.15	0.277	0.244	0.507
101	Dimethyl ether	C_2H_6O	115106	46.069	400.1	5.27	0.171	0.271	0.192
102	Methyl ethyl ether	C_3H_8O	540670	60.096	437.8	4.47	0.221	0.271	0.229
103	Methyl n-propyl ether	$C_4H_{10}O$	557175	74.123	476.3	3.77	0.276	0.263	0.264
104	Methyl isopropyl ether	$C_4H_{10}O$	598538	74.123	464.5	3.89	0.276	0.278	0.280
105	Methyl n-butyl ether	$C_5H_{12}O$	628284	88.150	510	3.31	0.329	0.257	0.335
106	Methyl isobutyl ether	$C_5H_{12}O$	625445	88.150	497	3.41	0.331	0.273	0.310
107	Methyl tert-butyl ether	$C_5H_{12}O$	1634044	88.150	497.1	3.41	0.329	0.272	0.264
108	Diethyl ether	$C_4H_{10}O$	60297	74.123	466.7	3.64	0.281	0.264	0.281
109	Ethyl propyl ether	$C_5H_{12}O$	628320	88.150	500.23	3.37	0.336	0.273	0.347
110	Ethyl isopropyl ether	$C_5H_{12}O$	625547	88.150	489	3.41	0.329	0.276	0.306
111	Methyl phenyl ether	C_7H_8O	100663	108.140	645.6	4.27	0.337	0.268	0.353
112	Diphenyl ether	$C_{12}H_{10}O$	101848	170.211	766.8	3.10	0.503	0.244	0.441
113	Formaldehyde	CH_2O	50000	30.026	408	6.59	0.115	0.223	0.282
114	Acetaldehyde	C_2H_4O	75070	44.053	466	5.57	0.154	0.221	0.292
115	1-Propanal	C_3H_6O	123386	58.080	504.4	4.92	0.204	0.239	0.256
116	1-Butanal	C_4H_8O	123728	72.107	537.2	4.32	0.258	0.250	0.278
117	1-Pentanal	$C_5H_{10}O$	110623	86.134	566.1	3.97	0.313	0.264	0.347
118	1-Hexanal	$C_6H_{12}O$	66251	100.161	591	3.46	0.369	0.260	0.387
119	1-Heptanal	$C_7H_{14}O$	111717	114.188	617	3.18	0.421	0.261	0.427
120	1-Octanal	$C_8H_{16}O$	124130	128.214	638.1	2.97	0.474	0.265	0.474
121	1-Nonanal	$C_9H_{18}O$	124196	142.241	658	2.74	0.527	0.264	0.514
122	1-Decanal	$C_{10}H_{20}O$	112312	156.268	674.2	2.60	0.580	0.269	0.582

TABLE 2-164   Critical Constants and Acentric Factors of Inorganic and Organic Compounds (Continued)

Cmpd. no.	Name	Formula	CAS no.	Mol. wt.	T_c, K	$P_c \times$ 1E-06 Pa	V_c, m³/Kmol	Z_c	Acentric factor
123	Acetone	C_3H_6O	67641	58.080	508.2	4.71	0.210	0.234	0.307
124	Methyl ethyl ketone	C_4H_8O	78933	72.107	535.5	4.12	0.267	0.247	0.320
125	2-Pentanone	$C_5H_{10}O$	107879	86.134	561.08	3.71	0.301	0.239	0.345
126	Methyl isopropyl ketone	$C_5H_{10}O$	563804	86.134	553	3.84	0.313	0.261	0.349
127	2-Hexanone	$C_6H_{12}O$	591786	100.161	587.05	3.31	0.369	0.250	0.395
128	Methyl isobutyl ketone	$C_6H_{12}O$	108101	100.161	571.4	3.27	0.369	0.254	0.389
129	3-Methyl-2-pentanone	$C_6H_{12}O$	565617	100.161	573	3.32	0.371	0.259	0.386
130	3-Pentanone	$C_5H_{10}O$	96220	86.134	560.95	3.70	0.336	0.267	0.340
131	Ethyl isopropyl ketone	$C_6H_{12}O$	565695	100.161	567	3.34	0.369	0.262	0.394
132	Diisopropyl ketone	$C_7H_{14}O$	565800	114.188	576	3.06	0.416	0.266	0.411
133	Cyclohexanone	$C_6H_{10}O$	108941	98.145	653	4.01	0.311	0.230	0.308
134	Methyl phenyl ketone	C_8H_8O	98862	120.151	709.5	3.85	0.386	0.252	0.365
135	Formic acid	CH_2O_2	64186	46.026	588	5.81	0.125	0.148	0.317
136	Acetic acid	$C_2H_4O_2$	64197	60.053	591.95	5.74	0.179	0.208	0.463
137	Propionic acid	$C_3H_6O_2$	79094	74.079	600.81	4.61	0.232	0.214	0.574
138	n-Butyric acid	$C_4H_8O_2$	107926	88.106	615.7	4.07	0.291	0.231	0.682
139	Isobutyric acid	$C_4H_8O_2$	79312	88.106	605	3.68	0.291	0.213	0.612
140	Benzoic acid	$C_7H_6O_2$	65850	122.123	751	4.47	0.347	0.248	0.603
141	Acetic anhydride	$C_4H_6O_3$	108247	102.090	606	3.97	0.290	0.229	0.450
142	Methyl formate	$C_2H_4O_2$	107313	60.053	487.2	5.98	0.173	0.255	0.254
143	Methyl acetate	$C_3H_6O_2$	79209	74.079	506.55	4.69	0.229	0.256	0.326
144	Methyl propionate	$C_4H_8O_2$	554121	88.106	530.6	4.03	0.284	0.259	0.349
145	Methyl n-butyrate	$C_5H_{10}O_2$	623427	102.133	554.5	3.48	0.340	0.257	0.378
146	Ethyl formate	$C_3H_6O_2$	109944	74.079	508.4	4.71	0.231	0.257	0.282
147	Ethyl acetate	$C_4H_8O_2$	141786	88.106	523.3	3.85	0.287	0.254	0.363
148	Ethyl propionate	$C_5H_{10}O_2$	105373	102.133	546	3.34	0.345	0.254	0.391
149	Ethyl n-butyrate	$C_6H_{12}O_2$	105544	116.160	571	2.94	0.403	0.249	0.399
150	n-Propyl formate	$C_4H_8O_2$	110747	88.106	538	4.03	0.286	0.257	0.310
151	n-Propyl acetate	$C_5H_{10}O_2$	109604	102.133	549.73	3.37	0.349	0.257	0.390
152	n-Butyl acetate	$C_6H_{12}O_2$	123864	116.160	579.15	3.11	0.389	0.251	0.410
153	Methyl benzoate	$C_8H_8O_2$	93583	136.150	693	3.59	0.436	0.272	0.421
154	Ethyl benzoate	$C_9H_{10}O_2$	93890	150.177	698	3.22	0.489	0.271	0.477
155	Vinyl acetate	$C_4H_6O_2$	108054	86.090	519.13	3.93	0.270	0.246	0.348
156	Methylamine	CH_5N	74895	31.057	430.05	7.41	0.154	0.319	0.279
157	Dimethylamine	C_2H_7N	124403	45.084	437.2	5.26	0.180	0.260	0.293
158	Trimethylamine	C_3H_9N	75503	59.111	433.25	4.10	0.254	0.289	0.210
159	Ethylamine	C_2H_7N	75047	45.084	456.15	5.59	0.202	0.298	0.283
160	Diethylamine	$C_4H_{11}N$	109897	73.138	496.6	3.67	0.301	0.268	0.300
161	Triethylamine	$C_6H_{15}N$	121448	101.192	535.15	3.04	0.389	0.266	0.316
162	n-Propylamine	C_3H_9N	107108	59.111	496.95	4.74	0.260	0.298	0.280
163	di-n-Propylamine	$C_6H_{15}N$	142847	101.192	550	3.11	0.401	0.273	0.446
164	Isopropylamine	C_3H_9N	75310	59.111	471.85	4.54	0.221	0.256	0.276
165	Diisopropylamine	$C_6H_{15}N$	108189	101.192	523.1	3.20	0.417	0.307	0.388
166	Aniline	C_6H_7N	62533	93.128	699	5.35	0.270	0.248	0.381
167	N-Methylaniline	C_7H_9N	100618	107.155	701.55	5.19	0.373	0.332	0.480
168	N,N-Dimethylaniline	$C_8H_{11}N$	121697	121.182	687.15	3.63	0.465	0.295	0.403
169	Ethylene oxide	C_2H_4O	75218	44.053	469.15	7.26	0.142	0.264	0.201
170	Furan	C_4H_4O	110009	68.075	490.15	5.55	0.218	0.297	0.205
171	Thiophene	C_4H_4S	110021	84.142	579.35	5.71	0.219	0.260	0.195
172	Pyridine	C_5H_5N	110861	79.101	619.95	5.64	0.254	0.278	0.239
173	Formamide	CH_3NO	75127	45.041	771	7.75	0.163	0.197	0.410
174	N,N-Dimethylformamide	C_3H_7NO	68122	73.095	649.6	4.37	0.262	0.212	0.312
175	Acetamide	C_2H_5NO	60355	59.068	761	6.57	0.215	0.223	0.419
176	N-Methylacetamide	C_3H_7NO	79163	73.095	718	5.00	0.267	0.224	0.437
177	Acetonitrile	C_2H_3N	75058	41.053	545.5	4.85	0.173	0.185	0.340
178	Propionitrile	C_3H_5N	107120	55.079	564.4	4.19	0.229	0.205	0.325
179	n-Butyronitrile	C_4H_7N	109740	69.106	582.25	3.79	0.278	0.217	0.371
180	Benzonitrile	C_7H_5N	100470	103.123	699.35	4.21	0.339	0.245	0.352
181	Methyl mercaptan	CH_4S	74931	48.109	469.15	7.23	0.145	0.268	0.158
182	Ethyl mercaptan	C_2H_6S	75081	62.136	499.15	5.49	0.206	0.273	0.188
183	n-Propyl mercaptan	C_3H_8S	107039	76.163	536.6	4.63	0.254	0.263	0.232
184	n-Butyl mercaptan	$C_4H_{10}S$	109795	90.189	570.1	3.97	0.307	0.257	0.272
185	Isobutyl mercaptan	$C_4H_{10}S$	513440	90.189	559	4.06	0.307	0.268	0.253
186	sec-Butyl mercaptan	$C_4H_{10}S$	513531	90.189	554	4.06	0.307	0.271	0.251
187	Dimethyl sulfide	C_2H_6S	75183	62.136	503.04	5.53	0.200	0.264	0.194
188	Methyl ethyl sulfide	C_3H_8S	624895	76.163	533	4.26	0.254	0.244	0.209
189	Diethyl sulfide	$C_4H_{10}S$	352932	90.189	557.15	3.96	0.320	0.273	0.294

TABLE 2-164 Critical Constants and Acentric Factors of Inorganic and Organic Compounds (*Concluded*)

Cmpd. no.	Name	Formula	CAS no.	Mol. wt.	T_c, K	$P_c \times$ 1E-06 Pa	V_c, m³/Kmol	Z_c	Acentric factor
190	Fluoromethane	CH₃F	593533	34.033	317.42	5.88	0.113	0.252	0.198
191	Chloromethane	CH₃Cl	74873	50.488	416.25	6.69	0.142	0.275	0.154
192	Trichloromethane	CHCl₃	67663	119.377	536.4	5.55	0.238	0.296	0.228
193	Tetrachloromethane	CCl₄	56235	153.822	556.35	4.54	0.274	0.270	0.191
194	Bromomethane	CH₃Br	74839	94.939	467	8.00	0.156	0.321	0.192
195	Fluoroethane	C₂H₅F	353366	48.060	375.31	5.01	0.164	0.263	0.218
196	Chloroethane	C₂H₅Cl	75003	64.514	460.35	5.46	0.155	0.221	0.206
197	Bromoethane	C₂H₅Br	74964	108.966	503.8	6.29	0.215	0.323	0.259
198	1-Chloropropane	C₃H₇Cl	540545	78.541	503.15	4.58	0.247	0.270	0.228
199	2-Chloropropane	C₃H₇Cl	75296	78.541	489	4.51	0.247	0.274	0.196
200	1,1-Dichloropropane	C₃H₆Cl₂	78999	112.986	560	4.24	0.292	0.266	0.253
201	1,2-Dichloropropane	C₃H₆Cl₂	78875	112.986	572	4.23	0.291	0.259	0.256
202	Vinyl chloride	C₂H₃Cl	75014	62.499	432	5.75	0.179	0.287	0.106
203	Fluorobenzene	C₆H₅F	462066	96.104	560.09	4.54	0.269	0.262	0.247
204	Chlorobenzene	C₆H₅Cl	108907	112.558	632.35	4.53	0.308	0.265	0.251
205	Bromobenzene	C₆H₅Br	108861	157.010	670.15	4.52	0.324	0.263	0.251
206	Air		132259100	28.951	132.45	3.79	0.092	0.318	0.000
207	Hydrogen	H₂	1333740	2.016	33.19	1.32	0.064	0.307	−0.215
208	Helium-4	He	7440597	4.003	5.2	0.23	0.058	0.305	−0.388
209	Neon	Ne	7440019	20.180	44.4	2.67	0.042	0.300	−0.038
210	Argon	Ar	7440371	39.948	150.86	4.90	0.075	0.292	0.000
211	Fluorine	F₂	7782414	37.997	144.12	5.17	0.067	0.287	0.053
212	Chlorine	Cl₂	7782505	70.905	417.15	7.79	0.124	0.279	0.073
213	Bromine	Br₂	7726956	159.808	584.15	10.28	0.135	0.286	0.128
214	Oxygen	O₂	7782447	31.999	154.58	5.02	0.074	0.287	0.020
215	Nitrogen	N₂	7727379	28.014	126.2	3.39	0.089	0.288	0.037
216	Ammonia	NH₃	7664417	17.031	405.65	11.30	0.072	0.241	0.253
217	Hydrazine	N₂H₄	302012	32.045	653.15	14.73	0.158	0.429	0.315
218	Nitrous oxide	N₂O	10024972	44.013	309.57	7.28	0.098	0.277	0.143
219	Nitric oxide	NO	10102439	30.006	180.15	6.52	0.058	0.252	0.585
220	Cyanogen	C₂N₂	460195	52.036	400.15	5.94	0.195	0.348	0.276
221	Carbon monoxide	CO	630080	28.010	132.92	3.49	0.095	0.300	0.048
222	Carbon dioxide	CO₂	124389	44.010	304.21	7.39	0.095	0.277	0.224
223	Carbon disulfide	CS₂	75150	76.143	552	8.04	0.160	0.280	0.118
224	Hydrogen fluoride	HF	7664393	20.006	461.15	6.49	0.069	0.117	0.383
225	Hydrogen chloride	HCl	7647010	36.461	324.65	8.36	0.082	0.253	0.134
226	Hydrogen bromide	HBr	10035106	80.912	363.15	8.46	0.100	0.280	0.069
227	Hydrogen cyanide	HCN	74908	27.026	456.65	5.35	0.139	0.195	0.407
228	Hydrogen sulfide	H₂S	7783064	34.082	373.53	9.00	0.099	0.287	0.096
229	Sulfur dioxide	SO₂	7446095	64.065	430.75	7.86	0.123	0.269	0.244
230	Sulfur trioxide	SO₃	7446119	80.064	490.85	8.19	0.127	0.255	0.423
231	Water	H₂O	7732185	18.015	647.13	21.94	0.056	0.228	0.343

All substances are listed in alphabetical order in Table 2-6a.

Compiled from Daubert, T. E., R. P. Danner, H. M. Sibul, and C. C. Stebbins, DIPPR Data Compilation of Pure Compound Properties, Project 801 Sponsor Release, July, 1993, Design Institute for Physical Property Data, AIChE, New York, NY; and from Ambrose, D. "Vapour-Liquid Critical Properties", Report Chem 107, National Physical Laboratory, Teddington, UK, October, 1979.

In order to ensure thermodynamic consistency, in almost all cases these properties are calculated from T_c and the vapor pressure and liquid density correlation coefficients listed in those tables. This means that there will be slight differences between the values listed here and those in the DIPPR tables. Most of the differences are less than 1%, and almost all the rest are less than the estimated accuracy of the quantity in question.

The atomic weights used, taken from *J. Phys. Chem. Ref. Data* **22**(6), 1993, are C = 12.011, H = 1.00794, O = 15.9994, N = 14.00674, S = 32.066, F = 18.9984, Cl = 35.4527, Br = 79.904, and I = 126.90447.

The value of the gas constant, R, used here is 8314.51 J/(kmol·K), as given by E. R. Cohen and B. N. Taylor in *J. Phys. Chem. Ref. Data* **17**, 1988. K − 273.15 = °C; 1.8 × K − 459.67 = °F; Pa × 9.869233E-06 = atm; Pa × 1.450377E-04 = psia; m³/kmol × (1E + 03/mol. wt.) = cm³/g; m³/kmol × (1.601846E + 01/mol wt) = ft³/lb.

COMPRESSIBILITIES

INTRODUCTION

The increasing ranges of pressure and temperature of interest to technology for an ever-increasing number of substances would necessitate additional tables in this subsection as well as in the subsection "Thermodynamic Properties." Space restrictions preclude this. Hence, in the present revision, an attempt was made to update the fluid-compressibility tables for selected fluids and to omit tables for other fluids. The reader is thus referred to the fourth edition for tables on miscellaneous gases at 0°C, acetylene, ammonia, ethane, ethylene, hydrogen-nitrogen mixtures, and methyl chloride. The reader is also reminded that compressibilities can be calculated from the pressure—volume (or density)—temperature tables of the subsection "Thermodynamic Properties."

UNITS CONVERSIONS

For this subsection, the following units conversions are applicable:

$$°R = \% \, K.$$

To convert bars to pounds-force per cubic inch, multiply by 14.504.
To convert bars to kilopascals, multiply by 1×10^2.

TABLE 2-165   Compressibility Factors for Air*

Temp., K	\multicolumn{14}{c}{Pressure, bar}
	1	5	10	20	40	60	80	100	150	200	250	300	400	500
75	0.0052	0.0260	0.0519	0.1036	0.2063	0.3082	0.4094	0.5099	0.7581	1.0025				
80		0.0250	0.0499	0.0995	0.1981	0.2958	0.3927	0.4887	0.7258	0.9588	1.1931	1.4139		
90	0.9764	0.0236	0.0471	0.0940	0.1866	0.2781	0.3686	0.4581	0.6779	0.8929	1.1098	1.3110	1.7161	2.1105
100	0.9797	0.8872	0.0453	0.0900	0.1782	0.2635	0.3498	0.4337	0.6386	0.8377	1.0395	1.2227	1.5937	1.9536
120	0.9880	0.9373	0.8660	0.6730	0.1778	0.2557	0.3371	0.4132	0.5964	0.7720	0.9530	1.1076	1.5091	1.7366
140	0.9927	0.9614	0.9205	0.8297	0.5856	0.3313	0.3737	0.4340	0.5909	0.7699	0.9114	1.0393	1.3202	1.5903
160	0.9951	0.9748	0.9489	0.8954	0.7803	0.6603	0.5696	0.5489	0.6340	0.7564	0.8840	1.0105	1.2585	1.4970
180	0.9967	0.9832	0.9660	0.9314	0.8625	0.7977	0.7432	0.7084	0.7180	0.7986	0.9000	1.0068	1.2232	1.4361
200	0.9978	0.9886	0.9767	0.9539	0.9100	0.8701	0.8374	0.8142	0.8061	0.8549	0.9311	1.0185	1.2054	1.3944
250	0.9992	0.9957	0.9911	0.9822	0.9671	0.9549	0.9463	0.9411	0.9450	0.9713	1.0152	1.0702	1.1990	1.3392
300	0.9999	0.9987	0.9974	0.9950	0.9917	0.9901	0.9903	0.9930	1.0074	1.0326	1.0669	1.1089	1.2073	1.3163
350	1.0000	1.0002	1.0004	1.0014	1.0038	1.0075	1.0121	1.0183	1.0377	1.0635	1.0947	1.1303	1.2116	1.3015
400	1.0002	1.0012	1.0025	1.0046	1.0100	1.0159	1.0229	1.0312	1.0533	1.0795	1.1087	1.1411	1.2117	1.2890
450	1.0003	1.0016	1.0034	1.0063	1.0133	1.0210	1.0287	1.0374	1.0614	1.0913	1.1183	1.1463	1.2090	1.2778
500	1.0003	1.0020	1.0034	1.0074	1.0151	1.0234	1.0323	1.0410	1.0650	1.0913	1.1183	1.1463	1.2051	1.2667
600	1.0004	1.0022	1.0039	1.0081	1.0164	1.0253	1.0340	1.0434	1.0678	1.0920	1.1172	1.1427	1.1947	1.2475
800	1.0004	1.0020	1.0038	1.0077	1.0157	1.0240	1.0321	1.0408	1.0621	1.0844	1.1061	1.1283	1.1720	1.2150
1000	1.0004	1.0018	1.0037	1.0068	1.0142	1.0215	1.0290	1.0365	1.0556	1.0744	1.0948	1.1131	1.1515	1.1889

*Calculated from values of pressure, volume (or density), and temperature in Vasserman, Kazavchinskii, and Rabinovich, *Thermophysical Properties of Air and Air Components,* Moscow, Nauka, 1966, and NBS-NSF Trans. TT 70-50095, 1971; and Vasserman and Rabinovich, *Thermophysical Properties of Liquid Air and Its Components,* Moscow, 1968, and NBS-NSF Trans. 69-55092, 1970.

TABLE 2-166   Compressibility Factors for Argon*

Temp., K	\multicolumn{12}{c}{Pressure, bar}
	1	5	10	20	40	60	80	100	200	300	400	500
100	0.9773	0.0183	0.0366	0.0729	0.1449	0.2162	0.2867	0.3567	0.6975	1.0267	1.3470	1.6932
150	0.9932	0.9647	0.9273	0.8447	0.6101	0.2249	0.2781	0.3324	0.5934	0.8387	1.0732	1.2995
200	0.9972	0.9857	0.9713	0.9419	0.8810	0.8208	0.7624	0.7121	0.6870	0.8360	1.0051	1.1982
250	0.9988	0.9935	0.9869	0.9741	0.9494	0.9263	0.9056	0.8877	0.8590	0.9207	1.0262	1.1479
300	0.9995	0.9969	0.9941	0.9884	0.9777	0.9686	0.9611	0.9552	0.9533	0.9950	1.0673	1.1786
400	1.0001	0.9997	0.9998	0.9999	1.0004	1.0018	1.0031	1.0056	1.0280	1.0656	1.1157	1.1976
500	1.0002	1.0007	1.0012	1.0034	1.0071	1.0113	1.0154	1.0205	1.0501	1.0874	1.1301	1.1997
600	1.0003	1.0012	1.0025	1.0046	1.0094	1.0143	1.0198	1.0250	1.0553	1.0904	1.1291	1.1933
800	1.0003	1.0012	1.0023	1.0050	1.0102	1.0151	1.0205	1.0258	1.0552	1.0830	1.1147	1.1707
1000	1.0002	1.0013	1.0022	1.0050	1.0096	1.0142	1.0193	1.0239	1.0484	1.0736	1.0999	1.1497

*Calculated from PVT values tabulated in Rabinovich (ed.), *Thermophysical Properties of Neon, Argon, Krypton and Xenon,* Standard Press, Moscow, 1976. This book was published in English translation by Hemisphere, New York, 1988 (604 pp.).

TABLE 2-167 Compressibility Factors for Carbon Dioxide*

Temp., °C	Pressure, bar											
	1	5	10	20	40	60	80	100	200	300	400	500
0	0.9933	0.9658	0.9294	0.8496								
50	0.9964	0.9805	0.9607	0.9195	0.8300	0.7264	0.5981	0.4239				
100	0.9977	0.9883	0.9764	0.9524	0.9034	0.8533	0.8022	0.7514	0.5891	0.6420		
150	0.9985	0.9927	0.9853	0.9705	0.9416	0.9131	0.8854	0.8590	0.7651	0.7623	0.8235	0.9098
200	0.9991	0.9953	0.9908	0.9818	0.9640	0.9473	0.9313	0.9170	0.8649	0.8619	0.8995	0.9621
250	0.9994	0.9971	0.9943	0.9886	0.9783	0.9684	0.9593	0.9511	0.9253	0.9294	0.9508	1.0096
300	0.9996	0.9982	0.9967	0.9936	0.9875	0.9822	0.9773	0.9733	0.9640	0.9746	1.0030	1.0464
350	0.9998	0.9991	0.9983	0.9964	0.9938	0.9914	0.9896	0.9882	0.9895	1.0053	1.0340	1.0734
400	0.9999	0.9997	0.9994	0.9989	0.9982	0.9979	0.9979	0.9984	1.0073	1.0266	1.0559	1.0928
450	1.0000	1.0000	1.0003	1.0005	1.0013	1.0023	1.0038	1.0056	1.0070	1.0412	1.0709	1.1067
500	1.0000	1.0004	1.0008	1.0015	1.0035	1.0056	1.0079	1.0107	1.0282	1.0522	1.0820	1.1165
600	1.0000	1.0007	1.0013	1.0030	1.0062	1.0093	1.0129	1.0168	1.0386	1.0648	1.0948	1.1277
700	1.0003	1.0010	1.0017	1.0036	1.0073	1.0161	1.0155	1.0198	1.0436	1.0707	1.1000	1.1318
800	1.0002	1.0009	1.0019	1.0040	1.0082	1.0122	1.0168	1.0212	1.0458	1.0731	1.1016	1.1324
900	1.0002	1.0009	1.0020	1.0041	1.0083	1.0128	1.0171	1.0221	1.0463	1.0726	1.1012	1.1303
1000	1.0002	1.0009	1.0021	1.0042	1.0084	1.0128	1.0172	1.0218	1.0460	1.0725	1.0725	1.1274

*Calculated from density-pressure-temperature data in Vukalovitch and Altunin, *Thermophysical Properties of Carbon Dioxide*, Atomizdat, Moscow, 1965, and Collet's, London, 1968, translation.

TABLE 2-168 Compressibility Factors for Carbon Monoxide*

Temp., K	Pressure, atm						
	1	4	7	10	40	70	100
200	0.9973	0.9893	0.9813	0.9734			
250	0.9989	0.9957	0.9926	0.9896	0.9632		
300	0.9997	0.9987	0.9977	0.9968	0.9907	0.9896	0.9935
350	1.0000	1.0002	1.0003	1.0005	1.0042	1.0112	1.0216
400	1.0002	1.0010	1.0017	1.0025	1.0042	1.0112	1.0216
450	1.0003	1.0014	1.0025	1.0035	1.0152	1.0285	1.0433
500	1.0004	1.0016	1.0029	1.0041	1.0172	1.0314	1.0469
600	1.0005	1.0018	1.0032	1.0045	1.0186	1.0332	1.0485
700	1.0005	1.0018	1.0032	1.0045	1.0183	1.0325	1.0470
800	1.0004	1.0017	1.0030	1.0044	1.0175	1.0309	1.0445
900	1.0004	1.0017	1.0029	1.0041	1.0166	1.0291	1.0418
1000	1.0004	1.0016	1.0027	1.0039	1.0156	1.0273	1.0391
1500	1.0003	1.0012	1.0021	1.0029	1.0115	1.0200	1.0286
2000	1.0002	1.0009	1.0016	1.0022	1.0088	1.0155	1.0221
2500	1.0002	1.0007	1.0013	1.0018	1.0071	1.0124	1.0178
3000	1.0002	1.0006	1.0010	1.0015	1.0059	1.0104	1.0148

*From Hilsenrath *et al., N.B.S. Circ.* 564, 1955. Some of the above values have been rounded to four decimal places. Values at 10-K increments below 1000 K and at 50 K increments for higher temperatures appear in the original, also for pressures below atmospheric.

TABLE 2-169 Compressibility Factors for Ethanol

Temp., K	Pressure, bar								
	0.1	0.5	1.013	10	20	50	100	250	500
300	0.0022	0.0023	0.0024	0.0229	0.0458	0.114	0.228	0.565	1.11
350				0.0215	0.0411	0.107	0.208	0.509	1.03
400	0.999	0.993	0.986	0.0204	0.0408	0.101	0.201	0.490	0.95
450	1.000	0.997	0.991	0.908		0.101	0.198	0.472	0.898
500	1.000	0.997	0.994	0.941	0.874	0.122	0.214	0.473	0.868
600	1.000	0.998	0.997	0.972	0.943		0.672	0.470	0.868
700	1.000	0.999	0.999	0.985	0.971	0.948	0.902	0.760	0.921
800	1.000	1.000	0.999	0.992	0.984	0.973	0.953	0.890	0.988
900	1.000	1.000	1.000	0.996	0.992	0.988	0.981	0.962	1.04
1000	1.000	1.000	1.000	0.998	0.997	0.993	0.990	1.002	1.08

Rounded and interpolated from Thermodynamics Research Center tables, Texas A&M University.

TABLE 2-170 Compressibility Factors for Ethylene

Pressure, bar	Temperature, K								
	110	150	200	250	300	350	400	450	500
1	0.0047	0.0038	0.9808	0.9902	0.9944	0.9966	0.9979	0.9986	0.9991
5	0.0237	0.0189	0.0162	0.9495	0.9717	0.9828	0.9894	0.9935	0.9959
10	0.0472	0.0378	0.0323	0.8946	0.9425	0.9659	0.9785	0.9867	0.9919
15	0.0710	0.0566	0.0484	0.8320	0.9121	0.9479	0.9679	0.9749	0.9876
20	0.0946	0.0754	0.0644	0.7578	0.8804	0.9299	0.9574	0.9734	0.9833
30	0.1418	0.1129	0.0963	0.0950	0.8122	0.8936	0.9357	0.9603	0.9754
40	0.1889	0.1504	0.1280	0.1251	0.7342	0.8560	0.9144	0.9477	0.9677
60	0.2831	0.2251	0.1910	0.1838	0.5235	0.7791	0.8730	0.9231	0.9541
80	0.3767	0.2994	0.2533	0.2410	0.3302	0.7023	0.9056	0.9009	0.9428
100	0.4702	0.3734	0.3150	0.2968	0.3480	0.6359	0.9220	0.8825	0.9321
150	0.7030	0.5567	0.4671	0.4324	0.4528	0.5842	0.7483	0.8523	0.9167
200	0.9337	0.7382	0.6161	0.5630	0.5641	0.6347	0.7499	0.8494	0.9184
250	1.1636	0.9179	0.7630	0.6904	0.6740	0.7110	0.7895	0.8710	0.9343
300	1.3917	1.0960	0.9075	0.8148	0.7816	0.7969	0.8479	0.9095	0.9631
400	1.8441	1.4475	1.1910	1.0565	0.9909	0.9726	0.9849	1.0142	1.0450
500	solid	1.7934	1.4679	1.2908	1.1932	1.1468	1.1304	1.1341	1.1436

Calculated from Jacobsen, R.T., M. Jahangiri, et al., *Ethylene*, Blackwell Sci. Publs., Oxford, 1988 (29 pp.).

TABLE 2-171 Compressibility Factors for Normal Hydrogen*

Temp., K	Pressure, bar											
	1	10	20	40	60	80	100	200	400	600	800	1000
20	0.0169	0.1680	0.3302	0.6430	0.9434	1.2346	1.5166	2.844				
40	0.9848	0.8340	0.6311	0.5240	0.6627	0.8118	0.9590	1.650	2.578	3.993	5.034	6.019
60	0.9955	0.9562	0.9169	0.8608	0.8498	0.8832	0.9432	1.347	2.158	2.902	3.598	4.263
80	0.9986	0.9776	0.9763	0.9655	0.9676	0.9842	1.0138	1.257	1.834	2.389	2.907	3.404
100	0.9998	0.9979	0.9976	1.0022	1.0133	1.0280	1.0528	1.225	1.659	2.095	2.512	2.902
200	1.0007	1.0066	1.0134	1.0275	1.0422	1.0575	1.0734	1.163	1.455	1.555	1.753	1.936
300	1.0005	1.0059	1.0117	1.0236	1.0357	1.0479	1.0603	1.124	1.253	1.383	1.510	1.636
400	1.0004	1.0048	1.0096	1.0192	1.0289	1.0386	1.0484	1.098	1.196	1.293	1.388	1.481
500	1.0004	1.0040	1.0080	1.0160	1.0240	1.0320	1.0400	1.080	1.159	1.236	1.311	1.385
600	1.0003	1.0034	1.0068	1.0136	1.0204	1.0272	1.0340	1.068	1.133	1.197	1.259	1.320
800	1.0002	1.0026	1.0052	1.0104	1.0156	1.0208	1.0259	1.051	1.100	1.147	1.193	1.237
1000	1.0002	1.0021	1.0042	1.0084	1.0126	1.0168	1.0209	1.041	1.080	1.117	1.153	1.187
2000	1.0009	1.0013	1.0023	1.0044	1.0065	1.0086	1.0107	1.021	1.040	1.057	1.073	1.088

*Calculated from PVT tables of McCarty, Hord, and Roder, NBS Monogr. 168, 1981.

TABLE 2-172 Compressibility Factors for KLEA 60

Temp., K	Pressure, bar							Z_{sat}	P_{sat}
	1	5	10	15	20	25	30		
250	0.9687							0.9494	2.08
260	0.9780							0.9315	3.11
270	0.9803							0.9098	4.49
280	0.9824	0.9099						0.8839	6.30
290	0.9848	0.9199						0.8538	8.62
300	0.9867	0.9284	0.8459					0.8175	11.55
310	0.9872	0.9359	0.8637	0.7800				0.7756	15.19
320	0.9884	0.9425	0.8790	0.8066				0.7261	19.66
330	0.9894	0.9484	0.8908	0.8299	0.7577	0.6700		0.6666	25.10
340	0.9905	0.9537	0.9026	0.8488	0.7888	0.7184	0.6305		
350	0.9920	0.9582	0.9139	0.8663	0.8145	0.7570	0.6908		
Z_{sat}	0.9712	0.9022	0.8361	0.7777	0.7224	0.6677	0.6118		
T_{sat}	234.0	273.1	295.0	309.5	320.7	329.8	337.6		

Converted and interpolated from "Thermodynamic Properties of KLEA 60," British units, © ICI Chemicals and Polymers, 1993 (20 pp.). Reproduced by permission. KLEA 60 is R32/125/134a (20/40/40 wt %).

TABLE 2-173   Compressibility Factors for KLEA 61

Temp., K	\multicolumn{7}{c}{Pressure, bar}							Z_{sat}	P_{sat}
	1	5	10	15	20	25	30		
250	0.9746							0.9381	2.46
260	0.9773							0.9172	3.63
270	0.9798							0.8920	5.18
280	0.9787	0.9067						0.8622	7.19
290	0.9838	0.9185						0.8272	9.75
300	0.9854	0.9270	0.8431					0.7868	12.21
310	0.9868	0.9348	0.8615	0.7755				0.7377	16.88
320	0.9881	0.9416	0.8772	0.8042	0.7148			0.6801	21.68
330	0.9892	0.9481	0.8909	0.8280	0.7518	0.6659		0.6087	27.50
340	0.9903	0.9529	0.9027	0.8484	0.7934	0.7174	0.6312		
350	0.9917	0.9577	0.9131	0.8653	0.8134	0.7565	0.6916		
Z_{sat}	0.9686	0.8944	0.8237	0.7602	0.7003	0.6399	0.5780		
T_{sat}	230.0	269.0	290.9	305.5	316.7	325.9	333.7		

Converted and interpolated from "Thermodynamic Properties of KLEA 61," British units, © ICI Chemicals and Polymers, 1993 (23 pp.). Reproduced by permission. KLEA 61 is R32/125/134a (10/70/20 wt %).

TABLE 2-174   Compressibility Factors for KLEA 66

Temp., K	\multicolumn{7}{c}{Pressure, bar}							Z_{sat}	P_{sat}
	1	5	10	15	20	25	30		
250	0.974							0.9541	1.89
260	0.9772							0.9374	2.84
270	0.9796							0.9172	4.12
280	0.9838	0.9089						0.8931	5.81
290	0.9858	0.9209						0.8645	7.98
300	0.9872	0.9287	0.8461					0.8328	10.73
310	0.9883	0.9359	0.8663					0.7920	14.15
320	0.9896	0.9431	0.8786	0.8056				0.7462	18.37
330	0.9907	0.9490	0.8910	0.8292	0.7551			0.6918	23.50
340	0.9917	0.9540	0.9035	0.8492	0.7878	0.7147		0.6255	29.73
350	0.9926	0.9588	0.9137	0.8659	0.8127	0.7542	0.6843		
Z_{sat}	0.9719	0.9044	0.8397	0.7827	0.7289	0.6759	0.6220		
T_{sat}	236.1	275.5	297.6	312.1	323.5	332.6	340.4		

Converted and interpolated from "Thermodynamic properties of KLEA 66," British units, © ICI Chemicals and Polymers, 1993 (20 pp.). Reproduced by permission. KLEA 66 is R32/125/134a (23/25/52 wt %).

TABLE 2-175   Compressibility Factors for Krypton*

Temp., K	\multicolumn{12}{c}{Pressure, bar}											
	1	5	10	20	40	60	80	100	200	300	400	500
150	0.9837	0.9155	0.0310	0.0618	0.1227	0.1829	0.2423	0.3012	0.5875	0.8636	1.1315	1.3932
200	0.9933	0.9648	0.9278	0.8459	0.6039	0.1870	0.2393	0.2903	0.5313	0.7568	0.9730	1.1820
250	0.9966	0.9841	0.9635	0.9265	0.8468	0.7605	0.6680	0.5810	0.5785	0.7461	0.9197	1.0891
300	0.9982	0.9899	0.9800	0.9595	0.9197	0.8807	0.8437	0.8097	0.7337	0.7954	0.9302	1.0627
350	0.9989	0.9949	0.9897	0.9793	0.9522	0.9415	0.9250	0.9110	0.8774	0.8992	0.9799	1.0664
400	0.9993	0.9967	0.9933	0.9867	0.9746	0.9635	0.9539	0.9459	0.9323	0.9570	1.0150	1.0910
450	0.9998	0.9985	0.9969	0.9939	0.9886	0.9838	0.9800	0.9774	0.9663	1.0011	1.0543	1.1142
500	0.9998	0.9992	0.9984	0.9970	0.9942	0.9921	0.9910	0.9906	1.0019	1.0311	1.0732	1.1258
600	1.0000	1.0003	1.0005	1.0012	1.0025	1.0043	1.0064	1.0091	1.0301	1.0618	1.1000	1.1431
800	1.0002	1.0010	1.0020	1.0041	1.0079	1.0122	1.0170	1.0214	1.0475	1.0779	1.1112	1.1147
1000	1.0002	1.0013	1.0023	1.0045	1.0091	1.0135	1.0184	1.0230	1.0486	1.0767	1.1063	1.1369

*Calculated from PVT values tabulated in Rabinovich (ed.), *Thermophysical Properties of Neon, Argon, Krypton and Xenon*, Standards Press, Moscow, 1976. This book was published in English translation by Hemisphere, New York, 1988 (604 pp.).

TABLE 2-176   Compressibility Factors for Methane (R50)*

Temp., K	Pressure, bar											
	1	5	10	20	40	60	80	100	200	300	400	500
150	0.9854	0.9225	0.8275	0.0714	0.1411	0.2093	0.2763	0.3423	0.6599	0.9623	1.2537	1.5363
200	0.9936	0.9676	0.9339	0.8599	0.6784	0.3559	0.3172	0.3618	0.6141	0.8568	1.0887	1.3122
250	0.9965	0.9838	0.9680	0.9352	0.8682	0.8020	0.7386	0.6854	0.6699	0.8554	1.0359	1.2155
300	0.9983	0.9915	0.9830	0.9667	0.9343	0.9047	0.8783	0.8556	0.8580	0.9154	1.0432	1.1829
350	0.9991	0.9954	0.9911	0.9825	0.9662	0.9520	0.9401	0.9306	0.9527	0.9800	1.0723	1.1804
400	0.9995	0.9977	0.9953	0.9912	0.9835	0.9772	0.9726	0.9696	0.9779	1.0245	1.0986	1.1859
450	0.9997	0.9989	0.9979	0.9963	0.9935	0.9917	0.9911	0.9916	1.0098	1.0528	1.1152	1.1899
500	0.9999	0.9997	0.9995	0.9995	0.9996	1.0005	1.0022	1.0048	1.0285	1.0699	1.1248	1.1899
600	1.0000	1.0009	1.0020	1.0039	1.0081	1.0125	1.0171	1.0217	1.0540	1.0969	1.1470	1.2019
800	1.0003	1.0017	1.0034	1.0068	1.0130	1.0197	1.0263	1.0330	1.0678	1.1068	1.1496	1.1951
1000	1.0004	1.0014	1.0035	1.0071	1.0141	1.0207	1.0274	1.0342	1.0678	1.1033	1.1400	1.1790

*Calculated from PVT values tabulated in Goodwin, NBS Tech. Note 653, 1974, for temperatures up to 500 K, and from PVT values tabulated in Zhuravlev. *Thermophysical Properties of Gaseous and Liquid Methane,* Standartov, Moscow, 1969, and NBS-NSF transl. TT 70-50097, 1970.

TABLE 2-177   Compressibility Factors for Methanol

Temp., K	Pressure, bar												
	0.1	0.5	1.0133	10	20	50	100	150	200	250	300	400	500
200	0.0002	0.0011	0.0022	0.0219	0.0438	0.1091	0.2174	0.3250	0.4319	0.5381	0.6437	0.8531	1.6030
250	0.0002	0.0009	0.0019	0.0185	0.0370	0.0923	0.1837	0.2743	0.3643	0.4535	0.5422	0.7176	0.8909
300	0.9792	0.0008	0.0017	0.0164	0.0327	0.0813	0.1617	0.2413	0.3201	0.3981	0.4755	0.6284	0.7791
350	0.9844	0.9713	0.9551	0.0150	0.0298	0.0742	0.1473	0.2193	0.2904	0.3606	0.4301	0.5671	0.7016
400	0.9872	0.9795	0.9722	0.0142	0.0283	0.0702	0.1386	0.2056	0.2714	0.3362	0.4000	0.5253	0.6478
450	0.9890	0.9835	0.9792	0.9145	0.7989	0.0701	0.1366	0.2007	0.2629	0.3238	0.3834	0.4997	0.6128
500	0.9903	0.9859	0.9828	0.9525	0.9081	0.6799	0.1505	0.2110	0.2699	0.3271	0.3829	0.4912	0.5959
600	0.9922	0.9889	0.9867	0.9756	0.9643	0.9042	0.7629	0.6275	0.5255	0.4921	0.5010	0.5606	0.6358
700	0.9934	0.9907	0.9889	0.9816	0.9778	0.9541	0.8932	0.8392	0.8027	0.7797	0.7675	0.7713	0.7993
800	0.9964	0.9920	0.9904	0.9838	0.9818	0.9711	0.9411	0.9156	0.9025	0.8994	0.9026	0.9205	0.9485

Goodwin, R.D., *J. Phys. Chem. Ref. Data,* **16** (4), 799, 1987.

TABLE 2-178   Compressibility Factors for Neon*

Temp., K	Pressure, bar											
	1	5	10	20	40	60	80	100	200	300	400	500
50	0.9913	0.9472	0.9083	0.8013	0.3810	0.4398	0.4984	0.5850	0.9864	1.3659	1.7289	2.0794
100	0.9993	0.9970	0.9949	0.9913	0.9854	0.9245	0.9864	0.9930	1.0795	1.2197	1.3796	1.5473
150	1.0002	1.0017	1.0036	1.0078	1.0162	1.0262	1.0375	1.0497	1.1235	1.2131	1.3113	1.4150
200	1.0003	1.0023	1.0049	1.0100	1.0204	1.0318	1.0427	1.0551	1.1191	1.1909	1.2655	1.3422
250	1.0001	1.0022	1.0045	1.0097	1.0198	1.0295	1.0403	1.0502	1.1057	1.1633	1.2223	1.2822
300	1.0000	1.0020	1.0041	1.0091	1.0181	1.0277	1.0369	1.0469	1.0961	1.1476	1.1997	1.2520
400	1.0000	1.0017	1.0036	1.0074	1.0151	1.0216	1.0301	1.0376	1.0771	1.1172	1.1575	1.1981
500	1.0000	1.0014	1.0029	1.0058	1.0124	1.0188	1.0252	1.0316	1.0641	1.0963	1.1291	1.1621
600	1.0000	1.0012	1.0024	1.0049	1.0107	1.0160	1.0214	1.0267	1.0542	1.0814	1.1091	1.1369
800	1.0000	1.0009	1.0018	1.0043	1.0081	1.0123	1.0163	1.0206	1.0413	1.0622	1.0829	1.1039
1000	1.0000	1.0007	1.0014	1.0034	1.0068	1.0098	1.0132	1.0165	1.0330	1.0500	1.0670	1.0836

*Calculated from PVT values tabulated in Rabinovich (ed.), *Thermophysical Properties of Neon, Argon, Krypton and Xenon,* Standards Press, Moscow, 1976. This book was published in English translation by Hemisphere, New York, 1988 (604 pp.).

TABLE 2-179 Compressibility Factors for Nitrogen*

Temp., K	Pressure, bar											
	1	5	10	20	40	60	80	100	200	300	400	500
70	0.0057	0.0287	0.0573	0.1143	0.2277	0.3400	0.4516	0.5623	1.1044	1.6308	Solid	Solid
80	0.9593	0.0264	0.0528	0.1053	0.2093	0.3122	0.4140	0.5148	1.0061	1.4797	1.9396	2.3879
90	0.9722	0.0251	0.0500	0.0996	0.1975	0.2938	0.3888	0.4826	0.9362	1.3700	1.7890	2.1962
100	0.9798	0.8910	0.0487	0.0966	0.1905	0.2823	0.3720	0.4605	0.8840	1.2852	1.6707	2.0441
120	0.9883	0.9397	0.8732	0.7059	0.1975	0.2822	0.3641	0.4438	0.8188	1.1684	1.5015	1.8223
140	0.9927	0.9635	0.9253	0.8433	0.6376	0.4251	0.4278	0.4799	0.7942	1.0996	1.3920	1.6726
160	0.9952	0.9766	0.9529	0.9042	0.8031	0.7017	0.6304	0.6134	0.8107	1.0708	1.3275	1.5762
180	0.9967	0.9846	0.9690	0.9381	0.8782	0.8125	0.7784	0.7530	0.8550	1.0669	1.2893	1.5105
200	0.9978	0.9897	0.9791	0.9592	0.9212	0.8882	0.8621	0.8455	0.9067	1.0760	1.2683	1.4631
250	0.9992	0.9960	0.9924	0.9857	0.9741	0.9655	0.9604	0.9589	1.0048	1.1143	1.2501	1.3962
300	0.9998	0.9990	0.9983	0.9971	0.9964	0.9973	1.0000	1.0052	1.0559	1.1422	1.2480	1.3629
350	1.0001	1.0007	1.0011	1.0029	1.0069	1.0125	1.0189	1.0271	1.0810	1.1560	1.2445	1.3405
400	1.0002	1.0011	1.0024	1.0057	1.0125	1.0199	1.0283	1.0377	1.0926	1.1609	1.2382	1.3216
450	1.0003	1.0018	1.0033	1.0073	1.0153	1.0238	1.0332	1.0430	1.0973	1.1606	1.2303	1.3043
500	1.0004	1.0020	1.0040	1.0081	1.0167	1.0257	1.0350	1.0451	1.0984	1.1575	1.2213	1.2881
600	1.0004	1.0021	1.0040	1.0084	1.0173	1.0263	1.0355	1.0450	1.0951	1.1540	1.2028	1.2657
800	1.0004	1.0017	1.0036	1.0074	1.0157	1.0237	1.0320	1.0402	1.0832	1.1264	1.1701	1.2140
1000	1.0003	1.0015	1.0034	1.0067	1.0136	1.0205	1.0275	1.0347	1.0714	1.1078	1.1449	1.1814

*Computed from pressure-volume-temperature tables in the Vasserman monographs referenced under Table 2-165.

TABLE 2-180 Compressibility Factors for Oxygen*

Temp., K	Pressure, bar											
	1	5	10	20	40	60	80	100	200	300	400	500
75	0.0043	0.0213	0.0425	0.0849	0.1693	0.2533	0.3368	0.4200	0.8301	1.2322	1.6278	2.0175
80	0.0041	0.0203	0.0405	0.0811	0.1616	0.2418	0.3214	0.4007	0.7912	1.1738	1.5495	1.9196
90	0.0038	0.0188	0.0375	0.0750	0.1494	0.2233	0.2966	0.3696	0.7281	1.0780	1.4211	1.7580
100	0.9757	0.0177	0.0354	0.0705	0.1404	0.2096	0.2783	0.3464	0.6798	1.0040	1.3206	1.6309
120	0.9855	0.9246	0.8367	0.0660	0.1302	0.1935	0.2558	0.3173	0.6148	0.8999	1.1762	1.4456
140	0.9911	0.9535	0.9034	0.7852	0.1334	0.1940	0.2527	0.3099	0.5815	0.8374	1.0832	1.3214
160	0.9939	0.9697	0.9379	0.8689	0.6991	0.3725	0.2969	0.3378	0.5766	0.8058	1.0249	1.2364
180	0.9960	0.9793	0.9579	0.9134	0.8167	0.7696	0.5954	0.5106	0.6043	0.8025	0.9990	1.1888
200	0.9970	0.9853	0.9705	0.9399	0.8768	0.8140	0.7534	0.6997	0.6720	0.8204	0.9907	1.1623
250	0.9987	0.9938	0.9871	0.9736	0.9477	0.9237	0.9030	0.8858	0.8563	0.9172	1.0222	1.1431
300	0.9994	0.9968	0.9941	0.9884	0.9771	0.9676	0.9597	0.9542	0.9560	0.9972	1.0689	1.1572
350	0.9998	0.9990	0.9979	0.9961	0.9919	0.9890	0.9870	0.9870	1.0049	1.0451	1.1023	1.1722
400	1.0000	1.0000	1.0000	1.0000	1.0003	1.0011	1.0022	1.0045	1.0305	1.0718	1.1227	1.1816
450	1.0002	1.0007	1.0015	1.0024	1.0048	1.0074	1.0106	1.0152	1.0445	1.0859	1.1334	1.1859
500	1.0002	1.0011	1.0022	1.0038	1.0075	1.0115	1.0161	1.0207	1.0523	1.0927	1.1380	1.1866
600	1.0003	1.0014	1.0024	1.0052	1.0102	1.0153	1.0207	1.0266	1.0582	1.0961	1.1374	1.1803
800	1.0003	1.0014	1.0026	1.0055	1.0109	1.0164	1.0219	1.0271	1.0565	1.0888	1.1231	1.1582
1000	1.0003	1.0013	1.0026	1.0053	1.0101	1.0149	1.0198	1.0253	1.0507	1.0783	1.1072	1.1369

*Calculated from pressure-volume-temperature tables in the Vasserman monographs listed under Table 2-165.

TABLE 2-181 Compressibility Factors for Refrigerant 32*

Temp., K	Pressure, bar									Z_{sat}	P_{sat}
	1	5	10	15	20	25	30	40	50		
230	0.9656									0.9453	1.54
240	0.9711									0.9278	2.40
250	0.9755									0.9062	3.60
260	0.9791	0.8865								0.8811	5.22
270	0.9819	0.9036								0.8522	7.34
280	0.9844	0.9180	0.8210							0.8194	10.07
290	0.9864	0.9285	0.8476							0.7822	13.51
300	0.9880	0.9376	0.8686	0.7899						0.7401	17.76
310	0.9894	0.9453	0.8358	0.8197	0.7439					0.6922	22.95
320	0.9904	0.9518	0.8998	0.8436	0.7812	0.7089				0.6370	29.21
330	0.9914	0.9573	0.9118	0.8628	0.8102	0.7518	0.6851			0.5719	36.72
340	0.9923	0.9619	0.9203	0.8790	0.8338	0.7846	0.7316	0.6021		0.4905	45.66
350	0.9932	0.9655	0.9296	0.8932	0.8534	0.8115	0.7671	0.6675	0.5312	0.3702	56.35
Z_{sat}	0.9595	0.8843	0.8202	0.7670	0.7191	0.6722	0.6303	0.5427	0.4467		
T_{sat}	221.2	258.8	279.8	293.8	304.6	313.5	321.1	333.9	344.3		

*Converted and interpolated from British units shown in *Thermodynamic properties of KLEA 32*, ICI Chemicals and Polymers, 1993. Reproduced by permission.

TABLE 2-182   Compressibility Factors for Refrigerant 123

Temp., °C	1	2.5	5	7.5	10	12.5	15	17.5	20	22.5	25	Z_{sat}	Psat
40	0.9639											0.9427	1.54
50	0.9682											0.9294	2.13
60	0.9717	0.9248										0.9134	2.96
70	0.9745	0.9327										0.8950	3.78
80	0.9766	0.9401										0.8727	4.90
100	0.9804	0.9501	0.9197	0.8355								0.8262	7.87
120	0.9839	0.9591	0.9146	0.8667	0.8140							0.7640	12.01
140	0.9861	0.9650	0.9282	0.8915	0.8503	0.8023	0.7479	0.6916				0.6890	17.59
160	0.9886	0.9714	0.9406	0.9077	0.8747	0.8398	0.8026	0.7600	0.7134	0.6553		0.5820	24.92
180	0.9908	0.9762	0.9518	0.9254	0.8970	0.8709	0.8402	0.8072	0.7712	0.7346	0.6841	0.3926	34.54
200	0.9924	0.9806	0.9602	0.9388	0.9174	0.8931	0.8688	0.8422	0.8163	0.7882	0.7539	—	—
225	0.9938	0.9846	0.9692	0.9526	0.9378	0.9170	0.8972	0.8800	0.8566	0.8401	0.8157	—	—
250	0.9954	0.9885	0.9	0.9651	0.9528	0.9382	0.9229	0.9101	0.8930	0.8771	0.8581	—	—
Z_{sat}	0.9575	0.9210	0.9730	0.8292	0.7947	0.7654	0.7229	0.7110	0.6564	0.6206	0.5821	—	—
T	27.5	55.4	80.8	97.9	111.1	122.0	131.4	139.7	147.2	154.0	160.2	—	—

Dashes indicate inaccessible states; blanks indicate no available data.

TABLE 2-183   Compressibility Factors for Refrigerant 124

Temp., °C	1	2.5	5	7.5	10	12.5	15	17.5	20	22.5	25	Z_{sat}	Psat
−20												0.9562	0.72
−10												0.9431	1.10
0	0.9573											0.9284	1.63
10	0.9641											0.9243	2.34
20	0.9693											0.8920	3.27
30	0.9736	0.9313										0.8828	4.45
40	0.9675	0.9396	0.8728									0.8427	5.93
50	0.9798	0.9473	0.8889	0.8229								0.8151	7.75
60	0.9820	0.9534	0.9017	0.8462								0.7803	9.96
80	0.9854	0.9633	0.9226	0.8820	0.8366		0.7251					0.7024	15.74
100	0.9880	0.9700	0.9370	0.9040	0.8710	0.8314	0.7918	0.7463	0.6950	0.6380		0.5955	23.75
120	0.9899	0.9749	0.9478	0.9206	0.8935	0.8634	0.8329	0.8022	0.7682	0.7285	0.6878	0.3912	34.70
140	0.9917	0.9794	0.9575	0.9357	0.9138	0.8884	0.8641	0.8391	0.8105	0.7896	0.7647		
160		0.9825	0.9645	0.9464	0.9247	0.9061	0.8868	0.8644	0.8489	0.8285	0.7868		
180			0.9690	0.9536	0.9382	0.9213	0.9056	0.8857	0.8634	0.8574	0.8379		
200				0.9601	0.9471	0.9338	0.9211	0.9042	0.8951	0.8783	0.8647		
225					0.9589	0.9488	0.9391	0.9223	0.9160	0.9040	0.8947		
250					0.9650	0.9573	0.9443	0.9412	0.9333	0.9252	0.9174		
Z_{sat}	0.9468	0.9071	0.8605	0.8185	0.7830	0.7488	0.7157	0.6825	0.6484	0.6279	0.5788		
T	−12.4	11.9	34.0	48.7	60.2	69.6	77.8	85.0	91.4	96.3	102.6		

Dashes indicate inaccessible states; blanks indicate no available data.

TABLE 2-184   Compressibility Factors for Refrigerant 134a

Temp., °C	1	5	10	15	20	25	30	40	50	Zsat	Psat
−10	0.9622									0.9316	2.005
0	0.9710									0.9119	2.926
10	0.9752									0.8888	4.144
20	0.9778	0.8819								0.8621	5.716
30	0.9817	0.8973								0.8314	7.701
40	0.9839	0.9098	0.8005							0.7963	10.17
50	0.9857	0.9206	0.8280							0.7560	13.18
60	0.9872	0.9296	0.8449	0.7361						0.7098	16.82
70	0.9886	0.9376	0.8678	0.7917	0.6853					0.6562	21.17
80	0.9897	0.9442	0.8828	0.8137	0.7327	0.6290				0.5911	26.38
90	0.9908	0.9495	0.8954	0.8390	0.7682	0.6860	0.5832			0.5054	32.45
100	0.9916	0.9543	0.9062	0.8555	0.7965	0.7335	0.6557			0.3462	39.72
110	0.9920	0.9592	0.9151	0.8630	0.8144	0.7630	0.7046	0.5732	0.4530	—	—
120	0.9924	0.9638	0.9235	0.8802	0.8386	0.7915	0.7418	0.6240	0.4885	—	—
130	0.9927	0.9673	0.9308	0.8949	0.8553	0.8165	0.7716	0.677	0.5645	—	—
140	0.9929	0.9691	0.9370	0.9040	0.8694	0.8350	0.7964	0.7160	0.6303	—	—
150	0.9931	0.9727	0.9428	0.8877	0.8817	0.8495	0.8173	0.7480	0.6783	—	—
satn.	0.9567	0.8741	0.7989	0.7017	0.6704	0.6094	0.5415	0.4442	—	—	—
sat. T	−26.37	15.74	39.39	55.23	67.49	77.57	86.20	100.35	—	—	—

Dashes indicate inaccessible states; blanks indicate no available data.

TABLE 2-185 Compressibility Factors for Water Substance (fps units)*

Pressure, lb/in² abs.	Temp., °F																		
	400	600	800	1000	1200	1400	1600	1800	2000	2200	2400	2600	2800	3000	3200	3400	3600	3800	4000
10	0.9965	0.9959	0.9992	0.9995	0.9999	0.9999	0.9999	1.0000	1.0000	1.0000	1.0001	1.0006	1.0012	1.0024	1.0053	1.0084	1.0145	1.0211	1.0332
15	0.9943	0.9972	0.9986	0.9993	0.9997	0.9998	0.9999	0.9999	1.0000	1.0000	1.0001	1.0004	1.0012	1.0022	1.0042	1.0072	1.0124	1.0188	1.0295
20	0.9930	0.9970	0.9981	0.9991	0.9995	0.9996	0.9998	0.9999	1.0000	1.0000	1.0001	1.0003	1.0011	1.0020	1.0036	1.0065	1.0112	1.0173	1.0269
40	0.9861	0.9940	0.9967	0.9981	0.9990	0.9994	0.9996	0.9998	0.9999	0.9999	1.0001	1.0003	1.0010	1.0018	1.0028	1.0054	1.0090	1.0139	1.0214
60	0.9788	0.9910	0.9951	0.9973	0.9984	0.9991	0.9994	0.9997	0.9999	0.9999	1.0001	1.0003	1.0009	1.0018	1.0024	1.0048	1.0080	1.0120	1.0186
80	0.9714	0.9878	0.9935	0.9963	0.9979	0.9987	0.9992	0.9996	0.9998	0.9999	1.0001	1.0003	1.0008	1.0016	1.0023	1.0044	1.0073	1.0108	1.0170
100	0.9469	0.9848	0.9919	0.9954	0.9974	0.9985	0.9990	0.9995	0.9998	0.9999	1.0001	1.0004	1.0007	1.0015	1.0022	1.0042	1.0067	1.0099	1.0157
150	0.9435	0.9770	0.9879	0.9931	0.9960	9.9976	0.9985	0.9993	0.9997	0.9998	1.0001	1.0004	1.0006	1.0014	1.0021	1.0039	1.0059	1.0087	1.0137
200	0.9216	0.9690	0.9839	0.9908	0.9947	0.9968	0.9980	0.9991	0.9996	0.9998	1.0001	1.0005	1.0007	1.0015	1.0021	1.0037	1.0055	1.0080	1.0126
400		0.9356	0.9675	0.9817	0.9893	0.9935	0.9960	0.9982	0.9992	0.9998	1.0002	1.0007	1.0011	1.0017	1.0023	1.0033	1.0049	1.0070	1.0105
600		0.8989	0.9509	0.9725	0.9839	0.9904	0.9942	0.9973	0.9988	0.9997	1.0002	1.0008	1.0014	1.0019	1.0026	1.0034	1.0048	1.0066	1.0097
800		0.8586	0.9336	0.9633	0.9790	0.9872	0.9925	0.9964	0.9985	0.9996	1.0003	1.0010	1.0016	1.0022	1.0029	1.0036	1.0049	1.0065	1.0094
1,000		0.8138	0.9162	0.9540	0.9733	0.9841	0.9905	0.9955	0.9981	0.9994	1.0004	1.0012	1.0019	1.0025	1.0032	1.0039	1.0052	1.0066	1.0092
1,500		0.6702	0.8695	0.9305	0.9600	0.9764	0.9859	0.9932	0.9971	0.9992	1.0007	1.0017	1.0026	1.0033	1.0040	1.0048	1.0059	1.0072	1.0096
2,000			0.8188	0.9067	0.9468	0.9687	0.9813	0.9900	0.9958	0.9990	1.0010	1.0023	1.0034	1.0042	1.0049	1.0058	1.0068	1.0082	1.0104
4,000			0.5608	0.8060	0.8942	0.9392	0.9647	0.9836	0.9930	0.9989	1.0024	1.0050	1.0069	1.0082	1.0093	1.0106	1.0118	1.0132	1.0149
6,000				0.7042	0.8442	0.9121	0.9497	0.9771	0.9907	0.9991	1.0048	1.0081	1.0110	1.0128	1.0139	1.0152	1.0165	1.0179	1.0195
8,000				0.6185	0.8003	0.8883	0.9371	0.9714	0.9895	1.0004	1.0075	1.0118	1.0152	1.0172	1.0188	1.0204	1.0216	1.0229	1.0242
10,000				0.5699	0.7657	0.8693	0.9274	0.9668	0.9890	1.0025	1.0105	1.0158	1.0196	1.0220	1.0240	1.0258	1.0271	1.0284	1.0298

*Calculated by P. E. Liley from various steam tables for the lower temperatures and from Paper B-11 by P. H. Kesselman and Yu. I. Blank, 7th. Int. Conf. Properties of Steam, Tokyo, 1968, for the higher temperatures.

TABLE 2-186 Compressibility Factors of Water Substance (SI units)*

Temperature, K	Pressure, bar																				
	1	5	10	15	20	25	30	40	50	60	80	100	150	200	250	300	400	500	600	800	1000
400	0.990	0.003	0.006	0.009	0.012	0.014	0.017	0.023	0.029	0.035	0.046	0.058	0.086	0.114	0.143	0.171	0.227	0.282	0.336	0.445	0.552
450	0.993	0.003	0.006	0.009	0.012	0.014	0.016	0.022	0.027	0.033	0.043	0.054	0.080	0.107	0.134	0.159	0.206	0.255	0.304	0.402	0.498
500	0.996	0.980	0.958	0.930	0.901	0.878	0.016	0.021	0.026	0.031	0.042	0.052	0.077	0.102	0.127	0.152	0.201	0.249	0.297	0.390	0.482
550	0.997	0.985	0.969	0.956	0.939	0.922	0.904	0.865	0.822	0.773	0.042	0.052	0.077	0.102	0.126	0.150	0.181	0.198	0.289	0.378	0.464
600	0.998	0.990	0.979	0.970	0.961	0.948	0.935	0.910	0.885	0.858	0.798	0.726	0.082	0.107	0.131	0.155	0.201	0.246	0.290	0.375	0.457
650	0.999	0.992	0.984	0.977	0.968	0.959	0.958	0.937	0.919	0.902	0.864	0.824	0.702	0.514	0.177	0.183	0.221	0.260	0.303	0.383	0.460
700	1.000	0.994	0.988	0.984	0.976	0.967	0.966	0.952	0.941	0.929	0.900	0.876	0.800	0.716	0.618	0.503	0.326	0.316	0.340	0.406	0.476
750	1.000	0.996	0.991	0.988	0.981	0.975	0.971	0.961	0.955	0.945	0.927	0.907	0.856	0.801	0.743	0.682	0.557	0.465	0.435	0.456	0.509
800	1.000	0.997	0.993	0.991	0.985	0.982	0.976	0.970	0.966	0.957	0.945	0.929	0.892	0.853	0.813	0.773	0.693	0.620	0.568	0.538	0.561
850	1.000	0.997	0.995	0.992	0.989	0.984	0.981	0.977	0.973	0.967	0.957	0.946	0.917	0.889	0.860	0.831	0.775	0.715	0.679	0.631	0.629
900	1.000	0.998	0.997	0.993	0.992	0.989	0.986	0.982	0.979	0.974	0.965	0.958	0.936	0.915	0.893	0.872	0.830	0.792	0.760	0.714	0.700
950	1.000	0.998	0.997	0.994	0.994	0.993	0.991	0.985	0.983	0.980	0.973	0.967	0.950	0.933	0.916	0.901	0.867	0.839	0.816	0.780	0.761
1000	1.000	0.999	0.998	0.995	0.995	0.994	0.993	0.990	0.987	0.985	0.978	0.973	0.960	0.948	0.935	0.923	0.900	0.878	0.859	0.831	0.816
1200	1.000	1.000	0.999	0.998	0.998	0.997	0.997	0.995	0.994	0.994	0.992	0.990	0.986	0.982	0.975	0.968	0.961	0.957	0.949	0.942	0.937
1400	1.000	1.000	1.000	1.000	1.000	1.000	1.000	0.999	0.998	0.998	0.998	0.997	0.996	0.995	0.995	0.994	0.993	0.992	0.994	0.996	0.998
1600	1.000	1.000	1.000	1.000	1.000	1.000	1.000	1.000	1.000	1.000	1.000	1.000	1.001	1.002	1.002	1.004	1.006	1.009	1.012	1.015	1.020
1800	1.001	1.001	1.001	1.000	1.000	1.000	1.000	1.000	1.000	1.001	1.002	1.003	1.003	1.004	1.005	1.008	1.011	1.014	1.017	1.021	1.031
2000	1.003	1.002	1.002	1.002	1.002	1.002	1.002	1.002	1.002	1.003	1.003	1.004	1.004	1.006	1.008	1.011	1.014	1.018	1.021	1.032	1.043

*Calculated by P. E. Liley from various steam tables for the lower temperatures and from Pap. B-11 by P. H. Kesselman and Yu. I. Blank, 7th Internal Conference on the Properties of Steam, Tokyo, 1968, for the higher temperatures.

TABLE 2-187 Compressibility Factors for Xenon*

Temperature, K	Pressure, bar											
	1	5	10	20	40	60	80	100	200	300	400	500
200	0.9831	0.9088	0.0293	0.0584	0.1162	0.1733	0.2300	0.2861	0.5601	0.8253	1.0833	1.3356
250	0.9911	0.9545	0.9052	0.7887	0.1114	0.1642	0.2158	0.2663	0.5074	0.7355	0.9546	1.1670
300	0.9949	0.9736	0.9465	0.8885	0.7517	0.5492	0.2794	0.3016	0.5021	0.6997	0.8886	1.0707
350	0.9967	0.9834	0.9669	0.9322	0.8473	0.7840	0.7039	0.6249	0.5645	0.7124	0.8706	1.0269
400	0.9977	0.9892	0.9183	0.9562	0.9128	0.8696	0.8278	0.7888	0.6916	0.7642	0.8850	1.0148
450	0.9989	0.9928	0.9856	0.9714	0.9429	0.9163	0.8911	0.8679	0.7335	0.8331	0.9187	1.0224
500	0.9982	0.9951	0.9902	0.9810	0.9623	0.9452	0.9293	0.9156	0.8774	0.8953	0.9572	1.0412
600	0.9996	0.9979	0.9957	0.9917	0.9841	0.9772	0.9715	0.9667	0.9596	0.9791	1.0211	1.0799
800	1.0000	0.9998	1.0002	1.0004	1.0012	1.0020	1.0034	1.0054	1.0213	1.0476	1.0818	1.1222
1000	1.0000	1.0004	1.0015	1.0031	1.0144	1.0101	1.0133	1.0172	1.0394	1.0669	1.0979	1.1331

*Calculated from PVT values tabulated in Rabinovich (ed.), *Thermophysical Properties of Neon, Argon, Krypton and Xenon*, Standards Press, Moscow, 1976. This book was published in English translation by Hemisphere, New York, 1988 (604 pp.).

TABLE 2-188 Compressibilities of Liquids*

At the constant temperature T, the compressibility $\beta = (1/\overline{V}_0)(dV/dP)$. In general as P increases, β decreases rapidly at first and then slowly; the change of β with T is large at low pressures but very small at pressures above 1000 to 2000 megabars. 1 megabar = 0.987 atm. = 10^6 dynes/cm² based upon the older usage, 1 bar = 1 dyne/cm². The use of the bar as a pressure unit is not encouraged.

Substance	Temp., °C	Pressure, megabars	Compressibility per megabar $\beta \times 10^6$	Substance	Temp., °C	Pressure, megabars	Compressibility per megabar $\beta \times 10^6$	Substance	Temp., °C	Pressure, megabars	Compressibility per megabar $\beta \times 10^6$
Acetone	14	23	111	Ethyl acetate	20	400	75	Methyl alcohol	15	23	103
Acetone	20	500	61	alcohol	14	23	100	alcohol	20	200	95
Acetone	20	1,000	52	alcohol	20	500	63	alcohol	20	400	80
Acetone	40	12,000	9	alcohol	20	1,000	54	alcohol	20	500	65
Amyl alcohol	14	23	88	alcohol	20	12,000	8	alcohol	20	1,000	54
alcohol, iso.	20	200	84	bromide	20	200	100	alcohol	20	12,000	8
alcohol, iso.	20	400	70	bromide	20	400	82	Nitric acid	0	17	32
alcohol, n	20	500	61	bromide	20	500	70	Oils:			
alcohol, n	20	1,000	45	bromide	20	1,000	54	Almond	15	5	53
alcohol, n	20	12,000	8	bromide	20	12,000	8	Castor	15	5	46
alcohol, n	40	12,000	8	chloride	15	23	151	Linseed	15	5	51
Benzene	17	5	89	chloride	20	500	102	Olive	15	5	55
Benzene	20	200	77	chloride	20	1,000	66	Rapeseed	20		59
Benzene	20	400	67	chloride	20	12,000	8	Phosphorus trichloride	10	250	71
Bromine	20	200	56	ether	25	23	188	trichloride	20	500	63
Bromine	20	400	51	ether	20	500	84	trichloride	20	1,000	47
Butyl alcohol, iso	18	8	97	ether	20	1,000	61	trichloride	20	12,000	8
alcohol, iso	20	200	81	ether	20	12,000	10	Propyl alcohol (n)	20	200	77
alcohol, iso	20	400	64	iodide	20	200	81	alcohol (n)	20	400	67
alcohol, iso	20	500	56	iodide	20	400	69	alcohol (n?)	20	500	65
alcohol, iso	20	1,000	46	iodide	20	500	64	alcohol (n?)	20	1,000	47
alcohol, iso	20	12,000	8	iodide	20	1,000	50	alcohol (n?)	20	12,000	7
Carbon bisulfide	16	21	86	iodide	20	12,000	8	Toluene	20	200	74
bisulfide	20	500	57	Gallium	30	300	3.97	Toluene	20	400	64
bisulfide	20	1,000	48	Glycerol	15	5	22	Turpentine	20		74
bisulfide	20	12,000	6	Hexane	20	200	117	Water	20	13	49
tetrachloride	20	200	86	Hexane	20	400	91	Water	20	200	43
tetrachloride	20	400	73	Kerosene	20	500	55	Water	20	400	41
Chloroform	20	200	83	Kerosene	20	1,000	45	Water	20	500	39
Chloroform	20	400	70	Kerosene	20	12,000	8	Water	40	500	38
Dichloroethylsulfide	32	1,000	54	Mercury	20	300	3.95	Water	40	1,000	33
Dichloroethylsulfide	32	2,000	54	Mercury	22	500	3.97	Water	40	12,000	9
Ethyl acetate	13	23	103	Mercury	22	1,000	3.91	Xylene, meta	20	200	69
acetate	20	200	90	Mercury	22	12,000	2.37	meta	20	400	60

* *Smithsonian Tables*, Table 106.

Scott (*Cryogenic Engineering*, Van Nostrand, Princeton, NJ, 1959) gives data for liquid nitrogen (p. 283), oxygen (p. 276), and hydrogen (p. 303). For a convenient index to the high-pressure work of Bridgman, see *American Institute of Physics Handbook*, p. 2-163, McGraw-Hill, New York, 1957.

TABLE 2-189   Compressibilities of Solids

Many data on the compressibility of solids obtained prior to 1926 are contained in Gruneisen, *Handbuch der Physik*, vol. 10, Springer, Berlin, 1926, pp. 1–52; also available as translation, NASA RE 2-18-59W, 1959. See also Tables 271, 273, 276, 278, and other material in *Smithsonian Physical Tables*, 9th ed., 1954. For a review of high-pressure work to 1946, see Bridgman, *Rev. Mod. Phys.*, **18**, 1 (1946).

LATENT HEATS

UNITS CONVERSIONS

For this subsection, the following units conversions are applicable:

$$°F = \tfrac{9}{5}°C + 32.$$

To convert calories per gram-mole to British thermal units per pound-mole, multiply by 1.799; to convert calories per gram to British thermal units per pound, multiply by 1.799.

To convert millimeters of mercury to pounds-force per square inch, multiply by 1.934×10^{-2}.

TABLE 2-190 Heats of Fusion and Vaporization of the Elements and Inorganic Compounds*

Unless stated otherwise, the values have been taken from the compilations by K. K. Kelley on "Heats of Fusion of Inorganic Compounds," U.S. Bur. Mines Bull. 393 (1936), and "The Free Energies of Vaporization and Vapor Pressures of Inorganic Substances," U.S. Bur. Mines Bull. 383 (1935).

Substance	mp, °C	Heat of fusion,[a,b] cal/mole	bp at 1 atm, °C	Heat of vaporization,[a,k] cal/mole
Aluminum				
Al	660.0	2,550	2057	61,020
Al_2Br_6	97.5	5,420	256.4	10,920
Al_2Cl_6	192.5	16,960	180.2[c]	26,750[c]
$AlF_3 \cdot 3NaF$	1000	16,380		
Al_2I_6	191.0	7,960	385.5	15,360
Al_2O_3	2045	(26,000)	3000	
Antimony				
Sb	630.5	4,770	1440	46,670
$SbBr_3$	97	3,510		
$SbCl_3$	73.4	3,030	219	10,360
$SbCl_5$	4	2,400	172[d]	11,570
Sb_4O_6	655	(27,000)	1425	17,820
Sb_4S_6	546	11,200		
Argon				
A	−189.3	290	−185.8	1,590
Arsenic				
As	814	(6,620)	610[c]	31,000[c]
$AsBr_3$	31	2,810		
$AsCl_3$	−16	2,420	122	7,570
AsF_5	−80.7	2,800	−52.8	4,980
As_4O_6	313	8,000	457.2	14,300
Barium				
Ba	704	(1,400)[e]	1638	35,670
$BaBr_2$	847	6,000		
$BaCl_2$	960	5,370		
BaF_2	1287	3,000		
$Ba(NO_3)_2$	595	(5,980)		
$Ba_3(PO_4)_2$	1730	18,600		
$BaSO_4$	1350	9,700		
Beryllium				
Be	1280	2,500[e]		
Bismuth				
Bi	271.3	2,505	1420	
$BiBr_3$			461	18,020
$BiCl_3$	224	2,600	441	17,350
Bi_2O_3	817	6,800		
Bi_2S_5	747	8,900		
Boron				
BBr_3			91.3	7,300
BCl_3			12.5	5,680
BF_3	−128	480	−100.9	4,620
B_2H_6	−165.5		−92.4	3,685
B_3H_{10}	−119.8		16	6,470
B_5H_9	−46.9		58	7,700
B_5H_{11}			67	8,500
$B_{10}H_{14}$	99.7	7,800	f	11,600
B_2H_5Br	−104		16	6,230
$B_3N_3H_6$	−58		50.4	7,670
Bromine				
Br_2	−7.2	2,580	58.0	7,420
BrF_5	−61.3	1,355	40.4	7,470
Cadmium				
Cd	320.9	1,460	765	23,870
$CdBr_2$	568	(5,000)		
$CdCl_2$	568	5,300	967	29,860
CdF_2	1110	(5,400)		
CdI_2	387	3,660	796	25,400
CdO			1559[c]	53,820[c]
$CdSO_4$	1000	4,790		
Calcium				
Ca	851	2,230	1487	36,580
$CaBr_2$	730	4,180		
$CaCO_3$	1282	(12,700)		
$CaCl_2$	782	6,100		
CaF_2	1392	4,100		
$Ca(NO_3)_2$	561	5,120		
CaO	2707	(12,240)		
$CaO \cdot Al_2O_3 \cdot 2SiO_2$	1550	29,400		
$CaO \cdot MgO \cdot 2SiO_2$	1392	(18,200)		
$CaO \cdot SiO_2$	1512	13,400		
$CaSO_4$	1297	6,700		
Carbon				
C (graphite)	3600	11,000[e]		
CBr_4	90	1,050		
CCl_4	−24.0	644	77	7,280
CF_4			−127.9	3,110
CH_4	−182.5	224	−161.4	2,040
C_2N_2	−27.8	1,938[u]	−21.1	5,576[u]
CNBr	52			11,010[c]
CNCl	−5	2,240	13	6,300

Substance	mp, °C	Heat of fusion,[a,b] cal/mole	bp at 1 atm, °C	Heat of vaporization,[a,b] cal/mole
Carbon (Cont.)				
CNF			−72.8	5,780[c]
CNI			141	13,980[c]
CO	−205.0	200	−191.5	1,444
CO_2	−57.5	1,900	−78.4[c]	6,030[c,r]
COS	−138.8	1,129[k]	−50.2	4,423[k]
$COCl_2$			8.0	5,990
CS_2	−112.0	1,049[l]		
Cerium				
Ce	775	2,120		
Cesium				
Cs	28.4	500	690	16,320
CsBr			1300	35,990
CsCl	642	3,600	1300	35,690
CsF	715	(2,450)	1251	34,330
CsI			1280	35,930
$CsNO_3$	407	3,250		
Chlorine				
Cl_2	−101.0	1,531[m]	−34.1	4,878[m]
ClF			−101	
ClF_3			11.3	5,890
Cl_2O			2.0	6,280
ClO_2			10.9	7,100
Cl_2O_7			79	8,480
Chromium				
Cr	1550	3,930	2475	
CrO_2Cl_2			117	8,250
Cobalt				
Co	1490	3,660		
$CoCl_2$	727	7,390	1050	27,170
Copper				
Cu	1083.0	3,110	2595	72,810
Cu_2Br_2			1355	16,310
Cu_2Cl_2	430	4,890	1490	11,920
CuI			1336	15,940
$Cu_2(CN)_2$	473	(5,400)		
Cu_2O	1230	(13,400)		
CuO	1447	2,820		
Cu_2S	1127	5,500		
Fluorine				
F_2	−223		−188.2	1,640
F_2O			−144.8	2,650
Gallium				
Ga	29.8	1,336	2071	
Germanium				
Ge	959	(8,300)		
GeH_4	−165		−89.1	3,580
Ge_2H_6	−109		31.4	5,900
Ge_3H_8	−105.6		110.6	7,550
$GeHCl_3$	−71		75[g]	8,000
$GeBr_4$	26.1		189	8,560
$GeCl_4$	−49.5		84	7,030
$Ge(CH_3)_4$	−88		44	6,460
Gold				
Au	1063.0	3,030	2966	81,800
Helium				
He	−271.4		−268.4	22
Hydrogen				
H_2	−259.2	28	−252.7	216
HBr	−86.9	575	−66.7	4,210
HCl	−114.2	476	−85.0	3,860
HCN	−13.2	2,009[l]	25.7	6,027[l]
HF	−83.0	1,094	33.3	7,460
$(HF)_6$			51.2	5,020
HI	−50.8	686		
H_2O	0.0	1,436	100.0	9,729[h,q]
$H_2O (= D_2O)$	3.8	1,501[s]	101.4	9,945[c,q]
H_2O_2	−2	2,520[c]	158	10,270
HNO_3	−47	600		
H_3PO_2	17.4	2,310		
H_3PO_3	74	3,070		
H_3PO_4	42.4	2,520		
$H_4P_2O_6$	55	8,300		
H_2S	−85.5	568[t]	−60.3	4,463[t]
H_2S_2	−87.6	1,805		
H_2SO_4	10.5	2,360		
H_2Se			−41.3	4,880
H_2SeO_4	58	3,450		
H_2Te	−48.9	1,670	−2.2	5,650
Indium				
In	156.4	781		

*See also subsection "Thermodynamic Properties."

TABLE 2-190   Heats of Fusion and Vaporization of the Elements and Inorganic Compounds (Continued)

Substance	mp, °C	Heat of fusion,[a,b] cal/mole	bp at 1 atm, °C	Heat of vaporization,[a,b] cal/mole
Iodine				
I_2	113.0	3,650	183	10,390
$ICl(\alpha)$	17.2	2,660		
$ICl(\beta)$	13.9	2,270		
IF_7			4[c]	7,460[c]
Iron				
Fe	1530	3,560	2735	84,600
$FeCl_2$	677	7,800	1026	30,210
Fe_2Cl_6	304	20,590	319	12,040
$Fe(CO)_5$	−21	3,250	105	9,000
FeO	1380	(7,700)		
FeS	1195	5,000		
Krypton				
Kr	−157	360[e]	152.9	2,310[e]
Lead				
Pb	327.4	1,224	1744	42,060
$PbBr_2$	488	4,290	914	27,700
$PbCl_2$	498	5,650	954	29,600
PbF_2	824	1,860	1293	38,300
PbI_2	412	5,970	872	24,850
$PbMoO_4$	1065	(25,800)		
PbO	890	2,820	1472	51,310
PbS	1114	4,150	1281	(50,000)
$PbSO_4$	1087	9,600		
$PbWO_4$	1123	(15,200)		
Lithium				
Li	179	1,100	1372	32,250
$LiBO_2$	845	(5,570)		
LiBr	552	2,900	1310	35,420
LiCl	614	3,200	1382	35,960
LiF	847	(2,360)	1681	50,970
LiI	440	(1,420)	1171	40,770
LiOH	462	2,480		
Li_2MoO_4	705	4,200		
$LiNO_3$				
Li_2SiO_3	1177	7,210		
Li_4SiO_4	1249	7,430		
Li_2SO_4	857	3,040		
Li_2WO_4	742	(6,700)		
Magnesium				
Mg	650	2,160	1107	32,520
$MgBr_2$	711	8,300		
$MgCl_2$	712	8,100	1418	32,690
MgF_2	1221	5,900		
MgO	2642	18,500		
$Mg_3(PO_4)_2$	1184	(11,300)		
$MgSiO_3$	1524	14,700		
$MgSO_4$	1127	3,500		
$MgZn_2$	589	(8,270)		
Manganese				
Mn	1220	3,450	2152	55,150
$MnCl_2$	650	7,340	1190	29,630
$MnSiO_3$	1274	(8,200)		
$MnTiO_3$	1404	(7,960)		
Mercury				
Hg	−38.9	557	361	13,980
$HgBr_2$	241	3,960	319	14,080
$HgCl_2$	277	4,150	304	14,080
HgI_2	250	4,500	354	14,260
$HgSO_4$	850	(1,440)		
Molybdenum				
Mo	2622	(6,660)	(4800)	(128,000)
MoF_6	17	2,500	36	6,000
MoO_3	745	(2,500)	1151	
Neon				
Ne	−248.5	77	−246.0	440[e]
Nickel				
Ni	1455	4,200	2730	87,300
$NiCl_2$			987[c]	48,360[c]
$Ni(CO)_4$			42.5	7,000
Ni_2S	645	(2,980)		
Ni_3S_2	790	5,800		
Nitrogen				
N_2	−210.0	172	−195.8	1,336
NF_3	−129.0			3,000
NH_3	−77.7	1,352[n]	−33.4	5,581[n]
NH_4CNS	146	(4,700)		
NH_4NO_3	169.6	1,460		
N_2O	−90.8	1,563	−88.5	3,950
NO	−163.6	550	−151.7	3,307
N_2O_4	−13	5,540	30	7,040
N_2O_5			32.4	13,800[c]
NOCl			−6.4	6,140
Osmium				
OsF_8			47.4	6,840
OsO_4 (yellow)	56	4,060	130	9,450
OsO_4 (white)	42	2,340		
Oxygen				
O_2	−218.9	106	−183.0	1,629
O_3			−111	2,880
Palladium				
Pd	1554	4,120		
Phosphorus				
P_4 (yellow)	44.2	615	280	12,520
P_4 (violet)			417[c]	25,600[c]
P_4 (black)			453[c]	33,100
PCl_3			74.2	7,280
PH_3	−133.8	270[o]	−87.7	3,489[o]
P_4O_6	23.8	3,360	174	10,380
$P_4O_{10}(\alpha)$	569	17,080	591	20,670
$P_4O_{10}(\beta)$			358[c]	
$POCl_3$	1.1	3,110	105.1	8,380
P_2S_3			508	
Platinum				
Pt	1773.5	4,700	(4400)	(107,000)
Potassium				
K	63.5	574	776	18,920
KBO_2	947	(5,700)		
KBr	742	5,000	1383	37,060
KCl	770	6,410	1407	38,840
KCN	623	(3,500)		
KCNS	179	2,250		
K_2CO_3	897	7,800		
K_2CrO_4	984	6,920		
$K_2Cr_2O_7$	398	8,770		
KF	857	6,500		
KI	682	4,100	1324	34,690
K_2MoO_4	922	(4,000)		
KNO_3	338	2,840		
KOH	360	(2,000)	1327	30,850
KPO_3	817	2,110		
K_3PO_4	1340	8,900		
$K_4P_2O_7$	1092	14,000		
K_2SO_4	1074	8,100		
K_2TiO_3	810	(10,600)		
K_2WO_4	927	(4,400)		
Praseodymium				
Pr	932	2,700		
Radon				
Rn	−71		−61.8	4,010
Rhenium				
Re	(3000)			
Re_2O_7	296	15,340	362.4	18,060
Re_2O_8	147	3,800		
Rubidium				
Rb	39.1	525	679	18,110
RbBr	677	3,700	1352	37,120
RbCl	717	4,400	1381	36,920
RbF	833	4,130	1408	39,510
RbI	638	2,990	1304	35,960
$RbNO_3$	305	1,340		
Selenium				
Se_2	217	1,220	753	25,490
Se_6			736	20,600
SeF_6			−45.8[c]	6,350[c]
SeO_2			317[c]	20,900
$SeOCl_2$	10	1,010	168	
Silicon				
Si	1427	9,470	2290	
$SiCl_4$	−67.6	1,845	56.8	6,860
Si_2Cl_6	−1		139	
Si_3Cl_8			211.4	12,340
$(SiCl_3)_2O$	−33		135.6	8,820
SiF_4			−94.8[c]	6,130[c]
Si_2F_6	−18.5	3,900	−18.9[c]	10,400[c]
SiF_3Cl	−138		−70.1	4,460
SiF_2Cl_2	−144		−31.5	5,080
SiH_4	−185		−111.6	2,960
Si_2H_6	−132.5		−14.3	5,110
Si_3H_8	−117		53.1	6,780
Si_4H_{10}	−93.5		100	8,890
SiH_3Br	−93.8		2.4	5,650
SiH_2Br_2	−70.0		70.5	6,840
$SiHCl_3$	−126.5		31.8	6,360
$(SiH_3)_3N$	−105.6		48.7	6,850
$(SiH_3)_2O$	−144		−15.4	5,350
SiO_2 (quartz)	1470	3,400	2230	
SiO_2 (cristobalite)	1700	2,100		
Silver				
Ag	960.5	2,700	2212	60,720
AgBr	430	2,180		
AgCl	455	3,155	1564	42,520
AgCN	350	2,750		
AgI	557	2,250	1506	34,450
$AgNO_3$	209	2,755		
Ag_2S	842	3,360		
Ag_2SO_4	657	(4,300)		
Sodium				
Na	97.7	630	914	23,120
$NaBO_2$	966	8,660		

TABLE 2-190   Heats of Fusion and Vaporization of the Elements and Inorganic Compounds (*Concluded*)

Substance	mp, °C	Heat of fusion,[a,b] cal/mole	bp at 1 atm, °C	Heat of vaporization,[a,b] cal/mole
Sodium (*Cont.*)				
NaBr	747	6,140	1392	37,950
NaCl	800	7,220	1465	40,810
NaClO$_3$	255	5,290		
NaCN	562	(4,400)	1500	37,280
NaCNS	323	4,450		
Na$_2$CO$_3$	854	7,000		
NaF	992	7,000	1704	53,260
NaI	662	5,240		
Na$_2$MoO$_4$	687	3,600		
NaNO$_3$	310	3,760		
NaOH	322	2,000	1378	
½Na$_2$O·½Al$_2$O$_3$·3SiO$_2$	1107	13,150		
NaPO$_3$	988	(5,000)		
Na$_4$P$_2$O$_7$	970	(13,700)		
Na$_2$S	920	(1,200)		
Na$_2$SiO$_3$	1087	10,300		
Na$_2$Si$_2$O$_5$	884	8,460		
Na$_2$SO$_4$	884	5,830		
Na$_2$WO$_4$	702	5,800		
Strontium				
Sr	757	2,190	1384	33,610
SrBr$_2$	643	4,780		
SrCl$_2$	872	4,100		
SrF$_2$	1400	4,260		
Sr$_3$(PO$_4$)$_2$	1770	18,500		
Sulfur				
S (rhombic)	112.8		444.6	2,200
S (monoclinic)	119.2			
S$_2$Cl$_2$			138	8,720
SF$_6$			−63.5[c]	5,600[c]
SO$_2$	−75.5	1,769[p]	−5.0	5,960[p]
SO$_3$(α)	17	2,060	44.8	10,190
SO$_3$(β)	32.4	2,890		
SO$_3$(γ)	62.2	6,310		
SOBr$_2$			139.5	9,920
SOCl$_2$			75.4	7,600
SO$_2$Cl$_2$			69.2	7,760
Tellurium				
Te	453	3,230	1090	
TeCl$_4$			392	16,830
TeF$_6$			−38.6[c]	6,700[c]

Substance	mp, °C	Heat of fusion,[a,b] cal/mole	bp at 1 atm, °C	Heat of vaporization,[a,b] cal/mole
Thallium				
Tl	302.5	1,030	1457	38,810
TlBr	460	5,990	819	23,800
TlCl	427	4,260	807	24,420
Tl$_2$CO$_3$	273	4,400		
TlI	440	3,125	823	25,030
TlNO$_3$	207	2,290		
Tl$_2$S	449	3,000		
Tl$_2$SO$_4$	632	5,500		
Tin				
Sn$_4$	231.8	1,720	2270	68,000
SnBr$_2$	232	(1,700)		
SnBr$_4$	30	3,000		
SnCl$_2$	247	3,050	623	20,740
SnCl$_4$	−33.2	2,190	113	8,330
Sn(CH$_3$)$_4$			78.3	7,320
SnH$_4$	−149.8		−52.3	4,420
SnI$_4$	143.5	(4,300)		
Titanium				
TiBr$_4$	38.2	(2,060)		
TiCl$_4$	−23	2,240	136	8,350
TiO$_2$	1825	(11,400)		
Tungsten				
W	3390	(8,400)	(5900)	(176,000)
WF$_6$	−0.4	1,800	17.3	6,350
Uranium				
UF$_6$			55.1[c]	9,990[c]
Xenon				
Xe	−111.5	740	−108.0	3,110
Zinc				
Zn	419.5	1,595	907	27,430
ZnCl$_2$	283	(5,500)	732	28,710
Zn(C$_2$H$_5$)$_2$			118	8,960
ZnO	1975	4,470		
ZnS	1645	(9,000)		
Zirconium				
ZrBr$_4$			357[c]	25,800[c]
ZrCl$_4$			311[c]	25,290[c]
ZrI$_4$			431[c]	29,030[c]
ZrO$_2$	2715	20,800		

[a] Values in parentheses are uncertain.
[b] For the freezing point or the normal boiling point unless otherwise stated.
[c] Sublimation.
[d] Decomposes at about 75°C; value obtained by extrapolation.
[e] Bichowsky and Rossini, "Thermochemistry of the Chemical Substances," Reinhold, New York (1936).
[f] Decomposes before the normal boiling point is reached.
[g] Decomposes at about 40°C; value obtained by extrapolation.
[h] See also pp. 2-304 through 2-307 on steam table.
[i] Giauque and Ruehrwein, *J. Am. Chem. Soc.*, **61** (1939): 2626.
[j] Giauque and Egan, *J. Chem. Phys.*, **5** (1937): 45.

[k] Kemp and Giauque, *J. Am. Chem. Soc.*, **59** (1937): 79.
[l] Brown and Manov, *J. Am. Chem. Soc.*, **59** (1937): 500.
[m] Giauque and Powell, *J. Am. Chem. Soc.* **61** (1939): 1970.
[n] Overstreet and Giauque, *J. Am. Chem. Soc* **59** (1937): 254.
[o] Stephenson and Giauque, *J. Chem. Phys.*, **5** (1937): 149.
[p] Giauque and Stephenson, *J. Am. Chem. Soc.*, **60** (1938): 1389.
[q] Osborne, Stimson, and Ginnings, *Bur. Standards J. Research*, **23**, 197 (1939): 261.
[r] Miles and Menzies, *J. Am. Chem. Soc.*, **58** (1936): 1067.
[s] Long and Kemp, *J. Am. Chem. Soc.*, **58** (1936): 1829.
[t] Giauque and Blue, *J. Am. Chem. Soc.*, **58** (1936): 831.
[u] Ruehrwein and Giauque, *J. Am. Chem. Soc.*, **61** (1939): 2940.

TABLE 2-191   Heats of Fusion of Miscellaneous Materials

Material	mp, °C	Heat of fusion, cal/g
Alloys		
30.5 Pb + 69.5 Sn	183	17
36.9 Pb + 63.1 Sn	179	15.5
63.7 Pb + 36.3 Sn	177.5	11.6
77.8 Pb + 22.2 Sn	176.5	9.54
1 Pb + 9 Sn	236	28
24 Pb + 27.3 Sn + 48.7 Bi	98.8	6.85
25.8 Pb + 14.7 Sn + 52.4 Bi + 7 Cd	75.5	8.4
Silicates		
Anorthite (CaAl$_2$Si$_2$O$_8$)		100
Orthoclase (KAlSi$_3$O$_8$)		100
Microcline (KAlSi$_3$O$_8$)		83
Wollastonite (CaSiO$_3$)		100
Malacolite (Ca$_8$MgSi$_4$O$_{12}$)		94
Diopside (CaMgSi$_2$O$_4$)		100
Olivine (Mg$_2$SiO$_4$)		130
Fayalite (Fe$_2$SiO$_4$)		85
Spermaceti	43.9	37.0
Wax (bees')	61.8	42.3

TABLE 2-192 Heats of Fusion of Organic Compounds

The values for the hydrocarbons are from the tables of the American Petroleum Institute Research Project 44 at the National Bureau of Standards, with some from Parks and Huffman, *Ind. Eng. Chem.*, **23**, 1138 (1931).

The values for the nonhydrocarbon compounds were recalculated from data in *International Critical Tables*, vol. 5.

Hydrocarbon compounds	Formula	mp, °C	Heat of fusion, cal/g	Hydrocarbon compounds	Formula	mp, °C	Heat of fusion, cal/g
Paraffins				Aromatics—(*Cont.*)			
Methane	CH_4	−182.48	14.03	1-Methyl-3-ethylbenzene	C_9H_{12}	−95.55	15.14
Ethane	C_2H_6	−183.23	22.712	1-Methyl-4-ethylbenzene	C_9H_{12}	−62.350	25.29
Propane	C_3H_8	−187.65	19.100	1,2,3-Trimethylbenzene	C_9H_{12}	−25.375	16.64
n-Butane	C_4H_{10}	−138.33	19.167	1,2,4-Trimethylbenzene	C_9H_{12}	−43.80	24.54
2-Methylpropane	C_4H_{10}	−159.60	18.668	1,3,5-Trimethylbenzene	C_9H_{12}	−44.720	18.97
n-Pentane	C_5H_{12}	−129.723	27.874	Naphthalene	$C_{10}H_8$	+80.0	36.0
2-Methylbutane	C_5H_{12}	−159.890	17.076	Camphene	$C_{10}H_{12}$	+51	57
2,2-Dimethylpropane	C_5H_{12}	−16.6	10.786	Durene	$C_{10}H_{14}$	+79.3	37.4
n-Hexane	C_6H_{14}	−95.320	36.138	Isodurene	$C_{10}H_{14}$	−24.0	23.0
2-Methylpentane	C_6H_{14}	−153.680	17.407	Prehnitene	$C_{10}H_{14}$	−7.7	20.0
2,2-Dimethylbutane	C_6H_{14}	−99.73	1.607	p-Cymene	$C_{10}H_{14}$	−68.9	17.1
2,3-Dimethylbutane	C_6H_{14}	−128.41	2.251	n-Butyl benzene	$C_{10}H_{14}$	−88.5	19.5
n-Heptane	C_7H_{16}	−90.595	33.513	tert-Butyl benzene	$C_{10}H_{14}$	−58.1	14.9
2-Methylhexane	C_7H_{16}	−118.270	21.158	β-Methyl naphthalene	$C_{11}H_{10}$	+34.1	20.1
3-Ethylpentane	C_7H_{16}	−118.593	22.555	Diphenyl	$C_{12}H_{10}$	+68.6	28.8
2,2-Dimethylpentane	C_7H_{16}	−123.790	13.982	Hexamethyl benzene	$C_{12}H_{18}$	+165.5	30.4
2,4-Dimethylpentane	C_7H_{16}	−119.230	15.968	Diphenyl methane	$C_{13}H_{12}$	+25.2	26.4
3,3-Dimethylpentane	C_7H_{16}	−134.46	16.856	Anthracene	$C_{14}H_{10}$	+216.5	38.7
2,2,3-Trimethylbutane	C_7H_{16}	−24.96	5.250	Phenanthrene	$C_{14}H_{10}$	+96.3	25.0
n-Octane	C_8H_{18}	−56.798	43.169	Tolane	$C_{14}H_{10}$	+60	28.7
2-Methylheptane	C_8H_{18}	−109.04	21.458	Stilbene	$C_{14}H_{12}$	+124	40.0
3-Methylheptane	C_8H_{18}	−120.50	23.795	Dibenzil	$C_{14}H_{14}$	+51.4	30.7
4-Methylheptane	C_8H_{18}	−120.955	22.692	Triphenyl methane	$C_{19}H_{16}$	+92.1	21.1
2,2-Dimethylhexane	C_8H_{18}	−121.18	24.226	Alkyl cyclohexanes			
2,5-Dimethylhexane	C_8H_{18}	−91.200	26.903	Cyclohexane	C_6H_{12}	+6.67	7.569
3,3-Dimethylhexane	C_8H_{18}	−126.10	14.9	Methylcyclohexane	C_7H_{14}	−126.58	16.429
2-Methyl-3-ethylpentane	C_8H_{18}	−114.960	23.690	Alkyl cyclopentanes			
3-Methyl-3-ethylpentane	C_8H_{18}	−90.870	22.657	Cyclopentane	C_5H_{10}	−93.80	2.068
2,2,3-Trimethylpentane	C_8H_{18}	−112.27	18.061	Methylcyclopentane	C_6H_{12}	−142.445	19.68
2,2,4-Trimethylpentane	C_8H_{18}	−107.365	19.278	Ethylcyclopentane	C_7H_{14}	−138.435	11.10
2,3,3-Trimethylpentane	C_8H_{18}	−100.70	3.204	1,1-Dimethylcyclopentane	C_7H_{14}	−69.73	3.36
2,3,4-Trimethylpentane	C_8H_{18}	−109.210	19.392	cis-1,2-Dimethylcyclopentane	C_7H_{14}	−53.85	3.87
2,2,3,3-Tetramethylbutane	C_8H_{18}	+100.69	14.900	trans-1,2-Dimethylcyclopentane	C_7H_{14}	−117.57	15.68
n-Nonane	C_9H_{20}	−53.9	41.2	trans-1,3-Dimethylcyclopentane	C_7H_{14}	−133.680	17.93
n-Decane	$C_{10}H_{22}$	−30.0	48.3	Monoolefins			
n-Undecane	$C_{11}H_{24}$	−25.9	34.1	Ethene (Ethylene)	C_2H_4	−169.15	28.547
n-Dodecane	$C_{12}H_{26}$	−9.6	51.3	Propene (Propylene)	C_3H_6	−185.25	17.054
Eicosane	$C_{20}H_{42}$	+36.4	52.0	1-Butene	C_4H_8	−185.35	16.393
Pentacosane	$C_{25}H_{52}$	+53.3	53.6	cis-2-Butene	C_4H_8	−138.91	31.135
Tritriacontane	$C_{33}H_{68}$	+71.1	54.0	trans-2-Butene	C_4H_8	−105.55	41.564
Aromatics				2-Methylpropene (isobutene)	C_4H_8	−140.35	25.265
Benzene	C_6H_6	+5.533	30.100	1-Pentene	C_5H_{10}	−165.27	16.82
Methylbenzene (Toluene)	C_7H_8	−94.991	17.171	cis-2-pentene	C_5H_{10}	−151.363	24.239
Ethylbenzene	C_8H_{10}	−94.950	20.629	trans-2-pentene	C_5H_{10}	−140.235	26.536
o-Xylene	C_8H_{10}	−25.187	30.614	2-Methyl-1-butene	C_5H_{10}	−137.560	26.879
m-Xylene	C_8H_{10}	−47.872	26.045	3-Methyl-1-butene	C_5H_{10}	−168.500	18.009
p-Xylene	C_8H_{10}	+13.263	38.526	2-Methyl-2-butene	C_5H_{10}	−133.780	25.738
n-Propylbenzene	C_9H_{12}	−99.500	16.97	Acetylenes			
Isopropylbenzene	C_9H_{12}	−96.028	19.22	Acetylene	C_2H_2	−81.5	23.04
1-Methyl-2-ethylbenzene	C_9H_{12}	−80.833	21.13	2-Butyne (dimethylacetylene)	C_4H_6	−132.23	40.808

Nonhydrocarbon compounds	Formula	mp, °C	Heat of fusion, cal/g	Nonhydrocarbon compounds	Formula	mp, °C	Heat of fusion, cal/g
Acetic acid	$C_2H_4O_2$	16.7	46.68	Butyl alcohol (n-)	$C_4H_{10}O$	−89.2	29.93
Acetone	C_3H_6O	−95.5	23.42	(t-)	$C_4H_{10}O$	25.4	21.88
Acrylic acid	$C_3H_4O_2$	12.3	37.03	Butyric acid (n-)	$C_4H_8O_2$	−5.7	30.04
Allo-cinnamic acid	$C_9H_8O_2$	68	27.35				
Aminobenzoic acid (o-)	$C_7H_7NO_2$	145	35.48	Capric acid (n-)	$C_{10}H_{20}O_2$	31.99	38.87
(m-)	$C_7H_7NO_2$	179.5	38.03	Caprylic acid (n-)	$C_8H_{16}O_2$	16.3	35.40
(p-)	$C_7H_7NO_2$	188.5	36.46	Carbazole	$C_{12}H_9N$	243	42.05
Amyl alcohol	$C_5H_{12}O$	−78.9	26.65	Carbon tetrachloride	CCl_4	−22.8	41.57
Anethole	$C_{10}H_{12}O$	22.5	25.80	Carvoxime (d-)	$C_{10}H_{15}NO$	71.5	23.29
Aniline	$C_6H_5NH_2$	−6.3	27.09	(l-)	$C_{10}H_{15}NO$	71	23.41
Anthraquinone	$C_{14}H_8O_2$	284.8	37.48	(dl-)	$C_{10}H_{15}NO$	91	24.61
Apiol	$C_{12}H_{14}O_4$	29.5	25.80	Cetyl alcohol	$C_{16}H_{34}O$	49.27	33.80
Azobenzene	$C_{12}H_{10}N_2$	67.1	28.91	Chloracetic acid (α-)	$C_2H_3ClO_2$	61.2	31.06
Azoxybenzene	$C_{12}H_{10}N_2O$	36	21.62	(β-)	$C_2H_3ClO_2$	56	35.12
				Chloral alcoholate	$C_4H_7Cl_3O_2$	9	24.03
Benzil	$C_{14}H_{10}O_2$	95.2	22.15	hydrate	$C_2H_3Cl_3O_2$	47.4	33.18
Benzoic acid	$C_7H_6O_2$	122.45	33.90	Chloroaniline (p-)	C_6H_6ClN	71	37.15
Benzophenone	$C_{13}H_{10}O$	47.85	23.53	Chlorobenzoic acid (o-)	$C_7H_5ClO_2$	140.2	39.30
Benzylaniline	$C_{13}H_{13}N$	32.37	21.86	(m-)	$C_7H_5ClO_2$	154.25	36.41
Bromocamphor	$C_{10}H_{15}BrO$	78	41.57	(p-)	$C_7H_5ClO_2$	239.7	49.21
Bromochlorbenzene (o-)	C_6H_4BrCl	−12.6	15.41	Chloronitrobenzene (m-)	$C_6H_4ClNO_2$	44.4	29.38
(m-)	C_6H_4BrCl	−21.2	15.29	(p-)	$C_6H_4ClNO_2$	83.5	31.51
(p-)	C_6H_4BrCl	64.6	23.41	Cinnamic acid	$C_9H_8O_2$	133	36.50
Bromoiodobenzene (o-)	C_6H_4BrI	21	12.18	anhydride	$C_{18}H_{14}O_3$	48	28.14
(m-)	C_6H_4BrI	9.3	10.27	Cresol (p-)	C_7H_8O	34.6	26.28
(p-)	C_6H_4BrI	90.1	16.60	Crotonic acid (α-)	$C_4H_6O_2$	72	25.32
Bromol hydrate	$C_2H_3Br_3O_2$	46	16.90	(cis-)	$C_4H_6O_2$	71.2	34.90
Bromophenol (p-)	C_6H_5BrO	63.5	20.50	Cyanamide	CH_2N_2	44	49.81
Bromotoluene (p-)	C_7H_7Br	28	20.86	Cyclohexanol	$C_6H_{12}O$	25.46	4.19

TABLE 2-192 Heats of Fusion of Organic Compounds (*Concluded*)

Nonhydrocarbon compounds	Formula	mp, °C	Heat of fusion, cal/g	Nonhydrocarbon compounds	Formula	mp, °C	Heat of fusion, cal/g
Dibromobenzene (o-)	$C_6H_4Br_2$	1.8	12.78	Naphthol (α-)	$C_{10}H_8O$	95.0	38.94
(m-)	$C_6H_4Br_2$	-6.9	13.38	(β-)	$C_{10}H_8O$	120.6	31.30
(p-)	$C_6H_4Br_2$	86	20.55	Naphthylamine (α-)	$C_{10}H_9N$	50	22.34
Dibromophenol (2, 4-)	$C_6H_4Br_2O$	12	13.97	Nitroaniline (o-)	$C_6H_6N_2O_2$	71.2	27.88
Dichloroacetic acid	$C_2H_2Cl_2O_2$	-4(?)	14.21	(m-)	$C_6H_6N_2O_2$	114.0	40.97
Dichlorobenzene (o-)	$C_6H_4Cl_2$	-16.7	21.02	(p-)	$C_6H_6N_2O_2$	147.3	36.46
(m-)	$C_6H_4Cl_2$	-24.8	20.55	Nitrobenzene	$C_6H_5NO_2$	5.85	22.52
(p-)	$C_6H_4Cl_2$	53.13	29.67	Nitrobenzoic acid (o-)	$C_7H_5NO_4$	145.8	40.06
Dihydroxybenzene (o-)	$C_6H_6O_2$	104.3	49.40	(m-)	$C_7H_5NO_4$	141.1	27.59
(m-)	$C_6H_6O_2$	109.65	46.20	(p-)	$C_7H_5NO_4$	239.2	52.80
(p-)	$C_6H_6O_2$	172.3	58.77	Nitronaphthalene	$C_{10}H_7NO_2$	56.7	25.44
Di-iodobenzene (o-)	$C_6H_4I_2$	23.4	10.15	Nitrophenol (o-)	$C_6H_5NO_3$	45.13	26.76
(m-)	$C_6H_4I_2$	34.2	11.54				
(p-)	$C_6H_4I_2$	129	16.20	Palmitic acid	$C_{16}H_{32}O_2$	61.82	39.18
Dimethyl tartrate (dl-)	$C_6H_{10}O_6$	87	35.12	Paraldehyde	$C_6H_{12}O_3$	10.5	25.02
(d-)	$C_6H_{10}O_6$	49	21.50	Pelargic acid (n-) (β-)	$C_9H_{18}O_2$		39.04
pyrone	$C_7H_8O_2$	132	56.14	Pelargonic acid (n-) (α-)	$C_9H_{18}O_2$	12.35	30.63
Dinitrobenzene (o-)	$C_6H_4N_2O_4$	116.93	32.25	Phenol	C_6H_6O	40.92	29.03
(m-)	$C_6H_4N_2O_4$	89.7	24.70	Phenylacetic acid	$C_8H_8O_2$	76.7	25.44
(p-)	$C_6H_4N_2O_4$	173.5	39.99	Phenylhydrazine	$C_6H_8N_2$	19.6	36.31
Dinitrotoluene (2, 4-)	$C_7H_6N_2O_4$	70.14	26.40	Propyl ether (n)	$C_6H_{14}O$	-126.1	20.66
Dioxane	$C_4H_8O_2$	11.0	34.85				
Diphenyl amine	$C_{12}H_{11}N$	52.98	25.23	Quinone	$C_6H_4O_2$	115.7	40.85
Elaidic acid	$C_{18}H_{34}O_2$	44.4	52.08	Stearic acid	$C_{18}H_{36}O_2$	68.82	47.54
Ethyl acetate	$C_4H_8O_2$	83.8	28.43	Succinic anhydride	$C_4H_4O_3$	119	48.74
alcohol	C_2H_6O	-114.4	25.76	Succinonitrile	$C_4H_4N_2$	54.5	11.71
Ethylene dibromide	$C_2H_4Br_2$	10.012	13.52				
Ethyl ether	$C_4H_{10}O$	-116.3	23.54	Tetrachloroxylene (o-)	$C_8H_6Cl_4$	86	21.02
				(p-)	$C_8H_6Cl_4$	95	22.10
Formic acid	CH_2O_2	8.40	58.89	Thiophene	C_4H_4S	-39.4	14.11
				Thiosinamine	$C_4H_8N_2S$	77	33.45
Glutaric acid	$C_5H_8O_4$	97.5	37.39	Thymol	$C_{10}H_{14}O$	51.5	27.47
Glycerol	$C_3H_8O_3$	18.07	47.49	Toluic acid (o-)	$C_8H_8O_2$	103.7	35.40
Glycol, ethylene	$C_2H_6O_2$	-11.5	43.26	(m-)	$C_8H_8O_2$	108.75	27.59
				(p-)	$C_8H_8O_2$	179.6	39.90
Hydrazo benzene	$C_{12}H_{12}N_2$	134	22.89	Toluidine (p-)	C_7H_9N	43.3	39.90
Hydrocinnamic acid	$C_9H_{10}O_2$	48	28.14	Tribromophenol (2, 4, 6-)	$C_6H_3Br_3O$	93	13.38
Hydroxyacetanilide	$C_8H_9NO_2$	91.3	33.59	Trichloroacetic acid	$C_2HCl_3O_2$	57.5	8.60
				Trinitroglycerol	$C_3H_5N_3O_9$	12.3	23.02
Iodotoluene (p-)	C_7H_7I	34	18.75	Trinitrotoluene (2, 4, 6-)	$C_7H_5N_3O_6$	80.83	22.34
Isopropyl alcohol	C_3H_8O	-88.5	21.08	Tristearin	$C_{57}H_{110}O_6$	70.8, 54.5	45.63
ether	$C_6H_{14}O$	-86.8	25.79				
				Undecylic acid (α-) (n-)	$C_{11}H_{22}O_2$	28.25	32.20
Lauric acid (n-)	$C_{12}H_{24}O_2$	43.22	43.72	(β-) (n-)	$C_{11}H_{22}O_2$		42.91
Levulinic acid	$C_5H_8O_3$	33	18.97	Urethane	$C_3H_7NO_2$	48.7	40.85
Menthol (l-) (α)	$C_{10}H_{20}O$	43.5	18.63	Veratrol	$C_8H_{10}O_2$	22.5	27.45
Methyl alcohol	CH_4O	-97.8	23.7				
Myristic acid	$C_{14}H_{28}O_2$	53.86	47.49	Xylene dibromide (o-)	$C_8H_8Br_2$	95	24.25
Methyl cinnamate	$C_{10}H_{10}O_2$	36	26.53	(m-)	$C_8H_8Br_2$	77	21.45
fumarate	$C_6H_8O_4$	102	57.93	dichloride (o-)	$C_8H_8Cl_2$	55	29.03
oxalate	$C_4H_6O_4$	54.35	42.64	(m-)	$C_8H_8Cl_2$	34	26.64
phenylpropiolate	$C_{10}H_8O_2$	18	22.86	(p-)	$C_8H_8Cl_2$	100	32.73
succinate	$C_6H_{10}O_4$	19.5	35.72				

TABLE 2-193 Heats of Vaporization of Inorganic and Organic Compounds

Cmpd. no.	Name	Formula	Mol wt	CAS no.	C1 ×1E-07	C2	C3	C4	T_{min} K	ΔH_t at T_{min} ×1E-07	T_{max} K	ΔH_v at T_{max}
1	Methane	CH$_4$	16.043	74828	1.0194	0.26087	−0.14694	0.22154	90.69	0.8724	190.56	0
2	Ethane	C$_2$H$_6$	30.070	74840	2.1091	0.60646	−0.55492	0.32799	90.35	1.7879	305.32	0
3	Propane	C$_3$H$_8$	44.097	74986	2.9209	0.78237	−0.77319	0.39246	85.47	2.4787	369.83	0
4	n-Butane	C$_4$H$_{10}$	58.123	106978	3.6238	0.8337	−0.82274	0.39613	134.86	2.8684	425.12	0
5	n-Pentane	C$_5$H$_{12}$	72.150	109660	3.9109	0.38681	0	0	143.42	3.3968	469.7	0
6	n-Hexane	C$_6$H$_{14}$	86.177	110543	4.4544	0.39002	0	0	177.83	3.7647	507.6	0
7	n-Heptane	C$_7$H$_{16}$	100.204	142825	5.0014	0.38795	0	0	182.57	4.2619	540.2	0
8	n-Octane	C$_8$H$_{18}$	114.231	111659	5.5180	0.38467	0	0	216.38	4.5598	568.7	0
9	n-Nonane	C$_9$H$_{20}$	128.258	111842	6.0370	0.38522	0	0	219.66	5.0545	594.6	0
10	n-Decane	C$_{10}$H$_{22}$	142.285	124185	6.6126	0.39797	0	0	243.51	5.4168	617.7	0
11	n-Undecane	C$_{11}$H$_{24}$	156.312	1120214	7.2284	0.40607	0	0	247.57	5.9240	639	0
12	n-Dodecane	C$_{12}$H$_{26}$	170.338	112403	7.7337	0.40681	0	0	263.57	6.2802	658	0
13	n-Tridecane	C$_{13}$H$_{28}$	184.365	629505	8.4339	0.4257	0	0	267.76	6.8015	675	0
14	n-Tetradecane	C$_{14}$H$_{30}$	198.392	629594	9.0539	0.44467	0	0	279.01	7.2002	693	0
15	n-Pentadecane	C$_{15}$H$_{32}$	212.419	629629	9.6741	0.45399	0	0	283.07	7.6728	708	0
16	n-Hexadecane	C$_{16}$H$_{34}$	226.446	544763	10.1560	0.45726	0	0	291.31	8.0225	723	0
17	n-Heptadecane	C$_{17}$H$_{36}$	240.473	629787	10.4730	0.4374	0	0	295.13	8.3699	736	0
18	n-Octadecane	C$_{18}$H$_{38}$	254.500	593453	10.9690	0.44327	0	0	301.31	8.7246	747	0
19	n-Nonadecane	C$_{19}$H$_{40}$	268.527	629925	11.6740	0.45865	0	0	305.04	9.2185	758	0
20	n-Eicosane	C$_{20}$H$_{42}$	282.553	112958	12.8600	0.50351	0.32986	−0.42184	309.58	9.5933	768	0
21	2-Methylpropane	C$_4$H$_{10}$	58.123	75285	3.1667	0.3855	0	0	113.54	2.7927	408.14	0
22	2-Methylbutane	C$_5$H$_{12}$	72.150	78784	3.7700	0.3952	0	0	113.25	3.3720	460.43	0
23	2,3-Dimethylbutane	C$_6$H$_{14}$	86.177	79298	4.1404	0.38124	0	0	145.19	3.6328	499.98	0
24	2-Methylpentane	C$_6$H$_{14}$	86.177	107835	4.2780	0.384	0	0	119.55	3.8495	497.5	0
25	2,3-Dimethylpentane	C$_7$H$_{16}$	100.204	565593	4.6536	0.37579	0	0	160	4.0747	537.35	0
26	2,3,3-Trimethylpentane	C$_8$H$_{18}$	114.231	560214	4.9910	0.383	0	0	172.22	4.3530	573.5	0
27	2,2,4-Trimethylpentane	C$_8$H$_{18}$	114.231	540841	4.7721	0.37992	0	0	165.78	4.1565	543.96	0
28	Ethylene	C$_2$H$_4$	28.054	74851	2.8694	0.3746	0	0	104	1.6025	282.34	0
29	Propylene	C$_3$H$_6$	42.081	115071	3.2300	0.8375	−0.9216	0.5012	87.89	2.4031	365.57	0
30	1-Butene	C$_4$H$_8$	56.108	106989	3.4190	0.3747	0	0	87.8	2.9582	419.95	0
31	cis-2-Butene	C$_4$H$_8$	56.108	590181	3.3754	0.3754	0	0	134.26	2.9773	435.58	0
32	trans-2-Butene	C$_4$H$_8$	56.108	624646	3.3320	0.3736	0	0	167.62	2.7684	428.63	0
33	1-Pentene	C$_5$H$_{10}$	70.134	109671	3.7740	0.37647	0	0	107.93	3.4166	464.78	0
34	1-Hexene	C$_6$H$_{12}$	84.161	592416	4.3236	0.3788	0	0	133.39	3.8483	504.03	0
35	1-Heptene	C$_7$H$_{14}$	98.188	592767	4.8120	0.3685	0	0	154.27	4.2478	537.29	0
36	1-Octene	C$_8$H$_{16}$	112.215	111660	5.3980	0.3835	0	0	171.45	4.7013	566.65	0
37	1-Nonene	C$_9$H$_{18}$	126.242	124118	5.9940	0.3953	0	0	191.78	5.1366	593.25	0
38	1-Decene	C$_{10}$H$_{20}$	140.269	872059	6.4898	0.39187	0	0	206.89	5.5289	616.4	0
39	2-Methylpropene	C$_4$H$_8$	56.108	115117	3.2720	0.383	0	0	132.81	2.8262	417.9	0
40	2-Methyl-1-butene	C$_5$H$_{10}$	70.134	563462	3.9091	0.39866	0	0	135.58	3.4072	465	0
41	2-Methyl-2-butene	C$_5$H$_{10}$	70.134	513359	3.9121	0.3634	0	0	139.39	3.4437	471	0
42	1,2-Butadiene	C$_4$H$_6$	54.092	590192	3.5220	0.395	0	0	136.95	3.0540	452	0
43	1,3-Butadiene	C$_4$H$_6$	54.092	106990	3.2580	0.373	0	0	164.25	2.7155	425.17	0
44	2-Methyl-1,3-butadiene	C$_5$H$_8$	68.119	78795	3.9310	0.425	0	0	127.27	3.4529	484	0
45	Acetylene	C$_2$H$_2$	26.038	74862	2.3795	0.375	0	0	192.4	1.6488	308.32	0
46	Methylacetylene	C$_3$H$_4$	40.065	74997	3.2775	0.3997	0	0	170.45	2.6297	402.39	0
47	Dimethylacetylene	C$_4$H$_6$	54.092	503173	3.8650	0.3737	0	0	240.91	2.9557	473.2	0
48	3-Methyl-1-butyne	C$_5$H$_8$	68.119	598232	3.7920	0.3565	0	0	183.45	3.1681	463.2	0
49	1-Pentyne	C$_5$H$_8$	68.119	627190	3.9540	0.3512	0	0	167.45	3.4025	481.2	0
50	2-Pentyne	C$_5$H$_8$	68.119	627214	4.4158	0.44347	0	0	163.83	3.7321	519	0
51	1-Hexyne	C$_6$H$_{10}$	82.145	693027	4.5740	0.3698	0	0	141.25	4.0640	516.2	0
52	2-Hexyne	C$_6$H$_{10}$	82.145	764352	4.9110	0.4392	0	0	183.65	4.1067	549	0

No.	Name	Formula										
53	3-Hexyne	C_6H_{10}	928494	82.145	4.8080	0.436	0	0	170.05	4.0831	544	0
54	1-Heptyne	C_7H_{12}	628717	96.172	5.0514	0.41163	0	0	192.22	4.2470	559	0
55	1-Octyne	C_8H_{14}	629050	110.199	5.6306	0.4148	0	0	193.55	4.7663	585	0
56	Vinylacetylene[1]	C_4H_4	689974	52.076	3.6490	0.4	0.043	0	173.15	2.9876	454	0
57	Cyclopentane	C_5H_{10}	287923	70.134	3.8900	0.361	0	0	179.28	3.3292	511.76	0
58	Methylcyclopentane	C_6H_{12}	96377	84.161	4.3600	0.38531	0	0	130.73	3.9118	532.79	0
59	Ethylcyclopentane	C_7H_{14}	1640897	98.188	4.8288	0.37809	0	0	134.71	4.3604	569.52	0
60	Cyclohexane	C_6H_{12}	110827	84.161	4.4940	0.3974	0	0	279.69	3.3977	553.58	0
61	Methylcyclohexane	C_7H_{14}	108872	98.188	4.7534	0.39461	0	0	146.58	4.2295	572.19	0
62	1,1-Dimethylcyclohexane	C_8H_{16}	590669	112.215	5.0402	0.4036	0	0	239.66	4.0862	591.15	0
63	Ethylcyclohexane	C_8H_{16}	1678917	112.215	5.3832	0.41763	0	0	161.84	4.7318	609.15	0
64	Cyclopentene	C_5H_8	142290	68.119	3.8107	0.3543	0	0	138.13	3.4046	507	0
65	1-Methylcyclopentene	C_6H_{10}	693590	82.145	4.3541	0.36805	0	0	146.62	3.8769	542	0
66	Cyclohexene	C_6H_{10}	110838	82.145	4.4405	0.37479	0	0	169.67	3.8791	560.4	0
67	Benzene	C_6H_6	71432	78.114	4.7500	0.45238	0.0534	-0.1181	278.68	3.4909	562.16	0
68	Toluene	C_7H_8	108883	92.141	5.0144	0.3859	0	0	178.18	4.3670	591.8	0
69	o-Xylene	C_8H_{10}	95476	106.167	5.5330	0.377	0	0	247.98	4.5826	630.33	0
70	m-Xylene	C_8H_{10}	108383	106.167	5.4600	0.3726	0	0	225.3	4.6097	617.05	0
71	p-Xylene	C_8H_{10}	106423	106.167	5.3740	0.3656	0	0	286.41	4.2761	616.23	0
72	Ethylbenzene	C_8H_{10}	100414	106.167	5.4640	0.392	0	0	178.15	4.7811	617.2	0
73	Propylbenzene	C_9H_{12}	103651	120.194	5.7663	0.3956	-8.9129E-03	0	215.03	5.0574	574.54	2.4695E+07
74	1,2,4-Trimethylbenzene	C_9H_{12}	95636	120.194	5.9126	0.35632	0	0	229.33	5.0621	649.13	0
75	Isopropylbenzene	C_9H_{12}	98928	120.194	5.7950	0.3956	0	0	177.14	5.0869	631.1	0
76	1,3,5-Trimethylbenzene	C_9H_{12}	108678	120.194	6.0380	0.37999	0	0	228.42	5.1010	637.36	0
77	p-Isopropyltoluene	$C_{10}H_{14}$	99576	134.221	6.3314	0.40289	0	0	205.25	5.4387	653.15	0
78	Naphthalene	$C_{10}H_8$	91203	128.174	7.0510	0.4612	0	0	353.43	5.2508	748.35	0
79	Biphenyl	$C_{12}H_{10}$	92524	154.211	7.5736	0.3975	0	0	342.2	6.0420	789.26	0
80	Styrene	C_8H_8	100425	104.152	5.7260	0.4055	0	0	242.54	4.7128	636	0
81	m-Terphenyl	$C_{18}H_{14}$	92068	230.309	10.1230	0.3767	0	0	360	8.4070	924.85	0
82	Methanol	CH_4O	67561	32.042	5.2390	0.3682	0	0	175.47	4.4900	512.64	0
83	Ethanol	C_2H_6O	64175	46.069	5.6900	0.3359	0	0	159.05	5.0245	513.92	0
84	1-Propanol	C_3H_8O	71238	60.096	6.3300	0.3575	0.2915	0	146.95	5.6460	536.78	0
85	1-Butanol	$C_4H_{10}O$	71363	74.123	6.7390	0.173	0	0	184.51	6.0575	563.05	0
86	2-Butanol	$C_4H_{10}O$	78922	74.123	7.2560	0.4774	0	0	158.45	6.1383	536.05	0
87	2-Propanol	C_3H_8O	67630	60.096	6.3080	0.3921	0	0	185.28	5.2307	508.3	0
88	2-Methyl-2-propanol	$C_4H_{10}O$	75650	74.123	7.7320	0.5645	0	0	298.97	4.6703	506.21	0
89	1-Pentanol	$C_5H_{12}O$	71410	88.150	8.3100	0.511	0	0	195.56	6.7533	586.15	0
90	2-Methyl-1-butanol	$C_5H_{12}O$	137326	88.150	7.7839	0.45313	0	0	203	6.3619	565	0
91	3-Methyl-1-butanol	$C_5H_{12}O$	123513	88.150	8.0815	0.50185	0	0	155.95	6.8999	577.2	0
92	1-Hexanol	$C_6H_{14}O$	111273	102.177	8.5980	0.513	0	0	228.55	6.7623	611.35	0
93	1-Heptanol	$C_7H_{16}O$	111706	116.203	9.6900	0.572	0	0	239.15	7.3822	631.9	0
94	Cyclohexanol	$C_6H_{12}O$	108930	100.161	9.2440	0.64825	0	0	296.6	6.2273	650	0
95	Ethylene glycol	$C_2H_6O_2$	107211	62.068	8.2900	0.4266	0	0	260.15	6.8461	719.7	0
96	1,2-Propylene glycol	$C_3H_8O_2$	57556	76.095	8.0700	0.295	0	0	213.15	7.1374	626	0
97	Phenol	C_6H_6O	108952	94.113	7.3060	0.4246	0	0	314.06	5.6577	694.25	0
98	o-Cresol	C_7H_8O	95487	108.140	7.1979	0.40317	0	0	304.19	5.7135	697.55	0
99	m-Cresol	C_7H_8O	108394	108.140	8.0082	0.45314	0	0	285.39	6.3326	705.85	0
100	p-Cresol	C_7H_8O	106445	108.140	8.4942	0.50234	0	0	307.93	6.3649	704.65	0
101	Dimethyl ether	C_2H_6O	115106	46.069	2.9940	0.3505	0	0	131.65	2.6032	400.1	0
102	Methyl ethyl ether	C_3H_8O	540670	60.096	3.5300	0.376	0	0	160	2.9751	437.8	0
103	Methyl n-propyl ether	$C_4H_{10}O$	557175	74.123	3.9795	0.3729	0	0	133.97	3.5184	476.3	0
104	Methyl isopropyl ether	$C_4H_{10}O$	598538	74.123	3.9305	0.3711	0	0	127.93	3.4876	464.5	0
105	Methyl-n-butyl ether	$C_5H_{12}O$	628284	88.150	4.5328	0.3824	0	0	157.48	3.9358	510	0
106	Methyl isobutyl ether	$C_5H_{12}O$	625445	88.150	4.2678	0.37995	0	0	150	3.7232	497	0
107	Methyl tert-butyl ether	$C_5H_{12}O$	1634044	88.150	4.2024	0.37826	0	0	164.55	3.6096	497.1	0

TABLE 2-193 Heats of Vaporization of Inorganic and Organic Compounds (*Continued*)

Cmpd. no.	Name	Formula	CAS no.	Mol wt	C1 ×1E-07	C2	C3	C4	T_{min} K	ΔH_t at T_{min} ×1E-07	T_{max} K	ΔH_v at T_{max}
108	Diethyl ether	C₄H₁₀O	60297	74.123	4.0600	0.3868	0	0	156.85	3.4651	466.7	0
109	Ethyl propyl ether	C₅H₁₂O	625320	88.150	5.4380	0.60624	0	0	145.65	4.4140	500.23	0
110	Ethyl isopropyl ether	C₅H₁₂O	625547	88.150	4.2580	0.37221	0	0	140	3.7556	489	0
111	Methyl phenyl ether	C₇H₈O	100663	108.140	5.8662	0.37127	0	0	235.65	4.9560	645.6	0
112	Diphenyl ether	C₁₂H₁₀O	101848	170.211	6.8243	0.30877	0	0	300.03	5.8546	766.8	0
113	Formaldehyde	CH₂O	50000	30.026	3.0760	0.2954	0	0	181.15	2.5863	408	0
114	Acetaldehyde	C₂H₄O	75070	44.053	4.6070	0.62	0	0	150.15	3.6199	466	0
115	1-Propanal	C₃H₆O	123386	58.080	4.1492	0.36751	0	0	170	3.5675	504.4	0
116	1-Butanal	C₄H₈O	123728	72.107	4.6403	0.3849	0	0	176.75	3.9797	537.2	0
117	1-Pentanal	C₅H₁₀O	110623	86.134	5.1478	0.37541	0	0	182	4.4502	566.1	0
118	1-Hexanal	C₆H₁₂O	66251	100.161	5.6661	0.38533	0	0	217.15	4.7495	591	0
119	1-Heptanal	C₇H₁₄O	111717	114.188	6.1299	0.37999	0	0	229.8	5.1353	617	0
120	1-Octanal	C₈H₁₆O	124130	128.214	6.8347	0.41039	0	0	246	5.5966	638.1	0
121	1-Nonanal	C₉H₁₈O	124196	142.241	7.3363	0.41735	0	0	255.15	5.9779	658	0
122	1-Decanal	C₁₀H₂₀O	112312	156.268	7.9073	0.4129	0	0	267.15	6.4201	674.2	0
123	Acetone	C₃H₆O	67641	58.080	4.2150	0.3397	0	0	178.45	3.6390	508.2	0
124	Methyl ethyl ketone	C₄H₈O	78933	72.107	4.6220	0.355	0	0	186.48	3.9704	535.5	0
125	2-Pentanone	C₅H₁₀O	107879	86.134	5.1740	0.39422	0	0	196.29	4.3663	561.08	0
126	Methyl isopropyl ketone	C₅H₁₀O	563804	86.134	5.1400	0.3858	0	0	250	4.0753	553	0
127	2-Hexanone	C₆H₁₂O	591786	100.161	5.6770	0.3817	0	0	217.35	4.7584	587.05	0
128	Methyl isobutyl ketone	C₆H₁₂O	108101	100.161	5.4000	0.383	0	0	189.15	4.6294	571.4	0
129	3-Methyl-2-pentanone	C₆H₁₂O	565617	100.161	5.1130	0.3395	0	0	167.15	4.5480	573	0
130	3-Pentanone	C₅H₁₀O	96220	86.134	5.2359	0.40465	0	0	234.18	4.2075	560.95	0
131	Ethyl isopropyl ketone	C₆H₁₂O	565695	100.161	5.3580	0.40616	0	0	200	4.5154	567	0
132	Diisopropyl ketone	C₇H₁₄O	565800	114.188	5.5980	0.3774	0	0	204.81	4.7426	576	0
133	Cyclohexanone	C₆H₁₀O	108941	98.145	5.5500	0.3538	0	0	242	4.7114	653	0
134	Methyl phenyl ketone	C₈H₈O	98862	120.151	6.6104	0.37425	0	0	292.81	5.4166	709.5	0
135	Formic acid	CH₂O₂	64186	46.026	2.3700	1.999	-5.1503	3.331	281.45	1.9532	588	0
136	Acetic acid	C₂H₄O₂	64197	60.053	2.0265	0.11911	-1.3487	1.4227	289.81	2.3185	591.95	0
137	Propionic acid	C₃H₆O₂	79094	74.079	2.7290	0.06954	-1.0423	1.1152	252.45	2.9964	600.81	0
138	n-Butyric acid	C₄H₈O₂	107926	88.106	7.4996	2.333	-3.8644	2.016	267.95	4.1566	615.7	0
139	Isobutyric acid	C₄H₈O₂	79312	88.106	4.4967	1.1615	-2.4573	1.5823	227.15	3.6179	605	0
140	Benzoic acid[2]	C₇H₆O₂	65850	122.123	10.1900	0.478	0	0	395.45	7.1277	751	0
141	Acetic anhydride	C₄H₆O₃	108247	102.090	6.3520	0.3986	0	0	200.15	5.4139	606	0
142	Methyl formate	C₂H₄O₂	107313	60.053	4.1030	0.3625	0	0	174.15	3.4644	487.2	0
143	Methyl acetate	C₃H₆O₂	79209	74.079	4.4920	0.3685	0	0	175.15	3.8418	506.55	0
144	Methyl propionate	C₄H₈O₂	554121	88.106	5.0080	0.3959	0	0	185.65	4.2231	530.6	0
145	Methyl n-butyrate	C₅H₁₀O₂	623427	102.133	5.3781	0.39523	0	0	187.35	4.5694	554.5	0
146	Ethyl formate	C₃H₆O₂	109944	74.079	4.5909	0.4123	0	0	193.55	3.7679	508.4	0
147	Ethyl acetate	C₄H₈O₂	141786	88.106	4.9330	0.3847	0	0	189.6	4.1490	523.3	0
148	Ethyl propionate	C₅H₁₀O₂	105373	102.133	5.3325	0.401	0	0	199.25	4.4449	546	0
149	Ethyl n-butyrate	C₆H₁₂O₂	105544	116.160	5.6419	0.37985	0	0	175.15	4.9090	571	0
150	n-Propyl formate	C₄H₈O₂	110747	88.106	4.9687	0.4025	0	0	180.25	4.2162	538	0
151	n-Propyl acetate	C₅H₁₀O₂	109604	102.133	5.4327	0.407	0	0	178.15	4.6322	549.73	0
152	n-Butyl acetate	C₆H₁₂O₂	123864	116.160	5.7800	0.3935	0	0	199.65	4.8943	579.15	0
153	Methyl benzoate	C₈H₈O₂	93583	136.150	6.9650	0.4061	0	0	260.75	5.7500	693	0
154	Ethyl benzoate	C₉H₁₀O₂	93890	150.177	6.3400	0.2911	0	0	238.45	5.6137	698	0
155	Vinyl acetate	C₄H₆O₂	108054	86.090	4.7700	0.3765	0	0	180.35	4.0619	519.13	0
156	Methylamine	CH₅N	74895	31.057	3.8580	0.404	0	0	179.69	3.1006	430.05	0
157	Dimethylamine	C₂H₇N	124403	45.084	4.0900	0.42005	0	0	180.96	3.2678	437.2	0
158	Trimethylamine	C₃H₉N	75503	59.111	3.3050	0.354	0	0	156.08	2.8216	433.25	0

No.	Name	Formula	No.	Mol. wt.									
159	Ethylamine	C_2H_7N	75047	45.084	4.2750	0.5857	-0.332	0.169	192.15	3.2955	456.15	0	0
160	Diethylamine	$C_4H_{11}N$	109897	73.138	4.6133	0.42628	0	0	223.35	3.5761	496.6	0	0
161	Triethylamine	$C_6H_{15}N$	121448	101.192	4.6640	0.3663	0	0	158.45	4.1011	535.15	0	0
162	n-Propylamine	C_3H_9N	107108	59.111	4.4488	0.39494	0	0	188.36	3.6857	496.95	0	0
163	di-n-Propylamine	$C_6H_{15}N$	142847	101.192	5.4280	0.3665	0	0	210.15	4.5500	550	0	0
164	Isopropylamine	C_3H_9N	75310	59.111	4.4041	0.43325	0	0	177.95	3.5574	471.85	0	0
165	Diisopropylamine	$C_6H_{15}N$	108189	101.192	5.0070	0.4362	0	0	176.85	4.1823	523.1	0	0
166	Aniline	C_6H_7N	62533	93.128	7.1950	0.458	0	0	267.13	5.7710	699	0	0
167	N-Methylaniline	C_7H_9N	100618	107.155	6.3360	0.3104	0	0	216.15	5.6961	701.55	0	0
168	N,N-Dimethylaniline	$C_8H_{11}N$	121697	121.182	6.7900	0.4053	0	0	275.6	5.5162	687.15	0	0
169	Ethylene oxide	C_2H_4O	75218	44.053	3.6652	0.37878	0	0	160.65	3.1271	469.15	0	0
170	Furan	C_4H_4O	110009	68.075	4.0050	0.3995	0	0	196.29	3.2647	490.15	0	0
171	Thiophene	C_4H_4S	110021	84.142	4.5793	0.38557	0	0	234.94	3.7472	579.35	0	0
172	Pyridine	C_5H_5N	110861	79.101	5.1740	0.38865	0	0	231.51	4.3144	619.95	0	0
173	Formamide[3]	CH_3NO	75127	45.041	7.3580	0.3564	0	0	275.7	6.2844	771	0	0
174	N,N-Dimethylformamide	C_3H_7NO	68122	73.095	5.9217	0.37996	0	0	212.72	5.0931	649.6	0	0
175	Acetamide	C_2H_5NO	60355	59.068	8.1070	0.42	0	0	353.15	6.2386	761	0	0
176	N-Methylacetamide	C_3H_7NO	79163	73.095	7.3402	0.38974	0	0	301.15	5.9384	718	0	0
177	Acetonitrile	C_2H_3N	75058	41.053	4.3511	0.34765	0	0	229.32	3.5996	545.5	0	0
178	Propionitrile	C_3H_5N	107120	55.079	4.9348	0.41873	0	0	180.26	4.2005	564.4	0	0
179	n-Butyronitrile	C_4H_7N	109740	69.106	5.2200	0.165	-0.539	0.6692	161.25	4.7223	582.25	0	0
180	Benzonitrile	C_7H_5N	100470	103.123	6.2615	0.35427	0	0	260.4	5.3091	699.35	0	0
181	Methyl mercaptan	CH_4S	74931	48.109	3.4448	0.37427	0	0	150.18	2.9825	469.95	0	0
182	Ethyl mercaptan	C_2H_6S	75081	62.136	3.8440	0.37534	0	0	125.26	3.3489	499.15	0	0
183	n-Propyl mercaptan	C_3H_8S	107039	76.163	4.4782	0.41073	0	0	159.95	3.8723	536.6	0	0
184	n-Butyl mercaptan	$C_4H_{10}S$	109795	90.189	4.9702	0.41199	0	0	157.46	4.3505	570.1	0	0
185	Isobutyl mercaptan	$C_4H_{10}S$	513440	90.189	4.7420	0.40535	0	0	128.31	4.2664	559	0	0
186	sec-Butyl mercaptan	$C_4H_{10}S$	513531	90.189	4.6432	0.399	0	0	133.02	4.1614	554	0	0
187	Dimethyl sulfide	C_2H_6S	75183	62.136	3.8690	0.3694	0	0	174.88	3.3042	503.04	0	0
188	Methyl ethyl sulfide	C_3H_8S	624895	76.163	4.4740	0.4097	0	0	167.23	3.3844	533	0	0
189	Diethyl sulfide	$C_4H_{10}S$	352932	90.189	4.7182	0.3643	0	0	169.2	4.1353	557.15	0	0
190	Fluoromethane	CH_3F	593533	34.033	2.4708	0.37014	0	0	131.35	2.0276	317.42	0	0
191	Chloromethane	CH_3Cl	74873	50.488	2.9745	0.353	0	0	175.43	2.4520	416.25	0	0
192	Trichloromethane	$CHCl_3$	67663	119.377	4.1860	0.3584	0	0	209.63	3.5047	536.4	0	0
193	Tetrachloromethane	CCl_4	56235	153.822	4.3252	0.37688	0	0	250.33	3.4528	556.35	0	0
194	Bromomethane	CH_3Br	74839	94.939	3.1690	0.3015	0	0	179.47	2.7379	467	0	0
195	Fluoroethane	C_2H_5F	353366	48.060	2.7617	0.32162	0	0	129.95	2.4089	375.31	0	0
196	Chloroethane	C_2H_5Cl	75003	64.514	3.5240	0.3652	0	0	134.8	3.1052	460.35	0	0
197	Bromoethane	C_2H_5Br	74964	108.966	3.9004	0.38012	0	0	154.55	3.3933	503.8	0	0
198	1-Chloropropane	C_3H_7Cl	540545	78.541	3.9890	0.37956	0	0	150.35	3.4862	503.15	0	0
199	2-Chloropropane	C_3H_7Cl	75296	78.541	3.8871	0.38043	0	0	155.97	3.3586	489	0	0
200	1,1-Dichloropropane	$C_3H_6Cl_2$	78999	112.986	4.7740	0.39204	0	0	200	4.0147	560	0	0
201	1,2-Dichloropropane	$C_3H_6Cl_2$	78875	112.986	4.6750	0.36529	0	0	172.71	4.0997	572	0	0
202	Vinyl chloride	C_2H_3Cl	75014	62.499	3.4125	0.4513	0	0	119.36	2.9491	432	0	0
203	Fluorobenzene	C_6H_5F	462066	96.104	4.5820	0.3717	0	0	230.94	3.7605	560.09	0	0
204	Chlorobenzene	C_6H_5Cl	108907	112.558	5.1480	0.36614	0	0	227.95	4.3707	632.35	0	0
205	Bromobenzene	C_6H_5Br	108861	157.010	5.5520	0.37694	0	0	242.43	4.6875	670.15	0	0
206	Air		132259100	28.951	0.8474	0.3822	-1.817	1.447	59.15	0.6759	132.45	0	0
207	Hydrogen	H_2	1333740	2.016	0.1013	0.698	-2.6954	1.7098	13.95	0.0913	33.19	0	0
208	Helium-4	He	7440597	4.003	0.0125	1.3038	0	0	2.2	0.0097	5.2	0	0
209	Neon	Ne	7440019	20.180	0.2389	0.3494	0	0	24.56	0.1803	44.4	0	0
210	Argon	Ar	7440371	39.948	0.8731	0.3526	0	0	83.78	0.6561	150.86	0	0
211	Fluorine	F_2	7782414	37.997	0.8876	0.34072	-0.9001	0.453	53.48	0.7578	144.12	0	0
212	Chlorine	Cl_2	7782505	70.905	0.3680	0.8458	0	0	172.12	2.2878	417.15	0	0
213	Bromine	Br_2	7726956	159.808	4.0000	0.351	0	0	265.85	3.2323	554.15	0	0

TABLE 2-193 Heats of Vaporization of Inorganic and Organic Compounds (*Concluded*)

Cmpd. no.	Name	Formula	CAS no.	Mol wt	C1 ×1E-07	C2	C3	C4	T_{min} K	ΔH_v at T_{min} ×1E-07	T_{max} K	ΔH_v at T_{max}
214	Oxygen	O_2	7782447	31.999	0.9008	0.4542	−0.4096	0.3183	54.36	0.7742	154.58	0
215	Nitrogen	N_2	7727379	28.014	0.7491	0.40406	−0.317	0.27343	63.15	0.6024	126.2	0
216	Ammonia	NH_3	7664417	17.031	3.1523	0.3914	−0.2289	0.2309	195.41	2.5298	405.65	0
217	Hydrazine	N_2H_4	302012	32.045	5.9794	0.9424	−1.398	0.8862	274.69	4.5238	653.15	0
218	Nitrous oxide	N_2O	10024972	44.013	2.3215	0.384		0	182.3	1.6502	309.57	0
219	Nitric oxide	NO	10102439	30.006	2.1310	0.4056	0	0	109.5	1.4578	180.15	0
220	Cyanogen	C_2N_2	460195	52.036	3.3840	0.3707	0	0	245.25	2.3803	400.15	915280
221	Carbon monoxide	CO	630080	28.010	0.8555	0.4921	−0.326	0.2231	68.13	0.6517	132.5	0
222	Carbon dioxide	CO_2	124389	44.010	2.1730	0.382	−0.4339	0.42213	216.58	1.5202	304.21	0
223	Carbon disulfide	CS_2	75150	76.143	3.4960	0.2986	0	0	161.11	3.1537	552	0
224	Hydrogen fluoride	HF	7664393	20.006	13.4510	13.36	−23.383	10.785	277.56	0.7104	461.15	0
225	Hydrogen chloride	HCl	7647010	36.461	2.2093	0.3466	0	0	158.97	1.7498	324.65	0
226	Hydrogen bromide	HBr	10035106	80.912	2.4850	0.39	0	0	185.15	1.8817	363.15	0
227	Hydrogen cyanide[2]	HCN	74908	27.026	3.3490	0.2053	0	0	259.83	2.8176	456.65	0
228	Hydrogen sulfide	H_2S	7783064	34.082	2.5676	0.37358	0	0	187.68	1.9782	373.53	0
229	Sulfur dioxide	SO_2	7446095	64.065	3.6760	0.4	0	0	197.67	2.8753	430.75	0
230	Sulfur trioxide	SO_3	7446119	80.064	7.3370	0.5647	0	0	289.95	4.4303	490.85	0
231	Water	H_2O	7732185	18.015	5.2053	0.3199	−0.212	0.25795	273.16	4.4733	647.13	0

All substances are listed in alphabetical order in Table 2-6*a*.
Compiled from Daubert, T. E.; R. P. Danner, H. M. Sibul, and C. C. Stebbins, DIPPR Data Compilation of Pure Compound Properties, Project 801 Sponsor Release, July, 1993, Design Institute for Physical Property Data, AIChE, New York, NY; and from Thermodynamics Research Center, "Selected Values of Properties of Hydrocarbons and Related Compounds," Thermodynamics Research Center Hydrocarbon Project, Texas A&M University, College Station, Texas (extant 1994).
Temperatures are expressed in kelvins; heats of vaporization, in J/kmol.
J/kmol × 2.390E−04 = cal/gmol; J/kmol × 4.302106E−04 = Btu/lbmol.
The heat of vaporization equation used is $\Delta H_v = C1 \times (1 - T_r)^{C2 + C3 \times T_r + C4 \times T_r \times T_r}$. T_r is the reduced temperature, T/T_c.
[1] Coefficients are hypothetical; compound *decomposes violently* on heating.
[2] For the monomer.
[3] Equation coefficients are hypothetical above the decomposition temperature.

SPECIFIC HEATS OF PURE COMPOUNDS

UNITS CONVERSIONS

For this subsection, the following units conversions are applicable:

$°F = \frac{9}{5} °C + 32$

$°F = 1.8 K$

To convert calories per gram-kelvin to British thermal units per pound-degree Rankine, multiply by 1.0; to convert calories per gram-mole-kelvin to British thermal units per pound-mole-degree Rankine, multiply by 1.0.

To convert kilojoules per kilogram-kelvin to British thermal units per pound-degree Rankine, multiply by 0.2388.

ADDITIONAL REFERENCES

Additional data are contained in the subsection "Thermodynamic Properties." Data on water are also contained in that subsection. Additional tables for water are found in Eng. Sci. Data Item 68008, 251 Regent Street, London, England, which contains about 5000 values from 1 to 1000 bar, 0 to 1500°C.

TABLE 2-194   Heat Capacities of the Elements and Inorganic Compounds*

Substance	State†	Heat capacity at constant pressure $(T = K; 0°C = 273.1 K)$, cal/deg mol	Range of temperature, K	Uncertainty, %
Aluminum[1]				
Al	c	$4.80 + 0.00322T$	273–931	1
	l	7.00	931–1273	5
AlBr$_3$	c	$18.74 + 0.01866T$	273–370	3
	l	29.5	370–407	5
AlCl$_3$	c	$13.25 + 0.02800T$	273–465	3
	l	31.2	465–504	3
AlCl$_3$·6H$_2$O	c	76	288–327	?
AlF$_3$	c	19.3	288–326	?
AlF$_3$·3½H$_2$O	c	50.5	288–326	?
AlF$_3$·3NaF	c	$38.63 + 0.04760T - 449200/T^2$	273–1273	2
	l	142	1273–1373	?
AlI$_3$	c	$16.88 + 0.02266T$	273–464	3
	l	28.8	464–480	5
Al$_2$O$_3$	c	$22.08 + 0.008971T - 522500/T^2$	273–1973	3
Al$_2$O$_3$·SiO$_2$	c, sillimanite	$40.79 + 0.004763T - 992800/T^2$	273–1573	3
	c, disthene	$41.81 + 0.005283T - 1211000/T^2$	273–1673	2
	c, andalusite	$43.96 + 0.001923T - 1086000/T^2$	273–1573	3
3Al$_2$O$_3$·2SiO$_2$	c, mullite	$59.65 + 0.0670T$	273–576	5
4Al$_2$O$_3$·3SiO$_2$	c	$113.2 + 0.0652T$	273–575	3
Al$_2$(SO$_4$)$_3$	c	63.5	273–373	?
Al$_2$(SO$_4$)$_3$·18H$_2$O	c	235	288–325	?
Antimony				
Sb	c	$5.51 + 0.00178T$	273–903	2
	l	7.15	903–1273	5
SbBr$_3$	c	$17.2 + 0.0293T$	273–370	?
SbCl$_3$	c	$10.3 + 0.0511T$	273–346	?
Sb$_2$O$_3$	c	$19.1 + 0.0171T$	273–929	?
Sb$_2$O$_4$	c	$22.6 + 0.0162T$	273–1198	?
Sb$_2$S$_3$	c	$24.2 + 0.0132T$	273–821	?
Argon[2]				
A	g	4.97	All	0
Arsenic				
As	c	$5.17 + 0.00234T$	273–1168	5
AsCl$_3$	l	31.9	286–371	?
As$_2$O$_3$	c	$8.37 + 0.0486T$	273–548	?
As$_2$S$_3$	c	25.8	293–373	?
Barium				
BaCl$_2$	c	$17.0 + 0.00334T$	273–1198	?
BaCl$_2$·H$_2$O	c	28.2	273–307	?
BaCl$_2$·2H$_2$O	c	37.3	273–307	?
Ba(ClO$_3$)$_2$·H$_2$O	c	51	289–320	?
BaCO$_3$	c, α	$17.26 + 0.0131T$	273–1083	5
	c, β	30.0	1083–1255	15
BaMoO$_4$	c	34	273–297	?
Ba(NO$_3$)$_2$	c	39.8	285–371	?
BaSO$_4$	c	$21.35 + 0.0141T$	273–1323	5
Beryllium[3,4]				
Be	c	$4.698 + 0.001555T - 121000/T^2$	273–1173	1
BeO	c	$8.69 + 0.00365T - 313000/T^2$	273–1175	5
BeO·Al$_2$O$_3$	c	25.4	273–373	?
BeSO$_4$	c	20.8	273–373	?

*From Kelley, U.S. Bur. Mines Bull. 371, 1934. For a revision see Kelley, U.S. Bur. Mines Bull. 477, 1948. Data for many elements and compounds are given by Johnson (ed.), WADD-TR-60-56, 1960, for cryogenic temperatures. Tabulated data for gases can be obtained from many of the references cited in the "Thermodynamic Properties" subsection and other tables in this section. Thinh, Duran, et al., *Hydrocarbon Process.*, **50**, 98 (January 1971), review previous equation fits and give newer fits for 408 hydrocarbons and related compounds. Later publications include Duran, Thinh, et al., *Hydrocarbon Process.*, **55**, 153 (August 1976); Thompson, *J. Chem. Eng. Data*, **22**(4), 431 (1977); and Passut and Danner, *Ind. Eng. Chem. Process Des. Dev.*, **11**, 543 (1972); **13**, 193 (1974).

†The symbols in this column have the following meaning; *c*, crystal; *l*, liquid; *g*, gas; *gls*, glass.

Substance	State†	Heat capacity at constant pressure $(T = \text{K}; 0°\text{C} = 273.1 \text{ K})$, cal/deg mol	Range of temperature, K	Uncertainty, %
Bismuth[4]				
Bi	c	$5.38 + 0.00260T$	273–544	3
	l	7.60	544–1273	3
Bi_2O_3	c	$23.27 + 0.01105T$	273–777	2
Bi_2S_3	c	30.4	284–372	?
Boron				
B	c	$1.54 + 0.00440T$	273–1174	5
B_2O_3	gls	$5.14 + 0.0320T$	273–513	3
	gls	30.4	513–623	3
BN	c	$1.61 + 0.00400T$	273–1173	5
Bromine				
Br_2	g	9.00	300–2000	5
Cadmium				
Cd	c	$5.46 + 0.002466T$	273–594	1
	l	7.13	594–973	5
CdO	c	$9.65 + 0.00208T$	273–2086	?
CdS	c	$12.9 + 0.00090T$	273–1273	?
$CdSO_4 \cdot 8/3H_2O$	c	51.3	293	?
Calcium				
Ca	c	$5.31 + 0.00333T$	273–673	2
	c	$6.29 + 0.00140T$	673–873	2
$CaCl_2$	c	$16.9 + 0.00386T$	273–1055	?
$CaCO_3$	c	$19.68 + 0.01189T - 307600/T^2$	273–1033	3
CaF_2	c	$14.7 + 0.00380T$	273–1651	?
$CaMg(CO_3)_2$	c	40.1	299–372	?
$CaMoO_4$	c	33	273–297	?
CaO	c	$10.00 + 0.00484T - 108000/T^2$	273–1173	2
$Ca(OH)_2$	c	21.4	276–373	?
$CaO \cdot Al_2O_3 \cdot 2SiO_2$	c, anorthite	$63.13 + 0.01500T - 1537000/T^2$	273–1673	1
	gls	$67.41 + 0.01048T - 1874000/T^2$	273–973	1
$CaO \cdot MgO \cdot 2SiO_2$	c, diopside	$54.46 + 0.005746T - 1500000/T^2$	273–1573	1
	gls	$51.68 + 0.009724T - 1308000/T^2$	273–973	1
$CaO \cdot SiO_2$	c, wollastonite	$27.95 + 0.002056T - 745600/T^2$	273–1573	1
	c, pseudowollastonite	$25.48 + 0.004132T - 488100/T^2$	273–1673	1
	gls	$23.16 + 0.009672T - 487100/T^2$	273–973	1
CaP_2O_6	c	39.5	287–371	?
$CaSO_4$	c	$18.52 + 0.02197T - 156800/T^2$	273–1373	5
$CaSO_4 \cdot 2H_2O$	c	46.8	282–373	?
$CaWO_4$	c	27.9	292–322	?
Carbon[5]				
C	c, graphite	$2.673 + 0.002617T - 116900/T^2$	273–1373	2
	c, diamond	$2.162 + 0.003059T - 130300/T^2$	273–1313	3
CH_4	g	$5.34 + 0.0115T$	273–1200	2
CO^6	g	$6.60 + 0.00120T$	273–2500	1½
CO_2	g	$10.34 + 0.00274T - 195500/T^2$	273–1200	1½
CS_2	l	18.4	293	?
Cerium				
Ce	c	$5.88 + 0.00123T$	273–908	?
CeO_2	c	15.1	273–373	?
$Ce_2(MoO_4)_3$	c	96	273–297	?
$Ce_2(SO_4)_3$	c	66.4	273–373	?
$Ce_2(SO_4)_3 \cdot 5H_2O$	c	131.6	273–319	?
Cesium				
Cs	c	$1.96 + 0.0182T$	273–301	3
	l	8.00	302	3
	g	4.97	All	0
CsBr	c	$12.6 + 0.00259T$	273–909	?
CsCl	c	$11.7 + 0.00309T$	273–752	?
CsF	c	$11.3 + 0.00285T$	273–957	?
CsI	c	$11.6 + 0.00268T$	273–894	?
Chlorine				
Cl_2	g	$8.28 + 0.00056T$	273–2000	1½
Chromium[4]				
Cr	c	$4.84 + 0.00295T$	273–1823	5
	l	9.70	1823–1923	10
$CrCl_3$	c	23	286–319	?
Cr_2O_3	c	$26.0 + 0.00400T$	273–2263	?
CrSb	c	$12.3 + 0.00120T$	273–1383	?
$CrSb_2$	c	$19.2 + 0.00184T$	273–949	?
$Cr_2(SO_4)_3$	c	67.4	273–373	?
Cobalt[4]				
Co	c	$5.12 + 0.00333T$	273–1763	5
	l	8.40	1763–1873	5
$CoAs_2 \cdot CoS_2$	c	32.9	283–373	?
CoSb	c	$11.7 + 0.00156T$	273–1464	?
Co_2Sn	c	$15.83 + 0.00950T$	273–903	2
CoS	c	$10.6 + 0.00251T$	273–1373	?
$CoSO_4 \cdot 7H_2O$	c	96	286–303	?

Substance	State†	Heat capacity at constant pressure (T = K; 0°C = 273.1 K), cal/deg mol	Range of temperature, K	Uncertainty, %
Copper[7]				
Cu	c	$5.44 + 0.001462T$	273–1357	1
	l	7.50	1357–1573	3
CuAl	c	$9.88 + 0.00500T$	273–733	2
CuAl$_2$	c	$16.78 + 0.00366T$	273–773	2
Cu$_3$Al	c	$19.61 + 0.01054T$	273–775	2
CuI	c	$12.1 + 0.00286T$	273–675	?
CuI$_2$	c	20.1	274–328	?
CuO	c	$10.87 + 0.003576T - 150600/T^2$	273–810	2
CuO·SiO$_2$·H$_2$O	c	29	293–323	?
CuS	c	$10.6 + 0.00264T$	273–1273	?
Cu$_2$S	c, α	$9.38 + 0.0312T$	273–376	3
	c, β	20.9	376–1173	2
CuS·FeS	c	24	292–321	?
Cu$_2$Sb	c	$13.73 + 0.01350T$	273–573	2
Cu$_2$Sb	c	$21.79 + 0.00900T$	273–693	2
Cu$_2$Se	c, α	20.85	273–383	5
	c, β	20.35	383–488	5
Cu$_3$Si	c	$20.3 + 0.00587T$	273–1135	?
CuSO$_4$	c	24.1	282	?
CuSO$_4$·H$_2$O	c	31.3	282	?
CuSO$_4$·3H$_2$O	c	49.0	282	?
CuSO$_4$·5H$_2$O	c	67.2	282	?
Fluorine[8]				
F$_2$	g	$6.50 + 0.00100T$	300–3000	5
Gallium				
Ga$_2$O$_3$	c	$18.2 + 0.0252T$	273–923	?
Ga$_2$(SO$_4$)$_3$	c	62.4	273–373	?
Germanium[4]				
Ge	c			
Gold				
Au	c	$5.61 + 0.00144T$	273–1336	2
	l	7.00	1336–1573	5
AuSb$_2$	c, α	$17.12 + 0.00465T$	273–628	1
	$c, \beta\gamma$	$11.47 + 0.01756T$	628–713	?
Helium[9]				
He	g	4.97	All	0
Hydrogen[10]				
H	g	4.97	All	0
H$_2$	g	$6.62 + 0.00081T$	273–2500	2
HBr	g	$6.80 + 0.00084T$	273–2000	2
HCl	g	$6.70 + 0.00084T$	273–2000	1½
HI	g	$6.93 + 0.00083T$	273–2000	2
H$_2$O	l	See Tables 2-355 through 2-357		?
	g	$8.22 + 0.00015T + 0.00000134T^2$	300–2500	?
H$_2$S	g	$7.20 + 0.00360T$	300–600	8
H$_2$S$_2$O$_7$	c	27	281	?
	l	58	308	?
Indium				
In	c			
Iodine				
I$_2$	g	9.00	300–2000	5
Iridium				
Ir	c	$5.50 + 0.00148T$	273–1873	1
Iron[4]				
Fe	c, α	$4.13 + 0.00638T$	273–1041	3
	c, β	$6.12 + 0.00336T$	1041–1179	3
	c, γ	8.40	1179–1674	5
	c, δ	10.0	1674–1803	5
	l	8.15	1803–1873	5
FeAs$_2$	c	17.8	283–373	?
Fe$_3$C	c	$25.17 + 0.00223T$	273–1173	10
FeCO$_3$	c	22.7	293–368	?
FeO	c	$12.62 + 0.001492T - 76200/T^2$	273–1173	2
Fe$_2$O$_3$	c	$24.72 + 0.01604T - 423400/T^2$	273–1097	2
Fe$_3$O$_4$	c	$41.17 + 0.01882T - 979500/T^2$	273–1065	2
Fe$_2$O$_3$·3H$_2$O	c	47.8	286–373	?
FeS	c, α	$2.03 + 0.0390T$	273–411	5
	c, β	$12.05 + 0.00273T$	411–1468	3
FeS$_2$	c	$10.7 + 0.01336T$	273–773	?
FeSi	c	$10.54 + 0.00458T$	273–903	2
Fe$_2$SiO$_4$	c	$33.57 + 0.01907T - 879700/T^2$	273–1161	2
FeSO$_4$	c	22	293–373	?
Fe$_2$(SO$_4$)$_3$	c	66.2	273–373	?
FeSO$_4$·4H$_2$O	c	63.6	282	?
FeSO$_4$·7H$_2$O	c	96	291–319	?
Krypton				
Kr	g	4.97	All	0

TABLE 2-194   Heat Capacities of the Elements and Inorganic Compounds (*Continued*)

Substance	State†	Heat capacity at constant pressure (T = K; 0°C = 273.1 K), cal/deg mol	Range of temperature, K	Uncertainty, %
Lanthanum				
La	c	$5.91 + 0.00100T$	273–1009	?
La_2O_3	c	$22.6 + 0.00544T$	273–2273	?
$La_2(MoO_4)_3$	c	86	273–307	?
$La_2(SO_4)_3$	c	66.9	273–373	?
$La_2(SO_4)_3 \cdot 9H_2O$	c	152	273–319	?
Lead[4]				
Pb	c	$5.77 + 0.00202T$	273–600	2
	l	6.8	600–1273	5
$Pb_3(AsO_4)_2$	c	65.5	286–370	?
PbB_2O_4	c	26.5	288–371	?
PbB_4O_7	c	41.4	289–371	?
$PbBr_2$	c	$18.13 + 0.00310T$	273–761	2
	l	27.4	761–860	10
$PbCl_2$	c	$15.88 + 0.00835T$	273–771	2
	l	27.2	771–851	10
$2PbCl_2 \cdot NH_4Cl$	c	53.1	293	?
$PbCO_3$	c	21.1	286–320	?
$PbCrO_4$	c	29.1	292–323	?
PbF_2	c	$16.5 + 0.00412T$	273–1091	?
PbI_2	c	$18.66 + 0.00293T$	273–648	2
	l	32.3	648–776	20
$PbMoO_4$	c	30.4	292–322	?
$Pb(NO_3)_2$	c	36.4	286–320	?
PbO	c	$10.33 + 0.00318T$	273–544	2
PbO_2	c	$12.7 + 0.00780T$	273–?	?
$Pb_3P_2O_7$	c	48.3	284–371	?
PbS	c	$10.63 + 0.00401T$	273–873	3
$PbSO_4$	c	26.4	293–372	?
PbS_2O_3	c	29	293–373	?
$PbWO_4$	c	35	273–297	?
Lithium				
Li	c	$0.68 + 0.0180T$	273–459	10
	g	4.97	All	0
LiBr	c	$11.5 + 0.00302T$	273–825	?
$LiBr \cdot H_2O$	c	22.6	278–318	?
LiCl	c	$11.0 + 0.00339T$	273–887	?
$LiCl \cdot H_2O$	c	23.6	279–360	?
LiF	c	$8.20 + 0.00520T$	273–1117	?
LiI	c	$12.5 + 0.00208T$	273–723	?
$LiI \cdot H_2O$	c	23.6	277–359	?
$LiI \cdot 2H_2O$	c	32.9	277–345	?
$LiI \cdot 3H_2O$	c	43.2	277–347	?
$LiNO_3$	c	$9.17 + 0.0360T$	273–523	5
	l	26.8	523–575	5
Magnesium[4]				
Mg	c	$6.20 + 0.00133T - 67800/T^2$	273–923	1
	l	7.4	923–1048	10
MgAg	c	$10.58 + 0.00412T$	273–905	2
Mg_4Al_3	c	$34.4 + 0.0198T$	273–736	?
MgAu	c	$11.3 + 0.00189T$	273–1433	?
Mg_2Au	c	$16.2 + 0.00451T$	273–1073	?
Mg_3Au	c	$21.2 + 0.00614T$	273–1103	?
$MgCl_2$	c	$17.3 + 0.00377T$	273–991	?
$MgCl_2 \cdot 6H_2O$	c	77.1	292–342	?
$MgCO_3$	c	16.9	290	?
$MgCu_2$	c	$14.96 + 0.00776T$	273–903	3
Mg_2Cu	c	$15.5 + 0.00652T$	273–843	?
$MgNi_2$	c	$15.87 + 0.00692T$	273–903	2
MgO	c	$10.86 + 0.001197T - 208700/T^2$	273–2073	2
$MgO \cdot Al_2O_3$	c	28	288–319	?
$MgO \cdot SiO_2$	c, amphibole	$25.60 + 0.004380T - 674200/T^2$	273–1373	1
	c, pyroxene	$23.35 + 0.008062T - 558800/T^2$	273–773	1
	gls	$23.30 + 0.007734T - 542000/T^2$	273–973	1
$6MgO \cdot MgCl_2 \cdot 8B_2O_3$	c, α	$58.7 + 0.408T$	273–538	5
	c, β	$107.2 + 0.2876T$	538–623	5
$Mg(OH)_2$	c	18.2	292–323	?
Mg_3Sb_2	c	$28.2 + 0.00560T$	273–1234	?
Mg_2Si	c	$15.4 + 0.00415T$	273–1343	?
$MgSO_4$	c	26.7	296–372	?
$MgSO_4 \cdot H_2O$	c	33	282	?
$MgSO_4 \cdot 6H_2O$	c	80	282	?
$MgSO_4 \cdot 7H_2O$	c	89	291–319	?

TABLE 2-194 **Heat Capacities of the Elements and Inorganic Compounds** (Continued)

Substance	State†	Heat capacity at constant pressure (T = K; 0°C = 273.1 K), cal/deg mol	Range of temperature, K	Uncertainty, %
Manganese				
Mn	c, α	$3.76 + 0.00747T$	273–1108	5
	c, β	$5.06 + 0.00395T$	1108–1317	5
	c, γ	$4.80 + 0.00422T$	1317–1493	5
	l	11.0	1493–1673	10
$MnCl_2$	c	$16.2 + 0.00520T$	273–923	?
$MnCO_3$	c	$7.79 + 0.0421T + 0.0000090T^2$	273–773	?
MnO	c	$7.43 + 0.01038T - 0.00000362T^2$	273–1923	?
Mn_2O_3	c	$10.33 + 0.0530T - 0.0000257T^2$	273–1173	?
Mn_3O_4	c	$19.25 + 0.0538T - 0.0000209T^2$	273–1773	?
MnO_2	c	$1.92 + 0.0471T - 0.0000297T^2$	273–773	?
$Mn_2O_3 \cdot H_2O$	c	31	291–322	?
MnS	c	$10.21 + 0.00656T - 0.00000242T^2$	273–1883	?
$MnSO_4$	c	27.5	293–373	?
$MnSO_4 \cdot 5H_2O$	c	78	290–319	?
Mercury[11]				
Hg	l	6.61	273–630	1
	g	4.97	All	0
Hg_2	g	9.00	300–2000	5
$HgCl$	c	$11.05 + 0.00370T$	273–798	?
$HgCl_2$	c	$15.3 + 0.0103T$	273–553	?
$Hg(CN)_2$	c	25	285–319	?
HgI	c	$11.4 + 0.00461T$	273–563	?
HgI_2	c, α	$17.4 + 0.004001T$	273–403	3
	c, β	20.2	403–523	3
HgO	c	11.5	278–371	?
HgS	c	$10.9 + 0.00365T$	273–853	?
Hg_2SO_4	c	31.0	273–307	?
Molybdenum				
Mo	c	$5.69 + 0.00188T - 50300/T^2$	273–1773	5
MoO_3	c	$15.1 + 0.0121T$	273–1068	?
MoS_2	c	$19.7 + 0.00315T$	273–729	?
Neon[12]				
Ne	g	4.97	All	0
Nickel[4]				
Ni	c, α	$4.26 + 0.00640T$	273–626	2
	c, β	$6.99 + 0.000905T$	626–1725	5
	l	8.55	1725–1903	10
NiO	c	$11.3 + 0.00215T$	273–1273	?
NiS	c	$9.25 + 0.00640T$	273–597	3
Ni_2Si	c	$15.8 + 0.00329T$	273–1582	?
$NiSi$	c	$10.0 + 0.00312T$	273–1273	?
Ni_3Sn	c	$20.78 + 0.0102T$	273–904	2
$NiSO_4$	c	33.4	293–373	?
$NiSO_4 \cdot 6H_2O$	c	82	291–325	?
$NiTe$	c	$11.00 + 0.00433T$	273–700	2
Nitrogen[13]				
N_2	g	$6.50 + 0.00100T$	300–3000	3
NH_3	g	$6.70 + 0.00630T$	300–800	1½
NH_4Br	c	22.8	274–328	?
NH_4Cl	c, α	$9.80 + 0.0368T$	273–457	5
	c, β	$5.0 + 0.0340T$	457–523	5
NH_4I	c	17.8	273–328	?
NH_4NO_3	c	31.8	273–293	?
$(NH_4)_2SO_4$	c	51.6	275–328	?
NO	g	$8.05 + 0.000233T - 156300/T^2$	300–5000	2
Osmium				
Os	c	$5.686 + 0.000875T$	273–1877	1
Oxygen[14]				
O_2	g	$8.27 + 0.000258T - 187700/T^2$	300–5000	1
Palladium				
Pd	c	$5.41 + 0.00184T$	273–1822	2
Phosphorus				
P	c, yellow	5.50	273–317	5
	c, red	$0.21 + 0.0180T$	273–472	10
	l	6.6	317–373	10
PCl_3	l	28.7	284–371	?
P_4O_{10}	c	$15.72 + 0.1092T$	273–631	2
	g	73.6	631–1371	3
Platinum[4]				
Pt	c	$5.92 + 0.00116T$	273–1873	1
Potassium				
K	c	$5.24 + 0.00555T$	273–336	5
	l	7.7	336–373	5

TABLE 2-194   Heat Capacities of the Elements and Inorganic Compounds (*Continued*)

Substance	State†	Heat capacity at constant pressure (T = K; 0°C = 273.1 K), cal/deg mol	Range of temperature, K	Uncertainty, %
Potassium—(*Cont.*)				
K	g	4.97	All	0
K_2	g	9.00	300–2000	5
$KAsO_3$	c	25.3	290–372	?
KBO_2	c	$12.6 + 0.0126T$	273–1220	?
$K_2B_4O_7$	c	51.3	290–372	?
KBr	c	$11.49 + 0.00360T$	273–543	2
KCl	c	$10.93 + 0.00376T$	273–1043	2
$KClO_3$	c	25.7	289–371	?
$KClO_4$	c	26.3	287–318	?
$2KCl \cdot CuCl_2 \cdot 2H_2O$	c	63	292–323	?
$2KCl \cdot PtCl_4$	c	55	286–319	?
$2KCl \cdot SnCl_4$	c	54.5	292–323	?
$2KCl \cdot ZnCl_2$	c	43.4	279–319	?
$2KCN \cdot Zn(CN)_2$	c	57.4	277–319	?
K_2CO_3	c	29.9	296–372	?
K_2CrO_4	c	35.9	289–371	?
$K_2Cr_2O_7$	c	$42.80 + 0.0410T$	273–671	5
	l	96.9	671–757	5
KF	c	$10.8 + 0.00284T$	273–1129	?
$K_4Fe(CN)_6$	c	80.1	273–319	?
$K_4Fe(CN)_6 \cdot 3H_2O$	c	114.5	273–310	?
KH_2AsO_4	c	32	289–319	?
KH_2PO_4	c	28.3	290–320	?
$KHSO_4$	c	30	292–324	?
$KMnO_4$	c	28	287–318	?
KNO_3	c	$6.42 + 0.0530T$	273–401	10
	c	28.8	401–611	5
	l	29.5	611–683	10
$K_2O \cdot Al_2O_3 \cdot 3SiO_2$	c, orthoclase	$69.26 + 0.00821T - 2331000/T^2$	273–1373	1½
	gls, orthoclase	$69.81 + 0.01053 - 2403000/T^2$	273–1373	1½
	c, microcline	$65.65 + 0.01102T - 1748000/T^2$	273–1373	1½
	gls, microcline	$64.83 + 0.014438T - 1641000/T^2$	273–1373	1½
$K_4P_2O_7$	c	63.1	290–371	?
K_2SO_4	c	33.1	287–371	?
$K_2S_2O_3$	c	37	293–373	?
$K_2SO_4 \cdot Al_2(SO_4)_3 \cdot 24H_2O$	c	352	292–322	?
$K_2SO_4 \cdot Cr_2(SO_4)_3 \cdot 24H_2O$	c	324	292–324	?
$K_2SO_4 \cdot MgSO_4 \cdot 6H_2O$	c	106	292–323	?
$K_2SO_4 \cdot NiSO_4 \cdot 6H_2O$	c	107	289–319	?
$K_2SO_4 \cdot ZnSO_4 \cdot 6H_2O$	c	120	293–317	?
Prometheum				
Pr	c			
Radon				
Rn	g	4.97	All	0
Rhenium				
Re	c	$6.30 + 0.00053T$	273–2273	?
Rhodium				
Rh	c	$5.40 + 0.00219T$	273–1877	2
Rubidium				
Rb	c	$3.27 + 0.0131T$	273–312	2
	l	7.85	312–373	5
RbBr	c	$11.6 + 0.00255T$	273–954	?
RbCl	c	$11.5 + 0.00249T$	273–987	?
Rb_2CO_3	c	28.4	291–320	?
RbF	c	$11.3 + 0.00256T$	273–1048	?
RbI	c	$11.6 + 0.00263T$	273–913	?
Scandium				
Sc_2O_3	c	21.1	273–373	?
$Sc_2(SO_4)_3$	c	62.0	273–373	?
Selenium				
Se	c	$4.53 + 0.00550T$	273–490	2
	l	8.35	490–570	3
Silicon				
Si	c	$5.74 + 0.000617T - 101000/T^2$	273–1174	2
SiC	c	$8.89 + 0.00291T - 284000/T^2$	273–1629	2
$SiCl_4$	l	32.4	293–373	?
SiO_2	c, quartz, α	$10.87 + 0.008712T - 241200/T^2$	273–848	1
	c, quartz, β	$10.95 + 0.00550T$	848–1873	3½
	c, cristobalite, α	$3.65 + 0.0240T$	273–523	2½
	c, cristobalite, β	$17.09 + 0.000454T - 897200/T^2$	523–1973	2
	gls	$12.80 + 0.00447T - 302000/T^2$	273–1973	3½
Silver[4]				
Ag	c	$5.60 + 0.00150T$	273–1234	1
	l	8.2	1234–1573	3

TABLE 2-194   Heat Capacities of the Elements and Inorganic Compounds (*Continued*)

Substance	State†	Heat capacity at constant pressure (T = K; 0°C = 273.1 K), cal/deg mol	Range of temperature, K	Uncertainty, %
Silver—(*Cont.*)				
Ag_3Al	c	$22.56 + 0.00570T$	273–902	2
Ag_2Al	c	$16.85 + 0.00450T$	273–903	2
$AgAl_{12}$	c	$58.62 + 0.0575T$	273–768	5
$AgBr$	c	$8.58 + 0.0141T$	273–703	6
	l	14.9	703–836	5
$AgCl$	c	$9.60 + 0.00929T$	273–728	2
	l	14.05	728–806	5
$AgCNO$	c	18.7	273–353	?
AgI	c, α	$8.58 + 0.0141T$	273–423	6
$AgNO_3$	c, α	$18.83 + 0.0160T$	273–433	2
	c, β	25.7	433–482	5
	l	30.2	482–541	5
Ag_3PO_4	c	37.5	293–325	?
Ag_2S	c, α	18.8	273–448	5
	c, β	21.8	448–597	5
Ag_3Sb	c	$19.53 + 0.0160T$	273–694	5
Ag_2Se	c, α	20.2	273–406	5
	c, β	20.4	406–460	5
Sodium[15]				
Na	c	$5.01 + 0.00536T$	273–371	1½
	l	7.50	371–451	2
	g	4.97	All	0
$NaBO_2$	c	$10.4 + 0.0199T$	273–1239	?
$Na_2B_4O_7$	c	47.9	289–371	?
$Na_2B_4O_7 \cdot 10H_2O$	c	147	292–323	?
$NaBr$	c	$11.74 + 0.00233T$	273–543	2
$NaCl$	c	$10.79 + 0.00420T$	273–1074	2
	l	15.9	1073–1205	3
$NaClO_3$	c	$9.48 + 0.0468T$	273–528	3
	l	31.8	528–572	5
$NaCNO$	c	13.1	273–353	?
Na_2CO_3	c	28.9	288–371	?
NaF	c	$10.4 + 0.00289T$	273–1261	?
$Na_2HPO_4 \cdot 7H_2O$	c	86.6	275–307	?
$Na_2HPO_4 \cdot 12H_2O$	c	133.4	275–307	?
NaI	c	$12.5 + 0.00162T$	273–936	?
$NaNO_3$	c	$4.56 + 0.0580T$	273–583	5
	l	37.2	583–703	10
$Na_2O \cdot Al_2O_3 \cdot 3SiO_2$	c, albite	$63.78 + 0.01171T - 1678000/T^2$	273–1373	1
	gls	$61.25 + 0.01768T - 1545000/T^2$	273–1173	1
$NaPO_3$	c	22.1	290–319	?
$Na_4P_2O_7$	c	60.7	290–371	?
Na_2SO_4	c	32.8	289–371	?
$Na_2S_2O_3$	c	34.9	273–307	?
$Na_2S_2O_3 \cdot 5H_2O$	c	86.2	273–307	?
Sodium-potassium alloys[15]	l			
Strontium				
$SrBr_2$	c	$18.1 + 0.00311T$	273–923	?
$SrBr_2 \cdot H_2O$	c	28.9	277–370	?
$SrBr_2 \cdot 6H_2O$	c	82.1	276–327	?
$SrCl_2$	c	$18.2 + 0.00244T$	273–1143	?
$SrCl_2 \cdot H_2O$	c	28.7	276–365	?
$SrCl_2 \cdot 2H_2O$	c	38.3	277–366	?
$SrCO_3$	c	21.8	281–371	?
SrI_2	c	$18.6 + 0.00304T$	273–783	?
$SrI_2 \cdot H_2O$	c	28.5	276–363	?
$SrI_2 \cdot 2H_2O$	c	39.1	275–336	?
$SrI_2 \cdot 6H_2O$	c	84.9	275–333	?
$SrMoO_4$	c	37	273–297	?
$Sr(NO_3)_2$	c	38.3	290–320	?
$SrSO_4$	c	26.2	293–369	?
Sulfur[16]				
S	c, rhombic	$3.63 + 0.00640T$	273–368	3
	c, monoclinic	$4.38 + 0.00440T$	368–392	3
S_2	g	$8.58 + 0.00030T$	300–2500	5
S_2Cl_2	l	27.5	273–332	?
SO_2	g	$7.70 + 0.00530T - 0.00000083T^2$	300–2500	2½
Tantalum				
Ta	c	$5.91 + 0.00099T$	273–1173	2
Tellurium				
Te	c	$5.19 + 0.00250T$	273–600	3
Thallium				
Tl	c, α	$5.32 + 0.00385T$	273–500	1
	c, β	8.12	500–576	1

TABLE 2-194 Heat Capacities of the Elements and Inorganic Compounds (*Concluded*)

Substance	State†	Heat capacity at constant pressure (T = K; 0°C = 273.1 K), cal/deg mol	Range of temperature, K	Uncertainty, %
Thallium—(*Cont.*)				
Tl	l	7.12	576–773	3
TlBr	c	$12.53 + 0.00100T$	273–733	10
	l	16.0	733–800	10
TlCl	c	$12.56 + 0.00088T$	273–700	5
	l	14.2	700–803	10
Thorium				
Th	c	6.40	273–373	?
ThO_2	c	$14.6 + 0.00507T$	273–1273	?
$Th(SO_4)_2$	c	41.2	273–373	?
Tin[4]				
Sn	c	$5.05 + 0.00480T$	273–504	2
	l	6.6	504–1273	10
SnAu	c	$11.79 + 0.00233T$	273–581	1
$SnCl_2$	c	$16.2 + 0.00926T$	273–520	?
$SnCl_4$	l	38.4	286–371	?
SnO	c	$9.40 + 0.00362T$	273–1273	?
SnO_2	c	$13.94 + 0.00565T - 252000/T^2$	273–1373	?
SnPt	c	$11.49 + 0.00190T$	273–1318	1
SnS	c	$12.1 + 0.00165T$	273–1153	?
SnS_2	c	$20.5 + 0.00400T$	273–873	?
Titanium				
Ti	c	$8.91 + 0.00114T - 433000/T^2$	273–713	3
$TiCl_4$	l	35.7	285–372	?
TiO_2	c	$11.81 + 0.00754T - 41900/T^2$	273–713	3
Tungsten				
W	c	$5.65 + 0.00866$	273–2073	1
WO_3	c	$16.0 + 0.00774T$	273–1550	?
Uranium				
U	c	6.64	273–372	?
U_3O_8	c	59.8	276–314	?
Vanadium				
V	c	$5.57 + 0.00097T$	273–1993	?
Xenon				
Xe	g	4.97	All	0
Zinc[4]				
Zn	c	$5.25 + 0.00270T$	273–692	1
	l	$7.59 + 0.00055T$	692–1122	3
$ZnCl_2$	c	$15.9 + 0.00800T$	273–638	?
ZnO	c	$11.40 + 0.00145T - 182400/T^2$	273–1573	1
ZnS	c	$12.81 + 0.00095T - 194600/T^2$	273–1173	5
ZnSb	c	$11.5 + 0.00313T$	273–810	?
$ZnSO_4$	c	28	293–373	?
$ZnSO_4 \cdot H_2O$	c	34.7	282	?
$ZnSO_4 \cdot 6H_2O$	c	80.8	282	?
$ZnSO_4 \cdot 7H_2O$	c	100.2	273–307	?
Zirconium				
ZrO_2	c	$11.62 + 0.01046T - 177700/T^2$	273–1673	5
$ZrO_2 \cdot SiO_2$	c	26.7	297–372	?

[1] See also Table 2-195. Data to 298 K are also given by Scott, *Cryogenic Engineering*, Van Nostrand, Princeton, N.J., 1959.
[2] For liquid and gas data, see Johnson (ed.), WADD-TR-60-56, 1960.
[3] Stalder, NACA Tech. Note 4141, 1957 (Fig. 5), gives data from 400 to 2600°R.
[4] See also Table 2-195.
[5] For data from 400 to 5500°R see Stalder, NACA Tech. Note 4141, 1975 (Fig. 4).
[6] For solid, liquid, and gas data, see Johnson (ed.), WADD-TR-60-56, 1960.
[7] For data from 400 to 2350°R see Stalder, NACA Tech. Note 4141, 1957.
[8] For solid, liquid, and gas data, see Johnson (ed.), WADD-TR-60-56, 1960.
[9] For liquid and gas data, see Johnson (ed.), WADD-TR-60-56, 1960.
[10] For solid, liquid, and gas data, see Johnson (ed.), WADD-TR-60-56, 1960.
[11] See also Table 2-195; Douglas, Ball, et al., *Bur. Stand. J. Res.*, **46** (1951): 334; Busey and Giaque, *J. Am. Chem. Soc.*, **75** (1953): 806; Sheldon, ASME Pap. 49-A-30, 1949.
[12] For solid, liquid, and gas data, see Johnson (ed.), WADD-TR-56-60, 1960.
[13] For solid, liquid, and gas data, see Johnson (ed.), WADD-TR-56-60, 1960.
[14] For solid, liquid, and gas data, see Johnson (ed.), WADD-TR-56-60, 1960. Ozone: For liquid see Brabets and Waterman, *J. Chem. Phys.*, **28** (1958): 1212.
[15] For data on liquid Na-K alloys to 1500°F and for liquid Na to 1460°F, see Lubarsky and Kaufman, NACA Rep. 1270, 1956.
[16] See also Evans and Wagman, *Bur. Stand. J. Res.* **49** (1952): 141; Gratch, OTS PB 124957, 1950; Guthrie, Scott, et al., *J. Am. Chem. Soc.*, **76** (1954): 1488.

TABLE 2-195 Specific Heat [kJ/(kg·K)] of Selected Elements

Symbol	Temperature, K														
	4	6	8	10	20	40	60	80	100	200	250	300	400	600	800
Al	0.00026	0.00050	0.00088	0.00140	0.0089	0.0775	0.214	0.357	0.481	0.797	0.859	0.902	0.949	1.042	1.134
Be	0.00008			0.00028	0.0014				0.195	1.109	1.537	1.840	2.191	2.605	2.823
Bi	0.00054	0.00220	0.00541	0.01040	0.0340	0.0729	0.092	0.102	0.109	0.120	0.121	0.122	0.123	0.142	0.136
Cr	0.00016	0.00029	0.00050	0.00081	0.0021	0.0107	0.059	0.127	0.190	0.382	0.424	0.450	0.501	0.565	0.611
Co	0.00036	0.00059	0.00085	0.00121	0.0048	0.0404	0.110	0.184	0.234	0.376	0.406	0.426	0.451	0.509	0.543
Cu	0.00011	0.00024	0.00048	0.00086	0.0076	0.059	0.137	0.203	0.254	0.357	0.377	0.386	0.396	0.431	0.448
Ge			0.00037	0.00081	0.0129	0.0619	0.108	0.153	0.192	0.286	0.305	0.323	0.343	0.364	0.377
Au	0.00018	0.00047	0.00126	0.00255	0.0163	0.0569	0.084	0.100	0.109	0.124	0.127	0.129	0.131	0.136	0.141
Ir				0.00032	0.0021				0.090	0.122	0.128	0.131	0.133	0.140	0.146
Fe	0.00038	0.00061	0.00090	0.00127	0.0039	0.0276	0.086	0.154	0.216	0.384	0.422	0.450	0.491	0.555	0.692
Pb	0.00075	0.00242	0.00747	0.01350	0.0531	0.0944	0.108	0.114	0.118	0.125	0.127	1.129	0.132	0.142	
Mg	0.00034	0.00080	0.00155	0.00172	0.0148	0.138	0.336	0.513	0.648	0.929	0.985	1.005	1.082	1.177	1.263
Hg	0.00417	0.01420	0.01820	0.02250	0.0515	0.0895	0.107	0.116	0.121	0.136	0.141	0.139	0.136	0.135	0.104
Mo	0.00011	0.00019	0.00032	0.00050	0.0029	0.0236	0.061	0.105	0.140	0.223	0.241	0.248	0.261	0.280	0.292
Ni	0.00054	0.00086	0.00121	0.00178	0.0058	0.0380	0.103	0.173	0.232	0.383	0.416	0.444	0.490	0.590	0.530
Pt	0.00019	0.00028	0.00067	0.00112	0.0077	0.0382	0.069	0.088	0.101	0.127	0.132	0.134	0.136	0.140	0.146
Ag	0.00016	0.00035	0.00093	0.00186	0.0159	0.0778	0.133	0.166	0.187	0.225	0.232	0.236	0.240	0.251	0.264
Sn	0.00024	0.00127	0.00423	0.00776	0.0400	0.108	0.149	0.173	0.189	0.214	0.220	0.222	0.245	0.257	0.257
Zn	0.00011	0.00029	0.00096	0.00250	0.0269	0.123	0.205	0.258	0.295	0.366	0.380	0.389	0.404	0.435	0.479

TABLE 2-196 Heat Capacities of Inorganic and Organic Liquids

Cmpd. no.	Name	Formula	CAS no.	Mol wt	C1	C2	C3	C4	C5	T_{min} K	C_p at T_{min} ×1E−05	T_{max} K	C_p at T_{max} ×1E−05
1	Methane (eqn. 2)	CH_4	74828	16.043	6.5708E+01	3.8883E+04	−2.5795E+02	6.1407E+02	1	90.69	0.5361	190	14.9780
2	Ethane (eqn. 2)	C_2H_6	74840	30.070	4.4009E+01	8.9718E+04	9.1877E+02	−1.8860E+03	0	92	0.6855	290	1.2444
3	Propane (eqn. 2)	C_3H_8	74986	44.097	6.2983E+01	1.1363E+05	6.3321E+02	−8.7346E+02	0	85.47	0.8488	360	2.6079
4	n-Butane (eqn. 2)	C_4H_{10}	106978	58.123	6.4730E+01	1.6184E+05	9.8341E+02	−1.4315E+03	0	134.86	1.1380	420	5.0822
5	n-Pentane	C_5H_{12}	109660	72.150	1.5908E+05	−2.7050E+02	9.9537E−01	0	0	143.42	1.4076	390	2.0498
6	n-Hexane	C_6H_{14}	110543	86.177	1.7212E+05	−1.8378E+02	8.8734E−01	0	0	177.83	1.6750	460	2.7534
7	n-Heptane (eqn. 2)	C_7H_{16}	142825	100.204	6.1260E+01	3.1441E+05	1.8246E+03	−2.5479E+03	0	182.57	1.9989	520	4.0657
8	n-Octane	C_8H_{18}	111659	114.231	2.2483E+05	−1.8663E+02	9.5891E−01	0	0	216.38	2.2934	460	3.4189
9	n-Nonane	C_9H_{20}	111842	128.258	3.8308E+05	−1.1398E+02	2.7101E+00	0	0	219.66	2.6348	325	2.9990
10	n-Decane	$C_{10}H_{22}$	124185	142.285	2.7862E+05	−1.9791E+02	1.0737E+00	0	0	243.51	2.9409	460	4.1478
11	n-Undecane	$C_{11}H_{24}$	1120214	156.312	2.9398E+05	−1.1498E+02	9.6936E−01	0	0	247.57	3.2493	433.42	4.2624
12	n-Dodecane	$C_{12}H_{26}$	112403	170.338	5.0821E+05	−1.3687E+02	3.1015E+00	0	0	263.57	3.6292	330	3.9429
13	n-Tridecane	$C_{13}H_{28}$	629505	184.365	3.5018E+05	−1.0470E+02	1.0022E+00	0	0	267.76	3.9400	508.62	5.5619
14	n-Tetradecane	$C_{14}H_{30}$	629594	198.392	3.5314E+05	2.9130E+01	8.6116E−01	0	0	279.01	4.2831	526.73	6.0741
15	n-Pentadecane	$C_{15}H_{32}$	629629	212.419	3.4691E+05	2.1954E+02	6.5632E−01	0	0	283.07	4.6165	543.84	6.6042
16	n-Hexadecane	$C_{16}H_{34}$	544763	226.446	3.7035E+05	2.3147E+02	6.8632E−01	0	0	291.31	4.9602	560.01	7.1521
17	n-Heptadecane	$C_{17}H_{36}$	629787	240.473	3.7697E+05	3.4782E+02	5.7895E−01	0	0	295.13	5.3005	575.3	7.6869
18	n-Octadecane	$C_{18}H_{38}$	593453	254.500	3.9943E+05	3.7464E+02	5.8156E−01	0	0	301.31	5.6511	589.86	8.2276
19	n-Nonadecane	$C_{19}H_{40}$	629925	268.527	3.4257E+05	7.6208E+02	2.0481E−01	0	0	305.04	5.9409	603.05	8.7663
20	n-Eicosane	$C_{20}H_{42}$	112958	282.553	3.5272E+05	8.0732E+02	2.1220E−01	0	0	309.58	6.2299	616.93	9.3154
21	2-Methylpropane	C_4H_{10}	75285	58.123	1.7237E+05	−1.7839E+03	1.4759E+01	−4.7909E+02	5.8050E−05	113.54	0.9961	380	2.0725
22	2-Methylbutane	C_5H_{12}	78784	72.150	1.0630E+05	1.4600E+02	−2.9200E−01	1.5100E−03	0	113.25	1.2328	310	1.7048
23	2,3-Dimethylbutane	C_6H_{14}	79298	86.177	1.2945E+05	1.8500E+02	6.0800E+00	0	0	145.19	1.4495	331.13	2.0224
24	2-Methylpentane	C_6H_{14}	107835	86.177	1.4222E+05	−4.7830E+02	7.3900E+00	0	0	119.55	1.4706	333.41	2.0842
25	2,3-Dimethylpentane[1]	C_7H_{16}	565593	100.204	1.4642E+05	5.9200E+02	6.0400E+00	0	0	90	1.5664	380	2.5613
26	2,3,3-Trimethylpentane	C_8H_{18}	560214	114.231	3.8862E+05	−1.4395E+03	3.2187E+00	2.1734E−03	0	280	2.3791	320	2.5757
27	2,2,4-Trimethylpentane	C_8H_{18}	540841	114.231	9.5275E+04	6.9670E+02	−1.3765E+00	0	0	165.78	1.8285	520	3.9095
28	Ethylene	C_2H_4	74851	28.054	2.4739E+05	−4.4280E+03	4.0936E+01	−1.6970E−01	2.6816E−04	103.97	0.7013	252.7	0.9758
29	Propylene	C_3H_6	115071	42.081	1.1720E+05	−3.8632E+02	1.2348E+00	0	0	87.89	0.9279	298.15	1.1178
30	1-Butene	C_4H_8	106989	56.108	1.3589E+05	−4.7739E+02	2.1835E+00	−2.2230E−03	0	87.8	1.0930	300	1.2917
31	cis-2-Butene	C_4H_8	590181	56.108	1.2669E+05	−6.5470E+02	−6.4000E+00	2.9120E−03	0	134.26	1.1340	350	1.5022
32	trans-2-Butene	C_4H_8	624646	56.108	1.1276E+05	−1.0470E+02	5.2140E+00	0	0	167.62	1.0986	274.03	1.2322
33	1-Pentene	C_5H_{10}	109671	70.134	1.5467E+05	−4.2600E+02	1.9640E+00	−1.8038E−03	0	107.93	1.2930	310	1.5761
34	1-Hexene	C_6H_{12}	592416	84.161	1.9263E+05	−5.7116E+02	2.4004E+00	−1.9758E−03	0	133.39	1.5446	336.63	1.9700
35	1-Heptene	C_7H_{14}	592767	98.188	1.5997E+05	−1.5670E+02	3.4300E+00	1.5222E−03	0	154.27	1.7955	330	2.3032
36	1-Octene	C_8H_{16}	111660	112.215	3.7930E+05	−2.1175E+03	8.2362E+00	−9.0093E−03	0	171.45	2.1295	315	2.4793
37	1-Nonene	C_9H_{18}	124118	126.242	2.5875E+05	−3.5450E+02	1.3126E+00	0	0	191.78	2.3904	420.02	3.4142
38	1-Decene	$C_{10}H_{20}$	872059	140.269	3.1950E+05	−5.7621E+02	1.7085E+00	0	0	206.89	2.7343	443.75	4.0027
39	2-Methylpropene	C_4H_8	115117	56.108	8.7680E+04	2.1710E+02	−9.1530E+00	2.2660E−03	0	132.81	1.0568	343.15	1.4596
40	2-Methyl-1-butene[2]	C_5H_{10}	563462	70.134	1.4951E+05	−2.4763E+02	9.1849E+00	0	0	135.58	1.3282	304.31	1.5921
41	2-Methyl-2-butene[2]	C_5H_{10}	513359	70.134	1.5160E+05	−2.6672E+02	9.0847E−01	0	0	139.39	1.3207	311.71	1.5673
42	1,2-Butadiene	C_4H_6	590192	54.092	1.3515E+05	−3.1114E+02	9.7007E+00	−1.5230E−04	0	136.95	1.1034	290	1.2279
43	1,3-Butadiene	C_4H_6	106990	54.092	1.2886E+05	−3.2310E+02	1.0150E+00	3.2000E−05	0	165	1.0333	350	1.4148
44	2-Methyl-1,3-butadiene	C_5H_8	78795	68.119	1.4148E+05	−2.8870E+02	1.0910E+00	0	0	130.32	1.2239	307.2	1.5575
45	Acetylene	C_2H_2	74862	26.038	2.011E+05	−1.1988E+03	3.0027E+00	0	0	192.4	0.8061	250	0.8808
46	Methylacetylene	C_3H_4	74997	40.065	7.9791E+04	8.9490E+01	0	0	0	200	0.9769	249.94	1.0216
47	Dimethylacetylene	C_4H_6	503173	54.092	8.8153E+04	1.2416E+02	0	0	0	240.91	1.1806	300.13	1.2542
48	3-Methyl-1-butyne	C_5H_8	598232	68.119	1.0520E+05	1.9110E+02	0	0	0	200	1.4342	299.49	1.6243
49	1-Pentyne	C_5H_8	627190	68.119	8.6200E+04	2.5660E+02	0	0	0	200	1.3752	313.33	1.6660
50	2-Pentyne	C_5H_8	627214	68.119	6.8671E+04	2.4666E+02	0	0	0	200	1.1800	329.27	1.4989

51	1-Hexyne	C_6H_{10}	693027	82.145	9.3000E+04	3.3600E+02	0	0	0	200	1.5820	344.48	2.0530
52	2-Hexyne	C_6H_{10}	764352	82.145	9.4860E+04	2.5415E+02	0	0	0	300	1.7110	357.67	1.8576
53	3-Hexyne	C_6H_{10}	928494	82.145	8.2795E+04	2.8340E+02	0	0	0	300	1.6781	354.35	1.8322
54	1-Heptyne	C_7H_{12}	628717	96.172	8.5122E+04	4.0247E+02	0	0	0	192.22	1.6248	372.93	2.3522
55	1-Octyne	C_8H_{14}	629050	110.199	9.1748E+04	4.7140E+02	0	0	0	193.55	1.8299	399.35	2.8000
56	Vinylacetylene[3]	C_4H_4	688974	52.076	6.8720E+04	1.3500E+02	0	0	0	200	0.9572	278.25	1.0628
57	Cyclopentane	C_5H_{10}	287923	70.134	1.2253E+05	-4.0380E+02	1.7344E+00	-1.0975E-03	0	179.28	0.9956	322.4	1.3584
58	Methylcyclopentane	C_6H_{12}	96377	84.161	1.5592E+05	-4.9000E+02	2.1383E+00	-1.5585E-03	0	130.73	1.2492	366.48	1.8682
59	Ethylcyclopentane	C_7H_{14}	1640897	98.188	1.7852E+05	-5.1835E+02	2.3255E+00	-1.6518E-03	0	134.71	1.4678	301.82	1.8767
60	Cyclohexane	C_6H_{12}	110827	84.161	-2.2060E+05	3.1183E+03	-9.4216E+00	1.0687E-02	0	279.69	1.4836	400	2.0323
61	Methylcyclohexane	C_7H_{14}	108872	98.188	1.3134E+05	-6.3100E+01	8.1151E-01	0	0	146.58	1.3955	320	1.9435
62	1,1-Dimethylcyclohexane	C_8H_{16}	590669	112.215	1.3450E+05	8.7650E+00	6.4738E-01	0	0	239.66	1.8321	392.7	2.6309
63	Ethylcyclohexane	C_8H_{16}	1678917	112.215	1.3236E+05	7.2740E+01	1.1430E+00	0	0	161.84	1.6109	404.95	2.6798
64	Cyclopentene	C_5H_8	142290	68.119	1.2538E+05	-3.4970E+02		0	0	138.13	0.9888	317.38	1.2953
65	1-Methylcyclopentene	C_6H_{10}	693390	82.145	5.3271E+04	-3.3792E+02		0	0	200	1.1885	348.64	1.6760
66	Cyclohexene	C_6H_{10}	110838	82.145	1.0585E+05	-6.0000E+01	6.8000E-01	0	0	169.67	1.1525	356.12	1.7072
67	Benzene	C_6H_6	71432	78.114	1.2944E+05	-1.6950E+02	6.4781E-01	0	0	278.68	1.3251	353.24	1.5040
68	Toluene	C_7H_8	108883	92.111	1.1011E+05	1.5230E+02	6.0500E-01	0	0	178.18	1.3007	500	2.0774
69	o-Xylene	C_8H_{10}	95476	106.167	3.6500E+04	1.0175E+03	-2.6300E+00	3.0200E-03	0	248	1.7315	415	2.2166
70	m-Xylene	C_8H_{10}	108383	106.167	1.7555E+05	-2.9950E+02	1.0880E+00	0	0	225.3	1.6330	360	2.0873
71	p-Xylene	C_8H_{10}	106423	106.167	-3.5500E+04	1.2872E+03	-2.5990E+00	2.4260E-03	0	286.41	1.7697	600	3.2520
72	Ethylbenzene	C_8H_{10}	100414	106.167	1.3316E+05	4.4507E+01	3.9645E-01	0	0	178.15	1.5367	409.35	2.1781
73	Propylbenzene	C_9H_{12}	103651	120.194	2.3477E+05	-8.0022E+02	3.4037	-3.1739E-03	0	173.59	1.8182	370	2.4389
74	1,2,4-Trimethylbenzene	C_9H_{12}	95636	120.194	1.7880E+05	-1.2847E+02	8.3741E-01	0	0	229.33	1.9338	350	2.3642
75	Isopropylbenzene	C_9H_{12}	98828	120.194	1.8290E+05	-1.7400E+02	9.1200E-01	0	0	177.14	1.8069	500	3.2390
76	1,3,5-Trimethylbenzene	C_9H_{12}	108678	120.194	1.4560E+05	1.9700E+01	6.2260E-01	0	0	228.42	1.8503	350	2.3121
77	p-Isopropyltoluene	$C_{10}H_{14}$	99876	134.221	2.9500E+04	2.4870E+02	1.5700E-01	0	0	205.25	2.0452	450.28	2.9550
78	Naphthalene	$C_{10}H_8$	91203	128.174	1.2177E+05	5.2750E+02	0	0	0	353.43	2.1623	491.14	2.8888
79	Biphenyl	$C_{12}H_{10}$	92524	154.211	1.1334E+05	4.2930E+02	0	0	0	342.2	2.6868	533.37	3.5075
80	Styrene	C_8H_8	100425	104.152	1.9567E+05	2.9020E+02	-6.0510E-01	1.3567E-03	0	242.54	1.6749	418.31	2.2816
81	m-Terphenyl	$C_{18}H_{14}$	92068	230.309		5.9407E+02	0	0	0	360	4.0954	650	5.8182
82	Methanol	CH_4O	67561	32.042	1.0580E+05	-3.6223E+02	9.3790E-01	0	0	175.47	0.7112	400	1.1097
83	Ethanol	C_2H_6O	64175	46.069	1.0264E+05	-1.3963E+02	-3.0341E-02	2.0386E-03	0	159.05	0.8787	390	1.6450
84	1-Propanol	C_3H_8O	71238	60.096	1.5876E+05	-6.3500E+02	1.9690E+00	0	0	146.95	1.0797	400	2.1980
85	1-Butanol	$C_4H_{10}O$	71363	74.123	1.9120E+05	-7.3040E+02	2.2998E+00	0	0	184.51	1.3473	390.81	2.5701
86	2-Propanol	C_3H_8O	67630	60.096	2.0670E+05	-1.0204E+03	3.2900E+00	0	0	185.28	1.2762	372.7	2.8340
87	2-Butanol	$C_4H_{10}O$	78922	74.123	7.2355E+05	-9.0950E+02	3.6662E+01	-6.6395E-02	0	158.45	1.1189	480	2.8122
88	2-Methyl-2-propanol	$C_4H_{10}O$	75650	74.123	-9.2546E+05	7.8949E+03	-1.7661E+01	1.3617E-02	4.4064E-05	298.96	2.2016	460	2.9455
89	1-Pentanol	$C_5H_{12}O$	71410	88.150	2.0120E+05	-6.5130E+02	2.2750E+00	0	0	200.14	1.6198	389.15	2.9227
90	2-Methyl-1-butanol	$C_5H_{12}O$	137326	88.150	8.2937E+04	4.5598E+02	0.0000E+00	0	0	250	1.9793	401.85	2.6778
91	3-Methyl-1-butanol	$C_5H_{12}O$	123513	88.150	-5.3777E+04	8.3342E+02	0	0	0	295.52	2.0729	350	2.5542
92	1-Hexanol	$C_6H_{14}O$	111273	102.177	4.8466E+05	-2.7613E+02	6.5555E+00	0	0	228.55	1.9599	320	2.7233
93	1-Heptanol	$C_7H_{16}O$	111706	116.203	4.3790E+05	-2.0947E+02	5.2090E+00	0	0	239.15	2.3487	370	3.7597
94	Cyclohexanol	$C_6H_{12}O$	108930	100.161	-4.0000E+04	8.5300E+02	0	0	0	296.6	2.1300	434	3.3020
95	Ethylene glycol	$C_2H_6O_2$	107211	62.068	3.5540E+04	4.3678E+02	-1.8486E-01	0	0	260.15	1.3666	493.15	2.0598
96	1,2-Propylene glycol	$C_3H_8O_2$	57556	76.095	5.8080E+04	4.4520E+02	0	0	0	213.15	1.5297	460.75	2.6321
97	Phenol	C_6H_6O	108952	94.113	1.0172E+05	3.1761E+02	0	0	0	314.06	2.0147	425	2.3670
98	o-Cresol	C_7H_8O	95487	108.140	-1.5515E+05	3.1480E+03	-8.0367E+00	7.2540E-03	0	304.2	2.3297	400	2.5243
99	m-Cresol	C_7H_8O	108394	108.140	-2.4670E+05	3.2568E+03	-7.4202E+00	6.0467E-03	0	285.39	2.1895	400	2.5578
100	p-Cresol	C_7H_8O	106445	108.140	2.5999E+05	-1.1123E+03	4.9270E+00	-5.4367E-03	0	307.93	2.2740	400	2.5794

TABLE 2-196   Heat Capacities of Inorganic and Organic Liquids (*Continued*)

Cmpd. no.	Name	Formula	CAS no.	Mol wt	C1	C2	C3	C4	C5	T_{min} K	C_p at T_{min} ×1E−05	T_{max} K	C_p at T_{max} ×1E−05
101	Dimethyl ether	C_2H_6O	115106	46.069	1.1010E+05	−1.5747E+02	5.1853E−01	0	0	131.65	0.9836	250	1.0314
102	Methyl ethyl ether	C_3H_8O	540670	60.096	1.2977E+05	−3.3196E+02	1.3369E+00	0	0	218.9	1.2356	325.35	1.7030
103	Methyl-n-propyl ether	$C_4H_{10}O$	557175	74.123	1.4411E+05	−1.0209E+02	5.8113E−01	0	0	133.97	1.4086	312.2	1.6888
104	Methyl isopropyl ether	$C_4H_{10}O$	598538	74.123	1.4344E+05	−1.5407E+02	7.2550E−01	0	0	127.93	1.3560	310	1.6540
105	Methyl-n-butyl ether	$C_5H_{12}O$	628284	88.150	1.7785E+05	−1.7157E+02	7.4379E−01	0	0	157.48	1.6928	343.35	2.0663
106	Methyl isobutyl ether	$C_5H_{12}O$	625445	88.150	5.1380E+04	4.5040E+02	0	0	0	300	1.8650	370	2.1803
107	Methyl tert-butyl ether	$C_5H_{12}O$	1634044	88.150	1.4012E+05	−9.0000E+00	5.6300E−01	0	0	164.55	1.5388	328.35	1.9786
108	Diethyl ether	$C_4H_{10}O$	60297	74.123	4.4400E+04	1.3010E+03	−5.5000E+00	8.7630E−03	0	156.92	1.4698	460	3.3202
109	Ethyl propyl ether	$C_5H_{12}O$	628320	88.150	1.0368E+05	7.2630E+02	−2.6047E+00	4.0957E−03	0	145.65	1.6686	320	2.0358
110	Ethyl isopropyl ether	$C_5H_{12}O$	625547	88.150	1.0625E+05	2.9215E+02	0	0	0	298.15	1.9335	326.15	2.0153
111	Methyl phenyl ether	C_7H_8O	100663	108.140	1.5094E+05	9.3455E+01	0	0	0	298.15	1.9978	484.2	2.5153
112	Diphenyl ether	$C_{12}H_{10}O$	101848	170.211	1.3416E+05	4.4767E+02	2.3602E−01	0	0	300.03	2.6847	570	3.8933
113	Formaldehyde[4]	CH_2O	50000	30.026	6.1900E+04	2.8300E+01	0	0	0	204	0.6767	234	0.6852
114	Acetaldehyde	C_2H_4O	75070	44.053	1.1510E+05	−4.3300E+02	1.4250E+00	0	0	150.15	0.8221	294	1.1097
115	1-Propanal	C_3H_6O	123386	58.080	9.9306E+04	1.1573E+02	0	0	0	200	1.2245	328.75	1.3735
116	1-Butanal	C_4H_8O	123728	72.107	6.5682E+04	1.3291E+03	−7.1579E+00	1.2755E−02	0	176.75	1.4741	300	1.6459
117	1-Pentanal	$C_5H_{10}O$	110623	86.134	1.1205E+05	2.5778E+02	0	0	0	200	1.6361	376.15	2.0901
118	1-Hexanal	$C_6H_{12}O$	66251	100.161	1.1770E+05	3.2952E+02	0	0	0	217.15	1.8926	401.45	2.4999
119	1-Heptanal	$C_7H_{14}O$	111717	114.188	2.2236E+05	−1.0517E+02	6.5074E−01	0	0	229.8	2.3256	381.25	2.7685
120	1-Octanal	$C_8H_{16}O$	124130	128.214	1.3065E+05	4.6361E+02	0	0	0	246	2.4470	447.15	3.3795
121	1-Nonanal	$C_9H_{18}O$	124196	142.241	1.3682E+05	5.3129E+02	0	0	0	255.15	2.7238	468.15	3.8554
122	1-Decanal	$C_{10}H_{20}O$	112312	156.268	1.5046E+05	5.8663E+02	0	0	0	267.15	3.0718	488.15	4.3682
123	Acetone	C_3H_6O	67641	58.080	1.3560E+05	−1.7700E+02	2.8370E−01	6.8900E−04	0	178.45	1.1696	329.44	1.3271
124	Methyl ethyl ketone	C_4H_8O	78933	72.107	1.3230E+05	−2.0087E+02	−9.5970E−01	1.9533E−03	0	186.48	1.4905	373.15	1.7511
125	2-Pentanone	$C_5H_{10}O$	107879	86.134	1.9459E+05	−2.6396E+02	7.6808E−01	0	0	196.29	1.7239	375.46	2.0380
126	Methyl isopropyl ketone	$C_5H_{10}O$	563804	86.134	1.8361E+05	−2.6885E+02	8.6080E−01	0	0	181.15	1.6316	367.55	2.0108
127	2-Hexanone	$C_6H_{12}O$	591786	100.161	2.7249E+05	−7.9070E+02	2.5834E+00	−2.0040E−03	0	220.87	2.0228	382.62	2.3590
128	Methyl isobutyl ketone	$C_6H_{12}O$	108101	100.161	1.2492E+05	3.0410E+02	0	0	0	298.15	2.1559	390	2.4352
129	3-Methyl-2-pentanone	$C_6H_{12}O$	565617	100.161	9.9815E+04	3.4672E+02	0	0	0	298.15	2.0319	390.55	2.3523
130	3-Pentanone	$C_5H_{10}O$	96220	86.134	1.9302E+05	−1.7643E+02	5.6690E−01	0	0	234.18	1.8279	375.14	2.0661
131	Ethyl isopropyl ketone	$C_6H_{12}O$	565695	100.161	8.3630E+04	3.9900E+02	0	0	0	298.15	2.0259	425	2.5320
132	Diisopropyl ketone	$C_7H_{14}O$	565800	114.188	1.7927E+05	2.8370E+01	5.3750E−01	0	0	204.81	2.0763	410	2.8126
133	Cyclohexanone	$C_6H_{10}O$	108941	98.145	1.0980E+05	2.6150E+02	0	0	0	290	1.8563	486.5	2.3702
134	Methyl phenyl ketone	C_8H_8O	98862	120.151	7.2692E+04	3.3783E+02	3.5572E−01	0	0	298.2	2.0506	532.12	3.5318
135	Formic acid	CH_2O_2	64186	46.026	7.8060E+04	7.1540E+01	0	0	0	281.45	0.9820	380	1.0525
136	Acetic acid	$C_2H_4O_2$	64197	60.053	1.3964E+05	−3.2080E+02	8.9550E−01	0	0	289.81	1.2213	391.05	1.5159
137	Propionic acid	$C_3H_6O_2$	79094	74.079	2.1366E+05	−7.0270E+02	1.6605E+00	0	0	252.45	1.4209	414.32	2.0756
138	n-Butyric acid	$C_4H_8O_2$	107926	88.106	2.3770E+05	−7.4640E+02	1.8290E+00	0	0	267.95	1.6902	436.42	2.6031
139	Isobutyric acid	$C_4H_8O_2$	79312	88.106	1.2754E+05	−6.5350E+01	8.2867E−01	0	0	270	1.7031	427.65	2.5114
140	Benzoic acid	$C_7H_6O_2$	65850	122.123	−5.4800E+03	6.4712E+02	0	0	0	395.45	2.5042	450	2.8572
141	Acetic anhydride	$C_4H_6O_3$	108247	102.090	3.6600E+04	5.1100E+02	0	0	0	250	1.6435	350	2.1545
142	Methyl formate	$C_2H_4O_2$	107313	60.053	1.3020E+05	−3.9600E+02	1.2100E+00	0	0	174.15	1.2991	304.9	1.2195
143	Methyl acetate	$C_3H_6O_2$	79209	74.079	6.1260E+04	2.7090E+02	0	0	0	253.4	1.7179	373.4	1.6241
144	Methyl propionate	$C_4H_8O_2$	554121	88.106	7.1140E+04	3.3550E+02	0	0	0	300	1.8678	390	2.0198
145	Methyl-n-butyrate	$C_5H_{10}O_2$	623427	102.133	1.0293E+05	1.2910E+02	6.2516E−01	0	0	277.25	1.3684	415.87	2.6474
146	Ethyl formate	$C_3H_6O_2$	109944	74.079	8.0000E+04	2.2360E+02	0	0	0	254.2	1.6068	374.2	1.6367
147	Ethyl acetate	$C_4H_8O_2$	141786	88.106	2.2623E+05	−6.2480E+02	1.4720E+00	0	0	189.6	1.9562	350.21	1.8796
148	Ethyl propionate	$C_5H_{10}O_2$	105373	102.133	7.6330E+04	4.0010E+02	1.4720E+00	0	0	298.15	1.9562	410	2.4037

No.	Name	Formula	CAS No.	Mol. wt.	C1	C2	C3	C4	C5				
149	Ethyl-n-butyrate	C$_6$H$_{12}$O$_2$	105544	116.160	8.2434E+04	4.2245E+02	2.0992E-01	0	0	285.5	2.2015	428.25	3.0185
150	n-Propyl formate	C$_4$H$_8$O$_2$	110747	88.106	7.5700E+04	3.2610E+02	0	0	0	298.15	1.7293	398.15	2.0554
151	n-Propyl acetate	C$_5$H$_{10}$O$_2$	109604	102.133	8.3400E+04	3.8410E+02	0	0	0	274.7	1.8891	404.7	2.3885
152	n-Butyl acetate	C$_6$H$_{12}$O$_2$	123864	116.160	1.1730E+05	3.5220E+02	0	0	0	289.58	1.9929	429.58	2.6860
153	Methyl benzoate	C$_8$H$_8$O$_2$	93583	136.150	1.1950E+05	2.9400E+02	0	0	0	260.75	1.9616	472.65	2.5846
154	Ethyl benzoate	C$_9$H$_{10}$O$_2$	93890	150.177	1.2450E+05	3.7060E+02	0	0	0	238.45	2.1287	486.55	3.0482
155	Vinyl acetate	C$_4$H$_6$O$_2$	108054	86.090	1.3630E+05	-1.0617E+02	7.5175E-01	0	0	259.56	1.5939	389.35	2.0892
156	Methylamine	CH$_5$N	74895	31.057	9.2520E+04	3.7450E+01	0	0	0	179.69	0.9925	266.82	1.0251
157	Dimethylamine	C$_2$H$_7$N	124403	45.084	2.1487E+05	3.7872E+03	-1.3781E-01	1.6924E-02	0	180.96	1.1947	298.15	1.3779
158	Trimethylamine	C$_3$H$_9$N	75503	59.111	1.3605E+05	-2.8800E+02	9.9130E-01	0	0	156.08	1.1525	276.02	1.3208
159	Ethylamine	C$_2$H$_7$N	75047	45.084	1.2170E+05	-3.8993E+01	0	0	0	192.15	1.2919	289.73	1.3300
160	Diethylamine	C$_4$H$_{11}$N	109897	73.138	1.0133E+05	2.4318E+02	0	0	0	223.35	1.5564	328.6	1.8124
161	Triethylamine	C$_6$H$_{15}$N	121448	101.192	1.1148E+05	3.6813E+02	0	0	0	200	1.8511	361.92	2.4471
162	n-Propylamine	C$_3$H$_9$N	107108	59.111	1.3953E+05	7.8000E+02	0	0	0	188.36	1.5422	340	1.6605
163	di-n-Propylamine	C$_6$H$_{15}$N	142847	101.192	4.9120E+04	5.6224E+02	0	0	0	277.9	2.0537	407.9	2.7846
164	Isopropylamine	C$_3$H$_9$N	75310	59.111	-3.2469E+04	1.9771E+03	0	0	0	177.95	1.4621	320	1.6671
165	Diisopropylamine	C$_6$H$_{15}$N	108189	101.192	9.8434E+04	4.2904E+02	-7.0145E+00	8.6913E-03	0	275	2.1642	357.05	2.5162
166	Aniline	C$_6$H$_7$N	62533	93.128	1.4150E+05	1.7190E+02	0	0	0	267.13	1.8723	457.15	2.1076
167	N-Methylaniline	C$_7$H$_9$N	100618	107.155	1.2850E+05	1.0020E+02	0	0	0	216.15	1.6763	469.02	2.5777
168	N,N-Dimethylaniline	C$_8$H$_{11}$N	121697	121.182	4.1860E+04	5.2750E+02	3.7400E-01	0	0	343.58	2.2310	513.58	3.1277
169	Ethylene oxide	C$_2$H$_4$O	75218	44.053	1.4471E+05	-7.5587E+02	2.8261E+00	-3.0640E-03	0	160.65	0.8303	283.85	0.8693
170	Furan	C$_4$H$_4$O	110009	68.075	1.1437E+05	-2.1569E+02	7.2691E-01	0	0	187.55	0.9949	304.5	1.1609
171	Thiophene	C$_4$H$_4$S	110021	84.142	8.1350E+04	1.2980E+02	-3.9000E-03	0	0	234.94	1.1163	357.31	1.2723
172	Pyridine	C$_5$H$_5$N	110861	79.101	1.0785E+05	-3.4787E+01	-3.9565E-01	0	0	231.51	1.2100	388.41	1.5403
173	Formamide[5]	CH$_3$NO	75127	45.041	6.3400E+04	1.5060E+02	0	0	0	292	1.0738	493	1.3765
174	N,N-Dimethylformamide	C$_3$H$_7$NO	68122	73.095	1.4790E+05	-1.0600E+02	3.8400E-01	0	0	273.82	1.4767	466.44	1.8200
175	Acetamide	C$_2$H$_5$NO	60355	59.068	1.0230E+05	1.2870E+02	0	0	0	354.15	1.4788	571	1.7579
176	N-Methylacetamide	C$_3$H$_7$NO	79163	73.095	6.2600E+04	2.4340E+02	0	0	0	359	1.4998	538.5	1.9367
177	Acetonitrile	C$_2$H$_3$N	75058	41.053	9.7582E+04	-1.2220E+02	3.4085E-01	0	0	229.32	0.8748	354.75	0.9713
178	Propionitrile	C$_3$H$_5$N	107120	55.079	1.1819E+05	-1.2098E+02	4.2075E-01	0	0	180.26	1.1005	370.5	1.3112
179	n-Butyronitrile	C$_4$H$_7$N	109740	69.106	1.0400E+05	1.7400E+02	0	0	0	161.25	1.3206	390.75	1.7199
180	Benzonitrile	C$_7$H$_5$N	100470	103.123	7.6900E+04	3.1420E+02	0	0	0	260.4	1.5872	464.15	2.2274
181	Methyl mercaptan	CH$_4$S	74931	48.109	1.1530E+05	-2.6323E+02	6.0412E-01	0	0	150.18	0.8939	298.15	0.9052
182	Ethyl mercaptan	C$_2$H$_6$S	75081	62.136	1.3467E+05	-2.3439E+02	5.9656E-01	0	0	125.26	1.1467	315.25	1.2007
183	n-Propyl mercaptan	C$_3$H$_8$S	107039	76.163	1.6733E+05	-3.1910E+02	8.1270E-01	0	0	159.95	1.3708	340.87	1.5299
184	n-Butyl mercaptan	C$_4$H$_{10}$S	109795	90.189	2.3219E+05	-8.0435E+02	2.7063E+00	-2.3017E-03	0	157.46	1.6365	390	1.9359
185	Isobutyl mercaptan	C$_4$H$_{10}$S	513440	90.189	1.7336E+05	-2.1732E+02	7.0933E-01	0	0	128.31	1.5715	361.64	1.8754
186	sec-Butyl mercaptan[2]	C$_4$H$_{10}$S	513531	90.189	1.9789E+05	-4.9154E+02	1.7219E+00	-1.2499E-03	0	133.02	1.6003	370	1.8844
187	Dimethyl sulfide	C$_2$H$_6$S	75183	62.136	1.4695E+05	-3.8006E+02	1.2035E+00	-8.4787E-04	0	174.88	1.1276	310.48	1.1959
188	Methyl ethyl sulfide	C$_3$H$_8$S	624895	76.163	1.6124E+05	-2.8861E+02	7.8179E-01	0	0	167.23	1.3484	339.8	1.5344
189	Diethyl sulfide	C$_4$H$_{10}$S	352932	90.189	2.3852E+05	-1.0384E+02	4.0587E+00	-4.4691E-03	0	181.95	1.5703	322.08	1.7579
190	Fluoromethane[2]	CH$_3$F	593533	34.033	7.4746E+04	-1.3232E+02	5.3772E-01	0	0	140	0.6676	220	0.7166
191	Chloromethane	CH$_3$Cl	74873	50.488	9.6910E+04	-2.0790E+02	3.7456E-01	4.8800E-04	0	175.43	0.7460	373.15	0.9684
192	Trichloromethane	CHCl$_3$	67663	119.377	1.2485E+05	-1.6634E+02	-4.3209E-01	0	0	233.15	1.0956	366.48	1.2192
193	Tetrachloromethane	CCl$_4$	56235	153.822	-7.5270E+05	8.9661E+03	-3.0394E-01	3.4550E-02	0	250.33	1.2763	388.71	1.6374
194	Bromomethane	CH$_3$Br	74839	94.939	1.2973E+05	-5.9654E+02	2.1600E+00	-2.4234E-03	0	184.45	0.7798	276.71	0.7870
195	Fluoroethane	C$_2$H$_5$F	353366	48.060	8.3303E+04	6.5454E+01	0	0	0	200	0.9639	281.48	1.0173
196	Chloroethane	C$_2$H$_5$Cl	75003	64.514	1.2790E+05	-3.4515E+02	9.1500E-01	0	0	134.8	0.9800	340	1.1632
197	Bromoethane	C$_2$H$_5$Br	74964	108.966	9.4364E+04	-1.0912E+02	4.4032E-01	0	0	160	0.8818	320	1.0453

TABLE 2-196 Heat Capacities of Inorganic and Organic Liquids (*Concluded*)

Cmpd. no.	Name	Formula	CAS no.	Mol wt	C1	C2	C3	C4	C5	T_{min} K	C_p at T_{min} ×1E-05	T_{max} K	C_p at T_{max} ×1E-05
198	1-Chloropropane	C_3H_7Cl	540545	78.541	9.6344E+04	1.1752E+02	0	0	0	230	1.2337	319.67	1.3391
199	2-Chloropropane	C_3H_7Cl [1]	75296	78.541	6.9362E+04	2.1501E+02	0	0	0	200	1.1236	308.85	1.3577
200	1,1-Dichloropropane	$C_3H_6Cl_2$	78999	112.986	7.0010E+04	2.6660E+02	0	0	0	280	1.4466	420	1.8198
201	1,2-Dichloropropane	$C_3H_6Cl_2$	78875	112.986	1.1094E+05	8.3496E+00	4.7218E-01	0	0	286	1.5195	429	2.0142
202	Vinyl chloride	C_2H_3Cl	75014	62.499	-1.0320E+04	3.2280E+02	0	0	0	200	0.5424	400	1.1880
203	Fluorobenzene	C_6H_5F	462066	96.104	-9.9120E+05	1.1734E+04	-4.0669E+01	4.7333E-02	0	239.99	1.3675	319.99	1.5018
204	Chlorobenzene	C_6H_5Cl	108907	112.558	-1.3075E+06	1.5333E+04	-5.3974E+01	6.3483E-02	0	227.95	1.3617	360	1.8101
205	Bromobenzene	C_6H_5Br	108861	157.010	1.2160E+05	-9.4500E+00	3.5800E-01	0	0	293.15	1.4960	495.08	2.0467
206	Air		132259100	28.951	-2.1446E+05	9.1851E+03	-1.0612E+02	4.1616E-01	0	75	0.5307	115	0.7132
207	Hydrogen (eqn. 2)	H_2	1333740	2.016	6.6653E+03	6.7659E+03	-1.2363E+02	4.7827E+02	3.2129E+03	13.95	0.1262	32	1.3122
208	Helium-4[6]	He	7440597	4.003	3.8722E+05	-4.6557E+05	2.1180E+05	-4.2494E+04	1.3841E+00	2.2	0.1087	4.6	0.2965
209	Neon	Ne	7440019	20.180	1.0341E+06	-1.3887E+05	7.1540E+03	-1.6255E+02	0	24.56	0.3666	40	0.6980
210	Argon	Ar	7440371	39.948	1.3439E+05	-1.9894E+05	1.1043E+01	0	0	83.78	0.4523	135	0.6708
211	Fluorine	F_2	7782414	37.997	-9.4585E+04	7.5299E+03	-1.3960E+02	1.1301E+00	-3.3241E-03	58	0.5541	98	0.5966
212	Chlorine	Cl_2	7782505	70.905	6.3936E+04	4.6350E+01	-1.6230E+01	0	0	172.12	0.6711	239.12	0.6574
213	Bromine	Br_2	7726956	159.808	3.7570E+04	3.3850E+02	-6.7000E-01	-9.2382E-01	2.7963E-03	265.9	0.7755	305.37	0.7541
214	Oxygen	O_2	7782447	31.999	1.7543E+05	-6.1523E+03	1.1392E+02	-2.2182E+00	7.4902E-03	54.36	0.5365	142	0.9066
215	Nitrogen	N_2	7727379	28.014	2.8197E+05	-1.2281E+04	2.4800E+02	-2.6510E+03	0	63.15	0.5593	112	0.7960
216	Ammonia (eqn. 2)	NH_3	7664417	17.031	1.6289E+01	8.0925E+04	7.9940E+02	0	0	203.15	0.7575	401.15	4.1847
217	Hydrazine	N_2H_4	302012	32.045	7.9815E+04	5.0929E+01	4.3379E-02	0	0	274.69	0.9708	653.15	1.3158
218	Nitrous oxide	N_2O	10024972	44.013	6.7556E+04	5.4373E+01	0	0	0	182.3	0.7747	200	0.7843
219	Nitric oxide	NO	10102439	30.006	-2.9796E+06	7.6600E+04	-6.5259E+02	1.8879E+00	0	109.5	0.6229	150	1.9909
220	Cyanogen	C_2N_2	460195	52.036	3.1322E+06	-2.4320E+04	4.8844E+01	0	0	245.25	1.0557	300	2.3216
221	Carbon monoxide (eqn. 2)	CO	630080	28.010	6.5429E+01	2.8723E+04	-8.4739E+02	1.9596E+03	0	68.15	0.5912	132	6.4799
222	Carbon dioxide	CO_2	124389	44.010	-8.3043E+06	1.0437E+05	-4.3333E+02	6.0052E-01	2.0080E-06	220	0.7827	290	1.6603
223	Carbon disulfide	CS_2	75150	76.143	8.5600E+04	-1.2200E+02	5.6050E-01	-1.4520E-03	0	161.11	0.7577	552	1.3125
224	Hydrogen fluoride	HF	7664393	20.006	6.2520E+04	-2.2302E+02	6.2970E-01	0	0	189.79	0.4288	292.67	0.5119
225	Hydrogen chloride	HCl	7647010	36.461	4.7300E+04	9.0000E+01	0	0	0	165	0.6215	185	0.6395
226	Hydrogen bromide	HBr	10035106	80.912	5.7720E+04	9.9000E+00	0	0	0	185.15	0.5955	206.45	0.5976
227	Hydrogen cyanide	HCN	74908	27.026	9.5398E+04	-1.9752E+02	3.8830E-01	0	0	259.83	0.7029	298.85	0.7105
228	Hydrogen sulfide (eqn. 2)	H_2S	7783064	34.082	6.4666E+01	4.9354E+04	2.2493E+01	-1.6230E+03	0	187.68	0.6733	370	4.9183
229	Sulfur dioxide	SO_2	7446095	64.065	8.5743E+04	5.7443E+00	0	0	0	197.67	0.8688	350	0.8775
230	Sulfur trioxide	SO_3	7446119	80.064	2.5809E+05	0.0000E+00	0	0	0	303.15	2.5809	303.15	2.5809
231	Water	H_2O	7732185	18.015	2.7637E+05	-2.0901E+03	8.1250E+00	-1.4116E-02	9.3701E-06	273.16	0.7615	533.15	0.8939

All substances are listed in alphabetical order in Table 2-6a.
Compiled from Daubert, T. E., R. P. Danner, H. M. Sibul, and C. C. Stebbins, DIPPR Data Compilation of Pure Compound Properties, Project 801 Sponsor Release, July, 1993, Design Institute for Physical Properties Data, AIChE, New York, NY; and from Thermodynamics Research Center, "Selected Values of Properties of Hydrocarbons and Related Compounds," Thermodynamics Research Center Hydrocarbon Project, Texas A&M University, College Station, Texas (extant 1994).

Temperatures are expressed in kelvins; liquid heat capacities are in J/kmol·K.
J/(kmol·K) × 2.390E−04 = cal/(gmol·°C); J/(kmol·K) × 2.390059E−04 = Btu/(lbmol·°F).
Equation 1, heat capacity $= C1 + C2 \times T + C3 \times T^2 + C4 \times T^3 + C5 \times T^4$, should be used except as otherwise specified.
Equation 2 is heat capacity $= C1^2/t + C2 - (2 \times C1 \times C3)t - (C1 \times C4)t^2 - (C3^2)t^3 - (C3 \times C4/2)t^4 - (C4^2/5)t^5$. $t = (1 - T_r)$ and T_r is the reduced temperature, T/T_c.

[1] Coefficients are hypothetical.
[2] For the saturated heat capacity.
[3] Coefficients are hypothetical; compound *decomposes violently* on heating.
[4] Coefficients are hypothetical and are based on predicted data.
[5] Coefficients are hypothetical.
[6] Exhibits superfluid properties below 2.2 K.

Coefficients are for the monomer and are hypothetical above 473 K.

TABLE 2-197 Specific Heats of Organic Solids
Recalculated from *International Critical Tables*, vol. 5, pp. 101–105

Compound	Formula	Temperature, °C	sp ht, cal/g °C
Acetic acid	$C_2H_4O_2$	−200 to +25	$0.330 + 0.00080t$
Acetone	C_3H_6O	−210 to −80	$0.540 + 0.0156t$
Aminobenzoic acid (o-)	$C_7H_7NO_2$	85 to mp	$0.254 + 0.00136t$
(m-)	$C_7H_7NO_2$	120 to mp	$0.253 + 0.00122t$
(p-)	$C_7H_7NO_2$	128 to mp	$0.287 + 0.00088t$
Aniline	C_6H_7N		0.741
Anthracene	$C_{14}H_{10}$	50	0.308
		100	0.350
		150	0.382
Anthraquinone	$C_{14}H_8O_2$	0 to 270	$0.258 + 0.00069t$
Apiol	$C_{12}H_{14}O_4$	10	0.299
Azobenzene	$C_{12}H_{10}N_2$	28	0.330
Benzene	C_6H_6	−250	0.0399
		−225	0.0908
		−200	0.124
		−150	0.170
		−100	0.227
		−50	0.299
		0	0.375
Benzoic acid	$C_7H_6O_2$	20 to mp	$0.287 + 0.00050t$
Benzophenone	$C_{13}H_{10}O$	−150	0.115
		−100	0.172
		−50	0.220
		0	0.275
		+20	0.303
Betol	$C_{17}H_{12}O_3$	−150	0.129
		−100	0.167
		0	0.248
		+50	0.308
Bromoiodobenzene (o-)	C_6H_4BrI	−50 to 0	$0.143 + 0.00025t$
(m-)	C_6H_4BrI	−75 to −15	0.143
(p-)	C_6H_4BrI	−40 to 50	$0.116 + 0.00032t$
Bromonaphthalene (β-)	$C_{10}H_7Br$	41	0.260
Bromophenol	C_6H_5BrO	32	0.263
Camphene	$C_{10}H_{16}$	35	0.380
Capric acid	$C_{10}H_{20}O_2$	8	0.695
Caprylic acid	$C_8H_{16}O_2$	−2	0.628
Carbon tetrachloride	CCl_4	−240	0.013
		−200	0.081
		−160	0.131
		−120	0.162
		−80	0.182
		−40	0.201
Cerotic acid	$C_{27}H_{54}O_2$	15	0.387
Chloral alcoholate	$C_4H_7Cl_3O_2$	78	0.509
hydrate	$C_2H_3Cl_3O_2$	32	0.213
Chloroacetic acid	$C_2H_3ClO_2$	60	0.363
Chlorobenzoic acid (o-)	$C_7H_5ClO_2$	80 to mp	$0.228 + 0.00084t$
(m-)	$C_7H_5ClO_2$	94 to mp	$0.232 + 0.00073t$
(p-)	$C_7H_5ClO_2$	180 to mp	$0.242 + 0.00055t$
Chlorobromobenzene (o-)	C_6H_4BrCl	−34	0.192
(m-)	C_6H_4BrCl	−52	0.150
(p-)	C_6H_4BrCl	−40	0.150
Crotonic acid	$C_4H_6O_2$	38 to 70	$0.520 + 0.00020t$
Cyamelide	$C_3H_3N_3O_3$	40	0.263
Cyanamide	CH_2N_2	20	0.547
Cyanuric acid	$C_3H_3N_3O_3$	40	0.318
Dextrin	$(C_6H_{10}O_5)x$	0 to 90	$0.291 + 0.00096t$
Dextrose	$C_6H_{12}O_6$	−250	0.016
		−200	0.077
		−100	0.160
		0	0.277
		20	0.300
Dibenzyl	$C_{14}H_{14}$	28	0.363
Dibromobenzene (o-)	$C_6H_4Br_2$	−36	0.248
(m-)	$C_6H_4Br_2$	−25	0.134
(p-)	$C_6H_4Br_2$	−50 to +50	$0.139 + 0.00038t$
Dichloroacetic acid	$C_2H_2Cl_2O_2$		0.406
Dichlorobenzene (o-)	$C_6H_4Cl_2$	−48.5	0.185
(m-)	$C_6H_4Cl_2$	−52	0.186
(p-)	$C_6H_4Cl_2$	−50 to +53	$0.219 + 0.0021t$
Dicyandiamide	$C_2H_4N_4$	0 to 204	0.456

TABLE 2-197   Specific Heats of Organic Solids (*Continued*)

Recalculated from *International Critical Tables*, vol. 5, pp. 101–105

Compound	Formula	Temperature, °C	sp ht, cal/g °C
Dihydroxybenzene (*o-*)	$C_6H_6O_2$	−163 to mp	$0.278 + 0.00098t$
(*m-*)	$C_6H_6O_2$	−160 to mp	$0.269 + 0.00118t$
(*p-*)	$C_6H_6O_2$	−250	0.025
		−240	0.038
		−220	0.061
		−200	0.081
		−150 to mp	$0.268 + 0.00093t$
Di-iodobenzene (*o-*)	$C_6H_4I_2$	−50 to +15	$0.109 + 0.00026t$
(*m-*)	$C_6H_4I_2$	−52 to −42	$0.100 + 0.00026t$
(*p-*)	$C_6H_4I_2$	−50 to +80	$0.101 + 0.00026t$
Dimethyl oxalate	$C_4H_6O_4$	10 to 50	$0.212 + 0.0044t$
Dimethylpyrene	$C_7H_8O_2$	50	0.368
Dinitrobenzene (*o-*)	$C_6H_4N_2O_4$	−160 to mp	$0.252 + 0.00083t$
(*m-*)	$C_6H_4N_2O_4$	−160 to mp	$0.248 + 0.00077t$
(*p-*)	$C_6H_4N_2O_4$	119 to mp	$0.259 + 0.00057t$
Diphenyl	$C_{12}H_{10}$	40	0.385
Diphenylamine	$C_{12}H_{11}N$	26	0.337
Dulcitol	$C_6H_{14}O_6$	20	0.282
Erythritol	$C_4H_{10}O_4$	60	0.351
Ethyl alcohol	C_2H_6O (crystalline)	−190	0.232
		−180	0.248
		−160	0.282
		−140	0.318
		−130	0.376
	(vitreous)	−190	0.260
		−180	0.296
		−175	0.380
		−170	0.399
Ethylene glycol	$C_2H_6O_2$	−190 to −40	$0.366 + 0.00110t$
Formic acid	CH_2O_2	−22	0.387
		0	0.430
Glutaric acid	$C_5H_8O_4$	20	0.299
Glycerol	$C_3H_8O_3$	−265	0.009
		−260	0.022
		−250	0.047
		−220	0.085
		−200	0.115
		−100	0.217
		0	0.330
Hexachloroethane	C_2Cl_6	25	0.174
Hexadecane	$C_{16}H_{34}$		0.495
Hydroxyacetanilide	$C_8H_9NO_2$	41 to mp	$0.249 + 0.00154t$
Iodobenzene	C_6H_5I	40	0.191
Isopropyl alcohol	C_3H_8O	−200 to −160	$0.051 + 0.00165t$
Lactose	$C_{12}H_{22}O_{11}$	20	0.287
	$C_{12}H_{22}O_{11} \cdot H_2O$	20	0.299
Lauric acid	$C_{12}H_{24}O_2$	−30 to +40	$0.430 + 0.000027t$
Levoglucosane	$C_6H_{10}O_5$	40	0.607
Levulose	$C_6H_{12}O_6$	20	0.275
Malonic acid	$C_3H_4O_4$	20	0.275
Maltose	$C_{12}H_{22}O_{11}$	20	0.320
Mannitol	$C_6H_{14}O_6$	0 to 100	$0.313 + 0.00025t$
Melamine	$C_3H_6N_6$	40	0.351
Myristic acid	$C_{14}H_{28}O_2$	0 to 35	$0.381 + 0.00545t$
Naphthalene	$C_{10}H_8$	−130 to mp	$0.281 + 0.00111t$
Naphthol (*α-*)	$C_{10}H_8O$	50 to mp	$0.240 + 0.00147t$
(*β-*)	$C_{10}H_8O$	61 to mp	$0.252 + 0.00128t$
Naphthylamine (*α-*)	$C_{10}H_9N$	0 to 50	$0.270 + 0.0031t$
Nitroaniline (*o-*)	$C_6H_6N_2O_2$	−160 to mp	$0.269 + 0.000920t$
(*m-*)	$C_6H_6N_2O_2$	−160 to mp	$0.275 + 0.000946t$
(*p-*)	$C_6H_6N_2O_2$	−160 to mp	$0.276 + 0.001000t$
Nitrobenzoic acid (*o-*)	$C_7H_5NO_4$	−163 to mp	$0.256 + 0.00085t$
(*m-*)	$C_7H_5NO_4$	66 to mp	$0.258 + 0.00091t$
(*p-*)	$C_7H_5NO_4$	−160 to mp	$0.247 + 0.00077t$
Nitronaphthalene	$C_{10}H_7NO_2$	0 to 55	$0.236 + 0.00215t$

TABLE 2-197 Specific Heats of Organic Solids (*Concluded*)

Recalculated from *International Critical Tables*, vol. 5, pp. 101–105

Compound	Formula	Temperature, °C	sp ht, cal/g °C
Oxalic acid	$C_2H_2O_4$	−200 to +50	$0.259 + 0.00076t$
	$C_2H_2O_4 \cdot 2H_2O$	−200	0.117
		−100	0.239
		0	0.338
		+50	0.385
		100	0.416
Palmitic acid	$C_{16}H_{32}O_2$	−180	0.167
		−140	0.208
		−100	0.251
		−50	0.306
		0	0.382
		+20	0.430
Phenol	C_6H_6O	14 to 26	0.561
Phthalic acid	$C_8H_6O_4$	20	0.232
Picric acid	$C_6H_3N_3O_7$	−100	0.165
		0	0.240
		+50	0.263
		100	0.297
		120	0.332
Propionic acid	$C_3H_6O_2$	−33	0.726
Propyl alcohol (*n*-)	C_3H_8O	−200	0.170
		−175	0.363
		−150	0.471
		−130	0.497
Pyrotartaric acid	$C_6H_8O_4$	20	0.301
Quinhydrone	$C_{12}H_{10}O_4$	−250	0.017
		−225	0.061
		−200	0.098
		−100	0.191
		0	0.256
Quinone	$C_6H_4O_2$	−250	0.031
		−225	0.082
		−200	0.113
		−150 to mp	$0.282 + 0.00083t$
Salol	$C_{13}H_{10}O_3$	32	0.289
Stearic acid	$C_{18}H_{36}O_2$	15	0.399
Succinic acid	$C_4H_6O_4$	0 to 160	$0.248 + 0.00153t$
Sucrose	$C_{12}H_{22}O_{11}$	20	0.299
Sugar (cane)	$C_{12}H_{22}O_{11}$	22 to 51	0.301
Tartaric acid	$C_4H_6O_6$	36	0.287
Tartaric acid	$C_4H_6O_6 \cdot H_2O$	−150	0.112
		−100	0.170
		−50	0.231
		0	0.308
		+50	0.366
Tetrachloroethylene	C_2Cl_4	−40 to 0	$0.198 + 0.00018t$
Tetryl	$C_7H_5N_5O_8$	−100	0.182
		−50	0.199
		0	0.212
		+100	0.236
1 Tetryl + 1 picric acid	$C_{13}H_8N_8O_{15}$	−100 to +100	$0.253 + 0.00072t$
1 Tetryl + 2 TNT	$C_{21}H_{15}N_{11}O_{20}$	−100	0.172
		0	0.280
		+50	0.325
Thymol	$C_{10}H_{14}O$	0 to 49	$0.315 + 0.0031t$
Toluic acid (*o*-)	$C_8H_8O_2$	54 to mp	$0.277 + 0.00120t$
(*m*-)	$C_8H_8O_2$	54 to mp	$0.239 + 0.00195t$
(*p*-)	$C_8H_8O_2$	130 to mp	$0.271 + 0.00106t$
Toluidine (*p*-)	C_7H_9N	0	0.337
		20	0.387
		40	0.440
Trichloroacetic acid	$C_2HCl_3O_2$	solid	0.459
Trimethyl carbinol	$C_4H_{10}O$	−4	0.559
Trinitrotoluene	$C_7H_5N_3O_6$	−100	0.170
		−50	0.253
		0	0.311
		+100	0.385
Trinitroxylene	$C_8H_7N_3O_6$	−185 to +23	0.241
		20 to 50	0.423
Triphenylmethane	$C_{19}H_{16}$	0 to 91	$0.189 + 0.0027t$
Urea	CH_4N_2O	20	0.320

TABLE 2-198 Heat Capacities of Inorganic and Organic Compounds in the Ideal Gas State

Cmpd. no.	Name	Formula	CAS no.	Mol. wt.	C1 ×1E-05	C2 ×1E-05	C3 ×1E-03	C4 ×1E-05	C5	T_{min} K	C_p at T_{min} ×1E-05	T_{max} K	C_p at T_{max} ×1E-05
1	Methane	CH$_4$	74828	16.043	0.3330	0.7993	2.0869	0.4160	991.96	50	0.3330	1500	0.8890
2	Ethane	C$_2$H$_6$	74840	30.070	0.4033	1.3422	1.6555	0.7322	752.87	200	0.4256	1500	1.4562
3	Propane	C$_3$H$_8$	74986	44.097	0.5192	1.9245	1.6265	1.1680	723.6	200	0.5632	1500	2.0556
4	n-Butane	C$_4$H$_{10}$	106978	58.123	0.7134	2.4300	1.6300	1.5033	730.42	200	0.7673	1500	2.6602
5	n-Pentane	C$_5$H$_{12}$	109660	72.150	0.8805	3.0110	1.6502	1.8920	747.6	200	0.9404	1500	3.2927
6	n-Hexane	C$_6$H$_{14}$	110543	86.177	1.0440	3.5230	1.6946	2.3690	761.6	200	1.1117	1500	3.8620
7	n-Heptane	C$_7$H$_{16}$	142825	100.204	1.2015	4.0010	1.6766	2.7400	756.4	200	1.2828	1500	4.4283
8	n-Octane	C$_8$H$_{18}$	111659	114.231	1.3554	4.4310	1.6356	3.0540	746.4	200	1.4529	1500	4.9764
9	n-Nonane	C$_9$H$_{20}$	111842	128.258	1.5175	4.9150	1.6448	3.4700	749.6	200	1.6257	1500	5.5407
10	n-Decane	C$_{10}$H$_{22}$	124185	142.285	1.6720	5.3530	1.6141	3.7820	742	200	1.7967	1500	6.0932
11	n-Undecane	C$_{11}$H$_{24}$	1120214	156.312	1.9529	6.0998	1.7087	4.1302	775.4	200	2.0594	1500	6.8342
12	n-Dodecane	C$_{12}$H$_{26}$	112403	170.338	2.1295	6.6330	1.7155	4.5161	777.5	200	2.2442	1500	7.4325
13	n-Tridecane	C$_{13}$H$_{28}$	629505	184.365	2.1496	7.3045	1.6695	4.9998	741.02	200	2.3156	1500	8.0251
14	n-Tetradecane	C$_{14}$H$_{30}$	629594	198.392	2.3082	7.8678	1.6823	5.4486	743.1	200	2.4864	1500	8.6225
15	n-Pentadecane	C$_{15}$H$_{32}$	629629	212.419	2.4679	8.4212	1.6865	5.8537	743.6	200	2.6586	1500	9.2209
16	n-Hexadecane	C$_{16}$H$_{34}$	544763	226.446	2.6283	8.9733	1.6912	6.2640	744.41	200	2.8312	1500	9.8182
17	n-Heptadecane	C$_{17}$H$_{36}$	629787	240.473	2.7878	9.5247	1.6935	6.6651	744.57	200	3.0034	1500	10.4160
18	n-Octadecane	C$_{18}$H$_{38}$	593453	254.500	2.9502	10.0340	0.7711	-4.3012	916.73	200	3.1800	1500	11.0160
19	n-Nonadecane	C$_{19}$H$_{40}$	629925	268.527	3.1062	10.5750	0.7679	-4.5661	-912.03	200	3.3533	1500	11.6130
20	n-Eicosane	C$_{20}$H$_{42}$	112958	282.553	3.2481	11.0900	1.6360	7.4500	-726.27	200	3.5235	1500	12.2110
21	2-Methylpropane	C$_4$H$_{10}$	75285	58.123	0.6549	2.4776	1.5870	1.5750	-706.99	200	0.7218	1500	2.6656
22	2-Methylbutane	C$_5$H$_{12}$	78784	72.150	0.7460	3.2650	1.5450	1.9230	666.7	200	0.8546	1500	3.3792
23	2,3-Dimethylbutane	C$_6$H$_{14}$	79298	86.177	0.7772	4.0320	1.5440	2.5080	-649.95	200	0.9363	1500	4.0353
24	2-Methylpentane	C$_6$H$_{14}$	107835	86.177	0.9030	3.8010	1.6020	2.4530	-691.6	200	1.0192	1500	3.9617
25	2,3-Dimethylpentane	C$_7$H$_{16}$	565593	100.204	0.8544	4.5772	1.5181	2.9740	641.01	200	1.0550	1500	4.5983
26	2,3,3-Trimethylpentane	C$_8$H$_{18}$	560214	114.231	0.9820	5.4020	1.5310	3.4930	639.9	200	1.2194	1500	5.3754
27	2,2,4-Trimethylpentane	C$_8$H$_{18}$	540841	114.231	1.1390	5.2860	1.5940	3.3510	677.94	200	1.3139	1500	5.3769
28	Ethylene	C$_2$H$_4$	74851	28.054	0.3338	0.9479	1.5960	0.5510	740.8	60	0.3338	1500	1.0987
29	Propylene	C$_3$H$_6$	115071	42.081	0.4339	1.5200	1.4250	0.7860	623.9	130	0.4388	1500	1.6836
30	1-Butene	C$_4$H$_8$	106989	56.108	0.5998	2.0846	1.5884	1.2940	707.3	200	0.6547	1500	2.2853
31	cis-2-Butene	C$_4$H$_8$	590181	56.108	0.5765	2.1150	1.6299	1.2872	739.1	200	0.6199	1500	2.2715
32	trans-2-Butene	C$_4$H$_8$	624646	56.108	0.6592	2.0700	1.6733	1.2510	742.2	200	0.7004	1500	2.2904
33	1-Pentene	C$_5$H$_{10}$	109671	70.134	0.7595	2.5525	1.5820	1.6660	713	200	0.8273	1500	2.8467
34	1-Hexene	C$_6$H$_{12}$	592416	84.161	0.9180	3.0220	1.5742	2.0320	715	200	0.9995	1500	3.4088
35	1-Heptene	C$_7$H$_{14}$	592767	98.188	1.0775	3.4900	1.5705	2.4030	717.4	200	1.1723	1500	3.9706
36	1-Octene	C$_8$H$_{16}$	111660	112.215	1.2355	3.9570	1.5640	2.7669	718.17	200	1.3440	1500	4.5322
37	1-Nonene	C$_9$H$_{18}$	124118	126.242	1.3950	4.4255	1.5624	3.1370	719.6	200	1.5168	1500	5.0938
38	1-Decene	C$_{10}$H$_{20}$	872059	140.269	1.7573	5.1710	1.7664	3.6210	803.02	200	1.8333	1500	5.8682
39	2-Methylpropene	C$_4$H$_8$	115117	56.108	0.6125	2.0660	1.5450	1.2057	676	200	0.6763	1500	2.2814
40	2-Methyl-1-butene	C$_5$H$_{10}$	563462	70.134	0.8703	2.5556	1.7757	1.7636	807.82	200	0.9060	1500	2.8923
41	2-Methyl-2-butene	C$_5$H$_{10}$	513359	70.134	0.8192	2.6038	1.7593	1.7195	800.93	200	0.8559	1500	2.8709
42	1,2-Butadiene	C$_4$H$_6$	590192	54.092	0.5750	1.6476	1.5270	0.9900	677.3	200	0.6269	1500	1.9202
43	1,3-Butadiene	C$_4$H$_6$	106990	54.092	0.5095	1.7050	1.5324	1.3370	685.6	200	0.5756	1500	1.9555
44	2-Methyl-1,3-butadiene	C$_5$H$_8$	78795	68.119	0.6527	2.2993	1.4943	1.5164	-647.15	200	0.7508	1500	2.5571
45	Acetylene	C$_2$H$_2$	74862	26.038	0.3199	0.5424	1.5940	0.4325	607.1	200	0.3566	1500	0.7575
46	Methylacetylene	C$_3$H$_4$	74997	40.065	0.4478	1.0917	1.5508	0.6750	658.2	200	0.4882	1500	1.3293
47	Dimethylacetylene	C$_4$H$_6$	503173	54.092	0.6534	1.6179	1.7837	1.0242	821.4	200	0.6721	1500	1.9148
48	3-Methyl-1-butyne	C$_5$H$_8$	598232	68.119	0.8274	2.1377	1.7550	1.5149	782	200	0.8646	1500	2.5255
49	1-Pentyne	C$_5$H$_8$	627190	68.119	0.7530	2.0905	1.5307	1.3780	672.8	200	0.8276	1500	2.4754
50	2-Pentyne	C$_5$H$_8$	627214	68.119	0.7074	2.2229	1.5570	1.3125	690.78	200	0.7700	1500	2.5052
51	1-Hexyne	C$_6$H$_{10}$	693027	82.145	0.9129	2.5577	1.5290	1.7370	683	200	1.0004	1500	3.0371
52	2-Hexyne	C$_6$H$_{10}$	764352	82.145	1.0360	3.0090	2.1160	2.1060	902.4	300	1.2215	1500	3.1894
53	3-Hexyne	C$_6$H$_{10}$	928494	82.145	0.9376	3.0150	1.9057	1.9960	817	300	1.1909	1500	3.1889

No.	Name	Formula	CAS No.	Mol. wt.									
55	1-Octyne	C$_8$H$_{14}$	629050	110.199	1.2307	3.4942	1.5280	2.4617	694.81	200	1.3448	1500	4.1604
56	Vinylacetylene	C$_4$H$_4$	689974	52.076	0.5598	1.2141	1.6102	0.8908	−710.4	200	0.5967	1500	1.5590
57	Cyclopentane	C$_5$H$_{10}$	287923	70.134	0.4160	3.0140	1.4617	1.8095	−668.8	100	0.4165	1500	2.9298
58	Methylcyclopentane	C$_6$H$_{12}$	96377	84.161	0.6646	3.5070	1.5592	2.3526	727.13	200	0.7510	1500	3.5495
59	Ethylcyclopentane	C$_7$H$_{14}$	1640897	98.188	0.8205	4.0342	1.5670	2.6697	715.52	200	0.9272	1500	4.1472
60	Cyclohexane	C$_6$H$_{12}$	110827	84.161	0.4320	3.7350	1.1920	1.6350	−530.1	100	0.4366	1500	3.6516
61	Methylcyclohexane	C$_7$H$_{14}$	108872	98.188	0.9227	4.1150	1.6504	2.9006	779.48	200	0.9953	1500	4.3180
62	1,1-Dimethylcyclohexane	C$_8$H$_{16}$	590669	112.215	1.0776	4.6718	1.6540	3.3397	792.5	200	1.1535	1500	4.9543
63	Ethylcyclohexane	C$_8$H$_{16}$	1678917	112.215	1.1059	4.6306	1.6628	3.2290	781.1	200	1.1875	1500	4.9184
64	Cyclopentene	C$_5$H$_8$	142290	68.119	0.4807	2.5159	1.5803	1.7454	718.37	150	0.4918	1500	2.5619
65	1-Methylcyclopentene	C$_6$H$_{10}$	693890	82.145	0.6941	3.0209	1.6903	2.1209	781.56	200	0.7464	1500	3.1496
66	Cyclohexene	C$_6$H$_{10}$	110838	82.145	0.5817	3.1717	1.5435	2.1273	701.62	150	0.5978	1500	3.2132
67	Benzene	C$_6$H$_6$	71432	78.114	0.4442	2.3205	1.4946	1.7213	−678.15	200	0.5340	1500	2.4169
68	Toluene	C$_7$H$_8$	108883	92.141	0.5814	2.8630	1.4406	1.8980	−650.43	200	0.7016	1500	3.0029
69	o-Xylene	C$_8$H$_{10}$	95476	106.167	0.8521	3.2954	1.4944	2.1150	−675.8	200	0.9643	1500	3.5965
70	m-Xylene	C$_8$H$_{10}$	108383	106.167	0.7568	3.3924	1.4960	2.2470	−675.9	200	0.8759	1500	3.5920
71	p-Xylene	C$_8$H$_{10}$	106423	106.167	0.7512	3.3970	1.4928	2.2470	−675.1	200	0.8710	1500	3.5923
72	Ethylbenzene	C$_8$H$_{10}$	100414	106.167	0.7844	3.3990	1.5390	2.4200	−702	200	0.8012	1500	3.6147
73	Propylbenzene (eqn. 3)	C$_9$H$_{12}$	103651	120.194	−21.4827	3.8070	54701	−0.001713	0	200	1.0802	1500	4.1537
74	1,2,4-Trimethylbenzene	C$_9$H$_{12}$	95636	120.194	1.0106	3.8314	1.5010	2.3950	678.3	200	1.1354	1500	4.1854
75	Isopropylbenzene	C$_9$H$_{12}$	98828	120.194	1.0810	3.7932	1.7505	3.0027	794.8	200	1.1480	1500	4.1808
76	1,3,5-Trimethylbenzene	C$_9$H$_{12}$	108678	120.194	0.9154	3.9270	1.4980	2.5090	676.9	200	1.0474	1500	4.1807
77	p-Isopropyltoluene	C$_{10}$H$_{14}$	99876	134.218	1.3186	4.3036	1.7734	3.2570	811.9	200	1.3825	1500	4.7952
78	Naphthalene	C$_{10}$H$_8$	91203	128.174	0.6805	3.5494	1.4262	2.5984	650.1	200	0.8454	1500	3.7359
79	Biphenyl	C$_{12}$H$_{10}$	92524	154.211	0.9060	4.2634	1.4553	3.1550	661.2	200	1.0913	1500	4.5581
80	Styrene	C$_8$H$_8$	100425	104.152	0.8930	2.1503	0.7720	0.9990	2442	100	0.8931	1500	3.2416
81	m-Terphenyl	C$_{18}$H$_{14}$	92068	230.309	1.6397	6.0125	1.6902	5.1314	757.5	298.15	2.4618	1500	6.6678
82	Methanol	CH$_4$O	67561	32.042	0.3925	0.8790	1.9165	0.5365	896.7	200	0.3980	1500	1.0533
83	Ethanol	C$_2$H$_6$O	64175	46.069	0.4920	1.4577	1.6628	0.9390	744.7	200	0.5224	1500	1.6576
84	1-Propanol	C$_3$H$_8$O	71238	60.096	0.6190	2.0213	1.6293	1.2956	727.4	200	0.6665	1500	2.2458
85	1-Butanol	C$_4$H$_{10}$O	71363	74.123	0.7454	2.5907	1.6073	1.7320	712.4	200	0.8162	1500	2.8509
86	2-Butanol	C$_4$H$_{10}$O	78922	74.123	0.8202	2.5220	1.6010	1.5864	−704.15	200	0.8890	1500	2.8513
87	2-Propanol	C$_3$H$_8$O	67630	60.096	0.5723	1.9100	1.4210	1.2155	626	150	0.5924	1500	2.1792
88	2-Methyl-2-propanol	C$_4$H$_{10}$O	75650	74.123	0.7704	2.5390	1.5502	1.6690	−679.3	200	0.8567	1500	2.8508
89	1-Pentanol	C$_5$H$_{12}$O	71410	88.150	0.9060	3.0620	1.6054	2.1150	−717.97	200	0.9890	1500	3.4133
90	2-Methyl-1-butanol	C$_5$H$_{12}$O	137326	88.150	1.0890	2.1850	0.8530	1.4000	2906	298.15	1.3247	1500.1	3.4718
91	3-Methyl-1-butanol	C$_5$H$_{12}$O	123513	88.150	1.1060	2.2100	0.8760	1.2200	2940	298.15	1.3213	1200.15	3.1770
92	1-Hexanol	C$_6$H$_{14}$O	111273	102.177	1.0625	3.5210	1.5835	2.4620	715.75	200	1.1607	1500	3.9726
93	1-Heptanol	C$_7$H$_{16}$O	111706	116.203	1.2215	3.9910	1.5800	2.8350	717.7	200	1.3330	1500	4.5346
94	Cyclohexanol	C$_6$H$_{12}$O	108930	100.161	0.9043	2.5771	0.7882	1.3068	1952.2	200	0.9648	1500	3.8251
95	Ethylene glycol	C$_2$H$_6$O$_2$	107211	62.068	0.8200	1.2780	1.6980	0.9290	−754	200	0.8481	1000.15	1.8521
96	1,2-Propylene glycol	C$_3$H$_8$O$_2$	57556	76.095	2.0114	0.8082	1.8656	−2.4404	279.98	298.15	1.0218	1000.15	2.1175
97	Phenol	C$_6$H$_6$O	108952	94.113	0.4340	2.4450	1.1520	1.5120	−507	100	0.4401	1500	2.6045
98	o-Cresol	C$_7$H$_8$O	95487	108.140	0.7988	2.8530	1.4765	2.0420	−664.7	200	0.9158	1500	3.2163
99	m-Cresol	C$_7$H$_8$O	108394	108.140	0.7515	2.0900	0.6666	1.2120	2214	200	0.8701	1500	3.2075
100	p-Cresol	C$_7$H$_8$O	106445	108.140	0.7384	2.9080	1.4559	2.0910	−650.42	200	0.8707	1500	3.2102
101	Dimethyl ether	C$_2$H$_6$O	115106	46.069	0.5148	1.4420	1.6034	0.7747	725.4	200	0.5436	1500	1.6581
102	Methyl ethyl ether	C$_3$H$_8$O	540670	60.096	0.6868	1.9959	1.5534	1.1168	692.04	200	0.7396	1500	2.2931
103	Methyl-n-propyl ether	C$_4$H$_{10}$O	557175	74.123	0.9215	2.3943	1.6936	1.4896	797.79	298	1.1251	1200	2.6391
104	Methyl isopropyl ether	C$_4$H$_{10}$O	598538	74.123	0.8923	2.4765	1.6960	1.5598	791.4	200	0.9280	1200	2.8696
105	Methyl-n-butyl ether	C$_5$H$_{12}$O	628284	88.150	0.8205	3.0869	1.3864	1.7886	613.87	300	1.3300	1200	3.1994
106	Methyl isobutyl ether	C$_5$H$_{12}$O	623445	88.150	0.7284	3.1713	1.3520	1.8948	585.14	300	1.3200	1200	3.1987

TABLE 2-198 Heat Capacities of Inorganic and Organic Compounds in the Ideal Gas State (Continued)

Cmpd. no.	Name	Formula	CAS no.	Mol. wt.	C1 ×1E-05	C2 ×1E-05	C3 ×1E-03	C4 ×1E-05	C5	T_{min} K	C_p at T_{min} ×1E-05	T_{max} K	C_p at T_{max} ×1E-05
107	Methyl tert-butyl ether	$C_5H_{12}O$	1634044	88.150	0.9933	3.0667	1.7426	2.0764	795.59	200	1.0394	1500	3.4321
108	Diethyl ether	$C_4H_{10}O$	60297	74.123	0.8621	2.5510	1.5413	1.4370	-688.9	200	0.9316	1500	2.9244
109	Ethyl propyl ether	$C_5H_{12}O$	628320	88.150	1.1320	2.9400	1.8270	2.0550	-852	298.15	1.3538	1500	3.4535
110	Ethyl isopropyl ether	$C_5H_{12}O$	625547	88.150	1.0953	3.0032	1.7988	2.1311	817.35	298.15	1.3620	1200	3.2289
111	Methyl phenyl ether	C_7H_8O	100663	108.140	0.7637	2.9377	1.6051	2.1700	751.2	300	1.1302	1200	3.0226
112	Diphenyl ether	$C_{12}H_{10}O$	101848	170.211	1.0985	4.3412	1.6222	3.6455	743.62	300	1.7298	1200	4.5143
113	Formaldehyde	CH_2O	50000	30.026	0.3327	0.4954	1.8666	0.2808	934.9	50	0.3327	1500	0.7113
114	Acetaldehyde	C_2H_4O	75070	44.053	0.4451	1.0687	1.6141	0.6135	737.8	200	0.4660	1500	1.2994
115	1-Propanal	C_3H_6O	123386	58.080	0.7174	1.9140	2.0144	1.1708	930.6	200	0.7266	1500	2.1149
116	1-Butanal	C_4H_8O	123728	72.107	0.8966	2.3731	1.9754	1.5866	904.13	298.15	0.9119	1500	2.6775
117	1-Pentanal	$C_5H_{10}O$	110623	86.134	1.0743	2.8363	1.9559	2.0146	890.44	200	1.0960	1500	3.2404
118	1-Hexanal	$C_6H_{12}O$	66251	100.161	1.2320	2.2146	0.8400	1.2190	2205	200	1.2672	1500	3.7314
119	1-Heptanal	$C_7H_{14}O$	111717	114.188	1.4040	2.5907	0.8315	1.3120	2201	200	1.4479	1500	4.2863
120	1-Octanal	$C_8H_{16}O$	124130	128.214	1.6088	4.2180	1.9126	3.2780	869	200	1.6504	1500	4.9286
121	1-Nonanal	$C_9H_{18}O$	124196	142.241	1.7347	4.5115	1.7120	3.3256	810.96	200	1.8005	1500	5.4439
122	1-Decanal	$C_{10}H_{20}O$	112312	156.268	1.9641	5.1412	1.8989	4.1278	862.51	200	2.0192	1500	6.0539
123	Acetone	C_3H_6O	67641	58.080	0.5704	1.6320	1.6070	0.9650	731.5	200	0.6049	1500	1.8820
124	Methyl ethyl ketone	C_4H_8O	78933	72.107	0.7840	2.1032	1.5488	1.1855	693	200	0.8397	1500	2.4816
125	2-Pentanone	$C_5H_{10}O$	107879	86.134	0.9005	2.7085	1.6592	1.8012	743.96	200	0.9591	1500	3.0797
126	Methyl isopropyl ketone	$C_5H_{10}O$	563804	86.134	1.5914	1.7640	1.2076	-407.4000	10.503	300	1.1291	1500	2.9991
127	2-Hexanone	$C_6H_{12}O$	591786	100.161	1.0940	1.8070	0.6990	0.4740	1772	200	1.1815	1200	3.3207
128	Methyl isobutyl ketone	$C_6H_{12}O$	108101	100.161	1.2270	2.1950	0.8420	1.1910	2460	298.15	1.4755	1500.15	3.6532
129	3-Methyl-2-pentanone	$C_6H_{12}O$	565617	100.161	1.0028	3.3169	1.6900	2.3000	770.7	300	1.3604	1200	3.4275
130	3-Pentanone	$C_5H_{10}O$	96220	86.134	0.9690	2.4907	1.4177	1.3010	646.7	200	1.0536	1500	3.0358
131	Ethyl isopropyl ketone	$C_6H_{12}O$	565695	100.161	1.2400	3.2000	1.9670	2.3460	896	298.15	1.4479	1200	3.4234
132	Diisopropyl ketone	$C_7H_{14}O$	565800	114.188	1.0869	4.0540	1.7802	2.9786	791.6	300	1.5102	1500	4.3093
133	Cyclohexanone	$C_6H_{10}O$	108941	98.145	0.5776	3.3535	1.2202	5.7700	586.92	200	0.7321	1500	3.4870
134	Methyl phenyl ketone	C_8H_8O	98862	120.151	0.8540	2.3340	0.8310	0.7730	2227	298.15	1.1313	1500	3.2797
135	Formic acid[1]	CH_2O_2	64186	46.026	0.3381	0.7593	1.1925	0.3180	550	50	0.3381	1500	0.9933
136	Acetic acid[2]	$C_2H_4O_2$	64197	60.053	0.4020	1.3675	1.2620	0.7003	569.7	50	0.4020	1500	1.5756
137	Propionic acid[2]	$C_3H_6O_2$	79094	74.079	0.6959	1.7778	1.7098	1.2654	-763.78	298.15	0.8938	1500	2.1248
138	n-Butyric acid[2]	$C_4H_8O_2$	107926	88.106	1.4880	1.3522	1.1460	-678.0000	6.98	298.15	1.1533	1200.1	2.4716
139	Isobutyric acid[2]	$C_4H_8O_2$	79312	88.106	0.7469	2.4356	1.7150	1.8484	757.75	298.15	1.0427	1200	2.5383
140	Benzoic acid	$C_7H_6O_2$	65850	122.123	0.7759	2.6455	1.7925	2.2382	835.59	200	0.8126	1500	2.9712
141	Acetic anhydride	$C_4H_6O_3$	108247	102.090	0.7130	2.2220	1.6203	1.6760	746.5	200	0.7665	1500	2.5675
142	Methyl formate	$C_2H_4O_2$	107313	60.053	0.5060	1.2190	1.6370	0.8940	743	250	0.5888	1500	1.5109
143	Methyl acetate	$C_3H_6O_2$	79209	74.079	0.5550	1.7820	1.2600	0.8530	562	298	0.8489	1500	2.0754
144	Methyl propionate	$C_4H_8O_2$	554121	88.106	0.7765	2.4420	1.7140	1.8180	716	300	1.1242	1200	2.5276
145	Methyl n-butyrate	$C_5H_{10}O_2$	623427	102.133	0.8940	2.9100	1.5700	2.0730	678.3	298	1.3461	1500	3.0766
146	Ethyl formate	$C_3H_6O_2$	109944	74.079	0.5370	1.8860	1.2070	0.8640	496	100	0.5412	1500	2.1485
147	Ethyl acetate	$C_4H_8O_2$	141786	88.106	0.9981	2.0931	2.0226	1.8030	928.05	200	1.0126	1500	2.6594
148	Ethyl propionate	$C_5H_{10}O_2$	105373	102.133	0.9370	2.8290	1.6480	2.1550	724.7	300	1.3377	1200	3.0569
149	Ethyl n-butyrate	$C_6H_{12}O_2$	105544	116.160	1.1150	3.3910	1.6705	2.5180	733.6	298	1.5583	1200	3.6213
150	n-Propyl formate	$C_4H_8O_2$	110747	88.106	0.8710	2.4470	1.9254	1.8880	-821.3	298.15	1.1022	1500	2.7484
151	n-Propyl acetate	$C_5H_{10}O_2$	109604	102.133	1.7994	1.7530	1.1960	-4.1200	108.2	298.15	1.3594	1200	3.2276
152	n-Butyl acetate	$C_6H_{12}O_2$	123864	116.160	1.1684	3.7690	1.9560	2.8180	811.2	300	1.5358	1200	3.6724
153	Methyl benzoate	$C_8H_8O_2$	93583	136.150	0.9396	2.5590	0.8250	1.3600	3000	300	1.2586	1200	3.3569
154	Ethyl benzoate	$C_9H_{10}O_2$	93890	150.177	1.0944	4.1794	0.8838	-1.6900	-1183.1	300	1.4598	1200	4.2540
155	Vinyl acetate	$C_4H_6O_2$	108054	86.090	0.5360	2.1190	1.1980	1.1470	510	100	0.5404	1500	2.3750
156	Methylamine	CH_5N	74895	31.057	0.4100	1.0578	1.7080	0.6836	735	150	0.4136	1500	1.2388
157	Dimethylamine	C_2H_7N	124403	45.084	0.5565	1.6384	1.7341	1.0899	793.04	200	0.5812	1500	1.8585
158	Trimethylamine	C_3H_9N	75503	59.111	0.7107	1.5051	0.7966	0.8454	2187.6	200	0.7439	1500	2.4322
159	Ethylamine	C_2H_7N	75047	45.084	0.5940	1.6180	1.8120	1.0780	820	200	0.6139	1500	1.8528

No.	Name	Formula	CAS No.	Mol. wt.	C1	C2	C3	C4	C5	Tmin	Cp@Tmin	Tmax	Cp@Tmax
160	Diethylamine	$C_4H_{11}N$	109897	73.138	0.9102	2.6740	1.7190	1.7926	794.94	200	0.9502	1500	3.0519
161	Triethylamine	$C_6H_{15}N$	121448	101.192	1.2766	2.5559	0.8094	1.4829	2231.7	200	1.3278	1500	4.2046
162	n-Propylamine	C_3H_9N	107108	59.111	0.7608	2.1049	1.7256	1.3936	789.03	200	0.7933	1500	2.4353
163	di-n-Propylamine	$C_6H_{15}N$	142847	101.192	1.2114	2.6127	0.7896	1.6903	2394.4	300	1.5900	1500	4.2484
164	Isopropylamine	C_3H_9N	75310	59.111	0.6855	2.1876	1.5831	1.3855	691.76	200	0.7510	1500	2.4540
165	Diisopropylamine	$C_6H_{15}N$	108189	101.192	1.1384	2.5747	0.7384	1.6200	2143	300	1.5995	1500	4.1941
166	Aniline	C_6H_7N	62533	93.128	0.6533	2.5192	1.4608	1.8870	-653.1	200	0.7705	1500	2.8047
167	N-Methylaniline	C_7H_9N	100618	107.155	0.7796	3.0280	1.5203	2.3250	699.8	300	1.2602	1500	3.3641
168	N,N-Dimethylaniline	$C_8H_{11}N$	121697	121.182	0.8742	2.7204	0.7242	1.1300	1949	300	1.3903	1500	3.8844
169	Ethylene oxide	C_2H_4O	75218	44.053	0.3346	1.2116	1.6084	0.8241	737.3	50	0.3346	1500	1.3297
170	Furan	C_4H_4O	110009	68.075	0.3727	1.6606	1.5112	1.3145	686	200	0.4376	1500	1.7940
171	Thiophene	C_4H_4S	110021	84.142	0.4040	1.6270	1.4564	1.3212	649	200	0.4884	1500	1.8097
172	Pyridine	C_5H_5N	110861	79.101	0.4413	2.0830	1.4783	1.5330	676.8	200	0.5220	1500	2.2194
173	Formamide	CH_3NO	75127	45.041	0.3822	0.9300	1.8450	0.6900	850	150	0.3833	1500	1.1203
174	N,N-Dimethylformamide	C_3H_7NO	68122	73.095	0.7220	1.7830	1.5320	1.3100	762	200	0.7594	1500	2.2596
175	Acetamide	C_2H_5NO	60355	59.068	0.3420	1.2940	1.0750	0.6400	502	100	0.3448	1500	1.4997
176	N-Methylacetamide	C_3H_7NO	70163	73.095	0.6116	2.0090	1.7683	1.3302	835.5	300	0.7698	1500	2.2209
177	Acetonitrile	C_2H_3N	75058	41.053	0.4191	0.8876	1.5818	0.5032	699.8	100	0.4192	1500	1.1285
178	Propionitrile	C_3H_5N	107120	55.079	0.5357	1.4617	1.5530	0.9120	678.2	200	0.5832	1500	1.7235
179	n-Butyronitrile	C_4H_7N	109740	69.106	0.6906	1.9996	1.5494	1.3146	675	200	0.7607	1500	2.3273
180	Benzonitrile	C_7H_5N	100470	103.123	0.7186	2.2700	1.4669	1.6930	-680.77	200	0.8053	1500	2.6706
181	Methyl mercaptan	CH_4S	74931	48.109	0.4146	0.8307	1.5890	0.4612	716.7	200	0.4329	1500	1.0781
182	Ethyl mercaptan	C_2H_6S	75081	62.136	0.5576	1.3617	1.5221	0.8073	687.5	200	0.5970	1500	1.6729
183	n-Propyl mercaptan	C_3H_8S	107039	76.163	0.7474	1.9523	1.6310	1.2112	750.92	200	0.7848	1500	2.3216
184	n-Butyl mercaptan	$C_4H_{10}S$	109795	90.189	0.9248	2.7795	1.6837	1.5974	758.68	200	0.9714	1500	3.1008
185	Isobutyl mercaptan	$C_4H_{10}S$	513440	90.189	0.9142	2.4513	1.6265	1.6157	745.8	200	0.9660	1500	2.9005
186	sec-Butyl mercaptan	$C_4H_{10}S$	513531	90.189	0.9237	2.5166	1.6109	1.5641	739.2	200	0.9763	1500	2.9615
187	Dimethyl sulfide	C_2H_6S	75183	62.136	0.6037	1.3747	1.6410	0.7988	-743.5	200	0.6298	1500	1.6949
188	Methyl ethyl sulfide	C_3H_8S	624895	76.163	0.7508	1.9577	1.6424	1.1949	749.19	273.16	0.9004	1500	2.3178
189	Diethyl sulfide	$C_4H_{10}S$	352932	90.189	0.9429	2.6863	1.7624	1.6752	-798.3	200	0.9794	1500	3.0338
190	Fluoromethane	CH_3F	593533	34.033	0.3329	0.7399	1.8639	0.4608	891.16	50	0.3329	1500	0.9024
191	Chloromethane	CH_3Cl	74873	50.488	0.3409	0.7246	1.7230	0.4480	780.5	150	0.3424	1500	0.9097
192	Trichloromethane	$CHCl_3$	67663	119.377	0.3942	0.6573	0.9280	0.4930	399.6	100	0.4048	1500	1.0063
193	Tetrachloromethane	CCl_4	56235	153.822	0.3758	0.7054	0.5121	0.4850	236.1	100	0.4730	1500	1.0662
194	Bromomethane	CH_3Br	74839	94.939	0.3377	0.7150	1.5780	0.4175	691.4	100	0.3378	1500	0.9107
195	Fluoroethane	C_2H_5F	353366	48.060	0.4437	1.3119	1.6422	0.8544	738.77	100	0.4726	1500	1.5008
196	Chloroethane	C_2H_5Cl	75003	64.514	0.4568	1.2967	1.5992	0.8590	708.8	100	0.4569	1500	1.5112
197	Bromoethane	C_2H_5Br	74964	108.966	0.4719	1.2787	1.5957	0.8517	703.87	200	0.5089	1500	1.5121
198	1-Chloropropane	C_3H_7Cl	540545	78.541	0.6210	1.8430	1.6290	1.2337	724	200	0.6674	1500	2.1126
199	2-Chloropropane	C_3H_7Cl	75296	78.541	0.6181	1.8023	1.5438	1.1893	685.93	200	0.6768	1500	2.1023
200	1,1-Dichloropropane	$C_3H_6Cl_2$	78999	112.986	0.7145	1.7344	1.5240	1.2230	674.2	150	0.7268	1500	2.1609
201	1,2-Dichloropropane	$C_3H_6Cl_2$	78875	112.986	0.7866	1.7429	1.7157	1.2627	765.1	200	0.8217	1500	2.1894
202	Vinyl chloride	C_2H_3Cl	75014	62.499	0.4236	0.8735	1.6492	0.6556	739.07	200	0.4457	1500	1.1423
203	Fluorobenzene	C_6H_5F	462066	96.104	0.6265	2.1646	1.5640	1.7278	-724.29	200	0.6914	1500	2.4736
204	Chlorobenzene	C_6H_5Cl	108907	112.558	0.8011	2.3100	2.1570	2.0460	-897.6	200	0.8219	1500	2.5327
205	Bromobenzene	C_6H_5Br	108861	157.010	0.7210	2.0640	1.6504	1.6870	765.3	200	0.7679	1500	2.4628
206	Air		132259100	28.951	0.2896	0.0939	3.0120	0.0758	1484	50	0.2896	1500	0.3496
207	Hydrogen[3]	H_2	1333740	2.016	0.2762	0.0956	2.4660	0.0376	567.6	250	0.2843	1500	0.3225
208	Helium-4 (eqn 2)	He	7440597	4.003	0.2079	0	0	0	0	100	0.2079	1500	0.2079
209	Neon	Ne	7440019	20.180	0.2079	0	0	0	0	100	0.2079	1500	0.2079
210	Argon	Ar	7440371	39.948	0.2079	0	0	0	0	100	0.2079	1500	0.2079
211	Fluorine	F_2	7782414	37.997	0.2912	0.1013	1.4530	0.0941	662.91	50	0.2912	1500	0.3812

TABLE 2-198 Heat Capacities of Inorganic and Organic Compounds in the Ideal Gas State (Concluded)

Cmpd. no.	Name	Formula	CAS no.	Mol wt.	C1 ×1E-05	C2 ×1E-05	C3 ×1E-03	C4 ×1E-05	C5	T_{min} K	C_p at T_{min} ×1E-05	T_{max} K	C_p at T_{max} ×1E-05
212	Chlorine	Cl_2	7782505	70.905	0.2914	0.0918	0.9490	0.1003	425	50	0.2914	1500	0.3793
213	Bromine	Br_2	7726956	159.808	0.3011	0.0801	0.7514	0.1078	314.6	100	0.3090	1500	0.3794
214	Oxygen	O_2	7782447	31.999	0.2910	0.1004	2.5265	0.0936	1153.8	50	0.2910	1500	0.3653
215	Nitrogen	N_2	7727379	28.014	0.2911	0.0861	1.7016	0.0010	909.79	50	0.2911	1500	0.3484
216	Ammonia	NH_3	7664417	17.031	0.3343	0.4898	2.0360	0.2256	882	100	0.3343	1500	0.6647
217	Hydrazine	N_2H_4	302012	32.045	0.3871	0.8576	1.7228	0.5664	733.53	200	0.4070	1500	1.0571
218	Nitrous oxide	N_2O	10024972	44.013	0.2934	0.3236	1.1238	0.2177	479.4	100	0.2948	1500	0.5828
219	Nitric oxide (eqn 2)	NO	10102439	30.006	0.3498	-3.5320E-04	7.7290E-05	-5.7357E-10	1.4526E-08	100	0.3217	1500	0.3586
220	Cyanogen	C_2N_2	460195	52.036	0.3545	0.5015	1.0570	0.4520	-396	100	0.3648	1500	0.8100
221	Carbon monoxide	CO	630080	28.010	0.2911	0.0877	3.0851	0.0846	1538.2	60	0.2911	1500	0.3521
222	Carbon dioxide	CO_2	124389	44.010	0.2937	0.3454	1.4280	0.2640	588	50	0.2937	5000	0.6335
223	Carbon disulfide	CS_2	75150	76.143	0.3010	0.3338	0.8960	0.2893	374.7	100	0.3100	1500	0.6148
224	Hydrogen fluoride	HF	7664393	20.006	0.2913	0.0933	2.9050	0.0020	1326	50	0.2913	1500	0.3224
225	Hydrogen chloride	HCl	7647010	36.461	0.2916	0.0905	2.0938	-0.0011	120	50	0.2914	1500	0.3406
226	Hydrogen bromide	HBr	10035106	80.912	0.2912	0.0953	2.1420	0.0157	1400	50	0.2912	1500	0.3479
227	Hydrogen cyanide	HCN	74908	27.026	0.3013	0.3171	1.6102	0.2179	626	100	0.3014	1500	0.5522
228	Hydrogen sulfide	H_2S	7783064	34.082	0.3329	0.2609	0.9134	-0.1798	949.4	100	0.3329	1500	0.5143
229	Sulfur dioxide	SO_2	7446095	64.065	0.3338	0.2586	0.9328	0.1088	423.7	100	0.3354	1500	0.5695
230	Sulfur trioxide	SO_3	7446119	80.064	0.3341	0.4968	0.8732	0.2856	393.74	100	0.3408	1500	0.7967
231	Water	H_2O	7732185	18.015	0.3336	0.2679	2.6105	0.0890	1169	100	0.3336	2273.15	0.5276

All substances are listed in alphabetical order in Table 2-6a.

Compiled from Daubert, T. E., R. P. Danner, H. M. Sibul, and C. C. Stebbins, DIPPR Data Compilation of Pure Compound Properties, Project 801 Sponsor Release, July, 1993, Design Institute for Physical Property Data, AIChE, New York, NY; and from Thermodynamics Research Center, "Selected Values of Properties of Hydrocarbons and Related Compounds," Thermodynamics Research Center Hydrocarbon Project, Texas A&M University, College Station, Texas (extant 1994).

Temperatures are expressed in kelvins; heat capacities, in J/kmol·K.

$J/(kmol \cdot K) \times 2.390E{-}04 = cal/(gmol \cdot °C)$; $J/(kmol \cdot K) \times 2.390059E{-}04 = Btu/(lbmol \cdot °F)$.

Use heat capacity $= C1 + C2 \left[\dfrac{C3}{T} \bigg/ \sinh \left(\dfrac{C3}{T} \right) \right]^2 + C4 \left[\dfrac{C5}{T} \bigg/ \cosh \left(\dfrac{C5}{T} \right) \right]^2$ unless otherwise specified.

Equation 2 is heat capacity $= C1 + C2 \times T + C3 \times T^2 + C4 \times T^3 + C5 \times T^4$.

Equation 3 is heat capacity $= C1 + C2 \times \ln T + C3/T + C4 \times T$.

[1] For the monomer. Monomer and dimer are in equilibrium below 600 K.

[2] For the monomer.

[3] For equilibrium mixture of *ortho* and *para* hydrogen.

TABLE 2-199 C_p/C_v: Ratios of Specific Heats of Gases at 1-atm Pressure*

Compound	Formula	Temperature, °C	Ratio of specific heats, $(\gamma) = C_p/C_v$	Compound	Formula	Temperature, °C	Ratio of specific heats, $(\gamma) = C_p/C_v$
Acetaldehyde	C_2H_4O	30	1.14	Hydrogen (Cont.)			
Acetic acid	$C_2H_4O_2$	136	1.15	iodide	HI	20–100	1.40
Acetylene	C_2H_2	15	1.26	sulfide	H_2S	15	1.32
		−71	1.31			−45	1.30
Air		925	1.36			−57	1.29
		17	1.403				
		−78	1.408	Iodine	I_2	185	1.30
		−118	1.415	Isobutane	C_4H_{10}	15	1.11
Ammonia	NH_3	15	1.310				
Argon	A	15	1.668	Krypton	Kr	19	1.68
		−180	1.76 (?)				
		0–100	1.67	Mercury	Hg	360	1.67
				Methane	CH_4	600	1.113
Benzene	C_6H_6	90	1.10			300	1.16
Bromine	Br_2	20–350	1.32			15	1.31
						−80	1.34
Carbon dioxide	CO_2	15	1.304			−115	1.41
		−75	1.37	Methyl acetate	$C_3H_6O_2$	15	1.14
disulfide	CS_2	100	1.21	alcohol	CH_4O	77	1.203
monoxide	CO	15	1.404	ether	C_2H_6O	6–30	1.11
		−180	1.41	Methylal	$C_3H_8O_2$	13	1.06
Chlorine	Cl_2	15	1.355			40	1.09
Chloroform	$CHCl_3$	100	1.15				
Cyanogen	$(CN)_2$	15	1.256	Neon	Ne	19	1.64
Cyclohexane	C_6H_{12}	80	1.08	Nitric oxide	NO	15	1.400
						−45	1.39
Dichlorodifluormethane	CCl_2F_2	25	1.139			−80	1.38
				Nitrogen	N_2	15	1.404
Ethane	C_2H_6	100	1.19			−181	1.47
		15	1.22	Nitrous oxide	N_2O	100	1.28
		−82	1.28			15	1.303
Ethyl alcohol	C_2H_6O	90	1.13			−30	1.31
ether	$C_4H_{10}O$	35	1.08			−70	1.34
		80	1.086				
Ethylene	C_2H_4	100	1.18	Oxygen	O_2	15	1.401
		15	1.255			−76	1.415
		−91	1.35			−181	1.45
Helium	He	−180	1.660	Pentane (n-)	C_5H_{12}	86	1.086
Hexane (n-)	C_6H_{14}	80	1.08	Phosphorus	P	300	1.17
Hydrogen	H_2	15	1.410	Potassium	K	850	1.77
		−76	1.453				
		−181	1.597	Sodium	Na	750–920	1.68
bromide	HBr	20	1.42	Sulfur dioxide	SO_2	15	1.29
chloride	HCl	15	1.41				
		100	1.40	Xenon	Xe	19	1.66
cyanide	HCN	65	1.31				
		140	1.28				
		210	1.24				

*From *International Critical Tables*, vol. 5, pp. 80–82.

TABLE 2-200 Specific Heat Ratio, C_p/C_v, for Air

Temperature, K	Pressure, bar															
	1	10	20	40	60	80	100	150	200	250	300	400	500	600	800	1000
150	1.410	1.510	1.668	2.333	4.120	3.973	3.202	2.507	2.243	2.091	1.988	1.851	1.768	1.712	1.654	1.639
200	1.406	1.452	1.505	1.630	1.781	1.943	2.093	2.274	2.236	2.140	2.050	1.920	1.832	1.771	1.682	1.619
250	1.403	1.429	1.457	1.517	1.577	1.640	1.699	1.816	1.877	1.896	1.885	1.836	1.782	1.743	1.681	1.636
300	1.402	1.418	1.436	1.470	1.505	1.537	1.570	1.640	1.687	1.716	1.730	1.727	1.707	1.683	1.645	1.619
350	1.399	1.411	1.422	1.446	1.467	1.488	1.509	1.553	1.589	1.612	1.627	1.640	1.638	1.629	1.605	1.585
400	1.395	1.404	1.412	1.429	1.444	1.460	1.472	1.505	1.529	1.548	1.563	1.579	1.584	1.580	1.567	1.555
450	1.392	1.397	1.404	1.416	1.428	1.438	1.449	1.471	1.490	1.505	1.518	1.533	1.541	1.542	1.537	1.528
500	1.387	1.391	1.395	1.406	1.414	1.421	1.430	1.448	1.463	1.474	1.484	1.499	1.507	1.510	1.510	1.504
600	1.377	1.378	1.382	1.386	1.392	1.398	1.403	1.413	1.423	1.432	1.439	1.448	1.457	1.461	1.465	1.466
800	1.353	1.355	1.357	1.359	1.361	1.365	1.366	1.372	1.375	1.381	1.384	1.392	1.397	1.401	1.406	1.409
1000	1.336	1.337	1.338	1.339	1.342	1.343	1.343	1.345	1.348	1.350	1.354	1.358	1.361	1.365	1.368	1.372

Calculated from C_p, C_v values of Sychev, V. V., A. A. Vasserman, et al., "Thermodynamic Properties of Air," Standartov, Moscow, 1978 and Hemisphere, New York, 1988 (276 pp.).

SPECIFIC HEATS OF AQUEOUS SOLUTIONS

UNITS CONVERSIONS

For this subsection, the following units conversions are applicable:

$$°F = \tfrac{9}{5}\,°C + 32.$$

To convert calories per gram-degree Celsius to British thermal units per pound-degree Fahrenheit, multiply by 1.0.

TABLE 2-201   Acetic Acid (at 38°C)

Mole % acetic acid	0	6.98	30.9	54.5	100
Cal/g °C	1.0	0.911	0.73	0.631	0.535

TABLE 2-202   Ammonia

Mole % NH$_3$	Specific heat, cal/g °C			
	2.4°C	20.6°C	41°C	61°C
0	1.01	1.0	0.995	1.0
10.5	0.98	0.995	1.06	1.02
20.9	.96	.99	1.03	
31.2	.956	1.0		
41.4	.985			

TABLE 2-203   Aniline (at 20°C)

Mol % aniline	100	95	90.5	82.3	75.2
Cal/g °C		0.497	0.52	0.53	0.56	0.581

TABLE 2-204   Copper Sulfate

Composition	Temperature	Specific heat, cal/g °C
CuSO$_4$ + 50H$_2$O	12° to 15°C	0.848
CuSO$_4$ + 200H$_2$O	12° to 14°C	.951
CuSO$_4$ + 400H$_2$O	13° to 17°C	.975

TABLE 2-205   Ethyl Alcohol

Mole % C$_2$H$_5$OH	Specific heat, cal/g °C		
	3°C	23°C	41°C
4.16	1.05	1.02	1.02
11.5	1.02	1.03	1.03
37.0	0.805	0.86	0.875
61.0	.67	.727	.748
100.0	.54	.577	.621

TABLE 2-206   Glycerol

Mole % C$_3$H$_5$(OH)$_3$	Specific heat, cal/g °C	
	15°C	32°C
2.12	0.961	0.960
4.66	.929	.924
11.5	.851	.841
22.7	.765	.758
43.9	.67	.672
100.0	.555	.576

TABLE 2-207   Hydrochloric Acid

Mole % HCl	Specific heat, cal/g °C				
	0°C	10°C	20°C	40°C	60°C
0.0	1.00				
9.09	0.72	0.72	0.74	0.75	0.78
16.7	.61	.605	.631	.645	.67
20.0	.58	.575	.591	.615	.638
25.9	.55				.61

ADDITIONAL REFERENCES

For additional data, see *International Critical Tables,* vol. 5, pp. 115–116, 122–125.

TABLE 2-208   Methyl Alcohol

Mole % CH$_3$OH	Specific heat, cal/g °C		
	5°C	20°C	40°C
5.88	1.02	1.0	0.995
12.3	0.975	0.982	.98
27.3	.877	.917	.92
45.8	.776	.811	.83
69.6	.681	.708	.726
100	.576	.60	.617

TABLE 2-209   Nitric Acid

% HNO$_3$ by Weight	Specific Heat at 20°C, cal/g °C
0	1.000
10	0.900
20	.810
30	.730
40	.675
50	.650
60	.640
70	.615
80	.575
90	.515

TABLE 2-210   Phosphoric Acid*

%H$_2$PO$_4$	C_p at 21.3°C cal/g °C	%H$_3$PO$_4$	C_p at 21.3°C cal/g °C
2.50	0.9903	50.00	0.6350
3.80	.9970	52.19	.6220
5.33	.9669	53.72	.6113
8.81	.9389	56.04	.5972
10.27	.9293	58.06	.5831
14.39	.8958	60.23	.5704
16.23	.8796	62.10	.5603
19.99	.8489	64.14	.5460
22.10	.8300	66.13	.5349
24.56	.8125	68.14	.5242
25.98	.8004	69.97	.5157
28.15	.7856	69.50	.5160
29.96	.7735	71.88	.5046
32.09	.7590	73.71	.4940
33.95	.7432	75.79	.4847
36.26	.7270	77.69	.4786
38.10	.7160	79.54	.4680
40.10	.7024	80.00	.4686
42.08	.6877	82.00	.4593
44.11	.6748	84.00	.4500
46.22	.6607	85.98	.4419
48.16	.6475	88.01	.4359
49.79	.6370	89.72	.4206

Z. Physik. Chem., A167, **42** (1933).

TABLE 2-211   Potassium Chloride

Mole % KCl	Specific heat, cal/g °C			
	6°C	20°C	33°C	40°C
0.99	0.945	0.947	0.947	0.947
3.85	.828	.831	.835	.837
5.66	.77	.775	.778	.775
7.41		.727		

TABLE 2-212   Potassium Hydroxide (at 19°C)

Mole % KOH	0	0.497	1.64	4.76	9.09
Cal/g °C	1.0	0.975	0.93	0.814	0.75

TABLE 2-213   Normal Propyl Alcohol

Mole % C$_3$H$_7$OH	Specific heat, cal/g °C		
	5°C	20°C	40°C
1.55	1.03	1.02	1.01
5.03	1.07	1.06	1.03
11.4	1.035	1.032	0.99
23.1	0.877	0.90	.91
41.2	.75	.78	.815
73.0	.612	.645	.708
100.0	.534	.57	.621

TABLE 2-214   Sodium Carbonate*

% Na$_2$CO$_3$ by weight	Temperature, °C			
	17.6	30.0	76.6	98.0
0.000	0.9992	0.9986	1.0098	1.0084
1.498	.9807			
2.000		.9786		
2.901	.9597			
4.000		.9594		
5.000	.9428		0.9761	
6.000		.9392		
8.000	.9183			
10.000	.9086		.9452	
13.790	.8924			
13.840		.8881		
20.000		.8631	.8936	
25.000			.8615	0.8911

*J. Chem. Soc. 3062–3079 (1931).

TABLE 2-215   Sodium Chloride

Mole % NaCl	Specific heat, cal/g °C			
	6°C	20°C	33°C	57°C
0.249		0.99		
.99	0.96	.97	0.97	
2.44	.91	.915	.915	0.923
9.09	.805	.81	.81	.82

TABLE 2-216   Sodium Hydroxide (at 20°C)

Mole % NaOH	0	0.5	1.0	9.09	16.7	28.6	37.5
Cal/g °C	1.0	0.985	0.97	0.835	0.80	0.784	0.782

TABLE 2-217   Sulfuric Acid*

%H$_2$SO$_4$	C_p at 20°C, cal/g °C	%H$_2$SO$_4$	C_p at 20°C, cal/g °C
0.34	0.9968	35.25	0.7238
0.68	.9937	37.69	.7023
1.34	.9877	40.49	.6770
2.65	.9762	43.75	.6476
3.50	.9688	47.57	.6153
5.16	.9549	52.13	.5801
9.82	.9177	57.65	.5420
15.36	.8767	64.47	.5012
21.40	.8339	73.13	.4628
22.27	.8275	77.91	.4518
23.22	.8205	81.33	.4481
24.25	.8127	82.49	.4467
25.39	.8041	84.48	.4408
26.63	.7945	85.48	.4346
28.00	.7837	89.36	.4016
29.52	.7717	91.81	.3787
30.34	.7647	94.82	.3554
31.20	.7579	97.44	.3404
33.11	.7422	100.00	.3352

*Vinal and Craig, *Bur. Standards J. Research,* **24,** 475 (1940).

TABLE 2-218   Zinc Sulfate

Composition	Temperature	Specific heat, cal/g °C
ZnSO$_4$ + 50H$_2$O	20° to 52°C	0.842
ZnSO$_4$ + 200H$_2$O	20° to 52°C	.952

SPECIFIC HEATS OF MISCELLANEOUS MATERIALS

TABLE 2-219 Specific Heats of Miscellaneous Liquids and Solids

Material	Specific heat, cal/g °C
Alumina	0.2 (100°C); 0.274 (1500°C)
Alundum	0.186 (100°C)
Asbestos	0.25
Asphalt	0.22
Bakelite	0.3 to 0.4
Brickwork	About 0.2
Carbon	0.168 (26° to 76°C)
	0.314 (40° to 892°C)
	0.387 (56° to 1450°C)
(gas retort)	0.204
(see under Graphite)	
Cellulose	0.32
Cement, Portland Clinker	0.186
Charcoal (wood)	0.242
Chrome brick	0.17
Clay	0.224
Coal	0.26 to 0.37
tar oils	0.34 (15° to 90°C)
Coal tars	0.35 (40°C); 0.45 (200°C)
Coke	0.265 (21° to 400°C)
	0.359 (21° to 800°C)
	0.403 (21° to 1300°C)
Concrete	0.156 (70° to 312°F); 0.219 (72° to 1472°F)
Cryolite	0.253 (16° to 55°C)
Diamond	0.147
Fireclay brick	0.198 (100°C); 0.298 (1500°C)
Fluorspar	0.21 (30°C)
Gasoline	0.53
Glass (crown)	0.16 to 0.20
(flint)	0.117
(pyrex)	0.20
(silicate)	0.188 to 0.204 (0 to 100°C)
	0.24 to 0.26 (0 to 700°C)
wool	0.157
Granite	0.20 (20° to 100°C)
Graphite	0.165 (26° to 76°C); 0.390 (56° to 1450°C)
Gypsum	0.259 (16° to 46°C)
Kerosene	0.47
Limestone	0.217
Litharge	0.055
Magnesia	0.234 (100°C); 0.188 (1500°C)
Magnesite brick	0.222 (100°C); 0.195 (1500°C)
Marble	0.21 (18°C)
Porcelain, fired Berlin	0.189 (60°C)
Porcelain, green Berlin	0.185 (60°C)
Porcelain, fired earthenware	0.186 (60°C)
Porcelain, green earthenware	0.181 (60°C)

TABLE 2-219 Specific Heats of Miscellaneous Liquids and Solids (*Concluded*)

Material	Specific heat, cal/g °C
Pyrex glass	0.20
Pyrites (copper)	0.131 (30°C)
Pyrites (iron)	0.136 (30°C)
Pyroxylin plastics	0.34 to 0.38
Quartz	0.17 (0°C); 0.28 (350°C)
Rubber (vulcanized)	0.415
Sand	0.191
Silica	0.316
Silica brick	0.202 (100°C); 0.195 (1500°C)
Silicon carbide brick	0.202 (100°C)
Silk	0.33
Steel	0.12
Stone	about 0.2
Stoneware (common)	0.188 (60°C)
Turpentine	0.42 (18°C)
Wood (Oak)	0.570
Woods, miscellaneous	0.45 to 0.65
Wool	0.325
Zirconium oxide	0.11 (100°C); 0.179 (1500°C)

TABLE 2-219a Oils (Animal, Vegetable, Mineral Oils)

$$C_p[\text{cal}/(g \cdot °C)] = A/\sqrt{d_4^{15}} + B(t - 15)$$

where d = density, g/cm^3.

°F = $\frac{9}{5}$ °C + 32; to convert calories per gram-degree Celsius to British thermal units per pound-degree Fahrenheit, multiply by 1.0; to convert grams per cubic centimeter to pounds per cubic foot, multiply by 62.43.

Oils	A	B
Castor	0.500	0.0007
Citron	(0.438 at 54°C)	
Fatty drying	0.440	0.0007
non-drying	0.450	0.0007
semidrying	0.445	0.0007
oils (except castor)	0.450	0.0007
Naphthene base	0.405	0.0009
Olive	(0.47 at 7°C)	
Paraffin base	0.425	0.0009
Petroleum oils	0.415	0.0009

HEATS AND FREE ENERGIES OF FORMATION

UNITS CONVERSIONS

°F = $\frac{9}{5}$ °C + 32; to convert kilocalories per gram-mole to British thermal units per pound-mole, multiply by 1.799×10^{-3}.

TABLE 2-220 Heats and Free Energies of Formation of Inorganic Compounds

The values given in the following table for the heats and free energies of formation of inorganic compounds are derived from (a) Bichowsky and Rossini, "Thermochemistry of the Chemical Substances," Reinhold, New York, 1936; (b) Latimer, "Oxidation States of the Elements and Their Potentials in Aqueous Solution," Prentice-Hall, New York, 1938; (c) the tables of the American Petroleum Institute Research Project 44 at the National Bureau of Standards; and (d) the tables of Selected Values of Chemical Thermodynamic Properties of the National Bureau of Standards. The reader is referred to the preceding books and tables for additional details as to methods of calculation, standard states, and so on.

Compound	State	Heat of formation ΔH (formation) at 25°C, kcal/mole	Free energy of formation ΔF (formation) at 25°C, kcal/mole	Compound	State	Heat of formation ΔH (formation) at 25°C, kcal/mole	Free energy of formation ΔF (formation) at 25°C, kcal/mole
Aluminum				Barium (Cont.)			
Al	c	0.00	0.00	BaF_2	c	−287.9	
$AlBr_3$	c	−123.4			aq, 1600	−284.6	−265.3
	aq	−209.5	−189.2	BaH_2	c	−40.8	−31.5
Al_4C_3	c	−30.8	−29.0	$Ba(HCO_3)_2$	aq	−459	−414.4
$AlCl_3$	c	−163.8		BaI_2	c	−144.6	
	aq, 600	−243.9	−209.5		aq, 400	−155.17	−158.52
AlF_3	c	−329		$Ba(IO_3)_2$	c	−264.5	
	aq	−360.8	−312.6		aq	−237.50	−198.35
AlI_3	c	−72.8		$BaMoO_4$	c	−370	
	aq	−163.4	−152.5	Ba_3N_2	c	−90.7	
AlN	c	−57.7	−50.4	$Ba(NO_2)_2$	c	−184.5	
$Al(NH_4)(SO_4)_2$	c	−561.19	−486.17		aq	−179.05	−150.75
$Al(NH_4)(SO_4)_2 \cdot 12H_2O$	c	−1419.36	−1179.26	$Ba(NO_3)_2$	c	−236.99	−189.94
$Al(NO_3)_3 \cdot 6H_2O$	c	−680.89	−526.32		aq, 600	−227.74	
$Al(NO_3)_3 \cdot 9H_2O$	c	−897.59		BaO	c	−133.0	
Al_2O_3	c, corundum	−399.09	−376.87	$Ba(OH)_2$	c	−225.9	
$Al(OH)_3$	c	−304.8	−272.9		aq, 400	−237.76	−209.02
$Al_2O_3 \cdot SiO_2$	c, sillimanite	−648.7		$BaO \cdot SiO_2$	c	−363	
$Al_2O_3 \cdot SiO_2$	c, disthene	−642.4		$Ba_3(PO_4)_2$	c	−992	
$Al_2O_3 \cdot SiO_2$	c, andalusite	−642.0		$BaPtCl_6$	c	−284.9	
$3Al_2O_3 \cdot 2SiO_2$	c, mullite	−1874		BaS	c	−111.2	
Al_2S_3	c	−121.6		$BaSO_3$	c	−282.5	
$Al_2(SO_4)_3$	c	−820.99	−739.53	$BaSO_4$	c	−340.2	−313.4
	aq	−893.9	−759.3	$BaWO_4$	c	−402	
$Al_2(SO_4)_3 \cdot 6H_2O$	c	−1268.15	−1103.39	Beryllium			
$Al_2(SO_4)_3 \cdot 18H_2O$	c	−2120		Be	c	0.00	0.00
Antimony				$BeBr_2$	c	−79.4	
Sb	c	0.00	0.00		aq	−142	−127.9
$SbBr_3$	c	−59.9		$BeCl_2$	c	−112.6	
$SbCl_3$	c	−91.3	−77.8		aq	−163.9	−141.4
$SbCl_5$	l	−104.8		BeI_2	c	−39.4	
SbF_3	c	−216.6			aq	−112	−103.4
SbI_3	c	−22.8		Be_3N_2	c	−134.5	−122.4
Sb_2O_3	c, I, orthorhombic	−165.4	−146.0	BeO	c	−145.3	−138.3
	c, II, octahedral	−166.6		$Be(OH)_2$	c	−215.6	
Sb_2O_4	c	−213.0	−186.6	BeS	c	−56.1	
Sb_2O_5	c	−230.0	−196.1	$BeSO_4$	c	−281	
Sb_2S_3	c, black	−38.2	−36.9		aq		−254.8
Arsenic				Bismuth			
As	c	0.00	0.00	Bi	c	0.00	0.00
$AsBr_3$	c	−45.9		$BiCl_3$	c	−90.5	−76.4
$AsCl_3$	l	−80.2	−70.5		aq	−101.6	
AsF_3	l	−223.76	−212.27	BiI_3	c	−24	
AsH_3	g	43.6	37.7		aq	−27	
AsI_3	c	−13.6		BiO	c	−49.5	−43.2
As_2O_3	c	−154.1	−134.8	Bi_2O_3	c	−137.1	−117.9
As_2O_5	c	−217.9	−183.9	$Bi(OH)_3$	c	−171.1	
As_2S_3	c	−20	−20	Bi_2S_3	c	−43.9	−39.1
	amorphous	−34.76		$Bi_2(SO_4)_3$	c	−607.1	
Barium				Boron			
Ba	c	0.00	0.00	B	c	0.00	0.00
$BaBr_2$	c	−180.38		BBr_3	l	−52.7	
	aq, 400	−185.67	−183.0		g	−44.6	−50.9
$BaCl_2$	c	−205.25		BCl_3	g	−94.5	−90.8
	aq, 300	−207.92	−196.5	BF_3	g	−265.2	−261.0
$Ba(ClO_3)_2$	c	−176.6		B_2H_6	g	7.5	19.9
	aq, 1600	−170.0	−134.4	BN	c	−32.1	−27.2
$Ba(ClO_4)_2$	c	−210.2		B_2O_3	c	−302.0	−282.9
	aq, 800		−155.3		gls	−297.6	−280.3
$Ba(CN)_2$	c	−48		$B(OH)_3$	c	−260.0	−229.4
$Ba(CNO)_2$	c	−212.1		B_2S_3	c	−56.6	
	aq		−180.7	Bromine			
$BaCN_2$	c	−63.6		Br_2	l	0.00	0.00
$BaCO_3$	c, witherite	−284.2	−271.4		g	7.47	0.931
$BaCrO_4$	c	−342.2		BrCl	g	3.06	−0.63

*For footnotes see end of table.

TABLE 2-220   Heats and Free Energies of Formation of Inorganic Compounds (*Continued*)

Compound	State[†]	Heat of formation[‡][§] ΔH (formation) at 25°C, kcal/mole	Free energy of formation[∥][¶] ΔF (formation) at 25°C, kcal/mole	Compound	State[†]	Heat of formation[‡][§] ΔH (formation) at 25°C, kcal/mole	Free energy of formation[∥][¶] ΔF (formation) at 25°C, kcal/mole
Cadmium				**Cesium** (*Cont.*)			
Cd	c	0.00	0.00	Cs_2CO_3	c	−271.88	
$CdBr_2$	c	−75.8	−70.7	CsF	c	−131.67	
	aq, 400	−76.6	−67.6		aq, 400	−140.48	−135.98
$CdCl_2$	c	−92.149	−81.889	CsH	c	−12	−7.30
	aq, 400	−96.44	−81.2	$CsHCO_3$	c	−230.6	
$Cd(CN)_2$	c	36.2			aq, 2000	−226.6	−210.56
$CdCO_3$	c	−178.2	−163.2	CsI	c	−83.91	
CdI_2	c	−48.40			aq, 400	−75.74	−82.61
	aq, 400	−47.46	−43.22	$CsNH_2$	c	−28.2	
Cd_3N_2	c	39.8		$CsNO_3$	c	−121.14	
$Cd(NO_3)_2$	aq, 400	−115.67	−71.05		aq, 400	−111.54	−96.53
CdO	c	−62.35	−55.28	Cs_2O	c	−82.1	
$Cd(OH)_2$	c	−135.0	−113.7	CsOH	c	−100.2	
CdS	c	−34.5	−33.6		aq, 200	−117.0	−107.87
$CdSO_4$	c	−222.23		Cs_2S	c	−87	
	aq, 400	−232.635	−194.65	Cs_2SO_4	c	−344.86	
Calcium					aq	−340.12	−316.66
Ca	c	0.00	0.00	**Chlorine**			
$CaBr_2$	c	−162.20		Cl_2	g	0.00	0.00
	aq, 400	−187.19	−181.86	ClF	g	−25.7	
CaC_2	c	−14.8	−16.0	ClO	g	33	
$CaCl_2$	c	−190.6	−179.8	ClO_2	g	24.7	29.5
	aq	−209.15	−195.36	ClO_3	g	37	
$CaCN_2$	c	−85		Cl_2O	g	18.20	22.40
$Ca(CN)_2$	c	−43.3		Cl_2O_7	g	63	
	aq		−54.0	**Chromium**			
$CaCO_3$	c, calcite	−289.5	−270.8	Cr	c	0.00	0.00
	c, aragonite	−289.54	−270.57	$CrBr_3$	aq		−122.7
$CaCO_3 \cdot MgCO_3$	c	−558.8		Cr_3C_2	c	−21.008	−21.20
CaC_2O_4	c	−332.2		Cr_4C	c	−16.378	−16.74
$Ca(C_2H_3O_2)_2$	c	−356.3		$CrCl_2$	c	−103.1	−93.8
	aq	−364.1	−311.3		aq		−102.1
CaF_2	c	−290.2		CrF_2	c	−152	
	aq	−286.5	−264.1	CrF_3	c	−231	
CaH_2	c	−46	−35.7	CrI_2	c	−63.7	
CaI_2	c	−128.49			aq		−64.1
	aq, 400	−156.63	−157.37	CrO_3	c	−139.3	
Ca_3N_2	c	−103.2	−88.2	Cr_2O_3	c	−268.8	−249.3
$Ca(NO_3)_2$	c	−224.05	−177.38	$Cr_2(SO_4)_3$	aq		−626.3
	aq, 400	−228.29		**Cobalt**			
$Ca(NO_3)_2 \cdot 2H_2O$	c	−367.95	−293.57	Co	c	0.00	0.00
$Ca(NO_3)_2 \cdot 3H_2O$	c	−439.05	−351.58	$CoBr_2$	c	−55.0	
$Ca(NO_3)_2 \cdot 4H_2O$	c	−509.43	−409.32		aq	−73.61	−61.96
CaO	c	−151.7	−144.3	Co_3C	c	9.49	7.08
$Ca(OH)_2$	c	−235.58	−213.9	$CoCl_2$	c	−76.9	−66.6
	aq, 800	−239.2	−207.9		aq, 400	−95.58	−75.46
$CaO \cdot SiO_2$	c, II, wollastonite	−377.9	−357.5	$CoCO_3$	c	−172.39	−155.36
	c, I, pseudo-wollastonite	−376.6	−356.6	CoF_2	aq	−172.98	−144.2
				CoI_2	c	−24.2	
CaS	c	−114.3	−113.1		aq	−43.15	−37.4
$CaSO_4$	c, insoluble form	−338.73	−311.9	$Co(NO_3)_2$	c	−102.8	
	c, soluble form α	−336.58	−309.8		aq	−114.9	−65.3
	c, soluble form β	−335.52	−308.8	CoO	c	−57.5	
$CaSO_4 \cdot \frac{1}{2}H_2O$	c	−376.13		Co_3O_4	c	−196.5	
$CaSO_4 \cdot 2H_2O$	c	−479.33	−425.47	$Co(OH)_2$	c	−131.5	−108.9
$CaWO_4$	c	−387		$Co(OH)_3$	c	−177.0	−142.0
Carbon				CoS	c	−22.3	−19.8
C	c, graphite	0.00	0.00	Co_2S_3	c	−40.0	
	c, diamond	0.453	0.685	$CoSO_4$	c	−216.6	
CO	g	−26.416	−32.808		aq, 400		−188.9
CO_2	g	−94.052	−94.260	**Columbium**			
Cerium				Cb	c	0.00	0.00
Ce	c	0.00	0.00	Cb_2O_5	c	−462.96	
CeN	c	−78.2	−70.8	**Copper**			
Cesium				Cu	c	0.00	0.00
Cs	c	0.00	0.00	CuBr	c	−26.7	−23.8
CsBr	c	−97.64		$CuBr_2$	c	−34.0	
	aq, 500	−91.39	−94.86		aq	−42.4	−33.25
CsCl	c	−106.31		CuCl	c	−31.4	−24.13
	aq, 400	−102.01	−101.61	$CuCl_2$	c	−48.83	
					aq, 400	−64.7	

TABLE 2-220   Heats and Free Energies of Formation of Inorganic Compounds (Continued)

Compound	State†	Heat of formation‡§ ΔH (formation) at 25°C, kcal/mole	Free energy of formation‖¶ ΔF (formation) at 25°C, kcal/mole	Compound	State†	Heat of formation‡§ ΔH (formation) at 25°C, kcal/mole	Free energy of formation‖¶ ΔF (formation) at 25°C, kcal/mole
Copper (Cont.)				Hydrogen (Cont.)			
CuClO$_4$	aq	−28.3	1.34	H$_2$CO$_3$	aq	−167.19	−149.0
Cu(ClO$_3$)$_2$	aq, 400		15.4	HF	g	−64.2	−64.7
Cu(ClO$_4$)$_2$	aq		−5.5		aq, 200	−75.75	
CuI	c	−17.8	−16.66	HI	g	6.27	0.365
CuI$_2$	c	−4.8			aq, 400	−13.47	−12.35
	aq	−11.9	−8.76	HIO	aq	−38	−23.33
Cu$_3$N	c	17.78		HIO$_3$	c	−56.77	
Cu(NO$_3$)$_2$	c	−73.1			aq	−54.8	−32.25
	aq, 200	−83.6	−36.6	HN$_3$	g	70.3	78.50
CuO	c	−38.5	−31.9	HNO$_3$	g	−31.99	−17.57
Cu$_2$O	c	−43.00	−38.13		l	−41.35	−19.05
Cu(OH)$_2$	c	−108.9	−85.5		aq, 400	−49.210	
CuS	c	−11.6	−11.69	HNO$_3$·H$_2$O	l	−112.91	−78.36
Cu$_2$S	c	−18.97	−20.56	HNO$_3$·3H$_2$O	l	−252.15	−193.70
CuSO$_4$	c	−184.7	−158.3	H$_2$O	g	−57.7979	−54.6351
	aq, 800	−200.78	−160.19		l	−68.3174	−56.6899
Cu$_2$SO$_4$	c	−179.6		H$_2$O$_2$	l	−45.16	−28.23
	aq		−152.0		aq, 200	−45.80	−31.47
Erbium				H$_3$PO$_2$	c	−145.5	
Er	c	0.00	0.00		aq	−145.6	−120.0
Er(OH)$_3$	c	−326.8		H$_3$PO$_3$	c	−232.2	
Fluorine					aq	−232.2	−204.0
F$_2$	g	0.00	0.00	H$_3$PO$_4$	c	−306.2	
F$_2$O	g	5.5	9.7		aq, 400	−309.32	−270.0
Gallium				H$_2$S	g	−4.77	−7.85
Ga	c	0.00	0.00		aq, 2000	−9.38	
GaBr$_3$	c	−92.4		H$_2$S$_2$	l	−3.6	
GaCl$_3$	c	−125.4		H$_2$SO$_3$	aq, 200	−146.88	−128.54
GaN	c	−26.2		H$_2$SO$_4$	l	−193.69	
Ga$_2$O	c	−84.3			aq, 400	−212.03	
Ga$_2$O$_3$	c	−259.9		H$_2$Se	g	20.5	17.0
Germanium					aq	18.1	18.4
Ge	c	0.00	0.00	H$_2$SeO$_3$	c	−126.5	
Ge$_3$N$_4$	c	−15.7			aq	−122.4	−101.36
GeO$_2$	c	−128.6		H$_2$SeO$_4$	c	−130.23	
Gold					aq, 400	−143.4	
Au	c	0.00	0.00	H$_2$SiO$_3$	c	−267.8	−247.9
AuBr	c	−3.4		H$_4$SiO$_4$	c	−340.6	
AuBr$_3$	c	−14.5		H$_2$Te	g	36.9	33.1
	aq	−11.0	24.47	H$_2$TeO$_3$	c	−145.0	−115.7
AuCl	c	−8.3			aq	−145.0	
AuCl$_3$	c	−28.3		H$_2$TeO$_4$	aq	−165.6	
	aq	−32.96	4.21	Indium			
AuI	c	0.2	−0.76	In	c	0.00	0.00
Au$_2$O$_3$	c	11.0	18.71	InBr$_3$	c	−97.2	
Au(OH)$_3$	c	−100.6			aq	−112.9	−97.2
Hafnium				InCl$_3$	c	−128.5	
Hf	c	0.00	0.00		aq	−145.6	−117.5
HfO$_2$	c	−271.1	−258.2	InI$_3$	c	−56.5	
Hydrogen					aq	−67.2	−60.5
H$_3$AsO$_3$	aq	−175.6	−153.04	InN	c	−4.8	
H$_3$AsO$_4$	c	−214.9		In$_2$O$_3$	c	−222.47	
	aq	−214.84	−183.93	Iodine			
HBr	g	−8.66	−12.72	I$_2$	c	0.00	0.00
	aq, 400	−28.80	−24.58		g	14.88	4.63
HBrO	aq	−25.4	−19.90	IBr	g	10.05	1.24
HBrO$_3$	aq	−11.51	5.00	ICl	g	4.20	−1.32
HCl	g	−22.063	−22.778	ICl$_3$	c	−21.8	−6.05
	aq, 400	−39.85	−31.330	I$_2$O$_5$	c	−42.5	
HCN	g	31.1	27.94	Iridium			
	aq, 100	24.2	26.55	Ir	c	0.00	0.00
HClO	aq, 400	−28.18	−19.11	IrCl	c	−20.5	−16.9
HClO$_3$	aq	−23.4	−0.25	IrCl$_2$	c	−40.6	−32.0
HClO$_4$	aq, 660	−31.4	−10.70	IrCl$_3$	c	−50.5	−46.5
HC$_2$H$_3$O$_2$	l	−116.2	−93.56	IrF$_6$	l	−130	
	aq, 400	−116.74	−96.8	IrO$_2$	c	−40.14	
H$_2$C$_2$O$_4$	c	−196.7		Iron			
	aq, 300	−194.6	−165.64	Fe	c, α	0.00	0.00
HCOOH	l	−97.8	−82.7	FeBr$_2$	c	−57.15	
	aq, 200	−98.0	−85.1		aq, 540	−78.7	−69.47

TABLE 2-220   Heats and Free Energies of Formation of Inorganic Compounds (*Continued*)

Compound	State†	Heat of formation‡§ ΔH (formation) at 25°C, kcal/mole	Free energy of formation‖¶ ΔF (formation) at 25°C, kcal/mole	Compound	State†	Heat of formation‡§ ΔH (formation) at 25°C, kcal/mole	Free energy of formation‖¶ ΔF (formation) at 25°C, kcal/mole
Iron (*Cont.*)				**Lithium** (*Cont.*)			
$FeBr_3$	aq	−95.5	−76.26	$LiC_2H_3O_2$	aq	−183.9	−160.00
Fe_3C	c	5.69	4.24	Li_2CO_3	c	−289.7	−269.8
$Fe(CO)_5$	l	−187.6			aq, 1900	−293.1	−267.58
$FeCO_3$	c, siderite	−172.4	−154.8	$LiCl$	c	−97.63	
$FeCl_2$	c	−81.9	−72.6		aq, 278	−106.45	−102.03
	aq	−100.0	−83.0	$LiClO_3$	aq	−87.5	−70.95
$FeCl_3$	c	−96.4		$LiClO_4$	aq	−106.3	−81.4
	aq, 2000	−128.5	−96.5	LiF	c	−145.57	
FeF_2	aq, 1200	−177.2	−151.7		aq, 400	−144.85	−136.40
FeI_2	c	−24.2		LiH	c	−22.9	
	aq	−47.7	−45	$LiHCO_3$	aq, 2000	−231.1	−210.98
FeI_3	aq	−49.7	−39.5	LiI	c	−65.07	
Fe_4N	c	−2.55	0.862		aq, 400	−80.09	−83.03
$Fe(NO_3)_2$	aq	−118.9	−72.8	$LiIO_3$	aq	−121.3	−102.95
$Fe(NO_3)_3$	aq, 800	−156.5	−81.3	Li_3N	c	−47.45	−37.33
FeO	c	−64.62	−59.38	$LiNO_3$	c	−115.350	
Fe_2O_3	c	−198.5	−179.1		aq, 400	−115.88	−96.95
Fe_3O_4	c	−266.9	−242.3	Li_2O	c	−142.3	
$Fe(OH)_2$	c	−135.9	−115.7	Li_2O_2	c	−151.9	−138.0
$Fe(OH)_3$	c	−197.3	−166.3		aq	−159	
$FeO·SiO_2$	c	−273.5		$LiOH$	c	−116.58	−106.44
Fe_2P	c	−13			aq, 400	−121.47	−108.29
$FeSi$	c	−19.0		$LiOH·H_2O$	c	−188.92	
FeS	c	−22.64	−23.23	$Li_2O·SiO_2$	gls	−374	
FeS_2	c, pyrites	−38.62	−35.93	Li_2Se	c	−84.9	
	c, marcasite	−33.0			aq	−95.5	−105.64
$FeSO_4$	c	−221.3	−195.5	Li_2SO_4	c	−340.23	−314.66
	aq, 400	−236.2	−196.4		aq, 400	−347.02	
$Fe_2(SO_4)_3$	aq, 400	−653.3	−533.4	$Li_2SO_4·H_2O$	c	−411.57	−375.07
$FeTiO_3$	c, ilmenite	−295.51	−277.06	**Magnesium**			
Lanthanum				Mg	c	0.00	0.00
La	c	0.00	0.00	$Mg(AsO_4)_2$	c	−731.3	
$LaCl_3$	c	−253.1			aq	−749	−630.14
	aq	−284.7		$MgBr_2$	c	−123.9	
La_3H_8	c	−160			aq, 400	−167.33	−156.94
LaN	c	−72.0	−64.6	$Mg(CN)_2$	aq	−39.7	−29.08
La_2O_3	c	−539		$MgCN_2$	c	−61	
LaS_2	c	−148.3		$Mg(C_2H_3O_2)_2$	aq	−344.6	−286.38
La_2S_3	c	−351.4		$MgCO_3$	c	−261.7	−241.7
$La_2(SO_4)_3$	aq	−972		$MgCl_2$	c	−153.220	−143.77
Lead					aq, 400	−189.76	
Pb	c	0.00	0.00	$MgCl_2·H_2O$	c	−230.970	−205.93
$PbBr_2$	c	−66.24	−62.06	$MgCl_2·2H_2O$	c	−305.810	−267.20
	aq	−56.4	−54.97	$MgCl_2·4H_2O$	c	−453.820	−387.98
$PbCO_3$	c, cerussite	−167.6	−150.0	$MgCl_2·6H_2O$	c	−597.240	−505.45
$Pb(C_2H_3O_2)_2$	c	−232.6		MgF_2	c	−263.8	
	aq, 400	−234.2	−184.40	MgI_2	c	−86.8	
PbC_2O_4	c	−205.3			aq, 400	−136.79	−132.45
$PbCl_2$	c	−85.68	−75.04	$MgMoO_4$	c	−329.9	
	aq	−82.5	−68.47	Mg_3N_2	c	−115.2	−100.8
PbF_2	c	−159.5	−148.1	$Mg(NO_3)_2$	c	−188.770	−140.66
PbI_2	c	−41.77	−41.47		aq, 400	−209.927	−160.28
$Pb(NO_3)_2$	c	−106.88		$Mg(NO_3)_2·2H_2O$	c	−336.625	
	aq, 400	−99.46	−58.3	$Mg(NO_3)_2·6H_2O$	c	−624.48	−496.03
PbO	c, red	−51.72	−45.53	MgO	c	−143.84	−136.17
	c, yellow	−50.86	−43.88	$MgO·SiO_2$	c	−347.5	−326.7
PbO_2	c	−65.0	−52.0	$Mg(OH)_2$	c, ppt.	−221.90	−200.17
Pb_3O_4	c	−172.4	−142.2		c, brucite	−223.9	−193.3
$Pb(OH)_2$	c	−123.0	−102.2	MgS	c	−84.2	
PbS	c	−22.38	−21.98		aq	−108	
$PbSO_4$	c	−218.5	−192.9	$MgSO_4$	c	−304.94	−277.7
Lithium					aq, 400	−325.4	−283.88
Li	c	0.00	0.00	$MgTe$	c	−25	
$LiBr$	c	−83.75		$MgWO_4$	c	−345.2	
	aq, 400	−95.40	−95.28	**Manganese**			
$LiBrO_3$	aq	−77.9	−65.70	Mn	c, α	0.00	0.00
Li_2C_2	c	−13.0		$MnBr_2$	c	−91	
$LiCN$	aq	−31.4	−31.35		aq	−106	−97.8
$LiCNO$	aq	−101.2	−94.12	Mn_3C	c	1.1	1.26

TABLE 2-220   Heats and Free Energies of Formation of Inorganic Compounds (Continued)

Compound	State†	Heat of formation‡§ ΔH (formation) at 25°C, kcal/mole	Free energy of formation‖¶ ΔF (formation) at 25°C, kcal/mole	Compound	State†	Heat of formation‡§ ΔH (formation) at 25°C, kcal/mole	Free energy of formation‖¶ ΔF (formation) at 25°C, kcal/mole
Manganese (Cont.)				**Nickel** (Cont.)			
$Mn(C_2H_3O_2)_2$	c	−270.3			aq, 400	−94.34	−74.19
	aq	−282.7	−227.2	NiF_2	c	−157.5	
$MnCO_3$	c	−211	−192.5		aq	−171.6	−142.9
MnC_2O_4	c	−240.9		NiI_2	c	−22.4	
$MnCl_2$	c	−112.0	−102.2		aq	−42.0	−36.2
	aq, 400	−128.9		$Ni(NO_3)_2$	c	−101.5	
MnF_2	aq, 1200	−206.1	−180.0		aq, 200	−113.5	−64.0
MnI_2	c	−49.8		NiO	c	−58.4	−51.7
	aq	−76.2	−73.3	$Ni(OH)_2$	c	−129.8	−105.6
Mn_5N_2	c	−57.77	−46.49	$Ni(OH)_3$	c	−163.2	
$Mn(NO_3)_2$	c	−134.9		NiS	c	−20.4	
	aq, 400	−148.0	−101.1	$NiSO_4$	c	−216	
$Mn(NO_3)_2.6H_2O$	c	−557.07	−441.2		aq, 200	−231.3	−187.6
MnO	c	−92.04	−86.77	**Nitrogen**			
MnO_2	c	−124.58	−111.49	N_2	g	0.00	0.00
Mn_2O_3	c	−229.5	−209.9	NF_3	g	−27	
Mn_3O_4	c	−331.65	−306.22	NH_3	g	−10.96	−3.903
$MnO.SiO_2$	c	−301.3	−282.1		aq, 200	−19.27	
$Mn(OH)_2$	c	−163.4	−143.1	NH_4Br	c	−64.57	
$Mn(OH)_3$	c	−221	−190		aq	−60.27	−43.54
$Mn_3(PO_4)_2$	c	−736		$NH_4C_2H_3O_2$	c	−148.1	
$MnSe$	c	−26.3	−27.5		aq, 400	−148.58	−108.26
MnS	c, green	−47.0	−48.0	NH_4CN	c	−0.7	
$MnSO_4$	c	−254.18	−228.41		aq	3.6	20.4
	aq, 400	−265.2		NH_4CNS	c	−17.8	
$Mn_2(SO_4)_3$	c	−635			aq	−12.3	4.4
	aq	−657		$(NH_4)_2CO_3$	aq	−223.4	−164.1
Mercury				$(NH_4)_2C_2O_4$	c	−266.3	
Hg	l	0.00	0.00		aq	−260.6	−196.2
$HgBr$	g	23	18	NH_4Cl	c	−75.23	−48.59
$HgBr_2$	c	−40.68	−38.8		aq, 400	−71.20	
	aq	−38.4	−9.74	NH_4ClO_4	c	−69.4	
$Hg(C_2H_3O_2)_2$	c	−196.3			aq	−63.2	−21.1
	aq	−192.5	−139.2	$(NH_4)_2CrO_4$	c	−276.9	
$HgCl_2$	c	−53.4	−42.2		aq	−271.3	−209.3
	aq	−50.3	−23.25	NH_4F	c	−111.6	
$HgCl$	g	19	14		aq	−110.2	−84.7
Hg_2Cl_2	c	−63.13		NH_4I	c	−48.43	
$Hg(CN)_2$	c	62.8			aq	−44.97	−31.3
	aq, 1110	66.25		NH_4NO_3	c	−87.40	
HgC_2O_4	c	−159.3			aq, 500	−80.89	
HgH	g	57.1	52.25	NH_4OH	aq	−87.59	
HgI_2	c, red	−25.3	−24.0	$(NH_4)_2S$	aq, 400	−55.21	−14.50
HgI	g	33	23	$(NH_4)_2SO_4$	c	−281.74	−215.06
Hg_2I_2	c	−28.88	−26.53		aq, 400	−279.33	−214.02
$Hg(NO_3)_2$	aq	−56.8	−13.09	N_2H_4	l	12.06	
$Hg_2(NO_3)_2$	aq	−58.5	−15.65	$N_2H_4.H_2O$	l	−57.96	
HgO	c, red	−21.6	−13.94	$N_2H_4.H_2SO_4$	c	−232.2	
	c, yellow ppt.	−20.8		N_2O	g	19.55	24.82
Hg_2O	c	−21.6	−12.80	NO	g	21.600	20.719
HgS	c, black	−10.7	−8.80	NO_2	g	7.96	12.26
$HgSO_4$	c	−166.6		N_2O_4	g	2.23	23.41
Hg_2SO_4	c	−177.34	−149.12	N_2O_5	c	−10.0	
Molybdenum				$NOBr$	l	11.6	19.26
Mo	c	0.00	0.00	$NOCl$	g	12.8	16.1
Mo_2C	c	4.36	2.91	**Osmium**			
Mo_2N	c	−8.3		Os	c	0.00	0.00
MoO_2	c	−130	−118.0	OsO_4	c	−93.6	−70.9
MoO_3	c	−180.39	−162.01		g	−80.1	−68.1
MoS_2	c	−56.27	−54.19	**Oxygen**			
MoS_3	c	−61.48	−57.38	O_2	g	0.00	0.00
Nickel				O_3	g	33.88	38.86
Ni	c	0.00	0.00	**Palladium**			
$NiBr_2$	c	−53.4		Pd	c	0.00	0.00
	aq	−72.6	−60.7	PdO	c	−20.40	
Ni_3C	c	9.2	8.88	**Phosphorus**			
$Ni(C_2H_3O_2)_2$	aq	−249.6	−190.1	P	c, white ("yellow")	0.00	0.00
$Ni(CN)_2$	aq	230.9	66.3		c, red ("violet")	−4.22	−1.80
$NiCl_2$	c	−75.0		P	g	150.35	141.88

TABLE 2-220   Heats and Free Energies of Formation of Inorganic Compounds (*Continued*)

Compound	State†	Heat of formation‡§ ΔH (formation) at 25°C, kcal/mole	Free energy of formation‖¶ ΔF (formation) at 25°C, kcal/mole	Compound	State†	Heat of formation‡§ ΔH (formation) at 25°C, kcal/mole	Free energy of formation‖¶ ΔF (formation) at 25°C, kcal/mole
Phosphorus (*Cont.*)				Potassium (*Cont.*)			
P_2	g	33.82	24.60	KNH_2	c	-28.25	
P_4	g	13.2	5.89	KNO_2	aq	-86.0	-75.9
PBr_3	l	-45		KNO_3	c	-118.08	-94.29
PBr_5	c	-60.6			aq, 400	-109.79	-93.68
PCl_3	g	-70.0	-65.2	K_2O	c	-86.2	
	l	-76.8	-63.3	$K_2O \cdot Al_2O_3 \cdot SiO_2$	c, leucite	-1379.6	
PCl_5	g	-91.0	-73.2		gls	-1368.2	
PH_3	g	2.21	-1.45	$K_2O \cdot Al_2O_3 \cdot SiO_2$	c, adularia	-1784.5	
PI_3	c	-10.9			c, microcline	-1784.5	
P_2O_5	c	-360.0			gls	-1747	
$POCl_3$	g	-138.4	-127.2	KOH	c	-102.02	
Platinum					aq, 400	-114.96	-105.0
Pt	c	0.00	0.00	K_3PO_3	aq	-397.5	
$PtBr_4$	c	-40.6		K_3PO_4	aq	-478.7	-443.3
	aq	-50.7		KH_2PO_4	c	-362.7	-326.1
$PtCl_2$	c	-34		K_2PtCl_4	c	-254.7	
$PtCl_4$	c	-62.6			aq	-242.6	-226.5
	aq	-82.3		K_2PtCl_6	c	-299.5	-263.6
PtI_4	c	-18			aq, 9400	-286.1	
$Pt(OH)_2$	c	-87.5	-67.9	K_2Se	c	-74.4	
PtS	c	-20.18	-18.55		aq	-83.4	-99.10
PtS_2	c	-26.64	-24.28	K_2SeO_4	aq	-267.1	-240.0
Potassium				K_2S	c	-121.5	
K	c	0.00	0.00		aq, 400	-110.75	-111.44
K_3AsO_3	aq	-323.0		K_2SO_3	c	-267.7	
K_3AsO_4	aq	-390.3	-355.7		aq	-269.7	-251.3
KH_2AsO_4	c	-271.2	-236.7	K_2SO_4	c	-342.65	-314.62
KBr	c	-94.06	-90.8		aq, 400	-336.48	-310.96
	aq, 400	-89.19	-92.0	$K_2SO_4 \cdot Al_2(SO_4)_3$	c	-1178.38	-1068.48
$KBrO_3$	c	-81.58	-60.30	$K_2SO_4 \cdot Al_2(SO_4)_3 \cdot 24H_2O$	c	-2895.44	-2455.68
	aq, 1667	-71.68		$K_2S_2O_6$	c	-418.62	
$KC_2H_3O_2$	c	-173.80		Rhenium			
	aq, 400	-177.38	-156.73	Re	c	0.00	0.00
KCl	c	-104.348	-97.76	ReF_6	g	-274	
	aq, 400	-100.164	-98.76	Rhodium			
$KClO_3$	c	-93.5	-69.30	Rh	c	0.00	0.00
	aq, 400	-81.34		RhO	c	-21.7	
$KClO_4$	c	-103.8	-72.86	Rh_2O	c	-22.7	
	aq, 400	-101.14		Rh_2O_3	c	-68.3	
KCN	c	-28.1		Rubidium			
	aq, 400	-25.3	-28.08	Rb	c	0.00	0.00
KCNO	c	-99.6		RbBr	c	-95.82	
	aq	-94.5	-90.85		g	-45.0	-52.50
KCNS	c	-47.0			aq, 500	-90.54	-93.38
	aq, 400	-41.07	-44.08	RbCN	aq	-25.9	
K_2CO_3	c	-274.01		Rb_2CO_3	c	-273.22	
	aq, 400	-280.90	-264.04		aq, 220	-282.61	-263.78
$K_2C_2O_4$	c	-319.9		RbCl	c	-105.06	-98.48
	aq, 400	-315.5	-293.1		g	-53.6	-57.9
K_2CrO_4	c	-333.4			aq, ∞	-101.06	-100.13
	aq, 400	-328.2	-306.3	RbF	c	-133.23	
$K_2Cr_2O_7$	c	-488.5			aq, 400	-139.31	-134.5
	aq, 400	-472.1	-440.9	$RbHCO_3$	c	-230.01	
KF	c	-134.50			aq, 2000	-225.59	-209.07
	aq, 180	-138.36	-133.13	RbI	c	-81.04	
$K_3Fe(CN)_6$	c	-48.4			g	-31.2	-40.5
	aq	-34.5			aq, 400	-74.57	-81.13
$K_4Fe(CN)_6$	c	-131.8		$RbNH_2$	c	-27.74	
	aq	-119.9		$RbNO_3$	c	-119.22	
KH	c	-10	-5.3		aq, 400	-110.52	-95.05
$KHCO_3$	c	-229.8		Rb_2O	c	-82.9	
	aq, 2000	-224.85	-207.71	Rb_2O_2	c	-107	
KI	c	-78.88	-77.37	RbOH	c	-101.3	
	aq, 500	-73.95	-79.76		aq, 200	-115.8	-106.39
KIO_3	c	-121.69	-101.87	Ruthenium			
	aq, 400	-115.18	-99.68	Ru	c	0.00	0.00
KIO_4	aq	-98.1		RuS_2	c	-46.99	-44.11
$KMnO_4$	c	-192.9	-169.1	Selenium			
	aq, 400	-182.5	-168.0	Se	c, I, hexagonal	0.00	0.00
K_2MoO_4	aq, 880	-364.2	-342.9				

TABLE 2-220 Heats and Free Energies of Formation of Inorganic Compounds (*Continued*)

Compound	State[†]	Heat of formation[‡][§] ΔH (formation) at 25°C, kcal/mole	Free energy of formation[‖][¶] ΔF (formation) at 25°C, kcal/mole
Selenium (*Cont.*)			
	c, II, red, monoclinic	0.2	
Se_2Cl_2	l	−22.06	−13.73
SeF_6	g	−246	−222
SeO_2	c	−56.33	
Silicon			
Si	c	0.00	0.00
$SiBr_4$	l	−93.0	
SiC	c	−28	−27.4
$SiCl_4$	l	−150.0	−133.9
	g	−142.5	−133.0
SiF_4	g	−370	−360
SiH_4	g	−14.8	−9.4
SiI_4	c	−29.8	
Si_3N_4	c	−179.25	−154.74
SiO_2	c, cristobalite, 1600° form	−202.62	
	c, cristobalite, 1100° form	−202.46	
	c, quartz	−203.35	−190.4
	c, tridymite	−203.23	
Silver			
Ag	c	0.00	0.00
AgBr	c	−23.90	−23.02
Ag_2C_2	c	84.5	
$AgC_2H_3O_2$	c	−95.9	
	aq	−91.7	−70.86
AgCN	c	33.8	38.70
Ag_2CO_3	c	−119.5	−103.0
$Ag_2C_2O_4$	c	−158.7	
AgCl	c	−30.11	−25.98
AgF	c	−48.7	
	aq, 400	−53.1	−47.26
AgI	c	−15.14	−16.17
$AgIO_3$	c	−42.02	−24.08
$AgNO_2$	c	−11.6	3.76
	aq	−2.9	9.99
$AgNO_3$	c	−29.4	−7.66
	aq, 6500	−24.02	−7.81
Ag_2O	c	−6.95	−2.23
Ag_2S	c	−5.5	−7.6
Ag_2SO_4	c	−170.1	−146.8
	aq	−165.8	−139.22
Sodium			
Na	c	0.00	0.00
Na_3AsO_3	aq, 500	−314.61	
Na_3AsO_4	c	−366	
	aq, 500	−381.97	−341.17
NaBr	c	−86.72	
	aq, 400	−86.33	−87.17
NaBrO	aq	−78.9	
$NaBrO_3$	aq, 400	−68.89	−57.59
$NaC_2H_3O_2$	c	−170.45	
	aq, 400	−175.450	−152.31
NaCN	c	−22.47	
	aq, 200	−22.29	−23.24
NaCNO	c	−96.3	
	aq	−91.7	−86.00
NaCNS	c	−39.94	
	aq, 400	−38.23	−39.24
Na_2CO_3	c	−269.46	−249.55
	aq, 1000	−275.13	−251.36
$NaCO_2NH_2$	c	−142.17	
$Na_2C_2O_4$	c	−313.8	
	aq, 600	−309.92	−283.42
NaCl	c	−98.321	−91.894
	aq, 400	−97.324	−93.92
$NaClO_3$	c	−83.59	
	aq, 400	−78.42	−62.84
$NaClO_4$	c	−101.12	
Sodium (*Cont.*)			
	aq, 476	−97.66	−73.29
Na_2CrO_4	c	−319.8	
	aq, 800	−323.0	−296.58
$Na_2Cr_2O_7$	aq, 1200	−465.9	−431.18
NaF	c	−135.94	−129.0
	aq, 400	−135.711	−128.29
NaH	c	−14	−9.30
$NaHCO_3$	c	−226.0	−202.66
	aq	−222.1	−202.87
NaI	c	−69.28	
	aq, ∞	−71.10	−74.92
$NaIO_3$	aq, 400	−112.300	−94.84
Na_2MoO_4	c	−364	
	aq	−358.7	−333.18
$NaNO_2$	c	−86.6	
	aq	−83.1	−71.04
$NaNO_3$	c	−111.71	−87.62
	aq, 400	−106.880	−88.84
Na_2O	c	−99.45	−90.06
Na_2O_2	c	−119.2	−105.0
$Na_2O \cdot SiO_2$	c	−383.91	−361.49
$Na_2O \cdot Al_2O_3 \cdot 3SiO_2$	c, natrolite	−1180	
$Na_2O \cdot Al_2O_3 \cdot 4SiO_2$	c	−1366	
NaOH	c	−101.96	−90.60
	aq, 400	−112.193	−100.18
Na_3PO_3	aq, 1000	−389.1	
Na_3PO_4	c	−457	
	aq, 400	−471.9	−428.74
Na_2PtCl_4	aq	−237.2	−216.78
Na_2PtCl_6	c	−272.1	
	aq	−280.9	
Na_2Se	c	−59.1	
	aq, 440	−78.1	−89.42
Na_2SeO_4	c	−254	
	aq, 800	−261.5	−230.30
Na_2S	c	−89.8	
	aq, 400	−105.17	−101.76
Na_2SO_3	c	−261.2	−240.14
	aq, 800	−264.1	−241.58
Na_2SO_4	c	−330.50	−302.38
	aq, 1100	−330.82	−301.28
$Na_2SO_4 \cdot 10H_2O$	c	−1033.85	−870.52
Na_2WO_4	c	−391	
	aq	−381.5	−345.18
Strontium			
Sr	c	0.00	0.00
$SrBr_2$	c	−171.0	
	aq, 400	−187.24	−182.36
$Sr(C_2H_3O_2)_2$	c	−358.0	
	aq	−364.4	−311.80
$Sr(CN)_2$	aq	−59.5	−54.50
$SrCO_3$	c	−290.9	−271.9
$SrCl_2$	c	−197.84	
	aq, 400	−209.20	−195.86
SrF_2	c	−289.0	
$Sr(HCO_3)_2$	aq	−459.1	−413.76
SrI_2	c	−136.1	
	aq, 400	−156.70	−157.87
Sr_3N_2	c	−91.4	−76.5
$Sr(NO_3)_2$	c	−233.2	
	aq, 400	−228.73	−185.70
SrO	c	−140.8	−133.7
$SrO \cdot SiO_2$	gls	−364	
SrO_2	c	−153.3	−139.0
Sr_2O	c	−153.6	
$Sr(OH)_2$	c	−228.7	
	aq, 800	−239.4	−208.27
$Sr_3(PO_4)_2$	c	−980	
	aq	−985	−881.54
SrS	c	−113.1	

TABLE 2-220   Heats and Free Energies of Formation of Inorganic Compounds (*Continued*)

Compound	State†	Heat of formation‡§ ΔH (formation) at 25°C, kcal/mole	Free energy of formation‖¶ ΔF (formation) at 25°C, kcal/mole	Compound	State†	Heat of formation‡§ ΔH (formation) at 25°C, kcal/mole	Free energy of formation‖¶ ΔF (formation) at 25°C, kcal/mole
Strontium (*Cont.*)				Tin			
	aq	−120.4	−109.78	Sn	c, II, tetragonal	0.00	0.00
SrSO₄	c	−345.3			c, III, "gray," cubic	0.6	1.1
	aq, 400	−345.0	−309.30	SnBr₂	c	−61.4	
SrWO₄	c	−393			aq	−60.0	−55.43
Sulfur				SnBr₄	c	−94.8	
S	c, rhombic	0.00	0.00		aq	−110.6	−97.66
	c, monoclinic	−0.071	0.023	SnCl₂	c	−83.6	
	l, λ	0.257	0.072		aq	−81.7	−68.94
	l, λμ equilibrium		0.071	SnCl₄	l	−127.3	−110.4
	g	53.25	43.57		aq	−157.6	−124.67
S₂	g	31.02	19.36	SnI₂	c	−38.9	
S₆	g	27.78	13.97		aq	−33.3	−30.95
S₈	g	27.090	12.770	SnO	c	−67.7	−60.75
S₂Br₂	l	−4		SnO₂	c	−138.1	−123.6
SCl₄	l	−13.7		Sn(OH)₂	c	−136.2	−115.95
S₂Cl₂	l	−14.2	−5.90	Sn(OH)₄	c	−268.9	−226.00
S₂Cl₄	l	−24.1		SnS	c	−18.61	
SF₆	g	−262	−237	Titanium			
SO	g	19.02	12.75	Ti	c	0.00	0.00
SO₂	g	−70.94	−71.68	TiC	c	−110	−109.2
SO₃	g	−94.39	−88.59	TiCl₄	l	−181.4	−165.5
	l	−103.03	−88.28	TiN	c	−80.0	−73.17
	c, α	−105.09	−88.22	TiO₂	c, III, rutil	−225.0	−211.9
	c, β	−105.92	−88.34		amorphous	−214.1	−201.4
	c, γ	−109.34	−88.98	Tungsten			
SO₂Cl₂	g	−82.04	−74.06	W	c	0.00	0.00
	l	−89.80	−75.06	WO₂	c	−130.5	−118.3
Tantalum				WO₃	c	−195.7	−177.3
Ta	c	0.00	0.00	WS₂	c	−84	
TaN	c	−51.2	−45.11	Uranium			
Ta₂O₅	c	−486.0	−453.7	U	c	0.00	0.00
Tellurium				UC₂	c	−29	
Te	c	0.00	0.00	UCl₃	c	−213	
TeBr₄	c	−49.3		UCl₄	c	−251	
TeCl₄	c	−77.4	−57.4	U₃N₄	c	−274	−249.6
TeF₆	g	−315	−292	UO₂	c	−256.6	−242.2
TeO₂	c	−77.56	−64.66	UO₂(NO₃)₂·6H₂O	c	−756.9	−617.8
Thallium				UO₃	c	−291.6	
Tl	c	0.00	0.00	U₃O₈	c	−845.1	
TlBr	c	−41.5	−39.43	Vanadium			
	aq	−28.0	−32.34	V	c	0.00	0.00
TlCl	c	−49.37	−44.46	VCl₂	c	−147	
	aq	−38.4	−39.09	VCl₃	l	−187	
TlCl₃	c	−82.4		VCl₄	l	−165	
	aq	−91.0	−44.25	VN	c	−41.43	−35.08
TlF	aq	−77.6	−73.46	V₃O₂	c	−195	
TlI	c	−31.1	−31.3	V₂O₃	c	−296	−277
	aq	−12.7	−20.09	V₂O₄	c	−342	−316
TlNO₃	c	−58.2	−36.32	V₂O₅	c	−373	−342
	aq	−48.4	−34.01	Zinc			
Tl₂O	c	−43.18		Zn	c	0.00	0.00
Tl₂O₃	c	−120		ZnSb	c	−3.6	−3.88
TlOH	c	−57.44	−45.54	ZnBr₂	c	−77.0	−72.9
	aq	−53.9	−45.35		aq, 400	−93.6	
Tl₂S	c	−22		Zn(C₂H₃O₂)₂	c	−259.4	
Tl₂SO₄	c	−222.8	−197.79		aq, 400	−269.4	−214.4
	aq, 800	−214.1	−191.62	Zn(CN)₂	c	17.06	
Thorium				ZnCO₃	c	−192.9	−173.5
Th	c	0.00	0.00	ZnCl₂	c	−99.9	−88.8
ThBr₄	c	−281.5			aq, 400	−115.44	
	aq	−352.0	−295.31	ZnF₂	aq	−192.9	−166.6
ThC₂	c	−45.1		ZnI₂	c	−50.50	−49.93
ThCl₄	c	−335			aq	−61.6	
	aq	−392	−322.32	Zn(NO₃)₂	aq, 400	−134.9	−87.7
ThI₄	aq	−292.0	−246.33	ZnO	c, hexagonal	−83.36	−76.19
Th₃N₄	c	−309.0	−282.3	ZnO·SiO₂	c	−282.6	
ThO₂	c	−291.6	−280.1	Zn(OH)₂	c, rhombic	−153.66	
Th(OH)₄	c, "soluble"	−336.1		ZnS	c, wurtzite	−45.3	−44.2
Th(SO₄)₂	c	−632		ZnSO₄	c	−233.4	
	aq	−668.1	−549.2		aq, 400	−252.12	−211.28

TABLE 2-220   Heats and Free Energies of Formation of Inorganic Compounds *(Concluded)*

Compound	State†	Heat of formation‡§ ΔH (formation) at 25°C, kcal/mole	Free energy of formation‖¶ ΔF (formation) at 25°C, kcal/mole	Compound	State†	Heat of formation‡§ ΔH (formation) at 25°C, kcal/mole	Free energy of formation‖¶ ΔF (formation) at 25°C, kcal/mole
Zirconium				Zirconium *(Cont.)*			
Zr	c	0.00	0.00	ZrO_2	c, monoclinic	−258.5	−244.6
ZrC	c	−29.8	−34.6	$Zr(OH)_4$	c	−411.0	
$ZrCl_4$	c	−268.9		$ZrO(OH)_2$	c	−357	−307.6
ZrN	c	−82.5	−75.9				

† The physical state is indicated as follows: *c*, crystal (solid); *l*, liquid; *g*, gas; *gls*, glass or solid supercooled liquid; *aq*, in aqueous solution. A number following the symbol *aq* applies only to the values of the heats of formation (not to those of free energies of formation); and indicates the number of moles of water per mole of solute; when no number is given, the solution is understood to be dilute. For the free energy of formation of a substance in aqueous solution, the concentration is always that of the hypothetical solution of unit molality.

‡ The increment in heat content, ΔH, in the reaction of forming the given substance from its elements in their standard states. When ΔH is negative, heat is evolved in the process, and, when positive, heat is absorbed.

§ The heat of solution in water of a given solid, liquid, or gaseous compound is given by the difference in the value for the heat of formation of the given compound in the solid, liquid, or gaseous state and its heat of formation in aqueous solution. The following two examples serve as an illustration of the procedure: (1) For NaCl(*c*) and NaCl(*aq*, 400H$_2$O), the values of ΔH(formation) are, respectively, −98.321 and −97.324 kg-cal per mole. Subtraction of the first value from the second gives $\Delta H = 0.998$ kg-cal per mole for the reaction of dissolving crystalline sodium chloride in 400 moles of water. When this process occurs at a constant pressure of 1 atm, 0.998 kg-cal of energy are absorbed. (2) For HCl(*g*) and HCl(*aq*, 400H$_2$O), the values for ΔH(formation) are, respectively, −22.06 and −39.85 kg-cal per mole. Subtraction of the first from the second gives $\Delta H = -17.79$ kg-cal per mole for the reaction of dissolving gaseous hydrogen chloride in 400 moles of water. At a constant pressure of 1 atm, 17.79 kg-cal of energy are evolved in this process.

‖ The increment in the free energy, ΔF, in the reaction of forming the given substance in its standard state from its elements in their standard states. The standard states are: for a gas, fugacity (approximately equal to the pressure) of 1 atm; for a pure liquid or solid, the substance at a pressure of 1 atm; for a substance in aqueous solution, the hypothetical solution of unit molality, which has all the properties of the infinitely dilute solution except the property of concentration.

¶ The free energy of solution of a given substance from its normal standard state as a solid, liquid, or gas to the hypothetical one molal state in aqueous solution may be calculated in a manner similar to that described in footnote § for calculating the heat of solution.

HEATS OF COMBUSTION

TABLE 2-221   Enthalpies and Gibbs Energies of Formation, Entropies, and Net Enthalpies of Combustion of Inorganic and Organic Compounds at 298.15 K

Cmpd. no.	Name	Formula	CAS no.	Mol wt	Ideal gas enthalpy of formation, J/kmol × 1E-07	Ideal gas Gibbs energy of formation, J/kmol × 1E-07	Ideal gas entropy, J/(kmol·K) × 1E-05	Standard net enthalpy of combustion, J/kmol × 1E-09
1	Methane	CH_4	74828	16.043	−7.4520	−5.0490	1.8627	−0.8026
2	Ethane	C_2H_6	74840	30.070	−8.3820	−3.1920	2.2912	−1.4286
3	Propane	C_3H_8	74986	44.097	−10.4680	−2.4390	2.7020	−2.0431
4	*n*-Butane	C_4H_{10}	106978	58.123	−12.5790	−1.6700	3.0991	−2.6573
5	*n*-Pentane	C_5H_{12}	109660	72.150	−14.6760	−0.8813	3.4945	−3.2449
6	*n*-Hexane	C_6H_{14}	110543	86.177	−16.6940	−0.0066	3.8874	−3.8551
7	*n*-Heptane	C_7H_{16}	142825	100.204	−18.7650	0.8165	4.2798	−4.4647
8	*n*-Octane	C_8H_{18}	111659	114.231	−20.8750	1.6000	4.6723	−5.0742
9	*n*-Nonane	C_9H_{20}	111842	128.258	−22.8740	2.4980	5.0640	−5.6846
10	*n*-Decane	$C_{10}H_{22}$	124185	142.285	−24.9460	3.3180	5.4570	−6.2942
11	*n*-Undecane	$C_{11}H_{24}$	1120214	156.312	−27.0430	4.1160	5.8493	−6.9036
12	*n*-Dodecane	$C_{12}H_{26}$	112403	170.338	−29.0720	4.9810	6.2415	−7.5137
13	*n*-Tridecane	$C_{13}H_{28}$	629505	184.365	−31.1770	5.7710	6.6337	−8.1229
14	*n*-Tetradecane	$C_{14}H_{30}$	629594	198.392	−33.2440	6.5990	7.0259	−8.7328
15	*n*-Pentadecane	$C_{15}H_{32}$	629629	212.419	−35.3110	7.4260	7.4181	−9.3424
16	*n*-Hexadecane	$C_{16}H_{34}$	544763	226.446	−37.4170	8.2160	7.8102	−9.9515
17	*n*-Heptadecane	$C_{17}H_{36}$	629787	240.473	−39.4450	9.0830	8.2023	−10.5618
18	*n*-Octadecane	$C_{18}H_{38}$	593453	254.500	−41.5120	9.9100	8.5945	−11.1715
19	*n*-Nonadecane	$C_{19}H_{40}$	629925	268.527	−43.5790	10.7400	8.9866	−11.7812
20	*n*-Eicosane	$C_{20}H_{42}$	112958	282.553	−45.6460	11.5700	9.3787	−12.3908
21	2-Methylpropane	C_4H_{10}	75285	58.123	−13.4180	−2.0760	2.9539	−2.6490
22	2-Methylbutane	C_5H_{12}	78784	72.150	−15.3700	−1.4050	3.4374	−3.2395
23	2,3-Dimethylbutane	C_6H_{14}	79298	86.177	−17.6800	−0.3125	3.6592	−3.8476
24	2-Methylpentane	C_6H_{14}	107835	86.177	−17.4550	−0.5338	3.8089	−3.8492
25	2,3-Dimethylpentane	C_7H_{16}	565593	100.204	−19.4100	0.5717	4.1455	−4.4608
26	2,3,3-Trimethylpentane	C_8H_{18}	560214	114.231	−21.8450	1.8280	4.2702	−5.0688
27	2,2,4-Trimethylpentane	C_8H_{18}	540841	114.231	−22.4010	1.3940	4.2296	−5.0653
28	Ethylene	C_2H_4	74851	28.054	5.2510	6.8440	2.1920	−1.3230
29	Propylene	C_3H_6	115071	42.081	1.9710	6.2150	2.6660	−1.9257

TABLE 2-221 Enthalpies and Gibbs Energies of Formation, Entropies, and Net Enthalpies of Combustion of Inorganic and Organic Compounds (*Continued*)

Cmpd. no.	Name	Formula	CAS no.	Mol wt	Ideal gas enthalpy of formation, J/kmol × 1E-07	Ideal gas Gibbs energy of formation, J/kmol × 1E-07	Ideal gas entropy, J/(kmol·K) × 1E-05	Standard net enthalpy of combustion, J/kmol × 1E-09
30	1-Butene	C_4H_8	106989	56.108	−0.0540	7.0270	3.0775	−2.5408
31	*cis*-2-Butene	C_4H_8	590181	56.108	−0.7400	6.5360	3.0120	−2.5339
32	*trans*-2-Butene	C_4H_8	624646	56.108	−1.1000	6.3160	2.9650	−2.5303
33	1-Pentene	C_5H_{10}	109671	70.134	−2.1300	7.8450	3.4699	−3.1296
34	1-Hexene	C_6H_{12}	592416	84.161	−4.2000	8.7390	3.8389	−3.7394
35	1-Heptene	C_7H_{14}	592767	98.188	−6.2800	9.4830	4.2549	−4.3489
36	1-Octene	C_8H_{16}	111660	112.215	−8.3600	10.3000	4.6469	−4.9606
37	1-Nonene	C_9H_{18}	124118	126.242	−10.4000	11.1500	5.0399	−5.5684
38	1-Decene	$C_{10}H_{20}$	872059	140.269	−12.4700	11.9800	5.4319	−6.1781
39	2-Methylpropene	C_4H_8	115117	56.108	−1.7100	5.8080	2.9309	−2.5242
40	2-Methyl-1-butene	C_5H_{10}	563462	70.134	−3.5300	6.6680	3.3950	−3.1159
41	2-Methyl-2-butene	C_5H_{10}	513359	70.134	−4.1800	6.0450	3.3860	−3.1088
42	1,2-Butadiene	C_4H_6	590192	54.092	16.2300	19.8600	2.9300	−2.4617
43	1,3-Butadiene	C_4H_6	106990	54.092	10.9240	14.9720	2.7889	−2.4090
44	2-Methyl-1,3-butadiene	C_5H_8	78795	68.119	7.5730	14.5896	3.1564	−2.9842
45	Acetylene	C_2H_2	74862	26.038	22.8200	21.0680	2.0081	−1.2570
46	Methylacetylene	C_3H_4	74997	40.065	18.4900	19.3840	2.4836	−1.8487
47	Dimethylacetylene	C_4H_6	503173	54.092	14.5700	18.4900	2.8330	−2.4189
48	3-Methyl-1-butyne	C_5H_8	598232	68.119	13.8000	20.7200	3.1890	−3.0460
49	1-Pentyne	C_5H_8	627190	68.119	14.4400	21.0300	3.2980	−3.0510
50	2-Pentyne	C_5H_8	627214	68.119	12.5100	19.0700	3.3084	−3.0291
51	1-Hexyne	C_6H_{10}	693027	82.145	12.3700	21.8500	3.6940	−3.6610
52	2-Hexyne	C_6H_{10}	764352	82.145	10.5000	19.9000	3.7200	−3.6400
53	3-Hexyne	C_6H_{10}	928494	82.145	10.6000	19.9000	3.7600	−3.6400
54	1-Heptyne	C_7H_{12}	628717	96.172	10.3000	22.7000	4.0850	−4.2717
55	1-Octyne	C_8H_{14}	629050	110.199	8.2300	23.5000	4.4780	−4.8815
56	Vinylacetylene	C_4H_4	689974	52.076	30.4600	30.6000	2.7940	−2.3620
57	Cyclopentane	C_5H_{10}	287923	70.134	−7.7030	3.8850	2.9290	−3.0709
58	Methylcyclopentane	C_6H_{12}	96377	84.161	−10.6200	3.6300	3.3990	−3.6741
59	Ethylcyclopentane	C_7H_{14}	1640897	98.188	−12.6900	4.4800	3.7830	−4.2839
60	Cyclohexane	C_6H_{12}	110827	84.161	−12.3300	3.1910	2.9728	−3.6560
61	Methylcyclohexane	C_7H_{14}	108872	98.188	−15.4800	2.7330	3.4430	−4.2571
62	1,1-Dimethylcyclohexane	C_8H_{16}	590669	112.215	−18.1000	3.5229	3.6501	−4.8639
63	Ethylcyclohexane	C_8H_{16}	1678917	112.215	−17.1500	3.9550	3.8260	−4.8705
64	Cyclopentene	C_5H_8	142290	68.119	3.3100	11.0500	2.9127	−2.9393
65	1-Methylcyclopentene	C_6H_{10}	693890	82.145	−0.3800	10.3800	3.2640	−3.5340
66	Cyclohexene	C_6H_{10}	110838	82.145	−0.4600	10.7700	3.1052	−3.5320
67	Benzene	C_6H_6	71432	78.114	8.2880	12.9600	2.6930	−3.1360
68	Toluene	C_7H_8	108883	92.141	5.0170	12.2200	3.2099	−3.7340
69	*o*-Xylene	C_8H_{10}	95476	106.167	1.9080	12.2000	3.5383	−4.3330
70	*m*-Xylene	C_8H_{10}	108383	106.167	1.7320	11.8760	3.5854	−4.3318
71	*p*-Xylene	C_8H_{10}	106423	106.167	1.8030	12.1400	3.5223	−4.3330
72	Ethylbenzene	C_8H_{10}	100414	106.167	2.9920	13.0730	3.6063	−4.3450
73	Propylbenzene	C_9H_{12}	103651	120.194	0.7910	13.8090	3.9843	−4.9542
74	1,2,4-Trimethylbenzene	C_9H_{12}	95636	120.194	−1.3800	11.7100	3.9610	−4.9307
75	Isopropylbenzene	C_9H_{12}	98828	120.194	0.4000	13.7900	3.8600	−4.9510
76	1,3,5-Trimethylbenzene	C_9H_{12}	108678	120.194	−1.5900	11.8100	3.8560	−4.9291
77	*p*-Isopropyltoluene	$C_{10}H_{14}$	99876	134.221	−2.9000	13.3520	4.2630	−5.5498
78	Naphthalene	$C_{10}H_8$	91203	128.174	15.0580	22.4080	3.3315	−4.9809
79	Biphenyl	$C_{12}H_{10}$	92524	154.211	18.2420	28.0230	3.9367	−6.0317
80	Styrene	C_8H_8	100425	104.152	14.7400	21.3900	3.4510	−4.2190
81	*m*-Terphenyl	$C_{18}H_{14}$	92068	230.309	27.6600	42.3000	5.2630	−9.0530
82	Methanol	CH_4O	67561	32.042	−20.0940	−16.2320	2.3988	−0.6382
83	Ethanol	C_2H_6O	64175	46.069	−23.4950	−16.7850	2.8064	−1.2350
84	1-Propanol	C_3H_8O	71238	60.096	−25.5200	−15.9900	3.2247	−1.8438
85	1-Butanol	$C_4H_{10}O$	71363	74.123	−27.4600	−15.0300	3.6148	−2.4560
86	2-Butanol	$C_4H_{10}O$	78922	74.123	−29.2900	−16.9600	3.6469	−2.4408
87	2-Propanol	C_3H_8O	67630	60.096	−27.2700	−17.3470	3.0920	−1.8300
88	2-Methyl-2-propanol	$C_4H_{10}O$	75650	74.123	−31.2400	−17.7600	3.2630	−2.4239
89	1-Pentanol	$C_5H_{12}O$	71410	88.150	−29.8737	−14.6022	4.0250	−3.0605
90	2-Methyl-1-butanol	$C_5H_{12}O$	137326	88.150	−30.2085	−14.6709	3.9351	−3.0620
91	3-Methyl-1-butanol	$C_5H_{12}O$	123513	88.150	−30.2100	−14.5000	3.8770	−3.0623
92	1-Hexanol	$C_6H_{14}O$	111273	102.177	−31.6500	−13.4400	4.4010	−3.6766
93	1-Heptanol	$C_7H_{16}O$	111706	116.203	−33.6400	−12.5300	4.7919	−4.2860
94	Cyclohexanol	$C_6H_{12}O$	108930	100.161	−28.6200	−10.9500	3.2770	−3.4639
95	Ethylene glycol	$C_2H_6O_2$	107211	62.068	−38.7500	−30.2600	2.2350	−1.0590
96	1,2-Propylene glycol	$C_3H_8O_2$	57556	76.095	−42.1500	−30.4000	3.5200	−1.6476

TABLE 2-221   Enthalpies and Gibbs Energies of Formation, Entropies, and Net Enthalpies of Combustion of Inorganic and Organic Compounds (*Continued*)

Cmpd. no.	Name	Formula	CAS no.	Mol wt	Ideal gas enthalpy of formation, J/kmol × 1E-07	Ideal gas Gibbs energy of formation, J/kmol × 1E-07	Ideal gas entropy, J/(kmol·K) × 1E-05	Standard net enthalpy of combustion, J/kmol × 1E-09
97	Phenol	C_6H_6O	108952	94.113	−9.6399	−3.2637	3.1481	−2.9210
98	*o*-Cresol	C_7H_8O	95487	108.140	−12.8570	−3.5430	3.5259	−3.5280
99	*m*-Cresol	C_7H_8O	108394	108.140	−13.2300	−4.0190	3.5604	−3.5278
100	*p*-Cresol	C_7H_8O	106445	108.140	−12.5350	−3.1660	3.5075	−3.5226
101	Dimethyl ether	C_2H_6O	115106	46.069	−18.4100	−11.2800	2.6670	−1.3284
102	Methyl ethyl ether	C_3H_8O	540670	60.096	−21.6400	−11.7100	3.0881	−1.9314
103	Methyl *n*-propyl ether	$C_4H_{10}O$	557175	74.123	−23.8200	−11.1000	3.5200	−2.5174
104	Methyl isopropyl ether	$C_4H_{10}O$	598538	74.123	−25.2000	−12.1800	3.4160	−2.5311
105	Methyl *n*-butyl ether	$C_5H_{12}O$	628284	88.150	−25.8100	−10.1700	3.9010	−3.1282
106	Methyl isobutyl ether	$C_5H_{12}O$	625445	88.150	−26.6000	−10.7000	3.8100	−3.1220
107	Methyl tert-butyl ether	$C_5H_{12}O$	1634044	88.150	−28.3500	−11.7500	3.5780	−3.1049
108	Diethyl ether	$C_4H_{10}O$	60297	74.123	−25.2100	−12.2100	3.4230	−2.5035
109	Ethyl propyl ether	$C_5H_{12}O$	628320	88.150	−27.2200	−11.5200	3.8810	−3.1200
110	Ethyl isopropyl ether	$C_5H_{12}O$	625547	88.150	−28.5800	−12.6400	3.8000	−3.1030
111	Methyl phenyl ether	C_7H_8O	100663	108.140	−6.7900	2.2700	3.6100	−3.6072
112	Diphenyl ether	$C_{12}H_{10}O$	101848	170.211	5.2000	17.5000	4.1300	−5.8939
113	Formaldehyde	CH_2O	50000	30.026	−10.8600	−10.2600	2.1866	−0.5268
114	Acetaldehyde	C_2H_4O	75070	44.053	−16.6200	−13.3100	2.6420	−1.1045
115	1-Propanal	C_3H_6O	123386	58.080	−18.6300	−12.4600	3.0440	−1.6857
116	1-Butanal	C_4H_8O	123728	72.107	−20.7000	−11.6300	3.4365	−2.3035
117	1-Pentanal	$C_5H_{10}O$	110623	86.134	−22.7800	−10.7100	3.8289	−2.9100
118	1-Hexanal	$C_6H_{12}O$	66251	100.161	−24.8600	−10.0050	4.2214	−3.5200
119	1-Heptanal	$C_7H_{14}O$	111717	114.188	−26.9400	−9.1910	4.6138	−4.1360
120	1-Octanal	$C_8H_{16}O$	124130	128.214	−29.0200	−8.3770	5.0063	−4.7400
121	1-Nonanal	$C_9H_{18}O$	124196	142.241	−31.0900	−7.5530	5.3988	−5.3500
122	1-Decanal	$C_{10}H_{20}O$	112312	156.268	−33.1700	−6.7390	5.7912	−5.9590
123	Acetone	C_3H_6O	67641	58.080	−21.5700	−15.1300	2.9540	−1.6590
124	Methyl ethyl ketone	C_4H_8O	78933	72.107	−23.9000	−14.7000	3.3940	−2.2680
125	2-Pentanone	$C_5H_{10}O$	107879	86.134	−25.9200	−13.8300	3.7860	−2.8796
126	Methyl isopropyl ketone	$C_5H_{10}O$	563804	86.134	−26.2400	−13.9000	3.6990	−2.8770
127	2-Hexanone	$C_6H_{12}O$	591786	100.161	−27.9826	−13.0081	4.1786	−3.4900
128	Methyl isobutyl ketone	$C_6H_{12}O$	108101	100.161	−28.0000	−13.5000	4.0700	−3.4900
129	3-Methyl-2-pentanone	$C_6H_{12}O$	565617	100.161	−28.1000	−12.9000	4.1200	−3.4900
130	3-Pentanone	$C_5H_{10}O$	96220	86.134	−25.7900	−13.4400	3.7000	−2.8804
131	Ethyl isopropyl ketone	$C_6H_{12}O$	565695	100.161	−28.6100	−13.3000	4.0690	−3.4860
132	Diisopropyl ketone	$C_7H_{14}O$	565800	114.188	−31.1400	−13.2000	4.5700	−4.0950
133	Cyclohexanone	$C_6H_{10}O$	108941	98.145	−22.6100	−8.6620	3.2200	−3.2990
134	Methyl phenyl ketone	C_8H_8O	98862	120.151	−8.6700	−0.1364	3.8450	−3.9730
135	Formic acid	CH_2O_2	64186	46.026	−37.8600	−35.1000	2.4870	−0.2115
136	Acetic acid	$C_2H_4O_2$	64197	60.053	−43.2800	−37.4600	2.8250	−0.8146
137	Propionic acid	$C_3H_6O_2$	79094	74.079	−45.3500	−36.6700	3.2300	−1.3950
138	*n*-Butyric acid	$C_4H_8O_2$	107926	88.106	−47.5800	−36.0000	3.6200	−2.0077
139	Isobutyric acid	$C_4H_8O_2$	79312	88.106	−48.4100	−36.2100	3.4120	−2.0004
140	Benzoic acid	$C_7H_6O_2$	65850	122.123	−29.4100	−21.4200	3.6900	−3.0951
141	Acetic anhydride	$C_4H_6O_3$	108247	102.090	−57.2500	−47.3400	3.8990	−1.6750
142	Methyl formate	$C_2H_4O_2$	107313	60.053	−35.2400	−29.5000	2.8520	−0.8924
143	Methyl acetate	$C_3H_6O_2$	79209	74.079	−41.1900	−32.4200	3.1980	−1.4610
144	Methyl propionate	$C_4H_8O_2$	554121	88.106	−42.7500	−31.1000	3.5960	−2.0780
145	Methyl *n*-butyrate	$C_5H_{10}O_2$	623427	102.133	−45.0700	−30.5300	3.9880	−2.6860
146	Ethyl formate	$C_3H_6O_2$	109944	74.079	−38.8300	−30.3100	3.2820	−1.5070
147	Ethyl acetate	$C_4H_8O_2$	141786	88.106	−44.4500	−32.8000	3.5970	−2.0610
148	Ethyl propionate	$C_5H_{10}O_2$	105373	102.133	−46.3600	−31.9300	4.0250	−2.6740
149	Ethyl *n*-butyrate	$C_6H_{12}O_2$	105544	116.160	−48.5500	−31.2200	4.4170	−3.2840
150	*n*-Propyl formate	$C_4H_8O_2$	110747	88.106	−40.7600	−29.3600	3.6780	−2.0410
151	*n*-Propyl acetate	$C_5H_{10}O_2$	109604	102.133	−46.4800	−32.0400	4.0230	−2.6720
152	*n*-Butyl acetate	$C_6H_{12}O_2$	123864	116.160	−48.5600	−31.2600	4.4250	−3.2800
153	Methyl benzoate	$C_8H_8O_2$	93583	136.150	−28.7900	−18.1000	4.1400	−3.7720
154	Ethyl benzoate	$C_9H_{10}O_2$	93890	150.177	−32.6000	−19.0500	4.5500	−4.4100
155	Vinyl acetate	$C_4H_6O_2$	108054	86.090	−31.4900	−22.7900	3.2800	−1.9500
156	Methylamine	CH_5N	74895	31.057	−2.2970	3.2070	2.4330	−0.9751
157	Dimethylamine	C_2H_7N	124403	45.084	−1.8450	6.8390	2.7296	−1.6146
158	Trimethylamine	C_3H_9N	75503	59.111	−2.4310	9.8990	2.8700	−2.2449
159	Ethylamine	C_2H_7N	75047	45.084	−4.7150	3.6160	2.8480	−1.5874
160	Diethylamine	$C_4H_{11}N$	109897	73.138	−7.1420	7.3080	3.5220	−2.8003
161	Triethylamine	$C_6H_{15}N$	121448	101.192	−9.5800	11.4100	4.0540	−4.0405
162	*n*-Propylamine	C_3H_9N	107108	59.111	−7.0500	4.1700	3.2420	−2.1650

TABLE 2-221   Enthalpies and Gibbs Energies of Formation, Entropies, and Net Enthalpies of Combustion of Inorganic and Organic Compounds (*Continued*)

Cmpd. no.	Name	Formula	CAS no.	Mol wt	Ideal gas enthalpy of formation, J/kmol × 1E-07	Ideal gas Gibbs energy of formation, J/kmol × 1E-07	Ideal gas entropy, J/(kmol·K) × 1E-05	Standard net enthalpy of combustion, J/kmol × 1E-09
163	di-*n*-Propylamine	$C_6H_{15}N$	142847	101.192	−11.6000	8.6800	4.2900	−4.0189
164	Isopropylamine	C_3H_9N	75310	59.111	−8.3800	3.1920	3.1240	−2.1566
165	Diisopropylamine	$C_6H_{15}N$	108189	101.192	−15.0000	5.7900	4.1200	−3.9900
166	Aniline	C_6H_7N	62533	93.128	8.7100	16.6800	3.1980	−3.2390
167	*N*-Methylaniline	C_7H_9N	100618	107.155	8.8000	20.2000	3.4100	−3.9000
168	*N,N*-Dimethylaniline	$C_8H_{11}N$	121697	121.182	10.0500	24.7728	3.6600	−4.5250
169	Ethylene oxide	C_2H_4O	75218	44.053	−5.2630	−1.3230	2.4299	−1.2180
170	Furan	C_4H_4O	110009	68.075	−3.4800	0.0823	2.6714	−1.9959
171	Thiophene	C_4H_4S	110021	84.142	11.5440	12.6620	2.7865	−2.4352
172	Pyridine	C_5H_5N	110861	79.101	14.0370	19.0490	2.8278	−2.6721
173	Formamide	CH_3NO	75127	45.041	−19.2200	−14.7100	2.4857	−0.5021
174	*N,N*-Dimethylformamide	C_3H_7NO	68122	73.095	−19.1700	−8.8400	3.2600	−1.7887
175	Acetamide	C_2H_5NO	60355	59.068	−23.8300	−15.9600	2.7220	−1.0741
176	*N*-Methylacetamide	C_3H_7NO	79163	73.095	−24.0000	−13.5000	3.2000	−1.7100
177	Acetonitrile	C_2H_3N	75058	41.053	7.4040	9.1868	2.4329	−1.1904
178	Propionitrile	C_3H_5N	107120	55.079	5.1800	9.7495	2.8614	−1.8007
179	*n*-Butyronitrile	C_4H_7N	109740	69.106	3.4058	10.8658	3.2543	−2.4148
180	Benzonitrile	C_7H_5N	100470	103.123	21.8823	26.0872	3.2104	−3.5224
181	Methyl mercaptan	CH_4S	74931	48.109	−2.2900	−0.9800	2.5500	−1.1517
182	Ethyl mercaptan	C_2H_6S	75081	62.136	−4.6300	−0.4814	2.9610	−1.7366
183	*n*-Propyl mercaptan	C_3H_8S	107039	76.163	−6.7500	0.2583	3.3650	−2.3458
184	*n*-Butyl mercaptan	$C_4H_{10}S$	109795	90.189	−8.7800	1.1390	3.7520	−2.9554
185	Isobutyl mercaptan	$C_4H_{10}S$	513440	90.189	−9.6900	0.5982	3.6280	−2.9490
186	sec-Butyl mercaptan	$C_4H_{10}S$	513531	90.189	−9.6600	0.5120	3.6670	−2.9490
187	Dimethyl sulfide	C_2H_6S	75183	62.136	−3.7240	0.7302	2.8585	−1.7449
188	Methyl ethyl sulfide	C_3H_8S	624895	76.163	−5.9600	1.1470	3.3320	−2.3531
189	Diethyl sulfide	$C_4H_{10}S$	352932	90.189	−8.3470	1.7780	3.6800	−2.9607
190	Fluoromethane	CH_3F	593533	34.033	−23.4300	−21.0400	2.2273	−0.5219
191	Chloromethane	CH_3Cl	74873	50.488	−8.1960	−5.8440	2.3418	−0.6754
192	Trichloromethane	$CHCl_3$	67663	119.377	−10.2900	−7.0100	2.9560	−0.3800
193	Tetrachloromethane	CCl_4	56235	153.822	−9.5810	−5.3540	3.0991	−0.2653
194	Bromomethane	CH_3Br	74839	94.939	−3.7700	−2.8190	2.4580	−0.7054
195	Fluoroethane	C_2H_5F	353366	48.060	−26.4400	−21.2300	2.6440	−1.1270
196	Chloroethane	C_2H_5Cl	75003	64.514	−11.2260	−6.0499	2.7578	−1.2849
197	Bromoethane	C_2H_5Br	74964	108.966	−6.3600	−2.5820	2.8730	−1.2850
198	1-Chloropropane	C_3H_7Cl	540545	78.541	−13.3180	−5.2610	3.1547	−1.8670
199	2-Chloropropane	C_3H_7Cl	75296	78.541	−14.4770	−6.1360	3.0594	−1.8630
200	1,1-Dichloropropane	$C_3H_6Cl_2$	78999	112.986	−15.0800	−6.5200	3.4480	−1.7200
201	1,2-Dichloropropane	$C_3H_6Cl_2$	78875	112.986	−16.2800	−8.0180	3.5480	−1.7070
202	Vinyl chloride	C_2H_3Cl	75014	62.499	2.8450	4.1950	2.7354	−1.1780
203	Fluorobenzene	C_6H_5F	462066	96.104	−11.6566	−6.9036	3.0263	−2.8145
204	Chlorobenzene	C_6H_5Cl	108907	112.558	5.1090	9.8290	3.1403	−2.9760
205	Bromobenzene	C_6H_5Br	108861	157.010	10.5018	13.8532	3.2439	−3.0192
206	Air		132259100	28.951	0	0	1.9900	0
207	Hydrogen	H_2	1333740	2.016	0	0	1.3057	0
208	Helium-4	He	7440597	4.003	0	0	1.2604	−0.2418
209	Neon	Ne	7440019	20.180	0	0	1.4622	0
210	Argon	Ar	7440371	39.948	0	0	1.5474	0
211	Fluorine	F_2	7782414	37.997	0	0	2.0268	0
212	Chlorine	Cl_2	7782505	70.905	0	0	2.2297	0
213	Bromine	Br_2	7726956	159.808	3.0910	0.3140	2.4535	0
214	Oxygen	O_2	7782447	31.999	0	0	2.0504	0
215	Nitrogen	N_2	7727379	28.014	0	0	1.9150	0
216	Ammonia	NH_3	7664417	17.031	−4.5898	−1.6400	1.9266	−0.3168
217	Hydrazine	N_2H_4	302012	32.045	9.5353	15.9170	2.3861	−5.3420
218	Nitrous oxide	N_2O	10024972	44.013	8.2050	10.4160	2.1985	−0.0820
219	Nitric oxide	NO	10102439	30.006	9.0250	8.6570	2.1060	−0.0902
220	Cyanogen	C_2N_2	460195	52.036	30.9072	29.7553	2.4146	−1.0961
221	Carbon monoxide	CO	630080	28.010	−11.0530	−13.7150	1.9756	−0.2830
222	Carbon dioxide	CO_2	124389	44.010	−39.3510	−39.4370	2.1368	0
223	Carbon disulfide	CS_2	75150	76.143	11.6900	6.6800	2.3790	−1.0769
224	Hydrogen fluoride	HF	7664393	20.006	−27.3300	−27.5400	1.7367	0.1524
225	Hydrogen chloride	HCl	7647010	36.461	−9.2310	−9.5300	1.8679	−0.0286
226	Hydrogen bromide	HBr	10035106	80.912	−3.6290	−5.3340	1.9859	−0.0690
227	Hydrogen cyanide	HCN	74908	27.026	13.5143	12.4725	2.0172	−0.6233

TABLE 2-221 Enthalpies and Gibbs Energies of Formation, Entropies, and Net Enthalpies of Combustion of Inorganic and Organic Compounds (*Concluded*)

Cmpd. no.	Name	Formula	CAS no.	Mol wt	Ideal gas enthalpy of formation, J/kmol × 1E-07	Ideal gas Gibbs energy of formation, J/kmol × 1E-07	Ideal gas entropy, J/(kmol·K) × 1E-05	Standard net enthalpy of combustion, J/kmol × 1E-09
228	Hydrogen sulfide	H_2S	7783064	34.082	−2.0630	−3.3440	2.0560	−0.5180
229	Sulfur dioxide	SO_2	7446095	64.065	−29.6840	−30.0120	2.4810	0
230	Sulfur trioxide	SO_3	7446119	80.064	−39.5720	−37.0950	2.5651	0.0989
231	Water	H_2O	7732185	18.015	−24.1814	−22.8590	1.8872	0

All substances are listed in alphabetical order in Table 2-6a.
Compiled from Daubert, T. E., R. P. Danner, H. M. Sibul, and C. C. Stebbins, DIPPR Data Compilation of Pure Compound Properties, Project 801 Sponsor Release, July, 1993, Design Institute for Physical Property Data, AIChE, New York, NY; and from Thermodynamics Research Center, "Selected Values of Properties of Hydrocarbons and Related Compounds," Thermodynamics Research Center Hydrocarbon Project, Texas A&M University, College Station, Texas (extant 1994).
   The compounds are considered to be formed from the elements in their standard states at 298.15 K and 101,325 P_a. These include C (graphite) and S (rhombic).
Enthalpy of combustion is the net value for the compound in its standard state at 298.15K and 101,325 Pa.
Products of combustion are taken to be CO_2 (gas), H_2O (gas), F_2 (gas), Cl_2 (gas), Br_2 (gas), I_2 (gas), SO_2 (gas), N_2 (gas), H_3PO_4 (solid), and SiO_2 (crystobalite).
J/kmol × 2.390E-04 = cal/gmol; J/kmol × 4.302106E-04 = Btu/lbmol.
J/(kmol·K) × 2.390E-04 = cal/(gmol·°C); J/(kmol·K) × 2.390059E-04 = Btu/(lbmol·°F).

TABLE 2-222 Ideal Gas Sensible Enthalpies, $h_T - h_{298}$ (kJ/kgmol), of Combustion Products

Temperature, K	CO	CO_2	H	OH	H_2	N	NO	NO_2	N_2	N_2O	O	O_2	SO_2	H_2O
200	−2858	−3414	−2040	−2976	−2774	−2040	−2951	−3495	−2857	−3553	−2186	−2868	−3736	−3282
240	−1692	−2079	−1209	−1756	−1656	−1209	−1743	−2104	−1692	−2164	−1285	−1703	−2258	−1948
260	−1110	−1383	−793	−1150	−1091	−793	−1142	−1392	−1110	−1438	−840	−1118	−1496	−1279
280	−529	−665	−377	−546	−522	−378	−543	−672	−528	−692	−398	−533	−718	−609
298.15	0	0	0	0	0	0	0	0	0	0	0	0	0	0
300	54	69	38	55	53	38	55	68	54	72	41	54	74	62
320	638	823	454	654	630	454	652	816	636	854	478	643	881	735
340	1221	1594	870	1251	1209	870	1248	1571	1219	1654	913	1234	1702	1410
360	1805	2382	1285	1847	1791	1286	1845	2347	1802	2470	1346	1828	2538	2088
380	2389	3184	1701	2442	2373	1701	2442	3130	2386	3302	1777	2425	3387	2769
400	2975	4003	2117	3035	2959	2117	3040	3927	2971	4149	2207	3025	4250	3452
420	3563	4835	2532	3627	3544	2533	3638	4735	3557	5010	2635	3629	5126	4139
440	4153	5683	2948	4219	4131	2949	4240	5557	4143	5884	3063	4236	6015	4829
460	4643	6544	3364	4810	4715	3364	4844	6392	4731	6771	3490	4847	6917	5523
480	5335	7416	3779	5401	5298	3780	5450	7239	5320	7670	3918	5463	7831	6222
500	5931	8305	4196	5992	5882	4196	6059	8099	5911	8580	4343	6084	8758	6925
550	7428	10572	5235	7385	6760	5235	7592	10340	7395	10897	5402	7653	11123	8699
600	8942	12907	6274	8943	8811	6274	9144	12555	8894	13295	6462	9244	13544	10501
650	10477	15303	7314	10423	10278	7314	10716	14882	10407	15744	7515	10859	16022	12321
700	12023	17754	8353	11902	11749	8353	12307	17250	11937	18243	8570	12499	18548	14192
750	13592	20260	9392	13391	13223	9329	13919	19671	13481	20791	9620	14158	21117	16082
800	15177	22806	10431	14880	14702	10431	15548	22136	15046	23383	10671	15835	23721	18002
850	16781	25398	11471	16384	16186	11471	17195	24641	16624	26014	11718	17531	26369	19954
900	18401	28030	12510	17888	17676	12510	18858	27179	18223	28681	12767	19241	29023	21938
950	20031	30689	13550	19412	19175	13550	20537	29749	19834	31381	13812	20965	31714	23954
1000	21690	33397	14589	20935	20680	14589	22229	32344	21463	34110	14860	22703	34428	26000
1100	25035	38884	16667	24024	23719	16667	25653	37605	24760	39647	16950	26212	39914	30191
1200	28430	44473	18746	27160	26797	18746	29120	42946	28109	45274	19039	29761	45464	34506
1300	31868	50148	20824	30342	29918	20824	32626	48351	31503	50976	21126	33344	51069	38942
1400	35343	55896	22903	33569	33082	22903	36164	53808	34936	56740	23212	36957	56718	43493
1500	38850	61705	24982	36839	36290	24982	39729	59309	38405	62557	25296	40599	62404	48151
1600	42385	67569	27060	40151	39541	27060	43319	64846	41904	68420	27381	44256	68123	52908
1700	45945	73480	29139	43502	42835	29139	46929	70414	45429	74320	29464	47958	73870	57758
1800	49526	79431	31217	46889	46169	31218	50557	76007	48978	80254	31547	51673	79642	62693
1900	53126	85419	33296	50310	49541	33296	54201	81624	52548	86216	33630	55413	85436	67706
2000	56744	91439	35375	53762	52951	35375	57859	87259	56137	92203	35713	59175	91250	72790
2100	60376	97488	37453	57243	56397	37454	61530	92911	59742	98212	37796	62961	97081	77941
2200	64021	103562	39532	60752	59876	39534	65212	98577	63361	104240	39878	66769	102929	83153
2300	67683	109660	41610	64285	63387	41614	68904	104257	66995	110284	41962	70600	108792	88421
2400	71324	115779	43689	67841	66928	43695	72606	109947	70640	116344	44045	74453	114669	93741
2500	74985	121917	45768	71419	70498	45777	76316	115648	74296	122417	46130	78328	120559	99108
2600	78673	128073	47846	75017	74096	47860	80034	121357	77963	128501	48216	82224	126462	104520
2700	82369	134246	49925	78633	77720	49945	83759	127075	81639	134596	50303	86141	132376	109973
2800	86074	140433	52004	82267	81369	52033	87491	132799	85323	140701	52391	90079	138302	115464
2900	89786	146636	54082	85918	85043	54124	91229	138530	89015	146814	54481	94036	144238	120990
3000	93504	152852	56161	89584	88740	56218	94973	144267	92715	152935	56574	98013	150184	126549
3500	112185	184109	66554	108119	107555	66769	113768	173020	111306	183636	67079	118165	180057	154768
4000	130989	215622	75947	126939	126874	77532	132671	201859	130027	214453	77675	188705	210145	183552
4500	149895	247354	87340	145991	146660	88614	151662	230756	148850	245348	88386	159572	240427	212764
5000	168890	279283	97733	165246	166876	100111	170730	259692	167763	276299	99222	180749	270893	242313

Converted and usually rounded off from JANAF Thermochemical Tables, NSRDS-NBS-37, 1971 (1141 pp.)

TABLE 2-223 Ideal Gas Entropies, $s°$, kJ/kgmol·K, of Combustion Products

Temperature, K	CO	CO_2	H	OH	H_2	N	NO	NO_2	N_2	N_2O	O	O_2	SO_2	H_2O
200	186.0	200.0	106.4	171.6	119.4	145.0	198.7	225.9	180.0	205.6	152.2	193.5	233.0	175.5
240	191.3	206.0	110.1	177.1	124.5	148.7	204.1	232.2	185.2	211.9	156.2	198.7	239.9	181.4
260	193.7	208.8	111.8	179.5	126.8	150.4	206.6	235.0	187.6	214.8	158.0	201.1	242.8	184.1
280	195.3	211.5	113.3	181.8	129.2	151.9	208.8	237.7	189.8	217.5	159.7	203.3	245.8	186.6
298.15	197.7	213.8	114.7	183.7	130.7	153.3	210.8	240.0	191.6	220.0	161.1	205.1	248.2	188.8
300	197.8	214.0	114.8	183.9	130.9	153.4	210.9	240.3	191.8	220.2	161.2	205.3	248.5	189.0
320	199.7	216.5	116.2	185.9	132.8	154.8	212.9	242.7	193.7	222.7	162.6	207.2	251.1	191.2
340	201.5	218.8	117.4	187.7	134.5	156.0	214.7	245.0	195.5	225.2	163.9	209.0	253.6	193.3
360	203.2	221.0	118.6	189.4	136.2	157.2	216.4	247.2	197.2	227.5	165.2	210.7	256.0	195.2
380	204.7	223.2	119.7	191.0	137.7	158.3	218.0	249.3	198.7	229.7	166.3	212.5	258.2	197.1
400	206.2	225.3	120.8	192.5	139.2	159.4	219.5	251.3	200.2	231.9	167.4	213.8	260.4	198.8
420	207.7	227.3	121.8	194.0	140.6	160.4	221.0	253.2	201.5	234.0	168.4	215.3	262.5	200.5
440	209.0	229.3	122.8	195.3	141.9	161.4	222.3	255.1	202.9	236.0	169.4	216.7	264.6	202.0
460	210.4	231.2	123.7	196.6	143.2	162.3	223.7	257.0	204.2	238.0	170.4	218.0	266.6	203.6
480	211.6	233.1	124.6	197.9	144.5	163.1	225.0	258.8	205.5	239.9	171.3	219.4	268.5	205.1
500	212.8	234.9	125.5	199.1	145.7	164.0	226.3	260.6	206.7	241.8	172.2	220.7	270.5	206.5
550	215.7	239.2	127.5	201.8	148.6	166.0	229.1	264.7	209.4	246.2	174.2	223.7	274.9	210.5
600	218.3	243.3	129.3	204.4	151.1	167.8	231.9	268.8	212.2	250.4	176.1	226.5	279.2	213.1
650	220.8	247.1	131.0	206.8	153.4	169.4	234.4	272.6	214.6	254.3	177.7	229.1	283.1	215.9
700	223.1	250.8	132.5	209.0	155.6	171.0	236.8	276.0	216.9	258.0	179.3	231.5	286.9	218.7
750	225.2	255.4	133.9	211.1	157.6	172.5	239.0	279.3	219.0	261.5	180.7	233.7	290.4	221.3
800	227.3	257.5	135.2	213.0	159.5	173.8	241.1	282.5	221.0	264.8	182.1	235.9	293.8	223.8
850	229.2	260.6	136.4	214.8	161.4	175.1	243.0	285.5	223.0	268.0	183.4	237.9	297.0	226.2
900	231.1	263.6	137.7	216.5	163.1	176.3	245.0	288.4	224.8	271.1	184.6	239.9	300.1	228.5
950	232.8	266.5	138.8	218.1	164.7	177.4	246.8	291.3	226.5	274.0	185.7	241.8	303.0	230.6
1000	234.5	269.3	139.9	219.7	166.2	178.5	248.4	293.9	228.2	276.8	186.8	243.6	305.8	232.7
1100	237.7	274.5	141.9	222.7	169.1	180.4	251.8	298.9	231.3	282.1	188.8	246.9	311.0	236.7
1200	240.7	279.4	143.7	225.4	171.8	182.2	254.8	303.6	234.2	287.0	190.6	250.0	315.8	240.5
1300	243.4	283.9	145.3	228.0	174.3	183.9	257.6	307.9	236.9	291.5	192.3	252.9	320.3	244.0
1400	246.0	288.2	146.9	230.3	176.6	185.4	260.2	311.9	239.5	295.8	193.8	255.6	324.5	247.4
1500	248.4	292.2	148.3	232.6	178.8	186.9	262.7	315.7	241.9	299.8	195.3	258.1	328.4	250.6
1600	250.7	296.0	149.6	234.7	180.9	188.2	265.0	319.3	244.1	303.6	196.6	260.4	332.1	253.7
1700	252.9	299.6	150.9	236.8	182.9	189.5	267.2	322.7	246.3	307.2	197.9	262.7	335.6	256.6
1800	254.9	303.0	152.1	238.7	184.8	190.7	269.3	325.9	248.3	310.6	199.1	264.8	338.9	259.5
1900	256.8	306.2	153.2	240.6	186.7	191.8	271.3	328.9	250.2	313.8	200.2	266.8	342.0	262.2
2000	258.7	309.3	154.3	242.3	188.4	192.9	273.1	331.8	252.1	316.9	201.3	268.7	345.0	264.8
2100	260.5	312.2	155.3	244.0	190.1	193.9	274.9	334.5	253.8	319.8	202.3	270.6	347.9	267.3
2200	262.2	315.1	156.3	245.7	191.7	194.8	276.6	337.2	255.5	322.6	203.2	272.4	350.6	269.7
2300	263.8	317.8	157.2	247.2	193.3	195.8	278.3	339.7	257.1	325.3	204.2	274.1	353.2	272.0
2400	265.4	320.4	158.1	248.7	194.8	196.7	279.8	342.1	258.7	327.9	205.0	275.7	355.7	274.3
2500	266.9	322.9	158.9	250.2	196.2	197.5	281.4	344.5	260.2	330.4	205.9	277.3	358.1	276.5
2600	268.3	325.3	159.7	251.6	197.7	198.3	282.8	346.7	261.6	332.7	206.7	278.8	360.4	278.6
2700	269.7	327.6	160.5	253.0	199.0	199.1	284.2	348.9	263.0	335.0	207.5	280.3	362.6	280.7
2800	271.0	329.9	161.3	254.3	200.3	199.9	285.6	350.9	264.3	337.3	208.3	281.7	364.8	282.7
2900	272.3	332.1	162.0	255.6	201.6	200.6	286.9	352.9	265.6	339.4	209.0	283.1	366.9	284.6
3000	273.6	334.2	162.7	256.8	202.9	201.3	288.2	354.9	266.9	341.5	209.7	284.4	368.9	286.5
3500	279.4	343.8	165.9	262.5	208.7	204.6	294.0	363.8	272.6	350.9	212.9	290.7	378.1	295.2
4000	284.4	352.2	168.7	267.6	213.8	207.4	299.0	371.5	277.6	359.2	215.8	296.2	386.1	302.9
4500	288.8	359.7	171.1	272.1	218.5	210.1	303.5	378.3	282.1	366.5	218.3	301.1	393.3	309.8
5000	292.8	366.4	173.3	276.1	222.8	212.5	307.5	384.4	286.0	373.0	220.6	305.5	399.7	316.0

Usually rounded off from JANAF Thermochemical Tables, NSRDS-NBS-37, 1971 (1141 pp.). Equilibrium constants can be calculated by combining $\Delta h_f°$ values from Table 2-221, $h_T - h_{298}$ from Table 2-222, and $s°$ values from the above, using the formula $\ln k_p = -\Delta G/(RT)$, where $\Delta G = \Delta h_f° + (h_T - h_{298}) - T s°$.

HEATS OF SOLUTION

TABLE 2-224 Heats of Solution of Inorganic Compounds in Water

Heat evolved, in kilogram-calories per gram formula weight, on solution in water at 18°C. Computed from data in Bichowsky and Rossini, *Thermochemistry of Chemical Substances*, Reinhold, New York, 1936.

Substance	Dilution°	Formula	Heat, kg-cal/ g-mole	Substance	Dilution°	Formula	Heat, kg-cal/ g-mole
Aluminum bromide	aq	$AlBr_3$	+85.3	Calcium—(*Cont.*)			
chloride	600	$AlCl_3$	+77.9	bromide	∞	$CaBr_2$	+24.86
	600	$AlCl_3 \cdot 6H_2O$	+13.2		∞	$CaBr_2 \cdot 6H_2O$	-0.9
fluoride	aq	AlF_3	+31	chloride	∞	$CaCl_2$	+4.9
	aq	$AlF_3 \cdot \frac{1}{2}H_2O$	+19.0		∞	$CaCl_2 \cdot H_2O$	+12.3
	aq	$AlF_3 \cdot 3\frac{1}{2}H_2O$	-1.7		∞	$CaCl_2 \cdot 2H_2O$	+12.5
iodide	aq	AlI_3	+89.0		∞	$CaCl_2 \cdot 4H_2O$	+2.4
sulfate	aq	$Al_2(SO_4)_3$	+126		∞	$CaCl_2 \cdot 6H_2O$	-4.11
	aq	$Al_2(SO_4)_3 \cdot 6H_2O$	+56.2	formate	400	$Ca(CHO_2)_2$	+0.7
	aq	$Al_2(SO_4)_3 \cdot 18H_2O$	+6.7	iodide	∞	CaI_2	+28.0
Ammonium bromide	aq	NH_4Br	-4.45		∞	$CaI_2 \cdot 8H_2O$	+1.8
chloride	∞	NH_4Cl	-3.82	nitrate	∞	$Ca(NO_3)_2$	+4.1
chromate	aq	$(NH_4)_2CrO_4$	-5.82		∞	$Ca(NO_3)_2 \cdot H_2O$	+0.7
dichromate	600	$(NH_4)_2Cr_2O_7$	-12.9		∞	$Ca(NO_3)_2 \cdot 2H_2O$	-3.2
iodide	aq	NH_4I	-3.56		∞	$Ca(NO_3)_2 \cdot 3H_2O$	-4.2
nitrate	∞	NH_4NO_3	-6.47		∞	$Ca(NO_3)_2 \cdot 4H_2O$	-7.99
perborate	aq	$NH_4BO_3 \cdot H_2O$	-9.0	phosphate, mono-	aq	$Ca(H_2PO_4)_2 \cdot H_2O$	-0.6
sulfate	∞	$(NH_4)_2SO_4$	-2.75	dibasic	aq	$CaHPO_4 \cdot 2H_2O$	-1
sulfate, acid	800	NH_4HSO_4	+0.56	sulfate	∞	$CaSO_4$	+5.1
sulfite	aq	$(NH_4)_2SO_3$	-1.2		∞	$CaSO_4 \cdot \frac{1}{2}H_2O$	+3.6
	aq	$(NH_4)_2SO_3 \cdot H_2O$	-4.13		∞	$CaSO_4 \cdot 2H_2O$	-0.18
Antimony fluoride	aq	SbF_3	-1.7	Chromous chloride	aq	$CrCl_2$	+18.6
iodide	aq	SbI_3	-0.8			$CrCl_2 \cdot 3H_2O$	+5.3
Arsenic acid	aq	H_3AsO_4	-0.4			$CrCl_2 \cdot 4H_2O$	+2.0
				iodide	aq	CrI_2	+5.7
Barium bromate	∞	$Ba(BrO_3)_2 \cdot H_2O$	-15.9	Cobaltous bromide	aq	$CoBr_2$	+18.4
bromide	∞	$BaBr_2$	+5.3		aq	$CoBr_2 \cdot 6H_2O$	-1.25
	∞	$BaBr_2 \cdot H_2O$	-0.8	chloride	400	$CoCl_2$	+18.5
	∞	$BaBr_2 \cdot 2H_2O$	-3.87		400	$CoCl_2 \cdot 2H_2O$	+9.8
chlorate	∞	$Ba(ClO_3)_2$	-6.7		400	$CoCl_2 \cdot 6H_2O$	-2.9
	∞	$Ba(ClO_3)_2 \cdot H_2O$	-10.6	iodide	aq	CoI_2	+18.8
chloride	∞	$BaCl_2$	+2.4	sulfate	400	$CoSO_4$	+15.0
	∞	$BaCl_2 \cdot H_2O$	-2.17		400	$CoSO_4 \cdot 6H_3O$	-1.4
	∞	$BaCl_2.2H_2O$	-4.5		400	$CoSO_4 \cdot 7H_2O$	-3.6
cyanide	aq	$Ba(CN)_2$	+1.5	Cupric acetate	aq	$Cu(C_2H_3O_2)_2$	+2.4
	aq	$Ba(CN)_2 \cdot H_2O$	-2.4	formate	aq	$Cu(CHO_2)_2$	+0.5
	aq	$Ba(CN)_2 \cdot 2H_2O$	-4.9	nitrate	200	$Cu(NO_3)_2$	+10.3
iodate	∞	$Ba(IO_3)_2$	-9.1		200	$Cu(NO_3)_2 \cdot 3H_2O$	-2.6
	∞	$Ba(IO_3)_2 \cdot H_2O$	-11.3		200	$Cu(NO_3)_2 \cdot 6H_2O$	-10.7
iodide	∞	BaI_2	+10.5	sulfate	800	$CuSO_4$	+15.9
	∞	$BaI_2 \cdot H_2O$	+2.7			$CuSO_4 \cdot H_2O$	+9.3
	∞	$BaI_2 \cdot 2H_2O$	+0.14			$CuSO_4 \cdot 3H_2O$	+3.65
	∞	$BaI_2 \cdot 2\frac{1}{2}H_2O$	-0.58			$CuSO_4 \cdot 5H_2O$	-2.85
	∞	$BaI_2 \cdot 7H_2O$	-6.61	Cuprous sulfate	aq	Cu_2SO_4	+11.6
nitrate	∞	$Ba(NO_3)_2$	-10.2	Ferric chloride	1000	$FeCl_3$	+31.7
perchlorate	∞	$Ba(ClO_4)_2$	-2.8		1000	$FeCl_3 \cdot 2\frac{1}{2}H_2O$	+21.0
	∞	$Ba(ClO_4)_2 \cdot 3H_2O$	-10.5		1000	$FeCl_3 \cdot 6H_2O$	+5.6
sulfide	∞	BaS	+7.2	nitrate	800	$Fe(NO_3)_3 \cdot 9H_2O$	-9.1
Beryllium bromide	aq	$BeBr_2$	+62.6	Ferrous bromide	aq	$FeBr_2$	+18.0
chloride	aq	$BeCl_2$	+51.1	chloride	400	$FeCl_2$	+17.9
iodide	aq	BeI_2	+72.6		400	$FeCl_2 \cdot 2H_2O$	+8.7
sulfate	aq	$BeSO_4$	+18.1		400	$FeCl_2 \cdot 4H_2O$	+2.7
	aq	$BeSO_4 \cdot H_2O$	+13.5	iodide	aq	FeI_2	+23.3
	aq	$BeSO_4 \cdot 2H_2O$	+7.9	sulfate	400	$FeSO_4$	+14.7
	aq	$BeSO_4 \cdot 4H_2O$	+1.1		400	$FeSO_4 \cdot H_2O$	+7.35
Bismuth iodide	aq	BiI_3	+3		400	$FeSO_4 \cdot 4H_2O$	+1.4
Boric acid	aq	H_3BO_3	-5.4		400	$FeSO_4 \cdot 7H_2O$	-4.4
Cadmium bromide	400	$CdBr_2$	+0.4				
	400	$CdBr_2 \cdot 4H_2O$	-7.3	Lead acetate	400	$Pb(C_2H_3O_2)_2$	+1.4
chloride	400	$CdCl_2$	+3.1		400	$Pb(C_2H_3O_2)_2 \cdot 3H_2O$	-5.9
	400	$CdCl_2 \cdot H_2O$	+0.6	bromide	aq	$PbBr_2$	-10.1
	400	$CdCl_2 \cdot 2\frac{1}{2}H_2O$	-3.00	chloride	aq	$PbCl_2$	-3.4
nitrate	400	$Cd(NO_3)_2 \cdot H_2O$	+4.17	formate	aq	$Pb(CHO_2)_2$	-6.9
	400	$Cd(NO_3)_2 \cdot 4H_2O$	-5.08	nitrate	400	$Pb(NO_3)_2$	-7.61
sulfate	400	$CdSO_4$	+10.69	Lithium bromide	∞	$LiBr$	+11.54
	400	$CdSO_4 \cdot H_2O$	+6.05		∞	$LiBr \cdot H_2O$	+5.30
	400	$CdSO_4 \cdot 2\frac{2}{3}H_2O$	+2.51		∞	$LiBr \cdot 2H_2O$	+2.05
Calcium acetate	∞	$Ca(C_2H_3O_2)_2$	+7.6		∞	$LiBr \cdot 3H_2O$	-1.59
	∞	$Ca(C_2H_3O_2)_2 \cdot H_2O$	+6.5	chloride	∞	$LiCl$	+8.66

°The numbers represent moles of water used to dissolve 1 g formula weight of substance; ∞ means "infinite dilution"; and *aq* means "aqueous solution of unspecified dilution."

TABLE 2-224   Heats of Solution of Inorganic Compounds in Water (Continued)

Substance	Dilution°	Formula	Heat, kg-cal/ g-mole	Substance	Dilution°	Formula	Heat, kg-cal/ g-mole
Lithium—(Cont.)				Phosphoric acid, ortho-	400	H_3PO_4	+2.79
	∞	$LiCl·H_2O$	+4.45		400	$H_3PO_4·\frac{1}{2}H_2O$	−0.1
	∞	$LiCl·2H_2O$	+1.07	pyro-	aq	$H_4P_2O_7$	+25.9
	∞	$LiCl·3H_2O$	−1.98		aq	$H_4P_2O_7·1\frac{1}{2}H_2O$	+4.65
fluoride	∞	LiF	−0.74	Potassium acetate	∞	$KC_2H_3O_2$	+3.55
hydroxide	∞	$LiOH$	+4.74	aluminum sulfate	600	$KAl(SO_4)_2$	+48.5
	∞	$LiOH·\frac{1}{8}H_2O$	+4.39		600	$KAl(SO_4)_2·3H_2O$	+26.6
	∞	$LiOH·H_2O$	+9.6			$KAl(SO_4)_2·12H_2O$	−10.1
iodide	∞	LiI	+14.92	bicarbonate	2000	$KHCO_3$	−5.1
	∞	$LiI·\frac{1}{2}H_2O$	+10.08	bromate	∞	$KBrO_3$	−10.13
	∞	$LiI·H_2O$	+6.93	bromide	∞	KBr	−5.13
	∞	$LiI·2H_2O$	+3.43	carbonate	∞	K_2CO_3	+6.58
	∞	$LiI·3H_2O$	−0.17			$K_2CO_3·\frac{1}{2}H_2O$	+4.25
nitrate	∞	$LiNO_3$	+0.466			$K_2CO_3·1\frac{1}{2}H_2O$	−0.43
	∞	$LiNO_3·3H_2O$	−7.87	chlorate	∞	$KClO_3$	−10.31
sulfate	∞	Li_2SO_4	+6.71	chloride	∞	KCl	−4.404
	∞	$Li_2SO_4·H_2O$	+3.77	chromate	2185	K_2CrO_4	−4.9
				chrome sulfate	600	$KCr(SO_4)_2$	+55
Magnesium bromide	∞	$MgBr_2$	+43.7			$KCr(SO_4)_2·H_2O$	+42
	∞	$MgBr_2·H_2O$	+35.9			$KCr(SO_4)_2·2H_2O$	+33
	∞	$MgBr_2·6H_2O$	+19.8			$KCr(SO_4)_2·6H_2O$	+7
chloride	∞	$MgCl_2$	+36.3			$KCr(SO_4)_2·12H_2O$	−9.5
	∞	$MgCl_2·2H_2O$	+20.8	cyanide	200	KCN	−3.0
	∞	$MgCl_2·4H_2O$	+10.5	dichromate	1600	$K_2Cr_2O_7$	−17.8
	∞	$MgCl_2·6H_2O$	+3.4	fluoride	∞	KF	+3.96
iodide	∞	MgI_2	+50.2		∞	$KF·2H_2O$	−1.85
nitrate	∞	$Mg(NO_3)_2·6H_2O$	−3.7		∞	$KF·4H_2O$	−6.05
phosphate	aq	$Mg_3(PO_4)_2$	+10.2	hydrosulfide	∞	KHS	+0.86
sulfate	∞	$MgSO_4$	+21.1		∞	$KHS·\frac{1}{4}H_2O$	+1.21
	∞	$MgSO_4·H_2O$	+14.0	hydroxide	∞	KOH	+12.91
	∞	$MgSO_4·2H_2O$	+11.7		∞	$KOH·\frac{3}{4}H_2O$	+4.27
	∞	$MgSO_4·4H_2O$	+4.9		∞	$KOH·H_2O$	+3.48
	∞	$MgSO_4·6H_2O$	+0.55		∞	$KOH·7H_2O$	+0.86
	∞	$MgSO_4·7H_2O$	−3.18	iodate	∞	KIO_3	−6.93
sulfide	aq	MgS	+25.8	iodide	∞	KI	−5.23
Manganic nitrate	400	$Mn(NO_3)_2$	+12.9	nitrate	∞	KNO_3	−8.633
	400	$Mn(NO_3)_2·3H_2O$	−3.9	oxalate	400	$K_2C_2O_4$	−4.6
	400	$Mn(NO_3)_2·6H_2O$	−6.2			$K_2C_2O_4·H_2O$	−7.5
sulfate	aq	$Mn_2(SO_4)_3$	+22	perchlorate	∞	$KClO_4$	−12.94
Manganous acetate	aq	$Mn(C_2H_3O_2)_2$	+12.2	permanganate	400	$KMnO_4$	−10.4
	aq	$Mn(C_2H_3O_2)_2·4H_2O$	+1.6	phosphate, dihydrogen	aq	KH_2PO_4	+4.7
bromide	aq	$MnBr_2$	+15	pyrosulfite	aq	$K_2S_2O_5$	−11.0
	aq	$MnBr_2·H_2O$	+14.4		aq	$K_2S_2O_5·\frac{1}{2}H_2O$	−10.22
	aq	$MnBr_2·4H_2O$	+16.1	sulfate	∞	K_2SO_4	−6.32
chloride	400	$MnCl_2$	+16.0	sulfate, acid	800	$KHSO_4$	−3.10
	400	$MnCl_2·2H_2O$	+8.2	sulfide	∞	K_2S	−11.0
	400	$MnCl_2·4H_2O$	+1.5	sulfite	aq	K_2SO_3	+1.8
formate	aq	$Mn(CHO_2)_2$	+4.3		aq	$K_2SO_3·H_2O$	+1.37
	aq	$Mn(CHO_2)_2·2H_2O$	−2.9	thiocyanate	∞	$KCNS$	−6.08
iodide	aq	MnI_2	+26.2	thionate, di-	aq	$K_2S_2O_6$	−13.0
	aq	$MnI_2·H_2O$	+24.1	thiosulfate	∞	$K_2S_2O_3$	−4.5
	aq	$MnI_2·2H_2O$	+22.7				
	aq	$MnI_2·4H_2O$	+19.9	Silver acetate	aq	$AgC_2H_3O_2$	−5.4
	aq	$MnI_2·6H_2O$	+21.2	nitrate	200	$AgNO_3$	−4.4
sulfate	400	$MnSO_4$	+13.8	Sodium acetate	∞	$NaC_2H_3O_2$	+4.085
	400	$MnSO_4·H_2O$	+11.9		∞	$NaC_2H_3O_2·3H_2O$	−4.665
	400	$MnSO_4·7H_2O$	−1.7	arsenate	500	Na_3AsO_4	+15.6
Mercuric acetate	aq	$Hg(C_2H_3O_2)_2$	−4.0		500	$Na_3AsO_4·12H_2O$	−12.61
bromide	aq	$HgBr_2$	−2.4	bicarbonate	1800	$NaHCO_3$	−4.1
chloride	aq	$HgCl_2$	−3.3	borate, tetra-	900	$Na_2B_4O_7$	+10.0
nitrate	aq	$Hg(NO_3)_2·\frac{1}{2}H_2O$	−0.7		900	$Na_2B_4O_7·10H_2O$	−16.8
Mercurous nitrate	aq	$Hg_2(NO_3)_2·2H_2O$	−11.5	bromide	∞	$NaBr$	−0.58
					∞	$NaBr·2H_2O$	−4.57
Nickel bromide	aq	$NiBr_2$	+19.0	carbonate	∞	Na_2CO_3	+5.57
	aq	$NiBr_2·3H_2O$	+0.2		∞	$Na_2CO_3·H_2O$	+2.19
Nickel chloride	800	$NiCl_2$	+19.23		∞	$Na_2CO_3·7H_2O$	−10.81
	800	$NiCl_2·2H_2O$	+10.4		∞	$Na_2CO_3·10H_2O$	−16.22
	800	$NiCl_2·4H_2O$	+4.2	chlorate	∞	$NaClO_3$	−5.37
	800	$NiCl_2·6H_2O$	−1.15	chloride	∞	$NaCl$	−1.164
iodide	aq	NiI_2	+19.4	chromate	800	Na_2CrO_4	+2.50
nitrate	200	$Ni(NO_3)_2$	+11.8		800	$Na_2CrO_4·4H_2O$	−7.52
	200	$Ni(NO_3)_2·6H_2O$	−7.5		800	$Na_2CrO_4·10H_2O$	−16.0
sulfate	200	$NiSO_4$	+15.1	cyanide	200	$NaCN$	−0.37
	200	$NiSO_4·7H_2O$	−4.2		200	$NaCN·\frac{1}{2}H_2O$	−0.92

TABLE 2-224   Heats of Solution of Inorganic Compounds in Water (Concluded)

Substance	Dilution°	Formula	Heat, kg-cal/ g-mole	Substance	Dilution°	Formula	Heat, kg-cal/ g-mole
Sodium—(Cont.)				Sodium—(Cont.)			
	200	$NaCN \cdot 2H_2O$	−4.41	thionate, di-	aq	$Na_2S_2O_6$	−5.80
fluoride	∞	NaF	−0.27		aq	$Na_2S_2O_6 \cdot 2H_2O$	−11.86
hydrosulfide	∞	$NaHS$	+4.62	Sodium thiosulfate	aq	$Na_2S_2O_3$	+2.0
	∞	$NaHS \cdot 2H_2O$	−1.49		aq	$Na_2S_2O_3 \cdot 5H_2O$	−11.30
Sodium hydroxide	∞	$NaOH$	+10.18	Stannic bromide	aq	$SnBr_4$	+15.5
	∞	$NaOH \cdot \tfrac{1}{2}H_2O$	+8.17	Stannous bromide	aq	$SnBr_2$	−1.6
	∞	$NaOH \cdot \tfrac{2}{3}H_2O$	+7.08	iodide	aq	SnI_2	−5.8
	∞	$NaOH \cdot \tfrac{3}{4}H_2O$	+6.48	Strontium acetate	∞	$Sr(C_2H_3O_2)_2$	+6.2
	∞	$NaOH \cdot H_2O$	+5.17		∞	$Sr(C_2H_3O_2)_2 \cdot \tfrac{1}{2}H_2O$	+5.9
iodide	∞	NaI	+1.57	bromide	∞	$SrBr_2$	+16.4
	∞	$NaI \cdot 2H_2O$	−3.89		∞	$SrBr_2 \cdot H_2O$	+9.25
metaphosphate	600	$NaPO_3$	+3.97		∞	$SrBr_2 \cdot 2H_2O$	+6.5
nitrate	∞	$NaNO_3$	−5.05		∞	$SrBr_2 \cdot 4H_2O$	+0.4
nitrite	aq	$NaNO_2$	−3.6		∞	$SrBr_2 \cdot 6H_2O$	−6.1
perchlorate	∞	$NaClO_4$	−4.15	chloride	∞	$SrCl_2$	+11.54
phosphate di	1600	Na_2HPO_4	+5.21		∞	$SrCl_2 \cdot H_2O$	+6.4
tri-	1600	Na_3PO_4	+13		∞	$SrCl_2 \cdot 2H_2O$	+2.95
phosphate di	1600	$Na_3PO_4 \cdot 12H_2O$	−15.3		∞	$SrCl_2 \cdot 6H_2O$	−7.1
di-	1600	$Na_2HPO_4 \cdot 2H_2O$	−0.82	iodide	∞	SrI_2	+20.7
	1600	$Na_2HPO_4 \cdot 7H_2O$	−12.04		∞	$SrI_2 \cdot H_2O$	+12.65
	1600	$Na_2HPO_4 \cdot 12H_2O$	−23.18		∞	$SrI_2 \cdot 2H_2O$	+10.4
phosphite, mono-	600	NaH_2PO_3	+0.90		∞	$SrI_2 \cdot 6H_2O$	−4.5
	600	$NaH_2PO_3 \cdot 2\tfrac{1}{2}H_2O$	−5.29	nitrate	∞	$Sr(NO_3)_2$	−4.8
di-	800	Na_2HPO_3	+9.30		∞	$Sr(NO_3)_2 \cdot 4H_2O$	−12.4
	800	$Na_2HPO_3 \cdot 5H_2O$	−4.54	sulfate	∞	$SrSO_4$	+0.5
pyrophosphate	1600	$Na_4P_2O_7$	+11.9	Sulfuric acid, pyro-	∞	$H_2S_2O_7$	−18.08
	1600	$Na_4P_2O_7 \cdot 10H_2O$	−11.7	Zinc acetate	400	$Zn(C_2H_3O_2)_2$	+9.8
di-	1200	$Na_2H_2P_2O_7$	−2.2		400	$Zn(C_2H_3O_2)_2 \cdot H_2O$	+7.0
	1200	$Na_2H_2P_2O_7 \cdot 6H_2O$	−14.0		400	$Zn(C_2H_3O_2)_2 \cdot 2H_2O$	+3.9
sulfate	∞	Na_2SO_4	+0.28	bromide	400	$ZnBr_2$	+15.0
	∞	$Na_2SO_4 \cdot 10H_2O$	−18.74	chloride	400	$ZnCl_2$	+15.72
sulfate, acid	800	$NaHSO_4$	+1.74	iodide	aq	ZnI_2	+11.6
	800	$NaHSO_4 \cdot H_2O$	+0.15	nitrate	400	$Zn(NO_3)_2 \cdot 3H_2O$	−5
sulfide	∞	Na_2S	+15.2		400	$Zn(NO_3)_2 \cdot 6H_2O$	−6.0
	∞	$Na_2S \cdot 4\tfrac{1}{2}H_2O$	+0.09	sulfate	400	$ZnSO_4$	+18.5
	∞	$Na_2S \cdot 5H_2O$	−6.54		400	$ZnSO_4 \cdot H_2O$	+10.0
	∞	$Na_2S \cdot 9H_2O$	−16.65		400	$ZnSO_4 \cdot 6H_2O$	−0.8
sulfite	∞	Na_2SO_3	+2.8		400	$ZnSO_4 \cdot 7H_2O$	−4.3
	∞	$Na_2SO_3 \cdot 7H_2O$	−11.1				
thiocyanate	∞	$NaCNS$	−1.83				

NOTE: To convert kilocalories per gram-mole to British thermal units per pound-mole, multiply by 1.799×10^{-3}.

TABLE 2-225 Heats of Solution of Organic Compounds in Water (at Infinite Dilution and Approximately Room Temperature)

Recalculated and rearranged from *International Critical Tables*, vol. 5, pp. 148–150. (g·cal)/(g·mol) = Btu/(lb·mol) × 1.799.

Solute	Heat of Solution, G-cal/g-mole Solute°	Solute	Heat of Solution, G-cal/g-mole Solute°
Acetic acid (solid), $C_2H_4O_2$	−2,251	Oxalic acid, $C_2H_2O_4$	−2,290
Acetylacetone, $C_5H_8O_2$	−641	(2H_2O)	−8,485
Acetylurea, $C_3H_6N_2O_2$	−6,812	Phenol (solid), C_6H_6O	−2,605
Aconitic acid, $C_6H_6O_6$	−4,206	Phthalic acid, $C_8H_6O_4$	−4,871
Ammonium benzoate, $C_7H_9NO_2$	−2,700	Picric acid, $C_6H_3N_3O_7$	−7,098
picrate	−8,700	Piperic acid, $C_{12}H_{10}O_4$	−10,492
succinate (n-)	−3,489	Piperonylic acid, $C_8H_6O_4$	−9,106
Aniline, hydrochloride, C_6H_8ClN	−2,732	Potassium benzoate	−1,506
Barium picrate	−4,708	citrate	2,820
Benzoic acid, $C_7H_6O_2$	−6,501	tartrate (n-) (0.5 H_2O)	−5,562
Camphoric acid, $C_{10}H_{16}O_4$	−502	Pyrogallol, $C_6H_6O_3$	−3,705
Citric acid, $C_6H_8O_7$	−5,401	Pyrotartaric acid	−5,019
Dextrin, $C_{12}H_{20}O_{10}$	268	Quinone	−3,991
Fumaric acid, $C_4H_4O_4$	−5,903	Raffinose, $C_{18}H_{32}O_{16}$ (5H_2O)	−9,703
Hexamethylenetetramine, $C_6H_{12}N_4$	4,780	Resorcinol, $C_6H_6O_2$	−3,960
Hydroxybenzamide (m-), $C_7H_7NO_2$	−4,161	Silver malonate (n-)	−9,799
(m-), (HCl)	−7,003	Sodium citrate (tri-)	5,270
(o-), $C_7H_7NO_2$	−4,340	picrate	−6,441
(p-)	−5,392	potassium tartrate	−1,817
Hydroxybenzoic acid (o-), $C_7H_6O_3$	−6,350	(4H_2O)	−12,342
(p-), $C_7H_6O_3$	−5,781	succinate (n-)	2,390
Hydroxybenzyl alcohol (o-), $C_7H_8O_2$	−3,203	(6H_2O)	−10,994
Inulin, $C_{36}H_{62}O_{31}$	−96	tartrate (n-)	−1,121
Isosuccinic acid, $C_4H_6O_4$	−3,420	(2H_2O)	−5,882
Itaconic acid, $C_5H_6O_4$	−5,922	Strontium picrate	7,887
Lactose, $C_{12}H_{22}O_{11} \cdot H_2O$	−3,705	(6H_2O)	−14,412
Lead picrate	−7,098	Succinic acid, $C_4H_6O_4$	−6,405
(2H_2O)	−13,193	Succinimide, $C_4H_5NO_2$	−4,302
Magnesium picrate	14,699	Sucrose, $C_{12}H_{22}O_{11}$	−1,319
(8H_2O)	−15,894	Tartaric acid (d-)	−3,451
Maleic acid, $C_4H_4O_4$	−4,441	Thiourea, CH_4N_2S	−5,330
Malic acid, $C_4H_6O_5$	−3,150	Urea, CH_4N_2O	−3,609
Malonic acid, $C_3H_4O_4$	−4,493	acetate	−8,795
Mandelic acid, $C_8H_2O_3$	−3,090	formate	−7,194
Mannitol, $C_6H_{14}O_6$	−5,260	nitrate	−10,803
Menthol, $C_{10}H_{20}O$	0	oxalate	−17,806
Nicotine dihydrochloride, $C_{10}H_{16}Cl_2N_2$	6,561	Vanillic acid	−5,160
Nitrobenzoic acid (m-), $C_7H_5NO_4$	−5,593	Vanillin	−5,210
(o-), $C_7H_5NO_4$	−5,306	Zinc picrate	−11,496
(p-), $C_7H_5NO_4$	−8,891	(8H_2O)	−15,894
Nitrophenol (m-), $C_6H_5NO_3$	−5,210		
(o-), $C_6H_5NO_3$	−6,310		
(p-), $C_6H_5NO_3$	−4,493		

°+ denotes heat evolved, and − denotes heat absorbed. All values are positive unless otherwise noted. The data in the *International Critical Tables* were calculated by E. Anderson.

THERMODYNAMIC PROPERTIES

EXPLANATION OF TABLES

The following subsection presents information on the thermodynamic properties of a number of fluids. In some cases transport properties are also included.

Notation

c_p = specific heat
e = specific internal energy
h = enthalpy
k = thermal conductivity
p = pressure
s = specific entropy
t = temperature
T = absolute temperature
u = specific internal energy
μ = viscosity
v = specific volume
f = subscript denoting saturated liquid
g = subscript denoting saturated vapor

UNITS CONVERSIONS

For this subsection, the following units conversions are applicable:

c_p, specific heat: To convert kilojoules per kilogram-kelvin to British thermal units per pound–degree Fahrenheit, multiply by 0.23885.

e, internal energy: To convert kilojoules per kilogram to British thermal units per pound, multiply by 0.42992.

g, gravity acceleration: To convert meters per second squared to feet per second squared, multiply by 3.2808.

h, enthalpy: To convert kilojoules per kilogram to British thermal units per pound, multiply by 0.42992.

k, thermal conductivity: To convert watts per meter-kelvin to British thermal unit–feet per hour–square foot–degree Fahrenheit, multiply by 0.57779.

p, pressure: To convert bars to kilopascals, multiply by 1×10^2; to convert bars to pounds-force per square inch, multiply by 14.504; and to convert millimeters of mercury to pounds-force per square inch, multiply by 0.01934.

s, entropy: to convert kilojoules per kilogram-kelvin to British thermal units per pound–degree Rankine, multiply by 0.23885.

t, temperature: $°F = \frac{9}{5} °C + 32$.

T, absolute temperature: $°R = \frac{9}{5} K$.

u, internal energy: to convert kilojoules per kilogram to British thermal units per pound, multiply by 0.42992.

μ, viscosity: to convert pascal-seconds to pound-force–seconds per square foot, multiply by 0.020885; to convert pascal-seconds to c_p, multiply by 1000.

v, specific volume: to convert cubic meters per kilogram to cubic feet per pound, multiply by 16.018.

ρ, density: to convert kilograms per cubic meter to pounds per cubic foot, multiply by 0.062428.

ADDITIONAL REFERENCES

Bretsznajder, *Prediction of Transport and Other Physical Properties of Fluids,* Pergamon, New York, 1971. D'Ans and Lax, *Handbook for Chemists and Physicists* (in German), 3 vols., Springer-Verlag, Berlin. *Engineering Data Book,* Natural Gas Processors Suppliers Association, Tulsa, Okla. Ganic, Hartnett, and Rohsenow, *Handbook of Heat Transfer,* 2d ed., McGraw-Hill, New York, 1984. Gray, *American Institute of Physics Handbook,* 3d ed., McGraw-Hill, New York, 1972. Kay and Laby, *Tables of Physical and Chemical Constants,* Longman, London, various editions and dates. *Landolt-Börnstein Tables,* many volumes and dates, Springer-Verlag, Berlin. Lange, *Handbook of Chemistry,* McGraw-Hill, New York, various editions and dates. Partington, *Advanced Treatise on Physical Chemistry,* 5 vols., Longman, London, 1950. Raznjevic, *Handbook of Thermodynamic Tables and Charts,* McGraw-Hill, New York, 1976 and other editions. Reynolds, *Thermodynamic Properties in SI,* Department of Mechanical Engineering, Stanford University, 1979. Stephan and Lucas, *Viscosity of Dense Fluids,* Plenum, New York and London, 1979. *Selected Values of Properties of Chemical Compounds* and *Selected Values of the Properties of Hydrocarbons and Related Compounds,* Thermodynamics Research Center, Texas A&M University, College Station, loose-leaf, intermittent publication. Vargaftik, *Tables of the Thermophysical Properties of Gases and Liquids,* Wiley, New York, 1975. Vargaftik, Filippov, Tarzimanov, and Totskiy, *Thermal Conductivity of Liquids and Gases* (in Russian), Standartov, Moscow, 1978. Weast, *Handbook of Chemistry and Physics,* Chemical Rubber Co., Boca Raton, FL, annually.

TABLE 2-226 Thermophysical Properties of Saturated Acetone

Temperature, K	Pressure, bar	v_f, m³/kg	v_g, m³/kg	h_f, kJ/kg	h_g, kJ/kg	s_f, kJ/kg·K	s_g, kJ/kg·K	c_{pf}, kJ/kg·K	μ_f, 10^{-6} Pa·s	k_f, W/m·K	Pr
300	0.318	0.001 261	1.415	−67	466	−0.213	1.561				
310	0.482	0.001 285	0.942	−46	476	−0.144	1.540				
320	0.710	0.001 309	0.645	−22	490	−0.068	1.531				
329.3[b]	1.013	0.001 333	0.456	0	506	0	1.537	2.29	232	0.141	3.77
330	1.040	0.001 335	0.448	2	506	0.003	1.521	2.29	231	0.141	3.75
340	1.52	0.001 359	0.311	25	509	0.075	1.514	2.33	212	0.137	3.61
350	2.04	0.001 383	0.237	51	529	0.150	1.516	2.38	200	0.132	3.61
360	2.74	0.001 408	0.179	78	543			2.43	187	0.128	3.55
370	3.60	0.001 435	0.138	103	554			2.48	176	0.124	3.52
380	4.52	0.001 464	0.110	127	566			2.53	165	0.119	3.51
390	5.87	0.001 495	0.0854	151	577			2.59	153	0.115	3.45
400	7.31	0.001 528	0.0684	184	588			2.65	141	0.111	3.37
410	8.94	0.001 564	0.0556	207	598			2.73	130	0.107	3.32
420	10.82	0.001 604	0.0454	231	608			2.82	119	0.103	3.26
430	13.64	0.001 647	0.0356	256	618			2.92	109	0.099	3.21
440	16.37	0.001 695	0.0292	281	625			3.03	99	0.095	3.16
450	19.42	0.001 748	0.0240	308	632			3.15	90	0.092	3.08
460	22.79	0.001 81	0.0199	337	637			3.29	80	0.088	2.99
470	27.52	0.001 88	0.0159	365	641			3.45	71	0.083	2.95
480	32.52	0.001 98	0.0130	396	638			3.76	64	0.077	3.13
490	37.73	0.002 15	0.0091								
500	43.08	0.002 46	0.0063								
508.2[c]	47.61	0.003 67	0.0037								

b = normal boiling point; c = critical point
P, v, h, and s interpolated and converted from *Heat Exchanger Design Handbook*, vol. 5, Hemisphere, Washington, DC, 1983 and reproduced in Beaton, C. F. and
G. F. Hewitt, *Physical Property Data for the Design Engineer*, Hemisphere, New York, 1989 (394 pp.). Other values compiled by P. E. Liley
An enthalpy-pressure diagram to 1000 psia, 250–500 °F appears in *J. Chem. Eng. Data* **7**, 1 (1962): 75–78.

TABLE 2-227 Saturated Acetylene*

Temperature, K	Pressure, bar	v_{cond}, m³/kg	v_g, m³/kg	h_{cond}, kJ/kg	h_g, kJ/kg	s_{cond}, kJ/(kg·K)	s_g, kJ/(kg·K)
162.0	0.101		5.081	158	983	2.967	8.062
169.3	0.203		2.644	173	994	3.039	7.889
173.9	0.304		1.805	182	999	3.095	7.797
180.0	0.507		1.116	194	1007	3.161	7.672
184.3	0.709		0.810	203	1011	3.216	7.596
189.1	1.013		0.5780	214	1015	3.272	7.511
192.4[t]	1.283		0.4617	221	1018	3.312	7.455
192.4[t]	1.283	0.00164	0.4617	378	1018	4.127	7.455
200.9	2.027	0.00165	0.3011	411	1027	4.296	7.362
209.4	3.040	0.00169	0.2074	445	1035	4.461	7.280
221.5	5.066	0.00174	0.1264	493	1046	4.684	7.180
230.4	7.093	0.00179	0.0907	528	1052	4.837	7.111
240.7	10.13	0.00186	0.0635	565	1058	4.990	7.037
253.2	15.20	0.00195	0.0420	602	1061	5.133	6.947
263.0	20.27	0.00204	0.0309	628	1061	5.231	6.878
271.6	25.33	0.00213	0.0240	654	1060	5.326	6.822
278.9	30.40	0.00223	0.0193	680	1057	5.414	6.767
284.9	35.46	0.00232	0.0159	704	1051	5.494	6.716
290.4	40.53	0.00242	0.0133	727	1041	5.576	6.658
300.0	50.66	0.00270	0.0093	778	1017	5.737	6.534
307.8	60.80	0.00335	0.0061	850	968	5.965	6.351
308.7[c]	62.47	0.00434	0.0043	908	908	6.158	6.158

*Values recalculated into SI units from those of Din. *Thermodynamic Functions of Gases*, vol. 2, Butterworth, London,
1956. Above the solid line the condensed phase is solid; below the line it is liquid. t = triple point; c = critical point.

TABLE 2-228   Saturated Air*

$T,$ K	$P_f,$ bar	$P_g,$ bar	$v_f,$ m^3/kg	$v_g,$ m^3/kg	$h_f,$ kJ/kg	$h_g,$ kJ/kg	$s_f,$ kJ/(kg·K)	$s_g,$ kJ/(kg·K)	$c_{pf},$ kJ/(kg·K)	$\mu_f,$ 10^{-4} Pa·s	$k_f,$ W/(m·K)
60			1.040.–3	5.55	−159.2	59.7	2.528	6.255		3.25	0.180
62			1.050.–3	3.73	−155.2	61.7	2.585	6.164		2.98	0.176
64	0.123	0.071	1.060.–3	2.57	−151.4	63.6	2.641	6.080		2.75	0.173
66	0.174	0.104	1.070.–3	1.82	−147.8	65.5	2.696	6.002		2.54	0.169
68	0.239	0.147	1.080.–3	1.313	−144.2	67.4	2.747	5.929		2.36	0.166
70	0.323	0.205	1.089.–3	0.968	−140.6	69.2	2.797	5.862	1.817	2.21	0.163
72	0.429	0.280	1.101.–3	0.728	−137.1	71.0	2.847	5.799	1.827	2.07	0.160
74	0.560	0.376	1.113.–3	0.556	−133.5	72.8	2.895	5.740	1.838	1.95	0.156
76	0.721	0.495	1.125.–3	0.431	−129.9	74.5	2.941	5.685	1.849	1.84	0.152
78	0.915	0.644	1.136.–3	0.339	−126.3	76.2	2.988	5.634	1.861	1.74	0.148
80	1.146	0.825	1.146.–3	0.270	−122.6	77.8	3.034	5.585	1.873	1.65	0.145
82	1.420	1.043	1.160.–3	0.217	−118.8	79.4	3.079	5.540	1.885	1.58	0.142
84	1.741	1.305	1.173.–3	0.177	−115.0	80.9	3.123	5.496	1.898	1.51	0.139
86	2.114	1.614	1.187.–3	0.145	−111.2	82.3	3.167	5.454	1.912	1.44	0.135
88	2.544	1.976	1.201.–3	0.120	−107.4	83.6	3.209	5.414	1.927	1.38	0.132
90	3.036	2.397	1.216.–3	0.1002	−103.5	84.8	3.251	5.376	1.944	1.32	0.128
92	3.596	2.884	1.231.–3	0.0843	−99.5	85.9	3.293	5.340	1.962	1.27	0.125
94	4.229	3.441	1.247.–3	0.0713	−95.5	87.0	3.335	5.304	1.982	1.23	0.121
96	4.940	4.075	1.265.–3	0.0607	−91.5	87.9	3.376	5.270	2.003	1.18	0.117
98	5.736	4.792	1.283.–3	0.0520	−87.5	88.7	3.416	5.236	2.027	1.14	0.114
100	6.621	5.599	1.302.–3	0.0447	−83.3	89.3	3.456	5.204	2.053	1.10	0.110
105	9.265	8.056	1.355.–3	0.0312	−72.8	90.2	3.553	5.124	2.137	1.02	0.102
110	12.59	11.22	1.418.–3	0.0222	−61.9	90.1	3.649	5.045	2.264	0.95	0.093
115	16.68	15.21	1.495.–3	0.0159	−50.3	88.4	3.747	4.964	2.477	0.87	0.084
120	21.61	20.14	1.596.–3	0.0115	−37.5	84.8	3.850	4.877	2.916	0.75	0.076
125	27.43	26.14	1.757.–3	0.0081	−22.0	78.2	3.969	4.776	4.585	0.42	0.067
130	34.16	33.32	2.075.–3	0.0054	0.4	66.1	4.136	4.644			
132.55^c		37.69	3.196.–3	0.0032	37.4	37.4	4.410	4.410	∞		∞

*Liquid properties extracted or converted from Vasserman and Rabinovich, *Thermophysical Properties of Liquid Air and Its Components,* Moscow, 1968, and NBS-NSF transl. TT 69-55092, 1970. Copyrighted material. Reproduced by permission. Vapor properties extracted or converted from Vasserman, Kazavchinskii, and Rabinovich, *Thermophysical Properties of Air and Its Components,* Nauka, Moscow, 1966, and NBS-NSF transl. TT 70-50095, 1971. Copyrighted material. Reproduced by permission. Note that on pages 150–151 of the TT 69-55092 publication certain values of TT 70-50095 were adjusted. As a complete retabulation was not given, the tables here are based upon the two separate publications, as indicated. See also Table 2-235 for the argon-oxygen-nitrogen equilibrium data. c = critical point. The notation 1.040.–3 signifies 1.040×10^{-3}.

TABLE 2-229 Thermophysical Properties of Compressed Air*

Pressure, bar		80	90	100	120	140	160	180	200	220	240	260	280	300
1	v		0.251	0.281	0.340	0.399	0.457	0.515	0.537	0.631	0.688	0.746	0.803	0.861
	h		87.9	98.3	118.8	139.1	159.3	179.5	199.7	219.8	239.9	260.0	280.2	300.3
	s	Mix	5.650	5.759	5.946	6.103	6.238	6.357	6.463	6.559	6.647	6.727	6.802	6.871
	C_p		1.044	1.032	1.020	1.014	1.010	1.008	1.007	1.006	1.006	1.006	1.006	1.007
	μ		0.064	0.071	0.085	0.097	0.109	0.121	0.133	0.144	0.154	0.165	0.175	0.185
	k		0.0084	0.0093	0.0112	0.0129	0.0147	0.0164	0.0181	0.0198	0.0214	0.0231	0.0247	0.0263
5	v	0.00115	0.00122	0.0509	0.0646	0.0773	0.0895	0.102	0.114	0.125	0.137	0.149	0.160	0.172
	h	−122.3	−103.3	90.6	113.6	135.3	156.4	177.1	197.7	218.1	238.5	258.8	279.1	299.4
	s	3.031	3.250	5.246	5.455	5.623	5.763	5.885	5.994	6.092	6.180	6.262	6.337	6.406
	C_p	1.868	1.941	1.212	1.107	1.065	1.045	1.033	1.025	1.020	1.017	1.015	1.013	1.013
	μ	1.794	1.163	0.077	0.087	0.098	0.110	0.122	0.134	0.145	0.155	0.165	0.175	0.185
	k	0.146	0.128	0.0103	0.0119	0.0135	0.0151	0.0168	0.0185	0.0201	0.0217	0.0234	0.0250	0.0265
10	v	0.00115	0.00121	0.00130	0.0298	0.0370	0.0436	0.0499	0.0561	0.0621	0.0681	0.0741	0.0800	0.0859
	h	−122.0	−103.1	−83.2	106.2	130.2	152.5	174.1	195.2	216.1	236.7	257.3	277.8	298.3
	s	3.028	3.246	3.452	5.214	5.398	5.548	5.675	5.786	5.885	5.975	6.058	6.134	6.204
	C_p	1.863	1.932	2.041	1.270	1.146	1.093	1.065	1.049	1.038	1.031	1.026	1.023	1.201
	μ	1.816	1.177	0.838	0.089	0.101	0.112	0.124	0.135	0.146	0.156	0.166	0.176	0.186
	k	0.146	0.128	0.111	0.0126	0.0141	0.0157	0.0173	0.0189	0.0205	0.0221	0.0237	0.0253	0.0268
20	v	0.00114	0.00121	0.00129	0.0116	0.0167	0.0206	0.0241	0.0274	0.0306	0.0337	0.0368	0.0398	0.0428
	h	−121.3	−102.5	−82.9	85.2	118.5	144.3	167.7	190.1	211.9	233.2	254.3	275.2	296.0
	s	3.022	3.239	3.442	4.882	5.140	5.312	5.450	5.568	5.672	5.765	5.849	5.927	5.998
	C_p	1.853	1.916	2.010	2.237	1.390	1.215	1.141	1.101	1.076	1.061	1.050	1.042	1.037
	μ	1.859	1.205	0.857	0.098	0.106	0.116	0.127	0.137	0.148	0.158	0.168	0.178	0.187
	k	0.147	0.130	0.112	0.0152	0.0157	0.0169	0.0182	0.0197	0.0212	0.0228	0.0243	0.0258	0.0273
40	v	0.00114	0.00120	0.00128	0.00153	0.0058	0.0090	0.0114	0.0131	0.0148	0.0165	0.0182	0.0198	0.0214
	h	−120.0	−101.4	−82.2	−39.8	83.6	125.3	154.3	179.7	203.5	226.3	248.5	270.2	291.7
	s	3.011	3.225	3.424	3.807	4.745	5.025	5.196	5.330	5.444	5.543	5.632	5.712	5.786
	C_p	1.834	1.886	1.958	2.432	3.193	1.610	1.335	1.221	1.159	1.122	1.097	1.081	1.068
	μ	1.943	1.261	0.896	0.516	0.132	0.129	0.135	0.144	0.154	0.163	0.172	0.182	0.191
	k	0.149	0.132	0.115	0.0814	0.0460	0.0201	0.0206	0.0217	0.0229	0.0242	0.0256	0.0270	0.0284
60	v	0.00113	0.00119	0.00126	0.00147	0.00222	0.00505	0.00687	0.00833	0.00963	0.0108	0.0120	0.0131	0.0142
	h	−118.6	−100.3	−81.4	−40.8	22.8	90.0	132.6	163.9	191.1	216.1	240.0	263.1	285.6
	s	3.000	3.211	3.407	3.773	4.260	4.798	5.020	5.174	5.298	5.404	5.497	5.581	5.657
	C_p	1.818	1.860	1.915	2.205	4.808	2.338	1.594	1.361	1.249	1.186	1.146	1.119	1.100
	μ	2.028	1.318	0.936	0.559	0.277	0.153	0.149	0.154	0.161	0.169	0.178	0.186	0.195
	k	0.150	0.134	0.117	0.0861	0.0480	0.0360	0.0238	0.0240	0.0248	0.0258	0.0270	0.0283	0.0296
80	v	0.00113	0.00119	0.00126	0.00145	0.00188	0.00327	0.00480	0.00601	0.00706	0.00803	0.00894	0.00981	0.0107
	h	−117.2	−99.1	−80.4	−41.3	9.0	78.4	125.3	158.7	187.1	212.9	237.3	260.8	283.7
	s	2.989	3.198	3.391	3.745	4.138	4.597	4.875	5.051	5.186	5.299	5.396	5.484	5.562
	C_p	1.802	1.838	1.881	2.078	2.992	3.029	1.887	1.510	1.342	1.250	1.194	1.156	1.130
	μ	2.12	1.38	0.977	0.597	0.356	0.194	0.167	0.166	0.170	0.177	0.184	0.191	0.200
	k	0.152	0.134	0.120	0.0901	0.0599	0.0420	0.0278	0.0268	0.0269	0.0276	0.0286	0.0296	0.0308
100	v	0.00112	0.00118	0.00125	0.00142	0.00174	0.00252	0.00366	0.00467	0.00556	0.00637	0.00713	0.00785	0.00855
	h	−115.8	−97.8	−79.4	−41.3	3.9	61.7	111.8	148.8	179.4	206.7	232.2	256.4	279.9
	s	2.978	3.186	3.376	3.721	4.076	4.457	4.753	4.949	5.095	5.214	5.315	5.406	5.486
	C_p	1.789	1.818	1.852	1.992	2.506	2.874	2.114	1.650	1.431	1.311	1.239	1.191	1.158
	μ	2.21	1.44	1.02	0.631	0.405	0.249	0.193	0.181	0.181	0.185	0.191	0.198	0.205
	k	0.154	0.137	0.122	0.0936	0.0669	0.0500	0.0327	0.0299	0.0293	0.0295	0.0302	0.0311	0.0320
150	v	0.00111	0.00116	0.00122	0.00137	0.00158	0.00194	0.00247	0.00309	0.00369	0.00425	0.00478	0.00529	0.00578
	h	−112.2	−94.5	−76.6	−40.1	0.5	45.2	89.5	129.2	163.2	193.4	221.0	247.0	271.8
	s	2.954	3.157	3.342	3.673	3.988	4.287	4.548	4.757	4.919	5.051	5.161	5.257	5.343
	C_p	1.789	1.818	1.852	1.992	2.506	2.874	2.114	1.650	1.431	1.311	1.239	1.267	1.220
	μ	2.44	1.60	1.13	0.709	0.490	0.349	0.266	0.229	0.215	0.211	0.212	0.215	0.220
	k	0.157	1.142	0.127	0.101	0.0785	0.0588	0.0455	0.0389	0.0360	0.0348	0.0346	0.0349	0.0354
200	v	0.00110	0.00115	0.00120	0.00133	0.00150	0.00174	0.00206	0.00245	0.00287	0.00328	0.00368	0.00407	0.00446
	h	−108.5	−91.2	−73.6	−38.0	0.2	40.2	79.8	117.6	152.2	183.6	212.5	239.6	265.5
	s	2.930	3.130	3.312	3.634	3.931	4.198	4.432	4.631	4.796	4.932	5.048	5.149	5.238
	C_p	1.733	1.747	1.761	1.809	1.905	1.988	1.953	1.814	1.643	1.501	1.396	1.321	1.266
	μ	2.70	1.78	1.25	0.782	0.561	0.420	0.331	0.279	0.253	0.241	0.236	0.235	0.237
	k	0.161	0.146	0.132	0.107	0.0868	0.0691	0.0559	0.0476	0.0429	0.0405	0.0393	0.0389	0.0389
250	v	0.00109	0.00114	0.00119	0.00130	0.00144	0.00162	0.00186	0.00214	0.00244	0.00276	0.00307	0.00338	0.00368
	h	−104.8	−87.6	−70.3	−35.4	1.3	38.9	75.8	111.7	145.6	177.1	206.6	234.3	260.8
	s	2.909	3.106	3.285	3.601	3.886	4.138	4.355	4.544	4.706	4.843	4.961	5.064	5.155
	C_p	1.712	1.722	1.733	1.767	1.824	1.854	1.831	1.748	1.635	1.522	1.427	1.353	1.297
	μ	2.96	1.97	1.39	0.855	0.625	0.476	0.385	0.327	0.292	0.272	0.262	0.257	0.256
	k	0.165	0.150	0.137	0.113	0.0935	0.0769	0.0641	0.0552	0.0495	0.0460	0.0441	0.0430	0.0426

*For sources, units, and remarks, see Table 2-228. v = specific volume, m³/kg; h = specific enthalpy, kJ/kg; s = specific entropy, kJ/(kg·K); c_p = specific heat at constant pressure, kJ/(kg·K); μ = viscosity, 10^{-4} Pa·s; and k = thermal conductivity, W/(m·K). For specific heat ratio, see Table 2-200; for Prandtl number, see Table 2-369.

						Temperature, K						
350	400	450	500	600	800	1000	1200	1400	1600	1800	2000	2500
1.005	1.148	1.292	1.436	1.723	2.297	2.872	3.446	4.020	4.594	5.168	5.743	7.200
350.7	401.2	452.1	503.4	607.5	822.5	1046.8	1278	1515	1764	2017	2279	3011
7.026	7.161	7.282	7.389	7.579	7.888	8.138	8.349	8.531	8.695	8.844	8.983	9.308
1.009	1.014	1.021	1.030	1.051	1.099	1.141	1.175	1.207	1.248	1.286	1.337	1.665
0.208	0.230	0.251	0.270	0.306	0.370	0.424	0.473	0.527	0.584	0.637	0.689	0.818
0.0301	0.0336	0.0371	0.0404	0.0466	0.0577	0.0681	0.0783	0.0927	0.106	0.120	0.137	0.222
0.201	0.230	0.259	0.288	0.345	0.460	0.575	0.690	0.805	0.920	1.034	1.149	1.438
350.0	400.8	451.8	503.2	607.4	822.6	1046.9	1279	1516	1764	2017	2278	2981
6.563	6.698	6.818	6.927	7.116	7.426	7.676	7.887	8.069	8.233	8.382	8.520	8.832
1.014	1.017	1.024	1.032	1.053	1.100	1.142	1.175	1.208	1.248	1.285	1.326	1.516
0.208	0.230	0.251	0.270	0.306	0.370	0.425	0.473	0.527	0.584	0.637	0.689	0.818
0.0303	0.0338	0.0372	0.0405	0.0467	0.0578	0.0681	0.0783	0.0927	0.106	0.120	0.136	0.195
0.101	0.115	0.130	0.144	0.173	0.231	0.288	0.345	0.403	0.460	0.518	0.575	0.720
349.2	400.2	451.4	502.9	607.3	822.7	1047.2	1279	1516	1765	2018	2279	2974
6.361	6.497	6.618	6.727	6.917	7.226	7.477	7.688	7.870	8.034	8.183	8.321	8.630
1.019	1.021	1.027	1.034	1.055	1.100	1.142	1.175	1.208	1.248	1.284	1.324	1.481
0.209	0.231	0.252	0.271	0.306	0.370	0.425	0.473	0.527	0.584	0.637	0.689	0.817
0.0305	0.0340	0.0374	0.0407	0.0469	0.0579	0.0682	0.0784	0.0927	0.106	0.120	0.135	0.187
0.0503	0.0577	0.0650	0.0723	0.0868	0.116	0.145	0.173	0.202	0.231	0.260	0.288	0.360
347.7	399.1	450.7	502.4	607.2	823.0	1047.7	1280	1517	1766	2019	2279	2970
6.158	6.295	6.417	6.526	6.716	7.027	0.277	7.489	7.671	7.835	7.984	8.121	8.428
1.030	1.029	1.033	1.039	1.057	1.102	1.143	1.176	1.209	1.249	1.284	1.322	1.456
0.210	0.232	0.253	0.272	0.307	0.371	0.425	0.474	0.527	0.584	0.637	0.689	0.817
0.0309	0.0344	0.0377	0.0410	0.0471	0.0581	0.0685	0.0787	0.0928	0.106	0.120	0.135	0.181
0.0252	0.0290	0.0327	0.0364	0.0438	0.0583	0.0728	0.0872	0.102	0.116	0.130	0.145	0.181
344.6	397.0	449.2	501.5	606.9	823.7	1048.8	1281	1519	1768	2021	2281	2969
5.950	6.090	6.212	6.323	6.515	6.826	7.077	7.289	7.473	7.636	7.785	7.922	8.229
1.051	1.044	1.044	1.049	1.063	1.105	1.145	1.177	1.210	1.249	1.284	1.322	1.438
0.213	0.235	0.255	0.274	0.309	0.372	0.426	0.474	0.527	0.584	0.637	0.689	0.817
0.0318	0.0351	0.0384	0.0416	0.0476	0.0584	0.0687	0.0789	0.0928	0.106	0.120	0.135	0.177
0.0169	0.0194	0.0220	0.0245	0.0294	0.0392	0.0489	0.0585	0.0681	0.0776	0.0872	0.0968	0.1207
340.4	394.0	447.1	500.6	606.8	824.3	1050.0	1283	1521	1770	2023	2284	2969
5.824	5.967	6.091	6.202	6.396	6.708	6.960	7.172	7.355	7.520	7.669	7.806	8.112
1.072	1.059	1.055	1.057	1.069	1.108	1.147	1.178	1.210	1.249	1.286	1.322	1.430
0.217	0.237	0.257	0.275	0.310	0.373	0.427	0.475	0.527	0.584	0.637	0.689	0.817
0.0328	0.0359	0.0391	0.0422	0.0481	0.0588	0.0690	0.0790	0.0929	0.106	0.120	0.134	0.176
0.0127	0.0147	0.0166	0.0185	0.0223	0.0296	0.0369	0.0442	0.0513	0.0585	0.0657	0.0729	0.0908
339.0	393.1	446.5	499.8	606.7	825.1	1051.1	1284	1522	1772	2025	2285	2971
5.733	5.878	6.004	6.116	6.311	6.624	6.877	7.089	7.273	7.437	7.586	7.723	8.029
1.091	1.073	1.066	1.065	1.075	1.111	1.149	1.180	1.210	1.249	1.286	1.322	1.426
0.220	0.240	0.259	0.278	0.312	0.374	0.428	0.475	0.527	0.584	0.637	0.689	0.817
0.0337	0.0368	0.0398	0.0428	0.0486	0.0592	0.0693	0.0793	0.0929	0.106	0.120	0.134	0.175
0.0102	0.0118	0.0134	0.0149	0.0180	0.0239	0.0298	0.0356	0.0413	0.0470	0.0528	0.0584	0.0729
336.5	391.3	445.3	499.0	606.6	825.8	1052.4	1286	1524	1774	2027	2288	2972
5.661	5.807	5.935	6.048	6.244	6.559	6.812	7.024	7.208	7.373	7.522	7.659	7.964
1.110	1.087	1.076	1.073	1.080	1.114	1.151	1.181	1.211	1.250	1.288	1.323	1.423
0.224	0.243	0.262	0.280	0.314	0.375	0.429	0.477	0.527	0.584	0.637	0.689	0.817
0.0347	0.0376	0.0405	0.0434	0.0491	0.0595	0.0696	0.0795	0.0930	0.106	0.120	0.134	0.175
0.00695	0.00806	0.00914	0.0102	0.0123	0.0163	0.0202	0.0241	0.0279	0.0317	0.0356	0.0394	0.0490
330.9	387.5	442.9	497.5	606.6	827.8	1055.5	1290	1529	1779	2033	2294	2977
5.525	5.677	5.807	5.922	6.121	6.439	6.693	6.906	7.092	7.256	7.405	7.543	7.848
1.151	1.117	1.099	1.092	1.093	1.121	1.155	1.184	1.213	1.252	1.290	1.325	1.418
0.235	0.252	0.270	0.286	0.318	0.379	0.431	0.478	0.527	0.584	0.637	0.689	
0.0374	0.0398	0.0424	0.0451	0.0504	0.0605	0.0703	0.0801	0.0932	0.106	0.120	0.133	
0.00534	0.00620	0.00702	0.00783	0.00940	0.0125	0.0154	0.0184	0.0212	0.0241	0.0269	0.0298	0.0370
326.5	384.5	440.9	496.6	607.0	829.9	1058.7	1294	1533	1783	2038	2299	2982
5.426	5.581	5.715	5.831	6.033	6.353	6.608	6.822	7.009	7.173	7.323	7.460	7.765
1.184	1.141	1.119	1.108	1.104	1.128	1.160	1.187	1.214	1.254	1.292	1.326	1.415
0.248	0.262	0.278	0.293	0.324	0.382	0.434	0.481	0.528	0.585	0.638		
0.0400	0.0420	0.0423	0.0467	0.0517	0.0614	0.0711	0.0808	0.0934	0.106	0.120		
0.00440	0.00509	0.00576	0.00642	0.00770	0.0102	0.0126	0.0149	0.0172	0.0195	0.0218	0.0241	0.0298
323.2	382.3	439.6	496.0	607.6	832.2	1062.0	1298	1538	1789	2043	2304	2988
5.348	5.506	5.641	5.760	5.963	6.286	6.542	6.757	6.944	7.108	7.258	7.396	7.701
1.208	1.161	1.135	1.121	1.115	1.135	1.164	1.190	1.216	1.256	1.294	1.328	1.414
0.262	0.273	0.286	0.301	0.329	0.386	0.437	0.483	0.528	0.585			
0.0429	0.0443	0.0462	0.0484	0.0531	0.0624	0.0718	0.0814	0.0937	0.106			

TABLE 2-229   Thermophysical Properties of Compressed Air (*Concluded*)

Pressure, bar		Temperature, K												
		80	90	100	120	140	160	180	200	220	240	260	280	300
300	v	0.00108	0.00112	0.00117	0.00127	0.00139	0.00155	0.00173	0.00195	0.00219	0.00243	0.00269	0.00294	0.00318
	h	−101.0	−84.0	−67.0	−32.4	3.1	39.2	74.5	109.0	142.0	173.2	202.7	230.8	257.7
	s	2.888	3.083	3.260	3.572	3.849	4.090	4.298	4.480	4.637	4.773	4.891	4.995	5.088
	C_p	1.694	1.703	1.713	1.740	1.769	1.777	1.751	1.689	1.607	1.518	1.438	1.370	1.316
	μ	3.24	2.18	1.53	0.932	0.687	0.529	0.433	0.370	0.329	0.303	0.288	0.280	0.276
	k	0.168	0.154	0.141	0.118	0.0996	0.0836	0.0710	0.0619	0.0555	0.0514	0.0487	0.0471	0.0462
400	v		0.00110	0.00114	0.00123	0.00133	0.00145	0.00158	0.00173	0.00189	0.00206	0.00224	0.00242	0.00260
	h		−76.6	−59.8	−25.9	8.3	42.4	75.8	108.5	140.1	170.5	199.7	227.8	254.8
	s		3.042	3.216	3.523	3.788	4.016	4.214	4.386	4.537	4.669	4.786	4.890	4.983
	C_p		1.674	1.686	1.704	1.702	1.685	1.654	1.607	1.550	1.490	1.431	1.378	1.331
	μ		2.63	1.86	1.10	0.802	0.631	0.500	0.446	0.397	0.364	0.341	0.325	0.316
	k		0.161	0.149	0.127	0.110	0.0946	0.0823	0.0729	0.0660	0.0610	0.0574	0.0550	0.0533
500	v		0.00109	0.00112	0.00120	0.00128	0.00138	0.00148	0.00160	0.00173	0.00186	0.00199	0.00213	0.00227
	h		−69.0	−52.3	−18.7	14.4	47.4	79.8	111.4	142.0	171.7	200.5	228.4	255.4
	s		3.005	3.177	3.482	3.743	3.966	4.151	4.317	4.463	4.593	4.708	4.811	4.905
	C_p		1.655	1.670	1.686	1.667	1.644	1.598	1.557	1.509	1.461	1.415	1.371	1.331
	μ		3.13	2.24	1.31	0.924	0.710	0.0560	0.512	0.459	0.420	0.391	0.370	0.356
	k		0.167	0.156	0.135	0.119	0.104	0.0916	0.0822	0.0749	0.0694	0.0653	0.0622	0.0599
600	v								0.00151	0.00161	0.00172	0.00183	0.00194	0.00205
	h								116.0	146.1	175.3	203.6	231.2	258.1
	s								2.263	4.406	4.533	4.646	4.749	4.842
	C_p								1.525	1.480	1.438	1.398	1.361	1.327
	μ									0.516	0.472	0.439	0.414	0.396
	k								0.0903	0.0828	0.0769	0.0724	0.0689	0.0662
800	v									0.00147	0.00155	0.00163	0.00171	0.00179
	h									157.4	185.9	213.7	240.3	267.3
	s									4.318	4.442	4.553	4.653	4.745
	C_p									1.445	1.406	1.372	1.342	1.314
	μ											0.529	0.497	0.473
	k									0.0964	0.0901	0.0850	0.0809	0.0776
1000	v											0.00151	0.00157	0.00163
	h											226.4	253.2	279.5
	s											4.482	4.582	4.672
	C_p											1.355	1.327	1.303
	μ													0.546
	k											0.0961	0.0916	0.0878

Temperature, K												
350	400	450	500	600	800	1000	1200	1400	1600	1800	2000	2500
0.00379	0.00437	0.00493	0.00548	0.00656	0.00864	0.0107	0.0126	0.0145	0.0164	0.0183	0.0202	0.0250
320.9	380.9	438.9	495.9	608.5	834.5	1065.3	1302	1542	1794	2049	2310	2993
5.283	5.443	5.580	5.700	5.906	6.230	6.488	6.703	6.891	7.056	7.206	7.344	7.648
1.226	1.176	1.148	1.133	1.124	1.140	1.168	1.193	1.217	1.257	1.298	1.330	1.413
0.276	0.284	0.296	0.308	0.335	0.390	0.440	0.485	0.529				
0.0457	0.0466	0.0481	0.0501	0.0544	0.0634	0.0726	0.0820	0.0940				
0.00304	0.00348	0.00390	0.00432	0.00514	0.00673	0.00826	0.00977	0.0111	0.0126	0.0140	0.0155	0.0190
319.1	380.0	439.0	496.8	611.0	839.4	1072.0	1310	1552	1804	2059	2321	3004
5.181	5.344	5.483	5.605	5.813	6.142	6.401	6.618	6.808	6.972	7.123	7.261	7.566
1.246	1.195	1.166	1.149	1.138	1.151	1.176	1.199	1.222	1.258	1.301	1.333	1.412
0.307	0.308	0.315	0.325	0.348	0.398	0.446	0.490					
0.0513	0.0512	0.0521	0.0535	0.0571	0.0653	0.0740	0.0832					
0.00262	0.00296	0.00330	0.00364	0.00430	0.00558	0.00683	0.00804	0.00911	0.0103	0.0114	0.0126	0.0154
319.9	381.3	440.8	499.1	614.3	844.6	1078.8	1318	1561	1814	2070	2332	3015
5.255	5.267	5.408	5.531	5.741	6.072	6.333	6.550	6.743	6.907	7.058	7.196	7.501
1.255	1.206	1.176	1.159	1.148	1.159	1.183	1.205	1.226	1.265	1.306	1.337	1.412
0.338	0.333	0.336	0.343	0.361	0.407	0.452	0.495					
0.0568	0.0557	0.0560	0.0569	0.0598	0.0672	0.0755	0.0844					
0.00234	0.00262	0.00290	0.00318	0.00374	0.00481	0.00586	0.00689	0.00776	0.00873	0.00970	0.0107	0.0130
322.6	384.2	444.0	502.6	618.5	850.1	1085.5	1326	1570	1824	2080	2343	3026
5.041	5.205	5.346	5.470	5.681	6.014	6.277	6.495	6.690	6.854	7.005	7.144	7.449
1.258	1.211	1.182	1.166	1.154	1.166	1.189	1.210	1.231	1.267	1.310	1.341	1.412
0.370	0.359	0.358	0.361	0.375	0.416	0.459	0.501					
0.0620	0.0602	0.0598	0.0603	0.0625	0.0691	0.0770	0.0857					
0.00200	0.00221	0.00242	0.00263	0.00304	0.00385	0.00465	0.00544	0.00608	0.00681	0.00754	0.00826	0.0101
331.6	393.8	453.4	512.3	625.8	862.0	1099.3	1341	1588	1844	2101	2365	3049
4.943	5.108	5.250	5.374	5.586	5.922	6.136	6.407	6.605	6.769	6.921	7.060	7.366
1.257	1.216	1.188	1.172	1.161	1.175	1.198	1.219	1.240	1.275	1.318	1.347	1.412
0.432	0.411	0.402	0.399	0.405	0.436	0.474	0.512					
0.0718	0.0688	0.0673	0.0669	0.0679	0.0730	0.0800	0.0881					
0.00180	0.00196	0.00213	0.00230	0.00262	0.00328	0.00392	0.00455	0.00507	0.00565	0.00624	0.00681	0.00825
343.4	405.1	465.3	524.4	641.2	875.1	1113.3	1356	1606	1863	2121	2386	3071
4.869	5.034	5.176	5.300	5.513	5.850	6.115	6.337	6.539	6.703	6.856	6.995	7.302
1.254	1.217	1.192	1.175	1.164	1.179	1.204	1.225	1.248	1.283	1.325	1.354	1.413
0.494	0.463	0.446	0.438	0.435	0.456	0.489	0.524					
0.0810	0.0768	0.0744	0.0733	0.0732	0.0768	0.0830	0.0906					

TABLE 2-230   Enthalpy and Psi Functions for Ideal-Gas Air*

T, K	h, kJ/kg	Ψ	T, K	h, kJ/kg	Ψ	T, K	h, kJ/kg	Ψ
200	200.0	−0.473	650	659.8	1.339	1200	1278	2.376
210	210.0	−0.400	660	670.5	1.364	1220	1301	2.406
220	220.0	−0.329	670	681.1	1.388	1240	1325	2.435
230	230.1	−0.262	680	691.8	1.412	1260	1349	2.463
240	240.1	−0.197	690	702.5	1.436	1280	1372	2.491
250	250.1	−0.135	700	713.3	1.459	1300	1396	2.519
260	260.1	−0.076	710	724.0	1.482	1320	1420	2.547
270	270.1	−0.018	720	734.8	1.505	1340	1444	2.574
280	280.1	0.037	730	745.6	1.528	1360	1467	2.601
290	290.2	0.090	740	756.4	1.550	1380	1491	2.627
300	300.2	0.142	750	767.3	1.572	1400	1515	2.653
310	310.3	0.191	760	778.2	1.594	1420	1539	2.679
320	320.3	0.240	770	789.1	1.615	1440	1563	2.705
330	330.4	0.286	780	800.0	1.637	1460	1587	2.730
340	340.4	0.332	790	811.0	1.658	1480	1612	2.755
350	350.5	0.376	800	821.9	1.679	1500	1636	2.779
360	360.6	0.419	810	832.9	1.699	1520	1660	2.803
370	370.7	0.461	820	844.0	1.720	1540	1684	2.827
380	380.8	0.502	830	855.0	1.740	1560	1709	2.851
390	390.9	0.541	840	866.1	1.760	1580	1738	2.875
400	401.0	0.580	850	877.2	1.780	1600	1758	2.898
410	411.2	0.618	860	888.3	1.800	1620	1782	2.921
420	421.3	0.655	870	899.4	1.819	1640	1806	2.944
430	431.5	0.691	880	910.6	1.838	1660	1831	2.966
440	441.7	0.727	890	921.8	1.857	1680	1855	2.988
450	451.8	0.761	900	933.0	1.876	1700	1880	3.010
460	462.1	0.795	910	944.2	1.895	1720	1905	3.032
470	472.3	0.829	920	955.4	1.914	1740	1929	3.054
480	482.5	0.861	930	966.7	1.932	1760	1954	3.075
490	492.8	0.893	940	978.0	1.950	1780	1979	3.096
500	503.1	0.925	950	989.3	1.969	1800	2003	3.117
510	513.4	0.956	960	1000.6	1.987	1820	2028	3.138
520	523.7	0.986	970	1011.9	2.004	1840	2053	3.158
530	534.0	1.016	980	1023.3	2.022	1860	2078	3.178
540	544.4	1.045	990	1034.7	2.039	1880	2102	3.198
550	554.8	1.074	1000	1046.1	2.057	1900	2127	3.218
560	565.2	1.102	1020	1068.9	2.091	1920	2152	3.238
570	575.6	1.130	1040	1091.9	2.125	1940	2177	3.258
580	586.1	1.158	1060	1114.9	2.158	1960	2202	3.277
590	596.5	1.185	1080	1138.0	2.190	1980	2227	3.296
600	607.0	1.211	1100	1161.1	2.223	2000	2252	3.215
610	617.5	1.238	1120	1184.3	2.254	2050	2315	3.362
620	628.1	1.264	1140	1207.6	2.285	2100	2377	3.408
630	638.6	1.289	1160	1230.9	2.316	2150	2440	3.453
640	649.2	1.314	1180	1254.3	2.346	2200	2504	3.496

*Values rounded off from Chappell and Cockshutt, Nat. Res. Counc. Can. Rep. NRC LR 759 (NRC No. 14300), 1974. This source tabulates values of seven thermodynamic functions at 1-K increments from 200 to 2200 K in SI units and at other increments for two other unit systems. An earlier report (NRC LR 381, 1963) gives a more detailed description of an earlier fitting from 200 to 1400 K. In the above table h = specific enthalpy, kJ/kg, and $\Psi_2 - \Psi_1 = \log_{10}(P_2/P_1)_s$ for an isentrope. In terms of the Keenan and Kaye function ϕ, $\Psi = (\log_{10} e/R) \cdot \phi$.

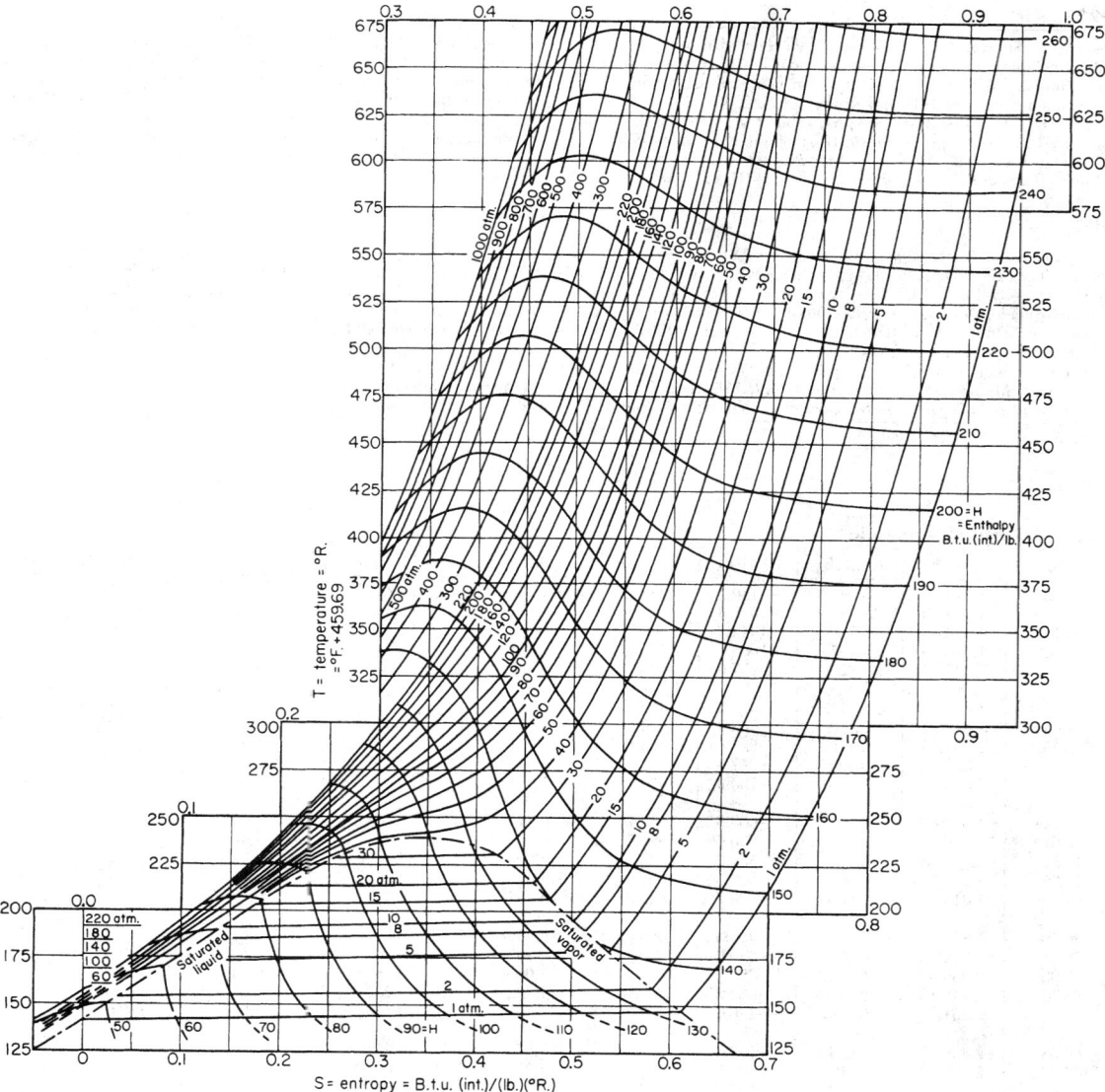

FIG. 2-5 Temperature-entropy diagram for air. [*Landsbaum, Dadds, Stevens, et al.,* Am. Inst. Chem. Eng. J., **1**(3), 303 (1955). *Reproduced by permission of the authors and of the editor, American Institute of Chemical Engineers.*]

TABLE 2-231   Air

Other tables include Stewart, R. B., S. G. Penoncello, et al., University of Idaho CATS report, 85-5, 1985 (0.1–700 bar, 85–750 K), and a revision is in process of publication. Tables including reactions with hydrocarbons include Gordon, S., NASA Techn. Paper 1907, 4 vols., 1982. See also Gupta, R. N., K-P. Lee, et al., NASA RP 1232, 1990 (89 pp.) and RP 1260, 1991 (75 pp.). Analytic expressions for high temperatures were given by Matsuzaki, R., *Jap. J. Appl. Phys.*, **21**, 7 (1982): 1009–1013 and Japanese National Aerospace Laboratory report NAL TR 671, 1981 (45 pp.). Functions from 1500 to 15000 K were tabulated by Hilsenrath, J. and M. Klein, AEDC-TR-65-58 = AD 612 301, 1965 (333 pp.). Tables from 10000 to 10,000,000 K were authored by Gilmore, F. R., Lockheed rept. 3-27-67-1, vol 1., 1967 (340 pp.), also published as *Radiative Properties of Air*, IFI/Plenum, New York, 1969 (648 pp.). Saturation and superheat tables and a chart to 7000 psia, 660°R appear in Stewart, R. B., R. T. Jacobsen, et al., *Thermodynamic Properties of Refrigerants*, ASHRAE, Atlanta, GA, 1986 (521 pp.). For specific heat, thermal conductivity, and viscosity see *Thermophysical Properties of Refrigerants*, ASHRAE, 1993.

AIR, MOIST

An ASHRAE publication, *Thermodynamic Properties of Dry Air and Water and S. I. Psychrometric Charts*, 1983 (360 pp.), extensively reviews moist air properties. Gandiduson, P., *Chem. Eng.*, Oct. 29, 1984 gives on page 118 a nomograph from 50 to 120°F, while equations in SI units were given by Nelson, B., *Chem. Eng. Progr.* **76**, 5 (May 1980): 83–85. Liley, P. E., *2000 Solved Problems in M.E. Thermodynamics*, McGraw-Hill, New York, 1989, gives four simple equations with which most calculations can be made. Devres, Y.O., *Appl. Energy* **48** (1994): 1–18 gives equations with which three known properties can be used to determine four others. Klappert, M. T. and G. F. Schilling, Rand RM-4244-PR = AD 604 856, 1984 (40 pp.) gives tables from 100 to 270 K, while programs from −60 to 2°F are given by Sando, F. A., *ASHRAE Trans.*, **96**, 2 (1990): 299–308.

Viscosity references include Kestin, J. and J. H. Whitelaw, *Int. J. Ht. Mass Transf.* **7**, 11 (1964): 1245–1255; Studnokov, E. L., *Inz.-Fiz. Zhur.* **19**, 2 (1970): 338–340; Hochramer, D. and F. Munczak, *Setzb. Ost. Acad. Wiss II* **175**, 10 (1966): 540–550. For thermal conductivity see, for instance, Mason, E. A. and L. Monchick, *Humidity and Moisture Control in Science and Industry*, Reinhold, New York, 1965 (257–272).

TABLE 2-232   Saturated Ammonia*

T, K	P, bar	v_f, m³/kg	v_g, m³/kg	h_f, kJ/kg	h_g, kJ/kg	s_f, kJ/(kg·K)	s_g, kJ/(kg·K)	c_{pf}, kJ/(kg·K)	μ_f, 10^{-4} Pa·s	k_f, W/(m·K)
195.5^t	0.0608	1.327.−3	15.648	−1110.1	380.1	4.203	11.827	4.73	4.25	0.715
200	0.0865	1.372.−3	11.237	−1088.8	388.5	4.311	11.698	4.61	4.07	0.709
210	0.1775	1.394.−3	5.729	−1044.1	406.7	4.529	11.438	4.38	3.69	0.685
220	0.3381	1.417.−3	3.135	−1000.6	424.1	4.731	11.207	4.35	3.34	0.661
230	0.6044	1.442.−3	1.822	−957.0	440.7	4.925	11.002	4.38	3.02	0.638
240	1.0226	1.468.−3	1.115	−912.9	456.2	5.113	10.817	4.43	2.73	0.615
250	1.6496	1.495.−3	0.712	−868.2	470.6	5.294	10.650	4.48	2.45	0.592
260	2.5529	1.524.−3	0.472	−823.1	483.8	5.471	10.498	4.54	2.20	0.569
270	3.8100	1.551.−3	0.324	−777.3	495.6	5.643	10.358	4.60	1.97	0.546
280	5.5077	1.589.−3	0.228	−730.9	506.0	5.811	10.228	4.66	1.76	0.523
290	7.741	1.626.−3	0.165	−683.8	514.7	5.975	10.108	4.73	1.58	0.500
300	10.61	1.666.−3	0.121	−636.0	521.5	6.135	9.994	4.82	1.41	0.477
310	14.24	1.710.−3	0.091	−587.2	526.1	6.293	9.885	4.91	1.26	0.454
320	18.72	1.760.−3	0.069	−537.5	528.2	6.448	9.779	5.02	1.13	0.431
330	24.20	1.815.−3	0.053	−486.7	527.5	6.602	9.675	5.17	1.02	0.408
340	30.79	1.878.−3	0.0410	−434.3	523.3	6.755	9.571	5.37	0.92	0.385
350	38.64	1.952.−3	0.0319	−380.0	515.1	6.908	9.465	5.64	0.83	0.361
360	47.90	2.039.−3	0.0249	−323.2	501.8	7.063	9.354	6.04	0.75	0.337
370	58.74	2.148.−3	0.0194	−262.6	481.9	7.222	9.235	6.68	0.69	0.313
380	71.35	2.291.−3	0.0149	−196.5	452.7	7.391	9.100	7.80	0.61	0.286
390	85.98	2.499.−3	0.0113	−120.9	408.1	7.578	8.935	10.3	0.50	0.254
400	103.0	2.882.−3	0.0077	−23.5	329.0	7.813	8.694	21.	0.39	0.21
405.4^c	113.0	4.255.−3	0.0043	142.7	142.7	8.216	8.216	∞	0.25	∞

*P, v, h, and s values condensed from *ASHRAE Handbook, 1981: Fundamentals*. Copyright 1981 by the American Society of Heating, Refrigerating and Air Conditioning Engineers, Inc., and reproduced by permission of the copyright owner. c_p, μ, and k values are interpolated and converted from *Thermophysical Properties of Refrigerants*, ASHRAE, New York, 1976. t = triple point; c = critical point. The notation 1.327.−3 signifies 1.327×10^{-3}. At 195.5 K, the viscosity of the saturated liquid is 4.25×10^{-4} Pa·s.

Most recent tabulations of ammonia properties are based upon the extensive tabulation to 5000 bar, 750 K of Haar, L. and J. S. Gallagher, *J. Phys. Chem. Ref. Data*, **7**, 3 (1978): 635–792, which does, however, neglect dissociation. For tables to 70,000 psia, 920°F, see Stewart, R. B., R. T. Jacobsen, et al., *Thermodynamic Properties of Refrigerants*, ASHRAE, Atlanta, GA, 1986 (521 pp.). A chart in fps units corresponding with these tables appears on page 17.34 of the ASHRAE 1989 *Fundamentals Handbook*.

Simmons, A. L., C. E. Miller III, et al., *Tables and Charts of Equilibrium Thermodynamic Properties of Ammonia for Temperatures from 500 to 50000 K*, NASA SP 3099, 1976 (255 pp.), tabulates ρ, h, s, c_p, c_v, Z, and so on, from 0.01 to 400 bar and also 18 species of decomposition products.

The 1993 ASHRAE *Handbook—Fundamentals* (SI ed.) gives material for integral degrees Celsius with temperatures on the ITS 90 scale for saturation temperatures from −77.66 to 132.22 °C. The same diagram reproduced here appears in that source.

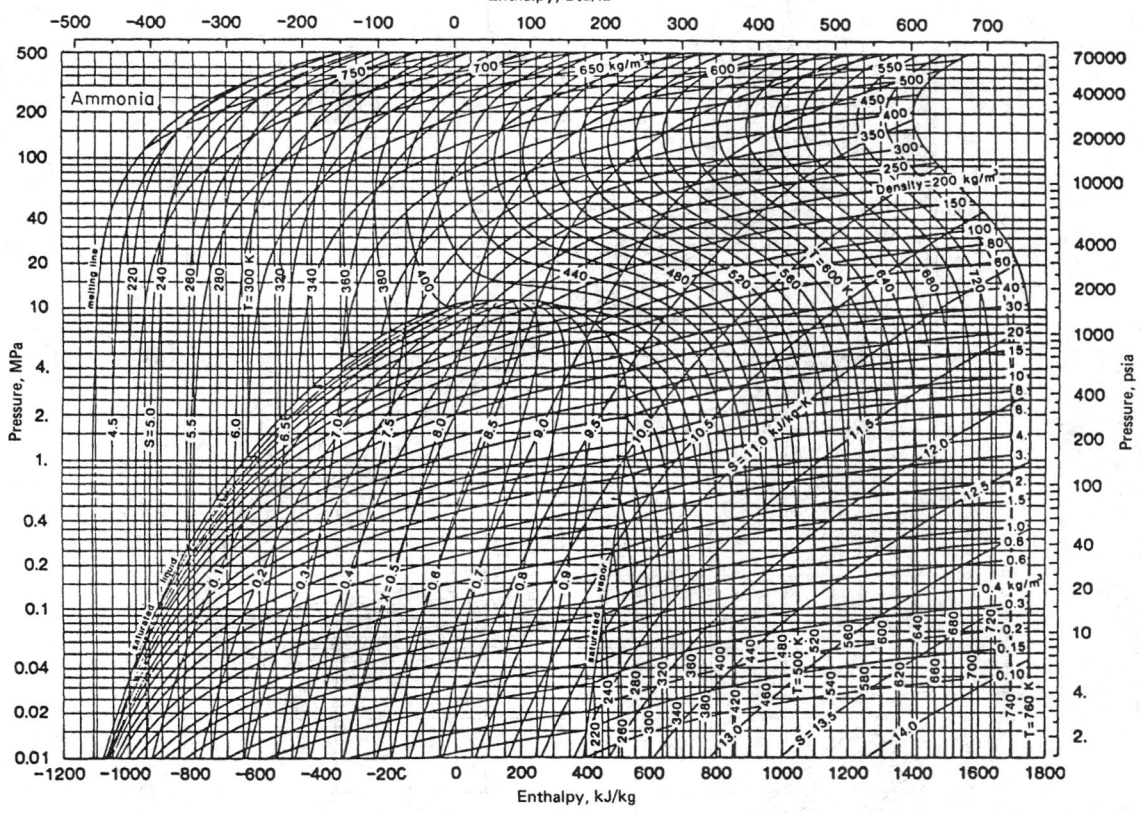

FIG. 2-6 Enthalpy–log-pressure diagram for ammonia. 1 MPa = 10 bar. (*Copyright 1981 by the American Society of Heating, Refrigerating and Air-Conditioning Engineers and reproduced by permission of the copyright owner.*)

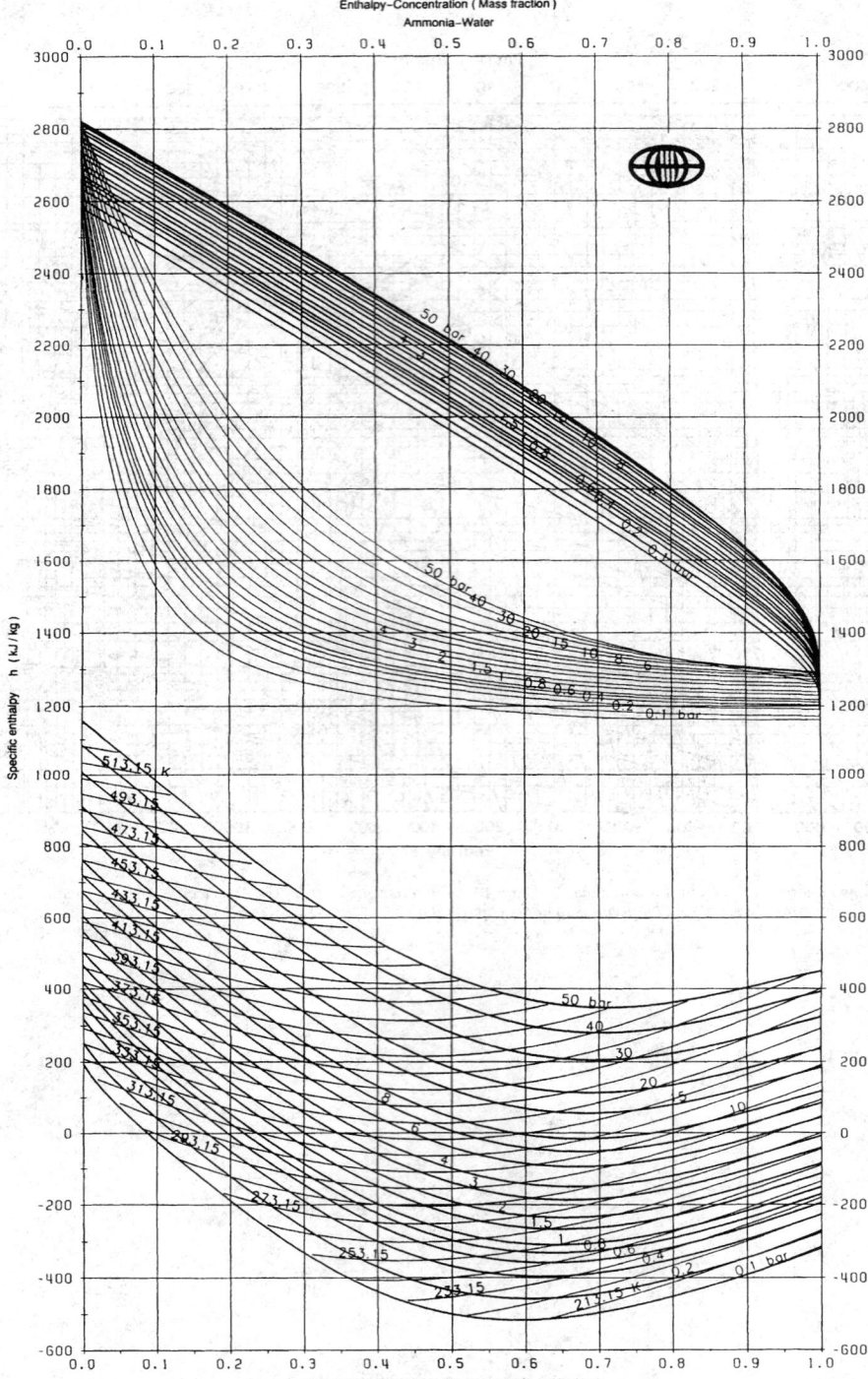

FIG. 2-7 Enthalpy-concentration diagram for aqueous ammonia. From *Thermodynamic and Physical Properties NH₃–H₂O*, Int. Inst. Refrigeration, Paris, France, 1994 (88 pp.). Reproduced by permission. In order to determine equilibrium compositions, draw a vertical from any liquid composition on any boiling line (the lowest plots) to intersect the appropriate auxiliary curve (the intermediate curves). A horizontal then drawn from this point to the appropriate dew line (the upper curves) will establish the vapor composition. The Int. Inst. Refrigeration publication also gives extensive *P-v-x* tables from −50 to 316°C. Other sources include Park, Y. M. and Sonntag, R. E., *ASHRAE Trans.*, **96**, 1 (1990): 150–159 (*x, h, s*, tables, 360 to 640 K); Ibrahim, O. M. and S. A. Klein, *ASHRAE Trans.*, **99**, 1 (1993): 1495–1502 (Eqs., 0.2 to 110 bar, 293 to 413 K); Smolen, T. M., D. B. Manley, et al., *J. Chem. Eng. Data*, **36** (1991): 202–208 (*p-x* correlation, 0.9 to 450 psia, 293–413 K); Ruiter, J. P., *Int. J. Refrig.*, **13** (1990): 223–236 gives ten subroutines for computer calculations.

TABLE 2-233 Saturated Argon (R740)*

T, K	P, bar	v_f, m³/kg	v_g, m³/kg	h_f, kJ/kg	h_g, kJ/kg	s_f, kJ/(kg·K)	s_g, kJ/(kg·K)	c_{pf}, kJ/(kg·K)	μ_f, 10^{-4} Pa·s	k_f, W/(m·K)
10		5.646.−4		0.20		0.0266		0.083		
20		5.666.−4		2.20		0.1559		0.306		
30		5.707.−4		6.12		0.3129		0.466		
40		5.763.−4		11.30		0.4610		0.560		
50		5.831.−4		17.26		0.5937		0.627		
60		5.912.−4		23.85		0.7138		0.687		
70	0.082	6.008.−4	2.1800	31.08	229.08	0.8250	3.415	0.752		
80	0.406	6.125.−4	0.3918	39.07	232.88	0.9316	3.364	0.836		
83.8ᵗ	0.687	6.178.−4	0.2434	42.34	235.06	0.9720	3.280	0.877		
83.8ᵗ	0.687	7.068.−4	0.2434	71.88	235.06	1.333	3.280	1.050	2.93	0.134
85	0.790	7.107.−4	0.2145	73.16	235.55	1.348	3.258	1.058	2.81	0.132
87.3	1.013	7.174.−4	0.1710	75.61	236.39	1.375	3.216	1.073	2.60	0.128
90	1.338	7.269.−4	0.1327	78.55	237.37	1.403	3.168	1.091	2.40	0.124
95	2.137	7.440.−4	0.0864	84.15	238.91	1.462	3.091	1.124	2.08	0.116
100	3.247	7.628.−4	0.0588	89.85	240.20	1.520	3.023	1.158	1.82	0.109
110	6.665	8.064.−4	0.0299	101.83	241.66	1.632	2.903	1.229	1.46	0.096
115	9.107	8.322.−4	0.0221	108.11	241.78	1.685	2.848	1.274	1.32	0.090
120	12.13	8.618.−4	0.0166	114.62	241.33	1.738	2.794	1.336	1.21	0.084
125	15.81	8.965.−4	0.0126	121.50	240.30	1.792	2.743	1.427	1.12	0.078
130	20.23	9.620.−4	0.0096	128.79	238.41	1.846	2.690	1.550	1.01	0.072
135	25.49	9.906.−4	0.0074	136.76	234.60	1.902	2.633	1.752	0.89	0.066
140	31.68	1.061.−3	0.0056	145.58	230.74	1.961	2.570		0.75	0.060
145	38.93	1.172.−3	0.0041	155.73	223.09	2.026	2.490		0.60	0.054
150	47.39	1.468.−3	0.0026	174.64	204.35	2.133	2.331		0.45	
150.9	48.98	1.867.−3	0.0019	189.94	189.94	2.201	2.201		0.28	∞

*Values extracted and in some cases rounded off from those cited in Rabinovich (ed.), *Thermophysical Properties of Neon, Argon, Krypton and Xenon*, Standards Press, Moscow, 1976. This source contains values for the compressed state for pressures up to 1000 bar, etc. *t* = triple point. Above the solid line the condensed phase is solid; below it, it is liquid. The notation 5.646.−4 signifies 5.646×10^{-4}. At 83.8 K, the viscosity of the saturated liquid is 2.93×10^{-4} Pa·s = 0.000293 Ns/m². This book was published in English translation by Hemisphere, New York, 1988 (604 pp.).

TABLE 2-234 Thermodynamic Properties of Compressed Argon*

T, K						Pressure, bar					
	1	100	200	300	400	500	600	700	800	900	1000
100 v	0.2035	7.420.−4	7.255.−4	7.120.−4	7.006.−4	6.907.−4	6.819.−4	6.050.−4	6.009.−4	5.976.−4	5.935.−4
h	243.4	93.6	97.9	102.5	107.2	112.0	116.8	91.1	96.1	101.0	106.0
s	3.299	1.494	1.464	1.438	1.414	1.393	1.372	1.037	1.026	1.016	1.007
200 v	0.4151	2.96.−3	1.430.−5	1.159.−3	1.045.−3	9.778.−4	9.312.−4	8.962.−4	8.683.−4	8.454.−4	8.260.−4
h	296.4	250.2	217.1	209.1	207.9	209.2	211.9	215.1	218.9	223.0	227.4
s	3.667	2.538	2.276	2.173	2.112	2.068	2.033	2.004	1.979	1.957	1.936
300 v	0.6241	5.96.−3	2.976.−3	2.071.−3	1.666.−3	1.443.−3	1.304.−3	1.207.−3	1.136.−3	1.081.−3	1.037.−3
h	348.6	330.9	316.3	306.6	301.4	299.3	299.2	300.5	302.7	305.6	310.0
s	3.879	2.872	2.686	2.572	2.493	2.435	2.389	2.352	2.320	2.293	2.269
400 v	0.8326	8.37.−3	4.279.−3	2.957.−3	2.322.−3	1.955.−3	1.719.−3	1.557.−3	1.435.−3	1.344.−3	1.271.−3
h	400.7	391.3	383.6	378.4	375.2	373.8	373.8	374.8	376.6	379.2	382.0
s	4.028	3.048	2.881	2.780	2.707	2.651	2.603	2.565	2.533	2.505	3.480
500 v	1.0409	1.062.−2	5.464.−3	3.772.−3	2.940.−3	2.448.−3	2.124.−3	1.899.−3	1.730.−3	1.607.−3	1.506.−3
h	452.8	447.7	444.3	442.0	440.9	440.6	441.4	422.9	444.7	447.1	449.9
s	4.145	3.174	3.018	2.924	2.854	2.801	2.755	2.718	2.685	2.658	2.633
600 v	1.2489	1.280.−2	6.589.−3	4.539.−3	3.525.−3	2.922.−3	2.522.−3	2.238.−3	2.023.−3	1.866.−3	1.736.−3
h	504.9	502.4	501.6	501.4	501.8	503.0	504.6	506.6	508.7	511.2	513.9
s	4.240	3.274	3.122	3.031	2.966	2.914	2.870	2.834	2.801	2.774	2.750
700 v	1.4569	1.495.−2	7.686.−3	5.281.−3	4.088.−3	3.377.−3	2.906.−3	2.570.−3	2.317.−3	2.123.−3	1.966.−3
h	556.9	556.5	556.9	558.0	559.8	561.8	564.2	566.9	569.6	527.5	575.3
s	4.320	3.356	3.207	3.118	3.054	3.005	2.963	2.928	2.897	2.870	2.845
800 v	1.6659	1.708.−2	8.768.−3	6.011.−3	4.640.−3	3.822.−3	3.280.−3	2.893.−3	2.603.−3	2.376.−3	2.196.−3
h	609.9	609.8	611.0	612.9	615.2	618.1	621.2	624.5	627.8	631.3	634.8
s	4.389	3.427	3.279	3.191	3.129	3.081	3.039	3.005	2.975	2.948	2.924
900 v	1.8739	1.920.−2	9.841.−3	6.732.−3	5.183.−3	4.259.−3	3.646.−3	3.209.−3	2.881.−3	2.626.−3	2.423.−3
h	661.0	662.7	664.6	667.2	670.1	673.3	676.8	680.7	684.4	688.3	692.3
s	4.451	3.490	3.342	3.255	3.193	3.145	3.105	3.071	3.042	3.016	2.992
1000 v	2.0819	2.131.−2	1.091.−2	7.448.−3	5.723.−3	4.692.−3	4.008.−3	3.520.−3	3.156.−3	2.872.−3	2.645.−3
h	713.1	715.4	717.9	720.9	724.3	727.8	731.5	735.6	739.8	744.1	748.5
s	4.506	3.545	3.398	3.312	3.250	3.203	3.163	3.129	3.100	3.074	3.051

*Values extracted and in some cases rounded off from those cited in Rabinovich (ed.), *Thermophysical Properties of Neon, Argon, Krypton and Xenon*, Standards Press, Moscow, 1976. v = specific volume, m³/kg; h = specific enthalpy, kJ/kg; s = specific entropy, kJ/(kg·K). This source contains an exhaustive tabulation of values. The notation 7.420.−4 signifies 7.420×10^{-4}. This book was published in English translation by Hemisphere, New York, 1988 (604 pp.). The 1993 ASHRAE *Handbook—Fundamentals* (SI ed.) has a thermodynamic chart for pressures from 1 to 2000 bar, temperatures from 90 to 700 K. Saturation and superheat tables and a chart to 50,000 psia, 1220 °R appear in Stewart, R. B., R. T. Jacobsen, et al., *Thermodynamic Properties of Refrigerants*, ASHRAE, Atlanta, GA, 1986 (521 pp.). For specific heat, thermal conductivity, and viscosity see *Thermophysical Properties of Refrigerants*, ASHRAE, 1993.

Extensive tables for 10 properties from 0.9–100 bar, 86–400 K are given by Jacques, A., Fermi Accelerator Lab., Batavia, IL, rept TM 1517, 1988 (201 pp.). In Hilsenrath, J., C. G. Messina, et al., AEDC-TR-66-248 = AD 644 081, 1966 (121 pp.), thermodynamic properties and chemical composition from 2400 to 35,000 K are tabulated. See also Drellishak, K. S. et al., AEDC-TDR-63-146, 1963; AEDC-TDR-64-12 = AD 427839, 1964.

TABLE 2-235 Liquid-Vapor Equilibrium Data for the Argon-Nitrogen-Oxygen System[*]

Liquid mole fraction		Vapor mole fraction			Temper-ature,°R	Relative volatility			Pressure activity coefficient			Enthalpy, Btu/(lb·mol)		Heat capacity, Btu/(lb·mol·°R)	
N₂/N₂ + O₂	Ar	N₂	Ar	O₂		N₂/Ar	N₂/O₂	Ar/O₂	N₂	Ar	O₂	Liquid	Vapor	Liquid	Vapor
						Pressure, 1 atm									
0.	0.	0.	0.	1.0000	162.4	2.575	4.010	1.557	1.118	1.165	0.999	−1841.	1093.	13.2	7.406
0.	0.01	0.	0.0154	0.9845	162.3	2.581	4.007	1.553	1.117	1.161	1.000	−1844.	1087.	13.1	7.374
0.	0.02	0.	0.0306	0.9694	162.2	2.586	4.004	1.548	1.115	1.158	1.000	−1847.	1082.	13.1	7.342
0.	0.03	0.	0.0456	0.9544	162.1	2.592	4.001	1.544	1.113	1.155	1.000	−1850.	1076.	13.1	7.311
0.	0.04	0.	0.0603	0.9397	162.0	2.597	3.998	1.540	1.112	1.151	1.001	−1852.	1071.	13.0	7.281
0.	0.05	0.	0.0748	0.9253	161.9	2.602	3.995	1.535	1.110	1.148	1.001	−1855.	1066.	13.0	7.251
0.	0.07	0.	0.1031	0.8970	161.7	2.613	3.989	1.526	1.107	1.142	1.002	−1860.	1056.	12.9	7.192
0.	0.10	0.	0.1439	0.8561	161.5	2.629	3.979	1.513	1.103	1.132	1.003	−1868.	1041.	12.9	7.107
0.	0.20	0.	0.2687	0.7313	160.7	2.682	3.941	1.469	1.091	1.104	1.010	−1893.	997.	12.6	6.847
0.	0.40	0.	0.4796	0.5204	159.4	2.786	3.852	1.382	1.076	1.058	1.034	−1938.	924.	11.9	6.406
0.	0.60	0.	0.6605	0.3395	158.5	2.888	3.746	1.297	1.075	1.026	1.072	−1978.	862.	11.3	6.026
0.	0.80	0.	0.8293	0.1707	157.7	2.991	3.632	1.214	1.087	1.008	1.127	−2015.	807.	10.7	5.669
0.	0.90	0.	0.9136	0.0865	157.5	3.042	3.572	1.174	1.099	1.003	1.162	−2032.	779.	10.4	5.491
0.10	0.	0.3135	0.	0.6865	157.7	2.621	4.111	1.568	1.103	1.168	1.012	−1834.	1060.	13.2	7.410
0.10	0.01	0.3095	0.0119	0.6786	157.6	2.626	4.106	1.563	1.102	1.164	1.012	−1837.	1057.	13.1	7.386
0.10	0.02	0.3056	0.0237	0.6707	157.6	2.631	4.100	1.558	1.100	1.161	1.012	−1839.	1053.	13.1	7.361
0.10	0.03	0.3017	0.0354	0.6630	157.6	2.636	4.095	1.554	1.099	1.157	1.013	−1842.	1049.	13.1	7.337
0.10	0.04	0.2978	0.0470	0.6553	157.6	2.641	4.090	1.549	1.098	1.154	1.013	−1844.	1045.	13.1	7.313
0.10	0.05	0.2939	0.0585	0.6476	157.5	2.645	4.085	1.544	1.096	1.151	1.013	−1846.	1042.	13.0	7.289
0.10	0.07	0.2863	0.0812	0.6325	157.5	2.655	4.074	1.534	1.094	1.144	1.014	−1851.	1034.	13.0	7.242
0.10	0.10	0.2752	0.1145	0.6103	157.4	2.669	4.058	1.520	1.090	1.135	1.015	−1858.	1024.	12.9	7.173
0.10	0.20	0.2399	0.2207	0.5394	157.2	2.717	4.003	1.473	1.080	1.106	1.022	−1882.	990.	12.5	6.951
0.10	0.40	0.1759	0.4170	0.4072	157.0	2.812	3.887	1.382	1.070	1.061	1.045	−1926.	928.	11.9	6.540
0.10	0.60	0.1169	0.6036	0.2795	156.9	2.906	3.766	1.296	1.072	1.029	1.082	−1969.	871.	11.3	6.147
0.10	0.80	0.0595	0.7933	0.1471	156.9	3.001	3.640	1.213	1.086	1.009	1.134	−2009.	813.	10.7	5.746
0.10	0.90	0.0303	0.8937	0.0762	157.1	3.048	3.576	1.173	1.099	1.004	1.166	−2029.	783.	10.4	5.534
0.20	0.	0.5095	0.	0.4905	154.0	2.641	4.155	1.573	1.085	1.171	1.026	−1814.	1035.	13.2	7.422
0.20	0.01	0.5042	0.0096	0.4861	154.0	2.646	4.149	1.568	1.084	1.168	1.026	−1816.	1032.	13.2	7.402
0.20	0.02	0.4990	0.0192	0.4818	154.0	2.651	4.143	1.563	1.083	1.164	1.026	−1819.	1029.	13.1	7.382
0.20	0.03	0.4938	0.0288	0.4775	154.1	2.655	4.137	1.558	1.082	1.161	1.027	−1821.	1027.	13.1	7.362
0.20	0.04	0.4886	0.0383	0.4731	154.1	2.660	4.131	1.553	1.081	1.158	1.027	−1824.	1024.	13.1	7.342
0.20	0.05	0.4834	0.0477	0.4688	154.1	2.665	4.125	1.548	1.080	1.154	1.027	−1826.	1021.	13.0	7.322
0.20	0.07	0.4732	0.0666	0.4602	154.1	2.674	4.112	1.538	1.078	1.148	1.028	−1831.	1016.	13.0	7.283
0.20	0.10	0.4580	0.0946	0.4474	154.2	2.688	4.094	1.523	1.075	1.139	1.030	−1839.	1008.	12.9	7.224
0.20	0.20	0.4083	0.1866	0.4051	154.3	2.735	4.032	1.474	1.068	1.110	1.036	−1863.	981.	12.6	7.031
0.20	0.40	0.3123	0.3680	0.3197	154.8	2.829	3.907	1.381	1.062	1.064	1.058	−1911.	930.	11.9	6.648
0.20	0.60	0.2162	0.5550	0.2288	155.4	2.921	3.779	1.294	1.068	1.032	1.093	−1958.	877.	11.3	6.252
0.20	0.80	0.1144	0.7602	0.1254	156.2	3.009	3.647	1.212	1.086	1.011	1.140	−2004.	819.	10.7	5.817
0.20	0.90	0.0593	0.8744	0.0663	156.7	3.052	3.580	1.173	1.099	1.006	1.169	−2026.	787.	10.4	5.574
0.40	0.	0.7333	0.	0.2667	148.7	2.629	4.124	1.569	1.050	1.187	1.065	−1748.	997.	13.3	7.452
0.40	0.01	0.7279	0.0070	0.2651	148.8	2.634	4.119	1.564	1.049	1.183	1.065	−1751.	996.	13.3	7.437
0.40	0.02	0.7226	0.0140	0.2635	148.8	2.640	4.114	1.558	1.049	1.179	1.065	−1754.	994.	13.2	7.422
0.40	0.03	0.7172	0.0210	0.2619	148.9	2.645	4.108	1.553	1.048	1.176	1.066	−1757.	992.	13.2	7.407
0.40	0.04	0.7118	0.0280	0.2602	148.9	2.650	4.103	1.548	1.048	1.172	1.066	−1760.	991.	13.2	7.392
0.40	0.05	0.7064	0.0350	0.2586	149.0	2.656	4.098	1.543	1.047	1.169	1.066	−1763.	989.	13.1	7.377
0.40	0.07	0.6956	0.0491	0.2553	149.1	2.667	4.087	1.533	1.047	1.162	1.066	−1770.	986.	13.1	7.347
0.40	0.10	0.6794	0.0703	0.2503	149.3	2.683	4.072	1.517	1.046	1.152	1.067	−1779.	981.	13.0	7.301
0.40	0.20	0.6244	0.1426	0.2331	149.9	2.737	4.018	1.468	1.044	1.122	1.071	−1810.	964.	12.6	7.145
0.40	0.40	0.5075	0.2977	0.1948	151.2	2.841	3.907	1.375	1.048	1.074	1.088	−1871.	928.	11.9	6.811
0.40	0.60	0.3743	0.4776	0.1482	152.8	2.939	3.788	1.289	1.062	1.039	1.116	−1932.	885.	11.3	6.423
0.40	0.80	0.2121	0.7010	0.0870	154.8	3.025	3.657	1.209	1.084	1.016	1.154	−1991.	829.	10.7	5.944
0.40	0.90	0.1141	0.8382	0.0477	155.9	3.063	3.586	1.171	1.099	1.008	1.177	−2020.	794.	10.4	5.651
0.60	0.	0.8569	0.	0.1431	144.9	2.575	3.993	1.551	1.024	1.218	1.126	−1663.	970.	13.6	7.483
0.60	0.01	0.8521	0.0056	0.1424	145.0	2.582	3.991	1.546	1.024	1.214	1.125	−1667.	969.	13.5	7.471
0.60	0.02	0.8472	0.0111	0.1416	145.1	2.589	3.988	1.541	1.024	1.210	1.125	−1672.	968.	13.5	7.459
0.60	0.03	0.8424	0.0167	0.1409	145.1	2.595	3.985	1.536	1.024	1.206	1.124	−1676.	967.	13.4	7.446
0.60	0.04	0.8375	0.0224	0.1402	145.2	2.602	3.983	1.531	1.024	1.202	1.124	−1680.	966.	13.4	7.434
0.60	0.05	0.8326	0.0280	0.1395	145.3	2.609	3.980	1.526	1.023	1.198	1.123	−1684.	965.	13.4	7.421
0.60	0.07	0.8227	0.0394	0.1380	145.5	2.622	3.975	1.516	1.023	1.190	1.122	−1692.	963.	13.3	7.396
0.60	0.10	0.8076	0.0566	0.1357	145.7	2.642	3.966	1.501	1.023	1.179	1.121	−1704.	960.	13.2	7.357
0.60	0.20	0.7557	0.1163	0.1280	146.5	2.707	3.937	1.454	1.025	1.145	1.120	−1744.	948.	12.8	7.224
0.60	0.40	0.6391	0.2507	0.1102	148.4	2.833	3.867	1.365	1.036	1.090	1.126	−1824.	923.	12.0	6.926
0.60	0.60	0.4938	0.4191	0.0871	150.6	2.945	3.777	1.283	1.056	1.049	1.143	−1903.	888.	11.3	6.556
0.60	0.80	0.2961	0.6500	0.0539	153.5	3.037	3.662	1.206	1.083	1.021	1.169	−1978.	837.	10.7	6.055
0.60	0.90	0.1647	0.8047	0.0306	155.2	3.071	3.592	1.170	1.098	1.010	1.185	−2014.	801.	10.4	5.723
0.80	0.	0.9384	0.	0.0616	142.0	2.501	3.811	1.524	1.013	1.273	1.214	−1570.	949.	14.0	7.514
0.80	0.01	0.9340	0.0047	0.0613	142.0	2.509	3.811	1.519	1.013	1.268	1.212	−1575.	948.	13.9	7.503
0.80	0.02	0.9296	0.0094	0.0610	142.1	2.517	3.812	1.515	1.013	1.263	1.210	−1580.	947.	13.9	7.492
0.80	0.03	0.9252	0.0142	0.0607	142.2	2.525	3.812	1.510	1.013	1.257	1.209	−1585.	947.	13.8	7.481
0.80	0.04	0.9207	0.0189	0.0604	142.3	2.533	3.813	1.506	1.013	1.258	1.207	−1590.	946.	13.8	7.470
0.80	0.05	0.9162	0.0237	0.0601	142.4	2.540	3.813	1.501	1.013	1.247	1.205	−1595.	945.	13.7	7.459
0.80	0.07	0.9071	0.0334	0.0595	142.6	2.556	3.814	1.492	1.012	1.238	1.202	−1606.	944.	13.6	7.437
0.80	0.10	0.8934	0.0481	0.0586	142.8	2.580	3.814	1.478	1.013	1.224	1.197	−1621.	942.	13.5	7.403
0.80	0.20	0.8452	0.0994	0.0554	143.8	2.658	3.814	1.435	1.015	1.181	1.185	−1671.	934.	13.0	7.284
0.80	0.40	0.7339	0.2177	0.0483	146.0	2.809	3.797	1.352	1.028	1.112	1.173	−1772.	916.	12.2	7.013
0.80	0.60	0.5869	0.3740	0.0391	148.7	2.943	3.751	1.275	1.051	1.062	1.173	−1870.	889.	11.4	6.662
0.80	0.80	0.3690	0.6059	0.0252	152.2	3.045	3.662	1.202	1.082	1.026	1.184	−1964.	843.	10.7	6.153
0.80	0.90	0.2117	0.7736	0.0147	154.5	3.078	3.595	1.168	1.098	1.013	1.193	−2007.	806.	10.3	5.790

[*]Calculated values, from Wilson, Silverberg, and Zellner, USAF Aero Propulsion Laboratory. Rep. APL TDR 64-64 (AD 603 151), 1964. Relative volatility = $\alpha_{i-j} = (y_i/x_i)(x_j/y_j)$, where x = liquid composition, y = vapor composition. Pressure activity coefficient = $y_i P/x_i p^0$, where p^0 = vapor pressure. These data were confirmed by the analyses of Bender, *Cryogenics*, **13**, 11 (1973); and by Elshayal and Lu, *J. Chem. Eng. Data*, **16**, 31 (1971). See also Armstrong, G. T., J. M. Goldstein, et al., *J. Res. N.B.S.*, **55**, 5 (1955): 265–277; Bender, E., *Cryogenics*, **13**, 1 (1973): 11–18; Elshayal, I. M. and B. C-y Lu, *J. Chem. Eng. Data*, **16**, 1 (1971): 31–37; Funada, I., S. Yoshimura, et al., *Advan. Cryog. Engng.*, **7** (1982): 893–901; and Hwang, S-C., *Fluid Phase Equila.*, **37** (1987): 153–167.

TABLE 2-235   Liquid-Vapor Equilibrium Data for the Argon-Nitrogen-Oxygen System (Continued)

Liquid mole fraction		Vapor mole fraction			Temper- ature,°R	Relative volatility			Pressure activity coefficient			Enthalpy, Btu/ (lb·mol)		Heat capacity, Btu/(lb·mol·°R)	
N$_2$/N$_2$ + O$_2$	Ar	N$_2$	Ar	O$_2$		N$_2$/Ar	N$_2$/O$_2$	Ar/O$_2$	N$_2$	Ar	O$_2$	Liquid	Vapor	Liquid	Vapor
Pressure, 1 atm (Cont.)															
0.90	0.	0.9709	0.	0.0291	140.6	2.459	3.710	1.509	1.015	1.311	1.271	−1522.	939.	14.2	7.530
0.90	0.01	0.9667	0.0044	0.0289	140.7	2.468	3.712	1.504	1.014	1.305	1.268	−1527.	938.	14.2	7.519
0.90	0.02	0.9624	0.0088	0.0288	140.8	2.476	3.714	1.500	1.014	1.299	1.265	−1533.	938.	14.1	7.509
0.90	0.03	0.9581	0.0133	0.0286	140.9	2.485	3.716	1.496	1.013	1.293	1.263	−1538.	937.	14.1	7.498
0.90	0.04	0.9538	0.0177	0.0285	141.0	2.493	3.718	1.491	1.013	1.287	1.260	−1544.	937.	14.0	7.488
0.90	0.05	0.9494	0.0222	0.0284	141.1	2.502	3.720	1.487	1.013	1.281	1.257	−1550.	936.	14.0	7.477
0.90	0.07	0.9407	0.0312	0.0281	141.3	2.519	3.724	1.478	1.013	1.270	1.252	−1561.	935.	13.8	7.456
0.90	0.10	0.9274	0.0450	0.0276	141.6	2.545	3.729	1.465	1.012	1.254	1.245	−1578.	934.	13.7	7.423
0.90	0.20	0.8808	0.0931	0.0261	142.6	2.629	3.743	1.424	1.014	1.204	1.226	−1633.	928.	13.2	7.310
0.90	0.40	0.7723	0.2048	0.0229	144.9	2.793	3.755	1.344	1.026	1.126	1.200	−1745.	912.	12.3	7.050
0.90	0.60	0.6262	0.3552	0.0186	147.8	2.938	3.733	1.270	1.050	1.069	1.190	−1853.	889.	11.4	6.708
0.90	0.80	0.4018	0.5860	0.0122	151.7	3.048	3.660	1.201	1.081	1.029	1.192	−1956.	846.	10.7	6.197
0.90	0.90	0.2339	0.7589	0.0072	154.2	3.082	3.597	1.167	1.098	1.014	1.197	−2004.	809.	10.3	5.822
0.97	0.	0.9916	0.	0.0084	139.8	2.429	3.638	1.498	1.018	1.342	1.316	−1488.	933.	14.4	7.541
0.97	0.01	0.9874	0.0042	0.0084	139.9	2.438	3.641	1.494	1.018	1.335	1.313	−1494.	932.	14.4	7.531
0.97	0.02	0.9832	0.0085	0.0083	140.0	2.447	3.644	1.489	1.017	1.329	1.309	−1500.	932.	14.3	7.520
0.97	0.03	0.9790	0.0127	0.0083	140.1	2.456	3.647	1.485	1.017	1.322	1.306	−1505.	931.	14.2	7.510
0.97	0.04	0.9748	0.0170	0.0083	140.2	2.465	3.650	1.481	1.016	1.315	1.302	−1511.	931.	14.2	7.500
0.97	0.05	0.9705	0.0213	0.0082	140.3	2.474	3.653	1.477	1.016	1.309	1.299	−1517.	930.	14.1	7.489
0.97	0.07	0.9619	0.0300	0.0081	140.5	2.492	3.658	1.468	1.015	1.296	1.293	−1529.	929.	14.0	7.469
0.97	0.10	0.9488	0.0432	0.0080	140.8	2.519	3.667	1.456	1.014	1.278	1.284	−1547.	928.	13.9	7.437
0.97	0.20	0.9032	0.0893	0.0076	141.8	2.608	3.691	1.415	1.014	1.224	1.257	−1606.	923.	13.3	7.327
0.97	0.40	0.7965	0.1969	0.0066	144.2	2.780	3.722	1.339	1.025	1.137	1.220	−1725.	910.	12.3	7.073
0.97	0.60	0.6513	0.3433	0.0054	147.2	2.934	3.718	1.267	1.049	1.075	1.202	−1841.	889.	11.4	6.737
0.97	0.80	0.4236	0.5728	0.0036	151.3	3.050	3.658	1.199	1.081	1.031	1.198	−1951.	847.	10.7	6.226
0.97	0.90	0.2489	0.7489	0.0021	153.9	3.084	3.597	1.167	1.098	1.015	1.200	−2002.	810.	10.3	5.844
1.00	0.	1.0000	0.	0.	139.4	2.416	3.607	1.493	1.021	1.357	1.338	−1473.	930.	14.5	7.546
1.00	0.01	0.9959	0.0041	0.0000	139.5	2.425	3.611	1.489	1.020	1.350	1.334	−1479.	929.	14.4	7.535
1.00	0.02	0.9917	0.0083	0.0000	139.6	2.434	3.614	1.485	1.019	1.343	1.330	−1485.	929.	14.4	7.525
1.00	0.03	0.9875	0.0125	0.0000	139.7	2.443	3.617	1.480	1.019	1.336	1.326	−1491.	929.	14.3	7.515
1.00	0.04	0.9833	0.0167	0.0000	139.8	2.452	3.621	1.476	1.018	1.329	1.322	−1497.	928.	14.3	7.505
1.00	0.05	0.9791	0.0209	0.0000	139.9	2.462	3.624	1.472	1.018	1.322	1.318	−1503.	928.	14.2	7.495
1.00	0.07	0.9705	0.0295	0.0000	140.1	2.480	3.630	1.464	1.017	1.309	1.311	−1516.	927.	14.1	7.474
1.00	0.10	0.9576	0.0424	0.0000	140.4	2.507	3.640	1.452	1.016	1.290	1.301	−1534.	926.	13.9	7.443
1.00	0.20	0.9122	0.0878	0.0000	141.5	2.598	3.668	1.412	1.015	1.232	1.271	−1595.	921.	13.4	7.333
1.00	0.40	0.8063	0.1937	0.0000	143.9	2.774	3.708	1.337	1.025	1.142	1.230	−1716.	909.	12.4	7.082
1.00	0.60	0.6616	0.3384	0.	147.0	2.932	3.712	1.266	1.048	1.077	1.208	−1836.	888.	11.4	6.749
1.00	0.80	0.4327	0.5674	0.	151.1	3.050	3.657	1.199	1.080	1.032	1.201	−1949.	848.	10.7	6.838
1.00	0.90	0.2553	0.7447	0.	153.8	3.085	3.598	1.166	1.098	1.016	1.201	−2001.	811.	10.3	5.853
Pressure, 4 atm															
0.	0.	0.	0.	1.0000	190.6	2.020	2.776	1.375	0.987	1.111	1.000	−1466.	1224.	13.4	8.122
0.	0.01	0.	0.0137	0.9863	190.5	2.023	2.775	1.372	0.986	1.109	1.001	−1470.	1218.	13.4	8.094
0.	0.02	0.	0.0272	0.9728	190.4	2.026	2.773	1.369	0.985	1.106	1.001	−1473.	1212.	13.4	8.065
0.	0.03	0.	0.0405	0.9595	190.3	2.030	2.772	1.366	0.985	1.104	1.002	−1477.	1207.	13.4	8.037
0.	0.04	0.	0.0537	0.9463	190.2	2.033	2.770	1.362	0.984	1.102	1.002	−1480.	1201.	13.4	8.010
0.	0.05	0.	0.0668	0.9332	190.1	2.037	2.768	1.359	0.983	1.100	1.003	−1484.	1196.	13.3	7.982
0.	0.07	0.	0.0924	0.9076	189.9	2.043	2.765	1.353	0.982	1.096	1.004	−1490.	1185.	13.3	7.929
0.	0.10	0.	0.1299	0.8701	189.6	2.053	2.759	1.344	0.980	1.089	1.005	−1500.	1169.	13.3	7.850
0.	0.20	0.	0.2471	0.7529	188.8	2.086	2.738	1.313	0.974	1.070	1.013	−1533.	1121.	13.1	7.605
0.	0.40	0.	0.4548	0.5452	187.5	2.148	2.688	1.251	0.967	1.038	1.034	−1593.	1037.	12.7	7.166
0.	0.60	0.	0.6410	0.3590	186.5	2.207	2.627	1.190	0.967	1.016	1.066	−1647.	964.	12.3	6.771
0.	0.80	0.	0.8190	0.1811	185.8	2.264	2.560	1.131	0.976	1.003	1.110	−1698.	896.	11.9	6.389
0.	0.90	0.	0.9084	0.0916	185.5	2.292	2.524	1.101	0.983	1.000	1.137	−1723.	862.	11.8	6.196
0.10	0.	0.2393	0.	0.7607	186.2	2.063	2.831	1.372	0.994	1.120	1.020	−1452.	1193.	13.7	8.150
0.10	0.01	0.2363	0.0116	0.7521	186.1	2.066	2.828	1.369	0.994	1.117	1.021	−1455.	1188.	13.6	8.126
0.10	0.02	0.2334	0.0230	0.7436	186.1	2.069	2.825	1.365	0.993	1.115	1.021	−1459.	1184.	13.6	8.102
0.10	0.03	0.2305	0.0344	0.7351	186.0	2.072	2.822	1.362	0.992	1.113	1.021	−1462.	1180.	13.6	8.078
0.10	0.04	0.2277	0.0457	0.7267	186.0	2.075	2.820	1.359	0.991	1.110	1.022	−1465.	1175.	13.6	8.054
0.10	0.05	0.2248	0.0569	0.7183	186.0	2.078	2.817	1.356	0.990	1.108	1.022	−1469.	1171.	13.5	8.031
0.10	0.07	0.2192	0.0792	0.7017	185.9	2.083	2.811	1.349	0.988	1.104	1.023	−1475.	1163.	13.5	7.984
0.10	0.10	0.2109	0.1120	0.6772	185.8	2.092	2.802	1.340	0.986	1.097	1.024	−1485.	1150.	13.4	7.915
0.10	0.20	0.1843	0.2174	0.5983	185.5	2.119	2.772	1.308	0.979	1.077	1.030	−1517.	1110.	13.2	7.691
0.10	0.40	0.1353	0.4150	0.4497	185.2	2.173	2.707	1.246	0.971	1.044	1.049	−1578.	1037.	12.8	7.269
0.10	0.60	0.0896	0.6045	0.3058	185.0	2.224	2.638	1.186	0.971	1.020	1.078	−1637.	968.	12.4	6.860
0.10	0.80	0.0452	0.7961	0.1588	185.1	2.273	2.564	1.128	0.978	1.006	1.117	−1693.	900.	12.0	6.444
0.10	0.90	0.0229	0.8957	0.0814	185.2	2.296	2.526	1.100	0.985	1.002	1.141	−1720.	865.	11.8	6.226
0.20	0.	0.4170	0.	0.5831	182.4	2.094	2.861	1.366	0.997	1.128	1.041	−1431.	1166.	13.8	8.189
0.20	0.01	0.4126	0.0099	0.5776	182.4	2.097	2.857	1.363	0.996	1.125	1.041	−1434.	1162.	13.8	8.167
0.20	0.02	0.4082	0.0198	0.5721	182.4	2.100	2.854	1.359	0.995	1.123	1.042	−1438.	1159.	13.8	8.146
0.20	0.03	0.4038	0.0297	0.5666	182.4	2.102	2.851	1.356	0.994	1.120	1.042	−1441.	1155.	13.8	8.125
0.20	0.04	0.3994	0.0395	0.5611	182.4	2.105	2.847	1.353	0.993	1.118	1.042	−1445.	1152.	13.7	8.104
0.20	0.05	0.3951	0.0493	0.5557	182.4	2.107	2.844	1.349	0.992	1.116	1.042	−1448.	1149.	13.7	8.083
0.20	0.07	0.3864	0.0688	0.5448	182.4	2.112	2.837	1.343	0.990	1.111	1.043	−1455.	1142.	13.7	8.041
0.20	0.10	0.3735	0.0979	0.5286	182.5	2.120	2.827	1.333	0.988	1.104	1.044	−1465.	1132.	13.6	7.978
0.20	0.20	0.3316	0.1932	0.4752	182.6	2.145	2.792	1.301	0.982	1.083	1.049	−1498.	1099.	13.3	7.771
0.20	0.40	0.2506	0.3808	0.3686	183.0	2.194	2.720	1.240	0.974	1.049	1.065	−1562.	1035.	12.9	7.363
0.20	0.60	0.1706	0.5714	0.2580	183.6	2.239	2.645	1.181	0.974	1.024	1.090	−1625.	971.	12.4	6.944
0.20	0.80	0.0883	0.7742	0.1376	184.3	2.281	2.568	1.126	0.980	1.008	1.124	−1687.	904.	12.0	6.497
0.20	0.90	0.0452	0.8834	0.0715	184.8	2.301	2.528	1.099	0.986	1.003	1.145	−1717.	867.	11.8	6.255

TABLE 2-235   Liquid-Vapor Equilibrium Data for the Argon-Nitrogen-Oxygen System (*Concluded*)

Liquid mole fraction		Vapor mole fraction			Temper-ature,°R	Relative volatility			Pressure activity coefficient			Enthalpy, Btu/(lb-mol)		Heat capacity, Btu/(lb-mol·°R)	
$N_2/N_2 + O_2$	Ar	N_2	Ar	O_2		N_2/Ar	N_2/O_2	Ar/O_2	N_2	Ar	O_2	Liquid	Vapor	Liquid	Vapor
						Pressure, 4 atm (*Cont.*)									
0.40	0.	0.6560	0.	0.3441	176.3	2.124	2.859	1.346	0.992	1.146	1.090	−1372.	1121.	14.1	8.277
0.40	0.01	0.6506	0.0077	0.3417	176.4	2.127	2.856	1.343	0.991	1.143	1.090	−1376.	1119.	14.0	8.260
0.40	0.02	0.6453	0.0155	0.3393	176.4	2.129	2.853	1.340	0.991	1.140	1.090	−1380.	1117.	14.0	8.242
0.40	0.03	0.6400	0.0232	0.3369	176.5	2.132	2.850	1.337	0.990	1.138	1.090	−1384.	1115.	14.0	8.224
0.40	0.04	0.6347	0.0310	0.3345	176.6	2.135	2.846	1.333	0.990	1.135	1.090	−1387.	1112.	13.9	8.206
0.40	0.05	0.6293	0.0387	0.3320	176.6	2.137	2.843	1.330	0.989	1.133	1.090	−1391.	1110.	13.9	8.188
0.40	0.07	0.6186	0.0543	0.3271	176.8	2.143	2.837	1.324	0.988	1.128	1.090	−1399.	1106.	13.9	8.153
0.40	0.10	0.6025	0.0778	0.3197	177.0	2.151	2.827	1.315	0.986	1.120	1.090	−1410.	1099.	13.8	8.098
0.40	0.20	0.5482	0.1575	0.2943	177.7	2.176	2.794	1.284	0.982	1.098	1.091	−1449.	1077.	13.5	7.915
0.40	0.40	0.4348	0.3259	0.2393	179.2	2.223	2.725	1.226	0.978	1.061	1.100	−1525.	1029.	13.0	7.527
0.40	0.60	0.3104	0.5141	0.1756	180.9	2.264	2.652	1.171	0.979	1.033	1.116	−1600.	975.	12.5	7.095
0.40	0.80	0.1684	0.7334	0.0982	182.9	2.297	2.573	1.120	0.985	1.013	1.139	−1674.	910.	12.0	6.597
0.40	0.90	0.0882	0.8595	0.0523	184.1	2.309	2.531	1.096	0.989	1.006	1.153	−1710.	872.	11.8	6.312
0.60	0.	0.8076	0.	0.1924	171.7	2.120	2.798	1.320	0.986	1.173	1.154	−1296.	1086.	14.2	8.369
0.60	0.01	0.8023	0.0064	0.1913	171.8	2.123	2.796	1.317	0.986	1.170	1.153	−1301.	1084.	14.2	8.354
0.60	0.02	0.7971	0.0127	0.1902	171.9	2.127	2.794	1.314	0.985	1.167	1.152	−1305.	1083.	14.1	8.338
0.60	0.03	0.7918	0.0192	0.1891	172.0	2.130	2.792	1.311	0.985	1.164	1.152	−1310.	1082.	14.1	8.322
0.60	0.04	0.7865	0.0256	0.1879	172.1	2.133	2.790	1.308	0.985	1.161	1.151	−1315.	1080.	14.1	8.306
0.60	0.05	0.7812	0.0321	0.1868	172.2	2.137	2.788	1.305	0.984	1.159	1.150	−1319.	1079.	14.0	8.289
0.60	0.07	0.7704	0.0451	0.1845	172.4	2.143	2.784	1.299	0.984	1.153	1.149	−1329.	1076.	14.0	8.257
0.60	0.10	0.7542	0.0649	0.1810	172.6	2.153	2.778	1.290	0.983	1.144	1.148	−1343.	1072.	13.9	8.208
0.60	0.20	0.6980	0.1332	0.1688	173.7	2.184	2.757	1.262	0.981	1.119	1.143	−1389.	1056.	13.6	8.038
0.60	0.40	0.5739	0.2848	0.1413	175.9	2.239	2.708	1.209	0.980	1.076	1.141	−1482.	1021.	13.0	7.666
0.60	0.60	0.4261	0.4667	0.1073	178.5	2.283	2.648	1.160	0.984	1.043	1.145	−1573.	976.	12.5	7.228
0.60	0.80	0.2413	0.6963	0.0625	181.6	2.311	2.576	1.115	0.989	1.019	1.155	−1661.	916.	12.0	6.690
0.60	0.90	0.1293	0.8367	0.0340	183.4	2.317	2.533	1.093	0.992	1.009	1.161	−1704.	876.	11.8	6.367
0.80	0.	0.9152	0.	0.0848	168.0	2.095	2.699	1.288	0.986	1.216	1.239	−1209.	1057.	14.3	8.462
0.80	0.01	0.9102	0.0055	0.0843	168.1	2.099	2.699	1.286	0.986	1.212	1.237	−1215.	1056.	14.2	8.447
0.80	0.02	0.9052	0.0110	0.0839	168.2	2.103	2.699	1.283	0.986	1.209	1.235	−1220.	1055.	14.2	8.432
0.80	0.03	0.9001	0.0165	0.0834	168.3	2.107	2.698	1.280	0.985	1.205	1.234	−1226.	1054.	14.2	8.417
0.80	0.04	0.8950	0.0221	0.0829	168.4	2.112	2.698	1.278	0.985	1.201	1.232	−1232.	1053.	14.1	8.402
0.80	0.05	0.8899	0.0277	0.0825	168.5	2.116	2.698	1.275	0.984	1.198	1.230	−1237.	1052.	14.1	8.387
0.80	0.07	0.8795	0.0390	0.0815	168.7	2.124	2.698	1.270	0.984	1.190	1.227	−1249.	1050.	14.0	8.356
0.80	0.10	0.8638	0.0562	0.0801	169.1	2.136	2.697	1.263	0.983	1.180	1.222	−1266.	1047.	14.0	8.309
0.80	0.20	0.8087	0.1162	0.0751	170.3	2.175	2.693	1.238	0.981	1.148	1.208	−1322.	1037.	13.7	8.148
0.80	0.40	0.6826	0.2535	0.0638	173.1	2.244	2.674	1.192	0.983	1.095	1.188	−1434.	1012.	13.1	7.786
0.80	0.60	0.5231	0.4274	0.0496	176.4	2.295	2.636	1.149	0.988	1.055	1.176	−1543.	976.	12.5	7.347
0.80	0.80	0.3077	0.6624	0.0299	180.3	2.323	2.576	1.109	0.993	1.024	1.171	−1647.	920.	12.0	6.777
0.80	0.90	0.1684	0.8150	0.0166	182.7	2.325	2.535	1.090	0.994	1.012	1.169	−1698.	880.	11.8	6.420
0.90	0.	0.9596	0.	0.0404	166.3	2.077	2.641	1.271	0.990	1.245	1.292	−1163.	1044.	14.3	8.509
0.90	0.01	0.9547	0.0051	0.0402	166.4	2.081	2.641	1.269	0.990	1.241	1.289	−1169.	1043.	14.3	8.494
0.90	0.02	0.9497	0.0103	0.0399	166.5	2.086	2.642	1.267	0.989	1.236	1.287	−1175.	1042.	14.2	8.479
0.90	0.03	0.9448	0.0155	0.0397	166.7	2.091	2.643	1.264	0.989	1.232	1.284	−1181.	1042.	14.2	8.464
0.90	0.04	0.9397	0.0208	0.0395	166.8	2.095	2.644	1.262	0.988	1.228	1.281	−1188.	1041.	14.2	8.449
0.90	0.05	0.9347	0.0260	0.0393	166.9	2.100	2.645	1.260	0.988	1.224	1.279	−1194.	1040.	14.1	8.434
0.90	0.07	0.9245	0.0367	0.0388	167.2	2.109	2.646	1.255	0.987	1.215	1.274	−1206.	1039.	14.1	8.404
0.90	0.10	0.9090	0.0529	0.0381	167.5	2.122	2.649	1.248	0.986	1.203	1.267	−1225.	1036.	14.0	8.358
0.90	0.20	0.8546	0.1096	0.0358	168.8	2.166	2.653	1.225	0.984	1.167	1.246	−1287.	1028.	13.7	8.198
0.90	0.40	0.7287	0.2407	0.0305	171.8	2.242	2.651	1.182	0.985	1.107	1.215	−1409.	1007.	13.1	7.840
0.90	0.60	0.5659	0.4102	0.0239	175.3	2.299	2.627	1.143	0.990	1.062	1.193	−1527.	975.	12.5	7.400
0.90	0.80	0.3387	0.6467	0.0146	179.7	2.328	2.575	1.106	0.995	1.027	1.179	−1640.	922.	12.0	6.818
0.90	0.90	0.1873	0.8045	0.0082	182.4	2.329	2.536	1.089	0.995	1.013	1.173	−1694.	882.	11.8	6.446
0.97	0.	0.9882	0.	0.0118	165.2	2.062	2.598	1.260	0.995	1.269	1.334	−1130.	1035.	14.3	8.541
0.97	0.01	0.9834	0.0050	0.0117	165.3	2.067	2.599	1.257	0.994	1.264	1.330	−1136.	1035.	14.3	8.527
0.97	0.02	0.9784	0.0099	0.0116	165.5	2.072	2.601	1.255	0.994	1.259	1.327	−1143.	1034.	14.2	8.512
0.97	0.03	0.9735	0.0149	0.0116	165.6	2.077	2.602	1.253	0.993	1.254	1.324	−1149.	1033.	14.2	8.497
0.97	0.04	0.9685	0.0200	0.0115	165.7	2.082	2.604	1.251	0.992	1.250	1.321	−1156.	1033.	14.2	8.482
0.97	0.05	0.9635	0.0250	0.0114	165.8	2.087	2.605	1.248	0.992	1.245	1.317	−1163.	1032.	14.2	8.467
0.97	0.07	0.9534	0.0353	0.0113	166.1	2.097	2.608	1.244	0.991	1.236	1.311	−1176.	1031.	14.1	8.437
0.97	0.10	0.9380	0.0509	0.0111	166.5	2.111	2.612	1.237	0.989	1.222	1.302	−1195.	1029.	14.0	8.391
0.97	0.10	0.9380	0.0509	0.0111	166.5	2.111	2.612	1.237	0.989	1.222	1.302	−1195.	1029.	14.0	8.391
0.97	0.20	0.8840	0.1056	0.0104	167.9	2.158	2.624	1.216	0.987	1.182	1.276	−1261.	1022.	13.7	8.229
0.97	0.40	0.7584	0.2327	0.0089	170.9	2.240	2.633	1.176	0.986	1.116	1.235	−1390.	1003.	13.1	7.877
0.97	0.60	0.5940	0.3990	0.0070	174.6	2.302	2.620	1.138	0.991	1.067	1.206	−1516.	974.	12.5	7.436
0.97	0.80	0.3597	0.6361	0.0043	179.3	2.332	2.574	1.104	0.996	1.030	1.185	−1635.	923.	12.0	6.846
0.97	0.90	0.2003	0.7972	0.0024	182.1	2.331	2.537	1.088	0.996	1.014	1.176	−1692.	883.	11.8	6.464
1.00	0.	1.0000	0.	0.	164.7	2.057	2.580	1.254	0.997	1.280	1.352	−1115.	1032.	14.3	8.555
1.00	0.01	0.9951	0.0049	0.0000	164.9	2.061	2.581	1.252	0.997	1.275	1.349	−1122.	1031.	14.3	8.541
1.00	0.02	0.9902	0.0098	0.0000	165.0	2.066	2.583	1.250	0.996	1.270	1.346	−1129.	1031.	14.3	8.526
1.00	0.03	0.9853	0.0147	0.0000	165.1	2.071	2.584	1.248	0.995	1.265	1.342	−1136.	1030.	14.2	8.511
1.00	0.04	0.9803	0.0197	0.0000	165.3	2.076	2.586	1.246	0.995	1.260	1.338	−1142.	1029.	14.2	8.496
1.00	0.05	0.9753	0.0247	0.0000	165.4	2.081	2.588	1.243	0.994	1.255	1.335	−1149.	1029.	14.2	8.481
1.00	0.07	0.9653	0.0347	0.0000	165.7	2.091	2.591	1.239	0.993	1.245	1.328	−1163.	1028.	14.1	8.451
1.00	0.10	0.9499	0.0501	0.0000	166.1	2.106	2.596	1.233	0.991	1.231	1.319	−1183.	1026.	14.0	8.405
1.00	0.20	0.8960	0.1040	0.0000	167.4	2.154	2.610	1.212	0.988	1.189	1.289	−1250.	1019.	13.7	8.247
1.00	0.40	0.7706	0.2294	0.0000	170.6	2.239	2.626	1.173	0.987	1.120	1.244	−1382.	1002.	13.1	7.892
1.00	0.60	0.6055	0.3945	0.	174.4	2.303	2.617	1.136	0.992	1.069	1.211	−1511.	973.	12.5	7.451
1.00	0.80	0.3684	0.6316	0.	179.1	2.333	2.574	1.103	0.996	1.030	1.188	−1633.	923.	12.0	6.858
1.00	0.90	0.2058	0.7942	0.	182.0	2.332	2.537	1.088	0.996	1.015	1.177	−1691.	884.	11.8	6.471

TABLE 2-236 Thermodynamic Properties of the International Standard Atmosphere*

Z, m	T, K	P, bar	ρ, kg/m³	g, m/s²	M	a, m/s	μ, Pa·s	k, W/(m·K)	λ, m	H, m
0	288.15	1.01325	1.2250	9.80665	28.964	340.29	1.79.−5	2.54.−5	6.63.−8	0
1,000	281.65	0.89876	1.1117	9.8036	28.964	336.43	1.76.−5	2.49.−5	7.31.−8	1,000
2,000	275.15	0.79501	1.0066	9.8005	28.964	332.53	1.73.−5	2.43.−5	8.07.−8	2,999
3,000	268.66	0.70121	0.90925	9.7974	28.964	328.58	1.69.−5	2.38.−5	8.94.−8	2,999
4,000	262.17	0.61660	0.81935	9.7943	28.964	324.59	1.66.−5	2.33.−5	9.92.−8	3,997
5,000	255.68	0.54048	0.73643	9.7912	28.964	320.55	1.63.−5	2.28.−5	1.10.−7	4,996
6,000	249.19	0.47217	0.66011	9.7882	28.964	316.45	1.59.−5	2.22.−5	1.23.−7	5,994
7,000	242.70	0.41105	0.59002	9.7851	28.964	312.31	1.56.−5	2.17.−5	1.38.−7	6,992
8,000	236.22	0.35651	0.52579	9.7820	28.964	308.11	1.53.−5	2.12.−5	1.55.−7	7,990
9,000	229.73	0.30800	0.46706	9.7789	28.964	303.85	1.49.−5	2.06.−5	1.74.−7	8,987
10,000	223.25	0.26499	0.41351	9.7759	28.964	299.53	1.46.−5	2.01.−5	1.97.−7	9,984
15,000	216.65	0.12111	0.19476	9.7605	28.964	295.07	1.42.−5	1.95.−5	4.17.−7	14,965
20,000	216.65	0.05529	0.08891	9.7452	28.964	295.07	1.42.−5	1.95.−5	9.14.−7	19,937
25,000	221.55	0.02549	0.04008	9.7300	28.964	298.39	1.45.−5	1.99.−5	2.03.−6	24,902
30,000	226.51	0.01197	0.01841	9.7147	28.964	301.71	1.48.−5	2.04.−5	4.42.−6	29,859
40,000	250.35	2.87.−3	4.00.−3	9.6844	28.964	317.19	1.60.−5	2.23.−5	2.03.−5	39,750
50,000	270.65	8.00.−4	1.03.−3	9.6542	28.964	329.80	1.70.−5	2.40.−5	7.91.−5	49,610
60,000	247.02	2.20.−4	3.10.−4	9.6241	28.964	315.07	1.58.−5	2.21.−5	2.62.−4	59,439
70,000	219.59	5.22.−5	8.28.−5	9.5942	28.964	297.06	1.44.−5	1.98.−5	9.81.−4	69,238
80,000	198.64	1.05.−5	1.85.−5	9.5644	28.964	282.54	1.32.−5	1.80.−5	4.40.−3	79,006
90,000	186.87	1.84.−6	3.43.−6	9.5348	28.95				2.37.−2	88,744
100,000	195.08	3.20.−7	5.60.−7	9.5052	28.40				0.142	98,451
150,000	634.39	4.54.−9	2.08.−9	9.3597	24.10				33	146,542
200,000	854.56	8.47.−10	2.54.−10	9.2175	21.30				240	193,899
250,000	941.33	2.48.−10	6.07.−11	9.0785	19.19				890	240,540
300,000	976.01	8.77.−11	1.92.−11	8.9427	17.73				2600	286,480
400,000	995.83	1.45.−11	2.80.−12	8.6799	15.98				1.6.+4	376,320
500,000	999.24	3.02.−12	5.22.−13	8.4286	14.33				7.7.+4	463,540
600,000	999.85	8.21.−13	1.14.−13	8.1880	11.51				2.8.+5	548,252
800,000	999.99	1.70.−13	1.14.−14	7.7368	5.54				1.4.+6	710,574
1,000,000	1000.00	7.51.−14	3.56.−15	7.3218	3.94				3.1.+6	864,071

*Extracted from *U.S. Standard Atmosphere, 1976, National Oceanic and Atmospheric Administration*, National Aeronautics and Space Administration and the U.S. Air Force, Washington, 1976. Z = geometric altitude, T = temperature, P = pressure, g = acceleration of gravity, M = molecular weight, a = velocity of sound, μ = viscosity, k = thermal conductivity, λ = mean free path, ρ = density, and H = geopotential altitude. The notation 1.79.−5 signifies 1.79×10^{-5}.

TABLE 2-237 Saturated Benzene*

T, K	P, bar	v_f, m³/kg	v_g, m³/kg	h_f, kJ/kg	h_g, kJ/kg	s_f, kJ/(kg·K)	s_g, kJ/(kg·K)	c_{pf}, kJ/(kg·K)	μ_f, 10^{-4} Pa·s	k_f, W/(m·K)
290	0.0860	1.133×10^{-3}	3.569	371.1	810.3	2.172	3.686	1.719	6.75	0.147
300	0.1382	1.147×10^{-3}	2.292	388.3	820.4	2.229	3.670	1.746	5.80	0.144
310	0.2139	1.162×10^{-3}	1.525	405.9	830.8	2.286	3.657	1.774	5.14	0.141
320	0.3206	1.176×10^{-3}	1.046	423.8	841.5	2.344	3.650	1.804	4.52	0.138
330	0.4665	1.192×10^{-3}	7.379×10^{-1}	442.1	852.4	2.400	3.643	1.836	3.95	0.135
340	0.6615	1.207×10^{-3}	5.332×10^{-1}	460.8	863.6	2.455	3.641	1.868	3.55	0.132
350	0.9162	1.224×10^{-3}	3.938×10^{-1}	479.6	875.0	2.510	3.641	1.890	3.23	0.129
360	1.2419	1.241×10^{-3}	2.965×10^{-1}	498.7	886.7	2.564	3.642	1.920	2.99	0.126
370	1.6517	1.259×10^{-3}	2.233×10^{-1}	518.1	898.6	2.617	3.646	1.950	2.72	0.123
380	2.1588	1.277×10^{-3}	1.767×10^{-1}	537.7	910.6	2.669	3.651	1.989	2.46	0.120
390	2.7774	1.297×10^{-3}	1.393×10^{-1}	557.6	922.9	2.592	3.657	2.030	2.24	0.117
400	3.5228	1.318×10^{-3}	1.112×10^{-1}	577.9	935.2	2.644	3.665	2.070	2.05	0.114
410	4.4091	1.340×10^{-3}	8.972×10^{-2}	598.6	947.8	2.823	3.674	2.110	1.88	0.111
420	5.4540	1.363×10^{-3}	7.309×10^{-2}	619.7	960.4	2.873	3.684	2.160	1.73	0.107
430	6.6739	1.388×10^{-3}	6.003×10^{-2}	641.3	973.0	2.924	3.695	2.210	1.60	0.104
440	8.0861	1.415×10^{-3}	4.965×10^{-2}	663.5	985.6	2.974	3.706	2.260	1.48	0.101
450	9.7088	1.444×10^{-3}	4.131×10^{-2}	686.3	998.2	3.025	3.718	2.320	1.37	0.098
460	11.451	1.475×10^{-3}	3.455×10^{-2}	709.7	1010.7	3.075	3.730	2.380	1.28	0.095
470	13.660	1.510×10^{-3}	2.901×10^{-2}	733.8	1022.9	3.126	3.742	2.450	1.10	0.092
480	16.028	1.548×10^{-3}	2.441×10^{-2}	758.6	1034.9	3.179	3.753	2.519	1.12	0.089
490	18.685	1.591×10^{-3}	2.059×10^{-2}	784.3	1046.4	3.230	3.765	2.590	1.05	0.086
500	21.651	1.640×10^{-3}	1.736×10^{-2}	810.9	1057.3	3.284	3.777	2.670	0.98	0.083
510	24.952	1.697×10^{-3}	1.462×10^{-2}	838.5	1067.5	3.336	3.785	2.750	0.91	
520	28.613	1.765×10^{-3}	1.226×10^{-2}	867.2	1076.6	3.391	3.794	2.839	0.84	
530	32.669	1.849×10^{-3}	1.020×10^{-2}	897.2	1084.3	3.446	3.800	2.941	0.77	
540	37.161	2.126×10^{-3}	8.349×10^{-3}	928.8	1089.5	3.504	3.802		0.70	
550	42.144	2.258×10^{-3}	6.616×10^{-3}	963.2	1090.4	3.565	3.797		0.65	
560	47.696	2.512×10^{-3}	4.696×10^{-3}	1007.3	1077.6	3.642	3.769		0.60	
562.2	48.979	3.290×10^{-3}	3.290×10^{-3}	1043.0	1043.0	3.706	3.706			

*Converted from a tabulation by Counsell, Lawrenson, and Lees, Nat. Phys. Lab. Teddington (U.K.) Rep. Chem. 52, 1976. Another tabulation by Kesselman et al., in Vargaftik (ed.), *Tables on the Thermophysical Properties of Liquids and Gases*, Hemisphere, Washington and London, 1975, shows some differences. The notation 1.133.−6 signifies 1.133×10^{-6}. Other tables are given by Goodwin, R. D., *J. Phys. Chem. Ref. Data*, **17**, 4 (1988): 1541–1636.

TABLE 2-238 Saturated Bromine*

T, K	P, bar	v_f, m³/kg	v_g, m³/kg	h_f, kJ/kg	h_g, kJ/kg	s_f, kJ/(kg·K)	s_g, kJ/(kg·K)	c_{pf}, kJ/(kg·K)	μ_f, 10^{-4} Pa·s	k_f, W/(m·K)
260	0.042	3.106.−4	3.195	−147.2	51.8	0.903	1.669	0.486	13.4	0.131
280	0.124	3.168.−4	1.169	−138.9	56.2	0.933	1.629	0.479	11.5	0.127
300	0.310	3.232.−4	0.5002	−131.6	60.6	0.956	1.597	0.475	9.3	0.122
320	0.680	3.311.−4	0.2425	−124.2	64.8	0.978	1.570	0.473	7.8	0.118
340	1.330	3.385.−4	0.1309	−112.3	71.1	1.004	1.539	0.471	6.7	0.114
360	2.384	3.464.−4	0.0767	−108.6	73.1	1.026	1.531	0.470	5.7	0.109
380	4.010	3.550.−4	0.0477	−100.6	76.9	1.048	1.515	0.471	5.0	0.104
400	6.390	3.647.−4	0.0311	−93.4	80.6	1.063	1.501	0.475	4.5	0.099
420	9.730	3.752.−4	0.0211	−85.8	84.0	1.084	1.488	0.480	4.0	0.094
440	14.25	3.885.−4	0.0148	−77.7	87.1	1.103	1.477	0.489	3.7	0.089
460	20.17	4.023.−4	0.0107	−69.0	89.9	1.122	1.467	0.503	3.3	0.084
480	27.75	4.179.−4	0.00786	−59.7	92.2	1.142	1.457	0.527	3.1	0.079
500	37.21	4.378.−4	0.00589	−49.3	94.0	1.161	1.448	0.595	2.8	0.073
520	48.81	4.623.−4	0.00445	−37.7	95.0	1.183	1.438	0.710	2.6	0.066
540	62.80	4.938.−4	0.00337	−24.0	94.8	1.207	1.428	0.860	2.5	0.059
560	79.41	5.368.−4	0.00251	−7.1	92.5	1.237	1.414	1.063	2.3	0.050
580	98.90	6.250.−4	0.00167	18.8	82.5	1.280	1.390	2.31	2.2	0.035
584.2^c	103.4	8.475.−4	0.00085	64.8	64.8	1.356	1.356	∞	2.1	∞

*Reproduced or converted from a tabulation by Seshadri, Viswanath, and Kuloor, *Ind. J. Technol.*, 6 (1970): 191–198. c = critical point.

TABLE 2-239 Saturated Normal Butane (R600)*

T, K	P, bar	v_f, m³/kg	v_g, m³/kg	h_f, kJ/kg	h_g, kJ/kg	s_f, kJ/(kg·K)	s_g, kJ/(kg·K)	c_{pf}, kJ/(kg·K)	μ_f, 10^{-4} Pa·s	k_f, W/(m·K)
134.9^t	6.7.−6	1.360.−3	28630	0.00	494.21	2.3056	5.9702	1.946	15.8	0.181
140	1.7.−5	1.369.−3	11635	9.95	499.96	2.3778	5.8779	1.953	14.4	0.179
150	8.7.−5	1.387.−3	2470	29.44	511.39	2.5121	5.7251	1.970	12.0	0.175
160	3.5.−4	1.405.−3	654	49.10	523.13	2.6389	5.6016	1.985	9.94	0.171
170	1.17.−3	1.424.−3	207	68.94	535.16	2.7592	5.5017	2.001	8.26	0.167
180	3.37.−3	1.443.−3	76.4	88.97	547.48	2.8738	5.4211	2.018	6.87	0.163
190	8.53.−3	1.463.−3	31.8	109.22	560.07	2.9835	5.3564	2.035	5.71	0.160
200	1.94.−2	1.484.−3	14.7	129.71	572.93	3.0887	5.3048	2.055	4.83	0.156
210	4.05.−2	1.505.−3	7.39	150.45	586.06	3.1900	5.2643	2.077	4.15	0.152
220	7.81.−2	1.528.−3	4.00	171.49	599.42	3.2879	5.2331	2.101	3.61	0.148
230	0.1411	1.551.−3	2.31	192.83	613.02	3.3828	5.2097	2.128	3.18	0.144
240	0.2408	1.575.−3	1.40	214.50	626.83	3.4749	5.1929	2.158	2.83	0.140
250	0.3915	1.601.−3	0.893	236.52	640.82	3.5647	5.1818	2.192	2.55	0.136
260	0.6100	1.628.−3	0.592	258.92	654.97	3.6523	5.1755	2.231	2.31	0.132
270	0.9155	1.656.−3	0.406	281.72	669.24	3.7380	5.1732	2.274	2.10	0.128
280	1.3297	1.686.−3	0.286	309.94	683.60	3.8220	5.1744	2.323	1.93	0.124
290	1.8765	1.718.−3	0.207	328.62	697.99	3.9046	5.1783	2.377	1.77	0.120
300	2.5811	1.752.−3	0.1533	352.77	712.36	3.9860	5.1846	2.437	1.62	0.116
310	3.4706	1.790.−3	0.1156	377.46	726.67	4.0663	5.1928	2.503	1.47	0.113
320	4.5731	1.830.−3	0.0885	402.71	740.84	4.1458	5.2025	2.577	1.34	0.109
330	5.9179	1.874.−3	0.0687	428.61	754.80	4.2248	5.2132	2.657	1.21	0.105
340	7.5354	1.923.−3	0.0539	455.25	768.49	4.3035	5.2248	2.746	1.08	0.101
350	9.4573	1.978.−3	0.0427	482.74	781.79	4.3822	5.2367	2.842	0.97	0.097
360	11.72	2.041.−3	0.0340	511.22	794.60	4.4613	5.2485	2.947	0.87	0.093
370	14.35	2.114.−3	0.0272	540.88	806.72	4.5412	5.2597	3.062	0.78	0.089
380	17.40	2.200.−3	0.0218	571.94	817.86	4.6225	5.2696	3.20	0.69	0.085
390	20.90	2.307.−3	0.0174	604.76	827.56	4.7058	5.2771	3.34	0.62	0.081
400	24.92	2.447.−3	0.0138	639.85	834.95	4.7922	5.2800	3.50	0.55	0.077
410	29.54	2.652.−3	0.0106	678.30	838.10	4.8842	5.2740	3.69	0.49	0.074
420	34.86	3.048.−3	0.0075	723.89	830.34	4.9903	5.2437	3.84	0.44	0.072
425.2^c	37.96	4.405.−3	0.0044	783.50	783.50	5.1290	5.1290	∞		∞

*Values rounded and reproduced or converted from Goodwin, NBSIR 79-1621, 1979. t = triple point; c = critical point. The notation 6.7.−6 signifies 6.7×10^{-6}.

TABLE 2-240   Superheated Normal Butane*

P, bar		150	200	250	300	350	400	450	500	600	700
					Temperature, K						
1.013	v	0.00139	0.00148	0.00160	0.4106	0.4847	0.5575	0.6297	0.7013	0.8440	0.9861
	h	29.6	129.8	236.6	718.9	810.7	913.1	1026.0	1149.0	1423	1730
	s	2.512	3.088	3.564	5.334	5.616	5.889	6.155	6.414	6.913	7.386
5	v	0.00139	0.00148	0.00160	0.00175	0.0909	0.1078	0.1238	0.1393	0.1693	0.1988
	h	30.0	130.2	237.0	352.9	798.5	904.3	1019.3	1143.7	1420	1728
	s	2.511	3.088	3.563	3.985	5.363	5.645	5.916	6.178	6.680	7.155
10	v	0.00139	0.00148	0.00160	0.00175	0.00198	0.0502	0.0593	0.0677	0.0835	0.0987
	h	30.6	130.8	237.4	353.3	482.7	891.9	1010.3	1136.8	1415	1725
	s	2.510	3.087	3.562	3.983	4.382	5.524	5.803	6.069	6.575	7.052
20	v	0.00138	0.00148	0.00160	0.00174	0.00196	0.0205	0.0268	0.0318	0.0406	0.0487
	h	31.7	131.8	238.4	354.0	482.6	860.0	990.1	1122.0	1406	1718
	s	2.509	3.085	3.560	3.980	4.376	5.364	5.670	5.948	6.464	6.945
30	v	0.00138	0.00148	0.00160	0.00174	0.00195	0.00240	0.0156	0.0198	0.0263	0.0320
	h	32.8	132.9	239.3	354.7	482.6	637.3	965.5	1105.9	1396	1711
	s	2.507	3.082	3.557	3.976	4.370	4.783	5.570	5.866	6.394	6.880
40	v	0.00138	0.00148	0.00159	0.00173	0.00194	0.00234	0.0097	0.0137	0.0192	0.0237
	h	33.9	134.0	240.3	355.4	482.7	633.6	932.2	1088.1	1387	1705
	s	2.505	3.080	3.555	3.973	4.365	4.768	5.468	5.797	6.341	6.832
50	v	0.00138	0.00148	0.00159	0.00173	0.00193	0.00229	0.00549	0.0101	0.0149	0.0188
	h	35.0	135.0	241.3	356.2	428.8	631.0	877.0	1068.2	1377	1699
	s	2.503	3.078	3.552	3.970	4.360	4.755	5.329	5.734	6.297	6.792
60	v	0.00138	0.00148	0.00159	0.00172	0.00192	0.00255	0.00352	0.00764	0.0121	0.0155
	h	36.2	136.1	242.3	356.9	483.1	629.1	825.1	1046.4	1367	1692
	s	2.501	3.076	3.550	3.967	4.355	4.745	5.204	5.673	6.258	6.759
80	v	0.00138	0.00147	0.00158	0.00172	0.00190	0.00219	0.00286	0.00482	0.00868	0.0114
	h	38.4	138.3	244.2	358.5	483.7	626.5	798.1	1001.5	1347	1680
	s	2.498	3.072	3.545	3.960	4.346	4.727	5.130	5.559	6.191	6.704
100	v	0.00138	0.00147	0.00158	0.00171	0.00188	0.00214	0.00264	0.00368	0.00669	0.00901
	h	40.6	140.4	246.2	360.1	484.5	624.9	787.9	971.3	1329	1668
	s	2.495	3.069	3.540	3.954	4.337	4.712	5.095	5.310	6.134	6.658
200	v	0.00137	0.00146	0.00156	0.00167	0.00178	0.00200	0.00225	0.00258	0.00349	0.00460
	h	51.9	151.3	257.9	368.8	490.3	624.4	773.3	933.7	1270	1623
	s	2.478	3.049	3.518	3.927	4.301	4.660	5.010	5.348	5.960	6.849
300	v	0.00136	0.00145	0.00154	0.00164	0.00176	0.00191	0.00209	0.00231	0.00284	0.00345
	h	63.2	162.2	266.7	378.3	498.0	629.2	773.4	928.4	1255	1603
	s	2.462	3.032	3.498	3.903	4.273	4.623	4.962	5.288	5.884	6.419
400	v	0.00136	0.00144	0.00152	0.00162	0.00173	0.00185	0.00200	0.00217	0.00255	0.00298
	h	74.5	173.3	277.4	388.2	506.8	636.2	778.0	930.2	1253	1600
	s	2.447	3.015	3.479	3.882	4.248	4.593	4.927	5.247	5.836	6.366
500	v	0.00136	0.00143	0.00151	0.00160	0.00170	0.00181	0.00193	0.00207	0.00240	0.00272
	h	85.8	184.4	288.1	398.4	516.3	644.5	784.8	935.3	1256	1599
	s	2.432	2.999	3.461	3.863	4.226	4.569	4.898	5.215	5.799	6.328

*Converted and rounded from tables of Goodwin, NBSIR 79-1621, 1979.

Saturation and superheat tables and a diagram to 100 bar, 580 K are given by Reynolds, W. C., *Thermodynamic Properties in S.I.*, Stanford Univ. publ., 1979 (173 pp.). For material to 10,000 psia, 640°F, see Stewart, R. B., R. T. Jacobsen, et al., *Thermodynamic Properties of Refrigerants*, ASHRAE, Atlanta, GA, 1986 (521 pp.). For specific heat, thermal conductivity, and viscosity, see *Thermophysical Properties of Refrigerants*, ASHRAE, 1993.

TABLE 2-241   Saturated Carbon Dioxide*

T, K	P, bar	v_f, m³/kg	v_g, m³/kg	h_f, kJ/kg	h_g, kJ/kg	s_f, kJ/(kg·K)	s_g, kJ/(kg·K)	c_{pf}, kJ/(kg·K)	μ_f, 10^{-4} Pa·s	k_f, W/(m·K)
216.6	5.180	8.484.−4	0.0712	386.3	731.5	2.656	4.250	1.707		0.182
220	5.996	8.574.−4	0.0624	392.6	733.1	2.684	4.232	1.761		0.178
225	7.357	8.710.−4	0.0515	401.8	735.1	2.723	4.204			0.171
230	8.935	8.856.−4	0.0428	411.1	736.7	2.763	4.178	1.879	1.64	0.164
235	10.75	9.011.−4	0.0357	420.5	737.9	2.802	4.152			0.160
240	12.83	9.178.−4	0.0300	430.2	738.9	2.842	4.128	1.933	1.45	0.156
245	15.19	9.358.−4	0.0253	440.1	739.4	2.882	4.103			0.148
250	17.86	9.554.−4	0.0214	450.3	739.6	2.923	4.079	1.992	1.28	0.140
255	20.85	9.768.−4	0.0182	460.8	739.4	2.964	4.056			0.134
260	24.19	1.000.−3	0.0155	471.6	738.7	3.005	4.032	2.125	1.14	0.128
270	32.03	1.056.−3	0.0113	494.4	735.6	3.089	3.981	2.410	1.02	0.116
275	36.59	1.091.−3	0.0097	506.5	732.8	3.132	3.954			0.109
280	41.60	1.130.−3	0.0082	519.2	729.1	3.176	3.925	2.887	0.91	0.102
290	53.15	1.241.−3	0.0058	547.6	716.9	3.271	3.854	3.724	0.79	0.088
300	67.10	1.470.−3	0.0037	585.4	690.2	3.393	3.742		0.60	0.074
304.2^c	73.83	2.145.−3	0.0021	636.6	636.6	3.558	3.558	∞	0.31	∞

*c = critical point. The notation 8.484.−4 signifies 8.484×10^{-4}.

The 1993 ASHRAE *Handbook—Fundamentals* (SI ed.) gives material for integral degrees Celsius with temperatures on the LPTS 68 scale for saturation temperatures from −56.57 to 30.98 degrees Celsius. The thermodynamic diagram from 4 to 1000 bar extends to 420°C.

Saturation and superheat tables and a chart to 15,000 psia, 840°F appear in Stewart, R. B., R. T. Jacobsen, et al., *Thermodynamic Properties of Refrigerants*, ASHRAE, Atlanta, GA, 1986 (521 pp.). For specific heat, thermal conductivity, and viscosity, see ASHRAE *Thermophysical Properties of Refrigerants*, 1993.

Saturation and superheat tables and a diagram to 200 bar, 1000 K are given by Reynolds, W. C., *Thermodynamic Properties in S.I.*, Stanford Univ. publ., 1979 (173 pp.). Holste, J. C., D. M. Bailey, et al., *Energy Progr.*, **6**, 2 (1986): 125–130, give properties mainly in the range 0–100 bar, 200–450 K for the superheated vapor. Compare these with Angus, S., B. Armstrong, et al., *International Tables of the Fluid State—Carbon Dioxide*, Pergamon, Oxford, 1976 (377 pp.). In Miller, C. E. III and S. E. Wilder, NASA SP 3097, 1976 (489 pp.), many properties and decomposition products are tabulated for pressures from 10^{-7} to 10^4 atm., 100–25,000 K. For the range to 50 kb, 400–2100 K, see Bottinga, Y. and P. Richet, *Amer. J. Sci.*, **281** (1981): 615–660.

TABLE 2-242 Superheated Carbon Dioxide*

P, bar		300	350	400	450	500	600	700	800	900	1000
						Temperature, K					
1	v	0.5639	0.6595	0.7543	0.8494	0.9439	1.1333	1.3324	1.5115	1.7005	1.8894
	h	809.3	853.1	899.1	947.1	997.0	1102	1212	1327	1445	1567
	s	4.860	4.996	5.118	5.231	5.337	5.527	5.697	5.850	5.990	6.120
5	v	0.1106	0.1304	0.1498	0.1691	0.1882	0.2264	0.2645	0.3024	0.3403	0.3782
	h	805.5	850.3	897.0	945.5	995.8	1101	1211	1326	1445	1567
	s	4.548	4.686	4.810	4.925	50.31	5.222	5.392	5.546	5.685	5.814
10	v	0.0539	0.0642	0.0742	0.0841	0.0938	0.1131	0.1322	0.1513	0.1703	0.1893
	h	800.7	846.9	894.4	943.5	994.1	1100	1211	1326	1445	1567
	s	4.405	4.548	4.674	4.790	4.897	5.089	5.260	5.414	5.555	5.683
20	v	0.0255	0.0311	0.0364	0.0416	0.0466	0.0564	0.0661	0.0757	0.0853	0.0948
	h	790.2	839.8	888.3	939.4	990.8	1098	1209	1325	1444	1567
	s	4.249	4.402	4.534	4.653	4.762	4.955	5.127	5.282	5.423	5.551
30	v	0.0159	0.0201	0.0238	0.0274	0.0309	0.0375	0.0441	0.0505	0.0570	0.0633
	h	778.5	832.4	882.8	935.2	987.3	1096	1208	1324	1444	1566
	s	4.144	4.341	4.447	4.569	4.679	4.876	5.049	5.204	5.346	5.474
40	v	0.0110	0.0146	0.0175	0.0203	0.0230	0.0281	0.0331	0.0379	0.0428	0.0476
	h	764.9	824.6	878.3	931.1	984.3	1094	1205	1323	1443	1566
	s	4.055	4.239	4.380	4.507	4.619	4.818	4.993	5.148	5.291	5.419
50	v	0.0080	0.0112	0.0138	0.0161	0.0183	0.0224	0.0265	0.0304	0.0343	0.0382
	h	748.2	816.3	872.6	926.9	981.1	1091	1205	1322	1443	1566
	s	3.968	4.179	4.330	4.457	4.572	4.773	4.948	5.104	5.247	5.377
60	v	0.0058	0.0090	0.0113	0.0133	0.0151	0.0187	0.0221	0.0254	0.0286	0.0318
	h	726.9	807.7	866.9	922.7	977.8	1089	1204	1321	1442	1565
	s	3.878	4.126	4.314	4.416	4.532	4.736	4.912	5.069	5.212	5.341
80	v		0.0062	0.0081	0.0097	0.0112	0.0140	0.0166	0.0191	0.0216	0.0240
	h		788.4	855.1	914.2	971.3	1085	1201	1320	1441	1565
	s		4.029	4.208	4.347	4.468	4.675	4.854	5.011	5.155	5.286
100	v		0.0045	0.0062	0.0076	0.0089	0.0111	0.0133	0.0153	0.0173	0.0193
	h		766.2	843.0	905.7	964.9	1081	1198	1318	1440	1564
	s		3.936	4.144	4.290	4.417	4.627	4.808	4.967	5.111	5.241
150	v		0.0023	0.0038	0.0049	0.0058	0.0074	0.0089	0.0103	0.0117	0.0130
	h		704.5	811.9	884.8	949.4	1072	1192	1314	1437	1562
	s		3.716	4.005	4.177	4.313	4.536	4.722	4.884	5.030	5.162
200	v		0.0017	0.0027	0.0035	0.0043	0.0056	0.0067	0.0078	0.0088	0.0099
	h		670.0	785.2	865.2	934.9	1063	1186	1310	1435	1561
	s		3.591	3.894	4.088	4.234	4.468	4.668	4.824	4.970	5.104
300	v			0.0017	0.0023	0.0029	0.0038	0.0046	0.0053	0.0060	0.0067
	h			745.3	834.0	910.6	1047	1176	1303	1431	1559
	s			3.747	3.956	4.118	4.367	4.573	4.743	4.886	5.021
400	v			0.0015	0.0018	0.0022	0.0029	0.0035	0.0041	0.0047	0.0052
	h			728.1	814.6	893.3	1035	1168	1298	1428	1558
	s			3.633	3.867	4.033	4.292	4.497	4.671	4.824	4.960
500	v				0.0016	0.0018	0.0024	0.0029	0.0034	0.0038	0.0043
	h				803.5	881.9	1027	1162	1294	1426	1557
	s				3.805	3.970	4.234	4.443	4.620	4.774	4.913

*Interpolated and rounded from Vukalovich and Altunin, *Thermophysical Properties of Carbon Dioxide*, Atomizdat, Moscow, 1965; and Collett, England, 1968.

TABLE 2-243 Saturated Carbon Monoxide*

T, K	P, bar	v_f, m³/kg	v_g, m³/kg	h_f, kJ/kg	h_g, kJ/kg	s_f, kJ/(kg·K)	s_g, kJ/(kg·K)
81.62	1.01	1.268.–3	0.0666	150.25	365.30	3.005	5.640
83.36	1.52	1.295.–3	0.0631	158.56	368.07	3.104	5.559
88.25	2.03	1.317.–3	0.0606	165.00	370.00	3.178	5.501
96.16	4.05	1.385.–3	0.0547	182.76	374.21	3.368	5.359
101.51	6.08	1.440.–3	0.0513	195.0	375.98	3.489	5.271
105.69	8.12	1.489.–3	0.0318	204.8	376.6	3.580	5.206
109.17	10.13	1.535.–3	0.0253	213.2	376.6	3.656	5.152
116.08	15.20	1.651.–3	0.0163	231.0	374.5	3.807	5.043
121.48	20.27	1.778.–3	0.0116	246.3	370.2	3.918	4.948
125.97	25.33	1.936.–3	0.0085	261.2	363.6	4.041	4.854
129.84	30.40	2.168.–3	0.0063	277.6	313.15	4.161	4.747
132.91ᶜ	34.96	3.337.–3	0.0033				

*Pressure and volume values converted, and enthalpy and entropy values reproduced, from Hust and Stewart, NBS Tech. Note 202, 1963. This source gives values at and above 72.373 K at closer pressure intervals. c = critical point. The notation 1.268.–3 signifies 1.268×10^{-3}.

Goodwin, R. D., *J. Phys. Chem. Ref. Data*, **14**, 4 (1985): 849–932, gives properties to 1000 bar, 68–1000 K.

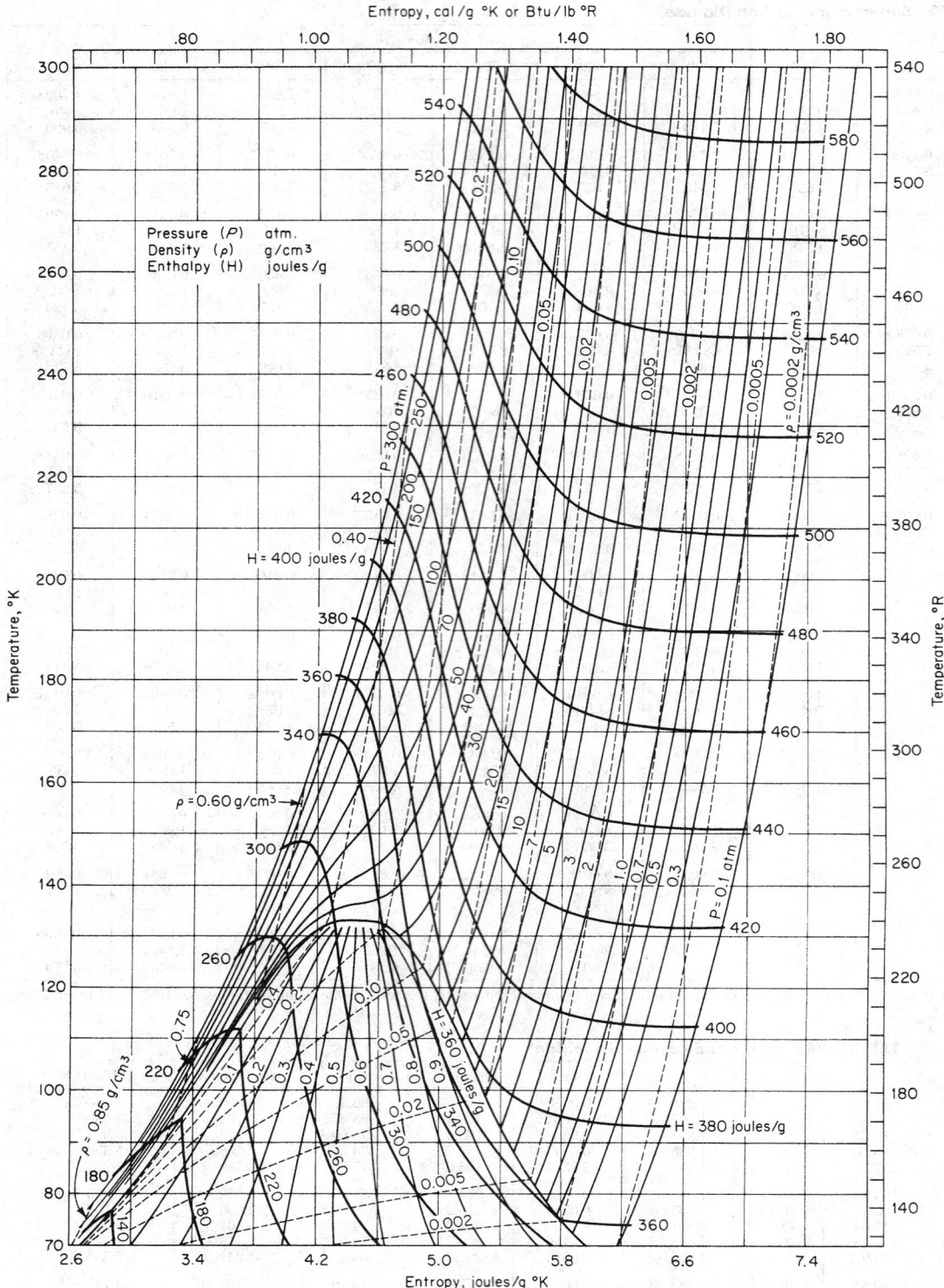

FIG. 2-8 Temperature-entropy diagram for carbon monoxide. Pressure P, in atmospheres; density ρ, in grams per cubic centimeter; enthalpy H, in joules per gram. (*From Hust and Stewart, NBS Tech. Note 202, 1963.*)

TABLE 2-244   Thermophysical Properties of Saturated Carbon Tetrachloride

T, K	P, bar	v_f, m³/kg	v_g, m³/kg	h_f, kJ/kg	h_g, kJ/kg	s_f, kJ/(kg·K)	s_g, kJ/(kg·K)	c_{pf}, kJ/(kg·K)	μ_f, 10^{-6} Pa·s	k_f, W/(m·K)	Pr
280	0.064	0.000 619	2.414	205.5	420.7	1.018	1.787	0.835	1042	0.1043	8.34
290	0.105	0.000 625	1.495	212.9	425.7	1.042	1.775	0.844	892	0.1020	7.38
300	0.165	0.000 633	0.971	220.9	430.9	1.068	1.768	0.853	774	0.0998	6.62
310	0.251	0.000 641	0.669	228.8	436.1	1.095	1.764	0.863	679	0.0975	6.01
320	0.370	0.000 649	0.463	236.9	441.3	1.121	1.760	0.874	603	0.0952	5.54
330	0.531	0.000 657	0.3306	246.0	446.4	1.149	1.756	0.885	539	0.0930	5.13
340	0.743	0.000 666	0.2407	254.5	451.5	1.174	1.754	0.897	486	0.0907	4.81
350	1.017	0.000 674	0.1802	263.1	456.6	1.199	1.752	0.910	441	0.0884	4.54
360	1.361	0.000 684	0.1370	271.8	461.7	1.224	1.751	0.924	402	0.0861	4.31
370	1.795	0.000 694	0.1053	280.8	466.6	1.248	1.751	0.939	368	0.0839	4.12
380	2.327	0.000 704	0.0820	289.7	471.5	1.272	1.750	0.954	338	0.0816	3.95
390	2.970	0.000 715	0.0651	298.1	475.8	1.295	1.751	0.970	311	0.0794	3.80
400	3.735	0.000 727	0.0525	307.9	481.2	1.319	1.752	0.987	287	0.0771	3.67
410	4.642	0.000 739	0.0426	317.1	485.8	1.341	1.753	1.010	265	0.0749	3.57
420	5.700	0.000 753	0.0350	326.0	490.4	1.363	1.754	1.034	246	0.0726	3.50
430	6.927	0.000 766	0.02899	335.2	494.9	1.384	1.756	1.060	227	0.0704	3.42
440	8.342	0.000 780	0.02413	344.3	499.2	1.405	1.757	1.094	211	0.0682	3.38
450	9.958	0.000 796	0.02020	353.6	503.4	1.426	1.759	1.141	195	0.0660	3.37
460	11.792	0.000 801	0.01692	363.1	507.3	1.446	1.760	1.207	180	0.0638	3.36
470	13.869	0.000 834	0.01425	372.8	511.1	1.467	1.761	1.240	167	0.0666	3.36
480	16.21	0.000 856	0.01205	382.6	514.6	1.487	1.762	1.278	156	0.0594	3.36
490	18.83	0.000 880	0.01011	392.0	517.5	1.507	1.763	1.320	145	0.0511	3.35
500	21.77	0.000 858	0.00858	402.5	520.2	1.526	1.762	1.375	133	0.0549	3.35
510	25.02	0.000 945	0.00722	412.9	522.6	1.546	1.761	1.44			
520	28.68	0.000 987	0.00607	424.3	524.2	1.568	1.760	1.52			
530	32.71	0.001 041	0.00500	436.4	524.5	1.590	1.756				
540	37.18	0.001 121	0.00400	448.3	522.7	1.614	1.749				
550	44.12	0.001 248	0.00309	463.4	518.2	1.638	1.738				
556.4^c	45.60	0.001 792	0.00179	494.4	494.4	1.692	1.692				

c = critical point. Base points: h_f = 200 at 273.15 K = 0°C = h_A − 300 kJ/kg; s_f = 1.000 at 273.15 K = 0°C = s_A − 4.000 kJ/(kg·K). Values mostly rounded and converted from Altunin, V. V., V. Z. Geller, et al., *Thermophysical Properties of Freons*, vol. 9, Hemisphere, Washington, DC, 1987 (243 pp.). Some irregularities exist in these data.

TABLE 2-245   Saturated Carbon Tetrafluoride (R14)*

T, K	P, bar	v_f, m³/kg	v_g, m³/kg	h_f, kJ/kg	h_g, kJ/kg	s_f, kJ/(kg·K)	s_g, kJ/(kg·K)	c_{pf}, kJ/(kg·K)	μ_f, 10^{-4} Pa·s	k_f, W/(m·K)
100	0.0089	5.370.−4	10.77	495.8	648.4	5.487	7.003	0.887		0.136
110	0.0286	5.515.−4	3.648	502.7	652.9	5.556	6.919	0.887		0.128
120	0.0924	5.668.−4	1.228	510.4	657.1	5.624	6.847	0.890		0.119
130	0.2986	5.834.−4	0.4051	518.8	661.1	5.691	6.786	0.896		0.111
140	0.6901	6.018.−4	0.1855	527.7	664.8	5.757	6.736	0.904	3.56	0.104
150	1.4074	6.225.−4	0.0951	537.2	668.3	5.822	6.696	0.922	3.28	0.097
160	2.598	6.460.−4	0.0532	549.4	671.4	5.885	6.662	0.975	3.03	0.089
170	4.426	6.733.−4	0.0318	557.6	674.0	5.947	6.629	1.031	2.80	0.081
180	7.067	7.055.−4	0.0200	568.2	676.1	6.007	6.607	1.104	2.59	0.072
190	10.702	7.449.−4	0.0131	579.3	677.4	6.066	6.583	1.203	2.39	0.064
200	15.531	7.957.−4	0.0087	591.0	677.8	6.124	6.558	1.334	2.19	0.057
210	21.794	8.674.−4	0.0058	603.5	676.4	6.182	6.536	1.506	2.01	0.049
220	29.269	9.931.−4	0.0036	618.5	671.4	6.233	6.490	1.73	1.85	0.042
227.5^c	37.45	1.598.−3	0.0016	646.9	646.9	6.371	6.371	∞		∞

P, v, h,* and *s* values interpolated, extrapolated, and converted from Oguchi, *Reito,* **52 (1977): 869–889. *c* = critical point. The notation 5.370.−4 signifies 5.370×10^{-4}.
   Equations and constants approximated to ASHRAE tables are given by Mecaryk, K. and M. Masaryk, *Heat Recovery Systems and CHP,* **11,** 2–3 (1991). The 1993 ASHRAE *Handbook—Fundamentals* (S.I. ed.) contains a saturation table from −140 to −45.65°C and an enthalpy–log-pressure diagram from 0.1 to 300 bar, −140 to 300°C. For properties to 1000 bar from 90 to 420 K, see Rublo, R. G., J. A. Zollweg, et al., *J. Chem. Eng. Data,* **36** (1991): 171–184. Saturation and superheat tables and a diagram to 80 bar, 600 K are given by Reynolds, W. C., *Thermodynamic Properties in S.I.,* Stanford Univ. publ., 1979 (173 pp.). Chari, Ph.D. thesis, University of Michigan, 1960, presents saturation-temperature tables in fps units for 1°F increments from −270 to −51°F. Thermodynamic and transport properties, equations, and computer code and tables at constant entropy from 89 to 845 K are given by Hunt, J. L. and Boney, L. R., NASA TN D-7181, 1973 (105 pp.), largely based upon the Chari data.

TABLE 2-246 Saturated Cesium*

T, K	P, bar	v_f, m³/kg	v_g, m³/kg	h_f, kJ/kg	h_g, kJ/kg	s_f, kJ/(kg·K)	s_g, kJ/(kg·K)	c_{pf}, kJ/(kg·K)
301.6^m	2.66.−9	5.444.−4	7.01.+7	74.6	637.6	0.696	2.563	0.245
400	3.83.−6	5.615.−4	6.54.+4	98.5	651.9	0.765	2.148	0.240
500	3.11.−4	5.800.−4	1001	122.0	666.1	0.817	1.905	0.232
600	5.65.−3	5.999.−4	65.63	144.9	678.4	0.859	1.748	0.224
700	0.0440	6.215.−4	9.671	167.0	688.9	0.893	1.638	0.219
800	0.2029	6.443.−4	2.353	188.7	698.3	0.922	1.559	0.217
900	0.6620	6.689.−4	0.796	210.6	707.3	0.975	1.500	0.222
1000	1.693	6.954.−4	0.335	233.2	716.4	0.972	1.455	0.231
1200	6.790	7.628.−4	0.097	281.1	736.1	1.015	1.394	0.248
1500	27.6	8.84.−4	0.029	358.8	772.2	1.072	1.345	0.275

*Converted from tables in Vargaftik, *Tables of the Thermophysical Properties of Liquids and Gases*, Nauka, Moscow, 1972, and Hemisphere, Washington, 1975. m = melting point. The notation 2.66.−9 signifies 2.66×10^9.

Many of the Vargaftik values also appear in Ohse, R. W., *Handbook of Thermodynamic and Transport Properties of Alkali Metals*, Blackwell Sci. Pubs., Oxford, 1985 (1020 pp.). This source contains superheat data.

Saturation and superheat tables and a diagram to 30 bar, 1550 K are given by Reynolds, W. C., *Thermodynamic Properties in S.I.*, Stanford Univ. publ., 1979 (173 pp.). For a Mollier diagram from 0.1 to 327 psia, 1300–2700°R, see Weatherford, W. D., J. C. Tyler, et al., WADD-TR-61-96, 1961.

An extensive review of properties of the solid and the saturated liquid was given by Alcock, C. B., M. W. Chase, et al., *J. Phys. Chem. Ref. Data,* **23,** 3 (1994): 385–497.

TABLE 2-247 Thermophysical Properties of Saturated Chlorine

T, °C	P, bar	v_f, m³/kg	v_g, m³/kg	h_f, kJ/kg	h_g, kJ/kg	s_f, kJ/(kg·K)	s_g, kJ/(kg·K)	c_{pf}, kJ/(kg·K)	c_{pg}, kJ/(kg·K)	μ_f, 10^{-6} Pa·s	μ_g, 10^{-6} Pa·s	k_f, W/(m·K)	k_g, W/(m·K)	Pr_f	Pr_g
−50	0.475	0.000 623	0.5448	221.5	518.2	1.7650	3.0946	0.9454	0.476	565	10.3	0.1684	0.0061	3.17	0.809
−40	0.773	0.000 634	0.3481	231.0	522.2	1.8074	3.0562	0.9474	0.484	520	10.8	0.1650	0.0065	2.99	0.815
−30	1.203	0.000 645	0.2314	240.6	526.1	1.8480	3.0223	0.9496	0.497	483	11.4	0.1613	0.0069	2.85	0.820
−20	1.802	0.000 656	0.1593	250.3	529.9	1.8869	2.9921	0.9520	0.513	452	11.9	0.1573	0.0074	2.74	0.826
−10	2.608	0.000 668	0.1134	260.0	533.9	1.9243	2.9649	0.9547	0.532	422	12.4	0.1527	0.0078	2.64	0.841
0	3.664	0.000 681	0.0829	269.7	537.4	1.9604	2.9402	0.9579	0.554	393	13.0	0.1478	0.0083	2.55	0.864
10	5.014	0.000 695	0.0619	279.4	540.5	1.9953	2.9177	0.9618	0.579	368	13.5	0.1427	0.0088	2.48	0.888
20	6.702	0.000 710	0.0471	289.2	543.3	2.0291	2.8924	0.9667	0.607	348	14.1	0.1378	0.0093	2.45	0.918
30	8.774	0.000 726	0.0364	299.0	545.7	2.0622	2.8777	0.9728	0.638	333	14.7	0.1327	0.0099	2.44	0.950
40	11.27	0.000 744	0.0286	308.8	548.0	2.0946	2.8593	0.9816	0.674	318	15.2	0.1282	0.0104	2.43	0.985
50	14.25	0.000 763	0.02276	318.6	549.8	2.1264	2.8417	0.9968	0.720	304	15.8	0.1230	0.0110	2.46	1.034
60	17.76	0.000 784	0.01827	329.1	551.2	2.1578	2.8245	1.022	0.786	290	16.4	0.1171	0.0117	2.53	1.107
70	21.85	0.000 808	0.01481	340.0	552.1	2.1892	2.8074	1.054	0.885	278	17.1	0.1122	0.0126	2.61	1.201
80	26.65	0.000 834	0.01202	351.4	552.5	2.2207	2.7900	1.124	1.017	267	17.9	0.1050	0.0137	2.85	1.331
90	32.17	0.000 865	0.00972	364.1	552.4	2.2528	2.7714	1.253	1.205	256	18.7	0.0986	0.0149	3.26	1.510
100	38.44	0.000 901	0.00789	377.8	551.0	2.2860	2.7502	1.418	1.434	247	19.5	0.0916	0.0163	3.82	1.700
110	45.54	0.000 956	0.00639	391.3	548.8	2.3207	2.7317	1.632	1.696	238	20.6	0.0850	0.0178	4.57	1.96
120	53.57	0.001 016	0.00508	407.1	543.7	2.3590	2.7064	1.991	1.960	230	22.2	0.0775	0.0195	5.61	2.23
130	62.68	0.001 121	0.00392	426.1	535.0	2.4032	2.6733								
140	72.84	0.001 335	0.00282	451.1	517.3	2.4595	2.6198								
144^c	77.10	0.001 77	0.00177	483.1	483.1	2.5365	2.5365								

c = critical point.
Values interpolated and converted from Martin, J. J., 1977 (private communication), and from *Heat Exchanger Design Handbook*, vol. 5, Hemisphere, Washington, DC, 1983. Values of Ziegler, *Chem.-Ing.-Tech.*, **22** (1950): 229, apparently were also used in Landolt-Bornstein, **IVa**, (1967): 238–239, and in Ullmans *Enzyklopädie der technische Chemie*, 9, Verlag Chemie, Weinheim, 1975 (317–372).

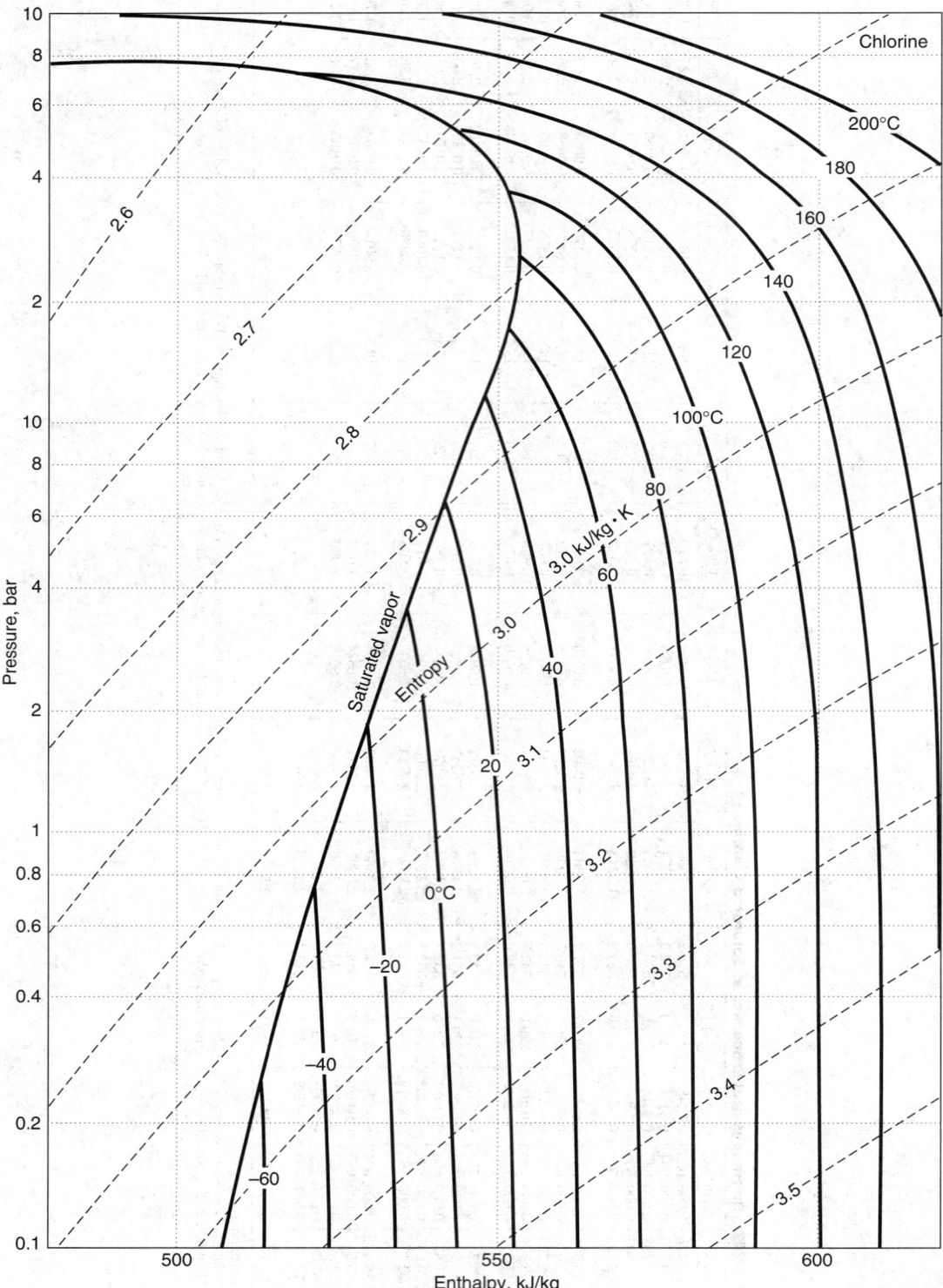

FIG. 2-9   Enthalpy–log-pressure diagram for chlorine.

TABLE 2-248   Saturated Chloroform (R20)

T, K	P, bar	v_f, m³/kg	v_g, m³/kg	h_f, kJ/kg	h_g, kJ/kg	s_f, kJ/(kg·K)	s_g, kJ/(kg·K)	c_{pf}, kJ/(kg·K)	μ_f, 10^{-6} Pa·s	k_f, W/(m·K)	Pr_f
280	0.115	0.000 660	1.689	−46.0	219.5	−0.165	0.798		748	0.120	
300	0.293	0.000 678	0.714	−32.6	230.6	−0.105	0.773		587	0.114	
320	0.620	0.000 695	0.358	−13.4	241.1	−0.041	0.754		468	0.109	
340	1.224	0.000 715	0.190	5.2	252.1	0.015	0.741		381	0.103	
360	2.255	0.000 739	0.107	23.3	263.0	0.065	0.731	1.03	319	0.095	3.35
380	3.830	0.000 765	0.0653	41.7	273.7	0.114	0.725	1.07	273	0.0921	3.17
400	6.039	0.000 795	0.0425	61.4	284.2	0.165	0.722	1.11	237	0.0863	3.04
420	9.058	0.000 822	0.0288	82.8	294.2	0.217	0.721	1.15	206	0.0808	2.93
440	13.39	0.000 871	0.0195	106.1	303.6	0.270	0.719	1.21	177	0.0750	2.86
460	18.80	0.000 921	0.0137	131.6	311.2	0.325	0.716	1.32	155	0.0694	2.95
480	26.00	0.000 980	0.00962	157.4	316.5	0.380	0.711	1.43	129.6	0.0641	2.89
500	34.66	0.001 059	0.00673	186.2	320.8	0.436	0.706	1.59	105.5	0.0584	2.87
520	44.68	0.001 193	0.00467	219.6	321.3	0.499	0.694		81.2	0.0518	
530	50.44	0.001 328	0.00359	242.7	315.7	0.540	0.678		67.7	0.0461	
536.6^c	54.72	0.002 00	0.00200	284.1	284.1	0.602	0.602				

c = critical point. $h_f = s_f = 0$ at n.b.p., 334.5 K.
P, v, h, and s interpolated from Altunin, V. V., V. Z. Geller, et al., *Thermophysical Properties of Freons*, U.S.S.R. N.S.R.D.S. series, vol. 9., Hemisphere.

TABLE 2-249   Saturated Decane*

T, K	P, bar	v_f, m³/kg	v_g, m³/kg	h_f, kJ/kg	h_g, kJ/kg	s_f, kJ/(kg·K)	s_g, kJ/(kg·K)	c_{pf}, kJ/(kg·K)	μ_f, 10^{-4} Pa·s	k_f, W/(m·K)
243.5^m	0.00001	1.319.−3	20750	418.1	812.5	2.561	4.092	2.119	25.0	0.149
260	0.00006	1.334.−3	3300.	452.7	836.3	2.699	4.120	2.109	16.6	0.144
280	0.00042	1.356.−3	443.	495.3	866.9	2.856	4.158	2.155	11.3	0.139
300	0.00197	1.381.−3	88.74	539.0	899.2	3.007	4.200	2.217	8.2	0.134
320	0.00720	1.410.−3	22.73	584.0	933.2	3.153	4.246	2.286	6.5	0.129
340	0.02155	1.442.−3	8.883	631.1	968.9	3.303	4.296		5.2	0.124
360	0.05522	1.478.−3	3.763	680.1	1006.2	3.443	4.350		4.16	0.119
380	0.1248	1.515.−3	1.750	730.7	1045.0	3.581	4.408		3.52	0.116
400	0.2549	1.552.−3	0.892	782.0	1085.0	3.712	4.469		2.98	0.110
420	0.4789	1.591.−3	0.490	835.6	1126.2	3.842	4.534		2.54	
440	0.8387	1.632.−3	0.290	889.6	1168.4	3.968	4.602		2.23	
447.3	1.0133	1.650.−3	0.243	909.4	1184.0	4.014	4.627		2.09	
460	1.3852	1.682.−3	0.178	944.5	1211.4	4.089	4.670			
480	2.1745	1.735.−3	0.115	1002.6	1255.2	4.213	4.739			
500	3.2690	1.797.−3	0.0759	1062.7	1299.4	4.335	4.808			
520	4.733	1.868.−3	0.0525	1124.5	1344.4	4.456	4.879			
540	6.633	1.952.−3	0.0369	1190.1	1389.5	4.573	4.949			
560	9.062	2.067.−3	0.0248	1256.1	1432.2	4.698	5.011			
580	12.16	2.255.−3	0.0154	1318.5	1468.1	4.802	5.060			
600	16.12	2.588.−3	0.0093	1384.5	1495.6	4.913	5.098			
617.5^c	20.97	4.238.−3	0.0042	1483.2	1483.2	5.073	5.073			

*Values converted from Das and Kuloor, *Ind. J. Technol.*, **5** (1967): 75. m = melting point; c = critical point. The notation 1.319.−3 signifies 1.319×10^{-3}.

TABLE 2-250   Saturated Normal Deuterium*

T, K	P, bar	v_f, m³/kg	v_g, m³/kg	h_f, kJ/kg	h_g, kJ/kg	s_f, kJ/(kg·K)	s_g, kJ/(kg·K)
18.71	0.1709	0.005752	2.232	−161.1	158.6	4.54	21.62
19	0.1944	0.005771	1.988	−160.0	159.1	4.68	21.48
20	0.2944	0.005840	1.365	−152.8	163.9	4.97	20.81
21	0.4297	0.005914	0.968	−145.9	167.6	5.30	20.23
22	0.6072	0.005993	0.705	−138.7	170.6	5.63	19.69
23	0.8344	0.00608	0.5256	−131.4	173.0	5.95	19.18
24	1.1192	0.00617	0.3995	−123.8	174.6	6.26	18.70
25	1.4694	0.00627	0.3088	−116.1	175.5	6.57	18.23
26	1.8932	0.00638	0.2421	−108.2	175.7	6.87	17.79
27	2.3989	0.00650	0.1921	−100.2	175.1	7.16	17.36
28	2.995	0.00663	0.1540	−92.0	173.8	7.44	16.94
29	3.690	0.00678	0.1246	−83.6	171.7	7.72	16.52
30	4.493	0.00694	0.1015	−74.9	168.7	8.00	16.12
31	5.412	0.00713	0.0831	−65.9	165.0	8.27	15.72
32	6.457	0.00735	0.0683	−56.5	160.3	8.54	15.32
33	7.455	0.00761	0.0563	−46.4	154.7	8.83	14.92
34	8.962	0.00793	0.0465	−35.5	148.0	9.12	14.52
35	10.44	0.00834	0.0382	−23.2	140.0	9.45	14.11
36	12.09	0.00890	0.0311	−8.6	130.1	9.82	13.67
37	13.91	0.00976	0.0249	10.0	117.1	10.28	13.17
38	15.92	0.01158	0.0185	39.7	95.0	11.01	12.47
38.34[c]	16.65	0.01433	0.0143	69.2	69.2	11.76	11.76

*Condensed and converted from tables of Prydz, NBS Rep. 9276, 1967. c = critical point.
For equations and T-s and Z charts from 0.1 to 100 atm, 20–300 K, see also Prydz, R. and K. D. Timmerhaus, *Advan. Cryog. Eng.*, **13** (1968): 384–396.

TABLE 2-251   Saturated Deuterium Oxide*

T, K	P, bar	v_f, m³/kg	v_g, m³/kg	h_f, kJ/kg	h_g, kJ/kg	s_f, kJ/(kg·K)	s_g, kJ/(kg·K)
277.0[t]	0.00668	9.047.−4	172.2	0.0	2320.9	0.000	8.380
278.2	0.00720	9.045.−4	160.4	5.0	2322.5	0.0188	8.351
283.2	0.01030	9.042.−4	114.1	25.9	2330.9	0.0920	8.233
288.2	0.01449	9.043.−4	82.48	46.9	2339.3	0.166	8.122
293.2	0.02011	9.047.−4	60.45	67.8	2347.6	0.239	8.016
298.2	0.02758	9.054.−4	44.88	88.7	2356.0	0.311	7.915
303.2	0.03730	9.063.−4	33.71	109.6	2364.0	0.382	7.818
308.2	0.04990	9.075.−4	25.59	130.5	2372.3	0.450	7.725
313.2	0.06598	9.091.−4	19.66	151.5	2380.7	0.518	7.637
318.2	0.08638	9.108.−4	15.24	172.4	2388.6	0.585	7.550
323.2	0.1120	9.127.−4	11.93	193.3	2396.6	0.650	7.468
333.2	0.1831	9.170.−4	7.52	234.7	2413.3	0.776	7.315
353.2	0.4439	9.274.−4	3.27	318.4	2445.1	1.020	7.042
373.2	0.9646	9.403.−4	1.58	402.0	2474.8	1.253	6.807
398.2	2.2427	9.599.−4	0.72	507.5	2509.6	1.527	6.555
423.2	4.653	9.835.−4	0.362	612.5	2541.8	1.781	6.341
448.2	8.806	1.012.−3	0.198	718.8	2569.4	2.020	6.149
473.2	15.46	1.044.−3	0.115	826.8	2585.7	2.256	5.973
498.2	25.52	1.082.−3	0.0704	938.5	2597.0	2.483	5.812
523.2	39.99	1.133.−3	0.0447	1055.2	2598.7	2.707	5.658
548.2	60.04	1.200.−3	0.0290	1177.4	2587.0	2.930	5.501
573.2	86.97	1.276.−3	0.0191	1306.7	2555.6	3.153	5.332
598.2	122.4	1.392.−3	0.0124	1445.6	2492.4	3.356	5.132
623.2	168.3	1.596.−3	0.0075	1607.1	2366.5	3.631	4.850
644.7[c]	218.4	2.950.−3	0.0030				

*Extracted or converted from values in Kazavchinskii, Kesselman, et al., *Thermophysical Properties of Heavy Water*, Moscow and Leningrad, 1963; NBS-NSF transl. 70-50094, 1971. t = triple point; c = critical point. The notation 9.047.−4 signifies 9.047×10^{-4}.
Hill, P. G., MacMillan, R. D. and others give extensive tables for 0–1000 bar, 4–800°C in Atomic Energy of Canada, Chalk River rept. AECL-7531, 1981 (196 pp.). See also *J. Phys. Chem. Ref. Data*, **11**, 1 (1982): 1–14; **19**, 5 (1990): 1233–1274.

TABLE 2-252   Deuterium Oxide Gas at 1-kg/cm³ Pressure

T, K	400	450	500	550	600	650	700	750
v, m³/kg	1.676	1.895	2.112	2.322	2.535	2.747	2.960	3.172
h, kJ/kg	2525	2619	2712	2807	2904	3002	3102	3205
s, kJ/(kg·K)	6.931	7.151	7.349	7.529	7.697	7.855	8.003	8.153

TABLE 2-253   Saturated Diphenyl*

T, K	P, bar	v_f, m³/kg	v_g, m³/kg	h_f, kJ/kg	h_g, kJ/kg	s_f, kJ/(kg·K)	s_g, kJ/(kg·K)	c_{pf}, kJ/(kg·K)	μ_f, 10⁻⁴ Pa·s	k_f, W/(m·K)
343	0.0010	1.010.−3	252.5	0.0	444.2	0.000	1.298	1.760	15.0	0.139
350	0.0016	1.014.−3	156.1	13.0	444.2	0.036	1.266	1.782	13.5	0.138
360	0.0029	1.021.−3	85.0	30.0	446.7	0.084	1.236	1.813	11.7	0.136
370	0.0049	1.030.−3	49.9	47.2	449.7	0.130	1.213	1.844	10.3	0.135
380	0.0064	1.037.−3	29.9	65.0	454.5	0.178	1.200	1.875	9.1	0.133
390	0.0129	1.046.−3	18.3	82.7	462.7	0.224	1.194	1.906	8.1	0.132
400	0.0200	1.054.−3	11.7	99.3	461.2	0.273	1.202	1.936	7.3	0.130
420	0.0432	1.072.−3	5.84	139.9	499.0	0.358	1.228	1.998	6.0	0.127
440	0.0879	1.092.−3	3.021	180.3	532.4	0.451	1.267	2.060	5.0	0.125
460	0.1694	1.112.−3	1.652	222.7	569.7	0.545	1.378	2.122	4.3	0.122
480	0.3112	1.132.−3	0.9594	267.6	611.6	0.652	1.367	2.184	3.7	0.119
500	0.5218	1.154.−3	0.4452	314.9	651.8	0.746	1.424	2.246	3.3	0.116
520	0.8375	1.177.−3	0.3652	361.5	687.8	0.824	1.477	2.308	2.7	0.113
540	1.290	1.204.−3	0.2261	404.5	723.8	0.915	1.529	2.370	2.4	0.110
560	1.941	1.230.−3	0.1447	457.2	762.7	1.032	1.582	2.432	2.2	0.107
580	2.818	1.258.−3	0.0977	522.3	801.7	1.125	1.635	2.494	1.90	0.105
600	3.926	1.291.−3	0.0685	563.7	842.4	1.223	1.688	2.556	1.71	0.102
620	5.408	1.326.−3	0.0504	630.4	886.4	1.316	1.740	2.618	1.54	0.099
640	7.328	1.366.−3	0.0381	689.1	930.9	1.375	1.748	2.680	1.39	0.096
660	9.572	1.412.−3	0.0301	745.9	977.1	1.457	1.791	2.741	1.24	0.093
680	12.05	1.465.−3	0.0236	802.8	1024.9	1.585	1.856	2.803	1.10	0.090
700	15.21	1.529.−3	0.0186	860.1	1073.1	1.663	1.951	2.865	0.97	0.087
720	19.14	1.56.−3	0.0147	917.5	1116.7	1.746	2.003	2.93		
740	23.93	1.70.−3	0.0113	975.2	1152.8	1.822	2.058	3.00		
760	28.71	1.95.−3	0.0085	1033.1	1182.5	1.901	2.099			
780	34.83	2.16.−3	0.0058	1091.2	1163.0	1.977	2.107			
800	42.46	3.18.−3	0.0032	1148.4	1148.4	2.047	2.047			

*Interpolated by P. E. Liley from the Landolt-Börnstein band IVa, p. 557, 1967 tables based on *Technical Data on Fuel*, British National Committee, World Energy Conference, London.

TABLE 2-254   Saturated Ethane (R170)*

T, K	P, bar	v_f, m³/kg	v_g, m³/kg	h_f, kJ/kg	h_g, kJ/kg	s_f, kJ/(kg·K)	s_g, kJ/(kg·K)	c_{pf}, kJ/(kg·K)	μ_f, 10⁻⁴ Pa·s	k_f, W/(m·K)
90.4ᵗ	1.131.−5	1.534.−3	21945	176.8	769.4	2.560	9.113	2.260	14.19	0.215
100	1.110.−4	1.546	2484.5	198.7	782.4	2.790	8.627	2.274	9.37	0.208
110	7.467.−3	1.573	407.0	221.5	795.0	3.008	8.222	2.284	6.57	0.201
120	3.545.−3	1.615	93.61	244.4	807.2	3.207	7.897	2.292	4.89	0.194
130	1.291.−2	1.644	27.83	267.4	819.3	3.391	7.637	2.302	3.81	0.187
140	3.831.−2	1.675	10.08	290.5	831.4	3.562	7.426	2.316	3.07	0.180
150	9.672.−2	1.708	4.263	313.7	843.5	3.722	7.254	2.333	2.55	0.174
160	0.2146	1.743	2.039	337.2	855.6	3.873	7.113	2.355	2.17	0.167
170	0.4290	1.780	1.075	360.9	867.6	4.017	6.998	2.383	1.88	0.160
180	0.7874	1.819	0.6139	384.9	879.4	4.154	6.901	2.417	1.65	0.153
190	1.347	1.862	0.3738	409.3	890.8	4.285	6.819	2.458	1.47	0.147
200	2.174	1.908	0.2395	434.2	901.7	4.412	6.750	2.508	1.33	0.140
210	3.340	1.958	0.1602	459.7	911.9	4.535	6.689	2.568	1.21	0.133
220	4.922	2.014	0.1109	485.9	921.4	4.655	6.635	2.640	1.11	0.126
230	7.004	2.076	0.0789	512.8	929.6	4.773	6.585	2.730	1.03	0.119
240	9.670	2.148	0.0573	540.8	936.6	4.890	6.539	2.843	0.96	0.112
250	13.01	2.231	0.0423	569.9	941.9	5.006	6.493	2.991	0.82	0.106
260	17.12	2.330	0.0316	600.7	945.4	5.123	6.449	3.214	0.73	0.099
270	22.10	2.452	0.0237	633.6	946.4	5.233	6.392	3.511	0.64	0.092
280	28.06	2.613	0.0177	669.3	943.6	5.370	6.350	4.011	0.55	0.085
290	35.14	2.847	0.0129	709.8	934.7	5.502	6.278	5.089	0.44	0.078
300	43.54	3.295	0.0087	761.6	910.8	5.669	6.166	9.919	0.31	0.067
305.3ᶜ	48.71	4.891	0.0048	841.2	841.2	5.919	5.919	∞		

*Values reproduced or converted from Goodwin, Roder, and Straty, NBS Tech. Note 684, 1976. t = triple point; c = critical point. The notation 1.131.−5 signifies 1.131×10^{-5}.

TABLE 2-255   Superheated Ethane*

P, bar		100	150	200	250	300	350	400	450	500	600	700
1.013	v	0.00156	0.00171	0.5310	0.6725	0.8118	0.9500	1.0877	1.2250	1.3622	1.6360	1.9096
	h	198.9	313.8	909.3	984.7	1068.3	1161.5	1265.3	1379.8	1504.6	1783	2097
	s	2.790	3.722	6.993	7.330	7.634	7.921	8.198	8.467	8.730	9.237	9.720
5	v	0.00156	0.00171	0.00191	0.1288	0.1595	0.1890	0.2178	0.2464	0.2747	0.3308	0.3867
	h	199.4	314.3	434.5	973.3	1060.3	1155.6	1260.7	1376.1	1501.5	1781	2096
	s	2.789	3.720	4.411	6.858	7.175	7.468	7.748	8.020	8.284	8.793	9.227
10	v	0.00156	0.00171	0.00190	0.0590	0.0765	0.0923	0.1073	0.1220	0.1365	0.1650	0.1933
	h	200.0	314.9	435.0	956.5	1050.0	1148.2	1255.0	1371.5	1497.9	1777	2094
	s	2.788	3.719	4.408	6.618	6.959	7.262	7.547	7.821	8.087	8.598	9.083
20	v	0.00156	0.00170	0.00190	0.00222	0.0346	0.0438	0.0521	0.0599	0.0674	0.0822	0.0966
	h	201.3	316.1	435.9	569.8	1026.1	1132.3	1243.3	1362.4	1490.5	1774	2090
	s	2.785	3.715	4.404	4.999	6.710	7.038	7.334	7.614	7.884	8.399	8.886
40	v	0.00155	0.00170	0.00189	0.00219	0.0118	0.0193	0.0244	0.0288	0.0329	0.0407	0.0482
	h	203.9	318.5	437.9	569.9	947.9	1096.2	1218.6	1343.8	1475.9	1764	2083
	s	2.780	3.709	4.394	4.982	6.309	6.770	7.097	7.391	7.670	8.194	8.686
60	v	0.00155	0.00170	0.00188	0.00217	0.00290	0.0109	0.0132	0.0185	0.0215	0.0270	0.0321
	h	206.5	321.0	439.8	570.3	738.1	1050.9	1192.0	1324.8	1461.2	1754	2077
	s	2.775	3.702	4.385	4.966	5.574	6.557	6.934	7.247	7.535	8.068	8.564
80	v	0.00155	0.00169	0.00188	0.00215	0.00273	0.00667	0.0106	0.0134	0.0158	0.0201	0.0459
	h	209.1	323.4	441.9	570.9	728.1	993.8	1163.6	1305.5	1446.7	1745	2070
	s	2.769	3.696	4.377	4.951	5.522	6.345	6.800	7.135	7.432	7.975	8.476
100	v	0.00155	0.00169	0.00187	0.00213	0.00263	0.00465	0.00791	0.0104	0.0124	0.0160	0.0193
	h	211.7	325.8	443.9	571.8	722.7	924.4	1134.7	1286.3	1432.4	1736	2064
	s	2.764	3.690	4.368	4.938	5.486	6.166	6.682	7.040	7.348	7.900	8.406
150	v	0.00155	0.00168	0.00185	0.00209	0.00247	0.00328	0.00488	0.00655	0.00805	0.0107	0.0130
	h	218.1	332.0	449.2	574.6	716.4	887.4	1075.2	1242.3	1399.3	1715	2050
	s	2.752	3.674	4.348	4.907	5.423	5.955	6.457	6.851	7.182	7.758	8.274
200	v	0.00154	0.00167	0.00184	0.00205	0.00237	0.00291	0.00383	0.00495	0.00605	0.00806	0.00986
	h	224.6	338.2	454.7	578.2	714.8	870.5	1041.7	1210.2	1327.3	1697	2038
	s	2.738	3.660	4.329	4.880	5.377	5.863	6.320	6.717	7.059	7.651	8.176
300	v	0.00153	0.00166	0.00181	0.00200	0.00225	0.00259	0.00307	0.00367	0.00433	0.00563	0.00686
	h	237.6	350.6	465.9	586.8	715.9	860.9	1014.9	1175.5	1338.7	1671	2019
	s	2.715	3.632	4.294	4.833	5.309	5.757	6.168	6.547	6.891	7.496	8.032
400	v	0.00153	0.00165	0.00179	0.00195	0.00216	0.00244	0.00276	0.00316	0.00361	0.00454	0.00545
	h	250.6	363.2	477.6	596.6	723.7	861.6	1008.3	62.5	1322.7	1657	2008
	s	2.692	3.605	4.262	4.793	5.257	5.688	6.080	6.443	6.780	7.388	7.930
500	v	0.00152	0.00163	0.00176	0.00192	0.00210	0.00232	0.00258	0.00288	0.00322	0.00392	0.00465
	h	263.5	375.8	489.3	607.1	732.0	866.5	1009.3	1159.3	1316.9	1650	2003
	s	2.670	3.580	4.234	4.758	5.213	5.634	6.015	6.369	6.00	7.306	7.851

*Converted and rounded off from the tables of Goodwin, Roder, and Straty, NBS Tech. Note 684, 1976. v = specific volume, m³/kg; h = specific enthalpy, kJ/kg; s = specific entropy, kJ/(kg·K).

Saturation and superheat tables and a diagram to 300 bar, 580 K are given by Reynolds, W. C., *Thermodynamic Properties in S.I.*, Stanford Univ. publ., 1979 (173 pp.). Saturation and superheat tables and a chart to 10,000 psia, 640°F appear in Stewart, R. B., R. T. Jacobsen, et al., *Thermodynamic Properties of Refrigerants*, ASHRAE, Atlanta, GA, 1986 (521 pp.). For specific heat, thermal conductivity, and viscosity, see *Thermophysical Properties of Refrigerants*, ASHRAE, 1993. The 1993 ASHRAE *Handbook—Fundamentals* (SI ed.) contains a thermodynamic diagram from 0.1 to 700 bar for temperatures to 600 K.

TABLE 2-256 Saturated Ethanol

T, K	P, bar	v_f, m³/kg	v_g, m³/kg	h_f, kJ/kg	h_g, kJ/kg	s_f, kJ/(kg·K)	s_g, kJ/(kg·K)	c_{pf}, kJ/(kg·K)	μ_f, 10⁻⁶ Pa·s	k_f, W/(m·K)	Pr_f
250	0.0027	0.001 184						2.113	295	0.177	35.2
260	0.0059	0.001 196						2.167	229	0.175	28.4
270	0.0128	0.001 208						2.227	193	0.173	24.8
280	0.025	0.001 220						2.294	156	0.171	20.9
290	0.048	0.001 233						2.369	127	0.170	17.7
300	0.088	0.001 246						2.45	104	0.168	15.2
310	0.151	0.001 260						2.54	86	0.165	13.2
320	0.253	0.001 274						2.64	72	0.162	11.7
330	0.406	0.001 288						2.75	61	0.159	10.6
340	0.632	0.001 304						2.86	52	0.157	9.5
350	0.956	0.001 318	0.7656	199.9	1161.9			2.99	45.0	0.155	8.7
360	1.409	0.001 337	0.5052	230.1	1178.4			3.12	39.0	0.153	8.0
370	2.023	0.001 357	0.3555	262.2	1193.9			3.27	34.2	0.151	7.4
380	2.837	0.001 379	0.2556	295.1	1208.4			3.42	30.0	0.149	6.9
390	3.897	0.001 403	0.1873	329.1	1221.5			3.58	26.1	0.147	6.3
400	5.251	0.001 430	0.1398	364.2	1233.6			3.74	22.7	0.145	5.9
410	6.954	0.001 461	0.1058	400.8	1244.2			3.99	20.0	0.144	5.5
420	9.063	0.001 495	0.0812	435.7	1254.2			4.26	17.6	0.142	5.3
430	11.64	0.001 532	0.0631	472.2	1262.3			4.55	15.3	0.140	5.0
440	14.72	0.001 574	0.0493	512.7	1269.2			4.88	13.9	0.139	4.9
450	18.33	0.001 623	0.0389	557.2	1274.2			5.23	12.5	0.137	4.8
460	22.61	0.001 682	0.0308	605.0	1275.5						
470	27.66	0.001 752	0.0243	653.7	1271.1						
480	33.55	0.001 832	0.0193	704.5	1262.3						
490	40.39	0.001 950	0.0148	757.7	1250.2						
500	48.28	0.002 091	0.0110	818.9	1232.7						
510	57.32										
516.3ᶜ	63.90										

c = critical point.
Values interpolated and converted from *Heat Exchanger Design Handbook,* vol. 5, Hemisphere, Washington, DC, 1983, and from various literature sources.

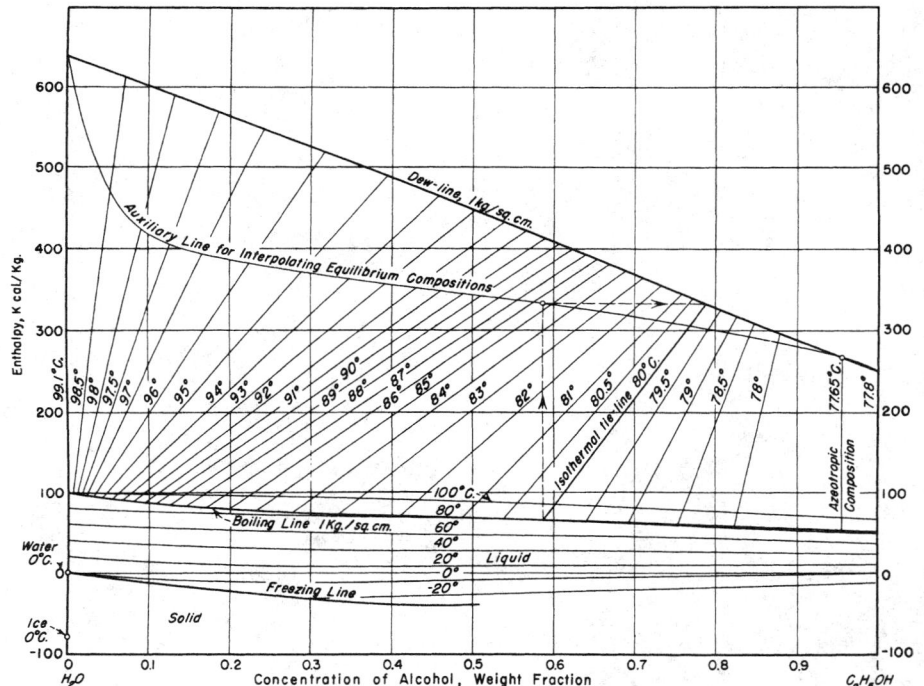

FIG. 2-10 Enthalpy-concentration diagram for aqueous ethyl alcohol. Reference states: Enthalpies of liquid water and ethyl alcohol at 0°C are zero. NOTE: In order to interpolate equilibrium compositions, a vertical may be erected from any liquid composition on the boiling line and its intersection with the auxiliary line determined. A horizontal from this intersection will establish the equilibrium vapor composition on the dew line. (*Bosnjakovic,* Technische Thermodynamik, T. Steinkopff, Leipzig, 1935.)

TABLE 2-257   Saturated Ethylene (Ethene—R1150)

Temperature, K	Pressure, bar	v_f, m³/kg	v_g, m³/kg	h_f, kJ/kg	h_g, kJ/kg	s_f, kJ/(kg·K)	s_g, kJ/(kg·K)	c_{pf}, kJ/(kg·K)
104.0^t	0.00123	0.001 527	251.36	−323.81	244.36	−1.9901	3.4730	2.497
110	0.00334	0.001 545	97.57	−309.54	251.47	−1.8571	3.2431	2.500
120	0.01380	0.001 576	25.75	−284.17	263.23	−1.6362	2.9255	2.539
130	0.04456	0.001 609	8.62	−259.13	274.87	−1.4358	2.6717	2.465
140	0.1191	0.001 644	3.46	−234.80	286.28	−1.2554	2.4663	2.405
150	0.2747	0.001 681	1.5977	−210.90	297.37	−1.0908	2.2977	2.377
160	0.5636	0.001 721	0.8232	−187.12	308.00	−0.9378	2.1566	2.377
170	1.0526	0.001 763	0.4625	−163.23	318.04	−0.7935	2.0375	2.395
180	1.8207	0.001 810	0.2784	−139.05	327.35	−0.6559	1.9352	2.427
190	2.9574	0.001 861	0.1770	−114.46	335.79	−0.5244	1.7812	2.472
200	4.560	0.001 918	0.1177	−89.33	343.21	−0.3967	1.7659	2.531
210	6.730	0.001 981	0.0810	−63.52	349.41	−0.2730	1.6932	2.608
220	9.575	0.002 054	0.0573	−36.84	354.18	−0.1515	1.6258	2.711
230	13.206	0.002 139	0.0413	−9.04	357.17	−0.0314	1.5609	2.852
240	17.742	0.002 241	0.0302	20.23	357.90	0.0088	1.4957	3.055
250	23.307	0.002 369	0.02222	51.55	355.37	0.2114	1.4276	3.372
260	30.046	0.002 541	0.01624	85.91	348.68	0.3397	1.3503	3.945
270	38.132	0.002 804	0.01152	125.79	333.71	0.4819	1.3054	5.40
280	47.834	0.003 442	0.00720	183.40	292.83	0.6803	1.0711	20.0
282.3^c	50.403	0.004 669	0.00467	234.55	234.55	0.8585	0.8585	

t = triple point; c = critical point. $h_f = s_f = 0$ at 233.15 K = −40°C.

Converted from Jacobsen, R. T., M. Jahangiri, et al., *Ethylene—Intl. Thermodyn. Tables of the Fluid State—10*, Blackwell Sci. Publ., Oxford, U.K., 1988 (299 pp.). Saturation and superheat tables and a diagram to 100 bar, 460 K are given by Reynolds, W. C., *Thermodynamic properties in S.I.*, Stanford Univ. publ., 1979 (173 pp.). Saturation and superheat tables and a chart to 6000 psia, 360°F appear in *Thermodynamic Properties of Refrigerants*, ASHRAE, Atlanta, GA, 1986 (521 pp.). For specific heat, thermal conductivity, and viscosity, see *Thermophysical Properties of Refrigerants*, ASHRAE, 1993.

The 1993 ASHRAE *Handbook—Fundamentals* (SI ed.) has a thermodynamic chart for pressures from 0.1 to 400 bar and temperatures up to 460 K.

TABLE 2-258 Compressed Ethylene

Pressure, bar	Temperature, K							
	110	125	150	175	200	225	250	275
v (m³/kg)	0.001 545	0.001 592	0.001 681	0.5036	0.5814	0.6580	0.7337	0.8091
1 h (kJ/kg)	−309.4	−271.4	−210.8	324.8	357.0	389.9	424.0	459.7
s (kJ/kg·K)	−1.858	−1.534	−1.091	2.091	2.264	2.419	2.562	2.698
v (m³/kg)	0.001 544	0.001 591	0.001 680	0.001 785	0.001 917	0.1240	0.1407	0.1569
5 h (kJ/kg)	−308.9	−271.0	−210.4	−150.8	−89.3	378.4	415.0	452.3
s (kJ/kg·K)	−1.859	−1.535	−1.093	−0.726	−0.397	1.907	2.061	2.203
v (m³/kg)	0.001 543	0.001 591	0.001 679	0.001 783	0.001 914	0.05643	0.06672	0.07525
10 h (kJ/kg)	−308.3	−270.4	−209.8	−150.3	−89.0	361.2	402.4	442.3
s (kJ/kg·K)	−1.860	−1.537	−1.095	−0.728	−0.400	1.646	1.820	1.973
v (m³/kg)	0.001 542	0.001 589	0.001 676	0.001 780	0.001 908	0.002 084	0.02810	0.03405
20 h (kJ/kg)	−307.1	−269.2	−208.7	−149.4	−88.2	−23.0	370.3	419.7
s (kJ/kg·K)	−1.863	−1.540	−1.098	−0.733	−0.406	−0.099	1.520	1.708
v (m³/kg)	0.001 541	0.001 588	0.001 674	0.001 776	0.001 903	0.002 072	0.002 347	0.01978
30 h (kJ/kg)	−305.9	−268.0	−207.6	−148.4	−87.5	−22.8	50.5	390.7
s (kJ/kg·K)	−1.866	−1.543	−1.102	−0.737	−0.412	−0.107	0.201	1.508
v (m³/kg)	0.001 540	0.001 587	0.001 672	0.001 773	0.001 897	0.002 062	0.002 318	0.01163
40 h (kJ/kg)	−304.7	−266.8	−206.5	−147.4	−86.7	−22.5	49.1	344.7
s (kJ/kg·K)	−1.869	−1.546	−1.106	−0.741	−0.418	−0.115	0.186	1.284
v (m³/kg)	0.001 539	0.001 585	0.001 670	0.001 770	0.001 892	0.002 052	0.002 293	0.002 846
50 h (kJ/kg)	−303.5	−265.7	−205.4	−146.4	−85.9	−22.2	48.1	139.8
s (kJ/kg·K)	−1.872	−1.550	−1.110	−0.746	−0.423	−0.123	0.173	0.521
v (m³/kg)	0.001 538	0.001 584	0.001 668	0.001 767	0.001 887	0.002 043	0.002 270	0.002 723
60 h (kJ/kg)	−302.3	−264.5	−204.2	−145.4	−85.1	−21.8	47.4	132.3
s (kJ/kg·K)	−1.875	−1.553	−1.113	−0.750	−0.428	−0.130	0.161	0.484
v (m³/kg)	0.001 535	0.001 581	0.001 664	0.001 761	0.001 877	0.002 025	0.002 232	0.002 585
80 h (kJ/kg)	−299.8	−262.1	−202.0	−143.4	−83.5	−20.9	46.5	124.1
s (kJ/kg·K)	−1.881	−1.559	−1.120	−0.759	−0.439	−0.145	0.139	0.434
v (m³/kg)	0.001 533	0.001 579	0.001 660	0.001 754	0.001 867	0.002 009	0.002 199	0.002 495
100 h (kJ/kg)	−297.4	−259.7	−199.7	−141.2	−81.8	−19.9	46.1	119.6
s (kJ/kg·K)	−1.887	−1.565	−1.127	−0.767	−0.449	−0.158	0.120	0.400
v (m³/kg)	0.001 528	0.001 571	0.001 650	0.001 740	0.001 846	0.001 973	0.002 136	0.002 356
150 h (kJ/kg)	−291.3	−253.7	−194.0	−136.0	−77.3	−16.7	46.6	114.4
s (kJ/kg·K)	−1.901	−1.580	−1.145	−0.787	−0.473	−0.188	0.079	0.337
v (m³/kg)	0.001 522	0.001 565	0.001 641	0.001 727	0.001 826	0.001 943	0.002 086	0.002 268
200 h (kJ/kg)	−285.3	−247.7	−188.3	−130.7	−72.5	−13.0	48.6	113.2
s (kJ/kg·K)	−1.914	−1.595	−1.161	−0.806	−0.495	−0.215	0.045	0.291
v (m³/kg)	0.001 517	0.001 559	0.001 633	0.001 715	0.001 809	0.001 918	0.002 046	0.002 203
250 h (kJ/kg)	−279.2	−241.7	−182.5	−125.2	−67.6	−8.9	51.4	113.9
s (kJ/kg·K)	−1.928	−1.610	−1.177	−0.824	−0.516	−0.240	0.015	0.253
v (m³/kg)	0.001 512	0.001 552	0.001 625	0.001 704	0.001 793	0.001 895	0.002 012	0.002 151
300 h (kJ/kg)	−273.0	−235.7	−174.5	−119.6	−62.5	−4.4	54.9	115.9
s (kJ/kg·K)	−1.942	−1.623	−1.192	−0.841	−0.536	−0.262	−0.012	0.220
v (m³/kg)	0.001 503	0.001 542	0.001 609	0.001 683	0.001 765	0.001 855	0.001 957	0.002 072
400 h (kJ/kg)	−260.8	−223.6	−164.8	−108.3	−51.9	5.1	63.0	122.1
s (kJ/kg·K)	−1.968	−1.650	−1.221	−0.873	−0.572	−0.303	−0.059	0.166
v (m³/kg)	0.001 499	0.001 531	0.001 596	0.001 665	0.001 740	0.001 823	0.001 913	0.002 01
500 h (kJ/kg)	−246.9	−211.4	−152.9	−96.8	−40.9	15.3	72.3	130.1
s (kJ/kg·K)	−1.978	−1.676	−1.249	−0.906	−0.605	−0.339	−0.099	0.121

Converted from Jacobsen, R. T., M. Jahangiri, et al., *Ethylene—Intl. Thermodyn. Tables of the Fluid State—10*, Blackwell Sci. Publ., Oxford, 1988 (299 pp.). $s_f = h_f = 0$ at 233.15 K = −40°C.

TABLE 2-258 Compressed Ethylene (*Concluded*)

Pressure, bar		300	325	350	375	400	425	450
					Temperature, K			
1	v (m³/kg)	0.8842	0.9591	1.0339	1.1084	1.1830	1.2575	1.3319
	h (kJ/kg)	497.3	536.9	578.8	622.8	668.9	717.2	767.9
	s (kJ/kg·K)	2.829	2.956	3.079	3.201	3.320	3.437	3.553
5	v (m³/kg)	0.1728	0.1884	0.2039	0.2193	0.2346	0.2499	0.2650
	h (kJ/kg)	491.0	531.5	574.1	618.6	665.2	713.9	764.9
	s (kJ/kg·K)	2.338	2.467	2.593	2.716	2.836	2.954	3.071
10	v (m³/kg)	0.08380	0.09207	0.1002	0.1081	0.1160	0.1238	0.1316
	h (kJ/kg)	482.8	542.5	568.0	613.3	660.5	709.7	761.1
	s (kJ/kg·K)	2.113	2.247	2.375	2.500	2.622	2.742	2.859
20	v (m³/kg)	0.03914	0.04379	0.04823	0.05257	0.05675	0.06088	0.06491
	h (kJ/kg)	465.0	509.8	555.5	602.4	650.9	701.2	753.5
	s (kJ/kg·K)	1.866	2.009	2.144	2.274	2.399	2.521	2.640
30	v (m³/kg)	0.02404	0.02763	0.03090	0.03400	0.03700	0.03990	0.04270
	h (kJ/kg)	444.7	493.8	542.3	591.2	641.2	692.6	745.9
	s (kJ/kg·K)	1.696	1.853	1.996	2.131	2.261	2.387	2.508
40	v (m³/kg)	0.01630	0.01947	0.02220	0.02473	0.02710	0.02938	0.03160
	h (kJ/kg)	420.6	476.3	528.3	579.5	631.2	688.9	738.2
	s (kJ/kg·K)	1.550	1.728	1.882	2.023	2.157	2.286	2.409
50	v (m³/kg)	0.01140	0.01451	0.01697	0.01916	0.02119	0.02311	0.02495
	h (kJ/kg)	390.4	456.9	513.4	567.5	621.1	675.1	730.5
	s (kJ/kg·K)	1.404	1.617	1.784	1.933	2.072	2.207	2.330
60	v (m³/kg)	0.007 757	0.01116	0.01347	0.01546	0.01725	0.01892	0.02052
	h (kJ/kg)	347.8	435.1	497.7	555.2	610.9	666.3	722.9
	s (kJ/kg·K)	1.230	1.510	1.696	1.854	1.999	2.135	2.263
80	v (m³/kg)	0.003 672	0.006 864	0.009 136	0.01085	0.01237	0.01374	0.01502
	h (kJ/kg)	238.7	382.8	463.5	529.4	590.1	648.5	707.6
	s (kJ/kg·K)	0.832	1.295	1.534	1.717	1.874	2.016	2.151
100	v (m³/kg)	0.003 094	0.004 698	0.006 596	0.008 163	0.009 492	0.01068	0.01177
	h (kJ/kg)	210.2	330.2	427.5	503.1	569.3	630.9	692.7
	s (kJ/kg·K)	0.715	1.098	1.387	1.596	1.768	1.918	2.059
150	v (m³/kg)	0.002 684	0.003 223	0.004 040	0.004 983	0.005 914	0.006 765	0.007 578
	h (kJ/kg)	188.8	272.7	361.4	445.5	521.6	592.1	658.1
	s (kJ/kg·K)	0.596	0.864	1.126	1.359	1.556	1.722	1.878
200	v (m³/kg)	0.002 508	0.002 840	0.003 292	0.003 838	0.004 445	0.005 058	0.005 664
	h (kJ/kg)	181.7	255.0	332.4	410.6	487.0	560.1	629.5
	s (kJ/kg·K)	0.529	0.763	0.992	1.208	1.406	1.580	1.742
250	v (m³/kg)	0.002 397	0.002 644	0.003 024	0.003 327	0.003 743	0.004 190	0.004 648
	h (kJ/kg)	179.2	247.7	319.1	392.1	465.5	538.2	608.5
	s (kJ/kg·K)	0.480	0.698	0.910	1.111	1.301	1.476	1.639
300	v (m³/kg)	0.002 317	0.002 517	0.002 578	0.003 037	0.003 351	0.003 690	0.004 042
	h (kJ/kg)	179.1	244.6	312.6	382.2	452.9	524.1	593.9
	s (kJ/kg·K)	0.440	0.670	0.850	1.043	1.226	1.398	1.558
400	v (m³/kg)	0.002 203	0.002 352	0.002 522	0.002 711	0.002 919	0.003 413	0.003 382
	h (kJ/kg)	182.7	244.9	308.8	374.3	441.3	509.8	578.3
	s (kJ/kg·K)	0.377	0.576	0.764	0.946	1.119	1.285	1.442
500	v (m³/kg)	0.002 122	0.002 245	0.002 379	0.002 524	0.002 678	0.002 847	0.003 022
	h (kJ/kg)	189.1	250.7	312.8	376.1	440.4	505.8	572.0
	s (kJ/kg·K)	0.326	0.523	0.707	0.882	1.048	1.206	1.358

TABLE 2-259   Saturated Fluorine*

T, K	P, bar	v_f, m³/kg	v_g, m³/kg	h_f, kJ/kg	h_g, kJ/kg	s_f, kJ/(kg·K)	s_g, kJ/(kg·K)	c_{pf}, kJ/(kg·K)	μ_f, 10^{-4} Pa·s	k_f, W/(m·K)
53.5[t]	0.0025	5.866.−4	46.2	−158.6	40.9	1.602	5.314	1.446	8.8	0.186
55	0.0041	5.898.−4	17.1	−153.5	42.0	1.642	5.235	1.442	8.0	0.184
60	0.0155	6.005.−4	8.46	−149.1	45.8	1.768	5.004	1.437	6.0	0.177
65	0.0477	6.119.−4	2.93	−141.8	49.6	1.885	4.816	1.442	4.7	0.170
70	0.1230	6.240.−4	1.24	−134.4	53.2	1.995	4.666	1.450	3.8	0.162
75	0.276	6.369.−4	0.583	−127.0	56.8	2.097	4.540	1.460	3.19	0.154
80	0.555	6.508.−4	0.309	−119.5	60.1	2.194	4.433	1.474	2.71	0.146
85	1.019	6.657.−4	0.176	−111.9	63.3	2.285	4.342	1.498	2.33	0.137
90	1.740	6.819.−4	0.108	−104.3	66.1	2.372	4.262	1.535	2.00	0.129
95	2.802	6.997.−4	0.069	−96.5	68.6	2.455	4.191	1.555	1.76	0.120
100	4.280	7.193.−4	0.0466	−88.6	70.7	2.535	4.127	1.585	1.53	0.112
105	6.280	7.412.−4	0.0323	−80.5	72.4	2.612	4.068	1.630	1.36	0.103
110	8.885	7.659.−4	0.0231	−72.2	73.6	2.688	4.012	1.692	1.21	0.095
115	12.20	7.948.−4	0.0168	−63.6	74.1	2.763	3.959	1.782	1.08	0.087
120	16.33	8.283.−4	0.0125	−54.5	73.9	2.837	3.906	1.888	0.96	0.080
125	21.37	8.696.−4	0.0093	−44.9	72.7	2.912	3.864	2.05	0.86	0.073
130	27.48	9.223.−4	0.0069	−34.5	70.2	2.989	3.795	2.33	0.74	0.066
135	34.72	9.963.−4	0.0051	−22.7	65.6	3.073	3.727	2.90	0.63	0.070
140	43.47	1.119.−3	0.0036	−8.4	56.9	3.170	3.636	3.64	0.49	0.105
144.3[c]	52.15	1.743.−3	0.0017	23.9	23.9	3.388	3.388	∞		∞

*Values reproduced or converted from Prydz and Straty, NBS Tech. Note 392, rev., September 1973. *t* = triple point; *c* = critical point. The notation 5.866.−4 signifies 5.866×10^{-4}.

TABLE 2-260   Fluorine Gas at Atmospheric Pressure*

T, K	84.95	90	100	120	140	160	180	200	220	240	260	280	300
v, m³/kg	0.1776	0.1892	0.2118	0.2562	0.3002	0.3439	0.3874	0.4309	0.4744	0.5176	0.5610	0.6043	0.6476
h, kJ/kg	63.22	67.30	75.27	90.96	106.53	122.06	137.62	153.2	169.0	184.9	201.0	217.2	233.7
s, kJ/(kg·K)	4.342	4.390	4.474	4.616	4.737	4.840	4.932	5.014	5.090	5.158	5.221	5.282	5.340

*Extracted from Prydz and Straty, NBS Tech. Note 392, 1970. This source is recommended for other pressures and temperatures. Other information is contained in *J. Chem. Phys.*, **53** (1970): 2359; and *J. Res. NBS*, **74A** (1970): 499, 661, 747.

TABLE 2-261   Flutec

Proprietary name for a series of fluorocarbons produced by the Imperial Smelting Corp., Avonmouth, Bristol, UK. Bulletins of thermodynamic properties include PP1 (C_6F_{14}), PP2 (C_7F_{14}), PP3 (C_8F_{16}), PP5 ($C_{10}F_{18}$), PP9 ($C_{11}F_{20}$), and PP50, usually for 0.1–100 kg/m², 0–500°C. See also Green, S. W., *Chem. & Ind.* (1969): 63–67.

TABLE 2-262   Halon

A series of fire-extinguishing fluids. Halon 1211 is produced by ICI, and Halon 1301, by duPont, the latter issuing a bulletin with thermodynamic properties and a diagram for the range 0.6–600 psia, −160–460°F.

TABLE 2-263   Saturated Helium³*

T, K	P, bar	v_f, m³/kg	v_g, m³/kg	h_f, kJ/kg	h_g, kJ/kg	s_f, kJ/(kg·K)	s_g, kJ/(kg·K)
1.0	0.0122	0.01222	1.72	−5.69	6.75	2.28	14.72
1.1	0.0182	0.01224	1.33	−5.49	7.40	2.65	14.34
1.2	0.0274	0.01227	1.02	−5.26	8.03	2.95	14.02
1.3	0.0370	0.01231	0.805	−5.01	8.65	3.20	13.70
1.4	0.0517	0.01236	0.649	−4.75	9.27	3.40	13.41
1.5	0.0659	0.01241	0.526	−4.47	9.88	3.60	13.13
1.6	0.0871	0.01247	0.437	−4.17	10.46	3.80	12.88
1.7	0.107	0.01254	0.363	−3.84	11.04	3.91	12.53
1.8	0.137	0.01262	0.308	−3.47	11.60	4.01	12.38
1.9	0.163	0.01271	0.260	−3.07	12.15	4.13	12.14
2.0	0.202	0.01282	0.222	−2.64	12.68	4.26	11.91
2.1	0.237	0.01294	0.189	−2.17	13.19	4.40	11.69
2.2	0.284	0.01308	0.164	−1.55	13.67	4.55	11.47
2.3	0.326	0.01324	0.142	−0.99	14.13	4.71	11.25
2.4	0.385	0.01343	0.124	−0.34	14.57	4.87	11.04
2.5	0.438	0.01365	0.109	0.36	14.98	5.03	10.84
2.6	0.508	0.01390	0.096	1.16	15.37	5.20	10.64
2.7	0.576	0.01419	0.085	2.01	15.89	5.38	10.41
2.8	0.653	0.01456	0.074	2.96	16.40	5.57	10.17
2.9	0.732	0.01497	0.064	4.01	16.37	5.77	9.92
3.0	0.803	0.01549	0.055	5.28	16.32	6.00	9.66
3.1	0.907	0.01614	0.047	6.70	16.20	6.24	9.34
3.2	1.023	0.01720	0.039	8.44	15.98	6.54	8.90
3.3	1.128	0.01902	0.028	10.66	14.50	6.96	8.35
3.32[c]	1.165	0.02394	0.024	13.25	13.25	7.50	7.50

*Converted and smoothed from a tabulation of Gibbons and Nathan, USAF Rep. AFML-TR-67-175, 1967. *c* = critical point. Kelly, D. P. and W. K. Haubach, in AEC R&D rept. MLM 1161, 1963 (56 pp.), give a comprehensive graphical comparison of the properties of He³ and He⁴.

TABLE 2-264   Saturated Helium[4]

Temperature, K	Pressure, bar	v_f, m³/kg	v_g, m³/kg	h_f, kJ/kg	h_g, kJ/kg	s_f, kJ/(kg·K)	s_g, kJ/(kg·K)	c_{pf}, kJ/(kg·K)	c_{pg}, kJ/(kg·K)
0.8	1.475.−5	0.00689	1125.9	0.0019	19.42	0.0047	23.94	0.022	5.210
0.9	5.379.−5	0.00689	347.1	0.0054	19.94	0.0087	21.86	0.050	5.230
1.0	1.557.−4	0.00689	133.0	0.0127	20.44	0.0163	20.44	0.100	5.262
1.1	3.800.−4	0.00689	59.8	0.0268	20.95	0.0296	18.82	0.185	5.305
1.2	8.148.−4	0.00689	30.4	0.0518	21.44	0.0510	17.67	0.318	5.360
1.3	0.00158	0.00689	16.93	0.0932	21.92	0.0836	16.70	0.511	5.424
1.4	0.00282	0.00689	10.17	0.1579	22.40	0.1308	15.86	0.780	5.496
1.5	0.00472	0.00689	6.49	0.2543	22.87	0.1962	15.13	1.138	5.574
1.6	0.00746	0.00688	4.35	0.3923	23.32	0.2839	14.49	1.602	5.654
1.7	0.01128	0.00688	3.04	0.5836	23.77	0.3981	13.93	2.193	5.736
1.8	0.01638	0.00688	2.1993	0.8422	24.20	0.5437	13.43	2.938	5.818
1.9	0.02299	0.00687	1.6420	1.186	24.63	0.7270	12.98	3.893	5.898
2.0	0.03129	0.00686	1.2601	1.642	25.04	0.9578	12.58	5.187	5.975
2.1	0.04141	0.00685	0.9921	2.261	25.45	1.256	12.23	7.244	6.046
2.2	0.05335	0.00684	0.7994	3.090	25.85	1.638	11.92	4.222	6.111
2.3	0.06730	0.00685	0.6566	3.418	26.24	1.780	11.65	2.685	6.170
2.4	0.08354	0.00687	0.5470	3.678	26.63	1.886	11.40	2.375	6.228
2.5	0.01023	0.00690	0.4608	3.922	27.00	1.980	11.17	2.284	6.285
2.6	0.1237	0.00693	0.3923	4.161	27.37	2.068	10.96	2.320	6.344
2.7	0.1481	0.00695	0.3367	4.408	27.72	2.155	10.76	2.351	6.406
2.8	0.1755	0.00699	0.2913	4.662	28.06	2.240	10.57	2.403	6.470
2.9	0.2063	0.00703	0.2537	4.923	28.38	2.324	10.39	2.486	6.540
3.0	0.2405	0.00707	0.2223	5.195	28.69	2.408	10.22	2.597	6.616
3.1	0.2784	0.00713	0.1958	5.483	28.98	2.494	10.05	2.740	6.700
3.2	0.3201	0.00717	0.1728	5.787	29.26	2.581	9.90	2.896	6.792
3.3	0.3659	0.00723	0.1542	6.108	29.52	2.670	9.747	3.061	6.897
3.4	0.4159	0.00728	0.1376	6.448	29.76	2.780	9.600	3.273	7.015
3.5	0.4704	0.00735	0.1232	6.806	29.97	2.852	9.458	3.413	7.150
3.6	0.5296	0.00742	0.1107	7.184	30.17	2.946	9.318	3.601	7.305
3.7	0.5935	0.00749	0.0997	7.581	30.34	3.042	9.181	3.801	7.484
3.8	0.6625	0.00758	0.0900	7.998	30.48	3.140	9.046	4.017	7.694
3.9	0.7366	0.00766	0.0814	8.437	30.60	3.239	8.911	4.254	7.942
4.0	0.8162	0.00776	0.0738	8.899	30.68	3.341	8.776	4.519	8.238
4.1	0.9014	0.00786	0.0669	9.387	30.73	3.444	8.641	4.820	8.641
4.2	0.9923	0.00797	0.0606	9.901	30.74	3.551	8.504	5.170	9.033
4.3	1.089	0.00810	0.0550	10.45	30.71	3.661	8.363	5.587	9.58
4.4	1.193	0.00824	0.0499	11.02	30.62	3.775	8.218	6.097	10.29
4.5	1.303	0.00841	0.0452	11.64	30.47	3.893	8.067	6.742	11.22
4.6	1.419	0.00860	0.0408	12.31	30.24	4.018	7.906	7.590	12.50
4.7	1.543	0.00881	0.0367	13.04	29.91	4.151	7.732	8.763	14.37
4.8	1.674	0.00907	0.0329	13.85	29.45	4.296	7.539	10.51	17.32
4.9	1.813	0.00941	0.0291	14.76	28.80	4.458	7.327	13.38	22.64
5.0	1.960	0.00986	0.0252	15.85	27.83	4.649	7.041	19.02	34.93
5.1	2.116	0.01056	0.0207	17.26	26.08	4.898	6.624	34.60	95.84
5.195ᶜ	2.275	0.01436	0.0145						

c = critical pt.

From Arp, V. D. and R. D. McCarty, N.I.S.T. TN 1334, 1989 (142 pp.).

TABLE 2-265 Superheated Helium*

P, bars		0	100	200	300	Temp., °C 400	500	600	800	1000
1	v	5.677	7.754	9.831	11.908	13.985	16.063	18.140	22.294	26.448
	h	0.327	519.6	1039	1558	2078	2597	3116	4155	5193
	s	0.0116	1.620	2.853	3.849	4.684	5.403	6.035	7.106	7.993
5	v	1.138	1.553	1.968	2.384	2.799	3.215	3.630	4.461	5.291
	h	1.636	520.9	1040	1560	2079	2598	3117	4156	5194
	s	−3.343	−1.723	−0.490	0.506	1.341	2.060	2.692	3.763	4.650
10	v	0.5704	0.780	0.986	1.193	1.401	1.609	1.816	2.232	2.647
	h	3.272	522.5	1042	1561	2080	2600	3119	4157	5196
	s	−4.782	−3.162	−1.929	−0.934	−0.098	0.621	1.252	2.323	3.211
20	v	0.2867	0.3904	0.4942	0.5979	0.7017	0.9093	1.1169	1.3245	1.8435
	h	6.544	525.8	1045	1564	2083	2603	3122	4160	5199
	s	−6.221	−4.601	−3.368	−2.373	−1.537	−0.818	−0.187	0.884	1.771
50	v	0.1164	0.1579	0.1993	0.2408	0.2822	0.3257	0.3652	0.4481	0.5311
	h	16.360	535.5	1055	1574	2093	2612	3131	4169	5207
	s	−8.121	−6.501	−5.268	−4.273	−3.438	−2.719	−2.088	−1.017	−0.130
100	v	0.0597	0.0803	0.1010	0.1217	0.1424	0.1631	0.1838	0.2252	0.2666
	h	37.720	551.7	1071	1590	2108	2627	3146	4184	5222
	s	−9.555	−7.936	−6.703	−5.709	−4.874	−4.155	−3.524	−2.454	−1.567
150	v	0.0407	0.0545	0.0682	0.0820	0.0958	0.1095	0.1233	0.1509	0.1785
	h	49.080	567.9	1087	1605	2124	2643	3161	4199	5236
	s	−10.391	−8.773	−7.541	−6.546	−5.712	−4.994	−4.363	−3.293	−2.407
200	v	0.0312	0.0416	0.0518	0.0622	0.0725	0.0828	0.0931	0.1137	0.1344
	h	65.440	584.1	1103	1621	2140	2658	3176	4213	5250
	s	−10.983	−9.635	−8.134	−7.139	−6.306	−5.588	−4.957	−3.888	−3.002

*Extracted from Tsederberg, Popov, et al., *Thermodynamic and Thermophysical Properties of Helium*, Atomizdat, Moscow, 1969, and NBS-NSF TT 50096, 1971. Copyright material. Reproduced by permission. This source contains entries for many more temperatures and pressures than can be reproduced here. v = volume, m³/kg; h = enthalpy, kJ/kg; s = entropy, kJ/(kg·K).

The 1993 ASHRAE *Handbook—Fundamentals* (SI ed.) has a thermodynamic chart for pressures from 0.1 to 50 bar and temperatures from 2.5 to 15 K. Saturation and superheat tables to 9000 psia, 800°R; and a chart to 700 psia, 40°R appear in Stewart, R. B., R. T. Jacobsen, et al., *Thermodynamic Properties of Refrigerants*, ASHRAE, Atlanta, GA, 1986 (521 pp.). For specific heat, thermal conductivity, and viscosity, see *Thermophysical Properties of Refrigerants*, ASHRAE, 1993. A useful compilation of properties is given by Betts, D. S., *Cryogenics*, **16**, 1 (1976): 3–16. A 32-term equation of state for the range up to 20,000 bar, 2–1500 K is given by McCarty, R. D. and V. D. Arp, *Advan. Cryog. Engng.*, **35** (1990): 1465–1475.

TABLE 2-266 Helium⁴ Gas at Atmospheric Pressure*

T, K	4.224	5	10	20	30	40	50	75	100	200	300	400	500	600	800	1000
v, m³/kg	0.0591	0.0834	0.1612	0.4094	0.6161	0.8218	1.0273	1.5403	2.053	4.102	6.154	8.191	10.24	12.31	16.40	20.50
h, kJ/kg	30.30	36.18	64.91	117.95	170.24	222.4	274.4	404.4	534.2	1054	1573	2092	2612	3131	4170	5208
s, kJ/(kg·K)	8.327	9.614	13.369	17.321	19.442	20.94	22.10	24.21	25.71	29.30	31.41	32.90	34.06	35.01	36.50	37.66

*From McCarty, NBS Rep. 9762, 1970. Reproduced by permission. The source contains values for further temperatures and for other functions, usually to additional significant figures.

TABLE 2-267 Saturated n-Heptane*

T, K	P, bar	v_f, m³/kg	v_g, m³/kg	h_f, kJ/kg	h_g, kJ/kg	s_f, kJ/(kg·K)	s_g, kJ/(kg·K)	c_{pf}, kJ/(kg·K)	μ_f, 10^{-4} Pa·s	k_f, W/(m·K)
182.6ᵗ		1.292.–3		284.1		2.260		2.025	39.4	0.150
200	0.00002	1.316.–3		319.4	722.6	2.441	4.457	2.011	21.0	0.148
220	0.00019	1.344.–3		359.7	757.1	2.636	4.442	2.026	12.6	0.145
240	0.00133	1.374.–3		400.5	791.4	2.814	4.443	2.063	8.52	0.142
250	0.00303	1.389.–3		421.3	808.3	2.899	4.447	2.088	7.23	0.140
260	0.00635	1.405.–3		442.3	824.9	2.981	4.453	2.117	6.52	0.137
270	0.01316	1.422.–3		463.6	841.2	3.061	4.460	2.147	5.46	0.135
280	0.02347	1.440.–3		485.2	857.8	3.140	4.471	2.180	4.83	0.132
290	0.03997	1.457.–3		507.2	874.8	3.217	4.485	2.216	4.29	0.129
300	0.06674	1.475.–3	3.744	529.6	891.9	3.293	4.501	2.252	3.85	0.126
310	0.1070	1.494.–3	2.412	552.3	908.9	3.367	4.517	2.291	3.48	0.123
320	0.1656	1.514.–3	1.596	575.4	926.0	3.441	4.537	2.329	3.17	0.121
330	0.2461	1.534.–3	1.101	598.8	943.3	3.513	4.557	2.370	2.89	0.119
340	0.3614	1.555.–3	0.7650	622.8	961.2	3.584	4.579	2.412	2.66	0.116
350	0.5130	1.578.–3	0.5510	647.0	979.1	3.655	4.604	2.454	2.45	0.114
360	0.712	1.601.–3	0.4058	671.9	997.5	3.725	4.629	2.500	2.24	0.111
370	0.967	1.625.–3	0.3036	697.1	1016.1	3.794	4.656	2.548	2.04	0.109
371.6	1.013	1.629.–3	0.2904	701.9	1019.8	3.805	4.660	2.556	2.01	0.108
380	1.289	1.651.–3	0.2308	723.9	1035.4	3.864	4.684	2.60	1.86	0.107
390	1.689	1.678.–3	0.1781	750.4	1054.2	3.932	4.711	2.65	1.71	0.105
400	2.180	1.708.–3	0.1388	777.2	1073.2	4.000	4.740	2.70	1.58	0.103
420	3.471	1.775.–3	0.0734					2.81	1.35	0.099
440	5.268	1.853.–3	0.0576					2.93	1.15	0.095
460	7.691	1.954.–3	0.0389					3.05	0.97	0.091
480	10.92	2.065.–3	0.0265					3.19	0.82	0.087
500	15.10	2.235.–3	0.0178					3.38	0.67	0.080
520	20.43	2.52.–3						3.7		
540.1ᶜ	27.35	4.3.–3	0.0043							

*Values of P and v interpolated and converted from tables in Vargaftik, *Handbook of Thermophysical Properties of Gases and Liquids,* Hemisphere, Washington, and McGraw-Hill, New York, 1975. Values of h and s calculated from API tables published by the Thermodynamics Research Center, Texas A&M University, College Station. t = triple point; c = critical point.

Saturation and superheat tables and a diagram to 200 bar, 680 K are given by Reynolds, W. C., *Thermodynamic Properties in S.I.,* Stanford Univ. publ., 1979 (173 pp.).

TABLE 2-268 Hexane

Saturation and superheat tables and a diagram to 100 bar, 680 K, are given by Reynolds, W. C., *Thermodynamic Properties in S.I.,* Stanford Univ. publ., 1979 (173 pp.).

TABLE 2-269 Saturated Hydrazine

Temperature, K	Pressure, bar	v_f, m³/kg	v_g, m³/kg	h_f, kJ/kg	h_g, kJ/kg	s_f, kJ/(kg·K)	s_g, kJ/(kg·K)
386.6	1.013	0.001 053	0.9833	−105.9	65.4	0.5994	1.0426
390	1.135	0.001 060	0.8850	−104.8	66.0	0.6029	1.0409
400	1.560	0.001 081	0.6579	−101.4	68.2	0.6120	1.0360
410	2.102	0.001 104	0.4994	−97.6	70.6	0.6211	1.0314
420	2.786	0.001 127	0.3850	−93.9	73.0	0.6300	1.0275
440	4.732	0.001 178	0.2355	−86.1	77.6	0.6492	1.0212
460	7.610	0.001 235	0.1500	−76.9	82.1	0.6707	1.0163
480	11.76	0.001 299	0.1005	−67.1	86.6	0.6916	1.0118
500	17.42	0.001 374	0.0690	−57.3	90.8	0.7124	1.0086
520	29.59	0.001 460	0.0407	−47.8	94.6	0.7320	1.0058
540	34.75	0.001 563	0.0353	−36.0	97.7	0.7566	1.0042
560	47.09	0.001 681	0.0263	−25.2	101.2	0.7762	1.0020
580	62.44	0.001 835	0.0196	−12.4	103.6	0.8002	1.0002
600	81.17	0.002 045	0.0142	5.2	104.2	0.8335	0.9988
620	102.7	0.002 320	0.0106	23.2	103.6	0.8671	0.9967
640	128.1	0.002 86	0.0074	45.9	98.1	0.9035	0.9906
653^c	146.9	0.004 33	0.0043	83.7	83.7	0.9715	0.9715

Converted from E. F. Fricke, Republic Aviation Co. rept. F-5028-101. c = critical point.

TABLE 2-270 Saturated n-Hydrogen*

T, K	P, bar	v_f, m³/kg	v_g, m³/kg	h_f, kJ/kg	h_g, kJ/kg	s_f, kJ/(kg·K)	s_g, kJ/(kg·K)	c_{pf}, kJ/(kg·K)	μ_f, 10⁻⁴ Pa·s	k_f, W/(m·K)
13.95^t	0.072	0.01298	7.974	218.3	667.4	14.079	46.635	6.36	0.255	0.073
14	0.074	0.01301	7.205	219.6	669.3	14.173	46.301	6.47	0.248	0.075
15	0.127	0.01316	4.488	226.4	678.2	14.640	44.763	6.91	0.218	0.083
16	0.204	0.01332	2.954	233.8	686.7	15.104	43.418	7.36	0.194	0.089
17	0.314	0.01348	2.032	241.6	694.7	15.568	42.227	7.88	0.175	0.093
18	0.461	0.01366	1.449	249.9	702.1	16.032	41.158	8.42	0.159	0.095
19	0.654	0.01387	1.064	258.8	708.8	16.498	40.188	8.93	0.146	0.097
20	0.901	0.01407	0.8017	268.3	714.8	16.966	39.299	9.45	0.135	0.098
21	1.208	0.01430	0.6177	278.4	720.2	17.440	38.485	10.13	0.125	0.100
22	1.585	0.01455	0.4828	289.2	724.4	17.919	37.710	10.82	0.116	0.101
23	2.039	0.01483	0.3829	300.8	727.6	18.405	36.973	11.69	0.108	0.101
24	2.579	0.01515	0.3072	313.3	729.8	18.901	36.266	12.52	0.101	0.101
25	3.213	0.01551	0.2489	326.7	730.7	19.408	35.579	13.44	0.094	0.100
26	3.950	0.01592	0.2032	341.2	730.2	19.929	34.900	14.80	0.088	0.098
27	4.800	0.01639	0.1667	357.0	728.0	20.473	34.221	16.17	0.082	0.096
28	5.770	0.01696	0.1370	374.3	723.7	21.041	33.524	18.48	0.076	0.094
29	6.872	0.01765	0.1125	393.6	716.6	21.650	32.795	22.05	0.070	0.091
30	8.116	0.01854	0.0919	415.4	705.9	22.315	32.002	26.59	0.065	0.087
31	9.510	0.01977	0.0738	441.3	689.7	23.075	31.091	36.55	0.058	0.086
32	11.07	0.02174	0.0571	474.7	663.2	24.032	29.926	65.37	0.051	0.092
33.18^c	13.13	0.03182	0.0318	565.4	565.4	26.680	26.680	∞		∞

*Values extracted and occasionally rounded off from McCarty, Hord, and Roder, NBS Monogr. 168, 1981. t = triple point; c = critical point.

TABLE 2-271   Compressed *n*-Hydrogen*

Pressure, bar		Temperature, K									
		15	20	30	40	50	60	80	100	150	200
0.1	v	6.076	8.176	12.333	16.473	20.606	24.736	32.991	41.244	61.870	82.495
	h	679.2	731.6	835.5	938.9	1042.3	1146	1356	1575	2172	2826
	s	46.02	49.04	53.25	56.23	58.53	60.43	63.45	65.89	70.68	74.46
1	v	0.0131	0.0141	1.196	1.625	2.046	2.463	3.295	4.123	6.190	8.254
	h	227.3	268.3	826.0	932.7	1037.9	1143	1354	1574	2172	2826
	s	14.62	16.96	43.56	46.63	48.98	50.89	53.93	56.38	61.17	64.96
5	v	0.0131	0.0140	0.2006	0.3039	0.3958	0.4839	0.6553	0.8238	1.241	1.655
	h	231.7	272.1	775.0	903.4	1017.6	1128	1345	1568	2170	2826
	s	14.57	16.88	35.80	39.52	42.07	44.07	47.20	49.68	54.66	58.31
10	v	0.0130	0.0138	0.0181	0.1376	0.1895	0.2366	0.3255	0.4116	0.6221	0.8303
	h	237.2	277.0	412.1	861.8	991.1	1109	1334	1560	2167	2826
	s	14.50	16.77	22.09	35.95	38.85	40.99	44.23	46.75	51.63	55.44
20	v	0.0129	0.0136	0.0167	0.0521	0.0866	0.1135	0.1611	0.2057	0.3129	0.4179
	h	248.2	286.9	406.5	752.0	934.7	1070	1312	1546	2163	2826
	s	14.37	16.58	21.33	31.07	35.19	37.67	41.15	43.76	48.71	52.55
40	v		0.0133	0.0155	0.0216	0.0376	0.0533	0.0796	0.1033	0.1586	0.2119
	h		307.3	413.5	589.3	823.5	997	1271	1521	2155	2826
	s		16.26	20.50	25.49	30.73	33.91	37.87	40.65	45.75	49.64
60	v		0.0130	0.0147	0.0182	0.0254	0.0351	0.0532	0.0697	0.1073	0.1433
	h		328.0	427.2	570.1	757.0	940	1237	1499	2149	2828
	s		15.98	19.95	24.03	28.19	31.54	35.82	38.76	43.99	47.92
80	v		0.0127	0.0142	0.0167	0.0211	0.0273	0.0406	0.0531	0.0818	0.1090
	h		348.9	443.5	572.3	732.8	905	1210	1482	2146	2831
	s		15.74	19.53	23.21	26.78	29.93	34.34	37.37	42.72	46.69
100	v		0.0125	0.0138	0.0158	0.0190	0.0233	0.0335	0.0434	0.0666	0.0885
	h		369.8	461.1	581.5	727.4	888	1192	1469	2144	2835
	s		15.53	19.19	22.63	25.88	28.80	33.19	36.28	41.73	45.73
200	v		0.0117	0.0125	0.0136	0.0150	0.0167	0.0207	0.0253	0.0368	0.0480
	h		474.4	556.1	658.7	776.9	908	1182	1458	2156	2869
	s		14.71	17.99	20.93	23.56	25.94	29.88	32.97	38.59	42.72
400	v			0.0113	0.0119	0.0126	0.0134	0.0151	0.0171	0.0225	0.0279
	h			751.0	841.9	945.4	1059	1303	1560	2249	2973
	s			16.59	19.20	21.50	23.58	27.07	29.94	35.48	39.67
600	v			0.0106	0.0110	0.0115	0.0120	0.0131	0.0144	0.0178	0.0214
	h			941.5	1027	1124	1231	1463	1709	2385	3107
	s			15.68	18.14	20.29	22.24	25.57	28.31	33.74	37.92
800	v				0.0104	0.0107	0.0111	0.0120	0.0130	0.0155	0.0181
	h				1209	1302	1405	1628	1870	2535	3255
	s				17.35	19.43	21.30	24.50	27.20	32.54	36.70
1000	v				0.0099	0.0102	0.0106	0.0112	0.0120	0.0140	0.0160
	h				1387	1478	1578	1796	2032	2692	3403
	s				16.72	18.75	20.58	23.70	26.33	31.63	35.75

*Values extracted and sometimes rounded off from the tables of McCarty, Hord, and Roder, NBS Monogr. 168, 1981. This source contains an exhaustive tabulation of property values for both the normal and the para forms of hydrogen. v = specific volume, m^3/kg; h = specific enthalpy, kJ/kg; s = specific entropy, $kJ/(kg \cdot K)$.

The 1993 ASHRAE *Handbook—Fundamentals* (SI ed.) has a thermodynamic diagram for 0.1 to 500 bar for temperatures up to 100 K. Tables and a Mollier chart from 10^{-4} to 1000 atm, 300–20,000 K are given by Kubin, R. F. and L. L. Presley, NASA SP 3002, 1964. Liebenberg, D. H., R. L. Mills, and others, in LA-6645-MS, 1977 (26 pp.), give properties from 75 to 307 K for pressures from 2 to 20 kbar. See also Baker, J. R. and H. F. Swift, *J. Appl. Phys.*, **43**, 3 (1972): 950–953. An extensive collection of data for H_2, D_2, T_2, and so on below 30 K is given by Souers, P. C., UCRL 52628, 1979 (91 pp.); and for temperatures below 40 K by Roder, H. M., G. E. Childs, et al., NBS TN 641, 1973 (114 pp.).

Saturation and superheat tables to 10,000 psia, 900°R and a chart to 180°R appear in Stewart, R. B., R. T. Jacobsen, et al., *Thermodynamic Properties of Refrigerants*, ASHRAE, Atlanta, GA, 1986 (521 pp.). For viscosity, thermal conductivity, and specific heat, see *Thermophysical Properties of Refrigerants*, ASHRAE, 1993.

					Temperature, K					
250	300	350	400	450	500	600	700	800	900	1000
103.12	123.23	144.35	164.97	185.60	206.22	247.46	288.70	329.94	371.18	412.43
3517	4227	4945	5668	6393	7118	8571	10028	11493	12969	14458
77.53	80.13	82.34	84.27	85.98	87.51	90.15	92.40	94.36	96.10	97.66
10.32	12.38	14.44	16.50	18.57	20.63	24.75	28.88	33.00	37.13	41.25
3517	4227	4946	5669	6393	7118	8571	10029	11494	12969	14459
68.03	70.63	72.85	74.78	76.48	78.01	80.66	82.91	84.86	86.60	88.17
2.069	2.482	2.895	3.307	3.720	4.132	4.957	5.782	6.607	7.432	8.257
3518	4229	4948	5671	6396	7121	8574	10032	11497	12973	14462
61.39	63.99	66.21	68.14	69.84	71.37	74.02	76.27	78.23	79.96	81.53
1.038	1.245	1.451	1.658	1.864	2.070	2.483	2.896	3.308	3.720	4.133
3519	4231	4951	5674	6399	7125	8578	10036	11501	12977	14467
58.52	61.12	63.34	65.28	66.98	68.51	71.16	73.41	75.37	77.10	78.67
0.522	0.6259	0.7294	0.8328	0.9361	1.040	1.246	1.452	1.658	1.865	2.071
3522	4235	4956	5680	6406	7132	8586	10044	11509	12985	14475
55.65	58.26	60.48	62.41	64.12	65.65	68.30	70.55	72.51	74.24	75.81
0.2644	0.3166	0.3685	0.4204	0.4721	0.5238	0.6271	0.7303	0.8335	0.9366	1.040
3527	4244	4967	5692	6419	7146	8601	10059	11525	13002	14492
52.76	55.38	57.61	59.55	61.26	62.79	65.44	67.69	69.65	71.39	72.95
0.1786	0.2136	0.2483	0.2829	0.3174	0.3519	0.4209	0.4897	0.5585	0.6273	0.6961
3533	4253	4978	5705	6432	7160	8616	10075	11542	13018	14508
51.05	53.69	55.92	57.86	59.58	61.11	63.76	66.02	67.97	70.51	71.28
0.1357	0.1621	0.1882	0.2142	0.2401	0.2660	0.3177	0.3694	0.4120	0.4726	0.5242
3540	4263	4989	5718	6446	7174	8631	10091	11558	13035	14525
49.84	52.49	54.73	56.67	58.39	59.92	62.57	64.83	66.79	68.52	70.09
0.1099	0.1312	0.1521	0.1730	0.1937	0.2145	0.2559	0.2972	0.3385	0.3798	0.4211
3547	4273	5001	5731	6460	7189	8647	10107	11574	13051	14542
48.89	51.55	53.79	55.74	57.46	59.00	61.65	63.90	65.87	67.60	69.17
0.0588	0.0695	0.0801	0.0905	0.1001	0.1114	0.1321	0.1528	0.1734	0.1941	0.2147
3594	4329	5064	5798	6531	7263	8724	10187	11656	13134	14625
45.94	48.62	50.89	52.85	54.58	56.12	58.78	61.04	63.00	64.74	66.31
0.0334	0.0388	0.0441	0.0493	0.0545	0.0597	0.0701	0.0804	0.0908	0.1011	0.1114
3716	4458	5202	5943	6681	7416	8883	10349	11820	13300	14792
42.98	45.68	47.97	49.95	51.69	53.24	55.91	58.17	60.14	61.88	63.45
0.0249	0.0285	0.0321	0.0355	0.0390	0.0425	0.0494	0.0562	0.0631	0.0700	0.0768
3854	4600	5349	6095	6836	7574	9045	10513	11985	13466	14958
41.24	43.95	46.26	48.26	50.00	51.56	54.24	56.50	58.47	60.21	61.78
0.0207	0.0234	0.0260	0.0286	0.0312	0.0338	0.0390	0.0441	0.0492	0.0543	0.0594
4003	4748	5501	6249	6993	7734	9207	10677	12150	13631	15124
40.03	42.73	45.05	47.05	48.81	50.37	53.05	55.32	57.29	59.03	60.60
0.0181	0.0202	0.0223	0.0244	0.0265	0.0286	0.0327	0.0367	0.0408	0.0449	0.0490
4156	4898	5654	6405	7151	7893	9370	10842	12316	13797	15289
39.10	41.79	44.12	46.12	47.88	49.45	52.14	54.41	56.38	58.12	59.69

TABLE 2-272 Saturated *para*-Hydrogen*

T, K	P, bar	v_f, m³/kg	v_g, m³/kg	h_f, kJ/kg	h_g, kJ/kg	s_f, kJ/(kg·K)	s_g, kJ/(kg·K)	c_{pf}, kJ/(kg·K)	μ_f, 10^{-4} Pa·s	k_f, W/(m·K)
13.8^t	0.070	0.0130	7.97	−308.9	140.3	4.97	37.52	6.37	0.255	0.073
14	0.079	0.0130	7.20	−307.6	142.1	5.06	37.19	6.47	0.248	0.075
15	0.134	0.0132	4.49	−300.9	151.1	5.53	36.65	6.91	0.218	0.082
16	0.216	0.0133	2.96	−293.4	159.6	5.99	34.31	7.36	0.194	0.089
17	0.329	0.0135	2.03	−285.6	167.6	6.45	33.11	7.88	0.175	0.092
18	0.482	0.0137	1.45	−277.3	175.0	6.92	32.05	8.42	0.159	0.095
19	0.682	0.0139	1.07	−268.4	181.7	7.38	31.08	8.93	0.146	0.097
20	0.935	0.0141	0.802	−258.9	187.7	7.85	30.19	9.45	0.135	0.098
21	1.250	0.0143	0.618	−248.8	193.0	8.32	29.37	10.13	0.125	0.100
22	1.634	0.0146	0.483	−237.9	197.3	8.80	28.60	10.82	0.116	0.101
23	2.096	0.0148	0.383	−226.3	200.5	9.29	27.86	11.69	0.108	0.101
24	2.645	0.0152	0.307	−213.9	202.7	9.78	27.15	12.52	0.101	0.100
25	3.288	0.0155	0.249	−200.4	203.6	10.29	26.46	13.44	0.094	0.099
26	4.035	0.0159	0.203	−185.9	203.1	10.81	25.79	14.81	0.088	0.098
27	4.892	0.0164	0.167	−170.2	200.9	11.36	25.11	16.18	0.082	0.096
28	5.88	0.0170	0.137	−152.9	196.5	11.93	24.41	18.5	0.076	0.094
29	6.98	0.0177	0.113	−133.6	189.5	12.54	23.68	22.1	0.070	0.091
30	8.23	0.0185	0.092	−111.7	178.8	13.20	22.89	26.6	0.065	0.087
31	9.63	0.0198	0.074	−85.8	162.6	13.96	21.98	36.6	0.058	0.088
32	11.20	0.0217	0.057	−52.4	136.1	14.92	20.81	65.4	0.051	0.092
33^c	12.93	0.0318	0.032	38.3	38.3	17.56	17.56	∞		∞

*Values extracted and occasionally rounded off from McCarty, Hord, and Roder, NBS *Monogr.* 168, 1981. *t* = triple point; *c* = critical point.
Saturation and superheat tables to 12,000 psia, 900°R and a chart to 180°R appear in Stewart, R. B., R. T. Jacobsen, et al., *Thermodynamic Properties of Refrigerants,* ASHRAE, Atlanta, GA, 1986 (521 pp.). For specific heat, thermal conductivity, and viscosity, see *Thermophysical Properties of Refrigerants,* ASHRAE, 1993. The 1993 ASHRAE *Handbook—Fundamentals* (SI ed.) has a thermodynamic chart for pressures from 0.1 to 1000 bar for temperatures up to 100 K.

TABLE 2-273 Saturated Hydrogen Peroxide*

T, K	P, bar	v_f, m³/kg	v_g, m³/kg	h_f, kJ/kg	h_g, kJ/kg	s_f, kJ/(kg·K)	s_g, kJ/(kg·K)	c_{pf}, kJ/(kg·K)	μ_f, 10^{-4} Pa·s	k_f, W/(m·K)
273	0.0004	0.00068	1672	−5577	−4027	2.990	8.662	1.45	18.0	0.483
300	0.0031	0.00069	235	−5510	−3995	3.224	8.269	1.48	11.3	0.481
350	0.0564	0.00072	15.1	−5376	−3933	3.631	7.758	1.54	4.3	0.474
400	0.4521	0.00076	2.12	−5238	−3878	4.032	7.440	1.61	2.2	0.464
450	2.143	0.00081	0.487	−5091	−3820	4.346	7.172	1.68	1.3	0.453
500	7.126	0.00088	0.155	−4945	−3777	4.656	6.992	1.75	0.89	0.443
550	18.56	0.00095	0.0605	−4794	−3745	4.941	6.846	1.82	0.65	0.431
600	40.75	0.00107	0.0268	−4635	−3731	5.209	6.720	1.90	0.50	0.416
650	79.27	0.00125	0.0125	−4463	−3746	5.485	6.582			
700	141.7	0.00171	0.0048	−4195	−3860	5.682	6.339			
708.5^c	155.3	0.00284	0.0028	−4012	−4012	5.732	5.732			

*Values reproduced or converted from a tabulation by Tsykalo and Tabachnikov in V. A. Rabinovich (ed.), *Thermophysical Properties of Gases and Liquids,* Standartov, Moscow, 1968; NBS-NSF transl. TT 69-55091, 1970. The reader may be reminded that very pure hydrogen peroxide is very difficult to obtain owing to its decomposition or instability. *c* = critical point. The FMC Corp., Philadelphia, PA tech. bull. 67, 1969 (100 pp.) contains an enthalpy-pressure diagram to 3000 psia, 1100 K.

TABLE 2-274 Hydrogen Sulfide

West, J. R., *Chem. Eng. Progr.,* **44,** 4 (1948): 207–292 gives tables and a chart for the range 1–90 atm., −76 to 1300°F while properties from 10 to 330 bar, 300 to 500 K were tabulated by Lui, C-H., D. M. Bailey, et al., *Hydroc. Proc.,* **65,** 7 (July 1986): 41–43.

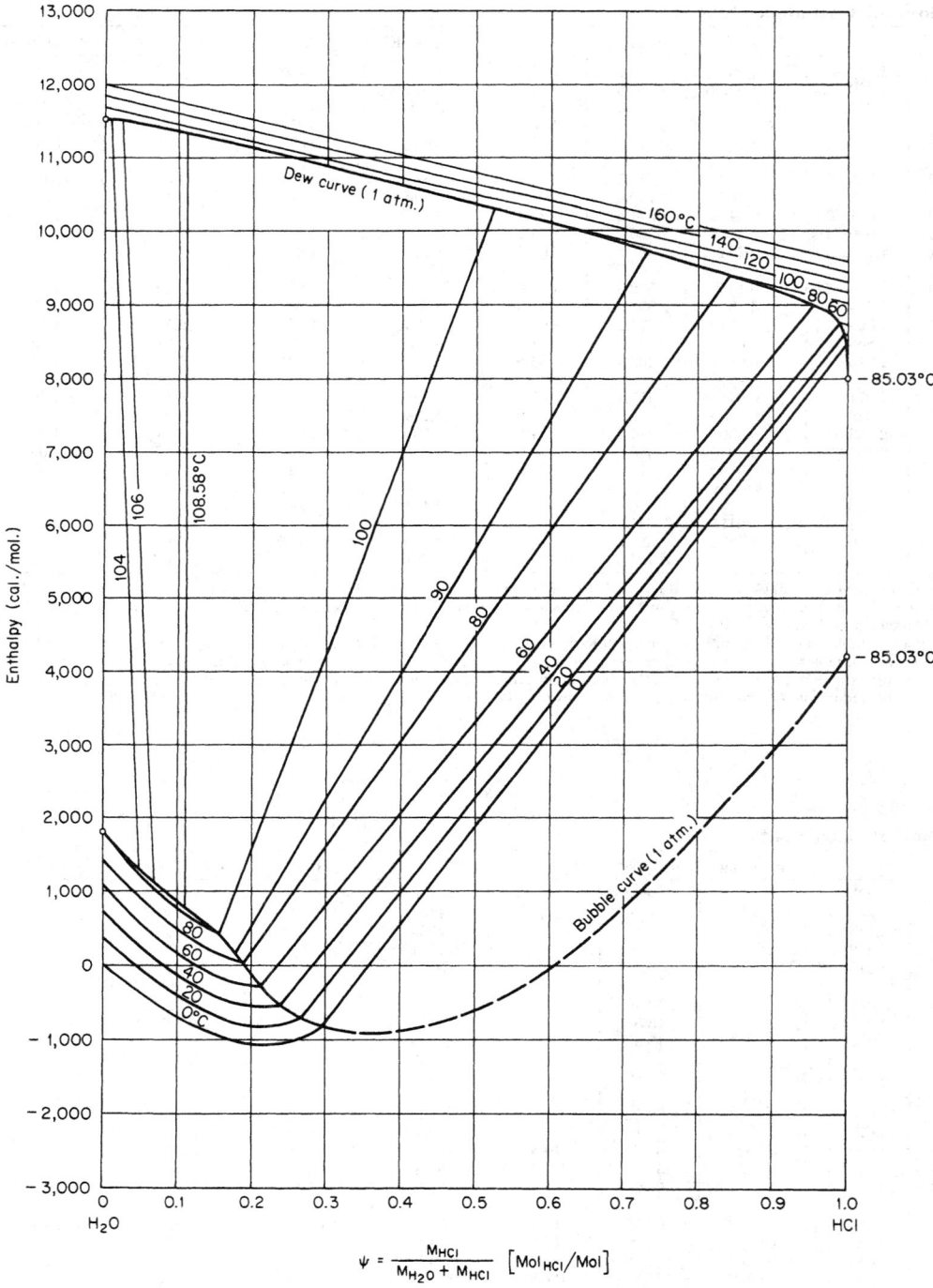

$$\psi = \frac{M_{HCl}}{M_{H_2O} + M_{HCl}} \quad \left[Mol_{HCl}/Mol\right]$$

FIG. 2-11 Enthalpy-concentration diagram for aqueous hydrogen chloride at 1 atm. Reference states: enthalpy of liquid water at 0°C is zero; enthalpy of pure saturated HCl vapor at 1 atm (−85.03°C) is 8000 kcal/mol. NOTE: It should be observed that the weight basis includes the vapor, which is particularly important in the two-phase region. Saturation values may be read at the ends of the tie lines. [*Van Nuys*, Trans. Am. Inst. Chem. Eng., **39,** *663 (1943).*]

TABLE 2-275   Saturated Isobutane (R600a)*

T, K	P, bar	v_f, m³/kg	v_g, m³/kg	h_f, kJ/kg	h_g, kJ/kg	s_f, kJ/(kg·K)	s_g, kJ/(kg·K)	c_{pf}, kJ/(kg·K)	μ_f, 10^{-4} Pa·s	k_f, W/(m·K)
113.6ᵗ	1.9.−7	1.349.−3	8.60.+6	0.0	485.3	1.863	6.136			
120	9.3.−7	1.360.−3	1.84.+6	11.0	491.1	1.957	5.957	1.78		
140	4.8.−5	1.396.−3	4210	46.0	510.1	2.226	5.541	1.87		0.163
160	8.2.−4	1.435.−3	278.2	82.1	530.8	2.467	5.272	1.93		0.158
180	0.0070	1.476.−3	36.66	119.5	533.0	2.688	5.097	1.99	9.46	0.149
200	0.0369	1.520.−3	7.723	158.5	576.7	2.893	4.984	2.05	6.06	0.142
220	0.1374	1.568.−3	2.265	199.0	601.5	3.086	4.916	2.12	4.21	0.134
240	0.3989	1.621.−3	0.8432	241.4	627.4	3.270	4.878	2.19	3.11	0.127
260	0.9600	1.680.−3	0.3738	285.8	654.2	3.446	4.863	2.28	2.40	0.120
270	1.4081	1.712.−3	0.2617	308.8	667.7	3.532	4.861	2.33	2.14	0.117
280	2.0020	1.746.−3	0.1882	332.3	681.3	3.617	4.863	2.39	1.93	0.113
290	2.7686	1.784.−3	0.1385	356.4	694.9	3.700	4.867	2.46	1.75	0.110
300	3.7365	1.824.−3	0.1040	381.1	708.4	3.783	4.874	2.53	1.59	0.106
310	4.934	1.868.−3	0.0794	406.4	721.7	3.865	4.882	2.61	1.46	0.102
320	6.392	1.916.−3	0.0614	432.4	734.8	3.946	4.891	2.70	1.35	0.099
330	8.140	1.971.−3	0.0481	459.2	747.7	4.028	4.902	2.81	1.25	0.095
340	10.21	2.032.−3	0.0380	486.9	760.0	4.109	4.912	2.92	1.15	0.092
350	12.64	2.103.−3	0.0301	515.7	771.8	4.191	4.923	3.04	1.05	0.088
360	15.46	2.187.−3	0.0240	545.6	782.7	4.273	4.932	3.17	0.95	0.083
370	18.72	2.289.−3	0.0190	577.1	792.3	4.357	4.939	3.31	0.85	0.080
380	22.48	2.420.−3	0.0150	610.6	799.8	4.444	4.942	3.45	0.75	0.076
390	26.82	2.604.−3	0.0115	647.1	803.7	4.536	4.937	3.62	0.63	0.071
400	31.86	2.920.−3	0.0083	689.6	799.6	4.639	4.915	3.85	0.51	0.065
408.0ᶜ	36.55	4.464.−3	0.0045	752.5	752.5	4.791	4.791	∞		∞

*Values reproduced or converted from Goodwin, NBSIR 79-1612, 1979. t = triple point; c = critical point. The notation 1.9.−7 signifies 1.9×10^{-7}.
Slightly different values for the range 0.5 to 34.5 bar, 250–404 K appear in Waxman, M. and J. S. Gallagher, *J. Chem. Eng. Data,* **28,** (1983): 224–241. This source also contains superheat tables for 1–400 bar, 250–600 K.
Saturation and superheat tables and a diagram to 200 bar, 600 K are given by Reynolds, W. C., *Thermodynamic properties in S.I.,* Stanford Univ. publ., 1979 (173 pp.). Saturation and superheat tables and a chart to 10,000 psia, 640°F appear in Stewart, R. B., R. T. Jacobsen, et al., *Thermodynamic Properties of Refrigerants,* ASHRAE, Atlanta, GA, 1986 (521 pp.). For specific heat, thermal conductivity, and viscosity, see *Thermophysical Properties of Refrigerants,* ASHRAE, 1993. Equations and data for thermal conductivity and viscosity are given by Nieuwoldt, J. C., B. LeNeindre, et al., *J. Chem. Eng. Data,* **32,** (1987): 1–8.

TABLE 2-276   Saturated Krypton*

T, K	P, bar	v_f, m³/kg	v_g, m³/kg	h_f, kJ/kg	h_g, kJ/kg	s_f, kJ/(kg·K)	s_g, kJ/(kg·K)	c_{pf}, kJ/(kg·K)	μ_f, 10^{-4} Pa·s	k_f, W/(m·K)
10		3.235.−4		0.22		0.0256		0.070		
20		3.246.−4		1.59		0.1141		0.188		
30		3.265.−4		3.84		0.2034		0.247		
40		3.288.−4		6.49		0.2791		0.276		
50		3.313.−4		9.37		0.3431		0.295		
60		3.341.−4		12.40		0.3982		0.311		
70		3.372.−4		15.57		0.4471		0.327		
80		3.407.−4		18.97		0.4925		0.345		
90		3.446.−4		22.58		0.5353		0.366		
100		3.492.−4		26.42		0.5765		0.389		
110		3.544.−4		30.52		0.6165		0.414		
115.76		3.579.−4		33.18		0.6390		0.427		
115.76	0.732	4.090.−4	0.1529	52.78	161.8	0.8095	1.751	0.547		
119.76	1.013	4.143.−4	0.1136	54.99	162.6	0.8279	1.726	0.545		
120	1.032	4.146.−4	0.1116	55.09	162.6	0.8291	1.724	0.544	3.72	0.0900
130	2.112	4.284.−4	0.0578	60.55	164.1	0.8724	1.669	0.542	3.16	0.0828
140	3.878	4.440.−4	0.0330	66.02	165.3	0.9124	1.622	0.546	2.64	0.0756
150	6.552	4.619.−4	0.0201	71.58	166.1	0.9499	1.580	0.559	2.20	0.0688
160	10.37	4.831.−4	0.0130	77.34	166.4	0.9859	1.543	0.587	1.87	0.0625
170	15.57	5.091.−4	0.0086	83.48	166.0	1.022	1.507	0.641	1.54	0.0558
180	22.41	5.423.−4	0.0059	90.26	164.6	1.058	1.472	0.734	1.28	0.0494
190	31.20	5.882.−4	0.0040	98.19	161.8	1.098	1.433	0.905	1.05	0.0433
200	42.23	6.641.−4	0.0026	108.40	156.0	1.147	1.386	1.515	0.80	0.0348
209.39	54.96	1.098.−3	0.0011	133.90	133.9	1.262	1.262	∞		∞

*Values extracted and in some cases rounded off from those cited in Rabinovich (ed.), *Thermophysical Properties of Neon, Argon, Krypton and Xenon,* Standards Press, Moscow, 1976. This source contains values for the compressed state for pressures up to 1000 bar, etc. The notation 3.235.−4 signifies 3.235×10^{-4}. This book was published in English translation by Hemisphere, New York, 1988 (604 pp.).

TABLE 2-277 Compressed Krypton*

Temperature, K		Pressure, bar											
		1	10	20	40	60	80	100	200	400	600	800	1000
100	v	3.49.−4	3.49.−4	3.49.−4	3.48.−4	3.48.−4	3.47.−4	3.47.−4	3.45.−4	3.42.−4	3.39.−4	3.36.−4	3.33.−4
	h	26.42	26.69	26.99	27.59	28.18	28.78	29.38	32.38	38.42	44.47	50.52	56.57
	s	0.5765	0.5760	0.5755	0.5745	0.5735	0.5724	0.5714	0.5667	0.5580	0.5503	0.5432	0.5366
200	v	0.1971	0.0184	8.39.−3	3.00.−3	6.19.−4	5.94.−4	5.76.−4	5.27.−4	4.83.−4	4.58.−4	4.41.−4	4.28.−4
	h	183.1	179.3	174.5	159.4	105.6	104.4	103.6	102.7	105.3	109.7	114.9	120.3
	s	1.859	1.618	1.533	1.405	1.129	1.116	1.106	1.073	1.037	1.013	0.993	0.977
300	v	0.2971	0.0292	0.0143	6.84.−3	4.37.−3	3.14.−3	2.41.−3	1.09.−3	6.92.−4	5.94.−4	5.44.−4	5.13.−4
	h	208.1	206.3	204.2	200.0	195.7	191.2	186.6	166.8	155.1	155.2	158.2	162.3
	s	1.961	1.728	1.654	1.575	1.525	1.485	1.451	1.333	1.239	1.196	1.169	1.149
400	v	0.3966	0.0394	0.0196	9.67.−3	6.37.−3	4.73.−3	3.75.−3	1.85.−3	1.01.−3	7.79.−4	6.76.−4	6.14.−4
	h	233.0	231.9	230.7	228.3	225.9	223.6	221.3	211.4	199.7	196.8	197.9	200.8
	s	2.032	1.802	1.730	1.657	1.612	1.579	1.552	1.463	1.368	1.317	1.284	1.259
500	v	0.4960	0.0495	0.0247	0.0123	8.20.−3	6.15.−3	4.91.−3	2.49.−3	1.33.−3	9.81.−4	8.22.−4	7.29.−4
	h	257.8	257.1	256.3	254.9	253.3	251.9	250.5	244.5	236.9	234.2	234.7	237.4
	s	2.088	1.858	1.788	1.716	1.673	1.642	1.617	1.537	1.451	1.400	1.365	1.340
600	v	0.5953	0.0596	0.0298	0.0149	9.96.−3	7.49.−3	6.01.−3	3.07.−3	1.64.−3	1.18.−3	9.67.−4	8.44.−4
	h	282.7	282.2	281.7	280.7	279.7	278.8	277.9	274.2	269.6	268.1	269.1	271.7
	s	2.133	1.904	1.834	1.763	1.721	1.691	1.667	1.591	1.511	1.462	1.428	1.403
700	v	0.6946	0.0696	0.0348	0.0175	0.0117	8.80.−3	7.07.−3	3.62.−3	1.93.−3	1.38.−3	1.11.−3	9.56.−4
	h	307.5	307.2	306.9	306.2	305.6	305.1	304.5	302.2	299.8	299.6	301.1	304.0
	s	2.171	1.942	1.873	1.803	1.761	1.732	1.708	1.634	1.557	1.511	1.478	1.453
800	v	0.7939	0.0795	0.0399	0.0200	0.0134	0.0101	8.11.−3	4.16.−3	2.21.−3	1.57.−3	1.25.−3	1.07.−3
	h	332.3	332.2	331.9	331.6	331.2	330.9	330.5	329.3	328.6	329.4	331.5	334.6
	s	2.204	1.975	1.906	1.837	1.795	1.766	1.743	1.671	1.596	1.551	1.518	1.494
900	v	0.8931	0.0895	0.0448	0.0225	0.0151	0.0114	9.13.−3	4.68.−3	2.48.−3	1.75.−3	1.39.−3	1.18.−3
	h	357.1	357.0	356.9	356.8	356.6	356.4	356.3	355.8	356.3	358.0	360.7	364.2
	s	2.233	2.005	1.936	1.866	1.825	1.796	1.773	1.702	1.628	1.584	1.553	1.528
1000	v	0.9924	0.0994	0.0498	0.0250	0.0168	0.0126	0.0102	5.20.−3	2.74.−3	1.93.−3	1.53.−3	1.29.−3
	h	381.9	381.9	381.9	381.8	381.8	381.8	381.8	381.9	383.4	385.9	389.0	392.8
	s	2.260	2.031	1.962	1.893	1.852	1.823	1.800	1.729	1.657	1.614	1.583	1.559

*Values extracted and in some cases rounded off from those cited in Rabinovich (ed.), *Thermophysical Properties of Neon, Argon, Krypton and Xenon*, Standards Press, Moscow, 1976. This source contains an exhaustive tabulation of values. v = specific volume, m³/kg; h = specific enthalpy, kJ/kg; s = specific entropy, kJ/(kg·K). The notation 3.49.−4 signifies 3.49×10^{-4}. This book was published in English translation by Hemisphere, New York, 1988 (604 pp.).

TABLE 2-278 Saturated Lithium*

T, K	P, bar	v_f, m³/kg	v_g, m³/kg	h_f, kJ/kg	h_g, kJ/kg	s_f, kJ/(kg·K)	s_g, kJ/(kg·K)	c_{pf}, kJ/(kg·K)
453.7^m	1.78.−13	1.912.−3		1703	24259	6.776	56.492	4.30
500	8.21.−12	1.946.−3		1905	24390	7.199	52.169	4.34
600	4.18.−9	1.988.−3		2334	24674	7.983	45.216	4.23
700	3.51.−7	2.028.−3	2.40.+7	2697	24869	8.633	40.307	4.19
800	9.57.−6	2.070.−3	9.94.+5	3174	25162	9.192	36.678	4.17
900	1.24.−4	2.114.−3	8.55.+4	3590	25341	9.682	33.850	4.16
1000	9.60.−4	2.160.−3	1.22.+4	4006	25477	10.120	31.591	4.16
1200	0.0204	2.262.−3	669.3	4835	25654	10.876	28.225	4.14
1400	0.1794	2.370.−3	86.06	5668	25778	11.518	25.882	4.19
1500	0.4269	2.433.−3	38.17	6088	25845	11.808	24.979	4.20

*Converted from tables in Vargaftik, *Tables of the Thermophysical Properties of Liquids and Gases*, Nauka, Moscow, 1972, and Hemisphere, Washington, 1975. m = melting point. The notation 1.78.−13 signifies 1.78×10^{-13}.

Many of the Vargaftik values also appear in Ohse, R. W., *Handbook of Thermodynamic and Transport Properties of Alkali Metals*, Blackwell Sci. Pubs., Oxford, 1985 (1020 pp.). This source contains superheat data.

Saturation and superheat tables and a diagram to 14 bar, 2200 K are given by Reynolds, W. C., *Thermodynamic properties in S.I.*, Stanford Univ. publ., 1979 (173 pp.). For a Mollier diagram from 0.1 to 140 psia, 2100–3600°R, see Weatherford, P. M., J. C. Tyler, et al., WADD-TR-61-96, 1961.

An extensive review of properties of the solid and the saturated liquid was given by Alcock, C. B., M. W. Chase, et al., *J. Phys. Chem. Ref. Data*, **23**, 3 (1994): 385–497.

TABLE 2-279 Lithium Bromide—Water Solutions

Ruiter, J. P., *Rev. Int. Froid = Int. J. Refrig.*, **13** (1990): 223–236 gives subroutines for computer calculations. See also ASHRAE *Handbook—Fundamentals*.

TABLE 2-280   Saturated Mercury*

T, K	P, bar	$v_f \times 10^5$, m³/kg	v_g, m³/kg	h_f, kJ/kg	h_g, kJ/kg	h_{fg}, kJ/kg	s_f, kJ/(kg·K)	s_g, kJ/(kg·K)
203.15	$2.298 \cdot 10^{-11}$	7.26239	$3.665 \cdot 10^9$	33.131	342.637	309.506	0.32434	1.84787
213.15	$1.288 \cdot 10^{-10}$	7.27570	$6.862 \cdot 10^8$	34.567	343.674	309.107	0.33124	1.78142
223.15	$6.169 \cdot 10^{-10}$	7.28900	$1.499 \cdot 10^8$	35.997	344.710	308.713	0.33780	1.72123
233.15	$2.580 \cdot 10^{-9}$	7.30231	$3.746 \cdot 10^7$	37.422	345.746	308.324	0.34404	1.66647
243.15	$9.573 \cdot 10^{-9}$	7.31563	$1.053 \cdot 10^7$	38.842	346.782	307.940	0.35001	1.61647
253.15	$3.198 \cdot 10^{-8}$	7.32896	$3.281 \cdot 10^6$	40.258	347.819	307.561	0.35571	1.57065
263.15	$9.736 \cdot 10^{-8}$	7.34229	$1.120 \cdot 10^6$	41.668	348.855	307.187	0.36118	1.52852
273.15	$2.728 \cdot 10^{-7}$	7.35563	$4.150 \cdot 10^5$	43.074	349.891	306.817	0.36642	1.48967
283.15	$7.101 \cdot 10^{-7}$	7.36898	$1.653 \cdot 10^5$	44.476	350.927	306.451	0.37146	1.45375
293.15	$1.729 \cdot 10^{-6}$	7.38234	$7.026 \cdot 10^4$	45.874	351.964	306.090	0.37631	1.42045
303.15	$3.968 \cdot 10^{-6}$	7.39572	$3.167 \cdot 10^4$	47.268	353.000	305.732	0.38099	1.38951
313.15	$8.626 \cdot 10^{-6}$	7.40911	$1.505 \cdot 10^4$	48.659	354.036	305.377	0.38550	1.36068
323.15	$1.786 \cdot 10^{-5}$	7.42252	$7.501 \cdot 10^3$	50.046	355.072	305.026	0.38986	1.33378
333.15	$3.356 \cdot 10^{-5}$	7.43594	$3.905 \cdot 10^3$	51.430	356.108	304.678	0.39408	1.30862
343.15	$6.724 \cdot 10^{-5}$	7.44938	$2.115 \cdot 10^3$	52.810	357.145	304.335	0.39816	1.28505
353.15	$1.232 \cdot 10^{-4}$	7.46285	$1.188 \cdot 10^3$	54.188	358.181	303.993	0.40212	1.26292
363.15	$2.182 \cdot 10^{-4}$	7.47633	$6.899 \cdot 10^2$	55.563	359.217	303.654	0.40596	1.24213
373.15	$3.745 \cdot 10^{-4}$	7.48984	413.0	56.936	360.253	303.317	0.40969	1.22255
383.15	$6.247 \cdot 10^{-4}$	7.50337	254.2	58.306	361.289	302.983	0.41331	1.20408
393.15	$1.015 \cdot 10^{-3}$	7.51693	153.6	59.674	362.326	302.652	0.41684	1.18665
403.15	$1.608 \cdot 10^{-3}$	7.53052	103.9	61.039	363.362	302.323	0.42027	1.17017
413.15	$2.491 \cdot 10^{-3}$	7.55415	68.75	62.403	364.397	301.994	0.42361	1.15456
423.15	$3.778 \cdot 10^{-3}$	7.55780	46.43	63.765	365.433	301.668	0.42687	1.13978
433.15	$5.618 \cdot 10^{-3}$	7.57148	31.96	65.125	366.469	301.344	0.43004	1.12575
443.15	$8.204 \cdot 10^{-3}$	7.58520	22.39	66.484	367.504	301.020	0.43314	1.11242
453.15	$1.178 \cdot 10^{-2}$	7.59897	15.95	67.842	368.539	300.697	0.43617	1.09975
463.15	$1.664 \cdot 10^{-2}$	7.61277	11.54	69.198	369.574	300.376	0.43913	1.08768
473.15	$2.315 \cdot 10^{-2}$	7.62662	8.469	70.553	370.609	300.056	0.44203	1.07619
483.15	$3.177 \cdot 10^{-2}$	7.64051	6.301	71.908	371.642	299.734	0.44486	1.06524
493.15	$4.304 \cdot 10^{-2}$	7.65444	4.748	73.261	372.676	299.415	0.44763	1.05478
503.15	$5.758 \cdot 10^{-2}$	7.66843	3.621	74.614	373.708	299.094	0.45035	1.04479
513.15	$7.614 \cdot 10^{-2}$	7.68247	2.793	75.967	374.740	298.773	0.45301	1.03525
523.15	$9.959 \cdot 10^{-2}$	7.69656	2.176	77.319	375.771	298.452	0.45562	1.02611
533.15	0.12892	7.71071	1.7132	78.671	376.800	298.129	0.45818	1.01737
543.15	0.16527	7.72491	1.3613	80.023	377.829	297.806	0.46069	1.00899
553.15	0.20993	7.73918	1.0912	81.375	378.855	297.480	0.46316	1.00095
563.15	0.26435	7.75351	0.88213	82.728	379.880	297.152	0.46558	0.99324
573.15	0.33015	7.7679	0.71874	84.080	380.904	296.824	0.46796	0.98584
583.15	0.40910	7.7823	0.59002	85.434	381.925	296.491	0.47030	0.97893
593.15	0.50320	7.7969	0.48779	86.788	382.944	296.156	0.47260	0.97190
603.15	0.61460	7.8115	0.40600	88.143	383.960	295.817	0.47487	0.96532
613.15	0.74567	7.8262	0.34008	89.499	384.973	295.474	0.47709	0.95899
623.15	0.89896	7.8409	0.28660	90.856	385.984	295.128	0.47929	0.95289
633.15	1.0772	7.8558	0.24291	92.215	386.991	294.776	0.48145	0.94702
643.15	1.2834	7.8707	0.20702	93.575	387.994	294.419	0.48358	0.94135
653.15	1.5207	7.8858	0.17735	94.937	388.994	294.057	0.48568	0.93589
663.15	1.9725	7.9008	0.15269	96.300	389.989	293.689	0.48774	0.93061
673.15	2.1024	7.9160	0.13207	97.666	390.980	293.314	0.48978	0.92552
683.15	2.454	7.9313	0.11476	99.033	391.966	292.933	0.49180	0.92059
693.15	2.852	7.9467	0.10014	100.403	392.947	292.544	0.49378	0.91583
703.15	3.299	7.9622	0.08775	101.775	393.923	292.148	0.49574	0.91123
713.15	3.801	7.9778	0.07719	103.150	394.893	291.743	0.49768	0.90677
723.15	4.362	7.9935	0.06815	104.528	395.858	291.330	0.49959	0.90245
733.15	4.986	8.0094	0.06039	105.908	396.816	290.908	0.50148	0.89827
743.15	5.679	8.0252	0.05369	107.292	397.767	290.475	0.50335	0.89422
753.15	6.446	8.0413	0.04789	108.679	398.711	290.032	0.50519	0.89029
763.15	7.292	8.0574	0.04285	110.069	399.649	289.580	0.50702	0.88647
773.15	8.222	8.074	0.03846	111.463	400.579	289.116	0.50882	0.88277
783.15	9.242	8.090	0.03462	112.861	401.501	288.640	0.51061	0.87917
793.15	10.358	8.106	0.03124	114.262	402.415	288.153	0.51238	0.87568
803.15	11.576	8.123	0.02827	115.668	403.321	287.653	0.51412	0.87228
813.15	12.901	8.140	0.02565	117.078	404.218	287.140	0.51586	0.86898
823.15	14.340	8.157	0.02333	118.492	405.106	286.614	0.51757	0.86576
833.15	15.899	8.174	0.02126	119.911	405.985	286.074	0.51927	0.86263

*From Vukalovich, Ivanov, Fokin, and Yakovlev, *Thermophysical Properties of Mercury*, Standartov, Moscow, 1971. For the saturated liquid the specific volume at 203.15 K is 7.26239×10^{-5} m³/kg, etc. All the tabular values for 203.15 K, 213.15 K, 223.15 K, and 233.15 K represent a metastable equilibrium between the subcooled liquid and the saturated vapor.

Saturation and superheat tables and a diagram to 100 bar, 1600 K are given by Reynolds, W. C., *Thermodynamic properties in S.I.*, Stanford Univ. publ., 1979 (173 pp.). For a Mollier diagram from 1 to 8200 psia and 2700°R, see Weatherford, W. D., J. C. Tyler, et al., WADD-TR-61-96, 1961.

TABLE 2-280   Saturated Mercury* (*Concluded*)

T, K	P, bar	$v_f \times 10^5$, m³/kg	v_g, m³/kg	h_f, kJ/kg	h_g, kJ/kg	h_{fg}, kJ/kg	s_f, kJ/(kg·K)	s_g, kJ/(kg·K)
843.15	17.584	8.191	0.019426	121.335	406.855	285.520	0.52095	0.85959
853.15	19.403	8.209	0.017785	122.763	407.715	284.952	0.52262	0.85662
863.15	21.36	8.226	0.016317	124.197	408.565	284.368	0.52427	0.85372
873.15	23.46	8.244	0.015000	125.636	409.405	283.769	0.52591	0.85090
883.15	25.72	8.262	0.013815	127.080	410.235	283.155	0.52753	0.84815
893.15	28.14	8.280	0.012748	128.530	411.054	282.524	0.52914	0.84546
903.15	30.72	8.298	0.011784	129.986	411.861	281.875	0.53074	0.84284
913.15	33.47	8.316	0.010911	131.448	412.658	281.210	0.53232	0.84028
923.15	36.41	8.335	0.010120	132.915	413.444	280.529	0.53389	0.83777
933.15	39.53	8.353	0.009401	134.389	414.218	279.829	0.53545	0.83533
943.15	42.85	8.372	0.008746	135.869	414.980	279.111	0.53700	0.83294
953.15	46.36	8.391	0.008150	137.356	415.731	278.375	0.53854	0.83060
963.15	50.09	8.410	0.007604	138.850	416.469	277.619	0.54006	0.82831
973.15	54.03	8.430	0.007105	140.350	417.195	276.845	0.54158	0.82606
983.15	58.20	8.450	0.006648	141.858	417.909	276.051	0.54308	0.82387
993.15	62.59	8.468	0.006228	143.372	418.610	275.238	0.54458	0.82172
1003.15	67.22	8.488	0.005842	144.894	419.298	274.404	0.54607	0.81961
1013.15	72.10	8.508	0.005487	146.424	419.974	273.550	0.54754	0.81754
1023.15	77.22	8.529	0.005159	147.961	420.636	272.675	0.54901	0.81552
1033.15	82.60	8.550	0.004856	149.506	421.286	271.780	0.55047	0.81353
1043.15	88.25	8.570	0.004576	151.059	421.923	270.864	0.55192	0.81158
1053.15	94.17	8.590	0.004317	152.619	422.546	269.927	0.55336	0.80966
1063.15	100.37	8.612	0.004077	154.188	423.156	268.968	0.55479	0.80778
1073.15	106.85	8.632	0.003854	155.766	423.752	267.986	0.55621	0.80593

TABLE 2-281   Saturated Methane*

T, K	P, bar	v_f, m³/kg	v_g, m³/kg	h_f, kJ/kg	h_g, kJ/kg	s_f, kJ/(kg·K)	s_g, kJ/(kg·K)	c_{pf}, kJ/(kg·K)	μ_f, 10^{-4} Pa·s	k_f, W/(m·K)
90.7^t	0.117	2.215.−3	3.976	216.4	759.9	4.231	10.225	3.288	2.02	0.225
95	0.198	2.244.−3	2.463	232.5	769.0	4.406	10.034	3.318	1.71	0.215
100	0.345	2.278.−3	1.479	246.3	776.9	4.556	9.862	3.369	1.56	0.206
105	0.565	2.316.−3	0.940	263.2	785.7	4.719	9.710	3.425	1.33	0.197
110	0.884	2.353.−3	0.625	280.1	794.5	4.882	9.558	3.478	1.22	0.189
115	1.325	2.396.−3	0.430	297.7	802.5	5.035	9.436	3.525	1.09	0.181
120	1.919	2.438.−3	0.306	315.3	810.4	5.188	9.314	3.570	0.98	0.173
125	2.693	2.487.−3	0.223	333.5	817.3	5.332	9.062	3.620	0.89	0.165
130	3.681	2.536.−3	0.167	351.7	824.1	5.476	8.810	3.679	0.81	0.158
135	4.912	2.594.−3	0.127	370.6	829.5	5.614	8.871	3.755	0.73	0.150
140	6.422	2.652.−3	0.098	389.5	834.8	5.751	8.932	3.849	0.66	0.143
145	8.246	2.722.−3	0.077	409.5	844.4	5.885	8.891	3.965	0.61	0.136
150	10.41	2.792.−3	0.061	429.4	853.9	6.019	8.849	4.101	0.56	0.129
155	12.97	2.882.−3	0.049	450.8	848.5	6.151	8.725	4.27	0.51	0.122
160	15.94	2.971.−3	0.039	472.1	843.0	6.283	8.601	4.47	0.46	0.115
165	19.39	3.095.−3	0.032	495.4	840.0	6.417	8.513	4.75	0.42	0.108
170	23.81	3.218.−3	0.026	518.6	837.0	6.551	8.424	5.16	0.38	0.101
175	27.81	3.419.−3	0.020	545.8	827.6	6.697	8.315	5.89	0.34	0.094
180	32.86	3.619.−3	0.016	572.9	818.1	6.843	8.205	7.27	0.30	0.088
185	38.59	3.979.−3	0.012	605.4	797.7	7.017	8.049	11.1	0.25	0.085
190	45.20	4.900.−3	0.008	661.6	750.7	7.293	7.762	70.	0.19	0.090
190.6^c	45.99	6.233.−3	0.006	704.4	704.4	7.516	7.516	∞	0.17	∞

*Values reproduced or converted from Goodwin, NBS Tech. Note 653, 1974. *t* = triple point; *c* = critical point. The notation 2.215.−3 signifies 2.215×10^{-3}.

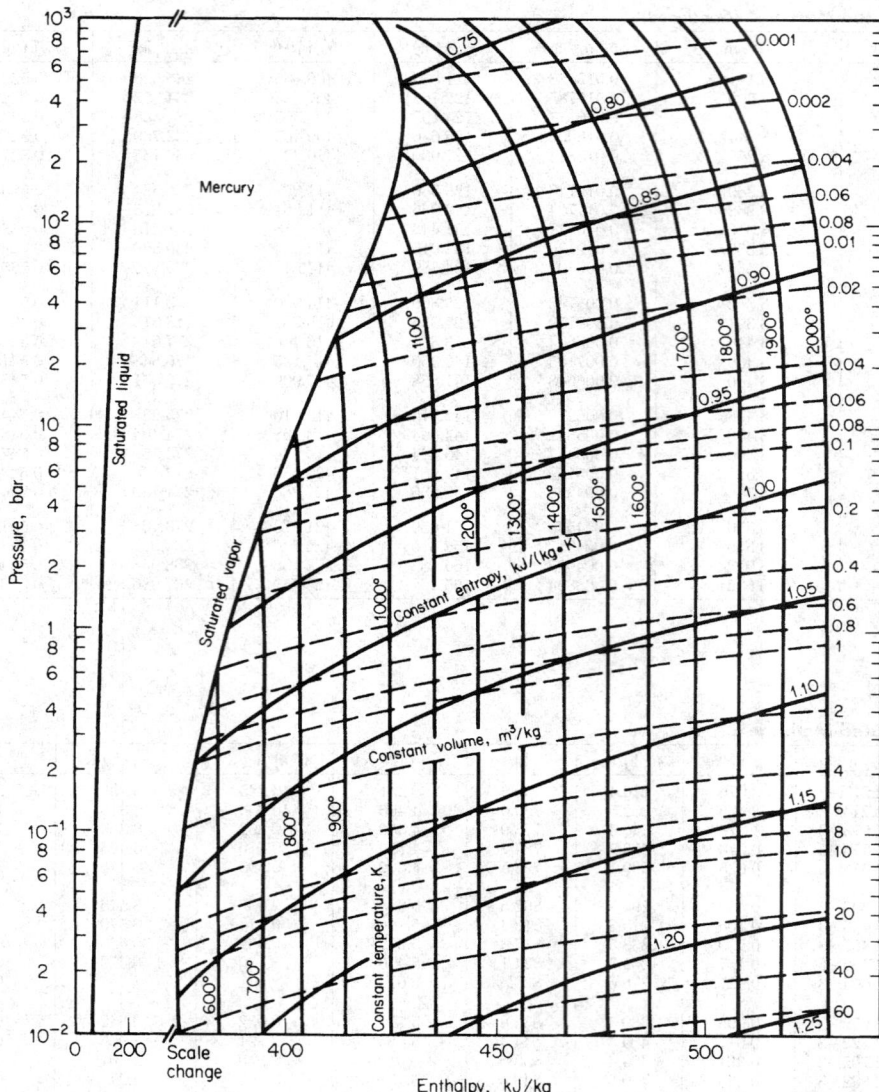

FIG. 2-12 Enthalpy-log-pressure diagram for mercury. (*Drawn from tabular data in footnote reference to Table 2-280.*)

TABLE 2-282   Superheated Methane*

P, bar		100	150	200	250	300	350	400	450	500
						Temperature, K				
1	v	0.00228	0.7661	1.0299	1.2915	1.5521	1.8122	2.0719	2.3669	2.5911
	h	246.4	879.0	984.3	1090.4	1199.8	1314.8	1437.4	1568.8	1708.9
	s	4.555	10.152	10.757	11.230	11.629	11.983	12.310	12.618	12.914
5	v	0.00228	0.1434	0.2006	0.2549	0.3083	0.3611	0.4136	0.4657	0.5181
	h	247.0	865.0	976.1	1084.7	1195.5	1311.5	1434.7	1566.6	1706.9
	s	4.553	9.256	9.896	10.381	10.785	11.142	11.471	11.781	12.066
10	v	0.00227	0.0643	0.0968	0.1254	0.1528	0.1798	0.2063	0.2327	0.2590
	h	247.8	843.6	965.5	1077.9	1190.6	1307.9	1432.0	1564.1	1705.3
	s	4.549	8.797	9.501	10.002	10.414	10.775	11.106	11.417	11.715
20	v	0.00227	0.00277	0.0446	0.0606	0.0751	0.0891	0.1027	0.1162	0.1295
	h	249.4	429.8	941.9	1063.6	1180.7	1300.6	1426.5	1560.3	1702.1
	s	4.542	6.003	9.059	9.603	10.030	10.400	10.736	11.050	11.349
40	v	0.00226	0.00274	0.0176	0.0281	0.0363	0.0438	0.0510	0.0579	0.0648
	h	252.5	430.8	879.3	1032.9	1160.5	1286.0	1415.7	1552.1	1696.0
	s	4.528	5.973	8.465	9.155	9.621	10.008	10.354	10.674	10.978
60	v	0.00226	0.00271	0.00615	0.0173	0.0234	0.0287	0.0338	0.0386	0.0432
	h	255.7	432.2	734.0	999.8	1140.0	1271.7	1405.1	1544.2	1690.0
	s	4.515	5.946	7.623	8.847	9.359	9.765	10.121	10.440	10.756
80	v	0.00225	0.00268	0.00411	0.0119	0.0171	0.0213	0.0252	0.0289	0.0324
	h	258.9	433.8	660.5	964.4	1119.7	1257.7	1394.9	1536.6	1684.4
	s	4.502	5.920	7.209	8.590	9.158	9.584	9.951	10.283	10.595
100	v	0.00224	0.00266	0.00375	0.00888	0.0133	0.0169	0.0201	0.0231	0.0260
	h	262.1	435.5	644.5	928.5	1099.6	1244.2	1385.2	1529.4	1679.0
	s	4.489	5.897	7.090	8.364	8.991	9.437	9.814	10.153	10.469
150	v	0.00223	0.00261	0.00337	0.00555	0.00852	0.0111	0.0134	0.0155	0.0175
	h	270.2	440.7	630.2	860.0	1054.1	1213.1	1362.8	1513.0	1667.0
	s	4.458	5.843	6.930	7.953	8.664	9.155	9.555	9.907	10.233
200	v	0.00221	0.00256	0.00318	0.00447	0.00644	0.00837	0.0101	0.0118	0.0133
	h	278.3	446.5	626.5	825.0	1019.8	1187.2	1343.8	1498.9	1656.9
	s	4.429	5.796	6.829	7.719	8.426	8.944	9.362	9.727	10.060
300	v	0.00218	0.00249	0.00296	0.00369	0.00474	0.00593	0.00708	0.00818	0.00924
	h	294.7	459.6	629.2	804.4	982.9	1153.6	1316.8	1478.5	1642.2
	s	4.373	5.714	6.690	7.471	8.122	8.649	9.085	9.465	9.811
400	v		0.00244	0.00282	0.00336	0.00406	0.00486	0.00569	0.00560	0.00729
	h		473.8	637.7	802.4	970.1	1137.8	1303.0	1467.7	1634.7
	s		5.645	6.588	7.323	7.935	8.451	8.893	9.280	9.633
500	v		0.00239	0.00272	0.00315	0.00368	0.00428	0.00492	0.00555	0.00616
	h		488.8	648.9	807.7	969.0	1132.8	1297.8	1464.2	1633.2
	s		5.584	6.507	7.215	7.802	8.307	8.748	9.139	9.496

*Converted and rounded off from the tables of Goodwin, NBS Tech. Note 654, 1974. v = specific volume, m^3/kg; h = specific enthalpy, kJ/kg; s = specific entropy, kJ/(kg·K).

For a thermodynamic diagram from 0.1 to 400 bar and 620°C, see the 1993 ASHRAE *Handbook—Fundamentals* (SI ed.).

Saturation and superheat tables and a chart to 6000 psia, 680°F appear in Stewart, R. B., R. T. Jacobsen, et al., *Thermodynamic Properties of Refrigerants*, ASHRAE, Atlanta, GA, 1986 (521 pp.). For specific heat, thermal conductivity, and viscosity, see *Thermophysical Properties of Refrigerants*, ASHRAE, 1993. See also Friend, D. G., J. F. Ely, et al., *J. Phys. Chem. Ref. Data.* **18,** 2 (1989): 583–638.

TABLE 2-283   Thermophysical Properties of Saturated Methanol

Pressure, bar	Temp., K	v_f, m³/kg	v_g, m³/kg	h_f, kJ/kg	h_g, kJ/kg	s_f, kJ/(kg·K)	s_g, kJ/(kg·K)	c_{pf}, kJ/(kg·K)	μ_f, 10^{-6} Pa·s	k_f, W/(m·K)	Pr_f
4×10^{-6t}	175.6	0.001 057	1700000	0.0	1303.1	2.8114	10.2328			0.204	7.75
0.1	288.4	0.001 257	7.309	261.0	1440.3	3.9383	8.0281	2.531	625	0.196	6.84
0.2	301.7	0.001 276	3.801	293.9	1455.4	4.0493	7.9032	2.554	525	0.193	5.55
0.5	320.7	0.001 307	1.599	345.0	1476.2	4.2117	7.7386	2.669	401	0.189	4.83
1.013	337.7	0.001 336	0.819	391.7	1492.1	4.3516	7.6104	2.777	329		
1.5	348.0	0.001 356	0.5632	421.0	1500.3	4.4361	7.5379	2.845	288	0.186	4.41
2.0	356.0	0.001 371	0.4276	444.2	1505.8	4.5014	7.4836	2.894	268	0.184	4.22
2.5	362.5	0.001 385	0.3443	463.6	1509.8	4.5536	7.4398	2.946	242	0.182	3.92
3.0	368.0	0.001 396	0.2893	479.8	1512.4	4.5992	7.4051	2.984	227	0.181	3.74
4.0	377.1	0.001 417	0.2188	507.8	1515.9	4.6728	7.3474	3.050	204	0.179	3.48
5	384.5	0.001 434	0.17569	529.7	1517.4	4.7307	7.2992	3.117	187	0.178	3.27
6	390.8	0.001 450	0.14683	549.6	1518.4	4.7836	7.2624	3.176	174	0.177	3.12
8	401.3	0.001 479	0.11015	582.7	1518.0	4.8678	7.1988	3.265	156	0.175	2.91
10	409.8	0.001 504	0.08783	610.3	1516.1	4.9366	7.1471	3.349	141	0.173	2.73
15	426.3	0.001 560	0.05761	665.8	1507.9	5.0708	7.0461	3.540	117	0.171	2.42
20	438.9	0.001 611	0.04224	710.5	1553.8	5.1744	6.9677	3.72	102	0.169	2.25
25	449.3	0.001 666	0.03290	749.0	1486.4	5.2605	6.9017	3.91	92	0.167	2.15
30	458.2	0.001 710	0.02661	783.8	1474.7	5.3355	6.8435	4.12	84	0.165	2.10
40	472.9	0.001 814	0.01863	846.7	1450.1	5.4650	6.7409	4.67	72	0.160	2.10
50	484.9	0.001 934	0.01373	905.2	1423.2	5.5793	6.6475	5.55	63	0.154	2.27
60	495.1	0.002 086	0.01032	963.3	1391.8	5.6889	6.5543				
80	508.1	0.002 507	0.00642	1065.3	1318.7	5.8803	6.3791				
80.95^c	512.6	0.003 715	0.00372	1186.8	1186.8	6.0979	6.0979				

t = triple point; c = critical point. v, h, s, and c_p interpolated and converted from Goodwin, R. D., *J. Phys. Chem. Ref. Data*, **16**, 4 (1987): 799–891.

TABLE 2-284 Thermodynamic Properties of Compressed Methanol

Pressure, bar		200	250	300	350	400	450	500	550	600
						Temperature, K				
0.1	v (m³/kg)	0.001137	0.001203	7.630	8.942	10.23	11.56	12.84		15.45
	h (kJ/kg)	57.5	169.8	1456.7	1529.5	1607.5	1691.5	1781.7	1878.1	1980.2
	s (kJ/kg·K)	3.096	3.597	8.081	8.305	8.514	8.711	8.901	9.085	9.263
0.5	v (m³/kg)	0.001137	0.001202	0.001274	1.764	2.033	2.296	2.558	2.818	3.078
	h (kJ/kg)	57.5	169.9	290.5	1522.7	1603.0	1687.9	1778.9	1875.3	1977.4
	s (kJ/kg·K)	3.096	3.597	4.038	7.877	8.091	8.291	8.482	8.666	8.844
1.013	v (m³/kg)	0.001137	0.001202	0.001274	0.8560	0.9958	1.1283	1.2843	1.3870	1.5157
	h (kJ/kg)	57.6	169.9	290.5	1514.0	1598.7	1685.1	1795.4	1873.5	1975.8
	s (kJ/kg·K)	3.096	3.597	4.038	7.675	7.902	8.105	8.117	8.482	8.660
10	v (m³/kg)	0.001136	0.001201	0.001272	0.001357	0.001474	0.1068	0.1236	0.1381	0.1519
	h (kJ/kg)	58.4	170.7	291.2	427.4	578.8	1638.1	1751.5	1858.0	1965.2
	s (kJ/kg·K)	3.095	3.596	4.036	4.451	4.857	7.427	7.667	7.870	8.056
15	v (m³/kg)	0.001136	0.001201	0.001272	0.001356	0.001472	0.0673	0.0806	0.0911	0.1007
	h (kJ/kg)	58.8	171.1	291.6	427.7	578.9	1601.9	1735.6	1849.4	1960.2
	s (kJ/kg·K)	3.094	3.595	4.035	4.450	4.856	7.253	7.536	7.752	7.946
20	v (m³/kg)	0.001135	0.001200	0.001271	0.001355	0.001469	0.0466	0.0589	0.0675	0.0751
	h (kJ/kg)	59.2	171.6	292.0	428.1	579.0	1565.3	1717.7	1840.0	1954.8
	s (kJ/kg·K)	3.094	3.595	4.035	4.449	4.854	7.087	7.431	7.664	7.864
30	v (m³/kg)	0.001134	0.001199	0.001269	0.001355	0.001465	0.001659	0.0367	0.0436	0.0492
	h (kJ/kg)	60.1	172.4	292.9	428.8	579.4	751.3	1675.4	1818.7	1942.7
	s (kJ/kg·K)	3.092	3.593	4.036	4.447	4.851	5.264	7.253	7.526	7.743
40	v (m³/kg)	0.001133	0.001198	0.001268	0.001350	0.001461	0.001649	0.0251	0.0314	0.0361
	h (kJ/kg)	61.0	173.3	293.7	429.5	579.8	750.4	1623.0	1794.2	1928.8
	s (kJ/kg·K)	3.091	3.592	4.032	4.445	4.849	5.258	7.088	7.414	7.650
50	v (m³/kg)	0.001133	0.001197	0.001266	0.001348	0.001457	0.001637	0.0176	0.0239	0.0282
	h (kJ/kg)	61.9	174.2	294.5	430.2	580.2	749.7	1556.4	1766.7	1913.4
	s (kJ/kg·K)	3.090	3.591	4.030	4.443	4.846	5.579	6.912	7.314	7.570
60	v (m³/kg)	0.001131	0.001196	0.001265	0.001346	0.001453	0.001628	0.0120	0.0188	0.0228
	h (kJ/kg)	62.8	175.0	295.3	430.9	580.6	749.1	1461.8	1736.1	1896.6
	s (kJ/kg·K)	3.089	3.589	4.029	4.442	4.843	5.248	6.692	7.220	7.500
75	v (m³/kg)	0.001130	0.001194	0.001263	0.001343	0.001448	0.001614	0.002084	0.01359	0.0174
	h (kJ/kg)	64.1	176.3	296.6	431.9	581.2	748.3	982.1	1683.9	1869.1
	s (kJ/kg·K)	3.087	3.587	4.027	4.439	4.839	5.241	5.718	7.081	7.405
100	v (m³/kg)	0.001128	0.001191	0.001259	0.001337	0.001439	0.001595	0.001952	0.	0.01188
	h (kJ/kg)	66.3	178.5	298.6	433.8	582.4	747.5	964.8	1572.9	1818.8
	s (kJ/kg·K)	3.084	3.584	4.023	4.435	4.833	5.230	5.673	6.829	7.261
150	v (m³/kg)	0.001125	0.001186	0.001252	0.001328	0.001423	0.001562	0.001825		0.006513
	h (kJ/kg)	70.7	182.8	302.8	437.4	584.9	746.8	948.4	1248.8	1704.3
	s (kJ/kg·K)	3.078	3.578	4.016	4.426	4.822	5.211	5.622	6.302	6.997
200	v (m³/kg)	0.001121	0.001182	0.001246	0.001317	0.001408	0.001535	0.001751	0.002314	0.004091
	h (kJ/kg)	75.1	187.2	307.0	441.2	587.8	747.0	939.9	1223.5	1583.5
	s (kJ/kg·K)	3.071	3.571	4.009	4.418	4.811	5.194	5.587	6.125	6.752
300	v (m³/kg)	0.001113	0.001172	0.001234	0.001302	0.001384	0.001492	0.001656	0.001957	0.002600
	h (kJ/kg)	83.9	195.9	315.4	448.9	593.8	749.4	932.0	1173.4	1443.5
	s (kJ/kg·K)	3.060	3.559	3.996	4.403	4.791	5.166	5.537	5.996	6.466
400	v (m³/kg)	0.001107	0.001164	0.001223	0.001288	0.001363	0.001459	0.001593	0.001808	0.002182
	h (kJ/kg)	92.7	204.7	324.0	456.9	600.5	753.4	929.6	1154.4	1388.1
	s (kJ/kg·K)	3.048	3.548	3.983	4.388	4.774	5.142	5.500	5.926	6.335
500	v (m³/kg)	0.001101	0.001156	0.001213	0.001274	0.001345	0.001431	0.001546	0.001716	0.001980
	h (kJ/kg)	101.5	213.4	332.6	465.1	607.7	758.4	930.4	1145.6	1360.6
	s (kJ/kg·K)	3.037	3.536	3.971	4.375	4.757	5.121	5.470	5.880	6.254

Converted and interpolated from Goodwin, R. D., *J. Phys. Chem. Ref. Data,* **16,** 4 (1987): 799–891. These extensive tables extend to 700 bar for temperatures from 175.6 to 800 K. Another extensive compilation is by deReuck, K. M. and R. J. B. Craven, *Methanol,* C.R.C. Press, 1993 (320 pp.).

Equations and diagrams to 30 bar, 200°C are given by Eicholz, H. D., S. Schulz, et al., *Kalte u Klim.,* no. 9, (1981) 322–331. For pressures to 1040 bar, 298–489 K, see Machado, J. R. S. and W. B. Street, *J.Chem.Eng.Data,* **28** (1983): 218–223; to 2800 bar from 273 to 333 K, see Sun, T., S. N. Biswas, et al., *J. Chem. Eng. Data,* **33** (1988): 395–398. Dissociation was considered by Yerlett, T. K. and C. J. Wormald, *J. Chem. Thermo.,* **18** (1986): 719–726, and by Kazarnovskii, Ya. S. and E. V. Pavlova, *Russ. J. Phys. Chem.,* **56,** 6 (1982): 847–851.

TABLE 2-285 Saturated Methyl Chloride*

T, K	P, bar	v_f, m³/kg	v_g, m³/kg	h_f, kJ/kg	h_g, kJ/kg	s_f, kJ/(kg·K)	s_g, kJ/(kg·K)	c_{pf}, kJ/(kg·K)	μ_f, 10^{-4} Pa·s	k_f, W/(m·K)
175	0.0117	8.84.−4	27.90	274.5	764.3	3.529	6.328	1.469		
180	0.0165	8.91.−4	19.85	280.9	767.7	3.570	6.274	1.472		
185	0.0233	8.97.−4	14.12	287.5	771.0	3.603	6.222	1.475		
190	0.0327	9.04.−4	10.12	294.5	774.3	3.647	6.172	1.477		
195	0.0462	9.10.−4	7.208	301.7	777.5	3.684	6.124	1.480		
200	0.0653	9.17.−4	5.137	309.0	780.7	3.722	6.080	1.483	4.44	0.241
205	0.0919	9.25.−4	3.835	316.3	783.9	3.756	6.038	1.486	4.27	0.236
210	0.1315	9.33.−4	2.656	323.7	787.0	3.791	5.998	1.489	4.11	0.232
215	0.181	9.40.−4	1.975	331.0	790.1	3.825	5.961	1.492	3.96	0.228
220	0.243	9.48.−4	1.505	338.4	793.2	3.859	5.928	1.496	3.82	0.224
225	0.319	9.56.−4	1.168	345.7	796.3	3.892	5.896	1.500	3.69	0.219
230	0.417	9.65.−4	0.911	353.1	799.3	3.925	5.866	1.504	3.57	0.215
235	0.539	9.73.−4	0.718	360.5	802.3	3.957	5.845	1.508	3.46	0.211
240	0.688	9.81.−4	0.572	368.0	805.3	3.988	5.822	1.513	3.35	0.207
245	0.866	9.89.−4	0.462	375.6	808.2	4.019	5.786	1.518	3.25	0.202
250	1.076	9.98.−4	0.377	383.2	811.1	4.050	5.762	1.523	3.16	0.198
255	1.328	10.08.−4	0.311	390.7	814.0	4.080	5.740	1.528	3.08	0.194
260	1.627	10.18.−4	0.257	398.3	816.8	4.110	5.720	1.533	3.00	0.190
265	1.970	10.27.−4	0.215	406.0	819.4	4.139	5.699	1.539	2.92	0.186
270	2.364	10.36.−4	0.1807	413.7	822.0	4.168	5.680	1.546	2.85	0.182
275	2.830	10.46.−4	0.1524	421.5	824.4	4.197	5.662	1.554	2.78	0.177
280	3.347	10.57.−4	0.1301	429.4	826.8	4.225	5.644	1.565	2.72	0.173
285	3.936	10.68.−4	0.1115	437.3	829.0	4.253	5.628	1.574	2.66	0.169
290	4.612	10.79.−4	0.0960	445.2	831.2	4.280	5.612	1.583	2.61	0.165
295	5.361	10.91.−4	0.0830	453.2	833.2	4.308	5.597	1.594	2.56	0.160
300	6.189	11.03.−4	0.0723	461.2	835.2	4.334	5.581	1.605	2.51	0.156
305	7.110	11.15.−4	0.0632	469.3	837.0	4.361	5.567	1.617	2.46	0.152
310	8.111	11.27.−4	0.0556	477.4	838.8	4.388	5.553	1.631	2.42	0.148
315	9.243	11.40.−4	0.0489	485.6	840.5	4.414	5.540	1.644	2.37	0.143
320	10.47	11.55.−4	0.0433	493.8	841.9	4.440	5.527	1.658	2.33	0.139
325	11.78	11.70.−4	0.0386	502.1	843.3	4.465	5.516		2.30	0.135
330	13.27	11.86.−4	0.0343	510.4	844.5	4.491	5.504		2.27	0.131
340	16.52	12.17.−4	0.0282	518.8	846.4	4.542	5.481		2.12	0.124
350	20.53	12.54.−4	0.0228	538.3	847.5	4.592	5.457		1.99	0.117
360	25.29	12.97.−4	0.0186	562.9	847.6	4.643	5.434		1.87	0.110
370	30.74	13.47.−4	0.0151	581.6	845.9	4.694	5.398		1.77	0.103
380	36.99	14.11.−4	0.0117	602.8	842.6	4.747	5.382		1.67	0.095
390	44.05	14.67.−4	0.0096	622.9	837.4	4.805	5.358		1.59	0.086
400	52.29	15.66.−4	0.0075	643.6	826.4	4.870	5.323		1.51	0.075
405	56.6	16.48.−4	0.0063	663.2	819.1	4.904	5.289			
410	61.5	17.97.−4	0.0052	677.3	807.1	4.954	5.256			
415	67.4	21.10.−4	0.0038	714.1	778.6	5.025	5.200			
416^c	69.0	27.40.−4	0.0027	749.3	749.3	5.116	5.116			

*Interpolated by P. E. Liley from the Landolt-Börnstein band IVa, p. 677, 1967 tables by Steinle/Dienemann. c = critical point. The notation 8.84.−4 signifies 8.84×10^{-4}.

TABLE 2-286 Saturated Neon*

T, K	P, bar	v_f, m³/kg	v_g, m³/kg	h_f, kJ/kg	h_g, kJ/kg	s_f, kJ/(kg·K)	s_g, kJ/(kg·K)	c_{pf}, kJ/(kg·K)	μ_f, 10^{-4} Pa·s	k_f, W/(m·K)
10		6.654.−4		0.75		0.0992		0.278		
20		6.823.−4		6.78		0.4906		0.945		
24.6^m		6.696.−4		11.96		0.7257		1.345		
24.6^m	0.434	8.012.−4	0.2266	28.22	117.0	1.388	5.006	1.802	1.57	0.146
26	0.718	8.172.−4	0.1429	30.90	118.1	1.494	4.846	1.868	1.37	0.132
28	1.321	8.413.−4	0.0817	34.75	119.3	1.634	4.653	1.955	1.16	0.124
30	2.238	8.687.−4	0.0501	38.80	120.1	1.771	4.483	2.052	1.00	0.115
32	3.552	9.001.−4	0.0323	43.06	120.6	1.905	4.329	2.163	0.84	0.106
34	5.352	9.370.−4	0.0217	47.57	120.6	2.036	4.184	2.302	0.71	0.097
36	7.728	9.820.−4	0.0149	52.34	119.9	2.166	4.043	2.506	0.59	0.088
38	10.78	1.039.−3	0.0104	57.52	118.4	2.297	3.900	2.825	0.48	0.078
40	14.62	1.116.−3	0.0073	63.33	115.8	2.435	3.749	3.436	0.38	0.069
42	19.39	1.232.−3	0.0050	69.82	111.8	2.582	3.582	5.26	0.31	0.059
44	25.22	1.538.−3	0.0031	80.83	103.0	2.812	3.316	25.0	0.25	
44.4^c	26.53	2.070.−3	0.0021	92.50	92.5	3.062	3.062	∞		∞

*Values extracted and in some cases rounded off from those cited in Rabinovich (ed.), *Thermophysical Properties of Neon, Argon, Krypton and Xenon*, Standards Press, Moscow, 1976. m = melting point; c = critical point. The notation 6.654.−4 signifies 6.654×10^{-3}. This source contains values for the compressed state up to 1000 bar, etc. This book was published in English translation by Hemisphere, New York 1988 (604 pp.).

Saturation and superheat tables and a diagram to 200 bar, 320 K are given by Reynolds, W. C., *Thermodynamic Properties in S.I.*, Stanford Univ. publ., 1979 (173 pp.). Saturation and superheat tables to 60,000 psia, 900°R and a chart to 4000 psia, 560°R appear in Stewart, R. B., R. T. Jacobsen, et al., *Thermodynamic Properties of Refrigerants*, ASHRAE, Atlanta, GA, 1986 (521 pp.). For specific heat, thermal conductivity, and viscosity, see *Thermophysical Properties of Refrigerants*, ASHRAE, 1993.

TABLE 2-287 Compressed Neon*

Temperature, K		1	10	20	40	60	80	100	200	400	600	800	1000
						Pressure, bar							
100	v	0.4117	0.0410	0.0204	0.0102	6.76.−3	5.08.−3	4.09.−3	2.22.−3	1.42.−3	1.18.−3	1.06.−3	9.74.−4
	h	195.4	194.0	192.4	189.4	186.6	184.0	181.6	174.1	173.3	180.0	189.2	199.2
	s	6.129	5.168	4.869	4.556	4.363	4.221	4.106	3.739	3.386	3.197	3.066	2.964
200	v	0.8243	0.0828	0.0416	0.0210	0.0142	0.0107	8.69.−3	4.61.−3	2.61.−3	1.95.−3	1.63.−3	1.43.−3
	h	298.5	298.4	298.4	298.2	298.2	298.2	298.3	299.4	304.6	312.4	321.8	332.1
	s	6.844	5.893	5.605	5.315	5.143	5.020	4.924	4.620	4.308	4.124	3.994	3.893
300	v	1.236	0.1241	0.0624	0.0315	0.0212	0.0160	0.0129	6.77.−3	3.71.−3	2.69.−3	2.18.−3	1.87.−3
	h	401.6	401.8	402.2	402.8	403.5	404.1	404.9	408.8	417.8	427.8	438.5	449.7
	s	7.262	6.312	6.026	5.739	5.570	5.450	5.357	5.065	4.769	4.593	4.469	4.372
400	v	1.648	0.1654	0.0830	0.0418	0.0281	0.0212	0.0171	8.88.−3	4.77.−3	3.40.−3	2.72.−3	2.30.−3
	h	504.6	505.0	505.4	506.4	507.4	508.3	509.3	514.4	525.3	536.7	548.4	560.2
	s	7.558	6.609	6.323	6.037	5.896	5.750	5.657	5.369	5.078	4.907	4.785	4.690
500	v	2.060	0.2066	0.1036	0.0521	0.0350	0.0264	0.0213	0.0110	5.82.−3	4.10.−3	3.24.−3	2.73.−3
	h	607.6	608.1	608.6	609.7	610.8	611.9	613.0	618.8	630.7	642.9	655.2	667.5
	s	7.788	6.839	6.553	6.267	6.100	5.981	5.889	5.601	5.313	5.144	5.023	4.929
600	v	2.472	0.2478	0.1242	0.0625	0.0419	0.0316	0.0254	0.0130	6.85.−3	4.80.−3	3.77.−3	3.15.−3
	h	710.6	711.1	711.7	712.9	714.1	715.3	716.5	722.5	735.0	747.8	760.5	773.2
	s	7.975	7.027	6.741	6.455	6.288	6.169	6.077	5.791	5.504	5.335	5.215	5.122
700	v	2.884	0.2890	0.1449	0.0728	0.0487	0.0367	0.0295	0.0151	7.89.−3	5.49.−3	4.29.−3	3.57.−3
	h	813.5	814.1	814.7	816.0	817.2	818.5	819.7	826.0	838.9	851.9	865.0	878.1
	s	8.134	7.186	6.900	6.614	6.447	6.328	6.236	5.950	5.664	5.496	5.376	5.284
800	v	3.296	0.3302	0.1655	0.0831	0.0556	0.0419	0.0336	0.0172	8.92.−3	6.18.−3	4.81.−3	3.98.−3
	h	916.5	917.1	917.7	919.0	920.3	921.6	922.9	929.3	942.4	955.7	969.0	982.2
	s	8.272	7.323	7.038	6.752	6.585	6.466	6.374	6.088	5.802	5.634	5.515	5.423
900	v	3.708	0.3714	0.1861	0.0934	0.0625	0.0470	0.0378	0.0192	9.96.−3	6.87.−3	5.32.−3	4.40.−3
	h	1020	1020	1021	1022	1023	1025	1026	1033	1046	1059	1073	1086
	s	8.393	7.444	7.159	6.873	6.706	6.588	6.496	6.210	5.924	5.756	5.637	5.545
1000	v	4.120	0.4126	0.2067	0.1037	0.0693	0.0522	0.0419	0.0213	0.0110	7.56.−3	5.84.−3	4.81.−3
	h	1123	1123	1124	1125	1126	1128	1129	1136	1149	1163	1176	1190
	s	8.502	7.553	7.267	6.982	6.815	6.696	6.604	6.318	6.032	5.856	5.746	5.654

*Values extracted and in some cases rounded off from those cited in Rabinovich (ed.), *Thermophysical Properties of Neon, Argon, Krypton and Xenon*, Standards Press, Moscow, 1976. This source contains an exhaustive tabulation of values. v = specific volume, m³/kg; h = specific enthalpy, kJ/kg; s = specific entropy, kJ/(kg·K). The notation 6.76.−3 signifies 6.76×10^{-3}. This book was published in English translation by Hemisphere, New York, 1988 (604 pp.).

TABLE 2-288 Saturated Nitrogen (R728)*

T, K	P, bar	v_f, 10^{-3} m³/kg	v_g, m³/kg	h_f, kJ/kg	h_g, kJ/kg	s_f, kJ/(kg·K)	s_g, kJ/(kg·K)	c_{pf}, kJ/(kg·K)	μ_f, 10^{-4} Pa·s	k_f, W/(m·K)
63.15^t	0.1253	1.155	1477	−148.5	64.1	2.459	5.826	1.928		0.170
65	0.1743	1.165	1091	−144.9	65.8	2.516	5.757	1.930	2.74	0.160
70	0.3859	1.193	525.6	−135.2	70.5	2.657	5.595	1.937	2.17	0.151
75	0.7609	1.224	281.8	−125.4	74.9	2.789	5.460	1.948	1.77	0.141
77.35	1.0133	1.239	216.9	−120.8	76.8	2.849	5.404	1.955	1.60	0.136
80	1.369	1.258	164.0	−115.6	78.9	2.913	5.345	1.964	1.48	0.132
85	2.287	1.297	101.7	−105.7	82.3	3.032	5.244	1.989	1.27	0.123
90	3.600	1.340	66.28	−95.6	85.0	3.147	5.152	2.028	1.10	0.114
95	5.398	1.390	44.87	−85.2	86.8	3.256	5.067	2.086	0.97	0.105
100	7.775	1.447	31.26	−74.5	87.7	3.363	4.985	2.176	0.87	0.097
105	10.83	1.514	22.23	−63.8	87.4	3.469	4.904	2.319	0.79	0.088
110	14.67	1.597	15.98	−51.4	85.6	3.575	4.820	2.566	0.71	0.080
115	19.40	1.714	11.47	−38.1	81.8	3.687	4.729	3.063	0.60	0.071
120	25.15	1.892	8.031	−21.4	74.3	3.821	4.619		0.48	0.063
125	32.05	2.324	5.016	5.1	57.2	4.024	4.444		0.32	0.052
126.25^c	33.96	3.289	3.289	34.8	34.8	4.252	4.252	∞		∞

*Reproduced and converted from Vasserman and Rabinovich, *Thermophysical Properties of Liquid Air and Its Components*, Standartov, Moscow, 1968; and Israel Program for Scientific Translations, TT 69-55092, 1970. t = triple point; c = critical point.

Other extensive tables are given by Angus, S., *International Thermodynamic Tables of the Fluid State—6. Nitrogen*, Pergamon, 1977 (244 pp.); Hanley, H. J. M., R. D. McCarty, et al., *J. Phys. Chem. Ref. Data*, **3** (1974): 979–1019.

Saturation and superheat tables to 30,000 psia and a chart to 10,000 psia, all to 860°R, appear in Stewart, R. B., R. T. Jacobsen, et al., *Thermodynamic Properties of Refrigerants*, ASHRAE, Atlanta, GA, 1986 (521 pp.). For specific heat, thermal conductivity, and viscosity, see *Thermophysical Properties of Refrigerants*, ASHRAE, 1993.

The 1993 ASHRAE *Handbook—Fundamentals* (SI ed.) has a thermodynamic chart for pressures from 0.1 to 800 bar and temperatures from 80 to 500 K.

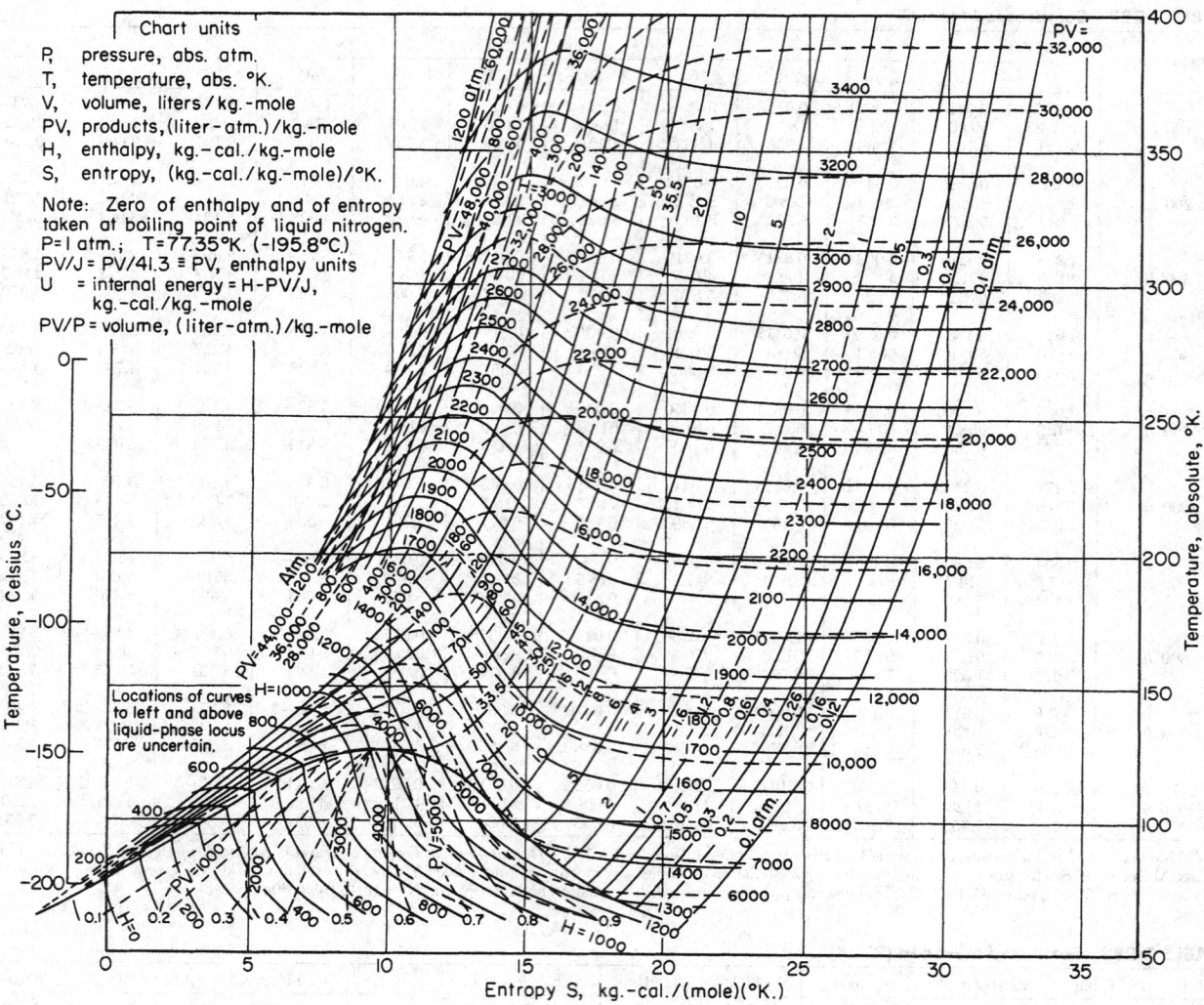

FIG. 2-13 Temperature-entropy diagram for nitrogen. Section of *T-S* diagram for nitrogen by E. S. Burnett, 1950. (*Reprinted from U.S. Bur. Mines Rep. Invest. 4729.*)

TABLE 2-289 Thermophysical Properties of Nitrogen (R728) at Atmospheric Pressure

T (K)	77.4[b]	80	100	120	140	160	180	200	220	240
v (m³/kg)	0.2164	0.2252	0.2871	0.3474	0.4071	0.4664	0.5255	0.5845	0.6434	0.7023
h (kJ/kg)	76.7	80.0	101.9	123.1	144.2	165.2	186.1	207.0	227.8	248.7
s (kJ/kg·K)	5.403	5.446	5.690	5.884	6.046	6.186	6.309	6.419	6.519	6.609
c_p (kJ/kg·K)	1.341	1.196	1.067	1.056	1.050	1.047	1.045	1.043	1.043	1.042
Z	0.9545	0.9610	0.9801	0.9801	0.9883	0.9927	0.9952	0.9967	0.9984	0.9990
$\bar{v}_s$ (m/s)	172	177	202	222	240	257	273	288	302	316
η (10⁻⁶ Pa·s)	5.0	5.2	6.7	8.0	9.3	10.6	11.8	12.9	14.0	15.0
k (W/m·K)	0.0074	0.0077	0.0098	0.0117	0.0136	0.0154	0.0171	0.0187	0.0203	0.0218
N_{Pr}	0.913	0.811	0.728	0.727	0.723	0.721	0.720	0.719	0.718	0.717

T (K)	260	280	300	320	340	360	380	400	420	440
v (m³/kg)	0.7611	0.8199	0.8786	0.9371	0.9960	1.0546	1.1134	1.1719	1.2305	1.2892
h (kJ/kg)	269.5	290.3	311.2	332.0	352.8	373.7	394.5	411.5	436.3	457.3
s (kJ/kg·K)	6.693	6.770	6.842	6.909	6.972	7.032	7.088	7.142	7.193	7.242
c_p (kJ/kg·K)	1.042	1.041	1.041	1.042	1.042	1.043	1.044	1.045	1.047	1.048
Z	0.9994	0.9997	0.9998	0.9999	1.0000	1.0001	1.0002	1.0002	1.0002	1.0003
$\bar{v}_s$ (m/s)	329	341	359	365	376	387	397	408	417	427
η (10⁻⁶ Pa·s)	16.0	17.0	17.9	18.8	19.7	20.5	21.4	22.2	23.0	23.8
k (W/m·K)	0.0232	0.0247	0.0260	0.0273	0.0286	0.0299	0.0311	0.0324	0.0336	0.0347
N_{Pr}	0.717	0.716	0.716	0.717	0.717	0.717	0.717	0.717	0.717	0.717

T (K)	460	480	500	600	700	800	900	1000	1500	2000
v (m³/kg)	1.3481	1.4065	1.4654	1.758	2.052	2.344	2.636	2.931	4.396	5.862
h (kJ/kg)	478.3	499.3	520.4	626.9	735.6	846.6	960.0	1075.7	1680.5	2313.5
s (kJ/kg·K)	7.288	7.333	7.376	7.570	7.738	7.886	8.019	8.141	8.630	8.995
c_p (kJ/kg·K)	1.051	1.053	1.056	1.075	1.098	1.122	1.146	1.167	1.244	1.284
Z	1.0003	1.0004	1.0004	1.000	1.000	1.000	1.000	1.001	1.001	1.001
$\bar{v}_s$ (m/s)	437	446	455	496	534	568	601	631	765	879
η (10⁻⁶ Pa·s)	24.5	25.3	26.0	29.5	32.8	35.9	38.8	41.6		
k (W/m·K)	0.0359	0.0371	0.0383	0.0440	0.0496	0.0551	0.0606	0.658		
N_{Pr}	0.718	0.718	0.718	0.722	0.726	0.730	0.734	0.737		

b = normal boiling point.

TABLE 2-290 Saturated Nitrogen Tetroxide

Pressure, bar	Temperature, K	v_f, m³/kg	v_g, m³/kg	M_f	M_g
1.0133	299.32	0.000 694	0.2996	91.857	79.157
2	309.57	0.000 711	0.1630	91.886	76.503
4	326.66	0.000 733	0.0876	91.766	73.538
6	337.43	0.000 749	0.0608	91.625	71.748
8	345.45	0.000 762	0.0469	91.488	70.480
10	351.88	0.000 774	0.0382	91.346	69.483
15	364.09	0.000 800	0.0262	90.979	67.742
20	373.17	0.000 822	0.0199	90.601	66.547
30	386.57	0.000 863	0.0133	89.823	64.997
40	396.52	0.000 903	0.0098	89.018	64.099
50	404.50	0.000 945	0.00761	88.191	63.532
60	411.20	0.000 993	0.00607	87.344	63.181
80	422.07	0.001 129	0.00394	85.602	62.959
100	430.76	0.001 577	0.00209	83.817	63.366

Condensed from McCarty, R. D., H-U. Steurer, et al., NBS IR 86 - 3054, 1986 (106 pp.). M = mol wt for the reaction N_2O_4 A $2NO_2$ A $2NO + O_2$. No derived thermodynamic functions were tabulated due to unduly large differences in literature values, but 92 references are given.

TABLE 2-291 Saturated Nitrous Oxide

Temp., °F	Pressure, psia	v_f, ft³/lb$_m$	v_g, ft³/lb$_m$	h_f, Btu/lb$_m$	h_g, Btu/lb$_m$	s_f, Btu/lb$_m$·°R	s_g, Btu/lb$_m$·°R
−127.2	14.70	0.01310	5.069	0.0	161.7	0.0000	0.4864
−100	33.68	0.01358	2.374	11.7	165.9	0.0304	0.4591
−80	56.79	0.01398	1.463	20.8	168.8	0.0534	0.4433
−60	90.29	0.01444	0.939	30.2	171.5	0.0782	0.4315
−40	136.68	0.01495	0.648	40.3	173.7	0.1044	0.4222
−20	198.62	0.01555	0.450	50.6	175.3	0.1296	0.4133
0	278.97	0.01625	0.316	60.5	176.2	0.1518	0.4036
20	380.88	0.01711	0.227	70.2	176.2	0.1718	0.3928
40	507.51	0.01819	0.164	80.3	175.0	0.1920	0.3815
60	662.69	0.01968	0.117	91.9	172.1	0.2145	0.3687
80	851.5	0.0222	0.0792	105.7	165.0	0.2382	0.3480
90	961.0	0.0247	0.0611	114.7	157.5	0.2523	0.3302
97.6[c]	1052.2	0.0354	0.0354	136.4	136.4	0.2890	0.2890

Rounded and condensed from Couch, E. J. and K. A. Kobe, Univ. Texas Rep., Cont. DAI-23-072-ORD-685, June 1, 1956. c = critical point.

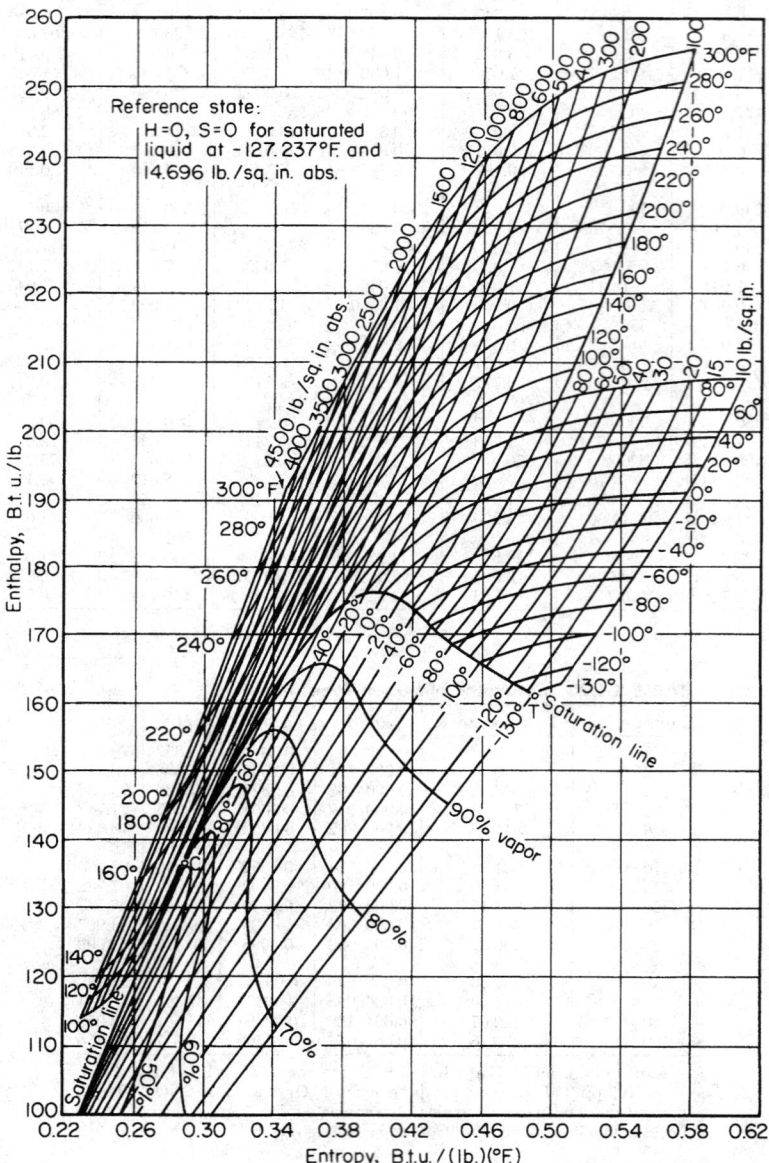

FIG. 2-14 Mollier diagram for nitrous oxide. (*Fig. 9, Univ. Texas Rep., Cont. DAI-23-072-ORD-685, June 1, 1956, by Couch and Kobe. Reproduced by permission.*) Some irregularity in the compressibility factors from 80 to 160 atm, 50 to 100°C exists (*Couch, private communication, 1967*). See Couch et al., *J. Chem. Eng. Data*, **6**, (1961) for PVT data.

TABLE 2-292 Nonane*

T, K	P, bar	v_f, m³/kg	v_g, m³/kg	h_f, kJ/kg	h_g, kJ/kg	s_f, kJ/(kg·K)	s_g, kJ/(kg·K)	c_{pf}, kJ/(kg·K)	μ_f, 10⁻⁴ Pa·s	k_f, W/(m·K)
219.7ᵗ	2.6.−6			358.4		2.424		2.07	33.5	0.150
220	2.7.−6			359.2		2.427		2.07	33.0	0.150
240	3.74.−5			400.6		2.607		2.08	17.9	0.145
260	2.97.−4			442.2	828.7	2.774	4.210	2.10	12.1	0.140
280	1.61.−3			484.8	859.4	2.932	4.243	2.16	8.7	0.134
300	6.40.−3	1.404.−3	30.35	528.6	891.7	3.083	4.282	2.22	6.53	0.129
320	0.0203	1.436.−3	10.19	573.8	925.6	3.229	4.324	2.30	5.13	0.123
340	0.0547	1.471.−3	4.00	622.0	961.1	3.370	4.368		4.16	0.118
360	0.1279	1.508.−3	1.80	671.3	998.2	3.511	4.419		3.44	0.112
380	0.2678	1.548.−3	0.894	722.5	1036.5	3.650	4.476		2.91	0.107
400	0.513	1.591.−3	0.485	776.7	1076.0	3.788	4.536		2.50	0.101
420	0.911	1.637.−3	0.286	833.3	1116.6	3.927	4.601		2.18	0.096
440	1.521	1.690.−3	0.161	890.2	1157.1	4.053	4.660			0.092
460	2.401	1.748.−3	0.104	950.3	1199.2	4.186	4.727			0.089
480	3.639	1.815.−3	0.069	1012.1	1241.3	4.316	4.794			0.085
500	5.309	1.895.−3	0.045	1076.2	1282.9	4.444	4.857			0.082
520	7.437	2.00.−3	0.030	1141.3	1324.5	4.569	4.921			
540	10.20	2.13.−3	0.021	1207.7	1363.8	4.691	3.980			
560	13.76	2.35.−3	0.013	1275.4	1338.7	4.811	5.029			
580	18.02	2.78.−3	0.008	1342.9	1318.1	4.927	5.056			
594.6ᶜ	22.90	4.23.−3	0.004	1305.2	1305.2	5.032	5.032			

*Values of p and v interpolated and converted from tables in Vargaftik, *Handbook of Thermophysical Properties of Gases and Liquids*, Hemisphere, Washington, and McGraw-Hill, New York, 1975. Values of h and s calculated from API tables published by Texas A&M University, College Station. t = triple point; c = critical point.

TABLE 2-293 Octane*

T, K	P, bar	v_f, m³/kg	v_g, m³/kg	h_f, kJ/kg	h_g, kJ/kg	s_f, kJ/(kg·K)	s_g, kJ/(kg·K)	c_{pf}, kJ/(kg·K)	μ_f, 10⁻⁵ Pa·s	k_f, W/(m·K)
216.4ᵗ	1.49.−5			365.9		2.487		2.033	2.25	0.149
220	2.41.−5			373.2		2.520		2.035	2.01	0.148
240	2.18.−4	1.353.−3	700	414.1	811.4	2.698	4.207	2.059	1.24	0.143
260	0.0014	1.368.−3	125	455.8	842.1	2.865	4.259	2.105	0.87	0.138
280	0.0061	1.384.−3	31.9	498.4	873.5	3.023	4.312	2.165	0.65	0.133
300	0.0207	1.420.−3	10.7	542.4	906.2	3.175	4.366	2.231	0.504	0.128
320	0.0575	1.457.−3	4.01	589.8	939.8	3.325	4.419		0.405	0.123
340	0.1384	1.495.−3	1.752	637.9	974.6	3.471	4.461		0.334	0.118
360	0.3000	1.536.−3	0.844	687.1	1010.4	3.611	4.509		0.282	0.112
380	0.5856	1.582.−3	0.448	737.7	1047.3	3.747	4.562		0.244	0.107
400	1.0507	1.632.−3	0.252	790.1	1084.8	3.881	4.617		0.200	0.102
420	1.758	1.685.−3	0.155	843.1	1123.6	4.010	4.677		0.167	0.099
440	2.797	1.747.−3	0.100	897.5	1162.5	4.137	4.740		0.143	0.095
460	4.246	1.818.−3	0.066	954.8	1202.0	4.264	4.802		0.121	0.091
480	6.201	1.904.−3	0.045	1013.5	1241.8	4.388	4.864		0.103	0.087
500	8.785	2.013.−3	0.031	1072.8	1281.2	4.508	4.924		0.086	0.083
520	12.15	2.16.−3	0.021	1136.0	1318.6	4.629	4.980		0.072	
540	16.46	2.37.−3	0.014	1201.5	1352.4	4.749	5.028		0.058	
560	21.98	2.81.−3	0.008	1276.7	1370.4	4.880	5.048		0.044	
568.8ᶜ	24.97	4.26.−3	0.004	1331.7	1331.7	4.977	4.977			

*Values of p and v interpolated and converted from tables in Vargaftik, *Handbook of Thermophysical Properties of Gases and Liquids*, Hemisphere, Washington, and McGraw-Hill, New York, 1975. Values of h and s calculated from API tables published by Texas A&M University, College Station. t = triple point; c = critical point. Saturation and superheat tables and a diagram to 100 bar, 680 K are given by Reynolds, W. C., *Thermodynamic Properties in S.I.*, Stanford Univ. publ., 1979 (173 pp.).

TABLE 2-294 Saturated Oxygen (R732)*

T, K	P, bar	v_f, 10^{-3} m³/kg	v_g, 10^{-3} m³/kg	h_f, kJ/kg	h_g, kJ/kg	s_f, kJ/(kg·K)	s_g, kJ/(kg·K)	c_{pf}, kJ/(kg·K)	μ_f, 10^{-4} Pa·s	k_f, W/(m·K)
54.35^t	0.0015	0.776	93980	−189.8	48.9	2.156	6.548			
55	0.0018	0.778	77920	−188.9	49.5	2.172	6.507			
60	0.0073	0.790	21240	−181.1	53.8	2.308	6.223			
65	0.0233	0.802	7200	−173.3	58.1	2.432	5.992			
70	0.0624	0.816	2894	−165.5	62.4	2.545	5.801			
75	0.1448	0.827	1330	−159.2	66.6	2.631	5.642	1.570	3.04	0.170
80	0.3003	0.845	680.7	−149.7	70.8	2.754	5.510	1.589	2.54	0.164
85	0.5677	0.862	379.7	−141.7	74.9	2.849	5.397	1.607	2.16	0.157
90	0.9943	0.880	227.1	−133.7	78.8	2.940	5.301	1.625	1.88	0.151
90.18	1.0133	0.881	223.2	−133.4	78.9	2.943	5.297	1.626	1.87	0.151
95	1.634	0.899	143.9	−125.4	82.4	3.045	5.216	1.645	1.66	0.144
100	2.547	0.920	95.46	−117.1	85.7	3.113	5.141	1.672	1.51	0.138
105	3.794	0.944	65.81	−108.6	88.5	3.196	5.073	1.706	1.34	0.131
110	5.443	0.970	46.81	−99.9	90.8	3.276	5.009	1.752	1.20	0.125
115	7.559	0.998	34.15	−90.0	92.6	3.354	4.950	1.814	1.07	0.118
120	10.21	1.031	25.42	−81.6	93.6	3.432	4.892	1.896	0.97	0.111
125	13.48	1.070	19.21	−71.8	93.9	3.510	4.836	2.004	0.86	0.103
130	17.44	1.116	14.67	−61.5	93.3	3.588	4.779	2.148	0.78	0.096
135	22.19	1.170	11.25	−50.6	91.6	3.667	4.720	2.341	0.70	0.088
140	27.82	1.237	8.612	−38.9	88.4	3.748	4.657	2.629	0.60	0.080
145	34.45	1.332	6.499	−25.9	82.9	3.833	4.583	3.141	0.52	0.072
150	42.23	1.487	4.705	−10.8	73.1	3.928	4.487	3.935		
154.77^c	50.87	2.464	2.464	35.2	35.2	4.219	4.219	∞		∞

*Reproduced and converted from Vasserman and Rabinovich, *Thermophysical Properties of Liquid Air and Its Components*, Standartov, Moscow, 1968; and Israel Program for Scientific Translations, TT 69-55092, 1970. t = triple point; c = critical point.

Other tables are given by Sytchev, V. V., A. A. Vasserman, et al., *Thermodynamic Properties of Oxygen*, Hemisphere, New York, 1987 (307 pp.); Stewart, R. B., R. T. Jacobsen, et al., *J. Phys. Chem. Ref. Data*, **20**, 5 (1991): 917–1021; For fps units, see *Thermodynamic Properties of Refrigerants*, ASHRAE, Atlanta, GA, 1986 (521 pp.). For specific heat, thermal conductivity, and viscosity, see *Thermophysical Properties of Refrigerants*, ASHRAE, 1993. See also Roder, H. M., *Transport Properties of Oxygen*, NASA Ref. Publ. 1102, 1983 (87 pp.); Laesecke, A., K. Krauss, et al., *J. Phys. Chem. Ref. Data*, **19**, 5 (1990): 1089–1122.

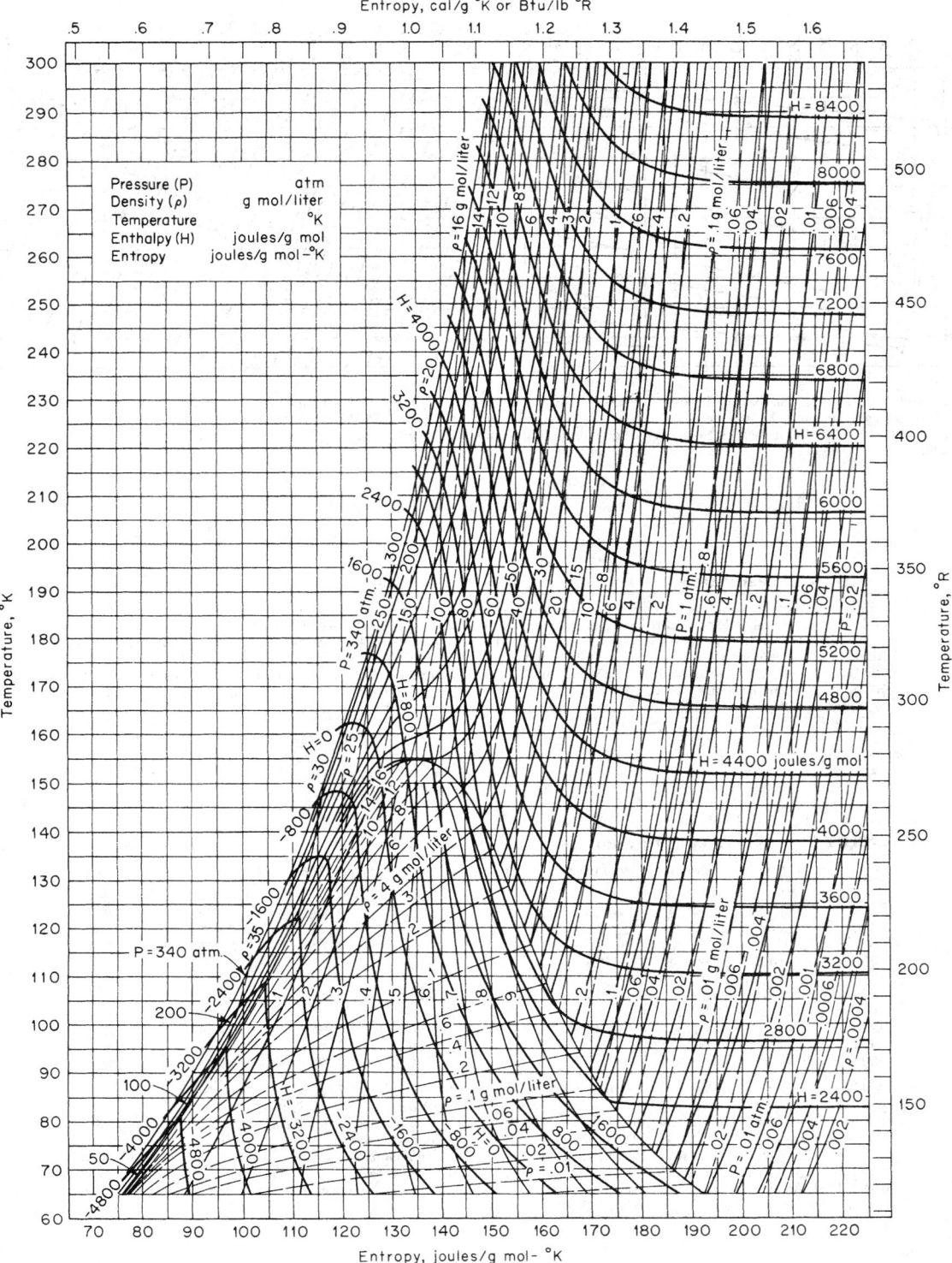

FIG. 2-15 Temperature-entropy chart for oxygen. Pressure *P*, atm; density ρ, (g·mol)/L; temperature, K; enthalpy *H*, J/(g·mol); entropy, J/(g·mol·K). (*NBS Chart D-56. Reproduced by permission.*)

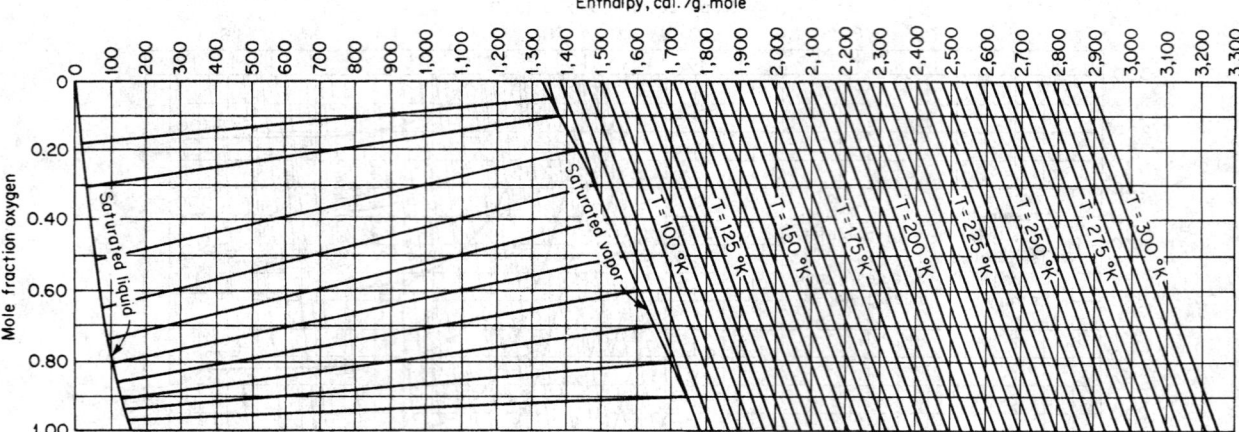

Enthalpy, cal./g. mole

FIG. 2-16   Enthalpy-concentration diagram for oxygen-nitrogen mixture at 1 atm. Reference states: Enthalpies of liquid oxygen and liquid nitrogen at the normal boiling point of nitrogen are zero. (*Dodge,* Chemical Engineering Thermodynamics, *McGraw-Hill, New York, 1944.*) Wilson, Silverberg, and Zellner, AFAPL TDR 64-64 (AD 603151), 1964, p. 314, present extensive vapor-liquid equilibrium data for the three-component system argon-nitrogen-oxygen as well as for binary systems including oxygen-nitrogen.

TABLE 2-295   Pentane

Canjar and Manning (*Thermodynamic Properties and Reduced Correlations for Gases,* Gulf, Houston, 1967) give extensive tables and an enthalpy–log-pressure diagram, based upon Brydon, Walen, and Canjar [*Chem. Eng. Prog. Symp. Ser.,* **49,** 7, (1951): 151–157]. For isopentane, Arnold, Liou, and Eldridge [*J. Chem. Eng. Data,* **10,** 88 (1965)] used the Benedict-Webb-Rubin equation to generate information to 600°F and 60 atm. Das and Kuloor used the same equation in *Ind. J. Technol.,* **5,** 46 (1967) to calculate information up to 1500 K and 1000 atm. Saturation and superheat tables and a diagram to 200 bar, 600 K are given by Reynolds, W. C., *Thermodynamic Properties in S.I.,* Stanford Univ. publ., 1979 (173 pp.). For equations, see Grigoryev, B. A., Yu. L. Rastorguyev, et al., *Int. J. Thermophys.,* **11,** 3 (1990): 487–502.

TABLE 2-296   Saturated Potassium*

T, K	P, bar	v_f, m³/kg	v_g, m³/kg	h_f, kJ/kg	h_g, kJ/kg	s_f, kJ/(kg·K)	s_g, kJ/(kg·K)	c_{pf}, kJ/(kg·K)
336.4^m	1.37.–9	0.001208		93.8	2327	1.928	8.567	0.822
400	1.84.–7	0.001229	4.64.+6	145.5	2342	2.068	7.559	0.805
500	3.13.–5	0.001266	3.39.+4	225.1	2390	2.246	6.576	0.785
600	9.26.–4	0.001304	3164	302.7	2433	2.388	5.937	0.771
700	0.01022	0.001346	142.3	379.4	2468	2.506	5.490	0.762
800	0.06116	0.001389	26.75	455.5	2498	2.608	5.161	0.761
1000	0.7322	0.001488	2.691	609.7	2552	2.780	4.722	0.792
1200	3.913	0.001605	0.584	773.5	2610	2.929	4.459	0.846
1400	12.44	0.001742	0.207	948.0	2679	3.063	4.299	0.899
1500	20.0	0.001816	0.132	1040.0	2718	3.123	4.209	0.924

*Converted from tables in Vargaftik, *Tables of the Thermophysical Properties of Liquids and Gases,* Nauka, Moscow, 1972; and Hemisphere, Washington, 1975. m = melting point. The notation 1.37.–9 signifies 1.37×10^{-9}.

Many of the Vargaftik values also appear in Ohse, R. W., *Handbook of Thermodynamic and Transport Properties of Alkali Metals,* Blackwell Sci. Pubs., Oxford, 1985 (1020 pp.). This source contains superheat data. Saturation and superheat tables and a diagram to 30 bar, 1650 K are given by Reynolds, W. C., *Thermodynamic Properties in S.I.,* Stanford Univ. publ., 1979 (173 pp.). For a Mollier diagram from 0.1 to 250 psia, 1300 to 2700°R, see Weatherford, W. D., J. C. Tyler, et al., WADD-TR-61-96, 1961. An extensive review of properties of the solid and the saturated liquid is given by Alcock, C. B., M. W. Chase, et al., *J. Phys. Chem. Ref. Data,* **23,** 3 (1994): 385–497.

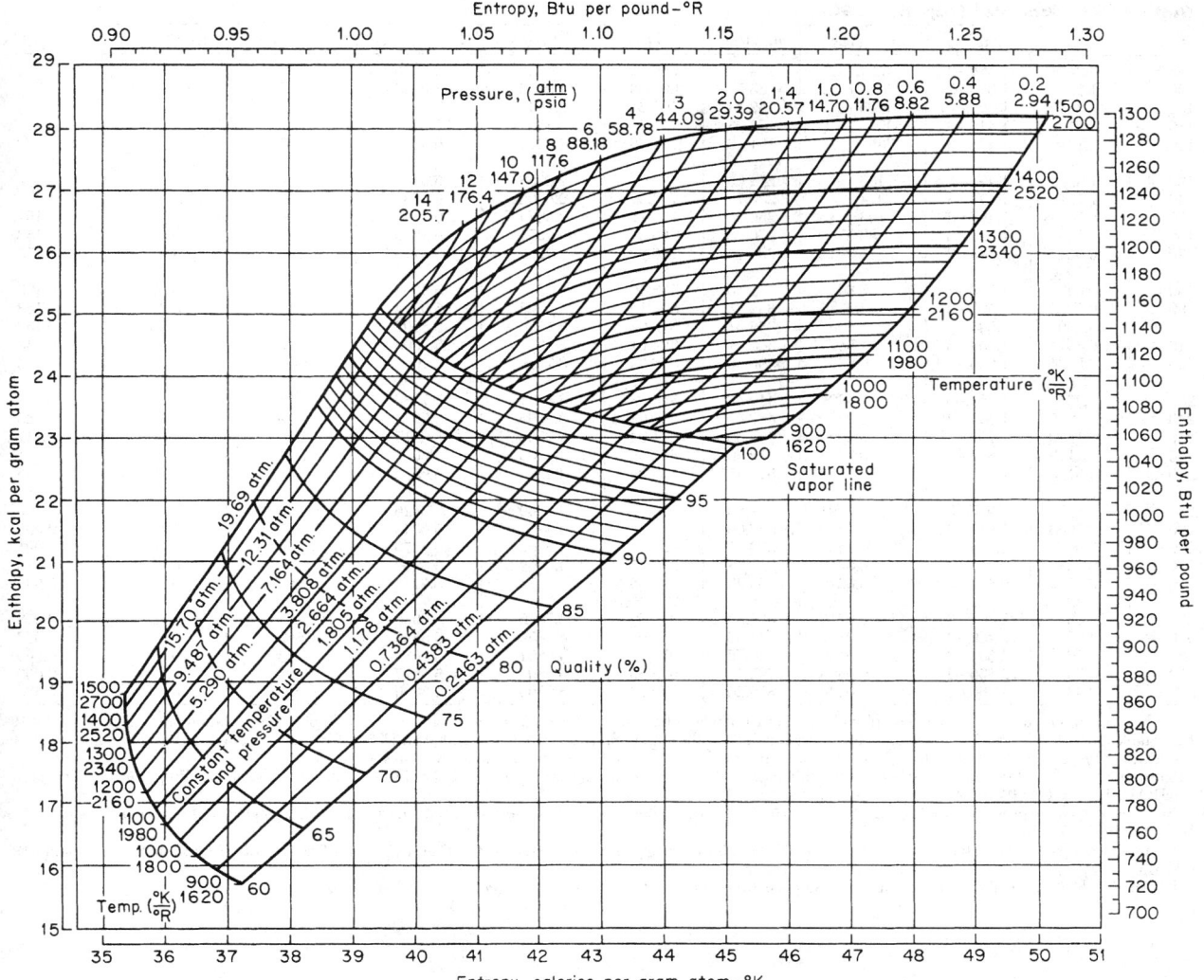

FIG. 2-17 Mollier diagram for potassium. Basis: enthalpy = 0.0 cal/g atom at 298 K; entropy = 15.8 cal/(g atom·K) at 298 K. (*Aerojet-General Rep. AGN8194, vol. 2, 1967. Reproduced by permission.*)

TABLE 2-297   Saturated Propane (R290)*

T, K	P, bar	v_f, m³/kg	v_g, m³/kg	h_f, kJ/kg	h_g, kJ/kg	s_f, kJ/(kg·K)	s_g, kJ/(kg·K)	c_{pf}, kJ/(kg·K)	μ_f, 10^{-4} Pa·s	k_f, W/(m·K)
85.5^t	3.0.−9	1.364.−3	5.37.+7	124.92	690.02	1.8738	8.3548	1.92		
90	1.5.−8	1.373.−3	1.12.+7	133.56	693.58	1.9723	8.0953	1.92		
100	3.2.−7	1.392.−3	5.85.+5	152.74	702.23	2.1743	7.6163	1.93		
110	3.9.−6	1.412.−3	53275	172.03	711.71	2.3581	7.2377	1.94		
120	3.1.−5	1.432.−3	7350	191.46	721.78	2.5271	6.9343	1.95		
130	1.8.−4	1.453.−3	1400	211.03	732.27	2.6838	6.6885	1.96		
140	7.7.−4	1.475.−3	344	230.77	743.07	2.8300	6.4881	1.98		
150	2.74.−3	1.497.−3	103	250.67	754.12	2.9674	6.3237	2.00	6.61	0.191
160	8.22.−3	1.521.−3	36.8	270.78	765.37	3.0971	6.1886	2.02	5.54	0.183
170	0.0214	1.545.−3	15.0	291.10	776.80	3.2202	6.0775	2.04	4.67	0.175
180	0.0495	1.570.−3	6.84	311.66	788.40	3.3377	5.9862	2.07	3.97	0.166
190	0.1035	1.597.−3	3.43	332.48	800.15	3.4503	5.9114	2.10	3.27	0.158
200	0.1993	1.625.−3	1.868	353.61	812.03	3.5586	5.8502	2.13	2.98	0.150
210	0.3574	1.654.−3	1.087	375.07	824.01	3.6631	5.8005	2.16	2.65	0.143
220	0.6031	1.686.−3	0.669	396.90	836.04	3.7645	5.7603	2.20	2.36	0.136
230	0.9661	1.719.−3	0.432	419.16	848.08	3.8631	5.7280	2.25	2.07	0.129
240	1.4800	1.754.−3	0.290	442.07	860.07	3.9605	5.7022	2.29	1.86	0.123
250	2.1819	1.792.−3	0.2020	465.58	871.94	4.0563	5.6817	2.34	1.69	0.117
260	3.1118	1.833.−3	0.1445	489.70	883.62	4.1505	5.6656	2.41	1.53	0.111
270	4.3120	1.878.−3	0.1059	514.45	895.02	4.2433	5.6528	2.48	1.40	0.106
280	5.8278	1.927.−3	0.0791	539.88	906.03	4.3349	5.6426	2.56	1.29	0.100
290	7.7063	1.982.−3	0.0600	566.06	916.54	4.4257	5.6343	2.65	1.19	0.096
300	9.9973	2.044.−3	0.0461	593.11	926.41	4.5160	5.6270	2.76	1.10	0.091
310	12.75	2.115.−3	0.0357	621.18	935.45	4.6062	5.6200	2.89	0.93	0.086
320	16.03	2.200.−3	0.0279	650.49	943.38	4.6971	5.6124	3.06	0.82	0.082
330	19.88	2.301.−3	0.0218	681.37	949.79	4.7896	5.6030	3.28	0.72	0.078
340	24.36	2.430.−3	0.0170	714.38	953.92	4.8850	5.5896	3.62	0.62	0.073
350	29.56	2.607.−3	0.0130	750.52	954.23	4.9861	5.5681	4.23	0.52	0.069
360	35.55	2.896.−3	0.0095	792.50	946.56	5.0997	5.5277	5.98	0.40	0.066
369.8^c	42.42	4.566.−3	0.0046	879.20	879.20	5.3300	5.3300	∞	0.29	∞

*Values converted and mostly rounded off from those of Goodwin, NBSIR 77-860, 1977. t = triple point; c = critical point. The notation 3.0.−9 signifies 3.0×10^{-9}. Later tables for the same temperature range for saturation and for the superheat state from 0.1 to 1000 bar, 85.5 to 600 K, were published by Younglove, B. A. and J. F. Ely, *J. Phys. Chem. Ref. Data,* **16,** 4 (1987): 685–721, but the lower temperature saturation tables contain some errors.

Saturation and superheat tables and a chart to 10,000 psia, 800°F appear in Stewart, R. B., R. T. Jacobsen, et al., *Thermodynamic Properties of Refrigerants,* ASHRAE, Atlanta, GA, 1986 (521 pp.).

For thermodynamic properties for 0.1 to 1000 bar, 100 to 700 K, see Sychev, V. V., A. A. Vasserman, et al., *Thermodynamic Properties of Propane,* Hemisphere, New York, NY, 1991 (275 pp.).

Saturation and superheat tables and a diagram to 200 bar, 600 K are given by Reynolds, W. C., *Thermodynamic Properties in S.I.,* Stanford Univ. publ., 1979 (173 pp.).

For specific heat, thermal conductivity, and viscosity, see *Thermophysical Properties of Refrigerants,* ASHRAE, 1993.

TABLE 2-298 Saturated Propylene (Propene, R1270)

T, K	P, bar	v_f, m³/kg	v_g, m³/kg	h_f, kJ/kg	h_g, kJ/kg	s_f, kJ/(kg·K)	s_g, kJ/(kg·K)	c_{pf}, kJ/(kg·K)	μ_f, 10⁻⁶ Pa·s	k_f, W/(m·K)	Pr
87.9^t	9.54.−9	0.001 301	1.82.+7	−290.1	279.2	−1.923	4.554				
90	2.05.−8	0.001 305	8.66.+6	−285.1	281.1	−1.867	4.424				
100	4.81.−7	0.001 325	411 165	−265.4	290.2	−1.659	3.897	1.695	2017	0.214	15.98
110	6.08.−6	0.001 346	35 753	−247.7	299.6	−1.490	3.488	1.760	1526	0.209	12.85
120	4.88.−5	0.001 367	4 856	−229.8	309.3	−1.335	3.158	1.820	1185	0.204	10.57
130	2.77.−4	0.001 389	927.0	−211.4	319.3	−1.187	2.895	1.875	941	0.198	8.91
140	1.20.−3	0.001 411	230.91	−192.4	329.4	−1.046	2.681	1.923	735	0.193	7.32
150	4.17.−3	0.001 434	71.043	−172.9	339.8	−0.912	2.506	1.964	587	0.188	6.13
160	0.0122	0.001 458	25.903	−153.1	350.4	−0.784	2.363	1.996	478	0.183	5.21
170	0.0309	0.001 483	10.842	−133.1	361.2	−0.663	2.245	2.020	397	0.178	4.50
180	0.0697	0.001 508	5.080	−112.7	372.1	−0.547	2.147	2.044	334.5	0.173	3.95
190	0.1425	0.001 535	2.613	−92.2	383.1	−0.436	2.066	2.067	286.1	0.168	3.52
200	0.2686	0.001 563	1.452	−71.4	394.2	−0.329	1.999	2.094	244.9	0.162	3.17
210	0.4727	0.001 593	0.860	−50.3	405.3	−0.226	1.943	2.128	212.7	0.157	2.88
220	0.7849	0.001 624	0.538	−28.8	416.3	−0.127	1.896	2.162	187.0	0.152	2.66
225.5^b	1.0133	0.001 642	0.4241	−16.9	422.2	−0.073	1.874	2.182	175.0	0.149	2.56
230	1.2401	0.001 657	0.3515	−7.0	427.1	−0.030	1.857	2.199	166.2	0.147	2.49
240	1.8775	0.001 693	0.2388	15.3	437.8	0.064	1.825	2.243	149.2	0.142	2.36
250	2.7401	0.001 732	0.1674	38.0	448.2	0.157	1.797	2.298	135.0	0.137	2.26
260	3.8737	0.001 774	0.1206	61.3	458.2	0.247	1.774	2.369	123.0	0.131	2.22
270	5.3269	0.001 820	0.0888	85.2	467.8	0.336	1.753	2.418	112.9	0.126	2.17
280	7.1499	0.001 872	0.0666	109.9	476.9	0.425	1.735	2.494	106.6	0.121	2.20
290	9.3954	0.001 929	0.0507	135.3	485.3	0.512	1.719	2.584	100.0	0.116	2.23
300	12.12	0.001 995	0.0390	161.6	492.8	0.600	1.704	2.693	93.0	0.112	2.24
310	15.38	0.002 071	0.0303	189.0	499.3	0.688	1.688	2.842	85.6	0.109	2.23
320	19.23	0.002 162	0.0236	217.7	504.3	0.776	1.672	3.007	77.8	0.104	2.25
330	23.75	0.002 273	0.0184	248.2	507.4	0.867	1.652	3.335	69.6	0.097	2.39
340	29.01	0.002 418	0.0142	280.9	507.6	0.961	1.627	3.723	61.0	0.090	2.52
350	35.12	0.002 628	0.0107	317.6	502.8	1.062	1.592	4.669		0.082	
360	42.20	0.003 038	0.0075	364.1	486.0	1.188	1.527				
365.6^c	46.65	0.004 476	0.0045	433.3	433.3	1.374	1.374				

t = triple point; b = normal boiling point; c = critical point. The notation 9.54.−9 signifies 9.54×10^{-9}. $h_f = s_f = 0$ at 233.15 K = −40°C. Converted from Angus, S., B. Armstrong, et al., *Intnl. Thermodynamic Properties of the Fluid State—7. Propylene (Propene) R1270*, Pergamon Press, Oxford, 1980 (401 pp.).

TABLE 2-299 Compressed Propylene (Propene, R1270)

Pressure, bar		225	250	275	300	325	350	375	400	425	450
						Temperature, K					
1	v (m³/kg)	0.00164	0.4817	0.5334	0.5846	0.6354	0.6858	0.7361	0.7861	0.8361	0.8859
	h (kJ/kg)	−17.9	455.5	491.2	529.1	569.1	611.4	656.0	702.8	751.9	803.2
	s (kJ/kg·K)	−0.0779	2.0169	2.1530	2.2847	2.4128	2.5380	2.6610	2.7821	2.9010	3.0183
10	v (m³/kg)	0.00164	0.00173	0.00184	0.04986	0.05670	0.06295	0.06889	0.07460	0.08014	0.08557
	h (kJ/kg)	−17.2	38.5	97.6	501.2	547.0	593.2	640.8	689.7	740.6	793.3
	s (kJ/kg·K)	−0.0810	0.1535	0.3788	1.7631	1.9653	2.0466	2.1774	2.3042	2.4273	2.5480
20	v (m³/kg)	0.00163	0.00172	0.00184	0.00198	0.02324	0.02773	0.03149	0.03489	0.03803	0.04104
	h (kJ/kg)	−16.3	39.3	98.1	161.5	512.7	568.2	621.5	673.5	726.5	781.5
	s (kJ/kg·K)	−0.0841	0.1497	0.3736	0.5941	1.6920	1.8567	2.0024	2.1381	2.2674	2.3923
40	v (m³/kg)	0.00163	0.00172	0.00182	0.00196	0.00216	0.00256		0.01465	0.01682	0.01872
	h (kJ/kg)	−14.4	40.8	99.0	161.3	230.0	313.3		633.6	695.4	755.6
	s (kJ/kg·K)	−0.0908	0.1419	0.3638	0.5806	0.8001	1.0466		1.9256	2.0758	2.2131
60	v (m³/kg)	0.00162	0.00171	0.00181	0.00194	0.00211	0.00240		0.00743	0.00944	0.01126
	h (kJ/kg)	−12.6	42.3	100.1	161.5	228.2	303.8		575.4	656.7	726.1
	s (kJ/kg·K)	−0.0970	0.1345	0.3546	0.5684	0.7816	1.0055		1.7272	1.9250	2.0832
80	v (m³/kg)	0.00162	0.00170	0.00180	0.00192	0.00208	0.00231		0.00402	0.00605	0.00757
	h (kJ/kg)	−10.7	44.0	101.3	162.0	227.2	299.0		499.7	607.7	693.3
	s (kJ/kg·K)	−0.1031	0.1274	0.3458	0.5570	0.7657	0.9781		1.5107	1.7795	1.9693
100	v (m³/kg)	0.00161	0.00169	0.00179	0.00190	0.00204	0.00224	0.00256	0.00316	0.00426	0.00551
	h (kJ/kg)	−8.8	45.6	102.6	162.7	226.7	296.1	373.5	464.9	567.8	660.1
	s (kJ/kg·K)	−0.1091	0.1202	0.3374	0.5466	0.7514	0.9570	1.1704	1.4061	1.6456	1.8669
150	v (m³/kg)	0.00160	0.00167	0.00176	0.00186	0.00198	0.00214	0.00234	0.00262	0.00300	0.00354
	h (kJ/kg)	−4.1	49.9	106.1	165.0	227.0	292.8	362.9	438.5	519.5	604.6
	s (kJ/kg·K)	−0.1236	0.1038	0.3180	0.5228	0.7214	0.9163	1.1100	1.3049	1.5021	1.6958
200	v (m³/kg)	0.00159	0.00166	0.00174	0.00183	0.00194	0.00207	0.00222	0.00242	0.00266	0.00296
	h (kJ/kg)	0.8	54.4	110.0	167.9	228.7	292.3	359.2	429.9	504.4	581.4
	s (kJ/kg·K)	−0.1371	0.0884	0.3004	0.5021	0.6963	0.8852	1.0701	1.2521	1.4319	1.6086
300	v (m³/kg)	0.00157	0.00163	0.00170	0.00178	0.00187	0.00197	0.00208	0.00221	0.00236	0.00253
	h (kJ/kg)	10.9	63.8	118.5	175.2	234.2	295.5	359.3	425.7	494.4	565.6
	s (kJ/kg·K)	−0.1625	0.0601	0.2688	0.4660	0.6549	0.8367	1.0127	1.1839	1.3507	1.5133
400	v (m³/kg)	0.00155	0.00161	0.00167	0.00174	0.00182	0.00190	0.00199	0.00209	0.00220	0.00232
	h (kJ/kg)	21.2	73.6	127.7	183.6	241.5	301.5	363.7	428.0	494.4	562.9
	s (kJ/kg·K)	−0.1863	0.0347	0.2407	0.4351	0.6207	0.7985	0.9705	1.1362	1.2972	1.4536
500	v (m³/kg)	0.00153	0.00159	0.00165	0.00171	0.00178	0.00185	0.00193	0.00201	0.00210	0.00220
	h (kJ/kg)	31.6	83.7	137.3	192.6	249.8	309.0	370.1	433.3	498.7	565.4
	s (kJ/kg·K)	−0.2082	0.0112	0.2155	0.4080	0.5910	0.7664	0.9351	1.0981	1.2559	1.4092

Converted and interpolated from Angus, S., B. Armstrong, et al., *International Thermodynamic Tables of the Fluid State—7. Propylene*, Pergamon, Oxford, 1980 (401 pp.).

The 1993 ASHRAE *Handbook—Fundamentals* (SI ed.) has a thermodynamic chart for pressures from 0.1 to 1000 bar for temperatures up to 580 K. Saturation and superheat tables and a diagram to 30,000 psia, 580°F appear in Stewart, R. B., R. T. Jacobsen, et al., *Thermodynamic Properties of Refrigerants*, ASHRAE, Atlanta, GA, 1986 (521 pp.). For specific heat, thermal conductivity, and viscosity, see *Thermophysical Properties of Refrigerants*, ASHRAE, 1993.

TABLE 2-300 Saturated Refrigerant 11*

T, K	P, bar	v_f, m³/kg	v_g, m³/kg	h_f, kJ/kg	h_g, kJ/kg	s_f, kJ/(kg·K)	s_g, kJ/(kg·K)	c_{pf}, kJ/(kg·K)	μ_f, 10^{-4} Pa·s	k_f, W/(m·K)
200	0.0043	5.901.−4	28.06	−14.37	186.30	−0.0651	0.9431	0.815	1.674	0.115
220	0.0417	6.061.−4	6.272	−8.20	195.89	−0.0361	0.8925	0.828	1.142	0.110
240	0.0768	6.225.−4	1.882	4.97	205.85	0.0210	0.8581	0.842	0.831	0.104
260	0.2215	6.398.−4	0.703	21.01	216.06	0.0851	0.8353	0.856	0.635	0.098
270	0.3514	6.491.−4	0.458	29.53	221.23	0.1172	0.8272	0.863	0.563	0.095
280	0.5364	6.587.−4	0.309	38.25	226.40	0.1489	0.8209	0.870	0.504	0.093
290	0.7917	6.688.−4	0.216	47.10	231.58	0.1799	0.8160	0.878	0.454	0.090
300	1.1341	6.794.−4	0.154	56.06	236.73	0.2102	0.8124	0.887	0.413	0.087
310	1.5821	6.908.−4	0.113	65.10	241.83	0.2397	0.8099	0.897	0.377	0.084
320	2.1556	7.027.−4	0.0847	74.22	246.88	0.2686	0.8081	0.907	0.346	0.081
330	2.876	7.156.−4	0.0645	83.42	251.84	0.2967	0.8071	0.917	0.320	0.079
340	3.764	7.293.−4	0.0500	92.72	256.69	0.3243	0.8065	0.928	0.297	0.076
350	4.845	7.442.−4	0.0392	102.12	261.40	0.3513	0.8064	0.939	0.276	0.073
360	6.142	7.603.−4	0.0311	111.64	265.95	0.3778	0.8065	0.950	0.259	0.070
380	9.487	7.974.−4	0.0201	131.12	274.40	0.4298	0.8069	0.975	0.229	0.065
400	14.02	8.435.−4	0.0134	151.38	281.69	0.4808	0.8066	1.004	0.203	0.059
420	19.98	9.042.−4	0.0090	172.76	287.20	0.5317	0.8041	1.04	0.169	0.053
440	27.65	9.930.−4	0.0059	196.01	289.72	0.5840	0.7970	1.09	0.131	0.048
460	37.36	1.167.−3	0.0036	223.85	285.36	0.6435	0.7773	1.19	0.084	0.037
471.2ᶜ	44.09	1.799.−3	0.0018	258.70	258.70	0.7162	0.7162	∞	0.033	∞

°Values reproduced or converted from Table 1, p. 17.75, *ASHRAE Handbook, 1981: Fundamentals*, American Society of Heating, Refrigerating and Air-Conditioning Engineers, Atlanta, 1981. Copyright material. Reproduced by permission of the copyright owner. *c* = critical point. The notation 5.901.−4 signifies 5.901×10^{-4}. The 1993 ASHRAE *Handbook—Fundamentals* (SI ed.) gives material for integral degrees Celsius with temperatures on the ITS 90 scale. For experimental isochores for the compressed liquid from 12 to 301 bar, 254 to 453 K, see Blanke, W. and R. Weiss, *PTB Bericht W 30*, Braunschweig, Germany, 1992 (54 pp.). Equations and constants approximated to 1985 ASHRAE tables are given by Mecarik, K. and M. Masaryk, *Heat Recovery Systems and CHP*, **11**, 2/3 (1991): 193–197. For tables and a chart to 3000 psia, 460°F, see Stewart, R. B., R. T. Jacobsen, et al., *Thermodynamic Properties of Refrigerants*, ASHRAE, Atlanta, GA, 1986 (521 pp.). For similar material to 80 bar, 650 K, see Reynolds, W. C., *Thermodynamic Properties in S.I.*, Stanford Univ. publ., 1979 (173 pp.). For specific heat at constant pressure, thermal conductivity, and viscosity in both SI and fps units, see Liley, P. E., *Thermophysical Properties of Refrigerants*, ASHRAE, Atlanta, GA, 1993.

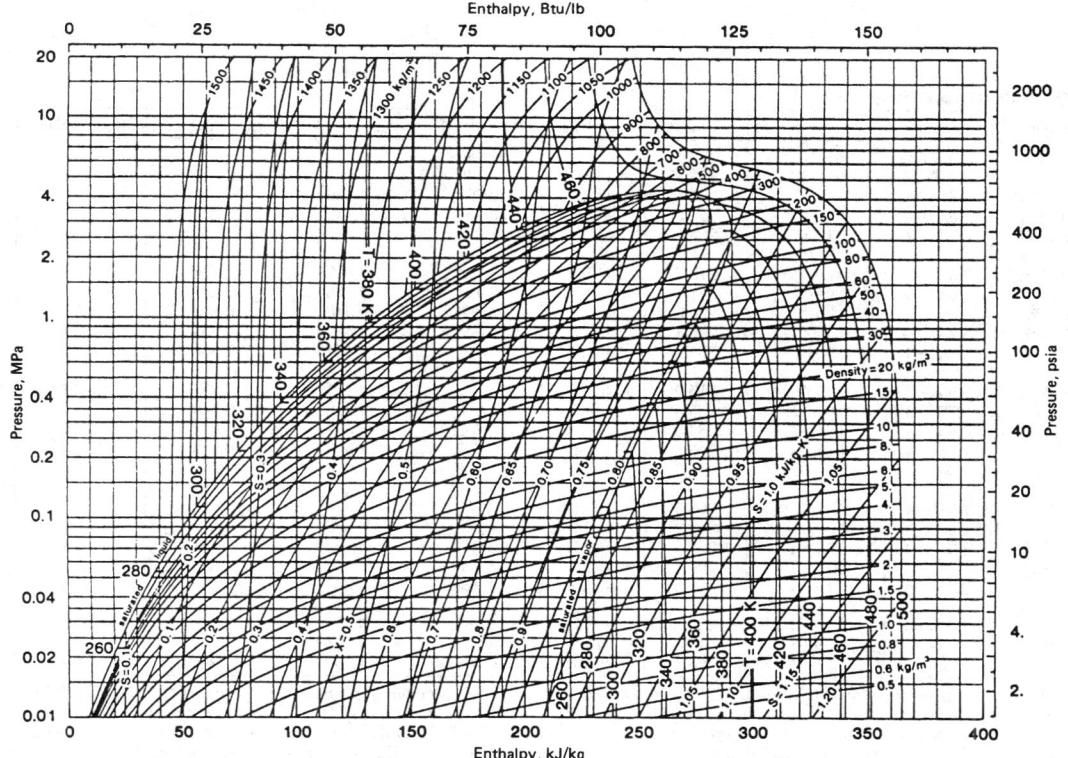

FIG. 2-18 Enthalpy–log-pressure diagram for Refrigerant 11. 1 MPa = 10 bar. (*Copyright 1981 by the American Society of Heating, Refrigerating and Air-Conditioning Engineers and reproduced by permission of the copyright owner.*) This chart, redrawn with a different zero point and temperatures in Celsius, appears in ASHRAE *Handbook—Fundamentals* (SI ed.), Atlanta, GA, 1993.

TABLE 2-301   Saturated Refrigerant 12*

T, K	P, bar	v_f, m³/kg	v_g, m³/kg	h_f, kJ/kg	h_g, kJ/kg	s_f, kJ/(kg·K)	s_g, kJ/(kg·K)	c_{pf}, kJ/(kg·K)	μ_f, 10^{-4} Pa·s	k_f, W/(m·K)
150	0.00091	5.767.−4	179.12	294.6	496.0	3.492	4.835	0.808	18.9	0.123
160	0.00305	5.849.−4	36.05	302.3	500.2	3.543	4.780	0.817	15.1	0.119
170	0.00871	5.926.−4	13.40	310.3	504.5	3.591	4.734	0.827	12.1	0.116
180	0.02178	6.024.−4	5.666	318.3	508.9	3.637	4.696	0.836	9.69	0.113
190	0.04877	6.118.−4	2.665	326.5	513.5	3.681	4.665	0.845	7.94	0.109
200	0.0996	6.217.−4	1.370	334.8	518.1	3.724	4.640	0.855	6.64	0.105
210	0.1879	6.139.−4	0.7589	343.2	522.7	3.765	4.620	0.864	5.65	0.102
220	0.3317	6.431.−4	0.4476	351.8	527.4	3.805	4.603	0.873	4.88	0.098
230	0.5531	6.549.−4	0.2784	360.6	531.1	3.844	4.590	0.882	4.26	0.094
240	0.8781	6.675.−4	0.1811	369.5	536.8	3.881	4.579	0.891	3.77	0.090
250	1.3359	6.810.−4	0.1225	378.0	541.5	3.918	4.570	0.902	3.37	0.087
260	1.959	6.970.−4	0.08559	387.7	546.1	3.954	4.563	0.913	3.03	0.083
270	2.784	7.112.−4	0.06147	397.0	550.7	3.989	4.558	0.926	2.75	0.080
280	3.825	7.282.−4	0.04543	406.5	555.1	4.023	4.554	0.942	2.52	0.076
290	5.184	7.470.−4	0.03888	416.1	559.4	4.057	4.551	0.959	2.31	0.072
300	6.840	7.678.−4	0.02582	426.0	563.5	4.090	4.548	0.979	2.14	0.069
310	8.860	7.912.−4	0.01992	436.0	567.3	4.122	4.546	1.005	2.00	0.065
320	11.29	8.173.−4	0.01553	446.2	570.9	4.154	4.543	1.041	1.86	0.061
330	14.17	8.478.−4	0.01218	456.8	574.0	4.186	4.541	1.093	1.74	0.058
340	17.58	8.840.−4	0.00957	467.8	576.5	4.218	4.538	1.166	1.60	0.054
350	21.57	9.286.−4	0.00750	479.4	578.2	4.250	4.533	1.264	1.45	0.050
360	26.19	9.868.−4	0.00582	492.1	578.7	4.285	4.525	1.39	1.28	0.046
370	31.56	1.072.−3	0.00439	506.4	577.2	4.322	4.514	1.55	1.06	0.041
380	37.76	1.237.−3	0.00305	524.7	571.2	4.369	4.900		0.75	
385^c	41.31	1.876.−3	0.00188	551.1	551.1	4.437	4.437	∞	0.31	∞

*P, v, h, and s data interpolated from Perelshteyn (ed.), *Tables and Diagrams of the Thermodynamic Properties of Refrigerants 12, 13, and 22*, Moscow, 1971. c_p, μ, and k data interpolated and converted from *Thermophysical Properties of Refrigerants*, American Society of Heating, Refrigerating and Air-Conditioning Engineers, New York, 1976. c = critical point. The notation 5.767.−4 signifies 5.767×10^{-4}.

TABLE 2-302   Saturated Refrigerant 13*

T, K	P, bar	v_f, m³/kg	v_g, m³/kg	h_f, kJ/kg	h_g, kJ/kg	s_f, kJ/(kg·K)	s_g, kJ/(kg·K)	c_{pf}, kJ/(kg·K)	μ_f, 10^{-4} Pa·s	k_f, W/(m·K)
91	3.817.−6	5.367.−4	19557	238.1	424.9	3.080	5.133			
100	3.418.−5	5.448.−4	2392	243.8	429.0	3.140	4.990			
110	2.563.−4	5.538.−4	347.0	251.1	433.1	3.205	4.860			
120	0.00137	5.635.−4	70.25	258.2	437.2	3.267	4.759			
130	0.00571	5.739.−4	18.15	265.8	441.3	3.327	4.677			
140	0.01895	5.850.−4	5.865	273.7	455.3	3.385	4.610			
150	0.05250	5.969.−4	2.2617	281.7	449.3	3.441	4.558	0.826	6.83	0.114
160	0.1258	6.095.−4	1.0019	290.0	453.5	3.494	4.516	0.845	5.60	0.109
170	0.2680	6.231.−4	0.4962	298.4	457.6	3.545	4.482	0.865	4.59	0.104
180	0.5186	6.380.−4	0.2689	307.1	461.8	3.594	4.454	0.884	3.83	0.099
190	0.9269	6.536.−4	0.1567	315.9	465.9	3.642	4.431	0.898	3.26	0.093
200	1.5507	6.709.−4	9.69.−2	325.0	469.9	3.688	4.413	0.910	2.82	0.088
210	2.456	6.899.−4	6.28.−2	334.3	473.8	3.732	4.397	0.924	2.48	0.083
220	3.712	7.110.−4	4.24.−2	343.8	477.5	3.777	4.385	0.943	2.20	0.078
230	5.396	7.346.−4	2.95.−2	353.6	481.0	3.820	4.374	0.972	1.97	0.072
240	7.589	7.615.−4	2.11.−2	363.5	484.1	3.862	4.364	1.014	1.79	0.067
250	10.37	7.928.−4	1.53.−2	373.9	486.1	3.903	4.355	1.072	1.63	0.062
260	13.85	8.302.−4	1.13.−2	384.7	489.1	3.944	4.346	1.151	1.50	0.057
270	18.13	8.769.−4	8.28.−3	396.2	490.5	3.986	4.336	1.255	1.34	0.051
280	23.32	9.320.−4	6.10.−3	408.8	490.6	4.029	4.323	1.386	1.14	0.045
290	29.57	1.035.−3	4.34.−3	423.6	488.3	4.080	4.303	1.549	0.87	0.038
300	37.05	1.284.−3	2.60.−3	445.3	477.5	4.151	4.257	1.75	0.52	
302.0^c	38.70	1.808.−3	1.81.−3	463.1	463.1	4.209	4.209	∞	0.29	∞

*P, v, h, and s data interpolated from Perelshteyn (ed.), *Tables and Diagrams of the Thermodynamic Properties of Refrigerants 12, 13 and 22*, Moscow, 1971. c_p, μ, and k data interpolated and converted from *Thermophysical Properties of Refrigerants*, American Society of Heating, Refrigerating and Air-Conditioning Engineers, New York, 1976. c = critical point. The notation 3.817.−6 signifies 3.817×10^{-6}. The 1993 ASHRAE *Handbook—Fundamentals* (SI ed.) contains a table at closer temperature increments and also an enthalpy–log-pressure diagram from 0.1 to 70 bar, −100 to 240°C. Equations and constants approximated to 1985 ASHRAE tables are given by Mecarik, K. and M. Masaryk, *Heat Recovery Systems and CHP*, **11**, 2/3 (1991): 193–197. Saturation and superheat tables and a diagram to 60 bar, 600 K are given by Reynolds, W. C., *Thermodynamic Properties in SI*, Stanford Univ. publ., 1979 (173 pp.). For tables and a chart to 1000 psia, 520°F, see Stewart, R. B., R. T. Jacobsen, et al., *Thermodynamic Properties of Refrigerants*, ASHRAE, Atlanta, GA, 1986 (521 pp.). For specific heat, thermal conductivity, and viscosity, see *Thermophysical Properties of Refrigerants*, ASHRAE, 1993.

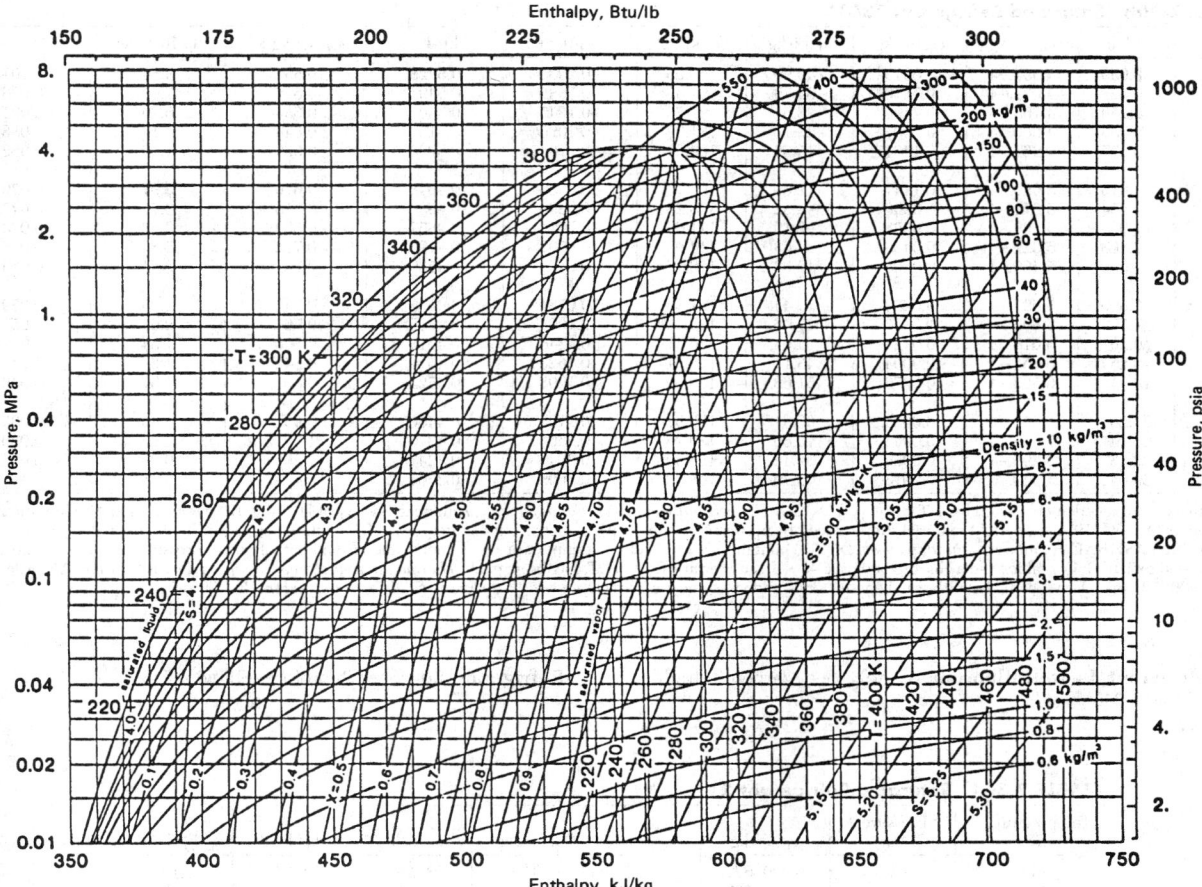

FIG. 2-19 Enthalpy–log-pressure diagram for Refrigerant 12. 1 MPa = 10 bar. (*Copyright 1981 by the American Society of Heating, Refrigerating and Air-Conditioning Engineers and reproduced by permission of the copyright owner.*) This chart, redrawn for integral Celsius temperatures with a different zero point, appears on p. 17.4 of the 1993 ASHRAE *Handbook—Fundamentals* (SI ed.). This handbook gives material for integral degrees Celsius with temperatures on the ITS 90 scale. For experimental isochores for the compressed liquid from 10 to 302 bar, 122 to 462 K, see Blanke, W. and R. Weiss, *PTB Bericht W30*, Braunschweig, Germany, 1992 (54 pp.). Equations and constants approximated to 1985 ASHRAE tables are given by Mecarik, K. and M. Masaryk, *Heat Recovery Systems and CHP*, **11**, 2/3 (1991): 193–197. Tables at 2°C increments to 240°C, 50 bar are given by Watson, J. T. R., *Thermophysical Properties of Refrigerant 12*, H.M.S.O., Edinburgh, Scotland, 1975 (183 pp.). Saturation and superheat tables and a diagram to 40 bar, 620 K are given by Reynolds, W. C., *Thermodynamic Properties in S.I.*, Stanford Univ., 1979 (173 pp.). Tables and a chart to 1100 psia, 480°F are given by Stewart, R. B., R. T. Jacobsen, et al., *Thermodynamic Properties of Refrigerants*, ASHRAE, Atlanta, GA 1986 (521 pp.). For specific heat, thermal conductivity, and viscosity, see *Thermophysical Properties of Refrigerants*, ASHRAE, 1993.

TABLE 2-303   Saturated Refrigerant 13B1*

T, K	P, bar	v_f, m³/kg	v_g, m³/kg	h_f, kJ/kg	h_g, kJ/kg	s_f, kJ/(kg·K)	s_g, kJ/(kg·K)	c_{pf}, kJ/(kg·K)	μ_f, 10^{-4} Pa·s	k_f, W/(m·K)
170	0.059	4.594.−4	1.6015	−40.90	90.95	−0.2033	0.5723	0.597	9.54	0.101
180	0.127	4.677.−4	0.7840	−34.75	94.37	−0.1682	0.5491	0.618	7.60	0.096
190	0.250	4.765.−4	0.4190	−28.51	97.83	−0.1345	0.5305	0.634	6.20	0.091
200	0.455	4.860.−4	0.2407	−22.17	101.32	−0.1020	0.5154	0.648	5.13	0.086
210	0.777	4.961.−4	0.1467	−15.68	104.82	−0.0704	0.5033	0.663	4.33	0.082
215.4	1.013	5.020.−4	0.1147	−12.09	106.70	−0.0536	0.4978	0.670	3.97	0.079
220	1.254	5.071.−4	0.0940	−9.02	108.28	−0.0396	0.4936	0.676	3.71	0.077
230	1.933	5.190.−4	0.0628	−2.19	111.68	−0.0094	0.4857	0.690	3.22	0.073
240	2.863	5.321.−4	0.0433	4.83	114.99	0.0202	0.4793	0.703	2.83	0.068
250	4.096	5.466.−4	0.0308	12.03	118.16	0.0494	0.4739	0.721	2.51	0.063
260	5.690	5.627.−4	0.0224	19.44	121.16	0.0781	0.4693	0.742	2.25	0.059
270	7.703	5.809.−4	0.0166	27.06	123.93	0.1064	0.4652	0.767	2.04	0.054
280	10.20	6.018.−4	0.0124	34.94	126.41	0.1345	0.4612	0.800	1.84	0.049
290	13.25	6.264.−4	0.0094	43.11	128.51	0.1625	0.4570	0.842	1.69	0.045
300	16.91	6.562.−4	0.0072	51.68	130.09	0.1908	0.4522	0.891	1.57	0.040
310	21.28	6.940.−4	0.0055	60.81	130.97	0.2197	0.4460	0.951	1.45	0.035
320	26.44	7.458.−4	0.0041	70.80	130.76	0.2503	0.4376	1.09	1.26	0.030
330	32.48	8.295.−4	0.0030	82.42	128.59	0.2845	0.4245	1.29	0.99	0.026
340.2^c	39.64	1.344.−3	0.0013	108.70	108.70	0.3605	0.3605	∞	0.35	∞

*Values reproduced or converted from Table 4, p. 17.83, *ASHRAE Handbook, 1981: Fundamentals,* American Society of Heating, Refrigerating and Air-Conditioning Engineers, Atlanta, 1981. Copyright material. Reproduced by permission of the copyright owner. c = critical point. The notation 4.594.−4 signifies 4.594×10^{-4}.

The 1993 ASHRAE *Handbook—Fundamentals* (SI ed.) contains a table at closer temperature increments and also an enthalpy–log-pressure diagram from 0.1 to 35 bar, −80 to 220°C. For tables and a chart to 500 psia, 480°F, see Stewart, R. B., R. T. Jacobsen, et al., *Thermodynamic Properties of Refrigerants,* ASHRAE, Atlanta, GA, 1986 (521 pp.). For specific heat, thermal conductivity, and viscosity, see *Thermophysical Properties of Refrigerants,* ASHRAE, 1993.

Refrigerant 14 (tetrafluoromethane)   See Carbon Tetrafluoride (Table 2-245).

Refrigerant 20   See Chloroform (Table 2-248).

TABLE 2-304   Saturated Refrigerant 21

Temperature, K	Pressure, bar	v_f, m³/kg	v_g, m³/kg	h_f, kJ/kg	h_g, kJ/kg	s_f, kJ/(kg·K)	s_g, kJ/(kg·K)
250	0.2415	0.000 677	0.8292	16.6	274.8	0.0687	1.1015
260	0.3953	0.000 687	0.5247	26.5	279.9	0.1076	1.0820
270	0.6200	0.000 698	0.3455	36.6	284.9	0.1454	1.0653
280	0.9364	0.000 709	0.2355	46.7	290.0	0.1824	1.0511
290	1.3682	0.000 722	0.1654	57.1	295.0	0.2186	1.0389
300	1.9417	0.000 735	0.1192	67.7	300.0	0.2543	1.0286
310	2.6849	0.000 748	0.0879	78.4	304.8	0.2894	1.0196
320	3.6279	0.000 763	0.0661	89.5	309.5	0.3242	1.0119
330	4.8022	0.000 778	0.0505	100.7	314.1	0.3586	1.0051
340	6.2409	0.000 794	0.0391	112.3	318.4	0.3927	0.9989
350	7.978	0.000 812	0.0307	124.1	322.4	0.4266	0.9932
360	10.049	0.000 830	0.0243	136.2	326.1	0.4602	0.9877
370	12.489	0.000 850	0.0194	148.6	329.3	0.4935	0.9820
380	15.337	0.000 870	0.0155	161.2	331.9	0.5264	0.9758
390	18.630	0.000 893	0.0125	173.9	333.8	0.5587	0.9688
400	22.41	0.000 918	0.01011	186.4	334.8	0.5896	0.9605
410	26.72	0.000 944	0.00820	198.3	334.7	0.6180	0.9506
420	31.60	0.000 972	0.00672	208.7	333.7	0.6418	0.9394
430	37.10	0.001 002	0.00564	216.4	332.4	0.6587	0.9286
440	43.26	0.001 034	0.00491	221.1	332.3	0.6682	0.9208

Reproduced and rounded from unpublished Center for Applied Thermodynamic Studies, Moscow ID report, 1981. For a thermodynamic diagram to 350 bar, 370°C, see Rombusch, U. K., *Allgem. Warme.,* **11,** 3 (1962).

TABLE 2-305 Saturated Refrigerant 22*

T, K	P, bar	v_f, m³/kg	v_g, m³/kg	h_f, kJ/kg	h_g, kJ/kg	s_f, kJ/(kg·K)	s_g, kJ/(kg·K)	c_{pf}, kJ/(kg·K)	μ_f, 10^{-4} Pa·s	k_f, W/(m·K)
150	0.0017	6.209.−4	83.40	268.2	547.3	3.355	5.215	1.059		0.161
160	0.0054	6.293.−4	28.20	278.2	552.1	3.430	5.141	1.058		0.156
170	0.0150	6.381.−4	10.85	288.3	557.0	3.494	5.075	1.057	0.770	0.151
180	0.0369	6.474.−4	4.673	298.7	561.9	3.551	5.013	1.058	0.647	0.146
190	0.0821	6.573.−4	2.225	308.6	566.8	3.605	4.963	1.060	0.554	0.141
200	0.1662	6.680.−4	1.145	318.8	571.6	3.657	4.921	1.065	0.481	0.136
210	0.3116	6.794.−4	0.6370	329.1	576.5	3.707	4.885	1.071	0.424	0.131
220	0.5470	6.917.−4	0.3772	339.7	581.2	3.756	4.854	1.080	0.378	0.126
230	0.9076	7.050.−4	0.2352	350.6	585.9	3.804	4.828	1.091	0.340	0.121
240	1.4346	7.195.−4	0.1532	361.7	590.5	3.852	4.805	1.105	0.309	0.117
250	2.174	7.351.−4	0.1037	373.0	594.9	3.898	4.785	1.122	0.282	0.112
260	3.177	7.523.−4	0.07237	384.5	599.0	3.942	4.768	1.143	0.260	0.107
270	4.497	7.733.−4	0.05187	396.3	603.0	3.986	4.752	1.169	0.241	0.102
280	6.192	7.923.−4	0.03803	408.2	606.6	4.029	4.738	1.193	0.225	0.097
290	8.324	8.158.−4	0.02838	420.4	610.0	4.071	4.725	1.220	0.211	0.092
300	10.956	8.426.−4	0.02148	432.7	612.8	5.113	4.713	1.257	0.198	0.087
310	14.17	8.734.−4	0.01643	445.5	615.1	4.153	4.701	1.305	0.185	0.082
320	18.02	9.096.−4	0.01265	458.6	616.7	4.194	4.688	1.372	0.175	0.077
330	22.61	9.535.−4	9.753.−3	472.4	617.3	4.235	4.674	1.460	0.167	0.072
340	28.03	1.010.−3	7.479.−3	487.2	616.5	4.278	4.658	1.573	0.151	0.067
350	34.41	1.086.−3	5.613.−3	503.7	613.3	4.324	4.637	1.718	0.130	0.062
360	41.86	1.212.−3	4.036.−3	523.7	605.5	4.378	4.605	1.897	0.106	
369.3^c	49.89	2.015.−3	2.015.−3	570.0	570.0	4.501	4.501	∞		

*P, v, h, and s data interpolated from Perelshteyn (ed.), *Tables and Diagrams of the Thermodynamic Properties of Refrigerants 12, 13 and 22*, Moscow, 1971. c_p, μ, and k data interpolated and converted from *Thermophysical Properties of Refrigerants*, American Society of Heating, Refrigerating and Air-Conditioning Engineers, New York, 1976. c = critical point. The notation 6.209.−4 signifies 6.209×10^{-4}. The 1993 ASHRAE *Handbook—Fundamentals* (SI ed.) gives a saturation table from −150 to 96.14°C and an enthalpy–log-pressure diagram from 0.1 to 150 bar, −60 to 200°C. For experimental isochores for the compressed liquid from 12 to 297 bar, 120 to 378 K, see Blanke, W. and R. Weiss, *PTB Bericht W 30*, Braunschweig, Germany, 1992 (54 pp.). Equations and constants approximated to 1985 ASHRAE tables are given by Mecarik, K. and M. Masaryk, *Heat Recovery Systems and CHP*, **11** 2/3 (1991): 193–197. Saturation and superheat tables and a diagram to 100 bar, 620 K are given by Reynolds, W. C., *Thermodynamic Properties in S.I.*, Stanford Univ. publ., 1979 (179 pp.). For tables and a chart to 2000 psia, 480°F, see Stewart, R. B., R. T. Jacobsen, et al., *Thermodynamic Properties of Refrigerants*, ASHRAE, Atlanta, GA, 1986 (521 pp.). For specific heat, thermal conductivity, and viscosity, see *Thermophysical Properties of Refrigerants*, ASHRAE, 1993.

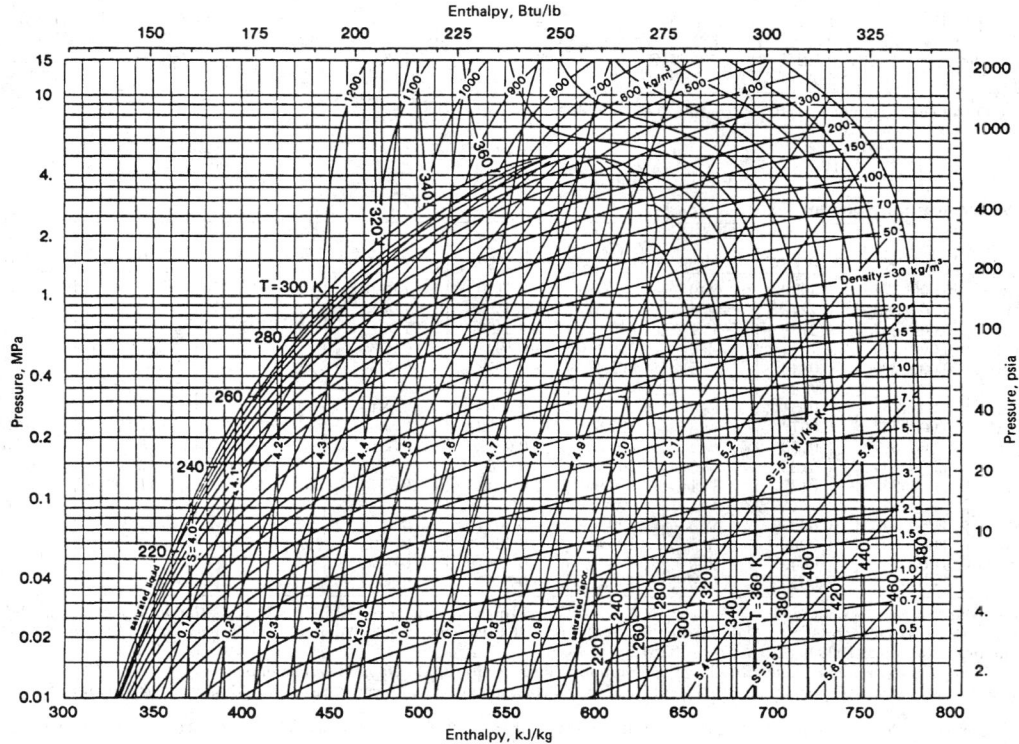

FIG. 2-20 Enthalpy–log-pressure diagram for Refrigerant 22. 1 MPa = 10 bar. (*Copyright 1981 by the American Society of Heating, Refrigerating and Air-Conditioning Engineers and reproduced by permission of the copyright owner.*)

TABLE 2-306 Thermophysical Properties of Compressed R22

Pressure, bar	Property	Temperature, K							
		275	300	325	350	375	400	425	450
1	c_p (kJ/kg·K)	0.639	0.653	0.689	0.714	0.739	0.758	0.781	0.806
	μ (10^{-6} Pa·s)	11.8	12.8	13.9	14.9	15.8	16.7	17.7	18.7
	k (W/m·K)	0.0091	0.0106	0.0121	0.0136	0.0151	0.0166	0.0181	0.0196
	Pr	0.829	0.793	0.787	0.782	0.773	0.762	0.765	0.769
5	c_p (kJ/kg·K)	0.725	0.728	0.744	0.759	0.766	0.775	0.791	0.816
	μ (10^{-6} Pa·s)	11.8	12.8	13.8	15.0	16.2	17.0	18.0	18.8
	k (W/m·K)	0.0096	0.0107	0.0123	0.0138	0.0153	0.0170	0.0184	0.0199
	Pr	0.887	0.871	0.852	0.839	0.803	0.775	0.773	0.771
10	c_p (kJ/kg·K)	1.166	0.847	0.810	0.799	0.797	0.803	0.814	0.828
	μ (10^{-6} Pa·s)	211	13.7	14.4	15.1	16.1	17.1	18.1	19.0
	k (W/m·K)	0.0954	0.0121	0.0128	0.0144	0.0160	0.0175	0.0190	0.0205
	Pr	2.58	0.959	0.901	0.838	0.802	0.785	0.775	0.767
20	c_p (kJ/kg·K)	0.164	1.237		0.949	0.889	0.865	0.858	0.859
	μ (10^{-6} Pa·s)	211	159		16.5	17.3	18.0	18.8	19.6
	k (W/m·K)	0.0963	0.0849		0.0157	0.0172	0.0184	0.0199	0.0214
	Pr	2.55	2.32		0.997	0.894	0.846	0.811	0.787
40	c_p (kJ/kg·K)	1.152	1.217	1.359		1.373	1.089	0.996	0.956
	μ (10^{-6} Pa·s)	218	164	123		20.7	20.5	20.7	21.2
	k (W/m·K)	0.0980	0.0872	0.0767		0.0219	0.0210	0.0217	0.0233
	Pr	2.56	2.29	2.18		1.30	1.063	0.950	0.870
60	c_p (kJ/kg·K)	1.142	1.191	1.311	1.460		1.767	1.221	1.089
	μ (10^{-6} Pa·s)	221	170	128	94.6		24.7	24.2	23.9
	k (W/m·K)	0.0993	0.0889	0.0786			0.0305	0.0287	0.0268
	Pr	2.54	2.28	2.14			1.431	1.030	0.971
80	c_p (kJ/kg·K)	1.132	1.177	1.277	1.444	1.861		1.396	1.262
	μ (10^{-6} Pa·s)	226	175	133	101	73.6		29.9	27.7
	k (W/m·K)	0.1003	0.0904	0.0803	0.0690	0.0523		0.0374	0.0337
	Pr	2.55	2.28	2.12	2.12	2.62		1.12	1.04
100	c_p (kJ/kg·K)	1.122	1.154	1.247	1.361	1.564	2.073	1.923	1.471
	μ (10^{-6} Pa·s)	230	179	138	108	83.2	55.5	37.6	32.4
	k (W/m·K)	0.1013	0.0916	0.0817	0.0716	0.0607	0.0504	0.0421	0.0378
	Pr	2.55	2.26	2.11	2.04	2.14	2.28	1.72	1.26

Some values are approximate as significant differences exist in the literature.

TABLE 2-307 Saturated Refrigerant 23

Temp., K	Pressure, bar	v_f, m³/kg	v_g, m³/kg	h_f, kJ/kg	h_g, kJ/kg	s_f, kJ/(kg·K)	s_g, kJ/(kg·K)	c_{pf}, kJ/(kg·K)	μ_f, 10^{-6} Pa·s	k_f, W/(m·K)	Pr
180	0.510	0.000 678	0.4088	−66.0	181.1	−0.3179	1.0549				
190	0.950	0.000 693	0.2279	−54.4	185.3	−0.2554	1.0062				
191.1[b]	1.013	0.000 695	0.2139	−53.1	185.7	−0.2485	1.0011				
200	1.652	0.000 710	0.1353	−42.6	189.1	−0.1948	0.9635				
210	2.709	0.000 729	0.0845	−30.3	192.4	−0.1353	0.9254				
220	4.298	0.000 751	0.0551	−17.5	195.4	−0.0764	0.8913				
230	6.312	0.000 777	0.0372	−4.3	197.8	−0.0182	0.8602	0.710	170.1	0.105	1.15
240	9.091	0.000 807	0.0259	9.4	199.6	0.0392	0.8314	1.043	150.4	0.098	1.56
250	12.69	0.000 844	0.0183	23.6	200.7	0.0957	0.8042	1.289	131.2	0.091	1.85
260	17.25	0.000 889	0.0132	38.1	200.9	0.1512	0.7773	1.497	113.0	0.084	2.00
270	22.94	0.000 948	0.0095	53.5	199.8	0.2071	0.7493				
280	29.98	0.001 031	0.0068	70.5	196.4	0.2665	0.7162				
290	38.68	0.001 169	0.0046	92.0	188.1	0.3387	0.6698				
299.1[c]	48.36	0.001 905	0.0019	143.0	143.0	0.5062	0.5062				

b = normal boiling point; c = critical point. $h_f = s_f = 0$ at 233.15 K = −40°C. Interpolated and converted from ASHRAE *Handbook—Fundamentals*, 1993. Experimental *P*–ρ–*T* data from 95 to 413 K reported in *J. Phys. Chem.*, **89** (1985): 4637–4646 were used by Rubio, R. G., J. A. Zollweg, et al., *J. Chem. Eng. Data*, **36**, (1991): 171–184, to calculate properties up to 1000 bar from 126 to 332° K.

The 1993 ASHRAE *Handbook—Fundamentals* (SI ed.) gives a saturation table from −100 to 25.92°C and an enthalpy–log-pressure diagram from 0.1 to 80 bar, −100 to 280°C. Equations and constants approximated to the 1985 ASHRAE tables are given by Mecaryk, K. and M. Masaryk, *Heat Recovery Systems and CHP*, **11**, 2/3 (1991): 193–197.

For an enthalpy–log-pressure diagram from 0.005 to 200 bar, −140 to 180°C, see Morsy, T. E., *Kaltetechnik—Klimat.*, **18**, 9 (1966): 347–349. Saturation and superheat tables and a diagram to 100 bar, 600 K are given by Reynolds, W. C., *Thermodynamic Properties in S.I.*, Stanford Univ. publ., 1979 (173 pp.).

For tables and a chart to 1000 psia, 560°F, see Stewart, R. B., R. T. Jacobsen, et al., *Thermodynamic Properties of Refrigerants*, ASHRAE, Atlanta, GA, 1986 (21 pp.).

For specific heat, thermal conductivity, and viscosity, see *Thermophysical Properties of Refrigerants*, ASHRAE, 1993.

TABLE 2-308 Thermophysical Properties of Saturated Difluoromethane (R32)

Temp., K	Pressure, bar	v_f, m³/kg	v_g, m³/kg	h_f, kJ/kg	h_g, kJ/kg	s_f, kJ/(kg·K)	s_g, kJ/(kg·K)	c_{pf}, kJ/(kg·K)	c_{pg}, kJ/(kg·K)	μ_f, 10^{-6} Pa·s	μ_g, 10^{-6} Pa·s	k_f, W/(m·K)	k_g, W/(m·K)	Pr_f	Pr_g
200	0.2960	7.845.−4	1.0580	−52.340	351.160	−0.2418	1.7757								
210	0.5440	8.025.−4	0.5990	−37.750	356.880	−0.1652	1.7098								
220	0.9384	8.208.−4	0.3593	−20.690	361.970	−0.0909	1.6485	1.557	0.799	283.8	10.30				
230	1.5345	8.402.−4	0.2261	−4.984	366.773	−0.0215	1.5948	1.580	0.839	247.8	10.37				
240	2.3963	8.611.−4	0.1483	10.947	371.129	0.0459	1.5468	1.613	0.894	220.4	10.46				
250	3.5966	8.842.−4	0.1005	27.1778	374.971	0.1117	1.5030	1.642	0.963	198.1	10.66	0.1646	0.0097	1.98	1.06
260	5.2160	9.096.−4	0.07020	43.786	378.224	0.1763	1.4624	1.682	1.043	177.9	10.95	0.1562	0.0106	1.92	1.08
270	7.3423	9.376.−4	0.05009	60.849	380.786	0.2397	1.4427	1.730	1.138	159.1	11.31	0.1487	0.0115	1.85	1.12
280	10.070	9.696.−4	0.03643	78.456	382.525	0.3029	1.3886	1.786	1.244	141.8	11.70	0.1403	0.0125	1.81	1.16
290	13.502	1.006.−3	0.02687	96.713	383.262	0.3654	1.3534	1.863	1.375	126.1	12.21	0.1308	0.0136	1.80	1.23
300	17.749	1.049.−3	0.02001	115.754	382.737	0.4283	1.3182	1.955	1.560	112.0	12.82	0.1228	0.0149	1.78	1.34
310	22.931	1.100.−3	0.01497	135.801	380.576	0.4919	1.2815	2.084	1.810	98.8	13.71	0.1155	0.0165	1.78	1.50
320	29.186	1.166.−3	0.01117	157.212	376.163	0.5574	1.2415	2.282	2.16	86.1	14.4	0.1073	0.0184	1.83	1.69
330	36.675	1.243.−3	0.00822	180.724	368.357	0.6264	1.1950	2.620	2.62	75.1	15.3	0.0990	0.0205	1.99	1.96
340	45.603	1.394.−3	0.00581	208.262	354.460	0.7047	1.1347	3.560	4.21	65.4	17.3		0.0236		3.10
350	56.336	0.00317		274.640	337.933	0.8927	1.0735								
351.4^c	57.927	0.00237	0.00237	286.675	286.675	0.9269	0.9269								

c = critical point. The notation 7.845.−4 signifies 7.845×10^{-4}. P, v, T, h, s, and c_p converted and extrapolated from Defibaugh, D. R. G. Morrison, et al., *J. Chem. Eng. Data*, **39** (1994): 333–340. Saturated liquid and vapor viscosities from smooth curve fits of Oliveira, C. M. B. P. and W. A. Wakeham, *Int. J. Thermophys.*, **14**, 6 (1993): 1131–1143. Thermal conductivity values based upon papers by Geller, V. Z. and M. E. Perlaitis, and by Gross, *Proc. 10th Symp. Thermophys. Props.*, Boulder, CO. 1994.

The 1993 ASHRAE Handbook—Fundamentals (SI ed.) gives a saturation table to 78.41°C and a diagram to 200 bar, 200°C.

TABLE 2-309 Specific Heat at Constant Pressure, Thermal Conductivity, Viscosity, and Prandtl Number of R32 Gas

Temp., K	Property	P, bar 1	5	10	15
250	c_p (kJ/kg·K)	0.805			
	μ (10^{-6} Pa·s)	10.55			
	k (W/m·K)	0.0094			
	Pr	0.908			
260	c_p (kJ/kg·K)	0.810	1.025		
	μ (10^{-6} Pa·s)	11.00	10.96		
	k (W/m·K)	0.0100	0.0104		
	Pr	0.890	1.080		
270	c_p (kJ/kg·K)	0.818	0.991		
	μ (10^{-6} Pa·s)	11.42	11.37		
	k (W/m·K)	0.0107	0.0111		
	Pr	0.873	1.015		
280	c_p (kJ/kg·K)	0.825	0.969	1.238	
	μ (10^{-6} Pa·s)	11.82	11.77	11.68	
	k (W/m·K)	0.0116	0.0118	0.0125	
	Pr	0.860	0.967	1.157	
290	c_p (kJ/kg·K)	0.837	0.959	1.161	
	μ (10^{-6} Pa·s)	12.28	12.22	12.17	
	k (W/m·K)	0.0121	0.0125	0.0131	
	Pr	0.849	0.938	1.079	
300	c_p (kJ/kg·K)	0.849	0.951	1.118	1.370
	μ (10^{-6} Pa·s)	12.70	12.69	12.66	12.62
	k (W/m·K)	0.0128	0.0132	0.0138	0.0144
	Pr	0.842	0.914	1.026	1.201

Temp., K	Property	P, bar 1	5	10	15	20	25	30	40
310	c_p (kJ/kg·K)	0.861	0.945	1.084	1.279	1.560			
	μ (10^{-6} Pa·s)	13.12	13.10	13.08	13.07	13.09			
	k (W/m·K)	0.0135	0.0139	0.0144	0.0150	0.0159			
	Pr	0.878	0.891	0.985	1.114	1.284			
320	c_p (kJ/kg·K)	0.873	0.944	1.059	1.207	1.400	1.704		
	μ (10^{-6} Pa·s)	13.54	13.54	13.55	13.56	13.60	13.75		
	k (W/m·K)	0.0142	0.0146	0.0150	0.0156	0.0164	0.0173		
	Pr	0.836	0.875	0.957	1.049	1.161	1.354		
330	c_p (kJ/kg·K)	0.885	0.942	1.038	1.158	1.301	1.508	1.837	
	μ (10^{-6} Pa·s)	13.96	13.96	13.98	14.01	14.15	14.28	14.52	
	k (W/m·K)	0.0148	0.0152	0.0156	0.0162	0.0169	0.0177	0.0187	
	Pr	0.834	0.865	0.930	1.001	1.089	1.217	1.426	
340	c_p (kJ/kg·K)	0.897	0.937	1.020	1.135	1.242	1.388	1.612	2.488
	μ (10^{-6} Pa·s)	14.38	14.40	14.43	14.47	14.53	14.65	14.85	16.00
	k (W/m·K)	0.0155	0.0159	0.0163	0.0168	0.0175	0.0182	0.0190	0.0217
	Pr	0.832	0.849	0.903	0.978	1.031	1.117	1.260	1.834
350	c_p (kJ/kg·K)	0.910	0.934	1.004	1.118	1.200	1.308	1.440	1.914
	μ (10^{-6} Pa·s)	14.80	14.82	14.84	14.87	14.92	15.06	15.21	16.16
	k (W/m·K)	0.0162	0.0165	0.0169	0.0174	0.0180	0.0186	0.0194	0.0216
	Pr	0.831	0.839	0.882	0.955	0.995	1.060	1.130	1.432

Some values read from charts may be approximate. c_p values interpolated and converted from *Thermodynamic Properties of KLEA 32*, I.C.I., 1993 (47 pp.). Viscosity interpolated from Takahashi, M., C. Yokoyama, et al., *Proc. 14th Symp. Thermophys. Props.*, Japan, 1993 (pp. 427–430). Thermal conductivities are taken from Geller, V. Z. and M. E. Perlaitis, and from Gross, *Proc. 10th Symp. Thermophys. Props.*, Boulder, CO, 1994.

TABLE 2-310 Saturated SUVA MP 39

Temp., °C	P_f, bar	P_g, bar	v_f, m³/kg	v_g, m³/kg	h_f, kJ/kg	h_g, kJ/kg	s_f, kJ/(kg·K)	s_g, kJ/(kg·K)	c_{pf}, kJ/(kg·K)	μ_f, 10^{-6} Pa·s	k_f, W/(m·K)	Pr_f
−40	0.733	0.533	0.000 712	0.3778	154.0	385.0	0.8188	1.8244	1.078	351	0.1209	3.13
−30	1.155	0.871	0.000 728	0.2391	164.9	390.6	0.8647	1.8059	1.109	323	0.1154	3.06
−20	1.748	1.361	0.000 744	0.1576	176.2	396.3	0.9099	1.7907	1.137	291	0.1107	2.99
−10	2.553	2.043	0.000 762	0.1075	188.6	401.8	0.9577	1.7781	1.165	266	0.1057	2.93
0	3.615	2.965	0.000 781	0.0755	200.0	407.3	1.0000	1.7675	1.197	241	0.1012	2.85
10	4.984	4.177	0.000 803	0.0544	212.7	412.6	1.0454	1.7587	1.233	221	0.0967	2.82
20	6.712	5.733	0.000 826	0.0399	225.3	417.6	1.0884	1.7510	1.277	202	0.0922	2.80
30	8.857	7.697	0.000 851	0.0298	238.3	422.2	1.1316	1.7439	1.329	186	0.0877	2.83
40	11.475	10.133	0.000 878	0.0225	252.0	426.5	1.1752	1.7372	1.392	170	0.0830	2.85
50	14.628	13.112	0.000 909	0.0172	266.4	430.1	1.2194	1.7304	1.468	157	0.0781	2.95
60	18.378	16.711	0.000 944	0.01313	281.6	433.0	1.2647	1.7228	1.564	143	0.0737	3.04
70	22.79	21.01	0.000 988	0.01005	297.9	434.9	1.3118	1.7138	1.652	131	0.0684	3.16
80	27.92	26.12	0.001 028	0.00764	315.9	435.4	1.3616	1.7022	1.802	122	0.0631	3.48
90	33.83	32.13	0.001 084	0.00570	336.2	433.5	1.4163	1.6858	1.958	115	0.0577	3.90
100	40.53	39.22	0.001 140	0.00403	361.4	426.9	1.4820	1.6584	2.16	110	0.0533	4.46
108.0^c	46.04	46.04	0.001 96	0.00196	397	397						

c = critical point. SUVA MP 39 = R401A = $CHClF_2$ (R22) 53% wt + CH_3CHF_2 (R 152a) 13% wt + $CHClFCF_3$ (R124) 34% wt, near-azeotropic blend. Some values read from charts are approximate. Material used by permission of DuPont Fluoroproducts.

TABLE 2-311 SUVA MP 39 at Atmospheric Pressure

Temp., °C	−27.01	−20	0	20	40	60	80	100	120	140
v (m³/kg)	0.2102	0.2167	0.2351	0.2534	0.2715	0.2896	0.3076	0.3256	0.3435	0.3613
h (kJ/kg)	351.7	396.9	410.4	424.5	439.2	454.4	470.3	486.6	503.5	521.2
s (kJ/kg·K)	1.8009	1.8193	1.8706	1.9204	1.9689	2.0161	2.0623	2.1073	2.1513	2.1943
c_p (kJ/kg·K)	0.648	0.669	0.698	0.727	0.757	0.787	0.811	0.836	0.859	0.883
μ (10^{-6} Pa·s)	10.17	10.43	11.18	11.93	12.68	13.42	14.17	14.89	15.61	16.32
k (W/m·K)	0.00878	0.00921	0.01041	0.01161	0.01282	0.01404	0.01536	0.01668	0.01796	0.01929
Pr	0.750	0.758	0.750	0.749	0.749	0.748	0.748	0.748	0.747	0.747
Z	0.9829	0.9852	0.9906	0.9949	0.9979	1.0005	1.0025	1.0043	1.0056	1.0060

For composition see footnote to Table 2-310. Some values read from charts are approximate. Material used by permission of DuPont Fluoroproducts.

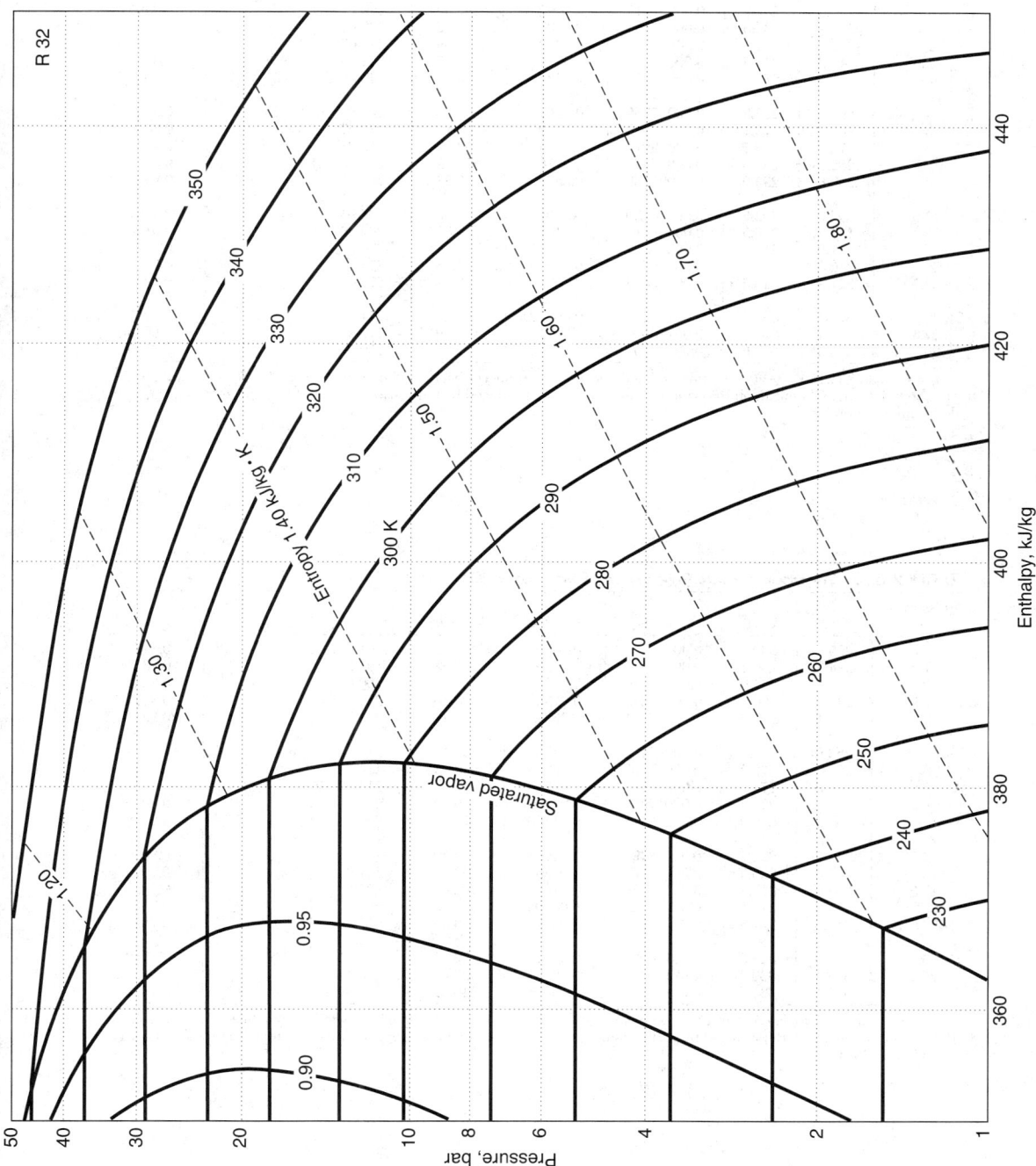

FIG. 2-21 Enthalpy–log-pressure diagram for Refrigerant 32.

TABLE 2-312   Thermodynamic Properties of Saturated KLEA 60

Pressure, bar	T_f, K	T_g, K	v_f, m³/kg	v_g, m³/kg	h_f, kJ/kg	h_g, kJ/kg	s_f, kJ/(kg·K)	s_g, kJ/(kg·K)
1	227.3	234.0	0.000 7118	0.2097	−7.80	229.64		0.9965
1.5	236.1	242.5	0.000 7263	0.1433	3.92	235.02		0.9833
2	242.8	249.1	0.000 7381	0.1093	12.89	239.07		0.9744
2.5	248.3	254.5	0.000 7483	0.0885	20.27	242.35		0.9679
3	253.0	259.1	0.000 7573	0.0744	26.57	245.08		0.9629
4	260.7	266.8	0.000 7735	0.0564	37.23	249.54		0.9552
5	267.3	273.1	0.000 7880	0.0442	46.12	253.07		0.9496
6	272.9	278.5	0.000 8012	0.0384	53.84	255.24		0.9450
8	282.1	287.5	0.000 8254	0.0286	67.02	260.70		0.9378
10	289.8	295.0	0.000 8480	0.0228	78.23	263.86		0.9318
12.5	297.9	302.8	0.000 8750	0.01802	90.50	266.95		0.9257
15	304.8	309.5	0.000 9017	0.01481	101.51	269.12		0.9190
17.5	311.0	315.4	0.000 9290	0.01247	111.64	270.58		0.9128
20	316.5	320.7	0.000 9613	0.01069	121.18	271.46		0.9065
22.5	321.4	325.5	0.000 9884	0.00928	130.31	271.79		0.8999
25	326.1	329.8	0.001 023	0.00828	139.17	271.63		0.8927
27.5	330.4	333.9	0.001 063	0.00717	147.89	270.97		0.8850
30	334.5	337.6	0.001 115	0.00635	156.58	269.81		0.8765

$h_f = s_f = 0$ at 233.15 K = −40°C. Converted and interpolated from *Thermodynamic Properties of Klea 60* (British units, 20 pp.), copyright ICI Chemicals and Polymers Limited, 1993. Reproduced by permission. T_f = bubble point temperature; T_g = dew point temperature.

TABLE 2-313   Thermodynamic Properties of Saturated KLEA 61

Pressure, bar	T_f, K	T_g, K	v_f, m³/kg	v_g, m³/kg	h_f, kJ/kg	h_g, kJ/kg	s_f, kJ/(kg·K)	s_g, kJ/(kg·K)
1	225.6	230.0	0.000 6852	0.1800	−9.45	191.64		0.8433
1.5	234.3	238.5	0.000 6994	0.1230	2.52	196.90		0.8341
2	241.8	245.0	0.000 7110	0.0937	9.72	200.88		0.8282
2.5	246.4	250.4	0.000 7211	0.0758	16.59	204.10		0.8245
3	251.1	254.9	0.000 7301	0.0637	22.47	206.80		0.8215
4	258.9	262.6	0.000 7463	0.04831	32.43	211.22		0.8172
5	265.4	269.0	0.000 7607	0.03888	40.76	214.74		0.8141
6	270.9	274.4	0.000 7740	0.03249	48.00	217.65		0.8123
8	280.2	283.4	0.000 7985	0.02435	59.82	222.21		0.8080
10	287.8	290.9	0.000 8214	0.01936	70.98	225.63		0.8048
12.5	295.8	298.7	0.000 8491	0.01528	82.59	228.80		0.8010
15	302.8	305.5	0.000 8768	0.01251	93.02	231.08		0.7971
17.5	308.8	311.4	0.000 9053	0.01049	102.67	232.64		0.7929
20	314.3	316.7	0.000 9353	0.00896	111.79	233.60		0.7882
22.5	319.3	321.5	0.000 9680	0.00774	120.55	233.99		0.7829
25	323.9	325.9	0.001 005	0.00674	129.11	233.85		0.7769
27.5	328.1	330.0	0.001 048	0.00590	137.62	233.16		0.7700
30	332.1	333.7	0.001 102	0.00518	146.21	231.84		0.7619

Converted and interpolated from *Thermodynamic Properties of Klea 61* (British units, 20 pp.), copyright ICI Chemicals and Polymers Limited, 1993. Reproduced by permission. T_f = bubble-point temperature; T_g = dew-point temperature. $h_f = s_f = 0$ at 233.15 K = −40 °C.

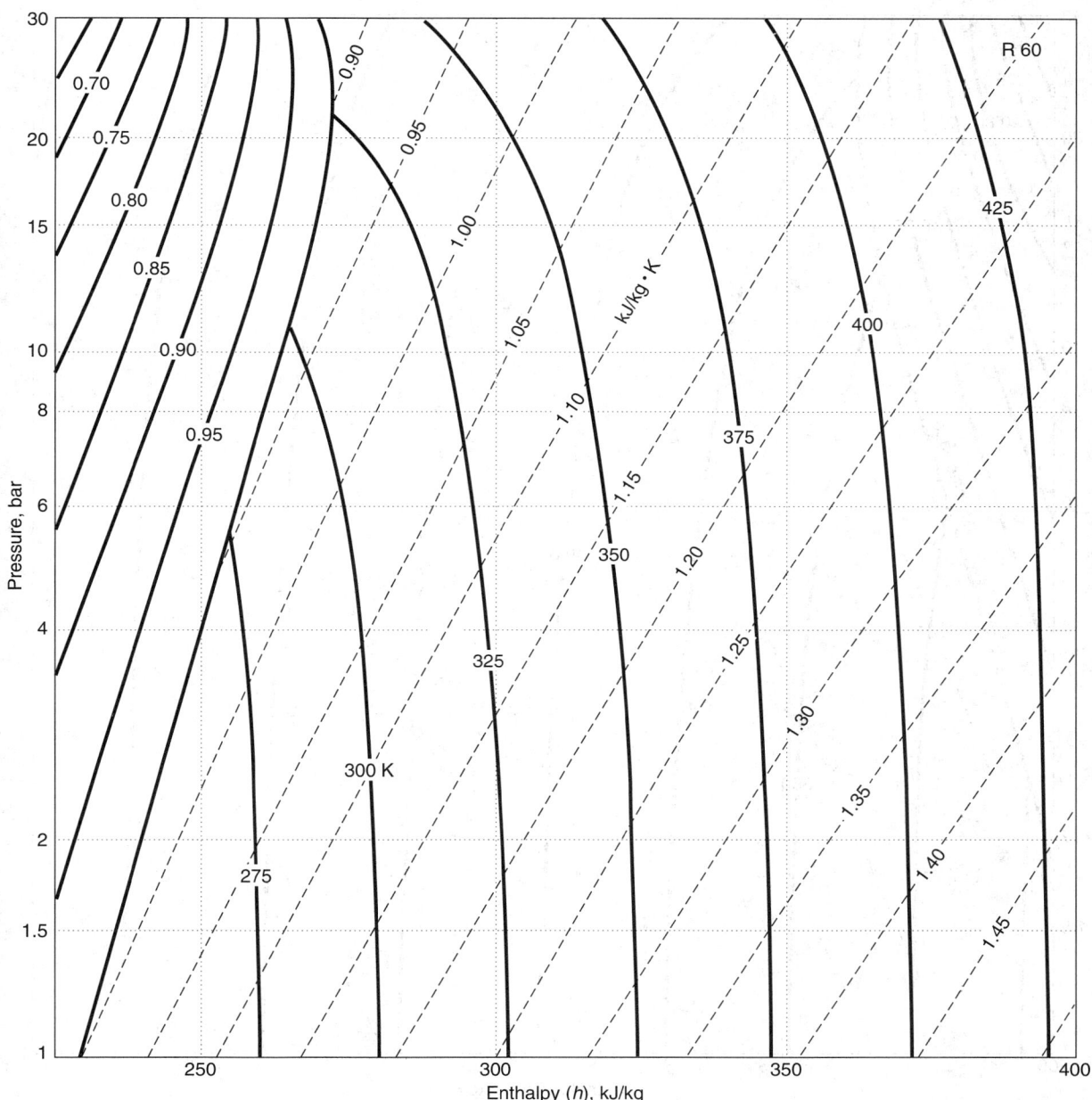

FIG. 2-22   Enthalpy–log-pressure diagram for KLEA 60.

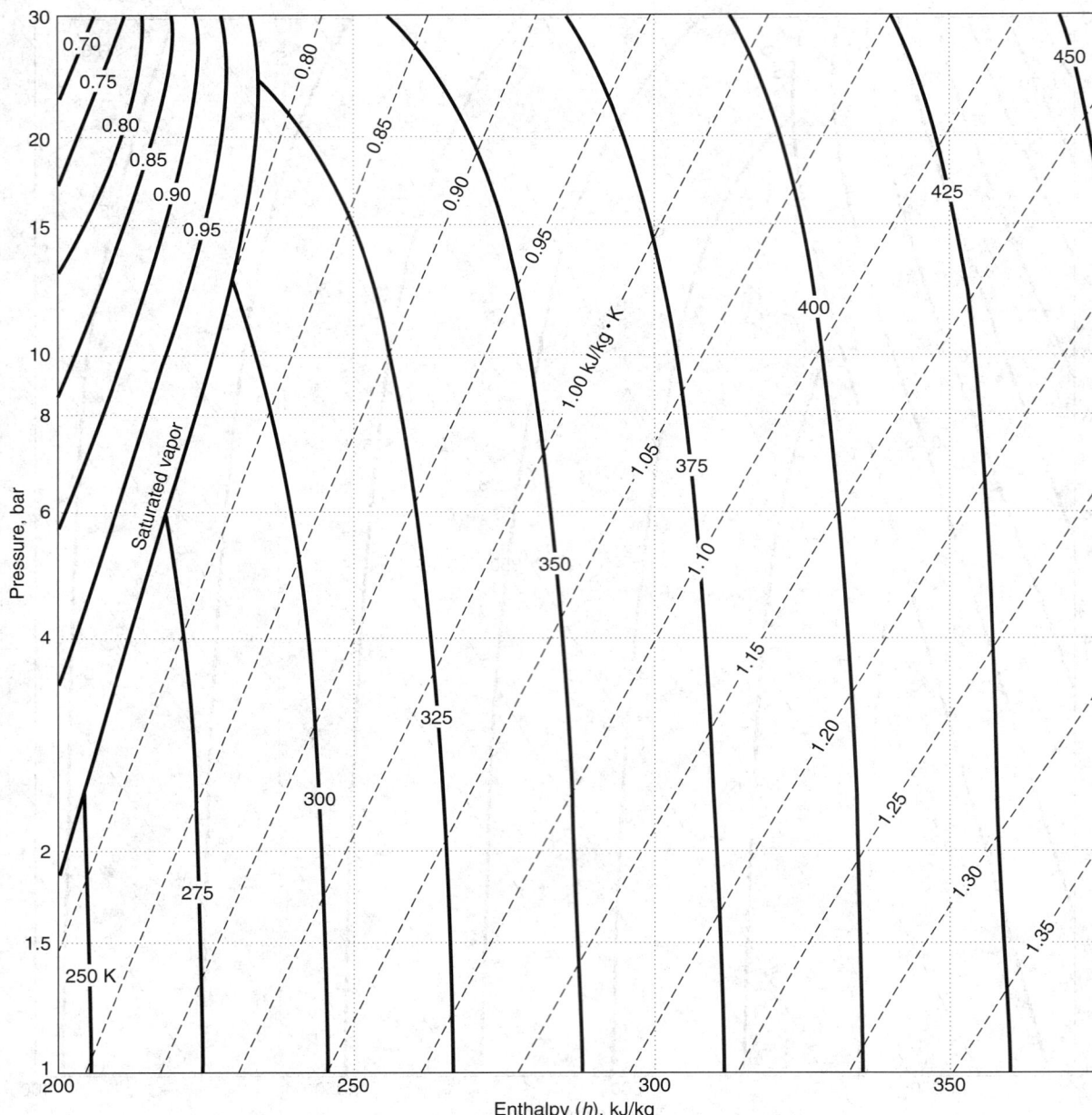

FIG. 2-23   Enthalpy–log-pressure diagram for KLEA 61.

TABLE 2-314 Saturated SUVA HP 62

Temp., °C	P_f, bar	P_g, bar	v_f, m³/kg	v_g, m³/kg	h_f, kJ/kg	h_g, kJ/kg	s_f, kJ/(kg·K)	s_g, kJ/(kg·K)	c_{pf}, kJ/(kg·K)	μ_f, 10^{-6} Pa·s	k_f, W/(m·K)	Pr_f
−50	0.852	0.821	0.000 761	0.2244	133.1	337.3	0.7318	1.6487		370	0.0970	
−40	1.367	1.325	0.000 779	0.1434	145.6	343.8	0.7862	1.6380		318		
−30	2.095	2.041	0.000 799	0.0953	159.9	350.3	0.8460	1.6301	1.220	276	0.0868	3.88
−20	3.087	3.018	0.000 820	0.0656	172.8	356.5	0.8975	1.6245	1.260	238	0.0834	3.60
−10	4.404	4.321	0.000 843	0.0463	186.1	362.6	0.9487	1.6202	1.302	207	0.0801	3.37
0	6.111	6.013	0.000 868	0.03338	200.0	368.3	1.0000	1.6188	1.351	181	0.0767	3.19
10	8.278	8.165	0.000 898	0.02444	214.5	373.6	1.0515	1.6138	1.412	158	0.0733	3.04
20	10.977	10.851	0.000 933	0.01809	229.9	378.3	1.1038	1.6106	1.489	138	0.0698	2.94
30	14.287	14.150	0.000 977	0.01348	246.2	382.2	1.1574	1.6065	1.592	122	0.0663	2.93
40	18.292	18.148	0.001 037	0.01003	263.8	385.0	1.2130	1.6005	1.753	106	0.0624	2.98
50	23.08	22.94	0.001 122	0.00739	283.2	386.1	1.2723	1.5910	2.09	91	0.0583	3.26
60	28.75	28.63	0.001 261	0.00527	305.8	384.2	1.3389	1.5742		76	0.0535	
70	35.58			0.00285	339.8	375.9				61		
72.1^c	37.32	37.32	0.002 06	0.00206	361	361						

c = critical point. SUVA HP 62 = CHF_2CF_3 (R125) 44% wt + CH_3CF_3 (R143a) 52% wt + CH_2FCF_3 (R134a) 4% wt, near-azeotropic blend. Material used by permission of DuPont Fluoroproducts. Some values read from charts may be approximate.

TABLE 2-315 SUVA HP 62 at Atmospheric Pressure

Temp., °C	−45.63	−40	−20	0	20	40	60	80	100	120
v (m³/kg)	0.1866	0.1921	0.2100	0.2278	0.2455	0.2630	0.2805	0.2980	0.3153	0.3325
h (kJ/kg)	336.0	344.4	359.9	376.2	393.1	410.9	429.3	448.4	468.2	488.7
s (kJ/kg·K)	1.6599	1.6636	1.7274	1.7891	1.8491	1.9076	1.9646	2.0203	2.0747	2.1278
c_p (kJ/kg·K)	0.732	0.738	0.781	0.821	0.860	0.897	0.933	0.967	1.000	1.032
μ (10^{-6} Pa·s)	9.47	9.68	10.45	11.22	11.99	12.76	13.53	14.30	15.07	15.84
k (W/m·K)	0.00860	0.00932	0.01059	0.01186	0.01313	0.01440	0.01568	0.01695	0.01827	0.01949
Pr	0.806	0.767	0.771	0.777	0.785	0.795	0.805	0.816	0.827	0.839
Z	0.9755	0.9800	0.9867	0.9919	0.9961	0.9989	1.0014	1.0037	1.0050	1.0060

v, h, and s from DuPont bull. T—HP62—SI, June 1993 (17 pp.). c_p and k from DuPont bull. ART 18, June 1993 (37 pp.). Some values read from charts may be approximate. Material used by permission of DuPont Fluoroproducts.

TABLE 2-316 Thermodynamic Properties of Saturated KLEA 66

Pressure, bar	T_f, K	T_g, K	v_f, m³/kg	v_g, m³/kg	h_f, kJ/kg	h_g, kJ/kg	s_f, kJ/(kg·K)	s_g, kJ/(kg·K)
0.69	221.46	228.77	0.000 7122	0.31325	−16.16	241.25		1.0729
1	228.89	236.05	0.000 7237	0.22131	−5.89	245.91		1.0580
1.5	237.69	244.69	0.000 7382	0.15140	6.22	251.38		1.0430
2	244.45	251.33	0.000 7501	0.11537	15.51	255.52		1.0330
2.5	249.99	256.76	0.000 7600	0.09104	23.12	258.95		1.0258
3	254.36	261.39	0.000 7695	0.07855	29.62	261.63		1.0201
4	262.60	269.14	0.000 7857	0.05964	40.60	266.16		1.0114
5	269.12	275.51	0.000 8001	0.04806	49.74	269.74		1.0055
6	274.70	280.98	0.000 8133	0.04021	57.69	272.68		0.9993
8	284.03	290.08	0.000 8375	0.03022	71.22	277.25		0.9913
10	291.74	297.56	0.000 8599	0.02410	82.73	280.64		0.9834
12.5	299.87	305.44	0.000 8867	0.01910	95.32	283.74		0.9770
15	306.87	312.18	0.000 9131	0.01571	106.59	285.93		0.9701
17.5	313.05	318.10	0.000 9400	0.01324	116.97	287.28		0.9633
20	318.60	323.40	0.000 9680	0.01137	126.73	288.26		0.9564
22.5	323.7	328.2	0.000 9981	0.00988	136.0	288.6		0.9493
25	328.3	332.5	0.001 032	0.00883	145.1	288.4		0.9418
27.5	332.7	336.6	0.001 072	0.00766	153.9	287.8		0.9338
30	336.7	340.4	0.001 125	0.00703	162.7	286.6		0.9251

Converted and interpolated from *Thermodynamic Properties of Klea 66* (British units, 22 pp.), copyright ICI Chemicals and Polymers Limited, 1993. Reproduced by permission. T_f = bubble-point temperature; T_g = dew-point temperature. $h_f = s_f = 0$ at 233.15 K = −40°C.

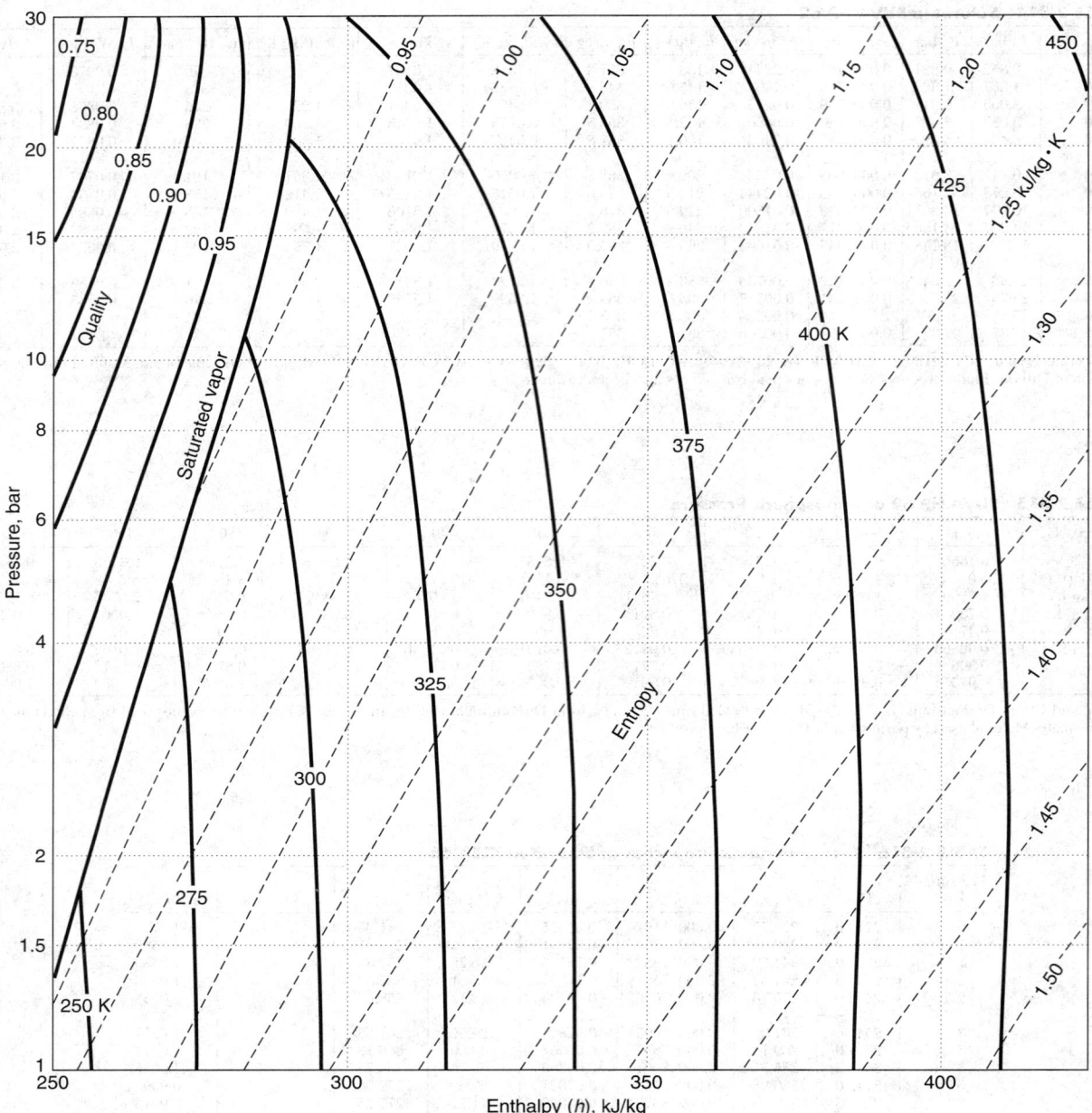

FIG. 2-24   Enthalpy–log-pressure diagram for KLEA 66.

TABLE 2-317 Saturated SUVA MP 66

Temp., °C	P_f, bar	P_g, bar	v_f, m³/kg	v_g, m³/kg	h_f, kJ/kg	h_g, kJ/kg	s_f, kJ/(kg·K)	s_g, kJ/(kg·K)	c_{pf}, kJ/(kg·K)	μ_f, 10^{-6} Pa·s	k_f, W/(m·K)	Pr_f
−40	0.788	0.585	0.000 710	0.3498	153.8	386.0	0.8184	1.8291	1.078	349	0.1209	3.11
−30	1.239	0.952	0.000 725	0.2224	164.8	391.6	0.8643	1.8100	1.109	313	0.1154	3.01
−20	1.872	1.479	0.000 740	0.1471	176.0	397.1	0.9095	1.7940	1.137	282	0.1106	2.90
−10	2.726	2.212	0.000 758	0.1008	188.6	402.6	0.9577	1.7807	1.165	257	0.1057	2.83
0	3.850	3.198	0.000 778	0.0710	200.0	407.8	1.0000	1.7694	1.197	236	0.1012	2.79
10	5.297	4.491	0.000 801	0.05124	212.6	412.9	1.0450	1.7598	1.233	217	0.0967	2.77
20	7.120	6.146	0.000 827	0.03771	225.1	417.7	1.0879	1.7512	1.277	198	0.0922	2.74
30	9.379	8.229	0.000 858	0.02818	238.2	422.1	1.1311	1.7433	1.329	181	0.0877	2.74
40	12.133	10.808	0.000 895	0.02131	251.9	426.1	1.1747	1.7357	1.392	168	0.0830	2.82
50	15.444	13.955	0.000 939	0.01625	266.3	429.4	1.2190	1.7278	1.468	151	0.0781	2.84
60	19.378	17.750	0.000 994	0.01244	281.6	431.9	1.2645	1.7191	1.564	139	0.0737	2.95
70	24.00	22.28	0.001 066	0.00951	298.1	433.4	1.3120	1.7088	1.652	127	0.0684	3.07
80	29.37	27.64	0.001 164	0.00721	316.3	433.2	1.3625	1.6956	1.802	116	0.0631	3.31
90	35.55	33.96	0.001 313	0.00534	337.2	430.4	1.4187	1.6768			0.0577	
100	42.30										0.0533	
106.1^c	46.82	46.82	0.001 95	0.00195	389	389						

c = critical point. SUVA MP 66 = R401 = $CHClF_2$ (R22) 61% wt + CH_3CHF_2 (R152a) 11% wt + $CHClFCF_3$ (R124) 28% wt, near-azeotropic blend. Material used by permission of DuPont Fluoroproducts. Some values read from charts are approximate.

TABLE 2-318 SUVA MP 66 at Atmospheric Pressure

Temp., °C	−28.63^b	−20	0	20	40	60	80	100	120	140
v (m³/kg)	0.2086	0.2177	0.2362	0.2545	0.2727	0.2908	0.3089	0.3269	0.3449	0.3629
h (kJ/kg)	392.2	397.9	411.2	425.1	439.6	454.6	470.1	486.2	502.7	519.4
s (kJ/kg·K)	1.8081	1.8299	1.8804	1.9295	1.9772	2.0237	2.0690	2.1132	2.1564	2.1986
c_p (kJ/kg·K)	0.641	0.652	0.688	0.716	0.744	0.771	0.796	0.822	0.844	0.866
μ (10^{-6} Pa·s)	9.78	10.43	11.18	11.93	12.68	13.42	14.17	14.89	15.61	16.32
k (W/m·K)	0.00817	0.00921	0.01041	0.01161	0.01282	0.01404	0.01536	0.01668	0.01796	0.01929
Pr	0.767	0.738	0.737	0.736	0.735	0.735	0.734	0.734	0.733	0.733
Z	0.9652	0.9730	0.9783	0.9822	0.9852	0.9876	0.9896	0.9912	0.9925	0.9937

v, h, and s from DuPont bull. T—MP 66—SI, Jan. 1993 (17 pp.). c_p, μ, and k from DuPont bull. ART 10, Jan. 1993 (27 pp.). Some values read from charts may be approximate. Material used by permission of DuPont Fluoroproducts. b = normal boiling point.

TABLE 2-319 Saturated SUVA HP 80

Temp., °C	P_f, bar	P_g, bar	v_f, m³/kg	v_g, m³/kg	h_f, kJ/kg	h_g, kJ/kg	s_f, kJ/(kg·K)	s_g, kJ/(kg·K)	c_{pf}, kJ/(kg·K)	μ_f, 10^{-6} Pa·s	k_f, W/(m·K)	Pr_f
−50	0.962	0.872	0.000 679	0.2033	139.6	334.1	0.7578	1.6327		377	0.0970	
−40	1.520	1.403	0.000 695	0.1303	150.8	339.9	0.8070	1.6206		317		
−30	2.305	2.156	0.000 713	0.0869	163.1	345.6	0.8584	1.6110	1.193	283	0.0880	3.84
−20	3.370	3.188	0.000 733	0.0598	174.9	351.1	0.9053	1.6034	1.217	247	0.0849	3.54
−10	4.776	4.560	0.000 757	0.0423	187.6	356.4	0.9541	1.5972	1.236	215	0.0813	3.27
0	6.588	6.336	0.000 785	0.03060	200.0	361.3	1.0000	1.5919	1.253	188	0.0778	3.03
10	8.877	8.592	0.000 819	0.02248	213.0	365.9	1.0461	1.5870	1.286	165	0.0743	2.86
20	11.720	11.404	0.000 860	0.01671	226.7	369.8	1.0927	1.5820	1.340	146	0.0708	2.76
30	15.195	14.855	0.000 911	0.01250	241.2	373.1	1.1403	1.5762	1.412	128	0.0672	2.69
40	19.388	19.034	0.000 977	0.00936	256.8	375.4	1.1897	1.5690	1.512	113	0.0634	2.70
50	24.39	24.04	0.001 070	0.00696	273.9	376.2	1.2420	1.5589	1.64	98	0.0593	2.71
60	30.30	29.97	0.001 212	0.00505	293.6	374.6	1.2998	1.5433	1.81	83	0.0551	2.79
70										68		
75.5^c	41.35	41.35	0.001 850	0.00185	340	340						

c = critical point. SUVA HP 80 = R402 = CHF_2CF_3 (R125) 60% wt + $CH_3CH_2CH_3$ (R290) 2% wt + $CHClF_2$ (R22) 38% wt, near-azeotropic blend. Material used by permission of DuPont Fluoroproducts. Some values, read from charts, may be approximate.

TABLE 2-320   SUVA HP 80 at Atmospheric Pressure

Temp., °C	−46.95[b]	−40	−20	0	20	40	60	80	100	120
v (m³/kg)	0.1768	0.1827	0.1996	0.2164	0.2331	0.2497	0.2663	0.2828	0.2992	0.3155
h (kJ/kg)	335.9	340.5	354.3	368.6	383.5	398.7	414.9	431.4	448.5	466.1
s (kJ/kg·K)	1.6286	1.6490	1.7055	1.7599	1.8124	1.8633	1.9128	1.9610	2.0081	2.0541
c_p (kJ/kg·K)	0.648	0.654	0.687	0.721	0.749	0.779	0.807	0.836	0.863	0.890
μ (10⁻⁶ Pa·s)	9.42	9.69	10.45	11.22	11.99	12.75	13.52	14.29	15.06	15.82
k (W/m·K)	0.00888	0.00932	0.01059	0.01186	0.01313	0.01440	0.01568	0.01695	0.01822	0.01949
Pr	0.687	0.680	0.678	0.681	0.685	0.690	0.696	0.703	0.713	0.722
Z	0.9673	0.9697	0.9758	0.9804	0.9840	0.9868	0.9892	0.9910	0.9923	0.9932

b = normal boiling pt. v, h, and s from DuPont bull. T—HP 80—SI, Jan. 1993 (17 pp.). c_p, μ, and k from DuPont bull. ART 18, June 1993 (37 pp.). Some values read from charts may be approximate. Material used by permission of DuPont Fluoroproducts.

TABLE 2-321   Saturated SUVA HP 81

Temp., °C	P_f, bar	P_g, bar	v_f, m³/kg	v_g, m³/kg	h_f, kJ/kg	h_g, kJ/kg	s_f, kJ/(kg·K)	s_g, kJ/(kg·K)	c_{pf}, kJ/(kg·K)	μ_f, 10⁻⁶ Pa·s	k_f, W/(m·K)	Pr_f
−50	0.883	0.787	0.000 687	0.2425	140.3	351.7	0.7606	1.7122		383	0.1031	
−40	1.403	1.273	0.000 702	0.1548	151.4	357.2	0.8092	1.6957		333	0.0983	
−30	2.135	1.967	0.000 719	0.1028	163.3	362.7	0.8589	1.6820	1.178	290	0.0941	3.63
−20	3.132	2.923	0.000 739	0.0707	174.9	368.0	0.9054	1.6706	1.191	253	0.0900	3.35
−10	4.451	4.198	0.000 761	0.0499	187.8	373.0	0.9550	1.6611	1.204	223	0.0863	3.11
0	6.153	5.852	0.000 787	0.03610	200.0	377.8	1.0000	1.6528	1.221	195	0.0818	2.91
10	8.307	7.959	0.000 817	0.02656	212.7	382.2	1.0450	1.6451	1.288	173	0.0790	2.82
20	10.984	10.591	0.000 854	0.01980	226.0	386.0	1.0905	1.6376	1.313	151	0.0753	2.63
30	14.261	13.827	0.000 899	0.01490	240.1	389.3	1.1367	1.6299	1.37	137	0.0715	2.49
40	18.216	17.750	0.000 955	0.01125	255.1	391.5	1.1842	1.6211	1.75	122	0.0676	3.16
50	22.93	22.45	0.001 030	0.00848	271.4	392.8	1.2339	1.6104	2.07	106	0.0633	3.47
60	28.50	28.03	0.001 136	0.00632	289.5	392.2	1.2873	1.5961		91	0.0586	
70	35.01	34.60	0.001 307	0.00456	299.6	390.9	1.3164	1.5866		75	0.0544	
80												
82.6[c]	44.45	44.45	0.001 88	0.00188	351	351						

c = critical point. SUVA HP 81 = R402 (38/2/60) = CHF_2CF_3 (R125) 38% wt + $CH_3CH_2CH_3$ (R290) 2% wt + $CHClF_2$ (R22) 60% wt, near-azeotropic blend. Material used by permission of DuPont Fluoroproducts. Some values read from charts may be approximate.

TABLE 2-322   SUVA HP 81 at Atmospheric Pressure

Temp., °C	−44.87[b]	−40	−20	0	20	40	60	80	100	120
v (m³/kg)	0.1903	0.1960	0.2142	0.2322	0.2500	0.2678	0.2856	0.3032	0.3209	0.3386
h (kJ/kg)	354.7	357.7	370.8	384.6	398.8	413.6	428.9	444.7	461.0	477.7
s (kJ/kg·K)	1.7032	1.7169	1.7711	1.8232	1.8735	1.9222	1.9696	2.0158	2.0607	2.1047
c_p (kJ/kg·K)	1.187	1.177	1.169	1.159	1.149	1.143	1.134	1.128	1.124	1.120
μ (10⁻⁶ Pa·s)	10.16	10.33	11.10	11.86	12.62	13.39	14.15	14.78	15.54	16.30
k (W/m·K)	0.00739	0.00768	0.00902	0.01036	0.01170	0.01304	0.01438	0.01572	0.01706	0.01840
Pr	1.632	1.583	1.439	1.327	1.239	1.174	1.124	1.061	1.024	0.992
Z	0.9622	0.9703	0.9766	0.9811	0.9843	0.9870	0.9894	0.9909	0.9926	0.9940

b = normal boiling point. v, h, and s from DuPont bull. T—HP 81—SI, Jan. 1993 (17 pp.). c_p, μ, and k from DuPont bull. ART 18, June 1993 (37 pp.). Some values, read from charts, may be approximate. Material used by permission of DuPont Fluoroproducts.

TABLE 2-323 Saturated Refrigerant 113*

T, K	P, bar	v_f, m³/kg	v_g, m³/kg	h_f, kJ/kg	h_g, kJ/kg	s_f, kJ/(kg·K)	s_g, kJ/(kg·K)	c_{pf}, kJ/(kg·K)	μ_f, 10⁻⁴ Pa·s	k_f, W/(m·K)
240	0.0233	5.908.−4	4.548	5.70	171.97	0.0241	0.7169	0.845	17.9	0.087
250	0.0435	5.986.−4	2.537	14.19	178.06	0.0587	0.7142	0.877	14.8	0.084
260	0.0767	6.066.−4	1.492	22.83	184.22	0.0926	0.7134	0.895	12.3	0.083
270	0.1290	6.150.−4	0.9189	31.65	190.46	0.1259	0.7141	0.916	10.⪯	0.081
280	0.2076	6.237.−4	0.5893	40.63	196.75	0.1585	0.7161	0.933	8.9	0.079
290	0.3217	6.328.−4	0.3917	49.77	203.08	0.1906	0.7192	0.946	7.6	0.077
300	0.4817	6.442.−4	0.2687	59.07	209.44	0.2221	0.7233	0.958	6.6	0.075
310	0.6999	6.522.−4	0.1895	68.51	215.80	0.2530	0.7281	0.971	5.9	0.073
320	0.9897	6.626.−4	0.1370	78.09	222.17	0.2833	0.7336	0.983	5.2	0.071
330	1.3657	6.737.−4	0.1012	87.80	228.53	0.3131	0.7396	0.992	4.7	0.069
340	1.8347	6.854.−4	0.0762	97.64	234.86	0.3424	0.7460	1.000	4.2	0.066
350	2.4406	6.979.−4	0.0584	107.58	241.16	0.3711	0.7528	1.013	3.8	0.065
360	3.174	7.112.−4	0.0454	117.65	247.41	0.3993	0.7598	1.029	3.4	0.062
370	4.062	7.255.−4	0.0357	127.82	253.59	0.4270	0.7669	1.042	3.2	0.060
380	5.123	7.411.−4	0.0284	138.11	259.70	0.4542	0.7742	1.059	2.9	0.058
390	6.379	7.580.−4	0.0229	148.52	265.71	0.4810	0.7815	1.084	2.7	0.056
400	7.849	7.767.−4	0.0185	159.07	271.59	0.5075	0.7888	1.109	2.46	0.054
410	9.556	7.975.−4	0.0151	169.78	277.31	0.5336	0.7958	1.14	2.28	0.052
420	11.52	8.211.−4	0.0124	180.69	282.83	0.5595	0.8027	1.18	2.10	0.050
430	13.78	8.483.−4	0.0102	191.85	288.09	0.5853	0.8091	1.22	1.93	0.047
440	16.35	8.806.−4	0.0083	203.35	292.98	0.6112	0.8149	1.27	1.75	0.045
450	19.26	9.201.−4	0.0068	215.31	297.38	0.6375	0.8198	1.32	1.58	0.042
460	22.56	9.713.−4	0.0055	227.97	301.03	0.6645	0.8234	1.38	1.33	0.039
470	26.29	1.044.−3	0.0044	241.79	303.41	0.6933	0.8244	1.45	1.07	0.035
480	30.52	1.174.−3	0.0032	258.16	303.00	0.7264	0.8198	1.54	0.77	0.031
487.5ᶜ	34.11	1.754.−3	0.0018	288.10	288.10	0.7828	0.7828	∞	0.30	∞

*Values reproduced or converted from Table 8, p. 17.91, *ASHRAE Handbook, 1981: Fundamentals,* American Society of Heating, Refrigerating and Air-Conditioning Engineers, Atlanta, 1981. Copyright material. Reproduced by permission of the copyright owner. *c* = critical point. The notation 5.908.−4 signifies 5.908×10^{-4}. The 1993 ASHRAE *Handbook—Fundamentals* (SI ed.) gives a saturation table from −30 to 214.4°C and an enthalpy–log-pressure diagram from 0.1 to 60 bar, 0 to 260°C. Equations and constants approximated to the 1985 ASHRAE tables were given by Mecaryk, K. and M. Masaryk, *Heat Recovery Systems and CHP,* **11,** 2/3 (1991): 193–197. For experimental isochores for the compressed liquid from 21 to 304 bar, 266 to 453 K, see Blanke, W. and R. Weiss, PTB Bericht W 30, Braunschweig, Germany, 1992 (54 pp.).

For tables to 300 bar, 460 K, see Geller, V. Z. and V. A. Rabinovich (ed.), *Thermophysical Properties of Substances and Materials,* Standartov, Moscow, **7** (1973): 135–154. Mastroianni, M. J., R. F. Stahl, et al., *J. Chem. Eng. Data,* **23,** 2 (1978): 113–118 give a diagram to 1000 psia, 600°F. Tables and a diagram to 800 psia, 520°F are given by Stewart, R. B., R. T. Jacobsen, et al., *Thermodynamic Properties of Refrigerants,* ASHRAE, Atlanta, GA, 1986 (521 pp.). For specific heat, thermal conductivity, and viscosity, see *Thermophysical Properties of Refrigerants,* ASHRAE, 1993.

TABLE 2-324 Saturated Refrigerant 114*

T, K	P, bar	v_f, m³/kg	v_g, m³/kg	h_f, kJ/kg	h_g, kJ/kg	s_f, kJ/(kg·K)	s_g, kJ/(kg·K)	c_{pf}, kJ/(kg·K)	μ_f, 10⁻⁴ Pa·s	k_f, W/(m·K)
190	0.0058	6.326.−4	15.823	−42.58	125.78	−0.2091	0.6794	0.765	23.9	0.093
200	0.0137	6.344.−4	7.094	−31.87	131.01	−0.1542	0.6648	0.787	18.2	0.090
210	0.029	6.366.−4	3.465	−21.48	136.41	−0.1035	0.6541	0.810	14.3	0.088
220	0.059	6.391.−4	1.822	−11.37	141.95	−0.0565	0.6466	0.831	11.5	0.085
230	0.109	6.421.−4	1.021	−1.50	147.61	−0.0126	0.6419	0.854	9.4	0.082
240	0.190	6.457.−4	0.604	8.18	153.36	0.0286	0.6393	0.877	7.9	0.080
250	0.317	6.500.−4	0.375	17.74	159.18	0.0676	0.6387	0.900	6.61	0.077
260	0.505	6.554.−4	0.2431	27.22	165.05	0.1047	0.6396	0.923	5.66	0.075
270	0.773	6.619.−4	0.1633	36.71	170.95	0.1405	0.6418	0.946	4.96	0.072
280	1.143	6.700.−4	0.1132	46.27	176.85	0.1751	0.6452	0.967	4.30	0.069
290	1.636	6.799.−4	0.0807	55.95	182.75	0.2090	0.6494	0.991	3.80	0.067
300	2.279	6.918.−4	0.0590	65.79	188.61	0.2422	0.6543	1.015	3.35	0.064
310	3.096	7.060.−4	0.0440	75.79	194.44	0.2748	0.6598	1.038	3.02	0.061
320	4.116	7.224.−4	0.0334	85.92	200.19	0.3067	0.6657	1.062	2.69	0.059
330	5.366	7.412.−4	0.0257	96.16	205.84	0.3379	0.6719	1.087	2.48	0.056
340	6.877	7.624.−4	0.0201	106.49	211.37	0.3685	0.6781	1.111	2.27	0.054
350	8.683	7.863.−4	0.0158	116.96	216.71	0.3984	0.6843	1.136	2.07	0.051
360	10.82	8.135.−4	0.0125	127.63	221.82	0.4280	0.6903	1.160	1.91	0.048
370	13.32	8.453.−4	0.0099	138.60	226.57	0.4575	0.6957	1.185	1.76	0.045
380	16.24	8.836.−4	0.0079	149.99	230.84	0.4872	0.7002	1.210	1.59	0.042
390	19.62	9.324.−4	0.0062	162.01	234.36	0.5176	0.7032	1.236	1.39	0.038
400	23.52	1.001.−3	0.0048	175.03	236.61	0.5496	0.7036	1.261	1.17	0.034
410	28.00	1.118.−3	0.0035	190.13	236.20	0.5857	0.6980	1.5	0.87	0.030
419.0ᶜ	32.61	1.795.−3	0.0018	219.90	219.90	0.6559	0.6559	∞	0.34	∞

*Values reproduced or converted from Table 9, p. 17.93, *ASHRAE Handbook, 1981: Fundamentals,* American Society of Heating, Refrigerating and Air-Conditioning Engineers, Atlanta, 1981. Copyright material. Reproduced by permission of the copyright owner. *c* = critical point. The notation 6.326.−4 signifies 6.326×10^{-4}. The 1993 ASHRAE *Handbook—Fundamentals* (SI ed.) gives a saturation table from −80 to 145.88°C and an enthalpy–log-pressure diagram from 0.1 to 100 bar, −20 to 220°C. Equations and constants approximated to the 1985 ASHRAE tables were given by Mecaryk, K. and M. Masaryk, *Heat Recovery Systems and CHP.,* **11,** 2/3 (1991): 193–197.

Saturation and superheat tables and a diagram to 60 bar, 540 K are given by Reynolds, W. C., *Thermodynamic Properties in S.I.,* Stanford Univ. publ., 1979 (179 pp.). Tables and a chart to 1500 psia, 480°F are given by Stewart, R. B., R. T. Jacobsen, et al., *Thermodynamic Properties of Refrigerants,* ASHRAE, Atlanta, GA, 1986 (521 pp.). For specific heat, thermal conductivity, and viscosity, see *Thermophysical Properties of Refrigerants,* ASHRAE, 1993.

TABLE 2-325 Saturated Refrigerant 115*

Temp., °F	Pressure, lb/in² abs.	Volume, ft³/lb		Enthalpy, Btu/lb		Entropy, Btu/(lb)(°F)	
		Liquid	Vapor	Liquid	Vapor	Liquid	Vapor
−100	2.327	0.00966	10.57	−13.07	45.83	−0.0335	0.1302
−80	4.573	0.00986	5.624	−8.78	48.39	−0.0219	0.1286
−60	8.306	0.01009	3.218	−4.43	50.96	−0.0108	0.1278
−40	14.13	0.01033	1.953	0.00	53.53	0.0000	0.1275
−20	22.74	0.01060	1.245	4.50	56.07	0.0104	0.1277
0	34.94	0.01090	0.8257	9.09	58.56	0.0206	0.1282
20	51.59	0.01123	0.5657	13.76	61.00	0.0305	0.1290
40	73.65	0.01161	0.3979	18.54	63.35	0.0401	0.1298
60	102.1	0.01204	0.2857	23.45	65.60	0.0496	0.1308
80	138.1	0.01255	0.2081	28.54	67.71	0.0591	0.1317
100	182.7	0.01316	0.1530	33.85	69.63	0.0686	0.1325
120	237.3	0.01393	0.1125	39.50	71.24	0.0782	0.1330
140	303.2	0.01496	0.0817	45.67	72.36	0.0884	0.1329
160	382.0	0.01664	0.0567	52.76	72.42	0.0996	0.1314
170	427.0	0.01838	0.0444	56.56	71.33	0.1055	0.1290
175.89ᶜ	457.6	0.0261	0.0261	64.30	64.30	0.1175	0.1175

*Unpublished data of General Chemicals Division, Allied Chemical Company. Used by permission. *c* = critical temperature. No material in SI units appears in the 1993 ASHRAE *Handbook—Fundamentals* (SI ed.). Tables and a chart to 50 ata, 200°C are given by Mathias, H. and H. J. Loffler, Techn. Univ. Berlin rept., 1966 (42 pp.). A chart to 1500 psia, 500°F was given by Mears, W. H., E. Rosenthal, et al., *J. Chem. Eng. Data,* **11,** 3 (1966): 338–343.

TABLE 2-326   Thermodynamic Properties of Refrigerant 123

Pressure, bar	Temp., K	v_f, m³/kg	v_g, m³/kg	h_f, kJ/kg	h_g, kJ/kg	s_f, kJ/(kg·K)	s_g, kJ/(kg·K)	c_{pf}, kJ/(kg·K)	k_f, W/(m·K)	μ_f, 10⁻⁶ Pa·s	Pr_f
0.1	249.49	0.000 6315	1.3430	13.25	198.51	0.0548	0.7977	0.849	0.0908	798.7	7.46
0.5	282.87	0.000 6664	0.2993	41.72	218.53	0.1610	0.7863	0.923	0.0811	503.7	5.73
1.0	300.62	0.000 6862	0.1567	58.62	229.20	0.2195	0.7869	1.000	0.0759	409.8	5.40
1.013	300.99	0.000 6868	0.1546	58.99	229.43	0.2208	0.7870	1.001	0.0758	408.1	5.39
1.5	312.25	0.000 7008	0.1070	70.51	236.52	0.2582	0.7892	1.038	0.0726	361.5	5.17
2.0	321.18	0.000 7126	0.08139	79.90	241.76	0.2877	0.7917	1.063	0.0696	329.6	5.03
2.5	328.50	0.000 7230	0.06546	87.76	246.20	0.3118	0.7942	1.079	0.0678	306.1	4.87
3.0	334.79	0.000 7323	0.05525	94.59	249.96	0.3323	0.7965	1.091	0.0660	287.4	4.75
4.0	345.29	0.000 7490	0.03836	106.16	256.17	0.3661	0.8006	1.108	0.0630	259.7	4.57
5.0	353.95	0.000 7640	0.03358	115.83	261.17	0.3935	0.8042	1.120	0.0605	239.1	4.43
6	361.41	0.000 7779	0.02799	124.23	265.34	0.4168	0.8073	1.130			
8	373.92	0.000 8038	0.02090	138.48	272.04	0.4551	0.8124	1.148			
10	384.19	0.000 8280	0.01675	150.35	277.18	0.4860	0.8162	1.168			
15	404.54	0.000 8874	0.01062	174.49	286.01	0.5462	0.8218	1.234			
20	420.30	0.000 9512	0.00751	194.19	291.01	0.5928	0.8232	1.345			
25	433.33	0.001 030	0.00549	212.00	293.05	0.6334	0.8203	1.559			
30	444.10	0.001 136	0.00408	228.26	291.27	0.6692	0.8112	2.005			
36.68°	456.83	0.001 818	0.00182	264.54	264.54	0.7393	0.7393				

$h_f = s_f = 0$ at −40°C = 233.15 K. s_f, s_g, c_{pf} units: kJ/kg·K. Interpolated and converted from 1993 ASHRAE *Handbook—Fundamentals* (SI ed.) saturation table from −40 to 183.68°C. This source also contains an enthalpy–log-pressure diagram from 0.1 to 200 bar, −40 to 320°C.

TABLE 2-327   Saturated Refrigerant 124

Temp., °C	Pressure, bar	v_f, m³/kg	v_g, m³/kg	h_f, kJ/kg	h_g, kJ/kg	s_f, kJ/(kg·K)	s_g, kJ/(kg·K)
−40	0.2680	0.000 644	0.5173	159.1	334.9	0.8384	1.5927
−30	0.4499	0.000 655	0.3185	169.3	340.6	0.8813	1.5856
−20	0.7197	0.000 668	0.2049	179.5	346.2	0.9222	1.5808
−10	1.1044	0.000 681	0.1369	189.7	351.8	0.9616	1.5777
0	1.6348	0.000 696	0.0945	200.0	357.4	1.0000	1.5762
10	2.3447	0.000 711	0.06703	210.5	363.0	1.0376	1.5760
20	3.2710	0.000 728	0.04867	221.3	368.5	1.0747	1.5768
30	4.4529	0.000 747	0.03604	282.3	373.9	1.1115	1.5785
40	5.9320	0.000 768	0.02713	243.7	379.2	1.1480	1.5808
50	7.7521	0.000 791	0.02069	255.4	384.4	1.1843	1.5836
60	9.9599	0.000 818	0.01594	267.5	389.3	1.2207	1.5864
70	12.605	0.000 849	0.01236	280.1	393.9	1.2572	1.5890
80	15.742	0.000 887	0.00961	293.2	398.0	1.2942	1.5909
90	19.432	0.000 935	0.00744	307.0	401.3	1.3318	1.5915
100	23.749	0.000 999	0.00569	321.9	403.4	1.3710	1.5894
110	28.787	0.001 098	0.00420	338.4	403.0	1.4133	1.5820
120	34.702	0.001 338	0.00269	360.6	394.9	1.4685	1.5558
122.5ᶜ	36.340	0.001 810	0.00181	378.5	378.5		

c = critical point.

Bull. T—124—SI, Jan. 1993 (28 pp.). Used by permission of DuPont Fluoroproducts. The 1993 ASHRAE *Handbook— Fundamentals* (SI ed.) gives a saturation table to 122.47°C and a diagram to 200 bar, 320°C.

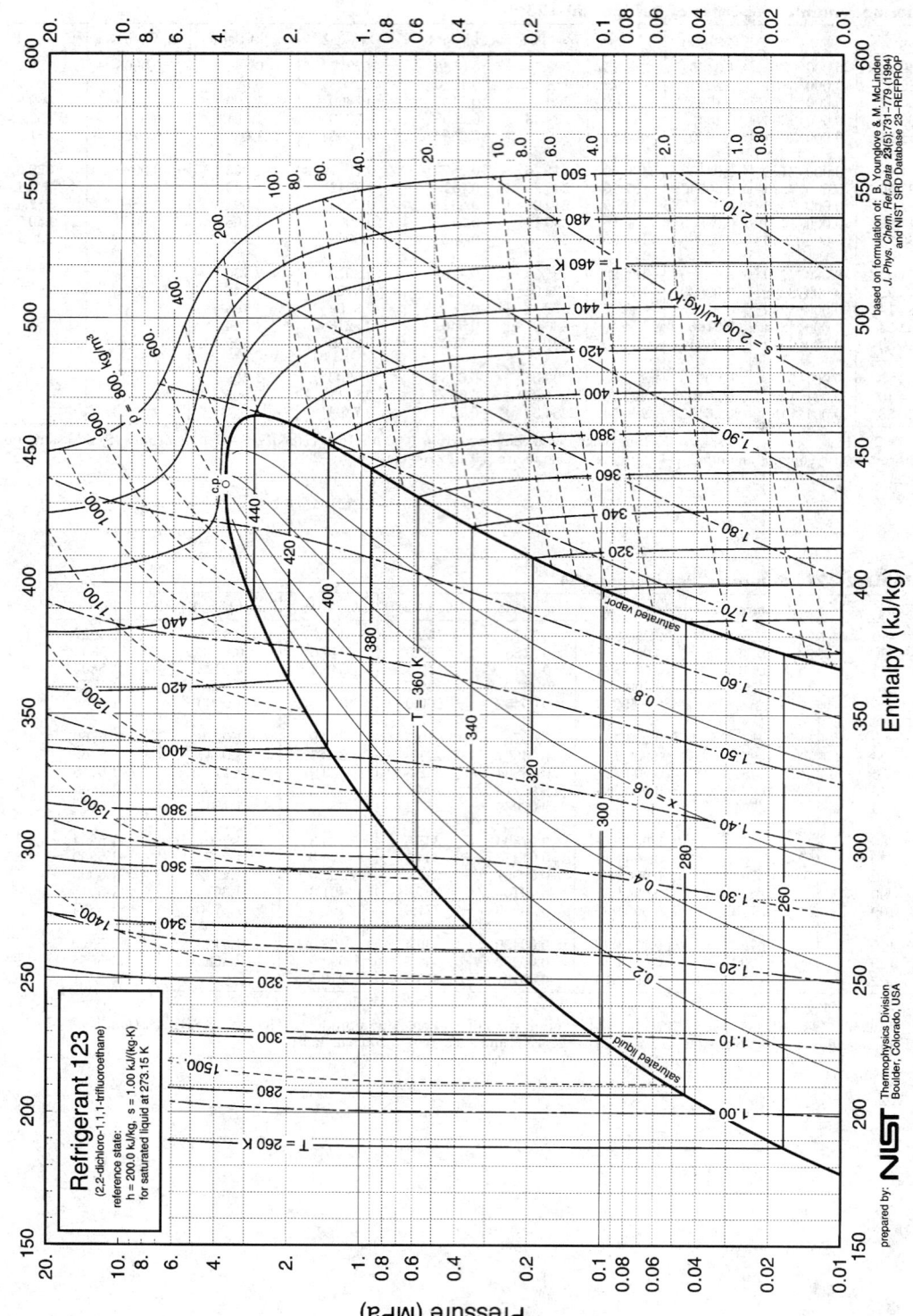

FIG. 2-25 Enthalpy–log-pressure diagram for Refrigerant 123.

TABLE 2-328 Thermophysical Properties of Saturated Refrigerant 125

Temp., K	Pressure, bar	v_f, m³/kg	v_g, m³/kg	h_f, kJ/kg	h_g, kJ/kg	s_f, kJ/(kg·K)	s_g, kJ/(kg·K)	c_{pf}, kJ/(kg·K)	μ_f, 10⁻⁶ Pa·s
172.5°	0.035	0.000591	3.48						
180	0.064	0.000599	1.958						
190	0.133	0.000611	0.986						
200	0.257	0.000624	0.5312	−30.5	140.8	−0.1386	0.7183		644
210	0.465	0.000638	0.3057	−23.3	146.7	−0.1024	0.7067		531
220	0.794	0.000653	0.1854	−13.9	152.6	−0.0604	0.6981		445.8
224.9†	1.013	0.000660	0.1475	−8.8	155.5	−0.0386	0.6948		411.2
230	1.290	0.000669	0.1175	−3.4	158.4	−0.0147	0.6919	1.077	379.6
240	2.005	0.000686	0.0775	7.7	164.2	0.0324	0.6875	1.139	326.6
250	3.000	0.000705	0.0527	19.3	170.0	0.0800	0.6847	1.184	282.8
260	4.336	0.000725	0.0369	31.3	175.3	0.1274	0.6831	1.221	245.8
270	6.078	0.000749	0.0264	43.7	180.5	0.1743	0.6822	1.257	213.6
280	8.298	0.000776	0.0193	56.5	185.4	0.2206	0.6819	1.299	185.3
290	11.068	0.000809	0.0143	69.7	189.9	0.2666	0.6815	1.356	159.7
300	14.476	0.000848	0.0106	83.5	194.2	0.3126	0.6805	1.437	136.3
310	18.62	0.000898	0.0079	98.1	196.9	0.3597	0.6774	1.57	115.0
320	23.63	0.000969	0.0059	113.9	198.5	0.4079	0.6726	1.82	95.4
330	29.65	0.001088	0.0041	132.2	197.4	0.4621	0.6639		
339.4‡	35.95	0.00175	0.0018	169.0	169.0	0.5699	0.5699		

° = triple point; † = normal boiling point; ‡ = critical point. Converted, extrapolated and interpolated from 1993 ASHRAE *Handbook—Fundamentals* (SI ed.) $h_f = s_f = 0$ at 233.15 K = −40°C. This source also contains an enthalpy–log-pressure diagram from 0.3 to 100 bar, −65 to 175°C. An apparently identical diagram but a different saturation table is contained in Duarte-Garza, H.A., Hwang, C.A. et al., ASHRAE Trans., **99**, 2 (1993): 649–664. R124: The 1993 ASHRAE *Handbook—Fundamentals* (SI ed.) contains a saturation table from −60 to 122.47°C.

TABLE 2-329 Thermophysical Properties of Refrigerant 134a

Pressure, bar	Temp., K	v_f, m³/kg	v_g, m³/kg	h_f, kJ/kg	h_g, kJ/kg	s_f, kJ/(kg·K)	s_g, kJ/(kg·K)	c_{pf}, kJ/(kg·K)	μ_f, 10⁻⁶ Pa·s	k_f, W/(m·K)	Pr_f
0.0039t	169.85	0.0006285	35.263	−76.68	186.50	−0.3830	1.1665	1.147	2187		
0.5	232.69	0.0007062	0.3692	−0.57	225.27	−0.0025	0.9669	1.242	506	0.1121	5.61
0.6	236.22	0.0007113	0.3015	3.85	227.52	0.0161	0.9636	1.248	480	0.1105	5.42
0.8	242.04	0.0007199	0.2375	11.15	231.19	0.0467	0.9560	1.258	438	0.1078	5.12
1.0	246.80	0.0007272	0.1924	17.14	234.15	0.0713	0.9507	1.267	408	0.1056	4.90
1.013	247.03	0.0007276	0.1902	17.50	234.33	0.0728	0.9503	1.268	406	0.1054	4.89
1.5	256.03	0.0007421	0.1312	28.96	239.86	0.1181	0.9419	1.288	358.7	0.1013	4.56
2.0	263.09	0.0007543	0.0999	38.13	244.14	0.1533	0.9364	1.306	326.6	0.0980	4.35
2.5	268.88	0.0007648	0.0806	45.75	247.60	0.1819	0.9326	1.322	303.2	0.0954	4.20
3.0	273.82	0.0007743	0.0677	52.33	250.50	0.2059	0.9297	1.337	285.1	0.0931	4.09
4.0	282.08	0.0007912	0.0512	63.50	255.22	0.2458	0.9256	1.363	257.7	0.0893	3.93
5	288.89	0.0008063	0.04116	72.87	258.99	0.2784	0.9232	1.387	237.5	0.0861	3.83
6	294.72	0.0008203	0.03434	81.04	262.09	0.3062	0.9208	1.410	221.6	0.0835	3.74
8	304.47	0.0008460	0.02565	95.00	267.01	0.3522	0.9171	1.454	197.6	0.0790	3.64
10	312.53	0.0008703	0.02035	106.86	270.74	0.3901	0.9144	1.497	179.5	0.0753	3.57
12	319.47	0.0008938	0.01675	117.34	273.65	0.4227	0.9120	1.541	165.1	0.0721	3.53
14	325.57	0.0009170	0.01414	126.80	275.92	0.4515	0.9095	1.589	153.0	0.0693	3.51
16	330.11	0.0009362	0.01247	134.00	277.40	0.4729	0.9073	1.631	144.3	0.0672	3.50
18	336.04	0.0009555	0.01059	143.68	279.01	0.5013	0.9041	1.698	133.2	0.0645	3.51
20	340.63	0.0009894	0.00931	151.39	279.95	0.5236	0.9010	1.764	124.8	0.0623	3.53
25	350.73	0.0010585	0.00695	169.30	280.64	0.5738	0.8913	1.987	106.6	0.0577	3.67
30	359.37	0.001144	0.00528	185.05	278.32	0.6212	0.8807	2.418	90.4	0.0538	4.06
35	366.89	0.001270	0.00399	203.19	273.52	0.6657	0.8574				
40	373.50	0.001606	0.00255	229.24	257.12	0.7292	0.8038				
40.56c	374.18	0.001948	0.00195	241.22	241.22	0.7620	0.7620				

t = triple point, c = critical point. $h_f = s_f = 0$ at −40°C = 233.15 K. T, v, h, and s interpolated and converted from *Refrigerant 134a—Thermodynamic and Physical Properties*, Int. Inst. Refrig., Paris, France, 1992 (28 pp.).

Other properties from this source and from Oliveira, C. M. B. P. and W. A. Wakeham, *Int. J. Thermophys.*, **14**, 1 (1993): 33–44; Krauss, R., J. Luettmer-Strathmann, et al., *Int. J. Thermophys.*, **14**, 4 (1993): 951–988; ASHRAE *Handbook—Fundamentals*, Atlanta, GA, 1993; ICI KLEA 134a bulletin, 1993 (43 pp.); and *R134a—Thermodynamic and Physical Properties*, Int. Inst. Refrig., Paris, France, 1992 (28 pp.).

Papers giving polynomial curve fits and similar simple equations include Cleland, A. C., *Rev. Int. Froid = Int. J. Refrig.*, **17**, 4 (1994): 245–249; Dobrokhotov, A., A. Grebenkov, et al., *Proc. 14th Japan Symp. Thermophys. Props.*, (1993): 271–274; Huber, M. L. and J. F. Ely, *Rev. Int. Froid = Int. J. Refrig.*, **17**, 1 (1994): 18–31 (includes extensive list of vapor pressure and liquid density sources for many refrigerants); Modic, J., *Proc. 11th Int. Symp. Htg, Refrig and Air-Condg.*, Zagreb, (1991): 174–185; and Kabelac, S., *Int. J. Refrig.*, **14**, (1991): 217–222. The 1993 ASHRAE *Handbook—Fundamentals* (SI ed.) gives saturation data for integral degrees Celsius with temperatures on the ITS 90 scale from −103.03°C to 101.03°C. The thermodynamic diagram from 0.1 to 200 bar extends to 320°C.

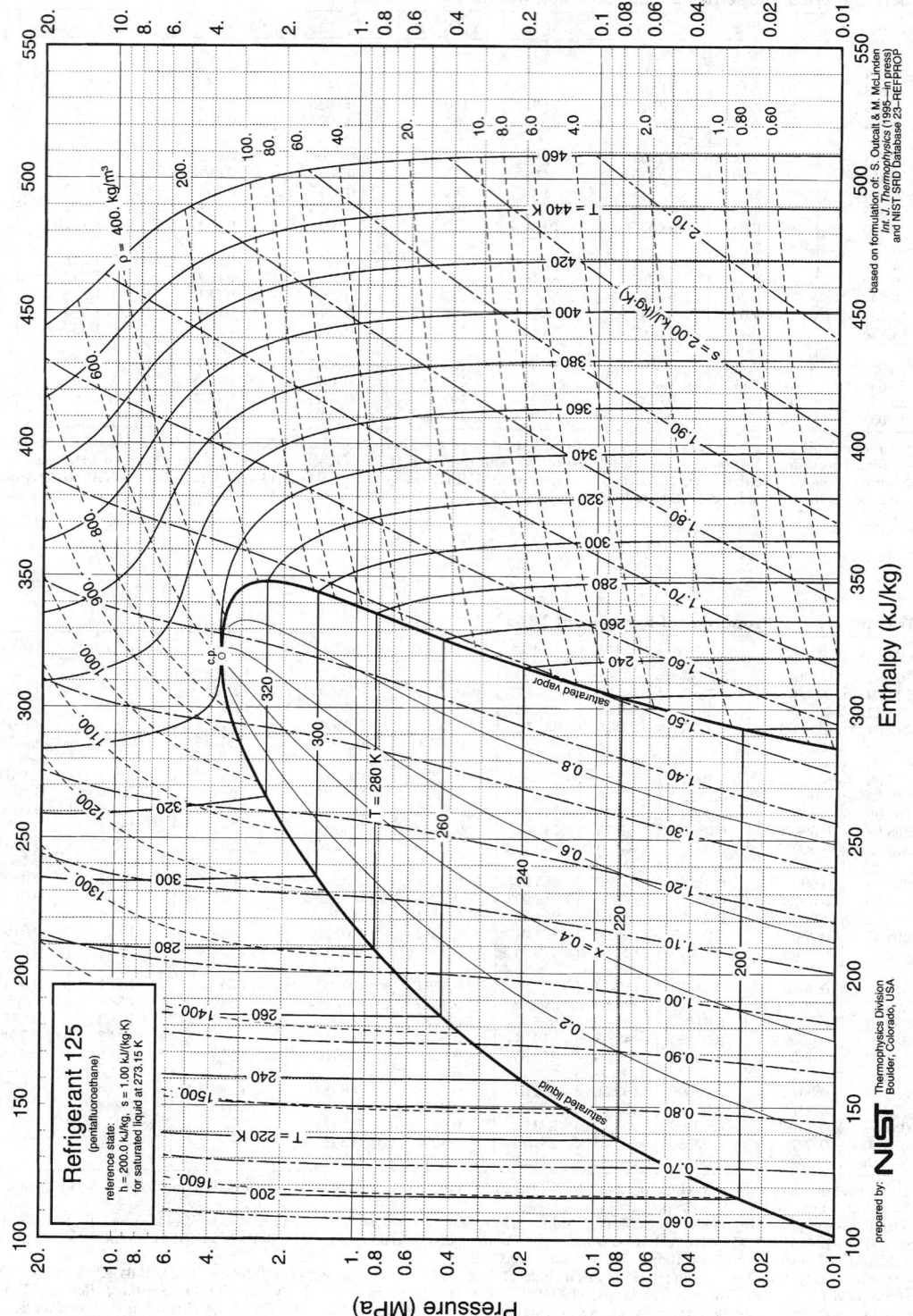

FIG. 2-26 Enthalpy–log-pressure diagram for Refrigerant 125.

TABLE 2-330 Thermophysical Properties of Compressed Gaseous Refrigerant 134a

Temp., K		Pressure, bar						
		0	1	2	3	4	5	6
230	c_p (kJ/kg·K)		—	—	—	—	—	—	—
	μ (10⁻⁶ Pa·s)		—	—	—	—	—	—	—
	k (W/m·K)		—	—	—	—	—	—	—
	Pr		—	—	—	—	—	—	—
240	c_p (kJ/kg·K)		—	—	—	—	—	—	—
	μ (10⁻⁶ Pa·s)		—	—	—	—	—	—	—
	k (W/m·K)		—	—	—	—	—	—	—
	Pr		—	—	—	—	—	—	—
250	c_p (kJ/kg·K)	0.7437	0.7953	—	—	—	—	—
	μ (10⁻⁶ Pa·s)	10.11	10.15	—	—	—	—	—
	k (W/m·K)	0.0096	0.0097	—	—	—	—	—
	Pr	0.783	0.797	—	—	—	—	—
260	c_p (kJ/kg·K)	0.7627	0.8048	—	—	—	—	—
	μ (10⁻⁶ Pa·s)	10.47	10.51	—	—	—	—	—
	k (W/m·K)	0.0105	0.0107	—	—	—	—	—
	Pr	0.761	0.790	—	—	—	—	—
270	c_p (kJ/kg)	0.7813	0.8158	0.8557	—	—	—	—
	μ (10⁻⁶ Pa·s)	10.84	10.88	10.94	—	—	—	—
	k (W/m·K)	0.0117	0.0118	0.0118	—	—	—	—
	Pr	0.724	0.761	0.793	—	—	—	—
280	c_p (kJ/kg·K)	0.7996	0.8283	0.8604		—	—	—
	μ (10⁻⁶ Pa·s)	11.22	11.26	11.29		—	—	—
	k (W/m·K)	0.0122	0.0123	0.0123		—	—	—
	Pr	0.735	0.757	0.790		—	—	—
290	c_p (kJ/kg·K)	0.8176	0.8412	0.8673	0.8938	0.9335		—
	μ (10⁻⁶ Pa·s)	11.62	11.65	11.68	11.71	11.74		—
	k (W/m·K)	0.0130	0.0131	0.0131	0.0133			—
	Pr	0.731	0.748	0.773	0.787			—
300	c_p (kJ/kg·K)	0.8354	0.8556	0.8771	0.8972	0.9277	0.9606	0.9976
	μ (10⁻⁶ Pa·s)	12.05	12.06	12.08	12.10	12.15		
	k (W/m·K)	0.0139	0.0139	0.0140	0.0141	0.0142	0.0143	
	Pr	0.730	0.742	0.757	0.770	0.792	0.816	
310	c_p (kJ/kg·K)	0.8530	0.8703	0.8875	0.9046	0.9292	0.9546	0.9827
	μ (10⁻⁶ Pa·s)	12.44	12.45	12.47	12.49	12.52	12.54	
	k (W/m·K)	0.0145	0.0146	0.0147	0.0148	0.0149	0.0150	0.0152
	Pr	0.730	0.742	0.753	0.763	0.781	0.798	
320	c_p (kJ/kg·K)	0.8703	0.8843	0.8993	0.9163	0.9356	0.9548	0.9750
	μ (10⁻⁶ Pa·s)	12.83	12.84	12.86	12.88	12.90	12.93	12.97
	k (W/m·K)	0.0153	0.0153	0.0154	0.0155	0.0156	0.0157	0.0158
	Pr	0.730	0.740	0.751	0.761	0.774	0.786	0.800
330	c_p (kJ/kg·K)	0.8874	0.8996	0.9114	0.9268	0.9398	0.9569	0.9750
	μ (10⁻⁶ Pa·s)	13.22	13.23	13.25	13.27	13.29	13.32	13.35
	k (W/m·K)	0.0160	0.0161	0.0161	0.0162	0.0163	0.0164	0.0165
	Pr	0.729	0.739	0.750	0.759	0.766	0.777	0.789
340	c_p (kJ/kg·K)	0.9042	0.9152	0.9262	0.9372	0.9502	0.9632	0.9770
	μ (10⁻⁶ Pa·s)	13.61	13.62	13.64	13.66	13.68	13.70	13.73
	k (W/m·K)	0.0169	0.0169	0.0169	0.0170	0.0170	0.0171	0.0171
	Pr	0.728	0.738	0.748	0.755	0.765	0.772	0.780
350	c_p (kJ/kg·K)	0.9208	0.9307	0.9406	0.9505	0.9607	0.9695	0.9830
	μ (10⁻⁶ Pa·s)	13.98	13.99	14.01	14.03	14.05	14.07	14.10
	k (W/m·K)	0.0175	0.0176	0.0176	0.0177	0.0177	0.0178	0.0179
	Pr	0.730	0.740	0.749	0.754	0.763	0.767	0.774

Dashes indicate unavailable states; blanks indicate no data.

Note that "profound differences" presently exist in the transport properties of R134a according to *Chemistry International,* **16**(6), 233, Nov. 1994, and **18**(2) 44–47, 1996.

TABLE 2-330 Thermophysical Properties of Compressed Gaseous R134a (*Concluded*)

Temp., K		Pressure, bar						
		8	10	12.5	15	17.5	20	22.5
300	c_p (kJ/kg·K)	—	—	—	—	—	—	—
	μ (10^{-6} Pa·s)	—	—	—	—	—	—	—
	k (W/m·K)	—	—	—	—	—	—	—
	Pr	—	—	—	—	—	—	—
310	c_p (kJ/kg·K)	1.053	—	—	—	—	—	—
	μ (10^{-6} Pa·s)	—	—	—	—	—	—	—
	k (W/m·K)	0.0155	—	—	—	—	—	—
	Pr	—	—	—	—	—	—	—
320	c_p (kJ/kg·K)	1.028	1.097	—	—	—	—	—
	μ (10^{-6} Pa·s)	13.05	13.13	—	—	—	—	—
	k (W/m·K)	0.0161	—	—	—	—	—	—
	Pr	0.833	—	—	—	—	—	—
330	c_p (kJ/kg·K)	1.015	1.065	1.151	1.276	—	—	—
	μ (10^{-6} Pa·s)	13.41	13.49	13.64	13.86	—	—	—
	k (W/m·K)	0.0168	0.0171	0.0177	0.0184	—	—	—
	Pr	0.810	0.840	0.887	0.961	—	—	—
340	c_p (kJ/kg·K)	1.008	1.049	1.107	1.187	1.319		—
	μ (10^{-6} Pa·s)	13.79	13.86	13.98	14.17		—	—
	k (W/m·K)	0.0174	0.0177	0.181	0.0187		—	
	Pr	0.799	0.821	0.855	0.899		—	
350	c_p (kJ/kg·K)	1.008	1.040	1.086	1.148	1.225	1.340	1.525
	μ (10^{-6} Pa·s)	14.15	14.22	14.34	14.49		14.97	
	k (W/m·K)	0.0181	0.0183	0.0186	0.0192	0.0198	0.0205	0.0215
	Pr	0.788	0.828	0.837	0.866			

TABLE 2-331 Refrigerant 141b

The 1993 ASHRAE *Handbook—Fundamentals* (SI ed.) gives saturation data to 150°C and a diagram to 20 bar, 150°C. For equation of state including decomposition, see Weber, L. A., paper 69, *Proc. 18th Int. Congr. Refrig.*, Montreal, 1991.

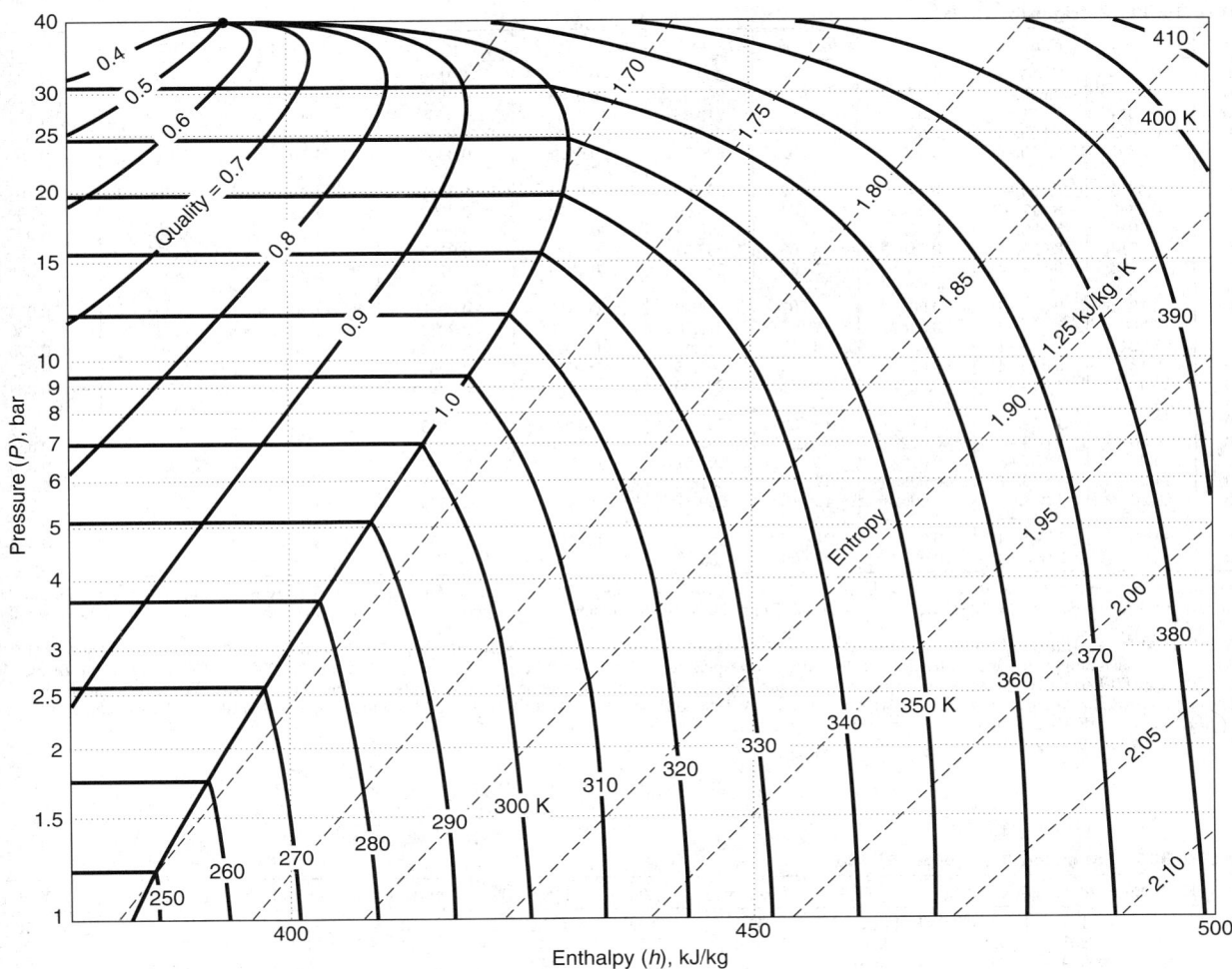

FIG. 2-27   Enthalpy–log-pressure diagram for Refrigerant 134a.

TABLE 2-332 Refrigerant 142b*

T, K	P, bar	v_f, m³/kg	v_g, m³/kg	h_f, kJ/kg	h_g, kJ/kg	s_f, kJ/(kg·K)	s_g, kJ/(kg·K)	c_{pf}, kJ/(kg·K)	μ_f, 10^{-4} Pa·s	k_f, W/(m·K)
200	0.0380	7.505.−4	4.337	−24.36	200.49	−0.1123	1.0119			0.123
210	0.0728	7.626.−4	2.374	−17.48	206.82	−0.0788	0.9893	1.15		0.118
220	0.1314	7.751.−4	1.373	−10.21	213.28	−0.0450	0.9708	1.17		0.114
230	0.2252	7.883.−4	0.833	−2.52	219.82	−0.0109	0.9558	1.18		0.111
240	0.3691	8.019.−4	0.527	9.92	229.74	0.0414	0.9387	1.19	0.517	0.109
250	0.5815	8.164.−4	0.346	14.32	233.05	0.0592	0.9341	1.21	0.466	0.103
260	0.8846	8.317.−4	0.234	23.54	239.66	0.0952	0.9264	1.22	0.422	0.099
270	1.3046	8.480.−4	0.162	33.32	246.18	0.1320	0.9204	1.24	0.385	0.095
280	1.8714	8.653.−4	0.115	43.68	252.57	0.1695	0.9155	1.26	0.355	0.091
290	2.6184	8.843.−4	0.0838	54.60	258.77	0.2076	0.9116	1.28	0.329	0.088
300	3.583	9.047.−4	0.0619	66.07	264.69	0.2462	0.9082	1.30	0.305	0.084
310	4.803	9.273.−4	0.0464	78.07	270.26	0.2851	0.9051	1.32	0.285	0.080
320	6.324	9.525.−4	0.0353	90.55	275.40	0.3243	0.9020	1.34	0.267	0.075
330	8.187	9.810.−4	0.0271	103.45	280.01	0.3634	0.8985		0.241	0.072
340	10.44	1.014.−3	0.0210	116.71	283.99	0.4024	0.8943		0.216	0.068
350	13.13	1.052.−3	0.0164	130.30	287.23	0.4409	0.8893		0.192	0.064
360	16.30	1.099.−3	0.0129	144.18	289.61	0.4791	0.8831			0.060
370	20.01	1.157.−3	0.0102	158.45	291.01	0.5170	0.8753			0.056
380	24.29	1.235.−3	0.0080	173.45	291.22	0.5557	0.8656			0.052
390	29.20	1.348.−3	0.0062	190.16	289.77	0.5974	0.8528			0.048
400	34.78	1.541.−3	0.0046	212.57	284.04	0.6521	0.8307			0.044
410ᶜ	41.5	2.300.−3	0.0023	255.00	255.00					∞

*Values reproduced and converted from Table 10, p. 17.95, *ASHRAE Handbook, 1981: Fundamentals*, American Society of Heating, Refrigerating and Air-Conditioning Engineers, Atlanta, 1981. Copyright material. Reproduced by permission of the copyright owner. c = critical point. The notation 7.505.−4 signifies 7.505×10^{-4}.

The 1993 ASHRAE *Handbook—Fundamentals* (SI ed.) gives material for integral degrees Celsius with temperatures on the IPTS 68 scale from −50 to 125°C. The thermodynamic diagram from 0.1 to 35 bar extends to 180°C. For experimental isochores for the compressed liquid from 6 to 298 bar, 147 to 432 K, see Blanke, W. and R. Weiss, *PTB Bericht W 30*, Braunschweig, Germany, 1992 (54 pp.). Tables and a diagram to 500 psia, 400°F are given in Stewart, R. B., R. T. Jacobsen, et al., *Thermodynamic Properties of Refrigerants*, ASHRAE, Atlanta, GA, 1986 (521 pp.). For specific heat, thermal conductivity, and viscosity, see *Thermophysical Properties of Refrigerants*, ASHRAE, 1993.

TABLE 2-333 Saturated Refrigerant R143a*

Temp., K	Pressure, bar	v_f, m³/kg	v_g, m³/kg	h_f, kJ/kg	h_g, kJ/kg	s_f, kJ/(kg·K)	s_g, kJ/(kg·K)	c_{pf}, kJ/(kg·K)	μ_f, 10^{-4} Pa·s	k_f, W/(m·K)
161.82ᵗ	0.01124	0.000752	14.22	53.2	320.5	0.3181	1.970	1.188	6.011	0.1416
170	0.02497	0.000764	6.709	63.0	325.5	0.3774	1.922	1.215	5.366	0.1403
180	0.05914	0.000778	2.991	75.3	331.8	0.4474	1.872	1.235	4.692	0.1372
190	0.126	0.000793	1.474	87.7	338.1	0.5147	1.832	1.252	4.121	0.1331
200	0.2458	0.000809	0.7898	100.3	344.5	0.5792	1.800	1.268	3.636	0.1281
210	0.4455	0.000826	0.4532	113.1	350.8	0.6415	1.774	1.287	3.221	0.1226
220	0.7586	0.000845	0.2754	126.1	357.1	0.7018	1.752	1.308	2.864	0.1168
225.92	1.01325	0.000856	0.2098	133.9	360.8	0.7367	1.741	1.323	2.676	0.1133
230	1.225	0.000865	0.1755	139.3	363.3	0.7604	1.734	1.333	2.556	0.1108
240	1.89	0.000886	0.1164	152.8	369.4	0.8176	1.720	1.362	2.288	0.1047
250	2.806	0.000910	0.07975	166.6	375.3	0.8736	1.708	1.394	2.055	0.09862
260	4.027	0.000936	0.05617	180.8	380.9	0.9287	1.698	1.431	1.852	0.09272
270	5.613	0.000966	0.04045	195.3	386.2	0.983	1.690	1.475	1.673	0.08682
280	7.629	0.000999	0.02964	210.3	391.1	1.037	1.682	1.527	1.496	0.08098
290	10.14	0.001038	0.02200	225.9	395.4	1.091	1.675	1.593	1.334	0.07521
300	13.23	0.001084	0.01646	242.1	399.0	1.144	1.668	1.679	1.192	0.06951
310	16.98	0.001140	0.01234	259.2	401.6	1.199	1.659	1.804	1.067	0.06381
320	21.48	0.001214	0.009182	277.4	402.7	1.255	1.647	2.006	0.9569	0.05803
330	26.85	0.001321	0.006678	297.5	401.0	1.315	1.629	2.421	0.8321	0.05202
340	33.25	0.001514	0.004520	321.8	393.4	1.385	1.595	4.021	0.6560	0.04480
346.75ᶜ	38.32	0.002311	0.002311	360.6	360.6	1.471	1.471	—	—	

*Values calculated from NIST Thermodynamic Properties of Refrigerants and Refrigerant Mixtures Database (REFPROP, Version 5). Thermodynamic properties are from 32-term MBWR equation of state; transport properties are from extended corresponding states model. t = triple point; c = critical point.

TABLE 2-334 Saturated Refrigerant R152a*

Temp., K	Pressure, bar	v_f, m³/kg	v_g, m³/kg	h_f, kJ/kg	h_g, kJ/kg	s_f, kJ/(kg·K)	s_g, kJ/(kg·K)	c_{pf}, kJ/(kg·K)	μ_f, 10^{-4} Pa·s	k_f, W/(m·K)
154.56t	0.000641	0.000839	303.6	14.0	419.8	0.1130	2.738	1.492	10.85	0.1932
160	0.001297	0.000846	155.2	22.2	423.5	0.1647	2.673	1.500	9.614	0.1894
170	0.004145	0.000859	51.59	37.2	430.6	0.2560	2.570	1.510	8.058	0.1822
180	0.01141	0.000873	19.82	52.4	437.9	0.3425	2.484	1.517	6.940	0.1753
190	0.02775	0.000887	8.588	67.6	445.3	0.4247	2.413	1.525	6.012	0.1685
200	0.06088	0.000902	4.110	82.9	452.8	0.5032	2.353	1.535	5.236	0.1618
210	0.1224	0.000918	2.138	98.3	460.4	0.5784	2.303	1.547	4.582	0.1552
220	0.2284	0.000935	1.193	113.8	468.0	0.6507	2.261	1.562	4.028	0.1487
230	0.4004	0.000952	0.7064	129.5	475.6	0.7205	2.225	1.580	3.556	0.1423
240	0.6647	0.000971	0.4397	145.5	483.1	0.7881	2.195	1.600	3.153	0.1361
249.12	1.01325	0.000989	0.2961	160.2	489.8	0.8481	2.171	1.622	2.834	0.1304
250	1.053	0.000991	0.2855	161.6	490.5	0.8538	2.169	1.624	2.805	0.1299
260	1.603	0.001012	0.1922	178.0	497.7	0.9178	2.147	1.651	2.500	0.1239
270	2.354	0.001035	0.1334	194.7	504.7	0.9805	2.129	1.681	2.231	0.1179
280	3.354	0.001060	0.09500	211.7	511.4	1.042	2.112	1.716	1.995	0.1121
290	4.650	0.001087	0.06916	229.1	517.8	1.103	2.098	1.756	1.789	0.1064
300	6.297	0.001118	0.05126	246.9	523.8	1.162	2.085	1.803	1.607	0.1008
310	8.351	0.001152	0.03857	265.2	529.3	1.222	2.073	1.859	1.447	0.09526
320	10.87	0.001190	0.02935	284.1	534.1	1.281	2.062	1.928	1.305	0.08986
330	13.92	0.001235	0.02251	303.7	538.2	1.340	2.051	2.015	1.180	0.08457
340	17.57	0.001289	0.01735	324.2	541.3	1.400	2.038	2.131	1.069	0.07939
350	21.90	0.001355	0.01335	345.7	542.9	1.460	2.024	2.299	1.005	0.07336
360	27.00	0.001440	0.01020	368.8	542.5	1.523	2.006	2.573	0.9225	0.06666
370	32.97	0.001563	0.007603	394.3	538.5	1.591	1.980	3.143	0.8191	0.06032
380	39.97	0.001785	0.005274	425.4	526.2	1.671	1.936	5.407	0.6638	0.05176
386.41c	45.17	0.002717	0.002717	477.3	477.3	1.778	1.778	—	—	—

*Values calculated from NIST Thermodynamic Properties of Refrigerants and Refrigerant Mixtures Database (REFPROP, Version 5). Thermodynamic properties are from 32-term MBWR equation of state; transport properties are from extended corresponding states model. t = triple point; c = critical point.

TABLE 2-335 Saturated Refrigerant 216*

Temp., °F	Pressure, lb/in² abs.	Volume, ft³/lb		Enthalpy, Btu/lb		Entropy, Btu/(lb)(°F)	
		Liquid	Vapor	Liquid	Vapor	Liquid	Vapor
−40	0.339	0.00927	59.957	0.000	62.415	0.0000	0.1487
−20	0.713	0.00942	29.749	4.778	65.276	0.0111	0.1487
0	1.382	0.00958	15.986	9.541	68.208	0.0217	0.1493
20	2.497	0.00974	9.184	14.298	71.199	0.0318	0.1504
40	4.247	0.00992	5.582	19.056	74.239	0.0415	0.1520
60	6.862	0.01010	3.558	23.821	77.319	0.0509	0.1538
80	10.612	0.01030	2.361	28.598	80.429	0.0599	0.1559
100	15.797	0.01050	1.6215	33.391	83.559	0.0686	0.1582
120	22.753	0.01073	1.1462	38.205	86.701	0.0770	0.1607
140	31.845	0.01097	0.8304	43.049	89.845	0.0852	0.1632
160	43.468	0.01124	0.6142	47.930	92.981	0.0931	0.1658
180	58.046	0.01153	0.4623	52.861	96.099	0.1009	0.1685
200	76.033	0.01186	0.3529	57.857	99.186	0.1085	0.1712
220	97.913	0.01223	0.2725	62.939	102.225	0.1161	0.1739
240	124.21	0.01266	0.2121	68.132	105.196	0.1235	0.1765
260	155.50	0.01317	0.1660	73.474	108.066	0.1309	0.1790
280	192.40	0.01378	0.1300	79.015	110.789	0.1384	0.1813
300	235.63	0.01458	0.1013	84.835	113.282	0.1460	0.1834
320	286.03	0.01570	0.0776	91.089	115.373	0.1539	0.1851
340	344.81	0.01764	0.0565	98.234	116.538	0.1628	0.1856
355.98c	399.45	0.02771	0.0277	110.248	110.248	0.1773	0.1773

*From published data, Chemicals Division, Union Carbide Corporation. Used by permission. The paper describing these data is by Shank, *ASHRAE J.*, **7** (1965): 94–101. c = critical temperature.

The 1993 ASHRAE *Handbook—Fundamentals* (SI ed.) gives material for integral degrees Celsius with temperatures on the ITS 90 scale −118.59 to 113.26°C. The thermodynamic diagram from 0.1 to 30 bar extends to 180°C. For tables and a diagram to 400 psia, 360°F, see Stewart, R. B., R. T. Jacobsen, et al., *Thermodynamic Properties of Refrigerants*, ASHRAE, Atlanta, GA, 1986 (521 pp.). For specific heat and viscosity, see *Thermophysical Properties of Refrigerants*, ASHRAE, 1993. Thermal conductivity data as a function of pressure and temperature are reported by Krauss, R. and K. Stephan, *Proc. 12th Symp. Thermophys. Props.*, Boulder, CO, 1994.

TABLE 2-336 Saturated Refrigerant 245*

T, K	P, bar	v_f, m³/kg	v_g, m³/kg	h_f, kJ/kg	h_g, kJ/kg	s_f, kJ/(kg·K)	s_g, kJ/(kg·K)
172	0.0034	6.46.−4	31.49	−63.4	133.8	−0.3131	0.8327
180	0.0076	6.57.−4	14.63	−55.9	138.7	−0.2707	0.8099
190	0.0190	6.70.−4	6.20	−46.2	145.1	−0.2182	0.7885
200	0.0425	6.83.−4	2.91	−36.0	151.7	−0.1666	0.7725
210	0.0870	6.97.−4	1.48	−25.7	158.5	−0.1157	0.7612
220	0.1654	7.11.−4	0.822	−14.8	165.4	−0.0654	0.7539
230	0.2946	7.25.−4	0.475	−3.6	172.5	−0.0156	0.7500
240	0.4958	7.40.−4	0.292	8.0	179.6	0.0337	0.7487
250	0.7946	7.55.−4	0.192	19.9	186.8	0.0824	0.7497
260	1.2204	7.72.−4	0.125	32.3	194.0	0.1305	0.7525
270	1.806	7.89.−4	0.0862	44.9	201.1	0.1781	0.7567
280	2.584	8.08.−4	0.0611	57.9	208.3	0.2249	0.7621
290	3.600	8.30.−4	0.0443	71.1	215.3	0.2711	0.7683
300	4.888	8.53.−4	0.0327	84.6	222.2	0.3161	0.7751
310	6.491	8.80.−4	0.0246	98.4	228.9	0.3614	0.7822
320	8.456	9.11.−4	0.0186	112.6	235.3	0.4057	0.7893
330	10.83	9.48.−4	0.0143	127.1	241.4	0.4497	0.7960
340	13.67	9.93.−4	0.0111	142.1	246.9	0.4937	0.8018
350	17.04	0.00105	0.0084	157.2	251.5	0.5382	0.8060
360	21.02	0.00113	0.0063	174.7	254.8	0.5844	0.8071
370	25.71	0.00125	0.0045	193.6	255.2	0.6349	0.8013
375	28.46	0.00137	0.0036	205.2	252.5	0.6649	0.7953
380.1^c	31.37	0.00204	0.0020	231.8	231.8	0.7341	0.7341

*Values converted from tables of Shank, *Thermodynamic Properties of UCON 245 Refrigerant*, Union Carbide Corporation, New York, 1966. See also Shank, *J. Chem. Eng. Data*, **12**, 474–480 (1967). c = critical point. The notation 6.46.−4 signifies 6.46×10^{-4}.

TABLE 2-337 Refrigerant C 318*

T, K	P, bar	v_f, m³/kg	v_g, m³/kg	h_f, kJ/kg	h_g, kJ/kg	s_f, kJ/(kg·K)	s_g, kJ/(kg·K)	c_{pf}, kJ/(kg·K)	μ_f, 10^{-4} Pa·s	k_f, W/(m·K)
200	0.0216	5.507.−4	3.810	353.5	498.0	3.909	4.560			
210	0.0449	5.593.−4	1.931	361.0	500.1	3.947	4.564			
220	0.0875	5.683.−4	1.038	369.2	502.2	3.984	4.569			
230	0.1608	5.778.−4	0.588	377.6	504.4	4.022	4.574	0.98	11.7	0.088
240	0.2810	5.879.−4	0.349	386.4	510.9	4.060	4.578	1.00	9.55	0.085
250	0.466	5.988.−4	0.2166	395.6	517.4	4.097	4.584	1.02	7.90	0.082
260	0.741	6.106.−4	0.1401	405.2	524.0	4.133	4.592	1.03	6.63	0.078
270	1.133	6.234.−4	0.0938	415.1	530.7	4.172	4.599	1.05	5.64	0.075
280	1.672	6.375.−4	0.0647	425.8	537.3	4.210	4.609	1.07	4.85	0.071
290	2.392	6.529.−4	0.0458	436.2	543.9	4.247	4.618	1.09	4.22	0.068
300	3.325	6.694.−4	0.0332	447.3	550.4	4.284	4.626	1.12	3.70	0.065
310	4.522	6.893.−4	0.0245	458.7	556.9	4.322	4.638	1.15	3.20	0.061
320	6.007	7.115.−4	0.0184	470.5	563.3	4.359	4.648	1.18	2.94	0.058
330	7.826	7.365.−4	0.0139	482.7	569.4	4.396	4.659	1.23	2.66	0.054
340	10.018	7.666.−4	0.0106	495.2	575.4	4.433	4.669	1.27	2.33	0.051
350	12.632	8.034.−4	0.0082	508.1	581.0	4.469	4.678	1.32	2.00	0.048
360	15.71	8.508.−4	0.0062	521.5	585.8	4.507	4.685	1.39		
370	19.33	9.172.−4	0.0047	535.6	589.9	4.544	4.691			
380	23.59	1.031.−3	0.0033	551.4	591.5	4.585	4.691			
388.5^c	27.83	1.613.−3	0.0016	577.2	577.2	4.651	4.651			

*Values of P, v, h, and s interpolated, extrapolated, and converted from tables of Oguchi, *Reito*, **52** (1977): 869–889. Values of c_p, μ, and k interpolated and converted from tables in *Thermophysical Properties of Refrigerants*, American Society of Heating, Refrigerating and Air-Conditioning Engineers, New York, 1976. c = critical point. Saturation and superheat tables and a diagram to 80 bar, 580 K are given by Reynolds, W. C., *Thermodynamic Properties in S.I.*, Stanford Univ. publ., 1979 (173 pp.). For equations, see Cipollone, R., *ASHRAE Trans.*, **97**, 2 (1991): 262–267.

TABLE 2-338 Saturated Refrigerant 500*

T, K	P, bar	v_f, m³/kg	v_g, m³/kg	h_f, kJ/kg	h_g, kJ/kg	s_f, kJ/(kg·K)	s_g, kJ/(kg·K)	c_{pf}, kJ/(kg·K)	μ_f, 10⁻⁴ Pa·s	k_f, W/(m·K)
200	0.1219	6.966.−4	1.360	−29.56	185.87	−0.1363	0.9408	1.044	6.11	0.113
210	0.2258	7.090.−4	0.766	−21.03	191.25	−0.0948	0.9161	1.018	5.15	0.109
220	0.3936	7.222.−4	0.457	−12.17	196.63	−0.0536	0.8955	0.997	4.42	0.106
230	0.6511	7.361.−4	0.286	−2.97	201.96	−0.0130	0.8782	0.987	3.85	0.102
240	1.0291	7.509.−4	0.187	6.58	207.23	0.0277	0.8638	0.987	3.42	0.098
250	1.5632	7.668.−4	0.1261	16.50	212.40	0.0680	0.8517	0.997	3.04	0.094
260	2.2932	7.839.−4	0.0879	26.78	217.45	0.1082	0.8415	1.017	2.74	0.090
270	3.2624	8.024.−4	0.0628	37.44	222.35	0.1481	0.8329	1.048	2.48	0.086
280	4.5172	8.226.−4	0.0459	48.48	227.06	0.1878	0.8257	1.089	2.26	0.082
290	6.1064	8.450.−4	0.0342	59.91	231.56	0.2275	0.8194	1.140	2.08	0.078
300	8.0809	8.699.−4	0.0259	71.76	235.79	0.2671	0.8139	1.201	1.92	0.074
310	10.49	8.981.−4	0.0198	84.05	239.69	0.3067	0.8088	1.273	1.77	0.070
320	13.40	9.306.−4	0.0154	96.83	243.19	0.3464	0.8038	1.355	1.63	0.066
330	16.86	9.690.−4	0.0119	110.17	246.14	0.3864	0.7985	1.447	1.48	0.062
340	20.93	1.016.−3	0.0093	124.20	248.36	0.4271	0.7922	1.550	1.34	0.058
350	25.70	1.077.−3	0.0072	139.18	249.47	0.4689	0.7841	1.663		
360	31.25	1.162.−3	0.0055	155.66	248.71	0.5135	0.7721	1.919		
370	37.72	1.307.−3	0.0040	175.59	244.26	0.5650	0.7509	2.07		
378.6ᶜ	44.26	2.012.−3	0.0020	219.50	219.50	0.6729	0.6729	∞		

*Values reproduced and converted from Table 12, p. 17.99, *ASHRAE Handbook, 1981: Fundamentals*, American Society of Heating, Refrigerating and Air-Conditioning Engineers, Atlanta, 1981. Copyright material. Reproduced by permission of the copyright owner. *c* = critical point. The notation 6.966.−4 signifies 6.966 × 10⁻⁴.

The 1993 ASHRAE *Handbook—Fundamentals* (SI ed.) gives material for integral degrees Celsius with temperatures on the IPTS 68 scale from −70 to 105.60°C. The thermodynamic diagram from 0.1 to 70 bar extends to 240°C. Equations and constants approximated to the 1985 ASHRAE tables were given by Mecaryk, K. and M. Masaryk, *Heat Recovery Systems and CHP*, **11**, 2/3 (1991): 193–197. Saturation and superheat tables and a diagram to 80 bar, 560 K are given by Reynolds, W. C., *Thermodynamic Properties in S.I.*, Stanford Univ. publ., 1979 (173 pp.). Tables and a chart to 1000 psia, 480°F are given by Stewart, R. B., R. T. Jacobsen, et al., *Thermodynamic Properties of Refrigerants*, ASHRAE, Atlanta, GA, 1986 (521 pp.). Specific heat and viscosity appear in *Thermophysical Properties of Refrigerants*, ASHRAE, 1993.

TABLE 2-339 Saturated Refrigerant 502*

T, K	P, bar	v_f, m³/kg	v_g, m³/kg	h_f, kJ/kg	h_g, kJ/kg	s_f, kJ/(kg·K)	s_g, kJ/(kg·K)	c_{pf}, kJ/(kg·K)	μ_f, 10⁻⁴ Pa·s	k_f, W/(m·K)
200	0.2274	6.381.−4	0.646	−29.04	153.34	−0.1337	0.7782	1.018	5.72	0.103
210	0.4098	6.507.−4	0.374	−20.83	158.42	−0.0937	0.7599	1.036	4.85	0.099
220	0.6965	6.640.−4	0.228	−12.15	163.49	−0.0534	0.7449	1.055	4.23	0.095
230	1.1251	6.783.−4	0.146	−2.99	168.50	−0.0128	0.7328	1.075	3.71	0.091
240	1.7392	6.938.−4	0.0969	6.66	173.42	0.0280	0.7228	1.097	3.28	0.087
250	2.5867	7.105.−4	0.0665	16.78	178.20	0.0691	0.7148	1.120	2.94	0.083
260	3.7188	7.289.−4	0.0470	27.36	182.81	0.1102	0.7082	1.144	2.65	0.079
270	5.1893	7.492.−4	0.0340	38.36	187.21	0.1514	0.7027	1.170	2.41	0.075
280	7.0530	7.720.−4	0.0251	49.77	191.35	0.1923	0.6980	1.197	2.18	0.072
290	9.3660	7.979.−4	0.0188	61.55	195.16	0.2330	0.6937	1.225	1.99	0.068
300	12.19	8.280.−4	0.0143	73.68	198.56	0.2734	0.6896	1.254	1.79	0.064
310	15.57	8.637.−4	0.0109	86.17	201.43	0.3134	0.6852	1.285	1.59	0.060
320	19.60	9.081.−4	0.0084	99.06	203.57	0.3532	0.6798	1.317	1.40	0.056
330	24.35	9.666.−4	0.0064	112.53	204.62	0.3933	0.6723	1.351	1.23	0.052
340	29.95	1.053.−3	0.0048	127.13	203.71	0.4351	0.6604	1.386	1.07	0.048
350	36.62	1.220.−3	0.0033	145.44	197.82	0.4859	0.6355	1.422	0.93	0.044
355.3ᶜ	40.75	1.786.−3	0.0018	174.00	174.00	0.5634	0.5634			

*Values reproduced and converted from Table 13, p. 17.101, *ASHRAE Handbook, 1981: Fundamentals*, American Society of Heating, Refrigerating and Air-Conditioning Engineers, Atlanta, 1981. Copyright material. Reproduced by permission of the copyright owner. *c* = critical point. The notation 6.381.−4 signifies 6.381 × 10⁻⁴.

The 1993 ASHRAE *Handbook—Fundamentals* (SI ed.) gives material for integral degrees Celsius with temperatures on the IPTS 68 scale from −70 to 82.2°C. The thermodynamic diagram from 0.1 to 80 bar extends to 180°C. Equations and constants approximated to 1985 ASHRAE tables are given by Mecaryk, K., and M. Masaryk, *Heat Recovery Systems and CHP*, **11**, 2/3 (1991): 193–197. Saturation and superheat tables and a diagram to 20 bar, 515 K are given by Reynolds, W. C., *Thermodynamic Properties in S.I.*, Stanford Univ. publ., 1979 (173 pp.). Tables and a chart to 1000 psia, 400°F appear in Stewart, R. B., R. T. Jacobsen, et al., *Thermodynamic Properties of Refrigerants*, ASHRAE, Atlanta, GA, 1986 (521 pp.). For specific heat and viscosity, see *Thermophysical Properties of Refrigerants*, ASHRAE, 1993.

TABLE 2-340   Saturated Refrigerant 503*

T, K	P, bar	v_f, m³/kg	v_g, m³/kg	h_f, kJ/kg	h_g, kJ/kg	s_f, kJ/(kg·K)	s_g, kJ/(kg·K)	c_{pf}, kJ/(kg·K)	μ_f, 10⁻⁴ Pa·s	k_f, W/(m·K)
150	0.0750	6.384.−4	1.894	−89.60	111.02	−0.4694	0.8681	0.482	6.12	0.128
160	0.1798	6.478.−4	0.837	−79.73	115.40	−0.4057	0.8139	0.554	5.05	0.123
170	0.3828	6.585.−4	0.414	−69.55	119.70	−0.3441	0.7691	0.620	4.16	0.116
180	0.7395	6.700.−4	0.224	−59.08	123.84	−0.2844	0.7318	0.682	3.43	0.111
190	1.3187	6.850.−4	0.130	−48.36	127.77	−0.2267	0.7003	0.747	2.94	0.105
200	2.1999	7.014.−4	0.0803	−37.45	131.45	−0.1710	0.6735	0.817	2.56	0.099
210	3.4713	7.204.−4	0.0520	−26.36	134.84	−0.1173	0.6503	0.896	2.25	0.094
220	5.2281	7.426.−4	0.0350	−15.10	137.87	−0.0656	0.6298	0.988	1.98	0.088
230	7.5713	7.687.−4	0.0242	− 3.65	140.49	−0.0155	0.6112	1.017	1.73	0.082
240	10.61	8.001.−4	0.0172	8.07	142.58	0.0334	0.5939	1.227	1.52	0.076
250	14.46	8.386.−4	0.0124	20.22	143.98	0.0817	0.5767	1.382	1.33	0.070
260	19.25	8.874.−4	0.0090	33.10	144.38	0.1305	0.5585	1.57	1.17	0.065
270	25.13	9.526.−4	0.0064	47.22	143.23	0.1816	0.5373	1.79	1.03	0.059
280	32.27	1.050.−3	0.0045	63.64	139.25	0.2384	0.5085	2.03	0.91	0.054
290	40.87	1.264.−3	0.0028	86.41	127.51	0.3131	0.4548	2.35		
292.6ᶜ	43.57	1.773.−3	0.0018	110.20	110.20	0.3864	0.3864	∞		∞

*P, v, h, and s values reproduced and converted from Table 14, p. 17.103, *ASHRAE Handbook, 1981: Fundamentals,* American Society of Heating, Refrigerating and Air-Conditioning Engineers, Atlanta, 1981. Copyright material. Reproduced by permission of the copyright owner. c_p, μ, and k values interpolated and converted from *Thermophysical Properties of Refrigerants,* American Society of Heating, Refrigerating and Air-Conditioning Engineers, New York, 1976. c = critical point. The notation 6.384.−4 signifies 6.384×10^{-4}.

Saturation and superheat tables and a diagram to 80 bar, 600 K are given by Reynolds, W. C., *Thermodynamic Properties in S.I.,* Stanford Univ. publ., 1979 (173 pp.). Tables and a chart to 1000 psia, 460°F are given by Stewart, R. B., R. T. Jacobsen, et al., *Thermodynamic Properties of Refrigerants,* ASHRAE, Atlanta, GA, 1986 (521 pp.). For specific heat and viscosity see *Thermophysical Properties of Refrigerants,* ASHRAE, 1993. The 1993 ASHRAE *Handbook—Fundamentals* (SI ed.) gives material for integral degrees Celsius with temperatures on the IPTS 68 scale for saturation conditions from −125 to 19.50°C. The thermodynamic diagram from 0.1 to 80 bar extends to 220°C.

TABLE 2-341   Saturated Refrigerant 504*

Temp., °F	Pressure, lb/in² abs.	Volume, ft³/lb		Enthalpy, Btu/lb		Entropy, Btu/(lb)(°F)	
		Liquid	Vapor	Liquid	Vapor	Liquid	Vapor
−120	2.964	0.01095	15.31	−21.48	86.69	−0.0565	0.2609
−100	6.042	0.01119	7.874	−16.39	89.31	−0.0420	0.2519
−80	11.34	0.01146	4.372	−11.12	91.84	−0.0277	0.2435
−60	19.85	0.01175	2.585	−5.65	94.25	−0.0137	0.2362
−40	32.76	0.01206	1.609	0.00	96.50	0.0000	0.2299
−20	51.44	0.01242	1.045	5.85	98.58	0.0135	0.2244
0	77.41	0.01282	0.7029	11.91	100.45	0.0269	0.2195
20	112.3	0.01328	0.4859	18.22	102.09	0.0401	0.2150
40	158.0	0.01379	0.3431	24.81	103.44	0.0533	0.2107
60	216.2	0.01443	0.2458	31.78	104.41	0.0667	0.2065
80	289.2	0.01522	0.1773	39.25	104.85	0.0804	0.2020
100	379.1	0.01629	0.1274	47.43	104.49	0.0948	0.1968
120	488.3	0.01783	0.0893	56.78	102.72	0.1107	0.1899
140	618.1	0.02083	0.0578	69.97	97.70	0.1322	0.1784
150	692.2	0.02597	0.0394	76.96	89.76	0.1432	0.1642

*Unpublished data of Allied Chemical Company, 1970. Used by permission.

TABLE 2-342   Thermodynamic Properties of Refrigerant 507*

Temp., K	Pressure, bar	v_f, m³/kg	v_g, m³/kg	h_f, kJ/kg	h_g, kJ/kg	s_f, kJ/(kg·K)	s_g, kJ/(kg·K)
230.5	1.013	0.000 574	0.1280	−3.1	143.3	−0.015	0.620
240	1.59	0.000 602	0.0826	10.3	150.2	0.042	0.623
250	2.42	0.000 627	0.0546	22.6	154.5	0.092	0.619
260	3.54	0.000 658	0.0377	37.6	159.0	0.149	0.617
270	4.95	0.000 695	0.0270	51.6	163.8	0.202	0.618
280	6.70	0.000 738	0.0198	64.7	169.0	0.250	0.620
290	8.85	0.000 787	0.0148	77.2	174.6	0.295	0.634
300	11.52	0.000 839	0.0112	89.4	180.3	0.336	0.640
310	14.74	0.000 903	0.0084	101.6	185.4	0.378	0.648
320	18.76	0.001 006	0.0062	115.7	188.6	0.422	0.649
330	23.65	0.001 221	0.0042	135.5	189.3	0.481	0.641
340	29.57	0.001 618	0.0025	161.7	179.9	0.557	0.611
341.5ᶜ	32.67	0.001 97	0.0020	172.7	172.7	0.590	0.590

*Azeotropic mixture of R152a and R218. $h_f = s_f = 0$ at 233.15 K = −40°C. Interpolated, extrapolated and converted from Lavrenchenko, G. K., M. G. Khmelnuk, et al., *Int. J. Refrig.,* **17,** 7 (1994): 461. Some values are tentative. This source also gives a ln *P–h* diagram from 0.6 to 30 bar, −50 to 70°C. Differences exist between the published diagram and tables. c = critical point.

TABLE 2-343 Saturated Rubidium*

T, K	P, bar	v_f, m³/kg	v_g, m³/kg	h_f, kJ/kg	h_g, kJ/kg	s_f, kJ/(kg·K)	s_g, kJ/(kg·K)	c_{pf}, kJ/(kg·K)
312.7ᵐ	2.46.−9	6.75.−4		118.7	1036	0.998	3.932	0.379
400	1.69.−6	6.98.−4	2.3.+5	151.6	1057	1.091	3.355	0.375
500	1.73.−4	7.22.−4	2790	188.8	1078	1.174	2.953	0.369
600	0.0037	7.46.−4	156.6	225.4	1096	1.241	2.692	0.362
700	0.0317	7.73.−4	20.75	261.3	1111	1.296	2.511	0.357
800	0.1584	8.10.−4	4.662	296.8	1124	1.343	2.378	0.353
1000	1.467	8.65.−4	0.605	367.6	1150	1.422	2.205	0.360
1200	6.466	9.40.−4	0.159	440.1	1179	1.490	2.104	0.385
1400	18.6	1.03.−3						
1500	28.5	1.08.−3						

*Converted from tables in Vargaftik, *Tables of the Thermophysical Properties of Liquids and Gases,* Nauka, Moscow, 1972, and Hemisphere, Washington, 1975. *m* = melting point. The notation 2.46.−9 signifies 2.46×10^{-9}.

Many of the Vargaftik values also appear in Ohse, R. W., *Handbook of Thermodynamic and Transport Properties of Alkali Metals,* Blackwell Sci. Pubs., Oxford, 1985 (1020 pp.). This source contains superheat data.

Saturation and superheat tables and a diagram to 40 bar, 1600 K are given by Reynolds, W. C., *Thermodynamic Properties in S.I.,* Stanford Univ. publ., 1979 (173 pp.).

For a Mollier diagram from 0.1 to 320 psia, 1200 to 2700°R, see Weatherford, W. D., J. C. Tyler, et al., WADD-TR-61-96, 1961. An extensive review of properties of the solid and the saturated liquid was given by Alcock, C. B., M. W. Chase, et al., *J. Phys. Chem. Ref. Data,* **23**, 3 (1994): 385–497.

TABLE 2-344 Thermophysical Properties of Saturated Seawater

Temp., °C	Pressure, bar	v, (m³/kg)10³	c_p, kJ/(kg·K)	μ, Ns/m²	k, W/(m·K)	N_{Pr}	$10^5\kappa$, 1/bar
0	0.005993	1.000158	4.000	0.001884	0.560	13.46	5.06
1	0.006438	1.000099	4.000	0.001827	0.563	12.98	5.02
2	0.006916	1.000077	4.000	0.001772	0.565	12.55	4.98
3	0.007427	1.000033	4.000	0.001720	0.567	12.13	4.95
4	0.007970	1.000025	4.001	0.001669	0.569	11.74	4.92
5	0.008548	1.000033	4.001	0.001620	0.571	11.35	4.89
6	0.009163	1.000057	4.001	0.001574	0.574	10.97	4.86
7	0.009816	1.000096	4.002	0.001529	0.576	10.62	4.83
8	0.010511	1.000149	4.002	0.001486	0.578	10.29	4.80
9	0.011248	1.000261	4.002	0.001445	0.580	9.97	4.78
10	0.01203	1.000298	4.003	0.001405	0.582	9.70	4.76
11	0.01286	1.000392	4.003	0.001367	0.584	9.37	4.74
12	0.01374	1.000500	4.003	0.001330	0.586	9.09	4.72
13	0.01467	1.000620	4.004	0.001294	0.588	8.81	4.70
14	0.01566	1.000727	4.004	0.001259	0.590	8.54	4.68
15	0.01671	1.000899	4.005	0.001226	0.592	8.29	4.66
16	0.01781	1.001055	4.005	0.001195	0.594	8.06	4.65
17	0.01898	1.001224	4.006	0.001165	0.595	7.82	4.63
18	0.02022	1.001404	4.006	0.001136	0.597	7.62	4.62
19	0.02153	1.001595	4.007	0.001107	0.599	7.41	4.60
20	0.02291	1.001796	4.007	0.001080	0.600	7.21	4.59
21	0.02437	1.002009	4.007	0.001054	0.602	7.02	4.57
22	0.02591	1.002232	4.008	0.001029	0.604	6.82	4.56
23	0.02753	1.002465	4.008	0.001005	0.605	6.66	4.55
24	0.02924	1.002708	4.009	0.000981	0.607	6.48	4.54
25	0.03104	1.002961	4.009	0.000958	0.608	6.31	4.53
26	0.03294	1.003224	4.009	0.000936	0.609	6.16	4.52
27	0.03494	1.003496	4.010	0.000915	0.611	6.01	4.51
28	0.03705	1.003778	4.010	0.000895	0.612	5.86	4.50
29	0.03926	1.004069	4.011	0.000875	0.614	5.72	4.49
30	0.04159	1.004369	4.011	0.000855	0.615	5.58	4.48

$\kappa = (-1/V)(\partial v/\partial p)_T \cdot 10^5$. Thus, at 0°C, the compressibility is 5.06×10^{-5}/bar.

For further information see, for instance, Bromley, LeR. A., *J. Chem. Eng. Data,* **12**, 2 (1967): 202–206; **13**, 1 (1968): 60–62 and **13**, 3: 399–402; **15**, 2 (1970): 246–253; and *A.I.Ch.E.J.,* **20**, 2 (1974): 326–335.

Thermal conductivity data sources include Castelli, V. J., E. M. Stanley, et al., *Deep Sea Res.,* **211** (1974): 311–318; Levy, F. L., *Int. J. Refrig.,* **5**, 3 (1982): 155–159.

For velocity of sound, see, for instance, U.S. Naval Oceanographic Office SP 58, 1962 (50 pp.). More recent information is contained in UNESCO technical papers. See *Marine Science* No. 38, 1981 (6 pp.) and No. 44, 1983 (53 pp.).

For sea ice properties, see Fukusako, S., *Int. J. Thermophys.,* **11**, 2 (1990): 353–372.

TABLE 2-345 Saturated Sodium

Temp., K	Pressure, bar	v_f, m³/kg	v_g, m³/kg	h_f, kJ/kg	h_g, kJ/kg	s_f, kJ/(kg·K)	s_g, kJ/(kg·K)	c_{pf}, kJ/(kg·K)	c_{pg}, kJ/(kg·K)	μ_f, 10⁻⁶ Pa·s	μ_g, 10⁻⁶ Pa·s	k_f, W/(m·K)	k_g, W/(m·K)	Pr_f	Pr_g
371	1.59.–10	0.001 078	8.54.+9	207	4739	2.259	14.475	1.383		688		89.4		0.0106	
400	1.80.–9	0.001 088	8.08.+8	247	4757	2.920	14.195	1.372	0.86	599		87.2		0.0094	
500	8.99.–7	0.001 115	1.99.+6	382	4817	3.222	12.092	1.334	1.25	415		80.1		0.0069	
600	5.57.–5	0.001 144	38022	514	4872	3.462	10.745	1.301	1.80	321		73.7		0.0057	
700	0.00105	0.001 174	2320	642	4921	3.661	10.631	1.277	2.28	264		68.0		0.0050	
800	0.00941	0.001 208	291.5	769	4966	3.830	9.076	1.260	2.59	227	19.6	62.9	0.0343	0.0045	1.48
900	0.05147	0.001 242	58.8	895	5007	3.978	8.547	1.252	2.72	201	20.6	58.3	0.0406	0.0043	1.38
1000	0.1995	0.001 280	16.6	1020	5044	4.110	8.134	1.252	2.70	181	23.0	54.2	0.0455	0.0042	1.36
1100	0.6016	0.001 323	5.95	1146	5079	4.230	7.805	1.261	2.62	166	25.3	50.5	0.0492	0.0042	1.35
1154.7	1.013	0.001 347	3.89	1215	5097	4.290	7.652	1.271	2.56	159	26.5	48.7	0.0522	0.0041	1.30
1200	1.50	0.001 366	2.54	1273	5111	4.340	7.538	1.279	2.51	153	27.5	47.2	0.0547	0.0041	1.26
1300	3.26	0.001 416	1.24	1402	5140	4.444	7.319	1.305	2.43	143	29.9	44.0	0.0570	0.0042	1.27
1400	6.30	0.001 471	0.676	1534	5168	4.542	7.138	1.340	2.39	135	32.2	41.1	0.0592	0.0044	1.30
1500	11.13	0.001 531	0.400	1671	5193	4.636	6.984	1.384	2.36	128	34.6	38.2		0.0046	
1600	18.28	0.001 597	0.253	1812	5217	4.727	6.855	1.437	2.34	122	37.1	35.4		0.0050	
1700	28.28	0.001 675	0.168	1959	5238	4.816	6.745	1.500	2.41	117		32.6		0.0054	
1800	41.61	0.001 761	0.117	2113	5256	4.904	6.650	1.574	2.46	112		29.7		0.0059	
1900	58.70	0.001 862	0.084	2274	5268	4.992	6.568	1.661	2.53	108		26.6		0.0067	
2000	79.91	0.001 984	0.063	2444	5273	5.079	6.494	1.764	2.66	104		23.2		0.0079	
2100	105.5	0.002 174	0.0472	2625	5265			1.926	2.91						
2200	135.7	0.002 320	0.0361	2822	5241			2.190	3.40						
2300	170.6	0.002 584	0.0275	3047	5188			2.690	4.47						
2400	210.3	0.002 985	0.0203	3331	5078			4.012	8.03						
2500	254.7	0.004 19	0.0098	3965	4617			39.3	417.						
2503.7ᶜ	256.4	0.004 57	0.0046	4294	4294										

c = critical point.

s_f values converted from Cordfunke, E. H. P. and R. J. M. Konings, *Thermochemical Data for Reactor Materials and Fission Products*, North Holland Elsevier, NY, 1990. s_g determined as $s_f + (h_g − h_f)/T$. μ_g and k_g values estimated by P. E. Liley. All other values are from Fink, J. K. and L. Leibowitz, Argonne Nat. Lab Rept. ANL/RE-95-2, 1995. The Fink and Leibowitz work also appeared in *High Temp. Materials Sci.*, **35**, 65–103, 1996. Saturation and superheat tables and a diagram to 14 bar; 1700 K are given by Reynolds, W. C., *Thermodynamic Properties in S.I.*, Stanford Univ. publ., 1979 (173 pp.). For a Mollier diagram for 0.1–150 psia, 1500–2700°R, see Weatherford, P. M., J. C. Tyler, et al., WADD-TR-61-96, 1961.

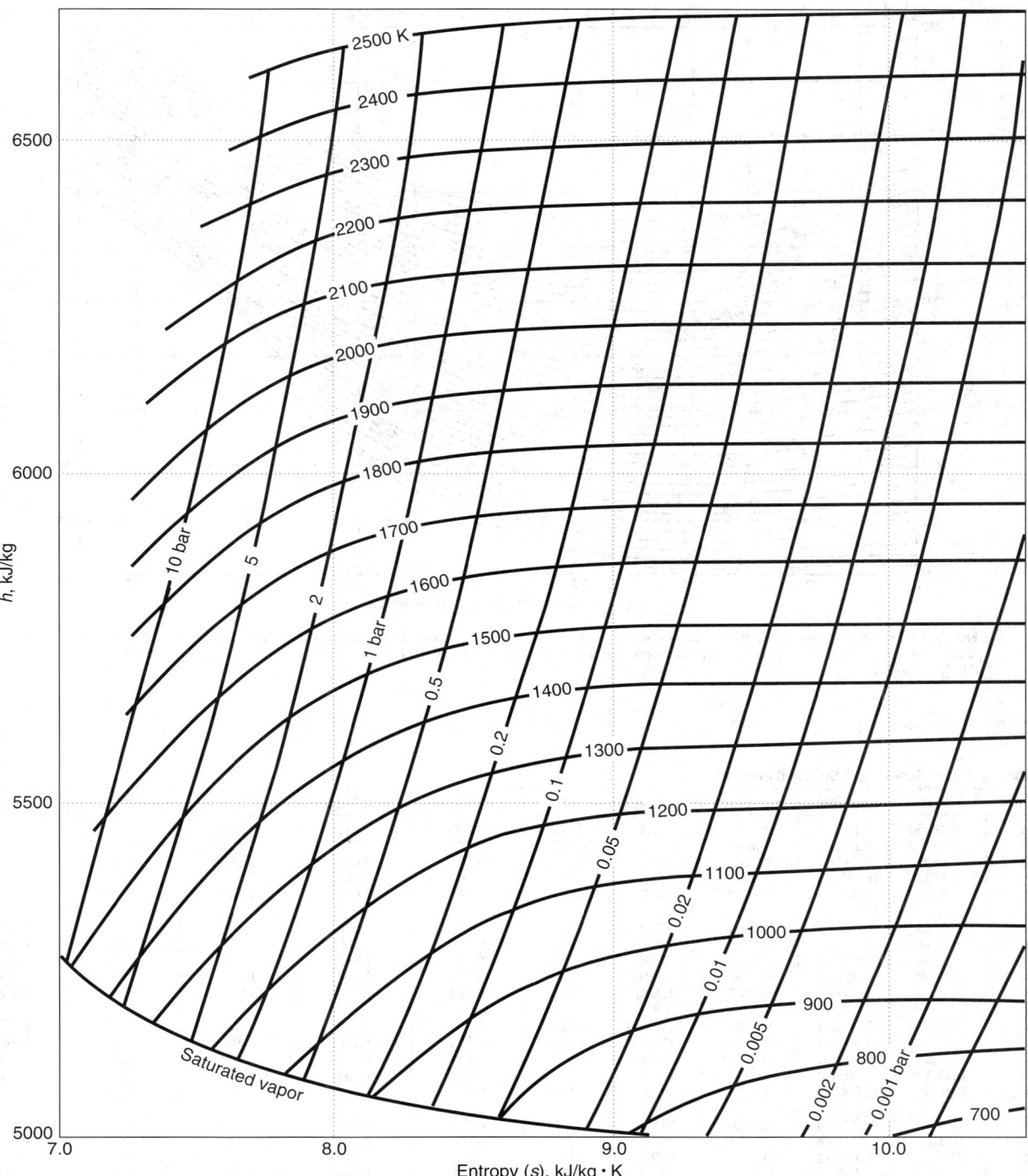

FIG. 2-28   Mollier Diagram for Sodium. Drawn from the Vargaftik et al. values in Ohse, R. W., *Handbook of Thermodynamic and Transport Properties of Alkali Metals,* Blackwell Sci. Pubs., Oxford, UK, 1985. These values are identical with those of Vargaftik, N. B., *Handbook of Thermophysical Properties of Gases and Liquids,* Moscow, 1972, and the Hemisphere translation, pp. 19. An apparent discontinuity exists between the superheat values and the saturation values, not reproduced here. For a Mollier diagram in f.p.s. units from 0.1 to 150 psia, 1500 to 2700°R, see Fig. 3-36, p. 3-232 of the 6th edition of this handbook. An extensive review of properties of the solid and the saturated liquid was given by Alcock, C. B., Chase, M. W. et al., *J. Phys. Chem. Ref. Data,* **23**(3), 385–497, 1994.

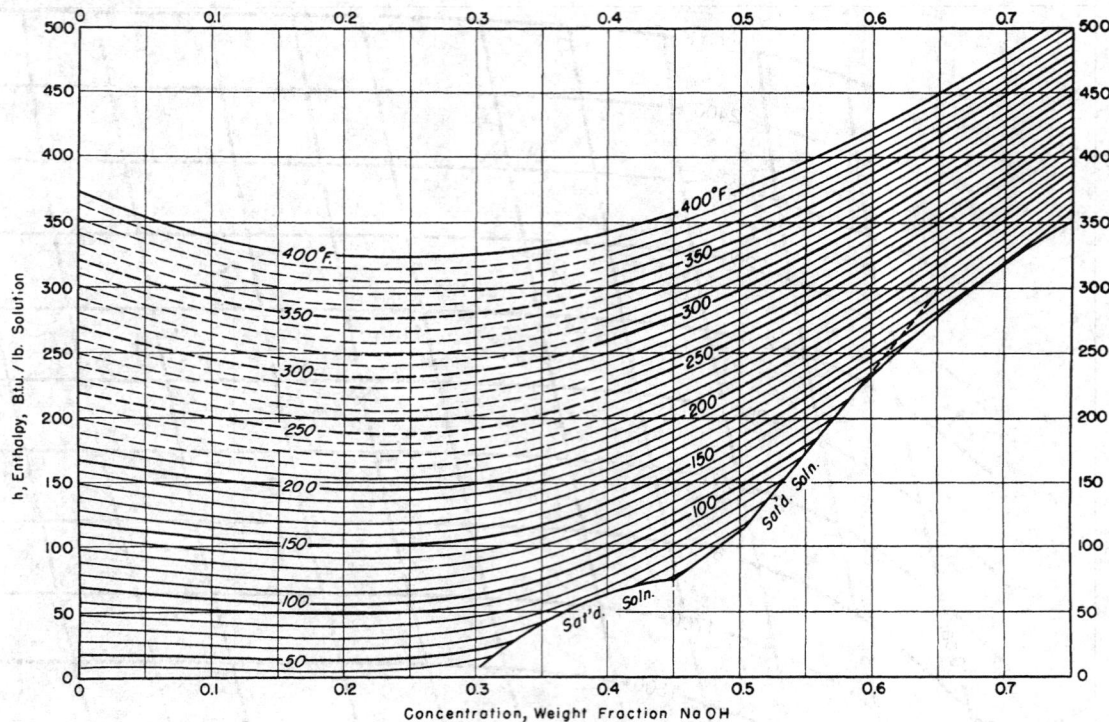

FIG. 2-29 Enthalpy-concentration diagram for aqueous sodium hydroxide at 1 atm. Reference states: enthalpy of liquid water at 32°F and vapor pressure is zero; partial molal enthalpy of infinitely dilute NaOH solution at 64°F and 1 atm is zero. [*McCabe*, Trans. Am. Inst. Chem. Eng., *31, 129 (1935)*.]

TABLE 2-346 Saturated Sulfur Dioxide*

T, K	P, bar	v_f, m³/kg	v_g, m³/kg	h_f, kJ/kg	h_g, kJ/kg	s_f, kJ/(kg·K)	s_g, kJ/(kg·K)	c_{pf}, kJ/(kg·K)	μ_f, 10^{-4} Pa·s	k_f, W/(m·K)
200	0.02056	6.189.−4	12.602	7.4	433.3	0.033	2.212	1.280	12.3	
210	0.04569	6.284.−4	5.946	9.1	446.1	0.041	2.159	1.284	10.6	
220	0.09997	6.384.−4	2.876	28.6	453.8	0.123	2.075	1.288	8.37	
230	0.1844	6.488.−4	1.605	43.5	459.5	0.198	2.001	1.293	7.03	
240	0.3202	6.596.−4	0.9602	56.5	464.5	0.254	1.952	1.299	5.97	
250	0.5430	6.707.−4	0.5864	70.0	469.7	0.308	1.906	1.308	5.11	0.262
260	0.8778	6.819.−4	0.3745	85.1	474.5	0.363	1.865	1.317	4.39	0.243
270	1.3634	6.938.−4	0.2479	99.8	479.3	0.425	1.827	1.328	3.78	0.224
280	2.0402	7.057.−4	0.1699	114.8	484.3	0.473	1.793	1.343	3.30	0.206
290	2.9574	7.184.−4	0.1197	129.2	488.5	0.523	1.763	1.363	2.87	0.190
300	4.1675	7.312.−4	0.08647	143.1	492.5	0.568	1.732	1.389	2.51	0.174
310	5.7372	7.447.−4	0.06366	157.1	496.3	0.612	1.706	1.422	2.19	0.162
320	7.8226	7.590.−4	0.04707	170.1	498.9	0.649	1.678	1.459	1.91	0.151
330	10.301	7.847.−4	0.03572	183.0	501.2	0.690	1.654	1.499	1.67	0.139
340	13.229	8.066.−4	0.02792	196.0	502.5	0.731	1.633	1.546	1.46	0.128
350	16.759	8.303.−4	0.02209	211.2	502.9	0.781	1.614	1.603	1.27	0.117
360	21.01	8.571.−4	0.01755	223.7	503.1	0.817	1.593	1.68	1.11	0.108
370	26.01	8.877.−4	0.01399	239.9	502.9	0.862	1.573	1.75	0.96	0.098
380	31.92	9.236.−4	0.01110	257.9	502.7	0.910	1.555	1.84	0.84	0.089
390	38.76	9.671.−4	0.00877	277.7	500.7	0.962	1.534	1.97	0.73	0.081
400	46.67	1.023.−3	0.00685	300.2	496.7	1.020	1.511	2.12	0.63	0.072
410	55.80	1.098.−3	0.00559	326.2	489.5	1.083	1.481		0.53	0.064
420	66.19	1.235.−3	0.00387	355.6	474.1	1.155	1.436		0.44	0.055
425.1ᶜ	78.81	1.906.−3	0.00191	423.6	423.6	1.304	1.304			

*Values interpolated and converted from tables of Kang, McKetta, et al., Bur. Eng. Res. Repr. 59, University of Texas, Austin, 1961. See also *J. Chem. Eng. Data,* **6** (1961): 220–227; and *Am. Inst. Chem. Eng. J.,* **7** (1961): 418. *c* = critical point. The notation 6.189.−4 signifies 6.189 × 10⁻⁴. The AIChE publication contains a Mollier diagram to 4500 psia, 480°F, while the reprint contains saturation and superheat tables.

TABLE 2-347   Thermodynamic Properties of Saturated Sulfur Hexafluoride (SF₆)*

Temp., K	Pressure, bar	v_f, m³/kg	v_g, m³/kg	h_f, kJ/kg	h_g, kJ/kg	s_f, kJ/(kg·K)	s_g, kJ/(kg·K)	c_{pf}, kJ/(kg·K)	c_{pg}, kJ/(kg·K)
222.4	2.200	0.0005389	0.05428	−57.55	59.08	−0.2310	0.2935		0.579
225	2.470	0.0005429	0.04861	−54.41	60.49	−0.2171	0.2936		0.583
230	3.045	0.0005507	0.03978	−48.51	63.12	−0.1913	0.2940		0.592
235	3.710	0.0005588	0.03286	−42.67	65.68	−0.1663	0.2947		0.602
240	4.475	0.0005675	0.02737	−36.87	68.18	−0.1421	0.2956		0.613
245	5.346	0.0005768	0.02296	−31.13	70.61	−0.1186	0.2966		0.626
250	6.332	0.0005866	0.01939	−25.47	72.96	−0.0960	0.2977		0.640
255	7.442	0.0005971	0.01647	−19.87	75.22	−0.0741	0.2988		0.656
260	8.684	0.0006085	0.01406	−14.33	77.39	−0.0528	0.2999		0.674
265	10.07	0.0006207	0.01205	−8.85	79.44	−0.0323	0.3009		0.695
270	11.60	0.0006341	0.01035	−3.41	81.38	−0.0123	0.3017		0.720
275	13.30	0.0006488	0.00892	2.00	83.19	0.0071	0.3024		0.748
280	15.18	0.0006652	0.00769	7.42	84.84	0.0262	0.3027		0.783
285	17.25	0.0006836	0.00663	12.88	86.30	0.0451	0.3027		0.827
290	19.52	0.0007047	0.00571	18.45	87.52	0.0639	0.3021	0.409	0.882
295	22.01	0.0007295	0.00490	24.20	88.45	0.0829	0.3008	0.631	0.941
300	24.75	0.0007594	0.00418	30.22	89.00	0.1025	0.2984	0.870	1.070
305	27.76	0.000798	0.00352	36.75	88.97	0.1233	0.2945	1.17	1.26
310	31.05	0.000851	0.00291	44.05	88.06	0.1462	0.2881	1.63	1.63
315	34.67	0.000949	0.00228	53.98	85.22	0.1769	0.2761	2.48	2.40
318.7	37.79	0.001372	0.00137	71.74	71.74	0.2317	0.2317	∞	∞

*See also Oda, A., M. Uematsu, et al., *Bull. JSME*, **26**, 219 (1983): 1590–1596. Ulybin, S.A., *Thermodynamic Properties of Sulfur Hexafluoride*, Moscow, 1977 (53 pp.). For thermal conductivity to 500 bar, see Rastorguev, Yu L., B. A. Grigorev, et al., *Teploenergetika* **24**, 6 (1977): 78–81 and Bakulin, S. S. and S. A. Ulybin, *Teplofiz. Vysok. Temp.*, **16**, 1 (1978): 59–66. For viscosity to 400 bar, see Grigorev, B. A., A. S. Keramidi, et al., *Teploenergetika*, **24**, 9 (1977): 85–87; and Ulybin, S. A. and V. I. Makanushkin, *Teplofiz. Vysok. Temp.*, **15**, 6 (1977): 1195–1201.

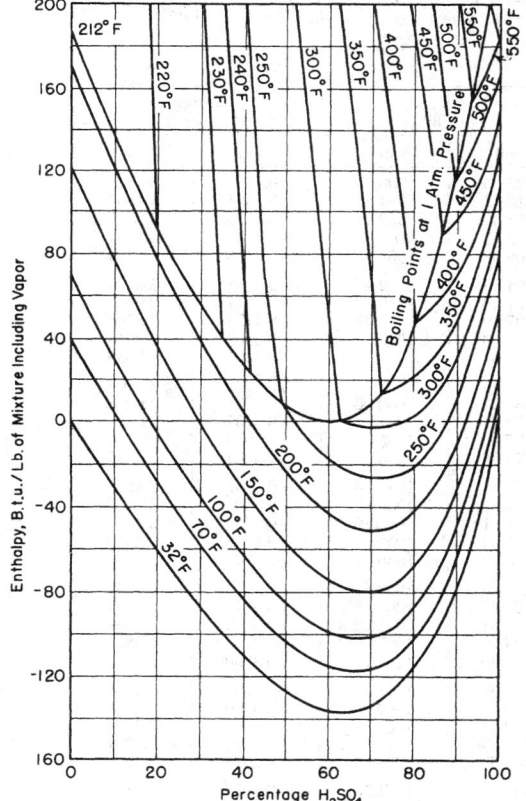

FIG. 2-30   Enthalpy-concentration diagram for aqueous sulfuric acid at 1 atm. Reference states: enthalpies of pure-liquid components at 32°F and vapor pressures are zero. NOTE: It should be observed that the weight basis includes the vapor, which is particularly important in the two-phase region. The upper ends of the tie lines in this region are assumed to be pure water. (*Hougen and Watson, Chemical Process Principles, part I, Wiley, New York, 1943.*)

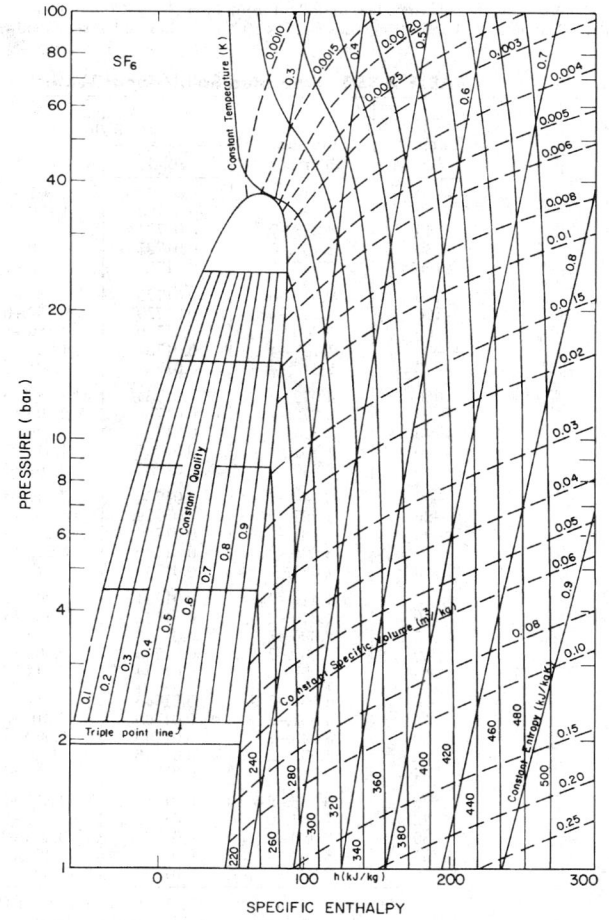

FIG. 2-31   Enthalpy–log-pressure diagram for sulfur hexafluoride.

TABLE 2-348 Saturated SUVA AC 9000

DuPont bulletin T–AC–9000–SI, 1994 (16 pp.) gives tables and a chart to 100 bar, 235°C. With a stated composition of 23% wt CH_2F_2 (R23), 25% wt CHF_2CF_3 (R125), and 52% wt CH_2FCH_3 (R134a) this is apparently identical to KLEA 66, to which the reader is referred.

TABLE 2-349 Saturated Toluene*

T, K	P, bar	v_f, m³/kg	v_g, m³/kg	h_f, kJ/kg	h_g, kJ/kg	s_f, kJ/(kg·K)	s_g, kJ/(kg·K)	c_{pf}, kJ/(kg·K)	μ_f, 10^{-4} Pa·s	k_f, W/(m·K)
270	0.0076	1.127.–3	34.9	316.7	745.7	2.236	3.825	1.64	8.02	0.141
280	0.0139	1.138.–3	19.1	333.0	756.1	2.295	3.806	1.66	6.96	0.138
290	0.0246	1.150.–3	10.6	349.6	766.8	2.353	3.792	1.68	6.10	0.136
300	0.0418	1.162.–3	6.46	366.5	777.8	2.410	3.782	1.71	5.41	0.133
310	0.0682	1.175.–3	4.08	383.7	789.2	2.467	3.776	1.74	4.83	0.131
320	0.1072	1.188.–3	2.67	401.3	800.9	2.522	3.771	1.78	4.34	0.128
330	0.1633	1.201.–3	1.80	419.6	812.9	2.577	3.771	1.81	3.93	0.126
340	0.2416	1.215.–3	1.25	437.4	825.2	2.632	3.772	1.84	3.58	0.124
350	0.3480	1.230.–3	0.891	456.0	837.8	2.686	3.777	1.88	3.28	0.121
360	0.4894	1.245.–3	0.698	475.1	850.7	2.739	3.783	1.92	3.01	0.119
370	0.6736	1.261.–3	0.481	494.6	863.8	2.792	3.791	1.96	2.78	0.117
380	0.9090	1.277.–3	0.364	514.4	877.2	2.846	3.801	2.01	2.56	0.114
390	1.2049	1.294.–3	0.279	534.7	890.9	2.898	3.811	2.05	2.37	0.112
400	1.5713	1.312.–3	0.218	555.4	904.8	2.950	3.824	2.09	2.19	0.110
420	2.5589	1.350.–3	0.137	598.1	933.1	3.054	3.852	2.17	1.89	0.105
440	3.965	1.393.–3	9.00.–2	642.3	962.0	3.156	3.883	2.24	1.64	0.101
460	5.892	1.443.–3	6.11.–2	688.1	991.3	3.258	3.917	2.31		0.096
480	8.451	1.499.–3	4.26.–2	735.5	1021.1	3.358	3.953	2.38		0.091
500	11.76	1.567.–3	3.03.–2	784.4	1051.3	3.457	3.989	2.45		0.086
520	15.96	1.651.–3	2.19.–2	834.9	1081.4	3.554	4.027	2.53		0.082
540	21.99	1.761.–3	1.58.–2	887.3	1109.6	3.651	4.062	2.65		0.078
560	27.65	1.919.–3	1.13.–2	942.8	1132.1	3.750	4.088	2.82		0.074
580	35.56	2.213.–3	7.59.–3	1005.6	1142.3	3.857	4.093			
590	40.16	2.650.–3	5.28.–3	1050.2	1128.1	3.932	4.063			
591.8^c	41.04	3.432.–3	3.43.–3	1084.9	1084.9	3.989	3.989			

*Values converted and mostly rounded off from the tables of Counsell, Lawrenson, and Lees, Nat. Phys. Lab., Teddington (U.K.) Rep. Chem. 52, 1976. c = critical point. The notation 1.127.–6 signifies 1.127×10^{-6}. For other tables, see Goodwin, R. D., *J. Phys. Chem. Ref. Data*, **18**, 4 (1989): 1565–1636.

TABLE 2-350 Saturated Solid/Vapor Water*

Temp., °F	Pressure, lb/in² abs.	Volume, ft³/lb		Enthalpy, Btu/lb		Entropy, Btu/(lb)(°F)	
		Solid	Vapor	Solid	Vapor	Solid	Vapor
−160	4.949.–8	0.01722	3.607.+9	−222.05	990.38	−0.4907	3.5549
−150	1.620.–7	0.01723	1.139.+9	−218.82	994.80	−0.4801	3.4387
−140	4.928.–7	0.01724	3.864.+8	−215.49	999.21	−0.4695	3.3301
−130	1.403.–6	0.01725	1.400.+8	−212.08	1003.63	−0.4590	3.2284
−120	3.757.–6	0.01726	5.386.+7	−208.58	1008.05	−0.4485	3.1330
−110	9.517.–6	0.01728	2.189.+7	−204.98	1012.47	−0.4381	3.0434
−100	2.291.–5	0.01729	9.352.+6	−201.28	1016.89	−0.4277	2.9591
−90	5.260.–5	0.01730	4.186.+6	−197.49	1021.31	−0.4173	2.8796
−80	1.157.–4	0.01731	1.955.+6	−193.60	1025.73	−0.4069	2.8045
−70	2.443.–4	0.01732	9.501.+5	−189.61	1030.15	−0.3965	2.7336
−60	4.972.–4	0.01734	4.788.+5	−185.52	1034.58	−0.3862	2.6664
−50	9.776.–4	0.01735	2.496.+5	−181.34	1039.00	−0.3758	2.6028
−45	1.354.–3	0.01736	1.824.+5	−179.21	1041.21	−0.3707	2.5723
−40	1.861.–3	0.01737	1.343.+5	−177.06	1043.42	−0.3655	2.5425
−35	2.540.–3	0.01737	9.961.+4	−174.88	1045.63	−0.3604	2.5135
−30	3.440.–3	0.01738	7.441.+4	−172.68	1047.84	−0.3552	2.4853
−25	4.627.–3	0.01739	5.596.+4	−170.46	1050.05	−0.3501	2.4577
−20	6.181.–3	0.01739	4.237.+4	−168.21	1052.26	−0.3449	2.4308
−15	8.204.–3	0.01740	3.228.+4	−165.94	1054.47	−0.3398	2.4046
−10	1.082.–2	0.01741	2.475.+4	−163.65	1056.67	−0.3347	2.3791
−5	1.419.–2	0.01741	1.909.+4	−161.33	1058.88	−0.3295	2.3541
0	1.849.–2	0.01742	1.481.+4	−158.98	1061.09	−0.3244	2.3297
5	2.396.–2	0.01743	1.155.+4	−156.61	1063.29	−0.3193	2.3039
10	3.087.–2	0.01744	9.060.+3	−154.22	1065.50	−0.3142	2.2827
15	3.957.–2	0.01744	7.144.+3	−151.80	1067.70	−0.3090	2.2600
16	4.156.–2	0.01745	6.817.+3	−151.32	1068.14	−0.3080	2.2555
18	4.581.–2	0.01745	6.210.+3	−150.34	1069.02	−0.3060	2.2466
20	5.045.–2	0.01745	5.662.+3	−149.36	1069.90	−0.3039	2.2378
22	5.552.–2	0.01746	5.166.+3	−148.38	1070.38	−0.3019	2.2291
24	6.105.–2	0.01746	4.717.+3	−147.39	1071.66	−0.2998	2.2205
26	6.708.–2	0.01746	4.311.+3	−146.40	1072.53	−0.2978	2.2119
28	7.365.–2	0.01746	3.943.+3	−145.40	1073.41	−0.2957	2.2034
30	8.080.–2	0.01747	3.608.+3	−144.40	1074.29	−0.2937	2.1950
31	8.461.–2	0.01747	3.453.+3	−143.90	1074.73	−0.2927	2.1908
32	8.858.–2	0.01747	3.305.+3	−143.40	1075.16	−0.2916	2.1867

*Condensed from *Fundamentals*, American Society of Heating, Refrigerating and Air-Conditioning Engineers, 1967 and 1972. Reproduced by permission. The validity of many standard reference tables has been critically reviewed by Jancso, Pupezin, and van Hook, *J. Phys. Chem.*, **74** (1970): 2984. This source is recommended for further study. The notation 4.949.–8, 3.607.+9, etc., means 4.949×10^{-8}, 3.607×10^9, etc.

TABLE 2-351   Saturated Water Substance—Temperature (fps units)*

Temp., °F	Pressure, lb/in² abs.	Volume, ft³/lb		Enthalpy, Btu/lb		Entropy, Btu/(lb)(°F)	
		Liquid	Vapor	Liquid	Vapor	Liquid	Vapor
32.018	0.08865	0.016022	3302.4	0.000	1075.5	0.0000	2.1872
35	0.09991	0.016020	2948.1	3.002	1076.8	0.0061	2.1767
40	0.12163	0.016019	2445.8	8.027	1079.0	0.0162	2.1594
45	0.14744	0.016020	2037.8	13.044	1081.2	0.0262	2.1426
50	0.17796	0.016023	1704.8	18.054	1083.4	0.0361	2.1262
55	0.21392	0.016027	1432.0	23.059	1085.6	0.0458	2.1102
60	0.25611	0.016033	1207.6	28.060	1087.7	0.0555	2.0946
65	0.30545	0.016041	1022.1	33.057	1089.9	0.0651	2.0794
70	0.36292	0.016050	868.4	38.052	1092.1	0.0745	2.0645
75	0.42964	0.016060	740.3	43.045	1094.3	0.0839	2.0500
80	0.50683	0.016072	633.3	48.037	1096.4	0.0932	2.0359
85	0.59583	0.016085	543.6	53.027	1098.6	0.1024	2.0221
90	0.69813	0.016099	468.1	58.018	1100.8	0.1115	2.0086
95	0.81534	0.016114	404.4	63.008	1102.9	0.1206	1.9954
100	0.94294	0.016130	350.4	67.999	1105.1	0.1295	1.9825
110	1.2750	0.016165	265.39	77.98	1109.3	0.1472	1.9577
120	1.6927	0.016204	203.26	87.97	1113.6	0.1646	1.9339
130	2.2230	0.016247	157.33	97.96	1117.8	0.1817	1.9112
140	2.8892	0.016293	122.98	107.89	1122.0	0.1985	1.8895
150	3.7184	0.016343	97.07	117.95	1126.1	0.2150	1.8686
160	4.7414	0.016395	77.27	127.96	1130.2	0.2313	1.8487
170	5.9926	0.016451	62.06	137.97	1134.2	0.2473	1.8295
180	7.5110	0.016510	50.225	148.00	1138.2	0.2631	1.8111
190	9.340	0.016572	40.957	158.04	1142.1	0.2787	1.7934
200	11.526	0.016637	33.639	168.09	1146.0	0.2940	1.7764
210	14.123	0.016705	27.816	178.15	1149.7	0.3091	1.7600
212	14.696	0.016719	26.799	180.17	1150.5	0.3121	1.7568
220	17.186	0.016775	23.148	188.23	1153.4	0.3241	1.7442
230	20.779	0.016849	19.381	198.33	1157.1	0.3388	1.7290
240	24.968	0.016926	16.321	208.45	1160.6	0.3533	1.7142
250	29.825	0.017066	13.819	218.59	1164.0	0.3677	1.7000
260	35.427	0.017089	11.762	228.76	1167.4	0.3819	1.6862
270	41.856	0.017175	10.060	238.95	1170.6	0.3960	1.6729
280	49.200	0.017264	8.644	249.17	1173.8	0.4098	1.6599
290	57.550	0.01736	7.4603	259.4	1167.8	0.4236	1.6473
300	67.005	0.01745	6.4658	269.7	1179.7	0.4372	1.6351
320	89.643	0.01766	4.9138	290.4	1185.2	0.4640	1.6116
340	117.992	0.01787	3.7878	311.3	1190.1	0.4902	1.5892
360	153.01	0.01811	2.9573	332.3	1194.4	0.5161	1.5678
380	195.73	0.01836	2.3353	353.6	1198.0	0.5416	1.5473
400	247.26	0.01864	1.8630	375.1	1201.0	0.5667	1.5274
420	308.78	0.01894	1.4997	396.9	1203.1	0.5915	1.5080
440	381.54	0.01926	1.2169	419.0	1204.4	0.6161	1.4890
460	466.87	0.01961	0.99424	441.5	1204.8	0.6405	1.4704
480	566.15	0.02000	0.81717	464.5	1204.1	0.6648	1.4518
500	680.86	0.02043	0.67492	487.9	1202.2	0.6890	1.4333
520	812.53	0.02091	0.55956	512.0	1199.0	0.7133	1.4146
540	962.79	0.02146	0.46513	536.8	1194.3	0.7378	1.3954
560	1133.38	0.02207	0.38714	562.4	1187.7	0.7625	1.3757
580	1326.17	0.02279	0.32216	589.1	1179.0	0.7876	1.3550
600	1543.2	0.02364	0.26747	617.1	1167.7	0.8134	1.3330
620	1786.9	0.02466	0.22081	646.9	1153.2	0.8403	1.3092
640	2059.9	0.02595	0.18021	679.1	1133.7	0.8686	1.2821
660	2365.7	0.02768	0.14431	714.9	1107.0	0.8995	1.2498
680	2708.6	0.03037	0.11117	758.5	1068.5	0.9365	1.2086
700	3094.3	0.03662	0.07519	825.2	991.7	0.9924	1.1359
702	3135.5	0.03824	0.06997	835.0	979.7	1.0006	1.1210
704	3177.2	0.04108	0.06300	854.2	956.2	1.0169	1.1046
705.47	3208.2	0.05078	0.05078	906.0	906.0	1.0612	1.0612

*Extracted and condensed from 1967 ASME Steam Tables. Copyright reserved. Reproduced by permission.

TABLE 2-352 Saturated Water Substance—Temperature (SI units)

Temp, K	Pressure, bar°	Volume, m³/kg Condensed†	Volume, m³/kg Vapor	Enthalpy, kJ/kg Condensed†	Enthalpy, kJ/kg Vapor	Entropy, kJ/(kg·K) Condensed†	Entropy, kJ/(kg·K) Vapor	Specific heat, C_p, kJ/(kg·K) Condensed†	Specific heat, C_p, kJ/(kg·K) Vapor	Viscosity, Ns/m² Condensed†	Viscosity, Ns/m² Vapor	Thermal conductivity, W/(m·K) Condensed†	Thermal conductivity, W/(m·K) Vapor	Prandtl no. Condensed†	Prandtl no. Vapor	Surface tension, N/m Condensed†	Temp, K
150	6.30,-11	1.073,-3	9.55,+9	-539.6	2273	-2.187	16.54	1.155				3.73					150
160	7.72,-10	1.074,-3	9.62,+8	-525.7	2291	-2.106	15.49	1.233				3.52					160
170	7.29,-9	1.076,-3	1.08,+8	-511.7	2310	-2.026	14.57	1.311				3.34					170
180	5.38,-8	1.077,-3	1.55,+7	-497.8	2328	-1.947	13.76	1.389				3.18					180
190	3.23,-7	1.078,-3	2.72,+6	-483.8	2347	-1.868	13.03	1.467				3.04					190
200	1.62,-6	1.079,-3	5.69,+5	-467.5	2366	-1.789	12.38	1.545				2.91					200
210	7.01,-6	1.081,-3	1.39,+5	-451.2	2384	-1.711	11.79	1.623				2.79					210
220	2.65,-5	1.082,-3	3.83,+4	-435.0	2403	-1.633	11.20	1.701				2.69					220
230	8.91,-5	1.084,-3	1.18,+4	-416.3	2421	-1.555	10.79	1.779				2.59					230
240	3.72,-4	1.085,-3	4.07,+3	-400.1	2440	-1.478	10.35	1.857				2.50					240
250	7.59,-4	1.087,-3	1.52,+3	-381.5	2459	-1.400	9.954	1.935				2.42					250
255	1.23,-3	1.087,-3	956.4	-369.8	2468	-1.361	9.768	1.974				2.38					255
260	1.96,-3	1.088,-3	612.2	-360.5	2477	-1.323	9.590	2.013				2.35					260
265	3.06,-3	1.089,-3	400.4	-351.2	2486	-1.281	9.461	2.052				2.31					265
270	4.69,-3	1.090,-3	265.4	-339.6	2496	-1.296	9.255	2.091				2.27					270
273.15	6.11,-3	1.091,-3	206.3	-333.5	2502	-1.221	9.158	2.116				2.26					273.15
273.15	0.00611	1.000,-3	206.3	0.0	2502	0.000	9.158	4.217	1.854	1750,-6	8.02,-6	0.569	0.0182	12.99	0.815	0.0755	273.15
275	0.00697	1.000,-3	181.7	7.8	2505	0.028	9.109	4.211	1.855	1652,-6	8.09,-6	0.574	0.0183	12.22	0.817	0.0753	275
280	0.00990	1.000,-3	130.4	28.8	2514	0.104	8.980	4.198	1.858	1422,-6	8.29,-6	0.582	0.0186	10.26	0.825	0.0748	280
285	0.01387	1.000,-3	99.4	49.8	2523	0.178	8.857	4.189	1.861	1225,-6	8.49,-6	0.590	0.0189	8.81	0.833	0.0743	285
290	0.01917	1.001,-3	69.7	70.7	2532	0.251	8.740	4.184	1.864	1080,-6	8.69,-6	0.598	0.0193	7.56	0.841	0.0737	290
295	0.02617	1.002,-3	51.94	91.6	2541	0.323	8.627	4.181	1.868	959,-6	8.89,-6	0.606	0.0195	6.62	0.849	0.0727	295
300	0.03531	1.003,-3	39.13	112.5	2550	0.393	8.520	4.179	1.872	855,-6	9.09,-6	0.613	0.0196	5.83	0.857	0.0717	300
305	0.04712	1.005,-3	27.90	133.4	2559	0.462	8.417	4.178	1.877	769,-6	9.29,-6	0.620	0.0201	5.20	0.865	0.0709	305
310	0.06221	1.007,-3	22.93	154.3	2568	0.530	8.318	4.178	1.882	695,-6	9.49,-6	0.628	0.0204	4.62	0.873	0.0700	310
315	0.08132	1.009,-3	17.82	175.2	2577	0.597	8.224	4.179	1.888	631,-6	9.69,-6	0.634	0.0207	4.16	0.883	0.0692	315
320	0.1053	1.011,-3	13.98	196.1	2586	0.649	8.151	4.180	1.895	577,-6	9.89,-6	0.640	0.0210	3.77	0.894	0.0683	320
325	0.1351	1.013,-3	11.06	217.0	2595	0.727	8.046	4.182	1.903	528,-6	10.09,-6	0.645	0.0213	3.42	0.901	0.0675	325
330	0.1719	1.016,-3	8.82	237.9	2604	0.791	7.962	4.184	1.911	489,-6	10.29,-6	0.650	0.0217	3.15	0.908	0.0666	330
335	0.2167	1.018,-3	7.09	258.8	2613	0.854	7.881	4.186	1.920	453,-6	10.49,-6	0.655	0.0220	2.88	0.916	0.0658	335
340	0.2713	1.021,-3	5.74	279.8	2622	0.916	7.804	4.188	1.930	420,-6	10.69,-6	0.660	0.0223	2.66	0.925	0.0649	340
345	0.3372	1.024,-3	4.683	300.7	2630	0.977	7.729	4.191	1.941	389,-6	10.89,-6	0.665	0.0226	2.45	0.933	0.0641	345
350	0.4163	1.027,-3	3.846	321.7	2639	1.038	7.657	4.195	1.954	365,-6	11.09,-6	0.668	0.0230	2.29	0.942	0.0632	350
355	0.5100	1.030,-3	3.180	342.7	2647	1.097	7.588	4.199	1.968	343,-6	11.29,-6	0.671	0.0233	2.14	0.951	0.0623	355
360	0.6209	1.034,-3	2.645	363.7	2655	1.156	7.521	4.203	1.983	324,-6	11.49,-6	0.674	0.0237	2.02	0.960	0.0614	360
365	0.7514	1.038,-3	2.212	384.7	2663	1.214	7.456	4.209	1.999	306,-6	11.69,-6	0.677	0.0241	1.91	0.969	0.0605	365
370	0.9040	1.041,-3	1.861	405.8	2671	1.271	7.394	4.214	2.017	289,-6	11.89,-6	0.679	0.0245	1.80	0.978	0.0595	370
373.15	1.0133	1.044,-3	1.679	419.1	2676	1.307	7.356	4.217	2.029	279,-6	12.02,-6	0.680	0.0248	1.76	0.984	0.0589	373.15
375	1.0815	1.045,-3	1.574	426.8	2679	1.328	7.333	4.220	2.036	274,-6	12.09,-6	0.681	0.0249	1.70	0.987	0.0586	375
380	1.2869	1.049,-3	1.337	448.0	2687	1.384	7.275	4.226	2.057	260,-6	12.29,-6	0.683	0.0254	1.61	0.995	0.0576	380
385	1.5233	1.053,-3	1.142	469.2	2694	1.439	7.218	4.232	2.080	248,-6	12.49,-6	0.685	0.0258	1.53	1.004	0.0566	385
390	1.794	1.058,-3	0.980	490.4	2702	1.494	7.163	4.239	2.104	237,-6	12.69,-6	0.686	0.0263	1.47	1.013	0.0556	390
400	2.455	1.067,-3	0.731	532.9	2716	1.605	7.058	4.256	2.158	217,-6	13.05,-6	0.688	0.0272	1.34	1.033	0.0536	400
410	3.302	1.077,-3	0.553	575.6	2729	1.708	6.959	4.278	2.221	200,-6	13.42,-6	0.688	0.0282	1.24	1.054	0.0515	410
420	4.370	1.088,-3	0.425	618.6	2742	1.810	6.865	4.302	2.291	185,-6	13.79,-6	0.688	0.0293	1.16	1.075	0.0494	420
430	5.699	1.099,-3	0.331	661.8	2753	1.911	6.775	4.331	2.369	173,-6	14.14,-6	0.685	0.0304	1.09	1.10	0.0472	430

440	7.333	1.110,–3	0.261	705.3	2764	2.011	6.689	4.36	2.46	162.–6	14.50.–6	0.682	0.0317	1.04	1.12	0.0451	440
450	9.319	1.123,–3	0.208	749.2	2773	2.109	6.607	4.40	2.56	152.–6	14.85.–6	0.678	0.0331	0.99	1.14	0.0429	450
460	11.71	1.137,–3	0.167	793.5	2782	2.205	6.528	4.44	2.68	143.–6	15.19.–6	0.673	0.0346	0.95	1.17	0.0407	460
470	14.55	1.152,–3	0.136	838.2	2789	2.301	6.451	4.48	2.79	136.–6	15.54.–6	0.667	0.0363	0.92	1.20	0.0385	470
480	17.90	1.167,–3	0.111	883.4	2795	2.395	6.377	4.53	2.94	129.–6	15.88.–6	0.660	0.0381	0.89	1.23	0.0362	480
490	21.83	1.184,–3	0.0922	929.1	2799	2.479	6.312	4.59	3.10	124.–6	16.23.–6	0.651	0.0401	0.87	1.25	0.0339	490
500	26.40	1.203,–3	0.0766	975.6	2801	2.581	6.233	4.66	3.27	118.–6	16.59.–6	0.642	0.0423	0.86	1.28	0.0316	500
510	31.66	1.222,–3	0.0631	1023	2802	2.673	6.163	4.74	3.47	113.–6	16.95.–6	0.631	0.0447	0.85	1.31	0.0293	510
520	37.70	1.244,–3	0.0525	1071	2801	2.765	6.093	4.84	3.70	108.–6	17.33.–6	0.621	0.0475	0.84	1.35	0.0269	520
530	44.58	1.268,–3	0.0445	1119	2798	2.856	6.023	4.95	3.96	104.–6	17.72.–6	0.608	0.0506	0.85	1.39	0.0245	530
540	52.38	1.294,–3	0.0375	1170	2792	2.948	5.953	5.08	4.27	101.–6	18.1.–6	0.594	0.0540	0.86	1.43	0.0221	540
550	61.19	1.323,–3	0.0317	1220	2784	3.039	5.882	5.24	4.64	97.–6	18.6.–6	0.580	0.0583	0.87	1.47	0.0197	550
560	71.08	1.355,–3	0.0269	1273	2772	3.132	5.808	5.43	5.09	94.–6	19.1.–6	0.563	0.0637	0.90	1.52	0.0173	560
570	82.16	1.392,–3	0.0228	1328	2757	3.225	5.733	5.68	5.67	91.–6	19.7.–6	0.548	0.0698	0.94	1.59	0.0150	570
580	94.51	1.433,–3	0.0193	1384	2737	3.321	5.654	6.00	6.40	88.–6	20.4.–6	0.528	0.0767	0.99	1.68	0.0128	580
590	108.3	1.482,–3	0.0163	1443	2717	3.419	5.569	6.41	7.35	84.–6	21.5.–6	0.513	0.0841	1.05	1.84	0.0105	590
600	123.5	1.541,–3	0.0137	1506	2682	3.520	5.480	7.00	8.75	81.–6	22.7.–6	0.497	0.0929	1.14	2.15	0.0084	600
610	137.3	1.612,–3	0.0115	1573	2641	3.627	5.318	7.85	11.1	77.–6	24.1.–6	0.467	0.103	1.30	2.60	0.0063	610
620	159.1	1.705,–3	0.0094	1647	2588	3.741	5.259	9.35	15.4	72.–6	25.9.–6	0.444	0.114	1.52	3.46	0.0045	620
625	169.1	1.778,–3	0.0085	1697	2555	3.805	5.191	10.6	18.3	70.–6	27.0.–6	0.430	0.121	1.65	4.20	0.0035	625
630	179.7	1.856,–3	0.0075	1734	2515	3.875	5.115	12.6	22.1	67.–6	28.0.–6	0.412	0.130	2.0	4.8	0.0026	630
635	190.9	1.935,–3	0.0066	1783	2466	3.950	5.025	16.4	27.6	64.–6	30.0.–6	0.392	0.141	2.7	6.0	0.0015	635
640	202.7	2.075,–3	0.0057	1841	2401	4.037	4.912	26	42	59.–6	32.0.–6	0.367	0.155	4.2	9.6	0.0008	640
645	215.2	2.351,–3	0.0045	1931	2292	4.223	4.732	90		54.–6	37.0.–6	0.331	0.178	12	26	0.0001	645
647.3‡	221.2	3.170,–3	0.0032	2107	2107	4.443	4.443	∞	∞	45.–6	45.0.–6	0.238	0.238	∞	∞	0.0000	647.3‡

*1 bar = 10⁵ N/m².

*1 bar = 10^5 N/m².

†Above the solid line, the condensed phase is solid; below it, liquid.

‡Critical temperature.

NOTE: The notations 6.30,–11, 1.073,–3, 9.55,+9, etc. signify 6.30×10^{-11}, 1.073×10^{-3}, 955×10^{9}, etc.

Tables 2-351 and 2-352 are provided for general use. Tables to higher precision are available over certain ranges and for various properties. The most current internationally accepted tables are found in Haar, L., J. S. Gallagher, and G. S. Kell, *NBS/NRC Steam Tables*, Hemisphere, Washington, DC, 1984 (320 pp.). These do not tabulate certain properties at saturation states. A revised release on the IAPWS Skeleton Tables 1985 for the thermodynamic properties of ordinary water substance, Sept. 1993 (15 pp.) is apparently the latest international publication. In *J. Phys. Chem. Ref. Data* **17**, 4 (1988): 1439–1540. H. Sato, M. Uematsu, and others review existing steam tables and present the 1985 formulation of skeleton tables. Property codes and programs include Cheng, S. C. and C. Nguyen, *Modeling and Simulation on Microcomputers 1989* (R. W. Allen, ed.), S.C.S. Intl. San Diego, 1989 (pp. 138–141); Garland, W. J. and B. J. Hand, *Nucl. Engng. & Des.*, **113**, (1989): 21–34; Dickey, D. S., *Chem. Eng.* **98**, 9 (1991): 207–8 and **98**, 11: 235–6; Muneer, T. and S. M. Scott, *Proc. Inst. Mech. Eng.*, **205**, (1991): 25–29; and *Energy Convsn. Mgmt.*, **31**, 4 (1991): 315–325. Useful pictorial representations of 20 properties as a function of both temperature (to 800°C) and pressure (to 1000 bar) are given by Grigull, U., J. Bach, et al., *Wärme- u. Stoff.*, **1** (1968): 202–213. Property equations for the saturated liquid for the range 0–300°C are given by Charters, W. W. S. and H. A. Sadafi, *Rev. Int. Froid*, **10**, (Mar. 1987): 105–6. Gordon, S., NASA Tech. Paper 1906, 1982 gives detailed tables for ice from 0 K. Ice and snow properties are reviewed by Fukusako, S., *Int. J. Thermophys*, **11**, 2 (1990): 353–372. See also Wagner, W., A. Saul, et al., *J. Phys. Chem. Ref. Data*, **23**, 3 (1994): 515–525, and Table 2-358.

TABLE 2-353 Saturated Liquid Water—Miscellaneous Properties

Temperature, °C	$10^4\,\beta$	$10^4\,k_T/\text{bar}$	$10^4\,k_s/\text{bar}$	v_s, m/s	μ_f, 10^{-6} Pa·s	c_p, kJ/kg·K	k, W/m·K	Pr, bar	σ, N/m
0	−0.681	0.50885	0.50855	1402.4	1.793	4.2176	0.567	13.32	0.07565
1	−0.501	0.50509	0.50493	1407.4	1.732	4.2140	0.569	12.83	0.07551
2	−0.327	0.50151	0.50143	1412.2	1.675	4.2107	0.570	12.37	0.07537
3	−0.160	0.49808	0.49806	1417.0	1.621	4.2077	0.572	11.93	0.07522
4	0.003	0.49481	0.49481	1421.6	1.569	4.2048	0.573	11.51	0.07508
5	0.160	0.49169	0.49167	1426.2	1.520	4.2022	0.575	11.11	0.07494
6	0.312	0.48871	0.48865	1430.6	1.474	4.1999	0.577	10.73	0.07480
7	0.460	0.48587	0.48573	1434.9	1.429	4.1977	0.578	10.38	0.07465
8	0.604	0.48315	0.48291	1439.1	1.387	4.1956	0.580	10.04	0.07451
9	0.744	0.48056	0.48019	1443.3	1.346	4.1938	0.581	9.72	0.07436
10	0.880	0.47809	0.47757	1447.3	1.308	4.1921	0.5828	9.41	0.07422
11	1.012	0.47573	0.47504	1451.2	1.271	4.1906	0.5844	9.11	0.07407
12	1.141	0.47347	0.47260	1455.0	1.236	4.1892	0.5859	8.84	0.07393
13	1.267	0.47133	0.47024	1458.7	1.202	4.1879	0.5875	8.57	0.07378
14	1.389	0.46928	0.46797	1462.4	1.170	4.1867	0.5891	8.32	0.07364
15	1.509	0.46733	0.46578	1465.9	1.139	4.1856	0.5906	8.07	0.07349
16	1.626	0.46548	0.46366	1469.4	1.110	4.1847	0.5922	7.84	0.07334
17	1.740	0.46371	0.46162	1472.7	1.081	4.1838	0.5937	7.62	0.07319
18	1.852	0.46203	0.45966	1476.0	1.054	4.1830	0.5953	7.41	0.07304
19	1.961	0.46043	0.45776	1479.2	1.028	4.1823	0.5968	7.20	0.07289
20	2.068	0.45892	0.45593	1482.3	1.003	4.1817	0.5983	7.01	0.07274
21	2.173	0.45748	0.45417	1485.3	0.979	4.1812	0.5999	6.82	0.07259
22	2.275	0.45612	0.45248	1488.3	0.955	4.1807	0.6014	6.64	0.07244
23	2.376	0.45484	0.45084	1491.2	0.933	4.1802	0.6029	6.47	0.07228
24	2.475	0.45362	0.44927	1493.9	0.911	4.1798	0.6044	6.30	0.07213
25	2.572	0.45247	0.44776	1496.7	0.891	4.1795	0.6059	6.15	0.07198
26	2.667	0.45139	0.44630	1499.3	0.871	4.1792	0.6074	5.99	0.07182
27	2.761	0.45038	0.44490	1501.9	0.852	4.1790	0.6089	5.85	0.07167
28	2.852	0.44943	0.44355	1504.3	0.833	4.1788	0.6104	5.70	0.07151
30	3.032	0.44771	0.44102	1509.1	0.798	4.1785	0.6133	5.44	0.07120
32	3.206	0.44622	0.43869	1513.6	0.765	4.1783	0.6162	5.19	0.07089
34	3.375	0.44496	0.43655	1517.8	0.734	4.1782	0.6190	4.95	0.07058
36	3.539	0.44390	0.43459	1521.7	0.705	4.1783	0.6218	4.74	0.07025
38	3.698	0.44305	0.43280	1525.4	0.679	4.1784	0.6246	4.54	0.06992
40	3.853	0.44239	0.43118	1528.9	0.653	4.1786	0.6273	4.35	0.06960
42	4.004	0.44192	0.42972	1532.1	0.629	4.1789	0.6299	4.17	0.06927
44	4.152	0.44162	0.42842	1535.0	0.607	4.1792	0.6315	4.02	0.06894
46	4.296	0.44149	0.42726	1537.7	0.586	4.1797	0.6351	3.86	0.06861
48	4.438	0.44153	0.42624	1540.3	0.566	4.1801	0.6375	3.71	0.06828
50	4.576	0.44173	0.42535	1542.6	0.547	4.1807	0.6400	3.57	0.06795
55	4.910	0.44290	0.42370	1547.4	0.5043	4.1824	0.6457	3.267	0.06710
60	5.231	0.44496	0.42281	1551.0	0.4668	4.1844	0.6511	3.000	0.06624
65	5.539	0.44788	0.42262	1553.4	0.4338	4.1869	0.6561	2.768	0.06537
70	5.837	0.45162	0.42309	1554.8	0.4045	4.1897	0.6607	2.565	0.06449
75	6.128	0.45614	0.42418	1555.1	0.3784	4.1929	0.6649	2.386	0.06359
80	6.411	0.46143	0.42587	1554.4	0.3550	4.1965	0.6686	2.228	0.06268
85	6.689	0.46748	0.42812	1552.9	0.3340	4.2005	0.6721	2.088	0.06176
90	6.962	0.47429	0.43093	1550.5	0.3150	4.2050	0.6753	1.962	0.06083
95	7.233	0.48185	0.43429	1547.2	0.2979	4.2102	0.6779	1.850	0.05988
100	7.501	0.49019	0.43819	1543.1	0.2823	4.2164	0.6800	1.756	0.05892

Values mostly from Aleksandrov, A. A. and M. S. Trakhtenhertz, *Thermophysical Properties of Water at Atmospheric Pressure,* Standartov, Moscow, 1977 (99 pp.).

TABLE 2-354 Thermodynamic Properties of Compressed Steam*

Temperature, K		Pressure, bar									
		0.1	0.5	1	5	10	20	40	60	80	100
350	v	16.12	1.027.−3	1.027.−3	1.027.−3	1.027.−3	1.026.−3	1.025.−3	1.024.−3	1.023.−3	1.023.−3
	h	2644	321.7	231.8	322.1	322.5	323.3	324.9	326.4	328.1	329.7
	s	8.327	1.037	1.037	1.037	1.037	1.036	1.035	1.034	1.032	1.031
400	v	18.44	3.67	1.827	1.067.−3	1.067.−3	1.066.−3	1.065.−3	1.064.−3	1.063.−3	1.061.−3
	h	2739	2735	2730	533.1	533.4	534.1	535.4	536.8	538.2	539.6
	s	8.581	7.831	7.502	1.601	1.600	1.599	1.597	1.595	1.593	1.592
450	v	20.75	4.14	2.063	0.410	1.124.−3	1.123.−3	1.121.−3	1.119.−3	1.118.−3	1.116.−3
	h	2835	2833	2830	2804	749.0	749.8	750.8	751.9	753.0	754.1
	s	8.811	8.061	7.736	6.949	2.110	2.107	2.105	2.102	2.099	2.097
500	v	23.07	4.61	2.298	0.452	0.221	0.104	1.201.−3	1.198.−3	1.196.−3	1.193.−3
	h	2932	2931	2929	2912.4	2891.2	2839.4	975.9	976.3	976.8	977.3
	s	9.012	8.261	7.944	7.177	6.823	6.422	2.578	2.575	2.571	2.567
600	v	27.7	5.53	2.76	0.548	0.271	0.133	0.0630	0.0396	0.0276	0.0201
	h	3131	3130	3129	3120	3109	3087	3036	2976	2906	2820
	s	9.374	8.630	8.309	7.560	7.223	6.875	6.590	6.224	5.997	5.775
700	v	32.3	6.46	3.23	0.643	0.319	0.158	0.0769	0.0500	0.0346	0.0283
	h	3335	3335	3334	3328	3322	3307	3278	3247	3214	3179
	s	9.692	8.946	8.625	7.877	7.550	7.215	6.864	6.644	6.431	6.334
800	v	36.9	7.38	3.69	0.736	0.367	0.182	0.0889	0.0589	0.0436	0.0343
	h	3547	3546	3546	3542	3537	3526	3506	3485	3464	3442
	s	9.971	9.228	8.908	8.161	7.837	7.507	7.151	6.965	6.809	6.685
900	v	41.5	8.31	4.15	0.829	0.414	0.206	0.102	0.0674	0.0501	0.0398
	h	3765	3765	3764	3761	3757	3750	3737	3719	3704	3688
	s	10.228	9.485	9.165	8.420	8.097	7.770	7.462	7.237	7.092	6.975
1000	v	46.2	9.23	4.615	0.921	0.460	0.229	0.114	0.0758	0.0564	0.0449
	h	3990	3990	3990	3987	3984	3978	3967	3955	3944	3935
	s	10.466	9.723	9.402	8.659	8.336	8.011	7.682	7.486	7.345	7.233
1500	v	69.2	13.9	6.92	1.385	0.692	0.341	0.1730	0.1153	0.0865	0.0692
	h	5231	5228	5227	5225	5224	5221	5217	5212	5207	5203
	s	11.47	10.77	10.40	9.66	9.34	9.015	8.693	8.503	8.368	8.262
2000	v	93.0	18.6	9.26	1.850	0.925	0.462	0.231	0.1543	0.1157	0.0926
	h	6832	6734	6706	6662	6649	6639	6629	6623	6619	6616
	s	12.38	11.58	11.25	10.48	10.15	9.828	9.503	9.313	9.178	9.073
2500	v	123.7	24.0	11.90	2.35	1.171	0.583	0.291	0.1942	0.1457	0.1166
	h	10417	9330	9046	8621	8504	8413	8342	8307	8285	8269
	s	13.95	12.73	12.28	11.35	10.80	10.62	10.26	10.06	9.920	9.810

*v = specific volume, m³/kg; h = specific enthalpy, kJ/kg; s = specific entropy, kJ/(kg·K). The notation 1.027.−3 signifies 1.027×10^{-3}.

TABLE 2-354   Thermodynamic Properties of Compressed Steam (*Concluded*)

Temperature, K		Pressure, bar										
		150	200	250	300	400	500	600	700	800	900	1000
350	v	1.020.−3	1.018.−3	1.016.−3	1.014.−3	1.009.−3	1.005.−3	1.002.−3	9.977.−4	9.937.−4	9.900.−4	9.865.−4
	h	333.7	337.7	341.7	344.7	353.8	361.8	369.7	377.7	385.7	393.7	401.7
	s	1.028	1.025	1.022	1.019	1.013	1.007	1.001	0.996	0.991	0.985	0.979
400	v	1.059.−3	1.056.−3	1.053.−3	1.050.−3	1.045.−3	1.041.−3	1.035.−3	1.031.−3	1.027.−3	1.022.−3	1.018.−3
	h	543.1	546.5	550.1	553.5	560.6	567.8	574.9	582.1	589.3	596.5	603.8
	s	1.587	1.583	1.578	1.574	1.565	1.557	1.549	1.541	1.533	1.526	1.518
450	v	1.112.−3	1.108.−3	1.105.−3	1.101.−3	1.094.−3	1.088.−3	1.082.−3	1.076.−3	1.070.−3	1.065.−3	1.059.−3
	h	756.8	759.5	762.3	765.2	771.0	776.9	783.0	789.6	795.3	801.6	807.9
	s	2.088	2.082	2.076	2.070	2.060	2.049	2.039	2.029	2.019	2.010	2.002
500	v	1.187.−3	1.181.−3	1.175.−3	1.170.−3	1.160.−3	1.151.−3	1.142.−3	1.134.−3	1.126.−3	1.119.−3	1.112.−3
	h	978.8	980.3	981.9	983.7	987.4	991.5	995.9	1000.5	1005.3	1010.3	1015.4
	s	2.558	2.549	2.541	2.533	2.517	2.502	2.488	2.474	2.461	2.449	2.437
600	v	1.519.−3	1.483.−3	1.454.−3	1.428.−3	1.392.−3	1.362.−3	1.337.−3	1.315.−3	1.296.−3	1.280.−3	1.265.−3
	h	1499	1489	1479	1472	1462	1456	1452	1449	1447	1447	1447
	s	3.501	3.469	3.443	3.419	3.379	3.346	3.316	3.290	3.266	3.244	3.223
700	v	1.724.−2	1.157.−2	7.986.−3	5.416.−3	2.630.−3	2.038.−3	1.831.−3	1.716.−3	1.639.−3	1.589.−3	1.536.−3
	h	3082	2965	2821	2635	2233	2084	2021	1986	1962	1946	1931
	s	6.037	5.770	5.494	5.179	4.554	4.308	4.192	4.116	4.058	4.012	3.972
800	v	2.195.−2	1.575.−2	1.201.−2	9.512.−3	6.391.−3	4.576.−3	3.496.−3	2.866.−3	2.484.−3	2.239.−3	2.072.−3
	h	3386	3325	3261	3193	3047	2895	2734	2648	2567	2508	2465
	s	6.444	6.252	6.086	5.934	5.654	5.397	5.175	4.998	4.864	4.761	4.701
900	v	2.590.−2	1.899.−2	1.483.−2	1.207.−2	8.619.−3	6.581.−3	5.257.−3	4.348.−3	3.704.−3	3.454.−3	2.907.−3
	h	3649	3609	3568	3526	3440	3354	3269	3188	3113	3049	2995
	s	6.755	6.587	6.449	6.327	6.119	5.940	5.780	5.637	5.510	5.399	5.305
1000	v	2.954.−2	2.186.−2	1.726.−2	1.420.−2	1.038.−2	8.102.−3	6.605.−3	5.557.−3	4.792.−3	4.212.−3	3.763.−3
	h	3904	3874	3845	3816	3756	3697	3640	3584	3532	3482	3435
	s	7.023	6.867	6.741	6.633	6.453	6.302	6.172	6.055	5.951	5.856	5.727
1500	v	0.0461	0.0346	0.0277	0.0231	0.0173	0.0139	0.0116	0.00993	0.00871	0.00776	0.00700
	h	5202	5198	5186	5180	5171	5157	5144	5133	5120	5108	5095
	s	8.074	7.936	7.827	7.738	7.597	7.484	7.391	7.310	7.239	7.176	7.118
2000	v	0.0619	0.0465	0.0372	0.0311	0.0234	0.0188	0.0157	0.0135	0.0119	0.0106	0.0096
	h	6613	6610	6608	6605	6599	6595	6590	6585	6581	6577	6574
	s	8.883	8.748	8.642	8.555	8.418	8.310	8.222	8.147	8.082	8.024	7.971
2500	v	0.0778	0.0584	0.0468	0.0391	0.0294	0.0236	0.0197	0.0170	0.0149	0.0133	0.0120
	h	8269	8269	8269	8268	8267	8265	8261	2856	8250	8244	8240
	s	9.610	9.468	9.358	9.270	9.129	9.020	8.930	8.854	8.788	8.730	8.677

TABLE 2-355 Density, Specific Heats at Constant Volume and at Constant Pressure and Velocity of Sound for Compressed Water, 1–1000 bar, 0–150°C

P, bar	0°C (ITS-90) density, kg/m³	C_p, kJ/(kg·K)	C_v, kJ/(kg·K)	w, m/s	10°C (ITS-90) density, kg/m³	C_p, kJ/(kg·K)	C_v, kJ/(kg·K)	w, m/s	20°C (ITS-90) density, kg/m³	C_p, kJ/(kg·K)	C_v, kJ/(kg·K)	w, m/s	30°C (ITS-90) density, kg/m³	C_p, kJ/(kg·K)	C_v, kJ/(kg·K)	w, m/s
1	999.702	4.1923	4.1877	1447.3	998.207	4.1812	4.1538	1482.3	995.650	4.1774	4.1148	1509.1	992.217	4.1775	4.0715	1528.9
50	1002.03	4.174	4.168	1455	1000.44	4.166	4.137	1491	997.82	4.164	4.099	1517	994.36	4.166	4.058	1537
100	1004.38	4.156	4.149	1464	1002.69	4.151	4.119	1499	1000.02	4.151	4.084	1526	996.52	4.154	4.044	1546
150	1006.71	4.139	4.130	1472	1004.93	4.137	4.103	1507	1002.19	4.139	4.069	1534	998.66	4.142	4.031	1554
200	1009.01	4.123	4.112	1480	1007.13	4.124	4.087	1516	1004.34	4.127	4.055	1543	1000.77	4.131	4.018	1563
250	1011.28	4.108	4.095	1489	1009.32	4.110	4.071	1524	1006.47	4.115	4.041	1551	1002.87	4.121	4.005	1571
300	1013.53	4.093	4.078	1497	1011.48	4.098	4.056	1532	1008.57	4.104	4.027	1559	1004.94	4.110	3.993	1579
400	1017.97	4.065	4.046	1513	1015.74	4.074	4.027	1548	1012.72	4.083	4.001	1576	1009.03	4.091	3.969	1596
500	1022.31	4.040	4.016	1529	1019.92	4.052	3.999	1565	1016.79	4.063	3.976	1592	1013.03	4.072	3.946	1612
600	1026.57	4.018	3.988	1545	1024.02	4.032	3.974	1581	1020.79	4.044	3.952	1608	1016.97	4.055	3.924	1628
800	1034.85	3.979	3.937	1577	1031.99	3.996	3.926	1613	1028.56	4.011	3.908	1640	1024.62	4.023	3.884	1660
1000	1042.83	3.948	3.892	1609	1039.68	3.967	3.884	1644	1036.06	3.982	3.869	1671	1032.00	3.995	3.847	1692

P, bar	40°C (ITS-90) density, kg/m³	C_p, kJ/(kg·K)	C_v, kJ/(kg·K)	w, m/s	50°C (ITS-90) density, kg/m³	C_p, kJ/(kg·K)	C_v, kJ/(kg·K)	w, m/s	60°C (ITS-90) density, kg/m³	C_p, kJ/(kg·K)	C_v, kJ/(kg·K)	w, m/s	70°C (ITS-90) density, kg/m³	C_p, kJ/(kg·K)	C_v, kJ/(kg·K)	w, m/s
1	988.036	4.1799	4.0248	1542.6	983.197	4.1840	3.9755	1551.0	977.766	4.1896	3.9246	1554.8	971.791	4.1967	3.8727	1554.5
50	990.16	4.169	4.012	1551	985.33	4.173	3.964	1560	979.92	4.179	3.915	1564	973.98	4.186	3.864	1564
100	992.31	4.158	4.000	1560	987.48	4.163	3.953	1568	982.09	4.169	3.905	1573	976.18	4.176	3.855	1573
150	994.44	4.147	3.988	1568	989.61	4.152	3.943	1577	984.23	4.158	3.895	1582	978.35	4.165	3.846	1582
200	996.54	4.137	3.976	1577	991.71	4.142	3.932	1586	986.36	4.148	3.885	1590	980.51	4.155	3.838	1591
250	998.62	4.126	3.965	1585	993.80	4.132	3.922	1594	988.46	4.139	3.876	1599	982.63	4.146	3.829	1600
300	1000.68	4.117	3.954	1594	995.86	4.123	3.911	1603	990.53	4.129	3.867	1608	984.74	4.136	3.821	1609
400	1004.74	4.098	3.932	1610	999.92	4.105	3.892	1620	994.62	4.111	3.849	1625	988.87	4.118	3.805	1627
500	1008.72	4.080	3.911	1627	1003.90	4.087	3.873	1637	998.62	4.094	3.832	1642	992.92	4.101	3.789	1644
600	1012.62	4.063	3.892	1643	1007.80	4.071	3.855	1653	1002.54	4.078	3.815	1659	996.88	4.085	3.774	1662
800	1020.21	4.033	3.854	1676	1015.38	4.041	3.821	1686	1010.15	4.048	3.784	1693	1004.56	4.054	3.745	1696
1000	1027.53	4.005	3.820	1707	1022.69	4.013	3.789	1718	1017.48	4.020	3.754	1726	1011.94	4.027	3.717	1730

Prepared by H. Sato. Keio University, Japan. Oct. 1994. Based upon "An equation of state for the thermodynamic properties of water in the liquid phase including the metastable state," from *Properties of Water and Steam*," *Proc. 11th Int. Conf. Props. Steam* (M. Pichal and O. Sifner, eds.), Hemisphere, New York, 1990 (551 pp.).

TABLE 2-355 Density, Specific Heats at Constant Pressure and at Constant Volume and Velocity of Sound for Compressed Water, 1–1000 bar, 0–150°C (*Concluded*)

80°C (ITS-90)

P, bar	density, kg/m³	C_p, kJ/(kg·K)	C_v, kJ/(kg·K)	w, m/s
1	965.309	4.2056	3.8206	1550.5
50	967.54	4.195	3.813	1560
100	969.79	4.184	3.805	1569
150	972.00	4.174	3.797	1579
200	974.20	4.164	3.789	1588
250	976.36	4.154	3.782	1597
300	978.50	4.144	3.774	1607
400	982.71	4.126	3.759	1625
500	986.82	4.108	3.745	1643
600	990.83	4.092	3.731	1661
800	998.62	4.061	3.704	1696
1000	1006.08	4.033	3.678	1731

90°C (ITS-90)

P, bar	density, kg/m³	C_p, kJ/(kg·K)	C_v, kJ/(kg·K)	w, m/s
1	958.348	4.2164	3.7689	1543.1
50	960.64	4.205	3.762	1553
100	962.94	4.194	3.755	1563
150	965.21	4.184	3.748	1572
200	967.45	4.173	3.741	1582
250	969.67	4.163	3.734	1592
300	971.85	4.153	3.727	1601
400	976.15	4.135	3.714	1620
500	980.34	4.117	3.701	1639
600	984.43	4.100	3.688	1658
800	992.34	4.068	3.663	1694
1000	999.92	4.039	3.639	1730

100°C (ITS-90)

P, bar	density, kg/m³	C_p, kJ/(kg·K)	C_v, kJ/(kg·K)	w, m/s
1	950.927	4.2296	3.7181	1532.5
50	953.28	4.218	3.712	1543
100	955.65	4.206	3.706	1553
150	957.99	4.195	3.699	1563
200	960.30	4.184	3.693	1573
250	962.57	4.174	3.687	1583
300	964.82	4.164	3.681	1593
400	969.21	4.144	3.669	1613
500	973.50	4.126	3.657	1632
600	977.69	4.108	3.645	1652
800	985.76	4.076	3.621	1690
1000	993.47	4.046	3.598	1727

110°C (ITS-90)

P, bar	density, kg/m³	C_p, kJ/(kg·K)	C_v, kJ/(kg·K)	w, m/s
1	943.059	4.2453	3.6684	1519.0
50	945.50	4.233	3.663	1530
100	947.95	4.221	3.657	1540
150	950.36	4.209	3.652	1551
200	952.74	4.198	3.646	1561
250	955.09	4.187	3.641	1572
300	957.40	4.176	3.635	1582
400	961.92	4.155	3.624	1603
500	966.32	4.136	3.613	1623
600	970.61	4.118	3.602	1644
800	978.87	4.084	3.579	1683
1000	986.73	4.053	3.556	1723

120°C (ITS-90)

P, bar	density, kg/m³	C_p, kJ/(kg·K)	C_v, kJ/(kg·K)	w, m/s
1	934.749	4.2639	3.6201	1502.8
50	937.28	4.251	3.615	1514
100	939.83	4.238	3.610	1525
150	942.33	4.225	3.605	1536
200	944.79	4.213	3.600	1547
250	947.22	4.201	3.595	1558
300	949.61	4.190	3.590	1569
400	954.27	4.168	3.580	1591
500	958.81	4.148	3.569	1612
600	963.22	4.128	3.558	1634
800	971.68	4.092	3.536	1676
1000	979.72	4.059	3.512	1717

130°C (ITS-90)

P, bar	density, kg/m³	C_p, kJ/(kg·K)	C_v, kJ/(kg·K)	w, m/s
1	925.997	4.2859	3.5733	1484.1
50	928.64	4.271	3.569	1496
100	931.29	4.257	3.564	1508
150	933.90	4.244	3.560	1519
200	936.46	4.231	3.555	1531
250	938.97	4.218	3.550	1543
300	941.45	4.206	3.545	1554
400	946.28	4.183	3.535	1577
500	950.96	4.161	3.525	1600
600	955.51	4.140	3.514	1623
800	964.20	4.101	3.491	1667
1000	972.44	4.066	3.466	1710

140°C (ITS-90)

P, bar	density, kg/m³	C_p, kJ/(kg·K)	C_v, kJ/(kg·K)	w, m/s
1	916.797	4.3114	3.5279	1463.0
50	919.57	4.296	3.524	1475
100	922.34	4.280	3.520	1488
150	925.06	4.266	3.516	1501
200	927.73	4.251	3.511	1513
250	930.35	4.238	3.506	1525
300	932.92	4.224	3.501	1538
400	937.94	4.199	3.491	1562
500	942.79	4.176	3.480	1586
600	947.48	4.153	3.469	1610
800	956.43	4.112	3.444	1658
1000	964.88	4.073	3.417	1704

150°C (ITS-90)

P, bar	density, kg/m³	C_p, kJ/(kg·K)	C_v, kJ/(kg·K)	w, m/s
1	907.143	4.3408	3.4848	1439.8
50	910.06	4.324	3.481	1453
100	912.97	4.307	3.477	1467
150	915.82	4.291	3.473	1480
200	918.61	4.276	3.469	1493
250	921.34	4.261	3.464	1507
300	924.03	4.246	3.459	1520
400	929.24	4.219	3.448	1546
500	934.27	4.194	3.436	1572
600	939.13	4.169	3.423	1598
800	948.36	4.124	3.395	1648
1000	957.04	4.081	3.364	1698

TABLE 2-356 Specific Heat and Other Thermophysical Properties of Water Substance*

Pressure, bar		Temperature, K														
		300	350	400	450	500	600	700	800	900	1000	1200	1400	1600	1800	2000
1	μ	8.57-4	3.70-4	1.32-5	1.52-5	1.73-5	2.15-5	2.57-5	2.98-5	3.39-5	3.78-5	4.48-5	5.06-5	5.65-5	6.19-5	6.70-5
	c_p	4.18	4.19	1.99	1.97	1.98	2.02	2.09	2.15	2.22	2.29	2.43	2.58	2.73	3.02	3.79
	k	0.614	0.668	0.0268	0.0311	0.0358	0.0464	0.0581	0.0710	0.0843	0.0981	0.13	0.16	0.21	0.33	0.57
	Pr	5.81	2.32	0.980	0.967	0.955	0.936	0.920	0.906	0.891	0.881	0.83	0.80	0.75	0.57	0.45
5	μ	8.57-4	3.70-4	2.17-4	1.49-5	1.72-5	2.15-5	2.57-5	2.98-5	3.39-5	3.78-5	4.45-5	5.06-5	5.65-5	6.19-5	6.70-5
	c_p	4.18	4.19	4.26	2.21	2.10	2.07	2.11	2.16	2.23	2.29	2.43	2.58	2.73	2.98	3.40
	k	0.614	0.668	0.689	0.0335	0.0369	0.0469	0.0585	0.0713	0.0846	0.0984	0.13	0.16	0.20	0.28	0.43
	Pr	5.82	2.32	1.34	0.983	0.973	0.947	0.925	0.907	0.892	0.881	0.83	0.81	0.77	0.65	0.53
10	μ	8.57-4	3.70-4	2.17-4	1.51-4	1.71-4	2.15-5	2.58-5	2.99-5	3.39-5	3.78-5	4.45-5	5.06-5	5.65-5	6.19-5	6.70-5
	c_p	4.18	4.19	4.25	4.39	2.29	2.13	2.13	2.18	2.24	2.30	2.44	2.58	2.73	2.95	3.29
	k	0.615	0.668	0.689	0.677	0.0380	0.0474	0.0590	0.0717	0.0851	0.0988	0.13	0.16	0.20	0.26	0.39
	Pr	5.82	2.32	1.34	0.981	1.028	0.963	0.931	0.908	0.892	0.881	0.84	0.82	0.78	0.70	0.57
20	μ	8.56-4	3.71-4	2.18-4	1.51-4	1.68-4	2.15-5	2.59-5	3.00-5	3.40-5	3.79-5	4.46-5	5.06-5	5.65-5	6.19-5	6.70-5
	c_p	4.17	4.19	4.25	4.39	2.84	2.26	2.19	2.21	2.26	2.32	2.45	2.59	2.73	2.92	3.21
	k	0.616	0.669	0.689	0.679	0.0402	0.0485	0.0599	0.0726	0.0859	0.0996	0.13	0.16	0.20	0.25	0.36
	Pr	5.80	2.32	1.34	0.979	1.19	0.999	0.946	0.912	0.893	0.881	0.84	0.82	0.79	0.72	0.60
40	μ	8.55-4	3.71-4	2.18-4	1.52-4	1.19-4	2.15-5	2.61-5	3.02-5	3.42-5	3.80-5	4.47-5	5.07-5	5.65-5	6.19-5	6.70-5
	c_p	4.17	4.19	4.25	4.38	4.65	2.60	2.32	2.28	2.30	2.34	2.46	2.59	2.73	2.90	3.14
	k	0.617	0.671	0.690	0.680	0.644	0.516	0.0620	0.0744	0.0877	0.101	0.13	0.16	0.19	0.24	0.33
	Pr	5.78	2.31	1.34	0.977	0.862	1.08	0.975	0.924	0.895	0.881	0.84	0.82	0.80	0.73	0.63
60	μ	8.54-4	3.72-4	2.19-4	1.53-4	1.20-4	2.14-5	2.63-5	3.04-5	3.43-5	3.82-5	4.48-5	5.07-5	5.66-5	6.19-5	6.70-5
	c_p	4.16	4.18	4.24	4.37	4.63	3.11	2.47	2.35	2.34	2.37	2.48	2.60	2.73	2.89	3.11
	k	0.619	0.672	0.692	0.682	0.646	0.0561	0.0645	0.0764	0.0895	0.103	0.13	0.16	0.19	0.24	0.32
	Pr	5.74	2.31	1.34	0.976	0.859	1.19	1.008	0.934	0.899	0.879	0.84	0.82	0.81	0.74	0.65
80	μ	8.53-4	3.72-4	2.19-4	1.53-4	1.20-4	2.14-5	2.66-5	3.06-5	3.45-5	3.83-5	4.48-5	5.08-5	5.66-5	6.19-5	6.70-5
	c_p	4.16	4.18	4.24	4.36	4.62	3.88	2.65	2.43	2.39	2.40	2.49	2.61	2.73	2.88	3.09
	k	0.620	0.674	0.693	0.684	0.648	0.0628	0.0672	0.0785	0.0914	0.105	0.13	0.16	0.10	0.24	0.31
	Pr	5.72	2.31	1.34	0.975	0.856	1.33	1.046	0.946	0.902	0.877	0.84	0.83	0.81	0.74	0.66
100	μ	8.52-4	3.73-4	2.20-4	1.53-4	1.21-4	2.14-5	2.69-5	3.08-5	3.47-5	3.85-5	4.49-5	5.08-5	5.66-5	6.19-5	6.70-5
	c_p	4.15	4.17	4.23	4.35	4.60	5.22	2.85	2.52	2.44	2.44	2.50	2.62	2.73	2.88	3.08
	k	0.622	0.675	0.694	0.685	0.651	0.0730	0.0704	0.0807	0.0934	0.107	0.13	0.16	0.19	0.24	0.31
	Pr	5.69	2.31	1.34	0.975	0.853	1.74	1.088	0.960	0.905	0.876	0.84	0.83	0.81	0.74	0.67
150	μ	8.51-4	3.74-4	2.22-4	1.56-4	1.22-4	8.22-5	2.72-5	3.12-5	3.51-5	3.89-5	4.52-5	5.09-5	5.67-5	6.19-5	6.70-5
	c_p	4.14	4.16	4.22	4.34	4.54		3.55	2.74	2.57	2.53	2.54	2.65	2.75	2.88	3.06
	k	0.624	0.678	0.699	0.693	0.657	0.520	0.079	0.086	0.098	0.110	0.14	0.16	0.19	0.23	
	Pr	5.64	2.30	1.34	0.974	0.842		1.22	0.994	0.916	0.891	0.84	0.83	0.82	0.76	
200	μ	8.50-4	3.75-4	2.24-4	1.57-4	1.23-4	8.32-5	2.80-5	3.17-5	3.54-5	3.93-5	4.54-5	5.11-5	5.67-5		
	c_p	4.12	4.15	4.21	4.32	4.51		4.67	3.04	2.71	2.62	2.57	2.67	2.76	2.88	3.05
	k	0.627	0.681	0.702	0.697	0.661	0.525	0.095	0.095	0.104	0.113	0.14	0.16	0.19		
	Pr	5.59	2.29	1.34	0.974	0.833		1.38	1.014	0.925	0.903	0.84	0.83	0.82		
250	μ	8.49-4	3.76-4	2.26-4	1.59-4	1.23-4	8.41-5	2.89-5	3.24-5	3.59-5	3.98-5	4.56-5	5.12-5	5.68-5		
	c_p	4.12	4.14	4.20	4.30	4.49	5.90	6.16	3.40	2.86	2.71	2.61	2.69	2.77	2.89	3.04
	k	0.629	0.683	0.705	0.701	0.672	0.537	0.112	0.103	0.110	0.119	0.136	0.16	0.19		
	Pr	5.57	2.28	1.34	0.974	0.826	0.924	1.590	1.070	0.940	0.910	0.85	0.84	0.82		
300	μ	8.49-4	3.77-4	2.28-4	1.60-4	1.24-4	8.50-5	3.7-5	3.4-5	3.64-5	4.02-5	4.59-5	5.14-5	5.68-5		
	c_p	4.10	4.13	4.19	4.29	4.44	5.60	10.20	3.82	3.03	2.81	2.65	2.72	2.78	2.90	3.04
	k	0.629	0.685	0.708	0.704	0.675	0.548	0.173	0.113	0.113	0.123	0.14				
	Pr	5.53	2.27	1.34	0.973	0.820	0.859	2.18	1.149	0.976	0.917	0.87				
400	μ	8.49-4	3.80-4	2.30-4	1.62-4	1.26-4	8.64-5	5.3-5	3.6-5	3.8-5	4.1-5	4.6-5	5.17-5			
	c_p	4.08	4.12	4.16	4.26	4.42	5.31	13.20	4.86	3.39	3.01	2.70	2.77	2.81	2.91	3.04
	k	0.631	0.689	0.714	0.710	0.676	0.567	0.327	0.145	0.129	0.134	0.15				
	Pr	5.49	2.26	1.34	0.971	0.817	0.799	2.14	1.207	0.999	0.926					
500	μ	8.50-4	3.82-4	2.31-4	1.64-4	1.28-4	8.83-5	5.8-5	4.0-5	4.0-5	4.2-5	4.7-5				
	c_p	4.06	4.10	4.15	4.23	4.38	5.08	8.44	5.70	3.90	3.21	2.77	2.81	2.84	2.92	3.04
	k	0.634	0.695	0.719	0.717	0.693	0.583	0.378	0.186	0.147	0.145					
	Pr	5.44	2.25	1.33	0.971	0.814	0.773	1.30	1.225	1.061	0.932					

TABLE 2-356 Specific Heat and Other Thermophysical Properties of Water Substance (*Concluded*)

Pressure, bar		Temperature, K															
		300	350	400	450	500	600	700	800	900	1000	1200	1400	1600	1800	2000
600	μ	8.51.−4	3.85.−4	2.32.−4	1.66.−4	1.30.−4	9.17.−5	6.5.−5	4.4.−5	4.2.−5	4.4.−5					
	c_p	4.04	4.08	4.13	4.20	4.33	4.92	6.93	6.83	4.19	3.38					
	k	0.639	0.699	0.725	0.725	0.700	0.597	0.420	0.239	0.170	0.159					
	Pr	5.38	2.24	1.32	0.970	0.812	0.755	1.073	1.175	1.035	0.935	2.87	2.86	2.86	2.92	3.04
700	μ	8.52.−4	3.87.−4	2.33.−4	1.69.−4	1.33.−4	9.50.−5	6.9.−5	4.9.−5	4.5.−5	4.6.−5					
	c_p	4.01	4.07	4.12	4.17	4.29	4.78	6.12	6.26	4.62	3.59					
	k	0.644	0.706	0.730	0.732	0.707	0.614	0.442	0.279	0.198	0.177					
	Pr	5.33	2.23	1.32	0.970	0.810	0.739	1.047	1.098	1.010	0.935	2.94	2.91	2.88	2.93	3.05
800	μ	8.53.−4	3.90.−4	2.34.−4	1.72.−4	1.36.−4	9.82.−5	7.3.−5	5.4.−5	4.8.−5	4.8.−5					
	c_p	3.99	4.05	4.10	4.15	4.26	4.67	5.60	6.09	4.77	3.75					
	k	0.648	0.709	0.735	0.736	0.714	0.625	0.478	0.320	0.228	0.193					
	Pr	5.28	2.23	1.31	0.970	0.808	0.725	0.855	1.028	1.003	0.933	3.01	2.96	2.91	2.95	3.05
900	μ	8.54.−4	3.93.−4	2.35.−4	1.74.−4	1.38.−4	1.00.−4	7.6.−5	5.8.−5	5.1.−5	5.0.−5					
	c_p	3.98	4.03	4.08	4.13	4.23	4.57	5.29	5.86	4.85	3.86					
	k	0.651	0.713	0.738	0.742	0.724	0.636	0.496	0.351	0.260	0.210					
	Pr	5.23	2.22	1.30	0.969	0.806	0.712	0.810	0.968	0.950	0.919	3.08	3.00	2.94	2.97	3.06
1000	μ	8.56.−4	3.96.−4	2.36.−4	1.76.−4	1.40.−4	1.02.−4	7.9.−5	6.2.−5	5.4.−5	5.1.−5					
	c_p	3.97	4.02	4.06	4.11	4.20	4.47	5.08	5.51	4.88	3.96					
	k	0.653	0.717	0.743	0.747	0.731	0.650	0.516	0.372	0.288	0.228					
	Pr	5.19	2.22	1.30	0.968	0.804	0.701	0.778	0.918	0.900	0.886	3.16	3.05	2.97	2.98	3.07

°μ = viscosity, Ns/m²; c_p = specific heat at constant pressure, kJ/(kg·K); k = thermal conductivity, W/(m·K); Pr = Prandtl number.

TABLE 2-357 Thermodynamic Properties of Water Substance along the Melting Line

P, bar	T, °C	$10^3 v_f$, m³/kg	h_f, kJ/kg	s_f, kJ/kg·K	c_{pf}, kJ/kg·K	c_{melt}, kJ/kg·K	$10^6\alpha_f$, K⁻¹	$10^6 K_{f,T}$ bar⁻¹
6.117.–5^t	0.0100	1.00021	0	0	4.219	3.969	−67.42	50.90
1.01325	0.0026	1.00016	0.0719	−0.0001	4.218	3.970	−67.17	50.88
50	−0.3618	0.99770	3.5140	−0.0054	4.196	3.997	−54.92	50.30
100	−0.7410	0.99523	6.9794	−0.0110	4.174	4.023	−42.52	49.73
150	−1.1249	0.99278	10.3964	−0.0167	4.152	4.047	−30.24	49.17
200	−1.5166	0.99037	13.7648	−0.0225	4.132	4.070	−18.05	48.63
250	−1.9151	0.98798	17.0843	−0.0285	4.112	4.092	−5.93	48.11
300	−2.3206	0.98562	20.3547	−0.0347	4.092	4.113	6.12	47.59
400	−3.1532	0.98098	26.7472	−0.0474	4.056	4.150	30.09	46.61
500	−4.0156	0.97643	32.9403	−0.0607	4.022	4.184	53.97	45.68
600	−4.909	0.97196	38.932	−0.0747	3.992	4.215	77.87	44.80
800	−6.790	0.96326	50.300	−0.1046	3.937	4.270	126.18	43.19
1000	−8.803	0.95493	60.836	−0.1371	3.893	4.320	175.98	41.74

Condensed from U. Grigull, Private communication, January 18, 1995.

Materials prepared at Technical University München, Germany by U. Grigull and S. Marek. For a table as a function of temperature, see Grigull, U. and S. Marek, *Warme u. Stoff.*, **30** (1994): 1–8.

t = the triple point (at 6.117×10^{-5} bar, 0.01°C); $v_f = 0.0010021$ m³/kg; $\alpha_f = -67.42 \times 10^{-6}$/K.

Other equations for properties are given by Jones, F. E. and G. L. Harris, *J. Res. N.I.S.T.*, **97**, 3 (1992): 335–340, and by Wagner, W. and A. Pruss, *J. Phys. Chem. Ref. Data*, **22,** 3 (1993): 783–787. Steam tables include Walker, W. A., U.S. Naval Ordn. Lab. rept. NOLTR NOLTR-66-217 = AD 651105 (0–1000 bar, 0–150°C), 1967 (72 pp.); Grigull, U., J. Straub, et al., *Steam Tables in S.I. Units* (0.01–1000 bar, 0–1000°C), Springer-Verlag, Berlin, 1990 (133 pp.); Tseng, C. M., T. A. Hamp, et al., Atomic Energy of Canada rept. (30 props, sat liq & vap., 1–220 bar), AECL-5910 1977 (90 pp.). For dissociation, see e.g., Knonicek, V., *Rozpr. Cesko Acad Ved., Rada techn ved* (0.01–100 bar, 1000–5000 K). **77,** 1 (1967). The proceedings of the 10th international conference on the properties of steam were edited by Sytchev, V. V. and A. A. Aleksandrov, Plenum, NY, 1984; and for the 11th conference by Pichal, M. and O. Sifner, Hemisphere, 1989 (550 pp.).

For electrical conductivity, see e.g., Marshall, W. L., *J. Chem. Eng. Data*, **32** (1987): 221–226.

TABLE 2-358   Saturated Xenon*

T, K	P, bar	v_f, m³/kg	v_g, m³/kg	h_f, kJ/kg	h_g, kJ/kg	s_f, kJ/(kg·K)	s_g, kJ/(kg·K)	c_{pf}, kJ/(kg·K)	μ_f, 10⁻⁴ Pa·s	k_f, W/(m·K)
10		2.642.−4		0.19		0.0236		0.058		
20		2.650.−4		1.21		0.0901		0.133		
30		2.661.−4		2.74		0.1510		0.164		
40		2.675.−4		4.47		0.2003		0.178		
50		2.689.−4		6.31		0.2410		0.186		
60		2.704.−4		8.21		0.2755		0.191		
80		2.737.−4		12.14		0.3319		0.202		
100		2.776.−4		16.30		0.3783		0.214		
120		2.820.−4		20.81		0.4197		0.231		
140		2.874.−4		25.67		0.4581		0.251		
160		2.941.−4		30.94		0.4946		0.270		
161.4^m		2.946.−4		31.30		0.4969		0.271		
161.4^m	0.816	3.372.−4	0.1219	48.98	145.5	0.6072	1.206	0.350		
170	1.336	3.439.−4	0.0776	52.01	146.5	0.6253	1.181	0.349	4.50	0.0707
180	2.218	3.523.−4	0.0487	55.52	147.5	0.6452	1.156	0.349	3.99	0.0663
190	3.480	3.615.−4	0.0321	59.04	148.3	0.6641	1.134	0.352	3.51	0.0622
200	5.212	3.715.−4	0.0220	62.61	148.9	0.6820	1.113	0.357	3.09	0.0582
210	7.504	3.828.−4	0.0156	66.25	149.2	0.6994	1.095	0.365	2.71	0.0542
220	10.45	3.955.−4	0.0113	70.00	149.4	0.7163	1.077	0.379	2.39	0.0506
230	14.16	4.100.−4	0.0084	73.91	149.2	0.7330	1.060	0.400	2.09	0.0468
240	18.72	4.271.−4	0.0063	78.05	148.5	0.7498	1.044	0.432	1.83	0.0429
250	24.25	4.476.−4	0.0047	82.54	147.5	0.7671	1.027	0.482	1.60	0.0393
260	30.87	4.730.−4	0.0036	87.52	145.7	0.7855	1.009	0.560	1.38	0.0355
270	38.69	5.079.−4	0.0027	93.30	142.8	0.8058	0.989	0.685	1.18	0.0313
280	47.86	5.689.−4	0.0019	100.6	138.0	0.8308	0.964	0.995	0.95	0.0275
289.7^c	58.21	9.091.−4	0.0009	120.0	120.0	0.8962	0.896	∞		∞

*Values extracted and in some cases rounded off from those cited in Rabinovich (ed.), *Thermophysical Properties of Neon, Argon, Krypton and Xenon*, Standards Press, Moscow, 1976. This source contains values for the compressed state for pressures up to 1000 bar, etc. m = melting point; c = critical point. The notation 2.642.−4 signifies 2.642×10^{-4}. This book was published in English translation by Hemisphere, New York, 1988 (604 pp.).

TABLE 2-359 Compressed Xenon*

T, K		Pressure, bar									
	1	100	200	300	400	500	600	700	800	900	1000
100 v	2.776.−4	2.764.−4	2.752.−4	2.742.−4	2.731.−4	2.721.−4	2.711.−4	2.702.−4	2.693.−4	2.684.−4	2.675.−4
h	16.30	18.84	21.40	23.95	26.50	29.05	31.59	34.13	36.67	39.21	41.74
s	0.3783	0.3762	0.3742	0.3723	0.3704	0.3686	0.3669	0.3652	0.3636	0.3621	0.3802
200 v	0.1245	3.623.−4	3.547.−4	3.484.−4	3.430.−4	3.383.−4	3.342.−4	3.304.−4	3.270.−4	3.240.−4	3.211.−4
h	151.8	64.22	66.14	68.19	70.34	72.56	74.83	77.13	79.46	81.81	84.18
s	1.228	0.6727	0.6643	0.6570	0.6505	0.6446	0.6391	0.6340	0.6292	0.6247	0.6204
300 v	0.1890	5.729.−4	4.769.−4	4.431.−4	4.220.−4	4.068.−4	3.955.−4	3.862.−4	3.783.−4	3.716.−4	3.657.−4
h	168.0	106.4	101.6	101.3	102.0	103.3	104.9	106.7	108.5	110.6	112.8
s	1.294	0.8401	0.8073	0.7908	0.7789	0.7691	0.7608	0.7540	0.7477	0.7424	0.7370
400 v	0.2527	1.998.−3	8.759.−4	6.452.−4	5.604.−4	5.141.−4	4.839.−4	4.622.−4	4.457.−4	4.325.−4	4.217.−4
h	183.9	164.2	145.4	137.4	134.7	134.1	134.5	135.5	136.8	138.3	140.0
s	1.340	1.012	0.9330	0.8945	0.8730	0.8581	0.8467	0.8373	0.8292	0.8220	0.8162
500 v	0.3163	2.899.−3	1.389.−3	9.449.−4	7.577.−4	6.593.−4	5.986.−4	5.570.−4	5.268.−4	5.038.−4	4.859.−4
h	199.8	187.8	177.1	169.4	165.1	163.0	162.3	162.4	163.1	164.3	165.7
s	1.375	1.065	1.004	0.9664	0.9409	0.9228	0.9088	0.8975	0.8881	0.8801	0.8731
600 v	0.3798	3.673.−3	1.823.−3	1.240.−3	9.699.−4	8.206.−4	7.273.−4	6.636.−4	6.172.−4	5.820.−4	5.545.−4
h	215.7	207.4	200.3	194.8	191.1	188.9	187.9	187.6	188.0	188.8	189.9
s	1.404	1.101	1.047	1.013	0.9885	0.9700	0.9555	0.9435	0.9334	0.9247	0.9172
700 v	0.4432	4.397.−3	2.217.−3	1.513.−3	1.175.−3	9.815.−4	8.583.−4	7.734.−4	7.115.−4	6.642.−4	6.268.−4
h	231.5	225.6	220.6	216.7	213.8	212.2	211.3	211.1	211.3	212.0	213.1
s	1.428	1.129	1.078	1.047	1.023	1.006	0.9916	0.9797	0.9695	0.9606	0.9528
800 v	0.5066	5.093.−3	2.587.−3	1.769.−3	1.370.−3	1.137.−3	9.870.−4	8.824.−4	8.057.−4	7.469.−4	7.005.−4
h	247.4	243.0	239.5	236.7	234.8	233.6	233.0	232.9	233.3	234.0	235.0
s	1.450	1.152	1.103	1.073	1.052	1.035	1.021	1.009	0.9988	0.9901	0.9823
900 v	0.5700	5.773.−3	2.944.−3	2.014.−3	1.557.−3	1.288.−3	1.112.−3	9.893.−4	8.989.−4	8.289.−4	7.737.−4
h	263.2	260.1	257.5	255.7	254.4	253.6	253.4	253.7	254.2	254.9	256.1
s	1.468	1.172	1.125	1.096	1.075	1.058	1.045	1.033	1.023	1.015	1.007
1000 v	0.6333	6.441.−3	3.291.−3	2.252.−3	1.738.−3	1.435.−3	1.235.−3	1.094.−3	9.899.−4	9.097.−4	8.461.−4
h	279.1	276.8	275.1	273.9	273.2	272.9	273.0	273.4	274.1	275.1	276.2
s	1.485	1.190	1.143	1.115	1.095	1.079	1.065	1.054	1.044	1.036	1.028

*Values extracted and in some cases rounded off from those cited in Rabinovich (ed.), *Thermophysical Properties of Neon, Argon, Krypton and Xenon.* Standards Press, Moscow, 1976. This source contains an exhaustive tabulation of values. v = specific volume, m^3/kg; h = specific enthalpy, kJ/kg; s = specific entropy, $kJ/(kg \cdot K)$. The notation 2.776.−4 signifies 2.776×10^{-4}. This book was published in English translation by Hemisphere, New York, 1988 (604 pp.).

TABLE 2-360   Surface Tension (N/m) of Saturated Liquid Refrigerants*

R no.	Temperature, °C								
	−50	−25	0	25	50	75	100	125	150
11	0.0279	0.0244	0.0210	0.0178	0.0146	0.0116	0.0087	0.0060	0.0036
12	0.0188	0.0152	0.0118	0.0085	0.0055	0.0029	0.0007	—	—
13	0.0092	0.0056	0.0025	0.0002	—	—	—	—	—
22	0.0197	0.0156	0.0117	0.0081	0.0047	0.0018	—	—	—
23	0.0115	0.0065	0.0025		—	—	—	—	—
32				0.0069	0.0032	0.0002	—	—	—
113	—	0.0231	0.0201	0.0172	0.0144	0.0118	0.0092	0.0067	0.0045
114			0.0138	0.0109	0.0082	0.0056	0.0033	0.0012	—
115			0.0075	0.0047	0.0022	—	—	—	—
134a	0.0192	0.0154	0.0117	0.0082	0.0050	0.0021	0.0000	—	—
142b	0.0213	0.0178	0.0145	0.0113	0.0083	0.0055	0.0029		—
152a	0.0201	0.0166	0.0132	0.0100	0.0068	0.0038	0.0011		—
170	0.0100	0.0051	0.0032	0.0005	—	—	—		—
290			0.0101	0.0082	0.0041	0.0016	—	—	—
C318	—	0.0143	0.0113	0.0085	0.0048	0.0033	0.0011	—	—
502	0.0159	0.0121	0.0086	0.0054	0.0026		—	—	—
503	0.0094	0.0053	0.0018	—	—	—	—	—	—
600		0.0180	0.0150	0.0122	0.0094	0.0068	0.0043	0.0020	0.0001
600a			0.0132	0.0101	0.0073	0.0047	0.0024	0.0005	—
718	—	—	0.0755	0.0720	0.0680	0.0636	0.0590	0.0540	0.0488
744		0.0096	0.0044	0.0005	—	—	—	—	—
1150	0.0100	0.0055	0.0013		—	—	—	—	—
1270	0.0171	0.0136	0.0102	0.0070	0.0041	0.0014	—	—	—

*Dashes indicate inaccessible states; blanks indicate no available data.

Values and equations were given by Srinivasan, K., *Can. J. Chem. Eng.* (27 liquids), **68** (1990): 493; Lielmezs, J. and T. A. Herrick, *Chem. Eng. J.* (34 liquids), **32** (1986): 165–169. Somayajulu, G. R., *Int. J. Thermophys.* (64 liquids); **9**, 4 (1988): 559–566; Ibrahim, N. and S. Murad, *Chem. Eng. Commun.* (29 polar liquids), **79** (1979): 165–174; Yaws, C. L.; Morachevsky, A. G. and I. B. Sladkov, *Physico-Chemical Properties of Molecular Inorganic Compounds* (200 compounds), Khimiya, Leningrad, 1987, Jasper, J., *J. Phys. Chem. Ref. Data* (2200 compounds), **1**, (1972): 841–1009; and Vargaftik, N. B., B. N. Volkov, et al., *J. Phys. Chem. Ref. Data* (water), **12**, 3 (1983): 817–820. See also Escobedo, J. and Mansoori, G. R., *AIChE J.*, **42**(5), May 1996: 1425–1433.

TABLE 2-361   Velocity of Sound (m/s) in Gaseous Refrigerants at Atmospheric Pressure*

R. no.	Temperature, °C								
	−50	−25	0	25	50	75	100	125	150
11	—	—	—	141	147	153	158	163	168
12	—	136	143	150	156	162	168	173	179
13	142	150	157	164	170	176	182	188	193
14	158	166	173	180	187	194	200	206	212
22	—	166	174	182	189	196	202	208	215
23	179	188	197	205	212	220	227	234	240
32									
113	—	—	—	—	121	126	131	135	140
114	—	—	—	120	126	131	136	141	146
134a		146	154	162	169	175	180	186	192
170	272	286	299	311	323	334	344	355	364
290	—	227	238	249	258	268	277	286	294
600	—	—	200	210	220	228	237	245	252
600a	—	—	201	211	221	229	237	246	253
718	—	—	—	—	—	—	473	490	505
744		248	258	269	279	288	297	307	316
1150	290	305	318	330	341	352	363	373	384
1270	—	235	246	257	267	277	286	295	303

*Dashes indicate inaccessible states; blanks indicate no available data.

TABLE 3-362   Velocity of Sound (m/s) in Saturated Liquid Refrigerants*

R. no.	Temperature, °C								
	−50	−25	0	25	50	75	100	125	150
11	933	843	772	705	639	569	493	408	323
12	829	695	564	434			—	—	—
13	602	444	302		—	—	—	—	—
14	182	—	—	—	—	—	—	—	—
22	899	790	682	571	446	319	—	—	—
23		538	348	191	—		—	—	—
32							—	—	—
113	—	871	786	700	633				
114	853	726	623	540	453	371	284	183	—
115			454	346	255		—	—	—
134a	858	743	626	517	387	262	105	—	—
290	1210	982	884	719	551	367	—	—	—
600	1290	1163	1031	896	759	609	477	325	142
600a	1205	1078	947	812	661	528	378	208	—
718	—	—	1402	1495	1542	1554	1543	1514	1468
744		751	525	272	—		—	—	—
1150	874	644	372						
1270	1184	1022	859	694	524	335			

*Dashes indicate inaccessible states; blanks indicate no available data.

TRANSPORT PROPERTIES

INTRODUCTION

Extensive tables of the viscosity and thermal conductivity of air and of water or steam for various pressures and temperatures are given with the thermodynamic-property tables. The thermal conductivity and the viscosity for the saturated-liquid state are also tabulated for many fluids along with the thermodynamic-property tables earlier in this section.

UNITS CONVERSIONS

For this subsection the following units conversions are applicable:

Diffusivity: to convert square centimeters per second to square feet per hour, multiply by 3.8750; to convert square meters per second to square feet per hour, multiply by 38,750.

Pressure: to convert bars to pounds-force per square inch, multiply by 14.504.

Temperature: °F = ⅑ °C + 32; °R = ⅑ K.

Thermal conductivity: to convert watts per meter-kelvin to British thermal unit–feet per hour–square foot–degree Fahrenheit, multiply by 0.57779; and to convert British thermal unit–feet per hour–square foot–degree Fahrenheit to watts per meter-kelvin, multiply by 1.7307.

Viscosity: to convert pascal-seconds to centipoises, multiply by 1000.

ADDITIONAL REFERENCES

An extensive coverage of the general pressure and temperature variation of thermal conductivity is given in the monograph by Vargaftik,

Filippov, Tarzimanov, and Totskiy, *Thermal Conductivity of Liquids and Gases* (in Russian), Standartov, Moscow, 1978, now published in English translation by CRC Press, Miami, FL.

For a similar work on viscosity, see Stephan and Lucas, *Viscosity of Dense Fluids*, Plenum, New York and London, 1979. Tables and polynomial fits for refrigerants in both the gaseous and the liquid state are contained in *ASHRAE Thermophysical Properties of Refrigerants,* American Society of Heating, Refrigerating and Ventilating Engineers, Atlanta, GA, 1993. Other sources for viscosity include Fischer & Porter Co. catalog 10-A-94, "Fluid densities and viscosities," 1953 (200 industrial fluids in 48 pp.) and van Velzen, D., R. L. Cardozo et al., EURATOM Ispra, Italy rept. 4735 e, 1972 (160 pp.). Liquid viscosity, 314 cpds, is summarized in *I&EC Fundtls.,* 11 (1972): 20–26. Five hundred forty-nine binary and ternary systems are discussed in Skubla, P., *Coll. Czech. Chem. Commun.,* 46 (1981): 303–339.

See also Duhne, C. R., *Chem. Eng.* (NY), 86, 15 (July 16, 1979): 83–91 (equations and 326 liquids); and Rao, K. V. K., *Chem. Eng.* (NY), 90, 11 (May 30, 1983): 90–91 (nomograph, 87 liquids). For rheology, non-Newtonian behavior, and the like, see, for instance, Barnes, H., *The Chem. Engr.* (UK), (June 24, 1993): 17–23; Hyman, W. A., *I&EC Fundtls.,* 16 (1976): 215–218; and Ferguson, J. and Z. Kemblowski, *Applied Fluid Rheology*, Elsevier, 1991 (325 pp.). Other sources for thermal conductivity include Ho, C. Y., R. W. Powell et al., *J. Phys. Chem. Ref. Data,* 1 (1972) and 3, suppl. 1 (1974); Childs, Ericks et al., N.B.S. Monogr. 131, 1973; Jamieson, D. T., J. B. Irving et al., *Liquid Thermal Conductivity*, H.M.S.O., Edinburgh, Scotland, 1975 (220 pp.).

TABLE 2-363 Transport Properties of Selected Gases at Atmospheric Pressure*

Substance	Thermal conductivity, W/(m·K) Temperature, K					Viscosity, 10^{-4} Pa·s Temperature, K					Prandtl number, dimensionless Temperature, K			
	250	300	400	500	600	250	300	400	500	600	250	300	400	500
Acetone	0.0080	0.0115	0.0201	0.0310			0.077	0.101	0.128	0.156				
Acetylene	0.0162	0.0213	0.0332	0.0452	0.0561		0.104	0.135	0.164					
Ammonia	0.0197	0.0246	0.0364	0.0506	0.0656	0.085	0.102	0.139	0.175	0.211		0.91	0.87	0.86
Argon	0.0152	0.0177	0.0223	0.0264	0.0301	0.195	0.229	0.289	0.343	0.390	0.669	0.668	0.666	0.663
Benzene	0.0077	0.0104	0.0195	0.0335	0.0524		0.076	0.101	0.127	0.154				
Bromine	0.0038	0.0048	0.0067											
Butane	0.0117	0.0160	0.0264	0.0377			0.076	0.101	0.125	0.151		0.805	0.820	
CO_2	0.0129	0.0166	0.0244	0.0323	0.0403	0.126	0.150	0.196	0.239	0.278	0.793	0.778	0.752	0.734
CCl_4	0.0053	0.0067	0.0099	0.0126			0.101	0.131	0.162	0.191				
Chlorine	0.0071	0.0089	0.0124	0.0156	0.0190		0.136	0.178	0.218	0.259				
Deuterium	0.122	0.141	0.176			0.111	0.126	0.153	0.178	0.201	0.817	0.773	0.746	0.746
Ethane	0.0156	0.0218	0.0360	0.0516	0.0685	0.079	0.094	0.123	0.148	0.171	0.812	0.796	0.769	0.750
Ethylene	0.0152	0.0214	0.0342	0.0491	0.0653	0.087	0.103	0.135	0.162	0.187	0.671	0.668	0.663	0.661
Helium	0.134	0.150	0.180	0.211	0.247	0.176	0.199	0.243	0.284	0.322				
Heptane	0.0082	0.0120	0.0214	0.0325	0.0447		0.080	0.099	1.116					
Hydrogen	0.156	0.182	0.221	0.256	0.291	0.080	0.090	0.109	0.126	0.143	0.71	0.71	0.71	0.71
Methane	0.0277	0.0343	0.0484	0.0671	0.0948	0.095	0.112	0.142	0.170	0.195	0.742	0.739	0.737	0.736
Nitrogen	0.0222	0.0260	0.0325	0.0386	0.0441	0.156	0.180	0.223	0.261	0.295	0.721	0.714	0.708	0.707
Oxygen	0.0225	0.0267	0.0343	0.0412	0.0480	0.179	0.207	0.258	0.306	0.348				
Pentane	0.0107	0.0152	0.0250	0.0362										
Propane	0.0129	0.0183	0.0295	0.0417		0.069	0.082	0.108	0.131		0.810	0.774	0.788	0.826
Propylene	0.0114	0.0168	0.0226	0.0430	0.0580	0.073	0.087	0.115	0.141		0.860	0.797	0.762	
R 11		0.0078	0.0119			0.094	0.110	0.144				0.814	0.761	
R 12	0.0072	0.0097	0.0151	0.0208		0.108	0.126	0.162			0.827	0.781	0.745	0.708
R 13	0.0091	0.0121	0.0185	0.0248		0.123	0.145	0.190			0.796	0.766	0.759	0.757
R 21		0.0088	0.0135	0.0181		0.100	0.115	0.154				0.779	0.773	
R 22	0.0080	0.0109	0.0170	0.0230	0.0290	0.109	0.129	0.168			0.820	0.771	0.760	
SO_2	0.0078	0.0096	0.0143	0.0200	0.0256		0.129	0.175	0.217	0.256				

*An approximate interpolation scheme is to plot the logarithm of the viscosity or the thermal conductivity versus the logarithm of the absolute temperature. At 250 K the viscosity of gaseous argon is to be read as 1.95×10^{-5} Pa·s = 0.0000195 Ns/m².

TABLE 2-364   Viscosities of Gases: Coordinates for Use with Fig. 2-32*

Gas	X	Y	$\mu \times 10^7$ p	Ref.	Gas	X	Y	$\mu \times 10^7$ p	Ref.
Acetic acid	7.0	14.6	825 (50°C)	1	Hydrogen–sulfur dioxide				4
Acetone	8.4	13.2	735	1	10% H_2, 90% SO_2	8.7	18.1	1259 (17)	
Acetylene	9.3	15.5	1017	1	20% H_2, 80% SO_2	8.6	18.2	1277 (17)	
Air	10.4	20.4	1812	1	50% H_2, 50% SO_2	8.9	18.3	1332 (17)	
Ammonia	8.4	16.0	1000	1	80% H_2, 20% SO_2	9.7	17.7	1306 (17)	
Amylene (β)	8.6	12.2	676	1	Hydrogen bromide	8.4	21.6	1843	1
Argon	9.7	22.6	2215	1	Hydrogen chloride	8.5	19.2	1425	1
Arsine	8.6	20.0	1576	1	Hydrogen cyanide	7.1	14.5	737	1
Benzene	8.7	13.2	746	1	Hydrogen iodide	8.5	21.5	1830	1
Bromine	8.8	19.4	1495	1	Hydrogen sulfide	8.4	18.0	1265	1
Butane (n)	8.6	13.2	735	1	Iodine	8.7	18.7	1730 (100)	1
Butane (iso)	8.6	13.2	744	1	Krypton	9.4	24.0	2480	1
Butyl acetate (iso)	5.7	16.3	778	1	Mercury	7.4	24.9	4500 (200)	1
Butylene (α)	8.4	13.5	761	1	Mercuric bromide	8.5	19.0	2253	1
Butylene (β)	8.7	13.1	746	1	Mercuric chloride	7.7	18.7	2200 (200)	1
Butylene (iso)	8.3	13.9	786	1	Mercuric iodide	8.4	18.0	2045 (200)	1
Butyl formate (iso)	6.6	16.0	840	1	Mesitylene	9.5	10.2	660 (50)	1
Cadmium	7.8	22.5	5690 (500)	1	Methane	9.5	15.8	1092	1
Carbon dioxide	8.9	19.1	1463	1	Methane (deuterated)	9.5	17.6	1290	1
Carbon disulfide	8.5	15.8	990	1	Methanol	8.3	15.6	935	1
Carbon monoxide	10.5	20.0	1749	1	Methyl acetate	8.4	14.0	870 (50)	1
Carbon oxysulfide	8.2	17.9	1220	1	Methyl acetylene	8.9	14.3	867	1
Carbon tetrachloride	8.0	15.3	966	1	3-Methyl-1-butene	8.0	13.3	716	1
Chlorine	8.8	18.3	1335	1	Methyl butyrate (iso)	6.6	15.8	824	1
Chloroform	8.8	15.7	1000	1	Methyl bromide	8.1	18.7	1327	1
Cyanogen	8.2	16.2	1002	1	Methyl chloride	8.5	16.5	1062	1
Cyclohexane	9.0	12.2	701	1	3-Methylene-1-butene	8.0	13.3	716	1
Cyclopropane	8.3	14.7	870	1	Methylene chloride	8.5	15.8	989	1
Deuterium	11.0	16.2	1240	1	Methyl formate	5.1	18.0	923	6
Diethyl ether	8.8	12.7	730	1	Neon	11.1	25.8	3113	1
Dimethyl ether	9.0	15.0	925	1	Nitric oxide	10.4	20.8	1899	1
Diphenyl ether	8.6	10.4	610 (50)	1	Nitrogen	10.6	20.0	1766	1
Diphenyl methane	8.0	10.3	605 (50)	1	Nitrous oxide	9.0	19.0	1460	1
Ethane	9.0	14.5	915	1	Nonane (n)	9.2	8.9	554 (50)	1
Ethanol	8.2	14.5	835	1	Octane (n)	8.8	9.8	586 (50)	1
Ethyl acetate	8.4	13.4	743	1	Oxygen	10.2	21.6	2026	1
Ethyl chloride	8.5	15.6	978	1	Pentane (n)	8.5	12.3	668	1
Ethylene	9.5	15.2	1010	1	Pentane (iso)	8.9	12.1	685	1
Ethyl propionate	12.0	12.4	890	1	Phosphene	8.8	17.0	1150	1
Fluorine	7.3	23.8	2250	2	Propane	8.9	13.5	800	1
Freon-11	8.6	16.2	1298 (93)	3	Propanol (n)	8.4	13.5	770	1
Freon-12	9.0	17.4	1496 (93)	3	Propanol (iso)	8.4	13.6	774	1
Freon-14	9.5	20.4	1716	5	Propyl acetate	8.0	14.3	797	1
Freon-21	9.0	16.7	1389 (93)	3	Propylene	8.5	14.4	840	1
Freon-22	9.0	17.7	1554 (93)	3	Pyridine	8.6	13.3	830 (50)	1
Freon-113	11.0	14.0	1166 (93)	3	Silane	9.0	16.8	1148	1
Freon-114	9.4	16.4	1364 (93)	3	Stannic chloride	9.1	16.0	1330 (100)	1
Helium	11.3	20.8	1946	1	Stannic bromide	9.0	16.7	142 (100)	1
Heptane (n)	8.6	10.6	618 (50)	1	Sulfur dioxide	8.4	18.2	1250	1
Hexane (n)	8.4	12.0	644	1	Thiazole	10.0	14.4	958	1
Hydrogen	11.3	12.4	880	1	Thiophene	8.3	14.2	901 (50)	1
Hydrogen-helium				1	Toluene	8.6	12.5	686	1
10% H_2, 90% He	11.0	20.5	1780 (0)		2,2,3-Trimethylbutane	10.0	10.4	691 (50)	1
25% H_2, 75% He	11.0	19.4	1603 (0)		Trimethylethane	8.0	13.0	686	1
40% H_2, 60% He	10.7	18.4	1431 (0)		Water	8.0	16.0	1250 (100)	1
60% H_2, 40% He	10.8	16.7	1227 (0)		Xenon	9.3	23.0	2255	
81% H_2, 19% He	10.5	15.0	1016 (0)		Zinc	8.0	22.0	5250 (500)	1

*Viscosity at 20°C unless otherwise indicated. From Beerman, *Meas. Control* (June 1982): 154–157.

References:

1. I. F. Golubev, *Viscosity of Gases and Gas Mixtures*, Moscow 1959; transl. U.S. Department of Commerce, Clearinghouse for Federal Scientific and Technical Information, Springfield, Va., TT 70-50022, ISPT Cat. No. 5680, Table 4, Jerusalem 1970.

2. R. H. Perry and C. H. Chilton, *Chemical Engineers' Handbook*, 5th ed., McGraw-Hill, New York, 1973, pp. 3-210, 3-211.

3. Ibid., Table 3-282, p. 3-210.

4. By interpolation of data in Ref. 1.

5. *Thermophysical Properties of Refrigerants*, American Society of Heating, Refrigerating and Air-Conditioning Engineers, New York.

6. N. A. Lange, *Handbook of Chemistry*, 4th ed., Handbook Publishers, Sandusky, Ohio, 1941.

For another alignment chart for 165 hydrocarbons from −100 to 500°C, see Sastry, R. C. and A. Satyanarayan, *Chem. Industry Devs.* (July 1978): 11–14.

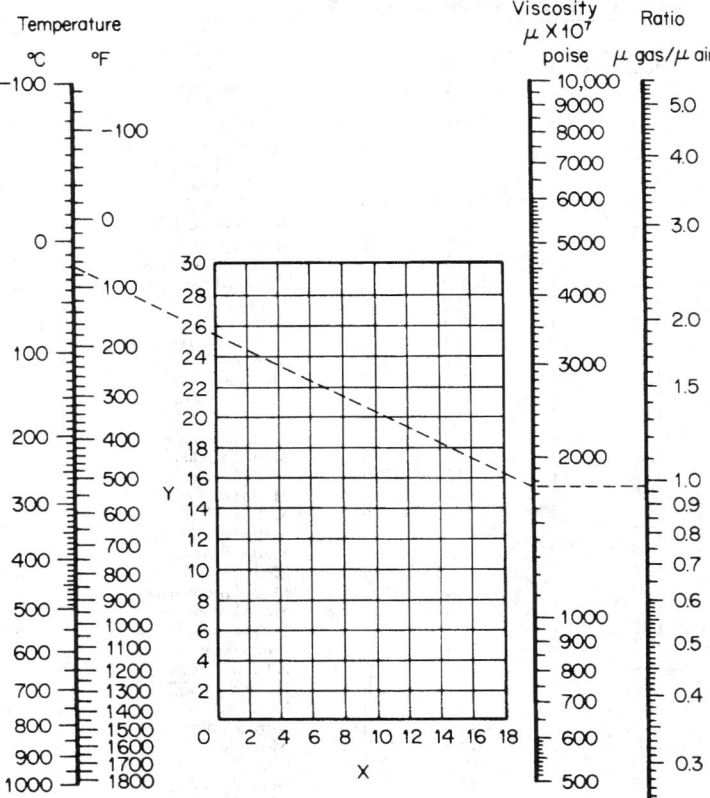

FIG. 2-32 Nomograph for determining (*a*) absolute viscosity of a gas as a function of temperature near ambient pressure and (*b*) relative viscosity of a gas compared with air. For coordinates see Table 2-364. To convert poises to pascal-seconds, multiply by 0.1. [*From Beerman, Meas. Control, 154–157 (June 1982).*]

TABLE 2-365   Viscosities of Liquids: Coordinates for Use with Fig. 2-33

Liquid	X	Y	Liquid	X	Y
Acetaldehyde	15.2	4.8	Freon-113	12.5	11.4
Acetic acid, 100%	12.1	14.2	Glycerol, 100%	2.0	30.0
Acetic acid, 70%	9.5	17.0	Glycerol, 50%	6.9	19.6
Acetic anhydride	12.7	12.8	Heptane	14.1	8.4
Acetone, 100%	14.5	7.2	Hexane	14.7	7.0
Acetone, 35%	7.9	15.0	Hydrochloric acid, 31.5%	13.0	16.6
Acetonitrile	14.4	7.4	Iodobenzene	12.8	15.9
Acrylic acid	12.3	13.9	Isobutyl alcohol	7.1	18.0
Allyl alcohol	10.2	14.3	Isobutyric acid	12.2	14.4
Allyl bromide	14.4	9.6	Isopropyl alcohol	8.2	16.0
Allyl iodide	14.0	11.7	Isopropyl bromide	14.1	9.2
Ammonia, 100%	12.6	2.0	Isopropyl chloride	13.9	7.1
Ammonia, 26%	10.1	13.9	Isopropyl iodide	13.7	11.2
Amyl acetate	11.8	12.5	Kerosene	10.2	16.9
Amyl alcohol	7.5	18.4	Linseed oil, raw	7.5	27.2
Aniline	8.1	18.7	Mercury	18.4	16.4
Anisole	12.3	13.5	Methanol, 100%	12.4	10.5
Arsenic trichloride	13.9	14.5	Methanol, 90%	12.3	11.8
Benzene	12.5	10.9	Methanol, 40%	7.8	15.5
Brine, CaCl₂, 25%	6.6	15.9	Methyl acetate	14.2	8.2
Brine, NaCl, 25%	10.2	16.6	Methyl acrylate	13.0	9.5
Bromine	14.2	13.2	Methyl i-butyrate	12.3	9.7
Bromotoluene	20.0	15.9	Methyl n-butyrate	13.2	10.3
Butyl acetate	12.3	11.0	Methyl chloride	15.0	3.8
Butyl acrylate	11.5	12.6	Methyl ethyl ketone	13.9	8.6
Butyl alcohol	8.6	17.2	Methyl formate	14.2	7.5
Butyric acid	12.1	15.3	Methyl iodide	14.3	9.3
Carbon dioxide	11.6	0.3	Methyl propionate	13.5	9.0
Carbon disulfide	16.1	7.5	Methyl propyl ketone	14.3	9.5
Carbon tetrachloride	12.7	13.1	Methyl sulfide	15.3	6.4
Chlorobenzene	12.3	12.4	Napthalene	7.9	18.1
Chloroform	14.4	10.2	Nitric acid, 95%	12.8	13.8
Chlorosulfonic acid	11.2	18.1	Nitric acid, 60%	10.8	17.0
Chlorotoluene, ortho	13.0	13.3	Nitrobenzene	10.6	16.2
Chlorotoluene, meta	13.3	12.5	Nitrogen dioxide	12.9	8.6
Chlorotoluene, para	13.3	12.5	Nitrotoluene	11.0	17.0
Cresol, meta	2.5	20.8	Octane	13.7	10.0
Cyclohexanol	2.9	24.3	Octyl alcohol	6.6	21.1
Cyclohexane	9.8	12.9	Pentachloroethane	10.9	17.3
Dibromomethane	12.7	15.8	Pentane	14.9	5.2
Dichloroethane	13.2	12.2	Phenol	6.9	20.8
Dichloromethane	14.6	8.9	Phosphorus tribromide	13.8	16.7
Diethyl ketone	13.5	9.2	Phosphorus trichloride	16.2	10.9
Diethyl oxalate	11.0	16.4	Propionic acid	12.8	13.8
Diethylene glycol	5.0	24.7	Propyl acetate	13.1	10.3
Diphenyl	12.0	18.3	Propyl alcohol	9.1	16.5
Dipropyl ether	13.2	8.6	Propyl bromide	14.5	9.6
Dipropyl oxalate	10.3	17.7	Propyl chloride	14.4	7.5
Ethyl acetate	13.7	9.1	Propyl formate	13.1	9.7
Ethyl acrylate	12.7	10.4	Propyl iodide	14.1	11.6
Ethyl alcohol, 100%	10.5	13.8	Sodium	16.4	13.9
Ethyl alcohol, 95%	9.8	14.3	Sodium hydroxide, 50%	3.2	25.8
Ethyl alcohol, 40%	6.5	16.6	Stannic chloride	13.5	12.8
Ethyl benzene	13.2	11.5	Succinonitrile	10.1	20.8
Ethyl bromide	14.5	8.1	Sulfur dioxide	15.2	7.1
2-Ethyl butyl acrylate	11.2	14.0	Sulfuric acid, 110%	7.2	27.4
Ethyl chloride	14.8	6.0	Sulfuric acid, 100%	8.0	25.1
Ethyl ether	14.5	5.3	Sulfuric acid, 98%	7.0	24.8
Ethyl formate	14.2	8.4	Sulfuric acid, 60%	10.2	21.3
2-Ethyl hexyl acrylate	9.0	15.0	Sulfuryl chloride	15.2	12.4
Ethyl iodide	14.7	10.3	Tetrachloroethane	11.9	15.7
Ethyl propionate	13.2	9.9	Thiophene	13.2	11.0
Ethyl propyl ether	14.0	7.0	Titanium tetrachloride	14.4	12.3
Ethyl sulfide	13.8	8.9	Toluene	13.7	10.4
Ethylene bromide	11.9	15.7	Trichloroethylene	14.8	10.5
Ethylene chloride	12.7	12.2	Triethylene glycol	4.7	24.8
Ethylene glycol	6.0	23.6	Turpentine	11.5	14.9
Ethylidene chloride	14.1	8.7	Vinyl acetate	14.0	8.8
Fluorobenzene	13.7	10.4	Vinyl toluene	13.4	12.0
Formic acid	10.7	15.8	Water	10.2	13.0
Freon-11	14.4	9.0	Xylene, ortho	13.5	12.1
Freon-12	16.8	5.6	Xylene, meta	13.9	10.6
Freon-21	15.7	7.5	Xylene, para	13.9	10.9
Freon-22	17.2	4.7			

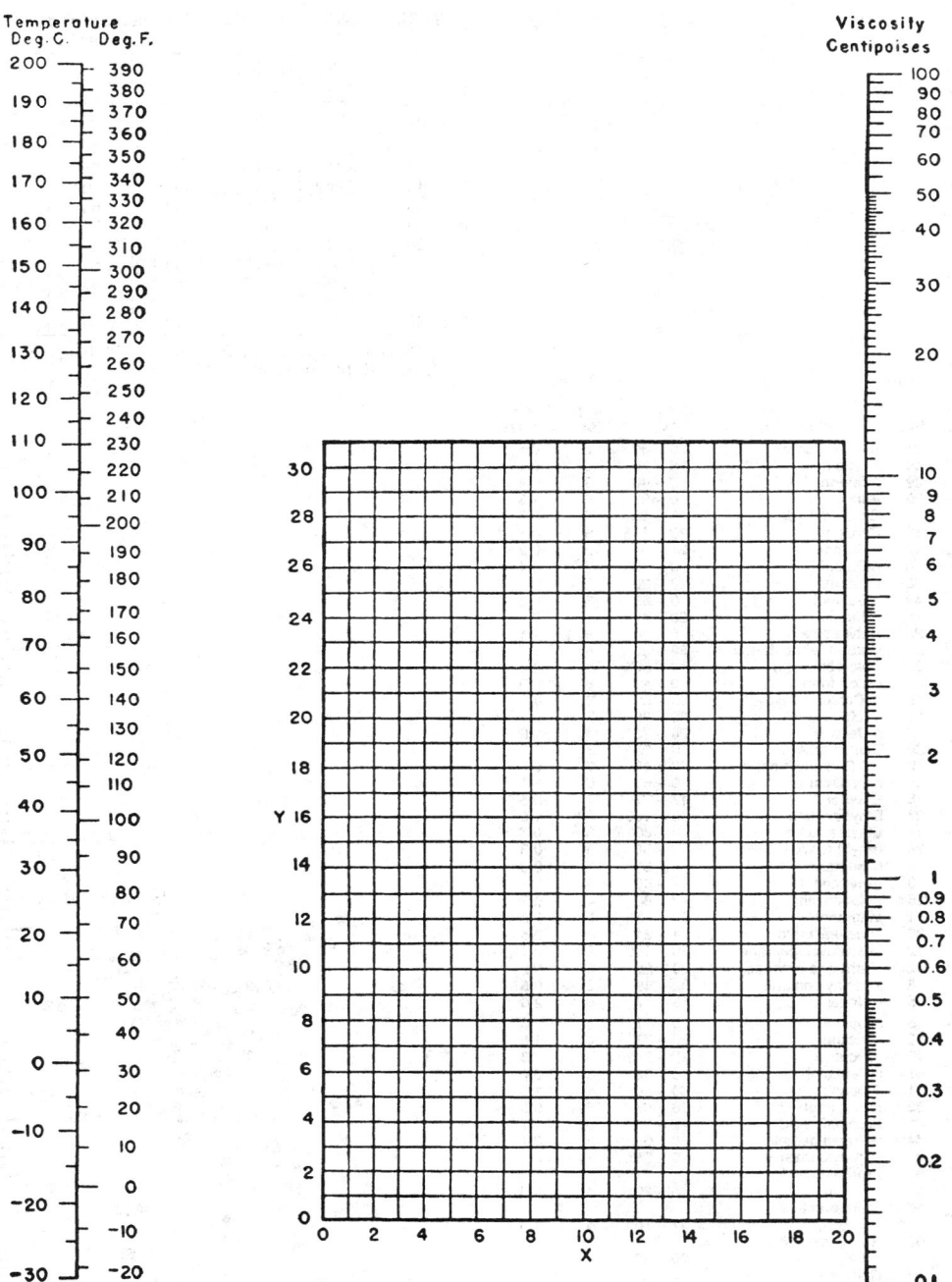

FIG. 2-33 Nomograph for viscosities of liquids at 1 atm. For coordinates see Table 2-365. To convert centipoises to pascal-seconds, multiply by 0.001.

TABLE 2-367

TABLE 2-366   Viscosity of Sucrose Solutions*

Viscosity in centipoises

Temp., °C	Percentage sucrose by weight			Temp., °C	Percentage sucrose by weight		
	20	40	60		20	40	60
0	3.818	14.82		50	0.974	2.506	14.06
5	3.166	11.60		55	0.887	2.227	11.71
10	2.662	9.830	113.9	60	0.811	1.989	9.87
15	2.275	7.496	74.9	65	0.745	1.785	8.37
20	1.967	6.223	56.7	70	0.688	1.614	7.18
25	1.710	5.206	44.02	75	0.637	1.467	6.22
30	1.510	4.398	34.01	80	0.592	1.339	5.42
35	1.336	3.776	26.62	85	0.552	1.226	4.75
40	1.197	3.261	21.30	90		1.127	4.17
45	1.074	2.858	17.24	95		1.041	3.73

International Critical Tables, vol. 5, p. 23. Bingham and Jackson, *Bur. Standards Bull.* **14** (1919): 59.

No.	Compound	Range, °C	Exp pt	% Avg abs devi
31	Acetaldehyde	0 – 30	2	0.4
40	Acetic acid	18 – 79	5	0.1
29	Acetone	0 – 40	3	0.4
20	Aniline	0 – 93	6	1.3
2	Benzaldehyde	16 – 68	4	0.7
13	Benzene	20 – 116	5	0.6
47	n-Butane	−21 – 0	2	0.2
16	n-Butanol	0 – 102	7	0.4
48	i-Butanol	0 – 20	2	0.6
49	s-Butanol	0 – 65	3	0.2
18	t-Butanol	20 – 77	3	2.5
28	Butyl acetate	0 – 38	4	1.1
23	Carbon tetrachloride	−20 – 25	4	1.1
25	Chlorobenzene	−40 – 80	6	0.5
42	Chloroform	0 – 40	3	1.3
19	m-Cresol	20 – 80	2	0.2
11	Cyclohexane	20 – 38	2	0.4
7	n-Decane	20 – 76	5	1.0
30	Diethyl ether	0 – 25	3	0.7
9	2,3-Dimethylbutane	32 – 49	2	0.3
38	n-Dodecane	0 – 63	6	1.1
15	Ethanol	17 – 77	7	0.5
27	Ethyl acetate	6 – 160	9	1.6
14	Ethylbenzene	0 – 80	3	0.7
24	Ethyl bromide	0 – 30	3	0.3
4	n-Heptane	0 – 60	3	0.3
3	n-Hexane	17 – 40	5	1.4
26	Iodobenzene	−20 – 80	4	0.7
41	Methanol	20 – 62	6	0.2
39	Methyl acetate	4 – 21	2	0.4
12	Methylcyclopentane	20 – 38	2	0.1
22	Methylene chloride	−20 – 20	3	0.9
8	2-Methylpentane	32 – 49	2	0.5
6	n-Nonane	16 – 77	5	2.0
5	n-Octane	0 – 77	3	0.3
10	i-Octane	20 – 77	4	1.6
17	n-Octanol	20 – 212	4	1.3
43	n-Pentanol	0 – 38	3	0.4
1	Propane	−60 – 50	4	1.6
32	n-Propanol	16 – 66	6	0.6
21	Propionic acid	12 – 30	3	1.7
44	n-Propyl acetate	0 – 38	2	0.1
33	i-Propylbenzene	0 – 38	4	0.2
45	Refrigerant-11, CFCl₃	0 – 20	4	0.5
46	Refrigerant-12, CF₂Cl₂	−42 – 70	4	1.8
50	Refrigerant-13, CF₃Cl	−20 – 20	4	1.7
51	Refrigerant-22, CHF₂Cl	−65 – 60	7	2.5
52	Refrigerant-113, C₂F₃Cl₃	0 – 20	4	0.7
53	Refrigerant-114, C₂F₄Cl₂	−25 – 20	4	0.8
54	Refrigerant-142, C₂H₃F₂Cl	0 – 90	2	0.9
34	Toluene	−70 – 200	5	1.8
55	n-Tridecane	20 – 228	5	2.4
36	m-Xylene	0 – 208	5	1.0
35	o-Xylene	0 – 208	5	0.7
37	p-Xylene	0 – 251	5	1.4

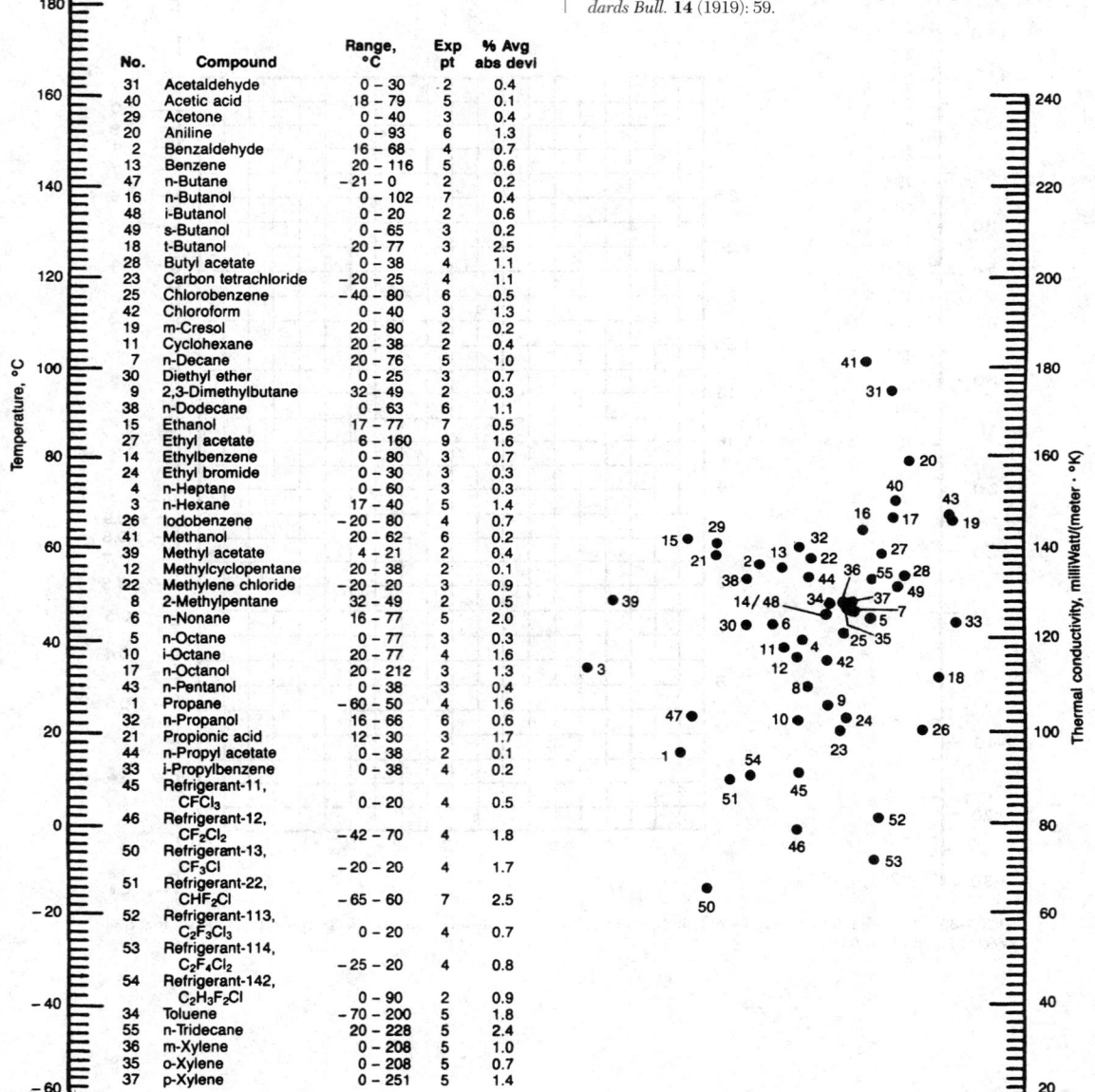

FIG. 2-34 and TABLE 2-367   Nomograph (*right*) for thermal conductivity of organic liquids.

TABLE 2-368   Prandtl Number of Air*

Temperature, K	Pressure, bar											
	1	5	10	20	30	40	50	60	70	80	90	100
80	mix	2.31	2.32	2.35	2.37	2.40	2.42	2.45	2.48	2.51	2.54	2.57
90	0.796	1.76	1.77	1.78	1.79	1.81	1.82	1.83	1.85	1.87	1.89	1.91
100	0.786	0.872	1.54	1.53	1.53	1.53	1.53	1.53	1.53	1.54	1.54	1.55
120	0.773	0.813	0.89	1.44	1.65	1.54	1.48	1.43	1.40	1.38	1.36	1.34
140	0.763	0.782	0.82	0.94	1.20	1.59	2.14	2.43	2.07	1.78	1.62	1.52
160	0.754	0.765	0.78	0.84	0.92	1.03	1.13	1.25	1.37	1.65	1.83	1.72
180	0.745	0.754	0.763	0.792	0.830	0.876	0.932	1.00	1.07	1.14	1.20	1.25
200	0.738	0.743	0.749	0.766	0.788	0.812	0.841	0.87	0.90	0.95	0.97	1.00
240	0.724	0.727	0.729	0.737	0.746	0.756	0.767	0.78	0.80	0.81	0.81	0.82
280	0.710	0.711	0.713	0.717	0.721	0.726	0.731	0.737	0.742	0.75	0.75	0.76
300	0.705	0.707	0.708	0.712	0.715	0.717	0.721	0.725	0.728	0.732	0.737	0.742
350	0.699	0.699	0.699	0.701	0.703	0.705	0.707	0.709	0.711	0.712	0.714	0.716
400	0.694	0.694	0.694	0.695	0.696	0.697	0.698	0.699	0.700	0.701	0.703	0.704
450	0.691	0.691	0.691	0.691	0.692	0.692	0.693	0.693	0.694	0.695	0.695	0.696
500	0.689	0.689	0.689	0.689	0.689	0.690	0.690	0.690	0.690	0.691	0.691	0.691
600	0.690	0.690	0.690	0.689	0.689	0.689	0.689	0.689	0.689	0.690	0.690	0.690
700	0.696	0.696	0.695	0.695	0.695	0.695	0.695	0.695	0.695	0.695	0.695	0.695
800	0.705	0.704	0.704	0.704	0.704	0.703	0.703	0.703	0.703	0.702	0.702	0.702
900	0.709	0.709	0.708	0.708	0.708	0.708	0.708	0.708	0.708	0.708	0.708	0.708
1000	0.711	0.711	0.711	0.711	0.711	0.711	0.710	0.710	0.710	0.710	0.709	0.709

*Compiled by P. E. Liley from tables of specific heat at constant pressure, thermal conductivity, and viscosity given in SI units for integral kelvin temperatures and pressures in bars by Vasserman. *Thermophysical Properties of Air and Its Components* and *Thermophysical Properties of Liquid Air and Its Components*. Nauka, Moscow, and in translated form by the National Bureau of Standards, Washington. The number of significant figures given above reflects the similar numbers appearing for the constituent properties in the source references. While reasonable agreement occurs for atmospheric pressure with some other works, the fragmentary data available for the saturated, etc., states show large deviations.

TABLE 2-369   Prandtl Number of Liquid Refrigerants*

Refrigerant	No.	Temperature, K										
		180	200	220	240	260	280	300	320	340	360	380
Trichlorofluoromethane	11		11.9	8.64	6.73	5.33	4.74	4.18				
Dichlorodifluoromethane	12	7.00	5.25	4.27	3.65	3.27	3.08	3.04	3.19	3.44	4.00	
Chlorotrifluoromethane	13		2.96	2.67	2.69	3.05	3.57		—	—	—	—
Bromotrifluoromethane	13B1	4.80	3.75	3.27	2.94	2.83	3.03	3.61	4.52			
Dichlorofluoromethane	21			5.72	4.50	3.87	3.48	3.25	3.16	3.17		
Chlorodifluoromethane	22	4.68	3.76	3.23	2.93	2.79	2.77	2.87	3.18	3.54		—
Methyl chloride	40			2.53	2.42	2.40	2.45	2.60	2.85			
Trichlorotrifluoroethane	113	—	—	—					7.04	6.23	5.61	5.18
Dichlorotetrafluoroethane	114	25.7	15.13	11.18	8.59	6.94	5.77	5.06	4.78	4.82		
Chloropentafluoroethane	115	—	7.85	6.16	5.21	4.67	4.40	4.46	4.90		—	—
Ethane	170	2.55	2.29	2.22	2.40	2.70			—	—	—	—
Propane	290	5.28	4.46	3.88	3.44	3.16	3.02	3.16				
Octafluorocyclobutane	C318	—	—	—	11.2	8.74	7.35	6.37	5.87	5.96		—
Dichlorodifluoromethane/difluoroethane	500		5.78	4.23	3.40	3.13	3.01	3.13	3.35	3.72		—
Chlorodifluoromethane/chloropentafluoroethane	502		5.73	4.71	4.13	3.81				—	—	—
Trifluoromethane/chlorotrifluoroethane	503	2.10	2.09	2.24	2.43	2.89			—	—	—	—
Methylene fluoride/chloropentafluoroethane	504		4.90	3.60	3.04	2.79	2.69	2.85	3.30	—	—	—
Butane	600	8.35	6.19	5.20	4.44	3.83	3.44	3.22	3.07	3.02		
Isobutane (2-methyl propane)	600a		8.26	6.36	5.18	4.49	3.93	3.66	3.53	3.53	3.77	4.68
Ammonia	717	—			1.97	1.76	1.54	1.40	1.29	1.24	1.25	1.34
Water	718	—	—	—	—	—	10.3	5.69	3.65	2.60	1.99	1.59
Ethylene	1150	1.85	1.74	1.78	2.07	2.70	4.4		—	—	—	—
Propylene	1270	3.80	2.24	1.88	1.71	1.71	1.88	2.24	3.91	4.73		—

*Dashes indicate inaccessible states. Average uncertainty is about 20 percent. Values derived from formulations for thermal conductivity, specific heat at constant pressure, and viscosity contained in *Thermophysical Properties of Refrigerants*. American Society of Heating, Refrigerating and Air-Conditioning Engineers, New York, 1976. For further details see M. W. Johnson, M.S.M.E. thesis, Purdue University, West Lafayette, Ind., 1976.

TABLE 2-370 Thermophysical Properties of Miscellaneous Saturated Liquids

Substance	Property	Temperature, °C															
		−50	−40	−30	−20	−10	0	10	20	30	40	50	60	70	80	90	100
Acetaldehyde	ρ (kg/m³)	863	852	840	828	816	804	794	783								
	c_p (kJ/kg·K)	2.05	2.08	2.11	2.14	2.17	2.20	2.24	2.28								
	μ (10⁻⁶Pa·s)	460	404	358	321	290	263	241	222								
	k (W/m·K)	0.211	0.206	0.200	0.195	0.189	0.184	0.182	0.180								
	Pr	4.47	4.08	3.78	3.52	3.33	3.14	2.97	2.81								
Acetic acid	ρ (kg/m³)								1049	1039	1028	1018	1006	995	984	972	960
	c_p (kJ/kg·K)								2.031								
	μ (10⁻⁶Pa·s)								1210	1102	1010	795	600				
	k (W/m·K)								0.173	0.170	0.168	0.167	0.165	0.163	0.161		
	Pr								14.2								
Aniline	ρ (kg/m³)	—	—	—	—	—	1039	1030	1022	1013	1005	996	987	978	969	960	951
	c_p (kJ/kg·K)	—	—	—	—	—	2.024	2.047	2.071	2.093	2.113	2.132	2.17	2.20	2.23	2.27	2.32
	μ (10⁻⁶Pa·s)	—	—	—	—	—	10200	6500	4400	3160	2370	1850	1510	1270	1090	935	825
	k (W/m·K)	—	—	—	—	—	0.186	0.184	0.182	0.180	0.177	0.174	0.171	0.169	0.168	0.167	0.167
	Pr	—	—	—	—	—	111	72	50	36.7	28.3	22.7	19.2	16.5	14.5	12.7	11.5
Butanol	ρ (kg/m³)	845	841	837	833	829	825	817	810	803	797	791	784	776	768	760	753
	c_p (kJ/kg·K)	1.947	1.996	2.046	2.100	2.153	2.202	2.262	2.345	2.437	2.524	2.621					
	μ (10⁻⁶Pa·s)	34700	22400	14700	10300	7400	5190	3870	2950	2300	1780	1410	1140	930	760	630	535
	k (W/m·K)	0.175	0.174	0.173	0.172	0.171	0.170	0.168	0.167	0.166	0.165	0.164	0.163	0.162	0.161	0.160	0.159
	Pr	3860	2570	1740	1260	930	670	120	41	33.8	27.2	22.5					
Carbon disulfide	ρ (kg/m³)	1362	1348	1334	1320	1306	1292	1278	1263								
	c_p (kJ/kg·K)	0.988	0.989	0.990	0.991	0.993	0.996	1.004	1.017								
	μ (10⁻⁶Pa·s)	630	580	535	496	463	435	405	375	350	330						
	k (W/m·K)	0.194	0.190	0.186	0.182	0.178	0.174	0.170	0.166	0.161	0.158	0.156	0.154	0.152	0.150		
	Pr	3.21	3.02	2.85	2.70	2.58	2.49	2.39	2.30								
Cyclohexane	ρ (kg/m³)	—	—	—	—	—	—	789	779	769	759	750	740	731	721		
	c_p (kJ/kg·K)	—	—	—	—	—	—	2.068	2.081	2.094	2.106	2.119					
	μ (10⁻⁶Pa·s)	—	—	—	—	—	—	1175	980	820	710	605	540				
	k (W/m·K)	—	—	—	—	—	—	0.122	0.120	0.119	0.118	0.117	0.116	0.114	0.112		
	Pr	—	—	—	—	—	—	19.9	17.0	14.4	12.7	11.0					
Ethanol	ρ (kg/m³)						806	798	789	781	776	763	754	745	735	725	716
	c_p (kJ/kg·K)	2.01	2.04	2.08	2.13	2.19	2.27	2.35	2.43	2.52	2.62	2.73	2.83	2.93	3.03	3.19	3.30
	μ (10⁻⁶Pa·s)	6400	4790	3650	2825	2220	1770	1470	1200	1000	835	700	590	500	435	370	314
	k (W/m·K)	0.188	0.186	0.184	0.181	0.179	0.177	0.175	0.173	0.171	0.168	0.165	0.162	0.159	0.156	0.153	0.151
	Pr	68.4	52.5	41.3	33.2	27.2	22.7	19.7	16.9	14.7	13.0	11.6	10.3	9.2	8.4	7.7	6.9
Ethyl acetate	ρ (kg/m³)				947	935	924	912	901	888	876	863	851	838	825	811	797
	c_p (kJ/kg·K)								2.01								
	μ (10⁻⁶Pa·s)	1090					580	510	455	400	370	345	310	280	250	230	220
	k (W/m·K)								0.145	0.142	0.139	0.136	0.133	0.130	0.127	0.123	0.119
	Pr								6.3								
Ethylamine	ρ (kg/m³)	761	750	739	729	718	707	695	683	671	658	646	633	620	607		
	c_p (kJ/kg·K)	2.95	2.97	2.98	3.00	3.01	3.03										
	μ (10⁻⁶Pa·s)	580	500	435	390	350	320										
	k (W/m·K)	0.204	0.201	0.199	0.196	0.194	0.191										
	Pr	8.39	7.39	6.51	5.97	5.43	5.08										
Ethyl ether	ρ (kg/m³)	790	780	769	758	747	736	725	714	702	689	676	666	653	640	625	611
	c_p (kJ/kg·K)	2.135	2.156	2.179	2.205	2.233	2.265	2.299	2.332	2.36	2.39	2.43	2.47	2.51			
	μ (10⁻⁶Pa·s)	550	470	410	365	330	290	265	233	214	197	181	166	153	140	129	118
	k (W/m·K)	0.159	0.155	0.151	0.147	0.144	0.140	0.139	0.134	0.129	0.125	0.120	0.116	0.112			
	Pr	7.39	6.54	5.92	5.48	5.12	4.69	4.38	4.05	3.92	3.77	3.67	3.54	3.43			
Ethyl iodide	ρ (kg/m³)																
	c_p (kJ/kg·K)				0.656	0.663	0.670	0.677	0.684	0.691	0.698	0.705	0.712	0.718	0.724		
	μ (10⁻⁶Pa·s)							730	655	590	539	495	455	420	390		
	k (W/m·K)							0.092	0.090	0.088	0.086	0.085	0.083	0.081	0.080		
	Pr							5.37	4.98	4.63	4.30	4.11	3.90	3.72	3.53		
Ethylene glycol	ρ (kg/m³)						1127	1120	1113	1106	1099	1092	1085	1077	1070	1063	1056
	c_p (kJ/kg·K)						2.272	2.327	2.381	2.431	2.484	2.536	2.586	2.636	2.685	2.734	2.779
	μ (10⁻⁶Pa·s)						57000	33300	20200	13400	9100	7070	4000	3450	3000	2440	2000
	k (W/m·K)						0.254	0.255	0.256	0.258	0.259	0.260					
	Pr						510	305	190	126	87.3	69.0					
Formic acid	ρ (kg/m³)						1241	1231	1220	1209	1196	1184	1170	1156	1140	1124	1108
	c_p (kJ/kg·K)																
	μ (10⁻⁶Pa·s)							2260	1800	1470	1220	1030	890	780	680	615	550
	k (W/m·K)						0.265	0.261	0.257	0.257	0.253	0.250	0.246	0.243	0.240	0.236	0.232
	Pr																

TABLE 2-370 Thermophysical Properties of Miscellaneous Saturated Liquids (*Concluded*)

Temperature, °C

Substance	Property	−50	−40	−30	−20	−10	0	10	20	30	40	50	60	70	80	90	100
Gasoline	ρ (kg/m^3)				784	775	767	759	751	743	735	721	717	708	699	690	681
	c_p(kJ/kg·K)				1.88	1.92	1.97	2.02	2.06	2.11	2.15	2.20	2.25	2.30	2.35	2.41	2.46
	μ (10^{-6}Pa·s)	1710	1400	1170	990	850	735	645	530	464	410	367	330	298	270	246	225
	k (W/m·K)	0.131	0.128	0.125	0.123	0.121	0.120	0.118	0.116	0.114	0.112	0.110	0.108	0.106	0.104	0.102	0.100
	Pr				15.1	13.5	12.1	11.0	9.41	8.59	7.87	7.34	6.88	6.47	6.10	5.81	5.54
Glycerol	ρ (kg/m^3)	—	—	—	—	—	1276	1270	1260	1254	1248	1242					
	c_p(kJ/kg·K)								2.393	2.406	2.457	2.504	2.548	2.588	2.625	2.657	2.686
	μ (10^{-6}Pa·s)						1.2.+7	4.0.+6	1.5.+6								
	k (W/m·K)								0.284	0.285	0.287	0.288	0.289	0.291	0.293	0.294	0.295
	Pr						12650										
Kerosine	ρ (kg/m^3)						781	774	767	760	754	748	742				
	c_p(kJ/kg·K)						1.91	1.96	2.02	2.07	2.13	2.18	2.23	2.28	2.32	2.35	2.38
	μ (10^{-6}Pa·s)	1150	725	500	360	275	215	173	149	126	108	95	83	73	66	60	55
	k (W/m·K)						0.140	0.139	0.139	0.138	0.138	0.137	0.137				
	Pr						2.93	2.44	2.17	1.89	1.67	1.51	1.35				
Methanol	ρ (kg/m^3)									783	774	766	756	746	736	725	711
	c_p(kJ/kg·K)	2.30	2.32	2.35	2.37	2.40	2.42	2.45	2.47	2.49	2.52	2.55	2.65	2.78	2.94	3.13	3.30
	μ (10^{-6}Pa·s)	2305	1800	1410	1170	975	820	692	590	510	455	400	355	315	271	240	218
	k(W/m·K)	0.225	0.222	0.219	0.216	0.212	0.209	0.206	0.203	0.199	0.195	0.192	0.189	0.187	0.184	0.182	0.180
	Pr	23.6	18.8	15.1	12.9	11.0	9.53	8.23	7.18	6.38	5.88	5.31	4.98	4.68	4.34	4.13	3.99
Methyl formate	ρ (kg/m^3)	1069	1056	1043	1030	1017	1003	989	975	960	944	929	913	897	880	863	845
	c_p(kJ/kg·K)	1.84	1.86	1.88	1.90	1.92	1.95	1.99	2.03	2.08							
	μ (10^{-6}Pa·s)	830	711	618	544	481	430	380	345	315							
	k (W/m·K)	0.217	0.213	0.209	0.205	0.200	0.195	0.191	0.186	0.180							
	Pr	7.04	6.21	5.56	5.04	4.62	4.30	3.96	3.77	3.64							
Oil, castor	ρ (kg/m^3)																
	c_p(kJ/kg·K)																
	μ (10^{-6}Pa·s)								2,420,000	986,000	451,000	231,000	125,000	74,000	43,000		
	k (W/m·K)						0.182	0.181	0.180	0.179	0.178	0.177	0.176		0.175	0.174	0.17
	Pr																
Oil, olive	ρ (kg/m^3)								914								
	c_p(kJ/kg·K)								1.633								
	μ (10^{-6}Pa·s)						138,000	84,000	52,000	36,300	24,500	17,000	12,400				
	k (W/m·K)						0.170	0.169	0.168	0.167	0.166	0.166	0.165		0.165	0.164	0.164
	Pr								810								
Pentane	ρ (kg/m^3)	693	684	674	665	656	646	636	626	616	606	596	585	574	562	550	538
	c_p(kJ/kg·K)	2.060	2.084	2.110	2.137	2.167	2.206	2.239	2.273								
	μ (10^{-6}Pa·s)	489	428	379	339	307	279	254	234	209	190	175	161	148	137	124	113
	k (W/m·K)	0.142	0.139	0.136	0.132	0.128	0.125	0.122	0.119	0.115	0.112	0.108	0.105	0.101	0.098	0.095	0.091
	Pr	7.14	6.42	5.88	5.49	5.20	4.92	4.66	4.47								
Propanol	ρ (kg/m^3)	849					819	811	814	796	788	779	770	761	752	747	743
	c_p(kJ/kg·K)	1.955					2.219										
	μ (10^{-6}Pa·s)	20,200	13,500	9500	6900	5110	3900	2900	2245	1720	1400	1130	921	760	630	508	447
	k (W/m·K)	0.167	0.166	0.165							0.171	0.169	0.168	0.167	0.165	0.164	0.162
	Pr	236															
Sulfuric acid	ρ (kg/m^3)								1834								
	c_p(kJ/kg·K)								1.382								
	μ (10^{-6}Pa·s)						48,400	35,200	25,400	15,700	11,500	8820	7220	6090	5190		
	k (W/m·K)						0.314										
	Pr																
Toluene	ρ (kg/m^3)	932	923	913	904	895	886	876	867	858	848	839	829	820	810	800	790
	c_p(kJ/kg·K)	1.514	1.535	1.556	1.579	1.602	1.633	1.652	1.675	1.701	1.73	1.76	1.80	1.83	1.87	1.92	1.97
	μ (10^{-6}Pa·s)	2120	1670	1345	1100	915	770	670	590	520	470	420	380	355	325	295	270
	k (W/m·K)	0.152	0.149	0.147	0.144	0.142	0.139	0.137	0.134	0.132	0.129	0.126	0.124	0.122	0.119	0.117	0.114
	Pr	21.1	17.8	14.2	12.1	10.3	9.0	8.1	7.4	6.7	6.3	5.9	5.5	5.3	5.1	4.8	4.7
Turpentine	ρ (kg/m^3)																
	c_p(kJ/kg·K)						1.72	1.76	1.80			1.93					
	μ (10^{-6}Pa·s)						2250	1780	1490	1270	1070	925	820	730	675		
	k (W/m·K)						0.130	0.129	0.128	0.127	0.126	0.125					
	Pr						29.8	24.3	20.9	18.4	16.1	14.3					

TABLE 2-371 Diffusivities of Pairs of Gases and Vapors (1 atm)

D_v in cm²/s

Substance	Temp., °C	Air	A	H$_2$	O$_2$	N$_2$	CO$_2$	N$_2$O	CH$_4$	C$_2$H$_6$	C$_2$H$_4$	n-C$_4$H$_{10}$	i-C$_4$H$_{10}$	Ref.
Acetic acid	0	0.1064		0.416			0.0716							8
Acetone	0	.109		.361										6, 16
n-Amyl alcohol	0	.0589		.235			.0422							8
sec-Amyl alcohol	30	.072												5
Amyl butyrate	0	.040												8
Amyl formate	0	.0543												8
i-Amyl formate	0	.058												8
Amyl isobutyrate	0	.0419		.171										8
Amyl propionate	0	.046		.1914			.0347							8
Aniline	0	.0610												8
	30	.075												5
Anthracene	0	.0421												8
Argon	20					0.194								18
Benzene	0	.077		.306	0.0797		.0528							8, 15
Benzidine	0	.0298												8
Benzyl chloride	0	.066												8
n-Butyl acetate	0	.058												8
i-Butyl acetate	0	.0612		.2364			.0425							8
n-Butyl alcohol	0	.0703		.2716			.0476							8
	30	.088												5
i-Butyl alcohol	0	.0727		.2771			.0483							8
Butyl amine	0	.0821												8
i-Butyl amine	0	.0853												8
i-Butyl butyrate	0	.0468		.185			.0327							8
i-Butyl formate	0	.0705												8
i-Butyl isobutyrate	0	.0457		.191			.0364							8
i-Butyl proprionate	0	.0529		.203			.0366							8
i-Butyl valerate	0	.0424		.173			.0308							8
Butyric acid	0	.067		.264			.0476							8
i-Butyric acid	0	.0679		.271			.0471							8
Cadmium	0					.17								13
Caproic acid	0	.050												8
i-Caproic acid	0	.0513												8
Carbon dioxide	0	.138		.550	.139			0.096	0.153					8
	20					.163								19
	25							.0996°	.00215†					1, 9
	500†				.9									18
Carbon disulfide	0	.0892		.369			.063							8
Carbon monoxide	0			.651	.185		.137				0.116			8
	450†			1.0										18
Carbon tetrachloride	0			.293	0.0636									16, 17
Chlorobenzene	30	.075												5
Chloroform	0	.091												6
Chloropicrin	25	.088												10
m-Chlorotoluene	0	.054												8
o-Chlorotoluene	0	.059												8
p-Chlorotoluene	0	.051												8
Cyanogen chloride	0	.111												10
Cyclohexane	15		0.0719	.319	.0744	.0760								3
	45	.086												6
n-Decane	90			.306		.0841								3
Diethylamine	0	.0884												8
2,3-Dimethyl butane	15		.0657	.301	.0753	.0751								3
Diphenyl	0	.0610												8
n-Dodecane	126			.308		.0813								3
Ethane	0			.459										8
Ethanol	0			.377			.0686							20
Ether (diethyl)	0	.0778		.298			.0546							7, 8
Ethyl acetate	0	.0715		.273			.0487							8
	30	.089												5
Ethyl alcohol	0	.102		.375			.0685							8
Ethyl benzene	0	.0658												8
Ethyl n-butyrate	0	.0579		.224			.0407							8
Ethyl i-butyrate	0	.0591		.229			.0413							8
Ethylene	0			.486										8
Ethyl formate	0	.0840		.337			.0573							8
Ethyl propionate	0	.068		.236			.0450							4, 8
Ethyl valerate	0	.0512		.205			.0367							8
Eugenol	0	.0377												8
Formic acid	0	.1308		.510			.0874							8
Helium	0		.641											8
	20					.705								19
n-Heptane	38								.066§					
n-Hexane	15		.0663	.290	.0753	.0757								3
Hexyl alcohol	0	.0499		.200			.0351							8
Hydrogen	0	.611			.697	.674	.550	.535	.625	0.459	0.486	0.272	0.277	8
	25						.646			.537	.726			2
	500				4.2									18

TABLE 2-371 Diffusivities of Pairs of Gases and Vapors (1 atm) *(Concluded)*

D_v in cm^2/s

Substance	Temp., °C	Air	A	H$_2$	O$_2$	N$_2$	CO$_2$	N$_2$O	CH$_4$	C$_2$H$_6$	C$_2$H$_4$	n-C$_4$H$_{10}$	i-C$_4$H$_{10}$	Ref.
Hydrogen cyanide	0	.173												10
Hydrogen peroxide	60	.188												11
Iodine	0	.07				.070								8, 12, 14
Mercury	0	.112		.53		.13								8, 12, 13
Mesitylene	0	.056												8
Methane	500				1.1									18
Methyl acetate	0	.084		.333			.0567							8
Methyl alcohol	0	.132		.506			.0879							8
Methyl butyrate	0	.0633		.242			.0446							8
Methyl i-butyrate	0	.0639		.257			.0451							8
Methyl cyclopentane	15		.0731	.318	0.0742	.0758								3
Methyl formate	0	.0872												8
Methyl propionate	0	.0735		.295			.0528							8
Methyl valerate	0	0.0569												8
Naphthalene	0	.0513												8
Nitrogen	0				0.181									8
	25						0.165			0.148	0.163	0.0960	0.0908	2
Nitrous oxide	0			0.535			.096							8
n-Octane	0	.0505												8
	30		0.0642	.271	.0705	0.0710								3
Oxygen	0	.178		.697		.181	.139							8
Phosgene	0	.095												10
Propionic acid	0	.0829		.330			.0588							8
Propyl acetate	0	.067												8
n-Propyl alcohol	0	.085		.315			.0577							8
i-Propyl alcohol	0	.0818												8
	30	.101												5
n-Propyl benzene	0	.0481												8
i-Propyl benzene	0	.0489												8
i-Propyl bromide	0	.085												8
n-Propyl bromide	0	.0902												8
Propyl butyrate	0	.0530		.206			.0364							8
Propyl formate	0	.0712		.281			.0490							8
n-Propyl iodide	0	.079												8
i-Propyl iodide	0	.0802												8
n-Propyl isobutyrate	0	.0549		.212			.0388							8
i-Propyl isobutyrate	0	.059												8
Propyl propionate	0	.057		.212			.0395							8
Propyl valerate	0	.0466		.189			.0341							8
Safrol	0	.0434												8
i-Safrol	0	.0455												8
Sulfur hexafluoride	25			.418										2
Toluene	0	.076	.071											4, 8
	30	.088												5
Trimethyl carbinol	0	.087												8
2,2,4-Trimethyl pentane	30		.0618	.288	.0688	.0705								3
2,2,3-Trimethyl heptane	90			.270		.0684								3
n-Valeric acid	0	.050												8
i-Valeric acid	0	.0544		.212			.0376							8
Water	0	.220		.75			.138							8, 20
	450				1.3									18

° 320 mm Hg.
† 40 atm.
‡ Also at other temperatures.
§ Strong function of concentration.

References
[1] Amdur, Irvine, Mason, and Ross, *J. Chem. Phys.,* **20,** 436 (1952).
[2] Boyd, Stein, Steingrimsson, and Rumpel, *J. Chem. Phys.,* **19,** 548 (1951).
[3] Cummings and Ubbelohde, *J. Chem. Soc. (London),* 1953, p. 3751.
[4] Fairbanks and Wilke, *Ind. Eng. Chem.,* **42,** 471 (1950).
[5] Gilliland, *Ind. Eng. Chem.,* **26,** 681 (1934).
[6] Gorynnova and Kuvskinskii, *Zhur. Tekh. Fiz.,* **18,** 1421 (1948).
[7] Hansen, Dissertation, Jena, 1907.
[8] "International Critical Tables," vol. 5, p. 62.
[9] Jeffries and Drickamer, *J. Chem. Phys.,* **22,** 436 (1954).
[10] Klotz and Miller, *J. Am. Chem. Soc.,* **69,** 2557 (1947).
[11] McMurtrie and Keyes, *J. Am. Chem. Soc.,* **70,** 3755 (1948).
[12] Mullaly and Jacques, *Phil. Mag.,* **48,** 6, 1105 (1924).
[13] Spier, *Physica,* **6** (1939): 453; **7,** 381 (1940).
[14] Topley and Whytlaw-Gray, *Phil. Mag.,* **4,** 873 (1927).
[15] Trautz and Ludwig, *Ann. Physik,* **5,** 5, 887 (1930).
[16] Trautz and Muller, *Ann. Physik,* **22,** 353 (1935).
[17] Trautz and Ries, *Ann. Physik,* **8,** 163 (1931).
[18] Walker and Westenberg, *J. Chem. Phys.,* **32,** 136 (1960).
[19] Westenberg and Walker, *J. Chem. Phys.,* **26,** 1753 (1957).
[20] Winkelmann, *Wied. Ann.,* **22,** 152 (1884); **23,** 203 (1884); **26,** 105 (1885); **33,** 445 (1888); **36,** 92 (1889).

In this table are a representative selection of diffusion coefficients. The subsection "Prediction and Correlation of Physical Properties" should be consulted for estimation techniques. As general references, the works by Hirschfelder, Curtiss, and Bird, *Molecular Theory of Gases and Liquids,* Wiley, New York, 1964; Chapman and Cowling, *The Mathematical Theory of Non-Uniform Gases,* Cambridge, New York, 1970; Reid and Sherwood, *The Properties of Gases and Liquids,*

McGraw-Hill, New York, 1964; and Bretsznajder, *Prediction of Transport and Other Physical Properties of Fluids,* Pergamon, New York, 1971, may be found useful. The most exhaustive recent compilation for gases is by Mason and Marrero, *J. Phys. Chem. Ref. Data,* **1** (1972). Unfortunately, the Mason and Marrero work cites only equations and equation constants and not direct tabulations. For these, the Landolt-Börnstein series is suggested.

TABLE 2-372   Diffusivities in Liquids (25°C)

Dilute solutions and 1 atm unless otherwise noted; use $D_L\mu/T$ = constant to estimate effect of temperature; ° indicates that reference gives effect of concentration.

Solute	Solvent	$D_L \times 10^5$, sq cm/sec	Estimated possible, error, ± %1	Ref.
Acetal°	Ethanol	1.25	5	11
Acetamide°	Ethanol	0.68	5	11
Acetamide°	Water	1.19	3	11
Acetic acid	Acetone	3.31		4
Acetic acid	Benzene	2.11		1, 4
Acetic acid	Carbon tetrachloride	1.49		4
Acetic acid	Ethylene glycol	0.13		4
Acetic acid	Toluene	2.26		4
Acetic acid°	Water	1.24	3	11
Acetonitrile	Water	1.66	5	11
Acetylene	Water	1.78, 2.11		1, 24
Allyl alcohol°	Ethanol	1.06	5	11
Allyl alcohol	Water	1.19	6	11
Ammonia°	Water	1.7, 2.0, 2.3		1, 11
i-Amyl alcohol°	Ethanol	0.87	5	11
i-Amyl alcohol	Water	1.0	8	11, 25
Benzene	Carbon tetrachloride	1.53		7
Benzene (50 mole %)	*n*-Decane	1.72		26
Benzene (50 mole %)	2,4-Dimethyl pentane	2.49		26
Benzene (50 mole %)	*n*-Dodecane	1.40		26
Benzene (50 mole %)	*n*-Heptane	2.47		26
Benzene (50 mole %)	*n*-Hexadecane	0.96		26
Benzene (50 mole %)	*n*-Octadecane	0.86		26
Benzoic acid	Acetone	2.62		4
Benzoic acid	Benzene	1.38		4
Benzoic acid	Carbon tetrachloride	0.91		4
Benzoic acid	Ethylene glycol	0.043		4
Benzoic acid	Toluene	1.49		4
Bromine	Benzene	2.7		11
Bromine	Carbon disulfide	4.1		11
Bromine	Water	1.3		11
Bromobenzene	Benzene	2.30		25
Bromoform°	Acetone	2.90		11
Bromoform	*i*-Amyl alcohol	0.53		11
Bromoform	Ethanol	1.08	5	11
Bromoform°	Ethyl ether	3.62		11
Bromoform	Methanol	2.20		23
Bromoform	*n*-Propanol	0.94		11
n-Butanol	Water	0.96	5	1, 11, 18, 25
Caffeine	Water	0.63	6	11
Carbon dioxide	Ethanol	4.0	6	11
Carbon dioxide	Water	1.96	1	1, 3, 5, 20, 24, 28
Carbon disulfide (50 mole %, 200 atm.)	*n*-Butanol	3.57		14
Carbon disulfide (50 mole %, 200 atm.)	*i*-Butanol	2.42		14
Carbon disulfide (50 mole %, 218 atm.)	Chlorobenzene	3.00		14
Carbon disulfide (50 mole %, 200 atm.)	2,4-Dimethyl pentane	3.63		14
Carbon disulfide (50 mole %, 100 atm.)	*n*-Heptane	3.0		14
Carbon disulfide (50 mole %, 50 atm.)	Methyl cyclohexane	3.5		14
Carbon disulfide (50 mole %, 200 atm.)	*n*-Octane	3.10		14
Carbon disulfide (50 mole %)	Toluene	2.06		14
Carbon tetrachloride	Benzene	2.04	3	7, 9
Carbon tetrachloride°	Cyclohexane	1.49	2	9, 10°
Carbon tetrachloride	Decalin	0.776	2	9
Carbon tetrachloride	Dioxane	1.02	2	9
Carbon tetrachloride°	Ethanol	1.50	2	9, 10°
Carbon tetrachloride	*n*-Heptane	3.17	2	9
Carbon tetrachloride	Kerosene	0.961	2	9
Carbon tetrachloride	Methanol	2.30	2	9
Carbon tetrachloride	*i*-Octane	2.57	2	9
Carbon tetrachloride	Tetralin	0.735	2	9
Chloral°	Ethanol	0.68	5	11
Chloral hydrate	Water	0.77	7	11

TABLE 2-372   Diffusivities in Liquids (25°C) (*Continued*)

Dilute solutions and 1 atm unless otherwise noted; use $D_L \mu / T$ = constant to estimate effect of temperature; ° indicates that reference gives effect of concentration.

Solute	Solvent	$D_L \times 10^5$, sq cm/sec	Estimated possible, error, $\pm$ %1	Ref.
Chlorine	Water	1.44	4	1, 28
Chlorobenzene	Benzene	2.66		25
Chloroform	Benzene	2.50	6	1, 25
Chloroform	Ethanol	1.38	3	11
Cinnamic acid	Acetone	2.41		4
Cinnamic acid	Benzene	1.12		4
Cinnamic acid	Carbon tetrachloride	0.76		4
Cinnamic acid	Toluene	2.41		4
1,1'-Dichloropropanol	Water	1.0	6	11
Dicyanodiamide°	Water	1.18	4	11
Diethyl ether	Benzene	2.73		25
Diethyl ether	Water	0.85		2
2,4-Dimethyl pentane (50 mole %)	n-Dodecane	1.44		26
2,4-Dimethyl pentane (50 mole %)	n-Hexadecane	0.88		26
Ethanol°	Water	1.28	4	1, 7, 9,° 11,° 22
Ethyl acetate	Ethyl benzoate	0.94		6
Ethylene dichloride	Benzene	2.8		1, 25
Formic acid	Acetone	3.77		4
Formic acid	Benzene	2.28		4
Formic acid	Carbon tetrachloride	1.89		4
Formic acid	Ethylene glycol	0.094		4
Formic acid	Toluene	2.65		
Formic acid	Water	1.37	10	11
Glucose	Water	0.69	6	11
Glycerol	i-Amyl alcohol	0.12		11
Glycerol	Ethanol	0.56		11
Glycerol°	Water	0.94	6	1, 11°
n-Heptane (50 mole %)	n-Dodecane	1.58		26
n-Heptane (50 mole %)	n-Hexadecane	1.00		26
n-Heptane (50 mole %)	n-Octadecane	0.92		26
n-Heptane (50 mole %)	n-Tetradecane	1.29		26
Hexamethylene tetramine	Water	0.67		11
Hydrogen chloride°	Water	3.10	3	4, 11,° 12°
Hydrogen	Water	5.85 (4.4)		1, 11, 24(?)
Hydrogen sulfide	Water	1.61		1
Hydroquinone°	Ethanol	0.53	5	11
Hydroquinone°	Water	0.88, 1.12		2, 11°
Iodine	Acetic acid	1.13		11
Iodine	Anisole	1.25		11
Iodine	Benzene	1.98		9, 19, 23
Iodine	Bromobenzene	1.25	10	4, 11, 19
Iodine	Carbon disulfide	3.2		11, 19, 23
Iodine	Carbon tetrachloride	1.45	8	9, 11, 19
Iodine	Chloroform	2.30	3	11, 23
Iodine	Cyclohexane	1.80		4
Iodine	Dioxane	1.07		9
Iodine°	Ethanol	1.30		4, 11°
Iodine	Ethyl acetate	2.2		11, 19
Iodine	Ethyl ether	3.61		11
Iodine	Ethylene bromide	0.93		11
Iodine	n-Heptane	3.4, 2.5		9, 11, 19
Iodine	n-Hexane	4.15		4, 9
Iodine	Mesitylene	1.49		9
Iodine	Methanol	1.74		19
Iodine	Methyl cyclohexane	2.1		4
Iodine	n-Octane	2.76		4
Iodine	Tetrabromoethane	2.0		11
Iodine	n-Tetradecane	0.96		4
Iodine	Toluene	2.1		11
Iodine	m-Xylene	1.82		9, 11
Iodobenzene	Ethanol	1.09	3	11
Lactose°	Water	0.49	5	11
Maltose°	Water	0.48	5	11
Mannitol°	Water	0.65	5	11
Methanol	Water	1.6		1, 7, 11
Nicotine°	Water	0.60	8	11
Nitric acid°	Water	2.98	2	11
Nitrobenzene	Carbon tetrachloride	1.00		7
Nitrogen	Water	1.9		1, 24
Nitrous oxide	Water	1.8		1, 11
Oxalic acid°	Water	1.61	2	11

TABLE 2-372 Diffusivities in Liquids (25°C) (*Concluded*)

Dilute solutions and 1 atm unless otherwise noted; use $D_L\mu/T$ = constant to estimate effect of temperature; ° indicates that reference gives effect of concentration.

Solute	Solvent	$D_L \times 10^5$, sq cm/sec	Estimated possible, error, ± %1	Ref.
Oxygen	Glycerol°-water (106 poise)	0.24		13
Oxygen	Sucrose°-water (125 poise)	0.25		13
Oxygen	Water	2.5	20	1, 3, 15, 21, 24
Pentaerythritol°	Water	0.77	4	11
Phenol	*i*-Amyl alcohol	0.2		11
Phenol	Benzene	1.68		1
Phenol	Carbon disulfide	3.7		11
Phenol	Chloroform	2.0		11
Phenol	Ethanol	0.89		11
Phenol	Ethyl ether	3.9		11
n-Propanol	Water	1.1		1, 7, 11
Pyridine°	Ethanol	1.24	3	11
Pyridine	Water	0.76	7	11
Pyrogallol	Water	0.74	7	11
Raffinose°	Water	0.41	4	11
Resorcinol°	Ethanol	0.46	5	11
Resorcinol°	Water	0.87	4	11
Saccharose°	Water	0.49	4	11
Stearic acid°	Ethanol	0.65	5	11
Succinic acid°	Water	0.94		11
Sucrose	Water	0.56	6	2, 27
Sulfur dioxide	Water	1.7		15, 17
Sulfuric acid°	Water	1.97	3	11
Tartaric acid°	Water	0.80	10	11
1,1,2,2-Tetrabromoethane	1,1,2,2-Tetra-chloroethane	0.61	4	11
Toluene	*n*-Decane	2.09		4
Toluene	*n*-Dodecane	1.38		4
Toluene	*n*-Heptane	3.72		4
Toluene	*n*-Hexane	4.21		4
Toluene	*n*-Tetradecane	1.02		4
Urea	Ethanol	0.73		11
Urea	Water	1.37	2	8, 11
Urethane	Water	1.06		11, 25
Water	Glycerol	0.021		16

References

[1] Arnold, *J. Am. Chem. Soc.*, **52**, 3937 (1930).
[2] Calvet, *J. Chim. Phys.*, **44**, 47 (1947).
[3] Carlson, *J. Am. Chem. Soc.*, **33**, 1027 (1911).
[4] Chang and Wilke, *J. Phys. Chem.*, **59**, 592 (1955).
[5] Davidson and Cullen, *Trans. Inst. Chem. Eng.*, **35**, 51 (1957).
[6] Dummer, *Z. Anorg. Chem.*, **109**, 31 (1949).
[7] Gerlach, *Ann. Phys. (Leipzig)*, **10**, 437 (1931).
[8] Gosting and Akeley, *J. Am. Chem. Soc.*, **74**, 2058 (1952).
[9] Hammond and Stokes, *Trans. Faraday Soc.*, **49**, 890 (1953); **49**, 886 (1953).
[10] Hammond and Stokes, *Trans. Faraday Soc.*, **52**, 781 (1956).
[11] *International Critical Tables*, vol. 5, p. 63.
[12] James, Hollingshead, and Gordon, *J. Chem. Phys.*, **7**, 89 (1939); **7**, 836 (1939).
[13] Jordon, Ackermann, and Berger, *J. Am. Chem. Soc.*, **78**, 2979 (1956).
[14] Koeller and Drickamer, *J. Chem. Phys.*, **21**, 575 (1953).
[15] Kolthoff and Miller, *J. Am. Chem. Soc.*, **63**, 1013 (1941).

TABLE 2-373 Thermal Conductivities of Some Building and Insulating Materials*

$$k = \text{Btu}/(\text{h·ft}^2)(°\text{F/ft})$$

Material	Apparent density ρ, lb/ft³ at room temperature	t, °C	k	Material	Apparent density ρ, lb/ft³ at room temperature	t, °C	k
Aerogel, silica, opacified	8.5	120	0.013	Cotton wool	5	30	0.024
		290	.026	Cork board	10	30	.025
Asbestos-cement boards	120	20	.43	Cork (regranulated)	8.1	30	.026
Asbestos sheets	55.5	51	.096	(ground)	9.4	30	.025
Asbestos slate	112	0	.087	Diatomaceous earth powder, coarse	20.0	38	.036
	112	60	.114	(Note 2)	20.0	871	.082
Asbestos	29.3	−200	.043	fine (Note 2)	17.2	204	.040
	29.3	0	.090		17.2	871	.074
	36	0	.087	molded pipe covering (Note 2)	26.0	204	.051
	36	100	.111		26.0	871	.088
	36	200	.120	4 vol. calcined earth and 1 vol. cement,			
	36	400	.129	poured and fired (Note 2)	61.8	204	.16
	43.5	−200	.090		61.8	871	.23
	43.5	0	.135	Dolomite	167	50	1.0
Aluminum foil (7 air spaces per 2.5 in.)	0.2	38	.025	Ebonite			0.10
		177	.038	Enamel, silicate	38		0.5–0.75
Ashes, wood		0–100	.041	Felt, wool	20.6	30	0.03
Asphalt	132	20	.43	Fiber insulating board	14.8	21	.028
Boiler scale (Note 1)				Fiber, red	80.5	20	.27
Bricks:				(with binder, baked)		20–97	.097
Alumina (92–99% Al₂O₃ by wt.) fused		427	1.8	Gas carbon		0–100	2.0
Alumina (64–65% Al₂O₃ by wt.)		1315	2.7	Glass			0.2–0.73
(See also Bricks, fire clay)	115	800	0.62	Borosilicate type	139	30–75	0.63
	115	1100	.63	Window glass			0.3–0.61
Building brick work		20	.4	Soda glass			0.3–0.44
Carbon	96.7		3.0	Granite			1.0–2.3
Chrome brick (32% Cr₂O₃ by wt.)	200	200	.67	Graphite, longitudinal		20	95
	200	650	.85	powdered, through 100 mesh	30	40	0.104
	200	1315	1.0	Gypsum (molded and dry)	78	20	.25
Diatomaceous earth, natural, across strata				Hair felt (perpendicular to fibers)	17	30	.021
(Note 2)	27.7	204	0.051	Ice	57.5	0	1.3
	27.7	871	.077	Infusorial earth, see diatomaceous earth			
Diatomaceous, natural, parallel to strata				Kapok	0.88	20	0.020
(Note 2)	27.7	204	.081	Lampblack	10	40	.038
	27.7	871	.106	Lava			.49
Diatomaceous earth, molded and fired	38	204	.14	Leather, sole	62.4		.092
(Note 2)	38	871	.18	Limestone (15.3 vol. % H₂O)	103	24	.54
Diatomaceous earth and clay, molded and				Linen		30	.05
fired (Note 2)	42.3	204	.14	Magnesia (powdered)	49.7	47	.35
	42.3	871	.19	Magnesia (light carbonate)	13	21	.034
Diatomaceous earth, high burn, large pores				Magnesium oxide (compressed)	49.9	20	.32
(Note 3)	37	200	.13	Marble			1.2–1.7
	37	1000	.34	Mica (perpendicular to planes)		50	0.25
Fire clay (Missouri)		200	.58	Mill shavings			0.033–0.05
		600	.85	Mineral wool	9.4	30	0.0225
		1000	.95		19.7	30	.024
		1400	1.02	Paper			.075
Kaolin insulating brick (Note 3)	27	500	0.15	Paraffin wax		0	.14
	27	1150	.26	Petroleum coke		100	3.4
Kaolin insulating firebrick (Note 4)	19	200	.050			500	2.9
	19	760	.113	Porcelain		200	0.88
Magnesite (86.8% MgO, 6.3% Fe₂O₃, 3%				Portland cement, see concrete		90	.17
CaO, 2.6% SiO₂ by wt.)	158	204	2.2	Pumice stone		21–66	.14
	158	650	1.6	Pyroxylin plastics			.075
	158	1200	1.1	Rubber (hard)	74.8	0	.087
Silicon carbide brick, recrystallized	129	600	10.7	(para)		21	.109
(Note 3)	129	800	9.2	(soft)		21	0.075–0.092
	129	1000	8.0	Sand (dry)	94.6	20	0.19
	129	1200	7.0	Sandstone	140	40	1.06
	129	1400	6.3	Sawdust	12	21	0.03
Calcium carbonate, natural	162	30	1.3	Scale (Note 1)			
White marble			1.7	Silk	6.3		.026
Chalk	96		0.4	varnished		38	.096
Calcium sulfate (4H₂O), artificial	84.6	40	.22	Slag, blast furnace		24–127	.064
plaster (artificial)	132	75	.43	Slag wool	12	30	.022
(building)	77.9	25	.25	Slate		94	.86
Cambric (varnished)		38	.091	Snow	34.7	0	.27
Carbon, gas		0–100	2.0	Sulfur (monoclinic)		100	0.09–0.097
Carbon stock	94	−184	0.55	(rhombic)		21	0.16
		0	3.6	Wall board, insulating type	14.8	21	.028
Cardboard, corrugated			0.037	Wall board, stiff paste board	43	30	.04
				Wood shavings	8.8	30	.034

TABLE 2-373   Thermal Conductivities of Some Building and Insulating Materials* (Concluded)

$$k = Btu/(h \cdot ft^2)(°F/ft)$$

Material	Apparent density ρ, lb/ft³ at room temperature	t, °C	k	Material	Apparent density ρ, lb/ft³ at room temperature	t, °C	k
Celluloid	87.3	30	.12	Wood (across grain):			
Charcoal flakes	11.9	80	.043	Balsa	7–8	30	0.025–0.03
	15	80	.051	Oak	51.5	15	0.12
Clinker (granular)		0–700	.27	Maple	44.7	50	.11
Coke, petroleum		100	3.4	Pine, white	34.0	15	.087
		500	2.9	Teak	40.0	15	.10
Coke, petroleum (20–100 mesh)	62	400	0.55	White fir	28.1	60	.062
Coke (powdered)		0–100	.11	Wood (parallel to grain):			
Concrete (cinder)			.20	Pine	34.4	21	.20
(stone)			.54	Wool, animal	6.9	30	.021
(1:4 dry)			.44				

*Marks, *Mechanical Engineers' Handbook,* 4th ed., McGraw-Hill, New York, 1941. *International Critical Tables,* McGraw-Hill, 1929, and other sources.
Note 1: B. Kamp [*Z. tech. Physik,* **12**, 30 (1931)] shows the effect of increased porosity in decreasing thermal conductivity of boiler scale. Partridge [University of Michigan, *Eng. Research Bull.,* **15**, 1930] has published a 170-page treatise on Formation and Properties of Boiler Scale.
Note 2: Townshend and Williams, *Chem. & Met.,* **39**, 219 (1932).
Note 3: Norton, *Refractories,* 2d ed., McGraw-Hill, New York, 1942.
Note 4: Norton, private communication.

TABLE 2-374   Thermal-Conductivity-Temperature Table for Metals*

Thermal conductivities tabulated in watts per meter-kelvin

Substance	Temperature, K														
	10	20	40	60	80	100	200	300	400	500	600	800	1000	1200	1400
Alumina	7	32	121	174	160	125	55	36	26	20	16	10	8	7	6
Aluminum	38,000	13,500	2,300	850	380	300	237	273	240	237	232	220	93	99	105
Antimony	470	230	110	80	60	48	32	26	22	20					
Beryllium oxide	47	196	810	1,400	1,650	1,490	480	272	196	146	111	70	47	33	25
Bismuth	240	100	45	31	24	22	18	16	14	12					
Boron	165	305	400	327	230	170	45	25	15	12					
Cadmium	900	250	150	120	110	110	105	104	101	99					
Chromium	400	570	450	250	180	158	111	90	87	85	81	71	65	62	61
Cobalt	250	450	380	250	190	160	120	100	85	70					
Constantan	4	9	16	18	19	20	23	25	27	30					
Copper	19,000	10,700	2,100	850	570	483	413	398	392	388	383	371	357	342	
Gallium	2,200	640	250	200	170	140	100	85							
Gold	2,800	1,500	520	380	350	345	327	315	312	309	304	292	278	262	
Graphite†	27	108	135	81	54	39	15	10	7	5	4	3	3	2	2
Graphite‡	81	420	1,630	2,980	4,290	4,980	3,250	2,000	1,460	1,140	930	680	530	440	370
Hastelloy	1	3	4	5	6	7	9	10	11	13					
Inconel	2	4	8	10	11	11	14	15							
Iridium	1,300	1,900	750	360	230	172	147	145	143	140					
Iron	710	1,000	560	270	170	132	94	80	69	61	55	43	33	28	31
Lead	175	57	43	42	41	40	37	35	34	33	31	19	22	24	26
Magnesium	1,200	1,300	620	290	190	169	159	156	153	151	149	146	84	98	112
Magnesium oxide	1,100	3,100	2,200	950	460	260	75	48	36	27	21	13	10	8	7
Manganese	2	2	4	5	5	6	7	8	9	9					
Manganin	2	4	9	11	13	13	17	22	28	34	40				
Mercury	54	40	35	33	33	32	32	8	10	11	12	13	14		
Molybdenum	150	280	350	250	210	179	143	138	134	130	126	118	112	105	100
Nickel	2,600	1,700	570	290	200	158	106	91	80	72	66	67	72	76	80
Nylon	0.04	0.10	0.17	0.20	0.23	0.25	0.28	0.30							
Palladium	1,200	610	160	100	88	80	78	78	78	80					
Platinum	1,200	490	130	92	82	79	75	73	72	72	72	73	78	78	81
PTFE§	0.94	1.43	1.94	2.1	2.15	2.16	2.20	2.25	2.3	2.5					
Pyrex	0.12	0.20	0.33	0.42	0.51	0.57	0.88	1.1	1.6	2.1					
Quartz	1,200	480	82	40	30										
Rhodium	2,900	3,900	1,000	370	250	190	160	150	145	140					
Rubber			0.13	0.15	0.16	0.17	0.20	0.22	0.24	0.25					
Selenium (axis)	140	57	25	15	10	8	6	4	3	2					
Silica								1.34	1.52	1.70	1.87	2.22	2.60		
Silver	16,500	5,200	1,100	630	500	430	425	424	420	413	405	389	374	358	
Tantalum	108	146	88	68	62	59	58	57	58	58	59	59	60	61	62
Tellurium	300	93	29	17	13	11	6	4	3	3					
Tin		320	130	101	90	84	72	67	62	60					
Titanium	14	28	39	37	33	31	26	21	20	20	19				
Tungsten			880	330	310	280	190	180	170	150	140				
Uranium				20	22	23	26	28	30	32					
Zinc				150	135	130	123	120	116	110	110				
Zirconium	100	110	59	42	38	34	25	23	22	21	21				

* Especially at low temperatures, the thermal conductivity can often be markedly reduced by even small traces of impurities. This table, for the highest-purity specimens available, should thus be used with caution in applications with commercial materials. From Perry, *Engineering Manual,* 3d ed., McGraw-Hill, New York, 1976. A more detailed table appears as Section 5.5.6 in the *Heat Exchanger Design Handbook,* Hemisphere Pub. Corp., Washington, DC, 1983.
† Parallel to basal plane.
‡ Perpendicular to basal plane.
§ Also known as Teflon, etc.

TABLE 2-375   Thermal Conductivity of Chromium Alloys*

$$k = Btu/(h \cdot ft^2)(°F/ft)$$

American Iron and Steel Institute Type No.	k at 212°F	k at 932°F
301, 302, 302B, 303, 304, 316†	9.4	12.4
308	8.8	12.5
309, 310	8.0	10.8
321, 347	9.3	12.8
403, 406, 410, 414, 416†	14.4	16.6
430, 430F†	15.1	15.2
442	12.5	14.2
501, 502†	21.2	19.5

*Table 3-322 is based on information from manufacturers.
† Shelton and Swanger (National Bureau of Standards), *Trans. Am. Soc. Steel Treat.*, **21**, 1061–1078 (1933).

TABLE 2-376   Thermal Conductivity of Some Alloys at High Temperature*

	Thermal conductivity, Btu/(ft)(hr)(°R)					
°R	Kovar	Advance	Monel	Hastelloy A	Inconel	Nichrome V
500	7.8		9.0	5.6	6.0	5.5
600	8.3	11.4	10.2	6.2	6.5	6.1
700	8.6	12.6	11.2	6.8	7.0	6.7
800	8.7	13.9	12.3	7.3	7.6	7.3
900	8.7	15.1	13.4	7.8	8.1	7.8
1000	8.9	16.4	14.4	8.4	8.6	8.4
1100	9.2	17.6	15.4	9.0	9.1	9.0
1200	9.5	18.8	16.5	9.5	9.7	9.5
1300	9.8	20.0	17.6	10.1	10.2	10.1
1400	10.2	21.2	18.7	10.7	10.8	10.7
1500	10.5	22.5	19.8	11.3	11.3	11.3
1600	10.8	23.8	20.8	11.8	11.8	11.9
1700	11.1	25.0	21.9	12.3	12.4	12.4
1800	11.3	26.2	23.0	12.9	13.0	13.0
1900	11.5	27.4	24.0	13.4	13.6	13.5
2000	11.8	28.7	25.1	14.0	14.0	14.1
2100	12.1	30.0	26.1	14.6	14.5	14.7
2200	12.3		27.2	15.1	15.0	15.3

*Silverman, *J. Metals*, **5**, 631 (1953). Copyright American Institute of Mining, Metallurgical and Petroleum Engineers, Inc.

TABLE 2-377   Thermal Conductivities of Some Materials for Refrigeration and Building Insulation*

$$k = Btu/(h \cdot ft^2)(°F/ft) \text{ at approximately room temperature}$$

Material	Apparent density, lb/cu ft . room temp.	k
Soft flexible materials in sheet form:		
Chemically treated wood fiber	2.2	0.023
Eel grass between paper	3.4–4.6	0.021–0.022
Felted cattle hair	11–13	0.022
Flax fibers between paper	4.9	0.023
Hair and asbestos fibers, felted	7.8	0.023
Insulating hair, and jute	6.1–6.3	0.022–0.023
Jute and asbestos fibers, felted	10.0	0.031
Loose materials:		
Cork, regranulated, fine particles	8–9	.025
Charcoal, 6 mesh	15.2	.031
Diatomaceous earth, powdered	10.6	.026
Glass wool, curled	4–10	.024
Gypsum in powdered form	26–34	0.043–0.05
Mineral wool, fibrous	6	0.0217
	10	.0225
	14	.0233
	18	.0242
Sawdust	12	.034
Wood shavings, from planer	8.8	.034
Semiflexible materials in sheet form:		
Flax fiber	13.0	.026
Semirigid materials in board form:		
Corkboard	7.0	.0225
	10.6	.025
Mineral wool, block, with binder	16.7	.031
Stiff fibrous materials in sheet form:		
Wood pulp	16.2–16.9	.028
Sugar-cane fiber	13.2–14.8	.028
Cellular gypsum	8	.029
	12	.037
	18	.049
	24	.064
	30	.083

*Abstracted from *U.S. Bur. Standards Letter Circ.* **227**, Apr. 19, 1927.

TABLE 2-378   Thermal Conductivities of Insulating Materials at High Temperatures*

$$k = Btu/(h \cdot ft^2)(°F/ft)$$

Material	For temperatures, °F up to	Mean temperatures, °F									
		100	200	300	400	500	600	800	1000	1500	2000
Laminated asbestos felt (approx. 40 laminations per in)	700	0.033	0.037	0.040	0.044	0.048					
Laminated asbestos felt (approx. 20 laminations per in)	500	.045	.050	.055	.060	.065					
Corrugated asbestos (4 plies per in)	300	.050	.058	.069							
85% magnesia (density, 13 lb/ft³)	600	.034	.036	.038	.040						
Diatomaceous earth, asbestos and bonding material	1600	.045	.047	.049	.050	.053	0.055	0.060	0.065		
Diatomaceous earth brick	1600	.054	.056	.058	.060	.063	.065	.069	.073		
Diatomaceous earth brick	2000	.127	.130	.133	.137	.140	.143	.150	.158	0.176	
Diatomaceous earth brick	2500	.128	.131	.135	.139	.143	.148	.155	.163	.183	0.203
Diatomaceous earth powder (density, 18 lb/ft³)		.039	.042	.044	.048	.051	.054	.061	.068		
Rock wool		.030	.034	.039	.044	.050	.057				

Asbestos cement, 1.2; 85% magnesia cement, 0.05; asbestos and rock wool cement, 0.075 approx.
*Marks, "Mechanical Engineers' Handbook," 4th ed., McGraw-Hill, New York, 1941.

TABLE 2-379   Thermal Conductivities of Insulating Materials at Moderate Temperatures (Nusselt)*

$k = \text{Btu}/(\text{h} \cdot \text{ft}^2)(\text{°F/ft})$

Material	Weight, lb/cu ft	Temperatures, °F						
		32	100	200	300	400	600	800
Asbestos	36.0	0.087	0.097	0.110	0.117	0.121	0.125	0.130
Burned infusorial earth for pipe coverings	12.5	.043	.046	.052	.057	.062	.073	.085
Insulating composition (loose)	25.0	.040	.046	.050	.053	.055		
Cotton	5.0	.032	.035	.039				
Silk hair	9.1	.026	.030	.034				
Silk	6.3	.025	.028	.034				
Wool	8.5	.022	.027	.033				
Pulverized cork	10.0	.021	.026	.032				
Infusorial earth (loose)	22.0	.035	.039	.045	.047	.050	.053	

*Marks, *Mechanical Engineers' Handbook,* 4th ed., McGraw-Hill, New York, 1941.

TABLE 2-380   Thermal Conductivities of Insulating Materials at Low Temperatures (Gröber)*

$k = \text{Btu}/(\text{h} \cdot \text{ft}^2)(\text{°F/ft})$

Material	Weight, lb/cu ft	Temperatures, °F				
		32	−50	−100	−200	−300
Asbestos	44.0	0.135	0.132	0.130	0.125	0.100
Asbestos	29.0	.0894	.0860	.0820	.0720	.0545
Cotton	5.0	.0325	.0302	.0276	.0235	.0198
Silk	6.3	.0290	.0256	.0235	.0196	.0155

*Marks, *Mechanical Engineers' Handbook,* 4th ed., McGraw-Hill, New York, 1941.

TABLE 2-381   Thermal Diffusivity (m²/s) of Selected Elements*

Element	Temperature, K									
	20	40	60	80	100	200	400	600	800	1000
Aluminum	0.50	0.012	0.0014	4.4. − 4	2.3. − 4	1.1. − 4	9.4. − 5	8.4. − 5	7.4. − 5	6.6. − 5
Beryllium					0.0036	1.5. − 4	4.0. − 5	2.6. − 5	2.1. − 5	1.7. − 5
Chromium	0.038	0.0037	5.9. − 4	2.0. − 4	1.2. − 4	4.1. − 5	2.6. − 5	2.0. − 5	1.7. − 5	1.4. − 5
Copper	0.16	0.0040	6.9. − 4	3.1. − 4	2.2. − 4	1.3. − 4	1.1. − 4	1.0. − 4	9.0. − 5	9.0. − 5
Gold	0.005	4.5. − 4	2.3. − 4	1.8. − 4	1.5. − 4	1.3. − 4	1.2. − 4	1.2. − 4	1.1. − 4	9.8. − 5
Iridium	0.046				8.4. − 5	5.6. − 5	4.8. − 5	4.4. − 5	4.1. − 5	3.5. − 5
Iron	0.043	3.2. − 3	4.9. − 4	1.6. − 4	8.2. − 5	3.1. − 5	1.8. − 5	1.3. − 5	1.1. − 5	1.0. − 5
Lead	9.3. − 5	3.9. − 4	3.3. − 5	3.1. − 4	2.9. − 5	2.6. − 5	2.3. − 5	2.0. − 5	1.3. − 5	1.5. − 5
Molybdenum	0.0095	0.0014	4.0. − 4	2.0. − 4	1.3. − 4	6.3. − 5	5.1. − 5	4.5. − 5	4.2. − 5	3.8. − 5
Nickel	0.033	0.0017	3.1. − 4	1.3. − 4	8.0. − 5	3.1. − 5	1.9. − 5	1.3. − 5	1.4. − 5	1.5. − 5
Platinum	0.0029	1.6. − 4	6.3. − 5	4.3. − 5	3.6. − 5	2.7. − 5	2.5. − 5	2.5. − 5	2.5. − 5	2.5. − 5
Silver	0.031	0.0013	4.5. − 4	2.8. − 4	2.3. − 4	1.8. − 4	1.7. − 4	1.6. − 4	1.5. − 4	1.4. − 4
Zinc	0.0046	3.1. − 4	1.0. − 4	7.0. − 5	5.5. − 5	4.7. − 5	3.9. − 5	3.4. − 5	1.8. − 5	2.2. − 5

*Tables for up to 24 temperatures for 47 elements appear in the *Handbook of Heat Transfer,* 2d ed., McGraw-Hill, New York, 1984. The notation 3.2. − 4 signifies 2.3×10^{-4}.

TABLE 2-382 Thermophysical Properties of Selected Nonmetallic Solid Substances

Material	Density, kg/m^3	Emissivity	Specific heat, $kJ/(kg \cdot K)$	Thermal conductivity, $W/(m \cdot K)$	Thermal diffusivity, $m^2/s \times 10^6$
Alumina	3975		0.765	36	11.9
Asphalt	2110		0.920	0.06	0.03
Bakelite	1300		1.465	1.4	0.74
Beryllia	3000	0.82	1.030	270	88
Brick	1925	0.93	0.835	0.72	0.45
Brick, fireclay	2640	0.93	0.960	1.0	0.39
Carbon, amorphous	1950	0.86	0.724	1.6	1.13
Clay	1460	0.91	0.880	1.3	1.01
Coal	1350	0.80	1.26	0.26	0.15
Cotton	80		1.30	0.06	0.58
Diamond	3500		0.509	2300	1290
Granite	2630		0.775	2.79	1.37
Hardboard	1000		1.38	0.15	0.11
Magnesite	3025	0.38	1.13	4.0	1.2
Magnesia	3635	0.72	0.943	48	14
Oak	770	0.90	2.38	0.18	0.10
Paper	930	0.83	1.34	0.011	0.01
Pine	525	0.84	2.75	0.12	0.54
Plaster board	800	0.91		0.17	
Plywood	540		1.22	0.12	0.18
Pyrex	2250	0.92	0.835	1.4	0.74
Rubber	1150	0.92	2.00	0.2	0.09
Rubber, foam	70	0.90		0.03	
Salt		0.34	0.854	7.1	
Sandstone	2150	0.59	0.745	2.9	1.8
Silica		0.79	0.743	1.3	
Sapphire	3975	0.48	0.765	46	15
Silicon carbide	3160	0.86	0.675	490	230
Soil	2050	0.38	1.84	0.52	0.14
Teflon	2200	0.92	0.35	0.26	0.34
Thoria	4160	0.28	0.71	14	4.7
Urethane foam	70		1.05	0.03	0.36
Vermiculite	120		0.84	0.06	0.60

NOTE: Difficulties of accurately characterizing many of the specimens mean that many of the values presented here must be regarded as being of order of magnitude only. For some materials, actual measurement may be the only way to obtain data of the required accuracy. To convert kilograms per cubic meter to pounds per cubic foot, multiply by 0.062428; to convert kilojoules per kilogram-kelvin to British thermal units per pound-degree Fahrenheit, multiply by 0.23885.

PREDICTION AND CORRELATION OF PHYSICAL PROPERTIES

INTRODUCTION

In the absence of reliable experimental data, the methods presented here provide physical property estimates that are sufficiently accurate for many engineering applications. These techniques have been selected on the basis of accuracy, generality, and, in most cases, simplicity; they are divided into 11 categories: (1) pure component constants: critical properties, normal freezing and boiling temperatures, acentric factor, radius of gyration, dipole moment, and van der Waals area and volume; (2) vapor pressure; (3) ideal gas thermal properties: heat capacity and enthalpy, Gibbs energy, and entropy of formation; (4) enthalpy of vaporization and fusion; (5) solid and liquid heat capacity; (6) vapor, liquid, and solid density; (7) vapor and liquid viscosity; (8) vapor and liquid thermal conductivity; (9) vapor and liquid diffusivity; (10) surface tension; and (11) flammability properties: flash point, flammability limits, and autoignition temperature. The definition of the property and limitations and accuracy of each method of correlation or prediction are given for each property. Numerical examples are included for many of the methods. Equation symbols are listed under "Nomenclature," and literature citations, indicated by superscript numbers, follow the nomenclature under "References." Essentially all of the methods are derived from work on the *American Petroleum Institute Technical Data Book*[23] (hydrocarbon compounds and their mixtures), the AIChE Design Institute for Physical Property Data (DIPPR) *Data Prediction Manual*[22] (nonhydrocarbon compounds and their mixtures), and the DIPPR Data Compilation Project[24].

UNITS

Applicable dimensional units are shown individually with each equation. The International Metric System (SI) is used when feasible; otherwise commonly used U.S. engineering units are employed. The reader is referred to Sec. 1 for unit conversion factors.

Nomenclature

Symbol	Definition	SI units	U.S. customary units
B	Second virial coefficient	m³/kmol	ft³/lbmol
C_p	Heat capacity at constant pressure	J/(kmol·K)	Btu/(lbm °F)
C_v	Heat capacity at constant volume	kg/m³	lbm/ft³
d	Mass density	m²/s	ft²/s
D	Diffusivity		
F	Conductivity factor in Eq. (2-136)		
k	Thermal conductivity	W/(m·K)	Btu/(h·ft °F)
K	Watson/UOP characterization factor = $(1.8T_b)^{1/3}$/rel den		
$\ln$	Denotes natural logarithm		
M	Molecular weight		
$MeABP$	Mean average boiling point		
N	Carbon number		
N_A	Avogadro's number		
P	Pressure	Pa	lbf/in²
P^{sat}	Vapor pressure	Pa	lbf/in²
$[P]$	Parachor		
$rel\ den$	Relative density at 15°C and 0.1 MPa		
R	Universal gas constant; e.g., 8314 Pa·m³/kmol K		
$\overline{R}$	Radius of gyration		
S	Absolute entropy	J/kg	Btu/lbm
T	Absolute temperature	K	R
V	Molar volume	m³/kmol	ft³/lbmole
x	Mole fraction of component in liquid phase		
y	Mole fraction of component in vapor phase		
Z	Compressibility factor = PV/RT		

Greek symbols

ΔG_f	Gibbs energy of formation	J/kg	Btu/lbm
ΔH_f	Enthalpy (heat) of formation	J/kg	Btu/lbm
ΔH_{fus}	Enthalpy (heat) of fusion	J/kg	Btu/lbm
ΔH_V	Enthalpy (heat) of vaporization	J/kg	Btu/lbm
ΔS_f	Entropy of formation	J/(kg·K)	Btu/lbm
ΔS_{fus}	Entropy of fusion	J/(kg·K)	Btu/lbm
ΔZ_V	Difference of vapor and liquid compressibility factors defined in Eq. (2-55)		
Θ	T_b/T_c		
λ	Dipole moment		
μ	Absolute viscosity	Pa·s	lbm/(ft·s)
ν	Kinematic viscosity	m²/s	ft²/s
ρ	Molar density	kmol/m³	lb·mole/ft³
σ	Surface tension	N/m	dyne/cm
ϕ	Volume fraction		
ω	Acentric factor		

Superscripts

r	At reference condition
(0)	Simple spherical molecule, corresponding states
(1)	Correction factor, corresponding states
$\circ$	Of the ideal gas
$'$	At atmospheric pressure

Subscripts

b	At normal boiling temperature
bp	At bubble point
c	At critical point
G	Of the gas/vapor
HI	Upper limit
i	Component index, of the ith component
j	Component index, of the jth component
L	Of the liquid
m	Of the mixture
mc	Mixture correspondence
mlt	At melting temperature
o	Of organic component
pc	Pseudocritical quantity
r	Reduced quantity
RA	Rackett parameter
sat	Saturated
S	Of the solid
V	Of the vapor
w	Of water

REFERENCES

1. Affens, W. A., *J. Chem. Eng. Data*, **11** (1966): 197.
2. Ambrose, D., *National Physical Laboratory Reports Chem. 92, 98, and 107*, Teddington, Middlesex, United Kingdom, 1978, 1979, and 1980.
3. Angus, S., B. Armstrong, and K. M. de Reuck, *Carbon Dioxide, International Thermodynamic Tables of the Fluid State*, Vol. 3. IUPAC, Pergamon Press, Elmsford, NY, 1976.
4. Antoine, C., *Comptes Rendus*, **107** (1888): 681, 836.
5. Baroncini, C., P. Di Filippo, G. Latini, and M. Pacetti, *Int. J. of Thermophysics*, **2** (1981): 21.
6. Benson, S. W., *Thermochemical Kinetics*, 2d ed., John Wiley and Sons, New York, 1976.
7. Benson, S. W., F. R. Cruickshank, D. M. Golden, G. R. Haugen, H. E. O'Neal, A. S. Rodgers, R. Shaw, and R. Walsh, *Chem. Rev.*, **69** (1969): 279.
8. Bondi, A., *Physical Properties of Molecular Crystals, Liquids, and Glasses*, John Wiley and Sons, New York, 1968.
9. Brock, J. R., and R. B. Bird, *Am. Inst. Chem. Eng. J.*, **1** (1955): 174.
10. Brokaw, R. S., *NASA Technical Note D-4496*, 1968.
11. Brokaw, R. S., *Ind. Eng. Chem. Process Des. Dev.*, **8** (1969): 240.
12. Bromley, L. A., *Thermal Conductivity of Gases at Moderate Pressures*, University of California Radiation Laboratory, Report No. UCRL-1852, Berkeley, CA (1952).
13. Bromley, L. A., and C. R. Wilke, *Ind. Eng. Chem.*, **43** (1951): 1641.
14. Caldwell, C. S., and A. L. Babb, *J. Phys. Chem.*, **60** (1956): 51.
15. Chase, M. W., Jr., C. A. Davies, J. R. Downey, Jr., D. J. Frurip, R. A. McDonald, and A. N. Syverud, "JANAF Thermochemical Tables," *J. Phys. Chem. Ref. Data*, **14**, suppl. 1 (1985).
16. Chickos, J. S., C. M. Braton, D. G. Hesse, and J. F. Liebman, *J. Org. Chem.*, **56** (1991): 927.
17. Chueh, C. F., and A. C. Swanson, *Chem. Eng. Prog.*, **69**, 7 (1973): 83.
18. Chueh, P. L., and J. M. Prausnitz, *AIChE J.*, **13** (1967): 1099.
19. Chueh, P. L., and J. M. Prausnitz, *AIChE J.*, **13** (1967): 1107.
20. Constantinou, L., and R. Gani, *AIChE J.*, **40** (1994): 1697.
21. Cox, J. D., D. D. Wagman, and V. A. Medvedev, *CODATA Key Values for Thermodynamics*, Hemisphere Publishing Corp., New York, 1989.
22. Danner, R. P., and T. E. Daubert, *Manual for Predicting Chemical Process Design Data*, Design Institute for Physical Property Data, AIChE, New York, extant 1989.
23. Daubert, T. E., and R. P. Danner, *Technical Data Book - Petroleum Refining*, 5th ed., American Petroleum Institute, Washington, DC, extant 1994.
24. Daubert, T. E., R. P. Danner, H. M. Sibul, and C. C. Stebbins, *Physical and Thermodynamic Properties of Pure Chemicals: Data Compilation*, Taylor & Francis, Bristol, PA, extant 1994.
25. Dean, D. E., and L. I. Stiel, *AIChE J.*, **11**, (1965): 526.
26. De Soria, M. L. G., J. L. Zurita, M. A. Postigo, and M. Katz, *J. Colloid Interface Sci.*, **103** (1985): 354.
27. Dixon, S. L., and P. C. Jurs, *J. Comput. Chem.*, **13** (1992): 492.
28. Fairbanks, D. F., and C. R. Wilke, *Ind. Eng. Chem.*, **42** (1950): 471.
29. Fedors, R. F., *AIChE J.*, **25** (1979): 202.
30. Fuller, E. N., P. D. Shettler, and J. C. Giddings, *Ind. Eng. Chem.*, **58** (1966): 19.
31. Gambill, W. R., *Chem. Eng.*, **66**, 15 (1959): 181.
32. Gilliland, E. R., *Ind. Eng. Chem.*, **26** (1934): 681.
33. Gmehling, J., and P. Rasmussen, *Ind. Eng. Chem. Fund.*, **21** (1982): 186, 321.
34. Haggenmacher, J. E., *J. Am. Chem. Soc.*, **68** (1946): 1633.
35. Hankinson, R. W., and G. H. Thomson, *AIChE J.*, **25** (1979): 653.
36. Harrison, B. K., and W. H. Seaton, *Ind. Eng. Chem. Res.*, **27** (1988): 1536.
37. Hayduk, W., and W. H. Laudie, *AIChE J.*, **20** (1974): 611.
38. High, M. S., and R. P. Danner, *Ind. Eng. Chem. Research*, **26** (1987): 1395.
39. High, M. S., and C. T. Siegel, Private Communication, Dept. Chem. Eng., Penn State Univ., 1985.
40. Hirschfelder, J. O., C. F. Curtiss, and R. B. Bird, *Molecular Theory of Gases and Liquids*, John Wiley and Sons, New York, 1954.
41. Hurst, J. E., and B. K. Harrison, *Chem. Eng. Comm.*, **112** (1992): 21.
42. Jamieson, D. T., and E. H. Hastings, *Thermal Conductivity*, Proceedings of the Eighth Conference, C. Y. Ho and R. E. Taylor, ed., Plenum Press, New York, 1969.
43. Jasper, J. J., *J. Phys. Chem. Ref. Data*, **1** (1972): 841.
44. Joback, K. G., *A Unified Approach to Physical Property Estimation Using Multivariate Statistical Techniques*, M.S. Thesis, Massachusetts Institute of Technology, Cambridge, MA, 1984.
45. Jossi, J. A., L. I. Stiel, and G. Thodos, *AIChE J.*, **8** (1962): 59.
46. Kanitkar, D., and G. Thodos, *Can. J. Chem. Eng.*, **47** (1969): 427.
47. Kay, W. B., *Ind. Eng. Chem.*, **28** (1936): 1014.
48. Kendall, J., and K. P. Monroe, *J. Am. Chem. Soc.*, **39** (1917): 1787.
49. King, C. J., L. Hsueh, and K. Mao, *J. Chem. Eng. Data*, **10** (1965): 348.
50. Kopp, H., *Ann. Chem. Pharm. (Leibig)*, **126** (1863): 362.
51. Kouzel, B., *Hydrocarbon Processing*, **44**, 3 (1965): 120.
52. Kreglewski, A., and W. B. Kay, *J. Phys. Chem.*, **73** (1969): 3359.
53. Landolt-Bornstein, *Numerical Data*, Springer-Verlag, New York, 1961+.
54. Lee, B. I., J. H. Erbar, and W. C. Edmister, *AIChE J.*, **19** (1973): 349.
55. Lee, B. I., and M. G. Kesler, *AIChE J.*, **21** (1975): 510.
56. Leffler, J., and J. T. Cullinan, *Ind. Eng. Chem. Fundam.*, **9** (1970): 88.
57. Lenoir, J. M., *Petroleum Refiner*, **36** (1957): 162.
58. Letsou, A., and L. I. Stiel, *AIChE J.*, **19** (1973): 409.
59. Li, C. C., *AIChE J.*, **22** (1976): 927.
60. Li, C. C., *Can. J. Chem. Eng.*, **49** (1971): 709; Errata **50** (1972): 152.
61. Lindsay, A. L., and L. A. Bromley, *Ind. Eng. Chem.*, **42** (1950): 1508.
62. Lu, B. C. Y., *Chem. Eng.*, **66**, 9 (1959): 137.
63. Lydersen, A. L., R. A. Greenkorn, and O. A. Hougen, *Univ. Wisconsin Coll. Eng. Exp. Stn. Rept. 3*, Madison, WI, 1955.
64. Lyman, W. J., W. F. Reehl, and D. H. Rosenblatt, "Handbook of Chemical Property Estimation Methods", McGraw Hill, New York, 1982.
65. Macleod, D. B., *Trans. Faraday Soc.*, **19** (1923): 38.
66. Maczek, A. O. S., and P. Gray, *Trans. Faraday Soc.*, **65** (1969): 1473.
67. Maxwell, J. B., and L. S. Bonnell, *Vapor Pressure Charts for Petroleum Engineers*, Esso Res. and Engrg. Co., Linden, NJ, 1955.
68. McClellan, A. L., *Tables of Experimental Dipole Moments*, Freeman, San Francisco, 1963, and Rahara, El Cerrito, CA, 1974 and 1989.
69. Miller, D. G., *Ind. Eng. Chem. Fundam.*, **2** (1963): 78.
70. Misic, D., and G. Thodos, *AIChE J.*, **7** (1961): 264.
71. Misic, D., and G. Thodos, *J. Chem. Eng. Data*, **9** (1963): 540.
72. Missenard, A., *Conductivité Thermique des Solides, Liquides, Gaz et de Leurs Mélanges*, Éditions Eyrolles, Paris (1965).
73. Missenard, A., *C. R. Acad. Sc. Paris*, **260**, Group 7 (1965): 5521.
74. Missenard, A., *Rev. Gen. Thermodyn.*, **101** (1970): 649.
75. Myers, K. H., *Thermodynamic and Transport Property Prediction Methods for Organometallic Compounds*, M.S. Thesis, The Pennsylvania State University, University Park, PA, 1990.
76. Nokay, R., *Chem. Eng.*, **66**, 4 (1959): 147.
77. Othmer, D. F., P. W. Maurer, C. S. Molinary, and R. C. Kowalaski, *Ind. Eng. Chem.*, **49** (1957): 125.
78. Othmer, D. F., and E. S. Yu, *Ind. Eng. Chem.*, **60** (1968): 22.
79. Pachaiyappan, V., S. H. Ibrahim, and N. R. Kuloor, *Chem. Eng.*, **74** (1967): 140.
80. Pailhes, F., *Fluid Phase Equil.*, **41** (1988): 97.
81. Peng, D. Y., and D. B. Robinson, *Ind. Eng. Chem. Fundam.*, **15** (1976): 59.
82. Pitzer, K. S., D. Z. Kippman, R. F. Curl, Jr., C. M. Huggins, and D. E. Petersen, *Am. Chem. Soc.*, **77** (1955): 3433.
83. Quale, O. R., *Chem Rev.*, **53** (1953): 439.
84. Rackett, H. G., *J. Chem. Eng. Data*, **15** (1970): 514.
85. Reichenberg, D., *AIChE J.*, **21** (1975): 181.
86. Reid, R. C., J. M. Prausnitz, and B. E. Poling, *The Properties of Gases and Liquids*, 4th ed., McGraw-Hill, New York, 1987.
87. Rice, O. K., *J. Phys. Chem.*, **64** (1960): 976.
88. Riedel, L., *Chem. Ingr. Tech.*, **21** (1949): 349.
89. Riedel, L., *Chem. Ingr. Tech.*, **26** (1954): 83.
90. Riedel, L., *Chem. Ingr. Tech.*, **26** (1954): 679.
91. Riedel, L., *Chem. Ingr. Tech.*, **26** (1954): 679; **28** (1956): 557.
92. Ritter, R. B., J. M. Lenoir, and J. L. Schweppe, *Petrol. Refiner*, **37**, 11 (1958): 225.
93. Ruzicka, V., and E. S. Domalski, *J. Phys. Chem. Ref. Data*, **22** (1993): 597.
94. Ruzicka, V., and E. S. Domalski, *J. Phys. Chem. Ref. Data*, **22** (1993): 619.
95. Seaton, W. H., *J. Hazard Mat.*, **27** (1991): 169.
96. Shebeko, N. Y., A. Y. Korolchenko, A. V. Ivanov, and E. N. Alekhina, *Soviet Chem. Ind.*, **16** (1984): 1371.
97. Shebeko, Y. N., A. V. Ivanov, and T. M. Dmitrieva, *Soviet Chem. Ind.*, **15** (1983): 311.
98. Soave, G., *Chem. Eng. Science*, **27** (1972): 1197.
99. Spencer, C. F., and R. P. Danner, *J. Chem. Eng. Data*, **17** (1972): 236.
100. Spencer, C. F., and R. P. Danner, *J. Chem. Eng. Data*, **18** (1973): 230.
101. Stiel, L. I., and G. Thodos, *AIChE J.*, **7** (1961): 611.
102. Stiel, L. I., and G. Thodos, *AIChE J.*, **10** (1964): 26.
103. Stiel, L. I., and G. Thodos, *AIChE J.*, **10** (1964): 275.
104. Stull, D. R., F. F. Westrum, Jr., and G. C. Sinke, *The Chemical Thermodynamics of Organic Compounds*, John Wiley and Sons, New York, 1969.
105. Stuper, A. J., W. E. Brugger, and P. C. Jurs, *Computer Assisted Studies of Chemical Structure and Biological Function*, John Wiley and Sons, New York, 1979.
106. Sugden, S., *J. Chem. Soc.*, **125** (1924): 32.
107. Sugden, S., *J. Chem. Soc.*, **125** (1924): 1177.
108. Suzuki, T., K. Ohtaguchi, and K. Kodie, *J. Chem. Eng. Japan*, **25** (1992): 606.
109. Takahashi, S., *J. Chem. Eng. Japan*, **7** (1974): 417.
110. Tamura, M., M. Kurata, and H. Odani, *Bull. Chem. Soc. Japan*, **28** (1955): 83.

111. Tang, Y. P., and D. M. Himmelblau, *AIChE J.*, **11** (1965): 54.
112. Tarakad, R. R., and R. P. Danner, *AIChE J.*, **23** (1977): 944.
113. Thinh, T.-P., J.-L. Duran, and R. S. Romalho, *Ind. Eng. Chem. Process Des. Dev.*, **10** (1971): 576.
114. Thomson, G. H., K. P. Brobst, and R. W. Hankinson, *AIChE J.*, **28** (1982): 671.
115. *TRC Thermodynamic Tables—Hydrocarbons,* Thermodynamics Research Center, The Texas A&M University System, College Station, TX, extant 1994.
116. *TRC Thermodynamic Tables—Non-Hydrocarbons,* Thermodynamics Research Center, The Texas A&M University System, College Station, TX, extant 1994.
117. Tsonopoulos, C., *AIChE J.*, **20** (1974): 263.
118. Tsonopoulos, C., *AIChE J.*, **21** (1975): 827.
119. Tsonopoulos, C., *AIChE J.*, **24** (1978): 1112.
120. Tyn, M. T., and W. F. Calus, *Processing*, **21,** 4 (1975): 16.
121. Umesi, N. O., *Correlating Diffusion Coefficients in Dilute Liquid Mixtures,* M.S. Thesis, Penn State University, University Park, PA, 1980.
122. van Velzen, D., R. L. Cardozo, and H. Langenkamp, *Ind. Eng. Chem. Fundam.*, **11** (1972): 20.
123. Vargaftik, N. B., *Tables on the Thermophysical Properties of Liquids and Gases,* 2d ed., Hemisphere Publishing Corp., Washington, DC, 1975.
124. Vargaftik, N. B., L. P. Filippov, A. A. Tarzimanov, and E. E. Totskii, *Handbook of Thermal Conductivity of Liquids and Gases,* CRC Press, Boca Raton, FL, 1994.
125. Wagner, W., *Cryogenics*, **13** (1973): 470.
126. Wassiljewa, A., *Physik. Z.*, **5** (1904): 737.
127. Watson, K. M., *Ind. Eng. Chem.*, **35** (1943): 398.
128. Wilke, C. R., *Chem. Eng. Prog.*, **46** (1950): 95.
129. Wilke, C. R., and P. Chang, *AIChE J.*, **1** (1955): 264.
130. Winterfeld, P. H., L. E. Scriven, and H. T. Davis, *AIChE J.*, **24** (1978): 1010.
131. Wright, W. A., *ASLE Trans.*, **10** (1967): 349.
132. Yoon, P., and G. Thodos, *AIChE J.*, **16** (1970): 300.

PURE COMPONENT CONSTANTS

Basic pure component constants required to characterize components or mixtures for calculation of other properties include the melting point, normal boiling point, critical temperature, critical pressure, critical volume, critical compressibility factor, acentric factor, and several other characterization properties. This section details for each property the method of calculation for an accurate technique of prediction for each category of compound, and it references other accurate techniques for which space is not available for inclusion.

Critical Temperature   The critical temperature of a compound is the temperature above which a liquid phase cannot be formed, no matter what the pressure on the system. The critical temperature is important in determining the phase boundaries of any compound and is a required input parameter for most phase equilibrium thermal property or volumetric property calculations using analytic equations of state or the theorem of corresponding states. Critical temperatures are predicted by various empirical methods according to the type of compound or mixture being considered.

For **pure hydrocarbons,** the method of Ambrose[2] is the most accurate and will also be useful for predicting critical pressure and volume. Equation (2-1) requires only the normal boiling point, T_b, and the molecular structure of the compound.

$$T_c = T_b \left[1 + \frac{1}{1.242 + \sum \Delta_T - 0.023\Delta(\text{Platt no.})} \right] \quad (2\text{-}1)$$

T_c and T_b are the critical and normal boiling temperatures, respectively, expressed in kelvins. Values of Δ_T from Table 2-383 are summed for each part of the molecule to yield $\sum \Delta_T$ (e.g., for isobutane, $3 \times$ —$CH_3 + 1 \times$ >CH—). Δ(Platt no.) is equal to the Platt number of any alkyl chains in the molecule minus the Platt number of the *n*-alkane with the same number of carbon atoms. The Platt number is defined as the number of pairs of carbon atoms that are separated by three carbon-carbon bonds and is an indicator of branching (e.g., for 2,2,3-trimethylpentane, the Platt number is 5). The Platt number of an *n*-alkane is the number of carbon atoms minus three (e.g., for *n*-octane, the Platt number is five). Errors in T_c average about 4 K for paraffins to C_{20} and other hydrocarbons to C_{14}.

Equation (2-2), another somewhat simpler method for estimating the critical temperature of pure hydrocarbons only, is the method of Nokay[76] and requires the normal boiling point, the relative density, and the compound family.

$$\log T_c = A + B \log_{10} (\text{rel den}) + C \log T_b \quad (2\text{-}2)$$

T_c and T_b are the critical and normal boiling temperatures, respectively, expressed in kelvins. The relative density (rel den) of the liquid at 15°C is 0.1 MPa. The regression constants A, B, and C are tabulated by family in Table 2-384. Errors average about 3 K.

For **pure nonhydrocarbon organics,** the most accurate method for prediction of critical temperature for all compound groups is also the Ambrose[2] method. Equation (2-1) applies to all nonhydrocarbon compounds except perfluorocarbons, where the constant 1.242 is replaced by 1.570. For compounds containing any of C, H, O, N, S, or halogens up to C_{13} and ranging in critical temperature from 228–790 K, the average error is about 6 K.

Alternate methods for nonhydrocarbon organics are the first order method of Lydersen[63] with an average error of 9 K although the method of Ambrose is considerably better for alcohols and ketones.

Equation (2-3) is the Lydersen equation for critical temperature and requires only the normal boiling point and the molecular structure for solution.

$$T_c = \frac{T_b}{[0.567 + \sum \Delta_T - (\sum \Delta_T)^2]} \quad (2\text{-}3)$$

Contributions to $\sum \Delta_T$ are given in Table 2-385.

For **pure inorganic compounds,** the method of Gambill[31] was modified to yield Eq. (2-4) and only requires the normal boiling point as input.

$$T_c = 1.64 T_b \quad (2\text{-}4)$$

Although this equation was tested with available experimental data (38 compounds), it can only be considered a rough approximation. **Inorganic-organic** and **inorganic-halide** compounds are predicted better by replacing the constant 1.64 in Eq. (2-4) by 1.55.

For both **hydrocarbons** and **nonhydrocarbon organic defined mixtures,** the method of Li[60] is used with a relatively simple volumetric average mixing rule as shown in Eq. (2-5) to calculate the true critical temperature.

$$T_{cm} = \sum_j \left(\frac{y_j V_{c_j}}{\sum_i y_i V_{c_i}} \right) T_{c_j} \quad (2\text{-}5)$$

T_{cm} is the mixture critical temperature in K. V_c is the critical volume of a component, m³/kmole. The mole fraction of a component is y. The mixture contains i components.

For hydrocarbon systems, the average error is about 3 K, while for systems containing nonhydrocarbons, the average error is 15 K with the highest errors occurring where simple gases are present. The method of Chueh and Prausnitz[18] yields average errors slightly lower than the method of Li, but it is computationally more complex.

Critical Pressure   The critical pressure of a compound is the vapor pressure of the compound at the critical temperature. Below the critical temperature, any compound above its vapor pressure will be a liquid. The critical pressure is required for calculations discussed in the part of the section on critical temperature.

For **pure hydrocarbons,** the method of Ambrose[2] is the most accurate. Equation (2-6) requires only the molecular weight (M) and the molecular structure of the compound.

$$P_c = \frac{0.101325M}{[0.339 + \sum \Delta_P - 0.026\Delta(\text{Platt no.})]^2} \quad (2\text{-}6)$$

P_c is the critical pressure, MPa. Values of Δ_P from Table 2-383 are summed for each part of the molecule to yield $\sum \Delta_P$. Calculation of the Platt number is discussed under "Critical Temperature." Errors in P_c average 0.07 MPa and are less reliable for compounds with 12 or more carbon atoms.

TABLE 2-383 Group Increments for the Ambrose Method

Group description	Δ_T	Δ_P	Δ_V	Group description	Δ_T	Δ_P	Δ_V
—CH₃	0.138	0.2260	55.1	**Aromatic Compounds (Cont.)**			
CH₂	0.138	0.2260	55.1		0.468	0.8840	222
CH—	0.095	0.2200	47.1				
C	0.018	0.1960	38.1		0.468	0.8840	222
=CH₂	0.113	0.1935	45.1				
=CH—	0.113	0.1935	45.1		0.418	0.8040	222
=C<	0.070	0.1875	37.1				
=C=	0.088	0.1610	35.1		0.368	0.7240	222
≡CH	0.038	0.1410	35.1				
≡C—	0.038	0.1410	35.1		0.220	0.5150	148
Ring Increments				in fused ring°			
CH₂	0.090	0.1820	44.5	**Nonring Increments**			
CH—	0.090	0.1820	44.5	—OH	Use Eq. (a), below	Use Eq. (b), below	
CH— in fused ring°	0.030	0.1820	44.5	—O—	0.138	0.160	
C	0.090	0.1820	44.5	CO	0.220	0.282	
=CH—	0.075	0.1495	37.0	—CHO	0.220	0.220	
=C<	0.075	0.1495	37.0	—COOH	0.578	0.450	
=C=	0.060	0.1170	29.5	—CO—O—OC—	1.156	0.900	
Aromatic Compounds				—CO—O—	0.330	0.470	
	0.458	0.9240	222	—NO₂	0.370	0.420	
				—NH₂	0.208	0.095	
	0.448	0.8940	222	—NH—	0.208	0.135	
				N	0.088	0.170	
	0.488	0.9440	222	—CN	0.423	0.360	
				—S—	0.105	0.270	
	0.488	0.9440	222	—SH	0.090	0.270	
				—Si—	0.138	0.461	
	0.438	0.8640	222	—SiH—	0.371	0.507	
				—SiH₃	0.195	—	
	0.478	0.9140	222	—Si—O—	0.159	0.725	
	0.428	0.8340	222	[—Si—O—]cyclic	0.131	0.663	

TABLE 2-383   Group Increments for the Ambrose Method (Concluded)

Group description	Δ_T	Δ_P	Δ_V	Group description	Δ_T	Δ_P	Δ_V
Double Bond	−0.050	−0.065		—Cl	0.080	0.318	
Triple Bond	−0.200	−0.170		—Br	0.080	0.600	
—F	0.180 (1st) 0.055	0.223		—I		0.850	
—Cl	0.110 (1st) 0.055	0.318		The single or first substituent			
—Br	0.110 (1st) 0.055	0.500		on an aromatic ring	0.010	0	
Aromatic Compounds				The second or subsequent ring			
Corrections				substituents	0.030	0.020	
benzene	0.448	0.924		Each pair of ring substituents			
pyridine	0.448	0.850		in ortho positions with respect			
—OH	0.198	−0.025		to each other	−0.040	−0.050	
C_4H_4 (fused ring)	0.220	0.515		If one of the ortho pair is —OH	−0.080	0	
—F	0.080	0.183					

*Group contributions for $>$CH and $<$ in fused rings have been calculated from minimal data and may be less reliable than the other values.

Substituent increments do not apply to halogens.
Eq. (a): Δ_T (—OH) $= 0.87 − 0.11n + 0.003n^2$
Eq. (b): Δ_P (—OH) $= 0.100 − 0.013n$
n = carbon number of compound. For branched alcohols, an effective carbon number can be determined by interpolation between the normal alcohol of the same carbon number and the immediately lower normal alcohol using normal boiling points. (See Example 1.)

TABLE 2-384   Nokay Equation

Type compound	A	B	C
Paraffin	1.35940	0.43684	0.56224
Naphthene	0.65812	−0.07165	0.81196
Olefin	1.09534	0.27749	0.65563
Acetylene	0.74673	0.30381	0.79987
Diolefin	0.1384	−0.39618	0.99481
Aromatic	1.0615	0.22732	0.66929

Example 1   Estimate the critical temperature and critical pressure of 2-butanol using the Ambrose method, Eqs. (2-1) and (2-6). The experimental normal boiling point is 372.7 K.

$$T_c = T_b \left[1 + \frac{1}{1.242 + \sum \Delta_T − 0.023\Delta \text{ (Platt no.)}} \right]$$

$$P_c = \frac{0.101325M}{[0.339 + \sum \Delta_P − 0.026\Delta \text{ (Platt no.)}]^2}$$

Determine group contributions from Table 2-383 for a structure $CH_3CHOHCH_2CH_3$:

Group	Number of groups	Δ_T	Δ_P
— CH$_3$	2	0.138	0.2260
$>$CH$_2$	1	0.138	0.2260
$>$CH$^-$	1	0.095	0.2200
— OH	1	(a)	(b)

$\Delta_T°$(—OH) $= 0.87 − 0.11(n) + 0.003 \, n^2 = 0.558$
$\Delta_P°$(—OH) $= 0.100 − 0.013(n) = 0.0597$

To determine n, the normal boiling points of 1-propanol (370.3 K) and 1-butanol (390.9 K) are required:

$$n = 3 + \left(\frac{372.7 − 370.3}{390.9 − 370.3} \right)(4 − 3) = 3.1$$

The Platt number of the compound is determined by substituting a CH_3 group for the OH.

$$CH_3—CH(CH_3) − CH_2CH_3, \text{ Platt no.} = 2$$

The Platt number of the n-alkane of the same carbon number is $5 − 3 = 2$. Thus, Δ(Platt no.) $= 2 − 2 = 0$. Therefore:

$$\sum \Delta_T = 2(.138) + .138 + .095 + .558 = 1.067$$

$$T_c = 372.7 \left[1 + \frac{1}{1.242 + 1.067 − 0} \right]$$

$$T_c = 534.1 \text{ K}$$

An accurate experimental value is 536.05 K.

$$\sum \Delta_P = 2(.226) + .226 + .220 + .0597 = 0.9577$$

$$P_c = \frac{.101325(74.12)}{(.339 + .9577 − 0)^2}$$

$$P_c = 4.467 \text{ MPa}$$

An accurate experimental value is 4.179 MPa.

For **pure nonhydrocarbon organics,** the simplest accurate method for prediction of critical pressure is the method of Lydersen.[63] Equation (2-7) requires the molecular weight (M) and the molecular structure of the compound.

$$P_c = \frac{0.101325M}{(0.34 + \sum \Delta_P)^2} \qquad (2-7)$$

P_c is the critical pressure, MPa. Values of Δ_P from Table 2-385 are summed to yield $\sum \Delta_P$. The average error in P_c is about 0.2 MPa when tested on compounds ranging in carbon number from C_1 to C_{10}.

Example 2   Estimate the critical temperature and critical pressure of 2-butanol, which has an experimental normal boiling point of 372.7 K. Use the Lydersen method, Eqs. (2-3) and (2-7).

The structure of 2-butanol is $CH_3CHOHCH_2CH_3$. Determine group contributions from Table 2-385.

Group	Number of groups	Δ_T	Δ_P
— CH$_3$	2	0.020	0.227
— CH$_2$ —	1	0.020	0.227
— CH —	1	0.012	0.210
— OH	1	0.082	0.06
		$\sum$ 0.154	0.951

$$T_c = \frac{T_b}{0.567 + \sum \Delta_T − (\sum \Delta_T)^2} = \frac{372.7}{.567 + .154 − (.154)^2}$$

$$T_c = 534.5 \text{ K}$$

TABLE 2-385 Group Increments for the Lydersen Method

Group description	Incremental contributions			Group description	Incremental contributions		
	Δ_T	Δ_P	Δ_V		Δ_T	Δ_P	Δ_V
Nonring Elements				**Oxygen Increments (*Cont.*)**			
—CH₃, —CH₂—	0.020	0.227	0.055				
					0.085	(0.4)	0.080
—CH—	0.012	0.210	0.051	—COOH (acid)			
					0.047	0.47	0.080
				—COO— (ester)			
					(0.02)	(0.12)	(0.011)
—C—	0.00	0.210	0.041	=O (except for combinations above)			
=CH₂, =CH	0.018	0.198	0.045	**Halogen Increments**			
				—F	0.018	0.224	0.018
				—Cl	0.017	0.320	0.049
=C—, =C=	0.00	0.198	0.036	—Br	0.010	(0.50)	(0.070)
				—I	0.012	(0.83)	(0.095)
≡CH, ≡C—	0.005	0.153	(0.036)	**Nitrogen Increments**			
				—NH₂	0.031	0.095	0.028
Ring Increments					0.031	0.135	(0.037)
—CH₂—	0.013	0.184	0.0445	—NH (nonring)			
					(0.024)	(0.09)	(0.027)
—CH	0.012	0.192	0.046	—NH (ring)			
					0.014	0.17	(0.042)
—C—	(−0.007)	0.154	(0.031)	—N— (nonring)			
					(0.007)	(0.13)	(0.032)
=C—, =C=	0.011	0.154	0.036	—N— (ring)			
=CH	0.011	0.154	0.037	—CN	(0.060)	(0.36)	(0.080)
				—NO₂	(0.055)	(0.42)	(0.078)
⬡	0.066	0.924		**Sulfur Increments**			
				—SH	0.015	0.27	0.055
				—S— (nonring)	0.015	0.27	0.055
				—S— (ring)	(0.008)	(0.24)	(0.045)
				=S	(0.003)	(0.24)	(0.047)
Oxygen Increments				**Organometallic Increments**			
—OH (alcohols)	0.082	0.06	(0.018)	—Si—	0.026	0.468	—
—OH (phenols)	0.031	(−0.02)	(0.003)				
—O— (nonring)	0.021	0.16	0.020	—SiH	0.040	0.513	—
—O— (ring)	(0.014)	(0.12)	(0.008)	—SiH₃	0.027	—	—
—C=O (nonring)	0.040	0.29	0.060	—Si—O—	0.025	0.730	—
—C=O (ring)	(0.033)	(0.2)	(0.050)	[—Si—O—] *cyclic*	0.027	0.668	—
HC=O (aldehyde)	0.048	0.33	0.073				

Values in parentheses are based on too few experimental points to be reliable.

The accurate experimental critical temperature is 536.05 K.

$$P_c = \frac{0.101325M}{(0.34 + \sum \Delta_P)^2} = \frac{(0.101325)(74.12)}{(0.34 + 0.951)^2}$$

$$P_c = 4.506 \text{ MPa}$$

The accurate experimental critical pressure is 4.179 MPa. No known method is available to predict the critical pressure of inorganic compounds.

For both **hydrocarbon** and **nonhydrocarbon organic defined mixtures,** the method of Kreglewski and Kay[52] is recommended. The critical temperature, critical pressure, and acentric factor of each compound and the critical temperature of the mixture must be known or predicted from the methods of this section.

$$P_{cm} = P_{pc} + P_{pc} \left[5.808 + 4.93 \left(\sum_{i=1}^{n} x_i \omega_i \right) \right] \left[\frac{T_{cm} - T_{pc}}{T_{pc}} \right] \quad (2\text{-}8)$$

Use of Eq. (2-8) requires the pseudocritical properties defined by Eqs. (2-9) and (2-10)

$$T_{pc} = \sum_{i=1}^{n} x_i T_{c_i} \quad (2\text{-}9)$$

$$P_{pc} = \sum_{i=1}^{n} x_i P_{c_i} \quad (2\text{-}10)$$

Each component i of the mixture must have available its T_c, P_c, and ω. T_{cm} can be predicted from Eq. (2-5). For hydrocarbon systems, average errors in predicted critical pressures are about 0.2 MPa, except when organic gases are present and errors are unacceptably large. Errors for nonhydrocarbon organics not including inorganic gases average 0.5 MPa.

Critical Volume The critical volume of a compound is the volume occupied by a set mass of a compound at its critical temperature and pressure. While useful in itself, the critical volume is extensively used in equations for estimating volumetric fractions.

For **pure hydrocarbons,** two methods are quite accurate. The Ambrose[2] method used for T_c and P_c is also used for critical volume. Eq. (2-11) only requires the molecular structure of the compound.

$$V_c = 10^{-3}(40 + \sum \Delta_V) \quad (2\text{-}11)$$

V_c is the critical volume, m^3/kmole. Values of Δ_V are given in Table 2-383.

The average error for hydrocarbons of twelve or less carbon atoms is about 0.01 m^3/kmole.

The Riedel method[90] requires the critical temperature (T_c), critical pressure (P_c), and acentric factor (ω) of the compound as given by Eqs. (2-12) and (2-13). If the gas constant is in Pa·m^3/kmole·K, the critical volume will be in m^3/kmole.

$$V_c = \frac{RT_c}{P_c[3.72 + 0.26(\alpha - 7.00)]} \quad (2\text{-}12)$$

$$\alpha = 5.811 + 4.919\omega \quad (2\text{-}13)$$

The average error for paraffins up to C_{18} and other hydrocarbon families up to C_{11} is about 0.015 m^3/kmole. If estimated values of T_c and P_c are used, errors may be higher.

For **pure nonhydrocarbon organics,** the method of Fedors,[29] which requires only molecular structure, is the most accurate. Equation (2-14) shows the method to depend only on the molecular structure. The resulting V_c will be in m^3/kmole.

$$V_c = 0.0266 + \sum \Delta_V \quad (2\text{-}14)$$

Values for Δ_V are given in Table 2-386. The average error for compounds up to C_7 is about 0.007 m^3/kmole, although a maximum error of 0.03 m^3/kmole has been noted. No experimental data above C_7 are available for comparison.

Example 3 Estimate the critical volume of 2-butanol. Use the method of Fedors, Eq. (2-14).

$$V_c = 0.0266 + \sum \Delta_V$$

The molecular formula is $C_4H_{10}O$. Using the atomic contribution values from Table 2-386.

$$\sum \Delta_V = 4(.034426) + 10(.009172) + 1(.018000)$$
$$\sum \Delta_V = 0.2474$$
$$V_c = 0.0266 + 0.2474 = 0.2740 \text{ m}^3/\text{kmole}$$

The accurate experimental critical volume is 0.2690 m^3/kmole.

TABLE 2-386 Atomic and Structural Contributions for the Fedors Method

Atomic increments		Structural increments	
Atom	Δ_V	Feature	Δ_V
C	0.034426	Cl	0.052801
H	0.009172	Br	0.071774
O	0.020291	I	0.096402
O (alcohols)	0.018000	S	0.050866
N	0.048855	3-membered ring	−0.105824
N (amines)	0.047422	4-membered ring	−0.017247
F	0.022242	5-membered ring	−0.039126
		6-membered ring	−0.039508
Si	0.086174	double bond	+0.005028
Si$_{\text{siloxane}}$	0.126483	triple bond	+0.000797
Si$_{\text{cyclic siloxane}}$	0.126483	ring attached directly to another ring	+0.035524

The method of Lydersen[63] may also be used for prediction of critical volume, but it is not so accurate as the method of Fedors. Equation (2-15) depends only on molecular structure and gives a critical volume in m^3/kmole.

$$V_c = 0.040 + \sum \Delta_V \quad (2\text{-}15)$$

Group contributions for Δ_V are given in Table 2-385. Errors average about 0.01 m^3/kmole.

There is no known method for predicting the critical volume of inorganic compounds.

For both **hydrocarbon** and **nonhydrocarbon organic defined mixtures,** the method of Chueh and Prausnitz[9] is useful. For hydrocarbon systems, the mixing rule is shown by Eq. (2-16) for binaries and by Eq. (2-17) for multicomponents. Equations (2-18) through (2-20) give the input parameters. The mixture critical volume V_{cm} is a function of the pure component critical volumes. The constant C is zero for hydrocarbon systems and 0.1559 for systems containing a nonhydrocarbon gas.

$$V_{cm} = \phi_1 V_{c1} + \phi_2 V_{c2} + 2\phi_1\phi_2 v_{12} \quad (2\text{-}16)$$

$$V_{cm} = \sum_{i}^{n} \sum_{j}^{n} \phi_i \phi_j v_{ij} \quad (i \neq j) \quad (2\text{-}17)$$

$$\phi_j = \frac{x_j V_{cj}^{2/3}}{\sum_{i=1}^{n} x_i V_{ci}^{2/3}} \quad (2\text{-}18)$$

$$v_{ij} = \frac{V_{ij}(V_{ci} + V_{cj})}{2.0} \quad (2\text{-}19)$$

$$V_{ij} = -1.4684\left(\left|\frac{V_{ci} - V_{cj}}{V_{ci} + V_{cj}}\right|\right) + C \quad (2\text{-}20)$$

Errors average about 10 percent for systems containing hydrocarbons.

Systems containing only organics or organics and gases may give very high errors. A specialized modification of the method (Chueh and Prausnitz[18]) is available for binary mixtures containing organics but may give errors over 20 percent.

Critical Compressibility Factor The critical compressibility factor of a compound is calculated from the experimental or predicted values of the critical properties by the definition, Eq. (2-21).

$$Z_c = \frac{P_c V_c}{RT_c} \quad (2\text{-}21)$$

Critical compressibility factors are used as characterization parameters in corresponding states methods (especially those of Lydersen) to predict volumetric and thermal properties. The factor varies from about 0.23 for water to 0.26–0.28 for most hydrocarbons to slightly above 0.30 for light gases.

Normal Freezing Temperature (Melting Point) The melting point is the temperature at which melting occurs at atmospheric pressure. In most cases, measurements are made in air, making values slightly lower than if the measurements were made in vacuum. Impurities can cause a substantial decrease in the measured melting point. The melting point is very slightly higher than the triple point temperature—the temperature at which equilibrium exists between solid, liquid, and vapor—for a pure compound. For practical purposes, the two temperatures are equal. Reliable methods for predicting melting points have not until recently been advanced. Constantinou and

Gani[20] derived a group contribution method that shows promise. Initial evaluations show average errors of 5 to 10 percent (10–30 K) on a wide variety of compounds, but larger errors can occur. It is recommended that several compounds of known melting point in the same or a similar family be predicted in order to estimate the probable error.

Normal Boiling Temperature   The normal boiling temperature (point) is the temperature at which the vapor pressure equals exactly 101,325 Pa (1 atmosphere). Caution should be taken in using values from older references, where the temperature may be reported for the prevailing pressure (0.95–0.97 atm) rather than at 1 atmosphere. If at least two values of vapor pressure very close to 1 atmosphere are available, the normal boiling point can be interpolated or extrapolated on a plot of $\log P^{sat}$ vs. $1/T$. The section on vapor pressure discusses this in more detail.

Various methods are available for estimation of the normal boiling point of organic compounds. Lyman et al.[64] review and give calculational procedures for the methods of Meissner, Miller, and Lydersen/Forman-Thodos. A more recent method that has been determined to be more accurate is the method of Pailhes,[80] which requires one experimental vapor pressure point and Lydersen group contributions for critical temperature and critical pressure (Table 2-385).

$$T_b = T \frac{\log P_c + (1 - \Theta) \log (1/p)}{\log P_c} \qquad (2\text{-}22)$$

where T and p = the one low pressure vapor pressure point, in K and atm, respectively

$\Theta = T_c/T_b$ calculated from Eq. (2-3) using Lydersen contributions

P_c = critical pressure calculated from Eq. (2-7) using Lydersen contributions

A recent study of the method on a wide variety of complex organics shows an overall average error of less than 2 percent (~10 K). If no vapor pressure point is available, the new group contribution method of Constantinou and Gani[20] discussed under the section on melting point gives an overall average error of about 4 percent (~20 K) and may be useful. The method of Miller (Lyman et al.[64]), which requires only the molecular structure, has also been found to be relatively accurate for organics.

Example 4   Estimate the normal boiling point of 2-butanol. One vapor pressure point of 0.802 psia at 100°F is available.
Use the Pailhes method, Eq. (2-22).

$$T_B = T \frac{\log P_c + (1 - \theta) \log (1/p)}{\log P_c}$$

From Example 2, $P_c = 4.506$ MPa; $\theta = T_c/T_b = 1.434$; $T = 100°F = 310.9$ K; and $p = 0.802$ psia $= 5528$ Pa.

$$T_b = (310.9) \frac{\log (4.506 \times 10^6) + (1 - 1.434) \log (1/5528)}{\log (4.506 \times 10^6)}$$

$$T_b = (310.9)(1.244) = 386.8 \text{ K}$$

An accepted experimental normal boiling point is 372.7 K.
Note the error here is 3.8 percent, a value above the average. If the vapor pressure point available would have been closer to one atmosphere, the error would have been much lower.

Acentric Factor   The acentric factor of a compound (ω) is primarily a measure of the shape of a molecule, though it also measures a molecule's polarity. It is calculated from the reduced vapor pressure (P_r^{sat}) at a reduced temperature of 0.7 by the definition, Eq. (2-23).

$$\omega = -\log (P_r^{sat})_{T_r = 0.7} - 1.000 \qquad (2\text{-}23)$$

Critical temperature and pressure are required and can be estimated from the methods of this section. Vapor pressure is predicted by the methods of the next section. Experimental values should be used if available. The acentric factor is used as a third parameter with T_c and P_c in Pitzer-type corresponding states methods to predict volumetric properties and cubic equations of state such as the Redlich-Kwong-Soave and Peng-Robinson equations. For simple spherical molecules, the acentric factor is essentially zero, rising as branching and molecu-

lar weight increases. For compounds of similar size and shape the acentric factor increases slightly with increasing polarity.

For mixtures, the acentric factor is usually taken as a simple molar average value of the n components of the mixture.

$$\omega = \sum_{i=1}^{n} x_i \omega_i \qquad (2\text{-}24)$$

Miscellaneous Characterizing Constants   The radius of gyration ($\overline{R}$) is a simultaneous size-shape factor varying with the manner in which mass is distributed about the center of gravity of the molecule. For planar molecules, the radius of gyration is

$$\overline{R} = \sqrt{\frac{(AB)^{1/2} N_A}{M}} \qquad (2\text{-}25)$$

For three-dimensional molecules, it is

$$\overline{R} = \sqrt{\frac{(ABC)^{1/3} 2\pi N_A}{M}} \qquad (2\text{-}26)$$

AB and ABC are the products of the principal moments of inertia. Moments of inertia are calculated from bond angles and bond lengths. Many values are given by Landolt-Bornstein.[53] N_A is Avogadro's number, and M is the molecular weight of the molecule. Stuper et al.[105] give a computerized method for prediction of the radius of gyration.

The dipole moment (λ) of a molecule is the first moment of the electric charge density of a molecule. Paraffins have dipole moments of zero, while dipole moments of almost all hydrocarbons are small. McClellan[68] lists many dipole moments. The computer method of Dixon and Jurs[27] is the most useful method for predicting dipole moments. Lyman et al.[64] give other methods of calculation.

The van der Waals volume and area are characterizing parameters relating molecular configurations. Bondi[8] describes group contribution methods for their calculation.

VAPOR PRESSURE

Vapor pressure is the most important of the basic thermodynamic properties affecting liquids and vapors. The vapor pressure is the pressure exerted by a pure component at equilibrium at any temperature when both liquid and vapor phases exist and thus extends from a minimum at the triple point temperature to a maximum at the critical temperature, the critical pressure. This section briefly reviews methods for both correlating vapor pressure data and for predicting vapor pressure of pure compounds. Except at very high total pressures (above about 10 MPa), there is no effect of total pressure on vapor pressure. If such an effect is present, a correction, the Poynting correction, can be applied. The pressure exerted above a solid-vapor mixture may also be called vapor pressure but is normally only available as experimental data for common compounds that sublime.

Correlation Methods   Vapor pressure is correlated as a function of temperature by numerous methods mainly derived from the Clapeyron equation discussed in the section on enthalpy of vaporization. The classic simple equation used for correlation of low to moderate vapor pressures is the Antoine[4] equation (2-27).

$$\ln P^{sat} = A + \frac{B}{T + C} \qquad (2\text{-}27)$$

A, B, and C are regression constants for the specific compound.

The Antoine equation does not fit data accurately much above the normal boiling point. Thus, as regression by computer is now standard, more accurate expressions applicable to the critical point have become usable. The entire DIPPR Compilation[24] is regressed with the modified Riedel[89] equation (2-28) with constants available for over 1500 compounds.

$$\ln P^{sat} = A + \frac{B}{T} + C \ln T + DT^E \qquad (2\text{-}28)$$

A, B, C, and D are regression constants and E is an exponent equal to 1, 2, or 6 depending on which regression gives the most accurate fit of the data.

For purposes of the *API Technical Data Book* (Daubert and Danner[23]), another modified Riedel equation (2-29) was chosen and found to fit hydrocarbon data well over the entire pressure range. Coefficients are given for several hundred hydrocarbons.

$$\ln P^{\text{sat}} = A + \frac{B}{T} + C \ln T + DT^2 + \frac{E}{T^2} \qquad (2\text{-}29)$$

Both equations (2-28) and (2-29) are also extrapolatable above the critical temperature where necessary for thermodynamic calculations.

The other modern equation used for correlation is the modified and linearized Wagner[124] equation (2-30), which has the advantage that it will match critical data exactly, although it cannot be extrapolated above the critical point. The equation is also included with coefficients to several hundred compounds in the *Technical Data Book—Petroleum Refining*.

$$\ln P_r = aX_1 + bX_2 + cX_3 + dX_4 \qquad (2\text{-}30)$$

where $\quad X_1 = \dfrac{1 - T_r}{T_r}, X_2 = \dfrac{(1 - T_r)^{1.5}}{T_r}, X_3 = \dfrac{(1 - T_r)^{2.6}}{T_r}, X_4 = \dfrac{(1 - T_r)^5}{T_r}$

$$P_r = P^{\text{sat}}/P_c \qquad T_r = T^{\text{sat}}/T_c$$

Both Riedel and Wagner regressions usually fit data within a few tenths of a percent over the entire range between the triple point and the critical point.

Prediction Methods   Two methods have gained almost universal acceptance for prediction of the vapor pressure of **pure hydrocarbons.**

The method of Lee and Kesler[55] is the preferred method if the critical temperature and the critical pressure of the hydrocarbon is known or can be reasonably predicted by the methods of the first section. The corresponding states method is shown in equation (2-31) with the simple fluid and correction terms to be calculated from equations (2-32) and (2-33), respectively, for any T_r.

$$\ln P_r^{\text{sat}} = (\ln P_r^{\text{sat}})^{(0)} + \omega(\ln P_r^{\text{sat}})^{(1)} \qquad (2\text{-}31)$$

$$(\ln P_r^{\text{sat}})^{(0)} = 5.92714 - 6.09648/T_r - 1.28862 \ln T_r + 0.169347T_r^6 \qquad (2\text{-}32)$$

$$(\ln P_r^{\text{sat}})^{(1)} = 15.2518 - 15.6875/T_r - 13.4721 \ln T_r + 0.43577T_r^6 \qquad (2\text{-}33)$$

The method is applicable at reduced temperatures above 0.30 or the freezing point, whichever is higher, and below the critical point. The method is most reliable when $0.5 < T_r < 0.95$, where errors in prediction average 3.5 percent when experimental critical properties are known. Errors are higher for predicted criticals. The method is useful when solved iteratively with Eq. (2-23) to predict the acentric factor.

Example 5   Estimate the vapor pressure of 1-butene at 98°C. Use Eq. (2-31):

$$\ln P_r^{\text{sat}} = (\ln P_r^{\text{sat}})^{(0)} + \omega(\ln P_r^{\text{sat}})^{(1)}$$

Pure component properties of 1-butene are $T_c = 146.4°C$, $P_c = 4.02$ MPa, and $\omega = 0.1867$.

$$T_r = \frac{371.1}{419.5} = 0.885$$

From Eq. (2-32): $(\ln P_r^{\text{sat}})^{(0)} = -0.7227$

From Eq. (2-33): $(\ln P_r^{\text{sat}})^{(1)} = -0.6190$

$$\ln P_r^{\text{sat}} = -.7227 + (.1867)(-0.6190)$$

$$\ln P_r^{\text{sat}} = -0.8383$$

$$P_r^{\text{sat}} = 0.4325$$

$$P^{\text{sat}} = P_r^{\text{sat}}P_c = (.4325)(4.02) = 1.74 \text{ MPa}$$

An experimental value is 1.72 MPa.

When criticals cannot be estimated with reasonable accuracy, the method of Maxwell and Bonnell[67] is recommended. The normal boiling point and the specific gravity at 60°F (15.5°C) are required inputs. According to what vapor pressure range is expected, the vapor pressure is calculated from Eqs. (2-34), (2-35), or (2-36). If the wrong range is selected, the procedure will need to be repeated.

For $X > 0.0022$ ($P^{\text{sat}} < 2$ mm Hg):

$$\log P^{\text{sat}} = \frac{3000.538X - 6.761560}{43X - 0.987672} \qquad (2\text{-}34)$$

For $0.0013 \le X \le 0.0022$ (2 mm Hg $\le P^{\text{sat}} \le 760$ mm Hg):

$$\log P^{\text{sat}} = \frac{2663.129X - 5.994296}{95.76X - 0.972546} \qquad (2\text{-}35)$$

For $X < 0.0013$ ($P^{\text{sat}} > 760$ mm Hg):

$$\log P^{\text{sat}} = \frac{2770.085X - 6.412631}{36X - 0.989679} \qquad (2\text{-}36)$$

X is calculated from Eq. (2-37) and T'_b is calculated from Eq. (2-38). Iterative calculation may be required.

$$X = \frac{\dfrac{T'_b}{T} - 0.0002867(T'_b)}{748.1 - 0.2145(T'_b)} \qquad (2\text{-}37)$$

$$T_b - T'_b = 2.5f(K - 12) \log \frac{P^{\text{sat}}}{760} \qquad (2\text{-}38)$$

where T_b = normal boiling point, °R
    T'_b = normal boiling point corrected to $K = 12$, °R
    T = absolute temperature, °R
    f = correction factor. For all subatmospheric vapor pressures and for all substances having normal boiling points greater than 400°F, $f = 1$. For substances having normal boiling points less than 200°F, $f = 0$. For superatmospheric vapor pressures of substances having normal boiling points between 200°F and 400°F, f is given by $(T_b - 659.7)/200$
    K = Watson characterization factor, $T_b^{1/3}$/sp gr

Evaluation of the method for pure hydrocarbons shows errors averaging 8 percent for vapor pressures above 1 mm Hg and 30 percent below 1 mm Hg. The method is also usable for narrow boiling (range up to 50°F) **undefined hydrocarbon mixtures** with the only change being that the mean average boiling point replaces the normal boiling point in all calculations.

Example 6   Estimate the vapor pressure of tetralin at 150°C (302°F). Its normal boiling point is 207.6°C (405.7°F) and its Watson characterization factor is 9.78.

Use the Maxwell-Bonnell method, Eq. (2-35).

$$\log P^{\text{sat}} = \frac{2663.129X - 5.994296}{95.76X - 0.972546}$$

At 150°C with a normal boiling point of 207.6°C, a vapor pressure between 2 and 760 mm Hg would be expected.

Assume $T'_b = T_b = 405.7°F$ as a first trial. From Eq. (2-37):

$$X = \frac{(865.7/762) - 0.0002867(865.7)}{748.1 - 0.2145(865.7)}$$

$$X = 0.001579$$

$$\log P^{\text{sat}} = \frac{2663.129(0.001579) - 5.994296}{95.76(0.001579) - 0.972546} = 2.179$$

$$P^{\text{sat}} = 151.0 \text{ mm Hg}$$

Use Eq. (2-38) to calculate the correction.

$$T_b - T'_b = 2.5f(K - 12) \log \frac{P^{\text{sat}}}{760} = 2.5(1)(9.48 - 12) \log \frac{151}{760}$$

$$T_b - T'_b = 4.4°$$

Thus, for the second trial, $T'_b = 405.7 - 4.4 = 401.3°F$.

Using Eq. (2-37), recalculate X.

$$X = \frac{(861.3/762) - 0.0002867(861.3)}{748.1 - 0.2145(861.3)} = 0.001568$$

From Eq. (2-35), $\log P^{\text{sat}} = 2.214$; $P^{\text{sat}} = 163.7$ mm Hg. Use Eq. (2-38) to recalculate the correction.

$$T_b - T'_b = 2.5(1)(9.48 - 12) \log \frac{163.7}{760} = 4.2°$$

Thus, $T'_b = 401.5°F$. If greater accuracy is desired, carry out a third trial. (An experimental vapor pressure is 161.8 mm Hg.)

For **nonhydrocarbon organics,** vapor pressures above 15 kPa for compounds of known or estimable normal boiling point are predicted using the method of Riedel[89] given by Eq. (2-39).

$$\log P_r^o = \phi(T_r) - (\alpha - 7)\psi(T_r) \qquad (2\text{-}39)$$

Correlation functions $\phi(T_r)$, $\psi(T_r)$, and $\zeta(T_r)$ are given by Eqs. (2-40), (2-41), and (2-42), respectively.

$$\phi(T_r) = 0.118\zeta(T_r) - 7\log_{10} T_r \qquad (2\text{-}40)$$

$$\psi(T_r) = 0.0364\zeta(T_r) - \log_{10} T_r \qquad (2\text{-}41)$$

$$\zeta(T_r) = 36/T_r + 96.7\log_{10} T_r - 35 - T_r^6 \qquad (2\text{-}42)$$

The Riedel α is calculated from Eq. (2-43).

$$\alpha = \frac{0.136\zeta(T_b) + \log_{10} P_c - 5.01}{0.0364\zeta(T_b) - \log_{10} T_b} \qquad (2\text{-}43)$$

Critical properties, if not available, can be estimated from the methods of the previous section. T_r is the reduced temperature at the temperature of interest, while T_{r_b} is the reduced temperature at the normal boiling point.

The method is accurate within 2 to 3 percent above 15 kPa, while errors increase to 10–30 percent at lower pressures. Care should be taken not to use the method below the freezing point temperature.

Example 7   Estimate the vapor pressure of thiophene at 500 K. Pure component properties are $T_c = 579.4$ K, $P_c = 5.694$ MPa, and $T_b = 357.5$ K. Use the Riedel method, Eq. (2-39).

$$\log P_r^{sat} = -\phi(T_r) - (\alpha - 7)\,\psi\,(T_r)$$

$$T_{r_b} = \frac{357.5}{579.4} = 0.6170 \qquad T_r = \frac{500}{579.4} = 0.8629$$

From Eq. (2-42):

$$\zeta(T_{r_b}) = \frac{(36)}{(0.6170)} + 96.7\log(0.6170) - 35 - (0.6170)^6 = 3.01$$

From Eq. (2-43):

$$\alpha = \frac{0.136(3.01) + \log(5.694 \times 10^6) - 5.01}{(0.0364)(3.01) - \log(0.6170)}$$

$$\alpha = 6.749$$

From Eq. (2-42), calculate $\zeta(T_r)$ and then calculate $\phi(T_r)$ and $\psi(T_r)$ from Eqs. (2-40) and (2-41).

$$\log P_r^{sat} = -0.461 - (6.749 - 7)(0.068)$$

$$\log P_r^{sat} = -0.444$$

$$P_r^{sat} = 0.3598$$

$$P^{sat} = P_r^{sat}P_c = (.3598)(5.694) = 2.049 \text{ MPa}$$

An experimental value is 2.037 MPa.

For **nonhydrocarbon organics** for which normal boiling points are unknown or expected vapor pressures are below 15 kPa, the reference substance method of Othmer and Yu[78] as given by Eq. (2-44) is recommended.

$$\log P^{sat} = m\log P_w^{sat} + C \qquad (2\text{-}44)$$

The vapor pressure of water P_w^{sat} may be calculated by Eq. (2-45).

$$\log P_w^{sat} = 31.51 - \frac{3.1298 \times 10^3}{T} - 7.1385\log T + 1.757 \times 10^{-6}T^2 \quad (2\text{-}45)$$

with temperatures in K and vapor pressures in Pa.

Values of the compound specific constants m and c were originally derived by Othmer et al. and greatly expanded to over 600 common organics by Danner and Daubert.[22] If constants are not available but any two vapor pressure data points are available, the constants m and C can be calculated using Eqs. (2-46) and (2-47).

$$m = \frac{\log P_1^{sat} - \log P_2^{sat}}{\log P_{w_1}^{sat} - \log P_{w_2}^{sat}} \qquad (2\text{-}46)$$

$$C = \log P_1^{sat} - m\log P_{w_1}^{sat} \qquad (2\text{-}47)$$

where the subscripts 1 and 2 refer to the two reference temperatures T_1 and T_2.

Average errors at low pressures for compounds with tabulated m and C are within a few percent. When values of m and C are calculated from only two vapor pressure points, the method should be used only for interpolation and limited extrapolation. The method is usable from about 220 K (so long as it is above the freezing point of the compound) to the critical point of water (about 647 K).

Example 8   Estimate the vapor pressure of acetaldehyde at 0°C. Two vapor pressure points are 20.0 kPa at 256.55 K and 107.6 kPa at 294.85 K.

Use Eq. (2-44), determining parameters from Eqs. (2-45), (2-46), and (2-47).

$$\log P^{sat} = m\log P_w^{sat} + C$$

at $T_1 = 256.55$ K, $P_1^{sat} = 2.00 \times 10^4$ Pa, $\log P_1^{sat} = 4.3010$

at $T_2 = 294.95$ K, $P_2^{sat} = 1.067 \times 10^5$ Pa, $\log P_2^{sat} = 5.0282$

Use Eq. (2-45) to calculate the vapor pressure of water at T_1 and T_2.

$$\log P_{\omega_1}^{sat} = 2.2282, \quad P_{\omega_1}^{sat} = 169.12 \text{ Pa}$$

$$\log P_{\omega_2}^{sat} = 3.4213, \quad P_{\omega_2}^{sat} = 2637.9 \text{ Pa}$$

From Eq. (2-46):

$$m = \frac{\log P_1^{sat} - \log P_2^{sat}}{\log P_{w_1}^{sat} - \log P_{w_2}^{sat}} = \frac{4.3010 - 5.0282}{2.2282 - 3.4213} = 0.6093$$

From Eq. (2-47):

$$C = \log P_1^{sat} - m\log P_{w_1}^{sat} = 4.3010 - 0.6093(2.2282)$$

$$C = 2.9434$$

at

$$T = 273.15 \text{ K } (0°C)$$

$$\log P_w^{sat} = 31.51 - \frac{3129.8}{273.15} - 7.1385\log 273.15 + 1.757 \times 10^{-6}(273.15)^2$$

$$\log P_w^{sat} = 2.7907$$

$$\log P^{sat} = (0.6093)(2.7907) + 2.9434$$

$$\log P^{sat} = 4.6438$$

$$P^{sat} = 44,030 \text{ Pa} = 44.0 \text{ kPa}$$

IDEAL GAS THERMAL PROPERTIES

A substance is in the ideal gas state when the volume of its molecules is a zero fraction of the total volume taken up by the substance and when the individual molecules are far enough apart from each other so that there is no interaction between them. Although this only occurs at infinite volume and zero pressure, in practice, ideal gas properties can be used for gases up to a pressure of two atmospheres with little loss of accuracy. Thermal properties of ideal gas mixtures may be obtained by mole-fraction averaging the pure component values.

Heat Capacity, C_p^o   Heat capacity is defined as the amount of energy required to change the temperature of a unit mass or mole one degree; typical units are J/kg·K or J/kmol·K. There are many sources of ideal gas heat capacities in the literature; e.g., Daubert et al.,[24] Daubert and Danner,[23] JANAF thermochemical tables,[15] TRC thermodynamic tables,[115,116] and Stull et al.[104] If C_p^o values are not in the preceding sources, there are several estimation techniques that require only the molecular structure. The methods of Thinh et al.[113] and Benson et al.[6,7] are the most accurate but are also somewhat complicated to use. The equation of Harrison and Seaton[36] for C_p^o between 300 and 1500 K is almost as accurate and easy to use:

$$C_p^o = a_1 + a_2C + a_3H + a_4O + a_5N + a_6S + a_7F + a_8Cl$$

$$+ a_9I + a_{10}Br + a_{11}Si + a_{12}Al + a_{13}B + a_{14}P + a_{15}E \quad (2\text{-}48)$$

where
C_p^o = ideal gas heat capacity, J/mol K
a_1–a_{15} = constant parameters obtained from Table 2-387 as a function of temperature
C = number of carbon atoms in the molecule
H = number of hydrogen atoms in the molecule
O = number of oxygen atoms in the molecule
N = number of nitrogen atoms in the molecule
S = number of sulfur atoms in the molecule
F = number of fluorine atoms in the molecule
Cl = number of chlorine atoms in the molecule
I = number of iodine atoms in the molecule
Br = number of bromine atoms in the molecule
Si = number of silicon atoms in the molecule
Al = number of aluminum atoms in the molecule
B = number of boron atoms in the molecule
P = number of phosphorus atoms in the molecule
E = number of atoms in the molecule excluding the 13 atom-types listed above

TABLE 2-387   Values of the Constant Parameters a_1–a_{15} in Eq. (2-48) at Different Temperatures

Temp., K	Parameter a														
	1	2	3	4	5	6	7	8	9	10	11	12	13	14	15
300	4.86	9.04	5.69	11.4	11.9	15.3	12.7	16.8	18.7	17.8	14.6	15.8	11.5	18.0	19.5
400	0.864	12.6	7.37	13.9	14.0	17.0	16.2	18.9	20.5	19.9	17.5	18.3	14.7	20.9	20.8
500	−1.85	15.5	8.89	15.7	16.0	19.4	17.9	20.2	22.1	21.2	19.6	20.0	17.0	21.6	21.7
600	−4.61	17.5	10.5	17.5	17.3	20.3	20.1	21.4	23.3	22.4	20.9	21.1	18.3	22.8	22.1
800	−7.49	20.1	13.1	19.4	19.4	22.3	21.5	22.4	25.0	23.4	23.2	22.3	20.8	23.0	23.0
1000	−8.53	21.6	15.2	20.4	20.4	22.9	22.4	22.8	25.4	23.8	23.9	22.8	22.2	23.4	23.3
1500	−7.37	23.9	17.9	20.6	21.1	22.5	22.1	22.6	24.6	23.0	24.1	23.2	24.2	24.2	23.3

Results and parameters may be interpolated between temperatures. Average errors are between 2 and 6 percent, with the higher errors at the lower temperatures.

Example 9   Using Eq. 2-48 to estimate the ideal gas heat capacity of acetone (C_3H_6O) at 600 K:

$$C_p^\circ = -4.61 + (17.5)(3) + (10.5)(6) + (17.5)(1) = 128.39 \text{ J/mol K}$$

Daubert et al.[24] report a value of 121.8 J/mol K.

Enthalpy of Formation   The ideal gas standard enthalpy (heat) of formation (ΔH_{f298}°) of a chemical compound is the increment of enthalpy associated with the reaction of forming that compound in the ideal gas state from the constituent elements in their standard states, defined as the existing phase at a temperature of 298.15 K and one atmosphere (101.3 kPa). Sources for data are Refs. 15, 23, 24, 104, 115, and 116. The most accurate, but again complicated, estimation method is that of Benson et al.[6,7] A compromise between complexity and accuracy is based on the additive atomic group-contribution scheme of Joback[44]; his original units of kcal/mol have been converted to kJ/mol by the conversion 1 kcal/mol = 4.1868 kJ/mol:

$$\Delta H_{f298}^\circ = 68.29 + \sum_{i=1}^{n} N_i \, \Delta_{Hi} \qquad (2\text{-}49)$$

where ΔH_{f298}° = enthalpy of formation at 298.15 K, kJ/mol
    n = number of different atomic groups contained in the molecule
    N_i = number of atomic groups i contained in the molecule
    Δ_{Hi} = numeric value of atomic group i obtained from Table 2-388.

Average expected errors are about 9 kJ/mol.
For **other temperatures:**

$$\Delta H_{fT}^\circ = \Delta H_{f298}^\circ + \int_{298}^{T} C_p^\circ \, dT \qquad (2\text{-}50)$$

See above for discussion of the ideal gas heat capacity (C_p°).

Example 10   The ΔH_{f298}° of 2-butanol is estimated using Table 2-388. The molecular groups are 2CH_3, 1CH_2, 1CH (all nonring), and 1OH (alcohol). Therefore:

$$\Delta H_{f298}^\circ = 68.29 + 2(-76.45) + (-20.64) + (29.89) + (-208.04) = -283.40 \text{ kJ/mol}$$

The value from Daubert et al.[24] is −292.9 kJ/mol.

Gibbs Energy of Formation   The ideal gas standard Gibbs energy of formation (ΔG_{f298}°) of a chemical compound is the increment of Gibbs energy associated with the reaction of forming that compound in the ideal gas state from the constituent elements in their standard state defined as the existing phase at a temperature of 298.15 K and one atmosphere (101.325 kPa). Refs. 15, 23, 24, 104, 115, and 116 are good sources of data. The additive atomic group-contribution scheme of Joback[44] may be used to estimate ΔG_{f298}°; his original units of kcal/mol have been converted to kJ/mol by the conversion 1 kcal/mol = 4.1868 kJ/mol:

$$\Delta G_{f298}^\circ = 53.88 + \sum_{i=1}^{n} N_i \, \Delta_{Gi} \qquad (2\text{-}51)$$

where ΔG_{f298}° = Gibbs energy of formation at 298.15 K, kJ/mol

n = number of different atomic groups contained in the molecule
    N_i = number of atomic groups i contained in the molecule
    Δ_{Gi} = numeric value of atomic group i obtained from Table 2-388

Average errors of 8 to 9 kJ/mol may be expected.

Example 11   The ΔG_{f298}° of phenol is estimated using Table 2-388. The molecular groups are

$$5 \quad =CH$$

$$1 \quad =C- \text{ (both ring)}$$

$$1 \quad OH \text{ (phenol)}$$

Therefore,

$$\Delta G_{f298}^\circ = 53.88 + 5(11.30) + (54.05) + (-197.37) = -32.94 \text{ kJ/mol}$$

The value from Daubert et al.[24] is −32.64 kJ/mol.
For **other temperatures,** the exact Eq. (2-52) may be used at temperature T (K):

$$\Delta G_{fT}^\circ = \Delta H_{fT}^\circ - T \, \Delta S_{fT}^\circ \qquad (2\text{-}52)$$

where ΔG_{fT}° = Gibbs energy of formation at T, kJ/mol
    ΔH_{fT}° = enthalpy of formation at T, kJ/mol (see above)
    ΔS_{fT}° = entropy of formation at T, kJ/mol K (see below)

Entropy of Formation   The ideal gas standard entropy of formation (ΔS_{f298}°) of a chemical compound is the increment of entropy associated with the reaction of forming that compound in the ideal gas state from the constituent elements in their standard state defined as the existing phase at a temperature of 298.15 K and one atmosphere (101.325 kPa). Thus:

$$\Delta S_{f298}^\circ = S_{\text{compound}}^\circ - \sum_{i=1}^{n} N_i S_{\text{element } i}^r \qquad (2\text{-}53)$$

where ΔS_{f298}° = entropy of formation at 298.15 K and 1 atm, J/mol K
    $S_{\text{compound}}^\circ$ = ideal gas absolute entropy of the compound at 298.15 K and 1 atm, J/mol K
    n = number of different elements contained in the compound
    N_i = moles of element i contained in one mole of compound
    $S_{\text{element } i}^r$ = absolute entropy of element i in its standard state at 298.15 K and 1 atm, J/mol K.

Ideal gas absolute entropies of many compounds may be found in Daubert et al.,[24] Daubert and Danner,[23] JANAF Thermochemical Tables,[15] TRC Thermodynamic Tables,[115,116] and Stull et al.[104] Otherwise, the estimation method of Benson et al.[6,7] is reasonably accurate, with average errors of 1–2 J/mol K. Elemental standard-state absolute entropies may be found in Cox et al.[21] Values from this source for some common elements are listed in Table 2-389. ΔS_{f298}° may also be calculated from Eq. (2-52) if values for ΔH_{f298}° and ΔG_{f298}° are known.

TABLE 2-388 Atomic Group Contributions to Estimate ΔH°_{f298} and ΔG°_{f298}

	Δ_H	Δ_G		Δ_F	Δ_G
Nonring Increments			**Oxygen Increments (*Cont.*)**		
— CH₃	−76.45	−43.96	— CHO (aldehyde)	−162.03	−143.48
— CH₂ —	−20.64	8.42	— COOH (acid)	−426.72	−387.87
— CH	29.89	58.36	— COO — (ester)	−337.92	−301.95
— C —	82.23	116.02	═ O (except for above)	−247.61	−250.83
═ CH₂	−9.63	3.77	**Nitrogen Increments**		
			— NH₂	−22.02	14.07
═ CH	37.97	48.53	— NH (nonring)	53.47	89.39
═ C —	83.99	92.36	— NH (ring)	31.65	75.61
═ C ═	142.14	136.70	— N — (nonring)	123.34	163.16
≡ CH	79.30	77.71	— N ═ (nonring)	23.61	—
≡ C —	115.51	109.82	— N ═ (ring)	55.52	79.93
Ring Increments			═ NH	93.70	119.66
— CH₂ —	−26.80	−3.68	— CN	88.43	89.22
— CH	8.67	40.99	— NO₂	−66.57	−16.83
			Sulfur Increments		
			—SH	−17.33	−22.99
— C —	79.72	87.88	—S— (nonring)	41.87	33.12
			—S— (ring)	39.10	27.76
═ CH (aromatic or cyclic olefin)	2.09	11.30	**Halogen Increments**		
			—F	−251.92	−247.19
═ C — (aromatic or cyclic olefin)	46.43	54.05	—Cl	−71.55	−64.31
Oxygen Increments			—Br	−29.48	−38.06
— OH (alcohol)	−208.04	−189.20	—I	21.06	5.74
— OH (phenol)	−221.65	−197.37			
— O — (nonring)	−132.22	−105.00			
— O — (ring)	−138.16	−98.22			
— C ═ O (nonring)	−133.22	−120.50			
— C ═ O (ring)	−164.50	−126.27			

ENTHALPY OF VAPORIZATION AND FUSION

Enthalpy of Vaporization The enthalpy (heat) of vaporization ΔH_V is defined as the difference of the enthalpies of a unit mole or mass of a saturated vapor and saturated liquid of a pure component; i.e., at a temperature (below the critical temperature) and corresponding vapor pressure. ΔH_V is related to vapor pressure by the thermodynamically exact Clausius-Clapeyron equation:

$$\Delta H_v = -R \, \Delta Z_V \frac{d \ln P^{\text{sat}}}{d \, (1/T)} \tag{2-54}$$

where R = gas constant in energy units
$$\Delta Z_V = Z_G - Z_L \tag{2-55}$$
Z_G = compressibility factor of the saturated vapor
Z_L = compressibility factor of the saturated liquid
P^{sat} = vapor pressure
T = absolute temperature

If accurate Z_G and Z_L data are available, excellent ΔH_V values can be obtained by differentiating a vapor pressure correlation and using Eq. (2-54). If not, ΔZ_V may be esimated by Haggenmacher's equation[34]:

TABLE 2-389 **Standard-State Entropy of Elements at 298.15 K and 1 Atmosphere**

Element	State	Absolute entropy J/mol K
C	crystal (graphite)	5.74
H_2	gas	130.571
O_2	gas	205.043
N_2	gas	191.500
S	crystal (rhombic)	32.054
F_2	gas	202.682
Cl_2	gas	222.972
Br_2	liquid	152.21
I_2	crystal	116.14

$$\Delta Z_v = \left(1 - \frac{P_r}{T_{r^3}}\right)^{1/2} \qquad (2\text{-}56)$$

where P_r = reduced pressure = P/P_c
T_r = reduced temperature = T/T_c

However, Eq. (2-56) should be used only near or below the normal boiling point; even then, the accuracy of the resulting ΔH_V is significantly reduced.

The corresponding states approach suggested by Pitzer et al.[82] requires only the critical temperature and acentric factor of the compound. For a close approximation, an analytical representation of this method proposed by Reid et al.[86] for $0.6 < T_r < 1.0$ is:

$$\Delta H_V / RT_c = 7.08(1 - T_r)^{0.354} + 10.95\omega(1 - T_r)^{0.456} \qquad (2\text{-}57)$$

where ΔH_V = enthalpy of vaporization, kJ/mol
R = gas constant = 0.008314 kJ/mol K
T_c = critical temperature, K
T_r = reduced temperature, T/T_c
T = temperature, K
ω = acentric factor

Maximum errors are in the order of 8 percent.

Example 12 Estimate ΔH_V of Propionaldehyde at 350 K. The required properties from Daubert et al.[24] are T_c = 504.4 K and ω = 0.2559. T_r = 350.0/504.4 = 0.6939. Substituting in Eq. (2-57):

$$\Delta H_V / RT_c = (7.08)(1 - 0.6939)^{0.354} + (10.95)(0.2559)(1 - 0.6939)^{0.456} = 6.289$$

$$\Delta H_V = (6.289)(0.008314)(504.4) = 26.37 \text{ kJ/mol}$$

The reported value is 26.85 kJ/mol.[24]

The enthalpy of vaporization **at the normal boiling temperature** ΔH_{vb} (kJ/mol) can be estimated by an equation suggested by Riedel[90]:

$$\Delta H_{vb} = 1.093 \, RT_c \left[T_{br} \frac{(\ln [P_c / 101.325] - 1)}{0.930 - T_{br}} \right] \qquad (2\text{-}58)$$

where R = gas constant = 0.008314 kJ/mol K.
T_c = critical temperature, K
T_{br} = reduced normal boiling temperature = T_b/T_r
T_b = normal boiling temperature, K
P_c = critical pressure, kPa

Average errors are about 2 percent.

Example 13 Estimate ΔH_{Vb} of Ethyl Acetate. The required properties for ethyl acetate are from Daubert et al.[24]: T_c = 523.3 K, T_b = 350.2 K, and P_c = 3880.0 kPa. T_{br} = 350.2/523.3 = 0.6692. Substituting in Eq. (2-58):

$$\Delta H_{Vb} = (1.093)(0.008314)(523.3) \left[(0.6692) \left(\frac{(\ln [3880.0/101.325] - 1)}{0.930 - 0.6692} \right) \right]$$

$$= 32.28 \text{ kJ/mol}$$

The value from Daubert et al.[24] is 32.23 kJ/mol.

The enthalpy of vaporization decreases with temperature and is zero at the critical point. If the value of an enthalpy of vaporization ΔH_{v_1} is known at temperature T_1, this temperature dependency can be represented by the Watson relation[127] to calculate another enthalpy of vaporization ΔH_{v_2} at any other temperature T_2:

$$\Delta H_{v_2} = \Delta H_{v_1} \left(\frac{1 - T_{r_2}}{1 - T_{r_1}} \right)^{0.38} \qquad (2\text{-}59)$$

where $T_{r_{1,2}}$ = reduced temperature = T_1/T_c or T_2/T_c
$T_{1,2}$ = temperature, K
T_c = critical temperature, K

Equation (2-59) works best between the normal boiling and critical temperatures, producing values of engineering accuracy.

Example 14 Estimate ΔH_v of ethyl acetate at 450 K, using the normal boiling point values as a basis (see Example 13). ΔH_{v_1} = 32.23 kJ/mol, T_{r_1} = 0.6692, and T_{r_2} = 450.0/523.3 = 0.8599. Substituting in Eq. (2-59):

$$\Delta H_{v(450)} = 32.23 \left(\frac{1 - 0.8599}{1 - 0.6692} \right)^{0.38} = 23.25 \text{ kJ/mol}$$

A value of 23.16 kJ/mol is obtained from Daubert et al.[24]

Enthalpy of Fusion The enthalpy (heat) of fusion ΔH_{fus} is defined as the difference of the enthalpies of a unit mole or mass of a solid and liquid at its melting temperature and one atmosphere pressure of a pure component. There are no generally applicable estimation techniques that are very accurate. However, if the melting temperature is known, the atomic group contribution method of Chickos et al.[16] yields approximate results:

$$\Delta H_{fus} = T_{mlt} \Delta S_{fus} \qquad (2\text{-}60)$$

where ΔH_{fus} = enthalpy of fusion at the melting temperature, J/mol
T_{mlt} = melting temperature, K
ΔS_{fus} = $a + b$ = entropy of fusion at the melting temperature, J/mol K.

It should be noted that the methodology for a and b results in a ΔS_{fus} associated with the phase change from a solid at 0 K to the liquid at T_{mlt}. No entropy changes resulting from solid transitions are taken into account, and ΔS_{fus} for a substance that undergoes such a transition will be overestimated by this technique.

$$a = 35.19 \, N_R + 4.289 \, (N_{CH_2} - 3 \, N_R) \qquad (2\text{-}61)$$

where N_R = number of nonaromatic rings
N_{CH_2} = number of —CH_2— atomic groups in nonaromatic ring(s) required to form a cyclic paraffin of the same ring size(s) as contained in the molecule of interest.

Example: For [pentagon with O], N_{CH_2} = 5; a = 0 if there are no nonaromatic rings in the molecule of interest. If a nonaromatic ring in fact contains a —CH_2— atomic group, then no consideration of that group in the b term in Eq. (2-62) is required.

$$b = \sum_{i=1}^{n_g} (N_g)_i (\Delta_s)_i + \sum_{j=1}^{n_s} (N_s)_j (C_s)_j (\Delta_s)_j + \sum_{k=1}^{n_f} (N_f)_k (C_t)_k (\Delta_s)_k \qquad (2\text{-}62)$$

where n_g = number of different nonring or aromatic C-H atomic groups bonded to other carbon atoms in the molecule of interest
$(N_g)_i$ = number of C-H atomic groups i bonded to other carbon atoms in the molecule of interest
n_s = number of different nonring or aromatic C-H atomic groups bonded to at least one functional group or atom in the molecule of interest
$(N_s)_j$ = number of C-H atomic groups j bonded to at least one functional group or atom in the molecule of interest
n_f = number of different functional groups or atoms in the molecule of interest
$(N_f)_k$ = number of functional groups or atoms k in the molecule of interest
$(C_s)_j$ = coefficient for C-H atomic group j bonded to at least one functional group or atom in the molecule of interest; numeric values for C-H atomic groups are found in Table 2-390
C_t = coefficient for the functional group or atom k in the molecule of interest, where t = the total number of functional groups or atoms in the molecule of interest. *Exception:* Molecules containing any number of fluorine atoms are treated as having only one functional fluorine atom. Numeric values of C_1–C_4 are in Table 2-391 for functional groups or atoms
$(\Delta_s)_{i,j,k}$ = contribution of the atomic group or atom i, j, or k to the entropy of fusion, J/mol K. Numeric values for C-H atomic groups are in Table 2-390; values for functional groups or atoms are in Table 2-391

TABLE 2-390 C_s and Δ_s Values for C-H Atomic Groups to Estimate ΔH_{fus}

	C_s	Δ_s
Nonring		
—CH₃	1.0	18.33
—CH₂—	1.0	9.41
—CH (with two vertical bonds)	0.69	−16.19
—C— (with vertical bonds)	0.67	−38.70
=CH₂	1.0	14.56
=CH	3.23	4.85
=C—	1.0	−11.38
≡CH	1.0	10.88
≡C—	1.0	2.18
Aromatic		
CH	1.0	6.44
C— (bonded to paraffinic C)	1.0	−10.33
C— (bonded to olefinic C or non-C)	1.0	−4.27
C— (bonded to acetylenic C)	1.0	−2.51
Ring		
CH—	0.76	−15.98
C	1.0	−32.97
CH	0.62	−4.35
C—	0.86	−11.72
C or =C=	1.0	−5.36

Chickos et al.[16] report an average error of 2050 J/mol for monofunctional molecules and 3180 J/mol for multifunctional molecules when using their method to estimate ΔH_{fus}. Four **example** estimations are shown in Table 2-392.

SOLID AND LIQUID HEAT CAPACITY

The heat capacity is defined as the amount of energy required to change the temperature of a unit mass or mole one degree; typical units are J/kg·K or J/kmol·K.

Solid Heat Capacity Solid heat capacity increases with increasing temperature, with steep rises near the triple point for many compounds. When experimental data are available, a simple polynomial equation in temperature is often used to correlate the data. It should be noted that step changes in heat capacity occur if the compound undergoes crystalline state changes at different temperatures.

There are no reliable prediction methods for solid heat capacity as a function of temperature. However, the atomic element contribution method of Hurst and Harrison,[41] which is a modification of Kopp's Rule,[50] provides estimations at 298.15 K and is easy to use:

$$C_{pS} = \sum_{i=1}^{n} N_i \Delta_{Ei} \qquad (2\text{-}63)$$

where C_{pS} = solid heat capacity at 298.15 K, J/mol K
n = number of different atomic elements in the compound
N_i = number of atomic elements i in the compound
Δ_{Ei} = numeric value of the contribution of atomic element i found in Table 2-393

Average errors are in the 9–10 percent range.

Example 15 Estimate solid heat capacity of dibenzothiophene, $C_{12}H_8S$. The required atomic element contributions from Table 2-393 are: C = 10.89, H = 7.56, and S = 12.36. Substituting in Eq. (2-63):

$$C_{pS} = (12)(10.89) + (8)(7.56) + (1)(12.36) = 203.52 \text{ J/mol K}$$

Daubert et al.[24] report a value of 198.5 J/mol K.

Liquid Heat Capacity The two commonly used liquid heat capacities are either at constant pressure or at saturated conditions. There is negligible difference between them for most compounds up to a reduced temperature (temperature/critical temperature) of 0.7. Liquid heat capacity increases with increasing temperature, although a minimum occurs near the triple point for many compounds.

There are a number of reliable estimating techniques for obtaining **pure-component** liquid heat capacity as a function of temperature, including Ruzicka and Dolmalski,[93,94] Tarakad and Danner,[112] and Lee and Kesler.[55] These methods are somewhat complicated. The relatively simple atomic group contribution approach of Chueh and Swanson[17] for liquid heat capacity **at 293.15 K** is presented here:

$$C_{pL} = \sum_{i=1}^{n} N_i \Delta_{cpi} + 18.83m \qquad (2\text{-}64)$$

where C_{pL} = liquid heat capacity at 293.15 K, J/mol K.
n = number of different atomic groups in the compound
N_i = number of atomic groups i in the compound
Δ_{cpi} = numeric value of the contribution of atomic element i found in Table 2-394. The original units of cal/mol K have been converted to J/mol K by the conversion 1 cal/mol K = 4.184 J/mol K
m = number of carbon groups requiring an additional contribution, which are those that are joined by a single bond to a carbon group, which in turn is connected to a third carbon group by a double or triple bond. If a carbon group meets this criterion in more than one way, m should be increased by one for each of the ways. *Exceptions:* —CH₃ groups or carbon groups in a ring never require an additional contribution; and the *first* additional contribution for a —CH₂— group is 10.46 J/mol K rather than 18.83 J/mol K. However, if the —CH₂— group meets the criterion in a second way, the *second* additional contribution reverts to the 18.83 J/mol K value (see Example 17, below).

Errors should be less than 6 percent for all compounds except for acids, amines, and halides.

Example 16 Estimate the liquid heat capacity at 293.15 K of 2-butanol. The atomic groups are:

$$2 \quad \text{—CH}_3$$
$$1 \quad \text{—CH}_2\text{—}$$
$$1 \quad \text{—CHOH}$$

Substituting in Eq. (2-64) the atomic group contributions from Table 2-394 with $m = 0$:

$$C_{pL} = (2)(36.82) + (1)(30.38) + (1)(76.15) = 180.17 \text{ J/mol K}$$

The value from Daubert et al.[24] is 190.3 J/mol K.

TABLE 2-391 C_i and Δ_s Values for Functional Groups and Atoms to Estimate ΔH_{fus}

	C_1	C_2	C_3	C_4	Δ_s
—OH (alcohol)	1.0	12.6	18.9	26.4	1.13
—OH (phenol)	1.0	1.0	1.0	1.0	16.57
—O— (ether, nonring)	1.0	1.0	1.0	1.0	1.09
—O— (ether, ring)	1.0	1.0	1.0	1.0	1.34
—C=O (ketone, nonring)	1.0	1.0			3.14
C=O (ketone, ring)	1.0	1.0			−1.88
—CHO (aldehyde)	1.0				19.66
—COOH (acid)	1.0	1.83	1.88	1.72	14.90
—COO— (ester)	1.0	1.0	1.0	1.0	3.68
—NH₂ (aliphatic)	1.0	1.82			16.23
—NH₂ (aromatic)	1.0	1.0			15.48
—NH (nonring)	1.0	1.0			−2.18
—NH (ring)	1.0				1.84
—N— (nonring)	1.0				−15.90
N— (ring)	1.0	1.0			−17.07
=N— (ring)	1.0	1.0			1.67
=N— (aromatic)	1.0	1.0	1.0		7.32
—CN (nitrile)	1.0	1.4			9.62
—NO₂	1.0	1.0	1.0		17.36
—C(=O)NH₂	1.0	1.0			26.19
—C(=O)NH—	1.0	1.0			−0.42
—SH	1.0	1.0			17.99
—S— (nonring)	1.0			0.36	7.20
—S— (ring)	1.0	1.0			2.18
—SO₂⁻ (nonring)	1.0				3.26
—F (on —C)	1.0	1.0	1.0	1.0	14.73
—F (on =C)	1.0	1.0	1.0	1.0	13.01
—F (on ring C)	1.0	1.0	1.0	1.0	15.90
—Cl	1.0	2.0	2.0	1.93	8.37
—Br	1.0	1.0	1.0	0.82	17.95
—I	1.0	1.0			16.95

TABLE 2-392   Examples of Estimations of ΔH_{fus}, J/mol

Molecule	Melting temp., K	Atomic group	Contribution		
SH $(t = 1)$	155.4	cyclopentane	$(35.19)(1) + (4.289)[5 - (3)(1)]$	=	43.77
		CH— (ring)	$(1)(0.76)(-15.98)$	=	−12.14
		—SH	$(1)(1.0)(17.99)$ ΔS_{fus}	= =	17.99 49.62
			$\Delta H_{fus} = (49.62)(155.4)$ $\Delta H_{fus} = \text{(experimental)}^{16}$	= =	7710.9 7802
S $(t = 1)$	304.5	cyclopentane	$(35.19)(1) + (4.289)[5 - (3)(1)]$	=	43.77
		CH (aromatic)	$(4)(6.44)$	=	25.76
		C— (ring)	$(1)(-11.72)$	=	−11.72
		C— (ring)	$(1)(0.86)(-11.72)$	=	−10.08
		CH (ring)	$(1)(-4.35)$	=	−4.35
		CH (ring)	$(1)(0.62)(-4.35)$	=	−2.69
		— S — (ring)	$(1)(1.0)(2.18)$ ΔS_{fus}	= =	2.18 42.87
			$\Delta H_{fus} = (42.87)(304.5)$ $\Delta H_{fus} = \text{(experimental)}^{16}$	= =	13054 11823
F CH—COOH OH $(t = 3)$	363.0	CH (aromatic)	$(4)(6.44)$	=	25.76
		C— (aromatic)	$(1)(1.0)(-4.27)$	=	−4.27
		—F (on =C)	$(1)(1.0)(13.01)$	=	13.01
		C— (aromatic)	$(1)(-10.33)$	=	−10.33
		—CH	$(1)(0.69)(-16.19)$	=	−11.17
		—COOH	$(1)(1.88)(14.90)$	=	28.01
		— OH (alcohol)	$(1)(18.9)(1.13)$ ΔS_{fus}	= =	21.36 62.37
			$\Delta H_{fus} = (62.37)(363.0)$ $\Delta H_{fus} = \text{(experimental)}^{16}$	= =	22640 20959
H Br HO—CH$_2$—C—C—CH$_2$—OH Br H $(t = 4)$	338.2	—CH$_2$—	$(2)(1.0)\ 9.41$	=	18.82
		—CH	$(2)(0.69)(-16.19)$	=	−22.34
		— OH (alcohol)	$(2)(26.4)(1.13)$	=	59.66
		— Br	$(2)(0.82)(17.95)$ ΔS_{fus}	= =	29.44 85.58
			$\Delta H_{fus} = (85.58)(338.2)$ $\Delta H_{fus} = \text{(experimental)}^{16}$	= =	28943 29291

TABLE 2-393   Atomic Element Contributions to Estimate Solid Heat Capacity at 298.15 K

Atomic element	Δ_E	Atomic element	Δ_E	Atomic element	Δ_E
C	10.89	Ba	32.37	Mo	29.44
H	7.56	Be	12.47	Na	26.19
O	13.42	Ca	28.25	Ni	25.46
N	18.74	Co	25.71	Pb	31.60
S	12.36	Cu	26.92	Si	17.00
F	26.16	Fe	29.08	Sr	28.41
Cl	24.69	Hg	27.87	Ti	27.24
Br	25.36	K	28.78	V	29.36
I	25.29	Li	23.25	W	30.87
Al	18.07	Mg	22.69	Zr	26.82
B	10.10	Mn	28.06	All other	26.63

Example 17   Estimate liquid heat capacity at 293.15 K of 1,4-pentadiene, CH_2=CH—CH_2—CH=CH_2. The atomic groups are:

$$2 \quad =CH_2$$
$$2 \quad =CH$$
$$1 \quad —CH_2—$$

The —CH_2— group is twice joined by a single bond to a carbon group, which in turn is connected to a third carbon group by a double bond, and $m = 2$. However, by the second exception, the first additional contribution is 10.46 J/mol K rather than 18.83 J/mol K. Substituting in Eq. (2-64) the atomic group contributions from Table 2-394:

$$C_{pL} = (2)(21.76) + (2)(21.34) + (1)(30.38) + 10.46 + 18.83 = 145.87 \text{ J/mol K}$$

TABLE 2-394   Atomic Group Contributions to Estimate Liquid Heat Capacity at 293.15 K

	Δ_{cp}		Δ_{cp}
Nonring Increments		**Oxygen Increments (Cont.)**	
—CH_3	36.82	—CH_2OH	73.22
—CH_2—	30.38	—CHOH	76.15
—CH	20.92	—COH	111.29
—C—	7.36	—OH (except for above)	44.77
=CH_2	21.76	—ONO_2	119.24
=CH	21.34	**Nitrogen Increments**	
=C—	15.90	—NH_2	58.58
≡CH	24.69	—NH	43.93
≡C—	24.69	—N—	31.38
Ring Increments		—N= (ring)	18.83
—CH_2—	25.94	—CN	58.16
—CH	18.41	**Sulfur Increments**	
		—SH	44.77
		—S—	33.47
—C— or =C	12.13	**Halogen Increments**	
		—F	16.74
=CH	22.18	—Cl (first or second on a carbon)	35.98
Oxygen Increments		—Cl (third or fourth on a carbon)	25.10
—O—	35.15	—Br	37.66
—C=O	52.97	—I	35.98
—CHO (aldehyde)	52.97	**Hydrogen Increment**	
—COOH (acid)	79.91	—H (for formic acid, formates, hydrogen cyanide, etc.)	14.64
—COO— (ester)	60.67		

Daubert et al.[24] report a value of 145.6 J/mol K.

For **liquid mixtures,** the values of the pure components can be mole-fraction-averaged. This procedure neglects any heat of mixing effects.

DENSITY

Density is defined as the mass of a substance contained in a unit volume. In the SI system of units, the ratio of the density of a substance to the density of water at 15°C is known as its relative density, while the older term *specific gravity* is the ratio relative to water at 60°F. Various units of density, such as kg/m^3, lb-mass/ft^3, and g/cm^3, are commonly used. In addition, molar densities, or the density divided by the molecular weight, is often specified. This section briefly discusses methods of correlation of density as a function of temperature and presents the most common accurate methods for prediction of vapor, liquid, and solid density.

Correlation Methods   Vapor densities are not correlated as functions of temperature alone, as pressure and temperature are both important. At high temperatures and very low pressures, the ideal gas law can be applied; while at moderate temperature and low pressure, vapor density is usually correlated by the virial equation. Both methods will be discussed later.

Molar liquid density (ρ) is best correlated by an equation adopted from the Rackett predictor. The equation has the form of Eq. (2-65):

$$\rho = \frac{A}{B\left(1 - \frac{T}{C}\right)^D} \quad (2\text{-}65)$$

The regression constants A, B, and D are determined from the nonlinear regression of available data, while C is usually taken as the critical temperature. The liquid density decreases approximately linearly from the triple point to the normal boiling point and then nonlinearly to the critical density (the reciprocal of the critical volume). A few compounds such as water cannot be fit with this equation over the entire range of temperature. Liquid density data to be regressed should be at atmospheric pressure up to the normal boiling point, above which saturated liquid data should be used. Constants for 1500 compounds are given in the DIPPR compilation.[24]

Solid density data are sparse and usually only available over a narrow temperature range, for which the general decrease in density with temperature is approximately linear.

Vapor Density Prediction   A myriad of methods exist for prediction of vapor density as a function of temperature and pressure. This section will only present the most accurate and generally used methods.

For simple molecules at temperatures above the critical and at pressures no more than a few atmospheres, the ideal gas law, Eq. (2-66), may be used to estimate vapor density.

$$\rho = \frac{1}{V} = \frac{P}{RT} \quad (2\text{-}66)$$

At slightly higher pressures up to a reduced pressure of about 0.4, the truncated virial equation, Eq. (2-67), is commonly used for all types of organic fluids.

$$Z = \frac{PV}{RT} = 1 + \frac{B}{V} \quad (2\text{-}67a)$$

$$Z = \frac{P}{\rho RT} = 1 + B\rho \quad (2\text{-}67b)$$

Second virial coefficients, B, are a function of temperature and are available for about 1500 compounds in the DIPPR compilation.[24] The second virial coefficient can be regressed from experimental PVT data or can be reasonably and accurately predicted. Tsonopoulos[117] proposed a prediction method for nonpolar compounds that requires the critical temperature, critical pressure, and acentric factor. Equations (2-68) through (2-70) describe the method.

$$\frac{BP_c}{RT_c} = B_0 + \omega B_1 \quad (2\text{-}68)$$

$$B_0 = 0.1445 - \left(\frac{0.330}{T_r}\right) - \left(\frac{0.1385}{T_r^2}\right) - \left(\frac{0.0121}{T_r^3}\right) - \left(\frac{0.000607}{T_r^5}\right) \quad (2\text{-}69)$$

$$B_1 = 0.0637 + \left(\frac{0.0331}{T_r^2}\right) - \left(\frac{0.423}{T_r^3}\right) - \left(\frac{0.008}{T_r^8}\right) \quad (2\text{-}70)$$

For non-hydrogen-bonding polar compounds such as carbonyls and ethers, Tsonopoulos[117] recommends that Eq. (2-68) be expanded to a third term that is a function of the reduced dipole moment (μ_r) as described by Eqs. (2-71) through (2-73):

$$\frac{BP_c}{RT_c} = B_0 + \omega B_1 + B_2 \quad (2\text{-}71)$$

$$B_2 = -0.0002410\lambda_r - 4.308 \times 10^{-21}\lambda_r^8 \quad (2\text{-}72)$$

$$\lambda_r = \frac{10^5 \lambda_p^2 P_c}{T_c^2} \quad (2\text{-}73)$$

The dipole moment λ_p in Eq. (2-73) is in debyes, while P_c is in atm and T_c is in K. Units must be watched carefully. For hydrogen-bonding molecules, Eq. (2-71) can be used with a value of B_2 calculated by Eq. (2-74).

$$B_2 = \frac{a}{T_r^6} - \frac{b}{T_r^8} \quad (2\text{-}74)$$

Variables a and b are specific constants reported by Tsonopoulos[117] for some alcohols and water (e.g., methanol: a = 0.0878, b = 0.0560; and water: a = 0.0279, b = 0.0229). Tsonopoulos also gives specific prediction methods for haloalkanes[118] and water pollutants.[119]

Example 18   Estimate the molar volume of isobutane at 155°C and 1.0 MPa pressure. Properties of isobutane are $T_c = 135.0°C$, $P_c = 3.647$ MPa, and $\omega = 0.1170$.

$$T_r = \frac{155 + 273.1}{135 + 273.1} = 1.05 \qquad P_r = \frac{1.0}{3.647} = 0.274$$

Since reduced pressure is below 0.4, use virial equation (2-67a). Calculate B by the Tsonopoulos method, Eq. (2-68).

$$\frac{BP_c}{RT_c} = B_0 + \omega B_1$$

Using Eq. (2-69):

$$B_0 = 0.1445 - \left(\frac{0.330}{1.05}\right) - \left(\frac{0.1385}{1.05^2}\right) - \left(\frac{0.0121}{1.05^3}\right) - \left(\frac{0.000607}{1.05^5}\right)$$

$$B_0 = 0.1445 - 0.3143 - 0.1256 - 0.0105 - 0.0005 = -0.3064$$

Using Eq. (2-70):

$$B_1 = 0.0637 + \left(\frac{0.0331}{1.05^2}\right) - \left(\frac{0.423}{1.05^3}\right) - \left(\frac{0.008}{1.05^8}\right)$$

$$B_1 = 0.0637 + 0.0300 - 0.3654 - 0.0054 = -0.2771$$

$$B = \frac{[-0.3064 + (0.1770)(-0.2771)](8314)(408.1)}{(3.647 \times 10^6)} = -0.3307$$

$$\frac{PV}{RT} = 1 + \frac{B}{V}$$

$$\frac{10^6 V}{(8314)(428.1)} = 1 - \frac{0.3307}{V}$$

Trial and error or the quadratic formula can be used for the solution. If you opt for trial and error, start with the ideal gas value (B = 0) where V = 3.559 m^3/kmole [Eq. (2-66)].

Solving, V = 3.190 m^3/kmole.

For prediction of vapor density of **pure hydrocarbon and nonpolar gases,** the corresponding states method of Pitzer et al.[82] is the most accurate method, with errors of less than 1 percent except in the critical region where errors of up to 30 percent can occur. The method correlates the compressibility factor by Eq. (2-75), after which the density can be calculated by Eq. (2-75):

$$Z = Z^{(0)} + \omega Z^{(1)} \quad (2\text{-}75)$$

$$\rho = \frac{P}{ZRT} = \frac{1}{V} \quad (2\text{-}76)$$

$Z^{(0)}$ is the compressibility factor for the simple fluid, while $Z^{(1)}$ is the correction term for molecular acentricity, both of which are functions of T_r and P_r. Both plots and detailed tabulations of the functions are available in the *Technical Data Book*.[23] Critical temperature and pressure and the acentric factor from tabulations or as predicted are required. For hydrogen, T_c and P_c should be taken as 41.7 K and 2100 kPa, respectively. For approximate calculations, Figs. 2-35 and 2-36 should be used for calculating $Z^{(0)}$ and $Z^{(1)}$, respectively, for superheated vapors with $0.2 \leq P_r \leq 10$. $Z^{(0)}$ will approach 1, and $Z^{(1)}$ will approach 0 for $P_r < 0.2$. More accurately, Eq. (2-77) can be used for P_r between 0 and 0.2.

$$Z = 1 + \frac{P_r}{T_r}[(0.1445 + 0.073\omega) - (0.330 - 0.46\omega)T_r^{-1}$$
$$- (0.1385 + 0.50\omega)T_r^{-2} - (0.0121 + 0.097\omega)T_r^{-3} - 0.0073\omega T_r^{-8}] \quad (2\text{-}77)$$

Extension of the pressure range to $P_r = 14$ is available in the *Technical Data Book*. For saturated vapor densities, the values of $Z^{(0)}$ and $Z^{(1)}$ are tabulated as a

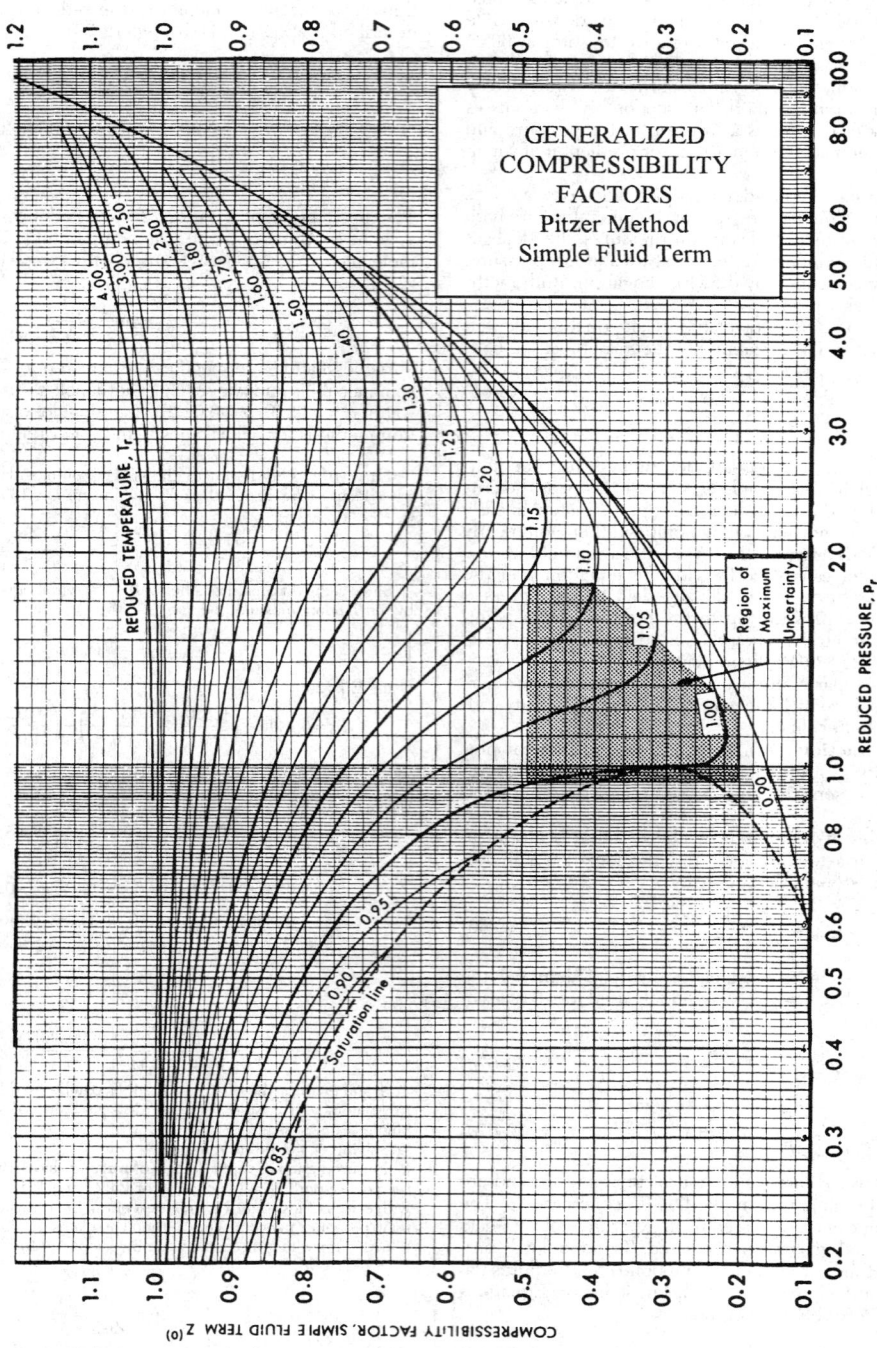

FIG. 2-35 Generalized compressibility factors—Pitzer Method, simple fluid term.

function of reduced pressure in Table 2-395. If the saturation temperature rather than the saturation pressure is known, the vapor pressure of the compound can be determined either from data or the vapor pressure prediction methods discussed earlier.

Example 19 Estimate the molar volume of isobutane at 155°C and 8.6 MPa pressure.

As high pressure, use Eq. (2-75) to calculate Z and then Eq. (2-76) to estimate the molar volume.

The properties of isobutane necessary are $T_c = 135.0°C$, $P_c = 3.647$ MPa, and $\omega = 0.1770$.

$$T_r = \frac{155 + 273.1}{135 + 273.1} = 1.05 \qquad P_r = \frac{8.6}{3.647} = 2.36$$

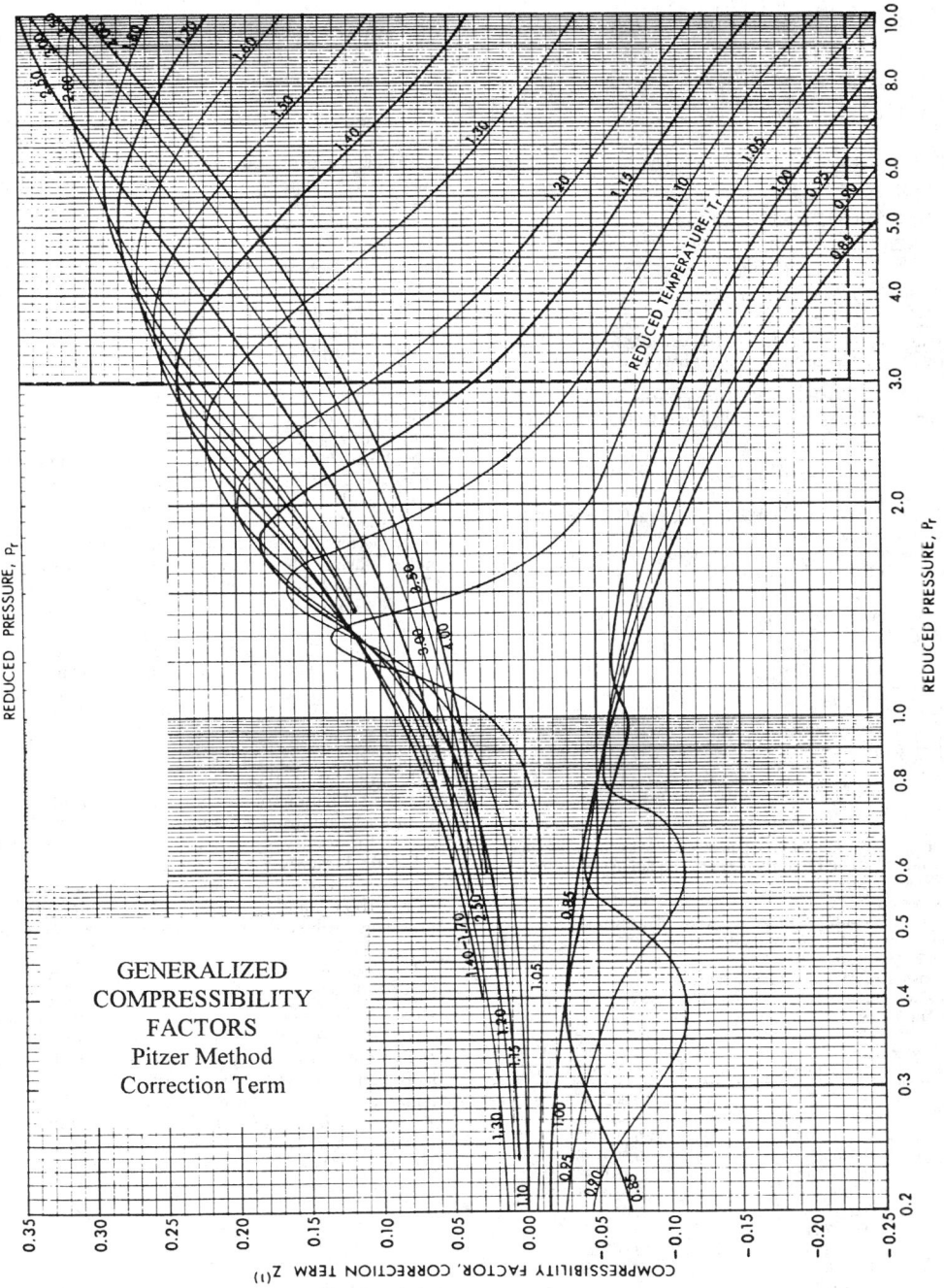

FIG. 2-36 Generalized compressibility factors—Pitzer Method, correction term.

TABLE 2-395 **Saturated Vapor Density Parameters**

P_r	$Z^{(0)}$	$Z^{(1)}$	P_r	$Z^{(0)}$	$Z^{(1)}$
1.00	0.291	−0.080	0.65	0.615	−0.069
0.99	0.35	−0.083	0.60	0.64	−0.063
0.98	0.38	−0.085	0.55	0.665	−0.056
0.97	0.40	−0.087	0.50	0.688	−0.049
0.96	0.41	−0.088	0.45	0.711	−0.041
0.95	0.42	−0.089	0.40	0.734	−0.033
0.94	0.43	−0.089	0.35	0.758	−0.025
0.92	0.45	−0.090	0.30	0.783	−0.018
0.90	0.47	−0.091	0.25	0.809	−0.012
0.85	0.50	−0.090	0.20	0.835	−0.008
0.80	0.53	−0.087	0.15	0.864	−0.005
0.75	0.56	−0.081	0.10	0.896	−0.002
0.70	0.59	−0.075	0.05	0.935	0.000

Using Fig. 2-35, $Z^{(0)} = 0.385$. Using Fig. 2-36, $Z^{(1)} = -0.063$.

$$Z = Z^{(0)} + \omega Z^{(1)}$$

$$Z = 0.385 + 0.1770(-0.063) = 0.374$$

$$V = \frac{ZRT}{P} = \frac{(0.374)(8314)(428.1)}{8.6 \times 10^6}$$

$$V = 0.155 \text{ m}^3/\text{kmol}$$

An experimental value for Z is 0.377.

Note that use of the Lee-Kesler fit [Eq. (2-78)] would give a slightly more accurate answer than the graphical method, and this fit is used for any computer applications.

Lee and Kesler[55] fit the entire Pitzer method to equations, rewriting the basic Eq. (2-75) with respect to a heavy reference fluid (*n*-octane) as shown by Eq. (2-78).

$$Z = Z^{(0)} + \frac{\omega}{\omega^{(h)}}(Z^{(h)} - Z^{(0)}) \qquad (2\text{-}78)$$

where h specifies the heavy reference fluid with an acentric factor of 0.3978.

The parameters in the equation are calculated for the simple fluid and the heavy reference fluid with an acentric factor of 0.3978. The parameters in the equation are calculated for the simple fluid and the heavy reference fluid from Eq. (2-79).

$$Z^{(i)} = \frac{P_r V_r}{T_r} = 1 + \frac{B}{V_r} + \frac{C}{V_r^2} + \frac{D}{V_r^5} + \frac{c_4}{T_r^3 V_r^2}\left[\beta + \frac{\gamma}{V_r^2}\right] \exp\left[\frac{-\gamma}{V_r^2}\right] \qquad (2\text{-}79)$$

$$B = b_1 - \left(\frac{b_2}{T_r}\right) - \left(\frac{b_3}{T_r^2}\right) - \left(\frac{b_4}{T_r^3}\right)$$

$$C = c_1 - \left(\frac{c_2}{T_r}\right) - \left(\frac{c_3}{T_r^3}\right)$$

$$D = d_1 + \left(\frac{d_2}{T_r}\right)$$

where $Z^{(i)} = Z^{(0)}$ when the constants in the equation correspond to the simple fluid and $Z^{(h)}$ when the constants in the equation correspond to the heavy reference fluid
  P = pressure, kPa
  P_c = critical pressure of the compound whose density is sought, kPa
  V = molar volume of the simple fluid or the heavy reference fluid, as the case may be, in m³/kmole.
  R = gas constant = 8.3140 m³ kPa/k mole K
  T_c = critical temperature of the compound whose density is sought, K
  T = temperature, K

Constant	Simple fluid	Heavy reference fluid
b_1	0.1181193	0.2026579
b_2	0.265728	0.331511
b_3	0.154790	0.027655
b_4	0.030323	0.203488
c_1	0.0236744	0.0313385
c_2	0.0186984	0.0503618
c_3	0.0	0.016901
c_4	0.042724	0.041577
$d_1 \times 10^4$	0.155488	0.48736
$d_2 \times 10^4$	0.623689	0.0740336
β	0.65392	1.226
γ	0.060167	0.03754

For **hydrocarbon and nonpolar gas mixtures,** the Pitzer pure component method can be used to predict vapor density by replacing the true critical properties with pseudocritical properties defined in Eqs. (2-80) and (2-81) by Kay.[47]

$$T_{pc} = \sum_{i=1}^{n} x_i T_{c_i} \qquad (2\text{-}80)$$

$$P_{pc} = \sum_{i=1}^{n} x_i P_{c_i} \qquad (2\text{-}81)$$

The mixture acentric factor, Eq. (2-82), can also be used.

$$\omega = \sum_{i=1}^{n} x_i \omega_i \qquad (2\text{-}82)$$

Errors in compressibility factors tabulated for over 6500 data points rarely exceed 2 percent except in the critical region, where 15 percent errors may be expected and 50 percent errors can occur. For mixtures near the critical point, special techniques are available as discussed in the sixth chapter of the *Technical Data Book*.

For **pure organic vapors,** the Lydersen et al.[63] corresponding states method is the most accurate technique for predicting compressibility factors and, hence, vapor densities. Critical temperature, critical pressure, and critical compressibility factor defined by Eq. (2-21) are used as input parameters. Figure 2-37 is used to predict the compressibility factor at $Z_c = 0.27$, and the result is corrected to the Z_c of the desired fluid using Eq. (2-83).

$$Z = Z_{@ Z_c = 0.27} + D_i(Z_c - 0.27) \qquad (2\text{-}83)$$

D_i is equal to D_a read from Fig. 2-38 if $Z_c > 0.27$; and D_i is equal to D_b read from Fig. 2-39 if $Z_c < 0.27$. At reduced temperatures less than 0.9, D_i can be taken as 0. The density is then calculated from Eq. (2-76). All families of organic compounds except mercaptans and carboxylic acids are predicted within an average deviation of 5 percent.

No specific mixing rules have been tested for predicting compressibility factors for **defined organic mixtures.** However, the Lydersen method using pseudocritical properties as defined in Eqs. (2-80), (2-81), and (2-82) in place of true critical properties will give a reasonable estimate of the compressibility factor and hence the vapor density.

Vapor densities for pure compounds can also be predicted by cubic equations of state. For hydrocarbons, relatively accurate Redlich-Kwong-type equations such as the Soave[98] and Peng-Robinson[81] equations are often used. Both require only T_c, P_c, and ω as inputs. For organic compounds, the Lee-Erbar-Edmister[54] equation (which requires the same input parameters) has been used with errors essentially equivalent to those determined for the Lydersen method. While analytical equations of state are not often used when only densities are required, values from equations of state are used as inputs to equation of state formulations for thermal and equilibrium properties.

Liquid Density Prediction   Methods for the prediction of pure saturated hydrocarbons and nonhydrocarbon organics, compressed hydrocarbon liquids, and defined and undefined hydrocarbon mixtures were evaluated. Only the most accurate and convenient methods are included here.

The most convenient method for predicting the saturated liquid density of both **pure hydrocarbons and pure organic liquids** is the method of Rackett[54] as modified by Spencer and Danner.[99] Equation (2-84) is used to calculate the saturated liquid molar density at any temperature using input parameters of T_c, P_c, and Z_{RA}. Z_{RA} is a parameter regressed from experimental data. Values for some common substances are given in Table 2-396. Extensive tabulations are given in the *Technical Data Book*[23] for hydrocarbons and nonhydrocarbons as well as organic and inorganic gases. Additional values are given in the *Data Prediction Manual*[22] for nonhydrocarbons.

$$\frac{1}{\rho_{\text{sat}}} = V_{\text{sat}} = \left(\frac{RT_c}{P_c}\right) Z_{RA}^n \qquad (2\text{-}84)$$

$$n = 1.0 + (1.0 - T_r)^{2/7}$$

Errors for hydrocarbons between the triple and critical points average about 0.7 percent, with organics averaging about 1.2 percent. The cor-

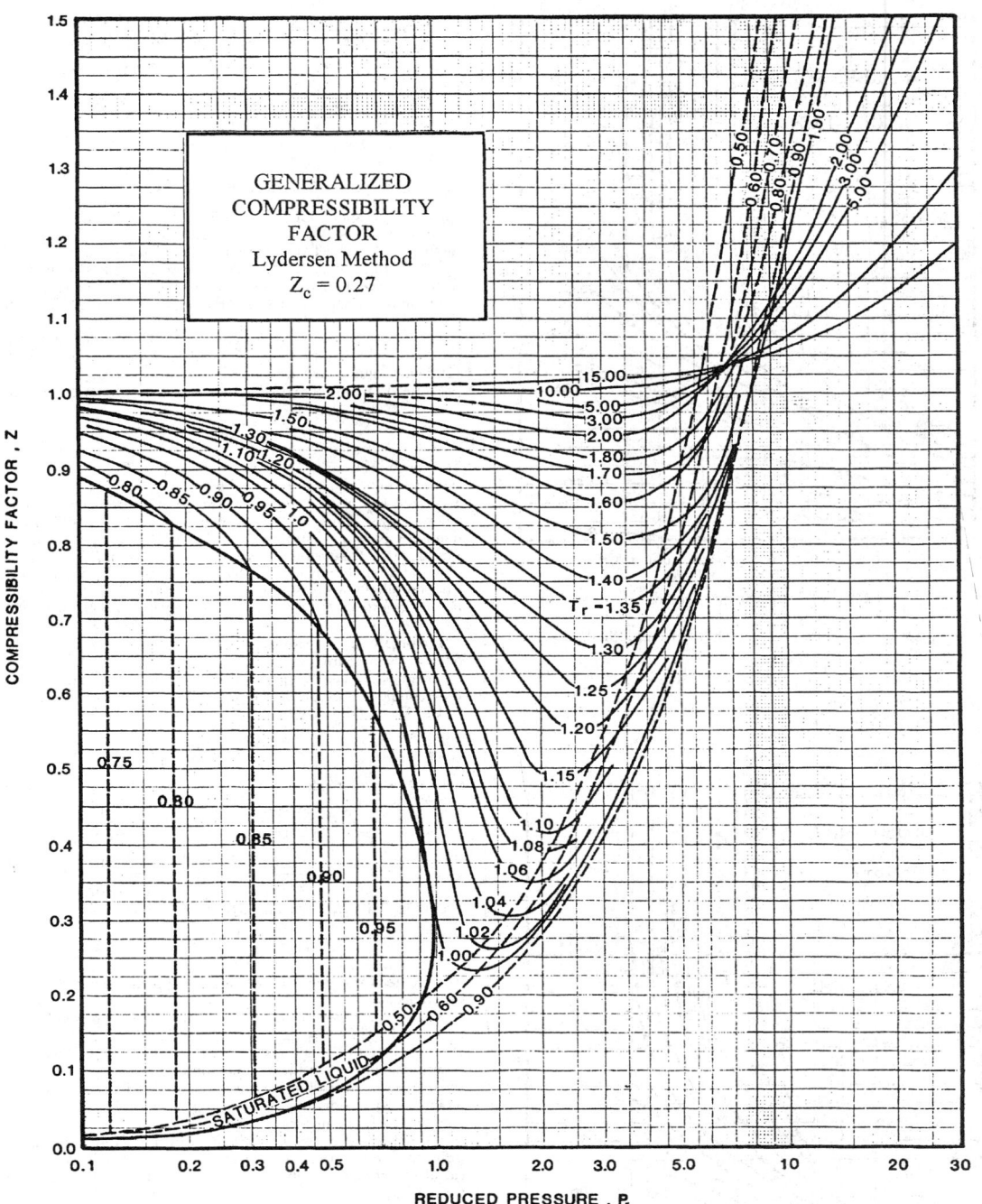

FIG. 2-37 Generalized compressibility factor—Lydersen Method, $Z_c = 0.27$.

relation is especially sensitive to the value of Z_{RA} near the critical point.

If no value of Z_{RA} is available or derivable, the critical compressibility factor can be used in Eq. (2-84) as originally proposed by Rackett. Use of Z_c increases the average error to about 3.0 percent.

Example 20 Estimate the density of saturated liquid propane at 0°C. Use Eq. (2-84).

$$\rho_{sat} = \left(\frac{P_c}{RT_c} \right) \left(\frac{1}{Z_{RA}^n} \right)$$

$$n = 1.0 + (1.0 - T_r)^{2/7}$$

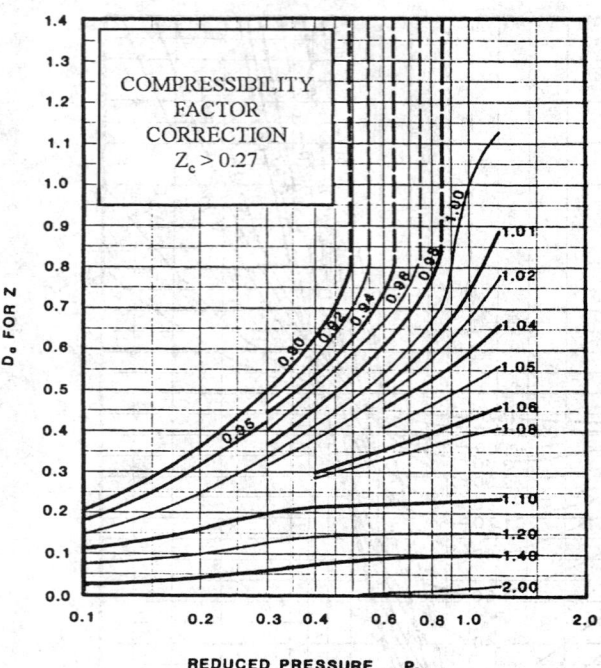

FIG. 2-38 Compressibility factor correction for Lydersen Method, $Z_c > 0.27$.

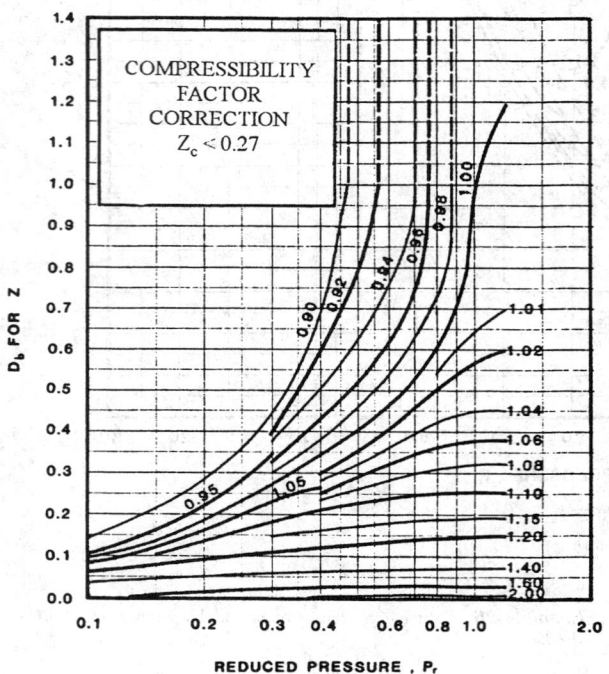

FIG. 2-39 Compressibility factor correction for Lydersen Method, $Z_c < 0.27$.

Pure component properties of propane are $M = 44.1$, $T_c = 96.7°C$, and $P_c = 4.246$ MPa. From Table 2-388, $Z_{RA} = 0.2763$.

$$T_r = \frac{273.1}{369.8} = 0.739, \quad n = 1.0 + (1.0 - 0.739)^{2/7} = 1.6813$$

$$n = 1.6813$$

$$\rho_{sat} = \frac{(4.246 \times 10^6)}{(8314)(369.8)} \left(\frac{1}{0.2763^{1.6813}} \right)$$

$$\rho_{sat} = 1.201 \text{ kmole/m}^3 = 529.6 \text{ kg/m}^3$$

An accepted experimental value is 531 kg/m³.

An alternate method with approximately the same accuracy as the Rackett method is the COSTALD method of Hankinson and Thomson.[35] The critical temperature, a characteristic volume near the critical volume, and an acentric factor optimized for vapor pressure prediction by the Soave[98] equation of state are required input parameters. The method is detailed in the *Technical Data Book*.[23]

Prediction of the density of **compressed pure liquid hydrocarbons** and their defined mixtures are readily and accurately predicted by the method of Lu.[62] One value of low pressure liquid density and the critical temperature and pressure are required to predict the density at higher pressures from Eq. (2-85).

$$\rho_2 = \left(\frac{C_2}{C_1} \right) \rho_1 \tag{2-85}$$

The constants C_1 and C_2 are both obtained from Fig. 2-40: C_1, usually from the saturated liquid line; and C_2, at the higher pressure. Errors should be less than 1 percent for pure hydrocarbons except at reduced temperatures above 0.95 where errors of up to 10 percent may occur. The method can be used for defined mixtures substituting pseudocritical properties for critical properties. For mixtures, the *Technical Data Book—Petroleum Refining* gives a more complex and accurate mixing rule than merely using the pseudocritical properties. The saturated low pressure value should be obtained from experiment or from prediction procedures discussed in this section for both pure and mixed liquids.

Example 21 Estimate the liquid density of *n*-nonane at 104.5°C and 6.893 MPa pressure.

A tabulated value of liquid density at 60°F (15.5°C) and 1 atm is 719.8 kg/m³. Pure component properties are $T_c = 321.5°C$ and $P_c = 2.288$ MPa.

Use Eq. (2-85) to correct a low pressure density.

$$\rho_2 = \left(\frac{C_2}{C_1} \right) \rho_1$$

where ρ_1 is a reference density. C_2 and C_1 are functions of T_r and P_r.

$$T_{r_1} = \frac{288.6}{594.6} = 0.485 \qquad T_{r_2} = \frac{377.6}{594.6} = 0.635$$

$$P_{r_1} = \frac{0.1013}{2.288} = 0.0443 \qquad P_{r_2} = \frac{6.893}{2.288} = 3.013$$

From Fig. 2-40, $C_1 = 1.08$ and $C_2 = 0.998$.

$$\rho_2 = \left(\frac{0.998}{1.08} \right)(719.8) = 665.1 \text{ kg/m}^3$$

An accepted experimental value at this temperature is 658.5 kg/m³, a 1 percent difference.

An analytical method for the prediction of compressed liquid densities was proposed by Thomson et al.[114] The method requires the saturated liquid density at the temperature of interest, the critical pressure, the critical temperature, an acentric factor (preferably the one optimized for vapor pressure data), and the vapor pressure at the temperature of interest. All properties not known experimentally may be estimated. Errors range from about 1 percent for hydrocarbons to 2 percent for nonhydrocarbons.

For prediction of the densities of a **defined liquid mixture at its bubble point** (ρ_{bp}), the method of Spencer and Danner[100] is the simplest. The density is calculated from Eq. (2-86) using inputs from Eqs. (2-87) and (2-88). For hydrocarbons, T_{mc} is calculated by Eqs. (2-89) through (2-92) if high accuracy is desired or by Eq. (2-93) for a less accurate answer.

$$\frac{1}{\rho_{bp}} = R \left(\sum_{i=1}^{n} x_i \frac{T_{c_i}}{P_{c_i}} \right) Z_{RA_m}^{[1 + (1 - T_r)^{2/7}]} \tag{2-86}$$

$$Z_{RA_m} = \sum_{i=1}^{n} x_i Z_{RA_i} \tag{2-87}$$

$$T_r = \frac{T}{T_{mc}} \tag{2-88}$$

$$T_{mc} = \sum_{i=1}^{n} \sum_{j=i}^{n} \phi_i \phi_j T_{c_{ij}} \tag{2-89}$$

TABLE 2-396   The Modified Rackett Equation Input Parameters for Calculating Pure Saturated Liquid Densities

Liquid	Z_{RA}	Liquid	Z_{RA}
Hydrocarbons		**Organics**	
Methane	0.2880	Acetic acid	0.2242
Ethane	0.2819	Methanol	0.2340
Propane	0.2763	Ethanol	0.2523
n-Butane	0.2730	2-Propanol	0.2508
2-Methylpropane (isobutane)	0.2760	Acetaldehyde	0.2387
n-Pentane	0.2685	Acetone	0.2448
n-Hexane	0.2637	Methyl ethyl ketone	0.2524
n-Heptane	0.2610	Methyl isobutyl ketone	0.2589
n-Octane	0.2569	Ethylamine	0.2640
n-Nonane	0.2555	Aniline	0.2607
n-Decane	0.2527	Methyl formate	0.2581
n-Dodecane	0.2471	Methyl acetate	0.2553
n-Tetradecane	0.2270	Ethyl acetate	0.2538
n-Hexadecane	0.2386	Ethyl acrylate	0.2583
n-Octadecane	0.2292	Methyl-n-butyl ether	0.2655
n-Eicosane	0.2281	Diethyl ether	0.2643
Cyclopentane	0.2709	Diisopropyl ether	0.2699
Methylcyclopentane	0.2712		
Cyclohexane	0.2729	**Halogen Compounds**	
Methylcyclohexane	0.2702	Methyl chloride	0.2679
Ethene (ethylene)	0.2813	Dichloromethane	0.2619
Propene (propylene)	0.2783	Chloroform	0.2751
1-Butene	0.2735	Tetrachloromethane	0.2721
cis-2-Butene	0.2705	Chlorobenzene	0.2650
trans-2-Butene	0.2722	Propionitrile	0.2156
2-Methylpropene (isobutylene)	0.2727		
1-Hexene	0.2654	**Inorganics**	
1,3-Butadiene	0.2713	Ammonia	0.2466
2-Methyl-1,3-butadiene	0.2680	Argon	0.2933
Ethyne (acetylene)	0.2707	Carbon dioxide	0.2729
Propyne (methylacetylene)	0.2703	Carbon disulfide	0.2850
Benzene	0.2696	Carbon monoxide	0.2898
Methylbenzene (toluene)	0.2645	Chlorine	0.2781
Ethylbenzene	0.2619	Hydrogen	0.3218
1,2-Dimethylbenzene (o-xylene)	0.2626	Hydrogen chloride	0.2673
1,3-Dimethylbenzene (m-xylene)	0.2594	Hydrogen sulfide	0.2818
1,4-Dimethylbenzene (p-xylene)	0.2590	Nitrogen	0.2893
Isopropylbenzene (cumene)	0.2616	Oxygen	0.2890
Biphenyl	0.2746	Sulfur dioxide	0.2667
Naphthalene	0.2611	Sulfur trioxide	0.2513

$$\phi_i = \frac{x_i V_{c_i}}{\sum_{j=1}^{n} x_j V_{c_j}} \qquad (2\text{-}90)$$

$$T_{c_{ij}} = \sqrt{T_{c_i} T_{c_j}} \,(1 - k_{ij}) \qquad (2\text{-}91)$$

$$k_{ij} = 1.0 - \left[\frac{\sqrt{V_{c_i}^{1/3} V_{c_j}^{1/3}}}{(V_{c_i}^{1/3} + V_{c_j}^{1/3})/2} \right]^3 \qquad (2\text{-}92)$$

$$T_{mc} = \sum_{i=1}^{n} x_i T_{c_i} \qquad (2\text{-}93)$$

Errors for binary hydrocarbon systems average about 2.5 percent except near the critical, where errors can approach 20 percent. If inorganic gases are included, errors average 4 percent except at high concentrations of carbon dioxide or hydrogen, where higher errors would be expected. If the simplified method for predicting T_{mc} is used, average errors of 5 to 7 percent should be expected for both binary hydrocarbon and nonhydrocarbon systems. No data are available to test the method for systems with more than two components.

A similarly accurate but slightly more complex method for prediction of densities of defined liquid hydrocarbon mixtures at their bubble points was published by Hankinson and Thomson[35] and was previously cited for prediction of pure liquid hydrocarbons.

For undefined hydrocarbon mixtures, the liquid density may be predicted at any temperature (T) from the mean average boiling point (MeABP) and the specific gravity (sp gr) by Eq. (2-94), adopted from Ritter et al.[92]

$$P = 62.3636 \left[(\text{sp gr})^2 - \frac{(1.2655)(\text{sp gr}) - 0.5098 + 8.011 \times 10^{-5} \text{MeABP}(T - 519.67)}{\text{MeABP}} \right]^{1/2} \qquad (2\text{-}94)$$

The density is calculated in lb_m/ft^3 if the temperatures are both in °R. Errors

average about 0.3 percent at atmospheric pressure. At high pressures, the liquid density of undefined hydrocarbon mixtures can be predicted from the low pressure value by the method of Wright[131] fully outlined in the *Technical Data Book*.[23]

Solid Density Prediction   The prediction of solid density is an inexact science and sometimes is taken as the liquid density at the triple point, although the solid density normally is higher than this value with a discontinuity at the triple point. Based on solid density data reviewed for the DIPPR compilation,[24] the solid density at the triple point can be estimated for organic compounds as 1.17 times the liquid density at the triple point. As liquid density at low temperatures varies little with temperature, the density of the liquid at the lowest estimable point above the triple point can be used with little degradation of the result. As solid density only decreases very slightly with increasing temperature and very little data on solid density as a function of temperature exist, no methods have been developed for predicting the solid density vs. temperature.

VISCOSITY

Viscosity is defined as the shear stress per unit area at any point in a confined fluid divided by the velocity gradient in the direction perpendicular to the direction of flow. If this ratio is constant with time at a given temperature and pressure for any species, the fluid is called a Newtonian fluid. This section is limited to Newtonian fluids, which include all gases and most nonpolymeric liquids and their mixtures. Most polymers, pastes, slurries, waxy oils, and some silicate esters are examples of non-Newtonian fluids.

The absolute viscosity (μ) is defined as the sheer stress at a point

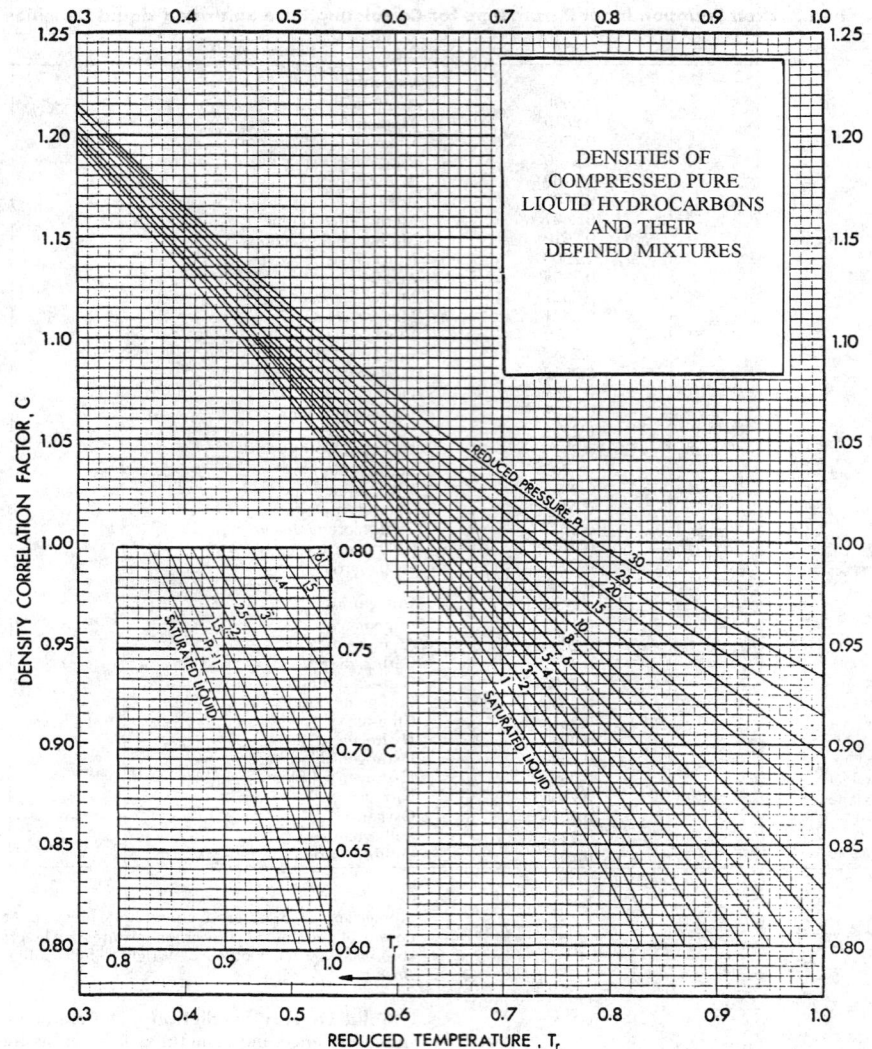

FIG. 2-40 Densities of compressed pure liquid hydrocarbons and their defined mixtures.

divided by the velocity gradient at that point. The most common unit is the poise (1 g/cm sec). The SI unit is the Pa·sec (1 kg/m sec). As many common fluids have viscosities in the hundredths of a poise the centipoise (cp) is often used. One centipoise is then equal to one mPa sec.

The kinematic viscosity (ν) is defined as the ratio of the absolute viscosity to density at the same temperature and pressure. The most common unit corresponding to the poise is the stoke (1 cm²/sec). The SI unit would be m²/sec.

Correlation Methods This section briefly discusses methods for correlating viscosities as a function of temperature and presents the most common accurate methods for prediction of vapor and liquid viscosity.

Vapor viscosity is accurately correlated as a function of temperature by Eq. (2-95).

$$\mu_v = \frac{AT^B}{1 + \dfrac{C}{T} + \dfrac{D}{T^2}} \qquad (2\text{-}95)$$

If data are available over a wide range, all four regression constants (A, B, C, and D) are usually used. Over narrow temperature ranges, only constants A and B are necessary.

Liquid viscosity is accurately correlated as a function of temperature by the modified Riedel equation previously discussed for correlation of vapor pressure and shown by Eq. (2-96).

$$\mu_\ell = \exp\left(A + \frac{B}{T} + C \ln(T) + DT^E\right) \qquad (2\text{-}96)$$

For most systems, only the first three terms are used. Only the first two terms are used for narrow ranges. If data are available in a wide range extending far above the normal boiling point, all four terms are used, with values of E varying in integers from −10 to 10 (excluding 0 and −1).

Constants for about 1500 compounds for both viscosities are available in the DIPPR compilation.[24]

Vapor Viscosity Methods for prediction of vapor viscosity abound such that only the most accurate and generally used methods are included.

For prediction of the vapor viscosity of **pure hydrocarbons at low pressure** (below T_r of 0.6), the method of Stiel and Thodos[101] is the most accurate. Only the molecular weight, the critical temperature, and the critical pressure are required. Equation (2-97) with values of N from Eqs. (2-98) and (2-99) is used.

$$\mu_v = 4.60 \times 10^{-4} \frac{NM^{1/2}P_c^{2/3}}{T_c^{1/6}} \tag{2-97}$$

$$N = 0.0003400T_r^{0.94} \quad \text{for } T_r \le 1.5 \tag{2-98}$$

$$N = 0.0001778(4.58T_r - 1.67)^{0.625} \quad \text{for } T_r > 1.5 \tag{2-99}$$

The resultant viscosity is in centipoise (mPa·sec) if T_c and P_c are given in K and Pa, respectively. This method can also be used for light nonhydrocarbon gases except for hydrogen where, special N's are required. For hydrocarbons below ten carbon atoms, average errors of about 3 percent can be expected, with errors increasing to 5–10 percent for heavier hydrocarbons.

Example 22   Estimate the vapor viscosity of propane at 101.3 kPa and 80°C.
Use Eq. (2-97).

$$\mu_v = 4.60 \times 10^{-4}N \frac{M^{1/2}P_c^{2/3}}{T_c^{1/6}}$$

$$T_c = 96.7°C, \quad P_c = 4.246 \text{ MPa}, \quad \text{and } M = 44.1.$$

To determine whether Eq. (2-98) or (2-99) should be used to calculate N, calculate T_r.

$$T_r = \frac{80 + 273.1}{96.7 + 273.1} = 0.955$$

Use Eq. (2-98): $N = 0.0003400T_r^{0.94}$

Thus, $N = 3.255 \times 10^{-4}$.

$$\mu_v = \frac{(4.60 \times 10^{-4})(3.255 \times 10^{-4})(44.1)^{1/2}(4.246 \times 10^6)^{2/3}}{(369.8)^{1/6}}$$

$$\mu_v = 0.0097 \text{ cp}$$

An experimental value of 0.0095 cp compares favorably.

For prediction of the vapor viscosity of **gaseous mixtures of hydrocarbons and nonhydrocarbon gases at low pressures** below a T_r of 0.6, the method of Bromley and Wilke[13] is recommended.

The mixing rule is given by Eq. (2-100) with the interaction parameter Q for each pair of components defined by Eq. (2-101).

$$\mu_m = \sum_{i=1}^{n} \frac{\mu_i}{1 + \sum_{\substack{j=1 \\ j \ne i}}^{n} \left(Q_{ij} \frac{x_j}{x_i} \right)} \tag{2-100}$$

$$Q_{ij} = \frac{1 + \left[\left(\frac{\mu_i}{\mu_j} \right)^{1/2} \left(\frac{M_j}{M_i} \right)^{1/4} \right]^2}{\sqrt{8} \left[1 + \frac{M_i}{M_j} \right]^{1/2}} \tag{2-101}$$

Errors, when tested against binary and multicomponent mixtures of both hydrocarbons and nonhydrocarbon gas mixtures, average about 3 percent.

For prediction of the vapor viscosity of **gaseous hydrocarbons and mixtures of hydrocarbons at high pressures** (not applicable to nonhydrocarbon gases) above a T_r of 0.6, low pressure values are calculated from Eq. (2-97) and/or (2-100) and then corrected for pressure by the method of Dean and Stiel[25] given by Eq. (2-102).

$$\mu - \mu_o = 5.0 \times 10^{-8} \frac{M^{1/2}P_c^{2/3}}{T_c^{1/6}} [\exp (1.439\rho_r) - \exp (-1.11\rho_r^{1.858})] \tag{2-102}$$

If critical pressure and critical temperature are given in Pa and K, respectively, viscosities in centipoise result. The variable μ_o is either the low pressure pure component or mixture viscosity according to whether a pure component or mixture is being considered. For mixtures, simple molar average pseudocritical temperature (Kay's rule), pressure, and density, and molar average molecular weight are used. The vapor density can be predicted by the methods previously discussed. Errors of above 5 percent are common for hydrocarbons and their mixtures. Experimental densities will reduce the errors slightly.

Example 23   Estimate the vapor viscosity of a mixture of propane and methane. Assume 60 mole percent methane and 40 mole percent propane at 125°C and 10.34 MPa total pressure. The low pressure viscosity is 0.0123 cp.
Use Eq. (2-102):

$$\mu - \mu_o = 5.0 \times 10^{-8} \frac{M^{1/2}P_c^{2/3}}{T_c^{1/6}} [\exp (1.439\rho_r) - \exp (-1.11\rho_r^{1.858})]$$

Properties of the pure components are:
Methane. $M = 16.04$, $T_c = -110.4°C$, $P_c = 4.593$ MPa
Propane. $M = 44.10$, $T_c = 96.7°C$, $P_c = 4.246$ MPa
For the mixture:
$T_{pc} = (0.60)(-110.4 + 273.1) + (0.40)(96.7 + 273.1) = 245.5$ K
$P_{pc} = (0.60)(4.593) + (0.40)(4.246) = 4.454$ MPa
$M_m = (0.60)(16.04) + (0.40)(44.10) = 27.26$
To calculate ρ_r, use Eq. (2-76) to calculate ρ using Pitzer corresponding states first to calculate Z. Then calculate $\rho_r = \rho/\rho_c = \rho V_{pc}$.
Additional properties required are:
Methane $\omega = 0.0115$, $V_c = 0.0986$ m³/kmole
Propane $\omega = 0.1523$, $V_c = 0.2002$ m³/kmole
For the mixture:
$\omega_m = (0.60)(0.0115) + (0.40)(0.1523) = 0.0678$
$V_{pc} = (0.60)(0.0986) + (0.40)(0.2002) = 0.1392$ m³/kmole

Using Eq. (2-58): $Z = Z^{(0)} + \omega Z^{(1)}$

$$T_r = \frac{398.1}{245.5} = 1.62 \qquad P_r = \frac{10.34}{4.454} = 2.32$$

From Fig. 2-35: $Z^{(0)} = 0.87$

From Fig. 2-36: $Z^{(0)} \cong 0.20$

$$Z = 0.87 + (0.0678)(.20) = 0.88$$

Hence: $\rho = \dfrac{P}{ZRT} = \dfrac{(10.34)}{(0.88)(8.314 \times 10^{-3})(398.1)}$

$$\rho = 3.55 \text{ kmole/m}^3$$

$$\rho_r = \rho V_{pc} = (3.55)(.1392) = 0.494$$

$$\mu - \mu_o = \frac{(5.0 \times 10^{-8})(27.26)^{1/2}(4.454 \times 10^6)^{2/3}}{(245.5)^{1/6}} [e^{1.439(.494)} - e^{-1.11(.494)^{1.858}}]$$

$$\mu - \mu_o = \frac{(5.08 \times 10^{-8})(5.22)(27,085)}{(2.503)} [2.0358 - 0.7413]$$

$$\mu - \mu_o = 0.0037$$

$$\mu = 0.0037 + 0.0123 = 0.0160 \text{ cp}$$

An experimental value of 0.0167 cp compares favorably.

For **pure nonhydrocarbon polar gases at low pressures,** the viscosity can be estimated by the method of Reichenberg[85] given by Eq. (2-103).

$$\mu = \frac{AT_r}{[1 + 0.36T_r(T_r - 1)]^{1/6}} \tag{2-103}$$

For organic compounds:

$$A = \frac{M^{1/2}T_c}{\sum n_i C_i} \tag{2-104}$$

For inorganic gases:

$$A = 1.6104 \times 10^{-10} \left[\frac{M^{1/2}P_c^{2/3}}{T_c^{1/6}} \right] \tag{2-105}$$

Viscosities are calculated in Pa sec (10^3 cp = 1 Pa sec) with T_c in K and P_c in Pa. Group contributions based on atomic structure for organic compounds necessary for Eq. (2-104) are tabulated in Table 2-397. Errors average about 5 percent for most organics, with slightly higher errors for inorganic gases. For pure nonhydrocarbon nonpolar gases, an alternate method is the method of Yoon and Thodos,[132] which requires the same input parameters as the Reichenberg method. The method, when evaluated for the *Technical Data Manual,* showed errors of about 3 percent for compounds with low dipole moments and requires special correlations for hydrogen and helium.

Example 24   Estimate the vapor viscosity of isopropyl alcohol at 251°C and atmospheric pressure.
Use Eq. (2-103) with A determined from Eq. (2-104).

$$\mu = \frac{AT_r}{[1 + 0.36T_r(T_r - 1)]^{1/6}}$$

$$A = \frac{M^{1/2}T_c}{\sum n_i C_i}$$

TABLE 2-397   Group Contribution Values for Eq. (2-104)

Group	Contribution $C_i \times 10^{-8}$ m·s·K/kg
—CH₃	0.904
＼CH₂ (nonring)	0.647
＼CH — (nonring)	0.267
＼C＼ (nonring)	−0.153
＝CH₂	0.768
＝CH — (nonring)	0.553
＼C＝ (nonring)	0.178
≡CH	0.741
≡C — (nonring)	0.524
＼CH₂ (ring)	0.691
＼CH — (ring)	0.116
＼C／ (ring)	0.023
＝CH — (ring)	0.590
＼C＝ (ring)	0.359
—F	0.446
—Cl	1.006
—Br	1.283
—OH (alcohols)	0.796
＼O (nonring)	0.359
＼C＝O (nonring)	1.202
—CHO (aldehydes)	1.402
—COOH (acids)	1.865
—COO— (esters) or HCOO— (formates)	1.341
—NH₂	0.971
＼NH (nonring)	0.368
＝N— (ring)	0.497
—CN	1.813
＼S (ring)	0.886

Pure component properties are $T_c = 508.3$ K and $M = 60.1$.

Isopropyl alcohol group contributions for $\sum n_i C_i$ are calculated using Table 2-397.

$$2 \quad —CH_3 \text{ groups} = (2)(.904 \times 10^8) +$$

$$1 \quad \diagdown CH— \text{ group} = (1)(.267 \times 10^8) +$$

$$1 \quad —OH \text{ group} = (1)(.796 \times 10^8) = 2.871 \times 10^8$$

$$T_r = \frac{524.1}{508.3} = 1.031$$

$$A = \frac{60.1^{1/2}(508.3)}{2.871} \times 10^8 = 1.373 \times 10^{-5} \text{ Pa sec}$$

$$\mu = \frac{(1.373 \times 10^{-5})(1.031)}{[1 + 0.36(1.031)(.031)]^{1/6}} = 1.413 \times 10^{-5} \text{ Pa sec}$$

(An experimental value of 1.380×10^{-5} Pa sec compares favorably.)

For **pure nonhydrocarbon polar and nonpolar gases at high pressure,** a method of prediction attributed to Stiel and Thodos[102] depending on the reduced density as a corrector to the low pressure gas viscosity (μ^o) takes various forms for polar gases according to the reduced density (ρ_r) as shown in Eqs. (2-106) through (2-109).

$$\rho_r \leq 0.1 \qquad (\mu - \mu^o)B = 1.656 \times 10^{-7}\rho_r^{1.111} \qquad (2\text{-}106)$$

$$0.1 \leq \rho_r \leq 0.9 \qquad (\mu - \mu^o)B = 6.07 \times 10^{-9}(9.045\rho_r + 0.63)^{1.739} \qquad (2\text{-}107)$$

$$0.9 \leq \rho_r < 2.2 \qquad \log\{4 - \log_{10}[(\mu - \mu^o)10^7 B]\} = 0.6439 - 0.1005\rho_r \qquad (2\text{-}108)$$

$$2.2 \leq \rho_r < 2.6 \qquad \log\{4 - \log_{10}[(\mu - \mu^o)10^7 B]\} = 0.6439 - 0.1005\rho_r$$
$$- 4.75 \times 10^{-4}(\rho_r^3 - 10.65)^2 \qquad (2\text{-}109)$$

$$B = 2173.424\, T_c^{1/6}\, M^{-1/2}\, P_c^{-2/3} \qquad (2\text{-}110)$$

For nonpolar gases, Jossi et al.[45] extended the method as shown in Eq. (2-111) for $0.1 < \rho_r < 3.0$.

$$[(\mu - \mu^o)B10^7 + 1]^{1/4} = 1.0230 + 0.23364\rho_r$$
$$+ 0.58533\rho_r^2 - 0.40758\rho_r^3 + 0.093324\rho_r^4 \qquad (2\text{-}111)$$

In all cases viscosities are in Pa sec with T_c in K and P_c in Pa. For nonpolars, errors are very small; while for polars, average errors reach 11 percent.

Example 25 Estimate the vapor viscosity of carbon dioxide at 350 K and a total pressure of 20 MPa. An experimental low pressure viscosity at 350 K is 1.7386×10^{-5} Pa sec.

Use Eq. (2-111) with B calculated by Eq. (2-110).

The pure component properties necessary are $T_c = 304.19$ K, $P_c = 7.3815$ MPa, $V_c = 0.094$ m³/kmole, and $M = 44.01$.

Using the Lee-Kesler form of the Pitzer method [Eq. (2-78)], $Z = 0.4983$.

$$\rho_r = \frac{V_c}{V} = \frac{V_c P}{ZRT} = \frac{(0.094)(2 \times 10^7)}{(0.4983)(8314)(350)} = 1.2965$$

$$B = \frac{2173.424 T_c^{1/6}}{M^{1/2} P_c^{2/3}} = \frac{(2173.424)(304.19)^{1/6}}{(44.01)^{1/2}(7.3815 \times 10^6)^{2/3}}$$

$$B = 2.2411 \times 10^{-2}$$

$$[(\mu - 1.7386 \times 10^{-5})B10^7 + 1]^{1/4} = 1.0230 + 0.23364(1.2965) + 0.58533(1.2965)^2$$
$$- 0.40758(1.2965)^3 + 0.093324(1.2965)^4$$

$$= 1.6852$$

$$(\mu - 1.7386 \times 10^{-5})(2.2411 \times 10^{-2})(10^7) + 1 = 8.0659$$

$$\mu - 1.7386 \times 10^{-5} = 3.1529 \times 10^{-5}$$

$$\mu = 4.89 \times 10^{-5} \text{ Pa sec}$$

An experimental value of 4.73×10^{-5} Pa sec compares favorably.

For both polar and nonpolar **nonhydrocarbon gaseous mixtures at low pressures,** the most accurate viscosity prediction method is the method of Brokaw.[10,11] The method is quite accurate but requires the dipole moment and the Stockmayer energy parameter (ε/k) for polar components as well as pure component viscosities, molecular weights, the normal boiling point, and the liquid molar volume at the normal boiling point. The *Technical Data Manual* should be consulted for the full method.

For nonpolar, nonhydrocarbon vapor mixtures at high pressures, the method of Dean and Stiel[25] [Eq. (2-102)] discussed earlier can be used. The accuracy of the method is excellent and dependent on the pure component viscosity values used as input parameters.

Liquid Viscosity The viscosity of both **pure hydrocarbon and pure nonhydrocarbon liquids** are most accurately predicted by the method of van Velzen et al.[122] The basic equation (2-112) depends on group contributions which are dependent on structure for the calculation of compound-specific constants B and T_o.

$$\log \mu = B\left(\frac{1}{T} - \frac{1}{T_o}\right) - 3.0 \qquad (2\text{-}112)$$

Resultant viscosities are in Pa sec. If the -3.0 on the right is deleted,

answers are in cp. T_o is calculated by Eq. (2-114) or (2-115) according to the value of an adjusted carbon number N° calculated by Eq. (2-113) using the actual carbon number N and group contributions from Table 2-398.

$$N^\circ = N + \sum_i \Delta N_i \qquad (2\text{-}113)$$

$$N^\circ \leq 20 \quad T_o = 28.86 + 37.439N^\circ - 1.3547N^{\circ 2} + 0.02076N^{\circ 3} \quad (2\text{-}114)$$

$$N^\circ > 20 \quad T_o = 8.164N^\circ + 238.59 \qquad (2\text{-}115)$$

TABLE 2-398 Group Contribution Values for Liquid Viscosity Prediction

Structures or functional group	ΔN_i	ΔB_i	Remarks
n-Alkanes	0	0	
Isoalkanes	$1.389 - 0.238N$	15.51	
Saturated hydrocarbons with two methyl groups in isoposition	$2.319 - 0.238N$	15.51	
n-Alkanes	$-0.152 - 0.042N$	$-44.94 + 5.410N^\circ$	
n-Alkadienes	$-0.304 - 0.084N$	$-44.94 + 5.410N^\circ$	
Isoalkanes	$1.237 - 0.084N$	$-36.01 + 5.410N^\circ$	
Isoalkadienes	$1.085 - 0.322N$	$-36.01 + 5.410N^\circ$	
Hydrocarbon with one double bond and two methyl groups in isoposition	$2.626 - 0.518N$	$-36.01 + 5.410N^\circ$	For any additional CH_3 groups in isoposition, increase ΔN by $1.389 - 0.238N$
Hydrocarbon with two double bonds and two methyl groups in isoposition	$2.474 - 0.560N$	$-36.01 + 5.410N^\circ$	For any additional CH_3 groups in isoposition, increase ΔN by $1.389 - 0.238N$
Cyclopentanes	$0.205 + 0.069N$	$-45.96 + 2.224N^\circ$	$N \leq 16$; not recommended for $N = 5,6$
	$3.971 - 0.172N$	$-339.67 + 23.135N^\circ$	$N \geq 16$
Cyclohexanes	1.48	$-272.85 + 25.041N^\circ$	$N < 17$; not recommended for $N = 6,7$
	$6.517 - 0.311N$	$-272.85 + 25.041N^\circ$	$N \geq 17$
Alkyl benzenes	0.60	$-140.04 + 13.869N^\circ$	$N < 16$; not recommended for $N = 6,7$[a,e,f]
	$3.055 - 0.161N$	$-140.04 + 13.869N^\circ$	$N \geq 16$[a,e,f]
Polyphenols	$-5.340 + 0.815N$	$-188.40 + 9.558N^\circ$	[a]
Alcohols			
Primary	$10.606 - 0.276N$	$-589.44 + 70.519N^\circ$	[b]
Secondary	$11.200 - 0.605N$	497.58	[b]
Tertiary	$11.200 - 0.605N$	928.83	[b]
Diols (correction)	See remarks	557.77	For ΔN, use alcohol contributions and add $N - 2.50$
Phenols (correction)	$16.17 - N$	213.68	[a,c,d]
——OH on side chain to aromatic ring (correction)	-0.16	213.68	
Acids	$6.795 + 0.365N$	$-249.12 + 22.449N^\circ$	$N < 11$, not recommended for $N = 1,2$
	10.71	$-249.12 + 22.449N^\circ$	$N \geq 11$
Isoacids	See remarks	$-249.12 + 22.449N^\circ$	Calculate ΔB as for straight-chain acid; calculate ΔN for straight-chain acid but reduce ΔN by 0.24 for each methyl group in isoposition
Acids with aromatic nucleus in structure (correction)	4.81	$-188.40 + 9.558N^\circ$	
Esters	$4.337 - 0.230N$	$-149.13 + 18.695N^\circ$	If hydrocarbon groups have isoconfiguration, see [e]
Esters with aromatic nucleus in structure (correction)	$-1.174 + 0.376N$	$-140.04 + 13.869N^\circ$	Add to values of ΔN, ΔB calculated for ester
Ketones	$3.265 - 0.122N$	$-117.21 + 15.781N^\circ$	If hydrocarbon groups have isoconfiguration, see [e]
Ketones with aromatic nucleus in structure (correction)	2.70	$-760.65 + 50.478N^\circ$	Add to values of ΔN, ΔB calculated for ketone
Ethers	$0.298 + 0.209N$	$-9.39 + 2.848N^\circ$	If hydrocarbon groups have isoconfiguration, see [e]
Aromatic ethers	$11.5 - N$	$-140.04 + 13.869N^\circ$	The ΔN value is not a correction to regular ether value, but the ΔB value is a correction to regular ether[e]
Amines			
Primary	$3.581 + 0.325N$	$25.39 + 8.744N^\circ$	If hydrocarbon groups have isoconfiguration, see [e]
Primary amine in side chain of aromatic compound (correction)	-0.16	0	Corrections to be added to amine calculation[e]
Secondary	$1.390 - 0.461N$	$25.39 + 8.744N^\circ$	If hydrocarbon groups have isoconfiguration, see [e]
Tertiary	3.27	$25.39 + 8.744N^\circ$	If hydrocarbon groups have isoconfiguration, see [e]
Primary amines with NH_2 group on aromatic nucleus	$15.04 - N$	0	The ΔN value is not a correction to regular amine value; to find ΔB, use primary amine value[a,c]
Secondary or tertiary amine with at least one aromatic group attached to amino nitrogen	[f]	[f]	
Nitro compounds			
1-Nitro	$7.812 - 0.236N$	$-213.14 + 18.330N^\circ$	
2-Nitro	5.84	$-213.4 + 18.330N^\circ$	Note alkene contribution is necessary
3-Nitro	5.56	$-338.01 + 25.086N^\circ$	
4-Nitro; 5-Nitro	5.36	$-338.01 + 25.086N^\circ$	
Aromatic nitrocompounds	$7.182 - 0.236N$	$-213.14 + 18.330N^\circ$	For aromatic correction, see [f]
Nitrile	$4.039 - 0.0103N$	$-241.66 + 27.937N^\circ$	

TABLE 2-398 Group Contribution Values for Liquid Viscosity Prediction (*Concluded*)

Structures or functional group	ΔN_i	ΔB_i	Remarks
Amines (*Cont.*)			
Isomethyl on nitrile	$-0.7228 + 0.1755N$	$286.26 - 31.009N°$	
Aromatic nitrile	$2.321 - 0.2357N$	$-26.063 - 11.516N°$	
Dinitrile	$10.452 - 1.1276N$	$3599.9 - 199.96N°$	
Halogenated compounds			
Fluoride	1.43	5.75	*e,f*
Chloride	3.21	-17.03	*e,f*
Bromide	4.39	$-101.97 + 5.954N°$	*e,f*
Iodide	5.76	-85.32	
Special configurations (corrections)			
$C(Cl)_x$	$1.91 - 1.459x$	-26.38	
——CCl——CCl——	0.96	0	
——$C(Br)_x$——	0.50	$81.34 - 86.850x$	
——CBr——CBr——	1.60	-57.73	
CF_3			
In alcohols	-3.93	341.68	
In other compounds	-3.93	25.55	
Aldehydes	3.38	$146.45 - 25.11N°$	
Aldehydes with an aromatic nucleus in structure (correction)	2.70	$-760.65 + 50.478N°$	
Anhydrides	$7.97 - 0.50N$	-33.50	
Anhydrides with an aromatic nucleus in structure (correction)	2.70	$-760.65 + 50.478N°$	
Amides	$13.12 + 1.49N$	$524.63 - 20.72N°$	
Amides with an aromatic nucleus in structure (correction)	2.70	$-760.65 + 50.478N°$	
Sulfide	$3.9965 - 0.1861N$	$-76.676 + 8.1403N°$	
Isomethyl on sulfide	0.1601	-25.026	

[a] For substitutions on an aromatic nucleus in more than one position, additional corrections are required:

Ortho	$\Delta N = 0.51$	$\Delta B = \begin{cases} -571.94 \text{ (with ——OH)} \\ 54.84 \text{ (without ——OH)} \end{cases}$	
Meta	$\Delta N = 0.11$	$\Delta B = 27.25$	
Para	$\Delta N = -0.04$	$\Delta B = 17.57$	

[b] For alcohols, if there is a methyl group in the isoposition, increase ΔN by 0.24 and ΔB by 94.23.

[c] If the compound has an aromatic ——OH or ——NH_2, or if there is an aromatic ether, use ΔN contribution in table but neglect other substituents on the ring such as halogen, CH_3, NO_2, and the like. For the calculation of ΔB, however, such substituents must be taken into account.

[d] For aromatic alcohols and compounds with an ——OH on a side chain, the alcohol contribution (primary, etc.) must be included. For example, *o*-chlorophenol:

$\Delta B = \Delta B$ (primary alcohol) $+ \Delta B$ (chlorine) $+ \Delta B$ (phenol) $+ \Delta B$ (ortho correction—see footnote *a*)

With $N° = 16.17$ (see footnote *c*):

$\Delta B = (-589.44 + 70.519 \times 16.17) + (-17.03) + (213.68) + (-571.94) = 175.56$

$B_a = 745.94 \quad B = B_a + \Delta B = 921.50$

2-Phenylethanol:

$N = 8; \Delta N = \Delta N$ (primary alcohol) $+ \Delta N$ (correction) $= [10.606 - (0.276)(8)] + (-0.16) = 8.24$

$N° = N + \Delta N = 8 + 8.24 = 16.24$

$\Delta B = \Delta B$ (primary alcohol) $+ \Delta B$ (correction) $= [-589.44 + (70.519)(16.24)] + 213.68 = 769.47$

$B_a = 747.43 \quad B = B_a + \Delta B = 1516.9$

[e] For esters, alkylbenzenes, halogenated hydrocarbons, and ketones: If the hydrocarbon chain has a methyl group in an isoposition, decrease ΔN by 0.24 and increase ΔB by 8.93 for each such grouping. For ethers and amines, decrease ΔN by 0.50 and increase ΔB by 8.93 for each isogroup.

[f] For alkylbenzenes, nitrobenzenes, halogenated benzenes and for secondary or tertiary amines where at least one aromatic group is connected to an amino nitrogen, add the following corrections for each aromatic nucleus. If $N < 16$, increase ΔN by 0.60; if $N \geq 16$, increase ΔN by $3.055 - 0.161N$ for each aromatic group. For any N, increase ΔB by $(-140.04 + 13.869N°)$.

From van Velzen et al. (122).

B is calculated by Eq. (2-116) using values of B_a calculated from Eq. (2-117) or (2-118) according to the value of $N°$ and group contributions from Table 2-398.

$$B = B_a + \sum_i \Delta B_i \tag{2-116}$$

$$N° \leq 20 \quad B_a = 24.79 + 66.885N° - 1.3173N°^2 - 0.00377N°^3 \tag{2-117}$$

$$N° > 20 \quad B_a = 530.59 + 13.740N° \tag{2-118}$$

The method should not be used for the first member of a homologous series or for temperatures much above the normal boiling point ($T_r \approx 0.75$). Errors for both hydrocarbons and nonhydrocarbons average 15 percent for a wide variety of compounds. Higher errors are noted for amines, diols, ethers, and fluorides. Table 2-398 gives ΔN and ΔB contributions for most common groups. Space prohibits examples for

each type of compound or inclusion of specialized cases that are more fully discussed in the *Technical Data Manual*.

Example 26 Estimate the liquid viscosity of *cis*-1,4-dimethylcyclohexane at 0°C.

Use Eq. (2-112) with $N°$ calculated from Eq. (2-113) and B calculated from Eq. (2-116).

Determine group contributions from Table 2-398.

		ΔN_i	ΔB_i
Cyclohexanes	(1)	1.48	$-272.85 + 25.041 N°$
n-Alkanes	(2)	0	0

$$N° = N + \sum_i n_i \Delta N_i$$

$$N° = 8 + (1)(1.48) + (2)(0) = 9.48$$

Use Eq. (2-114) to calculate T_o:

$$T_o = 28.86 + 37.439(9.48) - 1.3547(9.48)^2 + (.02076)(9.48)^3$$

$$T_o = 279.72$$

$$B = B_a + \sum_i n_i \Delta B_i$$

Use Eq. (2-117) to calculate B_a:

$$B_a = 24.79 + 66.885(9.48) - 1.3173(9.48)^2 - 0.00377(9.48)^3$$

$$B_a = 537.26$$

$$B = 537.26 + (1)[-272.85 + 25.041 \ (9.48)]$$

$$B = 501.80$$

$$\log \mu = B\left(\frac{1}{T} - \frac{1}{T_o}\right) - 3.0$$

$$\log \mu = 501.80\left(\frac{1}{273.15} - \frac{1}{279.72}\right) - 3.0$$

$$\log \mu = -2.9569$$

$$\mu = 0.001104 \text{ Pa sec} = 1.104 \text{ cp}$$

An experimental value at 0°C is 1.224 cp.

A mixing rule developed by Kendall and Monroe[48] is useful for determining the **liquid viscosity of defined hydrocarbon mixtures.** Equation (2-119) depends only on the pure component viscosities at the given temperature and pressure and the mixture composition.

$$\mu_m = \left(\sum_{i=1}^{n} x_i \mu_i^{1/3}\right)^3 \qquad (2\text{-}119)$$

For mixtures of the same chemical family, errors average less than 3 percent, while errors overall average 5–6 percent, with errors of mixed families averaging from 10–15 percent.

For estimating the liquid viscosity of **defined nonhydrocarbon mixtures,** a mixing rule shown by Eq. (2-120) was recommended by the *Technical Data Manual.*

$$\ln \mu_m = \sum_i x_i \ln \mu_i \qquad (2\text{-}120)$$

Errors average near 15 percent.

For interpolating viscosities of hydrocarbon mixtures within a limited range knowing viscosities at two temperatures, ASTM Procedure D341-89, including both charts and equations, is recommended. Several recommended methods for predicting the viscosity of undefined hydrocarbon mixtures, such as petroleum fractions and coal liquids, are presented and evaluated in the *Technical Data Book—Petroleum Refining.* In addition, several methods for determining the liquid viscosity of blends of hydrocarbon mixtures and a method for liquid viscosity of pure hydrocarbons blended with undefined mixtures are given.

The most accurate method for predicting the liquid viscosity of **hydrocarbons containing fewer than 20 carbon atoms at high pressure** is a corresponding states method developed by Graboski and included in the *Technical Data Book—Petroleum Refining.* Critical temperature, critical pressure, the acentric factor, and knowing or being able to calculate at least one viscosity at a reference temperature and pressure are required. Errors average about 5 percent. For high molecular weight hydrocarbons at high pressure, low pressure viscosities can be converted instead of using the method of Kouzel,[51] which requires a low pressure liquid viscosity as input. Compounds of more than 20 carbon atoms and their mixtures are treated in this way.

For predicting the **liquid viscosity of pure hydrocarbon mixtures at high temperatures,** the method of Letsou and Stiel[58] is available. Error analyses with only a small amount of data shows errors averaging 34 percent in the reduced temperature range of 0.76 to 0.98. Equation (2-121) defines the method with inputs of Eqs. (2-122) and (2-123).

$$\mu\left(2173.424 \ \frac{T_c^{1/6}}{M^{1/2}P_c^{2/3}}\right) = \mu^{(0)} + \omega\mu^{(1)} \qquad (2\text{-}121)$$

$$\mu^{(0)} = 1.5174 \times 10^{-5} - 2.135 \times 10^{-5}T_r + 7.5 \times 10^{-6}T_r^2 \qquad (2\text{-}122)$$

$$\mu^{(1)} = 4.2552 \times 10^{-5} - 7.674 \times 10^{-5}T_r + 3.4 \times 10^{-5}T_r^2 \qquad (2\text{-}123)$$

Results are in Pa sec with inputs of T_c in K and P_c in Pa.

VAPOR AND LIQUID THERMAL CONDUCTIVITY

Thermal conductivity describes the ease with which conductive heat can flow through a vapor, liquid, or solid layer of a substance. It is defined as the proportionality constant in Fourier's law of heat conduction in units of energy · length/time · area · temperature; e.g., W/m K.

Gases   For **pure component, low pressure** (<350 kPa) **hydrocarbon** gases, Misic and Thodos[70,71] recommend the following equations. For **methane** and **cyclic compounds** below reduced temperatures of 1.0:

$$k_G = 4.45 \times 10^{-7} \ T_r \frac{C_p}{\lambda} \qquad (2\text{-}124)$$

For these hydrocarbons above reduced temperatures of 1.0 and for other hydrocarbons at any temperature:

$$k_G = 10^{-7} \ (14.52 \ T_r - 5.14)^{2/3} \left(\frac{C_p}{\lambda}\right) \qquad (2\text{-}125)$$

In these equations,

$$\lambda = T_c^{1/6} \ M^{1/2} \left(\frac{101.325}{P_c}\right)^{2/3} \qquad (2\text{-}126)$$

where k_G = vapor thermal conductivity, W/m K
  T_r = reduced temperature, T/T_c
  T = temperature, K
  T_c = critical temperature, K
  C_p = heat capacity at constant pressure, J/kmol K
  M = molecular weight
  P_c = critical pressure, kPa

C_p may be assumed to be the ideal gas heat capacity, C_p^o. Average errors can be expected to be less than 5 percent.

Example 27   Estimate thermal conductivity for *n*-hexane. For *n*-hexane at 373.15 K and low pressure, the required properties from Daubert et al.[24] are: T_c = 507.6 K, M = 86.18, P_c = 3025.0 kPa, and C_p^o = 1.721 × 10⁵. T_r = 373.15/507.6 = 0.7351. Using Eq. (2-126):

$$\lambda = (507.6)^{1/6} \ (86.18)^{1/2} \left(\frac{101.325}{3025.0}\right)^{2/3}$$

$$\lambda = 2.724$$

Substituting into Eq. (2-125):

$$k_G = 10^{-7} \ [(14.52)(0.7351) - 5.14]^{2/3} \left(\frac{1.721 \times 10^5}{2.724}\right)$$

$$k_G = 0.01977 \text{ W/m K}$$

The reported value is 0.02025 W/m K.[24]

For **pure nonhydrocarbon** gases at low pressures (up through 1 atm), the following equations may be used at temperature T (K):

Monatomic gases[12]:
$$k_G = 2.5 \ \frac{\eta_G C_v}{M} \qquad (2\text{-}127)$$

Linear molecules[12]:
$$k_G = \frac{\eta_G}{M}\left(1.30 \ C_v + 14644.0 - \frac{2928.8}{T_r}\right) \qquad (2\text{-}128)$$

Nonlinear molecules[102]:
$$k_G = \frac{\eta_G}{M} \ (1.15 \ C_v + 16903.36) \qquad (2\text{-}129)$$

where k_G = vapor thermal conductivity, W/m K
  η_G = vapor viscosity, Pa·s
  C_v = heat capacity at constant volume, J/kmol K
  M = molecular weight
  T_r = reduced temperature, T/T_c
  T_c = critical temperature, K

C_v may be calculated as $C_p^o - R$, where C_p^o is the ideal gas heat capacity in J/kmol K and R is the gas constant, 8314 J/kmol K. Average errors are in the 8–10 percent range but may be higher for polar compounds. This method should not be used for molecules that associate; e.g., organic acids.

Example 28   Estimate thermal conductivity of carbon dioxide at 370 K and low pressure. For carbon dioxide (a linear molecule) at 370.0 K and low pressure, the required properties from Daubert et al.[24] are: T_c = 304.2 K, η_G = 1.828 × 10⁻⁵ Pa·s, M = 44.01, and C_p^o = 40520 J/kmol K. T_r = 370.0/304.2 = 1.2163, and C_v = 40520 − 8314 = 32206 J/kmol K. Substituting in Eq. (2-128):

$$k_G = \frac{1.828 \times 10^{-5}}{44.01}\left[(1.30)(32206) + 14644.0 - \frac{2928.8}{1.2163}\right]$$

$$k_G = 0.02247 \text{ W/m K}$$

A value of 0.0220 W/m K is obtained from Vargaftik et al.[124]

For pure gases **above atmospheric pressure,** the method of Stiel and Thodos[102] may be used:

$$k_G = k'_G + \frac{A \times 10^{-4}(e^{B\rho_r} + C)}{\left(\dfrac{T_c^{1/6}M^{1/2}}{P_c^{2/3}}\right)Z_c^5} \qquad (2\text{-}130)$$

$\rho_r < 0.5$	$A = 2.702$	$B = 0.535$	$C = -1.000$
$0.5 < \rho_r > 2.0$	$A = 2.528$	$B = 0.670$	$C = -1.069$
$2.0 < \rho_r > 2.8$	$A = 0.574$	$B = 1.155$	$C = 2.016$

where k_G = vapor thermal conductivity at the temperature T (K) and pressure P of interest, W/m K
k'_G = vapor thermal conductivity at T and atmospheric pressure, W/m K
ρ_r = reduced density = V_c/V
V_c = critical molar volume, m³/kmol
V = molar volume at T and P, m³/kmol
T_c = critical temperature, K
M = molecular weight
P_c = critical pressure, MPa
Z_c = critical compressibility factor = P_cV_c/RT_c
R = gas constant = 0.008314 MPa m³/kmol K

Errors in the range of 5–6 percent are typical with this method but may be higher for branched compounds.

Example 29 Estimate the thermal conductivity of carbon dioxide at 370 K and 10 MPa pressure. The required properties from Daubert et al.[24] are: $T_c = 304.2$ K, $M = 44.01$, $P_c = 7.383$ MPa, and $V_c = 0.0940$ m³/kmol; $Z_c = (7.383) \times (0.0940)/(0.008314)(304.2) = 0.274$. $k'_G = 0.0220$ W/m K[123], and $V = 0.22809$ m³/kmol³; $\rho_r = 0.0940/0.22809 = 0.4121$. Substituting in Eq. (2-130) using the equation constants A, B, and C for $\rho_r < 0.5$:

$$k_G = 0.0220 + \frac{(2.702 \times 10^{-4})(e^{(0.535)(0.4121)} - 1.000)}{\dfrac{(304.2)^{1/6}(44.01)^{1/2}}{(7.383)^{2/3}}(0.274)^5} = 0.0315 \text{ W/m K}$$

Vargaftik et al.[124] report a value of 0.0308 W/m K.

The thermal conductivity of **low pressure** (1 atm or less) **gas mixtures** can be determined from the relation of Wassiljewa[126]:

$$k_m = \sum_{i=1}^{n} \frac{y_i k_i}{\displaystyle\sum_{j-1}^{n} y_j A_{ij}} \qquad (2\text{-}131)$$

where k_m = mixture thermal conductivity, W/m K
n = number of components
$y_{i,j}$ = mole fraction of component i or j in the vapor mixture
k_i = thermal conductivity of pure component i at the temperature of interest

The binary interaction parameter A_{ij} is obtained by the method of Lindsay and Bromley[61]:

$$A_{ij} = \frac{1}{4}\left\{1 + \left[\frac{\mu_i}{\mu_j}\left(\frac{M_j}{M_i}\right)^{3/4}\left(\frac{T + S_i}{T + S_j}\right)\right]^{1/2}\right\}^2\left(\frac{T + S_{ij}}{T + S_i}\right) \qquad (2\text{-}132)$$

where $\mu_{i,j}$ = vapor viscosity of pure component i or j at the temperature T of interest and low pressure, Pa·s
$M_{i,j}$ = molecular weight of pure component i or j
T = temperature, K
$S_{ij} = S_{ji}$; see Eq. (2-133)
$C = 1.0$ except when either or both components i and j are very polar; then $C = 0.73$
$S_{i,j}$ = 79 K for helium, hydrogen, and neon; for all others, see Eq. (2-134)
$T_{bi,j}$ = normal boiling temperature of pure component i or j, K

$$S_{ji} = C(S_iS_j)^{1/2} \qquad (2\text{-}133)$$

$$S_{i,j} = 1.5T_{bi,j} \qquad (2\text{-}134)$$

Expected errors for this method are 4–5 percent. At **higher pressures,** a pressure correction using Eq. (2-130) may be used. The mixture is treated as a hypothetical pure component with mixture critical properties obtained via Eqs. (2-5), (2-8), and (2-17) and with the molecular weight being mole-averaged.

Example 30 Estimate thermal conductivity of a mixture of 0.23 mole fraction dimethylether (1) and 0.77 mole fraction methyl chloride (2) at

373.15 K and low pressure. The required pure component properties from Daubert et al.[24] are: $k_1 = 0.02504$ W/m K, $k_2 = 0.01587$ W/m K, $\mu_1 = 1.161 \times 10^{-5}$ Pa·s, $\mu_2 = 1.361 \times 10^{-5}$ Pa·s, $M_1 = 46.07$, $M_2 = 50.49$, $T_{b1} = 248.3$ K, and $T_{b2} = 248.9$ K. By Eq. (2-134), $S_1 = (1.5)(248.3) = 372.45$, and $S_2 = (1.5)(248.9) = 373.35$. By Eq. (2-133), $S_{12} = S_{21} = (1.0)[(372.45)(373.35)]^{1/2} = 372.90$. Substituting in Eq. (2-132):

$$A_{11} = A_{22} = 1.0$$

$$A_{12} = \frac{1}{4}\left\{1 + \left[\left(\frac{1.161 \times 10^{-5}}{1.361 \times 10^{-5}}\right)\left(\frac{50.49}{46.07}\right)^{3/4}\left(\frac{373.15 + 372.45}{373.15 + 373.35}\right)\right]^{1/2}\right\}^2$$
$$\times\left(\frac{373.15 + 372.90}{373.15 + 372.45}\right) = 0.956$$

$$A_{21} = \frac{1}{4}\left\{1 + \left[\left(\frac{1.361 \times 10^{-5}}{1.161 \times 10^{-5}}\right)\left(\frac{46.07}{50.49}\right)^{3/4}\left(\frac{373.15 + 373.35}{373.15 + 372.45}\right)\right]^{1/2}\right\}^2$$
$$\times\left(\frac{373.15 + 372.90}{373.15 + 373.35}\right) = 1.047$$

Using Eq. (2-131):

$$k_m = \frac{(0.23)(0.02504)}{(0.23)(1.0) + (0.77)(0.956)} + \frac{(0.77)(0.01587)}{(0.23)(1.047) + (0.77)(1.0)} = 0.01805 \text{ W/m K}$$

The experimental value is 0.01778 W/m K.[66]

Liquids For **pure component hydrocarbon** liquids at reduced temperatures between 0.25 and 0.8 and at pressures below 3.4 MPa, an equation based on the methods of Pachaiyappan et al.[79] and Riedel[88] may be used:

$$k_L = C\rho M^n\left[\frac{3 + 20(1 - T_r)^{2/3}}{3 + 20\left(1 - \dfrac{293.15}{T_c}\right)^{2/3}}\right] \qquad (2\text{-}135)$$

where k_L = liquid thermal conductivity, W/m K
M = molecular weight
ρ = molar density at 293.15 K, kmol/m³
T_r = reduced temperature, T/T_c
T = temperature, K
T_c = critical temperature, K

For unbranched, straight chain hydrocarbons, $n = 1.001$ and $C = 1.811 \times 10^{-4}$.

For branched and cyclic hydrocarbons, $n = 0.7717$ and $C = 4.407 \times 10^{-4}$.

Average errors are 5 percent when this equation is used. For pressures greater than 3.4 MPa, the thermal conductivity from Eq. (2-135) may be corrected by the technique suggested by Lenoir.[57] The correction factor is the ratio of conductivity factors F/F', where F is at the desired temperature and higher pressure, and F' is at the same temperature and lower pressure (usually atmospheric). The conductivity factors are calculated from:

$$F = 17.77 + 0.065P_r - 7.764T_r - \frac{2.054T_r^2}{e^{0.2P_r}} \qquad (2\text{-}136)$$

where T_r = reduced temperature as in Eq. (2-135)
P_r = reduced pressure, P/P_c
P = pressure, MPa
P_c = critical pressure, MPa

The average error in the pressure correction alone is typically 3 percent.

Example 31 Estimate thermal conductivity of n-octane at 373.15 K and pressures of 0.1 MPa and 20.0 MPa. The required properties from Daubert et al.[24] are: ρ at 293.15 K = 6.155 kmol/m³, $M = 114.2$, $T_c = 568.7$ K, and $P_c = 2.490$ MPa. $T_r = 373.15/568.7 = 0.6561$, $n = 1.001$, and $C = 1.811 \times 10^{-4}$. Substituting in Eq. (2-135) for the thermal conductivity at 0.1 MPa:

$$k_L = (1.811 \times 10^{-4})(6.155)(114.2)^{1.001}\left[\frac{3 + (20)(1 - 0.6561)^{2/3}}{3 + (20)\left(1 - \dfrac{293.15}{568.7}\right)^{2/3}}\right]$$

$$= 0.107 \text{ W/m K}$$

The reported value is 0.111 W/m K.[123] To correct to 20 MPa, $P_r = 20.0/2.490 = 8.032$, and $P'_r = 0.1/2.490 = 0.04016$. Substituting in Eq. (2-136) to calculate F and F':

$$F = 17.77 + (0.065)(8.032) - (7.764)(0.6561) - \frac{(2.054)(0.6561)^2}{e^{(0.2)(8.032)}}$$

$$F = 13.02$$

$$F' = 17.77 + (0.065)(0.04016) - (7.764)(0.6561) - \frac{(2.054)(0.6561)^2}{e^{(0.2)(0.04016)}}$$

$$F' = 11.80$$

Thus, the thermal conductivity of n-octane at 373.15 K and 20.0 MPa is estimated to be $(0.111)(13.02/11.80) = 0.122$ W/m K as compared with an experimental value of 0.121 W/m K.[123]

For pure component hydrocarbon liquids **above the normal boiling point and all pressures,** the method of Kanitkar and Thodos[46] is recommended:

$$k_L = \frac{\alpha\, e^{\beta\rho_r}}{\lambda} \qquad\qquad P < 10{,}000 \text{ kPa} \qquad (2\text{-}137)$$

$$k_L = \frac{2.596 \times 10^{-4}\, P_r^{1.6} + \alpha\, e^{\beta\rho_r}}{\lambda} \qquad P > 10{,}000 \text{ kPa} \qquad (2\text{-}138)$$

where k_L = liquid thermal conductivity at the temperature T (K) and pressure P (kPa) of interest, W/m K

$\alpha = 0.0112\, \beta^{-3.322}$ (2-139)

$\beta = 0.40 + 0.986 e^{-0.64\lambda}$ (2-140)

$$\lambda = T_c^{1/6}\, M^{1/2} \left(\frac{101.325}{P_c}\right)^{2/3} \qquad (2\text{-}141)$$

ρ_r = reduced density = V_c/V
V_c = critical molar volume, m³/kmol
V = molar volume at T and P, m³/kmol
T_c = critical temperature, K
M = molecular weight
P_c = critical pressure, kPa
P_r = reduced pressure, P/P_c

Average errors can be expected to be in the order of 10 percent.

Example 32   Estimate thermal conductivity of n-octane at 473.15 K and 15,000 kPa. The required properties from Daubert et al.[24] are: $T_c = 568.7$ K, $M = 114.2$, $P_c = 2490.0$ kPa, and $V_c = 0.4860$ m³/kmol. The specific volume at 473.15 K and 15,000 kPa is 0.001717 m³/kg[123]; $V = (0.001717)(114.2) = 0.1961$ m³/kmol. Thus, $P_r = 15000/2490 = 6.024$, and $\rho_r = 0.4860/0.1961 = 2.478$. Substituting in Eqs. (2-141), (2-140), and (2-139):

$$\lambda = (568.7)^{1/6}\, (114.2)^{1/2} \left(\frac{101.325}{2490}\right)^{2/3} = 3.639$$

$$\beta = 0.40 + 0.986 e^{-(0.64)(3.639)} = 0.496$$

$$\alpha = (0.0112)(0.496)^{-3.322} = 0.115$$

Using Eq. (2-138):

$$k_L = \frac{(2.596 \times 10^{-4})(6.024)^{1.6} + 0.115 e^{(0.496)(2.478)}}{3.639} = 0.109 \text{ W/m K}$$

The reported value is 0.106 W/m K.[123]

The thermal conductivity of **pure component nonhydrocarbon** liquids may be estimated by the method of Baroncini et al.,[5] with a modification by Myers[75] for silicon compounds, at reduced temperatures between 0.3 and 0.8 and at pressures below 3.5 MPa:

$$k_L = \left(\frac{abc}{m}\right)\left(\frac{(1 - T_r)^{0.38}}{T_r^{1/6}}\right) \qquad (2\text{-}142)$$

where k_L = liquid thermal conductivity, W/m K
T_r = reduced temperature, T/T_c
T = temperature, K
T_c = critical temperature, K
a = constant parameter
b = constant parameter = function of normal boiling temperature T_b
c = constant paramerer = function of critical temperature T_c, K
m = constant parameter = function of molecular weight M

Values of a, b, c, and m for various compound classes are found in Table 2-399. Average errors are about 8 percent.

Example 33   Estimate thermal conductivity of n-butanol. The properties required to estimate the liquid thermal conductivity of n-butanol at 360.0 K and 0.1 MPa from Daubert et al.[24] are: $T_c = 563.0$ K, $M = 74.12$, and $T_b = $

TABLE 2-399   Values of Constant Parameters in Eq. (2-142) for Various Compound Classes

Compound class	a	b	c	m
Acids°	0.00319	$T_b^{6/5}$	$T_c^{-1/6}$	$M^{1/2}$
Alcohols, phenols	0.00339	$T_b^{6/5}$	$T_c^{-1/6}$	$M^{1/2}$
Esters†	0.0415	$T_b^{6/5}$	$T_c^{-1/6}$	M
Ethers	0.0385	$T_b^{6/5}$	$T_c^{-1/6}$	M
Halides‡	0.494	1.0	$T_c^{1/6}$	$M^{1/2}$
Refrigerants R20–R23	0.562	1.0	$T_c^{1/6}$	$M^{1/2}$
Ketones	0.00383	$T_b^{6/5}$	$T_c^{-1/6}$	$M^{1/2}$
Alkoxysilanes	0.00482	$T_b^{6/5}$	$T_c^{-1/6}$	$M^{1/2}$
Alkyl-(aryl)-chlorosilanes	0.6510	1.0	$T_c^{1/6}$	$M^{1/2}$

° Do not use for formic, myristic, or oleic acids.
† Do not use for butyl stearate.
‡ Do not use for refrigerants R20–R23 ($CHCl_3$, $CHFCl_2$, $CHClF_2$, or CHF_3).

390.8 K. $T_r = 360.0/563.0 = 0.6394$. From Table 2-399, $a = 0.00339$, $b = (T_b)^{6/5} = (390.8)^{6/5} = 1289.3$, $c = (T_c)^{-1/6} = (563.0)^{-1/6} = 0.3480$, and $m = M^{1/2} = (74.12)^{1/2} = 8.609$. Substituting in Eq. (2-142):

$$k_L = \left[\frac{(0.00339)(1289.3)(0.3480)}{8.609}\right]\left[\frac{(1 - 0.6394)^{0.38}}{(0.6394)^{1/6}}\right] = 0.1292 \text{ W/m K}$$

The reported value is 0.1429 W/m K.[123]

For pure component nonhydrocarbon liquids **for which Eq. (2-142) is not applicable,** the method of Missenard[72,73] may be used at temperature T (K) and below pressures of 3.5 MPa:

$$k_L = k_L^r \left[\frac{3 + 20\,(1 - T_r)^{2/3}}{3 + 20\left(1 - \dfrac{273.15}{T_c}\right)^{2/3}}\right] \qquad (2\text{-}143)$$

$$k_L^r = 2.656 \times 10^{-7}\, \frac{(T_b\rho^r)^{1/2}\, C_p^r}{M^{1/2}\, N^{1/4}} \qquad (2\text{-}144)$$

where k_L = liquid thermal conductivity, W/m K
k_L^r = liquid thermal conductivity at 273.15 K, W/m K
T_r = reduced temperature, T/T_c
T_c = critical temperature, K
T_b = normal boiling temperature, K
ρ^r = molar density at 273.15 K, kmol/m³
C_p^r = molar heat capacity at 273.15 K, J/kmol K
M = molecular weight
N = number of atoms in the molecule

Errors in the order of 8 percent can be expected.

Example 34   Estimate the thermal conductivity of n-propionaldehyde (CH_3CH_2CHO) at 318.15 K and low pressure (0.1 MPa); the necessary properties from Daubert et al.[24] are: $T_c = 504.4$ K, $T_b = 321.1$ K, $\rho^r = 14.11$ kmol/m³, $C_p^r = 1.309 \times 10^5$ J/kmol K, and $M = 58.08$. $N = 10$, and $T_r = 318.15/504.4 = 0.6307$. Substituting in Eq. (2-144):

$$k_L^r = (2.656 \times 10^{-7})\, \frac{[(321.1)\,(14.11)]^{1/2}\,(1.309 \times 10^5)}{(58.08)^{1/2}\,(10)^{1/4}} = 0.1727 \text{ W/m K}$$

Using Eq. (2-143):

$$k_L = 0.1727 \left[\frac{3 + 20\,(1 - 0.6307)^{2/3}}{3 + 20\,(1 - 273.15/504.4)^{2/3}}\right] = 0.1542 \text{ W/m K}$$

A value from the literature is 0.1541 W/m K.[24]

For **pressures greater than 3.5 MPa,** the correction factor suggested by Missenard[74] may be used to obtain the thermal conductivity of pure component nonhydrocarbon liquids. Thus:

$$k_L = k_L'\left[0.98 + 0.0079\, P_r T_r^{1.4} + 0.63\, T_r^{1.2}\left(\frac{P_r}{30 + P_r}\right)\right] \qquad (2\text{-}145)$$

where k_L = liquid thermal conductivity at the desired temperature T (K) and pressure P (MPa), W/m K
k_L' = liquid thermal conductivity at T and pressure of 0.1 MPa, W/m K
P_r = reduced pressure, P/P_c
P_c = critical pressure, MPa
T_r = reduced temperature, T/T_c
T_c = critical temperature, K

Average errors are in the range of 5–20 percent.

Example 35   Estimate thermal conductivity of n-butanol. The required properties at 360 K and 15 MPa from Daubert et al.[24] are: $T_c = 563.0$ K and

$P_c = 4.423$ MPa. $T_r = 360.0/563.0 = 0.6394$ and $P_r = 15.0/4.423 = 3.391$. $k'_L = 0.1429$ W/m K.[123] Substituting in Eq. (2-145):

$$k_L = 0.1429 \left[0.98 + (0.0079)(3.391)(0.6394)^{1.4} + (0.63)(0.6394)^{1.2} \left(\frac{3.391}{30 + 3.391} \right) \right]$$

$$= 0.1474 \text{ W/m K}$$

Vargaftik[123] reports a value of 0.1494 W/m K.

For both aqueous and nonaqueous **liquid mixtures,** the method of Li[59] is suggested for pressures below 3.5 MPa:

$$k_m = \sum_{i=1}^{n} \sum_{j=1}^{n} \phi_i \, \phi_j \, k_{ij} \qquad (2\text{-}146)$$

$$\phi_i = \frac{x_i \, V_i}{\sum_{j=1}^{n} x_j \, V_j} \qquad (2\text{-}147)$$

$$k_{ij} = \frac{2}{(1/k_i) + (1/k_j)} \qquad (2\text{-}148)$$

where k_m = mixture liquid thermal conductivity at temperature T (K), W/m K
  n = number of components
  $x_{i,j}$ = mole fraction of component i or j in the liquid mixture
  $V_{i,j}$ = liquid molar volume of pure component i or j at temperature T, m³/kmol
  $k_{i,j}$ = liquid thermal conductivity of pure component i or j at temperature T, W/m K

Expected errors are in the 4–6 percent range. At **pressures greater than 3.5 MPa,** a pressure correction using Eq. (2-145) may be used. The mixture is treated as a hypothetical pure component with mixture critical properties obtained via Eqs. (2-5) and (2-8).

Example 36   Estimate thermal conductivity of a mixture of 0.302 mole fraction diethyl ether (1) and 0.698 mole fraction methanol (2) at 273.15 K and 0.1 MPa. The required properties from Daubert et al.[24] are: $V_1 = 0.1007$ m³/kmol, $V_2 = 0.03942$ m³/kmol, $k_1 = 0.1383$ W/m K, and $k_2 = 0.2069$ W/m K. Substituting in Eq. (2-148):

$$k_{12} = \frac{2}{1/0.1383 + 1/0.2069} = 0.1658$$

Then, using Eq. (2-147) to obtain ϕ_1 and ϕ_2:

$$\phi_1 = \frac{(0.302)(0.1007)}{(0.302)(0.1007) + (0.698)(0.03942)} = \frac{0.03041}{0.05793} = 0.525$$

$$\phi_2 = \frac{(0.698)(0.03942)}{0.05793} = 0.475$$

Substituting in Eq. (2-146):

$$k_m = (0.525)^2(0.1383) + 2(0.525)(0.475)(0.1658) + (0.475)^2(0.2069) = 0.167 \text{ W/m K}$$

Jamieson and Hastings[42] report a value of 0.173 W/m K for this mixture.

DIFFUSIVITY

Diffusion is the molecular transport of mass without flow. The diffusivity (D) or diffusion coefficient is the proportionality constant between the diffusion and the concentration gradient causing diffusion. It is usually defined by Fick's first law for one-dimensional, binary component diffusion for molecular transport without turbulence shown by Eq. (2-149)

$$\frac{N_1}{A} = -D_{12} \frac{dC_1}{dL} \qquad (2\text{-}149)$$

The molar flow of species 1 (N_1) per unit area (A) is directly proportional to the change in concentration of species 1 (C_1) per distance diffused (L). The usual units of diffusivity are m²/sec.

In chemical engineering, the primary application of the diffusivity is to calculate the Schmidt number $(\mu/\rho D)$ used to correlate mass transfer properties. This number is also used in reaction rate calculations involving transport to and away from catalyst surfaces.

Experimental diffusion coefficients are scarce and not highly accurate, especially in the liquid phase, leading to prediction methods with marginal accuracy. However, use of the values predicted are generally suitable for engineering calculations. At concentrations above about 10 mole percent, predicted values should be used with caution. Diffusivities in liquids are 10^4–10^5 times lower than those in gases.

Gas Diffusivity   For prediction of the gas diffusivity of **binary hydrocarbon-hydrocarbon gas systems at low pressures** (below about 500 psia [3.5 MPa]) the method of Gilliland[32] given by Eq. (2-150) is recommended.

$$D_{12} = \frac{0.1014 T^{1.5} \left(\frac{1}{M_1} + \frac{1}{M_2} \right)^{0.5}}{P(V_1^{1/3} + V_2^{1/3})^2} \qquad (2\text{-}150)$$

Component 1 is the solute, while component 2 is the solvent. Units of T, P, and V are °R, psia, and cm³/gmole, respectively. Diffusivity is then in ft²/hr. The molar volumes V_1 and V_2 at the normal boiling point are estimated by Tyn and Calus,[120] Eq. (2-151).

$$V_i = 0.285 V_{c_i}^{1.048} \qquad (2\text{-}151)$$

The method gives average errors of less than 4 percent.

For prediction of the gas diffusivity of **binary air-hydrocarbon or nonhydrocarbon gas mixtures at low pressures,** the method of Fuller et al.[30] given by Eq. (2-152) is recommended.

$$D_{12} = \frac{0.1013 T^{1.75} \left(\frac{1}{M_1} + \frac{1}{M_2} \right)^{0.5}}{P[(\sum v_1)^{1/3} + (\sum v_2)^{1/3}]^2} \qquad (2\text{-}152)$$

Units of T and P are K and Pa, respectively, with the resulting diffusivity in m²/sec. All v_i are group contribution values for the subscript component summed over atoms, groups, and structural features given in Table 2-400.

For air-hydrocarbon systems, average deviations do not exceed 9 percent. For general nonhydrocarbon gas systems, the average deviation is about 6 percent.

Example 37   Estimate the diffusivity of benzene vapor diffusing into air at 30°C and 96.5 kPa total pressure.
  Use Eq. (2-152).

$$D_{12} = \frac{0.1013 T^{1.75} \left(\frac{1}{M_1} + \frac{1}{M_2} \right)^{0.5}}{P[(\sum v_1)^{1/3} + (\sum v_2)^{1/3}]^2}$$

$T = 303.1$ K, $M_1 = 78.1$, $M_2 = 28.86$, and $P = 96,500$ Pa.

Now consult Table 2-400. For benzene (C_6H_6):

$$\sum v_1 = 6(16.5) + 6(1.98) - 20.2$$

$$\sum v_1 = 90.68$$

For air, $\sum v_2 = 20.1$:

$$D_{12} = \frac{(.01013)303.1^{1.75} \left(\frac{1}{78.1} + \frac{1}{28.86} \right)^{0.5}}{(96,500)(90.68^{1/3} + 20.1^{1/3})^2}$$

$$D_{12} = 9.68 \times 10^{-6} \text{ m}^2/\text{sec}$$

This value is within 1 percent of an available experimental value.

TABLE 2-400   Atomic Diffusion Volumes for Use in Eq. (2-153)

Atomic and structural diffusion-volume increments v			
C	16.5	Cl	(19.5)
H	1.98	S	(17.0)
O	5.481	Aromatic ring	−20.2
N	(5.69)	Heterocyclic ring	−20.2
Diffusion volumes for simple molecules			
H_2	7.07	CO	18.9
D_2	6.70	CO_2	26.9
He	2.88	N_2O	35.9
N_2	17.9	NH_3	14.9
O_2	16.6	H_2O	12.7
Air	20.1	CCl_2F_2	(114.8)
Ar	16.1	SF_6	(69.7)
Kr	22.8	Cl_2	(37.7)
Xe	(37.9)	Br_2	(67.2)
		SO_2	(41.1)

Parentheses indicate that the value listed is based only on a few data points.

TABLE 2-401 Parameters for Eq. (2-153)

P_r	$(D_{12}P)_R$	A	B	C	E
0.1	1.01	0.38042	1.52267	0.0	
0.2	1.01	0.067433	2.16794	0.0	
0.3	1.01	0.098371	2.42910	0.0	
0.4	1.01	0.137610	2.77605	0.0	
0.5	1.01	0.175081	2.98256	0.0	
0.6	1.01	0.216376	3.11384	0.0	
0.8	1.01	0.314051	0.50264	0.0	
1.0	1.02	0.385736	3.07773	0.141211	13.45454
1.2	1.02	0.514553	3.54744	0.278407	14.00000
1.4	1.02	0.599184	3.61216	0.372683	10.00900
1.6	1.02	0.557725	3.41882	0.504894	8.57519
1.8	1.03	0.593007	3.18415	0.678469	10.37483
2.0	1.03	0.696001	3.3760	0.0665702	11.21674
2.5	1.04	0.790770	3.27984	0.0	
3.0	1.05	0.502100	2.39031	0.602907	6.19043
4.0	1.06	0.837452	3.23513	0.0	
5.0	1.07	0.890390	3.13001	0.0	

An alternate method for gas diffusivity of binary gas mixtures at low pressures is the method of Hirschfelder et al.[40] The method requires several molecular parameters and, when evaluated, gives an average absolute error of about 10 percent. The method is discussed in detail in the *Data Prediction Manual*.

For predicting the diffusivity of *binary gas mixtures at high pressures*, the method of Takahashi,[109] Eq. (2-153), applies.

$$D_{12} = 1.013 \times 10^5 \frac{D'_{12}}{P}(D_{12}P)_R(1 - AT_r^{-B})(1 - CT_r^{-E}) \qquad (2\text{-}153)$$

D'_{12} is the low pressure diffusivity at the temperature of interest. $(D_{12}P)_R$ is a reduced diffusivity pressure product at infinite reduced temperature; and A, B, C, and E are constants. All are a function of P_r tabulated in Table 2-401. Component 1 is the diffusing species, while component 2 is the concentrated species. Critical properties are for the solvent. The pressure is given in Pa. The diffusivity is in m^2/sec. Errors from evaluation average near 15 percent.

Example 38 Estimate the diffusivity of hydrogen (1) in nitrogen (2) at 60°C and 17.23 MPa. A value of the low pressure diffusivity obtained using Eq. (2-152) is $D'_{12} = 9.2 \times 10^{-5}$ m^2/sec. Use Eq. (2-153):

$$D_{12} = 1.013 \times 10^5 \frac{D'_{12}}{P}(D_{12}P)_R(1 - AT_r^{-B})(1 - CT_r^{-E})$$

For nitrogen, $T_c = 126.2$ K and $P_c = 3.394$ MPa.

$$T_r = \frac{333.2}{126.2} = 2.640 \qquad P_r = \frac{17.23}{3.394} = 5.077$$

From Table 2-401:

$$(D_{12}P)_R = 1.07, \quad A = 0.89039, \quad B = 3.1300, \quad C = 0$$

$$D_{12} = 1.013 \times 10^5 \left(\frac{9.2 \times 10^{-5}}{1.723 \times 10^7}\right)(1.07)[1 - .89039(2.64)^{-3.13}]$$

$$D_{12} = 5.54 \times 10^{-7} \ m^2/sec$$

An available experimental value is 4.89×10^{-7} m^2/sec.

For prediction of gas phase diffusion coefficients in **multicomponent hydrocarbon/nonhydrocarbon gas systems,** the method of Wilke[125] shown in Eq. (2-154) is used.

$$D_{im} = \frac{1 - y_i}{\sum\limits_{j \neq i}^{j} y_j/D_{ij}} \qquad (2\text{-}154)$$

This mixing rule is used to determine the diffusivity of any component in a $j + 1$ component mixture and requires binary diffusivities of component i with all other components. It has been estimated that errors are about 5 percent greater than the greatest error in the binary diffusivities. Fairbanks and Wilke,[28] using the same Eq. (2-154), made the same recommendation with essentially the same errors.

For prediction of the diffusivity of a **dilute dissolved gas (hydrocarbon or nonhydrocarbon) in a liquid,** the standard method is that of Wilke and Chang[129] shown by Eq. (2-155).

$$D_{12} = 1.1728 \times 10^{-16} \frac{T(\chi_2 M_2)^{1/2}}{\mu_2 V_1^{0.6}} \qquad (2\text{-}155)$$

Component 1 is the diffusing gas, while component 2 is the solvent. The solvent viscosity μ_2 in Pa sec, the solute molar volume at the normal boiling point V_1 in $m^3/kmole$, and the solvent association parameter χ_2 multiplied by the solvent

molecular weight are required input parameters. For the common solvents, χ decreases to 1 as polarity decreases, with values of 2.6 for water, 1.9 for methanol, 1.5 for ethanol, and 1.0 for less polar solvents. When tested with both hydrocarbon and nonhydrocarbon systems, average errors are about 25 percent—not excessive, considering the magnitude of the diffusivity.

Example 39 Estimate the infinite dilution diffusivity of propane (1) in chlorobenzene (2) at 25°C. Use Eq. (2-113):

$$D_{12} = 1.1728 \times 10^{-16} \frac{T(\chi_2 M_2)^{1/2}}{\mu_2 V_1^{0.6}}$$

For chlorobenzene, $M_2 = 112.56$, $\chi_2 = 1.0$, μ_2 @ 25°C $= 7.548 \times 10^{-4}$ Pa sec, and $V_1 = 0.0745$ $m^3/kmole$.

$$D_{12} = \frac{(1.1728 \times 10^{-16})(298.15)(1.0 \times 112.56)^{1/2}}{(7.548 \times 10^{-4})(0.0745)^{0.6}}$$

$$D_{12} = 2.33 \times 10^{-9} \ m^2/sec$$

An experimental value of 2.77×10^{-9} m^2/sec is available.

Liquid Diffusivity Liquid diffusivities are in general not as accurately predicted as vapor diffusivities, and specialized methods have been developed. References to each method determined to be accurate are given, but only the most common methods will be presented.

For predicting liquid diffusivities of **binary nonpolar liquid systems at high solute dilution,** Umesi[120] developed a method that only depends on the viscosity of the solvent (2) and the radius of gyration of the solvent (2) and the solute (1). The *Technical Data Book—Petroleum Refining* gives the method and values of the radii of gyration for common hydrocarbons. Errors average 16 percent but reach 30 percent at times.

For predicting diffusivities in **binary polar or associating liquid systems at high solute dilution,** the method of Wilke and Chang[129] defined in Eq. (2-156) can be utilized. The Tyn and Calus equation (2-152) can be used to determine the molar volume of the solute at the normal boiling point. Errors average 20 percent, with occasional errors of 35 percent. The method is not considered to be accurate above a solute concentration of 5 mole percent.

For **concentrated binary nonpolar liquid systems** (more than 5 mole percent solute), the diffusivity can be estimated by a molar average mixing rule developed by Caldwell and Babb,[14] Eq. (2-156).

$$D_{1m} = x_1 D_{21} + (1 - x_1)D_{12} \qquad (2\text{-}156)$$

D_{21} and D_{12} are dilute solution binary diffusivities. Errors depend on the procedure used to determine the dilute solution diffusivities.

For **multicomponent nonpolar liquid systems,** Leffler and Cullinan[56] developed a mixing rule, Eq. (2-157).

$$D_{1m}\mu_m = (D_{12}\mu_2)^{x_2}(D_{13}\mu_3)^{x_3} \ldots \qquad (2\text{-}157)$$

The diffusivity of solute 1 in the mixture is related to the binary infinite dilution diffusivities for each of the other components calculated from Eq. (2-155) or the Umesi method. The viscosities are calculated by the methods in the previous section. Errors are not quantifiable, as little experimental data exist, although these errors would be related to those assumed for the binary pairs.

For **concentrated binary liquid nonhydrocarbon systems,** the method of Caldwell and Babb,[14] Eq. (2-156), has been modified by introduction of a thermodynamic correction term as shown in Eq. (2-158).

$$D_{1m}\mu_m = [x_1 D_{21} + (1 - x_1)D_{12}]\left(1 + \frac{d \ln \gamma_1}{d \ln x_1}\right) \qquad (2\text{-}158)$$

The activity coefficient (γ) based corrector is calculated using any applicable activity correlating equation such as the van Laar (slightly polar) or Wilson (more polar) equations. The average absolute error is 20 percent.

An alternate method for **binary concentrated liquid systems** where activity coefficients are not available or estimable is the method of Leffler and Cullinan[56] previously given in Eq. (2-156). Absolute errors average 25 percent.

For estimating the diffusivity of the **dilute solute (10 mole percent) in water,** the method of Hayduk and Laudie,[37] Eq. (2-159), applies.

$$D_{12} = \frac{8.621 \times 10^{-14}}{\mu_2^{1.14} V_1^{0.589}} \tag{2-159}$$

Component 1 is the solute, while component 2 is water. The molar volume of the solute in m^3/kmole is at the solute normal boiling point, while the viscosity of water in Pa sec is at the temperature of the system resulting in a diffusivity in m^2/sec. The average error is about 9 percent when tested on 36 experimental systems.

For estimating the diffusivity of a **dilute solute (<10 mole percent) in any solvent except water,** the method of King et al.,[49] Eq. (2-160), applies.

$$D_{12} = 4.4 \times 10^{-15} \frac{T}{\mu_2} \left(\frac{V_2}{V_1}\right)^{1/6} \left(\frac{\lambda_2}{\lambda_1}\right)^{1/2} \tag{2-160}$$

Component 1 is the solute, while component 2 is the solvent. The latent heats, λ, are at the normal boiling point, as are the molar volumes. Using T in K and μ in Pa sec yields a diffusivity in m^2/sec. The average error is 21 percent when tested on 237 experimental systems.

For prediction of the **liquid diffusivity of a solute in a pair of mixed solvents,** the method of Tang and Himmelblau,[111] Eq. (2-161), is recommended.

$$\ln (D_{1m}\mu_m^{1/2}) = x_2 \ln (D_{12}^\circ \mu_2^{1/2}) + x_3 \ln (D_{13}^\circ \mu_3^{1/2}) \tag{2-161}$$

The solute 1 is dissolved in a solvent pair of 2 and 3. D° are infinite dilution binary diffusivities estimated by the proper method discussed previously. The mixture viscosity can be predicted by methods of the previous section. The average absolute error when tested on 40 systems is 25 percent. The method gives higher errors if the solute is gaseous.

SURFACE TENSION

The molecules in a gas-liquid interface are in tension and tend to contract to a minimum surface area. This tension may be quantified by the surface tension, which is defined as the force in the plane of the surface per unit length. Jasper[43] has made a critical evaluation of experimental surface tension data for approximately 2200 pure chemicals. He correlates surface tension σ (mN/m = dyn/cm) with temperature T (°C) over a specified temperature range as

$$\sigma = a - bT \tag{2-162}$$

where a and b are listed for most of the substances. To obtain values at a higher temperature than the upper temperature limit indicated by Jasper, the following expression may be used:

$$\sigma = d(1 - T_r)^e \tag{2-163}$$

where d and e are determined such that σ and $d\sigma/dT$ at the upper temperature limit T_{HI} have the same values when calculated from both Eqs. (2-162) and (2-163):

$$e = \frac{b(T_c - T_{HI})}{a - b \, T_{HI}} \tag{2-164}$$

$$d = (a - b \, T_{HI})(1 - T_{HI})^{-e} \tag{2-165}$$

where T_c = critical temperature, °C
       T_r = reduced temperature = $(T + 273.15)/(T_c + 273.15)$

Eq. (2-163) correctly predicts that the surface tension becomes zero at the critical point.[87]

For **nonpolar, nonhydrocarbon** chemicals not found in Jasper,[43] use can be made of the corresponding states approach of Brock and Bird[9] as modified by Miller[69] at temperature T (K):

$$\sigma = 4.601 \times 10^{-4} P_c^{2/3} T_c^{1/3} Q (1 - T_r)^{11/9} \tag{2-166}$$

$$Q = 0.1207 \left[1 + \frac{T_{br}(\ln P_c - 11.5261)}{1 - T_{br}}\right] - 0.281 \tag{2-167}$$

where σ = surface tension, mN/m
       P_c = critical pressure, Pa

T_c = critical temperature, K
T_r = reduced temperature, T/T_c
T_b = normal boiling temperature, K
T_{br} = reduced normal boiling temperature, T_b/T_c

Errors are usually less than 5 percent.

Example 40   Estimate surface tension for ethyl mercaptan. The required properties from Daubert et al.[24] are $P_c = 5.49 \times 10^6$ Pa, $T_c = 499.15$ K, and $T_b = 308.15$ K. For $T = 303.15$ K, $T_r = 303.15/499.15 = 0.6073$, and $T_{br} = 308.15/499.15 = 0.6173$. Substituting into Eqs. (2-167) and (2-166):

$$Q = 0.1207 \left[1 + \frac{0.6173 \, (\ln 5.49 \times 10^6 - 11.5261)}{1 - 0.6173}\right] - 0.281$$

$$Q = 0.6170$$

$$\sigma = 4.601 \times 10^{-4}(5.49 \times 10^6)^{2/3}(499.15)^{1/3}(0.6170)(1 - 0.6073)^{11/9}$$

$$\sigma = 22.36 \text{ mN/m}$$

The reported experimental value is 22.68 mN/m.[43]

For **hydrocarbon and polar** chemicals, the approach originally suggested by Macleod[65] as further developed by Sugden[106,107] can be used:

$$\sigma = \left\{\frac{[P]}{1000} (\rho_L - \rho_G)\right\}^4 \tag{2-168}$$

where σ = surface tension, mN/m
        ρ_L = liquid density, kmol/m^3
        ρ_G = vapor density, kmol/m^3

The temperature-independent parachor $[P]$ may be calculated by the additive scheme proposed by Quale.[83] The atomic group contributions for this method, with contributions for silicon, boron, and aluminum from Myers,[75] are shown in Table 2-402. At low pressures, where $\rho_L \gg \rho_G$, the vapor density term may be neglected. Errors using Eq. (2-168) are normally less than 5 to 10 percent.

Example 41   Estimate surface tension for isobutyric acid. For isobutyric acid, the liquid density from Daubert et al.[24] is 10.77 kmol/m^3 at 293.15 K. $[P]$ is determined from Table 2-402:

is made up of the two groups CH_3—$CH(CH_3)$— and —COOH. Therefore, $[P] = 133.3 + 73.8 = 207.1$. With Eq. (2-168), neglecting the vapor density,

$$\sigma = \left[\frac{207.1}{1000} (10.77)\right]^4 = 24.75 \text{ mN/m}$$

Jasper[43] quotes a value of 25.04 mN/m at 293.15 K.

In general, the surface tension of a **liquid mixture** is not a simple function of the pure component surface tensions because the composition of the mixture surface is not the same as the bulk. For **nonaqueous** solutions of n components, the method of Winterfeld, Scriven, and Davis[130] is applicable:

$$\sigma_m = \sum_{i=1}^{n} \sum_{j=1}^{n} \rho^2 \left(\frac{x_i}{\rho_{Li}}\right)\left(\frac{x_i}{\rho_{Li}}\right)(\sigma_i\sigma_j)^{1/2} \tag{2-169}$$

$$\frac{1}{\rho} = \sum_{i=1}^{n} \frac{x_i}{\rho_{Li}} \tag{2-170}$$

where σ_m = mixture surface tension, mN/m
        $x_{i,j}$ = mole fraction of component i or j in the liquid mixture
        $\rho_{Li,j}$ = pure component liquid density of component i or j, kmol/m^3
        $\sigma_{i,j}$ = pure component surface tension of component i or j, mN/m

Accuracies of 3–4 percent average deviation are typical when using this method.

Example 42   Estimate surface tension of a mixture. At 298.15 K, Daubert et al.[24] report the liquid density of n-pentane to be 8.617 kmol/m^3 and its surface tension to be 15.47 mN/m. From the same source, the corresponding values for dichloromethane are 15.52 kmol/m^3 and 27.22 mN/m. Using Eqs. (2-170) and (2-169) for a mixture of 0.1606 mole fraction n-pentane and 0.8394 mole fraction dichloromethane:

$$\frac{1}{\rho} = \frac{0.1606}{8.617} + \frac{0.8394}{15.52}$$

$$\rho = 13.75 \text{ kmol/}m^3$$

TABLE 2-402 Atomic Group Contributions for Calculation of the Parachor [P]

Atomic group	[P]	Atomic group	[P]
Carbon-hydrogen		**Special Groups (*Cont.*)**	
C	9.0	B	13.2
H	15.5	Al	34.9
CH_3—	55.5	F	26.1
(—CH_2—)$_n$		Cl	55.2
$n = 1 - 12$	40.0	Br	68.0
$n > 12$	40.3	I	90.3
CH_3—$CH(CH_3)$—	133.3		
CH_3—CH_2—$CH(CH_3)$—	171.9	**Ethylenic Bond**	
CH_3—CH_2—CH_2—$CH(CH_3)$—	211.7	Terminal°	19.1
CH_3—$CH(CH_3)$—CH_2—	173.3	2,3-position	17.7
CH_3—CH_2—$CH(C_2H_5)$—	209.5	3,4-position†	16.3
CH_3—$C(CH_3)_2$—	170.4		
CH_3—CH_2—$C(CH_3)_2$—	207.5	**Triple Bond**	40.6
CH_3—$CH(CH_3)$—$CH(CH_3)$—	207.9		
CH_3—$CH(CH_3)$—$C(CH_3)_2$—	243.5	**Ring Closure**	
C_6H_5—	189.6	3-membered	12.5
		4-membered	6.0
Special Groups		5-membered	3.0
H in OH	10.0	6-membered	0.8
H in HN	12.5	7-membered	4.0
O	19.8		
—OH	29.8	**=0 (ketone)**	
O_2 in acids, esters	54.8	3 carbon atoms	22.3
—COO—	63.8	4 carbon atoms	20.0
—COOH	73.8	5 carbon atoms	18.5
N	17.5	6 carbon atoms	17.3
—NH_2	42.5	7 carbon atoms	17.3
S	49.1	8 carbon atoms	15.1
P	40.5	9 carbon atoms	14.1
Si	30.3	10 carbon atoms	13.0
Si (silanes)	43.3	11 carbon atoms	12.6

° Use the value for double bonds in cyclic compound. Assume 3 double bonds for the aromatic ring.
† Use 16.3 for double bonds in the 3, 4 or higher positions.

$$\sigma_m = (13.75)^2 \left(\frac{0.1606}{8.617}\right)^2 (15.47)$$

$$+ 2(13.75)^2 \left(\frac{0.1606}{8.617}\right)\left(\frac{0.8394}{15.52}\right)[(15.47)(27.22)]^{1/2}$$

$$+ (13.75)^2 \left(\frac{0.8394}{15.52}\right)^2 (27.22)$$

$$\sigma_m = 23.89 \text{ mN/m}$$

De Soria et al.[26] give an experimental value of 24.24 mN/m for this mixture.

Surface tensions for **aqueous** solutions are more difficult to predict than those for nonaqueous mixtures because of the nonlinear dependence on mole fraction. Small concentrations of the organic material may significantly affect the mixture surface tension value. For many binary organic-water mixtures, the method of Tamura, Kurata, and Odani[110] may be used:

$$\sigma_m^{1/4} = \psi_w \sigma_w^{1/4} + \psi_o \sigma_o^{1/4} \qquad (2\text{-}171)$$

where
- σ_m = mixture surface tension, mN/m
- σ_w = surface tension of pure water, mN/m
- σ_o = surface tension of pure organic component, mN/m
- $\psi_o = 1 - \psi_w$ (2-172)

ψ_w is defined by the relation:

$$\log_{10} \frac{(\psi_w)^q}{(1 - \psi_w)} = \log_{10}\left[\frac{(x_w V_w)^q}{x_o V_o}(x_w V_w + x_o V_o)^{1-q}\right]$$

$$+ 44.1 \frac{q}{T}\left[\frac{\sigma_o V_o^{2/3}}{q} - \sigma_w V_w^{2/3}\right] \qquad (2\text{-}173)$$

where
- x_w = bulk mole fraction of pure water
- x_o = bulk mole fraction of pure organic component
- V_w = molar volume of pure water, m³/kmol
- V_o = molar volume of pure organic component, m³/kmol
- T = temperature, K
- q = constant depending upon the size and type of the organic component; see table:

Organic component	q	Example
Fatty acids, alcohols	Number of carbon atoms	Acetic acid: $q = 2$
Ketones	One less than the number of carbon atoms	Acetone: $q = 2$
Halogen derivatives of fatty acids	Number of carbon atoms times the ratio of the molar volume of the halogen derivative to the parent fatty acid	Chloroacetic acid

$$q = 2\, \frac{V(\text{chloroacetic acid})}{V(\text{acetic acid})}$$

Expected errors are less than 10 percent when q is less than 5 and within 20 percent when q is greater than 5.

Example 43 Estimate surface tension of a water-methanol mixture. Equation (2-171) can be used with a water-methanol mixture at 303.15 K when the methanol mole fraction is 0.122. From Jasper,[43] $\sigma_w = 71.40$ mN/m, and $\sigma_o = 21.73$ mN/m. The density of water (per Ref. 24) is 55.16 kmol/m³; $V_w = 0.01813$ m³/kmol. The density of methanol is 24.49 kmol/m³ (Ref. 24); $V_o = 0.04083$ m³/kmol. For methanol, $q = 1$. Using Eq. (2-173) to obtain ψ_w:

$$\log_{10} \frac{\psi_w}{1 - \psi_w} = \log_{10}\left[\frac{(0.878)(0.01813)}{(0.122)(0.04083)}\right]$$

$$+ \frac{44.1}{303.15}[(21.73)(0.04083)^{2/3} - (71.40)(0.01813)^{2/3}]$$

$$= 0.505 - 0.342 = 0.163$$

$$\frac{\psi_w}{1 - \psi_w} = 10^{0.163} = 1.455$$

$$\psi_w = \frac{1.455}{2.455} = 0.593$$

Using Eq. (2-172): $\psi_o = 1 - 0.593 = 0.407$. Substituting into Eq. (2-171):

$$\sigma_m^{1/4} = (0.593)(71.40)^{1/4} + (0.407)(21.73)^{1/4}$$

$$\sigma_m^{1/4} = 2.603$$

$$\sigma_m = 45.91 \text{ mN/m}$$

The reported experimental value is 46.1 mN/m.[110]

FLAMMABILITY PROPERTIES

Flash points, lower and upper flammability limits, and autoignition temperatures are the three properties used to indicate safe operating limits of temperature when processing organic materials. Prediction methods are somewhat erratic, but, together with comparisons with reliable experimental values for families or similar compounds, they are valuable in setting a conservative value for each of the properties. The DIPPR compilation includes evaluated values for over 1000 common organics. Detailed examples of most of the methods discussed are available in Danner and Daubert.[22]

The flash point is the lowest temperature at which a liquid gives off sufficient vapor to form an ignitable mixture with air near the surface of the liquid or within the vessel used. ASTM test methods include procedures using a closed cup (ASTM D56, ASTM D93, and ASTM D3828), which is preferred, and an open cup (ASTM D92 and ASTM D1310). When several values are available, the *lowest* temperature is usually taken in order to assure safe operation of the process.

The method of Shebeko et al.[96] is the preferred flash point prediction method. The formula of the compound, the system pressure, and vapor pressure data for the compound must be available or estimable. Equation (2-174) is the basic equation.

$$P^{\text{sat}} = \frac{P}{1 + 4.76(2\beta - 1)} = 0 \qquad (2\text{-}174)$$

$$\beta = N_C + N_S + \frac{(N_H - N_X)}{4} - \frac{N_O}{2}$$

N's are the numbers of **atoms** of carbon (C), sulfur (S), hydrogen (H), halogens (X), and oxygen (O) in the molecule. P is the total system pressure. P^{sat} is the vapor pressure of the compound at the flash point temperature.

If P^{sat} is available as a function of temperature, Eq. (2-174) can be solved directly for the flash point temperature. Otherwise, trial and error with a table of P^{sat} vs. T is required. Errors average about 5°C but may be as much as 15°C.

An alternate method for flash point prediction is the method of Gmehling and Rasmussen[33] and depends on the lower flammability limit (discussed later). Vapor pressure as a function of temperature is also required. The method is generally not as accurate as the preceding method as flammability limit errors are propagated. The authors have also extended the method to defined mixtures of organics.

The upper and lower flammability limits are the boundary-line mixtures of vapor or gas with air, which, if ignited, will just propagate flame and are given in terms of percent by volume of gas or vapor in the air. Each of these limits also has a temperature at which the flammability limits are reached. The temperature corresponding to the lower-limit partial vapor pressure should equal the flash point. The

temperature corresponding to the upper-limit partial vapor pressure is somewhat above the lower limit and is usually considerably below the autoignition temperature. Flammability limits are calculated at one atmosphere total pressure and normally are considered synonymous with explosive limits. Limits in oxygen rather than air are sometimes measured and available. Limits are generally reported at 298 K and 1 atm. If temperature or pressure are increased, the lower limit will decrease while the upper limit will increase, giving a wider range of compositions over which flame will propagate.

The most generally applicable method for prediction of the property is the method of Seaton,[95] which depends only on the molecular structure of the molecule and utilizes second order (Benson-type) groups to construct the molecule. Equation (2-175) sums the groups' number of each type group (n_i) to get both the upper and lower limits.

$$z_u \quad \text{or} \quad z_l = \frac{\sum (n_i f_i)}{\sum \left(\dfrac{n_i f_i}{g_i} \right)} \qquad (2\text{-}175)$$

Two sets of f_i and g_i are given in the article for each second-order group to cover both upper (u) and lower (l) limits (z) in volume percent units. A study of this method for about 80 organic compounds in 14 families shows absolute errors of 0.15 percent and 2.3 percent for the lower and upper limits, respectively. The upper limit prediction should not be used for ethers.

Alternate group contribution methods dependent only on molecular structure are the method of Shebeko et al.[97] modified by High and Siegel[39] for lower flammability limit and the method of High and Danner[38] for upper flammability limit. Both methods are detailed by Danner and Daubert.[22] A study comparing these methods with the Seaton method shows slightly higher absolute errors of 0.23 percent and 2.9 percent for the lower and upper limits, respectively. The upper limit prediction should not be used for ethers. Both methods are recommended to be used only for qualitative guidance. Lower flammability limits can also be back-calculated from a known flash point by the method of Gmehling and Rasmussen[33] discussed earlier.

The autoignition temperature is the minimum temperature for a substance to initiate self-combustion in air in the absence of a spark or flame. The temperature is no lower than and is generally considerably higher than the temperature corresponding to the upper flammability limit. Large differences can occur in reported values determined by different procedures. The lowest reasonable value should be accepted in order to assure safety. Values are also sometimes given in oxygen rather than in air.

Values for hydrocarbons other than alkynes and alkadienes can be predicted by the method of Suzuki et al.[108] The best model includes the descriptors T_c, P_c, the parachor, the molecular surface area (which can be approximated by the van der Waals area), and the zero-order connectivity index. Excluding alkynes and alkadienes, a study for 58 alkanes, aromatics, and cycloalkanes showed an average deviation from experimental values of about 30 K.

Another method of estimating autoignition temperatures is to compare values for a compound with other members of its homologous series on a plot vs. carbon number as the temperature decreases and carbon number increases. Affens[1] gives a formal procedure for such estimation.

Section 3

Mathematics*

Bruce A. Finlayson, Ph.D., *Rehnberg Professor and Chair, Department of Chemical Engineering, University of Washington: Member, National Academy of Engineering (Numerical methods and all general material; section editor)*

James F. Davis, Ph.D., *Professor of Chemical Engineering, Ohio State University (Intelligent Systems)*

Arthur W. Westerberg, Ph.D., *Swearingen University Professor of Chemical Engineering, Carnegie Mellon University: Member, National Academy of Engineering (Optimization)*

Yoshiyuki Yamashita, Ph.D., *Associate Professor of Chemical Engineering, Tohoku University, Sendai, Japan (Intelligent Systems)*

* The contributions of William F. Ames (retired), Georgia Institute of Technology; Arthur E. Hoerl (deceased), University of Delaware; and M. Zuhair Nashed, University of Delaware, to material that was used from the sixth edition is gratefully acknowledged.

DIMENSIONAL ANALYSIS

PROCESS SIMULATION

INTELLIGENT SYSTEMS IN PROCESS ENGINEERING

GENERAL REFERENCES: The list of references for this section is selected to provide a broad perspective on classical and modern mathematical methods that are useful in chemical engineering. The references supplement and extend the treatment given in this section. Also included are selected references to important areas of mathematics that are not covered in the *Handbook* but that may be useful for certain areas of chemical engineering, e.g., additional topics in numerical analysis and software, optimal control and system theory, linear operators, and functional-analysis methods. Readers interested in brief summaries of theory, together with many detailed examples and solved problems on various topics of college mathematics and mathematical methods for engineers, are referred to the Schaum's Outline Series in Mathematics, published by the McGraw-Hill Book Comapny.

1. Abramowitz, M., and I. A. Stegun. *Handbook of Mathematical Functions,* National Bureau of Standards, Washington, D.C. (1964).
2. Action, F. S. *Numerical Methods That Work,* Math. Assoc. of Am. (1990).
3. Adey, R. A., and C. A. Brebbia. *Basic Computational Techniques for Engineers,* Wiley, New York (1983).
4. Akai, T. *Applied Numerical Methods for Engineers,* Wiley, New York (1994).
5. Akin, J. E. *Finite Element Anaysis for Undergraduates,* Academic, New York (1986).
6. Alder, H., N. Karmarker, M. Resende, and G. Veigo. *Mathematical Programming* **44** (1989): 297–335.
7. American Institute of Chemical Engineers. "Advanced Simulators Migrate to PCs," *Chem. Eng. Prog.* **90** (Oct. 1994): 13–14.
8. American Institute of Chemical Engineers. "CEP Software Directory," *Chem. Engn. Prog.* (Dec. 1994).
9. Ames, W. F. *Nonlinear Partial Differential Equations in Engineering,* Academic, New York (1965).
10. ———. *Nonlinear Ordinary Differential Equations in Transport Processes,* Academic, New York (1968).
11. ———. *Numerical Methods for Partial Differential Equations,* 2d ed., Academic, New York (1977).
12. ———. *Ind. Eng. Chem. Fund.* **8** (1969): 522–536.
13. Amundson, N. R. *Mathematical Methods in Chemical Engineering,* Prentice Hall, Englewood Cliffs, NJ (1966).
14. Anderson, E. et al. *LAPACK Users' Guide,* SIAM (1992).
15. Antsaklis, P. J., and K. M. Passino (eds.). *An Introduction to Intelligent and Autonomous Control,* Kluwer Academic Publishers (1993).
16. Aris, R. *The Mathematical Theory of Diffusion and Reaction in Permeable Catalysis,* vols. 1 and 2, Oxford University Press, Oxford (1975).
17. ———. *Mathematical Modelling Techniques,* Pitman, London (1978).
18. ———. *Vectors, Tensors, and the Basic Equations of Fluid Mechanics,* Prentice Hall (1962).
19. ——— and N. Amundson. *Mathematical Methods in Chemical Engineering,* vols. 1 and 2, Prentice Hall, Englewood Cliffs, NJ (1973).
20. Arya, J. C., and R. W. Lardner. *Algebra and Trigonometry with Applications,* Prentice Hall, Englewood Cliffs, NJ (1983).
21. Ascher, U., J. Christiansen, and R. D. Russell. *Math. Comp.* **33** (1979): 659–679.
22. Atkinson, K. E. *An Introduction to Numerical Analysis,* Wiley, New York (1978).
23. Atkinson, K. E. *A Survey of Numerical Methods for the Solution of Fredholm Integral Equations of the Second Kind,* SIAM, Philadelphia (1976).
24. Badiru, A. B. *Expert Systems Applications in Engineering and Manufacturing,* Prentice Hall, Englewood Cliffs, NJ (1992).
25. Baird, D. C. *Experimentation: An Introduction to Measurement Theory and Experiment Design,* 3d ed., Prentice Hall, Engelwood Cliffs, NJ (1995).
26. Baker, C. T. H. *The Numerical Treatment of Integral Equations,* Oxford University Press, New York (1977).
27. Barker, V. A. (ed.). *Sparse Matrix Techniques—Copenhagen 1976,* Lecture Notes in Mathematics 572, Springer-Verlag, New York (1977).
28. Beckenbach, E. F., and R. E. Bellman. *Inequalities,* 3d printing, Springer-Verlag, Berlin (1971).
29. Becker, E. B., G. F. Carey, and J. T. Oden: *Finite Elements: An Introduction,* Prentice Hall, Englewood Cliffs, NJ (1981).
30. Bellman, R. E., and K. L. Cooke. *Differential-Difference Equations,* Academic, New York (1972).
31. Bender, E. A. *An Introduction to Mathematical Modeling,* Wiley, New York (1978).
32. Bender, C. M., and Orszag, S. A., *Advanced Mathematical Methods for Scientists and Engineers,* McGraw-Hill (1978).
33. Ben-Israel, A., and T. N. E. Greville. *Generalized Inverses: Theory and Applications,* Wiley-Interscience, New York (1974).
34. Boas, R. P. Jr. *Am. Math. Mon.* **84** (1977): 237–258.
35. Bodewig, E. *Matrix Calculus,* 2d ed., Interscience, New York (1959).
36. Bogacki, M. B., Alejski, K., and Szymanowski, J. *Comp. Chem. Eng.* **13** (1989): 1081–1085.
37. Book, D. L. *Finite-Difference Techniques for Vectorized Fluid Dynamics Calculations,* Springer-Verlag, New York (1981).
38. Boor, C. de. *A Practical Guide to Splines,* Springer-Verlag, New York (1978).
39. Botha, J. F., and G. F. Pinder. *Fundamental Concepts in the Numerical Solution of Differential Equations,* Wiley, New York (1983).
40. Box, G. E. P., Hunter, W. G., and Hunter, J. S. *Statistics for Experimenters,* Wiley, New York (1978).
41. Boyce, W. E., and R. C. Di Prima. *Elementary Differential Equations and Boundary Value Problems,* 5th ed., Wiley, New York (1992).
42. Bradley, S. P., A. C. Hax, and T. L. Magnante. *Applied Mathematical Programming,* Addison-Wesley, Reading, MA (1977).
43. Brand, L. *Differential and Difference Equations,* Wiley, New York (1966).
44. Braun, M. *Differential Equations and Their Applications: An Introduction to Applied Mathematics,* 4th ed., Springer-Verlag, New York (1993).
45. Brebbia, C. A., and J. Dominguez, *Boundary Elements—An Introductory Course,* Computational Mechanics Publications, Southhampton (1988).
46. Brent, R. *Algorithms for Minimization without Derivatives,* Prentice Hall, Englewood Cliffs, NJ (1973).
47. Brigham, E. *The Fast Fourier Transform and its Application,* Prentice Hall, Englewood Cliffs, NJ (1988).
48. Bronshtein, I. N., and K. A. Semendyayev (K. A. Hirsch, trans.). *Handbook of Mathematics,* Van Nostrand (1985).
49. Brown, David C., and B. Chandrasekaran. *Design Problem Solving: Knowledge Structures and Control Strategies,* Pitman, London; and Morgan Kaufman, San Mateo, CA (1989).
50. Broyden, C. G. *J. Inst. Math. Applic.* **6** (1970): 76.
51. Brujin, N. G. de. *Asymptotic Methods in Analysis,* Dover, New York (1981).
52. Bryson, A. E., and Y-C Ho. *Applied Optimal Control,* Hemisphere Publishing, Washington, DC (1975).
53. Buck, R. C. *Advanced Calculus,* 3d ed., McGraw-Hill, New York, 1978.
54. Bulsari, A. B. (ed.). *Neural Networks for Chemical Engineers,* Elsevier Science Publishers, Amsterdam (1995).
55. Bunch, J. R., and D. J. Rose (ed.). *Sparse Matrix Computations,* Academic, New York (1976).
56. Burden, R. L., J. D. Faires, and A. C. Reynolds. *Numerical Analysis,* 5th ed., Prindle, Weber & Schmidt, Boston (1993).
57. Byrd, P., and M. Friedman. *Handbook of Elliptic Integrals for Scientists and Engineers,* 2d ed., Springer-Verlag, New York (1971).
58. Byrne, G. A., and P. R. Ponzi. *Comp. Chem. Eng.* **12** (1988): 377–382.
59. Carnahan, B., H. Luther, and J. Wilkes. *Applied Numerical Methods,* Wiley, New York (1969).
60. Carnahan, B., and J. O. Wilkes. "Numerical Solution of Differential Equations—An Overview" in *Foundations of Computer-Aided Chemical Process Design,* AIChE, New York (1981).
61. Carrier, G., and C. Pearson. *Partial Differential Equations: Theory and Technique,* 2d ed., Academic, New York (1988).
62. Carrier, G. F., and C. E. Pearson. *Ordinary Differential Equations,* SIAM (1991).
63. Carslaw, H. S. *The Theory of Fourier Series and Integrals,* 3d ed., Dover, New York (1930).
64. ——— and J. Jaeger. *Operational Methods in Applied Mathematics,* 2d ed., Clarendon Press, Oxford (1948).
65. Chamberlain, R. M., C. Lemarechal, H. C. Pedersen, and M. J. D. Powell. "The Watchdog Technique for Forcing Convergence in Algorithms for Constrained Optimization," *Math. Prog. Study* **16** (1982).
66. Chan, T. F. C., and H. B. Keller. *SIAM J. Sci. Stat. Comput.* **3** (1982): 173–194.
67. Chang, M. W., and B. A. Finlayson. *Int. J. Num. Methods Eng.* **15** (1980): 935–942.
68. Char, B. W., K. O. Geddes, G. H. Gonnet, B. L. Leong, M. B. Monagan, and S. M. Watt. *Maple V. Language Reference Manual,* Springer-Verlag, Berlin (1991).

69. Chatterjee, S., and B. Price. *Regression Analysis by Example,* 2d ed., Wiley, New York (1991).
70. Cheney, E. W., and D. Kincaid. *Numerical Mathematics and Computing,* Brooks/Cole, Monterey, CA (1980).
71. Churchill, R. V. *Operational Mathematics,* 3d ed., McGraw-Hill, New York (1972).
72. ——— and J. W. Brown. *Fourier Series and Boundary Value Problems,* 4th ed., McGraw-Hill, New York (1987).
73. ———, J. W. Brown, and R. V. Verhey. *Complex Variables and Applications,* 4th ed., McGraw-Hill, New York (1984).
74. Clarke, F. H. *Optimization and Nonsmooth Analysis,* Wiley, New York (1983).
75. Cochran, J. A. *The Analysis of Linear Integral Equations,* McGraw-Hill, New York (1972).
76. Collatz, L. *The Numerical Treatment of Differential Equations,* 3d ed., Springer-Verlag, Berlin and New York (1960).
77. Conte, S. D., and C. de Boor. *Elementary Numerical Analysis: An Algorithmic Approach,* 3d ed., McGraw-Hill, New York (1980).
78. Cooper, L., and D. Steinberg. *Methods and Applications of Linear Programming,* Saunders, Philadelphia (1974).
79. Courant, R., and D. Hilbert. *Methods of Mathematical Physics,* Interscience, New York (1953, 1962).
80. Crandall, S. *Engineering Analysis,* McGraw-Hill, New York (1956).
81. Creese, T. M., and R. M. Haralick. *Differential Equations for Engineers,* McGraw-Hill, New York (1978).
82. Cropley, J. B. "Heuristic Approach to Complex Kinetics," pp. 292–302 in *Chemical Reaction Engineering—Houston,* ACS Symposium Series 65, American Chemical Society, Washington, DC (1978).
83. Cuvelier, C., A. Segal, and A. A. van Steenhoven. *Finite Element Methods and Navier-Stokes Equations,* Reidel, Dordrecht (1986).
84. Davidon, W. C. "Variable Metric Methods for Minimization," AEC R&D Report ANL-5990, rev. (1959).
85. Davis, M. E. *Numerical Methods and Modeling for Chemical Engineers,* Wiley, New York (1984).
86. Davis, P. J. *Interpolation and Approximation,* Dover, New York (1980).
87. ——— and P. Rabinowitz. *Methods of Numerical Integration,* 2d ed., Academic, New York (1984).
88. Denn, M. M. *Stability of Reaction and Transport Processes,* Prentice Hall, Englewood Cliffs, NJ (1974).
89. Dennis, J. E., and J. J. More. *SIAM Review* **21** (1977): 443.
90. Dimian, A. *Chem. Eng. Prog.* **90** (Sept. 1994): 58–66.
91. Doherty, M. F., and J. M. Ottino. *Chem. Eng. Sci.* **43** (1988): 139–183.
92. Dongarra, J. J., J. R. Bunch, C. B. Moler, and G. W. Stewart. *LINPACK Users Guide,* Society for Industrial and Applied Mathematics, Philadelphia (1979).
93. Draper, N. R., and H. Smith. *Applied Regression Analysis,* 2d ed., Wiley, New York (1981).
94. Dubois, D., H. Prade, and R. R. Yager (eds.) *Readings in Fuzzy Sets for Intelligent Systems,* Morgan Kaufmann (1993).
95. Duff, I. S. (ed.). *Sparse Matrices and Their Uses,* Academic, New York (1981).
96. Duff, I. S. *Direct Methods for Sparse Matrices,* Oxford, Charendon Press (1986).
97. Duffy, D. G. *Transform Methods for Solving Partial Differential Equations,* CRC Press (1994).
98. Dym, C. L., and E. S. Ivey. *Principles of Mathematical Modeling,* Academic, New York (1980).
99. Edgar, T. F., and D. M. Himmelblau. *Optimization of Chemical Processes,* McGraw-Hill (1988).
100. Eisenstat, S. C. *SIAM J. Sci. Stat. Comp.* **2** (1981): 1–4.
101. Eisenstat, S. C., M. H. Schultz, and A. H. Sherman. *SIAM J. Sci. Stat. Comput.* **2** (1981): 225–237.
102. Elich, J., and C. J. Elich. *College Alegebra with Calculator Applications,* Addison-Wesley, Boston (1982).
103. Ferguson, N. B., and B. A. Finlayson. *A. I. Ch. E. J.* **20** (1974): 539–550.
104. Finlayson, B. A. *The Method of Weighted Residuals and Variational Principles,* Academic, New York (1972).
105. Finlayson, B., L. T. Biegler, I. E. Grossmann, and A. W. Westerberg. "Mathematics in Chemical Engineering," *Ullmann's Encyclopedia of Industrial Chemistry,* Vol. B1, VCH, Weinheim (1990).
106. Finlayson, B. A. *Nonlinear Analysis in Chemical Engineering,* McGraw-Hill, New York (1980).
107. Finlayson, B. A. *Numerical Methods for Problems with Moving Fronts,* Ravenna Park Publishing, Seattle (1992).
108. Fisher, R. C., and A. D. Ziebur. *Integrated Algebra, Trigonometry, and Analytic Geometry,* 4th ed., Prentice Hall, Englewood Cliffs, NJ (1982).
109. Fletcher, R. *Computer J.* **13** (1970): 317.
110. Fletcher, R. *Practical Methods of Optimization,* Wiley, New York (1987).
111. Forsythe, G. E., M. A. Malcolm, and C. B. Moler. *Computer Methods for Mathematical Computations,* Prentice Hall, Englewood Cliffs, NJ (1977).
112. Forsyth, G., and C. B. Moler. *Computer Solution of Linear Algbraic Systems,* Prentice Hall, Englewood Cliffs (1967).
113. Fourer, R., D. M. Gay, and B. W. Kerninghan. *Management Science* **36** (1990): 519–554.
114. Friedman, N. A. *Calculus and Mathematical Models,* Prindle, Weber & Schmidt, Boston (1979).
115. Gantmacher, F. R. *Applications of the Theory of Matrices,* Interscience, New York (1959).
116. Garbow, B. S., J. M. Boyle, J. J. Dongarra, and C. B. Moler: *Matrix Eigensystem Routines—EISPACK Guide Extensions,* Springer-Verlag, Berlin and New York (1977).
117. Gear, G. W. *Numerical Initial Value Problems in Ordinary Differential Equations,* Prentice Hall, Englewood Cliffs, NJ (1971).
118. Gellert, W., H. Küstner, M. Hellwich, H. Kästner (ed.). *The VNR Concise Encyclopedia of Mathematics,* Van Nostrand Reinhold Co., New York (1975).
119. Gill, P., and W. Murray. *Math. Prog.* **14** (1978): 349.
120. Gill, P. E., W. Murray, and M. Wright. *Practical Optimization,* Academic, New York (1981).
121. Goldberg, D. E. *Genetic Algorithms in Search, Optimization and Machine Learning,* Addison-Wesley (1989).
122. Goldfarb, D. *Math. Comp.* **24** (1970): 23.
123. ——— and A. Idnani. *Math. Prog.* **27** (1983): 1.
124. ——— and M. J. Todd. "Linear Programming," Chapter II in *Optimization* (G. L. Nemhauser, A. H. G. Rinnoy Kan, and M. J. Todd, eds.), North Holland, Amsterdam (1989).
125. Gottlieb, D., and S. A. Orszag. *Numerical Analysis of Spectral Methods: Theory and Applications,* SIAM, Philadelphia (1977).
126. Gradshteyn, I. S., and I. M. Ryzhik. *Tables of Integrals, Series, and Products,* Academic, New York (1980).
127. Greenberg, M. M. *Foundations of Applied Mathematics,* Prentice Hall, Englewood Cliffs, NJ (1978).
128. Groetsch, C. W. *Generalized Inverses of Linear Operators,* Marcel Dekker, New York (1977).
129. ———. *Elements of Applicable Functional Analysis,* Marcel Dekker, New York (1980).
130. Gunzburger, M. D. *Finite Element Methods for Viscous Incompressible Flows,* Academic, New York (1989).
131. Gustafson, R. D., and P. D. Frisk. *Plane Trigonometry,* Brooks/Cole, Monterey, CA (1982).
132. Haberman, R. *Mathematical Models,* Prentice Hall, Englewood Cliffs, NJ (1977).
133. Hageman, L. A., and D. M. Young. *Applied Iterative Methods,* Academic, New York (1981).
134. Hamburg, M. *Statistical Analysis for Decision Making,* 2d ed., Harcourt, New York (1977).
135. Hamming, R. W. *Numerical Methods for Scientists and Engineers,* 2d ed., McGraw-Hill, New York (1973).
136. Han, S-P. *J. Opt. Theo. Applics.* **22** (1977): 297.
137. Hanna, R. *Fourier Series and Integrals of Boundary Value Problems,* Wiley, New York (1982).
138. Hanna, O. T., and O. C. Sandall. *Computational Methods in Chemical Engineering,* Prentice Hall, Upper Saddle River, NJ (1994).
139. Hardy, G. H., J. E. Littlewood, and G. Polya. *Inequalities,* 2d ed., Cambridge University Press, Cambridge (1952).
140. Haykin, S. *Neural Networks: A Comprehensive Foundation,* Macmillan, New York (1994).
141. Henrici, P. *Applied and Computational Complex Analysis,* Wiley, New York (1974).
142. Hestenes, M. R. *Conjugate Gradient Methods in Optimization,* Springer-Verlag (1980).
143. Hildebrand, F. B. *Introduction to Numerical Analysis,* 2d ed., McGraw-Hill, New York (1974).
144. ———. *Advanced Calculus for Applications,* 2d ed., Prentice Hall, Englewood Cliffs, NJ (1976).
145. Hill, J. M. *Differential Equations and Group Methods for Scientists and Engineers,* CRC Press (1992).
146. Hille, E. *Ordinary Differential Equations in the Complex Domain,* Wiley (1976).
147. Hille, E. *Methods in Classical and Functional Analysis,* Addison-Wesley, Reading, MA (1972).
148. Hindmarsh, A. C. *ACM SIGNUM Newsletter* **15** (1980): 10–11.
149. Hindmarsh, A. C. "GEARB: Solution of Ordinary Differential Equations Having Banded Jacobian," UCID-30059, Rev. 1 Computer Documentation, Lawrence Livermore Laboratory, University of California (1975).
150. Hornbeck, R. W. *Numerical Methods,* Prentice Hall, Englewood Cliffs, NJ (1975).
151. Hougen, O. A., R. M. Watson, and R. A. Ragatz. Part II, "Thermodynamics," in *Chemical Process Principles,* 2d ed., Wiley, New York (1959).
152. Householder, A. S. *The Theory of Matrices in Numerical Analysis,* Dover, New York (1979).

153. ———. *Numerical Treatment of a Single Nonlinear Equation,* McGraw-Hill, New York, (1970) and Dover, New York (1980).

154. Houstis, E. N., W. F. Mitchell, and T. S. Papatheodoros. *Int. J. Num. Methods Engn.* **19** (1983): 665–704.

155. Isaacson, E., and H. B. Keller. *Analysis of Numerical Methods,* Wiley, New York (1966).

156. Jeffreys, H., and B. Jeffreys. *Methods of Mathematical Physics,* 3d ed., Cambridge University Press, London (1972).

157. Jennings, A., and J. J. McKeown. *Matrix Computations for Engineers and Scientists,* Wiley, New York (1992).

158. Johnson, R. E., and F. L. Kiokemeister. *Calculus with Analytic Geometry,* 4th ed., Allyn and Bacon, Boston (1969).

159. Joseph, D. D., M. Renardy, and J. C. Saut. *Arch. Rational Mech. Anal.* **87** (1985): 213–251.

160. Juncu, G., and R. Mihail. *Comp. Chem. Eng.* **13** (1989): 259–270.

161. Kalos, M. H., and P. A. Whitlock. *Monte Carlo Methods,* vol. I, Wiley, New York (1986).

162. Kantorovich, L. V., and G. P. Akilov. *Functional Analysis,* 2d ed., Pergamon, Oxford (1982).

163. Kaplan, W. *Advanced Calculus,* 2d ed., Addison-Wesley, Reading, MA (1973).

164. Kardestuncer, H., and D. H. Norrie (ed.). *Finite Element Handbook,* McGraw-Hill (1987).

165. Karmarker, N. *Combinatorica* **4** (1984): 373–395.

166. Keedy, M. L., and M. L. Bittinger. *Trigonometry: Triangles and Functions,* 3d ed., Addison-Wesley, New York (1983).

167. Keller, H. B. *Numerical Methods for Two-Point Boundary-Value Problems,* Blaisdell, New York (1972).

168. Kemeny, J. G., J. L. Snell, and G. L. Thompson. *Introduction to Finite Mathematics,* 3d ed., Prentice Hall, Englewood Cliffs, NJ (1975).

169. Kendall, M. G., A. Stuart, J. K. Ord, and A. O'Hogan. *Advanced Theory of Statistics,* Halsted, New York (1994).

170. Kevorkian, J., and J. D. Cole: *Perturbation Methods in Applied Mathematics,* Springer-Verlag, New York (1981).

171. Kincaid, D. R., and D. M. Young. "Survey of Iterative Methods," in *Encyclopedia of Computer Science and Technology,* Marcel Dekker, New York (1979).

172. Krantz, S. G. *Function Theory of Several Complex Variables,* 2d ed., Wadsworth and Brooks, New York (1992).

173. Kreyszig, E. *Advanced Engineering Mathematics,* 7th ed., Wiley, New York (1993).

174. ———. *Introductory Functional Analysis with Applications,* Wiley, New York (1978).

175. Krieger, J. H. *Chem. Eng. News* **73** (Mar. 27, 1995): 50–61.

176. Kubicek, M., and M. Marek. *Computational Methods in Bifurcation Theory and Dissipative Structures,* Springer-Verlag, Berlin (1983).

177. Kuhn, H. W., and A. W. Tucker. "Nonlinear Programming" in Neyman, J. (ed.), *Proc. Second Berkeley Symp. Mathematical Statistics and Probability* (1951): 402–411.

178. Kuipers, B. *Qualitative Reasoning: Modeling and Simulation with Incomplete Knowledge,* MIT Press, Boston (1994).

179. Kyrala, A. *Applied Functions of a Complex Variable,* Interscience, New York (1972).

180. Lagerstrom, P. A. *Matched Asymptotic Expansions: Ideas and Techniques,* Springer-Verlag (1988).

181. Lambert, J. D. *Computational Methods in Ordinary Differential Equations,* Wiley, New York (1973).

182. Lanczos, C. *J. Math. Phys.* **17** (1938): 123–199.

183. Lanczos, C. *Applied Analysis,* Prentice Hall, Englewood Cliffs, NJ (1956).

184. Lapidus, L., and G. F. Pinder. *Numerical Solution of Partial Differential Equations in Science and Engineering,* Interscience, New York (1982).

185. Lapidus, L., and J. Seinfeld. *Numerical Solution of Ordinary Differential Equations,* Academic, New York (1971).

186. Lapin, L. L. *Statistics for Modern Business Decisions,* 2d ed., Harcourt, New York (1982).

187. Lau, H. T. *A Numerical Library in C for Scientists and Engineers,* CRC Press (1995).

188. Lawrence, J. D. *A Catalog of Special Plane Curves,* Dover, New York (1972).

189. Lawson, C. L., and R. J. Hanson. *Solving Least Squares Problems,* Prentice Hall, Englewood Cliffs, NJ (1974).

190. Lebedev, N. N. *Special Functions and Their Applications,* Dover, New York (1972).

191. Leithold, L. *College Algebra and Trigonometry,* Addison-Wesley (1989).

192. LeVeque, R. J. *Numerical Methods for Conservation Laws,* Birkhäuser, Basel (1992).

193. Levy, H. *Analytic Geometry,* Harcourt, Brace & World, New York (1969).

194. Lin, C. C., and L. A. Segel. *Mathematics Applied to Deterministic Problems in the Natural Sciences,* Macmillan, New York (1974).

195. Linz, P. *Analytical and Numerical Methods for Volterra Equations,* SIAM Publications, Philadelphia (1985).

196. Liusternik, L. A., and V. J. Sobolev. *Elements of Functional Analysis,* 3d ed., Wiley, New York (1974).

197. Luke, Y. L. *Mathematical Functions and Their Applications,* Academic, New York (1975).

198. Luyben, W. L. *Process Modeling, Simulation and Control for Chemical Engineers,* 2d ed., McGraw-Hill, New York (1990).

199. MacDonald, W. B., A. N. Hrymak, and S. Treiber. "Interior Point Algorithms for Refinery Scheduling Problems" in *Proc. 4th Annual Symp. Process Systems Engineering* (Aug. 5–9, 1991): III.13.1–16.

200. Mackerle, J., and C. A. Brebbia (eds.). *Boundary Element Reference Book,* Springer Verlag, Berlin-Heidelberg, New York and Tokyo (1988).

201. Mah, R. S. H. *Chemical Process Structures and Information Flows,* Butterworths (1990).

202. Mansfield, R. *Trigonometry with Applications,* Wadsworth, New York (1972).

203. Margenau, H., and G. M. Murphy. *The Mathematics of Physics and Chemistry,* 2d ed., Van Nostrand, Princeton, NJ (1956).

204. Martin, R. H. Jr. *Ordinary Differential Equations,* McGraw-Hill, New York (1983).

205. Mavrovouniotis, Michael L. (ed.). *Artificial Intelligence in Process Engineering,* Academic, Boston (1990).

206. McIntosh, A. *Fitting Linear Models: An Application of Conjugate Gradient Algorithms,* Springer-Verlag, New York (1982).

207. McCormick, G. P. *Nonlinear Programming: Theory, Algorithms, and Applications,* Wiley, New York (1983).

208. *McGraw-Hill Encyclopedia of Science and Technology,* McGraw-Hill, New York (1971).

209. Mei, C. C. *Mathematical Analysis in Engineering,* Cambridge (1995).

210. Mitchell, A. R., and R. Wait. *The Finite Element Method in Partial Differential Equations,* Wiley, New York (1977).

211. Mood, A. M., R. A. Graybill, and D. C. Boes. *Introduction to the Theory of Statistics,* 3d ed., McGraw-Hill, New York (1974).

212. Morse, P. M., and H. Feshbach. *Methods of Theoretical Physics,* vols. I and II, McGraw-Hill, New York (1953).

213. Morton, K. W., and D. F. Mayers. *Numerical Solution of Partial Differential Equations,* Cambridge (1995).

214. Nayfeh, A. H. *Perturbation Methods,* Wiley, New York (1973).

215. ———. *Introduction to Perturbation Techniques,* Wiley, New York (1981).

216. Naylor, A. W., and G. R. Sell. *Linear Operator Theory in Engineering and Science,* Springer-Verlag, New York (1982).

217. Oberhettinger, F. *Fourier Expansions: A Collection of Formulas,* Academic, New York (1973).

218. Ogunnaike, B. A., and W. H. Ray. *Process Dynamics, Modeling, and Control,* Oxford University Press (1994).

219. Ortega, J. M. *Numerical Analysis: A Second Course,* SIAM (1990).

220. Pao, C. V. *Nonlinear Parabolic and Elliptic Equations,* Plenum (1992).

221. Peaceman, D. W. *Fundamentals of Numerical Reservoir Simulation,* Elsevier, Amsterdam (1977).

222. Pearson, Carl E. (ed.). *Handbook of Applied Mathematics,* 2d ed., Van Nostrand Reinhold Co., New York (1983).

223. Perlmutter, D. *Stability of Chemical Reactors,* Prentice Hall, Englewood Cliffs, NJ (1972).

224. Petzold, L. R. "A Description of DASSL: A Differential-Algebraic System Solver," Sandia National Laboratory Report SAND82-8637; also in Stepleman, R. S. et al., eds. *IMACS Trans. on Scientific Computing,* vol. 1, pp. 65–68.

225. Pike, R. W. *Optimization for Engineering Systems,* Van Nostrand Reinhold (1986).

226. Pontelides, C. C., D. Gritsis, K. R. Morison, and R. W. H. Sargent. *Comp. Chem. Eng.* **12** (1988): 449–454.

227. Poulain, C. A., and B. A. Finlayson. *Int. J. Num. Methods Fluids* **17** (1993): 839–859.

228. Powell, M. J. D. "A Fast Algorithm for Nonlinearly Constrained Optimization Calculations," *Lecture Notes in Mathematics* **630** (1977).

229. Powers, D. L. *Boundary Value Problems,* Academic, New York (1972).

230. Prenter, P. M. *Splines and Variational Methods,* Wiley, New York (1975).

231. Press, W. H., B. P. Flannery, S. A. Teukolsky, and W. T. Vetterling. *Numerical Recipes,* Cambridge University Press, Cambridge (1986).

232. Quantrille, T. E., and Y. A. Liu. *Artificial Intelligence in Chemical Engineering,* Academic Press, San Diego (1991).

233. Quarteroni, A., and A. Valli. *Numerical Approximation of Partial Differential Equations,* Springer-Verlag (1994).

234. Råde, L., and B. Westergren. β *Mathematics Handbook,* 2d ed., Chartwell-Bratt, Lund, Sweden (1990).

235. Rainville, E. D. *Special Functions,* Chelsea Publishing Company, New York (1972).

236. Rainville, E. D., and P. E. Bedient. *Elementary Differential Equations,* 7th ed., Macmillan, New York (1989).

237. Rall, L. B. *Computational Solution of Nonlinear Operator Equations,* Wiley, New York (1969) and Dover, New York (1981).
238. Ralston, A., and A. Rabinowitz. *A First Course in Numerical Analysis,* 2d ed., McGraw-Hill, New York (1978).
239. Ramirez, W. F. *Computational Methods for Process Simulations,* Butterworths, Boston (1989).
240. Rauch, J. *Partial Differential Equations,* Springer-Verlag (1991).
241. Reddy, J. N., and D. K. Gartling. *The Finite Element Method in Heat Transfer and Fluid Dynamics,* CRC Press (1994).
242. Reklaitis, G. V. *Introduction to Material and Energy Balances,* Wiley (1983).
243. Rekalitis, G. V., and H. D. Spriggs. *Proceedings of the First International Conference on Foundations of Computer-Aided Operations,* Elsevier Science Publishers, Inc., New York (1987).
244. Reklaitis, G. V., A. Ravindran, and K. M. Ragsdell. *Engineering Optimization Methods and Applications,* Wiley, New York (1983).
245. Rhee, H.-K., R. Aris, and N. R. Amundson. *First-Order Partial Differential Equations,* vol. I., Prentice Hall, Englewood Cliffs, NJ (1986).
246. ———. *Matrix Computations and Mathematical Software,* McGraw-Hill, New York (1981).
247. ———. *Numerical Methods, Software, and Analysis,* 2d ed., Academic, New York (1993).
248. Rich, E., and K. Kevin. *Artificial Intelligence,* 2d ed., McGraw-Hill, New York (1991).
249. Riggs, J. B. *An Introduction to Numerical Methods for Chemical Engineers,* Texas Tech Univ. Press, Lubbock, TX (1994).
250. Rippin, D. W. T., J. C. Hale, and J. F. Davis (ed.). *Proceedings of the Second International Conference on Foundations of Computer-Aided Operations,* CACHE Corporation, Austin, TX (1994).
251. Ritchmyer, R., and K. Morton. *Difference Methods for Initial-Value Problems,* 2d ed., Interscience, New York (1967).
252. Saaty, T. L., and J. Bram. *Nonlinear Mathematics,* McGraw-Hill, New York (1964) and Dover, New York (1981).
253. Schiesser, W. E. *The Numerical Method of Lines,* Academic Press (1991).
254. Schittkowski, K. *Num. Math.* **38** (1982): 83.
255. Seader, J. D. *Computer Modeling of Chemical Processes,* AIChE Monog. Ser. No. 15 (1985).
256. Seborg, D. E., T. F. Edgar, and D. A. Mellichamp. *Process Dynamics and Control,* Wiley, New York (1989).
257. Shampine, L. *Numerical Solution of Ordinary Differential Equations,* Chapman & Hall (1994).
258. Shapiro, S. C., D. Eckroth. et al (ed.). *Encyclopedia of Artificial Intelligence,* Wiley, New York (1987).
259. Shanno, D. F. *Math. Comp.* **24** (1970): 647.
260. Shenk, A. *Calculus and Analytic Geometry,* Goodyear Publishing Company, Santa Monica, CA (1977).
261. Shockley, J. E. *Calculus and Analytic Geometry,* Saunders, Philadelphia (1982).
262. Siirola, J. J., I. E. Grossmann, and G. Stephanopoulos. *Proceedings of the Second International Conference on Foundations of Computer-Aided Design,* Elsevier Science Publishers, Inc., New York (1990).
263. Simmons, G. F. *Differential Equations,* McGraw-Hill, New York (1972).
264. Simmonds, J. G. *A Brief on Tensor Analysis,* Springer-Verlag (1994).
265. Sincich, T., and Mendenhall, W. *Statistics for Engineering and the Sciences,* 4th ed., Prentice Hall, Englewood Cliffs, NJ (1995).
266. Sincovec, R. F. *Math. Comp.* **26** (1972): 893–895.
267. Smith, J., J. L. Siemienivich, and I. Gladwell. "A Comparison of Old and New Methods for Large Systems of Ordinary Differential Equations Arising from Parabolic Partial Differential Equations," Num. Anal. Rep. Department of Engineering, no. 13, University of Manchester, England (1975).
268. Smith, W. K. *Analytic Geometry,* Macmillan (1972).
269. Sobel, M. A., and N. Lerner. *College Algebra,* Prentice Hall, Englewood Cliffs, NJ (1983).
270. Sod, G. A. *Numerical Methods in Fluid Dynamics,* Cambridge Univ. Press (1985).
271. Sokolnikoff, I. S., and Sokolnikoff, E. S. *Higher Mathematics for Engineers and Physicists,* McGraw-Hill, New York (1941).
272. Spiegel, M. R. *Applied Differential Equations,* 3d ed., Prentice Hall, Englewood Cliffs, NJ (1981).
273. Stakgold, I. *Green's Functions and Boundary Value Problems,* Interscience, New York (1979).
274. Stein, S. K. *Calculus and Analytic Geometry,* 3d ed., McGraw-Hill, New York (1982).
275. Stephanopoulos, G., and J. F. Davis (eds.). *Artificial Intelligence in Process Engineering,* CACHE Monograph Series, CACHE, Austin (1990–1992).
276. Stephanopoulos, G., and H. Chonghun. "Intelligent Systems in Process Engineering: A Review," *Proceedings of PSE '94,* Korea (1994).
277. Stillwell, J. C. *Elements of Algebra,* CRC Press, New York (1994).
278. Stoer, J., and R. Bulirsch. *Introduction to Numerical Analysis,* Springer, New York (1993).
279. Strang, G. *Linear Algebra and Its Applications,* 2d ed., Academic, New York (1980).
280. Strang, G. *Introduction to Linear Algebra,* Wellesley-Cambridge, Cambridge, MA (1993).
281. ——— and G. Fix. *An Analysis of the Finite Element Method,* Prentice Hall, Englewood Cliffs, NJ (1973).
282. Swokowski, E. W. *Calculus with Analytic Geometry,* 2d ed., Prindle, Weber & Schmidt, Boston (1981).
283. Taylor, A. E., and D. C. Lay. *Introduction to Functional Analysis,* 2d ed., Wiley, New York (1980).
284. Umeda, T., and A. Ichikawa. *I&EC Proc. Design Develop.* **10** (1971): 229.
285. Vasantharajan, S., and L. T. Biegler. *Computers and Chemical Engineering* **12** (1988): 1087.
286. Vemuri, V., and W. Karplus. *Digital Computer Treatment of Partial Differential Equations,* Prentice Hall, Englewood Cliffs, NJ (1981).
287. Vichnevetsky, R. *Computer Methods for Partial Differential Equations,* vols. 1 and 2, Prentice Hall, Englewood Cliffs, NJ (1981, 1982).
288. Villadsen, J. V., and M. L. Michelsen. *Solution of Differential Equation Models by Polynomial Approximation,* Prentice Hall, Englewood Cliffs, NJ (1978).
289. Villadsen, J., and W. E. Stewart. *Chem. Eng. Sci.* **22** (1967): 1483–1501.
290. Walas, S. M. *Modeling with Differential Equations in Chemical Engineering,* Butterworth-Heinemann, Stoneham, MA (1991).
291. Weisberg, S. *Applied Linear Regression,* 2d ed., Wiley, New York (1985).
292. Weld, D. S., and J. de Kleer (ed.). *Readings in Qualitative Reasoning About Physical Systems,* Morgan Kaufman, San Mateo, CA (1990).
293. Westerberg, A. W., H. P. Hutchison, R. L. Motard, and P. Winter. *Process Flowsheeting,* Cambridge University Press, London (1979).
294. Westerberg, A. W., and H. H. Chien (ed.). *Proceedings of the Second International Conference on Foundations of Computer-Aided Design,* CACHE Corporation, Austin, TX (1984).
295. Westerberg, A. W. "Optimization" in A. K. Sunol, D. W. T. Rippin, G. V. Reklaitis, O. Hortacsu (eds.), *Batch Processing Systems Engineering: Current Status and Future Directions,* vol. 143, NATO ASI Series F, Springer, Berlin (1995).
296. Whipkey, K. L., and M. N. Whipkey. *The Power of Calculus,* 3d ed., Wiley, New York (1979).
297. Wilkinson, J. H. *The Algebraic Eigenvalue Problem,* Clarendon Press, Oxford (1988).
298. Williams, G. *Computational Linear Algebra with Models,* 2d ed., Allyn and Bacon, Boston (1981).
299. Wolfram, S. *Mathematica,* Addison-Wesley, New York (1988).
300. Wouk, A. *A Course of Applied Functional Analysis,* Interscience, New York (1979).
301. Wylie, C. R. *Advanced Engineering Mathematics,* 5th ed., McGraw-Hill, New York (1982).
302. Young, D. M. *Iterative Solution for Large Linear Systems,* Academic, New York (1971).
303. Zienkiewicz, O. C., and R. L. Taylor. *The Finite Element Method,* McGraw-Hill, London (1989).
304. ——— and K. Morgan. *Finite Elements and Approximations,* Wiley, New York (1983).

REFERENCES FOR GENERAL AND SPECIFIC TOPICS

Advanced engineering mathematics:
Upper undergraduate level, 19, 80, 127, 144, 156, 173, 194, 203, 209, 301. Graduate level, 79, 127, 212, 273. Mathematical tables, mathematical dictionaries, and handbooks of mathematical functions and formulas, 1, 28, 48, 57, 118, 126, 188, 208, 217, 222, 234. Mathematical modeling of physical phenomena, 17, 19, 31, 44, 98, 132, 194, 273. Mathematical theory of reaction, diffusion, and transport processes, 10, 16, 19, 88, 223. Mathematical methods in chemical engineering, 13, 15, 61, 85, 104, 106, 138, 239, 249, 288. Inequalities, 28, 126, 139, 290.

Vector and tensor analysis, 18, 163, 173, 264.
Special functions in physics and engineering, 190, 197, 235.
Green's functions and applications, 75, 127, 273.
Perturbation and asymptotic methods in applied mathematics, 170, 215, 216.
Approximation theory and interpolation, 86, 87.
Functional analysis; linear operators, 129, 147, 162, 174, 196, 216, 226, 283, 300.
Generalized inverses and least-squares problems, 33, 128, 189.

MATHEMATICS

GENERAL

The basic problems of the sciences and engineering fall broadly into three categories:

1. *Steady state problems.* In such problems the configuration of the system is to be determined. This solution does not change with time but continues indefinitely in the same pattern, hence the name "steady state." Typical chemical engineering examples include steady temperature distributions in heat conduction, equilibrium in chemical reactions, and steady diffusion problems.

2. *Eigenvalue problems.* These are extensions of equilibrium problems in which critical values of certain parameters are to be determined in addition to the corresponding steady-state configurations. The determination of eigenvalues may also arise in propagation problems. Typical chemical engineering problems include those in heat transfer and resonance in which certain boundary conditions are prescribed.

3. *Propagation problems.* These problems are concerned with predicting the subsequent behavior of a system from a knowledge of the initial state. For this reason they are often called the transient (time-varying) or unsteady-state phenomena. Chemical engineering examples include the transient state of chemical reactions (kinetics), the propagation of pressure waves in a fluid, transient behavior of an adsorption column, and the rate of approach to equilibrium of a packed distillation column.

The mathematical treatment of engineering problems involves four basic steps:

1. *Formulation.* The expression of the problem in mathematical language. That translation is based on the appropriate physical laws governing the process.

2. *Solution.* Appropriate mathematical operations are accomplished so that logical deductions may be drawn from the mathematical model.

3. *Interpretation.* Development of relations between the mathematical results and their meaning in the physical world.

4. *Refinement.* The recycling of the procedure to obtain better predictions as indicated by experimental checks.

Steps 1 and 2 are of primary interest here. The actual details are left to the various subsections, and only general approaches will be discussed.

The formulation step may result in algebraic equations, difference equations, differential equations, integral equations, or combinations of these. In any event these mathematical models usually arise from statements of physical laws such as the laws of mass and energy conservation in the form.

Input of conserved quantity – output of conserved quantity
    + conserved quantity produced
        = accumulation of conserved quantity

Rate of input of conserved quantity – rate of output of
    conserved quantity + rate of conserved quantity produced
        = rate of accumulation of conserved quantity

These statements may be abbreviated by the statement

Input – output + production = accumulation

When the basic physical laws are expressed in this form, the formulation is greatly facilitated. These expressions are quite often given the names, "material balance," "energy balance," and so forth. To be a little more specific, one could write the law of conservation of energy in the steady state as

Rate of energy in – rate of energy out + rate of energy produced = 0

Many general laws of the physical universe are expressible by differential equations. Specific phenomena are then singled out from the infinity of solutions of these equations by assigning the individual initial or boundary conditions which characterize the given problem. In mathematical language one such problem, the equilibrium problem,

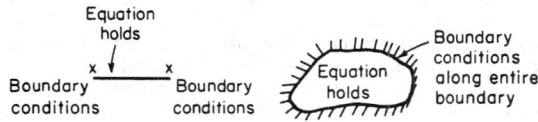

FIG. 3-1   Boundary conditions.

is called a boundary-value problem (Fig. 3-1). Schematically, the problem is characterized by a differential equation plus an open region in which the equation holds and, on the boundaries of the region, by certain conditions (boundary conditions) that are dictated by the physical problem. The solution of the equation must satisfy the differential equation inside the region and the prescribed conditions on the boundary.

In mathematical language, the propagation problem is known as an initial-value problem (Fig. 3-2). Schematically, the problem is characterized by a differential equation plus an open region in which the equation holds. The solution of the differential equation must satisfy the initial conditions plus any "side" boundary conditions.

The description of phenomena in a "continuous" medium such as a gas or a fluid often leads to partial differential equations. In particular, phenomena of "wave" propagation are described by a class of partial differential equations called "hyperbolic," and these are essentially different in their properties from other classes such as those that describe equilibrium ("elliptic") or diffusion and heat transfer ("parabolic"). Prototypes are:

1. *Elliptic.*   Laplace's equation

$$\frac{\partial^2 u}{\partial x^2} + \frac{\partial^2 u}{\partial y^2} = 0$$

Poisson's equation

$$\frac{\partial^2 u}{\partial x^2} + \frac{\partial^2 u}{\partial y^2} = g(x,y)$$

These do not contain the variable t (time) explicitly; accordingly, their solutions represent equilibrium configurations. Laplace's equation corresponds to a "natural" equilibrium, while Poisson's equation corresponds to an equilibrium under the influence of an external force of density proportional to $g(x, y)$.

2. *Parabolic.*   The heat equation

$$\frac{\partial u}{\partial t} = \frac{\partial^2 u}{\partial x^2} + \frac{\partial^2 u}{\partial y^2}$$

describes nonequilibrium or propagation states of diffusion as well as heat transfer.

3. *Hyperbolic.*   The wave equation

$$\frac{\partial^2 u}{\partial t^2} = \frac{\partial^2 u}{\partial x^2} + \frac{\partial^2 u}{\partial y^2}$$

describes wave propagation of all types when the assumption is made that the wave amplitude is small and that interactions are linear.

The solution phase has been characterized in the past by a concentration on methods to obtain analytic solutions to the mathematical

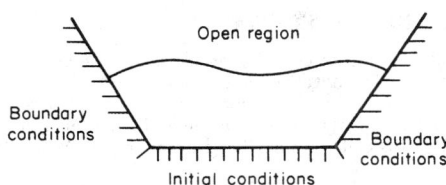

FIG. 3-2   Propagation problem.

equations. These efforts have been most fruitful in the area of the linear equations such as those just given. However, many natural phenomena are nonlinear. While there are a few nonlinear problems that can be solved analytically, most cannot. In those cases, numerical methods are used. Due to the widespread availability of software for computers, the engineer has quite good tools available.

Numerical methods almost never fail to provide an answer to any particular situation, but they can never furnish a general solution of any problem.

The mathematical details outlined here include both analytic and numerical techniques useful in obtaining solutions to problems.

Our discussion to this point has been confined to those areas in which the governing laws are well known. However, in many areas, information on the governing laws is lacking. Interest in the application of statistical methods to all types of problems has grown rapidly since World War II. Broadly speaking, statistical methods may be of use whenever conclusions are to be drawn or decisions made on the basis of experimental evidence. Since statistics could be defined as the technology of the scientific method, it is primarily concerned with the first two aspects of the method, namely, the performance of experiments and the drawing of conclusions from experiments. Traditionally the field is divided into two areas:

1. *Design of experiments.* When conclusions are to be drawn or decisions made on the basis of experimental evidence, statistical techniques are most useful when experimental data are subject to errors. The design of experiments may then often be carried out in such a fashion as to avoid some of the sources of experimental error and make the necessary allowances for that portion which is unavoidable. Second, the results can be presented in terms of probability statements which express the reliability of the results. Third, a statistical approach frequently forces a more thorough evaluation of the experimental aims and leads to a more definitive experiment than would otherwise have been performed.

2. *Statistical inference.* The broad problem of statistical inference is to provide measures of the uncertainty of conclusions drawn from experimental data. This area uses the theory of probability, enabling scientists to assess the reliability of their conclusions in terms of probability statements.

Both of these areas, the mathematical and the statistical, are intimately intertwined when applied to any given situation. The methods of one are often combined with the other. And both in order to be successfully used must result in the numerical answer to a problem—that is, they constitute the means to an end. Increasingly the numerical answer is being obtained from the mathematics with the aid of computers.

MISCELLANEOUS MATHEMATICAL CONSTANTS

Numerical values of the constants that follow are approximate to the number of significant digits given.

$\pi = 3.1415926536$	Pi
$e = 2.7182818285$	Napierian (natural) logarithm base
$\gamma = 0.5772156649$	Euler's constant
$\ln \pi = 1.1447298858$	Napierian (natural) logarithm of pi, base e
$\log \pi = 0.4971498727$	Briggsian (common logarithm of pi, base 10
Radian $= 57.2957795131°$	
Degree $= 0.0174532925$ rad	
Minute $= 0.0002908882$ rad	
Second $= 0.0000048481$ rad	

THE REAL-NUMBER SYSTEM

The natural numbers, or counting numbers, are the positive integers: 1, 2, 3, 4, 5, ... The negative integers are $-1, -2, -3, ...$

A number in the form a/b, where a and b are integers, $b \neq 0$, is a rational number. A real number that cannot be written as the quotient of two integers is called an irrational number, e.g., $\sqrt{2}$, $\sqrt{3}$, $\sqrt{5}$, π, e, $\sqrt[3]{2}$.

There is a one-to-one correspondence between the set of real numbers and the set of points on an infinite line (coordinate line).

Order among Real Numbers; Inequalities

$a > b$ means that $a - b$ is a positive real number.
If $a < b$ and $b < c$, then $a < c$.
If $a < b$, then $a \pm c < b \pm c$ for any real number c.
If $a < b$ and $c > 0$, then $ac < bc$.
If $a < b$ and $c < 0$, then $ac > bc$.
If $a < b$ and $c < d$, then $a + c < b + d$.
If $0 < a < b$ and $0 < c < d$, then $ac < bd$.
If $a < b$ and $ab > 0$, then $1/a > 1/b$.
If $a < b$ and $ab < 0$, then $1/a < 1/b$.

Absolute Value
For any real number x, $|x| = \begin{cases} x & \text{if } x \geq 0 \\ -x & \text{if } x < 0 \end{cases}$

Properties
If $|x| = a$, where $a > 0$, then $x = a$ or $x = -a$.
$|x| = |-x|$; $-|x| \leq x \leq |x|$; $|xy| = |x| \, |y|$.
If $|x| < c$, then $-c < x < c$, where $c > 0$.
$||x| - |y|| \leq |x + y| \leq |x| + |y|$.
$\sqrt{x^2} = |x|$.

Proportions
If $\dfrac{a}{b} = \dfrac{c}{d}$, then $\dfrac{a+b}{b} = \dfrac{c+d}{d}$, $\dfrac{a-b}{b} = \dfrac{c-d}{d}$, $\dfrac{a-b}{a+b} = \dfrac{c-d}{c+d}$.

Indeterminants

Form	Example	
$(\infty)(0)$	xe^{-x}	$x \to \infty^+$
0^0	x^x	$x \to 0^+$
∞^0	$(\tan x)^{\cos x}$	$x \to \frac{1}{2}\pi^-$
1^∞	$(1+x)^{1/x}$	$x \to 0^+$
$\infty - \infty$	$\sqrt{x+1} - \sqrt{x-1}$	$x \to \infty$
$\dfrac{0}{0}$	$\dfrac{\sin x}{x}$	$x \to 0$
$\dfrac{\infty}{\infty}$	$\dfrac{e^x}{x}$	$x \to \infty$

Limits of the type $0/\infty$, $\infty/0$, 0^∞, $\infty \cdot \infty$, $(+\infty) + (+\infty)$, and $(-\infty) + (-\infty)$ are *not* indeterminate forms.

Integral Exponents (Powers and Roots) If m and n are positive integers and a, b are numbers or functions, then the following properties hold:

$$a^{-n} = 1/a^n \qquad a \neq 0$$
$$(ab)^n = a^n b^n$$
$$(a^n)^m = a^{nm}, \qquad a^n a^m = a^{n+m}$$
$$\sqrt[n]{a} = a^{1/n} \quad \text{if} \quad a > 0$$
$$\sqrt[m]{\sqrt[n]{a}} = \sqrt[mn]{a}, a > 0$$
$$a^{m/n} = (a^m)^{1/n} = \sqrt[n]{a^m}, a > 0$$
$$a^0 = 1 \ (a \neq 0)$$
$$0^a = 0 \ (a \neq 0)$$

Infinity (∞) is not a real number. It is possible to extend the real-number system by adjoining to it "∞" and "$-\infty$," and within the extended system, certain operations involving $+\infty$ or $-\infty$ are possible. For example, if $0 < a < 1$, then $a^\infty = \lim_{x \to \infty} a^x = 0$, whereas if $a > 1$, then $a^\infty = \infty$, $\infty^a = \infty$ $(a > 0)$, $\infty^a = 0$ $(a < 0)$.

Care should be taken in the case of roots and fractional powers of a product; e.g., $\sqrt{xy} \neq \sqrt{x}\sqrt{y}$ if x and y are negative. This rule applies if one is careful about the domain of the functions involved; so $\sqrt{xy} = \sqrt{x}\sqrt{y}$ if $x > 0$ and $y > 0$.

Given any number $b > 0$, there is a unique function $f(x)$ defined for all real numbers x such that (1) $f(x) = b^x$ for all rational x; (2) f is increasing if $b > 1$, constant if $b = 1$, and decreasing if $0 < b < 1$. This function is called the exponential function b^x. For any $b > 0$, $f(x) = b^x$ is a continuous function. Also with $a,b > 0$ and x,y any real numbers, we have

$$(ab)^x = a^x b^x$$
$$b^x b^y = b^{x+y}$$
$$(b^x)^y = b^{xy}$$

The exponential function with base b can also be defined as the inverse of the logarithmic function. The most common exponential function in applications corresponds to choosing b the transcendental number e.

Logarithms $\log ab = \log a + \log b$, $a > 0$, $b > 0$
$$\log a^n = n \log a$$
$$\log (a/b) = \log a - \log b$$
$$\log \sqrt[n]{a} = (1/n) \log a$$

The common logarithm (base 10) is denoted $\log a$ or $\log_{10} a$. The natural logarithm (base e) is denoted $\ln a$ (or in some texts $\log_e a$).

Roots   If a is a real number, n is a positive integer, then x is called the nth root of a if $x^n = a$. The number of nth roots is n, but not all of them are necessarily real. The principal nth root means the following: (1) if $a > 0$ the principal nth root is the unique positive root, (2) if $a < 0$, and n odd, it is the unique negative root, and (3) if $a < 0$ and n even, it is any of the complex roots. In cases (1) and (2), the root can be found on a calculator by taking $y = \ln a/n$ and then $x = e^y$. In case (3), see the section on complex variables.

PROGRESSIONS

Arithmetic Progression

$$\sum_{k=0}^{n-1} (a + kd) = na + \frac{1}{2} n(n-1)d = \frac{n}{2} (a + \ell)$$

where ℓ is the last term, $\ell = a + (n-1)d$.

Geometric Progression

$$\sum_{k=1}^{n} ar^{k-1} = \frac{a(r^n - 1)}{r - 1} \qquad (r \neq 1)$$

Arithmetic-Geometric Progression

$$\sum_{k=0}^{n-1} (a + kd)r^k = \frac{a - [a + (n-1)d]r^n}{1 - r} + \frac{dr(1 - r^{n-1})}{(1 - r)^2} \qquad (r \neq 1)$$

$$\sum_{k=1}^{n} k^5 = \frac{1}{12} n^2 (n+1)^2 (2n^2 + 2n - 1)$$

$$\sum_{k=1}^{n} (2k - 1) = n^2$$

$$\sum_{k=1}^{n} (2k - 1)^2 = \frac{1}{3} n(4n^2 - 1)$$

$$\sum_{k=1}^{n} (2k - 1)^3 = n^2(2n^2 - 1)$$

$$\gamma = \lim_{n \to \infty} \left(\sum_{m=1}^{n} \frac{1}{m} - \ln n \right) = 0.577215$$

ALGEBRAIC INEQUALITIES

Arithmetic-Geometric Inequality   Let A_n and G_n denote respectively the arithmetic and the geometric means of a set of positive numbers $a_1, a_2, \ldots, a_n$. The $A_n \geq G_n$, i.e.,

$$\frac{a_1 + a_2 + \cdots + a_n}{n} \geq (a_1 a_2 \cdots a_n)^{1/n}$$

The equality holds only if all of the numbers a_i are equal.

Carleman's Inequality   The arithmetic and geometric means just defined satisfy the inequality

$$\sum_{r=1}^{n} G_r \leq neA_n$$

or, equivalently,

$$\sum_{r=1}^{n} (a_1 a_2 \cdots a_r)^{1/r} \leq neA_n$$

where e is the best possible constant in this inequality.

Cauchy-Schwarz Inequality   Let $a = (a_1, a_2, \ldots, a_n)$, $b = (b_1, b_2, \ldots, b_n)$, where the a_i's and b_i's are real or complex numbers. Then

$$\left| \sum_{k=1}^{n} a_k \overline{b}_k \right|^2 \leq \left(\sum_{k=1}^{n} |a_k|^2 \right) \left(\sum_{k=1}^{n} |b_k|^2 \right)$$

The equality holds if, and only if, the vectors a, b are linearly dependent (i.e., one vector is scalar times the other vector).

Minkowski's Inequality   Let $a_1, a_2, \ldots, a_n$ and $b_1, b_2, \ldots, b_n$ be any two sets of complex numbers. Then for any real number $p > 1$,

$$\left(\sum_{k=1}^{n} |a_k + b_k|^p \right)^{1/p} \leq \left(\sum_{k=1}^{n} |a_k|^p \right)^{1/p} + \left(\sum_{k=1}^{n} |b_k|^p \right)^{1/p}$$

Hölder's Inequality   Let $a_1, a_2, \ldots, a_n$ and $b_1, b_2, \ldots, b_n$ be any two sets of complex numbers, and let p and q be positive numbers with $1/p + 1/q = 1$. Then

$$\left| \sum_{k=1}^{n} a_k \overline{b}_k \right| \leq \left(\sum_{k=1}^{n} |a_k|^p \right)^{1/p} \left(\sum_{k=1}^{n} |b_k|^q \right)^{1/q}$$

The equality holds if, and only if, the sequences $|a_1|^p, |a_2|^p, \ldots, |a_n|^p$ and $|b_1|^q, |b_2|^q, \ldots, |b_n|^q$ are proportional and the argument (angle) of the complex numbers $a_k \overline{b}_k$ is independent of k. This last condition is of course automatically satisfied if $a_1, \ldots, a_n$ and $b_1, \ldots, b_n$ are positive numbers.

Lagrange's Inequality   Let $a_1, a_2, \ldots, a_n$ and $b_1, b_2, \ldots, b_n$ be real numbers. Then

$$\left(\sum_{k=1}^{n} a_k b_k \right)^2 = \left(\sum_{k=1}^{n} a_k^2 \right) \left(\sum_{k=1}^{n} b_k^2 \right) - \sum_{1 \leq k \leq j \leq n} (a_k b_j - a_j b_k)^2$$

Example   Two chemical engineers, John and Mary, purchase stock in the same company at times $t_1, t_2, \ldots, t_m$, when the price per share is respectively $p_1, p_2, \ldots, p_n$. Their methods of investment are different, however: John purchases x shares each time, whereas Mary invests P dollars each time (fractional shares can be purchased). Who is doing better?

While one can argue intuitively that the average cost per share for Mary does not exceed that for John, we illustrate a mathematical proof using inequalities. The average cost per share for John is equal to

$$\frac{\text{Total money invested}}{\text{Number of shares purchased}} = \frac{x \sum_{i=1}^{n} p_i}{nx} = \frac{1}{n} \sum_{i=1}^{n} p_i$$

The average cost per share for Mary is

$$\frac{nP}{\sum_{i=1}^{n} \dfrac{P}{p_i}} = \frac{n}{\sum_{i=1}^{n} \dfrac{1}{p_i}}$$

Thus the average cost per share for John is the arithmetic mean of $p_1, p_2, \ldots, p_n$, whereas that for Mary is the harmonic mean of these n numbers. Since the harmonic mean is less than or equal to the arithmetic mean for any set of positive numbers and the two means are equal only if $p_1 = p_2 = \cdots = p_n$, we conclude that the average cost per share for Mary is less than that for John if two of the prices p_i are distinct. One can also give a proof based on the Cauchy-Schwarz inequality. To this end, define the vectors

$$a = (p_1^{-1/2}, p_2^{-1/2}, \ldots, p_n^{-1/2}) \qquad b = (p_1^{1/2}, p_2^{1/2}, \ldots, p_n^{1/2})$$

Then $a \cdot b = 1 + \cdots + 1 = n$, and so by the Cauchy-Schwarz inequality

$$(a \cdot b)^2 = n^2 \leq \sum_{i=1}^{n} \frac{1}{p_i} \sum_{i=1}^{n} p_i$$

with the equality holding only if $p_1 = p_2 = \cdots = p_n$. Therefore

$$\frac{n}{\sum_{i=1}^{n} \dfrac{1}{p_i}} \leq \frac{\sum_{i=1}^{n} p_i}{n}$$

MENSURATION FORMULAS

Let A denote areas and V, volumes, in the following.

PLANE GEOMETRIC FIGURES WITH STRAIGHT BOUNDARIES

Triangles (see also "Plane Trigonometry") $A = \frac{1}{2}bh$ where $b =$ base, $h =$ altitude.

Rectangle $A = ab$ where a and b are the lengths of the sides.

Parallelogram (opposite sides parallel) $A = ah = ab \sin \alpha$ where a, b are the lengths of the sides, h the height, and α the angle between the sides. See Fig. 3-3.

Rhombus (equilateral parallelogram) $A = \frac{1}{2}ab$ where a, b are the lengths of the diagonals.

Trapezoid (four sides, two parallel) $A = \frac{1}{2}(a + b)h$ where the lengths of the parallel sides are a and b, and $h =$ height.

Quadrilateral (four-sided) $A = \frac{1}{2}ab \sin \theta$ where a, b are the lengths of the diagonals and the acute angle between them is θ.

Regular Polygon of n Sides   See Fig. 3-4.

$$A = \frac{1}{4} nl^2 \cot \frac{180°}{n} \quad \text{where } l = \text{length of each side}$$

$$R = \frac{l}{2} \csc \frac{180°}{n} \quad \text{where } R \text{ is the radius of the circumscribed circle}$$

$$r = \frac{l}{2} \cot \frac{180°}{n} \quad \text{where } r \text{ is the radius of the inscribed circle}$$

$$\beta = \frac{360°}{n}$$

$$\theta = \frac{(n-2)180°}{n}$$

$$l = 2r \tan \frac{\beta}{2} = 2R \sin \frac{\beta}{2}$$

Inscribed and Circumscribed Circles with Regular Polygon of n Sides   Let $l =$ length of one side.

Figure	n	Area	Radius of circumscribed circle	Radius of inscribed circle
Equilateral triangle	3	$0.4330\, l^2$	$0.5774\, l$	$0.2887\, l$
Square	4	$1.0000\, l^2$	$0.7071\, l$	$0.5000\, l$
Pentagon	5	$1.7205\, l^2$	$0.8507\, l$	$0.6882\, l$
Hexagon	6	$2.5981\, l^2$	$1.0000\, l$	$0.8660\, l$
Octagon	8	$4.8284\, l^2$	$1.3065\, l$	$1.2071\, l$
Decagon	10	$7.6942\, l^2$	$1.6180\, l$	$1.5388\, l$
Dodecagon	12	$11.1962\, l^2$	$1.8660\, l$	$1.9318\, l$

Radius r of Circle Inscribed in Triangle with Sides a, b, c

$$r = \sqrt{\frac{(s-a)(s-b)(s-c)}{s}} \quad \text{where } s = \frac{1}{2}(a + b + c)$$

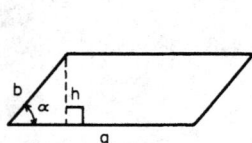

FIG. 3-3   Parallelogram.

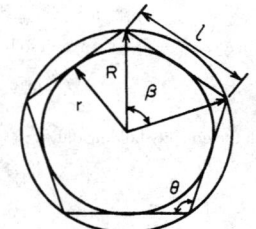

FIG. 3-4   Regular polygon.

Radius R of Circumscribed Circle

$$R = \frac{abc}{4\sqrt{s(s-a)(s-b)(s-c)}}$$

Area of Regular Polygon of n Sides Inscribed in a Circle of Radius r

$$A = (nr^2/2) \sin (360°/n)$$

Perimeter of Inscribed Regular Polygon

$$P = 2nr \sin (180°/n)$$

Area of Regular Polygon Circumscribed about a Circle of Radius r

$$A = nr^2 \tan (180°/n)$$

Perimeter of Circumscribed Regular Polygon

$$P = 2nr \tan \frac{180°}{n}$$

PLANE GEOMETRIC FIGURES WITH CURVED BOUNDARIES

Circle (Fig. 3-5)   Let
$C =$ circumference
$r =$ radius
$D =$ diameter
$A =$ area
$S =$ arc length subtended by θ
$l =$ chord length subtended by θ
$H =$ maximum rise of arc above chord, $r - H = d$
$\theta =$ central angle (rad) subtended by arc S
$C = 2\pi r = \pi D \quad (\pi = 3.14159 \ldots)$
$S = r\theta = \frac{1}{2}D\theta$
$l = 2\sqrt{r^2 - d^2} = 2r \sin (\theta/2) = 2d \tan (\theta/2)$

$$d = \frac{1}{2} \sqrt{4r^2 - l^2} = \frac{1}{2} l \cot \frac{\theta}{2}$$

$$\theta = \frac{S}{r} = 2 \cos^{-1} \frac{d}{r} = 2 \sin^{-1} \frac{l}{D}$$

$$A \text{ (circle)} = \pi r^2 = \frac{1}{4}\pi D^2$$

$$A \text{ (sector)} = \frac{1}{2}rS = \frac{1}{2}r^2\theta$$

$$A \text{ (segment)} = A \text{ (sector)} - A \text{ (triangle)} = \frac{1}{2}r^2(\theta - \sin \theta)$$

$$= r^2 \cos^{-1} \frac{r - H}{r} - (r - H) \sqrt{2rH - H^2}$$

Ring (area between two circles of radii r_1 and r_2)   The circles need not be concentric, but one of the circles must enclose the other.

$$A = \pi(r_1 + r_2)(r_1 - r_2) \qquad r_1 > r_2$$

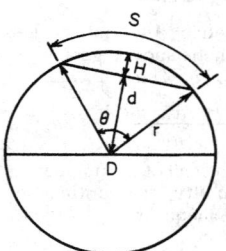

FIG. 3-5   Circle.

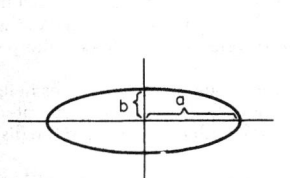

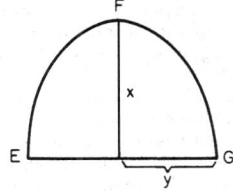

FIG. 3-6 Ellipse.   **FIG. 3-7** Parabola.

Ellipse (Fig. 3-6)   Let the semiaxes of the ellipse be a and b

$$A = \pi ab$$
$$C = 4aE(k)$$

where $e^2 = 1 - b^2/a^2$ and $E(e)$ is the complete elliptic integral of the second kind,

$$E(e) = \frac{\pi}{2}\left[1 - \left(\frac{1}{2}\right)^2 e^2 + \cdots\right]$$

[an approximation for the circumference $C = 2\pi\sqrt{(a^2 + b^2)/2}$].

Parabola   (Fig. 3-7)

$$\text{Length of arc } EFG = \sqrt{4x^2 + y^2} + \frac{y^2}{2x}\ln\frac{2x + \sqrt{4x^2 + y^2}}{y}$$

$$\text{Area of section } EFG = \frac{4}{3}xy$$

Catenary (the curve formed by a cord of uniform weight suspended freely between two points A, B; Fig. 3-8)

$$y = a\cosh(x/a)$$

Length of arc between points A and B is equal to $2a\sinh(L/a)$. Sag of the cord is $D = a\cosh(L/a) - 1$.

SOLID GEOMETRIC FIGURES WITH PLANE BOUNDARIES

Cube   Volume $= a^3$; total surface area $= 6a^2$; diagonal $= a\sqrt{3}$, where $a =$ length of one side of the cube.

Rectangular Parallelepiped   Volume $= abc$; surface area $= 2(ab + ac + bc)$; diagonal $= \sqrt{a^2 + b^2 + c^2}$, where a, b, c are the lengths of the sides.

Prism   Volume $=$ (area of base) $\times$ (altitude); lateral surface area $=$ (perimeter of right section) $\times$ (lateral edge).

Pyramid   Volume $= \frac{1}{3}$ (area of base) $\times$ (altitude); lateral area of regular pyramid $= \frac{1}{2}$ (perimeter of base) $\times$ (slant height) $= \frac{1}{2}$ (number of sides) (length of one side) (slant height).

Frustum of Pyramid   (formed from the pyramid by cutting off the top with a plane)

$$V = \frac{1}{3}(A_1 + A_2 + \sqrt{A_1 \cdot A_2})h$$

where $h =$ altitude and A_1, A_2 are the areas of the base; lateral area of a regular figure $= \frac{1}{2}$ (sum of the perimeters of base) $\times$ (slant height).

Volume and Surface Area of Regular Polyhedra with Edge l

Type of surface	Name	Volume	Surface area
4 equilateral triangles	Tetrahedron	$0.1179\,l^3$	$1.7321\,l^2$
6 squares	Hexahedron (cube)	$1.0000\,l^3$	$6.0000\,l^2$
8 equilateral triangles	Octahedron	$0.4714\,l^3$	$3.4641\,l^2$
12 pentagons	Dodecahedron	$7.6631\,l^3$	$20.6458\,l^2$
20 equilateral triangles	Icosahedron	$2.1817\,l^3$	$8.6603\,l^2$

SOLIDS BOUNDED BY CURVED SURFACES

Cylinders (Fig. 3-9) $V =$ (area of base) $\times$ (altitude); lateral surface area $=$ (perimeter of right section) $\times$ (lateral edge).

Right Circular Cylinder $V = \pi$ (radius)$^2 \times$ (altitude); lateral surface area $= 2\pi$ (radius) $\times$ (altitude).

Truncated Right Circular Cylinder

$$V = \pi r^2 h; \text{ lateral area} = 2\pi rh$$
$$h = \frac{1}{2}(h_1 + h_2)$$

Hollow Cylinders   Volume $= \pi h(R^2 - r^2)$, where r and R are the internal and external radii and h is the height of the cylinder.

Sphere   (Fig. 3-10)

$$V\text{ (sphere)} = \frac{4}{3}\pi R^3, \frac{1}{6}\pi D^3$$

$$V\text{ (spherical sector)} = \frac{2}{3}\pi R^2 h = \frac{1}{6}\pi h_1(3r_2^2 + h_1^2)$$

$$V\text{ (spherical segment of one base)} = \frac{1}{6}\pi h_1(3r_2^2 + h_1^2)$$

$$V\text{ (spherical segment of two bases)} = \frac{1}{6}\pi h_2(3r_1^2 + 3r_2^2 + h_2^2)$$

$$A\text{ (sphere)} = 4\pi R^2 = \pi D^2$$

$$A\text{ (zone)} = 2\pi Rh = \pi Dh$$

A (lune on the surface included between two great circles, the inclination of which is θ radians) $= 2R^2\theta$.

Cone $V = \frac{1}{3}$ (area of base) $\times$ (altitude).

Right Circular Cone $V = (\pi/3)\,r^2 h$, where h is the altitude and r is the radius of the base; curved surface area $= \pi r\sqrt{r^2 + h^2}$, curved surface of the frustum of a right cone $= \pi(r_1 + r_2)\sqrt{h^2 + (r_1 - r_2)^2}$, where r_1, r_2 are the radii of the base and top, respectively, and h is the altitude; volume of the frustum of a right cone $= \pi(h/3)(r_1^2 + r_1 r_2 + r_2^2) = h/3(A_1 + A_2 + \sqrt{A_1 A_2})$, where $A_1 =$ area of base and $A_2 =$ area of top.

Ellipsoid $V = (\frac{4}{3})\pi abc$, where a, b, c are the lengths of the semiaxes.

Torus   (obtained by rotating a circle of radius r about a line whose distance is $R > r$ from the center of the circle)

$$V = 2\pi^2 Rr^2 \qquad \text{Surface area} = 4\pi^2 Rr$$

Prolate Spheroid   (formed by rotating an ellipse about its major axis [$2a$])

$$\text{Surface area} = 2\pi b^2 + 2\pi(ab/e)\sin^{-1} e \qquad V = \frac{4}{3}\pi ab^2$$

where a, b are the major and minor axes and $e =$ eccentricity ($e < 1$).

Oblate Spheroid (formed by the rotation of an ellipse about its minor axis [$2b$])   Data as given previously.

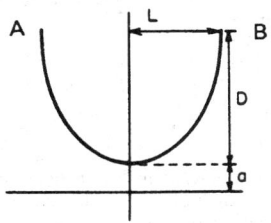

FIG. 3-8 Catenary.

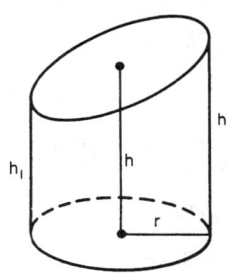

FIG. 3-9 Cylinder.

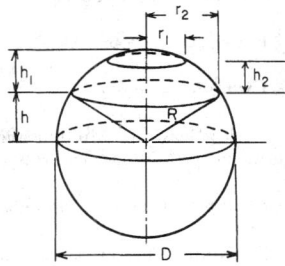

FIG. 3-10 Sphere.

$$\text{Surface area} = 2\pi a^2 + \pi\,\frac{b^2}{e}\ln\frac{1+e}{1-e} \qquad V = \tfrac{4}{3}\pi a^2 b$$

MISCELLANEOUS FORMULAS

See also "Differential and Integral Calculus."
Volume of a Solid Revolution   (the solid generated by rotating a plane area about the x axis)

$$V = \pi \int_a^b [f(x)]^2\,dx$$

where $y = f(x)$ is the equation of the plane curve and $a \le x \le b$.
Area of a Surface of Revolution

$$S = 2\pi \int_a^b y\,ds$$

where $ds = \sqrt{1 + (dy/dx)^2}\,dx$ and $y = f(x)$ is the equation of the plane curve rotated about the x axis to generate the surface.
Area Bounded by $f(x)$, the x Axis, and the Lines $x = a$, $x = b$

$$A = \int_a^b f(x)\,dx \qquad [f(x) \ge 0]$$

Length of Arc of a Plane Curve
If $y = f(x)$,

$$\text{Length of arc } s = \int_a^b \sqrt{1 + \left(\frac{dy}{dx}\right)^2}\,dx$$

If $x = g(y)$,

$$\text{Length of arc } s = \int_c^d \sqrt{1 + \left(\frac{dx}{dy}\right)^2}\,dy$$

If $x = f(t)$, $y = g(t)$,

$$\text{Length of arc } s = \int_{t_0}^{t_1} \sqrt{\left(\frac{dx}{dt}\right)^2 + \left(\frac{dy}{dt}\right)^2}\,dt$$

In general, $(ds)^2 = (dx)^2 + (dy)^2$.
Theorems of Pappus (for volumes and areas of surfaces of revolution)

1.   If a plane area is revolved about a line which lies in its plane but does not intersect the area, then the volume generated is equal to the product of the area and the distance traveled by the area's center of gravity.
2.   If an arc of a plane curve is revolved about a line that lies in its plane but does not intersect the arc, then the surface area generated by the arc is equal to the product of the length of the arc and the distance traveled by its center of gravity.

These theorems are useful for determining volumes V and surface areas S of solids of revolution if the centers of gravity are known. If S and V are known, the centers of gravity may be determined.

IRREGULAR AREAS AND VOLUMES

Irregular Areas   Let $y_0, y_1, \ldots, y_n$ be the lengths of a series of equally spaced parallel chords and h be their distance apart. The area of the figure is given approximately by any of the following:

$$A_T = (h/2)[(y_0 + y_n) + 2(y_1 + y_2 + \cdots + y_{n-1})] \qquad \text{(trapezoidal rule)}$$

$$A_s = (h/3)[(y_0 + y_n) + 4(y_1 + y_3 + y_5 + \cdots + y_{n-1})$$
$$+ 2(y_2 + y_4 + \cdots + y_{n-2})] \quad (n \text{ even, Simpson's rule})$$

The greater the value of n, the greater the accuracy of approximation.
Irregular Volumes   To find the volume, replace the y's by cross-sectional areas A_j and use the results in the preceding equations.

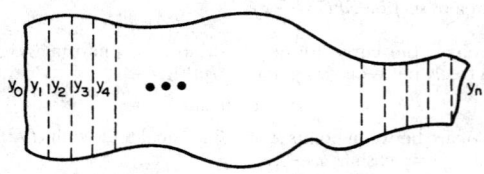

FIG. 3-11   Irregular area.

ELEMENTARY ALGEBRA

REFERENCES: 20, 102, 108, 191, 269, 277.

OPERATIONS ON ALGEBRAIC EXPRESSIONS

An algebraic expression will here be denoted as a combination of letters and numbers such as

$$3ax - 3xy + 7x^2 + 7x^{3/2} - 2.8xy$$

Addition and Subtraction   Only like terms can be added or subtracted in two algebraic expressions.

Example $(3x + 4xy - x^2) + (3x^2 + 2x - 8xy) = 5x - 4xy + 2x^2$.

Example $(2^x + 3xy - 4x^{1/2}) + (3^x + 6x - 8xy) = 2^x + 3^x + 6x - 5xy - 4x^{1/2}$.

Multiplication   Multiplication of algebraic expressions is term by term, and corresponding terms are combined.

Example $(2x + 3y - 2xy)(3 + 3y) = 6x + 9y + 9y^2 - 6xy^2$.

Division   This operation is analogous to that in arithmetic.

Example   Divide $3e^{2x} + e^x + 1$ by $e^x + 1$.

$$
\begin{array}{r}
\text{Dividend} \\
\text{Divisor } e^x + 1\,\big|\,3e^{2x} + e^x + 1\;\underline{3e^x - 2}\text{ quotient} \\
\underline{3e^{2x} + 3e^x} \\
-2e^x + 1 \\
\underline{-2e^x - 2} \\
+ 3 \text{ (remainder)}
\end{array}
$$

Therefore, $3e^{2x} + e^x + 1 = (e^x + 1)(3e^x - 2) + 3$.

Operations with Zero   All numerical computations (except division) can be done with zero: $a + 0 = 0 + a = a$; $a - 0 = a$; $0 - a = -a$; $(a)(0) = 0$; $a^0 = 1$ if $a \ne 0$; $0/a = 0$, $a \ne 0$. $a/0$ and $0/0$ have no meaning.
Fractional Operations

$$-\frac{x}{y} = -\left(\frac{-x}{-y}\right) = \frac{x}{-y} = \frac{-x}{y}; \quad \frac{x}{y} = \frac{-x}{-y}; \quad \frac{x}{y} = \frac{ax}{ay}, \text{ if } a \ne 0.$$

$$\frac{x}{y} \pm \frac{z}{y} = \frac{x \pm z}{y}; \quad \left(\frac{x}{y}\right)\left(\frac{z}{t}\right) = \frac{xz}{yt}; \quad \frac{x/y}{z/t} = \left(\frac{x}{y}\right)\left(\frac{t}{z}\right) = \frac{xt}{yz}$$

Factoring   That process of analysis consisting of reducing a given expression into the product of two or more simpler expressions called *factors*. Some of the more common expressions are factored here:

(1) $(x^2 - y^2) = (x - y)(x + y)$

(2) $x^2 + 2xy + y^2 = (x + y)^2$

(3) $x^2 + ax + b = (x + c)(x + d)$ where $c + d = a$, $cd = b$

(4) $by^2 + cy + d = (ey + f)(gy + h)$ where $eg = b$, $fg + eh = c$, $fh = d$

(5) $x^2 + y^2 + z^2 + 2yz + 2xz + 2xy = (x + y + z)^2$

(6) $x^2 - y^2 - z^2 - 2yz = (x - y - z)(x + y + z)$

(7) $x^2 + y^2 + z^2 - 2xy - 2xz + 2yz = (x - y - z)^2$

(8) $x^3 - y^3 = (x - y)(x^2 + xy + y^2)$

(9) $(x^3 + y^3) = (x + y)(x^2 - xy + y^2)$

(10) $(x^4 - y^4) = (x - y)(x + y)(x^2 + y^2)$

(11) $x^5 + y^5 = (x + y)(x^4 - x^3y + x^2y^2 - xy^3 + y^4)$

(12) $x^n - y^n = (x - y)(x^{n-1} + x^{n-2}y + x^{n-3}y^2 + \cdots + y^{n-1})$

Laws of Exponents

$(a^n)^m = a^{nm}$; $a^{n+m} = a^n \cdot a^m$; $a^{n/m} = (a^n)^{1/m}$; $a^{n-m} = a^n/a^m$; $a^{1/m} = \sqrt[m]{a}$; $a^{1/2} = \sqrt{a}$; $\sqrt{x^2} = |x|$ (absolute value of x). For $x > 0$, $y > 0$, $\sqrt{xy} = \sqrt{x}\sqrt{y}$; for $x > 0$ $\sqrt{x^m} = x^{m/n}$; $\sqrt[n]{1/x} = 1/\sqrt{x}$

THE BINOMIAL THEOREM

If n is a positive integer,

$$(a + b)^n = a^n + na^{n-1}b + \frac{n(n-1)}{2!} a^{n-2}b^2$$

$$+ \frac{n(n-1)(n-2)}{3!} a^{n-3}b^3 + \cdots + b^n = \sum_{j=0}^{n} \binom{n}{j} a^{n-j}b^j$$

where $\binom{n}{j} = \frac{n!}{j!(n-j)!}$ = number of combinations of n things taken j at a time. $n! = 1 \cdot 2 \cdot 3 \cdot 4 \cdots n$, $0! = 1$.

Example Find the sixth term of $(x + 2y)^{12}$. The sixth term is obtained by setting $j = 5$. It is

$$\binom{12}{5} x^{12-5}(2y)^5 = 792x^7(2y)^5$$

Example $\sum_{j=0}^{14} \binom{n}{j} = (1 + 1)^{14} = 2^{14}$.

If n is not a positive integer, the sum formula no longer applies and an infinite series results for $(a + b)^n$. The coefficients are obtained from the first formulas in this case.

Example $(1 + x)^{1/2} = 1 + \frac{1}{2}x - \frac{1}{2} \cdot \frac{1}{4}x^2 + \frac{1}{2} \cdot \frac{1}{4} \cdot \frac{3}{6}x^3 \cdots$ (convergent for $x^2 < 1$).

Additional discussion is under "Infinite Series."

PROGRESSIONS

An arithmetic progression is a succession of terms such that each term, except the first, is derivable from the preceding by the addition of a quantity d called the common difference. All arithmetic progressions have the form a, $a + d$, $a + 2d$, $a + 3d$, ... With a = first term, l = last term, d = common difference, n = number of terms, and s = sum of the terms, the following relations hold:

$$l = a + (n - 1)d = -\frac{d}{2} + \sqrt{2ds + \left(a - \frac{d}{2}\right)^2}$$

$$= \frac{s}{n} + \frac{(n-1)}{2} d$$

$$s = \frac{n}{2} [2a + (n-1)d] = \frac{n}{2} (a + l) = \frac{n}{2} [2l - (n-1)d]$$

$$a = l - (n-1)d = \frac{s}{n} - \frac{(n-1)d}{2} = \frac{2s}{n} - l$$

$$d = \frac{l - a}{n - 1} = \frac{2(s - an)}{n(n-1)} = \frac{2(nl - s)}{n(n-1)}$$

$$n = \frac{l - a}{d} + 1 = \frac{2s}{l + a} = \frac{2l + d + \sqrt{(2l + d)^2 - 8ds}}{2d}$$

The **arithmetic mean or average** of two numbers a, b is $(a + b)/2$; of n numbers $a_1, \ldots, a_n$ is $(a_1 + a_2 + \cdots + a_n)/n$.

A **geometric progression** is a succession of terms such that each term, except the first, is derivable from the preceding by the multiplication of a quantity r called the common ratio. All such progressions have the form a, ar, ar^2, ..., ar^{n-1}. With a = first term, l = last term, r = ratio, n = number of terms, s = sum of the terms, the following relations hold:

$$l = ar^{n-1} = \frac{[a + (r - 1)s]}{r} = \frac{(r - 1)sr^{n-1}}{r^n - 1}$$

$$s = \frac{a(r^n - 1)}{r - 1} = \frac{a(1 - r^n)}{1 - r} = \frac{rl - a}{r - 1} = \frac{lr^n - l}{r^n - r^{n-1}}$$

$$a = \frac{l}{r^{n-1}} = \frac{(r - 1)s}{r^n - 1} \quad r = \frac{s - a}{s - l} \quad \log r = \frac{\log l - \log a}{n - 1}$$

$$n = \frac{\log l - \log a}{\log r} + 1 = \frac{\log[a + (r - 1)s] - \log a}{\log r}$$

The geometric mean of two nonnegative numbers a, b is $\sqrt{ab}$; of n numbers is $(a_1 a_2 \ldots a_n)^{1/n}$.

Example Find the sum of $1 + \frac{1}{2} + \frac{1}{4} + \cdots + \frac{1}{64}$. Here $a = 1$, $r = \frac{1}{2}$, $n = 7$. Thus

$$s = \frac{\frac{1}{2}(\frac{1}{64}) - 1}{\frac{1}{2} - 1} = 127/64$$

$$s = a + ar + ar^2 + \cdots + ar^{n-1} = \frac{a}{1 - r} - \frac{ar^n}{1 - r}$$

If $|r| < 1$, then $\lim_{n \to \infty} s = \frac{a}{1 - r}$

which is called the sum of the infinite geometric progression.

Example The present worth (PW) of a series of cash flows C_k at the end of year k is

$$\text{PW} = \sum_{k=1}^{n} \frac{C_k}{(1 + i)^k}$$

where i is an assumed interest rate. (Thus the present worth always requires specification of an interest rate.) If all the payments are the same, $C_k = R$, the present worth is

$$\text{PW} = R \sum_{k=1}^{n} \frac{1}{(1 + i)^k}$$

This can be rewritten as

$$\text{PW} = \frac{R}{1 + i} \sum_{k=1}^{n} \frac{1}{(1 + i)^{k-1}} = \frac{R}{1 + i} \sum_{j=0}^{n-1} \frac{1}{(1 + i)^j}$$

This is a geometric series with $r = 1/(1 + i)$ and $a = R/(1 + i)$. The formulas above give

$$\text{PW} (=s) = \frac{R}{i} \frac{(1 + i)^n - 1}{(1 + i)^n}$$

The same formula applies to the value of an annuity (PW) now, to provide for equal payments R at the end of each of n years, with interest rate i.

A progression of the form a, $(a + d)r$, $(a + 2d)r^2$, $(a + 3d)r^3$, etc., is a combined arithmetic and geometric progression. The sum of n such terms is

$$s = \frac{a - [a + (n - 1)d]r^n}{1 - r} + \frac{rd(1 - r^{n-1})}{(1 - r)^2}$$

If $|r| < 1$, $\lim_{n \to \infty} s = \frac{a}{1 - r} + rd/(1 - r)^2$.

The non-zero numbers a, b, c, etc., form a harmonic progression if their reciprocals $1/a$, $1/b$, $1/c$, etc., form an arithmetic progression.

Example   The progression 1, $\frac{1}{3}$, $\frac{1}{5}$, $\frac{1}{7}$, . . . , $\frac{1}{31}$ is harmonic since 1, 3, 5, 7, . . . , 31 form an arithmetic progression.

The **harmonic mean** of two numbers a, b is $2ab/(a + b)$.

PERMUTATIONS, COMBINATIONS, AND PROBABILITY

Each separate arrangement of all or a part of a set of things is called a **permutation.** The number of permutations of n things taken r at a time, written

$$P(n, r) = \frac{n!}{(n-r)!} = n(n-1)(n-2) \cdots (n-r+1)$$

Example   The permutations of a, b, c two at a time are ab, ac, ba, ca, cb, and bc. The formula is $P(3,2) = 3!/1! = 6$. The permutations of a, b, c three at a time are abc, bac, cab, acb, bca, and cba.

Each separate selection of objects that is possible irrespective of the order in which they are arranged is called a combination. The number of combinations of n things taken r at a time, written $C(n, r) = n!/[r!(n-r)!]$.

Example   The combinations of a, b, c taken 2 at a time are ab, ac, bc; taken 3 at a time is abc.

An important relation is $r! \, C(n, r) = P(n, r)$.

If an event can occur in p ways and fail to occur in q ways, all ways being equally likely, the *probability* of its occurrence is $p/(p + q)$, and that of its failure $q/(p + q)$.

Example   Two dice may be thrown in 36 separate ways. What is the probability of throwing such that their sum is 7? Seven may arise in 6 ways: 1 and 6, 2 and 5, 3 and 4, 4 and 3, 5 and 2, 6 and 1. The probability of shooting 7 is $\frac{1}{6}$.

THEORY OF EQUATIONS

Linear Equations   A linear equation is one of the first degree (i.e., only the first powers of the variables are involved), and the process of obtaining definite values for the unknown is called solving the equation. Every linear equation in one variable is written $Ax + B = 0$ or $x = -B/A$. Linear equations in n variables have the form

$$a_{11}x_1 + a_{12}x_2 + \cdots + a_{1n}x_n = b_1$$
$$a_{21}x_1 + a_{22}x_2 + \cdots + a_{2n}x_n = b_2$$
$$\vdots$$
$$a_{m1}x_1 + a_{m2}x_2 + \cdots + a_{mn}x_n = b_m$$

The solution of the system may then be found by elimination or matrix methods if a solution exists (see "Matrix Algebra and Matrix Computations").

Quadratic Equations   Every quadratic equation in one variable is expressible in the form $ax^2 + bx + c = 0$. $a \neq 0$. This equation has two solutions, say, x_1, x_2, given by

$$\left.\begin{array}{c} x_1 \\ x_2 \end{array}\right\} = \frac{-b \pm \sqrt{b^2 - 4ac}}{2a}$$

If a, b, c are real, the discriminant $b^2 - 4ac$ gives the character of the roots. If $b^2 - 4ac > 0$, the roots are real and unequal. If $b^2 - 4ac < 0$, the roots are complex conjugates. If $b^2 - 4ac = 0$ the roots are **real** and **equal.**

Two quadratic equations in two variables can in general be solved only by numerical methods (see "Numerical Analysis and Approximate Methods"). If one equation is of the first degree, the other of the second degree, a solution may be obtained by solving the first for one unknown. This result is substituted in the second equation and the resulting quadratic equation solved.

Cubic Equations   A cubic equation, in one variable, has the form $x^3 + bx^2 + cx + d = 0$. Every cubic equation having complex coefficients

has three complex roots. If the coefficients are real numbers, then at least one of the roots must be real. The cubic equation $x^3 + bx^2 + cx + d = 0$ may be reduced by the substitution $x = y - (b/3)$ to the form $y^3 + py + q = 0$, where $p = \frac{1}{3}(3c - b^2)$, $q = \frac{1}{27}(27d - 9bc + 2b^3)$. This equation has the solutions $y_1 = A + B$, $y_2 = -\frac{1}{2}(A + B) + (i\sqrt{3}/2)(A - B)$, $y_3 = -\frac{1}{2}(A + B) - (i\sqrt{3}/2)(A - B)$, where $i^2 = -1$, $A = \sqrt[3]{-q/2 + \sqrt{R}}$, $B = \sqrt[3]{-q/2 - \sqrt{R}}$, and $R = (p/3)^3 + (q/2)^2$. If b, c, d are all real and if $R > 0$, there are one real root and two conjugate complex roots; if $R = 0$, there are three real roots, of which at least two are equal; if $R < 0$, there are three real unequal roots. If $R < 0$, these formulas are impractical. In this case, the roots are given by $y_k = \mp 2 \sqrt{-p/3} \cos [(\phi/3) + 120k]$, $k = 0, 1, 2$ where

$$\phi = \cos^{-1} \sqrt{\frac{q^2/4}{-p^3/27}}$$

and the upper sign applies if $q > 0$, the lower if $q < 0$.

Example $x^3 + 3x^2 + 9x + 9 = 0$ reduces to $y^3 + 6y + 2 = 0$ under $x = y - 1$. Here $p = 6$, $q = 2$, $R = 9$. Hence $A = \sqrt[3]{2}$, $B = \sqrt[3]{-4}$. The desired roots in y are $\sqrt[3]{2} - \sqrt[3]{-4}$ and $-\frac{1}{2}(\sqrt[3]{2} - \sqrt[3]{4}) \pm (i\sqrt{3}/2)(\sqrt[3]{2} + \sqrt[3]{4})$. The roots in x are $x = y - 1$.

Example $y^3 - 7y + 7 = 0$. $p = -7$, $q = 7$, $R < 0$. Hence

$$x_k = -\sqrt{\frac{28}{3}} \cos\left(\frac{\phi}{3} + 120k\right)$$

where

$$\phi = \sqrt{\frac{27}{28}}, \frac{\phi}{3} = 3°37'52''.$$

The roots are approximately -3.048916, 1.692020, and 1.356897.

Example   Many equations of state involve solving cubic equations for the compressibility factor Z. For example, the Redlich-Kwong-Soave equation of state requires solving

$$Z^3 - Z^2 + cZ + d = 0, \quad d < 0$$

where c and d depend on critical constants of the chemical species. In this case, only positive solutions, $Z > 0$, are desired.

Quartic Equations   See Ref. 118.

General Polynomials of the nth Degree   Denote the general polynomial equation of degree n by

$$P(x) = a_0 x^n + a_1 x^{n-1} + \cdots + a_{n-1}x + a_n = 0$$

If $n > 4$, there is no formula which gives the roots of the general equation. For fourth and higher order (even third order), the roots can be found numerically (see "Numerical Analysis and Approximate Methods"). However, there are some general theorems that may prove useful.

Remainder Theorems   When $P(x)$ is a polynomial and $P(x)$ is divided by $x - a$ until a remainder independent of x is obtained, this remainder is equal to $P(a)$.

Example $P(x) = 2x^4 - 3x^2 + 7x - 2$ when divided by $x + 1$ (here $a = -1$) results in $P(x) = (x + 1)(2x^3 - 2x^2 - x + 8) - 10$ where -10 is the remainder. It is easy to see that $P(-1) = -10$.

Factor Theorem   If $P(a)$ is zero, the polynomial $P(x)$ has the factor $x - a$. In other words, if a is a root of $P(x) = 0$, then $x - a$ is a factor of $P(x)$.

If a number a is found to be a root of $P(x) = 0$, the division of $P(x)$ by $(x - a)$ leaves a polynomial of degree one less than that of the original equation, i.e., $P(x) = Q(x)(x - a)$. Roots of $Q(x) = 0$ are clearly roots of $P(x) = 0$.

Example $P(x) = x^3 - 6x^2 + 11x - 6 = 0$ has the root $+3$. Then $P(x) = (x - 3)(x^2 - 3x + 2)$. The roots of $x^2 - 3x + 2 = 0$ are 1 and 2. The roots of $P(x)$ are therefore 1, 2, 3.

Fundamental Theorem of Algebra   Every polynomial of degree n has exactly n real or complex roots, counting multiplicities.

Every polynomial equation $a_0 x^n + a_1 x^{n-1} + \cdots + a_n = 0$ with *rational coefficients* may be rewritten as a polynomial, of the same degree, with *integral coefficients* by multiplying each coefficient by the least common multiple of the denominators of the coefficients.

Example   The coefficients of $\frac{3}{2}x^4 + \frac{7}{3}x^3 - \frac{5}{6}x^2 + 2x - \frac{1}{6} = 0$ are rational numbers. The least common multiple of the denominators is $2 \times 3 = 6$. Therefore, the equation is equivalent to $9x^4 + 14x^3 - 5x^2 + 12x - 1 = 0$.

Upper Bound for the Real Roots   Any number that exceeds all the roots is called an upper bound to the real roots. If the coefficients of a polynomial equation are all of like sign, there is no positive root. Such equations are excluded here since zero is the upper bound to the real roots. If the coefficient of the highest power of $P(x) = 0$ is negative, replace the equation by $-P(x) = 0$.

If in a polynomial $P(x) = c_0 x^n + c_1 x^{n-1} + \cdots + c_{n-1}x + c_n = 0$, with $c_0 > 0$, the first negative coefficient is preceded by k coefficients which are positive or zero, and if G denotes the greatest of the numerical values of the negative coefficients, then each real root is less than $1 + \sqrt[k]{G/c_0}$.

A lower bound to the negative roots of $P(x) = 0$ may be found by applying the rule to $P(-x) = 0$.

Example $P(x) = x^7 + 2x^5 + 4x^4 - 8x^2 - 32 = 0$. Here $k = 5$ (since 2 coefficients are zero), $G = 32$, $c_0 = 1$. The upper bound is $1 + \sqrt[5]{32} = 3$. $P(-x) = -x^7 - 2x^5 + 4x^4 - 8x^2 - 32 = 0$. $-P(-x) = x^7 + 2x^5 - 4x^4 + 8x^2 + 32 = 0$. Here $k = 3$, $G = 4$, $c_0 = 1$. The lower bound is $-(1 + \sqrt[3]{4}) \approx -2.587$. Thus all real roots r lie in the range $-2.587 < r < 3$.

Descartes Rule of Signs   The number of positive real roots of a polynomial equation with real coefficients either is equal to the number v of its variations in sign or is less than v by a positive even integer. The number of negative roots of $P(x) = 0$ either is equal to the number of variations of sign of $P(-x)$ or is less than that number by a positive even integer.

Example $P(x) = x^4 + 3x^3 + x - 1 = 0$. $v = 1$; so $P(x)$ has one positive root. $P(-x) = x^4 - 3x^3 - x - 1$. Here $v = 1$; so $P(x)$ has one negative root. The other two roots are complex conjugates.

Example $P(x) = x^4 - x^2 + 10x - 4 = 0$. $v = 3$; so $P(x)$ has three or one positive roots. $P(-x) = x^4 - x^2 - 10x - 4$. $v = 1$; so $P(x)$ has exactly one negative root.

Numerical methods are often used to find the roots of polynomials. A detailed discussion of these techniques is given under "Numerical Analysis and Approximate Methods."

Determinants   Consider the system of two linear equations

$$a_{11}x_1 + a_{12}x_2 = b_1$$
$$a_{21}x_1 + a_{22}x_2 = b_2$$

If the first equation is multiplied by a_{22} and the second by $-a_{12}$ and the results added, we obtain

$$(a_{11}a_{22} - a_{21}a_{12})x_1 = b_1 a_{22} - b_2 a_{12}$$

The expression $a_{11}a_{22} - a_{21}a_{12}$ may be represented by the symbol

$$\begin{vmatrix} a_{11} & a_{12} \\ a_{21} & a_{22} \end{vmatrix} = a_{11}a_{22} - a_{21}a_{12}$$

This symbol is called a determinant of second order. The value of the square array of n^2 quantities a_{ij}, where $i = 1, \ldots, n$ is the row index, $j = 1, \ldots, n$ the column index, written in the form

$$|A| = \begin{vmatrix} a_{11} & a_{12} & a_{13} \cdots a_{1n} \\ a_{21} & a_{22} & \ldots a_{2n} \\ \vdots & & \\ a_{n1} & a_{n2} & a_{n3} \cdots a_{nn} \end{vmatrix}$$

is called a determinant. The n^2 quantities a_{ij} are called the elements of the determinant. In the determinant $|A|$ let the ith row and jth column be deleted and a new determinant be formed having $n - 1$ rows and columns. This new determinant is called the minor of a_{ij} denoted M_{ij}.

Example $\begin{vmatrix} a_{11} & a_{12} & a_{13} \\ a_{21} & a_{22} & a_{23} \\ a_{31} & a_{32} & a_{33} \end{vmatrix}$ The minor of a_{23} is $M_{23} = \begin{vmatrix} a_{11} & a_{12} \\ a_{31} & a_{32} \end{vmatrix}$

The cofactor A_{ij} of the element a_{ij} is the signed minor of a_{ij} determined by the rule $A_{ij} = (-1)^{i+j}M_{ij}$. The *value* of $|A|$ is obtained by forming any of the equivalent expressions $\sum_{j=1}^{n} a_{ij}A_{ij}$, $\sum_{i=1}^{n} a_{ij}A_{ij}$, where the elements a_{ij} must be taken from a single row or a single column of A.

Example

$$\begin{vmatrix} a_{11} & a_{12} & a_{13} \\ a_{21} & a_{22} & a_{23} \\ a_{31} & a_{32} & a_{33} \end{vmatrix} = a_{31}A_{31} + a_{32}A_{32} + a_{33}A_{33}$$

$$= a_{31}\begin{vmatrix} a_{12} & a_{13} \\ a_{22} & a_{23} \end{vmatrix} - a_{32}\begin{vmatrix} a_{11} & a_{13} \\ a_{21} & a_{23} \end{vmatrix} + a_{33}\begin{vmatrix} a_{11} & a_{12} \\ a_{21} & a_{22} \end{vmatrix}$$

In general, A_{ij} will be determinants of order $n - 1$, but they may in turn be expanded by the rule. Also,

$$\sum_{j=1}^{n} a_{ji}A_{jk} = \sum_{j=1}^{n} a_{ij}A_{jk} = \begin{cases} |A| & i = k \\ 0 & i \neq k \end{cases}$$

Fundamental Properties of Determinants
1.   The value of a determinant $|A|$ is not changed if the rows and columns are interchanged.
2.   If the elements of one row (or one column) of a determinant are all zero, the value of $|A|$ is zero.
3.   If the elements of one row (or column) of a determinant are multiplied by the same constant factor, the value of the determinant is multiplied by this factor.
4.   If one determinant is obtained from another by interchanging any two rows (or columns), the value of either is the negative of the value of the other.
5.   If two rows (or columns) of a determinant are identical, the value of the determinant is zero.
6.   If two determinants are identical except for one row (or column), the sum of their values is given by a single determinant obtained by adding corresponding elements of dissimilar rows (or columns) and leaving unchanged the remaining elements.

Example

$$\begin{vmatrix} 3 & 2 \\ 1 & 5 \end{vmatrix} + \begin{vmatrix} 4 & 2 \\ 7 & 5 \end{vmatrix} = 13 + 6 = 19 \quad \text{Directly}$$

$$\begin{vmatrix} 7 & 2 \\ 8 & 5 \end{vmatrix} = 35 - 16 = 19 \quad \text{By rule 6}$$

7.   The value of a determinant is not changed if to the elements of any row (or column) are added a constant multiple of the corresponding elements of any other row (or column).
8.   If all elements but one in a row (or column) are zero, the value of the determinant is the product of that element times its cofactor.

The evaluation of determinants using the definition is quite laborious. The labor can be reduced by applying the fundamental properties just outlined.

The solution of n linear equations (not all b_i zero)

$$a_{11}x_1 + a_{12}x_2 + \cdots + a_{1n}x_n = b_1$$
$$a_{21}x_1 + a_{22}x_2 + \cdots + a_{2n}x_n = b_2$$
$$\vdots \qquad\qquad\qquad \vdots$$
$$a_{n1}x_1 + a_{n2}x_2 + \cdots + a_{nn}x_n = b_n$$

where $|A| = \begin{vmatrix} a_{11} & \cdots & a_{1n} \\ a_{21} & \cdots & a_{2n} \\ \vdots & & \\ a_{n1} & \cdots & a_{nn} \end{vmatrix} \neq 0$

has a unique solution given by $x_1 = |B_1|/|A|$, $x_2 = |B_2|/|A|$, $\ldots$, $x_n = |B_n|/|A|$, where B_k is the determinant obtained from A by replacing its kth column by $b_1, b_2, \ldots, b_n$. This technique is called **Cramer's rule.** It requires more labor than the method of elimination and should not be used for computations.

ANALYTIC GEOMETRY

REFERENCES: 108, 188, 193, 260, 261, 268, 274, 282.

Analytic geometry uses algebraic equations and methods to study geometric problems. It also permits one to visualize algebraic equations in terms of geometric curves, which frequently clarifies abstract concepts.

PLANE ANALYTIC GEOMETRY

Coordinate Systems The basic concept of analytic geometry is the establishment of a one-to-one correspondence between the points of the plane and number pairs (x, y). This correspondence may be done in a number of ways. The rectangular or cartesian coordinate system consists of two straight lines intersecting at right angles (Fig. 3-12). A point is designated by (x, y), where x (the abscissa) is the distance of the point from the y axis measured parallel to the x axis, positive if to the right, negative to the left. y (ordinate) is the distance of the point from the x axis, measured parallel to the y axis, positive if above, negative if below the x axis. The **quadrants** are labeled 1, 2, 3, 4 in the drawing, the coordinates of points in the various quadrants having the depicted signs. Another common coordinate system is the polar coordinate system (Fig. 3-13). In this system the position of a point is designated by the pair (r, θ), $r = \sqrt{x^2 + y^2}$ being the distance to the origin $0(0,0)$ and θ being the angle the line r makes with the positive x axis (polar axis). To change from polar to rectangular coordinates, use $x = r \cos \theta$ and $y = r \sin \theta$. To change from rectangular to polar coordinates, use $r = \sqrt{x^2 + y^2}$ and $\theta = \tan^{-1}(y/x)$ if $x \neq 0$; $\theta = \pi/2$ if $x = 0$. The distance between two points (x_1, y_1), (x_2, y_2) is defined by $d = \sqrt{(x_1 - x_2)^2 + (y_1 - y_2)^2}$ in rectangular coordinates or by $d = \sqrt{r_1^2 + r_2^2 - 2r_1r_2 \cos(\theta_1 - \theta_2)}$ in polar coordinates. Other coordinate systems are sometimes used. For example, on the surface of a sphere latitude and longitude prove useful.

The Straight Line (Fig. 3-14) The slope m of a straight line is the tangent of the inclination angle θ made with the positive x axis. If (x_1, y_1) and (x_2, y_2) are any two points on the line, slope $= m = (y_2 - y_1)/(x_2 - x_1)$. The slope of a line parallel to the x axis is zero; parallel to the y axis, it is undefined. Two lines are parallel if and only if they have the same slope. Two lines are perpendicular if and only if the product of their slopes is -1 (the exception being that case when the lines are parallel to the coordinate axes). Every equation of the type $Ax + By + C = 0$ represents a straight line, and every straight line has an equation of this form. A straight line is determined by a variety of conditions:

Given conditions	Equation of line
(1) Parallel to x axis	$y =$ constant
(2) Parallel y axis	$x =$ constant
(3) Point (x_1, y_1) and slope m	$y - y_1 = m(x - x_1)$
(4) Intercept on y axis $(0, b)$, m	$y = mx + b$
(5) Intercept on x axis $(a, 0)$, m	$y = m(x - a)$
(6) Two points (x_1, y_1), (x_2, y_2)	$y - y_1 = \dfrac{y_2 - y_1}{x_2 - x_1}(x - x_1)$
(7) Two intercepts $(a, 0)$, $(0, b)$	$x/a + y/b = 1$

The angle β a line with slope m_1 makes with a line having slope m_2 is given by $\tan \beta = (m_2 - m_1)/(m_1m_2 + 1)$. A line is determined if the length and direction of the perpendicular to it (the normal) from the origin are given (see Fig. 3-15). Let p = length of the perpendicular and α the angle that the perpendicular makes with the positive x axis. The equation of the line is $x \cos \alpha + y \sin \alpha = p$. The equation of a line perpendicular to a given line of slope m and passing through a point (x_1, y_1) is $y - y_1 = -(1/m)(x - x_1)$. The distance from a point (x_1, y_1) to a line with equation $Ax + by + C = 0$ is

$$d = \frac{|Ax_1 + By_1 + C|}{\sqrt{A^2 + B^2}}$$

Example If it is known that centigrade C and Fahrenheit F are linearly related and when C = 0°, F = 32°; C = 100°, F = 212°, find the equation relating C and F and that point where C = F. By using the two-point form, the equation is

$$F - 32 = \frac{212 - 32}{100 - 0}(C - 0)$$

or $F = \frac{9}{5}C + 32$. Equivalently

$$C - 0 = \frac{100 - 0}{212 - 32}(F - 32)$$

or $C = \frac{5}{9}(F - 32)$. Letting C = F, we have from either equation F = C = −40.

Occasionally some nonlinear algebraic equations can be reduced to linear equations under suitable substitutions or changes of variables. In other words, certain curves become the graphs of lines if the scales or coordinate axes are appropriately transformed.

Example Consider $y = bx^n$. $B = \log b$. Taking logarithms $\log y = n \log x + \log b$. Let $Y = \log y$, $X = \log x$, $B = \log b$. The equation then has the form $Y = nX + B$, which is a linear equation. Consider $k = k_0 \exp(-E/RT)$, taking logarithms $\log_e k = \log_e k_0 - E/(RT)$. Let $Y = \log_e k$, $B = \log_e k_0$, and $m = -E/R$, $X = 1/T$, and the result is $Y = mX + B$. Next consider $y = a + bx^n$. If the substitution $t = x^n$ is made, then the graph of y is a straight line versus t.

Asymptotes The limiting position of the tangent to a curve as the point of contact tends to an infinite distance from the origin is called an **asymptote.** If the equation of a given curve can be expanded in a Laurent power series such that

$$f(x) = \sum_{k=0}^{n} a_k x^k + \sum_{k=0}^{n} \frac{b_k}{x^k}$$

and

$$\lim_{x \to \infty} f(x) = \sum_{k=0}^{n} a_k x^k$$

then the equation of the asymptote is $y = \sum_{k=0}^{n} a_k x^k$. If $n = 1$, then the asymptote is (in general oblique) a line. In this case, the equation of the asymptote may be written as

$$y = mx + b \qquad m = \lim_{x \to \infty} f'(x)$$

$$b = \lim_{x \to \infty} [f(x) - xf'(x)]$$

Geometric Properties of a Curve When the Equation Is Given The analysis of the properties of an equation is facilitated by the investigation of the equation by using the following techniques:

1. *Points of maximum, minimum, and inflection.* These may be investigated by means of the calculus.

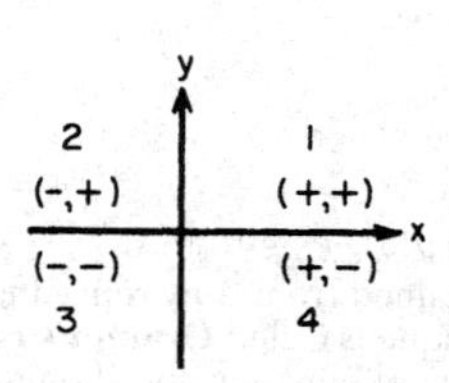

FIG. 3-12 Rectangular coordinates.

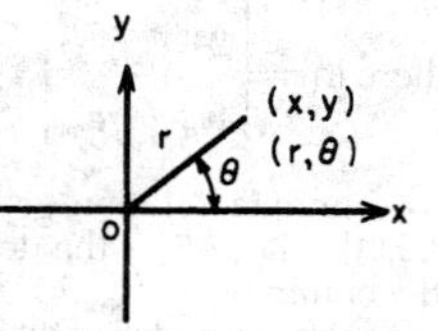

FIG. 3-13 Polar coordinates.

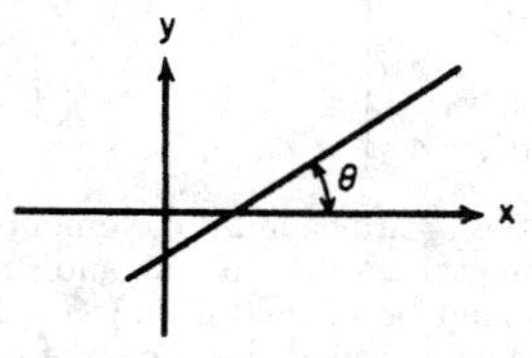

FIG. 3-14 Straight line.

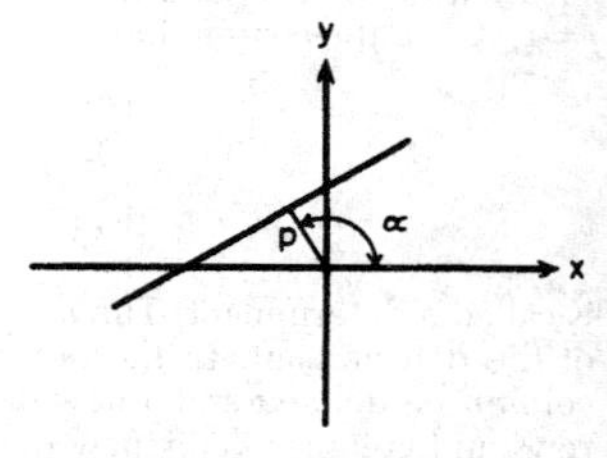

FIG. 3-15 Determination of line.

2.   *Symmetry.*   Let $F(x, y) = 0$ be the equation of the curve.

Condition on $F(x, y)$	Symmetry
$F(x, y) = F(-x, y)$	With respect to y axis
$F(x, y) = F(x, -y)$	With respect to x axis
$F(x, y) = F(-x, -y)$	With respect to origin
$F(x, y) = F(y, x)$	With respect to the line $y = x$

3.   *Extent.*   Only real values of x and y are considered in obtaining the points (x, y) whose coordinates satisfy the equation. The extent of them may be limited by the condition that negative numbers do not have real square roots.

4.   *Intercepts.*   Find those points where the curves of the function cross the coordinate axes.

5.   *Asymptotes.*   See preceding discussion.

6.   *Direction at a point.*   This may be found from the derivative of the function at a point. This concept is useful for distinguishing among a family of similar curves.

Example $y^2 = (x^2 + 1)/(x^2 - 1)$ is symmetric with respect to the x and y axis, the origin, and the line $y = x$. It has the vertical asymptotes $x = \pm 1$. When $x = 0$, $y^2 = -1$; so there are no y intercepts. If $y = 0$, $(x^2 + 1)/(x^2 - 1) = 0$; so there are no x intercepts. If $|x| < 1$, y^2 is negative; so $|x| > 1$. From $x^2 = (y^2 + 1)/(y^2 - 1)$, $y = \pm 1$ are horizontal asymptotes and $|y| > 1$. As $x \to 1^+$, $y \to +\infty$; as $x \to +\infty$, $y \to +1$. The graph is given in Fig. 3-16.

Conic Sections   The curves included in this group are obtained from plane sections of the cone. They include the circle, ellipse, parabola, hyperbola, and degeneratively the point and straight line. A **conic** is the locus of a point whose distance from a fixed point called the **focus** is in a constant ratio to its distance from a fixed line, called the directrix. This ratio is the eccentricity e. If $e = 0$, the conic is a circle; if $0 < e < 1$, the conic is an ellipse; if $e = 1$, the conic is a parabola; if $e > 1$, the conic is a hyperbola. Every conic section is representable by an equation of second degree. Conversely, every equation of second degree in two variables represents a conic. The general equation of the second degree is $Ax^2 + Bxy + Cy^2 + Dx + Ey + F = 0$. Let Δ be defined as the determinant

$$\Delta = \begin{vmatrix} 2A & B & D \\ B & 2C & E \\ D & E & 2F \end{vmatrix}$$

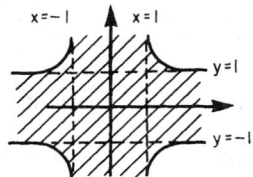

FIG. 3-16   Graph of $y^2 = (x^2 + 1)/(x^2 - 1)$

The table characterizes the curve represented by the equation.

	$B^2 - 4AC < 0$	$B^2 - 4AC = 0$	$B^2 - 4AC > 0$
$\Delta \neq 0$	$A\Delta < 0$ $A \neq C$, an ellipse $A\Delta < 0$ $A = C$, a circle $A\Delta > 0$, no locus	Parabola	Hyperbola
$\Delta = 0$	Point	2 parallel lines if $Q = D^2 + E^2 - 4(A + C)F > 0$ 1 straight line if $Q = 0$, no locus if $Q < 0$	2 intersecting straight lines

Example $3x^2 + 4xy - 2y^2 + 3x - 2y + 7 = 0$.

$$\Delta = \begin{vmatrix} 6 & 4 & 3 \\ 4 & -4 & -2 \\ 3 & -2 & 14 \end{vmatrix} = -596 \neq 0, B^2 - 4AC = 40 > 0$$

The curve is therefore a hyperbola.

To translate the axes to a new origin at (h, k), substitute for x and y in the original equation $x + h$ and $y + k$. Translation of the axes can always be accomplished to eliminate the linear terms in the second-degree equation in two variables having no xy term.

Example $x^2 + y^2 + 2x - 4y + 2 = 0$. Rewrite this as $x^2 + 2x + 1 + y^2 - 4y + 4 - 5 + 2 = 0$ or $(x + 1)^2 + (y - 2)^2 = 3$. Let $u = x + 1$, $v = y - 2$. Then $u^2 + v^2 = 3$. The axis has been translated to the new origin $(-1, 2)$.

The type of curve determined by a specific equation of the second degree can also be easily determined by reducing it to a standard form by translation and/or rotation. In the case in which the equation has no xy term, the procedure is merely to complete the squares of the terms in x and y separately.

To rotate the axes through an angle α, substitute for x the quantity $x \cos \alpha - y \sin \alpha$ and for y the quantity $x \sin \alpha + y \cos \alpha$. A rotation of the axes through $\alpha = \frac{1}{2} \cot^{-1} (A - C)/B$ will eliminate the cross-product term in the general second-degree equation.

Example   Consider $3x^2 + 2xy + y^2 - 2x + 3y = 7$. A rotation of axes through $\alpha = \frac{1}{2} \cot^{-1} 1 = 22\frac{1}{2}°$ eliminates the cross-product term.

The following tabulation gives the form of the more common equations.

Polar equation	Type of curve
(1) $r = a$	Circle
(2) $r = 2a \cos \theta$	Circle
(3) $r = 2a \sin \theta$	Circle
(4) $r^2 - 2br \cos (\theta - \beta) + b^2 - a^2 = 0$	Circle at (b, β), radius a
(5) $r = \dfrac{ke}{1 - e \cos \theta}$	$e = 1$ parabola $0 < e < 1$ ellipse $e > 1$ hyperbola

Some common equations in parametric form are given below.

(1) $(x - h)^2 + (y - k)^2 = a^2$	$x = h + a \cos \theta$ $y = k + a \sin \theta$	Circle (Fig. 3-23) Parameter is angle θ.
(2) $\dfrac{(x - h)^2}{a^2} + \dfrac{(y - k)^2}{b^2} = 1$	$x = h + a \cos \phi$ $y = k + a \sin \phi$	Ellipse (Fig. 3-20) Parameter is angle ϕ.
(3) $z^2 + y^2 = a^2$	$x = \dfrac{-at}{\sqrt{t^2 + 1}}$ $y = \dfrac{a}{\sqrt{t^2 + 1}}$	Circle Parameter is $t = \dfrac{dy}{dx} = $ slope of tangent at (x, y).
(4) $y = a \cosh \dfrac{x}{a}$	$x = a \sinh^{-1} \dfrac{s}{a}$ $y^2 = a^2 + s^2$	Catenary (Fig. 3-24; such as hanging cable under gravity) Parameter $s = $ arc length from $(0, a)$ to (x, y). See Fig. 3-24.
(5) Cycloid	$x = a(\phi - \sin \phi)$ $y = a(1 - \cos \phi)$	

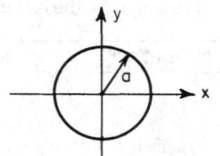

FIG. 3-17   Circle center $(0,0)$ $r = a$.

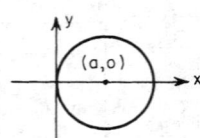

FIG. 3-18   Circle center $(a,0)$ $r = 2a \cos \theta$.

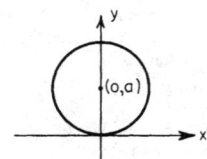

FIG. 3-19   Circle center $(0,a)$ $r = 2a \sin \theta$.

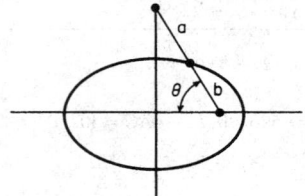

FIG. 3-20   Ellipse, $0 < e < 1$.

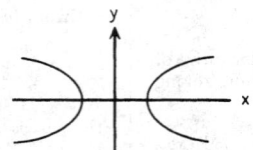

FIG. 3-21   Hyperbola, $e > 1$, $r = ke/(1 - e \cos \theta)$.

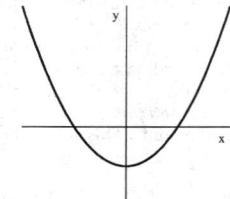

FIG. 3-22   Parabola, $e = 1$.

Circle at (b, β), radius a: $r^2 - 2br \cos (\theta - \beta) + b^2 - a^2 = 0$.

Graphs of Polar Equations   The equation $r = 0$ corresponds to $x = 0$, $y = 0$ regardless of θ. The same point may be represented in several different ways; thus the point $(2, \pi/3)$ or $(2, 60°)$ has the following representations: $(2, 60°)$, $(2, -300°)$. These are summarized in $(2, 60° + n \, 360°)$, $n = 0, \pm 1, \pm 2$, or in radian measure $[2, (\pi/3) + 2n\pi]$, $n = 0, \pm 1, \pm 2$. Plotting of polar equations can be facilitated by the following steps:
1.   Find those points where r is a maximum or minimum.
2.   Find those values of θ where $r = 0$, if any.
3.   Symmetry: The curve is symmetric about the origin if the equation is unchanged when θ is replaced by $\theta \pm \pi$, symmetric about the x axis if the equation is unchanged when θ is replaced by $-\theta$, and symmetric about the y axis if the equation is unchanged when θ is replaced by $\pi - \theta$.

Parametric Equations   It is frequently useful to write the equations of a curve in terms of an auxiliary variable called a parameter. For example, a circle of radius a, center at $(0, 0)$, can be written in the equivalent form $x = a \cos \theta$, $y = a \sin \phi$ where θ is the parameter. Similarly, $x = a \cos \phi$, $y = b \sin \phi$ are the parametric equations of the ellipse $x^2/a^2 + y^2/b^2 = 1$ with parameter ϕ.

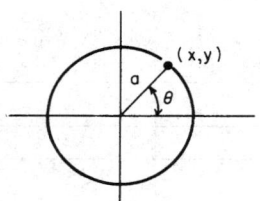

FIG. 3-23   Circle.

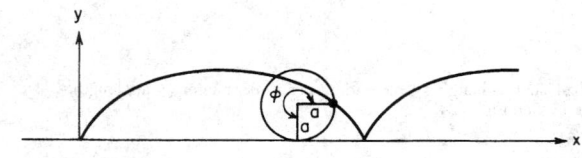

FIG. 3-24   Cycloid.

SOLID ANALYTIC GEOMETRY

Coordinate Systems   The commonly used coordinate systems are three in number. Others may be used in specific problems (see Ref. 212). The **rectangular** (cartesian) system (Fig. 3-25) consists of mutually orthogonal axes x, y, z. A triple of numbers (x, y, z) is used to represent each point. The **cylindrical** coordinate system $(r, \theta, z;$ Fig. 3-26) is frequently used to locate a point in space. These are essentially the polar coordinates (r, θ) coupled with the z coordinate. As

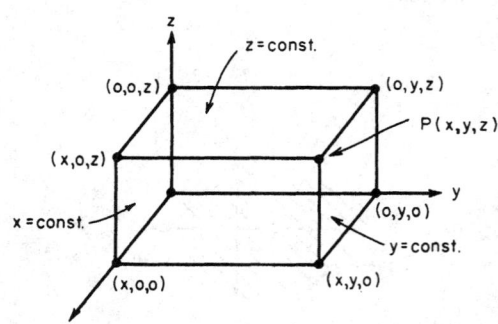

FIG. 3-25   Cartesian coordinates.

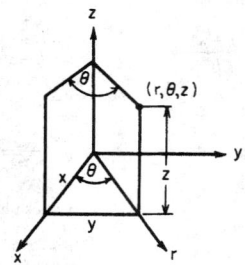

FIG. 3-26   Cylindrical coordinates.

before, $x = r \cos\theta$, $y = r \sin\theta$, $z = z$ and $r^2 = x^2 + y^2$, $y/x = \tan\theta$. If r is held constant and θ and z are allowed to vary, the locus of (r, θ, z) is a right circular cylinder of radius r along the z axis. The locus of $r = C$ is a circle, and $\theta = $ constant is a plane containing the z axis and making an angle θ with the xz plane. Cylindrical coordinates are convenient to use when the problem has an axis of symmetry.

The **spherical** coordinate system is convenient if there is a point of symmetry in the system. This point is taken as the origin and the coordinates (ρ, ϕ, θ) illustrated in Fig. 3-27. The relations are $x = \rho \sin\phi \cos\theta$, $y = \rho \sin\phi \sin\theta$, $z = \rho \cos\phi$, and $r = \rho \sin\phi$. $\theta = $ constant is a plane containing the z axis and making an angle θ with the xz plane. $\phi = $ constant is a cone with vertex at 0. $\rho = $ constant is the surface of a sphere of radius ρ, center at the origin 0. Every point in the space may be given spherical coordinates restricted to the ranges $0 \leq \phi \leq \pi$, $\rho \geq 0$, $0 \leq \theta < 2\pi$.

Lines and Planes   The distance between two points (x_1, y_1, z_1), (x_2, y_2, z_2) is $d = \sqrt{(x_1 - x_2)^2 + (y_1 - y_2)^2 + (z_1 - z_2)^2}$. There is nothing in the geometry of three dimensions quite analogous to the slope of a line in the plane case. Instead of specifying the direction of a line by a trigonometric function evaluated for one angle, a trigonometric function evaluated for three angles is used. The angles α, β, γ that a line segment makes with the positive x, y, and z axes, respectively, are called the **direction angles** of the line, and $\cos\alpha$, $\cos\beta$, $\cos\gamma$ are called the **direction cosines**. Let (x_1, y_1, z_1), (x_2, y_2, z_2) be on the line. Then $\cos\alpha = (x_2 - x_1)/d$, $\cos\beta = (y_2 - y_1)/d$, $\cos\gamma = (z_2 - z_1)/d$, where $d = $ the distance between the two points. Clearly $\cos^2\alpha + \cos^2\beta + \cos^2\gamma = 1$. If two lines are specified by the direction cosines $(\cos\alpha_1, \cos\beta_1, \cos\gamma_1)$, $(\cos\alpha_2, \cos\beta_2, \cos\gamma_2)$, then the angle θ between the lines is $\cos\theta = \cos\alpha_1 \cos\alpha_2 + \cos\beta_1 \cos\beta_2 + \cos\gamma_1 \cos\gamma_2$. Thus the lines are perpendicular if and only if $\theta = 90°$ or $\cos\alpha_1 \cos\alpha_2 + \cos\beta_1 \cos\beta_2 + \cos\gamma_1 \cos\gamma_2 = 0$. The equation of a line with direction cosines $(\cos\alpha, \cos\beta, \cos\gamma)$ passing through (x_1, y_1, z_1) is $(x - x_1)/\cos\alpha = (y - y_1)/\cos\beta = (z - z_1)/\cos\gamma$.

The equation of every plane is of the form $Ax + By + Cz + D = 0$. The numbers

$$\frac{A}{\sqrt{A^2 + B^2 + C^2}}, \quad \frac{B}{\sqrt{A^2 + B^2 + C^2}}, \quad \frac{C}{\sqrt{A^2 + B^2 + C^2}}$$

are direction cosines of the normal lines to the plane. The plane through the point (x_1, y_1, z_1) whose normals have these as direction cosines is $A(x - x_1) + B(y - y_1) + C(z - z_1) = 0$.

Example   Find the equation of the plane through $(1, 5, -2)$ perpendicular to the line $(x + 9)/7 = (y - 3)/-1 = z/8$. The numbers $(7, -1, 8)$ are called **direction numbers**. They are a constant multiple of the direction cosines. $\cos\alpha = 7/114$, $\cos\beta = -1/114$, $\cos\gamma = 8/114$. The plane has the equation $7(x - 1) - 1(y - 5) + 8(z + 2) = 0$ or $7x - y + 8z + 14 = 0$.

The distance from the point (x_1, y_1, z_1) to the plane $Ax + By + Cz + D = 0$ is

$$d = \frac{|Ax_1 + By_1 + Cz_1 + D|}{\sqrt{A^2 + B^2 + C^2}}$$

Space Curves   Space curves are usually specified as the set of points whose coordinates are given parametrically by a system of equations $x = f(t)$, $y = g(t)$, $z = h(t)$ in the parameter t.

Example   The equation of a straight line in space is $(x - x_1)/a = (y - y_1)/b = (z - z_1)/c$. Since all these quantities must be equal (say, to t), we may write $x = x_1 + at$, $y = y_1 + bt$, $z = z_1 + ct$, which represent the parametric equations of the line.

Example   The equations $z = a \cos\beta t$, $y = a \sin\beta t$, $z = bt$, a, β, b positive constants, represent a circular helix.

Surfaces   The locus of points (x, y, z) satisfying $f(x, y, z) = 0$, broadly speaking, may be interpreted as a surface. The simplest surface is the **plane.** The next simplest is a **cylinder,** which is a surface generated by a straight line moving parallel to a given line and passing through a given curve.

Example   The parabolic cylinder $y = x^2$ (Fig. 3-28) is generated by a straight line parallel to the z axis passing through $y = x^2$ in the plane $z = 0$.

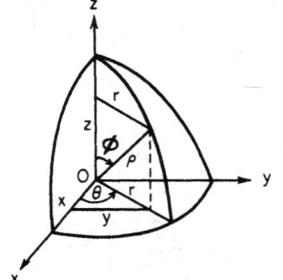

FIG. 3-27   Spherical coordinates.

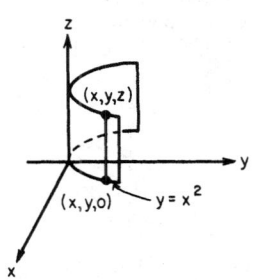

FIG. 3-28   Parabolic cylinder.

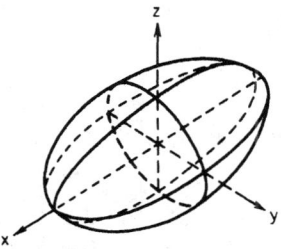

FIG. 3-29   Ellipsoid. $\dfrac{x^2}{a^2} + \dfrac{y^2}{b^2} + \dfrac{z^2}{c^2} = 1$ (sphere if $a = b = c$)

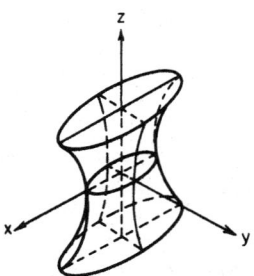

FIG. 3-30   Hyperboloid of one sheet. $\dfrac{x^2}{a^2} + \dfrac{y^2}{b^2} - \dfrac{z^2}{c^2} = 1$

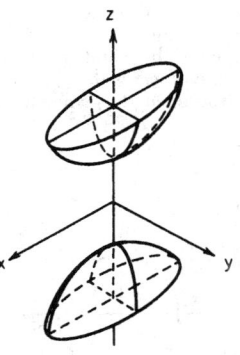

FIG. 3-31   Hyperboloid of two sheets. $\dfrac{x^2}{a^2} + \dfrac{y^2}{b^2} - \dfrac{z^2}{c^2} = -1$

A surface whose equation is a quadratic in the variables x, y, and z is called a **quadric surface.** Some of the more common such surfaces are tabulated and pictured in Figs. 3-29 to 3-37.

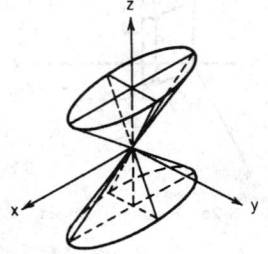

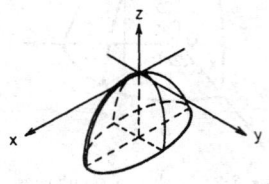

FIG. 3-33   Elliptic paraboloid.

FIG. 3-32   Cone. $\dfrac{x^2}{a^2} + \dfrac{y^2}{b^2} + \dfrac{z^2}{c^2} = 0$

$\dfrac{x^2}{a^2} + \dfrac{y^2}{b^2} + 2z = 0$

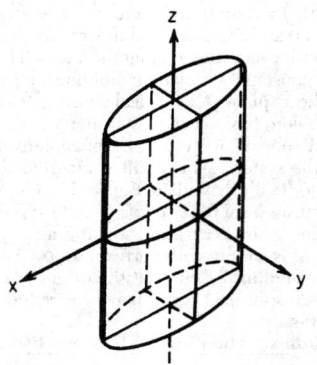

FIG. 3-35   Elliptic cylinder. $\dfrac{x^2}{a^2} + \dfrac{y^2}{b^2} = 1$

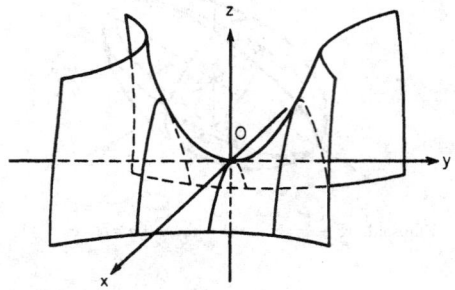

FIG. 3-34   Hyperbolic paraboloid. $\dfrac{x^2}{a^2} - \dfrac{y^2}{b^2} + 2z = 0$

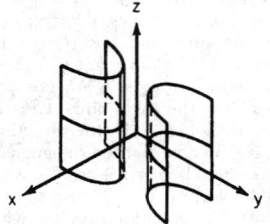

FIG. 3-36   Hyperbolic cylinder.

$\dfrac{x^2}{a^2} - \dfrac{y^2}{b^2} = 1$

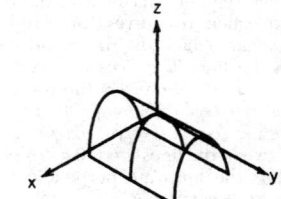

FIG. 3-37   Parabolic cylinder.

$y^2 + 2ax = 0$

PLANE TRIGONOMETRY

REFERENCES: 20, 108, 131, 158, 166, 202.

ANGLES

An angle is generated by the rotation of a line about a fixed center from some initial position to some terminal position. If the rotation is clockwise, the angle is negative; if it is counterclockwise, the angle is positive. Angle size is unlimited. If α, β are two angles such that $\alpha + \beta = 90°$, they are complementary; they are supplementary if $\alpha + \beta = 180°$. Angles are most commonly measured in the sexagesimal system or by radian measure. In the first system there are 360 degrees in one complete revolution; one degree = $\frac{1}{90}$ of a right angle. The degree is subdivided into 60 minutes; the minute is subdivided into 60 seconds. In the radian system one radian is the angle at the center of a circle subtended by an arc whose length is equal to the radius of the circle. Thus 2π rad = $360°$; 1 rad = $57.29578°$; $1° = 0.01745$ rad; 1 min = 0.00029089 rad. The advantage of radian measure is that it is *dimensionless*. The quadrants are conventionally labeled as Fig. 3-38 shows.

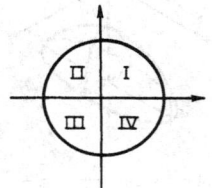

FIG. 3-38   Quadrants.

FUNCTIONS OF CIRCULAR TRIGONOMETRY

The trigonometric functions of angles are the ratios between the various sides of the reference triangles shown in Fig. 3-39 for the various quadrants. Clearly $r = \sqrt{x^2 + y^2} \geq 0$. The fundamental functions (see Figs. 3-40, 3-41, 3-42) are

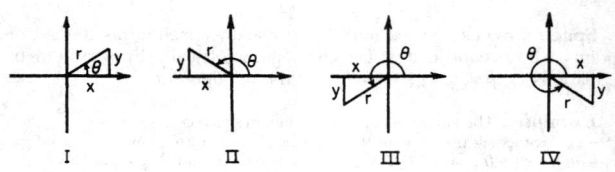

FIG. 3-39   Triangles.

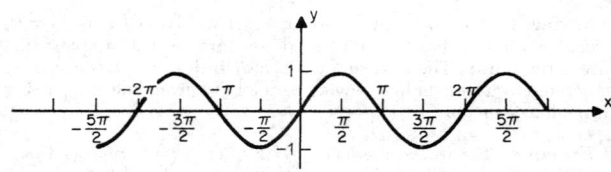

FIG. 3-40   Graph of $y = \sin x$.

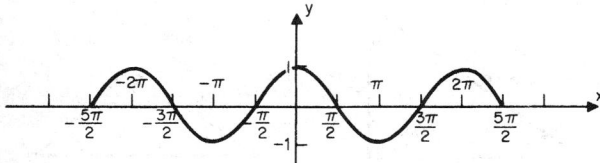

FIG. 3-41   Graph of $y = \cos x$.

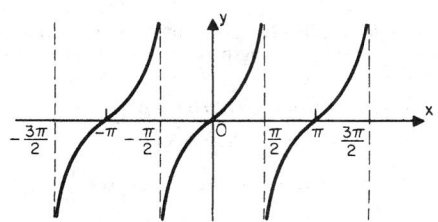

FIG. 3-42   Graph of $y = \tan x$.

Plane Trigonometry

Sine of $\theta = \sin \theta = y/r$	Secant of $\theta = \sec \theta = r/x$
Cosine of $\theta = \cos \theta = x/r$	Cosecant of $\theta = \csc \theta = r/y$
Tangent of $\theta = \tan \theta = y/x$	Cotangent of $\theta = \cot \theta = x/y$

Magnitude and Sign of Trigonometric Functions $0 \le \theta \le 360°$

Function	0° to 90°	90° to 180°	180° to 270°	270° to 360°
$\sin \theta$	$+0$ to $+1$	$+1$ to $+0$	-0 to -1	-1 to -0
$\csc \theta$	$+\infty$ to $+1$	$+1$ to $+\infty$	$-\infty$ to -1	-1 to $-\infty$
$\cos \theta$	$+1$ to 0	-0 to -1	-1 to -0	$+0$ to $+1$
$\sec \theta$	$+1$ to $+\infty$	$-\infty$ to -1	-1 to $-\infty$	$+\infty$ to $+1$
$\tan \theta$	$+0$ to $+\infty$	$-\infty$ to -0	$+0$ to $+\infty$	$-\infty$ to -0
$\cot \theta$	$+\infty$ to $+0$	-0 to $-\infty$	$+\infty$ to $+0$	-0 to $-\infty$

Values of the Trigonometric Functions for Common Angles

$\theta°$	θ, rad	$\sin \theta$	$\cos \theta$	$\tan \theta$
0	0	0	1	0
30	$\pi/6$	$1/2$	$\sqrt{3}/2$	$\sqrt{3}/3$
45	$\pi/4$	$\sqrt{2}/2$	$\sqrt{2}/2$	1
60	$\pi/3$	$\sqrt{3}/2$	$1/2$	$\sqrt{3}$
90	$\pi/2$	1	0	$+\infty$

If $90° \le \theta \le 180°$, $\sin \theta = \sin (180° - \theta)$; $\cos \theta = -\cos (180° - \theta)$; $\tan \theta = -\tan (180° - \theta)$. If $180° \le \theta \le 270°$, $\sin \theta = -\sin (270° - \theta)$; $\cos \theta = -\cos (270° - \theta)$; $\tan \theta = \tan (270° - \theta)$. If $270° \le \theta \le 360°$, $\sin \theta = -\sin (360° - \theta)$; $\cos \theta = \cos (360° - \theta)$; $\tan \theta = -\tan (360° - \theta)$. The reciprocal properties may be used to find the values of the other functions.

If it is desired to find the angle when a function of it is given, the procedure is as follows: There will in general be two angles between 0° and 360° corresponding to the given value of the function.

Given $(a > 0)$	Find an acute angle θ_0 such that	Required angles are
$\sin \theta = +a$	$\sin \theta_0 = a$	θ_0 and $(180° - \theta_0)$
$\cos \theta = +a$	$\cos \theta_0 = a$	θ_0 and $(360° - \theta_0)$
$\tan \theta = +a$	$\tan \theta_0 = a$	θ_0 and $(180° + \theta_0)$
$\sin \theta = -a$	$\sin \theta_0 = a$	$180° + \theta_0$ and $360° - \theta_0$
$\cos \theta = -a$	$\cos \theta_0 = a$	$180° - \theta_0$ and $180° + \theta_0$
$\tan \theta = -a$	$\tan \theta_0 = a$	$180° - \theta_0$ and $360° - \theta_0$

Relations between Functions of a Single Angle $\sec \theta = 1/\cos \theta$; $\csc \theta = 1/\sin \theta$, $\tan \theta = \sin \theta/\cos \theta = \sec \theta/\csc \theta = 1/\cot \theta$; $\sin^2 \theta + \cos^2 \theta = 1$; $1 + \tan^2 \theta = \sec^2 \theta$; $1 + \cot^2 \theta = \csc^2 \theta$. For $0 \le \theta \le 90°$ the following results hold:

$$\sin \theta = \cos \theta/\cot \theta = \sqrt{1 - \cos^2 \theta} = \cos \theta \tan \theta$$

$$= \frac{\tan \theta}{\sqrt{1 + \tan^2 \theta}} = \frac{1}{\sqrt{1 + \cot^2 \theta}} = 2 \sin \left(\frac{\theta}{2}\right) \cos \left(\frac{\theta}{2}\right)$$

and $\cos \theta = \sqrt{1 - \sin^2 \theta} = \dfrac{1}{\sqrt{1 + \tan^2 \theta}}$

$$= \frac{\cot \theta}{\sqrt{1 + \cot^2 \theta}} = \frac{\sin \theta}{\tan \theta} = \cos^2 \left(\frac{\theta}{2}\right) - \sin^2 \left(\frac{\theta}{2}\right)$$

The cofunction property is very important. $\cos \theta = \sin (90° - \theta)$, $\sin \theta = \cos (90° - \theta)$, $\tan \theta = \cot (90° - \theta)$, $\cot \theta = \tan (90° - \theta)$, etc.

Functions of Negative Angles $\sin (-\theta) = -\sin \theta$, $\cos (-\theta) = \cos \theta$, $\tan (-\theta) = -\tan \theta$, $\sec (-\theta) = \sec \theta$, $\csc (-\theta) = -\csc \theta$, $\cot (-\theta) = -\cot \theta$.

Identities

Sum and Difference Formulas   Let x, y be two angles. $\sin (x \pm y) = \sin x \cos y \pm \cos x \sin y$; $\cos (x \pm y) = \cos x \cos y \mp \sin x \sin y$; $\tan (x \pm y) = (\tan x \pm \tan y)/(1 \mp \tan x \tan y)$; $\sin x \pm \sin y = 2 \sin \frac{1}{2}(x \pm y) \cos \frac{1}{2}(x \mp y)$; $\cos x + \cos y = 2 \cos \frac{1}{2}(x + y) \cos \frac{1}{2}(x - y)$; $\cos x - \cos y = -2 \sin \frac{1}{2}(x + y) \sin \frac{1}{2}(x - y)$; $\tan x \pm \tan y = [\sin (x \pm y)]/(\cos x \cos y)$; $\sin^2 x - \sin^2 y = \cos^2 y - \cos^2 x = \sin (x + y) \sin (x - y)$; $\cos^2 x - \sin^2 y = \cos^2 y - \sin^2 x = \cos (x + y) \cos (x - y)$; $\sin (45° + x) = \cos (45° - x)$; $\sin (45° - x) = \cos (45° + x)$; $\tan (45° \pm x) = \cot (45° \mp x)$. $A \cos x + B \sin x = \sqrt{A^2 + B^2} \sin (\alpha + x) = \sqrt{A^2 + B^2} \cos (\beta - x)$ where $\tan \alpha = A/B$, $\tan \beta = B/A$; both α and β are positive acute angles.

Multiple and Half Angle Identities   Let $x =$ angle, $\sin 2x = 2 \sin x \cos x$; $\sin x = 2 \sin \frac{1}{2}x \cos \frac{1}{2}x$; $\cos 2x = \cos^2 x - \sin^2 x = 1 - 2 \sin^2 x = 2 \cos^2 x - 1$. $\tan 2x = (2 \tan x)/(1 - \tan^2 x)$; $\sin 3x = 3 \sin x - 4 \sin^3 x$; $\cos 3x = 4 \cos^3 x - 3 \cos x$. $\tan 3x = (3 \tan x - \tan^3 x)/(1 - 3 \tan^2 x)$; $\sin 4x = 4 \sin x \cos x - 8 \sin^3 x \cos x$; $\cos 4x = 8 \cos^4 x - 8 \cos^2 x + 1$.

$$\sin \left(\frac{x}{2}\right) = \sqrt{\tfrac{1}{2}(1 - \cos x)}$$

$$\cos \left(\frac{x}{2}\right) = \sqrt{\tfrac{1}{2}(1 + \cos x)}$$

$$\tan \left(\frac{x}{2}\right) = \sqrt{\frac{1 - \cos x}{1 + \cos x}} = \frac{\sin x}{1 + \cos x} = \frac{1 - \cos x}{\sin x}$$

Relations between Three Angles Whose Sum Is 180°   Let x, y, z be the angles.

$$\sin x + \sin y + \sin z = 4 \cos \left(\frac{x}{2}\right) \cos \left(\frac{y}{2}\right) \cos \left(\frac{z}{2}\right)$$

$$\cos x + \cos y + \cos z = 4 \sin \left(\frac{x}{2}\right) \sin \left(\frac{y}{2}\right) \sin \left(\frac{z}{2}\right) + 1$$

$$\sin x + \sin y - \sin z = 4 \sin \left(\frac{x}{2}\right) \sin \left(\frac{y}{2}\right) \cos \left(\frac{z}{2}\right)$$

$\sin^2 x + \sin^2 y + \sin^2 z = 2 \cos x \cos y \cos z + 2$; $\tan x + \tan y + \tan z = \tan x \tan y \tan z$; $\sin 2x + \sin 2y + \sin 2z = 4 \sin x \sin y \sin z$.

INVERSE TRIGONOMETRIC FUNCTIONS

$y = \sin^{-1} x = \arcsin x$ is the angle y whose sine is x.

Example $y = \sin^{-1} \frac{1}{2}$, y is 30°.

The complete solution of the equation $x = \sin y$ is $y = (-1)^n \sin^{-1} x + n(180°)$, $-\pi/2 \le \sin^{-1} x \le \pi/2$ where $\sin^{-1} x$ is the principal value of the angle whose sine is x. The range of principal values of the $\cos^{-1} x$ is $0 \le \cos^{-1} x \le \pi$ and $-\pi/2 \le \tan^{-1} x \le \pi/2$. If these restrictions are allowed to hold, the following formulas result:

$$\sin^{-1} x = \cos^{-1} \sqrt{1-x^2} = \tan^{-1} \frac{x}{\sqrt{1-x^2}} = \cot^{-1} \frac{\sqrt{1-x^2}}{x}$$

$$= \sec^{-1} \frac{1}{\sqrt{1-x^2}} = \csc^{-1} \frac{1}{x} = \frac{\pi}{2} - \cos^{-1} x$$

$$\cos^{-1} x = \sin^{-1} \sqrt{1-x^2} = \tan^{-1} \frac{\sqrt{1-x^2}}{x}$$

$$= \cot^{-1} \frac{x}{\sqrt{1-x^2}} = \sec^{-1} \frac{1}{x}$$

$$= \csc^{-1} \frac{1}{\sqrt{1-x^2}} = \frac{\pi}{2} - \sin^{-1} x$$

$$\tan^{-1} x = \sin^{-1} \frac{x}{\sqrt{1-x^2}} = \cos^{-1} \frac{1}{\sqrt{1+x^2}}$$

$$= \cot^{-1} \frac{1}{x} = \sec^{-1} \sqrt{1+x^2} = \csc^{-1} \frac{\sqrt{1+x^2}}{x}$$

RELATIONS BETWEEN ANGLES AND SIDES OF TRIANGLES

Solutions of Triangles (Fig. 3-43)   Let a, b, c denote the sides and α, β, γ the angles opposite the sides in the triangle. Let $2s = a + b + c$, A = area, r = radius of the inscribed circle, R = radius of the circumscribed circle, and h = altitude. In any triangle $\alpha + \beta + \gamma = 180°$.

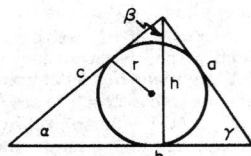

FIG. 3-43   Triangle.

Law of Sines $\sin \alpha/a = \sin \beta/b = \sin \gamma/c$.
Law of Tangents

$$\frac{a+b}{a-b} = \frac{\tan \frac{1}{2}(\alpha+\beta)}{\tan \frac{1}{2}(\alpha-\beta)}; \frac{b+c}{b-c} = \frac{\tan \frac{1}{2}(\beta+\gamma)}{\tan \frac{1}{2}(\beta-\gamma)}; \frac{a+c}{a-c} = \frac{\tan \frac{1}{2}(\alpha+\gamma)}{\tan \frac{1}{2}(\alpha-\gamma)}$$

Law of Cosines $a^2 = b^2 + c^2 - 2bc \cos \alpha$; $b^2 = a^2 + c^2 - 2ac \cos \beta$; $c^2 = a^2 + b^2 - 2ab \cos \gamma$.
Other Relations   In this subsection, where appropriate, two more formulas can be generated by replacing a by b, b by c, c by a, α by β, β by γ, and γ by α. $\cos \alpha = (b^2 + c^2 - a^2)/2bc$; $a = b \cos \gamma + c = \cos \beta$; $\sin \alpha = (2/bc) \sqrt{s(s-a)(s-b)(s-c)}$;

$$\sin\left(\frac{\alpha}{2}\right) = \sqrt{\frac{(s-b)(s-c)}{bc}}; \cos\left(\frac{\alpha}{2}\right) = \sqrt{\frac{s(s-a)}{bc}}; A = \frac{1}{2} bh$$

$$= \frac{1}{2} ab \sin \gamma = \frac{a^2 \sin \beta \sin \gamma}{2 \sin \alpha} = \sqrt{s(s-a)(s-b)(s-c)} = rs$$

where $r = \sqrt{\dfrac{(s-a)(s-b)(s-c)}{s}}$

$R = a/(2 \sin \alpha) = abc/4A$; $h = c \sin a = a \sin \gamma = 2rs/b$.

Example $a = 5$, $b = 4$, $\alpha = 30°$. Use the law of sines. $0.5/5 = \sin \beta/4$, $\sin \beta = \frac{2}{5}$, $\beta = 23°35'$, $\gamma = 126°25'$. So $c = \sin 126°25'/\frac{1}{10} = 10(.8047) = 8.05$.

The relations given here suffice to solve any triangle. One method for each triangle is given.

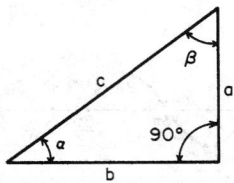

FIG. 3-44   Right triangle.

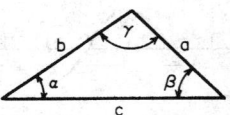

FIG. 3-45   Oblique triangle.

Right Triangle (Fig. 3-44)   Given one side and any acute angle α or any two sides, the remaining parts can be obtained from the following formulas:

$$a = \sqrt{(c+b)(c-b)} = c \sin \alpha = b \tan \alpha$$

$$b = \sqrt{(c+a)(c-a)} = c \cos \alpha = a \cot \alpha$$

$$c = \sqrt{a^2 \pm b^2}, \sin \alpha = \frac{a}{c}, \cos \alpha = \frac{b}{c}, \tan \alpha = \frac{a}{b}, \beta = 90° - \alpha$$

$$A = \frac{1}{2} ab = \frac{a^2}{2 \tan \alpha} = \frac{b^2 \tan \alpha}{2} = \frac{c^2 \sin 2\alpha}{4}$$

Oblique Triangles (Fig. 3-45)   There are four possible cases.
1.   Given b, c and the included angles α,

$$\frac{1}{2}(\beta + \gamma) = 90° - \frac{1}{2}\alpha; \tan \frac{1}{2}(\beta - \gamma) = \frac{b-c}{b+c} \tan \frac{1}{2}(\beta + \gamma)$$

$$\beta = \frac{1}{2}(\beta + \gamma) + \frac{1}{2}(\beta - \gamma); \gamma = \frac{1}{2}(\beta + \gamma) - \frac{1}{2}(\beta - \gamma); a = \frac{b \sin \alpha}{\sin \beta}$$

2.   Given the three sides a, b, c, $s = \frac{1}{2}(a + b + c)$;

$$r = \sqrt{\frac{(s-a)(s-b)(s-c)}{s}}$$

$$\tan \frac{1}{2} \alpha = \frac{r}{s-a}; \tan \frac{1}{2} \beta = \frac{r}{s-b}; \tan \frac{1}{2} \gamma = \frac{r}{s-c}$$

3.   Given any two sides a, c and an angle opposite one of them α, $\sin \gamma = (c \sin \alpha)/a$; $\beta = 180° - a - \gamma$; $b = (a \sin \beta)/(\sin \alpha)$. There may be two solutions here. γ may have two values γ_1, γ_2; $\gamma_1 < 90°$, $\gamma_2 = 180° - \gamma_1 > 90°$. If $\alpha + \gamma_2 > 180°$, use only γ_1. This case may be *impossible* if $\sin \gamma > 1$.
4.   Given any side c and two angles α and β, $\gamma = 180° - \alpha - \beta$; $a = (c \sin \alpha)/(\sin \gamma)$; $b = (c \sin \beta)/(\sin \gamma)$.

HYPERBOLIC TRIGONOMETRY

The hyperbolic functions are certain combinations of exponentials e^x and e^{-x}.

$$\cosh x = \frac{e^x + e^{-x}}{2}; \sinh x = \frac{e^x - e^{-x}}{2}; \tanh x = \frac{\sinh x}{\cosh x} = \frac{e^x - e^{-x}}{e^x + e^{-x}}$$

$$\coth x = \frac{e^x + e^{-x}}{e^x - e^{-x}} = \frac{1}{\tanh x} = \frac{\cosh x}{\sinh x}; \operatorname{sech} x = \frac{1}{\cosh x} = \frac{2}{e^x + e^{-x}};$$

$$\operatorname{csch} x = \frac{1}{\sinh x} = \frac{2}{e^x - e^{-x}}$$

Fundamental Relationships $\sinh x + \cosh x = e^x$; $\cosh x - \sinh x = e^{-x}$; $\cosh^2 x - \sinh^2 x = 1$; $\operatorname{sech}^2 x + \tanh^2 x = 1$; $\coth^2 x - \operatorname{csch}^2 x = 1$; $\sinh 2x = 2 \sinh x \cosh x$; $\cosh 2x = \cosh^2 x + \sinh^2 x = 1 + 2 \sinh^2 x = 2 \cosh^2 x - 1$. $\tanh 2x = (2 \tanh x)/(1 + \tanh^2 x)$; $\sinh (x \pm y) = \sinh x \cosh y \pm \cosh x \sinh y$; $\cosh (x \pm y) = \cosh x \cosh y \pm \sinh x \sinh y$; $2 \sinh^2 x/2 = \cosh x - 1$; $2 \cosh^2 x/2 = \cosh x + 1$; $\sinh (-x) = -\sinh x$; $\cosh (-x) = \cosh x$; $\tanh (-x) = -\tanh x$.
When $u = a \cosh x$, $v = a \sinh x$, then $u^2 - v^2 = a^2$; which is the equation for a hyperbola. In other words, the hyperbolic functions in the parametric equations $u = a \cosh x$, $v = a \sinh x$ have the same relation to the hyperbola $u^2 - v^2 = a^2$ that the equations $u = a \cos \theta$, $v = a \sin \theta$ have to the circle $u^2 + v^2 = a^2$.

Inverse Hyperbolic Functions   If $x = \sinh y$, then y is the inverse hyperbolic sine of x written $y = \sinh^{-1} x$ or arcsinh x. $\sinh^{-1} x = \log_e (x + \sqrt{x^2 + 1})$

$$\cosh^{-1} x = \log_e (x + \sqrt{x^2 + 1}); \tanh^{-1} x = \frac{1}{2} \log_e \frac{1 + x}{1 - x};$$

$$\coth^{-1} x = \frac{1}{2} \log_e \frac{x + 1}{x - 1}; \operatorname{sech}^{-1} x = \log_e \left(\frac{1 + \sqrt{1 - x^2}}{x} \right);$$

$$\operatorname{csch}^{-1} = \log_e \left(\frac{1 + \sqrt{1 + x^2}}{x} \right)$$

Magnitude of the Hyperbolic Functions $\cosh x \geq 1$ with equality only for $x = 0$; $-\infty < \sinh x < \infty$; $-1 < \tanh x < 1$. $\cosh x \sim e^x/2$ as $x \to \infty$; $\sinh x \to e^x/2$ as $x \to \infty$.

APPROXIMATIONS FOR TRIGONOMETRIC FUNCTIONS

For small values of θ (θ measured in radians) $\sin \theta \approx \theta$, $\tan \theta \approx \theta$; $\cos \theta \approx 1 - (\theta^2/2)$. The following relations actually hold: $\sin \theta < \theta < \tan \theta$; $\cos \theta < \sin \theta/\theta < 1$; $\theta \sqrt{1 - \theta^2} < \sin \theta < \theta$; $\cos \theta < \theta/\tan \theta < 1$;

$$\theta \left(1 - \frac{\theta^2}{2} \right) < \sin \theta < \theta \text{ and } \theta < \tan \theta < \frac{\theta}{\sqrt{1 - \theta^2}}$$

The behavior ratio of the functions as $\theta \to 0$ is given by the following:

$$\lim_{\theta \to 0} \sin \theta/\theta = 1; \sin \theta/\tan \theta = 1.$$

DIFFERENTIAL AND INTEGRAL CALCULUS

REFERENCES: 114, 158, 260, 261, 274, 282, 296. See also "General References: References for General and Specific Topics—Advanced Calculus." For computer evaluations of the calculus described here, see Refs. 68, 299.

DIFFERENTIAL CALCULUS

An Example of Functional Notation   Suppose that a storage warehouse of 16,000 ft^3 is required. The construction costs per square foot are \$10, \$3, and \$2 for walls, roof, and floor respectively. What are the minimum cost dimensions? Thus, with h = height, x = width, and y = length, the respective costs are

$$\begin{aligned}
\text{Walls} &= 2 \times 10hy + 2 \times 10hx = 20h(y + x) \\
\text{Roof} &= 3xy \\
\text{Floor} &= 2xy \\
\text{Total cost} &= 2xy + 3xy + 20h(x + y) = 5xy + 20h(x + y) \quad (3\text{-}1)
\end{aligned}$$

and the restriction

$$\text{Total volume} = xyh \quad (3\text{-}2)$$

Solving for h from Eq. (3-2),

$$h = \text{volume}/xy = 16,000/xy \quad (3\text{-}3)$$

$$\text{Cost} = 5xy + \frac{320,000}{xy}(y + x) = 5xy + 320,000 \left(\frac{1}{x} + \frac{1}{y} \right) \quad (3\text{-}4)$$

In this form it can be shown that the minimum cost will occur for $x = y$; therefore

$$\text{Cost} = 5x^2 + 640,000 \, (1/x)$$

By evaluation, the smallest cost will occur when $x = 40$.

$$\text{Cost} = 5(1600) + 640,000/40 = \$24,000$$

The dimensions are then $x = 40$ ft, $y = 40$ ft, $h = 16,000/(40 \times 40) = 10$ ft. Symbolically, the original cost relationship is written

$$\text{Cost} = f(x, y, h) = 5xy + 20h(y + x)$$

and the volume relation

$$\text{Volume} = g(x, y, h) = xyh = 16,000$$

In terms of the derived general relationships (3-1) and (3-2), x, y, and h are **independent variables**—cost and volume, **dependent variables.** That is, the cost and volume become fixed with the specification of dimensions. However, corresponding to the given restriction of the problem, relative to volume, the function $g(x, y, z) = xyh$ becomes a constraint function. In place of three independent and two dependent variables the problem reduces to two independent (volume has been constrained) and two dependent as in functions (3-3) and (3-4). Further, the requirement of minimum cost reduces the problem to three dependent variables (x, y, h) and no **degrees of freedom,** that is, freedom of independent selection.

Limits   The limit of function $f(x)$ as x approaches a (a is finite or else x is said to increase without bound) is the number N.

$$\lim_{x \to a} f(x) = N$$

This states that $f(x)$ can be calculated as close to N as desirable by making x sufficiently close to a. This does not put any restriction on $f(x)$ when $x = a$. Alternatively, for any given positive number ε, a number δ can be found such that $0 < |a - x| < \delta$ implies that $|N - f(x)| < \varepsilon$.

The following operations with limits (when they exist) are valid:

$$\lim_{x \to a} bf(x) = b \lim_{x \to a} f(x)$$

$$\lim_{x \to a} [f(x) + g(x)] = \lim_{x \to a} f(x) + \lim_{x \to a} g(x)$$

$$\lim_{x \to a} [f(x)g(x)] = \lim_{x \to a} f(x) \cdot \lim_{x \to a} g(x)$$

$$\lim_{x \to a} \frac{f(x)}{g(x)} = \frac{\lim_{x \to a} f(x)}{\lim_{x \to a} g(x)} \quad \text{if} \quad \lim_{x \to a} g(x) \neq 0$$

Continuity   A function $f(x)$ is continuous at the point $x = a$ if

$$\lim_{h \to 0} [f(a + h) - f(a)] = 0$$

Rigorously, it is stated $f(x)$ is continuous at $x = a$ if for any positive ε there exists a $\delta > 0$ such that $|f(a + h) - f(a)| < \varepsilon$ for all x with $|x - a| < \delta$. For example, the function $(\sin x)/x$ is not continuous at $x = 0$ and therefore is said to be discontinuous. Discontinuities are classified into three types:

1.   Removable $y = \sin x/x$ at $x = 0$
2.   Infinite $y = 1/x$ at $x = 0$
3.   Jump $y = 10/(1 + e^{1/x})$ at $x = 0^+$ $y = 0^+$
         $x = 0$ $y = 0$
         $x = 0^-$ $y = 10$

Derivative   The function $f(x)$ has a derivative at $x = a$, which can be denoted as $f'(a)$, if

$$\lim_{h \to 0} \frac{f(a + h) - f(a)}{h}$$

exists. This implies continuity at $x = a$. Conversely, a function may be continuous but not have a derivative. The derivative function is

$$f'(x) = \frac{df}{dx} = \lim_{h \to 0} \frac{f(x + h) - f(x)}{h}$$

Differentiation   Define $\Delta y = f(x + \Delta x) - f(x)$. Then dividing by Δx

$$\frac{\Delta y}{\Delta x} = \frac{f(x + \Delta x) - f(x)}{\Delta x}$$

Call
$$\lim_{\Delta x \to 0} \frac{\Delta y}{\Delta x} = \frac{dy}{dx}$$

then
$$\frac{dy}{dx} = \lim_{\Delta x \to 0} \frac{f(x + \Delta x) - f(x)}{\Delta x}$$

Example   Find the derivative of $y = \sin x$.
$$\frac{dy}{dx} = \lim_{\Delta x \to 0} \frac{\sin(x + \Delta x) - \sin(x)}{\Delta x}$$
$$= \lim_{\Delta x \to 0} \frac{\sin x \cos \Delta x + \sin \Delta x \cos x - \sin x}{\Delta x}$$
$$= \lim_{\Delta x \to 0} \frac{\sin x(\cos \Delta x - 1)}{\Delta x} + \lim_{\Delta x \to 0} \frac{\sin \Delta x \cos x}{\Delta x}$$
$$= \cos x \text{ since } \lim_{\Delta x \to 0} \frac{\sin \Delta x}{\Delta x} = 1$$

Differential Operations   The following differential operations are valid: $f, g, \ldots$ are differentiable functions of x, c is a constant; e is the base of the natural logarithms.

$$\frac{dc}{dx} = 0 \tag{3-5}$$

$$\frac{dx}{dx} = 1 \tag{3-6}$$

$$\frac{d}{dx}(f + g) = \frac{df}{dx} + \frac{dg}{dx} \tag{3-7}$$

$$\frac{d}{dx}(f \times g) = f\frac{dg}{dx} + g\frac{df}{dx} \tag{3-8}$$

$$\frac{dy}{dx} = \frac{1}{dx/dy} \quad \text{if} \quad \frac{dx}{dy} \neq 0 \tag{3-9}$$

$$\frac{d}{dx}f^n = nf^{n-1}\frac{df}{dx} \tag{3-10}$$

$$\frac{d}{dx}\left(\frac{f}{g}\right) = \frac{g(df/dx) - f(dg/dx)}{g^2} \tag{3-11}$$

$$\frac{df}{dx} = \frac{df}{dv} \times \frac{dv}{dx} \quad \text{(chain rule)} \tag{3-12}$$

$$\frac{df^g}{dx} = gf^{g-1}\frac{df}{dx} + f^g \ln f \frac{dg}{dx} \tag{3-13}$$

$$\frac{da^x}{dx} = (\ln a)\, a^x \tag{3-14}$$

Example   Derive dy/dx for $x^2 + y^3 = x + xy + A$.

Here
$$\frac{d}{dx}x^2 + \frac{d}{dx}y^3 = \frac{d}{dx}x + \frac{d}{dx}xy + \frac{d}{dx}A$$
$$2x + 3y^2\frac{dy}{dx} = 1 + y + x\frac{dy}{dx} + 0$$

by rules (3-10), (3-10), (3-6), (3-8), and (3-5) respectively.

Thus
$$\frac{dy}{dx} = \frac{2x - 1 - y}{x - 3y^2}$$

Differentials

$$de^x = e^x\, dx \tag{3-15a}$$
$$d(a^x) = a^x \log a\, dx \tag{3-15b}$$
$$d \ln x = (1/x)\, dx \tag{3-16}$$
$$d \log x = (\log e/x)dx \tag{3-17}$$
$$d \sin x = \cos x\, dx \tag{3-18}$$

$$d \cos x = -\sin x\, dx \tag{3-19}$$
$$d \tan x = \sec^2 x\, dx \tag{3-20}$$
$$d \cot x = -\csc^2 x\, dx \tag{3-21}$$
$$d \sec x = \tan x \sec x\, dx \tag{3-22}$$
$$d \csc x = -\cot x \csc x\, dx \tag{3-23}$$
$$d \sin^{-1} x = (1 - x^2)^{-1/2}\, dx \tag{3-24}$$
$$d \cos^{-1}x = -(1 - x^2)^{-1/2}\, dx \tag{3-25}$$
$$d \tan^{-1} x = (1 + x^2)^{-1}\, dx \tag{3-26}$$
$$d \cot^{-1} x = -(1 + x^2)^{-1}\, dx \tag{3-27}$$
$$d \sec^{-1} x = x^{-1}(x^2 - 1)^{-1/2}\, dx \tag{3-28}$$
$$d \csc^{-1} x = -x^{-1}(x^2 - 1)^{-1/2}\, dx \tag{3-29}$$
$$d \sinh x = \cosh x\, dx \tag{3-30}$$
$$d \cosh x = \sinh x\, dx \tag{3-31}$$
$$d \tanh x = \operatorname{sech}^2 x\, dx \tag{3-32}$$
$$d \coth x = -\operatorname{csch}^2 x\, dx \tag{3-33}$$
$$d \operatorname{sech} x = -\operatorname{sech} x \tanh x\, dx \tag{3-34}$$
$$d \operatorname{csch} x = -\operatorname{csch} x \coth x\, dx \tag{3-35}$$
$$d \sinh^{-1} x = (x^2 + 1)^{-1/2}\, dx \tag{3-36}$$
$$d \cosh^{-1} x = (x^2 - 1)^{-1/2}\, dx \tag{3-37}$$
$$d \tanh^{-1} x = (1 - x^2)^{-1}\, dx \tag{3-38}$$
$$d \coth^{-1} x = -(x^2 - 1)^{-1}\, dx \tag{3-39}$$
$$d \operatorname{sech}^{-1} x = -(1/x)(1 - x^2)^{-1/2}\, dx \tag{3-40}$$
$$d \operatorname{csch}^{-1} x = -x^{-1}(x^2 + 1)^{-1/2}\, dx \tag{3-41}$$

Example   Find dy/dx for $y = \sqrt{x} \cos(1 - x^2)$.

$$\frac{dy}{dx} = \sqrt{x}\frac{d}{dx}\cos(1 - x^2) + \cos(1 - x^2)\frac{d}{dx}\sqrt{x} \quad \text{Using (3-8)}$$
$$\frac{d}{dx}\cos(1 - x^2) = -\sin(1 - x^2)\frac{d}{dx}(1 - x^2) \quad (3\text{-}19)$$
$$= -\sin(1 - x^2)(0 - 2x) \quad (3\text{-}5),\,(3\text{-}10)$$
$$\frac{d\sqrt{x}}{dx} = \frac{1}{2}x^{-1/2} \quad (3\text{-}10)$$
$$\frac{dy}{dx} = 2x^{3/2}\sin(1 - x^2) + \frac{1}{2}x^{-1/2}\cos(1 - x^2)$$

Example   Find the derivative of $\tan x$ with respect to $\sin x$.

$$v = \sin x$$
$$y = \tan x$$
$$\frac{d \tan x}{d \sin x} = \frac{dy}{dv} = \frac{dy}{dx}\frac{dx}{dv} \quad \text{Using (3-12)}$$
$$= \frac{d \tan x}{dx}\frac{1}{\dfrac{d \sin x}{dx}} \quad (3\text{-}9)$$
$$= \sec^2 x/\cos x \quad (3\text{-}18),\,(3\text{-}20)$$

Very often in experimental sciences and engineering functions and their derivatives are available only through their numerical values. In particular, through measurements we may know the values of a function and its derivative only at certain points. In such cases the preceding operational rules for derivatives, including the chain rule, can be applied numerically.

Example   Given the following table of values for differentiable functions f and g; evaluate the following quantities:

x	$f(x)$	$f'(x)$	$g(x)$	$g'(x)$
1	3	1	4	-4
3	0	2	4	7
4	-2	10	3	6

$$\frac{d}{dx}[f(x) + g(x)]|_{x=4} = f'(4) + g'(4) = 10 + 6 = 16$$

$$\left(\frac{f}{g}\right)'(1) = \frac{f'(1)g(1) - f(1)g'(1)}{[g(1)]^2} = \frac{1 \cdot 4 - 3(-4)}{(-4)^2} = \frac{16}{16} = 1$$

Higher Differentials   The first derivative of $f(x)$ with respect to x is denoted by f' or df/dx. The derivative of the first derivative is called the second derivative of $f(x)$ with respect to x and is denoted by f'', $f^{(2)}$, or d^2f/dx^2; and similarly for the higher-order derivatives.

Example   Given $f(x) = 3x^3 + 2x + 1$, calculate all derivative values at $x = 3$.

$$\frac{df(x)}{dx} = 9x^2 + 2 \qquad x = 3, f'(3) = 9(9) + 2 = 83$$

$$\frac{d^2f(x)}{dx^2} = 18x \qquad x = 3, f''(3) = 18(3) = 54$$

$$\frac{d^3f(x)}{dx^3} = 18 \qquad x = 3, f'''(3) = 18$$

$$\frac{d^nf(x)}{dx^n} = 0 \qquad \text{for } n \geq 4$$

If $f'(x) > 0$ on (a, b), then f is increasing on (a, b). If $f'(x) < 0$ on (a, b), then f is decreasing on (a, b).

The graph of a function $y = f(x)$ is concave up if f' is increasing on (a, b); it is concave down if f' is decreasing on (a, b).

If $f''(x)$ exists on (a, b) and if $f''(x) > 0$, then f is concave up on (a, b). If $f''(x) < 0$, then f is concave down on (a, b).

An inflection point is a point at which a function changes the direction of its concavity.

Indeterminate Forms: L'Hospital's Theorem   Forms of the type $0/0$, ∞/∞, $0 \times \infty$, etc., are called indeterminates. To find the limiting values that the corresponding functions approach, L'Hospital's theorem is useful: If two functions $f(x)$ and $g(x)$ both become zero at $x = a$, then the limit of their quotient is equal to the limit of the quotient of their separate derivatives, if the limit exists or is $+ \infty$ or $- \infty$.

Example   Find $\lim\limits_{n \to 0} \dfrac{\sin x}{x}$.

Here

$$\lim_{x \to 0} \frac{\sin x}{x} = \lim_{x \to 0} \frac{d \sin x}{dx} = \lim_{x \to 0} \frac{\cos x}{1} = 1$$

Example   Find $\lim\limits_{x \to \infty} \dfrac{(1.1)^x}{x^{1000}}$.

$$\lim_{x \to \infty} \frac{(1.1)^x}{x^{1000}} = \lim_{x \to \infty} \frac{d(1.1)^x}{dx^{1000}} = \lim_{x \to \infty} \frac{(\ln 1.1)(1.1)^x}{1000x^{999}}$$

Obviously $\lim\limits_{x \to \infty} \dfrac{1.1^x}{x^{1000}} = \infty$ since repeated application of the rule will reduce the denominator to a finite number 1000! while the numerator remains infinitely large.

Example   Find $\lim\limits_{x \to \infty} x^3 e^{-x}$.

$$\lim_{x \to \infty} x^3 e^{-x} = \lim_{x \to \infty} \frac{x^3}{e^x} = \lim_{x \to \infty} \frac{6}{e^x} = 0$$

Example   Find $\lim\limits_{x \to 0} (1 - x)^{1/x}$.

Let

$$y = (1 - x)^{1/x}$$

$$\ln y = (1/x) \ln (1 - x)$$

$$\lim_{x \to 0} (\ln y) = \lim_{x \to 0} \frac{\ln(1 - x)}{x} = -1$$

Therefore,

$$\lim_{x \to 0} y = e^{-1}$$

Partial Derivative   The abbreviation $z = f(x, y)$ means that z is a function of the two variables x and y. The derivative of z with respect to x, treating y as a constant, is called the partial derivative with respect to x and is usually denoted as $\partial z/\partial x$ or $\partial f(x, y)/\partial x$ or simply f_x. Partial differentiation, like full differentiation, is quite simple to apply. Conversely, the solution of partial differential equations is appreciably more difficult than that of differential equations.

Example   Find $\partial z/\partial x$ and $\partial z/\partial y$ for $z = ye^{x^2} + xe^y$.

$$\frac{\partial z}{\partial x} = y\frac{\partial e^{x^2}}{\partial x} + e^y\frac{\partial x}{\partial x} \qquad \frac{\partial z}{\partial y} = e^{x^2}\frac{\partial y}{\partial y} + x\frac{\partial e^y}{\partial y}$$

$$= 2xye^{x^2} + e^y \qquad\qquad = e^{x^2} + xe^y$$

Order of Differentiation   It is generally true that the order of differentiation is immaterial for any number of differentiations or variables provided the function and the appropriate derivatives are continuous. For $z = f(x, y)$ it follows:

$$\frac{\partial^3 f}{\partial y^2 \, \partial x} = \frac{\partial^3 f}{\partial y \, \partial x \, \partial y} = \frac{\partial^3 f}{\partial x \, \partial y^2}$$

General Form for Partial Differentiation
1. Given $f(x, y) = 0$ and $x = g(t)$, $y = h(t)$.

Then $\dfrac{df}{dt} = \dfrac{\partial f}{\partial x}\dfrac{dx}{dt} + \dfrac{\partial f}{\partial y}\dfrac{dy}{dt}$

$$\frac{d^2f}{dt^2} = \frac{\partial^2 f}{\partial x^2}\left(\frac{dx}{dt}\right)^2 + 2\frac{\partial^2 f}{\partial x \, \partial y}\frac{dx}{dt}\frac{dy}{dt} + \frac{\partial^2 f}{\partial y^2}\left(\frac{dy}{dt}\right)^2 + \frac{\partial f}{\partial x}\frac{d^2x}{dt^2}$$

$$+ \frac{\partial f}{\partial y}\frac{d^2y}{dt^2}$$

Example   Find df/dt for $f = xy$, $x = \rho \sin t$, $y = \rho \cos t$.

$$\frac{df}{dt} = \frac{\partial(xy)}{\partial x}\left(\frac{d\rho \sin t}{dt}\right) + \frac{\partial(xy)}{\partial y}\left(\frac{d\rho \cos t}{dt}\right)$$

$$= y(\rho \cos t) + x(-\rho \sin t)$$

$$= \rho^2 \cos^2 t - \rho^2 \sin^2 t$$

2. Given $f(x, y) = 0$ and $x = g(t, s)$, $y = h(t, s)$.

Then

$$\frac{\partial f}{\partial t} = \frac{\partial f}{\partial x}\frac{\partial x}{\partial t} + \frac{\partial f}{\partial y}\frac{\partial y}{\partial t}$$

$$\frac{\partial f}{\partial s} = \frac{\partial f}{\partial x}\frac{\partial x}{\partial s} + \frac{\partial f}{\partial y}\frac{\partial y}{\partial x}$$

Differentiation of Composite Function

Rule 1.   Given $f(x, y) = 0$, then $\dfrac{dy}{dx} = -\dfrac{\partial f/\partial x}{\partial f/\partial y}\left(\dfrac{\partial f}{\partial y} \neq 0\right)$.

Rule 2.   Given $f(u) = 0$ where $u = g(x)$, then

$$\frac{df}{dx} = f'(u)\frac{du}{dx}$$

$$\frac{d^2f}{dx^2} = f''(u)\left(\frac{du}{dx}\right)^2 + f'(u)\frac{d^2u}{dx^2}$$

Example   Find df/dx for $f = \sin^2 u$ and $u = \sqrt{1 - x^2}$.

$$\frac{df}{dx} = \frac{d \sin^2 u}{du}\frac{d\sqrt{1 - x^2}}{dx}$$

$$= 2 \sin u \cos u\left(\frac{1}{2}\right)(-2x)(1 - x^2)^{-1/2}$$

$$= -2\frac{\sqrt{1 - u^2}}{u} \sin u \cos u$$

Rule 3.   Given $f(u) = 0$ where $u = g(x, y)$, then

$$\frac{\partial f}{\partial x} = f'(u)\frac{\partial u}{\partial x} \qquad \frac{\partial f}{\partial y} = f'(u)\frac{\partial u}{\partial y}$$

$$\frac{\partial^2 f}{\partial x^2} = f'' \left(\frac{\partial u}{\partial x}\right)^2 + f' \frac{\partial^2 u}{\partial x^2}$$

$$\frac{\partial^2 f}{\partial x \, \partial y} = f'' \frac{\partial u}{\partial x} \frac{\partial u}{\partial y} + f' \frac{\partial^2 u}{\partial x \, \partial y}$$

$$\frac{\partial^2 f}{\partial y^2} = f'' \left(\frac{\partial u}{\partial y}\right)^2 + f' \frac{\partial^2 u}{\partial y^2}$$

MULTIVARIABLE CALCULUS APPLIED TO THERMODYNAMICS

Many of the functional relationships needed in thermodynamics are direct applications of the rules of multivariable calculus. This section reviews those rules in the context of the needs of themodynamics. These ideas were expounded in one of the classic books on chemical engineering thermodynamics.[151]

State Functions   State functions depend only on the state of the system, not on past history or how one got there. If z is a function of two variables, x and y, then $z(x,y)$ is a state function, since z is known once x and y are specified. The differential of z is

$$dz = M \, dx + N \, dy$$

The line integral

$$\int_C (M \, dx + N \, dy)$$

is independent of the path in x-y space if and only if

$$\frac{\partial M}{\partial y} = \frac{\partial N}{\partial x} \qquad (3\text{-}42)$$

The total differential can be written as

$$dz = \left(\frac{\partial z}{\partial x}\right)_y dx + \left(\frac{\partial z}{\partial y}\right)_x dy \qquad (3\text{-}43)$$

and the following condition guarantees path independence.

$$\frac{\partial}{\partial y}\left(\frac{\partial z}{\partial x}\right)_y = \frac{\partial}{\partial x}\left(\frac{\partial z}{\partial y}\right)_x$$

or

$$\frac{\partial^2 z}{\partial y \, \partial x} = \frac{\partial^2 z}{\partial x \, \partial y} \qquad (3\text{-}44)$$

Example   Suppose z is constant and apply Eq. (3-43).

$$\left[0 = \left(\frac{\partial z}{\partial x}\right)_y dx + \left(\frac{\partial z}{\partial y}\right)_x dy \right]_z$$

Rearrangement gives

$$\left(\frac{\partial z}{\partial x}\right)_y = -\left(\frac{\partial y}{\partial x}\right)_z \left(\frac{\partial z}{\partial y}\right)_x = -\frac{(\partial y/\partial x)_z}{(\partial y/\partial z)_x} \qquad (3\text{-}45)$$

Alternatively, divide Eq. (3-43) by dy when holding some other variable w constant to obtain

$$\left(\frac{\partial z}{\partial y}\right)_w = \left(\frac{\partial z}{\partial x}\right)_v \left(\frac{\partial x}{\partial y}\right)_w + \left(\frac{\partial z}{\partial y}\right)_x \qquad (3\text{-}46)$$

Also divide both numerator and denominator of a partial derivative by dw while holding a variable y constant to get

$$\left(\frac{\partial z}{\partial x}\right)_y = \frac{(\partial z/\partial w)_y}{(\partial x/\partial w)_y} = \left(\frac{\partial z}{\partial w}\right)_y \left(\frac{\partial w}{\partial x}\right)_y \qquad (3\text{-}47)$$

Themodynamic State Functions   In thermodynamics, the state functions include the internal energy, U; enthalpy, H; and Helmholtz and Gibbs free energies, A and G, respectively, defined as follows:

$$H = U + p \, V$$
$$A = U - TS$$
$$G = H - TS = U + pV - TS = A + pV$$

S is the entropy, T the absolute temperature, p the pressure, and V the volume. These are also state functions, in that the entropy is specified once two variables (like T and p) are specified, for example. Likewise,

V is specified once T and p are specified; it is therefore a state function.

All applications are for closed systems with constant mass. If a process is reversible and only p-V work is done, the first law and differentials can be expressed as follows:

$$dU = T \, dS - p \, dV$$
$$dH = T \, dS + V \, dp$$
$$dA = -S \, dT - p \, dV$$
$$dG = -S \, dT + V \, dp$$

Alternatively, if the internal energy is considered a function of S and V, then the differential is:

$$dU = \left(\frac{\partial U}{\partial S}\right)_V dS + \left(\frac{\partial U}{\partial V}\right)_S dV$$

This is the equivalent of Eq. (3-43) and gives the following definitions.

$$T = \left(\frac{\partial U}{\partial S}\right)_V, \quad p = -\left(\frac{\partial U}{\partial V}\right)_S$$

Since the internal energy is a state function, then Eq. (3-44) must be satisfied.

$$\frac{\partial^2 U}{\partial V \, \partial S} = \frac{\partial^2 U}{\partial S \, \partial V}$$

This is

$$\left(\frac{\partial T}{\partial V}\right)_S = -\left(\frac{\partial p}{\partial S}\right)_V$$

This is one of the Maxwell relations, and the other Maxwell relations can be derived in a similar fashion by applying Eq. (3-44).

$$\left(\frac{\partial T}{\partial p}\right)_S = \left(\frac{\partial V}{\partial S}\right)_p$$
$$\left(\frac{\partial S}{\partial V}\right)_T = \left(\frac{\partial p}{\partial T}\right)_V$$
$$\left(\frac{\partial S}{\partial p}\right)_T = -\left(\frac{\partial V}{\partial T}\right)_p$$

In process simulation it is necessary to calculate enthalpy as a function of state variables. This is done using the following formulas, derived from the above relations by considering S and H as functions of T and p.

$$dH = C_p \, dT + \left[V - T\left(\frac{\partial V}{\partial T}\right)_p\right] dp$$

Enthalpy differences are then given by the following formula.

$$H(T_2, p_2) - H(T_1, p_1) = \int_{T_1}^{T_2} C_p(T, p_1) \, dT + \int_{p_1}^{p_2} \left[V - T\left(\frac{\partial V}{\partial T}\right)_p\right]\Bigg|_{T_2, p} dp$$

The same manipulations can be done for internal energy as a function of T and V.

$$dU = C_V \, dT - \left[p + T\frac{(\partial V/\partial T)_p}{(\partial V/\partial p)_T}\right] dV$$

Partial Derivatives of All Thermodynamic Functions   The various partial derivatives of the thermodynamic functions can be classified into six groups. In the general formulas below, the variables U, H, A, G or S are denoted by Greek letters, while the variables V, T, or p are denoted by Latin letters.

Type I   (3 possibilities plus reciprocals)

$$\text{General: } \left(\frac{\partial a}{\partial b}\right)_c; \quad \text{Specific: } \left(\frac{\partial p}{\partial T}\right)_V$$

Eq. (3-45) gives

$$\left(\frac{\partial p}{\partial T}\right)_V = -\left(\frac{\partial V}{\partial T}\right)_p \left(\frac{\partial p}{\partial V}\right)_T = -\frac{(\partial V/\partial T)_p}{(\partial V/\partial p)_T}$$

Type II   (30 possibilities)

$$\text{General: } \left(\frac{\partial \alpha}{\partial b}\right)_c; \quad \text{Specific: } \left(\frac{\partial G}{\partial T}\right)_V$$

The differential for G gives

$$\left(\frac{\partial G}{\partial T}\right)_V = -S + V\left(\frac{\partial p}{\partial T}\right)_V$$

Using the other equations for U, H, A, or S gives the other possibilities.

Type III   (15 possibilities plus reciprocals)

General: $\left(\dfrac{\partial a}{\partial b}\right)_\alpha$; Specific: $\left(\dfrac{\partial V}{\partial T}\right)_S$

First expand the derivative using Eq. (3-45).

$$\left(\frac{\partial V}{\partial T}\right)_S = -\left(\frac{\partial S}{\partial T}\right)_V\left(\frac{\partial V}{\partial S}\right)_T = -\frac{(\partial S/\partial T)_V}{(\partial S/\partial V)_T}$$

Then evaluate the numerator and denominator as type II derivatives.

$$\left(\frac{\partial V}{\partial T}\right)_S = -\frac{\dfrac{C_V}{T}}{-\left(\dfrac{\partial V}{\partial T}\right)_p\left(\dfrac{\partial p}{\partial V}\right)_T} = \frac{C_V}{T}\frac{\left(\dfrac{\partial V}{\partial p}\right)_T}{\left(\dfrac{\partial V}{\partial T}\right)_p}$$

These derivatives are of importance for reversible, adiabatic processes (such as in an ideal turbine or compressor), since then the entropy is constant. An example is the Joule-Thomson coefficient.

$$\left(\frac{\partial T}{\partial p}\right)_H = \frac{1}{C_p}\left[-V + T\left(\frac{\partial V}{\partial T}\right)_p\right]$$

Type IV   (30 possibilities plus reciprocals)

General: $\left(\dfrac{\partial \alpha}{\partial \beta}\right)_c$; Specific: $\left(\dfrac{\partial G}{\partial A}\right)_p$

Use Eq. (3-47) to introduce a new variable.

$$\left(\frac{\partial G}{\partial A}\right)_p = \left(\frac{\partial G}{\partial T}\right)_p\left(\frac{\partial T}{\partial A}\right)_p = \frac{(\partial G/\partial T)_p}{(\partial A/\partial T)_p}$$

This operation has created two type II derivatives; by substitution we obtain

$$\left(\frac{\partial G}{\partial A}\right)_p = \frac{S}{S + p\,(\partial V/\partial T)_p}$$

Type V   (60 possibilities)

General: $\left(\dfrac{\partial \alpha}{\partial b}\right)_\beta$; Specific: $\left(\dfrac{\partial G}{\partial p}\right)_A$

Start from the differential for dG. Then we get

$$\left(\frac{\partial G}{\partial p}\right)_A = -S\left(\frac{\partial T}{\partial p}\right)_A + V$$

The derivative is type III and can be evaluated by using Eq. (3-45).

$$\left(\frac{\partial G}{\partial p}\right)_A = S\frac{(\partial A/\partial p)_T}{(\partial A/\partial T)_p} + V$$

The two type II derivatives are then evaluated.

$$\left(\frac{\partial G}{\partial p}\right)_A = \frac{Sp\,(\partial V/\partial p)_T}{S + p\,(\partial V/\partial T)_p} + V$$

These derivatives are also of interest for free expansions or isentropic changes.

Type VI   (30 possibilities plus reciprocals)

General: $\left(\dfrac{\partial \alpha}{\partial \beta}\right)_\gamma$; Specific: $\left(\dfrac{\partial G}{\partial A}\right)_H$

We use Eq. (3-47) to obtain two type V derivatives.

$$\left(\frac{\partial G}{\partial A}\right)_H = \frac{(\partial G/\partial T)_H}{(\partial A/\partial T)_H}$$

These can then be evaluated using the procedures for Type V derivatives.

INTEGRAL CALCULUS

Indefinite Integral   If $f'(x)$ is the derivative of $f(x)$, an antiderivative of $f'(x)$ is $f(x)$. Symbolically, the indefinite integral of $f'(x)$ is

$$\int f'(x)\,dx = f(x) + c$$

where c is an arbitrary constant to be determined by the problem. By virtue of the known formulas for differentiation the following relationships hold (a is a constant):

$$\int (du + dv + dw) = \int du + \int dv + \int dw \tag{3-48}$$

$$\int a\,dv = a\int dv \tag{3-49}$$

$$\int v^n\,dv = \frac{v^{n+1}}{n+1} + c\ (n \neq -1) \tag{3-50}$$

$$\int \frac{dv}{v} = \ln|v| + c \tag{3-51}$$

$$\int a^v\,dv = \frac{a^v}{\ln a} + c \tag{3-52}$$

$$\int e^v\,dv = e^v + c \tag{3-53}$$

$$\int \sin v\,dv = -\cos v + c \tag{3-54}$$

$$\int \cos v\,dv = \sin v + c \tag{3-55}$$

$$\int \sec^2 v\,dv = \tan v + c \tag{3-56}$$

$$\int \csc^2 v\,dv = -\cot v + c \tag{3-57}$$

$$\int \sec v \tan v\,dv = \sec v + c \tag{3-58}$$

$$\int \csc v \cot v\,dv = -\csc v + c \tag{3-59}$$

$$\int \frac{dv}{v^2 + a^2} = \frac{1}{a}\tan^{-1}\frac{v}{a} + c \tag{3-60}$$

$$\int \frac{dv}{\sqrt{a^2 - v^2}} = \sin^{-1}\frac{v}{a} + c \tag{3-61}$$

$$\int \frac{dv}{v^2 - a^2} = \frac{1}{2a}\ln\left|\frac{v-a}{v+a}\right| + c \tag{3-62}$$

$$\int \frac{dv}{\sqrt{v^2 \pm a^2}} = \ln|v + \sqrt{v^2 \pm a^2}| + c \tag{3-63}$$

$$\int \sec v\,dv = \ln(\sec v + \tan v) + c \tag{3-64}$$

$$\int \csc v\,dv = \ln(\csc v - \cot v) + c \tag{3-65}$$

Example   Derive $\int a^v\,dv = (a^v/\ln a) + c$. By reference to the differentiation formula $da^v/dv = a^v \ln a$, or in the more usable form $d(a^v/\ln a) = a^v\,dv$, let $f' = a^v\,dv$; then $f = a^v/\ln a$ and hence $\int a^v\,dv = (a^v/\ln a) + c$.

Example   Find $\int (3x^2 + e^x - 10)\,dx$ using Eq. (3-48). $\int (3x^2 + e^x - 10)\,dx = 3\int x^2\,dx + \int e^x\,dx - 10\int dx = x^3 + e^x - 10x + c$ (by Eqs. 3-50, 3-53).

Example   Find $\int \dfrac{7x\,dx}{2 - 3x^2}$. Let $v = 2 - 3x^2$; $dv = -6x\,dx$

Thus

$$\int \frac{7x\,dx}{2 - 3x^2} = 7\int \frac{x\,dx}{2 - 3x^2} = -\frac{7}{6}\int \frac{-6x\,dx}{2 - 3x^2}$$

$$= -\frac{7}{6}\int \frac{dv}{v}$$

$$= -\frac{7}{6}\ln|v| + c$$

$$= -\frac{7}{6}\ln|2 - 3x^2| + c$$

Example—Constant of Integration By definition the derivative of x^3 is $3x^2$, and x^3 is therefore the integral of $3x^2$. However, if $f = x^3 + 10$, it follows that $f' = 3x^2$, and $x^3 + 10$ is therefore also the integral of $3x^2$. For this reason the constant c in $\int 3x^2\, dx = x^3 + c$ must be determined by the problem conditions, i.e., the value of f for a specified x.

Methods of Integration In practice it is rare when generally encountered functions can be directly integrated. For example, the integrand in $\int \sqrt{\sin x}\, dx$ which appears quite simple has no elementary function whose derivative is $\sqrt{\sin x}$. In general, there is no explicit way of determining whether a particular function can be integrated into an elementary form. As a whole, integration is a trial-and-error proposition which depends on the effort and ingenuity of the practitioner. The following are general procedures which can be used to find the elementary forms of the integral when they exist. When they do not exist or cannot be found either from tabled integration formulas or directly, the only recourse is series expansion as illustrated later. Indefinite integrals cannot be solved numerically unless they are redefined as definite integrals (see "Definite Integral"), i.e., $F(x) = \int f(x)\, dx$, indefinite, whereas $F(x) = \int_a^x f(t)\, dt$, definite.

Direct Formula Many integrals can be solved by transformation in the integrand to one of the forms given previously.

Example Find $\int x^2 \sqrt{3x^3 + 10}\, dx$. Let $v = 3x^3 + 10$ for which $dv = 9x^2\, dx$. Thus

$$\int x^2 \sqrt{3x^3 + 10}\, dx = \int (3x^3 + 10)^{1/2}\, (x^2\, dx)$$

$$= \frac{1}{9} \int (3x^3 + 10)^{1/2} (9x^2\, dx)$$

$$= \frac{1}{9} \int v^{1/2}\, dv$$

$$= \frac{1}{9} \frac{v^{3/2}}{3/2} + c \quad \text{[by Eq. (3-50)]}$$

$$= \frac{2}{27}(3x^3 + 10)^{3/2} + c$$

Trigonometric Substitution This technique is particularly well adapted to integrands in the form of radicals. For these the function is transformed into a trigonometric form. In the latter form they may be more easily recognizable relative to the identity formulas. These functions and their transformations are

$$\sqrt{x^2 - a^2} \quad \text{Let } x = a \sec \theta$$
$$\sqrt{x^2 + a^2} \quad \text{Let } x = a \tan \theta$$
$$\sqrt{a^2 - x^2} \quad \text{Let } x = a \sin \theta$$

Example Find $\int \dfrac{\sqrt{4 - 9x^2}}{x^2}\, dx$. Let $x = \dfrac{2}{3} \sin \theta$; then $dx = \dfrac{2}{3} \cos \theta\, d\theta$.

$$3 \int \frac{\sqrt{(2/3)^2 - x^2}}{x^2}\, dx = 3 \int \frac{2/3\sqrt{1 - \sin^2 \theta}}{(2/3)^2 \sin^2 \theta} \left(\frac{2}{3} \cos \theta\, d\theta \right)$$

$$= 3 \int \frac{\cos^2 \theta}{\sin^2 \theta}\, d\theta$$

$$= 3 \int \cot^2 \theta\, d\theta$$

$$= -3 \cot \theta - 3\theta + c \text{ by trigonometric transform}$$

$$= -\frac{\sqrt{4 - 9x^2}}{x} - 3 \sin^{-1} \frac{3}{2} x + c \text{ in terms of } x$$

Algebraic Substitution Functions containing elements of the type $(a + bx)^{1/n}$ are best handled by the algebraic transformation $y^n = a + bx$.

Example Find $\int \dfrac{x\, dx}{(3 + 4x)^{1/4}}$. Let $3 + 4x = y^4$; then $4dx = 4y^3\, dy$ and

$$\int \frac{x\, dx}{(3 + 4x)^{1/4}} = \int \frac{\dfrac{y^4 - 3}{4}\, y^3\, dy}{y}$$

$$= \frac{1}{4} \int y^2(y^4 - 3)\, dy$$

$$= \frac{1}{4} \frac{y^7}{7} - \frac{3}{4} \frac{y^3}{3} + c$$

$$= \frac{1}{28} (3 + 4x)^{7/4} - \frac{1}{4} (3 + 4x)^{3/4} + c$$

General The number of possible transformations one might use are unlimited. No specific overall rules can be given. Success in handling integration problems depends primarily upon experience and ingenuity. The following example illustrates the extent to which alternative approaches are possible.

Example Find $\int \dfrac{dx}{e^x - 1}$. Let $e^x = y$; then $e^x\, dx = dy$ or $dx = 1/y\, dy$.

$$\int \frac{dx}{e^x - 1} = \int \frac{(1/y)\, dy}{y - 1} = \int \frac{dy}{y^2 - y} = \ln \frac{y - 1}{y} = \ln \frac{e^x - 2}{e^x}$$

Partial Fractions Rational functions are of the type $f(x)/g(x)$ where $f(x)$ and $g(x)$ are polynomial expressions of degrees m and n respectively. If the degree of f is higher than g, perform the algebraic division—the remainder will then be at least one degree less than the denominator. Consider the following types:

Type 1 Reducible denominator to linear unequal factors. For example,

$$\frac{1}{x^3 - x^2 - 4x + 4} = \frac{1}{(x + 2)(x - 2)(x - 1)}$$

$$= \frac{A}{x + 2} + \frac{B}{x - 2} + \frac{C}{x - 1}$$

$$= \frac{A(x - 2)(x - 1) + B(x + 2)(x - 1) + C(x + 2)(x - 2)}{(x + 2)(x - 2)(x - 1)}$$

$$= \frac{x^2(A + B + C) + x(-3A + B) + (2A - 2B - 4C)}{(x + 2)(x - 2)(x - 1)}$$

Equate coefficients and solve for A, B, and C.

$$A + B + C = 0$$
$$-3A + B = 0$$
$$2A - 2B - 4C = 1$$
$$A = \tfrac{1}{12},\ B = \tfrac{1}{4},\ C = -\tfrac{1}{3}$$

$$\frac{1}{x^3 - x^2 - 4x + 4} = \frac{1}{12(x + 2)} + \frac{1}{4(x - 2)} - \frac{1}{3(x - 1)}$$

Hence

$$\int \frac{dx}{x^3 - x^2 - 4x + 4} = \int \frac{dx}{12(x + 2)} + \int \frac{dx}{4(x - 2)} - \int \frac{dx}{3(x - 1)}$$

Parts An extremely useful formula for integration is the relation

$$d(uv) = u\, dv + v\, du$$

and

$$uv = \int u\, dv + \int v\, du$$

or

$$\int u\, dv = uv - \int v\, du$$

No general rule for breaking an integrand can be given. Experience alone limits the use of this technique. It is particularly useful for trigonometric and exponential functions.

Example Find $\int xe^x\, dx$. Let

$$u = x \quad \text{and} \quad dv = e^x\, dx$$
$$du = dx \qquad\qquad v = e^x$$

Therefore
$$\int xe^x\,dx = xe^x - \int e^x\,dx$$
$$= xe^x - e^x + c$$

Example   Find $\int e^x \sin x\,dx$. Let

$$u = e^x \qquad\qquad dv = \sin x\,dx$$
$$du = e^x\,dx \qquad\qquad v = -\cos x$$

$$\int e^x \sin x\,dx = -e^x \cos x + \int e^x \cos x\,dx$$

Again
$$u = e^x \qquad\qquad dv = \cos x\,dx$$
$$du = e^x\,dx \qquad\qquad v = \sin x$$

$$\int e^x \sin x\,dx = -e^x \cos x + e^x \sin x - \int e^x \sin x\,dx + c$$
$$= (e^x/2)(\sin x - \cos x) + \frac{c}{2}$$

Series Expansion   When an explicit function cannot be found, the integration can sometimes be carried out by a series expansion.

Example   Find $\int e^{-x^2}\,dx$. Since

$$e^{-x^2} = 1 - x^2 + \frac{x^4}{2!} - \frac{x^6}{3!} + \cdots$$

$$\int e^{-x^2}\,dx = \int dx - \int x^2\,dx + \int \frac{x^4}{2!}\,dx - \int \frac{x^6}{3!}\,dx + \cdots$$

$$= x - \frac{x^3}{3} + \frac{x^5}{5.2!} - \frac{x^7}{7.3!} + \cdots \qquad \text{for all } x$$

Definite Integral   The concept and derivation of the definite integral are completely different from those for the indefinite integral. These are by definition different types of operations. However, the formal operation $\int$ as it turns out treats the integrand in the same way for both.

Consider the function $f(x) = 10 - 10e^{-2x}$. Define $x_1 = a$ and $x_n = b$, and suppose it is desirable to compute the area between the curve and the coordinate axis $y = 0$ and bounded by $x_1 = a$, $x_n = b$. Obviously, by a sufficiently large number of rectangles this area could be approximated as closely as desired by the formula

$$\sum_{i=1}^{n-1} f(\xi_i)(x_{i+1} - x_i) = f(\xi_1)(x_2 - a) + f(\xi_2)(x_3 - x_2)$$
$$+ \cdots + f(\xi_{n-1})(b - x_{n-1}) \qquad x_{i-1} \le \xi_{i-1} \le x_i$$

The definite integral of $f(x)$ is defined as

$$\int_a^b f(x)\,dx = \lim_{n \to \infty} \sum_{i=1}^{n} f(\xi_i)(x_{i+1} - x_i)$$

where the points $x_1, x_2, \ldots, x_n$ are equally spaced. For a rigorous definition of the definite integral the references should be consulted.

Thus, the value of a definite integral depends on the limits a, b, and any selected variable coefficients in the function but not on the dummy variable of integration x. Symbolically

$$F(x) = \int f(x)\,dx \qquad \text{indefinite integral where } dF/dx = f(x)$$

or $$F(a, b) = \int_a^b f(x)\,dx \qquad \text{definite integral}$$

$$F(\alpha) = \int_a^b f(x, \alpha)\,dx$$

There are certain restrictions of the integration definition, "The function $f(x)$ must be continuous in the finite interval (a, b) with at most a finite number of finite discontinuities," which must be observed before integration formulas can be generally applied. Two of these restrictions give rise to so-called **improper integrals** and require special handling. These occur when
1.   The limits of integration are not both finite, i.e., $\int_0^\infty e^{-x}\,dx$.
2.   The function becomes infinite within the interval of integration, i.e.,

$$\int_0^1 \frac{1}{\sqrt{x}}\,dx$$

Techniques for determining when integration is valid under these conditions are available in the references. However, the following simplified rules will, in general, serve as a guide for most practical applications.

Rule 1   For the integral

$$\int_0^\infty \frac{\phi(x)}{x^n}\,dx$$

if $\phi(x)$ is bounded, the integral will converge for $n > 1$ and not converge for $n \le 1$.

It is easily seen that $\int_0^\infty e^{-x}\,dx$ converges by noting $1/x^2 > 1/e^x > 0$ for large x.

Rule 2   For the integral

$$\int_a^b \frac{\phi(x)}{(a-x)^n}\,dx,$$

if $\phi(x)$ is bounded, the integral will converge for $n < 1$ and diverge for $n \ge 1$. Thus

$$\int_0^1 \frac{1}{\sqrt{x}}\,dx$$

will converge (exist) since $\frac{1}{2} = n < 1$.

Properties   The fundamental theorem of calculus states

$$\int_a^b f(x)\,dx = F(b) - F(a)$$

where $$dF(x)/dx = f(x)$$

Other properties of the definite integral are

$$\int_a^b c[f(x)\,dx] = c \int_a^b f(x)\,dx$$

$$\int_a^b [f_1(x) + f_2(x)]\,dx = \int_a^b f_1(x)\,dx + \int_a^b f_2(x)\,dx$$

$$\int_a^b f(x)\,dx = -\int_b^a f(x)\,dx$$

$$\int_a^b f(x)\,dx = \int_a^c f(x)\,dx + \int_c^b f(x)\,dx$$

$$\int_a^b f(x)\,dx = (b - a)f(\xi) \quad \text{for some } \xi \text{ in } (a, b)$$

$$\frac{\partial}{\partial b} \int_a^b f(x)\,dx = f(b)$$

$$\frac{\partial}{\partial a} \int_a^b f(x)\,dx = -f(a)$$

$$\frac{dF(\alpha)}{d\alpha} = \int_a^b \frac{\partial f(x, \alpha)}{\partial \alpha}\,dx \quad \text{if } a \text{ and } b \text{ are constant}$$

$$\int_a^b dx \int_c^d f(x, \alpha)\,d\alpha = \int_c^d d\alpha \int_a^b f(x, \alpha)\,dx \qquad (3\text{-}66)$$

when $F(x) = \int_{a(x)}^{b(x)} f(x, y)\,dy$

the Leibniz rule gives

$$\frac{dF}{dx} = \frac{db}{dx} f[x, b(x)] - \frac{da}{dx} f[x, a(x)] + \int_{a(x)}^{b(x)} \frac{\partial f}{\partial x}\,dy$$

Example   Find $\int_0^{\pi/2} \sin x\,dx$.

$$\int_0^{\pi/2} \sin x\,dx = [-\cos x]_0^{\pi/2} = -\left(\cos \frac{\pi}{2} - \cos 0\right) = 1$$

since $$-d\cos x/dx = \sin x$$

Example   Find $\int_0^2 \dfrac{dx}{(x-1)^2}$. Direct application of the formula would yield the incorrect value

$$\int_0^2 \frac{dx}{(x-1)^2} = \left[-\frac{1}{x-1}\right]_0^2 = -2$$

It should be noted that $f(x) = 1/(x-1)^2$ becomes unbounded as $x \to 1$ and by Rule 2 the integral diverges and hence is said not to exist.

Methods of Integration   All the methods of integration available for the indefinite integral can be used for definite integrals. In addition, several others are available for the latter integrals and are indicated below.

Change of Variable   This substitution is basically the same as previously indicated for indefinite integrals. However, for definite integrals, the limits of integration must also be changed: i.e., for $x = \phi(t)$,

$$\int_a^b f(x)\,dx = \int_{t_0}^{t_1} f[\phi(t)]\phi'(t)\,dt$$

where $t = t_0$ when $x = a$
           $t = t_1$ when $x = b$

Example   Find $\int_0^4 \sqrt{16 - x^2}\,dx$. Let

$$x = 4 \sin\theta \qquad (x = 0, \theta = 0)$$
$$dx = 4 \cos\theta\,d\theta \qquad (x = 4, \theta = \pi/2)$$

Then $\int_0^4 \sqrt{16 - x^2}\,dx = 16\int_0^{\pi/2} \cos^2\theta\,d\theta = 16[\tfrac{1}{2}\theta + \tfrac{1}{4}\sin 2\theta]_0^{\pi/2} = 4\pi$

Differentiation   Here the application of the general rules for differentiating under the integral sign may be useful.

Example   Find

$$\phi(\alpha) = \int_0^\infty \frac{e^{-\alpha x}\sin x}{x}\,dx \quad (\alpha > 0)$$

Since this is a continuous function of α, it may be differentiated under the integral sign

$$\frac{d\phi}{d\alpha} = -\int_0^\infty e^{-\alpha x}\sin x\,dx$$

$$= -1/(1 + \alpha^2)$$

$$\phi(\alpha) = -\tan^{-1}\alpha + c$$

and since $\phi(\alpha) \to 0$ as $\alpha \to \infty$,

$$c = \pi/2$$
$$\phi(\alpha) = -\tan^{-1}\alpha + \pi/2$$

Integration   It is sometimes useful to generate a double integral to solve a problem. By this approach, the fundamental theorem indicated by Eq. (3-66) can be used.

Example   Find $\int_0^1 \dfrac{x^b - x^\alpha}{\ln x}\,dx$

Consider $\int_0^1 x^\alpha\,dx = \dfrac{1}{\alpha + 1} \quad (\alpha > -1)$

Then multiplying both sides by $d\alpha$ and integrating between a and b,

$$\int_a^b d\alpha \int_0^1 x^\alpha\,dx = \int_a^b \frac{d\alpha}{\alpha + 1} = \ln\left|\frac{b+1}{a+1}\right|$$

But also

$$\int_a^b d\alpha \int_0^1 x^\alpha\,dx = \int_0^1 dx \int_a^b x^\alpha\,d\alpha = \int_0^1 \frac{x^b - x^\alpha}{\ln x}\,dx$$

Therefore $\int_0^1 \dfrac{x^b - x^\alpha}{\ln x}\,dx = \ln\left|\dfrac{b+1}{a+1}\right|$

Complex Variable   Certain definite integrals can be evaluated by the technique of complex variable integration. This is described in the references for "Complex Variables."

Numerical   Because of the property of definite integrals another method for obtaining their solution is available which cannot be applied to indefinite integrals. This involves a numerical approximation based on the previously outlined summation definition:

$$\lim_{n\to\infty}\sum_1^{n-1} f(\xi_i)(x_{i+1} - x_i) = \int_a^b f(x)\,dx$$

where $x_1 = a$ and $x_n = b$

Examples of this procedure are given in the subsection "Numerical Analysis and Approximate Methods."

INFINITE SERIES

REFERENCES: 53, 126, 127, 163. For asymptotic series and asymptotic methods, see Refs. 51, 127.

DEFINITIONS

A succession of numbers or terms that are formed according to some definite rule is called a sequence. The indicated sum of the terms of a sequence is called a series. A series of the form $a_0 + a_1(x - c) + a_2(x - c)^2 + \cdots + a_n(x - c)^n + \cdots$ is called a power series.

Consider the sum of a finite number of terms in the geometric series (a special case of a power series).

$$S_n = a + ar + ar^2 + ar^3 + \cdots + ar^{n-1} \tag{3-67}$$

For any number of terms n, the sum equals

$$S_n = a\frac{1 - r^n}{1 - r}$$

In this form, the geometric series is assumed finite.

In the form of Eq. (3-67), it can further be defined that the terms in the series be nonending and therefore an infinite series.

$$S = a + ar + ar^2 + \cdots + ar^n + \cdots \tag{3-68}$$

However, the defined sum of the terms [Eq. (3-67)]

$$S_n = a\frac{1 - r^n}{1 - r} \qquad r \neq 1$$

while valid for any finite value of r and n now takes on a different interpretation. In this sense it is necessary to consider the limit of S_n as n increases indefinitely:

$$S = \lim_{n\to\infty} S_n$$

$$= a\lim_{n\to\infty}\frac{1 - r^n}{1 - r}$$

For this, it is stated the infinite series converges if the limit of S_n approaches a fixed finite value as n approaches infinity. Otherwise, the series is **divergent.**

On this basis an analysis of

$$S = a\lim_{n\to\infty}\frac{1 - r^n}{1 - r}$$

shows that if r is less than 1 but greater than -1, the infinite series is convergent. For values outside of the range $-1 < r < 1$, the series is divergent because the sum is not defined. The range $-1 < r < 1$ is called the **region of convergence.** (We assume $a \neq 0$.)

Consider the divergence of Eq. (3-68) when $r = -1$ and $+1$. For the former case $r = -1$,

$$S = a + a(-1) + a(-1)^2 + a(-1)^3 + \cdots + a(-1)^n + \cdots$$
$$= a - a + a - a + a - \cdots$$

and for which

$$S = a \lim_{n \to \infty} \frac{1 - r^n}{1 - r}$$

$$= a \lim_{n \to \infty} \frac{1 - (-1)^n}{1 + 1} \quad \text{undefined limit (if } a \neq 0\text{)}$$

Since the limit sum does not exist, the series is divergent. This is defined as a **bounded or oscillating divergent series.** Similarly for the value $r = +1$,

$$S = a + a(1) + a(1)^2 + a(1)^3 + \cdots + a(1)^n + \cdots$$

$$S = a + a + a + a + \cdots + a + \cdots \qquad (a \neq 0)$$

The series is also divergent but defined as an **unbounded divergent series.**

There are also two types of convergent series. Consider the new series

$$S = 1 - \frac{1}{2} + \frac{1}{3} - \frac{1}{4} + \cdots + (-1)^{n+1} \frac{1}{n} + \cdots \qquad (3\text{-}69)$$

It can be shown that the series (3-69) does converge to the value $S = \log 2$. However, if each term is replaced by its absolute value, the series becomes unbounded and therefore divergent (unbounded divergent):

$$S = 1 + \frac{1}{2} + \frac{1}{3} + \frac{1}{4} + \frac{1}{5} + \cdots \qquad (3\text{-}70)$$

In this case the series (3-69) is defined as a **conditionally convergent** series. If the replacement series of absolute values also converges, the series is defined to **converge absolutely.**

Series (3-69) is further defined as an **alternating series,** while series (3-70) is referred to as a **positive series.**

OPERATIONS WITH INFINITE SERIES

1. The convergence or divergence of an infinite series is unaffected by the removal of a finite number of finite terms. This is a trivial theorem but useful to remember, especially when using the comparison test to be described in the subsection "Tests for Convergence and Divergence."

2. If a series is conditionally convergent, its sums can be made to have any arbitrary value by a suitable rearrangement of the series; it can in fact be made divergent or oscillatory (Riemann's theorem). This seemingly paradoxical theorem can be illustrated by the following example.

Example $S = 1 - \dfrac{1}{2} + \dfrac{1}{3} - \dfrac{1}{4} + \dfrac{1}{5} - \dfrac{1}{6} + \cdots$

The series is rearranged so that each positive term is followed by two negative terms:

$$t = 1 - \frac{1}{2} - \frac{1}{4} + \frac{1}{3} - \frac{1}{6} - \frac{1}{8} + \frac{1}{5} - \frac{1}{10} - \frac{1}{12} + \cdots$$

Define t_{3n} for the first $3n$ terms in the series

$$t_{3n} = \left(1 - \frac{1}{2}\right) - \frac{1}{4} + \left(\frac{1}{3} - \frac{1}{6}\right) - \frac{1}{8} + \cdots + \left(\frac{1}{2n-1} - \frac{1}{4n-2}\right) - \frac{1}{4n}$$

$$= \frac{1}{2} - \frac{1}{4} + \frac{1}{6} - \frac{1}{8} + \cdots + \frac{1}{4n-2} - \frac{1}{4n}$$

$$= \frac{1}{2}\left(1 - \frac{1}{2} + \frac{1}{3} - \frac{1}{4} + \cdots + \frac{1}{2n-1} - \frac{1}{2n}\right)$$

$$= \frac{1}{2} S_{2n}$$

where S_{2n} is the sum of the first $2n$ terms of the original series. Thus

$$\lim_{n \to \infty} t_{3n} = \lim_{n \to \infty} \frac{1}{2} S_{2n}$$

$$t = \frac{1}{2} S$$

and since $\lim t_{3n+2} = \lim t_{3n+1} = \lim t_{3n}$, it follows the sum of the series t is $(\frac{1}{2}) S$. Hence a rearrangement of the terms of an alternating series alters the sum of the series.

3. A series of positive terms, if convergent, has a sum independent of the order of its terms; but if divergent, it remains divergent however its terms are rearranged.

4. An oscillatory series can always be made to converge by grouping the terms in brackets.

Example   Consider the series

$$1 - \frac{1}{2} + \frac{2}{3} - \frac{3}{4} + \frac{4}{5} - \frac{5}{6} + \cdots$$

which oscillates between the values 0.306 and 1.306. However, the series

$$\left(1 - \frac{1}{2}\right) + \left(\frac{2}{3} - \frac{3}{4}\right) + \left(\frac{4}{5} - \frac{5}{6}\right) + \cdots = \frac{1}{2} - \frac{1}{12} - \frac{1}{30} - \frac{1}{56} - \cdots \cong 0.306 \cdots$$

and

$$1 - \left(\frac{1}{2} - \frac{2}{3}\right) - \left(\frac{3}{4} - \frac{4}{5}\right) - \left(\frac{5}{6} - \frac{6}{7}\right) + \cdots = 1 + \frac{1}{6} + \frac{1}{20} + \frac{1}{42} + \cdots = 1.306 \cdots$$

5. A power series can be inverted, provided the first-degree term is not zero. Given

$$y = b_1 x + b_2 x^2 + b_3 x^3 + b_4 x^4 + b_5 x^5 + b_6 x^6 + b_7 x^7 + \cdots$$

then

$$x = B_1 y + B_2 y^2 + B_3 y^3 + B_4 y^4 + B_5 y^5 + B_6 y^6 + B_7 y^7 + \cdots$$

where $B_1 = 1/b_1$
$B_2 = -b_2/b_1^3$
$B_3 = (1/b_1^5)(2b_2^2 - b_1 b_3)$
$B_4 = (1/b_1^7)(5b_1 b_2 b_3 - b_1^2 b_4 - 5b_2^3)$

Additional coefficients are available in the references.

6. Two series may be added or subtracted term by term provided each is a convergent series. The joint sum is equal to the sum (or difference) of the individuals.

7. The sum of two divergent series can be convergent. Similarly, the sum of a convergent series and a divergent series must be divergent.

Example   Given

$$\sum_{n=1}^{\infty} \left(\frac{1+n}{n^2}\right) = \frac{2}{1} + \frac{3}{4} + \frac{4}{9} + \frac{5}{16} + \cdots \qquad \text{(a divergent series)}$$

$$\sum_{n=1}^{\infty} \left(\frac{1-n}{n^2}\right) = -\frac{1}{4} - \frac{2}{9} - \frac{3}{16} + \cdots \qquad \text{(a divergent series)}$$

However,
$$\sum \left(\frac{1+n}{n^2}\right) + \sum \left(\frac{1-n}{n^2}\right) = \sum \left(\frac{1+n+1-n}{n^2}\right)$$

$$= 2 \sum \frac{1}{n^2} \quad \text{(convergent)}$$

8. A power series may be integrated term by term to represent the integral of the function within an interval of the region of convergence. If $f(x) = a_0 + a_1 x + a_2 x^2 + \cdots$, then

$$\int_{x_1}^{x_2} f(x) \, dx = \int_{x_1}^{x_2} a_0 \, dx + \int_{x_1}^{x_2} a_1 x \, dx + \int_{x_1}^{x_2} a_2 x^2 \, dx + \cdots$$

9. A power series may be differentiated term by term and represents the function $df(x)/dx$ within the same region of convergence as $f(x)$.

TESTS FOR CONVERGENCE AND DIVERGENCE

In general, the problem of determining whether a given series will converge or not can require a great deal of ingenuity and resourcefulness. There is no all-inclusive test which can be applied to all series. As the only alternative, it is necessary to apply one or more of the developed theorems in an attempt to ascertain the convergence or divergence of the series under study. The following defined tests are given in relative order of effectiveness. For examples, see references on advanced calculus.

1. *Comparison Test.*   A series will converge if the absolute value of each term (with or without a finite number of terms) is less than the corresponding term of a known convergent series. Similarly, a positive series is divergent if it is termwise larger than a known divergent series of positive terms.

2. *nth-Term Test.*   A series is divergent if the nth term of the series does not approach zero as n becomes increasingly large.

3. *Ratio Test.*   If the absolute ratio of the $(n + 1)$ term divided by the nth term as n becomes unbounded approaches

   a. A number less than 1, the series is absolutely convergent
   b. A number greater than 1, the series is divergent
   c. A number equal to 1, the test is inconclusive

4. *Alternating-Series Leibniz Test.*   If the terms of a series are alternately positive and negative and never increase in value, the absolute series will converge, provided that the terms tend to zero as a limit.

5. *Cauchy's Root Test.*   If the nth root of the absolute value of the nth term, as n becomes unbounded, approaches

   a. A number less than 1, the series is absolutely convergent
   b. A number greater than 1, the series is divergent
   c. A number equal to 1, the test is inconclusive

6. *Maclaurin's Integral Test.*   Suppose $\sum a_n$ is a series of positive terms and f is a continuous decreasing function such that $f(x) \geq 0$ for $1 \leq x < \infty$ and $f(n) = a_n$. Then the series and the improper integral $\int_1^\infty f(x)\,dx$ either both converge or both diverge.

SERIES SUMMATION AND IDENTITIES

Sums for the First n Numbers to Integer Powers

$$\sum_{j=1}^{n} j = \frac{n(n + 1)}{2} = 1 + 2 + 3 + 4 + \cdots + n$$

$$\sum_{j=1}^{n} j^2 = \frac{n(n + 1)(2n + 1)}{6} = 1^2 + 2^2 + 3^2 + 4^2 + \cdots + n^2$$

$$\sum_{j=1}^{n} j^3 = \frac{n^2(n + 1)^2}{4} = 1^3 + 2^3 + 3^3 + \cdots + n^3$$

$$\sum_{j=1}^{n} j^4 = \frac{n(n + 1)(2n + 1)(3n^2 + 3n - 1)}{30} = 1^4 + 2^4 + 3^4 + \cdots + n^4$$

Arithmetic Progression

$$\sum_{k=1}^{n} [a + (k - 1)d] = a + (a + d) + (a + 2d)$$
$$+ (a + 3d) + \cdots + [a + (n - 1)]d$$
$$= na + \frac{1}{2}n(n - 1)d$$

Geometric Progression

$$\sum_{j=1}^{n} ar^{j-1} = a + ar + ar^2 + ar^3 + \cdots + ar^{n-1}$$
$$= a\,\frac{1 - r^n}{1 - r} \qquad r \neq 1$$

Harmonic Progression

$$\sum_{k=0}^{n} \frac{1}{a + kd} = \frac{1}{a} + \frac{1}{a + d} + \frac{1}{a + 2d} + \frac{1}{a + 3d} + \frac{1}{a + 4d} + \cdots + \frac{1}{a + nd}$$

The reciprocals of the terms of the arithmetic-progression series are called harmonic progression. No general summation formulas are available for this series.

Binomial Series

$$(x + y)^n = x^n + nx^{n-1}y + \frac{n(n - 1)}{2!}x^{n-2}y^2$$
$$+ \frac{n(n - 1)(n - 2)}{3!}x^{n-3}y^3 + \cdots + \frac{n!}{(n - r)!r!}x^{n-r}y^r + \cdots + y^n$$

$$(1 \pm x)^n = 1 \pm nx + \frac{n(n - 1)}{2!}x^2 \pm \frac{n(n - 1)(n - 2)}{3!}x^3 + \cdots \quad (x^2 < 1)$$

Taylor's Series

$$f(x + h) = f(h) + xf'(h) + \frac{x^2}{2!}f''(h) + \frac{x^3}{3!}f'''(h) + \cdots$$

or $\quad f(x) = f(x_0) + f'(x_0)(x - x_0) + \dfrac{f''(x_0)}{2!}(x - x_0)^2 + \dfrac{f'''(x_0)}{3!}(x - x_0)^3 + \cdots$

Example   Find a series expansion for $f(x) = \ln(1 + x)$ about $x_0 = 0$.

$$f'(x) = (1 + x)^{-1}, \quad f''(x) = -(1 + x)^{-2}, \quad f'''(x) = 2(1 + x)^{-3}, \text{ etc.}$$

thus $\qquad f(0) = 0, \quad f'(0) = 1, \quad f''(0) = -1, \quad f'''(1) = 2, \text{ etc.}$

$$\ln(x + 1) = x - \frac{x^2}{2} + \frac{x^3}{3} - \frac{x^4}{4} + \cdots + (-1)^{n+1}\frac{x^n}{n} + \cdots$$

which converges for $-1 < x \leq 1$.

Maclaurin's Series

$$f(x) = f(0) + xf'(0) + \frac{x^2}{2!}f''(0) + \frac{x^3}{3!}f'''(0) + \cdots$$

This is simply a special case of Taylor's series when h is set to zero.

Exponential Series

$$e^x = 1 + x + \frac{x^2}{2!} + \frac{x^3}{3!} + \cdots + \frac{x^n}{n!} + \cdots -\infty < x < \infty$$

Logarithmic Series

$$\ln x = \frac{x - 1}{x} + \frac{1}{2}\left(\frac{x - 1}{x}\right)^2 + \frac{1}{3}\left(\frac{x - 1}{x}\right)^3 + \cdots \qquad (x > \tfrac{1}{2})$$

$$\ln x = 2\left[\left(\frac{x - 1}{x + 1}\right) + \frac{1}{3}\left(\frac{x - 1}{x + 1}\right)^3 + \cdots \right] \qquad (x > 0)$$

Trigonometric Series°

$$\sin x = x - \frac{x^3}{3!} + \frac{x^5}{5!} - \frac{x^7}{7!} + \cdots \qquad -\infty < x < \infty$$

$$\cos x = 1 - \frac{x^2}{2!} + \frac{x^4}{4!} - \frac{x^6}{6!} + \cdots \qquad -\infty < x < \infty$$

$$\sin^{-1} x = x + \frac{x^3}{6} + \frac{1}{2} \cdot \frac{3}{4} \cdot \frac{x^5}{5} + \frac{1}{2} \cdot \frac{3}{4} \cdot \frac{5}{6} \cdot \frac{x^7}{7} + \cdots \qquad (x^2 < 1)$$

$$\tan^{-1} x = x - \frac{1}{3}x^3 + \frac{1}{5}x^5 - \frac{1}{7}x^7 + \cdots \qquad (x^2 < 1)$$

Taylor Series   The Taylor series for a function of two variables, expanded about the point (x_0, y_0), is

$$f(x, y) = f(x_0, y_0) + \frac{\partial f}{\partial x}\bigg|_{x_0, y_0}(x - x_0) + \frac{\partial f}{\partial y}\bigg|_{x_0, y_0}(y - y_0)$$
$$+ \frac{1}{2!}\left[\frac{\partial^2 f}{\partial x^2}\bigg|_{x_0, y_0}(x - x_0)^2 + 2\frac{\partial^2 f}{\partial x\partial y}\bigg|_{x_0, y_0}(x - x_0)(y - y_0)\right.$$
$$\left. + \frac{\partial^2 f}{\partial y^2}\bigg|_{x_0, y_0}(y - y_0)^2\right] + \cdots$$

Partial Sums of Infinite Series, and How They Grow   Calculus textbooks devote much space to tests for convergence and divergence of series that are of little practical value, since a convergent

---

° tan x series has awkward coefficients and should be computed as $\left[(\text{sign})\dfrac{\sin x}{\sqrt{1 - \sin^2 x}}\right]$.

series either converges rapidly, in which case almost any test (among those presented in the preceding subsections) will do; or it converges slowly, in which case it is not going to be of much use unless there is some way to get at its sum without adding up an unreasonable number of terms. To find out, as accurately as possible, how fast a convergent series converges and how fast a divergent series diverges, see Ref. 34.

COMPLEX VARIABLES

REFERENCES: *General.* 73, 163, 172, 179. *Applied and computational complex analysis.* 141, 146, 179.

Numbers of the form $z = x + iy$, where x and y are real, $i^2 = -1$, are called complex numbers. The numbers $z = x + iy$ are representable in the plane as shown in Fig. 3-46. The following definitions and terminology are used:
1. Distance $OP = r = $ modulus of z written $|z|$. $|z| = \sqrt{x^2 + y^2}$.
2. x is the real part of z.
3. y is the imaginary part of z.
4. The angle θ, $0 \le \theta < 2\pi$, measured counterclockwise from the positive x axis to OP is the argument of z. $\theta = \arctan y/x = \arcsin y/r = \arccos x/r$ if $x \ne 0$, $\theta = \pi/2$ if $x = 0$ and $y > 0$.
5. The numbers r, θ are the polar coordinates of z.
6. $\bar{z} = x - iy$ is the complex conjugate of z.

ALGEBRA

Let $z_1 = x_1 + iy_1$, $z_2 = x_2 + iy_2$.
  Equality $z_1 = z_2$ if and only if $x_1 = x_2$ and $y_1 = y_2$.
  Addition $z_1 + z_2 = (x_1 + x_2) + i(y_1 + y_2)$.
  Subtraction $z_1 - z_2 = (x_1 - x_2) + i(y_1 - y_2)$.
  Multiplication $z_1 \cdot z_2 = (x_1 x_2 - y_1 y_2) + i(x_1 y_2 + x_2 y_1)$.

  Division $z_1/z_2 = \dfrac{x_1 x_2 + y_1 y_2}{x_2^2 + y_2^2} + i\dfrac{x_2 y_1 - x_1 y_2}{x_2^2 + y_2^2}$, $z_2 \ne 0$.

SPECIAL OPERATIONS

$z\bar{z} = x^2 + y^2 = |z|^2$; $\overline{z_1 \pm z_2} = \bar{z}_1 \pm \bar{z}_2$; $\overline{\bar{z}_1} = z_1$; $\overline{z_1 z_2} = \bar{z}_1 \bar{z}_2$; $|z_1 \cdot z_2| = |z_1| \cdot |z_2|$; $\arg(z_1 \cdot z_2) = \arg z_1 + \arg z_2$; $\arg(z_1/z_2) = \arg z_1 - \arg z_2$; $i^{4n} = 1$ for n any integer; $i^{2n} = -1$ where n is any odd integer; $z + \bar{z} = 2x$; $z - \bar{z} = 2iy$.
Every complex quantity can be expressed in the form $x + iy$.

TRIGONOMETRIC REPRESENTATION

By referring to Fig. 3-46, there results $x = r \cos \theta$, $y = r \sin \theta$ so that $z = x + iy = r(\cos \theta + i \sin \theta)$, which is called the polar form of the complex number. $\cos \theta + i \sin \theta = e^{i\theta}$. Hence $z = x + iy = re^{i\theta}$. $\bar{z} = x - iy = re^{-i\theta}$. Two important results from this are $\cos \theta = (e^{i\theta} + e^{-i\theta})/2$ and $\sin \theta = (e^{i\theta} - e^{-i\theta})/2i$. Let $z_1 = r_1 e^{i\theta_1}$, $z_2 = r_2 e^{i\theta_2}$. This form is convenient for multiplication for $z_1 z_2 = r_1 r_2 e^{i(\theta_1 + \theta_2)}$ and for division for $z_1/z_2 = (r_1/r_2) e^{i(\theta_1 - \theta_2)}$, $z_2 \ne 0$.

POWERS AND ROOTS

If n is a positive integer, $z^n = (re^{i\theta})^n = r^n e^{in\theta} = r^n(\cos n\theta + i \sin n\theta)$.
  If n is a positive integer,

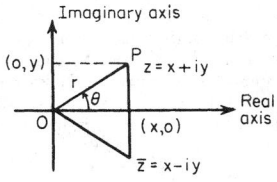

FIG. 3-46 Complex plane.

$$z^{1/n} = r^{1/n} e^{i[(\theta + 2k\pi)/n]} = r^{1/n}\left[\cos\left(\frac{\theta + 2k\pi}{n}\right) + i \sin\left(\frac{\theta + 2k\pi}{n}\right)\right]$$

and selecting values of $k = 0, 1, 2, 3, \ldots, n - 1$ give the n distinct values of $z^{1/n}$. The n roots of a complex quantity are uniformly spaced around a circle, with radius $r^{1/n}$, in the complex plane in a symmetric fashion.

Example Find the three cube roots of -8. Here $r = 8$, $\theta = \pi$. The roots are $z_0 = 2(\cos \pi/3 + i \sin \pi/3) = 1 + i\sqrt{3}$, $z_1 = 2(\cos \pi + i \sin \pi) = -2$, $z_2 = 2(\cos 5\pi/3 + i \sin 5\pi/3) = 1 - i\sqrt{3}$.

ELEMENTARY COMPLEX FUNCTIONS

Polynomials A polynomial in z, $a_n z^n + a_{n-1} z^{n-1} + \cdots + a_0$, where n is a positive integer, is simply a sum of complex numbers times integral powers of z which have already been defined. Every polynomial of degree n has precisely n complex roots provided each multiple root of multiplicity m is counted m times.
  Exponential Functions The exponential function e^z is defined by the equation $e^z = e^{x+iy} = e^x \cdot e^{iy} = e^x(\cos y + i \sin y)$. Properties: $e^0 = 1$; $e^{z_1} \cdot e^{z_2} = e^{z_1 + z_2}$; $e^{z_1}/e^{z_2} = e^{z_1 - z_2}$; $e^{z + 2k\pi i} = e^z$.
  Trigonometric Functions $\sin z = (e^{iz} - e^{-iz})/2i$; $\cos z = (e^{iz} + e^{-iz})/2$; $\tan z = \sin z/\cos z$; $\cot z = \cos z/\sin z$; $\sec z = 1/\cos z$; $\csc z = 1/\sin z$. Fundamental identities for these functions are the same as their real counterparts. Thus $\cos^2 z + \sin^2 z = 1$, $\cos(z_1 \pm z_2) = \cos z_1 \cos z_2 \mp \sin z_1 \sin z_2$, $\sin(z_1 \pm z_2) = \sin z_1 \cos z_2 \pm \cos z_1 \sin z_2$. The sine and cosine of z are periodic functions of period 2π; thus $\sin(z + 2\pi) = \sin z$. For computation purposes $\sin z = \sin(x + iy) = \sin x \cosh y + i \cos x \sinh y$, where $\sin x$, $\cosh y$, etc., are the real trigonometric and hyperbolic functions. Similarly, $\cos z = \cos x \cosh y - i \sin x \sinh y$. If $x = 0$ in the results given, $\cos iy = \cosh y$, $\sin iy = i \sinh y$.

Example Find all solutions of $\sin z = 3$. From previous data $\sin z = \sin x \cosh y + i \cos x \sinh y = 3$. Equating real and imaginary parts $\sin x \cosh y = 3$, $\cos x \sinh y = 0$. The second equation can hold for $y = 0$ or for $x = \pi/2, 3\pi/2, \ldots$. If $y = 0$, $\cosh 0 = 1$ and $\sin x = 3$ is impossible for real x. Therefore, $x = \pm\pi/2, \pm 3\pi/2, \ldots \pm(2n + 1)\pi/2$, $n = 0, \pm 1, \pm 2, \ldots$. However, $\sin 3\pi/2 = -1$ and $\cosh y \ge 1$. Hence $x = \pi/2, 5\pi/2, \ldots$. The solution is $z = [(4n + 1)\pi]/2 + i \cosh^{-1} 3$, $n = 0, 1, 2, 3, \ldots$.

Example Find all solutions of $e^z = -i$. $e^z = e^x(\cos y + i \sin y) = -i$. Equating real and imaginary parts gives $e^x \cos y = 0$, $e^x \sin y = -1$. From the first $y = \pm\pi/2$, $\pm 3\pi/2, \ldots$. But $e^x > 0$. Therefore, $y = 3\pi/2, 7\pi/2, -\pi/2, \ldots$. Then $x = 0$. The solution is $z = i[(4n + 3)\pi]/2$.

Two important facets of these functions should be recognized. First, the $\sin z$ is *unbounded;* and, second, e^z takes *all* complex values *except* 0.
  Hyperbolic Functions $\sinh z = (e^z - e^{-z})/2$; $\cosh z = (e^z + e^{-z})/2$; $\tanh z = \sinh z/\cosh z$; $\coth z = \cosh z/\sinh z$; $\text{csch } z = 1/\sinh z$; $\text{sech } z = 1/\cosh z$. Identities are: $\cosh^2 z - \sinh^2 z = 1$; $\sinh(z_1 + z_2) = \sinh z_1 \cosh z_2 + \cosh z_1 \sinh z_2$; $\cosh(z_1 + z_2) = \cosh z_1 \cosh z_2 + \sinh z_1 \sinh z_2$; $\cosh z + \sinh z = e^z$; $\cosh z - \sinh z = e^{-z}$. The hyperbolic sine and hyperbolic cosine are periodic functions with the imaginary period $2\pi i$. That is, $\sinh(z + 2\pi i) = \sinh z$.
  Logarithms The logarithm of z, $\log z = \log |z| + i(\theta + 2n\pi)$, where $\log |z|$ is taken to the base e and θ is the principal argument of z, that is, the particular argument lying in the interval $0 \le \theta < 2\pi$. The logarithm of z is infinitely many valued. If $n = 0$, the resulting logarithm is called the principal value. The familiar laws $\log z_1 z_2 = \log z_1 + \log z_2$, $\log z_1/z_2 = \log z_1 - \log z_2$, $\log z^n = n \log z$ hold for the principal value.

Example $\log(1+i) = \log\sqrt{2} + i\left(\dfrac{\pi}{4} + 2n\pi\right)$.

General powers of z are defined by $z^\alpha = e^{\alpha \log z}$. Since $\log z$ is infinitely many valued, so too is z^α unless α is a rational number.

DeMoivre's formula can be derived from properties of e^z.

$$z^n = r^n (\cos\theta + i\sin\theta)^n = r^n (\cos n\theta + i\sin n\theta)$$

Thus $(\cos\theta + i\sin\theta)^n = \cos n\theta + i\sin n\theta$

Example $i^i = e^{i\log i} = e^{i[\log|i| + i(\pi/2 + 2n\pi)]} = e^{-(\pi/2 + 2n\pi)}$. Thus i^i is real with principal value $(n=0) = e^{-\pi/2}$.

Example $(\sqrt{2})^{1+i} = e^{(1+i)\log\sqrt{2}} = e^{\log\sqrt{2}} \cdot e^{i\log\sqrt{2}} = \sqrt{2} \cdot (\cos\log\sqrt{2} + i\sin\log\sqrt{2}) = \sqrt{2}[\cos(0.3466) + i\sin(0.3466)]$.

Inverse Trigonmetric Functions $\cos^{-1} z = -i\log(z \pm \sqrt{z^2 - 1})$; $\sin^{-1} z = -i\log(iz \pm \sqrt{1 - z^2})$; $\tan^{-1} z = \dfrac{i}{2}\log\left(\dfrac{i+z}{i-z}\right)$. These functions are infinitely many valued.

Inverse Hyperbolic Functions $\cosh^{-1} z = \log(z \pm \sqrt{z^2 - 1})$; $\sinh^{-1} z = \log(z \pm \sqrt{z^2 + 1})$; $\tanh^{-1} z = \dfrac{1}{2}\log\left(\dfrac{1+z}{1-z}\right)$.

COMPLEX FUNCTIONS (ANALYTIC)

In the real-number system a greater than $b(a > b)$ and b less than $c(b < c)$ define an order relation. These relations have no meaning for complex numbers. The absolute value is used for ordering. Some important relations follow: $|z| \geq x$; $|z| \geq y$; $|z_1 + z_2| \leq |z_1| + |z_2|$; $|z_1 - z_2| \geq ||z_1| - |z_2||$; $|z| \geq (|x| + |y|)/\sqrt{2}$. Parts of the complex plane, commonly called **regions** or **domains**, are described by using inequalities.

Example $|z - 3| \leq 5$. This is equivalent to $\sqrt{(x-3)^2 + y^2} \leq 5$, which is the set of all points within and on the circle, centered at $x = 3$, $y = 0$ of radius 5.

Example $|z - 1| \leq x$ represents the set of all points inside and on the parabola $2x = y^2 + 1$ or, equivalently, $2x \geq y^2 + 1$.

Functions of a Complex Variable   If $z = x + iy$, $w = u + iv$ and if for each value of z in some region of the complex plane one or more values of w are defined, then w is said to be a function of z, $w = f(z)$. Some of these functions have already been discussed, e.g., $\sin z$, $\log z$. All functions are reducible to the form $w = u(x, y) + iv(x, y)$, where u, v are real functions of the real variables x and y.

Example $z^3 = (x + iy)^3 = x^3 + 3x^2(iy) + 3x(iy)^2 + (iy)^3 = (x^3 - 3xy^2) + i(3x^2y - y^3)$.

Example $\cos z = \cos x \cosh y - i\sin x \sinh y$.

Differentiation   The *derivative* of $w = f(z)$ is

$$\frac{dw}{dz} = \lim_{\Delta z \to 0} \frac{f(z + \Delta z) - f(z)}{\Delta z}$$

and for the derivative to exist the limit must be the same no matter how Δz approaches zero. If w_1, w_2 are differentiable functions of z, the following rules apply:

$$\frac{d(w_1 \pm w_2)}{dz} = \frac{dw_1}{dz} \pm \frac{dw_2}{dz} \qquad \frac{d(w_1 w_2)}{dz} = w_2 \frac{dw_1}{dz} + w_1 \frac{dw_2}{dz}$$

$$\frac{d(w_1/w_2)}{dz} = \frac{w_2(dw_1/dz) - w_1(dw_2/dz)}{w_2^2}$$

and

$$\frac{dw_1^n}{dz} = nw_1^{n-1}\frac{dw_1}{dz}$$

For $w = f(z)$ to be differentiable, it is necessary that $\partial u/\partial x = \partial v/\partial y$ and

$\partial v/\partial x = -\partial u/\partial y$. The last two equations are called the Cauchy-Riemann equations. The derivative

$$\frac{dw}{dz} = \frac{\partial u}{\partial x} + i\frac{\partial v}{\partial x} = \frac{\partial v}{\partial y} - i\frac{\partial u}{\partial y}$$

If $f(z)$ possesses a derivative at z_o and at every point in some neighborhood of z_0, then $f(z)$ is said to be analytic at z_0. If the Cauchy-Riemann equations are satisfied and

$$u, v, \frac{\partial u}{\partial x}, \frac{\partial u}{\partial y}, \frac{\partial v}{\partial x}, \frac{\partial v}{\partial y}$$

are continuous in a region of the complex plane, then $f(z)$ is analytic in that region.

Example $w = z\bar{z} = x^2 + y^2$. Here $u = x^2 + y^2$, $v = 0$. $\partial u/\partial x = 2x$, $\partial u/\partial y = 2y$, $\partial v/\partial x = \partial v/\partial y = 0$. These are continuous everywhere, but the Cauchy-Riemann equations hold only at the origin. Therefore, w is nowhere analytic, but it is differentiable at $z = 0$ only.

Example $w = e^z = e^x \cos y + ie^x \sin y$. $u = e^x \cos y$, $v = e^x \sin y$. $\partial u/\partial x = e^x \cos y$, $\partial u/\partial y = -e^x \sin y$, $\partial v/\partial x = e^x \sin y$, $\partial v/\partial y = e^x \cos y$. The continuity and Cauchy-Riemann requirements are satisfied for all finite z. Hence e^z is analytic (except at ∞) and $dw/dz = \partial u/\partial x + i(\partial v/\partial x) = e^z$.

Example $w = \dfrac{1}{z} = \dfrac{x - iy}{x^2 + y^2} = \dfrac{x}{x^2 + y^2} - i\dfrac{y}{x^2 + y^2}$

It is easy to see that dw/dz exists except at $z = 0$. Thus $1/z$ is analytic except at $z = 0$.

Singular Points   If $f(z)$ is analytic in a region except at certain points, those points are called singular points.

Example $1/z$ has a singular point at zero.

Example $\tan z$ has singular points at $z = \pm(2n + 1)(\pi/2)$, $n = 0, 1, 2, \ldots$.

The derivatives of the common functions, given earlier, are the same as their real counterparts.

Example $(d/dz)(\log z) = 1/z$, $(d/dz)(\sin z) = \cos z$.

Harmonic Functions   Both the *real* and the *imaginary* parts of any analytic function $f = u + iv$ satisfy Laplace's equation $\partial^2\phi/\partial x^2 + \partial^2\phi/\partial y^2 = 0$. A function which possesses continuous second partial derivatives and satisfies Laplace's equation is called a harmonic function.

Example $e^z = e^x \cos y + ie^x \sin y$. $u = e^x \cos y$, $\partial u/\partial x = e^x \cos y$, $\partial^2 u/\partial x^2 = e^x \cos y$, $\partial u/\partial y = -e^x \sin y$, $\partial^2 u/\partial y^2 = -e^x \cos y$. Clearly $\partial^2 u/\partial x^2 + \partial^2 u/\partial y^2 = 0$. Similarly, $v = e^x \sin y$ is also harmonic.

If $w = u + iv$ is analytic, the curves $u(x, y) = c$ and $v(x, y) = k$ intersect at right angles, if $w'(z) \neq 0$.

Example $z^3 = (x^3 - 3xy^2) + i(3x^2y - y^3)$. Set $u = x^3 - 3xy^2 = c$, $v = 3x^2y - y^3 = k$. By implicit differentiation there results, respectively, $dy/dx = (x^2 - y^2)/2xy$, $dy/dx = 2xy/(y^2 - x^2)$, which are clearly negative reciprocals, the condition for perpendicularity.

Integration   In much of the work with complex variables a simple extension of integration called line or curvilinear integration is of fundamental importance. Since any complex line integral can be expressed in terms of real line integrals, we define only real line integrals. Let $F(x,y)$ be a real, continuous function of x and y and c be any continuous curve of finite length joining the points A and B (Fig. 3-47). $F(x,y)$ is not related to the curve c. Divide c up into n segments, Δs_i, whose projection on the x axis is Δx_i and on the y axis is Δy_i. Let (ε_i, η_i) be the coordinates of an arbitrary point on Δs_i. The limits of the sums

$$\lim_{\Delta s_i \to 0} \sum_{i=1}^{n} F(\varepsilon_i, \eta_i)\,\Delta s_i = \int_c F(x, y)\,ds$$

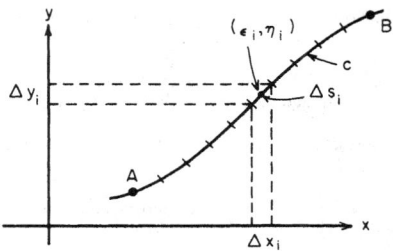

FIG. 3-47   Line integral.

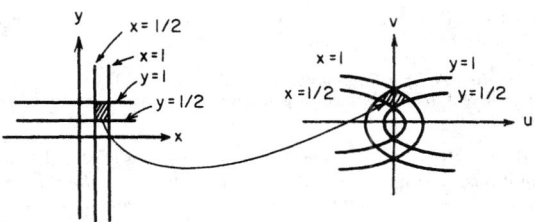

FIG. 3-48   Conformal transformation.

$$\lim_{\Delta s_i \to 0} \sum_{i=1}^{n} F(\varepsilon_i, \eta_i)\, \Delta x_i = \int_c F(x, y)\, dx$$

$$\lim_{\Delta s_i \to 0} \sum_{i=1}^{n} F(\varepsilon_i, \eta_i)\, \Delta y_i = \int_c F(x, y)\, dy$$

are known as line integrals. Much of the initial strangeness of these integrals will vanish if it be observed that the ordinary definite integral $\int_a^b f(x)\,dx$ is just a line integral in which the curve c is a line segment on the x axis and $F(x, y)$ is a function of x alone. The evaluation of line integrals can be reduced to evaluation of ordinary integrals.

Example $\int_c y(1 + x)\, dy$, where c: $y = 1 - x^2$ from $(-1, 0)$ to $(1, 0)$. Clearly $y = 1 - x^2$, $dy = -2x\, dx$. Thus $\int_c y(1+x)\, dy = -2 \int_{-1}^{1} (1 - x^2)(1 + x)x\, dx = -\frac{8}{15}$.

Example $\int_c x^2 y\, ds$, c is the square whose vertices are $(0, 0)$, $(1, 0)$, $(1, 1)$, $(0, 1)$. $ds = \sqrt{dx^2 + dy^2}$. When $dx = 0$, $ds = dy$. From $(0, 0)$ to $(1, 0)$, $y = 0$, $dy = 0$. Similar arguments for the other sides give

$$\int x^2 y\, ds = \int_0^1 0.x^2\, dx + \int_0^1 y\, dy + \int_1^0 x^2\, dx + \int_1^0 0.y\, dy = \frac{1}{2} - \frac{1}{3} = \frac{1}{6}$$

Let $f(z)$ be any function of z, analytic or not, and c any curve as above. The complex integral is calculated as $\int_c f(z)\, dz = \int_c (u\, dx - v\, dy) + i \int_c (v\, dx + u\, dy)$, where $f(z) = u(x, y) + iv(x, y)$. Properties of line integrals are the same as those for ordinary integrals. That is, $\int_c [f(z) \pm g(z)]\, dz = \int_c f(z)\, dz \pm \int_c g(z)\, dz$; $\int_c kf(z)\, dz = k \int_c f(z)\, dz$ for any constant k, etc.

Example $\int_c (x^2 + iy)\, dz$ along c: $y = x$, 0 to $1 + i$. This becomes

$$\int_c (x^2 + iy)\, dz = \int_c (x^2\, dx - y\, dy)$$

$$+ i \int_c (y\, dx + x^2\, dy) = \int_0^1 x^2\, dx - \int_0^1 x\, dx + i \int_0^1 x\, dx + i \int_0^1 x^2\, dx = -\frac{1}{6} + 5i/6$$

Conformal Mapping   Every function of a complex variable $w = f(z) = u(x, y) + iv(x, y)$ transforms the x, y plane into the u, v plane in some manner. A conformal transformation is one in which angles between curves are preserved in *magnitude* and *sense*. Every analytic function, except at those points where $f'(z) = 0$, is a conformal transformation. See Fig. 3-48.

Example $w = z^2$. $u + iv = (x^2 - y^2) + 2ixy$ or $u = x^2 - y^2$, $v = 2xy$. These are the transformation equations between the (x, y) and (u, v) planes. Lines parallel to the x axis, $y = c_1$ map into curves in the u, v plane with parametric equations $u = x^2 - c_1^2$, $v = 2c_1x$. Eliminating x, $u = (v^2/4c_1^2) - c_1^2$, which represents a family of parabolas with the origin of the w plane as focus, the line $v = 0$ as axis and opening to the right. Similar arguments apply to $x = c_2$.

The principles of complex variables are useful in the solution of a variety of applied problems. See the references for additional information.

DIFFERENTIAL EQUATIONS

REFERENCES: *Ordinary Differential Equations:* Elementary level, 41, 44, 62, 81, 204, 236, 263. Intermediate level, 30, 43, 144. Theory and Advanced topics, 252. Applications, 9, 263. Partial Differential Equations: Elementary level and solution methods, 9, 41, 61, 72, 144, 156, 229. Theory and advanced level, 79, 220, 240.
    See also "Numerical Analysis and Approximate Methods" and "General References: References for General and Specific Topics—Advanced Engineering Mathematics" for additional references on topics in ordinary and partial differential equations.

The natural laws in any scientific or technological field are not regarded as precise and definitive until they have been expressed in mathematical form. Such a form, often an equation, is a relation between the quantity of interest, say, product yield, and independent variables such as time and temperature upon which yield depends. When it happens that this equation involves, besides the function itself, one or more of its derivatives it is called a differential equation.

Example   The homogeneous bimolecular reaction $A + B \xrightarrow{k} C$ is characterized by the differential equation $dx/dt = k(a - x)(b - x)$, where a = initial concentration of A, b = initial concentration of B, and $x = x(t)$ = concentration of C as a function of time t.

Example   The differential equation of heat conduction in a moving fluid with velocity components v_x, v_y is

$$\frac{\partial u}{\partial t} + v_x \frac{\partial u}{\partial x} + v_y \frac{\partial u}{\partial y} = \frac{K}{\rho c_p} \left(\frac{\partial^2 u}{\partial x^2} + \frac{\partial^2 u}{\partial y^2} \right)$$

where $u = u(x, y, t)$ = temperature, K = thermal conductivity, ρ = density, and c_p = specific heat at constant pressure.

ORDINARY DIFFERENTIAL EQUATIONS

When the function involved in the equation depends upon only one variable, its derivatives are ordinary derivatives and the differential equation is called an ordinary differential equation. When the function depends upon several independent variables, then the equation is called a partial differential equation. The theories of ordinary and partial differential equations are quite different. In almost every respect the latter is more difficult.

Whichever the type, a differential equation is said to be of nth order if it involves derivatives of order n but no higher. The equation in the first example is of first order and that in the second example of second order. The degree of a differential equation is the power to which the derivative of the highest order is raised after the equation has been cleared of fractions and radicals in the dependent variable and its derivatives.

A relation between the variables, involving no derivatives, is called a solution of the differential equation if this relation, when substituted in the equation, satisfies the equation. A solution of an ordinary differential equation which includes the maximum possible number of "arbitrary" constants is called the **general solution.** The maximum number of "arbitrary" constants is exactly equal to the order of the differential equation. If any set of specific values of the constants is chosen, the result is called a **particular solution.**

Example   The general solution of $(d^2x/dt^2) + k^2x = 0$ is $x = A \cos kt + B \sin kt$, where A, B are arbitrary constants. A particular solution is $x = \frac{1}{2} \cos kt + 3 \sin kt$.

In the case of some equations still other solutions exist called singular solutions. A **singular solution** *is any solution of the differential equation which is not included in the general solution.*

Example $y = x(dy/dx) - \frac{1}{4}(dy/dx)^2$ has the general solution $y = cx - \frac{1}{4}c^2$, where c is an arbitrary constant; $y = x^2$ is a singular solution, as is easily verified.

ORDINARY DIFFERENTIAL EQUATIONS OF THE FIRST ORDER

Equations with Separable Variables   Every differential equation of the first order and of the first degree can be written in the form $M(x, y)\, dx + N(x, y)\, dy = 0$. If the equation can be transformed so that M does not involve y and N does not involve x, then the variables are said to be separated. The solution can then be obtained by **quadrature,** which means that $y = \int f(x)\, dx + c$, which may or may not be expressible in simpler form.

Example   Two liquids A and B are boiling together in a vessel. Experimentally it is found that the ratio of the rates at which A and B are evaporating at any time is proportional to the ratio of the amount of A (say, x) to the amount of B (say, y) still in the liquid state. This physical law is expressible as $(dy/dt)/(dx/dt) = ky/x$ or $dy/dx = ky/x$, where k is a proportionality constant. This equation may be written $dy/y = k(dx/x)$, in which the variables are separated. The solution is $\ln y = k \ln x + \ln c$ or $y = cx^k$.

Exact Equations   The equation $M(x, y)\, dx + N(x, y)\, dy = 0$ is exact if and only if $\partial M/\partial y = \partial N/\partial x$. In this case there exists a function $w = f(x, y)$ such that $\partial f/\partial x = M$, $\partial f/\partial y = N$, and $f(x, y) = C$ is the required solution. $f(x, y)$ is found as follows: treat y as though it were constant and evaluate $\int M(x, y)\, dx$. Then treat x as though it were constant and evaluate $\int N(x, y)\, dy$. The sum of all unlike terms in these two integrals (including no repetitions) is $f(x, y)$.

Example $(2xy - \cos x)\, dx + (x^2 - 1)\, dy = 0$ is exact for $\partial M/\partial y = 2x$, $\partial N/\partial x = 2x$. $\int M\, dx = \int (2xy - \cos x)\, dx = x^2y - \sin x$, $\int N\, dy = \int (x^2 - 1)\, dy = x^2y - y$. The solution is $x^2y - \sin x - y = C$, as may easily be verified.

Linear Equations   A differential equation is said to be linear when it is of first degree in the dependent variable and its derivatives. The general linear first-order differential equation has the form $dy/dx + P(x)y = Q(x)$. Its general solution is

$$y = e^{-\int P\, dx}\left[\int Qe^{\int P\, dx}\, dx + C \right]$$

Example   A tank initially holds 200 gal of a salt solution in which 100 lb is dissolved. Six gallons of brine containing 4 lb of salt run into the tank per minute. If mixing is perfect and the output rate is 4 gal/min, what is the amount A of salt in the tank at time t? The differential equation of A is $dA/dt + [1/(100 + t)]A = 4$. Its general solution is $A = 2(100 + t) + C/(100 + t)$. At $t = 0$, $A = 100$; so the particular solution is $A = 2(100 + t) - 10^4/(100 + t)$.

ORDINARY DIFFERENTIAL EQUATIONS OF HIGHER ORDER

The higher-order differential equations, especially those of order 2, are of great importance because of physical situations describable by them.

Equation $y^{(n)} = f(x)$ Such a differential equation can be solved by n integrations. The solution will contain n arbitrary constants.

Linear Differential Equations with Constant Coefficients and Right-Hand Member Zero (Homogeneous)   The solution of $y'' + ay' + by = 0$ depends upon the nature of the roots of the characteristic equation $m^2 + am + b = 0$ obtained by substituting the trial solution $y = e^{mx}$ in the equation.

Distinct Real Roots   If the roots of the characteristic equation

are distinct real roots, r_1 and r_2, say, the solution is $y = Ae^{r_1x} + Be^{r_2x}$, where A and B are arbitrary constants.

Example $y'' + 4y' + 3 = 0$. The characteristic equation is $m^2 + 4m + 3 = 0$. The roots are -3 and -1, and the general solution is $y = Ae^{-3x} + Be^{-x}$.

Multiple Real Roots   If $r_1 = r_2$, the solution of the differential equation is $y = e^{r_1x}(A + Bx)$.

Example $y'' + 4y + 4 = 0$. The characteristic equation is $m^2 + 4m + 4 = 0$ with roots -2 and -2. The solution is $y = e^{-2x}(A + Bx)$.

Complex Roots   If the characteristic roots are $p \pm iq$, then the solution is $y = e^{px}(A \cos qx + B \sin qx)$.

Example   The differential equation $My'' + Ay' + ky = 0$ represents the vibration of a linear system of mass M, spring constant k, and damping constant A. If $A < 2\sqrt{kM}$, the roots of the characteristic equation

$$Mm^2 + Am + k = 0 \text{ are complex} - \frac{A}{2M} \pm i\sqrt{\frac{k}{M} - \left(\frac{A}{2M}\right)^2}$$

and the solution is $y = e^{-(At/2M)}$

$$\left\{ c_1 \cos\left(\sqrt{\frac{k}{M} - \left(\frac{A}{2M}\right)^2}\right)t + c_2 \sin\left(\sqrt{\frac{k}{M} - \left(\frac{A}{2M}\right)^2}\right)t \right\}$$

This solution is oscillatory, representing undercritical damping.

All these results generalize to homogeneous linear differential equations with constant coefficients of order higher than 2. These equations (especially of order 2) have been much used because of the ease of solution. Oscillations, electric circuits, diffusion processes, and heat-flow problems are a few examples for which such equations are useful.

Second-Order Equations: Dependent Variable Missing   Such an equation is of the form

$$F\left(x, \frac{dy}{dx}, \frac{d^2y}{dx^2}\right) = 0$$

It can be reduced to a first-order equation by substituting $p = dy/dx$ and $dp/dx = d^2y/dx^2$.

Second-Order Equations: Independent Variable Missing   Such an equation is of the form

$$F\left(y, \frac{dy}{dx}, \frac{d^2y}{dx^2}\right) = 0$$

Set

$$\frac{dy}{dx} = p, \quad \frac{d^2y}{dx^2} = p\frac{dp}{dy}$$

The result is a first-order equation in p,

$$F\left(y, p, p\frac{dp}{dy}\right) = 0$$

Example   The capillary curve for one vertical plate is given by

$$\frac{d^2y}{dx^2} = \frac{4y}{c^2}\left[1 + \left(\frac{dy}{dx}\right)^2\right]^{3/2}$$

Its solution by this technique is

$$x + \sqrt{c^2 - y^2} - \sqrt{c^2 - h_0^2} = \frac{c}{2}\left(\cosh^{-1}\frac{c}{y} - \cosh^{-1}\frac{c}{h_0}\right)$$

where c, h_0 are physical constants.

Example   The equation governing chemical reaction in a porous catalyst in plane geometry of thickness L is

$$D\frac{d^2c}{dx^2} = k\, f(c), \quad \frac{dc}{dx}(0) = 0, \quad c(L) = c_0$$

where D is a diffusion coefficient, k is a reaction rate parameter, c is the concentration, $k\, f(c)$ is the rate of reaction, and c_0 is the concentration at the boundary. Making the substitution gives

$$p\frac{dp}{dc} = \frac{k}{D}f(c)$$

Integrating gives

$$\frac{p^2}{2} = \frac{k}{D} \int_{c(0)}^{c} f(c)\, dc$$

If the reaction is very fast, $c(0) \approx 0$ and the average reaction rate is related to $p(L)$. See Ref. 106. This variable is given by

$$p(L) = \left[\frac{2k}{D} \int_0^{c_0} f(c)\, dc\right]^{1/2}$$

Thus, the average reaction rate can be calculated without solving the complete problem.

Linear Nonhomogeneous Differential Equations

Linear Differential Equations Right-Hand Member f(x) ≠ 0 Again the specific remarks for $y'' + ay' + by = f(x)$ apply to differential equations of similar type but higher order. We shall discuss two general methods.

Method of Undetermined Coefficients Use of this method is limited to equations exhibiting both constant coefficients and particular forms of the function $f(x)$. In most cases $f(x)$ will be a sum or product of functions of the type constant, x^n (n a positive integer), e^{mx}, $\cos kx$, $\sin kx$. When this is the case, the solution of the equation is $y = H(x) + P(x)$, where $H(x)$ is a solution of the homogeneous equations found by the method of the preceding subsection and $P(x)$ is a particular integral found by using the following table subject to these conditions: (1) When $f(x)$ consists of the sum of several terms, the appropriate form of $P(x)$ is the sum of the particular integrals corresponding to these terms individually. (2) When a term in any of the trial integrals listed is already a part of the homogeneous solution, the indicated form of the particular integral is multiplied by x.

Form of Particular Integral

If $f(x)$ is	Then $P(x)$ is
a (constant)	A (constant)
ax^n	$A_n x^n + A_{n-1} x^{n-1} + \cdots A_1 x + A_0$
ae^{rx}	Be^{rx}
$\left.\begin{array}{l} c\cos kx \\ d\sin kx \end{array}\right\}$	$A\cos kx + B\sin kx$
$\left.\begin{array}{l} gx^n e^{rx}\cos kx \\ hx^n e^{rx}\sin kx \end{array}\right\}$	$(A_n x^n + \cdots + A_0)e^{rx}\cos kx + (B_n x^n + \cdots + B_0)e^{rx}\sin kx$

Since the form of the particular integral is known, the constants may be evaluated by substitution in the differential equation.

Example $y'' + 2y' + y = 3e^{2x} - \cos x + x^3$. The characteristic equation is $(m+1)^2 = 0$ so that the homogeneous solution is $y = (c_1 + c_2 x)e^{-x}$. To find a particular solution we use the trial solution from the table, $y = a_1 e^{2x} + a_2 \cos x + a_3 \sin x + a_4 x^3 + a_5 x^2 + a_6 x + a_7$. Substituting this in the differential equation collecting and equating like terms, there results $a_1 = 1/3$, $a_2 = 0$, $a_3 = -1/2$, $a_4 = 1$, $a_5 = -6$, $a_6 = 18$, and $a_7 = -24$. The solution is $y = (c_1 + c_2 x)e^{-x} + 1/3 e^{2x} - 1/2 \sin x + x^3 - 6x^2 + 18x - 24$.

Method of Variation of Parameters This method is applicable to any linear equation. The technique is developed for a second-order equation but immediately extends to higher order. Let the equation be $y'' + a(x)y' + b(x)y = R(x)$ and let the solution of the homogeneous equation, found by some method, be $y = c_1 f_1(x) + c_2 f_2(x)$. It is now assumed that a particular integral of the differential equation is of the form $P(x) = u f_1 + v f_2$ where u, v are functions of x to be determined by two equations. One equation results from the requirement that $u f_1 + v f_2$ satisfy the differential equation, and the other is a degree of freedom open to the analyst. The best choice proves to be

$$u'f_1 + v'f_2 = 0 \qquad \text{and} \qquad u'f_1' + v'f_2' = R(x)$$

Then

$$u' = \frac{du}{dx} = -\frac{f_2}{f_1 f_2' - f_2 f_1'} R(x)$$

$$v' = \frac{dv}{dx} = \frac{f_1}{f_1 f_2' - f_2 f_1'} R(x)$$

and since f_1, f_2, and R are known u, v may be found by direct integration.

Example $(1 - x^2)\dfrac{d^2 y}{dx^2} - \dfrac{1}{x}\dfrac{dy}{dx} = x$. The homogeneous equation

$$(1 - x^2)\frac{d^2 y}{dx^2} - \frac{1}{x}\frac{dy}{dx} = 0$$

reduces to

$$\frac{dp}{p} = \frac{dx}{x(1 - x^2)}$$

when we set $dy/dx = p$. Upon integrating twice, $y = c_1\sqrt{x^2 - 1} + c_2$ is the homogeneous solution. Now assume that the particular solution has the form $y = u\sqrt{x^2 - 1} + v$. The equations for u and v become

$$u' = du/dx = \sqrt{x^2 - 1}$$

$$v' = \frac{dv}{dx} = 1 - x^2$$

so that

$$u = \frac{1}{2}[x\sqrt{x^2 - 1} - \ln(x + \sqrt{x^2 - 1})] \qquad \text{and} \qquad v = x - x^3/3.$$

The complete solution is

$$y = c_1\sqrt{x^2 - 1} + c_2 + \frac{x}{2} - \frac{x^3}{6} - \frac{1}{2}\sqrt{x^2 - 1}\ln(x + \sqrt{x^2 - 1}).$$

Perturbation Methods If the ordinary differential equation has a parameter that is small and is not multiplying the highest derivative, perturbation methods can give solutions for small values of the parameter.

Example Consider the differential equation for reaction and diffusion in a catalyst; the reaction is second order: $c'' = ac^2$, $c'(0) = 0$, $c(1) = 1$. The solution is expanded in the following Taylor series in a.

$$c(x, a) = c_0(x) + ac_1(x) + a^2 c_2(x) + \cdots$$

The goal is to find equations governing the functions $\{c_i(x)\}$ and solve them. Substitution into the equations gives the following equations:

$$c_0''(x) + a\, c_1''(x) + a^2 c_2''(x) + \cdots = a[c_0(x) + ac_1(x) + a^2 c_2(x) + \cdots]^2$$

$$c_0'(0) + ac_1'(0) + a^2 c_2'(0) + \cdots = 0$$

$$c_0(1) + ac_1(1) + a^2 c_2(1) + \cdots = 1$$

Like terms in powers of a are collected to form the individual problems.

$$c_0'' = 0, \quad c_0'(0) = 0, \quad c_0(1) = 1$$

$$c_1'' = c_0^2, \quad c_1'(0) = 0, \quad c_1(1) = 0$$

$$c_2'' = 2c_0 c_1, \quad c_2'(0) = 0, \quad c_2(1) = 0$$

The solution proceeds in turn.

$$c_0(x) = 1, \quad c_1(x) = \frac{(x^2 - 1)}{2}, \quad c_2(x) = \frac{5 - 6x^2 + x^4}{12}$$

SPECIAL DIFFERENTIAL EQUATIONS (SEE REF. 1)

Euler's Equation The linear equation $x^n y^{(n)} + a_1 x^{n-1} y^{(n-1)} + \cdots + a_{n-1} xy' + a_n y = R(x)$ can be reduced to a linear equation with constant coefficients by the change of variable $x = e^t$. To solve the homogeneous equation substitute $y = x^r$ into it, cancel the powers of x, which are the same for all terms, and solve the resulting polynomial for r. In case of multiple or complex roots there results the form $y = x^r(\log x)^r$ and $y = x^\alpha[\cos(\beta\log x) + i\sin(\beta\log x)]$.

Example Solve $x^2 y'' - 2y = 0$. By setting $y = x^r$, $x^r[r(r-1) - 2] = 0$. The roots of $r^2 - r - 2 = 0$ are $r = 2, -1$. The general solution is $y = Ax^2 + B/x$.

The equation $(ax + b)^n y^{(n)} + a_1(ax + b)^{n-1} y^{(n-1)} + \cdots + a_n y = R(x)$ can be reduced to the Euler form by the substitution $ax + b = z$. It may be treated without change of variable, the homogeneous equation having solutions of the form $y = (ax + b)^r$.

Bessel's Equation The linear equation $x^2(d^2 y/dx^2) + (1 - 2\alpha) x(dy/dx) + [\beta^2\gamma^2 x^{2\gamma} + (\alpha^2 - p^2\gamma^2)]y = 0$ is the general Bessel equation. By series methods, not to be discussed here, this equation can be shown to have the solution

$$y = Ax^\alpha J_p(\beta x^\gamma) + Bx^\alpha J_{-p}(\beta x^\gamma) \qquad p \text{ not an integer or zero}$$

$$y = Ax^\alpha J_p(\beta x^\gamma) + Bx^\alpha Y_p(\beta x^\gamma) \qquad p \text{ an integer}$$

where
$$J_p(x) = \left(\frac{x}{2}\right)^p \sum_{k=0}^{\infty} \frac{(-1)^k (x/2)^{2k}}{k!\Gamma(p+k+1)}$$

$$J_{-p}(x) = \left(\frac{x}{2}\right)^{-p} \sum_{k=0}^{\infty} \frac{(-1)^k (x/2)^{2k}}{k!\Gamma(k+1-p)} \quad p \text{ not an integer}$$

$$\Gamma(n) = \int_0^{\infty} x^{n-1} e^{-x}\, dx \quad n > 0$$

is the gamma function. For p an integer
$$J_p(x) = \left(\frac{x}{2}\right)^p \sum_{k=0}^{\infty} \frac{(-1)^k (x/2)^{2k}}{k!(p+k)!}$$

(Bessel function of the first kind of order p)

$$Y_p(x) = \frac{[J_p(x)\cos(p\pi) - J_{-p}(x)]}{\sin(p\pi)}$$

(replace right-hand side by limiting value if P is an integer or zero).

The series converge for all x. Much of the importance of Bessel's equation and Bessel functions lies in the fact that the solutions of numerous linear differential equations can be expressed in terms of them.

Example $d^2y/dx^2 + [9x - (63/4x^2)]y = 0$. In general form this is $x^2(d^2y/dx^2) + (9x^3 - 6\frac{3}{4})y = 0$. Thus $\alpha = \frac{1}{2}$, $\gamma = \frac{3}{2}$; $\beta = 2$, $p = \frac{5}{3}$. The solution is (since $p \neq$ integer) $y = Ax^{1/2}J_{5/3}(2x^{3/2}) + Bx^{1/2}J_{-5/3}(2x^{3/2})$. Tables are available for the evaluation of many of these functions.

Example The heat flow through a wedge-shaped fin is characterized by the equation $x^2(d^2y/dx^2) + x(dy/dx) - a^2xy = 0$, where $y = T - T_{\text{air}}$, α is a combination of physical constants, and x = distance from fin end. By comparing this with the standard equation, there results $\alpha = 0, p = 0, \gamma = \frac{1}{2}, \beta^2 = -4a^2$ or $\beta = 2ai$. The solution is $y = AJ_0(2ai\sqrt{x}) + BY_0(2ai\sqrt{x})$.

Legendre's Equation The Legendre equation $(1 - x^2)y'' - 2xy' + n(n + 1)y = 0$, $n \geq 0$, has the solution $y = Au_n(x) + Bv_n(x)$ for n not an integer where

$$u_n(x) = 1 - \frac{n(n+1)}{2!}x^2 + \frac{n(n-2)(n+1)(n+3)}{4!}x^4$$
$$- \frac{n(n-2)(n-4)(n+1)(n+3)(n+5)}{6!}x^6 + \cdots$$

$$v_n(x) = x - \frac{(n-1)(n+2)}{3!}x^3 + \frac{(n-1)(n-3)(n+2)(n+4)}{5!}x^5 - \cdots$$

If n is an even integer or zero, u_n is a polynomial in x. If n is an odd integer, then v_n is a polynomial. The interval of convergence for the series is $-1 < x < 1$. If n is an integer, set

$$P_n(x) = \frac{u_n(x)}{u_n(1)} \text{ (n even or zero)}, \quad P_n = \frac{v_n(x)}{v_n(1)} \text{ (n odd)}$$

The polynomials P_n are the so-called Legendre polynomials, $P_0(x) = 1$, $P_1(x) = x$, $P_2(x) = \frac{1}{2}(3x^2 - 1)$, $P_3(x) = \frac{1}{2}(5x^3 - 3x)$, ...

Laguerre's Equation The Laguerre equation $x(d^2y/dx^2) + (c-x)(dy/dx) - ay = 0$ is satisfied by the confluent hypergeometric function. See Refs. 1 and 173.

Hermite's Equation The Hermite equation $y'' - 2xy' + 2ny = 0$ is satisfied by the Hermite polynomial of degree n, $y = AH_n(x)$ if n is a positive integer or zero. $H_0(x) = 1$, $H_1(x) = 2x$, $H_2(x) = 4x^2 - 2$, $H_3(x) = 8x^3 - 12x$, $H_4(x) = 16x^4 - 48x^2 + 12$, $H_{r+1}(x) = 2xH_r(x) - 2rH_{r-1}(x)$.

Example $y'' - 2xy' + 6y = 0$. Here $n = 3$; so $y = AH_3 = A(8x^3 - 12x)$ is a solution.

Chebyshev's Equation The equation $(1 - x^2)y'' - xy' + n^2y = 0$ for n a positive integer or zero is satisfied by the nth Chebyshev polynomial $y = AT_n(x)$. $T_0(x) = 1$, $T_1(x) = x$, $T_2(x) = 2x^2 - 1$, $T_3(x) = 4x^3 - 3x$, $T_4(x) = 8x^4 - 8x^2 + 1$; $T_{r+1}(x) = 2xT_r(x) - T_{r-1}(x)$.

Example $(1 - x^2)y'' - xy' + 36y = 0$. Here $n = 6$. A solution is $y = T_6(x) = 2xT_5(x) - T_4(x) = 2x(2xT_4 - T_3) - T_4 = 32x^6 - 48x^4 + 18x^2 - 1$. Further details on these special equations and others can be found in the literature.

PARTIAL DIFFERENTIAL EQUATIONS

The analysis of situations involving two or more independent variables frequently results in a partial differential equation.

Example The equation $\partial T/\partial t = K(\partial^2 T/\partial x^2)$ represents the unsteady one-dimensional conduction of heat.

Example The equation for the unsteady transverse motion of a uniform beam clamped at the ends is

$$\frac{\partial^4 y}{\partial x^4} + \frac{\rho}{EI}\frac{\partial^2 y}{\partial t^2} = 0$$

Example The expansion of a gas behind a piston is characterized by the simultaneous equations

$$\frac{\partial u}{\partial t} + u\frac{\partial u}{\partial x} + \frac{c^2}{\rho}\frac{\partial \rho}{\partial x} = 0 \quad \text{and} \quad \frac{\partial \rho}{\partial t} + u\frac{\partial \rho}{\partial x} + \rho\frac{\partial u}{\partial x} = 0$$

Example The heating of a diathermanous solid is characterized by the equation $\alpha(\partial^2\theta/\partial x^2) + \beta e^{-\gamma x} = \partial\theta/\partial t$.

The partial differential equation $\partial^2 f/\partial x \,\partial y = 0$ can be solved by two integrations yielding the solution $f = g(x) + h(y)$, where $g(x)$ and $h(y)$ are arbitrary differentiable functions. This result is an example of the fact that the general solution of partial differential equations involves arbitrary functions in contrast to the solution of ordinary differential equations, which involve only arbitrary constants. A number of methods are available for finding the general solution of a partial differential equation. In most applications of partial differential equations the general solution is of limited use. In such applications the solution of a partial differential equation must satisfy both the equation and certain auxiliary conditions called **initial** and/or **boundary** conditions, which are dictated by the problem. Examples of these include those in which the wall temperature is a fixed constant $T(x_0) = T_0$, there is no diffusion across a nonpermeable wall, and the like. In ordinary differential equations these auxiliary conditions allow definite numbers to be assigned to the constants of integration. In partial differential equations the boundary conditions demand that the arbitrary functions resulting from integration assume specific forms. Except for a few cases (some first-order equations, D'Alembert's solution of the wave equation, and others) a procedure which first determines the arbitrary functions and then specializes them to fit the boundary conditions is usually not feasible. A more fruitful attack is to determine directly a set of particular solutions and then combine them so that the boundary conditions are satisfied. The only area in which much analysis has been accomplished is for linear homogeneous partial differential equations. Such equations have the property that if $f_1, f_2, \ldots, f_n, \ldots$ are individually solutions, then the function $f = \sum_{i=1}^{\infty} f_i$ is also a solution, provided the series converges and is differentiable up to the order (termwise) of the equation.

Partial Differential Equations of Second and Higher Order Many of the applications to scientific problems fall naturally into partial differential equations of second order, although there are important exceptions in elasticity, vibration theory, and elsewhere.

A second-order differential equation can be written as

$$a\frac{\partial^2 u}{\partial x^2} + b\frac{\partial^2 u}{\partial x \partial y} + c\frac{\partial^2 u}{\partial y^2} = f$$

where $a, b, c,$ and f depend upon $x, y, u, \partial u/\partial x$, and $\partial u/\partial y$. This equation is hyperbolic, parabolic, or elliptic, depending on whether the discriminant $b^2 - 4ac$ is >0, $=0$, or <0, respectively. Since $a, b, c,$ and f depend on the solution, the type of equation can be different at different x and y locations. If the equation is hyperbolic, discontinuities can be propagated. See Refs. 11, 79, 105, 159, and 192.

Phenomena of **propagation** such as vibrations are characterized by equations of "hyperbolic" type which are essentially different in their properties from other classes such as those which describe equilibrium (elliptic) or unsteady diffusion and heat transfer (parabolic). Prototypes are as follows:

Elliptic Laplace's equation $\partial^2 u/\partial x^2 + \partial^2 u/\partial y^2 = 0$ and Poisson's equation $\partial^2 u/\partial x^2 + \partial^2 u/\partial y^2 = g(x, y)$ do not contain the variable time explicitly and consequently represent equilibrium configurations. Laplace's equation is satisfied by static electric or magnetic potential at points free from electric charges or magnetic poles. Other important functions satisfying Laplace's equation are the velocity potential of the irrotational motion of an incompressible fluid, used in hydrodynamics; the steady temperature at points in a homogeneous solid, and the steady state of diffusion through a homogeneous body. The gravitational potential V at points occupied by mass of density d satisfies Poisson's equation $\partial^2 V/\partial x^2 + \partial^2 V/\partial y^2 + \partial^2 V/\partial z^2 = -4\pi d$.

Parabolic   The heat equation $\partial T/\partial t = \partial^2 T/\partial x^2 + \partial^2 T/\partial y^2$ represents nonequilibrium or unsteady states of heat conduction and diffusion.

Hyperbolic   The wave equation $\partial^2 u/\partial t^2 = c^2(\partial^2 u/\partial x^2 + \partial^2 u/\partial y^2)$ represents wave propagation of many varied types.

Quasilinear first-order differential equations are like

$$a \frac{\partial u}{\partial x} + b \frac{\partial u}{\partial y} = f$$

where a, b, and f depend on x, y, and u, with $a^2 + b^2 \neq 0$. This equation can be solved using the method of characteristics, which writes the solution in terms of a parameter s, which defines a path for the characteristic.

$$\frac{dx}{ds} = a, \quad \frac{dy}{ds} = b, \quad \frac{du}{ds} = f$$

These equations are integrated from some initial conditions. For a specified value of s, the value of x and y shows the location where the solution is u. The equation is semilinear if a and b depend just on x and y (and not u), and the equation is linear if a, b, and f all depend on x and y, but not u. Such equations give rise to shock propagation, and conditions have been derived to deduce the presence of shocks, Ref. 245. For further information, see Refs. 79, 159, 192, and 245.

An example of a linear hyperbolic equation is the advection equation for flow of contaminants when the x and y velocity components are u and v, respectively.

$$\frac{\partial c}{\partial t} + u \frac{\partial c}{\partial x} + v \frac{\partial c}{\partial y} = 0$$

The equations for flow and adsorption in a packed bed or chromatography column give a quasilinear equation.

$$\phi \frac{\partial c}{\partial t} + \phi\, u \frac{\partial c}{\partial x} + (1 - \phi) \frac{df}{dc} \frac{\partial c}{\partial t} = 0$$

Here $n = f(c)$ is the relation between concentration on the adsorbent and fluid concentration.

The solution of problems involving partial differential equations often revolves about an attempt to reduce the partial differential equation to one or more ordinary differential equations. The solutions of the ordinary differential equations are then combined (if possible) so that the boundary conditions as well as the original partial differential equation are simultaneously satisfied. Three of these techniques are illustrated.

Similarity Variables   The physical meaning of the term "similarity" relates to internal similitude, or self-similitude. Thus, similar solutions in boundary-layer flow over a horizontal flat plate are those for which the horizontal component of velocity u has the property that two velocity profiles located at different coordinates x differ only by a scale factor. The mathematical interpretation of the term similarity is a transformation of variables carried out so that a reduction in the number of independent variables is achieved. There are essentially two methods for finding similarity variables, "separation of variables" (not the classical concept) and the use of "continuous transformation groups." The basic theory is available in Ames (see the references).

Example   The equation $\partial \theta/\partial x = (A/y)(\partial^2 \theta/\partial y^2)$ with the boundary conditions $\theta = 0$ at $x = 0$, $y > 0$; $\theta = 0$ at $y = \infty$, $x > 0$; $\theta = 1$ at $y = 0$, $x > 0$ represents the nondimensional temperature θ of a fluid moving past an infinitely wide flat plate immersed in the fluid. Turbulent transfer is neglected, as is molecular transport except in the y direction. It is now assumed that the equation and the boundary conditions can be satisfied by a solution of the form $\theta = f(y/x^n) = f(u)$, where $\theta = $

0 at $u = \infty$ and $\theta = 1$ at $u = 0$. The purpose here is to replace the independent variables x and y by the single variable u when it is hoped that a value of n exists which will allow x and y to be completely eliminated in the equation. In this case since $u = y/x^n$, there results after some calculation $\partial \theta/\partial x = -(nu/x)(d\theta/du)$, $\partial^2 \theta/\partial y^2 = (1/x^{2n})(d^2\theta/du^2)$, and when these are substituted in the equation, $-(1/x)nu(d\theta/du) = (1/x^{3n})(A/u)(d^2\theta/du^2)$. For this to be a function of u only, choose $n = \frac{1}{3}$. There results $(d^2\theta/du^2) + (u^2/3A)(d\theta/du) = 0$. Two integrations and use of the boundary conditions for this ordinary differential equation give the solution

$$\theta = \int_u^\infty \exp\left(-u^3/9A\right) du \Big/ \int_0^\infty \exp\left(-u^3/9A\right) du$$

Group Method   The type of transformation can be deduced using group theory. For a complete exposition, see Refs. 9, 12, and 145; a shortened version is in Ref. 106. Basically, a similarity transformation should be considered when one of the independent variables has no physical scale (perhaps it goes to infinity). The boundary conditions must also simplify (and combine) since each transformation leads to a differential equation with one fewer independent variable.

Example   A similarity variable is found for the problem

$$\frac{\partial c}{\partial t} = \frac{\partial}{\partial x}\left(D(c) \frac{\partial c}{\partial x}\right), \quad c(0,t) = 1, \quad c(\infty,t) = 0, \quad c(x,0) = 0$$

Note that the length dimension goes to infinity, so that there is no length scale in the problem statement; this is a clue to try a similarity transformation. The transformation examined here is

$$\bar{t} = a^\alpha t, \quad \bar{x} = a^\beta x, \quad \bar{c} = a^\gamma c$$

With this substitution, the equation becomes

$$a^{\alpha - \gamma} \frac{\partial \bar{c}}{\partial \bar{t}} = a^{2\beta - \gamma} \frac{\partial}{\partial \bar{x}}\left[D(a^{-\gamma}\bar{c}) \frac{\partial \bar{c}}{\partial \bar{x}}\right]$$

Group theory says a system is conformally invariant if it has the same form in the new variables; here, that is

$$\gamma = 0, \quad \alpha - \gamma = 2\beta - \gamma, \quad \text{or } \alpha = 2\beta$$

The invariants are

$$\eta = \frac{x}{t^\delta}, \quad \delta = \frac{\beta}{\alpha}$$

and the solution is

$$c(x, t) = f(\eta)t^{\gamma/\alpha}$$

We can take $\gamma = 0$ and $\delta = \beta/\alpha = \frac{1}{2}$. Note that the boundary conditions combine because the point $x = \infty$ and $t = 0$ give the same value of η and the conditions on c at $x = \infty$ and $t = 0$ are the same. We thus make the transformation

$$\eta = \frac{x}{\sqrt{4D_0 t}}, \quad c(x, t) = f(\eta)$$

The use of the 4 and D_0 makes the analysis below simpler. The result is

$$\frac{d}{d\eta}\left[D(c) \frac{df}{d\eta}\right] + 2\eta \frac{df}{d\eta} = 0, \quad f(0) = 1, \quad f(\infty) = 0$$

Thus, we solve a two-point boundary value problem instead of a partial differential equation. When the diffusivity is constant, the solution is the error function, a tabulated function.

$$c(x,t) = 1 - \text{erf } \eta = \text{erfc } \eta$$

$$\text{erf } \eta = \int_0^\eta e^{-\xi^2} d\xi \Big/ \int_0^\infty e^{-\xi^2} d\xi$$

Separation of Variables   This is a powerful, well-utilized method which is applicable in certain circumstances. It consists of assuming that the solution for a partial differential equation has the form $U = f(x)g(y)$. If it is then possible to obtain an ordinary differential equation on one side of the equation depending only on x and on the other side only on y, the partial differential equation is said to be separable in the variables x, y. If this is the case, one side of the equation is a function of x alone and the other of y alone. The two can be equal only if each is a constant, say λ. Thus the problem has again been reduced to the solution of ordinary differential equations.

Example   Laplace's equation $\partial^2 V/\partial x^2 + \partial^2 V/\partial y^2 = 0$ plus the boundary conditions $V(0, y) = 0$, $V(l, y) = 0$, $V(x, \infty) = 0$, $V(x, 0) = f(x)$ represents the steady-state potential in a thin plate (in z direction) of infinite extent in the y direction and of width l in the x direction. A potential $f(x)$ is impressed (at $y = 0$) from $x = 0$ to $x = 1$, and the sides are grounded. To obtain a solution of this boundary-value problem assume $V(x, y) = f(x)g(y)$. Substitution in the differential equation yields $f''(x)g(y) + f(x)g''(y) = 0$, or $g''(y)/g(y) = -f''(x)/f(x) = \lambda^2$ (say). This system becomes $g''(y) - \lambda^2 g(y) = 0$ and $f''(x) + \lambda^2 f(x) = 0$. The solutions of these ordinary differential equations are respectively $g(y) = Ae^{\lambda y} + Be^{-\lambda y}$, $f(x) = C \sin \lambda x + D \cos \lambda x$. Then $f(x)g(y) = (Ae^{\lambda y} + Be^{-\lambda y})(C \sin \lambda x + D \cos \lambda x)$. Now $V(0, y) = 0$ so that $f(0)g(y) = (Ae^{\lambda y} + Be^{-\lambda y}) D \equiv 0$ for all y. Hence $D = 0$. The solution then has the form $\sin \lambda x (Ae^{\lambda y} + Be^{-\lambda y})$ where the multiplicative constant C has been eliminated. Since $V(l, y) = 0$, $\sin \lambda l (Ae^{\lambda y} + Be^{-\lambda y}) \equiv 0$. Clearly the bracketed function of y is not zero, for the solution would then be the identically zero solution. Hence $\sin \lambda l = 0$ or $\lambda_n = n\pi/l$, $n = 1, 2, \ldots$ where $\lambda_n = n$th eigenvalue.

The solution now has the form $\sin (n\pi x/l)(Ae^{n\pi y/l} + Be^{-n\pi y/l})$. Since $V(x, \infty) = 0$, A must be taken to be zero because e^y becomes arbitrarily large as $y \to \infty$. The solution then reads $B_n \sin (n\pi x/l)e^{-n\pi y/l}$, where B_n is the multiplicative constant. The differential equation is linear and homogeneous so that $\sum_{n=1}^{\infty} B_n e^{-n\pi y/l} \sin (n\pi x/l)$ is also a solution. Satisfaction of the last boundary condition is ensured by taking

$$B_n = \frac{2}{l} \int_0^l f(x) \sin (n\pi x/l)\, dx = \text{Fourier sine coefficients of } f(x)$$

Further, convergence and differentiability of this series are established quite easily. Thus the solution is

$$V(x, y) = \sum_{n=1}^{\infty} B_n e^{-n\pi y/l} \sin \frac{n\pi x}{l}$$

Example The diffusion problem

$$\frac{\partial c}{\partial t} = \frac{\partial}{\partial x}\left(D(c)\frac{\partial c}{\partial x}\right), \quad c(0, t) = 1, \quad c(\infty, t) = 0, \quad c(x, 0) = 0$$

can be solved by separation of variables. First transform the problem so that the boundary conditions are homogeneous (having zeroes on the right-hand side). Let

$$c(x, t) = 1 - x + u(x, t)$$

Then $u(x, t)$ satisfies

$$\frac{\partial u}{\partial t} = D \frac{\partial^2 u}{\partial x^2}, \quad u(x, 0) = x - 1, \quad u(0, t) = 0, \quad u(1, t) = 0$$

Assume a solution of the form $u(x, t) = X(x) T(t)$, which gives

$$X \frac{dT}{dt} = D\, T \frac{d^2 X}{dx^2}$$

Since both sides are constant, this gives the following ordinary differential equations to solve.

$$\frac{1}{D\,T}\frac{dT}{dt} = -\lambda, \quad \frac{1}{X}\frac{d^2 X}{dx^2} = -\lambda$$

The solution of these is

$$T = A\, e^{-\lambda Dt}, \quad X = B \cos \sqrt{\lambda}\, x + E \sin \sqrt{\lambda}\, x$$

The combined solution for $u(x,t)$ is

$$u = A\, (B \cos \sqrt{\lambda}\, x + E \sin \sqrt{\lambda}\, x)\, e^{-\lambda Dt}$$

Apply the boundary condition that $u(0,t) = 0$ to give $B = 0$. Then the solution is

$$u = A\, (\sin \sqrt{\lambda}\, x)e^{-\lambda Dt}$$

where the multiplicative constant E has been eliminated. Apply the boundary condition at $x = L$.

$$0 = A\, (\sin \sqrt{\lambda}\, L)e^{-\lambda Dt}$$

This can be satisfied by choosing $A = 0$, which gives no solution. However, it can also be satisfied by choosing λ such that

$$\sin \sqrt{\lambda}\, L = 0, \quad \sqrt{\lambda}\, L = n\,\pi$$

Thus

$$\lambda = \frac{n^2 \pi^2}{L^2}$$

The combined solution can now be written as

$$u = A\left(\frac{\sin n\pi x}{L}\right)e^{-n^2 \pi^2 Dt/L^2}$$

Since the initial condition must be satisfied, we use an infinite series of these functions.

$$u = \sum_{n=1}^{\infty} A_n\left(\frac{\sin n\pi x}{L}\right)e^{-n^2 \pi^2 Dt/L^2}$$

At $t = 0$, we satisfy the initial condition.

$$x - 1 = \sum_{n=1}^{\infty} A_n\left(\frac{\sin n\pi x}{L}\right)$$

This is done by multiplying the equation by

$$\frac{\sin m\pi x}{L}$$

and integrating over x: $0 \to L$. (This is the same as minimizing the mean-square error of the initial condition.) This gives

$$\frac{A_m L}{2} = \int_0^L (x - 1) \sin m\pi x\, dx$$

which completes the solution.

Integral-Transform Method A number of integral transforms are used in the solution of differential equations. Only one, the Laplace transform, will be discussed here [for others, see "Integral Transforms (Operational Methods)"]. The one-sided Laplace transform indicated by $L[f(t)]$ is defined by the equation $L[f(t)] = \int_0^{\infty} f(t)e^{-st}\, dt$. It has numerous important properties. The ones of interest here are $L[f'(t)] = sL[f(t)] - f(0)$; $L[f''(t)] = s^2 L[f(t)] - sf(0) - f'(0)$; $L[f^{(n)}(t)] = s^n L[f(t)] - s^{n-1}f(0) - s^{n-2}f'(0) - \cdots - f^{(n-1)}(0)$ for ordinary derivatives. For partial derivatives an indication of which variable is being transformed avoids confusion. Thus, if

$$y = y(x, t), \quad L_t\left[\frac{\partial y}{\partial t}\right] = sL[y(x, t)] - y(x, 0)$$

whereas

$$L_t\left[\frac{\partial y}{\partial x}\right] = \frac{dL_t[y(x, t)]}{dx}$$

since $L[y(x, t)]$ is "really" only a function of x. Otherwise the results are similar. These facts coupled with the linearity of the transform, i.e., $L[af(t) + bg(t)] = aL[f(t)] + bL[g(t)]$, make it a useful device in solving some linear differential equations. Its use reduces the solution of ordinary differential equations to the solution of algebraic equations for $L[y]$. The solution of partial differential equations is reduced to the solution of ordinary differential equations. In both situations the inverse transform must be obtained either from tables, of which there are several, or by use of complex inversion methods.

Example The equation $\partial c/\partial t = D(\partial^2 c/\partial x^2)$ represents the diffusion in a semi-infinite medium, $x \geq 0$. Under the boundary conditions $c(0, t) = c_0$, $c(x, 0) = 0$ find a solution of the diffusion equation. By taking the Laplace transform of both sides with respect to t,

$$\int_0^{\infty} e^{-st}\frac{\partial^2 c}{\partial x^2}\, dt = \frac{1}{D}\int_0^{\infty} e^{-st}\frac{\partial c}{\partial t}\, dt$$

or

$$\frac{d^2 F}{dx^2} = (1/D)sF - c(x, 0) = \frac{sF}{D}$$

where $F(x, s) = L_t[c(x, t)]$. Hence

$$\frac{d^2 F}{dx^2} - \left(\frac{s}{D}\right)F = 0$$

The other boundary condition transforms into $F(0, s) = c_0/s$. Finally the solution of the ordinary differential equation for F subject to $F(0, s) = c_0/s$ and F remains finite as $x \to \infty$ is $F(x, s) = (c_0/s)e^{-\sqrt{s/D}x}$. Reference to a table shows that the function having this as its Laplace transform is

$$c(x, t) = c_0\left[1 - \frac{2}{\sqrt{\pi}}\int_0^{x/2\sqrt{Dt}} e^{-u^2}\, du\right]$$

Matched-Asymptotic Expansions Sometimes the coefficient in front of the highest derivative is a small number. Special perturbation techniques can then be used, provided the proper scaling laws are found. See Refs. 32, 170, and 180.

DIFFERENCE EQUATIONS

REFERENCES: 30, 43.

Certain situations are such that the independent variable does not vary continuously but has meaning only for discrete values. Typical illustrations occur in the stagewise processes found in chemical engineering such as distillation, staged extraction systems, and absorption columns. In each of these the operation is characterized by a finite between-stage change of the dependent variable in which the independent variable is the integral number of the stage. The importance of difference equations is twofold: (1) to analyze problems of the type described and (2) to obtain approximate solutions of problems which lead, in their formulation, to differential equations. In this subsection only problems of analysis are considered; the application to approximate solutions is considered under "Numerical Analysis and Approximate Methods."

ELEMENTS OF THE CALCULUS OF FINITE DIFFERENCES

Let $y = f(x)$ be defined for discrete equidistant values of x, which will be denoted by x_n. The corresponding value of y will be written $y_n = f(x_n)$. The first forward difference of $f(x)$ denoted by $\Delta f(x) = f(x + h) - f(x)$ where $h = x_n - x_{n-1} = $ interval length.

Example   Let $f(x) = x^2$. Then $\Delta f(x) = (x + h)^2 - x^2 = 2hx + h^2$.

The second forward difference is obtained by taking the difference of the first; thus $\Delta\Delta f(x) = \Delta^2 f(x) = \Delta f(x + h) - \Delta f(x) = f(x + 2h) - 2f(x + h) + f(x)$.

Example $f(x) = x^2$, $\Delta^2 f(x) = \Delta[\Delta f(x)] = \Delta 2hx + \Delta h^2 = 2h(x + h) - 2hx + h^2 - h^2 = 2h^2$.

Similarly the nth forward difference is defined by the relation $\Delta^n f(x) = \Delta[\Delta^{n-1} f(x)]$. Other difference relations are also quite useful. Some of these are $\nabla f(x) = f(x) - f(x - h)$, which is called the backward difference, and $\delta f(x) = f[x + (h/2)] - f[x - (h/2)]$, called the central difference. Some properties of the operator Δ are quite important. If C is any constant, $\Delta C = 0$; if $f(x)$ is any function of period h, $\Delta f(x) = 0$ (in fact, periodic functions of period h play the same role here as constants do in the differential calculus); $\Delta[f(x) + g(x)] = \Delta f(x) + \Delta g(x)$; $\Delta^m[\Delta^n f(x)] = \Delta^{m+n} f(x)$; $\Delta[f(x)g(x)] = f(x) \Delta g(x) + g(x + h) \Delta f(x)$

$$\Delta\left[\frac{f(x)}{g(x)}\right] = \frac{g(x) \Delta f(x) - f(x) \Delta g(x)}{g(x)g(x + h)}$$

Example $\Delta(x \sin x) = x\Delta \sin x + \sin (x + h) \Delta x = 2x \sin (h/2) \cos [x + (h/2)] + h \sin (x + h)$.

DIFFERENCE EQUATIONS

A difference equation is a relation between the differences and the independent variable, $\phi(\Delta^n y, \Delta^{n-1}y, \ldots, \Delta y, y, x) = 0$, where ϕ is some given function. The general case in which the interval between the successive points is any real number h, instead of 1, can be reduced to that with interval size 1 by the substitution $x = hx'$. Hence all further difference-equation work will assume the interval size between successive points is 1.

Example $f(x + 1) - (\alpha + 1)f(x) + \alpha f(x - 1) = 0$. Common notation usually is $y_x = f(x)$. This equation is then written $y_{x+1} - (\alpha + 1)y_x + \alpha y_{x-1} = 0$.

Example $y_{x+2} + 2y_{x+1} + y_x = x^2$.

Example $y_{x+1} - y_x = 2^x$.

The order of the difference equation is the difference between the largest and smallest arguments when written in the form of the second example. The first and second examples are both of order 2, while the third example is of order 1. A linear difference equation involves no products or other nonlinear functions of the dependent variable and its differences. The first and third examples are linear, while the second example is nonlinear.

A solution of a difference equation is a relation between the variables which satisfies the equation. If the difference equation is of order n, the general solution involves n arbitrary constants. The techniques for solving difference equations resemble techniques used for differential equations.

Equation $\Delta^n y = a$ The solution of $\Delta^n y = a$, where a is a constant, is a polynomial of degree n plus an arbitrary periodic function of period 1. That is, $y = (ax^n/n!) + c_1 x^{n-1} + c_2 x^{n-2} + \cdots + c_n + f(x)$, where $f(x + 1) = f(x)$.

Example $\Delta^3 y = 6$. The solution is $y = x^3 + c_1 x^2 + c_2 x + c_3 + f(x)$; c_1, c_2, c_3 are arbitrary constants, and $f(x)$ is an arbitrary periodic function of period 1.

Equation $y_{x+1} - y_x = \phi(x)$ This equation states that the first difference of the unknown function is equal to the given function $\phi(x)$. The solution by analogy with solving the differential equation $dy/dx = \phi(x)$ by integration is obtained by "finite integration" or summation. When there are only a finite number of data points, this is easily accomplished by writing $y_x = y_0 + \sum_{t=1}^{x} \phi(t - 1)$, where the data points are numbered from 1 to x. This is the only situation considered here.

Examples   If $\phi(x) = 1$, $y_x = x$. If $\phi(x) = x$, $y_x = [x(x - 1)]/2$. If $\phi(x) = a^x$, $a \neq 0$, $y_x = a^x/(a - 1)$. In all cases $y_0 = 0$.

Other examples may be evaluated by using summation, that is, $y_2 = y_1 + \phi(1)$, $y_3 = y_2 + \phi(2) = y_1 + \phi(1) + \phi(2)$, $y_4 = y_3 + \phi(3) = y_1 + \phi(1) + \phi(2) + \phi(3), \ldots, y_x = y_1 + \sum_{t=1}^{x-1} \phi(t)$.

Example $y_{x+1} - ry_x = 1$, r constant, $x > 0$ and $y_0 = 1$. $y_1 = 1 + r$, $y_2 = 1 + r + r^2, \ldots, y_x = 1 + r + \cdots + r^x = (1 - r^{x+1})/(1 - r)$ for $r \neq 1$ and $y_x = 1 + x$ for $r = 1$.

Linear Difference Equations   The linear difference equation of order n has the form $P_n y_{x+n} + P_{n-1} y_{x+n-1} + \cdots + P_1 y_{x+1} + P_0 y_x = Q(x)$ with $P_n \neq 0$ and $P_0 \neq 0$ and P_j; $j = 0, \ldots, n$ are functions of x.

Constant Coefficient and $Q(x) = 0$ (Homogeneous)   The solution is obtained by trying a solution of the form $y_x = c\beta^x$. When this trial solution is substituted in the difference equation, a polynomial of degree n results for β. If the solutions of this polynomial are denoted by $\beta_1, \beta_2, \ldots, \beta_n$ then the following cases result: (1) if all the β_j's are real and unequal, the solution is $y_x = \sum_{j=1}^{n} c_j \beta_j^x$, where the $c_1, \ldots, c_n$ are arbitrary constants; (2) if the roots are real and repeated, say, β_j has multiplicity m, then the partial solution corresponding to β_j is $\beta_j^x(c_1 + c_2 x + \cdots + c_m x^{m-1})$; (3) if the roots are complex conjugates, say, $a + ib = pe^{i\theta}$ and $a - ib = pe^{-i\theta}$, the partial solution corresponding to this pair is $p^x(c_1 \cos \theta x + c_2 \sin \theta x)$; and (4) if the roots are multiple complex conjugates, say, $a + ib = pe^{i\theta}$ and $a - ib = pe^{-i\theta}$ are m-fold, then the partial solution corresponding to these is $p^x[(c_1 + c_2 x + \cdots + c_m x^{m-1}) \cos \theta x + (d_1 + d_2 x + \cdots + d_m x^{m-1}) \sin \theta x]$.

Example   The equation $y_{x+1} - (\alpha + 1)y_x + \alpha y_{x-1} = 0$, $y_0 = c_0$ and $y_{m+1} = x_{m+1}/k$ represents the steady-state composition of transferable material in the raffinate stream of a staged countercurrent liquid-liquid extraction system. Clearly y is a function of the stage number x. α is a combination of system constants. By using the trial solution $y_x = c\beta^x$, there results $\beta^2 - (\alpha + 1)\beta + \alpha = 0$, so that $\beta_1 = 1$, $\beta_2 = \alpha$. The general solution is $y_x = c_1 + c_2 \alpha^x$. By using the side conditions, $c_1 = c_0 - c_2$, $c_2 = (y_{m+1} - c_0)/(\alpha^{m+1} - 1)$. The desired solution is $(y_x - c_0)/(y_{m+1} - c_0) = (\alpha^x - 1)/(\alpha^{m+1} - 1)$.

Example $y_{x+3} - 3y_{x+2} + 4y_x = 0$. By setting $y_x = c\beta^x$, there results $\beta^3 - 3\beta^2 + 4 = 0$ or $\beta_1 = -1$, $\beta_2 = 2$, $\beta_3 = 2$. The general solution is $y_x = c_1(-1)^x + 2^x(c_2 + c_3 x)$.

Example $y_{x+1} - 2y_x + 2y_{x-1} = 0$. $\beta_1 = 1 + i$, $\beta_2 = 1 - i$. $p = \sqrt{1 + 1} = \sqrt{2}$, $\theta = \pi/4$. The solution is $y_x = 2^{x/2}[c_1 \cos (x\pi/4) + c_2 \sin (x\pi/4)]$.

Constant Coefficients and $Q(x) \neq 0$ (Nonhomogeneous)   In this case the general solution is found by first obtaining the homogeneous solution, say, y_x^H and adding to it any particular solution with $Q(x) \neq 0$, say, y_x^P. There are several means of obtaining the particular solution.

Method of Undetermined Coefficients If $Q(x)$ is a product or linear combination of products of the functions e^{bx}, a^x, x^p (p a positive integer or zero) $\cos cx$ and $\sin cx$, this method may be used. The "families" $[a^x]$, $[e^{bx}]$, $[\sin cx, \cos cx]$ and $[x^p, x^{p-1}, \ldots, x, 1]$ are defined for each of the above functions in the following way: The family of a term f_x is the set of all functions of which f_x and all operations of the form a^{x+y}, $\cos c(x+y)$, $\sin c(x+y)$, $(x+y)^p$ on f_x and their linear combinations result in. The technique involves the following steps: (1) Solve the homogeneous system. (2) Construct the family of each term. (3) If the family has no representative in the homogeneous solution, assume y_x^P is a linear combination of the families of each term and determine the constants so that the equation is satisfied. (4) If a family has a representative in the homogeneous solution, multiply each member of the family by the smallest integral power of x for which all such representatives are removed and revert to step 3.

Example $y_{x+1} - 3y_x + 2y_{x-1} = 1 + a^x$. $a \neq 0$. The homogeneous solution is $y_x^H = c_1 + c_2 2^x$. The family of 1 is 1 and of a^x is a^x. However, 1 is a solution of the homogeneous system. Therefore, try $y_x^P = Ax + Ba^x$. Substituting in the equation there results

$$y_x = c_1 + c_2 2^x - x + \frac{a}{(a-1)(a-2)} a^x \, a \neq 1, a \neq 2$$

If $a = 1$, $y_x = c_1 + c_2 2^x - 2x$. If $a = 2$, $y_x = c_1 + c_2 2^x - x + x2^x$.

Example The family of $x^2 3^x$ is $[x^2 3^x, x3^x, 3^x]$.

Method of Variation of Parameters This technique is applicable to general linear difference equations. It is illustrated for the second-order system $y_{x+2} + Ay_{x+1} + By_x = \phi(x)$. Assume that the homogeneous solution has been found by some technique and write $y_x^H = c_1 u_x + c_2 v_x$. Assume that a particular solution $y_x^P = D_x u_x + E_x v_x$. E_x and D_x can be found by solving the equations:

$$E_{x+1} - E_x = \frac{u_{x+1} \phi(x)}{u_{x+1} v_{x+2} - u_{x+2} v_{x+1}}$$

$$D_{x+1} - D_x = \frac{v_{x+1} \phi(x)}{v_{x+1} u_{x+2} - v_{x+2} u_{x+1}}$$

by summation. The general solution is then $y_x = y_x^P + y_x^H$.

Variable Coefficients The method of variation of parameters applies equally well to the linear difference equation with variable coefficients. Techniques are therefore needed to solve the homogeneous system with variable coefficients.

Equation $y_{x+1} - a_x y_x = 0$ By assuming that this equation is valid for $x \geq 0$ and $y_0 = c$, the solution is $y_x = c \prod_{n=1}^{x} a_{n-1}$.

Example $y_{x+1} + \dfrac{x+2}{x+1} y_x = 0$. The solution is

$$y_x = c \prod_{n=1}^{x} \left(-\frac{n+1}{n} \right) = c(-1)^x \cdot \frac{2}{1} \cdot \frac{3}{2} \cdots \frac{x+1}{x} = (-1)^x c(x+1)$$

Example $y_{x+1} - xy_x = 0$. The solution is $y_x = c(x-1)!$

Reduction of Order If one homogeneous solution, say, u_x, can be found by inspection or otherwise, an equation of lower order can be obtained by the substitution $v_x = y_x/u_x$. The resultant equation must be satisfied by $v_x = $ constant or $\Delta v_x = 0$. Thus the equation will be of reduced order if the new variable $U_x = \Delta(y_x/u_x)$ is introduced.

Example $(x+2)y_{x+2} - (x+3)y_{x+1} + y_x = 0$. By observation $u_x = 1$ is a solution. Set $U_x = \Delta y_x = y_{x+1} - y_x$. There results $(x+2)U_{x+1} - U_x = 0$, which is of degree one lower than the original equation. The complete solution for y_x is finally

$$y_x = c_0 \sum_{n=0}^{x} \frac{1}{n!} + c_1$$

Factorization If the difference equation can be factored, then the general solution can be obtained by solving two or more successive equations of lower order. Consider $y_{x+2} + A_x y_{x+1} + B_x y_x = \phi(x)$. If there exists a_x, b_x such that $a_x + b_x = -A_x$ and $a_x b_x = B_x$, then the difference equation may be written $y_{x+2} - (a_x + b_x) y_{x+1} + a_x b_x y_x = \phi(x)$. First solve $U_{x+1} - b_x U_x = \phi(x)$ and then $y_{x+1} - a_x y_x = U_x$.

Example $y_{x+2} - (2x+1)y_{x+1} + (x^2 + x)y_x = 0$. Set $a_x = x$, $b_x = x + 1$. Solve $u_{x+1} - (x+1)u_x = 0$ and then $y_{x+1} - xy_x = u_x$.

Substitution If it is possible to rearrange a difference equation so that it takes the form $af_{x+2}y_{x+2} + bf_{x+1}y_{x+1} + cf_x y_x = \phi(x)$ with a, b, c constants, then the substitution $u_x = f_x y_x$ reduces the equation to one with constant coefficients.

Example $(x+2)^2 y_{x+2} - 3(x+1)^2 y_{x+1} + 2x^2 y_x = 0$. Set $u_x = x^2 y_x$. The equation becomes $u_{x+2} - 3u_{x+1} + 2u_x = 0$, which is linear and easily solved by previous methods.

The substitution $u_x = y_x/f_x$ reduces $af_x f_{x+1} y_{x+2} + bf_x f_{x+2} y_{x+1} + cf_{x+1} f_{x+2} y_x = \phi(x)$ to an equation with constant coefficients.

Example $x(x+1)y_{x+2} + 3x(x+2)y_{x+1} - 4(x+1)(x+2)y_x = x$. Set $u_x = y_x/f_x = y_x/x$. Then $y_x = xu_x$, $y_{x+1} = (x+1)u_{x+1}$ and $y_{x+2} = (x+2)u_{x+2}$. Substitution in the equation yields $x(x+1)(x+2)u_{x+2} + 3x(x+2)(x+1)u_{x+1} - 4x(x+1)(x+2)u_x = x$ or $u_{x+2} + 3u_{x+1} - 4u_x = 1/(x+1)(x+2)$, which is a linear equation with constant coefficients.

Nonlinear Difference Equations: Riccati Difference Equation The Riccati equation $y_{x+1}y_x + ay_{x+1} + by_x + c = 0$ is a nonlinear difference equation which can be solved by reduction to linear form. Set $y = z + h$. The equation becomes $z_{x+1}z_x + (h+a)z_{x+1} + (h+b)z_x + h^2 + (a+b)h + c = 0$. If h is selected as a root of $h^2 + (a+b)h + c = 0$ and the equation is divided by $z_{x+1}z_x$ there results $[(h+b)/z_{x+1}] + [(h+a)/z_x] + 1 = 0$. This is a linear equation with constant coefficients. The solution is

$$y_x = h + \cfrac{1}{c\left[-\dfrac{a+h}{b+h} \right]^x - \cfrac{1}{(a+h) + (b+h)}}$$

Example This equation is obtained in distillation problems, among others, in which the number of theoretical plates is required. If the relative volatility is assumed to be constant, the plates are theoretically perfect, and the molal liquid and vapor rates are constant, then a material balance around the nth plate of the enriching section yields a Riccati difference equation.

INTEGRAL EQUATIONS

REFERENCES: 75, 79, 105, 195, 273. See also "Numerical Analysis and Approximate Methods."

An integral equation is any equation in which the unknown function appears under the sign of integration and possibly outside the sign of integration. If derivatives of the dependent variable appear elsewhere in the equation, the equation is said to be integrodifferential.

CLASSIFICATION OF INTEGRAL EQUATIONS

Volterra integral equations have an integral with a variable limit. The Volterra equation of the second kind is

$$u(x) = f(x) + \lambda \int_a^x K(x, t)u(t)\, dt$$

whereas a Volterra equation of the first kind is

$$u(x) = \lambda \int_a^x K(x, t)u(t)\, dt$$

Equations of the first kind are very sensitive to solution errors so that they present severe numerical problems. Volterra equations are similar to initial value problems.

A Fredholm equation of the second kind is

$$u(x) = f(x) + \lambda \int_a^b K(x, t)u(t)\, dt$$

whereas a Fredholm equation of the first kind is

$$u(x) = \int_a^b K(x, t)u(t)\, dt$$

The limits of integration are fixed, and these problems are analogous to boundary value problems.

An eigenvalue problem is a homogeneous equation of the second kind, and solutions exist only for certain λ.

$$u(x) = \lambda \int_a^b K(x, t)u(t)\, dt$$

See Refs. 105 and 195 for further information and existence proofs.

If the unknown function u appears in the equation in any way except to the first power, the integral equation is said to be nonlinear. The equation $u(x) = f(x) + \int_a^b K(x, t)[u(t)]^{3/2}\, dt$ is nonlinear. The differential equation $du/dx = g(x, u)$ is equivalent to the nonlinear integral equation $u(x) = c + \int_a^x g[t, u(t)]\, dt$.

An integral equation is said to be singular when either one or both of the limits of integration become infinite or if $K(x, t)$ becomes infinite for one or more points of the interval under discussion.

Example $u(x) = x + \int_0^\infty \cos\ (xt)u(t)\, dt$ and $f(x) = \int_0^x \dfrac{u(t)}{x-t}\, dt$ are both singular. The kernel of the first equation is $\cos\ (xt)$, and that of the second is $(x-t)^{-1}$.

RELATION TO DIFFERENTIAL EQUATIONS

The Leibniz rule (see "Integral Calculus") can be used to show the equivalence of the initial-value problem consisting of the second-order differential equation $d^2y/dx^2 + A(x)(dy/dx) + B(x)y = f(x)$ together with the prescribed initial conditions $y(a) = y_0$, $y'(a) = y_0'$ to the integral equation.

$$y(x) = \int_a^x K(x, t)y(t)\, dt + F(x)$$

where $K(x, t) = (t-x)[B(t) - A'(t)] - A(t)$

and $F(x) = \int_a^x (x-t)f(t)\, dt + [A(a)y_0 + y_0'](x-a) + y_0$

This integral equation is a **Volterra equation** of the second kind. Thus the initial-value problem is equivalent to a Volterra integral equation of the second kind.

Example $d^2y/dx^2 + x^2(dy/dx) + xy = x$, $y(0) = 1$, $y'(0) = 0$. Here $A(x) = x^2$, $B(x) = x$, $f(x) = x$. The equivalent integral equation is $y(x) = \int_0^x K(x, t)y(t)\, dt + F(x)$ where $K(x, t) = t(x-t) - t^2$ and $F(x) = \int_0^x (x-t)t\, dt + 1 = x^3/6 + 1$. Combining these $y(x) = \int_0^x t[x - 2t]y(t)\, dt + x^3/6 + 1$.

Eigenvalue problems can also be related. For example, the problem $(d^2y/dx^2) + \lambda y = 0$ with $y(0) = 0$, $y(a) = 0$ is equivalent to the integral equation $y(x) = \lambda \int_0^a K(x, t)y(t)\, dt$, where $K(x, t) = (t/a)(a - x)$ when $t < x$ and $K(x, t) = (x/a)(a - t)$ when $t > x$. The differential equation may be recovered from the integral equation by differentiating the integral equation by using the Leibniz rule.

METHODS OF SOLUTION

In general, the solution of integral equations is not easy, and a few exact and approximate methods are given here. Often numerical methods must be employed, as discussed in "Numerical Solution of Integral Equations."

Equations of Convolution Type   The equation $u(x) = f(x) + \lambda \int_0^x K(x - t)u(t)\, dt$ is a special case of the linear integral equation of the second kind of Volterra type. The integral part is the convolution integral discussed under "Integral Transforms (Operational Methods)"; so the solution can be accomplished by Laplace transforms; $L[u(x)] = L[f(x)] + \lambda L[u(x)]L[K(x)]$ or

$$L[u(x)] = \frac{L[f(x)]}{1 - \lambda L[K(x)]}, \quad u(x) = L^{-1}\left[\frac{L[f(x)]}{1 - \lambda L[K(x)]}\right]$$

Equations of the type considered here occur quite frequently in practice in what can be called "cause-and-effect" systems.

Example   In a certain linear system, the effect $E(t)$ due to a cause $C = \lambda E$ at time τ is a function only of the elapsed time $t - \tau$. If the system has the activity level 1 at time $t < 0$, the cause λE and effect (E) relation is given by the integral equation $E(t) = 1 + \lambda \int_0^t K(t - \tau)E(\tau)\, d\tau$. Let $K(t - \tau) = t - \tau$. Then $E(t) = 1 + \lambda \int_0^t (t - \tau)E(\tau)\, d\tau$. By using the transform method

$$E(t) = L^{-1}\left[\frac{L[1]}{1 - \lambda L[K(t)]}\right] = L^{-1}\left[\frac{1/p}{1 - \lambda/p^2}\right] = L^{-1}\left[\frac{p}{p^2 - \lambda}\right] = \cosh\sqrt{\lambda}\, t$$

Method of Successive Approximations   Consider the equation $y(x) = f(x) + \lambda \int_a^b K(x, t)y(t)\, dt$. In this method a unique solution is obtained in sequence form as follows: Substitute in the right-hand member of the equation $y_0(t)$ for $y(t)$. Upon integration there results $y_1(t) = f(x) + \lambda \int_a^b K(x, t)y_0(t)\, dt$. Continue in like manner by replacing y_0 by y_1, y_1 by y_2, etc. A series of functions $y_0(x)$, $y_1(x)$, $y_2(x)$, . . . are obtained which satisfy the equations

$$y_n(x) = f(x) + \lambda \int_a^b K(x, t)y_{n-1}(t)\, dt$$

Then $y_n(x) = f(x) + \lambda \int_a^b K(x, t)f(t)\, dt + \lambda^2 \int_a^b K(x, t) \int_a^b K(t, t_1)f(t_1)\, dt_1\, dt + \lambda^3 \int_a^b K(x, t) \int_a^b K(t, t_1) \int_a^b K(t_1, t_2)f(t_2)\, dt_2\, dt_1\, dt + \cdots + R_n$, where R_n is the remainder, and

$$|R_n| \le |\lambda^n| \binom{\max.\ y_0}{a \le x \le b} M^n(b - a)^n$$

where M = maximum value of $|K|$ in the rectangle $a \le t \le b$, $a \le x \le b$. If $|\lambda|M(b - a) < 1$, $\lim\limits_{n \to \infty} R_n = 0$. Then $y_n(x) \to y(x)$, which is the unique solution.

Example   Consider the equation $y(x) = 1 + \lambda \int_0^1 (1 - 3xt)y(t)\, dt$.

$$y(x) = 1 + \lambda \int_0^1 (1 - 3xt)\, dt + \lambda^2 \int_0^1 (1 - 3xt) \int_0^1 (1 - 3tt_1)\, dt_1\, dt + \cdots$$

$$= 1 + \lambda\left(1 - \frac{3}{2}x\right) + \lambda^2 \frac{1}{4} + \frac{1}{4}\lambda^3\left(1 - \frac{3}{2}x\right) + \frac{\lambda^4}{16} + \frac{1}{16}\lambda^5\left(1 - \frac{3}{2}x\right) + \cdots$$

$$= \left(1 + \frac{\lambda^2}{4} + \frac{\lambda^4}{16} + \cdots\right)\left(1 + \lambda\left(1 - \frac{3}{2}x\right)\right)$$

$$= \frac{1 + \lambda(1 - \frac{3}{2}x)}{1 - \frac{1}{4}\lambda^2}, \quad |\lambda| < 2$$

Example $dy/dx = x^2 + y$, $x_0 = 0$, $y_0 = 1$. This problem is equivalent to the integral equation $y = 1 + \int_0^x (x^2 + y)\, dx$. Let the initial approximation for y be 1. Then

$$y^{(1)} = 1 + \int_0^x (x^2 + 1)\, dx = 1 + x + \frac{x^3}{3}$$

$$y^{(2)} = 1 + \int_0^x [x^2 + y^{(1)}]\, dx = 1 + \int_0^x \left[x^2 + 1 + x + \frac{x^3}{3}\right]\, dx$$

$$= 1 + x + \frac{x^2}{2} + \frac{x^3}{3} + \frac{x^4}{12}, \text{ etc.}$$

INTEGRAL TRANSFORMS (OPERATIONAL METHODS)

REFERENCES: 63, 64, 71, 72, 97, 137, 217.

The term "operational method" implies a procedure of solving differential and difference equations by which the boundary or initial conditions are automatically satisfied in the course of the solution. The technique offers a very powerful tool in the applications of mathematics, but it is limited to linear problems.

Most integral transforms are special cases of the equation $g(s) = \int_a^b f(t)K(s, t)\, dt$ in which $g(s)$ is said to be the transform of $f(t)$ and $K(s, t)$ is called the kernel of the transform. A tabulation of the more important kernels and the interval (a, b) of applicability follows. The first three transforms are considered here.

Name of transform	(a, b)	$K(s, t)$
Laplace	$(0, \infty)$	e^{-st}
Fourier	$(-\infty, \infty)$	$\dfrac{1}{\sqrt{2\pi}}\, e^{-ist}$
Fourier cosine	$(0, \infty)$	$\sqrt{\dfrac{2}{\pi}}\, \cos st$
Fourier sine	$(0, \infty)$	$\sqrt{\dfrac{2}{\pi}}\, \sin st$
Mellin	$(0, \infty)$	t^{s-1}
Hankel	$(0, \infty)$	$tJ_\nu(st),\ \nu \ge -\tfrac{1}{2}$

LAPLACE TRANSFORM

The Laplace transform of a function $f(t)$ is defined by $F(s) = L\{f(t)\} = \int_0^\infty e^{-st}f(t)\, dt$, where s is a complex variable. Note that the transform is an improper integral and therefore may not exist for all continuous functions and all values of s. We restrict consideration to those values of s and those functions f for which this improper integral converges.

The function $L[f(t)] = g(s)$ is called the direct transform, and $L^{-1}[g(s)] = f(t)$ is called the inverse transform. Both the direct and the inverse transforms are tabulated for many often-occurring functions. In general,

$$L^{-1}[g(s)] = \frac{1}{2\pi i} \int_{\alpha - i\infty}^{\alpha + i\infty} e^{st}g(s)\, ds$$

and to evaluate this integral requires a knowledge of complex variables, the theory of residues, and contour integration.

A function is said to be piecewise continuous on an interval if it has only a finite number of finite (or jump) discontinuities. A function f on $0 < t < \infty$ is said to be of exponential growth at infinity if there exist constants M and α such that $|f(t)| \le Me^{\alpha t}$ for sufficiently large t.

Sufficient Conditions for the Existence of Laplace Transform
Suppose f is a function which is (1) piecewise continuous on every finite interval $0 < t < T$, (2) of exponential growth at infinity, and (3) $\int_0^\delta |f(t)|\, dt$ exist (finite) for every finite $\delta > 0$. Then the Laplace transform of f exists for all complex numbers s with sufficiently large real part.

Note that condition 3 is automatically satisfied if f is assumed to be piecewise continuous on every finite interval $0 \le t < T$. The function $f(t) = t^{-1/2}$ is not piecewise continuous on $0 \le t \le T$ but satisfies conditions 1 to 3.

Let Λ denote the class of all functions on $0 < t < \infty$ which satisfy conditions 1 to 3.

Example   Let $f(t)$ be the Heaviside step function at $t = t_0$; i.e., $f(t) = 0$ for $t \le t_0$, and $f(t) = 1$ for $t > t_0$. Then

$$L\{f(t)\} = \int_{t_0}^\infty e^{-st}\, dt = \lim_{T \to \infty} \int_{t_0}^T e^{-st}\, dt = \lim_{T \to \infty} \frac{1}{s}(e^{-st_0} - e^{-sT}) = \frac{e^{-st_0}}{s}$$

provided $s > 0$.

Example   Let $f(t) = e^{at}$, $t \ge 0$, where a is a real number. Then $L\{e^{at}\} = \int_0^\infty e^{-(s-a)t}\, dt = 1/(s - a)$, provided Re $s > a$.

Properties of the Laplace Transform

1.   The Laplace transform is a linear operator: $L\{af(t) + bg(t)\} = aL\{f(t)\} + bL\{g(t)\}$ for any constants a, b and any two functions f and g whose Laplace transforms exist.

2.   The Laplace transform of a real-valued function is real for real s. If $f(t)$ is a complex-valued function, $f(t) = u(t) + iv(t)$, where u and v are real, then $L\{f(t)\} = L\{u(t)\} + iL\{v(t)\}$. Thus $L\{u(t)\}$ is the real part of $L\{f(t)\}$, and $L\{v(t)\}$ is the imaginary part of $L\{f(t)\}$.

3.   The Laplace transform of a function in the class Λ has derivatives of all orders, and $L\{t^k f(t)\} = (-1)^k d^k F(s)/ds^k$, $k = 1, 2, 3, \ldots$.

Example $\int_0^\infty e^{-st} \sin at\, dt = \dfrac{a}{s^2 + a^2}$, $s > 0$. By property 3, $\dfrac{2as}{(s^2 + a^2)^2} = \int_0^\infty e^{-st}\, t \sin at\, dt = L\{t \sin at\}$.

Example   By applying property 3 with $f(t) = 1$ and using the preceding results, we obtain

$$L\{t^k\} = (-1)^k \frac{d^k}{ds^k}\left(\frac{1}{s}\right) = \frac{k!}{s^{k+1}}$$

provided Re $s > 0$; $k = 1, 2, \ldots$. Similarly, we obtain

$$L\{t^k e^{at}\} = (-1)^k \frac{d^k}{ds^k}\left(\frac{1}{s - a}\right) = \frac{k!}{(s - a)^{k+1}}$$

4.   Frequency-shift property (or, equivalently, the transform of an exponentially modulated function). If $F(s)$ is the Laplace transform of a function $f(t)$ in the class Λ, then for any constant a, $L\{e^{at} f(t)\} = F(s - a)$.

Example $L\{te^{-at}\} = \dfrac{1}{(s + a)^2}$, $s > 0$.

5.   Time-shift property. Let $u(t - a)$ be the unit step function at $t = a$. Then $L\{f(t - a)u(t - a)\} = e^{-as}F(s)$.

6.   Transform of a derivative. Let f be a differentiable function such that both f and f' belong to the class Λ. Then $L\{f'(t)\} = sF(s) - f(0)$.

7.   Transform of a higher-order derivative. Let f be a function which has continuous derivatives up to order n on $(0, \infty)$, and suppose that f and its derivatives up to order n belong to the class Λ. Then $L\{f^{(j)}(t)\} = s^j F(s) - s^{j-1}f(0) - s^{j-2}f'(0) - \cdots - sf^{(j-2)}(0) - f^{(j-1)}(0)$ for $j = 1, 2, \ldots, k$.

Example $L\{f''(t)\} = s^2 L\{f(t)\} - sf(0) - f'(0)$
$L\{f'''(t)\} = s^3 L\{f(t)\} - s^2 f(0) - sf'(0) - f''(0)$

Example   Solve $y'' + y = 2e^t$, $y(0) = y'(0) = 2$. $L[y''] = -y'(0) - sy(0) + s^2 L[y] = -2 - 2s + s^2 L[y]$. Thus

$$-2 - 2s + s^2 L[y] + L[y] = 2L[e^t] = \frac{2}{s - 1}$$

$$L[y] = \frac{2s^2}{(s - 1)(s^2 + 1)} = \frac{1}{s - 1} + \frac{s}{s^2 + 1} + \frac{1}{s^2 + 1}$$

Hence $y = e^t + \cos t + \sin t$.

A short table (Table 3-1) of very common Laplace transforms and inverse transforms follows. The references include more detailed tables. NOTE: $\Gamma(n + 1) = \int_0^\infty x^n e^{-x}\, dx$ (gamma function); $J_n(t) =$ Bessel function of the first kind of order n.

8. $L\left[\displaystyle\int_a^t f(t)\, dt\right] = \dfrac{1}{s} L[f(t)] + \dfrac{1}{s} \int_a^0 f(t)\, dt$

Example   Find $f(t)$ if $L[f(t)] = \dfrac{1}{s^2}\left[\dfrac{1}{s^2 - a^2}\right]$ $L\left[\dfrac{1}{a} \sinh at\right] = \dfrac{1}{s^2 - a^2}$

Therefore $f(t) = \displaystyle\int_0^t \left[\int_0^t \frac{1}{a} \sinh at\, dt\right] dt = \frac{1}{a^2}\left[\frac{\sinh at}{a} - t\right]$

TABLE 3-1   Laplace Transforms

$f(t)$	$g(s)$	$f(t)$	$g(s)$
1	$1/s$	$e^{-at}(1-at)$	$\dfrac{s}{(s+a)^2}$
t^n, ($n\,a$ + integer)	$\dfrac{n!}{s^{n+1}}$	$\dfrac{t\sin at}{2a}$	$\dfrac{s}{(s^2+a^2)^2}$
t^n, $n \neq$ + integer	$\dfrac{\Gamma(n+1)}{s^{n+1}}$	$\dfrac{1}{2a^2}\sin at \sinh at$	$\dfrac{s}{s^4+4a^4}$
$\cos at$	$\dfrac{s}{s^2+a^2}$	$\cos at \cosh at$	$\dfrac{s^2}{s^4+4a^4}$
$\sin at$	$\dfrac{a}{s^2+a^2}$	$\dfrac{1}{2a}(\sinh at + \sin at)$	$\dfrac{s^2}{s^4-a^4}$
$\cosh at$	$\dfrac{s}{s^2-a^2}$	$\tfrac{1}{2}(\cosh at + \cos at)$	$\dfrac{s^2}{s^4-a^4}$
$\sinh at$	$\dfrac{a}{s^2-a^2}$	$\dfrac{\sin at}{t}$	$\tan^{-1}\dfrac{a}{s}$
e^{-at}	$\dfrac{1}{s+a}$	$J_0(at)$	$\dfrac{1}{\sqrt{s^2+a^2}}$
$e^{-bt}\cos at$	$\dfrac{s+b}{(s+b)^2+a^2}$	$na^n\dfrac{J_n(at)}{t}$	$\dfrac{1}{\sqrt{s^2+a^2}-s^n}$
$e^{-bt}\sin a$	$\dfrac{a}{(s+b)^2+a^2}$	$J_0(2\sqrt{at})$	$\dfrac{1}{s}e^{-a/s}$

9. $L\left[\dfrac{f(t)}{t}\right] = \displaystyle\int_s^\infty g(s)\,ds$ $L\left[\dfrac{f(t)}{t^k}\right] = \underbrace{\int_s^\infty \cdots \int_s^\infty}_{k\text{ integrals}} g(s)(ds)^k$

Example $L\left[\dfrac{\sin at}{t}\right] = \displaystyle\int_s^\infty L[\sin at]\,ds = \int_s^\infty \dfrac{a\,ds}{s^2+a^2} = \cot^{-1}\dfrac{s}{a}$

10. The unit step function

$$u(t-a) = \begin{cases} 0 & t<a \\ 1 & t>a \end{cases} \qquad L[u(t-a)] = e^{-as}/s$$

11. The unit impulse function is

$$\delta(a) = u'(t-a) = \begin{cases} \infty & \text{at } t=a \\ 0 & \text{elsewhere} \end{cases} \qquad L[u'(t-a)] = e^{-as}$$

12. $L^{-1}[e^{-as}g(s)] = f(t-a)u(t-a)$ (second shift theorem).

13. If $f(t)$ is periodic of period b, i.e., $f(t+b)=f(t)$, then

$$L[f(t)] = \left[\dfrac{1}{1-e^{-bs}}\right]\int_0^b e^{-st}f(t)\,dt$$

Example   The partial differential equations relating gas composition to position and time in a gas chromatograph are $\partial y/\partial n + \partial x/\partial \theta = 0$, $\partial y/\partial n = x - y$, where $x = mx'$, $n = (k_G aP/G_m)h$, $\theta = (mk_G a/\rho_B)t$ and G_M = molar velocity, y = mole fraction of the component in the gas phase, ρ_B = bulk density, h = distance from the entrance, P = pressure, k_G = mass-transfer coefficient, and m = slope of the equilibrium line. These equations are equivalent to $\partial^2 y/\partial n\ \partial\theta + \partial y/\partial n +$ $\partial y/\partial \theta = 0$, where the boundary conditions considered here are $y(0, \theta) = 0$ and $x(n, 0) = y(n, 0) + (\partial y/\partial n)(n, 0) = \delta(0)$ (see property 11). The problem is conveniently solved by using the Laplace transform of y with respect to n; write $g(s, \theta) = \int_0^\infty e^{-ns}y(n, \theta)\,dn$. Operating on the partial differential equation gives $s(dg/d\theta) - (\partial y/\partial\theta)(0, \theta) + sg - y(0, \theta) + dg/d\theta = 0$ or $(s+1)(dg/d\theta) + sg = (\partial y/\partial\theta)$ $(0, \theta) + y(0, \theta) = 0$. The second boundary condition gives $g(s, 0) + sg(s, 0) -$ $y(0, 0) = 1$ or $g(s, 0) + sg(s, 0) = 1$ ($L[\delta(0)] = 1$). A solution of the ordinary differential equation for g consistent with this second condition is

$$g(s, \theta) = \dfrac{1}{s+1}e^{-s\theta/(s+1)}$$

Inversion of this transform gives the solution $y(n, \theta) = e^{-(n+\theta)}I_0(2\sqrt{n\theta})$ where

I_0 = zero-order Bessel function of an imaginary argument. For large u, $I_n(u) \sim e^u/\sqrt{2\pi u}$. Hence for large n,

$$y(n, \theta) \sim \dfrac{\exp[-(\sqrt{\theta}-\sqrt{n})^2]}{2\pi^{1/2}(n\theta)^{1/4}}$$

or for sufficiently large n, the peak concentration occurs near $\theta = n$.

Other applications of Laplace transforms are given under "Differential Equations."

CONVOLUTION INTEGRAL

The convolution integral (faltung) of two functions $f(t)$, $r(t)$ is $x(t) = f(t)°r(t) = \int_0^t f(\tau)r(t-\tau)\,d\tau$.

Example $t°\sin t = \displaystyle\int_0^t \tau\sin(t-\tau)\,d\tau = t - \sin t.$

$$L[f(t)]L[h(t)] = L[f(t)°h(t)]$$

z-TRANSFORM

See Refs. 198, 218, and 256. The z-transform is useful when data is available at only discrete points. Let

$$f°(t) = f(t_k)$$

be the value of f at the sample points

$$t_k = k\,\Delta t, \quad k = 0, 1, 2, \ldots$$

Then the function $f°(t)$ is

$$f°(t) = \sum_{k=0}^\infty f(t_k)\,\delta(t-t_k)$$

Take the Laplace transform of this.

$$g°(s) = L[f°(t)] = \sum_{k=0}^\infty f(t_k)\,e^{-stk} = \sum_{k=0}^\infty f(t_k)\,e^{-s\Delta tk}$$

For convenience, replace $e^{-s\Delta t}$ by z and call $g°(z)$ the z-transform of $f°(t)$.

$$g°(z) = \sum_{k=0}^\infty f(t_k)\,z^{-k}$$

The z-transform is used in process control when the signals are at intervals of Δt. A brief table (Table 3-2) is provided here.

The z-transform can also be used to solve difference equations, just like the Laplace transform can be used to solve differential equations.

TABLE 3-2   z-Transforms

$f(k)$	$g°(z)$
$1(k)$	$\dfrac{1}{1-z^{-1}}$
$k\,\Delta t$	$\dfrac{\Delta t\,z^{-1}}{(1-z^{-1})^2}$
$(k\,\Delta t)^{n-1}$	$\displaystyle\lim_{a\to 0}(-1)^{n-1}\dfrac{\partial^{n-1}}{\partial a^{n-1}}\left(\dfrac{1}{1-e^{-a\Delta t}z^{-1}}\right)$
$\sin a\,k\,\Delta t$	$\dfrac{z^{-1}\sin c\,\Delta t}{(1-2z^{-1}\cos a\,\Delta t + z^{-2})}$
$\cos a\,k\,\Delta t$	$\dfrac{1-z^{-1}\cos a\,\Delta t}{(1-2z^{-1}\cos a\,\Delta t + z^{-2})}$
$e^{-ak\Delta t}$	$\dfrac{1}{1-e^{-a\Delta t}z^{-1}}$
$e^{-bk\Delta t}\cos a\,k\,\Delta t$	$\dfrac{1-z^{-1}e^{-b\Delta t}\cos a\,\Delta t}{1-2z^{-1}e^{-b\Delta t}\cos a\,\Delta t + z^{-2}e^{-2b\Delta t}}$
$\dfrac{1}{b}e^{-bk\Delta t}\sin a\,k\,\Delta t$	$\dfrac{1}{b}\dfrac{z^{-1}e^{-b\Delta t}\sin a\,\Delta t}{1-2z^{-1}e^{-b\Delta t}\cos a\,\Delta t + z^{-2}e^{-2b\Delta t}}$

Example The difference equation for $y(k)$ is

$$y(k) + a_1 y(k-1) + a_2 y(k-2) = b_1 u(k)$$

Take the z-transform

$$(1 + a_1 z^{-1} + a_2 z^{-2})\, y^\circ(z) = u^\circ(z)$$

Then

$$y^\circ(z) = \frac{u^\circ(z)}{1 + a_1 z^{-1} + a_2 z^{-2}}$$

The inverse transform must be found, usually from a table of inverse transforms.

FOURIER TRANSFORM

The Fourier transform is given by

$$F[f(t)] = \frac{1}{\sqrt{2\pi}} \int_{-\infty}^{\infty} f(t) e^{-ist}\, dt = g(s)$$

and its inverse by

$$F^{-1}[g(s)] = \frac{1}{\sqrt{2\pi}} \int_{-\infty}^{\infty} g(s) e^{ist}\, dt = f(t)$$

In brief, the condition for the Fourier transform to exist is that $\int_{-\infty}^{\infty} |f(t)|\, dt < \infty$, although certain functions may have a Fourier transform even if this is violated.

Example The function $f(t) = \begin{cases} 1 - a \le t \le a \\ 0 \text{ elsewhere} \end{cases}$ has $F[f(t)] = \int_{-a}^{a} e^{-ist}\, dt =$

$$\int_{0}^{a} e^{ist}\, dt + \int_{0}^{a} e^{-ist}\, dt = 2 \int_{0}^{a} \cos st\, dt = \frac{2 \sin sa}{s}$$

Properties of the Fourier Transform Let $F[f(t)] = g(s)$; $F^{-1}[g(s)] = f(t)$.

1. $F[f^{(n)}(t)] = (is)^n F[f(t)]$.
2. $F[af(t) + bh(t)] = aF[f(t)] + bF[h(t)]$.
3. $F[f(-t)] = g(-s)$.
4. $F[f(at)] = \dfrac{1}{a} g\left(\dfrac{s}{a}\right)$, $a > 0$.
5. $F[e^{-iwt}f(t)] = g(s + w)$.
6. $F[f(t + t_1)] = e^{ist_1} g(s)$.
7. $F[f(t)] = G(is) + G(-is)$ if $f(t) = f(-t)$ (f even)
   $F[f(t)] = G(is) - G(-is)$ if $f(t) = -f(-t)$ (f odd)

where $G(s) = L[f(t)]$. This result allows the use of the Laplace-transform tables to obtain the Fourier transforms.

Example Find $F[e^{-a|t|}]$ by property 7. $e^{-a|t|}$ is even. So $L[e^{-at}] = 1/(s + a)$. Therefore, $F[e^{-a|t|}] = 1/(is + a) + 1/(-is + a) = 2a/(s^2 + a^2)$.

Tables of this transform may be found in *Higher Transcendental Functions*, vols. I, II, and III, A. Erdelyi, et al., McGraw-Hill, New York, 1953–1955.

FOURIER COSINE TRANSFORM

The Fourier cosine transform is given by

$$F_c[f(t)] = g(s) = \sqrt{\frac{2}{\pi}} \int_{0}^{\infty} f(t) \cos st\, dt$$

and its inverse by

$$F_c^{-1}[g(s)] = f(t) = \sqrt{\frac{2}{\pi}} \int_{0}^{\infty} g(s) \cos st\, ds$$

The Fourier sine transform F_s is obtainable by replacing the cosine by the sine in these integrals.

Example $F_c[f(t)]$, $f(t) = \begin{cases} 1 & 0 < t < a \\ 0 & a < t < \infty \end{cases}$ $F_c[f(t)] = \sqrt{\dfrac{2}{\pi}} \int_{0}^{a} \cos st\, dt = \sqrt{\dfrac{2}{\pi}} \dfrac{\sin as}{s}$

Properties of the Fourier Cosine Transform $F_c[f(t)] = g(s)$.

1. $F_c[af(t) + bh(t)] = aF_c[f(t)] + bF_c[h(t)]$.
2. $F_c[f(at)] = (1/a)g(s/a)$.
3. $F_c[f(at) \cos bt] = \dfrac{1}{2a}\left[g\left(\dfrac{s+b}{a}\right) + g\left(\dfrac{s-b}{a}\right) \right]$, $a, b > 0$.
4. $F_c[t^{-2n}f(t)] = (-1)^n\, (d^{2n}g/ds^{2n})$.
5. $F_c[t^{2n+1}f(t)] = (-1)^n\, (d^{2n+1}/ds^{2n+1})\, F_s[f(t)]$.

A short table (Table 3-3) of Fourier cosine transforms follows.

TABLE 3-3   Fourier Transforms

$f(t)$		$\dfrac{g(s)}{\sqrt{2/\pi}}$
t	$0 < t < 1$	$\dfrac{1}{s^2}[2 \cos s - 1 - \cos 2s]$
$2 - t$	$1 < t < 2$	
0	$2 < t < \infty$	$\pi^{1/2}(2s)^{-1/2}$
$t^{-1/2}$		
0	$0 < t < a$	
$(t-a)^{-1/2}$	$a < t < \infty$	$\pi^{1/2}(2s)^{-1/2}[\cos as - \sin as]$
$(t^2 + a^2)^{-1}$		$\tfrac{1}{2}\pi a^{-1} e^{-as}$
e^{-at}	$a > 0$	$\dfrac{a}{s^2 + a^2}$
e^{-at^2}	$a > 0$	$\tfrac{1}{2}\pi^{1/2}a^{-1/2}e^{-s^2/4a}$
$\dfrac{\sin at}{t}$	$a > 0$	$\begin{cases} \pi/2 & s < a \\ \pi/4 & s = a \\ 0 & s > a \end{cases}$

Example The temperature θ in the semi-infinite rod $0 \le x < \infty$ is determined by the differential equation $\partial\theta/\partial t = k(\partial^2\theta/\partial x^2)$ and the condition $\theta = 0$ when $t = 0$, $x \ge 0$; $\partial\theta/\partial x = -\mu = $ constant when $x = 0$, $t > 0$. By using the Fourier cosine transform a solution may be found as

$$\theta(x, t) = \frac{2\mu}{\pi} \int_{0}^{\infty} \frac{\cos px}{p} (1 - e^{-kp^2 t})\, dp.$$

MATRIX ALGEBRA AND MATRIX COMPUTATIONS

REFERENCES: General (textbooks that cover at an introductory level a variety of topics that constitute a core of numerical methods for practicing engineers), 2, 3, 4, 22, 56, 59, 70, 77, 133, 135, 143, 150, 155, 219. Numerical solution of nonlinear equations, 153, 171, 237, 302. Numerical solution of ordinary differential equations, 76, 117, 127, 185, 257. Numerical solution of integral equations, 23, 26, 129, 162. Numerical solution of partial differential equations, 11, 76, 127, 133, 155, 210, 251, 286, 287, 213, 233, 253. Spline functions and applications, 38, 56, 70, 230. Finite elements and applications, 5, 29, 83, 130, 164, 210, 241, 281, 287, 303, 304. Fast Fourier transforms, 47, 56, 135, 238. Software, 187, 231.

MATRIX ALGEBRA

Matrices   A rectangular array of mn quantities, arranged in m rows and n columns

$$A = (a_{ij}) = \begin{bmatrix} a_{11} & \cdots & a_{1n} \\ a_{21} & \cdots & a_{2n} \\ \vdots & & \\ a_{m1} & \cdots & a_{mn} \end{bmatrix}$$

is called a matrix. The elements a_{ij} may be real or complex. The notation a_{ij} means the element in the ith row and jth column, i is called the row index, j the column index. If $m = n$ the matrix is said to be square and of order n. A matrix, even if it is square, does not have a numerical value, as a determinant does. However, if the matrix A is square, a determinant can be formed which has the same elements as the matrix A. This is called the determinant of the matrix and is written det (A) or $|A|$. If A is square and det $(A) \neq 0$, A is said to be nonsingular; if det $(A) = 0$, A is said to be singular. A matrix A has rank r if and only if it has a nonvanishing determinant of order r and no nonvanishing determinant of order $> r$.

Equality of Matrices   Let $A = (a_{ij})$, $B = (b_{ij})$. Two matrices A and B are *equal* ($=$) if and only if they are identical; that is, they have the same number of rows and the same number of columns and equal corresponding elements ($a_{ij} = b_{ij}$ for all i and j).

Addition and Subtraction   The operations of addition ($+$) and subtraction ($-$) of two or more matrices are possible if and only if they have the same number of rows and columns. Thus $A \pm B = (a_{ij} \pm b_{ij})$; i.e., addition and subtraction are of corresponding elements.

Transposition   The matrix obtained from A by interchanging the rows and columns of A is called the transpose of A, written A' or A^T.

Example $A = \begin{bmatrix} 1 & 3 & 4 \\ 2 & 1 & 6 \end{bmatrix}$, $\quad A^T = \begin{bmatrix} 1 & 2 \\ 3 & 1 \\ 4 & 6 \end{bmatrix}$

Note that $(A^T)^T = A$.

Multiplication   Let $A = (a_{ij})$, $i = 1, \ldots, m_1$; $j = 1, \ldots, m_2$. $B = (b_{ij})$, $i = 1, \ldots, n_1$, $j = 1, \ldots, n_2$. The product AB is defined if and only if the number of columns of A (m_2) equals the number of rows of $B(n_1)$, i.e., $n_1 = m_2$. For two such matrices the product $P = AB$ is defined by summing the element by element products of a row of A by a column of B. This is the row by column rule. Thus

$$p_{ij} = \sum_{k=1}^{n_1} a_{ik} b_{kj}$$

The resulting matrix has m_1 rows and n_2 columns.

Example $\begin{bmatrix} 3 & 2 \\ 1 & 1 \\ 5 & 4 \end{bmatrix} \begin{bmatrix} 0 & 1 & 5 & 6 \\ -2 & 0 & 1 & 3 \end{bmatrix} = \begin{bmatrix} -4 & 3 & 17 & 24 \\ -2 & 1 & 6 & 9 \\ -8 & 5 & 29 & 42 \end{bmatrix}$

It is helpful to remember that the element p_{ij} is formed from the ith row of the first matrix and the jth column of the second matrix. The matrix product is not commutative. That is, $AB \neq BA$ in general.

Inverse of a Matrix   A square matrix A is said to have an inverse if there exists a matrix B such that $AB = BA = I$, where I is the identity matrix of order n.

$$\begin{bmatrix} 1 & 0 & \ldots & & 0 \\ 0 & 1 & & & \\ \vdots & & 1 & 0 \\ 0 & & \ldots & 0 & 1 \end{bmatrix}$$

The inverse B is a square matrix of the order of A, designated by A^{-1}. Thus $AA^{-1} = A^{-1}A = I$. A square matrix A has an inverse if and only if A is nonsingular.

Certain relations are important:

(1) $(AB)^{-1} = B^{-1}A^{-1}$

(2) $(AB)^T = B^T A^T$

(3) $(A^{-1})^T = (A^T)^{-1}$

(4) $(ABC)^{-1} = C^{-1}B^{-1}A^{-1}$

Scalar Multiplication   Let c be any real or complex number. Then $cA = (ca_{ij})$.

Adjugate Matrix of a Matrix   Let A_{ij} denote the cofactor of the element a_{ij} in the determinant of the matrix A. The matrix B^T where $B = (A_{ij})$ is called the adjugate matrix of A written adj $A = B^T$. The elements b_{ij} are calculated by taking the matrix A, deleting the ith row and jth column, and calculating the determinant of the remaining matrix times $(-1)^{i+j}$. Then $A^{-1} = $ adj $A/|A|$. This definition may be used to calculate A^{-1}. However, it is very laborious and the inversion is usually accomplished by numerical techniques shown under "Numerical Analysis and Approximate Methods."

Example   Let $A = \begin{bmatrix} 3 & 0 & -1 \\ -1 & 2 & 1 \\ 3 & 6 & 3 \end{bmatrix}$ Form $B = (A_{ij})$, $B = \begin{bmatrix} 0 & 6 & -12 \\ -6 & 12 & -18 \\ 2 & -2 & 6 \end{bmatrix}$

adj $A = B^T = \begin{bmatrix} 0 & -6 & 2 \\ 6 & 12 & -2 \\ -12 & -18 & 6 \end{bmatrix}$; $|A| = 12$

$$A^{-1} = \frac{\text{adj } A}{|A|} = \begin{bmatrix} 0 & -\frac{1}{2} & \frac{1}{6} \\ \frac{1}{2} & 1 & -\frac{1}{6} \\ -1 & -\frac{3}{2} & \frac{1}{2} \end{bmatrix}$$

Linear Equations in Matrix Form   Every set of n nonhomogeneous linear equations in n unknowns

$$a_{11}x_1 + a_{12}x_2 + \cdots + a_{1n}x_n = b_1$$
$$a_{21}x_1 + a_{22}x_2 + \cdots + a_{2n}x_n = b_2$$
$$\vdots \qquad\qquad \vdots$$
$$a_{n1}x_1 + a_{n2}x_2 + \cdots + a_{nn}x_n = b_n$$

can be written in matrix form as $AX = B$, where $A = (a_{ij})$, $X^T = [x_1 \cdots x_n]$, and $B^T = [b_1 \cdots b_n]$. The solution for the unknowns is $X = A^{-1}B$.

Special Square Matrices

1. A triangular matrix is a matrix all of whose elements above or below the main diagonal (set of elements $a_{11}, \ldots, a_{nn}$) are zero.

If A is triangular, det $(A) = a_{11} \cdot a_{22} \ldots a_{nn}$.

2. A diagonal matrix is one such that all elements both above and below the main diagonal are zero (i.e., $a_{ij} = 0$ for all $i \neq j$). If all diagonal elements are equal, the matrix is called scalar. If A is diagonal, $A = (a_{ij})$, $A^{-1} = (1/a_{ij})$.

3. If $a_{ij} = a_{ji}$ for all i and j (i.e., $A = A^T$), the matrix is symmetric.

4. If $a_{ij} = -a_{ji}$ for $i \neq j$ but the a_{ij} are not all zero, the matrix is skew.

5. If $a_{ij} = -a_{ji}$ for all i and j (i.e., $a_{ii} = 0$), the matrix is skew symmetric.

6. If $A^T = A^{-1}$, the matrix A is orthogonal.

7. If the matrix $A^\circ = (\bar{a}_{ij})^T$, $\bar{a}_{ij} = $ complex conjugate of a_{ij}, A° is the hermitian conjugate of A.

8. If $A = A^{-1}$, A is involutory.

9. If $A = A^\circ$, A is hermitian.

10. If $A = -A^\circ$, A is skew hermitian.

11. If $A^{-1} = A^\circ$, A is unitary.

If A is any matrix, then AA^T and A^TA are square symmetric matrices, usually of different order.

Example   Let $A = \begin{bmatrix} 5 & 1 & 3 & 0 \\ 3 & 4 & 1 & 5 \\ 2 & -2 & 0 & 1 \end{bmatrix}$, $A^T = \begin{bmatrix} 5 & 3 & 2 \\ 1 & 4 & -2 \\ 3 & 1 & 0 \\ 0 & 5 & 1 \end{bmatrix}$

$$AA^T = \begin{bmatrix} 35 & 22 & 8 \\ 22 & 51 & 3 \\ 8 & 3 & 9 \end{bmatrix}, \quad A^TA = \begin{bmatrix} 38 & 13 & 18 & 17 \\ 13 & 21 & 7 & 18 \\ 18 & 7 & 10 & 5 \\ 17 & 18 & 5 & 26 \end{bmatrix}$$

Using a program such as MATLAB, these are easily calculated.

Matrix Calculus

Differentiation   Let the elements of $A = [a_{ij}(t)]$ be differentiable functions of t. Then $\dfrac{dA}{dt} = \left[\dfrac{da_{ij}(t)}{dt} \right]$.

Example $A = \begin{bmatrix} \sin t & \cos t \\ -\cos t & \sin t \end{bmatrix}, \dfrac{dA}{dt} = \begin{bmatrix} \cos t & -\sin t \\ \sin t & \cos t \end{bmatrix}$.

Integration   The integral $\int A \, dt = [\int a_{ij}(t) \, dt]$.

Example $A = \begin{bmatrix} t & 2 \\ t^2 & e^t \end{bmatrix}, \int A \, dt = \begin{bmatrix} t^2/2 & 2t \\ t^3/3 & e^t \end{bmatrix}$.

The matrix $B = A - \lambda I$ is called the characteristic (eigen) matrix of A. Here A is square of order n, λ is a scalar parameter, and I is the $n \times n$ identity. $\det B = \det (A - \lambda I) = 0$ is the characteristic (eigen) equation for A. The characteristic equation is always of the same degree as the order of A. The roots of the characteristic equation are called the eigenvalues of A.

Example $A = \begin{bmatrix} 1 & 2 \\ 3 & 8 \end{bmatrix}, B = \begin{bmatrix} 1 & 2 \\ 3 & 8 \end{bmatrix} - \begin{bmatrix} \lambda & 0 \\ 0 & \lambda \end{bmatrix} = \begin{bmatrix} 1-\lambda & 2 \\ 3 & 8-\lambda \end{bmatrix}$

is the characteristic matrix and $f(\lambda) = \det (B) = \det (A - \lambda I) = (1 - \lambda)(8 - \lambda) - 6 = 2 - 9\lambda + \lambda^2 = 0$ is the characteristic equation. The eigenvalues of A are the roots of $\lambda^2 - 9\lambda + 2 = 0$, which are $(9 \pm \sqrt{73})/2$.

A nonzero matrix X_i, which has one column and n rows, called a column vector satisfying the equation

$$(A - \lambda I)X_i = 0$$

and associated with the ith characteristic root λ_i is called an eigenvector.

Vector and Matrix Norms   To carry out error analysis for approximate and iterative methods for the solutions of linear systems, one needs notions for vectors in R^n and for matrices that are analogous to the notion of length of a geometric vector. Let R^n denote the set of all vectors with n components, $x = (x_1, \ldots, x_n)$. In dealing with matrices it is convenient to treat vectors in R^n as columns, and so $x = (x_1, \ldots, x_n)^T$; however, we shall here write them simply as row vectors. A norm on R^n is a real-valued function f defined on R^n with the following properties:

1. $f(x) \geq 0$ for all $x \in R^n$.
2. $f(x) = 0$ if and only if $x = (0, 0, \ldots, 0)$.
3. $f(ax) = |a| f(x)$ for all real numbers a and $x \in R^n$.
4. $f(x + y) \leqq f(x) + f(y)$ for all $x, y \in R^n$.

The usual notation for a norm is $f(x) = \|x\|$.
The norm of a matrix is

$$\kappa(A) \equiv \|A\| \, \|A^{-1}\|$$

where

$$\|A\| = \sup_{x \neq 0} \frac{\|A x\|}{\|x\|} = \max_k \sum_{j=1}^{n} |a_{jk}|$$

The norm is useful when doing numerical calculations. If the computer's floating-point precision is 10^{-6}, then $\kappa = 10^6$ indicates an ill-conditioned matrix. If the floating-point precision is 10^{-12} (double precision), then a matrix with $\kappa = 10^{12}$ may be ill-conditioned. Two other measures are useful and are more easily calculated:

$$\text{Ratio} = \frac{\max_k |a_{kk}^{(k)}|}{\min_k |a_{kk}^{(k)}|}, \quad V = \frac{|\det A|}{\alpha_1 \alpha_2 \ldots \alpha_n}, \quad \alpha_i = (a_{i1}^2 + a_{i2}^2 + \cdots + a_{in}^2)^{1/2}$$

where $a_{kk}^{(k)}$ are the diagonal elements of the LU decomposition.

MATRIX COMPUTATIONS

The principal topics in linear algebra involve systems of linear equations, matrices, vector spaces, linear transformations, eigenvalues and eigenvectors, and least-squares problems. The calculations are routinely done on a computer.

LU Factorization of a Matrix   To every $m \times n$ matrix A there exists a permutation matrix P, a lower triangular matrix L with unit diagonal elements, and an $m \times n$ (upper triangular) echelon matrix U such that $PA = LU$. The Gauss elimination is in essence an algorithm to determine U, P, and L. The permutation matrix P may be needed since it may be necessary in carrying out the Gauss elimination to interchange two rows of A to produce a (nonzero) pivot, such as if we start with

$$A = \begin{bmatrix} 0 & 2 \\ 1 & 6 \end{bmatrix}$$

If A is a square matrix and if principal submatrices of A are all nonsingular, then we may choose P as the identity in the preceding factorization and obtain $A = LU$. This factorization is unique if L is normalized (as assumed previously), so that it has unit elements on the main diagonal.

Solution of $Ax = b$ by Using LU Factorization   Suppose that the indicated system is compatible and that $A = LU$ (the case $PA = LU$ is similarly handled and amounts to rearranging the equations). Let $z = Ux$. Then $Ax = LUx = b$ implies that $Lz = b$. Thus to solve $Ax = b$ we first solve $Lz = b$ for z and then solve $Ux = z$ for x. This procedure does not require that A be invertible and can be used to determine all solutions of a compatible system $Ax = b$. Note that the systems $Lz = b$ and $Ux = z$ are both in triangular forms and thus can be easily solved.

The LU decomposition is essentially a Gaussian elimination, arranged for maximum efficiency (Ref. 112). The chief reason for doing an LU decomposition is that it takes fewer multiplications than would be needed to find an inverse. Also, once the LU decomposition has been found, it is possible to solve for multiple right-hand sides with little increase in work. The multiplication count for an $n \times n$ matrix and m right-hand sides is

$$\text{operation count} = \frac{1}{3} n^3 - \frac{1}{3} n + mn^2$$

If an inverse is desired, it can be calculated by solving for the LU decomposition and then solving n problems with right-hand sides consisting of all zeroes except one entry. Thus $4n^2/3 - n/3$ multiplications are required for the inverse. The determinant is given by

$$\text{Det } \mathbf{A} = \prod_{i=1}^{n} a_{ii}^{(i)}$$

where $a_{ii}^{(i)}$ are the diagonal elements obtained in the LU decomposition.

A tridiagonal matrix is one in which the only nonzero entries lie on the main diagonal and the diagonal just above and just below the main diagonal. The set of equations can be written as

$$a_i x_{i-1} + b_i x_i + c_i x_{i+1} = d_i$$

The LU decomposition is

$$b_1 = b_1$$
$$\text{for } k=2,n \text{ do}$$
$$a'_k = \frac{a_k}{b'_{k-1}}, \quad b'_k = b_k - \frac{a_k}{b'_{k-1}} c_{k-1}$$
$$\text{enddo}$$
$$d'_1 = d_1$$
$$\text{for } k=2,n \text{ do}$$
$$d'_k = d_k - a'_k d'_{k-1}$$
$$\text{enddo}$$
$$x_n = d'_n/b'_n$$
$$\text{for } k=n-1,1 \text{ do}$$
$$x_k = \frac{d'_k - c_k x_{k+1}}{b'_k}$$
$$\text{enddo}$$

The operation count for an $n \times n$ matrix with m right-hand sides is

$$2(n - 1) + m(3n - 2)$$

If

$$|b_i| > |a_i| + |c_i|$$

no pivoting is necessary, and this is true for many boundary-value problems and partial-differential equations.

Sparse matrices are ones in which the majority of the elements are zero. If the structure of the matrix is exploited, the solution time on a computer is greatly reduced. See Refs. 27, 55, 95, 96, 101, and 246. The conjugate gradient method is one method for solving sparse matrix problems, since it only involves multiplication of a matrix times a vector. Thus the sparseness of the matrix is easy to exploit. The conjugate gradient method is an iterative method that converges for sure in n iterations where the matrix is an $n \times n$ matrix. See Refs. 142 and 206. The singular value decomposition is useful when the matrix is singular or nearly singular (see Ref. 231).

Matrix methods, in particular finding the rank of the matrix, can be used to find the number of independent reactions in a reaction set. If the stoichiometric numbers for the reactions and molecules are put in the form of a matrix, the rank of the matrix gives the number of independent reactions (see Ref. 13).

Pivoting in Gauss Elimination   It might seem that the Gauss elimination completely disposes of the problem of finding solutions of linear systems, and theoretically it does. In practice, however, things are not so simple.

Example   Assume three-decimal floating arithmetic (i.e., only the three most significant digits of any number are retained), and solve the following system by Gauss elimination:

$$0.000100x_1 + 1.00x_2 = 1.00$$
$$1.00x_1 + 1.00x_2 = 2.00$$

We obtain

$$0.100 \times 10^{-3}x_1 + 0.100 \times 10^1 x_2 = 0.100 \times 10^1$$
$$-0.100 \times 10^5 x_2 = -0.100 \times 10^5$$

so that $x_2 = 1.00$ and $x_1 = 0.00$.

We check our solution by computing the residual vector $\mathbf{r} = b - Ax$:

$$r_1 = 0.100 \times 10^1 - 0.100 \times 10^{-3}x_1 - 0.100 \times 10^1 x_2 = 0.00$$
$$r_2 = 0.200 \times 10^1 - 0.100 \times 10^1 x_1 - 0.100 \times 10^1 x_2 = 0.100 \times 10^1$$

The fact that $r_2 = 1$ indicates that our "solution" is not very good. Indeed the exact solution of the system is $x_1 = 1.00010$ and $x_2 = 0.99990$, so the result computed by Gauss elimination is pretty bad.

Now reverse the order of the equations (that is, pivot) and solve

$$0.100 \times 10^1 x_1 + 0.100 \times 10^1 x_2 = 0.200 \times 10^1$$
$$0.100 \times 10^1 x_2 = 0.100 \times 10^1$$

so that $x_2 = 1.00$ and $x_1 = 1.00$. In this case the residual vector is $r_1 = 0.00$ and $r_2 = 0.100 \times 10^{-3}$, a considerable improvement over the previous result. In fact, the solution is as good as one could hope for by using three-digit arithmetic.

The moral of the preceding example is that the order of equations can make a large difference in how good an answer is obtained. It should be clear that the poor results in the first case are caused by having the large multiplier $(0.100 \times 10^1)/(0.100 \times 10^{-3})$, which resulted from dividing by a relatively small a_{11}. It is not enough just to avoid zero "pivots"; one must also avoid using pivots that are relatively small.

This magnification of errors can be reduced if we arrange that the pivot at any stage is larger in magnitude than any remaining element in the column. If this is done, the multipliers will then be less than or equal to 1 in magnitude. Gauss elimination modified in this manner is called pivotal condensation or partial pivoting. This is routinely done by computer programs.

NUMERICAL APPROXIMATIONS TO SOME EXPRESSIONS

APPROXIMATION IDENTITIES

For the following relationships the sign $\cong$ means approximately equal to, when X is small:

Approximation	Approximation
$\dfrac{1}{1 \pm X} \cong 1 \mp X$	$\sqrt{1 \pm X} \cong 1 \pm \dfrac{X}{2}$
$\dfrac{1 + Y}{1 \mp X} \cong 1 + Y \pm X$	$(1 \pm X)^{-n} \cong 1 \mp nX$

Approximation	Approximation
$(1 \pm X)^n \cong 1 \pm nX$	$(1 \pm X)^{-1/2} \cong 1 \mp \dfrac{X}{2}$
$(a \pm X)^2 \cong a^2 \pm 2aX$	$e^z \cong 1 + X$
$\sin X \cong X (X \text{ rad})$	$\tan X \cong X$
$\sqrt{Y(Y + X)} \cong \dfrac{2Y + X}{2}$	$\sqrt{Y^2 + X^2} \cong Y + \dfrac{X^2}{2Y}\left(\dfrac{X}{Y}\text{ small}\right)$

NUMERICAL ANALYSIS AND APPROXIMATE METHODS

REFERENCES: General (textbooks that cover at an introductory level a variety of topics that constitute a core of numerical methods for practicing engineers), 2, 3, 4, 22, 56, 59, 70, 77, 133, 135, 143, 150, 155, 219. Numerical solution of nonlinear equations, 153, 171, 237, 302. Numerical solution of ordinary differential equations, 76, 117, 127, 185, 257. Numerical solution of integral equations, 23, 26, 129, 162. Numerical solution of partial differential equations, 11, 76, 127, 133, 155, 210, 251, 286, 287, 213, 233, 253. Spline functions and applications, 38, 56, 70, 230. Finite elements and applications, 5, 29, 83, 130, 164, 210, 241, 281, 287, 303, 304. Fast Fourier transforms, 47, 56, 135, 238. Software, 187, 231.

INTRODUCTION

The goal of approximate and numerical methods is to provide convenient techniques for obtaining useful information from mathematical formulations of physical problems. Often this mathematical statement is not solvable by analytical means. Or perhaps analytic solutions are available but in a form that is inconvenient for direct interpretation

numerically. In the first case it is necessary either to attempt to approximate the problem satisfactorily by one which will be amenable to analysis, to obtain an approximate solution to the original problem by numerical means, or to use the two techniques in combination.

Numerical techniques therefore do not yield exact results in the sense of the mathematician. Since most numerical calculations are inexact, the concept of error is an important feature. The error associated with an approximate value is defined as

$$\text{True value} = \text{approximate value} + \text{error}$$

The four sources of error are as follows:

1. *Gross errors.* These result from unpredictable human, mechanical, or electrical mistakes.

2. *Round-off errors.* These are the consequence of using a number specified by m correct digits to approximate a number which requires more than m digits for its exact specification. For example, approximate the irrational number $\sqrt{2}$ by 1.414. Such errors are often

present in experimental data, in which case they may be called inherent errors, due either to empiricism or to the fact that the computer dictates the number of digits. Such errors may be especially damaging in areas such as matrix inversion or the numerical solution of partial differential equations when the number of algebraic operations is extremely large.

3. *Truncation errors.* These errors arise from the substitution of a finite number of steps for an infinite sequence of steps which would yield the exact result. To illustrate this error consider the infinite series for $e^{-x} \cdot e^{-x} = 1 - x + x^2/2 - x^3/6 + E_T(x)$, where E_T is the truncation error, $E_T = (1/24)e^{-\varepsilon}x^4$, $0 < \varepsilon < x$. If x is positive, ε is also positive. Hence $e^{-\varepsilon} < 1$. The approximation $e^{-x} \approx 1 - x + x^2/2 - x^3/6$ is in error by a positive amount smaller than $(1/24)x^4$.

4. *Inherited errors.* These arise as a result of errors occurring in the previous steps of the computational algorithm.

The study of errors in a computation is related to the theory of probability. In what follows a relation for the error will be given in certain instances.

NUMERICAL SOLUTION OF LINEAR EQUATIONS

See the section entitled "Matrix Algebra and Matrix Computation."

NUMERICAL SOLUTION OF NONLINEAR EQUATIONS IN ONE VARIABLE

Special Methods for Polynomials   Consider a polynomial equation of degree n:

$$P(x) = a_0x^n + a_1x^{n-1} + a_2x^{n-2} + \cdots + a_{n-1}x + a_n = 0 \qquad (3\text{-}71)$$

with real coefficients. $P(x)$ has exactly n roots, which may be real or complex. If all the coefficients of $P(x)$ are integers, then any rational root, say, r/s (r, s integers, having no common divisors) of $P(x)$, must be such that r is an integral divisor of a_n and s is an integral divisor of a_0. Further, any polynomial with rational coefficients may be converted into one with integral coefficients by multiplying by the lowest common multiple of the denominators of the coefficients.

Example $3x^4 - \frac{5}{3}x^2 + \frac{1}{5}x - 2 = 0$. The lowest common multiple of the denominator is 15. Thus multiplying by 15 (which does not change the roots) gives $45x^4 - 25x^2 + 3x - 30 = 0$. The only possible rational roots r/s are such that r may have the values $\pm30, \pm15, \pm10, \pm6, \pm5, \pm3, \pm2, \pm1$. s may have the values $\pm45, \pm15, \pm9, \pm5, \pm3, \pm1$. The possible rational roots may then be formed from all possible quotients, having no common factor.

In addition to these results, one can obtain an upper and lower bound for the real roots by the following device: If $a_0 > 0$ in Eq. (3-71) and if in Eq. (3-71) the first negative coefficient is preceded by k coefficients which are positive or zero, and if G is the greatest of the absolute values of the negative coefficients, then each real root is less than $1 + \sqrt[k]{G/a_0}$.

Example $P(x) = x^5 + 3x^4 - 7x^2 - 40x + 2 = 0$. Here $a_0 = 1$, $G = 40$, and $k = 3$ since we must supply 0 as the coefficient for x^3. Thus $1 + \sqrt[3]{40} \approx 4.42$ is an upper bound for the real roots.

A lower bound to the real roots may be found by applying the criterion to the equation $P(-x)$.

Example $P(-x) = -x^5 + 3x^4 - 7x^2 + 40x + 2 = 0$, which is equivalent to $x^5 - 3x^4 + 7x^2 - 40x - 2 = 0$ since a_0 must be +. Then $a_0 = 1$, $G = 40$, and $k = 1$. Hence $-(1 + 40) = -41$ is a lower bound. Thus all real roots $-41 < r < 4.42$.

One last result is helpful in getting an estimate of how many positive and negative real roots there are.

Descartes' Rule   The number of positive real roots of a polynomial with real coefficients is either equal to the number of changes in sign v or is less than v by a positive even integer. The number of negative roots of $f(x)$ is either equal to the number of variations of sign of $f(-x)$ or is less than this by a positive even integer.

Example $f(x) = x^4 - 13x^2 + 4x - 2 = 0$ has three changes in sign; therefore, there are either three or one positive roots. $f(-x) = x^4 - 13x^2 - 4x - 2$ has one change in sign. Therefore, there is one negative root.

General Methods for Nonlinear Equations in One Variable

Successive Substitutions   Let $f(x) = 0$ be the nonlinear equation to be solved. If this is rewritten as $x = F(x)$, then an iterative scheme can be set up in the form $x_{k+1} = F(x_k)$. To start the iteration an initial guess must be obtained graphically or otherwise. The convergence or divergence of the procedure depends upon the method of writing $x = F(x)$, of which there will usually be several forms. However, if a is a root of $f(x) = 0$, and if $|F'(a)| < 1$, then for any initial approximation sufficiently close to a, the method converges to a. This process is called first order because the error in x_{k+1} is proportional to the first power of the error in x_k for large k.

Example $f(x) = x^3 - x - 1 = 0$. A rough plot shows a real root of approximately 1.3. The equation can be written in the form $x = F(x)$ in several ways such as $x = x^3 - 1$, $x = 1/(x^2 - 1)$, and $x = (1 + x)^{1/3}$. In the first case $F'(x) = 3x^2 = 5.07$ at $x = 1.3$, in the second $F(1.3) = -5.46$, and only in the third case is $F'(1.3) < 1$. Hence only the third iterative process has a chance to converge. This is illustrated in the following table.

Step k	$x = x^3 - 1$	$x = 1/(x^2 - 1)$	$x = (1+x)^{1/3}$
0	1.3	1.3	1.3
1	1.197	1.4493	1.32
2	0.7150	0.9088	1.3238
3	-0.6345	-5.742	1.3245
4			1.3247

Another way of writing the equation is $x_{k+1} = x_k + \beta f(x_k)$. The choice of β is made such that $|1 + \beta\, df/dx(a)| < 1$. Convergence is guaranteed by the theorem given for simultaneous equations.

Methods of Perturbation   Let $f(x) = 0$ be the equation. In general, the iterative relation is

$$x_{k+1} = x_k - [f(x_k)/a_k]$$

where the iteration begins with x_0 as an initial approximation and α_k as some functional.

Newton-Raphson Procedure   This variant chooses $\alpha_k = f'(x_k)$ where $f' = df/dx$ and geometrically consists of replacing the graph of $f(x)$ by the tangent line at $x = x_k$ in each successive step. If $f'(x)$ and $f''(x)$ have the same sign throughout an interval $a \leq x \leq b$ containing the solution, with $f(a)$, $f(b)$ of opposite signs, then the process converges starting from any x_0 in the interval $a \leq x \leq b$. The process is second order.

Example $f(x) = x - 1 + \dfrac{(0.5)^x - 0.5}{0.3}$

$$f'(x) = 1 - 2.3105[0.5]^x$$

An approximate root (obtained graphically) is 2.

Step k	x_k	$f(x_k)$	$f'(x_k)$
0	2	0.1667	0.4224
1	1.6054	0.0342	0.2407
2	1.4632	0.0055	0.1620

Method of False Position   This variant is commenced by finding x_0 and x_1 such that $f(x_0)$, $f(x_1)$ are of opposite signs. Then $\alpha_1 = $ slope of secant line joining $[x_0, f(x_0)]$ and $[x_1, f(x_1)]$ so that

$$x_2 = x_1 - \frac{x_1 - x_0}{f(x_1) - f(x_0)} f(x_1)$$

In each following step α_k is the slope of the line joining $[x_k, f(x_k)]$ to the most recently determined point where $f(x_j)$ has the opposite sign from that of $f(x_k)$. This method is of first order. If one uses the most recently determined point (regardless of sign), the method is a secant method.

Method of Wegstein This is a variant of the method of successive substitutions which forces and/or accelerates convergence. The iterative procedure $x_{k+1} = F(x_k)$ is revised by setting $\hat{x}_{k+1} = F(x_k)$ and then taking $x_{k+1} = qx_k + (1 - q)\hat{x}_{k+1}$, where q is a suitably chosen number which may be taken as constant throughout or may be adjusted at each step. Wegstein found that suitable q's are:

Behavior of successive substitution process	Range of optimum q
Oscillatory convergence	$0 < q < \frac{1}{2}$
Oscillatory divergence	$\frac{1}{2} < q < 1$
Monotonic convergence	$q < 0$
Monotonic divergence	$1 < q$

At each step q may be calculated to give a locally optimum value by setting

$$q = \frac{\hat{x}_{k+1} - \hat{x}_k}{\hat{x}_{k+1} - 2\hat{x}_{k+1} + \hat{x}_{k-1}}$$

The Wegstein method is a secant method applied to $g(x) \equiv x - F(x)$.

Numerical Solution of Simultaneous Nonlinear Equations The techniques illustrated here will be demonstrated for two simultaneous equations $f(x, y) = 0$, $g(x, y) = 0$. They immediately generalize to more than two simultaneous equations.

Method of Successive Substitutions Write a system of equations as

$$\alpha_i = f_i(\alpha), \quad \text{or } \alpha = \mathbf{f}(\alpha)$$

The following theorem guarantees convergence. Let α be the solution to $\alpha_i = f_i(\alpha)$. Assume that given $h > 0$, there exists a number $0 < \mu < 1$ such that

$$\sum_{j=1}^{n} \left| \frac{\partial f_i}{\partial x_j} \right| \leq \mu \quad \text{for} \quad |x_i - \alpha_i| < h, \quad i = 1, \dots, n$$

$$x_i^{k+1} = f_i(x_i^k)$$

Then $x_i^k \to \alpha_i$

as k increases (see Ref. 106).

Newton-Raphson Method To solve the set of equations

$$F_i(x_1, x_2, \dots, x_n) = 0, \quad \text{or } F_i(\{x_j\}) = 0, \quad \text{or } F_i(\mathbf{x}) = 0$$

one uses a truncated Taylor series to give

$$0 = F_i(\{\mathbf{x}^k\}) + \sum_{j=1}^{n} \left. \frac{\partial F_i}{\partial x_j} \right|_{\mathbf{x}^k} (x_j^{k+1} - x_j^k)$$

Thus one solves iteratively from one point to another.

$$\sum_{j=1}^{n} A_{ij}^k (x_j^{k+1} - x_j^k) = -F_i(\{\mathbf{x}^k\})$$

where $A_{ij}^k = \left. \frac{\partial F_i}{\partial x_j} \right|_{\mathbf{x}^k}$

This method requires solution of sets of linear equations until the functions are zero to some tolerance or the changes of the solution between iterations is small enough. Convergence is guaranteed provided the norm of the matrix $\mathbf{A}$ is bounded, $\mathbf{F}(\mathbf{x})$ is bounded for the initial guess, and the second derivative of $\mathbf{F}(\mathbf{x})$ with respect to all variables is bounded. See Refs. 106 and 155.

Example $f(x, y) = 4x^2 + 6x - 4xy + 2y^2 - 3$
$g(x, y) = 2x^2 - 4xy + y^2$

By plotting one of the approximate roots is found to be $x_0 = 0.4$, $y_0 = 0.3$. At this point there results $\partial f/\partial x = 8$, $\partial f/\partial y = -0.4$, $\partial g/\partial x = 0.4$, and $\partial g/\partial y = -1$.

$$8(x^{k+1} - x^k) - 0.4(y^{k+1} - y^k) = +0.26$$
$$0.4(x^{k+1} - x^k) - 1(y^{k+1} - y^k) = -0.07$$

The first few iteration steps are as follows:

Step k	x_k	y_k	$f(x_k, y_k)$	$g(x_k, y_k)$
0	0.4	0.3	−0.26	0.07
1	0.43673	0.24184	0.078	0.0175
2	0.42672	0.25573	−0.0170	−0.007
3	0.42925	0.24943	0.0077	0.0010

Method of Continuity (Homotopy) In the case of n equations in n unknowns, when n is large, determining the approximate solution may involve considerable effort. In such a case the method of continuity is admirably suited for use on digital computers. It consists basically of the introduction of an extra variable into the n equations

$$f_i(x_1, x_2, \dots, x_n) = 0 \qquad i = 1, \dots, n \qquad (3\text{-}72)$$

and replacing them by

$$f_i(x_1, x_2, \dots, x_n, \lambda) = 0 \qquad i = 1, \dots, n \qquad (3\text{-}73)$$

where λ is introduced in such a way that the functions (3-73) depend in a simple way upon λ and reduce to an easily solvable system for $\lambda = 0$ and to the original equations (3-72) for $\lambda = 1$. A system of ordinary differential equations, with independent variable λ, is then constructed by differentiating Eqs. (3-73) with respect to λ. There results

$$\sum_{j=1}^{n} \frac{\partial f_i}{\partial x_j} \frac{dx_j}{d\lambda} + \frac{\partial f_i}{\partial \lambda} = 0 \qquad (3\text{-}74)$$

where $x_1, \dots, x_n$ are considered as functions of λ. Equations (3-74) are integrated, with initial conditions obtained from Eqs. (3-73) with $\lambda = 0$, from $\lambda = 0$ to $\lambda = 1$. If the solution can be continued to $\lambda = 1$, the values of $x_1, \dots, x_n$ for $\lambda = 1$ will be a solution of the original equations. If the integration becomes infinite, the parameter λ must be introduced in a different fashion. Integration of the differential equations (which are usually nonlinear in λ) may be accomplished by using techniques described under "Numerical Solution of Ordinary Differential Equations."

Other Methods Other methods can be found in the literature. See Ref. 66.

INTERPOLATION AND FINITE DIFFERENCES

Linear Interpolation If a function $f(x)$ is approximately linear in a certain range, then the ratio

$$\frac{f(x_1) - f(x_0)}{x_1 - x_0} = f[x_0, x_1]$$

is approximately independent of x_0, x_1 in the range. The linear approximation to the function $f(x)$, $x_0 < x < x_1$ then leads to the interpolation formula

$$f(x) \approx f(x_0) + (x - x_0)f[x_0, x_1]$$

$$\approx f(x_0) + \frac{x - x_0}{x_1 - x_0} [f(x_1) - f(x_0)]$$

$$\approx \frac{1}{x_1 - x_0} [(x_1 - x)f(x_0) - (x_0 - x)f(x_1)]$$

Example Find cosh 0.83 by linear interpolation given cosh 0.8 and cosh 0.9.

x_i	$f(x_i)$	$x_i - 0.83$
0.8	1.33743	−0.03
0.9	1.43309	+0.07

$$f(0.83) \approx 1/0.10[(0.07)(1.33743) - (-0.03)(1.43309)]$$
$$f(0.83) \approx 1.36613$$

Since the true five-place value is 1.36468, it is seen that here linear interpolation gives three significant figures.

Divided Differences of Higher Order and Higher-Order Interpolation The first-order divided difference $f[x_0, x_1]$ was defined previously. Divided differences of second and higher order are defined iteratively by

$$f[x_0, x_1, x_2] = \frac{f[x_1, x_2] - f[x_0, x_1]}{x_2 - x_0}$$

$$\vdots$$

$$f[x_0, x_1, \ldots, x_k] = \frac{f[x_1, \ldots, x_k] - f[x_0, x_1, \ldots, x_{k-1}]}{x_k - x_0}$$

and a convenient form for computational purposes is

$$f[x_0, x_1, \ldots, x_k] = \sum_{j=0}^{k'} \frac{f(x_j)}{(x_j - x_0)(x_j - x_1) \cdots (x_j - x_k)}$$

for any $k \geq 0$, where the $'$ means that the term $(x_j - x_j)$ is omitted in the denominator. For example,

$$f[x_0, x_1, x_2] = \frac{f(x_0)}{(x_0 - x_1)(x_0 - x_2)} + \frac{f(x_1)}{(x_1 - x_0)(x_1 - x_2)} + \frac{f(x_2)}{(x_2 - x_0)(x_2 - x_1)}$$

If the accuracy afforded by a linear approximation is inadequate, a generally more accurate result may be based upon the assumption that $f(x)$ may be approximated by a polynomial of degree 2 or higher over certain ranges. This assumption leads to Newton's fundamental interpolation formula with divided differences

$$f(x) \approx f(x_0) + (x - x_0)f[x_0, x_1] + (x - x_0)(x - x_1)f[x_0, x_1, x_2]$$

$$+ \cdots + (x - x_0)(x - x_1) \cdots (x - x_{n-1})f[x_0, x_1, \ldots, x_n] + E_n(x)$$

where

$$E_n(x) = \text{error} = \frac{1}{(n+1)!} f^{n+1}(\varepsilon)\pi(x)$$

where minimum $(x_0, \ldots, x) < \varepsilon <$ maximum $(x_0, x_1, \ldots, x_n, x)$ and $\pi(x) = (x - x_0)(x - x_1) \cdots (x - x_n)$. In order to use the previous equation most effectively one may first form a divided-difference table. For example, for third-order interpolation the difference table is

$$
\begin{array}{l|l}
x_0 & f(x_0) \\
 & \qquad \searrow f[x_0, x_1] \\
x_1 & f(x_1) \qquad\qquad\qquad \searrow f[x_0, x_1, x_2] \\
 & \qquad \searrow f[x_1, x_2] \qquad\qquad\qquad\qquad \searrow f[x_0, x_1, x_2, x_3] \\
x_2 & f(x_2) \qquad\qquad\qquad \searrow f[x_1, x_2, x_3] \\
 & \qquad \searrow f[x_2, x_3] \\
x_3 & f(x_3)
\end{array}
$$

where each entry is given by taking the difference between diagonally adjacent entries to the left, divided by the abscissas corresponding to the ordinates intercepted by the diagonals passing through the calculated entry.

Equally Spaced Forward Differences If the ordinates are *equally spaced*, i.e., $x_j - x_{j-1} = \Delta x$ for all j, then the first differences are denoted by $\Delta f(x_0) = f(x_1) - f(x_0)$ or $\Delta y_0 = y_1 - y_0$, where $y = f(x)$. The differences of these first differences, called second differences, are denoted by $\Delta^2 y_0, \Delta^2 y_1, \ldots, \Delta^2 y_n$. Thus

$$\Delta^2 y_0 = \Delta y_1 - \Delta y_0 = y_2 - y_1 - y_1 + y_0 = y_2 - 2y_1 + y_0$$

And in general

$$\Delta^j y_0 = \sum_{n=0}^{j} (-1)^n \binom{j}{n} y_{j-n}$$

where $\binom{j}{n} = \frac{j!}{n!(j-n)!} = $ binomial coefficients.

If the ordinates are equally spaced,

$$x_{n+1} - x_n = \Delta x$$
$$y_n = y(x_n)$$

then the first and second differences are denoted by

$$\Delta y_n = y_{n+1} - y_n$$
$$\Delta^2 y_n = \Delta y_{n+1} - \Delta y_n = y_{n+2} - 2y_{n+1} + y_n$$

A new variable is defined

$$\alpha = \frac{x_\alpha - x_0}{\Delta x}$$

and the finite interpolation formula through the points $y_0, y_1, \ldots, y_n$ is written as follows:

$$y_\alpha = y_0 + \alpha \, \Delta y_0 + \frac{\alpha(\alpha - 1)}{2!} \Delta^2 y_0 + \cdots + \frac{\alpha(\alpha - 1) \cdots (\alpha - n + 1)}{n!} \Delta^n y_0$$

Keeping only the first two terms gives a straight line through (x_0, y_0)–(x_1, y_1); keeping the first three terms gives a quadratic function of position going through those points plus (x_2, y_2). The value $\alpha = 0$ gives $x = x_0$; $\alpha = 1$ gives $x = x_1$, and so on.

Equally Spaced Backward Differences Backward differences are defined by

$$\nabla y_n = y_n - y_{n-1}$$
$$\nabla^2 y_n = \nabla y_n - \nabla y_{n-1} = y_n - 2y_{n-1} + y_{n-2}$$

The interpolation polynomial of order n through the points $y_0, y_{-1}, \ldots, y_{-n}$ is

$$y_\alpha = y_0 + \alpha \, \nabla y_0 + \frac{\alpha(\alpha + 1)}{2!} \nabla^2 y_0 + \cdots + \frac{\alpha(\alpha + 1) \cdots (\alpha + n - 1)}{n!} \nabla^n y_0$$

The value of $\alpha = 0$ gives $x = x_0$; $\alpha = -1$ gives $x = x_{-1}$, and so on. Alternatively, the interpolation polynomial of order n through the points $y_1, y_0, y_{-1}, \ldots, y_{-n}$ is

$$y_\alpha = y_1 + (\alpha - 1) \, \nabla y_1 + \frac{\alpha(\alpha - 1)}{2!} \nabla^2 y_1$$

$$+ \cdots + \frac{(\alpha - 1)\alpha(\alpha + 1) \cdots (\alpha + n - 2)}{n!} \nabla^n y_1$$

Now $\alpha = 1$ gives $x = x_1$; $\alpha = 0$ gives $x = x_0$.

Central Differences The central difference denoted by

$$\delta f(x) = f\left(x + \frac{h}{2}\right) - f\left(x - \frac{h}{2}\right)$$

$$\delta^2 f(x) = \delta^{n-1} f\left(x + \frac{h}{2}\right) - \delta^{n-1} f\left(x - \frac{h}{2}\right)$$

is useful for calculating at the interior points of tabulated data.

Lagrange Interpolation Formulas A global polynomial is defined over the entire region of space

$$P_m(x) = \sum_{j=0}^{m} c_j x^j$$

This polynomial is of degree m (highest power is x^m) and order $m + 1$ ($m + 1$ parameters $\{c_j\}$). If we are given a set of $m + 1$ points

$$y_1 = f(x_1), \, y_2 = f(x_2), \ldots, y_{m+1} = f(x_{m+1})$$

then Lagrange's formula gives a polynomial of degree m that goes through the $m + 1$ points:

$$P_m(x) = \frac{(x - x_2)(x - x_3) \cdots (x - x_{m+1})}{(x_1 - x_2)(x_1 - x_3) \cdots (x_1 - x_{m+1})} \, y_1$$

$$+ \frac{(x - x_1)(x - x_3) \cdots (x - x_{m+1})}{(x_2 - x_1)(x_2 - x_3) \cdots (x_2 - x_{m+1})} \, y_2 + \cdots$$

$$+ \frac{(x - x_1)(x - x_2) \cdots (x - x_{m+1})}{(x_{m+1} - x_1)(x_{m+1} - x_2) \cdots (x_{m+1} - x_m)} \, y_{m+1}$$

Note that each coefficient of y_j is a polynomial of degree m that vanishes at the points $\{x_j\}$ (except for one value of j) and takes the value of 1.0 at that point:

$$P_m(x_j) = y_j, \quad j = 1, 2, \ldots, m + 1$$

If the function $f(x)$ is known, the error in the approximation is, per Ref. 14,

$$|\text{error}(x)| \leq \frac{|x_{m+1} - x_1|^{m+1}}{(m+2)!} \max_{x_1 \leq x \leq x_{m+1}} |f^{(m+2)}(x)|$$

The evaluation of $P_m(x)$ at a point other than at the defining points can be made with Neville's algorithm (Ref. 231). Let P_1 be the value at x of the unique function passing through the point (x_1, y_1); or $P_1 = y_1$. Let

P_{12} be the value at x of the unique polynomial passing through the points x_1 and x_2. Likewise, $P_{ijk\cdots r}$ is the unique polynomial passing through the points $x_i, x_j, x_k, \ldots x_r$. Then use the table

$$
\begin{array}{llll}
x_1 & y_1 = P_1 & & \\
& & P_{12} & \\
x_2 & y_2 = P_2 & & P_{123} \\
& & P_{23} & & P_{1234} \\
x_3 & y_3 = P_3 & & P_{234} \\
& & P_{34} & \\
x_4 & y_4 = P_4 & &
\end{array}
$$

These entries are defined using

$$P_{i(i+1)\cdots(i+m)} = \frac{(x - x_{i+m}) P_{i(i+1)\cdots(i+m-1)} + (x_i - x) P_{(i+1)(i+2)\cdots(i+m)}}{x_i - x_{i+m}}$$

For example, consider P_{1234}. The terms on the right-hand side involve P_{123} and P_{234}. The "parents," P_{123} and P_{234}, already agree at points 2 and 3. Here $i = 1$, $m = 3$; thus, the parents agree at $x_{i+1}, \ldots, x_{i+m-1}$ already. The formula makes $P_{i(i+1)\cdots(i+m)}$ agree with the function at the additional points x_{i+m} and x_i. Thus, $P_{i(i+1)\cdots(i+m)}$ agrees with the function at all the points $\{x_i, x_{i+1}, \ldots x_{i+m}\}$.

NUMERICAL DIFFERENTIATION

Numerical differentiation should be avoided whenever possible, particularly when data are empirical and subject to appreciable observation errors. Errors in data can affect numerical derivatives quite strongly; i.e., differentiation is a roughening process. When such a calculation must be made, it is usually desirable first to smooth the data to a certain extent.

Use of Interpolation Formula If the data are given over equidistant values of the independent variable x, an interpolation formula such as the Newton formula (see Refs. 143 and 185) may be used and the resulting formula differentiated analytically. If the independent variable is not at equidistant values, then Lagrange's formulas must be used. By differentiating three- and five-point Lagrange interpolation formulas the following differentiation formulas result for equally spaced tabular points:

Three-Point Formulas Let x_0, x_1, x_2 be the three points.

$$f'(x_0) = \frac{1}{2h} [-3f(x_0) + 4f(x_1) - f(x_2)] + \frac{h^2}{3} f'''(\varepsilon)$$

$$f'(x_1) = \frac{1}{2h} [-f(x_0) + f(x_2)] - \frac{h^2}{6} f'''(\varepsilon)$$

$$f'(x_2) = \frac{1}{2h} [f(x_0) - 4f(x_1) + 3f(x_2)] + \frac{h^2}{3} f'''(\varepsilon)$$

where the last term is an error term $\min_j x_j < \varepsilon < \max_j x_j$.

Smoothing Techniques These techniques involve the approximation of the tabular data by a least-squares fit of the data by using some known functional form, usually a polynomial (for the concept of least squares see "Statistics"). In place of approximating $f(x)$ by a single least-squares polynomial of degree n over the entire range of the tabulation, it is often desirable to replace each tabulated value by the value taken on by a least-squares polynomial of degree n relevant to a subrange of $2M + 1$ points centered, when possible, at the point for which the entry is to be modified. Thus each smoothed value replaces a tabulated value. Let $f_j = f(x_j)$ be the tabular points and $y_j =$ smoothed values.

First-Degree Least Squares with Three Points

$$y_0 = \tfrac{1}{6}[5f_0 + 2f_1 - f_2]$$
$$y_1 = \tfrac{1}{3}[f_0 + f_1 + f_2]$$
$$y_2 = \tfrac{1}{6}[-f_0 + 2f_1 + 5f_2]$$

Second-Degree Least Squares with Five Points For five evenly spaced points $x_{-2}, x_{-1}, x_0, x_1,$ and x_2 (separated by distance h) and their ordinates $f_{-2}, f_{-1}, f_0, f_1,$ and f_2, assume a parabola is fit by least squares. Then the derivative at the center point is

$$f'_0 = 1/10h \, [-2f_{-2} - f_{-1} + f_1 + 2f_2]$$

If derivatives are required at end points, with all points and ordinates to one side, the derivatives are

$$f'_0 = 1/20h \, [-21f_0 + 13f_1 + 17f_2 - 9f_3]$$
$$f'_1 = 1/20h \, [-11f_0 + 3f_1 + 7f_2 + f_3]$$
$$f'_0 = 1/20h \, [21f_0 - 13f_1 - 17f_2 + 9f_{-1}]$$
$$f'_{-1} = 1/20h \, [\, 11f_0 - 3f_{-1} - 7f_{-2} - f_{-1}]$$

Numerical Derivatives The results given above can be used to obtain numerical derivatives when solving problems on the computer, in particular for the Newton-Raphson method and homotopy methods. Suppose one has a program, subroutine, or other function evaluation device that will calculate f given x. One can estimate the value of the first derivative at x_0 using

$$\frac{df}{dx}\bigg|_{x_0} \approx \frac{f[x_0(1+\varepsilon)] - f[x_0]}{\varepsilon}$$

(a first-order formula) or

$$\frac{df}{dx}\bigg|_{x_0} \approx \frac{f[x_0(1+\varepsilon)] - f[x_0(1-\varepsilon)]}{2\varepsilon}$$

(a second-order formula). The value of ε is important; a value of 10^{-6} is typical, but smaller or larger values may be necessary depending on the computer precision and the application. One must also be sure that the value of x_0 is not zero and use a different increment in that case.

NUMERICAL INTEGRATION (QUADRATURE)

A multitude of formulas have been developed to accomplish numerical integration, which consists of computing the value of a definite integral from a set of numerical values of the integrand.

Newton-Cotes Integration Formulas (Equally Spaced Ordinates) for Functions of One Variable The definite integral $\int_a^b f(x)\, dx$ is to be evaluated.

Trapezoidal Rule This formula consists of subdividing the interval $a \le x \le b$ into n subintervals a to $a + h$, $a + h$ to $a + 2h, \ldots$ and replacing the graph of $f(x)$ by the result of joining the ends of adjacent ordinates by line segments. If $f_j = f(x_j) = f(a + jh), f_0 = f(a), f_n = f(b)$, the integration formula is

$$\int_a^b f(x)\, dx = \frac{h}{2} [f_0 + 2f_1 + 2f_2 + \cdots + 2f_{n-1} + f_n] + E_n$$

where $\quad |E_n| = \frac{nh^3}{12} |f''(\varepsilon)| = \frac{(b-a)^3}{12n^2} |f''(\varepsilon)| \qquad a < \varepsilon < b$

This procedure is not of high accuracy. However, if $f''(x)$ is continuous in $a < x < b$, the error goes to zero as $1/n^2$, $n \to \infty$.

Parabolic Rule (Simpson's Rule) This procedure consists of subdividing the interval $a < x < b$ into $n/2$ subintervals, each of length $2h$, where n is an *even* integer. By using the notation as above the integration formula is

$$\int_a^b f(x)\, dx = \frac{h}{3} [f_0 + 4f_1 + 2f_2 + 4f_3 + \cdots$$
$$+ 4f_{n-3} + 2f_{n-2} + 4f_{n-1} + f_n] + E_n$$

where $\quad |E_n| = \frac{nh^5}{180} |f^{(\text{IV})}(\varepsilon)| = \frac{(b-a)^5}{180n^4} |f^{(\text{IV})}(\varepsilon)| \qquad a < \varepsilon < b$

This method approximates $f(x)$ by a parabola on each subinterval. This rule is generally more accurate than the trapezoidal rule. It is the most widely used integration formula.

Gaussian Quadrature Gaussian quadrature provides a highly accurate formula based on irregularly spaced points, but the integral needs to be transformed onto the interval 0 to 1.

$$x = a + (b-a)u, \quad dx = (b-a)du$$

$$\int_a^b f(x)\, dx = (b-a) \int_0^1 f(u)\, du$$

$$\int_0^1 f(u)\,du = \sum_{i=1}^m W_i f(u_i)$$

The quadrature is exact when f is a polynomial of degree $2m-1$ in x. Because there are m weights and m Gauss points, we have $2m$ parameters that are chosen to exactly represent a polynomial of degree $2m-1$, which has $2m$ parameters. The Gauss points and weights are given in the table.

Gaussian Quadrature Points and Weights

m	u_i	W_i
2	0.21132 48654	0.50000 00000
	0.78867 51346	0.50000 00000
3	0.11270 16654	0.27777 77778
	0.50000 00000	0.44444 44445
	0.88729 83346	0.27777 77778
4	0.06943 18442	0.17392 74226
	0.33000 94783	0.32607 25774
	0.66999 05218	0.32607 25774
	0.93056 81558	0.17392 74226
5	0.04691 00771	0.11846 34425
	0.23076 53450	0.23931 43353
	0.50000 00000	0.28444 44444
	0.76923 46551	0.23931 43353
	0.95308 99230	0.11846 34425

Example   Calculate the value of the following integral.

$$I = \int_0^1 e^{-x} \sin x\,dx \qquad (3\text{-}75)$$

Using the Gaussian quadrature formulas gives the following values for various values of m. Clearly, three internal points, requiring evaluation of the integrand at only three points, gives excellent results.

m	I
1	0.908185
2	0.910089
3	0.909336367
4	0.909330666
5	0.909330674

Romberg's Method   Romberg's method uses extrapolation techniques to improve the answer (Ref. 231). If we let I_1 be the value of the integral obtained using interval size $h = \Delta x$, and I_2 be the value of I obtained when using interval size $h/2$, and I_0 the true value of I, then the error in a method is approximately h^m, or

$$I_1 \approx I_0 + ch^m$$

$$I_2 \approx I_0 + c\left(\frac{h}{2}\right)^m$$

Replacing the $\approx$ by an equality (an approximation) and solving for c and I_0 gives

$$I_0 = \frac{2^m I_2 - I_1}{2^m - 1}$$

This process can also be used to obtain $I_1, I_2, \ldots$, by halving h each time, and then calculating new estimates from each pair, calling them $J_1, J_2, \ldots$; that is, in the formula above, replace I_0 with J_1. The formulas are reapplied for each pair of J to obtain $K_1, K_2, \ldots$ The process continues until the required tolerance is obtained.

$$
\begin{array}{cccc}
I_1 & I_2 & I_3 & I_4 \\
 & J_1 & J_2 & J_3 \\
 & & K_1 & K_2 \\
 & & & L_1
\end{array}
$$

Romberg's method is most useful for a low-order method (small m) because significant improvement is then possible.

Example   Evaluate the same integral (3-75) using the trapezoid rule and then apply the Romberg method. To achieve four-digit accuracy, any result from J_2 through L_1 are suitable, even though the base results (I_1–I_4) are not that close.

$I_1 = 0.967058363$	$I_2 = 0.923704741$	$I_3 = 0.912920511$	$I_4 = 0.910227902$
	$J_1 = 0.909253534$	$J_2 = 0.909325768$	$J_3 = 0.909330366$
		$K_1 = 0.909349846$	$K_2 = 0.909331898$
			$L_1 = 0.909325916$

Singularities   When the integrand has singularities, a variety of techniques can be tried. The integral may be divided into one part that can be integrated analytically near the singularity and another part that is integrated numerically. Sometimes a change of argument allows analytical integration. Series expansion might be helpful, too. When the domain is infinite, it is possible to use Gauss-Legendre or Gauss-Hermite quadrature. Also a transformation can be made (Ref. 26). For example, let $u = 1/x$ and then

$$\int_a^b f(x)\,dx = \int_{1/b}^{1/a} \frac{1}{u^2} f\left(\frac{1}{u}\right) du \qquad ab > 0$$

Two-Dimensional Formula   Two-dimensional integrals can be calculated by breaking down the integral into one-dimensional integrals.

$$\int_a^b \int_{g_1(x)}^{g_2(x)} f(x, y)\,dx\,dy = \int_a^b G(x)\,dx$$

$$G(x) = \int_{g_1(x)}^{g_2(x)} f(x, y)\,dy$$

Gaussian quadrature can also be used in two dimensions, provided the integration is on a square or can be transformed to one. (Domain transformations might be used to convert the domain to a square.)

$$\int_0^1 \int_0^1 f(x, y)\,dx\,dy = \sum_{i=1}^{mx} W_{xi} \sum_{j=1}^{my} W_{yj} f(x_i, y_j)$$

NUMERICAL SOLUTION OF ORDINARY DIFFERENTIAL EQUATIONS AS INITIAL VALUE PROBLEMS

A differential equation for a function that depends on only one variable, often time, is called an ordinary differential equation. The general solution to the differential equation includes many possibilities; the boundary or initial conditions are needed to specify which of those are desired. If all conditions are at one point, then the problem is an initial value problem and can be integrated from that point on. If some of the conditions are available at one point and others at another point, then the ordinary differential equations become two-point boundary value problems, which are treated in the next section. Initial value problems as ordinary differential equations arise in control of lumped parameter models, transient models of stirred tank reactors, and in all models where there are no spatial gradients in the unknowns.

A higher-order differential equation

$$y^{(n)} + F(y^{(n-1)}, y^{(n-2)}, \ldots, y', y) = 0$$

with initial conditions

$$G_i(y^{(n-1)}(0), y^{(n-2)}(0), \ldots, y(0), y(0)) = 0, \quad i = 1, \ldots, n$$

can be converted into a set of first-order equations using

$$y_i \equiv y^{(i-1)} = \frac{d^{(i-1)}y}{dt^{(i-1)}} = \frac{d}{dt} y^{(i-2)} = \frac{dy_{i-1}}{dt}$$

The higher-order equation can be written as a set of first-order equations.

$$\frac{dy_1}{dt} = y_2$$

$$\frac{dy_2}{dt} = y_3$$

$$\frac{dy_3}{dt} = y_4$$

$$\cdots$$

$$\frac{dy_n}{dt} = -F(y_{n-1}, y_{n-2}, \ldots, y_2, y_1)$$

The initial conditions would have to be specified for variables $y_1(0), \ldots, y_n(0)$, or equivalently $y(0), \ldots, y^{(n-1)}(0)$. The set of equations is then written as

$$\frac{d\mathbf{y}}{dt} = \mathbf{f}(\mathbf{y}, t)$$

All the methods in this section are described for a single equation; the methods apply to multiple equations. See Refs. 106 and 185 for more details.

Euler's method is first-order.

$$y^{n+1} = y^n + \Delta t\, f(y^n)$$

and errors are proportional to Δt. The second-order Adams-Bashforth method is

$$y^{n+1} = y^n + \frac{\Delta t}{2}\,[3\,f(y^n) - f(y^{n-1})]$$

Errors are proportional to Δt^2, and high-order methods are available. Notice that the higher-order explicit methods require knowing the solution (or the right-hand side) evaluated at times in the past. Since these were calculated to get to the current time, this presents no problem except for starting the problem. Then it may be necessary to use Euler's method with a very small step size for several steps in order to generate starting values at a succession of time points. The error terms, order of the method, function evaluations per step, and stability limitations are listed in Ref. 106. The advantage of the high-order Adams-Bashforth method is that it uses only one function evaluation per step yet achieves high-order accuracy. The disadvantage is the necessity of using another method to start.

Runge-Kutta methods are explicit methods that use several function evaluations for each time step. Runge-Kutta methods are traditionally written for $f(t, y)$. The first-order Runge-Kutta method is Euler's method. A second-order Runge-Kutta method is

$$y^{n+1} = y^n + \frac{\Delta t}{2}\,[f^n + f(t^n + \Delta t, y^n + \Delta t\, f^n)]$$

while the midpoint scheme is also a second-order Runge-Kutta method.

$$y^{n+1} = y^n + \Delta t\, f\left(t^n + \frac{\Delta t}{2}, y^n + \frac{\Delta t}{2}\, f^n\right)$$

A popular fourth-order Runge-Kutta method is the Runge-Kutta-Feldberg formulas (Ref. 111), which have the property that the method is fourth-order but achieves fifth-order accuracy. The popular integration package RKF45 is based on this method.

$$k_1 = \Delta t\, f(t^n, y^n)$$

$$k_2 = \Delta t\, f\left(t^n + \frac{\Delta t}{4}, y^n + \frac{k_1}{4}\right)$$

$$k_3 = \Delta t\, f\left(t^n + \frac{3}{8}\,\Delta t, y^n + \frac{3}{32}\, k_1 + \frac{9}{32}\, k_2\right)$$

$$k_4 = \Delta t\, f\left(t^n + \frac{12}{13}\,\Delta t, y^n + \frac{1932}{2197}\, k_1 - \frac{7200}{2197}\, k_2 + \frac{7296}{2197}\, k_3\right)$$

$$k_5 = \Delta t\, f\left(t^n + \Delta t, y^n + \frac{439}{216}\, k_1 - 8k_2 + \frac{3680}{513}\, k_3 - \frac{845}{4104}\, k_4\right)$$

$$k_6 = \Delta t\, f\left(t^n + \frac{\Delta t}{2}, y^n - \frac{8}{27}\, k_1 + 2k_2 - \frac{3544}{2565}\, k_3 + \frac{1859}{4104}\, k_4 - \frac{11}{40}\, k_5\right)$$

$$y^{n+1} = y^n + \frac{25}{216}\, k_1 + \frac{1408}{2565}\, k_3 + \frac{2197}{4104}\, k_4 - \frac{1}{5}\, k_5$$

$$z^{n+1} = y^n + \frac{16}{135}\, k_1 + \frac{6656}{12825}\, k_3 + \frac{28561}{56430}\, k_4 - \frac{9}{50}\, k_5 + \frac{2}{55}\, k_6$$

The value of $y^{n+1} - z^{n+1}$ is an estimate of the error in y^{n+1} and can be used in step-size control schemes.

Usually one would use a high-order method to achieve high accuracy. The Runge-Kutta-Feldberg method is popular because it is high order and does not require a starting method (as does an Adams-Bashforth method). However, it does require four function evaluations per time step, or four times as many as a fourth-order Adams-Bashforth method. For problems in which the function evaluations are a significant portion of the calculation time, this might be important. Given the speed of present-day computers and the widespread availability of microcomputers (which can be run while you are doing something else, if need be), the efficiency of the methods is most important only for very large problems that are going to be solved many times. For other problems, the most important criterion for choosing a method is probably the time the user spends setting up the problem.

The stability limits for the explicit methods are based on the largest eigenvalue of the linearized system of equations (see Ref. 106). For linear problems, the eigenvalues do not change, so that the stability and oscillation limits must be satisfied for *every* eigenvalue of the matrix **A**. When solving nonlinear problems, the equations are linearized about the solution at the local time, and the analysis applies for small changes in time, after which a new analysis about the new solution must be made. Thus, for nonlinear problems, the eigenvalues keep changing, and the largest stable time step changes, too. The stability limits are:

Euler method, $\lambda\,\Delta t \leq 2$
Runge-Kutta, 2nd order, $\lambda\,\Delta t < 2$
Runge-Kutta-Feldberg, $\lambda\,\Delta t < 3.0$

Richardson extrapolation can be used to improve the accuracy of a method. Suppose we step forward one step Δt with a pth-order method. Then redo the problem, this time stepping forward from the same initial point, but in two steps of length $\Delta t/2$, thus ending at the same point. Call the solution of the one-step calculation y_1 and the solution of the two-step calculation y_2. Then an improved solution at the new time is given by

$$y = \frac{2^p y_2 - y_1}{2^p - 1}$$

This gives a good estimate provided Δt is small enough that the method is truly convergent with order p. This process can also be repeated in the same way Romberg's method was used for quadrature.

The error term in the various methods can be used to deduce a step size that will give a user-specified accuracy. Most packages today are based on a user-specified tolerance; the step-size is changed during the calculation to achieve that accuracy. The accuracy itself is not guaranteed, but it improves as the tolerance is decreased.

Implicit Methods By using different interpolation formulas involving y^{n+1}, it is possible to derive implicit integration methods. Implicit methods result in a nonlinear equation to be solved for y^{n+1} so that iterative methods must be used. The backward Euler method is a first-order method.

$$y^{n+1} = y^n + \Delta t\, f(y^{n+1})$$

Errors are proportional to Δt for small Δt. The trapezoid rule is a second-order method.

$$y^{n+1} = y^n + \frac{\Delta t}{2}\,[f(y^n) + f(y^{n+1})]$$

Errors are proportional to Δt^2 for small Δt. When the trapezoid rule is used with the finite difference method for solving partial differential equations, it is called the Crank-Nicolson method. The implicit methods are stable for any step size but do require the solution of a set of nonlinear equations, which must be solved iteratively. The set of equations can be solved using the successive substitution method or Newton-Raphson method. See Ref. 36 for an application to dynamic distillation problems.

The best packages for stiff equations (see below) use Gear's backward difference formulas. The formulas of various orders are, per Refs. 59 and 117,

(1) $y^{n+1} = y^n + \Delta t\, f(y^{n+1})$

(2) $y^{n+1} = \dfrac{4}{3} y^n - \dfrac{1}{3} y^{n-1} + \dfrac{2}{3} \Delta t\, f(y^{n+1})$

(3) $y^{n+1} = \dfrac{18}{11} y^n - \dfrac{9}{11} y^{n-1} + \dfrac{2}{11} y^{n-2} + \dfrac{6}{11} \Delta t\, f(y^{n+1})$

(4) $y^{n+1} = \dfrac{48}{25} y^n - \dfrac{36}{25} y^{n-1} + \dfrac{16}{25} y^{n-2} - \dfrac{3}{25} y^{n-3} + \dfrac{12}{25} \Delta t\, f(y^{n+1})$

(5) $y^{n+1} = \dfrac{300}{137} y^n - \dfrac{300}{137} y^{n-1} + \dfrac{200}{137} y^{n-2} - \dfrac{75}{137} y^{n-3} + \dfrac{12}{137} y^{n-4}$
$\qquad + \dfrac{60}{137} \Delta t\, f(y^{n+1})$

Stiffness   The concept of stiffness is described for a system of linear equations.

$$\frac{d\mathbf{y}}{dt} = \mathbf{A}\,\mathbf{y}$$

Let λ_i be the eigenvalues of the matrix $\mathbf{A}$ (Ref. 267). Then, per Ref. 181, the stiffness ratio is defined as

$$SR = \frac{\max_i |\mathrm{Re}\,(\lambda_i)|}{\min_i |\mathrm{Re}\,(\lambda_i)|}$$

$SR = 20$ is not stiff, $SR = 10^3$ is stiff, and $SR = 10^6$ is very stiff. If the problem is nonlinear, then the solution is expanded about the current state.

$$\frac{dy_i}{dt} = f_i\,[\mathbf{y}(t^n)] + \sum_{j=1}^{n} \frac{\partial f_i}{\partial y_j}\,[y_j - y_j(t^n)]$$

The question of stiffness then depends on the solution at the current time. Consequently nonlinear problems can be stiff during one time period and not stiff during another. While the chemical engineer may not actually calculate the eigenvalues, it is useful to know that they determine the stability and accuracy of the numerical scheme and the step size used.

Problems are stiff when the time constants for different phenomena have very different magnitudes. Consider flow through a packed bed reactor. The time constants for different phenomena are:
1. Time for device flow-through

$$t_{\text{flow}} = \frac{L}{u} = \frac{\phi A L}{Q}$$

where Q is the volumetric flow rate, A is the cross sectional area, L is the length of the packed bed, and ϕ is the void fraction;
2. Time for reaction

$$t_{r \times n} = \frac{1}{k}$$

where k is a rate constant (time^{-1});
3. Time for diffusion inside the catalyst

$$t_{\text{internal diffusion}} = \frac{\varepsilon R^2}{D_e}$$

where ε is the porosity of the catalyst, R is the catalyst radius, and D_e is the effective diffusion coefficient inside the catalyst;
4. Time for heat transfer is

$$t_{\text{internal heat transfer}} = \frac{R^2}{\alpha} = \frac{\rho_s C_s R^2}{k_e}$$

where ρ_s is the catalyst density, C_s is the catalyst heat capacity per unit mass, k_e is the effective thermal conductivity of the catalyst, and α is the thermal diffusivity. For example, in the model of a catalytic converter for an automobile (Ref. 103), the time constants for internal diffusion was 0.3 seconds; internal heat transfer, 21 seconds; and device flow-through, 0.003 seconds. The device flow-through is so fast that it might as well be instantaneous. The stiffness is approximately 7000.

Implicit methods must be used to integrate the equations. Alternatively, a quasistate model can be developed (Ref. 239).

Differential-Algebraic Systems   Sometimes models involve ordinary differential equations subject to some algebraic constraints. For example, the equations governing one equilibrium stage (as in a distillation column) are

$$M\frac{dx^n}{dt} = V^{n+1} y^{n+1} - L^n x^n - V^n y^n + L^{n-1} x^{n-1}$$
$$x^{n-1} - x^n = E^n(x^{n-1} - x^{\circ,n})$$
$$\sum_{i=1}^{N} x_i = 1$$

where x and y are the mole fraction in the liquid and vapor, respectively; L and V are liquid and vapor flow rates, respectively; M is the holdup; and the superscript is the stage number. The efficiency is E, and the concentration in equilibrium with the vapor is x°. The first equation is an ordinary differential equation for the mass of one component on the stage, while the third equation represents a constraint that the mass fractions add to one. This is a differential-algebraic system of equations.

Differential-algebraic equations can be written in the general notation

$$F\left(t, y, \frac{dy}{dt}\right) = 0$$

To solve the general problem using the backward Euler method, replace the nonlinear differential equation with the nonlinear algebraic equation for one step.

$$F\left(t, y^{n+1}, \frac{y^{n+1} - y^n}{\Delta t}\right) = 0$$

This equation must be solved for y^{n+1}. The Newton-Raphson method can be used, and if convergence is not achieved within a few iterations, the time step can be reduced and the step repeated. In actuality, the higher-order backward-difference Gear methods are used in DASSL (Ref. 224).

Differential-algebraic systems are more complicated than differential systems because the solution may not always be defined. Pontelides et al. (Ref. 226) introduced the term *index* to identify the possible problems. The index is defined as the minimum number of times the equations need to be differentiated with respect to time to convert the system to a set of ordinary differential equations. These higher derivatives may not exist, and the process places limits on which variables can be given initial values. Sometimes the initial values must be constrained by the algebraic equations (Ref. 226). For a differential-algebraic system modeling a distillation tower, Ref. 226 shows that the index depends on the specification of pressure for the column. Byrne and Ponzi (Ref. 58) also list several chemical engineering examples of differential-algebraic systems and solve one involving two-phase flow.

Computer Software   Efficient computer packages are available for solving ordinary differential equations as initial value problems. The packages are widely available and good enough that most chemical engineers use them and do not write their own. Here we discuss three of them: RKF45, LSODE, and EPISODE. In each of the packages, the user specifies the differential equation to be solved and a desired error criterion. The package then integrates in time and adjusts the step size to achieve the error criterion within the limitations imposed by stability.

A popular explicit, Runge-Kutta package is RKF45, developed by Forsythe et al. (Ref. 111). The method is based on the Runge-Kutta-Feldberg formulas. Notice there that an estimate of the truncation error at each step is available. Then the step size can be reduced until this estimate is below the user-specified tolerance. The method is thus automatic, and the user is assured of the results. Note, however, that the tolerance is set on the local truncation error, namely from one step to another, whereas the user is usually interested in the global truncation error, or the error after several steps. The global error is generally made smaller by making the tolerance smaller, but the absolute accuracy is not the same as the tolerance. If the problem is stiff, then very small step sizes are used; the computation becomes very lengthy. The RKF45 code discovers this and returns control to the user with a message indicating the problem is too hard to solve with RKF45.

A popular implicit package is LSODE, a version of Gear's method (Ref. 117) written by Alan Hindmarsh at Lawrence Livermore Laboratory (Ref. 148). In this package, the user specifies the differential equation to be solved and the tolerance desired. Now the method is implicit and therefore stable for any step size. The accuracy may not be acceptable, however, and sets of nonlinear equations must be solved. Thus, in practice the step size is limited but not nearly so much as in the Runge-Kutta methods. In these packages, both the step size and order of the method are adjusted by the package. Suppose we are calculating with a kth order method. The truncation error is determined by the $(k + 1)$th order derivative. This is estimated using difference formulas and the values of the right-hand sides at previous times. An estimate is also made for the kth and $(k + 2)$th derivative. Then it is possible to estimate the error in a $(k - 1)$th order method, a kth order method, and a $(k + 1)$th order method. Furthermore, the step size needed to satisfy the tolerance with each of these methods can be determined. Then we can choose the method and step size for the next step that achieves the biggest step, with appropriate adjustments due to the different work required for each order. The package generally starts with a very small step size and a first-order method, the backward Euler method. Then it integrates along, adjusting the order up (and later down) depending on the error estimates. The user is thus assured that the local truncation error meets the tolerance. There is a further difficulty, since the set of nonlinear equations must be solved. Usually a good guess of the solution is available, since the solution is evolving in time and past history can be extrapolated. Thus, the Newton-Raphson method will usually converge. The package protects itself, though, by only doing three iterations. If convergence is not reached within this many iterations, then the step size is reduced and the calculation is redone for that time step. The convergence theorem for the Newton-Raphson method (p. 3-50) indicates that the method will converge if the step size is small enough. Thus the method is guaranteed to work. Further economies are possible. The Jacobian needed in the Newton-Raphson method can be fixed over several time steps. Then, if the iteration does not converge, the Jacobian can be reevaluated at the current time-step. If the iteration still does not converge, then the step-size is reduced and a new Jacobian is evaluated. Also the successive substitution method can be used, which is even faster, except that it may not converge. However, it, too, will converge if the time step is small enough.

Comparisons of the methods and additional details are provided for chemical engineering problems by Refs. 59 and 106. Generally, the Runge-Kutta methods give extremely good accuracy, especially when the step size is kept small for stability reasons. When the computation time is comparable for LSODE and RKF45, the RKF45 package generally gives much more accurate results. The RKF45 package is unsuitable, however, for many chemical reactor problems because they are so stiff. Generally, though, standard packages must have a high-order explicit method (usually a version of Runge-Kutta) and a multi-step, implicit method (usually a version of GEAR, EPISODE, or LSODE). The package DASSL (Ref. 224) uses similar principles to solve the differential-algebraic systems.

The software described here is available by electronic mail over the Internet. Sending the message

```
send index to
netlib@ornl.gov
```

will retrieve an index and descriptions of how to obtain the software.

Stability, Bifurcations, Limit Cycles   Some aspects of this subject involve the solution of nonlinear equations; other aspects involve the integration of ordinary differential equations; applications include chaos and fractals as well as unusual operation of some chemical engineering equipment. Ref. 176 gives an excellent introduction to the subject and the details needed to apply the methods. Ref. 66 gives more details of the algorithms. A concise survey with some chemical engineering examples is given in Ref. 91. Bifurcation results are closely connected with stability of the steady states, which is essentially a transient phenomenon.

Sensitivity Analysis   When solving differential equations, it is frequently necessary to know the solution as well as the sensitivity of the solution to the value of a parameter. Such information is useful when doing parameter estimation (to find the best set of parameters for a model) and for deciding if a parameter needs to be measured accurately. See Ref. 105.

ORDINARY DIFFERENTIAL EQUATIONS-BOUNDARY VALUE PROBLEMS

Diffusion problems in one dimension lead to boundary value problems. The boundary conditions are applied at two different spatial locations: at one side the concentration may be fixed and at the other side the flux may be fixed. Because the conditions are specified at two different locations, the problems are not initial value in character. It is not possible to begin at one position and integrate directly because at least one of the conditions is specified somewhere else and there are not enough conditions to begin the calculation. Thus, methods have been developed especially for boundary value problems.

Shooting Methods   The first method is one that utilizes the techniques for initial value problems but allows for an iterative calculation to satisfy all the boundary conditions. Consider the nonlinear boundary value problem

$$\frac{d^2y}{dx^2} = f\left(x, y, \frac{dy}{dx}\right), \quad y(0) = \alpha, \quad y(1) = \beta$$

Convert this second-order equation into two first-order equations along with the boundary conditions written to include a parameter s to represent the unknown value of $v(0) = dy/dx(0)$.

$$\frac{dy}{dx} = v, \quad \frac{dv}{dx} = f(x, y, v), \quad y(0) = \alpha, \quad v(0) = s$$

The parameter s is chosen so that the last boundary condition is satisfied: $y(1) = \beta$. Define the function

$$\chi(s) = y(1, s) - \beta$$

and iterate on s to make $\chi(s) = 0$. Note that the condition at $x = 0$ is satisfied for any s, the differential equation is satisfied by the integration routine, and only the last boundary condition is yet to be satisfied. Both successive substitution and the Newton-Raphson methods can be used. The technique can be used when the boundary conditions are more general and convergence can be proved (see Refs. 106 and 167). Computer software exists: the IMSL program DTPTB uses DVERK, which employs Runge-Kutta integration to integrate the ordinary differential equations (Ref. 55).

Finite Difference Method   To apply the finite difference method, we first spread grid points through the domain. Figure 3-49 shows a uniform mesh of n points (nonuniform meshes are possible, too). The unknown, here $c(x)$, at a grid point x_i is assigned the symbol $c_i = c(x_i)$. The finite difference method can be derived easily by using a Taylor expansion of the solution about this point. Expressions for the derivatives are:

$$\frac{dc}{dx}\bigg|_i = \frac{c_{i+1} - c_i}{\Delta x} - \frac{d^2c}{dx^2}\bigg|_i \frac{\Delta x}{2} + \cdots, \quad \frac{dc}{dx}\bigg|_i = \frac{c_i - c_{i-1}}{\Delta x} + \frac{d^2c}{dx^2}\bigg|_i \frac{\Delta x}{2} + \cdots$$

$$\frac{dc}{dx}\bigg|_i = \frac{c_{i+1} - c_{i-1}}{2\Delta x} - \frac{d^3c}{dx^3}\bigg|_i \frac{\Delta x^2}{3!}$$

The truncation error in the first two expressions is proportional to Δx, and the methods are said to be first-order. The truncation error in the third expression is proportional to Δx^2, and the method is said to be second-order. Usually the last equation is used to insure the best accuracy. The finite difference representation of the second derivative is:

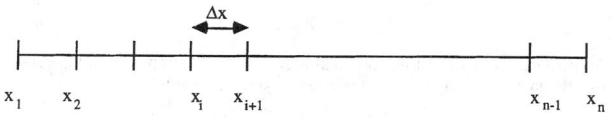

FIG. 3-49   Finite difference mesh; Δx uniform.

$$\frac{d^2c}{dx^2}\Big|_i = \frac{c_{i+1} - 2c_i + c_{i-1}}{\Delta x^2} - \frac{d^4c}{dx^4}\Big|_i \frac{2\Delta x^2}{4!} + \cdots$$

The truncation error is proportional to Δx^2. To solve a differential equation, it is evaluated at a point i and then these expressions are inserted for the derivatives.

Example Consider the equation for convection, diffusion, and reaction in a tubular reactor.

$$\frac{1}{\text{Pe}} \frac{d^2c}{dx^2} - \frac{dc}{dx} = \text{Da } R(c)$$

The finite difference representation is

$$\frac{1}{\text{Pe}} \frac{c_{i+1} - 2c_i + c_{i-1}}{\Delta x^2} - \frac{c_{i+1} - c_{i-1}}{2\Delta x} = \text{Da } R(c_i)$$

This equation is written for $i = 2$ to $n - 1$, or the internal points. The equations would then be coupled but would also involve the values of c_1 and c_n, as well. These are determined from the boundary conditions.

If the boundary condition involves a derivative, it is important that the derivatives be evaluated using points that exist. Three possibilities exist:

$$\frac{dc}{dx}\Big|_1 = \frac{c_2 - c_1}{\Delta x}$$

$$\frac{dc}{dx}\Big|_1 = \frac{-3c_1 + 4c_2 - c_3}{2\Delta x}$$

The third alternative is to add a false point, outside the domain, as $c_0 = c(x = -\Delta x)$.

$$\frac{dc}{dx}\Big|_1 = \frac{c_2 - c_0}{2\Delta x}$$

Since this equation introduces a new variable, c_0, another equation is needed and is obtained by writing the finite difference equation for $i = 1$, too.

The sets of equations can be solved using the Newton-Raphson method. The first form of the derivative gives a tridiagonal system of equations, and the standard routines for solving tridiagonal equations suffice. For the other two options, some manipulation is necessary to put them into a tridiagonal form (see Ref. 105).

Frequently, the transport coefficients, such as diffusion coefficient or thermal conductivity, depend on the dependent variable, concentration, or temperature, respectively. Then the differential equation might look like

$$\frac{d}{dx}\left(D(c) \frac{dc}{dx}\right) = 0$$

This could be written as two equations.

$$-\frac{dJ}{dx} = 0 \qquad J = -D(c) \frac{dc}{dx}$$

Because the coefficient depends on c, the equations are more complicated. A finite difference method can be written in terms of the fluxes at the midpoints, $i + 1/2$.

$$-\frac{J_{i+1/2} - J_{i-1/2}}{\Delta x} = 0 \qquad J_{i+1/2} = -D(c_{i+1/2}) \frac{c_{i+1} - c_i}{\Delta x}$$

These are combined to give the complete equation.

$$\frac{D(c_{i+1/2})(c_{i+1} - c_i) - D(c_{i-1/2})(c_i - c_{i-1})}{\Delta x^2} = 0$$

This represents a set of nonlinear algebraic equations that can be solved with the Newton-Raphson method. However, in this case, a viable iterative strategy is to evaluate the transport coefficients at the last value and then solve

$$\frac{D(c_{i+1/2}^k)(c_{i+1}^{k+1} - c_i^{k+1}) - D(c_{i-1/2}^k)(c_i^{k+1} - c_{i-1}^{k+1})}{\Delta x^2} = 0$$

The advantage of this approach is that it is easier to program than a full Newton-Raphson method. If the transport coefficients do not vary radically, then the method converges. If the method does not converge, then it may be necessary to use the full Newton-Raphson method.

There are three common ways to evaluate the transport coefficient at the midpoint. The first one uses the transport coefficient evaluated at the average value of the solutions on either side.

$$D(c_{i+1/2}) \approx D\left[\frac{1}{2}(c_{i+1} + c_i)\right]$$

The truncation error of this approach is Δx^2 (Ref. 106). The second approach uses the average of the transport coefficients on either side.

$$D(c_{i+1/2}) \approx \frac{1}{2}[D(c_{i+1}) + D(c_i)]$$

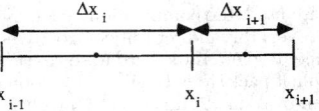

FIG. 3-50 Finite difference grid with variable spacing.

The truncation error of this approach is also Δx^2 (Ref. 106). The third approach uses an "upstream" transport coefficient.

$$D(c_{i+1/2}) \approx D(c_{i+1}), \quad \text{when } D(c_{i+1}) > D(c_i)$$

$$D(c_{i+1/2}) \approx D(c_i), \quad \text{when } D(c_{i+1}) < D(c_i)$$

This approach is used when the transport coefficients vary over several orders of magnitude, and the "upstream" direction is defined as the one in which the transport coefficient is larger. The truncation error of this approach is only Δx (Refs. 106 and 107), but this approach is useful if the numerical solutions show unrealistic oscillations.

If the grid spacing is not uniform, the formulas must be revised. The notation is shown in Fig. 3-50. The finite-difference form of the equations is then

$$-\frac{J_{i+1/2} - J_{i-1/2}}{1/2(\Delta x_i + \Delta x_{i+1})} = 0 \qquad J_{i+1/2} = -D_{i+1/2} \frac{c_{i+1} - c_i}{\Delta x_{i+1}}, \quad J_{i-1/2} = -D_{i-1/2} \frac{c_i - c_{i-1}}{\Delta x_i}$$

If average diffusion coefficients are used, then the finite difference equation is as follows.

$$\frac{1}{\Delta x_{i+1} + \Delta x_i}\left[\frac{1}{\Delta x_{i+1}}(D_{i+1} + D_i)(c_{i+1} - c_i) - \frac{1}{\Delta x_i}(D_i + D_{i-1})(c_i - c_{i-1})\right] = 0$$

Rigorous error bounds are discussed for linear ordinary differential equations solved with the finite difference method by Isaacson and Keller (Ref. 107). Computer software exists to solve two-point boundary value problems. The IMSL routine DVCPR uses the finite difference method with a variable step size (Ref. 247). Finlayson (Ref. 106) gives FDRXN for reaction problems.

Example A reaction diffusion problem is solved with the finite difference method.

$$\frac{d^2c}{dx^2} = \phi^2 c, \quad \frac{dc}{dx}(0) = 0, \quad c(1) = 1$$

The solution is derived for $\phi = 2$. It is solved several times, first with two intervals and three points (at $x = 0, 0.5, 1$), then with four intervals, then with eight intervals. The reason is that when an exact solution is not known, one must use several Δx and see that the solution converges as Δx approaches zero. With two intervals, the equations are as follows. The points are $x_1 = 0$, $x_2 = 0.5$, and $x_3 = 1.0$; and the solution at those points are c_1, c_2, and c_3, respectively. A false boundary is used at $x_0 = -0.5$.

$$\frac{c_0 - c_2}{2\Delta x} = 0, \quad \frac{c_0 - 2c_1 + c_2}{\Delta x^2} - \phi^2 c_1 = 0, \quad \frac{c_1 - 2c_2 + c_3}{\Delta x^2} - \phi^2 c_2 = 0, \quad c_3 = 1$$

The solution is $c_1 = 0.2857$, $c_2 = 0.4286$, and $c_3 = 1.0$. Since the solution is only an approximation and approaches the exact solution only as Δx approaches zero, it is necessary to find out if Δx is small enough to be considered zero. This is done by solving the problem again with more grid points. The value of concentration at $x = 0$ takes the following values for different Δx. These values are extrapolated using the Richardson extrapolation technique to give $c(0) = 0.265826$. Using this value as the best estimate of the exact solution, the errors in the solution are tabulated versus Δx. Clearly the errors go as Δx^2 (decreasing by a factor of 4 when Δx decreases by a factor of 2), thus validating the solution. The exact solution (given below) is 0.265802.

$n - 1$	Δx	$c(0)$
2	0.5	0.285714
4	0.25	0.271043
8	0.125	0.267131

$n - 1$	Δx	Error in $c(0)$
2	0.5	0.01989
4	0.25	0.00521
8	0.125	0.00130

Finite Difference Methods Solved with Spreadsheets A convenient way to solve the finite difference equations for simple

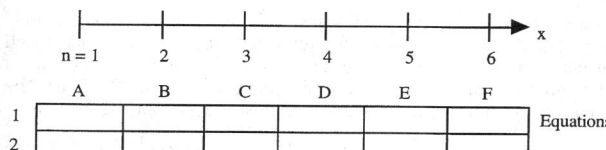

FIG. 3-51   Finite difference method using spreadsheets.

problems is to use a computer spreadsheet. The equations for the problem solved in the example can be cast into the following form

$$c_1 = \frac{2c_2}{2 + \phi^2 \Delta x^2}$$

$$c_i = \frac{c_{i+1} + c_{i-1}}{2 + \phi^2 \Delta x^2}$$

$$c_{n+1} = 1$$

Let us solve the problem using 6 nodes, or 5 intervals. Then the connection between the cell in the spreadsheet and the nodal value is shown in Fig. 3-51. The following equations are placed into the various cells.

```
A1: = 2*B1/(2.+(phi*dx)**2)
B1: = (A1 + C1)/(2.+(phi*dx)**2)
F1: = 1.
```

The equation in cell B1 is copied into cells C1 though E1. Then turn on the iteration scheme in the spreadsheet and watch the solution converge. Whether or not convergence is achieved can depend on how you write the equations, so some experimentation may be necessary. Theorems for convergence of the successive substitution method are useful in this regard.

Orthogonal Collocation   The orthogonal collocation method has found widespread application in chemical engineering, particularly for chemical reaction engineering. In the collocation method, the dependent variable is expanded in a series of orthogonal polynomials, and the differential equation is evaluated at certain collocation points. The collocation points are the roots to an orthogonal polynomial, as first used by Lanczos (Refs. 182 and 183). A major improvement was proposed by Villadsen and Stewart (Refs. 288 and 289), who proposed that the entire solution process be done in terms of the solution at the collocation points rather than the coefficients in the expansion. This method is especially useful for reaction-diffusion problems that frequently arise when modeling chemical reactors. It is highly efficient when the solution is smooth, but the finite difference method is preferred when the solution changes steeply in some region of space. See Ref. 105 for comparisons.

Galerkin Finite Element Method   In the finite element method, the domain is divided into elements and an expansion is made for the solution on each finite element. In the Galerkin finite element method an additional idea is introduced: the Galerkin method is used to solve the equation. The Galerkin method is explained before the finite element basis set is introduced, using the equations for reaction and diffusion in a porous catalyst pellet.

$$\frac{d^2c}{dx^2} = \phi^2 R(c)$$

$$\frac{dc}{dx}(0) = 0, \quad c(1) = 1$$

The unknown solution is expanded in a series of known functions $\{b_i(x)\}$ with unknown coefficients $\{a_i\}$.

$$c(x) = \sum_{i=1}^{NT} a_i b_i(x)$$

The trial solution is substituted into the differential equation to obtain the residual.

$$\text{Residual} = \sum_{i=1}^{NT} a_i \frac{d^2 b_i}{dx^2} - \phi^2 R\left[\sum_{i=1}^{NT} a_i b_i(x)\right]$$

The residual is then made orthogonal to the set of basis functions.

$$\int_0^1 b_j(x)\left\{\sum_{i=1}^{NT} a_i \frac{d^2 b_i}{dx^2} - \phi^2 R\left[\sum_{i=1}^{NT} a_i b_i(x)\right]\right\} dx = 0 \qquad j = 1, \ldots, NT$$

This is the process that makes the method a Galerkin method. The basis for the orthogonality condition is that a function that is made orthogonal to each member of a complete set is then zero. The residual is being made orthogonal, and if the basis functions are complete and you use infinitely many of them, then the residual is zero. Once the residual is zero, the problem is solved.

This equation is integrated by parts to give the following equation

$$-\sum_{i=1}^{NT} \int_0^1 \frac{db_j}{dx} \frac{db_i}{dx} dx a_i = \phi^2 \int_0^1 b_j(x) R\left[\sum_{i=1}^{NT} a_i b_i(x)\right] dx$$

$$j = 1, \ldots, NT - 1 \quad (3\text{-}76)$$

This equation defines the Galerkin method and a solution that satisfies this equation (for all $j = 1, \ldots, \infty$) is called a weak solution. For an approximate solution, the equation is written once for each member of the trial function, $j = 1, \ldots, NT - 1$, and the boundary condition is applied.

$$\sum_{i=1}^{NT} a_i b_i(1) = c_B$$

The Galerkin finite element method results when the Galerkin method is combined with a finite element trial function. The domain is divided into elements separated by nodes, as in the finite difference method. The solution is approximated by a linear (or sometimes quadratic) function of position within the element. These approximations are substituted into Eq. (3-76) to provide the Galerkin finite element equations. The element integrals are defined as

$$B_{JI}^e = -\frac{1}{\Delta x_e} \int_0^1 \frac{dN_J}{du} \frac{dN_I}{du} du, \quad F_J^e = \phi^2 \Delta x_e \int_0^1 N_J(u) R\left[\sum_{I=1}^{NP} c_I^e N_I(u)\right] du$$

and the entire method can be written in the following compact notation:

$$\sum_e B_{JI}^e c_I^e = \sum_e F_J^e$$

The matrices for various terms are given in the table. This equation can also be written in the form

$$\mathbf{AAc} = \mathbf{f}$$

where the matrix $\mathbf{AA}$ is sparse; if linear elements are used, the matrix is tridiagonal. Once the solution is found, the solution at any point can be recovered from

$$c^e(u) = c_{I=1}^e (1 - u) + c_{I=2}^e u$$

for linear elements.

Element Matrices for Galerkin Method with Linear Shape Functions

$$N_1 = 1 - u, \; N_2 = u, \; \frac{dN_1}{du} = -1, \; \frac{dN_2}{du} = 1$$

$$\int_0^1 \frac{dN_J}{du} \frac{dN_I}{du} du = \begin{pmatrix} 1 & -1 \\ -1 & 1 \end{pmatrix}, \int_0^1 N_J \frac{dN_I}{du} du = \begin{pmatrix} -\frac{1}{2} & \frac{1}{2} \\ -\frac{1}{2} & \frac{1}{2} \end{pmatrix}$$

$$\int_0^1 N_J N_I du = \begin{pmatrix} \frac{1}{3} & \frac{1}{6} \\ \frac{1}{6} & \frac{1}{3} \end{pmatrix}, \int_0^1 N_J du = \begin{pmatrix} \frac{1}{2} \\ \frac{1}{2} \end{pmatrix}, \int_0^1 N_J u \, du = \begin{pmatrix} \frac{1}{6} \\ \frac{1}{3} \end{pmatrix}$$

Example   Solve the specified problem when $\phi = 2$, the rate expression is linear, $R(c) = c$, and the boundary condition is 1.0. The Galerkin finite element method is used with $\Delta x = 0.33333$. The element nodes are at $x = 0$, 0.3333, 0.6667, and 1.0. The solution at $x = 1.0$ is $c_4 = 1.0$. The Galerkin equations for one element are obtained from the table.

$$-\begin{pmatrix} 1 & -1 \\ -1 & 1 \end{pmatrix} = \phi^2 \Delta x^2 \begin{pmatrix} \frac{1}{3} & \frac{1}{6} \\ \frac{1}{6} & \frac{1}{3} \end{pmatrix}$$

When these are summed over all elements the result is

$$
\begin{bmatrix}
-1 & 1 & 0 & 0 \\
1 & -1-1 & 1 & 0 \\
0 & 1 & -1-1 & 1 \\
0 & 0 & 0 & 1
\end{bmatrix}
\begin{bmatrix}
c_1 \\ c_2 \\ c_3 \\ c_4
\end{bmatrix}
= \frac{4}{9}
\begin{bmatrix}
\frac{1}{3} & \frac{1}{6} & 0 & 0 \\
\frac{1}{6} & \frac{1}{3}+\frac{1}{3} & \frac{1}{6} & 0 \\
0 & \frac{1}{6} & \frac{1}{3}+\frac{1}{3} & \frac{1}{6} \\
0 & 0 & 0 & 0
\end{bmatrix}
\begin{bmatrix}
c_1 \\ c_2 \\ c_3 \\ c_4
\end{bmatrix}
+
\begin{bmatrix}
0 \\ 0 \\ 0 \\ 1
\end{bmatrix}
$$

After rearrangement this is

$$
\begin{bmatrix}
{}^{31}\!/_{27} & {}^{-25}\!/_{27} & 0 \\
{}^{-25}\!/_{27} & {}^{62}\!/_{27} & {}^{-25}\!/_{27} \\
0 & {}^{-25}\!/_{27} & {}^{62}\!/_{27}
\end{bmatrix}
\begin{bmatrix}
c_1 \\ c_2 \\ c_3
\end{bmatrix}
=
\begin{bmatrix}
0 \\ 0 \\ {}^{25}\!/_{27}
\end{bmatrix}
$$

The solution is $c_1 = 0.2560$, $c_2 = 0.3174$, $c_3 = 0.5312$, and $c_4 = 1$. The exact solution is derived using the section entitled "Ordinary Differential Equations: Linear Differential Equations with Constant Coefficients."

$$
c = \frac{e^{2x} + e^{-2x}}{e^2 + e^{-2}} = \frac{\cosh(2x)}{\cosh(2)}
$$

The values of the exact solution at the same finite element nodes are $c_1 = 0.2658$, $c_2 = 0.3271$, $c_3 = 0.5392$, and $c_4 = 1$, indicating that the three-element finite element solution is accurate within 3 percent. When the exact solution is not known, the problem must be solved several times, each with a different number of elements, so that convergence is seen as the number of elements increases.

Cubic B-Splines   Cubic B-splines can also be used to solve differential equations (Refs. 105 and 266).

Adaptive Meshes   In many two-point boundary value problems, the difficulty in the problem is the formation of a boundary layer region, or a region in which the solution changes very dramatically. In such cases, it is prudent to use small mesh spacing there, either with the finite difference method or the finite element method. If the region is known *a priori*, small mesh spacings can be assumed at the boundary layer. If the region is not known, though, other techniques must be used. These techniques are known as adaptive mesh techniques. The mesh size is made small where some property of the solution is large. For example, if the truncation error of the method is nth order, then the nth-order derivative of the solution is evaluated and a small mesh is used where it is large. Alternatively, the residual (the differential equation with the numerical solution substituted into it) can be used as a criterion. See Refs. 21 and 107. It is also possible to define the error that is expected from a method one order higher and one order lower. Then a decision about whether to increase or decrease the order of the method can be made, taking into account the relative work of the different orders. This provides a method of adjusting both the mesh spacing (Δx, or sometimes called h) and the degree of polynomial (p). Such methods are called h-p methods.

Singular Problems and Infinite Domains   If the solution being sought has a singularity, it may be difficult to find a good numerical solution. Sometimes even the location of the singularity may not be known (Ref. 11). One method of solving such problems is to refine the mesh near the singularity, relying on the better approximation due to a smaller Δx. Another approach is to incorporate the singular trial function into the approximation. Thus, if the solution approaches $f(x)$ as x goes to zero and $f(x)$ becomes infinite, one may define a new variable $u(x) = y(x) - f(x)$ and derive an equation for u. The differential equation is more complicated, but the solution is better near the singularity. See Refs. 39 and 231.

Sometimes the domain is semi-infinite, as in boundary layer flow. The domain can be transformed from the x domain ($0-\infty$) to the η domain ($1-0$) using the transformation $\eta = \exp(-x)$. Another approach is to use a variable mesh, perhaps with the same transformation. For example, use $\eta = \exp(-\beta x)$ and a constant mesh size in η; the value of β is found experimentally. Still another approach is to solve on a finite mesh in which the last point is far enough away that its location does not influence the solution (Ref. 59). A location that is far enough away must be found by trial and error.

NUMERICAL SOLUTION OF INTEGRAL EQUATIONS

In this subsection is considered a method of solving numerically the Fredholm integral equation of the second kind:

$$
u(x) = f(x) + \lambda \int_a^b k(x, t)u(t)\, dt \qquad \text{for } u(x) \tag{3-77}
$$

The method discussed arises because a definite integral can be closely approximated by any of several numerical integration formulas (each of which arises by approximating the function by some polynomial over an interval). Thus the definite integral in Eq. (3-77) can be replaced by an integration formula, and Eq. (3-77) may be written

$$
u(x) = f(x) + \lambda(b - a)\left[\sum_{i=1}^{n} c_i k(x, t_i)u(t_i)\right] \tag{3-78}
$$

where $t_1, \ldots, t_n$ are points of subdivision of the t axis, $a \le t \le b$, and the c's are coefficients whose values depend upon the type of numerical integration formula used. Now Eq. (3-78) must hold for all values of x, $a \le x \le b$; so it must hold for $x = t_1$, $x = t_2$, $\ldots$, $x = t_n$. Substituting for x successively $t_1, t_2, \ldots, t_n$ and setting $u(t_i) = u_i$, $f(t_i) = f_i$, we get n linear algebraic equations for the n unknowns $u_1, \ldots, u_n$. That is,

$$
u_i = f_i + (b - a)[c_1 k(t_i, t_1)u_1 + c_2 k(t_i, t_2)u_2
$$
$$
+ \cdots + c_n k(t_i, t_n)u_n] \qquad i = 1, 2, \ldots, n
$$

These u_j may be solved for by the methods under "Numerical Solution of Linear Equations and Associated Problems" and substituted into Eq. (3-78) to yield an approximate solution for Eq. (3-77).

Example   Solve numerically $u(x) = x + \frac{1}{3}\int_0^1 (t + x)u(t)\, dt$. In this example $a = 0$, $b = 1$. Take $n = 3$, $t_1 = 0$, $t_2 = \frac{1}{2}$, $t_3 = 1$. Then Eq. (3-78) takes the form (for which we have used the parabolic rule)

$$
u(x) = x + (\tfrac{1}{3})\frac{\frac{1}{2}}{3}[(t_1 + x)u(t_1) + 4(t_2 + x)u(t_2) + (t_3 + x)u(t_3)]
$$
$$
= x + (1/18)[(t_1 + x)u(t_1) + 4(t_2 + x)u(t_2) + (t_3 + x)u(t_3)]
$$

This must hold for all x, $0 \le x \le 1$. Here $t_1 = 0$, $t_2 = \frac{1}{2}$, and $t_3 = 1$. Evaluate at $x = t_i$.

$$
u(t_1) = t_1 + \tfrac{1}{18}[2t_1 u(t_1) + 4(t_2 + t_1)u(t_2) + (t_3 + t_1)u(t_3)]
$$
$$
u(t_2) = t_2 + \tfrac{1}{18}[(t_1 + t_2)u(t_1) + 4(2t_2)u(t_2) + (t_3 + t_2)u(t_3)]
$$
$$
u(t_3) = t_3 + \tfrac{1}{18}[(t_1 + t_3)u(t_1) + 4(t_2 + t_3)u(t_2) + 2t_3 u(t_3)]
$$

By setting in the values of t_1, t_2, t_3 and $u(t_i) = u_i$,

$$
18u_1 - 2u_2 - u_3 = 0
$$
$$
-u_1 + 28u_2 - 3u_3 = 18
$$
$$
-u_1 - 6u_2 + 16u_3 = 18
$$

with the solution $u_1 = {}^{12}\!/_{71}$, $u_2 = {}^{57}\!/_{71}$, $u_3 = {}^{102}\!/_{71}$. Thus

$$
u(x) = x + \tfrac{1}{18}[x\,{}^{12}\!/_{71} + 4(\tfrac{1}{2} + x)\,{}^{57}\!/_{71} + (1 + x)\,{}^{102}\!/_{71}]
$$
$$
= {}^{90}\!/_{71}x + {}^{12}\!/_{71}
$$

Because of the work involved in solving large systems of simultaneous linear equations it is desirable that only a small number of u's be computed. Thus the gaussian integration formulas are useful because of the economy they offer. See references on numerical solutions of integral equations.

Solutions for Volterra equations are done in a similar fashion, except that the solution can proceed point by point, or in small groups of points depending on the quadrature scheme. See Refs. 105 and 195. There are methods that are analogous to the usual methods for integrating differential equations (Runge-Kutta, predictor-corrector, Adams methods, etc.). Explicit methods are fast and efficient until the time step is very small to meet the stability requirements. Then implicit methods are used, even though sets of simultaneous algebraic equations must be solved. The major part of the calculation is the evaluation of integrals, however, so that the added time to solve the algebraic equations is not excessive. Thus, implicit methods tend to be preferred (Ref. 195). Volterra equations of the first kind are not well posed, and small errors in the solution can have disastrous consequences. The boundary element method uses Green's functions and integral equations to solve differential equations (Refs. 45 and 200).

MONTE CARLO SIMULATIONS

Some physical problems, such as those involving interaction of molecules, are usually formulated as integral equations. Monte Carlo methods are especially well-suited to their solution. This section cannot give a comprehensive treatment of such methods, but their use in calculating the value of an integral will be illustrated. Suppose we wish to calculate the integral

$$G = \int_{\Omega_0} g(x)f(x)\, dx$$

where the distribution function $f(x)$ satisfies:

$$f(x) \geq 0, \quad \int_{\Omega_0} f(x)\, dx = 1$$

The distribution function $f(x)$ can be taken as constant; for example, $1/\Omega_0$. We choose variables $x_1, x_2, \ldots, x_N$ randomly from $f(x)$ and form the arithmetic mean

$$G_N = \frac{1}{N} \sum_i g(x_i)$$

The quantity G_N is an estimation of G, and the fundamental theorem of Monte Carlo guarantees that the expected value of G_N is G, if G exists (Ref. 161). The error in the calculation is given by

$$\varepsilon = \frac{\sigma_1}{N^{1/2}}$$

where σ_1^2 is calculated from

$$\sigma_1^2 = \int_{\Omega_0} g^2(x)\, f(x)\, dx - G^2$$

Thus the number of terms needed to achieve a specified accuracy can be calculated once an estimate of σ_1^2 is known.

$$N = \frac{\sigma_1^2}{\varepsilon}$$

Various methods, such as influence sampling, can be used to reduce the number of calculations needed (Ref. 161).

NUMERICAL SOLUTION OF PARTIAL DIFFERENTIAL EQUATIONS

Parabolic Equations in One Dimension   By combining the techniques applied to initial value problems and boundary value problems it is possible to easily solve parabolic equations in one dimension. The method is often called the method of lines. It is illustrated here using the finite difference method, but the Galerkin finite element method and the orthogonal collocation method can also be combined with initial value methods in similar ways. The analysis is done by example.

Example   Consider the diffusion equation, with boundary and initial conditions.

$$\frac{\partial c}{\partial t} = D \frac{\partial^2 c}{\partial x^2}$$

$$c(x, 0) = 0$$

$$c(0, t) = 1, \quad c(1, t) = 0$$

We denote by c_i the value of $c(x_i, t)$ at any time. Thus, c_i is a function of time, and differential equations in c_i are ordinary differential equations. By evaluating the diffusion equation at the ith node and replacing the derivative with a finite difference equation, the following working equation is derived for each node i, $i = 2, \ldots, n$ (see Fig. 3-52).

$$\frac{dc_i}{dt} = D \frac{c_{i+1} - 2c_i + c_{i-1}}{\Delta x^2}$$

This can be written in the general form of a set of ordinary differential equations by defining the matrix $\mathbf{AA}$.

$$\frac{d\mathbf{c}}{dt} = \mathbf{AA}\mathbf{c}$$

This set of ordinary differential equations can be solved using any of the standard methods, and the stability of the integration of these equations is governed by the largest eigenvalue of $\mathbf{AA}$. If Euler's method is used for integration, the time step is limited by

$$\Delta t \leq \frac{2}{|\lambda|_{\max}}$$

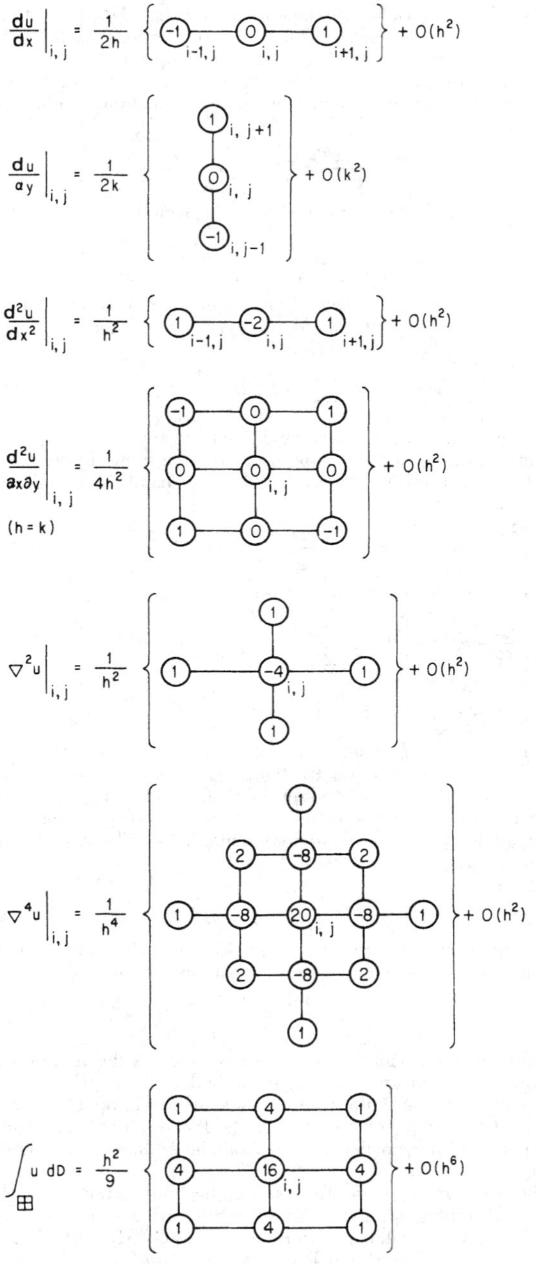

FIG. 3-52   Computational molecules. $h = \Delta x = \Delta y$.

whereas, if the Runge-Kutta-Feldberg method is used, the 2 in the numerator is replaced by 3.0. The largest eigenvalue of $\mathbf{AA}$ is bounded by Gerschgorin's Theorem (Ref. 155, p. 135).

$$|\lambda|_{\max} \leq \max_{2 < j < n} \sum_{i=2}^{n} |\mathbf{AA}_{ji}| = \frac{4D}{\Delta x^2}$$

This gives the well-known stability limit

$$\Delta t \frac{D}{\Delta x^2} \leq \frac{1}{2}$$

The smallest eigenvalue is independent of Δx (it is $D\pi^2/L^2$) so that the ratio of largest to smallest eigenvalue is proportional to $1/\Delta x^2$. Thus, the problem becomes stiff as Δx approaches zero (Ref. 106).

Another way to study the stability of explicit equations is to use the positivity theorem. For Euler's method, the equations can be written in the form

$$\frac{c_i^{n+1} - c_i^n}{\Delta t} = D\,\frac{c_{i+1}^n - 2c_i^n + c_{i-1}^n}{\Delta x^2}$$

where $c_i^n = c(x_i, t^n)$. Then the new value is given by

$$c_i^{n+1} = \frac{D\Delta t}{\Delta x^2}\,c_{i+1}^n + \left(1 - 2\,\frac{D\Delta t}{\Delta x^2}\right)c_i^n + \frac{D\Delta t}{\Delta x^2}\,c_{i-1}^n$$

Theorem. If $c_i^{n+1} = Ac_{i+1}^n + Bc_i^n + Cc_{i-1}^n$ and A, B, and C are positive and $A + B + C \le 1$, then the scheme is stable and the errors die out. Here the theorem requires

$$\left(1 - 2\,\frac{D\Delta t}{\Delta x^2}\right) > 0$$

which gives the same stability condition (Ref. 106).

Implicit methods can also be used. Write a finite difference form for the time derivative and average the right-hand sides, evaluated at the old and new time.

$$\frac{c_i^{n+1} - c_i^n}{\Delta t} = D(1 - \theta)\,\frac{c_{i+1}^n - 2c_i^n + c_{i-1}^n}{\Delta x^2} + D\theta\,\frac{c_{i+1}^{n+1} - 2c_i^{n+1} + c_{i-1}^{n+1}}{\Delta x^2}$$

Now the equations are of the form

$$-\frac{D\Delta t\theta}{\Delta x^2}\,c_{i+1}^{n+1} + \left[1 + 2\,\frac{D\Delta t\theta}{\Delta x^2}\right]c_i^{n+1} - \frac{D\Delta t\theta}{\Delta x^2}\,c_{i-1}^{n+1}$$
$$= c_i^n + \frac{D\Delta t(1 - \theta)}{\Delta x^2}\,(c_{i+1}^n - 2c_i^n + c_{i-1}^n)$$

and require solving a set of simultaneous equations, which have a tridiagonal structure. Using $\theta = 0$ gives the Euler method (as above), $\theta = 0.5$ gives the Crank-Nicolson method, and $\theta = 1$ gives the backward Euler method. The Crank-Nicolson method is also the same as applying the trapezoid rule to do the integration. The stability limit is given by

$$\frac{D\Delta t}{\Delta x^2} \le \frac{0.5}{1 - 2\theta}$$

If the Δt satisfies the following equation, then the solution will not oscillate from node to node (a numerical artifact). See Ref. 106.

$$\frac{D\Delta t}{\Delta x^2} \le \frac{0.25}{1 - \theta}$$

Other methods can be used in space, such as the finite element method, the orthogonal collocation method, or the method of orthogonal collocation on finite elements (see Ref. 106). Spectral methods employ Chebyshev polynomials and the Fast Fourier Transform and are quite useful for hyperbolic or parabolic problems on rectangular domains (Ref. 125).

Packages exist that use various discretizations in the spatial direction and an integration routine in the time variable. PDECOL uses B-splines for the spatial direction and various GEAR methods in time (Ref. 247). PDEPACK and DSS (Ref. 247) use finite differences in the spatial direction and GEARB in time (Ref. 66). REACOL (Ref. 106) uses orthogonal collocation in the radial direction and LSODE in the axial direction, while REACFD uses finite difference in the radial direction; both codes are restricted to modeling chemical reactors.

Elliptic Equations Elliptic equations can be solved with both finite difference and finite element methods. One-dimensional elliptic problems are two-point boundary value problems. Two- and three-dimensional elliptic problems are often solved with iterative methods when the finite difference method is used and direct methods when the finite element method is used. So there are two aspects to consider: how the equations are discretized to form sets of algebraic equations and how the algebraic equations are then solved.

The prototype elliptic problem is steady-state heat conduction or diffusion,

$$k\left(\frac{\partial^2 T}{\partial x^2} + \frac{\partial^2 T}{\partial y^2}\right) = Q$$

possibly with a heat generation term per unit volume, Q. The boundary conditions taken here are $T = f(x, y)$ on the boundary (S) with f a known function. Illustrations are given for constant thermal conductivity k while Q is a known function of position. The finite difference formulation is given using the following nomenclature:

$$T_{i,j} = T(i\Delta x, j\Delta y)$$

The finite difference formulation is then (see Fig. 3-52)

$$\frac{T_{i+1,j} - 2T_{i,j} + T_{i-1,j}}{\Delta x^2} + \frac{T_{i,j+1} - 2T_{i,j} + T_{i,j-1}}{\Delta y^2} = Q_{i,j} \qquad (3\text{-}79)$$

$$T_{i,j} = f(x_i, y_j) \text{ on } S$$

If the boundary is parallel to a coordinate axis any derivative is evaluated as in the section on boundary value problems, using either a one-sided, centered difference or a false boundary. If the boundary is more irregular and not parallel to a coordinate line then more complicated expressions are needed and the finite element method may be the better method.

Equation (3-79) is rewritten in the form

$$2\left(1 + \frac{\Delta x^2}{\Delta y^2}\right)T_{i,j} = T_{i+1,j} + T_{i-1,j} + \frac{\Delta x^2}{\Delta y^2}\,(T_{i,j+1} + T_{i,j-1}) - \Delta x^2\,\frac{Q_{i,j}}{k}$$

The relaxation method solves this equation iteratively.

$$2\left(1 + \frac{\Delta x^2}{\Delta y^2}\right)T_{i,j}^* = T_{i+1,j}^s + T_{i-1,j}^{s+1} + \frac{\Delta x^2}{\Delta y^2}\,(T_{i,j+1}^s + T_{i,j-1}^{s+1}) - \Delta x^2\,\frac{Q_{i,j}}{k}$$

$$T_{i,j}^{s+1} = T_{i,j}^s + \beta(T_{i,j}^* - T_{i,j}^s)$$

If $\beta = 1$, this is the Gauss-Seidel method. If $\beta > 1$, it is overrelaxation; if $\beta < 1$ it is underrelaxation. The value of β may be chosen empirically, $0 < \beta < 2$, but it can be selected theoretically for simple problems like this (Refs. 106 and 221). In particular, these equations can be programmed in a spreadsheet and solved using the iteration feature, provided the boundaries are all rectangular.

The alternating direction method can be used for elliptic problems by using sequences of iteration parameters (Refs. 106 and 221). The method is well suited to transient problems as well.

These are the classical iterative techniques. Recently preconditioned conjugate gradient methods have been developed (see Ref. 100). In these methods, a series of matrix multiplications are done iteration by iteration; and the steps lend themselves to the efficiency available in parallel computers. In the multigrid method, the problem is solved on several grids, each more refined than the previous one. As one iterates between the solutions on the different grids, one converges to the solution of the algebraic equations. See Juncu and Mihail (Ref. 68) for a chemical engineering application.

The Galerkin finite element method (FEM) is useful for solving elliptic problems and is particularly effective when the domain or geometry is irregular. As an example, cover the domain with triangles and define a trial function on each triangle. The trial function takes the value 1.0 at one corner and 0.0 at the other corners and is linear in between. See Fig. 3-53. These trial functions on each triangle are pieced together to give a trial function on the whole domain. General treatments of the finite element method are available (see references). The steps in the solution method are similar to those described for boundary value problems, except now the problems are much bigger so that the numerical analysis must be done very carefully to be efficient. Most engineers, though, just use a finite element program without generating it. There are three major caveats that must be addressed. The first one is that the solution is dependent on the mesh laid down, and the only way to assess the accuracy of the solution is to solve the problem with a more refined mesh. The second concern is that the solution obeys the shape of the trial function inside the element. Thus, if linear functions are used on triangles, a three-dimensional view of the solution, plotting the solution versus x and y, consists of a series of triangular planes joined together at the edges, as in a geodesic dome. The third caveat is that the Galerkin finite element method is applied to both the differential equations and the boundary conditions. Computer programs are usually quite general and may allow the user to specify boundary conditions that are not realistic. Also, natural boundary conditions are satisfied if no other boundary condition (ones involving derivatives) is set at a node. Thus, the user of finite element codes must be very clear what boundary conditions and differential equations are built into the computer code. When the problem is nonlinear, the Newton-Raphson method is used to iterate from an initial guess. Nonlinear problems lead to complicated integrals to evaluate, and they are usually evaluated using Gaussian quadrature.

One nice feature of the finite element method is the use of natural boundary conditions. It may be possible to solve the problem on a domain that is shorter than needed to reach some limiting condition (such as at an outflow boundary). The externally applied flux is still applied at the shorter domain, and the solution *inside* the truncated domain is still valid. Examples are given in Refs. 67 and 107. The effect of this is to allow solutions in domains that are smaller, thus saving computation time and permitting the solution in semi-infinite domains.

A general purpose package for general two-dimensional domains and rectangular three-dimensional rectangular domains is ELLPACK (Ref. 247). This package allows choice of a variety of methods: finite difference, Hermite collocation, spline Galerkin, collocation, as well as others. Comparisons of the various methods are available (Ref. 154). The program FISHPAK solves the Helmholtz equation in multiple dimensions when the domain is separable (since fast methods like FFT are used). See Ref. 247.

Hyperbolic Equations   The most common situation yielding hyperbolic equations involves unsteady phenomena with convection. Two typical equations are the convective diffusive equation

$$\frac{\partial c}{\partial t} + u \frac{\partial c}{\partial x} = D \frac{\partial^2 c}{\partial x^2}$$

and the chromatography equation (Ref. 245)

$$\phi \frac{\partial c}{\partial t} + \phi u \frac{\partial c}{\partial x} + (1 - \phi) \frac{df}{dc} \frac{\partial c}{\partial t} = 0$$

where ϕ is the void fraction and $f(c)$ gives the equilibrium relation between the concentration in the fluid phase and the concentration in the solid phase. If the diffusion coefficient is zero, the convective diffusion equation is hyperbolic. If D is small, the phenomenon may be essentially hyperbolic, even though the equations are parabolic. Thus the numerical methods for hyperbolic equations may be useful even for parabolic equations.

Equations for several methods are given here, as taken from the book by Finlayson (Ref. 107). If the convective term is treated with a centered difference expression, the solution exhibits oscillations from node to node, and these only go away if a very fine grid is used. The

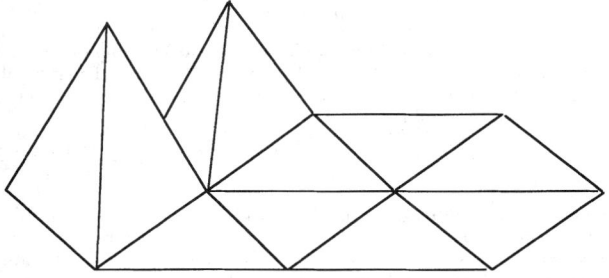

FIG. 3-53   Trial functions for Galerkin finite element method: linear polynomial on triangle.

simplest way to avoid the oscillations with a hyperbolic equation is to use upstream derivatives. If the flow is from left to right, this would give

$$\frac{dc_i}{dt} + u \frac{c_i - c_{i-1}}{\Delta x} = D \frac{c_{i+1} - 2c_i + c_{i-1}}{\Delta x^2}$$

$$\frac{d}{dt} \left[\phi c_i + (1 - \phi) f(c_i) \right] + \phi u_i \frac{c_i - c_{i-1}}{\Delta x} = 0$$

(See Ref. 227 for the reason the equation is written in this form.)

The effect of using upstream derivatives is to add artificial or numerical diffusion to the model. This can be ascertained by rearranging the finite difference form of the convective diffusion equation

$$\frac{dc_i}{dt} + u \frac{c_{i+1} - c_{i-1}}{2\Delta x} = \left(D + \frac{u\Delta x}{2} \right) \frac{c_{i+1} - 2c_i + c_{i-1}}{\Delta x^2}$$

Thus the diffusion coefficient has been changed from

$$D \text{ to } D + \frac{u\Delta x}{2}$$

Another method often used for hyperbolic equations is the MacCormack method. This method has two steps, and it is written here for the convective diffusion equation.

$$c_i^{\circ n+1} = c_i^n - \frac{u\Delta t}{\Delta x} (c_{i+1}^n - c_i^n) + \frac{D\Delta t}{\Delta x^2} (c_{i+1}^n - 2c_i^n + c_{i-1}^n)$$

$$c_i^{n+1} = \frac{1}{2} (c_i^n + c_i^{\circ n+1}) - \frac{u\Delta t}{2\Delta x} (c_i^{\circ n+1} - c_{i-}^{\circ n+1})$$

$$+ \frac{D\Delta t}{2\Delta x^2} (c_{i+1}^{\circ n+1} - 2c_i^{\circ n+1} + c_{i-1}^{\circ n+1})$$

The concentration profile is steeper for the MacCormack method than for the upstream derivatives, but oscillations can still be present. The flux-corrected transport method can be added to the MacCormack method. A solution is obtained both with the upstream algorithm and the MacCormack method and then they are combined to add just enough diffusion to eliminate the oscillations without smoothing the solution too much. The algorithm is complicated and lengthy but well worth the effort (Refs. 37, 107, and 270).

Stability conditions can be constructed in terms of $Co = u\Delta t/\Delta x$ and $r = D\Delta t/\Delta x^2$ by using Fourier analysis (Ref. 107). All the methods require

$$Co = \frac{u\Delta t}{\Delta x} \leq 1$$

where Co is the Courant number. How much Co should be less than one depends on the method and on $r = D\Delta t/\Delta x^2$. For example, the upstream method requires $Co \leq 1 - 2r$. The MacCormack method depends less on r and is stable for most Co as long as $r \leq 0.5$. Each of these methods is trying to avoid oscillations that would disappear if the mesh were fine enough. For the steady convective diffusion equation, these oscillations *do not* occur provided

$$\frac{u\Delta x}{2D} \leq 1$$

For large velocity u, the Δx must be small to meet this condition. An alternative is to use a small Δx in regions where the solution changes drastically. Since these regions change in time, it is necessary that the elements or grid points move. The criteria to move the grid points can be quite complicated, and typical methods are reviewed in Ref. 107. Similar considerations apply to the nonlinear chromatography problem (Ref. 227). See especially Ref. 192.

Parabolic Equations in Two or Three Dimensions   Computations become much more lengthy when there are two or more spatial dimensions. For example, we may have the unsteady heat conduction equation

$$\rho C_p \frac{\partial T}{\partial t} = k \left(\frac{\partial^2 T}{\partial x^2} + \frac{\partial^2 T}{\partial y^2} \right) - Q$$

In the finite difference method an explicit technique would evaluate the right-hand side at the nth time level.

$$\rho C_p \frac{T_{i,j}^{n+1} - T_{i,j}^n}{\Delta t} = \frac{k}{\Delta x^2}(T_{i+1,j}^n - 2T_{i,j}^n + T_{i-1,j}^n)$$

$$+ \frac{k}{\Delta y^2}(T_{i,j+1}^n - 2T_{i,j}^n + T_{i,j-1}^n) - Q$$

When $Q = 0$ and $\Delta x = \Delta y$, the time step limit can be found using the positivity rule.

$$\Delta t \leq \frac{\Delta x^2 \rho C_p}{4k} \quad \text{or} \quad \frac{\Delta x^2}{4D}$$

These time steps are smaller than for one-dimensional problems. For three dimensions, the limit is

$$\Delta t \leq \frac{\Delta x^2}{6D}$$

To avoid such small time steps, which become smaller as Δx decreases, an implicit method could be used. This leads to large, sparse matrices rather than convenient tridiagonal matrices. These can be solved, but the alternating direction method is also useful (Ref. 221). This reduces a problem on an $n \times n$ grid to a series of $2n$ one-dimensional problems on an n grid.

SPLINE FUNCTIONS

Splines are functions that match given values at the points $x_1, \ldots, x_{NT}$ and have continuous derivatives up to some order at the knots, or the points $x_2, \ldots, x_{NT-1}$. Cubic splines are most common; see Ref. 38. The function is represented by a cubic polynomial within each interval (x_i, x_{i+1}) and has continuous first and second derivatives at the knots. Two more conditions can be specified arbitrarily. These are usually the second derivatives at the two end points, which are commonly taken as zero; this gives the natural cubic splines.

Take $y_i = y(x_i)$ at each of the points x_i, and let $\Delta x_i = x_{i+1} - x_i$. Then, in the interval (x_i, x_{i+1}), the function is represented as a cubic polynomial.

$$C_i(x) = a_{0i} + a_{1i}x + a_{2i}x^2 + a_{3i}x^3$$

The interpolating function takes on specified values at the knots and has continuous first and second derivatives at the knots. Within the ith interval, the function is

$$C_i(x) = C_i(x_i) + C_i'(x_i)(x - x_i) + C_i''(x_i)\frac{(x - x_i)}{2}$$

$$+ [C_i''(x_{i+1}) - C_i''(x_i)]\frac{(x - x_i)^3}{6\Delta x_i}$$

where $C_i(x_i) = y_i$. The second derivative $C_i''(x_i) = y_i''$ is found by solving the following tridiagonal system of equations:

$$y_{i-1}''\Delta x_{i-1} + y_i''2(\Delta x_{i-1} + \Delta x_i) + y_{i+1}''\Delta x_i = 6\left(\frac{y_i - y_{i-1}}{\Delta x_{i-1}} - \frac{y_{i+1} - y_i}{\Delta x_i}\right)$$

Since the continuity conditions apply only for $i = 2, \ldots, NT - 1$, we have only $NT - 2$ conditions for the NT values of y_i''. Two additional conditions are needed, and these are usually taken as the value of the second derivative at each end of the domain, y_1'', y_{NT}''. If these values are zero, we get the natural cubic splines; they can also be set to achieve some other purpose, such as making the first derivative match some desired condition at the two ends. With these values taken as zero in the natural cubic spline, we have a $NT - 2$ system of tridiagonal equations, which is easily solved. Once the second derivatives are known at each of the knots, the first derivatives are given by

$$y_i' = \frac{y_{i+1} - y_i}{\Delta x_i} - y_i''\frac{\Delta x_i}{3} - y_{i+1}''\frac{\Delta x_i}{6}$$

The function itself is then known within each element.

FAST FOURIER TRANSFORM (REF. 231)

Suppose a signal $y(t)$ is sampled at equal intervals

$$y_n = y(n\Delta), \qquad n = \ldots, -2, -1, 0, 1, 2, \ldots$$

Δ = sampling rate (e.g., number of samples per second)

The Fourier transform and inverse transform are

$$Y(\omega) = \int_{-\infty}^{\infty} y(t)e^{i\omega t}\,dt$$

$$y(t) = \frac{1}{2\pi}\int_{-\infty}^{\infty} Y(\omega)e^{-i\omega t}\,dt$$

The Nyquist critical frequency or critical angular frequency is

$$f_c = \frac{1}{2\Delta}, \qquad \omega_c = \frac{\pi}{\Delta}$$

If a function $y(t)$ is bandwidth-limited to frequencies smaller than f_c, such as

$$Y(\omega) = 0 \qquad \text{for } \omega > \omega_c$$

then the function is completely determined by its samples y_n. Thus, the entire information content of a signal can be recorded by sampling at a rate $\Delta^{-1} = 2f_c$. If the function is *not* bandwidth-limited, then aliasing occurs. Once a sample rate Δ is chosen, information corresponding to frequencies greater than f_c is simply aliased into that range. The way to detect this in a Fourier transform is to see if the transform approaches zero at $\pm f_c$; if not, aliasing has occurred, and a higher sampling rate is needed.

Next, suppose we have N samples, where N is even

$$y_k = y(t_k) \qquad t_k = k\Delta \qquad k = 0,1,2, \ldots, N-1$$

and the sample rate is Δ. With only N values $\{y_k\}$, it is not possible to determine the complete Fourier transform $Y(\omega)$. We calculate the value $Y(\omega_n)$ at the discrete points

$$\omega_n = \frac{2\pi n}{N\Delta}, \qquad n = -\frac{N}{2}, \ldots, 0, \ldots, \frac{N}{2}$$

$$Y_n = \sum_{k=0}^{N-1} y_k e^{2\pi ikn/N}$$

$$Y(\omega_n) = \Delta Y_n$$

The discrete inverse Fourier transform is

$$y_k = \frac{1}{N}\sum_{n=0}^{N-1} Y_n e^{-2\pi ikn/N}$$

The fast Fourier transform (FFT) is used to calculate the Fourier transform as well as the inverse Fourier transform. A discrete Fourier transform of length N can be written as the sum of two discrete Fourier transforms, each of length $N/2$.

$$Y_k = Y_k^e + W^k Y_k^o$$

Here Y_k is the kth component of the Fourier transform of y, and Y_k^e is the kth component of the Fourier transform of the even components of $\{y_j\}$ and is of length $N/2$. Similarly, Y_k^o is the kth component of the Fourier transform of the odd components of $\{y_j\}$ and is of length $N/2$. W is a constant, which is taken to the kth power.

$$W = e^{2\pi i/N}$$

Since Y_k has N components, while Y_k^e and Y_k^o have $N/2$ components, Y_k^e and Y_k^o are repeated once to give N components in the calculation of Y_k. This decomposition can be used recursively. Thus, Y_k^e is split into even and odd terms of length $N/4$.

$$Y_k^e = Y_k^{ee} + W^k Y_k^{eo}$$

$$Y_k^o = Y_k^{oe} + W^k Y_k^{oo}$$

This process is continued until there is only one component. For this reason, the number N is taken as a power of 2. The vector $\{y_j\}$ is filled with zeroes, if need be, to make $N = 2^p$ for some p. For the computer program, see Ref. 26. The standard Fourier transform takes N^2 operations to calculation, whereas the fast Fourier transform takes only $N \log_2 N$. For large N, the difference is significant; at $N = 100$ it is a factor of 15, but for $N = 1000$ it is a factor of 100.

The discrete Fourier transform can also be used for differentiating a function, and this is used in the spectral method for solving differential equations. Suppose we have a grid of equidistant points

$$x_n = n\Delta x, \quad n = 0, 1, 2, \ldots, 2N - 1, \qquad \Delta x = \frac{L}{2N}$$

The solution is known at each of these grid points $\{Y(x_n)\}$. First the Fourier transform is taken.

$$y_k = \frac{1}{2N} \sum_{n=0}^{2N-1} Y(x_n) e^{-2ik\pi x_n / L}$$

The inverse transformation is

$$Y(x) = \frac{1}{L} \sum_{k=-N}^{N} y_k e^{2ik\pi x / L}$$

Differentiate this to get

$$\frac{dY}{dx} = \frac{1}{L} \sum_{k=-N}^{N} y_k \frac{2\pi i k}{L} e^{2ik\pi x / L}$$

Thus at the grid points

$$\left. \frac{dY}{dx} \right|_n = \frac{1}{L} \sum_{k=-N}^{N} y_k \frac{2\pi i k}{L} e^{2ik\pi x_n / L}$$

The process works as follows. From the solution at all grid points the Fourier transform is obtained using FFT, $\{y_k\}$. Then this is multiplied by $2\pi i k / L$ to obtain the Fourier transform of the derivative.

$$y_k = y_k \frac{2\pi i k}{L}$$

Then the inverse Fourier transform is taken using FFT, giving the value of the derivative at each of the grid points.

$$\left. \frac{dY}{dx} \right|_n = \frac{1}{L} \sum_{k=-N}^{N} y_k e^{2ik\pi x_n / L}$$

OPTIMIZATION

INTRODUCTION

Optimization should be viewed as a tool to aid in decision making. Its purpose is to aid in the selection of better values for the decisions that can be made by a person in solving a problem. To formulate an optimization problem, one must resolve three issues. First, one must have a representation of the artifact that can be used to determine how the artifact performs in response to the decisions one makes. This representation may be a mathematical model or the artifact itself. Second, one must have a way to evaluate the performance—an objective function—which is used to compare alternative solutions. Third, one must have a method to search for the improvement. This section concentrates on the third issue, the methods one might use. The first two items are difficult ones, but discussing them at length is outside the scope of this section.

Example optimization problems are: (1) determining the optimal thickness of pipe insulation; (2) finding the best equipment sizes and operating schedules for the design of a new batch process to make a given slate of products; (3) choosing the best set of operating conditions for a set of experiments to determine the constants in a kinetic model for a given reaction; (4) finding the amounts of a given set of ingredients one should use for making a carbon rod to be used as an electrode in an arc welder.

For the first problem, one will usually write a mathematical model of how insulation of varying thicknesses restricts the loss of heat from a pipe. Evaluation requires that one develop a cost model for the insulation (a capital cost in dollars) and the heat that is lost (an operating cost in dollars/year). Some method is required to permit these two costs to be compared, such as a present worth analysis. Finally, if the model is simple enough, the method one can use is to set the derivative of the evaluation function to zero with respect to wall thickness to find candidate points for its optimal thickness. For the second problem, selecting a best operating schedule involves discrete decisions, which will generally require models that have integer variables.

It may not be possible to develop a mathematical model for the fourth problem if not enough is known to characterize the performance of a rod versus the amounts of the various ingredients used in its manufacture. The rods may have to be manufactured and judged by ranking the rods relative to each other, perhaps based partially or totally on opinions. Pattern search methods have been devised to attack problems in this class.

In this section assume a mathematical model is possible for the problem to be solved. The model may be encoded in a subroutine and be known only implicitly, or the equations may be known explicitly. A general form for such an optimization problem is

$$\min F = F(z), \text{ such that } h(z) = 0 \text{ and } g(z) \leq 0$$

where F represents a specified objective function that is to be minimized. Functions h and g represent equality and inequality constraints that must be satisfied at the final problem solution.

Variables z are used to model such things as flows, mole fractions, physical properties, temperatures, and sizes. The objective function F is generally assumed to be a scalar function, one which represents such things as cost, net present value, safety, or flexibility. Sometimes several objective functions are specified (e.g., minimizing cost while maximizing reliability); these are commonly combined into one function, or else one is selected for the optimization while the others are specified as constraints. Equations $h(z) = 0$ are typically algebraic equations, linear or nonlinear, when modeling steady-state processes, or algebraic coupled with ordinary and/or partial differential equations when optimizing time-varying processes. Inequalities $g(z) \leq 0$ put limits on the values variables can take, such as a minimum and maximum temperature, or they restrict one pressure to be greater than another.

An important issue is how to solve large problems that occur in distributed systems. The optimization of distributed systems is discussed in Refs. 52, 120, 244, and 285. For further reading on optimization, readers are directed to Refs. 120 and 244 as well as introductory texts on optimization applied to chemical engineering (Refs. 99 and 225). The material in this section is part of a more advanced treatment (Ref. 295).

Packages   There are a number of packages available for optimization, some of which are listed here.

1. *Frameworks*
- *GAMS.* This framework is commercially available. It provides a uniform language to access several different optimization packages, many of them listed below. It will convert the model as expressed in "GAMS" into the form needed to run the package chosen.
- *AMPL.* This framework is by Fourier and coworkers (Ref. 113) at Northwestern University. It is well suited for constructing complex models.
- *ASCEND.* This framework is by Westerberg and coworkers (Ref. 295) at Carnegie-Mellon University. It features an object-oriented modeling language and is well suited for constructing complex models.
2. *Algebraic optimization with equality and inequality constraints*
- *SQP.* A package by Biegler at Carnegie-Mellon University.
- *MINOS5.4.* A package available from Stanford Research Institute (affiliated with Stanford University). This package is the state of the art for mildly nonlinear programming problems.
- *GRG.* A package from Lasdon at the University of Texas, Dept. of Management Science.
3. *Linear programming.* Most current commercial codes for lin-

ear programming extend the Simplex algorithm, and they can typically handle problems with up to 15,000 constraints.

- *MPSX.* From IBM
- *SCICONIC.* From the company of that name
- *MINOS5.4*
- *Cplex.* A package by R. Bixby at Rice University and Cplx, Inc.

CONDITIONS FOR OPTIMALITY

Local Minimum Point for Unconstrained Problems Consider the following unconstrained optimization problem:

$$\text{Min}_u \{F(u) \mid u \in \mathbf{R}^r\}$$

If F is continuous and has continuous first and second derivatives, it is necessary that F is stationary with respect to all variations in the independent variables u at a point $\hat{u}$, which is proposed as a minimum to F; that is,

$$\frac{\partial F}{\partial u_i} = 0, \quad i = 1, 2, \ldots, r \quad \text{or} \quad \nabla_u F = 0 \quad \text{at } u = \hat{u} \quad (3\text{-}80)$$

These are only necessary conditions, as point $\hat{u}$ may be a minimum, maximum, or saddle point.

Sufficient conditions are that any local move away from the optimal point $\hat{u}$ gives rise to an increase in the objective function. Expand F in a Taylor series locally around the candidate point $\hat{u}$ up to second-order terms:

$$F(u) = F(\hat{u}) + \nabla_u F^T\big|_{\hat{u}} (u - \hat{u}) + \frac{1}{2}(u - \hat{u})^T \nabla_{uu}^2 F\big|_{\hat{u}} (u - \hat{u}) + \cdots$$

If $\hat{u}$ satisfies necessary conditions [Eq. (3-80)], the second term disappears in this last line. Sufficient conditions for the point to be a local minimum are that the matrix of second partial derivatives $\nabla_{uu}^2 F$ is positive definite. This matrix is symmetric, so all of its eigenvalues are real; to be positive definite, they must all be greater than zero.

Constrained Derivatives—Equality Constrained Problems Consider minimizing the objective function F written in terms of n variables z and subject to m equality constraints $h(z) = 0$, or

$$\text{Min}_z \{F(z) \mid h(z) = 0, z \in \mathbf{R}^n, h{:}\mathbf{R}^n \to \mathbf{R}^m\} \quad (3\text{-}81)$$

The point $\hat{z}$ is tested to see if it could be a minimum point. It is necessary that F be stationary for all infinitesimal moves for z that satisfy the equality constraints. Linearize the m equality constraints around $\hat{z}$, getting

$$h(\hat{z} + \Delta z) = h(\hat{z}) + \nabla_z h^T\big|_{\hat{z}} \Delta z \quad (3\text{-}82)$$

where $\Delta z = z - \hat{z}$. There are m constraints here, so m of the variables are dependent, leaving $r = n - m$ independent variables. Partition the variables Δz into a set of m dependent variables Δx and $r = n - m$ independent variables Δu. Equation (3-82), rearranged and then rewritten in terms of these variables, becomes

$$\Delta h = \nabla_x h^T\big|_{\hat{z}} \Delta x + \nabla_u h^T\big|_{\hat{z}} \Delta u = 0$$

This enables the solution for Δx. Linearize the objective function $F(z)$ in terms of the partitioned variables

$$\Delta F = \nabla_x F^T\big|_{\hat{z}} \Delta x + \nabla_u F^T\big|_{\hat{z}} \Delta u$$

and substitute for Δx.

$$\Delta F = \{\nabla_x F^T - \nabla_u F^T[\nabla_x h^T]^{-1} \nabla_u h^T\}_{\hat{z}} \Delta u$$

$$= \left\{\frac{dF}{du}\right\}_{\Delta h = 0}^T \Delta u = \sum_{i=1}^r \left\{\frac{dF}{du_i}\right\}_{\Delta h = 0} \Delta u_i$$

There is one term for each Δu_i in the row vector which is in the curly braces {}. These terms are called **constrained derivatives,** which tells how the object function changes when the independent variables u_i are changed while keeping the constraints satisfied (by varying the dependent variables x_i).

Necessary conditions for optimality are that these constrained derivatives are zero; that is,

$$\left\{\frac{dF}{du_i}\right\}_{\Delta h = 0} = 0, \quad i = 1, 2, \ldots, r$$

Equality Constrained Problems—Lagrange Multipliers Form a scalar function, called the Lagrange function, by adding each of the equality constraints multiplied by an arbitrary multiplier to the objective function.

$$L(x, u, \lambda) = F(x, u) + \sum_{i=1}^m \lambda_i h_i(x, u) = F(x, u) + \lambda^T h(x, u)$$

At any point where the functions $h(z)$ are zero, the Lagrange function equals the objective function.

Next differentiate L with respect to variables x, u, and λ.

$$\nabla_x L^T\big|_{\hat{z}} = \nabla_x F^T\big|_{\hat{z}} + \lambda^T \nabla h_x^T\big|_{\hat{z}} \qquad = 0^T \quad (3\text{-}83)$$

$$\nabla_u L^T\big|_{\hat{z}} = \nabla_u F^T\big|_{\hat{z}} + \lambda^T \nabla h_u^T\big|_{\hat{z}} \qquad = 0^T \quad (3\text{-}84)$$

$$\nabla_\lambda L^T\big|_{\hat{z}} = h^T(x, u) \qquad\qquad = 0^T$$

Solve Eq. (3-83) for the Lagrange multipliers

$$\lambda^T = -\nabla_x F^T[\nabla h_x^T]^{-1} \quad (3\text{-}85)$$

and then eliminate these multipliers from Eq. (3-84).

$$\nabla_u L^T = \nabla_u F^T - \nabla_x F^T[\nabla h_x^T]\nabla h_u^T = 0^T$$

$\nabla_u L$ is equal to the constrained derivatives for the problem, which should be zero at the solution to the problem. Also, these stationarity conditions very neatly provide the necessary conditions for optimality of an equality-constrained problem.

Lagrange multipliers are often referred to as shadow prices, adjoint variables, or dual variables, depending on the context. Suppose the variables are at an optimum point for the problem. Perturb the variables such that only constraint h_i changes. We can write

$$\Delta L = \Delta F + \lambda_i \Delta h_i = 0$$

which is zero because, as just shown, the Lagrange function is at a stationary point at the optimum. Solving for the change in the objective function:

$$\Delta F = -\lambda_i \Delta h_i$$

The multiplier tells how the optimal value of the objective function changes for this small change in the value of a constraint while holding all the other constraints at zero. It is for this reason that they are often called shadow prices.

Equality- and Inequality-Constrained Problems—Kuhn-Tucker Multipliers Next a point is tested to see if it is an optimum one when there are inequality constraints. The problem is

$$\text{Min} \{F(z) \mid h(z) = 0, g(z) \leq 0, z \in \mathbf{R}^n, F{:}\mathbf{R}^n \to \mathbf{R}^1, h{:}\mathbf{R}^n \to \mathbf{R}^m, g{:}\mathbf{R}^n \to \mathbf{R}^p\}$$

The Lagrange function here is similar to that used above.

$$L(z, \lambda, \mu) \equiv F(z) + \lambda^T h(z) + \mu^T g(z)$$

Each of the inequality constraints $g_i(z)$ multiplied by what is called a Kuhn-Tucker multiplier μ_i is added to form the Lagrange function. The necessary conditions for optimality, called the Karush-Kuhn-Tucker conditions for inequality-constrained optimization problems, are

$$\nabla_z L = \nabla_z F\big|_{\hat{z}} + \nabla_z h\big|_{\hat{z}} \lambda + \nabla_z g\big|_{\hat{z}} \mu = 0$$

$$\nabla_\lambda L = h(z) = 0$$

$$g(z) \leq 0$$

$$\mu_i g_i(z) = 0, \quad i = 1, 2, \ldots, p \quad (3\text{-}86)$$

$$\mu_i \geq 0, \quad i = 1, 2, \ldots, p$$

Conditions in Eq. (3-86), called complementary slackness conditions, state that either the constraint $g_i(z) = 0$ and/or its corresponding multiplier μ_i is zero. If constraint $g_i(z)$ is zero, it is behaving like an equality constraint, and its multiplier μ_i is exactly the same as a Lagrange multiplier for an equality constraint. If the constraint is away from zero, it is not a part of the problem and should not affect it. Setting its multiplier to zero removes it from the problem.

As the goal is to minimize the objective function, releasing the constraint into the feasible region must not decrease the objective function. Using the shadow price argument above, it is evident that the multiplier must be nonnegative (Ref. 177).

Sufficiency conditions to assure that a Kuhn-Tucker point is a local minimum point require one to prove that the objective function will increase for any feasible move away from such a point. To carry out such a test, one has to generate the matrix of second derivatives of the Lagrange function with respect to all the variables z evaluated at $\hat{z}$. The test is seldom done, as it requires too much work.

STRATEGIES OF OPTIMIZATION

The theory just covered tells if a candidate point is or is not the optimum point, but how is the candidate point found? The simplest strategy is to place a grid of points throughout the feasible space, evaluating the objective function at every grid point. If the grid is fine enough, then the point yielding the highest value for the objective function can be selected as the optimum. Twenty variables gridded over only ten points would take place over 10^{20} points in our grid, and, at one nanosecond per evaluation, it would take in excess of four thousand years to carry out these evaluations.

Most strategies limit themselves to finding a local minimum point in the vicinity of the starting point for the search. Such a strategy will find the global optimum only if the problem has a single minimum point or a set of "connected" minimum points. A "convex" problem has only a global optimum.

Pattern Search   Suppose the optimization problem is to find the right mix of a given set of ingredients and the proper baking temperature and time to make the best cake possible. A panel of judges can be formed to judge the cakes; assume they are only asked to rank the cakes and that they can do that task in a consistent manner. Our approach will be to bake several cakes and ask the judges to rank them. For this type of problem, pattern-search methods can be used to find the better conditions for manufacturing the product. We shall only describe the ideas behind this approach. Details on implementing it can be found in Ref. 284.

The complex method is one such pattern search method (see Fig. 3-54). First, form a "complex" of at least $r + 1$ ($r = 2$ and 4 points are used in Fig. 3-54) different points at which to bake the cakes by picking a range of suitable values for the r independent variables for the baking process. Bake the cakes and then ask the judges to identify the worst cake.

For each independent variable, form the average value at which it was run in the complex. Draw a line from the coordinates of the worst cake through the average point—called the centroid—and continue on that line a distance that is twice that between these two points. This point will be the next test point. First decide if it is feasible. If so, bake the cake and discover if it leads to a cake that is better than the worst cake from the last set of cakes. If it is not feasible or it is not better, then return half the distance toward the average values from the last

test and try again. If it is better, toss out the worst point of the last test and replace it with this new one. Again, ask the judges to find the worst cake. Continue as above until the cakes are all the same quality in the most recent test. It might pay to restart at this point, stopping finally if the restart leads to no improvement. The method takes large steps if the steps are being successful in improving the recipe. It collapses onto a set of points quite close to each other otherwise. The method works reasonably well, but it requires one to bake lots of cakes.

The following strategies are all examples of Generalized Reduced Gradient (GRG) methods.

Optimization of Unconstrained Objective   Assume the objective function F is a function of independent variables u_i, $i = 1 \cdots r$. A computer program, given the values for the independent variables, can calculate F and its derivatives with respect to each u_i. Assume that F is well approximated as an as-yet-unknown quadratic function in u.

$$F \approx a + b^T u + \frac{1}{2} u^T Q u$$

where a is a scalar; b, a vector; and Q, an $r \times r$ symmetric positive definite matrix. The gradient of the approximate function is

$$\nabla_u F = b + Q u$$

Setting the gradient to zero allows an estimate for its minimum.

$$u = -Q^{-1} b \qquad (3\text{-}87)$$

Initially, Q and b are not known and the calculation proceeds as follows: b contains r unknown coefficients and Q another $r(r + 1)/2$. To estimate b and Q, the computer code is used repeatedly, getting r equations each time—namely

$$(\nabla_u F)(1) = b + Q u(1)$$
$$(\nabla_u F)(2) = b + Q u(2)$$
$$\cdots$$
$$(\nabla_u F)(t) = b + Q u(t) \qquad (3\text{-}88)$$

As soon as there are as many independent equations as there are unknown coefficients, these *linear* equations are solved for b and Q. A proper choice of the points $u(i)$ guarantees getting independent equations to solve here.

Given b and Q, Eq. (3-87) provides a new estimate for u as a candidate minimum point. The subroutine is used again to obtain the gradient of F at this point. If the gradient is essentially zero, the calculations stop, since a point has been found that satisfies the necessary conditions for optimality. If not, the equations are written in the form of Eq. (3-88) for this new point, adding them to the set while removing the oldest set of equations. The new set of equations for b and Q are solved, and the calculations continue until a minimum point is found. If removal of the oldest equations from the set in Eq. (3-88) leads to a singular set of equations, then different equations have to be selected for removal. Alternatively, all the older equations can be kept, with the new ones added to the top of the list. Pivoting can be done by proceeding down the list until a nonsingular set of equations is found. Then the older equations are used only if necessary. Also, since only one set of equations is being replaced, clever methods are available to find the solution to the equations with much less work than is required to solve the set of equations the first time (Refs. 89 and 259).

Quadratic Fit for the Equality Constrained Case   Next consider solving a problem of the form of Eq. (3-82). For each iteration k:

1.   Enter with values provided for variables $u(k)$.

2.   Given values for $u(k)$, solve equations $h(x, u) = 0$ for $x(k)$. These will be m equations in m unknowns. If the equations are nonlinear, solving can be done using a variant of the Newton-Raphson method.

3.   Use Eq. (3-85) to solve for the Lagrange multipliers $\lambda(k)$. If the Newton-Raphson method (or any or several variants to it) is used to solve the equations, the jacobian matrix $\nabla_x^T h|_{z(k)}$ and its LU factors are already known so solving Eq. (3-85) requires very little effort.

4.   Substitute $\lambda(k)$ into Eq. (3-84), which in general will not be zero. The gradient $\nabla_u L(k)$ computed will be the constrained derivatives of F with respect to the independent variables $u(k)$.

5.   Return.

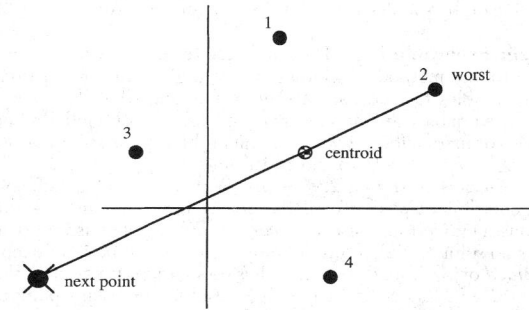

FIG. 3-54   Complex method, a pattern search optimization method.

The calculations begin with given values for the independent variables u and exit with the (constrained) derivatives of the objective function with respect to them. Use the routine described above for the unconstrained problem where a succession of quadratic fits is used to move toward the optimal point for an unconstrained problem. This approach is a form of the generalized reduced gradient (GRG) approach to optimizing, one of the better ways to carry out optimization numerically.

Inequality Constrained Problems To solve inequality constrained problems, a strategy is needed that can decide which of the inequality constraints should be treated as equalities. Once that question is decided, a GRG type of approach can be used to solve the resulting equality constrained problem. Solving can be split into two phases: phase 1, where the goal is to find a point that is feasible with respect to the inequality constraints; and phase 2, where one seeks the optimum while maintaining feasibility. Phase 1 is often accomplished by ignoring the objective function and using instead

$$F = \sum_{i=1}^{p} \left\{ \begin{array}{l} g_i^2(z) \text{ if } g_i(z) > 0) \\ 0 \text{ otherwise} \end{array} \right\}$$

until all the inequality constraints are satisfied.

Then at each point, check which of the inequality constraints are active, or exactly equal to zero. These can be placed into the active set and treated as equalities. The remaining can be put aside to be used only for testing. A step can then be proposed using the GRG algorithm. If it does not cause one to violate any of the inactive inequality constraints, the step is taken. Otherwise one can add the closest inactive inequality constraint to the active set. Finding the closet inactive equality will almost certainly require a line search in the direction proposed by the GRG algorithm.

When one comes to a stationary point, one has to test the active inequality constraints at that point to see if they should remain active. This test is done by examining the sign (they should be nonnegative if they are to remain active) of their respective Kuhn-Tucker multipliers. If any should be released, it has to be done carefully as the release of a constraint changes the multipliers for all the constraints. One can find oneself cycling through the testing to decide whether to release the constraints. A correct approach is to add slack variables s to the problem to convert the inequality constraints to equalities and then require the slack variables to remain positive. The multipliers associated with the inequalities $s \geq 0$ all behave independently, and their sign tells one directly to keep or release the constraints. In other words, simultaneously release all the slack variables that have multipliers strictly less than zero. If released, *the slack variables must be treated as a part of the set of independent variables* until one is well away from the associated constraints for this approach to work.

Successive Quadratic Programming (SQP) The above approach to finding the optimum is called a feasible path method, as it attempts at all times to remain feasible with respect to the equality and inequality constraints as it moves to the optimum. A quite different method exists called the Successive Quadratic Programming (SQP) method, which only requires one be feasible at the final solution. Tests that compare the GRG and SQP methods generally favor the SQP method so it has the reputation of being one of the best methods known for nonlinear optimization for the type of problems considered here.

Assume certain inequality constraints will be active at the final solution. The necessary conditions for optimality are

$$\nabla_z L(z, \mu, \lambda) = \nabla F + \nabla g_A \mu + \nabla h \lambda = 0, \quad g_A(z) = 0, \quad h(z) = 0$$

Then one can apply Newton's method to the necessary conditions for optimality, which are a set of simultaneous (non)linear equations. The Newton equations one would write are

$$\begin{bmatrix} \nabla_{zz} L(z(i), u(i), \lambda(i)) & \nabla g_A[z(i)] & \nabla h[z(i)] \\ \nabla g_A[z(i)]^T & 0 & 0 \\ \nabla h[z(i)]^T & 0 & 0 \end{bmatrix} \begin{bmatrix} \Delta z(i) \\ \Delta \mu(i) \\ \Delta \lambda(i) \end{bmatrix} = - \begin{bmatrix} \nabla_z L[z(i), \mu(i), \lambda(i)] \\ g_A[z(i)] \\ h[z(i)] \end{bmatrix}$$

A sufficient condition for a unique Newton direction is that the matrix of constraint derivatives is of full rank (linear independence

of constraints) and the Hessian matrix of the Lagrange function $[\nabla_{zz} L(z, \mu, \lambda)]$ *projected into the space of the linearized constraints* is positive definite. The linearized system actually represents the solution of the following quadratic programming problem:

$$\underset{\Delta z}{\text{Min}} \ \nabla F[z(i)]^T \Delta z + \frac{1}{2} \Delta z^T \nabla_{zz} L[z(i), \mu(i), \lambda(i)] \Delta z$$

subject to

$$g_A[z(i)] + \nabla g_A[z(i)]^T \Delta z = 0 \quad \text{and} \quad h[z(i)] + \nabla h[z(i)]^T \Delta z = 0$$

Reformulating the necessary conditions as a linear quadratic program has an interesting side effect. We can simply add linearizations of the inactive inequalities to the problem and let the active set be selected by the algorithm used to solve the linear quadratic program.

Problems with calculating second derivatives as well as maintaining positive definiteness of the Hessian matrix can be avoided by approximating this matrix by $B(i)$ using a quasi-Newton formula such as BFGS (Refs. 50, 84, 109, 110, 122, and 259). One maintains positive definiteness by skipping the update if it causes the matrix to lose this property. Here gradients of the Lagrange function are used to calculate the update formula (Refs. 136 and 228). The resulting quadratic program, which generates the search direction at each iteration i, becomes:

$$\underset{\Delta z}{\text{Min}} \ \nabla F[z(i)]^T \Delta z + \frac{1}{2} \Delta z^T B(i) \Delta z$$

subject to

$$g[z(i)] + \nabla g[z(i)]^T \Delta z \leq 0$$

$$h[z(i)] + \nabla h[z(i)]^T \Delta z = 0$$

This linear quadratic program will have a unique solution if $B(i)$ is kept positive definite. Efficient solution methods exist for solving it (Refs. 119 and 123).

Finally, to ensure convergence of this algorithm from poor starting points, a step size α is chosen along the search direction so that the point at the next iteration ($z^{i+1} = z^i + \alpha d$) is closer to the solution of the NLP (Refs. 65, 136, and 254).

These problems get very large as the Lagrange function involves all the variables in the problem. If one has a problem with 5000 variables z and the problem has only 10 degrees of freedom (i.e., the partitioning will select 4990 variables x and only 10 variables u), one is still faced with maintaining a matrix B that is 5000×5000. See Westerberg (Ref. 40) for references to this case.

Interior Point Algorithms for Linear Programming Problems There has been considerable excitement in the popular press about so-called interior point algorithms (Ref. 23) for solving extremely large linear programming problems. Computational demands for these algorithms grow less rapidly than for the Simplex algorithm, with a break-even point being a few thousand constraints. A key idea for an interior method is that one heads across the feasible region to locate the solution rather than around its edges as one does for the Simplex algorithm. This move is found by computing the direction of steepest descent for the objective function with respect to changing the slack variables. Variables u are computed in terms of the slack variables by using the inequality constraints. The direction of steepest descent is a function of the scaling of the variables used for the problem. See Refs. 6, 124, 199, and 295.

Linear Programming The combined term *linear programming* is given to any method for finding where a given linear function of several variables takes on an extreme value, and what that value is, when the variables are nonnegative and are constrained by linear equalities or inequalities. A very general problem consists of maximizing $f = \sum_{j=1}^{n} c_j z_j$ subject to the constraints $z_j \geq 0$ ($j = 1, 2, \ldots, n$) and $\sum_{j=1}^{n} a_{ij} z_j \leq b_i$ ($i = 1, 2, \ldots, m$). With S the set of all points whose coordinates z_j satisfy all the constraints, we must ask three questions: (1) Are the constraints *consistent*? If not, S is empty and there is no solution. (2) If S is not empty, does the function f become *unbounded* on S? If so, the problem has no solution. If not, then there is a point P of S that is optimal in the sense that if Q is any point of S then $f(Q) \leq f(P)$. (3) How can we find P?

The simplex algorithm, in a sense, prepares the problem before calculation in such a way that favorable answers to these questions are tentatively assumed for the given problem and can be guaranteed for the prepared problem. The calculations then reveal whether or not those assumptions are justified for the given problem. The simplex algorithm terminates automatically, yielding full information on the given problem and so-called dual problem. The dual of the general problem of linear programming is to minimize $d(\mu_1, \ldots, \mu_m) = \sum_{i=1}^{m} \mu_i b_i$ subject to $\mu_i \geq 0$ $(i = 1, 2, \ldots, m)$ and $\sum_{i=1}^{m} \mu_i a_{if} \geq c_j (j = 1, 2, \ldots, n)$. Let A be the matrix $[a_{ij}]$, $c = [c_j]$, $U = [\mu_i]$ be row vectors, and $B = [b_i]^T$, $Z = [z_j]^T$ be column vectors. In matrix form the original (primal) problem is to maximize $f(Z) = CZ$ subject to $Z \geq 0$, $AZ \leq B$. The dual is to minimize $d(U) = UB$ subject to $U \geq 0$, $UA \geq C$.

Example   Maximize $3z_1 + 4z_2$ subject to the constraints $z_1 \geq 0$, $z_2 \geq 0$, $2z_1 + 4z_2 \leq 8$, and $4z_1 + 3z_2 \leq 10$. The dual problem is to minimize $8\mu_1 + 10\mu_2$ subject to the constraints $\mu_1 \geq 0$, $\mu_2 \geq 0$, $2\mu_1 + 4\mu_2 \geq 3$, and $5\mu_1 + 3\mu_2 \geq 4$.

Simplex Method

1. *Original problem.* Let the column vector $[z_j]^T = z$ $(j = 1, 2, \ldots, n)$ and the row vector $[c_j] = c$. To maximize $f(z) = \sum_{j=1}^{n} c_j z_j = c^T z$ subject to the n constraints $z_j \geq 0$ $(j = 1, \ldots, n)$ and m further constraints $h_i: \sum_{j=1}^{n} a_{ij} z_j \,^{\circ}{}_i b_i$ $(i = 1, 2, \ldots, m)$ where $^{\circ}{}_i$ can be $\geq$ or $\leq$. If any $b_i \leq 0$, multiply h_i by -1; thus we may assume $b_i \geq 0$. We suppose the m constraints have been arranged so that $^{\circ}{}_i$ is $\geq$ for $i = 1, \ldots, g$; $^{\circ}{}_i$ is $=$ for $i = g + 1, \ldots, g + e$; $^{\circ}{}_i$ is $\leq$ for $i = g + e + 1, \ldots, g + e + l = m$.

2. *Adjusted original problem.* Introduce $m + g$ further variables

with associated constraints and coefficients for use in f. Thus, replacing j by $j + m$, f becomes $f(x) = \sum_{j=m+1}^{m+n} c_j z_j$ and constraints $z_j \geq 0$ and $h_i: \sum_{j=m+1}^{m+n} a_{ij} z_j \,^{\circ}{}_i b_i$, $i = 1, \ldots, m$.

3. *Prepared problem.* For $i = 1, \ldots, g$ replace h_i by H_i: $z_i + \sum_{j=m+1}^{m+n} a_{ij} z_j \, z_{m+n+i} = b_i$, define $c_i = -M (M > 0$ and "large") and $C_{m+n+i} = 0$, and add the constraints $z_i \geq 0$, $z_{m+n+i} \geq 0$. For $i = g + e + 1, \ldots, m$ replace h_i by H_i: $x_i + \sum_{j=m+1}^{m+n} a_{ij} z_j = b_i$, define $c_i = 0$, and adjoin $z_i \geq 0$. Let J run from 1 to $N = n + m + g$; put $Z = [z_i]^T$ and $j = m + 1, \ldots, m + n$. The new function to be maximized is $f(Z) = \sum_{J=1}^{N} c_j z_j$. Actually this is $f(Z) = -M \sum_{i=1}^{g+e} z_i + \sum_{j=m+1}^{vm+n} c_j z_j$, for all other coefficients are zero. Thus for $J = g + e + 1, \ldots, m$ and $m + n + 1, \ldots, N$ the variables z_j make no contribution to f. They are called slack variables, since they take up the slack permitted by the inequalities ($\leq$ and $\geq$) in h_i. Any variable z_j, $i = 1, \ldots, g + e$ whose value is not zero gives rise to a large negative term $-Mz_i$. Such a term will keep $f(Z)$ less than it would be with that $z_i = 0$. The effect of $c_i = -M(i = 1, \ldots, g + e)$ is to make it likely that the optimal solution will have the artificial variables $z_i = 0 (i = 1, \ldots, g + e)$.

The prepared problem now has the form—maximize $f(Z) = \sum_{j=1}^{N} c_j z_j$ subject to $z_j \geq 0$ and $H_i: \sum_{j=1}^{N} a_{ij} z_j = (i = 1, \ldots, m)$, where $b_i \geq 0$,

$$a_{i\beta} = \delta_{i\beta} = \begin{cases} 0 & i \neq \beta \\ 1 & i \neq \beta \end{cases} (\beta = 1, \ldots, m), \, a_{i,m+n+\beta} = -\delta_{i\beta} \, (\beta = 1, \ldots, g)$$

and a_{ij} came from h_i.

The set of feasible points S_P (points satisfying all constraints) for the prepared problem is not empty, and $f(Z)$ is bounded above on S_P.

STATISTICS

GENERAL REFERENCES: 69, 93, 134, 169, 186, 211, 265, 291.

INTRODUCTION

Statistics represents a body of knowledge which enables one to deal with quantitative data reflecting any degree of uncertainty. There are six basic aspects of applied statistics. These are:

1. Type of data
2. Random variables
3. Models
4. Parameters
5. Sample statistics
6. Characterization of chance occurrences

From these can be developed strategies and procedures for dealing with (1) estimation and (2) inferential statistics. The following has been directed more toward inferential statistics because of its broader utility.

Detailed illustrations and examples are used throughout to develop basic statistical methodology for dealing with a broad area of applications. However, in addition to this material, there are many specialized topics as well as some very subtle areas which have not been discussed. The references should be used for more detailed information.

Type of Data   In general, statistics deals with two types of data: counts and measurements. Counts represent the number of discrete outcomes, such as the number of defective parts in a shipment, the number of lost-time accidents, and so forth. Measurement data are treated as a continuum. For example, the tensile strength of a synthetic yarn theoretically could be measured to any degree of precision. A subtle aspect associated with count and measurement data is that some types of count data can be dealt with through the application of techniques which have been developed for measurement data alone. This ability is due to the fact that some simplified measurement statistics serve as an excellent approximation for the more tedious count statistics.

Random Variables   Applied statistics deals with quantitative data. In tossing a fair coin the successive outcomes would tend to be

different, with heads and tails occurring randomly over a period of time. Given a long strand of synthetic fiber, the tensile strength of successive samples would tend to vary significantly from sample to sample.

Counts and measurements are characterized as random variables, that is, observations which are susceptible to chance. Virtually all quantitative data are susceptible to chance in one way or another.

Models   Part of the foundation of statistics consists of the mathematical models which characterize an experiment. The models themselves are mathematical ways of describing the probability, or relative likelihood, of observing specified values of random variables. For example, in tossing a coin once, a random variable x could be defined by assigning to x the value 1 for a head and 0 for a tail. Given a fair coin, the probability of observing a head on a toss would be a .5, and similarly for a tail. Therefore, the mathematical model governing this experiment can be written as

x	$P(x)$
0	.5
1	.5

where $P(x)$ stands for what is called a probability function. This term is reserved for count data, in that probabilities can be defined for particular outcomes.

The probability function that has been displayed is a very special case of the more general case, which is called the binomial probability distribution.

For measurement data which are considered continuous, the term *probability density* is used. For example, consider a spinner wheel which conceptually can be thought of as being marked off on the circumference infinitely precisely from 0 up to, but not including, 1. In spinning the wheel, the probability of the wheel's stopping at a specified marking point at any particular x value, where $0 \leq x < 1$, is zero, for example, stopping at the value $x = \sqrt{.5}$. For the spinning wheel, the probability density function would be defined by $f(x) = 1$ for $0 \leq x < 1$. Graphically, this is shown in Fig. 3-55. The relative-probability

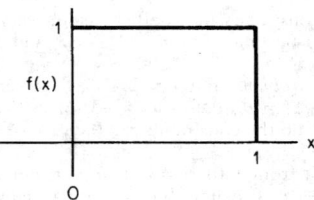

FIG. 3-55 Density function.

concept refers to the fact that density reflects the relative likelihood of occurrence; in this case, each number between 0 and 1 is equally likely.

For measurement data, probability is defined by the area under the curve between specified limits. A density function always must have a total area of 1.

Example For the density of Fig. 3-55 the

$$P[0 \leq x \leq .4] = .4$$
$$P[.2 \leq x \leq .9] = .7$$
$$P[.6 \leq x < 1] = .4$$

and so forth. Since the probability associated with any particular point value is zero, it makes no difference whether the limit point is defined by a closed interval ($\leq$ or $\geq$) or an open interval ($<$ or $>$).

Many different types of models are used as the foundation for statistical analysis. These models are also referred to as **populations.**

Parameters As a way of characterizing probability functions and densities, certain types of quantities called parameters can be defined. For example, the center of gravity of the distribution is defined to be the population mean, which is designated as μ. For the coin toss $\mu = .5$, which corresponds to the average value of x; i.e., for half of the time x will take on a value 0 and for the other half a value 1. The average would be .5. For the spinning wheel, the average value would also be .5.

Another parameter is called the **standard deviation,** which is designated as σ. The square of the standard deviation is used frequently and is called the popular **variance,** σ^2. Basically, the standard deviation is a quantity which measures the spread or dispersion of the distribution from its mean μ. If the spread is broad, then the standard deviation will be larger than if it were more constrained.

For specified probability and density functions, the respective means and variances are defined by the following:

Probability functions	Probability density functions
$E(x) = \mu = \sum_x x\, P(x)$	$E(x) = \mu = \int_x x\, f(x)\, dx$
$\text{Var}(x) = \sigma^2 = \sum_x (x - \mu)^2\, P(x)$	$\text{Var}(x) = \sigma^2 = \int_x (x - \mu)^2\, f(x)\, dx$

where $E(x)$ is defined to be the expected or average value of x.

Sample Statistics Many types of sample statistics will be defined. Two very special types are the **sample mean,** designated as $\bar{x}$, and the sample standard deviation, designated as s. These are, by definition, random variables. Parameters like μ and σ are not random variables; they are fixed constants.

Example In an experiment, six random numbers (rounded to four decimal places) were observed from the uniform distribution $f(x) = 1$ for $0 \leq x < 1$:

.1009
.3754
.0842
.9901
.1280
.6606

The sample mean corresponds to the arithmetic average of the observations, which will be designated as x_1 through x_6, where

$$\bar{x} = \frac{1}{n} \sum_{i=1}^{n} x_i \text{ with } n = 6$$
$$= (1/6)(.1009 + .3754 + \cdots + .6606)$$
$$= .3899$$

The sample standard deviation s is defined by the computation

$$s = \sqrt{\frac{\sum (x_i - \bar{x})^2}{n - 1}}$$
$$= \sqrt{\frac{n \sum x_i^2 - (\sum x_i)^2}{n(n - 1)}}$$

In effect, this represents the root of a statistical average of the squares. The divisor quantity $(n - 1)$ will be referred to as the degrees of freedom.

The value of $n - 1$ is used in the denominator because the deviations from the sample average must total zero, or

$$\sum (x_i - \bar{x}) = 0$$

Thus knowing $n - 1$ values of $x_i - \bar{x}$ permits calculation of the nth value of $x_i - \bar{x}$.

The sample value of the standard deviation for the data given is .3686. The following is a tabulation of the deviations $(x_i - \bar{x}_j)$ for the data:

x	$x - \bar{x}$
.1009	$-.2890$
.3754	$-.0145$
.0842	$-.3057$
.9901	.6002
.1280	$-.2619$
.6606	.2707
$\bar{x} = .3899$	$s = .3686$

In effect, the standard deviation quantifies the relative magnitude of the deviation numbers, i.e., a special type of "average" of the distance of points from their center. In statistical theory, it turns out that the corresponding variance quantities s^2 have remarkable properties which make possible broad generalities for sample statistics and therefore also their counterparts, the standard deviations.

For the corresponding population, the parameter values are $\mu = .50$ and $\sigma = .2887$. If, instead of using individual observations only, averages of six were reported, then the corresponding population parameter values would be $\mu = .50$ and $\sigma_{\bar{x}} = \sigma/\sqrt{6} = .1179$. The corresponding variance for an average will be written as $\text{Var}(\bar{x}) = \text{var}(x)/n$. In effect, the variance of an average is inversely proportional to the sample size n, which reflects the fact that sample averages will tend to cluster about μ much more closely than individual observations. This is illustrated in greater detail under "Measurement Data and Sampling Densities."

Characterization of Chance Occurrences To deal with a broad area of statistical applications, it is necessary to characterize the way in which random variables will vary by chance alone. The basic foundation for this characteristic is laid through a density called the gaussian, or normal, distribution.

Determining the area under the normal curve is a very tedious procedure. However, by standardizing a random variable that is normally distributed, it is possible to relate all normally distributed random variables to one table. The standardization is defined by the identity $z = (x - \mu)/\sigma$, where z is called the unit normal. Further, it is possible to standardize the sampling distribution of averages $\bar{x}$ by the identity $z = (\bar{x} - \mu)/(\sigma/\sqrt{n})$.

A remarkable property of the normal distribution is that, almost regardless of the distribution of x, sample averages $\bar{x}$ will approach the gaussian distribution as n gets large. Even for relatively small values of n, of about 10, the approximation in most cases is quite close. For example, sample averages of size 10 from the uniform distribution will have essentially a gaussian distribution.

Also, in many applications involving count data, the normal distribution can be used as a close approximation. In particular, the approximation is quite close for the binomial distribution within certain guidelines.

ENUMERATION DATA AND PROBABILITY DISTRIBUTIONS

Introduction   Many types of statistical applications are characterized by enumeration data in the form of counts. Examples are the number of lost-time accidents in a plant, the number of defective items in a sample, and the number of items in a sample that fall within several specified categories.

The sampling distribution of count data can be characterized through probability distributions. In many cases, count data are appropriately interpreted through their corresponding distributions. However, in other situations analysis is greatly facilitated through distributions which have been developed for measurement data. Examples of each will be illustrated in the following subsections.

Binomial Probability Distribution

Nature   Consider an experiment in which each outcome is classified into one of two categories, one of which will be defined as a success and the other as a failure. Given that the probability of success p is constant from trial to trial, then the probability of observing a specified number of successes x in n trials is defined by the binomial distribution. The sequence of outcomes is called a **Bernoulli process,**

Nomenclature
n = total number of trials
x = number of successes in n trials
p = probability of observing a success on any one trial
$\hat{p} = x/n$, the proportion of successes in n trials

Probability Law

$$P(x) = P\left(\frac{x}{n}\right) = \binom{n}{x} p^x (1-p)^{n-x} \qquad x = 0, 1, 2, \ldots, n$$

where $\binom{n}{x} = \dfrac{n!}{x!(n-x)!}$

Properties $E(x) = np$ $\mathrm{Var}(x) = np(1-p)$

$\qquad\qquad\quad E(\hat{p}) = p$ $\mathrm{Var}(\hat{p}) = p(1-p)/n$

Geometric Probability Distribution

Nature   Consider an experiment in which each outcome is classified into one of two categories, one of which will be defined as a success and the other as a failure. Given that the probability of success p is constant from trial to trial, then the probability of observing the first success on the xth trial is defined by the geometric distribution.

Nomenclature
p = probability of observing a success on any one trial
x = the number of trials to obtain the first success

Probability Law

$$P(x) = p(1-p)^{x-1} \qquad x = 1, 2, 3, \ldots$$

Properties

$$E(x) = 1/p \qquad \mathrm{Var}(x) = (1-p)/p^2$$

Poisson Probability Distribution

Nature   In monitoring a moving threadline, one criterion of quality would be the frequency of broken filaments. These can be identified as they occur through the threadline by a broken-filament detector mounted adjacent to the threadline. In this context, the random occurrences of broken filaments can be modeled by the Poisson distribution. This is called a Poisson process and corresponds to a probabilistic description of the frequency of defects or, in general, what are called arrivals at points on a continuous line or in time. Other examples include:
1.   The number of cars (arrivals) that pass a point on a high-speed highway between 10:00 and 11:00 A.M. on Wednesdays
2.   The number of customers arriving at a bank between 10:00 and 10:10 A.M.
3.   The number of telephone calls received through a switchboard between 9:00 and 10:00 A.M.
4.   The number of insurance claims that are filed each week

5.   The number of spinning machines that break down during 1 day at a large plant.

Nomenclature
x = total number of arrivals in a total length L or total period T
a = average rate of arrivals for a unit length or unit time
$\lambda = aL$ = expected or average number of arrivals for the total length L
$\lambda = aT$ = expected or average number of arrivals for the total time T

Probability Law   Given that a is constant for the total length L or period T, the probability of observing x arrivals in some period L or T is given by

$$P(x) = \frac{\lambda^x}{x!} e^{-\lambda} \qquad x = 0, 1, 2, \ldots$$

Properties $E(x) = \lambda$ $\mathrm{Var}(x) = \lambda$

Example   The number of broken filaments in a threadline has been averaging .015 per yard. What is the probability of observing exactly two broken filaments in the next 100 yd? In this example, $a = .015/\mathrm{yd}$ and $L = 100$ yd; therefore $\lambda = (.015)(100) = 1.5$:

$$P(x = 2) = \frac{(1.5)^2}{2!} e^{-1.5} = .2510$$

Example   A commercial item is sold in a retail outlet as a unit product. In the past, sales have averaged 10 units per month with no seasonal variation. The retail outlet must order replacement items 2 months in advance. If the outlet starts the next 2-month period with 25 items on hand, what is the probability that it will stock out before the end of the second month?

Given $a = 10/\mathrm{month}$, then $\lambda = 10 \times 2 = 20$ for the total period of 2 months:

$$P(x \geq 26) = \sum_{26}^{\infty} P(x) = 1 - \sum_{0}^{25} P(x)$$

$$\sum_{0}^{25} \frac{20^x}{x!} e^{-20} = e^{-20}\left[1 + \frac{20}{1} + \frac{20^2}{2!} + \cdots + \frac{20^{25}}{25!}\right]$$

$$= .887815$$

Therefore $P(x \geq 26) = .112185$ or roughly an 11 percent chance of a stockout.

Hypergeometric Probability Distribution

Nature   In an experiment in which one samples from a relatively small group of items, each of which is classified in one of two categories, A or B, the hypergeometric distribution can be defined. One example is the probability of drawing two red and two black cards from a deck of cards. The hypergeometric distribution is the analog of the binomial distribution when successive trials are not independent, i.e., when the total group of items is not infinite. This happens when the drawn items are not replaced.

Nomenclature
N = total group size
n = sample group size
X = number of items in the total group with a specified attribute A
$N - X$ = number of items in the total group with the other attribute B
x = number of items in the sample with a specified attribute A
$n - x$ = number of items in the sample with the other attribute B

	Population	Sample
Category A	X	x
Category B	$N - X$	$n - x$
Total	N	n

Probability Law

$$P(x) = \binom{N-X}{n-x}\binom{X}{x} \Big/ \binom{N}{n}$$

$$E(x) = \frac{nX}{N}$$

$$\mathrm{var}(x) = nP(1-P)\frac{N-n}{N-1}$$

Example   What is the probability that an appointed special committee of 4 has no female members when the members are randomly selected from a candidate group of 10 males and 7 females?

$$P(x=0) = \frac{\binom{10}{4}\binom{7}{0}}{\binom{17}{4}} = .0882$$

Example   A bin contains 300 items, of which 240 are good and 60 are defective. In a sample of 6 what is the probability of selecting 4 good and 2 defective items by chance?

$$P(x) = \frac{\binom{240}{4}\binom{60}{2}}{\binom{300}{6}} = .2478$$

Multinomial Distribution

Nature   For an experiment in which successive outcomes can be classified into two or more categories and the probabilities associated with the respective outcomes remain constant, then the experiment can be characterized through the multinomial distribution.

Nomenclature
n = total number of trials
k = total number of distinct categories
p_j = probability of observing category j on any one trial, $j = 1, 2, \ldots, k$
x_j = total number of occurrences in category j in n trials

Probability Law

$$P(x_1, x_2, \ldots, x_k) = \frac{n!}{x_1! x_2! \ldots x_k!} p_1^{x_1} p_2^{x_2} \cdots p_k^{x_k}$$

Example   In tossing a die 12 times, what is the probability that each face value will occur exactly twice?

$$p(2,2,2,2,2,2) = \frac{12!}{2!2!2!2!2!2!} \left(\frac{1}{6}\right)^2 \left(\frac{1}{6}\right)^2 \left(\frac{1}{6}\right)^2 \left(\frac{1}{6}\right)^2 \left(\frac{1}{6}\right)^2 \left(\frac{1}{6}\right)^2 = .003438$$

MEASUREMENT DATA AND SAMPLING DENSITIES

Introduction   The following example data are used throughout this subsection to illustrate concepts. Consider, for the purpose of illustration, that five synthetic-yarn samples have been selected randomly from a production line and tested for tensile strength on each of 20 production days. For this, assume that each group of five corresponds to a day, Monday through Friday, for a period of 4 weeks:

Monday 1	Tuesday 2	Wednesday 3	Thursday 4	Friday 5	Groups of 25 pooled
36.48	38.06	35.28	36.34	36.73	
35.33	31.86	36.58	36.25	37.17	
35.92	33.81	38.81	30.46	33.07	
32.28	30.30	33.31	37.37	34.27	
31.61	35.27	33.88	37.52	36.94	
$\bar{x}$ = 34.32	33.86	35.57	35.59	35.64	35.00
s = 2.22	3.01	2.22	2.92	1.85	2.40
6	7	8	9	10	
38.67	36.62	35.03	35.80	36.82	
32.08	33.05	36.22	33.16	36.49	
33.79	35.43	32.71	35.19	32.83	
32.85	36.63	32.52	32.91	32.43	
35.22	31.46	27.23	35.44	34.16	
$\bar{x}$ = 34.52	34.64	32.74	34.50	34.54	34.19
s = 2.60	2.30	3.46	1.36	2.03	2.35
11	12	13	14	15	
39.63	34.52	36.05	36.64	31.57	
34.38	37.39	35.36	31.18	36.21	
36.51	34.16	35.00	36.13	33.84	
30.00	35.76	33.61	37.51	35.01	
39.64	37.63	36.98	39.05	34.95	
$\bar{x}$ = 36.03	35.89	35.40	36.10	34.32	35.55
s = 4.04	1.59	1.25	2.96	1.75	2.42

Monday 16	Tuesday 17	Wednesday 18	Thursday 19	Friday 20	Groups of 25 pooled
37.68	35.97	33.71	35.61	36.65	
36.38	35.92	32.34	37.13	37.91	
38.43	36.51	33.29	31.37	42.18	
39.07	33.89	32.81	35.89	39.25	
33.06	36.01	37.13	36.33	33.32	
$\bar{x}$ = 36.92	35.66	33.86	35.27	37.86	35.91
s = 2.38	1.02	1.90	2.25	3.27	2.52

Pooled sample of 100: $\bar{x}$ = 35.16 s = 2.47

Even if the process were at steady state, tensile strength, a key property would still reflect some variation. Steady state, or stable operation of any process, has associated with it a characteristic variation. Superimposed on this is the testing method, which is itself a process with its own characteristic variation. The observed variation is a composite of these two variations.

Assume that the table represents "typical" production-line performance. The numbers themselves have been generated on a computer and represent random observations from a population with $\mu = 35$ and a population standard deviation $\sigma = 2.45$. The sample values reflect the way in which tensile strength can vary by chance alone. In practice, a production supervisor unschooled in statistics but interested in high tensile performance would be despondent on the eighth day and exuberant on the twentieth day. If the supervisor were more concerned with uniformity, the lowest and highest points would have been on the eleventh and seventeenth days.

An objective of statistical analysis is to serve as a guide in decision making in the context of normal variation. In the case of the production supervisor, it is to make a decision, with a high probability of being correct, that something has in fact changed the operation.

Suppose that an engineering change has been made in the process and five new tensile samples have been tested with the results:

36.81
38.34 $\bar{x}$ = 37.14
34.87 s = 1.85
39.58
36.12

In this situation, management would inquire whether the product has been improved by increased tensile strength. To answer this question, in addition to a variety of analogous questions, it is necessary to have some type of scientific basis upon which to draw a conclusion.

A scientific basis for the evaluation and interpretation of data is contained in the accompanying table descriptions. These tables characterize the way in which sample values will vary by chance alone in the context of individual observations, averages, and variances.

Table number	Designated symbol	Variable	Sampling distribution of
3-4	z	$\dfrac{x - \mu}{\sigma}$	Observations°
3-4	z	$\dfrac{\bar{x} - \mu}{\sigma/\sqrt{n}}$	Averages
3-5	t	$\dfrac{\bar{x} - \mu}{s/\sqrt{n}}$	Averages when σ is unknown°
3-6	χ^2	$(s^2/\sigma^2)(\text{df})$	Variances°
3-7	F	s_1^2/s_2^2	Ratio of two independent sample variances°

°When sampling from a gaussian distribution.

Normal Distribution of Observations   Many types of data follow what is called the gaussian, or bell-shaped, curve; this is especially true of averages. Basically, the gaussian curve is a purely mathematical function which has very special properties. However, owing to some mathematically intractable aspects primary use of the function is restricted to tabulated values.

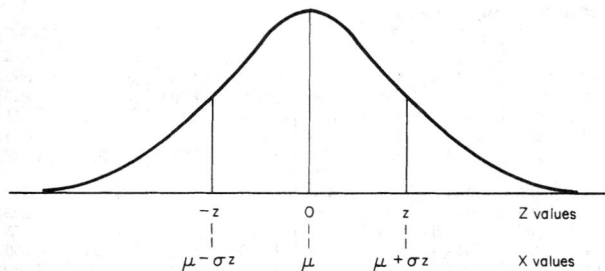

FIG. 3-56 Transformation of z values.

0 and 2 is .4772; therefore, the area between 1 and 2 is .4772 − .3413 = .1359.

Also, since the normal curve is symmetric, areas to the left can be determined in exactly the same way. For example, the area between −2 and +1 would include the area between −2 and 0, .4772 (the same as 0 to 2), plus the area between 0 and 1, .3413, or a total area of .8185.

Any types of observation which are applicable to the normal curve can be transformed to Z values by the relationship $z = (x − \mu)/\sigma$ and, conversely, Z values to x values by $x = \mu + \sigma z$, as shown in Fig. 3-56. For example, for tensile strength, with $\mu = 35$ and $\sigma = 2.45$, this would dictate $z = (x − 35)/2.45$ and $x = 35 + 2.45z$.

Example What proportion of tensile values will fall between 34 and 36?

$$z_1 = (34 − 35)/2.45 = −.41 \qquad z_2 = (36 − 35)/2.45 = .41$$

$$P[−.41 \leq z \leq .41] = .3182, \text{ or roughly 32 percent}$$

The value 0.3182 is interpolated from Table 3-4 using $z = \pm0.40$, $A = 0.3108$, and $z = \pm0.45$, $A = 0.3473$.

Example What midrange of tensile values will include 95 percent of the sample values? Since $P[−1.96 \leq z \leq 1.96] = .95$, the corresponding values of x are

$$x_1 = 35 − 1.96(2.45) = 30.2$$

$$x_2 = 35 + 1.96(2.45) = 39.8$$

or

$$P[30.2 \leq x \leq 39.8] = .95$$

Basically, the tabled values represent area (proportions or probability) associated with a scaling variable designated by Z in Fig. 3-56. The normal curve is centered at 0, and for particular values of Z, designated as z, the tabulated numbers represent the corresponding area under the curve between 0 and z. For example, between 0 and 1 the area is .3413. (Get this number from Table 3-4. The value of A includes the area on both sides of zero. Thus we want $A/2$. For $z = 1$, $A = 0.6827$, $A/2 = 0.3413$. For $z = 2$, $A/2 = 0.4772$.) The area between

TABLE 3-4   Ordinates and Areas between Abscissa Values −z and +z of the Normal Distribution Curve

z	X	Y	A	$1 − A$	z	X	Y	A	$1 − A$
0	μ	0.399	0.0000	1.0000	±1.50	$\mu \pm 1.50\sigma$	0.1295	0.8664	0.1336
±0.05	$\mu \pm 0.05\sigma$	.398	.0399	0.9601	±1.55	$\mu \pm 1.55\sigma$	.1200	.8789	.1211
± .10	$\mu \pm .10\sigma$	.397	.0797	.9203	±1.60	$\mu \pm 1.60\sigma$	.1109	.8904	.1096
± .15	$\mu \pm .15\sigma$	.394	.1192	.8808	±1.65	$\mu \pm 1.65\sigma$	.1023	.9011	.0989
± .20	$\mu \pm .20\sigma$	.391	.1585	.8415	±1.70	$\mu \pm 1.70\sigma$	.0940	.9109	.0891
± .25	$\mu \pm .25\sigma$	.387	.1974	.8026	±1.75	$\mu \pm 1.75\sigma$	.0863	.9199	.0801
± .30	$\mu \pm .30\sigma$	.381	.2358	.7642	±1.80	$\mu \pm 1.80\sigma$	.0790	.9281	.0719
± .35	$\mu \pm .35\sigma$	.375	.2737	.7263	±1.85	$\mu \pm 1.85\sigma$	.0721	.9357	.0643
± .40	$\mu \pm .40\sigma$	.368	.3108	.6892	±1.90	$\mu \pm 1.90\sigma$	.0656	.9446	.0574
± .45	$\mu \pm .45\sigma$	.361	.3473	.6527	±1.95	$\mu \pm 1.95\sigma$	.0596	.9488	.0512
± .50	$\mu \pm .50\sigma$	.352	.3829	.6171	±2.00	$\mu \pm 2.00\sigma$	.0540	.9545	.0455
± .55	$\mu + .55\sigma$	.343	.4177	.5823	±2.05	$\mu \pm 2.05\sigma$	.0488	.9596	.0404
± .60	$\mu \pm .60\sigma$	.333	.4515	.5485	±2.10	$\mu \pm 2.10\sigma$	.0440	.9643	.0357
± .65	$\mu \pm .65\sigma$	.323	.4843	.5157	±2.15	$\mu \pm 2.15\sigma$	.0396	.9684	.0316
± .70	$\mu \pm .70\sigma$	.312	.5161	.4839	±2.20	$\mu \pm 2.20\sigma$	.0335	.9722	.0278
± .75	$\mu \pm .75\sigma$	.301	.5467	.4533	±2.25	$\mu \pm 2.25\sigma$	.0317	.9756	.0244
± .80	$\mu \pm .80\sigma$	.290	.5763	.4237	±2.30	$\mu \pm 2.30\sigma$	.0283	.9786	.0214
± .85	$\mu \pm .85\sigma$	.278	.6047	.3953	±2.35	$\mu \pm 2.35\sigma$	.0252	.9812	.0188
± .90	$\mu \pm .90\sigma$	.266	.6319	.3681	±2.40	$\mu \pm 2.40\sigma$	.0224	.9836	.0164
± .95	$\mu \pm .95\sigma$	.254	.6579	.3421	±2.45	$\mu \pm 2.45\sigma$	.0198	.9857	.0143
±1.00	$\mu \pm 1.00\sigma$	.242	.6827	.3173	±2.50	$\mu \pm 2.50\sigma$	.0175	.9876	.0124
±1.05	$\mu \pm 1.05\sigma$	.230	.7063	.2937	±2.55	$\mu \pm 2.55\sigma$	.0154	.9892	.0108
±1.10	$\mu \pm 1.10\sigma$	.218	.7287	.2713	±2.60	$\mu \pm 2.60\sigma$	.0136	.9907	.0093
±1.15	$\mu \pm 1.15\sigma$	.206	.7499	.2501	±2.65	$\mu \pm 2.65\sigma$	.0119	.9920	.0080
±1.20	$\mu \pm 1.20\sigma$	.194	.7699	.2301	±2.70	$\mu \pm 2.70\sigma$	.0104	.9931	.0069
±1.25	$\mu \pm 1.25\sigma$	.183	.7887	.2113	±2.75	$\mu \pm 2.75\sigma$	.0091	.9940	.0060
±1.30	$\mu \pm 1.30\sigma$	.171	.8064	.1936	±2.80	$\mu \pm 2.80\sigma$	.0079	.9949	.0051
±1.35	$\mu \pm 1.35\sigma$	.160	.8230	.1770	±2.85	$\mu \pm 2.85\sigma$	.0069	.9956	.0044
±1.40	$\mu \pm 1.40\sigma$	.150	.8385	.1615	±2.90	$\mu \pm 2.90\sigma$	.0060	.9963	.0037
±1.45	$\mu \pm 1.45\sigma$	.139	.8529	.1471	±2.95	$\mu \pm 2.95\sigma$	.0051	.9968	.0032
±1.50	$\mu \pm 1.50\sigma$	.130	.8664	.1336	±3.00	$\mu \pm 3.00\sigma$	.0044	.9973	.0027
					±4.00	$\mu \pm 4.00\sigma$	.0001	.99994	.00006
					±5.00	$\mu \pm 5.00s$	.000001	.9999994	.0000006
±0.000	μ	0.3989	.0000	1.0000	±1.036	$\mu \pm 1.036\sigma$	0.2331	0.7000	0.3000
± .126	$\mu \pm 0.126\sigma$	.3958	.1000	0.9000	±1.282	$\mu \pm 1.282\sigma$	.1755	.8000	.2000
± .253	$\mu \pm .253\sigma$	.3863	.2000	.8000	±1.645	$\mu \pm 1.645\sigma$	.1031	.9000	.1000
± .385	$\mu \pm .385\sigma$	.3704	.3000	.7000	±1.960	$\mu + 1.960\sigma$	.0584	.9500	.0500
± .524	$\mu \pm .524\sigma$	.3477	.4000	.6000	±2.576	$\mu \pm 2.576\sigma$	.0145	.9900	.0100
± .674	$\mu \pm .674\sigma$	.3178	.5000	.5000	±3.291	$\mu \pm 3.291\sigma$	.0018	.9990	.0010
± .842	$\mu \pm .842\sigma$	.2800	.6000	.4000	±3.891	$\mu \pm 3.891\sigma$	.0002	.9999	.0001

Normal Distribution of Averages   An examination of the tensile-strength data previously tabulated would show that the range (largest minus the smallest) of tensile strength within days averages 5.72. The average range in $\bar{x}$ values within each week is 2.37, while the range in the four weekly averages is 1.72. This reflects the fact that averages tend to be less variable in a predictable way. Given that the variance of x is var $(x) = \sigma^2$, then the variance of $\bar{x}$ based on n observations is var $(\bar{x}) = \sigma^2/n$.

For averages of n observations, the corresponding relationship for the Z-scale relationship is

$$z = (\bar{x} - \mu/\sigma/\sqrt{n}) \qquad \text{or} \qquad \bar{x} = \mu + \frac{\sigma}{\sqrt{n}}z$$

Example   What proportion of daily tensile averages will fall between 34 and 36?

$$z_1 = (34 - 35)/(2.45/\sqrt{5}) = -.91 \qquad z_2 = (36 - 35)/(2.45/\sqrt{5}) = .91$$

$$P[-.91 \le z \le .91] = .6372, \text{ or roughly 64 percent}$$

Example   What midrange of daily tensile averages will include 95 percent of the sample values?

$$x_1 = 35 - 1.96(2.45/\sqrt{5}) = 32.85$$

$$x_2 = 35 + 1.96(2.45/\sqrt{5}) = 37.15$$

or

$$P[32.85 \le \bar{x} \le 37.15] = .95$$

Example   What proportion of weekly tensile averages will fall between 34 and 36?

$$z_1 = (34 - 35)/(2.45/\sqrt{25}) = -2.04$$

$$z_2 = (36 - 35)/(2.45/\sqrt{25}) = 2.04$$

$$P[-2.04 \le z \le 2.04] = .9586, \text{ or roughly 96 percent}$$

Distribution of Averages   The normal curve relies on a knowledge of σ, or in special cases, when it is unknown, s can be used with the normal curve as an approximation when $n > 30$. For example, with $n > 30$ the intervals $\bar{x} \pm s$ and $\bar{x} \pm 2s$ will include roughly 68 and 95 percent of the sample values respectively when the distribution is normal.

In applications sample sizes are usually small and σ unknown. In these cases, the t distribution can be used where

$$t = (\bar{x} - \mu)/(s/\sqrt{n}) \qquad \text{or} \qquad \bar{x} = \mu + ts/\sqrt{n}$$

The t distribution is also symmetric and centered at zero. It is said to be robust in the sense that even when the individual observations x are not normally distributed, sample averages of x have distributions which tend toward normality as n gets large. Even for small n of 5 through 10, the approximation is usually relatively accurate.

In reference to the tensile-strength table, consider the summary statistics $\bar{x}$ and s by days. For each day, the t statistic could be computed. If this were repeated over an extensive simulation and the resultant t quantities plotted in a frequency distribution, they would match the corresponding distribution of t values summarized in Table 3-5.

Since the t distribution relies on the sample standard deviation s, the resultant distribution will differ according to the sample size n. To designate this difference, the respective distributions are classified according to what are called the degrees of freedom and abbreviated as df. In simple problems, the df are just the sample size minus 1. In more complicated applications the df can be different. In general, degrees of freedom are the number of quantities minus the number of constraints. For example, four numbers in a square which must have row and column sums equal to zero have only one df, i.e., four numbers minus three constraints (the fourth constraint is redundant).

Example   For a sample size $n = 5$, what values of t define a midarea of 90 percent. For 4 df the tabled value of t corresponding to a midarea of 90 percent is 2.132; i.e., $P[-2.132 \le t \le 2.132] = .90$.

Example   For a sample size $n = 25$, what values of t define a midarea of 95 percent? For 24 df the tabled value of t corresponding to a midarea of 95 percent is 2.064; i.e., $P[-2.064 \le t \le 2.064] = .95$.

TABLE 3-5   Values of t

df	$t_{.40}$	$t_{.30}$	$t_{.20}$	$t_{.10}$	$t_{.05}$	$t_{.025}$	$t_{.01}$	$t_{.005}$
1	0.325	0.727	1.376	3.078	6.314	12.706	31.821	63.657
2	.289	.617	1.061	1.886	2.920	4.303	6.965	9.925
3	.277	.584	0.978	1.638	2.353	3.182	4.541	5.841
4	.271	.569	.941	1.533	2.132	2.776	3.747	4.604
5	.267	.559	.920	1.476	2.015	2.571	3.365	4.032
6	.265	.553	.906	1.440	1.943	2.447	3.143	3.707
7	.263	.549	.896	1.415	1.895	2.365	2.998	3.499
8	.262	.546	.889	1.397	1.860	2.306	2.896	3.355
9	.261	.543	.883	1.383	1.833	2.262	2.821	3.250
10	.260	.542	.879	1.372	1.812	2.228	2.764	3.169
11	.260	.540	.876	1.363	1.796	2.201	2.718	3.106
12	.259	.539	.873	1.356	1.782	2.179	2.681	3.055
13	.259	.538	.870	1.350	1.771	2.160	2.650	3.012
14	.258	.537	.868	1.345	1.761	2.145	2.624	2.977
15	.258	.536	.866	1.341	1.753	2.131	2.602	2.947
16	.258	.535	.865	1.337	1.746	2.120	2.583	2.921
17	.257	.534	.863	1.333	1.740	2.110	2.567	2.898
18	.257	.534	.862	1.330	1.734	2.101	2.552	2.878
19	.257	.533	.861	1.328	1.729	2.093	2.539	2.861
20	.257	.533	.860	1.325	1.725	2.086	2.528	2.845
21	.257	.532	.859	1.323	1.721	2.080	2.518	2.831
22	.256	.532	.858	1.321	1.717	2.074	2.508	2.819
23	.256	.532	.858	1.319	1.714	2.069	2.500	2.807
24	.256	.531	.857	1.318	1.711	2.064	2.492	2.797
25	.256	.531	.856	1.316	1.708	2.060	2.485	2.787
26	.256	.531	.856	1.315	1.706	2.056	2.479	2.779
27	.256	.531	.855	1.314	1.703	2.052	2.473	2.771
28	.256	.530	.855	1.313	1.701	2.048	2.467	2.763
29	.256	.530	.854	1.311	1.699	2.045	2.462	2.756
30	.256	.530	.854	1.310	1.697	2.042	2.457	2.750
40	.255	.529	.851	1.303	1.684	2.021	2.423	2.704
60	.254	.527	.848	1.296	1.671	2.000	2.390	2.660
120	.254	.526	.845	1.289	1.658	1.980	2.358	2.617
∞	.253	.524	.842	1.282	1.645	1.960	2.326	2.576

Above values refer to a single tail outside the indicated limit of t. For example, for 95 percent of the area to be between $-t$ and $+t$ in a two-tailed t distribution, use the values for $t_{0.025}$ or 2.5 percent for each tail.

Example   What is the sample value of t for the first day of tensile data?

$$\text{Sample } t = (34.32 - 35)/(2.22/\sqrt{5}) = -.68$$

Note that on the average 90 percent of all such sample values would be expected to fall within the interval ± 2.132.

t Distribution for the Difference in Two Sample Means

Population Variances Are Equal   The t distribution can be readily extended to the difference in two sample means when the respective populations have the same variance σ^2:

$$t = \frac{(\bar{x}_1 - \bar{x}_2) - (\mu_1 - \mu_2)}{s_p\sqrt{1/n_1 + 1/n_2}}$$

where s_p^2 is a pooled variance defined by

$$s_p^2 = \frac{(n_1 - 1)s_1^2 + (n_2 - 1)s_2^2}{(n_1 - 1) + (n_2 - 1)}$$

In this application, the t distribution has $(n_1 + n_2 - 2)$ df.

Population Variances Are Unequal   When population variances are unequal, an approximate t quantity can be used:

$$t = \frac{(\bar{x}_1 - \bar{x}_2) - (\mu_1 - \mu_2)}{\sqrt{a + b}}$$

with

$$a = s_1^2/n_1 \qquad b = s_2^2/n_2$$

and

$$df = \frac{(a + b)^2}{a^2/(n_1 - 1) + b^2/(n_2 - 1)}$$

Chi-Square Distribution   For some industrial applications, product uniformity is of primary importance. The sample standard deviation s is most often used to characterize uniformity. In dealing with this problem, the chi-square distribution can be used where $\chi^2 = (s^2/\sigma^2)$ (df). The chi-square distribution is a family of distributions which are defined by the degrees of freedom associated with the sample variance. For most applications, df is equal to the sample size minus 1.

The probability distribution function is

$$p(y) = y_0 y^{df-2} \exp\left[\frac{-(df)^2}{2}\right]$$

where y_0 is chosen such that the integral of $p(y)$ over all y is one.

In terms of the tensile-strength table previously given, the respective chi-square sample values for the daily, weekly, and monthly figures could be computed. The corresponding df would be 4, 24, and 99 respectively. These numbers would represent sample values from the respective distributions which are summarized in Table 3-6.

In a manner similar to the use of the t distribution, chi square can be interpreted in a direct probabilistic sense corresponding to a midarea of $(1 - \alpha)$:

$$P[\chi_1^2 \le (s^2/\sigma^2)(df) \le \chi_2^2] = 1 - \alpha$$

where χ_1^2 corresponds to a lower-tail area of $\alpha/2$ and χ_2^2 an upper-tail area of $\alpha/2$.

The basic underlying assumption for the mathematical derivation of chi square is that a random sample was selected from a normal distribution with variance σ^2. When the population is not normal but skewed, square probabilities could be substantially in error.

Example   On the basis of a sample size $n = 5$, what midrange of values will include the sample ratio s/σ with a probability of 90 percent?
Use Table 3-6 for 4 df and read $\chi_1^2 = 0.484$ for a lower tail area of 0.05/2, 2.5 percent, and read $\chi_2^2 = 11.1$ for an upper tail area of 97.5 percent.

$$P[.484 \le (s^2/\sigma^2)(4) \le 11.1] = .90$$

or

$$P[.35 \le s/\sigma \le 1.66] = .90$$

Example   On the basis of a sample size $n = 25$, what midrange of values will include the sample ratio s/σ with a probability of 90 percent?

$$P[12.4 \le (s^2/\sigma^2)(24) \le 39.4] = .90$$

or

$$P[.72 \le s/\sigma \le 1.28] = .90$$

This states that the sample standard deviation will be at least 72 percent and not more than 128 percent of the population variance 90 percent of the time. Conversely, 10 percent of the time the standard deviation will underestimate or overestimate the population standard deviation by the corresponding amount. Even for samples as large as 25, the relative reliability of a sample standard deviation is poor.

The chi-square distribution can be applied to other types of application which are of an entirely different nature. These include applications which are discussed under "Goodness-of-Fit Test" and "Two-Way Test for Independence of Count Data." In these applications, the mathematical formulation and context are entirely different, but they do result in the same table of values.

F Distribution   In reference to the tensile-strength table, the successive pairs of daily standard deviations could be ratioed and squared. These ratios of variance would represent a sample from a distribution called the F distribution or F ratio. In general, the F ratio is defined by the identity

$$F(\gamma_1, \gamma_2) = s_1^2/s_2^2$$

where γ_1 and γ_2 correspond to the respective df's for the sample variances. In statistical applications, it turns out that the primary area of interest is found when the ratios are greater than 1. For this reason, most tabled values are defined for an upper-tail area. However, defining F_2 to be that value corresponding to an upper-tail area of $\alpha/2$, then F_1 for a lower-tail area of $\alpha/2$ can be determined through the identity

TABLE 3-6   Percentiles of the χ^2 Distribution

df					Percent					
	0.5	1	2.5	5	10	90	95	97.5	99	99.5
1	0.000039	0.00016	0.00098	0.0039	0.0158	2.71	3.84	5.02	6.63	7.88
2	.0100	.0201	.0506	.1026	.2107	4.61	5.99	7.38	9.21	10.60
3	.0717	.115	.216	.352	.584	6.25	7.81	9.35	11.34	12.84
4	.207	.297	.484	.711	1.064	7.78	9.49	11.14	13.28	14.86
5	.412	.554	.831	1.15	1.61	9.24	11.07	12.83	15.09	16.75
6	.676	.872	1.24	1.64	2.20	10.64	12.59	14.45	16.81	18.55
7	.989	1.24	1.69	2.17	2.83	12.02	14.07	16.01	18.48	20.28
8	1.34	1.65	2.18	2.73	3.49	13.36	15.51	17.53	20.09	21.96
9	1.73	2.09	2.70	3.33	4.17	14.68	16.92	19.02	21.67	23.59
10	2.16	2.56	3.25	3.94	4.87	15.99	18.31	20.48	23.21	25.19
11	2.60	3.05	3.82	4.57	5.58	17.28	19.68	21.92	24.73	26.76
12	3.07	3.57	4.40	5.23	6.30	18.55	21.03	23.34	26.22	28.30
13	3.57	4.11	5.01	5.89	7.04	19.81	22.36	24.74	27.69	29.82
14	4.07	4.66	5.63	6.57	7.79	21.06	23.68	26.12	29.14	31.32
15	4.60	5.23	6.26	7.26	8.55	22.31	25.00	27.49	30.58	32.80
16	5.14	5.81	6.91	7.96	9.31	23.54	26.30	28.85	32.00	34.27
18	6.26	7.01	8.23	9.39	10.86	25.99	28.87	31.53	34.81	37.16
20	7.43	8.26	9.59	10.85	12.44	28.41	31.41	34.17	37.57	40.00
24	9.89	10.86	12.40	13.85	15.66	33.20	36.42	39.36	42.98	45.56
30	13.79	14.95	16.79	18.49	20.60	40.26	43.77	46.98	50.89	53.67
40	20.71	22.16	24.43	26.51	29.05	51.81	55.76	59.34	63.69	66.77
60	35.53	37.48	40.48	43.19	46.46	74.40	79.08	83.30	88.38	91.95
120	83.85	86.92	91.58	95.70	100.62	140.23	146.57	152.21	158.95	163.64

For large values of degrees of freedom the approximate formula

$$\chi_a^2 = n\left(1 - \frac{2}{9n} + z_a\sqrt{\frac{2}{9n}}\right)^3$$

where z_a is the normal deviate and n is the number of degrees of freedom, may be used. For example, $\chi_{.99}^2 = 60[1 - 0.00370 + 2.326(0.06086)]^3 = 60(1.1379)^3 = 88.4$ for the 99th percentile for 60 degrees of freedom.

$$F_1(\gamma_1, \gamma_2) = 1/F_2(\gamma_2, \gamma_1)$$

The F distribution, similar to the chi square, is sensitive to the basic assumption that sample values were selected randomly from a normal distribution.

Example For two sample variances with 4 df each, what limits will bracket their ratio with a midarea probability of 90 percent?

Use Table 3-7 with 4 df in the numerator and denominator and upper 5 percent points (to get both sides totaling 10 percent). The entry is 6.39. Thus:

$$P[1/6.39 \leq s_1^2/s_2^2 \leq 6.39] = .90$$

or

$$P[.40 \leq s_1/s_2 \leq 2.53] = .90$$

Confidence Interval for a Mean For the daily sample tensile-strength data with 4 df it is known that $P[-2.132 \leq t \leq 2.132] = .90$. This states that 90 percent of all samples will have sample t values which fall within the specified limits. In fact, for the 20 daily samples exactly 16 do fall within the specified limits (note that the binomial with $n = 20$ and $p = .90$ would describe the likelihood of exactly none through 20 falling within the prescribed limits—the sample of 20 is only a sample).

Consider the new daily sample (with $n = 5$, $\bar{x} = 37.14$, and $s = 1.85$) which was observed after a process change. In this case, the same probability holds. However, in this instance the sample value of t cannot be computed, since the new μ, under the process change, is not known. Therefore $P[-2.132 \leq (37.14 - \mu)/(1.85/\sqrt{5}) \leq 2.132] = .90$. In effect, this identity limits the magnitude of possible values for μ. The magnitude of μ can be only large enough to retain the t quantity above -2.132 and small enough to retain the t quantity below $+2.132$. This can be found by rearranging the quantities within the bracket; i.e., $P[35.78 \leq \mu \leq 38.90] = .90$. This states that we are 90 percent sure that the interval from 35.78 to 38.90 includes the unknown parameter μ.

In general,

$$P\left[\bar{x} - t\,\frac{s}{\sqrt{n}} \leq \mu \leq \bar{x} + t\,\frac{s}{\sqrt{n}}\right] = 1 - \alpha$$

where t is defined for an upper-tail area of $\alpha/2$ with $(n - 1)$ df. In this application, the interval limits $(\bar{x} + t\,s/\sqrt{n})$ are random variables which will cover the unknown parameter μ with probability $(1 - \alpha)$. The converse, that we are $100(1 - \alpha)$ percent sure that the parameter value is within the interval, is not correct. This statement defines a probability for the parameter rather than the probability for the interval.

Example What values of t define the midarea of 95 percent for weekly samples of size 25, and what is the sample value of t for the second week?

$$P[-2.064 \leq t \leq 2.064] = .95$$

and

$$(34.19 - 35)/(2.35/\sqrt{25}) = 1.72.$$

Example For the composite sample of 100 tensile strengths, what is the 90 percent confidence interval for μ?

Use Table 3-5 for $t_{.05}$ with df $\approx \infty$.

$$P\left[35.16 - 1.645\,\frac{2.47}{\sqrt{100}} < \mu < 35.16 + 1.645\,\frac{2.47}{\sqrt{100}}\right] = .90$$

or

$$P[34.75 \leq \mu \leq 35.57] = .90$$

Confidence Interval for the Difference in Two Population Means The confidence interval for a mean can be extended to include the difference between two population means. This interval is based on the assumption that the respective populations have the same variance σ^2:

$$(\bar{x}_1 - \bar{x}_2) - ts_p\sqrt{1/n_1 + 1/n_2} \leq \mu_1 - \mu_2 \leq (\bar{x}_1 - \bar{x}_2) + ts_p\sqrt{1/n_1 + 1/n_2}$$

Example Compute the 95 percent confidence interval based on the original 100-point sample and the subsequent 5-point sample:

$$s_p^2 = \frac{99(2.47)^2 + 4(1.85)^2}{103} = 5.997$$

or

$$s_p = 2.45$$

With 103 df and $\alpha = .05$, $t = \pm 1.96$ using $t_{.025}$ in Table 3-5. Therefore

$$(35.16 - 37.14) \pm 1.96(2.45)\sqrt{1/100 + 1/5} = -1.98 \pm 2.20$$

or

$$-4.18 \leq (\mu_1 - \mu_2) \leq .22$$

Note that if the respective samples had been based on 52 observations each rather than 100 and 5, the uncertainty factor would have been $\pm.94$ rather than the observed ±2.20. The interval width tends to be minimum when $n_1 = n_2$.

Confidence Interval for a Variance The chi-square distribution can be used to derive a confidence interval for a population variance σ^2 when the parent population is normally distributed. For a $100(1 - \alpha)$ percent confidence interval

$$\frac{(\mathrm{df})s^2}{\chi_2^2} \leq \sigma^2 \leq \frac{(\mathrm{df})s^2}{\chi_1^2}$$

where χ_1^2 corresponds to a lower-tail area of $\alpha/2$ and χ_2^2 to an upper-tail area of $\alpha/2$.

Example For the first week of tensile-strength samples compute the 90 percent confidence interval for σ^2 (df = 24, corresponding to $n = 25$, using 5 percent and 95 percent in Table 3-6):

$$\frac{24(2.40)^2}{36.4} \leq \sigma^2 \leq \frac{24(2.40)^2}{13.8}$$

$$3.80 \leq \sigma^2 \leq 10.02$$

or

$$1.95 \leq \sigma \leq 3.17$$

TESTS OF HYPOTHESIS

General Nature of Tests The general nature of tests can be illustrated with a simple example. In a court of law, when a defendant is charged with a crime, the judge instructs the jury initially to presume that the defendant is innocent of the crime. The jurors are then presented with evidence and counterargument as to the defendant's guilt or innocence. If the evidence suggests beyond a reasonable doubt that the defendant did, in fact, commit the crime, they have been instructed to find the defendant guilty; otherwise, not guilty. The burden of proof is on the prosecution.

Jury trials represent a form of decision making. In statistics, an analogous procedure for making decisions falls into an area of statistical inference called **hypothesis testing.**

Suppose that a company has been using a certain supplier of raw materials in one of its chemical processes. A new supplier approaches the company and states that its material, at the same cost, will increase the process yield. If the new supplier has a good reputation, the company might be willing to run a limited test. On the basis of the test results it would then make a decision to change suppliers or not. Good management would dictate that an improvement must be demonstrated (beyond a reasonable doubt) for the new material. That is, the burden of proof is tied to the new material. In setting up a test of hypothesis for this application, the initial assumption would be defined as a null hypothesis and symbolized as H_0. The null hypothesis would state that yield for the new material is no greater than for the conventional material. The symbol μ_0 would be used to designate the known current level of yield for the standard material and μ for the unknown population yield for the new material. Thus, the null hypothesis can be symbolized as H_0: $\mu \leq \mu_0$.

The alternative to H_0 is called the alternative hypothesis and is symbolized as H_1: $\mu > \mu_0$.

Given a series of tests with the new material, the average yield $\bar{x}$ would be compared with μ_0. If $\bar{x} < \mu_0$, the new supplier would be dismissed. If $\bar{x} > \mu_0$, the question would be: Is it sufficiently greater in the light of its corresponding reliability, i.e., beyond a reasonable doubt? If the confidence interval for μ included μ_0, the answer would be no, but if it did not include μ_0, the answer would be yes. In this simple application, the formal test of hypothesis would result in the same conclusion as that derived from the confidence interval. However, the utility of tests of hypothesis lies in their generality, whereas confidence intervals are restricted to a few special cases.

TABLE 3-7   F Distribution

Upper 5% Points ($F_{.95}$)

Degrees of freedom for denominator	Degrees of freedom for numerator																		
	1	2	3	4	5	6	7	8	9	10	12	15	20	24	30	40	60	120	∞
1	161	200	216	225	230	234	237	239	241	242	244	246	248	249	250	251	252	253	254
2	18.5	19.0	19.2	19.2	19.3	19.3	19.4	19.4	19.4	19.4	19.4	19.4	19.4	19.5	19.5	19.5	19.5	19.5	19.5
3	10.1	9.55	9.28	9.12	9.01	8.94	8.89	8.85	8.81	8.79	8.74	8.70	8.66	8.64	8.62	8.59	8.57	8.55	8.53
4	7.71	6.94	6.59	6.39	6.26	6.16	6.09	6.04	6.00	5.96	5.91	5.86	5.80	5.77	5.75	5.72	5.69	5.66	5.63
5	6.61	5.79	5.41	5.19	5.05	4.95	4.88	4.82	4.77	4.74	4.68	4.62	4.56	4.53	4.50	4.46	4.43	4.40	4.37
6	5.99	5.14	4.76	4.53	4.39	4.28	4.21	4.15	4.10	4.06	4.00	3.94	3.87	3.84	3.81	3.77	3.74	3.70	3.67
7	5.59	4.74	4.35	4.12	3.97	3.87	3.79	3.73	3.68	3.64	3.57	3.51	3.44	3.41	3.38	3.34	3.30	3.27	3.23
8	5.32	4.46	4.07	3.84	3.69	3.58	3.50	3.44	3.39	3.35	3.28	3.22	3.15	3.12	3.08	3.04	3.01	2.97	2.93
9	5.12	4.26	3.86	3.63	3.48	3.37	3.29	3.23	3.18	3.14	3.07	3.01	2.94	2.90	2.86	2.83	2.79	2.75	2.71
10	4.96	4.10	3.71	3.48	3.33	3.22	3.14	3.07	3.02	2.98	2.91	2.85	2.77	2.74	2.70	2.66	2.62	2.58	2.54
11	4.84	3.98	3.59	3.36	3.20	3.09	3.01	2.95	2.90	2.85	2.79	2.72	2.65	2.61	2.57	2.53	2.49	2.45	2.40
12	4.75	3.89	3.49	3.26	3.11	3.00	2.91	2.85	2.80	2.75	2.69	2.62	2.54	2.51	2.47	2.43	2.38	2.34	2.30
13	4.67	3.81	3.41	3.18	3.03	2.92	2.83	2.77	2.71	2.67	2.60	2.53	2.46	2.42	2.38	2.34	2.30	2.25	2.21
14	4.60	3.74	3.34	3.11	2.96	2.85	2.76	2.70	2.65	2.60	2.53	2.46	2.39	2.35	2.31	2.27	2.22	2.18	2.13
15	4.54	3.68	3.29	3.06	2.90	2.79	2.71	2.64	2.59	2.54	2.48	2.40	2.33	2.29	2.25	2.20	2.16	2.11	2.07
16	4.49	3.63	3.24	3.01	2.85	2.74	2.66	2.59	2.54	2.49	2.42	2.35	2.28	2.24	2.19	2.15	2.11	2.06	2.01
17	4.45	3.59	3.20	2.96	2.81	2.70	2.61	2.55	2.49	2.45	2.38	2.31	2.23	2.19	2.15	2.10	2.06	2.01	1.96
18	4.41	3.55	3.16	2.93	2.77	2.66	2.58	2.51	2.46	2.41	2.34	2.27	2.19	2.15	2.11	2.06	2.02	1.97	1.92
19	4.38	3.52	3.13	2.90	2.74	2.63	2.54	2.48	2.42	2.38	2.31	2.23	2.16	2.11	2.07	2.03	1.98	1.93	1.88
20	4.35	3.49	3.10	2.87	2.71	2.60	2.51	2.45	2.39	2.35	2.28	2.20	2.12	2.08	2.04	1.99	1.95	1.90	1.84
21	4.32	3.47	3.07	2.84	2.68	2.57	2.49	2.42	2.37	2.32	2.25	2.18	2.10	2.05	2.01	1.96	1.92	1.87	1.81
22	4.30	3.44	3.05	2.82	2.66	2.55	2.46	2.40	2.34	2.30	2.23	2.15	2.07	2.03	1.98	1.94	1.89	1.84	1.78
23	4.28	3.42	3.03	2.80	2.64	2.53	2.44	2.37	2.32	2.27	2.20	2.13	2.05	2.01	1.96	1.91	1.86	1.81	1.76
24	4.26	3.40	3.01	2.78	2.62	2.51	2.42	2.36	2.30	2.25	2.18	2.11	2.03	1.98	1.94	1.89	1.84	1.79	1.73
25	4.24	3.39	2.99	2.76	2.60	2.49	2.40	2.34	2.28	2.24	2.16	2.09	2.01	1.96	1.92	1.87	1.82	1.77	1.71
30	4.17	3.32	2.92	2.69	2.53	2.42	2.33	2.27	2.21	2.16	2.09	2.01	1.93	1.89	1.84	1.79	1.74	1.68	1.62
40	4.08	3.23	2.84	2.61	2.45	2.34	2.25	2.18	2.12	2.08	2.00	1.92	1.84	1.79	1.74	1.69	1.64	1.58	1.51
60	4.00	3.15	2.76	2.53	2.37	2.25	2.17	2.10	2.04	1.99	1.92	1.84	1.75	1.70	1.65	1.59	1.53	1.47	1.39
120	3.92	3.07	2.68	2.45	2.29	2.18	2.09	2.02	1.96	1.91	1.83	1.75	1.66	1.61	1.55	1.50	1.43	1.35	1.25
∞	3.84	3.00	2.60	2.37	2.21	2.10	2.01	1.94	1.88	1.83	1.75	1.67	1.57	1.52	1.46	1.39	1.32	1.22	1.00

Upper 1% Points ($F_{.99}$)

Degrees of freedom for denominator	Degrees of freedom for numerator																		
	1	2	3	4	5	6	7	8	9	10	12	15	20	24	30	40	60	120	∞
1	4052	5000	5403	5625	5764	5859	5928	5982	6023	6056	6106	6157	6209	6235	6261	6287	6313	6339	6366
2	98.5	99.0	99.2	99.2	99.3	99.3	99.4	99.4	99.4	99.4	99.4	99.4	99.4	99.5	99.5	99.5	99.5	99.5	99.5
3	34.1	30.8	29.5	28.7	28.2	27.9	27.7	27.5	27.3	27.2	27.1	26.9	26.7	26.6	26.5	26.4	26.3	26.2	26.1
4	21.2	18.0	16.7	16.0	15.5	15.2	15.0	14.8	14.7	14.5	14.4	14.2	14.0	13.9	13.8	13.7	13.7	13.6	13.5
5	16.3	13.3	12.1	11.4	11.0	10.7	10.5	10.3	10.2	10.1	9.89	9.72	9.55	9.47	9.38	9.29	9.20	9.11	9.02
6	13.7	10.9	9.78	9.15	8.75	8.47	8.26	8.10	7.98	7.87	7.72	7.56	7.40	7.31	7.23	7.14	7.06	6.97	6.88
7	12.2	9.55	8.45	7.85	7.46	7.19	6.99	6.84	6.72	6.62	6.47	6.31	6.16	6.07	5.99	5.91	5.82	5.74	5.65
8	11.3	8.65	7.59	7.01	6.63	6.37	6.18	6.03	5.91	5.81	5.67	5.52	5.36	5.28	5.20	5.12	5.03	4.95	4.86
9	10.6	8.02	6.99	6.42	6.06	5.80	5.61	5.47	5.35	5.26	5.11	4.96	4.81	4.73	4.65	4.57	4.48	4.40	4.31
10	10.0	7.56	6.55	5.99	5.64	5.39	5.20	5.06	4.94	4.85	4.71	4.56	4.41	4.33	4.25	4.17	4.08	4.00	3.91
11	9.65	7.21	6.22	5.67	5.32	5.07	4.89	4.74	4.63	4.54	4.40	4.25	4.10	4.02	3.94	3.86	3.78	3.69	3.60
12	9.33	6.93	5.95	5.41	5.06	4.82	4.64	4.50	4.39	4.30	4.16	4.01	3.86	3.78	3.70	3.62	3.54	3.45	3.36
13	9.07	6.70	5.74	5.21	4.86	4.62	4.44	4.30	4.19	4.10	3.96	3.82	3.66	3.59	3.51	3.43	3.34	3.25	3.17
14	8.86	6.51	5.56	5.04	4.70	4.46	4.28	4.14	4.03	3.94	3.80	3.66	3.51	3.43	3.35	3.27	3.18	3.09	3.00
15	8.68	6.36	5.42	4.89	4.56	4.32	4.14	4.00	3.89	3.80	3.67	3.52	3.37	3.29	3.21	3.13	3.05	2.96	2.87
16	8.53	6.23	5.29	4.77	4.44	4.20	4.03	3.89	3.78	3.69	3.55	3.41	3.26	3.18	3.10	3.02	2.93	2.84	2.75
17	8.40	6.11	5.19	4.67	4.34	4.10	3.93	3.79	3.68	3.59	3.46	3.31	3.16	3.08	3.00	2.92	2.83	2.75	2.65
18	8.29	6.01	5.09	4.58	4.25	4.01	3.84	3.71	3.60	3.51	3.37	3.23	3.08	3.00	2.92	2.84	2.75	2.66	2.57
19	8.19	5.93	5.01	4.50	4.17	3.94	3.77	3.63	3.52	3.43	3.30	3.15	3.00	2.92	2.84	2.76	2.67	2.58	2.49
20	8.10	5.85	4.94	4.43	4.10	3.87	3.70	3.56	3.46	3.37	3.23	3.09	2.94	2.86	2.78	2.69	2.61	2.52	2.42
21	8.02	5.78	4.87	4.37	4.04	3.81	3.64	3.51	3.40	3.31	3.17	3.03	2.88	2.80	2.72	2.64	2.55	2.46	2.36
22	7.95	5.72	4.82	4.31	3.99	3.76	3.59	3.45	3.35	3.26	3.12	2.98	2.83	2.75	2.67	2.58	2.50	2.40	2.31
23	7.88	5.66	4.76	4.26	3.94	3.71	3.54	3.41	3.30	3.21	3.07	2.93	2.78	2.70	2.62	2.54	2.45	2.35	2.26
24	7.82	5.61	4.72	4.22	3.90	3.67	3.50	3.36	3.26	3.17	3.03	2.89	2.74	2.66	2.58	2.49	2.40	2.31	2.21
25	7.77	5.57	4.68	4.18	3.86	3.63	3.46	3.32	3.22	3.13	2.99	2.85	2.70	2.62	2.53	2.45	2.36	2.27	2.17
30	7.56	5.39	4.51	4.02	3.70	3.47	3.30	3.17	3.07	2.98	2.84	2.70	2.55	2.47	2.39	2.30	2.21	2.11	2.01
40	7.31	5.18	4.31	3.83	3.51	3.29	3.12	2.99	2.89	2.80	2.66	2.52	2.37	2.29	2.20	2.11	2.02	1.92	1.80
60	7.08	4.98	4.13	3.65	3.34	3.12	2.95	2.82	2.72	2.63	2.50	2.35	2.20	2.12	2.03	1.94	1.84	1.73	1.60
120	6.85	4.79	3.95	3.48	3.17	2.96	2.79	2.66	2.56	2.47	2.34	2.19	2.03	1.95	1.86	1.76	1.66	1.53	1.38
∞	6.63	4.61	3.78	3.32	3.02	2.80	2.64	2.51	2.41	2.32	2.18	2.04	1.88	1.79	1.70	1.59	1.47	1.32	1.00

Interpolation should be performed using reciprocals of the degrees of freedom.

Test of Hypothesis for a Mean Procedure

Nomenclature

μ = mean of the population from which the sample has been drawn

σ = standard deviation of the population from which the sample has been drawn

μ_0 = base or reference level

H_0 = null hypothesis

H_1 = alternative hypothesis

α = significance level, usually set at .10, .05, or .01

t = tabled t value corresponding to the significance level α. For a two-tailed test, each corresponding tail would have an area of $\alpha/2$, and for a one-tailed test, one tail area would be equal to α. If σ^2 is known, then z would be used rather than the t.

$t = (\bar{x} - \mu_0)/(s/\sqrt{n})$ = sample value of the test statistic.

Assumptions

1. The n observations $x_1, x_2, \ldots, x_n$ have been selected randomly.

2. The population from which the observations were obtained is normally distributed with an unknown mean μ and standard deviation σ. In actual practice, this is a robust test, in the sense that in most types of problems it is not sensitive to the normality assumption when the sample size is 10 or greater.

Test of Hypothesis

1. Under the null hypothesis, it is assumed that the sample came from a population whose mean μ is equivalent to some base or reference designated by μ_0. This can take one of three forms:

Form 1	Form 2	Form 3
$H_0: \mu = \mu_0$	$H_0: \mu \le \mu_0$	$H_0: \mu \ge \mu_0$
$H_1: \mu \ne \mu_0$	$H_1: \mu > \mu_0$	$H_1: \mu < \mu_0$
Two-tailed test	Upper-tailed test	Lower-tailed test

2. If the null hypothesis is assumed to be true, say, in the case of a two-sided test, form 1, then the distribution of the test statistic t is known. Given a random sample, one can predict how far its sample value of t might be expected to deviate from zero (the midvalue of t) by chance alone. If the sample value of t does, in fact, deviate too far from zero, then this is defined to be sufficient evidence to refute the assumption of the null hypothesis. It is consequently rejected, and the converse or alternative hypothesis is accepted.

3. The rule for accepting H_0 is specified by selection of the α level as indicated in Fig. 3-57. For forms 2 and 3 the α area is defined to be in the upper or the lower tail respectively.

4. The decision rules for each of the three forms are defined as follows: If the sample t falls within the acceptance region, accept H_0 for lack of contrary evidence. If the sample t falls in the critical region, reject H_0 at a significance level of 100α percent.

Example

Application. In the past, the yield for a chemical process has been established at 89.6 percent with a standard deviation of 3.4 percent. A new supplier of raw materials will be used and tested for 7 days.

Procedure

1. The standard of reference is $\mu_0 = 89.6$ with a known $\sigma = 3.4$.

2. It is of interest to demonstrate whether an increase in yield is achieved with the new material; H_0 says it has not; therefore,

$$H_0: \mu \le 89.6 \qquad H_1: \mu > 89.6$$

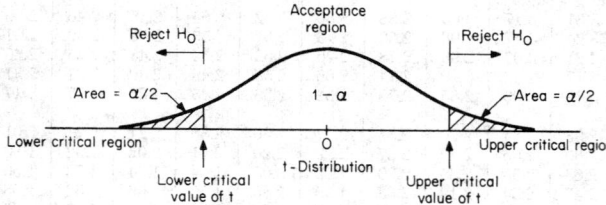

FIG. 3-57 Acceptance region.

3. Select $\alpha = .05$, and since σ is known (the new material would not affect the day-to-day variability in yield), the test statistic would be z with a corresponding critical value $cv(z) = 1.645$ (Table 3-5, df $= \infty$).

4. The decision rule:

Accept H_0 if sample $z < 1.645$

Reject H_0 if sample $z > 1.645$

5. A 7-day test was carried out, and daily yields averaged 91.6 percent with a sample standard deviation $s = 3.6$ (this is not needed for the test of hypothesis).

6. For the data sample $z = (91.6 - 89.6)/(3.4/\sqrt{7}) = 1.56$.

7. Since the sample $z < cv(z)$, accept the null hypothesis for lack of contrary evidence; i.e., an improvement has not been demonstrated beyond a reasonable doubt.

Example

Application. In the past, the break strength of a synthetic yarn has averaged 34.6 lb. The first-stage draw ratio of the spinning machines has been increased. Production management wants to determine whether the break strength has changed under the new condition.

Procedure

1. The standard of reference is $\mu_0 = 34.6$.

2. It is of interest to demonstrate whether a change has occurred; therefore,

$$H_0: \mu = 34.6 \qquad H_1: \mu \ne 34.6$$

3. Select $\alpha = .05$, and since with the change in draw ratio the uniformity might change, the sample standard deviation would be used, and therefore t would be the appropriate test statistic.

4. A sample of 21 ends was selected randomly and tested on an Instron with the results $\bar{x} = 35.55$ and $s = 2.041$.

5. For 20 df and a two-tailed α level of 5 percent, the critical values of t are given by ± 2.086 with a decision rule (Table 3-5, $t_{.025}$, df $= 20$):

Accept H_0 if $-2.086 <$ sample $t < 2.086$

Reject H_0 if sample $t < -2.086$ or > 2.086

6. For the data sample $t = (35.55 - 34.6)/(2.041/\sqrt{21}) = 2.133$.

7. Since $2.133 > 2.086$, reject H_0 and accept H_1. It has been demonstrated that an improvement in break strength has been achieved.

Two-Population Test of Hypothesis for Means

Nature Two samples were selected from different locations in a plastic-film sheet and measured for thickness. The thickness of the respective samples was measured at 10 close but equally spaced points in each of the samples. It was of interest to compare the average thickness of the respective samples to detect whether they were significantly different. That is, was there a significant variation in thickness between locations?

From a modeling standpoint statisticians would define this problem as a two-population test of hypothesis. They would define the respective sample sheets as two populations from which 10 sample thickness determinations were measured for each.

In order to compare populations based on their respective samples, it is necessary to have some basis of comparison. This basis is predicated on the distribution of the t statistic. In effect, the t statistic characterizes the way in which two sample means from two separate populations will tend to vary by chance alone when the population means and variances are equal. Consider the following:

Population 1		Population 2	
Normal	Sample 1	Normal	Sample 2
μ_1	n_1	μ_2	n_2
	$\bar{x}_1$		$\bar{x}_2$
σ_1^2	s_1^2	σ_2^2	s_2^2

Consider the hypothesis $\mu_1 = \mu_2$. If, in fact, the hypothesis is correct, i.e., $\mu_1 = \mu_2$ (under the condition $\sigma_1^2 = \sigma_2^2$), then the sampling distribution of $(\bar{x}_1 - \bar{x}_2)$ is predictable through the t distribution. The observed sample values then can be compared with the corresponding t distribution. If the sample values are reasonably close (as reflected through the α level), that is, $\bar{x}_1$ and $\bar{x}_2$ are not "too different" from each other on the basis of the t distribution, the null hypothesis would be accepted. Conversely, if they deviate from each other "too much" and the deviation is therefore not ascribable to chance, the conjecture would be questioned and the null hypothesis rejected.

Example

Application. Two samples were selected from different locations in a plastic-film sheet. The thickness of the respective samples was measured at 10 close but equally spaced points.

Procedure

1. Demonstrate whether the thicknesses of the respective sample locations are significantly different from each other; therefore,

$$H_0: \mu_1 = \mu_2 \qquad H_1: \mu_1 \neq \mu_2$$

2. Select $\alpha = .05$.
3. Summarize the statistics for the respective samples:

Sample 1		Sample 2	
1.473	1.367	1.474	1.417
1.484	1.276	1.501	1.448
1.484	1.485	1.485	1.469
1.425	1.462	1.435	1.474
1.448	1.439	1.348	1.452
$\bar{x}_1 = 1.434$	$s_1 = .0664$	$\bar{x}_2 = 1.450$	$s_2 = .0435$

4. As a first step, the assumption for the standard t test, that $\sigma_1^2 = \sigma_2^2$, can be tested through the F distribution. For this hypothesis, $H_0: \sigma_1^2 = \sigma_2^2$ would be tested against $H_1: \sigma_1^2 \neq \sigma_2^2$. Since this is a two-tailed test and conventionally only the upper tail for F is published, the procedure is to use the largest ratio and the corresponding ordered degrees of freedom. This achieves the same end result through one table. However, since the largest ratio is arbitrary, it is necessary to define the true α level as twice the value of the tabled value. Therefore, by using Table 3-7 with $\alpha = .05$ the corresponding critical value for $F(9,9) = 3.18$ would be for a true $\alpha = .10$. For the sample,

$$\text{Sample } F = (.0664/.0435)^2 = 2.33$$

Therefore, the ratio of sample variances is no larger than one might expect to observe when in fact $\sigma_1^2 = \sigma_2^2$. There is not sufficient evidence to reject the null hypothesis that $\sigma_1^2 = \sigma_2^2$.

5. For 18 df and a two-tailed α level of 5 percent the critical values of t are given by ± 2.101 (Table 3-5, $t_{0.025}$, $df = 18$).
6. The decision rule:

> Accept H_0 if $-2.101 \leq \text{sample } t \leq 2.101$
> Reject H_0 otherwise

7. For the sample the pooled variance estimate is given by

$$s_p^2 = \frac{9(.0664)^2 + 9(.0435)^2}{9+9} = \frac{(.0664)^2 + (.0435)^2}{2} = .00315$$

or

$$s_p = .056$$

8. The sample statistic value of t is

$$\text{Sample } t = \frac{1.434 - 1.450}{.056\sqrt{1/10 + 1/10}} = -.64$$

9. Since the sample value of t falls within the acceptance region, accept H_0 for lack of contrary evidence; i.e., there is insufficient evidence to demonstrate that thickness differs between the two selected locations.

Test of Hypothesis for Paired Observations

Nature In some types of applications, associated pairs of observations are defined. For example, (1) pairs of samples from two populations are treated in the same way, or (2) two types of measurements are made on the same unit. For applications of this type, it is not only more effective but necessary to define the random variable as the difference between the pairs of observations. The difference numbers can then be tested by the standard t distribution.

Examples of the two types of applications are as follows:

1. *Sample treatment*
   a. Two types of metal specimens buried in the ground together in a variety of soil types to determine corrosion resistance
   b. Wear-rate test with two different types of tractor tires mounted in pairs on n tractors for a defined period of time
2. *Same unit*
   a. Blood-pressure measurements made on the same individual before and after the administration of a stimulus
   b. Smoothness determinations on the same film samples at two different testing laboratories

Test of Hypothesis for Matched Pairs: Procedure

Nomenclature

d_i = sample difference between the ith pair of observations
s = sample standard deviation of differences
μ = population mean of differences
σ = population standard deviation of differences
μ_0 = base or reference level of comparison
H_0 = null hypothesis
H_1 = alternative hypothesis
α = significance level
t = tabled value with $(n-1)$ df
$t = (\bar{d} - \mu_0)/(s/\sqrt{n})$, the sample value of t

Assumptions

1. The n pairs of samples have been selected and assigned for testing in a random way.
2. The population of differences is normally distributed with a mean μ and variance σ^2. As in the previous application of the t distribution, this is a robust procedure, i.e., not sensitive to the normality assumption if the sample size is 10 or greater in most situations.

Test of Hypothesis

1. Under the null hypothesis, it is assumed that the sample came from a population whose mean μ is equivalent to some base or reference level designated by μ_0. For most applications of this type, the value of μ_0 is defined to be zero; that is, it is of interest generally to demonstrate a difference not equal to zero. The hypothesis can take one of three forms:

Form 1	Form 2	Form 3
$H_0: \mu = \mu_0$	$H_0: \mu \leq \mu_0$	$H_0: \mu \geq \mu_0$
$H_1: \mu \neq \mu_0$	$H_1: \mu > \mu_0$	$H_1: \mu < \mu_0$
Two-tailed test	Upper-tailed test	Lower-tailed test

2. If the null hypothesis is assumed to be true, say, in the case of a lower-tailed test, form 3, then the distribution of the test statistic t is known under the null hypothesis that limits $\mu = \mu_0$. Given a random sample, one can predict how far its sample value of t might be expected to deviate from zero by chance alone when $\mu = \mu_0$. If the sample value of t is too small, as in the case of a negative value, then this would be defined as sufficient evidence to reject the null hypothesis.
3. Select α.
4. The critical values or value of t would be defined by the tabled value of t with $(n-1)$ df corresponding to a tail area of α. For a two-tailed test, each tail area would be $\alpha/2$, and for a one-tailed test there would be an upper-tail or a lower-tail area of α corresponding to forms 2 and 3 respectively.
5. The decision rule for each of the three forms would be to reject the null hypothesis if the sample value of t fell in that area of the t distribution defined by α, which is called the critical region. Otherwise, the alternative hypothesis would be accepted for lack of contrary evidence.

Example

Application. Pairs of pipes have been buried in 11 different locations to determine corrosion on nonbituminous pipe coatings for underground use. One type includes a lead-coated steel pipe and the other a bare steel pipe.

Procedure

1. The standard of reference is taken as $\mu_0 = 0$, corresponding to no difference in the two types.
2. It is of interest to demonstrate whether either type of pipe has a greater corrosion resistance than the other. Therefore,

$$H_0: \mu = 0 \qquad H_1: \mu \neq 0$$

3. Select $\alpha = .05$. Therefore, with $n = 11$ the critical values of t with 10 df are defined by $t = \pm 2.228$ (Table 3.5, $t_{.025}$).
4. The decision rule:

> Accept H_0 if $-2.228 \leq \text{sample } t \leq 2.228$
> Reject H_0 otherwise

5. The sample of 11 pairs of corrosion determinations and their differences are as follows:

Soil type	Lead-coated steel pipe	Bare steel pipe	d = difference
A	27.3	41.4	−14.1
B	18.4	18.9	−0.5
C	11.9	21.7	−9.8
D	11.3	16.8	−5.5
E	14.8	9.0	5.8
F	20.8	19.3	1.5
G	17.9	32.1	−14.2
H	7.8	7.4	0.4
I	14.7	20.7	−6.0
J	19.0	34.4	−15.4
K	65.3	76.2	−10.9

6. The sample statistics:

$$\bar{d} = -6.245 \qquad s^2 = \frac{11 \sum d^2 - (\sum d)^2}{11 \times 10} = 52.56$$

or
$$s = 7.25$$

$$\text{Sample } t = (-6.245 - 0)/(7.25/\sqrt{11})$$
$$= -2.86$$

7. Since the sample t of −2.86 < tabled t of −2.228, reject H_0 and accept H_1; that is, it has been demonstrated that, on the basis of the evidence, lead-coated steel pipe has a greater corrosion resistance than bare steel pipe.

Example

Application. A stimulus was tested for its effect on blood pressure. Ten men were selected randomly, and their blood pressure was measured before and after the stimulus was administered. It was of interest to determine whether the stimulus had caused a significant increase in the blood pressure.

Procedure

1. The standard of reference was taken as $\mu_0 \le 0$, corresponding to no increase.
2. It was of interest to demonstrate an increase in blood pressure if in fact an increase did occur. Therefore,

$$H_0: \mu_0 \le 0 \qquad H_1: \mu_0 > 0$$

3. Select $\alpha = .05$. Therefore, with $n = 10$ the critical value of t with 9 df is defined by $t = 1.833$ (Table 3-5, $t_{.05}$, one-sided).
4. The decision rule:

Accept H_0 if sample $t < 1.833$

Reject H_0 if sample $t > 1.833$

5. The sample of 10 pairs of blood pressure and their differences were as follows:

Individual	Before	After	d = difference
1	138	146	8
2	116	118	2
3	124	120	−4
4	128	136	8
5	155	174	19
6	129	133	4
7	130	129	−1
8	148	155	7
9	143	148	5
10	159	155	−4

6. The sample statistics:

$$\bar{d} = 4.4 \qquad s = 6.85$$

$$\text{Sample } t = (4.4 - 0)/(6.85/\sqrt{10}) = 2.03$$

7. Since the sample $t = 2.03 >$ critical $t = 1.833$, reject the null hypothesis. It has been demonstrated that the population of men from which the sample was drawn tend, as a whole, to have an increase in blood pressure after the stimulus has been given. The distribution of differences d seems to indicate that the degree of response varies by individuals.

Test of Hypothesis for a Proportion

Nature Some types of statistical applications deal with counts and proportions rather than measurements. Examples are (1) the proportion of workers in a plant who are out sick, (2) lost-time worker accidents per month, (3) defective items in a shipment lot, and (4) preference in consumer surveys.

The procedure for testing the significance of a sample proportion follows that for a sample mean. In this case, however, owing to the nature of the problem the appropriate test statistic is Z. This follows from the fact that the null hypothesis requires the specification of the goal or reference quantity p_0, and since the distribution is a binomial proportion, the associated variance is $[p_0(1 - p_0)]n$ under the null hypothesis. The primary requirement is that the sample size n satisfy normal approximation criteria for a binomial proportion, roughly $np > 5$ and $n(1 - p) > 5$.

Test of Hypothesis for a Proportion: Procedure

Nomenclature

p = mean proportion of the population from which the sample has been drawn

p_0 = base or reference proportion

$[p_0(1 - p_0)]/n$ = base or reference variance

$\hat{p} = x/n$ = sample proportion, where x refers to the number of observations out of n which have the specified attribute

H_0 = assumption or null hypothesis regarding the population proportion

H_1 = alternative hypothesis

α = significance level, usually set at .10, .05, or .01

z = Tabled Z value corresponding to the significance level α. The sample sizes required for the z approximation according to the magnitude of p_0 are given in Table 3-5.

$z = (\hat{p} - p_0)/\sqrt{p_0(1 - p_0)/n}$, the sample value of the test statistic

Assumptions

1. The n observations have been selected randomly.
2. The sample size n is sufficiently large to meet the requirement for the Z approximation.

Test of Hypothesis

1. Under the null hypothesis, it is assumed that the sample came from a population with a proportion p_0 of items having the specified attribute. For example, in tossing a coin the population could be thought of as having an unbounded number of potential tosses. If it is assumed that the coin is fair, this would dictate $p_0 = 1/2$ for the proportional number of heads in the population. The null hypothesis can take one of three forms:

Form 1	Form 2	Form 3
$H_0: p = p_0$	$H_0: p \le p_0$	$H_0: p \ge p_0$
$H_1: p \ne p_0$	$H_1: p > p_0$	$H_1: p < p_0$
Two-tailed test	Upper-tailed test	Lower-tailed test

2. If the null hypothesis is assumed to be true, then the sampling distribution of the test statistic Z is known. Given a random sample, it is possible to predict how far the sample proportion x/n might deviate from its assumed population proportion p_0 through the Z distribution. When the sample proportion deviates too far, as defined by the significance level α, this serves as the justification for rejecting the assumption, that is, rejecting the null hypothesis.

3. The decision rule is given by

Form 1: Accept H_0 if lower critical $z <$ sample $z <$ upper critical z
Reject H_0 otherwise

Form 2: Accept H_0 if sample $z <$ upper critical z
Reject H_0 otherwise

Form 3: Accept H_0 if lower critical $z <$ sample z
Reject H_0 otherwise

Example

Application. A company has received a very large shipment of rivets. One product specification required that no more than 2 percent of the rivets have diameters greater than 14.28 mm. Any rivet with a diameter greater than this would be classified as defective. A random sample of 600 was selected and tested with a go–no-go gauge. Of these, 16 rivets were found to be defective. Is this sufficient evidence to conclude that the shipment contains more than 2 percent defective rivets?

Procedure

1.   The quality goal is $p \le .02$. It would be assumed initially that the shipment meets this standard; i.e., H_0: $p \le .02$.

2.   The assumption in step 1 would first be tested by obtaining a random sample. Under the assumption that $p \le .02$, the distribution for a sample proportion would be defined by the z distribution. This distribution would define an upper bound corresponding to the upper critical value for the sample proportion. It would be unlikely that the sample proportion would rise above that value if, in fact, $p \le .02$. If the observed sample proportion exceeds that limit, corresponding to what would be a very unlikely chance outcome, this would lead one to question the assumption that $p \le .02$. That is, one would conclude that the null hypothesis is false. To test, set

$$H_0: p \le .02 \qquad H_1: p > .02$$

3.   Select $\alpha = .05$.

4.   With $\alpha = .05$, the upper critical value of $Z = 1.645$ (Table 3-5, $t_{.05}$, df = ∞, one-sided).

5.   The decision rule:

$$\text{Accept } H_0 \text{ if sample } z < 1.645$$

$$\text{Reject } H_0 \text{ if sample } z > 1.645$$

6.   The sample z is given by

$$\text{Sample } z = \frac{(16/600) - .02}{\sqrt{(.02)(.98)/600}}$$

$$= 1.17$$

7.   Since the sample $z < 1.645$, accept H_0 for lack of contrary evidence; there is not sufficient evidence to demonstrate that the defect proportion in the shipment is greater than 2 percent.

Test of Hypothesis for Two Proportions

Nature   In some types of engineering and management-science problems, we may be concerned with a random variable which represents a proportion, for example, the proportional number of defective items per day. The method described previously relates to a single proportion. In this subsection two proportions will be considered.

A certain change in a manufacturing procedure for producing component parts is being considered. Samples are taken by using both the existing and the new procedures in order to determine whether the new procedure results in an improvement. In this application, it is of interest to demonstrate statistically whether the population proportion p_2 for the new procedure is less than the population proportion p_1 for the old procedure on the basis of a sample of data.

Test of Hypothesis for Two Proportions: Procedure

Nomenclature

p_1 = population 1 proportion
p_2 = population 2 proportion
n_1 = sample size from population 1
n_2 = sample size from population 2
x_1 = number of observations out of n_1 that have the designated attribute
x_2 = number of observations out of n_2 that have the designated attribute
$\hat{p}_1 = x_1/n_1$, the sample proportion from population 1
$\hat{p}_2 = x_2/n_2$, the sample proportion from population 2
α = significance level
H_0 = null hypothesis
H_1 = alternative hypothesis
z = tabled Z value corresponding to the stated significance level α
$z = \dfrac{\hat{p}_1 - \hat{p}_2}{\sqrt{\hat{p}_1(1 - \hat{p}_1)/n_1 + \hat{p}_2(1 - \hat{p}_2)/n_2}}$, the sample value of Z

Assumptions

1.   The respective two samples of n_1 and n_2 observations have been selected randomly.

2.   The sample sizes n_1 and n_2 are sufficiently large to meet the requirement for the Z approximation; i.e., $x_1 > 5$, $x_2 > 5$.

Test of Hypothesis

1.   Under the null hypothesis, it is assumed that the respective two samples have come from populations with equal proportions $p_1 = p_2$. Under this hypothesis, the sampling distribution of the corresponding Z statistic is known. On the basis of the observed data, if the resultant sample value of Z represents an unusual outcome, that is, if it falls within the critical region, this would cast doubt on the assumption of equal proportions. Therefore, it will have been demonstrated statistically that the population proportions are in fact not equal. The various hypotheses can be stated:

Form 1	Form 2	Form 3
H_0: $p_1 = p_2$	H_0: $p_1 \le p_2$	H_0: $p_1 \ge p_2$
H_1: $p_1 \ne p_2$	H_1: $p_1 > p_2$	H_1: $p_1 < p_2$
Two-tailed test	Upper-tailed test	Lower-tailed test

2.   The decision rule for form 1 is given by
Accept H_0 if lower critical z < sample z < upper critical z
Reject H_0 otherwise

Example

Application.   A change was made in a manufacturing procedure for component parts. Samples were taken during the last week of operations with the old procedure and during the first week of operations with the new procedure. Determine whether the proportional numbers of defects for the respective populations differ on the basis of the sample information.

Procedure

1.   The hypotheses are

$$H_0: p_1 = p_2 \qquad H_1: p_1 \ne p_2$$

2.   Select $\alpha = .05$. Therefore, the critical values of z are ± 1.96 (Table 3-4, $A = 0.9500$).

3.   For the samples, 75 out of 1720 parts from the previous procedure and 80 out of 2780 parts under the new procedure were found to be defective; therefore,

$$\hat{p}_1 = 75/1720 = .0436 \qquad \hat{p}_2 = 80/2780 = .0288$$

4.   The decision rule:
Accept H_0 if $-1.96 \le$ sample $Z \le 1.96$
Reject H_0 otherwise

5.   The sample statistic:

$$\text{Sample } z = \frac{.0436 - .0288}{\sqrt{(.0436)(.9564)/1720 + (.0288)(.9712)/2780}}$$

$$= 2.53$$

6.   Since the sample z of 2.53 > tabled z of 1.96, reject H_0 and conclude that the new procedure has resulted in a reduced defect rate.

Goodness-of-Fit Test

Nature   A standard die has six sides numbered from 1 to 6. If one were really interested in determining whether a particular die was well balanced, one would have to carry out an experiment. To do this, it might be decided to count the frequencies of outcomes, 1 through 6, in tossing the die N times. On the assumption that the die is perfectly balanced, one would expect to observe $N/6$ occurrences each for 1, 2, 3, 4, 5, and 6. However, chance dictates that exactly $N/6$ occurrences each will not be observed. For example, given a perfectly balanced die, the probability is only 1 chance in 65 that one will observe 1 outcome each, for 1 through 6, in tossing the die 6 times. Therefore, an outcome different from 1 occurrence each can be expected. Conversely, an outcome of six 3s would seem to be too unusual to have occurred by chance alone.

Some industrial applications involve the concept outlined here. The basic idea is to test whether or not a group of observations follows a preconceived distribution. In the case cited, the distribution is uniform; i.e., each face value should *tend* to occur with the same frequency.

Goodness-of-Fit Test: Procedure

Nomenclature   Each experimental observation can be classified into one of r possible categories or cells.

r = total number of cells
O_j = number of observations occurring in cell j
E_j = expected number of observations for cell j based on the pre-conceived distribution
N = total number of observations
f = degrees of freedom for the test. In general, this will be equal to $(r-1)$ minus the number of statistical quantities on which the E_j's are based (see the examples which follow for details).

Assumptions
1.   The observations represent a sample selected randomly from a population which has been specified.
2.   The number of expectation counts E_j within each category should be roughly 5 or more. If an E_j count is significantly less than 5, that cell should be pooled with an adjacent cell.

Computation for E_j   On the basis of the specified population, the probability of observing a count in cell j is defined by p_j. For a sample of size N, corresponding to N total counts, the expected frequency is given by $E_j = Np_j$.

Test Statistics: Chi Square

$$\chi^2 = \sum_{j=1}^{r} \frac{(O_j - E_j)^2}{E_j} \quad \text{with } f \text{ df}$$

Test of Hypothesis
1. H_0: The sample came from the specified theoretical distribution

   H_1: The sample did not come from the specified theoretical distribution

2.   For a stated level of α,

   Reject H_0 if sample χ^2 > tabled χ^2
   Accept H_0 if sample χ^2 < tabled χ^2

Example
Application   A production-line product is rejected if one of its characteristics does not fall within specified limits. The standard goal is that no more than 2 percent of the production should be rejected.
Computation
1.   Of 950 units produced during the day, 28 units were rejected.
2.   The hypotheses:

   H_0: the process is in control

   H_1: the process is not in control

3.   Assume that α = .05; therefore, the critical value of $\chi^2(1)$ = 3.84 (Table 3-6, 95 percent, df = 1). One degree of freedom is defined since $(r-1)$ = 1, and no statistical quantities have been computed for the data.
4.   The decision rule:

   Reject H_0 if sample χ^2 > 3.84

   Accept H_0 otherwise

5.   Since it is assumed that p = .02, this would dictate that in a sample of 950 there would be on the average $(.02)(950)$ = 19 defective items and 931 acceptable items:

Category	Observed O_j	Expectation $E_j = 950p_j$
Acceptable	922	931
Not acceptable	28	19
Total	950	950

$$\text{Sample } \chi^2 = \frac{(922 - 931)^2}{931} + \frac{(28 - 19)^2}{19}$$

$$= 4.35 \text{ with critical } \chi^2 = 3.84$$

6.   Conclusion. Since the sample value exceeds the critical value, it would be concluded that the process is not in control.

Example
Application   A frequency count of workers was tabulated according to the number of defective items that they produced. An unresolved question is whether the observed distribution is a Poisson distribution. That is, do observed and expected frequencies agree within chance variation?
Computation
1.   The hypotheses:
   H_0: there are no significant differences, in number of defective units, between workers
   H_1: there are significant differences

2.   Assume that α = .05.
3.   Test statistic:

No. of defective units	O_j		E_j	
0	8	} 10	2.06	} 8.70 pool
1	7		6.64	
2	9		10.73	
3	12		11.55	
4	9		9.33	
5	6		6.03	
6	3		3.24	
7	2		1.50	
8	0	} 6	.60	} 5.66 pool
9	1		.22	
≥ 10	0		.10	
Sum	52		52	

The expectation numbers E_j were computed as follows: For the Poisson distribution, $\lambda = E(x)$; therefore, an estimate of λ is the average number of defective units per worker, i.e., $\lambda = (1/52)(0 \times 3 + 1 \times 7 + \cdots + 9 \times 1) = 3.23$. Given this approximation, the probability of no defective units for a worker would be $(3.23)^0/0!)e^{-3.23} = .0396$. For the 52 workers, the number of workers producing no defective units would have an expectation $E = 52(0.0396) = 2.06$, and so forth.
The sample chi-square value is computed from

$$\chi^2 = \frac{(10 - 8.70)^2}{8.70} + \frac{(9 - 10.73)^2}{10.73} + \cdots + \frac{(6 - 5.66)^2}{5.66}$$

$$= .53$$

4.   The critical value of χ^2 would be based on four degrees of freedom. This corresponds to $(r-1) - 1$, since one statistical quantity λ was computed from the sample and used to derive the expectation numbers.
5.   The critical value of $\chi^2(4) = 9.49$ (Table 3-6) with α = .05; therefore, accept H_0.

Two-Way Test for Independence for Count Data

Nature   When individuals or items are observed and classified according to two different criteria, the resultant counts can be statistically analyzed. For example, a market survey may examine whether a new product is preferred and if it is preferred due to a particular characteristic.
Count data, based on a random selection of individuals or items which are classified according to two different criteria, can be statistically analyzed through the χ^2 distribution. The purpose of this analysis is to determine whether the respective criteria are dependent. That is, is the product preferred because of a particular characteristic?

Two-Way Test for Independence for Count Data: Procedure

Nomenclature
1.   Each observation is classified into each of two categories:
a.   The first one into 2, 3, . . . , or r categories
b.   The second one into 2, 3, . . . , or c categories
2. O_{ij} = number of observations (observed counts) in cell (i, j) with

$$i = 1, 2, \ldots, r$$
$$j = 1, 2, \ldots, c$$

3. N = total number of observations
4. E_{ij} = computed number for cell (i,j) which is an expectation based on the assumption that the two characteristics are independent
5. R_i = subtotal of counts in row i
6. C_j = subtotal of counts in column j
7. α = significance level
8. H_0 = null hypothesis
9. H_1 = alternative hypothesis
10. χ^2 = critical value of χ^2 corresponding to the significance level α and $(r-1)(c-1)$ df

11.   Sample $\chi^2 = \sum_{i,j}^{c,r} \frac{(O_{ij} - E_{ij})^2}{E_{ij}}$

Assumptions
1.   The observations represent a sample selected randomly from a large total population.

2. The number of expectation counts E_{ij} within each cell should be approximately 2 or more for arrays 3×3 or larger. If any cell contains a number smaller than 2, appropriate rows or columns should be combined to increase the magnitude of the expectation count. For arrays 2×2, approximately 4 or more are required. If the number is less than 4, the exact Fisher test should be used.

Test of Hypothesis   Under the null hypothesis, the classification criteria are assumed to be independent, i.e.,

H_0: the criteria are independent

H_1: the criteria are not independent

For the stated level of α,

Reject H_0 if sample χ^2 > tabled χ^2

Accept H_0 otherwise

Computation for E_{ij}   Compute E_{ij} across rows or down columns by using either of the following identities:

$$E_{ij} = C_j \left(\frac{R_i}{N} \right) \text{ across rows}$$

$$E_{ij} = R_i \left(\frac{C_j}{N} \right) \text{ down columns}$$

Sample χ^2 Value

$$\chi^2 = \sum_{i,j} \frac{(O_{ij} - E_{ij})^2}{E_{ij}}$$

In the special case of $r = 2$ and $c = 2$, a more accurate and simplified formula which does not require the direct computation of E_{ij} can be used:

$$\chi^2 = \frac{[|O_{11}O_{22} - O_{12}O_{21}| - \frac{1}{2}N]^2 N}{R_1 R_2 C_1 C_2}$$

Example

Application   A market research study was carried out to relate the subjective "feel" of a consumer product to consumer preference. In other words, is the consumer's preference for the product associated with the feel of the product, or is the preference independent of the product feel?

Procedure

1. It was of interest to demonstrate whether an association exists between feel and preference; therefore, assume

H_0: feel and preference are independent

H_1: they are not independent

2. A sample of 200 people was asked to classify the product according to two criteria:

a. Liking for this product
b. Liking for the feel of the product

		Like feel		
		Yes	No	R_i
Like product	Yes	114	13	= 127
	No	55	18	= 73
	C_j	169	31	200

3. Select $\alpha = .05$; therefore, with $(r - 1)(c - 1) = 1$ df, the critical value of χ^2 is 3.84 (Table 3-6, 95 percent).

4. The decision rule:

Accept H_0 if sample $\chi^2 < 3.84$

Reject H_0 otherwise

5. The sample value of χ^2 by using the special formula is

$$\text{Sample } \chi^2 = \frac{[|114 \times 18 - 13 \times 55| - 100]^2 200}{(169)(31)(127)(73)}$$

$$= 6.30$$

6. Since the sample χ^2 of 6.30 > tabled χ^2 of 3.84, reject H_0 and accept H_1. The relative proportionality of $E_{11} = 169(127/200) = 107.3$ to the observed 114 compared with $E_{22} = 31(73/200) = 11.3$ to the observed 18 suggests that when the consumer likes the feel, the consumer tends to like the product, and conversely for not liking the feel. The proportions $169/200 = 84.5$ percent and $127/200 = 63.5$ percent suggest further that there are other attributes of the product which tend to nullify the beneficial feel of the product.

LEAST SQUARES

When experimental data is to be fit with a mathematical model, it is necessary to allow for the fact that the data has errors. The engineer is interested in finding the parameters in the model as well as the uncertainty in their determination. In the simplest case, the model is a linear equation with only two parameters, and they are found by a least-squares minimization of the errors in fitting the data. Multiple regression is just linear least squares applied with more terms. Nonlinear regression allows the parameters of the model to enter in a nonlinear fashion. The following description of maximum likelihood applies to both linear and nonlinear least squares (Ref. 231). If each measurement point y_i has a measurement error Δy_i that is independently random and distributed with a normal distribution about the true model $y(x)$ with standard deviation σ_i, then the probability of a data set is

$$P = \prod_{i=1}^{N} \left\{ \exp \left[-\frac{1}{2} \left(\frac{y_i - y(x_i)}{\sigma_i} \right)^2 \right] \Delta y \right\}$$

Here, y_i is the measured value, σ_i is the standard deviation of the ith measurement, and Δy is needed to say a measured value $\pm \Delta y$ has a certain probability. Given a set of parameters (maximizing this function), the probability that this data set plus or minus Δy could have occurred is P. This probability is maximized (giving the maximum likelihood) if the negative of the logarithm is minimized.

$$\sum_{i=1}^{N} \left(\frac{y_i - y(x_i)}{\sqrt{2}\,\sigma_i} \right)^2 - N \log \Delta y$$

Since N, σ_i, and Δy are constants, this is the same as minimizing χ^2.

$$\chi^2 = \sum_{i=1}^{N} \left[\frac{y_i - y(x_i; a_1, \ldots, a_M)}{\sigma_i} \right]^2$$

with respect to the parameters $\{a_j\}$. Note that the standard deviations $\{\sigma_i\}$ of the measurements are expected to be known. The goodness of fit is related to the number of degrees of freedom, $\nu = N - M$. The probability that χ^2 would exceed a particular value $(\chi_0)^2$ is

$$P = 1 - Q \left(\frac{\nu}{2}, \frac{1}{2} \chi_0^2 \right)$$

where $Q(a, x)$ is the incomplete gamma function

$$Q(a, x) = \frac{1}{\Gamma(a)} \int_0^x e^{-t} t^{a-1}\, dt \qquad (a > 0)$$

and $\Gamma(a)$ is the gamma function

$$\Gamma(a) = \int_0^\infty t^{a-1} e^{-t}\, dt$$

Both functions are tabulated in mathematical handbooks (Ref. 1). The function P gives the goodness of fit. Call χ_0^2 the value of χ^2 at the minimum. Then $P > 0.1$ represents a believable fit; if $Q > 0.001$, it might be an acceptable fit; smaller values of Q indicate the model may be in error (or the σ_i are really larger.) A "typical" value of χ^2 for a moderately good fit is $\chi^2 \sim \nu$. Asymptotically for large ν, the statistic χ^2 becomes normally distributed with a mean ν and a standard deviation $\sqrt{(2\nu)}$ (Ref. 231).

If values σ_i are not known in advance, assume $\sigma_i = \sigma$ (so that its value does not affect the minimization of χ^2). Find the parameters by minimizing χ^2 and compute:

$$\sigma^2 = \sum_{i=1}^{N} \frac{[y_i - y(x_i)]^2}{N}$$

This gives some information about the errors (i.e., the variance and standard deviation of each data point), although the goodness of fit, P, cannot be calculated.

The minimization of χ^2 requires

$$\sum_{i=1}^{N} \left[\frac{y_i - y(x_i)}{\sigma_i^2} \right] \frac{\partial y(x_i; a_1, \ldots, a_M)}{\partial a_k} = 0, \quad k = 1, \ldots, M$$

Linear Least Squares   When the model is a straight line

$$\chi^2(a, b) = \sum_{i=1}^{N} \left[\frac{y_i - a - bx_i}{\sigma_i} \right]^2$$

Define $S = \sum_{i=1}^{N} \frac{1}{\sigma_i^2}$, $S_x = \sum_{i=1}^{N} \frac{x_i}{\sigma_i^2}$, $S_y = \sum_{i=1}^{N} \frac{y_i}{\sigma_i^2}$

$S_{xx} = \sum_{i=1}^{N} \frac{x_i^2}{\sigma_i^2}$, $S_{xy} = \sum_{i=1}^{N} \frac{x_i y_i}{\sigma_i^2}$, $t_i = \frac{1}{\sigma_i} \left(x_i - \frac{S_x}{S} \right)$, $S_{tt} = \sum_{i=1}^{N} t_i^2$

Then $b = \frac{1}{S_{tt}} \sum_{i=1}^{N} \frac{t_i y_i}{\sigma_i}$, $a = \frac{S_y - S_x b}{S}$, $\sigma_a^2 = \frac{1}{S}\left(1 + \frac{S_x^2}{SS_{tt}} \right)$, $\sigma_b^2 = \frac{1}{S_{tt}}$

$$\text{Cov}(a, b) = -\frac{S_x}{SS_{tt}}, \quad r_{ab} = \frac{\text{Cov}(a, b)}{\sigma_a \sigma_b}$$

We thus get the values of a and b with maximum likelihood as well as the variances of a and b. Using the value of χ^2 for this a and b, we can also calculate the goodness of fit, P. In addition, the linear correlation coefficient r is related by

$$\chi^2 = (1 - r^2) \sum_{i=1}^{N} (y_i - \overline{y})^2$$

Here $r = \dfrac{\sum\limits_{i=1}^{N} \dfrac{(x_i - \overline{x})(y_i - \overline{y})}{\sigma_i^2}}{\sqrt{\sum\limits_{i=1}^{N} \dfrac{(x_i - \overline{x})^2}{\sigma_i^2}} \sqrt{\sum\limits_{i=1}^{N} \dfrac{(y_i - \overline{y})^2}{\sigma_i^2}}}$

Values of r near 1 indicate a positive correlation; r near -1 means a negative correlation and r near zero means no correlation.

The form of the equations here is given to provide good accuracy when many terms are used and to provide the variances of the parameters. Another form of the equations for a and b is simpler, but is sometimes inaccurate unless many significant digits are kept in the calculations. The minimization of χ^2 when σ_i is the same for all i gives the following equations for a and b.

$$aN + b \sum_{i=1}^{N} x_i = \sum_{i=1}^{N} y_i$$

$$a \sum_{i=1}^{N} x_i + b \sum_{i=1}^{N} x_i^2 = \sum_{i=1}^{N} y_i x_i$$

The solution is $b = \dfrac{N \sum\limits_{i=1}^{N} y_i x_i - \sum\limits_{i=1}^{N} x_i \sum\limits_{i=1}^{N} y_i}{N \sum\limits_{i=1}^{N} x_i^2 - \left(\sum\limits_{i=1}^{N} x_i \right)^2}$

$$\overline{y} = \sum_{i=1}^{N} y_i \Big/ N, \quad \overline{x} = \sum_{i=1}^{N} x_i \Big/ N$$

$$a = \overline{y} - b\overline{x}$$

The value of χ^2 can be calculated from the formula

$$\chi^2 = \sum_{i=1}^{N} y_i^2 - a \sum_{i=1}^{N} y_i - b \sum_{i=1}^{N} y_i x_i$$

It is usually advisable to plot the observed pairs of y_i versus x_i to support the linearity assumption and to detect potential outliers. Suspected outliers can be omitted from the least-squares "fit" and then subsequently tested on the basis of the least-squares fit.

Example

Application.   Brenner (*Magnetic Method for Measuring the Thickness of Non-magnetic Coatings on Iron and Steel*, National Bureau of Standards, RP1081, March 1938) suggests an alternative way of measuring the thickness of nonmagnetic coatings of galvanized zinc on iron and steel. This procedure is based on a nondestructive magnetic method as a substitute for the standard destructive stripping method. A random sample of 11 pieces was selected and measured by both methods.

Nomenclature.   The calibration between the magnetic and the stripping methods can be determined through the model

$$y = a + bx + \varepsilon$$

where x = strip-method determination
y = magnetic-method determination

Sample data

Thickness, 10^{-5} In

Stripping method, x	Magnetic method, y
104	85
114	115
116	105
129	127
132	120
139	121
174	155
312	250
338	310
465	443
720	630

Computations.   The normal equations are defined by

$$na + \left(\sum x \right)b = \sum y$$
$$\left(\sum x \right)a + \left(\sum x^2 \right)b = \sum xy$$

For the sample

$$11a + 2743b = 2461$$
$$2743a + 1,067,143b = 952,517$$

with $\sum y^2 = 852,419$.

The solution to the normal equations is given by

$$a = 3.19960 \qquad b = .884362$$

The error sum of squares can be computed from the formula

$$\chi^2 = \sum y^2 - a \sum y - b \sum xy$$

if a sufficient number of significant digits is retained (usually six or seven digits are sufficient). Here

$$\chi^2 = 2175.14$$

If the normalized method is used in addition, the value of S_{tt} is $3.8314 \times 10^5/\sigma^2$, where σ^2 is the variance of the measurement of y. The values of a and b are, of course, the same. The variances of a and b are $\sigma_a^2 = 0.2532\sigma^2$, $\sigma_b^2 = 2.610 \times 10^{-6}\sigma^2$. The correlation coefficient is 0.996390, which indicates that there is a positive correlation between x and y. The small value of the variance for b indicates that this parameter is determined very well by the data. The residuals show no particular pattern, and the predictions are plotted along with the data in Fig. 3-58. If the variance of the measurements of y is known through repeated measurements, then the variance of the parameters can be made absolute.

Multiple Regression   A general linear model is one expressed as

$$y(x) = \sum_{k=1}^{M} a_k X_k(x)$$

where the parameters are $\{a_k\}$, and the expression is linear with respect to them, and $X_k(x)$ can be any (nonlinear) functions of x, not depending on the parameters $\{a_k\}$. Then:

$$\sum_{i=1}^{N} \frac{1}{\sigma_i^2} \left[y_i - \sum_{j=1}^{M} a_j X_j(x_i) \right] X_k(x_i) = 0, \quad k = 1, \ldots, M$$

This is rewritten as

$$\sum_{j=1}^{M} \left[\sum_{i=1}^{N} \frac{1}{\sigma_i^2} X_j(x_i) X_k(x_i) \right] a_j = \sum_{i=1}^{N} \frac{y_i}{\sigma_i^2} X_k(x_i)$$

or as

$$\sum_{j=1}^{M} \alpha_{kj} a_j = \beta_k$$

Solving this set of equations gives the parameters $\{a_j\}$, which maximize the likelihood. The variance of a_j is

$$\sigma^2(a_j) = C_{jj}$$

where $C_{jk} = \alpha_{jk}^{-1}$, or C is the inverse of α. The covariance of a_j and a_k

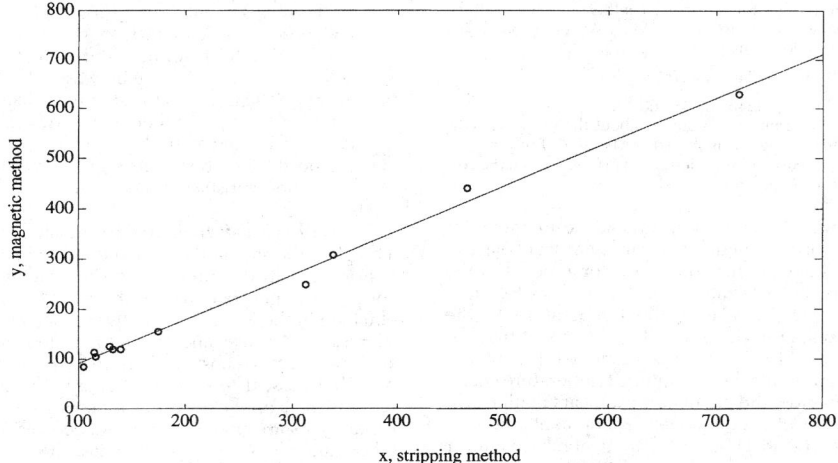

FIG. 3-58   Plot of data and correlating line.

is given by C_{jk}. If rounding errors affect the result, then we try to make the functions orthogonal. For example, using

$$X_k(x) = x^{k-1}$$

will cause rounding errors for a smaller M than

$$X_k(x) = P_{k-1}(x)$$

where P_{k-1} are orthogonal polynomials. If necessary, a singular value decomposition can be used.

Various global and piecewise polynomials can be used to fit the data. Most approximations are to be used with $M < N$. One can sometimes use more and more terms, and calculating the value of χ^2 for each solution. Then stop increasing M when the value of χ^2 no longer increases with increasing M.

Example

Application.   Merriman ("The Method of Least Squares Applied to a Hydraulic Problem," *J. Franklin Inst.*, 233–241, October 1877) reported on a study of stream velocity as a function of relative depth of the stream.

Sample data

Depth°	Velocity, y, ft/s
0	3.1950
.1	3.2299
.2	3.2532
.3	3.2611
.4	3.2516
.5	3.2282
.6	3.1807
.7	3.1266
.8	3.0594
.9	2.9759

°As a fraction of total depth.

Model.   Owing to the curvature of velocity with depth, a quadratic model was specified:

$$\text{Velocity} = \beta_0 + \beta_1 x_1 + \beta_2 x_2$$

where $x_2 = x_1^2$.

Normal equations.   The three normal equations are defined by

$$(n)\hat{\beta}_0 + (\textstyle\sum x_1)\hat{\beta}_1 + (\textstyle\sum x_2)\hat{\beta}_2 = \textstyle\sum y$$
$$(\textstyle\sum x_1)\hat{\beta}_0 + (\textstyle\sum x_1^2)\hat{\beta}_1 + (\textstyle\sum x_1 x_2)\hat{\beta}_2 = \textstyle\sum x_1 y$$
$$(\textstyle\sum x_2)\hat{\beta}_0 + (\textstyle\sum x_1 x_2)\hat{\beta}_1 + (\textstyle\sum x_2^2)\hat{\beta}_2 = \textstyle\sum x_2 y$$

For the sample data, the normal equations are

$$10\hat{\beta}_0 + 4.5\hat{\beta}_1 + 2.85\hat{\beta}_2 = 31.7616$$

$$4.5\hat{\beta}_0 + 2.85\hat{\beta}_1 + 2.025\hat{\beta}_2 = 14.08957$$
$$2.85\hat{\beta}_0 + 2.025\hat{\beta}_1 + 1.5333\hat{\beta}_2 = 8.828813$$

The algebraic solution to the simultaneous equations is

$$\hat{\beta}_0 = 3.19513 \qquad \hat{\beta}_1 = .4425 \qquad \hat{\beta}_2 = -.7653$$

The inverse of the product matrix

$$\alpha = \begin{pmatrix} 10 & 4.5 & 2.85 \\ 4.5 & 2.85 & 2.025 \\ 2.85 & 2.025 & 1.5333 \end{pmatrix}$$

is

$$\alpha^{-1} = \begin{pmatrix} .6182 & -2.5909 & 2.2727 \\ -2.5909 & 16.5530 & -17.0455 \\ 2.2727 & -17.0455 & 18.9394 \end{pmatrix}$$

The variances are then the diagonal elements of the inverse of matrix α (0.6182, 16.5530, 18.9394) times the variance of the measurement of y, σ_y^2. The value of χ^2 is 5.751×10^{-5}, the correlation coefficient $r = 0.99964$, and $\sigma = 0.002398$.

t values.   A sample t value can be computed for each regression coefficient j through the identity $t_j = \hat{\beta}_j / (\hat{\sigma}\sqrt{c_{jj}})$, where c_{jj} is the (j,j) element in the inverse. For the two variables x_1 and x_2,

Coefficient	c_{jj}	Sample t value
.4425	16.55	45.3
−.7653	18.94	−73.3

Computational note.   From a computational standpoint, it is usually advisable to define the variables in deviation units. For example, in the problem presented, let

$$x_1 = \text{depth} - \overline{\text{depth}}$$
$$= \text{depth} - .45$$

For expansion terms such as a square, define

$$x_2 = x_1^2 - \overline{x_1^2} \qquad (\overline{x_1^2} = .0825)$$

For the previous sample data,

Deviation units			
x_1	x_2	x_1	x_2
−.45	.12	.05	−.08
−.35	.04	.15	−.06
−.25	−.02	.25	−.02
−.25	−.02	.35	.04
−.15	−.06	.45	.12
−.05	−.08		

The resultant analysis-of-variance tables will remain exactly the same. However, the corresponding coefficient t value for the linear coefficient will usually be improved. This is an idiosyncrasy of regression modeling. With the coded data presented, the least-squares solution is given by

$$\hat{Y} = 3.17616 - .2462x_1 - .7653x_2$$

with a corresponding t value for $\hat{\beta}_1 = -.2462$ of $t = -63.63$.

When expansion terms are used but not expanded about the mean, the corresponding t values for the generating terms should not be used. For example, if $x_3 = x_1x_2$ is used rather than the correct expansion $(x_1 - \bar{x}_1)(x_2 - \bar{x}_2)$, then the corresponding t values for x_1 and x_2 should not be used.

Nonlinear Least Squares   There are no analytic methods for determining the most appropriate model for a particular set of data. In many cases, however, the engineer has some basis for a model. If the parameters occur in a nonlinear fashion, then the analysis becomes more difficult. For example, in relating the temperature to the elapsed time of a fluid cooling in the atmosphere, a model that has an asymptotic property would be the appropriate model (temp = $a + b \exp(-c\ \text{time})$, where a represents the asymptotic temperature corresponding to $t \to \infty$. In this case, the parameter c appears nonlinearly. The usual practice is to concentrate on model development and computation rather than on statistical aspects. In general, nonlinear regression should be applied only to problems in which there is a well-defined, clear association between the two variables; therefore, a test of hypothesis on the significance of the fit would be somewhat ludicrous. In addition, the generalization of the theory for the associate confidence intervals for nonlinear coefficients is not well developed.

The Levenberg-Marquardt method is used when the parameters of the model appear nonlinearly (Ref. 231). We still define

$$\chi^2(\mathbf{a}) = \sum_{i=1}^{N} \left[\frac{y_i - y(x_i; \mathbf{a})}{\sigma_i^2} \right]^2$$

and near the optimum represent χ^2 by

$$\chi^2(\mathbf{a}) = \chi_0^2 - \mathbf{d}^T \cdot \mathbf{a} + \frac{1}{2}\ \mathbf{a}^T \cdot \mathbf{D} \cdot \mathbf{a}$$

where $\mathbf{d}$ is an $M \times 1$ vector and $\mathbf{D}$ is an $M \times M$ matrix. We then calculate iteratively

$$\mathbf{D} \cdot (\mathbf{a}^{k+1} - \mathbf{a}^k) = -\nabla\chi^2(\mathbf{a}^k) \qquad (3\text{-}89)$$

The notation a_l^k means the lth component of $\mathbf{a}$ evaluated on the kth iteration. If $\mathbf{a}^k$ is a poor approximation to the optimum, we might use steepest descent instead.

$$\mathbf{a}^{k+1} - \mathbf{a}^k = -\text{constant} \times \nabla\chi^2(\mathbf{a}^k) \qquad (3\text{-}90)$$

and choose the constant somehow to decrease χ^2 as much as possible. The gradient of χ^2 is

$$\frac{\partial\chi^2}{\partial a_k} = -2 \sum_{i=1}^{N} \frac{y_i - y(x_i; \mathbf{a})}{\sigma_i^2} \frac{\partial y(x_i; \mathbf{a})}{\partial a_k} \qquad k = 1, 2, \ldots, M$$

The second derivative (in $\mathbf{D}$) is

$$\frac{\partial^2\chi^2}{\partial a_k \partial a_l} = 2 \sum_{i=1}^{N} \frac{1}{\sigma_i^2} \left\{ \frac{\partial y(x_i; \mathbf{a})}{\partial a_k} \frac{\partial y(x_i; \mathbf{a})}{\partial a_l} - [y_i - y(x_i; \mathbf{a})] \frac{\partial^2 y(x_i; \mathbf{a})}{\partial a_k \partial a_l} \right\}$$

Both Eq. (3-89) and Eq. (3-90) are included if we write

$$\sum_{l=1}^{M} \alpha'_{kl} (a_l^{k+1} - a_l^k) = \beta_k \qquad (3\text{-}91)$$

where $\alpha'_{kl} = \sum_{i=1}^{N} \frac{1}{\sigma_i^2} \frac{\partial y(x_i; \mathbf{a})}{\partial a_k} \frac{\partial y(x_i; \mathbf{a})}{\partial a_l} \qquad k \neq 1$

$$\alpha'_{kk} = \sum_{i=1}^{N} \frac{1}{\sigma_i^2} \left[\frac{\partial y(x_i; \mathbf{a})}{\partial a_k} \right]^2 (1 + \lambda)$$

$$\beta_k = \sum_{i=1}^{N} \frac{y_i - y(x_i; \mathbf{a})}{\sigma_i^2} \frac{\partial y(x_i; \mathbf{a})}{\partial a_k}$$

The second term in the second derivative is dropped because it is usually small [remember that y_i will be close to $y(x_i, \mathbf{a})$]. The Levenberg-Marquardt method then iterates as follows

1. Choose $\mathbf{a}$ and calculate $\chi^2(\mathbf{a})$.
2. Choose λ, say $\lambda = 0.001$.
3. Solve Eq. (3-91) for $\mathbf{a}^{k+1}$ and evaluate $\chi^2(\mathbf{a}^{k+1})$.
4. If $\chi^2(\mathbf{a}^{k+1}) \geq \chi^2(\mathbf{a}^k)$ then increase λ by a factor of, say, 10 and go back to step 3. This makes the step more like a steepest descent.
5. If $\chi^2(\mathbf{a}^{k+1}) < \chi^2(\mathbf{a}^k)$ then update $\mathbf{a}$, i.e., use $\underline{\mathbf{a}} = a^{k+1}$, decrease λ by a factor of 10, and go back to step 3.
6. Stop the iteration when the *decrease* in χ^2 from one step to another is not statistically meaningful, i.e., less than 0.1 or 0.01 or 0.001.
7. Set $\lambda = 0$ and compute the estimated covariance matrix: $\mathbf{C} = \alpha^{-1}$. This gives the standard errors in the fitted parameters $\mathbf{a}$.

For normally distributed errors the parameter region in which χ^2 = constant can give boundaries of the confidence limits. The value of $\mathbf{a}$ obtained in the Marquardt method gives the minimum $\chi^2_{\min}$. If we set $\chi^2 = \chi^2_{\min} + \Delta\chi$ for some $\Delta\chi$ and then look at contours in parameter space where χ_1^2 = constant then we have confidence boundaries at the probability associated with χ_1^2. For example, in a chemical reactor with radial dispersion the heat transfer coefficient and radial effective heat conductivity are closely connected: decreasing one and increasing the other can still give a good fit. Thus, the confidence boundaries may look something like Fig. 3-59. The ellipse defined by $\Delta\chi^2 = 2.3$ contains 68.3 percent of the normally distributed data. The curve defined by $\Delta\chi^2 = 6.17$ contains 95.4 percent of the data.

Example

Application.   Data were collected on the cooling of water in the atmosphere as a function of time.

Sample data

Time x	Temperature y
0	92.0
1	85.5
2	79.5
3	74.5
5	67.0
7	60.5
10	53.5
15	45.0
20	39.5

Model form.   On the basis of the nature of the data, an exponential model was selected initially to represent the trend $y = a + be^{cx}$. In this example, the resultant temperature would approach as an asymptotic (a with c negative) the wet-bulb temperature of the surrounding atmosphere. Unfortunately, this temperature was not reported.

Using a computer package in MATLAB gives the following results: $a = 33.54$, $b = 57.89$, $c = 0.11$. The value of χ^2 is 1.83. An alternative form of model is $y = a + b/(c + x)$. For this model the results were $a = 9.872$, $b = 925.7$, $c = 11.27$, and the value of χ^2 is 0.19. Since this model had a smaller value of χ^2, it might be the chosen one, but it is only a fit of the specified data and may not be generalized beyond that. Both curve fits give an equivalent plot. The second form is shown in Fig. 3-60.

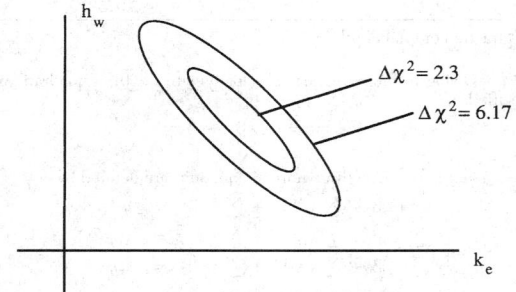

FIG. 3-59   Parameter estimation for heat transfer.

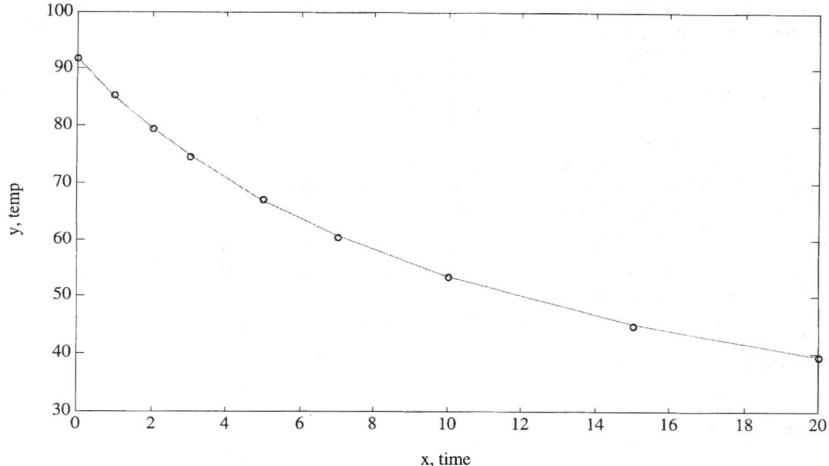

FIG. 3-60   Data in nonlinear regression example.

ERROR ANALYSIS OF EXPERIMENTS

Consider the problem of assessing the accuracy of a series of measurements. If measurements are for independent, identically distributed observations, then the errors are independent and uncorrelated. Then $\bar{y}$, the experimentally determined mean, varies about $E(y)$, the true mean, with variance σ^2/n, where n is the number of observations in $\bar{y}$. Thus, if one measures something several times today, and each day, and the measurements have the same distribution, then the variance of the means decreases with the number of samples in each day's measurement, n. Of course, other factors (weather, weekends) may make the observations on different days *not* distributed identically.

Consider next the problem of estimating the error in a variable that cannot be measured directly but must be calculated based on results of other measurements. Suppose the computed value Y is a linear combination of the measured variables $\{y_i\}$, $Y = \alpha_1 y_1 + \alpha_2 y_2 + \cdots$. Let the random variables $y_1, y_2, \ldots$ have means $E(y_1), E(y_2), \ldots$ and variances $\sigma^2(y_1), \sigma^2(y_2), \ldots$. The variable Y has mean

$$E(Y) = \alpha_1 E(y_1) + \alpha_2 E(y_2) + \cdots$$

and variance (Ref. 82)

$$\sigma^2(Y) = \sum_{i=1}^{n} \alpha_i^2 \sigma^2(y_i) + 2 \sum_{i=1}^{n} \sum_{j=i+1}^{n} \alpha_i \alpha_j \operatorname{Cov}(y_i, y_j)$$

If the variables are uncorrelated and have the same variance, then

$$\sigma^2(Y) = \left(\sum_{i=1}^{n} \alpha_i^2 \right) \sigma^2$$

Next suppose the model relating Y to $\{y_i\}$ is nonlinear, but the errors are small and independent of one another. Then a change in Y is related to changes in y_i by

$$dY = \frac{\partial Y}{\partial y_1} dy_1 + \frac{\partial Y}{\partial y_2} dy_2 + \cdots$$

If the changes are indeed small, then the partial derivatives are constant among all the samples. Then the expected value of the change, $E(dY)$, is zero. The variances are given by the following equation (Refs. 25 and 40):

$$\sigma^2(dY) = \sum_{i=1}^{N} \left(\frac{\partial Y}{\partial y_i} \right)^2 \sigma_i^2$$

Thus, the variance of the desired quantity Y can be found. This gives an independent estimate of the errors in measuring the quantity Y from the errors in measuring each variable it depends upon.

Example   Suppose one wants to measure the thermal conductivity of a solid (k). To do this, one needs to measure the heat flux (q), the thickness of the sample (d), and the temperature difference across the sample (ΔT). Each measurement has some error. The heat flux (q) may be the rate of electrical heat input ($\dot{Q}$) divided by the area (A), and both quantities are measured to some tolerance. The thickness of the sample is measured with some accuracy, and the temperatures are probably measured with a thermocouple to some accuracy. These measurements are combined, however, to obtain the thermal conductivity, and it is desired to know the error in the thermal conductivity. The formula is

$$k = \frac{d}{A \Delta T} \dot{Q}$$

The variance in the thermal conductivity is then

$$\sigma_k^2 = \left(\frac{k}{d} \right)^2 \sigma_d^2 + \left(\frac{k}{\dot{Q}} \right)^2 \sigma_{\dot{Q}}^2 + \left(\frac{k}{A} \right)^2 \sigma_A^2 + \left(\frac{k}{\Delta T} \right)^2 \sigma_{\Delta T}^2$$

FACTORIAL DESIGN OF EXPERIMENTS AND ANALYSIS OF VARIANCE

Statistically designed experiments consider, of course, the effect of primary variables, but they also consider the effect of extraneous variables and the interactions between variables, and they include a measure of the random error. Primary variables are those whose effect you wish to determine. These variables can be quantitative or qualitative. The quantitative variables are ones you may fit to a model in order to determine the model parameters (see the section "Least Squares"). Qualitative variables are ones you wish to know the effect of, but you do not try to quantify that effect other than to assign possible errors or magnitudes. Qualitative variables can be further subdivided into Type I variables, whose effect you wish to determine directly, and Type II variables, which contribute to the performance variability and whose effect you wish to average out. For example, if you are studying the effect of several catalysts on yield in a chemical reactor, each different type of catalyst would be a Type I variable because you would like to know the effect of each. However, each time the catalyst is prepared, the results are slightly different due to random variations; thus, you may have several batches of what purports to be the same catalyst. The variability between batches is a Type II variable. Since the ultimate use will require using different batches, you would like to know the overall effect including that variation, since knowing precisely the results from one batch of one catalyst might not be representative of the results obtained from all batches of the same catalyst. A randomized block design, incomplete block design, or Latin square design (Ref. 40), for example, all keep the effect of experimental error in the blocked variables from influencing the effect of the primary variables. Other uncontrolled variables are accounted for by introducing randomization in parts of the experimental design. To study all variables and their interaction requires a factorial design, involving all possible combinations of each variable, or a fractional factorial design, involving only a selected set. Statistical techniques are then used to determine which are the important variables, what are the important interactions, and what the error is in estimating these effects. The discussion here is only a brief overview of the excellent Ref. 40.

Suppose we have two methods of preparing some product and we wish to see which treatment is best. When there are only two treatments, then the sampling analysis discussed in the section "Two-Population Test of Hypothesis for Means" can be used to deduce if the means of the two treatments differ significantly. When there are more treatments, the analysis is more detailed. Suppose the experimental results are arranged as shown in the table: several measurements for each treatment. The goal is to see if the treatments differ significantly from each other; that is, whether their means are different when the samples have the same variance. The hypothesis is that the treatments are all the same, and the null hypothesis is that they are different. The statistical validity of the hypothesis is determined by an analysis of variance.

Estimating the Effect of Four Treatments

	Treatment			
	1	2	3	4
	—	—	—	—
	—	—	—	—
	—	—	—	—
				—
				—
Treatment average	—	—	—	—
Grand average		—		

The data for $k = 4$ treatments is arranged in the table. For each treatment, there are n_t experiments and the outcome of the ith experiment with treatment t is called y_{ti}. Compute the treatment average

$$\overline{y}_t = \frac{\sum_{i=1}^{n_t} y_{ti}}{n_t}$$

Also compute the grand average

$$\overline{y} = \frac{\sum_{t=1}^{k} n_t \overline{y}_t}{N}, \quad N = \sum_{t=1}^{k} n_t$$

Next compute the sum of squares of deviations from the average within the tth treatment

$$S_t = \sum_{i=1}^{n_t} (y_{ti} - \overline{y}_t)^2$$

Since each treatment has n_t experiments, the number of degrees of freedom is $n_t - 1$. Then the sample variances are

$$s_t^2 = \frac{S_t}{n_t - 1}$$

The within-treatment sum of squares is

$$S_R = \sum_{t=1}^{k} S_t$$

and the within-treatment sample variance is

$$s_R^2 = \frac{S_R}{N - k}$$

Now, if there is no difference between treatments, a second estimate of σ^2 could be obtained by calculating the variation of the treatment averages about the grand average. Thus compute the between-treatment mean square

$$s_T^2 = \frac{S_T}{k - 1}, \quad S_T = \sum_{t=1}^{k} n_t (\overline{y}_t - \overline{y})^2$$

Basically the test for whether the hypothesis is true or not hinges on a comparison of the within-treatment estimate s_R^2 (with $\nu_R = N - k$ degrees of freedom) with the between-treatment estimate s_T^2 (with $\nu_T = k - 1$ degrees of freedom). The test is made based on the F distribution for ν_R and ν_T degrees of freedom (Table 3-7).

Next consider the case that uses randomized blocking to eliminate the effect of some variable whose effect is of no interest, such as the batch-to-batch variation of the catalysts in the chemical reactor example. Suppose there are k treatments and n experiments in each treatment. The results from nk experiments can be arranged as shown in the block design table; within each block, the various treatments are applied in a random order. Compute the block average, the treatment average, as well as the grand average as before.

Block Design with Four Treatments and Five Blocks

Treatment	1	2	3	4	Block average
Block 1	—	—	—	—	—
Block 2	—	—	—	—	—
Block 3	—	—	—	—	—
Block 4	—	—	—	—	—
Block 5	—	—	—	—	—

The following quantities are needed for the analysis of variance table.

Name	Formula	dof
average	$S_A = nk\overline{y}^2$	1
blocks	$S_B = k \sum_{i=1}^{n} (\overline{y}_i - \overline{y})^2$	$n - 1$
treatments	$S_T = n \sum_{t=1}^{k} (\overline{y}_t - \overline{y})^2$	$k - 1$
residuals	$S_R = \sum_{t=1}^{k} \sum_{i=1}^{n} (y_{ti} - \overline{y}_i - \overline{y}_t + \overline{y})^2$	$(n-1)(k-1)$
total	$S = \sum_{t=1}^{k} \sum_{i=1}^{n} y_{ti}^2$	$N = nk$

The key test is again a statistical one, based on the value of

$$\frac{s_T^2}{s_R^2}, \quad s_T^2 = \frac{S_T}{k - 1}, \quad s_R^2 = \frac{S_R}{(n-1)(k-1)}$$

and the F distribution for ν_R and ν_T degrees of freedom (Table 3-7). The assumption behind the analysis is that the variations are linear (Ref. 40). There are ways to test this assumption as well as transformations to make if it is not true. Reference 40 also gives an excellent example of how the observations are broken down into a grand average, a block deviation, a treatment deviation, and a residual. For two-way factorial design in which the second variable is a real one rather than one you would like to block out, see Ref. 40.

To measure the effects of variables on a single outcome a factorial design is appropriate. In a two-level factorial design, each variable is considered at two levels only, a high and low value, often designated as a + and −. The two-level factorial design is useful for indicating trends, showing interactions, and it is also the basis for a fractional factorial design. As an example, consider a 2^3 factorial design with 3 variables and 2 levels for each. The experiments are indicated in the factorial design table.

Two-Level Factorial Design with Three Variables

	Variable		
Run	1	2	3
1	−	−	−
2	+	−	−
3	−	+	−
4	+	+	−
5	−	−	+
6	+	−	+
7	−	+	+
8	+	+	+

The main effects are calculated by calculating the difference between results from all high values of a variable and all low values of a variable; the result is divided by the number of experiments at each level. For example, for the first variable:

$$\text{Effect of variable } 1 = \frac{[(y_2 + y_4 + y_6 + y_8) - (y_1 + y_3 + y_5 + y_7)]}{4}$$

Note that all observations are being used to supply information on each of the main effects and each effect is determined with the precision of a fourfold replicated difference. The advantage of a one-at-a-time experiment is the gain in precision if the variables are additive and the measure of nonadditivity if it occurs (Ref. 40).

Interaction effects between variables 1 and 2 are obtained by calculating the difference between the results obtained with the high and low value of 1 at the low value of 2 compared with the results obtained with the high and low value 1 at the high value of 2. The 12-interaction is

$$12\text{-interaction} = \frac{[(y_4 - y_3 + y_8 - y_7) - (y_2 - y_1 + y_6 - y_5)]}{2}$$

The key step is to determine the errors associated with the effect of each variable and each interaction so that the significance can be determined. Thus, standard errors need to be assigned. This can be done by repeating the experiments, but it can also be done by using higher-order interactions (such as 123 interactions in a 2^4 factorial design). These are assumed negligible in their effect on the mean but can be used to estimate the standard error (see Ref. 40). Then, calculated effects that are large compared with the standard error are considered important, while those that are small compared with the standard error are considered to be due to random variations and are unimportant.

In a fractional factorial design one does only part of the possible experiments. When there are k variables, a factorial design requires 2^k experiments. When k is large, the number of experiments can be large; for $k = 5$, $2^5 = 32$. For a k this large, Box et al. (Ref. 82, p. 376) do a fractional factorial design. In the fractional factorial design with $k = 5$, only 16 experiments are done. Cropley (Ref. 82) gives an example of how to combine heuristics and statistical arguments in application to kinetics mechanisms in chemical engineering.

DIMENSIONAL ANALYSIS

Dimensional analysis allows the engineer to reduce the number of variables that must be considered to model experiments or correlate data. Consider a simple example in which two variables F_1 and F_2 have the units of force and two additional variables L_1 and L_2 have the units of length. Rather than having to deduce the relation of one variable on the other three, $F_1 = \text{fn } (F_2, L_1, L_2)$, dimensional analysis can be used to show that the relation must be of the form $F_1/F_2 = \text{fn } (L_1/L_2)$. Thus considerable experimentation is saved. Historically, dimensional analysis can be done using the Rayleigh method or the Buckingham pi method. This brief discussion is equivalent to the Buckingham pi method but uses concepts from linear algebra; see Ref. 13 for further information.

The general problem is posed as finding the minimum number of variables necessary to define the relationship between n variables. Let $\{Q_i\}$ represent a set of fundamental units, like length, time, force, and so on. Let $\{P_i\}$ represent the dimensions of a physical quantity P_i; there are n physical quantities. Then form the matrix α_{ij}

	$[P_1]$	$[P_2]$	$\cdots$	$[P_n]$
Q_1	α_{11}	α_{12}	$\cdots$	α_{1n}
Q_2	α_{21}	α_{22}	$\cdots$	α_{2n}
$\cdots$				
Q_m	α_{m1}	α_{m2}	$\cdots$	α_{mn}

in which the entries are the number of times each fundamental unit appears in the dimensions $[P_i]$. The dimensions can then be expressed as follows.

$$[P_i] = Q_1^{\alpha_{1i}} Q_2^{\alpha_{2i}} \cdots Q_m^{\alpha_{mi}}$$

Let m be the rank of the α matrix. Then $p = n - m$ is the number of dimensionless groups that can be formed. One can choose m variables $\{P_i\}$ to be the basis and express the other p variables in terms of them, giving p dimensionless quantities.

Example: Buckingham Pi Method—Heat-Transfer Film Coefficient It is desired to determine a complete set of dimensionless groups with which to correlate experimental data on the film coefficient of heat transfer between the walls of a straight conduit with circular cross section and a fluid flowing in that conduit. The variables and the dimensional constant believed to be involved and their dimensions in the engineering system are given below:

Film coefficient = $h = (F/L\theta T)$
Conduit internal diameter = $D = (L)$
Fluid linear velocity = $V = (L/\theta)$
Fluid density = $\rho = (M/L^3)$
Fluid absolute viscosity = $\mu = (M/L\theta)$
Fluid thermal conductivity = $k = (F/\theta T)$
Fluid specific heat = $c_p = (FL/MT)$
Dimensional constant = $g_c = (ML/F\theta^2)$

The matrix α in this case is as follows.

		h	D	V	ρ	μ	k	C_p	g_c
	F	1	0	0	0	0	1	1	-1
	M	0	0	0	1	1	0	-1	1
Q_j	L	-1	1	1	-3	-1	0	1	1
	θ	-1	0	-1	0	-1	-1	0	-2
	T	-1	0	0	0	0	-1	-1	0

Here $m \le 5$, $n = 8$, $p \ge 3$. Choose D, V, μ, k, and g_c as the primary variables. By examining the 5×5 matrix associated with those variables, we can see that its determinant is not zero, so the rank of the matrix is $m = 5$; thus, $p = 3$. These variables are thus a possible basis set. The dimensions of the other three variables h, ρ, and C_p must be defined in terms of the primary variables. This can be done by inspection, although linear algebra can be used, too.

$$[h] = D^{-1}k^{+1}; \text{ thus } \frac{h}{D^{-1}k} = \frac{hD}{k} \text{ is a dimensionless group}$$

$$[\rho] = \mu^1 V^{-1} D^{-1}; \text{ thus } \frac{\rho}{\mu^1 V^{-1} D^{-1}} = \frac{\rho V D}{\mu} \text{ is a dimensionless group}$$

$$[C_p] = k^{+1}\mu^{-1}; \text{ thus } \frac{C_p}{k^{+1}\mu^{-1}} = \frac{C_p \mu}{k} \text{ is a dimensionless group}$$

Thus, the dimensionless groups are

$$\frac{[P_i]}{Q_1^{\alpha_{1i}} Q_2^{\alpha_{2i}} \cdots Q_m^{\alpha_{mi}}} : \frac{hD}{k}, \frac{\rho V D}{\mu}, \frac{C_p \mu}{k}$$

The dimensionless group hD/k is called the Nusselt number, N_{Nu}, and the group $C_p\mu/k$ is the Prandtl number, N_{Pr}. The group $DV\rho/\mu$ is the familiar Reynolds number, N_{Re}, encountered in fluid-friction problems. These three dimensionless groups are frequently used in heat-transfer-film-coefficient correlations. Functionally, their relation may be expressed as

$$\phi(N_{\text{Nu}}, N_{\text{Pr}}, N_{\text{Re}}) = 0 \qquad (3\text{-}91)$$

or as

$$N_{\text{Nu}} = \phi_1(N_{\text{Pr}}, N_{\text{Re}})$$

It has been found that these dimensionless groups may be correlated well by an equation of the type

$$hD/k = K(c_p\mu/k)^a(DV\rho/\mu)^b$$

in which K, a, and b are experimentally determined dimensionless constants. However, any other type of algebraic expression or perhaps simply a graphical relation among these three groups that accurately fits the experimental data would be an equally valid manner of expressing Eq. (3-91).

Naturally, other dimensionless groups might have been obtained in the example by employing a different set of five repeating quantities that would not form a dimensionless group among themselves. Some of these groups may be found among those presented in Table 3-8. Such a complete set of three dimensionless groups might consist of Stanton, Reynolds, and Prandtl numbers or of Stanton, Peclet, and Prandtl numbers. Also, such a complete set different from that obtained in the preceding example will result from a multiplication of appropriate powers of the Nusselt, Prandtl, and Reynolds numbers. For such a set to be complete, however, it must satisfy the condition that each of the three dimensionless groups be independent of the other two.

TABLE 3-8   Dimensionless Groups in the Engineering System of Dimensions

Biot number	N_{Bi}	hL/k
Condensation number	N_{Co}	$(h/k)(\mu^2/\rho^2 g)^{1/3}$
Number used in condensation of vapors	N_{Cv}	$L^3\rho^2 g\lambda/k\mu\Delta t$
Euler number	N_{Eu}	$g_c(-dp)/\rho V^2$
Fourier number	N_{Fo}	$k\theta/\rho cL^2$
Froude number	N_{Fr}	V^2/Lg
Graetz number	N_{Gz}	wc/kL
Grashof number	N_{Gr}	$L^3\rho^2\beta g\Delta t/\mu^2$
Mach number	N_{Ma}	V/V_a
Nusselt number	N_{Nu}	hD/k
Peclet number	N_{Pe}	$DV\rho c/k$
Prandtl number	N_{Pr}	$c\mu/k$
Reynolds number	N_{Re}	$DV\rho/\mu$
Schmidt number	N_{Sc}	$\mu/\rho D_v$
Stanton number	N_{St}	$h/cV\rho$
Weber number	N_{We}	$LV^2\rho/\sigma g_c$

PROCESS SIMULATION

Classification   Process simulation refers to the activity in which mathematical models of chemical processes and refineries are modeled with equations, usually on the computer. The usual distinction must be made between steady-state models and transient models, following the ideas presented in the introduction to this section. In a chemical process, of course, the process is nearly always in a transient mode, at some level of precision, but when the time-dependent fluctuations are below some value, a steady-state model can be formulated. This subsection presents briefly the ideas behind steady-state process simulation (also called flowsheeting), which are embodied in commercial codes. The transient simulations are important for designing startup of plants and are especially useful for the operating of chemical plants.

Process Modules   The usual first step in process simulation is to perform a mass and energy balance for a chosen process. The most important aspect of the simulation is that the thermodynamic data of the chemicals be modeled correctly. The computer results of vapor-liquid equilibria, for example, must be checked against experimental data to insure their validity before using the data in more complicated computer calculations. At this first level of detail, it is not necessary to know the internal parameters for all the units, since what is desired is just the overall performance. For example, in a heat exchanger design, it suffices to know the heat duty, the total area, and the temperatures of the output streams; the details like the percentage baffle cut, tube layout, or baffle spacing can be specified later when the details of the proposed plant are better defined. Each unit operation is modeled by a subroutine, which is governed by equations (presented throughout this book). Some of the inputs to the units are known, some are specified by the user as design variables, and some are to be found using the simulation. It is important to know the number of degrees of freedom for each option of the unit operation, because at least that many parameters must be specified in order for the simulation to be able to calculate unit outputs. Sometimes the quantities the user would like to specify are targets, and parameters in the unit operation are to be changed to meet that target. This is not always possible, and the designer will have to adjust the parameters of the unit operation to achieve the desired target, possibly using the convergence tools discussed below. For example, in a reaction/separation system, if there is an impurity that must be purged, a common objective is to set the purge fraction so that the impurity concentration into the reactor is kept at some moderate value. Yet the solution techniques do not readily lend themselves to this connection, so convergence strategies must be employed.

Solution Strategies   Consider a chemical process consisting of a series of units, such as distillation towers, reactors, and so forth. If the feed to the process is known and the operating parameters of the unit operations are specified by the user, then one can begin with the first unit, take the process input, calculate the unit output, carry that output to the input of the next unit, and continue the process. In this way, one can simulate the entire process. However, if the process involves a recycle stream, as nearly all chemical processes do, then when the calculation is begun, it is discovered that the recycle stream is unknown. Thus the calculation cannot begin. This situation leads to the need for an iterative process: the flow rates, temperature, and pressure of the unknown recycle stream are guessed and the calculations proceed as before. When one reaches the end of the process, where the recycle stream is formed to return to the inlet, it is necessary to check to see if the recycle stream is the same as assumed. If not, an iterative procedure must be used to cause convergence. The techniques like Wegstein (see "Numerical Solution of Nonlinear Equations in One Variable") can be used to accelerate the convergence. When doing these iterations, it is useful to analyze the process using precedence ordering and tearing to minimize the number of recycle loops (Refs. 201, 242, 255, and 293). When the recycle loops interact with one another the iterations may not lead to a convergent solution.

The designer usually wants to specify stream flow rates or parameters in the process, but these may not be directly accessible. For example, the desired separation may be known for a distillation tower, but the simulation program requires the specification of the number of trays. It is left up to the designer to choose the number of trays that lead to the desired separation. In the example of the purge stream/reactor impurity, a controller module may be used to adjust the purge rate to achieve the desired reactor impurity. This further complicates the iteration process.

An alternative method of solving the equations is to solve them as simultaneous equations. In that case, one can specify the design variables and the desired specifications and let the computer figure out the process parameters that will achieve those objectives. It is possible to overspecify the system or give impossible conditions. However, the biggest drawback to this method of simulation is that large sets (10,000s) of algebraic equations must be solved simultaneously. As computers become faster, this is less of an impediment.

For further information, see Refs. 90, 175, 255, and 293. For information on computer software, see the Annual CEP Software Directory (Ref. 8) and other articles (Refs. 7 and 175).

INTELLIGENT SYSTEMS IN PROCESS ENGINEERING

REFERENCES: General, 232, 248, 258, 275, 276. Knowledge-Based Systems, 49, 232, 275. Neural Networks, 54, 140. Qualitative Simulation, 178, 292. Fuzzy Logic, 94. Genetic Algorithms, 121. Applications, 15, 24, 205, 232, 250, 262, 294.

Intelligent system is a term that refers to computer-based systems that include knowledge-based systems, neural networks, fuzzy logic and fuzzy control, qualitative simulation, genetic algorithms, natural language understanding, and others. The term is often associated with a variety of computer programming languages and/or features that are used as implementation media, although this is an imprecise use. Examples include object-oriented languages, rule-based languages, prolog, and lisp. The term *intelligent system* is preferred over the term *artificial intelligence*. The three intelligent-system technologies currently seeing the greatest amount of industrial application are knowledge-based systems, fuzzy logic, and artificial neural networks. These technologies are components of distributed systems. Mathematical models, conventional numeric and statistical approaches, neural networks, knowledge-based systems, and the like, all have their place in practical implementation and allow automation of tasks not well-treated by numerical algorithms.

Fundamentally, intelligent-system techniques are modeling techniques. They allow the encoding of qualitative models that draw upon experience and expertise, thereby extending modeling capacity beyond mathematical description. An important capability of intelligent system techniques is that they can be used not only to model physical behaviors but also decision-making processes. Decision processes reflect the selection, application, and interpretation of highly relevant pieces of information to draw conclusions about complex situations. Activity-specific decision processes can be expressed at a functional level, such as diagnosis, design, planning, and scheduling, or as their generic components, such as classification, abduction, and simulation. Decision process models address how information is organized and structured and then assimilated into active decisions.

Knowledge-Based Systems   Knowledge-based system (KBS) approaches capture the structural and information processing features of qualitative problem solving associated with sequential consideration, selection, and search. These technologies not only provide the means of capturing decision-making knowledge but also offer a medium for exploiting efficient strategies used by experts.

KBSs, then, are computer programs that model specific ways of organizing problem-specific fragments of knowledge and then searching through them by establishing appropriate relationships to reach correct conclusions. *Deliberation* is a general label for the algorithmic process for sorting through the knowledge fragments. The basic components of KBSs are knowledge representation (structure) and search. They are the programming mechanisms that facilitate the use and application of the problem-specific knowledge appropriate to solving the problem. Together they are used to form conclusions, decisions, or interpretations in a symbolic form. See Refs. 49, 232, and 275.

Qualitative simulation is a specific KBS model of physical processes that are not understood well enough to develop a physics-based numeric model. Corrosion, fouling, mechanical wear, equipment failure, and fatigue are not easily modeled, but decisions about them can be based on qualitative reasoning. See Refs. 178 and 292.

Qualitative description of physical behaviors require that each continuous variable space be quantized. Quantization is typically based on landmark values that are boundary points separating qualitatively distinct regions of continuous values. By using these qualitative quantity descriptions, dynamic relations between variables can be modeled as qualitative equations that represent the structure of the system. The solution to the equations represents the possible sequences of qualitative states as well as the explanations for changes in behaviors.

Building and explaining a complex model requires a unified view called an *ontology*. Methods of qualitative reasoning can be based on different viewpoints; the dominant viewpoints are device, process, and constraints. Behavior generation is handled with two approaches: (1) simulating successive states from one or more initial states, and (2) determining all possible state-to-state transitions once all possible states are determined.

Fuzzy Logic   Fuzzy logic is a formalism for mapping between numerical values and qualitative or linguistic interpretations. This is useful when it is difficult to define precisely such terms as "high" and "low," since there may be no fixed threshold. Fuzzy sets use the concept of degree of membership to overcome this problem. Degree of membership allows a descriptor to be associated with a range of numeric values but in varying degrees. A fuzzy set is explicitly defined by a degree of membership for each linguistic variable that is applicable, $m_A(x)$ where m_A is the degree of membership for linguistic variable A. For fuzzy sets, logical operators, such as complement (NOT), intersection (AND), and union (OR) are defined. The following are typical definitions.

$$\text{NOT: } m_{\text{NOT }A}(x) = 1 - m_A(x)$$

$$\text{AND: } m_{A \text{ AND } B}(x) = \min\left[m_A(x), m_B(x)\right]$$

$$\text{OR: } m_{A \text{ OR } B}(x) = \max\left[m_A(x), m_B(x)\right]$$

Using these operators, fuzzy inference mechanisms are then developed to manipulate rules that include fuzzy values. The largest difference between fuzzy inference and ordinary inference is that fuzzy inference allows "partial match" of input and produces an "interpolated" output. This technology is useful in control also. See Ref. 94.

Artificial Neural Networks   An artificial neural network (ANN) is a collection of computational units that are interconnected in a network. Knowledge is captured in the form of weights, and input-output mappings are produced by the interactions of the weights and the computational units. Each computational unit combines weighted inputs and generates an output base on an activation function. Typical activation functions are (1) specified limit, (2) sigmoid, and (3) gaussian. ANNs can be feedforward, with multiple layers of intermediate units, or feedback (sometimes called recurrent networks).

The ability to generalize on given data is one of the most important performance characteristics. With appropriate selection of training examples, an optimal network architecture, and appropriate training, the network can map a relationship between input and output that is complete but bounded by the coverage of the training data.

Applications of neural networks can be broadly classified into three categories:

1.   Numeric-to-numeric transformations are used as empirical mathematical models where the adaptive characteristics of neural networks learn to map between numeric sets of input-output data. In these modeling applications, neural networks are used as an alternative to traditional data regression schemes based on regression of plant data. Backpropagation networks have been widely used for this purpose.

2.   Numeric-to-symbolic transformations are used in pattern-recognition problems where the network is used to classify input data vectors into specific labeled classes. Pattern recognition problems include data interpretation, feature identification, and diagnosis.

3.   Symbolic-to-symbolic transformations are used in various symbolic manipulations, including natural language processing and rule-based system implementation. See Refs. 54 and 140.

Thermodynamics

Hendrick C. Van Ness, D.Eng., *Howard P. Isermann Department of Chemical Engineering, Rensselaer Polytechnic Institute; Fellow, American Institute of Chemical Engineers; Member, American Chemical Society*

Michael M. Abbott, Ph.D., *Howard P. Isermann Department of Chemical Engineering, Rensselaer Polytechnic Institute; Member, American Institute of Chemical Engineers*

Nomenclature and Units

Symbols are omitted that are correlation- or application-specific.

Symbol	Definition	SI units	U.S. customary units
A	Helmholtz energy	J	Btu
$\hat{a}_i$	Activity of species i in solution	Dimensionless	Dimensionless
B	2d virial coefficient, density expansion	cm^3/mol	cm^3/mol
C	3d virial coefficient, density expansion	cm^6/mol^2	cm^6/mol^2
D	4th virial coefficient, density expansion	cm^9/mol^3	cm^9/mol^3
B'	2d virial coefficient, pressure expansion	kPa^{-1}	kPa^{-1}
C'	3d virial coefficient, pressure expansion	kPa^{-2}	kPa^{-2}
D'	4th virial coefficient, pressure expansion	kPa^{-3}	kPa^{-3}
B_{ij}	Interaction 2d virial coefficient	cm^3/mol	cm^3/mol
C_{ijk}	Interaction 3d virial coefficient	cm^6/mol^2	cm^6/mol^2
C_P	Heat capacity at constant pressure	$J/(mol \cdot K)$	$Btu/(lb\ mol \cdot R)$
C_V	Heat capacity at constant volume	$J/(mol \cdot K)$	$Btu/(lb\ mol \cdot R)$
E_K	Kinetic energy	J	Btu
E_P	Gravitational potential energy	J	Btu
f_i	Fugacity of pure species i	kPa	psi
$\hat{f}_i$	Fugacity of species i in solution	kPa	psi
G	Molar or unit-mass Gibbs energy	J/mol or J/kg	Btu/lb mol or Btu/lbm
g	Acceleration of gravity	m/s^2	ft/s^2
$\mathfrak{g}$	$\equiv G^E/RT$		
H	Molar or unit-mass enthalpy	J/mol or J/kg	Btu/lb mol or Btu/lbm
K_i	Equilibrium K-value, y_i/x_i	Dimensionless	Dimensionless
K_j	Equilibrium constant for chemical reaction j	Dimensionless	Dimensionless
k_i	Henry's constant	kPa	psi
M	Molar or unit-mass value of any extensive thermodynamic property of a solution		
M_i	Molar or unit-mass value of any extensive property of pure species i		
$\overline{M}_i$	Partial molar property of species i in solution		
ΔM	Property change of mixing		
ΔM_j°	Standard property change of reaction j		
m	Mass	kg	lbm
$\dot{m}$	Mass flow rate	kg/s	lbm/s
n	Number of moles		
n_i	Number of moles of species i		
P	Absolute pressure	kPa	psi
P_c	Critical pressure	kPa	psi
P_i^{sat}	Saturation or vapor pressure of species i	kPa	psi
p_i	Partial pressure of species i in gas mixture ($\equiv y_i P$)	kPa	psi
Q	Heat	J	Btu
$\dot{Q}$	Rate of heat transfer	J/s	Btu/s
R	Universal gas constant	$J/(mol \cdot K)$	$Btu/(lb\ mol \cdot R)$

Symbol	Definition	SI units	U.S. customary units
S	Molar or unit-mass entropy	$J/(mol \cdot K)$ or $J/(kg \cdot K)$	$Btu/(lb\ mol \cdot R)$ or $Btu/(lb \cdot R)$
T	Absolute temperature	K	R
T_c	Critical temperature,	K	R
U	Molar or unit-mass internal energy	J/mol or J/kg	Btu/lb mol or Btu/lbm
u	Velocity	m/s	ft/s
V	Molar or unit-mass volume	m^3/mol or m^3/kg	$ft^3/lb\ mol$ or ft^3/lbm
W	Work	J	Btu
W_s	Shaft work for flow process	J	Btu
$\dot{W}_s$	Shaft power for flow process	J/s	Btu/s
x_i	Mole fraction in general or liquid-phase mole fraction of species i in solution	Dimensionless	Dimensionless
y_i	Vapor-phase mole fraction of species i in solution	Dimensionless	Dimensionless
Z	Compressibility factor	Dimensionless	Dimensionless
z	Elevation above a datum level	m	ft

	Superscripts		
E	Denotes excess thermodynamic property		
id	Denotes value for an ideal solution		
ig	Denotes value for an ideal gas		
l	Denotes liquid phase		
lv	Denotes phase transition from liquid to vapor		
R	Denotes residual thermodynamic property		
t	Denotes a total value of a thermodynamic property		
v	Denotes vapor phase		
∞	Denotes a value at infinite dilution		

	Subscripts		
C	Denotes a value for a colder heat reservoir		
c	Denotes a value for the critical state		
H	Denotes a value for a hotter heat reservoir		
r	Denotes a reduced value		
rev	Denotes a reversible process		

	Greek letters		
α, β	As superscripts, identify phases		
β_i	Volume expansivity, species i	K^{-1}	R^{-1}
ε_j	Reaction coordinate for reaction j	mol	lb mol
$\Gamma_i(T)$	Defined by Eq. (4-72)	J/mol	Btu/lb mol
γ	Heat-capacity ratio, C_P/C_V	Dimensionless	Dimensionless
γ_i	Activity coefficient of species i in solution	Dimensionless	Dimensionless
μ_i	Chemical potential of species i	J/mol	Btu/lb mol
$\nu_{i,j}$	Stoichiometric number of species i in reaction j	Dimensionless	Dimensionless
ρ	Molar density	$mols/m^3$	$lb\ moles/ft^3$
σ	As a subscript, denotes a heat reservoir		
Φ_i	Defined by Eq. (4-283)	Dimensionless	Dimensionless
ϕ_i	Fugacity coefficient of pure species i	Dimensionless	Dimensionless
$\hat{\phi}_i$	Fugacity coefficient of species i in solution	Dimensionless	Dimensionless
ω	Acentric factor	Dimensionless	Dimensionless

GENERAL REFERENCES: Abbott, M.M., and H.C. Van Ness, *Schaum's Outline of Theory and Problems of Thermodynamics,* 2d ed., McGraw-Hill, New York, 1989. Tester, J.W. and M. Modell, *Thermodynamics and its Applications,* 3d ed., Prentice-Hall, Englewood Cliffs, N.J., 1996. Prausnitz, J.M., R.N. Lichtenthaler, and E.G. de Azevedo, *Molecular Thermodynamics of Fluid-Phase Equilibria,* 2d ed., Prentice-Hall, Englewood Cliffs, N.J., 1986. Reid, R.C., J.M. Prausnitz, and B.E. Poling, *The Properties of Gases and Liquids,* 4th ed., McGraw-Hill, New York, 1987. Sandler, S.I., *Chemical and Engineering Thermodynamics,* 2d ed., Wiley, New York, 1989. Smith, J.M., H.C. Van Ness, and M.M. Abbott, *Introduction to Chemical Engineering Thermodynamics,* 5th ed., McGraw-Hill, New York, 1996. Van Ness, H.C., and M.M. Abbott, *Classical Thermodynamics of Nonelectrolyte Solutions: With Applications to Phase Equilibria,* McGraw-Hill, New York, 1982.

INTRODUCTION

Thermodynamics is the branch of science that embodies the principles of energy transformation in macroscopic systems. The general restrictions which experience has shown to apply to all such transformations are known as the *laws of thermodynamics*. These laws are primitive; they cannot be derived from anything more basic.

The first law of thermodynamics states that energy is conserved; that, although it can be altered in form and transferred from one place to another, the total quantity remains constant. Thus, the first law of thermodynamics depends on the concept of energy; but, conversely, energy is an *essential* thermodynamic function because it allows the first law to be formulated. This coupling is characteristic of the primitive concepts of thermodynamics.

The words *system* and *surroundings* are similarly coupled. A *system* is taken to be any object, any quantity of matter, any region, and so on, selected for study and set apart (mentally) from everything else, which is called the *surroundings*. The imaginary envelope which encloses the system and separates it from its surroundings is called the *boundary* of the system.

Attributed to this boundary are special properties which may serve either (1) to *isolate* the system from its surroundings, or (2) to provide for *interaction* in specific ways between system and surroundings. An isolated system exchanges neither matter nor energy with its surroundings. If a system is not isolated, its boundaries may permit exchange of matter or energy or both with its surroundings. If the exchange of matter is allowed, the system is said to be *open;* if only energy and not matter may be exchanged, the system is *closed* (but not isolated), and its mass is constant.

When a system is isolated, it cannot be affected by its surroundings. Nevertheless, changes may occur within the system that are detectable with such measuring instruments as thermometers, pressure gauges, and so on. However, such changes cannot continue indefinitely, and the system must eventually reach a final static condition of *internal equilibrium*.

For a closed system which interacts with its surroundings, a final static condition may likewise be reached such that the system is not only internally at equilibrium but also in *external equilibrium* with its surroundings.

The concept of equilibrium is central in thermodynamics, for associated with the condition of internal equilibrium is the concept of *state*. A system has an identifiable, reproducible state when all its *properties,* such as temperature T, pressure P, and molar volume V, are fixed. The concepts of *state* and *property* are again coupled. One can equally well say that the properties of a system are fixed by its state. Although the properties T, P, and V may be detected with measuring instruments, the existence of the *primitive* thermodynamic properties (see Postulates 1 and 3 following) is recognized much more indirectly. The number of properties for which values must be specified in order to fix the state of a system depends on the nature of the system and is ultimately determined from experience.

When a system is displaced from an equilibrium state, it undergoes a *process*, a change of state, which continues until its properties attain new equilibrium values. During such a process the system may be caused to interact with its surroundings so as to interchange energy in the forms of heat and work and so to produce in the system changes considered desirable for one reason or another. A process that proceeds so that the system is never displaced more than differentially from an equilibrium state is said to be *reversible*, because such a process can be reversed at any point by an infinitesimal change in external conditions, causing it to retrace the initial path in the opposite direction.

Thermodynamics finds its origin in experience and experiment, from which are formulated a few postulates that form the foundation of the subject. The first two deal with energy:

POSTULATE 1

There exists a form of energy, known as **internal energy,** *which for systems at internal equilibrium is an intrinsic property of the system, functionally related to its characteristic coordinates.*

POSTULATE 2
(FIRST LAW OF THERMODYNAMICS)

The **total** *energy of any system and its surroundings is conserved.*

Internal energy is quite distinct from such external forms as the kinetic and potential energies of macroscopic bodies. Although a macroscopic property characterized by the macroscopic coordinates T and P, internal energy finds its origin in the kinetic and potential energies of molecules and submolecular particles. In applications of the first law of thermodynamics, all forms of energy must be considered, including the internal energy. It is therefore clear that Postulate 2 depends on Postulate 1. For an isolated system, the first law requires that its energy be constant. For a closed (but not isolated) system, the first law requires that energy changes of the system be exactly compensated by energy changes in the surroundings. Energy is exchanged between such a system and its surroundings in two forms: heat and work.

Heat is energy crossing the system boundary under the influence of a temperature difference or gradient. A quantity of heat Q represents an amount of energy in transit between a system and its surroundings, and is not a property of the system. The convention with respect to sign makes numerical values of Q positive when heat is added to the system and negative when heat leaves the system.

Work is again energy in transit between a system and its surroundings, but resulting from the displacement of an external force acting on the system. Like heat, a quantity of work W represents an amount of energy, and is not a property of the system. The sign convention, analogous to that for heat, makes numerical values of W positive when work is done on the system by the surroundings and negative when work is done on the surroundings by the system.

When applied to closed (constant-mass) systems for which the only form of energy that changes is the internal energy, the first law of thermodynamics is expressed mathematically as

$$dU^t = dQ + dW \qquad (4\text{-}1)$$

where U^t is the total internal energy of the system. Note that dQ and dW, differential *quantities* representing energy exchanges between the system and its surroundings, serve to account for the energy change of the surroundings. On the other hand, dU^t is directly the differential *change* in internal energy of the system. Integration of Eq. (4-1) gives for a finite process

$$\Delta U^t = Q + W \qquad (4\text{-}2)$$

where ΔU^t is the finite change given by the difference between the final and initial values of U^t. The heat Q and work W are finite quantities of heat and work; they are not properties of the system nor functions of the thermodynamic coordinates that characterize the system.

POSTULATE 3

There exists a property called **entropy,** *which for systems at internal equilibrium is an intrinsic property of the system, functionally related to the measurable coordinates which characterize the system. For* **reversible** *processes, changes in this property may be calculated by the equation:*

$$dS^t = dQ_{\text{rev}}/T \qquad (4\text{-}3)$$

where S^t is the total entropy of the system and T is the absolute temperature of the system.

POSTULATE 4 (SECOND LAW OF THERMODYNAMICS)

The entropy change of any system and its surroundings, **considered together,** *resulting from any real process is positive, approaching zero when the process approaches reversibility.*

In the same way that the first law of thermodynamics cannot be formulated without the prior recognition of internal energy as a property, so also the second law can have no complete and quantitative expression without a prior assertion of the existence of entropy as a property.

The second law requires that the entropy of an isolated system either increase or, in the limit, where the system has reached an equilibrium state, remain constant. For a closed (but not isolated) system it requires that any entropy decrease in either the system or its surroundings be more than compensated by an entropy increase in the other part or that in the limit, where the process is reversible, the total entropy of the system plus its surroundings be constant.

The fundamental thermodynamic properties that arise in connection with the first and second laws of thermodynamics are internal energy and entropy. These properties, together with the two laws for which they are essential, apply to all types of systems. However, different types of systems are characterized by different sets of measurable coordinates or variables. The type of system most commonly

encountered in chemical technology is one for which the primary characteristic variables are temperature T, pressure P, molar volume V, and composition, not all of which are necessarily independent. Such systems are usually made up of fluids (liquid or gas) and are called *PVT* systems.

For closed systems of this kind, the work of a *reversible* process may always be calculated from

$$dW_{\text{rev}} = -P \, dV^t \qquad (4\text{-}4)$$

where P is the absolute pressure and V^t is the total volume of the system. This equation follows directly from the definition of mechanical work.

POSTULATE 5

The macroscopic properties of homogeneous PVT systems at internal equilibrium can be expressed as functions of temperature, pressure, and composition only.

This postulate imposes an idealization, and is the basis for all subsequent property relations for *PVT* systems. The *PVT* system serves as a satisfactory model in an enormous number of practical applications. In accepting this model one assumes that the effects of fields (e.g., electric, magnetic, or gravitational) are negligible and that surface and viscous-shear effects are unimportant.

Temperature, pressure, and composition are thermodynamic coordinates representing conditions imposed upon or exhibited by the system, and the functional dependence of the thermodynamic properties on these conditions is determined by experiment. This is quite direct for molar or specific volume V, which can be measured, and leads immediately to the conclusion that there exists an *equation of state* relating molar volume to temperature, pressure, and composition for any particular homogeneous *PVT* system. The equation of state is a primary tool in applications of thermodynamics.

Postulate 5 affirms that the other molar or specific thermodynamic properties of *PVT* systems, such as internal energy U and entropy S, are also functions of temperature, pressure, and composition. These molar or unit-mass properties, represented by the plain symbols V, U, and S, are independent of system size and are called *intensive*. Temperature, pressure, and the composition variables, such as mole fraction, are also intensive. Total-system properties (V^t, U^t, S^t) do depend on system size, and are *extensive*. For a system containing n moles of fluid, $M^t = nM$, where M is a molar property.

Applications of the thermodynamic postulates necessarily involve the abstract quantities internal energy and entropy. The solution of any problem in applied thermodynamics is therefore found through these quantities.

VARIABLES, DEFINITIONS, AND RELATIONSHIPS

Consider a single-phase closed system in which there are no chemical reactions. Under these restrictions the composition is fixed. If such a system undergoes a differential, reversible process, then by Eq. (4-1)

$$dU^t = dQ_{\text{rev}} + dW_{\text{rev}}$$

Substitution for dQ_{rev} and dW_{rev} by Eqs. (4-3) and (4-4) gives

$$dU^t = T \, dS^t - P \, dV^t$$

Although derived for a *reversible* process, this equation relates properties only and is valid for *any* change between equilibrium states in a closed system. It may equally well be written

$$d(nU) = T \, d(nS) - P \, d(nV) \qquad (4\text{-}5)$$

where n is the number of moles of fluid in the system and is constant for the special case of a closed, nonreacting system. Note that

$$n \equiv n_1 + n_2 + n_3 + \cdots = \sum_i n_i$$

where i is an index identifying the chemical species present. When U, S, and V represent *specific* (unit-mass) properties, n is replaced by m.

Equation (4-5) shows that for the single-phase, nonreacting, closed system specified,

$$nU = u(nS, nV)$$

Then

$$d(nU) = \left[\frac{\partial(nU)}{\partial(nS)} \right]_{nV,n} d(nS) + \left[\frac{\partial(nU)}{\partial(nV)} \right]_{nS,n} d(nV)$$

where the subscript n indicates that all mole numbers n_i (and hence n) are held constant. Comparison with Eq. (4-5) shows that

$$\left[\frac{\partial(nU)}{\partial(nS)} \right]_{nV,n} = T \qquad (4\text{-}6)$$

$$\left[\frac{\partial(nU)}{\partial(nV)} \right]_{nS,n} = -P \qquad (4\text{-}7)$$

Consider now an *open* system consisting of a single phase and assume that

$$nU = \mathcal{U}(nS, nV, n_1, n_2, n_3, \ldots)$$

Then

$$d(nU) = \left[\frac{\partial(nU)}{\partial(nS)}\right]_{nV,n} d(nS) + \left[\frac{\partial(nU)}{\partial(nV)}\right]_{nS,n} d(nV) + \sum_i \left[\frac{\partial(nU)}{\partial n_i}\right]_{nS,nV,n_j} dn_i$$

where the summation is over all species present in the system and subscript n_j indicates that all mole numbers are held constant except the ith. Let

$$\mu_i \equiv \left[\frac{\partial(nU)}{\partial n_i}\right]_{nS,nV,n_j}$$

Together with Eqs. (4-6) and (4-7), this definition allows elimination of all the partial differential coefficients from the preceding equation:

$$d(nU) = T\, d(nS) - P\, d(nV) + \sum_i \mu_i\, dn_i \qquad (4\text{-}8)$$

Equation (4-8) is the **fundamental property relation** for single-phase PVT systems, from which all other equations connecting properties of such systems are derived. The quantity μ_i is called the *chemical potential* of species i, and it plays a vital role in the thermodynamics of phase and chemical equilibria.

Additional property relations follow directly from Eq. (4-8). Since $n_i = x_i n$, where x_i is the mole fraction of species i, this equation may be rewritten:

$$d(nU) - T\, d(nS) + P\, d(nV) - \sum_i \mu_i\, d(x_i n) = 0$$

Upon expansion of the differentials and collection of like terms, this becomes

$$\left[dU - T\, dS + P\, dV - \sum_i \mu_i\, dx_i\right] n + \left[U - TS + PV - \sum_i x_i \mu_i\right] dn = 0$$

Since n and dn are independent and arbitrary, the terms in brackets must separately be zero. Then

$$dU = T\, dS - P\, dV + \sum_i \mu_i\, dx_i \qquad (4\text{-}9)$$

$$U = TS - PV + \sum_i x_i \mu_i \qquad (4\text{-}10)$$

Equations (4-8) and (4-9) are similar, but there is an important difference. Equation (4-8) applies to a system of n moles where n may vary; whereas Eq. (4-9) applies to a system in which n is unity and invariant. Thus Eq. (4-9) is subject to the constraint that $\sum_i x_i = 1$ or that $\sum_i dx_i = 0$. In this equation the x_i are not independent variables, whereas the n_i in Eq. (4-8) are.

Equation (4-10) dictates the possible combinations of terms that may be defined as additional primary functions. Those in common use are:

Enthalpy $\qquad\qquad H \equiv U + PV \qquad\qquad (4\text{-}11)$

Helmholtz energy $\qquad A \equiv U - TS \qquad\qquad (4\text{-}12)$

Gibbs energy $\qquad\quad G \equiv U + PV - TS = H - TS \quad (4\text{-}13)$

Additional thermodynamic properties are related to these and arise by arbitrary definition. Multiplication of Eq. (4-11) by n and differentiation yields the general expression:

$$d(nH) = d(nU) + P\, d(nV) + nV\, dP$$

Substitution for $d(nU)$ by Eq. (4-8) reduces this result to:

$$d(nH) = T\, d(nS) + nV\, dP + \sum_i \mu_i\, dn_i \qquad (4\text{-}14)$$

The total differentials of nA and nG are obtained similarly:

$$d(nA) = -nS\, dT - P\, d(nV) + \sum_i \mu_i\, dn_i \qquad (4\text{-}15)$$

$$d(nG) = -nS\, dT + nV\, dP + \sum_i \mu_i\, dn_i \qquad (4\text{-}16)$$

Equations (4-8) and (4-14) through (4-16) are equivalent forms of the fundamental property relation. Each expresses a property nU, nH,

and so on, as a function of a particular set of independent variables; these are the *canonical variables* for the property. The choice of which equation to use in a particular application is dictated by convenience. However, the Gibbs energy G is special, because of its unique functional relation to T, P, and the n_i, which are the variables of primary interest in chemical processing. A similar set of equations is developed from Eq. (4-9). This set also follows from the preceding set when $n = 1$ and $n_i = x_i$. The two sets are related exactly as Eq. (4-8) is related to Eq. (4-9). The equations written for $n = 1$ are, of course, less general. Furthermore, the interdependence of the x_i precludes those mathematical operations which depend on independence of these variables.

CONSTANT-COMPOSITION SYSTEMS

For 1 mole of a homogeneous fluid of constant composition Eqs. (4-8) and (4-14) through (4-16) simplify to:

$$dU = T\, dS - P\, dV \qquad (4\text{-}17)$$
$$dH = T\, dS + V\, dP \qquad (4\text{-}18)$$
$$dA = -S\, dT - P\, dV \qquad (4\text{-}19)$$
$$dG = -S\, dT + V\, dP \qquad (4\text{-}20)$$

Implicit in these are the following:

$$T = \left(\frac{\partial U}{\partial S}\right)_V = \left(\frac{\partial H}{\partial S}\right)_P \qquad (4\text{-}21)$$

$$-P = \left(\frac{\partial U}{\partial V}\right)_S = \left(\frac{\partial A}{\partial V}\right)_T \qquad (4\text{-}22)$$

$$V = \left(\frac{\partial H}{\partial P}\right)_S = \left(\frac{\partial G}{\partial P}\right)_T \qquad (4\text{-}23)$$

$$-S = \left(\frac{\partial A}{\partial T}\right)_V = \left(\frac{\partial G}{\partial T}\right)_P \qquad (4\text{-}24)$$

In addition, the common Maxwell equations result from application of the reciprocity relation for exact differentials:

$$\left(\frac{\partial T}{\partial V}\right)_S = -\left(\frac{\partial P}{\partial S}\right)_V \qquad (4\text{-}25)$$

$$\left(\frac{\partial T}{\partial P}\right)_S = \left(\frac{\partial V}{\partial S}\right)_P \qquad (4\text{-}26)$$

$$\left(\frac{\partial P}{\partial T}\right)_V = \left(\frac{\partial S}{\partial V}\right)_T \qquad (4\text{-}27)$$

$$\left(\frac{\partial V}{\partial T}\right)_P = -\left(\frac{\partial S}{\partial P}\right)_T \qquad (4\text{-}28)$$

In all these equations the partial derivatives are taken with composition held constant.

Enthalpy and Entropy as Functions of T and P At constant composition the molar thermodynamic properties are functions of temperature and pressure (Postulate 5). Thus

$$dH = \left(\frac{\partial H}{\partial T}\right)_P dT + \left(\frac{\partial H}{\partial P}\right)_T dP \qquad (4\text{-}29)$$

$$dS = \left(\frac{\partial S}{\partial T}\right)_P dT + \left(\frac{\partial S}{\partial P}\right)_T dP \qquad (4\text{-}30)$$

The obvious next step is to eliminate the partial-differential coefficients in favor of measurable quantities.

The *heat capacity at constant pressure* is defined for this purpose:

$$C_P \equiv \left(\frac{\partial H}{\partial T}\right)_P \qquad (4\text{-}31)$$

It is a property of the material and a function of temperature, pressure, and composition.

Equation (4-18) may first be divided by dT and restricted to constant pressure, and then be divided by dP and restricted to constant temperature, yielding the two equations:

$$\left(\frac{\partial H}{\partial T}\right)_P = T\left(\frac{\partial S}{\partial T}\right)_P$$

$$\left(\frac{\partial H}{\partial P}\right)_T = T\left(\frac{\partial S}{\partial P}\right)_T + V$$

In view of Eq. (4-31), the first of these becomes

$$\left(\frac{\partial S}{\partial T}\right)_P = \frac{C_P}{T} \qquad (4\text{-}32)$$

and in view of Eq. (4-28), the second becomes

$$\left(\frac{\partial H}{\partial P}\right)_T = V - T\left(\frac{\partial V}{\partial T}\right)_P \qquad (4\text{-}33)$$

Combination of Eqs. (4-29), (4-31), and (4-33) gives

$$dH = C_P\, dT + \left[V - T\left(\frac{\partial V}{\partial T}\right)_P\right] dP \qquad (4\text{-}34)$$

and in combination Eqs. (4-30), (4-32), and (4-28) yield

$$dS = \frac{C_P}{T}\, dT - \left(\frac{\partial V}{\partial T}\right)_P dP \qquad (4\text{-}35)$$

Equations (4-34) and (4-35) are general expressions for the enthalpy and entropy of homogeneous fluids *at constant composition* as functions of T and P. The coefficients of dT and dP are expressed in terms of measurable quantities.

Internal Energy and Entropy as Functions of T and V
Because V is related to T and P through an equation of state, V rather than P can serve as an independent variable. In this case the internal energy and entropy are the properties of choice; whence

$$dU = \left(\frac{\partial U}{\partial T}\right)_V dT + \left(\frac{\partial U}{\partial V}\right)_T dV \qquad (4\text{-}36)$$

$$dS = \left(\frac{\partial S}{\partial T}\right)_V dT + \left(\frac{\partial S}{\partial V}\right)_T dV \qquad (4\text{-}37)$$

The procedure now is analogous to that of the preceding section.
Define the *heat capacity at constant volume* by

$$C_V \equiv \left(\frac{\partial U}{\partial T}\right)_V \qquad (4\text{-}38)$$

It is a property of the material and a function of temperature, pressure, and composition.
Two relations follow immediately from Eq. (4-17):

$$\left(\frac{\partial U}{\partial T}\right)_V = T\left(\frac{\partial S}{\partial T}\right)_V$$

$$\left(\frac{\partial U}{\partial V}\right)_T = T\left(\frac{\partial S}{\partial V}\right)_T - P$$

As a result of Eq. (4-38) the first of these becomes

$$\left(\frac{\partial S}{\partial T}\right)_V = \frac{C_V}{T} \qquad (4\text{-}39)$$

and as a result of Eq. (4-27), the second becomes

$$\left(\frac{\partial U}{\partial V}\right)_T = T\left(\frac{\partial P}{\partial T}\right)_V - P \qquad (4\text{-}40)$$

Combination of Eqs. (4-36), (4-38), and (4-40) gives

$$dU = C_V\, dT + \left[T\left(\frac{\partial P}{\partial T}\right)_V - P\right] dV \qquad (4\text{-}41)$$

and Eqs. (4-37), (4-39), and (4-27) together yield

$$dS = \frac{C_V}{T}\, dT + \left(\frac{\partial P}{\partial T}\right)_V dV \qquad (4\text{-}42)$$

Equations (4-41) and (4-42) are general expressions for the internal energy and entropy of homogeneous fluids *at constant composition* as functions of temperature and molar volume. The coefficients of dT and dV are expressed in terms of measurable quantities.

Heat-Capacity Relations   In Eqs. (4-34) and (4-41) both dH and dU are exact differentials, and application of the reciprocity relation leads to

$$\left(\frac{\partial C_P}{\partial P}\right)_T = -T\left(\frac{\partial^2 V}{\partial T^2}\right)_P \qquad (4\text{-}43)$$

$$\left(\frac{\partial C_V}{\partial V}\right)_T = T\left(\frac{\partial^2 P}{\partial T^2}\right)_V \qquad (4\text{-}44)$$

Thus, the pressure or volume dependence of the heat capacities may be determined from PVT data. The temperature dependence of the heat capacities is, however, determined empirically and is often given by equations such as

$$C_P = \alpha + \beta T + \gamma T^2$$

Equations (4-35) and (4-42) both provide expressions for dS, which must be equal for the same change of state. Equating them and solving for dT gives

$$dT = \frac{T}{C_P - C_V}\left(\frac{\partial V}{\partial T}\right)_P dP + \frac{T}{C_P - C_V}\left(\frac{\partial P}{\partial T}\right)_V dV$$

However, at constant composition $T = T(P,V)$, and

$$dT = \left(\frac{\partial T}{\partial P}\right)_V dP + \left(\frac{\partial T}{\partial V}\right)_P dV$$

Equating coefficients of either dP or dV in these two expressions for dT gives

$$C_P - C_V = T\left(\frac{\partial V}{\partial T}\right)_P\left(\frac{\partial P}{\partial T}\right)_V \qquad (4\text{-}45)$$

Thus the *difference* between the two heat capacities may be determined from PVT data.

Division of Eq. (4-32) by Eq. (4-39) yields the *ratio* of these heat capacities:

$$\frac{C_P}{C_V} = \frac{(\partial S/\partial T)_P}{(\partial S/\partial T)_V} = \frac{(\partial S/\partial V)_P (\partial V/\partial T)_P}{(\partial S/\partial P)_V (\partial P/\partial T)_V}$$

Replacement of each of the four partial derivatives through the appropriate Maxwell relation gives finally

$$\gamma \equiv \frac{C_P}{C_V} = \left(\frac{\partial V}{\partial P}\right)_T\left(\frac{\partial P}{\partial V}\right)_S \qquad (4\text{-}46)$$

where γ is the symbol conventionally used to represent the heat-capacity ratio.

The Ideal Gas   The simplest equation of state is the ideal gas equation:

$$PV = RT$$

where R is a universal constant, values of which are given in Table 1-9. The following partial derivatives are obtained from the ideal gas equation:

$$\left(\frac{\partial P}{\partial T}\right)_V = \frac{R}{V} = \frac{P}{T} \qquad \left(\frac{\partial^2 P}{\partial T^2}\right)_V = 0$$

$$\left(\frac{\partial V}{\partial T}\right)_P = \frac{R}{P} = \frac{V}{T} \qquad \left(\frac{\partial^2 V}{\partial T^2}\right)_P = 0$$

$$\left(\frac{\partial P}{\partial V}\right)_T = -\frac{P}{V}$$

The general equations for constant-composition fluids derived in the preceding subsections reduce to very simple forms when the relations for an ideal gas are substituted into them:

$$\left(\frac{\partial U}{\partial V}\right)_T = \left(\frac{\partial H}{\partial P}\right)_T = 0$$

$$\left(\frac{\partial S}{\partial P}\right)_T = -\frac{R}{P} \qquad \left(\frac{\partial S}{\partial V}\right)_T = \frac{R}{V}$$

$$dU = C_V\, dT$$

$$dH = C_P\,dT$$

$$dS = \left(\frac{C_V}{T}\right)dT + \left(\frac{R}{V}\right)dV$$

$$dS = \left(\frac{C_P}{T}\right)dT - \left(\frac{R}{P}\right)dP$$

$$\left(\frac{\partial C_V}{\partial V}\right)_T = \left(\frac{\partial C_P}{\partial P}\right)_T = 0$$

$$C_P - C_V = R \qquad \gamma \equiv \frac{C_P}{C_V} = -\left(\frac{\partial \ln P}{\partial \ln V}\right)_S$$

These equations clearly show that for an ideal gas U, H, C_P, and C_V are functions of temperature only and are independent of P and V. The entropy of an ideal gas, however, is a function of both T and P or of both T and V.

SYSTEMS OF VARIABLE COMPOSITION

The composition of a system may vary because the system is open or because of chemical reactions even in a closed system. The equations developed here apply regardless of the cause of composition changes.

Partial Molar Properties   Consider a homogeneous fluid solution comprised of any number of chemical species. For such a PVT system let the symbol M represent the molar (or unit-mass) value of any extensive thermodynamic property of the solution, where M may stand in turn for U, H, S, and so on. A total-system property is then nM, where $n = \sum_i n_i$ and i is the index identifying chemical species. One might expect the solution property M to be related solely to the properties M_i of the pure chemical species which comprise the solution. However, no such generally valid relation is known, and the connection must be established experimentally for every specific system.

Although the chemical species which make up a solution do not in fact have separate properties of their own, a solution property may be arbitrarily apportioned among the individual species. Once an apportioning recipe is adopted, then the assigned property values are quite logically treated as though they were indeed properties of the species in solution, and reasoning on this basis leads to valid conclusions.

For a homogeneous PVT system, Postulate 5 requires that

$$nM = \mathcal{M}(T, P, n_1, n_2, n_3, \ldots)$$

The total differential of nM is therefore

$$d(nM) = \left[\frac{\partial(nM)}{\partial T}\right]_{P,n}dT + \left[\frac{\partial(nM)}{\partial P}\right]_{T,n}dP + \sum_i\left[\frac{\partial(nM)}{\partial n_i}\right]_{T,P,n_j}dn_i$$

where subscript n indicates that all mole numbers n_i are held constant, and subscript n_j signifies that all mole numbers are held constant except the ith. This equation may also be written

$$d(nM) = n\left(\frac{\partial M}{\partial T}\right)_{P,x}dT + n\left(\frac{\partial M}{\partial P}\right)_{T,x}dP + \sum_i\left[\frac{\partial(nM)}{\partial n_i}\right]_{T,P,n_j}dn_i$$

where subscript x indicates that all mole fractions are held constant. The derivatives in the summation are called *partial molar properties* $\overline{M}_i$; by definition,

$$\overline{M}_i \equiv \left[\frac{\partial(nM)}{\partial n_i}\right]_{T,P,n_j} \tag{4-47}$$

The basis for calculation of partial properties from solution properties is provided by this equation. Moreover, the preceding equation becomes

$$d(nM) = n\left(\frac{\partial M}{\partial T}\right)_{P,x}dT + n\left(\frac{\partial M}{\partial P}\right)_{T,x}dP + \sum_i \overline{M}_i\,dn_i \tag{4-48}$$

Important equations follow from this result through the relations:

$$d(nM) = n\,dM + M\,dn$$

$$dn_i = d(x_in) = x_i\,dn + n\,dx_i$$

Combining these expressions with Eq. (4-48) and collecting like terms gives

$$\left[dM - \left(\frac{\partial M}{\partial T}\right)_{P,x}dT - \left(\frac{\partial M}{\partial P}\right)_{T,x}dP - \sum_i \overline{M}_i\,dx_i\right]n + \left[M - \sum_i \overline{M}_i x_i\right]dn = 0$$

Since n and dn are independent and arbitrary, the terms in brackets must separately be zero; whence

$$dM = \left(\frac{\partial M}{\partial T}\right)_{P,x}dT + \left(\frac{\partial M}{\partial P}\right)_{T,x}dP + \sum_i \overline{M}_i\,dx_i \tag{4-49}$$

and

$$M = \sum_i x_i\overline{M}_i \tag{4-50}$$

Equation (4-49) is merely a special case of Eq. (4-48); however, Eq. (4-50) is a vital new relation. Known as the *summability equation*, it provides for the calculation of solution properties from partial properties. Thus, a solution property apportioned according to the recipe of Eq. (4-47) may be recovered simply by adding the properties attributed to the individual species, each weighted by its mole fraction in solution. The equations for partial molar properties are also valid for partial specific properties, in which case m replaces n and the x_i are mass fractions. Equation (4-47) applied to the definitions of Eqs. (4-11) through (4-13) yields the partial-property relations:

$$\overline{H}_i = \overline{U}_i + P\overline{V}_i$$
$$\overline{A}_i = \overline{U}_i - T\overline{S}_i$$
$$\overline{G}_i = \overline{H}_i - T\overline{S}_i$$

Pertinent examples on partial molar properties are presented in Smith, Van Ness, and Abbott (*Introduction to Chemical Engineering Thermodynamics*, 5th ed., Sec. 10.3, McGraw-Hill, New York, 1996).

Gibbs/Duhem Equation   Differentiation of Eq. (4-50) yields

$$dM = \sum_i x_i\,d\overline{M}_i + \sum_i \overline{M}_i\,dx_i$$

Since this equation and Eq. (4-49) are both valid in general, their right-hand sides can be equated, yielding

$$\left(\frac{\partial M}{\partial T}\right)_{P,x}dT + \left(\frac{\partial M}{\partial P}\right)_{T,x}dP - \sum_i x_i\,d\overline{M}_i = 0 \tag{4-51}$$

This general result, the Gibbs/Duhem equation, imposes a constraint on how the partial molar properties of any phase may vary with temperature, pressure, and composition. For the special case where T and P are constant:

$$\sum_i x_i\,d\overline{M}_i = 0 \qquad \text{(constant } T, P) \tag{4-52}$$

Symbol M may represent the molar value of any extensive thermodynamic property; for example, V, U, H, S, or G. When $M \equiv H$, the derivatives $(\partial H/\partial T)_P$ and $(\partial H/\partial P)_T$ are given by Eqs. (4-31) and (4-33). Equations (4-49), (4-50), and (4-51) then become

$$dH = C_P\,dT + \left[V - T\left(\frac{\partial V}{\partial T}\right)_{P,x}\right]dP + \sum_i \overline{H}_i\,dx_i \tag{4-53}$$

$$H = \sum_i x_i\overline{H}_i \tag{4-54}$$

$$C_P\,dT + \left[V - T\left(\frac{\partial V}{\partial T}\right)_{P,x}\right]dP - \sum_i x_i\,d\overline{H}_i = 0 \tag{4-55}$$

Similar equations are readily derived when M takes on other identities.

Equation (4-47), which defines a partial molar property, provides a general means by which partial property values may be determined. However, for a *binary* solution an alternative method is useful. Equation (4-50) for a binary solution is

$$M = x_1\overline{M}_1 + x_2\overline{M}_2 \tag{4-56}$$

Moreover, the Gibbs/Duhem equation for a solution at given T and P, Eq. (4-52), becomes

$$x_1\,d\overline{M}_1 + x_2\,d\overline{M}_2 = 0 \tag{4-57}$$

These two equations can be combined to give

$$\overline{M}_1 = M + x_2 \frac{dM}{dx_1} \qquad (4\text{-}58a)$$

$$\overline{M}_2 = M - x_1 \frac{dM}{dx_1} \qquad (4\text{-}58b)$$

Thus for a binary solution, the partial properties are given directly as functions of composition for given T and P. For multicomponent solutions such calculations are complex, and direct use of Eq. (4-47) is appropriate.

Partial Molar Gibbs Energy Implicit in Eq. (4-16) is the relation

$$\mu_i = \left[\frac{\partial(nG)}{\partial n_i} \right]_{T,P,n_j}$$

In view of Eq. (4-47), the chemical potential and the partial molar Gibbs energy are therefore identical:

$$\mu_i = \overline{G}_i \qquad (4\text{-}59)$$

The reciprocity relation for an exact differential applied to Eq. (4-16) produces not only the Maxwell relation, Eq. (4-28), but also two other useful equations:

$$\left(\frac{\partial \mu_i}{\partial P} \right)_{T,n} = \left[\frac{\partial(nV)}{\partial n_i} \right]_{T,P,n_j} = \overline{V}_i \qquad (4\text{-}60)$$

$$\left(\frac{\partial \mu_i}{\partial T} \right)_{P,n} = -\left[\frac{\partial(nS)}{\partial n_i} \right]_{T,P,n_j} = -\overline{S}_i \qquad (4\text{-}61)$$

In a solution of constant composition, $\mu_i = \mu(T,P)$; whence

$$d\mu_i \equiv d\overline{G}_i = \left(\frac{\partial \mu_i}{\partial T} \right)_{P,n} dT + \left(\frac{\partial \mu_i}{\partial P} \right)_{T,n} dP$$

or

$$d\overline{G}_i = -\overline{S}_i \, dT + \overline{V}_i \, dP \qquad (4\text{-}62)$$

Comparison with Eq. (4-20) provides an example of the parallelism that exists between the equations for a constant-composition solution and those for the corresponding partial properties. This parallelism exists whenever the solution properties in the parent equation are related linearly (in the algebraic sense). Thus, in view of Eqs. (4-17), (4-18), and (4-19):

$$d\overline{U}_i = T \, d\overline{S}_i - P \, d\overline{V}_i \qquad (4\text{-}63)$$
$$d\overline{H}_i = T \, d\overline{S}_i + \overline{V}_i \, dP \qquad (4\text{-}64)$$
$$d\overline{A}_i = -\overline{S}_i \, dT - P \, d\overline{V}_i \qquad (4\text{-}65)$$

Note that these equations hold only for species in a constant-composition solution.

The following equation is a mathematical identity:

$$d\left(\frac{nG}{RT} \right) \equiv \frac{1}{RT} d(nG) - \frac{nG}{RT^2} dT$$

Substitution for $d(nG)$ by Eq. (4-16) and for G by $H - TS$ (Eq. [4-13]) gives, after algebraic reduction,

$$d\left(\frac{nG}{RT} \right) = \frac{nV}{RT} dP - \frac{nH}{RT^2} dT + \sum_i \frac{\mu_i}{RT} dn_i \qquad (4\text{-}66)$$

Equation (4-66) is a useful alternative to the fundamental property relation given by Eq. (4-16). All terms in this equation have the units of moles; moreover, the enthalpy rather than the entropy appears on the right-hand side.

The Ideal Gas State and the Compressibility Factor The simplest equation of state for a PVT system is the ideal gas equation:

$$PV^{ig} = RT$$

where V^{ig} is the ideal-gas–state molar volume. Similarly, H^{ig}, S^{ig}, and G^{ig} are ideal gas–state values; that is, the molar enthalpy, entropy, and Gibbs energy values that a PVT system would have were the ideal gas equation the correct equation of state. These quantities provide reference values to which actual values may be compared. For example, the compressibility factor Z compares the true molar volume to the ideal gas molar volume as a ratio:

$$Z = \frac{V}{V^{ig}} = \frac{V}{RT/P} = \frac{PV}{RT}$$

Generalized correlations for the compressibility factor are treated in Sec. 2.

Residual Properties These quantities compare true and ideal gas properties through differences:

$$M^R \equiv M - M^{ig} \qquad (4\text{-}67)$$

where M is the molar value of an extensive thermodynamic property of a fluid in its actual state and M^{ig} is the corresponding value for the ideal gas state of the fluid at the same T, P, and composition. Residual properties depend on interactions *between* molecules and not on characteristics of individual molecules. Since the ideal gas state presumes the absence of molecular interactions, residual properties reflect deviations from ideality. Most commonly used of the residual properties are:

Residual volume	$V^R \equiv V - V^{ig}$
Residual enthalpy	$H^R \equiv H - H^{ig}$
Residual entropy	$S^R \equiv S - S^{ig}$
Residual Gibbs energy	$G^R \equiv G - G^{ig}$

SOLUTION THERMODYNAMICS

IDEAL GAS MIXTURES

An ideal gas is a model gas comprising imaginary molecules of zero volume that do not interact. Each chemical species in an ideal gas mixture therefore has its own private properties, uninfluenced by the presence of other species. The *partial pressure* of species i in a gas mixture is defined as

$$p_i = x_i P \qquad (i = 1, 2, \ldots, N)$$

where x_i is the mole fraction of species i. The sum of the partial pressures clearly equals the total pressure. *Gibbs' theorem* for a mixture of ideal gases may be stated as follows:

> The partial molar property, other than the volume, of a constituent species in an ideal gas mixture is equal to the corresponding molar property of the species as a pure ideal gas at the mixture temperature but at a pressure equal to its partial pressure in the mixture.

This is expressed mathematically for generic partial property $\overline{M}_i^{ig}$ by the equation

$$\overline{M}_i^{ig}(T, P) = M_i^{ig}(T, p_i) \quad (M \neq V) \qquad (4\text{-}68)$$

For those properties of an ideal gas that are independent of P, for example, U, H, and C_P, this becomes simply

$$\overline{M}_i^{ig} = M_i^{ig}$$

where M_i^{ig} is evaluated at the *mixture T and P*. Thus, for the enthalpy,

$$\overline{H}_i^{ig} = H_i^{ig} \qquad (4\text{-}69)$$

The entropy of an ideal gas *does* depend on pressure:

$$dS_i^{ig} = -R \, d \ln P \qquad (\text{constant } T)$$

Integration from p_i to P gives

$$S_i^{ig}(T, P) - S_i^{ig}(T, p_i) = -R \ln \frac{P}{p_i} = -R \ln \frac{P}{x_i P} = R \ln x_i$$

Whence

$$S_i^{ig}(T, p_i) = S_i^{ig}(T, P) - R \ln x_i$$

Substituting this result into Eq. (4-68) written for the entropy gives

$$\overline{S}_i^{ig} = S_i^{ig} - R \ln x_i \qquad (4\text{-}70)$$

where S_i^{ig} is evaluated at the mixture T and P.

For the Gibbs energy of an ideal gas mixture, $G^{ig} = H^{ig} - TS^{ig}$; the parallel relation for partial properties is

$$\overline{G}_i^{ig} = \overline{H}_i^{ig} - T\overline{S}_i^{ig}$$

In combination with Eqs. (4-69) and (4-70), this becomes

$$\overline{G}_i^{ig} = \overline{H}_i^{ig} - TS_i^{ig} + RT \ln x_i$$

or

$$\mu_i^{ig} \equiv \overline{G}_i^{ig} = G_i^{ig} + RT \ln x_i \qquad (4\text{-}71)$$

Elimination of G_i^{ig} from this equation is accomplished by Eq. (4-20), written for pure species i as:

$$dG_i^{ig} = V_i^{ig}\, dP = \frac{RT}{P}\, dP = RT\, d \ln P \qquad \text{(constant } T\text{)}$$

Integration gives

$$G_i^{ig} = \Gamma_i(T) + RT \ln P \qquad (4\text{-}72)$$

where $\Gamma_i(T)$, the integration constant for a given temperature, is a function of temperature only. Equation (4-71) now becomes

$$\mu_i^{ig} = \Gamma_i(T) + RT \ln x_i P \qquad (4\text{-}73)$$

FUGACITY AND FUGACITY COEFFICIENT

The chemical potential μ_i plays a vital role in both phase and chemical-reaction equilibria. However, the chemical potential exhibits certain unfortunate characteristics which discourage its use in the solution of practical problems. The Gibbs energy, and hence μ_i, is defined in relation to the internal energy and entropy, both primitive quantities for which absolute values are unknown. Moreover, μ_i approaches negative infinity when either P or x_i approaches zero. While these characteristics do not preclude the use of chemical potentials, the application of equilibrium criteria is facilitated by introduction of the *fugacity*, a quantity that takes the place of μ_i but which does not exhibit its less desirable characteristics.

The origin of the fugacity concept resides in Eq. (4-72), an equation valid only for pure species i in the ideal gas state. For a *real* fluid, an analogous equation is written:

$$G_i \equiv \Gamma_i(T) + RT \ln f_i \qquad (4\text{-}74)$$

in which a new property f_i replaces the pressure P. This equation serves as a partial definition of the *fugacity* f_i.

Subtraction of Eq. (4-72) from Eq. (4-74), both written for the same temperature and pressure, gives

$$G_i - G_i^{ig} = RT \ln \frac{f_i}{P}$$

According to the definition of Eq. (4-67), $G_i - G_i^{ig}$ is the residual Gibbs energy, G_i^R. The dimensionless ratio f_i/P is another new property called the *fugacity coefficient* ϕ_i. Thus,

$$G_i^R = RT \ln \phi_i \qquad (4\text{-}75)$$

where

$$\phi_i \equiv \frac{f_i}{P} \qquad (4\text{-}76)$$

The definition of fugacity is completed by setting the ideal-gas–state fugacity of pure species i equal to its pressure:

$$f_i^{ig} = P$$

Thus, for the special case of an ideal gas, $G_i^R = 0$, $\phi_i = 1$, and Eq. (4-72) is recovered from Eq. (4-74).

The definition of the fugacity of a species in solution is parallel to the definition of the pure-species fugacity. An equation analogous to the ideal gas expression, Eq. (4-73), is written for species i in a fluid mixture:

$$\mu_i \equiv \Gamma_i(T) + RT \ln \hat{f}_i \qquad (4\text{-}77)$$

where the partial pressure $x_i P$ is replaced by $\hat{f}_i$, the fugacity of species

i in solution. Since it is not a partial molar property, it is identified by a circumflex rather than an overbar.

Subtracting Eq. (4-73) from Eq. (4-77), both written for the same temperature, pressure, and composition, yields

$$\mu_i - \mu_i^{ig} = RT \ln \frac{\hat{f}_i}{x_i P}$$

Analogous to the defining equation for the residual Gibbs energy of a mixture, $G^R \equiv G - G^{ig}$, is the definition of a partial molar residual Gibbs energy:

$$\overline{G}_i^R \equiv \overline{G}_i - \overline{G}_i^{ig} = \mu_i - \mu_i^{ig}$$

Therefore

$$\overline{G}_i^R = RT \ln \hat{\phi}_i \qquad (4\text{-}78)$$

where by definition

$$\hat{\phi}_i \equiv \frac{\hat{f}_i}{x_i P} \qquad (4\text{-}79)$$

The dimensionless ratio $\hat{\phi}_i$ is called the *fugacity coefficient of species i in solution*.

Eq. (4-78) is the analog of Eq. (4-75), which relates ϕ_i to G_i^R. For an ideal gas, $\overline{G}_i^R$ is necessarily 0; therefore $\hat{\phi}_i^{ig} = 1$, and

$$\hat{f}_i^{ig} = x_i P$$

Thus, the fugacity of species i in an ideal gas mixture is equal to its partial pressure.

Pertinent examples are given in Smith, Van Ness, and Abbott (*Introduction to Chemical Engineering Thermodynamics*, 5th ed., Secs. 10.5–10.7, McGraw-Hill, New York, 1996).

FUNDAMENTAL RESIDUAL-PROPERTY RELATION

In view of Eq. (4-59), the fundamental property relation given by Eq. (4-66) may be written

$$d\left(\frac{nG}{RT}\right) = \frac{nV}{RT}\, dP - \frac{nH}{RT^2}\, dT + \sum_i \frac{\overline{G}_i}{RT}\, dn_i \qquad (4\text{-}80)$$

This equation is general, and may be written for the special case of an ideal gas:

$$d\left(\frac{nG^{ig}}{RT}\right) = \frac{nV^{ig}}{RT}\, dP - \frac{nH^{ig}}{RT^2}\, dT + \sum_i \frac{\overline{G}_i^{ig}}{RT}\, dn_i$$

Subtraction of this equation from Eq. (4-80) gives

$$d\left(\frac{nG^R}{RT}\right) = \frac{nV^R}{RT}\, dP - \frac{nH^R}{RT^2}\, dT + \sum_i \frac{\overline{G}_i^R}{RT}\, dn_i \qquad (4\text{-}81)$$

where the definitions $G^R \equiv G - G^{ig}$ and $\overline{G}_i^R \equiv \overline{G}_i - \overline{G}_i^{ig}$ have been imposed. Equation (4-81) is the *fundamental residual-property relation*. An alternative form follows by introduction of the fugacity coefficient as given by Eq. (4-78):

$$d\left(\frac{nG^R}{RT}\right) = \frac{nV^R}{RT}\, dP - \frac{nH^R}{RT^2}\, dT + \sum_i \ln \hat{\phi}_i\, dn_i \qquad (4\text{-}82)$$

These equations are of such generality that for practical application they are used only in restricted forms. Division of Eq. (4-82) by dP and restriction to constant T and composition leads to:

$$\frac{V^R}{RT} = \left[\frac{\partial (G^R/RT)}{\partial P}\right]_{T,x} \qquad (4\text{-}83)$$

Similarly, division by dT and restriction to constant P and composition gives

$$\frac{H^R}{RT} = -T\left[\frac{\partial (G^R/RT)}{\partial T}\right]_{P,x} \qquad (4\text{-}84)$$

Also implicit in Eq. (4-82) is the relation

$$\ln \hat{\phi}_i = \left[\frac{\partial (nG^R/RT)}{\partial n_i}\right]_{T,P,n_j} \qquad (4\text{-}85)$$

This equation demonstrates that $\ln \hat{\phi}_i$ is a partial property with respect to G^R/RT. The partial-property analogs of Eqs. (4-83) and (4-84) are therefore:

$$\left(\frac{\partial \ln \hat{\phi}_i}{\partial P}\right)_{Tx} = \frac{\overline{V}_i^R}{RT} \tag{4-86}$$

$$\left(\frac{\partial \ln \hat{\phi}_i}{\partial T}\right)_{Px} = -\frac{\overline{H}_i^R}{RT^2} \tag{4-87}$$

The partial-property relationship of $\ln \hat{\phi}_i$ to G^R/RT also means that the summability relation applies; thus

$$\frac{G^R}{RT} = \sum_i x_i \ln \hat{\phi}_i \tag{4-88}$$

THE IDEAL SOLUTION

The ideal gas is a useful model of the behavior of gases and serves as a standard to which real gas behavior can be compared. This is formalized by the introduction of residual properties. Another useful model is the *ideal solution*, which serves as a standard to which real solution behavior can be compared. This is formalized by introduction of *excess properties*.

The partial molar Gibbs energy of species i in an ideal gas mixture is given by Eq. (4-71). This equation takes on new meaning when G_i^{ig}, the Gibbs energy of pure species i in the ideal gas state, is replaced by G_i, the Gibbs energy of pure species i as it actually exists at the mixture T and P and in the same physical state (*real* gas, liquid, or solid) as the mixture. It then becomes applicable to species in real solutions; indeed, to liquids and solids as well as to gases. The ideal solution is therefore *defined* as one for which

$$\overline{G}_i^{id} \equiv G_i + RT \ln x_i \tag{4-89}$$

where superscript *id* denotes an ideal-solution property.

This equation is the basis for development of expressions for all other thermodynamic properties of an ideal solution. Equations (4-60) and (4-61), applied to an ideal solution with μ_i replaced by $\overline{G}_i$, can be written

$$\overline{V}_i^{id} = \left(\frac{\partial \overline{G}_i^{id}}{\partial P}\right)_{Tx} \quad \text{and} \quad \overline{S}_i^{id} = -\left(\frac{\partial \overline{G}_i^{id}}{\partial T}\right)_{Px}$$

Appropriate differentiation of Eq. (4-89) in combination with these relations and Eqs. (4-23) and (4-24) yields

$$\overline{V}_i^{id} = V_i \tag{4-90}$$

$$\overline{S}_i^{id} = S_i - R \ln x_i \tag{4-91}$$

Since $\overline{H}_i^{id} = \overline{G}_i^{id} + T\overline{S}_i^{id}$, substitutions by Eqs. (4-89) and (4-91) yield

$$\overline{H}_i^{id} = H_i \tag{4-92}$$

The summability relation, Eq. (4-50), written for the special case of an ideal solution, may be applied to Eqs. (4-89) through (4-92):

$$G^{id} = \sum_i x_i G_i + RT \sum_i x_i \ln x_i \tag{4-93}$$

$$V^{id} = \sum_i x_i V_i \tag{4-94}$$

$$S^{id} = \sum_i x_i S_i - R \sum_i x_i \ln x_i \tag{4-95}$$

$$H^{id} = \sum_i x_i H_i \tag{4-96}$$

A simple equation for the fugacity of a species in an ideal solution follows from Eq. (4-89). Written for the special case of species i in an ideal solution, Eq. (4-77) becomes

$$\mu_i^{id} \equiv \overline{G}_i^{id} = \Gamma_i(T) + RT \ln \hat{f}_i^{id}$$

When this equation and Eq. (4-74) are combined with Eq. (4-89), $\Gamma_i(T)$ is eliminated, and the resulting expression reduces to

$$\hat{f}_i^{id} = x_i f_i \tag{4-97}$$

This equation, known as the *Lewis/Randall rule*, applies to each species in an ideal solution at all conditions of T, P, and composition. It shows that the fugacity of each species in an ideal solution is proportional to its mole fraction; the proportionality constant is the fugacity of *pure* species i in the same physical state as the solution and at the

same T and P. Division of both sides of Eq. (4-97) by $x_i P$ and substitution of $\hat{\phi}_i^{id}$ for $\hat{f}_i^{id}/x_i P$ (Eq. [4-79]) and of ϕ_i for f_i/P (Eq. [4-76]) gives an alternative form:

$$\hat{\phi}_i^{id} = \phi_i \tag{4-98}$$

Thus, the fugacity coefficient of species i in an ideal solution is equal to the fugacity coefficient of *pure* species i in the same physical state as the solution and at the same T and P.

Ideal solution behavior is often approximated by solutions comprised of molecules not too different in size and of the same chemical nature. Thus, a mixture of isomers conforms very closely to ideal solution behavior. So do mixtures of adjacent members of a homologous series.

FUNDAMENTAL EXCESS-PROPERTY RELATION

The residual Gibbs energy and the fugacity coefficient are useful where experimental *PVT* data can be adequately correlated by equations of state. Indeed, if convenient treatment of all fluids by means of equations of state were possible, the thermodynamic-property relations already presented would suffice. However, *liquid* solutions are often more easily dealt with through properties that measure their deviations from ideal solution behavior, not from ideal gas behavior. Thus, the mathematical formalism of *excess* properties is analogous to that of the residual properties.

If M represents the molar (or unit-mass) value of any extensive thermodynamic property (e.g., V, U, H, S, G, and so on), then an excess property M^E is defined as the difference between the actual property value of a solution and the value it would have as an ideal solution at the same temperature, pressure, and composition. Thus,

$$M^E \equiv M - M^{id} \tag{4-99}$$

This definition is analogous to the definition of a residual property as given by Eq. (4-67). However, excess properties have no meaning for pure species, whereas residual properties exist for pure species as well as for mixtures. In addition, analogous to Eq. (4-99) is the partial-property relation,

$$\overline{M}_i^E = \overline{M}_i - \overline{M}_i^{id} \tag{4-100}$$

where $\overline{M}_i^E$ is a partial excess property. The fundamental excess-property relation is derived in exactly the same way as the fundamental residual-property relation and leads to analogous results. Equation (4-80), written for the special case of an ideal solution, is subtracted from Eq. (4-80) itself, yielding:

$$d\left(\frac{nG^E}{RT}\right) = \frac{nV^E}{RT} dP - \frac{nH^E}{RT^2} dT + \sum_i \frac{\overline{G}_i^E}{RT} dn_i \tag{4-101}$$

This is the *fundamental excess-property relation*, analogous to Eq. (4-81), the fundamental residual-property relation.

The excess Gibbs energy is of particular interest. Equation (4-77) may be written:

$$\overline{G}_i = \Gamma_i(T) + RT \ln \hat{f}_i$$

In accord with Eq. (4-97) for an ideal solution, this becomes

$$\overline{G}_i^{id} = \Gamma_i(T) + RT \ln x_i f_i$$

By difference

$$\overline{G}_i - \overline{G}_i^{id} = RT \ln \frac{\hat{f}_i}{x_i f_i}$$

The left-hand side is the partial excess Gibbs energy $\overline{G}_i^E$; the dimensionless ratio $\hat{f}_i/x_i f_i$ appearing on the right is called the *activity coefficient of species i in solution*, and is given the symbol γ_i. Thus, *by definition*,

$$\gamma_i \equiv \frac{\hat{f}_i}{x_i f_i} \tag{4-102}$$

and

$$\overline{G}_i^E = RT \ln \gamma_i \tag{4-103}$$

Comparison with Eq. (4-78) shows that Eq. (4-103) relates γ_i to $\overline{G}_i^E$ exactly as Eq. (4-78) relates $\hat{\phi}_i$ to $\overline{G}_i^R$. For an ideal solution, $\overline{G}_i^E = 0$, and therefore $\gamma_i = 1$.

An alternative form of Eq. (4-101) follows by introduction of the activity coefficient through Eq. (4-103):

$$d\left(\frac{nG^E}{RT}\right) = \frac{nV^E}{RT}\,dP - \frac{nH^E}{RT^2}\,dT + \sum_i \ln \gamma_i\,dn_i \qquad (4\text{-}104)$$

SUMMARY OF FUNDAMENTAL PROPERTY RELATIONS

For convenience, the three other fundamental property relations, Eqs. (4-16), (4-80), and (4-82), expressing the Gibbs energy and related properties as functions of T, P, and the n_i, are collected here:

$$d(nG) = nV\,dP - nS\,dT + \sum_i \mu_i\,dn_i \qquad (4\text{-}16)$$

$$d\left(\frac{nG}{RT}\right) = \frac{nV}{RT}\,dP - \frac{nH}{RT^2}\,dT + \sum_i \frac{\overline{G}_i}{RT}\,dn_i \qquad (4\text{-}80)$$

$$d\left(\frac{nG^R}{RT}\right) = \frac{nV^R}{RT}\,dP - \frac{nH^R}{RT^2}\,dT + \sum_i \ln \hat{\phi}_i\,dn_i \qquad (4\text{-}82)$$

These equations and Eq. (4-104) may also be written for the special case of 1 mole of solution by setting $n = 1$ and $n_i = x_i$. The x_i are then subject to the constraint that $\sum_i x_i = 1$.

If written for 1 mole of a *constant-composition* solution, they become:

$$dG = V\,dP - S\,dT \qquad (4\text{-}105)$$

$$d\left(\frac{G}{RT}\right) = \frac{V}{RT}\,dP - \frac{H}{RT^2}\,dT \qquad (4\text{-}106)$$

$$d\left(\frac{G^R}{RT}\right) = \frac{V^R}{RT}\,dP - \frac{H^R}{RT^2}\,dT \qquad (4\text{-}107)$$

$$d\left(\frac{G^E}{RT}\right) = \frac{V^E}{RT}\,dP - \frac{H^E}{RT^2}\,dT \qquad (4\text{-}108)$$

These equations are, of course, valid as a special case for a pure species; in this event they are written with subscript i affixed to the appropriate symbols.

The partial-property analogs of these equations are:

$$d\overline{G}_i = d\mu_i = \overline{V}_i\,dP - \overline{S}_i\,dT \qquad (4\text{-}109)$$

$$d\left(\frac{\overline{G}_i}{RT}\right) = d\left(\frac{\mu_i}{RT}\right) = \frac{\overline{V}_i}{RT}\,dP - \frac{\overline{H}_i}{RT^2}\,dT \qquad (4\text{-}110)$$

$$d\left(\frac{\overline{G}_i^R}{RT}\right) = d\ln \hat{\phi}_i = \frac{\overline{V}_i^R}{RT}\,dP - \frac{\overline{H}_i^R}{RT^2}\,dT \qquad (4\text{-}111)$$

$$d\left(\frac{\overline{G}_i^E}{RT}\right) = d\ln \gamma_i = \frac{\overline{V}_i^E}{RT}\,dP - \frac{\overline{H}_i^E}{RT^2}\,dT \qquad (4\text{-}112)$$

Finally, a Gibbs/Duhem equation is associated with each fundamental property relation:

$$V\,dP - S\,dT = \sum_i x_i\,d\mu_i \qquad (4\text{-}113)$$

$$\frac{V}{RT}dP - \frac{H}{RT^2}\,dT = \sum_i x_i\,d\left(\frac{\overline{G}_i}{RT}\right) \qquad (4\text{-}114)$$

$$\frac{V^R}{RT}dP - \frac{H^R}{RT^2}\,dT = \sum_i x_i\,d\ln \hat{\phi}_i \qquad (4\text{-}115)$$

$$\frac{V^E}{RT}dP - \frac{H^E}{RT^2}\,dT = \sum_i x_i\,d\ln \gamma_i \qquad (4\text{-}116)$$

This depository of equations stores an enormous amount of information. The equations themselves are so general that their direct application is seldom appropriate. However, by inspection one can write a vast array of relations valid for particular applications. For example, Eqs. (4-83) and (4-84) come directly from Eq. (4-107); Eqs. (4-86) and (4-87), from (4-111). Similarly, from Eq. (4-108),

$$\frac{V^E}{RT} = \left[\frac{\partial(G^E/RT)}{\partial P}\right]_{T,x} \qquad (4\text{-}117)$$

$$\frac{H^E}{RT} = -T\left[\frac{\partial(G^E/RT)}{\partial T}\right]_{P,x} \qquad (4\text{-}118)$$

and from Eq. (4-104)

$$\ln \gamma_i = \left[\frac{\partial(nG^E/RT)}{\partial n_i}\right]_{T,P,n_j} \qquad (4\text{-}119)$$

The last relation demonstrates that $\ln \gamma_i$ is a partial property with respect to G^E/RT. The partial-property analogs of Eqs. (4-117) and (4-118) follow from Eq. (4-112):

$$\left(\frac{\partial \ln \gamma_i}{\partial P}\right)_{T,x} = \frac{\overline{V}_i^E}{RT} \qquad (4\text{-}120)$$

$$\left(\frac{\partial \ln \gamma_i}{\partial T}\right)_{P,x} = -\frac{\overline{H}_i^E}{RT^2} \qquad (4\text{-}121)$$

Finally, an especially useful form of the Gibbs/Duhem equation follows from Eq. (4-116):

$$\sum_i x_i\,d\ln \gamma_i = 0 \qquad \text{(constant } T,P) \qquad (4\text{-}122)$$

Since $\ln \gamma_i$ is a partial property with respect to G^E/RT, the following form of the summability equation is valid:

$$\frac{G^E}{RT} = \sum_i x_i \ln \gamma_i \qquad (4\text{-}123)$$

The analogy between equations derived from the fundamental residual- and excess-property relations is apparent. Whereas the fundamental *residual*-property relation derives its usefulness from its direct relation to equations of state, the *excess*-property formulation is useful because V^E, H^E, and γ_i are all experimentally accessible. Activity coefficients are found from vapor/liquid equilibrium data, and V^E and H^E values come from mixing experiments.

PROPERTY CHANGES OF MIXING

If M represents a molar thermodynamic property of a homogeneous fluid solution, then by definition,

$$\Delta M \equiv M - \sum_i x_i M_i \qquad (4\text{-}124)$$

where ΔM is the property change of mixing, and M_i is the molar property of pure species i at the T and P of the solution and in the same physical state (gas or liquid). The summability relation, Eq. (4-50), may be combined with Eq. (4-124) to give

$$\Delta M = \sum_i x_i\,\overline{\Delta M_i} \qquad (4\text{-}125)$$

where by definition

$$\overline{\Delta M_i} \equiv \overline{M}_i - M_i \qquad (4\text{-}126)$$

All three quantities are for the same T, P, and physical state. Eq. (4-126) defines a partial molar property change of mixing, and Eq. (4-125) is the summability relation for these properties.

Each of Eqs. (4-93) through (4-96) is an expression for an ideal solution property, and each may be combined with the defining equation for an excess property (Eq. [4-99]), yielding

$$G^E = G - \sum_i x_i G_i - RT \sum_i x_i \ln x_i \qquad (4\text{-}127)$$

$$V^E = V - \sum_i x_i V_i \qquad (4\text{-}128)$$

$$S^E = S - \sum_i x_i S_i + R \sum_i x_i \ln x_i \qquad (4\text{-}129)$$

$$H^E = H - \sum_i x_i H_i \qquad (4\text{-}130)$$

In view of Eq. (4-124), these may be written

$$G^E = \Delta G - RT \sum_i x_i \ln x_i \qquad (4\text{-}131)$$

$$V^E = \Delta V \qquad (4\text{-}132)$$

$$S^E = \Delta S + R \sum_i x_i \ln x_i \qquad (4\text{-}133)$$

$$H^E = \Delta H \qquad (4\text{-}134)$$

where ΔG, ΔV, ΔS, and ΔH are the Gibbs energy change of mixing, the volume change of mixing, the entropy change of mixing, and the enthalpy change of mixing. For an ideal solution, each excess property is zero, and for this special case

$$\Delta G^{id} = RT \sum_i x_i \ln x_i \qquad (4\text{-}135)$$

$$\Delta V^{id} = 0 \qquad (4\text{-}136)$$

$$\Delta S^{id} = -R \sum_i x_i \ln x_i \qquad (4\text{-}137)$$

$$\Delta H^{id} = 0 \qquad (4\text{-}138)$$

Property changes of mixing and excess properties are easily calculated one from the other. The most commonly encountered property changes of mixing are the volume change of mixing ΔV and the enthalpy change of mixing ΔH, commonly called the *heat of mixing*. These properties are directly measurable and are identical to the corresponding excess properties.

Pertinent examples are given in Smith, Van Ness, and Abbott (*Introduction to Chemical Engineering Thermodynamics*, 5th ed., Sec. 11.4, McGraw-Hill, New York, 1996).

BEHAVIOR OF BINARY LIQUID SOLUTIONS

Property changes of mixing and excess properties find greatest application in the description of liquid mixtures at low reduced tempera-

tures, that is, at temperatures well below the critical temperature of each constituent species. The properties of interest to the chemical engineer are V^E ($\equiv \Delta V$), H^E ($\equiv \Delta H$), S^E, ΔS, G^E, and ΔG. The activity coefficient is also of special importance because of its application in phase-equilibrium calculations.

The behavior of binary liquid solutions is clearly displayed by plots of M^E, ΔM, and $\ln \gamma_i$ vs. x_1 at constant T and P. The volume change of mixing (or excess volume) is the most easily measured of these quantities and is normally small. However, as illustrated by Fig. 4-1, it is subject to individualistic behavior, being sensitive to the effects of molecular size and shape and to differences in the nature and magnitude of intermolecular forces.

The heat of mixing (excess enthalpy) and the excess Gibbs energy are also experimentally accessible, the heat of mixing by direct measurement and G^E (or $\ln \gamma_i$) indirectly as a product of the reduction of vapor/liquid equilibrium data. Knowledge of H^E and G^E allows calculation of S^E by Eq. (4-13) written for excess properties,

$$S^E = \frac{H^E - G^E}{T} \qquad (4\text{-}139)$$

with ΔS then given by Eq. (4-133).

Figure 4-2 displays plots of ΔH, ΔS, and ΔG as functions of composition for 6 binary solutions at 50°C. The corresponding excess properties are shown in Fig. 4-3; the activity coefficients, derived from Eq. (4-119), appear in Fig. 4-4. The properties shown here are insensitive to pressure, and for practical purposes represent solution properties at 50°C (122°F) and low pressure ($P \approx 1$ bar [14.5 psi]).

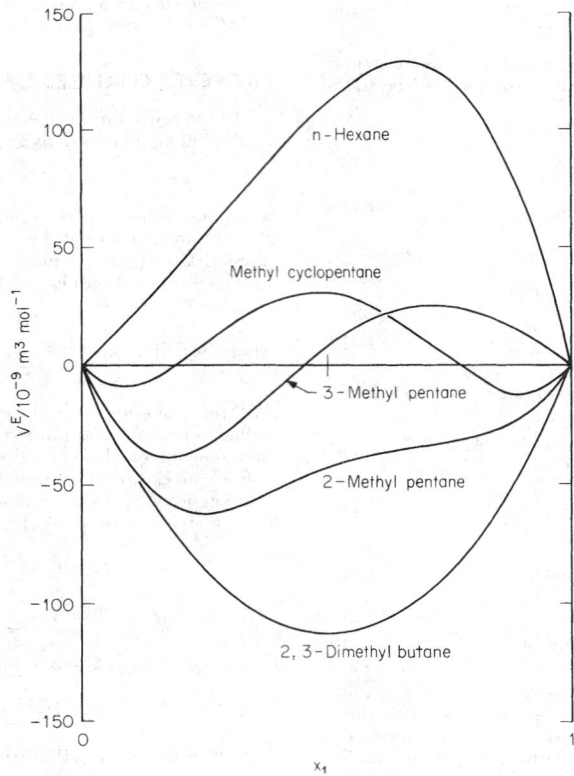

FIG. 4-1   Excess volumes at 25°C for liquid mixtures of cyclohexane(1) with some other C_6 hydrocarbons.

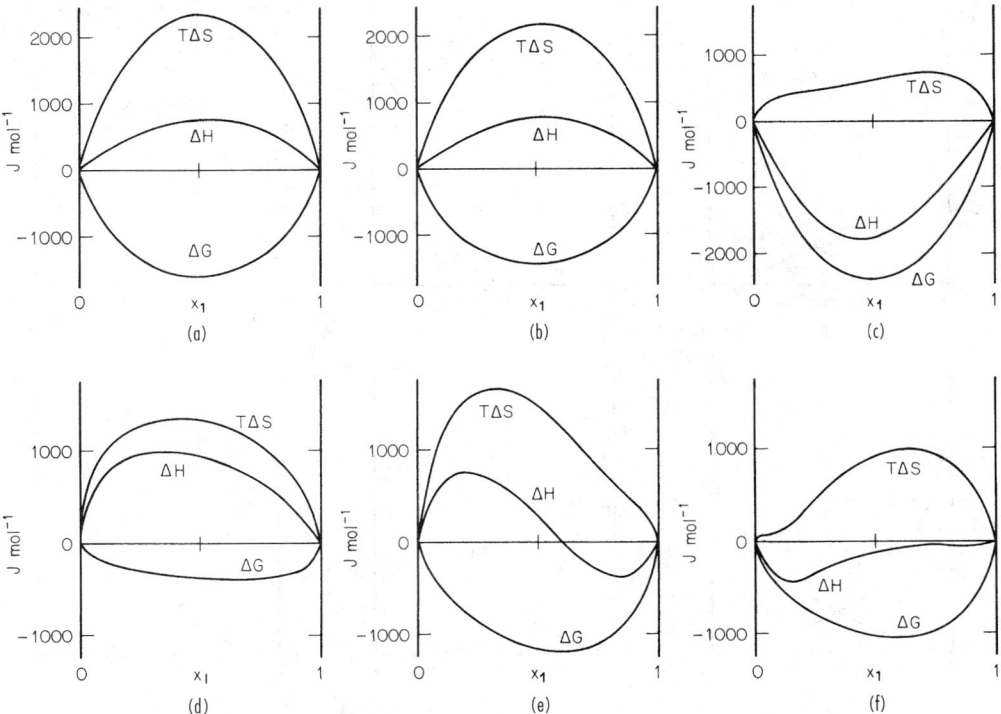

FIG. 4-2 Property changes of mixing at 50°C for 6 binary liquid systems: (*a*) chloroform(1)/*n*-heptane(2); (*b*) acetone(1)/methanol(2); (*c*) acetone(1)/chloroform(2); (*d*) ethanol(1)/*n*-heptane(2); (*e*) ethanol(1)/chloroform(2); (*f*) ethanol(1)/water(2).

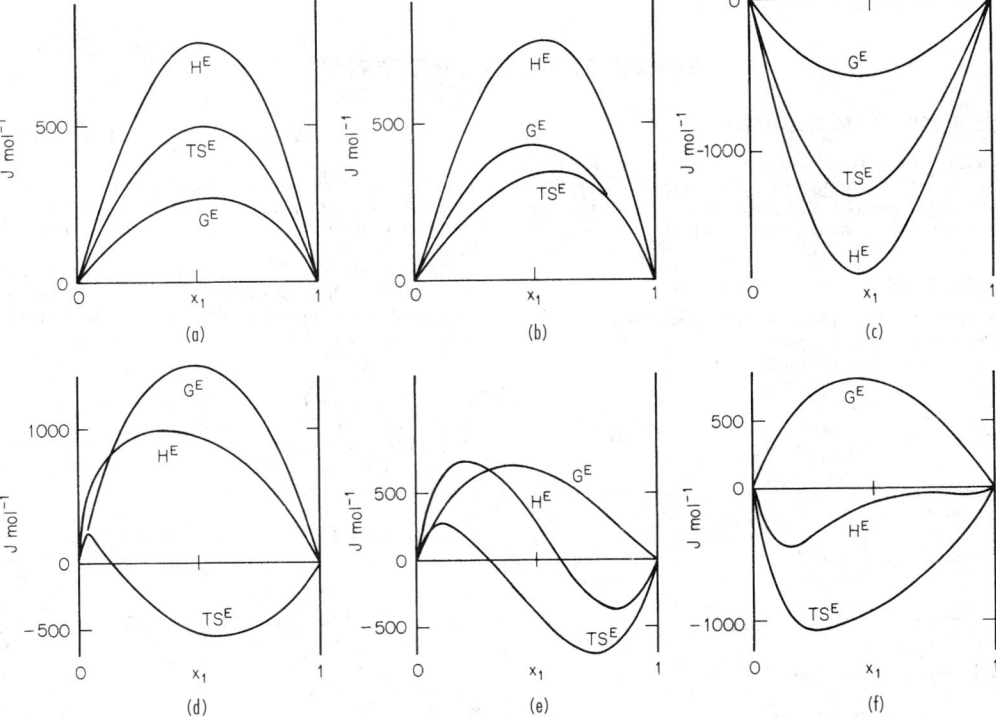

FIG. 4-3 Excess properties at 50°C for 6 binary liquid systems: (*a*) chloroform(1)/*n*-heptane(2); (*b*) acetone(1)/methanol(2); (*c*) acetone(1)/chloroform(2); (*d*) ethanol(1)/*n*-heptane(2); (*e*) ethanol(1)/chloroform(2); (*f*) ethanol(1)/water(2).

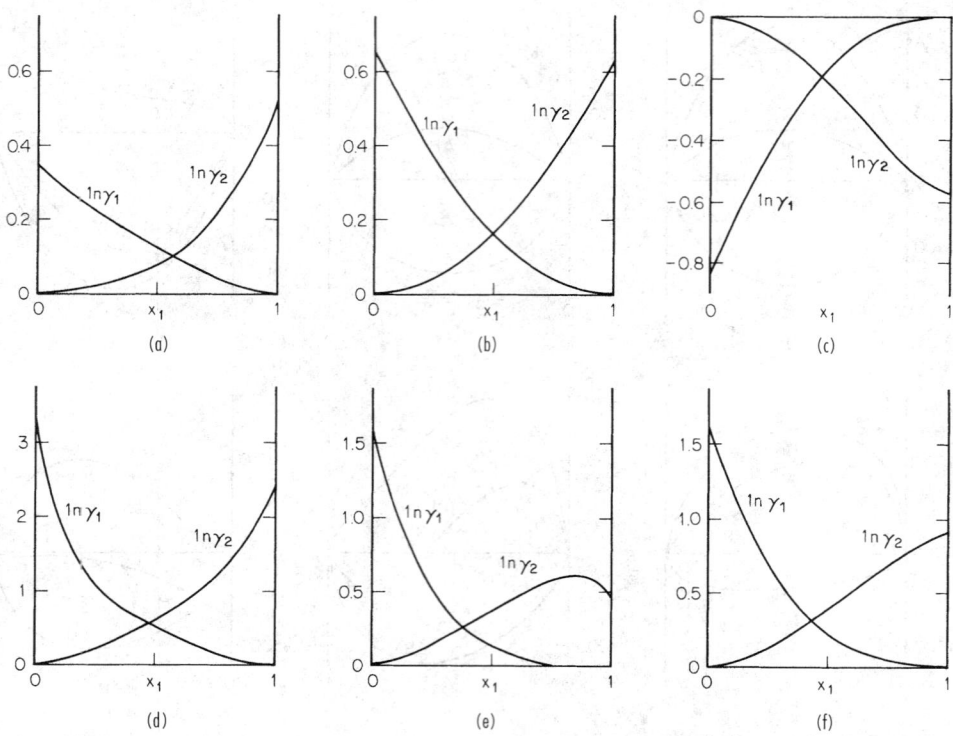

FIG. 4-4 Activity coefficients at 50°C for 6 binary liquid systems: (*a*) chloroform(1)/*n*-heptane(2); (*b*) acetone(1)/methanol(2); (*c*) acetone(1)/chloroform(2); (*d*) ethanol(1)/*n*-heptane(2); (*e*) ethanol(1)/chloroform(2); (*f*) ethanol(1)/water(2).

EVALUATION OF PROPERTIES

RESIDUAL-PROPERTY FORMULATIONS

The most satisfactory calculational procedure for thermodynamic properties of gases and vapors requires PVT data and ideal gas heat capacities. The primary equations are based on the concept of the ideal gas state and the definitions of residual enthalpy and residual entropy:

$$H = H^{ig} + H^R \qquad \text{and} \qquad S = S^{ig} + S^R$$

The enthalpy and entropy are simple sums of the ideal gas and residual properties, which are evaluated separately.

For the ideal gas state at constant composition,

$$dH^{ig} = C_P^{ig}\, dT$$

$$dS^{ig} = C_P^{ig}\,\frac{dT}{T} - R\frac{dP}{P}$$

Integration from an initial ideal gas *reference state* at conditions T_0 and P_0 to the ideal gas state at T and P gives:

$$H^{ig} = H_0^{ig} + \int_{T_0}^{T} C_P^{ig}\, dT$$

$$S^{ig} = S_0^{ig} + \int_{T_0}^{T} C_P^{ig}\,\frac{dT}{T} - R\ln\frac{P}{P_0}$$

Substitution into the equations for H and S yields

$$H = H_0^{ig} + \int_{T_0}^{T} C_P^{ig}\, dT + H^R \qquad (4\text{-}140)$$

$$S = S_0^{ig} + \int_{T_0}^{T} C_P^{ig}\,\frac{dT}{T} - R\ln\frac{P}{P_0} + S^R \qquad (4\text{-}141)$$

The reference state at T_0 and P_0 is arbitrarily selected, and the values assigned to H_0^{ig} and S_0^{ig} are also arbitrary. In practice, only *changes* in H and S are of interest, and the reference-state values ultimately cancel in their calculation.

The ideal-gas–state heat capacity C_P^{ig} is a function of T but not of P. For a mixture, the heat capacity is simply the molar average $\sum_i x_i C_P^{ig}$. Empirical equations giving the temperature dependence of C_P^{ig} are available for many pure gases, often taking the form

$$C_P^{ig} = A + BT + CT^2 + DT^{-2} \qquad (4\text{-}142)$$

where A, B, C, and D are constants characteristic of the particular gas, and either C or D is 0. Evaluation of the integrals $\int C_P^{ig}\, dT$ and $\int (C_P^{ig}/T)dT$ is accomplished by substitution for C_P^{ig}, followed by formal integration. For temperature limits of T_0 and T the results are conveniently expressed as follows:

$$\int_{T_0}^{T} C_P^{ig}\, dT = AT_0(\tau-1) + \frac{B}{2}\,T_0^2(\tau^2-1) + \frac{C}{3}\,T_0^3(\tau^3-1) + \frac{D}{T_0}\left(\frac{\tau-1}{\tau}\right) \qquad (4\text{-}143)$$

and

$$\int_{T_0}^{T}\frac{C_P^{ig}}{T}\, dT = A\ln\tau + \left[BT_0 + \left(CT_0^2 + \frac{D}{\tau^2 T_0^2}\right)\left(\frac{\tau+1}{2}\right)\right](\tau-1) \qquad (4\text{-}144)$$

where

$$\tau \equiv \frac{T}{T_0}$$

Equations (4-140) and (4-141) may sometimes be advantageously expressed in alternative form through use of mean heat capacities:

$$H = H_0^{ig} + \langle C_P^{ig} \rangle_H (T - T_0) + H^R \tag{4-145}$$

$$S = S_0^{ig} + \langle C_P^{ig} \rangle_S \ln \frac{T}{T_0} - R \ln \frac{P}{P_0} + S^R \tag{4-146}$$

where $\langle C_P^{ig} \rangle_H$ and $\langle C_P^{ig} \rangle_S$ are mean heat capacities specific respectively to enthalpy and entropy calculations. They are given by the following equations:

$$\langle C_P^{ig} \rangle_H = A + \frac{B}{2} T_0 (\tau + 1) + \frac{C}{3} T_0^2 (\tau^2 + \tau + 1) + \frac{D}{\tau T_0^2} \tag{4-147}$$

$$\langle C_P^{ig} \rangle_S = A + \left[BT_0 + \left(CT_0^2 + \frac{D}{\tau^2 T_0^2} \right) \left(\frac{\tau + 1}{2} \right) \right] \left(\frac{\tau - 1}{\ln \tau} \right) \tag{4-148}$$

LIQUID/VAPOR PHASE TRANSITION

When a differential amount of a pure liquid in equilibrium with its vapor in a piston-and-cylinder arrangement evaporates at constant temperature T and vapor pressure P_i^{sat}, Eq. (4-16) applied to the process reduces to $d(n_i G_i) = 0$, whence

$$n_i \, dG_i + G_i \, dn_i = 0$$

Since the system is closed, $dn_i = 0$ and, therefore, $dG_i = 0$; this requires the molar (or specific) Gibbs energy of the vapor to be identical with that of the liquid:

$$G_i^l = G_i^v \tag{4-149}$$

where G_i^l and G_i^v are the molar Gibbs energies of the individual phases.

If the temperature of a two-phase system is changed and if the two phases continue to coexist in equilibrium, then the vapor pressure must also change in accord with its temperature dependence. Since Eq. (4-149) holds throughout this change,

$$dG_i^l = dG_i^v$$

Substituting the expressions for dG_i^l and dG_i^v given by Eq. (4-16) yields

$$V_i^l \, dP_i^{sat} - S_i^l \, dT = V_i^v \, dP_i^{sat} - S_i^v \, dT$$

which upon rearrangement becomes

$$\frac{dP_i^{sat}}{dT} = \frac{S_i^v - S_i^l}{V_i^v - V_i^l} = \frac{\Delta S_i^{lv}}{\Delta V_i^{lv}}$$

The entropy change ΔS_i^{lv} and the volume change ΔV_i^{lv} are the changes which occur when a unit amount of a pure chemical species is transferred from phase l to phase v at constant temperature and pressure. Integration of Eq. (4-18) for this change yields the latent heat of phase transition:

$$\Delta H_i^{lv} = T \Delta S_i^{lv}$$

Thus, $\Delta S_i^{lv} = \Delta H_i^{lv}/T$, and substitution in the preceding equation gives

$$\frac{dP_i^{sat}}{dT} = \frac{\Delta H_i^{lv}}{T \Delta V_i^{lv}} \tag{4-150}$$

Known as the *Clapeyron equation*, this is an exact thermodynamic relation, providing a vital connection between the properties of the liquid and vapor phases. Its use presupposes knowledge of a suitable vapor pressure vs. temperature relation. Empirical in nature, such relations are approximated by the equation

$$\ln P^{sat} = A - \frac{B}{T} \tag{4-151}$$

where A and B are constants for a given species. This equation gives a rough approximation of the vapor-pressure relation for its entire temperature range. Moreover, it is an excellent basis for interpolation between values that are reasonably spaced.

The *Antoine equation*, which is more satisfactory for general use, has the form

$$\ln P^{sat} = A - \frac{B}{T + C} \tag{4-152}$$

A principal advantage of this equation is that values of the constants A, B, and C are readily available for a large number of species.

The accurate representation of vapor-pressure data over a wide temperature range requires an equation of greater complexity. The *Wagner equation*, one of the best, expresses the reduced vapor pressure as a function of reduced temperature:

$$\ln P_r^{sat} = \frac{A\tau + B\tau^{1.5} + C\tau^3 + D\tau^5}{1 - \tau} \tag{4-153}$$

where here

$$\tau \equiv 1 - T_r$$

and A, B, C, and D are constants. Values of the constants either for this equation or the Antoine equation are given for many species by Reid, Prausnitz, and Poling (*The Properties of Gases and Liquids*, 4th ed., App. A, McGraw-Hill, New York, 1987).

LIQUID-PHASE PROPERTIES

Given saturated-liquid enthalpies and entropies, the calculation of these properties for pure compressed liquids is accomplished by integration at constant temperature of Eqs. (4-34) and (4-35):

$$H_i = H_i^{sat} + \int_{P_i^{sat}}^{P} V_i (1 - \beta_i T) dP \tag{4-154}$$

$$S_i = S_i^{sat} - \int_{P_i^{sat}}^{P} \beta_i V_i \, dP \tag{4-155}$$

where the *volume expansivity* of species i at temperature T is

$$\beta_i \equiv \frac{1}{V_i} \left(\frac{\partial V_i}{\partial T} \right)_P \tag{4-156}$$

Since β_i and V_i are weak functions of pressure for liquids, they are usually assumed constant at the values for the saturated liquid at temperature T.

PROPERTIES FROM *PVT* CORRELATIONS

The empirical representation of the *PVT* surface for pure materials is treated later in this section. We first present general equations for evaluation of reduced properties from such representations.

Equation (4-83), applied to a pure material, may be written

$$d \left(\frac{G^R}{RT} \right) = \frac{V^R}{RT} dP \qquad \text{(constant } T\text{)}$$

Integration from zero pressure to arbitrary pressure P gives

$$\frac{G^R}{RT} = \int_0^P \frac{V^R}{RT} dP \qquad \text{(constant } T\text{)}$$

where at the lower limit G^R/RT is set equal to zero on the basis that the zero-pressure state is an ideal gas state. The residual volume is related directly to the compressibility factor:

$$V^R \equiv V - V^{ig} = \frac{ZRT}{P} - \frac{RT}{P} = (Z - 1) \frac{RT}{P}$$

whence

$$\frac{V^R}{RT} = \frac{Z - 1}{P} \tag{4-157}$$

Therefore

$$\frac{G^R}{RT} = \int_0^P (Z - 1) \frac{dP}{P} \qquad \text{(constant } T\text{)} \tag{4-158}$$

Differentiation of Eq. (4-158) with respect to temperature in accord with Eq. (4-84), gives

$$\frac{H^R}{RT} = -T \int_0^P \left(\frac{\partial Z}{\partial T} \right)_P \frac{dP}{P} \qquad \text{(constant } T\text{)} \tag{4-159}$$

Equation (4-13) written for residual properties becomes

$$\frac{S^R}{R} = \frac{H^R}{RT} - \frac{G^R}{RT} \tag{4-160}$$

In view of Eq. (4-75), Eqs. (4-158) and (4-160) may be expressed alternatively as

$$\ln \phi = \int_0^P (Z - 1) \frac{dP}{P} \qquad \text{(constant } T) \qquad (4\text{-}161)$$

and

$$\frac{S^R}{R} = \frac{H^R}{RT} - \ln \phi \qquad (4\text{-}162)$$

Values of Z and of $(\partial Z/\partial T)_P$ come from experimental PVT data, and the integrals in Eqs. (4-158), (4-159), and (4-161) may be evaluated by numerical or graphical methods. Alternatively, the integrals are expressed analytically when Z is given by an equation of state. Residual properties are therefore evaluated from PVT data or from an appropriate equation of state.

Pitzer's Corresponding-States Correlation A three-parameter corresponding-states correlation of the type developed by Pitzer, K.S. (*Thermodynamics*, 3d ed., App. 3, McGraw-Hill, New York, 1995) is described in Sec. 2. It has as its basis an equation for the compressibility factor:

$$Z = Z^0 + \omega Z^1 \qquad (4\text{-}163)$$

where Z^0 and Z^1 are each functions of reduced temperature T_r and reduced pressure P_r. The eccentric factor ω is defined by Eq. (2-23). The T_r and P_r dependencies of functions Z^0 and Z^1 are shown by Figs. 2-1 and 2-2. Generalized correlations are developed here for the residual enthalpy, residual entropy, and the fugacity coefficient.

Equations (4-161) and (4-159) are put into generalized form by substitution of the relationships

$$P = P_c P_r \qquad \qquad T = T_c T_r$$
$$dP = P_c \, dP_r \qquad dT = T_c \, dT_r$$

The resulting equations are:

$$\ln \phi = \int_0^{P_r} (Z - 1) \frac{dP_r}{P_r} \qquad (4\text{-}164)$$

and

$$\frac{H^R}{RT_c} = -T_r^2 \int_0^{P_r} \left(\frac{\partial Z}{\partial T_r} \right)_{P_r} \frac{dP_r}{P_r} \qquad (4\text{-}165)$$

The terms on the right-hand sides of these equations depend only on the upper limit P_r of the integrals and on the reduced temperature at which they are evaluated. Thus, values of $\ln \phi$ and H^R/RT_c may be determined once and for all at any reduced temperature and pressure from generalized compressibility factor data.

Substitution for Z in Eq. (4-164) by Eq. (4-163) yields

$$\ln \phi = \int_0^{P_r} (Z^0 - 1) \frac{dP_r}{P_r} + \omega \int_0^{P_r} Z^1 \frac{dP_r}{P_r}$$

This equation may be written in alternative form as

$$\ln \phi = \ln \phi^0 + \omega \ln \phi^1 \qquad (4\text{-}166)$$

where

$$\ln \phi^0 \equiv \int_0^{P_r} (Z^0 - 1) \frac{dP_r}{P_r}$$

$$\ln \phi^1 \equiv \int_0^{P_r} Z^1 \frac{dP_r}{P_r}$$

Since Eq. (4-166) may also be written

$$\phi = (\phi^0)(\phi^1)^\omega \qquad (4\text{-}167)$$

correlations may be presented for ϕ^0 and ϕ^1 as well as for their logarithms.

Differentiation of Eq. (4-163) yields

$$\left(\frac{\partial Z}{\partial T_r} \right)_{P_r} = \left(\frac{\partial Z^0}{\partial T_r} \right)_{P_r} + \omega \left(\frac{\partial Z^1}{\partial T_r} \right)_{P_r}$$

Substitution for $(\partial Z/\partial T_r)_{P_r}$ in Eq. (4-165) gives:

$$\frac{H^R}{RT_c} = -T_r^2 \int_0^{P_r} \left(\frac{\partial Z^0}{\partial T_r} \right)_{P_r} \frac{dP_r}{P_r} - \omega T_r^2 \int_0^{P_r} \left(\frac{\partial Z^1}{\partial T_r} \right)_{P_r} \frac{dP_r}{P_r}$$

Again, in alternative form,

$$\frac{H^R}{RT_c} = \frac{(H^R)^0}{RT_c} + \omega \frac{(H^R)^1}{RT_c} \qquad (4\text{-}168)$$

where

$$\frac{(H^R)^0}{RT_c} = -T_r^2 \int_0^{P_r} \left(\frac{\partial Z^0}{\partial T_r} \right)_{P_r} \frac{dP_r}{P_r}$$

$$\frac{(H^R)^1}{RT_c} = -T_r^2 \int_0^{P_r} \left(\frac{\partial Z^1}{\partial T_r} \right)_{P_r} \frac{dP_r}{P_r}$$

The residual entropy is given by Eq. (4-162), here written

$$\frac{S^R}{R} = \frac{1}{T_r} \left(\frac{H^R}{RT_c} \right) - \ln \phi \qquad (4\text{-}169)$$

Pitzer's original correlations for Z and the derived quantities were determined graphically and presented in tabular form. Since then, analytical refinements to the tables have been developed, with extended range and accuracy. The most popular Pitzer-type correlation is that of Lee and Kesler (*AIChE J.*, **21**, pp. 510–527 [1975]). These tables cover both the liquid and gas phases, and span the ranges $0.3 \leq T_r \leq 4.0$ and $0.01 \leq P_r \leq 10.0$. Shown by Figs. 4-5 and 4-6 are isobars of $-(H^R)^0/RT_c$ and $-(H^R)^1/RT_c$ with T_r as independent variable drawn from these tables. Figures 4-7 and 4-8 are the corresponding plots for $-\ln \phi^0$ and $-\ln \phi^1$. Figures 4-9 and 4-10 are isotherms of ϕ^0 and ϕ^1 with P_r as independent variable.

Although the Pitzer correlations are based on data for pure materials, they may also be used for the calculation of mixture properties. A set of recipes is required relating the parameters T_c, P_c, and ω for a mixture to the pure-species values and to composition. One such set is given by Eqs. (2-80) through (2-82) in Sec. 2, which define *pseudoparameters*, so called because the defined values of T_c, P_c, and ω have no physical significance for the mixture.

Alternative Property Formulations Direct application of Eqs. (4-159) and (4-161) can be made only to equations of state that are solvable for volume, that is, that are *volume explicit*. Most equations of state are in fact *pressure explicit*, and alternative equations are required.

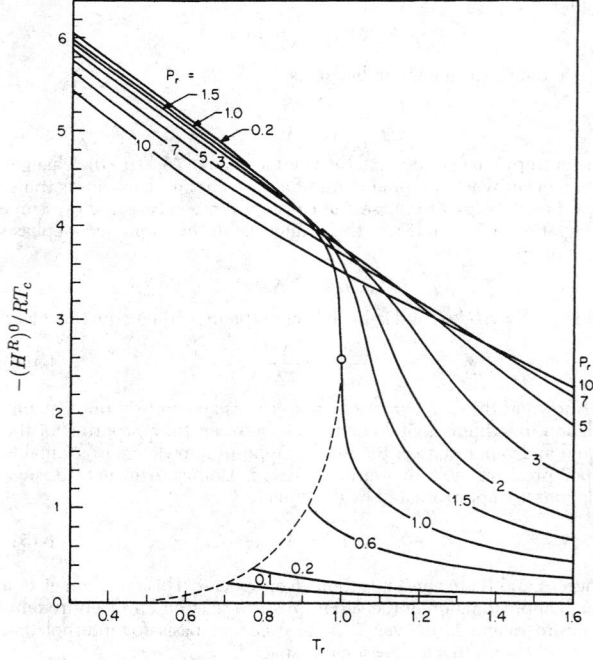

FIG. 4-5 Correlation of $-(H^R)^0/RT_c$, drawn from the tables of Lee and Kesler (*AIChE J.*, **21**, pp. 510–527 [1975]).

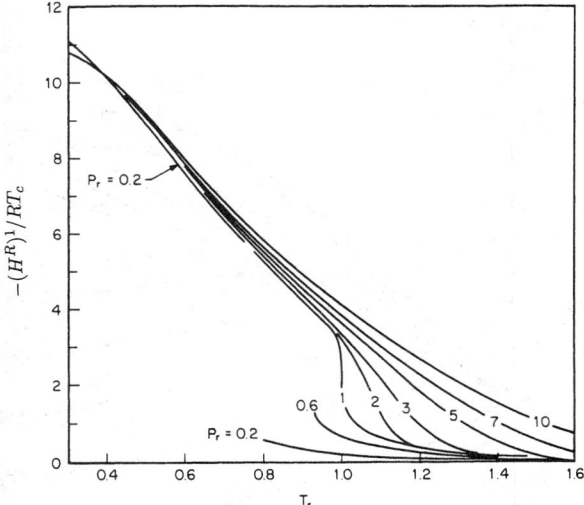

FIG. 4-6   Correlation of $-(H^R)^1/RT_c$, drawn from the tables of Lee and Kesler (*AIChE J.*, **21**, pp. 510–527 [1975]).

Equation (4-158) is converted through application of the general relation $PV = ZRT$. Differentiation at constant T gives

$$P\,dV + V\,dP = RT\,dZ \qquad \text{(constant } T)$$

which is readily transformed to

$$\frac{dP}{P} = \frac{dZ}{Z} - \frac{dV}{V} \qquad \text{(constant } T)$$

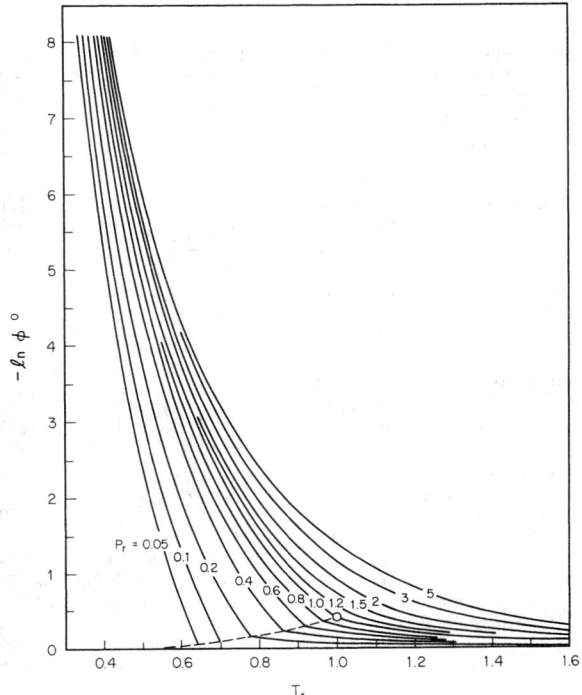

FIG. 4-7   Correlation of $[-\ln \phi^0]$ vs. T_r, drawn from the tables of Lee and Kesler (*AIChE J.*, **21**, pp. 510–527 [1975]).

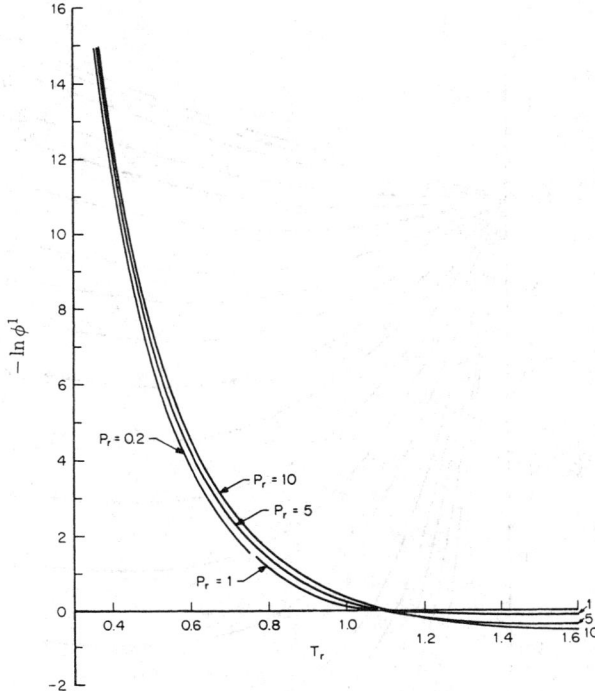

FIG. 4-8   Correlation of $[-\ln \phi^1]$ vs. T_r, drawn from the tables of Lee and Kesler (*AIChE J.*, **21**, pp. 510–527 [1975]).

Substitution into Eq. (4-158) leads to

$$\frac{G^R}{RT} = Z - 1 - \ln Z - \int_{\infty}^{V} (Z-1)\,\frac{dV}{V} \qquad (4\text{-}170)$$

The molar volume may be eliminated in favor of the molar density, $\rho = V^{-1}$, to give

$$\frac{G^R}{RT} = Z - 1 - \ln Z + \int_{0}^{\rho} (Z-1)\,\frac{d\rho}{\rho} \qquad (4\text{-}171)$$

For a pure material, Eq. (4-75) shows that $G^R/RT = \ln \phi$, in which case Eqs. (4-170) and (4-171) directly yield values of $\ln \phi$:

$$\ln \phi = Z - 1 - \ln Z - \int_{\infty}^{V} (Z-1)\,\frac{dV}{V} \qquad (4\text{-}172)$$

$$\ln \phi = Z - 1 - \ln Z + \int_{0}^{\rho} (Z-1)\,\frac{d\rho}{\rho} \qquad (4\text{-}173)$$

where subscript i is omitted for simplicity.

The corresponding equations for H^R are most readily found from Eq. (4-107) applied to a pure material. In view of Eqs. (4-75) and (4-157), this equation may be written

$$d \ln \phi = (Z-1)\,\frac{dP}{P} - \frac{H^R}{RT^2}\,dT$$

Division by dT and restriction to constant V gives, upon rearrangement,

$$\frac{H^R}{RT^2} = \frac{Z-1}{P}\left(\frac{\partial P}{\partial T}\right)_V - \left(\frac{\partial \ln \phi}{\partial T}\right)_V$$

Differentiation of $P = ZRT/V$ provides the first derivative on the right and differentiation of Eq. (4-172) provides the second. Substitution then leads to

$$\frac{H^R}{RT} = Z - 1 + T \int_{\infty}^{V} \left(\frac{\partial Z}{\partial T}\right)_V \frac{dV}{V} \qquad (4\text{-}174)$$

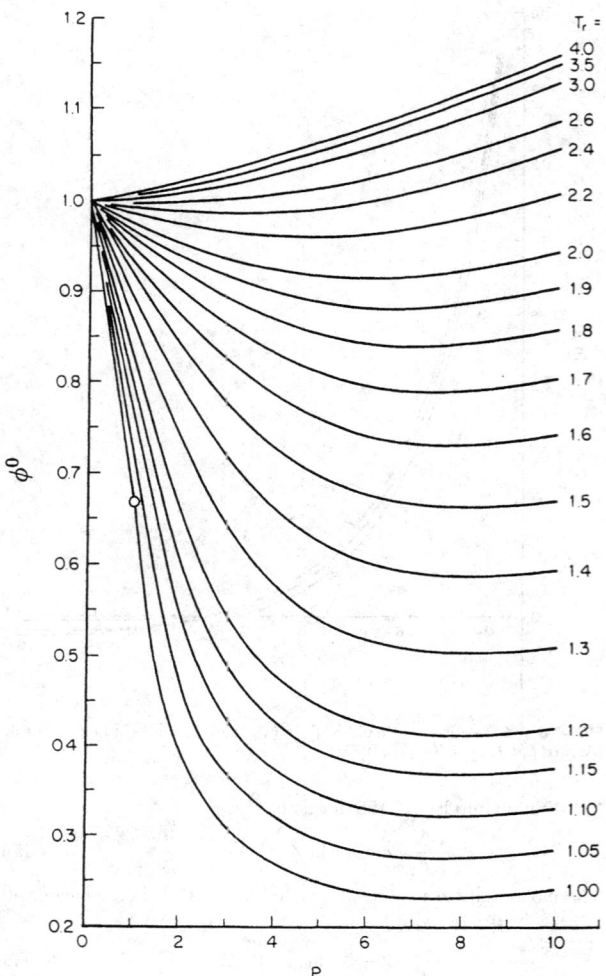

FIG. 4-9 Correlation of ϕ^0 vs. P_r, drawn from the tables of Lee and Kesler (*AIChE J.*, **21**, pp. 510–527 [1975]).

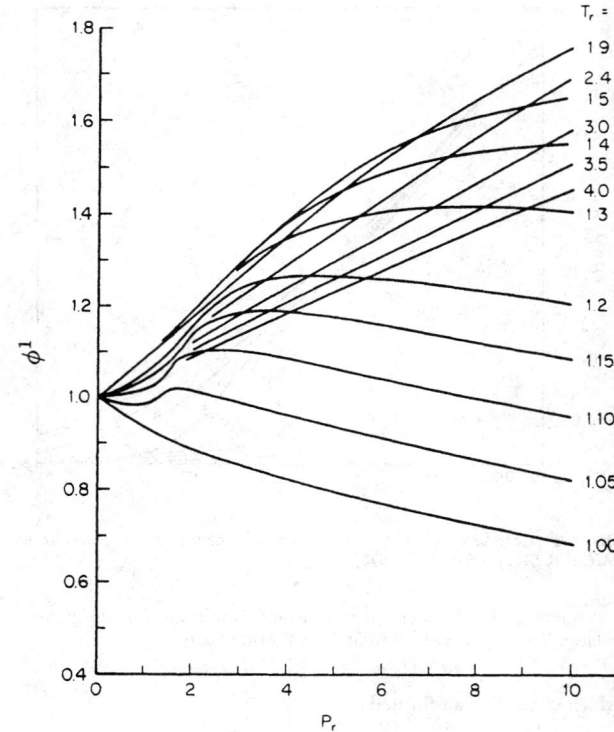

FIG. 4-10 Correlation of ϕ^1 vs. P_r, drawn from the tables of Lee and Kesler (*AIChE J.*, **21**, pp. 510–527 [1975]).

Alternatively,

$$\frac{H^R}{RT} = Z - 1 - T \int_0^\rho \left(\frac{\partial Z}{\partial T}\right)_\rho \frac{d\rho}{\rho} \tag{4-175}$$

As before, the residual entropy is found by Eq. (4-162).

In applications to equilibrium calculations, the fugacity coefficients of species in a mixture $\hat{\phi}_i$ are required. Given an expression for G^R/RT as determined from Eq. (4-158) for a constant-composition mixture, the corresponding recipe for $\ln \hat{\phi}_i$ is found through the partial-property relation

$$\ln \hat{\phi}_i = \left[\frac{\partial(nG^R/RT)}{\partial n_i}\right]_{T,P,n_j} \tag{4-85}$$

There are two ways to proceed: operate on the *result* of the integration of Eq. (4-158) in accord with Eq. (4-85) or apply Eq. (4-85) *directly* to Eq. (4-158), obtaining

$$\ln \hat{\phi}_i = \int_0^P (\bar{Z}_i - 1) \frac{dP}{P} \tag{4-176}$$

where $\bar{Z}_i$ is the partial compressibility factor, defined as

$$\bar{Z}_i \equiv \left[\frac{\partial(nZ)}{\partial n_i}\right]_{T,P,n_j} \tag{4-177}$$

Direct application of these results is possible only to equations of state explicit in volume. For pressure-explicit equations of state, alternative recipes are required. The basis is Eq. (4-82), which in view of Eq. (4-157) may be written

$$d\left(\frac{nG^R}{RT}\right) = \frac{n(Z-1)}{P} dP - \frac{nH^R}{RT^2} dT + \sum_i \ln \hat{\phi}_i \, dn_i$$

Division by dn_i and restriction to constant T, nV, and n_j $(j \neq i)$ leads to

$$\ln \hat{\phi}_i = \left[\frac{\partial(nG^R/RT)}{\partial n_i}\right]_{T,nV,n_j} - \frac{n(Z-1)}{P} \left(\frac{\partial P}{\partial n_i}\right)_{T,nV,n_j}$$

But $P = (nZ)RT/nV$, and therefore

$$\left(\frac{\partial P}{\partial n_i}\right)_{T,nV,n_j} = \frac{P}{nZ} \left[\frac{\partial(nZ)}{\partial n_i}\right]_{T,nV,n_j}$$

Combination of the last two equations gives

$$\ln \hat{\phi}_i = \left[\frac{\partial(nG^R/RT)}{\partial n_i}\right]_{T,nV,n_j} - \left(\frac{Z-1}{Z}\right)\left[\frac{\partial(nZ)}{\partial n_i}\right]_{T,nV,n_j} \tag{4-178}$$

Alternatively,

$$\ln \hat{\phi}_i = \left[\frac{\partial(nG^R/RT)}{\partial n_i}\right]_{T,\rho/n,n_j} - \left(\frac{Z-1}{Z}\right)\left[\frac{\partial(nZ)}{\partial n_i}\right]_{T,\rho/n,n_j} \tag{4-179}$$

These equations may either be applied to the results of integrations of Eqs. (4-170) and (4-171) or *directly* to Eqs. (4-170) and (4-171) as written for a mixture. In the latter case the following analogs of Eq. (4-176) are obtained:

$$\ln \hat{\phi}_i = -\int_\infty^V \left\{\left[\frac{\partial(nZ)}{\partial n_i}\right]_{T,nV,n_j} - 1\right\} \frac{dV}{V} - \ln Z \tag{4-180}$$

$$\ln \hat{\phi}_i = -\int_0^\rho \left\{\left[\frac{\partial(nZ)}{\partial n_i}\right]_{T,\rho/n,n_j} - 1\right\} \frac{d\rho}{\rho} - \ln Z \tag{4-181}$$

Virial Equations of State   The virial equation in *density* is an infinite-series representation of the compressibility factor Z in powers of molar density ρ (or reciprocal molar volume V^{-1}) about the real-gas state at zero density (zero pressure):

$$Z = 1 + B\rho + C\rho^2 + D\rho^3 + \cdots \qquad (4\text{-}182)$$

The density-series virial coefficients $B, C, D, \ldots$, depend on temperature and composition only. The composition dependencies are given by the exact recipes

$$B = \sum_i \sum_j y_i y_j B_{ij} \qquad (4\text{-}183)$$

$$C = \sum_i \sum_j \sum_k y_i y_j y_k C_{ijk} \qquad (4\text{-}184)$$

$$\text{and so on}$$

where y_i, y_j, and y_k are mole fractions for a gas mixture, with indices i, j, and k identifying species.

The coefficient B_{ij} characterizes a bimolecular interaction between molecules i and j, and therefore $B_{ij} = B_{ji}$. Two kinds of second virial coefficient arise: B_{ii} and B_{jj}, wherein the subscripts are the same $(i = j)$; and B_{ij}, wherein they are different $(i \neq j)$. The first is a virial coefficient for a pure species; the second is a mixture property, called a *cross coefficient*. Similarly for the third virial coefficients: C_{iii}, C_{jjj}, and C_{kkk} are for the pure species; and $C_{iij} = C_{iji} = C_{jii}$, and so on, are cross coefficients.

Although the virial equation itself is easily rationalized on empirical grounds, the "mixing rules" of Eqs. (4-183) and (4-184) follow rigorously from the methods of statistical mechanics. The temperature derivatives of B and C are given exactly by

$$\frac{dB}{dT} = \sum_i \sum_j y_i y_j \frac{dB_{ij}}{dT} \qquad (4\text{-}185)$$

$$\frac{dC}{dT} = \sum_i \sum_j \sum_k y_i y_j y_k \frac{dC_{ijk}}{dT} \qquad (4\text{-}186)$$

An alternative form of the virial equation expresses Z as an expansion in powers of pressure about the real-gas state at zero pressure (zero density):

$$Z = 1 + B'P + C'P^2 + D'P^3 + \cdots \qquad (4\text{-}187)$$

Equation (4-187) is the virial equation in *pressure*, and $B', C', D', \ldots$, are the pressure-series virial coefficients. Like the density-series coefficients, they depend on temperature and composition only. Moreover, the two sets of coefficients are related:

$$B' = \frac{B}{RT} \qquad (4\text{-}188)$$

$$C' = \frac{C - B^2}{(RT)^2} \qquad (4\text{-}189)$$

$$\text{and so on}$$

Application of an *infinite* series to practical calculations is, of course, impossible, and *truncations* of the virial equations are in fact employed. The degree of truncation is conditioned not only by the temperature and pressure but also by the availability of correlations or data for the virial coefficients. Values can usually be found for B (see Sec. 2), and often for C (see, e.g., De Santis and Grande, *AIChE J.*, **25**, pp. 931–938 [1979]), but rarely for higher-order coefficients. Application of the virial equations is therefore usually restricted to two- or three-term truncations. For pressures up to several bars, the two-term expansion in pressure, with B' given by Eq. (4-188), is usually preferred:

$$Z = 1 + \frac{BP}{RT} \qquad (4\text{-}190)$$

For supercritical temperatures, it is satisfactory to ever-higher pressures as the temperature increases. For pressures above the range where Eq. (4-190) is useful, but below the critical pressure, the virial expansion in density truncated to three terms is usually suitable:

$$Z = 1 + B\rho + C\rho^2 \qquad (4\text{-}191)$$

Equations for derived properties may be developed from each of these expressions. Consider first Eq. (4-190), which is explicit in volume. Equations (4-159), (4-161), and (4-176) are therefore applicable. Direct substitution for Z in Eq. (4-161) gives

$$\ln \phi = \frac{BP}{RT} \qquad (4\text{-}192)$$

Differentiation of Eq. (4-190) yields

$$\left(\frac{\partial Z}{\partial T}\right)_P = \left(\frac{dB}{dT} - \frac{B}{T}\right)\frac{P}{RT}$$

Whence, by Eq. (4-159)

$$\frac{H^R}{RT} = \frac{P}{R}\left(\frac{B}{T} - \frac{dB}{dT}\right) \qquad (4\text{-}193)$$

and by Eq. (4-162),

$$\frac{S^R}{R} = -\frac{P}{R}\frac{dB}{dT} \qquad (4\text{-}194)$$

Multiplication of Eq. (4-190) by n gives

$$nZ = n + (nB)\frac{P}{RT}$$

Differentiation in accord with Eq. (4-177) yields

$$\overline{Z}_i = 1 + \left[\frac{\partial(nB)}{\partial n_i}\right]_{T,n_j}\frac{P}{RT}$$

Whence, by Eq. (4-176),

$$\ln \hat{\phi}_i = \left[\frac{\partial(nB)}{\partial n_i}\right]_{T,n_j}\frac{P}{RT}$$

Equation (4-183) can be written

$$nB = \frac{1}{n}\sum_k \sum_l n_k n_l B_{kl}$$

from which, by differentiation,

$$\left[\frac{\partial(nB)}{\partial n_i}\right]_{T,n_j} = 2\sum_k y_k B_{ki} - B \qquad (4\text{-}195)$$

Whence

$$\ln \hat{\phi}_i = \left(2\sum_k y_k B_{ki} - B\right)\frac{P}{RT} \qquad (4\text{-}196)$$

Equation (4-191) is explicit in pressure, and Eqs. (4-173), (4-175), and (4-181) are therefore applicable. Direct substitution of Eq. (4-191) into Eq. (4-173) yields

$$\ln \phi = 2B\rho + \frac{3}{2}C\rho^2 - \ln Z \qquad (4\text{-}197)$$

Moreover,

$$\left(\frac{\partial Z}{\partial T}\right)_\rho = \frac{dB}{dT}\rho + \frac{dC}{dT}\rho^2$$

whence

$$\frac{H^R}{RT} = \left(B - T\frac{dB}{dT}\right)\rho + \left(C - \frac{T}{2}\frac{dC}{dT}\right)\rho^2 \qquad (4\text{-}198)$$

The residual entropy is given by Eq. (4-162).

Application of Eq. (4-181) provides an expression for $\ln \hat{\phi}_i$. First, from Eq. (4-191),

$$\left[\frac{\partial(nZ)}{\partial n_i}\right]_{T,\rho/n,n_j} = 1 + \left\{B + \left[\frac{\partial(nB)}{\partial n_i}\right]_{T,n_j}\right\}\rho + \left\{2C + \left[\frac{\partial(nC)}{\partial n_i}\right]_{T,n_j}\right\}\rho^2$$

Substitution into Eq. (4-181) gives, on integration,

$$\ln \hat{\phi}_i = \left\{B + \left[\frac{\partial(nB)}{\partial n_i}\right]_{T,n_j}\right\}\rho + \frac{1}{2}\left\{2C + \left[\frac{\partial(nC)}{\partial n_i}\right]_{T,n_j}\right\}\rho^2 - \ln Z$$

The mole-number derivative of nB is given by Eq. (4-195); the corresponding derivative of nC, similarly found from Eq. (4-184), is

$$\left[\frac{\partial(nC)}{\partial n_i}\right]_{T,n_j} = 3\sum_k \sum_l y_k y_l C_{kli} - 2C \qquad (4\text{-}199)$$

Finally,

$$\ln \hat{\phi}_i = 2\rho \sum_k y_k B_{ki} + \frac{3}{2} \rho^2 \sum_k \sum_l y_k y_l C_{kli} - \ln Z \qquad (4\text{-}200)$$

In a process calculation, T and P, rather than T and ρ (or T and V), are usually the favored independent variables. Application of Eqs. (4-197), (4-198), and (4-200) therefore requires prior solution of Eq. (4-191) for Z or ρ. Since $Z = P/\rho RT$, Eq. (4-191) may be written in two equivalent forms:

$$Z^3 - Z^2 - \left(\frac{BP}{RT}\right) Z - \frac{CP^2}{(RT)^2} = 0 \qquad (4\text{-}201)$$

or

$$\rho^3 + \left(\frac{B}{C}\right) \rho^2 + \left(\frac{1}{C}\right) \rho - \frac{P}{CRT} = 0 \qquad (4\text{-}202)$$

In the event that three real roots obtain for these equations, only the largest Z (smallest ρ) appropriate for the vapor phase has physical significance, because the virial equations are suitable only for vapors and gases.

Generalized Correlation for the Second Virial Coefficient Perhaps the most useful of all Pitzer-type correlations is the one for the second virial coefficient. The basic equation (see Eq. [2-68]) is

$$\frac{BP_c}{RT_c} = B^0 + \omega B^1 \qquad (4\text{-}203)$$

where for a pure material B^0 and B^1 are functions of reduced temperature only. Substitution for B by this expression in Eq. (4-190) yields

$$Z = 1 + (B^0 + \omega B^1) \frac{P_r}{T_r} \qquad (4\text{-}204)$$

By differentiation,

$$\left(\frac{\partial Z}{\partial T_r}\right)_{P_r} = P_r \left(\frac{dB^0/dT_r}{T_r} - \frac{B^0}{T_r^2}\right) + \omega P_r \left(\frac{dB^1/dT_r}{T_r} - \frac{B^1}{T_r^2}\right)$$

Substitution of these equations into Eqs. (4-164) and (4-165) and integration gives

$$\ln \phi = (B^0 + \omega B^1) \frac{P_r}{T_r} \qquad (4\text{-}205)$$

and

$$\frac{H^R}{RT_c} = P_r \left[B^0 - T_r \frac{dB^0}{dT_r} + \omega \left(B^1 - T_r \frac{dB^1}{dT_r} \right) \right] \qquad (4\text{-}206)$$

The residual entropy follows from Eq. (4-162):

$$\frac{S^R}{R} = -P_r \left(\frac{dB^0}{dT_r} + \omega \frac{dB^1}{dT_r}\right) \qquad (4\text{-}207)$$

In these equations, B^0 and B^1 and their derivatives are well represented by

$$B^0 = 0.083 - \frac{0.422}{T_r^{1.6}} \qquad (4\text{-}208)$$

$$B^1 = 0.139 - \frac{0.172}{T_r^{4.2}} \qquad (4\text{-}209)$$

$$\frac{dB^0}{dT_r} = \frac{0.675}{T_r^{2.6}} \qquad (4\text{-}210)$$

$$\frac{dB^1}{dT_r} = \frac{0.722}{T_r^{5.2}} \qquad (4\text{-}211)$$

Though limited to pressures where the two-term virial equation in pressure has approximate validity, this correlation is applicable to most chemical-processing conditions. As with all generalized correlations, it is least accurate for polar and associating molecules.

Although developed for pure materials, this correlation can be extended to gas or vapor mixtures. Basic to this extension is the mixing rule for second virial coefficients and its temperature derivative:

$$B = \sum_i \sum_j y_i y_j B_{ij} \qquad (4\text{-}183)$$

$$\frac{dB}{dT} = \sum_i \sum_j y_i y_j \frac{dB_{ij}}{dT} \qquad (4\text{-}185)$$

Values for the cross coefficients and their derivatives in these equations are provided by writing Eq. (4-203) in extended form:

$$B_{ij} = \frac{RT_{cij}}{P_{cij}} (B^0 + \omega_{ij} B^1) \qquad (4\text{-}212)$$

where B^0, B^1, dB^0/dT_r, and dB^1/dT_r are the same functions of T_r as given by Eqs. (4-208) through (4-211). Differentiation produces

$$\frac{dB_{ij}}{dT} = \frac{RT_{cij}}{P_{cij}} \left(\frac{dB^0}{dT} + \omega_{ij} \frac{dB^1}{dT}\right)$$

or

$$\frac{dB_{ij}}{dT} = \frac{R}{P_{cij}} \left(\frac{dB^0}{dT_{rij}} + \omega_{ij} \frac{dB^1}{dT_{rij}}\right) \qquad (4\text{-}213)$$

where $T_{rij} = T/T_{cij}$. The following are combining rules for calculation of ω_{ij}, T_{cij}, and P_{cij} as given by Prausnitz, Lichtenthaler, and de Azevedo (*Molecular Thermodynamics of Fluid-Phase Equilibria*, 2d ed., pp. 132 and 162, Prentice-Hall, Englewood Cliffs, N.J., 1986):

$$\omega_{ij} = \frac{\omega_i + \omega_j}{2} \qquad (4\text{-}214)$$

$$T_{cij} = (T_{ci} T_{cj})^{1/2} (1 - k_{ij}) \qquad (4\text{-}215)$$

$$P_{cij} = \frac{Z_{cij} RT_{cij}}{V_{cij}} \qquad (4\text{-}216)$$

with

$$Z_{cij} = \frac{Z_{ci} + Z_{cj}}{2} \qquad (4\text{-}217)$$

$$V_{cij} = \left(\frac{V_{ci}^{1/3} + V_{cj}^{1/3}}{2}\right)^3 \qquad (4\text{-}218)$$

In Eq. (4-215), k_{ij} is an empirical interaction parameter specific to an i-j molecular pair. When $i = j$ and for chemically similar species, $k_{ij} = 0$. Otherwise, it is a small (usually) positive number evaluated from minimal PVT data or in the absence of data set equal to zero.

When $i = j$, all equations reduce to the appropriate values for a pure species. When $i \neq j$, these equations define a set of interaction parameters having no physical significance. For a mixture, values of B_{ij} and dB_{ij}/dT from Eqs. (4-212) and (4-213) are substituted into Eqs. (4-183) and (4-185) to provide values of the mixture second virial coefficient B and its temperature derivative. Values of H^R and S^R for the mixture are then given by Eqs. (4-193) and (4-194), and values of $\ln \hat{\phi}_i$ for the component fugacity coefficients are given by Eq. (4-196).

Cubic Equations of State The simplest expressions that can (in principle) represent both the vapor- and liquid-phase volumetric behavior of pure fluids are equations cubic in molar volume. All such expressions are encompassed by the generic equation

$$P = \frac{RT}{V - b} - \frac{a(V - \eta)}{(V - b)(V^2 + \delta V + \varepsilon)} \qquad (4\text{-}219)$$

where parameters b, θ, δ, ε, and η can each depend on temperature and composition. Special cases are obtained by specification of values or expressions for the various parameters.

The modern development of cubic equations of state started in 1949 with publication of the Redlich/Kwong equation (Redlich and Kwong, *Chem. Rev.*, **44**, pp. 233–244 [1949]):

$$P = \frac{RT}{V - b} - \frac{a(T)}{V(V + b)} \qquad (4\text{-}220)$$

where

$$a(T) = \frac{a}{T^{1/2}}$$

and a and b are functions of composition only. This equation, like other cubic equations of state, has three volume roots, of which two may be complex. Physically meaningful values of V are always real, positive, and greater than the constant b. When $T > T_c$, solution for V at any positive value of P yields only one real positive root. When $T = T_c$, this is also true, except at the critical pressure, where there are three roots, all equal to V_c. For $T < T_c$, only one real positive root exists at high pressures, but for a range of lower pressures there are three real positive roots. Here, the middle root is of no significance; the smallest root is a liquid or liquidlike volume, and the largest root is a vapor or vaporlike volume. The volumes of saturated liquid and saturated vapor are given by the smallest and largest roots when P is the saturation vapor pressure P^{sat}.

The application of cubic equations of state to mixtures requires expression of the equation-of-state parameters as functions of composition. No exact theory like that for the virial coefficients prescribes this composition dependence, and empirical *mixing rules* provide approximate relationships. The mixing rules that have found general favor for the Redlich/Kwong equation are:

$$a = \sum_i \sum_j y_i y_j a_{ij} \qquad (4\text{-}221)$$

with $a_{ij} = a_{ji}$, and

$$b = \sum_i y_i b_i \qquad (4\text{-}222)$$

The a_{ij} are of two types: *pure-species parameters* (like subscripts) and *interaction parameters* (unlike subscripts). The b_i are parameters for the pure species.

Parameter evaluation may be accomplished with the equations

$$a_{ij} = \frac{0.42748 R^2 T_{cij}^{2.5}}{P_{cij}} \qquad (4\text{-}223)$$

$$b_i = \frac{0.08664 R T_{ci}}{P_{ci}} \qquad (4\text{-}224)$$

where Eqs. (4-215) through (4-218) provide for the calculation of the T_{cij} and P_{cij}.

Multiplication of the Redlich/Kwong equation (Eq. [4-220]) by V/RT leads to its expression in alternative form:

$$Z = \frac{1}{1-h} - \frac{a}{bRT^{1.5}} \left(\frac{h}{1+h} \right) \qquad (4\text{-}225)$$

Whence

$$Z - 1 = \frac{h}{1-h} - \frac{a}{bRT^{1.5}} \left(\frac{h}{1+h} \right) \qquad (4\text{-}226)$$

where

$$h = \frac{bP}{ZRT} \qquad (4\text{-}227)$$

Equations (4-170) and (4-174) in combination with Eq. (4-226) give

$$\frac{G^R}{RT} = Z - 1 - \ln(1-h)Z - \left(\frac{a}{bRT^{1.5}} \right) \ln(1+h) \qquad (4\text{-}228)$$

and

$$\frac{H^R}{RT} = Z - 1 - \left(\frac{3a}{2bRT^{1.5}} \right) \ln(1+h) \qquad (4\text{-}229)$$

Once a and b are determined by Eqs. (4-221) through (4-224), then for given T and P values of Z, G^R/RT, and H^R/RT are found by Eqs. (4-225), (4-228), and (4-229) and S^R/R by Eq. (4-160). The procedure requires initial solution of Eqs. (4-225) and (4-227) for Z and h.

The original Redlich/Kwong equation is rarely satisfactory for vapor/liquid equilibrium calculations, and equations have been developed specific to this purpose. The two most popular are the Soave/Redlich/Kwong (SRK) equation, a modification of the Redlich/Kwong equation (Soave, *Chem. Eng. Sci.*, **27**, pp. 1197–1203 [1972]), and the Peng/Robinson (PR) equation (Peng and Robinson, *Ind. Eng. Chem. Fundam.*, **15**, pp. 59–64 [1976]). Both equations are designed specifically to yield reasonable vapor pressures for pure fluids. Thus, there is no assurance that molar volumes calculated by these equations are more accurate than values given by the original Redlich/Kwong equation. Written for pure species i the SRK and PR equations are special cases of the following:

$$P = \frac{RT}{V_i - b_i} - \frac{a_i(T)}{(V_i + \varepsilon b_i)(V_i + \sigma b_i)} \qquad (4\text{-}230)$$

where

$$a_i(T) = \frac{\Omega_a \alpha(T_{ri}; \omega_i) R^2 T_{ci}^2}{P_{ci}}$$

$$b_i = \frac{\Omega_b R T_{ci}}{P_{ci}}$$

and ε σ, Ω_a, and Ω_b are equation-specific constants. For the Soave/Redlich/Kwong equation:

$$\alpha(T_{ri}; \omega_i) = [1 + (0.480 + 1.574 \omega_i - 0.176 \omega_i^2)(1 - T_{ri}^{1/2})]^2$$

For the Peng/Robinson equation:

$$\alpha(T_{ri}; \omega_i) = [1 + (0.37464 + 1.54226 \omega_i - 0.26992 \omega_i^2)(1 - T_{ri}^{1/2})]^2$$

Written for a mixture, Eq. (4-230) becomes

$$P = \frac{RT}{V - b} - \frac{a(T)}{(V + \varepsilon b)(V + \sigma b)} \qquad (4\text{-}231)$$

where a and b are mixture values, related to the a_i and b_i by *mixing rules*. Equation (4-170) applied to Eq. (4-231) leads to

$$\ln \hat{\phi}_i = \frac{\overline{b}_i}{b}(Z-1) - \ln \frac{(V-b)Z}{V} + \frac{a/bRT}{\varepsilon - \sigma} \left(1 + \frac{\overline{a}_i}{a} - \frac{\overline{b}_i}{b} \right) \ln \frac{V + \sigma b}{V + \varepsilon b} \qquad (4\text{-}232)$$

where a_i and b_i are *partial parameters* for species i, defined by

$$a_i = \left[\frac{\partial(na)}{\partial n_i} \right]_{T,n_j} \qquad (4\text{-}233)$$

and

$$b_i = \left[\frac{\partial(nb)}{\partial n_i} \right]_{T,n_j} \qquad (4\text{-}234)$$

These are general equations that do not depend on the particular *mixing rules* adopted for the composition dependence of a and b. The mixing rules given by Eqs. (4-221) and (4-222) can certainly be employed with these equations. However, for purposes of vapor/liquid equilibrium calculations, a special pair of mixing rules is far more appropriate, and will be introduced when these calculations are treated. Solution of Eq. (4-232) for fugacity coefficient $\hat{\phi}_i$ at given T and P requires prior solution of Eq. (4-231) for V, from which is found $Z = PV/RT$.

Benedict/Webb/Rubin Equation of State The BWR equation of state with Z as the dependent variable is written

$$Z = 1 + \left(B_0 - \frac{A_0}{RT} - \frac{C_0}{RT^3} \right) \rho + \left(b - \frac{a}{RT} \right) \rho^2$$
$$+ \frac{a\alpha}{RT} \rho^5 + \frac{c}{RT^3} \rho^2 (1 + \gamma \rho^2) \exp(-\gamma \rho^2) \qquad (4\text{-}235)$$

All eight parameters depend on composition; moreover, parameters C_0, b, and γ are for some applications treated as functions of T. By Eq. (4-171), the residual Gibbs energy is

$$\frac{G^R}{RT} = 2 \left(B_0 - \frac{A_0}{RT} - \frac{C_0}{RT^3} \right) \rho + \frac{3}{2} \left(b - \frac{a}{RT} \right) \rho^2 + \frac{6a\alpha}{5RT} \rho^5$$
$$+ \frac{c}{2\gamma RT^3} \left[(2\gamma^2 \rho^4 + \gamma \rho^2 - 2) \exp(-\gamma \rho^2) + 2 \right] - \ln Z \qquad (4\text{-}236)$$

With allowance for T dependence of C_0, b, and γ, Eq. (4-175) yields

$$\frac{H^R}{RT} = \left(B_0 - \frac{2A_0}{RT} - \frac{4C_0}{RT^3} + \frac{1}{RT^2} \frac{dC_0}{dT} \right) \rho$$
$$- \frac{1}{2} \left(T \frac{db}{dT} - 2b + \frac{3a}{RT} \right) \rho^2 + \frac{6a\alpha}{5RT} \rho^5$$
$$+ \frac{c}{2\gamma RT^3} \left[(2\gamma^2 \rho^4 - \gamma \rho^2 - 6) \exp(-\gamma \rho^2) + 6 \right]$$
$$- \frac{c}{2\gamma^2 RT^2} \frac{d\gamma}{dT} \left[(\gamma^2 \rho^4 + 2\gamma \rho^2 + 2) \exp(-\gamma \rho^2) - 2 \right] \qquad (4\text{-}237)$$

The residual entropy is given by Eq. (4-160).

Computation of $\ln \hat{\phi}_i$ is done via Eq. (4-181). The result is

$$\ln \hat{\phi}_i = \left(B_0 + \overline{B}_{0_i} - \frac{A_0 + \overline{A}_{0_i}}{RT} - \frac{C_0 + \overline{C}_{0_i}}{RT^3} \right) \rho$$
$$+ \frac{1}{2} \left(2b + \overline{b}_i - \frac{2a + \overline{a}_i}{RT} \right) \rho^2 + \left(\frac{4a\alpha + a\overline{\alpha}_i + \alpha\overline{a}_i}{5RT} \right) \rho^5$$
$$+ \frac{c}{2\gamma RT^3} \left\{ \left[\left(1 + \frac{\overline{\gamma}_i}{\gamma} \right) \gamma^2 \rho^4 + \left(\frac{2\overline{\gamma}_i}{\gamma} - \frac{\overline{c}_i}{c} \right) \rho^2 \right. \right.$$
$$\left. - 2 \left(1 + \frac{\overline{c}_i}{c} - \frac{\overline{\gamma}_i}{\gamma} \right) \right] \exp(-\gamma \rho^2) + 2 \left(1 + \frac{\overline{c}_i}{c} - \frac{\overline{\gamma}_i}{\gamma} \right) \right\} - \ln Z \qquad (4\text{-}238)$$

Here the quantities with overbars are *partial parameters* for species i, defined for arbitrary parameter π by

$$\bar{\pi}_i \equiv \left[\frac{\partial(n\pi)}{\partial n_i} \right]_{T,n_j} \qquad (4\text{-}239)$$

Application of these equations requires specific mixing rules. For example, if

$$\pi = \left(\sum_k y_k \pi_k^{1/r} \right)^r \qquad (4\text{-}240)$$

where r is a small integer, the recipe for $\bar{\pi}_i$ is

$$\bar{\pi}_i = \pi \left[r \left(\frac{\pi_i}{\pi} \right)^{1/r} - (r-1) \right] \qquad (4\text{-}241)$$

Specifically, if $r = 3$ for $\pi \equiv c$; then

$$\bar{c}_i = c \left[3 \left(\frac{c_i}{c} \right)^{1/3} - 2 \right]$$

where c_i is the parameter for pure i and c is the parameter for the mixture, given by

$$c = \left(\sum_k y_k c_k^{1/3} \right)^3$$

Equation-of-state examples are given in Smith, Van Ness, and Abbott (*Introduction to Chemical Engineering Thermodynamics*, 5th ed., Secs. 3.4–3.7 and 6.2–6.6, McGraw-Hill, New York, 1996).

EXPRESSIONS FOR THE EXCESS GIBBS ENERGY

In principle, equation-of-state procedures can be used for the calculation of liquid-phase as well as gas-phase properties, and much has been accomplished in the development of PVT equations of state suitable for both phases. However, a widely used alternative for the liquid phase is application of excess properties.

The excess property of primary importance for engineering calculations is the excess Gibbs energy G^E, because its canonical variables are T, P, and composition, the variables usually specified or sought in a design calculation. Knowing G^E as a function of T, P, and composition, one can in principle compute from it all other excess properties (see, for example, Eqs. [4-117] through [4-119]). As noted with respect to Fig. 4-1, the excess volume for liquid mixtures is usually small; the pressure dependence of G^E may then be safely ignored. Thus, the engineering efforts at describing G^E center on representing its composition and temperature dependence.

For binary systems *at constant T*, G^E is a function of just x_1, and the quantity most conveniently represented by an equation is G^E/x_1x_2RT. The simplest procedure is to express this quantity as a power series in x_1:

$$\frac{G^E}{x_1x_2RT} = a + bx_1 + cx_1^2 + \cdots \qquad \text{(constant } T)$$

An equivalent power series with certain advantages is known as the *Redlich/Kister expansion* (Redlich, Kister, and Turnquist, *Chem. Eng. Progr. Symp. Ser. No. 2*, **48**, pp. 49–61 [1952]):

$$\frac{G^E}{x_1x_2RT} = B + C(x_1 - x_2) + D(x_1 - x_2)^2 + \cdots$$

In application, different truncations of this series are appropriate. For each particular expression representing G^E/x_1x_2RT, specific expressions for $\ln \gamma_1$ and $\ln \gamma_2$ result from application of Eq. (4-119). When all parameters are zero, $G^E/RT = 0$, and the solution is ideal. If $C = D = \cdots = 0$, then

$$\frac{G^E}{x_1x_2RT} = B$$

where B is a constant for a given temperature. The corresponding equations for $\ln \gamma_1$ and $\ln \gamma_2$ are

$$\ln \gamma_1 = Bx_2^2 \qquad (4\text{-}242)$$

$$\ln \gamma_2 = Bx_1^2 \qquad (4\text{-}243)$$

The symmetrical nature of these relations is evident. The infinite-dilution values of the activity coefficients are $\ln \gamma_1^\infty = \ln \gamma_2^\infty = B$.

If $D = \cdots = 0$, then

$$\frac{G^E}{x_1x_2RT} = B + C(x_1 - x_2) = B + C(2x_1 - 1)$$

and in this case G^E/x_1x_2RT is linear in x_1. The substitutions, $B + C = A_{21}$ and $B - C = A_{12}$ transform this expression into the *Margules equation:*

$$G^E/x_1x_2RT = A_{21}x_1 + A_{12}x_2 \qquad (4\text{-}244)$$

Application of Eq. (4-119) yields

$$\ln \gamma_1 = x_2^2[A_{12} + 2(A_{21} - A_{12})x_1] \qquad (4\text{-}245)$$

$$\ln \gamma_2 = x_1^2[A_{21} + 2(A_{12} - A_{21})x_2] \qquad (4\text{-}246)$$

An alternative equation is obtained when the reciprocal quantity x_1x_2RT/G^E is expressed as a linear function of x_1:

$$\frac{x_1x_2}{G^E/RT} = B' + C'(x_1 - x_2) = B' + C'(2x_1 - 1)$$

This may also be written:

$$\frac{x_1x_2}{G^E/RT} = B'(x_1 + x_2) + C'(x_1 - x_2) = (B' + C')x_1 + (B' - C')x_2$$

The substitutions $B' + C' = 1/A'_{21}$ and $B' - C' = 1/A'_{12}$ produce

$$\frac{x_1x_2}{G^E/RT} = \frac{x_1}{A'_{21}} + \frac{x_2}{A'_{12}} = \frac{A'_{12}x_1 + A'_{21}x_2}{A'_{12}A'_{21}}$$

or

$$\frac{G^E}{x_1x_2RT} = \frac{A'_{12}A'_{21}}{A'_{12}x_1 + A'_{21}x_2} \qquad (4\text{-}247)$$

The activity coefficients implied by this equation are given by

$$\ln \gamma_1 = A'_{12} \left(1 + \frac{A'_{12}x_1}{A'_{21}x_2} \right)^{-2} \qquad (4\text{-}248)$$

$$\ln \gamma_2 = A'_{21} \left(1 + \frac{A'_{21}x_2}{A'_{12}x_1} \right)^{-2} \qquad (4\text{-}249)$$

These are known as the *van Laar equations*. When $x_1 = 0$, $\ln \gamma_1^\infty = A'_{12}$; when $x_2 = 0$, $\ln \gamma_2^\infty = A'_{21}$.

The Redlich/Kister expansion, the Margules equations, and the van Laar equations are all special cases of a very general treatment based on rational functions, that is, on equations for G^E given by ratios of polynomials (Van Ness and Abbott, *Classical Thermodynamics of Nonelectrolyte Solutions: With Applications to Phase Equilibria*, Sec. 5-7, McGraw-Hill, New York, 1982). Although providing great flexibility in the fitting of VLE data for binary systems, they are without theoretical foundation, with no rational basis for their extension to multicomponent systems. Nor do they incorporate an explicit temperature dependence for the parameters.

Modern theoretical developments in the molecular thermodynamics of liquid-solution behavior are often based on the concept of *local composition*, presumed to account for the short-range order and nonrandom molecular orientations that result from differences in molecular size and intermolecular forces. Introduced with the publication of a model of G^E behavior known as the Wilson equation (*J. Am. Chem. Soc.*, **86**, pp. 127–130 [1964]), it prompted the development of alternative local-composition models, most notably the NRTL (**N**on-**R**andom-**T**wo-**L**iquid) equation of Renon and Prausnitz (*AIChE J.*, **14**, pp. 135–144 [1968]) and the UNIQUAC (**UNI**versal **QUA**si-**C**hemical) equation of Abrams and Prausnitz (*AIChE J.*, **21**, pp. 116–128 [1975]). A further significant development, based on the UNIQUAC equation, is the UNIFAC method (**UNI**QUAC **F**unctional-group **A**ctivity **C**oefficients). Proposed by Fredenslund, Jones, and Prausnitz (*AIChE J.*, **21**, pp. 1086–1099 [1975]) and given detailed treatment by Fredenslund, Gmehling, and Rasmussen (*Vapor-Liquid Equilibrium Using UNIFAC*, Elsevier, Amsterdam, 1977), it provides for the calculation of activity coefficients from contributions of the various groups making up the molecules of a solution.

The Wilson equation, like the Margules and van Laar equations, contains just two parameters for a binary system (Λ_{12} and Λ_{21}), and is written:

$$\frac{G^E}{RT} = -x_1 \ln (x_1 + x_2\Lambda_{12}) - x_2 \ln (x_2 + x_1\Lambda_{21}) \tag{4-250}$$

$$\ln \gamma_1 = -\ln (x_1 + x_2\Lambda_{12}) + x_2 \left(\frac{\Lambda_{12}}{x_1 + x_2\Lambda_{12}} - \frac{\Lambda_{21}}{x_2 + x_1\Lambda_{21}} \right) \tag{4-251}$$

$$\ln \gamma_2 = -\ln (x_2 + x_1\Lambda_{21}) - x_1 \left(\frac{\Lambda_{12}}{x_1 + x_2\Lambda_{12}} - \frac{\Lambda_{21}}{x_2 + x_1\Lambda_{21}} \right) \tag{4-252}$$

whence

$$\ln \gamma_1^\infty = -\ln \Lambda_{12} + 1 - \Lambda_{21}$$

$$\ln \gamma_2^\infty = -\ln \Lambda_{21} + 1 - \Lambda_{12}$$

Both Λ_{12} and Λ_{21} must be positive numbers.

The NRTL equation contains three parameters for a binary system and is written:

$$\frac{G^E}{x_1 x_2 RT} = \frac{G_{21}\tau_{21}}{x_1 + x_2 G_{21}} + \frac{G_{12}\tau_{12}}{x_2 + x_1 G_{12}} \tag{4-253}$$

$$\ln \gamma_1 = x_2^2 \left[\tau_{21}\left(\frac{G_{21}}{x_1 + x_2 G_{21}}\right)^2 + \frac{G_{12}\tau_{12}}{(x_2 + x_1 G_{12})^2} \right] \tag{4-254}$$

$$\ln \gamma_2 = x_1^2 \left[\tau_{12}\left(\frac{G_{12}}{x_2 + x_1 G_{12}}\right)^2 + \frac{G_{21}\tau_{21}}{(x_1 + x_2 G_{21})^2} \right] \tag{4-255}$$

Here

$$G_{12} = \exp (-\alpha\tau_{12}) \qquad G_{21} = \exp (-\alpha\tau_{21})$$

and

$$\tau_{12} = \frac{b_{12}}{RT} \qquad \tau_{21} = \frac{b_{21}}{RT}$$

where α, b_{12}, and b_{21}, parameters specific to a particular pair of species, are independent of composition and temperature. The infinite-dilution values of the activity coefficients are given by the equations:

$$\ln \gamma_1^\infty = \tau_{21} + \tau_{12} \exp (-\alpha\tau_{12})$$

$$\ln \gamma_2^\infty = \tau_{12} + \tau_{21} \exp (-\alpha\tau_{21})$$

The local-composition models have limited flexibility in the fitting of data, but they are adequate for most engineering purposes. Moreover, they are implicitly generalizable to multicomponent systems without the introduction of any parameters beyond those required to describe the constituent binary systems. For example, the Wilson equation for multicomponent systems is written:

$$\frac{G^E}{RT} = -\sum_i x_i \ln \sum_j x_j \Lambda_{ij} \tag{4-256}$$

and

$$\ln \gamma_i = 1 - \ln \sum_j x_j \Lambda_{ij} - \sum_k \frac{x_k \Lambda_{ki}}{\sum_j x_j \Lambda_{kj}} \tag{4-257}$$

where $\Lambda_{ij} = 1$ for $i = j$, and so on. All indices in these equations refer to the same species, and all summations are over *all* species. For each ij pair there are two parameters, because $\Lambda_{ij} \neq \Lambda_{ji}$. For example, in a ternary system the three possible ij pairs are associated with the parameters Λ_{12}, Λ_{21}; Λ_{13}, Λ_{31}; and Λ_{23}, Λ_{32}.

The temperature dependence of the parameters is given by:

$$\Lambda_{ij} = \frac{V_j}{V_i} \exp \frac{-a_{ij}}{RT} \qquad (i \neq j) \tag{4-258}$$

where V_j and V_i are the molar volumes at temperature T of pure liquids j and i, and a_{ij} is a constant independent of composition and temperature. Thus the Wilson equation, like all other local-composition models, has built into it an *approximate* temperature dependence for the parameters. Moreover, all parameters are found from data for

binary (in contrast to multicomponent) systems. This makes parameter determination for the local-composition models a task of manageable proportions.

The UNIQUAC equation treats $g \equiv G^E/RT$ as comprised of two additive parts, a *combinatorial* term g^C, accounting for molecular size and shape differences, and a *residual* term g^R (not a residual property), accounting for molecular interactions:

$$g = g^C + g^R \tag{4-259}$$

Function g^C contains pure-species parameters only, whereas function g^R incorporates two *binary* parameters for each pair of molecules. For a multicomponent system,

$$g^C = \sum_i x_i \ln \frac{\Phi_i}{x_i} + 5 \sum_i q_i x_i \ln \frac{\theta_i}{\Phi_i} \tag{4-260}$$

$$g^R = -\sum_i q_i x_i \ln \left(\sum_j \theta_j \tau_{ji} \right) \tag{4-261}$$

where

$$\Phi_i \equiv \frac{x_i r_i}{\sum_j x_j r_j} \tag{4-262}$$

and

$$\theta_i \equiv \frac{x_i q_i}{\sum_j x_j q_j} \tag{4-263}$$

Subscript i identifies species, and j is a dummy index; all summations are over all species. Note that $\tau_{ji} \neq \tau_{ij}$; however, when $i = j$, then $\tau_{ii} = \tau_{jj} = 1$. In these equations r_i (a relative molecular volume) and q_i (a relative molecular surface area) are pure-species parameters. The influence of temperature on g enters through the interaction parameters τ_{ji} of Eq. (4-261), which are temperature dependent:

$$\tau_{ji} = \exp \frac{-(u_{ji} - u_{ii})}{RT} \tag{4-264}$$

Parameters for the UNIQUAC equation are therefore values of $(u_{ji} - u_{ii})$.

An expression for $\ln \gamma_i$ is found by application of Eq. (4-119) to the UNIQUAC equation for g (Eqs. [4-259] through [4-261]). The result is given by the following equations:

$$\ln \gamma_i = \ln \gamma_i^C + \ln \gamma_i^R \tag{4-265}$$

$$\ln \gamma_i^C = 1 - J_i + \ln J_i - 5q_i \left(1 - \frac{J_i}{L_i} + \ln \frac{J_i}{L_i} \right) \tag{4-266}$$

$$\ln \gamma_i^R = q_i \left(1 - \ln s_i - \sum_j \theta_j \frac{\tau_{ij}}{s_j} \right) \tag{4-267}$$

where in addition to Eqs. (4-263) and (4-264)

$$J_i = \frac{r_i}{\sum_j r_j x_j} \tag{4-268}$$

$$L_i = \frac{q_i}{\sum_j q_j x_j} \tag{4-269}$$

$$s_i = \sum_l \theta_l \tau_{li} \tag{4-270}$$

Again subscript i identifies species, and j and l are dummy indicies. Values for the parameters r_i, q_i, and $(u_{ij} - u_{jj})$ are given by Gmehling, Onken, and Arlt (*Vapor-Liquid Equilibrium Data Collection*, Chemistry Data Series, vol. I, parts 1–8, DECHEMA, Frankfurt/Main, 1974–1990).

The Wilson parameters Λ_{ij}, NRTL parameters G_{ij}, and UNIQUAC parameters τ_{ij} all inherit a Boltzmann-type T dependence from the origins of the expressions for G^E, but it is only approximate. Computations of properties sensitive to this dependence (e.g., heats of mixing and liquid/liquid solubility) are in general only qualitatively correct.

EQUILIBRIUM

CRITERIA

The equations developed in preceding sections are for *PVT* systems in states of internal equilibrium. The criteria for internal thermal and mechanical equilibrium are well known, and need not be discussed in detail. They simply require uniformity of temperature and pressure throughout the system. The criteria for phase and chemical-reaction equilibria are less obvious.

Consider a closed *PVT* system, either homogeneous or heterogeneous, of uniform *T* and *P*, which is in thermal and mechanical equilibrium with its surroundings, but which is not initially at internal equilibrium with respect to mass transfer or with respect to chemical reaction. Changes occurring in the system are then irreversible, and must necessarily bring the system closer to an equilibrium state. The first and second laws written for the entire system are

$$dU^t = dQ + dW$$

$$dS^t \geq \frac{dQ}{T}$$

Combination gives

$$dU^t - dW - T\, dS^t \leq 0$$

Since mechanical equilibrium is assumed,

$$dW = -P\, dV^t$$

Whence

$$dU^t + P\, dV^t - T\, dS^t \leq 0$$

The inequality applies to all incremental changes toward the equilibrium state, whereas the equality holds at the equilibrium state where any change is reversible.

Various constraints may be put on this expression to produce alternative criteria for the directions of irreversible processes and for the condition of equilibrium. For example, it follows immediately that

$$dU^t_{S^t, V^t} \leq 0$$

Alternatively, other pairs of properties may be held constant. The most useful result comes from fixing *T* and *P*, in which case

$$d(U^t + PV^t - TS^t)_{T,P} \leq 0$$

or

$$dG^t_{T,P} \leq 0$$

This expression shows that all irreversible processes occurring at constant *T* and *P* proceed in a direction such that the total Gibbs energy of the system decreases. Thus the equilibrium state of a closed system is the state with the minimum total Gibbs energy attainable at the given *T* and *P*. At the equilibrium state, differential variations may occur in the system at constant *T* and *P* without producing a change in *G^t*. This is the meaning of the equilibrium criterion

$$dG^t_{T,P} = 0 \qquad (4\text{-}271)$$

This equation may be applied to a closed, nonreactive, two-phase system. Each phase taken separately is an *open* system, capable of exchanging mass with the other, and Eq. (4-16) may be written for each phase:

$$d(nG)' = -(nS)'\, dT + (nV)'\, dP + \sum_i \mu_i'\, dn_i'$$

$$d(nG)'' = -(nS)''\, dT + (nV)''\, dP + \sum_i \mu_i''\, dn_i''$$

where the primes and double primes denote the two phases and the presumption is that *T* and *P* are uniform throughout the two phases. The change in the Gibbs energy of the two-phase system is the sum of these equations. When each total-system property is expressed by an equation of the form

$$nM = (nM)' + (nM)''$$

this sum is given by

$$d(nG) = (nV)\, dP - (nS)\, dT + \sum_i \mu_i'\, dn_i' + \sum_i \mu_i''\, dn_i''$$

If the two-phase system is at equilibrium, then application of Eq. (4-271) yields

$$dG^t_{T,P} \equiv d(nG)_{T,P} = \sum_i \mu_i'\, dn_i' + \sum_i \mu_i''\, dn_i'' = 0$$

Since the system is closed and without chemical reaction, material balances require that

$$dn_i'' = -dn_i'$$

Therefore

$$\sum_i (\mu_i' - \mu_i'')\, dn_i' = 0$$

Since the dn_i' are independent and arbitrary, it follows that

$$\mu_i' = \mu_i''$$

This is the criterion of two-phase equilibrium. It is readily generalized to multiple phases by successive application to pairs of phases. The general result is

$$\mu_i' = \mu_i'' = \mu_i''' = \cdots \qquad (4\text{-}272)$$

Substitution for each μ_i by Eq. (4-77) produces the equivalent result

$$\hat{f}_i' = \hat{f}_i'' = \hat{f}_i''' = \cdots \qquad (4\text{-}273)$$

These are the criteria of phase equilibrium applied in the solution of practical problems.

For the case of equilibrium with respect to chemical reaction within a single-phase closed system, combination of Eqs. (4-16) and (4-271) leads immediately to

$$\sum_i \mu_i\, dn_i = 0 \qquad (4\text{-}274)$$

For a system in which both phase and chemical-reaction equilibrium prevail, the criteria of Eqs. (4-272) and (4-274) are superimposed.

THE PHASE RULE

The *intensive* state of a *PVT* system is established when its temperature and pressure and the compositions of all phases are fixed. However, for equilibrium states these variables are not all independent, and fixing a limited number of them automatically establishes the others. This number of independent variables is given by the phase rule, and is called the *number of degrees of freedom* of the system. It is the number of variables which may be arbitrarily specified and which must be so specified in order to fix the *intensive* state of a system at equilibrium. This number is the difference between the number of variables needed to characterize the system and the number of equations that may be written connecting these variables.

For a system containing *N* chemical species distributed at equilibrium among π phases, the phase-rule variables are temperature and pressure, presumed uniform throughout the system, and *N* − 1 mole fractions in each phase. The number of these variables is $2 + (N - 1)\pi$. The masses of the phases are not phase-rule variables, because they have nothing to do with the intensive state of the system.

The equations that may be written connecting the phase-rule variables are:

1. Equation (4-272) for each species, giving $(\pi - 1)N$ phase-equilibrium equations.

2. Equation (4-274) for each independent chemical reaction, giving *r* equations.

The total number of independent equations is therefore $(\pi - 1)N + r$. In their fundamental forms these equations relate chemical potentials, which are functions of temperature, pressure, and composition, the phase-rule variables. Since the degrees of freedom of the system *F* is the difference between the number of variables and the number of equations,

$$F = 2 + (N - 1)\pi - (\pi - 1)N - r$$

or

$$F = 2 - \pi + N - r \qquad (4\text{-}275)$$

The number of independent chemical reactions r can be determined as follows:

1. Write *formation* reactions from the elements for each chemical compound present in the system.

2. Combine these reaction equations so as to eliminate from the set all elements not present as elements in the system. A systematic procedure is to select one equation and combine it with each of the other equations of the set so as to eliminate a particular element. This usually reduces the set by one equation for each element eliminated, though two or more elements may be simultaneously eliminated.

The resulting set of r equations is a complete set of independent reactions. More than one such set is often possible, but all sets number r and are equivalent.

Example 1: Application of the Phase Rule

a. For a system of two miscible nonreacting species in vapor/liquid equilibrium,

$$F = 2 - \pi + N - r = 2 - 2 + 2 - 0 = 2$$

The two degrees of freedom for this system may be satisfied by setting T and P, or T and y_1, or P and x_1, or x_1 and y_1, and so on, at fixed values. Thus, for equilibrium at a particular T and P, this state (if possible at all) exists only at one liquid and one vapor composition. Once the two degrees of freedom are used up, no further specification is possible that would restrict the phase-rule variables. For example, one cannot *in addition* require that the system form an azeotrope (assuming this possible), for this requires $x_1 = y_1$, an equation not taken into account in the derivation of the phase rule. Thus, the requirement that the system form an azeotrope imposes a special constraint and reduces the number of degrees of freedom to one.

b. For a gaseous system consisting of CO, CO_2, H_2, H_2O, and CH_4 in chemical-reaction equilibrium,

$$F = 2 - \pi + N - r = 2 - 1 + 5 - 2 = 4$$

The value of $r = 2$ is found from the formation reactions:

$$C + \tfrac{1}{2}O_2 \rightarrow CO$$
$$C + O_2 \rightarrow CO_2$$
$$H_2 + \tfrac{1}{2}O_2 \rightarrow H_2O$$
$$C + 2H_2 \rightarrow CH_4$$

Systematic elimination of C and O_2 from this set of chemical equations reduces the set to two. Three possible pairs of equations may result, depending on how the combination of equations is effected. Any *pair* of the following three equations represents a complete set of independent reactions, and all pairs are equivalent.

$$CH_4 + H_2O \rightarrow CO + 3H_2$$
$$CO + H_2O \rightarrow CO_2 + H_2$$
$$CH_4 + 2H_2O \rightarrow CO_2 + 4H_2$$

The result, $F = 4$, means that one is free to specify, for example, T, P, and two mole fractions in an equilibrium mixture of these five chemical species, provided nothing else is arbitrarily set. Thus, it cannot simultaneously be required that the system be prepared from specified amounts of particular constituent species.

Since the phase rule treats only the intensive state of a system, it applies to both closed and open systems. **Duhem's theorem,** on the other hand, is a rule relating to closed systems only: *For any closed system formed initially from given masses of prescribed chemical species, the equilibrium state is completely determined by any two properties of the system, provided only that the two properties are independently variable at the equilibrium state.* The meaning of *completely determined* is that both the intensive and extensive states of the system are fixed; not only are T, P, and the phase compositions established, but so also are the masses of the phases.

VAPOR/LIQUID EQUILIBRIUM

Vapor/liquid equilibrium (VLE) relationships (as well as other interphase equilibrium relationships) are needed in the solution of many engineering problems. The required data can be found by experiment, but such measurements are seldom easy, even for binary systems, and they become rapidly more difficult as the number of constituent species increases. This is the incentive for application of thermodynamics to the calculation of phase-equilibrium relationships.

The general VLE problem involves a multicomponent system of N constituent species for which the independent variables are T, P, $N-1$ liquid-phase mole fractions, and $N-1$ vapor-phase mole fractions. (Note that $\sum_i x_i = 1$ and $\sum_i y_i = 1$, where x_i and y_i represent liquid and vapor mole fractions respectively.) Thus there are $2N$ independent variables, and application of the phase rule shows that exactly N of these variables must be fixed to establish the intensive state of the system. This means that once N variables have been specified, the remaining N variables can be determined by simultaneous solution of the N equilibrium relations:

$$\hat{f}_i^l = \hat{f}_i^v \qquad (i = 1, 2, \ldots, N) \tag{4-276}$$

where superscripts l and v denote the liquid and vapor phases, respectively.

In practice, either T or P *and* either the liquid-phase or vapor-phase composition are specified, thus fixing $1 + (N-1) = N$ independent variables. The remaining N variables are then subject to calculation, provided that sufficient information is available to allow determination of all necessary thermodynamic properties.

Gamma/Phi Approach   For many VLE systems of interest the pressure is low enough that a relatively simple equation of state, such as the two-term virial equation, is satisfactory for the vapor phase. Liquid-phase behavior, on the other hand, may be conveniently described by an equation for the excess Gibbs energy, from which activity coefficients are derived. The fugacity of species i in the liquid phase is then given by Eq. (4-102), written

$$\hat{f}_i^l = \gamma_i x_i f_i$$

while the vapor-phase fugacity is given by Eq. (4-79), written

$$\hat{f}_i^v = \hat{\phi}_i^v y_i P$$

Equation (4-276) is now expressed as

$$\gamma_i x_i f_i = \hat{\phi}_i y_i P \qquad (i = 1, 2, \ldots, N) \tag{4-277}$$

The identifying superscripts l and v are omitted here with the understanding that γ_i and f_i are liquid-phase properties, whereas $\hat{\phi}_i$ is a vapor-phase property. Applications of Eq. (4-277) represent what is known as the *gamma/phi* approach to VLE calculations.

Evaluation of $\hat{\phi}_i$ is usually by Eq. (4-196), based on the two-term virial equation of state, but other equations, such as Eq. (4-200), are also applicable. The activity coefficient γ_i is evaluated by Eq. (4-119), which relates $\ln \gamma_i$ to G^E/RT as a partial property. Thus, what is required for the liquid phase is a relation between G^E/RT and composition. Equations in common use for this purpose have already been described.

The fugacity f_i of pure compressed liquid i must be evaluated at the T and P of the equilibrium mixture. This is done in two steps. First, one calculates the fugacity coefficient of saturated vapor $\phi_i^v = \phi_i^{sat}$ by an integrated form of Eq. (4-161), written for pure species i and evaluated at temperature T and the corresponding vapor pressure $P = P_i^{sat}$. Equation (4-276) written for pure species i becomes

$$f_i^v = f_i^l = f_i^{sat} \tag{4-278}$$

where f_i^{sat} indicates the value both for saturated liquid and for saturated vapor. The corresponding fugacity coefficient is

$$\phi_i^{sat} = \frac{f_i^{sat}}{P_i^{sat}} \tag{4-279}$$

This fugacity coefficient applies equally to saturated vapor and to saturated liquid at given temperature T. Equation (4-278) can therefore equally well be written

$$\phi_i^v = \phi_i^l \tag{4-280}$$

The second step is the evaluation of the change in fugacity of the liquid with a change in pressure to a value above or below P_i^{sat}. For this isothermal change of state from saturated liquid at P_i^{sat} to liquid at pressure P, Eq. (4-105) is integrated to give

$$G_i - G_i^{sat} = \int_{P_i^{sat}}^{P} V_i \, dP$$

Equation (4-74) is then written twice: for G_i and for G_i^{sat}. Subtraction provides another expression for $G_i - G_i^{sat}$:

$$G_i - G_i^{sat} = RT \ln \frac{f_i}{f_i^{sat}}$$

Equating the two expressions for $G_i - G_i^{sat}$ yields

$$\ln \frac{f_i}{f^{sat}} = \frac{1}{RT} \int_{P_i^{sat}}^{P} V_i \, dP$$

Since V_i, the liquid-phase molar volume, is a very weak function of P at temperatures well below T_c, an excellent approximation is often obtained when evaluation of the integral is based on the assumption that V_i is constant at the value for saturated liquid, V_i^l:

$$\ln \frac{f_i}{f_i^{sat}} = \frac{V_i^l (P - P_i^{sat})}{RT}$$

Substituting $f_i^{sat} = \phi_i^{sat} P_i^{sat}$ (Eq. [4-279]), and solving for f_i gives

$$f_i = \phi_i^{sat} P_i^{sat} \exp \frac{V_i^l (P - P_i^{sat})}{RT} \qquad (4\text{-}281)$$

The exponential is known as the *Poynting factor*.

Equation (4-277) may now be written

$$y_i P \Phi_i = x_i \gamma_i P_i^{sat} \qquad (i = 1, 2, \ldots, N) \qquad (4\text{-}282)$$

where

$$\Phi_i = \left(\frac{\hat{\phi}_i}{\phi_i^{sat}} \right) \exp \frac{-V_i^l (P - P_i^{sat})}{RT} \qquad (4\text{-}283)$$

If evaluation of ϕ_i^{sat} and $\hat{\phi}_i$ is by Eqs. (4-192) and (4-196), this reduces to

$$\Phi_i = \exp \left[\frac{P \overline{B}_i - P_i^{sat} B_{ii} - V_i^l (P - P_i^{sat})}{RT} \right] \qquad (4\text{-}284)$$

where $\overline{B}_i$ is given by Eq. (4-195):

$$\overline{B}_i \equiv \left[\frac{\partial (nB)}{\partial n_i} \right]_{T, n_j} = 2 \sum_k y_k B_{ki} - B \qquad (4\text{-}285)$$

with B evaluated by Eq. (4-183).

The N equations represented by Eq. (4-282) in conjunction with Eq. (4-284) may be used to solve for N unspecified phase-equilibrium variables. For a multicomponent system the calculation is formidable, but well suited to computer solution. The types of problems encountered for nonelectrolyte systems at low to moderate pressures (well below the critical pressure) are discussed by Smith, Van Ness, and Abbott (*Introduction to Chemical Engineering Thermodynamics*, 5th ed., McGraw-Hill, New York, 1996).

When Eq. (4-282) is applied to VLE for which the vapor phase is an ideal gas and the liquid phase is an ideal solution, it reduces to a very simple expression. For ideal gases, fugacity coefficients $\hat{\phi}_i$ and ϕ_i^{sat} are unity, and the right-hand side of Eq. (4-283) reduces to the Poynting factor. For the systems of interest here this factor is always very close to unity, and for practical purposes $\Phi_i = 1$. For ideal solutions, the activity coefficients γ_i are also unity. Equation (4-282) therefore reduces to

$$y_i P = x_i P_i^{sat} \qquad (i = 1, 2, \ldots, N)$$

an equation which expresses *Raoult's law*. It is the simplest possible equation for VLE, and as such fails to provide a realistic representation of real behavior for most systems. Nevertheless, it is useful as a standard of comparison.

When an appropriate correlating equation for G^E is not available, reliable estimates of activity coefficients may often be obtained from a group-contribution correlation. The Analytical Solution of Groups (ASOG) method (Kojima and Tochigi, *Prediction of Vapor-Liquid Equilibrium by the ASOG Method*, Elsevier, Amsterdam, 1979) and the UNIFAC method are both well developed. Additional references of interest include Hansen et al. (*Ind. Eng. Chem. Res.*, **30**, pp. 2352–2355 [1991]), Gmehling and Schiller (Ibid., **32**, pp. 178–193 [1993]); Larsen et al. (Ibid., **26**, pp. 2274–2286 [1987]); and Tochigi et al. (*J. Chem. Eng. Japan*, **23**, pp. 453–463 [1990]).

Data Reduction Correlations for G^E and the activity coefficients are based on VLE data taken at low to moderate pressures. The ASOG and UNIFAC group-contribution methods depend for validity on parameters evaluated from a large base of such data. The process

of finding a suitable analytic relation for g ($\equiv G^E/RT$) as a function of its independent variables T and x_1, thus producing a correlation of VLE data, is known as *data reduction*. Although g is in principle also a function of P, the dependence is so weak as to be universally and properly neglected. Given here is a brief description of the treatment of data taken for *binary* systems under *isothermal* conditions. A more comprehensive development is given by Van Ness (*J. Chem. Thermodyn.*, **27**, pp. 113–134 [1995]; *Pure & Appl. Chem.*, **67**, pp. 859–872 [1995]).

Presumed in all that follows is the existence of an equation inherently capable of representing correct values of G^E for the liquid phase as a function of x_1:

$$g \equiv G^E/RT = \mathcal{G}(x_1; \alpha, \beta, \ldots) \qquad (4\text{-}286)$$

where α, β, and so on, represent adjustable parameters.

The measured variables of binary VLE are x_1, y_1, T, and P. Experimental values of the activity coefficient of species i in the liquid are related to these variables by Eq. (4-282), written:

$$\gamma_i^\circ = \frac{y_i^\circ P^\circ}{x_i P_i^{sat}} \Phi_i \qquad (i = 1, 2) \qquad (4\text{-}287)$$

where Φ_i is given by Eq. (4-283), and the asterisks denote experimental values. A simple summability relation analogous to Eq. (4-123) defines an experimental value of g°:

$$g^\circ \equiv x_1 \ln \gamma_1^\circ + x_2 \ln \gamma_2^\circ \qquad (4\text{-}288)$$

Moreover, Eq. (4-122), the Gibbs/Duhem equation, may be written for experimental values in a binary system as

$$x_1 \frac{d \ln \gamma_1^\circ}{dx_1} + x_2 \frac{d \ln \gamma_2^\circ}{dx_1} = 0 \qquad (4\text{-}289)$$

Because experimental measurements are subject to systematic error, sets of values of $\ln \gamma_1^\circ$ and $\ln \gamma_2^\circ$ determined by experiment may not satisfy, that is, may not be *consistent* with, the Gibbs/Duhem equation. Thus, Eq. (4-289) applied to sets of experimental values becomes a test of the thermodynamic consistency of the data, rather than a valid general relationship.

Values of g given by the correlating equation, Eq. (4-286), are called *derived* values, and associated derived values of the activity coefficients are given by specialization of Eqs. (4-58):

$$\gamma_1 = \exp \left(g + x_2 \frac{dg}{dx_1} \right) \qquad (4\text{-}290)$$

$$\gamma_2 = \exp \left(g - x_1 \frac{dg}{dx_1} \right) \qquad (4\text{-}291)$$

These two equations may be combined to yield

$$\frac{dg}{dx_1} = \ln \frac{\gamma_1}{\gamma_2} \qquad (4\text{-}292)$$

This equation applies to *derived* property values. The corresponding *experimental* values are given by differentiation of Eq. (4-288):

$$\frac{dg^\circ}{dx_1} = x_1 \frac{d \ln \gamma_1^\circ}{dx_1} + \ln \gamma_1^\circ + x_2 \frac{d \ln \gamma_2^\circ}{dx_1} - \ln \gamma_2^\circ$$

or

$$\frac{dg^\circ}{dx_1} = \ln \frac{\gamma_1^\circ}{\gamma_2^\circ} + x_1 \frac{d \ln \gamma_1^\circ}{dx_1} + x_2 \frac{d \ln \gamma_2^\circ}{dx_1} \qquad (4\text{-}293)$$

Subtraction of Eq. (4-293) from Eq. (4-292) gives

$$\frac{dg}{dx_1} - \frac{dg^\circ}{dx_1} = \ln \frac{\gamma_1}{\gamma_2} - \ln \frac{\gamma_1^\circ}{\gamma_2^\circ} - \left(x_1 \frac{d \ln \gamma_1^\circ}{dx_1} + x_2 \frac{d \ln \gamma_2^\circ}{dx_1} \right)$$

The differences between like terms represent *residuals* between derived and experimental values. Defining these residuals as

$$\delta g \equiv g - g^\circ \qquad \text{and} \qquad \delta \ln \frac{\gamma_1}{\gamma_2} \equiv \ln \frac{\gamma_1}{\gamma_2} - \ln \frac{\gamma_1^\circ}{\gamma_2^\circ}$$

puts this equation into the form

$$\frac{d \delta g}{dx_1} = \delta \ln \frac{\gamma_1}{\gamma_2} - \left(x_1 \frac{d \ln \gamma_1^\circ}{dx_1} + x_2 \frac{d \ln \gamma_2^\circ}{dx_1} \right)$$

If a data set is reduced so as to make the δg residuals scatter about zero, then the derivative on the left is effectively zero, and the preceding equation becomes

$$\delta \ln \frac{\gamma_1}{\gamma_2} = x_1 \frac{d \ln \gamma_1^\circ}{dx_1} + x_2 \frac{d \ln \gamma_2^\circ}{dx_1} \qquad (4\text{-}294)$$

The right-hand side of this equation is exactly the quantity that Eq. (4-289), the Gibbs/Duhem equation, requires to be zero for consistent data. The residual on the left is therefore a direct measure of deviations from the Gibbs/Duhem equation. The extent to which values of this residual fail to scatter about zero measures the departure of the data from consistency with respect to this equation.

The data-reduction procedure just described provides parameters in the correlating equation for g that make the δg residuals scatter about zero. This is usually accomplished by finding the parameters that minimize the sum of squares of the residuals. Once these parameters are found, they can be used for the calculation of derived values of both the pressure P and the vapor composition y_1. Equation (4-282) is solved for y_iP and written for species 1 and for species 2. Adding the two equations gives

$$P = \frac{x_1 \gamma_1 P_1^{\text{sat}}}{\Phi_1} + \frac{x_2 \gamma_2 P_2^{\text{sat}}}{\Phi_2} \qquad (4\text{-}295)$$

whence by Eq. (4-282),

$$y_1 = \frac{x_1 \gamma_1 P_1^{\text{sat}}}{\Phi_1 P} \qquad (4\text{-}296)$$

These equations allow calculation of the *primary* residuals:

$$\delta P \equiv P - P^\circ \qquad \text{and} \qquad \delta y_1 \equiv y_1 - y_1^\circ$$

If the experimental values P° and y_1° are closely reproduced by the correlating equation for g, then these residuals, evaluated at the experimental values of x_1, scatter about zero. This is the result obtained when the data are thermodynamically consistent. When they are not, these residuals do not scatter about zero, and the correlation for g does not properly reproduce the experimental values P° and y_1°. Such a correlation is, in fact, unnecessarily divergent. An alternative is to process just the P-x_1 data; this is possible because the P-x_1-y_1 data set includes more information than necessary. Assuming that the correlating equation is appropriate to the data, one merely searches for values of the parameters α, β, and so on, that yield pressures by Eq. (4-295) that are as close as possible to the measured values. The usual procedure is to minimize the sum of squares of the residuals δP. Known as *Barker's method* (*Austral. J. Chem.*, **6**, pp. 207–210 [1953]), it provides the best possible fit of the experimental pressures. When the experimental data do not satisfy the Gibbs/Duhem equation, it cannot precisely represent the experimental y_1 values; however, it provides a better fit than does the procedure that minimizes the sum of the squares of the δg residuals.

Worth noting is the fact that Barker's method does not require experimental y_1° values. Thus the correlating parameters α, β, and so on, can be evaluated from a P-x_1 data subset. Common practice now is, in fact, to measure just such data. They are, of course, not subject to a test for consistency by the Gibbs/Duhem equation. The world's store of VLE data has been compiled by Gmehling et al. (*Vapor-Liquid Equilibrium Data Collection*, Chemistry Data Series, vol. I, parts 1–8, DECHEMA, Frankfurt am Main, 1979–1990).

Solute/Solvent Systems   The gamma/phi approach to VLE calculations presumes knowledge of the vapor pressure of each species at the temperature of interest. For certain binary systems species 1, designated the *solute*, is either unstable at the system temperature or is *supercritical* ($T > T_c$). Its vapor pressure cannot be measured, and its fugacity as a pure liquid at the system temperature f_1 cannot be calculated by Eq. (4-281).

Equations (4-282) and (4-283) are applicable to species 2, designated the *solvent*, but not to the solute, for which an alternative approach is required. Figure 4-11 shows a typical plot of the liquid-phase fugacity of the solute $\hat{f}_1$ vs. its mole fraction x_1 at constant temperature. Since the curve representing $\hat{f}_1$ does not extend all the way to $x_1 = 1$, the location of f_1, the liquid-phase fugacity of pure species 1, is not established. The tangent line at the origin, representing *Henry's*

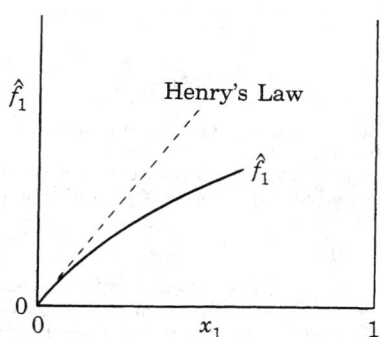

FIG. 4-11   Plot of solute fugacity $\hat{f}_1$ vs. solute mole fraction.

law, provides alternative information. The slope of the tangent line is *Henry's constant*, defined as

$$k_1 \equiv \lim_{x_1 \to 0} \frac{\hat{f}_1}{x_1} \qquad (4\text{-}297)$$

This is the definition of k_1 for temperature T and for a pressure equal to the vapor pressure of the pure solvent P_2^{sat}.

The activity coefficient of the solute at infinite dilution is

$$\lim_{x_1 \to 0} \gamma_1 = \lim_{x_1 \to 0} \frac{\hat{f}_1}{x_1 f_1} = \frac{1}{f_1} \lim_{x_1 \to 0} \frac{\hat{f}_1}{x_1}$$

In view of Eq. (4-297), this becomes $\gamma_1^\infty = k_1/f_1$, or

$$f_1 = \frac{k_1}{\gamma_1^\infty} \qquad (4\text{-}298)$$

where γ_1^∞ represents the infinite-dilution value of the activity coefficient of the solute. Since both k_1 and γ_1^∞ are evaluated at P_2^{sat}, this pressure also applies to f_1. However, the effect of P on a liquid-phase fugacity, given by a Poynting factor, is very small, and for practical purposes may usually be neglected. The activity coefficient of the solute, given by

$$\gamma_1 \equiv \frac{\hat{f}_1}{x_1 f_1} = \frac{y_1 P \hat{\phi}_1}{x_1 f_1}$$

then becomes

$$\gamma_1 = \frac{y_1 P \hat{\phi}_1 \gamma_1^\infty}{x_1 k_1}$$

For the solute, this equation takes the place of Eqs. (4-282) and (4-283). Solution for y_1 gives

$$y_1 = \frac{x_1 (\gamma_1/\gamma_1^\infty) k_1}{\hat{\phi}_1 P} \qquad (4\text{-}299)$$

For the solvent, species 2, the analog of Eq. (4-296) is

$$y_2 = \frac{x_2 \gamma_2 P_2^{\text{sat}}}{\Phi_2 P} \qquad (4\text{-}300)$$

Since $y_1 + y_2 = 1$,

$$P = \frac{x_1 (\gamma_1/\gamma_1^\infty) k_1}{\hat{\phi}_1} + \frac{x_2 \gamma_2 P_2^{\text{sat}}}{\Phi_2} \qquad (4\text{-}301)$$

Note that the same correlation that provides for the evaluation of γ_1 also allows evaluation of γ_1^∞.

There remains the problem of finding Henry's constant from the available VLE data. For equilibrium

$$\hat{f}_1 \equiv \hat{f}_1^l = \hat{f}_1^v = y_1 P \hat{\phi}_1$$

Division by x_1 gives

$$\frac{\hat{f}_1}{x_1} = P \hat{\phi}_1 \frac{y_1}{x_1}$$

Henry's constant is defined as the limit as $x_1 \to 0$ of the ratio on the left; therefore

$$k_1 = P_2^{sat} \hat{\phi}_1^{\infty} \lim_{x_1 \to 0} \frac{y_1}{x_1}$$

The limiting value of y_1/x_1 can be found by plotting y_1/x_1 vs. x_1 and extrapolating to zero.

K-Values A measure of how a given chemical species distributes itself between liquid and vapor phases is the equilibrium ratio:

$$K_i \equiv \frac{y_i}{x_i} \qquad (4\text{-}302)$$

Usually called simply a K-value, it adds nothing to thermodynamic knowledge of VLE. However, its use may make for computational convenience, allowing formal elimination of one set of mole fractions $\{y_i\}$ or $\{x_i\}$ in favor of the other. Moreover, it characterizes *lightness* of a constituent species. For a *light* species, tending to concentrate in the vapor phase, $K > 1$; for a *heavy* species, tending to concentrate in the liquid phase, $K < 1$.

Empirical correlations for K-values found in the older literature have little relation to thermodynamics. Their proper evaluation comes directly from Eq. (4-277):

$$K_i \equiv \frac{y_i}{x_i} = \frac{\gamma_i f_i}{\hat{\phi}_i P} \qquad (4\text{-}303)$$

When Raoult's law applies, this becomes $K_i = P_i^{sat}/P$. In general, K-values are functions of T, P, liquid composition, and vapor composition, making their direct and accurate correlation impossible. Those correlations that do exist are approximate and severely limited in application. The DePriester correlation, for example, gives K-values for light hydrocarbons (*Chem. Eng. Prog. Symp. Ser. No. 7*, **49**, pp. 1–43 [1953]).

Equation-of-State Approach Although the gamma/phi approach to VLE is in principle generally applicable to systems comprised of subcritical species, in practice it has found use primarily where pressures are no more than a few bars. Moreover, it is most satisfactory for correlation of constant-temperature data. A temperature dependence for the parameters in expressions for G^E is included only for the local-composition equations, and it is at best only approximate.

A generally applicable alternative to the gamma/phi approach results when both the liquid and vapor phases are described by the same equation of state. The defining equation for the fugacity coefficient, Eq. (4-79), may be applied to each phase:

Liquid: $\qquad f_i^l = \hat{\phi}_i^l x_i P$

Vapor: $\qquad f_i^v = \hat{\phi}_i^v y_i P$

Equation (4-276) now becomes

$$x_i \hat{\phi}_i^l = y_i \hat{\phi}_i^v \qquad (i = 1, 2, \dots, N) \qquad (4\text{-}304)$$

This introduces the compositions x_i and y_i into the equilibrium equations, but neither is explicit, because the $\hat{\phi}_i$ are functions, not only of T and P, but of composition. Thus Eq. (4-304) represents N complex relationships connecting T, P, the x_i, and the y_i, suitable for computer solution. Given an appropriate equation of state, one or another of Eqs. (4-178) through (4-181) provides for expression of the $\hat{\phi}_i$ as functions of T, P, and composition.

Because of inadequacies in empirical mixing rules, such as those given by Eqs. (4-221) and (4-222), the equation-of-state approach was long limited to systems exhibiting modest and well-behaved deviations from ideal solution behavior in the liquid phase; for example, to systems containing hydrocarbons and cryogenic fluids. However, the introduction by Wong and Sandler (*AIChE J.*, **38**, pp. 671–680 [1992]) of a new class of mixing rules for cubic equations of state has greatly expanded their useful application to VLE.

The Soave/Redlich/Kwong (SRK) and the Peng/Robinson (PR) equations of state, both expressed by Eqs. (4-230) and (4-231), were developed specifically for VLE calculations. The fugacity coefficients implicit in these equations are given by Eq. (4-232). When combined

with the theoretically based Wong/Sandler mixing rules for parameters a and b these equations provide the means for accurate correlation and prediction of VLE data.

The first of the Wong/Sandler mixing rules relates the difference in mixture quantities b and a/RT to the corresponding differences (identified by subscripts) for the pure species:

$$b - \frac{a}{RT} = \sum_p \sum_q x_p x_q E_{pq} \qquad (4\text{-}305)$$

where $\qquad E_{pq} \equiv \frac{1}{2} \left(b_p - \frac{a_p}{RT} + b_q - \frac{a_q}{RT} \right) (1 - k_{pq}) \qquad (4\text{-}306)$

Binary interaction parameters k_{pq} are determined for each pq pair ($p \neq q$) from experimental data. Note that $k_{pq} = k_{qp}$ and $k_{pp} = k_{qq} = 0$. Since the quantity on the left-hand side of Eq. (4-305) represents the second virial coefficient as predicted by Eq. (4-231), the basis for Eq. (4-305) lies in Eq. (4-183), which expresses the quadratic dependence of the mixture second virial coefficient on mole fraction.

The second Wong/Sandler mixing rule relates ratios of a/RT to b:

$$\frac{a}{bRT} = 1 - D \qquad (4\text{-}307)$$

where $\qquad D \equiv 1 + \frac{G^E}{cRT} - \sum_p x_p \frac{a_p}{b_p RT} \qquad (4\text{-}308)$

The quantity G^E/RT is given by an appropriate correlation for the excess Gibbs energy of the liquid phase, and is evaluated at the mixture composition, regardless of whether the mixture is liquid or vapor. The constant c is specific to the equation of state. The theoretical basis for these equations can be found in the literature (Wong and Sandler, op. cit.; *Ind. Eng. Chem. Res.*, **31**, pp. 2033–2039 [1992]; Eubank, et al., *Ind. Eng. Chem. Res.*, **34**, pp. 314–323 [1995]).

Elimination of a from Eq. (4-305) by Eq. (4-307) provides an expression for b:

$$b = \frac{1}{D} \sum_p \sum_q x_p x_q E_{pq} \qquad (4\text{-}309)$$

Mixture parameter a then follows from Eq. (4-307):

$$a = bRT(1 - D) \qquad (4\text{-}310)$$

Equations (4-233) and (4-234) may now be applied for the evaluation of partial parameters $\bar{a}_i$ and $\bar{b}_i$:

$$\bar{b}_i = \frac{1}{D} \left[2 \sum_j x_j E_{ij} - b \left(1 + \frac{\ln \gamma_i}{c} - \frac{a_i}{b_i RT} \right) \right] \qquad (4\text{-}311)$$

and $\qquad \bar{a}_i = bRT \left(\frac{a_i}{b_i RT} - \frac{\ln \gamma_i}{c} \right) + a \left(\frac{\bar{b}_i}{b} - 1 \right) \qquad (4\text{-}312)$

For pure species i, Eq. (4-232) reduces to

$$\ln \phi_i = Z_i - 1 - \ln \frac{(V_i - b_i)Z_i}{V_i} + \frac{a_i/b_i RT}{\varepsilon - \sigma} \ln \frac{V_i + \sigma b_i}{V_i + \varepsilon b_i} \qquad (4\text{-}313)$$

This equation may be applied separately to the liquid phase and to the vapor phase to yield the pure-species values ϕ_i^l and ϕ_i^v. For vapor/liquid equilibrium (Eq. [4-280]), these two quantities are equal. Given parameters a_i and b_i, the pressure P in Eq. (4-230) that makes these two values equal is P_i^{sat}, the equilibrium vapor pressure of pure species i as predicted by the equation of state.

The correlations for $\alpha(T_{ri}; \omega_i)$ that follow Eq. (4-230) are designed to provide values of a_i that yield pure-species vapor pressures which, on average, are in reasonable agreement with experiment. However, reliable correlations for P_i^{sat} as a function of temperature are available for many pure species. Thus when P_i^{sat} is known for a particular temperature, a_i should be evaluated so that the equation of state correctly predicts this known value. The procedure is to write Eq. (4-313) for each of the phases, combining the two equations in accord with Eq. (4-280), written

$$\ln \phi_i^l = \ln \phi_i^v$$

The resulting expression may be solved for a_i:

$$a_i = \frac{b_i RT(\varepsilon - \sigma)\left(\ln\dfrac{V_i^l - b_i}{V_i^v - b_i} + Z_i^v - Z_i^l\right)}{\ln\dfrac{(V_i^l + \sigma b_i)(V_i^v + \varepsilon b_i)}{(V_i^l + \varepsilon b_i)(V_i^v + \sigma b_i)}} \qquad (4\text{-}314)$$

where $Z_i^v = P_i^{sat} V_i^v/RT$ and $Z_i^l = P_i^{sat} V_i^l/RT$. Values of V_i^v and V_i^l come from solution of Eq. (4-230) for each phase with $P = P_i^{sat}$ at temperature T. Since a value of a_i is *required* for these calculations, an iterative procedure is implemented with an initial value for a_i from the appropriate correlation for $\alpha(T_{ri}; \omega_i)$.

The binary interaction parameters k_{pq} are evaluated from liquid-phase G^E correlations for binary systems. The most satisfactory procedure is to apply at infinite dilution the relation between a liquid-phase activity coefficient and its underlying fugacity coefficients, $\gamma_i^\infty = \hat{\phi}_i^\infty/\phi_i$. Rearrangement of the logarithmic form yields

$$\ln \hat{\phi}_i^\infty = \ln \gamma_i^\infty + \ln \phi_i \qquad (4\text{-}315)$$

where $\ln \gamma_i^\infty$ comes from the G^E correlation and $\ln \phi_i$ is given by Eq. (4-313) written for the liquid phase. Equation (4-315) supplies a value for $\ln \hat{\phi}_i^\infty$ which, when used with Eq. (4-232), ultimately (see following) leads to values for k_{pq}.

For a binary system comprised of species p and q, Eqs. (4-232), (4-312), and (4-315) may be written for species p at infinite dilution. The three resulting equations are then combined to yield

$$\frac{\bar{b}_p^\infty}{b_q} = \frac{\ln \gamma_p^\infty + \ln \phi_p - M_p}{Z_q - 1} \qquad (4\text{-}316)$$

where

$$M_p \equiv -\ln\frac{(V_q - b_q)Z_q}{V_q} + \frac{1}{\varepsilon - \sigma}\left(\frac{a_p}{b_p RT} - \frac{\ln \gamma_p^\infty}{c}\right)\ln\frac{V_q + \sigma b_q}{V_q + \varepsilon b_q} \qquad (4\text{-}317)$$

By Eq. (4-311) written for species p at infinite dilution in a pq binary,

$$\frac{\bar{b}_p^\infty}{b_q} = \frac{\dfrac{2E_{pq}}{b_q} - 1 - \dfrac{\ln \gamma_p^\infty}{c} + \dfrac{a_p}{b_p RT}}{1 - \dfrac{a_q}{b_q RT}} \qquad (4\text{-}318)$$

Equations (4-316) and (4-318) are set equal, E_{pq} is eliminated by Eq. (4-306), and k_{pq} is replaced by k_p, its infinite-dilution value at $x_p \to 0$. Solution for k_p then yields

$$k_p = 1 - \frac{\left(b_q - \dfrac{a_q}{RT}\right)\left(\dfrac{\ln \gamma_p^\infty + \ln \phi_p - M_p}{Z_q - 1}\right) + b_q\left(1 + \dfrac{\ln \gamma_p^\infty}{c} - \dfrac{a_p}{b_p RT}\right)}{b_p - \dfrac{a_p}{RT} + b_q - \dfrac{a_q}{RT}} \qquad (4\text{-}319)$$

where $\ln \phi_p$ comes from Eq. (4-313). All values in Eq. (4-319) are for the liquid phase at $P = P_q^{sat}$. The analogous equation for k_q, the infinite-dilution value of k_{pq} at $x_q \to 0$ is written

$$k_q = 1 - \frac{\left(b_p - \dfrac{a_p}{RT}\right)\left(\dfrac{\ln \gamma_q^\infty + \ln \phi_q - M_q}{Z_p - 1}\right) + b_p\left(1 + \dfrac{\ln \gamma_q^\infty}{c} - \dfrac{a_q}{b_q RT}\right)}{b_p - \dfrac{a_p}{RT} + b_q - \dfrac{a_q}{RT}} \qquad (4\text{-}320)$$

where M_q is given by an equation analogous to Eq. (4-317) but with subscripts reversed. All values in Eq. (4-320) are for the liquid phase at $P = P_p^{sat}$.

One advantage of this procedure is that k_p and k_q are found directly from the pure-species parameters a_p, a_q, b_p, and b_q. In addition, the required values of $\ln \gamma_p^\infty$ and $\ln \gamma_q^\infty$ can be found from experimental data for the pq binary system, independent of the correlating expression used for G^E.

A second advantage is that the procedure, applied for infinite dilution of each species, yields two values of k_{pq} from which a composition-dependent function can be generated, a simple linear relation proving fully satisfactory:

$$k_{pq} = k_p x_q + k_q x_p \qquad (4\text{-}321)$$

The two values k_p and k_q are usually not very different, and k_{pq} is not strongly composition dependent. Nevertheless, the quadratic dependence of $b - (a/RT)$ on composition indicated by Eq. (4-305) is not exactly preserved. Since this quantity is not a *true* second virial coefficient, only a value predicted by a cubic equation of state, a strict quadratic dependence is not required. Moreover, the composition-dependent k_{pq} leads to better results than does use of a constant value.

The equation-specific constants for the SRK and PR equations are given by the following table:

	SRK equation	PR equation
ε	0	-0.414214
σ	1	2.414214
Ω_a	0.42748	0.457235
Ω_b	0.08664	0.077796
c	0.69315	0.62323

Outlined below are the steps required for of a VLE calculation of vapor-phase composition and pressure, given the liquid-phase composition and temperature. A choice must be made of an equation of state. Only the Soave/Redlich/Kwong and Peng/Robinson equations, as represented by Eqs. (4-230) and (4-231), are considered here. These two equations usually give comparable results. A choice must also be made of a two-parameter correlating expression to represent the liquid-phase composition dependence of G^E for each pq binary. The Wilson, NRTL (with α fixed), and UNIQUAC equations are of general applicability; for binary systems, the Margules and van Laar equations may also be used. The equation selected depends on evidence of its suitability to the particular system treated. Reasonable estimates of the parameters in the equation must also be known at the temperature of interest. These parameters are directly related to infinite-dilution values of the activity coefficients for each pq binary.

Input information includes the known values of T and $\{x_i\}$, as well as the equation-of-state and G^E-expression parameters. Estimates are also needed of P and $\{y_i\}$, the quantities to be evaluated, and these require some preliminary calculations:

1. For the chosen equation of state (with appropriate values of ε, σ, and c), find values of b_i and preliminary values of a_i for each species from the information following Eq. (4-230).

2. If the vapor pressure P_i^{sat} for species i at temperature T is known, determine a new value for a_i by Eqs. (4-314) and (4-230).

3. Evaluate k_p and k_q by Eqs. (4-319) and (4-320) for each pq binary.

4. Although pressure P is to be determined, an estimate is required to permit any VLE calculations at all. A reasonable initial value is the sum of the pure-species vapor pressures, each weighted by its known liquid-phase mole fraction.

5. The vapor-phase composition is also to be determined, and it, too, is required to initiate calculations. Assuming both the liquid and vapor phases to be ideal solutions, Eqs. (4-98) and (4-304) combine to give

$$y_i = x_i \frac{\phi_i^l}{\phi_i^v}$$

Evaluation of the pure-species values ϕ_i^l and ϕ_i^v by Eq. (4-313) then provides values for y_i. Since these are not constrained to sum to unity, they should be normalized to yield an initial vapor-phase composition.

Given estimates for P and $\{y_i\}$ an iterative procedure can be initiated:

1. At the known *liquid-phase* composition, evaluate D by Eq. (4-308), b and a by Eqs. (4-309) and (4-310), and $\{\bar{b}_i\}$ and $\{\bar{a}_i\}$ by Eqs. (4-311) and (4-312).

2. Evaluate $\{\hat{\phi}_i^l\}$. The mixture volume V is determined from the equation of state, Eq. (4-231), applied to the liquid phase at the given composition, T, and P.

3. Repeat the two preceding items for the *vapor-phase* composition, thus evaluating $\{\hat{\phi}_i^v\}$.

4. Eq. (4-304) is now written

$$y_i = x_i \frac{\hat{\phi}_i^l}{\hat{\phi}_i^v}$$

The values of y_i so calculated are normalized by division by $\sum_i y_i$.

5. Recalculate the $\hat{\phi}_i^v$, and continue this iterative procedure until it converges to a fixed value for $\sum_i y_i$. This sum is appropriate to the pressure P for which the calculations have been made. Unless the sum is unity, the pressure is adjusted and the iteration process is repeated. Systematic adjustment of pressure P continues until $\sum_i y_i = 1$. The pressure and vapor compositions so found are the equilibrium values for the given temperature and liquid-phase composition as predicted by the equation of state.

A vast store of liquid-phase excess-property data for binary systems at temperatures near 30°C and somewhat higher is available in the literature. Effective use of these data to extend G^E correlations to higher temperatures is critical to the procedure considered here. The key relations are Eq. (4-118),

$$d\left(\frac{G^E}{RT}\right) = -\frac{H^E}{RT^2} \, dT \qquad \text{(constant } P,x)$$

and the excess-property analog of Eq. (4-31),

$$dH^E = C_P^E \, dT \qquad \text{(constant } P,x)$$

Integration of the first of these equations from T_0 to T gives

$$\frac{G^E}{RT} = \left(\frac{G^E}{RT}\right)_{T_0} - \int_{T_0}^{T} \frac{H^E}{RT^2} \, dT \qquad (4\text{-}322)$$

Similarly, the second equation may be integrated from T_1 to T:

$$H^E = H_1^E + \int_{T_1}^{T} C_P^E \, dT \qquad (4\text{-}323)$$

In addition, we may write

$$dC_P^E = \left(\frac{\partial C_P^E}{\partial T}\right)_{P,x} dT$$

Integration from T_2 to T yields

$$C_P^E = C_{P_2}^E + \int_{T_2}^{T} \left(\frac{\partial C_P^E}{\partial T}\right)_{P,x} dT$$

Combining this equation with Eqs. (4-322) and (4-323) leads to

$$\frac{G^E}{RT} = \left(\frac{G^E}{RT}\right)_{T_0} - \left(\frac{H^E}{RT}\right)_{T_1}\left(\frac{T_1}{T_0} - 1\right)\frac{T_1}{T}$$
$$- \frac{C_{P_2}^E}{R}\left[\ln\frac{T}{T_0} - \left(\frac{T}{T_0} - 1\right)\frac{T_1}{T}\right] - I \qquad (4\text{-}324)$$

where

$$I \equiv \int_{T_0}^{T} \frac{1}{RT^2} \int_{T_1}^{T} \int_{T_2}^{T} \left(\frac{\partial C_P^E}{\partial T}\right)_{P,x} dT \, dT \, dT$$

This general equation makes use of excess Gibbs-energy data at temperature T_0, excess enthalpy (heat-of-mixing) data at T_1, and excess heat-capacity data at T_2. Evaluation of the integral I requires information with respect to the temperature dependence of C_P^E. Because of the relative paucity of excess heat-capacity data, the most reasonable assumption is that this quantity is constant, independent of T. In this event, the integral is zero. and the closer T_0 and T_1 are to T, the less the influence of this assumption. When no information is available with respect to C_P^E, and excess enthalpy data are available at only a single temperature, the excess heat capacity must be assumed zero. In this case only the first two terms on the right-hand side of Eq. (4-324) are retained, and it more rapidly becomes imprecise as T increases.

Our primary interest in Eq. (4-324) is its application to binary systems at infinite dilution of one of the constituent species. For this pur-

pose, we divide Eq. (4-324) by the product $x_1 x_2$. For C_P^E independent of T (and thus with $I = 0$), it then becomes

$$\frac{G^E}{x_1 x_2 RT} = \left(\frac{G^E}{x_1 x_2 RT}\right)_{T_0} - \left(\frac{H^E}{x_1 x_2 RT}\right)_{T_1}\left(\frac{T_1}{T_0} - 1\right)\frac{T_1}{T}$$
$$- \frac{C_P^E}{x_1 x_2 R}\left[\ln\frac{T}{T_0} - \left(\frac{T}{T_0} - 1\right)\frac{T_1}{T}\right]$$

As shown by Smith, Van Ness and Abbott (*Introduction to Chemical Engineering Thermodynamics*, 5th ed., Chap. 11, McGraw-Hill, New York, 1996),

$$\left(\frac{G^E}{x_1 x_2 RT}\right)_{x_i = 0} \equiv \ln \gamma_i^\infty$$

The preceding equation may therefore be written

$$\ln \gamma_i^\infty = (\ln \gamma_i^\infty)_{T_0} - \left(\frac{H^E}{x_1 x_2 RT}\right)_{T_1, x_i = 0}\left(\frac{T_1}{T_0} - 1\right)\frac{T_1}{T}$$
$$- \left(\frac{C_P^E}{x_1 x_2 R}\right)_{x_i = 0}\left[\ln\frac{T}{T_0} - \left(\frac{T}{T_0} - 1\right)\frac{T_1}{T}\right] \qquad (4\text{-}325)$$

The methanol(1)/acetone(2) system serves as a specific example in conjunction with the Peng/Robinson equation of state. At a base temperature T_0 of 323.15 K (50°C), both VLE data (Van Ness and Abbott, *Int. DATA Ser., Ser. A, Sel. Data Mixtures,* **1978**, p. 67 [1978]) and excess enthalpy data (Morris, et al., *J. Chem. Eng. Data,* **20**, pp. 403–405 [1975]) are available. From the former,

$$(\ln \gamma_1^\infty)_{T_0} = 0.6281 \qquad \text{and} \qquad (\ln \gamma_2^\infty)_{T_0} = 0.6557$$

and from the latter

$$\left(\frac{H^E}{x_1 x_2 RT}\right)_{T_0, x_1 = 0} = 1.3636 \qquad \text{and} \qquad \left(\frac{H^E}{x_1 x_2 RT}\right)_{T_0, x_2 = 0} = 1.0362$$

The Margules equations (Eqs. [4-244], [4-245], and [4-246]) are well suited to this system, and the parameters for this equation are given as

$$A_{12} = \ln \gamma_1^\infty \qquad \text{and} \qquad A_{21} = \ln \gamma_2^\infty$$

This information allows prediction of VLE at 323.15 K and at the higher temperatures, 372.8, 397.7, and 422.6 K, for which measured VLE values are given by Wilsak, et al. (*Fluid Phase Equilibria,* **28**, pp. 13–37 [1986]). Values of $\ln \gamma_i^\infty$ and hence of the Margules parameters at the higher temperatures are given by Eq. (4-325) with $C_P^E = 0$. The pure-species vapor pressures in all cases are the measured values reported with the data sets. Results of these calculations are displayed in Table 4-1, where the parentheses enclose values from the gamma/phi approach as reported in the papers cited.

The results at 323.15 K (581.67 R) show both the suitability of the Margules equation for correlation of data for this system and the capability of the equation-of-state method to reproduce the data. Results for the three higher temperatures indicate the quality of predictions based only on vapor-pressure data for the pure species and on mixture data at 323.15 K (581.67 R). Extrapolations based on the same data to still higher temperatures can be expected to become progressively less accurate. When Eq. (4-325) can no longer be expected to produce reasonable values, better results are obtained for higher temperatures by assuming that the parameters, A_{12}, A_{21}, k_1, and k_2, do not change further at still-higher temperatures. This is also the course to be followed for extrapolation to supercritical temperatures.

Only the Wilson, NRTL, and UNIQUAC equations are suited to the treatment of multicomponent systems. For such systems, the parameters are determined for pairs of species exactly as for binary systems.

Examples treating the calculation of *VLE* are given in Smith, Van Ness, and Abbott (*Introduction to Chemical Engineering Thermodynamics,* 5th ed., Chap. 12, McGraw-Hill, New York, 1996).

LIQUID/LIQUID AND VAPOR/LIQUID/LIQUID EQUILIBRIA

Equation (4-273) is the basis for both liquid/liquid equilibria (LLE) and vapor/liquid/liquid equilibria (VLLE). Thus, for LLE with superscripts α and β denoting the two phases, Eq. (4-273) is written

TABLE 4-1 VLE Results for Methanol(1)/Acetone(2)

T, K	$\ln \gamma_1^\infty$	$\ln \gamma_2^\infty$	k_1	k_2	RMS δP, kPa	RMS % δP	RMS δy_1
323.15	0.6281	0.6557	0.1395	0.0955	0.08	0.12	
	(0.6281)	(0.6557)			(0.06)		
372.8	0.4465	0.5177	0.1432	0.1056	0.85	0.22	0.004
	(0.4607)	(0.5271)			(0.83)		(0.006)
397.7	0.3725	0.4615	0.1454	0.1118	2.46	0.32	0.014
	(0.3764)	(0.4640)			(1.39)		(0.013)
422.6	0.3072	0.4119	0.1480	0.1192	7.51	0.55	0.009
	(0.3079)	(0.3966)			(2.38)		(0.006)

$$\hat{f}_i^\alpha = \hat{f}_i^\beta \qquad (i = 1, 2, \ldots, N) \qquad (4\text{-}326)$$

Eliminating fugacities in favor of activity coefficients gives

$$x_i^\alpha \gamma_i^\alpha = x_i^\beta \gamma_i^\beta \qquad (i = 1, 2, \ldots, N) \qquad (4\text{-}327)$$

For most LLE applications, the effect of pressure on the γ_i can be ignored, and thus Eq. (4-327) constitutes a set of N equations relating equilibrium compositions to each other and to temperature. For a given temperature, solution of these equations requires a single expression for the composition dependence of G^E suitable for both liquid phases. Not all expressions for G^E suffice, even in principle, because some cannot represent liquid/liquid phase splitting. The UNIQUAC equation is suitable, and therefore prediction is possible by the UNIFAC method. A special table of parameters for LLE calculations is given by Magnussen, et al. (*Ind. Eng. Chem. Process Des. Dev.*, **20**, pp. 331–339 [1981]).

A comprehensive treatment of LLE is given by Sorensen, et al. (*Fluid Phase Equilibria*, **2**, pp. 297–309 [1979]; **3**, pp. 47–82 [1979]; **4**, pp. 151–163 [1980]). Data for LLE are collected in a three-part set compiled by Sorensen and Arlt (*Liquid-Liquid Equilibrium Data Collection*, Chemistry Data Series, vol. V, parts 1–3, DECHEMA, Frankfurt am Main, 1979–1980).

For vapor/liquid/liquid equilibria, Eq. (4-273) gives

$$\hat{f}_i^\alpha = \hat{f}_i^\beta = \hat{f}_i^v \qquad (i = 1, 2, \ldots, N) \qquad (4\text{-}328)$$

where α and β designate the two liquid phases. With activity coefficients applied to the liquid phases and fugacity coefficients to the vapor phase, the $2N$ equilibrium equations for subcritical VLLE are

$$\left.\begin{array}{l} x_i^\alpha \gamma_i^\alpha f_i^\alpha = y_i \hat{\phi}_i P \\ x_i^\beta \gamma_i^\beta f_i^\beta = y_i \hat{\phi}_i P \end{array}\right\} \qquad \text{(all } i) \qquad (4\text{-}329)$$

As for LLE, an expression for G^E capable of representing liquid/liquid phase splitting is required; as for VLE, a vapor-phase equation of state for computing the $\hat{\phi}_i$ is also needed.

CHEMICAL-REACTION STOICHIOMETRY

Consider a phase in which a chemical reaction occurs according to the equation

$$|\nu_1| A_1 + |\nu_2| A_2 + \cdots \rightarrow |\nu_3| A_3 + |\nu_4| A_4 + \cdots$$

where the $|\nu_i|$ are stoichiometric coefficients and the A_i stand for chemical formulas. The ν_i themselves are called *stoichiometric numbers,* and associated with them is a sign convention such that the value is positive for a product and negative for a reactant. More generally, for a system containing N chemical species, any or all of which can participate in r chemical reactions, the reactions can be represented by the equations:

$$0 = \sum_i \nu_{i,j} A_i \qquad (j = \text{I, II}, \ldots, r) \qquad (4\text{-}330)$$

where

$$\text{sign } (\nu_{i,j}) = \begin{cases} - \text{ for a reactant species} \\ + \text{ for a product species} \end{cases}$$

If species i does not participate in reaction j, then $\nu_{i,j} = 0$.

The stoichiometric numbers provide relations among the changes in mole numbers of chemical species which occur as the result of chemical reaction. Thus, for reaction j:

$$\frac{\Delta n_{1,j}}{\nu_{1,j}} = \frac{\Delta n_{2,j}}{\nu_{2,j}} = \cdots = \frac{\Delta n_{N,j}}{\nu_{N,j}} \qquad (4\text{-}331)$$

Since all of these terms are equal, they can be equated to the change in a single quantity ε_j, called the *reaction coordinate* for reaction j, thereby giving

$$\Delta n_{i,j} = \nu_{i,j} \Delta \varepsilon_j \qquad \begin{cases} i = 1, 2, \ldots, N \\ j = \text{I, II}, \ldots, r \end{cases} \qquad (4\text{-}332)$$

Since the total change in mole number Δn_i is just the sum of the changes $\Delta n_{i,j}$ resulting from the various reactions,

$$\Delta n_i = \sum_j \Delta n_{i,j} = \sum_j \nu_{i,j} \Delta \varepsilon_j \qquad (i = 1, 2, \ldots, N) \qquad (4\text{-}333)$$

If the initial number of moles of species i is n_{i^0} and if the convention is adopted that $\varepsilon_j = 0$ for each reaction in this initial state, then

$$n_i = n_{i_0} + \sum_j \nu_{i,j} \varepsilon_j \qquad (i = 1, 2, \ldots, N) \qquad (4\text{-}334)$$

Equation (4-334) is the basic expression of material balance for a closed system in which r chemical reactions occur. It shows for a reacting system that at most r mole number–related quantities ε_j are capable of independent variation. Note the absence of implied restrictions with respect to chemical-reaction equilibria; the reaction-coordinate formalism is merely an accounting scheme, valid for tracking the progress of each reaction to any arbitrary level of conversion. The reaction coordinate has units of moles. A change in ε_j of 1 mole signifies a *mole of reaction*, meaning that reaction j has proceeded to such an extent that the change in mole number of each reactant and product is equal to its stoichiometric number.

CHEMICAL-REACTION EQUILIBRIA

The general criterion of chemical-reaction equilibria is given by Eq. (4-274). For a system in which just a single reaction occurs, Eq. (4-334) becomes

$$n_i = n_{i_0} + \nu_i \varepsilon$$

whence

$$dn_i = \nu_i \, d\varepsilon$$

Substitution for dn_i in Eq. (4-274) leads to

$$\sum_i \nu_i \mu_i = 0 \qquad (4\text{-}335)$$

Generalization of this result to multiple reactions produces

$$\sum_i \nu_{i,j} \mu_i = 0 \qquad (j = \text{I, II}, \ldots, r) \qquad (4\text{-}336)$$

Standard Property Changes of Reaction A *standard* property change for the reaction

$$aA + bB \rightarrow lL + mM$$

is defined as the property change that occurs when a moles of A and b moles of B in their *standard states at temperature T* react to form l moles of L and m moles of M in their *standard states also at temperature T*. A *standard state* of species i is its real or hypothetical state as a pure species *at temperature T* and at a standard-state pressure P°. The standard property change of reaction j is given the symbol ΔM_j°, and its general mathematical definition is

$$\Delta M_j^\circ \equiv \sum_i \nu_{i,j} M_i^\circ \qquad (4\text{-}337)$$

For species present as gases in the actual reactive system, the standard state is the pure *ideal gas* at pressure $P°$. For liquids and solids, it is usually the state of pure real liquid or solid at $P°$. The standard-state pressure $P°$ is fixed at 100 kPa. Note that the standard states may represent different physical states for different species; any or all of the species may be gases, liquids, or solids.

The most commonly used standard property changes of reaction are

$$\Delta G_j° \equiv \sum_i v_{i,j} G_i° = \sum_i v_{i,j} \mu_i° \qquad (4\text{-}338)$$

$$\Delta H_j° \equiv \sum_i v_{i,j} H_i° \qquad (4\text{-}339)$$

$$\Delta C_{P_j}° \equiv \sum_i v_{i,j} C_{P_i}° \qquad (4\text{-}340)$$

The standard Gibbs-energy change of reaction $\Delta G_j°$ is used in the calculation of equilibrium compositions. The standard heat of reaction $\Delta H_j°$ is used in the calculation of the heat effects of chemical reaction, and the standard heat-capacity change of reaction is used for extrapolating $\Delta H_j°$ and $\Delta G_j°$ with T. Numerical values for $\Delta H_j°$ and $\Delta G_j°$ are computed from tabulated formation data, and $\Delta C_{P_j}°$ is determined from empirical expressions for the T dependence of the $C_{P_i}°$ (see, e.g., Eq. [4-142]).

Equilibrium Constants   For practical application, Eq. (4-336) must be reformulated. The initial step is elimination of the μ_i in favor of fugacities. Equation (4-74) for species i in its standard state is subtracted from Eq. (4-77) for species i in the equilibrium mixture, giving

$$\mu_i = G_i° + RT \ln \hat{a}_i \qquad (4\text{-}341)$$

where, by definition, $\hat{a}_i \equiv \hat{f}_i / f_i°$ and is called an *activity*. Substitution of this equation into Eq. (4-341) yields, upon rearrangement,

$$\sum_i [v_{i,j}(G_i° + RT \ln \hat{a}_i)] = 0$$

or

$$\sum_i (v_{i,j} G_i°) + RT \sum_i \ln \hat{a}_i^{v_{i,j}} = 0$$

or

$$\ln \prod_i \hat{a}_i^{v_{i,j}} = \frac{-\sum_i (v_{i,j} G_i°)}{RT}$$

The right-hand side of this equation is a function of temperature only for given reactions and given standard states. Convenience suggests setting it equal to $\ln K_j$; whence

$$\prod_i \hat{a}_i^{v_{i,j}} = K_j \qquad (\text{all } j) \qquad (4\text{-}342)$$

where

$$K_j \equiv \exp\left(\frac{-\Delta G_j°}{RT}\right) \qquad (4\text{-}343)$$

Quantity K_j is the chemical-reaction equilibrium constant for reaction j, and $\Delta G_j°$ is the corresponding standard Gibbs-energy change of reaction (see Eq. [4-338]). Although called a "constant," K_j is a function of T, but only of T.

The activities in Eq. (4-342) provide the connection between the *equilibrium* states of interest and the *standard* states of the constituent species, for which data are presumed available. The standard states are always at the equilibrium temperature. Although the standard state need not be the same for all species, for a *particular* species it must be the state represented by both $G_i°$ and the $f_i°$ upon which the activity $\hat{a}_i$ is based.

The application of Eq. (4-342) requires explicit introduction of composition variables. For gas-phase reactions this is accomplished through the fugacity coefficient:

$$\hat{a}_i \equiv \hat{f}_i / f_i° = y_i \hat{\phi}_i P / f_i°$$

However, the standard state for gases is the ideal gas state at the standard-state pressure, for which $f_i° = P°$. Therefore

$$\hat{a}_i = \frac{y_i \hat{\phi}_i P}{P°}$$

and Eq. (4-342) becomes

$$\prod_i (y_i \hat{\phi}_i)^{v_{i,j}} \left(\frac{P}{P°}\right)^{v_j} = K_j \qquad (\text{all } j) \qquad (4\text{-}344)$$

where $v_j \equiv \sum_i v_{i,j}$ and $P°$ is the standard-state pressure of 100 kPa, expressed in the same units used for P. The y_i may be eliminated in favor of equilibrium values of the reaction coordinates ε_j. Then, for fixed temperature Eqs. (4-344) relate the ε_j to P. In principle, specification of the pressure allows solution for the ε_j. However, the problem may be complicated by the dependence of the $\hat{\phi}_i$ on composition, that is, on the ε_j. If the equilibrium mixture is assumed an ideal solution, then each $\hat{\phi}_i$ becomes ϕ_i, the fugacity coefficient of pure species i at the mixture T and P. This quantity does not depend on composition and may be determined from experimental data, from a generalized correlation, or from an equation of state.

An important special case of Eq. (4-344) is obtained for gas-phase reactions when the phase can be assumed an ideal gas. In this event $\hat{\phi}_i = 1$, and

$$\prod_i (y_i)^{v_{i,j}} \left(\frac{P}{P°}\right)^{v_j} = K_j \qquad (\text{all } j) \qquad (4\text{-}345)$$

In the general case the evaluation of the $\hat{\phi}_i$ requires an iterative process. An initial step is to set the $\hat{\phi}_i$ equal to unity and to solve the problem by Eq. (4-345). This provides a set of y_i values, allowing evaluation of the $\hat{\phi}_i$ by, for example, Eq. (4-196), (4-200), or (4-231). Equation (4-344) can then be solved for a new set of y_i values, and the process continues to convergence.

For liquid-phase reactions, Eq. (4-342) is modified by introduction of the activity coefficient, $\gamma_i = \hat{f}_i / x_i f_i$, where x_i is the liquid-phase mole fraction. The activity is then

$$\hat{a}_i \equiv \frac{\hat{f}_i}{f_i°} = \gamma_i x_i \frac{f_i}{f_i°}$$

Both f_i and $f_i°$ represent fugacity of pure liquid i at temperature T, but at pressures P and $P°$, respectively. Except in the critical region, pressure has little effect on the properties of liquids, and the ratio $f_i/f_i°$ is often taken as unity. When this is not acceptable, this ratio is evaluated by the equation

$$\ln \frac{f_i}{f_i°} = \frac{1}{RT} \int_{P°}^{P} V_i \, dP \approx \frac{V_i(P - P°)}{RT}$$

When the ratio $f_i/f_i°$ is taken as unity, $\hat{a}_i = \gamma_i x_i$, and Eq. (4-342) becomes

$$\prod_i (\gamma_i x_i)^{v_{i,j}} = K_j \qquad (\text{all } j) \qquad (4\text{-}346)$$

Here the difficulty is to determine the γ_i, which depend on the x_i. This problem has not been solved for the general case. Two courses are open: the first is experiment; the second, assumption of solution ideality. In the latter case, $\gamma_i = 1$, and Eq. (4-346) reduces to

$$\prod_i (x_i)^{v_{i,j}} = K_j \qquad (\text{all } j) \qquad (4\text{-}347)$$

the *law of mass action*. The significant feature of Eqs. (4-345) and (4-347), the simplest expressions for gas- and liquid-phase reaction equilibrium, is that the temperature-, pressure-, and composition-dependent terms are distinct and separate.

Example 2: Single-Reaction Equilibrium   Consider the equilibrium state at 1,000 K and atmospheric pressure for the reaction

$$CO + H_2O \rightarrow CO_2 + H_2$$

Let the feed stream contain 3 mol CO, 1 mol H_2O, and 2 mol CO_2 for every mole of H_2 present. This initial constitution forms the *basis* for calculation, and for this single reaction, Eq. (4-334) becomes $n_i = n_{i_0} + v_i \varepsilon$. Whence

$$n_{CO} = 3 - \varepsilon$$
$$n_{H_2O} = 1 - \varepsilon$$
$$n_{CO_2} = 2 + \varepsilon$$
$$\underline{n_{H_2} = 1 + \varepsilon}$$
$$\sum n_i = 7$$

Each mole fraction is therefore given by $y_i = n_i/7$.

At 1,000 K, $\Delta G° = -2680$ J per mole of reaction; whence by Eq. (4-343)

$$K = \exp \frac{2680}{(8.314)(1000)} = 1.38$$

For the given conditions, the assumption of ideal gases is appropriate; Eq. (4-345) written for a single reaction (subscript j omitted) with $\nu = 0$ becomes

$$\prod_i y_i^{\nu_i} = \frac{\left(\dfrac{2+\varepsilon}{7}\right)\left(\dfrac{1+\varepsilon}{7}\right)}{\left(\dfrac{3-\varepsilon}{7}\right)\left(\dfrac{1-\varepsilon}{7}\right)} = K = 1.38$$

or

$$\frac{(2+\varepsilon)(1+\varepsilon)}{(3-\varepsilon)(1-\varepsilon)} = 1.38$$

whence

$$\varepsilon = 0.258$$

Thus, for the equilibrium mixture,

$n_{CO} = 2.74$ mol	$y_{CO} = 0.391$
$n_{H_2O} = 0.74$ mol	$y_{H_2O} = 0.106$
$n_{CO_2} = 2.26$ mol	$y_{CO_2} = 0.323$
$n_{H_2} = 1.26$ mol	$y_{H_2} = 0.180$
$\sum_i n_i = 7.00$ mol	$\sum_i y_i = 1.000$

The effect of temperature on the equilibrium constant follows from Eq. (4-106):

$$\frac{d(\Delta G_j°/RT)}{dT} = \frac{-\Delta H_j°}{RT^2} \qquad (4\text{-}348)$$

The total derivative is appropriate here because property changes of reaction are functions of temperature only. In combination with Eq. (4-343) this gives

$$\frac{d \ln K_j}{dT} = \frac{\Delta H_j°}{RT^2} \qquad (4\text{-}349)$$

For an endothermic reaction $\Delta H_j°$ is positive; for an exothermic reaction it is negative. The temperature dependence of $\Delta H_j°$ is given by

$$\frac{d\Delta H_j°}{dT} = \Delta C_{P_j}° \qquad (4\text{-}350)$$

Integration of Eq. (4-350) from reference temperature T_0 (usually 298.15 K) to temperature T gives

$$\Delta H° = \Delta H_0° + R \int_{T_0}^{T} \frac{\Delta C_P°}{R} dT \qquad (4\text{-}351)$$

where for simplicity subscript j has been suppressed. A convenient integrated form of Eq. (4-349) is

$$\ln K = \frac{-\Delta G°}{RT} = \frac{\Delta H_0° - \Delta G_0°}{RT} - \frac{\Delta H°}{RT} + \frac{1}{T}\int_{T_0}^{T} \frac{C_P°}{R} dT \qquad (4\text{-}352)$$

where $\Delta H°/RT$ is given by Eq. (4-351).

In the more extensive compilations of data, values of $\Delta G°$ and $\Delta H°$ for formation reactions are given for a wide range of temperatures, rather than just at the reference temperature of 298.15 K. (See in particular *TRC Thermodynamic Tables—Hydrocarbons* and *TRC Thermodynamic Tables—Non-hydrocarbons*, serial publications of the Thermodynamics Research Center, Texas A & M University System, College Station, Tex.; "The NBS Tables of Chemical Thermodynamic Properties," *J. Physical and Chemical Reference Data*, **11**, supp. 2 [1982]. Where data are lacking, methods of estimation are available; these are reviewed by Reid, Prausnitz, and Poling, *The Properties of Gases and Liquids*, 4th ed., Chap. 6, McGraw-Hill, New York, 1987. For an estimation procedure based on molecular structure, see Constantinou and Gani, *Fluid Phase Equilibria*, **103**, pp. 11–22 [1995]. (See also Sec. 2.)

Complex Chemical-Reaction Equilibria When the composition of an equilibrium mixture is determined by a number of simultaneous reactions, calculations based on equilibrium constants become complex and tedious. A more direct procedure (and one suitable for general computer solution) is based on minimization of the total Gibbs energy G^t in accord with Eq. (4-271). The treatment here is limited to gas-phase reactions for which the problem is to find the equilibrium composition for given T and P and for a given initial feed.

1. Formulate the constraining material-balance equations, based on conservation of the total number of atoms of each *element* in a system comprised of w elements. Let subscript k identify a particular atom, and define A_k as the total number of atomic masses of the kth element in the feed. Further, let a_{ik} be the number of atoms of the kth element present in each molecule of chemical species i. The material balance for element k is then

$$\sum_i n_i a_{ik} = A_k \qquad (k = 1, 2, \ldots, w) \qquad (4\text{-}353)$$

or

$$\sum_i n_i a_{ik} - A_k = 0 \qquad (k = 1, 2, \ldots, w)$$

2. Multiply each element balance by λ_k, a Lagrange multiplier:

$$\lambda_k \left(\sum_i n_i a_{ik} - A_k \right) = 0 \qquad (k = 1, 2, \ldots, w)$$

Summed over k, these equations give

$$\sum_k \lambda_k \left(\sum_i n_i a_{ik} - A_k \right) = 0$$

3. Form a function F by addition of this sum to G^t:

$$F = G^t + \sum_k \lambda_k \left(\sum_i n_i a_{ik} - A_k \right)$$

Function F is identical with G^t, because the summation term is zero. However, the partial derivatives of F and G^t with respect to n_i are different, because function F incorporates the constraints of the material balances.

4. The minimum value of both F and G^t is found when the partial derivatives of F with respect to n_i are set equal to zero:

$$\left(\frac{\partial F}{\partial n_i} \right)_{T,P,n_j} = \left(\frac{\partial G^t}{\partial n_i} \right)_{T,P,n_j} + \sum_k \lambda_k a_{ik} = 0$$

The first term on the right is the definition of the chemical potential; whence

$$\mu_i + \sum_k \lambda_k a_{ik} = 0 \qquad (i = 1, 2, \ldots, N) \qquad (4\text{-}354)$$

However, the chemical potential is given by Eq. (4-341); for gas-phase reactions and standard states as the pure ideal gases at $P°$, this equation becomes

$$\mu_i = G_i° + RT \ln \frac{\hat{f}_i}{P°}$$

If $G_i°$ is arbitrarily set equal to zero for all *elements* in their standard states, then for compounds $G_i° = \Delta G_{f_i}°$, the standard Gibbs-energy change of formation for species i. In addition, the fugacity is eliminated in favor of the fugacity coefficient by Eq. (4-79), $\hat{f}_i = y_i \hat{\phi}_i P$. With these substitutions, the equation for μ_i becomes

$$\mu_i = \Delta G_{f_i}° + RT \ln \frac{y_i \hat{\phi}_i P}{P°}$$

Combination with Eq. (4-354) gives

$$\Delta G_{f_i}° + RT \ln \frac{y_i \hat{\phi}_i P}{P°} + \sum_k \lambda_k a_{ik} = 0 \qquad (i = 1, 2, \ldots, N) \qquad (4\text{-}355)$$

If species i is an element, $\Delta G_{f_i}°$ is zero. There are N equilibrium equations (Eqs. [4-355]), one for each chemical species, and there are w material-balance equations (Eqs. [4-353]), one for each element—a total of $N + w$ equations. The unknowns in these equations are the n_i (note that $y_i = n_i / \sum_i n_i$), of which there are N, and the λ_k, of which there are w—a total of $N + w$ unknowns. Thus, the number of equations is sufficient for the determination of all unknowns.

Equation (4-355) is derived on the presumption that the $\hat{\phi}_i$ are known. If the phase is an ideal gas, then each $\hat{\phi}_i$ is unity. If the phase is an ideal solution, each $\hat{\phi}_i$ becomes ϕ_i, and can at least be estimated. For real gases, each $\hat{\phi}_i$ is a function of the y_i, the quantities being calculated. Thus an iterative procedure is indicated, initiated with each $\hat{\phi}_i$

set equal to unity. Solution of the equations then provides a preliminary set of y_i. For low pressures or high temperatures this result is usually adequate. Where it is not satisfactory, an equation of state with the preliminary y_i gives a new and more nearly correct set of $\hat{\phi}_i$ for use in Eq. (4-355). Then a new set of y_i is determined. The process is repeated to convergence. All calculations are well suited to computer solution.

In this procedure, the question of what chemical reactions are involved never enters directly into any of the equations. However, the choice of a set of species is entirely equivalent to the choice of a set of independent reactions among the species. In any event, a set of species or an equivalent set of independent reactions must always be assumed, and different assumptions produce different results.

Example 3: Minimization of Gibbs Energy Calculate the equilibrium compositions at 1,000 K and 1 bar of a gas-phase system containing the species CH_4, H_2O, CO, CO_2, and H_2. In the initial unreacted state there are present 2 mol of CH_4 and 3 mol of H_2O. Values of ΔG_f° at 1,000 K are

$$\Delta G_{f_{CH_4}}^\circ = 19{,}720 \text{ J/mol}$$
$$\Delta G_{f_{H_2O}}^\circ = -192{,}420 \text{ J/mol}$$
$$\Delta G_{f_{CO}}^\circ = -200{,}240 \text{ J/mol}$$
$$\Delta G_{f_{CO_2}}^\circ = -395{,}790 \text{ J/mol}$$

The required values of A_k are determined from the initial numbers of moles, and the values of a_{ik} come directly from the chemical formulas of the species. These are shown in the accompanying table.

	Element k		
	Carbon	Oxygen	Hydrogen
	A_k = no. of atomic masses of k in the system		
	$A_C = 2$	$A_O = 3$	$A_H = 14$
Species i	a_{ik} = no. of atoms of k per molecule of i		
CH_4	$a_{CH_4,C} = 1$	$a_{CH_4,O} = 0$	$a_{CH_4,H} = 4$
H_2O	$a_{H_2O,C} = 0$	$a_{H_2O,O} = 1$	$a_{H_2O,H} = 2$
CO	$a_{CO,C} = 1$	$a_{CO,O} = 1$	$a_{CO,H} = 0$
CO_2	$a_{CO_2,C} = 1$	$a_{CO_2,O} = 2$	$a_{CO_2,H} = 0$
H_2	$a_{H_2,C} = 0$	$a_{H_2,O} = 0$	$a_{H_2,H} = 2$

At 1 bar and 1,000 K the assumption of ideal gases is justified, and the $\hat{\phi}_i$ are all unity. Since $P = 1$ bar, Eq. (4-355) is written:

$$\frac{\Delta G_{f_i}^\circ}{RT} + \ln \frac{n_i}{\Sigma_i n_i} + \sum_k \frac{\lambda_k}{RT} a_{ik} = 0$$

The five equations for the five species then become:

$$CH_4: \quad \frac{19{,}720}{RT} + \ln \frac{n_{CH_4}}{\Sigma_i n_i} + \frac{\lambda_C}{RT} + \frac{4\lambda_H}{RT} = 0$$

$$H_2O: \quad \frac{-192{,}420}{RT} + \ln \frac{n_{H_2O}}{\Sigma_i n_i} + \frac{2\lambda_H}{RT} + \frac{\lambda_O}{RT} = 0$$

$$CO: \quad \frac{-200{,}240}{RT} + \ln \frac{n_{CO}}{\Sigma_i n_i} + \frac{\lambda_C}{RT} + \frac{\lambda_O}{RT} = 0$$

$$CO_2: \quad \frac{-395{,}790}{RT} + \ln \frac{n_{CO_2}}{\Sigma_i n_i} + \frac{\lambda_C}{RT} + \frac{2\lambda_O}{RT} = 0$$

$$H_2: \quad \ln \frac{n_{H_2}}{\Sigma_i n_i} + \frac{2\lambda_H}{RT} = 0$$

The three material-balance equations (Eq. [4-353]) are:

$$C: \quad n_{CH_4} + n_{CO} + n_{CO_2} = 2$$
$$H: \quad 4n_{CH_4} + 2n_{H_2O} + 2n_{H_2} = 14$$
$$O: \quad n_{H_2O} + n_{CO} + 2n_{CO_2} = 3$$

Simultaneous computer solution of these eight equations, with $RT = 8{,}314$ J/mol and

$$\sum_i n_i = n_{CH_4} + n_{H_2O} + n_{CO} + n_{CO_2} + n_{H_2}$$

produces the following results ($y_i = n_i / \Sigma_i n_i$):

$$y_{CH_4} = 0.0196 \qquad \frac{\lambda_C}{RT} = 0.7635$$
$$y_{H_2O} = 0.0980$$
$$y_{CO} = 0.1743 \qquad \frac{\lambda_O}{RT} = 25.068$$
$$y_{CO_2} = 0.0371$$
$$y_{H_2} = 0.6711 \qquad \frac{\lambda_H}{RT} = 0.1994$$
$$\sum_i y_i = 1.000$$

The values of λ_k/RT are of no significance, but are included to make the results complete.

THERMODYNAMIC ANALYSIS OF PROCESSES

Real irreversible processes can be subjected to thermodynamic analysis. The goal is to calculate the efficiency of energy use or production and to show how energy loss is apportioned among the steps of a process. The treatment here is limited to steady-state, steady-flow processes, because of their predominance in chemical technology.

CALCULATION OF IDEAL WORK

In any steady-state, steady-flow process *requiring* work, a minimum amount must be expended to bring about a specific change of state in the flowing fluid. In a process *producing* work, a maximum amount is attainable for a specific change of state in the flowing fluid. In either case, the limiting value obtains when the specific change of state is accomplished *completely reversibly*. The implications of this requirement are:

1. The process is internally reversible within the control volume.
2. Heat transfer external to the control volume is reversible.

The second item means that heat exchange between system and surroundings must occur at the temperature of the surroundings, presumed to constitute a heat reservoir at a constant and uniform temperature

T_σ. This may require Carnot engines or heat pumps internal to the system that provide for the reversible transfer of heat from the temperature of the flowing fluid to that of the surroundings. Since Carnot engines and heat pumps are cyclic, they undergo no net change of state.

The entropy change of the surroundings, found by integration of Eq. (4-3), is $\Delta S_\sigma = Q_\sigma/T_\sigma$; whence

$$Q_\sigma = T_\sigma \Delta S_\sigma \tag{4-356}$$

Since heat transfer with respect to the surroundings and with respect to the system are equal but of opposite sign, $Q_\sigma = -Q$. Moreover, the second law requires for a reversible process that the entropy changes of system and surroundings be equal but of opposite sign: $\Delta S_\sigma = -\Delta S^t$. Equation (4-356) can therefore be written $Q = T_\sigma \Delta S^t$. In terms of rates this becomes

$$\dot{Q} = T_\sigma \Delta (S\dot{m})_{fs} \tag{4-357}$$

where $\dot{Q}$ = rate of heat transfer with respect to the system
$\dot{m}$ = mass rate of flow of fluid

In addition, Δ denotes the difference between exit and entrance streams, and fs indicates that the term applies to all flowing streams.

The energy balance for a steady-state steady-flow process resulting from the first law of thermodynamics is

$$\Delta\left[\left(H + \frac{1}{2}u^2 + zg\right)\dot{m}\right]_{fs} = \dot{Q} + \dot{W}_s \tag{4-358}$$

where H = specific enthalpy of flowing fluid
u = velocity of flowing fluid
z = elevation of flowing fluid above datum level
g = local acceleration of gravity
W_s = shaft work

Eliminating $\dot{Q}$ in Eq. (4-358) by Eq. (4-357) gives

$$\Delta\left[\left(H + \frac{1}{2}u^2 + zg\right)\dot{m}\right]_{fs} = T_\sigma\Delta(S\dot{m})_{fs} + \dot{W}_s(\text{rev})$$

where $\dot{W}_s(\text{rev})$ indicates that the shaft work is for a completely reversible process. This work is called the *ideal work* $\dot{W}_{\text{ideal}}$. Thus

$$\dot{W}_{\text{ideal}} = \Delta\left[\left(H + \frac{1}{2}u^2 + zg\right)\dot{m}\right]_{fs} - T_\sigma\Delta(S\dot{m})_{fs} \tag{4-359}$$

In most applications to chemical processes, the kinetic- and potential-energy terms are negligible compared with the others; in this event Eq. (4-359) is written

$$\dot{W}_{\text{ideal}} = \Delta(H\dot{m})_{fs} - T_\sigma\Delta(S\dot{m})_{fs} \tag{4-360}$$

For the special case of a single stream flowing through the system, Eq. (4-360) becomes

$$\dot{W}_{\text{ideal}} = \dot{m}(\Delta H - T_\sigma\Delta S) \tag{4-361}$$

Division by $\dot{m}$ puts this equation on a unit-mass basis

$$W_{\text{ideal}} = \Delta H - T_\sigma\Delta S \tag{4-362}$$

A completely reversible processes is hypothetical, devised solely to find the ideal work associated with a given change of state. Its only connection with an actual process is that it brings about the same change of state as the actual process, allowing comparison of the actual work of a process with the work of the hypothetical reversible process.

Equations (4-359) through (4-362) give the work of a completely reversible process associated with given property changes in the flowing streams. When the same property changes occur in an actual process, the actual work $\dot{W}_s$ (or W_s) is given by an energy balance, and comparison can be made of the actual work with the ideal work. When $\dot{W}_{\text{ideal}}$ (or W_{ideal}) is positive, it is the *minimum work required* to bring about a given change in the properties of the flowing streams, and is smaller than $\dot{W}_s$. In this case a thermodynamic efficiency η_t is defined as the ratio of the ideal work to the actual work:

$$\eta_t(\text{work required}) = \frac{\dot{W}_{\text{ideal}}}{\dot{W}_s} \tag{4-363}$$

When $\dot{W}_{\text{ideal}}$ (or W_{ideal}) is negative, $|\dot{W}_{\text{ideal}}|$ is the *maximum work obtainable* from a given change in the properties of the flowing streams, and is larger than $|\dot{W}_s|$. In this case, the thermodynamic efficiency is defined as the ratio of the actual work to the ideal work:

$$\eta_t(\text{work produced}) = \frac{\dot{W}_s}{\dot{W}_{\text{ideal}}} \tag{4-364}$$

LOST WORK

Work that is wasted as the result of irreversibilities in a process is called *lost work* $\dot{W}_{\text{lost}}$, and is defined as the difference between the actual work of a process and the ideal work for the process. Thus, by definition,

$$W_{\text{lost}} \equiv W_s - W_{\text{ideal}} \tag{4-365}$$

In terms of rates this is written

$$\dot{W}_{\text{lost}} \equiv \dot{W}_s - \dot{W}_{\text{ideal}} \tag{4-366}$$

The actual work rate comes from Eq. (4-358)

$$\dot{W}_s = \Delta\left[\left(H + \frac{1}{2}u^2 + zg\right)\dot{m}\right]_{fs} - \dot{Q}$$

Subtracting the ideal work rate as given by Eq. (4-359) yields

$$\dot{W}_{\text{lost}} = T_\sigma\Delta(S\dot{m})_{fs} - \dot{Q} \tag{4-367}$$

For the special case of a single stream flowing through the control volume,

$$\dot{W}_{\text{lost}} = \dot{m}T_\sigma\Delta S - \dot{Q} \tag{4-368}$$

Division of this equation by $\dot{m}$ gives

$$W_{\text{lost}} = T_\sigma\Delta S - Q \tag{4-369}$$

where the basis is now a unit amount of fluid flowing through the control volume.

The total rate of entropy increase (in both system and surroundings) as a result of a process is

$$\dot{S}_{\text{total}} = \Delta(S\dot{m})_{fs} - \frac{\dot{Q}}{T_\sigma} \tag{4-370}$$

For a single stream, division by $\dot{m}$ provides an equation based on a unit amount of fluid flowing through the control volume:

$$S_{\text{total}} = \Delta S - \frac{Q}{T_\sigma} \tag{4-371}$$

Multiplication of Eq. (4-370) by T_σ gives

$$T_\sigma\dot{S}_{\text{total}} = T_\sigma\Delta(S\dot{m})_{fs} - \dot{Q}$$

Since the right-hand sides of this equation and of Eq. (4-367) are identical, it follows that

$$\dot{W}_{\text{lost}} = T_\sigma\dot{S}_{\text{total}} \tag{4-372}$$

For flow of a single stream on the basis of a unit amount of fluid, this becomes

$$W_{\text{lost}} = T_\sigma S_{\text{total}} \tag{4-373}$$

Since the second law of thermodynamics requires that

$$\dot{S}_{\text{total}} \geq 0 \qquad \text{and} \qquad S_{\text{total}} \geq 0$$

it follows that

$$\dot{W}_{\text{lost}} \geq 0 \qquad \text{and} \qquad W_{\text{lost}} \geq 0$$

When a process is completely reversible, the equality holds, and the lost work is zero. For irreversible processes the inequality holds, and the lost work, that is, the energy that becomes unavailable for work, is positive. The engineering significance of this result is clear: The greater the irreversibility of a process, the greater the rate of entropy production and the greater the amount of energy that becomes unavailable for work. Thus, every irreversibility carries with it a price.

ANALYSIS OF STEADY-STATE, STEADY-FLOW PROCESSES

Many processes consist of a number of steps, and lost-work calculations are then made for each step separately. Writing Eq. (4-372) for each step of the process and summing gives

$$\sum \dot{W}_{\text{lost}} = T_\sigma \sum \dot{S}_{\text{total}}$$

Dividing Eq. (4-372) by this result yields

$$\frac{\dot{W}_{\text{lost}}}{\sum \dot{W}_{\text{lost}}} = \frac{\dot{S}_{\text{total}}}{\sum \dot{S}_{\text{total}}}$$

Thus, an analysis of the lost work, made by calculation of the fraction that each individual lost-work term represents of the total lost work, is the same as an analysis of the rate of entropy generation, made by expressing each individual entropy-generation term as a fraction of the sum of all entropy-generation terms.

An alternative to the lost-work or entropy-generation analysis is a work analysis. This is based on Eq. (4-366), written

$$\sum \dot{W}_{\text{lost}} = \dot{W}_s - \dot{W}_{\text{ideal}} \tag{4-374}$$

For a work-requiring process, all of these work quantities are positive and $\dot{W}_s > \dot{W}_{\text{ideal}}$. The preceding equation is then expresed as

$$\dot{W}_s = \dot{W}_{\text{ideal}} + \sum \dot{W}_{\text{lost}} \tag{4-375}$$

A work analysis gives each of the individual work terms on the right as a fraction of $\dot{W}_s$.

TABLE 4-2 States and Values of Properties for the Process of Fig. 4-12*

Point	P, bar	T, K	Composition	State	H, J/mol	S, J/(mol · K)
1	55.22	300	Air	Superheated	12,046	82.98
2	1.01	295	Pure O_2	Superheated	13,460	118.48
3	1.01	295	91.48% N_2	Superheated	12,074	114.34
4	55.22	147.2	Air	Superheated	5,850	52.08
5	1.01	79.4	91.48% N_2	Saturated vapor	5,773	75.82
6	1.01	90	pure O_2	Saturated vapor	7,485	83.69
7	1.01	300	Air	Superheated	12,407	117.35

*Properties on the basis of Miller and Sullivan, U.S. Bur. Mines Tech. Pap. 424 (1928).

For a work-producing process, $\dot{W}_s$ and $\dot{W}_{ideal}$ are negative, and $|\dot{W}_{ideal}| > |\dot{W}_s|$. Equation (4-374) in this case is best written:

$$|\dot{W}_{ideal}| = |\dot{W}_s| + \sum \dot{W}_{lost} \qquad (4\text{-}376)$$

A work analysis here expresses each of the individual work terms on the right as a fraction of $|\dot{W}_{ideal}|$. A work analysis cannot be carried out in the case where a process is so inefficient that $\dot{W}_{ideal}$ is negative, indicating that the process should produce work, but $\dot{W}_s$ is positive, indicating that the process in fact requires work. A lost-work or entropy-generation analysis is always possible.

Example 4: Lost-Work Analysis Make a work analysis of a simple Linde system for the separation of air into gaseous oxygen and nitrogen, as depicted in Fig. 4-12. Table 4-2 lists a set of operating conditions for the numbered points of the diagram. Heat leaks into the column of 147 J/mol of entering air and into the exchanger of 70 J/mol of entering air have been assumed. Take $T_\sigma = 300$ K.

The basis for analysis is 1 mol of entering air, assumed to contain 79 mol % N_2 and 21 mol % O_2. By a material balance on the nitrogen, $0.79 = 0.9148\,x$; whence

$$x = 0.8636 \text{ mol of nitrogen product}$$
$$1 - x = 0.1364 \text{ mol of oxygen product}$$

Calculation of Ideal Work If changes in kinetic and potential energies are neglected, Eq. (4-360) is applicable. From the tabulated data,

$$\Delta(H\dot{m})_{fs} = (13,460)(0.1364) + (12,074)(0.8636) - (12,407)(1) = -144 \text{ J}$$
$$\Delta(S\dot{m})_{fs} = (118.48)(0.1364) - (114.34)(0.8636) - (117.35)(1) = -2.4453 \text{ J/K}$$

Thus, by Eq. (4-360),

$$\dot{W}_{ideal} = -144 - (300)(-2.4453) = 589.6 \text{ J}$$

Calculation of Actual Work of Compression For simplicity, the work of compression is calculated by the equation for an ideal gas in a three-stage reciprocating machine with complete intercooling and with isentropic compression in each stage. The work so calculated is assumed to represent 80 percent of the actual work. The following equation may be found in any number of textbooks on thermodynamics:

$$\dot{W}_s = \frac{n\gamma R T_1}{(0.8)(\gamma - 1)}\left[\left(\frac{P_2}{P_1}\right)^{(\gamma - 1)/n\gamma} - 1\right]$$

where n = number of stages, here taken as 3
  γ = ratio of heat capacities, here taken as 1.4
  T_1 = initial absolute temperature, 300 K
  P_2/P_1 = overall pressure ratio, 54.5
  R = universal gas constant, 8.314 J/(mol·K)

The efficiency factor of 0.8 is already included in the equation. Substitution of the remaining values gives

$$\dot{W}_s = \frac{(3)(1.4)(8.314)(300)}{(0.8)(0.4)}\left[(54.5)^{0.4/(3)(1.4)} - 1\right] = 15,171 \text{ J}$$

The heat transferred to the surroundings during compression as a result of intercooling and aftercooling to 300 K is found from the first law:

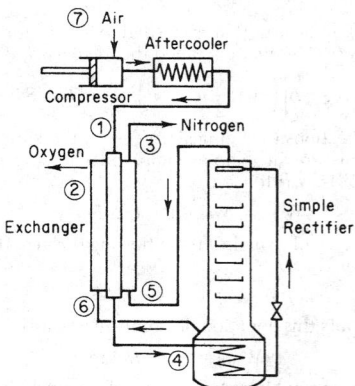

FIG. 4-12 Diagram of simple Linde system for air separation.

$$\dot{Q} = \dot{m}(\Delta H) - \dot{W}_s = (12,046 - 12,407) - 15,171 = -15,532 \text{ J}$$

Calculation of Lost Work Equation (4-367) may be applied to each of the major units of the process. For the compressor/cooler,

$$\dot{W}_{lost} = (300)[(82.98)(1) - (117.35)(1)] - (-15,532)$$
$$= 5,221.0 \text{ J}$$

For the exchanger,

$$\dot{W}_{lost} = (300)[(118.48)(0.1364) + (114.34)(0.8636) + (52.08)(1)$$
$$- (75.82)(0.8636) - (83.69)(0.1364) - (82.98)(1)] - 70$$
$$= 2,063.4 \text{ J}$$

Finally, for the rectifier,

$$\dot{W}_{lost} = (300)[(75.82)(0.8636) + (83.69)(0.1364) - (52.08)(1)] - 147$$
$$= 7,297.0 \text{ J}$$

Work Analysis Since the process requires work, Eq. (4-375) is appropriate for a work analysis. The various terms of this equation appear as entries in the following table, and are on the basis of 1 mol of entering air.

			% of $\dot{W}_s$
$\dot{W}_{ideal}$		589.6 J	3.9
$\dot{W}_{lost}$:	Compressor/cooler	5,221.0 J	34.4
$\dot{W}_{lost}$:	Exchanger	2,063.4 J	13.6
$\dot{W}_{lost}$:	Rectifier	7,297.0 J	48.1
$\dot{W}_s$		15,171.0 J	100.0

The thermodynamic efficiency of this process as given by Eq. (4-363) is only 3.9 percent. Significant inefficiencies reside with each of the primary units of the process.

Heat and Mass Transfer*

James G. Knudsen, Ph.D., *Professor Emeritus of Chemical Engineering, Oregon State University; Member, American Institute of Chemical Engineers, American Chemical Society; Registered Professional Engineer (Oregon). (Conduction and Convection; Condensation, Boiling; Section Coeditor)*

Hoyt C. Hottel, S.M., *Professor Emeritus of Chemical Engineering, Massachusetts Institute of Technology; Member, National Academy of Sciences, American Academy of Arts and Sciences, American Institute of Chemical Engineers, American Chemical Society, Combustion Institute. (Radiation)*

Adel F. Sarofim, Sc.D., *Lammot du Pont Professor of Chemical Engineering and Assistant Director, Fuels Research Laboratory, Massachusetts Institute of Technology; Member, American Institute of Chemical Engineers, American Chemical Society, Combustion Institute. (Radiation)*

Phillip C. Wankat, Ph.D., *Professor of Chemical Engineering, Purdue University; Member, American Institute of Chemical Engineers, American Chemical Society, International Adsorption Society. (Mass Transfer Section Coeditor)*

Kent S. Knaebel, Ph.D., *President, Adsorption Research, Inc.; Member, American Institute of Chemical Engineers, American Chemical Society, International Adsorption Society. Professional Engineer (Ohio). (Mass Transfer Section Coeditor)*

* The contribution to the section on Interphase Mass Transfer of Mr. William M. Edwards (editor of Sec. 14), who was an author for the sixth edition, is acknowledged.

Nomenclature and Units

Specialized heat transfer nomenclature used for radiative heat transfer is defined in the subsection "Heat Transmission by Radiation." Nomenclature for mass transfer is defined in the subsection "Mass Transfer."

Symbol	Definition	SI units	U.S. customary units
a	Proportionality coefficient	Dimensionless	Dimensionless
a_x	Cross-sectional area of a fin	m^2	ft^2
a'	Proportionality factor		
A	Area of heat transfer surface; A_i for inside; A_o for outside; A_m for mean; A_{avg} for average; A_1, A_2, and A_3 for points 1, 2, and 3 respectively; A_B for bare surface of finned tube; A_f for finned portion of tube; A_{uf} for external area of unfinned portion of finned tube; A_{of} for external area of finned tube before fins are attached, equals A_o; A_{oe} for effective area of finned surface; A_T for total external area of finned tube; A_d for surface area of dirt (scale) deposit	m^2	ft^2
b	Proportionality coefficient		
b'	Proportionality factor		
b_f	Height of fin	m	ft
B	Material constant $= 5D^{-0.5}$		
c_1, c_2, etc.	Constants of integration		
c, c_p	Specific heat at constant pressure; c_s for specific heat of solid; c_g for specific heat of gas	$J/(kg \cdot K)$	$Btu/(lb \cdot °F)$
C	Thermal conductance, equals kA/x, hA, or UA; C_1, C_2, C_3, C_n, thermal conductance of sections 1, 2, 3, and n respectively of a composite body	$J/(s \cdot K)$	$Btu/(h \cdot °F)$
C_r	Correlating constant; proportionality coefficient	Dimensionless	Dimensionless
d_m	Depth of divided solids bed	m	ft
D	Diameter; D_o for outside; D_i for inside; D_r for root diameter of finned tube	m	ft
D_c	Diameter of a coil or helix	m	ft
D_e	Equivalent diameter of a cross section, usually 4 times free area divided by wetted perimeter; D_w for equivalent diameter of window	m	ft
D_j	Diameter of a jacketed cylindrical vessel	m	ft
D_{otl}	Outside diameter of tube bundle	m	ft
D_p	Diameter of packing in a packed tube	m	ft
D_s	Inside diameter of heat-exchanger shell	m	ft
D_t	Solids-processing vessel diameter	m	ft
D_1, D_2	Diameter at points 1 and 2 respectively; inner and outer diameter of annulus respectively	m	ft
E_H	Eddy conductivity of heat	$J/(s \cdot m \cdot K)$	$Btu/(h \cdot ft \cdot °F)$
E_M	Eddy viscosity	$Pa \cdot s$	$lb/(ft \cdot h)$
f	Fanning friction factor; f_1 for inner wall and f_2 for outer wall of annulus; f_k for ideal tube bank; skin friction drag coefficient	Dimensionless	Dimensionless
F	Entrance factors		
F_a	Dry solids feed rate	$kg/(s \cdot m^2)$	$lb/(h \cdot ft^2)$
F_g	Gas volumetric flow rate	m^3 ($s \cdot m^2$ of bed area)	$ft^3/(h \cdot ft^2$ of bed area)
F_c	Fraction of total tubes in cross-flow; F_{bp} for fraction of cross-flow area available for bypass flow		
F_t	Factor, ratio of temperature difference across tube-side film to overall mean temperature difference	Dimensionless	Dimensionless
F_s	Factor, ratio of temperature difference across shell-side film to overall mean temperature difference	Dimensionless	Dimensionless
F_w	Factor, ratio of temperature difference across retaining wall to overall mean temperature difference between bulk fluids	Dimensionless	Dimensionless
F_D	Factor, ratio of temperature difference across combined dirt or scale films to overall mean temperature difference between bulk fluids	Dimensionless	Dimensionless
F_T	Temperature-difference correction factor		
g, g_L	Acceleration due to gravity	981 m/s^2	$(4.18)(10^8)$ ft/h^2
g_c	Conversion factor	1.0 $(kg \cdot m)/(N \cdot s^2)$	$(4.17)(10^8)(lb \cdot ft)/(lbf \cdot h^2)$
G	Mass velocity, equals $V\rho$ or W/S; G_v for vapor mass velocity	$kg/(m^2 \cdot s)$	$lb/(h \cdot ft^2)$
G_{max}	Mass velocity through minimum free area between rows of tubes normal to the fluid stream	$kg/(m^2 \cdot s)$	$lb/(h \cdot ft^2)$
G_{mf}	Minimum fluidizing mass velocity	$kg/(m^2 \cdot s)$	$lb/(h \cdot ft^2)$
h	Local individual coefficient of heat transfer, equals $dq/(dA)(\Delta T)$	$J/(m^2 \cdot s \cdot K)$	$Btu/(h \cdot ft^2 \cdot °F)$
h_{am}, h_{lm}	Film coefficient based on arithmetic-mean temperature difference and logarithmic-mean temperature difference respectively	$J/(m^2 \cdot s \cdot K)$	$Btu/(h \cdot ft^2 \cdot °F)$
h_b	Film coefficient delivered at base of fin	$J/(m^2 \cdot s \cdot K)$	$Btu/(h \cdot ft^2 \cdot °F)$
h_{cg}	Effective combined coefficient for simultaneous gas-vapor cooling and vapor condensation	$J/(m^2 \cdot s \cdot K)$	$Btu/(h \cdot ft^2 \cdot °F)$
$h_c + h_r$	Combined coefficient for conduction, convection, and radiation between surface and surroundings	$J/(m^2 \cdot s \cdot K)$	$Btu/(h \cdot ft^2 \cdot °F)$

Nomenclature and Units (*Continued*)

Symbol	Definition	SI units	U.S. customary units
h_{do}, h_{di}	Film coefficient for dirt or scale on outside or inside respectively of a surface	$J/(m^2 \cdot s \cdot K)$	$Btu/(h \cdot ft^2 \cdot {}^\circ F)$
h_f	Film coefficient for finned-tube exchangers based on total external surface	$J/(m^2 \cdot s \cdot K)$	$Btu/(h \cdot ft^2 \cdot {}^\circ F)$
h_{fi}	Effective outside film coefficient of a finned tube based on inside area	$J/(m^2 \cdot s \cdot K)$	$Btu/(h \cdot ft^2 \cdot {}^\circ F)$
h_{fo}	Film coefficient for air film of an air-cooled finned-tube exchanger based on external bare surface	$J/(m^2 \cdot s \cdot K)$	$Btu/(h \cdot ft^2 \cdot {}^\circ F)$
h_F, h_x	Effective film coefficient for dirt or scale on heat-transfer surface	$J/(m^2 \cdot s \cdot K)$	$Btu/(h \cdot ft^2 \cdot {}^\circ F)$
h_i, h_o	Film coefficient for heat transfer for inside and outside surface respectively	$J/(m^2 \cdot s \cdot K)$	$Btu/(h \cdot ft^2 \cdot {}^\circ F)$
h_k	Film coefficient for ideal tube bank; h_s for shell side of baffled exchanger; h_w for coefficient at liquid-vapor interface	$J/(m^2 \cdot s \cdot K)$	$Btu/(h \cdot ft^2 \cdot {}^\circ F)$
h_1	Condensing coefficient on top tube; h_N coefficient for N tubes in a vertical row	$J/(m^2 \cdot s \cdot K)$	$Btu/(h \cdot ft^2 \cdot {}^\circ F)$
h'	Film coefficient for enclosed spaces	$J/(m^2 \cdot s \cdot K)$	$Btu/(h \cdot ft^2 \cdot {}^\circ F)$
h_{lm}	Film coefficient based on log-mean temperature difference	$J/(m^2 \cdot s \cdot K)$	$Btu/(h \cdot ft^2 \cdot {}^\circ F)$
h_r	Heat-transfer coefficient for radiation	$J/(m^2 \cdot s \cdot K)$	$Btu/(h \cdot ft^2 \cdot {}^\circ F)$
h_T	Coefficient of total heat transfer by conduction, convection, and radiation between the surroundings and the surface of a body subject to unsteady-state heat transfer	$J/(m^2 \cdot s \cdot K)$	$Btu/(h \cdot ft^2 \cdot {}^\circ F)$
h_w	Equivalent coefficient of retaining wall, equals k/x	$J/(m^2 \cdot s \cdot K)$	$Btu/(h \cdot ft^2 \cdot {}^\circ F)$
j	Ordinate, Colburn j factor, equals $f/2$; j_H for heat transfer; j_{H1} for inner wall of annulus; j_{H2} for outer wall of annulus; j_k for heat transfer for ideal tube bank	Dimensionless	Dimensionless
J	Mechanical equivalent of heat	$1.0(N \cdot m)/J$	$778(ft \cdot lbf)/Btu$
J_b, J_c, J_l, J_r	Correction factors for baffle bypassing, baffle configuration, baffle leakage, and adverse temperature gradient respectively		
k	Thermal conductivity; k_1, k_2, k_3, thermal conductivities of bodies 1, 2, and 3	$J/(m \cdot s \cdot K)$	$(Btu \cdot ft)/(h \cdot ft^2 \cdot {}^\circ F)$
k_v	Thermal conductivity of vapor; k_1 for liquid thermal conductivity; k_s for thermal conductivity of solid	$J/(m \cdot s \cdot K)$	$(Btu \cdot ft)/(h \cdot ft^2 \cdot {}^\circ F)$
k_{avg}, k_m	Mean thermal conductivity	$J/(m \cdot s \cdot K)$	$(Btu \cdot ft)/(h \cdot ft^2 \cdot {}^\circ F)$
k_f	Thermal conductivity of fluid at film temperature	$J/(m \cdot s \cdot K)$	$(Btu \cdot ft)/(h \cdot ft^2 \cdot {}^\circ F)$
k_w	Thermal conductivity of retaining-wall material	$J/(m \cdot s \cdot K)$	$(Btu \cdot ft)/(h \cdot ft^2 \cdot {}^\circ F)$
K'	Property of non-Newtonian fluid		
l_c	Baffle cut; l_s for baffle spacing	m	ft
L	Length of heat-transfer surface	m	ft
L_o	Flow rate	kg/s	lb/h
L_u	Undisturbed length of path of fluid flow	m	ft
L_F	Thickness of dirt or scale deposit	m	ft
L_H	Depth of fluidized bed	m	ft
L_p	Diameter of agitator blade	m	ft
m	Ratio, term, or exponent as defined where used		
M	Molecular weight	kg/mol	lb/mol
M	Weight of fluid	kg	lb
n	Position ratio or number	Dimensionless	Dimensionless
n_t	Number of tubes in parallel in a heat exchanger		
n_r	Number of rows in a vertical plane		
n'	Flow-behavior index for nonnewtonian fluids		
n_b	Number of baffle-type coils		
N_r	Speed of agitator	rad/s	r/h
N	Number of tubes in a vertical row; or number of tubes in a bundle; N_b for number of baffles; N_T for total number of tubes in exchanger; N_c for number of tubes in one cross-flow section; N_{cw} for number of cross-flow rows in each window		
N_B	Biot number, $h_T \Delta x/k$		
N_d	Proportionality coefficient, dimensionless group	Dimensionless	Dimensionless
N_{Gr}	Grashof number, $L^3 \rho^2 g \beta \Delta t/\mu^2$		
N_{Nu}	Nusselt number, hD/k or hL/k		
N_{Pe}	Peclet number, DGc/k		
N_{Pr}	Prandtl number, $c\mu/k$		
N_{Re}	Reynolds number, DG/μ		
N_{St}	Stanton number, $N_{Nu}/N_{Re}N_{Pr}$		
N_{ss}	Number of sealing strips		
p	Pressure	kPa	lbf/ft^2 abs
p_f	Perimeter of a fin	m	ft
p, p'	Center-to-center spacing of tubes in tube bundle (tube pitch); p_n for tube pitch normal to flow; p_p for tube pitch parallel to flow	m	ft

Nomenclature and Units (*Continued*)

Symbol	Definition	SI units	U.S. customary units
Δp	Pressure of the vapor in a bubble minus saturation pressure of a flat liquid surface	kPa	lbf/ft^2 abs
P	Absolute pressure; P_c for critical pressure	kPa	lbf/ft^2
P'	Spacing between adjacent baffles on shell side of a heat exchanger (baffle pitch)	m	ft
$\Delta P_{bk}, \Delta P_{wk}$	Pressure drop for ideal-tube-bank cross-flow and ideal window respectively; ΔP_s for shell side of baffled exchanger	kPa	lbf/ft^2
q	Rate of heat flow, equals Q/θ	W, J/s	Btu/h
q'	Rate of heat generation	J/(s·m^3)	Btu/(h·ft^3)
$(q/A)_{max}$	Maximum heat flux in nucleate boiling	J/(s·m^2)	Btu/(h·ft^2)
Q	Quantity of heat; rate of heat transfer	J/s	Btu/h
Q	Quantity of heat; Q_T for total quantity	J	Btu
r	Radius; cylindrical and spherical coordinate; distance from midplane to a point in a body; r_1 for inner wall of annulus; r_2 for outer wall of annulus; r_i for inside radius of tube; r_m for distance from midplane or center of a body to the exterior surface of the body	m	ft
r_j	Inside radius	Dimensionless	Dimensionless
R	Thermal resistance, equals x/kA, $1/UA$, $1/hA$; R_1, R_2, R_3, R_n for thermal resistance of sections 1, 2, 3, and n of a composite body; R_T for sum of individual resistances of several resistances in series or parallel; R_{di} and R_{do} for dirt or scale resistance on inner and outer surface respectively	(s·K)/J	(h·°F)/Btu
R_j	Ratio of total outside surface of finned tube to area of tube having same root diameter		
S	Cross-sectional area; S_m for minimum cross-sectional area between rows of tubes, flow normal to tubes; S_{tb} for tube-to-baffle leakage area for one baffle; S_{sb} for shell-to-baffle area for one baffle; S_w for area for flow through window; S_{wg} for gross window area; S_{wt} for window area occupied by tubes	m^2	ft^2
S_r	Slope of rotary shell		
s	Specific gravity of fluid referred to liquid water		
t	Bulk temperature; temperature at a given point in a body at time θ	K	°F
t_1, t_2, t_n	Temperature at points 1, 2, and n in a system through which heat is being transferred	K	°F
t'	Temperature of surroundings	K	°F
t_1', t_2'	Inlet and outlet temperature respectively of hotter fluid	K	°F
t_1'', t_2''	Inlet and outlet temperature respectively of colder fluid	K	°F
t_b	Initial uniform bulk temperature of a body; bulk temperature of a flowing fluid	K	°F
t_H, t_L	High and low temperature respectively on tube side of a heat exchanger	K	°F
t_s	Surface temperature	K	°F
t_{sv}	Saturated-vapor temperature	K	°F
t_w	Wall temperature	K	°F
t_∞	Temperature of undisturbed flowing stream	K	°F
T_H, T_L	High and low temperature respectively on shell side of a heat exchanger	K	°F
T	Absolute temperature; T_b for bulk temperature; T_w for wall temperature; T_v for vapor temperature; T_c for coolant temperature; T_e for temperature of emitter; T_r for temperature of receiver	K	°R
$\Delta T, \Delta t$	Temperature difference; Δt_1, Δt_2, and Δt_3 temperature difference across bodies 1, 2, and 3 or at points 1, 2, and 3; ΔT_o, Δt_o for overall temperature difference; Δt_b for temperature difference between surface and boiling liquid	K	°F, °R
$\Delta t_{am}, \Delta t_{lm}$	Arithmetic- and logarithmic-mean temperature difference respectively	K	°F
Δt_{om}	Mean effective overall temperature difference	K	°F
$\Delta T_H, \Delta t_H$	Greater terminal temperature difference	K	°F
$\Delta T_L, \Delta t_L$	Lesser terminal temperature difference	K	°F
$\Delta T_m, \Delta t_m$	Mean temperature difference	K	°F
u	Velocity in x direction	m/s	ft/h
$u°$	Friction velocity	m/s	ft/h
U	Overall coefficient of heat transfer; U_o for outside surface basis; U' for overall coefficient between liquid-vapor interface and coolant	J/(s·m^2·K)	Btu/(h·ft^2·°F)
U_1, U_2	Overall coefficient of heat transfer at points 1 and 2 respectively	J/(s·m^2·K)	Btu/(h·ft^2·°F)
$U_{co}, U_{cv}, U_{ct}, U_{ra}$	Overall coefficients for divided solids processing by conduction, convection, contact, and radiation mechanism respectively	J/(s·m^2·K)	Btu/(h·ft^2·°F)
U_m	Mean overall coefficient of heat transfer	J/(s·m^2·K)	Btu/(h·ft^2·°F)
v	Velocity in y direction	m/s	ft/h
V_r	Volume of rotating shell	m^3	ft^3

Nomenclature and Units (*Concluded*)

Symbol	Definition	SI units	U.S. customary units
V	Velocity	m/s	ft/h
V', V_s	Velocity	m/s	ft/s
V_F	Face velocity of a fluid approaching a bank of finned tubes	m/s	ft/h
V_g, V_l	Specific volume of gas, liquid	m³/kg	ft³/lb
V'_{max}	Maximum velocity through minimum free area between rows of tubes normal to the fluid stream	m/s	ft/h
w	Velocity in z direction	m/s	ft/h
w	Flow rate	kg/s	lb/h
W	Total mass rate of flow; mass rate of vapor generated; W_F for total rate of vapor condensation in one tube	kg/s	lb/h
W_r	Weight rate of flow	kg/(s·tube)	lb/(h·tube)
W_1, W_o	Total mass rate of flow on tube side and shell side respectively of a heat exchanger	kg/s	lb/h
x_q	Vapor quality, x_i for inlet quality, x_o for outlet quality	kg/s	lb/h
x	Coordinate direction; length of conduction path; x_s for thickness of scale; x_1, x_2, and x_3 at positions 1, 2, and 3 in a body through which heat is being transferred	m	ft
X	Factor	Dimensionless	Dimensionless
y	Coordinate direction	m	ft
y^+	Wall distance	Dimensionless	Dimensionless
Y	Factor	Dimensionless	Dimensionless
z	Coordinate direction	m	ft
z_p	Distance (perimeter) traveled by fluid across fin	m	ft
Z_H	Ratio of sensible heat removed from vapor to total heat transferred	Dimensionless	Dimensionless

<table>
<tr><td colspan="4" align="center">Greek symbols</td></tr>
</table>

Symbol	Definition	SI units	U.S. customary units
α	Thermal diffusivity, equals $k/\rho c$; α_e for effective thermal diffusivity of powdered solids	m²/s	ft²/h
β	Volumetric coefficient of thermal expansion	K⁻¹	°F⁻¹
β'	Contact angle of a bubble	°	°
γ	Fluid consistency	kg/(s^{2-n'}·m)	lb/(ft·s^{2-n'})
Γ	Mass rate of flow of a falling film from a tube or surface per unit perimeter, equals $w/\pi D$ for vertical tube, $w/2L$ for horizontal tube	kg/(s·m)	lb/(h·ft)
δ_s	Correction factor, ratio of nonnewtonian to newtonian shear rates		
δ	Cell width	m	ft
δ_{sb}	Diametral shell-to-baffle clearance	m	ft
ϵ	Eddy diffusivity; ϵ_M for eddy diffusivity of momentum; ϵ_H for eddy diffusivity of heat	m²/s	ft²/h
ϵ_v	Fraction of voids in porous bed		
η	Fluidization efficiency		
θ	Time	s	h
θ_b	Baffle cut		
λ	Latent heat (enthalpy) of vaporization (condensation)	J/kg	Btu/lb
λ_m	Radius of maximum velocity	m	ft
μ	Viscosity; μ_w for viscosity at wall temperature; μ_b for viscosity at bulk temperature; μ_f for viscosity at film temperature; μ_G, μ_g, and μ_v for viscosity of gas or vapor; μ_L, μ_l for viscosity of liquid; μ_w for viscosity at wall; μ_l for viscosity of fluid at inner wall of annulus	Pa·s	lb/(h·ft)
ν	Kinematic viscosity	m²/s	ft²/h
ρ	Density; ρ_L, ρ_l for density of liquid; ρ_G, ρ_v for density of gas or vapor; ρ_s for density of solid	kg/m³	lb/ft³
σ	Surface tension between a liquid and its vapor	N/m	lbf/ft
Σ	Term indicating summation of variables		
τ	Shear stress τ_w for shear stress at the wall	N/m²	lbf/ft²
ϕ	Velocity-potential function		
ϕ_p	Particle sphericity		
Φ	Viscous-dissipation function		
ω	Angle of repose of powdered solid	rad	rad
Ω	Fin efficiency	Dimensionless	Dimensionless

GENERAL REFERENCES: Becker, *Heat Transfer*, Plenum, New York, 1986. Bejan, *Convection Heat Transfer*, Wiley, New York, 1984. Bird, Stewart, and Lightfoot, *Transport Phenomena*, Wiley, New York, 1960. Carslaw and Jaeger, *Conduction of Heat in Solids*, Clarendon Press, Oxford, 1959. Chapman, *Heat Transfer*, 2d ed., Macmillan, New York, 1967. Drew and Hoopes, *Advances in Chemical Engineering*, Academic, New York, vol. 1, 1956; vol. 2, 1958; vol. 5, 1964; vol. 6, 1966; vol. 7, 1968. Dusinberre, *Heat Transfer Calculations by Finite Differences*, International Textbook, Scranton, Pa., 1961. Eckert and Drake, *Heat and Mass Transfer*, 2d ed., McGraw-Hill, New York, 1959. Gebhart, *Heat Transfer*, McGraw-Hill, New York, 1961. Irvine and Hartnett, *Advances in Heat Transfer*, Academic, New York, vol. 1, 1964; vol. 2, 1965; vol. 3, 1966. Grigull and Sandner, *Heat Conduction*, Hemisphere Publishing, 1984. Jakob, *Heat Transfer*, Wiley, New York, vol. 1, 1949; vol. 2, 1957. Jakob and Hawkins, *Elements of Heat Transfer*, 3d ed., Wiley, New York, 1957. Kakac, Bergles, and Mayinger, *Heat Exchangers: Thermal Hydraulic Fundamentals and Design*, Hemisphere Publishing, Washington, 1981. Kakac and Yener, *Convective Heat Transfer*, Hemisphere Publishing, Washington, 1980. Kay, *An Introduction to Fluid Mechanics and Heat Transfer*, 2d ed., Cambridge University Press, Cambridge, England, 1963. Kays, *Convective Heat and Mass Transfer*, McGraw-Hill, New York, 1966. Kays and London, *Compact Heat Exchangers*, 3d ed., McGraw-Hill, New York, 1984. Kern, *Process Heat Transfer*, McGraw-Hill, New York, 1950. Kays and London, *Fluid Dynamics and Heat Transfer*, McGraw-Hill, New York, 1958. Kraus, *Analysis and Evaluation of Extended Surface Thermal Systems*, Hemisphere Publishing, Washington, 1982. Kutatladze, *A Concise Encyclopedia of Heat Transfer*, 1st English ed., Pergamon, New York, 1966. Lykov, *Heat and Mass Transfer in Capillary Porous Bodies*, translated from Russian, Pergamon, New York, 1966. McAdams, *Heat Transmission*, 3d ed., McGraw-Hill, New York, 1954. Mickley, Sherwood, and Reed, *Applied Mathematics in Chemical Engineering*, 2d ed., McGraw-Hill, New York, 1957. Rohsenow and Choi, *Heat, Mass, and Momentum Transfer*, Prentice-Hall, Englewood Cliffs, N.J., 1961. Schlünder (ed.), *Heat Exchanger Design Handbook*, Hemisphere Publishing, Washington, 1983 (Book 2; Chapter 2.4 [Conduction], Chapter 2.5 [Convection], Chapter 2.6 [Condensation], Chapter 2.7 [Boiling]). Skelland, *Non-Newtonian Flow and Heat Transfer*, Wiley, New York, 1967. Taborek and Bell, *Process Heat Exchanger Design*, Hemisphere Publishing, Washington, 1984. Taborek, Hewitt, and Afghan, *Heat Exchangers: Theory and Practice*, Hemisphere Publishing, Washington, 1983. TSederberg, *Thermal Conductivity of Liquids and Gases*, M.I.T., Cambridge, Mass., 1965. Welty, Wicks, and Wilson, *Fundamentals of Momentum, Heat and Mass Transfer*, 3d ed., Wiley, New York, 1984. Zenz and Othmer, *Fluidization and Fluid Particle Systems*, Reinhold, New York, 1960.

REFERENCES FOR DIFFUSIVITIES
1. Akita, *Ind. Eng. Chem. Fundam.*, **10**, 89 (1981).
2. Asfour and Dullien, *Chem. Eng. Sci.*, **41**, 1591 (1986).
3. Blanc, *J. Phys.*, **7**, 825 (1908).
4. Brokaw, *Ind. Eng. Chem. Process Des. and Dev.*, **8**, 2, 240 (1969).
5. Caldwell and Babb, *J. Phys. Chem.*, **60**, 51 (1956).
6. Catchpole and King, *Ind. Eng. Chem. Res.*, **33**, 1828 (1994).
7. Chen and Chen, *Chem. Eng. Sci.*, **40**, 1735 (1985).
8. Chung, Ajlan, Lee and Starling, *Ind. Eng. Chem. Res.*, **27**, 671 (1988).
9. Condon and Craven, *Aust. J. Chem.*, **25**, 695 (1972).
10. Cullinan, *AIChE J.*, **31**, 1740–1741 (1985).
11. Cullinan, *Can. J. Chem. Eng.*, **45**, 377–381 (1967).
12. Cussler, *AIChE J.*, **26**, 1 (1980).
13. Darken, *Trans. Am. Inst. Mining Met. Eng.*, **175**, 184 (1948).
14. Debenedetti and Reid, *AIChE J.*, **32**, 2034 (1986); see errata: *AIChE J.*, **33**, 496 (1987).
15. Elliott, R. W. and H. Watts, *Can. J. Chem.*, **50**, 31 (1972).
16. Erkey and Akgerman, *AIChE J.*, **35**, 443 (1989).
17. Ertl, Ghai, and Dullien, *AIChE J.*, **20**, 1, 1 (1974).
18. Fairbanks and Wilke, *Ind. Eng. Chem.*, **42**, 471 (1950).
19. Fuller, Schettler and Giddings, *Ind. Eng. Chem.*, **58**, 18 (1966).
20. Ghai, Ertl, and Dullien, *AIChE J.*, **19**, 5, 881 (1973).
21. Gordon, *J. Chem. Phys.*, **5**, 522 (1937).
22. Graham and Dranoff, *Ind. Eng. Chem. Fundam.*, **21**, 360–365 (1982).
23. Graham and Dranoff, *Ind. Eng. Chem. Fundam.*, **21**, 365–369 (1982).
24. Gurkan, *AIChE J.*, **33**, 175–176 (1987).
25. Hayduk and Laudie, *AIChE J.*, **20**, 3, 611 (1974).
26. Hayduk and Minhas, *Can. J. Chem. Eng.*, **60**, 195 (1982).
27. Hildebrand, *Science*, **174**, 490 (1971).
28. Hiss and Cussler, *AIChE J.*, **19**, 4, 698 (1973).
29. Jossi, Stiel, and Thodos, *AIChE J.*, **8**, 59 (1962).
30. Krishnamurthy and Taylor, *Chem. Eng. J.*, **25**, 47 (1982).
31. Lee and Thodos, *Ind. Eng. Chem. Fundam.*, **22**, 17–26 (1983).
32. Lee and Thodos, *Ind. Eng. Chem. Res.*, **27**, 992–997 (1988).
33. Lees and Sarram, *J. Chem. Eng. Data*, **16**, 1, 41 (1971).
34. Leffler and Cullinan, *Ind. Eng. Chem. Fundam.*, **9**, 84, 88 (1970).
35. Lugg, *Anal. Chem.*, **40**, 1072 (1968).
36. Marrero and Mason, *AIChE J.*, **19**, 498 (1973).
37. Mathur and Thodos, *AIChE J.*, **11**, 613 (1965).
38. Matthews and Akgerman, *AIChE J.*, **33**, 881 (1987).
39. Matthews, Rodden and Akgerman, *J. Chem. Eng. Data*, **32**, 317 (1987).
40. Olander, *AIChE J.*, **7**, 175 (1961).
41. Passut and Danner, *Chem. Eng. Prog. Symp. Ser.*, **140**, 30 (1974).
42. Perkins and Geankoplis, *Chem. Eng. Sci.*, **24**, 1035–1042 (1969).
43. Pinto and Graham, *AIChE J.*, **32**, 291 (1986).
44. Pinto and Graham, *AIChE J.*, **33**, 436 (1987).
45. Quale, *Chem. Rev.*, **53**, 439 (1953).
46. Rathbun and Babb, *Ind. Eng. Chem. Proc. Des. Dev.*, **5**, 273 (1966).
47. Reddy and Doraiswamy, *Ind. Eng. Chem. Fundam.*, **6**, 77 (1967).
48. Riazi and Whitson, *Ind. Eng. Chem. Res.*, **32**, 3081 (1993).
49. Robinson, Edmister, and Dullien, *Ind. Eng. Chem. Fundam.*, **5**, 75 (1966).
50. Rollins and Knaebel, *AIChE J.*, **37**, 470 (1991).
51. Siddiqi, Krahn, and Lucas, *J. Chem. Eng. Data*, **32**, 48 (1987).
52. Siddiqi and Lucas, *Can. J. Chem. Eng.*, **64**, 839 (1986).
53. Smith and Taylor, *Ind. Eng. Chem. Fundam.*, **22**, 97 (1983).
54. Sridhar and Potter, *AIChE J.*, **23**, 4, 590 (1977).
55. Stiel and Thodos, *AIChE J.*, **7**, 234 (1961).
56. Sun and Chen, *Ind. Eng. Chem. Res.*, **26**, 815 (1987).
57. Tanford, *Phys. Chem. of Macromolecule*, Wiley, New York, NY (1961).
58. Taylor and Webb, *Comput. Chem. Eng.*, **5**, 61 (1981).
59. Tyn and Calus, *J. Chem. Eng. Data*, **20**, 310 (1975).
60. Umesi and Danner, *Ind. Eng. Chem. Process Des. Dev.*, **20**, 662 (1981).
61. Van Geet and Adamson, *J. Phys. Chem.*, **68**, 2, 238 (1964).
62. Vignes, *Ind. Eng. Chem. Fundam.*, **5**, 184 (1966).
63. Wilke, *Chem. Eng. Prog.*, **46**, 2, 95 (1950).
64. Wilke and Chang, *AIChE J.*, **1**, 164 (1955).
65. Wilke and Lee, *Ind. Eng. Chem.*, **47**, 1253 (1955).

REFERENCES FOR DIFFUSIVITIES IN POROUS SOLIDS, TABLE 5-20
66. Ruthven, *Principles of Adsorption & Adsorption Processes*, Wiley, 1984.
67. Satterfield, *Mass Transfer in Heterogeneous Catalysis*, MIT Press, 1970.
68. Suzuki, *Adsorption Engineering*, Kodansha—Elsevier, 1990.
69. Yang, *Gas Separation by Adsorption Processes*, Butterworths, 1987.

REFERENCES FOR TABLES 5-21 TO 5-28
70. Bahmanyar, Chang-Kakoti, Garro, Liang, and Slater, *Chem. Engr. Rsch. Des.*, **68**, 74 (1990).
71. Beenackers and van Swaaij, *Chem. Engr. Sci.*, **48**, 3109 (1993).
72. Bird, Stewart and Lightfoot, *Transport Phenomena*, Wiley, 1960.
73. Blatt, Dravid, Michaels, and Nelson in Flinn (ed.), *Membrane Science and Technology*, **47**, Plenum, 1970.
74. Bolles and Fair, *Institution Chem. Eng. Symp. Ser.*, **56**, 3/35 (1979).
75. Bolles and Fair, *Chem. Eng.*, **89**(14), 109 (July 12, 1982).
76. Bravo and Fair, *Ind. Eng. Chem. Process Des. Dev.*, **21**, 162 (1982).
77. Bravo, Rocha and Fair, *Hydrocarbon Processing*, 91 (Jan. 1985).
78. Brian and Hales, *AIChE J.*, **15**, 419 (1969).
79. Calderbank and Moo-Young, *Chem. Eng. Sci.*, **16**, 39 (1961).
80. Chilton and Colburn, *Ind. Eng. Chem.*, **26**, 1183 (1934).
81. Cornell, Knapp, and Fair, *Chem. Engr. Prog.*, **56**(7), 68 (1960).
82. Cornet and Kaloo, *Proc. 3rd Int'l. Congr. Metallic Corrosion—Moscow*, **3**, 83 (1966).
83. Cussler, *Diffusion: Mass Transfer in Fluid Systems*, Cambridge, 1984.
84. Dwivedi and Upadhyay, *Ind. Eng. Chem. Process Des. Develop*, **16**, 1657 (1977).
85. Eisenberg, Tobias, and Wilke, *Chem. Engr. Prog. Symp. Sec.*, **51**(16), 1 (1955).
86. Elzinga and Banchero, *Chem. Engr. Progr. Symp. Ser.*, **55**(29), 149 (1959).
87. Fair, "Distillation" in Rousseau (ed.), *Handbook of Separation Process Technology*, Wiley, 1987.
88. Faust, Wenzel, Clump, Maus, and Andersen, *Principles of Unit Operations*, 2d ed., Wiley, 1980.
89. Frossling, *Gerlands Beitr. Geophys.*, **52**, 170 (1938).
90. Garner and Suckling, *AIChE J.*, **4**, 114 (1958).
91. Geankoplis, *Transport Processes and Unit Operations*, 3d ed., Prentice Hall, 1993.

92. Gibilaro, Davies, Cooke, Lynch, and Middleton, *Chem. Engr. Sci.,* **40,** 1811 (1985).
93. Gilliland and Sherwood, *Ind. Engr. Chem.,* **26,** 516 (1934).
94. Griffith, *Chem. Engr. Sci.,* **12,** 198 (1960).
95. Gupta and Thodos, *AIChE J.,* **9,** 751 (1963).
96. Gupta and Thodos, *Ind. Eng. Chem. Fundam.,* **3,** 218 (1964).
97. Harriott, *AIChE J.,* **8,** 93 (1962).
98. Hausen, *Verfahrenstech. Beih. Z. Ver. Dtsch. Ing.,* **4,** 91 (1943).
99. Heertjes, Holve, and Talsma, *Chem. Engr. Sci.,* **3,** 122 (1954).
100. Hines and Maddox, *Mass Transfer: Fundamentals and Applications,* Prentice Hall, 1985.
101. Hsiung and Thodos, *Int. J. Heat Mass Transfer,* **20,** 331 (1977).
102. Hsu, Sato, and Sage, *Ind. Engr. Chem.,* **46,** 870 (1954).
103. Hughmark, *Ind. Eng. Chem. Fundam.,* **6,** 408 (1967).
104. Johnson, Besic, and Hamielec, *Can. J. Chem. Engr.,* **47,** 559 (1969).
105. Johnstone and Pigford, *Trans. AIChE,* **38,** 25 (1942).
106. Kafesjian, Plank, and Gerhard, *AIChE J.,* **7,** 463 (1961).
107. Kelly and Swenson, *Chem. Eng. Prog.,* **52,** 263 (1956).
108. King, *Separation Processes,* 2d ed., McGraw-Hill (1980).
109. Kirwan, "Mass Transfer Principles" in Rousseau, *Handbook of Separation Process Technology,* Wiley, 1987.
110. Klein, Ward, and Lacey, "Membrane Processes—Dialysis and Electro-Dialysis" in Rousseau, *Handbook of Separation Process Technology,* Wiley, 1987.
111. Kohl, "Absorption and Stripping" in Rousseau, *Handbook of Separation Process Technology,* Wiley, 1987.
112. Kojima, Uchida, Ohsawa, and Iguchi, *J. Chem. Engng. Japan,* **20,** 104 (1987).
113. Koloini, Sopcic, and Zumer, *Chem. Engr. Sci.,* **32,** 637 (1977).
114. Lee, *Biochemical Engineering,* Prentice Hall, 1992.
115. Lee and Foster, *Appl. Catal.,* **63,** 1 (1990).
116. Lee and Holder, *Ind. Engr. Chem. Res.,* **34,** 906 (1995).
117. Levich, *Physicochemical Hydrodynamics,* Prentice Hall, 1962.
118. Levins and Gastonbury, *Trans. Inst. Chem. Engr.,* **50,** 32, 132 (1972).
119. Lim, Holder, and Shah, *J. Supercrit. Fluids,* **3,** 186 (1990).
120. Linton and Sherwood, *Chem. Eng. Prog.,* **46,** 258 (1950).
121. Ludwig, *Applied Process Design for Chemical and Petrochemical Plants,* 2d ed., vol. 2, Gulf Pub. Co., 1977.
122. McCabe, Smith, and Harriott, *Unit Operations of Chemical Engineering,* 5th ed., McGraw-Hill, 1993.
123. Nelson and Galloway, *Chem. Engr. Sci.,* **30,** 7 (1975).
124. Notter and Sleicher, *Chem. Eng. Sci.,* **26,** 161 (1971).
125. Ohashi, Sugawara, Kikuchi, and Konno, *J. Chem. Engr. Japan,* **14,** 433 (1981).
126. Onda, Takeuchi, and Okumoto, *J. Chem. Engr. Japan,* **1,** 56 (1968).
127. Pasternak and Gauvin, *AIChE J.,* **7,** 254 (1961).
128. Pasternak and Gauvin, *Can. J. Chem. Engr.,* **38,** 35 (April 1960).

129. Perez and Sandall, *AIChE J.,* **20,** 770 (1974).
130. Petrovic and Thodos, *Ind. Eng. Chem. Fundam.,* **7,** 274 (1968).
131. Pinczewski and Sideman, *Chem. Engr. Sci.,* **29,** 1969 (1974).
132. Prandtl, *Phys. Zeit.,* **29,** 487 (1928).
133. Prasad and Sirkar, *AIChE J.,* **34,** 177 (1988).
134. Rahman and Streat, *Chem. Engr. Sci.,* **36,** 293 (1981).
135. Ranz and Marshall, *Chem. Engr. Prog.,* **48,** 141, 173 (1952).
136. Reiss, *Ind. Eng. Chem. Process Des. Develop,* **6,** 486 (1967).
137. Riet, *Ind. Eng. Chem. Process Des. Dev.,* **18,** 357 (1979).
138. Rowe, *Chem. Engr. Sci.,* **30,** 7 (1975).
139. Rowe, Claxton, and Lewis, *Trans. Inst. Chem. Engr. London,* **43,** 14 (1965).
140. Ruckenstein and Rajagopalan, *Chem. Engr. Commun.,* **4,** 15 (1980).
141. Ruthven, *Principles of Adsorption & Adsorption Processes,* Wiley, 1984.
142. Satterfield, *AIChE J.,* **21,** 209 (1975).
143. Schluter and Deckwer, *Chem. Engr. Sci.,* **47,** 2357 (1992).
144. Schmitz, Steiff, and Weinspach, *Chem. Engng. Technol.,* **10,** 204 (1987).
145. Sherwood, Brian, Fisher, and Dresner, *Ind. Eng. Chem. Fundam.,* **4,** 113 (1965).
146. Sherwood, Pigford, and Wilke, *Mass Transfer,* McGraw-Hill, 1975.
147. Shulman, Ullrich, Proulx, and Zimmerman, *AIChE J.,* **1,** 253 (1955).
148. Shulman and Margolis, *AIChE J.,* **3,** 157 (1957).
149. Siegel, Sparrow, and Hallman, *Appl. Sci. Res. Sec. A.,* **7,** 386 (1958).
150. Sissom and Pitts, *Elements of Transport Phenomena,* McGraw-Hill, 1972.
151. Skelland, *Diffusional Mass Transfer,* Wiley (1974).
152. Skelland and Cornish, *AIChE J.,* **9,** 73 (1963).
153. Skelland and Moeti, *Ind. Eng. Chem. Res.,* **29,** 2258 (1990).
154. Skelland and Tedder, "Extraction—Organic Chemicals Processing" in Rousseau, *Handbook of Separation Process Technology,* Wiley, 1987, pp. 405–466.
155. Skelland and Wellek, *AIChE J.,* **10,** 491, 789 (1964).
156. Slater, "Rate Coefficients in Liquid-Liquid Extraction Systems" in Godfrey and Slater, *Liquid-Liquid Extraction Equipment,* Wiley, 1994, pp. 45–94.
157. Steinberger and Treybal, *AIChE J.,* **6,** 227 (1960).
158. Steiner, L., *Chem. Eng. Sci.,* **41,** 1979 (1986).
159. Taylor and Krishna, *Multicomponent Mass Transfer,* Wiley, 1993.
160. Tournie, Laguerie, and Couderc, *Chem. Engr. Sci.,* **34,** 1247 (1979).
161. Treybal, *Mass Transfer Operations,* 3d ed., McGraw-Hill, 1980.
162. Von Karman, *Trans. ASME,* **61,** 705 (1939).
163. Wakao and Funazkri, *Chem. Engr. Sci.,* **33,** 1375 (1978).
164. Wankat, *Equilibrium-Staged Separations,* Prentice Hall, 1988.
165. Wankat, *Rate-Controlled Separations,* Chapman-Hall, 1990.
166. Wilson and Geankoplis, *Ind. Eng. Chem. Fundam.,* **5,** 9 (1966).
167. Yagi and Yoshida, *Ind. Eng. Chem. Process Des. Dev.,* **14,** 488 (1975).
168. Yang, *Gas Separation by Adsorption Processes,* Butterworths, 1987.
169. Yoshida, Ramaswami, and Hougen, *AIChE J.,* **8,** 5 (1962).

HEAT TRANSFER

MODES OF HEAT TRANSFER

There are three fundamental types of heat transfer: conduction, convection, and radiation. All three types may occur at the same time, and it is advisable to consider the heat transfer by each type in any particular case.

Conduction is the transfer of heat from one part of a body to another part of the same body, or from one body to another in physical contact with it, without appreciable displacement of the particles of the body.

Convection is the transfer of heat from one point to another within a fluid, gas, or liquid by the mixing of one portion of the fluid with another. In natural convection, the motion of the fluid is entirely the result of differences in density resulting from temperature differences; in forced convection, the motion is produced by mechanical means. When the forced velocity is relatively low, it should be realized that "free-convection" factors, such as density and temperature difference, may have an important influence.

Radiation is the transfer of heat from one body to another, not in contact with it, by means of wave motion through space.

HEAT TRANSFER BY CONDUCTION

FOURIER'S LAW

Fourier's law is the fundamental differential equation for heat transfer by conduction:

$$dQ/d\theta = -kA(dt/dx) \qquad (5\text{-}1)$$

where $dQ/d\theta$ (quantity per unit time) is the rate of flow of heat, A is the area at right angles to the direction in which the heat flows, and $-dt/dx$ is the rate of change of temperature with the distance in the direction of the flow of heat, i.e., the temperature gradient. The factor k is called the thermal conductivity; it is a characteristic property of the material through which the heat is flowing and varies with temperature.

THREE-DIMENSIONAL CONDUCTION EQUATION

Equation (5-1) is used as a basis for derivation of the unsteady-state three-dimensional energy equation for **solids or static fluids:**

$$c\rho\,\frac{\partial t}{\partial \theta} = \frac{\partial}{\partial x}\left(k\,\frac{\partial t}{\partial x}\right) + \frac{\partial}{\partial y}\left(k\,\frac{\partial t}{\partial y}\right) + \frac{\partial}{\partial z}\left(k\,\frac{\partial t}{\partial z}\right) + q' \qquad (5\text{-}2)$$

where x, y, z are distances in the rectangular coordinate system and q' is the rate of heat generation (by chemical reaction, nuclear reaction, or electric current) in the solid per unit of volume. Solution of Eq. (5-2) with appropriate boundary and initial conditions will give the temperature as a function of time and location in the material. Equation (5-2) may be transformed into spherical or cylindrical coordinates to conform more closely to the physical shape of the system.

THERMAL CONDUCTIVITY

Thermal conductivity varies with temperature but not always in the same direction. The thermal conductivities for many materials, as a function of temperature, are given in Sec. 2. Additional and more comprehensive information may often be obtained from suppliers of the materials. Impurities, especially in metals, can give rise to variations in thermal conductivity of from 50 to 75 percent. In using thermal conductivities, engineers should remember that conduction is not the sole method of transferring heat and that, particularly with liquids and gases, radiation and convection may be much more important.

The thermal conductivity at a given temperature is a function of the apparent, or bulk, density. Thus, at 0°C (32°F), k for asbestos wool is 0.09 J/(m·s·K) [0.052 Btu/(hr·ft·°F)] when the bulk density is 400 kg/m³ (24.9 lb/ft³) and is 0.19 (0.111) for a density of 700 (43.6).

In determining the apparent thermal conductivities of **granular solids,** such as granulated cork or charcoal grains, Griffiths (Spec. Rep. 5, Food Investigation Board, H. M. Stationery Office, 1921) found that air circulates within the mass of granular solid. Under a certain set of conditions, the apparent thermal conductivity of a charcoal was 9 percent greater when the test section was vertical than when it was horizontal. When the apparent conductivity of a mixture of cellular or porous nonhomogeneous solid is determined, the observed temperature coefficient may be much larger than for the homogeneous solid alone, because heat is transferred not only by the mechanism of conduction but also by convection in the gas pockets and by radiation from surface to surface of the individual particles. If internal radiation is an important factor, a plot of the apparent conductivity as ordinate versus temperature should show a curve concave upward, since radiation increases with the fourth power of the absolute temperature. Griffiths noted that cork, slag, wool, charcoal, and wood fibers, when of good quality and dry, have thermal conductivities about 2.2 times that of still air, whereas a highly cellular form of rubber, 112 kg/m³ (7 lb/ft³), had a thermal conductivity only 1.6 times that of still air. In measuring the apparent thermal conductivity of diathermanous substances such as quartz (especially when exposed to radiation emitted at high temperatures), it should be remembered that a part of the heat is transmitted by radiation.

Bridgman [*Proc. Am. Acad. Arts Sci.*, **59**, 141 (1923)] showed that the thermal conductivity of **liquids** is increased by only a few percent under a pressure of 100,330 kPa (1000 atm). The thermal conductivity of some liquids varies with temperature through a maximum. It is often necessary for the engineer to estimate thermal conductivities; methods are indicated in Sec. 2.

Equation (5-2) considers the thermal conductivity to be variable. If k is expressed as a function of temperature, Eq. (5-2) is nonlinear and difficult to solve analytically except for certain special cases. Usually in complicated systems numerical solution by means of computer is possible. A complete review of heat conduction has been given by Davis and Akers [*Chem. Eng.*, **67**(4), 187, (5), 151 (1960)] and by Davis [*Chem. Eng.*, **67**(6), 213, (7), 135 (8), 137 (1960)].

STEADY-STATE CONDUCTION

For steady flow of heat, the term $dQ/d\theta$ in Eq. (5-1) is constant and may be replaced by Q/θ or q. Likewise, in Eq. (5-2) the term $\partial t/\partial \theta$ is zero. Hence, for constant thermal conductivity, Eq. (5-2) may be expressed as

$$\nabla^2 t = (q'/k) \qquad (5\text{-}3)$$

One-Dimensional Conduction Many heat-conduction problems may be formulated into a one-dimensional or pseudo-one-dimensional form in which only one space variable is involved. Forms of the conduction equation for rectangular, cylindrical, and spherical coordinates are, respectively,

$$\frac{\partial^2 t}{\partial x^2} = -\frac{q'}{k} \qquad (5\text{-}4a)$$

$$\frac{1}{r}\frac{d}{dr}\left(r\,\frac{dt}{dr}\right) = -\frac{q'}{k} \qquad (5\text{-}4b)$$

$$\frac{1}{r^2}\frac{d}{dr}\left(r^2\,\frac{dt}{dr}\right) = -\frac{q'}{k} \qquad (5\text{-}4c)$$

These are second-order differential equations which upon integration become, respectively,

$$t = -(q'x^2/2k) + c_1 x + c_2 \qquad (5\text{-}5a)$$

$$t = -(q'r^2/4k) + c_1 \ln r + c_2 \qquad (5\text{-}5b)$$

$$t = -(q'r^2/6k) - (c_1/r) + c_2 \qquad (5\text{-}5c)$$

Constants of integration c_1 and c_2 are determined by the boundary conditions, i.e., temperatures and temperature gradients at known locations in the system.

For the case of a solid surface exposed to surroundings at a different temperature and for a finite surface coefficient, the **boundary condition** is expressed as

$$h_T(t_s - t') = -k(dt/dx)_{\text{surf}} \qquad (5\text{-}6)$$

Inspection of Eqs. (5-5a), (5-5b), and (5-5c) indicates the form of temperature profile for various conditions and geometries and also reveals the effect of the heat-generation term q' upon the temperature distributions.

In the **absence of heat generation,** one-dimensional steady-state conduction may be expressed by integrating Eq. (5-1):

$$q\int_{x_1}^{x_2}\frac{dx}{A} = -\int_{t_1}^{t_2} k\,dt \qquad (5\text{-}7)$$

Area A must be known as a function of x. If k is constant, Eq. (5-7) is expressed in the integrated form

$$q = kA_{\text{avg}}(t_1 - t_2)/(x_2 - x_1) \qquad (5\text{-}8)$$

where

$$A_{\text{avg}} = \frac{1}{x_2 - x_1}\int_{x_1}^{x_2}\frac{dx}{A} \qquad (5\text{-}9)$$

Examples of values of A_{avg} for various functions of x are shown in the following table.

Area proportional to	A_{avg}
Constant	$A_1 = A_2$
x	$\dfrac{A_2 - A_1}{\ln (A_2/A_1)}$
x^2	$\sqrt{A_2 A_1}$

Usually, thermal conductivity k is not constant but is a function of temperature. In most cases, over the ranges of values used the relation is linear. Integration of Eq. (5-7), with k linear in t, gives

$$q\int_{x_1}^{x_2}\frac{dx}{A} = k_{\text{avg}}(t_1 - t_2) \qquad (5\text{-}10)$$

where k_{avg} is the arithmetic-average thermal conductivity between temperatures t_1 and t_2. This average probably gives results which are correct within the precision of the data in the majority of cases, though a special integration can be made whenever k is known to be greatly different from linear in temperature.

Conduction through Several Bodies in Series Figure 5-1 illustrates diagrammatically the temperature gradients accompanying the steady conduction of heat in series through three solids.

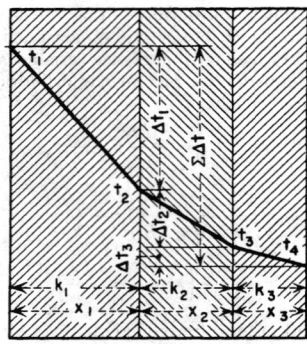

FIG. 5-1　Temperature gradients for steady heat conduction in series through three solids.

Since the heat flow through each of the three walls must be the same,

$$q = (k_1 A_1\,\Delta t_1/x_1) = (k_2 A_2\,\Delta t_2/x_2) = (k_3 A_3\,\Delta t_3/x_3) \qquad (5\text{-}11)$$

Since, by definition, individual thermal resistance

$$R = x/kA \qquad (5\text{-}12)$$

then

$$\Delta t_1 = qR_1 \quad \Delta t_2 = qR_2 \quad \Delta t_3 = qR_3 \qquad (5\text{-}13)$$

Adding the individual temperature drops, noting that q is uniform,

$$q(R_1 + R_2 + R_3) = \Delta t_1 + \Delta t_2 + \Delta t_3 = \sum \Delta t \qquad (5\text{-}14)$$

or

$$q = \sum \Delta t/R_T = (t_1 - t_4)/R_T \qquad (5\text{-}15)$$

where R_T is the overall resistance and is the sum of the individual resistances in series, then

$$R_T = R_1 + R_2 + \cdots + R_n \qquad (5\text{-}16)$$

When a wall is constructed of several layers of solids, the joints at adjacent layers may not perfectly exclude air spaces, and these additional resistances should not be overlooked.

Conduction through Several Bodies in Parallel　For n resistances in parallel, the rates of heat flow are additive:

$$q = \Delta t/R_1 + \Delta t/R_2 + \cdots + \Delta t/R_n \qquad (5\text{-}17a)$$

$$q = \left(\frac{1}{R_1} + \frac{1}{R_2} + \cdots + \frac{1}{R_n}\right)\Delta t \qquad (5\text{-}17b)$$

$$q = (C_1 + C_2 + \cdots + C_n)\Delta t = \sum C\,\Delta t \qquad (5\text{-}17c)$$

where R_1 to R_n are the individual resistances and C_1 to C_n are the individual conductances; $C = kA/x$.

Several Bodies in Series with Heat Generation　The simple Fourier type of equation indicated by Eq. (5-15) may *not* be used when heat generation occurs in one of the bodies in the series. In this case, Eq. (5-5a), (5-5b), or (5-5c) must be solved with appropriate boundary conditions.

Example 1: Steady-State Conduction with Heat Generation
A plate-type nuclear fuel element, consisting of a uranium-zirconium alloy $(3.2)(10^{-3})$ m (0.125 in) thick clad on each side with a $(6.4)(10^{-4})$-m- (0.025-in-) thick layer of zirconium, is cooled by water under pressure at 200°C (400°F), the heat-transfer coefficient being 42,600 J/(m²·s·K) [7500 Btu/(h·ft²·°F)]. If the temperature at the center of the fuel must not exceed 570°C (1050°F), determine the maximum rate of heat generation in the fuel. The zirconium and zirconium alloy have a thermal conductivity of 21 J/(m·s·K) [12 Btu/(h·ft²)(°F/ft)].

Solution.　Equation (5-4a) may be integrated for each material. The heat generation is zero in the cladding, and its value for the fuel may be determined from the integrated equations. Let $x = 0$ at the midplane of the fuel. Then $x_1 = (1.6)(10^{-3})$ m (0.0625 in) at the cladding-fuel interface and $x_2 = (2.2)(10^{-3})$ m (0.0875 in) at the cladding-water interface. Let the subscripts c, f refer to cladding and fuel respectively.

The boundary conditions are:

For fuel, at $x = 0$, $t = 570$°C (1050°F), $dt/dx = 0$ (this follows if the temperature is finite at the midplane).

For fuel and cladding, at $x = x_1$, $t_f = t_c$,

$$k_f(dt/dx) = k_c(dt/dx)$$

For cladding, at $x = x_2$,

$$t_c - 400 = -(k_c/42{,}600)(dt/dx)$$

For the fuel, the first integration of Eq. (10-4a) gives

$$dt_f/dx = -(q'/k_f)x + c_1$$

which gives $c_1 = 0$ when the boundary condition is applied. Thus the second integration gives

$$t_f = -(q'/2k_f)x^2 + c_2$$

from which c_2 is determined to be 570 (1050) upon application of the boundary condition. Thus the temperature profile in the fuel is

$$t_f = -(q'/2k_f)x^2 + 570$$

The temperature profile in the *cladding* is obtained by integrating Eq. (10-4a) twice with $q' = 0$. Hence

$$(dt_c/dx) = c_1 \qquad \text{and} \qquad t_c = c_1 x + c_2$$

There are now three unknowns, c_1, c_2, and q', and three boundary conditions by which they can be determined.
At $x = x_1$,

$$q'x_1^2/2k_f + 570 = c_1 x_1 + c_2 - k_f q'x_1/k_f = k_c c_1$$

At $x = x_2$,

$$c_1 x_2 + c_2 - 200 = -(k_c/42{,}600)c_1$$

From which　$q' = (2.53)(10^9)$ J/(m³·s)[$(2.38)(10^8)$Btu/(h·ft³)]
$$c_1 = -(1.92)(10^5)$$
$$c_2 = 724$$

Two-Dimensional Conduction　If the temperature of a material is a function of two space variables, the two-dimensional conduction equation is (assuming constant k)

$$\partial^2 t/\partial x^2 + \partial^2 t/\partial y^2 = -q'/k \qquad (5\text{-}18)$$

When q' is zero, Eq. (5-18) reduces to the familiar Laplace equation. The analytical solution of Eq. (10-18) as well as of Laplace's equation is possible for only a few boundary conditions and geometric shapes. Carslaw and Jaeger (*Conduction of Heat in Solids,* Clarendon Press, Oxford, 1959) have presented a large number of analytical solutions of differential equations applicable to heat-conduction problems. Generally, graphical or numerical **finite-difference methods** are most frequently used. Other numerical and relaxation methods may be found in the general references in the "Introduction." The methods may also be extended to three-dimensional problems.

UNSTEADY-STATE CONDUCTION

When temperatures of materials are a function of both time and space variables, more complicated equations result. Equation (5-2) is the three-dimensional unsteady-state conduction equation. It involves the rate of change of temperature with respect to time $\partial t/\partial\theta$. Solutions to most practical problems must be obtained through the use of digital computers. Numerous articles have been published on a wide variety of transient conduction problems involving various geometrical shapes and boundary conditions.

One-Dimensional Conduction　The one-dimensional transient conduction equations are (for constant physical properties)

$$\partial t/\partial\theta = \alpha(\partial^2 t/\partial x^2) + q'/c\rho \qquad \text{(rectangular coordinates)} \qquad (5\text{-}19a)$$

$$\frac{\partial t}{\partial\theta} = \frac{\alpha}{r}\frac{\partial}{\partial r}\left(r\frac{\partial t}{\partial r}\right) + \frac{q'}{c\rho} \qquad \text{(cylindrical coordinates)} \qquad (5\text{-}19b)$$

$$\frac{\partial t}{\partial\theta} = \frac{\alpha}{r^2}\frac{\partial}{\partial r}\left(r^2\frac{\partial t}{\partial r}\right) + \frac{q'}{c\rho} \qquad \text{(spherical coordinates)} \qquad (5\text{-}19c)$$

These equations have been solved analytically for solid slabs, cylinders, and spheres. The solutions are in the form of infinite series, and usually the results are plotted as curves involving four ratios [Gurney and Lurie, *Ind. Eng. Chem.,* **15,** 1170 (1923)] defined as follows with $q' = 0$:

$$Y = (t' - t)/(t' - t_b) \qquad X = k\theta/\rho c r_m^2 \qquad (5\text{-}20a,b)$$

$$m = k/h_T r_m \qquad n = r/r_m \qquad (5\text{-}20c,d)$$

Since each ratio is dimensionless, any consistent units may be employed in any ratio. The significance of the symbols is as follows: t' = temperature of the surroundings; t_b = initial uniform temperature of the body; t = temperature at a given point in the body at the time θ measured from the start of the heating or cooling operations; k = uniform thermal conductivity of the body; ρ = uniform density of the body; c = specific heat of the body; h_T = coefficient of total heat transfer between the surroundings and the surface of the body expressed as heat transferred per unit time per unit area of the surface per unit difference in temperature between surroundings and surface; r = distance, in the direction of heat conduction, from the midpoint or midplane of the body to the point under consideration; r_m = radius of a sphere or cylinder, one-half of the thickness of a slab heated from both faces, the total thickness of a slab heated from one face and insulated perfectly at the other; and x = distance, in the direction of heat conduction, from the surface of a semi-infinite body (such as the surface of the earth) to the point under consideration. In making the integrations which lead to the curves shown, the following factors were assumed constant: c, h_T, k, r, r_m, t', x, and ρ.

The working curves are shown in Figs. 5-2 to 5-5 for **cylinders of infinite length, spheres, slabs of infinite faces,** and **semi-infinite solids** respectively, with Y plotted as ordinates on a logarithmic scale versus X as abscissas to an arithmetic scale, for various values of the ratios m and n. To facilitate calculations involving instantaneous rates of cooling or heating of the semi-infinite body, Fig. 5-5 shows also a curve of dY/dX versus X. Similar plots to a larger scale are given in McAdams,

Brown and Marco, Schack, and Stoever (see "Introduction: General References"). For a solid of infinite thickness (Fig. 5-5) and with $m = 0$,

$$Y = \frac{2}{\sqrt{\pi}} \int_0^z \exp(-z^2)\, dz \qquad (5\text{-}21)$$

where $z = 1/\sqrt{2}X$ and the "error integral" may be evaluated from standard mathematical tables.

Various numerical and graphical methods are used for unsteady-state conduction problems, in particular the Schmidt graphical method (*Foppls Festschrift*, Springer-Verlag, Berlin, 1924). These methods are very useful because any form of initial temperature distribution may be used.

Two-Dimensional Conduction The governing differential equation for two-dimensional transient conduction is

$$\frac{\partial t}{\partial \theta} = \alpha \left(\frac{\partial^2 t}{\partial x^2} + \frac{\partial^2 t}{\partial y^2} \right) + \frac{q'}{c\rho} \qquad (5\text{-}22)$$

McAdams (*Heat Transmission*, 3d ed., McGraw-Hill, New York, 1954) gives various forms of transient difference equations and methods of solving transient conduction problems. The availability of computers and a wide variety of computer programs permits virtually routine solution of complicated conduction problems.

Conduction with Change of Phase A special type of transient problem (the Stefan problem) involves conduction of heat in a material when freezing or melting occurs. The liquid-solid interface moves with time, and in addition to conduction, latent heat is either generated or absorbed at the interface. Various problems of this type are discussed by Bankoff [in Drew et al. (eds.), *Advances in Chemical Engineering*, vol. 5, Academic, New York, 1964].

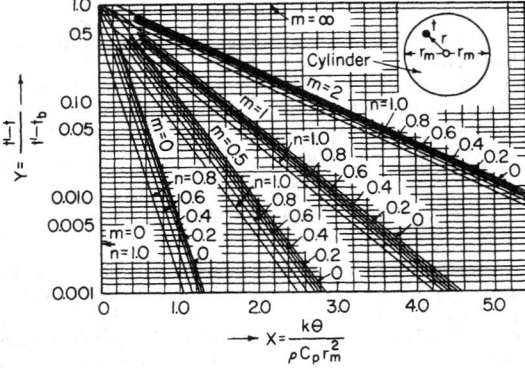

FIG. 5-2 Heating and cooling of a solid cylinder having an infinite ratio of length to diameter.

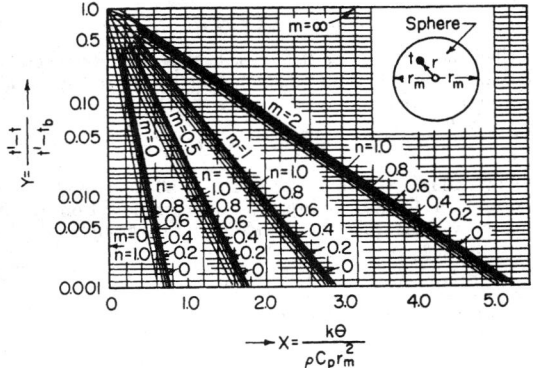

FIG. 5-3 Heating and cooling of a solid sphere.

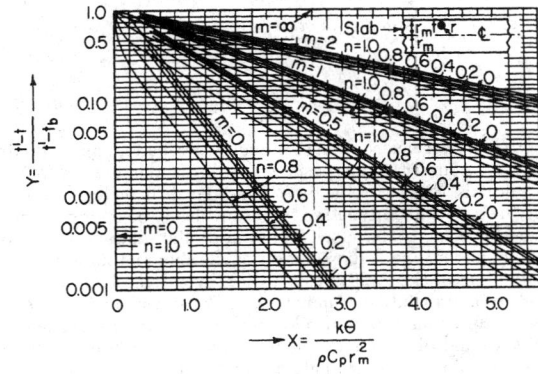

FIG. 5-4 Heating and cooling of a solid slab having a large face area relative to the area of the edges.

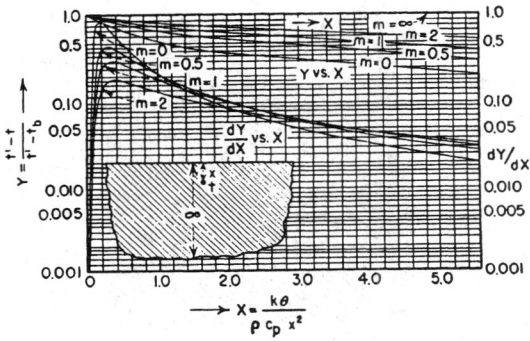

FIG. 5-5 Heating and cooling of a solid of infinite thickness, neglecting edge effects. (This may be used as an approximation in the zone near the surface of a body of finite thickness.)

HEAT TRANSFER BY CONVECTION

COEFFICIENT OF HEAT TRANSFER

In many cases of heat transfer involving either a liquid or a gas, convection is an important factor. In the majority of heat-transfer cases met in industrial practice, heat is being transferred from one fluid through a solid wall to another fluid. Assume a hot fluid at a temperature t_1 flowing past one side of a metal wall and a cold fluid at t_7 flowing past the other side to which a scale of thickness x_s adheres. In such a case, the conditions obtaining at a given section are illustrated diagrammatically in Fig. 5-6.

For turbulent flow of a fluid past a solid, it has long been known that, in the immediate neighborhood of the surface, there exists a relatively quiet zone of fluid, commonly called the **film.** As one approaches the wall from the body of the flowing fluid, the flow tends to become less turbulent and develops into laminar flow immediately adjacent to the wall. The film consists of that portion of the flow which is essentially in laminar motion (the laminar sublayer) and through which heat is transferred by molecular conduction. The resistance of the laminar layer to heat flow will vary according to its thickness and can range from 95 percent of the total resistance for some fluids to about 1 percent for other fluids (liquid metals). The turbulent core and the buffer layer between the laminar sublayer and turbulent core each offer a **resistance to heat transfer** which is a function of the turbulence and the thermal properties of the flowing fluid. The relative temperature difference across each of the layers is dependent upon their resistance to heat flow.

The Energy Equation   A complete energy balance on a flowing fluid through which heat is being transferred results in the energy equation (assuming constant physical properties):

$$c\rho \left(\frac{\partial t}{\partial \theta} + u\,\frac{\partial t}{\partial x} + v\,\frac{\partial t}{\partial y} + w\,\frac{\partial t}{\partial z} \right)$$

$$= k \left(\frac{\partial^2 t}{\partial x^2} + \frac{\partial^2 t}{\partial y^2} + \frac{\partial^2 t}{\partial z^2} \right) + q' + \Phi \quad (5\text{-}23)$$

where Φ is the term accounting for energy dissipation due to fluid viscosity and is significant in high-speed gas flow and in the flow of highly viscous liquids. Except for the time term, the left-hand terms of Eq. (5-23) are the so-called **convective terms** involving the energy carried by the fluid by virtue of its velocity. Therefore, the solution of the equation is dependent upon the solution of the momentum equations of flow. Solutions of Eq. (5-23) exist only for several simple flow cases and geometries and mainly for laminar flow. For turbulent flow the difficulties of expressing the fluid velocity as a function of space and time coordinates and of obtaining reliable values of the effective thermal conductivity of the flowing fluid have prevented solution of the equation unless simplifying assumptions and approximations are made.

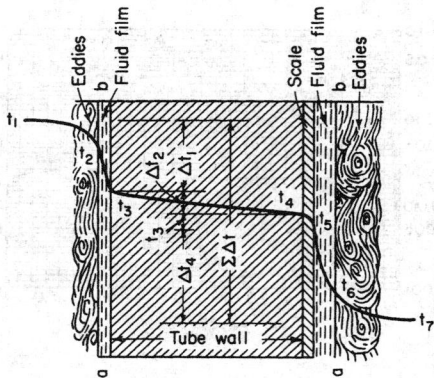

FIG. 5-6   Temperature gradients for a steady flow of heat by conduction and convection from a warmer to a colder fluid separated by a solid wall.

Individual Coefficient of Heat Transfer   Because of the complicated structure of a turbulent flowing stream and the impracticability of measuring thicknesses of the several layers and their temperatures, the **local rate of heat transfer** between fluid and solid is defined by the equations

$$dq = h_i\,dA_i\,(t_1 - t_3) = h_o\,dA_o\,(t_5 - t_7) \quad (5\text{-}24)$$

where h_i and h_o are the local heat-transfer coefficients inside and outside the wall, respectively, and temperatures are defined by Fig. 5-6.

The definition of the heat-transfer coefficient is arbitrary, depending on whether *bulk-fluid temperature, centerline temperature,* or *some other reference temperature* is used for t_1 or t_7. Equation (5-24) is an expression of Newton's law of cooling and incorporates all the complexities involved in the solution of Eq. (5-23). The **temperature gradients** in both the fluid and the adjacent solid at the fluid-solid interface may also be related to the heat-transfer coefficient:

$$dq = h_i\,dA_i\,(t_1 - t_3) = \left(-k\,\frac{dt}{dx}\right)_{\text{fluid}} = \left(-k\,\frac{dt}{dx}\right)_{\text{solid}} \quad (5\text{-}25)$$

Equation (5-25) holds for the liquid *only* if laminar flow exists immediately adjacent to the solid surface. The integration of Eq. (5-24) will give

$$A_i = \int_{\text{in}}^{\text{out}} \frac{dq}{h_i\,\Delta t_i} \quad \text{or} \quad A_o = \int_{\text{in}}^{\text{out}} \frac{dq}{h_o\,\Delta t_o} \quad (5\text{-}26)$$

which may be evaluated only if the quantities under the integral can be expressed in terms of a single variable. If q is a linear function of Δt and h is constant, then Eq. (5-26) gives

$$q = \frac{hA(\Delta t_{\text{in}} - \Delta t_{\text{out}})}{\ln\,(\Delta t_{\text{in}}/\Delta t_{\text{out}})} \quad (5\text{-}27)$$

where the Δt factor is the **logarithmic-mean temperature difference** between the wall and the fluid.

Frequently experimental data report average heat-transfer coefficients based upon an arbitrarily defined temperature difference, the two most common being

$$q = \frac{h_{lm}A(\Delta t_{\text{in}} - \Delta t_{\text{out}})}{\ln(\Delta t_{\text{in}}/\Delta t_{\text{out}})} \quad (5\text{-}28a)$$

$$q = \frac{h_{am}A(\Delta t_{\text{in}} + \Delta t_{\text{out}})}{2} \quad (5\text{-}28b)$$

where h_{lm} and h_{am} are average heat-transfer coefficients based upon the logarithmic-mean temperature difference and the arithmetic-average temperature difference, respectively.

Overall Coefficient of Heat Transfer   In testing commercial heat-transfer equipment, it is not convenient to measure tube temperatures (t_3 or t_4 in Fig. 5-6), and hence the overall performance is expressed as an overall coefficient of heat transfer U based on a convenient area dA, which may be dA_i, dA_o, or an average of dA_i and dA_o; whence, by definition,

$$dq = U\,dA\,(t_1 - t_7) \quad (5\text{-}29)$$

U is called the "overall coefficient of heat transfer," or merely the "overall coefficient." The rate of conduction through the tube wall and scale deposit is given by

$$dq = \frac{k\,dA_{\text{avg}}(t_3 - t_4)}{x} = h_d\,dA_d(t_4 - t_5) \quad (5\text{-}30)$$

Upon eliminating t_3, t_4, t_5 from Eqs. (5-24), (5-29), and (5-30), the complete expression for the **steady rate of heat flow** from one fluid through the wall and scale to a second fluid, as illustrated in Fig. 5-6, is

$$dq = \frac{t_1 - t_7}{\dfrac{1}{h_i\,dA_i} + \dfrac{x}{k\,dA_{\text{avg}}} + \dfrac{1}{h_d\,dA_d} + \dfrac{1}{h_o\,dA_o}} = U\,dA\,(t_1 - t_7) \quad (5\text{-}31)\degree$$

° Normally, dirt and scale resistance must be considered on both sides of the tube wall. The area dA is any convenient reference area.

Representation of Heat-Transfer Film Coefficients   There are two general methods of expressing film coefficients: (1) dimensionless relations and (2) dimensional equations.

The dimensionless relations are usually indicated in either of two forms, each yielding identical results. The preferred form is that suggested by Colburn [*Trans. Am. Inst. Chem. Eng.*, **29**, 174–210 (1933)]. It relates, primarily, three dimensionless groups: the Stanton number h/cG, the Prandtl number $c\mu/k$, and the Reynolds number DG/μ. For more accurate correlation of data (at Reynolds number <10,000), two additional dimensionless groups are used: ratio of length to diameter L/D and ratio of viscosity at wall (or surface) temperature to viscosity at bulk temperature. Colburn showed that the product of the Stanton number and the two-thirds power of the Prandtl number (and, in addition, power functions of L/D and μ_w/μ for Reynolds number <10,000) is approximately equal to half of the Fanning friction factor $f/2$. This product is called the **Colburn j factor.** Since the Colburn type of equation relates heat transfer and fluid friction, it has greater utility than other expressions for the heat-transfer coefficient.

The classical (and perhaps more familiar) form of dimensionless expressions relates, primarily, the Nusselt number hD/k, the Prandtl number $c\mu/k$, and the Reynolds number DG/μ, and viscosity-ratio modifications (for Reynolds number <10,000) also apply.

The **dimensional equations** are usually expansions of the dimensionless expressions in which the terms are in more convenient units and in which all numerical factors are grouped together into a single numerical constant. In some instances, the combined physical properties are represented as a linear function of temperature, and the dimensional equation resolves into an equation containing only one or two variables.

NATURAL CONVECTION

Natural convection occurs when a solid surface is in contact with a fluid of different temperature from the surface. Density differences provide the body force required to move the fluid. Theoretical analyses of natural convection require the simultaneous solution of the coupled equations of motion and energy. Details of theoretical studies are available in several general references (Brown and Marco, *Introduction to Heat Transfer*, 3d ed., McGraw-Hill, New York, 1958; and Jakob, *Heat Transfer*, Wiley, New York, vol. 1, 1949; vol. 2, 1957) but have generally been applied successfully to the simple case of a vertical plate. Solution of the motion and energy equations gives temperature and velocity fields from which heat-transfer coefficients may be derived. The general type of equation obtained is the so-called **Nusselt equation:**

$$\frac{hL}{k} = a\left(\frac{L^3\rho^2 g\beta\,\Delta t}{\mu^2}\,\frac{c\mu}{k}\right)^m \tag{5-32a}$$

$$N_{\text{Nu}} = a(N_{\text{Gr}}N_{\text{Pr}})^m \tag{5-32b}$$

Nusselt Equation for Various Geometries   Natural-convection coefficients for various bodies may be predicted from Eq. (5-32). The various numerical values of a and m have been determined experimen-

tally and are given in Table 5-1. Fluid properties are evaluated at $t_f = (t_s + t')/2$. For **vertical plates and cylinders** and $1 < N_{\text{Pr}} < 40$, Kato, Nishiwaki, and Hirata [*Int. J. Heat Mass Transfer*, **11**, 1117 (1968)] recommend the relations

$$N_{\text{Nu}} = 0.138N_{\text{Gr}}^{0.36}(N_{\text{Pr}}^{0.175} - 0.55) \tag{5-33a}$$

for $N_{\text{Gr}} > 10^9$, and

$$N_{\text{Nu}} = 0.683N_{\text{Gr}}^{0.25}N_{\text{Pr}}^{0.25}[N_{\text{Pr}}/(0.861 + N_{\text{Pr}})]^{0.25} \tag{5-33b}$$

for $N_{\text{Gr}} < 10^9$.

Simplified Dimensional Equations   Equation (5-32) is a dimensionless equation, and any consistent set of units may be used. Simplified dimensional equations have been derived for air, water, and organic liquids by rearranging Eq. (5-32) into the following form by collecting the fluid properties into a single factor:

$$h = b(\Delta t)^m L^{3m-1} \tag{5-34}$$

Values of b in SI and U.S. customary units are given in Table 5-1 for air, water, and organic liquids.

Simultaneous Loss by Radiation   The heat transferred by radiation is often of significant magnitude in the loss of heat from surfaces to the surroundings because of the diathermanous nature of atmospheric gases (air). It is convenient to represent radiant-heat transfer, for this case, as a **radiation film coefficient** which is added to the film coefficient for convection, giving the combined coefficient for convection and radiation ($h_c + h_r$). In Fig. 5-7 values of the film coefficient for radiation h_r are plotted against the two surface temperatures for emissivity = 1.0.

Table 5-2 shows values of ($h_c + h_r$) from single horizontal oxidized pipe surfaces.

Enclosed Spaces   The rate of heat transfer across an enclosed space is calculated from a special coefficient h' based upon the temperature difference between the two surfaces, where $h' = (q/A)/(t_{s1} - t_{s2})$. The value of $h'L/k$ may be predicted from Eq. (5-32) by using the values of a and m given in Table 5-3.

For **vertical enclosed cells** 10 in high and up to 2-in gap width, Landis and Yanowitz (*Proc. Third Int. Heat Transfer Conf.*, Chicago, 1966, vol. II, p. 139) give

$$\left(\frac{q}{A}\right)\frac{\delta}{k\,\Delta t} = 0.123(\delta/L)^{0.84}(N_{\text{Gr}}N_{\text{Pr}})^{0.28} \tag{5-35}$$

for $2 \times 10^3 < N_{\text{Gr}}N_{\text{Pr}}(\delta/L)^3 < 10^7$, where q/A is the uniform heat flux and Δt is the temperature difference at $L/2$. Equation (5-35) is applicable for air, water, and silicone oils.

For **horizontal annuli** Grugal and Hauf (*Proc. Third Int. Heat Transfer Conf.*, Chicago, 1966, vol. II, p 182) report

$$\frac{h\delta}{k} = \left(0.2 + 0.145\frac{\delta}{D_1}N_{\text{Gr}}\right)^{0.25}\exp\left(-0.02\frac{\delta}{D_1}\right) \tag{5-36}$$

for $0.55 < \delta/D_1 < 2.65$, where N_{Gr} is based upon gap width δ and D_1 is the core diameter of the annulus.

TABLE 5-1   Values of a, m, and b for Eqs. (5-32) and (5-34)

Configuration	$Y = N_{\text{Gr}}N_{\text{Pr}}$	a	m	b, air at 21°C	b, air at 70°F	b, water at 21°C	b, water at 70°F	b, organic liquid at 21°C	b, organic liquid at 70°F
Vertical surfaces	$<10^4$	1.36	⅕						
L = vertical dimension < 3 ft	$10^4 < Y < 10^9$	0.59	¼	1.37	0.28	127	26	59	12
	$>10^9$	0.13	⅓	1.24	0.18				
Horizontal cylinder	$<10^{-5}$	0.49	0						
L = diameter < 8 in	$10^{-5} < Y < 10^{-3}$	0.71	1/25						
	$10^{-3} < Y < 1$	1.09	1/10						
	$1 < Y < 10^4$	1.09	⅕						
	$10^4 < Y < 10^9$	0.53	¼	1.32	0.27				
	$>10^9$	0.13	⅓	1.24	0.18				
Horizontal flat surface	$10^5 < Y < 2 \times 10^7$(FU)	0.54	¼	1.86	0.38				
	$2 \times 10^7 < Y < 3 \times 10^{10}$(FU)	0.14	⅓						
	$3 \times 10^5 < Y < 3 \times 10^{10}$(FD)	0.27	¼	0.88	0.18				

NOTE: FU = facing upward; FD = facing downward. b in SI units is given in °C column; b in U.S. customary units, in °F column.

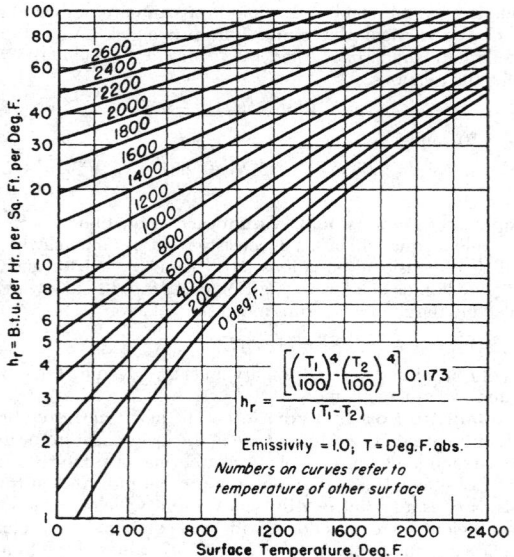

FIG. 5-7 Radiation coefficients of heat transfer h_r. To convert British thermal units per hour-square foot-degrees Fahrenheit to joules per square meter–second–kelvins, multiply by 5.6783; °C = (°F − 32)/1.8.

FORCED CONVECTION

Forced-convection heat transfer is the most frequently employed mode of heat transfer in the process industries. Hot and cold fluids, separated by a solid boundary, are pumped through the heat-transfer equipment, the rate of heat transfer being a function of the physical properties of the fluids, the flow rates, and the geometry of the system. Flow is generally turbulent, and the flow duct varies in complexity from circular tubes to baffled and extended-surface heat exchangers. Theoretical analyses of forced-convection heat transfer have been limited to relatively simple geometries and laminar flow. Analyses of turbulent-flow heat transfer have been based upon some mechanistic model and have not generally yielded relationships which were suitable for design purposes. Usually for complicated geometries only empirical relationships are available, and frequently these are based upon limited data and special operating conditions. Heat-transfer coefficients are strongly influenced by the mechanics of flow occurring during forced-convection heat transfer. Intensity of turbulence, entrance conditions, and wall conditions are some of the factors which must be considered in detail as greater accuracy in prediction of coefficients is required.

Analogy between Momentum and Heat Transfer The interrelationship of momentum transfer and heat transfer is obvious from examining the equations of motion and energy. For constant fluid properties, the equations of motion must be solved before the energy equation is solved. If fluid properties are not constant, the equations are coupled, and their solutions must proceed simultaneously. Considerable effort has been directed toward deriving some simple relationship between momentum and heat transfer. The methodology has been to use easily observed velocity profiles to obtain a measure of the diffusivity of momentum in the flowing stream. The analogy between heat and momentum is invoked by assuming that diffusion of heat and diffusion of momentum occur by essentially the same mechanism so that a relatively simple relationship exists between the diffusion coefficients. Thus, the diffusivity of momentum is used to predict temperature profiles and thence by Eq. (10-25) to predict the heat-transfer coefficient.

The analogy has been reasonably successful for simple geometries and for fluids of very low Prandtl number (liquid metals). For high-Prandtl-number fluids the **empirical analogy of Colburn** [*Trans. Am. Inst. Chem. Eng.*, **29**, 174 (1933)] has been very successful. A j factor for momentum transfer is defined as $j = f/2$, where f is the friction factor for the flow. The j factor for heat transfer is assumed to be equal to the j factor for momentum transfer

$$j = h/cG(c\mu/k)^{2/3} \qquad (5\text{-}37)$$

More involved analyses for **circular tubes** reduce the equations of motion and energy to the form

$$\frac{\tau g_c}{\rho} = -\frac{(\nu + \varepsilon_M)du}{dy} \qquad (5\text{-}38a)$$

$$\frac{q/A}{c\rho} = -\frac{(\alpha + \varepsilon_H)dt}{dy} \qquad (5\text{-}38b)$$

where ε_H is the eddy diffusivity of heat and ε_M is the eddy diffusivity of momentum. The units of diffusivity are L^2/θ. The eddy viscosity is $E_M = \rho\varepsilon_M$, and the eddy conductivity of heat is $E_H = \varepsilon_H c\rho$. Values of ε_M are determined via Eq. (5-38a) from experimental velocity-distribution data. By assuming $\varepsilon_H/\varepsilon_M$ = constant (usually unity), Eq. (10-38b) is solved to give the temperature distribution from which the heat-transfer coefficient may be determined. The major difficulties in solving Eq. (5-38b) are in accurately defining the thickness of the various flow layers (laminar sublayer and buffer layer) and in obtaining a suitable relationship for prediction of the eddy diffusivities. For assistance in predicting eddy diffusivities, see Reichardt (NACA Tech. Memo 1408, 1957) and Strunk and Chao [*Am. Inst. Chem. Eng. J.*, **10**, 269 (1964)].

Internal and External Flow Two main types of flow are considered in this subsection: internal or conduit flow, in which the fluid completely fills a closed stationary duct, and external or immersed flow, in which the fluid flows past a stationary immersed solid. With **internal flow**, the heat-transfer coefficient is theoretically infinite at the location where heat transfer begins. The local heat-transfer coefficient rapidly decreases and becomes constant, so that after a certain length the average coefficient in the conduit is independent of length. The local coefficient may follow an irregular pattern, however, if obstructions or turbulence promoters are present in the duct. For **immersed flow**, the local coefficient is again infinite at the point where heating begins, after which it decreases and may show various irregularities depending upon the configuration of the body. Usually in this instance the local coefficient never becomes constant as flow proceeds downstream over the body.

When heat transfer occurs during immersed flow, the rate is dependent upon the configuration of the body, the position of the body, the proximity of other bodies, and the flow rate and turbulence of the

TABLE 5-2 Values of $(h_c + h_r)$*

Btu/(h·ft²·°F from pipe to room)
For horizontal bare standard steel pipe of various sizes in a room at 80°F

Nominal pipe diameter, in	Temperature difference, °F														
	30	50	100	150	200	250	300	350	400	450	500	550	600	650	700
1	2.16	2.26	2.50	2.73	3.00	3.29	3.60	3.95	4.34	4.73	5.16	5.60	6.05	6.51	6.98
3	1.97	2.05	2.25	2.47	2.73	3.00	3.31	3.69	4.03	4.43	4.85	5.26	5.71	6.19	6.66
5		1.95	2.15	2.36	2.61	2.90	3.20	3.54	3.90						
10	1.80	1.87	2.07	2.29	2.54	2.82	3.12	3.47	3.84						

*Bailey and Lyell [*Engineering*, **147**, 60 (1939)] give values for $(h_c + h_r)$ up to Δt_s of 1000°F. °C = (°F − 32)/1.8; 5.6783 Btu/(h·ft²·°F) = J/(m²·s·/K).

TABLE 5-3   Values of a and m for Eq. (5-32)

Configuration	$N_{Gr}N_{Pr}(\delta/L)^3$	a	m
Vertical spaces	2×10^4 to 2×10^5	$0.20\,(\delta/L)^{-5/36}$	¼
	2×10^5 to 10^7	$0.071\,(\delta/L)^{1/9}$	⅓
Horizontal spaces	10^4 to 3×10^5	$0.21\,(\delta/L)^{-1/4}$	¼
	3×10^5 to 10^7	0.075	⅓

δ = cell width, L = cell length.

stream. The heat-transfer coefficient varies over the immersed body, since both the thermal and the momentum boundary layers vary in thickness. Relatively simple relationships are available for simple configurations immersed in an infinite flowing fluid. For complicated configurations and assemblages of bodies such as are found on the shell side of a heat exchanger, little is known about the local heat-transfer coefficient; empirical relationships giving average coefficients are all that are usually available. Research that has been conducted on local coefficients in complicated geometries has not been extensive enough to extrapolate into useful design relationships.

Laminar Flow   Normally, laminar flow occurs in closed ducts when $N_{Re} < 2100$ (based on equivalent diameter $D_e = 4 \times$ free area ÷ perimeter). Laminar-flow heat transfer has been subjected to extensive theoretical study. The energy equation has been solved for a variety of boundary conditions and geometrical configurations. However, true laminar-flow heat transfer very rarely occurs. Natural-convection effects are almost always present, so that the assumption that molecular conduction alone occurs is not valid. Therefore, empirically derived equations are most reliable.

Data are most frequently correlated by the Nusselt number $(N_{Nu})_{lm}$ or $(N_{NU})_{am}$, the Graetz number $N_{Gz} = (N_{Re}N_{Pr}D/L)$, and the Grashof (natural-convection effects) number N_{Gr}. Some correlations consider only the variation of viscosity with temperature, while others also consider density variation. Theoretical analyses indicate that for very long tubes $(N_{Nu})_{lm}$ approaches a limiting value. Limiting Nusselt numbers for various closed ducts are shown in Table 5-4.

Circular Tubes   For **horizontal tubes** and **constant wall temperature**, several relationships are available, depending on the Graetz number. For $0.1 < N_{Gz} < 10^4$, Hausen's [*Allg. Waermetech.*, **9**, 75 (1959)], the following equation is recommended.

$$(N_{Nu})_{lm} = 3.66 + \frac{0.19 N_{Gz}^{0.8}}{1 + 0.117 N_{Gz}^{0.467}}\left(\frac{\mu_b}{\mu_w}\right)^{0.14} \quad (5\text{-}39)$$

For $N_{Gz} > 100$, the Sieder-Tate relationship [*Ind. Eng. Chem.*, **28**, 1429 (1936)] is satisfactory for small diameters and Δt's:

$$(N_{Nu})_{am} = 1.86 N_{Gz}^{1/3}\,(\mu_b/\mu_w)^{0.14} \quad (5\text{-}40)$$

A more general expression covering all diameters and Δt's is obtained by including an additional factor $0.87(1 + 0.015N_{Gr}^{1/3})$ on the right side of Eq. (5-40). The diameter should be used in evaluating N_{Gr}. An equation published by Oliver [*Chem. Eng. Sci.*, **17**, 335 (1962)] is also recommended.

TABLE 5-4   Values of Limiting Nusselt Number in Laminar Flow in Closed Ducts

Configuration	Limiting Nusselt number $N_{Gr} < 4.0$	
	Constant wall temperature	Constant heat flux
Circular tube	3.66	4.36
Concentric annulus		Eq. (10-42)
Equilateral triangle		3.00
Rectangles		
Aspect ratio:		
1.0 (square)	2.89	3.63
0.713		3.78
0.500	3.39	4.11
0.333		4.77
0.25		5.35
0 (parallel planes)	7.60	8.24

For laminar flow in **vertical tubes** a series of charts developed by Pigford [*Chem. Eng. Prog. Symp. Ser.* 17, **51**, 79 (1955)] may be used to predict values of h_{am}.

Annuli   Approximate heat-transfer coefficients for laminar flow in annuli may be predicted by the equation of Chen, Hawkins, and Solberg [*Trans. Am. Soc. Mech. Eng.*, **68**, 99 (1946)]:

$$(N_{Nu})_{am} = 1.02 N_{Re}^{0.45} N_{Pr}^{0.5}\left(\frac{D_e}{L}\right)^{0.4}\left(\frac{D_2}{D_1}\right)^{0.8}\left(\frac{\mu_b}{\mu_1}\right)^{0.14} N_{Gr}^{0.05} \quad (5\text{-}41)$$

Limiting Nusselt numbers for **slug-flow annuli** may be predicted (for constant heat flux) from Trefethen (*General Discussions on Heat Transfer*, London, ASME, New York, 1951, p. 436):

$$(N_{Nu})_{lm} = \frac{8(m-1)(m^2-1)^2}{4m^4 \ln m - 3m^4 + 4m^2 - 1} \quad (5\text{-}42)$$

where $m = D_2/D_1$. The Nusselt and Reynolds numbers are based on the equivalent diameter, $D_2 - D_1$.

Limiting Nusselt numbers for laminar flow in annuli have been calculated by Dwyer [*Nucl. Sci. Eng.*, **17**, 336 (1963)]. In addition, theoretical analyses of laminar-flow heat transfer in concentric and eccentric annuli have been published by Reynolds, Lundberg, and McCuen [*Int. J. Heat Mass Transfer*, **6**, 483, 495 (1963)]. Lee [*Int. J. Heat Mass Transfer*, **11**, 509 (1968)] presented an analysis of turbulent heat transfer in entrance regions of concentric annuli. Fully developed local Nusselt numbers were generally attained within a region of 30 equivalent diameters for $0.1 < N_{Pr} < 30$, $10^4 < N_{Re} < 2 \times 10^5$, $1.01 < D_2/D_1 < 5.0$.

Parallel Plates and Rectangular Ducts   The limiting Nusselt number for parallel plates and flat rectangular ducts is given in Table 5-4. Norris and Streid [*Trans. Am. Soc. Mech. Eng.*, **62**, 525 (1940)] report for constant wall temperature

$$(N_{Nu})_{lm} = 1.85 N_{Gz}^{1/3} \quad (5\text{-}43)$$

for $N_{Gz} > 70$. Both Nusselt number and Graetz numbers are based on equivalent diameter. For large temperature differences it is advisable to apply the correction factor $(\mu_b/\mu_w)^{0.14}$ to the right side of Eq. (5-43).

For **rectangular ducts** Kays and Clark (Stanford Univ., Dept. Mech. Eng. Tech. Rep. 14, Aug. 6, 1953) published relationships for heating and cooling of air in rectangular ducts of various aspect ratios. For most **noncircular ducts** Eqs. (5-39) and (5-40) may be used if the equivalent diameter (= $4 \times$ free area/wetted perimeter) is used as the characteristic length. See also Kays and London, *Compact Heat Exchangers*, 3d ed., McGraw-Hill, New York, 1984.

Immersed Bodies   When flow occurs over immersed bodies such that the boundary layer is completely laminar over the whole body, laminar flow is said to exist even though the flow in the mainstream is turbulent. The following relationships are applicable to single bodies immersed in an infinite fluid and *are not valid for assemblages of bodies.*

In general, the average heat-transfer coefficient on immersed bodies is predicted by

$$N_{Nu} = C_r (N_{Re})^m (N_{Pr})^{1/3} \quad (5\text{-}44)$$

Values of C_r and m for various configurations are listed in Table 5-5. The characteristic length is used in both the Nusselt and the Reynolds numbers, and the properties are evaluated at the film temperature = $(t_w + t_\infty)/2$. The velocity in the Reynolds number is the undisturbed free-stream velocity.

Heat transfer from immersed bodies is discussed in detail by Eckert and Drake, Jakob, and Knudsen and Katz (see "Introduction: General References"), where equations for local coefficients and the effects of unheated starting length are presented. Equation (5-44) may also be expressed as

$$N_{St} N_{Pr}^{2/3} = C_r N_{Re}^{m-1} = f/2 \quad (5\text{-}45)$$

where f is the *skin-friction drag coefficient* (not the form drag coefficient).

Falling Films   When a liquid is distributed uniformly around the periphery at the top of a vertical tube (either inside or outside) and allowed to fall down the tube wall by the influence of gravity, the fluid

TABLE 5-5   Laminar-Flow Heat Transfer over Immersed Bodies [Eq. (5-44)]

Configuration	Characteristic length	N_{Re}	N_{Pr}	C_r	m
Flat plate parallel to flow	Plate length	10^3 to 3×10^5	>0.6	0.648	0.50
Circular cylinder axes perpendicular to flow	Cylinder diameter	1 – 4		0.989	0.330
		4 – 40		0.911	0.385
		40 – 4000	>0.6	0.683	0.466
		$4 \times 10^3 - 4 \times 10^4$		0.193	0.618
		$4 \times 10^4 - 2.5 \times 10^5$		0.0266	0.805
Non-circular cylinder, axis	Square, short diameter	$5 \times 10^3 - 10^5$		0.104	0.675
Perpendicular to flow, characteristic	Square, long diameter	$5 \times 10^3 - 10^5$		0.250	0.588
Length perpendicular to flow	Hexagon, short diameter	$5 \times 10^3 - 10^5$	>0.6	0.155	0.638
	Hexagon, long diameter	$5 \times 10^3 - 2 \times 10^4$		0.162	0.638
		$2 \times 10^4 - 10^5$		0.0391	0.782
Sphere°	Diameter	$1 - 7 \times 10^4$	0.6 – 400	0.6	0.50

°Replace N_{Nu} by $N_{Nu} - 2.0$ in Eq. (5-44).

does not fill the tube but rather flows as a thin layer. Similarly, when a liquid is applied uniformly to the outside and top of a horizontal tube, it flows in layer form around the periphery and falls off the bottom. In both these cases the mechanism is called gravity flow of liquid layers or falling films.

For the turbulent flow of **water** in layer form down the walls of **vertical tubes** the dimensional equation of McAdams, Drew, and Bays [*Trans. Am. Soc. Mech. Eng.*, **62**, 627 (1940)] is recommended:

$$h_{lm} = b\Gamma^{1/3} \qquad (5\text{-}46)$$

where $b = 9150$ (SI) or 120 (U.S. customary) and is based on values of $\Gamma = W_F/\pi D$ ranging from 0.25 to 6.2 kg/(m·s) [600 to 15,000 lb/(h·ft)] of wetted perimeter. This type of water flow is used in vertical vapor-in-shell ammonia condensers, acid coolers, cycle water coolers, and other process-fluid coolers.

The following dimensional equations may be used for **any liquid** flowing in layer form down **vertical surfaces:**

For $\quad \dfrac{4\Gamma}{\mu} > 2100 \quad h_{lm} = 0.01 \left(\dfrac{k^3 \rho^2 g}{\mu^2}\right)^{1/3} \left(\dfrac{c\mu}{k}\right)^{1/3} \left(\dfrac{4\Gamma}{\mu}\right)^{1/3} \quad (5\text{-}47a)$

For $\quad \dfrac{4\Gamma}{\mu} < 2100 \quad h_{am} = 0.50 \left(\dfrac{k^2 \rho^{4/3} cg^{2/3}}{L\mu^{1/3}}\right)^{1/3} \left(\dfrac{\mu}{\mu_w}\right)^{1/4} \left(\dfrac{4\Gamma}{\mu}\right)^{1/9} \quad (5\text{-}47b)$

Equation (5-47b) is based on the work of Bays and McAdams [*Ind. Eng. Chem.*, **29**, 1240 (1937)]. The significance of the term L is not clear. When $L = 0$, the coefficient is definitely not infinite. When L is large and the fluid temperature has not yet closely approached the wall temperature, it does not appear that the coefficient should necessarily decrease. Within the finite limits of 0.12 to 1.8 m (0.4 to 6 ft), this equation should give results of the proper order of magnitude.

For falling films applied to the **outside of horizontal tubes,** the Reynolds number rarely exceeds 2100. Equations may be used for falling films on the outside of the tubes by substituting $\pi D/2$ for L.

For **water** flowing over a **horizontal tube,** data for several sizes of pipe are roughly correlated by the dimensional equation of McAdams, Drew, and Bays [*Trans. Am. Soc. Mech. Eng.*, **62**, 627 (1940)].

$$h_{am} = b \, (\Gamma/D_0)^{1/3} \qquad (5\text{-}48)$$

where $b = 3360$ (SI) or 65.6 (U.S. customary) and Γ ranges from 0.94 to 4 kg/(m·s) [100 to 1000 lb/(h·ft)].

Falling films are also used for evaporation in which the film is both entirely or partially evaporated (juice concentration). This principle is also used in crystallization (freezing).

The advantage of high coefficient in falling-film exchangers is partially offset by the difficulties involved in distribution of the film, maintaining complete wettability of the tube, and pumping costs required to lift the liquid to the top of the exchanger.

Transition Region   Turbulent-flow equations for predicting heat transfer coefficients are usually valid only at Reynolds numbers greater than 10,000. The transition region lies in the range $2000 < N_{Re} < 10,000$. No simple equation exists for accomplishing a smooth mathematical transition from laminar flow to turbulent flow. Of the relationships proposed, Hausen's equation [*Z. Ver. Dtsch. Ing. Beih. Verfahrenstech.*, No.

4, 91 (1934)] fits both the laminar extreme and the fully turbulent extreme quite well.

$$(N_{Nu})_{am} = 0.116(N_{Re}^{2/3} - 125)N_{Pr}^{1/3} \left[1 + \left(\dfrac{D}{L}\right)^{2/3}\right] \left(\dfrac{\mu_b}{\mu_w}\right)^{0.14} \quad (5\text{-}49)$$

between 2100 and 10,000. It is customary to represent the probable magnitude of coefficients in this region by hand-drawn curves (Fig. 5-8). Equation (5-40) is plotted as a series of curves (j factor versus Reynolds number with L/D as parameters) terminating at Reynolds number = 2100. Continuous curves for various values of L/D are then hand-drawn from these terminal points to coincide tangentially with the curve for forced-convection, fully turbulent flow [Eq. (5-50c)].

Turbulent Flow

Circular Tubes   Numerous relationships have been proposed for predicting turbulent flow in tubes. For high-Prandtl-number fluids, relationships derived from the equations of motion and energy through the momentum-heat-transfer analogy are more complicated and no more accurate than many of the empirical relationships that have been developed.

For $N_{Re} > 10,000$, $0.7 < N_{Pr} < 170$, for properties based on the bulk temperature and for **heating,** the **Dittus-Boelter equation** [Boelter, Cherry, Johnson and Martinelli, *Heat Transfer Notes*, McGraw-Hill, New York (1965)] may be used:

$$N_{Nu} = 0.0243 \, N_{Re}^{0.8} \, N_{Pr}^{0.4} \, (\mu_b/\mu_w)^{0.14} \qquad (5\text{-}50a)$$

For **cooling,** the relationship is

$$N_{Nu} = 0.0265 \, N_{Re}^{0.8} \, N_{Pr}^{0.3} \, (\mu_b/\mu_w)^{0.14} \qquad (5\text{-}50b)$$

The **Colburn correlation** is

$$j_H = N_{St} N_{Pr}^{2/3} (\mu_w/\mu_b)^{0.14} = 0.023 N_{Re}^{-0.2} \qquad (5\text{-}50c)$$

In Eq. (5-50c), the viscosity-ratio factor may be neglected if properties are evaluated at the film temperature $(t_b + t_w)/2$.

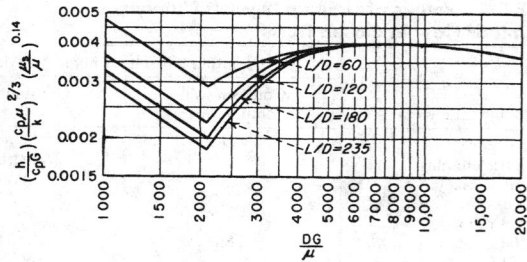

FIG. 5-8   Graphical representation of the Colburn j factor for the heating and cooling of fluids inside tubes. The curves for N_{Re} below 2100 are based on Eq. (5-40). L is the length of each pass in feet. The curve for N_{Re} above 10,000 is represented by Eq. (5-50c).

For the **transition** and **turbulent regions, including diameter to length effects,** Gnielinski [*Int. Chem. Eng.,* **16,** 359 (1976)] recommends a modification of an equation suggested by Petukhov and Popov [*High Temp.,* **1,** 69 (1963)]. This equation applies in the ranges $0 < D/L < 1$, $0.6 < N_{Pr} < 2000$, $2300 < N_{Re} < 10^6$.

$$N_{Nu} = \frac{(f/2)(N_{Re} - 1000) N_{Pr}}{1 + 12.7 (f/2)^{0.5}(N_{Pr}^{2/3} - 1)} \left(1 + \left(\frac{D}{L}\right)^{2/3}\right)\left(\frac{\mu_b}{\mu_w}\right)^{0.14} \quad (5\text{-}51a)$$

The Fanning friction f is determined by an equation recommended by Filonenko [*Teploenergetika,* **1,** 40 (1954)]

$$f = 0.25 (1.82 \log_{10} N_{Re} - 1.64)^{-2} \quad (5\text{-}51b)$$

Any other appropriate friction factor equation for smooth tubes may be used.

Approximate predictions for **rough pipes** may be obtained from Eq. (5-50c) if the right-hand term is replaced by $f/2$ for the rough pipe. For air, Nunner (*Z. Ver. Dtsch. Ing. Forsch.,* 1956, p. 455) obtains

$$\frac{(N_{Nu})_{rough}}{(N_{Nu})_{smooth}} = \frac{f_{rough}}{f_{smooth}} \quad (5\text{-}52)$$

Dippery and Sabersky [*Int. J. Heat Mass Transfer,* **6,** 329 (1963)] present a complete discussion of the influence of roughness on heat transfer in tubes.

Dimensional Equations for Various Conditions   For gases at ordinary pressures and temperatures based on $c\mu/k = 0.78$ and $\mu = (1.76)(10^{-5})$ Pa·s [0.0426 lb/(ft·h)]

$$h = bc\rho^{0.8}(V^{0.8}/D^{0.2}) \quad (5\text{-}53)$$

where $b = (3.04)(10^{-3})$ (SI) or $(1.44)(10^{-2})$ (U.S. customary). For air at atmospheric pressure

$$h = b(V^{0.8}/D^{0.2}) \quad (5\text{-}54)$$

where $b = 3.52$ (SI) or $(4.35)(10^{-4})$ (U.S. customary). For water [based on a temperature range of 5 to 104°C (40 to 220°F)]

$$h = 1057 (1.352 + 0.02t) (V^{0.8}/D^{0.2}) \quad (5\text{-}55a)$$

in SI units with $t = $ °C, or

$$h = 0.13(1 + 0.011t)(V^{0.8}/D^{0.2}) \quad (5\text{-}55b)$$

in U.S. customary units with $t = $ °F.

For organic liquids, based on $c = 2.092$ J/kg·K [0.5 Btu/(lb·°F)], $k = 0.14$ J/(m·s·K) [0.08 Btu/(h·ft·°F)], $\mu_b = (1)(10^{-3})$ Pa·s (1.0 cP), and $\rho = 810$ kg/m³ (50 lb/ft³),

$$h = b(V^{0.8}/D^{0.2}) \quad (5\text{-}56)$$

where $b = 423$ (SI) or $(5.22)(10^{-2})$ (U.S. customary). Within reasonable limits, coefficients for organic liquids are about one-third of the values obtained for water.

Entrance effects are usually not significant industrially if $L/D > 60$. Below this limit Nusselt recommended the conservative equation for $10 < L/D < 400$ and properties evaluated at bulk temperature

$$N_{Nu} = 0.036N_{Re}^{0.8} N_{Pr}^{1/3}(L/D)^{-0.054} \quad (5\text{-}57)$$

It is common to correlate entrance effects by the equation

$$h_m/h = 1 + F(D/L) \quad (5\text{-}58)$$

where h is predicted by Eq. (5-50a) or (5-50b), and h_m is the mean coefficient for the pipe in question. Values of F are reported by Boelter, Young, and Iverson [NACA Tech. Note 1451, 1948] and tabulated by Kays and Knudsen and Katz (see "Introduction: General References"). Selected values of F are as follows:

Fully developed velocity profile	1.4
Abrupt contraction entrance	6
90° right-angle bend	7
180° round bend	6

For **large temperature differences** different equations are necessary and usually are specifically applicable to either gases or liquids. Gambill (*Chem. Eng.,* Aug. 28, 1967, p. 147) provides a detailed review of high-flux heat transfers to gases. He recommends

$$N_{Nu} = \frac{0.021N_{Re}^{0.8}N_{Pr}^{0.4}}{(T_w/T_b)^{0.29 + 0.0019 (L/D)}} \quad (5\text{-}59)$$

for $10 < L/D < 240$, $110 < T_b < 1560$ K ($200 < T_b < 2800$°R), $1.1 < (T_w/T_b) < 8.0$, and properties evaluated at T_b. For liquids, Eq. (5-50c) is generally satisfactory.

Annuli   For diameter ratios $D_1/D_2 > 0.2$, Monrad and Pelton's equation [*Trans. Am. Inst. Chem. Eng.,* **38,** 593 (1942)] is recommended for either or both the inner and outer tube:

$$N_{Nu} = 0.020N_{Re}^{0.8} N_{Pr}^{1/3}(D_2/D_1)^{0.53} \quad (5\text{-}60a)$$

Equation (5-51a) may also be used for **smooth annuli** as follows:

$$\frac{(N_{Nu})_{ann}}{(N_{Nu})_{tube}} = \phi\left(\frac{D_1}{D_2}\right) \quad (5\text{-}60b)$$

The hydraulic diameter $D_2 - D_1$ is used in N_{Nu}, N_{Re}, and D/L is used for the annulus. The function on the right of Eq. (5-60b) is given by Petukhov and Roizen [*High Temp.,* **2,** 65 (1964)] as follows:

Inner tube heated	$0.86 (D_1/D_2)^{-0.16}$
Outer tube heated	$1 - 0.14 (D_1/D_2)^{0.6}$

If both tubes are heated, the function is the sum of the above two functions divided by $1 + D_1/D_2$ [Stephan, *Chem. Ing. Tech.,* **34,** 207 (1962)]. The Colburn form of relationship may be employed for the individual walls of the annulus by using the individual friction factor for each wall [see Knudsen, *Am. Inst. Chem. Eng. J.,* **8,** 566 (1962)]:

$$j_{H1} = (N_{St})_1 N_{Pr}^{2/3} = f_1/2 \quad (5\text{-}61a)$$
$$j_{H2} = (N_{St})_2 N_{Pr}^{2/3} = f_2/2 \quad (5\text{-}61b)$$

Rothfus, Monrad, Sikchi, and Heideger [*Ind. Eng. Chem.,* **47,** 913 (1955)] report that the friction factor f_2 for the outer wall bears the same relation to the Reynolds number for the outer portion of the annular stream $2(r_2^2 - \lambda_m)V\rho/r_2\mu$ as the friction factor for circular tubes does to the Reynolds number for circular tubes, where r_2 is the radius of the outer tube and λ_m is the position of maximum velocity in the annulus, estimated from

$$\lambda_m = \frac{r_2^2 - r_1^2}{\ln (r_2/r_1)^2} \quad (5\text{-}62)°$$

To calculate the friction factor f_1 for the inner tube use the relation

$$f_1 = \frac{f_2 r_2(\lambda_m - r_1^2)}{r_1(r_2^2 - \lambda_m)} \quad (5\text{-}63)$$

There have been several analyses of turbulent heat transfer in annuli: for example, Deissler and Taylor (NACA Tech. Note 3451, 1955), Kays and Leung [*Int. J. Heat Mass Transfer,* **6,** 537 (1963)], Lee [*Int. J. Heat Transfer,* **11,** 509 (1968)], Sparrow, Hallman and Siegel [*Appl. Sci. Res.,* **7A,** 37 (1958)], and Johnson and Sparrow [*Am. Soc. Mech. Eng. J. Heat Transfer,* **88,** 502 (1966)]. The reader is referred to these for details of the analyses.

For **annuli containing externally finned tubes** the heat-transfer coefficients are a function of the fin configurations. Knudsen and Katz (*Fluid Dynamics and Heat Transfer,* McGraw-Hill, New York, 1958) present relationships for transverse finned tubes, spined tubes, and longitudinal finned tubes in annuli.

Noncircular Ducts   Equations (5-50a) and (5-50b) may be employed for noncircular ducts by using the equivalent diameter $D_e = 4 \times$ free area per wetted perimeter. Kays and London (*Compact Heat Exchangers,* 3rd ed., McGraw-Hill, New York, 1984) give charts for various noncircular ducts encountered in compact heat exchangers.

Vibrations and pulsations generally tend to increase heat-transfer coefficients.

Example 2: Calculation of j Factors in an Annulus   Calculate the heat-transfer j factors for both walls of an annulus for the following condi-

---

° Equation (5-62) predicts the point of maximum velocity for *laminar* flow in annuli and is only an approximate equation for turbulent flow. Brighton and Jones [*Am. Soc. Mech. Eng. Basic Eng.,* **86,** 835 (1964)] and Macagno and McDougall [*Am. Inst. Chem. Eng. J.,* **12,** 437 (1966)] give more accurate equations for predicting the point of maximum velocity for turbulent flow.

tions: $D_1 = 0.0254$ m (1.0 in); $D_2 = 0.0635$ m (2.5 in); water at 15.6°C (60°F); $\mu/\rho = (1.124)(10^{-6})$ m²/s [$(1.21)(10^{-5})$ ft²/s]; velocity = 1.22 m/s (4 ft/s).

$$\lambda_m = \frac{0.0635^2 - 0.0254^2}{4 \ln (0.0635/0.0254)^2} = (4.621)(10^{-4}) \text{ m}^2 \ (0.716 \text{ in}^2)$$

$$\text{Re}_2 = \frac{2(r_2^2 - \lambda_m)V\rho}{r_2\mu} = \frac{2[0.0318^2 - (4.621)(10^{-4})(1.22)]}{(0.0318)(1.124)(10^{-6})} = (3.74)(10^4)$$

From Eq. (5-51b), $f_2 = 0.0055$. Hence

$$j_{H2} = (N_{St})_2 \, N_{Pr}^{2/3} = 0.00275$$

From Eq. (5-63),

$$f_1 = \frac{(0.0055)(0.0318)[(4.621)(10^{-4}) - 0.0127^2]}{(0.0127)[0.0318^2 - (4.621)(10^{-4})]} = 0.00754$$

from which $j_{H_1} = (N_{St})_1 N_{Pr}^{2/3} = 0.00377$.

These results indicate that for this system the heat-transfer coefficient on the inner tube is about 40 percent greater than on the outer tube.

Coils For flow *inside* **helical coils,** Reynolds number above 10,000, multiply the value of the film coefficient obtained from the applicable equation for straight tubes by the term $(1 + 3.5 \, D_i/D_c)$.

For flow inside helical coils, Reynolds number less than 10,000, substitute the term $(D_c/D_i)^{1/2}$ for (L/D_i) where the latter appears in the applicable equation for straight tubes (frequently as part of the Graetz number).

For **flat spiral (pancake) coils,** in which the ratio D_c/D_i varies for each turn, a different value of coefficient will be obtained for each turn; a weighted average based on length per turn is used.

For flow *outside* **helical coils** use the equation for flow normal to a bank of tubes, in-line flow.

Finned Tubes (Extended Surface) When the film coefficient on the outside of a metal tube is much lower than that on the inside, as when steam condensing in a pipe is being used to heat air, externally finned (or extended) heating surfaces are of value in increasing substantially the rate of heat transfer per unit length of tube. The data on extended heating surfaces, for the case of air flowing outside and at right angles to the axes of a bank of finned pipes, can be represented approximately by the dimensional equation derived from

$$h_f = b \, \frac{V_F^{0.6}}{D_0^{0.4}} \left(\frac{p'}{p' - D_0} \right)^{0.6} \tag{5-64}$$

where $b = 5.29$ (SI) or $(5.39)(10^{-3})$ (U.S. customary); h_f is the film coefficient of heat transfer on the air side; V_F is the face velocity of the air; p' is the center-to-center spacing, m, of the tubes in a row; and D_0 is the outside diameter, m, of the bare tube (diameter at the root of the fins).

In atmospheric air-cooled finned tube exchangers, the air-film coefficient from Eq. (5-64) is sometimes converted to a value based on outside bare surface as follows:

$$h_{fo} = h_f \, \frac{A_f + A_{uf}}{A_{of}} = h_f \, \frac{A_T}{A_o} \tag{5-65}$$

in which h_{fo} is the air-film coefficient based on external bare surface; h_f is the air-film coefficient based on total external surface; A_T is total external surface, and A_o is external bare surface of the unfinned tube; A_f is the area of the fins; A_{uf} is the external area of the unfinned portion of the tube; and A_{of} is area of tube before fins are attached.

Fin efficiency is defined as the ratio of the mean temperature difference from surface to fluid divided by the temperature difference from fin to fluid at the base or root of the fin. Graphs of fin efficiency for extended surfaces of various types are given by Gardner [*Trans. Am. Soc. Mech. Eng.*, **67**, 621 (1945)].

Heat-transfer coefficients for finned tubes of various types are given in a series of papers [*Trans. Am. Soc. Mech. Eng.*, **67**, 601 (1945)].

For flow of air normal to fins in the form of **short strips or pins,** Norris and Spofford [*Trans. Am. Soc. Mech. Eng.*, **64**, 489 (1942)] correlate their results for air by the dimensionless equation of Pohlhausen:

$$\frac{h_m}{c_p G_{max}} \left(\frac{c_p \mu}{k} \right)^{2/3} = 1.0 \left(\frac{z_p G_{max}}{\mu} \right)^{-0.5} \tag{5-66}$$

for values of $z_p G_{max}/\mu$ ranging from 2700 to 10,000.

For the general case, the treatment suggested by Kern (*Process Heat Transfer,* McGraw-Hill, New York, 1950, p. 512) is recommended. Because of the wide variations in fin-tube construction, it is convenient to convert all film coefficients to values based on the inside bare surface of the tube. Thus to convert the film coefficient based on outside area (finned side) to a value based on inside area Kern gives the following relationship:

$$h_{fi} = (\Omega A_f + A_o)(h_f/A_i) \tag{5-67}$$

in which h_{fi} is the effective outside film coefficient based on the inside area, h_f is the outside film coefficient calculated from the applicable equation for bare tubes, A_f is the surface area of the fins, A_o is the surface area on the outside of the tube which is not finned, A_i is the inside area of the tube, and Ω is the fin efficiency defined as

$$\Omega = (\tanh mb_f)/mb_f \tag{5-68}$$

in which

$$m = (h_f p_f/ka_x)^{1/2} \text{ m}^{-1} \ (\text{ft}^{-1}) \tag{5-69}$$

and b_f = height of fin. The other symbols are defined as follows: p_f is the perimeter of the fin, a_x is the cross-sectional area of the fin, and k is the thermal conductivity of the material from which the fin is made.

Fin efficiencies and fin dimensions are available from manufacturers. Ratios of finned to inside surface are usually available so that the terms A_f, A_o, and A_i may be obtained from these ratios rather than from the total surface areas of the heat exchangers.

Banks of Tubes For heating and cooling of fluids flowing normal to a bank of circular tubes at least 10 rows deep the following equations are applicable:

Colburn type:

$$\frac{h}{cG_{max}} \left(\frac{c\mu}{k} \right)^{2/3} = \frac{a}{(D_o G_{max}/\mu)^{0.4}} = j \tag{5-70}$$

Nusselt type:

$$\frac{hD}{k} = a \left(\frac{D_o G_{max}}{\mu} \right)^{0.6} \left(\frac{c\mu}{k} \right)^{1/3} \tag{5-71}$$

The dimensionless constant a in these equations varies depending upon conditions.

Conditions, Reynolds number > 3000	Value of a
Flow normal to apex of diamond, staggered arrangement	
No leakage	0.330
Normal leakage in baffled exchanger	0.198
Flow normal to flat side of diamond, not staggered (in-line) arrangement	
No leakage	0.260
Normal leakage in baffled exchanger	0.156

For Reynolds number less than 3000, Eq. (5-70) would give conservative results, but greater accuracy (if desired) may be obtained by using the following equation.

$$\frac{h}{cG_{max}} \left(\frac{c\mu}{k} \right)^{2/3} = \frac{a}{(D_o G_{max}/\mu)^m} = j \tag{5-72}$$

in which the constant a and exponent m are as follows:

Reynolds number	m	Tube pitch	Leakage	a
100–300	0.492	Staggered	None	0.695
			Normal	0.416
		In-line	None	0.548
			Normal	0.329
1–100	0.590	Staggered	None	1.086
			Normal	0.650
		In-line	None	0.855
			Normal	0.513

The following **dimensional equations** (5-73 to 5-77) are based on flow normal to a bank of staggered tubes without leakage. Multiply the values obtained for h by 0.6 for normal leakage and, in addition, by 0.79 for in-line (not staggered) tube arrangement.

$$h = b \frac{c^{1/3} k^{2/3} \rho^{0.6} V_{max}^{0.6}}{\mu^{0.267} D_0^{0.4}} \qquad (5\text{-}73)$$

where $b = 0.33$ (SI) or 0.261 (U.S. customary). For gases at ordinary pressures and temperatures, based on $c\mu/k = 0.78$; $\mu = (1.76)(10^{-5})$ Pa·s [0.0426 lb/(ft·h)],

$$h = bc \frac{G_{max}^{0.6}}{D_0^{0.4}} \qquad (5\text{-}74)$$

where $b = (4.82)(10^{-3})$ (SI) or 0.109 (U.S. customary). For air at atmospheric pressure

$$h = b \frac{V_{max}^{0.6}}{D_0^{0.4}} \qquad (5\text{-}75)$$

where $b = 5.33$ (SI) or $(5.44)(10^{-3})$ (U.S. customary). For water based on a temperature range 7 to 104°C (40 to 220°F)

$$h = 986(1.21 + 0.0121t) \frac{V_{max}^{0.6}}{D_0^{0.4}} \qquad (5\text{-}76a)$$

in SI units and t in °C.

$$h = 1.01(1 + 0.0067t) \frac{V_{max}^{0.6}}{D_0^{0.4}} \qquad (5\text{-}76b)$$

in U.S. customary units and t in °F. For organic liquids, based on $c = 2.22$ J/(kg·K) [0.53 Btu/(lb·°F)], $k = 0.14$ J/(m·s·K) [0.08 Btu/(h·ft·°F)], $\mu_b = (1)(10^{-3})$ Pa·s (1.0 cP), $\rho = 810$ kg/m^3 (50 lb/ft^3),

$$h = b \frac{V_{max}^{0.6}}{D_0^{0.4}} \qquad (5\text{-}77)$$

where $b = 400$ (SI) or 0.408 (U.S. customary).

JACKETS AND COILS OF AGITATED VESSELS

See Sec. 18.

NONNEWTONIAN FLUIDS

A wide variety of nonnewtonian fluids are encountered industrially. They may exhibit Bingham-plastic, pseudoplastic, or dilatant behavior and may or may not be thixotropic. For design of equipment to handle or process nonnewtonian fluids, the properties must usually be measured experimentally, since no generalized relationships exist to predict the properties or behavior of the fluids. Details of handling nonnewtonian fluids are described completely by Skelland (*Non-Newtonian Flow and Heat Transfer*, Wiley, New York, 1967). The generalized shear-stress rate-of-strain relationship for nonnewtonian fluids is given as

$$n' = \frac{d \ln (D \,\Delta P/4L)}{d \ln (8V/D)} \qquad (5\text{-}78)$$

as determined from a plot of shear stress versus velocity gradient.

For **circular tubes**, $N_{Gz} > 100$, $n' > 0.1$, and laminar flow

$$(N_{Nu})_{lm} = 1.75 \, \delta_s^{1/3} N_{Gz}^{1/3} \qquad (5\text{-}79)$$

where $\delta_s = (3n' + 1)/4n'$. When natural-convection effects are considered, Metzer and Gluck [*Chem. Eng. Sci.*, **12**, 185 (1960)] obtained the following for **horizontal tubes**:

$$(N_{Nu})_{lm} = 1.75 \, \delta_s^{1/3} \left[N_{Gz} + 12.6 \left(\frac{N_{Pr} N_{Gr} D}{L} \right)^{0.4} \right]^{1/3} \left(\frac{\gamma_b}{\gamma_w} \right)^{0.14} \qquad (5\text{-}80)$$

where properties are evaluated at the wall temperature, i.e., $\gamma = g_c K' 8^{n'-1}$ and $\tau_w = K'(8V/D)^{n'}$.

Metzner and Friend [*Ind. Eng. Chem.*, **51**, 879 (1959)] present relationships for turbulent heat transfer with nonnewtonian fluids. Relationships for heat transfer by natural convection and through laminar boundary layers are available in Skelland's book (op. cit.).

LIQUID METALS

Liquid metals constitute a class of heat-transfer media having Prandtl numbers generally below 0.01. Heat-transfer coefficients for liquid metals cannot be predicted by the usual design equations applicable to gases, water, and more viscous fluids with Prandtl numbers greater than 0.6. Relationships for predicting heat-transfer coefficients for liquid metals have been derived from solution of Eqs. (5-38a) and (5-38b). By the momentum-transfer-heat-transfer analogy, the eddy conductivity of heat is $kN_{Pr}(E_M/\mu) \approx k$ for small N_{Pr}. Thus in the solution of Eqs. (5-38a) and (5-38b) the knowledge of the thickness of various layers of flow is not critical. In fact, assumption of slug flow and constant conductivity ($=k$) across the duct gives reasonable values of heat-transfer coefficients for liquid metals.

For **constant heat flux:**

$$N_{Nu} = 5 + 0.025(N_{Re} N_{Pr})^{0.8} \qquad (5\text{-}81)$$

For **constant wall temperature:**

$$N_{Nu} = 7 + 0.025(N_{Re} N_{Pr})^{0.8} \qquad (5\text{-}82)$$

For $0.003 < N_{Pr} < 0.05$ and constant heat flux, Sleicher and Rouse [*Int. J. Heat Mass Transfer*, **18**, 677 (1975)] obtained the correlation

$$N_{Nu} = 6.3 + 0.0167 N_{Re}^{0.85} N_{Pr}^{0.93} \qquad (5\text{-}83)$$

For **parallel plates and annuli** with $D_2/D_1 < 1.4$ and uniform heat flux, Seban [*Trans. Am. Soc. Mech. Eng.*, **72**, 789 (1950)] obtained the equation

$$N_{Nu} = 5.8 + 0.020(N_{Re} N_{Pr})^{0.8} \qquad (5\text{-}84)$$

For annuli only, application of a factor of $0.70(D_2/D_1)^{0.53}$ is recommended for Eqs. (5-81) and (5-82). For more accurate semiempirical relationships for tubes, annuli, and rod bundles, refer to Dwyer [*Am. Inst. Chem. Eng. J.*, **9**, 261 (1963)].

Hsu [*Int. J. Heat Mass Transfer*, **7**, 431 (1964)] and Kalish and Dwyer [*Int. J. Heat Mass Transfer*, **10**, 1533 (1967)] discuss heat transfer to liquid metals flowing across **banks of tubes.** Hsu recommends the equations

$$N_{Nu} = 0.81 N_{Re} N_{Pr} (\phi/D)^{1/2} \quad \text{(for uniform heat flux)} \qquad (5\text{-}85)$$

$$N_{Nu} = 0.096 N_{Re} N_{Pr} (\phi/D)^{1/2} \quad \text{(for cosine surface temperature)} \qquad (5\text{-}86)$$

where the heat-transfer coefficient is based on the average circumferential temperature around the tubes, the Reynolds number is based on the superficial velocity through the tube bank, D is the tube outside diameter, and ϕ is a velocity potential function having the following values:

D/p'	ϕ/D square pitch	ϕ/D equilateral triangular pitch
0	2.00	2.00
0.1	2.02	2.02
0.2	2.07	2.06
0.3	2.16	2.15
0.4	2.30	2.27
0.5	2.52	2.45
0.6	2.84	2.71
0.7	3.34	3.11
0.8	4.23	3.80

Equations (5-85) and (5-86) are useful in calculating tube-surface temperatures.

Further information on liquid-metal heat transfer in tube banks is given by Hsu for spheres and elliptical rod bundles [*Int. J. Heat Mass Transfer*, **8**, 303 (1965)] and by Kalish and Dwyer for oblique flow across tube banks [*Int. J. Heat Mass Transfer*, **10**, 1533 (1967)]. For additional details of heat transfer with liquid metals for various systems see Dwyer (1968 ed., Na and Nak supplement to *Liquid Metals Handbook*) and Stein ("Liquid Metal Heat Transfer," in *Advances in Heat Transfer*, vol. 3, Academic, New York, 1966).

HEAT TRANSFER WITH CHANGE OF PHASE

In any operation in which a material undergoes a change of phase, provision must be made for the addition or removal of heat to provide for the latent heat of the change of phase plus any other sensible heating or cooling that occurs in the process. Heat may be transferred by any one or a combination of the three modes—conduction, convection, and radiation. The process involving change of phase involves mass transfer simultaneous with heat transfer.

CONDENSATION

Condensation Mechanisms   Condensation occurs when a saturated vapor comes in contact with a surface whose temperature is below the saturation temperature. Normally a film of condensate is formed on the surface, and the thickness of this film, per unit of breadth, increases with increase in extent of the surface. This is called **film-type condensation.**

Another type of condensation, called **dropwise,** occurs when the wall is not uniformly wetted by the condensate, with the result that the condensate appears in many small droplets at various points on the surface. There is a growth of individual droplets, a coalescence of adjacent droplets, and finally a formation of a rivulet. Adhesional force is overcome by gravitational force, and the rivulet flows quickly to the bottom of the surface, capturing and absorbing all droplets in its path and leaving dry surface in its wake.

Film-type condensation is more common and more dependable. Dropwise condensation normally needs to be promoted by introducing an impurity into the vapor stream. Substantially higher (6 to 18 times) coefficients are obtained for dropwise condensation of steam, but design methods are not available. Therefore, the development of equations for condensation will be for the film type only.

The physical properties of the liquid, rather than those of the vapor, are used for determining the film coefficient for condensation. Nusselt [Z. Ver. Dtsch. Ing., **60,** 541, 569 (1916)] derived theoretical relationships for predicting the film coefficient of heat transfer for condensation of a pure saturated vapor. A number of simplifying assumptions were used in the derivation.

The **Reynolds number** of the condensate film (falling film) is $4\Gamma/\mu$, where Γ is the weight rate of flow (loading rate) of condensate per unit perimeter kg/(s·m) [lb/(h·ft)]. The thickness of the condensate film for Reynolds number less than 2100 is $(3\mu\Gamma/\rho^2 g)^{1/3}$.

Condensation Coefficients

Vertical Tubes   For the following cases Reynolds number < 2100 and is calculated by using $\Gamma = W_F/\pi D$. The **Nusselt equation** for the heat-transfer coefficient for condensate films may be written in the following ways (using liquid physical properties and where L is the cooled length and Δt is $t_{sv} - t_s$):
Colburn type:

$$\frac{h}{cG}\frac{c\mu}{k} = \frac{5.35}{4\Gamma/\mu} \tag{5-87}$$

where $G = \dfrac{\Gamma}{(3\mu\Gamma/\rho^2 g)^{1/3}} = \left(\dfrac{W_F^2\rho^2 g}{29.6 D^2\mu}\right)^{1/3}$ kg/(s·m²) [lb/(h·ft²)]

Nusselt type:

$$\frac{hL}{k} = 0.943\left(\frac{L^3\rho^2 g\lambda}{k\mu\,\Delta t}\right)^{1/4} = 0.925\left(\frac{L^3\rho^2 g}{\mu\Gamma}\right)^{1/3} \tag{5-88}$$

Dimensional:

$$h = b(k^3\rho^2 D/\mu_b W_F)^{1/3} \tag{5-89}$$

where $b = 127$ (SI) or 756 (U.S. customary). For steam at atmospheric pressure, $k = 0.682$ J/(m·s·K) [0.394 Btu/(h·ft·°F)], $\rho = 960$ kg/m³ (60 lb/ft³), $\mu_b = (0.28)(10^{-3})$ Pa·s (0.28 cP),

$$h = b(D/W_F)^{1/3} \tag{5-90}$$

where $b = 2954$ (SI) or 6978 (U.S. customary). For organic vapors at normal boiling point, $k = 0.138$ J/(m·s·K) [0.08 Btu/(h·ft·°F)], $\rho = 720$ kg/m³ (45 lb/ft³), $\mu_b = (0.35)(10^{-3})$ Pa·s (0.35 cP),

$$h = b(D/W_F)^{1/3} \tag{5-91}$$

where $b = 457$ (SI) or 1080 (U.S. customary).
Horizontal Tubes   For the following cases Reynolds number < 2100 and is calculated by using $\Gamma = W_F/2L$.
Colburn type:

$$\frac{h}{cG}\frac{c\mu}{k} = \frac{4.4}{4\Gamma/\mu} \tag{5-92}$$

$$G = \frac{\Gamma}{(3\mu\Gamma/\rho^2 g)^{1/3}} = \left(\frac{W_F^2\rho^2 g}{12L^2\mu}\right)^{1/3} \text{ kg/(s·m²) [lb/(h·ft²)]}$$

Nusselt type:

$$\frac{hD}{k} = 0.73\left(\frac{D^3\rho^2 g\lambda}{k\mu\,\Delta t}\right)^{1/4} = 0.76\left(\frac{D^3\rho^2 g}{\mu\Gamma}\right)^{1/3} \tag{5-93}°$$

Dimensional:

$$h = b(k^3\rho^2 L/\mu_b W_F)^{1/3} \tag{5-94}$$

where $b = 205.4$ (SI) or 534 (U.S. customary). For steam at atmospheric pressure

$$h = b(L/W_F)^{1/3} \tag{5-95}$$

where $b = 2080$ (SI) or 4920 (U.S. customary). For organic vapors at normal boiling point

$$h = b(L/W_F)^{1/3} \tag{5-96}$$

where $b = 324$ (SI) or 766 (U.S. customary).

Figure 5-9 is a nomograph for determining coefficients of heat transfer for condensation of pure vapors.

Banks of Horizontal Tubes   ($N_{Re} < 2100$) In the idealized case of N tubes in a vertical row where the total condensate flows smoothly from one tube to the one beneath it, without splashing, and still in laminar flow on the tube, the mean condensing coefficient h_N for the entire row of N tubes is related to the condensing coefficient for the top tube h_1 by

$$h_N = h_1 N^{-1/4} \tag{5-97}$$

Dukler Theory   The preceding expressions for condensation are based on the classical Nusselt theory. It is generally known and conceded that the film coefficients for steam and organic vapors calculated by the Nusselt theory are conservatively low. Dukler [Chem. Eng. Prog., **55,** 62 (1959)] developed equations for velocity and temperature distribution in thin films on vertical walls based on expressions of Deissler (NACA Tech. Notes 2129, 1950; 2138, 1952; 3145, 1959) for the eddy viscosity and thermal conductivity near the solid boundary. According to the Dukler theory, three fixed factors must be known to establish the value of the average film coefficient: the terminal Reynolds number, the Prandtl number of the condensed phase, and a dimensionless group N_d defined as follows:

$$N_d = (0.250\mu_L^{1.173}\mu_G^{0.16})/(g^{2/3}D^2\rho_L^{0.553}\rho_G^{0.78}) \tag{5-98}$$

Graphical relationships of these variables are available in Document 6058, ADI Auxiliary Publications Project, Library of Congress, Washington. If rigorous values for condensing-film coefficients are desired, especially if the value of N_d in Eq. (5-98) exceeds $(1)(10^{-5})$, it is suggested that these graphs be used. For the case in which interfacial shear is zero, Fig. 5-10 may be used. It is interesting to note that, according to the Dukler development, there is no definite transition Reynolds number; deviation from Nusselt theory is less at low Reynolds numbers; and when the Prandtl number of a fluid is less

---

° If the vapor density is significant, replace ρ^2 with $\rho_l(\rho_l - \rho_v)$.

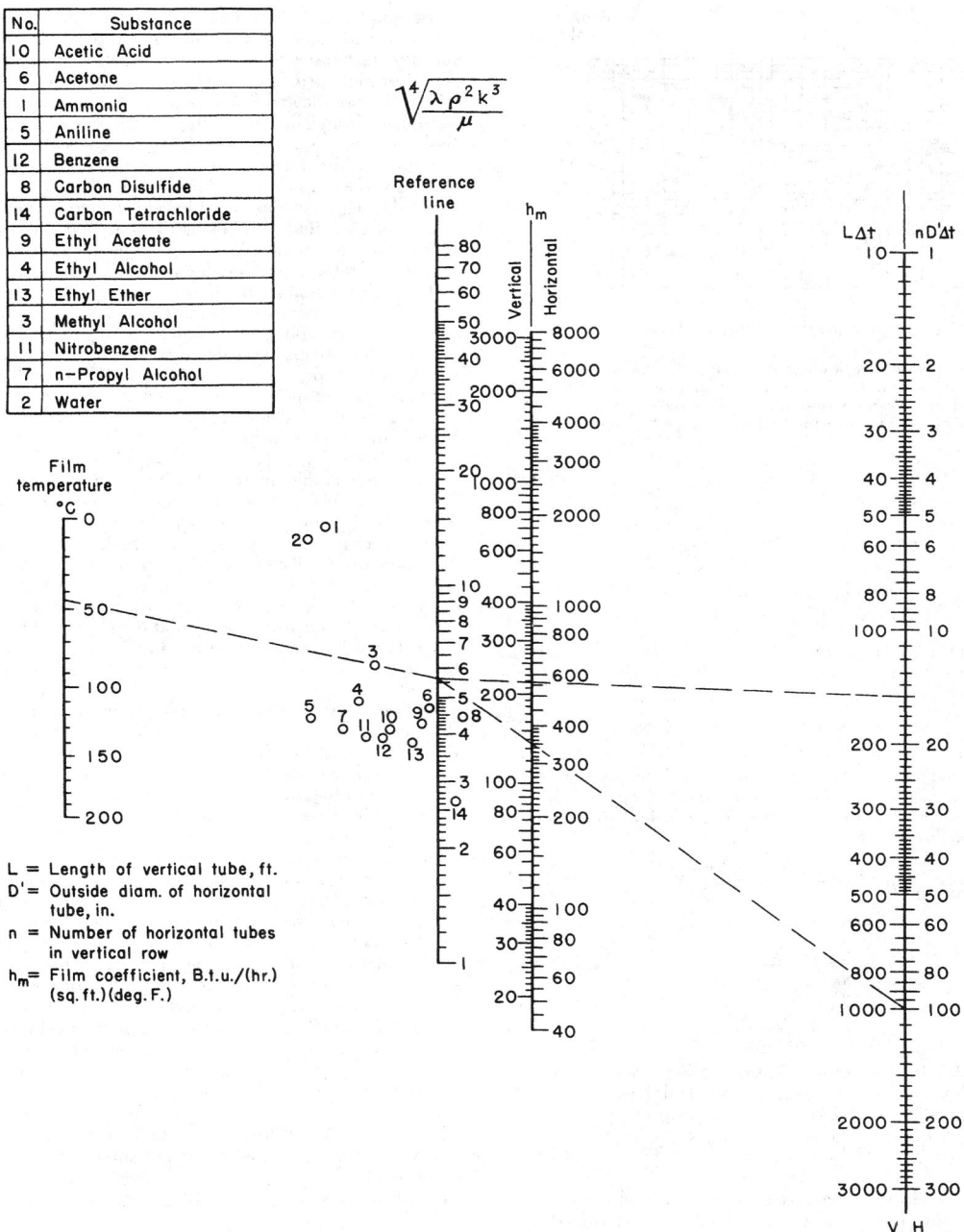

No.	Substance
10	Acetic Acid
6	Acetone
1	Ammonia
5	Aniline
12	Benzene
8	Carbon Disulfide
14	Carbon Tetrachloride
9	Ethyl Acetate
4	Ethyl Alcohol
13	Ethyl Ether
3	Methyl Alcohol
11	Nitrobenzene
7	n−Propyl Alcohol
2	Water

$$\sqrt[4]{\frac{\lambda \rho^2 k^3}{\mu}}$$

L = Length of vertical tube, ft.
D' = Outside diam. of horizontal tube, in.
n = Number of horizontal tubes in vertical row
h_m = Film coefficient, B.t.u./(hr.)(sq. ft.)(deg. F.)

FIG. 5-9   Chart for determining film coefficient h_m for film-type condensation of pure vapor, based on Eqs. 5-88 and 5-93. For vertical tubes multiply h_m by 1.2. If $4\Gamma/\mu_f$ exceeds 2100, use Fig. 5-10. $\sqrt[4]{\lambda\rho^2 k^3/\mu}$ is in U.S. customary units; to convert feet to meters, multiply by 0.3048; to convert inches to centimeters, multiply by 2.54; and to convert British thermal units per hour–square foot–degrees Fahrenheit to watts per square meter–kelvins, multiply by 5.6780.

than 0.4 (at Reynolds number above 1000), the predicted values for film coefficient are lower than those predicted by the Nusselt theory.

The Dukler theory is applicable for condensate films on horizontal tubes and also for falling films, in general, i.e., those not associated with condensation or vaporization processes.

Vapor Shear Controlling   For **vertical in-tube condensation** with vapor and liquid flowing cocurrently downward, if gravity controls, Figs. 5-9 and 5-10 may be used. If vapor shear controls, the Carpenter-Colburn correlation (*General Discussion on Heat Transfer*, London, 1951, ASME, New York, p. 20) is applicable:

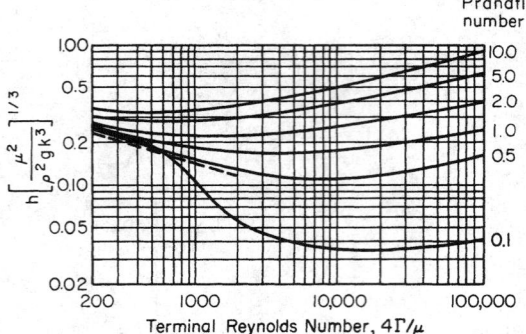

FIG. 5-10 Dukler plot showing average condensing-film coefficient as a function of physical properties of the condensate film and the terminal Reynolds number. (Dotted line indicates Nusselt theory for Reynolds number < 2100.) [*Reproduced by permission from* Chem. Eng. Prog., **55**, 64 *(1959).*]

$$h\mu_l/k_l\rho_l^{1/2} = 0.065(N_{Pr})^{1/2}F_{vc}^{1/2} \qquad (5\text{-}99a)$$

where
$$F_{vc} = fG_{vm}^2/2\rho_v \qquad (5\text{-}99b)$$

$$G_{vm} = \left(\frac{G_{vi}^2 + G_{vi}G_{vo} + G_{vo}^2}{3}\right)^{1/2} \qquad (5\text{-}99c)$$

and f is the Fanning friction factor evaluated at

$$(N_{Re})_{vm} = D_i G_{vm}/\mu_v \qquad (5\text{-}99d)$$

and the subscripts vi and vo refer to the vapor inlet and outlet, respectively. An alternative formulation, directly in terms of the friction factor, is

$$h = 0.065 \, (c\rho k f/2\mu \rho_v)^{1/2} G_{vm} \qquad (5\text{-}99e)$$

expressed in consistent units.

Another correlation for vapor-shear-controlled condensation is the Boyko-Kruzhilin correlation [*Int. J. Heat Mass Transfer*, **10**, 361 (1967)], which gives the mean condensing coefficient for a stream between inlet quality x_i and outlet quality x_o:

$$\frac{hD_i}{k_l} = 0.024 \left(\frac{D_i G_T}{\mu_l}\right)^{0.8}(N_{Pr})_l^{0.43}\frac{\sqrt{(\rho/\rho_m)_i} + \sqrt{(\rho/\rho_m)_o}}{2} \qquad (5\text{-}100a)$$

where G_T = total mass velocity in consistent units

$$\left(\frac{\rho}{\rho_m}\right)_i = 1 + \frac{\rho_l - \rho_v}{\rho_v}\,x_i \qquad (5\text{-}100b)$$

and
$$\left(\frac{\rho}{\rho_m}\right)_o = 1 + \frac{\rho_l - \rho_v}{\rho_v}\,x_o \qquad (5\text{-}100c)$$

For **horizontal in-tube condensation** at low flow rates Kern's modification (*Process Heat Transfer*, McGraw-Hill, New York, 1950) of the Nusselt equation is valid:

$$h_m = 0.761 \left[\frac{Lk_l^3\rho_l(\rho_l - \rho_v)g}{W_F\mu_l}\right]^{1/3} = 0.815\left[\frac{k_l^3\rho_l(\rho_l - \rho_v)g\lambda}{\pi\mu_l D_i \, \Delta t}\right]^{1/4} \qquad (5\text{-}101)$$

where W_F is the total vapor condensed in one tube and Δt is $t_{sv} - t_s$. A more rigorous correlation has been proposed by Chaddock [*Refrig. Eng.*, **65**(4), 36 (1957)]. Use consistent units.

At high condensing loads, with vapor shear dominating, tube orientation has no effect, and Eq. (5-100a) may also be used for horizontal tubes.

Condensation of pure vapors under laminar conditions in the presence of noncondensable gases, interfacial resistance, superheating, variable properties, and diffusion has been analyzed by Minkowycz and Sparrow [*Int. J. Heat Mass Transfer*, **9**, 1125 (1966)].

BOILING (VAPORIZATION) OF LIQUIDS

Boiling Mechanisms Vaporization of liquids may result from various mechanisms of heat transfer, singly or combinations thereof.

For example, vaporization may occur as a result of heat absorbed, by radiation and convection, at the surface of a pool of liquid; or as a result of heat absorbed by natural convection from a hot wall beneath the disengaging surface, in which case the vaporization takes place when the superheated liquid reaches the pool surface. Vaporization also occurs from falling films (the reverse of condensation) or from the flashing of liquids superheated by forced convection under pressure.

Pool boiling refers to the type of boiling experienced when the heating surface is surrounded by a relatively large body of fluid which is not flowing at any appreciable velocity and is agitated only by the motion of the bubbles and by natural-convection currents. Two types of pool boiling are possible: subcooled pool boiling, in which the bulk fluid temperature is below the saturation temperature, resulting in collapse of the bubbles before they reach the surface, and saturated pool boiling, with bulk temperature equal to saturation temperature, resulting in net vapor generation.

The general shape of the curve relating the heat-transfer coefficient to Δt_b, the temperature driving force (difference between the wall temperature and the bulk fluid temperature) is one of the few parametric relations that are reasonably well understood. The familiar boiling curve was originally demonstrated experimentally by Nukiyama [*J. Soc. Mech. Eng.* (*Japan*), **37**, 367 (1934)]. This curve points out one of the great dilemmas for boiling-equipment designers. They are faced with at least six heat-transfer regimes in pool boiling: natural convection (+), incipient nucleate boiling (+), nucleate boiling (+), transition to film boiling (−), stable film boiling (+), and film boiling with increasing radiation (+). The signs indicate the sign of the derivative $d(q/A)/d\,\Delta t_b$. In the transition to film boiling, heat-transfer rate *decreases* with driving force. The regimes of greatest commercial interest are the *nucleate-boiling* and *stable-film-boiling regimes*.

Heat transfer by **nucleate boiling** is an important mechanism in the vaporization of liquids. It occurs in the vaporization of liquids in kettle-type and natural-circulation reboilers commonly used in the process industries. High rates of heat transfer per unit of area (heat flux) are obtained as a result of bubble formation at the liquid-solid interface rather than from mechanical devices external to the heat exchanger. There are available several expressions from which reasonable values of the film coefficients may be obtained.

The boiling curve, particularly in the nucleate-boiling region, is significantly affected by the temperature driving force, the total system pressure, the nature of the boiling surface, the geometry of the system, and the properties of the boiling material. In the nucleate-boiling regime, heat flux is approximately proportional to the cube of the temperature driving force. Designers in addition must know the minimum Δt (the point at which nucleate boiling begins), the critical Δt (the Δt above which transition boiling begins), and the maximum heat flux (the heat flux corresponding to the critical Δt). For designers who do not have experimental data available, the following equations may be used.

Boiling Coefficients For the **nucleate-boiling coefficient** the Mostinski equation [*Teplenergetika*, **4**, 66 (1963)] may be used:

$$h = bP_c^{0.69}\left(\frac{q}{A}\right)^{0.7}\left[1.8\left(\frac{P}{P_c}\right)^{0.17} + 4\left(\frac{P}{P_c}\right)^{1.2} + 10\left(\frac{P}{P_c}\right)^{10}\right] \qquad (5\text{-}102)$$

where $b = (3.75)(10^{-5})$(SI) or $(2.13)(10^{-4})$ (U.S. customary), P_c is the critical pressure and P the system pressure, q/A is the heat flux, and h is the nucleate-boiling coefficient. The McNelly equation [*J. Imp. Coll. Chem. Eng. Soc.*, 7(18), (1953)] may also be used:

$$h = 0.225\left(\frac{qc_l}{A\lambda}\right)^{0.69}\left(\frac{Pk_l}{\sigma}\right)^{0.31}\left(\frac{\rho_l}{\rho_v} - 1\right)^{0.33} \qquad (5\text{-}103)$$

where c_l is the liquid heat capacity, λ is the latent heat, P is the system pressure, k_l is the thermal conductivity of the liquid, and σ is the surface tension.

An equation of the Nusselt type has been suggested by Rohsenow [*Trans. Am. Soc. Mech. Eng.*, **74**, 969 (1952)].

$$hD/k = C_r(DG/\mu)^{2/3}(c\mu/k)^{-0.7} \qquad (5\text{-}104a)$$

in which the variables assume the following form:

$$\frac{h\beta'}{k}\left[\frac{g_c\sigma}{g(\rho_L - \rho_v)}\right]^{1/2} = C_r\left[\frac{\beta'}{\mu}\left(\frac{g_c\sigma}{g(\rho_L - \rho_v)}\right)^{1/2}\frac{W}{A}\right]^{2/3}\left(\frac{c\mu}{k}\right)^{-0.7} \qquad (5\text{-}104b)$$

The coefficient C_r is not truly constant but varies from 0.006 to 0.015.° It is possible that the nature of the surface is partly responsible for the variation in the constant. The only factor in Eq. (5-104b) not readily available is the value of the contact angle β'.

Another Nusselt-type equation has been proposed by Forster and Zuber:†

$$N_{Nu} = 0.0015 N_{Re}^{0.62} N_{Pr}^{1/3} \qquad (5\text{-}105)$$

which takes the following form:

$$\frac{c\rho_L \sqrt{\pi\alpha}}{k\rho_v} \frac{W}{A} \left(\frac{2\sigma}{\Delta p}\right)^{1/2} \left(\frac{\rho_L}{\Delta p g_c}\right)^{1/4}$$
$$= 0.0015 \left[\frac{\rho_L}{\mu} \left(\frac{c\rho_L \,\Delta T \sqrt{\pi\alpha}}{\lambda\rho_v}\right)^2\right]^{0.62} \left(\frac{c\mu}{k}\right)^{1/2} \qquad (5\text{-}106)$$

where $\alpha = k/\rho c$ (all liquid properties)

Δp = pressure of the vapor in a bubble minus saturation pressure of a flat liquid surface

Equations (5-104b) and (5-106) have been arranged in dimensional form by Westwater.

The numerical constant may be adjusted to suit any particular set of data if one desires to use a certain criterion. However, surface conditions vary so greatly that deviations may be as large as ±25 percent from results obtained.

The **maximum heat flux** may be predicted by the Kutateladse-Zuber [*Trans. Am. Soc. Mech. Eng.*, **80**, 711 (1958)] relationship, using consistent units:

$$\left(\frac{q}{A}\right)_{max} = 0.18 g_c^{1/4} \rho_v \lambda \left[\frac{(\rho_l - \rho_v)\sigma g}{\rho_v^2}\right]^{1/4} \qquad (5\text{-}107)$$

Alternatively, Mostinski presented an equation which approximately represents the Cichelli-Bonilla [*Trans. Am. Inst. Chem. Eng.*, **41**, 755 (1945)] correlation:

$$\frac{(q/A)_{max}}{P_c} = b \left(\frac{P}{P_c}\right)^{0.35} \left(1 - \frac{P}{P_c}\right)^{0.9} \qquad (5\text{-}108)$$

where $b = 0.368$(SI) or 5.58 (U.S. customary); P_c is the critical pressure, Pa absolute; P is the system pressure; and $(q/A)_{max}$ is the maximum heat flux.

The lower limit of applicability of the nucleate-boiling equations is from 0.1 to 0.2 of the maximum limit and depends upon the magnitude of natural-convection heat transfer for the liquid. The best method of determining the lower limit is to plot two curves: one of h versus Δt for natural convection, the other of h versus Δt for nucleate boiling. The intersection of these two curves may be considered the lower limit of applicability of the equations.

These equations apply to single tubes or to flat surfaces in a large pool. In tube bundles the equations are only approximate, and designers must rely upon experiment. Palen and Small [*Hydrocarbon Process.*, **43**(11), 199 (1964)] have shown the effect of tube-bundle size on maximum heat flux.

$$\left(\frac{q}{A}\right)_{max} = b \frac{p}{D_o \sqrt{N_T}} \rho_v \lambda \left[\frac{g\sigma(\rho_l - \rho_v)}{\rho_v^2}\right]^{1/4} \qquad (5\text{-}109)$$

where $b = 0.43$ (SI) or 61.6 (U.S. customary), p is the tube pitch, D_o is the tube outside diameter, and N_T is the number of tubes (twice the number of complete tubes for U-tube bundles).

For **film boiling,** Bromley's [*Chem. Eng. Prog.*, **46**, 221 (1950)] correlation may be used:

$$h = b \left[\frac{k_v^3(\rho_l - \rho_v)\rho_v g}{\mu_v D_o \,\Delta t_b}\right]^{1/4} \qquad (5\text{-}110)$$

where $b = 4.306$ (SI) or 0.620 (U.S. customary). Katz, Myers, and Balekjian [*Pet. Refiner,* **34**(2), 113 (1955)] report boiling heat-transfer coefficients on finned tubes.

HEAT TRANSFER BY RADIATION

GENERAL REFERENCES: Much of the pertinent literature on radiative heat transfer has been surveyed in the following texts: Goody, *Atmospheric Radiation,* Clarendon Press, Oxford, 1964. Sparrow and Cess, *Radiation Heat Transfer,* Brooks/Cole Publishing Company, Belmont, Calif., 1966. Hottel and Sarofim, *Radiative Transfer,* McGraw-Hill, New York, 1967. Love, *Radiative Heat Transfer,* Merrill, Columbus, 1968. Siegel and Howell, *Thermal Radiation Heat Transfer,* NASA SP-164, GPO, Washington, 1968. Edwards, *Radiation Heat Transfer Notes,* Hemisphere Publishing Corp., 1981; Howell and Siegel, *Thermal Radiative Heat Transfer,* McGraw-Hill, 3d ed., 1992; Brewster, *Thermal Radiative Transfer and Properties,* Wiley, 1992; Modest, *Radiative Heat Transfer,* McGraw-Hill, 1993.

Additional sources are the *Journal of Applied Optics* and the *Journal of the Optical Society of America,* particularly for surface properties; the *Journal of Quantitative Spectroscopy and Radiative Transfer* for gas properties; the *Journal of Heat Transfer* and the *International Journal of Heat and Mass Transfer* for broad coverage; and the *Journal of the Institute of Energy* for applications to industrial furnaces.

Thermal radiation—electromagnetic energy in transport—is emitted within matter excited by temperature; it is absorbed in other matter at distances from the source which depend on the mean free path of the photons emitted. The ratio of the mean free path involved in an energy-transport process to a characteristic dimension of the system of interest determines the mathematical structure of the formulation. In molecular conduction this ratio is minute (unless the system or the density of matter is minute, which is the case of free molecular flow), and a differential equation of energy diffusion is involved. In gas radiation the ratio is generally large enough to give rise to an integral equation, with an unknown function inside the integral. Solids gener-

ally have small enough photon mean free paths (high enough absorption coefficients) for the radiation escaping through the surface to have originated close to the surface; radiative loss is then identifiable with its surface temperature, but an integral equation is still involved if all the surfaces of an enclosure filled with a diathermanous medium like air are not specified as to temperature or are not black.

Radiation differs from conduction and convection not only in mathematical structure but in its much higher sensitivity to temperature. It is of dominating importance in furnaces because of their temperature, and in cryogenic insulation because of the vacuum existing between particles. The temperature at which it accounts for roughly half of the total heat loss from a surface in air depends on such factors as surface emissivity and the convection coefficient. For pipes in free convection, this is room temperature; for fine wires of low emissivity it is above red heat. Gases at combustion-chamber temperatures lose more than 90 percent of their energy by radiation from the carbon dioxide, water vapor, and particulate matter.

NOMENCLATURE FOR RADIATIVE TRANSFER

Terms that are defined at specific places in the text are excluded.

a = effective energy fraction of blackbody spectrum in which a nongray gas absorbs.

A = area.

c = number concentration of particles in a cloud.

c_1, c_2 = first and second Planck-law constants.

° Reported by Westwater in Drew and Hoopes, *Advances in Chemical Engineering,* vol. I, Academic, New York, 1956, p. 15.

† Forster, *J. Appl. Phys.,* **25,** 1067 (1954); Forster and Zuber, *J. Appl. Phys.,* **25,** 474 (1954); Forster and Zuber, Conference on Nuclear Engineering, University of California, Los Angeles, 1955; excellent treatise on boiling of liquids by Westwater in Drew and Hoopes, *Advances in Chemical Engineering,* vol. I, Academic, New York, 1956.

C = axis-to-axis distance of separation of tubes.

C_b = mean specific heat of combustion products from base temperature T_o to leaving-gas temperature T_E.

C = cold-surface fraction of a furnace enclosure.

C_W = correction factor for pressure broadening of radiation from water vapor.

d = particle diameter.

D = tube diameter; characteristic dimension; dimensionless firing density.

D' = reduced firing density.

E = hemispherical emissive power of a blackbody.

f = fraction of blackbody radiation lying below λ.

f_v = volume fraction of space occupied by particles.

F = direct view factor; F_{ij}, fraction of isotropic radiation from A_i intercepted directly by A_j.

$\bar{F}$ = total view factor from black source to black sink, with allowance for refractory surfaces (subscripts identify source and sink).

$\mathscr{F}_{ij}$ = total view factor, radiation from i to j both directly and indirectly, expressed as fraction of blackbody radiation from A_i.

$\overline{gs}$ = direct-exchange area between gas volume and surface.

$\overline{GS}$ = total-exchange area between gas and surface; subscript R indicates allowance for radiatively adiabatic surfaces.

h = coefficient of convective heat transfer.

H = enthalpy of fuel plus air entering combustion chamber.

I = intensity, radiant-energy-flux density per unit solid angle of divergence.

ij = shorthand for $\overline{s_i s_j}$.

k = absorption or emission coefficient; or thermal conductivity.

K = constant defined in connection with Eq. (5-147).

L = mean beam length; L_0, at vanishingly small optical thickness; L_m, average value.

L = wall-loss group.

L_c = dimensionless convective loss.

L_o = dimensionless wall-opening loss.

L_r = dimensionless refractory-wall loss.

$\dot{m}$ = mass flow rate.

n = refractive index.

p = partial pressure, atm.; subscript c, CO_2; subscript w, water vapor.

P = total pressure atm.

q = heat-flux density, energy per time-area.

Q = heat flux, energy per time.

r = separating distance; or electrical resistivity; or refractory (radiatively adiabatic) surface.

$\overline{ss} \equiv AF$, direct-exchange area (subscripts identify surface zones).

$\overline{SS} \equiv A\mathscr{F}$, total-exchange area.

T = absolute temperature. Subscript 1 (or G), radiating surface (or gas); subscript E, exit-gas; subscript o, base temperature; subscript F, pseudoadiabatic flame temperature based on $\bar{C}_p$ averaged from T_o to T_E.

U = overall coefficient of heat transfer, gas convection to refractory wall to ambient air.

W = total leaving-flux density (also radiosity).

α = absorptivity or absorptance; α_{12}, absorptance of surface 1 for radiation from surface 2.

Δ = difference between radiating temperature and leaving-gas temperature divided by the pseudoadiabatic flame temperature T_F.

ε = emissivity or emittance.

η = thermal efficiency. Subscript G, gas-side; subscript 1, sink-side.

θ = polar angle.

λ = wavelength.

μm = micrometer (m^{-6}).

ρ = reflectance; ρ_s, specular reflectance.

σ = Stefan-Boltzmann constant.

τ = ratio of temperature to T_F. Subscript G, gas; subscript 1 sink; subscript o, base.

τ = transmittance.

Ω = solid angle.

ω = albedo of a surface.

NATURE OF THERMAL RADIATION

Consider a pencil of radiation, defined as all the rays passing through each of two small widely separated areas dA_1 and dA_2. The rays at dA_1 will have a solid angle of divergence $d\Omega_1$ equal to the apparent area of dA_2 viewed from dA_1, divided by the square of the separating distance. Let the normal to dA_1 make the angle θ_1 with the pencil. The flux density q (energy per time-area) normal to the beam and per unit solid angle of its divergence is called the intensity I, and the flux $d\dot{Q}_1$ (energy per time) through the area dA_1 (of apparent area $dA_1 \cos \theta_1$ normal to the beam) is therefore given by

$$d\dot{Q}_1 = dA_1(\cos \theta_1)q = I\, dA_1(\cos \theta_1)\, d\Omega_1 \qquad (5\text{-}111)$$

The intensity I along a pencil, in the absence of absorption or scatter, is constant (unless the beam passes into a medium of different refractive index n; then $I_1/n_1^2 = I_2/n_2^2$).

The emissive power° of a surface is the flux density (energy per time-surface area) due to emission from it throughout a hemisphere. If the intensity I of emission from a surface is independent of the angle of emission, Eq. (5-111) may be integrated to show that the surface emissive power is πI, though the emission is throughout 2π sr.

Blackbody Radiation Engineering calculations of thermal radiation from surfaces are best keyed to the radiation characteristics of the blackbody, or ideal radiator. The characteristic properties of a blackbody are that it absorbs all the radiation incident on its surface and that the quality and intensity of the radiation it emits are completely determined by its temperature. The total radiative flux throughout a hemisphere from a black surface of area A and absolute temperature T is given by the Stefan-Boltzmann law:

$$\dot{Q} = A\sigma T^4 \qquad \text{or} \qquad q = \sigma T^4 \qquad (5\text{-}112)$$

The Stefan-Boltzmann constant σ has the value $(0.1713)(10^{-8})$ Btu/$(\text{ft}^2 \cdot \text{h} \cdot °\text{R}^4)$; $(1.00)(10^{-8})$ CHU/$(\text{ft}^2 \cdot \text{h} \cdot \text{K}^4)$; $(4.88)(10^{-8})$ kcal/$(\text{m}^2 \cdot \text{h} \cdot \text{K}^4)$; $(1.356)(10^{-12})$ cal $(\text{cm}^2 \cdot \text{s} \cdot \text{K}^4)$; $(5.67)(10^{-12})$ W/$(\text{cm}^2 \cdot \text{K}^4)$; $(5.67)(10^{-8})$ W/$(\text{m}^2 \cdot \text{K}^4)$; or in terms of Planck constants, $c_1(\pi/c_2)^4/15$. From the definition of emissive power, σT^4 is the total emissive power of a blackbody, called E; the intensity I_B of blackbody emission is E/π or $\sigma T^4/\pi$.

The spectral distribution of energy flux from a black body is expressed by Planck's law:

$$E_\lambda\, d\lambda = (2\pi hc^2 n^2 \lambda^{-5})/(e^{hc/k\lambda T} - 1)\, d\lambda \qquad (5\text{-}113)$$

$$\equiv (n^2 c_1 \lambda^{-5})/(e^{c_2/\lambda T} - 1)\, d\lambda \qquad (5\text{-}114)$$

where $E_\lambda\, d\lambda$ is the hemispherical flux density lying in the wavelength range λ to $\lambda + d\lambda$; h is Planck's constant, $(6.6256)(10^{-27})$ erg·s; c is the velocity of light in vacuo, $(2.9979)(10^{10})$ cm/s; k is the Boltzmann constant, $(1.3805)(10^{-16})$ erg/K; λ is the wavelength measured in vacuo; and n is the refractive index of the emitter ($\lambda = n\lambda_m$, where λ_m is the wavelength measured in the medium; $E_\lambda\, d\lambda = E_{\lambda m}\, d\lambda_m$, where E_λ and $E_{\lambda m}$ are both measured in the medium; engineers commonly use E_λ). Equation (10-191) may be written

$$\frac{E_\lambda}{n^2 T^5} = \frac{c_1(\lambda T)^{-5}}{e^{c_2/\lambda T} - 1} \qquad (5\text{-}115)$$

The first and second Planck-law constants c_1 and c_2 are respectively $(3.740)(10^{-16})$ $(\text{J} \cdot \text{m}^2)$/s and $(1.4388)(10^{-2})$ m·K. The term E_λ/n^2T^5, clearly a function only of the product λT, is given in Fig. 5-11 which may be visualized as the monochromatic emissive power versus wavelength measured in vacuo of a black surface at 1 K discharging in vacuo.

The wavelength of maximum intensity is seen to be inversely proportional to the absolute temperature. The relation is known as **Wien's displacement law:** $\lambda_{max}T = (2.898)(10^{-3})$ m·K. This can be misleading, however, since the wavelength of maximum intensity depends on whether intensity is defined in terms of wavelength interval or frequency interval. More useful displacement laws refer to the value of λT corresponding to maximum energy per unit *fractional change* in wavelength or frequency [$(3.67)(10^{-3})$ m·K] or to the value of λT corresponding to half of the energy [$(4.11)(10^{-3})$ m·K]. Figure

---

° Variously called, in the literature, emittance, total hemispherical intensity, or radiant flux density.

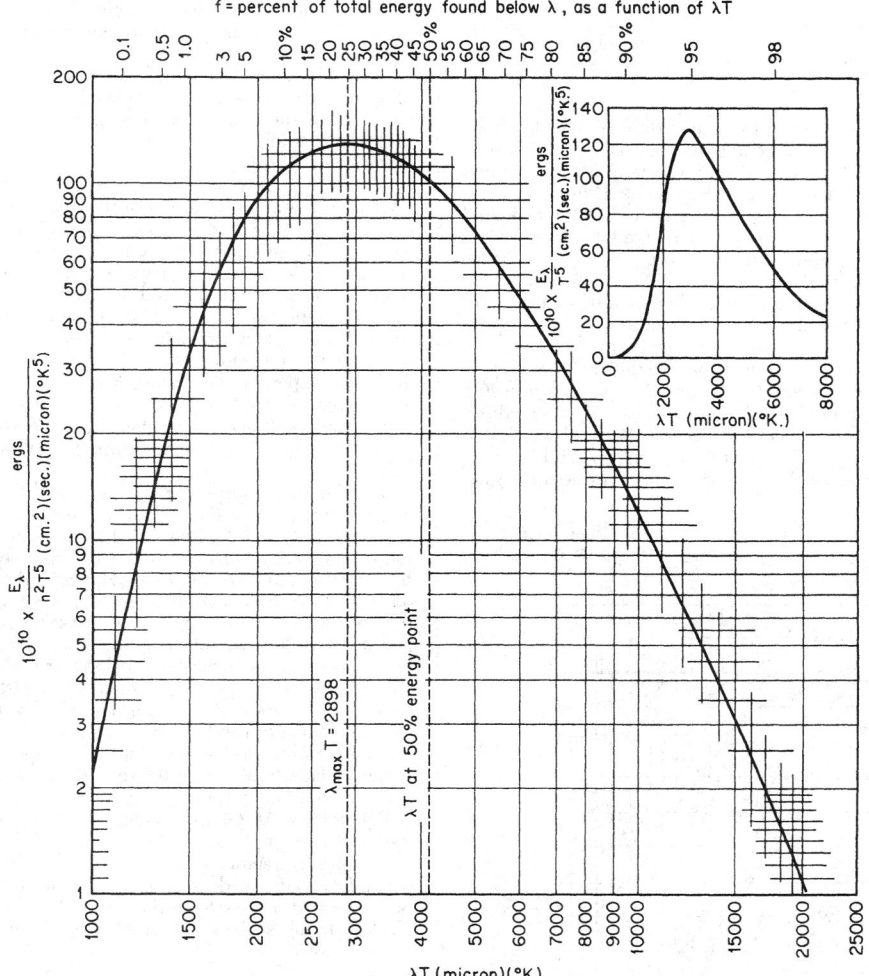

f = percent of total energy found below λ, as a function of λT

FIG. 5-11   Distribution of energy in the spectrum of a blackbody. To convert microns to micrometers, multiply by unity. To convert ergs per square centimeter–second–micron–K^5 to watts per square meter, per meter, per K^5, multiply by 10^{-3}.

5-11 carries, at the top, a scale giving the fraction f of the total energy in the spectrum that lies below λT. A generalization useful for identifying the spectral range of greatest interest in evaluations of radiative transfer is that roughly half of the energy from a black surface lies within the twofold range of λT geometrically centered on 3.67×10^{-3}, i.e., from $\lambda T = (3.67/\sqrt{2})(10^{-3})$ to $(3.67 \times \sqrt{2})(10^{-3})$ m·K.

One limiting form of the Planck equation, approached as $\lambda T \to 0$, is the **Wien equation** [Eqs. (5-113) and (5-114)] with the 1 missing in the denominator. The error is less than 1 percent when $\lambda T < (3)(10^{-3})$ m·K or when $T < 4800$ K if an optical pyrometer with red screen ($\lambda = 0.65\mu$m) is used.

RADIATIVE EXCHANGE BETWEEN SURFACES OF SOLIDS

Emittance and Absorptance   The ratio of the total radiating power of a real surface to that of a black surface at the same temperature is called the **emittance** of the surface (for a perfectly plane surface, the **emissivity**), designated by ε. Subscripts λ, θ, and n may be assigned to differentiate monochromatic, directional, and surface-normal values respectively from the total hemispherical value. If radi-

ation is incident on a surface, the fraction absorbed is called the **absorptance** (**absorptivity**), a term to which two subscripts may be appended, the first to identify the temperature of the surface and the second to identify the spectral energy distribution of the surface.

According to **Kirchhoff's law,** the emissivity and absorptivity of a surface *in surroundings at its own temperature* are the same for both monochromatic and total radiation. When the temperatures of the surface and its surroundings differ, the total emissivity and absorptivity of the surface often are found to be different, but, because absorptivity is substantially independent of irradiation density, the monochromatic emissivity and absorptivity of surfaces are for all practical purposes the same. The difference between total emissivity and absorptivity depends on the variation, with wavelength, of ε_λ and on the difference between the emitter temperature and the effective source temperature.

Consider radiative exchange between a body of area A_1 and temperature T_1 and black surroundings at T_2. The net interchange is given by

$$\dot{Q}_{1=2} = A_1 \int_0^\infty [\varepsilon_\lambda E_\lambda(T_1) - \alpha_\lambda E_\lambda(T_2)] \, d\lambda$$

$$= A_1(\varepsilon_1 \sigma T_1^4 - \alpha_{12} \sigma T_2^4) \qquad (5\text{-}116)$$

where

$$\varepsilon_1 = \int_0^1 \varepsilon_\lambda \, df_{\lambda T_1} \qquad (5\text{-}117)$$

and

$$\alpha_{12} = \int_0^1 \varepsilon_\lambda \, df_{\lambda T_2} \qquad (5\text{-}118)$$

The value of ε_1 (or α_{12}, the absorptivity of surface A_1 for blackbody radiation at T_2) is the area under a curve of ε_λ versus f, the latter read as a function of λT_1 (or λT_2) from the top ordinate of Fig. 5-11. For a gray surface, $\varepsilon_1 = \alpha_{12} = \varepsilon_\lambda$. A selective surface is one whose ε_λ changes dramatically with wavelength. If this change is unidirectional, ε_1 and α_{12} are, according to Eqs. (5-116) to (5-118), markedly different when the absolute-temperature ratio is far from 1; e.g., when $T_1 = 294$ K (530°R; ambient temperature), and $T_2 = 6000$ K (10,800°R; effective solar temperature), $\varepsilon_1 = 0.9$ and $\alpha_{12} = 0.1$ to 0.2 for a white paint, but ε_1 can be as low as 0.12 and α_{12} above 0.9 for a thin layer of copper oxide on bright aluminum.

The effect of radiation-source temperature on the low-temperature absorptivity of a number of additional materials is presented in Fig. 5-12. It will be noted that polished aluminum (curve 15) and anodized (surface-oxidized) aluminum (curve 13), representative of metals and nonmetals respectively, respond oppositely to a change in the temperature of the radiation source. The absorptance of surfaces for solar radiation may be read from the right of Fig. 10-45, if solar radiation is assumed to consist of blackbody radiation from a source at 5800 K (10,440°R).

Although values of emittance and absorptance depend in very complex ways on the real and imaginary components of the refractive index and on the geometrical structure of the surface layer, the generalizations that follow are possible.

Polished Metals

1. ε_λ in the infrared is governed by free-electron contributions, is quite low, and is a function of the resistivity-wavelength quotient r/λ (Fig. 5-13). For $\lambda > 8\mu$m, $\varepsilon_{\lambda,n}$ is approximately $0.0365 \sqrt{r/\lambda}$, where r is in ohm-meters and λ in micrometers (the Drude or Hagen-Rubens relation). At shorter wavelengths, bound-electron contributions become significant and ε_λ increases, sometimes exhibiting maxima; values of 0.4 to 0.8 are common in the visible spectrum (0.4 to 0.7 μm). ε_λ is approximately proportional to the square root of the absolute temperature ($\varepsilon_\lambda \propto \sqrt{r}$, and $r \propto T$) in the far infrared ($\lambda > 8\mu$m), is temperature-insensitive in the near infrared (0.7 to 1.5 μm), and decreases slightly as temperature increases in the visible.

2.   Total emittance is substantially proportional to absolute temperature; at moderate temperature, $\varepsilon_n = 0.058T\sqrt{rT}$, where T is in kelvin.

3.   The total absorptance of a metal at T_1 for radiation from a black or gray source at T_2 is equal to the emissivity evaluated at the geometric mean of T_1 and T_2. Figure 5-13 gives values of ε_λ, $\varepsilon_{\lambda,n}$, and their ratio as a function of r/λ (dashed lines); and total emissivities ε, ε_n and their ratio as a function of rT (solid lines). Although the figure is based on free-electron contributions to emissivity in the far infrared, the relations for total emissivity are remarkably good even at high temperatures. Unless extraordinary pains are taken to prevent oxidation, however, a metallic surface may exhibit several times the emittance or absorptance of a polished specimen. The emittance of iron and steel, for example, varies widely with degree of oxidation and roughness; clean metallic surfaces have an emittance of from 0.05 to 0.45 at ambient temperatures to 0.4 to 0.7 at high temperatures; oxidized and/or rough surfaces range from 0.6 to 0.95 at low temperatures to 0.9 to 0.95 at high temperatures.

Refractory Materials   Grain size and concentration of trace impurities are important.

1.   Most refractory materials have an ε_λ of 0.8 to 1.0 at wavelengths beyond 2 to 4 μm; ε_λ decreases rapidly toward shorter wavelengths for materials that are white in the visible but retains its high value for black materials such as FeO and Cr_2O_3. Small concentrations of FeO

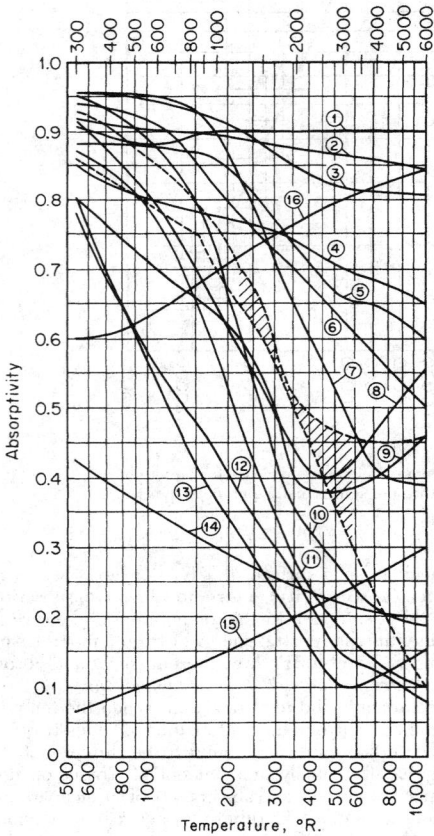

FIG. 5-12   Variation of absorptivity with temperature of radiation source. (1) Slate composition roofing. (2) Linoleum, red brown. (3) Asbestos slate. (4) Soft rubber, gray. (5) Concrete. (6) Porcelain. (7) Vitreous enamel, white. (8) Red brick. (9) Cork. (10) White dutch tile. (11) White chamotte. (12) MgO, evaporated. (13) Anodized aluminum. (14) Aluminum paint. (15) Polished aluminum. (16) Graphite. The two dashed lines bound the limits of data on gray paving brick, asbestos paper, wood, various cloths, plaster of paris, lithopone, and paper. To convert degrees Rankine to kelvins, multiply by $(5.556)(10^{-1})$.

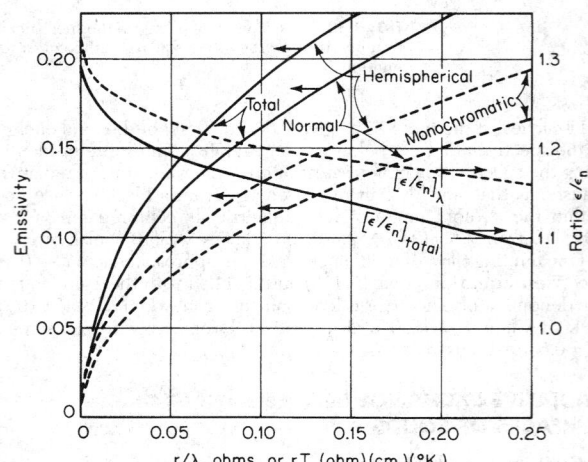

FIG. 5-13   Hemispherical and normal emissivities of metals and their ratio. Dashed lines: monochromatic (spectral) values versus r/λ. Solid lines: total values versus rT. To convert ohm-centimeter-kelvins to ohm-meter-kelvins, multiply by 10^{-2}.

and Cr_2O_3 or other colored oxides can cause marked increases in the emittance of materials that normally are white. The sensitivity of the emittance of refractory oxides to small additions of absorbing materials is demonstrated by the results of calculations, shown in Fig. 5-14, of the emittance of a semi-infinite absorbing-scattering medium as a function of its albedo: the ratio of the scatter coefficient to the sum of scatter and absorption coefficients. The results, pertinent to the radiative properties of fibrous materials, paints, oxide coatings, and refractories, show that when absorption accounts for only 0.5 percent (10 percent) of the total attenuation within the medium, the emittance is greater than 0.15 (0.5). ε_λ for refractory materials varies little with temperature, with the exception of some white oxides which at high temperatures become good emitters in the visible spectrum as a consequence of the induced electronic transitions.

2. Refractory materials generally have a total emittance which is high (0.7 to 1.0) at ambient temperatures and decreases with increase in temperature; a change from 1000 to 1570°C (1850 to 2850°F) may cause a decrease in ε of one-fourth to one-third.

3. The emittance and absorptance increase with increase in grain size over a grain-size range of 1 to 200 μm.

4. The ratio $\varepsilon/\varepsilon_n$ of hemispherical to normal emissivity of polished surfaces varies with refractive index n from 1 at $n = 1.0$ to 0.93 at $n = 1.5$ (common glass) and back to 0.96 at $n = 3$.

5. The ratio $\varepsilon/\varepsilon_n$ for a surface composed of particulate matter which scatters isotropically varies with ε from 1 when $\varepsilon = 1$ to 0.8 when $\varepsilon = 0.07$ (see Fig. 5-14).

6. The total absorptance shows a decrease with increase in temperature of the radiation source similar to the decrease in emittance with increase in the specimen temperature.

Figure 5-12 shows a regular variation of α_{12} with T_2. When T_2 is not very different from T_1, α_{12} may be expressed as $\varepsilon_1(T_2/T_1)^m$. It may be shown that Eq. (5-116) is then approximated by

$$\dot{Q}_{1,\text{net}} = \sigma A_1 \varepsilon_{av} \left(1 + \frac{m}{4}\right)(T_1^4 - T_2^4) \qquad (5\text{-}119)$$

where ε_{av} is evaluated at the arithmetic mean of T_1 and T_2. For metals m is about 0.5; for nonmetals it is small and negative.

Table 5-6, based on a critical evaluation of early data, is illustrative of the emittance of materials encountered in engineering practice; it shows the wide variation possible in the emissivity of a particular material due to variations in surface roughness and thermal pretreatment. (With few exceptions the values refer to emission normal to the surface; see above for conversion to hemispherical values.) More recent data support the range of emittance values given in Table 5-6 and their dependence on surface conditions. Extensive compilations of data are provided by Schmidt and Furthmann (*Mitt. Kaiser-Wilhelm-Inst. Eisenforsch.*, 109, 225), covering data to 1928; by Gubareff, Jansen, and Torborg [*Thermal Radiation Properties Survey*, Honeywell Research Center, Minneapolis, 1960), covering data to 1940; and by Goldsmith, Waterman, and Hirschhorn [*Thermophysical Properties of Matter*, Purdue University (Touloukian, Ed.), Plenum, 1970].

For opaque materials, the reflectance ρ is the complement of the absorptance. The directional distribution of the reflected radiation depends on the material, its degree of roughness or grain size, and, if a metal, its state of oxidation. Polished surfaces of homogeneous materials reflect specularly. In contrast, the intensity of the radiation reflected from a perfectly diffuse, or Lambert, surface is independent of direction. The directional distribution of reflectance of many oxidized metals, refractory materials, and natural products approximates that of a perfectly diffuse reflector. A better model, adequate for many calculational purposes, is achieved by assuming that the total reflectance ρ is the sum of diffuse and specular components ρ_D and ρ_S.

Black-Surface Enclosures

View Factor and Direct-Exchange Area When several surfaces are present, the need arises for evaluating a geometrical factor F, called the direct view factor. In the following discussion, restriction is to black surfaces, the intensity from which is independent of angle of emission. Define F_{12} as the fraction of the radiation leaving surface A_1 in all directions which is intercepted by surface A_2. Since the net interchange between A_1 and A_2 must be zero when their temperatures are alike, it follows that $A_1F_{12} = A_2F_{21}$. This product, having the dimensions of area, is called the direct-exchange area and is designated for brevity by $\overline{12}(\equiv\overline{21})$. It is sometimes designated $\overline{s_1s_2}$. Clearly, $\overline{11} + \overline{12} + \overline{13} + \cdots = A_1$; and when A_1 cannot "see" itself, $\overline{11} = 0$.

From Eq. (5-111) and the definition of F:

$$A_1F_{12} \equiv \overline{s_1s_2} \equiv \frac{Q_{1-2}}{E_1} = \int_{A_1}\int_{A_2} \frac{dA_1(\cos\theta_1)\,d\Omega_1}{\pi}$$

$$= \int_{A_1}\int_{A_2} \frac{dA_1(\cos\theta_1)\,dA_2\,(\cos\theta_2)}{\pi r^2} \qquad (5\text{-}120)$$

where $A(\cos\theta)$ is the projection of A normal to r, the line connecting dA_1 and dA_2. Values of $\overline{s_1s_2}$ (or of F_{12}) may be obtained by integrating either Eq. (5-120) or an equivalent contour integral (see Hottel and Sarofim, *Radiative Transfer*, McGraw-Hill, New York, 1967, chap. 2). Such values are given for opposed parallel disks or rectangles in Fig. 5-15. For rectangles of dimensions L_1 and L_2, $F \cong \sqrt{F_1F_2}$, where F_1 and F_2 are for squares of sides L_1 and L_2. The view factor for rectangles in perpendicular planes and having a common edge length x and ratios Y and Z of widths to common length is given by

$$F_{YZ} = \frac{1}{\pi Y}\left\{\begin{array}{l}\frac{1}{4}\ln\left[(1+Y^2)\left(\frac{Y^2}{1+Y^2}\right)^{Y^2}\left(\frac{Z^2}{Y^2+Z^2}\right)^{Z^2}\left(\frac{1+Z^2}{1+Y^2+Z^2}\right)^{1-Y^2-Z^2}\right] \\ + Y\tan^{-1}\frac{1}{Y} + Z\tan^{-1}\frac{1}{Z} - \sqrt{Y^2+Z^2}\tan^{-1}\frac{1}{\sqrt{Y^2+Z^2}}\end{array}\right\}$$

$$(5\text{-}121)$$

The direct-exchange area is given by

$$\overline{s_ys_z} = \frac{x^2}{\pi} \qquad (5\text{-}122)$$

When the maximum dimensions of each of two plane surfaces is small relative to their center-to-center separating distance r, Eq. (5-120) gives

$$\overline{12} = \frac{A_1(\cos\theta_1)A_2(\cos\theta_2)}{\pi r^2} \qquad (5\text{-}123)$$

and when, in addition, the normals to A_1 and A_2 are in a common plane,

$$\overline{12} = A_1A_2n_1n_2/\pi r^2 \qquad (5\text{-}124)$$

where n_1 is the normal-to-A_1 component of the distance to A_2. Equation (5-124) is, for example, in error only by +7 percent for the case of opposed squares separated by 3 times their side dimension. The view factors are given for finite coaxial coextensive cylinders in Fig. 5-16,

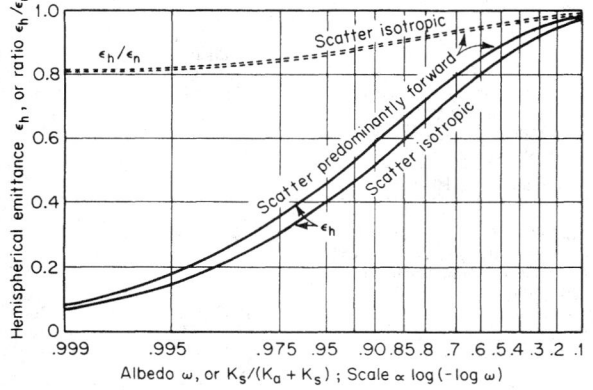

FIG. 5-14 Hemispherical emittance ε_h and the ratio of hemispherical to normal emittance $\varepsilon_h/\varepsilon_n$ for a semi-infinite absorbing-scattering medium.

TABLE 5-6   Normal Total Emissivity of Various Surfaces

A. Metals and Their Oxides

Surface	t, °F°	Emissivity°	Surface	t, °F°	Emissivity°
Aluminum			Sheet steel, strong rough oxide layer	75	0.80
Highly polished plate, 98.3% pure	440–1070	0.039–0.057	Dense shiny oxide layer	75	0.82
Polished plate	73	0.040	Cast plate:		
Rough plate	78	0.055	Smooth	73	0.80
Oxidized at 1110°F	390–1110	0.11–0.19	Rough	73	0.82
Aluminum-surfaced roofing	100	0.216	Cast iron, rough, strongly oxidized	100–480	0.95
Calorized surfaces, heated at 1110°F.			Wrought iron, dull oxidized	70–680	0.94
Copper	390–1110	0.18–0.19	Steel plate, rough	100–700	0.94–0.97
Steel	390–1110	0.52–0.57	High temperature alloy steels (see Nickel Alloys).		
Brass			Molten metal		
Highly polished:			Cast iron	2370–2550	0.29
73.2% Cu, 26.7% Zn	476–674	0.028–0.031	Mild steel	2910–3270	0.28
62.4% Cu, 36.8% Zn, 0.4% Pb, 0.3% Al	494–710	0.033–0.037	Lead		
82.9% Cu, 17.0% Zn	530	0.030	Pure (99.96%), unoxidized	260–440	0.057–0.075
Hard rolled, polished:			Gray oxidized	75	0.281
But direction of polishing visible	70	0.038	Oxidized at 390°F.	390	0.63
But somewhat attacked	73	0.043	Mercury	32–212	0.09–0.12
But traces of stearin from polish left on	75	0.053	Molybdenum filament	1340–4700	0.096–0.292
Polished	100–600	0.096	Monel metal, oxidized at 1110°F	390–1110	0.41–0.46
Rolled plate, natural surface	72	0.06	Nickel		
Rubbed with coarse emery	72	0.20	Electroplated on polished iron, then polished	74	0.045
Dull plate	120–660	0.22	Technically pure (98.9% Ni, + Mn), polished	440–710	0.07–0.087
Oxidized by heating at 1110°F	390–1110	0.61–0.59	Electroplated on pickled iron, not polished	68	0.11
Chromium; see Nickel Alloys for Ni-Cr steels	100–1000	0.08–0.26	Wire	368–1844	0.096–0.186
Copper			Plate, oxidized by heating at 1110°F	390–1110	0.37–0.48
Carefully polished electrolytic copper	176	0.018	Nickel oxide	1200–2290	0.59–0.86
Commercial, emeried, polished, but pits remaining	66	0.030	Nickel alloys		
Commercial, scraped shiny but not mirror-like	72	0.072	Chromnickel	125–1894	0.64–0.76
Polished	242	0.023	Nickelin (18–32 Ni; 55–68 Cu; 20 Zn), gray oxidized	70	0.262
Plate, heated long time, covered with thick oxide layer	77	0.78	KA-2S alloy steel (8% Ni; 18% Cr.), light silvery, rough, brown, after heating	420–914	0.44–0.36
Plate heated at 1110°F	390–1110	0.57	After 42 hr. heating at 980°F.	420–980	0.62–0.73
Cuprous oxide	1470–2010	0.66–0.54	NCT-3 alloy (20% Ni; 25% Cr.), brown, splotched, oxidized from service	420–980	0.90–0.97
Molten copper	1970–2330	0.16–0.13	NCT-6 alloy (60% Ni; 12% Cr), smooth, black, firm adhesive oxide coat from service	520–1045	0.89–0.82
Gold			Platinum		
Pure, highly polished	440–1160	0.018–0.035	Pure, polished plate	440–1160	0.054–0.104
Iron and steel			Strip	1700–2960	0.12–0.17
Metallic surfaces (or very thin oxide layer):			Filament	80–2240	0.036–0.192
Electrolytic iron, highly polished	350–440	0.052–0.064	Wire	440–2510	0.073–0.182
Polished iron	800–1880	0.144–0.377	Silver		
Iron freshly emeried	68	0.242	Polished, pure	440–1160	0.0198–0.0324
Cast iron, polished	392	0.21	Polished	100–700	0.0221–0.0312
Wrought iron, highly polished	100–480	0.28	Steel, see Iron.		
Cast iron, newly turned	72	0.435	Tantalum filament	2420–5430	0.194–0.31
Polished steel casting	1420–1900	0.52–0.56	Tin—bright tinned iron sheet	76	0.043 and 0.064
Ground sheet steel	1720–2010	0.55–0.61	Tungsten		
Smooth sheet iron	1650–1900	0.55–0.60	Filament, aged	80–6000	0.032–0.35
Cast iron, turned on lathe	1620–1810	0.60–0.70	Filament	6000	0.39
Oxidized surfaces:			Zinc		
Iron plate, pickled, then rusted red	68	0.612	Commercial, 99.1% pure, polished	440–620	0.045–0.053
Completely rusted	67	0.685	Oxidized by heating at 750°F.	750	0.11
Rolled sheet steel	70	0.657	Galvanized sheet iron, fairly bright	82	0.228
Oxidized iron	212	0.736	Galvanized sheet iron, gray oxidized	75	0.276
Cast iron, oxidized at 1100°F	390–1110	0.64–0.78			
Steel, oxidized at 1100°F	390–1110	0.79			
Smooth oxidized electrolytic iron	260–980	0.78–0.82			
Iron oxide	930–2190	0.85–0.89			
Rough ingot iron	1700–2040	0.87–0.95			

B. Refractories, Building Materials, Paints, and Miscellaneous

Surface	t, °F	Emissivity	Surface	t, °F	Emissivity
Asbestos			Carbon		
Board	74	0.96	T-carbon (Gebr. Siemens) 0.9% ash (this started with emissivity at 260°F. of 0.72, but on heating changed to values given)	260–1160	0.81–0.79
Paper	100–700	0.93–0.945			
Brick					
Red, rough, but no gross irregularities	70	0.93	Carbon filament	1900–2560	0.526
Silica, unglazed, rough	1832	0.80	Candle soot	206–520	0.952
Silica, glazed, rough	2012	0.85	Lampblack-waterglass coating	209–362	0.959–0.947
Grog brick, glazed	2012	0.75			
See Refractory Materials below.					

TABLE 5-6 Normal Total Emissivity of Various Surfaces (*Concluded*)

B. Refractories, Building Materials, Paints, and Miscellaneous

Surface	t, °F°	Emissivity°	Surface	t, °F°	Emissivity°
Same	260–440	0.957–0.952	Oil paints, sixteen different, all colors	212	0.92–0.96
Thin layer on iron plate	69	0.927	Aluminum paints and lacquers		
Thick coat	68	0.967	10% Al, 22% lacquer body, on rough or		
Lampblack, 0.003 in. or thicker	100–700	0.945	smooth surface	212	0.52
Enamel, white fused, on iron	66	0.897	26% Al, 27% lacquer body, on rough or		
Glass, smooth	72	0.937	smooth surface	212	0.3
Gypsum, 0.02 in. thick on smooth or			Other Al paints, varying age and Al		
blackened plate	70	0.903	content	212	0.27–0.67
Marble, light gray, polished	72	0.931	Al lacquer, varnish binder, on rough plate	70	0.39
Oak, planed	70	0.895	Al paint, after heating to 620°F.	300–600	0.35
Oil layers on polished nickel (lube oil)	68		Paper, thin		
Polished surface, alone		0.045	Pasted on tinned iron plate	66	0.924
+0.001-in. oil		0.27	On rough iron plate	66	0.929
+0.002-in. oil		0.46	On black lacquered plate	66	0.944
+0.005-in. oil		0.72	Plaster, rough lime	50–190	0.91
Infinitely thick oil layer		0.82	Porcelain, glazed	72	0.924
Oil layers on aluminum foil (linseed oil)			Quartz, rough, fused	70	0.932
Al foil	212	0.087†	Refractory materials, 40 different	1110–1830	
+1 coat oil	212	0.561	poor radiators		$\begin{bmatrix} 0.65 \\ 0.70 \end{bmatrix} - 0.75$
+2 coats oil	212	0.574			
Paints, lacquers, varnishes			good radiators		$\begin{bmatrix} 0.80 \\ 0.85 \end{bmatrix} - \begin{Bmatrix} 0.85 \\ 0.90 \end{Bmatrix}$
Snowhite enamel varnish or rough iron					
plate	73	0.906	Roofing paper	69	0.91
Black shiny lacquer, sprayed on iron	76	0.875	Rubber		
Black shiny shellac on tinned iron sheet	70	0.821	Hard, glossy plate	74	0.945
Black matte shellac	170–295	0.91	Soft, gray, rough (reclaimed)	76	0.859
Black lacquer	100–200	0.80–0.95	Serpentine, polished	74	0.900
Flat black lacquer	100–200	0.96–0.98	Water	32–212	0.95–0.963
White lacquer	100–200	0.80–0.95			

°When two temperatures and two emissivities are given, they correspond, first to first and second to second, and linear interpolation is permissible. °C = (°F − 32)/1.8.

†Although this value is probably high, it is given for comparison with the data by the same investigator to show the effect of oil layers. See Aluminum, Part A of this table.

and for an infinite plane parallel to a system of rows of parallel tubes as curves 1 and 3 of Fig. 5-17.

The exchange area between any two area elements of a sphere is independent of their relative shape and position and is simply the product of the areas divided by the area of the whole sphere; i.e., any spot on a sphere has equal views of all other spots.

For surfaces in two-dimensional systems (with third dimension infinite), A_1F_{12} per unit length in the third dimension may be obtained simply by evaluating, in a cross-sectional view, the sum of lengths of crossed strings from the ends of A_1 to the ends of A_2 less the sum of uncrossed strings from and to the same points, all divided by 2. The strings must be so drawn that all the flux from one surface to the other must cross each of a pair of crossed strings and neither of a pair of uncrossed ones. If one surface can see the other around both sides of an obstruction, two more pairs of strings are involved.

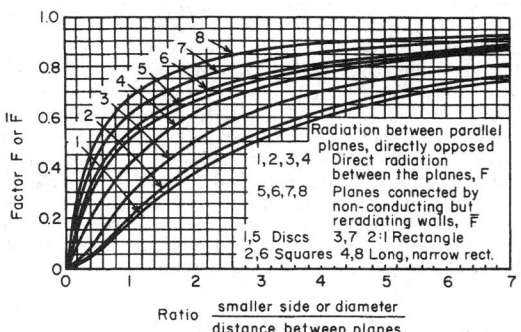

FIG. 5-15 Radiation between parallel planes, directly opposed.

Example 3: Calculation of View Factor   Evaluate the view factor between two parallel circular tubes long enough compared with their diameter D or their axis-to-axis separating distance C to make the problem two-dimensional. With reference to Fig. 5-18, the crossed-strings method yields, per unit of axial length,

$$A_1F_{12} = \frac{2(EFGH - HJ)}{2} = D\left\{\sin^{-1}\frac{D}{C} + \left[\left(\frac{C}{D}\right)^2 - 1\right]^{1/2} - \frac{C}{D}\right\}$$

Results for a large number of other cases are given by Hottel and Sarofim (op. cit., chap. 2) and Hamilton and Morgan (NACA-TN2836, December 1952). A comprehensive bibliography is provided by Siegel and Howell (*Thermal Radiation Heat Transfer*, McGraw-Hill, 1992).

The view factor F may often be evaluated from that for simpler configurations by the application of three principles: that of reciprocity, $A_iF_{ij} = A_jF_{ji}$; that of conservation, $\sum F_{ij} = 1$; and that due to Yamauti [*Res. Electrotech. Lab. (Tokyo)*, 148, 1924; 194, 1927; 250, 1929], showing that the exchange areas AF between two pairs of surfaces are equal when there is a one-to-one correspondence for all sets of symmetrically placed pairs of elements in the two surface combinations.

Example 4: Calculation of Exchange Area   The exchange area between the two squares 1 and 4 of Fig. 5-19 is to be evaluated. The following exchange areas may be obtained from Eq. 5-121. F for common-side rectangles: $\overline{13} = 0.24$, $\overline{24} = 2 \times 0.29 = 0.5S$, $\overline{(1+2)(3+4)} = 3 \times 0.32 = 0.96$. Expression of $\overline{(1+2)(3+4)}$ in terms of its components yields $\overline{(1+2)(3+4)} = \overline{13} + \overline{14} + \overline{23} + \overline{24}$. And by the Yamauti principle $\overline{14} = \overline{23}$, since for every pair of elements in 1 and 4 there is a corresponding pair in 2 and 3. Therefore,

$$\overline{14} = \frac{\overline{(1+2)(3+4)} - \overline{13} - \overline{24}}{2} = 0.07$$

Figure 5-16 may be used in the same way.

Non-Black-Surface Enclosures   In the following discussion we are concerned with enclosures containing gray sources and sinks, radiatively adiabatic surfaces, and no absorbing gas. The calculation of interchange between a source and a sink under conditions involving successive multiple reflections from other source-sink surfaces in the

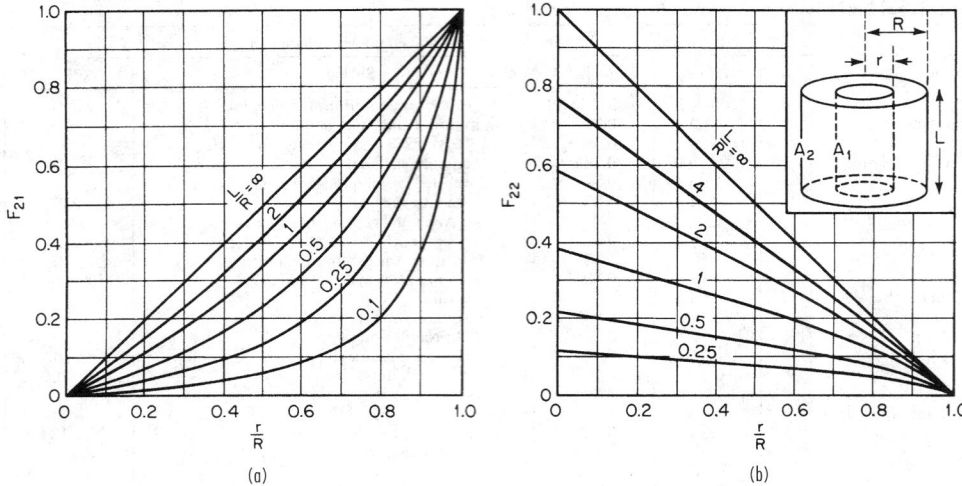

FIG. 5-16 View factors for a system of two concentric coaxial c to inner cylinder. (*b*) Inner surface of outer cylinder to itself.

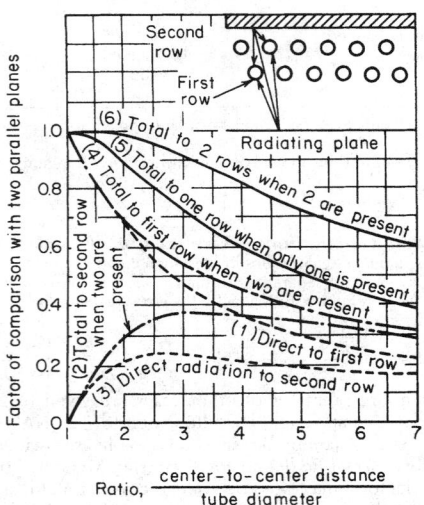

FIG. 5-17 Distribution of radiation to rows of tubes irradiated from one side. Dashed lines: direct view factor *F* from plane to tubes. Solid lines: total view factor $\bar{F}$ for black tubes backed by a refractory surface.

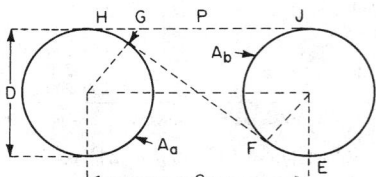

FIG. 5-18 Direct exchange between parallel circular tubes.

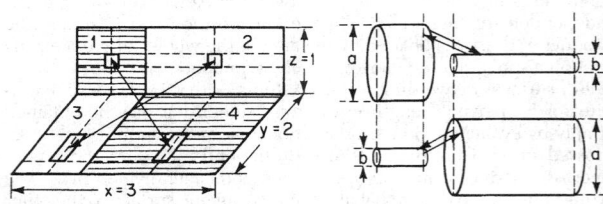

FIG. 5-19 Illustration of the Yamauti principle.

enclosure, as well as reradiation from refractory surfaces which are in radiative equilibrium, can become complicated.

Zone Method   Let a zone of a furnace enclosure be an area small enough to make all elements of itself have substantially equivalent "views" of the rest of the enclosure. (In a furnace containing a symmetry plane, parts of a single zone would lie on either side of the plane.) Zones are of two classes: source-sink surfaces, designated by numerical subscripts and having areas A_1, A_2, . . . , and emissivities ε_1, ε_2, . . . ; and surfaces at which the net radiant-heat flux is zero (fulfilled by the average refractory wall in which difference between internal convection and external loss is minute compared with incident radiation), designated by letter subscripts starting with *r*, and having areas A_r, A_s, ... It may be shown (see, for example, Hottel and Sarofim, op. cit., chap. 3) that the net radiation interchange between source-sink zones *i* and *j* is given by

$$\dot{Q}_{i=j} = A_i \mathscr{F}_{ij} \sigma\, T_i^4 - A_j \mathscr{F}_{ji} \sigma\, T_j^4 \qquad (5\text{-}125)$$

$\mathscr{F}_{ij}$ is called the **total view factor** from *i* to *j*, and the term $A_i\mathscr{F}_{ij}$, sometimes designated $\overline{S_i S_j}$, is called the **total interchange area** shared by areas A_i and A_j and depends on the shape of the enclosure and the emissivity and absorptivity of the source and sink zones. Restriction here is to gray source-sink zones, for which $A_i\mathscr{F}_{ij} = A_j\mathscr{F}_{ji}$; the more general case is treated elsewhere (Hottel and Sarofim, op. cit., chap. 5).

Evaluation of the $A\mathscr{F}$'s that characterize an enclosure involves solution of a system of radiation balances on the surfaces. If the assumption is made that all the zones of the enclosure are gray and emit and reflect diffusely,° then the direct-exchange area *ij*, as evaluated for the black-surface pair A_i and A_j, applies to emission *and* reflections between them. If at a surface the total leaving-flux density, emitted plus reflected, is denoted by *W* (and called by some the **radiosity** and by others the **exitance**), radiation balances take the form:

° So-called Lambert surfaces, which emit or reflect with an intensity independent of angle; approximately satisfied by most nonmetallic, tarnished, oxidized, or rough surfaces.

For source-sink j,

$$A_j \varepsilon_j E_j + \rho_j \sum_i (\overline{ij}) W_i = A_j W_j \qquad (5\text{-}126)$$

For adiabatic surface r,

$$\sum_i (\overline{ir}) W_i = A_r W_r \qquad (5\text{-}127)$$

where ρ is reflectance and the summation is over all surfaces in the enclosure. In matrix notation, Eq. (10-196) becomes, with source or sink zones represented by 1, 2, 3 . . . and adiabatic zones by r, s, t . . . ,

$$
\begin{bmatrix}
\overline{11} - \dfrac{A_1}{\rho_1} & \overline{12} & \overline{1r} & \overline{1s} \\[2mm]
\overline{12} & \overline{22} - \dfrac{A_2}{\rho_2} & \overline{2r} & \overline{2s} \\[2mm]
\overline{1r} & \overline{2r} & \overline{rr} - A_r & \overline{rs} \\[2mm]
\overline{1s} & \overline{2s} & \overline{rs} & \overline{ss} - A_s
\end{bmatrix}
\begin{bmatrix}
W_1 \\[2mm] W_2 \\[2mm] W_r \\[2mm] W_s
\end{bmatrix}
=
\begin{bmatrix}
-\dfrac{A_1 \varepsilon_1}{\rho_1} E_1 \\[2mm]
-\dfrac{A_2 \varepsilon_2}{\rho_2} E_2 \\[2mm]
0 \\[2mm] 0
\end{bmatrix}
$$

$$(5\text{-}128)$$

This represents a system of simultaneous equations equal in number to the number of rows of the square matrix. Each equation consists, on the left, of the sum of the products of the members of a row of the square matrix and the corresponding members of the W-column matrix and, on the right, of the member of that row in the third matrix. With this set of equations solved for W_i, the net flux at any surface A_i is given by

$$\dot{Q}_{i,\text{net}} = (A_i \varepsilon_i / \rho_i)(E_i - W_i) \qquad (5\text{-}129)$$

Refractory temperature is obtained from $W_r = E_r = \sigma T_r^4$.

The more general use of Eq. (5-128) is to obtain the set of total interchange areas $A\mathscr{F}$· which *constitute a complete description of the effect of shape, size, and emissivity on radiative flux, independent of the presence or absence of other transfer mechanisms.* It may be shown that

$$A_i \mathscr{F}_{ij} \equiv A_j \mathscr{F}_{ji} \equiv \overline{\overline{S_i S_j}} = \frac{A_i \varepsilon_i}{\rho_i}\left[\frac{A_j \varepsilon_j}{\rho_j}\left(-\frac{D'_{ij}}{D}\right) - \delta_{ij}\varepsilon_j\right] \qquad (5\text{-}130)$$

where D is the determinant of the square coefficient matrix in Eq. 5-128) and D'_{ij} is the cofactor of its ith row and jth column, or $(-1)^{i+j}$ times the minor of D formed by crossing out the ith row and ith column, and δ_{ij} is the Kronecker delta, 1 when $i = j$, otherwise 0.

As an example, consider radiation between two surfaces A_1 and A_2, which together form a complete enclosure. Equation (5-130) takes the form

$$A_1 \mathscr{F}_{12} = \left(\frac{A_1 \varepsilon_1}{\rho_1}\right)\left(\frac{A_2 \varepsilon_2}{\rho_2}\right)\left(\frac{\overline{12}}{\begin{vmatrix} \overline{11} - \dfrac{A_1}{\rho_1} & \overline{12} \\ \overline{12} & \overline{22} - \dfrac{A_2}{\rho_2} \end{vmatrix}}\right) \qquad (5\text{-}131)$$

Only one direct-view factor F_{12} or direct-exchange area $\overline{12}$ is needed because F_{11} equals $1 - F_{12}$ and F_{22} equals $1 - F_{21}$ equals $1 - F_{12}A_1/A_2$. Then $\overline{11}$ equals $A_1 - \overline{12}$ and $\overline{22}$ equals $A_2 - \overline{21}$. With these substitutions, Eq. (5-131) becomes

$$A_1 \mathscr{F}_{12} = \frac{A_1}{\dfrac{1}{F_{12}} + \dfrac{1}{\varepsilon_1} - 1 + \dfrac{A_1}{A_2}\left(\dfrac{1}{\varepsilon_2} - 1\right)} \qquad (5\text{-}132)$$

Special cases include
1. Parallel plates, large compared to clearance. Substitution of $F_{12} = 1$ and $A_1 = A_2$ gives

$$A_1 \mathscr{F}_{12} = \frac{A_1}{\dfrac{1}{\varepsilon_1} + \dfrac{1}{\varepsilon_2} - 1} \qquad (5\text{-}133)$$

2. Sphere of area A_1 concentric with surrounding sphere of area A_2. $F_{12} = 1$. Then

$$A_1 \mathscr{F}_{12} = \frac{A_1}{\dfrac{1}{\varepsilon_1} + \left(\dfrac{A_1}{A_2}\right)\left(\dfrac{1}{\varepsilon_2 - 1}\right)} \qquad (5\text{-}134)$$

3. Body of surface A_1 having no negative curvature, surrounded by very much larger surface A_2. $F_{12} = 1$ and $A_1/A_2 \to 0$. Then

$$\mathscr{F}_{12} = \varepsilon_1 \qquad (5\text{-}135)$$

Many furnace problems are adequately handled by dividing the enclosure into but two source-sink zones A_1 and A_2 and any number of no-flux zones A_r, A_s,. . . . For this case Eq. (5-130) yields

$$\frac{1}{A_1 \mathscr{F}_{12}}\left(\equiv \frac{1}{A_2 \mathscr{F}_{21}}\right) = \frac{1}{A_1}\left(\frac{1}{\varepsilon_1} - 1\right) + \frac{1}{A_2}\left(\frac{1}{\varepsilon_2} - 1\right) + \frac{1}{A_1 \overline{F}_{12}} \qquad (5\text{-}136)$$

where the expression $A_1 \overline{F}_{12}(\equiv A_2 \overline{F}_{21})$ represents the total interchange area for the limiting case of a black source and black sink (the refractory emissivity is of no moment). The factor $\overline{F}$ is known exactly for a few geometrically simple cases and may be approximated for others. If A_1 and A_2 are equal parallel disks, squares, or rectangles, connected by nonconducting but reradiating refractory walls, then $\overline{F}$ is given by Fig. 5-15, curves 5 to 8. If A_1 represents an infinite plane and A_2 is one or two rows of infinite parallel tubes in a parallel plane and if the only other surface is a refractory surface behind the tubes, $\overline{F}_{12}$ is given by curve 5 or 6 of Fig. 5-17.

If an enclosure may be divided into several radiant-heat sources or sinks A_1, A_2, etc., and the rest of the enclosure (reradiating refractory surface) may be lumped together as A_r at a uniform temperature T_r, then the total interchange area for zone pairs in the black system is given by

$$A_1 \overline{F}_{12}(\equiv A_2 \overline{F}_{21}) = \overline{12} + \frac{(\overline{1r})(\overline{r2})}{A_r - \overline{rr}} \qquad (5\text{-}137)$$

For the two-source-sink-zone system to which Eq. (5-136) applies, Eq. (5-137) simplifies to

$$A_1 \overline{F}_{12} = \overline{12} + 1/(1/\overline{1r} + 1/\overline{2r}) \qquad (5\text{-}138)$$

and if A_1 and A_2 each can see none of itself, there is further simplification to

$$A_1 \overline{F}_{12} = \overline{12} + \frac{1}{1/(A_1 - \overline{12}) + 1/(A_2 - \overline{12})}$$
$$= \frac{A_1 A_2 - (\overline{12})^2}{A_1 + A_2 - 2(\overline{12})} \qquad (5\text{-}139)$$

which necessitates the evaluation of but one geometrical factor F.

Equation (5-136) covers many of the problems of radiant-heat interchange between source and sink in a furnace enclosure. The error due to single zoning of source and sink is small even if the views of the enclosure from different parts of each zone are quite different, provided the emissivity is fairly high; the error in $\overline{F}$ is zero if it is obtainable from Fig. 5-15 or 5-17, small if Eq. (5-137) is used and the variation in temperature over the refractory is small. An approach to any desired accuracy can be made by use of Eqs. (5-126) and (5-130) with division of the surfaces into more zones.

From the definitions of F, $\overline{F}$, and $\mathscr{F}$ it is to be noted that

$$F_{11} + F_{12} + F_{13} + \cdots + F_{1r} + F_{1s} + \cdots = 1$$
$$\overline{F}_{11} + \overline{F}_{12} + \overline{F}_{13} + \cdots = 1$$
$$\mathscr{F}_{11} + \mathscr{F}_{12} + \mathscr{F}_{13} + \cdots = \varepsilon_1$$

Example 5: Radiation in a Furnace Chamber   A furnace chamber of rectangular parallelepipedal form is heated by the combustion of gas inside vertical radiant tubes lining the sidewalls. The tubes are of 0.127-m (5-in) outside diameter on 0.305-m (12-in) centers. The stock forms a continuous plane on the hearth. Roof and end walls are refractory. Dimensions are shown in Fig. 5-20. The radiant tubes and stock are gray bodies having emissivities of 0.8 and 0.9 respectively. What is the net rate of heat transmission to the stock by radiation when the mean temperature of the tube surface is 816°C (1500°F) and that of the stock is 649°C (1200°F)?

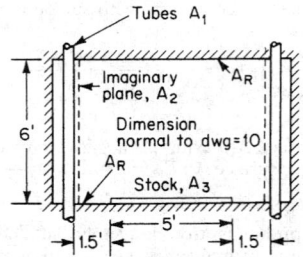

FIG. 5-20 Furnace-chamber cross section. To convert feet to meters, multiply by 0.3048.

This problem must be broken up into two parts, first considering the walls with their refractory-backed tubes. To imaginary planes A_2 of area 1.83 by 3.05 m (6 by 10 ft) and located parallel to and inside the rows of radiant tubes, the tubes emit radiation $\sigma T_1^4 A_1 \mathscr{F}_{12}$, which equals $\sigma T_1^4 A_2 \mathscr{F}_{21}$. To find $\mathscr{F}_{21}$, use Fig. 5-17, curve 5, from which $\overline{F}_{21} = 0.81$. Then from Eq. (10-200)

$$\mathscr{F}_{21} = \frac{1}{\left(\dfrac{1}{1} - 1\right) + \left(\dfrac{12}{5\pi}\right)\left(\dfrac{1}{0.8} - 1\right) + \dfrac{1}{0.81}} = 0.702$$

This amounts to saying that the system of refractory-backed tubes is equal in radiating power to a continuous plane A_2 replacing the tubes and refractory back of them, having a temperature equal to that of the tubes and an equivalent or effective emissivity of 0.702.

The new simplified furnace now consists of an enclosure formed by two 1.83- by 3.05-m (6- by 10-ft) radiating sidewalls (area A_2, emissivity 0.702), a 1.52- by 3.05-m (5- by 10-ft) receiving plane on the floor A_3, and refractory surfaces A_r to complete the enclosure (ends, roof, and floor side strips). The desired heat transfer is

$$\dot{Q}_{2 \rightleftarrows 3} = \sigma(T_1^4 - T_3^4) A_2 \mathscr{F}_{23}$$

To evaluate $\mathscr{F}_{23}$, start with the direct interchange factor F_{23}. $F_{23} = F$ from A_2 to $(A_3 + a$ strip of A_r alongside A_3, which has a common edge with A_2) minus F from A_2 to the strip only. These two F's may be evaluated from Fig. 5-20 and Eq. (5-121). For the first F, $Y = 1.83/3.05$, $Z = 1.98/3.05$, and $F = 0.239$; for the second F, $Y = 1.83/3.05$, $Z = 0.46/3.05$, and $F = 0.100$. Then $F_{23} = 0.239 - 0.10 = 0.139$. Now $\overline{F}$ may be evaluated. From Eq. 5-137 et seq.,

$$A_2 \overline{F}_{23} = \overline{23} \div \frac{1}{1/2r + 1/3r} \qquad \overline{F}_{23} = F_{23} + \frac{1}{(1/F_{2r}) + (A_2/A_3)(1/F_{3r})}$$

Since A_2 "sees" A_r, A_3, and some of itself (the plane opposite), $F_{2r} = 1 - F_{22} - F_{23}$. F_{22}, the direct interchange factor between parallel 1.83- by 3.05-m (6- by 10-ft) rectangles separated by 2.44 m (8 ft), may be taken as the geometric mean of the factors for 1.83-m (6-ft) squares separated by 2.44 m (8 ft) and for 3.05-m (10-ft) squares separated by 2.44 m (8 ft). These come from Fig. 5-15, curve 2, according to which $F_{22} = \sqrt{0.13 \times 0.255} = 0.182$. Then $F_{2r} = 1 - 0.182 - 0.139 = 0.679$. The other required direct factor is $F_{3r} = 1 - F_{32} = 1 - F_{23} A_2/A_3 = 1 - (0.139)(11.14)/4.65 = 0.666$. Then

$$\overline{F}_{23} = 0.139 + \frac{1}{(1/0.679) + (11.14/4.65)(1/0.666)} = 0.336$$

Having $\overline{F}_{23}$, we may now evaluate the factor $\mathscr{F}_{23}$:

$$\mathscr{F}_{23} = \frac{1}{(1/0.336) + [(1/0.702) - 1] + (11.14/4.65)[(1/0.9) - 1]} = 0.273$$

$$\dot{Q}_{net} = \sigma(T_1^4 - T_3^4) A_2 \mathscr{F}_{23} = 5.67(10.89^4 - 9.22^4)(11.15)(0.273)$$

$$= 118{,}000 \text{ J/s} \quad (402{,}000 \text{ Btu/h})$$

A result of interest is obtained by dividing the term $A_2 \mathscr{F}_{23}(11.15 \times 0.273$, or 3.04) by the actual area A_1 of the radiating tubes $(0.127\pi)(18.3)(2) = 14.6$ m^2 (157 ft^2). Thus $3.04/14.6 = 0.208$, which means that the net radiation from a tube to the stock is 20.8 percent as much as if the tube were black and completely surrounded by black stock.

Integral Formulation The zone method has the purpose of dodging the solution of an integral equation. If in Eq. (5-126) the zone on which the radiation balance is formulated is decreased to a differential element, that equation becomes

$$dA_j \varepsilon_j E_j + \rho_j \int \frac{dA_i \, dA_f (\cos \theta_i)(\cos \theta_j) W_i}{r^2} = dA_j W_j \qquad (5\text{-}140)$$

which is an integral equation with the unknown function W inside the integral. Integration is over the entire surface area. Exact solutions have been carried out for only a few simple cases. One of these is the evaluation of emittance of an isothermal spherical cavity, for which $dA_i \, dA_f(\cos \theta_i)(\cos \theta_j)/r^2$ in the integral of Eq. (5-140) becomes $dA_i \, dA_j/4\pi R^2$, where R is the sphere radius. For this special case W is, from Eq. (10-202), constant over the inner surface of the cavity and given by

$$W = \frac{\varepsilon E}{1 - \rho(1 - A_1/4\pi R^2)} \qquad (5\text{-}141)$$

where A_1 is the curved area of a hole in the sphere's surface. The ratio W/E is the effective emittance of the hole as sensed by a narrow-angle receiver viewing the cavity interior. If the material of construction of the cavity is a diffuse emitter and reflector and has an emissivity of 0.5 and the cavity is to appear at least 98 percent black, the curved area A_1 of the hole must be smaller than 2 percent of the total surface area of the sphere.

Enclosures of Surfaces That Are Not Diffuse Reflectors If no restriction that the surfaces be diffuse emitters and reflectors is imposed, Eq. (5-140) becomes much more complex. The W's are replaced by πI's and ε_j, I_i, and I_j all become functions of the angle of the leaving beam, and ρ_j goes inside the integral and becomes a function of angles of incidence and reflection. Seldom are such details of reflectance known. When they are and a solution is needed, the Monte Carlo method of tracing the history of a large number of beams emitted from random positions and in random initial directions is probably the best method of obtaining a solution. Another approach is possible, however, because of the tendency of most surfaces to fit a simpler reflection model. The total reflectance $\rho(\equiv 1 - \varepsilon)$ can be represented by the sum of a diffuse component ρ_D and a specular component ρ_S. For applications see Hottel and Sarofim (op. cit., chap. 5). The method yields the following relation for exchange between concentric spheres or infinite cylinders:

$$A_1 \mathscr{F}_{12} \equiv \overline{S_1 S_2}$$

$$= \frac{1}{1/A_1 \varepsilon_1 + (1/A_2)(1/\varepsilon_2 - 1) + [\rho_{s2}/(1 - \rho_{s2})](1/A_1 - 1/A_2)} \qquad (5\text{-}142)$$

When there is no specular reflectance, the third term in the denominator drops out, in agreement with Eqs. (5-134) and (5-135). When the reflectance is exclusively specular, the denominator becomes $1/A_1 \varepsilon_1 + \rho_{s2}/A_1(1 - \rho_{s2})$, easily derivable from first principles.

EMISSIVITIES OF COMBUSTION PRODUCTS

The radiation from a flame is due to radiation from burning soot particles of microscopic and submicroscopic dimensions, from suspended larger particles of coal, coke, or ash, and from the water vapor and carbon dioxide in the hot gaseous combustion products. The contribution of radiation emitted by the combustion process itself, so-called chemiluminescence, is relatively negligible. Common to these problems is the effect of the shape of the emitting volume on the radiative flux; this is considered first.

Mean Beam Lengths Evaluation of radiation from a nonisothermal volume is beyond the scope of this section (see Hottel and Sarofim, *Radiative Transfer*, McGraw-Hill, New York, 1967, chap. 11). Consider an isothermal gas confined within the volume bounded by the solid angle $d\Omega$ with vertex at dA and making the angle θ with the normal to dA. The ratio of the emission to dA from the gas to that from a blackbody at the gas temperature and filling the field of view $d\Omega$ is called the gas emissivity ε. Clearly, ε depends on the path length L through the volume to dA. A hemispherical volume radiating to a spot on the center of its base represents the only case in which L is independent of direction. Flux at that spot relative to hemispherical blackbody flux is thus an alternative way to visualize emissivity.

The flux density to a small area of interest on the envelope of an emitter volume of any shape can be matched by that at the base of a hemispherical volume of some radius L, which is called the mean beam length. It is found that although the ratio of L to a characteristic dimension D of the shape varies with opacity, the variation is small enough for most engineering purposes to permit use of a constant ratio L_M/D, where L_M is the average mean beam length. L_M can be defined to apply either to a spot on the envelope or to any finite portion of its area. An important limiting case is that of opacity approaching zero ($pD \to 0$, where p = partial pressure of the emitter constituent). For this case, L (called L_0) equals $4V/A$ (V = gas volume; A = bounding area) when interest is in radiation to the entire envelope. For the range of pD encountered in practice, the optimum value of L (now L_M) lies between 0.8 to 0.95 times L_0. For shapes not reported in Table 5-7, a factor of 0.88 (or $L_M = 0.88 L_0 = 3.5V/A$) is recommended.

Instead of using the average-mean-beam-length concept to approximate $A_1[\varepsilon(L_m)]$ (the flux per unit black emissive power from a gas volume partially bounded by a surface of area A_1), one may calculate the flux rigorously by integration, over the gas volume and over A_1, of the expression $4k\,dv\,\tau(r)\,dA\cos\theta/\pi r^2$. Here k is the emission coefficient of the gas, and $\tau(r)$ is the transmittance through the distance r between dv and dA. The result has the dimensions of area and, by analogy to $\overline{ss}$, is called $\overline{gs}_1$, the direct-exchange area between the gas zone and the surface zone (Hottel and Sarofim, op. cit., chap. 7). The use of $A_1[\varepsilon(L_m)]$ instead of $\overline{gs}_1$ is adequate when the problem is such that all the gas can be treated as a single zone in contact with A_1, and having a mean radiating temperature, but the $\overline{gs}$ concept is clearly useful if allowance is to be made for temperature variations within the gas.

Gaseous Combustion Products Radiation from water vapor and carbon dioxide occurs in spectral bands in the infrared. In magnitude it overshadows convection at furnace temperatures.

Carbon Dioxide The contribution ε_c to the emissivity of a gas containing CO_2 depends on gas temperature T_G, on the CO_2 partial pressure-beam length product $p_c L$ and, to a much lesser extent, on the total pressure P. Constants for use in evaluating ε_c at a total pressure of 101.3 kPa (1 atm) are given in Table 5-8 (more on this later). The gas absorptivity α_c equals the emissivity when the absorbing gas and the emitter are at the same temperature. When the emitter surface temperature is T_1, α_c is $(T_G/T_1)^{0.65}$ times ε_c, evaluated using Table 5-8 at T_1 instead of T_G and at $p_c L T_1/T_G$ instead of $p_c L$. Line broadening, due to

increases either in total pressure or in partial pressure of CO_2, makes a correction necessary. However, at a total pressure of 101.3 kPa (1 atm) the correction factor may be ignored, since it decreases with increase in temperature and is never more than 4 percent at temperatures above 1111 K(2000°R). Estimations of the correction in systems up to 1013.3 kPa (10 atm) are given by Hottel and Sarofim (op. cit., p. 228), and by Edwards [*J. Opt. Soc. Am.*, **50**, 617 (1960)] who in addition presents data on CO_2-band emission for use in calculations involving spectrally selective surfaces. The principal emission bands of CO_2 are at about 2.64 to 2.84, 4.13 to 4.5, and 13 to 17 µm.

Water Vapor The contribution ε_w to the emissivity of a gas containing H_2O depends on T_G and $p_w L$ and on total pressure P and partial pressure p_w. Table 5-8 gives constants for use in evaluating ε_w. Allowance for departure from the special pressure conditions is made by multiplying ε_w by a correction factor C_w read from Fig. 5-21 as a function of $(p_w + P)$ and $p_w L$. The absorptivity α_w of water vapor for blackbody radiation is ε_w evaluated from Table 5-8 but at T_1 instead of T_G and at $p_w L T_1/T_G$ instead of $p_w L$. Multiply by $(T_G/T_1)^{0.45}$.

The correction factor C_w still applies. Spectral data for water vapor, tabulated for 371 wavelength intervals from 1 to 40 µm, are also available [Ferriso, Ludwig, and Thompson, *J. Quant. Spectros. Radiat. Transfer*, **6**, 241–273 (1966)]. The principal emission is in bands at about 2.55 to 2.84, 5.6 to 7.6, and 12 to 25 µm.

Carbon Dioxide–Water-Vapor Mixtures When these gases are present together, the total radiation due to both is somewhat less than the sum of the separately calculated effects, because each gas is somewhat opaque to radiation from the other in the wavelength regions 2.7 and 15 µm.

Allowance for spectral overlap, the effect of pressure, and the effect of soot luminosity would make computation tedious. Table 5-8 gives constants for use in direct calculation, for H_2O/CO_2 mixtures, of the product $\varepsilon_G T$. The product term is used because it varies much less with T than does ε_G alone. Constants are given for mixtures, in nonradiating gases, of water vapor alone, CO_2 alone, and four p_w/p_c mixtures.

TABLE 5-7   Mean Beam Lengths for Volume Radiation

Shape	Characteristic dimension, D	L_0/D	L_M/D
Sphere	Diameter	0.67	0.63
Infinite cylinder	Diameter	1.0	0.94
Semi-infinite cylinder, radiating to:			
Center of base	Diameter	1.0	0.90
Entire base	Diameter	0.81	0.65
Right-circle cylinder, ht. = diam. radiating to:			
Center of base	Diameter	0.76	0.71
Whole surface	Diameter	0.67	0.60
Right-circle cylinder, ht. = 0.5 diam. radiating to:			
End	Diameter	0.47	0.43
Side	Diameter	0.52	0.46
Total surface	Diameter	0.50	0.45
Right-circle cylinder, ht. = 2 × diam. radiating to:			
End	Diameter	0.73	0.60
Side	Diameter	0.82	0.76
Total surface	Diameter	0.80	0.73
Infinite cylinder, half-circle cross section radiating to: middle of flats	Radius		1.26
Rectangular parallelepipeds:			
1:1:1 (cube)	Edge	0.67	0.60
1:1:4, radiating to:			
1 × 4 face	Shortest edge	0.90	0.82
1 × 1 face	Shortest edge	0.86	0.71
Whole surface	Shortest edge	0.89	0.81
1:2:6, radiating to:			
2 × 6 face	Shortest edge	1.18	
1 × 6 face	Shortest edge	1.24	
1 × 2 face	Shortest edge	1.18	
Whole surface	Shortest edge	1.2	
Infinite parallel planes	Clearance	2.00	1.76
Space outside bank of parallel tubes on equilateral triangular centers			
Tube diam. = clearance	Clearance 3.4	2.8	0.82
Tube diam. = ½ clearance	Clearance 4.45	3.8	0.85
Tube centers on squares, diam. = clearance	Clearance 4.1	3.5	0.85

TABLE 5-8 Emissivity ε_G of H_2O:CO_2 Mixtures

Limited range for furnaces, valid over 25-fold range of $p_{w+c}L$, 0.046–1.15 m atm (0.15–3.75 ft. atm)

p_w/p_c	0	½	1	2	3	∞
$\frac{p_w}{p_w+p_c}$	0	⅓(0.3–0.42)	½(0.42–0.5)	⅔(0.6–0.7)	¾(0.7–0.8)	1
	CO_2 only	corresponding to $(CH)_x$, covering coal, heavy oils, pitch	corresponding to $(CH_2)_x$, covering distillate oils, paraffins, olefines	corresponding to CH_4, covering natural gas and refinery gas	corresponding to $(CH_6)_x$, covering future high H_2 fuels	H_2O only

Constants b and n of Eq., $\varepsilon_G T = b(pL - 0.015)^n$, pL = m atm, T = K

T, K	b	n	b	n	b	n	b	n	b	n	b	n
1000	188	0.209	384	0.33	416	0.34	444	0.34	455	0.35	416	0.400
1500	252	0.256	448	0.38	495	0.40	540	0.42	548	0.42	548	0.523
2000	267	0.316	451	0.45	509	0.48	572	0.51	594	0.52	632	0.640

Constants b and n of Eq., $\varepsilon_G T = b(pL - 0.05)^n$, pL = ft. atm, T = °R

T, °R	b	n	b	n	b	n	b	n	b	n	b	n
1800	264	0.209	467	0.33	501	0.34	534	0.34	541	0.35	466	0.400
2700	335	0.256	514	0.38	555	0.40	591	0.42	600	0.42	530	0.523
3600	330	0.316	476	0.45	519	0.48	563	0.51	577	0.52	532	0.640

Full range, valid over 2000-fold range of $p_{w+c}L$, 0.005–10.0 m atm (0.016–32.0 ft. atm)
Constants of Eq., $\log_{10} \varepsilon_G T_G = a_0 + a_1 \log pL + a_2 \log^2 pL + a_3 \log^3 pL$

		pL = m atm, T = K					pL = ft. atm, T = °R				
$\frac{p_w}{p_c}$	$\frac{p_w}{p_w+p_c}$	T, K	a_0	a_1	a_2	a_3	T, °R	a_0	a_1	a_2	a_3
		1000	2.2661	0.1742	−0.0390	0.0040	1800	2.4206	0.2176	−0.0452	0.0040
0	0	1500	2.3954	0.2203	−0.0433	0.00562	2700	2.5248	0.2695	−0.0521	0.00562
		2000	2.4104	0.2602	−0.0651	−0.00155	3600	2.5143	0.3621	−0.0627	−0.00155
		1000	2.5754	0.2792	−0.0648	0.0017	1800	2.6691	0.3474	−0.0674	0.0017
½	⅓	1500	2.6451	0.3418	−0.0685	−0.0043	2700	2.7074	0.4091	−0.0618	−0.0043
		2000	2.6504	0.4279	−0.0674	−0.0120	3600	2.6686	0.4879	−0.0489	−0.0120
		1000	2.6090	0.2799	−0.0745	−0.0006	1800	2.7001	0.3563	−0.0736	−0.0006
1	½	1500	2.6862	0.3450	−0.0816	−0.0039	2700	2.7423	0.4561	−0.0756	−0.0039
		2000	2.7029	0.4440	−0.0859	−0.0135	3600	2.7081	0.5210	−0.0650	−0.0135
		1000	2.6367	0.2723	−0.0804	0.0030	1800	2.7296	0.3577	−0.0850	0.0030
2	⅔	1500	2.7178	0.3386	−0.0990	−0.0030	2700	2.7724	0.4384	−0.0944	−0.0030
		2000	2.7482	0.4464	−0.1086	−0.0139	3600	2.7461	0.5474	−0.0871	−0.0139
		1000	2.6432	0.2715	−0.0816	0.0052	1800	2.7359	0.3599	−0.0896	0.0052
3	¾	1500	2.7257	0.3355	−0.0981	0.0045	2700	2.7811	0.4403	−0.1051	0.0045
		2000	2.7592	0.4372	−0.1122	−0.0065	3600	2.7599	0.5478	−0.1021	−0.0065
		1000	2.5995	0.3015	−0.0961	0.0119	1800	2.6720	0.4102	−0.1145	0.0119
∞	1	1500	2.7083	0.3969	−0.1309	0.00123	2700	2.7238	0.5330	−0.1328	0.00123
		2000	2.7709	0.5099	−0.1646	−0.0165	3600	2.7215	0.6666	−0.1391	−0.0165

NOTE: $p_w/(p_w + p_c)$ of ⅓, ½, ⅔, and ¾ may be used to cover the ranges 0.2–0.4, 0.4–0.6, 0.6–0.7, and 0.7–0.8, respectively, with a maximum error in ε_G of 5 percent at pL = 6.5 m atm, less at lower pLs. Linear interpolation reduces the error generally to less than 1 percent. Linear interpolation or extrapolation on T introduces an error generally below 2 percent, less than the accuracy of the original data.

Four suffice, since a change halfway from one mixture ratio to the adjacent one changes the emissivity by a maximum of but 5 percent; linear interpolation may be used if considered necessary. The constants are given for three temperatures, adequate for linear interpolation since $\varepsilon_G T$ changes a maximum of one-sixth due to a change from one temperature base halfway to the adjacent one. The interpolation relation, with T_H and T_L representing the higher and lower base temperatures bracketing T, and with the brackets in the term $[A(x)]$ indicating that the parentheses refer not to a multiplier but to an argument, is

$$\overline{\varepsilon_G T_G} = \frac{[\varepsilon_G T_H(pL)](T_G - T_L) + [\varepsilon_G T_L(pL)](T_H - T_G)}{500} \tag{5-143}$$

Extrapolation to a temperature that is above the highest or below the lowest of the three base temperatures in Table 5-8 uses the same formulation, but one of its terms becomes negative.

The gas absorptivity may also be obtained from the constants for emissivities. The product $\overline{\alpha_{G_1} T_1}$ (gas absorptivity for black surface radiation), x (surface temperature), is $\varepsilon_G T_1$ evaluated at T_1 instead of T_G and at pLT_1/T_G instead of pL, then multiplied by $(T_G/T_1)^{0.5}$, or

$$\overline{\alpha_{G_1} T_1} = \left[\varepsilon_G T_1\left(\frac{pLT_1}{T_G}\right)\right]\left(\frac{T_G}{T_1}\right)^{0.5} \tag{5-144}$$

The exponent 0.5 is an adequate average of the exponents for the pure components. The interpolation relation for absorptivity is

$$\overline{\alpha_{G_1} T_1} = \left[\varepsilon_G T_H\left(\frac{pLT_H}{T_G}\right)\right]\left(\frac{T_G}{T_H}\right)^{0.5}\left(\frac{T_1 - T_L}{500}\right) + \left[\varepsilon_G T_L\left(\frac{pLT_L}{T_G}\right)\right]\left(\frac{T_G}{T_L}\right)^{0.5}\left(\frac{T_H - T_1}{500}\right) \tag{5-145}$$

The base temperature pair T_H and T_L can be different for evaluating ε_G and α_{G_1} if T_G and T_1 are far enough apart.

Example 6: Calculation of Gas Emissivity and Absorptivity This example will use only SI units, except that pressure will be in atm, not kPa. Flue gas containing 6 percent CO_2 and 11 percent H_2O vapor, wet basis, flows through a bank of tubes of 0.1016 in (4-in) outside diameter on equilateral 0.2032 m (8-in) triangular centers. In a section in which the gas and tube surface temperatures are 691°C (964 K) and 413°C (686 K), what are the emissivity and absorptivity of the gas? From Table 5-8, $L_m = (2.8)(0.01016) = 0.2845$ m (only SI units will be used in this example). $p = p_w + p_c = 0.17$ atm; $pL = 0.0484$ m atm, barely large enough to justify the short method, the top part of Table 5-8. p_w/p_c = 11/6, near enough to 2 to use col. 5. Since both T_G and T_1 are below the lowest T in the top part of the table, use the nearest pair, T_H = 1500 K and T_L = 1000 K. At T_H, $b = 540$, $n = 0.42$. $\varepsilon_G T_H = 540(0.0484 - 0.015)^{0.42} = 129.5$. At T_L,

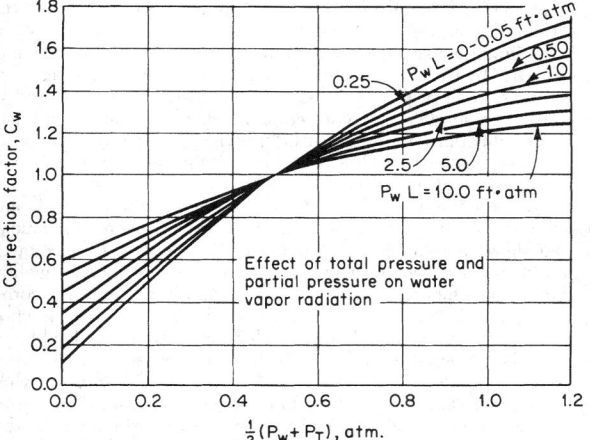

FIG. 5-21 Correction factor for converting emissivity of water vapor to values of P_w and P_T other than 0 to 1 atm respectively. To convert atmosphere-feet to kilopascal-meters, multiply by 30.89; to convert atmospheres to kilo-pascals, multiply by $(1.0133)(10^2)$.

$b = 444$, $n = 0.34$, $\overline{\epsilon_G T_L} = 444(0.0484 - 0.015)^{0.34} = 139.8$. From interpolation Eq. (5-143) (here extrapolation), $\epsilon_G T_G = [129.5 (964 - 1000) + 139.5 (1500 - 964)]/500 = 140.2$. For $\alpha_{G_1} T_1$ with $T_1 = T_H = 1500$, $pLT_H/T_G = 0.0753$. From Eq. (5-144), $\alpha_{GH} T_H = 540 (0.0753 - 0.015)^{0.42}(964/1500)^{0.5} = 133.1$. Similarly, $\alpha_{GL} T_L = 444 (0.0502 - 0.015)^{0.34} (964/1000)^{0.5} = 139.7$. From Eq. (5-145)

$$\overline{\alpha_{G_1} T_1} = \frac{133.1(686 - 1000) + 139.7(1500 - 686)}{500} = 143.8$$

Then $\epsilon_G = 140.2/964 = 0.145$ and $\alpha_{G_1} = 143.8/686 = 0.210$. If the longer method (the bottom part of Table 5-8) were used, $\epsilon_G = 0.141$ and $\alpha_{G_1} = 0.206$.

Other Gases Because of their practical importance, the emissivities of CO_2 and H_2O have been studied much more extensively than those of other gases, and the values summarized in the preceding paragraphs are based on extensive measurement of both total and integrated spectral values. Correction for pressure has reduced the disagreement among experimenters. A summary of the less adequate information on other gases appears in Table 5-9.

Flames and Particle Clouds

Luminous Flames Luminosity conventionally refers to soot radiation; it is important when combustion occurs under such conditions that the hydrocarbons in the flame are subject to heat in the absence of sufficient air well mixed on a molecular scale. Because soot parti-

cles are small relative to the wavelength of the radiation of interest [diameters $(2)(10^{-8})$ to $(1.4)(10^{-7})$ m (200 to 1400 Å)], the monochromatic emissivity ϵ_λ depends on the total particle volume per unit volume of space f_v regardless of particle size. It is given by

$$\epsilon_\lambda = 1 - e^{-Kf_v L/\lambda} \tag{5-146}$$

where L is the path length. Use of the perfect gas law and a material balance allows the restatement of the above to

$$\epsilon_\lambda = 1 - e^{-KPSL/\lambda T} \tag{5-147}$$

where P is the total pressure (atm) and S is the mole fraction of soot in the gas. S depends on the fractional conversion f_c of the fuel carbon to soot and is the mole fraction, wet basis, of carbon in gaseous form (CO_2, CO, CH_4, etc.) times $f_c/(1 - f_c)$ or, with negligible error, times f_c, which is a very small number. Evaluation of K is complex, and its numerical value depends somewhat on the age of the soot, the temperature at which it is formed, and its hydrogen content. It is recommended that $K = 0.526$ [K/atm].

The total emissivity of soot ϵ_s is obtained by integration over the wavelength spectrum, giving

$$\epsilon_s = 1 - \frac{15}{4}\left[\psi^{(3)}\left(1 + \frac{KPSL}{c_2}\right)\right], \tag{5-148}$$

where $\psi^{(3)}(x)$ is the pentagamma function of x. It may be shown that an excellent approximation to Eq. (5-148) is

$$\epsilon_s = 1 - [1 + 34.9SPL]^{-4} \tag{5-149}$$

where PL is in atm m. The error is less the lower ϵ_s, and is only 0.5 percent at $\epsilon_s = 0.5$ and 0.8 percent at 0.67.

There is at present no method of predicting soot concentration of a luminous flame analytically; reliance must be placed on experimental measurement on flames similar to that of interest. Visual observation is misleading; a flame so bright as to hide the wall behind it may be far from a "black" radiator. The chemical kinetics and fluid mechanics of soot burnout have not progressed far enough to evaluate the soot fraction f_c for relatively complex systems. Additionally, the soot in a combustion chamber is highly localized, and a mean value is needed for calculation of the radiative heat transfer performance of the chamber. On the basis of limited experience with fitting data to a model, the following procedure is recommended when total combustion chamber performance is being estimated: (1) When pitch, or a highly aromatic fuel, is burned, one percent of the fuel carbon appears as soot. This produces values of ϵ_s of 0.4–0.5 and ϵ_{G+s} of 0.6–0.7. These values are lower than some measurements on pitch flames, but the measurements are usually taken through the flame at points of high luminosity. (2) When No. 2 fuel oil is burned, 33 percent of the fuel carbon appears as soot (but that number varies greatly with burner design). (3) When natural gas is burned, any soot contribution to emissivity may be ignored. Admittedly, the numbers given should be functions of burner design and excess air, and they should be considered tentative, subject to change when good data show they are off target. The Inter-

TABLE 5-9 Total Emissivities of Some Gases

Temperature	1000°R			1600°R			2200°R			2800°R		
P_sL, (atm)(ft)	0.01	0.1	1.0	0.01	0.1	1.0	0.01	0.1	1.0	0.01	0.1	1.0
NH_3[a]	0.047	0.20	0.61	0.020	0.120	0.44	0.0057	0.051	0.25	(0.001)	(0.015)	(0.14)
SO_2[b]	0.020	0.13	0.28	0.013	0.090	0.32	0.0085	0.051	0.27	0.0058	0.043	0.20
CH_4[c]	0.020	0.060	0.15	0.023	0.072	0.194	0.022	0.070	0.185	0.019	0.059	0.17
CO[d]	0.011	0.031	0.061	0.022	0.057	0.10	0.022	0.050	0.080	(0.012)	(0.035)	(0.050)
NO[e]	0.0046	0.018	0.060	0.0046	0.021	0.070	0.0019	0.010	0.040	0.00078	0.004	0.025
HCl[f]	0.00022	0.00079	0.0020	0.00036	0.0013	0.0033	0.00037	0.0014	0.0036	0.00029	0.0010	0.0027

NOTE: Figures in this table are taken from plots in Hottel and Sarofim, *Radiative Transfer*, McGraw-Hill, New York, 1967, chap. 6. Values in parentheses are extrapolated. To convert degrees Rankine to kelvins, multiply by $(5.556)(10^{-1})$. To convert atmosphere-feet to kilopascal-meters, multiply by 30.89.
[a]Total-radiation measurements of Port (Sc.D. thesis in chemical engineering, MIT, 1940) at 1-atm total pressure, $L = 1.68$ ft, T to 2000°R.
[b]Calculations of Guerrieri (S.M. thesis in chemical engineering, MIT, 1932) from room-temperature absorption measurements of Coblentz (*Investigations of Infrared Spectra*, Carnegie Institution, Washington, 1905) with poor allowance for temperature.
[c]Band measurements of Lee and Happel [*Ind. Eng. Chem. Fundam.*, **3**, 167 (1964)] at T up to 2050°R plus calculations to extrapolate temperature to 3800°R.
[d]Total-radiation measurements of Ullrich (Sc.D. thesis in chemical engineering, MIT, 1953) at 1-atm total pressure, $L = 1.68$ ft, T to 2200°R.
[e]Calculations of Malkmus and Thompson [*J. Quant. Spectros. Radiat. Transfer*, **2**, 16 (1962)], to $T = 5400$°R and $PL = 30$ atm · ft.
[f]Calculations of Malkmus and Thompson [*J. Quant. Spectros. Radiat. Transfer*, **2**, 16 (1962)], to $T = 5400$°R and $PL = 300$ atm · ft.

national Flame Foundation has recorded data on many luminous flames from gas, oil, and coal (see *J. Inst. Energy,* formerly *J. Inst. Fuel,* 1956 to present).

Combined Soot, H₂O, and CO₂ Radiation The spectral overlap of H_2O and CO_2 radiation has been taken into account by the constants for obtaining ε_G. Additional overlap occurs when soot emissivity ε_s is added. If the emission bands of water vapor and CO_2 were randomly placed in the spectrum and soot radiation were gray, the combined emissivity would be ε_G plus ε_s minus an overlap correction $\varepsilon_G \varepsilon_s$. But monochromatic soot emissivity is higher the shorter the wavelength, and in a highly sooted flame at 1500 K half the soot emission lies below 2.5 μm where H_2O and CO_2 emission is negligible. Then the correction $\varepsilon_G \varepsilon_s$ must be reduced, and the following is recommended:

$$\varepsilon_{G+s} = \varepsilon_G + \varepsilon_s - M\varepsilon_G \varepsilon_s \qquad (5\text{-}150)$$

where M depends mostly on T_G and to a much less extent on optical density SPL. Values that have been calculated from this simple model can be represented with acceptable error by

$$M = 1.07 + 18\ SPL - 0.27 \left(\frac{T}{1000}\right) \qquad (5\text{-}151)$$

Clouds of Large Black Particles The emissivity ε_M of a cloud of particles with a perimeter large compared with wavelength λ is

$$\varepsilon_M = 1 - e^{-(a/v)L} \qquad (5\text{-}152)$$

where a/v is the projected area of the particles per unit volume of space. If the particles have no negative curvature (a particle can see none of itself) and are randomly oriented, $a = a'/4$, where a' is the actual surface area; and if the particles are uniform, $a/v = cA = cA'/4$ where A and A' are the projected and total areas of each particle and c is the number concentration of particles. For spherical particles, this gives

$$\varepsilon_M = 1 - e^{-(\pi/4)cd^2L} = 1 - e^{-1.5f_vL/d} \qquad (5\text{-}153)$$

As an example, consider heavy fuel oil ($CH_{1.5}$, specific gravity, 0.95) atomized to a surface mean particle diameter of d, burned with 20 percent excess air to produce coke-residue particles having the original drop diameter and suspended in combustion products at 1204°C (2200°F). The flame emissivity due to the particles along a path of L m will be, with d in micrometers,

$$\varepsilon_M = 1 - e^{-24.3L/d} \qquad (5\text{-}154)$$

With 200-μm particles and an L of 3.05 m (10 ft), the particle contribution to emissivity will be 0.31. Soot luminosity will increase this; particle burnout will decrease it.

Clouds of Nonblack Particles The correction for nonblackness of the particles is complicated by multiple scatter of the radiation reflected by each particle. The emissivity ε_M of a cloud of gray particles of individual surface emissivity ε_1 can be estimated by the use of Eq. (5-151), with its exponent multiplied by ε_1, if the optical thickness $(a/v)L$ does not exceed about 2. Modified Eq. (5-151) would predict an approach of ε_M to 1 as $L \to \infty$, an impossibility in a scattering system; the asymptotic value of ε_M can be read from Fig. 5-14 as ε_h, with albedo ω given by particle-surface reflectance $1 - \varepsilon_1$. Particles with a perimeter lying between 0.5 and 5 times the wavelength of interest can be handled with difficulty by use of the Mie equations (see Hottel and Sarofim, op. cit., chaps. 12 and 13).

Summation of Separate Contributions to Gas or Flame Emissivity Flame emissivity ε_{G+s} due to joint emission from gas and soot has already been treated. If massive-particle emissivity ε_M, such as from fly ash, coal char, or carbonaceous cenospheres from heavy fuel oil, are present, it is recommended that the total emissivity be approximated by

$$\varepsilon_{G+s} + \varepsilon_M - (\varepsilon_{G+s})(\varepsilon_M)$$

RADIATIVE EXCHANGE BETWEEN GASES OR SUSPENDED MATTER AND A BOUNDARY

Local Radiative Exchange The interchange rate $\dot{Q}$ between an isothermal gas mass at T_G and its isothermal black bounding surface of area A_1 is given by

$$\dot{Q} = A_1 \sigma(T_G^4 \varepsilon_G - T_1^4 \alpha_{G1}) \qquad (5\text{-}155)$$

Evaluation of α_{G1} is unnecessary when T_1 is less than one-half T_G; α_{G1} may then be assumed equal to ε_G.

If the bounding surface is gray rather than black, multiplication of Eq. (5-154) by surface emissivity ε_1 allows properly for reduction of the primary beams, gas-to-surface or surface-to-gas, but secondary reflections are ignored. The correction then lies between ε_1 and 1, and for most industrially important surfaces with $\varepsilon_1 > 0.8$ a value of $(1 + \varepsilon_1)/2$ is adequate. Rigorous allowance for this and other factors is presented later, e.g., Eq. (5-163).

If the bounding walls are mostly sink-type surfaces of area A_1 and temperature T_1, but in small part refractory surfaces of area A_r in radiative equilibrium at unknown temperature T_r, an energy balance on A_r is in principle necessary to determine T_r and the effect on energy flux. However, the total heat transfer to the sink may be visualized as corresponding to its having an effective area equal to its own plus a fraction x of that of the refractory, with the only temperatures involved being those of the gas and the heat sink. The fraction x varies from zero when the ratio of refractory to heat-sink surface is very high to unity when the ratio is very low and the value of ε_G is low. If A_r is small compared with A_1, a value for x of 0.7 may be used in the approximate method.

Long Exchanger This case, in which axial radiative flux is ignored, includes most radiatively modified heat exchangers of interest to chemical engineers. When the gas temperature transverse to the flow direction is reasonably uniform and the chamber is long compared with its mean hydraulic radius, the opposed upstream and downstream fluxes through the flow cross section will substantially cancel (hot combustion products through tubes or across tube banks, tunnel kilns, billet-reheating furnaces, Example 7). Under these conditions, the radiative contribution to local flux density q may be formulated in terms of local temperatures and beam lengths or exchange areas evaluated for a two-dimensional system infinite in the flow direction. The local flux density at the sink A_1 is then

$$q(T_G, T_1) = q_r(T_G, T_1) + h(T_G - T_1) \qquad (5\text{-}156)$$

where h is the local convective heat-transfer coefficient and $q_r(T_G, T_1)$ the radiation contribution calculated from T_G, T_1, ε_G, and ε_1 by using the approximate treatment in the preceding subsection or the more rigorous treatment in the following subsection. If $\dot{m}C_p$ is the hourly heat capacity of the gas stream, the temperature of which changes by dT_G over the sink-area increment dA_1, then

$$[q(T_G, T_1)]dA_1 = -\dot{m}C_p\, dT_G \qquad (5\text{-}157)$$

from which

$$A_1 = \dot{m} \int_{T_{G,\text{outlet}}}^{T_{G,\text{inlet}}} \frac{C_p\, dT_G}{q(T_G, T_1)} \qquad (5\text{-}158)$$

The area under a curve of C_p/q versus T_G or $1/q$ versus the specific enthalpy i may be used to solve for the area A_1 required to obtain a given outlet temperature or to obtain the outlet temperature given A_1. Three points generally suffice to determine the area under the curve within 10 percent.

Instead of using graphical integration, which can handle any complexity of variation of flux density q with T_G and T_1 along an interchanger flow path, one may evaluate a mean flux density based on mean gas and sink temperatures, based in turn on terminal temperatures. It has been found empirically that fair results are obtained by the use of a mean surface temperature equal to the arithmetic mean of the terminal surface temperatures and by the use of a mean gas temperature equal to the mean surface temperature plus the logarithmic mean of the temperature difference, gas to surface, at the two ends of the exchanger. When radiation dominates the transfer process, however, graphical integration is safer.

Example 7: Radiation in Gases Flue gas containing 6 percent carbon dioxide and 11 percent water vapor by volume (wet basis) flows through the convection bank of an oil tube still consisting of rows of 0.102-m (4-in) tubes on 0.203-m (8-in) centers, nine 7.62-m (25-ft) tubes in a row, the rows staggered to put the tubes on equilateral triangular centers. The flue gas enters at 871°C (1144 K, 1600°F) and leaves at 538°C (811 K, 1000°F). The oil flows in a countercurrent direction to the gas and rises from 316 to 427°C (600 to 800°F). Tube surface emissivity is 0.8. What is the average heat-input rate, due to gas radiation alone, per square meter of external tube area?

With each row of tubes there is associated $(0.203)(\sqrt{3}/2) = 0.176$ m $(0.577$ ft) of wall height, of area $[(0.203)(9)(2) + (7.62)(2)]0.176 - (9)(2)(\pi)(0.0508)^2 = 3.18$ m^2 $(34.2$ ft$^2)$. One row of tubes has an area of $(\pi)(0.102)(7.62)(9) = 22.0$ m^2 (236 ft^2). If the recommended factor of 0.7 on the refractory area is used, the effective area of the tubes is $[22.0 + (0.7)(3.18)]/22.0 = 1.10$ m^2/m^2 of actual area. The exact evaluation of the outside tube temperature from the known oil temperature would involve a knowledge of the oil-film coefficient, tube-wall resistance, and rate of heat flow into the tube, the evaluation usually involving trial and error. However, for the present purpose the temperature drop through the tube wall and oil film will be assumed to be $41.7°C$ ($75°F$), making the tube surface temperatures $357°C$ ($675°F$) and $468°C$ ($875°F$); the average is $412°C$ ($775°F$). The radiating gas temperature is

$$t_g = 412 + \frac{(871 - 468) - (538 - 357)}{2.3 \log [(871 - 468)/(538 - 357)]}$$

$$= 412 + 278 = 690°C \ (1274°F)$$

These temperatures, partial pressures, and dimensions were used in Example 6 to determine gas emissivity and absorptivity, $\varepsilon_G = 0.145$; $\alpha_{G1} = 0.210$. The approximate effective emissivity of the boundary is $(0.8 + 1)/2 = 0.9$. Then from Eq. (5-155), modified to allow for sink emissivity and for the presence of a small amount of refractory boundary,

$$\dot{Q}/A_1 = q = (0.9)(1.10)\sigma(T_G^4\varepsilon_G - T_1^4\alpha_G)$$

$$= (0.9)(1.10)(5.67)[(9.63)^4(0.145) - (6.85)^4(0.210)]$$

$$= 4405 \ \text{J/(m}^2 \ \text{tube area·s)} \ [1396, \ \text{Btu/(ft}^2 \ \text{tube area·h)}]$$

This is equivalent to a convection coefficient of 4405/278, or 15.85 W/(m^2)(K) which is of the order of magnitude expected of the convection coefficient itself. Radiation rapidly becomes dominant as the system temperature rises.

Total-Exchange Areas $\overline{SS}$ and $\overline{GS}$ The arguments leading to the development of the interchange factor $A_i\mathcal{F}_{ij}(\equiv \overline{S_iS_j})$ between surfaces apply to the case of absorption within the gas volume if in the evaluation of the direct-exchange areas allowance is made for attenuation of the radiant beam through the gas. This necessitates nothing more than redefinition, in Eqs. (5-126) to (5-130), of every term $\overline{ij}(\equiv \overline{S_iS_j} \equiv A_iF_{ij})$ to represent, per unit black emissive power, flux from A_i through an absorbing gas to A_j. This may be visualized as multiplication of A_iF_{ij} by the mean gas transmittance $T_{ij}(= 1 - \varepsilon_G$ for a gray gas). In a system containing an isothermal gas and source-sink boundaries of areas $A_1 \ldots A_n$, the total emission from A_1 per unit of its black emissive power is ε_1A_1, of which $\overline{S_1S_1} + \overline{S_1S_2} + \cdots + \overline{S_1S_n}$ is absorbed in the various source-sink surfaces by multiple reflections. The difference has been absorbed in the gas and is called the gas-surface total-exchange area $\overline{GS_1}$

$$\overline{GS_1} = A_1\varepsilon_1 - \sum_i \overline{S_1S_i} \tag{5-159}$$

Note that though $\overline{S_1S_i}$ is never used in calculating radiative exchange, its value is necessary for use of Eq. (5-159) to calculate $\overline{GS}$.

If the gas volume is not isothermal and is zoned, an additional magnitude, the gas-to-gas total-exchange area $\overline{G_iG_j}$, arises (see Hottel and Sarofim. *Radiative Transfer*, McGraw-Hill, New York, 1967, chap. 11). Space does not permit derivations of special cases; only the single-gas-zone system is treated here.

Single-Gas-Zone/Two-Surface-Zone Systems An enclosure consisting of but one isothermal gas zone and two gray surface zones can, properly specified, model so many industrially important radiation problems as to merit detailed presentation. One can evaluate the total radiation flux between any two of the three zones, including multiple reflection at all surfaces.

$$\dot{Q}_{G \Leftrightarrow 1} = \overline{GS_1}\sigma(T_G^4 - T_1^4)$$

$$\dot{Q}_{1 \Leftrightarrow 2} = \overline{S_1S_2}\sigma(T_1^4 - T_2^4)$$

The total-exchange area takes a relatively simple closed form, even when important allowance is made for gas radiation not being gray and when a reduction of the number of system parameters is introduced by assuming that one of the surface zones, if refractory, is radiatively adiabatic. Before allowance is made for these factors, the case of a gray gas enclosed by two source-sink surface zones will be presented. Modification of Eq. 5-130, as discussed in the first paragraph of this subsection, combined with the assumption that a single mean beam length applies to all transfers; that is, that there is but one gas transmittance $\tau(= 1 - \varepsilon_G)$, gives

$$\overline{S_1S_2} = \frac{A_1\varepsilon_1\varepsilon_2F_{12}}{1/\tau + \tau\rho_1\rho_2(1 - F_{12}/C_2) - \rho_1(1 - F_{12}) - \rho_2(1 - F_{21})} \tag{5-160}$$

$$\overline{S_1S_1} = \frac{A_1\varepsilon_1^2(F_{11} + \rho_2\tau(F_{12}/C_2 - 1))}{1/\tau + \tau\rho_1\rho_2(1 - F_{12}/C_2) - \rho_1(1 - F_{12}) - \rho_2(1 - F_{21})} \tag{5-161}$$

$$\overline{GS_1} = \frac{A_1\varepsilon_1\varepsilon_G(1/\tau + \rho_2(F_{12}/C_2 - 1))}{1/\tau + \tau\rho_1\rho_2(1 - F_{12}/C_2) - \rho_1(1 - F_{12}) - \rho_2(1 - F_{21})} \tag{5-162}$$

These three expressions suffice to formulate total-exchange areas for gas-enclosing arrangements which include, for example, the four cases illustrated in Table 5-10.

An additional surface arrangement of importance is a single-zone surface enclosing gas. With the gas assumed gray, the simplest derivation of $\overline{GS_1}$ is to note that the emission from surface A_1 per unit of its blackbody emissive power is $A_1\varepsilon_1$, of which the fractions ε_G and $(1 - \varepsilon_G)\varepsilon_1$ are absorbed by the gas and the surface, respectively, and the surface-reflected residue always repeats this distribution. Therefore,

$$\overline{GS}_{\substack{\text{single surface} \\ \text{zone surround-} \\ \text{ing gray gas}}} \equiv \overline{GS_1} = A_1\varepsilon_1\frac{\varepsilon_G}{\varepsilon_G + (1 - \varepsilon_G)\varepsilon_1} = \frac{A_1}{\frac{1}{\varepsilon_G} + \frac{1}{\varepsilon_1} - 1} \tag{5-163}$$

Alternatively, $\overline{GS_1}$ could be obtained from Case 1 of Table 5-10 by letting plane area A_1 approach 0, leaving A_2 as the sole surface zone.

Departure of gas from grayness has so marked an effect on radiative transfer that the subject will be presented prior to discussion of the systems covered by Table 5-10.

The Effect of Nongrayness of Gas on Total-Exchange Area A radiating gas departs from grayness in two ways: (1) Its transmittance τ through successive path lengths L_m due to surface reflection, instead of being constant, keeps increasing because at the wavelengths of high absorption the incremental absorption decreases with increasing pathlength; (2) Gas emissivity ε_G and absorptivity α_{G1} are not the same unless T_1 equals T_G. The total emissivity of a real gas, the spectral emissivity, and absorptivity ε_λ that varies in any way with λ can be expressed as the a-weighted mean of a suitable number of gray-gas emissivity or absorptivity terms $\varepsilon_{G,i}$ or $\alpha_{G,i}$, representing the gray-gas emissivity or absorptivity in the energy fractions a_i of the blackbody spectrum. Then

$$\varepsilon_G = \sum_0^n a_i\varepsilon_{G,i} = \sum_0^n a_i(1 - e^{-k_ipL}) \tag{5-164}$$

For simplicity, n should be as low as is consistent with small error. The retention of but two terms is feasible when one considers that if α_{G1} is so fitted that the first absorption and the second following surface reflection are correct, then further attenuation of the beam by successive surface reflections makes the errors in those absorptions decrease in importance. Let the gas be modeled as the sum of one gray gas plus a clear gas, with the gray gas occupying the energy fraction a of the blackbody spectrum and the clear gas the fraction $(1 - a)$. Then

$$[\varepsilon_G(pL)] = a(1 - e^{-kpL}) + (1 - a)0$$

$$[\varepsilon_G(2pL)] = a(1 - e^{-2kpL}) + (1 - a)0 \tag{5-165}$$

Solution of these gives

$$a = \frac{\varepsilon_G(pL)}{2 - \frac{\varepsilon_G(2pL)}{\varepsilon_G(pL)}}$$

$$kpL = -\ln\left[1 - \frac{\varepsilon_G(pL)}{a}\right] \tag{5-166}$$

Note that these values are specific to the subject problem in which the mean beam length is L_m, with ε_GS evaluated from basic data, such as Table 5-8. $(1 - e^{-kpL})$ in Eq. (5-165) represents the emissivity of a gray gas, which will be called $\varepsilon_{G,i}$. For later use, note that,

$$\varepsilon_{G,i} = \frac{\varepsilon_G(pL)}{a} \tag{5-167}$$

To allow for the difference between emissivity and absorptivity and combine them into a single emissivity-absorptivity term called effec-

TABLE 5-10 Total-Exchange Areas for Four Arrangements of Two-Zone-Surface Enclosures of a Gray Gas

A plane surface A_1 and a surface A_2 completing the enclosure	Infinite parallel planes	Concentric spherical or infinite cylindrical surface zones, A_1 inside	Two-surface-zone spherical surface, each zone one or more parts; or speckled enclosure, any shape
$F_{12} = 1$	$F_{12} = F_{21} = 1$	$F_{12} = 1; \quad F_{21} = A_1/A_2$	$F_{12} = F_{22} = C_2; \quad F_{21} = F_{11} = C_1$
$\dfrac{\overline{S_1S_2}}{A_1} = \dfrac{\epsilon_1\epsilon_2}{D_1}$	$\dfrac{\overline{S_1S_2}}{A_1} = \dfrac{\epsilon_1\epsilon_2}{D_2}$	$\dfrac{\overline{S_1S_2}}{A_1} = \dfrac{\epsilon_1\epsilon_2}{D_3}$	$\dfrac{\overline{S_1S_2}}{A_1} = \dfrac{\epsilon_1\epsilon_2 C_2}{D_4}$
$\dfrac{\overline{GS_1}}{A_1} = \dfrac{\epsilon_1\epsilon_G(1/\tau + \rho_2 A_1/A_2)}{D_1}$	$\dfrac{\overline{GS_1}}{A_1} = \dfrac{\epsilon_1\epsilon_G(1/\tau + \rho_2)}{D_2}$	$\dfrac{\overline{GS_1}}{A_1} = \dfrac{\epsilon_1\epsilon_G(1/\tau + \rho_2 A_1/A_2)}{D_3}$	$\dfrac{\overline{GS_1}}{A_1} = \dfrac{\epsilon_1\epsilon_G/\tau}{D_4}$
$\dfrac{\overline{GS_2}}{A_2} = \dfrac{\epsilon_2\epsilon_G(1/\tau + \rho_1 A_1/A_2)}{D_1}$	$\dfrac{\overline{S_1S_1}}{A_1} = \dfrac{\epsilon_1^2\rho_2\tau}{D_2}$	$\dfrac{\overline{GS_2}}{A_2} = \dfrac{\epsilon_2\epsilon_G(1/\tau + \rho_1 A_1/A_2)}{D_3}$	$\dfrac{\overline{S_1S_1}}{A_1} = \dfrac{\epsilon_1^2 C_1}{D_4}$
$\dfrac{\overline{S_1S_1}}{A_1} = \dfrac{\epsilon_1^2\tau\rho_2 A_1/A_2}{D_1}$	$D_2 \equiv 1/\tau - \tau\rho_1\rho_2$	$\dfrac{\overline{S_1S_1}}{A_1} = \dfrac{\epsilon_1^2\rho_2\tau A_1/A_2}{D_3}$	$D_4 \equiv 1/\tau - \rho_1 C_1 - \rho_2 C_2$
$D_1 \equiv \dfrac{1}{\tau} - \rho_2\left[1 - \dfrac{A_1}{A_2}(1-\tau\rho_1)\right]$		$D_3 \equiv \dfrac{1}{\tau} - \rho_2\left[1 - \dfrac{A_1}{A_2}(1-\tau\rho_1)\right]$	

tive emissivity $\epsilon_{G,e}$, one must first evaluate absorptivity α_{G1} using Eq. (5-161). Formulation of the net direct exchange can then be used to define $\epsilon_{G,e}$:

$$\sigma(\epsilon_G T_G^4 - \alpha_{G1}T_1^4) \equiv \sigma\epsilon_{G,e}(T_G^4 - T_1^4)$$

or

$$\epsilon_{G,e} = \frac{\epsilon_{G,e} - \alpha_{G1}(T_1^4/T_G^4)}{1 - (T_1/T_G)^4} \tag{5-168}$$

The emissivity and absorptivity of use in converting gray-gas total-exchange areas to real-gas values are $\epsilon_{G,e}$ and a_e, the latter obtained by using Eq. (5-166), except that $\epsilon_{G,e}(pL)$ replaces $\epsilon_G(pL)$; the same for $\epsilon_{G,e}(2pL)$. This means that, for the conversion, four terms will have to be formulated: $\epsilon_G(pL)$, $\epsilon_G(2pL)$, $\alpha_{G1}(pL)$, and $\alpha_{G1}(2pL)$. The gray emissivity term $\epsilon_{G,t}$ of Eq. (5-167) now becomes $\epsilon_{G,e}/a_e$.

Conversion of gray-gas total exchange areas $\overline{GS}$ and $\overline{SS}$ to their nongray form depends on the fact that the relation between radiative transfer and blackbody emissive power σT^4 is linear and proportional. The gray-gas-equivalent emissivity $\epsilon_{G,e}/a_e$ is applicable only to the energy fraction a_e of σT^4. In consequence, to convert $\overline{GS}$ or $\overline{SS}$ to its nongray form, wherever ϵ_G or τ appears in $\overline{GS}$ it must be replaced by $\epsilon_{G,e}/a_e$ or $(1 - \epsilon_{G,e}/a_e)$, respectively; the overall result is then multiplied by a_e. The converted $\overline{SS}$ is in two parts: The gray-gas contribution involves, as above, replacement of ϵ_G by $\epsilon_{G,e}/a_e$ and τ by $(1 - \epsilon_{G,e}/a_e)$, and the result multiplied by a_e; for the clear-gas contribution, ϵ_G is replaced by 0 and τ by 1, and the result is multiplied by $(1 - a_e)$ and added to the gray-gas contribution.

The simplest application of this simple gray-plus-clear model of gas radiation is the case of a single gas zone surrounded by a single surface zone, Eq. (5-163) for a gray gas. The gray-plus-clear model gives

$$\frac{\overline{GS_1}}{A_1} = \frac{a_e}{\dfrac{a_e}{\epsilon_{G,e}} + \dfrac{1}{\epsilon_1} - 1} \tag{5-169}$$

Example 8: Effective Gas Emissivity Methane is burned to completion with 20 percent excess air (air half-saturated with water vapor at 298 K (60°F), 0.0088 mols H_2O/mol dry air) in a furnace chamber of floor dimensions 3×10 m and height 5 m. The whole surface is a gray-energy sink of emissivity 0.8

at 1000 K. The surrounding gas is at 1500 K and is well stirred. Find the effective gas emissivity $\epsilon_{G,e}$, the weighting factor a_e, and the surface radiative flux density.

Solution: Combustion is $1CH_4 + 2 \times 1.2\ O_2 + 2 \times 1.2 \times 79/21\ N_2 + 2 \times 1.2 \times 100/21 \times 0.0088\ H_2O$, going to $1\ CO_2 + [2 + 2 \times 1.2 \times (100/21) \times 0.0088]\ H_2O + 0.4\ O_2 + 9.03\ N_2 = 12.53$ moles per mole of CH_4; $p_c + p_w = (1 + 2.1)/12.53 = 0.2474$ atm. The mean beam length, $L_m = 0.88 \times 4V/A_T = 0.88 \times 4(10 \times 3 \times 5)/[2 \times (10 \times 3 + 10 \times 5 + 3 \times 5)] = 2.779$ m. $pL_m = 0.2474 \times 2.779 = 0.6875$ m atm. From emissivity Table 5-8, $b(1500) = 540$; $n(1500) = 0.42$; $b(1000) = 444$; $n(1000) = 0.34$. $\epsilon_G(pL) = 540(0.6875 - 0.015)^{0.42}/1500 = 0.3047$; $\alpha_{G1}(pL) = 444(0.6875 \times 1000/1500 - 0.015)^{0.34}(1500/1000)^{0.5}/1000 = 0.4124$. Then $\epsilon_{G,e}(pL) = [0.3047 - 0.4124(1000/1500)^4]/[1 - (1000/1500)^4] = 0.2782$. Repeat all 3 computations for $pL = 2 \times 0.6875$ to give $\epsilon_G(2pL) = 0.4096$, $\alpha_{G1}(2pL) = 0.5250$, $\epsilon_{G,e}(2pL) = 0.3812$. Then $a_e = 0.2782/(2 - 0.3812/0.2782) = 0.4418$ and the emissivity substitute = $0.2782/0.4418 = 0.6297$. For a single enveloping surface zone, the total-exchange area comes from Eq. (5-169); $\overline{GS_1}/A_1 = a_e/(a_e/\epsilon_{G,e} + 1/\epsilon_1 - 1) = 0.4418/(0.4418/0.2782 + 1/0.8 - 1) = 0.2404$. The flux density is $Q/A = q = (\overline{GS_1}/A)\sigma(T_G^4 - T_1^4) = 0.2404 \times 56.7 \times [(1500/1000)^4 - (1000/1000)^4] = 55.37$ kW/m² (17,550 Btu/sq ft hr). (Note that allowing for average humidity in air adds 5 percent to H_2O and approximately 2 percent to gas emissivity.)

Total-exchange areas for the basic one-gas two-surface model [Eqs. (5-160) to (5-162)], used to evaluate the cases in Table 5-10, take the following form when converted by the above described procedure to their nongray form:

$$\overline{S_1S_2} = \frac{a_e A_1 F_{12}\epsilon_1\epsilon_2}{\begin{aligned}&1/(1 - \epsilon_{G,e}/a_e) + (1 - \epsilon_{G,e}/a_e)\rho_1\rho_2(1 - F_{12}C_2)\\ &\quad - \rho_1(1 - F_{12}) - \rho_2(1 - F_{21})\end{aligned}}$$

$$+ \frac{(1 - a_e)A_1 F_{12}\epsilon_1\epsilon_2}{1 + \rho_1\rho_2(1 - F_{12}C_2) - \rho_1(1 - F_{12}) - \rho_2(1 - F_{21})} \tag{5-170}$$

$$\overline{GS_1} = \frac{A_1\epsilon_1\epsilon_{G,e}(1/(1 - \epsilon_{G,e}/a_e) + \rho_2(F_{12}/C_2 - 1))}{\begin{aligned}&1/(1 - \epsilon_{G,e}/a_e) + (1 - \epsilon_{G,e}/a_e)\rho_1\rho_2(1 - F_{12}C_2)\\ &\quad - \rho_1(1 - F_{12}) - \rho_2(1 - F_{21})\end{aligned}} \tag{5-171}$$

Modification of Table 5-10 to make the total-exchange areas conform to the gray-plus-clear gas model is straightforward, following the instructions presented above. The results are given in Table 5-11.

TABLE 5-11   Conversion of Some Total-Exchange Areas to Their Gray-Plus-Clear Values

1. Plane slab A_1 and surface A_2, completing an enclosure of gas ($F_{12} = 1$)

$$\frac{\overline{S_1 S_2}}{A_1} = \frac{a \varepsilon_1 \varepsilon_2}{D_1} + \frac{(1-a)\varepsilon_1 \varepsilon_2}{1 - \rho_2(1 - \varepsilon_1 A_1/A_2)}$$

$$\left[D_1 = \frac{1}{1 - \varepsilon_G/a} - \rho_2 \left(1 - \left(\frac{A_1}{A_2} \right) \left(\varepsilon_1 + \frac{\rho_1 \varepsilon_G}{a} \right) \right) \right]$$

$$\frac{\overline{GS_1}}{A_1} = \frac{\varepsilon_1 \varepsilon_G (1/(1 - \varepsilon_G/a) + \rho_2 A_1/A_2)}{D_1}$$

$$\frac{\overline{GS_2}}{A_2} = \frac{\varepsilon_2 \varepsilon_G (1/(1 - \varepsilon_G/a) + \rho_1 A_1/A_2)}{D_1}$$

2. Infinite parallel planes, gas between ($F_{12} = F_{21} = 1$)

$$\frac{\overline{S_1 S_2}}{A_1} = \frac{a \varepsilon_1 \varepsilon_2}{D_2} + \frac{(1-a)\varepsilon_1 \varepsilon_2}{1 - \rho_1 \rho_2}$$

$$\left[D_2 = \frac{1}{1 - \varepsilon_G/a} - \left(1 - \frac{\varepsilon_G}{a} \right) \rho_1 \rho_2 \right]$$

$$\frac{\overline{GS_1}}{A_1} = \frac{\varepsilon_1 \varepsilon_G (1/(1 - \varepsilon_G/a) + \rho_2)}{D_2}$$

3. Concentric spherical or infinite cylindrical surface zones, A_1 inside ($F_{12} = 1$; $F_{21} = A_1/A_2$)

$$\frac{\overline{S_1 S_2}}{A_1} = \frac{a \varepsilon_1 \varepsilon_2}{D_3} + \frac{(1-a)\varepsilon_1 \varepsilon_2}{1 - \rho_2(1 - \varepsilon_1 A_1/A_2)}$$

$$D_3 = \frac{1}{(1 - \varepsilon_G/a)} - \rho_2 \left[1 - \left(\frac{A_1}{A_2} \right) \left(\varepsilon_1 + \frac{\rho_1 \varepsilon_G}{a} \right) \right]$$

$$\frac{\overline{GS_1}}{A_1} = \frac{\varepsilon_1 \varepsilon_G (1/(1 - \varepsilon_G/a) + \rho_2 A_1/A_2)}{D_3}$$

$$\frac{\overline{GS_2}}{A_2} = \frac{\varepsilon_2 \varepsilon_G (1/(1 - \varepsilon_G/a) + \rho_1 A_1/A_2)}{D_3}$$

4. Spherical enclosure of two surface zones *or* "speckled" A_1:A_2 enclosure ($F_{12} = F_{22} = C_2$; $F_{21} = F_{11} = C_1$)

$$\frac{\overline{S_1 S_2}}{A_1} = \frac{a \varepsilon_1 \varepsilon_2 C_2}{D_4} + \frac{(1-a)\varepsilon_1 \varepsilon_2 C_2}{1 - \rho_1 C_1 - \rho_2 C_2}$$

$$\left[D_4 = \frac{1}{(1 - \varepsilon_G/a)} - \rho_1 C_1 - \rho_2 C_2 \right]$$

$$\left[C_1 = \frac{A_1}{A_1 + A_2} \right]$$

$$\frac{\overline{GS_1}}{A_1} = \frac{\varepsilon_1 \varepsilon_G /(1 - \varepsilon_G/a)}{D_4}$$

Treatment of Refractory Walls Partially Enclosing a Radiating Gas   Another modification of the results in Table 5-10 becomes important when one of the surface zones is radiatively adiabatic; the need to find its temperature can be eliminated. If surface A_2, now called A_r, is radiatively adiabatic, its net radiative exchange with A_1 must equal its net exchange with the gas.

$$\overline{GS_r}(T_G^4 - T_r^4) = \overline{S_r S_1}(T_r^4 - T_1^4)$$

or

$$\frac{T_G^4 - T_r^4}{1/\overline{GS_r}} = \frac{T_r^4 - T_1^4}{1/\overline{S_r S_1}} = \frac{T_G^4 - T_1^4}{1/\overline{GS_r} + 1/\overline{S_r S_1}} \qquad (5\text{-}172)$$

The net flux from gas G is $\overline{GS_1} \sigma(T_G^4 - T_1^4) + \overline{GS_r}\sigma(T_G^4 - T_r^4)$ which, with replacement of the last term using Eq. (5-172), gives the single term

$$\sigma(T_G^4 - T_1^4)\left[\overline{GS_1} + \frac{1}{\dfrac{1}{\overline{GS_r}} + \dfrac{1}{\overline{S_r S_1}}} \right]$$

The bracketed term is called $(\overline{GS_1})_R$, the total exchange area from G to A_1 with assistance from a refractory surface. In summary,

$$\dot{Q}_{G \leftrightarrow 1} = (\overline{GS_1})_R \sigma(T_G^4 - T_1^4) = \left[\overline{GS_1} + \frac{1}{\dfrac{1}{\overline{GS_r}} + \dfrac{1}{\overline{S_r S_1}}} \right] \sigma(T_G^4 - T_1^4) \qquad (5\text{-}173)$$

Table 5-10 supplies the forms for the three terms needed to formulate $(\overline{GS_1})_R$, with A_r substituted for A_2. If, in addition, allowance is to be made for the gas not being gray, $\varepsilon_{G,e}$ and a_e are evaluated using values of the emissivity and absorptivity calculated using Table 5-8, and the procedure described in the previous subsection is followed with $\varepsilon_{G,e}/a_e$ replacing ε_G together with the addition of a clear-gas contribution, when $\overline{SS}$ is at issue. It is tempting to say that a surface A_2 (or A_r) could be made radiatively adiabatic simply by assigning its reflectance ρ a value of 1, making the terms in the brackets of Eq. (5-173) much easier to evaluate and the result much simpler. This is valid only if the gas is gray. If it is not, A_r is a net absorber of radiation in the spectral energy fraction a (or a_e) and a net emitter in the clear-gas fraction $(1 - a)$.

Conversion of $\overline{GS}$ to $(\overline{GS_1})_R$ will be carried out for two of the four cases of Table 5-10. Case 1 is an idealization of a metal-heating slab furnace or glass furnace, with its plane sink A_1 combining with refractory surface A_r to complete the enclosure. With insertion into Eq. (5-173) of $\overline{GS_1}$, $\overline{GS_r}$, and $\overline{S_rS_1}$ after converting each to its gray plus clear form, one obtains

$$\frac{(\overline{GS_1})_R}{A_1} = \frac{\varepsilon_G}{D_1}\left[\varepsilon_1\left(\rho_2\frac{C_1}{C_r} + \frac{1}{1 - \frac{\varepsilon_G}{a}}\right)\right.$$

$$\left. + \frac{\varepsilon_r}{\frac{1}{\rho_1 + (C_r/C_1)[1 - (\varepsilon_G/a)]} + \frac{\varepsilon_r + \rho_r\varepsilon_1(C_1/C_r)}{\frac{\varepsilon_1}{\varepsilon_G}\left[a\left(\varepsilon_r + \rho_r\varepsilon_1\frac{C_1}{C_r}\right)\right] + (1-a)D_1}}\right]$$

$$(5\text{-}174)$$

where $D_1 = 1/(1 - \varepsilon_G/a) - \rho_r\{1 - (C_1/C_r)[1 - \rho_1(1 - \varepsilon_G/a)]\}$.

Conversion of $(\overline{GS_1})_R$ to applicability to a gray gas comes by making a equal 1, producing the enormous simplification to

$$\left[\frac{(\overline{GS_1})_R}{A_1}\right]_{\substack{\text{gray}\\\text{gas}}} = \frac{1}{\frac{1 + C_1(1/\varepsilon_G - 2)}{1 - C_1\varepsilon_G} + \frac{\rho_1}{\varepsilon_1}} \qquad (5\text{-}175)$$

Note that the emissivity and reflectance of the refractory are without effect on $(\overline{GS_1})_R$ if the gas is gray.

The second conversion of $\overline{GS}$ to $(\overline{GS_1})_R$ will be Case 4 of Table 5-10, the two-surface-zone enclosure with computation simplified by assuming that the direct-view factor from any spot to a surface equals the fraction of the whole enclosure that the surface occupies (the speckled-furnace model). This case can be considered an idealization of many processing furnaces such as distilling and cracking coil furnaces, with parts of the enclosure tube-covered and part left refractory. (But the refractory under the tubes is not to be classified as part of the refractory zone.) Again, one starts with substitution into Eq. (5-173) of the terms $\overline{GS_1}$, $\overline{GS_r}$, and $\overline{S_rS_1}$ from Table 5-10, Case 4, with all terms first converted to their gray-plus-clear form. To indicate the procedure, one of the components, $\overline{S_rS_1}$, wil be formulated.

$$\frac{\overline{S_rS_1}}{A_1} = a\frac{C_r\varepsilon_1\varepsilon_r}{D_4} + (1-a)\frac{C_r\varepsilon_1\varepsilon_r}{1 - \rho_1C_1 - \rho_rC_r}$$

$$= \frac{C_r\varepsilon_1\varepsilon_r}{D_4}\left(1 + \frac{\varepsilon_G(1-a)(a-\varepsilon_G)}{1 - \rho_1C_1 - \rho_rC_r}\right)$$

With $D_4 = 1/(1 - \varepsilon_G/a) - \rho_1C_1 - \rho_rC_r$, the result of the full substitution simplifies to

$$\frac{(\overline{GS_1})_R}{A_1} = \frac{1}{C_1\left(\frac{1}{\varepsilon_G} - \frac{1}{a}\right) + \frac{1}{\varepsilon_1} + \frac{1/a - 1}{\varepsilon_1 + \varepsilon_r(C_r/C_1)}} \qquad (5\text{-}176)$$

For a gray gas ($a = 1$), the above becomes

$$\frac{(\overline{GS_1})_R}{A_1} = \frac{1}{C_1\left(\frac{1}{\varepsilon_G} - 1\right) + \frac{1}{\varepsilon_1}} \qquad (5\text{-}177)$$

Eq. (5-176) has wide applicability.

COMBUSTION CHAMBER HEAT TRANSFER

Treatment of radiative transfer in combustion chambers is available at varying levels of complexity, including allowance for temperature variation in both gas and refractory walls (Hottel and Sarofim,

Radiative Transfer, McGraw-Hill, New York, 1967, chap. 14). A less rigorous treatment suffices, however, for handling many problems. There are two limiting cases: the long chamber with gas temperature varying only in the direction of gas flow (already treated) and the compact chamber containing a gas or a flame to which can be assigned an effective or average radiating temperature. The latter will be considered.

Stirred-Chamber Model; Refractory Wall Loss Negligible What furnace engineers most need is a closed-form solution of the problem, theoretically sound in structure and therefore containing a minimum number of parameters and no empirical constants and, preferably, physically visualizable. They can then (1) correlate data on existing furnaces, (2) develop a performance equation for standard design, or (3) estimate performance of a new furnace type on which no data are available.

An equation representing an energy balance on a combustion chamber of two surface zones, a heat sink A_1 at temperature T_1, and a refractory surface A_r assumed radiatively adiabatic at T_r, is most simply solved if the total enthalpy input H is expressed as $\dot{m}\overline{C_p}(T_F - T_o)$; $\dot{m}$ is the mass rate of fuel plus air; and T_F is a pseudoadiabatic flame temperature based on a mean specific heat from base temperature T_o up to the gas exit temperature T_E rather than up to T_F. The heat transfer rate Q out of the gas is then $H - \dot{m}\overline{C_p}(T_E - T_o)$ or $\dot{m}\overline{C_p}(T_F - T_E)$. The energy balance, with ambient temperature taken as conventional base T_o, is

$$(\dot{Q} =)\dot{H} - \dot{m}\overline{C_p}(T_E - T_o) = (\overline{GS_1})_R\sigma(T_G^4 - T_1^4) + h_1A_1(T_G - T_1)$$
$$+ A_oF_o\sigma(T_G^4 - T_o^4) + UA_r(T_G - T_o) \quad (5\text{-}178)$$

To make the relation dimensionless, divide through by $(\overline{GS_1})_R\sigma T_F^4$ and let all temperatures, expressed as ratios to T_F, be called τs. For clarity, the terms are tabulated:

$$\dot{m}\overline{C_p}/(\overline{GS_1})_R\sigma T_F^3 = D, \text{ dimensionless firing density}$$
$$\text{l.h.s. term} = D(1 - \tau_E)$$
$$\text{1st r.h.s. term} = \tau_G^4 - \tau_1^4$$
$$(h_1A_1/(\overline{GS_1})_R\sigma T_F^3) = L_c, \text{ convection number (dimensionless)}$$
$$A_oF_o/(\overline{GS_1})_R = L_o, \text{ wall-openings loss number,}$$
$$\text{(dimensionless)}$$
$$UA_r/(\overline{GS_1})_R\sigma T_F^3 = L_r, \text{ refractory-wall loss number}$$
$$\text{(dimensionless)}$$

The equation then becomes

$$D(1 - \tau_E) = \tau_G^4 - \tau_1^4 + L_c(\tau_G - \tau_1) + L_o(\tau_G^4 - \tau_o^4) + L_r(\tau_G - \tau_o) \quad (5\text{-}179)$$

This equation has two unknowns (τ_G and τ_E), and an empirical relation between them is needed. Many have been tried, and one of the best is to assume that the excess of T_G over T_E expressed as a ratio to T_F (zero for a perfectly stirred chamber) is a constant $\Delta [\equiv (T_G - T_E)/T_F]$. Although Δ should vary with burner type, the effects of firing rate and percent excess air are small. In the absence of performance data on the kind of furnace under study, assume $\Delta = 300/T_F$, °R or $170/T_F$, K. The left side of Eq. (5-178) then becomes $D(1 - \tau_G + \Delta)$, and with coefficients of τ_G and τ_G^4 collected, the equation becomes

$$\tau_G^4 + \left(\frac{D + L_c + L_r}{1 + L_o}\right)\tau_G - \left(\frac{\tau_1^4 + L_c\tau_1 + L_o\tau_o^4 + L_r\tau_o + D(1 + \Delta)}{1 + L_o}\right) = 0$$
$$(5\text{-}180)$$

Though this is a quartic equation, it is capable of explicit solution because of the absence of second and third degree terms. Trial-and-error enters, however, because $(\overline{GS_1})_R$ and $\overline{C_p}$ are mild functions of T_G and related T_E, respectively, and a preliminary guess of T_G is necessary. An ambiguity can exist in interpretation of terms. If part of the enclosure surface consists of screen tubes over the chamber-gas exit to a convection section, radiative transfer to those tubes is included in the chamber energy balance, but convection is not, because it has no effect on chamber gas temperature.

With Eq. (5-180) solved, the gas-side efficiency η_G is $(1 - \tau_G + \Delta)/(1 - \tau_o)$. The sink-side efficiency η_1 is less by the amount $(L_o(\tau_G^4 - \tau_1^4) + L_r(\tau_G - \tau_o))/D(1 - \tau_o)$ and is also given by $[(\overline{GS_1})_R\sigma(T_G^4 - T_1^4) + h_1A_1(T_G - T_1)]/\dot{H}$. It must be remembered that the efficiency η_1

includes the losses through the wall from the backside of any wall–mounted heat sinks. Though the results must be considered approximations, depending as they do on the empirical Δ, the equation may be used to find the effect of firing rate, excess air, and air preheat on efficiency. With some performance data available, the small effect of various factors on Δ may be found.

The first term on the right side of Eq. (5-179) is so nearly dominant for most furnaces that consideration of the main features of chamber performance is clarified by ignoring the loss terms L_o and L_r or by assuming that they and L_c have a constant mean value. The relation of a modified chamber efficiency $\eta_G(1 - \tau_o)$ to a modified firing density $D/(1 - \tau_o)$ and to the normalized sink temperature $\tau = T_1/T_F$ is shown in Fig. 5-23, which is based on Eq. (5-178), with the radiative and convective transfer terms $(GS_1)_R\sigma(T_G^4 - T_1^4) + h_1A_1(T_G - T_1)$ replaced by a combined radiation/conduction term $(GS_1)_{R,c}\sigma(T_G^4 - T_1^4)$, where $(GS_1)_{R,c} = (GS_1)_R + h_1A_1/4\sigma T_{G1}^3$; T_{G1} is adequately approximated by the arithmetic mean of T_G and T_1.

Example 9: Radiation in a Furnace   Consider a furnace 3 m × 10 m × 5 m fired with methane and 20 percent excess air, at a methane firing rate of 2500 kg/hr. Two rows of 5-inch (0.127 m) tubes (outer diameter) are mounted on equilateral triangular centers, with center-to-center distance twice the tube diameter, on 60 percent of the interior surface of the chamber. The radiative properties of the gases for an enclosure of these dimensions, containing the same combustion products, have been estimated in Example 8 for a gas temperature of 1500 K and a sink temperature of 1000 K: 12.53 moles of combustion products are generated per mole of fuel, with a mean molar heat capacity between a base temperature of 298 K and the exit gas exit temperature T_E, adequately represented for this example by $MC_p = 7.01 + 0.875\,(T/1000)$ over a T_E range of 800 to 1600 K. The lower heating value of CH_4 is 191,760 cal/g mole. The air is preheated to 600°C, and has a mean MC_p of 7.31 cal/g mole. The alloy tube emissivity ε_1 is 0.7 and may be assumed gray; the mean tube surface temperature is 700°C. The convection coefficient, gas-to-tube plane and to refractory surface is 0.0170 kW/m² °C; h_{c+r} on the outside tube surface is 0.00050 kW/m °C with a k of 0.00050 kW/m °C and an assumed ε_r of 0.6; the walls are pierced by four 0.10-m × 0.23-m peepholes. The gas exit area, 1 m × 10 m, is tube-screen-covered.

What is the sink-side efficiency η_1, the gas exit temperature T_E, and the mean flux density through the tube surface?

Solution: Temporary basis—1 mole entering CH_4. Since no molal change occurs when CH_4 burns completely with half-saturated entering air 20 percent in excess of stoichiometric, the total number of moles produced equal 11.53 moles. Entering enthalpy = 191,760 + 11.53 × 7.31(600 − 25) = 240,220 Kcal/kg mol CH_4. H [of Eq. (5-178)] = 240,220 × (2500/16.04) = 37.44E6 Kcal/hr × 4.186/3600 = 43.54E3 Kw. $\dot{m} = [16.04 + 2 \times 1.2(100/21)(29 + 0.0088 \times 18.016)](2500/16.04) = 54,440$ kg/hr. Trial and error solution necessitates several sets of computations of T_G to check assumed T_E; only the last of these will be given. The first, to save time by using results attained elsewhere, assumes $T_G = 1500$ K; the resulting T_G is 369 K higher. The second set assumes $T_G = 2017$ K; the resulting T_G is 80 K lower. Linear interpolation indicates the third set should assume $T_G = 1934$ K. That set is presented: $T_E = 1934 − 170 = 1764$. $MC_p = 7.01 + 0.875 \times (1764/1000) = 8.554$ cal/(gmol)(K). $T_F = 240,220/(12.53 \times 8.554) +$

298 = 2539 K. $\overline{C}_p = (8.554/16.04)(4.186/3600) = 0.6201E-3$ kw-hr/(kg)(K). For ε_G, $pL_m = 0.247 \times 2.779 = 0.6875$ m atm. Use of Table 5-8, with $T_H = 2000$ K ($b = 572$, $n = 0.51$) and $T_L = 1500$ ($b = 540$, $n = 0.42$) gives $\varepsilon_G(pL) = [572(0.6875 − 0.015)^{0.51}(1934 − 1500) + 540(0.6725)^{0.42}(2000 − 1934)]/(500 × 1934) = 0.2409$. For α_{G1}, with $T_1 = 1000$, Table 5-8 gives $b = 444$, $n = 0.34$. $\alpha_{G1}(pL) = 444(0.6875 × 1000/1934 − 0.015)^{0.34}(1934/1000)^{0.5}/1000 = 0.4281$. $\varepsilon_{G,e}(pL) = [0.2409 − 0.4281(1000/1934)^4]/(1 − 1/1.934^4) = 0.2265$. $2pL_m = 2 \times 0.6875 = 1.375$ m atm. $\varepsilon_G(2pL) = [572(1.375 − 0.015)^{0.51} \times 434 + 540(1.36)^{0.42} \times 66]/(500 × 1934) = 0.3422$. $\alpha_{G1}(2pL) = 444 \times (1.375 × 1000/1934 − 0.015)^{0.34}(1.934)^{0.5}/1000 = 0.5459$. $\varepsilon_{G,e}(2pL) = [0.3422 − 0.5459(1/1.934)^4]/(1 − 1/1.934^4) = 0.3265$. $a_e = 0.2265/(2 − 0.3265/0.2265) = 0.4056$. Of all these, only $\varepsilon_{G,e}(pL)$ and a_e will be used from here on. From Eq. (5-176), $(GS_1)_R/A_1 = 1/[0.6(1/0.2265 − 1/0.4056) + 1/0.87 + (1/0.4056 − 1)/(0.87 + 0.6\,(0.4/0.6))] = 0.2879$. $A_1 = (3 × 10 + 3 × 5 + 10 × 5) × 2 = 190 × 0.6 = 114$ m². $(\overline{GS}_1)_R = 0.2879 × 114 = 32.82$ m². $(\overline{GS}_1)_R\sigma T_F^3 = 32.82 × (56.7E − 12) × 2539^3 = 30.46$ kw/K. $D = 54,440 \times (0.6201E-3)/30.46 = 1.1083$. $\Delta = 170/2539 = 0.06696$. $L_c = 0.017 × 114/30.46 = 0.0636$. $L_o = 4 × 0.1 × 0.23 × 0.335/32.82 = 9.4E − 4$. $L_r = 0.0012 × 46/30.46 = 0.001812$. In Eq. (5-180), the coefficient of τ_G equals $(1.1083 + 0.0636 + 0.0018)/1.00094 = 1.1726$. The constant in the equation equals $[(100/2539)^4 + 0.0636(1/2.539) + 0.00094 × 0.1174^4 + 0.00181(298/2539)] + 1.1083(1 + 170/2539)]/1.00094 = 1.2307$. The equation to solve is: $\tau_G^4 + 1.1726\tau_G − 1.2307 = 0$. Solution gives $\tau_G = 0.7620$; $T_G = 0.762 × 2539 = 1935$ K. $T_e = 1765$ K, only 1 K above value assumed for obtaining $\overline{C}_p$ and ε_G. $\eta_G = (1 − \tau_G)/(1 − \tau_o) = (1 − 1765/2539)/(1 − 298/2539) = 0.3454$. Sink-side efficiency η_1 0.3454 − $[0.00094(0.762^4 − 0.11744^4)] + 0.001812(0.762 − 0.1174)/1.1083(1 − 0.1174) = 0.344$, not including convection to screen tubes covering gas exit. $q_{plane\ of\ sink}/A_1 = 43,540 × 0.3439/114 = 131.3$ kw/m². $q_{tube\ surf} = 131.3\,(2D/2\pi D) = 41.8$ kw/m² × 3412 × 0.3048² = 13,300 Btu/(ft²)(hr). $T_E = 1765$ K = 1492°C = 2717°F.

In Fig. 5-22, the shaded areas indicate the operating regimes of a wide range of furnace types. Note the significant properties of the function presented. (1) As firing rate D' goes down, the efficiency rises and approaches $1 − \tau_1$ in the limit. (This conclusion is modified if wall losses are significant.) (2) Changes in sink temperature have little effect if $\tau_1 < 0.3$. (3) As the furnace walls approach complete coverage by a black sink [$C\varepsilon_1 \to 1$ in Eqs. (5-176) and (5-177)] and as convection becomes unimportant, the effect of flame emissivity on D becomes one of inverse proportionality; thus at very high firing rates at which efficiency approaches inverse proportionality to D, the efficiency of heat transfer varies directly as ε_G (gas-turbine chambers), but at low firing rates ε_G has relatively little effect. (4) When $C\varepsilon_1 << 1$ because of a nonblack sink or much refractory surface, the effect of changing flame emissivity is to produce a much less than proportional effect on heat flux.

The factor Δ, the allowance for imperfect stirring, must be estimated. Values in the range of 93 to 149°C (200 to 300°F) have been found to produce data correlation for a series of tests on marine boilers.

Equation (5-179) and Fig. 5-22 serve as a framework for correlating the performance of furnaces with flow patterns—plug flow, parabolic

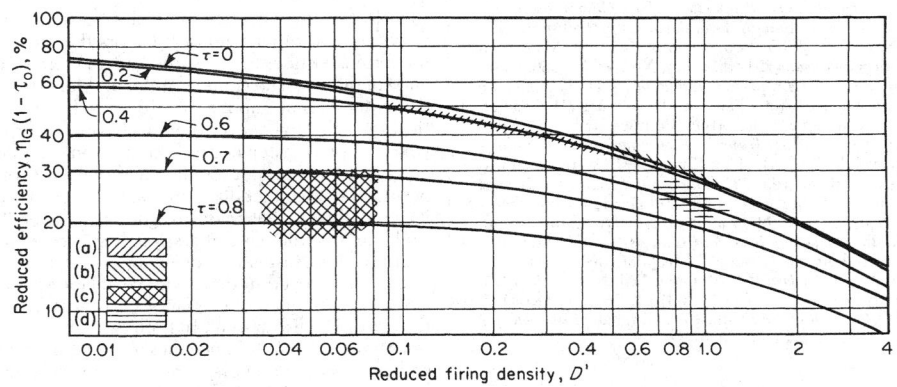

FIG. 5-22   Thermal performance of well-stirred furnace chambers; reduced efficiency as a function of reduced firing density D and reduced sink temperature τ_1. (*a*) Radiant section, oil tube stills, cracking coils. (*b*) Domestic boiler combustion chambers. (*c*) Open-hearth furnaces. (*d*) Soaking pits.

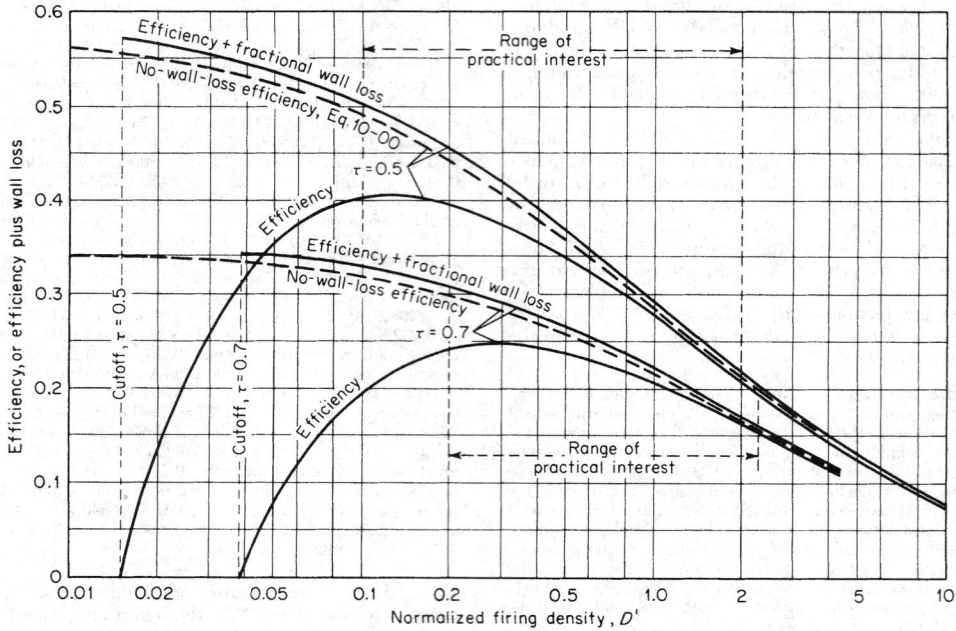

FIG. 5-23   Effect of wall-loss factor L on combustion-chamber performance; $L' = 0.02$, and $\tau = 0.5$ and 0.7.

profile, and recirculatory flow—which differ from the well-stirred model [Hottel and Sarofim, *Int. J. Mass Heat Transfer*, **8**, 1153 (1965)]. As expected, plug-flow furnaces show somewhat higher efficiency, mild-recirculation types somewhat lower efficiency, and strong-recirculation furnaces a performance closely similar to that of the well-stirred model.

If data on several furnaces of a single class are available, a similar treatment can lead to a partially empirical equation based on simplified rules for obtaining $(\overline{GS_1})_{R\mathscr{e}}$ or an effective Δ. Because Eq. (5-178) has a structure which covers a wide range of furnace types and has a sound theoretical basis, it provides safer structures of empirical design equations than many such equations available in the engineering literature.

MASS TRANSFER

GENERAL REFERENCES: Bird, Stewart, and Lightfoot, *Transport Phenomena*, Wiley, New York, 1960. Cussler, *Diffusion: Mass Transfer in Fluid Systems*, Cambridge University Press, Cambridge, 1984. Danner and Daubert, *Manual for Predicting Chemical Process Design Data*, AIChE, New York, 1983. Daubert and Danner, *Physical and Thermodynamic Properties of Pure Chemicals*, Taylor and Francis, Bristol, PA, 1989–1995. Fahien, *Fundamentals of Transport Phenomena*, McGraw-Hill, New York, 1983. Foust, Wenzel, Clump, Maus, and Andersen, *Principles of Unit Operations*, 2d ed., Wiley, New York, 1980. Gammon, Marsh, and Dewan, *Transport Properties and Related Thermodynamic Data of Binary Mixtures*, AIChE, New York, Part 1, 1993; Part 2, 1994. Geankoplis, *Transport Processes and Unit Operations*, 3d ed., Prentice Hall, Englewood Cliffs, NJ, 1993. Hines and Maddox, *Mass Transfer: Fundamentals and Applications*, Prentice Hall, Englewood Cliffs, NJ, 1985. Kirwan, "Mass Transfer Principles," Chap. 2 in Rousseau, R. W. (ed.), *Handbook of Separation Process Technology*, Wiley, New York, 1987. McCabe, Smith, and Harriott, *Unit Operations of Chemical Engineering*, 5th ed., McGraw-Hill, New York, 1993. Reid, Prausnitz, and Poling, *The Properties of Gases and Liquids*, 4th ed., McGraw-Hill, New York, 1987. Schwartzberg and Chao, *Food Technol.*, **36**(2), 73 (1982). Sherwood, Pigford, and Wilke, *Mass Transfer*, McGraw-Hill, New York, 1975. Skelland, *Diffusional Mass Transfer*, Wiley, New York, 1974. Taylor and Krishna, *Multicomponent Mass Transfer*, Wiley, New York, 1993. Treybal, *Mass-Transfer Operations*, 3d ed., McGraw-Hill, New York, 1980.

INTRODUCTION

This part of Sec. 5 provides a concise guide to solving problems in situations commonly encountered by chemical engineers. It deals with diffusivity and mass-transfer coefficient estimation and common flux equations, although material balances are also presented in typical coordinate systems to permit a wide range of problems to be formulated and solved.

Mass-transfer calculations involve transport properties, such as diffusivities, and other empirical factors that have been found to relate mass-transfer rates to measured "driving forces" in myriad geometries and conditions. The context of the problem dictates whether the fundamental or more applied coefficient should be used. One key distinction is that, whenever there is flow parallel to an interface through which mass transfer occurs, the relevant coefficient is an empirical combination of properties and conditions. Conversely, when diffusion occurs in stagnant media or in creeping flow without transverse velocity gradients, ordinary diffusivities may be suitable for solving the problem. In either case, it is strongly suggested to employ data, whenever available, instead of relying on correlations.

Units employed in diffusivity correlations commonly followed the cgs system. Similarly, correlations for mass transfer correlations used the cgs or English system. In both cases, only the most recent correlations employ SI units. Since most correlations involve other properties and physical parameters, often with mixed units, they are repeated here as originally stated. Common conversion factors are listed in Table 1-4.

Fick's First Law   This law relates flux of a component to its composition gradient, employing a constant of proportionality called a

Nomenclature and Units—Mass Transfer

Symbols	Definition	SI units	U.S. customary units
a	Effective interfacial mass transfer area per unit volume	m^2/m^3	ft^2/ft^3
A_{cs}	Cross-sectional area of vessel	m^2 or cm^2	ft^2
A'	Constant (see Table 5-28-I)		
a_p	See a		
c	Concentration $= P/RT$ for an ideal gas	mol/m^3 or mol/l or $gequiv/l$	$lbmol/ft^3$
c_i	Concentration of component $i = x_i c$ at gas-liquid interface	mol/m^3 or mol/l or $gequiv/l$	$lbmol/ft^3$
c_P	Specific heat	$kJ/(kg \cdot K)$	$Btu/(lb \cdot °F)$
d	Characteristic length	m or cm	ft
d_b	Bubble diameter	m	ft
d_c	Column diameter	m or cm	ft
d_{drop}	Sauter mean diameter	m	ft
d_{imp}	Impeller diameter	m	ft
d_{pore}	Pore diameter	m or cm	ft
$D_{A'A}$	Self-diffusivity $(= D_A$ at $x_A = 1)$	m^2/s or cm^2/s	ft^2/h
D_{AB}	Mutual diffusivity	m^2/s or cm^2/s	ft^2/h
D_{AB}^o	Mutual diffusivity at infinite dilution of A in B	m^2/s or cm^2/s	ft^2/h
D_{eff}	Effective diffusivity within a porous solid $= \varepsilon_p D/\tau$	m^2/s	ft^2/h
D_K	Knudson diffusivity for gases in small pores	m^2/s or cm^2/s	ft^2/h
D_L	Liquid phase diffusion coefficient	m^2/s	ft^2/h
D_S	Surface diffusivity	m^2/s or cm^2/s	ft^2/h
E	Energy dissipation rate/mass		
E_S	Activation energy for surface diffusion	J/mol or cal/mol	
f	Friction factor for fluid flow	Dimensionless	Dimensionless
F	Faraday's constant	96,487 Coulomb/gequiv	
g	Acceleration due to gravity	m/s^2	ft/h^2
g_c	Conversion factor	1.0	4.17×10^8 lb ft/[lbf·h²]
G	Gas-phase mass flux	$kg/(s \cdot m^2)$	$lb/(h \cdot ft^2)$
G_a	Dry air flux	$kg/(s \cdot m^2)$	$lb/(h \cdot ft^2)$
G_M	Molar gas-phase mass flux	$kmol/(s \cdot m^2)$	$(lbmol)/(h \cdot ft^2)$
h'	Heat transfer coefficient	$W/(m^2 \cdot K) = J/(s \cdot m^2 \cdot K)$	$Btu/(h \cdot ft^2 \cdot °F)$
h_T	Total height of tower packing	m	ft
H	Compartment height	m	ft
H	Henry's law constant	$kPa/(mole-fraction$ solute in liquid phase)	$(lbf/in^2)/(mole-fraction$ solute in liquid phase)
H'	Henry's law constant	$kPa/[kmol/(m^3$ solute in liquid phase)]	$(lbf/in^2)/[(lbmol)/(ft^3$ solute in liquid phase)] or $atm/[(lbmole)/(ft^3$ solute in liquid phase)]
H_G	Height of one transfer unit based on gas-phase resistance	m	ft
H_{OG}	Height of one overall gas-phase mass-transfer unit	m	ft
H_L	Height of one transfer unit based on liquid-phase resistance	m	ft
H_{OL}	Height of one overall liquid-phase mass-transfer unit	m	ft
HTU	Height of one transfer unit (general)	m	ft
j_D	Chilton-Colburn factor for mass transfer, Eq. (5-291)	Dimensionless	Dimensionless
j_H	Chilton-Colburn factor for heat transfer	Dimensionless	Dimensionless
j_M	See j_D		
$_mJ_A$	Mass flux of A by diffusion with respect to the mean mass velocity	$kmol/(m^2 \cdot s)$ or $mol/(cm^2 \cdot s)$	$lbmol/(ft^2 \cdot h)$
$_MJ_A$	Molar flux of A by diffusion with respect to mean molar velocity	$kmol/(m^2 \cdot s)$ or $mol/(cm^2 \cdot s)$	$lbmol/(ft^2 \cdot h)$
$_VJ_A$	Molar flux of A with respect to mean volume velocity	$kmol/(m^2 \cdot s)$	$lbmol/(ft^2 \cdot h)$
J_{Si}	Molar flux by surface diffusion	$kmol/(m^2 \cdot s)$ or $gmol/(cm^2 \cdot s)$	$lbmol/(ft^2 \cdot h)$
k	Boltzmann's constant	8.9308×10^{-10} gequiv ohm/s	
k	Film mass transfer coefficient	m/s or cm/s	ft/hr
k	Thermal conductivity	$(J \cdot m)/(s \cdot m^2 \cdot K)$	$Btu/(h \cdot ft \cdot °F)$
k'	Mass-transfer coefficient for dilute systems	$kmol/[(s \cdot m^2)(kmol/m^3)]$ or m/s	$lbmol/[(h \cdot ft^2)(lbmol/ft^3)]$ or ft/hr
k_G	Gas-phase mass-transfer coefficient for dilute systems	$kmol/[(s \cdot m^2)(kPa$ solute partial pressure)]	$lbmol/[(h \cdot ft^2)lbf/in^2$ solute partial pressure)]
k'_G	Gas-phase mass-transfer coefficient for dilute systems	$kmol/[(s \cdot m^2)(mole$ fraction in gas)]	$lbmol/[(h \cdot ft^2)(mole$ fraction in gas)]
$k_G a$	Volumetric gas-phase mass-transfer	$kmol/[(s \cdot m^3)(mole$ fraction)]	$(lbmol)/[(h \cdot ft^3)(mole$ fraction)]
$\hat{k}_G a$	Overall volumetric gas-phase mass-transfer coefficient for concentrated systems	$kmol/(s \cdot m^3)$	$lbmol/(h \cdot ft^3)$
$\hat{k}_L^o$	Liquid phase mass transfer coefficient for pure absorption (no reaction)	$kmol/(s \cdot m^2)$	$lbmol/(h \cdot ft^2)$
k_L	Liquid-phase mass-transfer coefficient for dilute systems	$kmol/[(s \cdot m^2)(mole-fraction$ solution in liquid)]	$(lbmol)/[(h \cdot ft^2)(mole-fraction$ solute in liquid)]
k'_L	Liquid-phase mass-transfer coefficient for dilute systems	$kmol/[(s \cdot m^2)(kmol/m^3)]$ or m/s	$(lbmol)/[(h \cdot ft^2)(lbmol/ft^3)]$ or ft/h
$\hat{k}_L$	Liquid-phase mass-transfer coefficient for concentrated systems	$kmol/(s \cdot m^2)$	$lbmol/(h \cdot ft^2)$
$k_L a$	Volumetric liquid-phase mass-transfer coefficient for dilute systems	$kmol/[(s \cdot m^3)(mole$ fraction)]	$(lbmol)/[(h \cdot ft^3)(mole$ fraction)]
K	Overall mass transfer coefficient	m/s or cm/s	ft/h
K	$\alpha/R =$ specific conductance	ohm/cm	
K_G	Overall gas-phase mass-transfer coefficient for dilute systems	$kmol/[(s \cdot m^2)(mole$ fraction)]	$(lbmol)/[(h \cdot ft^2)(mole$ fraction)]

Nomenclature and Units—Mass Transfer (*Continued*)

Symbols	Definition	SI units	U.S. customary units
$\hat{K}_G$	Overall gas-phase mass-transfer coefficient for concentrated systems	kmol/(s·m²)	lbmol/(h·ft²)
$K_G a$	Overall volumetric gas-phase mass-transfer dilute systems	kmol/[(s·m³)(mole-fraction solute in gas)]	(lbmol)/[(h·ft³)(mole-fraction solute in gas)]
$K'_G a$	Overall volumetric gas-phase mass-transfer dilute systems	kmol/[(s·m³)(kPa solute partial pressure)]	(lbmol)/[(h·ft³)(lbf/in² solute partial pressure)]
$(Ka)_H$	Overall enthalpy mass-transfer coefficient	kmol/[(s·m²)(mole fraction)]	lb/[(h·ft³)(lb water/lb dry air)]
K_L	Overall liquid-phase mass-transfer coefficient	kmol/[(s·m²)(mole fraction)]	(lbmol)/[(h·ft²)(mole fraction)]
$\hat{K}_L$	Liquid-phase mass-transfer coefficient for concentrated systems	kmol/(s·m²)	(lbmol)/(h·ft²)
$K_L a$	Overall volumetric liquid-phase mass-transfer coefficient for dilute systems	kmol/[(s·m³)(mole-fraction solute in liquid)]	(lbmol)/[(h·ft³)(mole-fraction solute in liquid)]
$\hat{K}_L a$	Overall volumetric liquid-phase mass-transfer coefficient for concentrated systems	kmol/(s·m³)	(lbmol)/(h·ft³)
L	Liquid-phase mass flux	kg/(s·m²)	lb/(h·ft²)
L_M	Molar liquid-phase mass flux	kmol/(s·m²)	(lbmol)/(h·ft²)
m	Slope of equilibrium curve = dy/dx (mole-fraction solute in gas)/(mole-fraction solute in liquid)	Dimensionless	Dimensionless
m	Molality of solute	mol/1000 g solvent	
M_i	Molecular weight of species i	kg/kmol or g/mol	lb/lbmol
M	Mass in a control volume V	kg or g	lb
$\|n_+\| \|n_-\|$	Valences of cationic and anionic species	Dimensionless	Dimensionless
n'	See Table 5-28-I	Dimensionless	Dimensionless
n_A	Mass flux of A with respect to fixed coordinates	kg/(s·m²)	lb/(h·ft²)
N	Impeller speed	Revolution/s	Revolution/min
N'	Number deck levels	Dimensionless	Dimensionless
N_A	Interphase mass-transfer rate of solute A per interfacial area with respect to fixed coordinates	kmol/(s·m²)	(lbmol)/(h·ft²)
N_c	Number of components	Dimensionless	Dimensionless
N_{Fr}	Froude Number ($d_{imp} N^2/g$)	Dimensionless	Dimensionless
N_{Gr}	Grashof number $\left(\dfrac{gx^3}{(\mu/\rho)^2}\left(\dfrac{\rho_\infty}{\rho_o}-1\right)\right)$	Dimensionless	Dimensionless
N_{OG}	Number of overall gas-phase mass-transfer units	Dimensionless	Dimensionless
N_{OL}	Number of overall liquid-phase mass-transfer units	Dimensionless	Dimensionless
NTU	Number of transfer units (general)	Dimensionless	Dimensionless
N_{Kn}	Knudson number = l/d_{pore}	Dimensionless	Dimensionless
N_{Pr}	Prandtl number ($c_p\mu/k$)	Dimensionless	Dimensionless
N_{Re}	Reynolds number (Gd/μ_G)	Dimensionless	Dimensionless
N_{Sc}	Schmidt number ($\mu_G/\rho_G D_{AB}$) or ($\mu_L/\rho_L D_L$)	Dimensionless	Dimensionless
N_{Sh}	Sherwood number ($\hat{k}_G RTd/D_{AB}p_T$)	Dimensionless	Dimensionless
N_{St}	Stanton number ($\hat{k}_G/G_M$) or ($\hat{k}_L/L_M$)	Dimensionless	Dimensionless
N_W	Weber number ($\rho_c N^2 d_{imp}^3/\sigma$)	Dimensionless	Dimensionless
p	Solute partial pressure in bulk gas	kPa	lbf/in²
$p_{B,M}$	Log mean partial pressure difference of stagnant gas B	Dimensionless	Dimensionless
p_i	Solute partial pressure at gas-liquid interface	kPa	lbf/in²
p_T	Total system pressure	kPa	lbf/in²
P	Pressure	Pa	lbf/in² or atm
P	Power	Watts	
P_c	Critical pressure	Pa	lbf/in² or atm
Per	Perimeter/area	m⁻¹	ft⁻¹
Q	Volumetric flow rate	m³/s	ft³/h
r_A	Radius of dilute spherical solute	Å	
R	Gas constant	8.314 J/mol K = 8.314 Pa m³/(mol K) = 82.057 atm cm³/mol K	10.73 ft³ psia/lbmol·h
R	Solution electrical resistance	ohm	
R_i	Radius of gyration of the component i molecule	Å	
s	Fractional surface-renewal rate	s⁻¹	h⁻¹
S	Tower cross-sectional area = $\pi d^2/4$	m²	ft²
t	Contact time	s	h
t_f	Formation time of drop	s	h
T	Temperature	K	°R
T_b	Normal boiling point	K	°R
T_c	Critical temperature	K	°R
T_r	Reduced temperature = T/T_c	Dimensionless	Dimensionless
u, v	Fluid velocity	m/s or cm/s	ft/h
u_o	Blowing or suction velocity	m/s	ft/h
u_∞	Velocity away from object	m/s	ft/h
u_L	Superficial liquid velocity in vertical direction	m/s	ft/h
v_s	Slip velocity	m/s	ft/h
v_T	Terminal velocity	m/s	ft/h
v_{TS}	Stokes law terminal velocity	m/s	ft/h
V	Packed volume in tower	m³	ft³
V	Control volume	m³ or cm³	ft³
V_b	Volume at normal boiling point	m³/kmol or cm³/mol	ft³/lbmol
V_i	Molar volume of i at its normal boiling point	m³/kmol or cm³/mol	ft³/lbmol
$\bar{v}_i$	Partial molar volume of i	m³/kmol or cm³/mol	ft³/lbmol

Nomenclature and Units—Mass Transfer (*Concluded*)

Symbols	Definition	SI units	U.S. customary units
V_{mli}	Molar volume of the liquid-phase component i at the melting point	m³/kmol or cm³/mol	ft³/lbmol
V_{tower}	Tower volume per area	m³/m²	ft³/ft²
w	Width of film	m	ft
x	Length along plate	m	ft
x	Mole-fraction solute in bulk-liquid phase	(kmol solute)/(kmol liquid)	(lbmol solute)/(lb mol liquid)
x_A	Mole fraction of component A	kmole A/kmole fluid	lbmole A/lb mole fluid
x^o	Mole-fraction solute in bulk liquid in equilibrium with bulk-gas solute concentration y	(kmol solute)/(kmol liquid)	(lbmol solute)/(lbmol liquid)
x_{BM}	Logarithmic-mean solvent concentration between bulk liquid and interface values	(kmol solvent)/(kmol liquid)	(lbmol solvent)/(lbmol liquid)
x_{BM}^o	Logarithmic-mean inert-solvent concentration between bulk-liquid value and value in equilibrium with bulk gas	(kmol solvent)/(kmol liquid)	(lbmol solvent)/(lbmol liquid)
x_i	Mole-fraction solute in liquid at gas-liquid interface	(kmol solute)/(kmol liquid)	(lbmol solute)/(lbmol liquid)
y	Mole-fraction solute in bulk-gas phase	(kmol solute)/(kmol gas)	(lbmol solute)/(lbmol gas)
y_{BM}	Logarithmic-mean inert-gas concentration (5-262)	(kmol inert gas)/(kmol gas)	(lbmol inert gas)/(lbmol gas)
y_{BM}^o	Logarithmic-mean inert-gas concentration	(kmol inert gas)/(kmol gas)	(lbmol inert gas)/(lbmol gas)
y_i	Mole fraction solute in gas at interface	(kmole solute)/(kmol gas)	(lbmol solute)/(lbmol gas)
y_i^o	Mole-fraction solute in gas at interface in equilibrium with the liquid-phase interfacial solute concentration x_i	(kmol solute)/(kmol gas)	(lbmol solute)/(lbmol gas)
z	Direction of unidimensional diffusion	m	ft

Greek symbols			
α	$1 + N_B/N_A$	Dimensionless	Dimensionless
α	Conductance cell constant (measured)	cm⁻¹	
β	$M_A^{1/2} P_c^{1/3}/T_c^{5/6}$	Dimensionless	Dimensionless
δ	Effective thickness of stagnant-film layer	m	ft
ε	Fraction of discontinuous phase in continuous phase for two-phase flow	Dimensionless	Dimensionless
ε	Void fraction available for gas flow or fractional gas holdup	m³/m³	ft³/ft³
ε_A	Characteristic Lennard-Jones energy	Dimensionless	Dimensionless
ε_{AB}	$(\varepsilon_A \varepsilon_B)^{1/2}$	Dimensionless	Dimensionless
γ_i	Activity coefficient of solute i	Dimensionless	Dimensionless
$\gamma_{\pm}$	Mean ionic activity coefficient of solute	Dimensionless	Dimensionless
λ_+, λ_-	Infinite dilution conductance of cation and anion	cm²/(gequiv·ohm)	
Λ	$1000\,K/C = \lambda_+ + \lambda_- = \Lambda_o + f(C)$	cm²/ohm gequiv	
Λ_o	Infinite dilution conductance	cm²/gequiv ohm	
μ_i	Dipole moment of i	Debeyes	
μ_i	Viscosity of pure i	cP or Pa s	lb/(h·ft)
μ_G	Gas-phase viscosity	kg/(s·m)	lb/(h·ft)
μ_L	Liquid-phase viscosity	kg/(s·m)	lb/(h·ft)
ν	Kinematic viscosity $= \rho/\mu$	m²/s	ft²/h
ρ	Density of A	kg/m³ or g/cm³	lb/ft³
ρ_c	Critical density of A	kg/m³ or g/cm³	lb/ft³
ρ_c	Density continuous phase	kg/m³	lb/ft³
ρ_G	Gas-phase density	kg/m³	lb/ft³
$\bar{\rho}_L$	Average molar density of liquid phase	kmol/m³	(lbmol)/ft³
ρ_p	Particle density	kg/m³ or g/cm³	lb/ft³
ρ_r	Reduced density $= \rho/\rho_c$	Dimensionless	Dimensionless
ψ_i	Parachor of component $i = V_i \sigma^{1/4}$		
ψ	Parameter, Table 5-28-G	Dimensionless	Dimensionless
ψ	Shape factor, Table 5-27-A	Dimensionless	Dimensionless
σ	Interfacial tension	dyn/cm	lbf/ft
σ_i	Characteristic length	Å	
σ_i	Surface tension of component i	dyn/cm	
σ_{AB}	Binary pair characteristic length $= (\sigma_A + \sigma_B)/2$	Å	
τ	Intraparticle tortuosity	Dimensionless	
ω	Pitzer's acentric factor $= -[1.0 + \log 10(P^o/P_c)]$	Dimensionless	Dimensionless
ω	Rotational velocity	Radians/s	
Ω	Diffusion collision integral $= f(kT/\varepsilon_{AB})$	Dimensionless	Dimensionless

Subscript			
A	Solute component in liquid or gas phase		
B	Inert-gas or inert-solvent component		
G	Gas phase		
m	Mean value		
L	Liquid phase		
super	Superficial velocity		

Superscript			
^{o}	At equilibrium		

diffusivity. It can be written in several forms, depending on the units and frame of reference. Three that are related but not identical are

$$_vJ_A = -D_{AB} \frac{dc_A}{dz} \approx _MJ_A = -cD_{AB} \frac{dx_A}{dz} \propto _mJ_A = -\rho D_{AB} \frac{dw_A}{dz} \quad (5\text{-}181)$$

The first equality (on the left-hand side) corresponds to the molar flux with respect to the volume average velocity, while the equality in the center represents the molar flux with respect to the molar average velocity and the one on the right is the mass flux with respect to the mass average velocity. These must be used with consistent flux expressions for fixed coordinates and for N_C components, such as:

$$N_A = _vJ_A + c_A \sum_{i=1}^{N_C} N_i \overline{V}_i = _MJ_A + x_A \sum_{i=1}^{N_C} N_i = \frac{_mJ_A + w_A \sum_{i=1}^{N_C} n_i}{M_A} \quad (5\text{-}182)$$

In each case, the term containing the summation accounts for *conveyance*, which is the amount of component A carried by the net flow in the direction of diffusion. Its impact on the total flux can be as much as 10 percent. In most cases it is much less, and it is frequently ignored. Some people refer to this as the "convective" term, but that conflicts with the other sense of convection which is promoted by flow perpendicular to the direction of flux.

Mutual Diffusivity, Mass Diffusivity, Interdiffusion Coefficient   Diffusivity is denoted by D_{AB} and is defined by Fick's first law as the ratio of the flux to the concentration gradient, as in Eq. (5-181). It is analogous to the thermal diffusivity in Fourier's law and to the kinematic viscosity in Newton's law. These analogies are flawed because both heat and momentum are conveniently defined with respect to fixed coordinates, irrespective of the direction of transfer or its magnitude, while mass diffusivity most commonly requires information about bulk motion of the medium in which diffusion occurs. For liquids, it is common to refer to the limit of infinite dilution of A in B using the symbol, D_{AB}°.

When the flux expressions are consistent, as in Eq. (5-182), the diffusivities in Eq. (5-181) are identical. As a result, experimental diffusivities are often measured under constant volume conditions but may be used for applications involving open systems. It turns out that the two versions are very nearly equivalent for gas-phase systems because there is negligible volume change on mixing. That is not usually true for liquids, however.

Self Diffusivity   Self-diffusivity is denoted by $D_{A'A}$ and is the measure of mobility of a species in itself; for instance, using a small concentration of molecules tagged with a radioactive isotope so they can be detected. Tagged and untagged molecules presumably do not have significantly different properties. Hence, the solution is ideal, and there are practically no gradients to "force" or "drive" diffusion. This kind of diffusion is presumed to be purely statistical in nature.

In the special case that A and B are similar in molecular weight, polarity, and so on, the self-diffusion coefficients of pure A and B will be approximately equal to the mutual diffusivity, D_{AB}. Second, when A and B are the less mobile and more mobile components, respectively, their self-diffusion coefficients can be used as rough lower and upper bounds of the mutual diffusion coefficient. That is, $D_{A'A} \leq D_{AB} \leq D_{B'B}$. Third, it is a common means for evaluating diffusion for gases at high pressure. Self-diffusion in liquids has been studied by many [Easteal *AIChE J.* **30,** 641 (1984), Ertl and Dullien, *AIChE J.* **19,** 1215 (1973), and Vadovic and Colver, *AIChE J.* **18,** 1264 (1972)].

Tracer Diffusivity   Tracer diffusivity, denoted by $D_{A'B}$ is related to both mutual and self-diffusivity. It is evaluated in the presence of a second component B, again using a tagged isotope of the first component. In the dilute range, tagging A merely provides a convenient method for indirect composition analysis. As concentration varies, tracer diffusivities approach mutual diffusivities at the dilute limit, and they approach self-diffusivities at the pure component limit. That is, at the limit of dilute A in B, $D_{A'B} \rightarrow D_{AB}^\circ$ and $D_{B'A} \rightarrow D_{B'B}$; likewise at the limit of dilute B in A, $D_{B'A} \rightarrow D_{BA}^\circ$ and $D_{A'B} \rightarrow D_{A'A}$.

Neither the tracer diffusivity nor the self-diffusivity has much practical value except as a means to understand ordinary diffusion and as

order-of-magnitude estimates of mutual diffusivities. Darken's equation [Eq. (5-220)] was derived for tracer diffusivities but is often used to relate mutual diffusivities at moderate concentrations as opposed to infinite dilution.

Mass-Transfer Coefficient   Denoted by k_c, k_x, K_x, and so on, the mass-transfer coefficient is the ratio of the flux to a concentration (or composition) difference. These coefficients generally represent rates of transfer that are much greater than those that occur by diffusion alone, as a result of convection or turbulence at the interface where mass transfer occurs. There exist several principles that relate that coefficient to the diffusivity and other fluid properties and to the intensity of motion and geometry. Examples that are outlined later are the film theory, the surface renewal theory, and the penetration theory, all of which pertain to idealized cases. For many situations of practical interest like investigating the flow inside tubes and over flat surfaces as well as measuring external flow through banks of tubes, in fixed beds of particles, and the like, correlations have been developed that follow the same forms as the above theories. Examples of these are provided in the subsequent section on mass-transfer coefficient correlations.

Problem Solving Methods   Most, if not all, problems or applications that involve mass transfer can be approached by a systematic course of action. In the simplest cases, the unknown quantities are obvious. In more complex (e.g., multicomponent, multiphase, multidimensional, nonisothermal, and/or transient) systems, it is more subtle to resolve the known and unknown quantities. For example, in multicomponent systems, one must know the fluxes of the components before predicting their effective diffusivities and vice versa. More will be said about that dilemma later. Once the known and unknown quantities are resolved, however, a combination of conservation equations, definitions, empirical relations, and properties are applied to arrive at an answer. Figure 5-24 is a flowchart that illustrates the primary types of information and their relationships, and it applies to many mass-transfer problems.

CONTINUITY AND FLUX EXPRESSIONS

Material Balances   Whenever mass-transfer applications involve equipment of specific dimensions, flux equations alone are inadequate to assess results. A material balance or continuity equation must also be used. When the geometry is simple, macroscopic balances suffice. The following equation is an overall mass balance for such a unit having N_m bulk-flow ports and N_n ports or interfaces through which diffusive flux can occur:

$$\frac{dM}{dt} = \sum_{i=1}^{N_m} m_i + \sum_{i=1}^{N_n} n_i A_{cs_i} \quad (5\text{-}183)$$

where M represents the mass in the unit volume V at any time t; m_i is the mass flow rate through the ith port; and n_i is the mass flux through the ith port, which has a cross-sectional area of A_{cs_i}. The corresponding balance equation for individual components includes a reaction term:

$$\frac{dM_j}{dt} = \sum_{i=1}^{N_m} m_{ij} + \sum_{i=1}^{N_n} n_{ij} A_{cs_i} + r_j V \quad (5\text{-}184)$$

For the jth component, $m_{ij} = m_i w_{ij}$ is the component mass flow rate in stream i; w_{ij} is the mass fraction of component j in stream i; and r_j is the net reaction rate (mass generation minus consumption) per unit volume V that contains mass M. If it is inconvenient to measure mass flow rates, the product of density and volumetric flow rate is used instead.

In addition, most situations that involve mass transfer require material balances, but the pertinent area is ambiguous. Examples are packed columns for absorption, distillation, or extraction. In such cases, flow rates through the discrete ports (nozzles) must be related to the mass-transfer rate in the packing. As a result, the mass-transfer rate is determined via flux equations, and the overall material balance incorporates the stream flow rates m_i and integrated fluxes. In such instances, it is common to begin with the most general, differential material balance equations. Then, by eliminating terms that are negligible, the simplest applicable set of equations remains to be solved.

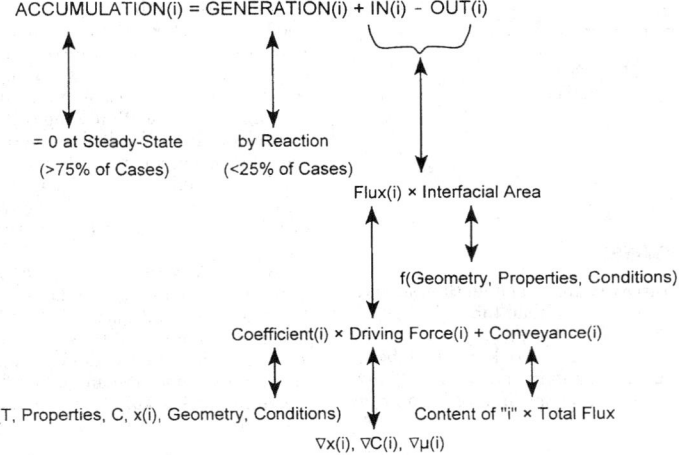

FIG. 5-24 Flowchart illustrating problem solving approach using mass-transfer rate expressions in the context of mass conservation.

Table 5-12 provides material balances for Cartesian, cylindrical, and spherical coordinates. The generic form applies over a unit cross-sectional area and constant volume:

$$\frac{\partial \rho_j}{\partial t} = -\nabla \cdot n_j + r_j \qquad (5\text{-}185a)$$

where $n_j = \rho v_j$. Applying Fick's law and expressing composition as concentration gives

$$\frac{\partial c_j}{\partial t} = -v \cdot \nabla c_j + D_j \nabla^2 c_j + r_j \qquad (5\text{-}185b)$$

Flux Expressions: Simple Integrated Forms of Fick's First Law Simplified flux equations that arise from Eqs. (5-181) and (5-182) can be used for unidimensional, steady-state problems with binary mixtures. The boundary conditions represent the compositions x_{A_L} and x_{A_R} at the left-hand and right-hand sides of a hypothetical layer having thickness Δz. The principal restriction of the following equations is that the concentration and diffusivity are assumed to be constant. As written, the flux is positive from left to right, as depicted in Fig. 5-25.

1. Equimolar counterdiffusion ($N_A = -N_B$)

$$N_A = {}_M J_A = -D_{AB} c \frac{dx_A}{dz} = \frac{D_{AB}}{\Delta z} c \,(x_{A_L} - x_{A_R}) \qquad (5\text{-}189)$$

2. Unimolar diffusion ($N_A \neq 0, N_B = 0$)

$$N_A = {}_M J_A + x_A N_A = \frac{D_{AB}}{\Delta z} c \ln \frac{1 - x_{A_R}}{1 - x_{A_L}} \qquad (5\text{-}190)$$

3. Steady state diffusion ($N_A \neq -N_B \neq 0$)

$$N_A = {}_M J_A + x_A(N_A + N_B) = \frac{N_A}{N_A + N_B} \frac{D_{AB}}{\Delta z} c \ln \frac{\dfrac{N_A}{N_A + N_B} - x_{A_R}}{\dfrac{N_A}{N_A + N_B} - x_{A_L}} \qquad (5\text{-}191)$$

The unfortunate aspect of the last relationship is that one must know a priori the ratio of the fluxes to determine the magnitudes. It is not possible to solve simultaneously the pair of equations that apply for components A and B because the equations are not independent.

Stefan-Maxwell Equations Following Eq. (5-182), a simple and intuitively appealing flux equation for applications involving N_c components is

$$N_i = -cD_{im} \nabla x_i + x_i \sum_{j=1}^{N_c} N_j \qquad (5\text{-}192)$$

In the late 1800s, the development of the kinetic theory of gases led to a method for calculating multicomponent gas diffusion (e.g., the flux of each species in a mixture). The methods were developed simultaneously by Stefan and Maxwell. The problem is to determine the diffusion coefficient D_{im}. The Stefan-Maxwell equations are simpler in principle since they employ binary diffusivities:

$$\nabla x_i = \sum_{j=1}^{N_c} \frac{1}{cD_{ij}} (x_i N_j - x_j N_i) \qquad (5\text{-}193)$$

If Eqs. (5-192) and (5-193) are combined, the multicomponent diffusion coefficient may be assessed in terms of binary diffusion coefficients [see Eq. (5-204)]. For gases, the values D_{ij} of this equation are approximately equal to the binary diffusivities for the ij pairs. The Stefan-Maxwell diffusion coefficients may be negative, and the method may be applied to liquids, even for electrolyte diffusion [Kraaijeveld, Wesselingh, and Kuiken, *Ind. Eng. Chem. Res.*, **33,** 750 (1994)]. Approximate solutions have been developed by linearization [Toor, H.L., *AIChE J.*, **10,** 448 and 460 (1964); Stewart and Prober, *Ind. Eng. Chem. Fundam.*, **3,** 224 (1964)]. Those differ in details but yield about the same accuracy. More recently, efficient algorithms for solving the equations exactly have been developed (see Taylor and Krishna, Krishnamurthy and Taylor, and Taylor and Webb).

TABLE 5-12 Continuity Equation in Various Coordinate Systems

Coordinate System	Equation	
Cartesian	$$\frac{\partial \rho_j}{\partial t} = -\left(\frac{\partial n_{x_j}}{\partial x} + \frac{\partial n_{y_j}}{\partial y} + \frac{\partial n_{z_j}}{\partial z} \right) + r_j$$	(5-186)
Cylindrical	$$\frac{\partial \rho_j}{\partial t} = -\left(\frac{1}{r} \frac{\partial r n_{r_j}}{\partial r} + \frac{1}{r} \frac{\partial n_{\theta_j}}{\partial \theta} + \frac{\partial n_{z_j}}{\partial z} \right) + r_j$$	(5-187)
Spherical	$$\frac{\partial \rho_j}{\partial t} = -\left(\frac{1}{r^2} \frac{\partial r^2 n_{r_j}}{\partial r} + \frac{1}{r \sin \theta} \frac{\partial n_{\theta_j} \sin \theta}{\partial \theta} + \frac{1}{r \sin \theta} \frac{\partial n_{\phi_j}}{\partial \phi} \right) + r_j$$	(5-188)

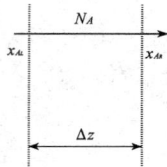

FIG. 5-25 Hypothetical film and boundary conditions.

DIFFUSIVITY ESTIMATION—GASES

Whenever measured values of diffusivities are available, they should be used. Typically, measurement errors are less than those associated with predictions by empirical or even semitheoretical equations. A few general sources of data are Sec. 2 of this handbook, Schwartzberg and Chao; Reid et al.; Gammon et al.; and Daubert and Danner. Many other more restricted sources are listed under specific topics later in this subsection.

Before using diffusivities from either data or correlations, it is a good idea to check their reasonableness with respect to values that have been commonly observed in similar situations. Table 5-13 is a compilation of several rules of thumb. These values are not authoritative; they simply represent guidelines based on experience.

Diffusivity correlations for gases are outlined in Table 5-14. Specific parameters for individual equations are defined in the specific text regarding each equation. References are given after Table 5-19. The errors reported for Eq. (5-194) through (5-197) were compiled by Reid et al., who compared the predictions with 68 experimental values of D_{AB}. Errors cited for Eqs. (5-198) to (5-202) were reported by the authors.

Binary Mixtures—Low Pressure—Nonpolar Components Many evaluations of correlations are available (Elliott and Watts; Lugg; Marrero and Mason). The differences in accuracy of the correlations are minor, and thus the major concern is ease of calculation. The Fuller-Schettler-Giddings equation is usually the simplest correlation to use and is recommended by Reid et al.

Chapman-Enskog (Bird et al.) and Wilke-Lee The inherent assumptions of these equations are quite restrictive (i.e., low density,

TABLE 5-13   Rules of Thumb for Diffusivities (See Cussler, Reid et al., Schwartzberg and Chao)

Continuous phase	D_i magnitude		D_i range		Comments
	m²/s	cm²/s	m²/s	cm²/s	
Gas at atmospheric pressure	10^{-5}	0.1	10^{-4}–10^{-6}	1–10^{-2}	Accurate theories exist, generally within ±10%; $D_i P \cong$ constant; $D_i \propto T^{1.66 \text{ to } 2.0}$
Liquid	10^{-9}	10^{-5}	10^{-8}–10^{-10}	10^{-4}–10^{-6}	Approximate correlations exist, generally within ±25%
Liquid occluded in solid matrix	10^{-10}	10^{-6}	10^{-8}–10^{-12}	10^{-4}–10^{-8}	Hard cell walls: $D_{eff}/D_i = 0.1$ to 0.2. Soft cell walls: $D_{eff}/D_i = 0.3$ to 0.9
Polymers and glasses	10^{-12}	10^{-8}	10^{-10}–10^{-14}	10^{-6}–10^{-10}	Approximate theories exist for dilute and concentrated limits; strong composition dependence
Solid	10^{-14}	10^{-10}	10^{-10}–10^{-34}	10^{-6}–10^{-30}	Approximate theories exist; strong temperature dependence

TABLE 5-14   Correlations of Diffusivities for Gases

Authors[*]	Equation		Error
	1. Binary Mixtures—Low Pressure—Nonpolar		
Chapman-Enskog	$D_{AB} = \dfrac{0.001858 T^{3/2} M_{AB}^{1/2}}{P \sigma_{AB}^2 \Omega_D}$	(5-194)	7.3%
Wilke-Lee [65]	$D_{AB} = \dfrac{(0.0027 - 0.0005\, M_{AB}^{1/2})\, T^{3/2} M_{AB}^{1/2}}{P \sigma_{AB}^2 \Omega_D}$	(5-195)	7.0%
Fuller-Schettler-Giddings [19]	$D_{AB} = \dfrac{0.001 T^{1.75} M_{AB}^{1/2}}{P\,[(\sum v)_A^{1/3} + (\sum v)_B^{1/3}]^2}$	(5-196)	5.4%
	2. Binary Mixtures—Low Pressure—Polar		
Brokaw [4]	$D_{AB} = \dfrac{0.001858 T^{3/2} M_{AB}^{1/2}}{P \sigma_{AB}^2 \Omega_D}$	(5-197)	9.0%
	3. Self-Diffusivity—High Pressure		
Mathur-Thodos [37]	$D_{AA} = \dfrac{10.7 \times 10^{-5} T_r}{\beta \rho_r} \; \{\rho_r \le 1.5\}$	(5-198)	5%
Lee-Thodos [31]	$D_{AA} = \dfrac{0.77 \times 10^{-5} T_r}{\rho_r \delta} \; \{\rho_r \le 1\}$	(5-199)	0.5%
Lee-Thodos [32]	$D_{AA} = \dfrac{(0.007094 G + 0.001916)^{2.5} T_r}{\delta},\; [\rho_r > 1,\, G < 1]$	(5-200)	17%
	4. Supercritical Mixtures		
Sun and Chen [56]	$D_{AB} = \dfrac{1.23 \times 10^{-10} T}{\mu^{0.799} V_{C_A}^{0.49}}$	(5-201)	5%
Catchpole and King [6]	$D_{AB} = 5.152\, D_c T_r \dfrac{(\rho^{-0.667} - 0.4510)\,(1 + M_A/M_B)\, R}{(1 + (V_{c_B}/V_{c_A})^{0.333})^2}$	(5-202)	10%

[*]References are listed on pages 5-7 and 5-8.

TABLE 5-15   Estimates for ε_i and σ_i (K, Å, atm, cm³, mol)

Critical point	$\varepsilon/k = 0.75\ T_c$	$\sigma = 0.841\ V_c^{1/3}$ or $2.44\ (T_c/P_c)^{1/3}$
Critical point	$\varepsilon/k = 65.3\ T_c z_c^{3.6}$	$\sigma = \dfrac{1.866\ V_c^{1/3}}{z_c^{1.2}}$
Normal boiling point	$\varepsilon/k = 1.15\ T_b$	$\sigma = 1.18\ V_b^{1/3}$
Melting point	$\varepsilon/k = 1.92\ T_m$	$\sigma = 1.222\ V_m^{1/3}$
Acentric factor	$\varepsilon/k = (0.7915 + 0.1693\ \omega)\ T_c$	$\sigma = (2.3551 - 0.087\ \omega)\left(\dfrac{T_c}{P_c}\right)^{1/3}$

NOTE: These values may not agree closely, so usage of a consistent basis is suggested (e.g., data at the normal boiling point).

spherical atoms), and the intrinsic potential function is empirical. Despite that, they provide good estimates of D_{AB} for many polyatomic gases and gas mixtures, up to about 1000 K and a maximum of 70 atm. The latter constraint is because observations for many gases indicate that $D_{AB}P$ is constant up to 70 atm.

The characteristic length is $\sigma_{AB} = (\sigma_A + \sigma_B)/2$ in Å. In order to estimate Ω_D for Eqs. (5-194) or (5-195), two empirical equations are available. The first is:

$$\Omega_D = (44.54T^{\circ -4.909} + 1.911T^{\circ -1.575})^{0.10} \qquad (5\text{-}203a)$$

where $T^\circ = kT/\varepsilon_{AB}$ and $\varepsilon_{AB} = (\varepsilon_A\ \varepsilon_B)^{1/2}$. Estimates for σ_i and ε_i are given in Table 5-15. This expression shows that Ω_D is proportional to temperature roughly to the −0.49 power at low temperatures and to the −0.16 power at high temperature. Thus, gas diffusivities are proportional to temperatures to the 2.0 power and 1.66 power, respectively, at low and high temperatures. The second is:

$$\Omega_D = \frac{A}{T^{\circ B}} + \frac{C}{\exp(DT^\circ)} + \frac{E}{\exp(FT^\circ)} + \frac{G}{\exp(HT^\circ)} \qquad (5\text{-}203b)$$

where $A = 1.06036$, $B = 0.15610$, $C = 0.1930$, $D = 0.47635$, $E = 1.03587$, $F = 1.52996$, $G = 1.76474$, and $H = 3.89411$.

Fuller-Schettler-Giddings   The parameters and constants for this correlation were determined by regression analysis of 340 experimental diffusion coefficient values of 153 binary systems. Values of $\sum v_i$ used in this equation are in Table 5-16.

Binary Mixtures—Low Pressure—Polar Components   The *Brokaw* correlation was based on the Chapman-Enskog equation, but σ_{AB° and Ω_{D° were evaluated with a modified Stockmayer potential for polar molecules. Hence, slightly different symbols are used. That potential model reduces to the Lennard-Jones 6-12 potential for interactions between nonpolar molecules. As a result, the method should yield accurate predictions for polar as well as nonpolar gas mixtures. Brokaw presented data for 9 relatively polar pairs along with the prediction. The agreement was good: an average absolute error of 6.4 percent, considering the complexity of some of the gas pairs [e.g., $(CH_3)_2O$ & CH_3Cl]. Despite that, Reid, *op. cit.*, found the average error was 9.0 percent for combinations of mixtures (including several polar-nonpolar gas pairs), temperatures and pressures. In this equation, Ω_D is calculated as described previously, and other terms are:

$$\Omega_{D^\circ} = \Omega_D + 0.19\ \delta_{AB}^2/T^\circ \qquad T^\circ = kT/\varepsilon_{AB^\circ}$$
$$\sigma_{AB^\circ} = (\sigma_{A^\circ}\ \sigma_{B^\circ})^{1/2} \qquad \sigma_{i^\circ} = [1.585\ V_{bi}/(1 + 1.3\ \delta_i^2)]^{1/3}$$
$$\delta_{AB} = (\delta_A\ \delta_B)^{1/2} \qquad \delta_i = 1.94 \times 10^3\ \mu_i^2/V_{bi}T_{bi}$$
$$\varepsilon_{AB^\circ} = (\varepsilon_{A^\circ}\varepsilon_{B^\circ})^{1/2} \qquad \varepsilon_{i^\circ}/k = 1.18\ (1 + 1.3\ \delta_i^2)T_{bi}$$

Self-Diffusivity—High Pressure   The criterion of high pressure is vague at best. For most "permanent" gases, such as the major constituents of air, it would mean $P > 70$ atm. For less volatile components, the criterion would be lower. At present, accurate prediction of mutual diffusion coefficients for dense gas mixtures is not possible. One major reason for this is the scarcity of data. Most high-pressure diffusion experiments have measured the self-diffusion coefficient. The general observation is that the product DP is near constant at low pressure, is not constant at high pressure, but rather decreases as pressure increases. In addition, although there are usually negligible composition effects on diffusivity of gases at low pressures, the effects are not negligible at high pressures.

Mathur-Thodos showed that for reduced densities less than unity, the product $D_{AA}\rho$ is approximately constant at a given temperature. Thus, by knowing the value of the product at low pressure, it is possible to estimate its value at a higher pressure. They found at higher pressures the density increases, but the product $D_{AA}\rho$ decreases rapidly. In their correlation, $\beta = M_A^{1/2}P_C^{1/3}/T_C^{5/6}$.

Lee-Thodos presented a generalized treatment of self-diffusivity for gases (and liquids). These correlations have been tested for more than 500 data points each. The average deviation of the first is 0.51 percent, and that of the second is 17.2 percent. $\delta = M_A^{1/2}/P_C^{1/2}V_C^{5/6}$, s/cm², and where $G = (X^\circ - X)/(X^\circ - 1)$, $X = \rho_r/T_r^{0.1}$, and $X^\circ = \rho_r/T_r^{0.1}$ evaluated at the solid melting point.

Lee and Thodos expanded their earlier treatment of self-diffusivity to cover 58 substances and 975 data points, with an average absolute deviation of 5.26 percent. Their correlation is too involved to repeat here, but those interested should refer to the original paper.

Supercritical Mixtures   *Debenedetti-Reid* showed that conventional correlations based on the Stokes-Einstein relation (for liquid phase) tend to overpredict diffusivities in the supercritical state. Nevertheless, they observed that the Stokes-Einstein group $D_{AB}\mu/T$ was constant. Thus, although no general correlation applies, only one data point is necessary to examine variations of fluid viscosity and/or temperature effects. They explored certain combinations of aromatic solids in SF_6 and CO_2.

Sun-Chen examined tracer diffusion data of aromatic solutes in alcohols up to the supercritical range and found their data correlated with average deviations of 5 percent and a maximum deviation of 17 percent for their rather limited set of data.

Catchpole-King examined binary diffusion data of near-critical fluids in the reduced density range of 1 to 2.5 and found that their data correlated with average deviations of 10 percent and a maximum deviation of 60 percent. They observed two classes of behavior. For the first, no correction factor was required ($R = 1$). That class was comprised of alcohols as solvents with aromatic or aliphatic solutes, or carbon dioxide as a solvent with aliphatics except ketones as solutes, or

TABLE 5-16   Atomic Diffusion Volumes for Use in Estimating D_{AB} by the Method of Fuller, Schettler, and Giddings

Atomic and Structural Diffusion–Volume Increments, v_i (cm³/mol)			
C	16.5	(Cl)	19.5
H	1.98	(S)	17.0
O	5.48	Aromatic ring	−20.2
(N)	5.69	Heterocyclic ring	−20.2

Diffusion Volumes for Simple Molecules, Σv_i (cm³/mol)			
H_2	7.07	CO	18.9
D_2	6.70	CO_2	26.9
He	2.88	N_2O	35.9
N_2	17.9	NH_3	14.9
O_2	16.6	H_2O	12.7
Air	20.1	(CCl_2F_2)	114.8
Ar	16.1	(SF_6)	69.7
Kr	22.8	(Cl_2)	37.7
(Xe)	37.9	(Br_2)	67.2
Ne	5.59	(SO_2)	41.1

Parentheses indicate that the value listed is based on only a few data points.

ethylene as a solvent with aliphatics except ketones and naphthalene as solutes. For the second class, the correction factor was $R = X^{0.17}$. The class was comprised of carbon dioxide with aromatics; ketones and carbon tetrachloride as solutes; and aliphatics (propane, hexane, dimethyl butane), sulfur hexafluoride, and chlorotrifluoromethane as solvents with aromatics as solutes. In addition, sulfur hexafluoride combined with carbon tetrachloride, and chlorotrifluoromethane combined with 2-propanone were included in that class. In all cases, $X = (1 + (V_{C^B}/V_{C^A})^{1/3})^2/(1 + M_A/M_B)$ was in the range of 1 to 10.

Low-Pressure/Multicomponent Mixtures These methods are outlined in Table 5-17. Stefan-Maxwell equations were discussed earlier. *Smith-Taylor* compared various methods for predicting multicomponent diffusion rates and found that Eq. (5-204) was superior among the effective diffusivity approaches, though none is very good. They also found that linearized and exact solutions are roughly equivalent and accurate.

Blanc provided a simple limiting case for dilute component i diffusing in a stagnant medium (i.e., $N \approx 0$), and the result, Eq. (5-205), is known as Blanc's law. The restriction basically means that the compositions of all the components, besides component i, are relatively large and uniform.

Wilke obtained solutions to the Stefan-Maxwell equations. The first, Eq. (5-206), is simple and reliable under the same conditions as Blanc's law. This equation applies when component i diffuses through a stagnant mixture. It has been tested and verified for diffusion of toluene in hydrogen + air + argon mixtures and for diffusion of ethyl propionate in hydrogen + air + argon mixtures (Fairbanks and Wilke). When the compositions vary far from one boundary to the other, Wilke recommends that the arithmetic average mole fractions be used. Wilke also suggested using the Stefan-Maxwell equation, which applies when the fluxes of two or more components are significant. In this situation, the mole fractions are arithmetic averages of the boundary conditions, and the solution requires iteration because the ratio of fluxes is not known a priori.

DIFFUSIVITY ESTIMATION—LIQUIDS

Many more correlations are available for diffusion coefficients in the liquid phase than for the gas phase. Most, however, are restricted to binary diffusion at infinite dilution D_{AB}° or to self-diffusivity $D_{A'A}$. This reflects the much greater complexity of liquids on a molecular level. For example, gas-phase diffusion exhibits negligible composition effects and deviations from thermodynamic ideality. Conversely, liquid-phase diffusion almost always involves volumetric and thermodynamic effects due to composition variations. For concentrations greater than a few mole percent of A and B, corrections are needed to obtain the true diffusivity. Furthermore, there are many conditions that do not fit any of the correlations presented here. Thus, careful consideration is needed to produce a reasonable estimate. Again, if diffusivity data are available at the conditions of interest, then they are strongly preferred over the predictions of any correlations.

Stokes-Einstein and Free-Volume Theories The starting point for many correlations is the Stokes-Einstein equation. This equation is derived from continuum fluid mechanics and classical thermodynamics for the motion of large spherical particles in a liquid.

For this case, the need for a molecular theory is cleverly avoided. The Stokes-Einstein equation is (Bird et al.)

$$D_{AB} = \frac{kT}{6\pi r_A \mu_B} \qquad (5\text{-}207)$$

where A refers to the solute and B refers to the solvent. This equation is applicable to very large unhydrated molecules ($M > 1000$) in low-molecular-weight solvents or where the molar volume of the solute is greater than 500 cm³/mol (Reddy and Doraiswamy; Wilke and Chang). Despite its intellectual appeal, this equation is seldom used "as is." Rather, the following principles have been identified: (1) The diffusion coefficient is inversely proportional to the size $r_A \approx V_A^{1/3}$ of the solute molecules. Experimental observations, however, generally indicate that the exponent of the solute molar volume is larger than one-third. (2) The term $D_{AB}\mu_B/T$ is approximately constant only over a 10-to-15 K interval. Thus, the dependence of liquid diffusivity on properties and conditions does not generally obey the interactions implied by that grouping. For example, Robinson et al. found that: $\ln D_{AB} \propto -1/T$. (3) Finally, pressure does not affect liquid-phase diffusivity much, since μ_B and V_A are only weakly pressure-dependent. Pressure does have an impact at very high levels.

Another advance in the concepts of liquid-phase diffusion was provided by Hildebrand, who adapted a theory of viscosity to self-diffusivity. He postulated that $D_{A'A} = B(V - V_{ms})/V_{ms}$, where $D_{A'A}$ is the self-diffusion coefficient, V is the molar volume, and V_{ms} is the molar volume at which fluidity is zero (i.e., the molar volume of the solid phase at the melting temperature). The difference $(V - V_{ms})$ can be thought of as the free volume, which increases with temperature; and B is a proportionality constant.

Ertl and Dullien [ibid.] found that Hildebrand's equation could not fit their data with B as a constant. They modified it by applying an empirical exponent n (a constant greater than unity) to the volumetric ratio. The new equation is not generally useful, however, since there is no means for predicting n. The theory does identify the free volume as an important physical variable, since $n > 1$ for most liquids implies that diffusion is more strongly dependent on free volume than is viscosity.

Dilute Binary Nonelectrolytes: General Mixtures These correlations are outlined in Table 5-18.

Wilke-Chang This correlation for D_{AB}° is one of the most widely used, and it is an empirical modification of the Stokes-Einstein equation. It is not very accurate, however, for water as the solute. Otherwise, it applies to diffusion of very dilute A in B. The average absolute error for 251 different systems is about 10 percent. ϕ_B is an association factor of solvent B that accounts for hydrogen bonding.

Component B	ϕ_B
Water	2.26
Methanol	1.9
Ethanol	1.5
Propanol	1.2
Others	1.0

The value of ϕ_B for water was originally stated as 2.6, although when the original data were reanalyzed, the empirical best fit was 2.26.

TABLE 5-17 Relationships for Diffusivities of Multicomponent Gas Mixtures at Low Pressure

Authors[*]	Equation	
Stefan-Maxwell, Smith and Taylor [53]	$D_{im} = \left[1 - x_i \left(\sum_{j=1}^{NC} N_j\right)\Big/N_i\right]\sum_{j=1}^{NC}\left[\left(x_j - \frac{x_i N_i}{N_i}\right)\Big/D_{ij}\right]$	(5-204)
Blanc [13]	$D_{im} = \left(\sum_{j=1}^{NC} \frac{x_j}{D_{ij}}\right)^{-1}$	(5-205)
Wilke [63]	$D_{im} = \left(\sum_{\substack{j=1 \\ j \neq i}}^{NC} \frac{x_j}{D_{ij}}\right)^{-1}$	(5-206)

[*]References are listed at the beginning of this subsection.

TABLE 5-18 Correlations for Diffusivities of Dilute, Binary Mixtures of Nonelectrolytes in Liquids

Authors[*]	Equation		Error
1. General Mixtures			
Wilke-Chang [64]	$D_{AB}^{\circ} = \dfrac{7.4 \times 10^{-8} \, (\phi_B M_B)^{1/2} \, T}{\mu_B \, V_A^{0.6}}$	(5-208)	20%
Tyn-Calus [59]	$D_{AB}^{\circ} = \dfrac{8.93 \times 10^{-8} \, (V_A/V_B^2)^{1/6} \, (\psi_B/\psi_A)^{0.6} \, T}{\mu_B}$	(5-209)	10%
Umesi-Danner [60]	$D_{AB}^{\circ} = \dfrac{2.75 \times 10^{-8} \, (R_B/R_A^{2/3}) \, T}{\mu_B}$	(5-210)	13%
Siddiqi-Lucas [52]	$D_{AB}^{\circ} = \dfrac{9.89 \times 10^{-8} \, V_B^{0.265} \, T}{V_A^{0.45} \, \mu_B^{0.907}}$	(5-211)	13%
2. Gases in Low Viscosity Liquids			
Sridhar-Potter [54]	$D_{AB}^{\circ} = D_{BB} \left(\dfrac{V_{c_B}}{V_{c_A}}\right)^{2/3} \left(\dfrac{V_B}{V_{ml_B}}\right)$	(5-212)	18%
Chen-Chen [7]	$D_{AB}^{\circ} = 2.018 \times 10^{-9} \, \dfrac{(\beta V_{c_B})^{2/3}(RT_{c_B})^{1/2}}{M_A^{1/6} \, (M_B V_{c_A})^{1/3}} \, (V_r - 1) \left(\dfrac{T}{T_{c_B}}\right)^{1/2}$	(5-213)	6%
3. Aqueous Solutions			
Hayduk-Laudie [25]	$D_{AW}^{\circ} = \dfrac{13.16 \times 10^{-5}}{\mu_w^{1.14} \, V_A^{0.589}}$	(5-214)	18%
Siddiqi-Lucas [52]	$D_{AW}^{\circ} = 2.98 \times 10^{-7} \, V_A^{-0.5473} \, \mu_w^{-1.026} \, T$	(5-215)	13%
4. Hydrocarbon Mixtures			
Hayduk-Minhas [26]	$D_{AB}^{\circ} = 13.3 \times 10^{-8} \, T^{1.47} \, \mu_B^{(10.2/V_A - 0.791)} \, V_A^{-0.71}$	(5-216)	5%
Matthews-Akgerman [38]	$D_{AB}^{\circ} = 32.88 \, M_A^{-0.61} V_D^{-1.04} \, T^{0.5} \, (V_B - V_D)$	(5-217)	5%
Riazi-Whitson [48]	$D_{AB} = 1.07 \, \dfrac{(\rho D_{AB})^{\circ}}{\rho} \left(\dfrac{\mu}{\mu^{\circ}}\right)^{-0.27 - 0.38\,\omega + (-0.05 + 0.1\,\omega)P_r}$	(5-218)	15%

[*]References are listed on pages 5-7 and 5-8.

Random comparisons of predictions with 2.26 versus 2.6 show no consistent advantage for either value, however. It has been suggested to replace the exponent of 0.6 with 0.7 and to use an association factor of 0.7 for systems containing aromatic hydrocarbons. These modifications, however, are not recommended by Umesi and Danner. Lees and Sarram present a comparison of the association parameters. The average absolute error for 87 different solutes in water is 5.9 percent.

Tyn-Calus   This correlation requires data in the form of molar volumes and parachors $\psi_i = V_i \sigma_i^{1/4}$ (a property which, over moderate temperature ranges, is nearly constant), measured at the same temperature (not necessarily the temperature of interest). The parachors for the components may also be evaluated at different temperatures from each other. Quale has compiled values of ψ_i for many chemicals. Group contribution methods are available for estimation purposes (Reid et al.). The following suggestions were made by Reid et al.: The correlation is constrained to cases in which $\mu_B < 30$ cP. If the solute is water or if the solute is an organic acid and the solvent is not water or a short-chain alcohol, dimerization of the solute A should be assumed for purposes of estimating its volume and parachor. For example, the appropriate values for water as solute at 25°C are $V_W = 37.4$ cm³/mol and $\psi_W = 105.2$ cm³g$^{1/4}$/s$^{1/2}$mol. Finally, if the solute is nonpolar, the solvent volume and parachor should be multiplied by $8\mu_B$.

Umesi-Danner   They developed an equation for nonaqueous solvents with nonpolar and polar solutes. In all, 258 points were involved in the regression. R_i is the radius of gyration in Å of the component molecule, which has been tabulated by Passut and Danner for 250 compounds. The average absolute deviation was 16 percent, compared with 26 percent for the Wilke-Chang equation.

Siddiqi-Lucas   In an impressive empirical study, these authors examined 1275 organic liquid mixtures. Their equation yielded an average absolute deviation of 13.1 percent, which was less than that for the Wilke-Chang equation (17.8 percent). Note that this correlation does not encompass aqueous solutions; those were examined and a separate correlation was proposed, which is discussed later.

Binary Mixtures of Gases in Low-Viscosity, Nonelectrolyte Liquids   *Sridhar-Potter* derived an equation for predicting gas diffusion through liquid by combining existing correlations. Hildebrand had postulated the following dependence of the diffusivity for a gas in a liquid: $D_{AB}^{\circ} = D_{B'B}(V_{cB}/V_{cA})^{2/3}$, where $D_{B'B}$ is the solvent self-diffusion coefficient and V_{ci} is the critical volume of component i, respectively. To correct for minor changes in volumetric expansion, Sridhar and Potter multiplied the resulting equation by V_B/V_{ml_B}, where V_{ml_B} is the molar volume of the liquid B at its melting point and $D_{B'B}$ can be estimated by the equation of Ertl and Dullien (see p. 5-50). Sridhar and Potter compared experimentally measured diffusion coefficients for twenty-seven data points of eleven binary mixtures. Their average absolute error was 13.5 percent, but Chen and Chen analyzed about 50 combinations of conditions and 3 to 4 replicates each and found an average error of 18 percent. This correlation does not apply to hydrogen and helium as solutes. However, it demonstrates the usefulness of self-diffusion as a means to assess mutual diffusivities and the value of observable physical property changes, such as molar expansion, to account for changes in conditions.

Chen-Chen   Their correlation was based on diffusion measurements of 50 combinations of conditions with 3 to 4 replicates each and exhibited an average error of 6 percent. In this correlation, $V_r = V_B/[0.9724 \, (V_{ml_B} + 0.04765)]$ and $V_{ml_B} =$ the liquid molar volume at the melting point, as discussed previously. Their association parameter β [which is different from the definition of that symbol in Eq. (5-219)] accounts for hydrogen bonding of the solvent. Values for acetonitrile and methanol are: β = 1.58 and 2.31, respectively.

Dilute Binary Mixtures of a Nonelectrolyte in Water The correlations that were suggested previously for general mixtures, unless specified otherwise, may also be applied to diffusion of miscellaneous solutes in water. The following correlations are restricted to the present case, however.

Hayduk-Laudie They presented a simple correlation for the infinite dilution diffusion coefficients of nonelectrolytes in water. It has about the same accuracy as the Wilke-Chang equation (about 5.9 percent). There is no explicit temperature dependence, but the 1.14 exponent on μ_w compensates for the absence of T in the numerator. That exponent was misprinted (as 1.4) in the original article and has been reproduced elsewhere erroneously.

Siddiqi-Lucas These authors examined 658 aqueous liquid mixtures in an empirical study. They found an average absolute deviation of 19.7 percent. In contrast, the Wilke-Chang equation gave 35.0 percent and the Hayduk-Laudie correlation gave 30.4 percent.

Dilute Binary Hydrocarbon Mixtures *Hayduk-Minhas* presented an accurate correlation for normal paraffin mixtures that was developed from 58 data points consisting of solutes from C_5 to C_{32} and solvents from C_5 to C_{16}. The average error was 3.4 percent for the 58 mixtures.

Matthews-Akgerman The free-volume approach of Hildebrand was shown to be valid for binary, dilute liquid paraffin mixtures (as well as self-diffusion), consisting of solutes from C_8 to C_{16} and solvents of C_6 and C_{12}. The term they referred to as the "diffusion volume" was simply correlated with the critical volume, as $V_D = 0.308$ V_c. We can infer from Table 5-15 that this is approximately related to the volume at the melting point as $V_D = 0.945 V_m$. Their correlation was valid for diffusion of linear alkanes at temperatures up to 300°C and pressures up to 3.45 MPa. Matthews et al. and Erkey and Akgerman completed similar studies of diffusion of alkanes, restricted to *n*-hexadecane and *n*-octane, respectively, as the solvents.

Riazi-Whitson They presented a generalized correlation in terms of viscosity and molar density that was applicable to both gases and liquids. The average absolute deviation for gases was only about 8 percent, while for liquids it was 15 percent. Their expression relies on the Chapman-Enskog correlation [Eq. (5-194)] for the low-pressure diffusivity and the Stiel-Thodos correlation for low-pressure viscosity:

$$\mu° = \frac{x_A \mu_A° M_A^{1/2} + x_B \mu_B° M_B^{1/2}}{x_A M_A^{1/2} + x_B M_B^{1/2}}$$

where $\mu_i° \xi_i = 3.4 \times 10^{-4} T_{r^i}^{0.94}$ for $T_{r^i} < 1.5$ or $\mu_i° \xi_i = 1.778 \times 10^{-4}$ (4.58 $T_{r^i} - 1.67)^{5/8}$ for $T_{r^i} > 1.5$. In these equations, $\xi_i = T_{c^i}^{1/6}/P_{c^i}^{2/3} M_i^{1/2}$, and units are in cP, atm, K, and mol. For dense gases or liquids, the Chung et al. or Jossi-Stiel-Thodos correlation may be used to estimate viscosity. The latter is:

$$(\mu - \mu°) \xi + 10^{-4} = (0.1023 + 0.023364 \, \rho_r$$
$$+ 0.058533 \, \rho_r^2 - 0.040758 \, \rho_r^3 + 0.093324 \, \rho_r^4)^4$$

where
$$\xi = \frac{(x_A T_{c_A} + x_B T_{c_B})^{1/6}}{(x_A M_A + x_B M_B)^{1/2} (x_A P_{c_A} + x_B P_{c_B})}$$

and
$$\rho_r = (x_A V_{c_A} + x_B V_{c_B})\rho.$$

Dilute Binary Mixtures of Nonelectrolytes with Water as the Solute *Olander* modified the Wilke-Chang equation to adapt it to the infinite dilution diffusivity of water as the solute. The modification he recommended is simply the division of the right-hand side of the Wilke-Chang equation by 2.3. Unfortunately, neither the Wilke-Chang equation nor that equation divided by 2.3 fit the data very well. A reasonably valid generalization is that the Wilke-Chang equation is accurate if water is very insoluble in the solvent, such as pure hydrocarbons, halogenated hydrocarbons, and nitro-hydrocarbons. On the other hand, the Wilke-Chang equation divided by 2.3 is accurate for solvents in which water is very soluble, as well as those that have low viscosities. Such solvents include alcohols, ketones, carboxylic acids,

and aldehydes. Neither equation is accurate for higher-viscosity liquids, especially diols.

Dilute Dispersions of Macromolecules in Nonelectrolytes The Stokes-Einstein equation has already been presented. It was noted that its validity was restricted to large solutes, such as spherical macromolecules and particles in a continuum solvent. The equation has also been found to predict accurately the diffusion coefficient of spherical latex particles and globular proteins. Corrections to Stokes-Einstein for molecules approximating spheroids is given by Tanford. Since solute-solute interactions are ignored in this theory, it applies in the dilute range only.

Hiss-Cussler Their basis is the diffusion of a small solute in a fairly viscous solvent of relatively large molecules, which is the opposite of the Stokes-Einstein assumptions. The large solvent molecules investigated were not polymers or gels but were of moderate molecular weight so that the macroscopic and microscopic viscosities were the same. The major conclusion is that $D_{AB}° \, \mu^{2/3}$ = constant at a given temperature and for a solvent viscosity from 5×10^{-3} to 5 Pa s or greater (5 to 5×10^3 cP). This observation is useful if $D_{AB}°$ is known in a given high-viscosity liquid (oils, tars, etc.). Use of the usual relation of $D_{AB}° \propto 1/\mu$ for such an estimate could lead to large errors.

Concentrated, Binary Mixtures of Nonelectrolytes Several correlations that predict the composition dependence of D_{AB} are summarized in Table 5-19. Most are based on known values of $D_{AB}°$ and $D_{BA}°$. In fact, a rule of thumb states that, for many binary systems, $D_{AB}°$ and $D_{BA}°$ bound the D_{AB} vs. x_A curve. Cullinan's equation predicts diffusivities even in lieu of values at infinite dilution, but requires accurate density, viscosity, and activity coefficient data.

Since the infinite dilution values $D_{AB}°$ and $D_{BA}°$ are generally unequal, even a thermodynamically ideal solution like $\gamma_A = \gamma_B = 1$ will exhibit concentration dependence of the diffusivity. In addition, non-ideal solutions require a thermodynamic correction factor to retain the true "driving force" for molecular diffusion, or the gradient of the chemical potential rather than the composition gradient. That correction factor is:

$$\beta_A = 1 + \frac{\partial \ln \gamma_A}{\partial \ln x_A} \qquad (5-219)$$

Caldwell-Babb Darken observed that solid-state diffusion in metallurgical applications followed a simple relation. His equation related the tracer diffusivities and mole fractions to the mutual diffusivity:

$$D_{AB} = (x_A D_B + x_B D_A) \beta_A \qquad (5-220)$$

Caldwell and Babb used virtually the same equation to evaluate the mutual diffusivity for concentrated mixtures of common liquids.

Van Geet and Adamson tested that equation for the *n*-dodecane (A) and *n*-octane (B) system and found the average deviation of D_{AB} from experimental values to be −0.68 percent. In addition, that equation was tested for benzene + bromobenzene, *n*-hexane + *n*-dodecane, benzene + CCl_4, octane + decane, heptane + cetane, benzene + diphenyl, and benzene + nitromethane with success. For systems that depart significantly from thermodynamic ideality, it breaks down, sometimes by a factor of eight. For example, in the binary systems acetone + CCl_4, acetone + chloroform, and ethanol + CCl_4, it is not accurate. Thus, it can be expected to be fairly accurate for nonpolar hydrocarbons of similar molecular weight but not for polar-polar mixtures. Siddiqi et al. found that this relation was superior to those of Vignes and Leffler and Cullinan for a variety of mixtures. Umesi and Danner found an average absolute deviation of 13.9 percent for 198 data points.

Rathbun-Babb suggested that Darken's equation could be improved by raising the thermodynamic correction factor β_A to a power, n, less than unity. They looked at systems exhibiting negative deviations from Raoult's law and found $n = 0.3$. Furthermore, for polar-nonpolar mixtures, they found $n = 0.6$. In a separate study, Siddiqi and Lucas followed those suggestions and found an average absolute error of 3.3 percent for nonpolar-nonpolar mixtures, 11.0 percent for polar-nonpolar mixtures, and 14.6 percent for polar-polar mixtures. Siddiqi et al. examined a few other mixtures and found that $n = 1$ was probably best. Thus, this approach is, at best, highly dependent on the type of components being considered.

TABLE 5-19 Correlations of Diffusivities for Concentrated, Binary Mixtures of Nonelectrolyte Liquids

Authors[*]	Equation	
Caldwell-Babb [5]	$D_{AB} = (x_A D_{BA}^\circ + x_B D_{AB}^\circ)\beta_A$	(5-221)
Rathbun-Babb [46]	$D_{AB} = (x_A D_{BA}^\circ + x_B D_{AB}^\circ)\beta_A^n$	(5-222)
Vignes [62]	$D_{AB} = D_{AB}^{\circ x_B} D_{BA}^{\circ x_A}\beta_A$	(5-223)
Leffler-Cullinan [34]	$D_{AB}\mu_{mix} = (D_{AB}^\circ \mu_B)^{x_B}(D_{BA}^\circ \mu_A)^{x_A}\beta_A$	(5-224)
Cussler [12]	$D_{AB} = D_0\left[1 + \dfrac{K}{x_A x_B}\left(\dfrac{\partial \ln x_A}{\partial \ln a_A} - 1\right)\right]^{-1/2}$	(5-225)
Cullinan [10]	$D_{AB} = \dfrac{kT}{2\pi\mu_{mix}(V/A)^{1/3}}\left[\dfrac{2\pi x_A x_B \beta_A}{1 + \beta_A(2\pi x_A x_B - 1)}\right]^{1/2}$	(5-226)
Asfour-Dullien [2]	$D_{AB} = \left(\dfrac{D_{AB}^\circ}{\mu_B}\right)^{x_B}\left(\dfrac{D_{BA}^\circ}{\mu_A}\right)^{x_A}\zeta\mu\beta_A$	(5-227)
Siddiqi-Lucas [52]	$D_{AB} = (C_B \overline{V}_B D_{AB}^\circ + C_A \overline{V}_A D_{BA}^\circ)\beta_A$	(5-228)

Relative errors for the correlations in this table are very dependent on the components of interest and are cited in the text.
[*]See pages 5-7 and 5-8 for references.

Vignes empirically correlated mixture diffusivity data for 12 binary mixtures. Later Ertl et al. evaluated 122 binary systems, which showed an average absolute deviation of only 7 percent. None of the latter systems, however, was very nonideal.

Leffler-Cullinan modified Vignes' equation using some theoretical arguments to arrive at Eq. (5-224), which the authors compared to Eq. (5-223) for the 12 systems mentioned above. The average absolute maximum deviation was only 6 percent. Umesi and Danner, however, found an average absolute deviation of 11.4 percent for 198 data points. For normal paraffins, it is not very accurate. In general, the accuracies of Eqs. (5-223) and (5-224) are not much different, and, since Vignes' is simpler to use, it is suggested. The application of either should be limited to nonassociating systems that do not deviate much from ideality ($0.95 < \beta_A < 1.05$).

Cussler studied diffusion in concentrated associating systems and has shown that, in associating systems, it is the size of diffusing clusters rather than diffusing solutes that controls diffusion. D_o is a reference diffusion coefficient discussed hereafter; a_A is the activity of component A; and K is a constant. By assuming that D_o could be predicted by Eq. (5-223) with $\beta = 1$, K was found to be equal to 0.5 based on five binary systems and validated with a sixth binary mixture. The limitations of Eq. (5-225) using D_o and K defined previously have not been explored, so caution is warranted. Gurkan showed that K should actually be closer to 0.3 (rather than 0.5) and discussed the overall results.

Cullinan presented an extension of Cussler's cluster diffusion theory. His method accurately accounts for composition and temperature dependence of diffusivity. It is novel in that it contains no adjustable constants, and it relates transport properties and solution thermodynamics. This equation has been tested for six very different mixtures by Rollins and Knaebel, and it was found to agree remarkably well with data for most conditions, considering the absence of adjustable parameters. In the dilute region (of either A or B), there are systematic errors probably caused by the breakdown of certain implicit assumptions (that nevertheless appear to be generally valid at higher concentrations).

Asfour-Dullien developed a relation for predicting alkane diffusivities at moderate concentrations that employs:

$$\zeta = \left(\frac{V_{fm}}{V_{f x_A} V_{f x_B}}\right)^{2/3}\frac{M_{x_A} M_{x_B}}{M_m}$$ (5-229)

where $V_{f x_i} = V_{fi}^{x_i}$; the fluid free volume is $V_{f^i} = V_i - V_{ml^i}$ for $i = A, B$, and m, in which V_{ml^i} is the molar volume of the liquid at the melting point and

$$V_{ml_m} = \left(\frac{x_A^2}{V_{ml_A}} + \frac{2 x_A x_B}{V_{ml_{AB}}} + \frac{x_B^2}{V_{ml_B}}\right)^{-1}$$

and

$$V_{ml_{AB}} = \left[\frac{V_{ml_A}^{1/3} + V_{ml_B}^{1/3}}{2}\right]^3$$

and μ is the mixture viscosity; M_m is the mixture mean molecular weight; and β_A is defined by Eq. (5-219). The average absolute error of this equation is 1.4 percent, while the Vignes equation and the Leffler-Cullinan equation give 3.3 percent and 6.2 percent, respectively.

Siddiqi-Lucas suggested that component volume fractions might be used to correlate the effects of concentration dependence. They found an average absolute deviation of 4.5 percent for nonpolar-nonpolar mixtures, 16.5 percent for polar-nonpolar mixtures, and 10.8 percent for polar-polar mixtures.

Binary Electrolyte Mixtures When electrolytes are added to a solvent, they dissociate to a certain degree. It would appear that the solution contains at least three components: solvent, anions, and cations. If the solution is to remain neutral in charge at each point (assuming the absence of any applied electric potential field), the anions and cations diffuse effectively as a single component, as for molecular diffusion. The diffusion of the anionic and cationic species in the solvent can thus be treated as a binary mixture.

Nernst-Haskell The theory of dilute diffusion of salts is well developed and has been experimentally verified. For dilute solutions of a single salt, the well-known Nernst-Haskell equation (Reid et al.) is applicable:

$$D_{AB}^\circ = \frac{RT}{F^2}\frac{\left|\dfrac{1}{n_+}\right| + \left|\dfrac{1}{n_-}\right|}{\dfrac{1}{\lambda_+^0} + \dfrac{1}{\lambda_-^0}} = 8.9304 \times 10^{-10}\, T\,\frac{\left|\dfrac{1}{n_+}\right| + \left|\dfrac{1}{n_-}\right|}{\dfrac{1}{\lambda_+^0} + \dfrac{1}{\lambda_-^0}}$$ (5-230)

where D_{AB}° = diffusivity based on molarity rather than normality of dilute salt A in solvent B, cm²/s.

The previous definitions can be interpreted in terms of ionic-species diffusivities and conductivities. The latter are easily measured and depend on temperature and composition. For example, the equivalent conductance Λ is commonly tabulated in chemistry handbooks as the limiting (infinite dilution) conductance Λ_o and at standard concentrations, typically at 25°C. $\Lambda = 1000\,K/C = \lambda_+ + \lambda_- = \Lambda_o + f(C)$, (cm²/ohm gequiv); $K = \alpha/R$ = specific conductance, (ohm cm)⁻¹; C = solution concentration, (gequiv/ℓ); α = conductance cell constant (measured), (cm⁻¹); R = solution electrical resistance, which is measured (ohm); and $f(C)$ = a complicated function of concentration. The resulting equation of the electrolyte diffusivity is

$$D_{AB} = \frac{|z_+| + |z_-|}{(|z_-|/D_+) + (|z_+|/D_-)}$$ (5-231)

where $|z_\pm|$ represents the magnitude of the ionic charge and where the cationic or anionic diffusivities are $D_\pm = 8.9304 \times 10^{-10}\, T\lambda_\pm/|z_\pm|$ cm²/s. The coefficient is $kN_0/F^2 = R/F^2$. In practice, the equivalent conductance of the ion pair of interest would be obtained and supplemented with conductances of permutations of those ions and one independent cation and anion. This would allow determination of all the ionic conductances and hence the diffusivity of the electrolyte solution.

Gordon Typically, as the concentration of a salt increases from infinite dilution, the diffusion coefficient decreases rapidly from D_{AB}°. As concentration is increased further, however, D_{AB} rises steadily, often becoming greater than D_{AB}°. Gordon proposed the following empirical equation, which is applicable up to concentrations of 2N:

$$D_{AB} = D_{AB}^\circ \frac{1}{C_B \overline{V}_B} \frac{\mu_B}{\mu} \left(1 + \frac{\ln \gamma_\pm}{\ln m}\right) \qquad (5\text{-}232)$$

where D_{AB}° is given by the Nernst-Haskell equation. References that tabulate $\gamma_\pm$ as a function of m, as well as other equations for D_{AB}, are given by Reid et al.

Multicomponent Mixtures No simple, practical estimation methods have been developed for predicting multicomponent liquid-diffusion coefficients. Several theories have been developed, but the necessity for extensive activity data, pure component and mixture volumes, mixture viscosity data, and tracer and binary diffusion coefficients have significantly limited the utility of the theories (see Reid et al.).

The generalized Stefan-Maxwell equations using binary diffusion coefficients are not easily applicable to liquids since the coefficients are so dependent on conditions. That is, in liquids, each D_{ij} can be strongly composition dependent in binary mixtures and, moreover, the binary D_{ij} is strongly affected in a multicomponent mixture. Thus, the convenience of writing multicomponent flux equations in terms of binary coefficients is lost. Conversely, they apply to gas mixtures because each D_{ij} is practically independent of composition by itself and in a multicomponent mixture (see Taylor and Krishna for details).

One particular case of multicomponent diffusion that has been examined is the dilute diffusion of a solute in a homogeneous mixture (e.g., of A in $B + C$). Umesi and Danner compared the three equations given below for 49 ternary systems. All three equations were equivalent, giving average absolute deviations of 25 percent.

Perkins-Geankoplis

$$D_{am}\,\mu_m^{0.8} = \sum_{\substack{j=1 \\ j \ne A}}^n x_j\,D_{Aj}^\circ\,\mu_j^{0.8} \qquad (5\text{-}233)$$

Cullinan This is an extension of Vignes' equation to multicomponent systems:

$$D_{am} = \prod_{\substack{j=1 \\ j \ne A}}^n (D_{Aj}^\circ)^{x_j} \qquad (5\text{-}234)$$

Leffler-Cullinan They extended their binary relation to an arbitrary multicomponent mixture, as follows:

$$D_{am}\,\mu_m = \prod_{\substack{j=1 \\ j \ne A}}^n (D_{Aj}^\circ\,\mu_j)^{x_j} \qquad (5\text{-}235)$$

where D_{Aj} is the dilute binary diffusion coefficient of A in j; D_{Am} is the dilute diffusion of A through m; x_j is the mole fraction; μ_j is the viscosity of component j; and μ_m is the mixture viscosity.

Akita Another case of multicomponent dilute diffusion of significant practical interest is that of gases in aqueous electrolyte solutions. Many gas-absorption processes use electrolyte solutions. Akita presents experimentally tested equations for this case.

Graham-Dranoff They studied multicomponent diffusion of electrolytes in ion exchangers. They found that the Stefan-Maxwell interaction coefficients reduce to limiting ion tracer diffusivities of each ion.

Pinto-Graham Pinto and Graham studied multicomponent diffusion in electrolyte solutions. They focused on the Stefan-Maxwell equations and corrected for solvation effects. They achieved excellent results for 1-1 electrolytes in water at 25°C up to concentrations of 4M.

DIFFUSION OF FLUIDS IN POROUS SOLIDS

Diffusion in porous solids is usually the most important factor controlling mass transfer in adsorption, ion exchange, drying, heterogeneous catalysis, leaching, and many other applications. Some of the applications of interest are outlined in Table 5-20. Applications of these equations are found in Secs. 16, 22, and 23.

Diffusion within the largest cavities of a porous medium is assumed to be similar to ordinary or bulk diffusion except that it is hindered by the pore walls (see Eq. 5-236). The tortuosity τ that expresses this hindrance has been estimated from geometric arguments. Unfortunately, measured values are often an order of magnitude greater than those estimates. Thus, the effective diffusivity D_{eff} (and hence τ) is normally determined by comparing a diffusion model to experimental measurements. The normal range of tortuosities for silica gel, alumina, and other porous solids is $2 \le \tau \le 6$, but for activated carbon, $5 \le \tau \le 65$.

In small pores and at low pressures, the mean free path ℓ of the gas molecule (or atom) is significantly greater than the pore diameter d_{pore}. Its magnitude may be estimated from

$$\ell = \frac{3.2\,\mu}{P}\left[\frac{RT}{2\pi M}\right]^{1/2}, \text{ in } m$$

As a result, collisions with the wall occur more frequently than with other molecules. This is referred to as the Knudsen mode of diffusion and is contrasted with ordinary or bulk diffusion, which occurs by intermolecular collisions. At intermediate pressures, both ordinary diffusion and Knudsen diffusion may be important [see Eqs. (5-239) and (5-240)].

For gases and vapors that adsorb on the porous solid, surface diffusion may be important, particularly at high surface coverage [see Eqs. (5-241) and (5-244)]. The mechanism of surface diffusion may be viewed as molecules hopping from one surface site to another. Thus, if adsorption is too strong, surface diffusion is impeded, while if adsorption is too weak, surface diffusion contributes insignificantly to the overall rate. Surface diffusion and bulk diffusion usually occur in parallel [see Eqs. (5-245) and (5-246)]. Although D_s is expected to be less than D_{eff}, the solute flux due to surface diffusion may be larger than that due to bulk diffusion if $\partial q_i/\partial z \gg \partial C_i/\partial z$. This can occur when a component is strongly adsorbed and the surface coverage is high. For all that, surface diffusion is not well understood. The references in Table 5-20 should be consulted for further details.

INTERPHASE MASS TRANSFER

Transfer of material between phases is important in most separation processes in which two phases are involved. When one phase is pure, mass transfer in the pure phase is not involved. For example, when a pure liquid is being evaporated into a gas, only the gas-phase mass transfer need be calculated. Occasionally, mass transfer in one of the two phases may be neglected even though pure components are not involved. This will be the case when the resistance to mass transfer is much larger in one phase than in the other. Understanding the nature and magnitudes of these resistances is one of the keys to performing reliable mass transfer. In this section, mass transfer between gas and liquid phases will be discussed. The principles are easily applied to the other phases.

Mass-Transfer Principles: Dilute Systems When material is transferred from one phase to another across an interface that separates the two, the resistance to mass transfer in each phase causes a concentration gradient in each, as shown in Fig. 5-26 for a gas-liquid interface. The concentrations of the diffusing material in the two phases immediately adjacent to the interface generally are unequal, even if expressed in the same units, but usually are assumed to be related to each other by the laws of thermodynamic equilibrium. Thus, it is assumed that the thermodynamic equilibrium is reached at the gas-liquid interface almost immediately when a gas and a liquid are brought into contact.

For systems in which the solute concentrations in the gas and liquid phases are dilute, the rate of transfer may be expressed by equations which predict that the rate of mass transfer is proportional to the difference between the bulk concentration and the concentration at the gas-liquid interface. Thus

$$N_A = k_G'(p - p_i) = k_L'(c_i - c) \qquad (5\text{-}248)$$

where N_A = mass-transfer rate, k_G' = gas-phase mass-transfer coefficient, k_L' = liquid-phase mass-transfer coefficient, p = solute partial pressure in

TABLE 5-20 Relations for Diffusion in Porous Solids

Mechanism	Equation		Applies to	References[°]
Bulk diffusion in pores	$D_{eff} = \dfrac{\varepsilon_p D}{\tau}$	(5-236)	Gases or liquids in large pores. $N_{K_n} = \ell/d_{pore} < 0.01$	[67]
Knudsen diffusion	$D_K = 48.5\, d_{pore}\left(\dfrac{T}{M}\right)^{1/2}$ in m²/s	(5-237)	Dilute (low pressure) gases in small pores. $N_{K_n} = \ell/d_{pore} > 10$	Geankoplis, [68, 69]
	$D_{Keff} = \dfrac{\varepsilon_p D_K}{\tau}$			
	$N_i = -D_K \dfrac{dC_i}{dz}$	(5-238)	" " " "	
Combined bulk and Knudsen diffusion	$D_{eff} = \left(\dfrac{1 - \alpha x_A}{D_{eff}} + \dfrac{1}{D_{Keff}}\right)^{-1}$	(5-239)	" " " " $N_A \neq N_B$	Geankoplis, [66, 69]
	$\alpha = 1 + \dfrac{N_B}{N_A}$			
	$D_{eff} = \left(\dfrac{1}{D_{eff}} + \dfrac{1}{D_{Keff}}\right)^{-1}$	(5-240)	$N_A = N_B$	
Surface diffusion	$J_{Si} = -D_{Seff}\,\rho_p\left(\dfrac{dq_i}{dz}\right)$	(5-241)	Adsorbed gases or vapors	[66, 68, 69]
	$D_{Seff} = \dfrac{\varepsilon_p D_S}{\tau}$	(5-242)	" " " "	
	$D_{S\theta} = \dfrac{D_{S\theta=0}}{(1-\theta)}$	(5-243)	θ = fractional surface coverage ≤ 0.6	
	$D_S = D_S'(q)\exp\left(\dfrac{-E_S}{RT}\right)$	(5-244)	" " " "	
Parallel bulk and surface diffusion	$J = -\left[D_{eff}\left(\dfrac{dp_i}{dz}\right) + D_{Seff}\,\rho_p\left(\dfrac{dq_i}{dz}\right)\right]$	(5-245)	" " " "	[68]
	$J = -D_{app}\left(\dfrac{dp_i}{dz}\right)$	(5-246)	" " " "	
	$D_{app} = D_{eff} + D_{Seff}\,\rho_p\left(\dfrac{dq_i}{dp_i}\right)$	(5-247)	" " " "	

[°]See pages 5-7 and 5-8 for references.

bulk gas, p_i = solute partial pressure at interface, c = solute concentration in bulk liquid, and c_i = solute concentration in liquid at interface.

The mass-transfer coefficients k_G' and k_L' by definition are equal to the ratios of the molal mass flux N_A to the concentration driving forces $(p - p_i)$ and $(c_i - c)$ respectively. An alternative expression for the rate of transfer in dilute systems is given by

$$N_A = k_G(y - y_i) = k_L(x_i - x) \tag{5-249}$$

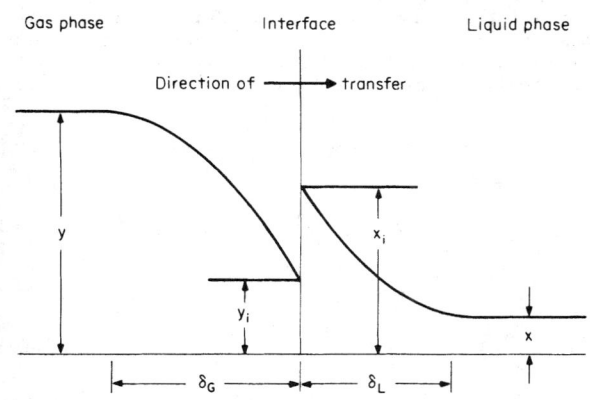

FIG. 5-26 Concentration gradients near a gas-liquid interface.

where N_A = mass-transfer rate, k_G = gas-phase mass-transfer coefficient, k_L = liquid-phase mass-transfer coefficient, y = mole-fraction solute in bulk-gas phase, y_i = mole-fraction solute in gas at interface, x = mole-fraction solute in bulk-liquid phase, and x_i = mole-fraction solute in liquid at interface.

The mass-transfer coefficients defined by Eqs. (5-248) and (5-249) are related to each other as follows:

$$k_G = k_G' p_T \tag{5-250}$$

$$k_L = k_L' \bar{\rho}_L \tag{5-251}$$

where p_T = total system pressure employed *during the experimental determinations* of k_G' values and $\bar{\rho}_L$ = average molar density of the liquid phase. The coefficient k_G is relatively independent of the total system pressure and therefore is more convenient to use than k_G', which is inversely proportional to the total system pressure.

The above equations may be used for finding the interfacial concentrations corresponding to any set of values of x and y provided the ratio of the individual coefficients is known. Thus

$$(y - y_i)/(x_i - x) = k_L/k_G = k_L'\bar{\rho}_L/k_G' p_T = L_M H_G/G_M H_L \tag{5-252}$$

where L_M = molar liquid mass velocity, G_M = molar gas mass velocity, H_L = height of one transfer unit based on liquid-phase resistance, and H_G = height of one transfer unit based on gas-phase resistance. The last term in Eq. (5-252) is derived from Eqs. (5-271) and (5-273).

Equation (5-252) may be solved graphically if a plot is made of the equilibrium vapor and liquid compositions and a point representing the bulk concentrations x and y is located on this diagram. A construction of this type is shown in Fig. 5-27, which represents a gas-absorption situation.

The interfacial mole fractions y_i and x_i can be determined by solving Eq. (5-252) simultaneously with the equilibrium relation $y_i^\circ = F(x_i)$ to obtain y_i and x_i. The rate of transfer may then be calculated from Eq. (5-249).

If the equilibrium relation $y_i^\circ = F(x_i)$ is sufficiently simple, e.g., if a plot of y_i° versus x_i is a straight line, not necessarily through the origin, the rate of transfer is proportional to the difference between the bulk concentration in one phase and the concentration (in that same phase) which would be in equilibrium with the bulk concentration in the second phase. One such difference is $y - y^\circ$, and another is $x^\circ - x$. In this case, there is no need to solve for the interfacial compositions, as may be seen from the following derivation.

The rate of mass transfer may be defined by the equation

$$N_A = K_G(y - y^\circ) = k_G(y - y_i) = k_L(x_i - x) = K_L(x^\circ - x) \quad (5\text{-}253)$$

where K_G = overall gas-phase mass-transfer coefficient, K_L = overall liquid-phase mass-transfer coefficient, y° = vapor composition in equilibrium with x, and x° = liquid composition in equilibrium with vapor of composition y. This equation can be rearranged to the formula

$$\frac{1}{K_G} = \frac{1}{k_G}\left(\frac{y - y^\circ}{y - y_i}\right) = \frac{1}{k_G} + \frac{1}{k_G}\left(\frac{y_i - y^\circ}{y - y_i}\right) = \frac{1}{k_G} + \frac{1}{k_L}\left(\frac{y_i - y^\circ}{x_i - x}\right)$$
$$(5\text{-}254)$$

in view of Eq. (5-252). Comparison of the last term in parentheses with the diagram of Fig. 5-27 shows that it is equal to the slope of the chord connecting the points (x, y°) and (x_i, y_i). If the equilibrium curve is a straight line, then this term is the slope m. Thus

$$1/K_G = (1/k_G + m/k_L) \quad (5\text{-}255)$$

When Henry's law is valid ($p_A = H x_A$ or $p_A = H' C_A$), the slope m can be computed according to the relationship

$$m = H/p_T = H' \rho_L/p_T \quad (5\text{-}256)$$

where m is defined in terms of mole-fraction driving forces compatible with Eqs. (5-249) through (5-255), i.e., with the definitions of k_L, k_G, and K_G.

If it is desired to calculate the rate of transfer from the overall concentration difference based on bulk-liquid compositions $(x^\circ - x)$, the appropriate overall coefficient K_L is related to the individual coefficients by the equation

$$1/K_L = (1/k_L + 1/m k_G) \quad (5\text{-}257)$$

Conversion of these equations to a k_G', k_L' basis can be accomplished readily by direct substitution of Eqs. (5-250) and (5-251).

Occasionally one will find k_L' or K_L' values reported in units (SI) of meters per second. The correct units for these values are kmol/

$[(\text{s}\cdot\text{m}^2)(\text{kmol/m}^3)]$, and Eq. (5-251) is the correct equation for converting them to a mole-fraction basis.

When k_G' and K_G' values are reported in units (SI) of kmol/[(s·m²)(kPa)], one must be careful in converting them to a mole-fraction basis by multiplying by the total pressure actually employed in the original experiments and *not* by the total pressure of the system to be designed. This conversion is valid for systems in which Dalton's law of partial pressures ($p = y p_T$) is valid.

Comparison of Eqs. (5-255) and (5-257) shows that for systems in which the equilibrium line is straight, the overall mass transfer coefficients are related to each other by the equation

$$K_L = m K_G \quad (5\text{-}258)$$

When the equilibrium curve is not straight, there is no strictly logical basis for the use of an overall transfer coefficient, since the value of m will be a function of position in the apparatus, as can be seen from Fig. 5-27. In such cases the rate of transfer must be calculated by solving for the interfacial compositions as described above.

Experimentally observed rates of mass transfer often are expressed in terms of overall transfer coefficients even when the equilibrium lines are curved. This procedure is empirical, since the theory indicates that in such cases the rates of transfer may not vary in direct proportion to the overall bulk concentration differences $(y - y^\circ)$ and $(x^\circ - x)$ at all concentration levels even though the rates may be proportional to the concentration difference in each phase taken separately, i.e., $(x_i - x)$ and $(y - y_i)$.

In most types of separation equipment such as packed or spray towers, the interfacial area that is effective for mass transfer cannot be accurately determined. For this reason it is customary to report experimentally observed rates of transfer in terms of transfer coefficients based on a unit volume of the apparatus rather than on a unit of interfacial area. Such volumetric coefficients are designated as $K_G a$, $k_L a$, etc., where a represents the interfacial area per unit volume of the apparatus. Experimentally observed variations in the values of these volumetric coefficients with variations in flow rates, type of packing, etc., may be due as much to changes in the effective value of a as to changes in k. Calculation of the overall coefficients from the individual volumetric coefficients is made by means of the equations

$$1/K_G a = (1/k_G a + m/k_L a) \quad (5\text{-}259)$$
$$1/K_L a = (1/k_L a + 1/m k_G a) \quad (5\text{-}260)$$

Because of the wide variation in equilibrium, the variation in the values of m from one system to another can have an important effect on the overall coefficient and on the selection of the type of equipment to use. For example, if m is large, the liquid-phase part of the overall resistance might be extremely large where k_L might be relatively small. This kind of reasoning must be applied with caution, however, since species with different equilibrium characteristics are separated under different operating conditions. Thus, the effect of changes in m on the overall resistance to mass transfer may partly be counterbalanced by changes in the individual specific resistances as the flow rates are changed.

Mass-Transfer Principles: Concentrated Systems When solute concentrations in the gas and/or liquid phases are large, the equations derived above for dilute systems no longer are applicable. The correct equations to use for concentrated systems are as follows:

$$N_A = \hat{k}_G(y - y_i)/y_{BM} = \hat{k}_L(x_i - x)/x_{BM}$$
$$= \hat{K}_G(y - y^\circ)/y_{BM}^\circ = \hat{K}_L(x^\circ - x)/x_{BM}^\circ \quad (5\text{-}261)$$

where ($N_B = 0$)

$$y_{BM} = \frac{(1 - y) - (1 - y_i)}{\ln[(1 - y)/(1 - y_i)]} \quad (5\text{-}262)$$

$$y_{BM}^\circ = \frac{(1 - y) - (1 - y^\circ)}{\ln[(1 - y)/(1 - y^\circ)]} \quad (5\text{-}263)$$

$$x_{BM} = \frac{(1 - x) - (1 - x_i)}{\ln[(1 - x)/(1 - x_i)]} \quad (5\text{-}264)$$

$$x_{BM}^\circ = \frac{(1 - x) - (1 - x^\circ)}{\ln[(1 - x)/(1 - x^\circ)]} \quad (5\text{-}265)$$

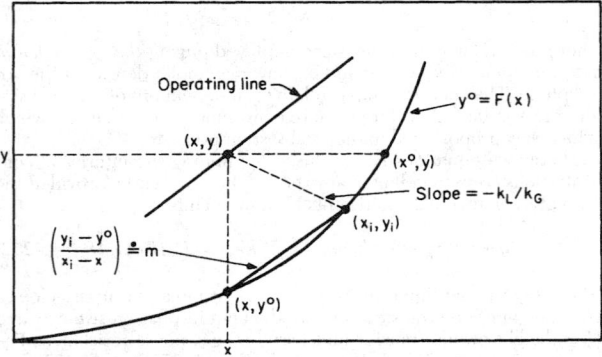

FIG. 5-27 Identification of concentrations at a point in a countercurrent absorption tower.

and where $\hat{k}_G$ and $\hat{k}_L$ are the gas-phase and liquid-phase mass-transfer coefficients for concentrated systems and $\hat{K}_G$ and $\hat{K}_L$ are the overall gas-phase and liquid-phase mass-transfer coefficients for concentrated systems. These coefficients are defined later in Eqs. (5-268) to (5-270).

The factors y_{BM} and x_{BM} arise from the fact that, in the diffusion of a solute through a second stationary layer of insoluble fluid, the resistance to diffusion varies in proportion to the concentration of the insoluble stationary fluid, approaching zero as the concentration of the insoluble fluid approaches zero. See Eq. (5-190).

The factors y_{BM}° and x_{BM}° cannot be justified on the basis of mass-transfer theory since they are based on overall resistances. These factors therefore are included in the equations by analogy with the corresponding film equations.

In dilute systems the logarithmic-mean insoluble-gas and nonvolatile-liquid concentrations approach unity, and Eq. (5-261) reduces to the dilute-system formula. For equimolar counter diffusion (e.g., binary distillation), these log-mean factors should be omitted. See Eq. (5-189).

Substitution of Eqs. (5-262) through (5-265) into Eq. (5-261) results in the following simplified formula:

$$
\begin{aligned}
N_A &= \hat{k}_G \ln\left[(1 - y_i)/(1 - y)\right] \\
&= \hat{K}_G \ln\left[(1 - y^\circ)/(1 - y)\right] \\
&= \hat{k}_L \ln\left[(1 - x)/(1 - x_i)\right] \\
&= \hat{K}_L \ln\left[(1 - x)/(1 - x^\circ)\right]
\end{aligned}
\tag{5-266}
$$

Note that the units of $\hat{k}_G$, $\hat{K}_G$, $\hat{k}_L$, and $\hat{K}_L$ are all identical to each other, i.e., kmol/(s·m²) in SI units.

The equation for computing the interfacial gas and liquid compositions in concentrated systems is

$$
\begin{aligned}
(y - y_i)/(x_i - x) &= \hat{k}_L y_{BM}/\hat{k}_G x_{BM} \\
&= L_M H_G y_{BM}/G_M H_L x_{BM} = k_L/k_G
\end{aligned}
\tag{5-267}
$$

This equation is identical to the one for dilute systems since $\hat{k}_G = k_G y_{BM}$ and $\hat{k}_L = k_L x_{BM}$. Note, however, that when $\hat{k}_G$ and $\hat{k}_L$ are given, the equation must be solved by trial and error, since x_{BM} contains x_i and y_{BM} contains y_i.

The overall gas-phase and liquid-phase mass-transfer coefficients for concentrated systems are computed according to the following equations:

$$
\frac{1}{\hat{K}_G} = \frac{y_{BM}}{y_{BM}^\circ}\frac{1}{\hat{k}_G} + \frac{x_{BM}}{y_{BM}^\circ}\frac{1}{\hat{k}_L}\left(\frac{y_i - y^\circ}{x_i - x}\right)
\tag{5-268}
$$

$$
\frac{1}{\hat{K}_L} = \frac{x_{BM}}{x_{BM}^\circ}\frac{1}{\hat{k}_L} + \frac{y_{BM}}{x_{BM}^\circ}\frac{1}{\hat{k}_G}\left(\frac{x^\circ - x_i}{y - y_i}\right)
\tag{5-269}
$$

When the equilibrium curve is a straight line, the terms in parentheses can be replaced by the slope m as before. In this case the overall mass-transfer coefficients for concentrated systems are related to each other by the equation

$$
\hat{K}_L = m\hat{K}_G(x_{BM}^\circ/y_{BM}^\circ)
\tag{5-270}
$$

All these equations reduce to their dilute-system equivalents as the inert concentrations approach unity in terms of mole fractions of inert concentrations in the fluids.

HTU (Height Equivalent to One Transfer Unit) Frequently the values of the individual coefficients of mass transfer are so strongly dependent on flow rates that the quantity obtained by dividing each coefficient by the flow rate of the phase to which it applies is more nearly constant than the coefficient itself. The quantity obtained by this procedure is called the height equivalent to one transfer unit, since it expresses in terms of a single length dimension the height of apparatus required to accomplish a separation of standard difficulty. The following relations between the transfer coefficients and the values of HTU apply:

$$
H_G = G_M/k_G a y_{BM} = G_M/\hat{k}_G a
\tag{5-271}
$$

$$
H_{OG} = G_M/K_G a y_{BM}^\circ = G_M/\hat{K}_G a
\tag{5-272}
$$

$$
H_L = L_M/k_L a x_{BM} = L_M/\hat{k}_L a
\tag{5-273}
$$

$$
H_{OL} = L_M/K_L a x_{BM}^\circ = L_M/\hat{K}_L a
\tag{5-274}
$$

The equations that express the addition of individual resistances in terms of HTUs, applicable to either dilute or concentrated systems, are

$$
H_{OG} = \frac{y_{BM}}{y_{BM}^\circ} H_G + \frac{mG_M}{L_M}\frac{x_{BM}}{y_{BM}^\circ} H_L
\tag{5-275}
$$

$$
H_{OL} = \frac{x_{BM}}{x_{BM}^\circ} H_L + \frac{L_M}{mG_M}\frac{y_{BM}}{x_{BM}^\circ} H_G
\tag{5-276}
$$

These equations are strictly valid only when m, the slope of the equilibrium curve, is constant, as noted previously.

NTU (Number of Transfer Units) The NTU required for a given separation is closely related to the number of theoretical stages or plates required to carry out the same separation in a stagewise or plate-type apparatus. For equimolal counterdiffusion, such as in a binary distillation, the number of overall gas-phase transfer units N_{OG} required for changing the composition of the vapor stream from y_1 to y_2 is

$$
N_{OG} = \int_{y_2}^{y_1} \frac{dy}{y - y^\circ}
\tag{5-277}
$$

When diffusion is in one direction only, as in the absorption of a soluble component from an insoluble gas,

$$
N_{OG} = \int_{y_2}^{y_1} \frac{y_{BM}^\circ \, dy}{(1 - y)(y - y^\circ)}
\tag{5-278}
$$

The total height of packing required is then

$$
h_T = H_{OG} N_{OG}
\tag{5-279}
$$

When it is known that H_{OG} varies appreciably within the tower, this term must be placed inside the integral in Eqs. (5-277) and (5-278) for accurate calculations of h_T. For example, the packed-tower design equation in terms of the overall gas-phase mass-transfer coefficient for absorption would be expressed as follows:

$$
h_T = \int_{y_2}^{y_1} \left[\frac{G_M}{K_G a y_{BM}^\circ}\right] \frac{y_{BM}^\circ \, dy}{(1 - y)(y - y^\circ)}
\tag{5-280}
$$

where the first term under the integral can be recognized as the HTU term. Convenient solutions of these equations for special cases are discussed later.

Definitions of Mass-Transfer Coefficients $\hat{k}_G$ and $\hat{k}_L$ The mass-transfer coefficient is defined as the ratio of the molal mass flux N_A to the concentration driving force. This leads to many different ways of defining these coefficients. For example, gas-phase mass-transfer rates may be defined as

$$
N_A = k_G(y - y_i) = k_G'(p - p_i) = \hat{k}_G(y - y_i)/y_{BM}
\tag{5-281}
$$

where the units (SI) of k_G are kmol/[(s·m²)(mole fraction)], the units of k_G' are kmol/[(s·m²)(kPa)], and the units of $\hat{k}_G$ are kmol/(s·m²). These coefficients are related to each other as follows:

$$
k_G = k_G y_{BM} = k_G' p_T y_{BM}
\tag{5-282}
$$

where p_T is the total system pressure (it is assumed here that Dalton's law of partial pressures is valid).

In a similar way, liquid-phase mass-transfer rates may be defined by the relations

$$
N_A = k_L(x_i - x) = k_L'(c_i - c) = \hat{k}_L(x_i - x)/x_{BM}
\tag{5-283}
$$

where the units (SI) of k_L are kmol/[(s·m²)(mole fraction)], the units of k_L' are kmol/[(s·m²)(kmol/m³)] or meters per second, and the units of $\hat{k}_L$ are kmol/(s·m²). These coefficients are related as follows:

$$
\hat{k}_L = k_L x_{BM} = k_L' \overline{\rho}_L x_{BM}
\tag{5-284}
$$

where $\overline{\rho}_L$ is the molar density of the liquid phase in units (SI) of kilomoles per cubic meter. Note that, for dilute solutions where $x_{BM} \doteq 1$, k_L and $\hat{k}_L$ will have identical numerical values. Similarly, for dilute gases $\hat{k}_G \doteq k_G$.

Simplified Mass-Transfer Theories In certain simple situations, the mass-transfer coefficients can be calculated from first principles. The film, penetration, and surface-renewal theories are attempts to extend these theoretical calculations to more complex situations. Although these theories are often not accurate, they are useful to provide a physical picture for variations in the mass-transfer coefficient.

For the special case of steady-state unidirectional diffusion of a component through an inert-gas film in an ideal-gas system, the rate of mass transfer is derived as

$$N_A = \frac{D_{AB}p_T}{RT\,\delta_G}\frac{(y-y_i)}{y_{BM}} = \frac{D_{AB}p_T}{RT\,\delta_G}\ln\frac{1-y_i}{1-y} \qquad (5\text{-}285)$$

where D_{AB} = the diffusion coefficient or "diffusivity," δ_G = the "effective" thickness of a stagnant-gas layer which would offer a resistance to molecular diffusion equal to the experimentally observed resistance, and R = the gas constant. [Nernst, *Z. Phys. Chem.*, **47**, 52 (1904); Whitman, *Chem. Mat. Eng.*, **29**, 149 (1923), and Lewis and Whitman, *Ind. Eng. Chem.*, **16**, 1215 (1924)].

The film thickness δ_G depends primarily on the hydrodynamics of the system and hence on the Reynolds number and the Schmidt number. Thus, various correlations have been developed for different geometries in terms of the following dimensionless variables:

$$N_{Sh} = \hat{k}_G RTd/D_{AB}p_T = f(N_{Re}, N_{Sc}) \qquad (5\text{-}286)$$

where N_{Sh} is the Sherwood number, N_{Re} (= Gd/μ_G) is the Reynolds number based on the characteristic length d appropriate to the geometry of the particular system; and N_{Sc} (= $\mu_G/\rho_G D_{AB}$) is the Schmidt number.

According to this analysis one can see that for gas-absorption problems, which often exhibit unidirectional diffusion, the most appropriate driving-force expression is of the form $(y - y_i)/y_{BM}$, and the most appropriate mass-transfer coefficient is therefore $\hat{k}_G$. This concept is to be found in all the key equations for the design of mass-transfer equipment.

The Sherwood-number relation for gas-phase mass-transfer coefficients as represented by the film diffusion model in Eq. (5-286) can be rearranged as follows:

$$N_{Sh} = (\hat{k}_G/G_M)N_{Re}N_{Sc} = N_{St}N_{Re}N_{Sc} = f(N_{Re}, N_{Sc}) \qquad (5\text{-}287)$$

where $N_{St} = \hat{k}_G/G_M = k'_G p_{BM}/G_M$ is known as the Stanton number. This equation can now be stated in the alternative functional forms

$$N_{St} = \hat{k}_G/G_M = g(N_{Re}, N_{Sc}) \qquad (5\text{-}288)$$

$$j_D = N_{St}\cdot N_{Sc}^{2/3} \qquad (5\text{-}289)$$

where j is the Chilton-Colburn "j factor" for mass transfer (discussed later).

The important point to note here is that the gas-phase mass-transfer coefficient $\hat{k}_G$ depends principally upon the transport properties of the fluid (N_{Sc}) and the hydrodynamics of the particular system involved (N_{Re}). It also is important to recognize that specific mass-transfer correlations can be derived only in conjunction with the investigator's particular assumptions concerning the numerical values of the effective interfacial area a of the packing.

The stagnant-film model discussed previously assumes a steady state in which the local flux across each element of area is constant; i.e., there is no accumulation of the diffusing species within the film. Higbie [*Trans. Am. Inst. Chem. Eng.*, **31**, 365 (1935)] pointed out that industrial contactors often operate with repeated brief contacts between phases in which the contact times are too short for the steady state to be achieved. For example, Higbie advanced the theory that in a packed tower the liquid flows across each packing piece in laminar flow and is remixed at the points of discontinuity between the packing elements. Thus, a fresh liquid surface is formed at the top of each piece, and as it moves downward, it absorbs gas at a decreasing rate until it is mixed at the next discontinuity. This is the basis of penetration theory.

If the velocity of the flowing stream is uniform over a very deep region of liquid (total thickness, $\delta_T \gg \sqrt{Dt}$), the time-averaged mass-transfer coefficient according to penetration theory is given by

$$k'_L = 2\sqrt{D_L/\pi t} \qquad (5\text{-}290)$$

where k'_L = liquid-phase mass-transfer coefficient, D_L = liquid-phase diffusion coefficient, and t = contact time.

In practice, the contact time t is not known except in special cases in which the hydrodynamics are clearly defined. This is somewhat similar to the case of the stagnant-film theory in which the unknown quantity is the thickness of the stagnant layer δ (in film theory, the liquid-phase mass-transfer coefficient is given by $k'_L = D_L/\delta$).

The penetration theory predicts that k'_L should vary by the square root of the molecular diffusivity, as compared with film theory, which predicts a first-power dependency on D. Various investigators have reported experimental powers of D ranging from 0.5 to 0.75, and the Chilton-Colburn analogy suggests a ⅔ power.

Penetration theory often is used in analyzing absorption with chemical reaction because it makes no assumption about the depths of penetration of the various reacting species, and it gives a more accurate result when the diffusion coefficients of the reacting species are not equal. When the reaction process is very complex, however, penetration theory is more difficult to use than film theory, and the latter method normally is preferred.

Danckwerts [*Ind. Eng. Chem.*, **42**, 1460 (1951)] proposed an extension of the penetration theory, called the surface renewal theory, which allows for the eddy motion in the liquid to bring masses of fresh liquid continually from the interior to the surface, where they are exposed to the gas for finite lengths of time before being replaced. In his development, Danckwerts assumed that every element of fluid has an equal chance of being replaced regardless of its age. The Danckwerts model gives

$$k'_L = \sqrt{Ds} \qquad (5\text{-}291)$$

where s = fractional rate of surface renewal.

Note that both the penetration and the surface-renewal theories predict a square-root dependency on D. Also, it should be recognized that values of the surface-renewal rate s generally are not available, which presents the same problems as do δ and t in the film and penetration models.

The predictions of correlations based on the film model often are nearly identical to predictions based on the penetration and surface-renewal models. Thus, in view of its relative simplicity, the film model normally is preferred for purposes of discussion or calculation. It should be noted that none of these theoretical models has proved adequate for making a priori predictions of mass-transfer rates in packed towers, and therefore empirical correlations such as those outlined later in Table 5-28. must be employed.

Mass-Transfer Correlations   Because of the tremendous importance of mass transfer in chemical engineering, a very large number of studies have determined mass-transfer coefficients both empirically and theoretically. Some of these studies are summarized in Tables 5-21 to 5-28. Each table is for a specific geometry or type of contactor, starting with flat plates, which have the simplest geometry (Table 5-21); then wetted wall columns (Table 5-22); flow in pipes and ducts (Table 5-23); submerged objects (Table 5-24); drops and bubbles (Table 5-25); agitated systems (Table 5-26); packed beds of particles for adsorption, ion exchange, and chemical reaction (Table 5-27); and finishing with packed bed two-phase contactors for distillation, absorption and other unit operations (Table 5-28). Graphical correlations for the Bolles and Fair correlation (Table 5-28-G) are in Figs. 5-28 to 5-30. Although extensive, these tables are not meant to be encyclopedic. For simple geometries, one may be able to determine a theoretical (T) form of the mass-transfer correlation. For very complex geometries, only an empirical (E) form can be found. In systems of intermediate complexity, semiempirical (S) correlations where the form is determined from theory and the coefficients from experiment are often useful. Although the major limitations and constraints in use are usually included in the tables, obviously many details cannot be included in this summary form. Readers are strongly encouraged to check the references before using the correlations in important situations. Note that even authoritative sources occasionally have typographical errors in the fairly complex correlation equations. Thus, it is a good idea to check several sources, including the original paper. The references will often include figures comparing the correlations with data. These figures are very useful since they provide a visual picture of the scatter in the data.

Since there are often several correlations that are applicable, how does one choose the correlation to use? First, the engineer must determine which correlations are closest to the current situation. This

TABLE 5-21   Mass Transfer Correlations for a Single Flat Plate or Disk—Transfer to or from Plate to Fluid

Situation	Correlation	Comments E = Empirical, S = Semiempirical, T = Theoretical	References*
A. Laminar, local, flat plate, forced flow	$N_{Sh,x} = \dfrac{k'x}{D} = 0.323(N_{Re,x})^{1/2}(N_{Sc})^{1/3}$ Coefficient 0.332 is a better fit.	[T] Low M.T. rates. Low mass-flux, constant property systems. $N_{Sh,x}$ is local k. Use with arithmetic difference in concentration. Coefficient 0.323 is Blasius' approximate solution. $N_{Re,x} = \dfrac{xu_\infty\rho}{\mu}$, x = length along plate	[100] p. 183 [108] p. 526 [146] p. 79 [150] p. 518 [151] p. 110
Laminar, average, flat plate, forced flow	$N_{Sh,avg} = \dfrac{k'_m L}{D} = 0.646(N_{Re,L})^{1/2}(N_{Sc})^{1/3}$ k'_m is mean mass-transfer coefficient for dilute systems.	$N_{Re,L} = \dfrac{Lu_\infty\rho}{\mu}$, 0.664 (Polhausen) is a better fit for $N_{Sc} > 0.6$, $N_{Re,x} < 3 \times 10^5$.	
j-factors	$j_D = j_H = \dfrac{f}{2} = 0.664(N_{Re,L})^{-1/2}$	[S] Analogy. $N_{Sc} = 1.0$, f = drag coefficient. j_D is defined in terms of k'_m.	[151] p. 271
B. Laminar, local, flat plate, blowing or suction and forced flow	$N_{Sh,x} = \dfrac{k'x}{D} = (\text{Slope})_{y=0}\,(N_{Re,x})^{1/2}(N_{Sc})^{1/3}$	[T] Blowing is positive. Other conditions as above. $\begin{array}{l}\dfrac{u_o}{u_\infty}\sqrt{N_{Re,x}} \quad 0.6 \;\; 0.5 \;\; 0.25 \; 0.0 \;\; -2.5 \\ \hline (\text{Slope})_{y=0} \;\; 0.01 \; 0.06 \; 0.17 \; 0.332 \; 1.64\end{array}$	[100] p. 185 [150] p. 271
C. Laminar, local, flat plate, natural convection vertical plate	$N_{Sh,x} = \dfrac{k'x}{D} = 0.508 N_{Sc}^{1/2}(0.952 + N_{Sc})^{-1/4} N_{Gr}^{1/4}$	[T] Low MT rates. Dilute systems, $\Delta\rho/\rho \ll 1$. $N_{Gr}N_{Sc} < 10^8$. Use with arithmetic concentration difference. x = length from plate bottom. $N_{Gr} = \dfrac{gx^3}{(\mu/\rho)^2}\left(\dfrac{\rho_\infty}{\rho_0} - 1\right)$	[151] p. 120
D. Laminar, stationary disk	$N_{Sh} = \dfrac{k'd_{\text{disk}}}{D} = \dfrac{8}{\pi}$	[T] Stagnant fluid. Use arithmetic concentration difference.	[146] p. 240
Laminar, spinning disk	$N_{Sh} = \dfrac{k'd_{\text{disk}}}{D} = 0.879 N_{Re}^{1/2} N_{Sc}^{1/3}$	[T] Asymptotic solution for large N_{Sc}. $N_{Re} < \sim 10^4$ $u = \omega d_{\text{disk/2}}$, ω = rotational speed, rad/s. Rotating disks are often used in electrochemical research.	[117] p. 60 [146] p. 240
E. Laminar, inclined, plate	$N_{Sh,avg} = 0.783 N_{Re,\text{film}}^{1/9} N_{Sc}^{1/3}\left(\dfrac{x^3\rho^2 g\sin\alpha}{\mu^2}\right)^{2/9}$ $N_{Sh,avg} = \dfrac{k'_m x}{D}$ $\delta_{\text{film}} = \left(\dfrac{3\mu Q}{w\rho g\sin\alpha}\right)^{1/3}$ = film thickness	[T] Constant-property liquid film with low mass-transfer rates. Use arithmetic concentration difference. $N_{Re,\text{film}} = \dfrac{4Q\rho}{\mu^2} < 2000$ w = width of plate, δ_f = film thickness, α = angle of inclination, x = distance from start soluble surface. Newtonian fluid. Solute does not penetrate past region of linear velocity profile. Differences between theory and experiment.	[151] p. 130 [146] p. 209
F. Turbulent, local flat plate, forced flow	$N_{Sh,x} = \dfrac{k'x}{D} = 0.0292 N_{Re,x}^{0.8}$	[S] Low mass-flux with constant property system. Use with arithmetic concentration difference. $N_{Sc} = 1.0$, $N_{Re,x} > 10^5$	[100] p. 191 [146] p. 201 [151] p. 221
Turbulent, average, flat plate, forced flow	$N_{Sh,avg} = \dfrac{k'L}{D} = 0.0365 N_{Re,L}^{0.8}$, average coefficient	Based on Prandtl's 1/7-power velocity law, $\dfrac{u}{u_\infty} = \left(\dfrac{y}{\delta}\right)^{1/7}$	
G. Laminar and turbulent, flat plate, forced flow	$j_D = j_H = \dfrac{f}{2} = 0.037 N_{Re,L}^{-0.2}$	Chilton-Colburn analogies, $N_{Sc} = 1.0$, (gases), f = drag coefficient. Corresponds to item 5-21-F and refers to same conditions. $8000 < N_{Re} < 300,000$. Can apply analogy, $j_D = f/2$, to entire plate (including laminar portion) if average values are used.	[100] p. 193 [109] p. 112 [146] p. 201 [151] p. 271

TABLE 5-21   Mass Transfer Correlations for a Single Flat Plate or Disk—Transfer to or from Plate to Fluid (*Concluded*)

Situation	Correlation	Comments E = Empirical, S = Semiempirical, T = Theoretical	References[*]
H. Laminar and turbulent, flat plate, forced flow	$N_{Sh,avg} = 0.037 N_{Sc}^{1/3}(N_{Re,L}^{0.8} - 15{,}500)$ to $N_{Re,L} = 320{,}000$ $N_{Sh,avg} = 0.037 N_{Sc}^{1/3}$ $\times \left(N_{Re,L}^{0.8} - N_{Re,Cr}^{0.8} + \dfrac{0.664}{0.037} N_{Re,Cr}^{1/2} \right)$ in range 3×10^5 to 3×10^6.	[E] Use arithmetic concentration difference. $N_{Sh,avg} = \dfrac{k'_m L}{D}$, $N_{Sc} > 0.5$ Entrance effects are ignored. $N_{Re,Cr}$ is transition laminar to turbulent.	[109] p. 112 [146] p. 201
I. Turbulent, local flat plate, natural convection, vertical plate Turbulent, average, flat plate, natural convection, vertical plate	$N_{Sh,x} = \dfrac{k'x}{D} = 0.0299 N_{Gr}^{2/5} N_{Sc}^{7/15}$ $\times (1 + 0.494 N_{Sc}^{2/3})^{-2/5}$ $N_{Sh,avg} = 0.0249 N_{Gr}^{2/5} N_{Sc}^{7/15} \times (1 + 0.494 N_{Sc}^{2/3})^{-2/5}$	[S] Low solute concentration and low transfer rates. Use arithmetic concentration difference. $N_{Gr} = \dfrac{gx^3}{(\mu/\rho)^2}\left(\dfrac{\rho_\infty}{\rho_0} - 1 \right)$ $N_{Gr} > 10^{10}$ Assumes laminar boundary layer is small fraction of total. $N_{Sh,avg} = \dfrac{k'_m L}{D}$	[151] p. 225
J. Turbulent, vertical plate	$N_{Sh,avg} = \dfrac{k'_m x}{D} = 0.327 N_{Re,film}^{2/9} N_{Sc}^{1/3} \left(\dfrac{x^3 \rho^2 g}{\mu^2} \right)^{2/9}$ $\delta_{film} = 0.172 \left(\dfrac{Q^2}{w^2 g} \right)^{1/3}$	[E] See 5-21-E for terms. $N_{Re,film} = \dfrac{4Q\rho}{w\mu^2} > 2360$ Solute remains in laminar sublayer.	[151] p. 229
K. Turbulent, spinning disk	$N_{Sh} = \dfrac{k' d_{disk}}{D} = 5.6 N_{Re}^{1.1} N_{Sc}^{1/3}$	[E] Use arithmetic concentration difference. $6 \times 10^5 < N_{Re} < 2 \times 10^6$ $120 < N_{Sc} < 1200$ $u = \omega d_{disk}/2$ where ω = rotational speed, radians/s. $N_{Re} = \rho \omega d^2/2\mu$.	[82] [146] p. 241
L. Mass transfer to a flat plate membrane in a stirred vessel	$N_{Sh} = \dfrac{k' d_{tank}}{D} = a N_{Re}^b N_{Sc}^c$ a depends on system. $a = 0.0443$ [73, 165]; b is often 0.65–0.70 [110]. If $N_{Re} = \dfrac{\omega d_{tank}^2 \rho}{\mu}$ $b = 0.785$ [73]. c is often 0.33 but other values have been reported [110].	[E] Use arithmetic concentration difference. ω = stirrer speed, radians/s. Useful for laboratory dialysis, R.O., U.F., and microfiltration systems.	[73] [110] p. 965 [165] p. 738

[*]See pages 5-7 and 5-8 for references.

involves recognizing the similarity of geometries, which is often challenging, and checking that the range of parameters in the correlation is appropriate. For example, the Bravo, Rocha, and Fair correlation for distillation with structured packings with triangular cross-sectional channels (Table 5-28-H) uses the Johnstone and Pigford correlation for rectification in vertical wetted wall columns (Table 5-22-D). Recognizing that this latter correlation pertains to a rather different application and geometry was a nontrivial step in the process of developing a correlation. If several correlations appear to be applicable, check to see if the correlations have been compared to each other and to the data. When a detailed comparison of correlations is not available, the following heuristics may be useful:

1.   Mass-transfer coefficients are derived from models. They must be employed in a similar model. For example, if an arithmetic concentration difference was used to determine k, that k should only be used in a mass-transfer expression with an arithmetic concentration difference.

2.   Semiempirical correlations are often preferred to purely empirical or purely theoretical correlations. Purely empirical correlations are dangerous to use for extrapolation. Purely theoretical correlations may predict trends accurately, but they can be several orders of magnitude off in the value of k.

3.   Correlations with broader data bases are often preferred.

4.   The analogy between heat and mass transfer holds over wider ranges than the analogy between mass and momentum transfer. Good heat transfer data (without radiation) can often be used to predict mass-transfer coefficients.

5.   More recent data is often preferred to older data, since end effects are better understood, the new correlation often builds on earlier data and analysis, and better measurement techniques are often available.

6.   With complicated geometries, the product of the interfacial area per volume and the mass-transfer coefficient is required. Correlations of ka_p or of HTU are more accurate than individual correlations of k and a_p since the measurements are simpler to determine the product ka_p or HTU.

7.   Finally, if a mass-transfer coefficient looks too good to be true, it probably is incorrect.

To determine the mass-transfer rate, one needs the interfacial area in addition to the mass-transfer coefficient. For the simpler geometries, determining the interfacial area is straightforward. For packed beds of particles a, the interfacial area per volume can be estimated as shown in Table 5-27-A. For packed beds in distillation, absorption, and so on in Table 5-28, the interfacial area per volume is included with the mass-transfer coefficient in the correlations for HTU. For agitated liquid-liquid systems, the interfacial area can be estimated

TABLE 5-22   Mass Transfer Correlations for Falling Films with a Free Surface in Wetted Wall Columns—Transfer between Gas and Liquid

Situation	Correlation	Comments E = Empirical, S = Semiempirical, T = Theoretical	References*
A. Laminar, vertical wetted wall column	$N_{Sh,avg} = \dfrac{k'_m x}{D} \approx 3.41 \dfrac{x}{\delta_{film}}$ (first term of infinite series) $\delta_{film} = \left(\dfrac{3\mu Q}{w\rho g}\right)^{1/3}$ = film thickness w = film width (circumference in column)	[T] Low rates M.T. Use with log mean concentration difference. Parabolic velocity distribution in films. $N_{Re,film} = \dfrac{4Q\rho}{w\mu} < 20$ Derived for flat plates, used for tubes if $r_{tube}\left(\dfrac{\rho g}{2\sigma}\right)^{1/2} > 3.0.\ \sigma$ = surface tension If $N_{Re,film} > 20$, surface waves and rates increase. An approximate solution $D_{apparent}$ can be used. Ripples are suppressed with a wetting agent good to N_{Re} = 1200.	[146] p. 78 [151] p. 137 [161] p. 50
B. Turbulent, vertical wetted wall column	$N_{Sh,avg} = \dfrac{k'_m d_t}{D} = 0.023 N_{Re}^{0.83} N_{Sc}^{0.44}$ A coefficient 0.0163 has also been reported using $N_{Re'}$, where $v = v$ of gas relative to liquid film.	[E] Use with log mean concentration difference for correlations in B and C. N_{Re} is for gas. N_{Sc} for vapor in gas. $2000 < N_{Re} \le 35,000, 0.6 \le N_{Sc} \le 2.5$. Use for gases, d_t = tube diameter.	[88] p. 266 [93] [100] p.181 [146] p. 211 [151] p. 265 [159] p. 212 [161] p. 71
C. Turbulent, vertical wetted wall column with ripples	$N_{Sh,avg} = \dfrac{k'_m d_t}{D} = 0.00814 N_{Re}^{0.83} N_{Sc}^{0.44}\left(\dfrac{4Q\rho}{w\mu}\right)^{0.15}$ $N_{Sh,avg} = \dfrac{k'_m d_t}{D} = 0.023 N_{Re}^{0.8} N_{Sc}^{1/3}$	[E] For gas systems with rippling. Fits B for $\left(\dfrac{4Q\rho}{w\mu}\right) = 1000$ $30 \le \left(\dfrac{4Q\rho}{w\mu}\right) < 1200$ [E] "Rounded" approximation to include ripples. Includes solid-liquid mass-transfer data to find ⅓ coefficient on N_{Sc}. May use $N_{Re}^{0.83}$. Use for liquids. See also Table 5-23.	[88] p. 266 [106] [146] p. 213
D. Rectification in vertical wetted wall column with turbulent vapor flow, Johnstone and Pigford correlation	$N_{Sh,avg} = \dfrac{k'_G d_{col} p_{BM}}{D_v p} = 0.0328(N'_{Re})^{0.77} N_{Sc}^{0.33}$ $3000 < N'_{Re} < 40,000,\ 0.5 < N_{Sc} < 3$ $N'_{Re} = \dfrac{d_{col} v_{rel}\rho_v}{\mu_v},\ v_{rel}$ = gas velocity relative to liquid film = $\dfrac{3}{2} u_{avg}$ in film	[E] Use logarithmic mean driving force at two ends of column. Based on four systems with gas-side resistance only. p_{BM} = logarithmic mean partial pressure of nondiffusing species B in binary mixture. p = total pressure Modified form is used for structured packings (See Table 5-28-H).	[105] [146] p. 214

*See pages 5-7 and 5-8 for references.

from the dispersed phase holdup and mean drop size correlations. Godfrey, Obi, and Reeve [*Chem. Engr. Prog.* **85,** 61 (Dec. 1989)] summarize these correlations. For many systems, $\bar{d}_{drop}/d_{imp}$ = (const)$N_{We}^{-0.6}$ where $N_{We} = \rho_c N^2 d_{imp}^3/\sigma$.

Effects of Total Pressure on $\hat{k}_G$ and $\hat{k}_L$ The influence of total system pressure on the rate of mass transfer from a gas to a liquid or to a solid has been shown to be the same as would be predicted from stagnant-film theory as defined in Eq. (5-285), where

$$\hat{k}_G = D_{AB} p_T / RT\, \delta_G \qquad (5\text{-}292)$$

Since the quantity $D_{AB} p_T$ is known to be relatively independent of the pressure, it follows that the rate coefficients $\hat{k}_G$, $k_G y_{BM}$, and $k'_G p_T y_{BM}$ ($= k'_G p_{BM}$) do not depend on the total pressure of the system, subject to the limitations discussed later.

Investigators of tower packings normally report $k'_G a$ values measured at very low inlet-gas concentrations, so that $y_{BM} = 1$, and at total pressures close to 100 kPa (1 atm). Thus, the correct rate coefficient for use in packed-tower designs involving the use of the driving force $(y - y_i)/y_{BM}$ is obtained by multiplying the reported $k'_G a$ values by the value of p_T employed in the actual test unit (e.g., 100 kPa) and *not* the total pressure of the system to be designed.

From another point of view one can correct the reported values of $k'_G a$ in kmol/[(s·m³)(kPa)], valid for a pressure of 101.3 kPa (1 atm), to some other pressure by dividing the quoted values of $k'_G a$ by the design pressure and multiplying by 101.3 kPa, i.e., ($k'_G a$ at design pressure p_T) = ($k'_G a$ at 1 atm) × 101.3/p_T.

One way to avoid a lot of confusion on this point is to convert the experimentally measured $k'_G a$ values to values of $\hat{k}_G a$ straightaway, before beginning the design calculations. A design based on the rate coefficient $\hat{k}_G a$ and the driving force $(y - y_i)/y_{BM}$ will be independent of the total system pressure with the following limitations: caution should be employed in assuming that $\hat{k}_G a$ is independent of total pressure for systems having significant vapor-phase nonidealities, for systems that operate in the vicinity of the critical point, or for total pressures higher than about 3040 to 4050 kPa (30 to 40 atm).

Experimental confirmations of the relative independence of $\hat{k}_G$ with respect to total pressure have been widely reported. Deviations do occur at extreme conditions. For example, Bretsznajder (*Prediction of Transport and Other Physical Properties of Fluids,* Pergamon Press, Oxford, 1971, p. 343) discusses the effects of pressure on the $D_{AB} p_T$ product and presents experimental data on the self-diffusion of CO_2 which show that the D-p product begins to decrease at a pressure of

TABLE 5-23   Mass-Transfer Correlations for Flow in Pipes and Ducts—Transfer is from Wall to Fluid

Situation	Correlation	Comments E = Empirical, S = Semiempirical, T = Theoretical	References[a]
A. Tubes, laminar, fully developed parabolic velocity profile, developing concentration profile, constant wall concentration	$N_{Sh} = \dfrac{k'd_t}{D} = 3.66 + \dfrac{0.0668(d_t/x)N_{Re}N_{Sc}}{1 + 0.04[(d_t/x)N_{Re}N_{Sc}]^{2/3}}$	[T] Use log mean concentration difference. For $\dfrac{x/d_t}{N_{Re}N_{Sc}} < 0.10$, $N_{Re} < 2100$. x = distance from tube entrance. Good agreement with experiment at values $10^4 > \dfrac{\pi}{4}\dfrac{d_t}{x}N_{Re}N_{Sc} > 10$	[98] [100] p. 176 [108] p. 525 [151] p. 159
B. Tubes, fully developed concentration profile	$N_{Sh} = \dfrac{k'd_t}{D} = 3.66$	[T] Subset of 5-23-A for fully developed concentration profile. $\dfrac{x/d_t}{N_{Re}N_{Sc}} > 0.1$	[98] [151] p. 165
C. Tubes, approximate solution	$N_{Sh,x} = \dfrac{k'd_t}{D} = 1.077\left(\dfrac{d_t}{x}\right)^{1/3}(N_{Re}N_{Sc})^{1/3}$ $N_{Sh,avg} = \dfrac{k'd_t}{D} = 1.615\left(\dfrac{d_t}{L}\right)^{1/3}(N_{Re}N_{Sc})^{1/3}$	[T] For arithmetic concentration difference. $\dfrac{W}{\rho Dx} > 400$ Leveque's approximation: Concentration BL is thin. Assume velocity profile is linear. High mass velocity. Fits liquid data well.	[151] p. 166
D. Tubes, laminar, uniform plug velocity, developing concentration profile, constant wall concentration	$N_{Sh,avg} = \dfrac{1}{2}\dfrac{d_t}{L}N_{Re}N_{Sc}\left[\dfrac{1-4\displaystyle\sum_{j=1}^{\infty}a_j^{-2}\exp\left(\dfrac{-2a_j^2(x/r_t)}{N_{Re}N_{Sc}}\right)}{1+4\displaystyle\sum_{j=1}^{\infty}a_j^{-2}\exp\left(\dfrac{-2a_j^2(x/r_t)}{N_{Re}N_{Sc}}\right)}\right]$ Graetz solution for heat transfer written for M.T.	[T] Use arithmetic concentration difference. Fits gas data well, for $\dfrac{W}{D\rho x} < 50$ (fit is fortuitous). $N_{Sh,avg} = (k'_m d_t)/D.$ $a_1 = 2.405$, $a_2 = 5.520$, $a_3 = 8.654$, $a_4 = 11.792$, $a_5 = 14.931$. Graphical solutions are in references.	[91] p. 443 [120] [151] p. 150
E. Laminar, fully developed parabolic velocity profile, constant mass flux at wall	$N_{Sh,x} = \left[\dfrac{11}{48} - \dfrac{1}{2}\displaystyle\sum_{j=1}^{\infty}\dfrac{\exp\left[-\lambda_j^2(x/r_t)/(N_{Re}N_{Sc})\right]}{C_j\lambda_j^4}\right]^{-1}$ | j	λ_j^2	c_j
1	25.68	7.630×10^{-3}
2	83.86	2.058×10^{-3}
3	174.2	0.901×10^{-3}
4	296.5	0.487×10^{-3}
5	450.9	0.297×10^{-3}		[T] Use log mean concentration difference. $N_{Re} < 2100$ $N_{Sh,x} = \dfrac{k'd_t}{D}$ $N_{Re} = \dfrac{vd_t\rho}{\mu}$	[149] [151] p. 167
F. Laminar, alternate	$N_{Sh} = 4.36 + \dfrac{0.023(d_t/L)N_{Re}N_{Sc}}{1 + 0.0012(d_t/L)N_{Re}N_{Sc}}$	[T] $N_{sh} = \dfrac{k'd_t}{D}$ Use log mean concentration difference. $N_{Re} < 2100$	[98] [100] p. 176
G. Laminar, fully developed concentration and velocity profile	$N_{Sh} = \dfrac{k'd_t}{D} = \dfrac{48}{11} = 4.3636$	[T] Use log mean concentration difference. $N_{Re} < 2100$	[98] [151] p. 167
H. Vertical tubes, laminar flow, forced and natural convection	$N_{Sh,avg} = 1.62N_{Gz}^{1/3}\left[1 \pm 0.0742\dfrac{(N_{Gr}N_{Sc}d/L)^{3/4}}{N_{Gz}}\right]^{1/3}$	[T] Approximate solution. Use minus sign if forced and natural convection oppose each other. $N_{Gz} = \dfrac{N_{Re}N_{Sc}d}{L}$ $N_{Gr} = \dfrac{g\Delta\rho d^3}{\rho v^2}$ Good agreement with experiment.	[140]
I. Tubes, laminar, RO systems	$N_{Sh,avg} = \dfrac{k'_m d_t}{D} = 1.632\left(\dfrac{ud_t^2}{DL}\right)^{1/3}$	Use arithmetic concentration difference. Thin concentration polarization layer, not fully developed. $N_{Re} < 2000$, L = length tube.	[73] [165] p. 738
J. Tubes and parallel plates, laminar RO	Graphical solutions for concentration polarization. Uniform velocity through walls.	[T]	[145] [165] p. 762

TABLE 5-23   Mass-Transfer Correlations for Flow in Pipes and Ducts—Transfer is from Wall to Fluid (*Continued*)

Situation	Correlation	Comments E = Empirical, S = Semiempirical, T = Theoretical	References°
K. Parallel plates, laminar, parabolic velocity, developing concentration profile, constant wall concentration	Graphical solution	[T] Low transfer rates.	[151] p. 176
L. 5-23-K, fully developed	$N_{Sh} = \dfrac{k'(2h)}{D} = 7.6$	[T] h = distance between plates. Use log mean concentration difference. $\dfrac{N_{Re}N_{Sc}}{x/(2h)} < 20$	[151] p. 177
M. Parallel plates, laminar, parabolic velocity, developing concentration profile, constant mass flux at wall	Graphical solution	[T] Low transfer rates.	[151] p. 176
N. 5-23-M, fully developed	$N_{Sh} = \dfrac{k'(2h)}{D} = 8.23$	[T] Use log mean concentration difference. $\dfrac{N_{Re}N_{Sc}}{x/(2h)} < 20$	[151] p. 177
O. Laminar flow, vertical parallel plates, forced and natural convection	$N_{Sh,avg} = 1.47 N_{Gz}^{1/3}\left[1 \pm 0.0989\dfrac{(N_{Gr}N_{Sc}\,h/L)^{3/4}}{N_{Gz}}\right]^{1/3}$	[T] Approximate solution. Use minus sign if forced and natural convection oppose each other. $N_{Gz} = \dfrac{N_{Re}N_{Sc}h}{L}$ $N_{Gr} = \dfrac{g\Delta\rho h^3}{\rho v^2}$ Good agreement with experiment.	[140]
P. Parallel plates, laminar, RO systems	$N_{Sh,avg} = \dfrac{k'(2H_p)}{D} = 2.354\left(\dfrac{uH_p^2}{DL}\right)^{1/3}$	Thin concentration polarization layer. Short tubes, concentration profile not fully developed. Use arithmetic concentration difference.	[73] [165] p. 738
Q. Tubes, turbulent	$N_{Sh,avg} = \dfrac{k'_m d_t}{D} = 0.023 N_{Re}^{0.83} N_{Sc}^{1/3}$	[E] Use with log mean concentration difference at two ends of tube. $2100 < N_{Re} < 35{,}000$ $0.6 < N_{Sc} < 3000$ From wetted wall column and dissolution data— see Table 5-22-B. Good fit for liquids.	[88] p. 266 [100] p. 181 [120] [161] p. 72
R. Tubes, turbulent	$N_{Sh,avg} = \dfrac{k'_m d_t}{D} = 0.023 N_{Re}^{0.83} N_{Sc}^{0.44}$	[E] Evaporation of liquids. Use with log mean concentration difference. See item above. Better fit for gases. $2000 < N_{Re} < 35{,}000$ $0.6 < N_{Sc} < 2.5$.	[93][100] p. 181 [109] p. 112 [146] p. 211
S. Tubes, turbulent	$N_{Sh} = \dfrac{k' d_t}{D} = 0.0096 N_{Re}^{0.913} N_{Sc}^{0.346}$	[E] $430 < N_{Sc} < 100{,}000$. Dissolution data. Use for high N_{Sc}.	[122] p. 668
T. Tubes, turbulent, smooth tubes, Reynolds analogy	$N_{Sh} = \dfrac{k' d_t}{D} = \left(\dfrac{f}{2}\right)N_{Re}N_{Sc}$ f = Fanning friction faction	[T] Use arithmetic concentration difference. N_{Sc} near 1.0 Turbulent core extends to wall. Of limited utility.	[91] p. 438 [100] p. 171 [151] p. 239 [159] p. 250
U. Tubes, turbulent, smooth tubes, Chilton-Colburn analogy	$j_D = j_H = \dfrac{f}{2}$ If $\dfrac{f}{2} = 0.023 N_{Re}^{-0.2}$, $j_D = \dfrac{N_{Sh}}{N_{Re}N_{Sc}^{1/3}} = 0.023 N_{Re}^{-0.2}$ $N_{Sh} = \dfrac{k' d_t}{D}$ $j_D = j_H = f(N_{Re},\text{ geometry and B.C.})$	[T] Use log-mean concentration difference. Relating j_D to $f/2$ approximate. N_{Pr} and N_{Sc} near 1.0. Low concentration. Results about 20% lower than experiment. $3\times10^4 < N_{Re} < 10^6$ [E] Good over wide ranges.	[72] p. 400, 647 [80][88] p. 269 [151] p. 264 [159] p. 251 [72] p. 647 [80]
V. Tubes, turbulent, smooth tubes, constant surface concentration, Prandtl analogy	$N_{Sh} = \dfrac{k' d_t}{D} = \dfrac{(f/2)N_{Re}N_{Sc}}{1 + 5\sqrt{f/2}(N_{Sc} - 1)}$ $\dfrac{f}{2} = 0.04 N_{Re}^{-0.25}$	[T] Use arithmetic concentration difference. Improvement over Reynolds analogy. Best for N_{Sc} near 1.0.	[100] p. 173 [132] [151] p. 241

°See pages 5-7 and 5-8 for references.

TABLE 5-23   Mass-Transfer Correlations for Flow in Pipes and Ducts—Transfer is from Wall to Fluid (*Concluded*)

Situation	Correlation	Comments E = Empirical, S = Semiempirical, T = Theoretical	References[*]
W. Tubes, turbulent, smooth tubes, Constant surface concentration, Von Karman analogy	$$N_{Sh} = \frac{(f/2)N_{Re}N_{Sc}}{1 + 5\sqrt{f/2}\left\{(N_{Sc}-1) + \ln\left[1 + \frac{5}{6}(N_{Sc}-1)\right]\right\}}$$ $$\frac{f}{2} = 0.04 N_{Re}^{-0.25}$$	[T] Use arithmetic concentration difference. $N_{Sh} = k'd_t/D$. Improvement over Prandtl, $N_{Sc} < 25$.	[100] p. 173 [151] p. 243 [159] p. 250 [162]
X. Tubes, turbulent, smooth tubes, constant surface concentration	For $0.5 < N_{Sc} < 10$: $$N_{Sh,avg} = 0.0097 N_{Re}^{9/10} N_{Sc}^{1/2}$$ $$\times (1.10 + 0.44 N_{Sc}^{-1/3} - 0.70 N_{Sc}^{-1/6})$$ For $10 < N_{Sc} < 1000$: $N_{Sh,avg}$ $$= \frac{0.0097 N_{Re}^{9/10} N_{Sc}^{1/2}(1.10 + 0.44 N_{Sc}^{-1/3} - 0.70 N_{Sc}^{-1/6})}{1 + 0.064 N_{Sc}^{1/2}(1.10 + 0.44 N_{Sc}^{-1/3} - 0.70 N_{Sc}^{-1/6})}$$ For $N_{Sc} > 1000$: $$N_{Sh,avg} = 0.0102 N_{Re}^{9/10} N_{Sc}^{1/3}$$	[S] Use arithmetic concentration difference. Based on partial fluid renewal and an infrequently replenished thin fluid layer for high N_{Sc}. Good fit to available data. $N_{Re} = \dfrac{u_{bulk}\, d_t}{\nu}$ $N_{Sh,avg} = \dfrac{k'_{avg}\, d_t}{D}$	[100] p. 179 [131]
Y. Turbulent flow, tubes	$$N_{St} = \frac{N_{Sh}}{N_{Pe}} = \frac{N_{Sh}}{N_{Re}N_{Sc}} = 0.0149 N_{Re}^{-0.12} N_{Sc}^{-2/3}$$	[E] Smooth pipe data. Data fits within 4% except at $N_{Sc} > 20{,}000$, where experimental data is underpredicted. $N_{Sc} > 100$, $10^5 > N_{Re} > 2100$	[124]
Z. Turbulent flow, noncircular ducts	Use correlations with $$d_{eq} = \frac{4 \text{ cross-sectional area}}{\text{wetted perimeter}}$$ Parallel plates: $$d_{eq} = 4\,\frac{2hw}{2w + 2h}$$	Can be suspect for systems with sharp corners.	[151] p. 289 [165] p. 738

[*]See pages 5-7 and 5-8 for references.

approximately 8100 kPa (80 atm). For reduced temperatures higher than about 1.5, the deviations are relatively modest for pressures up to the critical pressure. However, deviations are large near the critical point (see also p. 5-49). The effect of pressure on the gas-phase viscosity also is negligible for pressures below about 5060 kPa (50 atm).

For the liquid-phase mass-transfer coefficient $\hat{k}_L$, the effects of total system pressure can be ignored for all practical purposes. Thus, when using $\hat{k}_G$ and $\hat{k}_L$ for the design of gas absorbers or strippers, the primary pressure effects to consider will be those which affect the equilibrium curves and the values of m. If the pressure changes affect the hydrodynamics, then $\hat{k}_G$, $\hat{k}_L$, and a can all change significantly.

Effects of Temperature on $\hat{k}_G$ and $\hat{k}_L$ The Stanton-number relationship for gas-phase mass transfer in packed beds,

$$N_{St} = \hat{k}_G/G_M = g(N_{Re}, N_{Sc}) \tag{5-293}$$

indicates that for a given system geometry the rate coefficient $\hat{k}_G$ depends only on the Reynolds number and the Schmidt number. Since the Schmidt number for a gas is independent of temperature, the principal effect of temperature upon $\hat{k}_G$ arises from changes in the gas viscosity with changes in temperature. For normally encountered temperature ranges, these effects will be small owing to the fractional powers involved in Reynolds-number terms (see Tables 5-21 to 5-28). It thus can be concluded that for all practical purposes $\hat{k}_G$ is independent of temperature and pressure in the normal ranges of these variables.

For modest changes in temperature the influence of temperature upon the interfacial area a may be neglected. For example, in experiments on the absorption of SO_2 in water, Whitney and Vivian [*Chem. Eng. Prog.*, **45**, 323 (1949)] found no appreciable effect of temperature upon $k'_G a$ over the range from 10 to 50°C.

With regard to the liquid-phase mass-transfer coefficient, Whitney and Vivian found that the effect of temperature upon $k_L a$ could be explained entirely by variations in the liquid-phase viscosity and diffusion coefficient with temperature. Similarly, the oxygen-desorption data of Sherwood and Holloway [*Trans. Am. Inst. Chem. Eng.*, **36**, 39 (1940)] show that the influence of temperature upon H_L can be explained by the effects of temperature upon the liquid-phase viscosity and diffusion coefficients.

It is important to recognize that the effects of temperature on the liquid-phase diffusion coefficients and viscosities can be very large and therefore must be carefully accounted for when using $\hat{k}_L$ or H_L data. For liquids the mass-transfer coefficient $\hat{k}_L$ is correlated in terms of design variables by relations of the form

$$N_{St} = \hat{k}_L/L_M = f(N_{Re}, N_{Sc}) \tag{5-294}$$

A general relation for H_L which may be used as the basis for applying temperature corrections is as follows:

$$H_L = b N_{Re}^a N_{Sc}^{1/2} \tag{5-295}$$

where b is a proportionality constant and the exponent a may range from about 0.2 to 0.5 for different packings and systems. The liquid-phase diffusion coefficients may be corrected from a base temperature T_1 to another temperature T_2 by using the Einstein relation as recommended by Wilke [*Chem. Eng. Prog.*, **45**, 218 (1949)]:

$$D_2 = D_1(T_2/T_1)(\mu_1/\mu_2) \tag{5-296}$$

The Einstein relation can be rearranged to the following equation for relating Schmidt numbers at two temperatures:

$$N_{Sc2} = N_{Sc1}(T_1/T_2)(\rho_1/\rho_2)(\mu_2/\mu_1)^2 \tag{5-297}$$

TABLE 5-24 Mass Transfer Correlations for Flow Past Submerged Objects

Situation	Correlation	Comments E = Empirical, S = Semiempirical, T = Theoretical	References[*]
A. Single sphere	$N_{Sh} = \dfrac{k'_G p_{BLM} RT d_s}{PD} = \dfrac{2r}{r - r_s}$ $\begin{array}{c\|ccccc} r/r_s & 2 & 5 & 10 & 50 & \infty \text{ (asymptotic limit)} \\ \hline N_{Sh} & 4.0 & 2.5 & 2.22 & 2.04 & 2.0 \end{array}$	[T] Use with log mean concentration difference. r = distance from sphere, r_s, d_s = radius and diameter of sphere. No convection.	[151] p. 18
B. Single sphere, creeping flow with forced convection	$N_{Sh} = \dfrac{k'd}{D} = [4.0 + 1.21(N_{Re}N_{Sc})^{2/3}]^{1/2}$	[T] Use with log mean concentration difference. Average over sphere. Numerical calculations. $(N_{Re}N_{Sc}) < 10{,}000\ N_{Re} < 1.0$. Constant sphere diameter. Low mass-transfer rates.	[78][109] p. 114 [122] p. 671 [146] p. 214
	$N_{Sh} = \dfrac{k'd}{D} = a(N_{Re}N_{Sc})^{1/3}$ $a = 1.01, 1.0,$ or 0.991	[T] Fit to above ignoring molecular diffusion. $1000 < (N_{Re}N_{Sc}) < 10{,}000$.	[117] p. 80 146 p. 215
C. Single spheres, molecular diffusion, and forced convection, low flow rates	$N_{Sh} = 2.0 + A N_{Re}^{1/2} N_{Sc}^{1/3}$ $A = 0.5$ to 0.62	[E] Use with log mean concentration difference. Average over sphere. Frössling Eq. ($A = 0.552$), $2 \le N_{Re} \le 800$, $0.6 \le N_{sc} \le 2.7$. N_{Sh} lower than experimental at high N_{Re}.	[72] [89] p. 409, 647 [100], p. 194 109 p. 114 151 p. 276
	$A = 0.60.$	[E] Ranz and Marshall $2 \le N_{Re} \le 200$, $0.6 \le N_{sc} \le 2.5$. See also Table 5-27-L.	[72] p. 409, 647 [135] [146] p. 217 [151] p. 276
	$A = 0.95.$	[E] Liquids $2 \le N_{Re} \le 2{,}000$. Graph in Ref. 146, p. 217–218.	[90][91] p. 446 [146] p. 217
	$A = 0.95.$	[E] $100 \le N_{Re} \le 700$; $1{,}200 \le N_{Sc} \le 1525$.	[139][151] p. 276
	$A = 0.544.$	[E] Use with arithmetic concentration difference. $N_{Sc} = 1$; $50 \le N_{Re} \le 350$.	[102][151] p. 276
D. 5-24-C	$N_{Sh} = \dfrac{k'd_s}{D} = 2.0 + 0.575 N_{Re}^{1/2} N_{Sc}^{0.35}$	[E] Use with log mean concentration difference. $N_{Sc} \le 1$, $N_{Re} < 1$.	[94][151] p. 276
E. 5-24-C	$N_{Sh} = \dfrac{k'd_s}{D} = 2.0 + 0.552 N_{Re}^{0.53} N_{Sc}^{1/3}$	[E] Use with log mean concentration difference. $1.0 < N_{Re} \le 48{,}000$ Gases: $0.6 \le N_{Sc} \le 2.7$.	[91] p. 446
F. Single spheres, forced concentration, any flow rate	$N_{Sh} = \dfrac{k'_L d_s}{D} = 2.0 + 0.59 \left[\dfrac{E^{1/3} d_p^{4/3} \rho}{\mu} \right]^{0.57} N_{Sc}^{1/3}$ Energy dissipation rate per unit mass of fluid (ranges $570 < N_{Sc} < 1420$): $E = \left(\dfrac{C_{Dr}}{2} \right) \left(\dfrac{v_r^3}{d_p} \right) \dfrac{m^2}{s^3}$ $2 < \left(\dfrac{E^{1/3} d_p^{4/3} \rho}{\mu} \right) < 63{,}000$	[S] Correlates large amount of data and compares to published data. v_r = relative velocity between fluid and sphere, m/s. C_{Dr} = drag coefficient for single particle fixed in fluid at velocity v_r. See 5-27-G for calculation details and other applications.	[125]
G. Single spheres, forced convection, high flow rates, ignoring molecular diffusion	$N_{Sh} = \dfrac{k'd_s}{D} = 0.347 N_{Re}^{0.62} N_{Sc}^{1/3}$	[E] Use with arithmetic concentration difference. Liquids, $2000 < N_{Re} < 17{,}000$. High N_{Sc}, graph in Ref. 146, p. 217–218.	[91] p. 446 [157] [146] p. 217
	$N_{Sh} = \dfrac{k'd_s}{D} = 0.33 N_{Re}^{0.6} N_{Sc}^{1/3}$	[E] $1500 \le N_{Re} \le 12{,}000$.	[151] p. 276
	$N_{Sh} = \dfrac{k'd_s}{D} = 0.43 N_{Re}^{0.56} N_{Sc}^{1/3}$	[E] $200 \le N_{Re} \le 4 \times 10^4$, "air" $\le N_{Sc} \le$ "water."	[151] p. 276
	$N_{Sh} = \dfrac{k'd_s}{D} = 0.692 N_{Re}^{0.514} N_{Sc}^{1/3}$	[E] $500 \le N_{Re} \le 5000$.	[128] [151] p. 276
H. Single cylinders, perpendicular flow	$N_{Sh} = \dfrac{k'd_s}{D} = A N_{Re}^{1/2} N_{Sc}^{1/3}, A = 0.82$	[E] $100 < N_{Re} \le 3500$, $N_{Sc} = 1560$.	[151] p. 276
	$A = 0.74$	[E] $120 \le N_{Re} \le 6000$, $N_{Sc} = 2.44$.	[151] p. 276 [152]
	$A = 0.582$	[E] $300 \le N_{Re} \le 7600$, $N_{Sc} = 1200$.	
	$j_D = 0.600(N_{Re})^{-0.487}$	[E] Use with arithmetic concentration difference.	[151] p. 276
	$N_{Sh} = \dfrac{k'd_{cyl}}{D}$	$50 \le N_{Re} \le 50{,}000$; gases, $0.6 \le N_{Sc} \le 2.6$; liquids; $1000 \le N_{Sc} \le 3000$. Data scatter $\pm 30\%$.	[91] p. 450

TABLE 5-24 Mass Transfer Correlations for Flow Past Submerged Objects (*Concluded*)

Situation	Correlation	Comments E = Empirical, S = Semiempirical, T = Theoretical	References*
I. 5-24-H	Can use $j_D = j_H$. Graphical correlation.	[E] Used with linear concentration difference.	[72] p. 408, 647
		$N_{Sh} = \dfrac{k' d_s}{D}$	[146] p. 236 [151] p. 273
J. Rotating cylinder in an infinite liquid, no forced flow	$j'_D = \dfrac{k'}{v} N_{Sc}^{0.644} = 0.0791 N_{Re}^{-0.30}$	[E] Used with arithmetic concentration difference.	[85]
	Results presented graphically to $N_{Re} = 241{,}000$.	$112 < N_{Re} \le 100{,}000$. $835 < N_{Sc} < 11490$	[146] p. 238
	$N_{Re} = \dfrac{v d_{cyl} \mu}{\rho}$ where $v = \dfrac{\omega d_{cyl}}{2}$ = peripheral velocity	k' = mass-transfer coefficient, cm/s; ω = rotational speed, radian/s. Useful geometry in electrochemical studies.	
K. Oblate spheroid, forced convection	$j_D = \dfrac{N_{Sh}}{N_{Re} N_{Sc}^{1/3}} = 0.74 N_{Re}^{-0.5}$	[E] Used with arithmetic concentration difference.	[151] p. 284
	$N_{Re} = \dfrac{d_{ch} v \rho}{\mu}, d_{ch} = \dfrac{\text{total surface area}}{\text{perimeter normal to flow}}$	$120 \le N_{Re} \le 6000$; standard deviation 2.1%. Eccentricities between 1:1 (spheres) and 3:1.	[152]
	e.g., for cube with side length a, $d_{ch} = 1.27a$.		
	$N_{Sh} = \dfrac{k' d_{ch}}{D}$	Shape is often approximated by drops.	[109] p. 115
L. Other objects, including prisms, cubes, hemispheres, spheres, and cylinders; forced convection	$j_D = 0.692 N_{Re,p}^{-0.486}, N_{Re,p} = \dfrac{v d_{ch} \rho}{\mu}$	[E] Used with arithmetic concentration difference.	
	Terms same as in 5-24-J.	$500 \le N_{Re,p} \le 5000$. Turbulent. Agrees with cylinder and oblate spheroid results, $\pm 15\%$. Assumes molecular diffusion and natural convection are negligible.	[127, 128] [151] p. 285
M. Other objects, molecular diffusion limits	$N_{Sh} = \dfrac{k' d_{ch}}{D} = A$	[T] Use with arithmetic concentration difference. Hard to reach limits in experiments.	[109] p. 114
	Spheres and cubes $A = 2$, tetrahedrons $A = 2\sqrt{6}$ octahedron $2\sqrt{2}$.		
N. Shell side of microporous hollow fiber module for solvent extraction	$N_{Sh} = \beta [d_h (1 - \varphi)/L] N_{Re}^{0.6} N_{Sc}^{0.33}$	[E] Use with logarithmic mean concentration difference.	[133]
	$N_{Sh} = \dfrac{\overline{K} d_h}{D}$		
	$\beta = 5.8$ for hydrophobic membrane.	$N_{Re} = \dfrac{d_h v \rho}{\mu}, \overline{K}$ = overall mass-transfer coefficient	
	$\beta = 6.1$ for hydrophilic membrane.	d_h = hydraulic diameter $= \dfrac{4 \times \text{cross-sectional area of flow}}{\text{wetted perimeter}}$ φ = packing fraction of shell side. L = module length. Based on area of contact according to inside or outside diameter of tubes depending on location of interface between aqueous and organic phases. Can also be applied to gas-liquid systems with liquid on shell side.	

See Table 5-27 for flow in packed beds.
*See pp. 5-7 and 5-8 for references.

Substitution of this relation into Eq. (5-295) shows that for a given geometry the effect of temperature on H_L can be estimated as

$$H_{L2} = H_{L1} (T_1/T_2)^{1/2} (\rho_1/\rho_2)^{1/2} (\mu_2/\mu_1)^{1-a} \qquad (5\text{-}298)$$

In using these relations it should be noted that for equal liquid flow rates

$$H_{L2}/H_{L1} = (\hat{k}_L a)_1 / (\hat{k}_L a)_2 \qquad (5\text{-}299)$$

Effects of System Physical Properties on $\hat{k}_G$ and $\hat{k}_L$ When designing packed towers for nonreacting gas-absorption systems for which no experimental data are available, it is necessary to make corrections for differences in composition between the existing test data and the system in question. For example, the test data of Fellinger for ammonia-water absorption on various packings are frequently used as a base (see Table 5-28-B). In these tests it is estimated that $H_G = 0.9 H_{OG}$, so that one may wish to use these data as the basis for estimating H_G or $k_G a$ values for other systems. This may be done by taking H_G proportional to $N_{Sc}^{0.5}$ and $k_G a$ proportional to $N_{Sc}^{-0.5}$, based on a value of N_{Sc} for NH_3-air of 0.66 at 25°C. The coefficient k_G varies as the diffusivity D_{AB} to the 0.5 power. It should be noted, however, that

TABLE 5-25 Mass-Transfer Correlations for Drops and Bubbles

Conditions	Correlations	Comments E = Empirical, S = Semiempirical, T = Theoretical	References[o]
A. Single liquid drop in immiscible liquid, drop formation, discontinuous (drop) phase coefficient	$\hat{k}_{d,f} = A \left(\dfrac{\rho_d}{M_d}\right)_{av}\left(\dfrac{D_d}{\pi t_f}\right)^{1/2}$ $A = \dfrac{24}{7}$ (penetration theory) $A = 1.31$ (semiempirical value) $A = \left[\dfrac{24}{7}(0.8624)\right]$ (extension by fresh surface elements)	[T,S] Use arithmetic mole fraction difference. Fits some, but not all, data. Low mass transfer rate. M_d = mean molecular weight of dispersed phase; t_f = formation time of drop. $k_{L,d}$ = mean dispersed liquid phase M.T. coefficient kmole/[s · m² (mole fraction)].	[151] p. 399
B. 5-25-A	$\hat{k}_{df} = 0.0432$ $\times \dfrac{d_p}{t_f}\left(\dfrac{\rho_d}{M_d}\right)_{av}\left(\dfrac{u_o}{d_p g}\right)^{0.089}\left(\dfrac{d_p^2}{t_f D_d}\right)^{-0.334}\left(\dfrac{\mu_d}{\sqrt{\rho_d d_p\,\sigma g_c}}\right)^{-0.601}$	[E] Use arithmetic mole fraction difference. Based on 23 data points for 3 systems. Average absolute deviation 26%. Use with surface area of drop after detachment occurs. u_o = velocity through nozzle; σ = interfacial tension.	[151] p. 401 [154] p. 434
C. Single liquid drop in immiscible liquid, drop formation, continuous phase coefficient	$\hat{k}_{cf} = 4.6\left(\dfrac{\rho_c}{M_c}\right)_{av}\sqrt{\dfrac{D_c}{\pi t_f}}$	[T] Use arithmetic mole fraction difference. Based on rate of bubble growth away from fixed orifice. Approximately three times too high compared to experiments.	[151] p. 402
D. 5-25-C	$k_{L,c} = 0.386$ $\times\left(\dfrac{\rho_c}{M_c}\right)_{av}\left(\dfrac{D_c}{t_f}\right)^{0.5}\left(\dfrac{\rho_c \sigma g_c}{\Delta\rho g t_f \mu_c}\right)^{0.407}\left(\dfrac{g t_f^2}{d_p}\right)^{0.148}$	[E] Average absolute deviation 11% for 20 data points for 3 systems.	[151] p. 402 [154] p. 434
E. Single liquid drop in immiscible liquid, free rise or fall, discontinuous phase coefficient, stagnant drops	$k_{L,d,m} = \dfrac{-d_p}{6t}\left(\dfrac{\rho_d}{M_d}\right)_{av}\ln\left\{\dfrac{6}{\pi^2}\displaystyle\sum_{j=1}^{\infty}\dfrac{1}{j^2}\exp\left[\left(\dfrac{-D_d j^2\pi^2 t}{(d_p/2)^2}\right)\right]\right\}$	[T] Use with log mean mole fraction differences based on ends of column. t = rise time. No continuous phase resistance. Stagnant drops are likely if drop is very viscous, quite small, or is coated with surface active agent. $k_{L,d,m}$ = mean dispersed liquid M.T. coefficient.	[151] p. 404 [154] p. 435
F. 5-25-E	$\hat{k}_{L,d,m} = \dfrac{-d_p}{6t}\left(\dfrac{\rho_d}{M_d}\right)_{av}\ln\left[1 - \dfrac{\pi D_d^{1/2} t^{1/2}}{d_p/2}\right]$	[S] See 5-25-E. Approximation for fractional extractions less than 50%.	[151] p. 404 [154] p. 435
G. 5-25-E, continuous phase coefficient, stagnant drops, spherical	$N_{Sh} = \dfrac{k_{L,c,m}d_c}{D_c} = 0.74\left(\dfrac{\rho_c}{M_c}\right)_{av}N_{Re}^{1/2}(N_{Sc})^{1/3}$	[E] $N_{Re} = \dfrac{v_s d_p \rho_c}{\mu_c}$, special case Eq. (5-254). v_s = slip velocity between drop and continuous phase.	[151] p. 407 [152][154] p. 436
H. 5-25-E, oblate spheroid	$N_{Sh} = \dfrac{k_{L,c,m}d_3}{D_c} = 0.74\left(\dfrac{\rho_c}{M_c}\right)_{av}(N_{Re,3})^{1/2}(N_{Sc,c})^{1/3}$ $N_{Re,3} = \dfrac{v_s d_3 \rho_c}{\mu_c}$ v_s = slip velocity, $d_3 = \dfrac{\text{total drop surface area}}{\text{perimeter normal to flow}}$	[E] Used with log mean mole fraction. Differences based on ends of extraction column; 100 measured values ±2% deviation. Based on area oblate spheroid.	[151] p. 285, 406, 407
I. Single liquid drop in immiscible liquid, Free rise or fall, discontinuous phase coefficient, circulating drops	$k_{dr,circ} = -\dfrac{d_p}{6\theta}\ln\left[\dfrac{3}{8}\displaystyle\sum_{j=1}^{\infty}B_j^2\exp\left(-\dfrac{\lambda_j 64 D_d\theta}{d_p^2}\right)\right]$ Eigenvalues for Circulating Drop 	$k_d d_p/D_d$	λ_1	λ_2	λ_3	B_1	B_2	B_3
---	---	---	---	---	---	---
3.20	0.262	0.424		1.49	0.107	
10.7	0.680	4.92		1.49	0.300	
26.7	1.082	5.90	15.7	1.49	0.495	0.205
107	1.484	7.88	19.5	1.39	0.603	0.384
320	1.60	8.62	21.3	1.31	0.583	0.391
∞	1.656	9.08	22.2	1.29	0.596	0.386		[T] Use with arithmetic concentration difference. θ = drop residence time. A more complete listing of eigenvalues is given by Refs. 86 and 99. $k'_{L,d,circ}$ is m/s.	[86][99][151] p. 405 [161] p. 523
J. 5-25-I	$\hat{k}_{L,d,circ} = -\dfrac{d_p}{6\theta}\left(\dfrac{\rho_d}{M_d}\right)_{av}\ln\left[1 - \dfrac{R^{1/2}\pi D_d^{1/2}\theta^{1/2}}{d_p/2}\right]$	[E] Used with mole fractions for extraction less than 50%, $R \approx 2.25$.	[151] p. 405

TABLE 5-25 **Mass-Transfer Correlations for Drops and Bubbles** (*Continued*)

Conditions	Correlations	Comments E = Empirical, S = Semiempirical, T = Theoretical	References°
K. 5-25-I	$$N_{Sh} = \frac{\hat{k}_{L,d,\text{circ}} d_p}{D_d}$$ $$= 31.4\left(\frac{\rho_d}{M_f}\right)_{av}\left(\frac{4 D_d t}{d_p^2}\right)^{-0.34} N_{Sc,d}^{-0.125}\left(\frac{d_p v_s^2 \rho_c}{\sigma g_c}\right)^{-0.37}$$	[E] Used with log mean mole fraction difference. d_p = diameter of sphere with same volume as drop. $856 \le N_{Sc} \le 79{,}800$, $2.34 \le \sigma \le 4.8$ dynes/cm.	[154] p. 435 [155]
L. Liquid drop in immiscible liquid, free rise or fall, continuous phase coefficient, circulating single drops	$$N_{Sh,c} = \frac{k'_{L,c} d_p}{D_d}$$ $$= \left[2 + 0.463 N_{Re,\text{drop}}^{0.484} N_{Sc,c}^{0.339}\left(\frac{d_p g^{1/3}}{D_c^{2/3}}\right)^{0.072}\right] F$$ $$F = 0.281 + 1.615 K + 3.73 K^2 - 1.874 K$$ $$K = N_{Re,\text{drop}}^{1/8}\left(\frac{\mu_c}{\mu_d}\right)^{1/4}\left(\frac{\mu_c v_s}{\sigma g_c}\right)^{1/6}$$	[E] Used as an arithmetic concentration difference. $$N_{Re,\text{drop}} = \frac{d_p v_s \rho_c}{\mu_c}$$ Solid sphere form with correction factor *F*.	[103]
M. 5-25-L, circulating, single drop	$$N_{Sh} = \frac{k_{L,c} d_p}{D_c} = 0.6\left(\frac{\rho_c}{M_c}\right)_{av} N_{Re,\text{drop}}^{1/2} N_{Sc,c}^{1/2}$$	[E] Used as an arithmetic concentration difference. Low σ. $$N_{Re,\text{drop}} = \frac{d_p v_s \rho_c}{\mu_c}$$	[151] p. 407
N. 5-25-L, circulating swarm of drops	$$k_{L,c} = 0.725\left(\frac{\rho_c}{M_c}\right)_{av} N_{Re,\text{drop}}^{-0.43} N_{Sc,c}^{-0.58} v_s (1 - \phi_d)$$	[E] Used as an arithmetic concentration difference. Low σ, disperse-phase holdup of drop swarm. ϕ_d = volume fraction dispersed phase.	[151] p. 407 [154] p. 436
O. Liquid drops in immiscible liquid, free rise or fall, discontinuous phase coefficient, oscillating drops	$$N_{Sh} = \frac{k_{L,d,\text{osc}} d_p}{D_d}$$ $$= 0.32\left(\frac{\rho_d}{M_d}\right)_{av}\left(\frac{4 D_d t}{d_p^2}\right)^{-0.14} N_{Re,\text{drop}}^{0.68}\left(\frac{\sigma^3 g_c^3 \rho_c^2}{g \mu_c^4 \Delta\rho}\right)^{0.10}$$	[E] Used with a log mean mole fraction difference. Based on ends of extraction column. $$N_{Re,\text{drop}} = \frac{d_p v_s \rho_c}{\mu_c}$$ d_p = diameter of sphere with volume of drop. Average absolute deviation from data, 10.5%. $411 \le N_{Re} \le 3114$ Low interfacial tension (3.5–5.8 dynes), $\mu_c < 1.35$ centipoise.	[151] p. 406 [154] p. 435 [155]
P. 5-25-O	$$k_{L,d,\text{osc}} = \frac{0.00375 v_s}{1 + \mu_d/\mu_c}$$	[T] Use with log mean concentration difference. Based on end of extraction column. No continuous phase resistance. $k_{L,d,\text{osc}}$ in cm/s, v_s = drop velocity relative to continuous phase.	[146] p. 228 [151] p. 405
Q. Single liquid drop in immiscible liquid, range rigid to fully circulating	Rigid drops: $10^4 < N_{Pe,c} < 10^6$ $$N_{Sh,c,\text{rigid}} = \frac{k_c d_p}{D_c} = 2.43 + 0.774 N_{Re}^{0.5} N_{Sc}^{0.33}$$ $$+ 0.0103 N_{Re} N_{Sc}^{0.33}$$ Circulating drops: $10 < N_{Re} < 1200$, $190 < N_{Sc} < 241{,}000$, $10^3 < N_{Pe,c} < 10^6$ $$N_{Sh,c,\text{fully circular}} = \left[\frac{2}{\pi^{0.5}}\right] N_{Pe,c}^{0.5}$$ Drops in intermediate range: $$\frac{N_{Sh,c} - N_{Sh,c,\text{rigid}}}{N_{Sh,c,\text{fully circular}} - N_{Sh,c,\text{rigid}}} = 1 - \exp\left[-(4.18 \times 10^{-3}) N_{Pe,c}^{0.42}\right]$$	[E] Allows for slight effect of wake.	[156] p. 58 [158]
R. Coalescing drops in immiscible liquid, discontinuous phase coefficient	$$\hat{k}_{d,\text{coal}} = 0.173 \frac{d_p}{t_f}\left(\frac{\rho_d}{M_d}\right)_{av}\left(\frac{\mu_d}{\rho_d D_d}\right)^{-1.115}$$ $$\times \left(\frac{\Delta\rho g d_p^2}{\sigma g_c}\right)^{1.302}\left(\frac{v_s^2 t_f}{D_d}\right)^{0.146}$$	[E] Used with log mean mole fraction difference. 23 data points. Average absolute deviation 25%. t_f = formation time.	[151] p. 408
S. 5-25-R, continuous phase coefficient	$$\hat{k}_{c,\text{coal}} = 5.959 \times 10^{-4}\left(\frac{\rho}{M}\right)_{av}$$ $$\times \left(\frac{D_c}{t_f}\right)^{0.5}\left(\frac{\rho_d u_s^3}{g \mu_c}\right)^{0.332}\left(\frac{d_p^2 \rho_c \rho_d v_s^3}{\mu_d \sigma g_c}\right)^{0.525}$$	[E] Used with log mean mole fraction difference. 20 data points. Average absolute deviation 22%.	[151] p. 409

TABLE 5-25 Mass-Transfer Correlations for Drops and Bubbles (Concluded)

Conditions	Correlations	Comments E = Empirical, S = Semiempirical, T = Theoretical	References[*]
T. Single liquid drops in gas, gas side coefficient	$\dfrac{\hat{k}_g M_g d_p P}{D_{\text{gas}} \rho_g} = 2 + A N_{Re,g}^{1/2} N_{Sc,g}^{1/3}$ $A = 0.552$ or 0.60.	[E] Used for spray drying (arithmetic partial pressure difference). $N_{Re,g} = \dfrac{d_p \rho_g v_s}{\mu_g}$, v_s = slip velocity between drop and gas stream. Sometimes written with: $\dfrac{M_g P}{\rho_g} = RT$	[88] p. 489 [111] p. 388 [135]
U. Single water drop in air, liquid side coefficient	$k_L = 2 \left(\dfrac{D_L}{\pi t} \right)^{1/2}$, short contact times $k_L = 10 \dfrac{D_L}{d_p}$, long contact times	[T] Use arithmetic concentration difference. Penetration theory. t = contact time of drop. Gives plot for $k_G a$ also. Air-water system.	[111] p. 389
V. Single bubbles of gas in liquid, continuous phase coefficient, very small bubbles	$N_{Sh} = \dfrac{k_c' d_b}{D_c} = 1.0 (N_{Re} N_{Sc})^{1/3}$	[T] Solid-sphere Eq. (see Table 5-24-B). $d_b < 0.1$ cm, k_c' is average over entire surface of bubble.	[122] p. 673 [146] p. 214
W. 5-25-V, medium to large bubbles	$N_{Sh} = \dfrac{k_c' d_b}{D_c} = 1.13 (N_{Re} N_{Sc})^{1/2}$	[T] Use arithmetic concentration difference. Droplet equation: $d_b > 0.5$ cm.	[146] p. 231
X. 5-25-W	$N_{Sh} = \dfrac{k_c' d_b}{D_c} = 1.13 (N_{Re} N_{Sc})^{1/2} \left[\dfrac{d_b}{0.45 + 0.2 d_b} \right]$	[S] Use arithmetic concentration difference. Modification of above (W), $d_b > 0.5$ cm. $500 \le N_{Re} \le 8000$. No effect SAA for $d_p > 0.6$ cm.	[104][146] p. 231
Y. Rising small bubbles of gas in liquid, continuous phase	$N_{Sh} = \dfrac{k_c' d_b}{D_c} = 2 + 0.31 (N_{Gr})^{1/3} N_{Sc}^{1/3}$, $d_b < 0.25$ cm	[E] Use with arithmetic concentration difference. $N_{Ra} = \dfrac{d_b^3 |\rho_G - \rho_L| g}{\mu_L D_L}$ = Raleigh number Note that $N_{Ra} = N_{Gr} N_{Sc}$. Valid for single bubbles or swarms. Independent of agitation as long as bubble size is constant.	[79][91] p. 451 [109] p. 119 [161] p. 156
Z. 5-25-Y, large bubbles	$N_{Sh} = \dfrac{k_c' d_b}{D_c} = 0.42 (N_{Gr})^{1/3} N_{Sc}^{1/2}$, $d_b > 0.25$ cm $\dfrac{\text{Interfacial area}}{\text{volume}} = a = \dfrac{6 H_g}{d_b}$	[E] Use with arithmetic concentration difference. H_g = fractional gas holdup, volume gas/total volume. For large bubbles, k_c' is independent of bubble size and independent of agitation or liquid velocity. Resistance is entirely in liquid phase for most gas-liquid mass transfer.	[79][91] p. 452 [109] p. 119 [114] p. 249

See Table 5-26 for agitated systems.

[*]See pages 5-7 and 5-8 for references.

there is conflicting evidence concerning this exponent ($\frac{2}{3}$ versus $\frac{1}{2}$) as discussed by Yadav and Sharma [*Chem. Eng. Sci.*, **34**, 1423 (1979)].

The existing data indicate that $\hat{k}_L a$ is proportional to the square root of the solute-diffusion coefficient, and since the interfacial area a does not depend on D_L, it follows that $\hat{k}_L$ is proportional to $D_L^{0.5}$. An analysis of the design variables involved indicates that $\hat{k}_L$ should be proportional to $N_{Sc}^{-0.5}$ when the Reynolds number is held constant.

It should be noted that the influence of substituting solvents of widely differing viscosities upon the interfacial area a can be very large. One therefore should be cautious about extrapolating $\hat{k}_L a$ data to account for viscosity effects between different solvent systems.

Effects of High Solute Concentrations on $\hat{k}_G$ and $\hat{k}_L$ As discussed previously, the stagnant-film model indicates that $\hat{k}_G$ should be independent of y_{BM} and $\hat{k}_G$ should be inversely proportional to y_{BM}. The data of Vivian and Behrman [*Am. Inst. Chem. Eng. J.*, **11**, 656 (1965)] for the absorption of ammonia from an inert gas strongly suggest that the film model's predicted trend is correct. This is another indication that the most appropriate rate coefficient to use is $\hat{k}_G$ and the proper driving-force term is of the form $(y - y_i)/y_{BM}$.

The use of the rate coefficient $\hat{k}_L$ and the driving force $(x_i - x)/x_{BM}$ is believed to be appropriate. For many practical situations the liquid-phase solute concentrations are low, thus making this assumption unimportant.

Influence of Chemical Reactions on $\hat{k}_G$ and $\hat{k}_L$ When a chemical reaction occurs, the transfer rate may be influenced by the chemical reaction as well as by the purely physical processes of diffusion and convection within the two phases. Since this situation is common in gas absorption, gas absorption will be the focus of this discussion. One must consider the impacts of chemical equilibrium and reaction kinetics on the absorption rate in addition to accounting for the effects of gas solubility, diffusivity, and system hydrodynamics.

There is no sharp dividing line between pure physical absorption and absorption controlled by the rate of a chemical reaction. Most cases fall in an intermediate range in which the rate of absorption is limited both by the resistance to diffusion and by the finite velocity of the reaction. Even in these intermediate cases the equilibria between the various diffusing species involved in the reaction may affect the rate of absorption.

TABLE 5-26　Mass-Transfer Correlations for Particles, Drops, and Bubbles in Agitated Systems

Situation	Correlation	Comments E = Empirical, S = Semiempirical, T = Theoretical	References[a]
A. Solid particles suspended in agitated vessel containing vertical baffles, continuous phase coefficient	$$\frac{k'_{LT}d_p}{D} = 2 + 0.6N_{Re,T}^{1/2}N_{Sc}^{1/3}$$ Replace v_{slip} with v_T = terminal velocity. Calculate Stokes' law terminal velocity $$v_{Ts} = \frac{d_p^2 \lvert\rho_p - \rho_c\rvert g}{18\mu_c}$$ and correct: <table><tr><td>$N_{Re,Ts}$</td><td>1</td><td>10</td><td>100</td><td>1,000</td><td>10,000</td><td>100,000</td></tr><tr><td>v_T/v_{Ts}</td><td>0.9</td><td>0.65</td><td>0.37</td><td>0.17</td><td>0.07</td><td>0.023</td></tr></table> Approximate: $k'_L = 2k'_{LT}$	[S] Use log mean concentration difference. Modified Frossling equation: $$N_{Re,Ts} = \frac{v_{Ts}d_p\rho_c}{\mu_c}$$ (Reynolds number based on Stokes' law.) $$N_{Re,T} = \frac{v_T d_p\rho_c}{\mu_c}$$ (terminal velocity Reynolds number.) k'_L almost independent of d_p. Harriott suggests different correction procedures. Range k'_L/k'_{LT} is 1.5 to 8.0.	[97][146] p. 220 [87]
B. 5-26-A	Graphical comparisons experiments and correlations.	[E,S] For spheres. Includes transpiration effects and changing diameters.	[78][146] p. 222
C. Solid, neutrally buoyant particles, continuous phase coefficient	$$N_{Sh} = \frac{k'_L d_p}{D} = 2 + 0.47N_{Re,p}^{0.62}N_{Sc}^{0.36}\left(\frac{d_{imp}}{d_{tank}}\right)^{0.17}$$ Graphical comparisons are in Ref. 109, p. 116.	[E] Use log mean concentration difference. Density unimportant if particles are close to neutrally buoyant. [E] E = energy dissipation rate per unit mass fluid $$= \frac{Pg_c}{V_{tank}\rho_c}, P = \text{power}$$ $$N_{Re,p} = \frac{E^{1/3}d_p^{4/3}}{\nu}$$ Also used for drops. Geometric effect (d_{imp}/d_{tank}) is usually unimportant. Ref. 118 gives a variety of references on correlations.	[109] p. 115 [118] p. 132 [161] p. 523
D. 5-26-C, small particles	$N_{Sh} = 2 + 0.52N_{Re,p}^{0.52}N_{Sc}^{1/3}, N_{Re,p} < 1.0$	[E] Terms same as above.	[109] p. 116
E. Solid particles with significant density difference	$$N_{Sh} = \frac{k'_L d_p}{D} = 2 + 0.44\left(\frac{d_p v_{slip}}{\nu}\right)^{1/2}N_{Sc}^{0.38}$$	[E] Use log mean concentration difference. N_{Sh} standard deviation 11.1%. v_{slip} calculated by methods given in reference.	[118]
F. Small solid particles, gas bubbles or liquid drops, $d_p < 2.5$ mm	$$N_{Sh} = \frac{k'_L d_p}{D} = 2 + 0.31\left[\frac{d_p^3 \lvert\rho_p - \rho_c\rvert}{\mu_c D}\right]^{1/3}$$	[E] Use log mean concentration difference. $g = 9.80665$ m/s^2. Second term RHS is free-fall or rise term. For large bubbles, see Table 5-25-Z.	[79][91] p. 451 [114] p. 249
G. Highly agitated systems; solid particles, drops, and bubbles; continuous phase coefficient	$$k'_L N_{Sc}^{2/3} = 0.13\left[\frac{(P/V_{tank})\mu_c g_c}{\rho_c^2}\right]^{1/4}$$	[E] Use arithmetic concentration difference. Use when gravitational forces overcome by agitation. Up to 60% deviation. Correlation prediction is low (Ref. 118). (P/V_{tank}) = power dissipated by agitator per unit volume liquid.	[79][83] p. 231 [91] p. 452
H. Liquid drops in baffled tank with flat six-blade turbine	$$k'_c a = 2.621 \times 10^{-3}\frac{(ND)^{1/2}}{d_{imp}}$$ $$\times \phi^{0.304}\left(\frac{d_{imp}}{d_{tank}}\right)^{1.582}N_{Re}^{1.929}N_{Oh}^{1.025}$$	[E] Use arithmetic concentration difference. Studied for five systems. $N_{Re} = d_{imp}^2 N\rho_c/\mu_c, N_{Oh} = \mu_c/(\rho_c d_{imp}\sigma)^{1/2}$ ϕ = volume fraction dispersed phase. N = impeller speed (revolutions/time). For $d_{tank} = h_{tank}$, average absolute deviation 23.8%.	[154] p. 437
I. Liquid drops in baffled tank, low volume fraction dispersed phase	$$N_{Sh} = \frac{k'_c d_p}{D} = 1.237 \times 10^{-5}N_{Sc}^{1/3}N^{2/3}$$ $$\times N_{Fr}^{5/12}\left(\frac{d_{imp}}{d_p}\right)\left(\frac{d_p}{D_{tank}}\right)^{1/2}\left(\frac{\rho_d d_p^2}{\sigma}\right)^{5/4}\phi^{-1/2}$$ Stainless steel flat six-blade turbine. Tank had four baffles. Correlation recommended for $\phi \leq 0.06$ [Ref. 156] $a = 6\phi/d_{32}$, where d_{32} is Sauter mean diameter when 33% mass transfer has occurred.	[E] 180 runs, 9 systems, $\phi = 0.01$. k_c is time-averaged. Use arithmetic concentration difference. $N_{Re} = \left(\frac{d_{imp}^2 N_{Sc}}{\mu_c}\right), N_{Fr} = \left(\frac{d_{imp}N^2}{g}\right)$ d_p = particle or drop diameter; σ = interfacial tension, N/m; ϕ = volume fraction dispersed phase; a = interfacial volume, 1/m; and $k_c\alpha D_c^{2/3}$ implies rigid drops. Negligible drop coalescence. Average absolute deviation—19.71%. Graphical comparison given by Ref. 153.	[153, 156] p. 78

TABLE 5-26 Mass-Transfer Correlations for Particles, Drops, and Bubbles in Agitated Systems (*Concluded*)

Situation	Correlation	Comments E = Empirical, S = Semiempirical, T = Theoretical	References[°]
J. Gas bubble swarms in sparged tank reactors	$k'_L a \left(\dfrac{\nu}{g^2}\right)^{1/3} = C\left[\dfrac{P/V_L}{\rho(\nu g^4)^{1/3}}\right]^a \left[\dfrac{q_G}{V_L}\left(\dfrac{\nu}{g^2}\right)^{1/3}\right]^b$ Rushton turbines: $C = 7.94 \times 10^{-4}$, $a = 0.62$, $b = 0.23$. Intermig impellers: $C = 5.89 \times 10^{-4}$, $a = 0.62$, $b = 0.19$.	[E] Use arithmetic concentration difference. Done for biological system, O_2 transfer. $h_{tank}/D_{tank} = 2.1$; P = power, kW. V_L = liquid volume, m^3. q_G = gassing rate, m^3/s. $k'_L a = s^{-1}$. Since $a = m^2/m^3$, ν = kinematic viscosity, m^2/s. Low viscosity system. Better fit claimed with q_G/V_L than with u_G (see 5-26-K to O).	[143]
K. 5-26-J	$k'_L a = 2.6 \times 10^{-2}\left(\dfrac{P}{V_L}\right)^{0.4} u_G^{0.5}$	[E] Use arithmetic concentration difference. Ion free water $V_L < 2.6$, u_G = superficial gas velocity in m/s. $500 < P/V_L < 10{,}000$. P/V_L = watts/m^3, V_L = liquid volume, m^3.	[115, 137]
L. 5-26-J	$k'_L a = 2.0 \times 10^{-3}\left(\dfrac{P}{V_L}\right)^{0.7} u_G^{0.2}$	[E] Use arithmetic concentration difference. Water with ions. $0.002 < V_L < 4.4$, $500 < P/V_L < 10{,}000$. Same definitions as 5-26-J.	[115, 117]
M. 5-26-J, baffled tank with standard blade Rushton impeller	$k'_L a = 93.37\left(\dfrac{P}{V_L}\right)^{0.76} u_G^{0.45}$	[E] Air-water. Same definitions as 5-26-J. $0.005 < u_G < 0.025$, $3.83 < N < 8.33$, $400 < P/V_L < 7000$ $h = D_{tank} = 0.305$ or 0.610 m. V_G = gas volume, m^3, N = stirrer speed, rpm. Method assumes perfect liquid mixing.	[92, 115]
N. 5-26-M	$k'_L a \dfrac{d_{imp}^2}{D} = 7.57\left(\dfrac{\mu_{eff}}{\rho D}\right)^{0.5}\left(\dfrac{\mu_G}{\mu_{eff}}\right)^{0.694}$ $\times\left(\dfrac{d_{imp}^2 N \rho_L}{\mu_{eff}}\right)^{1.11}\left(\dfrac{u_G d}{\sigma}\right)^{0.447}$ d_{imp} = impeller diameter, m; D = diffusivity, m^2/s	[E] Use arithmetic concentration difference. CO_2 into aqueous carboxyl polymethylene. Same definitions as 5-26-M. μ_{eff} = effective viscosity from power law model, Pa·s. σ = surface tension liquid, N/m.	[115, 129]
O. 5-26-M, bubbles	$\dfrac{k'_L a d_{imp}^2}{D} = 0.060\left(\dfrac{d_{imp}^2 N \rho}{\mu_{eff}}\right)\left(\dfrac{d_{imp}^2 N^2}{g}\right)^{0.19}\left(\dfrac{\mu_{eff} u_G}{\sigma}\right)^{0.6}$	[E] Use arithmetic concentration difference. O_2 into aqueous glycerol solutions. O_2 into aqueous millet jelly solutions. Same definitions as 5-26-M.	[115, 167]
P. Gas bubble swarm in sparged stirred tank reactor with solids present	$\dfrac{k'_L a}{(k'_L a)_o} = 1 - 3.54(\varepsilon_s - 0.03)$ $300 \le P/V_{rx} < 10{,}000$ W/m^3, $0.03 \le \varepsilon_s \le 0.12$ $0.34 \le u_G \le 4.2$ cm/s, $5 < \mu_L < 75$ Pa·s	[E] Use arithmetic concentration difference. Solids are glass beads, $d_p = 320$ μm. ε_s = solids holdup m^3/m^3 liquid, $(k'_L a)_o$ = mass transfer in absence of solids. Ionic salt solution—noncoalescing.	[71, 144]
Q. 5-26-P	$\dfrac{k'_L a}{(k'_L a)_o} = 1 - \varepsilon_s$	[E] Use arithmetic concentration difference. Variety of solids, $d_p > 150$ μm (glass, amberlite, polypropylene). Tap water. Slope very different than item P. Coalescence may have occurred.	[71, 112]

See also Table 5-25.
[°]See pages 5-7 and 5-8 for references.

The gas-phase rate coefficient $\hat{k}_G$ is not affected by the fact that a chemical reaction is taking place in the liquid phase. If the liquid-phase chemical reaction is extremely fast and irreversible, the rate of absorption may be governed completely by the resistance to diffusion in the gas phase. In this case the absorption rate may be estimated by knowing only the gas-phase rate coefficient $\hat{k}_G$ or else the height of one gas-phase transfer unit $H_G = G_M/(\hat{k}_G a)$.

It should be noted that the highest possible absorption rates will occur under conditions in which the liquid-phase resistance is negligible and the equilibrium back pressure of the gas over the solvent is zero. Such situations would exist, for instance, for NH_3 absorption into an acid solution, for SO_2 absorption into an alkali solution, for vaporization of water into air, and for H_2S absorption from a dilute-gas stream into a strong alkali solution, provided there is a large excess of reagent in solution to consume all the dissolved gas. This is known as the gas-phase mass-transfer limited condition, when both the liquid-phase resistance and the back pressure of the gas equal zero. Even when the reaction is sufficiently reversible to allow a small back pres-

sure, the absorption may be gas-phase-controlled, and the values of $\hat{k}_G$ and H_G that would apply to a physical-absorption process will govern the rate.

The liquid-phase rate coefficient $\hat{k}_L$ is strongly affected by fast chemical reactions and generally increases with increasing reaction rate. Indeed, the condition for zero liquid-phase resistance $(m/\hat{k}_L)$ implies that either the equilibrium back pressure is negligible, or that $\hat{k}_L$ is very large, or both. Frequently, even though reaction consumes the solute as it is dissolving, thereby enhancing both the mass-transfer coefficient and the driving force for absorption, the reaction rate is slow enough that the liquid-phase resistance must be taken into account. This may be due either to an insufficient supply of a second reagent or to an inherently slow chemical reaction.

In any event the value of $\hat{k}_L$ in the presence of a chemical reaction normally is larger than the value found when only physical absorption occurs, $\hat{k}_L^0$. This has led to the presentation of data on the effects of chemical reaction in terms of the "reaction factor" or "enhancement factor" defined as

TABLE 5-27   Mass Transfer Correlations for Fixed and Fluidized Beds

Transfer is to or from particles.

Situation	Correlation	Comments E = Empirical, S = Semiempirical, T = Theoretical	References[a]
A. Heat or mass transfer in packed bed for gases and liquids (shape factor, Ψ)	$j_D = j_H = 0.91 \Psi N_{Re}^{-0.51}, 0.01 < N_{Re} < 50$ Equivalent $N_{Sh} = 0.91 \Psi N_{Re}^{0.49} N_{Sc}^{1/3}$. $j_D = j_H = 0.61 \Psi N_{Re}^{-0.41}, 50 < N_{Re} < 1000$ Equivalent $N_{Sh} = 0.61 \Psi N_{Re}^{0.59} N_{Sc}^{1/3}$. | particle	sphere	cylinder | |---|---|---| | 1.00 | 0.91 | 0.81 |	[E] Different constants and shape factors reported in other references. Evaluate terms at film temperature or composition. $N_{Sh} = \dfrac{k' d_s}{D}, j_D = \dfrac{N_{Sh}}{N_{Re} N_{Sc}^{1/3}}$ $N_{Re} = \dfrac{v_{super} \rho}{\mu \Psi a}, v_{super} = $ superficial velocity $a = \dfrac{\text{surface area}}{\text{volume}} = 6(1 - \varepsilon)/d_p$ For spheres, $d_p = $ diameter. For nonspherical: $d_p = 0.567 \sqrt{\text{Part. Surf. Area}}$ Results are from too-short beds—use with caution.	[100] p. 194 [169]
B. For gases, fixed and fluidized beds, Gupta and Thodos correlation	$j_H = j_D = \dfrac{2.06}{\varepsilon N_{Re}^{0.575}}, 90 \le N_{Re} \le A$ Equivalent: $N_{Sh} = \dfrac{2.06}{\varepsilon} N_{Re}^{0.425} N_{Sc}^{1/3}$ For other shapes: $\dfrac{\varepsilon j_D}{(\varepsilon j_D)_{sphere}} = 0.79$ (cylinder) or 0.71 (cube) Graphical results are available for N_{Re} from 1900 to 10,300.	[E] For spheres. $N_{Re} = \dfrac{v_{super} d_p \rho}{\mu}$ $A = 2453$ [Ref. 151], $A = 4000$ [Ref. 100]. For $N_{Re} > 1900, j_H = 1.05 j_D$. Heat transfer result is in absence of radiation. $N_{Sh} = \dfrac{k' d_s}{D}$	[95, 96] [100] p. 195 [151]
C. For gases, for fixed beds, Petrovic and Thodos correlation	$N_{Sh} = \dfrac{0.357}{\varepsilon} N_{Re}^{0.641} N_{Sc}^{1/3}$	[E] Packed spheres, deep beds, $3 < N_{Re} < 900$ can be extrapolated for $N_{Re} < 2000$. Corrected for axial dispersion with axial Peclet number = 2.0. Prediction is low at low N_{Re}. N_{Re} defined as in 5-27-A and B.	[130][141] p. 214 [163]
D. For gases and liquids, fixed and fluidized beds	$j_D = \dfrac{0.4548}{\varepsilon N_{Re}^{0.4069}}, 10 \le N_{Re} \le 2000$ $j_D = \dfrac{N_{Sh}}{N_{Re} N_{Sc}^{1/3}}, N_{Sh} = \dfrac{k' d_s}{D}$	[E] Packed spheres, deep bed. Average deviation $\pm 20\%, N_{Re} = d_p v_{super} \rho / \mu$. Can use for fluidized beds. $10 \le N_{Re} \le 4000$.	[85][91] p. 447
E. For gases, fixed beds	$j_D = \dfrac{0.499}{\varepsilon N_{Re}^{0.382}}$	[E] Data on sublimation of naphthalene spheres dispersed in inert beads. $0.1 < N_{Re} < 100, N_{Sc} = 2.57$. Correlation coefficient = 0.978.	[101]
F. For liquids, fixed bed, Wilson and Geankoplis correlation	$j_D = \dfrac{1.09}{\varepsilon N_{Re}^{2/3}}, 0.0016 < N_{Re} < 55$ $165 \le N_{Sc} \le 70,600, 0.35 < \varepsilon < 0.75$ Equivalent: $N_{Sh} = \dfrac{1.09}{\varepsilon} N_{Re}^{1/3} N_{Sc}^{1/3}$ $j_D = \dfrac{0.25}{\varepsilon N_{Re}^{0.31}}, 55 < N_{Re} < 1500, 165 \le N_{Sc} \le 10,690$ Equivalent: $N_{Sh} = \dfrac{0.25}{\varepsilon} N_{Re}^{0.69} N_{Sc}^{1/3}$	[E] Beds of spheres, $N_{Re} = \dfrac{d_p V_{super} \rho}{\mu}$ Deep beds. $N_{Sh} = \dfrac{k' d_s}{D}$	[91] p. 448 [100] p. 195 [151] p. 287 [166]

TABLE 5-27   Mass Transfer Correlations for Fixed and Fluidized Beds (*Continued*)

Situation	Correlation	Comments E = Empirical, S = Semiempirical, T = Theoretical	References[a]
G. For liquids, fixed beds, Ohashi et al. correlation	$N_{Sh} = \dfrac{k'd_s}{D} = 2 + 0.51\left(\dfrac{E^{1/3}d_p^{4/3}\rho}{\mu}\right)^{0.60} N_{Sc}^{1/3}$ E = Energy dissipation rate per unit mass of fluid $= 50(1-\varepsilon)\varepsilon^2 C_{Do}\left(\dfrac{v_r^3}{d_p}\right)$, m^2/s^3 $= \left[\dfrac{50(1-\varepsilon)C_D}{\varepsilon}\right]\left(\dfrac{v_{super}^3}{d_p}\right)$ General form: $N_{Sh} = 2 + K\left(\dfrac{E^{1/3}D_p^{4/3}\rho}{\mu}\right)^{\alpha} N_{Sc}^{\beta}$ applies to single particles, packed beds, two-phase tube flow, suspended bubble columns, and stirred tanks with different definitions of E.	[S] Correlates large amount of published data. Compares number of correlations, v_r = relative velocity, m/s. In packed bed, $v_r = v_{super}/\varepsilon$. C_{Do} = single particle drag coefficient at v_{super} calculated from $C_{Do} = AN_{Re_i}^{-m}$ | N_{Re}	A	m
---	---	---
0 to 5.8	24	1.0
5.8 to 500	10	0.5
>500	0.44	0 | Ranges for packed bed: $0.001 < N_{Re} < 1000$ $505 < N_{Sc} < 70600$ $0.2 < \dfrac{E^{1/3}d_p^{4/3}\rho}{\mu} < 4600$ Compares different situations versus general correlation. See also 5-24-F.	[125]
H. For liquids, fixed and fluidized beds	$\varepsilon j_D = \dfrac{1.1068}{N_{Re}^{0.72}}$, $1.0 < N_{Re} \le 10$ $\varepsilon j_D = \dfrac{N_{Sh}}{N_{Re}N_{Sc}^{1/3}}$, $N_{Sh} = \dfrac{k'd_s}{D}$	[E] Spheres: $N_{Re} = \dfrac{d_p v_{super}\rho}{\mu}$	[84][91] p. 448
I. For gases and liquids, fixed and fluidized beds, Dwivedi and Upadhyay correlation	$\varepsilon j_D = \dfrac{0.765}{N_{Re}^{0.82}} + \dfrac{0.365}{N_{Re}^{0.386}}$ Gases: $10 \le N_{Re} \le 15,000$. Liquids: $0.01 \le N_{Re} \le 15,000$.	[E] Deep beds of spheres, $j_D = \dfrac{N_{Sh}}{N_{Re}N_{Sc}^{1/3}}$ $N_{Re} = \dfrac{d_p v_{super}\rho}{\mu}$, $N_{Sh} = \dfrac{k'd_s}{D}$ Based on 20 gas studies and 17 liquid studies. Recommended instead of 5-27-D or F.	[84][100] p. 196
J. For gases and liquids, fixed bed	$j_D = 1.17N_{Re}^{-0.415}$, $10 \le N_{Re} \le 2500$ $j_D = \dfrac{k'}{v_{av}}\dfrac{p_{BM}}{P} N_{Sc}^{2/3}$ Comparison with other results are shown.	[E] Spheres: $N_{Re} = \dfrac{d_p v_{super}\rho}{\mu}$ Variation in packing that changes ε not allowed for. Extensive data referenced. $0.5 < N_{Sc} < 15,000$.	[146] p. 241
K. For liquids, fixed and fluidized beds, Rahman and Streat correlation	$N_{Sh} = \dfrac{0.86}{\varepsilon} N_{Re}N_{Sc}^{1/3}$, $2 \le N_{Re} \le 25$	[E] Can be extrapolated to $N_{Re} = 2000$. $N_{Re} = d_p v_{super}\rho/\mu$. Done for neutralization of ion exchange resin.	[134]
L. For liquids and gases, Ranz and Marshall correlation	$N_{Sh} = \dfrac{k'd}{D} = 2.0 + 0.6N_{Sc}^{1/3}N_{Re}^{1/2}$ $N_{Re} = \dfrac{d_p v_{super}\rho}{\mu}$	[E] Based on freely falling, evaporating spheres (see 5-24-C). Has been applied to packed beds. Prediction is low compared to experimental data for packed beds. Limit of 2.0 at low N_{Re} is too high. Not corrected for axial dispersion.	[135][141] p. 214 [163][168] p. 106
M. For liquids and gases, Wakao and Funazkri correlation	$N_{Sh} = 2.0 + 1.1N_{Sc}^{1/3}N_{Re}^{0.6}$, $3 < N_{Re} < 10,000$ $N_{Sh} = \dfrac{k'_{film}d_p}{D} t$ Graphical comparison with data shown by Refs. 141, p. 215, and 163.	[E] $N_{Re} = \dfrac{\rho_f v_{super}\rho}{\mu}$ Correlate 20 gas studies and 16 liquid studies. Corrected for axial dispersion with: $\dfrac{\varepsilon D_{axial}}{D} = 10 + 0.5N_{Sc}N_{Re}$ D_{axial} is axial dispersion coefficient.	[141] p. 214 [163] [165] p. 376 [168] p. 106

TABLE 5-27   Mass Transfer Correlations for Fixed and Fluidized Beds (*Concluded*)

Situation	Correlation	Comments E = Empirical, S = Semiempirical, T = Theoretical	References[*]
N. Liquid fluidized beds	$$N_{Sh} = \dfrac{\dfrac{2\xi/\varepsilon^m + \left[\dfrac{(2\xi/\varepsilon^m)(1-\varepsilon)^{1/2}}{[1-(1-\varepsilon)^{1/3}]^2} - 2\right]\tan h\,(\xi/\varepsilon^m)}{\dfrac{\xi/\varepsilon^m}{1-(1-\varepsilon^{1/2})} - \tan h\,(\xi/\varepsilon^m)}}{}$$ where $$\xi = \left[\frac{1}{(1-\varepsilon)^{1/3}} - 1\right]\frac{\alpha}{2} N_{Sc}^{1/3} N_{Re}^{1/2}$$ This simplifies to: $$N_{Sh} = \frac{\varepsilon^{1-2m}}{(1-\varepsilon)^{1/3}}\left[\frac{1}{(1-\varepsilon)^{1/3}} - 1\right]\frac{\alpha^2}{2} N_{Re} N_{Sc}^{2/3} \quad (N_{Re} < 0.1)$$	[S] Modification of theory to fit experimental data. For spheres, $m = 1$, $N_{Re} > 2$. $N_{Sh} = \dfrac{k_L' d_p}{D} \quad N_{Re} = \dfrac{V_{super} d_p \xi}{\mu}$ $m = 1$ for $N_{Re} > 2$; $m = 0.5$ for $N_{Re} < 1.0$; $\varepsilon =$ voidage; $\alpha =$ const. Best fit data is $\alpha = 0.7$. Comparison of theory and experimental ion exchange results in Ref. 113.	[113, 123, 138]
O. Liquid fluidized beds	$N_{Sh} = 0.250 N_{Re}^{0.023} N_{Ga}^{0.306}\left(\dfrac{\rho_s - \rho}{\rho}\right)^{0.282} N_{Sc}^{0.410} \quad (\varepsilon < 0.85)$ $N_{Sh} = 0.304 N_{Re}^{-0.057} N_{Ga}^{0.332}\left(\dfrac{\rho_s - \rho}{\rho}\right)^{0.297} N_{Sc}^{0.404} \quad (\varepsilon > 0.85)$ This can be simplified (with slight loss in accuracy at high ε) to $N_{Sh} = 0.245 N_{Ga}^{0.323}\left(\dfrac{\rho_s - \rho}{\rho}\right)^{0.300} N_{Sc}^{0.400}$	[E] Correlate amount of data from literature. Compare large number of published correlations. $N_{Sh} = \dfrac{k_L' d_p}{D},\ N_{Re} = \dfrac{d_p \rho v_{super}}{\mu}$ $N_{Ga} = \dfrac{d_p^3 \rho^2 g}{\mu^2},\ N_{Sc} = \dfrac{\mu}{\rho D}$ $1.6 < N_{Re} < 1320,\ 2470 < N_{Ga} < 4.42 \times 10^6$ $0.27 < \dfrac{\rho_s - \rho}{\rho} < 1.114,\ 305 < N_{Sc} < 1595$ Predicts very little dependence of N_{Sh} on velocity.	[160]
P. Liquid film flowing over solid particles with air present, trickle bed reactors, fixed bed	$N_{Sh} = \dfrac{k_L}{aD} = 1.8 N_{Re}^{1/2} N_{Sc}^{1/3},\ 0.013 < N_{Re} < 12.6$ two-phases, liquid trickle, no forced flow of gas. $N_{Sh} = 0.8 N_{Re}^{1/2} N_{Sc}^{1/3}$, one-phase, liquid only.	[E] $N_{Re} = \dfrac{L}{a\mu}$ $L =$ superficial liquid flow rate, kg/m²s. $a =$ surface area/col. volume, m²/m³. Irregular granules of benzoic acid, $0.29 \leq d_p \leq 1.45$ cm.	[142]
Q. Supercritical fluids in packed bed	$\dfrac{N_{Sh}}{(N_{Sc} N_{Gr})^{1/4}} = 0.1813\left(\dfrac{N_{Re}^2 N_{Sc}^{1/3}}{N_{Gr}}\right)^{1/4}(N_{Re}^{1/2} N_{Sc}^{1/3})^{3/4}$ $+ 1.2149\left\|\left(\dfrac{N_{Re}^2 N_{Sc}^{1/3}}{N_{Gr}}\right)^{3/4} - 0.01649\right\|^{1/3}$	[E] Natural and forced convection, $4 < N_{Re} < 135$.	[119]
R. Supercritical fluids in packed bed	$\dfrac{N_{Sh}}{(N_{Sc} N_{Gr})^{1/4}} = 0.5265\left(\dfrac{(N_{Re}^{1/2} N_{Sc}^{1/3})}{(N_{Sc} N_{Gr})^{1/4}}\right)^{1.6808}$ $+ 2.48\left\|\left(\dfrac{N_{Re}^2 N_{Sc}^{1/3}}{N_{Gr}}\right)^{0.6439} - 0.8768\right\|^{1.553}$	[E] Natural and forced convection. $0.3 < N_{Re} < 135$. Improvement of correlation in Q.	[116]

NOTE: For $N_{Re} < 3$ convective contributions which are not included may become important. Use with logarithmic concentration difference (integrated form) or with arithmetic concentration difference (differential form).

[*]See pages 5-7 and 5-8 for references.

$$\phi = \hat{k}_L / \hat{k}_L^0 \geq 1 \qquad (5\text{-}300)$$

where $\hat{k}_L$ = mass-transfer coefficient with reaction and $\hat{k}_L^0$ = mass-transfer coefficient for pure physical absorption.

It is important to understand that when chemical reactions are involved, this definition of $\hat{k}_L$ is based on the driving force defined as the difference between the concentration of *unreacted* solute gas at the interface and in the bulk of the liquid. A coefficient based on the total of both unreacted and reacted gas could have values *smaller* than the physical-absorption mass-transfer coefficient $\hat{k}_L^0$.

When liquid-phase resistance is important, particular care should be taken in employing any given set of experimental data to ensure that the equilibrium data used conform with those employed by the original author in calculating values of $\hat{k}_L$ or H_L. Extrapolation to widely different concentration ranges or operating conditions should be made with caution, since the mass-transfer coefficient $\hat{k}_L$ may vary in an unexpected fashion, owing to changes in the apparent chemical-reaction mechanism.

Generalized prediction methods for $\hat{k}_L$ and H_L do not apply when chemical reaction occurs in the liquid phase, and therefore one must use actual operating data for the particular system in question. A discussion of the various factors to consider in designing gas absorbers and strippers when chemical reactions are involved is presented by Astarita, Savage, and Bisio, *Gas Treating with Chemical Solvents*, Wiley (1983) and by Kohl and Riesenfeld, *Gas Purification*, 4th ed., Gulf (1985).

Effective Interfacial Mass-Transfer Area *a*   In a packed tower of constant cross-sectional area S the differential change in solute flow per unit time is given by

$$-d(G_M Sy) = N_A a\, dV = N_A aS\, dh \qquad (5\text{-}301)$$

TABLE 5-28 Mass Transfer Correlations for Packed Two-Phase Contactors—Absorption, Distillation, Cooling Towers, and Extractors (Packing Is Inert)

Situation	Correlations	Comments E = Empirical, S = Semiempirical, T = Theoretical	References[a]
A. Absorption, counter-current, liquid-phase coefficient H_L, Sherwood and Holloway correlation for random packings	$H_L = a_L \left(\dfrac{L}{\mu_L}\right)^n N_{Sc,L}^{0.5}$, L = lb/hr ft^2 $H_L = \dfrac{L_M}{\hat{k}_L a}$ L_M = lbmoles/hr ft^2, $\hat{k}_L$ = lbmoles/hr ft^2, a = ft^2/ft^3, μ_L in lb/(hr ft).	[E] From experiments on desorption of sparingly soluble gases from water. Graphs [Ref. 146], p. 606. Equation is dimensional. A typical value of n is 0.3 [Ref. 91] p. 633 has constants in kg, m, and s units for use in 5-28-A and B with $\hat{k}_G$ in kgmole/s m^2 and $\hat{k}_L$ in kgmole/s m^2 (kgmole/m^3). Constants for other packings are given by Refs. 121 p. 187 and 161, p. 239.	[121] p. 187 [122] p. 714 [146] p. 606 [164] p. 660

Ranges for 5-28-B (G and L)

Packing	a_G	b	c	G	L	a_L	n
				Raschig rings			
3/8 inch	2.32	0.45	0.47	200–500	500–1500	0.00182	0.46
1	7.00	0.39	0.58	200–800	400–500	0.010	0.22
1	6.41	0.32	0.51	200–600	500–4500		0.22
2	3.82	0.41	0.45	200–800	500–4500	0.0125	0.22
				Berl saddles			
1/2 inch	32.4	0.30	0.74	200–700	500–1500	0.0067	0.28
1/2	0.811	0.30	0.24	200–800	400–4500		0.28
1	1.97	0.36	0.40	200–800	400–4500	0.0059	0.28
1.5	5.05	0.32	0.45	200–1000	400–4500	0.0062	0.28

Range for 5-28-A is $400 < L < 15{,}000$ lb/hr ft^2

Situation	Correlations	Comments	References
B. Absorption counter-current, gas-phase coefficient H_G, for random packing	$H_G = \dfrac{a_G (G)^b N_{Sc,v}^{0.5}}{(L)^c}$, G = lb/hr ft^2 $H_G = \dfrac{G_M}{\hat{k}_G a}$ G_M = lbmoles/hr ft^2, $\hat{k}_G$ = lbmoles/hr ft^2.	[E] Based on ammonia-water-air data in Fellinger's 1941 MIT thesis. Curves: Refs. 121, p. 186 and 146 p. 607. Constants given in 5-28-A. The equation is dimensional.	[91] p. 633 [121] p. 189 [146] p. 607 [164] p. 660
C. Absorption, counter-current, gas-liquid individual coefficients and interfacial area, Shulman data for random packings	$\dfrac{k_G N_{Sc,v}^{2/3}}{G_M} = 1.195 \left[\dfrac{d_p G}{\mu_G (1 - \varepsilon_{Lo})}\right]^{-0.36}$ $\dfrac{\hat{k}_L d_p}{D_L} = 25.1 \left(\dfrac{d_p L}{\mu_L}\right)^{0.45} N_{Sc,L}^{0.5}$ Interfacial area a per volume given for Racshig rings and Berl saddles in graphical form by Refs. 78 and 121 p. 178, and in equation form by Ref. 161, p. 205. Liquid holdups are given by Refs. 161 (p. 206), 148, or 121, p. 174.	[E] Compared napthalene sublimation to aqueous absorption to obtain $\hat{k}_G$, a, and $\hat{k}_L$ separately. Raschig rings and Berl saddles. d_p = diameter of sphere with same surface area as packing piece. ε_{Lo} = operating void space = $\varepsilon - \phi_{Li}$, where ε = void fraction w/o liquid, and ϕ_{Li} = liquid holdup. Same definition as 5-28-A and B. Onda et al. correlation (5-28-D) is preferred. $G = \rho_G v_{super,gas}$	[76][123] p. 174, 186 [151, 158] [161] p. 203
D. Absorption and and distillation, counter-current, gas and liquid individual coefficients and wetted surface area, Onda et al. correlation for random packings	$\dfrac{k'_G RT}{a_p D_G} = A \left(\dfrac{G}{a_p \mu_G}\right)^{0.7} N_{Sc,G}^{1/3} (a_p d'_p)^{-2.0}$ $A = 5.23$ for packing $\geq 1/2$ inch (0.012 m) $A = 2.0$ for packing $< 1/2$ inch (0.012 m) k'_G = lbmoles/hr ft^2 atm [kg mol/s m^2 (N/m^2)] $k'_L \left(\dfrac{\rho_L}{\mu_L g}\right)^{1/3} = 0.0051 \left(\dfrac{L}{a_w \mu_L}\right)^{2/3} N_{Sc,L}^{-1/2} (a_p d'_p)^{0.4}$ k'_L = lbmoles/hr ft^2 (lbmoles/ft^3) [kgmoles/s m^2 (kgmoles/m^3)] $\dfrac{a_w}{a_p} = 1 - \exp\left\{\begin{array}{l} -1.45 \left(\dfrac{\sigma_c}{\sigma}\right)^{0.75} \left(\dfrac{L}{a_p \mu_L}\right)^{0.1} \\ \times \left(\dfrac{L^2 a_p}{\rho_L^2 g}\right)^{-0.05} \left(\dfrac{L}{\rho_L \sigma a_p}\right)^{0.2} \end{array}\right\}$ Critical surface tensions, $\sigma_C = 61$ (ceramic), 75 (steel), 33 (polyethylene), 40 (PVC), 56 (carbon) dynes/cm. Graphical comparison with data in Ref. 126.	[E] Gas absorption and desorption from water and organics plus vaporization of pure liquids for Raschig rings, saddles, spheres, and rods. d'_p = nominal packing size, a_p = dry packing surface area/volume, a_w = wetted packing surface area/volume. Equations are dimensionally consistent, so any set of consistent units can be used. σ = surface tension, dynes/cm. $4 < \dfrac{L}{a_w \mu_L} < 400$ $5 < \dfrac{G}{a_p \mu_G} < 1000$ Most data ± 20% of correlation, some ± 50%.	[76][88] p. 399 [111] p. 380 [126][159] p. 355

TABLE 5-28   Mass Transfer Correlations for Packed Two-Phase Contactors—Absorption, Distillation, Cooling Towers, and Extractors (Packing Is Inert) (Continued)

Situation	Correlations	Comments E = Empirical, S = Semiempirical, T = Theoretical	References[e]
E. Distillation and absorption, counter-current, random packings, modification of Onda correlation, Bravo and Fair correlation	Use Onda's correlations (5-28-D) for k'_G and k'_L. Calculate: $$H_G = \frac{G}{k'_G a_e P M_G}, \; H_L = \frac{L}{k'_L a_e \rho_L}$$ $$H_{OG} = H_G + \lambda H_L$$ where $$\lambda = \frac{m}{L_M/G_M}$$ Using $$a_e = 0.498 a_p \left(\frac{\sigma^{0.5}}{Z^{0.4}}\right)(N_{Ca,L} N_{Re,G})^{0.392}$$ where $$N_{Re,G} = \frac{6G}{a_p \mu_G},$$ $$N_{Ca,L} = \frac{L\mu_L}{\rho_L \sigma g_c} \text{ (dimensionless)}$$	[E] Use's Bolles & Fair (Ref. 75) data base to determine new effective area a_e to use with Onda et al. (Ref. 126) correlation. Same definitions as 5-28-D. P = total pressure, atm; M_G = gas, molecular weight; m = local slope of equilibrium curve; L_M/G_M = slope operating line; Z = height of packing in feet. Equation for a_e is dimensional. Fit to data for effective area quite good for distillation. Good for absorption at low values of $(N_{Ca,L} \times N_{Re,G})$, but correlation is too high at higher values of $(N_{Ca,L} \times N_{Re,G})$.	[76]
F. Absorption, co-current downward flow, random packings	Air-oxygen-water results correlated by $k'_L a = 0.12 E_L^{0.5}$. Extended to other systems. $$k'_L a = 0.12 E_L^{0.5} \left(\frac{D_L}{2.4 \times 10^5}\right)^{0.5}$$ $$E_L = \left(\frac{\Delta p}{\Delta L}\right)_{\text{2-phase}} v_L$$ $k'_L a = s^{-1}$ $D_L = cm/s$ $E_L = $ ft, lbf/s ft^3 v_L = superficial liquid velocity, ft/s $\frac{\Delta p}{\Delta L}$ = pressure loss in two-phase flow = lbf/ft^2 ft	[E] Based on oxygen transfer from water to air 77°F. Liquid film resistance controls. (D_{water} @ 77°F = 2.4×10^{-5}). Equation is dimensional. Data was for thin-walled polyethylene Raschig rings. Correlation also fit data for spheres. Fit ±25%. See Reiss for graph.	[136] [142] p. 217
	$k'_G a = 2.0 + 0.91 E_G^{2/3}$ for NH$_3$ $$E_g = \left(\frac{\Delta p}{\Delta L}\right)_{\text{2-phase}} v_g$$ v_g = superficial gas velocity, ft/s.	[E] Ammonia absorption into water from air at 70°F. Gas-film resistance controls. Thin-walled polyethylene Raschig rings and 1-inch Intalox saddles. Fit ±25%. See Reiss for fit. Terms defined as above.	[136]
G. Absorption, stripping, distillation, counter-current, H_L, and H_G, random packings, Cornell et al. correlation, and Bolles and Fair correlation	For Raschig rings, Berl saddles, and spiral tile: $$H_L = \frac{\phi C_{\text{flood}}}{3.28} N_{Sc,L}^{0.5} \left(\frac{Z}{3.05}\right)^{0.15}$$ $C_{\text{flood}} = 1.0$ if below 40% flood—otherwise, use Fig. 5-28. ϕ shown in Fig. 5-29 for different packings and sizes. Range $0.02 < \phi < 0.300$. $$H_G = \frac{A\psi(d'_{\text{col}})^m Z^{0.33} N_{Sc,G}^{0.5}}{\left[L\left(\frac{\mu_L}{\mu_{\text{water}}}\right)^{0.16}\left(\frac{\rho_{\text{water}}}{\rho_L}\right)^{1.25}\left(\frac{\sigma_{\text{water}}}{\sigma_L}\right)^{0.8}\right]^n}$$ $A = 0.017$ (rings) or 0.029 (saddles) d'_{col} = column diameter in m (if diameter > 0.6 m, use $d'_{\text{col}} = 0.6$) $m = 1.24$ (rings) or 1.11 (saddles) $n = 0.6$ (rings) or 0.5 (saddles) ψ is given in Fig. 5-30. Range: $25 < \psi < 190$ m.	[E] Z = packed height, m of each section with its own liquid distribution. The original work is reported in English units. Cornell et al. (Ref. 81) review early literature. Improved fit of Cornell's ϕ values given by Bolles and Fair (Refs. 74 and 75) and in Fig. 5-29. L = liquid rate, kg/(sm^2), $\mu_{\text{water}} = 1.0$ Pa · s, $\rho_{\text{water}} = 1000$ kg/m^3, $\sigma_{\text{water}} = 72.8$ mN/m (72.8 dynes/cm). H_G and H_L will vary from location to location. Design each section of packing separately.	[74, 75, 81] [100] p. 428 [111] p. 381 [151] p. 353 [164] p. 651

TABLE 5-28 Mass Transfer Correlations for Packed Two-Phase Contactors—Absorption, Distillation, Cooling Towers, and Extractors (Packing Is Inert) (*Concluded*)

Situation	Correlations	Comments E = Empirical, S = Semiempirical, T = Theoretical	References[a]
H. Distillation and absorption. Counter-current flow. Structured packings. Gauze-type with triangular flow channels, Bravo, Rocha, and Fair correlation	Equivalent channel: $d_{eq} = Bh\left[\dfrac{1}{B+2S} + \dfrac{1}{2S}\right]$ Use modified correlation for wetted wall column (See 5-22-D) $N_{Sh,v} = \dfrac{k'_v d_{eq}}{D_v} = 0.0338 N_{Re,v}^{0.8} N_{Sc,v}^{0.333}$ $N_{Re,v} = \dfrac{d_{eq}\rho_v(U_{v,\text{eff}} + U_{L,\text{eff}})}{\mu_v}$ where effective velocities $U_{v,\text{eff}} = \dfrac{U_{v,\text{super}}}{\varepsilon \sin\theta}$ $U_{L,\text{eff}} = \dfrac{3\Gamma}{2\rho_L}\left(\dfrac{\rho_L^2 g}{3\mu_L\Gamma}\right)^{0.333}, \Gamma = \dfrac{L}{\text{Per}}$ $\text{Per} = \dfrac{\text{Perimeter}}{\text{Area}} = \dfrac{4S+2B}{Bh}$ Calculate k'_L from penetration model (use time for liquid to flow distance s). $k'_L = 2(D_L U_{L,\text{eff}}/\pi S)^{1/2}$.	[T] Check of 132 data points showed average deviation 14.6% from theory. Johnstone and Pigford [Ref. 105] correlation (5-22-D) has exponent on N_{Re} rounded to 0.8. Assume gauze packing is completely wet. Thus, $a_{\text{eff}} = a_p$ to calculate H_G and H_L. Same approach may be used generally applicable to sheet-metal packings, but they will not be completely wet and need to estimate transfer area. L = liquid flux, kg/s m². G = vapor flux, kg/s m². Fit to data shown in Ref. 77. $H_G = \dfrac{G}{k'_v a_p \rho_v}, H_L = \dfrac{L}{k'_L a_p \rho_L}$	[77] [87] p. 310, 326 [159] p. 356, 362
I. High-voidage packings, cooling towers, splash-grid packings	$\dfrac{(Ka)_H V_{\text{tower}}}{L} = 0.07 + A'N'\left(\dfrac{L}{G_a}\right)^{-n'}$ A' and n' depend on deck type (Ref. 107), $0.060 \leq A' \leq 0.135$, $0.46 \leq n' \leq 0.62$. General form fits the graphical comparisons (Refs. 146 and 164).	[E] General form. G_a = lb dry air/hr ft². L = lb/h ft², N' = number of deck levels. $(Ka)_H$ = overall enthalpy transfer coefficient = lb/(h)(ft³)$\left(\dfrac{\text{lb water}}{\text{lb dry air}}\right)$ V_{tower} = tower volume, ft³/ft². If normal packings are used, use absorption mass-transfer correlations or Ref. 88, p. 452.	[107][121] p. 220 [146] p. 286 [164] p. 681
J. Liquid-liquid extraction, packed towers	Use k values for drops (Table 5-25). Enhancement due to packing is at most 20%. Packing decreases drop size and increases interfacial area.	[E]	[156] p. 79
K. Liquid-liquid extraction in Rotating-disc contactor (RDC)	$\dfrac{k_{c,\text{RDC}}}{k_c} = 1.0 + 2.44\left(\dfrac{N}{N_{Cr}}\right)^{2.5}$ $N_{Cr} = 7.6 \times 10^{-4}\left(\dfrac{\sigma}{d_{\text{drop}}\,\mu_c}\right)\left(\dfrac{H}{D_{\text{tank}}}\right)$ $\dfrac{k_{d,\text{RDC}}}{k_d} = 1.0 + 1.825\left(\dfrac{N}{N_{Cr}}\right)\dfrac{H}{D_{\text{tank}}}$	k_c, k_d are for drops (Table 5-25) N = impeller speed Breakage occurs when $N > N_{Cr}$. Maximum enhancement before breakage was factor of 2.0. H = compartment height, D_{tank} = tank diameter, σ = interfacial tension, N/m. Done in 0.152 and 0.600 m RDC.	[70][156] p. 79
L. Liquid-liquid extraction, stirred tanks	See Table 5-26-F, G, H, and I.	[E]	

[a]See pages 5-7 and 5-8 for references.

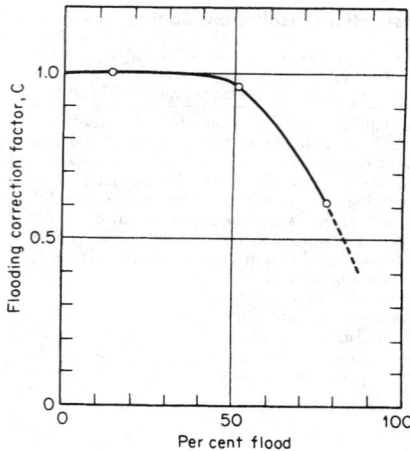

FIG. 5-28 Liquid-film correction factor (Table 5-28-G) for operation at high percent of flood. [*Cornell et al.*, Chem. Eng. Prog., **56**(8), 68 (1960).]

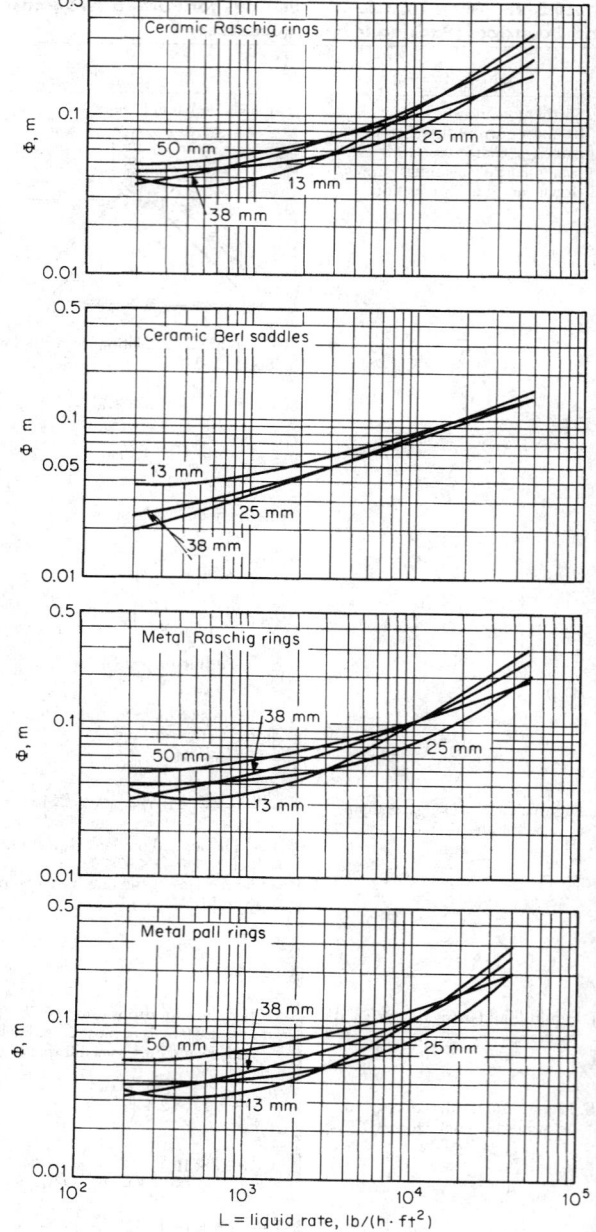

FIG. 5-29 H_l correlation for various packings (Table 5-28-G). To convert meters to feet, multiply by 3.281; to convert pounds per hour-square foot to kilograms per second-square meter, multiply by 0.001356; and to convert millimeters to inches, multiply by 0.0394. [*Bolles and Fair*, Inst. Chem. Eng. Symp. Ser., no. 56, 3.3/35 (1969).]

where a = interfacial area effective for mass transfer per unit of packed volume and V = packed volume. Owing to incomplete wetting of the packing surfaces and to the formation of areas of stagnation in the liquid film, the effective area normally is significantly less than the total external area of the packing pieces.

The effective interfacial area depends on a number of factors, as discussed in a review by Charpentier [*Chem. Eng. J.*, **11**, 161 (1976)]. Among these factors are (1) the shape and size of packing, (2) the packing material (for example, plastic generally gives smaller interfacial areas than either metal or ceramic), (3) the liquid mass velocity, and (4), for small-diameter towers, the column diameter.

Whereas the interfacial area generally increases with increasing liquid rate, it apparently is relatively independent of the superficial gas mass velocity below the flooding point. According to Charpentier's review, it appears valid to assume that the interfacial area is independent of the column height when specified in terms of unit packed volume (i.e., as a). Also, the existing data for chemically reacting gas-liquid systems (mostly aqueous electrolyte solutions) indicate that the interfacial area is independent of the chemical system. However, this situation may not hold true for systems involving large heats of reaction.

Rizzuti et al. [*Chem. Eng. Sci.*, **36**, 973 (1981)] examined the influence of solvent viscosity upon the effective interfacial area in packed columns and concluded that for the systems studied the effective interfacial area a was proportional to the kinematic viscosity raised to the 0.7 power. Thus, the hydrodynamic behavior of a packed absorber is strongly affected by viscosity effects. Surface-tension effects also are important, as expressed in the work of Onda et al. (see Table 5-28-D).

In developing correlations for the mass-transfer coefficients $\hat{k}_G$ and $\hat{k}_L$, the various authors have assumed different but internally compatible correlations for the effective interfacial area a. It therefore would be inappropriate to mix the correlations of different authors unless it has been demonstrated that there is a valid area of overlap between them.

Volumetric Mass-Transfer Coefficients $\hat{K}_G a$ and $\hat{K}_L a$ Experimental determinations of the individual mass-transfer coefficients $\hat{k}_G$ and $\hat{k}_L$ and of the effective interfacial area a involve the use of extremely difficult techniques, and therefore such data are not plentiful. More often, column experimental data are reported in terms of overall volumetric coefficients, which normally are defined as follows:

$$K'_G a = n_A/(h_T S p_T \Delta y^\circ_{1m}) \qquad (5\text{-}302)$$

and

$$K_L a = n_A/(h_T S \Delta x^\circ_{1m}) \qquad (5\text{-}303)$$

where $K'_G a$ = overall volumetric gas-phase mass-transfer coefficient, $K_L a$ = overall volumetric liquid-phase mass-transfer coefficient, n_A =

overall rate of transfer of solute A, h_T = total packed depth in tower, S = tower cross-sectional area, p_T = total system pressure employed during the experiment, and Δx°_{1m} and Δy°_{1m} are defined as

$$\Delta y^\circ_{1m} = \frac{(y - y^\circ)_1 - (y - y^\circ)_2}{\ln\left[(y - y^\circ)_1/(y - y^\circ)_2\right]} \qquad (5\text{-}304)$$

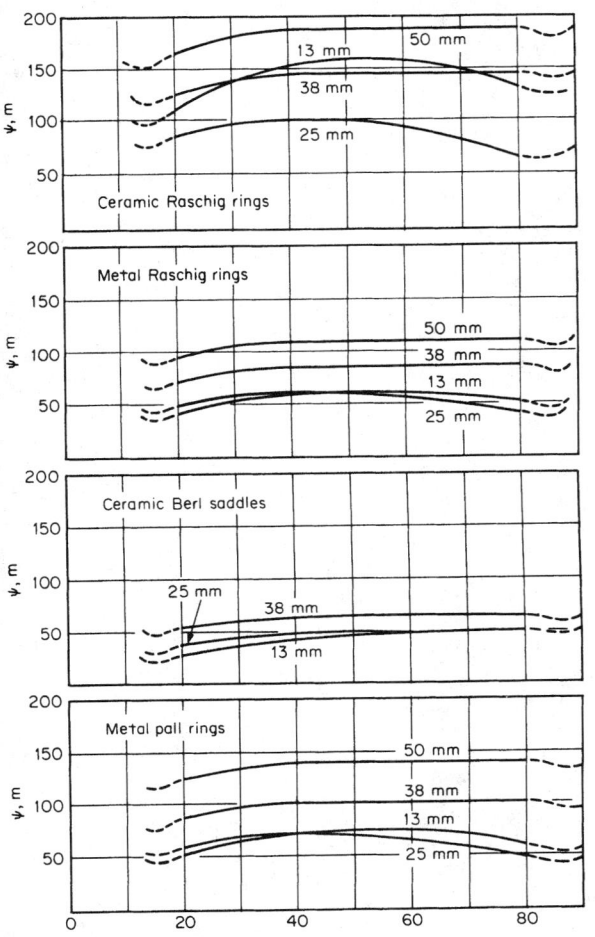

FIG. 5-30 H_g correlation for various packings (Table 5-28-G). To convert meters to feet, multiply by 3.281; to convert millimeters to inches, multiply by 0.03937. [*Bolles and Fair*, Inst. Chem. Eng. Symp. Ser., *no. 56, 3.3/35 (1979).*]

and

$$\Delta x_{1m}^{\circ} = \frac{(x^{\circ} - x)_2 - (x^{\circ} - x)_1}{\ln\left[(x^{\circ} - x)_2/(x^{\circ} - x)_1\right]} \qquad (5\text{-}305)$$

where subscripts 1 and 2 refer to the bottom and top of the tower respectively.

Experimental $K_G'a$ and K_La data are available for most absorption and stripping operations of commercial interest (see Sec. 15). The solute concentrations employed in these experiments normally are very low, so that $K_La \doteq \hat{K}_La$ and $K_G'ap_T \doteq \hat{K}_Ga$, where p_T is the total pressure employed in the actual experimental-test system. Unlike the individual gas-film coefficient $\hat{k}_Ga$, the overall coefficient $\hat{K}_Ga$ will

vary with the total system pressure except when the liquid-phase resistance is negligible (i.e., when either $m = 0$, or $\hat{k}_La$ is very large, or both).

Extrapolation of K_Ga data for absorption and stripping to conditions other than those for which the original measurements were made can be extremely risky, especially in systems involving chemical reactions in the liquid phase. One therefore would be wise to restrict the use of overall volumetric mass-transfer-coefficient data to conditions not too far removed from those employed in the actual tests. The most reliable data for this purpose would be those obtained from an operating commercial unit of similar design.

Experimental values of H_{OG} and H_{OL} for a number of distillation systems of commercial interest are also readily available. Extrapolation of the data or the correlations to conditions that differ significantly from those used for the original experiments is risky. For example, pressure has a major effect on vapor density and thus can affect the hydrodynamics significantly. Changes in flow patterns affect both mass-transfer coefficients and interfacial area.

Chilton-Colburn Analogy   When a fluid moves over either a liquid or a solid surface, the eddy motion that causes mass transfer also causes heat transfer and fluid friction owing to the transfer of thermal energy and momentum respectively. This close similarity among the mechanisms for the transfer of mass, heat, and momentum was brought out in the Reynolds analogy (see Table 5-23-T), which stated that the following dimensionless ratios are equal:

$$\hat{k}_G/G_M = h'/c_pG = f/2 \qquad (5\text{-}306)$$

where h' = heat-transfer coefficient, c_p = specific heat, G = mass flux, and f = friction factor.

Experimental data for mass transfer into gas streams agree approximately with Eq. (5-306) when the Schmidt number is close to unity and in smooth, straight tubes or along flat plates when the pressure drop is due entirely to skin friction against the surface. It does not, however, agree for cases involving "form" drag as well as skin friction. Also, it does not account for the mass-transfer resistance of the region of fluid near the liquid or solid boundary in which mass transfer occurs principally by molecular (as opposed to turbulent) motion.

Colburn [*Trans. Am. Inst. Chem. Eng.*, **29**, 174 (1933)] and Chilton and Colburn [*Ind. Eng. Chem.*, **26**, 1183 (1934)] showed empirically that the resistance of the laminar sublayer can be expressed by the following modification of the Reynolds analogy:

$$(\hat{k}_G/G_M)N_{Sc}^{2/3} = j_M = (h'/c_pG)N_{Pr}^{2/3} = j_H = f/2 \qquad (5\text{-}307)$$

for turbulent flow through straight tubes (see Table 5-23-U) and across plane surfaces (see Table 5-21-G), and

$$j_M = j_H \leq f/2 \qquad (5\text{-}308)$$

for turbulent flow around cylinders (see Table 5-24-I), where j_M = mass-transfer factor, j_H = heat-transfer factor, $N_{Pr} = c_p\mu/k$ = Prandtl number, and k = thermal conductivity; other symbols are as defined earlier.

On occasion one will find that heat-transfer-rate data are available for a system in which mass-transfer-rate data are not readily available. The Chilton-Colburn analogy provides a procedure for developing estimates of the mass-transfer rates based on heat-transfer data. Extrapolation of experimental j_M or j_H data obtained with gases to predict liquid systems (and vice versa) should be approached with caution, however. When pressure-drop or friction-factor data are available, one may be able to place an upper bound on the rates of heat and mass transfer, according to Eq. (5-308).

Fluid and Particle Dynamics*

James N. Tilton, Ph.D., P.E., *Senior Consultant, Process Engineering, E. I. du Pont de Nemours & Co.; Member, American Institute of Chemical Engineers; Registered Professional Engineer (Delaware)*

° The author acknowledges the contribution of B. C. Sakiadis, editor of this section in the sixth edition of the *Handbook*.

Nomenclature and Units*

In this listing, symbols used in this section are defined in a general way and appropriate SI units are given. Specific definitions, as denoted by subscripts, are stated at the place of application in the section. Some specialized symbols used in the section are defined only at the place of application. Some symbols have more than one definition; the appropriate one is identified at the place of application.

Symbol	Definition	SI units	U.S. customary units
a	Pressure wave velocity	m/s	ft/s
A	Area	m^2	ft^2
b	Wall thickness	m	in
b	Channel width	m	ft
c	Acoustic velocity	m/s	ft/s
c_f	Friction coefficient	Dimensionless	Dimensionless
C	Conductance	m^3/s	ft^3/s
Ca	Capillary number	Dimensionless	Dimensionless
C_0	Discharge coefficient	Dimensionless	Dimensionless
C_D	Drag coefficient	Dimensionless	Dimensionless
d	Diameter	m	ft
D	Diameter	m	ft
De	Dean number	Dimensionless	Dimensionless
D_{ij}	Deformation rate tensor components	1/s	1/s
E	Elastic modulus	Pa	lbf/in^2
$\dot{E}_v$	Energy dissipation rate	J/s	ft · lbf/s
Eo	Eotvos number	Dimensionless	Dimensionless
f	Fanning friction factor	Dimensionless	Dimensionless
f	Vortex shedding frequency	1/s	1/s
F	Force	N	lbf
F	Cumulative residence time distribution	Dimensionless	Dimensionless
Fr	Froude number	Dimensionless	Dimensionless
g	Acceleration of gravity	m/s^2	ft/s^2
G	Mass flux	kg/(m^2 · s)	lbm/(ft^2 · s)
h	Enthalpy per unit mass	J/kg	Btu/lbm
h	Liquid depth	m	ft
k	Ratio of specific heats	Dimensionless	Dimensionless
k	Kinetic energy of turbulence	J/kg	ft · lbf/lbm
K	Power law coefficient	kg/(m · s^{2-n})	lbm/(ft · s^{2-n})
l_v	Viscous losses per unit mass	J/kg	ft · lbf/lbm
L	Length	m	ft
$\dot{m}$	Mass flow rate	kg/s	lbm/s
M	Mass	kg	lbm
M	Mach number	Dimensionless	Dimensionless
M	Morton number	Dimensionless	Dimensionless
M_w	Molecular weight	kg/kgmole	lbm/lbmole
n	Power law exponent	Dimensionless	Dimensionless
N_b	Blend time number	Dimensionless	Dimensionless
N_D	Best number	Dimensionless	Dimensionless
N_P	Power number	Dimensionless	Dimensionless
N_Q	Pumping number	Dimensionless	Dimensionless
p	Pressure	Pa	lbf/in^2
q	Entrained flow rate	m^3/s	ft^3/s
Q	Volumetric flow rate	m^3/s	ft^3/s
Q	Throughput (vacuum flow)	Pa · m^3/s	lbf · ft^3/s
δQ	Heat input per unit mass	J/kg	Btu/lbm
r	Radial coordinate	m	ft
R	Radius	m	ft
R	Ideal gas universal constant	J/(kgmole · K)	Btu/(lbmole · R)
R_i	Volume fraction of phase i	Dimensionless	Dimensionless
Re	Reynolds number	Dimensionless	Dimensionless
s	Density ratio	Dimensionless	Dimensionless

Symbol	Definition	SI units	U.S. customary units
s	Entropy per unit mass	J/(kg · K)	Btu/(lbm · R)
S	Slope	Dimensionless	Dimensionless
S	Pumping speed	m^3/s	ft^3/s
S	Surface area per unit volume	l/m	l/ft
St	Strouhal number	Dimensionless	Dimensionless
t	Time	s	s
t	Force per unit area	Pa	lbf/in^2
T	Absolute temperature	K	R
u	Internal energy per unit mass	J/kg	Btu/lbm
u	Velocity	m/s	ft/s
U	Velocity	m/s	ft/s
v	Velocity	m/s	ft/s
V	Velocity	m/s	ft/s
V	Volume	m^3	ft^3
We	Weber number	Dimensionless	Dimensionless
$\dot{W}_s$	Rate of shaft work	J/s	Btu/s
δW_s	Shaft work per unit mass	J/kg	Btu/lbm
x	Cartesian coordinate	m	ft
y	Cartesian coordinate	m	ft
z	Cartesian coordinate	m	ft
z	Elevation	m	ft

<table>
<tr><td colspan="4" align="center">Greek symbols</td></tr>

Symbol	Definition	SI units	U.S. customary units
α	Velocity profile factor	Dimensionless	Dimensionless
α	Included angle	Radians	Radians
β	Velocity profile factor	Dimensionless	Dimensionless
β	Bulk modulus of elasticity	Pa	lbf/in^2
$\dot{\gamma}$	Shear rate	l/s	l/s
Γ	Mass flow rate per unit width	kg/(m · s)	lbm/(ft · s)
δ	Boundary layer or film thickness	m	ft
δ_{ij}	Kronecker delta	Dimensionless	Dimensionless
ϵ	Pipe roughness	m	ft
ϵ	Void fraction	Dimensionless	Dimensionless
ϵ	Turbulent dissipation rate	J/(kg · s)	ft · lbf/(lbm · s)
θ	Residence time	s	s
θ	Angle	Radians	Radians
λ	Mean free path	m	ft
μ	Viscosity	Pa · s	lbm/(ft · s)
ν	Kinematic viscosity	m^2/s	ft^2/s
ρ	Density	kg/m^3	lbm/ft^3
σ	Surface tension	N/m	lbf/ft
σ	Cavitation number	Dimensionless	Dimensionless
σ_{ij}	Components of total stress tensor	Pa	lbf/in^2
τ	Shear stress	Pa	lbf/in^2
τ	Time period	s	s
τ_{ij}	Components of deviatoric stress tensor	Pa	lbf/in^2
Φ	Energy dissipation rate per unit volume	J/(m^3 · s)	ft · lbf/(ft^3 · s)
ϕ	Angle of inclination	Radians	Radians
ω	Vorticity	1/s	1/s

* Note that with U.S. Customary units, the conversion factor g_c may be required to make equations in this section dimensionally consistent; $g_c = 32.17$ (lbm·ft)/lbf·s^2).

FLUID DYNAMICS

GENERAL REFERENCES: Batchelor, *An Introduction to Fluid Dynamics*, Cambridge University, Cambridge, 1967; Bird, Stewart, and Lightfoot, *Transport Phenomena*, Wiley, New York, 1960; Brodkey, *The Phenomena of Fluid Motions*, Addison-Wesley, Reading, Mass., 1967; Denn, *Process Fluid Mechanics*, Prentice-Hall, Englewood Cliffs, N.J., 1980; Landau and Lifshitz, *Fluid Mechanics*, 2d ed., Pergamon, 1987; Govier and Aziz, *The Flow of Complex Mixtures in Pipes*, Van Nostrand Reinhold, New York, 1972, Krieger, Huntington, N.Y., 1977; Panton, *Incompressible Flow*, Wiley, New York, 1984; Schlichting, *Boundary Layer Theory*, 8th ed., McGraw-Hill, New York, 1987; Shames, *Mechanics of Fluids*, 3d ed., McGraw-Hill, New York, 1992; Streeter, *Handbook of Fluid Dynamics*, McGraw-Hill, New York, 1971; Streeter and Wylie, *Fluid Mechanics*, 8th ed., McGraw-Hill, New York, 1985; Vennard and Street, *Elementary Fluid Mechanics*, 5th ed., Wiley, New York, 1975; Whitaker, *Introduction to Fluid Mechanics*, Prentice-Hall, Englewood Cliffs, N.J., 1968, Krieger, Huntington, N.Y., 1981.

NATURE OF FLUIDS

Deformation and Stress   A **fluid** is a substance which undergoes continuous deformation when subjected to a shear stress. Figure 6-1 illustrates this concept. A fluid is bounded by two large parallel plates, of area A, separated by a small distance H. The bottom plate is held fixed. Application of a force F to the upper plate causes it to move at a velocity U. The fluid continues to deform as long as the force is applied, unlike a solid, which would undergo only a finite deformation.

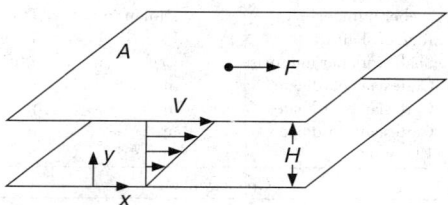

FIG. 6-1   Deformation of a fluid subjected to a shear stress.

The force is directly proportional to the area of the plate; the shear stress is $\tau = F/A$. Within the fluid, a linear velocity profile $u = Uy/H$ is established; due to the **no-slip condition,** the fluid bounding the lower plate has zero velocity and the fluid bounding the upper plate moves at the plate velocity U. The velocity gradient $\dot\gamma = du/dy$ is called the **shear rate** for this flow. Shear rates are usually reported in units of reciprocal seconds. The flow in Fig. 6-1 is a **simple shear flow.**

Viscosity   The ratio of shear stress to shear rate is the viscosity, μ.

$$\mu = \frac{\tau}{\dot\gamma} \qquad (6\text{-}1)$$

The SI units of viscosity are kg/(m · s) or Pa · s (pascal second). The cgs unit for viscosity is the poise; 1 Pa · s equals 10 poise or 1000 centipoise (cP) or 0.672 lbm/(ft · s). The terms *absolute viscosity* and *shear viscosity* are synonymous with the viscosity as used in Eq. (6-1). **Kinematic viscosity** $\nu \equiv \mu/\rho$ is the ratio of viscosity to density. The SI units of kinematic viscosity are m²/s. The cgs stoke is 1 cm²/s.

Rheology   In general, fluid flow patterns are more complex than the one shown in Fig. 6-1, as is the relationship between fluid deformation and stress. Rheology is the discipline of fluid mechanics which studies this relationship. One goal of rheology is to obtain **constitutive equations** by which stresses may be computed from deformation rates. For simplicity, fluids may be classified into rheological types in reference to the simple shear flow of Fig. 6-1. Complete definitions require extension to multidimensional flow. For more information, several good references are available, including Bird, Armstrong, and Hassager, (*Dynamics of Polymeric Liquids*, vol. 1: *Fluid Mechanics*, Wiley, New York, 1977); Metzner, ("Flow of Non-Newtonian Fluids" in Streeter, *Handbook of Fluid Dynamics*, McGraw-Hill, New York, 1971); and Skelland (*Non-Newtonian Flow and Heat Transfer*, Wiley, New York, 1967).

Fluids without any solidlike elastic behavior do not undergo any reverse deformation when shear stress is removed, and are called **purely viscous** fluids. The shear stress depends only on the rate of deformation, and not on the extent of deformation (strain). Those which exhibit both viscous and elastic properties are called **viscoelastic** fluids.

Purely viscous fluids are further classified into time-independent and time-dependent fluids. For time-independent fluids, the shear stress depends only on the instantaneous shear rate. The shear stress for time-dependent fluids depends on the past history of the rate of deformation, as a result of structure or orientation buildup or breakdown during deformation.

A **rheogram** is a plot of shear stress versus shear rate for a fluid in simple shear flow, such as that in Fig. 6-1. Rheograms for several types of time-independent fluids are shown in Fig. 6-2. The **Newtonian** fluid rheogram is a straight line passing through the origin. The slope of the line is the viscosity. For a Newtonian fluid, the viscosity is independent of shear rate, and may depend only on temperature and perhaps pressure. By far, the Newtonian fluid is the largest class of fluid of engineering importance. Gases and low molecular weight liquids are generally Newtonian. Newton's law of viscosity is a rearrangement of Eq. (6-1) in which the viscosity is a constant:

$$\tau = \mu\dot\gamma = \mu\,\frac{du}{dy} \qquad (6\text{-}2)$$

All fluids for which the viscosity varies with shear rate are **non-Newtonian fluids.** For non-Newtonian fluids the viscosity, defined as the ratio of shear stress to shear rate, is often called the **apparent viscosity** to emphasize the distinction from Newtonian behavior. Purely viscous, time-independent fluids, for which the apparent viscosity may be expressed as a function of shear rate, are called **generalized Newtonian fluids.**

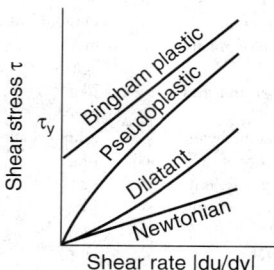

FIG. 6-2   Shear diagrams.

Non-Newtonian fluids include those for which a finite stress τ_y is required before continuous deformation occurs; these are called **yield-stress** materials. The **Bingham plastic** fluid is the simplest yield-stress material; its rheogram has a constant slope μ_∞, called the *infinite shear* viscosity.

$$\tau = \tau_y + \mu_\infty\dot\gamma \qquad (6\text{-}3)$$

Highly concentrated suspensions of fine solid particles frequently exhibit Bingham plastic behavior.

Shear-thinning fluids are those for which the slope of the rheogram decreases with increasing shear rate. These fluids have also been called *pseudoplastic,* but this terminology is outdated and discouraged. Many polymer melts and solutions, as well as some solids suspensions, are shear-thinning. Shear-thinning fluids without yield stresses typically obey a power law model over a range of shear rates.

$$\tau = K\dot\gamma^n \qquad (6\text{-}4)$$

The apparent viscosity is

$$\mu = K\dot\gamma^{n-1} \qquad (6\text{-}5)$$

The factor K is the consistency index or power law coefficient, and n is the power law exponent. The exponent n is dimensionless, while K is in units of $kg/(m \cdot s^{2-n})$. For shear-thinning fluids, $n < 1$. The power law model typically provides a good fit to data over a range of one to two orders of magnitude in shear rate; behavior at very low and very high shear rates is often Newtonian. Shear-thinning power law fluids with yield stresses are sometimes called *Herschel-Bulkley fluids*. Numerous other rheological model equations for shear-thinning fluids are in common use.

Dilatant, or shear-thickening, fluids show increasing viscosity with increasing shear rate. Over a limited range of shear rate, they may be described by the power law model with $n > 1$. Dilatancy is rare, observed only in certain concentration ranges in some particle suspensions (Govier and Aziz, pp. 33–34). Extensive discussions of dilatant suspensions, together with a listing of dilatant systems, are given by Green and Griskey (*Trans. Soc. Rheol.*, **12**[1], 13–25 [1968]); Griskey and Green (*AIChE J.*, **17**, 725–728 [1971]); and Bauer and Collins ("Thixotropy and Dilatancy," in Eirich, *Rheology*, vol. 4, Academic, New York, 1967).

Time-dependent fluids are those for which structural rearrangements occur during deformation at a rate too slow to maintain equilibrium configurations. As a result, shear stress changes with duration of shear. **Thixotropic** fluids, such as mayonnaise, clay suspensions used as drilling muds, and some paints and inks, show decreasing shear stress with time at constant shear rate. A detailed description of thixotropic behavior and a list of thixotropic systems is found in Bauer and Collins (ibid.).

Rheopectic behavior is the opposite of thixotropy. Shear stress increases with time at constant shear rate. Rheopectic behavior has been observed in bentonite sols, vanadium pentoxide sols, and gypsum suspensions in water (Bauer and Collins, ibid.) as well as in some polyester solutions (Steg and Katz, *J. Appl. Polym. Sci.*, **9**, 3, 177 [1965]).

Viscoelastic fluids exhibit elastic recovery from deformation when stress is removed. Polymeric liquids comprise the largest group of fluids in this class. A property of viscoelastic fluids is the *relaxation time,* which is a measure of the time required for elastic effects to decay. Viscoelastic effects may be important with sudden changes in rates of deformation, as in flow startup and stop, rapidly oscillating flows, or as a fluid passes through sudden expansions or contractions where accelerations occur. In many fully developed flows where such effects are absent, viscoelastic fluids behave as if they were purely viscous. In viscoelastic flows, normal stresses perpendicular to the direction of shear are different from those in the parallel direction. These give rise to such behaviors as the *Weissenberg effect,* in which fluid climbs up a shaft rotating in the fluid, and *die swell,* where a stream of fluid issuing from a tube may expand to two or more times the tube diameter.

A parameter indicating whether viscoelastic effects are important is the **Deborah number,** which is the ratio of the characteristic relaxation time of the fluid to the characteristic time scale of the flow. For small Deborah numbers, the relaxation is fast compared to the characteristic time of the flow, and the fluid behavior is purely viscous. For very large Deborah numbers, the behavior closely resembles that of an elastic solid.

Analysis of viscoelastic flows is very difficult. Simple constitutive equations are unable to describe all the material behavior exhibited by viscoelastic fluids even in geometrically simple flows. More complex constitutive equations may be more accurate, but become exceedingly difficult to apply, especially for complex geometries, even with advanced numerical methods. For good discussions of viscoelastic fluid behavior, including various types of constitutive equations, see Bird, Armstrong, and Hassager (*Dynamics of Polymeric Liquids,* vol. 1: *Fluid Mechanics,* vol. 2: *Kinetic Theory,* Wiley, New York, 1977); Middleman (*The Flow of High Polymers,* Interscience (Wiley) New York, 1968); or Astarita and Marrucci (*Principles of Non-Newtonian Fluid Mechanics,* McGraw-Hill, New York, 1974).

Polymer processing is the field which depends most on the flow of non-Newtonian fluids. Several excellent texts are available, including Middleman (*Fundamentals of Polymer Processing,* McGraw-Hill, New York, 1977) and Tadmor and Gogos (*Principles of Polymer Processing,* Wiley, New York, 1979).

There is a wide variety of instruments for measurement of Newtonian viscosity, as well as rheological properties of non-Newtonian fluids. They are described in Van Wazer, Lyons, Kim, and Colwell, (*Viscosity and Flow Measurement,* Interscience, New York, 1963); Coleman, Markowitz, and Noll (*Viscometric Flows of Non-Newtonian Fluids,* Springer-Verlag, Berlin, 1966); Dealy and Wissbrun (*Melt Rheology and its Role in Plastics Processing,* Van Nostrand Reinhold, 1990). Measurement of rheological behavior requires well-characterized flows. Such *rheometric* flows are thoroughly discussed by Astarita and Marrucci (*Principles of Non-Newtonian Fluid Mechanics,* McGraw-Hill, New York, 1974).

KINEMATICS OF FLUID FLOW

Velocity The term *kinematics* refers to the quantitative description of fluid motion or deformation. The rate of deformation depends on the distribution of velocity within the fluid. Fluid velocity **v** is a vector quantity, with three cartesian components v_x, v_y, and v_z. The velocity vector is a function of spatial position and time. A steady flow is one in which the velocity is independent of time, while in **unsteady** flow **v** varies with time.

Compressible and Incompressible Flow An incompressible flow is one in which the density of the fluid is constant or nearly constant. Liquid flows are normally treated as incompressible, except in the context of hydraulic transients (see following). Compressible fluids, such as gases, may undergo incompressible flow if pressure and/or temperature changes are small enough to render density changes insignificant. Frequently, compressible flows are regarded as flows in which the density varies by more than 5 to 10 percent.

Streamlines, Pathlines, and Streaklines These are curves in a flow field which provide insight into the flow pattern. *Streamlines* are tangent at every point to the local instantaneous velocity vector. A *pathline* is the path followed by a material element of fluid; it coincides with a streamline if the flow is steady. In unsteady flow the pathlines generally do not coincide with streamlines. *Streaklines* are curves on which are found all the material particles which passed through a particular point in space at some earlier time. For example, a streakline is revealed by releasing smoke or dye at a point in a flow field. For steady flows, streamlines, pathlines, and streaklines are indistinguishable. In two-dimensional incompressible flows, streamlines are contours of the **stream function.**

One-dimensional Flow Many flows of great practical importance, such as those in pipes and channels, are treated as one-dimensional flows. There is a single direction called the *flow direction;* velocity components perpendicular to this direction are either zero or considered unimportant. Variations of quantities such as velocity, pressure, density, and temperature are considered only in the flow direction. The fundamental conservation equations of fluid mechanics are greatly simplified for one-dimensional flows. A broader category of one-dimensional flow is one where there is only one nonzero velocity component, which depends on only one coordinate direction, and this coordinate direction may or may not be the same as the flow direction.

Rate of Deformation Tensor For general three-dimensional flows, where all three velocity components may be important and may vary in all three coordinate directions, the concept of deformation previously introduced must be generalized. The rate of deformation tensor D_{ij} has nine components. In Cartesian coordinates,

$$D_{ij} = \left(\frac{\partial v_i}{\partial x_j} + \frac{\partial v_j}{\partial x_i} \right) \quad (6\text{-}6)$$

where the subscripts i and j refer to the three coordinate directions. Some authors define the deformation rate tensor as one-half of that given by Eq. (6-6).

Vorticity The relative motion between two points in a fluid can be decomposed into three components: rotation, dilatation, and deformation. The rate of deformation tensor has been defined. Dilatation refers to the volumetric expansion or compression of the fluid, and vanishes for incompressible flow. Rotation is described by a tensor $\omega_{ij} = \partial v_i/\partial x_j - \partial v_j/\partial x_i$. The vector of vorticity given by one-half the curl of the velocity vector is another measure of rotation. In two-dimensional flow in the x-y plane, the vorticity ω is given by

$$\omega = \frac{1}{2}\left(\frac{\partial v_y}{\partial x} - \frac{\partial v_x}{\partial y}\right) \tag{6-7}$$

Here ω is the magnitude of the vorticity vector, which is directed along the z axis. An **irrotational** flow is one with zero vorticity. Irrotational flows have been widely studied because of their useful mathematical properties and applicability to flow regions where viscous effects may be neglected. Such flows without viscous effects are called **inviscid** flows.

Laminar and Turbulent Flow, Reynolds Number   These terms refer to two distinct types of flow. In *laminar flow,* there are smooth streamlines and the fluid velocity components vary smoothly with position, and with time if the flow is unsteady. The flow described in reference to Fig. 6-1 is laminar. In *turbulent flow,* there are no smooth streamlines, and the velocity shows chaotic fluctuations in time and space. Velocities in turbulent flow may be reported as the sum of a time-averaged velocity and a velocity fluctuation from the average. For any given flow geometry, a dimensionless **Reynolds number** may be defined for a Newtonian fluid as $\mathrm{Re} = LU\rho/\mu$ where L is a characteristic length. Below a critical value of Re the flow is laminar, while above the critical value a transition to turbulent flow occurs. The geometry-dependent critical Reynolds number is determined experimentally.

CONSERVATION EQUATIONS

Macroscopic and Microscopic Balances   Three postulates, regarded as laws of physics, are fundamental in fluid mechanics. These are *conservation of mass, conservation of momentum,* and *conservation of energy.* In addition, two other postulates, *conservation of moment of momentum* (angular momentum) and the *entropy inequality* (second law of thermodynamics) have occasional use. The conservation principles may be applied either to material systems or to control volumes in space. Most often, control volumes are used. The control volumes may be either of finite or differential size, resulting in either **algebraic** or **differential** conservation equations, respectively. These are often called **macroscopic** and **microscopic** balance equations.

Macroscopic Equations   An arbitrary control volume of finite size V_a is bounded by a surface of area A_a with an outwardly directed unit normal vector **n**. The control volume is not necessarily fixed in space. Its boundary moves with velocity **w**. The fluid velocity is **v**. Figure 6-3 shows the arbitrary control volume.

Mass balance   Applied to the control volume, the principle of conservation of mass may be written as (Whitaker, *Introduction to Fluid Mechanics,* Prentice-Hall, Englewood Cliffs, N.J., 1968, Krieger, Huntington, N.Y., 1981)

$$\frac{d}{dt}\int_{V_a}\rho\,dV + \int_{A_a}\rho(v-w)\cdot n\,dA = 0 \tag{6-8}$$

This equation is also known as the **continuity** equation.

Simplified forms of Eq. (6-8) apply to special cases frequently found in practice. For a control volume fixed in space with one inlet of area A_1 through which an incompressible fluid enters the control volume at an average velocity V_1, and one outlet of area A_2 through which fluid leaves at an average velocity V_2, as shown in Fig. 6-4, the continuity equation becomes

$$V_1A_1 = V_2A_2 \tag{6-9}$$

The average velocity across a surface is given by

$$V = (1/A)\int_A v\,dA$$

where v is the local velocity component perpendicular to the inlet surface. The volumetric flow rate Q is the product of average velocity and the cross-sectional area, $Q = VA$. The average **mass velocity** is $G = \rho V$. For steady flows through fixed control volumes with multiple inlets and/or outlets, conservation of mass requires that the sum of inlet mass flow rates equals the sum of outlet mass flow rates. For incompressible flows through fixed control volumes, the sum of inlet flow rates (mass or volumetric) equals the sum of exit flow rates, whether the flow is steady or unsteady.

Momentum Balance   Since momentum is a vector quantity, the momentum balance is a vector equation. Where gravity is the only body force acting on the fluid, the linear momentum principle, applied to the arbitrary control volume of Fig. 6-3, results in the following expression (Whitaker, ibid.).

$$\frac{d}{dt}\int_{V_a}\rho\mathbf{v}\,dV + \int_{A_a}\rho\mathbf{v}(\mathbf{v}-\mathbf{w})\cdot\mathbf{n}\,dA = \int_{V_a}\rho\mathbf{g}\,dV + \int_{A_a}\mathbf{t_n}\,dA \tag{6-10}$$

Here **g** is the gravity vector and $\mathbf{t_n}$ is the force per unit area exerted by the surroundings on the fluid in the control volume. The integrand of the area integral on the left-hand side of Eq. (6-10) is nonzero only on the entrance and exit portions of the control volume boundary. For the special case of steady flow at a mass flow rate $\dot{m}$ through a control volume fixed in space with one inlet and one outlet, (Fig. 6-4) with the inlet and outlet velocity vectors perpendicular to planar inlet and outlet surfaces, giving average velocity vectors $\mathbf{V}_1$ and $\mathbf{V}_2$, the momentum equation becomes

$$\dot{m}(\beta_2\mathbf{V}_2 - \beta_1\mathbf{V}_1) = -p_1\mathbf{A}_1 - p_2\mathbf{A}_2 + \mathbf{F} + M\mathbf{g} \tag{6-11}$$

where M is the total mass of fluid in the control volume. The factor β arises from the averaging of the velocity across the area of the inlet or outlet surface. It is the ratio of the area average of the square of velocity magnitude to the square of the area average velocity magnitude. For a uniform velocity, $\beta = 1$. For turbulent flow, β is nearly unity, while for laminar pipe flow with a parabolic velocity profile, $\beta = 4/3$. The vectors $\mathbf{A}_1$ and $\mathbf{A}_2$ have magnitude equal to the areas of the inlet and outlet surfaces, respectively, and are outwardly directed normal to the surfaces. The vector **F** is the force exerted on the fluid by the nonflow boundaries of the control volume. It is also assumed that the stress vector $\mathbf{t_n}$ is normal to the inlet and outlet surfaces, and that its magnitude may be approximated by the pressure p. Equation (6-11) may be generalized to multiple inlets and/or outlets. In such cases, the mass flow rates for all the inlets and outlets are not equal. A distinct flow rate $\dot{m}_i$ applies to each inlet or outlet i. To generalize the equation, $-p\mathbf{A}$ terms for each inlet and outlet, $-\dot{m}\beta\mathbf{V}$ terms for each inlet, and $\dot{m}\beta\mathbf{V}$ terms for each outlet are included.

Balance equations for angular momentum, or moment of momentum, may also be written. They are used less frequently than the lin-

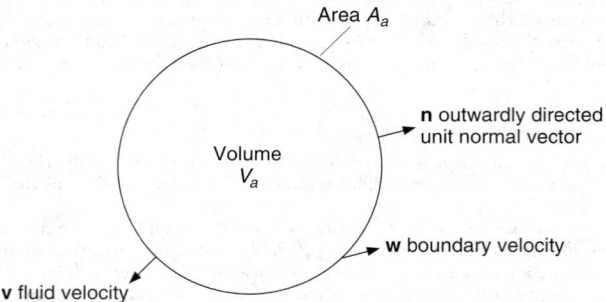

FIG. 6-3   Arbitrary control volume for application of conservation equations.

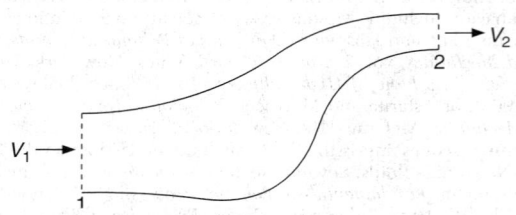

FIG. 6-4   Fixed control volume with one inlet and one outlet.

ear momentum equations. See Whitaker (*Introduction to Fluid Mechanics*, Prentice-Hall, Englewood Cliffs, N.J., 1968, Krieger, Huntington, N.Y., 1981; or Shames (*Mechanics of Fluids*, 3d ed., McGraw-Hill, New York, 1992).

Total Energy Balance   The total energy balance derives from the first law of thermodynamics. Applied to the arbitrary control volume of Fig. 6-3, it leads to an equation for the rate of change of the sum of internal, kinetic, and gravitational potential energy. In this equation, u is the internal energy per unit mass, v is the magnitude of the velocity vector $\mathbf{v}$, z is elevation, g is the gravitational acceleration, and $\mathbf{q}$ is the heat flux vector:

$$\frac{d}{dt}\int_{V_a}\rho\left(u+\frac{v^2}{2}+gz\right)dV + \int_{A_a}\rho\left(u+\frac{v^2}{2}+gz\right)(\mathbf{v}-\mathbf{w})\cdot\mathbf{n}\,dA$$

$$= \int_{A_a}(\mathbf{v}\cdot\mathbf{t_n})\,dA - \int_{A_a}(\mathbf{q}\cdot\mathbf{n})\,dA \quad (6\text{-}12)$$

The first integral on the right-hand side is the rate of work done on the fluid in the control volume by forces at the boundary. It includes both work done by moving solid boundaries and work done at flow entrances and exits. The work done by moving solid boundaries also includes that by such surfaces as pump impellers; this work is called **shaft work;** its rate is $\dot{W}_S$.

A useful simplification of the total energy equation applies to a particular set of assumptions. These are a control volume with fixed solid boundaries, except for those producing shaft work, steady state conditions, and mass flow at a rate $\dot{m}$ through a single planar entrance and a single planar exit (Fig. 6-4), to which the velocity vectors are perpendicular. As with Eq. (6-11), it is assumed that the stress vector $\mathbf{t_n}$ is normal to the entrance and exit surfaces and may be approximated by the pressure p. The **equivalent pressure,** $p + \rho gz$, is assumed to be uniform across the entrance and exit. The average velocity at the entrance and exit surfaces is denoted by V. Subscripts 1 and 2 denote the entrance and exit, respectively.

$$h_1 + \alpha_1\frac{V_1^2}{2} + gz_1 = h_2 + \alpha_2\frac{V_2^2}{2} + gz_2 - \delta Q - \delta W_S \quad (6\text{-}13)$$

Here, h is the enthalpy per unit mass, $h = u + p/\rho$. The shaft work per unit of mass flowing through the control volume is $\delta W_S = \dot{W}_S/\dot{m}$. Similarly, δQ is the heat input rate per unit of mass. The factor α is the ratio of the cross-sectional area average of the cube of the velocity to the cube of the average velocity. For a uniform velocity profile, $\alpha = 1$. In turbulent flow, α is usually assumed to equal unity; in turbulent pipe flow, it is typically about 1.07. For laminar flow in a circular pipe with a parabolic velocity profile, $\alpha = 2$.

Mechanical Energy Balance, Bernoulli Equation   A balance equation for the sum of kinetic and potential energy may be obtained from the momentum balance by forming the scalar product with the velocity vector. The resulting equation, called the *mechanical energy balance*, contains a term accounting for the dissipation of mechanical energy into thermal energy by viscous forces. The mechanical energy equation is also derivable from the total energy equation in a way that reveals the relationship between the dissipation and entropy generation. The macroscopic mechanical energy balance for the arbitrary control volume of Fig. 6-3 may be written, with p = thermodynamic pressure, as

$$\frac{d}{dt}\int_{V_a}\rho\left(\frac{v^2}{2}+gz\right)dV + \int_{A_a}\rho\left(\frac{v^2}{2}+gz\right)(\mathbf{v}-\mathbf{w})\cdot\mathbf{n}\,dA$$

$$= \int_{V_a}p\,\boldsymbol{\nabla}\cdot\mathbf{v}\,dV + \int_{A_a}(\mathbf{v}\cdot\mathbf{t_n})\,dA - \int_{V_a}\Phi\,dV \quad (6\text{-}14)$$

The last term is the rate of viscous energy dissipation to internal energy, $\dot{E}_v = \int_{V_a}\Phi\,dV$, also called the rate of viscous losses. These losses are the origin of frictional pressure drop in fluid flow. Whitaker and Bird, Stewart, and Lightfoot provide expressions for the dissipation function Φ for Newtonian fluids in terms of the local velocity gradients. However, when using macroscopic balance equations the local velocity field within the control volume is usually unknown. For such cases additional information, which may come from empirical correlations, is needed.

For the same special conditions as for Eq. (6-13), the mechanical energy equation is reduced to

$$\alpha_1\frac{V_1^2}{2} + gz_1 + \delta W_S = \alpha_2\frac{V_2^2}{2} + gz_2 + \int_{p_1}^{p_2}\frac{dp}{\rho} + l_v \quad (6\text{-}15)$$

Here $l_v = \dot{E}_v/\dot{m}$ is the energy dissipation per unit mass. This equation has been called the **engineering Bernoulli equation.** For an incompressible flow, Eq. (6-15) becomes

$$\frac{p_1}{\rho} + \alpha_1\frac{V_1^2}{2} + gz_1 + \delta W_S = \frac{p_2}{\rho} + \alpha_2\frac{V_2^2}{2} + gz_2 + l_v \quad (6\text{-}16)$$

The Bernoulli equation can be written for incompressible, inviscid flow along a streamline, where no shaft work is done.

$$\frac{p_1}{\rho} + \frac{V_1^2}{2} + gz_1 = \frac{p_2}{\rho} + \frac{V_2^2}{2} + gz_2 \quad (6\text{-}17)$$

Unlike the momentum equation (Eq. [6-11]), the Bernoulli equation is not easily generalized to multiple inlets or outlets.

Microscopic Balance Equations   Partial differential balance equations express the conservation principles at a point in space. Equations for mass, momentum, total energy, and mechanical energy may be found in Whitaker (ibid.), Bird, Stewart, and Lightfoot (*Transport Phenomena*, Wiley, New York, 1960), and Slattery (*Momentum, Heat and Mass Transfer in Continua*, 2d ed., Krieger, Huntington, N.Y., 1981), for example. These references also present the equations in other useful coordinate systems besides the cartesian system. The coordinate systems are fixed in inertial reference frames. The two most used equations, for mass and momentum, are presented here.

Mass Balance, Continuity Equation   The continuity equation, expressing conservation of mass, is written in cartesian coordinates as

$$\frac{\partial\rho}{\partial t} + \frac{\partial\rho v_x}{\partial x} + \frac{\partial\rho v_y}{\partial y} + \frac{\partial\rho v_z}{\partial z} = 0 \quad (6\text{-}18)$$

In terms of the **substantial derivative,** D/Dt,

$$\frac{D\rho}{Dt} \equiv \frac{\partial\rho}{\partial t} + v_x\frac{\partial\rho}{\partial x} + v_y\frac{\partial\rho}{\partial y} + v_z\frac{\partial\rho}{\partial z} = -\rho\left(\frac{\partial v_x}{\partial x} + \frac{\partial v_y}{\partial y} + \frac{\partial v_z}{\partial z}\right) \quad (6\text{-}19)$$

The substantial derivative, also called the **material derivative,** is the rate of change in a Lagrangian reference frame, that is, following a material particle. In vector notation the continuity equation may be expressed as

$$\frac{D\rho}{Dt} = -\rho\nabla\cdot\mathbf{v} \quad (6\text{-}20)$$

For incompressible flow,

$$\nabla\cdot\mathbf{v} = \frac{\partial v_x}{\partial x} + \frac{\partial v_y}{\partial y} + \frac{\partial v_z}{\partial z} = 0 \quad (6\text{-}21)$$

Stress Tensor   The stress tensor is needed to completely describe the stress state for microscopic momentum balances in multidimensional flows. The components of the stress tensor σ_{ij} give the force in the j direction on a plane perpendicular to the i direction, using a sign convention defining a positive stress as one where the fluid with the greater i coordinate value exerts a force in the positive i direction on the fluid with the lesser i coordinate. Several references in fluid mechanics and continuum mechanics provide discussions, to various levels of detail, of stress in a fluid (Denn; Bird, Stewart, and Lightfoot; Schlichting; Fung [*A First Course in Continuum Mechanics*, 2d. ed., Prentice-Hall, Englewood Cliffs, N.J., 1977]; Truesdell and Toupin [in Flügge, *Handbuch der Physik*, vol. 3/1, Springer-Verlag, Berlin, 1960]; Slattery [*Momentum, Energy and Mass Transfer in Continua*, 2d ed., Krieger, Huntington, N.Y., 1981]).

The stress has an isotropic contribution due to fluid pressure and dilatation, and a **deviatoric** contribution due to viscous deformation effects. The deviatoric contribution for a Newtonian fluid is the three-dimensional generalization of Eq. (6-2):

$$\tau_{ij} = \mu D_{ij} \quad (6\text{-}22)$$

The total stress is

$$\sigma_{ij} = (-p + \lambda\nabla\cdot\mathbf{v})\delta_{ij} + \tau_{ij} \quad (6\text{-}23)$$

The identity tensor δ_{ij} is zero for $i \neq j$ and unity for $i = j$. The coefficient λ is a material property related to the **bulk viscosity,** $\kappa = \lambda + 2\mu/3$. There is considerable uncertainty about the value of κ. Traditionally, Stokes' hypothesis, $\kappa = 0$, has been invoked, but the validity of this hypothesis is doubtful (Slattery, ibid.). For incompressible flow, the value of bulk viscosity is immaterial as Eq. (6-23) reduces to

$$\sigma_{ij} = -p\delta_{ij} + \tau_{ij} \qquad (6\text{-}24)$$

Similar generalizations to multidimensional flow are necessary for non-Newtonian constitutive equations.

Cauchy Momentum and Navier-Stokes Equations   The differential equations for conservation of momentum are called the **Cauchy momentum equations.** These may be found in general form in most fluid mechanics texts (e.g., Slattery [ibid.]; Denn; Whitaker; and Schlichting). For the important special case of an incompressible Newtonian fluid with constant viscosity, substitution of Eqs. (6-22) and (6-24) lead to the **Navier-Stokes equations,** whose three Cartesian components are

$$\rho\left(\frac{\partial v_x}{\partial t} + v_x\frac{\partial v_x}{\partial x} + v_y\frac{\partial v_x}{\partial y} + v_z\frac{\partial v_x}{\partial z}\right)$$
$$= -\frac{\partial p}{\partial x} + \mu\left(\frac{\partial^2 v_x}{\partial x^2} + \frac{\partial^2 v_x}{\partial y^2} + \frac{\partial^2 v_x}{\partial z^2}\right) + \rho g_x \quad (6\text{-}25)$$

$$\rho\left(\frac{\partial v_y}{\partial t} + v_x\frac{\partial v_y}{\partial x} + v_y\frac{\partial v_y}{\partial y} + v_z\frac{\partial v_y}{\partial z}\right)$$
$$= -\frac{\partial p}{\partial y} + \mu\left(\frac{\partial^2 v_y}{\partial x^2} + \frac{\partial^2 v_y}{\partial y^2} + \frac{\partial^2 v_y}{\partial z^2}\right) + \rho g_y \quad (6\text{-}26)$$

$$\rho\left(\frac{\partial v_z}{\partial t} + v_x\frac{\partial v_z}{\partial x} + v_y\frac{\partial v_z}{\partial y} + v_z\frac{\partial v_z}{\partial z}\right)$$
$$= -\frac{\partial p}{\partial z} + \mu\left(\frac{\partial^2 v_z}{\partial x^2} + \frac{\partial^2 v_z}{\partial y^2} + \frac{\partial^2 v_z}{\partial z^2}\right) + \rho g_z \quad (6\text{-}27)$$

In vector notation,

$$\rho\frac{D\mathbf{v}}{Dt} = \frac{\partial \mathbf{v}}{\partial t} + (\mathbf{v} \cdot \nabla)\mathbf{v} = -\nabla p + \mu\nabla^2\mathbf{v} + \rho\mathbf{g} \quad (6\text{-}28)$$

The pressure and gravity terms may be combined by replacing the pressure p by the equivalent pressure $P = p + \rho gz$. The left-hand side terms of the Navier-Stokes equations are the **inertial terms,** while the terms including viscosity μ are the **viscous terms.** Limiting cases under which the Navier-Stokes equations may be simplified include **creeping flows** in which the inertial terms are neglected, **potential flows** (inviscid or irrotational flows) in which the viscous terms are neglected, and **boundary layer** and **lubrication** flows in which certain terms are neglected based on scaling arguments. Creeping flows are described by Happel and Brenner (*Low Reynolds Number Hydrodynamics,* Prentice-Hall, Englewood Cliffs, N.J., 1965); potential flows by Lamb (*Hydrodynamics,* 6th ed., Dover, New York, 1945) and Milne-Thompson (*Theoretical Hydrodynamics,* 5th ed., Macmillan, New York, 1968); boundary layer theory by Schlichting (*Boundary Layer Theory,* 8th ed., McGraw-Hill, New York, 1987), and lubrication theory by Batchelor (*An Introduction to Fluid Dynamics,* Cambridge University, Cambridge, 1967) and Denn (*Process Fluid Mechanics,* Prentice-Hall, Englewood Cliffs, N.J., 1980).

Because the Navier-Stokes equations are first-order in pressure and second-order in velocity, their solution requires one pressure boundary condition and two velocity boundary conditions (for each velocity component) to completely specify the solution. The **no slip** condition, which requires that the fluid velocity equal the velocity of any bounding solid surface, occurs in most problems. Specification of velocity is a type of boundary condition sometimes called a *Dirichlet condition.* Often boundary conditions involve stresses, and thus velocity gradients, rather than the velocities themselves. Specification of velocity derivatives is a *Neumann boundary condition.* For example, at the boundary between a viscous liquid and a gas, it is often assumed that the liquid shear stresses are zero. In numerical solution of the Navier-

Stokes equations, Dirichlet and Neumann, or **essential** and **natural,** boundary conditions may be satisfied by different means.

Fluid statics, discussed in Sec. 10 of the *Handbook* in reference to pressure measurement, is the branch of fluid mechanics in which the fluid velocity is either zero or is uniform and constant relative to an inertial reference frame. With velocity gradients equal to zero, the momentum equation reduces to a simple expression for the pressure field, $\nabla p = \rho \mathbf{g}$. Letting z be directed vertically upward, so that $g_z = -g$ where g is the gravitational acceleration (9.806 m²/s), the pressure field is given by

$$dp/dz = -\rho g \qquad (6\text{-}29)$$

This equation applies to any incompressible or compressible static fluid. For an incompressible liquid, pressure varies linearly with depth. For compressible gases, p is obtained by integration accounting for the variation of ρ with z.

The **force exerted on a submerged planar surface** of area A is given by $F = p_c A$ where p_c is the pressure at the geometrical **centroid** of the surface. The **center of pressure,** the point of application of the net force, is always lower than the centroid. For details see, for example, Shames, where may also be found discussion of forces on **curved surfaces, buoyancy,** and **stability of floating bodies.**

Examples   Four examples follow, illustrating the application of the conservation equations to obtain useful information about fluid flows.

Example 1: Force Exerted on a Reducing Bend   An incompressible fluid flows through a reducing elbow (Fig. 6-5) situated in a horizontal plane. The inlet velocity V_1 is given and the pressures p_1 and p_2 are known. Selecting the inlet and outlet surfaces 1 and 2 as shown, the continuity equation Eq. (6-9) can be used to find the exit velocity $V_2 = V_1 A_1/A_2$. The mass flow rate is obtained by $\dot{m} = \rho V_1 A_1$.

Assume that the velocity profile is nearly uniform so that β is approximately unity. The force exerted on the fluid by the bend has x and y components; these can be found from Eq. (6-11). The x component gives

$$F_x = \dot{m}(V_{2x} - V_{1x}) + p_1 A_{1x} + p_2 A_{2x}$$

while the y component gives

$$F_y = \dot{m}(V_{2y} - V_{1y}) + p_1 A_{1y} + p_2 A_{2y}$$

The velocity components are $V_{1x} = V_1$, $V_{1y} = 0$, $V_{2x} = V_2 \cos\theta$, and $V_{2y} = V_2 \sin\theta$. The area vector components are $A_{1x} = -A_1$, $A_{1y} = 0$, $A_{2x} = A_2 \cos\theta$, and $A_{2y} = A_2 \sin\theta$. Therefore, the force components may be calculated from

$$F_x = \dot{m}(V_2 \cos\theta - V_1) - p_1 A_1 + p_2 A_2 \cos\theta$$
$$F_y = \dot{m}V_2 \sin\theta + p_2 A_2 \sin\theta$$

The force acting on the fluid is $\mathbf{F}$; the equal and opposite force exerted by the fluid on the bend is $-\mathbf{F}$.

Example 2: Simplified Ejector   Figure 6-6 shows a very simplified sketch of an ejector, a device that uses a high velocity primary fluid to pump another (secondary) fluid. The continuity and momentum equations may be

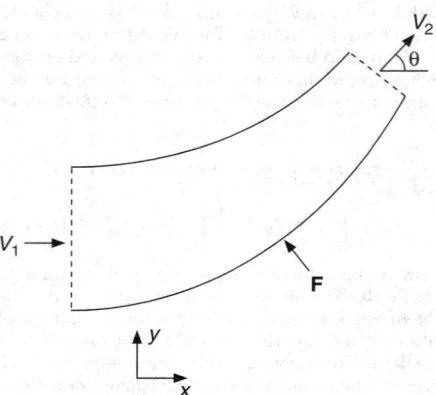

FIG. 6-5   Force at a reducing bend. $\mathbf{F}$ is the force exerted by the bend on the fluid. The force exerted by the fluid on the bend is $-\mathbf{F}$.

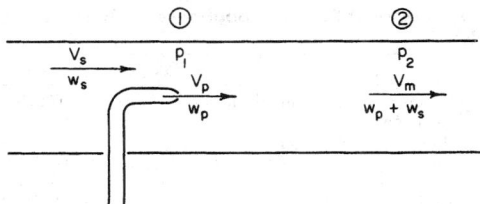

FIG. 6-6   Draft-tube ejector.

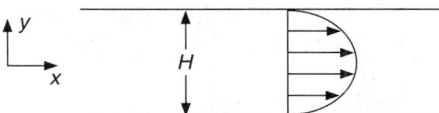

FIG. 6-8   Plane Poiseuille flow.

applied on the control volume with inlet and outlet surfaces 1 and 2 as indicated in the figure. The cross-sectional area is uniform, $A_1 = A_2 = A$. Let the mass flow rates and velocities of the primary and secondary fluids be $\dot{m}_p$, $\dot{m}_s$, V_p and V_s. Assume for simplicity that the density is uniform. Conservation of mass gives $\dot{m}_2 = \dot{m}_p + \dot{m}_s$. The exit velocity is $V_2 = \dot{m}_2/(\rho A)$. The principle momentum exchange in the ejector occurs between the two fluids. Relative to this exchange, the force exerted by the walls of the device are found to be small. Therefore, the force term F is neglected from the momentum equation. Written in the flow direction, assuming uniform velocity profiles, and using the extension of Eq. (6-11) for multiple inlets, it gives the pressure rise developed by the device:

$$(p_2 - p_1)A = (\dot{m}_p + \dot{m}_s)V_2 - \dot{m}_p V_p - \dot{m}_s V_s$$

Application of the momentum equation to ejectors of other types is discussed in Lapple (*Fluid and Particle Dynamics,* University of Delaware, Newark, 1951) and in Sec. 10 of the *Handbook.*

Example 3: Venturi Flowmeter   An incompressible fluid flows through the venturi flowmeter in Fig. 6-7. An equation is needed to relate the flow rate Q to the pressure drop measured by the manometer. This problem can be solved using the mechanical energy balance. In a well-made venturi, viscous losses are negligible, the pressure drop is entirely the result of acceleration into the throat, and the flow rate predicted neglecting losses is quite accurate. The inlet area is A and the throat area is a.

With control surfaces at 1 and 2 as shown in the figure, Eq. (6-17) in the absence of losses and shaft work gives

$$\frac{p_1}{\rho} + \frac{V_1^2}{2} = \frac{p_2}{\rho} + \frac{V_2^2}{2}$$

The continuity equation gives $V_2 = V_1 A/a$, and $V_1 = Q/A$. The pressure drop measured by the manometer is $p_1 - p_2 = (\rho_m - \rho)g\Delta z$. Substituting these relations into the energy balance and rearranging, the desired expression for the flow rate is found.

$$Q = \frac{1}{A} \sqrt{\frac{2(\rho_m - \rho)g\Delta z}{\rho[(A/a)^2 - 1]}}$$

Example 4: Plane Poiseuille Flow   An incompressible Newtonian fluid flows at a steady rate in the x direction between two very large flat plates, as shown in Fig. 6-8. The flow is laminar. The velocity profile is to be found. This example is found in most fluid mechanics textbooks; the solution presented here closely follows Denn.

This problem requires use of the microscopic balance equations because the velocity is to be determined as a function of position. The boundary conditions for this flow result from the no-slip condition. All three velocity components must be zero at the plate surfaces, $y = H/2$ and $y = -H/2$.

Assume that the flow is **fully developed,** that is, all velocity derivatives van-

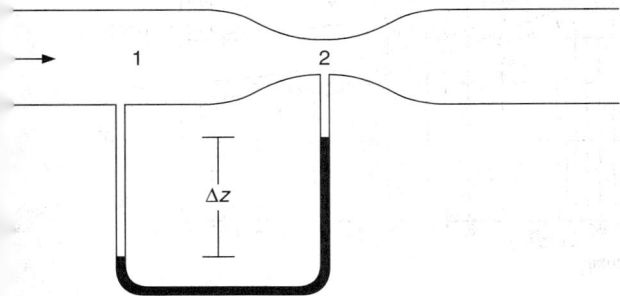

FIG. 6-7   Venturi flowmeter.

ish in the x direction. Since the flow field is infinite in the z direction, all velocity derivatives should be zero in the z direction. Therefore, velocity components are a function of y alone. It is also assumed that there is no flow in the z direction, so $v_z = 0$. The continuity equation Eq. (6-21), with $v_z = 0$ and $\partial v_x/\partial x = 0$, reduces to

$$\frac{dv_y}{dy} = 0.$$

Since $v_y = 0$ at $y = \pm H/2$, the continuity equation integrates to $v_y = 0$. This is a direct result of the assumption of fully developed flow.

The Navier-Stokes equations are greatly simplified when it is noted that $v_y = v_z = 0$ and $\partial v_x/\partial x = \partial v_x/\partial z = \partial v_x/\partial t = 0$. The three components are written in terms of the equivalent pressure P:

$$0 = -\frac{\partial P}{\partial x} + \mu \frac{\partial^2 v_x}{\partial y^2}$$

$$0 = -\frac{\partial P}{\partial y}$$

$$0 = -\frac{\partial P}{\partial z}$$

The latter two equations require that P is a function only of x, and therefore $\partial P/\partial x = dP/dx$. Inspection of the first equation shows one term which is a function only of x and one which is only a function of y. This requires that both terms are constant. The pressure gradient $-dP/dx$ is constant. The x-component equation becomes

$$\frac{d^2 v_x}{dy^2} = \frac{1}{\mu} \frac{dP}{dx}$$

Two integrations of the x-component equation give

$$v_x = \frac{1}{2\mu} \frac{dP}{dx} y^2 + C_1 y + C_2$$

where the constants of integration C_1 and C_2 are evaluated from the boundary conditions $v_x = 0$ at $y = \pm H/2$. The result is

$$v_x = \frac{H^2}{8\mu}\left(-\frac{dP}{dx}\right)\left[1 - \left(\frac{2y}{H}\right)^2\right]$$

This is a **parabolic** velocity distribution. The average velocity $V = (1/H)\int_{-H/2}^{H/2} v_x \, dy$ is

$$V = \frac{H^2}{12\mu}\left(-\frac{dP}{dx}\right)$$

This flow is one-dimensional, as there is only one nonzero velocity component, v_x, which, along with the pressure, varies in only one coordinate direction.

INCOMPRESSIBLE FLOW IN PIPES AND CHANNELS

Mechanical Energy Balance   The mechanical energy balance, Eq. (6-16), for **fully developed** incompressible flow in a straight circular pipe of constant diameter D reduces to

$$\frac{p_1}{\rho} + gz_1 = \frac{p_2}{\rho} + gz_2 + l_v \qquad (6\text{-}30)$$

In terms of the equivalent pressure, $P \equiv p + \rho gz$,

$$P_1 - P_2 = \rho l_v \qquad (6\text{-}31)$$

The pressure drop due to frictional losses l_v is proportional to pipe length L for fully developed flow and may be denoted as the (positive) quantity $\Delta P \equiv P_1 - P_2$.

Friction Factor and Reynolds Number   For a Newtonian fluid in a smooth pipe, dimensional analysis relates the frictional pressure drop per unit length $\Delta P/L$ to the pipe diameter D, density ρ, and average velocity V through two dimensionless groups, the **Fanning friction factor** f and the **Reynolds number** Re.

$$f \equiv \frac{D\Delta P}{2\rho V^2 L} \qquad (6\text{-}32)$$

$$\mathrm{Re} \equiv \frac{DV\rho}{\mu} \qquad (6\text{-}33)$$

For smooth pipe, the friction factor is a function only of the Reynolds number. In rough pipe, the relative roughness ϵ/D also affects the friction factor. Figure 6-9 plots f as a function of Re and ϵ/D. Values of ϵ for various materials are given in Table 6-1. The Fanning friction factor should not be confused with the Darcy friction factor used by Moody (*Trans. ASME*, **66**, 671 [1944]), which is four times greater. Using the momentum equation, the stress at the wall of the pipe may be expressed in terms of the friction factor:

$$\tau_w = f\,\frac{\rho V^2}{2} \qquad (6\text{-}34)$$

Laminar and Turbulent Flow Below a **critical Reynolds number** of about 2,100, the flow is laminar; over the range 2,100 < Re < 5,000 there is a transition to turbulent flow. For laminar flow, the Hagen-Poiseuille equation

$$f = \frac{16}{\mathrm{Re}}, \qquad \mathrm{Re} \le 2{,}100 \qquad (6\text{-}35)$$

may be derived from the Navier-Stokes equation and is in excellent agreement with experimental data. It may be rewritten in terms of volumetric flow rate, $Q = V\pi D^2/4$, as

$$Q = \frac{\pi \Delta P D^4}{128\mu L}, \qquad \mathrm{Re} \le 2{,}100 \qquad (6\text{-}36)$$

TABLE 6-1 Values of Surface Roughness for Various Materials*

Material	Surface roughness ϵ, mm
Drawn tubing (brass, lead, glass, and the like)	0.00152
Commercial steel or wrought iron	0.0457
Asphalted cast iron	0.122
Galvanized iron	0.152
Cast iron	0.259
Wood stove	0.183–0.914
Concrete	0.305–3.05
Riveted steel	0.914–9.14

* From Moody, *Trans. Am. Soc. Mech. Eng.*, **66**, 671–684 (1944); *Mech. Eng.*, **69**, 1005–1006 (1947). Additional values of ϵ for various types or conditions of concrete wrought-iron, welded steel, riveted steel, and corrugated-metal pipes are given in Brater and King, *Handbook of Hydraulics*, 6th ed., McGraw-Hill, New York, 1976, pp. 6-12–6-13. To convert millimeters to feet, multiply by 3.281×10^{-3}.

For turbulent flow in smooth tubes, the Blasius equation gives the friction factor accurately for a wide range of Reynolds numbers.

$$f = \frac{0.079}{\mathrm{Re}^{0.25}}, \qquad 4{,}000 < \mathrm{Re} < 10^5 \qquad (6\text{-}37)$$

The Colebrook formula (Colebrook, *J. Inst. Civ. Eng.* [*London*], **11**, 133–156 [1938–39]) gives a good approximation for the f-Re-(ϵ/D) data for rough pipes over the entire turbulent flow range:

$$\frac{1}{\sqrt{f}} = -4 \log\left[\frac{\epsilon}{3.7D} + \frac{1.256}{\mathrm{Re}\sqrt{f}} \right] \qquad \mathrm{Re} > 4{,}000 \qquad (6\text{-}38)$$

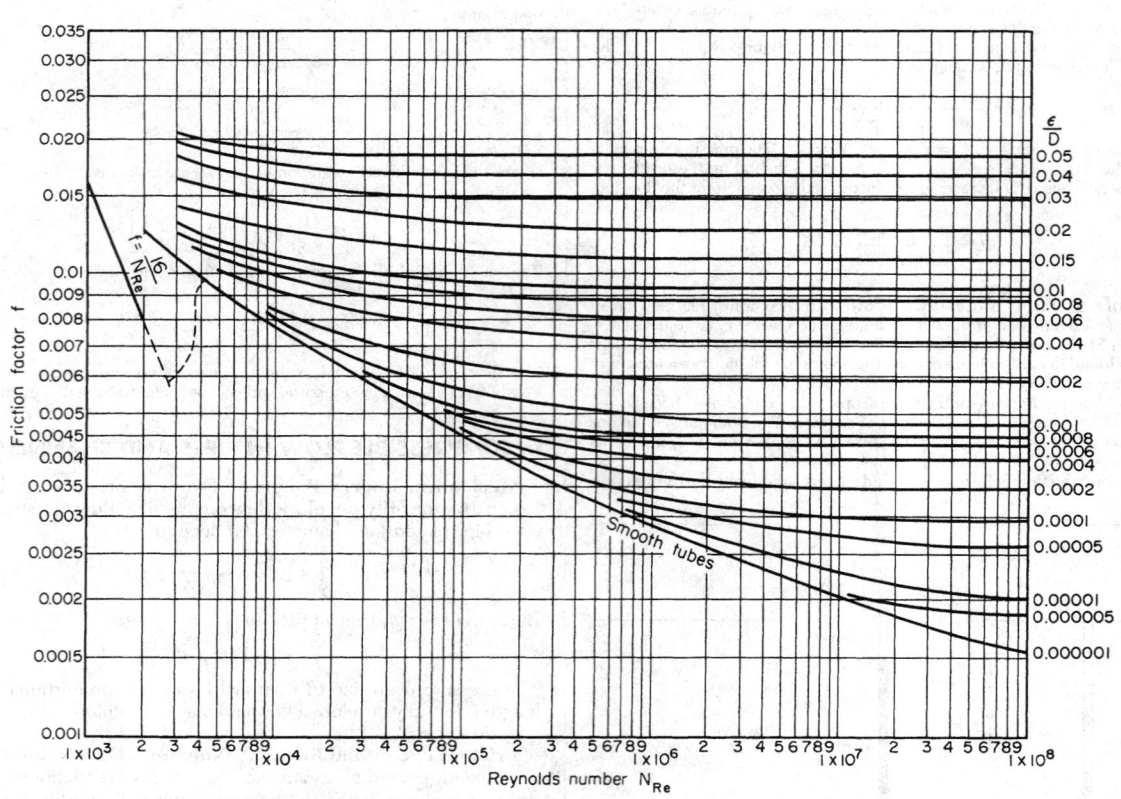

FIG. 6-9 Fanning Friction Factors. Reynolds number Re = $DV\rho/\mu$, where D = pipe diameter, V = velocity, ρ = fluid density, and μ = fluid viscosity. (*Based on Moody*, Trans. ASME, **66**, 671 [1944].)

An equation by Churchill (*Chem. Eng.*, **84**[24], 91–92 [Nov. 7, 1977]) for both smooth and rough tubes offers the advantage of being explicit in f:

$$\frac{1}{\sqrt{f}} = -4 \log \left[\frac{0.27\epsilon}{D} + (7/\text{Re})^{0.9} \right] \qquad \text{Re} > 4{,}000 \qquad (6\text{-}39)$$

In laminar flow, f is independent of ϵ/D. In turbulent flow, the friction factor for rough pipe follows the smooth tube curve for a range of Reynolds numbers (hydraulically smooth flow). For greater Reynolds numbers, f deviates from the smooth pipe curve, eventually becoming independent of Re. This region, often called *complete turbulence*, is frequently encountered in commercial pipe flows. The Reynolds number above which f becomes essentially independent of Re is (Davies, *Turbulence Phenomena*, Academic, New York, 1972, p. 37)

$$\text{Re} = \frac{20[3.2 - 2.46 \ln (\epsilon/D)]}{(\epsilon/D)} \qquad (6\text{-}40)$$

Roughness may also affect the transition from laminar to turbulent flow (Schlichting).

Common pipe flow problems include calculation of pressure drop given the flow rate (or velocity) and calculation of flow rate (or velocity) given pressure drop. When flow rate is given, the Reynolds number is first calculated to determine the flow regime, so that the appropriate relations between f and Re (or pressure drop and velocity or flow rate) are used. When pressure drop is given and the velocity is unknown, the Reynolds number and flow regime cannot be immediately determined. It is necessary to assume the flow regime and then verify by checking Re afterward. With experience, the initial guess for the flow regime will usually prove correct. When solving Eq. (6-38) for velocity when pressure drop is given, it is useful to note that the right-hand side is independent of velocity since $\text{Re}\sqrt{f} = D^{3/2}/\mu)\sqrt{\rho \Delta P/(2L)}$.

As Fig. 6-9 suggests, the friction factor is uncertain in the transition range $2{,}100 < \text{Re} < 4{,}000$ and a conservative choice should be made for design purposes.

Velocity Profiles   In laminar flow, the solution of the Navier-Stokes equation, corresponding to the Hagen-Poiseuille equation, gives the velocity v as a function of radial position r in a circular pipe of radius R in terms of the average velocity $V = Q/A$. The **parabolic** profile, with centerline velocity twice the average velocity, is shown in Fig. 6-10.

$$v = 2V\left(1 - \frac{r^2}{R^2}\right) \qquad (6\text{-}41)$$

In turbulent flow, the velocity profile is much more blunt, with most of the velocity gradient being in a region near the wall, described by a **universal** velocity profile. It is characterized by a **viscous sublayer**, a **turbulent core**, and a **buffer zone** in between.

Viscous sublayer

$$u_+ = y_+ \qquad \text{for} \qquad y_+ < 5 \qquad (6\text{-}42)$$

Buffer zone

$$u_+ = 5.00 \ln y_+ - 3.05 \qquad \text{for} \qquad 5 < y_+ < 30 \qquad (6\text{-}43)$$

Turbulent core

$$u_+ = 2.5 \ln y_+ + 5.5 \qquad \text{for} \qquad y_+ > 30 \qquad (6\text{-}44)$$

Here, $u_+ = v/u_\circ$ is the dimensionless, time-averaged axial velocity, $u_\circ =$

$$v = 2V\left(1 - \frac{r^2}{R^2}\right)$$

$$v_{\max} = 2V$$

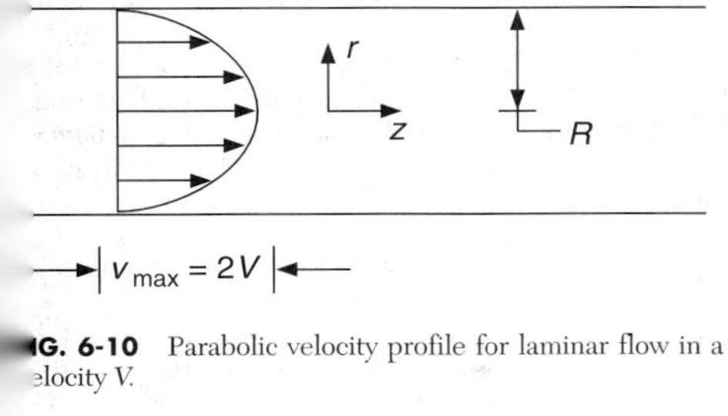

FIG. 6-10   Parabolic velocity profile for laminar flow in a pipe, with average velocity V.

$\sqrt{\tau_w/\rho}$ is the **friction velocity** and $\tau_w = f\rho V^2/2$ is the wall stress. The friction velocity is of the order of the root mean square velocity fluctuation perpendicular to the wall in the turbulent core. The dimensionless distance from the wall is $y_+ = yu_\circ \rho/\mu$. The universal velocity profile is valid in the wall region for any cross-sectional channel shape. For incompressible flow in constant diameter circular pipes, $\tau_w = \Delta P/4L$ where ΔP is the pressure drop in length L. In circular pipes, Eq. (6-44) gives a surprisingly good fit to experimental results over the entire cross section of the pipe, even though it is based on assumptions which are valid only near the pipe wall.

For rough pipes, the velocity profile in the turbulent core is given by

$$u_+ = 2.5 \ln y/\epsilon + 8.5 \qquad \text{for} \qquad y_+ > 30 \qquad (6\text{-}45)$$

when the dimensionless roughness $\epsilon_+ = \epsilon u_\circ \rho/\mu$ is greater than 5 to 10; for smaller ϵ_+, the velocity profile in the turbulent core is unaffected by roughness.

For velocity profiles in the transition region, see Patel and Head (*J. Fluid Mech.*, **38**, part 1, 181–201 [1969]) where profiles over the range $1{,}500 < \text{Re} < 10{,}000$ are reported.

Entrance and Exit Effects   In the entrance region of a pipe, some distance is required for the flow to adjust from upstream conditions to the fully developed flow pattern. This distance depends on the Reynolds number and on the flow conditions upstream. For a uniform velocity profile at the pipe entrance, the computed length in laminar flow required for the centerline velocity to reach 99 percent of its fully developed value is (Dombrowski, Foumeny, Ookawara and Riza, *Can. J. Chem. Engr.*, **71**, 472–476 [1993])

$$L_{\text{ent}}/D = 0.370 \exp (-0.148\text{Re}) + 0.0550\text{Re} + 0.260 \qquad (6\text{-}46)$$

In turbulent flow, the entrance length is about

$$L_{\text{ent}}/D = 40 \qquad (6\text{-}47)$$

The frictional losses in the entrance region are larger than those for the same length of fully developed flow. (See the subsection, "Frictional Losses in Pipeline Elements," following.) At the pipe exit, the velocity profile also undergoes rearrangement, but the exit length is much shorter than the entrance length. At low Re, it is about one pipe radius. At $\text{Re} > 100$, the exit length is essentially 0.

Residence Time Distribution   For laminar Newtonian pipe flow, the cumulative residence time distribution $F(\theta)$ is given by

$$F(\theta) = 0 \qquad \text{for} \qquad \theta < \frac{\theta_{\text{avg}}}{2}$$

$$F(\theta) = 1 - \frac{1}{4}\left(\frac{\theta_{\text{avg}}}{\theta}\right)^2 \qquad \text{for} \qquad \theta \geq \frac{\theta_{\text{avg}}}{2} \qquad (6\text{-}48)$$

where $F(\theta)$ is the fraction of material which resides in the pipe for less than time θ and θ_{avg} is the average residence time, $\theta = V/L$.

The residence time distribution in long transfer lines may be made narrower (more uniform) with the use of **flow inverters** or **static mixing elements.** These devices exchange fluid between the wall and central regions. Variations on the concept may be used to provide effective mixing of the fluid. See Godfrey ("Static Mixers," in Harnby, Edwards, and Nienow, *Mixing in the Process Industries*, 2d ed., Butterworth Heinemann, Oxford, 1992); Gretta and Smith (*Trans. ASME J. Fluids Eng.*, **115**, 255–263 [1993]); Kemblowski and Pustelnik (*Chem. Eng. Sci.*, **43**, 473–478 [1988]).

A theoretically derived equation for flow in helical pipe coils by Ruthven (*Chem. Eng. Sci.*, **26**, 1113–1121 [1971]; **33**, 628–629 [1978]) is given by

$$F(\theta) = 1 - \left(\frac{1}{4}\right)\left[\frac{\theta_{\text{avg}}}{\theta}\right]^{2.81} \qquad \text{for} \qquad 0.5 < \frac{\theta_{\text{avg}}}{\theta} < 1.63 \qquad (6\text{-}49)$$

and was substantially confirmed by Trivedi and Vasudeva (*Chem. Eng. Sci.*, **29**, 2291–2295 [1974]) for $0.6 < \text{De} < 6$ and $0.0036 < D/D_c < 0.097$ where $\text{De} = \text{Re}\sqrt{D/D_c}$ is the Dean number and D_c is the diameter of curvature of the coil. Measurements by Saxena and Nigam (*Chem. Eng. Sci.*, **34**, 425–426 [1979]) indicate that such a distribution will hold for $\text{De} > 1$. The residence time distribution for helical coils is narrower than for straight circular pipes, due to the secondary flow which exchanges fluid between the wall and center regions.

In turbulent flow, axial mixing is usually described in terms of turbulent diffusion or dispersion coefficients, from which cumulative residence time distribution functions can be computed. Davies (*Turbulence Phenomena*, Academic, New York, 1972, p. 93), gives $D_L = 1.01 v Re^{0.875}$ for the longitudinal dispersion coefficient. Levenspiel (*Chemical Reaction Engineering*, 2d ed., Wiley, New York, 1972, pp. 253–278) discusses the relations among various residence time distribution functions, and the relation between dispersion coefficient and residence time distribution.

Noncircular Channels   Calculation of frictional pressure drop in noncircular channels depends on whether the flow is laminar or turbulent, and on whether the channel is full or open. For **turbulent flow** in **ducts running full,** the **hydraulic diameter** D_H should be substituted for D in the friction factor and Reynolds number definitions, Eqs. (6-32) and (6-33). The hydraulic diameter is defined as **four times the channel cross-sectional area divided by the wetted perimeter.** For example, the hydraulic diameter for a circular pipe is $D_H = D$, for an annulus of inner diameter d and outer diameter D, $D_H = D - d$, for a rectangular duct of sides $a, b, D_H = ab/[2(a + b)]$. The **hydraulic radius** R_H is defined as **one-fourth** of the hydraulic diameter.

With the hydraulic diameter substituted for D in f and Re, Eqs. (6-37) through (6-40) are good approximations. Note that V appearing in f and Re is the actual average velocity $V = Q/A$; for noncircular pipes, it is **not** $Q/(\pi D_H^2/4)$. The pressure drop should be calculated from the friction factor for noncircular pipes. Equations relating Q to ΔP and D for circular pipes **may not be used** for noncircular pipes with D replaced by D_H because $V \neq Q/(\pi D_H^2/4)$.

Turbulent flow in noncircular channels is generally accompanied by secondary flows perpendicular to the axial flow direction (Schlichting). These flows may cause the pressure drop to be slightly greater than that computed using the hydraulic diameter method. For data on pressure drop in annuli, see Brighton and Jones (*J. Basic Eng.*, **86,** 835–842 [1964]); Okiishi and Serovy (*J. Basic Eng.*, **89,** 823–836 [1967]); and Lawn and Elliot (*J. Mech. Eng. Sci.*, **14,** 195–204 [1972]). For rectangular ducts of large aspect ratio, Dean (*J. Fluids Eng.*, **100,** 215–233 [1978]) found that the numerator of the exponent in the Blasius equation (6-37) should be increased to 0.0868. Jones (*J. Fluids Eng.*, **98,** 173–181 [1976]) presents a method to improve the estimation of friction factors for rectangular ducts using a modification of the hydraulic diameter–based Reynolds number.

The hydraulic diameter method does not work well for **laminar flow** because the shape affects the flow resistance in a way that cannot be expressed as a function only of the ratio of cross-sectional area to wetted perimeter. For some shapes, the Navier-Stokes equations have been integrated to yield relations between flow rate and pressure drop. These relations may be expressed in terms of **equivalent diameters** D_E defined to make the relations reduce to the second form of the Hagen-Poiseulle equation, Eq. (6-36); that is, $D_E \equiv (128 Q \mu L/\pi \Delta P)^{1/4}$. **Equivalent diameters are not the same as hydraulic diameters.** Equivalent diameters yield the correct relation between flow rate and pressure drop when substituted into Eq. (6-36), but not Eq. (6-35) because $V \neq Q/(\pi D_E/4)$. Equivalent diameter D_E is not to be used in the friction factor and Reynolds number; $f \neq 16/\text{Re}$ using the equivalent diameters defined in the following. This situation is, by arbitrary definition, opposite to that for the hydraulic diameter D_H used for turbulent flow.

Ellipse, semiaxes a and b (Lamb, *Hydrodynamics*, 6th ed., Dover, New York, 1945, p. 587):

$$D_E = \left(\frac{32a^3b^3}{a^2 + b^2} \right)^{1/4} \tag{6-50}$$

Rectangle, width a, height b (Owen, *Trans. Am. Soc. Civ. Eng.*, **119,** 1157–1175 [1954]):

$$D_E = \left(\frac{128ab^3}{\pi K} \right)^{1/4} \tag{6-51}$$

$a/b =$	1	1.5	2	3	4	5	10	∞
$K =$	28.45	20.43	17.49	15.19	14.24	13.73	12.81	12

Annulus, inner diameter D_1 outer diameter D_2 (Lamb, op. cit., p. 587):

$$D_E = \left\{ (D_2^2 - D_1^2) \left[D_2^2 + D_1^2 - \frac{D_2^2 - D_1^2}{\ln (D_2/D_1)} \right] \right\}^{1/4} \tag{6-52}$$

For isosceles triangles and regular polygons, see Sparrow (*AIChE J.*, **8,** 599–605 [1962]), Carlson and Irvine (*J. Heat Transfer*, **83,** 441–444 [1961]), Cheng (*Proc. Third Int. Heat Transfer Conf.*, New York, **1,** 64–76 [1966]), and Shih (*Can. J. Chem. Eng.*, **45,** 285–294 [1967]).

The critical Reynolds number for **transition from laminar to turbulent flow** in noncircular channels varies with channel shape. In rectangular ducts, $1{,}900 < \text{Re}_c < 2{,}800$ (Hanks and Ruo, *Ind. Eng. Chem. Fundam.*, **5,** 558–561 [1966]). In triangular ducts, $1{,}600 < \text{Re}_c < 1{,}800$ (Cope and Hanks, *Ind. Eng. Chem. Fundam.*, **11,** 106–117 [1972]; Bandopadhayay and Hinwood, *J. Fluid Mech.*, **59,** 775–783 [1973]).

Nonisothermal Flow   For nonisothermal flow of **liquids,** the friction factor may be increased if the liquid is being cooled or decreased if the liquid is being heated, because of the effect of temperature on viscosity near the wall. In shell and tube heat-exchanger design, the recommended practice is to first estimate f using the bulk mean liquid temperature over the tube length. Then, in laminar flow, the result is divided by $(\mu_a/\mu_w)^{0.23}$ in the case of cooling or $(\mu_a/\mu_w)^{0.38}$ in the case of heating. For turbulent flow, f is divided by $(\mu_a/\mu_w)^{0.11}$ in the case of cooling or $(\mu_a/\mu_w)^{0.17}$ in case of heating. Here, μ_a is the viscosity at the average bulk temperature and μ_w is the viscosity at the average wall temperature (Seider and Tate, *Ind. Eng. Chem.*, **28,** 1429–1435 [1936]). In the case of rough commercial pipes, rather than heat-exchanger tubing, it is common for flow to be in the "complete" turbulence regime where f is independent of Re. In such cases, the friction factor should not be corrected for wall temperature. If the liquid density varies with temperature, the average bulk density should be used to calculate the pressure drop from the friction factor. In addition, a (usually small) correction may be applied for acceleration effects by adding the term $G^2[(1/\rho_2) - (1/\rho_1)]$ from the mechanical energy balance to the pressure drop $\Delta P = P_1 - P_2$, where G is the mass velocity. This acceleration results from small compressibility effects associated with temperature-dependent density. Christiansen and Gordon (*AIChE J.*, **15,** 504–507 [1969]) present equations and charts for frictional loss in laminar nonisothermal flow of Newtonian and non-Newtonian liquids heated or cooled with constant wall temperature.

Frictional dissipation of mechanical energy can result in significant heating of fluids, particularly for very viscous liquids in small channels. Under adiabatic conditions, the bulk liquid temperature rise is given by $\Delta T = \Delta P/C_v \rho$ for incompressible flow through a channel of constant cross-sectional area. For flow of polymers, this amounts to about 4°C per 10 MPa pressure drop, while for hydrocarbon liquids it is about 6°C per 10 MPa. The temperature rise in laminar flow is highly nonuniform, being concentrated near the pipe wall where most of the dissipation occurs. This may result in significant viscosity reduction near the wall, and greatly increased flow or reduced pressure drop, and a flattened velocity profile. Compensation should generally be made for the heat effect when ΔP exceeds 1.4 MPa (203 psi) for adiabatic walls or 3.5 MPa (508 psi) for isothermal walls (Gerard, Steidler, and Appeldoorn, *Ind. Eng. Chem. Fundam.*, **4,** 332–339 [1969]).

Open Channel Flow   For flow in **open channels,** the data are largely based on experiments with water in turbulent flow, in channels of sufficient roughness that there is no Reynolds number effect. The hydraulic radius approach may be used to estimate a friction factor with which to compute friction losses. Under conditions of **uniform flow** where liquid depth and cross-sectional area do not vary significantly with position in the flow direction, there is a balance between gravitational forces and wall stress, or equivalently between frictional losses and potential energy change. The mechanical energy balance reduces to $l_v = g(z_1 - z_2)$. In terms of the friction factor and hydraulic diameter or hydraulic radius,

$$l_v = \frac{2fV^2L}{D_H} = \frac{fV^2L}{2R_H} = g(z_1 - z_2) \tag{6-53}$$

The hydraulic radius is the cross-sectional area divided by the wetted perimeter, where the wetted perimeter *does not include the free sur-*

face. Letting $S = \sin\theta =$ channel slope (elevation loss per unit length of channel, $\theta =$ angle between channel and horizontal), Eq. (6-53) reduces to

$$V = \sqrt{\frac{2gSR_H}{f}} \qquad (6\text{-}54)$$

The most often used friction correlation for open channel flows is due to Manning (*Trans. Inst. Civ. Engrs. Ireland,* **20,** 161 [1891]) and is equivalent to

$$f = \frac{29n^2}{R_H^{1/3}} \qquad (6\text{-}55)$$

where n is the channel roughness, with dimensions of $(\text{length})^{1/6}$. Table 6-2 gives roughness values for several channel types.

For gradual changes in channel cross section and liquid depth, and for slopes less than $10°$, the momentum equation for a rectangular channel of width b and liquid depth h may be written as a differential equation in the flow direction x.

$$\frac{dh}{dx}(1 - \text{Fr}) - \text{Fr}\left(\frac{h}{b}\right)\frac{db}{dx} = S - \frac{fV^2(b + 2h)}{2gbh} \qquad (6\text{-}56)$$

For a given fixed flow rate $Q = Vbh$, and channel width profile $b(x)$, Eq. (6-56) may be integrated to determine the liquid depth profile $h(x)$. The dimensionless Froude number is $\text{Fr} = V^2/gh$. When $\text{Fr} = 1$, the flow is **critical,** when $\text{Fr} < 1$, the flow is **subcritical,** and when $\text{Fr} > 1$, the flow is **supercritical.** Surface disturbances move at a wave velocity $c = \sqrt{gh}$; they cannot propagate upstream in supercritical flows. The **specific energy** E_{sp} is nearly constant.

$$E_{\text{sp}} = h + \frac{V^2}{2g} \qquad (6\text{-}57)$$

This equation is cubic in liquid depth. Below a minimum value of E_{sp} there are no real positive roots; above the minimum value there are two positive real roots. At this minimum value of E_{sp} the flow is critical; that is, $\text{Fr} = 1$, $V = \sqrt{gh}$, and $E_{\text{sp}} = (3/2)h$. Near critical flow conditions, wave motion and sudden depth changes called **hydraulic jumps** are likely. Chow (*Open Channel Hydraulics,* McGraw-Hill, New York, 1959), discusses the numerous surface profile shapes which may exist in nonuniform open channel flows.

For flow over a sharp-crested weir of width b and height L, from a liquid depth H, the flow rate is given approximately by

$$Q = \frac{2}{3}C_d b\sqrt{2g}(H - L)^{3/2} \qquad (6\text{-}58)$$

where $C_d \approx 0.6$ is a discharge coefficient. Flow through notched weirs is described under flow meters in Sec. 10 of the *Handbook.*

TABLE 6-2 Average Values of *n* for Manning Formula, Eq. (6-55)

Surface	n, m$^{1/6}$	n, ft$^{1/6}$
Cast-iron pipe, fair condition	0.014	0.011
Riveted steel pipe	0.017	0.014
Vitrified sewer pipe	0.013	0.011
Concrete pipe	0.015	0.012
Wood-stave pipe	0.012	0.010
Planed-plank flume	0.012	0.010
Semicircular metal flumes, smooth	0.013	0.011
Semicircular metal flumes, corrugated	0.028	0.023
Canals and ditches		
Earth, straight and uniform	0.023	0.019
Winding sluggish canals	0.025	0.021
Dredged earth channels	0.028	0.023
Natural-stream channels		
Clean, straight bank, full stage	0.030	0.025
Winding, some pools and shoals	0.040	0.033
Same, but with stony sections	0.055	0.045
Sluggish reaches, very deep pools, rather weedy	0.070	0.057

SOURCE: Brater and King, *Handbook of Hydraulics,* 6th ed., McGraw-Hill, New York, 1976, p. 7–22. For detailed information, see Chow, *Open-Channel Hydraulics,* McGraw-Hill, New York, 1959, pp. 110–123.

Non-Newtonian Flow For **isothermal laminar flow** of time-independent non-Newtonian liquids, integration of the Cauchy momentum equations yields the fully developed velocity profile and flow rate–pressure drop relations. For the **Bingham plastic** fluid described by Eq. (6-3), in a pipe of diameter D and a pressure drop per unit length $\Delta P/L$, the flow rate is given by

$$Q = \frac{\pi D^3 \tau_w}{32\mu_\infty}\left[1 - \frac{4\tau_y}{3\tau_w} + \frac{\tau_y^4}{3\tau_w^4}\right] \qquad (6\text{-}59)$$

where the wall stress is $\tau_w = D\Delta P/(4L)$. The velocity profile consists of a central nondeforming plug of radius $r_P = 2\tau_y/(\Delta P/L)$ and an annular deforming region. The velocity profile in the annular region is given by

$$v_z = \frac{1}{\mu_\infty}\left[\frac{\Delta P}{4L}(R^2 - r^2) - \tau_y(R - r)\right], \qquad r_P \leq r \leq R \qquad (6\text{-}60)$$

where r is the radial coordinate and R is the pipe radius. The velocity of the central, nondeforming plug is obtained by setting $r = r_P$ in Eq. (6-60). When Q is given and Eq. (6-59) is to be solved for τ_w and the pressure drop, multiple positive roots for the pressure drop may be found. The root corresponding to $\tau_w < \tau_y$ is physically unrealizable, as it corresponds to $r_P > R$ and the pressure drop is insufficient to overcome the yield stress.

For a **power law fluid,** Eq. (6-4), with constant properties K and n, the flow rate is given by

$$Q = \pi\left(\frac{\Delta P}{2KL}\right)^{1/n}\left(\frac{n}{1 + 3n}\right)R^{(1 + 3n)/n} \qquad (6\text{-}61)$$

and the velocity profile by

$$v_z = \left(\frac{\Delta P}{2KL}\right)^{1/n}\left(\frac{n}{1 + n}\right)[R^{(1 + n)/n} - r^{(1 + n)/n}] \qquad (6\text{-}62)$$

Similar relations for other non-Newtonian fluids may be found in Govier and Aziz and in Bird, Armstrong, and Hassager (*Dynamics of Polymeric Liquids,* vol. 1: *Fluid Mechanics,* Wiley, New York, 1977).

For steady-state laminar flow of any time-independent viscous fluid, at average velocity V in a pipe of diameter D, the Rabinowitsch-Mooney relations give a general relationship for the shear rate at the pipe wall.

$$\dot\gamma_w = \frac{8V}{D}\left(\frac{1 + 3n'}{4n'}\right) \qquad (6\text{-}63)$$

where n' is the slope of a plot of $D\Delta P/(4L)$ versus $8V/D$ on logarithmic coordinates,

$$n' = \frac{d\ln[D\Delta P/(4L)]}{d\ln(8V/D)} \qquad (6\text{-}64)$$

By plotting capillary viscometry data this way, they can be used directly for pressure drop design calculations, or to construct the rheogram for the fluid. For a pressure drop calculation, the flow rate and diameter determine the velocity, from which $8V/D$ is calculated and $D\Delta P/(4L)$ read from the plot. For a Newtonian fluid, $n' = 1$ and the shear rate at the wall is $\dot\gamma = 8V/D$. For a power law fluid, $n' = n$. To construct a rheogram, n' is obtained from the slope of the experimental plot at a given value of $8V/D$. The shear rate at the wall is given by Eq. (6-63) and the corresponding shear stress at the wall is $\tau_w = D\Delta P/(4L)$ read from the plot. By varying the value of $8V/D$, the shear rate versus shear stress plot can be constructed.

The generalized approach of Metzner and Reed (*AIChE J.,* **1,** 434 [1955]) for time-independent non-Newtonian fluids defines a modified Reynolds number as

$$\text{Re}_{\text{MR}} \equiv \frac{D^{n'}V^{2 - n'}\rho}{K'8^{n' - 1}} \qquad (6\text{-}65)$$

where K' satisfies

$$\frac{D\Delta P}{4L} = K'\left(\frac{8V}{D}\right)^{n'} \qquad (6\text{-}66)$$

With this definition, $f = 16/\text{Re}_{\text{MR}}$ is automatically satisfied at the value of $8V/D$ where K' and n' are evaluated. Equation (6-66) may be obtained by integration of Eq. (6-64) only when n' is a constant, as, for example, the cases of Newtonian and power law fluids. For Newtonian fluids, $K' = \mu$ and $n' = 1$; for power law fluids, $K' = K[(1 + 3n)/(4n)]^n$ and $n' = n$. For Bingham plastics, K' and n' are variable, given as a function of τ_w (Metzner, *Ind. Eng. Chem.*, **49**, 1429–1432 [1957]).

$$K = \tau_w^{1-n'}\left[\frac{\mu_\infty}{1 - 4\tau_y/3\tau_w + (\tau_y/\tau_w)^4/3}\right]^{n'} \quad (6\text{-}67)$$

$$n' = \frac{1 - 4\tau_y/(3\tau_w) + (\tau_y/\tau_w)^4/3}{1 - (\tau_y/\tau_w)^4} \quad (6\text{-}68)$$

For laminar flow of power law fluids in channels of noncircular cross section, see Schecter (*AIChE J.*, **7**, 445–448 [1961]), Wheeler and Wissler (*AIChE J.*, **11**, 207–212 [1965]), Bird, Armstrong, and Hassager (*Dynamics of Polymeric Liquids*, vol. 1: *Fluid Mechanics*, Wiley, New York, 1977), and Skelland (*Non-Newtonian Flow and Heat Transfer*, Wiley, New York, 1967).

Steady state, fully developed laminar flows of viscoelastic fluids in straight, constant-diameter pipes show no effects of viscoelasticity. The viscous component of the constitutive equation may be used to develop the flow rate–pressure drop relations, which apply downstream of the entrance region after viscoelastic effects have disappeared. A similar situation exists for time-dependent fluids.

The **transition to turbulent flow** begins at $\mathrm{Re}_{\mathrm{MR}}$ in the range of 2,000 to 2,500 (Metzner and Reed, *AIChE J.*, **1**, 434 [1955]). For Bingham plastic materials, K' and n' must be evaluated for the τ_w condition in question in order to determine $\mathrm{Re}_{\mathrm{MR}}$ and establish whether the flow is laminar. An alternative method for Bingham plastics is by Hanks (Hanks, *AIChE J.*, **9**, 306 [1963]; **14**, 691 [1968]; Hanks and Pratt, *Soc. Petrol. Engrs. J.*, **7**, 342 [1967]; and Govier and Aziz, pp. 213–215). The transition from laminar to turbulent flow is influenced by **viscoelastic** properties (Metzner and Park, *J. Fluid Mech.*, **20**, 291 [1964]) with the critical value of $\mathrm{Re}_{\mathrm{MR}}$ increased to beyond 10,000 for some materials.

For **turbulent flow of non-Newtonian fluids,** the design chart of Dodge and Metzner (*AIChE J.*, **5**, 189 [1959]), Fig. 6-11, is most widely used. For Bingham plastic materials in turbulent flow, it is generally assumed that stresses greatly exceed the yield stress, so that the friction factor–Reynolds number relationship for Newtonian fluids applies, with μ_∞ substituted for μ. This is equivalent to setting $n' = 1$ and $\tau_y/\tau_w = 0$ in the Dodge-Metzner method, so that $\mathrm{Re}_{\mathrm{MR}} = DV\rho/\mu_\infty$. Wilson and Thomas (*Can. J. Chem. Eng.*, **63**, 539–546 [1985]) give friction factor equations for turbulent flow of power law fluids and Bingham plastic fluids.

Power law fluids:

$$\frac{1}{\sqrt{f}} = \frac{1}{\sqrt{f_N}} + 8.2\,\frac{1-n}{1+n} + 1.77 \ln\left(\frac{1+n}{2}\right) \quad (6\text{-}69)$$

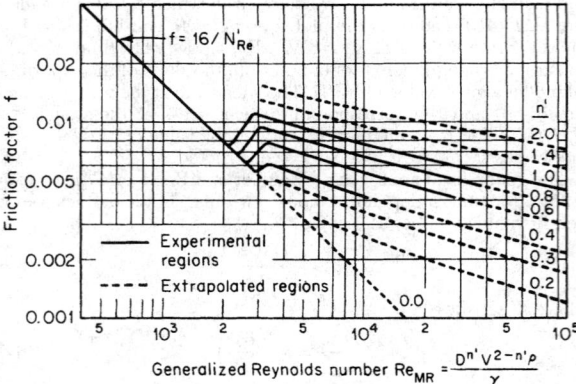

where f_N is the friction factor for Newtonian fluid evaluated at $\mathrm{Re} = DV\rho/\mu_{\mathrm{eff}}$ where the effective viscosity is

$$\mu_{\mathrm{eff}} = K\left(\frac{3n+1}{4n}\right)^{n-1}\left(\frac{8V}{D}\right)^{n-1} \quad (6\text{-}70)$$

Bingham fluids:

$$\frac{1}{\sqrt{f}} = \frac{1}{\sqrt{f_N}} + 1.77 \ln\left(\frac{(1-\xi)^2}{1+\xi}\right) + \xi(10 + 0.884\xi) \quad (6\text{-}71)$$

where f_N is evaluated at $\mathrm{Re} = DV\rho/\mu_\infty$ and $\xi = \tau_y/\tau_w$. Iteration is required to use this equation since $\tau_w = f\rho V^2/2$.

Drag reduction in turbulent flow can be achieved by adding soluble high molecular weight polymers in extremely low concentration to Newtonian liquids. The reduction in friction is generally believed to be associated with the viscoelastic nature of the solutions effective in the wall region. For a given polymer, there is a minimum molecular weight necessary to initiate drag reduction at a given flow rate, and a critical concentration above which drag reduction will not occur (Kim, Little and Ting, *J. Colloid Interface Sci.*, **47**, 530–535 [1974]). Drag reduction is reviewed by Hoyt (*J. Basic Eng.*, **94**, 258–285 [1972]); Little, et al. (*Ind. Eng. Chem. Fundam.*, **14**, 283–296 [1975]) and Virk (*AIChE J.*, **21**, 625–656 [1975]). At maximum possible drag reduction in smooth pipes,

$$\frac{1}{\sqrt{f}} = -19 \log\left(\frac{50.73}{\mathrm{Re}\sqrt{f}}\right) \quad (6\text{-}72)$$

or, approximately,

$$f = \frac{0.58}{\mathrm{Re}^{0.58}} \quad (6\text{-}73)$$

for $4{,}000 < \mathrm{Re} < 40{,}000$. The actual drag reduction depends on the polymer system. For further details, see Virk (ibid.).

Economic Pipe Diameter, Turbulent Flow The economic optimum pipe diameter may be computed so that the last increment of investment reduces the operating cost enough to produce the required minimum return on investment. For long cross-country pipelines, alloy pipes of appreciable length and complexity, or pipelines with control valves, detailed analyses of investment and operating costs should be made. Peters and Timmerhaus (*Plant Design and Economics for Chemical Engineers*, 4th ed., McGraw-Hill, New York, 1991) provide a detailed method for determining the economic optimum size. For pipelines of the lengths usually encountered in chemical plants and petroleum refineries, simplified selection charts are often adequate. In many cases there is an economic optimum velocity that is nearly independent of diameter, which may be used to estimate the economic diameter from the flow rate. For low-viscosity liquids in schedule 40 steel pipe, economic optimum velocity is typically in the range of 1.8 to 2.4 m/s (5.9 to 7.9 ft/s). For gases with density ranging from 0.2 to 20 kg/m³ (0.013 to 1.25 lbm/ft³), the economic optimum velocity is about 40 m/s to 9 m/s (131 to 30 ft/s). Charts and rough guidelines for economic optimum size do not apply to multiphase flows.

Economic Pipe Diameter, Laminar Flow Pipelines for the transport of high-viscosity liquids are seldom designed purely on the basis of economics. More often, the size is dictated by operability considerations such as available pressure drop, shear rate, or residence time distribution. Peters and Timmerhaus (ibid., Chap. 10) provide an economic pipe diameter chart for laminar flow. For non-Newtonian fluids, see Skelland (*Non-Newtonian Flow and Heat Transfer*, Chap. 7, Wiley, New York, 1967).

Vacuum Flow When gas flows under high vacuum conditions or through very small openings, the continuum hypothesis is no longer appropriate if the channel dimension is not very large compared to the mean free path of the gas. When the mean free path is comparable to the channel dimension, flow is dominated by collisions of molecules with the wall, rather than by collisions between molecules. An approximate expression based on Brown, et al. (*J. Appl. Phys.*, **17**, 802–813 [1946]) for the mean free path is

$$\lambda = \left(\frac{2\mu}{p}\right)\sqrt{\frac{8RT}{\pi M_w}} \quad (6\text{-}74)$$

The Knudsen number Kn is the ratio of the mean free path to the channel dimension. For pipe flow, $Kn = \lambda/D$. **Molecular flow** is characterized by $Kn > 1.0$; continuum viscous (laminar or turbulent) flow is characterized by $Kn < 0.01$. **Transition** or **slip** flow applies over the range $0.01 < Kn < 1.0$.

Vacuum flow is usually described with flow variables different from those used for normal pressures, which often leads to confusion. **Pumping speed** S is the actual volumetric flow rate of gas through a flow cross section. **Throughput** Q is the product of pumping speed and absolute pressure. In the SI system, Q has units of $Pa \cdot m^3/s$.

$$Q = Sp \tag{6-75}$$

The mass flow rate w is related to the throughput using the ideal gas law.

$$w = \frac{M_w}{RT} Q \tag{6-76}$$

Throughput is therefore proportional to mass flow rate. For a given mass flow rate, throughput is independent of pressure. The relation between throughput and pressure drop $\Delta p = p_1 - p_2$ across a flow element is written in terms of the **conductance** C. **Resistance** is the reciprocal of conductance. Conductance has dimensions of volume per time.

$$Q = C\Delta p \tag{6-77}$$

The conductance of a series of flow elements is given by

$$\frac{1}{C} = \frac{1}{C_1} + \frac{1}{C_2} + \frac{1}{C_3} + \cdots \tag{6-78}$$

while for elements in parallel,

$$C = C_1 + C_2 + C_3 + \cdots \tag{6-79}$$

For a vacuum pump of speed S_p withdrawing from a vacuum vessel through a connecting line of conductance C, the pumping speed at the vessel is

$$S = \frac{S_p C}{S_p + C} \tag{6-80}$$

Molecular Flow Under molecular flow conditions, conductance is independent of pressure. It is proportional to $\sqrt{T/M_w}$, with the proportionality constant a function of geometry. For fully developed pipe flow,

$$C = \frac{\pi D^3}{8L} \sqrt{\frac{RT}{M_w}} \tag{6-81}$$

For an orifice of area A,

$$C = 0.40A \sqrt{\frac{RT}{M_w}} \tag{6-82}$$

Conductance equations for several other geometries are given by Ryans and Roper (*Process Vacuum System Design and Operation*, Chap. 2, McGraw-Hill, New York, 1986). For a circular annulus of outer and inner diameters D_1 and D_2 and length L, the method of Guthrie and Wakerling (*Vacuum Equipment and Techniques*, McGraw-Hill, New York, 1949) may be written

$$C = 0.42K \frac{(D_1 - D_2)^2(D_1 + D_2)}{L} \sqrt{\frac{RT}{M_w}} \tag{6-83}$$

where K is a dimensionless constant with values given in Table 6-3.

For a short pipe of circular cross section, the conductance as calculated for an orifice from Eq. (6-82) is multiplied by a correction factor K which may be approximated as (Kennard, *Kinetic Theory of Gases*, McGraw-Hill, New York, 1938, pp. 306–308)

$$K = \frac{1}{1 + (L/D)} \qquad \text{for} \qquad 0 \le L/D \le 0.75 \tag{6-84}$$

$$K = \frac{1 + 0.8(L/D)}{1 + 1.90(L/D) + 0.6(L/D)^2} \qquad \text{for} \qquad L/D > 0.75 \tag{6-85}$$

For $L/D > 100$, the error in neglecting the end correction by using the fully developed pipe flow equation (6-81) is less than 2 percent. For rectangular channels, see Normand (*Ind. Eng. Chem.*, **40**, 783–787 [1948]).

Yu and Sparrow (*J. Basic Eng.*, **70**, 405–410 [1970]) give a theoretically derived chart for slot seals with or without a sheet located in or passing through the seal, giving mass flow rate as a function of the ratio of seal plate thickness to gap opening.

Slip Flow In the transition region between molecular flow and continuum viscous flow, the conductance for fully developed pipe flow is most easily obtained by the method of Brown, et al. (*J. Appl. Phys.*, **17**, 802–813 [1946]), which uses the parameter

$$X = \sqrt{\frac{8}{\pi}} \left(\frac{\lambda}{D}\right) = \left(\frac{2\mu}{p_m D}\right) \sqrt{\frac{RT}{M}} \tag{6-86}$$

where p_m is the arithmetic mean absolute pressure. A correction factor F, read from Fig. 6-12 as a function of X, is applied to the conductance for viscous flow.

$$C = F \frac{\pi D^4 p_m}{128\mu L} \tag{6-87}$$

For slip flow through **square channels,** see Milligan and Wilkerson (*J. Eng. Ind.*, **95**, 370–372 [1973]). For slip flow through **annuli,** see Maegley and Berman (*Phys. Fluids*, **15**, 780–785 [1972]).

The **pump-down time** θ for evacuating a vessel in the absence of air in-leakage is given approximately by

$$\theta = \left(\frac{V_t}{S_0}\right) \ln \left(\frac{p_1 - p_0}{p_2 - p_0}\right) \tag{6-88}$$

where V_t = volume of vessel plus volume of piping between vessel and pump; S_0 = system speed as given by Eq. (6-80), assumed independent of pressure; p_1 = initial vessel pressure; p_2 = final vessel pressure; and p_0 = lowest pump intake pressure attainable with the pump in question. See Dushman and Lafferty (*Scientific Foundations of Vacuum Technique*, 2d ed., Wiley, New York, 1962).

The amount of inerts which has to be removed by a pumping system after the pump-down stage depends on the in-leakage of air at the various fittings, connections, and so on. Air leakage is often correlated with system volume and pressure, but this approach introduces uncer-

TABLE 6-3 Constants for Circular Annuli

D_2/D_1	K	D_2/D_1	K
0	1.00	0.707	1.254
0.259	1.072	0.866	1.430
0.500	1.154	0.966	1.675

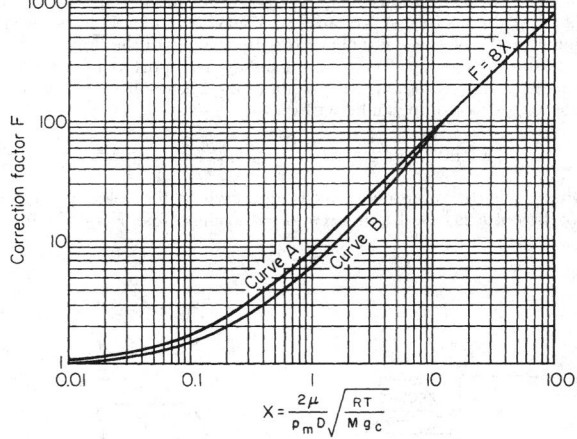

FIG. 6-12 Correction factor for Poiseuille's equation at low pressures. Curve *A*: experimental curve for glass capillaries and smooth metal tubes. (*From Brown, et al.*, J. Appl. Phys., **17**, *802 [1946].*) Curve *B*: experimental curve for iron pipe (*From Riggle, Courtesy of E. I. du Pont de Nemours & Co.*)

tainty because the number and size of leaks does not necessarily correlate with system volume, and leakage is sensitive to maintenance quality. Ryans and Roper (*Process Vacuum System Design and Operation*, McGraw-Hill, New York, 1986) present a thorough discussion of air leakage.

FRICTIONAL LOSSES IN PIPELINE ELEMENTS

The viscous or frictional loss term in the mechanical energy balance for most cases is obtained experimentally. For many common fittings found in piping systems, such as expansions, contractions, elbows and valves, data are available to estimate the losses. Substitution into the energy balance then allows calculation of pressure drop. A common error is to assume that pressure drop and frictional losses are equivalent. Equation (6-16) shows that in addition to frictional losses, other factors such as shaft work and velocity or elevation change influence pressure drop.

Losses l_v for incompressible flow in sections of straight pipe of constant diameter may be calculated as previously described using the Fanning friction factor:

$$l_v = \frac{\Delta P}{\rho} = \frac{2fV^2L}{D} \qquad (6\text{-}89)$$

where ΔP = drop in equivalent pressure, $P = p + \rho gz$, with p = pressure, ρ = fluid density, g = acceleration of gravity, and z = elevation. Losses in the fittings of a piping network are frequently termed *minor losses* or *miscellaneous losses*. These descriptions are misleading because in process piping fitting losses are often much greater than the losses in straight piping sections.

Equivalent Length and Velocity Head Methods Two methods are in common use for estimating fitting loss. One, the **equivalent length** method, reports the losses in a piping element as the length of straight pipe which would have the same loss. For turbulent flows, the equivalent length is usually reported as a number of diameters of pipe of the same size as the fitting connection; L_e/D is given as a fixed quantity, independent of D. This approach tends to be most accurate for a single fitting size and loses accuracy with deviation from this size. For laminar flows, L_e/D correlations normally have a size dependence through a Reynolds number term.

The other method is the **velocity head** method. The term $V^2/2g$ has dimensions of length and is commonly called a *velocity head*. Application of the Bernoulli equation to the problem of frictionless discharge at velocity V through a nozzle at the bottom of a column of liquid of height H shows that $H = V^2/2g$. Thus H is the liquid head corresponding to the velocity V. Use of the velocity head to scale pressure drops has wide application in fluid mechanics. Examination of the Navier-Stokes equations suggests that when the inertial terms dominate the viscous terms, pressure gradients are expected to be proportional to ρV^2 where V is a characteristic velocity of the flow.

In the velocity head method, the losses are reported as a number of velocity heads K. Then, the engineering Bernoulli equation for an incompressible fluid can be written

$$p_1 - p_2 = \alpha_2 \frac{\rho V_2^2}{2} - \alpha_1 \frac{\rho V_1^2}{2} + \rho g(z_2 - z_1) + K \frac{\rho V^2}{2} \qquad (6\text{-}90)$$

where V is the reference velocity upon which the velocity head loss coefficient K is based. For a section of straight pipe, $K = 4fL/D$.

Contraction and Entrance Losses For a **sudden contraction** at a sharp-edged entrance to a pipe or sudden reduction in cross-sectional area of a channel, as shown in Fig. 6-13a, the loss coefficient based on the downstream velocity V_2 is given for **turbulent flow** in Crane Co. Tech Paper 410 (1980) approximately by

$$K = 0.5 \left(1 - \frac{A_2}{A_1} \right) \qquad (6\text{-}91)$$

Example 5: Entrance Loss Water, $\rho = 1000$ kg/m³, flows from a large vessel through a sharp-edged entrance into a pipe at a velocity in the pipe of 2 m/s. The flow is turbulent. Estimate the pressure drop from the vessel into the pipe.

With $A_2/A_1 \sim 0$, the viscous loss coefficient is $K = 0.5$ from Eq. (6-91). The mechanical energy balance, Eq. (6-16) with $V_1 = 0$ and $z_2 - z_1 = 0$ and assuming uniform flow ($\alpha_2 = 1$) becomes

$$p_1 - p_2 = \frac{\rho V_2^2}{2} + 0.5 \frac{\rho V_2^2}{2} = 4{,}000 + 2{,}000 = 6{,}000 \text{ Pa}$$

Note that the total pressure drop consists of 0.5 velocity heads of frictional loss contribution, and 1 velocity head of velocity change contribution. The frictional contribution is a permanent loss of mechanical energy by viscous dissipation. The acceleration contribution is reversible; if the fluid were subsequently decelerated in a frictionless diffuser, a 4,000 Pa pressure rise would occur.

For a **trumpet-shaped** rounded entrance, with a radius of rounding greater than about 15 percent of the pipe diameter (Fig. 6-13b), the turbulent flow loss coefficient K is only about 0.1 (Vennard and Street, *Elementary Fluid Mechanics*, 5th ed., Wiley, New York, 1975, pp. 420–421). Rounding of the inlet prevents formation of the **vena contracta**, thereby reducing the resistance to flow.

For **laminar flow** the losses in sudden contraction may be estimated for area ratios $A_2/A_1 < 0.2$ by an equivalent additional pipe length L_e given by

$$L_e/D = 0.3 + 0.04\text{Re} \qquad (6\text{-}92)$$

where D is the diameter of the smaller pipe and Re is the Reynolds number in the smaller pipe. For laminar flow in the entrance to rectangular ducts, see Shah (*J. Fluids Eng.*, **100**, 177–179 [1978]) and Roscoe (*Philos. Mag.*, **40**, 338–351 [1949]). For creeping flow, Re < 1, of power law fluids, the entrance loss is approximately $L_e/D = 0.3/n$ (Boger, Gupta, and Tanner, *J. Non-Newtonian Fluid Mech.*, **4**, 239–248 [1978]). For viscoelastic fluid flow in circular channels with sudden contraction, a toroidal vortex forms upstream of the contraction plane. Such flows are reviewed by Boger (*Ann. Review Fluid Mech.*, **19**, 157–182 [1987]).

For creeping flow through **conical converging channels,** inertial acceleration terms are negligible and the viscous pressure drop $\Delta p = \rho l_v$ may be computed by integration of the differential form of the Hagen-Poiseuille equation Eq. (6-36), provided the angle of convergence is small. The result for a power law fluid is

$$\Delta p = K \left(\frac{3n+1}{4n} \right)^n \left(\frac{8V_2}{D_2} \right)^n \left\{ \frac{1}{6n \tan(\alpha/2)} \left[1 - \left(\frac{D_2}{D_1} \right)^{3n} \right] \right\} \qquad (6\text{-}93)$$

where D_1 = inlet diameter
D_2 = exit diameter
V_2 = velocity at the exit
α = total included angle

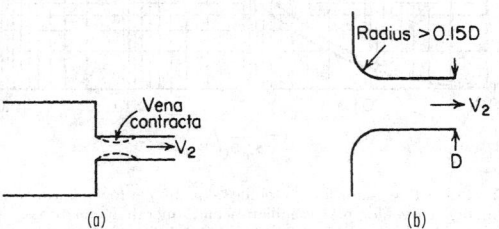

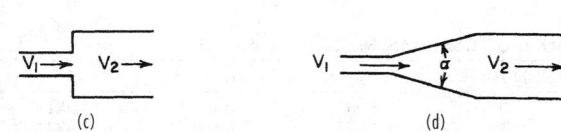

FIG. 6-13 Contractions and enlargements: (*a*) sudden contraction, (*b*) rounded contraction, (*c*) sudden enlargement, and (*d*) uniformly diverging duct.

Equation (6-93) agrees with experimental data (Kemblowski and Kiljanski, *Chem. Eng. J.* (*Lausanne*), **9**, 141–151 [1975]) for $\alpha < 11°$. For Newtonian liquids, Eq. (6-93) simplifies to

$$\Delta p = \mu \left(\frac{8V_2}{D_2} \right) \left\{ \frac{1}{6 \tan (\alpha/2)} \left[1 - \left(\frac{D_2}{D_1} \right)^3 \right] \right\} \qquad (6\text{-}94)$$

For creeping flow through rectangular or two-dimensional converging channels, the differential form of the Hagen-Poiseuille equation with equivalent diameter given by Eq. (6-49) may be used, provided the convergence is gradual.

Expansion and Exit Losses   For ducts of any cross section, the frictional loss for a **sudden enlargement** (Fig. 6-13c) with turbulent flow is given by the Borda-Carnot equation:

$$l_v = \frac{V_1^2 - V_2^2}{2} = \frac{V_1^2}{2} \left(1 - \frac{A_1}{A_2} \right)^2 \qquad (6\text{-}95)$$

where V_1 = velocity in the smaller duct
$\quad\quad\;\; V_2$ = velocity in the larger duct
$\quad\quad\;\; A_1$ = cross-sectional area of the smaller duct
$\quad\quad\;\; A_2$ = cross-sectional area of the larger duct

Equation (6-95) is valid for incompressible flow. For compressible flows, see Benedict, Wyler, Dudek, and Gleed (*J. Eng. Power*, **98**, 327–334 [1976]). For an infinite expansion, $A_1/A_2 = 0$, Eq. (6-95) shows that the **exit loss** from a pipe is 1 velocity head. This result is easily deduced from the mechanical energy balance Eq. (6-90), noting that $p_1 = p_2$. This exit loss is due to the dissipation of the discharged jet; there is no pressure drop at the exit.

For creeping Newtonian flow (Re < 1), the frictional loss due to a sudden enlargement should be obtained from the same equation for a sudden contraction (Eq. [6-92]). Note, however, that Boger, Gupta, and Tanner (ibid.) give an exit friction equivalent length of 0.12 diameter, increasing for power law fluids as the exponent decreases. For laminar flows at higher Reynolds numbers, the pressure drop is twice that given by Eq. (6-95). This results from the velocity profile factor α in the mechanical energy balance being 2.0 for the parabolic laminar velocity profile.

If the transition from a small to a large duct of any cross-sectional shape is accomplished by a **uniformly diverging duct** (see Fig. 6-13d) with a straight axis, the total frictional pressure drop can be computed by integrating the differential form of Eq. (6-89), $dl_v/dx = 2fV^2/D$ over the length of the expansion, provided the total angle α between the diverging walls is less than 7°. For angles between 7 and 45°, the loss coefficient should be estimated as 2.6 sin $(\alpha/2)$ times the loss coefficient for a sudden expansion; see Hooper (*Chem. Eng.*, Nov. 7, 1988). Gibson (*Hydraulics and Its Applications,* 5th ed., Constable, London 1952, p. 93) recommends multiplying the sudden enlargement loss by 0.13 for 5° < α < 7.5° and by $0.0110\alpha^{1.22}$ for 7.5° < α < 35°. For angles greater than 35 to 45°, the losses are normally considered equal to those for a sudden expansion, although in some cases the losses may be greater. Expanding flow through standard pipe reducers should be treated as sudden expansions.

Trumpet-shaped enlargements for **turbulent flow** designed for constant decrease in velocity head per unit length were found by Gibson (ibid., p. 95) to give 20 to 60 percent less frictional loss than straight taper pipes of the same length.

A special feature of expansion flows occurs when **viscoelastic** liquids are extruded through a die at a low Reynolds number. The extrudate may expand to a diameter several times greater than the die diameter, whereas for a Newtonian fluid the diameter expands only 10 percent. This phenomenon, called **die swell**, is most pronounced with short dies (Graessley, Glasscock, and Crawley, *Trans. Soc. Rheol.*, **14**, 519–544 [1970]). For velocity distribution measurements near the die exit, see Goulden and MacSporran (*J. Non-Newtonian Fluid Mech.*, **1**, 183–198 [1976]) and Whipple and Hill (*AIChE J.*, **24**, 664–671 [1978]). At high flow rates, the extrudate becomes distorted, suffering **melt fracture** at wall shear stresses greater than 10^5 N/m². This phenomenon is reviewed by Denn (*Ann. Review Fluid Mech.*, **22**, 13–34 [1990]). Ramamurthy (*J. Rheol.*, **30**, 337–357 [1986]) has found a dependence of apparent stick-slip behavior in melt fracture to be dependent on the material of construction of the die.

Fittings and Valves   For **turbulent flow,** the frictional loss for fittings and valves can be expressed by the equivalent length or velocity head methods. As fitting size is varied, K values are relatively more constant than L_e/D values, but since fittings generally do not achieve geometric similarity between sizes, K values tend to decrease with increasing fitting size. Table 6-4 gives K values for many types of fittings and valves.

Manufacturers of valves, especially control valves, express valve capacity in terms of a flow coefficient C_v, which gives the flow rate through the valve in gal/min of water at 60°F under a pressure drop of 1 lbf/in². It is related to K by

$$C_v = \frac{C_1 d^2}{\sqrt{K}} \qquad (6\text{-}96)$$

where C_1 is a dimensional constant equal to 29.9 and d is the diameter of the valve connections in inches.

For **laminar flow,** data for the frictional loss of valves and fittings are meager. (Beck and Miller, *J. Am. Soc. Nav. Eng.*, **56**, 62–83 [1944]; Beck, ibid., **56**, 235–271, 366–388, 389–395 [1944]; De Craene, *Heat. Piping Air Cond.*, **27**[10], 90–95 [1955]; Karr and Schutz, *J. Am. Soc. Nav. Eng.*, **52**, 239–256 [1940]; and Kittredge and Rowley, *Trans. ASME*, **79**, 1759–1766 [1957]). The data of Kittredge and Rowley indicate that K is constant for Reynolds numbers above 500 to 2,000, but increases rapidly as Re decreases below 500. Typical values for K for laminar flow Reynolds numbers are shown in Table 6-5.

Methods to calculate losses for **tee and wye junctions** for dividing and combining flow are given by Miller (*Internal Flow Systems*, 2d ed., Chap. 13, BHRA, Cranfield, 1990), including effects of Reynolds number, angle between legs, area ratio, and radius. Junctions with more than three legs are also discussed. The sources of data for the loss coefficient charts are Blaisdell and Manson (*U.S. Dept. Agric. Res. Serv. Tech. Bull.* 1283 [August 1963]) for combining flow and Gardel (*Bull. Tech. Suisses Romande*, **85**[9], 123–130 [1957]; **85**[10], 143–148 [1957]) together with additional unpublished data for dividing flow.

Miller (*Internal Flow Systems*, 2d ed., Chap. 13, BHRA, Cranfield, 1990) gives the most complete information on losses in **bends and curved pipes.** For turbulent flow in circular cross-section bends of constant area, as shown in Fig. 6-14a, a more accurate estimate of the loss coefficient K than that given in Table 6-4 is

$$K = K°C_{\text{Re}}C_oC_f \qquad (6\text{-}97)$$

where $K°$, given in Fig. 6-14b, is the loss coefficient for a smooth-walled bend at a Reynolds number of 10^6. The Reynolds number correction factor C_{Re} is given in Fig. 6-14c. For $0.7 < r/D < 1$ or for $K° < 0.4$, use the C_{Re} value for $r/D = 1$. Otherwise, if $r/D < 1$, obtain C_{Re} from

$$C_{\text{Re}} = \frac{K°}{K° + 0.2(1 - C_{\text{Re}, r/D = 1})} \qquad (6\text{-}98)$$

The correction C_o (Fig. 6-14d) accounts for the extra losses due to developing flow in the outlet tangent of the pipe, of length L_o. The total loss for the bend plus outlet pipe includes the bend loss K plus the straight pipe frictional loss in the outlet pipe $4fL_o/D$. Note that $C_o = 1$ for L_o/D greater than the termination of the curves on Fig. 6-14d, which indicate the distance at which fully developed flow in the outlet pipe is reached. Finally, the roughness correction is

$$C_f = \frac{f_{\text{rough}}}{f_{\text{smooth}}} \qquad (6\text{-}99)$$

where f_{rough} is the friction factor for a pipe of diameter D with the roughness of the bend, at the bend inlet Reynolds number. Similarly, f_{smooth} is the friction factor for smooth pipe. For Re > 10^6 and $r/D \geq 1$, use the value of C_f for Re = 10^6.

Example 6: Losses with Fittings and Valves   It is desired to calculate the liquid level in the vessel shown in Fig. 6-15 required to produce a discharge velocity of 2 m/s. The fluid is water at 20°C with $\rho = 1,000$ kg/m³ and $\mu = 0.001$ Pa · s, and the butterfly valve is at $\theta = 10°$. The pipe is 2-in Schedule 40, with an inner diameter of 0.0525 m. The pipe roughness is 0.046 mm. Assuming the flow is turbulent and taking the velocity profile factor $\alpha = 1$, the engineering Bernoulli equation Eq. (6-16), written between surfaces 1 and 2, where the

TABLE 6-4   Additional Frictional Loss for Turbulent Flow through Fittings and Valves[a]

Type of fitting or valve	Additional friction loss, equivalent no. of velocity heads, K
45° ell, standard[b,c,d,e,f]	0.35
45° ell, long radius[e]	0.2
90° ell, standard[b,c,e,f,g,h]	0.75
Long radius[b,c,d,e]	0.45
Square or miter[h]	1.3
180° bend, close return[b,c,e]	1.5
Tee, standard, along run, branch blanked off[e]	0.4
Used as ell, entering run[g,i]	1.0
Used as ell, entering branch[c,g,i]	1.0
Branching flow[i,j,k]	1[l]
Coupling[c,e]	0.04
Union[e]	0.04
Gate valve,[b,e,m] open	0.17
¾ open[n]	0.9
½ open[n]	4.5
¼ open[n]	24.0
Diaphragm valve,[o] open	2.3
¾ open[n]	2.6
½ open[n]	4.3
¼ open[n]	21.0
Globe valve,[e,m]	
Bevel seat, open	6.0
½ open[n]	9.5
Composition seat, open	6.0
½ open[n]	8.5
Plug disk, open	9.0
¾ open[n]	13.0
½ open[n]	36.0
¼ open[n]	112.0
Angle valve,[b,e] open	2.0
Y or blowoff valve,[b,m] open	3.0
Plug cock[p]	
$\theta = 5°$	0.05
$\theta = 10°$	0.29
$\theta = 20°$	1.56
$\theta = 40°$	17.3
$\theta = 60°$	206.0
Butterfly valve[p]	
$\theta = 5°$	0.24
$\theta = 10°$	0.52
$\theta = 20°$	1.54
$\theta = 40°$	10.8
$\theta = 60°$	118.0
Check valve,[b,e,m] swing	2.0[q]
Disk	10.0[q]
Ball	70.0[q]
Foot valve[e]	15.0
Water meter,[h] disk	7.0[r]
Piston	15.0[r]
Rotary (star-shaped disk)	10.0[r]
Turbine-wheel	6.0[r]

[a]Lapple, *Chem. Eng.*, **56**(5), 96–104 (1949), general survey reference.
[b]"Flow of Fluids through Valves, Fittings, and Pipe," Tech. Pap. 410, Crane Co., 1969.
[c]Freeman, *Experiments upon the Flow of Water in Pipes and Pipe Fittings,* American Society of Mechanical Engineers, New York, 1941.
[d]Giesecke, *J. Am. Soc. Heat. Vent. Eng.*, **32**, 461 (1926).
[e]*Pipe Friction Manual*, 3d ed., Hydraulic Institute, New York, 1961.
[f]Ito, *J. Basic Eng.*, **82**, 131–143 (1960).
[g]Giesecke and Badgett, *Heat. Piping Air Cond.*, **4**(6), 443–447 (1932).
[h]Schoder and Dawson, *Hydraulics*, 2d ed., McGraw-Hill, New York, 1934, p. 213.
[i]Hoopes, Isakoff, Clarke, and Drew, *Chem. Eng. Prog.*, **44**, 691–696 (1948).
[j]Gilman, *Heat. Piping Air Cond.*, **27**(4), 141–147 (1955).
[k]McNown, *Proc. Am. Soc. Civ. Eng.*, **79**, Separate 258, 1–22 (1953); discussion, ibid., **80**, Separate 396, 19–45 (1954). For the effect of branch spacing on junction losses in dividing flow, see Hecker, Nystrom, and Qureshi, *Proc. Am. Soc. Civ. Eng. J. Hydraul. Div.*, **103**(HY3), 265–279 (1977).
[l]This is pressure drcp (including friction loss) between run and branch, based on velocity in the mainstream before branching. Actual value depends on the flow split, ranging from 0.5 to 1.3 if mainstream enters run and from 0.7 to 1.5 if mainstream enters branch.
[m]Lansford, *Loss of Head in Flow of Fluids through Various Types of 1½-in. Valves,* Univ. Eng. Exp. Sta. Bull. Ser. 340, 1943.

pressures are both atmospheric and the fluid velocities are 0 and $V = 2$ m/s, respectively, and there is no shaft work, simplifies to

$$gZ = \frac{V^2}{2} + l_v$$

Contributing to l_v are losses for the entrance to the pipe, the three sections of straight pipe, the butterfly valve, and the 90° bend. Note that no exit loss is used because the discharged jet is outside the control volume. Instead, the $V^2/2$ term accounts for the kinetic energy of the discharging stream. The Reynolds number in the pipe is

$$Re = \frac{DV\rho}{\mu} = \frac{0.0525 \times 2 \times 1000}{0.001} = 1.05 \times 10^5$$

From Fig. 6-9 or Eq. (6-38), at $\epsilon/D = 0.046 \times 10^{-3}/0.0525 = 0.00088$, the friction factor is about 0.0054. The straight pipe losses are then

$$l_{v(sp)} = \left(\frac{4fL}{D}\right)\frac{V^2}{2}$$

$$= \left(\frac{4 \times 0.0054 \times (1+1+1)}{0.0525}\right)\frac{V^2}{2}$$

$$= 1.23\frac{V^2}{2}$$

The losses from Table 6-4 in terms of velocity heads K are $K = 0.5$ for the sudden contraction and $K = 0.52$ for the butterfly valve. For the 90° standard radius ($r/D = 1$), the table gives $K = 0.75$. The method of Eq. (6-94), using Fig. 6-14, gives

$$K = K^°C_{Re}C_oC_f$$

$$= 0.24 \times 1.24 \times 1.0 \times \left(\frac{0.0054}{0.0044}\right)$$

$$= 0.37$$

This value is more accurate than the value in Table 6-4. The value $f_{smooth} = 0.0044$ is obtainable either from Eq. (6-37) or Fig. 6-9.

The total losses are then

$$l_v = (1.23 + 0.5 + 0.52 + 0.37)\frac{V^2}{2} = 2.62\frac{V^2}{2}$$

and the liquid level Z is

$$Z = \frac{1}{g}\left(\frac{V^2}{2} + 2.62\frac{V^2}{2}\right) = 3.62\frac{V^2}{2g}$$

$$= \frac{3.62 \times 2^2}{2 \times 9.81} = 0.73 \text{ m}$$

Curved Pipes and Coils   For flow through curved pipe or coil, a secondary circulation perpendicular to the main flow called the **Dean effect** occurs. This circulation increases the friction relative to straight pipe flow and stabilizes laminar flow, delaying the transition Reynolds number to about

$$Re_{crit} = 2,100\left(1 + 12\sqrt{\frac{D}{D_c}}\right) \tag{6-100}$$

where D_c is the coil diameter. Equation (6-100) is valid for $10 < D_c/D < 250$. The **Dean number** is defined as

$$De = \frac{Re}{(D_c/D)^{1/2}} \tag{6-101}$$

In laminar flow, the friction factor for curved pipe f_c may be expressed in terms of the straight pipe friction factor $f = 16/Re$ as (Hart, *Chem. Eng. Sci.*, **43**, 775–783 [1988])

TABLE 6-5   Additional Frictional Loss for Laminar Flow through Fittings and Valves

Type of fitting or valve	Additional frictional loss expressed as K			
	Re = 1,000	500	100	50
90° ell, short radius	0.9	1.0	7.5	16
Gate valve	1.2	1.7	9.9	24
Globe valve, composition disk	11	12	20	30
Plug	12	14	19	27
Angle valve	8	8.5	11	19
Check valve, swing	4	4.5	17	55

SOURCE: From curves by Kittredge and Rowley, *Trans. Am. Soc. Mech. Eng.*, **79**, 1759–1766 (1957).

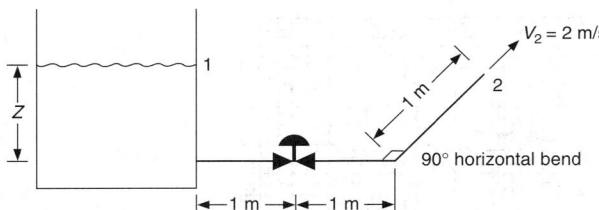

FIG. 6-15 Tank discharge example.

$$\frac{f_c}{f} = 1 + 0.090\left(\frac{\text{De}^{1.5}}{70 + \text{De}}\right) \qquad (6\text{-}102)$$

For turbulent flow, equations by Ito (*J. Basic Eng*, **81**, 123 [1959]) and Srinivasan, Nandapurkar, and Holland (*Chem. Eng.* [*London*] no. 218, CE113-CE119 [May 1968]) may be used, with probable accuracy of ±15 percent. Their equations are similar to

$$f_c = \frac{0.079}{\text{Re}^{0.25}} + \frac{0.0073}{\sqrt{(D_c/D)}} \qquad (6\text{-}103)$$

The pressure drop for flow in **spirals** is discussed by Srinivasan, et al. (loc. cit.) and Ali and Seshadri (*Ind. Eng. Chem. Process Des. Dev.*,

10, 328–332 [1971]). For friction loss in laminar flow through **semicircular ducts**, see Masliyah and Nandakumar (*AIChE J.*, **25**, 478–487 [1979]); for curved channels of **square cross section**, see Cheng, Lin, and Ou (*J. Fluids Eng.*, **98**, 41–48 [1976]).

For **non-Newtonian** (**power law**) **fluids** in coiled tubes, Mashelkar and Devarajan (*Trans. Inst. Chem. Eng.* (*London*), **54**, 108–114 [1976]) propose the correlation

$$f_c = (9.07 - 9.44n + 4.37n^2)(D/D_c)^{0.5}(\text{De}')^{-0.768+0.122n} \qquad (6\text{-}104)$$

where De' is a modified Dean number given by

$$\text{De}' = \frac{1}{8}\left(\frac{6n+2}{n}\right)^n \text{Re}_{\text{MR}}\sqrt{\frac{D}{D_c}} \qquad (6\text{-}105)$$

where Re_{MR} is the Metzner-Reed Reynolds number, Eq. (6-65). This correlation was tested for the range De' = 70 to 400, D/D_c = 0.01 to 0.135, and n = 0.35 to 1. See also Oliver and Asghar (*Trans. Inst. Chem. Eng.* [*London*], **53**, 181–186 [1975]).

Screens The pressure drop for incompressible flow across a screen of fractional free area α may be computed from

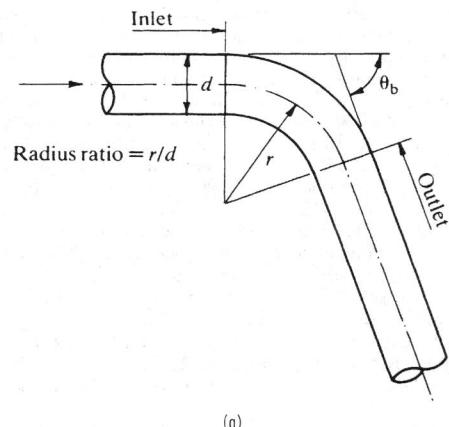

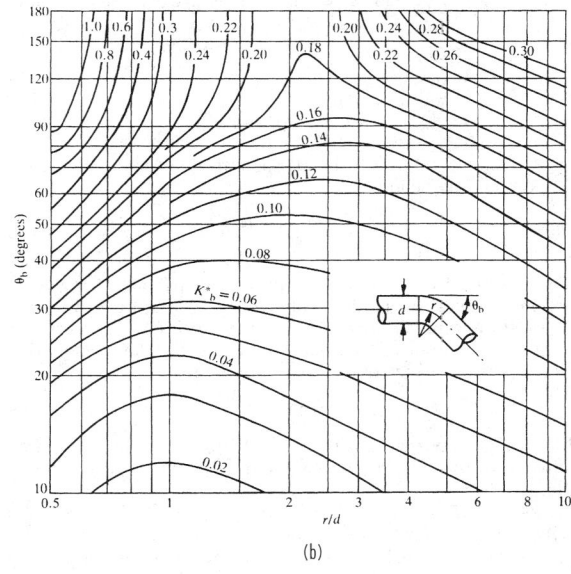

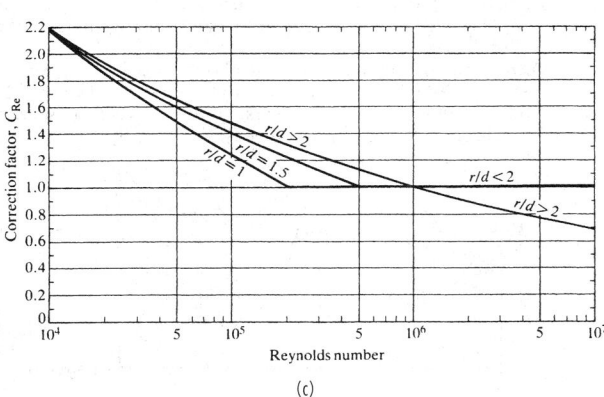

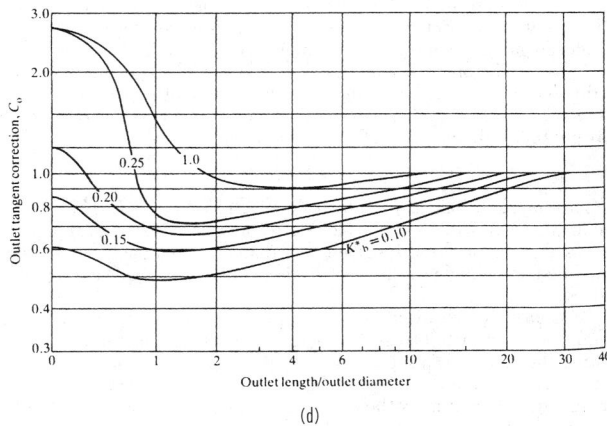

FIG. 6-14 Loss coefficients for flow in bends and curved pipes: (*a*) flow geometry, (*b*) loss coefficient for a smooth-walled bend at Re = 10⁶, (*c*) Re correction factor, (*d*) outlet pipe correction factor (*From D. S. Miller*, Internal Flow Systems, *2d. ed., BHRA, Cranfield, U.K., 1990.*)

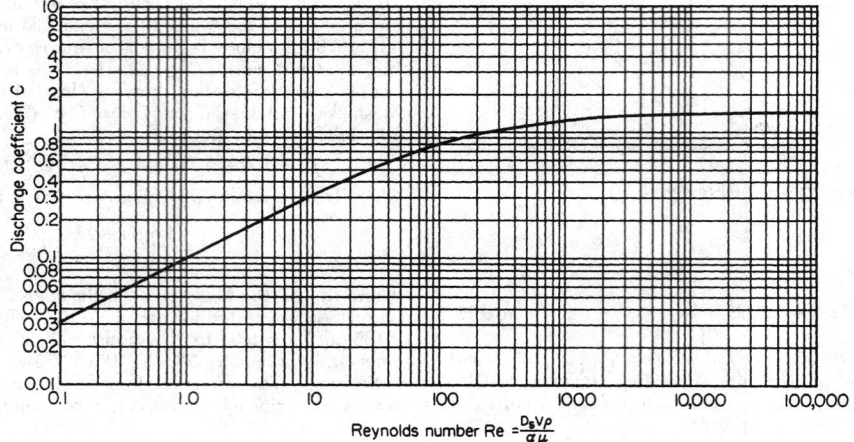

FIG. 6-16   Screen discharge coefficients, plain square-mesh screens. (*Courtesy of E. I. du Pont de Nemours & Co.*)

$$\Delta p = K \frac{\rho V^2}{2} \qquad (6\text{-}106)$$

where ρ = fluid density
V = superficial velocity based upon the gross area of the screen
K = velocity head loss

$$K = \left(\frac{1}{C^2}\right)\left(\frac{1 - \alpha^2}{\alpha^2}\right) \qquad (6\text{-}107)$$

The discharge coefficient for the screen C with aperture D_s is given as a function of screen Reynolds number $\text{Re} = D_s(V/\alpha)\rho/\mu$ in Fig. 6-16 for **plain square-mesh screens,** $\alpha = 0.14$ to 0.79. This curve fits most of the data within ± 20 percent. In the laminar flow region, Re < 20, the discharge coefficient can be computed from

$$C = 0.1\sqrt{\text{Re}} \qquad (6\text{-}108)$$

Coefficients greater than 1.0 in Fig. 6-16 probably indicate partial pressure recovery downstream of the minimum aperture, due to rounding of the wires.

Grootenhuis (*Proc. Inst. Mech. Eng.* [*London*], **A168,** 837–846 [1954]) presents data which indicate that for a series of screens, the total pressure drop equals the number of screens times the pressure drop for one screen, and is not affected by the spacing between screens or their orientation with respect to one another, and presents a correlation for frictional losses across plain square-mesh screens and sintered gauzes. Armour and Cannon (*AIChE J.,* **14,** 415–420 [1968]) give a correlation based on a packed bed model for plain, twill, and "dutch" weaves. For losses through monofilament fabrics see Pedersen (*Filtr. Sep.,* **11,** 586–589 [1975]). For screens **inclined at an angle** θ, use the normal velocity component V'

$$V' = V \cos \theta \qquad (6\text{-}109)$$

(Carothers and Baines, *J. Fluids Eng.,* **97,** 116–117 [1975]) in place of V in Eq. (6-106). This applies for Re > 500, $C = 1.26$, $\alpha \leq 0.97$ and $0 < \theta < 45°$, for square-mesh screens and diamond-mesh netting. Screens inclined at an angle to the flow direction also experience a tangential stress.

For **non-Newtonian** fluids in slow flow, friction loss across a square-woven or full-twill-woven screen can be estimated by considering the screen as a set of parallel tubes, each of diameter equal to the average minimal opening between adjacent wires, and length twice the diameter, without entrance effects (Carley and Smith, *Polym. Eng. Sci.,* **18,** 408–415 [1978]). For screen stacks, the losses of individual screens should be summed.

JET BEHAVIOR

A **free jet,** upon leaving an outlet, will entrain the surrounding fluid, expand, and decelerate. To a first approximation, total momentum is conserved as jet momentum is transferred to the entrained fluid. For practical purposes, a jet is considered free when its cross-sectional area is less than one-fifth of the total cross-sectional flow area of the region through which the jet is flowing (Elrod, *Heat. Piping Air Cond.,* **26**(3), 149–155 [1954]), and the surrounding fluid is the same as the jet fluid. A **turbulent jet** in this discussion is considered to be a free jet with Reynolds number greater than 2,000. Additional discussion on the relation between Reynolds number and turbulence in jets is given by Elrod (ibid.). Abramowicz (*The Theory of Turbulent Jets,* MIT Press, Cambridge, 1963) provides a thorough discourse on the theory of turbulent jets. Hussein, et al. (*J. Fluid Mech.,* **258,** 31–75 [1994]) give extensive velocity data for a free jet, as well as an extensive discussion of free jet experimentation and comparison of data with momentum conservation equations.

A turbulent free jet is normally considered to consist of four flow regions (Tuve, *Heat. Piping Air Cond.,* **25**(1), 181–191 [1953]; Davies, *Turbulence Phenomena,* Academic, New York, 1972) as shown in Fig. 6-17:

1.   Region of flow establishment—a short region whose length is about 6.4 nozzle diameters. The fluid in the conical core of the same length has a velocity about the same as the initial discharge velocity. The termination of this *potential core* occurs when the growing mixing or boundary layer between the jet and the surroundings reaches the centerline of the jet.

2.   A transition region that extends to about 8 nozzle diameters.

3.   Region of established flow—the principal region of the jet. In this region, the velocity profile transverse to the jet is self-preserving when normalized by the centerline velocity.

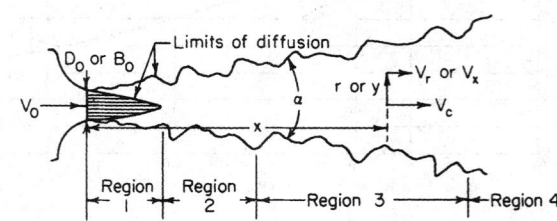

FIG. 6-17   Configuration of a turbulent free jet.

TABLE 6-6   Turbulent Free-Jet Characteristics
Where Both Jet Fluid and Entrained Fluid Are Air

Rounded-inlet circular jet

Longitudinal distribution of velocity along jet center line°†

$$\frac{V_c}{V_0} = K\frac{D_0}{x} \qquad \text{for } 7 < \frac{x}{D_0} < 100$$

$$K = 5 \qquad \text{for } V_0 = 2.5 \text{ to } 5.0 \text{ m/s}$$
$$K = 6.2 \qquad \text{for } V_0 = 10 \text{ to } 50 \text{ m/s}$$

Radial distribution of longitudinal velocity†

$$\log\left(\frac{V_c}{V_r}\right) = 40\left(\frac{r}{x}\right)^2 \qquad \text{for } 7 < \frac{x}{D_0} < 100$$

Jet angle°†

$$\alpha \simeq 20° \qquad \text{for } \frac{x}{D_0} < 100$$

Entrainment of surrounding fluid‡

$$\frac{q}{q_0} = 0.32\frac{x}{D_0} \qquad \text{for } 7 < \frac{x}{D_0} < 100$$

Rounded-inlet, infinitely wide slot jet

Longitudinal distribution of velocity along jet centerline‡

$$\frac{V_c}{V_0} = 2.28\left(\frac{B_0}{x}\right)^{0.5} \qquad \text{for } 5 < \frac{x}{B_0} < 2,000 \text{ and } V_0 = 12 \text{ to } 55 \text{ m/s}$$

Transverse distribution of longitudinal velocity‡

$$\log\left(\frac{V_c}{V_x}\right) = 18.4\left(\frac{y}{x}\right)^2 \qquad \text{for } 5 < \frac{x}{B_0} < 2,000$$

Jet angle‡

$$\alpha \text{ is slightly larger than that for a circular jet}$$

Entrainment of surrounding fluid‡

$$\frac{q}{q_0} = 0.62\left(\frac{x}{B_0}\right)^{0.5} \qquad \text{for } 5 < \frac{x}{B_0} < 2,000$$

°Nottage, Slaby, and Gojsza, *Heat, Piping Air Cond.*, **24**(1), 165–176 (1952).
†Tuve, *Heat, Piping Air Cond.*, **25**(1), 181–191 (1953).
‡Albertson, Dai, Jensen, and Rouse, *Trans. Am. Soc. Civ. Eng.*, **115**, 639–664 (1950), and Discussion, ibid., **115**, 665–697 (1950).

4.   A terminal region where the residual centerline velocity reduces rapidly within a short distance. For air jets, the residual velocity will reduce to less than 0.3 m/s, (1.0 ft/s) usually considered still air.

Several references quote a length of 100 nozzle diameters for the length of the established flow region. However, this length is dependent on initial velocity and Reynolds number.

Table 6-6 gives characteristics of **rounded-inlet circular jets** and **rounded-inlet infinitely wide slot jets** (aspect ratio > 15). The information in the table is for a homogeneous, incompressible air system under isothermal conditions. The table uses the following nomenclature:

B_0 = slot height
D_0 = circular nozzle opening
q = total jet flow at distance x
q_0 = initial jet flow rate
r = radius from circular jet centerline
y = transverse distance from slot jet centerline
V_c = centerline velocity
V_r = circular jet velocity at r
V_y = velocity at y

Witze (*Am. Inst. Aeronaut. Astronaut. J.*, **12**, 417–418 [1974]) gives equations for the centerline velocity decay of different types of subsonic and supersonic circular free jets. Entrainment of surrounding fluid in the region of flow establishment is lower than in the region of established flow (see Hill, *J. Fluid Mech.*, **51**, 773–779 [1972]). Data of Donald and Singer (*Trans. Inst. Chem. Eng.* [*London*], **37**, 255–267

[1959]) indicate that jet angle and the coefficients given in Table 6-6 depend upon the fluids; for a water system, the jet angle for a circular jet is 14° and the entrainment ratio is about 70 percent of that for an air system. Most likely these variations are due to Reynolds number effects which are not taken into account in Table 6-6. Rushton (*AIChE J.*, **26**, 1038–1041 [1980]) examined available published results for circular jets and found that the centerline velocity decay is given by

$$\frac{V_c}{V_0} = 1.41 \text{Re}^{0.135}\left(\frac{D_0}{x}\right) \qquad (6\text{-}110)$$

where $\text{Re} = D_0 V_0 \rho/\mu$ is the initial jet Reynolds number. This result corresponds to a jet angle tan $\alpha/2$ proportional to $\text{Re}^{-0.135}$.

Characteristics of **rectangular jets** of various aspect ratios are given by Elrod (*Heat., Piping, Air Cond.*, **26**[3], 149–155 [1954]). For **slot jets discharging into a moving fluid,** see Weinstein, Osterle, and Forstall (*J. Appl. Mech.*, **23**, 437–443 [1967]). **Coaxial jets** are discussed by Forstall and Shapiro (*J. Appl. Mech.*, **17**, 399–408 [1950]), and **double concentric jets** by Chigier and Beer (*J. Basic Eng.*, **86**, 797–804 [1964]). **Axisymmetric confined jets** are described by Barchilon and Curtet (*J. Basic Eng.*, **777–787** [1964]). **Restrained** turbulent jets of liquid discharging into air are described by Davies (*Turbulence Phenomena*, Academic, New York, 1972). These jets are inherently unstable and break up into drops after some distance. Lienhard and Day (*J. Basic Eng. Trans. AIME*, p. 515 [September 1970]) discuss the breakup of superheated liquid jets which flash upon discharge.

Density gradients affect the spread of a single-phase jet. A jet of lower density than the surroundings spreads more rapidly than a jet of the same density as the surroundings, and, conversely, a denser jet spreads less rapidly. Additional details are given by Keagy and Weller (*Proc. Heat Transfer Fluid Mech. Inst.*, ASME, pp. 89–98, June 22–24 [1949]) and Cleeves and Boelter (*Chem. Eng. Prog.*, **43**, 123–134 [1947]).

Few experimental data exist on **laminar jets** (see Gutfinger and Shinnar, *AIChE J.*, **10**, 631–639 [1964]). Theoretical analysis for velocity distributions and entrainment ratios are available in Schlichting and in Morton (*Phys. Fluids*, **10**, 2120–2127 [1967]).

Theoretical analyses of jet flows for power law **non-Newtonian fluids** are given by Vlachopoulos and Stournaras (*AIChE J.*, **21**, 385–388 [1975]), Mitwally (*J. Fluids Eng.*, **100**, 363 [1978]), and Sridhar and Rankin (*J. Fluids Eng.*, **100**, 500 [1978]).

FLOW THROUGH ORIFICES

Section 10 of this *Handbook* describes the use of orifice meters for flow measurement. In addition, **orifices** are commonly found within pipelines as flow-restricting devices, in perforated pipe distributing and return manifolds, and in perforated plates. Incompressible flow through an orifice in a pipeline, as shown in Fig. 6-18, is commonly described by the following equation for flow rate Q in terms of pressure drop across the orifice Δp, the orifice area A_o, the pipe cross-sectional area A, and the density ρ.

$$Q = C_o A_o \sqrt{\frac{2\Delta p/\rho}{[1 - (A_o/A)^2]}} \qquad (6\text{-}111)$$

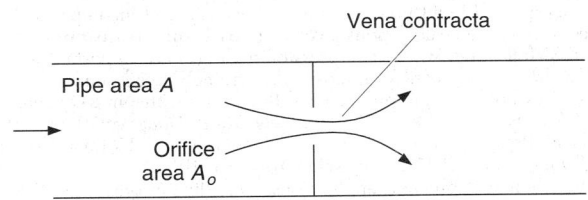

FIG. 6-18   Flow through an orifice.

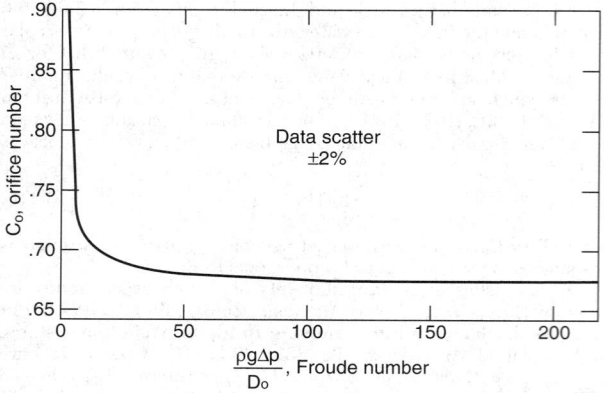

FIG. 6-19 Orifice coefficient vs. Froude number. (*Courtesy E. I. duPont de Nemours & Co.*)

The velocity of approach term $[1 - (A_o/A)^2]$ accounts for the kinetic energy approaching the orifice, while the **orifice coefficient** or **discharge coefficient** C_o accounts for the **vena contracta** effect which causes the fluid to accelerate to velocity greater than Q/A_o. The downstream pressure measurement corresponding to Δp in Eq. (6-111) is taken at the vena contracta. Downstream of the vena contracta, the velocity decelerates and some pressure recovery may be expected. Any pressure recovery is completed about 4 to 8 pipe diameters downstream of the orifice. As an approximation, the pressure recovery, expressed as a fraction of the orifice pressure drop, is approximately equal to the area ratio A_o/A. When the orifice discharges into a large chamber, instead of being installed within a pipe, there is negligible pressure recovery. Equation (6-111) may also be used for flow across a perforated plate with open area A_o and total area A.

The orifice coefficient has a value of about 0.62 at large Reynolds numbers (Re $= D_o V_o \rho/\mu > 20{,}000$), although values ranging from 0.60 to 0.70 are frequently used. At lower Reynolds numbers, the orifice coefficient varies with both Re and with the area or diameter ratio. See Sec. 10 for more details.

When liquids discharge vertically downward from orifices into gas, gravity increases the discharge coefficient. Figure 6-19 shows this effect, giving the discharge coefficient in terms of a modified Froude number, Fr $= \rho g \Delta p/D_o$.

The orifice coefficient deviates from its value for sharp-edged orifices when the orifice wall thickness exceeds about 75 percent of the orifice diameter. Some pressure recovery occurs within the orifice and the orifice coefficient increases. Pressure drop across **segmental orifices** is roughly 10 percent greater than that for concentric circular orifices of the same open area.

COMPRESSIBLE FLOW

Flows are typically considered **compressible** when the density varies by more than 5 to 10 percent. In practice compressible flows are normally limited to gases, supercritical fluids, and multiphase flows containing gases. Liquid flows are normally considered incompressible, except for certain calculations involved in **hydraulic transient** analysis (see following) where compressibility effects are important even for nearly incompressible liquids with extremely small density variations. Textbooks on compressible gas flow include Shapiro (*Dynamics and Thermodynamics of Compressible Fluid Flow*, vol. I and II, Ronald Press, New York [1953]) and Zucrow and Hofmann (*Gas Dynamics*, vol. I and II, Wiley, New York [1976]).

In chemical process applications, one-dimensional gas flows through nozzles or orifices and in pipelines are the most important applications of compressible flow. Multidimensional external flows are of interest mainly in aerodynamic applications.

Mach Number and Speed of Sound The **Mach number** $M = V/c$ is the ratio of fluid velocity, V, to the **speed of sound** or **acoustic velocity**, c. The speed of sound is the propagation velocity of infinitesimal pressure disturbances and is derived from a momentum balance. The compression caused by the pressure wave is adiabatic and frictionless, and therefore isentropic.

$$c = \sqrt{\left(\frac{\partial p}{\partial \rho}\right)_s} \qquad (6\text{-}112)$$

The derivative of pressure p with respect to density ρ is taken at constant entropy s. For an ideal gas,

$$\left(\frac{\partial p}{\partial \rho}\right)_s = \frac{kRT}{M_w}$$

where
   k = ratio of specific heats, C_p/C_v
   R = universal gas constant (8,314 J/kgmol K)
   T = absolute temperature
   M_w = molecular weight

Hence for an ideal gas,

$$c = \sqrt{\frac{kRT}{M_w}} \qquad (6\text{-}113)$$

Most often, the Mach number is calculated using the speed of sound evaluated at the local pressure and temperature. When $M = 1$, the flow is **critical** or **sonic** and the velocity equals the local speed of sound. For **subsonic** flow $M < 1$ while **supersonic** flows have $M > 1$. Compressibility effects are important when the Mach number exceeds 0.1 to 0.2. A common error is to assume that compressibility effects are always negligible when the Mach number is small. The proper assessment of whether compressibility is important should be based on relative density changes, not on Mach number.

Isothermal Gas Flow in Pipes and Channels Isothermal compressible flow is often encountered in long transport lines, where there is sufficient heat transfer to maintain constant temperature. Velocities and Mach numbers are usually small, yet compressibility effects are important when the total pressure drop is a large fraction of the absolute pressure. For an ideal gas with $\rho = p M_w/RT$, integration of the differential form of the momentum or mechanical energy balance equations, assuming a constant friction factor f over a length L of a channel of constant cross section and hydraulic diameter D_H, yields,

$$p_1^2 - p_2^2 = G^2 \frac{RT}{M_w}\left[\frac{4fL}{D_H} + 2\ln\left(\frac{p_1}{p_2}\right)\right] \qquad (6\text{-}114)$$

where the mass velocity $G = w/A = \rho V$ is the mass flow rate per unit cross-sectional area of the channel. The logarithmic term on the right-hand side accounts for the pressure change caused by acceleration of gas as its density decreases, while the first term is equivalent to the calculation of frictional losses using the density evaluated at the average pressure $(p_1 + p_2)/2$.

Solution of Eq. (6-114) for G and differentiation with respect to p_2 reveals a maximum mass flux $G_{max} = p_2\sqrt{M_w/(RT)}$ and a corresponding exit velocity $V_{2,max} = \sqrt{RT/M_w}$ and exit Mach number $M_2 = 1/\sqrt{k}$. This apparent **choking** condition, though often cited, is not physically meaningful for isothermal flow because at such high velocities, and high rates of expansion, isothermal conditions are not maintained.

Adiabatic Frictionless Nozzle Flow In process plant pipelines, compressible flows are usually more nearly adiabatic than isothermal. Solutions for adiabatic flows through frictionless nozzles and in channels with constant cross section and constant friction factor are readily available.

Figure 6-20 illustrates adiabatic discharge of a **perfect** gas through a frictionless nozzle from a large chamber where velocity is effectively zero. A perfect gas obeys the ideal gas law $\rho = p M_w/RT$ and also has constant specific heat. The subscript 0 refers to the **stagnation** conditions in the chamber. More generally, stagnation conditions refer to the conditions which would obtained by isentropically decelerating a gas flow to zero velocity. The minimum area section, or *throat*, of the nozzle is at the nozzle exit. The flow through the nozzle is isentropic

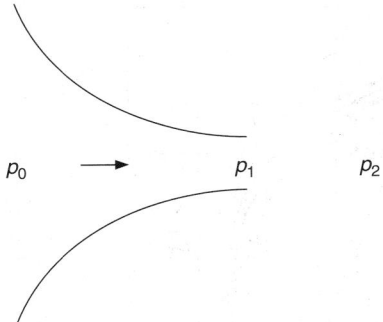

$p_0 \longrightarrow \qquad p_1 \qquad p_2$

FIG. 6-20   Isentropic flow through a nozzle.

because it is frictionless (reversible) and adiabatic. In terms of the exit Mach number M_1 and the upstream stagnation conditions, the flow conditions at the nozzle exit are given by

$$\frac{p_0}{p_1} = \left(1 + \frac{k-1}{2} M_1^2\right)^{k/(k-1)} \tag{6-115}$$

$$\frac{T_0}{T_1} = 1 + \frac{k-1}{2} M_1^2 \tag{6-116}$$

$$\frac{\rho_0}{\rho_1} = \left(1 + \frac{k-1}{2} M_1^2\right)^{1/(k-1)} \tag{6-117}$$

The mass velocity $G = w/A$, where w is the mass flow rate and A is the nozzle exit area, at the nozzle exit is given by

$$G = p_0 \sqrt{\frac{kM_w}{RT_0}} \frac{M_1}{\left(1 + \dfrac{k-1}{2} M_1^2\right)^{(k+1)/2(k-1)}} \tag{6-118}$$

These equations are consistent with the isentropic relations for a perfect gas $p/p_0 = (\rho/\rho_0)^k$, $T/T_0 = (p/p_0)^{(k-1)/k}$. Equation (6-116) is valid for adiabatic flows with or without friction; it does not require isentropic flow. However, Eqs. (6-115) and (6-117) do require isentropic flow.

The exit Mach number M_1 may not exceed unity. At $M_1 = 1$, the flow is said to be **choked, sonic,** or **critical.** When the flow is choked, the pressure at the exit is greater than the pressure of the surroundings into which the gas flow discharges. The pressure drops from the exit pressure to the pressure of the surroundings in a series of shocks which are highly nonisentropic. Sonic flow conditions are denoted by °; sonic exit conditions are found by substituting $M_1 = M_1^\circ = 1$ into Eqs. (6-115) to (6-118).

$$\frac{p^\circ}{p_0} = \left(\frac{2}{k+1}\right)^{k/(k-1)} \tag{6-119}$$

$$\frac{T^\circ}{T_0} = \frac{2}{k+1} \tag{6-120}$$

$$\frac{\rho^\circ}{\rho_0} = \left(\frac{2}{k+1}\right)^{1/(k-1)} \tag{6-121}$$

$$G^\circ = p_0 \sqrt{\left(\frac{2}{k+1}\right)^{(k+1)/(k-1)} \left(\frac{kM_w}{RT_0}\right)} \tag{6-122}$$

Note that under choked conditions, the exit velocity is $V = V^\circ = c^\circ = \sqrt{kRT^\circ/M_w}$, not $\sqrt{kRT_0/M_w}$. Sonic velocity must be evaluated at the exit temperature. For air, with $k = 1.4$, the critical pressure ratio p°/p_0 is 0.5285 and the critical temperature ratio $T^\circ/T_0 = 0.8333$. Thus, for air discharging from 300 K, the temperature drops by 50 K (90 R). This large temperature decrease results from the conversion of internal energy into kinetic energy and is reversible. As the discharged jet decelerates in the external stagnant gas, it recovers its initial enthalpy.

When it is desired to determine the discharge rate through a nozzle from upstream pressure p_0 to external pressure p_2, Equations (6-115) through (6-122) are best used as follows. The critical pressure is first determined from Eq. (6-119). If $p_2 > p^\circ$, then the flow is subsonic (subcritical, unchoked). Then $p_1 = p_2$ and M_1 may be obtained from Eq. (6-115). Substitution of M_1 into Eq. (6-118) then gives the desired mass velocity G. Eqs. (6-116) and (6-117) may be used to find the exit temperature and density. On the other hand, if $p_2 \le p^\circ$, then the flow is choked and $M_1 = 1$. Then $p_1 = p^\circ$, and the mass velocity is G° obtained from Eq. (6-122). The exit temperature and density may be obtained from Eqs. (6-120) and (6-121).

When the flow is choked, $G = G^\circ$ is independent of external downstream pressure. Reducing the downstream pressure will not increase the flow. The mass flow rate under choking conditions is directly proportional to the upstream pressure.

Example 7: Flow through Frictionless Nozzle   Air at p_0 and temperature $T_0 = 293$ K discharges through a frictionless nozzle to atmospheric pressure. Compute the discharge mass flux G, the pressure, temperature, Mach number, and velocity at the exit. Consider two cases: (1) $p_0 = 7 \times 10^5$ Pa absolute, and (2) $p_0 = 1.5 \times 10^5$ Pa absolute.

1. $p_0 = 7.0 \times 10^5$ Pa. For air with $k = 1.4$, the critical pressure ratio from Eq. (6-119) is $p^\circ/p_0 = 0.5285$ and $p^\circ = 0.5285 \times 7.0 \times 10^5 = 3.70 \times 10^5$ Pa. Since this is greater than the external atmospheric pressure $p_2 = 1.01 \times 10^5$ Pa, the flow is choked and the exit pressure is $p_1 = 3.70 \times 10^5$ Pa. The exit Mach number is 1.0, and the mass flux is equal to G° given by Eq. (6-118).

$$G^\circ = 7.0 \times 10^5 \times \sqrt{\left(\frac{2}{1.4+1}\right)^{(1.4+1)/(1.4-1)} \left(\frac{1.4 \times 29}{8314 \times 293}\right)} = 1{,}650 \text{ kg/m}^2 \cdot \text{s}$$

The exit temperature, since the flow is choked, is

$$T^\circ = \left(\frac{T^\circ}{T_0}\right) T_0 = \left(\frac{2}{1.4+1}\right) \times 293 = 244 \text{ K}$$

The exit velocity is $V = Mc = c^\circ = \sqrt{kRT^\circ/M_w} = 313$ m/s.

2. $p_0 = 1.5 \times 10^5$ Pa. In this case $p^\circ = 0.79 \times 10^5$ Pa, which is less than p_2. Hence, $p_1 = p_2 = 1.01 \times 10^5$ Pa. The flow is unchoked (subsonic). Equation (6-115) is solved for the Mach number.

$$\frac{1.5 \times 10^5}{1.01 \times 10^5} = \left(1 + \frac{1.4-1}{2} M_1^2\right)^{1.4/(1.4-1)}$$

$$M_1 = 0.773$$

Substitution into Eq. (6-118) gives G.

$$G = 1.5 \times 10^5 \times \sqrt{\frac{1.4 \times 29}{8{,}314 \times 293}}$$
$$\times \frac{0.773}{\left(1 + \left(\dfrac{1.4-1}{2}\right) \times 0.773^2\right)^{(1.4+1)/2(1.4-1)}} = 337 \text{ kg/m}^2 \cdot \text{s}$$

The exit temperature is found from Eq. (6-116) to be 261.6 K or −11.5°C. The exit velocity is

$$V = Mc = 0.773 \times \sqrt{\frac{1.4 \times 8314 \times 261.6}{29}} = 250 \text{ m/s}$$

Adiabatic Flow with Friction in a Duct of Constant Cross Section   Integration of the differential forms of the continuity, momentum, and total energy equations for a perfect gas, assuming a constant friction factor, leads to a tedious set of simultaneous algebraic equations. These may be found in Shapiro (*Dynamics and Thermodynamics of Compressible Fluid Flow,* vol. I, Ronald Press, New York, 1953) or Zucrow and Hofmann (*Gas Dynamics,* vol. I, Wiley, New York, 1976). Lapple's (*Trans. AIChE.,* **39,** 395–432 [1943]) widely cited graphical presentation of the solution of these equations contained a subtle error, which was corrected by Levenspiel (*AIChE J.,* **23,** 402–403 [1977]). Levenspiel's graphical solutions are presented in Fig. 6-21. These charts refer to the physical situation illustrated in Fig. 6-22, where a perfect gas discharges from stagnation conditions in a large chamber through an isentropic nozzle followed by a duct of length L. The resistance parameter is $N = 4fL/D_H$, where f = Fanning friction factor and D_H = hydraulic diameter.

The exit Mach number M_2 may not exceed unity. $M_2 = 1$ corresponds to choked flow; sonic conditions may exist only at the pipe exit. The mass velocity G° in the charts is the choked mass flux **for an isentropic nozzle** given by Eq. (6-118). For a pipe of finite length,

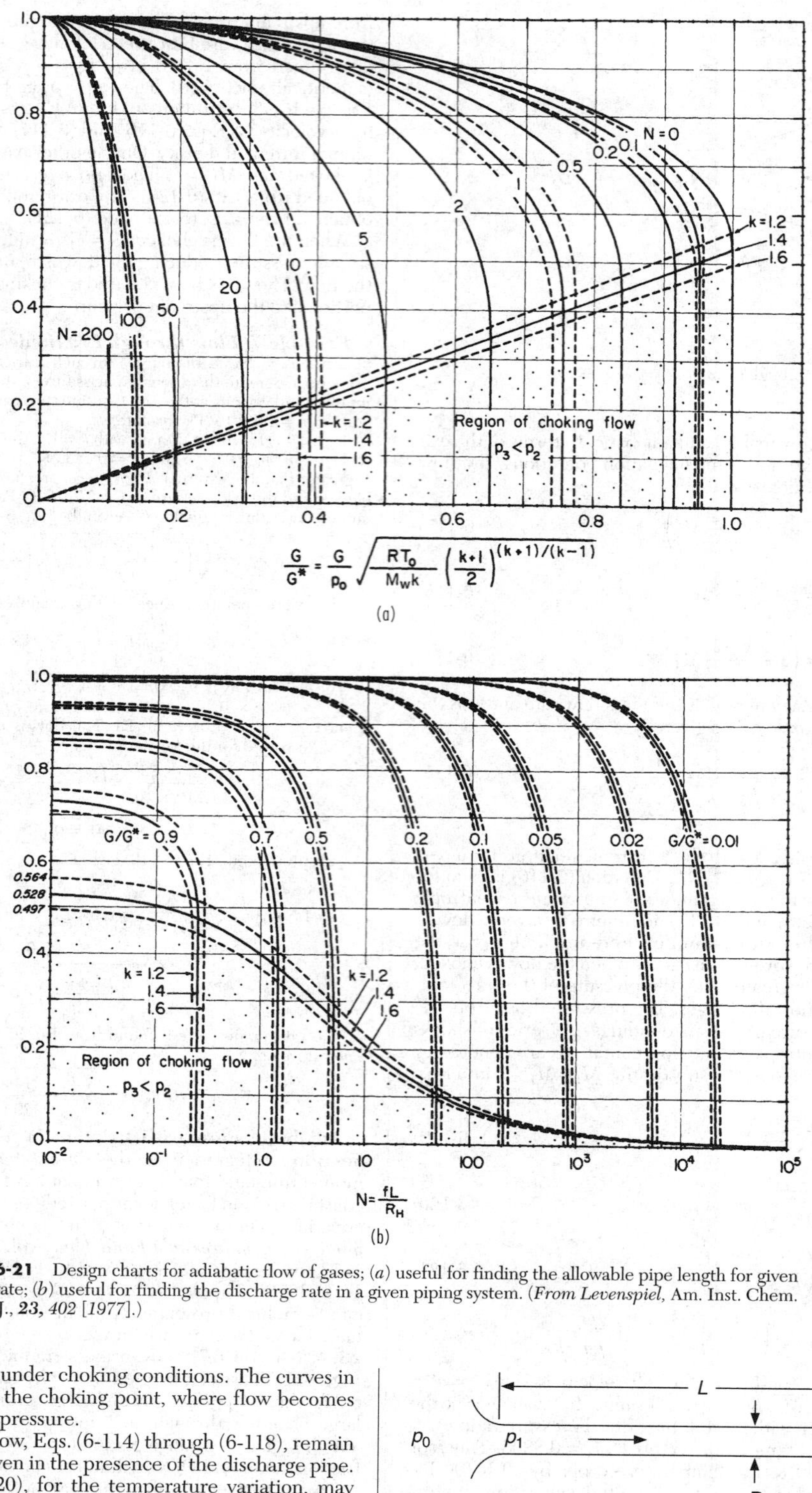

Figure (a):

$$\frac{G}{G^*} = \frac{G}{p_0}\sqrt{\frac{RT_0}{M_w k}}\left(\frac{k+1}{2}\right)^{(k+1)/(k-1)}$$

(a)

Figure (b):

$$N = \frac{fL}{R_H}$$

(b)

FIG. 6-21 Design charts for adiabatic flow of gases; (a) useful for finding the allowable pipe length for given flow rate; (b) useful for finding the discharge rate in a given piping system. (*From Levenspiel,* Am. Inst. Chem. Eng. J., **23,** 402 [1977].)

the mass flux is less than G^* under choking conditions. The curves in Fig. 6-21 become vertical at the choking point, where flow becomes independent of downstream pressure.

The equations for nozzle flow, Eqs. (6-114) through (6-118), remain valid for the nozzle section even in the presence of the discharge pipe. Equations (6-116) and (6-120), for the temperature variation, may also be used for the pipe, with M_2, p_2 replacing M_1, p_1 since they are valid for adiabatic flow, with or without friction.

The graphs in Fig. 6-21 are based on accurate calculations, but are

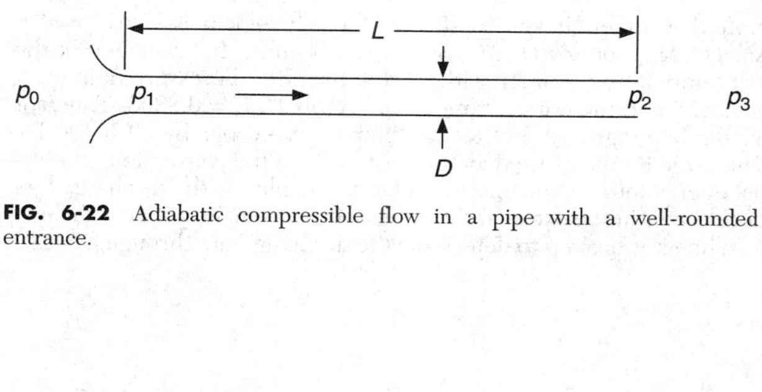

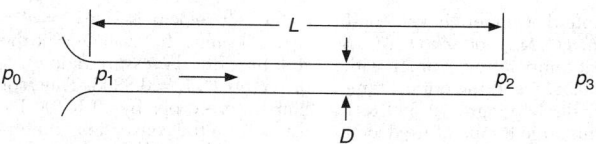

FIG. 6-22 Adiabatic compressible flow in a pipe with a well-rounded entrance.

difficult to interpolate precisely. While they are quite useful for rough estimates, precise calculations are best done using the equations for one-dimensional adiabatic flow with friction, which are suitable for computer programming. Let subscripts 1 and 2 denote two points along a pipe of diameter D, point 2 being downstream of point 1. From a given point in the pipe, where the Mach number is M, the additional length of pipe required to accelerate the flow to sonic velocity ($M = 1$) is denoted L_{max} and may be computed from

$$\frac{4fL_{max}}{D} = \frac{1 - M^2}{kM^2} + \frac{k+1}{2k} \ln \left(\frac{\frac{k+1}{2} M^2}{1 + \frac{k-1}{2} M^2} \right) \quad (6\text{-}123)$$

With L = length of pipe between points 1 and 2, the change in Mach number may be computed from

$$\frac{4fL}{D} = \left(\frac{4fL_{max}}{D} \right)_1 - \left(\frac{4fL_{max}}{D} \right)_2 \quad (6\text{-}124)$$

Eqs. (6-116) and (6-113), which are valid for adiabatic flow with friction, may be used to determine the temperature and speed of sound at points 1 and 2. Since the mass flux $G = \rho v = \rho c M$ is constant, and $\rho = PM_w/RT$, the pressure at point 2 (or 1) can be found from G and the pressure at point 1 (or 2).

The additional frictional losses due to pipeline fittings such as elbows may be added to the velocity head loss $N = 4fL/D_H$ using the same velocity head loss values as for incompressible flow. This works well for fittings which do not significantly reduce the channel cross-sectional area, but may cause large errors when the flow area is greatly reduced, as, for example, by restricting orifices. Compressible flow across restricting orifices is discussed in Sec. 10 of this *Handbook*. Similarly, elbows near the exit of a pipeline may choke the flow even though the Mach number is less than unity due to the nonuniform velocity profile in the elbow. For an abrupt contraction rather than rounded nozzle inlet, an additional 0.5 velocity head should be added to N. This is a reasonable approximation for G, but note that it allocates the additional losses to the pipeline, even though they are actually incurred in the entrance. It is an error to include one velocity head exit loss in N. The kinetic energy at the exit is already accounted for in the integration of the balance equations.

Example 8: Compressible Flow with Friction Losses Calculate the discharge rate of air to the atmosphere from a reservoir at 10^6 Pa gauge and 20°C through 10 m of straight 2-in Schedule 40 steel pipe (inside diameter = 0.0525 m), and 3 standard radius, flanged 90° elbows. Assume 0.5 velocity heads lost for the elbows.

For commercial steel pipe, with a roughness of 0.046 mm, the friction factor for fully rough flow is about 0.0047, from Eq. (6-38) or Fig. 6-9. It remains to be verified that the Reynolds number is sufficiently large to assume fully rough flow. Assuming an abrupt entrance with 0.5 velocity heads lost,

$$N = 4 \times 0.0047 \times \frac{10}{0.0525} + 0.5 + 3 \times 0.5 = 5.6$$

The pressure ratio p_3/p_0 is

$$\frac{1.01 \times 10^5}{(1 \times 10^6 + 1.01 \times 10^5)} = 0.092$$

From Fig. 6-21b at $N = 5.6$, $p_3/p_0 = 0.092$ and $k = 1.4$ for air, the flow is seen to be choked. At the choke point with $N = 5.6$ the critical pressure ratio p_2/p_0 is about 0.25 and $G/G°$ is about 0.48. Equation (6-122) gives

$$G° = 1.101 \times 10^6 \times \sqrt{\left(\frac{2}{1.4 + 1} \right)^{(1.4+1)/(1.4-1)} \frac{1.4 \times 29}{8,314 \times 293.15}} = 2,600 \text{ kg/m}^2 \cdot \text{s}$$

Multiplying by $G/G° = 0.48$ yields $G = 1,250 \text{ kg/m}^2 \cdot \text{s}$. The discharge rate is $w = GA = 1,250 \times \pi \times 0.0525^2/4 = 2.7 \text{ kg/s}$.

Before accepting this solution, the Reynolds number should be checked. At the pipe exit, the temperature is given by Eq. (6-120) since the flow is choked. Thus, $T_2 = T° = 244.6$ K. The viscosity of air at this temperature is about 1.6×10^{-5} Pa $\cdot$ s. Then

$$\text{Re} = \frac{DV\rho}{\mu} = \frac{DG}{\mu} = \frac{0.0525 \times 1,250}{1.6 \times 10^{-5}} = 4.1 \times 10^6$$

At the beginning of the pipe, the temperature is greater, giving greater viscosity

and a Reynolds number of 3.6×10^6. Over the entire pipe length the Reynolds number is very large and the fully rough flow friction factor choice was indeed valid.

Once the mass flux G has been determined, Fig. 6-21a or 6-21b can be used to determine the pressure at any point along the pipe, simply by reducing $4fL/D_H$ and computing p_2 from the figures, given G, instead of the reverse. Charts for calculation between two points in a pipe with known flow and known pressure at either upstream or downstream locations have been presented by Loeb (*Chem. Eng.*, **76**[5], 179–184 [1969]) and for known downstream conditions by Powley (*Can. J. Chem. Eng.*, **36**, 241–245 [1958]).

Convergent/Divergent Nozzles (De Laval Nozzles) During frictionless adiabatic one-dimensional flow with changing cross-sectional area A the following relations are obeyed:

$$\frac{dA}{A} = \frac{dp}{\rho V^2}(1 - M^2) = \frac{1 - M^2}{M^2} \frac{d\rho}{\rho} = -(1 - M^2) \frac{dV}{V} \quad (6\text{-}125)$$

Equation (6-125) implies that in converging channels, subsonic flows are accelerated and the pressure and density decrease. In diverging channels, subsonic flows are decelerated as the pressure and density increase. In subsonic flow, the converging channels act as nozzles and diverging channels as diffusers. In supersonic flows, the opposite is true. Diverging channels act as nozzles accelerating the flow, while converging channels act as diffusers decelerating the flow.

Figure 6-23 shows a converging/diverging nozzle. When p_2/p_0 is less than the critical pressure ratio ($p°/p_0$), the flow will be subsonic in the converging portion of the nozzle, sonic at the throat, and supersonic in the diverging portion. At the throat, where the flow is critical and the velocity is sonic, the area is denoted $A°$. The cross-sectional area and pressure vary with Mach number along the converging/diverging flow path according to the following equations for isentropic flow of a perfect gas:

$$\frac{A}{A°} = \frac{1}{M} \left[\frac{2}{k+1} \left(1 + \frac{k-1}{2} M^2 \right) \right]^{(k+1)/2(k-1)} \quad (6\text{-}126)$$

$$\frac{p_0}{p} = \left(1 + \frac{k-1}{2} M^2 \right)^{k/(k-1)} \quad (6\text{-}127)$$

The temperature obeys the adiabatic flow equation for a perfect gas,

$$\frac{T_0}{T} = 1 + \frac{k-1}{2} M^2 \quad (6\text{-}128)$$

Equation (6-128) does not require frictionless (isentropic) flow. The sonic mass flux through the throat is given by Eq. (6-122). With A set equal to the nozzle exit area, the exit Mach number, pressure, and temperature may be calculated. Only if the exit pressure equals the ambient discharge pressure is the ultimate expansion velocity reached in the nozzle. Expansion will be incomplete if the exit pressure exceeds the ambient discharge pressure; shocks will occur outside the nozzle. If the calculated exit pressure is less than the ambient discharge pressure, the nozzle is overexpanded and compression shocks within the expanding portion will result.

The shape of the converging section is a smooth trumpet shape similar to the simple converging nozzle. However, special shapes of the diverging section are required to produce the maximum supersonic exit velocity. Shocks result if the divergence is too rapid and excessive boundary layer friction occurs if the divergence is too shallow. See Liepmann and Roshko (*Elements of Gas Dynamics*, Wiley, New York, 1957, p. 284). If the nozzle is to be used as a thrust device, the diverg-

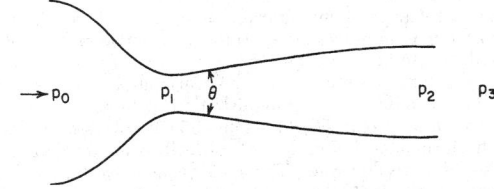

FIG. 6-23 Converging/diverging nozzle.

ing section can be conical with a total included angle of 30° (Sutton, *Rocket Propulsion Elements,* 2d ed., Wiley, New York, 1956). To obtain large exit Mach numbers, slot-shaped rather than axisymmetric nozzles are used.

MULTIPHASE FLOW

Multiphase flows, even when restricted to simple pipeline geometry, are in general quite complex, and several features may be identified which make them more complicated than single-phase flow. Flow pattern description is not merely an identification of laminar or turbulent flow. The relative quantities of the phases and the topology of the interfaces must be described. Because of phase density differences, vertical flow patterns are different from horizontal flow patterns, and horizontal flows are not generally axisymmetric. Even when phase equilibrium is achieved by good mixing in two-phase flow, the changing equilibrium state as pressure drops with distance, or as heat is added or lost, may require that interphase mass transfer, and changes in the relative amounts of the phases, be considered.

Wallis (*One-dimensional Two-phase Flow,* McGraw-Hill, New York, 1969) and Govier and Aziz present mass, momentum, mechanical energy, and total energy balance equations for two-phase flows. These equations are based on one-dimensional behavior for each phase. Such equations, for the most part, are used as a framework in which to interpret experimental data. Reliable prediction of multiphase flow behavior generally requires use of data or correlations. **Two-fluid modeling,** in which the full three-dimensional microscopic (partial differential) equations of motion are written for each phase, treating each as a continuum, occupying a volume fraction which is a continuous function of position, is a rapidly developing technique made possible by improved computational methods. For some relatively simple examples not requiring numerical computation, see Pearson (*Chem. Engr. Sci.,* **49,** 727–732 [1994]). Constitutive equations for two-fluid models are not yet sufficiently robust for accurate general-purpose two-phase flow computation, but may be quite good for particular classes of flows.

Liquids and Gases   For cocurrent flow of liquids and gases in vertical (upflow), horizontal, and inclined pipes, a very large literature of experimental and theoretical work has been published, with less work on countercurrent and cocurrent vertical downflow. Much of the effort has been devoted to predicting flow patterns, pressure drop, and volume fractions of the phases, with emphasis on fully developed flow. In practice, many two-phase flows in process plants are not fully developed.

The most reliable methods for fully developed gas/liquid flows use **mechanistic models** to predict flow pattern, and use different pressure drop and void fraction estimation procedures for each flow pattern. Such methods are too lengthy to include here, and are well suited to incorporation into computer programs; commercial codes for gas/liquid pipeline flows are available. Some key references for mechanistic methods for flow pattern transitions and flow regime–specific pressure drop and void fraction methods include Taitel and Dukler (*AIChE J.,* **22,** 47–55 [1976]), Barnea, et al. (*Int. J. Multiphase Flow,* **6,** 217–225 [1980]), Barnea (*Int. J. Multiphase Flow,* **12,** 733–744 [1986]), Taitel, Barnea, and Dukler (*AIChE J.,* **26,** 345–354 [1980]), Wallis (*One-dimensional Two-phase Flow,* McGraw-Hill, New York, 1969), and Dukler and Hubbard (*Ind. Eng. Chem. Fundam.,* **14,** 337–347 [1975]). For preliminary or approximate calculations, **flow pattern maps** and flow regime–independent empirical correlations, are simpler and faster to use. Such methods for horizontal and vertical flows are provided in the following.

In **horizontal pipe,** flow patterns for fully developed flow have been reported in numerous studies. Transitions between flow patterns are gradual, and subjective owing to the visual interpretation of individual investigators. In some cases, statistical analysis of pressure fluctuations has been used to distinguish flow patterns. Figure 6-24 (Alves, *Chem. Eng. Progr.,* **50,** 449–456 [1954]) shows seven flow patterns for horizontal gas/liquid flow. **Bubble flow** is prevalent at high ratios of liquid to gas flow rates. The gas is dispersed as bubbles which move at velocity similar to the liquid and tend to concentrate near the top of the pipe at lower liquid velocities. **Plug flow** describes a pattern in which alternate plugs of gas and liquid move along the upper

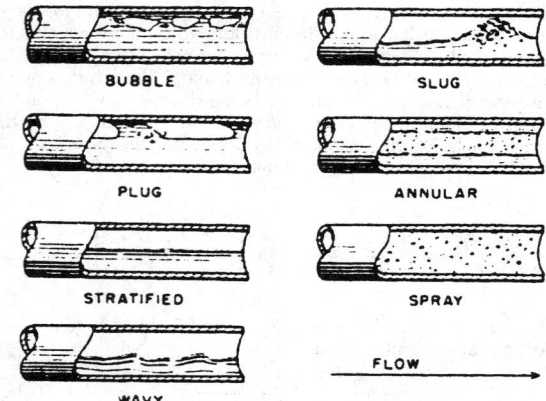

FIG. 6-24   Gas/liquid flow patterns in horizontal pipes. (*From Alves,* Chem. Eng. Progr., **50,** *449–456 [1954].)*

part of the pipe. In **stratified flow,** the liquid flows along the bottom of the pipe and the gas flows over a smooth liquid/gas interface. Similar to stratified flow, **wavy flow** occurs at greater gas velocities and has waves moving in the flow direction. When wave crests are sufficiently high to bridge the pipe, they form frothy slugs which move at much greater than the average liquid velocity. **Slug flow** can cause severe and sometimes dangerous vibrations in equipment because of impact of the high-velocity slugs against bends or other fittings. Slugs may also flood gas/liquid separation equipment.

In **annular flow,** liquid flows as a thin film along the pipe wall and gas flows in the core. Some liquid is entrained as droplets in the gas core. At very high gas velocities, nearly all the liquid is entrained as small droplets. This pattern is called **spray, dispersed,** or **mist flow.**

Approximate prediction of flow pattern may be quickly done using **flow pattern maps,** an example of which is shown in Fig. 6-25 (Baker, *Oil Gas J.,* **53**[12], 185–190, 192–195 [1954]). The Baker chart remains widely used; however, for critical calculations the mechanistic model methods referenced previously are generally preferred for their greater accuracy, especially for large pipe diameters and fluids with physical properties different from air/water at atmospheric pressure. In the chart,

$$\lambda = (\rho'_G \rho'_L)^{1/2} \tag{6-129}$$

$$\psi = \frac{1}{\sigma'}\left[\frac{\mu'_L}{(\rho'_L)^2}\right]^{1/3} \tag{6-130}$$

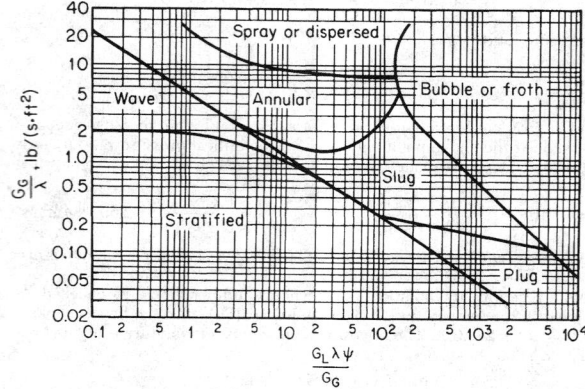

FIG. 6-25   Flow-pattern regions in cocurrent liquid/gas flow through horizontal pipes. To convert lbm/(ft²·s) to kg/(m²·s), multiply by 4.8824. (*From Baker,* Oil Gas J., **53**[12], 185–190, 192, 195 [1954].)

G_L and G_G are the liquid and gas mass velocities, μ'_L is the ratio of liquid viscosity to water viscosity, ρ'_G is the ratio of gas density to air density, ρ'_L is the ratio of liquid density to water density, and σ' is the ratio of liquid surface tension to water surface tension. The reference properties are at 20°C (68°F) and atmospheric pressure, water density 1,000 kg/m³ (62.4 lbm/ft³), air density 1.20 kg/m³ (0.075 lbm/ft³), water viscosity 0.001 Pa · s, (1.0 cp) and surface tension 0.073 N/m (0.0050 lbf/ft). The empirical parameters λ and ψ provide a crude accounting for physical properties. The Baker chart is dimensionally inconsistent since the dimensional quantity G_G/λ is plotted against a dimensionless one, $G_L\lambda\psi/G_G$, and so must be used with G_G in lbm/(ft² · s) units on the ordinate. To convert to kg/(m² · s), multiply by 4.8824.

Rapid approximate predictions of **pressure drop** for fully developed, incompressible horizontal gas/liquid flow may be made using the method of Lockhart and Martinelli (*Chem. Eng. Prog.*, **45**, 39–48 [1949]). First, the pressure drops that would be expected for each of the two phases as if flowing alone in single-phase flow are calculated. The Lockhart-Martinelli parameter X is defined in terms of the ratio of these pressure drops:

$$X = \left[\frac{(\Delta p/L)_L}{(\Delta p/L)_G}\right]^{1/2} \tag{6-131}$$

The two-phase pressure drop may be then be estimated from either of the single-phase pressure drops, using

$$\left(\frac{\Delta p}{L}\right)_{TP} = Y_L\left(\frac{\Delta p}{L}\right)_L \tag{6-132}$$

or

$$\left(\frac{\Delta p}{L}\right)_{TP} = Y_G\left(\frac{\Delta p}{L}\right)_G \tag{6-133}$$

where Y_L and Y_G are read from Fig. 6-26 as functions of X. The curve labels refer to the flow regime (laminar or turbulent) found for each of the phases flowing alone. The common turbulent-turbulent case is approximated well by

$$Y_L = 1 + \frac{20}{X} + \frac{1}{X^2} \tag{6-134}$$

Lockhart and Martinelli (ibid.) correlated pressure drop data from pipes 25 mm (1 in) in diameter or less within about ±50 percent. In general, the predictions are high for stratified, wavy, and slug flows and low for annular flow. The correlation can be applied to pipe diameters up to about 0.1 m (4 in) with about the same accuracy.

The **volume fraction,** sometimes called **holdup,** of each phase in two-phase flow is generally not equal to its volumetric flow rate fraction, because of velocity differences, or **slip,** between the phases. For each phase, denoted by subscript i, the relations among superficial velocity V_i, in situ velocity v_i, volume fraction R_i, total volumetric flow rate Q_i, and pipe area A are

$$Q_i = V_iA = v_iR_iA \tag{6-135}$$

$$v_i = \frac{V_i}{R_i} \tag{6-136}$$

The **slip velocity** between gas and liquid is $v_s = v_G - v_L$. For two-phase gas/liquid flow, $R_L + R_G = 1$. A very common mistake in practice is to assume that in situ phase volume fractions are equal to input volume fractions.

For fully developed incompressible horizontal gas/liquid flow, a quick estimate for R_L may be obtained from Fig. 6-27, as a function of the Lockhart-Martinelli parameter X defined by Eq. (6-131). Indications are that liquid volume fractions may be overpredicted for liquids more viscous than water (Alves, *Chem. Eng. Prog.*, **50**, 449–456 [1954]), and underpredicted for pipes larger than 25 mm diameter (Baker, *Oil Gas J.*, **53**[12], 185–190, 192–195 [1954]).

A method for predicting pressure drop and volume fraction for **non-Newtonian fluids** in annular flow has been proposed by Eisenberg and Weinberger (*AIChE J.*, **25**, 240–245 [1979]). Das, Biswas, and Matra (*Can. J. Chem. Eng.*, **70**, 431–437 [1993]) studied holdup in both horizontal and vertical gas/liquid flow with non-Newtonian liquids. Farooqi and Richardson (*Trans Inst. Chem. Engrs.*, **60**, 292–305, 323–333 [1982]) developed correlations for holdup and pressure drop for gas/non-Newtonian liquid horizontal flow. They used a modified Lockhart-Martinelli parameter for non-Newtonian liquid holdup. They found that two-phase pressure drop may actually be less than the single-phase liquid pressure drop with shear thinning liquids in laminar flow.

Pressure drop data for a 1-in **feed tee** with the liquid entering the run and gas entering the branch are given by Alves (*Chem. Eng. Progr.*, **50**, 449–456 [1954]). Pressure drop and division of two-phase **annular flow in a tee** are discussed by Fouda and Rhodes (*Trans. Inst. Chem. Eng. [London]*, **52**, 354–360 [1974]). Flow through tees can result in unexpected flow splitting. Further reading on gas/liquid flow through tees may be found in Mudde, Groen, and van den Akker (*Int. J. Multiphase Flow*, **19**, 563–573 [1993]); Issa and Oliveira (*Computers and Fluids*, **23**, 347–372 [1993]) and Azzopardi and Smith (*Int. J. Multiphase Flow*, **18**, 861–875 [1992]).

Results by Chenoweth and Martin (*Pet. Refiner*, **34**[10], 151–155 [1955]) indicate that single-phase data for **fittings** and **valves** can be used in their correlation for two-phase pressure drop. Smith, Murdock, and Applebaum (*J. Eng. Power*, **99**, 343–347 [1977]) evaluated existing correlations for two-phase flow of steam/water and other gas/liquid mixtures through sharp-edged **orifices** meeting ASTM standards for flow measurement. The correlation of Murdock (*J. Basic Eng.*, **84**, 419–433 [1962]) may be used for these orifices. See also Collins and Gacesa (*J. Basic Eng.*, **93**, 11–21 [1971]), for measurements with steam and water beyond the limits of this correlation.

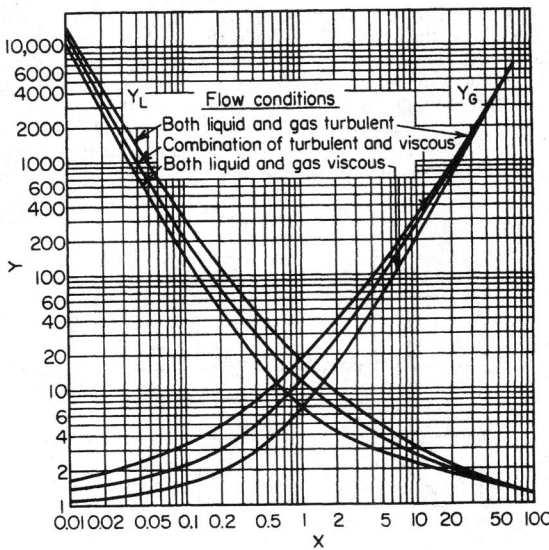

FIG. 6-26   Parameters for pressure drop in liquid/gas flow through horizontal pipes. (*Based on Lockhart and Martinelli*, Chem. Engr. Prog., **45**, 39 [1949].)

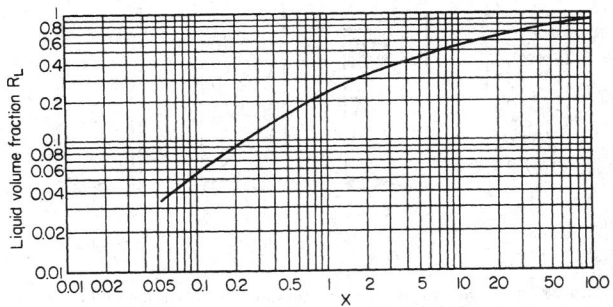

FIG. 6-27   Liquid volume fraction in liquid/gas flow through horizontal pipes. (*From Lockhart and Martinelli*, Eng. Prog., **45**, 39 [1949].)

For pressure drop and holdup in **inclined pipe** with upward or downward flow, see Beggs and Brill (*J. Pet. Technol.*, **25**, 607–617 [1973]); the mechanistic model methods referenced above may also be applied to inclined pipes. Up to 10° from horizontal, upward pipe inclination has little effect on holdup (Gregory, *Can. J. Chem. Eng.*, **53**, 384–388 [1975]).

For fully developed incompressible **cocurrent upflow** of gases and liquids in **vertical pipes**, a variety of flow pattern terminologies and descriptions have appeared in the literature; some of these have been summarized and compared by Govier, Radford, and Dunn (*Can. J. Chem. Eng.*, **35**, 58–70 [1957]). One reasonable classification of patterns is illustrated in Fig. 6-28.

In **bubble flow**, gas is dispersed as bubbles throughout the liquid, but with some tendency to concentrate toward the center of the pipe. In **slug flow**, the gas forms large **Taylor bubbles** of diameter nearly equal to the pipe diameter. A thin film of liquid surrounds the Taylor bubble. Between the Taylor bubbles are liquid slugs containing some bubbles. **Froth** or **churn flow** is characterized by strong intermittency and intense mixing, with neither phase easily described as continuous or dispersed. There remains disagreement in the literature as to whether churn flow is a real fully developed flow pattern or is an indication of large entry length for developing slug flow (Zao and Dukler, *Int. J. Multiphase Flow*, **19**, 377–383 [1993]; Hewitt and Jayanti, *Int. J. Multiphase Flow*, **19**, 527–529 [1993]).

Ripple flow has an upward-moving wavy layer of liquid on the pipe wall; it may be thought of as a transition region to **annular, annular mist**, or **film flow**, in which gas flows in the core of the pipe while an annulus of liquid flows up the pipe wall. Some of the liquid is entrained as droplets in the gas core. **Mist flow** occurs when all the liquid is carried as fine drops in the gas phase; this pattern occurs at high gas velocities, typically 20 to 30 m/s (66 to 98 ft/s).

The correlation by Govier, et al. (*Can. J. Chem. Eng.*, **35**, 58–70 [1957]), Fig. 6-29, may be used for quick estimate of flow pattern.

Slip, or relative velocity between phases, occurs for vertical flow as well as for horizontal. No completely satisfactory, flow regime–independent correlation for volume fraction or holdup exists for vertical flow. Two frequently used flow regime–independent methods are those by Hughmark and Pressburg (*AIChE J.*, **7**, 677 [1961]) and Hughmark (*Chem. Eng. Prog.*, **58**[4], 62 [April 1962]). **Pressure drop** in **upflow** may be calculated by the procedure described in Hughmark (*Ind. Eng. Chem. Fundam.*, **2**, 315–321 [1963]). The mechanistic, flow regime–based methods are advisable for critical applications.

For **upflow** in **helically coiled tubes**, the flow pattern, pressure drop, and holdup can be predicted by the correlations of Banerjee,

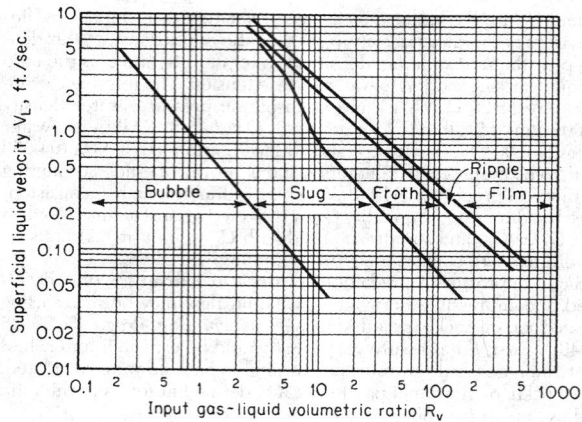

FIG. 6-29   Flow-pattern regions in cocurrent liquid/gas flow in upflow through vertical pipes. To convert ft/s to m/s, multiply by 0.3048. (*From Govier, Radford, and Dunn,* Can. J. Chem. Eng., **35**, 58–70 [1957].)

Rhodes, and Scott (*Can. J. Chem. Eng.*, **47**, 445–453 [1969]) and Akagawa, Sakaguchi, and Ueda (*Bull JSME*, **14**, 564–571 [1971]). Correlations for flow patterns in **downflow** in vertical pipe are given by Oshinowo and Charles (*Can. J. Chem. Eng.*, **52**, 25–35 [1974]) and Barnea, Shoham, and Taitel (*Chem. Eng. Sci.*, **37**, 741–744 [1982]). Use of **drift flux theory** for void fraction modeling in downflow is presented by Clark and Flemmer (*Chem. Eng. Sci.*, **39**, 170–173 [1984]). **Downward inclined** two-phase flow data and modeling are given by Barnea, Shoham, and Taitel (*Chem. Eng. Sci.*, **37**, 735–740 [1982]). Data for **downflow** in **helically coiled tubes** are presented by Casper (*Chem. Ing. Tech.*, **42**, 349–354 [1970]).

The entrance to a **drain** is flush with a horizontal surface, while the entrance to an **overflow** pipe is above the horizontal surface. When such pipes do not run full, considerable amounts of gas can be drawn down by the liquid. The amount of gas entrained is a function of pipe diameter, pipe length, and liquid flow rate, as well as the drainpipe outlet boundary condition. Extensive data on air entrainment and liquid head above the entrance as a function of water flow rate for pipe diameters from 43.9 to 148.3 mm (1.7 to 5.8 in) and lengths from about 1.22 to 5.18 m (4.0 to 17.0 ft) are reported by Kalinske (*Univ. Iowa Stud. Eng.*, Bull. 26, pp. 26–40 [1939–1940]). For heads greater than the critical, the pipes will run full with no entrainment. The critical head *h* for flow of water in drains and overflow pipes is given in Fig. 6-30. Kalinske's results show little effect of the height of protrusion of overflow pipes when the protrusion height is greater than about one pipe diameter. For conservative design, McDuffie (*AIChE J.*, **23**, 37–40 [1977]) recommends the following relation for minimum liquid height to prevent entrainment.

$$\mathrm{Fr} \leq 1.6 \left(\frac{h}{D}\right)^2 \tag{6-137}$$

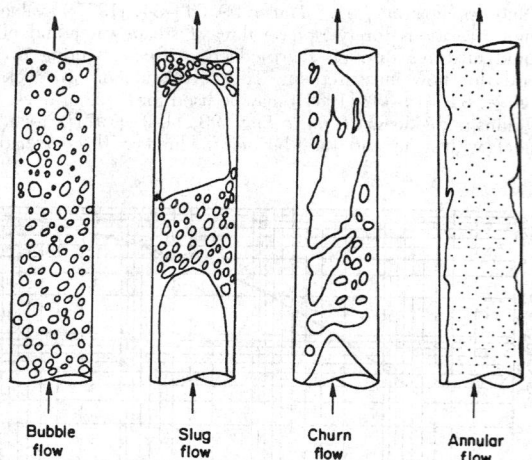

FIG. 6-28   Flow patterns in cocurrent upward vertical gas/liquid flow. (*From Taitel, Barnea, and Dukler,* AIChE J., **26**, 345–354 [1980]. Reproduced by permission of the American Institute of Chemical Engineers © 1980 AIChE. All rights reserved.)

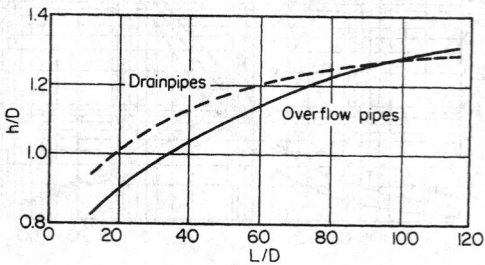

FIG. 6-30   Critical head for drain and overflow pipes. (*From Kalinske,* Univ. Iowa Stud. Eng., Bull. 26 [1939–1940].)

where the Froude number is defined by

$$\text{Fr} \equiv \frac{V_L}{\sqrt{g(\rho_L - \rho_G)D/\rho_L}} \qquad (6\text{-}138)$$

where g = acceleration due to gravity
V_L = liquid velocity in the drain pipe
ρ_L = liquid density
ρ_G = gas density
D = pipe inside diameter
h = liquid height

For additional information, see Simpson (*Chem. Eng.*, **75**(6), 192–214 [1968]). A critical Froude number of 0.31 to ensure vented flow is widely cited. Recent results (Thorpe, *3d Int. Conf. Multi-phase Flow*, The Hague, Netherlands, 18–20 May 1987, paper K2, and *4th Int. Conf. Multi-phase Flow*, Nice, France, 19–21 June 1989, paper K4) show hysteresis, with different critical Froude numbers for flooding and unflooding of drain pipes, and the influence of end effects. Wallis, Crowley, and Hagi (*Trans. ASME J. Fluids Eng.*, 405–413 [June 1977]) examine the conditions for horizontal discharge pipes to run full.

Flashing flow and **condensing flow** are two examples of multiphase flow with **phase change.** Flashing flow occurs when pressure drops below the bubble point pressure of a flowing liquid. A frequently used one-dimensional model for flashing flow through nozzles and pipes is the **homogeneous equilibrium model** which assumes that both phases move at the same in situ velocity, and maintain vapor/liquid equilibrium. It may be shown that a **critical flow** condition, analogous to sonic or critical flow during compressible gas flow, is given by the following expression for the mass flux G in terms of the derivative of pressure p with respect to mixture density ρ_m at constant entropy:

$$G_{\text{crit}} = \rho_m \sqrt{\left(\frac{\partial p}{\partial \rho_m}\right)_s} \qquad (6\text{-}139)$$

The corresponding acoustic velocity $\sqrt{(\partial p/\partial \rho_m)_s}$ is normally much less than the acoustic velocity for gas flow. The mixture density is given in terms of the individual phase densities and the **quality** (mass flow fraction vapor) x by

$$\frac{1}{\rho_m} = \frac{x}{\rho_G} + \frac{1-x}{\rho_L} \qquad (6\text{-}140)$$

Choked and unchoked flow situations arise in pipes and nozzles in the same fashion for homogeneous equilibrium flashing flow as for gas flow. For nozzle flow from stagnation pressure p_0 to exit pressure p_1, the mass flux is given by

$$G^2 = -2\rho_{m1}^2 \int_{p_0}^{p_1} \frac{dp}{\rho_m} \qquad (6\text{-}141)$$

The integration is carried out over an isentropic flash path: flashes at constant entropy must be carried out to evaluate ρ_m as a function of p. Experience shows that isenthalpic flashes provide good approximations unless the liquid mass fraction is very small. Choking occurs when G obtained by Eq. (6-141) goes through a maximum at a value of p_1 greater than the external discharge pressure. Equation (6-139) will also be satisfied at that point. In such a case the pressure at the nozzle exit equals the choking pressure and flashing shocks occur outside the nozzle exit.

For homogeneous flow in a pipe of diameter D, the differential form of the Bernoulli equation (6-15) rearranges to

$$\frac{dp}{\rho_m} + g\,dz + \frac{G^2}{\rho_m} d\,\frac{1}{\rho_m} + 2f \frac{dx'}{D} \frac{G^2}{\rho_m^2} = 0 \qquad (6\text{-}142)$$

where x' is distance along the pipe. Integration over a length L of pipe assuming constant friction factor f yields

$$G^2 = \frac{-\int_{p_1}^{p_2} \rho_m\, dp - g \int_{z_1}^{z_2} \rho_m^2\, dz}{\ln\,(\rho_{m1}/\rho_{m2}) + 2fL/D} \qquad (6\text{-}143)$$

Frictional pipe flow is not isentropic. Strictly speaking, the flashes must be carried out at constant $h + V^2/2 + gz$, where h is the enthalpy

per unit mass of the two-phase flashing mixture. The flash calculations are fully coupled with the integration of the Bernoulli equation; the velocity V must be known at every pressure p to evaluate ρ_m. Computational routines, employing the thermodynamic and material balance features of flowsheet simulators, are the most practical way to carry out such flashing flow calculations, particularly when multicomponent systems are involved. Significant simplification arises when the mass fraction liquid is large, for then the effect of the $V^2/2$ term on the flash splits may be neglected. If elevation effects are also negligible, the flash computations are decoupled from the Bernoulli equation integration. For many horizontal flashing flow calculations, this is satisfactory and the flash computatations may be carried out first, to find ρ_m as a function of p from p_1 to p_2, which may then be substituted into Eq. (6-143).

With flashes carried out along the appropriate thermodynamic paths, the formalism of Eqs. (6-139) through (6-143) applies to all homogeneous equilibrium compressible flows, including, for example, flashing flow, ideal gas flow, and nonideal gas flow. Equation (6-118), for example, is a special case of Eq. (6-141) where the quality $x = 1$ and the vapor phase is a perfect gas.

Various **nonequilibrium** and **slip flow** models have been proposed as improvements on the homogeneous equilibrium flow model. See, for example, Henry and Fauske (*Trans. ASME J. Heat Transfer*, 179–187 [May 1971]). Nonequilibrium and slip effects both increase computed mass flux for fixed pressure drop, compared to homogeneous equilibrium flow. For flow paths greater than about 100 mm, homogeneous equilibrium behavior appears to be the best assumption (Fischer, et al., *Emergency Relief System Design Using DIERS Technology*, AIChE, New York [1992]). For shorter flow paths, the best estimate may sometimes be given by linearly interpolating (as a function of length) between **frozen flow** (constant quality, no flashing) at 0 length and equilibrium flow at 100 mm.

In a series of papers by Leung and coworkers (*AIChE J.*, **32**, 1743–1746 [1986]; **33**, 524–527 [1987]; **34**, 688–691 [1988]; *J. Loss Prevention Proc. Ind.*, **2**[2], 78–86 [April 1989]; **3**(1), 27–32 [January 1990]; *Trans. ASME J. Heat Transfer*, **112**, 524–528, 528–530 [1990]; **113**, 269–272 [1991]) approximate techniques have been developed for homogeneous equilibrium calculations based on pseudo–equation of state methods for flashing mixtures.

Relatively less work has been done on **condensing flows.** Slip effects are more important for condensing than for flashing flows. Soliman, Schuster, and Berenson (*J. Heat Transfer*, **90**, 267–276 [1968]) give a model for condensing vapor in **horizontal pipe.** They assume the condensate flows as an annular ring. The Lockhart-Martinelli correlation is used for the frictional pressure drop. To this pressure drop is added an acceleration term based on homogeneous flow, equivalent to the $G^2 d(1/\rho_m)$ term in Eq. (6-142). Pressure drop is computed by integration of the incremental pressure changes along the length of pipe.

For **condensing vapor** in **vertical downflow,** in which the liquid flows as a thin annular film, the frictional contribution to the pressure drop may be estimated based on the gas flow alone, using the friction factor plotted in Fig. 6-31, where Re_G is the Reynolds number for the gas flowing alone (Bergelin, et al., *Proc. Heat Transfer Fluid Mech. Inst.*, ASME, June 22–24, 1949, pp. 19–28).

$$-\frac{dp}{dz} = \frac{2f_G' \rho_G V_G^2}{D} \qquad (6\text{-}144)$$

To this should be added the $G_G^2 d(1/\rho_G)/dx$ term to account for velocity change effects.

Gases and Solids The flow of gases and solids in **horizontal pipe** is usually classified as either **dilute phase** or **dense phase** flow. Unfortunately, there is no clear dilineation between the two types of flow, and the *dense phase* description may take on more than one meaning, creating some confusion (Knowlton, et al., *Chem. Eng. Progr.*, **90**(4), 44–54 [April 1994]). For dilute phase flow, achieved at low solids-to-gas weight ratios (loadings), and high gas velocities, the solids may be fully suspended and fairly uniformly dispersed over the pipe cross section (homogeneous flow), particularly for low-density or small particle size solids. At lower gas velocities, the solids may bounce along the bottom of the pipe. With higher loadings and lower gas velocities, the particles may settle to the bottom of the pipe, form-

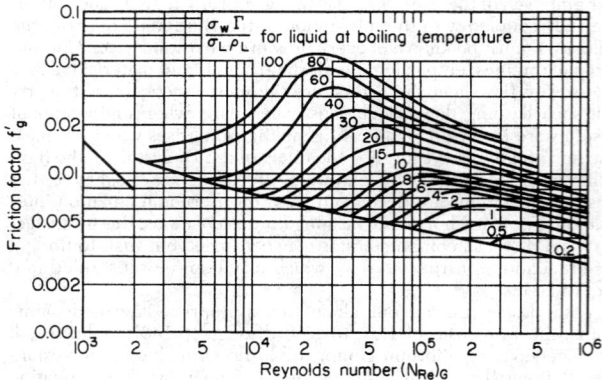

FIG. 6-31 Friction factors for condensing liquid/gas flow downward in vertical pipe. In this correlation $\Gamma/\rho L$ is in ft²/h. To convert ft²/h to m²/s, multiply by 0.00155. (*From Bergelin, et al., Proc. Heat Transfer Fluid Mech. Inst., ASME, 1949, p. 19.*)

ing dunes, with the particles moving from dune to dune. In dense phase conveying, solids tend to concentrate in the lower portion of the pipe at high gas velocity. As gas velocity decreases, the solids may first form dense moving strands, followed by slugs. Discrete plugs of solids may be created intentionally by timed injection of solids, or the plugs may form spontaneously. Eventually the pipe may become blocked. For more information on flow patterns, see Coulson and Richardson (*Chemical Engineering*, vol. 2, 2d ed., Pergamon, New York, 1968, p. 583); Korn (*Chem. Eng., 57*[3], 108–111 [1950]); Patterson (*J. Eng. Power*, **81**, 43–54 [1959]); Wen and Simons (*AIChE J., 5*, 263–267 [1959]); and Knowlton, et al. (*Chem. Eng. Progr., 90*[4], 44–54 [April 1994]).

For the **minimum velocity** required to prevent formation of dunes or settled beds in **horizontal flow**, some data are given by Zenz (*Ind. Eng. Chem. Fundam., 3*, 65–75 [1964]), who presented a correlation for the minimum velocity required to keep particles from depositing on the bottom of the pipe. This rather tedious estimation procedure may also be found in Govier and Aziz, who provide additional references and discussion on transition velocities. In practice, the actual conveying velocities used in systems with loadings less than 10 are generally over 15 m/s, (49 ft/s) while for high loadings (>20) they are generally less than 7.5 m/s (24.6 ft/s) and are roughly twice the actual solids velocity (Wen and Simons, *AIChE J., 5*, 263–267 [1959]).

Total **pressure drop** for **horizontal gas/solid** flow includes acceleration effects at the entrance to the pipe and frictional effects beyond the entrance region. A great number of correlations for pressure gradient are available, none of which is applicable to all flow regimes. Govier and Aziz review many of these and provide recommendations on when to use them.

For **upflow** of gases and solids in **vertical pipes,** the **minimum conveying velocity** for low loadings may be estimated as twice the terminal settling velocity of the largest particles. Equations for terminal settling velocity are found in the "Particle Dynamics" subsection, following. **Choking** occurs as the velocity is dropped below the minimum conveying velocity and the solids are no longer transported, collapsing into solid plugs (Knowlton, et al., *Chem. Eng. Progr., 90*[4], 44–54 [April 1994]). See Smith (*Chem. Eng. Sci., 33*, 745–749 [1978]) for an equation to predict the onset of choking.

Total **pressure drop** for vertical upflow of gases and solids includes acceleration and frictional affects also found in horizontal flow, plus potential energy or hydrostatic effects. Govier and Aziz review many of the pressure drop calculation methods and provide recommendations for their use. See also Yang (*AIChE J., 24*, 548–552 [1978]).

Drag reduction has been reported for low loadings of small diameter particles (<60 µm diameter), ascribed to damping of turbulence near the wall (Rossettia and Pfeffer, *AIChE J., 18*, 31–39 [1972]).

For **dense phase** transport in **vertical pipes** of small diameter, see

Sandy, Daubert, and Jones (*Chem. Eng. Prog., 66, Symp. Ser., 105*, 133–142 [1970]).

The **flow of bulk solids through restrictions and bins** is discussed in symposium articles (*J. Eng. Ind., 91*[2] [1969]) and by Stepanoff (*Gravity Flow of Bulk Solids and Transportation of Solids in Suspension,* Wiley, New York, 1969). Some problems encountered in discharge from bins include (Knowlton, et al., *Chem. Eng. Progr., 90*[4], 44–54 [April 1994]) flow stoppage due to **ratholing** or **arching, segregation** of fine and coarse particles, **flooding** upon collapse of ratholes, and poor **residence time distribution** when **funnel flow** occurs.

Solid/liquid or **slurry** flow may be divided roughly into two categories based on settling behavior (see Etchells in Shamlou, *Processing of Solid-Liquid Suspensions,* Chap. 12, Butterworth-Heinemann, Oxford, 1993). **Nonsettling** slurries are made up of very fine, highly concentrated, or neutrally buoyant particles. These slurries are normally treated as pseudohomogeneous fluids. They may be quite viscous and are frequently non-Newtonian. Slurries of particles that tend to settle out rapidly are called **settling slurries** or **fast-settling slurries.** While in some cases positively buoyant solids are encountered, the present discussion will focus on solids which are more dense than the liquid.

For **horizontal flow** of **fast-settling slurries,** the following rough description may be made (Govier and Aziz). Ultrafine particles, 10 µm or smaller, are generally fully suspended and the particle distributions are not influenced by gravity. Fine particles 10 to 100 µm (3.3×10^{-5} to 33×10^{-5} ft) are usually fully suspended, but gravity causes concentration gradients. Medium-size particles, 100 to 1000 µm, may be fully suspended at high velocity, but often form a moving deposit on the bottom of the pipe. Coarse particles, 1,000 to 10,000 µm, (0.0033 to .033 ft), are seldom fully suspended and are usually conveyed as a moving deposit. Ultracoarse particles larger than 10,000 µm (0.033 ft) are not suspended at normal velocities unless they are unusually light.

Figure 6-32, taken from Govier and Aziz, schematically indicates four flow pattern regions superimposed on a plot of pressure gradient vs. mixture velocity $V_M = V_L + V_S = (Q_L + Q_S)/A$ where V_L and V_S are the superficial liquid and solid velocities, Q_L and Q_S are liquid and solid volumetric flow rates, and A is the pipe cross-sectional area. V_{M4} is the transition velocity above which a bed exists in the bottom of the pipe, part of which is stationary and part of which moves by **saltation,** with the upper particles tumbling and bouncing over one another, often with formation of dunes. With a broad particle-size distribution, the finer particles may be fully suspended. Near V_{M4}, the pressure gra-

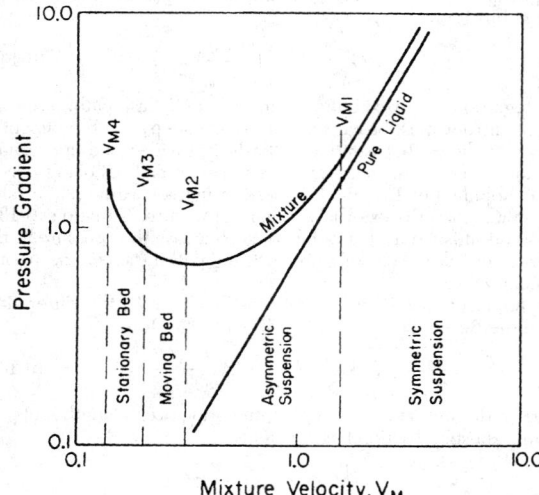

FIG. 6-32 Flow pattern regimes and pressure gradients in horizontal slurry flow. (*From Govier and Aziz, The Flow of Complex Mixtures in Pipes, Van Nostrand Reinhold, New York, 1972.*)

dient rapidly increases as V_M decreases. Above V_{M3}, the entire bed moves. Above V_{M2}, the solids are fully suspended; that is, there is no deposit, moving or stationary, on the bottom of the pipe. However, the concentration distribution of solids is asymmetric. This flow pattern is the most frequently used for fast-settling slurry transport. Typical mixture velocities are in the range of 1 to 3 m/s (3.3 to 9.8 ft/s). The minimum in the pressure gradient is found to be near V_{M2}. Above V_{M1}, the particles are symmetrically distributed, and the pressure gradient curve is nearly parallel to that for the liquid by itself.

The most important transition velocity, often regarded as the minimum transport or conveying velocity for settling slurries, is V_{M2}. The Durand equation (Durand, Minnesota Int. Hydraulics Conf., *Proc.,* 89, Int. Assoc. for Hydraulic Research [1953]; Durand and Condolios, *Proc. Colloq. On the Hyd. Transport of Solids in Pipes, Nat. Coal Board* [UK], Paper IV, 39–35 [1952]) gives the minimum transport velocity as

$$V_{M2} = F_L[2gD(s-1)]^{0.5} \qquad (6\text{-}145)$$

where g = acceleration of gravity
D = pipe diameter
$s = \rho_S/\rho_L$ = ratio of solid to liquid density
F_L = a factor influenced by particle size and concentration

Probably F_L is a function of particle Reynolds number and concentration, but Fig. 6-33 gives Durand's empirical correlation for F_L as a function of particle diameter and the input, feed volume fraction solids, $C_S = Q_S/(Q_S + Q_L)$. The form of Eq. (6-145) may be derived from turbulence theory, as shown by Davies (*Chem. Eng. Sci., 42,* 1667–1670 [1987]).

No single correlation for **pressure drop** in horizontal solid/liquid flow has been found satisfactory for all particle sizes, densities, concentrations, and pipe sizes. However, with reference to Fig. 6-32, the following simplifications may be considered. The minimum pressure gradient occurs near V_{M2} and for conservative purposes it is generally desirable to exceed V_{M2}. When V_{M2} is exceeded, a rough guide for pressure drop is 25 percent greater than that calculated assuming that the slurry behaves as a psuedohomogeneous fluid with the density of the mixture and the viscosity of the liquid. Above the transition velocity to symmetric suspension, V_{M1}, the pressure drop closely approaches the pseudohomogeneous pressure drop. The following

correlation by Spells (*Trans. Inst. Chem. Eng.* [*London*], **33**, 79–84 [1955]) may be used for V_{M1}:

$$V_{M1}^2 = 0.075 \left(\frac{DV_{M1}\rho_M}{\mu} \right)^{0.775} gD_S(s-1) \qquad (6\text{-}146)$$

where D = pipe diameter
D_S = particle diameter (such that 85 percent by weight of particles are smaller than D_S
ρ_M = the slurry mixture density
μ = liquid viscosity
$s = \rho_S/\rho_L$ = ratio of solid to liquid density

Between V_{M2} and V_{M1} the concentration of solids gradually becomes more uniform in the vertical direction. This transition has been modeled by several authors as a concentration gradient where turbulent diffusion balances gravitational settling. See, for example, Karabelas (*AIChE J.*, **23**, 426–434 [1977]).

Published correlations for pressure drop are frequently very complicated and tedious to use, may not offer significant accuracy advantages over the simple guide given here, and many of them are applicable only for velocities above V_{M2}. One which does include the effect of sliding beds is due to Gaessler (Doctoral Dissertation, Technische Hochschule, Karlsruhe, Germany [1967]; reproduced by Govier and Aziz, pp. 668–669). Turian and Yuan (*AIChE J.*, **23**, 232–243 [1977]; see also Turian and Oroskar, *AIChE J.*, **24**, 1144 [1978]) segregated a large body of data into four flow regime groups and developed empirical correlations for predicting pressure drop in each flow regime.

Pressure drop data for the flow of **paper stock** in pipes are given in the data section of *Standards of the Hydraulic Institute* (Hydraulic Institute, 1965). The flow behavior of fiber suspensions is discussed by Bobkowicz and Gauvin (*Chem. Eng. Sci.,* **22**, 229–241 [1967]), Bugliarello and Daily (*TAPPI,* **44**, 881–893 [1961]), and Daily and Bugliarello (*TAPPI,* **44**, 497–512 [1961]).

In **vertical flow** of fast-settling slurries, the in situ concentration of solids with density greater than the liquid will exceed the feed concentration $C = Q_S/(Q_S + Q_L)$ for upflow and will be smaller than C for downflow. This results from slip between the phases. The **slip velocity,** the difference between the in situ average velocities of the two phases, is roughly equal to the terminal settling velocity of the solids in the liquid. Specification of the slip velocity for a pipe of a given diameter, along with the phase flow rates, allows calculation of in situ volume fractions, average velocities, and holdup ratios by simple material balances. Slip velocity may be affected by particle concentration and by turbulence conditions in the liquid. **Drift-flux theory,** a framework incorporating certain functional forms for empirical expressions for slip velocity, is described by Wallis (*One-Dimensional Two-Phase Flow,* McGraw-Hill, New York, 1969). **Minimum transport velocity** for upflow for design purposes is usually taken as twice the particle settling velocity. **Pressure drop** in vertical pipe flow includes the effects of kinetic and potential energy (elevation) changes and friction. Rose and Duckworth (*The Engineer,* **227**[5,903], 392 [1969]; **227**[5,904], 430 [1969]; **227**[5,905], 478 [1969]; see also Govier and Aziz, pp. 487–493) have developed a calculation procedure including all these effects, which may be applied not only to vertical solid/liquid flow, but also to gas/solid flow and to horizontal flow.

For fast-settling slurries, ensuring conveyance is usually the key design issue while pressure drop is somewhat less important. For **nonsettling slurries** conveyance is not an issue, because the particles do not separate from the liquid. Here, viscous and rheological behavior, which control pressure drop, take on critical importance.

Fine particles, often at high concentration, form nonsettling slurries for which useful design equations can be developed by treating them as homogeneous fluids. These fluids are usually very viscous and often non-Newtonian. Shear-thinning and Bingham plastic behavior are common; dilatancy is sometimes observed. Rheology of such fluids must in general be empirically determined, although theoretical results are available for some very limited circumstances. Further discussion of both fast-settling and nonsettling slurries may be found in Shook (in Shamlou, *Processing of Solid-Liquid Suspensions,* Chap. 11, Butterworth-Heinemann, Oxford, 1993).

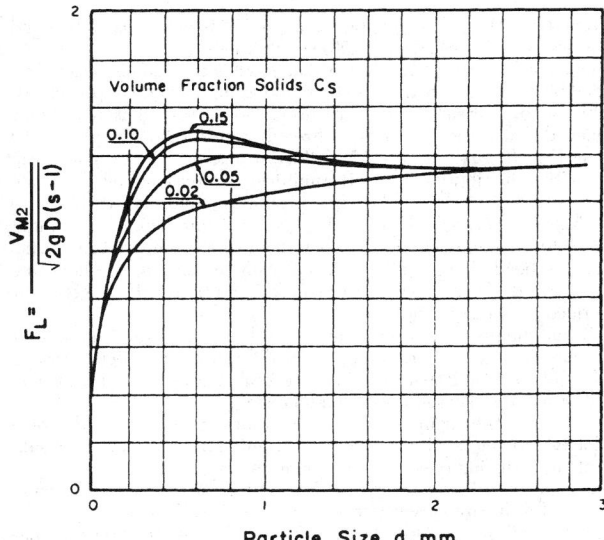

FIG. 6-33   Durand factor for minimum suspension velocity. (*From Govier and Aziz,* The Flow of Complex Mixtures in Pipes, *Van Nostrand Reinhold, New York, 1972.*)

FLUID DISTRIBUTION

Uniform fluid distribution is essential for efficient operation of chemical-processing equipment such as contactors, reactors, mixers, burners, heat exchangers, extrusion dies, and textile-spinning chimneys. To obtain optimum distribution, proper consideration must be given to flow behavior in the distributor, flow conditions upstream and downstream of the distributor, and the distribution requirements of the equipment. Even though the principles of fluid distribution have been well developed for more than three decades, they are frequently overlooked by equipment designers, and a significant fraction of process equipment needlessly suffers from maldistribution. In this subsection, guides for the design of various types of fluid distributors, taking into account only the flow behavior within the distributor, are given.

Perforated-Pipe Distributors   The simple perforated pipe or sparger (Fig. 6-34) is a common type of distributor. As shown, the flow distribution is uniform; this is the case in which pressure recovery due to kinetic energy or momentum changes, frictional pressure drop along the length of the pipe, and pressure drop across the outlet holes have been properly considered. In typical turbulent flow applications, inertial effects associated with velocity changes may dominate frictional losses in determining the pressure distribution along the pipe, unless the length between orifices is large. Application of the momentum or mechanical energy equations in such a case shows that the pressure inside the pipe increases with distance from the entrance of the pipe. If the outlet holes are uniform in size and spacing, the discharge flow will be biased toward the closed end. Disturbances upstream of the distributor, such as pipe bends, may increase or decrease the flow to the holes at the beginning of the distributor. When frictional pressure drop dominates the inertial pressure recovery, the distribution is biased toward the feed end of the distributor.

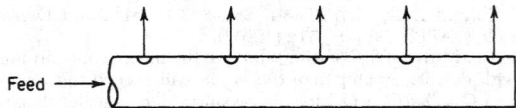

FIG. 6-34   Perforated-pipe distributor.

For turbulent flow, with roughly uniform distribution, assuming a constant friction factor, the combined effect of friction and inertial (momentum) pressure recovery is given by

$$\Delta p = \left(\frac{4fL}{3D} - 2K\right)\frac{\rho V_i^2}{2} \quad \text{(discharge manifolds)} \quad (6\text{-}147)$$

where Δp = net pressure drop over the length of the distributor
L = pipe length
D = pipe diameter
f = Fanning friction factor
V_i = distributor inlet velocity

The factor K would be 1 in the case of full momentum recovery, or 0.5 in the case of negligible viscous losses in the portion of flow which remains in the pipe after the flow divides at a takeoff point (Denn, pp. 126–127). Experimental data (Van der Hegge Zijnen, *Appl. Sci. Res.*, **A3**, 144–162 [1951–1953]; and Bailey, *J. Mech. Eng. Sci.*, **17**, 338–347 [1975]), while scattered, show that K is probably close to 0.5 for discharge manifolds. For inertially dominated flows, Δp will be negative. For **return manifolds** the recovery factor K is close to 1.0, and the pressure drop between the first hole and the exit is given by

$$\Delta p = \left(\frac{4fL}{3D} + 2K\right)\frac{\rho V_e^2}{2} \quad \text{(return manifolds)} \quad (6\text{-}148)$$

where V_e is the pipe exit velocity.

One means to obtain a desired uniform distribution is to make the average pressure drop across the holes Δp_o large compared to the pressure variation over the length of pipe Δp. Then, the relative variation in pressure drop across the various holes will be small, and so will be the variation in flow. When the area of an individual hole is

small compared to the cross-sectional area of the pipe, hole pressure drop may be expressed in terms of the discharge coefficient C_o and the velocity across the hole V_o as

$$\Delta p_o = \frac{1}{C_o^2}\frac{\rho V_o^2}{2} \quad (6\text{-}149)$$

Provided C_o is the same for all the holes, the *percent maldistribution*, defined as the percentage variation in flow between the first and last holes, may be estimated reasonably well for small maldistribution by (Senecal, *Ind. Eng. Chem.*, **49**, 993–997 [1957])

$$\text{Percent maldistribution} = 100\left(1 - \sqrt{\frac{\Delta p_o - |\Delta p|}{\Delta p_o}}\right) \quad (6\text{-}150)$$

This equation shows that for 5 percent maldistribution, the pressure drop across the holes should be about 10 times the pressure drop over the length of the pipe. For discharge manifolds with $K = 0.5$ in Eq. (6-147), and with $4fL/3D \ll 1$, the pressure drop across the holes should be 10 times the inlet velocity head, $\rho V_i^2/2$ for 5 percent maldistribution. This leads to a simple design equation.

Discharge manifolds, $4fL/3D \ll 1$, 5% maldistribution:

$$\frac{V_o}{V_i} = \frac{A_p}{A_o} = \sqrt{10}C_o \quad (6\text{-}151)$$

Here A_p = pipe cross-sectional area and A_o is the *total* hole area of the distributor. Use of large hole velocity to pipe velocity ratios promotes perpendicular discharge streams. In practice, there are many cases where the $4fL/3D$ term will be less than unity but not close to zero. In such cases, Eq. (6-151) will be conservative, while Eqs. (6-147), (6-149), and (6-150) will give more accurate design calculations. In cases where $4fL/(3D) > 2$, friction effects are large enough to render Eq. (6-151) nonconservative. When significant variations in f along the length of the distributor occur, calculations should be made by dividing the distributor into small enough sections that constant f may be assumed over each section.

For return manifolds with $K = 1.0$ and $4fL/(3D) \ll 1$, 5 percent maldistribution is achieved when hole pressure drop is 20 times the pipe exit velocity head.

Return manifolds, $4fL/3D \ll 1$, 5% maldistribution:

$$\frac{V_o}{V_e} = \frac{A_p}{A_o} = \sqrt{20}C_o \quad (6\text{-}152)$$

When $4fL/3D$ is not negligible, Eq. (6-152) is not conservative and Eqs. (6-148), (6-149), and (6-150) should be used.

One common misconception is that good distribution is always provided by high pressure drop, so that increasing flow rate improves distribution by increasing pressure drop. Conversely, it is mistakenly believed that turndown of flow through a perforated pipe designed using Eqs. (6-151) and (6-152) will cause maldistribution. However, when the distribution is nearly uniform, decreasing the flow rate decreases Δp and Δp_o in the same proportion, and Eqs. (6-151) and (6-152) are still satisfied, preserving good distribution independent of flow rate, as long as friction losses remain small compared to inertial (velocity head change) effects. Conversely, increasing the flow rate through a distributor with severe maldistribution will not generally produce good distribution.

Often, the pressure drop required for design flow rate is unacceptably large for a distributor pipe designed for uniform velocity through uniformly sized and spaced orifices. Several measures may be taken in such situations. These include the following:

1.   Taper the diameter of the distributor pipe so that the pipe velocity and velocity head remain constant along the pipe, thus substantially reducing pressure variation in the pipe.

2.   Vary the hole size and/or the spacing between holes to compensate for the pressure variation along the pipe. This method may be sensitive to flow rate and a distributor optimized for one flow rate may suffer increased maldistribution as flow rate deviates from design rate.

3.   Feed or withdraw from both ends, reducing the pipe flow velocity head and required hole pressure drop by a factor of 4.

The orifice discharge coefficient C_o is usually taken to be about 0.62. However, C_o is dependent on the ratio of hole diameter to pipe diameter, pipe wall thickness to hole diameter ratio, and pipe velocity to hole velocity ratio. As long as all these are small, the coefficient 0.62 is generally adequate.

Example 9: Pipe Distributor   A 3-in schedule 40 (inside diameter 7.793 cm) pipe is to be used as a distributor for a flow of 0.010 m³/s of water ($\rho = 1,000$ kg/m³, $\mu = 0.001$ Pa · s). The pipe is 0.7 m long and is to have 10 holes of uniform diameter and spacing along the length of the pipe. The distributor pipe is submerged. Calculate the required hole size to limit maldistribution to 5 percent, and estimate the pressure drop across the distributor.

The inlet velocity computed from $V_i = Q/A = 4Q/(\pi D^2)$ is 2.10 m/s, and the inlet Reynolds number is

$$\text{Re} = \frac{DV_i\rho}{\mu} = \frac{0.07793 \times 2.10 \times 1000}{0.001} = 1.64 \times 10^5$$

For commercial pipe with roughness $\epsilon = 0.046$ mm, the friction factor is about 0.0043. Approaching the last hole, the flow rate, velocity and Reynolds number are about one-tenth their inlet values. At Re = 16,400 the friction factor f is about 0.0070. Using an average value of f = 0.0057 over the length of the pipe, $4fL/3D$ is 0.068 and may reasonably be neglected so that Eq. (6-151) may be used. With $C_o = 0.62$,

$$\frac{V_o}{V_i} = \frac{A_p}{A_o} = \sqrt{10}C_o = \sqrt{10} \times 0.62 = 1.96$$

With pipe cross-sectional area $A_p = 0.00477$ m², the total hole area is 0.00477/1.96 = 0.00243 m². The area and diameter of each hole are then 0.00243/10 = 0.000243 m² and 1.76 cm. With V_o/V_i = 1.96, the hole velocity is 1.96 × 2.10 = 4.12 m/s and the pressure drop across the holes is obtained from Eq. (6-149).

$$\Delta p_o = \frac{1}{C_o^2}\frac{\rho V_o^2}{2} = \frac{1}{0.62^2} \times \frac{1000(4.12)^2}{2} = 22{,}100 \text{ Pa}$$

Since the hole pressure drop is 10 times the pressure variation in the pipe, the total pressure drop from the inlet of the distributor may be taken as approximately 22,100 Pa.

Further detailed information on pipe distributors may be found in Senecal (*Ind. Eng. Chem.*, **49**, 993–997 [1957]). Much of the information on tapered manifold design has appeared in the pulp and paper literature (Spengos and Kaiser, *TAPPI*, **46**[3], 195–200 [1963]; Madeley, *Paper Technology*, **9**[1], 35–39 [1968]; Mardon, et al., *TAPPI*, **46**[3], 172–187 [1963]; Mardon, et al., *Pulp and Paper Magazine of Canada*, **72**[11], 76–81 [November 1971]; Trufitt, *TAPPI*, **58**[11], 144–145 [1975]).

Slot Distributors   These are generally used in sheeting dies for extrusion of films and coatings and in air knives for control of thickness of a material applied to a moving sheet. A simple slotted pipe for turbulent flow conditions may give severe maldistribution because of nonuniform discharge velocity, but also because this type of design does not readily give perpendicular discharge (Koestel and Tuve, *Heat. Piping Air Cond.*, **20**[1], 153–157 [1948]; Senecal, *Ind. Eng. Chem.*, **49**, 993–997 [1957]; Koestel and Young, *Heat. Piping Air Cond.*, **23**[7], 111–115 [1951]). For slots in tapered ducts where the duct cross-sectional area decreases linearly to zero at the far end, the discharge angle will be constant along the length of the duct (Koestel and Young, ibid.). One way to ensure an almost perpendicular discharge is to have the ratio of the area of the slot to the cross-sectional area of the pipe equal to or less than 0.1. As in the case of perforated-pipe distributors, pressure variation within the slot manifold and pressure drop across the slot must be carefully considered.

In practice, the following methods may be used to keep the diameter of the pipe to a minimum consistent with good performance (Senecal, *Ind. Eng. Chem.*, **49**, 993–997 [1957]):

1.   Feed from both ends.

2.   Modify the cross-sectional design (Fig. 6-35); the slot is thus farther away from the influence of feed-stream velocity.

3.   Increase pressure drop across the slot; this can be accomplished by lengthening the lips (Fig. 6-35).

4.   Use screens (Fig. 6-35) to increase overall pressure drop across the slot.

Design considerations for air knives are discussed by Senecal (ibid.). Design procedures for extrusion dies when the flow is laminar,

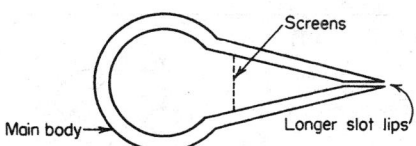

FIG. 6-35   Modified slot distributor.

as with highly viscous fluids, are presented by Bernhardt (*Processing of Thermoplastic Materials*, Rheinhold, New York, 1959, pp. 248–281).

Turning Vanes   In applications such as ventilation, the discharge profile from slots can be improved by turning vanes. The tapered duct is the most amenable for turning vanes because the discharge angle remains constant. One way of installing the vanes is shown in Fig. 6-36. The vanes should have a depth twice the spacing (*Heating, Ventilating, Air Conditioning Guide*, vol. 38, American Society of Heating, Refrigerating and Air-Conditioning Engineers, 1960, pp. 282–283) and a curvature at the upstream end of the vanes of a circular arc which is tangent to the discharge angle θ of a slot without vanes and perpendicular at the downstream or discharge end of the vanes (Koestel and Young, *Heat. Piping Air Cond.*, **23**[7], 111–115 [1951]). Angle θ can be estimated from

$$\cot \theta = \frac{C_d A_s}{A_d} \tag{6-153}$$

where A_s = slot area
A_d = duct cross-sectional area at upstream end
C_d = discharge coefficient of slot

Vanes may be used to improve velocity distribution and reduce frictional loss in bends, when the ratio of bend turning radius to pipe diameter is less than 1.0. For a miter bend with low-velocity flows, simple circular arcs (Fig. 6-37) can be used, and with high-velocity flows, vanes of special airfoil shapes are required. For additional details and references, see Ower and Pankhurst (*The Measurement of Air Flow*, Pergamon, New York, 1977, p. 102); Pankhurst and Holder (*Wind-Tunnel Technique*, Pitman, London, 1952, pp. 92–93); Rouse (*Engineering Hydraulics*, Wiley, New York, 1950, pp. 399–401); and Jorgensen (*Fan Engineering*, 7th ed., Buffalo Forge Co., Buffalo, 1970, pp. 111, 117, 118).

Perforated Plates and Screens   A nonuniform velocity profile in turbulent flow through channels or process equipment can be smoothed out to any desired degree by adding sufficient uniform resistance, such as perforated plates or screens across the flow channel, as shown in Fig. 6-38. Stoker (*Ind. Eng. Chem.*, **38**, 622–624 [1946]) provides the following equation for the effect of a uniform resistance on velocity profile:

$$\frac{V_{2,\max}}{V} = \sqrt{\frac{(V_{1,\max}/V)^2 + \alpha_2 - \alpha_1 + \alpha_2 K}{1 + K}} \tag{6-154}$$

Here, V is the area average velocity, K is the number of velocity heads of pressure drop provided by the uniform resistance, $\Delta p = \bar{K}\rho V^2/2$, and α is the velocity profile factor used in the mechanical energy bal-

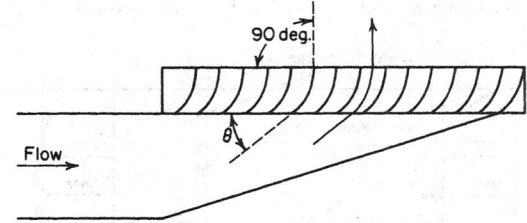

FIG. 6-36   Turning vanes in a slot distributor.

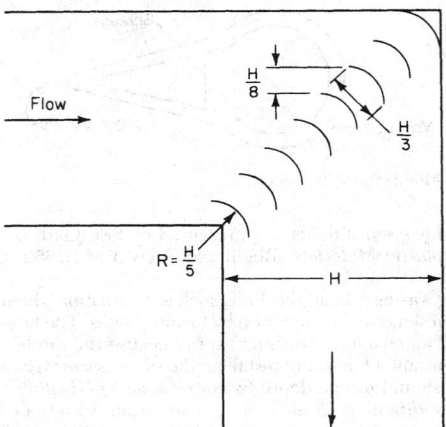

FIG. 6-37   Miter bend with vanes.

ance, Eq. (6-13). It is ratio of the area average of the cube of the velocity, to the cube of the area average velocity V. The shape of the exit velocity profile appears twice in Eq. (6-154), in $V_{2,max}/V$ and α_2. Typically, K is on the order of 10, and the desired exit velocity profile is fairly uniform so that $\alpha_2 \sim 1.0$ may be appropriate. Downstream of the resistance, the velocity profile will gradually reestablish the fully developed profile characteristic of the Reynolds number and channel shape. The screen or perforated plate open area required to produce the resistance K may be computed from Eqs. (6-107) or (6-111).

Screens and other flow restrictions may also be used to suppress stream swirl and turbulence (Loehrke and Nagib, *J. Fluids Eng.*, **98**, 342–353 [1976]). Contraction of the channel, as in a venturi, provides further reduction in turbulence level and flow nonuniformity.

Beds of Solids   A suitable depth of solids can be used as a fluid distributor. As for other types of distribution devices, a pressure drop of 10 velocity heads is typically used, here based on the superficial velocity through the bed. There are several substantial disadvantages to use of particle beds for flow distribution. Heterogeneity of the bed may actually worsen rather than improve distribution. In general, uniform flow may be found only downstream of the point in the bed where sufficient pressure drop has occurred to produce uniform flow. Therefore, inefficiency results when the bed also serves reaction or mass transfer functions, as in catalysts, adsorbents, or tower packings for gas/liquid contacting, since portions of the bed are bypassed. In the case of trickle flow of liquid downward through column packings, inlet distribution is critical since the bed itself is relatively ineffective in distributing the liquid. Maldistribution of flow through packed beds also arises when the ratio of bed diameter to particle size is less than 10 to 30.

Other Flow Straightening Devices   Other devices designed to produce uniform velocity or reduce swirl, sometimes with reduced pressure drop, are available. These include both commercial devices of proprietary design and devices discussed in the literature. For pipeline flows, see the references under flow inverters and static mixing elements previously discussed in the "Incompressible Flow in Pipes and Channels" subsection. For large area changes, as at the

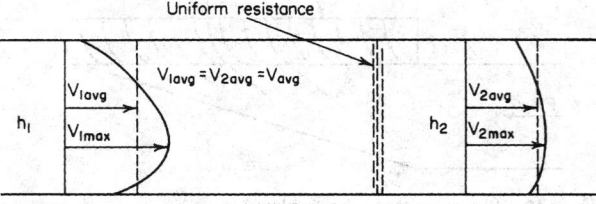

FIG. 6-38   Smoothing out a nonuniform profile in a channel.

entrance to a vessel, it is sometimes necessary to diffuse the momentum of the inlet jet discharging from the feed pipe in order to produce a more uniform velocity profile within the vessel. Methods for this application exist, but remain largely in the domain of proprietary, commercial design.

FLUID MIXING

Mixing of fluids is a discipline of fluid mechanics. Fluid motion is used to accelerate the otherwise slow processes of diffusion and conduction to bring about uniformity of concentration and temperature, blend materials, facilitate chemical reactions, bring about intimate contact of multiple phases, and so on. As the subject is too broad to cover fully, only a brief introduction and some references for further information are given here.

Several texts are available. These include Harnby, Edwards, and Nienow (*Mixing in the Process Industries,* 2d ed., Butterworths, London, 1992), Oldshue (*Fluid Mixing Technology,* McGraw-Hill, New York, 1983), Tatterson (*Fluid Mixing and Gas Dispersion in Agitated Tanks,* McGraw-Hill, New York, 1991), Uhl and Gray (*Mixing,* vols. I–III, Academic, New York, 1966, 1967, 1986), and Nagata (*Mixing: Principles and Applications,* Wiley, New York, 1975). A good overview of stirred tank agitation is given in the series of articles from *Chemical Engineering* (110–114, Dec. 8, 1975; 139–145, Jan. 5, 1976; 93–100, Feb. 2, 1976; 102–110, Apr. 26, 1976; 144–150, May 24, 1976; 141–148, July 19, 1976; 89–94, Aug. 2, 1976; 101–108, Aug. 30, 1976; 109–112, Sept. 27, 1976; 119–126, Oct. 25, 1976; 127–133, Nov. 8, 1976).

Process mixing is commonly carried out in pipeline and vessel geometries. The terms **radial mixing** and **axial mixing** are commonly used. Axial mixing refers to mixing of materials which pass a given point at different times, and thus leads to **backmixing.** For example, backmixing or axial mixing occurs in stirred tanks where fluid elements entering the tank at different times are intermingled. Mixing of elements initially at different axial positions in a pipeline is axial mixing. Radial mixing occurs between fluid elements passing a given point at the same time, as, for example, between fluids mixing in a pipeline tee.

Turbulent flow, by means of the chaotic eddy motion associated with velocity fluctuation, is conducive to rapid mixing and, therefore, is the preferred flow regime for mixing. **Laminar mixing** is carried out when high viscosity makes turbulent flow impractical.

Stirred Tank Agitation   Turbine impeller agitators, of a variety of shapes, are used for stirred tanks, predominantly in turbulent flow. Figure 6-39 shows typical stirred tank configurations and time-averaged flow patterns for axial flow and radial flow impellers. In order to prevent formation of a **vortex,** four vertical baffles are normally installed. These cause top-to-bottom mixing and prevent mixing-ineffective swirling motion.

For a given impeller and tank geometry, the impeller Reynolds number determines the flow pattern in the tank:

$$\mathrm{Re}_I = \frac{D^2 N \rho}{\mu} \qquad (6\text{-}155)$$

where D = impeller diameter, N = rotational speed, and ρ and μ are the liquid density and viscosity. Rotational speed N is typically reported in revolutions per minute, or revolutions per second in SI units. Radians per second are almost never used. Typically, $\mathrm{Re}_I > 10^4$ is required for fully turbulent conditions throughout the tank. A wide transition region between laminar and turbulent flow occurs over the range $10 < \mathrm{Re}_I < 10^4$.

The power P drawn by the impeller is made dimensionless in a group called the power number:

$$\mathrm{N}_P = \frac{P}{\rho N^3 D^5} \qquad (6\text{-}156)$$

Figure 6-40 shows power number vs. impeller Reynolds number for a typical configuration. The similarity to the friction factor vs. Reynolds number behavior for pipe flow is significant. In laminar flow, the power number is inversely proportional to Reynolds number, reflecting the dominance of viscous forces over inertial forces. In turbulent flow, where inertial forces dominate, the power number is nearly constant.

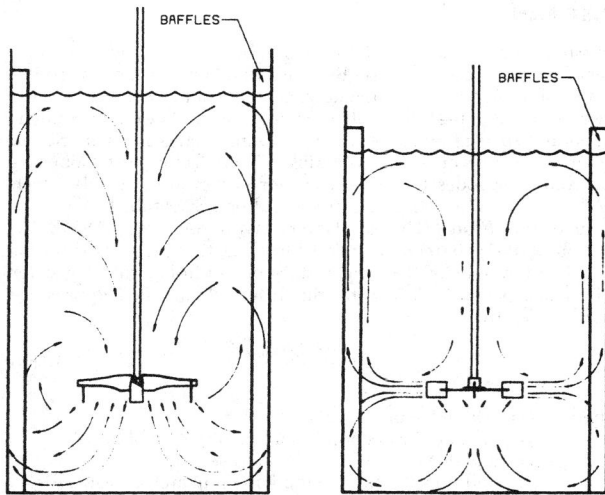

FIG. 6-39 Typical stirred tank configurations, showing time-averaged flow patterns for axial flow and radial flow impellers. (*From Oldshue,* Fluid Mixing Technology, *McGraw-Hill, New York, 1983.*)

Impellers are sometimes viewed as pumping devices; the total volumetric flow rate Q discharged by an impeller is made dimensionless in a pumping number:

$$N_Q = \frac{Q}{ND^3} \qquad (6\text{-}157)$$

Blend time t_b, the time required to achieve a specified maximum standard deviation of concentration after injection of a tracer into a stirred tank, is made dimensionless by multiplying by the impeller rotational speed:

$$N_b = t_b N \qquad (6\text{-}158)$$

Dimensionless pumping number and blend time are independent of Reynolds number under fully turbulent conditions. The magnitude of concentration fluctuations from the final well-mixed value in batch mixing decays exponentially with time.

The design of mixing equipment depends on the desired process result. There is often a tradeoff between operating cost, which depends mainly on power, and capital cost, which depends on agitator size and torque. For some applications bulk flow throughout the vessel is desired, while for others high local turbulence intensity is required. Multiphase systems introduce such design criteria as solids suspension and gas dispersion. In very viscous systems, helical ribbons, extruders, and other specialized equipment types are favored over turbine agitators.

Pipeline Mixing   Mixing may be carried out with **mixing tees, inline** or **motionless mixing elements,** or in empty pipe. In the latter case, large pipe lengths may be required to obtain adequate mixing. Coaxially injected streams require lengths on the order of 100 pipe diameters. Coaxial mixing in turbulent single-phase flow is characterized by the turbulent diffusivity (eddy diffusivity) D_E which determines the rate of radial mixing. Davies (*Turbulence Phenomena,* Academic, New York, 1972) provides an equation for D_E which may be rewritten as

$$D_E \sim 0.015 DV Re^{-0.125} \qquad (6\text{-}159)$$

where D = pipe diameter
V = average velocity
Re = pipe Reynolds number, $DV\rho/\mu$
ρ = density
μ = viscosity

Properly designed tee mixers, with due consideration given to main stream and injected stream momentum, are capable of producing high degrees of uniformity in just a few diameters. Forney (*Jet Injection for Optimum Pipeline Mixing,* in "Encyclopedia of Fluid Mechanics," vol. 2., Chap. 25, Gulf Publishing, 1986) provides a thorough discussion of tee mixing. Inline or motionless mixers are generally of

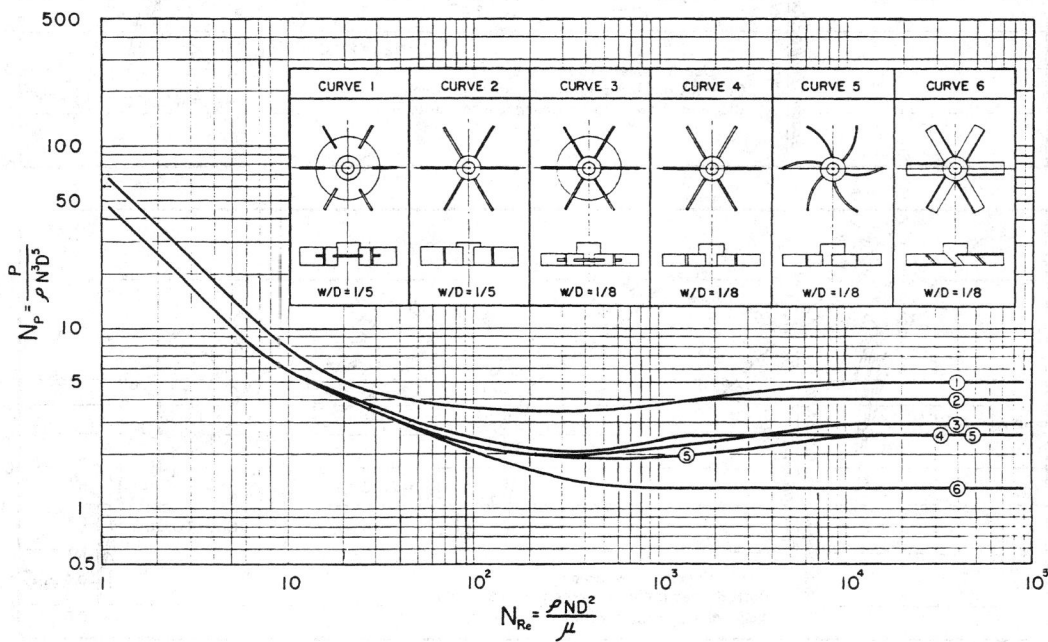

FIG. 6-40   Dimensionless power number in stirred tanks. (*Reprinted with permission from Bates, Fondy, and Corpstein,* Ind. Eng. Chem. Process Design Develop., *2, 310 [1963].*)

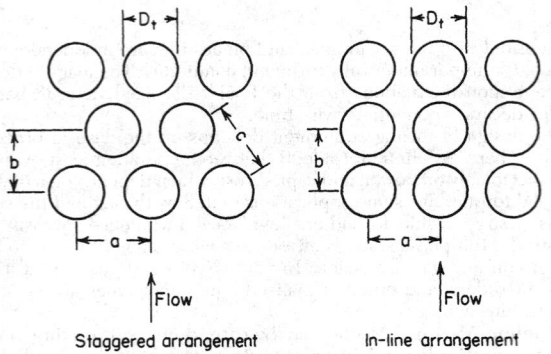

FIG. 6-41   Tube-bank configurations.

proprietary commercial design, and may be selected for viscous or turbulent, single or multiphase mixing applications. They substantially reduce required pipe length for mixing.

TUBE BANKS

Pressure drop across tube banks may not be correlated by means of a single, simple friction factor—Reynolds number curve, owing to the variety of tube configurations and spacings encountered, two of which are shown in Fig. 6-41. Several investigators have allowed for configuration and spacing by incorporating spacing factors in their friction factor expressions or by using multiple friction factor plots. Commercial computer codes for heat-exchanger design are available which include features for estimating pressure drop across tube banks.

Turbulent Flow   The correlation by Grimison (*Trans. ASME*, **59**, 583–594 [1937]) is recommended for predicting pressure drop for turbulent flow (Re ≥ 2,000) across staggered or in-line tube banks for tube spacings [(a/D_t), (b/D_t)] ranging from 1.25 to 3.0. The pressure drop is given by

$$\Delta p = \frac{4fN_r\rho V_{max}^2}{2} \tag{6-160}$$

where f = friction factor
  N_r = number of rows of tubes in the direction of flow
  ρ = fluid density
  V_{max} = fluid velocity through the minimum area available for flow.

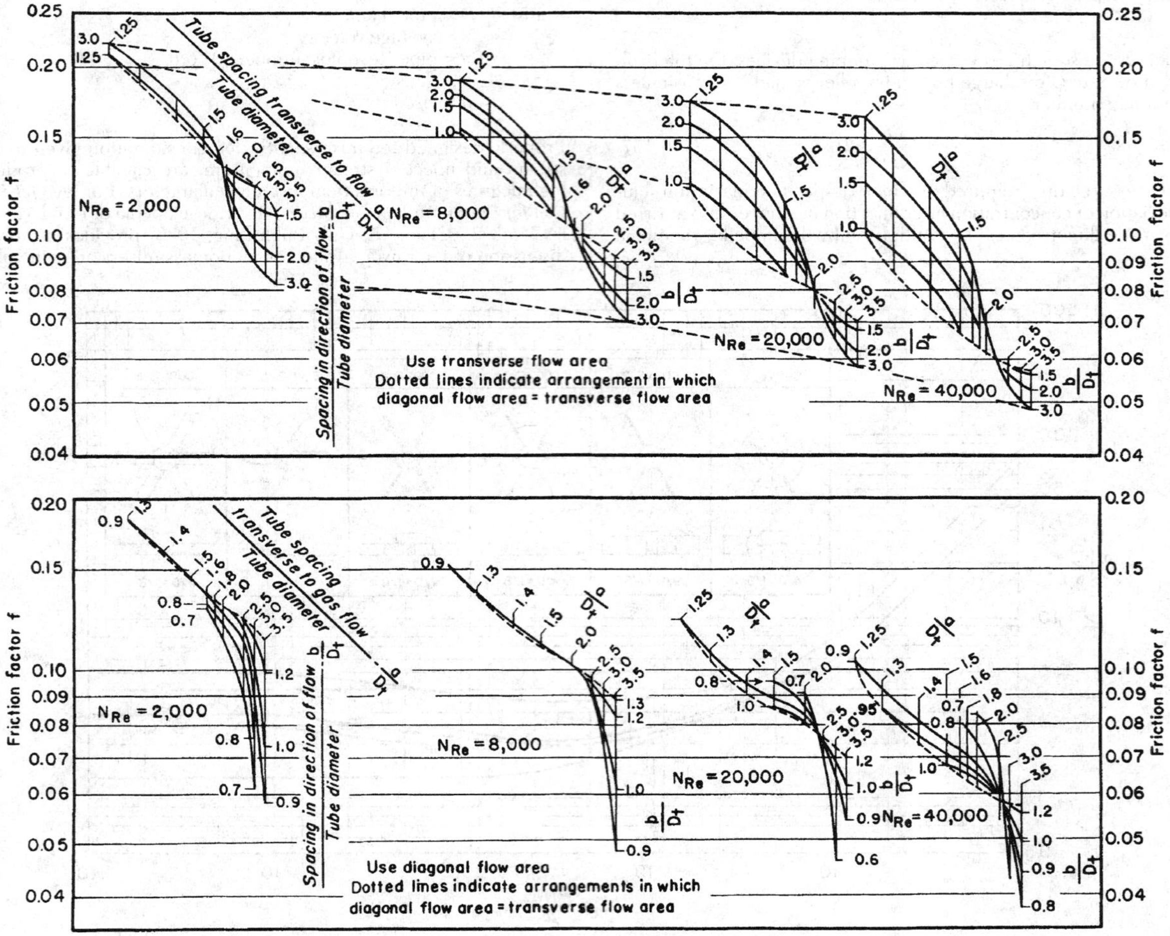

FIG. 6-42   Upper chart: Friction factors for staggered tube banks with minimum fluid flow area in transverse openings. Lower chart: Friction factors for staggered tube banks with minimum fluid flow area in diagonal openings. (*From Grimison*, Trans. ASME, **59**, 583 [1937].)

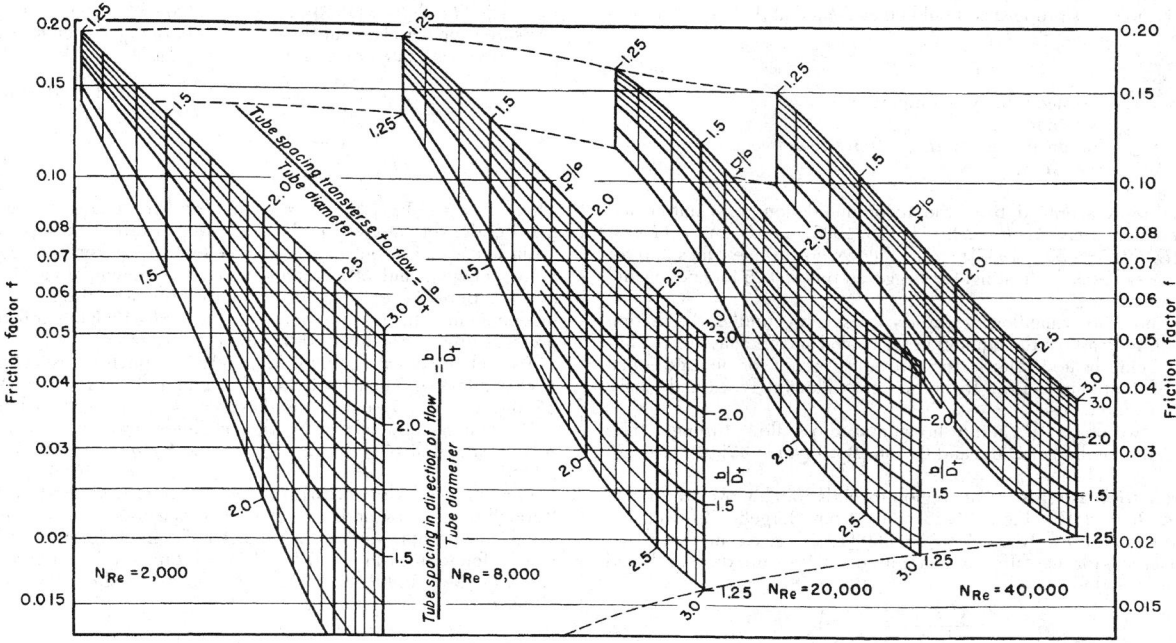

FIG. 6-43   Friction factors for in-line tube banks. (*From Grimison,* Trans. ASME, **59,** 583 [*1937*].)

For banks of **staggered tubes,** the friction factor for isothermal flow is obtained from Fig. (6-42). Each "fence" (group of parametric curves) represents a particular Reynolds number defined as

$$\text{Re} = \frac{D_t V_{max} \rho}{\mu} \qquad (6\text{-}161)$$

where D_t = tube outside diameter and μ = fluid viscosity. The numbers along each fence represent the transverse and inflow-direction spacings. The upper chart is for the case in which the minimum area for flow is in the transverse openings, while the lower chart is for the case in which the minimum area is in the diagonal openings. In the latter case, V_{max} is based on the area of the diagonal openings and N_r is the number of rows in the direction of flow minus 1. A critical comparison of this method with all the data available at the time showed an average deviation of the order of ±15 percent. (Boucher and Lapple, *Chem. Eng. Prog.,* **44,** 117–134 [1948]). For tube spacings greater than 3 tube diameters, the correlation by Gunter and Shaw (*Trans. ASME,* **67,** 643–660 [1945]) can be used as an approximation. As an **approximation,** the pressure drop can be taken as 0.72 velocity head (based on V_{max} per row of tubes for tube spacings commonly encountered in practice (Lapple, et al., *Fluid and Particle Mechanics,* University of Delaware, Newark, 1954).

For banks of **in-line tubes,** f for isothermal flow is obtained from Fig. 6-43. Average deviation from available data is on the order of ±15 percent. For tube spacings greater than $3D_t$, the charts of Gram, Mackey, and Monroe (*Trans. ASME,* **80,** 25–35 [1958]) can be used. As an **approximation,** the pressure drop can be taken as 0.32 velocity head (based on V_{max}) per row of tubes (Lapple, et al., *Fluid and Particle Mechanics,* University of Delaware, Newark, 1954).

For turbulent flow through **shallow** tube banks, the average friction factor per row will be somewhat greater than indicated by Figs. 6-42 and 6-43, which are based on 10 or more rows depth. A 30 percent increase per row for 2 rows, 15 percent per row for 3 rows and 7 percent per row for 4 rows can be taken as the maximum likely to be encountered (Boucher and Lapple, *Chem. Eng. Prog.,* **44,** 117–134 [1948]).

For a **single row of tubes,** the friction factor is given by Curve B in Fig. 6-44 as a function of tube spacing. This curve is based on

the data of several experimenters, all adjusted to a Reynolds number of 10,000. The values should be substantially independent of Re for 1,000 < Re < 100,000.

For **extended surfaces,** which include fins mounted perpendicularly to the tubes or spiral-wound fins, pin fins, plate fins, and so on, friction data for the specific surface involved should be used. For details, see Kays and London (*Compact Heat Exchangers,* 2d ed., McGraw-Hill, New York, 1964). If specific data are unavailable, the correlation by Gunter and Shaw (*Trans. ASME,* **67,** 643–660 [1945]) may be used as an approximation.

When a large temperature change occurs in a gas flowing across a

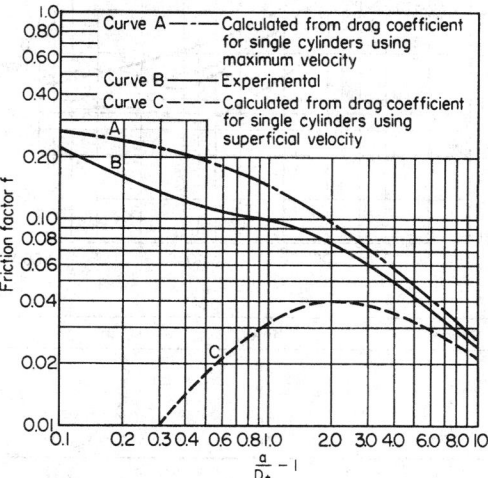

FIG. 6-44   Friction factors vs. transverse spacing for single row of tubes. (*From Boucher and Lapple,* Chem. Eng. Prog., **44,** 117 [1948].)

tube bundle, gas properties should be evaluated at the mean temperature

$$T_m = T_t + K\,\Delta T_{lm} \tag{6-162}$$

where T_t = average tube-wall temperature
K = constant
ΔT_{lm} = log-mean temperature difference between the gas and the tubes.

Values of K averaged from the recommendations of Chilton and Genereaux (*Trans. AIChE*, **29**, 151–173 [1933]) and Grimison (*Trans. ASME*, **59**, 583–594 [1937]) are as follows: for in-line tubes, 0.9 for cooling and −0.9 for heating; for staggered tubes, 0.75 for cooling and −0.8 for heating.

For nonisothermal flow of **liquids** across tube bundles, the friction factor is increased if the liquid is being cooled and decreased if the liquid is being heated. The factors previously given for nonisothermal flow of liquids in pipes ("Incompressible Flow in Pipes and Channels") should be used.

For two-phase gas/liquid horizontal cross flow through tube banks, the method of Diehl and Unruh (*Pet. Refiner*, **37**[10], 124–128 [1958]) is available.

Transition Region This region extends roughly over the range $200 < Re < 2,000$. Figure 6-45 taken from Bergelin, Brown, and Doberstein (*Trans. ASME*, **74**, 953–960 [1952]) gives curves for friction factor f_T for five different configurations. Pressure drop for liquid flow is given by

$$\Delta p = \frac{4 f_T N_r \rho V_{max}^2}{2}\left(\frac{\mu_s}{\mu_b}\right)^{0.14} \tag{6-163}$$

where N_r = number of major restrictions encountered in flow through the bank (equal to number of rows when minimum flow area occurs in transverse openings, and to number of rows minus 1 when it occurs in the diagonal openings); ρ = fluid density; V_{max} = velocity through minimum flow area; μ_s = fluid viscosity at tube-surface temperature and μ_b = fluid viscosity at average bulk temperature. This method gives the friction factor within about ±25 percent.

Laminar Region Bergelin, Colburn, and Hull (*Univ. Delaware*

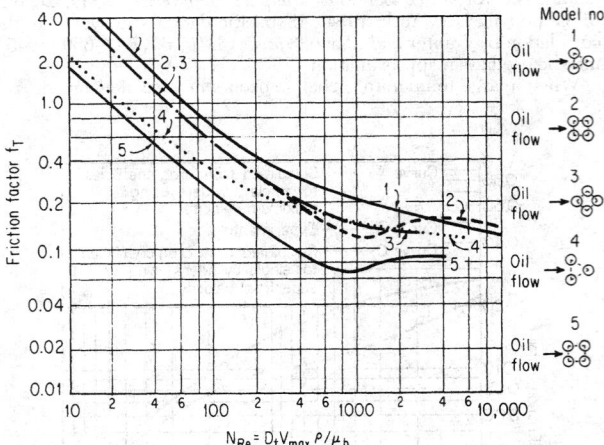

Model	Rows	D_t, in	Pitch/D_t
1	10	⅜	1.25
2	10	⅜	1.25
3	14	⅜	1.25
4	10	⅜	1.50
5	10	⅜	1.50

FIG. 6-45 Friction factors for transition region flow across tube banks. (Pitch is the minimum center-to-center tube spacing.) (*From Bergelin, Brown, and Doberstein*, Trans. ASME, **74**, 953 [1952].)

Eng. Exp. Sta. Bull., **2** [1950]) recommend the following equations for pressure drop with laminar flow ($Re_v < 100$) of liquids across banks of plain tubes with pitch ratios P/D_t of 1.25 and 1.50:

$$\Delta p = \frac{280 N_r}{Re_v}\left(\frac{D_t}{P}\right)^{1.6}\left(\frac{\mu_s}{\mu_b}\right)^m\left(\frac{\rho V_{max}^2}{2}\right) \tag{6-164}$$

$$m = \frac{0.57}{(Re_v)^{0.25}} \tag{6-165}$$

where $Re_v = D_v V_{max}\rho/\mu_b$; D_v = volumetric hydraulic diameter [(4 × free-bundle volume)/(exposed surface area of tubes)]; P = pitch (= a for in-line arrangements, = a or c [whichever is smaller] for staggered arrangements), and other quantities are as defined following Eq. (6-163). Bergelin, et al. (ibid.) show that pressure drop per row is independent of the number of rows in the bank with laminar flow. The pressure drop is predicted within about ±25 percent.

The validity of extrapolating Eq. (6-164) to pitch ratios larger than 1.50 is unknown. The correlation of Gunter and Shaw (*Trans. ASME*, **67**, 643–660 [1945]) may be used as an approximation in such cases.

For laminar flow of non-Newtonian fluids across tube banks, see Adams and Bell (*Chem. Eng. Prog.*, **64**, *Symp. Ser.*, **82**, 133–145 [1968]).

Flow-induced **tube vibration** occurs at critical fluid velocities through tube banks, and is to be avoided because of the severe damage that can result. Methods to predict and correct vibration problems may be found in Eisinger (*Trans. ASME J. Pressure Vessel Tech.*, **102**, 138–145 [May 1980]) and Chen (*J. Sound Vibration*, **93**, 439–455 [1984]).

BEDS OF SOLIDS

Fixed Beds of Granular Solids Pressure-drop prediction is complicated by the variety of granular materials and of their packing arrangement. For flow of a **single incompressible fluid** through an incompressible bed of granular solids, the pressure drop may be estimated by the correlation given in Fig. 6-46 (Leva, *Chem. Eng.*, **56**[5], 115–117 [1949]), or *Fluidization*, McGraw-Hill, New York, 1959). The modified friction factor and Reynolds number are defined by

$$f_m \equiv \frac{D_p \rho \phi_s^{3-n}\epsilon^3|\Delta p|}{2 G^2 L (1-\epsilon)^{3-n}} \tag{6-166}$$

$$Re' \equiv \frac{D_p G}{\mu} \tag{6-167}$$

where $-\Delta p$ = pressure drop
L = depth of bed
D_p = average particle diameter, defined as the diameter of a sphere of the same volume as the particle
ϵ = void fraction
n = exponent given in Fig. 6-46 as a function of Re'
ϕ_s = shape factor defined as the area of sphere of diameter D_p divided by the actual surface area of the particle
G = fluid superficial mass velocity based on the empty chamber cross section
ρ = fluid density
μ = fluid viscosity

As for any incompressible single-phase flow, the equivalent pressure $P = p + \rho gz$ where g = acceleration of gravity z = elevation, may be used in place of p to account for gravitational effects in flows with vertical components.

In creeping flow ($Re' < 10$),

$$f_m = \frac{100}{Re'} \tag{6-168}$$

At high Reynolds numbers the friction factor becomes nearly constant, approaching a value of the order of unity for most packed beds.

In terms of S, particle surface area per unit volume of bed,

$$D_p = \frac{6(1-\epsilon)}{\phi_s S} \tag{6-169}$$

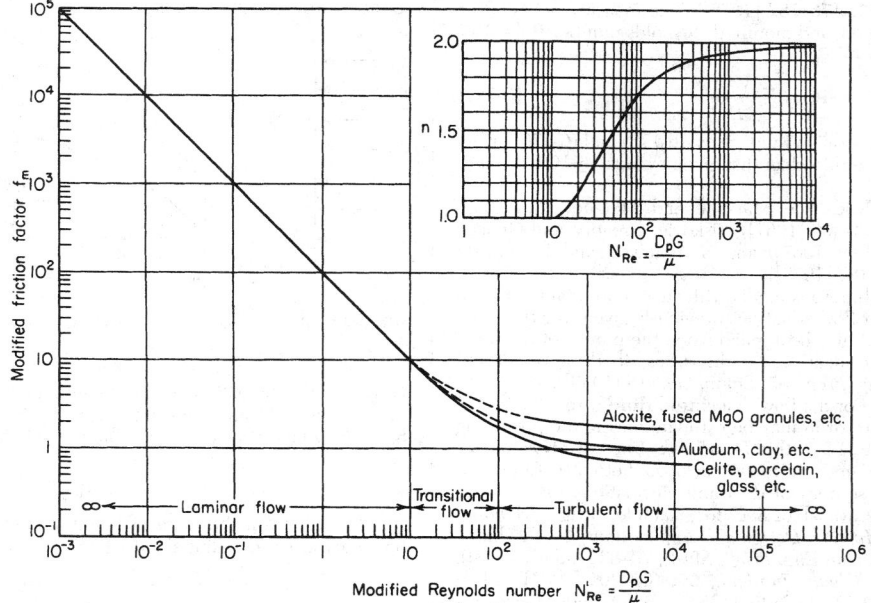

FIG. 6-46 Friction factor for beds of solids. (*From Leva,* Fluidization, *McGraw-Hill, New York, 1959, p. 49.*)

Porous Media Packed beds of granular solids are one type of the general class referred to as **porous media,** which include geological formations such as petroleum reservoirs and aquifers, manufactured materials such as sintered metals and porous catalysts, burning coal or char particles, and textile fabrics, to name a few. Pressure drop for incompressible flow across a porous medium has the same qualitative behavior as that given by Leva's correlation in the preceding. At low Reynolds numbers, viscous forces dominate and pressure drop is proportional to fluid viscosity and superficial velocity, and at high Reynolds numbers, pressure drop is proportional to fluid density and to the square of superficial velocity.

Creeping flow (Re$'$ <~ 1) through porous media is often described in terms of the **permeability** k and Darcy's Law:

$$\frac{-\Delta P}{L} = \frac{\mu}{k} V \tag{6-170}$$

where V = superficial velocity. The SI units for permeability are m^2. Creeping flow conditions generally prevail in geological porous media. For **multidimensional flows** through **isotropic** porous media, the superficial velocity $\mathbf{V}$ and pressure gradient ∇P vectors replace the corresponding one-dimensional variables in Eq. (6-170).

$$\nabla P = -\frac{\mu}{k} \mathbf{V} \tag{6-171}$$

For isotropic homogeneous porous media (uniform permeability and porosity), the pressure for creeping incompressible single phase-flow may be shown to satisfy the LaPlace equation:

$$\nabla^2 P = 0 \tag{6-172}$$

For **anisotropic** or **oriented** porous media, as are frequently found in geological media, permeability varies with direction and a **permeability tensor K,** with nine components K_{ij} giving the velocity component in the i direction due to a pressure gradient in the j direction, may be introduced. For further information, see Slattery (*Momentum, Energy and Mass Transfer in Continua,* Krieger, Huntington, New York, 1981, p. 193–218). See also Dullien (*Chem. Eng. J. [Laussanne],* **10,** 1,034 [1975]) for a review of pressure-drop methods in single-phase flow. Solutions for Darcy's law for several geometries of interest in petroleum reservoirs and aquifers, for both incompressible and compressible flows, are given in Craft and Hawkins (*Applied Petro-*

leum Reservoir Engineering, Prentice-Hall, Englewood Cliffs, N.J., 1959). See also Todd (*Groundwater Hydrology,* 2nd ed., Wiley, New York, 1980).

For granular solids of **mixed sizes** the average particle diameter may be calculated as

$$\frac{1}{D_p} = \sum_i \frac{x_i}{D_{p,i}} \tag{6-173}$$

where x_i = weight fraction of particles of size $D_{p,i}$.

For **isothermal compressible flow** of a gas with constant compressibility factor Z through a packed bed of granular solids, an equation similar to Eq. (6-114) for pipe flow may be derived:

$$p_1^2 - p_2^2 = \frac{2ZRG^2T}{M_w} \left[\ln \frac{v_2}{v_1} + \frac{2f_m L(1-\epsilon)^{3-n}}{\phi_s^{3-n} \epsilon^3 D_p} \right] \tag{6-174}$$

where p_1 = upstream absolute pressure
   p_2 = downstream absolute pressure
   R = gas constant
   T = absolute temperature
   M_w = molecular weight
   v_1 = upstream specific volume of gas
   v_2 = downstream specific volume of gas

For creeping flow of **power law** non-Newtonian fluids, the method of Christopher and Middleton (*Ind. Eng. Chem. Fundam.,* **4,** 422–426 [1965]) may be used:

$$-\Delta p = \frac{150HLV^n(1-\epsilon)^2}{D_p^2 \phi_s^2 \epsilon^3} \tag{6-175}$$

$$H = \frac{K}{12} \left(9 + \frac{3}{n} \right)^n \left[\frac{D_p^2 \phi_s^2 \epsilon^4}{(1-\epsilon)^2} \right]^{(1-n)/2} \tag{6-176}$$

where $V = G/\rho$ = superficial velocity, K, n = power law material constants, and all other variables are as defined in Eq. (6-166). This correlation is supported by data from Christopher and Middleton (ibid.), Gregory and Griskey (*AIChE J.,* **13,** 122–125 [1967]), Yu, Wen, and Bailie (*Can. J. Chem. Eng.,* **46,** 149–154 [1968]), Siskovic, Gregory, and Griskey (*AIChE J.,* **17,** 176–187 [1978]), Kemblowski and Mertl (*Chem. Eng. Sci.,* **29,** 213–223 [1974]), and Kemblowski and Dziuminski (*Rheol. Acta*, **17**, 176–187 [1978]). The measurements cover the range $n = 0.50$ to 1.60, and modified Reynolds number $Re' = 10^{-8}$ to 10, where

$$Re' = \frac{D_p V^{2-n} \rho}{H} \qquad (6-177)$$

For the case $n = 1$ (Newtonian fluid), Eqs. (6-175) and (6-176) give a pressure drop 25 percent less than that given by Eqs. (6-166) through (6-168).

For **viscoelastic fluids** see Marshall and Metzner (*Ind. Eng. Chem. Fundam.*, **6**, 393–400 [1967]), Siskovic, Gregory, and Griskey (*AIChE J.*, **13**, 122–125 [1967]) and Kemblowski and Dziubinski (*Rheol. Acta*, **17**, 176–187 [1978]).

For gas flow through porous media with small pore diameters, the slip flow and molecular flow equations previously given (see the "Vacuum Flow" subsection) may be applied when the pore is of the same or smaller order as the mean free path, as described by Monet and Vermeulen (*Chem. Eng. Prog.*, **55**, *Symp. Ser.*, **25** [1959]).

Tower Packings   For the flow of a **single fluid** through a bed of tower packing, pressure drop may be estimated using the preceding methods. See also Sec. 14 of this *Handbook*. For **countercurrent gas/liquid flow** in commercial tower packings, both structured and unstructured, several sources of data and correlations for pressure drop and flooding are available. See, for example, Strigle (*Random Packings and Packed Towers, Design and Applications,* Gulf Publishing, Houston, 1989; *Chem. Eng. Prog.*, **89**[8], 79–83 [August 1993]), Hughmark (*Ind. Eng. Chem. Fundam.*, **25**, 405–409 [1986]), Chen (*Chem. Eng. Sci.*, **40**, 2139–2140 [1985]), Billet and Mackowiak (*Chem. Eng. Technol.*, **11**, 213–217 [1988]), Krehenwinkel and Knapp (*Chem. Eng. Technol.*, **10**, 231–242 [1987]), Mersmann and Deixler (*Ger. Chem. Eng.*, **9**, 265–276 [1986]) and Robbins (*Chem. Eng. Progr.*, **87**[5], 87–91 [May 1991]). Data and correlations for flooding and pressure drop for structured packings are given by Fair and Bravo (*Chem. Eng. Progr.*, **86**[1], 19–29 [January 1990]).

Fluidized Beds   When gas or liquid flows upward through a vertically unconstrained bed of particles, there is a minimum fluid velocity at which the particles will begin to move. Above this minimum velocity, the bed is said to be *fluidized*. Fluidized beds are widely used, in part because of their excellent mixing and heat and mass transfer characteristics. See Sec. 17 of this *Handbook* for detailed information.

BOUNDARY LAYER FLOWS

Boundary layer flows are a special class of flows in which the flow far from the surface of an object is inviscid, and the effects of viscosity are manifest only in a thin region near the surface where steep velocity gradients occur to satisfy the no-slip condition at the solid surface. The thin layer where the velocity decreases from the inviscid, potential flow velocity to zero (relative velocity) at the solid surface is called the *boundary layer*. The thickness of the boundary layer is indefinite because the velocity asymptotically approaches the free-stream velocity at the outer edge. The boundary layer thickness is conventionally taken to be the distance for which the velocity equals 0.99 times the free-stream velocity. The boundary layer may be either laminar or turbulent. Particularly in the former case, the equations of motion may be simplified by scaling arguments. Schlichting (*Boundary Layer Theory*, 8th ed., McGraw-Hill, New York, 1987) is the most comprehensive source for information on boundary layer flows.

Flat Plate, Zero Angle of Incidence   For flow over a wide, thin flat plate at zero angle of incidence with a uniform free-stream velocity, as shown in Fig. 6-47, the **critical Reynolds number** at which the boundary layer becomes turbulent is normally taken to be

$$Re_x = \frac{xV\rho}{\mu} = 500,000 \qquad (6-178)$$

where V = free-stream velocity
ρ = fluid density
μ = fluid viscosity
x = distance from leading edge of the plate

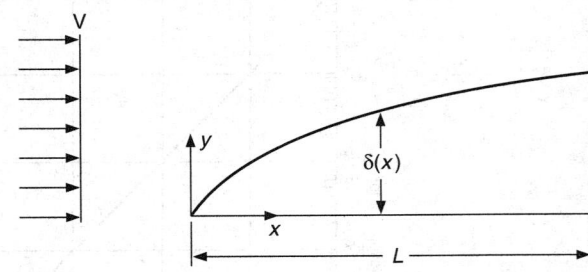

Uniform free-stream velocity

FIG. 6-47   Boundary layer on a flat plate at zero angle of incidence.

However, the transition Reynolds number depends on free-stream turbulence and may range from 3×10^5 to 3×10^6. The **laminar boundary layer** thickness δ is a function of distance from the leading edge:

$$\delta \approx 5.0 x Re_x^{-0.5} \qquad (6-179)$$

The total drag on the plate of length L and width b for a laminar boundary layer, including the drag on both surfaces, is:

$$F_D = 1.328 b L \rho V^2 Re_L^{-0.5} \qquad (6-180)$$

For **non-Newtonian power law fluids** (Acrivos, Shah, and Peterson, *AIChE J.*, **6**, 312–317 [1960]; Hsu, *AIChE J.*, **15**, 367–370 [1969]),

$$F_D = C b L \rho V^2 Re_L'^{-1/(1+n)} \qquad (6-181)$$

$n =$	0.2	0.3	0.4	0.5	0.6	0.7	0.8	0.9	1
$C =$	2.075	1.958	1.838	1.727	1.627	1.538	1.460	1.390	1.328

where $Re_L' = \rho V^{2-n} L^n / K$ and K and n are the power law material constants (see Eq. [6-4]).

For a **turbulent boundary layer,** the thickness may be estimated as

$$\delta \approx 0.37 x Re_x^{-0.2} \qquad (6-182)$$

and the total drag force on both sides of the plate of length L is

$$F_D = \left[\frac{0.455}{(\log Re_L)^{2.58}} - \frac{1,700}{Re_L} \right] \rho b L V^2 \qquad 5 \times 10^5 < Re_L < 10^9 \qquad (6-183)$$

Here the second term accounts for the laminar leading edge of the boundary layer and assumes that the critical Reynolds number is 500,000.

Cylindrical Boundary Layer   Laminar boundary layers on cylindrical surfaces, with flow parallel to the cylinder axis, are described by Glauert and Lighthill (*Proc. R. Soc.* [*London*], **230A**, 188–203 [1955]), Jaffe and Okamura (*Z. Angew. Math. Phys.*, **19**, 564–574 [1968]) and Stewartson (*Q. Appl. Math.*, **13**, 113–122 [1955]). For a turbulent boundary layer, the total drag may be estimated as

$$F_D = \bar{c_j} \pi r L \rho V^2 \qquad (6-184)$$

where r = cylinder radius, L = cylinder length, and the average friction coefficient is given by (White, *J. Basic Eng.*, **94**, 200–206 [1972])

$$\bar{c_j} = 0.0015 + \left[0.30 + 0.015 \left(\frac{L}{r} \right)^{0.4} \right] Re_L^{-1/3} \qquad (6-185)$$

for $Re_L = 10^6$ to 10^9 and $L/r < 10^6$.

Continuous Flat Surface   Boundary layers on continuous surfaces drawn through a stagnant fluid are shown in Fig. 6-48. Figure 6-48a shows the continuous flat surface (Sakiadis, *AIChE J.*, **7**, 26–28, 221–225, 467–472 [1961]). The critical Reynolds number for transition to turbulent flow may be greater than the 500,000 value for the finite flat-plate case discussed previously (Tsou, Sparrow, and Kurtz, *J. Fluid Mech.*, **26**, 145–161 [1966]). For a laminar boundary layer, the thickness is given by

$$\delta = 6.37 x Re_x^{-0.5} \qquad (6-186)$$

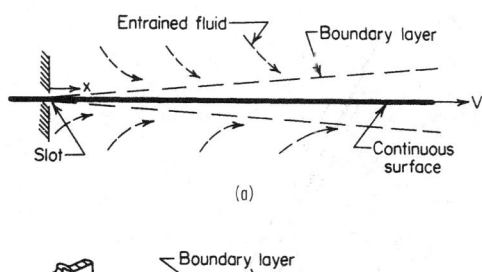

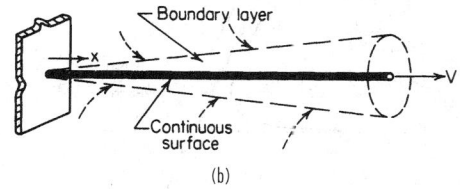

FIG. 6-48 Continuous surface: (*a*) continuous flat surface, (*b*) continuous cylindrical surface. (*From Sakiadis*, Am. Inst. Chem. Eng. J., **7**, *221, 467* [*1961*].)

and the total drag exerted on the two surfaces is

$$F_D = 1.776 bL\rho V^2 \text{Re}_L^{-0.5} \tag{6-187}$$

The total flow rate of fluid entrained by the surface is

$$q = 3.232 bLV \text{Re}_L^{-0.5} \tag{6-188}$$

The theoretical velocity field was experimentally verified by Tsou, Sparrow, and Goldstein (*Int. J. Heat Mass Transfer*, **10**, 219–235 [1967]) and Szeri, Yates, and Hai (*J. Lubr. Technol.*, **98**, 145–156 [1976]). For **non-Newtonian power law fluids** see Fox, Erickson, and Fan (*AIChE J.*, **15**, 327–333 [1969]).

For a turbulent boundary layer, the thickness is given by

$$\delta = 1.01 x \text{Re}_x^{-0.2} \tag{6-189}$$

and the total drag on both sides by

$$F_D = 0.056 bL\rho V^2 \text{Re}_L^{-0.2} \tag{6-190}$$

and the total entrainment by

$$q = 0.252 bLV \text{Re}_L^{-0.2} \tag{6-191}$$

When the laminar boundary layer is a significant part of the total length of the object, the total drag should be corrected by subtracting a calculated turbulent drag for the length of the laminar section and then adding the laminar drag for the laminar section. Tsou, Sparrow, and Goldstein (*Int. J. Heat Mass Transfer*, **10**, 219–235 [1967]) give an improved analysis of the turbulent boundary layer; their data indicate that Eq. (6-190) underestimates the drag by about 15 percent.

Continuous Cylindrical Surface The continuous surface shown in Fig. 6-48*b* is applicable, for example, for a wire drawn through a stagnant fluid (Sakiadis, *AIChE J.*, **7**, 26–28, 221–225, 467–472 [1961]). The critical-length Reynolds number for transition is $\text{Re}_x = 200{,}000$. The boundary layer thickness, total drag, and entrainment flow rate may be obtained from Fig. 6-49; the drag and entrainment rate are obtained from the *momentum area* Θ and *displacement area* Δ evaluated at $x = L$.

$$F_D = \rho V^2 \Theta \tag{6-192}$$

$$q = V\Delta \tag{6-193}$$

Further laminar boundary layer analysis is given by Crane (*Z. Angew. Math. Phys.*, **23**, 201–212 [1972]).

For a turbulent boundary layer, the total drag may be roughly estimated using Eqs. (6-184) and (6-185) for finite cylinders. Measured forces by Kwon and Prevorsek (*J. Eng. Ina.*, **101**, 73–79 [1979]) are greater than predicted this way.

The laminar boundary layer on **deforming continuous surfaces** with velocity varying with axial position is discussed by Vleggaar

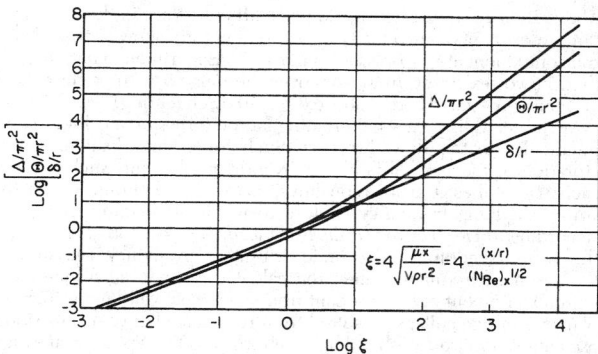

FIG. 6-49 Boundary layer parameters for continuous cylindrical surfaces. (*From Sakiadis*, Am. Inst. Chem. J., **7**, 467 [1961].)

(*Chem. Eng. Sci.*, **32**, 1517–1525 [1977]) and Crane (*Z. Angew. Math. Phys.*, **26**, 619–622 [1975]).

VORTEX SHEDDING

When fluid flows past objects or through orifices or similar restrictions, vortices may periodically be shed downstream. Objects such as smokestacks, chemical-processing columns, suspended pipelines, and electrical transmission lines can be subjected to damaging vibrations and forces due to the vortices, especially if the shedding frequency is close to a natural vibration frequency of the object. The shedding can also produce sound. See Krzywoblocki (*Appl. Mech. Rev.*, **6**, 393–397 [1953]) and Marris (*J. Basic Eng.*, **86**, 185–196 [1964]).

Development of a vortex street, or *von Kármán vortex street* is shown in Fig. 6-50. Discussions of the vortex street may be found in Panton (pp. 387–393). The Reynolds number is

$$\text{Re} = \frac{DV\rho}{\mu} \tag{6-194}$$

where D = diameter of cylinder or effective width of object
V = free-stream velocity
ρ = fluid density
μ = fluid viscosity

For flow past a cylinder, the vortex street forms at Reynolds numbers above about 40. The vortices initially form in the wake, the point of formation moving closer to the cylinder as Re is increased. At a Reynolds number of 60 to 100, the vortices are formed from eddies attached to the cylinder surface. The vortices move at a velocity slightly less than V. The frequency of vortex shedding f is given in terms of the Strouhal number, which is approximately constant over a wide range of Reynolds numbers.

$$\text{St} \equiv \frac{fD}{V} \tag{6-195}$$

For $40 < \text{Re} < 200$ the vortices are laminar and the Strouhal number has a nearly constant value of 0.2 for flow past a cylinder. Between $\text{Re} = 200$ and 400 the Strouhal number is no longer constant and the wake becomes irregular. Above about $\text{Re} = 400$ the vortices become turbulent, the wake is once again stable, and the Strouhal number remains constant at about 0.2 up to a Reynolds number of about 10^5.

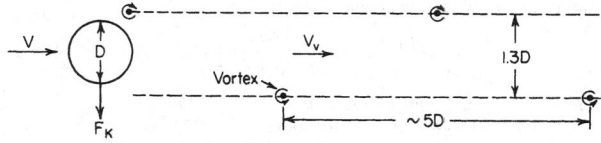

FIG. 6-50 Vortex street behind a cylinder.

Above $Re = 10^5$ the vortex shedding is difficult to see in flow visualization experiments, but velocity measurements still show a strong spectral component at $St = 0.2$ (Panton, p. 392). Experimental data suggest that the vortex street disappears over the range $5 \times 10^5 < Re < 3.5 \times 10^6$, but is reestablished at above 3.5×10^6 (Schlichting).

Vortex shedding exerts alternating lateral forces on a cylinder, perpendicular to the flow direction. Such forces may lead to severe vibration or mechanical failure of cylindrical elements such as heat-exchanger tubes, transmission lines, stacks, and columns when the vortex shedding frequency is close to resonant bending frequency. According to Den Hartog (*Proc. Nat. Acad. Sci.*, **40**, 155–157 [1954]), the vortex shedding and cylinder vibration frequency will shift to the resonant frequency when the calculated shedding frequency is within 20 percent of the resonant frequency. The well-known Tacoma Narrows bridge collapse resulted from resonance between a torsional oscillation and vortex shedding (Panton, p. 392). Spiral strakes are sometimes installed on tall stacks so that vortices at different axial positions are not shed simultaneously. The alternating lateral force F_K, sometimes called the *von Kármán force*, is given by (Den Hartog, *Mechanical Vibrations*, 4th ed., McGraw-Hill, New York, 1956, pp. 305–309):

$$F_K = C_K A \frac{\rho V^2}{2} \qquad (6\text{-}196)$$

where C_K = von Kármán coefficient
   A = projected area perpendicular to the flow
   ρ = fluid density
   V = free-stream fluid velocity

For a cylinder, $C_K = 1.7$. For a vibrating cylinder, the effective projected area exceeds, but is always less than twice, the actual cylinder projected area (Rouse, *Engineering Hydraulics*, Wiley, New York, 1950).

The following references pertain to discussions of vortex shedding in specific structures: steel stacks (Osker and Smith, *Trans. ASME*, **78**, 1381–1391 [1956]); Smith and McCarthy, *Mech. Eng.*, **87**, 38–41 [1965]); chemical-processing columns (Freese, *J. Eng. Ind.*, **81**, 77–91, [1959]); heat exchangers (Eisinger, *Trans. ASME J. Pressure Vessel Tech.*, **102**, 138–145 [May 1980]; Chen, *J. Sound Vibration*, **93**, 439–455 [1984]; Gainsboro, *Chem. Eng. Prog.*, **64**[3], 85–88 [1968]; "Flow-Induced Vibration in Heat Exchangers," *Symp. Proc.*, ASME, New York, 1970); suspended pipe lines (Baird, *Trans. ASME*, **77**, 797–804 [1955]); and suspended cable (Steidel, *J. Appl. Mech.*, **23**, 649–650 [1956]).

COATING FLOWS

In coating flows, liquid films are entrained on moving solid surfaces. For general discussions, see Ruschak (*Ann. Rev. Fluid Mech.*, **17**, 65–89 [1985]), Cohen and Gutoff (*Modern Coating and Drying Technology*, VCH Publishers, New York, 1992), and Middleman (*Fundamentals of Polymer Processing*, McGraw-Hill, New York, 1977). It is generally important to control the thickness and uniformity of the coatings.

In **dip coating,** or free withdrawal coating, a solid surface is withdrawn from a liquid pool, as shown in Fig. 6-51. It illustrates many of the features found in other coating flows, as well. Tallmadge and Gutfinger (*Ind. Eng. Chem.*, **59**[11], 19–34 [1967]) provide an early review of the theory of dip coating. The coating flow rate and film thickness are controlled by the withdrawal rate and the flow behavior in the meniscus region. For a withdrawal velocity V and an angle of inclination from the horizontal ϕ, the film thickness h may be estimated for low withdrawal velocities by

$$h \left(\frac{\rho g}{\sigma} \right)^{1/2} = \frac{0.944}{(1 - \cos \phi)^{1/2}} \, Ca^{2/3} \qquad (6\text{-}197)$$

where g = acceleration of gravity
   $Ca = \mu V/\sigma$ = capillary number
   μ = viscosity
   σ = surface tension

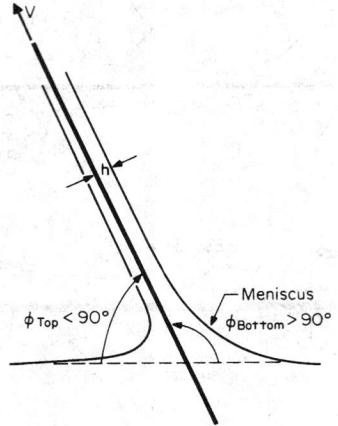

FIG. 6-51   Dip coating.

Equation (6-197) is asymptotically valid as $Ca \to 0$ and agrees with experimental data up to capillary numbers in the range of 0.01 to 0.03. In practice, where high production rates require high withdrawal speeds, capillary numbers are usually too large for Eq. (6-197) to apply. Approximate analytical methods for larger capillary numbers have been obtained by numerous investigators, but none appears wholly satisfactory, and some are based on questionable assumptions (Ruschak, *Ann. Rev. Fluid Mech.*, **17**, 65–89 [1985]). With the availability of high-speed computers and the development of the field of computational fluid dynamics, numerical solutions accounting for two-dimensional flow aspects, along with gravitational, viscous, inertial, and surface tension forces are now the most effective means to analyze coating flow problems.

Other common coating flows include premetered flows, such as **slide** and **curtain coating,** where the film thickness is an independent parameter that may be controlled within limits, and the curvature of the meniscus adjusts accordingly; the closely related **blade coating;** and **roll coating** and **extrusion coating.** See Ruschak (ibid.), Cohen and Gutoff (*Modern Coating and Drying Technology*, VCH Publishers, New York, 1992) and Middleman (*Fundamentals of Polymer Processing*, McGraw-Hill, New York, 1977). For dip coating of wires, see Taughy (*Int. J. Numerical Meth. Fluids*, **4**, 441–475 [1984]).

Many coating flows are subject to instabilities that lead to unacceptable coating defects. Three-dimensional flow instabilities lead to such problems as **ribbing.** Air entrainment is another common defect.

FALLING FILMS

Minimum Wetting Rate   The minimum liquid rate required for complete wetting of a vertical surface is about 0.03 to 0.3 kg/m · s for water at room temperature. The minimum rate depends on the geometry and nature of the vertical surface, liquid surface tension, and mass transfer between surrounding gas and the liquid. See Ponter, et al. (*Int. J. Heat Mass Transfer*, **10**, 349–359 [1967]; *Trans. Inst. Chem. Eng. [London]*, **45**, 345–352 [1967]), Stainthorp and Allen (*Trans. Inst. Chem. Eng. [London]*, **43**, 85–91 [1967]) and Watanabe, et al. (*J. Chem. Eng. [Japan]*, **8**[1], 75 [1975]).

Laminar Flow   For films falling down **vertical flat surfaces,** as shown in Fig. 6-52, or vertical tubes with small film thickness compared to tube radius, laminar flow conditions prevail for Reynolds numbers less than about 2,000, where the Reynolds number is given by

$$Re = \frac{4\Gamma}{\mu} \qquad (6\text{-}198)$$

where Γ = liquid mass flow rate per unit width of surface and μ = liq-

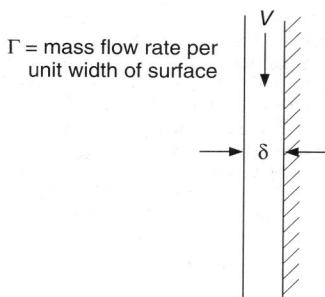

Γ = mass flow rate per unit width of surface

FIG. 6-52 Falling film.

uid viscosity. For a flat film surface, the following equations may be derived. The film thickness δ is

$$\delta = \left(\frac{3\Gamma\mu}{\rho^2 g}\right)^{1/3} \tag{6-199}$$

The average film velocity is

$$V = \frac{\Gamma}{\rho\delta} = \frac{g\rho\delta^2}{3\mu} \tag{6-200}$$

The downward velocity profile $u(x)$ where $x = 0$ at the solid surface and $x = \delta$ at the liquid/gas interface is given by

$$u = 1.5V\left[\frac{2x}{\delta} - \left(\frac{x}{\delta}\right)^2\right] \tag{6-201}$$

These equations assume that there is no drag force at the gas/liquid interface, such as would be produced by gas flow. For a flat surface, **inclined** at an angle θ with the horizontal, the preceding equations may be modified by replacing g by $g \sin \theta$. For films falling inside vertical tubes with film thickness up to and including the full pipe radius, see Jackson (*AIChE J.*, **1**, 231–240 [1955]).

These equations have generally given good agreement with experimental results for low-viscosity liquids (<0.005 Pa · s) (< 5 cp) whereas Jackson (ibid.) found film thicknesses for higher-viscosity liquids (0.01 to 0.02 Pa·s) (10 to 20 cp) were significantly less than predicted by Eq. (6-197). At Reynolds numbers of 25 or greater, **surface waves** will be present on the liquid film. West and Cole (*Chem. Eng. Sci.*, **22**, 1388–1389 [1967]) found that the surface velocity $u(x = \delta)$ is still within ±7 percent of that given by Eq. (6-201) even in wavy flow.

For laminar non-Newtonian film flow, see Bird, Armstrong, and Hassager (*Dynamics of Polymeric Liquids*, vol. 1: *Fluid Mechanics*, Wiley, New York, 1977, p. 215, 217), Astarita, Marrucci, and Palumbo (*Ind. Eng. Chem. Fundam.*, **3**, 333–339 [1964]) and Cheng (*Ind. Eng. Chem. Fundam.*, **13**, 394–395 [1974]).

Turbulent Flow   In turbulent flow, Re > 2,000, for vertical surfaces, the film thickness may be estimated to within ±25 percent using

$$\delta = 0.304\left(\frac{\Gamma^{1.75}\mu^{0.25}}{\rho^2 g}\right)^{1/3} \tag{6-202}$$

Replace g by $g \sin \theta$ for a surface inclined at angle θ to the horizontal. The average film velocity is $V = \Gamma/\rho\delta$.

Tallmadge and Gutfinger (*Ind. Eng. Chem.*, **59**[11], 19–34 [1967]) discuss prediction of drainage rates from liquid films on flat and cylindrical surfaces.

Effect of Surface Traction   If a drag is exerted on the surface of the film because of motion of the surrounding fluid, the film thickness will be reduced or increased, depending upon whether the drag acts with or against gravity. Thomas and Portalski (*Ind. Eng. Chem.*, **50**, 1081–1088 [1958]), Dukler (*Chem. Eng. Prog.*, **55**[10], 62–67 [1959]) and Kosky (*Int. J. Heat Mass Transfer*, **14**, 1220–1224 [1971]) have presented calculations of film thickness and film velocity. Film thickness data for falling water films with cocurrent and countercurrent air flow in pipes are given by Zhivaikin (*Int. Chem. Eng.*, **2**, 337–341 [1962]). Zabaras, Dukler, and Moalem-Maron (*AIChE J.*, **32**, 829–843

[1986]) and Zabaras and Dukler (*AIChE J.*, **34**, 389–396 [1988]) present studies of film flow in vertical tubes with both cocurrent and countercurrent gas flow, including measurements of film thickness, wall shear stress, wave velocity, wave amplitude, pressure drop, and flooding point for countercurrent flow.

Flooding   With countercurrent gas flow, a condition is reached with increasing gas rate for which flow reversal occurs and liquid is carried upward. The mechanism for this flooding condition has been most often attributed to waves either bridging the pipe or reversing direction to flow upward at flooding. However, the results of Zabaras and Dukler (ibid.) suggest that flooding may be controlled by flow conditions at the liquid inlet and that wave bridging or upward wave motion does not occur, at least for the 50.8-mm diameter pipe used for their study. Flooding mechanisms are still incompletely understood. Under some circumstances, as when the gas is allowed to develop its normal velocity profile in a "calming length" of pipe beneath the liquid draw-off, the gas superficial velocity at flooding will be increased, and increases with decreasing length of wetted pipe (Hewitt, Lacy, and Nicholls, *Proc. Two-Phase Flow Symp.*, University of Exeter, paper 4H, AERE-4 4614 [1965]). A bevel cut at the bottom of the pipe with an angle 30° from the vertical will increase the flooding velocity in small-diameter flow tubes at moderate liquid flow rates. If the gas approaches the tube from the side, the taper should be oriented with the point facing the gas entrance. Figures 6-53 and 6-54 give correlations for flooding in tubes with square and slant bottoms (courtesy Holmes, DuPont Co.) The superficial mass velocities of gas and liquid G_G and G_L, and the physical property parameters λ and ψ are the same as those defined for the Baker chart ("Multiphase Flow" subsection, Fig. 6-25). For tubes larger than 50 mm (2 in), flooding velocity appears to be relatively insensitive to diameter and the flooding curves for 1.98-in diameter may be used.

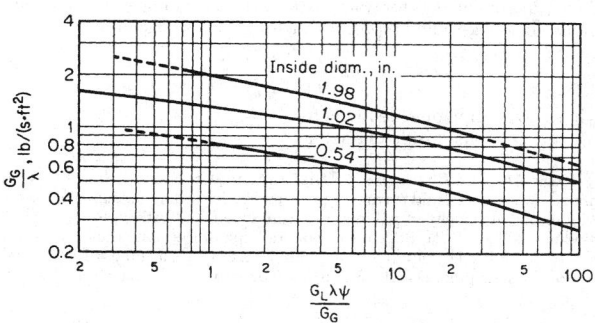

FIG. 6-53   Flooding in vertical tubes with square top and square bottom. To convert lbm/(ft²·s) to kg/(m²·s), multiply by 4.8824; to convert in to mm, multiply by 25.4. (*Courtesy of E. I. du Pont de Nemours & Co.*)

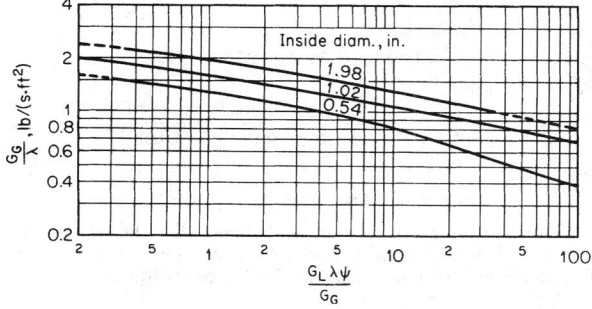

FIG. 6-54   Flooding in vertical tubes with square top and slant bottom. To convert lbm/(ft²·s) to kg/(m²·s), multiply by 4.8824; to convert in to mm, multiply by 25.4. (*Courtesy of E. I. du Pont de Nemours & Co.*)

HYDRAULIC TRANSIENTS

Many transient flows of liquids may be analyzed by using the full time-dependent equations of motion for incompressible flow. However, there are some phenomena that are controlled by the small compressibility of liquids. These phenomena are generally called **hydraulic transients.**

Water Hammer   When liquid flowing in a pipe is suddenly decelerated to zero velocity by a fast-closing valve, a pressure wave propagates upstream to the pipe inlet, where it is reflected; a pounding of the line commonly known as **water hammer** is often produced. For an instantaneous flow stoppage of a truly incompressible fluid in an inelastic pipe, the pressure rise would be infinite. Finite compressibility of the fluid and elasticity of the pipe limit the pressure rise to a finite value. The Joukowski formula gives the maximum pressure rise as

$$\Delta p = \rho a \Delta V \qquad (6\text{-}203)$$

where ρ = liquid density
ΔV = change in liquid velocity
a = pressure wave velocity

The wave velocity is given by

$$a = \frac{\sqrt{\beta/\rho}}{\sqrt{1 + (\beta/E)(D/b)}} \qquad (6\text{-}204)$$

where β = liquid bulk modulus of elasticity
E = elastic modulus of pipe wall
D = pipe inside diameter
b = pipe wall thickness

The numerator gives the wave velocity for perfectly rigid pipe, and the denominator corrects for wall elasticity. This formula is for thin-walled pipes; for thick-walled pipes, the factor D/b is replaced by

$$2\,\frac{(D_o^2 + D_i^2)}{(D_o^2 - D_i^2)}$$

where D_o = pipe outside diameter
D_i = pipe inside diameter

Example 10: Response to Instantaneous Valve Closing   Compute the wave speed and maximum pressure rise for instantaneous valve closing, with an initial velocity of 2.0 m/s, in a 4-in Schedule 40 steel pipe with elastic modulus 207×10^9 Pa. Repeat for a plastic pipe of the same dimensions, with $E = 1.4 \times 10^9$ Pa. The liquid is water with $\beta = 2.2 \times 10^9$ Pa and $\rho = 1,000$ kg/m³.

For the steel pipe, $D = 102.3$ mm, $b = 6.02$ mm, and the wave speed is

$$a = \frac{\sqrt{\beta/\rho}}{\sqrt{1 + (\beta/E)(D/b)}}$$

$$= \frac{\sqrt{2.2 \times 10^9/1000}}{\sqrt{1 + (2.2 \times 10^9/207 \times 10^9)(102.3/6.02)}}$$

$$= 1365 \text{ m/s}$$

The maximum pressure rise is

$$\Delta p = \rho a \Delta V$$

$$= 1,000 \times 1,365 \times 2.0 = 2.73 \times 10^6 \text{ Pa}$$

For the plastic pipe,

$$a = \frac{\sqrt{2.2 \times 10^9/1000}}{\sqrt{1 + (2.2 \times 10^9/1.4 \times 10^9)(102.3/6.02)}}$$

$$= 282 \text{ m/s}$$

$$\Delta p = \rho a \Delta V = 1,000 \times 282 \times 2.0 = 5.64 \times 10^5 \text{ Pa}$$

The maximum pressure surge is obtained when the valve closes in less time than the period τ required for the pressure wave to travel from the valve to the pipe inlet and back, a total distance of $2L$.

$$\tau = \frac{2L}{a} \qquad (6\text{-}205)$$

The pressure surge will be reduced when the time of flow stoppage exceeds the pipe period τ, due to cancellation between direct and reflected waves. Wood and Jones (*Proc. Am. Soc. Civ. Eng., J. Hydraul. Div.*, **99**, (HY1), 167–178 [1973]) present charts for reliable estimates of water-hammer pressure for different valve closure modes. Wylie and Streeter (*Hydraulic Transients*, McGraw-Hill, New York, 1978) describe several solution methods for hydraulic transients, including the method of characteristics, which is well suited to computer methods for accurate solutions. A rough approximation for the peak pressure for cases where the valve closure time t_c exceeds the pipe period τ is (Daugherty and Franzini, *Fluid Mechanics with Engineering Applications*, McGraw-Hill, New York, 1985):

$$\Delta p \approx \left(\frac{\tau}{t_c}\right)\rho a \Delta V \qquad (6\text{-}206)$$

Successive reflections of the pressure wave between the pipe inlet and the closed valve result in alternating pressure increases and decreases, which are gradually attenuated by fluid friction and imperfect elasticity of the pipe. Periods of reduced pressure occur while the reflected pressure wave is traveling from inlet to valve. Degassing of the liquid may occur, as may vaporization if the pressure drops below the vapor pressure of the liquid. Gas and vapor bubbles decrease the wave velocity. Vaporization may lead to what is often called *liquid column separation;* subsequent collapse of the vapor pocket can result in pipe rupture.

In addition to water hammer induced by changes in valve setting, including closure, numerous other hydraulic transient flows are of interest, as, for example (Wylie and Streeter, *Hydraulic Transients*, McGraw-Hill, New York, 1978), those arising from starting or stopping of pumps; changes in power demand from turbines; reciprocating pumps; changing elevation of a reservoir; waves on a reservoir; turbine governor hunting; vibration of impellers or guide vanes in pumps, fans, or turbines; vibration of deformable parts such as valves; draft-tube instabilities due to vortexing; and unstable pump or fan characteristics. Tube failure in heat exhangers may be added to this list.

Pulsating Flow   Reciprocating machinery (pumps and compressors) produces flow pulsations, which adversely affect flow meters and process control elements and can cause vibration and equipment failure, in addition to undesirable process results. Vibration and damage can result not only from the fundamental frequency of the pulse producer but also from higher harmonics. Multipiston double-acting units reduce vibrations. Pulsation dampeners are often added. Damping methods are described by M. W. Kellogg Co. (*Design of Piping Systems*, rev. 2d ed., Wiley, New York, 1965). For liquid phase pulsation damping, gas-filled surge chambers, also known as accumulators, are commonly used; see Wylie and Streeter (*Hydraulic Transients*, McGraw-Hill, New York, 1978).

Software packages are commercially available for simulation of hydraulic transients. These may be used to analyze piping systems to reveal unsatisfactory behavior, and they allow the assessment of design changes such as increases in pipe-wall thickness, changes in valve actuation, and addition of check valves, surge tanks, and pulsation dampeners.

Cavitation   Loosely regarded as related to water hammer and hydraulic transients because it may cause similar vibration and equipment damage, **cavitation** is the phenomenon of collapse of vapor bubbles in flowing liquid. These bubbles may be formed anywhere the local liquid pressure drops below the vapor pressure, or they may be injected into the liquid, as when steam is sparged into water. Local low-pressure zones may be produced by local velocity increases (in accordance with the Bernoulli equation; see the preceding "Conservation Equations" subsection) as in eddies or vortices, or near boundary contours; by rapid vibration of a boundary; by separation of liquid during water hammer; or by an overall reduction in static pressure, as due to pressure drop in the suction line of a pump.

Collapse of vapor bubbles once they reach zones where the pressure exceeds the vapor pressure can cause objectionable noise and vibration and extensive erosion or pitting of the boundary materials. The critical cavitation number at inception of cavitation, denoted σ_i, is useful in correlating equipment performance data:

$$\sigma_i = \frac{(p - p_v)}{\rho V^2/2} \qquad (6\text{-}207)$$

where p = static pressure in undisturbed flow
p_v = vapor pressure
ρ = liquid density
V = free-stream velocity of the liquid

The value of the cavitation number for incipient cavitation for a specific piece of equipment is a characteristic of that equipment. Cavitation numbers for various head forms of cylinders, for disks, and for various hydrofoils are given by Holl and Wislicenus (*J. Basic Eng.*, **83**, 385–398 [1961]) and for various surface irregularities by Arndt and Ippen (*J. Basic Eng.*, **90**, 249–261 [1968]), Ball (*Proc. ASCE J. Constr. Div.*, **89**(C02), 91–110 [1963]), and Holl (*J. Basic Eng.*, **82**, 169–183 [1960]). As a guide only, for blunt forms the cavitation number is generally in the range of 1 to 2.5, and for somewhat streamlined forms the cavitation number is in the range of 0.2 to 0.5. Critical cavitation numbers generally depend on a characteristic length dimension of the equipment in a way that has not been explained. This renders scale-up of cavitation data questionable.

For cavitation in flow through orifices, Fig. 6-55 (Thorpe, *Int. J. Multiphase Flow*, **16**, 1023–1045 [1990]) gives the critical cavitation number for inception of cavitation. To use this cavitation number in Eq. (6-207), the pressure p is the orifice backpressure downstream of the vena contracta after full pressure recovery, and V is the average velocity through the orifice. Fig. 6-55 includes data from Tullis and Govindarajan (*ASCE J. Hydraul. Div.*, **HY13**, 417–430 [1973]) modified to use the same cavitation number definition; their data also include critical cavitation numbers for 30.50- and 59.70-cm pipes (12.00- to 23.50-in). Very roughly, compared with the 15.40-cm pipe, the cavitation number is about 20 percent greater for the 30.50-cm (12.01-in) pipe and about 40 percent greater for the 59.70-cm (23.50-in) diameter pipe. Inception of cavitation appears to be related to release of dissolved gas and not merely vaporization of the liquid. For further discussion of cavitation, see Eisenberg and Tulin (Streeter, *Handbook of Fluid Dynamics*, Sec. 12, McGraw-Hill, New York, 1961).

TURBULENCE

Turbulent flow occurs when the Reynolds number exceeds a critical value above which laminar flow is unstable; the critical Reynolds number depends on the flow geometry. There is generally a transition regime between the critical Reynolds number and the Reynolds number at which the flow may be considered fully turbulent. The transition regime is very wide for some geometries. In turbulent flow, variables such as velocity and pressure fluctuate chaotically; statistical methods are used to quantify turbulence.

Time Averaging   In turbulent flows it is useful to define time-averaged and fluctuation values of flow variables such as velocity com-

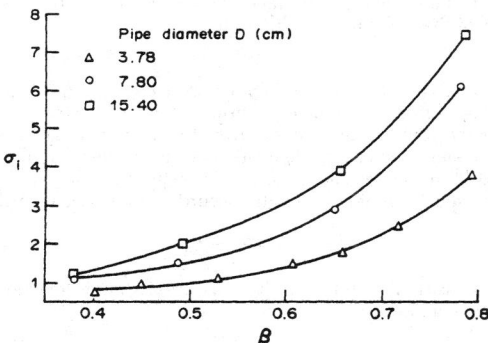

FIG. 6-55   Critical cavitation number vs. diameter ratio β. (*Reprinted from Thorpe, "Flow regime transitions due to cavitation in the flow through an orifice," Int. J. Multiphase Flow,* **16**, *1023–1045. Copyright © 1990, with kind permission from Elsevier Science, Ltd., The Boulevard, Langford Lane, Kidlington OX5 1GB, United Kingdom.*)

ponents. For example, the x-component velocity fluctuation v'_x is the difference between the actual instantaneous velocity v_x and the time-averaged velocity $\overline{v_x}$:

$$v'_x(x, y, z, t) = v_x(x, y, z, t) - \overline{v_x}(x, y, z) \qquad (6\text{-}208)$$

The actual and fluctuating velocity components are, in general, functions of the three spatial coordinates x, y, and z and of time t. The time-averaged velocity $\overline{v_x}$ is independent of time for a **stationary** flow. Nonstationary processes may be considered where averages are defined over time scales long compared to the time scale of the turbulent fluctuations, but short compared to longer time scales over which the time-averaged flow variables change due, for example, to time-varying boundary conditions. The time average over a time interval $2T$ centered at time t of a turbulently fluctuating variable $\zeta(t)$ is defined as

$$\overline{\zeta(t)} = \frac{1}{2T} \int_{t-T}^{t+T} \zeta(\tau) \, d\tau \qquad (6\text{-}209)$$

where τ = dummy integration variable. For stationary turbulence, $\overline{\zeta}$ does not vary with time.

$$\overline{\zeta} = \lim_{T \to \infty} \frac{1}{2T} \int_{t-T}^{t+T} \zeta(\tau) \, d\tau \qquad (6\text{-}210)$$

The time average of a fluctuation $\overline{\zeta'} = \overline{\zeta - \overline{\zeta}} = 0$. Fluctuation magnitudes are quantified by root mean squares.

$$\tilde{v}'_x = \sqrt{\overline{(v'_x)^2}} \qquad (6\text{-}211)$$

In **isotropic** turbulence, statistical measures of fluctuations are equal in all directions.

$$\tilde{v}'_x = \tilde{v}'_y = \tilde{v}'_z \qquad (6\text{-}212)$$

In **homogeneous** turbulence, turbulence properties are independent of spatial position. **The kinetic energy of turbulence** k is given by

$$k = \frac{1}{2}(\tilde{v}'^2_x + \tilde{v}'^2_y + \tilde{v}'^2_z) \qquad (6\text{-}213)$$

Turbulent velocity fluctuations ultimately dissipate their kinetic energy through viscous effects. Macroscopically, this energy dissipation requires pressure drop, or velocity decrease. The **energy dissipation rate** per unit mass is usually denoted ϵ. For steady flow in a pipe, the average energy dissipation rate per unit mass is given by

$$\epsilon = \frac{2fV^3}{D} \qquad (6\text{-}214)$$

where ρ = fluid density
f = Fanning friction factor
D = pipe inside diameter

When the continuity equation and the Navier-Stokes equations for incompressible flow are time averaged, equations for the time-averaged velocities and pressures are obtained which appear identical to the original equations (6-18 through 6-28), except for the appearance of additional terms in the Navier-Stokes equations. Called **Reynolds stress** terms, they result from the nonlinear effects of momentum transport by the velocity fluctuations. In each i-component ($i = x, y, z$) Navier-Stokes equation, the following additional terms appear on the right-hand side:

$$\sum_{j=1}^{3} \frac{\partial \tau^{(t)}_{ji}}{\partial x_j}$$

with j components also being x, y, z. The Reynolds stresses are given by

$$\tau^{(t)}_{ij} = -\rho \overline{v'_i v'_j} \qquad (6\text{-}215)$$

The Reynolds stresses are nonzero because the velocity fluctuations in different coordinate directions are correlated so that $\overline{v'_i v'_j}$ in general is nonzero.

Although direct numerical simulations under limited circumstances have been carried out to determine (unaveraged) fluctuating velocity fields, in general the solution of the equations of motion for turbulent flow is based on the time-averaged equations. This requires semiempirical models to express the Reynolds stresses in terms of time-averaged velocities. This is the **closure** problem of turbulence. In all but the simplest geometries, numerical methods are required.

Closure Models   Many closure models have been proposed. A few of the more important ones are introduced here. Many employ the Boussinesq approximation, simplified here for incompressible flow, which treats the Reynolds stresses as analogous to viscous stresses, introducing a scalar quantity called the turbulent or eddy viscosity μ_t.

$$-\rho \overline{v_i' v_j'} = \mu_t \left(\frac{\partial \overline{v_i}}{\partial x_j} + \frac{\partial \overline{v_j}}{\partial x_i} \right) \qquad (6\text{-}216)$$

An additional *turbulence pressure* term equal to $-\frac{2}{3}k\delta_{ij}$, where $k =$ turbulent kinetic energy and $\delta_{ij} = 1$ if $i = j$ and $\delta_{ij} = 0$ if $i \neq j$, is sometimes included in the right-hand side. To solve the equations of motion using the Boussinesq approximation, it is necessary to provide equations for the single scalar unknown μ_t (and k, if used) rather than the nine unknown tensor components $\tau_{ij}^{(t)}$. With this approximation, and using the effective viscosity $\mu_{eff} = \mu + \mu_t$, the time-averaged momentum equation is similar to the original Navier-Stokes equation, with time-averaged variables and μ_{eff} replacing the instantaneous variables and molecular viscosity. However, solutions to the time-averaged equations for turbulent flow are not identical to those for laminar flow because μ_{eff} is not a constant.

The universal turbulent velocity profile near the pipe wall presented in the preceding subsection "Incompressible Flow in Pipes and Channels" may be developed using the Prandtl mixing length approximation for the eddy viscosity,

$$\mu_t = \rho l_P^2 \left| \frac{d\overline{v_x}}{dy} \right| \qquad (6\text{-}217)$$

where l_P is the Prandtl mixing length. The turbulent core of the universal velocity profile is obtained by assuming that the mixing length is proportional to the distance from the wall. The proportionality constant is one of two constants adjusted to fit experimental data.

The Prandtl mixing length concept is useful for shear flows parallel to walls, but is inadequate for more general three-dimensional flows. A more complicated semiempirical model commonly used in numerical computations, and found in most commercial software for computational fluid dynamics (CFD; see the following subsection), is the k–ϵ model described by Launder and Spaulding (*Lectures in Mathematical Models of Turbulence*, Academic, London, 1972). In this model the eddy viscosity is assumed proportional to the ratio k^2/ϵ.

$$\mu_t = \rho C_\mu \frac{k^2}{\epsilon} \qquad (6\text{-}218)$$

where the value $C_\mu = 0.09$ is normally used. Semiempirical partial differential conservation equations for k and ϵ derived from the Navier-Stokes equations with simplifying closure assumptions are coupled with the equations of continuity and momentum:

$$\frac{\partial}{\partial t} (\rho k) + \frac{\partial}{\partial x_i} (\rho \overline{v_i} k)$$

$$= \frac{\partial}{\partial x_i} \left(\frac{\mu_t}{\sigma_k} \frac{\partial k}{\partial x_i} \right) + \mu_t \left(\frac{\partial \overline{v_i}}{\partial x_j} + \frac{\partial \overline{v_j}}{\partial x_i} \right) \frac{\partial \overline{v_i}}{\partial x_j} - \rho \epsilon \qquad (6\text{-}219)$$

$$\frac{\partial}{\partial t} (\rho \epsilon) + \frac{\partial}{\partial x_i} (\rho \overline{v_i} \epsilon)$$

$$= \frac{\partial}{\partial x_i} \left(\frac{\mu_t}{\sigma_\epsilon} \frac{\partial \epsilon}{\partial x_i} \right) + C_{1\epsilon} \frac{\epsilon \mu_t}{k} \left(\frac{\partial \overline{v_i}}{\partial x_j} + \frac{\partial \overline{v_j}}{\partial x_i} \right) \frac{\partial \overline{v_i}}{\partial x_j} - C_{2\epsilon} \frac{\rho \epsilon^2}{k} \qquad (6\text{-}220)$$

In these equations summations over repeated indices are implied. The values for the empirical constants $C_{1\epsilon} = 1.44$, $C_{2\epsilon} = 1.92$, $\sigma_k = 1.0$, and $\sigma_\epsilon = 1.3$ are widely accepted (Launder and Spaulding, *The Numerical Computation of Turbulent Flows*, Imperial Coll. Sci. Tech. London, NTIS N74-12066 [1973]). The k–ϵ model has proved reasonably accurate for many flows without highly curved streamlines or significant swirl. It usually underestimates flow separation and overestimates turbulence production by normal straining. The k–ϵ model is suitable for high Reynolds number flows. See Virendra, Patel, Rodi,

and Scheuerer (*AIAA J.*, **23**, 1308–1319 [1984]) for a review of low Reynolds number k–ϵ models.

More advanced models, more complex and computationally intensive, are being developed. For example, the renormalization group theory (Yakhot and Orszag, *J. Scientific Computing*, **1**, 1–51 [1986]; Yakhot, Orszag, Thangam, Gatski, and Speziale, *Phys. Fluids A*, **4**, 1510–1520 [1992]) modification of the k–ϵ model provides theoretical values of the model constants and provides substantial improvement in predictions of flows with stagnation, separation, normal straining, transient behavior such as vortex shedding, and relaminarization. Stress transport models provide equations for all nine Reynolds stress components, rather than introducing eddy viscosity. Algebraic closure equations for the Reynolds stresses are available, but are no longer in common use. Differential Reynolds stress models (e.g., Launder, Reece, and Rodi, *J. Fluid Mech.*, **68**, 537–566 [1975]) use differential conservation equations for all nine Reynolds stress components.

In **direct numerical simulation** of turbulent flows, the solution of the unaveraged equations of motion is sought. Due to the extreme computational intensity, solutions to date have been limited to relatively low Reynolds numbers (Re < about 10,000 to 20,000) in simple geometries such as channel flow. See, for example, Kim, Moin, and Moser (*J. Fluid Mech.*, **177**, 133 [1987]). Since computational grids must be sufficiently fine to resolve even the smallest eddies, the computational difficulty rapidly becomes prohibitive as Reynolds number increases. **Large eddy simulations** use models for subgrid turbulence while solving for larger-scale fluctuations.

Eddy Spectrum   The energy that produces and sustains turbulence is extracted from velocity gradients in the mean flow, principally through vortex stretching. At Reynolds numbers well above the critical value there is a wide spectrum of eddy sizes, often described as a cascade of energy from the largest down to the smallest eddies. The largest eddies are of the order of the equipment size. The smallest are those for which viscous forces associated with the eddy velocity fluctuations are of the same order as inertial forces, so that turbulent fluctuations are rapidly damped out by viscous effects at smaller length scales. Most of the turbulent kinetic energy is contained in the larger eddies, while most of the dissipation occurs in the smaller eddies. Large eddies, which extract energy from the mean flow velocity gradients, are generally anisotropic. At smaller length scales, the directionality of the mean flow exerts less influence, and **local isotropy** is approached. The range of eddy scales for which local isotropy holds is called the **equilibrium range.**

Davies (*Turbulence Phenomena*, Academic, New York, 1972) presents a good discussion of the spectrum of eddy lengths for well-developed isotropic turbulence. The smallest eddies, usually called *Kolmogorov eddies* (Kolmogorov, *Compt. Rend. Acad. Sci. URSS*, **30**, 301; **32**, 16 [1941]), have a characteristic velocity fluctuation $\tilde{v}_K'$ given by

$$\tilde{v}_K' = (\nu \epsilon)^{1/4} \qquad (6\text{-}221)$$

where $\nu =$ kinematic viscosity and $\epsilon =$ energy dissipation per unit mass. The size of the Kolmogorov eddy scale is

$$l_K = (\nu^3/\epsilon)^{1/4} \qquad (6\text{-}222)$$

The Reynolds number for the Kolmogorov eddy, $Re_K = l_K \tilde{v}_K'/\nu$, is equal to unity by definition. In the equilibrium range, which exists for well-developed turbulence and extends from the medium eddy sizes down to the smallest, the energy dissipation at the smaller length scales is supplied by turbulent energy drawn from the bulk flow and passed down the spectrum of eddy lengths according to the scaling rule

$$\epsilon = \frac{(\tilde{v}')^3}{l} \qquad (6\text{-}223)$$

which is consistent with Eqs. (6-221) and (6-222). For the medium, or energy-containing, eddy size,

$$\epsilon = \frac{(\tilde{v}_e')^3}{l_e} \qquad (6\text{-}224)$$

For turbulent pipe flow, the friction velocity $u_* = \sqrt{\tau_w/\rho}$ used earlier in describing the universal turbulent velocity profile may be used as an estimate for $\tilde{v}_e'$. Together with the Blasius equation for the friction factor from which ϵ may be obtained (Eq. 6-214), this provides an estimate for the energy-containing eddy size in turbulent pipe flow:

$$l_e = 0.05D\mathrm{Re}^{-1/8} \qquad (6\text{-}225)$$

where D = pipe diameter and Re = pipe Reynolds number. Similarly, the Kolmogorov eddy size is

$$l_K = 4D\mathrm{Re}^{-0.78} \qquad (6\text{-}226)$$

Most of the energy dissipation occurs on a length scale about 5 times the Kolmogorov eddy size. The characteristic fluctuating velocity for these energy-dissipating eddies is about 1.7 times the Kolmogorov velocity.

The eddy spectrum is normally described using Fourier transform methods; see, for example, Hinze (*Turbulence*, McGraw-Hill, New York, 1975), and Tennekes and Lumley (*A First Course in Turbulence*, MIT Press, Cambridge, 1972). The spectrum $E(\kappa)$ gives the fraction of turbulent kinetic energy contained in eddies of wavenumber between κ and $\kappa + d\kappa$, so that $k = \int_0^\infty E(\kappa)\, d\kappa$. The portion of the equilibrium range excluding the smallest eddies, those which are affected by dissipation, is the **inertial subrange.** The Kolmogorov law gives $E(\kappa) \propto \kappa^{-5/3}$ in the inertial subrange.

Several texts are available for further reading on turbulent flow, including Tennekus and Lumley (ibid.), Hinze (*Turbulence*, McGraw-Hill, New York, 1975), Landau and Lifshitz (*Fluid Mechanics*, 2d ed., Chap. 3, Pergamon, Oxford, 1987) and Panton (*Incompressible Flow*, Wiley, New York, 1984).

COMPUTATIONAL FLUID DYNAMICS

Computational fluid dynamics (CFD) emerged in the 1980s as a significant tool for fluid dynamics both in research and in practice, enabled by rapid development in computer hardware and software. Commercial CFD software is widely available. Computational fluid dynamics is the numerical solution of the equations of continuity and momentum (Navier-Stokes equations for incompressible Newtonian fluids) along with additional conservation equations for energy and material species in order to solve problems of nonisothermal flow, mixing, and chemical reaction.

Textbooks include Fletcher (*Computational Techniques for Fluid Dynamics*, vol. 1: *Fundamental and General Techniques*, and vol. 2: *Specific Techniques for Different Flow Categories*, Springer-Verlag, Berlin, 1988), Hirsch (*Numerical Computation of Internal and External Flows*, vol. 1: *Fundamentals of Numerical Discretization*, and vol. 2: *Computational Methods for Inviscid and Viscous Flows*, Wiley, New York, 1988), Peyret and Taylor (*Computational Methods for Fluid Flow*, Springer-Verlag, Berlin, 1990), Canuto, Hussaini, Quarteroni, and Zang (*Spectral Methods in Fluid Dynamics*, Springer-Verlag, Berlin, 1988), Anderson, Tannehill, and Pletcher (*Computational Fluid Mechanics and Heat Transfer*, Hemisphere, New York, 1984), and Patankar (*Numerical Heat Transfer and Fluid Flow*, Hemisphere, Washington, D.C., 1980).

A wide variety of numerical methods has been employed, but three basic steps are common to all CFD methods:

1. **Subdivision or discretization of the flow domain into cells or elements.** There are methods, called *boundary element methods,* in which the surface of the flow domain, rather than the volume, is discretized, but the vast majority of CFD work uses volume discretization. Discretization produces a set of **grid** lines or curves which define a **mesh** and a set of **nodes** at which the flow variables are to be calculated. The equations of motion are solved approximately on a domain defined by the grid. Curvilinear or **body-fitted** coordinate system grids may be used to ensure that the discretized domain accurately represents the true problem domain.

2. **Discretization of the governing equations.** In this step, the exact partial differential equations to be solved are replaced by approximate algebraic equations written in terms of the nodal values of the dependent variables. Among the numerous discretization methods, **finite difference, finite volume,** and **finite element** methods are the most common. The *finite difference* method estimates spatial derivatives in terms of the nodal values and spacing between nodes. The governing equations are then written in terms of the nodal unknowns at each interior node. *Finite volume* methods, related to finite difference methods, may be derived by a volume integration of the equations of motion, with application of the divergence theorem, reducing by one the order of the differential equations. Equivalently, macroscopic balance equations are written on each cell. *Finite element* methods are weighted residual techniques in which the unknown dependent variables are expressed in terms of **basis functions** interpolating among the nodal values. The basis functions are substituted into the equations of motion, resulting in error residuals which are multiplied by the weighting functions, integrated over the control volume, and set to zero to produce algebraic equations in terms of the nodal unknowns. Selection of the weighting functions defines the various finite element methods. For example, Galerkin's method uses the nodal interpolation basis functions as weighting functions. Each method also has its own method for implementing **boundary conditions.** The end result after discretization of the equations and application of the boundary conditions is a set of algebraic equations for the nodal unknown variables. Discretization in time is also required for the $\partial/\partial t$ time derivative terms in unsteady flow; finite differencing in time is often used. The discretized equations represent an approximation of the exact equations, and their solution gives an approximation for the flow variables. The accuracy of the solution improves as the grid is **refined;** that is, as the number of nodal points is increased.

3. **Solution of the algebraic equations.** For creeping flows, the algebraic equations are linear and a linear matrix equation is to be solved. Both direct and iterative solvers have been used. For most flows, the nonlinear inertial terms in the momentum equation are important and the algebraic discretized equations are therefore nonlinear. Solution yields the nodal values of the unknowns.

CFD solutions, especially for complex three-dimensional flows, generate very large quantities of solution data. Computer graphics have greatly improved the ability to examine CFD solutions and visualize flow.

CFD methods are used for incompressible and compressible, creeping, laminar and turbulent, Newtonian and non-Newtonian, and isothermal and nonisothermal flows. Chemically reacting flows, particularly in the field of combustion, have been simulated. Solution accuracy must be considered from several perspectives. These include convergence of the algorithms for solving the nonlinear discretized equations and convergence with respect to refinement of the mesh so that the discretized equations better approximate the exact equations and, in some cases, so that the mesh more accurately fits the true geometry. The possibility that steady-state solutions are unstable must always be considered. In addition to numerical sources of error, modeling errors are introduced in turbulent flow, where semiempirical closure models are used to solve time-averaged equations of motion, as discussed previously. Most commercial CFD codes include the k–ϵ turbulence model, which has been by far the most widely used. More accurate models, such as differential Reynolds stress and renormalization group theory models, are also becoming available. Significant solution error is known to result in some problems from inadequacy of the turbulence model. Closure models for nonlinear chemical reaction source terms may also contribute to inaccuracy. **Direct numerical simulation** and **large eddy simulation,** which involve solutions for velocity fluctuations, under limited conditions or with certain modeling assumptions, remain primarily research areas.

In its general sense, multiphase flow is not currently solvable by computational fluid dynamics. However, in certain cases reasonable solutions are possible. These include well-separated flows where the phases are confined to relatively well-defined regions separated by one or a few interfaces and flows in which a second phase appears as discrete particles of known size and shape whose motion may be approximately computed with drag coefficient formulations, or rigorously computed with refined meshes applying boundary conditions at the particle surface. **Two-fluid modeling,** in which the phases are treated as overlapping continua, with each phase occupying a volume fraction that is a continuous function of position (and time) is a useful approximation which is becoming available in commercial software. See Elghobashi and Abou-Arab (*J. Physics Fluids*, **26,** 931–938 [1983]) for a k–ϵ model for two-fluid systems.

Figure 6-56 gives an example CFD calculation for time-dependent flow past a square cylinder at a Reynolds number of 22,000 (Choudhury, et al., *Trans. ASME Fluids Div.*, Lake Tahoe, Nev. [1994]). The computation was done with an implementation of the renormalization group theory k–ϵ model. The series of contour plots of stream function shows a sequence in time over about 1 vortex-shedding period. The calculated Strouhal number (Eq. [6-195]) is 0.146, in excellent agreement with experiment, as is the time-averaged drag coefficient, $C_D = 2.24$. Similar computations for a circular cylinder at Re = 14,500 have given excellent agreement with experimental measurements for St and C_D (*Introduction to the Renormalization Group Method and Turbulence Modeling*, Fluent, Inc., 1993).

DIMENSIONLESS GROUPS

For purposes of data correlation, model studies, and scale-up, it is useful to arrange variables into dimensionless groups. Table 6-7 lists many of the dimensionless groups commonly found in fluid mechanics problems, along with their physical interpretations and areas of application. More extensive tabulations may be found in Catchpole and Fulford (*Ind. Eng. Chem.*, **58**[3], 46–60 [1966]) and Fulford and Catchpole (*Ind. Eng. Chem.*, **60**[3], 71–78 [1968]).

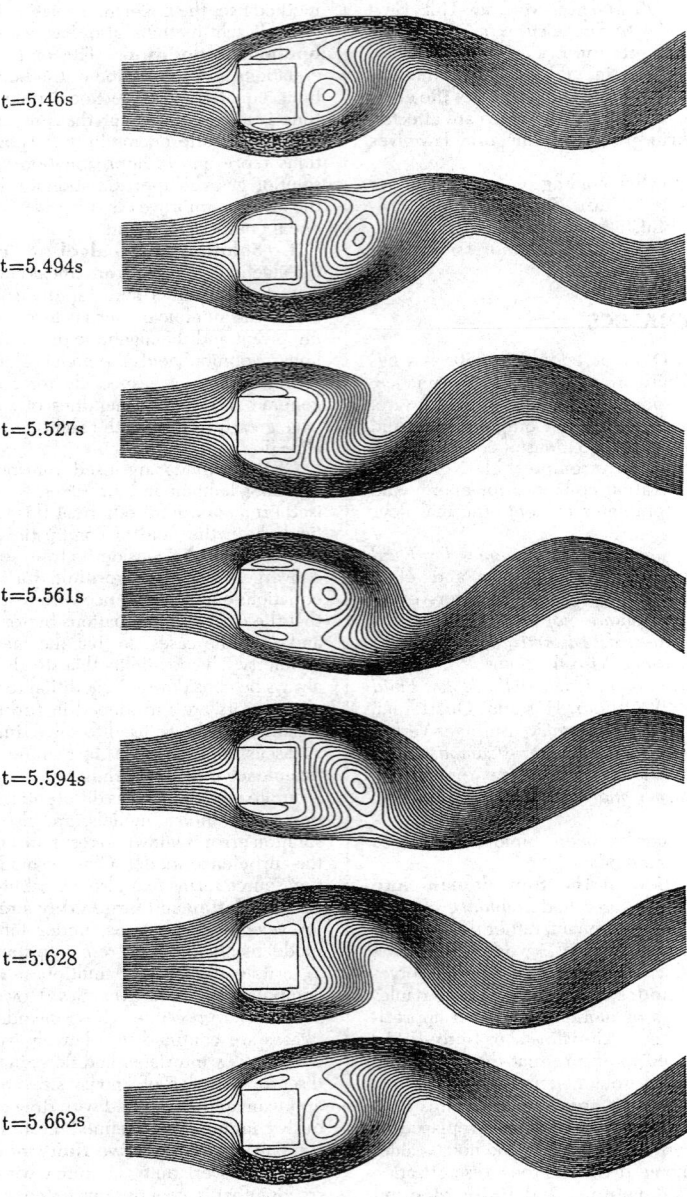

t=5.46s

t=5.494s

t=5.527s

t=5.561s

t=5.594s

t=5.628

t=5.662s

FIG. 6-56 Computational fluid dynamic simulation of flow over a square cylinder, showing one vortex shedding period. (*From Choudhury, et al.*, Trans. ASME Fluids Div., *TN-076 [1994]*.)

TABLE 6-7 Dimensionless Groups and Their Significance

Name	Symbol	Formula	Physical interpretation	Comments
Archimedes number	Ar	$\dfrac{gL^3(\rho_p - \rho)\rho}{\mu^2}$	$\dfrac{\text{inertial forces} \times \text{buoyancy forces}}{(\text{viscous forces})^2}$	Particle settling
Bingham number	Bm	$\dfrac{\tau_y L}{\mu_\infty V}$	$\dfrac{\text{yield stress}}{\text{viscous stress}}$	Flow of Bingham plastics = yield number, Y
Bingham Reynolds number	Re_B	$\dfrac{LV\rho}{\mu_\infty}$	$\dfrac{\text{inertial force}}{\text{viscous force}}$	Flow of Bingham plastics
Blake number	B	$\dfrac{V\rho}{\mu(1 - \epsilon)s}$	$\dfrac{\text{inertial force}}{\text{viscous force}}$	Beds of solids
Bond number	Bo	$\dfrac{(\rho_L - \rho_G)L^2 g}{\sigma}$	$\dfrac{\text{gravitational force}}{\text{surface-tension force}}$	Atomization = Eotvos number, Eo
Capillary number	Ca	$\dfrac{\mu V}{\sigma}$	$\dfrac{\text{viscous force}}{\text{surface-tension force}}$	Two-phase flows, free surface flows
Cauchy number	C	$\dfrac{\rho V^2}{\beta}$	$\dfrac{\text{inertial force}}{\text{compressibility force}}$	Compressible flow, hydraulic transients
Cavitation number	σ	$\dfrac{p - p_v}{\rho V^2/2}$	$\dfrac{\text{excess pressure above vapor pressure}}{\text{velocity head}}$	Cavitation
Dean number	D_e	$\dfrac{Re}{(Dc/D)^{1/2}}$	$\text{Reynolds number} \times \dfrac{\text{inertial force}}{\text{centrifugal force}}$	Flow in curved channels
Deborah number	De	$\lambda\omega$	$\dfrac{\text{fluid relaxation time}}{\text{flow characteristic time}}$	Viscoelastic flow
Drag coefficient	C_D	$\dfrac{F_D}{A\rho V^2/2}$	$\dfrac{\text{drag force}}{\text{projected area} \times \text{velocity head}}$	Flow around objects, particle settling
Elasticity number	El	$\dfrac{\lambda\mu}{\rho L^2}$	$\dfrac{\text{elastic force}}{\text{inertial force}}$	Viscoelastic flow
Euler number	Eu	$\dfrac{\Delta p}{\rho V^2}$	$\dfrac{\text{frictional pressure loss}}{2 \times \text{velocity head}}$	Fluid friction in conduits
Fanning friction factor	f	$\dfrac{D\Delta p}{2\rho V^2 L} = \dfrac{2\tau_w}{\rho V^2}$	$\dfrac{\text{wall shear stress}}{\text{velocity head}}$	Fluid friction in conduits Darcy friction factor = $4f$
Froude number	Fr	$\dfrac{V^2}{gL}$	$\dfrac{\text{inertial force}}{\text{gravity force}}$	Often defined as $\text{Fr} = V/\sqrt{gL}$
Densometric Froude number	Fr′	$\dfrac{\rho V^2}{(\rho_d - \rho)gL}$	$\dfrac{\text{inertial force}}{\text{gravity force}}$	or $\text{Fr}' = \dfrac{V}{\sqrt{(\rho_d - \rho)gL/\rho}}$
Hedstrom number	He	$\dfrac{L^2\tau_y\rho}{\mu_\infty^2}$	Bingham Reynolds number $\times$ Bingham number	Flow of Bingham plastics
Hodgson number	H	$\dfrac{V'\omega\Delta p}{\bar{q}\bar{p}}$	$\dfrac{\text{time constant of system}}{\text{period of pulsation}}$	Pulsating gas flow
Mach number	M	$\dfrac{V}{c}$	$\dfrac{\text{fluid velocity}}{\text{sonic velocity}}$	Compressible flow
Newton number				
Ohnesorge number	Z	$\dfrac{\mu}{(\rho L\sigma)^{1/2}}$	$\dfrac{\text{viscous force}}{(\text{inertial force} \times \text{surface tension force})^{1/2}}$	Atomization = $\dfrac{\text{Weber number}}{\text{Reynolds number}}$
Peclet number	Pe	$\dfrac{LV}{D}$	$\dfrac{\text{convective transport}}{\text{diffusive transport}}$	Heat, mass transfer, mixing
Pipeline parameter	Pn	$\dfrac{aV_o}{2gH}$	$\dfrac{\text{maximum water-hammer pressure rise}}{2 \times \text{static pressure}}$	Water hammer

TABLE 6-7 Dimensionless Groups and Their Significance (*Concluded*)

Name	Symbol	Formula	Physical interpretation	Comments
Power number	Po	$\dfrac{P}{\rho N^3 L^5}$	$\dfrac{\text{impeller drag force}}{\text{inertial force}}$	Agitation
Prandtl velocity ratio	v^+	$\dfrac{v}{(\tau_w/\rho)^{1/2}}$	velocity normalized by friction velocity	Turbulent flow near a wall, friction velocity $= \sqrt{\tau_w/\rho}$
Reynolds number	Re	$\dfrac{LV\rho}{\mu}$	$\dfrac{\text{inertial force}}{\text{viscous force}}$	
Strouhal number	St	$\dfrac{f'L}{V}$	vortex shedding frequency × characteristic flow time scale	Vortex shedding, von Karman vortex streets
Weber number	We	$\dfrac{\rho V^2 L}{\sigma}$	$\dfrac{\text{inertial force}}{\text{surface tension force}}$	Bubble, drop formation

Nomenclature		SI Units	Nomenclature		SI Units
a	Wave speed	m/s	P	Power	Watts
A	Projected area	m	$\overline{q}$	Average volumetric flow rate	m³/s
c	Sonic velocity	m/s	s	Particle area/particle volume	1/m
D	Diameter of pipe	m	v	Local fluid velocity	m/s
D_c	Diameter of curvature	m	V	Characteristic or average fluid velocity	m/s
D'	Diffusivity	m²/s	V'	System volume	m³
f'	Vortex shedding frequency	1/s	β	Bulk modulus	Pa
F_D	Drag force	N	ϵ	Void fraction	m³
g	Acceleration of gravity	m/s	λ	Fluid relaxation time	s
H	Static head	m	μ	Fluid viscosity	Pa · s
L	Characteristic length	m	μ_∞	Infinite shear viscosity (Bingham plastics)	Pa · s
N	Rotational speed	1/s	ρ	Fluid density	kg/m³
p	Pressure	Pa	ρ_G, ρ_L	Gas, liquid densities	kg/m³
p_v	Vapor pressure	Pa	ρ_d	Dispersed phase density	kg/m³
$\overline{p}$	Average static pressure	Pa	σ	Surface tension	N/m
Δp	Frictional pressure drop	Pa	ω	Characteristic frequency or reciprocal time scale of flow	1/s

PARTICLE DYNAMICS

GENERAL REFERENCES: Brodkey, *The Phenomena of Fluid Motions*, Addison-Wesley, Reading, Mass., 1967; Clift, Grace, and Weber, *Bubbles, Drops and Particles*, Academic New York, 1978; Govier and Aziz, *The Flow of Complex Mixtures in Pipes*, Van Nostrand Reinhold, New York, 1972, Krieger, Huntington, N.Y., 1977; Lapple, et al., *Fluid and Particle Mechanics*, University of Delaware, Newark, 1951; Levich, *Physicochemical Hydrodynamics*, Prentice-Hall, Englewood Cliffs, N.J., 1962; Orr, *Particulate Technology*, Macmillan, New York, 1966; Shook and Roco, *Slurry Flow*, Butterworth-Heinemann, Boston, 1991; Wallis, *One-dimensional Two-phase Flow*, McGraw-Hill, New York, 1969.

DRAG COEFFICIENT

Whenever relative motion exists between a particle and a surrounding fluid, the fluid will exert a drag upon the particle. In steady flow, the drag force on the particle is

$$F_D = \frac{C_D A_P \rho u^2}{2} \qquad (6\text{-}227)$$

where F_D = drag force
C_D = drag coefficient
A_P = projected particle area in direction of motion
ρ = density of surrounding fluid
u = relative velocity between particle and fluid

The drag force is exerted in a direction parallel to the fluid velocity. Equation (6-227) defines the **drag coefficient.** For some solid bodies, such as aerofoils, a lift force component perpendicular to the liquid velocity is also exerted. For free-falling particles, lift forces are generally not important. However, even spherical particles experience lift forces in shear flows near solid surfaces.

TERMINAL SETTLING VELOCITY

A particle falling under the action of gravity will accelerate until the drag force balances gravitational force, after which it falls at a constant **terminal** or **free-settling velocity** u_t, given by

$$u_t = \sqrt{\frac{2 g m_p (\rho_p - \rho)}{\rho \rho_p A_P C_D}} \qquad (6\text{-}228)$$

where g = acceleration of gravity
m_p = particle mass
ρ_p = particle density

and the remaining symbols are as previously defined.

Settling particles may undergo fluctuating motions owing to vortex shedding, among other factors. Oscillation is enhanced with increasing separation between the mass and geometric centers of the particle. Variations in mean velocity are usually less than 10 percent. The drag force on a particle fixed in space with fluid moving is somewhat lower than the drag force on a particle freely settling in a stationary fluid at the same relative velocity.

Spherical Particles   For spherical particles of diameter d_p, Eq. (6-228) becomes

$$u_t = \sqrt{\frac{4 g d_p (\rho_p - \rho)}{3 \rho C_D}} \qquad (6\text{-}229)$$

The drag coefficient for rigid spherical particles is a function of particle Reynolds number, $\text{Re}_p = d_p \rho u / \mu$ where μ = fluid viscosity, as shown in Fig. 6-57. At low Reynolds number, **Stokes' Law** gives

$$C_D = \frac{24}{\text{Re}_p} \qquad \text{Re}_p < 0.1 \qquad (6\text{-}230)$$

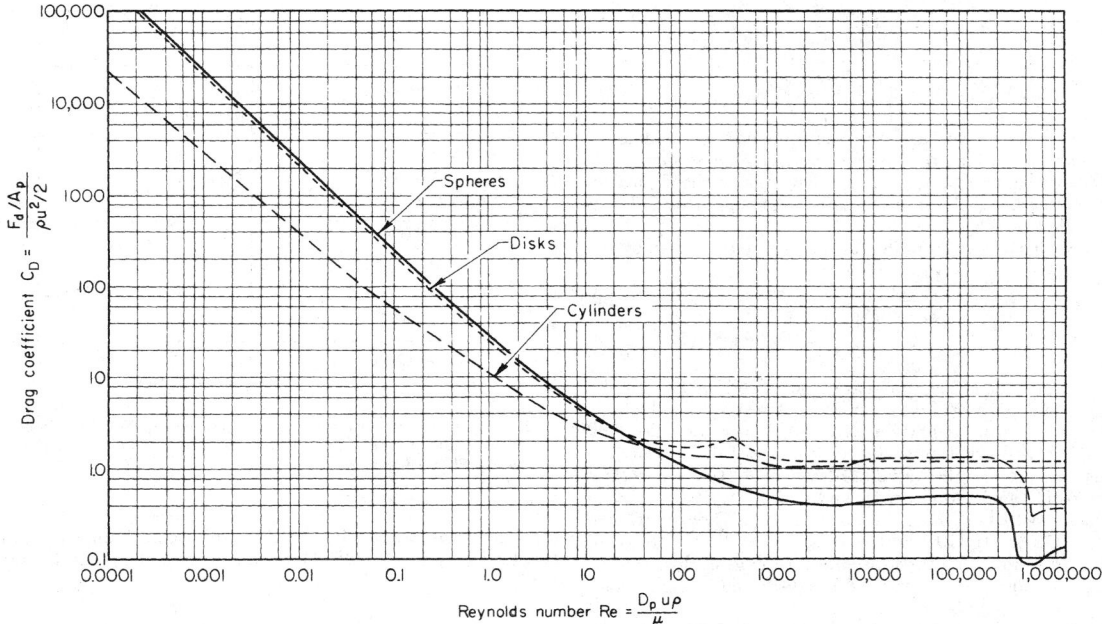

FIG. 6-57 Drag coefficients for spheres, disks, and cylinders: A_p = area of particle projected on a plane normal to direction of motion; C = overall drag coefficient, dimensionless; D_p = diameter of particle; F_d = drag or resistance to motion of body in fluid; Re = Reynolds number, dimensionless; u = relative velocity between particle and main body of fluid; μ = fluid viscosity; and ρ = fluid density. (*From Lapple and Shepherd,* Ind. Eng. Chem., **32**, *605 [1940].*)

which may also be written

$$F_D = 3\pi\mu u d_p \qquad \mathrm{Re}_p < 0.1 \qquad (6\text{-}231)$$

and gives for the terminal settling velocity

$$u_t = \frac{g d_p^2(\rho_p - \rho)}{18\mu} \qquad \mathrm{Re}_p < 0.1 \qquad (6\text{-}232)$$

In the **intermediate regime** ($0.1 < \mathrm{Re}_p < 1{,}000$), the drag coefficient may be estimated within 6 percent by

$$C_D = \left(\frac{24}{\mathrm{Re}_p}\right)\left(1 + 0.14\mathrm{Re}_p^{0.70}\right) \qquad 0.1 < \mathrm{Re}_p < 1{,}000 \qquad (6\text{-}233)$$

In the **Newton's Law** regime, which covers the range $1{,}000 < \mathrm{Re}_p < 350{,}000$, $C_D = 0.445$, within 13 percent. In this region, Eq. (6-227) becomes

$$u_t = 1.73 \sqrt{\frac{g d_p(\rho_p - \rho)}{\rho}} \qquad 1{,}000 < \mathrm{Re}_p < 350{,}000 \qquad (6\text{-}234)$$

Between about $\mathrm{Re}_p = 350{,}000$ and 1×10^6, the drag coefficient drops dramatically in a **drag crisis** owing to the transition to turbulent flow in the boundary layer around the particle, which delays aft separation, resulting in a smaller wake and less drag. Beyond $\mathrm{Re} = 1 \times 10^6$, the drag coefficient may be estimated from (Clift, Grace, and Weber):

$$C_D = 0.19 - \frac{8 \times 10^4}{\mathrm{Re}_p} \qquad \mathrm{Re}_p > 1 \times 10^6 \qquad (6\text{-}235)$$

Drag coefficients may be affected by turbulence in the free-stream flow; the drag crisis occurs at lower Reynolds numbers when the free stream is turbulent. Torobin and Guvin (*AIChE J.,* **7,** 615–619 [1961]) found that the drag crisis Reynolds number decreases with increasing free-stream turbulence, reaching a value of 400 when the relative turbulence intensity, defined as $\sqrt{u'}/\overline{U}_R$ is 0.4. Here $\sqrt{u'}$ is the rms fluctuating velocity and $\overline{U}_R$ is the relative velocity between the particle and the fluid.

For particles settling in **non-Newtonian** fluids, correlations are

given by Dallon and Christiansen (Preprint 24C, *Symposium on Selected Papers,* part III, 61st Ann. Mtg. AIChE, Los Angeles, Dec. 1–5, 1968) for spheres settling in shear-thinning liquids, and by Ito and Kajiuchi (*J. Chem. Eng. Japan,* **2**[1], 19–24 [1969]) and Pazwash and Robertson (*J. Hydraul. Res.,* **13,** 35–55 [1975]) for spheres settling in Bingham plastics. Beris, Tsamopoulos, Armstrong, and Brown (*J. Fluid Mech.,* **158** [1985]) present a finite element calculation for creeping motion of a sphere through a Bingham plastic.

Nonspherical Rigid Particles The drag on a nonspherical particle depends upon its shape and orientation with respect to the direction of motion. The orientation in free fall as a function of Reynolds number is given in Table 6-8.

The drag coefficients for **disks** (flat side perpendicular to the direction of motion) and for **cylinders** (infinite length with axis perpendicular to the direction of motion) are given in Fig. 6-57 as a function of Reynolds number. The effect of length-to-diameter ratio for cylinders in the Newton's law region is reported by Knudsen and Katz (*Fluid Mechanics and Heat Transfer,* McGraw-Hill, New York, 1958).

Pettyjohn and Christiansen (*Chem. Eng. Prog.,* **44,** 157–172 [1948]) present correlations for the effect of particle shape on free-settling velocities of **isometric particles.** For Re < 0.05, the terminal or free-settling velocity is given by

TABLE 6-8   Free-Fall Orientation of Particles

Reynolds number[*]	Orientation
0.1–5.5	All orientations are stable when there are three or more perpendicular axes of symmetry.
5.5–200	Stable in position of maximum drag.
200–500	Unpredictable. Disks and plates tend to wobble, while fuller bluff bodies tend to rotate.
500–200,000	Rotation about axis of least inertia, frequently coupled with spiral translation.

SOURCE: From Becker, *Can. J. Chem. Eng.,* **37,** 85–91 (1959).
[*]Based on diameter of a sphere having the same surface area as the particle.

$$u_t = K_1 \frac{gd_s^2(\rho_p - \rho)}{18\mu} \tag{6-236}$$

$$K_1 = 0.843 \log\left(\frac{\psi}{0.065}\right) \tag{6-237}$$

where ψ = sphericity, the surface area of a sphere having the same volume as the particle, divided by the actual surface area of the particle; d_s = equivalent diameter, equal to the diameter of the equivalent sphere having the same volume as the particle; and other variables are as previously defined.

In the **Newton's law region,** the terminal velocity is given by

$$u_t = \sqrt{\frac{4d_s(\rho_p - \rho)g}{3K_3\rho}} \tag{6-238}$$

$$K_3 = 5.31 - 4.88\psi \tag{6-239}$$

Equations (6-236) to (6-239) are based on experiments on cube-octahedrons, octahedrons, cubes, and tetrahedrons for which the sphericity ψ ranges from 0.906 to 0.670, respectively. See also Clift, Grace, and Weber. A graph of drag coefficient vs. Reynolds number with ψ as a parameter may be found in Brown, et al. (*Unit Operations,* Wiley, New York, 1950) and in Govier and Aziz.

For particles with $\psi < 0.67$, the correlations of Becker (*Can. J. Chem. Eng.,* **37,** 85–91 [1959]) should be used. Reference to this paper is also recommended for **intermediate region** flow. Settling characteristics of nonspherical particles are discussed by Clift, Grace, and Weber, Chaps. 4 and 6.

The terminal velocity of **axisymmetric particles in axial motion** can be computed from Bowen and Masliyah (*Can. J. Chem. Eng.,* **51,** 8–15 [1973]) for low–Reynolds number motion:

$$u_t = \frac{V'}{K_2} \frac{gD_s^2(\rho_p - \rho)}{18\mu} \tag{6-240}$$

$$K_2 = 0.244 + 1.035\Sigma - 0.712\Sigma^2 + 0.441\Sigma^3 \tag{6-241}$$

where D_s = diameter of sphere with perimeter equal to maximum particle projected perimeter
   V' = ratio of particle volume to volume of sphere with diameter D_s
   Σ = ratio of surface area of particle to surface area of a sphere with diameter D_s

and other variables are as defined previously.

Hindered Settling   When particle concentration increases, particle settling velocities decrease because of hydrodynamic interaction between particles and the upward motion of displaced liquid. The suspension viscosity increases. Hindered settling is normally encountered in sedimentation and transport of concentrated slurries. Below 0.1 percent volumetric particle concentration, there is less than a 1 percent reduction in settling velocity. Several expressions have been given to estimate the effect of particle volume fraction on settling velocity. Maude and Whitmore (*Br. J. Appl. Phys.,* **9,** 477–482 [1958]) give, for uniformly sized spheres,

$$u_t = u_{t0}(1 - c)^n \tag{6-242}$$

where u_t = terminal settling velocity
   u_{t0} = terminal velocity of a single sphere (infinite dilution)
   c = volume fraction solid in the suspension
   n = function of Reynolds number $Re_p = d_p u_{t0}\rho/\mu$ as given Fig. 6-58

In the Stokes' law region ($Re_p < 0.3$), $n = 4.65$ and in the Newton's law region ($Re_p > 1,000$), $n = 2.33$. Equation (6-242) may be applied to particles of any size in a polydisperse system, provided the volume fraction corresponding to all the particles is used in computing terminal velocity (Richardson and Shabi, *Trans. Inst. Chem. Eng.* [*London*], **38,** 33–42 [1960]). The concentration effect is greater for nonspherical and angular particles than for spherical particles (Steinour, *Ind. Eng. Chem.,* **36,** 840–847 [1944]). Theoretical developments for low–Reynolds number flow assemblages of spheres are given by Happel and Brenner (*Low Reynolds Number Hydrodynamics,* Prentice-

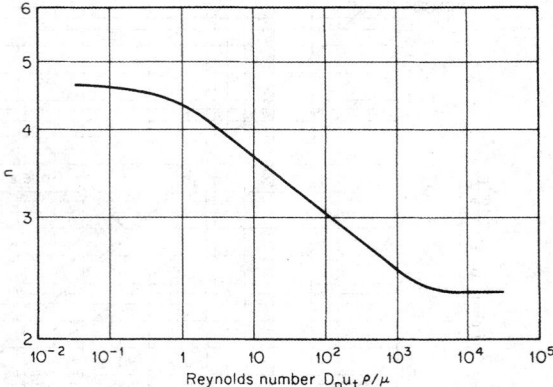

FIG. 6-58   Values of exponent n for use in Eq. (6-240). (*From Maude and Whitmore,* Br. J. Appl. Phys., **9,** 481 [1958]. *Courtesy of the Institute of Physics and the Physical Society.*)

Hall, Englewood Cliffs, N.J., 1965) and Famularo and Happel (*AIChE J.,* **11,** 981 [1965]) leading to an equation of the form

$$u_t = \frac{u_{t0}}{1 + \gamma c^{1/3}} \tag{6-243}$$

where γ is about 1.3. As particle concentration increases, resulting in interparticle contact, hindered settling velocities are difficult to predict. Thomas (*AIChE J.,* **9,** 310 [1963]) provides an empirical expression reported to be valid over the range $0.08 < u_t/u_{t0} < 1$:

$$\ln\left(\frac{u_t}{u_{t0}}\right) = -5.9c \tag{6-244}$$

Time-Dependent Motion   The time-dependent motion of particles is computed by application of Newton's second law, equating the rate of change of particle motion to the net force acting on the particle. Rotation of particles may also be computed from the net torque. For large particles moving through low-density gases, it is usually sufficient to compute the force due to fluid drag from the relative velocity and the drag coefficient computed for steady flow conditions. For two- and three-dimensional problems, the velocity appearing in the particle Reynolds number and the drag coefficient is the amplitude of the relative velocity. The drag force, not the relative velocity, is to be resolved into vector components to compute the particle acceleration components. Clift, Grace, and Weber (*Bubbles, Drops and Particles,* Academic, London, 1978) discuss the complexities that arise in the computation of transient drag forces on particles when the transient nature of the flow is important. Analytical solutions for the case of a single particle in creeping flow ($Re_p = 0$) are available. For example, the creeping motion of a spherical particle released from rest in a stagnant fluid is described by

$$\rho_p V \frac{dU}{dt} = g(\rho_p - \rho)V - 3\pi\mu d_p U - \frac{\rho}{2} V \frac{dU}{dt}$$

$$- \left(\frac{3}{2}\right) d_p^2 \sqrt{\pi\rho\mu} \int_0^t \frac{(dU/dt)_{t=s}\, ds}{\sqrt{t - s}} \tag{6-245}$$

Here, U = particle velocity, positive in the direction of gravity, and V = particle volume. The first term on the right-hand side is the net gravitational force on the particle, accounting for buoyancy. The second is the steady-state Stokes drag (Eq. 6-231). The third is the **added mass** or **virtual mass** term, which may be interpreted as the inertial effect of the fluid which is accelerated along with the particle. The volume of the added mass of fluid is half the particle volume. The last term, the **Basset force,** depends on the entire history of the transient motion, with past motions weighted inversely with the square root of elapsed time. Clift, et al. provide integrated solutions. In **turbulent flows,** particle velocity will closely follow fluid eddy velocities when (Clift et al.)

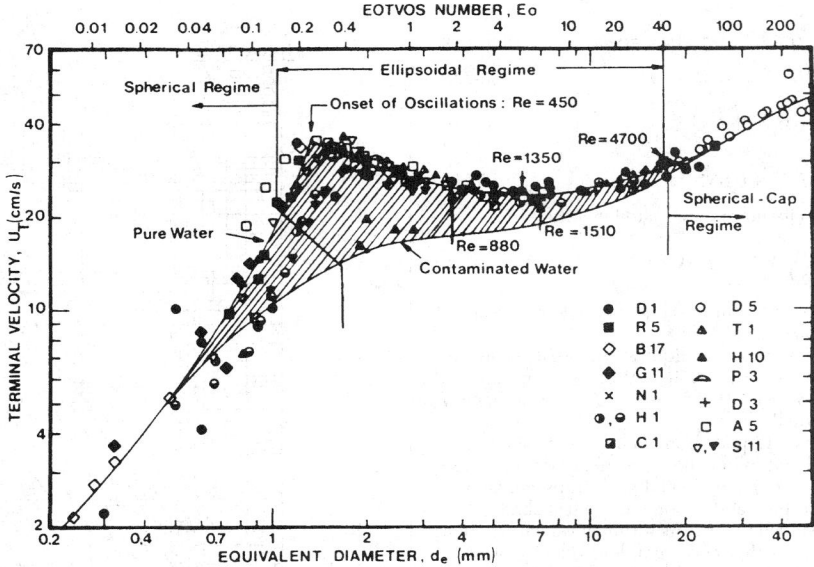

FIG. 6-59 Terminal velocity of air bubbles in water at 20°C. (*From Clift, Grace, and Weber,* Bubbles, Drops and Particles, *Academic, New York, 1978*).

$$\tau_0 \gg \frac{d_p^2[(2\rho_p/\rho) + 1]}{36\nu} \tag{6-246}$$

where τ_0 = oscillation period or eddy time scale, the right-hand side expression is the **particle relaxation time,** and ν = kinematic viscosity.

Gas Bubbles Fluid particles, unlike rigid solid particles, may undergo deformation and internal circulation. Figure 6-59 shows rise velocity data for air bubbles in stagnant water. In the figure, Eo = Eotvos number, $g(\rho_L - \rho_G)d_b^2/\sigma$, where ρ_L = liquid density, ρ_G = gas density, d_b = bubble diameter, and σ = surface tension. Small bubbles (<1-mm [0.04-in] diameter) remain spherical and rise in straight lines. The presence of surface active materials generally renders small bubbles rigid, and they rise roughly according to the drag coefficient and terminal velocity equations for spherical solid particles. Bubbles roughly in the range 2- to 8-mm (0.079- to 0.32-in) diameter assume flattened, ellipsoidal shape, and rise in a zig-zag or spiral pattern. This motion increases dissipation and drag, and the rise velocity may actually decrease with increasing bubble diameter in this region, characterized by rise velocities in the range of 20 to 30 cm/s (0.7 to 1.0 ft/s). Large bubbles, >8-mm (0.32-in) diameter, are greatly deformed, assuming a mushroomlike, spherical cap shape. These bubbles are unstable and may break into smaller bubbles. Carefully purified water, free of surface active materials, allows bubbles to freely circulate even when they are quite small. Under creeping flow conditions $Re_b = d_b u_r \rho_L/\mu_L < 1$, where u_r = bubble rise velocity and μ_L = liquid viscosity, the bubble rise velocity may be computed analytically from the Hadamard-Rybczynski formula (Levich, *Physicochemical Hydrodynamics,* Prentice-Hall, Englewood Cliffs, N.J., 1962, p. 402). When $\mu_G/\mu_L \ll 1$, which is normally the case, the rise velocity is 1.5 times the rigid sphere Stokes law velocity. However, in practice, most liquids, including ordinary distilled water, contain sufficient surface active materials to render small bubbles rigid. Larger bubbles undergo deformation in both purified and ordinary liquids; however, the variation in rise velocity for large bubbles with degree of purity is quite evident in Fig. 6-59. For additional discussion, see Clift, et al., Chap. 7. Figure 6-60 gives the drag coefficient as a function of bubble or drop Reynolds number for air bubbles in water and water drops in air, compared with the standard drag curve for rigid spheres. Information on bubble motion in **non-Newtonian** liquids may be found in Astarita and Apuzzo (*AIChE J.,* **11,** 815–820 [1965]); Calderbank, Johnson, and Loudon (*Chem. Eng. Sci.,* **25,** 235–256 [1970]); and Acharya, Mashelkar, and Ulbrecht (*Chem. Eng. Sci.,* **32,** 863–872 [1977]).

Liquid Drops in Liquids Very small liquid drops in immisicibile liquids behave like rigid spheres, and the terminal velocity can be approximated by use of the drag coefficient for solid spheres up to a Reynolds number of about 10 (Warshay, Bogusz, Johnson, and Kintner, *Can. J. Chem. Eng.,* **37,** 29–36 [1959]). Between Reynolds numbers of 10 and 500, the terminal velocity exceeds that for rigid spheres owing to internal circulation. In normal practice, the effect of drop phase viscosity is neglected. Grace, Wairegi, and Nguyen (*Trans. Inst. Chem. Eng.,* **54,** 167–173 [1976]; Clift, et al., op. cit., pp. 175–177) present a correlation for terminal velocity valid in the range

$$M < 10^{-3} \qquad Eo < 40 \qquad Re > 0.1 \tag{6-247}$$

where M = Morton number = $g\mu^4\Delta\rho/\rho^2\sigma^3$
  Eo = Eotvos number = $g\Delta\rho d^2/\sigma$
  Re = Reynolds number = $du\rho/\mu$
  $\Delta\rho$ = density difference between the phases
  ρ = density of continuous liquid phase
  d = drop diameter
  μ = continuous liquid viscosity
  σ = surface tension
  u = relative velocity

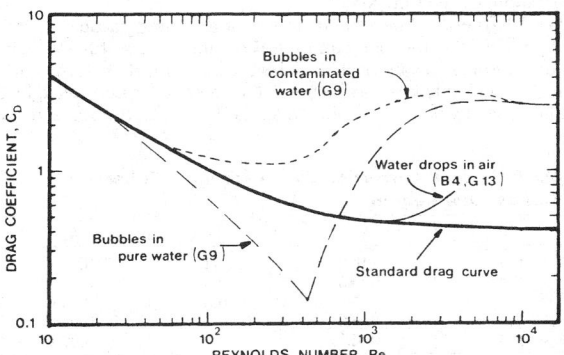

FIG. 6-60 Drag coefficient for water drops in air and air bubbles in water. Standard drag curve is for rigid spheres. (*From Clift, Grace, and Weber,* Bubbles, Drops and Particles, *Academic, New York, 1978.*)

The correlation is represented by

$$J = 0.94H^{0.757} \qquad (2 < H \le 59.3) \qquad (6\text{-}248)$$

$$J = 3.42H^{0.441} \qquad (H > 59.3) \qquad (6\text{-}249)$$

where

$$H = \frac{4}{3}\,\text{Eo}M^{-0.149}\left(\frac{\mu}{\mu_w}\right)^{-0.14} \qquad (6\text{-}250)$$

$$J = \text{Re}M^{0.149} + 0.857 \qquad (6\text{-}251)$$

Note that the terminal velocity may be evaluated explicitly from

$$u = \frac{\mu}{\rho d}\,M^{-0.149}(J - 0.857) \qquad (6\text{-}252)$$

In Eq. (6-250), μ = viscosity of continuous liquid and μ_w = viscosity of water, taken as 0.9 cP (0.0009 Pa · s).

For drop velocities in non-Newtonian liquids, see Mhatre and Kinter (*Ind. Eng. Chem.*, **51**, 865–867 [1959]); Marrucci, Apuzzo, and Astarita (*AIChE J.*, **16**, 538–541 [1970]); and Mohan, et al. (*Can. J. Chem. Eng.*, **50**, 37–40 [1972]).

Liquid Drops in Gases Liquid drops falling in stagnant gases appear to remain spherical and follow the rigid sphere drag relationships up to a Reynolds number of about 100. Large drops will deform, with a resulting increase in drag, and in some cases will shatter. The largest water drop which will fall in air at its terminal velocity is about 8 mm (0.32 in) in diameter, with a corresponding velocity of about 9 m/s (30 ft/s). Drops shatter when the Weber number defined as

$$\text{We} = \frac{\rho_G u^2 d}{\sigma} \qquad (6\text{-}253)$$

exceeds a critical value. Here, ρ_G = gas density, u = drop velocity, d = drop diameter, and σ = surface tension. A value of $\text{We}_c = 13$ is often cited for the critical Weber number.

Terminal velocities for water drops in air have been correlated by Berry and Prnager (*J. Appl. Meteorol.*, **13**, 108–113 [1974]) as

$$\text{Re} = \exp\left[-3.126 + 1.013 \ln N_D - 0.01912(\ln N_D)^2\right] \qquad (6\text{-}254)$$

for $2.4 < N_D < 10^7$ and $0.1 < \text{Re} < 3550$. The dimensionless group N_D (often called the *Best* number [Clift, et al.]) is given by

$$N_D = \frac{4\rho\,\Delta\rho g d^3}{3\mu^2} \qquad (6\text{-}255)$$

and is proportional to the similar Archimedes and Galileo numbers.

Figure 6-61 gives calculated settling velocities for solid spherical particles settling in air or water using the standard drag coefficient curve for spherical particles. For fine particles settling in air, the **Stokes-Cunningham correction** has been applied to account for particle size comparable to the mean free path of the gas. The correction is less than 1 percent for particles larger than 16 μm settling in air. Smaller particles are also subject to **Brownian motion.** Motion of particles smaller than 0.1 μm is dominated by Brownian forces and gravitational effects are small.

Wall Effects When the diameter of a settling particle is significant compared to the diameter of the container, the settling velocity is reduced. For rigid spherical particles settling with Re < 1, the correction given in Table 6-9 may be used. The factor k_w is multiplied by the settling velocity obtained from Stokes' law to obtain the corrected set-

TABLE 6-9 Wall Correction Factor for Rigid Spheres in Stokes' Law Region

β°	k_w	β	k_w
0.0	1.000	0.4	0.279
0.05	0.885	0.5	0.170
0.1	0.792	0.6	0.0945
0.2	0.596	0.7	0.0468
0.3	0.422	0.8	0.0205

SOURCE: From Haberman and Sayre, *David W. Taylor Model Basin Report* 1143, 1958.

°β = particle diameter divided by vessel diameter.

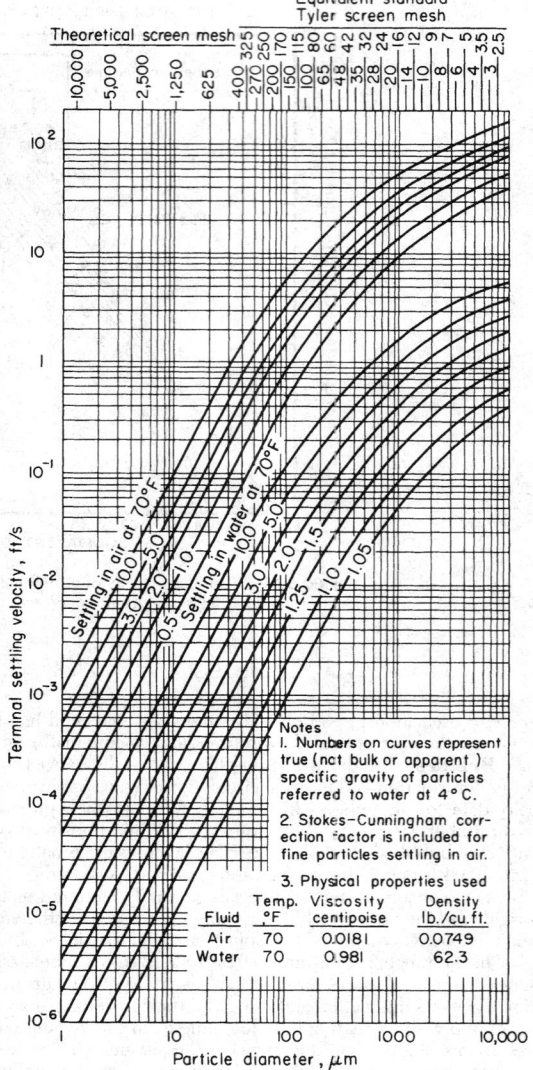

FIG. 6-61 Terminal velocities of spherical particles of different densities settling in air and water at 70°F under the action of gravity. To convert ft/s to m/s, multiply by 0.3048. (*From Lapple, et al.*, Fluid and Particle Mechanics, *University of Delaware, Newark, 1951, p. 292.*)

tling rate. For values of diameter ratio β = particle diameter/vessel diameter less than 0.05, $k_w = 1/(1 + 2.1\beta)$ (Zenz and Othmer, *Fluidization and Fluid-Particle Systems*, Reinhold, New York, 1960, pp. 208–209). In the range 100 < Re < 10,000, the computed terminal velocity for rigid spheres may be multiplied by k'_w to account for wall effects, where k'_w is given by (Harmathy, *AIChE J.*, **6**, 281 [1960])

$$k'_w = \frac{1 - \beta^2}{\sqrt{1 + \beta^4}} \qquad (6\text{-}256)$$

For gas bubbles in liquids, there is little wall effect for $\beta < 0.1$. For $\beta > 0.1$, see Uto and Kintner (*AIChE J.*, **2**, 420–424 [1956]), Maneri and Mendelson (*Chem. Eng. Prog.*, **64**, Symp. Ser., **82**, 72–80 [1968]), and Collins (*J. Fluid Mech.*, **28**, part 1, 97–112 [1967]).

Reaction Kinetics

Stanley M. Walas, Ph.D., *Professor Emeritus, Department of Chemical and Petroleum Engineering, University of Kansas; Fellow, American Institute of Chemical Engineers*

Nomenclature and Units

Following is a listing of typical nomenclature expressed in SI and U.S. customary units. Specific definitions and units are stated at the place of application in this section.

Symbol	Definition	SI units	U.S. customary units
A, B, C, . . .	Names of substances, or their concentrations		
A*	Free radical, as CH_3^*		
C_a	Concentration of substance A	kg mol/m³	lb mol/ft³
C^0	Initial mean concentration in vessel	kg mol/m³	lb mol/ft³
C_p	Heat capacity	kJ/(kg·K)	Btu/(lbm·°F)
CSTR	Continuous stirred tank reactor		
D, D_e, D_x	Dispersion coefficient	m²/s	ft²/s
D_{eff}	Effective diffusivity	m²/s	ft²/s
D_K	Knudsen diffusivity	m²/s	ft²/s
$E(t)$	Residence time distribution		
$E(t_r)$	Normalized residence time distribution		
f_a	C_a/C_{a0} or n_a/n_{a0}, fraction of A remaining unconverted		
$F(t)$	Age function of tracer		
ΔG	Gibbs energy change	kJ	Btu
Ha	Hatta number		
ΔH_r	Heat of reaction	kJ/kg mol	Btu/lb mol
K, K_e, K_y, K_ϕ	Chemical equilibrium constant		
k, k_c, k_p	Specific rate of reaction	Variable	Variable
L	Length of path in reactor	m	ft
n	Parameter of Erlang or Gamma distribution, or number of stages in a CSTR battery		
n_a	Number of mols of A present		
n_a'	Number of mols flowing per unit time; the prime (′) may be omitted when context is clear		
n_t	Total number of mols		
p_a	Partial pressure of substance A	kPₐ	psi
Pe	Peclet number for dispersion		
PFR	Plug flow reactor		
Q	Heat transfer	kJ	Btu
r	Radial position	m	ft
r_a	Rate of reaction of A per unit volume	Variable	Variable
R	Radius of cylindrical vessel	m	ft
Re	Reynolds number		
Sc	Schmidt number		

Symbol	Definition	SI units	U.S. customary units
t	Time	s	s
$\bar{t}$	Mean residence time	s	s
t_r	$t/\bar{t}$, reduced time		
TFR	Tubular flow reactor		
u	Linear velocity	m/s	ft/s
$u(t)$	Unit step input		
V	Volume of reactor contents	m³	ft³
V′	Volumetric flow rate	m³/s	ft³/s
V_r	Volume of reactor	m³	ft³
x	Axial position in a reactor	m	ft
x_a	$1 - f_a = 1 - C_a/C_{a0}$ or $1 - n_a/n_{a0}$, fraction of A converted		
z	x/L, normalized axial position		
	Greek letters		
β	r/R, normalized radial position		
$\gamma^3(t)$	Skewness of distribution		
$\delta(t)$	Unit impulse input, Dirac function		
ε	Fraction void space in a packed bed		
θ	$t/\bar{t}$, reduced time, fraction of surface covered by adsorbed species		
η	Effectiveness of porous catalyst		
$\Lambda(t)$	Intensity function		
μ	Viscosity	Pa·s	lbm/(ft·s)
ν	υ/ρ, kinematic viscosity	m²/s	ft²/s
π	Total pressure	Pa	psi
ρ	Density	kg/m³	lbm/ft³
ρ	r/R, normalized radial position in a pore		
$\sigma^2(t)$	Variance		
$\sigma^2(t_r)$	Normalized variance		
τ	$t/\bar{t}$, reduced time		
τ	Tortuosity		
φ	Thiele modulus		
ϕ_m	Modified Thiele modulus		
	Subscripts		
0	Subscript designating initial or inlet conditions, as in $C_{a0}, n_{a0}, V_0', \ldots$		

GENERAL REFERENCES
1. Aris, *Elementary Chemical Reactor Analysis*, Prentice-Hall, 1969.
2. Bamford and Tipper (eds.), *Comprehensive Chemical Kinetics*, Elsevier, 1969–date.
3. Boudart, *Kinetics of Chemical Processes*, Prentice-Hall, 1968.
4. Brotz, *Fundamentals of Chemical Reaction Engineering*, Addison-Wesley, 1965.
5. Butt, *Reaction Kinetics and Reactor Design*, Prentice-Hall, 1980.
6. Capello and Bielski, *Kinetic Systems: Mathematical Description of Kinetics in Solution*, Wiley, 1972.
7. Carberry, *Chemical and Catalytic Reaction Engineering*, McGraw-Hill, 1976.
8. Carberry and Varma (eds.), *Chemical Reaction and Reactor Engineering*, Dekker, 1987.
9. Chen, *Process Reactor Design*, Allyn & Bacon, 1983.
10. Cooper and Jeffreys, *Chemical Kinetics and Reactor Design*, Prentice-Hall, 1971.
11. Cremer and Watkins (eds.), *Chemical Engineering Practice*, vol. 8: *Chemical Kinetics*, Butterworths, 1965.
12. Denbigh and Turner, *Chemical Reactor Theory*, Cambridge, 1971.
13. Fogler, *Elements of Chemical Reaction Engineering*, Prentice-Hall, 1992.
14. Froment and Bischoff, *Chemical Reactor Analysis and Design*, Wiley, 1990.
15. Hill, *An Introduction to Chemical Engineering Kinetics and Reactor Design*, Wiley, 1977.
16. Holland and Anthony, *Fundamentals of Chemical Reaction Engineering*, Prentice-Hall, 1989.
17. Horak and Pasek, *Design of Industrial Chemical Reactors from Laboratory Data*, Heyden, 1978.
18. Kafarov, *Cybernetic Methods in Chemistry and Chemical Engineering*, Mir Publishers, 1976.
19. Laidler, *Chemical Kinetics*, Harper & Row, 1987.
20. Levenspiel, *Chemical Reaction Engineering*, Wiley, 1972.
21. Lewis (ed.), *Techniques of Chemistry*, vol. 4: *Investigation of Rates and Mechanisms of Reactions*, Wiley, 1974.
22. Naumann, *Chemical Reactor Design*, Wiley, 1987.
23. Panchenkov and Lebedev, *Chemical Kinetics and Catalysis*, Mir Publishers, 1976.
24. Petersen, *Chemical Reaction Analysis*, Prentice-Hall, 1965.
25. Rase, *Chemical Reactor Design for Process Plants: Principles and Case Studies*, Wiley, 1977.
26. Rose, *Chemical Reactor Design in Practice*, Elsevier, 1981.
27. Smith, *Chemical Engineering Kinetics*, McGraw-Hill, 1981.
28. Steinfeld, Francisco, and Hasse, *Chemical Kinetics and Dynamics*, Prentice-Hall, 1989.
29. Ulrich, *Guide to Chemical Engineering Reactor Design and Kinetics*, Ulrich, 1993.
30. Walas, *Reaction Kinetics for Chemical Engineers*, McGraw-Hill, 1959; reprint, Butterworths, 1989.
31. Walas, *Chemical Reaction Engineering Handbook of Solved Problems*, Gordon & Breach Publishers, 1995.
32. Westerterp, van Swaaij, and Beenackers, *Chemical Reactor Design and Operation*, Wiley, 1984.

REACTION KINETICS

INTRODUCTION

From an engineering viewpoint, reaction kinetics has these principal functions:

Establishing the chemical mechanism of a reaction
Obtaining experimental rate data
Correlating rate data by equations or other means
Designing suitable reactors

Specifying operating conditions, control methods, and auxiliary equipment to meet the technological and economic needs of the reaction process

Reactions can be classified in several ways. On the basis of *mechanism* they may be:
1. Irreversible
2. Reversible
3. Simultaneous
4. Consecutive

A further classification from the point of view of mechanism is with respect to the number of molecules participating in the reaction, the *molecularity:*
1. Unimolecular
2. Bimolecular and higher

Related to the preceding is the classification with respect to *order.* In the power law rate equation $r = kC_a^p C_b^q$, the exponent to which any particular reactant concentration is raised is called the order p or q with respect to that substance, and the sum of the exponents $p + q$ is the order of the reaction. At times the order is identical with the molecularity, but there are many reactions with experimental orders of zero or fractions or negative numbers. Complex reactions may not conform to any power law. Thus, there are reactions of:
1. Integral order
2. Nonintegral order
3. Non–power law; for instance, hyperbolic

With respect to *thermal conditions*, the principal types are:
1. Isothermal at constant volume
2. Isothermal at constant pressure
3. Adiabatic
4. Temperature regulated by heat transfer

According to the *phases* involved, reactions are:
1. Homogeneous, gaseous, liquid or solid
2. Heterogeneous:

Controlled by diffusive mass transfer
Controlled by chemical factors

A major distinction is between reactions that are:
1. Uncatalyzed
2. Catalyzed with homogeneous or solid catalysts

Equipment is also a basis for differentiation, namely:
1. Stirred tanks, single or in series
2. Tubular reactors, single or in parallel
3. Reactors filled with solid particles, inert or catalytic:

Fixed bed
Moving bed
Fluidized bed, stable or entrained

Finally, there are the *operating modes:*
1. Batch
2. Continuous flow
3. Semibatch or semiflow

Clearly, these groupings are not mutually exclusive. The chief distinctions are between homogeneous and heterogeneous reactions and between batch and flow reactions. These distinctions most influence the choice of equipment, operating conditions, and methods of design.

PRIMARY NOMENCLATURE

The participant A is identified by the subscript a. Thus, the concentration is C_a; the number of mols is n_a; the fractional conversion is x_a; the partial pressure is p_a; and the rate of decomposition is r_a. Capital letters are also used to represent concentration on occasion; thus, A instead of C_a. The flow rate in mol is n_a' but the prime (') is left off when the meaning is clear from the context. The volumetric flow rate is V'; reactor volume is V_r or simply V of batch reactors; the total pressure is π; and the temperature is T. The concentration is $C_a = n_a/V$ or n_a'/V'.

Throughout this section, equations are presented without specification of units. Use of any consistent unit set is appropriate.

SUMMARY

Basic kinetic relations of this section are summarized in Table 7-1.

TABLE 7-1   Basic Rate Equations

1. The *reference reaction* is

$$\nu_a A + \nu_b B + \cdots \rightarrow \nu_r R + \nu_s S + \cdots$$
$$\Delta\nu = \nu_r + \nu_s + \cdots - (\nu_a + \nu_b + \cdots)$$

2. *Stoichiometric balance* for any component *i*:

$$n_i = n_{i0} \pm \left(\frac{\nu_i}{\nu_a}\right)(n_{a0} - n_a)$$

$$\begin{cases} + \text{ for product (right-hand side, RHS)} \\ - \text{ for reactant (left-hand side, LHS)} \end{cases}$$

$$C_i = C_{i0} \pm \left(\frac{\nu_i}{\nu_a}\right)(C_{a0} - C_a), \qquad \text{at constant } T \text{ and } V \text{ only}$$

$$n_t = n_{t0} + \left(\frac{\Delta\nu}{\nu_a}\right)(n_{a0} - n_a)$$

3. *Law of mass action:*

$$r_a = -\frac{1}{V_r}\frac{dn_a}{dt} = kC_a^{\nu_a}C_b^{\nu_b}\cdots$$

$$= kC_a^{\nu_a}\left[C_{b0} - \left(\frac{\nu_b}{\nu_a}\right)(C_{a0} - C_a)\right]^{\nu_b}\cdots$$

$$r_a = kC_a^{\alpha}\left[C_{b0} - \left(\frac{\nu_b}{\nu_a}\right)(C_{a0} - C_a)\right]^{\beta}\cdots$$

where it is not necessarily true that $\alpha = \nu_{a'}, \beta = \nu_{b'}, \ldots$

4. At *constant volume*, $C_a = n_a/V_r$

$$kt = \int_{C_a}^{C_{a0}} \frac{1}{C_a^{\alpha}[C_{b0} - (\nu_b/\nu_a)(C_{a0} - C_a)]^{\beta}\cdots} dC_a$$

$$kt = \int_{n_a}^{n_{a0}} \frac{V_r^{-1+\alpha+\beta}}{n_a^{\alpha}[n_{b0} + (\nu_b/\nu_a)(n_{a0} - n_a)]^{\beta}\cdots} dn_a$$

Completed integrals for some values of α and β are in Table 7-4.

5. *Ideal gases at constant pressure:*

$$V_r = \frac{n_t RT}{P} = \frac{RT}{P}\left[n_{t0} + \frac{\Delta\nu}{\nu_a}(n_{a0} - n_a)\right]$$

$$r_a = kC_a^{\alpha}$$

$$kt = \left(\frac{RT}{P}\right)^{\alpha-1}\int_{n_a}^{n_{a0}} \frac{[n_{t0} + (\Delta\nu/\nu_a)(n_{a0} - n_a)]^{\alpha-1}}{n_a^{\alpha}} dn_a$$

6. *Temperature effect* on the specific rate:

$$k = k_\infty \exp\left(\frac{-E}{RT}\right) = \exp\left(a' - \frac{b'}{T}\right)$$

$E = $ energy of activation

7. *Simultaneous reactions.* The overall rate is the algebraic sum of the rates of the individual reactions. For example, take the three reactions:

$$A + B \xrightarrow{k_1} C + D \tag{1}$$
$$C + D \xrightarrow{k_2} A + B \tag{2}$$
$$A + C \xrightarrow{k_3} E \tag{3}$$

The rates are related by:

$$r_a = r_{a1} + r_{a2} + r_{a3} = k_1 C_a C_b - k_2 C_c C_d + k_3 C_a C_c$$
$$r_b = -r_d = k_1 C_a C_b - k_2 C_c C_d$$
$$r_c = k_1 C_a C_b + k_2 C_c C_d + k_3 C_a C_c$$
$$r_e = -k_3 C_a C_c$$

The number of independent rate equations is the same as the number of independent stoichiometric relations. In the present example, Reactions (1) and (2) are reversible reactions and are not independent. Accordingly, C_c and C_d, for example, can be eliminated from the equations for r_a and r_b which then become an integrable system. Usually only systems of linear differential equations with constant coefficients are solvable analytically.

8. *Mass transfer resistance:*

$C_{ai} = $ interfacial concentration of reactant A

$$r_a = -\frac{dC_a}{dt} = k_d(C_a - C_{ai}) = kC_{ai}^{\alpha} = k\left(C_a - \frac{r_a}{k_d}\right)^{\alpha}$$

$$kt = \int_{C_a}^{C_{a0}} \frac{1}{(C_a - r_a/k_d)^{\alpha}} dC_a$$

The relation between r_a and C_a must be established (numerically if need be) from the second line before the integration can be completed.

9. *Solid-catalyzed reactions.* Some Langmuir-Hinshelwood mechanisms for the reference reaction A + B → R + S (see also Tables 7.2, 7.3):

- Adsorption rate of A controlling:

$$r_a = -\frac{1}{V}\frac{dn_a}{dt} = kP_a\theta_v$$

$$\theta_v = 1\left/\left[1 + \frac{K_a}{K_e}\frac{P_rP_s}{P_b} + K_bP_b + K_rP_r + K_sP_s + K_lP_l\right]\right. \tag{1}$$

$K_e = P_rP_s/P_aP_b$ (equilibrium constant)

l is an adsorbed substance that is chemically inert.

- Surface reaction rate controlling:

$$r = kP_aP_b\theta_v^2$$

$$\theta_v = \frac{1}{1 + \sum K_jP_j}, \qquad \text{summation over all substances absorbed} \tag{2}$$

- Reaction $A_2 + B \rightarrow R + S$, with A_2 dissociated upon adsorption and with surface reaction rate controlling:

$$r_a = kP_aP_b\theta_v^3$$

$$\theta_v = \frac{1}{(1 + \sqrt{K_aP_a} + K_bP_b + \cdots)} \tag{3}$$

- At constant *P* and *T* the P_i are eliminated in favor of n_i and the total pressure by:

$$P_a = \frac{n_a}{n_t}P$$

$$P_i = \frac{n_i}{n_t}P = \frac{n_{i0} \pm (\nu_i/\nu_a)(n_{a0} - n_a)}{n_{t0} + (\Delta\nu/\nu_a)(n_{a0} - n_a)}P$$

$$\begin{cases} + \text{ for products, RHS} \\ - \text{ for reactants, LHS} \end{cases}$$

$$V = \frac{n_t RT}{P}$$

$$kt = \int_{n_a}^{n_{a0}} \frac{dn_a}{VP_aP_b\theta_v^2}, \qquad \text{for a Case (2) batch reaction} \tag{4}$$

10. A *continuously stirred tank reactor* (CSTR) battery
    Material balances:

$$n'_{a0} = n'_a + r_{a1}V_{r1}$$

$$n'_{aj-1} = n'_{aj} + r_{aj}V_{rj}, \qquad \text{for the } j\text{th stage}$$

For a first-order reaction, with $r_a = kC_a$:

$$\frac{C_{aj}}{C_{a0}} = \frac{1}{(1 + k_1\bar{t}_1)(1 + k_2\bar{t}_2)\cdots(1 + k_j\bar{t}_j)}$$

$$= \frac{1}{(1 + k\bar{t}_i)^j}$$

for *j* tanks in series with the same temperatures and residence times $\bar{t}_i = V_{ri}/V'_i$, where V' is the volumetric flow rate.

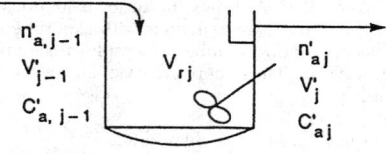

11. *Plug flow reactor* (PFR):

$$r_z = \frac{dn'_a}{dV_r} = kC_a^{\alpha}C_b^{\beta}\cdots$$

$$= k\left(\frac{n'_a}{V'}\right)^{\alpha}\left(\frac{n'_b}{V'}\right)^{\beta}\cdots$$

TABLE 7-1 Basic Rate Equations (*Concluded*)

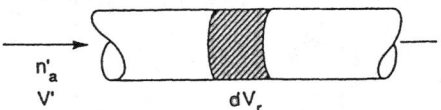

12. *Material and energy balances* for batch, CSTR, and PFR in Tables 7-5, 7-6, and 7-7.

13. *Notation.* A, B, R, S are participants in the reaction; the letters also are used to represent concentrations.

$$C_i = n_i/V_r \text{ or } n_i'/V', \text{ concentration}$$
n_i = mol of component i in the reactor
n_i' = molal flow rate of component i
V_r = volume of reactor
V' = volumetric flow rate
v_i = stoichiometric coefficient
r_i = rate of reaction of substance i [mol/(unit time)(unit volume)]
α, β = empirical exponents in a rate equation

SOURCE: Adapted from Walas, *Chemical Process Equipment Selection and Design*, Butterworth-Heinemann, 1990.

RATE EQUATIONS

RATE OF REACTION

The term *rate of reaction* means the rate of decomposition per unit volume,

$$r_a = -\frac{1}{V}\frac{dn_a}{dt}, \qquad \text{mol/(unit time) (unit volume)} \qquad (7\text{-}1)$$

$$= \frac{n_{a0}}{V}\frac{dx_a}{dt}, \qquad n_0 = n_{a0}(1 - x_a) \qquad (7\text{-}2)$$

where x_a is the fractional conversion of substance A. A rate of formation will have the opposite sign. The negative sign is required for the rate of decomposition to be a positive number. When the volume is constant,

$$r_a = -\frac{dC_a}{dt} \qquad \text{only at constant volume} \qquad (7\text{-}3)$$

Law of Mass Action The effect of concentration on the rate is isolated as

$$r_a = kf(C_a, C_b, \ldots) \qquad (7\text{-}4)$$

where the specific rate k is independent of concentration but does depend on temperature, catalysts, and other factors. The law of mass action states that the rate is proportional to the concentrations of the reactants. For the reaction

$$v_a A + v_b B + v_c C + \cdots \Rightarrow v_r R + v_s S + \cdots \qquad (7\text{-}5)$$

the rate equation is

$$r_a = -\frac{1}{V}\frac{dn_a}{dt} = kC_a^p C_b^q C_c^r \ldots \qquad (7\text{-}6)$$

$$\Rightarrow -\frac{dC_a}{dt} \qquad \text{at constant volume} \qquad (7\text{-}7)$$

The exponents ($p, q, r, \ldots$) are empirical, but they are identical with the stoichiometric coefficients ($v_a, v_b, v_c, \ldots$) when the stoichiometric equation truly represents the mechanism of reaction. The first group of exponents identifies the *order* of the reaction, the stoichiometric coefficients the *molecularity*.

Effect of Temperature The *Arrhenius equation* relates the specific rate to the absolute temperature,

$$k = k_0 \exp\left(\frac{-E}{RT}\right) \qquad (7\text{-}8)$$

$$= \exp\left(A - \frac{B}{T}\right) \qquad (7\text{-}9)$$

$$\ln k = A - \frac{B}{T} \qquad (7\text{-}10)$$

E is called the *activation energy* and k_0 the *preexponential factor.* When presumably accurate data deviate from linearity as stated by the

last equation, the reaction is believed to have a complex mechanism (Fig. 7-1g).

CONCENTRATION, MOLES, PARTIAL PRESSURE, AND MOLE FRACTION

Any property of a reacting system that changes regularly as the reaction proceeds can be formulated as a rate equation which should be convertible to the fundamental form in terms of concentration, Eq. (7-4). Examples are the rates of change of electrical conductivity, of pH, or of optical rotation. The most common other variables are partial pressure p_i and mole fraction N_i. The relations between these units are

$$n_i = VC_i = n_t N_i = \frac{n_t p_i}{\pi} \qquad (7\text{-}11)$$

where the subscript t denotes the total mol and π the total pressure. For ideal gases,

$$V = \frac{n_t RT}{\pi}$$

$$n_i = \frac{n_t RT}{\pi} C_i = \frac{n_i}{\pi} p_i = \frac{V}{RT} p_i = \frac{\pi V}{RT} N_i \qquad (7\text{-}12)$$

Other volume-explicit equations of state are sometimes required, such as the compressibility equation $V = zRT/P$ or the truncated virial equation $V = (1 + B'P)RT/P$. The quantities z and B' are not constants, so some kind of averaging will be required. More accurate equations of state are even more difficult to use but are not often justified for kinetic work.

Designate δ_a as the increase in the total mol per mol decrease of substance A according to the stoichiometric equation Eq. (7-5):

$$\delta_a = \frac{(v_r + v_s + \cdots) - (v_a + v_b + v_c + \cdots)}{v_a} \qquad (7\text{-}13)$$

The total number of mols present is

$$n_t = n_{t0} + \delta_a(n_{a0} - n_a) = n_{t0} + \delta_a x_a = n_{t0} + \delta_b x_b = \cdots \qquad (7\text{-}14)$$

Accordingly,

$$-\frac{dn_t}{dt} = \delta_a \frac{dn_a}{dt} = \delta_b \frac{dn_b}{dt} = \delta_c \frac{dn_c}{dt} = \cdots \qquad (7\text{-}15)$$

The various differentials are

$$dn_i = d(VC_i) = \frac{1}{RT}d(Vp_i) = d(n_t N_i) = \frac{n_t}{1 + \delta_i N_i}dN_i \qquad (7\text{-}16)$$

The rate equation

$$r_a = -\frac{1}{V}\frac{dC_a}{dt} = k_c C_a^\alpha \qquad (7\text{-}17)$$

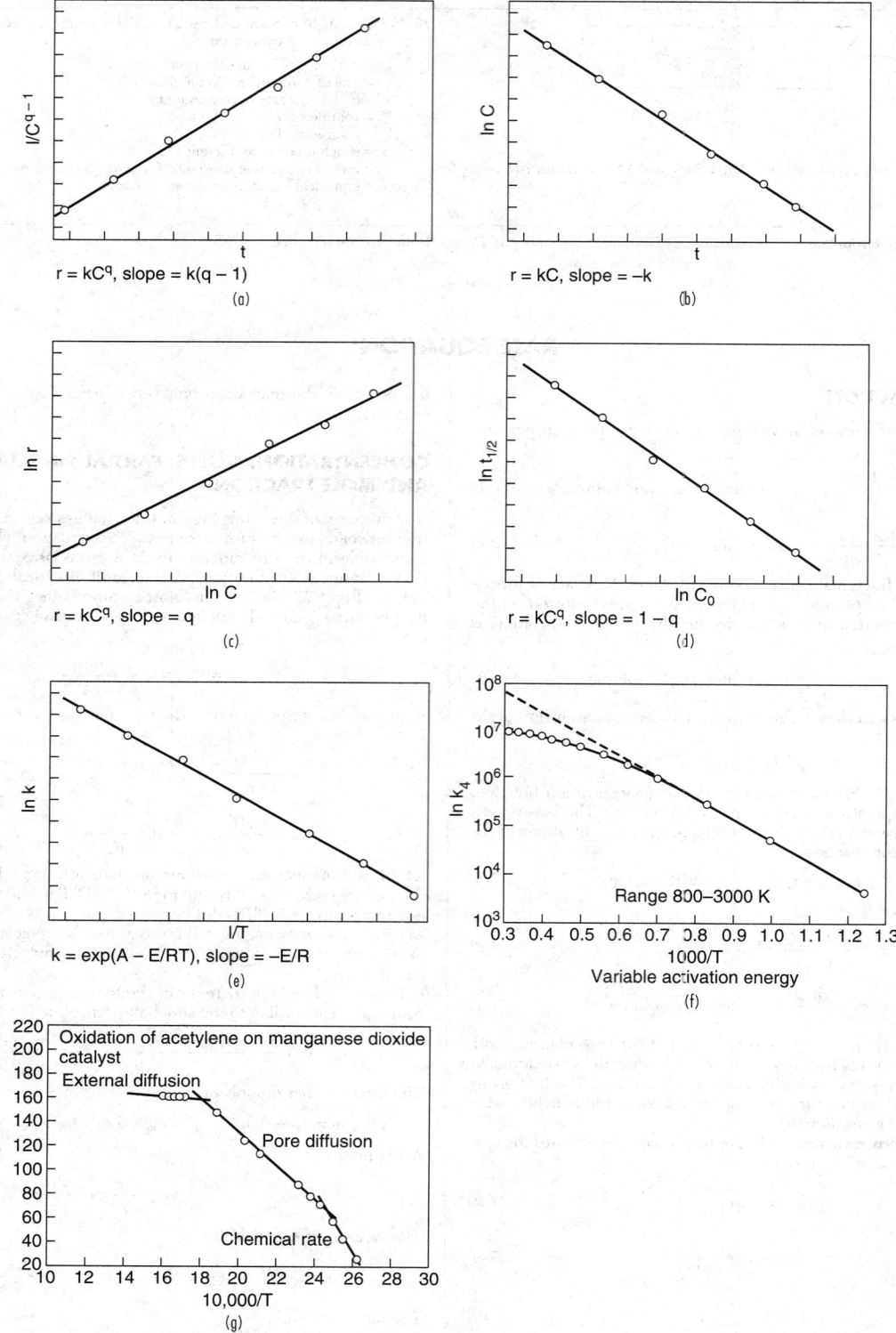

FIG. 7-1 Constants of the power law and Arrhenius equations by linearization: (*a*) integrated equation, (*b*) integrated first order, (*c*) differential equation, (*d*) half-time method, (*e*) Arrhenius equation, (*f*) variable activation energy, and (*g*) change of mechanism with temperature (*T* in K).

can be expressed in terms of pressure and mole fraction,

$$-\frac{1}{RTV}\frac{d(Vp_a)}{dt}=k_c\left(\frac{1}{RT}\right)^\alpha p_a^\alpha \qquad (7\text{-}18)$$

or at constant volume,

$$-\frac{dp_a}{dt}=k_c\left(\frac{1}{RT}\right)^{\alpha-1}p_a^\alpha=k_p p_a^\alpha \qquad (7\text{-}19)$$

where the specific rate in terms of partial pressure is

$$k_p=k_c\left(\frac{1}{RT}\right)^{\alpha-1} \qquad (7\text{-}20)$$

Typical Units of Specific Rates   For order α, typical units are:

k_c (L/g mol)$^{\alpha-1}\cdot$s^{-1}, and s^{-1} when first order
k_p (g mol)/L$\cdot$s$\cdot$atm$^\alpha$

Furthermore,

$$-\frac{n_t}{V(1+\delta_a N_a)}\frac{dN_a}{dt}=k_c\left(\frac{n_t}{V}\right)^\alpha N_a^\alpha$$

or

$$-\frac{dN_a}{dt}=k_c\left(\frac{n_t}{V}\right)^{\alpha-1}\left(\frac{1}{1+\delta_a N_a}\right)N_a^\alpha$$

$$=k_c\left(\frac{\pi}{RT}\right)^{\alpha-1}\left(\frac{1}{1+\delta_a N_a}\right)N_a^\alpha \qquad (7\text{-}21)$$

Various derivatives are evaluated in numerical Example 1.

Example 1: Rates of Change at Constant V or Constant P
Consider the ideal gas reaction $2A \Rightarrow B + 2C$ occurring at 800°R, starting with 5 lb mol of pure A at 10 atm. The rate equation is

$$r_a=-\frac{1}{V}\frac{dn_a}{dt}=700\,C_a^2 \text{ lb mol/(ft}^3\cdot\text{h)}$$

Evaluate the various rates of change at the time when the rate of reaction is $r_a = 0.1$ lb mol/(ft$^3\cdot$h) and the reaction proceeds at (1) constant volume, and (2) constant pressure.

$$r_a=-\frac{1}{V}\frac{dn_a}{dt}=700\,C_a^2=0.1 \text{ lb mol/(ft}^3\cdot\text{h)}$$

$$C_a=\sqrt{\frac{0.1}{700}}=0.01195 \text{ lb mol/ft}^3$$

$$V_0=\frac{n_{a0}RT}{\pi_0}=\frac{5(0.729)(800)}{10}=291.6 \text{ ft}^3$$

$$C_{a0}=\frac{n_{a0}}{V_0}=\frac{5}{291.6}=0.01715 \text{ lb mol/ft}^3$$

$$n_t=0.5(3n_{a0}-n_a) \text{ lb mol}$$

$$V=\frac{(3n_{a0}-n_a)RT}{2\pi_0}=29.16(15-n_a) \text{ ft}^3$$

$$\pi=\frac{n_t}{n_{t0}}\pi_0=3n_{a0}-n_a \text{ atm}$$

At constant volume, $n_a=V_0 C_a=291.6(0.01195)=3.4853$ lb mol

$$\frac{dn_a}{dt}=V_0\frac{dC_a}{dt}=-291.6(0.1)=-29.16 \text{ lb mol/h}$$

$$N_a=\frac{n_a}{n_t}=\frac{2n_a}{3n_{a0}-n_a}$$

$$\frac{dN_a}{dt}=\frac{6}{n_{a0}(3-n_a/n_{a0})^2}\left(\frac{dn_a}{dt}\right)=\frac{6}{5(3-3.4853/5)^2}(-29.16)$$

$$=-6.598 \text{ h}^{-1}$$

$$p_a=\frac{n_a RT}{V_0} \text{ atm}$$

$$\frac{dp_a}{dt}=\frac{RT}{V_0}\left(\frac{dn_a}{dt}\right)=\frac{0.729(800)}{291.6}(-29.16)=-58.32 \text{ atm/h}$$

$$\pi=\frac{n_t}{n_{t0}}\pi_0=\frac{3n_{a0}-n_a}{2n_{a0}}\pi_0=5\left(3-\frac{n_a}{n_{a0}}\right) \text{ atm}$$

$$\frac{d\pi}{dt}=-\frac{5}{n_{a0}}\frac{dn_a}{dt}=29.16 \text{ atm/h}$$

At constant pressure, $n_a=VC_a=29.16(15-n_a)(0.01195)=3.8768$ lb mol

$$\frac{dn_a}{dt}=-Vr_a=-324.4(0.1)=-32.44 \text{ lb mol/h}$$

since $V=29.16\,(15-3.8768)=324.4 \text{ ft}^3$

$$C_a=\frac{n_a}{V}=\frac{n_a}{29.16(15-n_a)} \text{ lb mol/ft}^3$$

$$\frac{dC_a}{dt}=\frac{1}{29.16}\left[\frac{15}{(15-n_a)^2}\right]\frac{dn_a}{dt}=\frac{1}{29.16}\left[\frac{15}{(15-3.8768)^2}\right](-32.44)$$

$$=0.1349 \text{ lb mol/(ft}^3\cdot\text{h)}$$

$$\frac{dN_a}{dt}=\frac{6n_{a0}}{(3n_{a0}-n_a)^2}\frac{dn_a}{dt}=\frac{30}{(15-3.8768)^2}(-32.44)=-7.8658 \text{ h}^{-1}$$

$$p_a=N_a\pi_0=\frac{2\pi_0 n_a}{3n_{a0}-n_a} \text{ atm}$$

$$\frac{dp_a}{dt}=\frac{6\pi_0}{(3n_{a0}-n_a)^2}\frac{dn_a}{dt}=\frac{(6)(10)(5)}{(15-3.8768)^2}(-32.44)=-78.66 \text{ atm/h}$$

$$V=\frac{(3n_{a0}-n_a)RT}{2\pi_0} \text{ ft}^3$$

$$\frac{dV}{dt}=\frac{RT}{2\pi_0}\left(-\frac{dn_a}{dt}\right)=\frac{0.729(800)}{20}(32.44)=945.95 \text{ ft}^3/\text{h}$$

$$\frac{dx_a}{dt}=\frac{d}{dt}\left(\frac{n_{a0}-n_a}{n_{a0}}\right)=-\frac{1}{n_{a0}}\frac{dn_a}{dt}=-\frac{1}{5}(-32.44)=6.488 \text{ h}^{-1}$$

SUMMARY

Rate	At constant V	At constant P
$dn_a dt$, lb mol/h	−29.16	−32.44
dN_a/dt, h^{-1}	−6.598	−7.866
dp_a/dt, atm/h	−58.32	−78.66
$d\pi/dt$, atm/h	29.16	0
dV/dt, ft^3/h	0	946.0
dx_a/dt, h^{-1}	5.832	6.488

REACTION TIME IN FLOW REACTORS

Flow reactors usually operate at nearly constant pressure, and thus at variable density when there is a change of moles of gas or of temperature. An *apparent residence time* is the ratio of reactor volume and the inlet volumetric flow rate,

$$\bar{t}_{app}=\frac{V_r}{V_0'} \qquad (7\text{-}22)$$

The *true residence time* is obtained by integration of the rate equation,

$$\bar{t}=\int\frac{dV_r}{V'}=\int\frac{dn}{V'r}=\int\frac{dn}{kV'(n/V')^q} \qquad (7\text{-}23)$$

The apparent time is readily evaluated and is popularly used to indicate the loading of a flow reactor.

A related concept is that of *space velocity*, which is a ratio of a flow rate at STP (usually 60°F, 1 atm) to the size of the reactor. The most common versions in typical units are:

GHSV (gas hourly space velocity) = (volumes of feed as gas at STP/h)/(volume of reactor or its content of catalyst) = SCFH gas feed/ft^3.

LHSV (liquid hourly space velocity) = (volume of liquid feed at 60°F/h)/(ft^3 of reactor) = SCFH liquid feed/ft^3.

WHSV (weight hourly space velocity) = (lb feed/h)/(lb catalyst).

It is usually advisable to spell out the units when the acronym is used, since the units are arbitrary.

CONSTANTS OF THE RATE EQUATION

The problem is to apply experimental data to find the constants of assumed rate equations, of which some of the simpler examples are:

$$r = -\frac{dC}{dt} = kC^q \tag{7-24}$$

or

$$r = -\frac{dC}{dt} = \exp\left(a - \frac{b}{t}\right)C^q \tag{7-25}$$

or

$$r_a = -\frac{dC_a}{dt} = kC_a^p C_b^q \tag{7-26}$$

Experimental data that are most easily obtained are of (C, t), (p, t), (r, t), or (C, T, t). Values of the rate are obtainable directly from measurements on a continuous stirred tank reactor (CSTR), or they may be obtained from (C, t) data by numerical means, usually by first curve fitting and then differentiating. When other properties are measured to follow the course of reaction—say, conductivity—those measurements are best converted to concentrations before kinetic analysis is started.

The most common ways of evaluating the constants are from linear rearrangements of the rate equations or their integrals. Figure 7-1 examines power law and Arrhenius equations, and Fig. 7-2 has some more complex cases.

From the Differential Equation   Linear regression can be applied with the differential equation to obtain constants. Taking logarithms of Eq. (7-25),

$$\ln r = a - \frac{b}{T} + q \ln C \tag{7-27}$$

The variables that are combined linearly are $\ln r$, $1/T$, and $\ln C$. Multilinear regression software can be used to find the constants, or only three sets of the data suitably spaced can be used and the constants found by simultaneous solution of three linear equations. For a linearized Eq. (7-26) the variables are logarithms of r, C_a, and C_b. The logarithmic form of Eq. (7-24) has only two constants, so the data can be plotted and the constants read off the slope and intercept of the best straight line.

From the Integrated Equation   The integral of Eq. (7-24) is

$$k = \frac{1}{t - t_0} \ln \frac{C_0}{C}, \quad \text{when } q = 1 \tag{7-28}$$

$$\frac{C_0^{q-1}}{(t - t_0)(q - 1)}\left[\left(\frac{C_0}{C}\right)^{q-1} - 1\right], \quad \text{when } q \neq 1 \tag{7-29}$$

A value of q is assumed and values of k are calculated for each data point. The correct value of q is chosen when the values of k are nearly constant or show no drift. This procedure is applicable for a rate equation of any complexity if it can be integrated. Eqs. (7-28) and (7-29) can also be put into linear form:

$$\ln\left(\frac{C_0}{C}\right) = k(t - t_0), \quad \text{when } q = 1 \tag{7-30}$$

$$\left(\frac{1}{C}\right)^{q-1} = \left(\frac{1}{C_0}\right)^{q-1} + k(q - 1)(t - t_0), \quad \text{when } q \neq 1 \tag{7-31}$$

When the plots are collinear, the correct value of k is found from the slope of the best straight line.

From Half-Times   The time by which one-half of the reactant has been converted is called the *half-time*. From Eq. (7-24),

$$kt_{1/2} = \ln 2, \quad q = 1$$

$$\frac{2^{q-1} - 1}{(q - 1)C_0^{q-1}} \quad q \neq 1 \tag{7-32}$$

When several sets of $(C_0, t_{1/2})$ are known, values of q are tried until one is found that makes all k values substantially the same. Alternatively, the constants may be found from a linearized plot,

$$\ln t_{1/2} = \ln \frac{2^{q-1} - 1}{(q - 1)k} + (1 - q)\ln C_0 \tag{7-33}$$

Complex Rate Equations   Complex rate equations may require individual treatment, although the examples in Fig. 7-2 are all linearizable. A perfectly general procedure is nonlinear regression. For instance, when $r = f(C, a, b, \ldots)$ where $(a, b, \ldots)$ are the constants to be found, the condition is

$$\sum [r_i - f(C_i, a, b, \ldots)]^2 \Rightarrow \text{Minimum} \tag{7-34}$$

and

$$\frac{\partial \Sigma}{\partial a} = \frac{\partial \Sigma}{\partial b} = \cdots = 0 \tag{7-35}$$

Much professional software is devoted to this problem. A diskette for sets of differential and algebraic equations with parameters to be found by this method is by Constantinides (*Applied Numerical Methods with Personal Computers*, McGraw-Hill, 1987).

The acquisition of kinetic data and parameter estimation can be at quite a sophisticated level, particularly for solid catalytic reactions: statistical design of experiments, refined equipment, computer monitoring of data acquisition, and statistical evaluation of the data. Two papers are devoted to this topic by Hofmann (in *Chemical Reaction Engineering, ACS Advances in Chemistry*, **109**, 519–534 [1972]; in de Lasa, ed., *Chemical Reactor Design and Technology*, Martinus Nijhoff, 1985, pp. 69–105).

MULTIPLE REACTIONS AND STOICHIOMETRIC BALANCES

Single Reaction   For the stoichiometric equation, Eq. (7-5), the relations between the conversions of the several participants are

$$\frac{x}{\nu_a} = \frac{n_{a0} - n_a}{\nu_a} = \frac{n_{b0} - n_b}{\nu_b} = \cdots = -\frac{n_{r0} - n_r}{\nu_r} = -\frac{n_{s0} - n_s}{\nu_s} = \cdots \tag{7-36}$$

$$C_a = \frac{n_a}{V} = \frac{n_{a0} - x}{V}, \quad C_b = \frac{n_{b0} - \nu_b x/\nu_a}{V}, \quad C_c = \frac{C_{c0} - \nu_c x/\nu_a}{V}, \text{ and so on} \tag{7-37}$$

Also,

$$C_b = C_{b0} - \frac{\nu_b}{\nu_a}(C_{a0} - C_a)$$

$$C_c = C_{c0} - \frac{\nu_c}{\nu_a}(C_{a0} - C_a), \text{ and so on} \tag{7-38}$$

Accordingly, the rate equation can be written in terms of the single dependent variable x; thus,

$$r_a = -\frac{1}{V}\frac{dn_a}{dt} = \frac{1}{V}\frac{dx}{dt}$$

$$= k\left(\frac{n_{a0} - x}{V}\right)^p\left(\frac{n_{b0} - \nu_b x/\nu_a}{V}\right)^q\left(\frac{n_{c0} - \nu_c x/\nu_a}{V}\right)^r \cdots \tag{7-39}$$

and in terms of concentrations,

$$r_a = -\frac{dC_a}{dt} = kC_a^p C_b^q C_c^r \cdots$$

$$= kC_a^p\left[C_{b0} - \frac{\nu_b}{\nu_a}(C_{a0} - C_a)\right]^q\left[C_{c0} - \frac{\nu_c}{\nu_a}(C_{a0} - C_a)\right]^r \cdots \tag{7-40}$$

Eq. (7-39) becomes integrable when V is properly expressed in terms of the composition of the system, and Eq. (7-40) can be integrated as it stands.

Multiple Reactions   When a substance participates in several reactions at the same time, its net rate of decomposition is the algebraic sum of its rates in the individual reactions. Identify the rates of the individual steps with subscripts, $(dC/dt)_1$, $(dC/dt)_2$, ... Take this case of three reactions,

$$A + B \xrightarrow{1} C$$

$$A + C \xrightarrow{2} D + E$$

$$D + E \xrightarrow{3} A + C$$

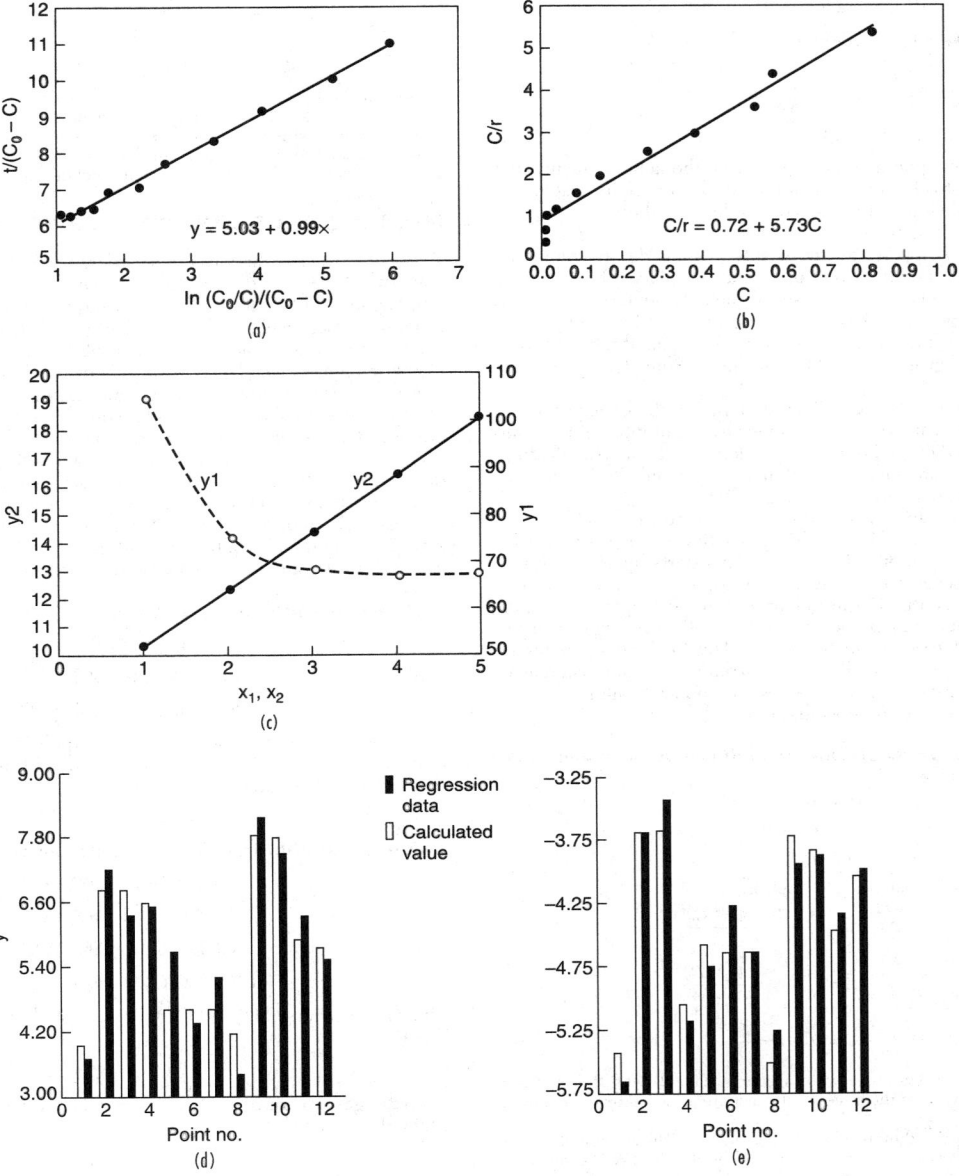

FIG. 7-2 Linear analysis of catalytic rate equations. (a), (b) Sucrose hydrolysis with an enzyme, $r = kM/(M + C)$. Data are (C, t) curve-fitted with a fourth-degree polynomial and differentiated for $r - (-dC/dt)$. Integrated equation,

$$\frac{t}{C_0 - C} = \frac{1}{k} + \frac{M}{k}\frac{\ln(C_0/C)}{C_0 - C}, \qquad k = 0.199, \, M = 4.98$$

Linearized rate equation,

$$\frac{C}{r} = \frac{M}{k} + \frac{C}{k}, \qquad M = 4.13$$

poor agreement. (c) For a solid catalyzed reaction, two possible equations in linear form are $y_1 = P_a/r = a + bp_a$ and $y_2 = P_a/\sqrt{r} = a + bp_a$, of which the second appears to fit. (d), (e) Hydrogenation of octenes, Hougen and Watson (*Chemical Process Principles*, Wiley, 1947, p. 943). The hyperbolic and power law fits are of about equal quality. Pressure in atm; r in lb mol/(ft^3·h).

$$y = \sqrt{\frac{P_u P_h}{r}} = a + bp_u + cp_s + dp_h$$

$$= 2.7655 + 1.5247p_u + 1.0092p_s + 1.1291p_h$$

$$\ln r = \ln k + a \ln P_u + b \ln p_s + c \ln p_h$$

$$= -4.059 + 0.469 \ln p_u - 0.2356 \ln p_s + 0.5997 \ln p_h$$

$$r = 0.0173 \, P_u^{0.469} \, p_s^{-0.2356} \, P_h^{0.5997}$$

The overall rates of the several participants are

$$r_a = r_{a1} + r_{a2} + r_{a3} = -k_1 C_a C_b - k_2 C_a C_c + k_3 C_d C_e$$

$$r_b = -k_1 C_a C_b$$

$$r_c = k_1 C_a C_b - k_2 C_a C_c + k_3 C_d C_e$$

$$r_d = r_e = k_2 C_a C_e - k_3 C_d C_e \qquad (7\text{-}41)$$

The number of independent rate equations is the same as the number of independent stoichiometric relations. In this example, reactions 2 and 3 are reversible and are not independent, so there are only two independent rate equations.

Some reactions apparently represented by single stoichiometric equations are in reality the result of several reactions, often involving short-lived intermediates. After a set of such elementary reactions is postulated by experience, intuition, and exercise of judgment, a rate equation is deduced and checked against experimental rate data. Several examples are given under "Mechanisms of Some Complex Reactions," following.

Stoichiometric Balances   The amounts of all participants in a group of reactions can be expressed in terms of a number of key components equal to the number of independent stoichiometric relations. The independent rate equations will then involve only those key components and will be, in principle, integrable.

For a single equation, Eqs. (7-36) and (7-37) relate the amounts of the several participants. For multiple reactions, the procedure for finding the concentrations of all participants starts by assuming that the reactions proceed consecutively. Key components are identified. Intermediate concentrations are identified by subscripts. The resulting concentration from a particular reaction is the starting concentration for the next reaction in the series. The final value carries no subscript. After the intermediate concentrations are eliminated algebraically, the compositions of the excess components will be expressible in terms of the key components.

Example 2: Analysis of Three Simultaneous Reactions   Consider the three reactions

$$A + 2B \overset{1}{\Rightarrow} 3C$$
$$A + C \overset{2}{\Rightarrow} 2D$$
$$C + D \overset{3}{\Rightarrow} 2E$$

with A, B, and C the key components. Apply Eq. (7-37),

$$A_0 - A_1 = \frac{B_0 - B}{2} = \frac{C_1 - C_0}{3}$$

$$A_1 - A = C_1 - C_2 = \frac{D_2 - D_0}{2}$$

$$C_2 - C = D_2 - D = \frac{E - E_0}{2} \qquad (7\text{-}42)$$

Elimination of the concentrations with subscripts 1 and 2 will find D and E in terms of A, B, and C, with the same results that are achieved by the following method.

This alternative procedure is called the *xyz* method. The amount of change by the first reaction is x, by the second y, and by the third z. For the same example,

$$A = A_0 - x - y$$
$$B = B_0 - 2x$$
$$C = C_0 + 3x - y - z$$
$$D = D_0 + 2y - z$$
$$E = E_0 + 2z \qquad (7\text{-}43)$$

Elimination of x, y, and z gives for the excess components:

$$D_0 - D = -3(A_0 - A) + 3(B_0 - B)$$
$$E_0 - E = 2(A_0 - A) - 4(B_0 - B) - 2(C_0 - C) \qquad (7\text{-}44)$$

The differential equations for the three key components become:

$$\frac{dA}{dt} = -k_1 AB^2 - k_2 AC$$

$$\frac{dB}{dt} = -2k_1 AB^2$$

$$\frac{dC}{dt} = 3k_1 AB^2 - k_2 AC - k_3 CD$$

$$= 3k_1 AB^2 - C\{k_2 A + k_3[D_0 + 3(A_0 - A) - 3(B_0 - B)]\} \qquad (7\text{-}45)$$

These equations will have to be solved numerically for A, B, and C as functions of time; then D and E can be found by algebra. Alternatively, five differential equations can be written and solved directly for the five participants as functions of time, thus avoiding the use of stoichiometric balances, although these are really involved in the formulation of the differential equations.

MECHANISMS OF SOME COMPLEX REACTIONS

The rates of many reactions are not represented by application of the law of mass action on the basis of their overall stoichiometric relations. They appear, rather, to proceed by a sequence of first- and second-order processes involving short-lived intermediates which may be new species or even unstable combinations of the reactants; for $2A + B \Rightarrow C$, the sequence could be $A + B \Rightarrow AB$ followed by $A + AB \Rightarrow C$.

Free radicals are molecular fragments having one or more unpaired electrons, usually short-lived (milliseconds) and highly reactive. They are detectable spectroscopically and some have been isolated. They occur as initiators and intermediates in such basic phenomena as oxidation, combustion, photolysis, and polymerization. The rate equation of a process in which they are involved is developed on the postulate that each free radical is at equilibrium or its net rate of formation is zero. Several examples of free radical and catalytic mechanisms will be cited, all possessing nonintegral power law or hyperbolic rate equations.

Phosgene Synthesis $CO + Cl_2 \Rightarrow COCl_2$, but with the sequence:

$$Cl_2 \Leftrightarrow 2Cl^\bullet$$
$$Cl^\bullet + CO \Leftrightarrow COCl^\bullet$$
$$COCl^\bullet + Cl_2 \Rightarrow COCl_2 + Cl^\bullet$$

Assuming the first two reactions to be in equilibrium, an expression is found for the concentration of $COCl^\bullet$ and when this is substituted into the third equation the rate becomes

$$r_{COCl_2} = k(CO)(Cl_2)^{3/2} \qquad (7\text{-}46)$$

Ozone and Chlorine   The assumed sequence is:

$$Cl_2 + O_3 \Rightarrow ClO^\bullet + ClO_2^\bullet$$
$$ClO_2^\bullet + O_3 \Rightarrow ClO_3^\bullet + O_2$$
$$ClO_3^\bullet + O_3 \Rightarrow ClO_2^\bullet + 2O_2$$
$$ClO_3^\bullet + ClO_3^\bullet \Rightarrow Cl_2 + 3O_2$$

The chain carriers $ClO_2^\bullet$ and $ClO_3^\bullet$ are assumed to attain steady state. Then,

$$r_{O_3} = k(Cl_2)^{1/2} (O_3)^{3/2} \qquad (7\text{-}47)$$

Hydrogen Bromide $H_2 + Br_2 \Rightarrow 2HBr$ (Bodenstein, 1906). The chain of reactions is:

$$Br_2 \overset{1}{\Rightarrow} 2Br^\bullet$$
$$Br^\bullet + H_2 \overset{2}{\Rightarrow} HBr$$
$$H^\bullet + Br_2 \overset{3}{\Rightarrow} HBR + Br^\bullet$$
$$H^\bullet + HBr \Rightarrow H_2 + Br^\bullet$$
$$Br^\bullet + Br^\bullet \Rightarrow Br_2$$

Assuming equilibrium for the concentrations of the free radicals, the rate equation becomes

$$\frac{d(HBr)}{dt} = k_1(Br^\bullet)(H_2) + k_2(H^\bullet)(Br_2) - k_3(H^\bullet)(HBr)$$

$$= \frac{k_1(H_2)(Br_2)^{3/2}}{k_2(Br_2) + k_3(HBr)} \qquad (7\text{-}48)$$

Enzyme Kinetics   The enzyme E and the reactant S are assumed to form a complex ES that then dissociates into product P and uncombined enzyme.

$$S + E \overset{1}{\underset{2}{\rightleftharpoons}} ES$$

$$ES \overset{3}{\Rightarrow} E + P$$

If equilibrium holds,

$$\frac{(S)(E)}{(ES)} = \frac{(S)[(E_0) - (ES)]}{(ES)} = K_m$$

where (E_0) is the total of the free and combined enzyme and K_m is a dissociation constant. Solve for (ES) and substitute into the rate equation,

$$r_p = \frac{d(P)}{dt} = k(ES) = \frac{k(E_0)(S)}{K_m + (S)} \qquad (7\text{-}49)$$

This hyperbolic equation is named after Michaelis and Menten (*Biochem. Zeit.*, **49**, 333 [1913]).

Chain Polymerization   The growth process of a polymer postulates a three-step mechanism:

1. An initiator I generates a free radical R·
2. The free radical reacts repeatedly with monomer by a process called *propagation*.
3. The free radical eventually disappears by some reaction, called *termination*. The stoichiometric equations are

$$I \overset{1}{\Rightarrow} 2R\text{·}$$

$$R\text{·} + M \overset{2}{\Rightarrow} RM\text{·}, \qquad \text{initiation}$$

$$RM\text{·} + M \overset{k_p}{\Rightarrow} RM_2^{\text{·}}$$

or $\qquad RM_n^{\text{·}} + M \overset{k_p}{\Rightarrow} RM_{n+1}^{\text{·}}, \qquad \text{propagation}$

$$RM_n^{\text{·}} + RM_m^{\text{·}} \overset{k_t}{\Rightarrow} R_2M_{n+m}$$

or $\qquad RM_n + RM_m, \qquad\qquad \text{termination}$

The rates of formation of the free radicals R· and M· reach steady states,

$$\frac{dR_n^{\text{·}}}{dt} = 2k_1(I) - k_2(R\text{·})(M) = 0$$

$$\frac{dM\text{·}}{dt} = k_2(R\text{·})(M) - 2k_t(M\text{·})^2 = 0$$

These equations are solved for $(R\text{·})$ and $(M\text{·})$ and substituted into the propagation equation. The rate of polymerization becomes

$$r_p = -\frac{dM}{dt} = k_p(M\text{·})(M) = k_p\left(\frac{k_1}{k_t}\right)^{1/2}(M)(I)^{1/2} \qquad (7\text{-}50)$$

Thus, the process of chain polymerization is first-order with respect to monomer and half-order with respect to initiator.

Solid Catalyzed Reaction   The pioneers were Langmuir (*J. Am. Chem. Soc.*, **40**, 1361 [1918]) and Hinshelwood (*Kinetics of Chemical Change*, Oxford, 1940). For a gas phase reaction $A + B \Rightarrow$ Products, catalyzed by a solid, the postulated mechanism consists of the following:

1. The reactants are first adsorbed on the surface, where they subsequently react and the product is desorbed.
2. The rate of adsorption is proportional to the partial pressure and to the fraction of uncovered surface ϑ_v.
3. The rate of desorption of A is proportional to the fraction ϑ_a of the surface covered by A.
4. Adsorptive equilibrium is maintained.
5. The rate of reaction between adsorbed species is proportional to their amounts on the surface.

The net rates of adsorption are:

$$r_a = k_a p_a \vartheta_v - k_{-a} \vartheta_a \Rightarrow 0$$

$$r_b = k_b p_b \vartheta_v - k_{-b} \vartheta_b \Rightarrow 0$$

Substitute $\vartheta_v = 1 - \vartheta_a - \vartheta_b$ and solve for the coverages:

$$\vartheta_a = \left(\frac{k_a}{k_{-a}}\right) p_a \vartheta_v = K_a p_a \vartheta_v$$

$$\vartheta_b = \left(\frac{k_b}{k_{-b}}\right) p_b \vartheta_v = K_b p_b \vartheta_v$$

$$\vartheta_v = \frac{1}{1 + K_a p_a + K_b p_b}$$

The rate of surface reaction is:

$$r = k\vartheta_a \vartheta_b = \frac{kK_a K_b p_a p_b}{(1 + K_a p_a + K_b p_b)^2} \qquad (7\text{-}51)$$

The linearized form can be used to find the constants,

$$y = \sqrt{\frac{p_a p_b}{r}} = \frac{1 + K_a P_a + K_b p_b}{\sqrt{kK_a K_b}}$$

More about this topic is presented later.

WITH DIFFUSION BETWEEN PHASES

When reactants are distributed between several phases, migration between phases ordinarily will occur: with gas/liquid, from the gas to the liquid; with fluid/solid, from the fluid to the solid; between liquids, possibly both ways because reactions can occur in either or both phases. The case of interest is at steady state, where the rate of mass transfer equals the rate of reaction in the destined phase. Take a hyperbolic rate equation for the reaction on a surface. Then,

$$r = r_d = r_s$$

$$= k_1(C - C_s) = \frac{k_2 C_s}{1 + k_3 C_s} = \frac{k_2(C - r/k_1)}{1 + k_3(C - r/k_1)} \qquad (7\text{-}52)$$

The unknown intermediate concentration C_s has been mathematically eliminated from the last term. In this case, r can be solved for explicitly, but that is not always possible with surface rate equations of greater complexity. The mass transfer coefficient k_1 is usually obtainable from correlations. When the experimental data are of (C, r) the other constants can be found by linear plotting.

CATALYSIS BY SOLIDS: LANGMUIR-HINSHELWOOD MECHANISM

A plausible mechanism of solid catalytic reactions is that the participants chemisorb on the surface and react while in the adsorbed state. The process of adsorption of A on an active site of the surface σ is represented by

$$A + \sigma \Rightarrow A\sigma$$

and the reaction between adsorbed molecules, for instance, by

$$A\sigma + B\sigma \Rightarrow C\sigma + D\sigma$$

Adsorptive Equilibrium   The fraction of the surface covered by A at equilibrium is

$$\vartheta_a = K_a p_a \vartheta_v \qquad (7\text{-}53)$$

$$\vartheta_v = \frac{1}{1 + K_a p_a + K_b p_b + K_c P_c + K_d p_d + \cdots} \qquad (7\text{-}54)$$

where terms may be added for adsorbed inerts that may be present, and analogous expressions for the other participants. The rate of reaction between species in adsorptive equilibrium is then

$$r = kp_a p_b \vartheta_v^2 \qquad (7\text{-}55)$$

Dissociation   A diatomic molecule A_2 may adsorb as atoms,

$$A_2 + 2\sigma \Rightarrow 2A\sigma$$

with the result,

$$\vartheta_a = \frac{\sqrt{K_a p_a}}{1 + \sqrt{K_a p_a} + K_b p_b + \cdots} = \sqrt{K_a p_a}\ \vartheta_v$$

and the rate of the reaction is

$$2A\sigma + B\sigma \Rightarrow \text{Products}$$

$$r = k'\vartheta_a^2 \vartheta_b = kp_a p_b \vartheta_v^3 \qquad (7\text{-}56)$$

Different Sites   When A and B adsorb on chemically different sites σ_1 and σ_2, the rate of the reaction $A + B \Rightarrow$ Unadsorbed products is

$$r = \frac{k p_a p_b}{(1 + K_a p_a)(1 + K_b p_b)} \tag{7-57}$$

Dual Sites   When the numbers of moles of reactants and products are unequal, $A \Leftrightarrow M + N$, the mechanism is assumed to be

$$A\sigma + \sigma \Leftrightarrow M\sigma + N\sigma$$

and the rate

$$r = k\left(\vartheta_a \vartheta_v - \frac{\vartheta_m \vartheta_n}{K}\right) = k\left(p_a - \frac{p_m p_n}{K}\right)\vartheta_v^2$$

$$= \frac{k(p_a - p_m p_n/K)}{(1 + K_a p_a + K_m p_m + K_n p_n)^2} \tag{7-58}$$

Reactant in the Gas Phase   When A in the gas phase reacts with adsorbed B,

$$A + B\sigma \Rightarrow \text{Products}$$

$$r = k p_a \vartheta_b = k p_a p_b \vartheta_v = \frac{k p_a p_b}{(1 + \Sigma K_i p_i)} \tag{7-59}$$

Chemical Equilibrium   When A is not in adsorptive equilibrium, it is assumed to be in chemical equilibrium, with $p_a^\circ = p_m p_n/K_e p_b$. This expression is substituted for p_a wherever it appears in the rate equation. Then

$$r = k p_a^\circ p_b \vartheta_v^2 = \frac{k p_m p_n/K_e}{(1 + K_a p_m p_n/K_e p_b + K_b p_b + K_m p_m + K_n p_n)^2} \tag{7-60}$$

All of these relations are brought together in the fundamental form

$$r = \frac{(\text{kinetic term})(\text{driving force})}{\text{adsorption term}} \tag{7-61}$$

Table 7-2 summarizes the cases when all substances are in adsorptive equilibrium and the surface reaction controls. In Table 7-3, substance A is not in adsorptive equilibrium, so its adsorption rate is controlling.

Details of the derivations of these and some other equations are presented by Yang and Hougen (*Chem. Eng. Prog.*, **46**, 146 [1950]),

Walas (*Reaction Kinetics for Chemical Engineers*, McGraw-Hill, 1959; Butterworths, 1989, pp. 153–164), and Rase (*Chemical Reactor Design for Process Plants*, vol. 1, Wiley, 1977, pp. 178–191).

All of the relations developed here assume that only one step is controlling. A more general case is that of the reaction $A \Rightarrow B$ with five steps controlling, namely

$r = k_1(p_{ag} - p_{ai})$	Diffusion of A to the surface
$r = k_2\left(p_{ai}\theta_v - \dfrac{\theta_a}{k_3}\right)$	Adsorption of A
$r = k_4\theta_a$	Surface reaction
$r = k_5\left(p_{bi}\theta_v - \dfrac{\theta_b}{k_6}\right)$	Desorption of B
$r = k_7(p_{bi} - p_{bg})$	Diffusion of B from the surface (7-62)

where $\theta_v = 1 - \theta_a - \theta_b$.

At steady state these rates are all the same. Upon elimination of the unmeasurable quantities p_{ai}, p_{bi}, ϑ_a, ϑ_b, and ϑ_v, the relation becomes

$$r = \frac{k_5(k_7 p_{bg} + r)}{k_7}\left(1 - \frac{r}{k_4}\right)$$
$$- k_5\left(p_{bg} + \frac{r}{k_7} + \frac{1}{k_6}\right)\left[1 - \frac{r}{k_4} - \frac{k_1 r}{k_1 p_{ag} - r}\left(\frac{1}{k_2} + \frac{1}{k_3 k_4}\right)\right] \tag{7-63}$$

Combinations of several adsorption and surface reaction steps are usually not felt to be necessary, since so many alternatives are available individually. Single steps in combination with diffusion to the surface are usually adequate, as in the case leading to Eq. (7-52).

Over the usual limited range of conditions, a power law rate equation often appears to be as satisfactory a fit of the data as a more complex Langmuir-Hinshelwood equation. The example of the hydrogenation of octenes is shown in Fig. 7-2d and 7-2e, and another case follows.

Example 3: Phosgene Synthesis   Rate data were obtained by Potter and Baron (*Chem. Eng. Prog.*, **47**, 478 [1951]) for the reaction CO (A) + Cl_2 (B_2) $\Rightarrow COCl_2$ (C) at 30.6. Three correlations of approximately equal statis-

TABLE 7-2   Surface-reaction Controlling (Adsorptive Equilibrium Maintained of All Participants)

Reaction	Special condition	Basic rate equation	Driving force	Adsorption term
1. $A \to M + N$	General case	$r = k\theta_a$	p_a	$1 + K_a p_a + K_m p_m + K_n p_n$
$A \to M + N$	Sparsely covered surface	$r = k\theta_a$	p_a	1
$A \to M + N$	Fully covered surface	$r = k\theta_a$	1	1
2. $A \rightleftharpoons M$		$r = k_1\theta_a - k_{-1}\theta_m$	$p_a - \dfrac{p_m}{K}$	$1 + K_a p_a + K_m p_m$
3. $A \rightleftharpoons M + N$	Adsorbed A reacts with vacant site	$r = k_1\theta_a\theta_v - k_{-1}\theta_m\theta_n$	$p_a - \dfrac{p_m p_n}{K}$	$(1 + K_a p_a + K_m p_m + K_n p_n)^2$
4. $A_2 \rightleftharpoons M$	Dissociation of A_2 upon adsorption	$r = k_1\theta_a^2 - k_{-1}\theta_m\theta_v$	$p_a - \dfrac{p_m}{K}$	$(1 + \sqrt{K_a p_a} + K_m p_m)^2$
5. $A + B \to M + N$	Adsorbed B reacts with A in gas but not	$r = k\theta_a\theta_b$	$p_a p_b$	$(1 + K_a p_a + K_b p_b + K_m p_m + K_n p_n)^2$
$A + B \to M + N$	with adsorbed A	$r = k p_a\theta_b$	$p_a p_b$	$1 + K_a p_a + K_b p_b + K_m p_m + K_n p_n$
6. $A + B \rightleftharpoons M$		$r = k_1\theta_a\theta_b - k_{-1}\theta_m\theta_v$	$p_a p_b - \dfrac{p_m}{K}$	$(1 + K_a p_a + K_b p_b + K_m p_m)^2$
7. $A + B \rightleftharpoons M + N$		$r = k_1\theta_a\theta_b - k_{-1}\theta_m\theta_n$	$p_a p_b - \dfrac{p_m p_n}{K}$	$(1 + K_a p_a + K_b p_b + K_m p_m + K_n p_n)^2$
8. $A_2 + B \rightleftharpoons M + N$	Dissociation of A_2 upon adsorption	$r = k_1\theta_a^2\theta_b - k_{-1}\theta_m\theta_n\theta_v$	$p_a p_b - \dfrac{p_m p_n}{K}$	$(1 + \sqrt{K_a p_a} + K_b p_b + K_m p_m + K_n p_n)^3$

NOTE: The rate equation is:

$$r = \frac{k\,(\text{driving force})}{\text{adsorption term}}$$

When an inert substance I is adsorbed, the term $K_i p_i$ is to be added to the adsorption term.
SOURCE: From Walas, *Reaction Kinetics for Chemical Engineers*, McGraw-Hill, 1959; Butterworths, 1989.

TABLE 7-3 Adsorption-rate Controlling (Rapid Surface Reaction)

Reaction	Special condition	Basic rate equation	Driving force	Adsorption term
1. $A \to M + N$		$r = kp_a\theta_v$	p_a	$1 + \dfrac{K_a p_m p_n}{K} + K_m p_m + K_n p_n$
2. $A \rightleftharpoons M$		$r = k\left(p_a\theta_v - \dfrac{\theta_a}{K_a}\right)$	$p_a - \dfrac{p_m}{K}$	$1 + \dfrac{K_a p_m}{K} + K_m p_m$
3. $A \rightleftharpoons M + N$		$r = k\left(p_a\theta_v - \dfrac{\theta_a}{K_a}\right)$	$p_a - \dfrac{p_m p_n}{K}$	$1 + \dfrac{K_a p_m p_n}{K} + K_m p_m + K_n p_n$
4. $A_2 \rightleftharpoons M$	Dissociation of A_2 upon adsorption	$r = k\left(p_a\theta_v^2 - \dfrac{\theta_a^2}{K_a}\right)$	$p_a - \dfrac{p_m}{K}$	$\left(1 + \sqrt{\dfrac{K_a p_m}{K}} + K_m p_m\right)^2$
5. $A + B \to M + N$	Unadsorbed A reacts with adsorbed B	$r = kp_a\theta_v$	p_a	$1 + \dfrac{K_a p_m p_n}{K p_b} + K_b p_b + K_m p_m + K_n p_n$
6. $A + B \rightleftharpoons M$		$r = k\left(p_a\theta_v - \dfrac{\theta_a}{K_a}\right)$	$p_a - \dfrac{p_m}{K p_b}$	$1 + \dfrac{K_a p m}{K p_b} + K_b p_b + K_m p_m$
7. $A + B \rightleftharpoons M + N$		$r = k\left(p_a\theta_v - \dfrac{\theta_a}{K_a}\right)$	$p_a - \dfrac{p_m p_n}{K p_b}$	$1 + \dfrac{K_a p_m p_n}{K p_b} + K_b p_b + K_m p_m + K_n p_n$
8. $A_2 + B \rightleftharpoons M + N$	Dissociation of A_2 upon adsorption	$r = k\left(p_a\theta_v^2 - \dfrac{\theta_a^2}{K_a}\right)$	$p_a - \dfrac{p_m p_n}{K p_b}$	$\left(1 + \sqrt{\dfrac{K_a p_m p_n}{K p_b}} + K_b p_b + K_m p_m + K_n p_n\right)^2$

NOTES: The rate equation is:

$$r = \frac{k \,(\text{driving force})}{\text{adsorption term}}$$

Adsorption rate of substance A is controlling in each case. When an inert substance I is adsorbed, the term $K_i p_i$ is to be added to the adsorption term.
SOURCE: From Walas, *Reaction Kinetics for Chemical Engineers*, McGraw Hill, 1959; Butterworths, 1989.

tical validity are:

1. $A\sigma + 2B\sigma \Rightarrow C + 3\sigma$

$$y = \left(\frac{p_a p_b}{r}\right)^{1/3} = 0.34(1 - 0.061p_a + 0.0032\sqrt{p_b} - 0.00046p_c)$$

2. $A\sigma + B_2\sigma \Rightarrow C + 2\sigma$

$$y = \left(\frac{p_a p_b}{r}\right)^{1/2} = 2.38(1 + 1.98p_b + 0.59p_c)$$

3. $r = 0.02 p_a^{1.33} p_b^{0.58} p_c^{-0.68}$

The data are partial pressures, atm and the rate r, g mol phosgene made/(h·g catalyst).

The first is ruled out because the constants physically cannot be negative. Although the other correlations are equally valid statistically, the Langmuir-Hinshelwood may be preferred to the power law form because it is more likely to be amenable to extrapolation.

CHEMICAL EQUILIBRIUM

The rate of a reversible reaction

$$a\text{A} + b\text{B} \underset{k_2}{\overset{k_1}{\rightleftharpoons}} c\text{C} + d\text{D}$$

may be written

$$r = k_1\left(C_a^a C_b^b - \frac{C_c^c C_d^d}{K_e}\right) \tag{7-64}$$

In terms of the compositions at equilibrium, the equilibrium constant is

$$K_e = \frac{C_{ce}^c C_{de}^d}{C_{ae}^a C_{be}^b}$$

With the aid of the stoichiometric "degree of advancement,"

$$\varepsilon = \frac{C_{a0} - C_a}{a} = \frac{C_{b0} - C_b}{b} = -\frac{C_{C0} - C_c}{c} = -\frac{C_{d0} - C_d}{d}$$

the equilibrium constant can be written in terms of a single variable. When several reactions occur simultaneously, each reaction is characterized by its own ε_i. When the K_es are known, the composition can be found by simultaneous solution of the several equations. The equilibrium composition of a mixture of known chemical species also can be found by a process of Gibbs energy minimization without the formulation of stoichiometric equations. Examples of the calculation of equilibria are in books on thermodynamics and in Walas (*Phase Equilibria in Chemical Engineering*, Butterworths, 1985).

The equilibrium constant depends on temperature according to

$$\frac{d \ln K}{dT} = \frac{\Delta H_r}{RT^2}$$

This is integrable to

$$\ln K = \ln K_{298} + \frac{1}{R}\int_{298}^T \frac{\Delta H_{r298} + \int_{298}^T \Delta Cp \, dT}{T^2} \, dT \tag{7-65}$$

where ΔH_r is the enthalpy change of reaction. Over a moderate temperature range, an adequate form of relation is

$$K = \exp\left(a + \frac{b}{t}\right) \tag{7-66}$$

Gaseous equilibria are expressed in terms of *fugacities* or fugacity coefficients. In terms of partial pressures, $p_i = y_i\pi$,

$$K_p = \frac{p_c^c p_d^d}{p_a^a p_b^b} = K_y \pi^{c+d-a-b} \tag{7-67}$$

Pressure affects the composition of an equilibrium mixture, but not the equilibrium constant itself.

Although the equilibrium constant can be evaluated in terms of kinetic data, it is usually found independently so as to simplify finding the other constants of the rate equation. With K_e known, the correct exponents of Eq. (7-64) can be found by choosing trial sets until k_1 comes out approximately constant. When the exponents are small integers or simple fractions, this process is not overly laborious.

Example 4: Reaction between Methane and Steam At 600°C the principal reactions between methane and steam are

$$
\begin{array}{cccccc}
CH_4 + H_2O \Leftrightarrow CO + 3H_2 & & CO + H_2O \Leftrightarrow CO_2 + H_2 \\
1-x \quad 5-x \quad x \quad 3x & & x-y \quad 5-x-y \quad y \quad 3x+y \quad \Sigma = 6+2x
\end{array}
$$

where $K_1 = 0.574$, $K_2 = 2.21$. Starting with 1 mol methane and 5 mol steam,

$$\frac{(x-y)(3x+y)^3}{(1-x)(5-x-y)(6+2x)^2} = 0.574, \qquad \frac{y(3x+y)}{(x-y)(5-x-y)} = 2.21$$

Simultaneous solution by the Newton-Raphson method yields $x = 0.9121$, $y = 0.6328$. Accordingly, the fractional compositions are:

$$CH_4 = \frac{(1-x)}{(6+2x)} = 0.0112$$

$$CO = \frac{(x-y)}{(6+2x)} = 0.0357$$

$$CO_2 = \frac{y}{(6+2x)} = 0.0809$$

$$H_2O = \frac{(5-x-y)}{(6+2x)} = 0.4416$$

$$H_2 = \frac{(3x+y)}{(6+2x)} = 0.4306$$

Approach to Equilibrium As equilibrium is approached the rate of reaction falls off, and the reactor size required to achieve a specified conversion goes up. At some point, the cost of increased reactor size will outweigh the cost of discarded or recycled unconverted material. No simple rule for an economic appraisal is really possible, but sometimes a basis of 95 percent of equilibrium conver-

sion is taken. For adiabatic operation, a certain approach to equilibrium temperature is common practice, say within 10 to 20°C (18 to 36°F), a number possibly based on experience with a particular process.

Example 5: Percent Approach to Equilibrium For a reversible reaction with rate equation $r = k[A^2 - (1 - A)^2/16]$, the size function kV_r/V' of a plug flow reactor will be found in terms of percent approach to equilibrium:

$$\frac{kV_r}{V'} = \int_A^1 \frac{dA}{A^2 - (1-A)^2/16}, \qquad A_{\text{equilib}} = 0.2000$$

Percent approach	70	90	95	98	99	99.5	100
A	0.440	0.280	0.240	0.216	0.208	0.204	0.2
kV_r/V'	1.319	3.053	4.309	6.090	7.600	9.315	∞

The volume escalates rapidly at high percent approaches.

INTEGRATION OF RATE EQUATIONS

In either batch or flow systems, many single-rate equations lead to integrands that are ratios of low-degree polynomials that can be integrated by inspection or with the briefest of integral tables. Some of the cases of frequent occurrence are summarized in Table 7-4. When the problem is to relate C and t, the constants are known, and the polynomials are of second degree or higher, numerical integration may save

TABLE 7-4   Some Isothermal Rate Equations and Their Integrals

1. $A \rightarrow$ Products:

$$-\frac{dA}{dt} = kA^q$$

$$\frac{A}{A_0} = \begin{cases} \exp\left[-k(t-t_0)\right], & q = 1 \\ \left[\dfrac{1}{1+(q-1)kA_0^{q-1}(t-t_0)}\right]^{1/(q-1)}, & q \neq 1 \end{cases}$$

2. $A + B \rightarrow$ Products:

$$-\frac{dA}{dt} = kAB = kA(A + B_0 - A_0)$$

$$k(t-t_0) = \frac{1}{B_0 - A_0} \ln \frac{A_0(A + B_0 - A_0)}{AB_0}$$

3. Reversible reaction $A \overset{k_1}{\underset{k_2}{\rightleftharpoons}} B$:

$$-\frac{dA}{dt} = k_1A - k_2(A_0 + B_0 - A) = (k_1 + k_2)A - k_2(A_0 + B_0)$$

$$(k_1 + k_2)(t - t_0) = \ln \frac{k_1A_0 - k_2B_0}{(k_1 + k_2)A - k_2(A_0 + B_0)}$$

4. Reversible reaction, second order, $A + B \overset{k_1}{\underset{k_2}{\rightleftharpoons}} R + S$:

$$-\frac{dA}{dt} = k_1AB - k_2RS = k_1A(A + B_0 - A_0)$$
$$\qquad - k_2(A_0 + R_0 - A)(A_0 + S_0 - A)$$
$$\qquad = \alpha A^2 + \beta A - \gamma$$
$$\alpha = k_1 - k_2$$
$$\beta = k_1(B_0 - A_0) + k_2(2A_0 + R_0 + S_0)$$
$$\gamma = k_2(A_0 + R_0)(A_0 + S_0)$$
$$q = \sqrt{\beta^2 + 4\alpha\gamma}$$

$$k(t - t_0) = \begin{cases} \dfrac{2\alpha A_0 + \beta}{2\alpha A + \beta}, & q = 0 \\ \dfrac{1}{q} \ln\left[\left(\dfrac{2\alpha A_0 + \beta - q}{2\alpha A_0 + \beta + q}\right)\left(\dfrac{2\alpha A + \beta + q}{2\alpha A + \beta - q}\right)\right], & q \neq 0 \end{cases}$$

5. The reaction $v_aA \rightarrow v_rR + v_sS$ between ideal gases at constant T and P:

$$-\frac{dn_a}{dt} = \frac{kn_a^\alpha}{V^{\alpha-1}}$$

$$V = n_t \frac{RT}{P} = \left[n_{t0} + \frac{\Delta v}{v_a}(n_{a0} - n_a)\right]\frac{RT}{P}$$

$$k(t - t_0) = \begin{cases} \displaystyle\int_{n_a}^{n_{a0}} \frac{V^{\alpha-1}}{n_a^\alpha} \, dr_{a}, & \text{in general} \\ \dfrac{RT}{P}\left[n_{t0} + \dfrac{\Delta v}{v_a}\left(\dfrac{1}{n_a} - \dfrac{1}{n_{a0}}\right)\right. \\ \left. \quad - \dfrac{\Delta v}{v_a}\ln\left(\dfrac{n_{a0}}{n_a}\right)\right], & \text{when } \alpha = 2 \end{cases}$$

6. Equations readily solvable by Laplace transforms. For example:

$$A \overset{k_1}{\underset{k_3}{\rightleftharpoons}} B \overset{k_2}{\rightarrow} C$$

Rate equations are

$$-\frac{dA}{dt} = k_1A - k_2B$$

$$-\frac{dB}{dt} = -k_1A + (k_2 + k_3)B$$

$$-\frac{dC}{dt} = -k_2B$$

Laplace transformations are made and rearranged to

$$(s + k_1)\overline{A} + k_3\overline{B} = A_0$$
$$-k_1\overline{A} + (s + k_2 + k_3)\overline{B} = B_0$$
$$-k_2\overline{B} + s\overline{C} = C_0$$

These linear equations are solved for the transforms as

$$D = s^2 + (k_1 + k_2 + k_3)s + k_1k_2$$

$$\overline{A} = \frac{A_0s + (k_2 + k_3)A_0 + k_3B_0}{D}$$

$$\overline{B} = \frac{B_0s + k_1(A_0 + B_0)}{D}$$

$$\overline{C} = \frac{k_2\overline{B} + C_0}{s}$$

Inversion of the transforms can be made to find the concentrations A, B, and C as functions of the time t.

SOURCE: Adapted from Walas, *Chemical Process Equipment Selection and Design*, Butterworth-Heinemann, 1990.

time and preserve reliability. Some 40 cases of integrations at constant volume are developed by Capellos and Bielski (*Kinetic Systems Mathematical Descriptions of Chemical Kinetics,* Wiley, 1972).

Sets of first-order rate equations are solvable by Laplace transform (Rodiguin and Rodiguina, *Consecutive Chemical Reactions,* Van Nostrand, 1964). The methods of linear algebra are applied to large sets of coupled first-order reactions by Wei and Prater (*Adv. Catal.,* **13,** 203 [1962]). Reactions of petroleum fractions are examples of this type.

Example 6: Laplace Transform Application   For the reaction

$$A \overset{1}{\Rightarrow} B \overset{2}{\underset{3}{\rightleftharpoons}} C$$

with $B_0 = C_0 = 0$

the rate equations are $\dfrac{dA}{dt} = k_1 A$

$$\frac{dB}{dt} = k_1 A - k_2 B + k_3(A_0 - A - B)$$

The transforms are $s\overline{A} - A_0 = -k_1 \overline{A}$

$$s\overline{B} = \frac{k_3 A_0}{s} + (k_1 - k_3)\,\overline{A} - (k_2 + k_3)\,\overline{B}$$

Explicitly, $\overline{A} = \dfrac{A_0}{s + k_1}$ $\overline{B} = \dfrac{A_0}{s + k_2 + k_3}\left(\dfrac{k_3}{s} + \dfrac{k_1 - k_3}{s + k_1}\right)$

A and *B* as functions of *t* are found by inversion with a table of *L-T* pairs.

When even second-order reactions are included in a group to be analyzed, individual integration methods may be needed. Three cases of coupled first- and second-order reactions will be touched on. All of them are amenable only with difficulty to the evaluation of specific rates from kinetic data. Numerical integrations are often necessary.

1. The reactions are $2A \overset{1}{\Rightarrow} B \overset{2}{\Rightarrow} C$. The partial solutions are

$$A = \frac{A_0}{(1 + 2k_1 A_0 t)}$$

$$\frac{dB}{dt} + k_2 B = \frac{k_1 A_0^2}{(1 + 2k_1 A_0 t)^2}$$

Although the differential equation is first-order linear, its integration requires evaluation of an infinite series of integrals of increasing difficulty.

2. The reactions are $A \overset{1}{\Rightarrow} B$ and $A + B \overset{2}{\Rightarrow} C$. After A is expressed in terms of B by elimination of *t*,

$$\frac{dB}{dt} = (k_1 - k_2 B)\left[A_0 - B_0 + B - 2k \ln\frac{B - k}{B + k}\right], \qquad k = \frac{k_1}{k_2}$$

but this cannot be integrated analytically.

3. For the reactions $A \overset{1}{\Rightarrow} B$, $2B \overset{2}{\Rightarrow} C$; $2A \overset{1}{\Rightarrow} B \overset{2}{\Rightarrow} C$; $2A \overset{1}{\Rightarrow} B$, $2B \overset{2}{\Rightarrow} C$; the rate equations are solved in terms of higher transcendental functions by Chien (*J. Am. Chem. Soc.,* **76,** 2256 [1948]). For the first case, with $B_0 = 0$:

$$A = \exp(-k_1 t)$$

$$B = A_0 \sqrt{\frac{\tau}{K}}\,\frac{iJ_1(\gamma) - \beta H_1^{(i)}(\gamma)}{J_0(\gamma) + \beta i H_0^{(i)}(\gamma)}$$

where $\tau = \exp(-k_1 t)$
    $K = k_1 k_2 A_0$
    $\gamma = 2i\sqrt{K\tau}$
    $\beta = iJ_1(\gamma)/H_1^{(i)}(\gamma)$

The notation of the Bessel functions is that of Jahnke and Emde (*Tables of Functions with Formulas and Curves,* Dover, 1945; Teubner, 1960).

IDEAL REACTORS

INTRODUCTION

A useful classification of kinds of reactors is in terms of their concentration distributions. The concentration profiles of certain limiting cases are illustrated in Fig. 7-3; namely, of batch reactors, continuously stirred tanks, and tubular flow reactors. Basic types of flow reactors are illustrated in Fig. 7-4. Many others, employing granular catalysts and for multiphase reactions, are illustrated throughout Sec. 23. The present material deals with the sizes, performances and heat effects of these ideal types. They afford standards of comparison.

In a batch reactor, all the reactants are loaded at once; the concentration then varies with time, but at any one time it is uniform throughout. Agitation serves to mix separate feeds initially and to enhance heat transfer. In a semibatch operation, some of the reactants are charged at once and the others are then charged gradually.

In an ideal continuously stirred tank reactor (CSTR), the conditions are uniform throughout and the condition of the effluent is the same as the condition in the tank. When a battery of such vessels is employed in series, the concentration profile is step-shaped if the abscissa is the total residence time or the stage number. The residence time of individual molecules varies exponentially from zero to infinity, as illustrated in Fig. 7-3e.

In another kind of ideal flow reactor, all portions of the feed stream have the same residence time; that is, there is no mixing in the axial direction but complete mixing radially. It is called a *plug flow reactor* (PFR), or a *tubular flow reactor* (TFR), because this flow pattern is characteristic of tubes and pipes. As the reaction proceeds, the concentration falls off with distance.

Often, complete mixing cannot be approached for economic reasons. Inactive or dead zones, bypassing, and limitations of energy input are common causes. Packed beds are usually predominantly used in plug flow reactors, but they may also have small mixing zones superimposed in series or in parallel. In tubular reactors for viscous fluids, laminar or non-Newtonian behavior gives rise to variations of residence time. Deviations from ideal behavior are analyzed at length in Sec. 23.

MATERIAL AND ENERGY BALANCES

These balances are based on the general conservation law,

$$\text{Input} + \text{Sources} = \text{Outputs} + \text{Sinks} + \text{Accumulation} \qquad (7\text{-}68)$$

The terms may be quantities or rates of flow of material or enthalpy. *Inputs* and *outputs* are streams that cross the vessel boundaries. A heat of reaction within the vessel is a *source*. A depletion of reactant in the vessel is a *sink*. *Accumulation* is the time derivative of the content of the reference quantity in the vessel; of the volume times the concentration, $\partial V_r C_a / \partial t$; or of the total enthalpy of the vessel contents, $\partial[W C_p (T - T_{\text{ref}})]/\partial t$.

BATCH REACTORS

Batch reactors are tanks, usually provided with agitation and some mode of heat transfer to maintain temperature within a desirable range. They are primarily employed for relatively slow reactions of several hours duration, since the downtime for filling and emptying large equipment may be an hour or so. Agitation maintains uniformity and improves heat transfer. Modes of heat transfer are illustrated in Figs. 23-1 and 23-2.

Except in the laboratory, batch reactors are mostly liquid phase. In semibatch operation, a gas of limited solubility may be fed in gradually as it is used up. Batch reactors are popular in practice because of their flexibility with respect to reaction time and to the kinds and quantities of reactions that they can process.

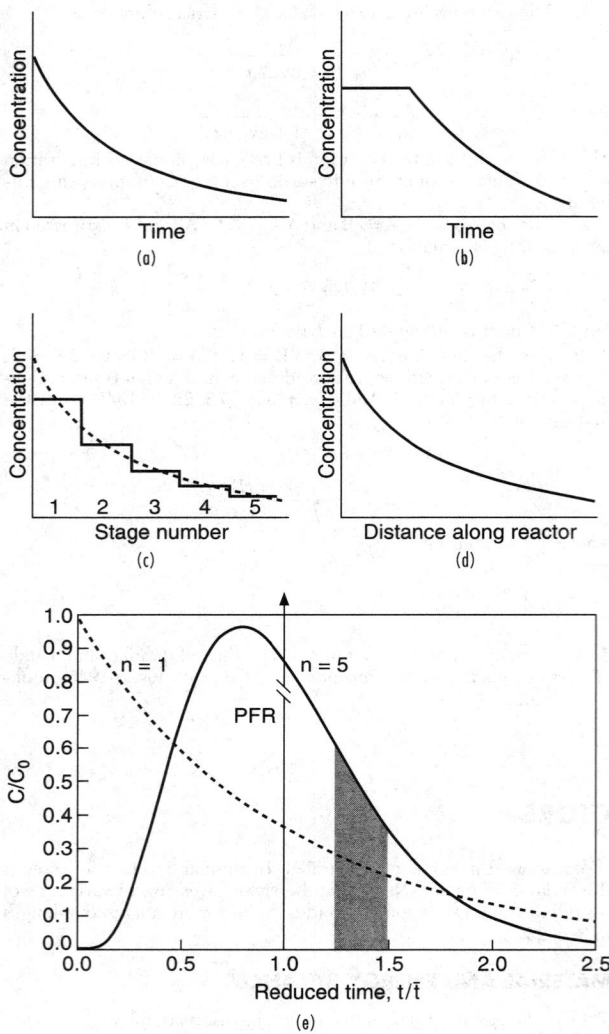

FIG. 7-3 Concentration profiles in batch and continuous flow: (a) batch time profile, (b) semibatch time profile, (c) five-stage distance profile, (d) tubular flow distance profile, (e) residence time distributions in single, five-stage, and PFR; the shaded area represents the fraction of the feed that has a residence time between the indicated abscissas.

Material and energy balances of a nonflow reactor are summarized in Table 7-5. Several batch operations are summarized in Fig. 7-5.

Daily Yield Say the downtime for filling and emptying a reactor is t_d and no reaction occurs during these periods. The reaction time t_r of a first-order reaction, for instance, is given by $kt_r = -\ln(1 - x)$. The daily yield with n batches per day will be

$$y = nV_r(C_0 - C) = \frac{24V_rC_0x}{t_r + t_d} = \frac{24V_rkC_0x}{-\ln(1-x) + kt_d} \tag{7-69}$$

Some conditions at which the daily yield is a maximum are

kt_d	0.01	0.1	0.5	5.0
x	0.13	0.45	0.68	0.88

Thus, the required conversion goes up as the downtime increases. Details are in Problem P2 of the "Solved Problems" subsection.

Filling and Emptying Periods Say the pumping rate is V', the full tank volume is V_{r1}, and the rate of reaction is $r = kC_a^q$. For $t \leq V_{r1}/V'$, the material balances with Eq. (7-68) are as follows.

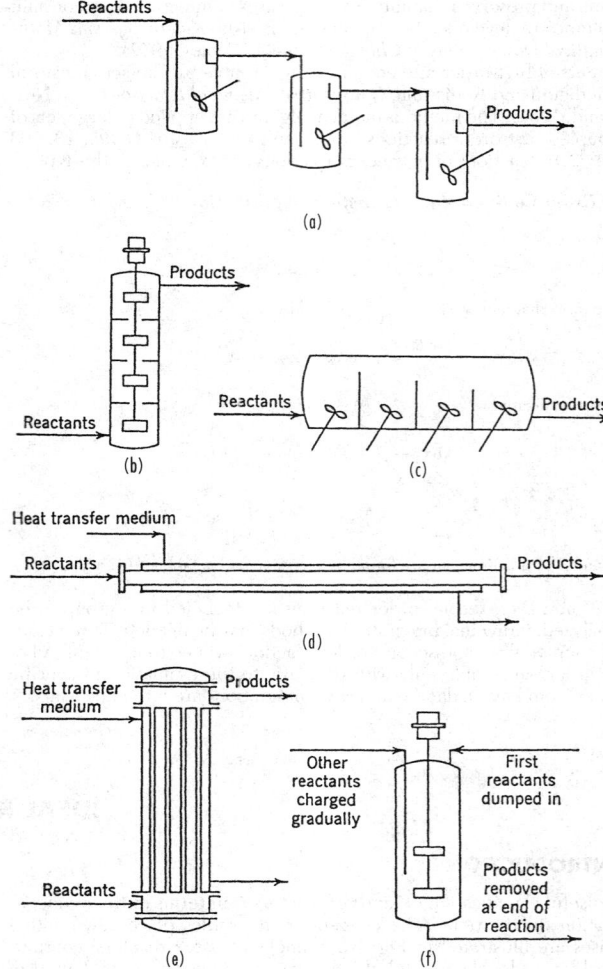

FIG. 7-4 Types of flow reactors: (a) stirred tank battery, (b) vertically staged, (c) compartmented, (d) single-jacketed tube, (e) shell and tube, (f) semiflow stirred tank.

Filling: $V_r = V't$, $C_a = C_{a0}$,   when $t = 0$,

$$V'C_{a0} = 0 + rV_r + \frac{d(V_rC_a)}{dt} = kV_rC_a^q + V'C_a + V't\frac{dC_a}{dt}$$

or $C_{a0} = C_a + k\left(\frac{V_r}{V'}\right)C_a^q + t\frac{dC_a}{dt}$ \hfill (7-70)

where the variables are separable.

Emptying: $V_r = V_{r1} - V't$, $C_a = C_{a1}$,   when $t = 0$,

$$0 = V'C_a + kV_rC_a^q + \frac{d(V_rC_a)}{dt} = V'C_a + kV_rC_a^q - V'C_a + V_r\frac{dC_a}{dt}$$

$$\frac{dC_a}{dt} = -kC_a^q \tag{7-71}$$

This is the same equation as for the full tank, but applies only for $t \leq V_{r1}/V'$. Figure 7-5e shows a complete batch cycle.

Optimum Operation of Reversible Reactions Often, equilibrium composition becomes less favorable and the rate of reaction becomes more favorable as the temperature increases, so a best condition may exist. If the temperature is adjusted at each composition to

TABLE 7-5 Material and Energy Balances of a Nonflow Reactor

Rate equations:

$$r_a = -\frac{1}{V_r}\frac{dn_a}{d\theta} = kC_a^\alpha = k\left(\frac{n_a}{V_r}\right)^\alpha \tag{1}$$

$$k = \exp\left(a' - \frac{b'}{T'}\right) \tag{2}$$

Heat of reaction:

$$\Delta H_r = \Delta H_{r298} + \int_{298}^{T}\Delta C_p\,dT \tag{3}$$

Rate of heat transfer:

$$Q' = UA(T_s - T) \tag{4}$$

(the simplest case is when UA and T_s are constant)

Enthalpy balance:

$$\frac{dT}{dn_a} = \frac{1}{\rho V_r \overline{C}_p}\left[\Delta H_r + \frac{UA(T_s - T)}{V_r k(n_a/V_r)^\alpha}\right] \tag{5}$$

$$\frac{dT}{dC_a} = \frac{1}{\rho \overline{C}_p}\left[\Delta H_r + \frac{UA(T_s - T)}{V_r k C_a^\alpha}\right] \tag{6}$$

$$T = T_0 \quad \text{when} \quad C_a = C_{a0} \tag{7}$$

$$\overline{C}_p = \frac{1}{\rho V_r}\sum n_i C_{pi} \tag{8}$$

Solve Eq. (6) to find $T = f(C_a)$; combine Eqs. (1) and (2) and integrate as

$$\theta = \int_{C_a}^{C_{a0}}\frac{1}{C_a^\alpha \exp[a' - b'/f(C_a)]}\,dC_a \tag{9}$$

SOURCE: Adapted from Walas, *Chemical Process Equipment Selection and Design*, Butterworth-Heinemann, 1990.

make the rate a maximum, then a minimum reactor size or maximum conversion will result. Take the first-order reversible process,

$$r = k_1(1 - x) - k_2 x$$

$k_1 = A_1 \exp(-B_1/T)$ and $k_2 = A_2 \exp(-B_2/T)$. The condition

$$(\partial r/\partial T)_x = 0$$

leads to

$$T = \frac{B_1 - B_2}{\ln\dfrac{A_1 B_1(1 - x)}{A_2 B_2 x}} \tag{7-72}$$

which tells what the temperature must be at each fractional conversion for the minimum reactor size. Practically, it may be difficult to vary the temperature of a batch reactor in this way, but the operation may be more nearly feasible with a CSTR battery or a PFR. Figure 7-5f shows an example of such a temperature profile for a batch reactor.

CONTINUOUS STIRRED TANK REACTORS (CSTR)

Flow reactors are used for greater production rates when the reaction time is comparatively short, when uniform temperature is desired, when labor costs are high. CSTRs are used singly or in multiple units in series, in either separate vessels or single, compartmented shells.

Material and energy balances are based on the conservation law, Eq. (7-69). In the operation of liquid phase reactions at steady state, the input and output flow rates are constant so the holdup is fixed. The usual control of the discharge is on the liquid level in the tank. When the mixing is adequate, concentration and temperature are uniform, and the effluent has these same properties. The steady state material balance on a reactant A is

$$V_0'C_{a0} = V'C_a + V_r r_a \tag{7-73}$$

Changes in density because of reaction or temperature changes are often small enough to be ignored. Then the volumetric flow rate is uniform and the balance becomes

$$C_{a0} = C_a + \bar{t}r_a \tag{7-74}$$

where the residence time is

$$\bar{t} = \frac{V_r}{V'} \tag{7-75}$$

A useful rearrangement,

$$r_a = \frac{C_{a0} - C_a}{\bar{t}} \tag{7-76}$$

emphasizes how CSTR measurements can provide data for the development of rate equations without integrating them.

During startup or discharge the material balance becomes

$$V_0'C_{a0} = V'C_a + V_r r_a + \frac{d(V_r C_a)}{dt} \tag{7-77}$$

where the reactor volume V_r is a known function of time.

For a power law rate equation at steady state,

$$C_{a0} = C_a + k\bar{t}C_a^q \tag{7-78}$$

A summary of material and energy balances is in Table 7-6.

For each vessel of a series,

$$C_{a,\,n-1} = C_{an} + \bar{t}_n r_{an} \tag{7-79}$$

The set of equations for all stages can be solved in succession, starting with the inlet to the first stage as C_{a0}.

Example 7: A Four-Stage Unit   When the material balances are $C_{n-1} = C_n + 1.5[C_n/(0.2 + C_n)]^2$ and $C_0 = 2$, the successive outlet concentrations are found by *RootSolver* to be 0.985, 0.580, 0.389, and 0.281.

The simplest problem is when all of the stages have the same $k\bar{t}$; then one of the three variables ($k\bar{t}$, n, or C_{an}) can be found when the others are specified. For first-order reactions,

$$\frac{C_{an}}{C_{a0}} = \frac{1}{(1 + k_1\bar{t}_1)(1 + k_2\bar{t}_2)\cdots(1 + k_n\bar{t}_n)} \tag{7-80}$$

$$\Rightarrow \frac{1}{(1 + k\bar{t}_{\text{total}}/n)^n} \tag{7-81}$$

for identical stages.

For multiple reactions, material balances are required for each stoichiometry.

Example 8: Consecutive Reactions   Take the reaction $A \xrightarrow{1} B \xrightarrow{2} C$, with $B_0 = C_0 = 0$. Define $\vartheta = k_1\bar{t}A_0$, $\alpha = 1/(1 + k_1\bar{t})$, $\beta = 1/(1 + k_2\bar{t})$. Then by setting up successive material balances, equations for the effluent from the nth stage are derived as

$$A_n = A_0\alpha^n \qquad B_n = \frac{\alpha\beta\vartheta}{\alpha - \beta}(\alpha^n - \beta^n) \qquad C_n = A_0 - A_n - B_n \tag{7-82}$$

When $n \Rightarrow \infty$, this equation reduces to

$$\frac{B}{A_0} = \frac{k_1}{k_2 - k_1}[\exp(-k_1\bar{t}) - \exp(-k_2\bar{t})] \tag{7-83}$$

This is the equation for a plug flow reactor. It can be derived directly from the rate equations with the aid of Laplace transforms. The sequences of second-order reactions of Figs. 7-5a and 7-5c required numerical integrations.

When CSTRs are operated in series, the sum of the reactor volumes drops off sharply with the number of stages. An economical number often is only 3 to 6, since the benefit of reduced volume may be outweighed by the increased cost of multiple agitators, pumps, and controls. When all stages are in a single shell, the economics are more favorable to large numbers of stages, but the single-shell arrangements lose some of the flexibility of the multiple-tank designs.

Example 9: Comparison of Batch and CSTR Volumes   For a first-order reaction, the ratio of n-stage CSTR and batch volumes is

$$\text{Ratio} = \frac{n[(C_0/C)^{1/n} - 1]}{\ln(C_0/C)}$$

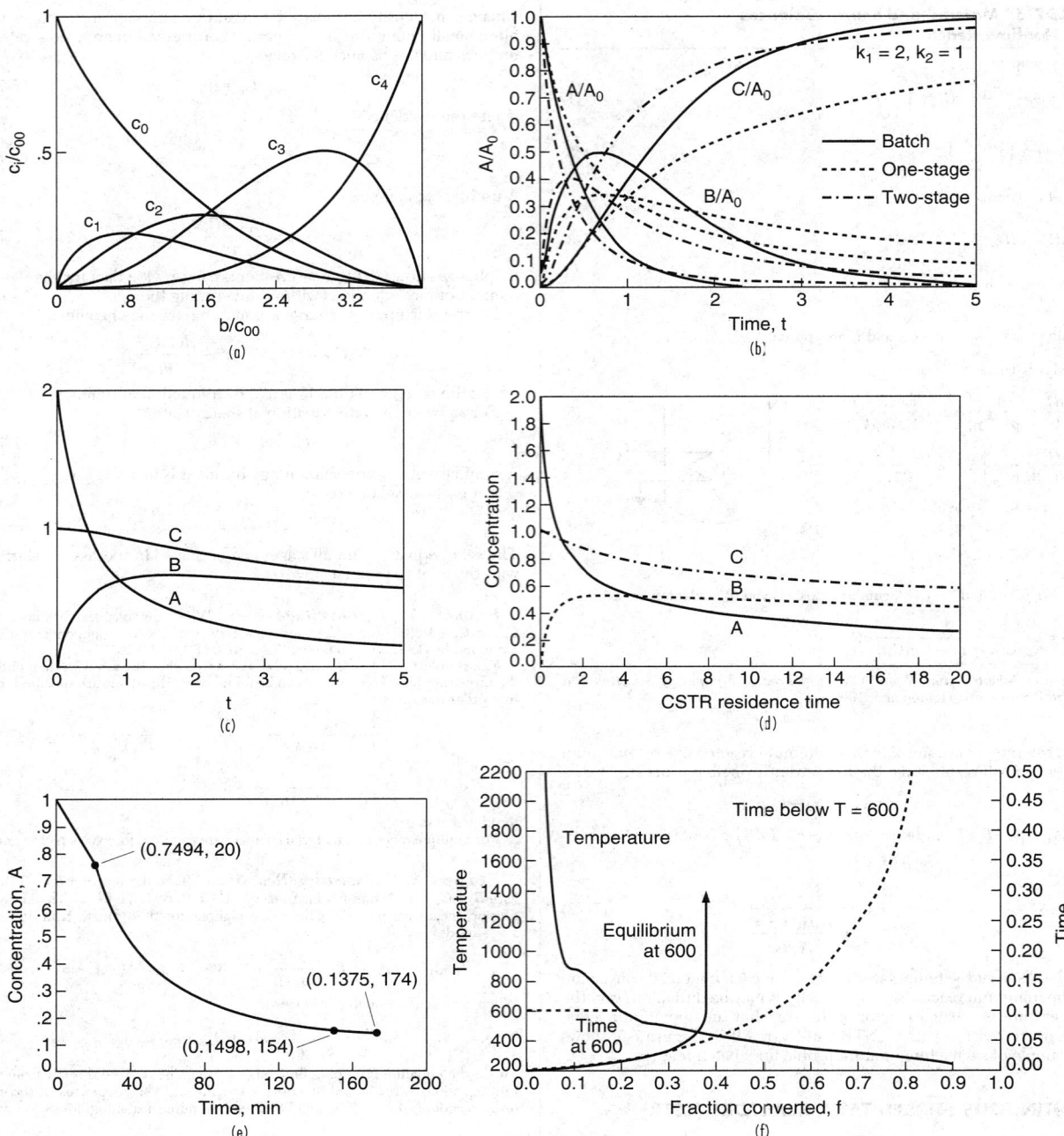

FIG. 7-5   Some batch operations, the P-code refers to detailed solutions in Walas (*Chemical Reaction Engineering Handbook of Solved Problems*, Gordon & Breach, 1995). (*a*) Methane chlorination in batch reactor or PFR; abscissa is ratio of chlorine to methane; P4.03.20. (*b*) Product yields of $A \Rightarrow B \Rightarrow C$ with $k_1 = 2$ and $k_2 = 1$ in batch reactor and CSTR; P4.04.60. (*c*) The reactions $2A \Rightarrow B$ and $2B + 2C \Rightarrow D$ with $k_1 = 1.0$ in batch reactor; P4.04.46. (*d*) Same as *c* but in CSTR. (*e*) Fill for 20 min, react, and discharge for 20 min with $r_a = 0.03[C_a(0.2 + C_a) - 0.04(1 - C_a)^2]$; P4.09.18. (*f*) Best temperature profile for a reversible reaction, $r = k_1(1 - x) - k_2 x$; when $t = 0.1$, $x = 0.37$ at 600 R, and $x = 0.51$ with optimum temperature profile; $x = 0.81$ when $t = 0.5$ and final temperature is 250 R; the full range is impractical; P4.11.02.

TABLE 7-6 Material and Energy Balance of a CSTR

The sketch identifies the nomenclature.

Mean residence time:

$$\bar{t} = \frac{V_r}{V'} \tag{1}$$

Temperature dependence:

$$k = \exp\left(a' - \frac{b'}{T}\right) \tag{2}$$

Rate equation:

$$r_a = kC_a^\alpha = kC_{a0}^\alpha(1-x)^\alpha, \qquad x = \frac{(C_{a0} - C_a)}{C_{a0}} \tag{3}$$

Material balance:

$$C_{a0} = C_a + k\bar{t}C_a \tag{4}$$

$$x = k\bar{t}C_{a0}^{\alpha-1}(1-x)^\alpha \tag{5}$$

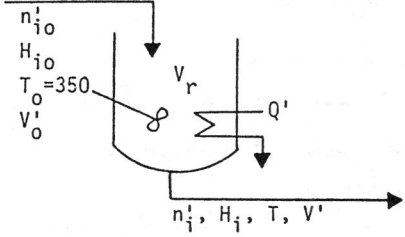

Enthalpy balance:

$$\sum n_i'H_i - \sum n_{i0}'H_{i0} = Q' - \Delta H_r(n_{a0}' - n_a') \tag{6}$$

$$H_i = \int_{298}^{T} C_{pi}\, dT \tag{7}$$

$$\Delta H_r = \Delta H_{r298} + \int_{298}^{T} \Delta C_p\, dT \tag{8}$$

For the reaction $aA + bB \rightarrow rR + sS$:

$$\Delta C_p = rC_{pr} + sC_{ps} - aC_{pa} - bC_{pb} \tag{9}$$

When the heat capacities are equal and constant, the heat balance is:

$$\overline{C}_p\rho V'(T - T_0) = Q' - \Delta H_{r298}V'(C_{a0} - C_a) \tag{10}$$

SOURCE: Adapted from Walas, *Chemical Process Equipment Selection and Design*, Butterworth-Heinemann, 1990.

Some values are

n	C_0/C		
	2	10	20
1	2.89	3.91	6.34
5	1.07	1.27	1.37

The ratio goes up sharply as the conversion increases and down sharply as the number of stages increases. For higher-order reactions the numbers are of comparable magnitudes.

Different Sizes Ordinarily, it is most economical to make all stages of a CSTR battery the same size. For a first-order reaction the resulting total volume is a minimum for a specified performance, but not so for other orders. Take a two-stage battery:

$$\bar{t}_1 + \bar{t}_2 = \frac{C_0 - C_1}{kC_1^q} + \frac{C_1 - C_2}{kC_2^q}$$

With C_0 and C_2 specified, the condition for a minimum is

$$\frac{\partial(\bar{t}_1 + \bar{t}_2)}{\partial C_1} = 0$$

and

$$C_1^{q+1} + C_2^q[(q-1)C_1 - qC_0] = 0$$

when $q = 1$, $C_1 = \sqrt{C_0 C_2}$, and $\bar{t}_1 = \bar{t}_2$

For higher orders $\bar{t}_1 \neq \bar{t}_2$ and the sum is less than twice the sum of equal stages, although usually not much different from that sum. As an example, when the second-order reaction between benzoquinone and cyclopentadiene is done in a three-stage unit, the reactor sizes are 3.25, 4.68, and 6.27, totaling 14.20, as compared to 14.56 with three equal stages (where consistent units are used). Details are in Walas (*Chemical Reaction Engineering Handbook of Solved Problems*, p. 4.11.15, Gordon & Breach, 1995).

Selectivity A significant respect in which CSTRs may differ from batch (or PFR) reactors is in the product distribution of complex reactions. However, each particular set of reactions must be treated individually to find the superiority. For the consecutive reactions $A \Rightarrow B \Rightarrow C$, Fig. 7-5b shows that a higher peak value of B is reached in batch reactors than in CSTRs; as the number of stages increases the batch performance is approached.

TUBULAR AND PACKED BED FLOW REACTORS

Tubular reactors are made up of one or more tubes in parallel, each of less than approximately 100-mm (3.94-in) diameter. With fluids of normal viscosity, plug flow exists in tubes of this size, with all molecules having essentially the same residence time. In packed beds of larger diameters, large-scale convection may be inhibited to such an extent that plug flow is also approached. Continuous gas phase reactions are predominantly done in such units, as are many liquid phase processes. Immiscible liquids are best handled in stirred tanks, although in-line mixers can facilitate such reactions in pipes. Reaction times are mostly short, made feasible by elevated temperatures. In such large-scale operations as oil cracking, the tubes may be several hundred meters long in a trombonelike arrangement. Temperature control is by heat transfer through the walls or by cold-shot injection. Shell-and-tube arrangements can provide large amounts of heat transfer. Product distribution of complex reactions is like that of batch reactors, but different from that of CSTRs.

Material and energy balances of a plug flow reactor are summarized in Table 7-7.

For convenience, the loading on a flow reactor is expressed as a size of reactor per unit of flow rate, say V_r/V', and is labeled the *space velocity*. Some of the units in practical use are stated in the Introduction. How the actual residence time is calculated when the density of flow varies is illustrated in Table 7-8.

Tubular flow reactors operate at nearly constant pressure. How the differential material balance is integrated for a number of second-order reactions will be explained. When n_a is the molal flow rate of reactant A the *flow reactor equation* is

$$-dn_a = n_{a0}dx = -V'dC_a = r_a dV_r \tag{7-84}$$

or

$$V_r = \int_{n_a}^{n_{a0}} \frac{dn_a}{r_a} \tag{7-85}$$

The equation is rendered integrable by application of the stoichiometry of the reaction, the ideal gas law, and, for instance, the power law for rate of reaction. Some details are shown in Table 7-9.

Frictional Pressure Drop Usually this does not have a significant effect on the reactor size, except perhaps when the flow is two-phase. Some approximate relations will be cited that are adequate for pressure-drop calculations of homogeneous flow reactions in pipelines. The pressure drop is given by

$$-dP = \frac{f\rho u^2}{2gD}\, dL \tag{7-86}$$

A good approximation to the friction factor in the turbulent flow range is

$$f = 0.046(\text{Re})^{-0.2} = 0.044\left(\frac{\mu D}{W}\right)^{0.2} \tag{7-87}$$

The mass flow rate is

$$W = 0.7854D^2\rho u$$

TABLE 7-7 Material and Energy Balances of a Plug Flow Reactor (PFR)

The balances are made over a differential volume dV_r of the reactor. Rate equation:

$$dV_r = \frac{-dn'_a}{r_a} \tag{1}$$

$$= -\frac{1}{k}\left(\frac{V'}{n'_a}\right)^\alpha dn'_a \tag{2}$$

$$= -\exp\left(-a' + \frac{b'}{T}\right)\left(\frac{n'_t RT}{Pn'_a}\right)^\alpha dn'_a \tag{3}$$

Enthalpy balance:

$$\Delta H_r = \Delta H_{r298} + \int_{298}^T \Delta C_p \, dT \tag{4}$$

$$dQ = U(T_s - T)\,dA_p = \frac{4U}{D}(T_s - T)\,dV_r$$

$$= -\frac{4U(T_s - T)}{Dr_a}dn'_a \tag{5}$$

$$dQ + \Delta H_r \, dn'_a = \sum n_i dH_i = \sum n_i C_{pi} dT \tag{6}$$

$$\frac{dT}{dn'_a} = \frac{\Delta H_r - 4U(T_s - T)/Dr_a}{\sum n_i c_{pi}} = f(T, T_s, n'_a) \tag{7}$$

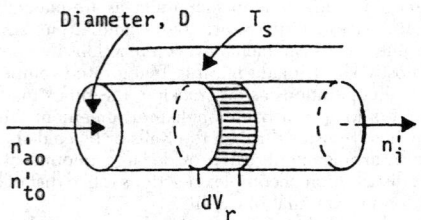

At constant T_s, Eq. (7) may be integrated numerically to yield the temperature as a function of the number of moles

$$T = \phi(n'_a) \tag{8}$$

Then the reactor volume is found by integration

$$V_r = \int_{n'_a}^{n'_{a0}} \frac{1}{\exp\left[a' - b'/\phi(n'_a)\right]\left[Pn'_a/n'_t R\phi(n'_a)\right]^\alpha}\,dn'_a \tag{9}$$

Adiabatic process:

$$dQ = 0 \tag{10}$$

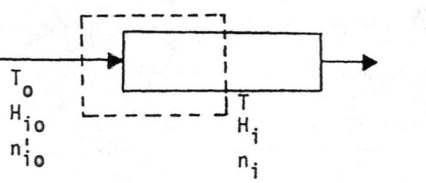

The balance around one end of the reactor is

$$\sum n_{i0}H_{i0} - \sum H_{r0}(n'_{a0} - n'_a) = \sum n_i H_i = \sum n_i \int C_{pi}\,dT \tag{11}$$

With reference temperature at T_0, enthalpies $H_{i0} = 0$

$$\Delta H_{r0} = \Delta H_{r298} + \int_{298}^{T_0} \Delta C_p \, dT \tag{12}$$

Substituting Eq. (12) into Eq. (10)

$$\left[-\Delta H_{r298} + \int_{298}^{T_0} \Delta C_p \, dT\right](n'_{a0} - n'_a) = \sum n_i \int_{T_0}^T C_{pi}\,dT \tag{13}$$

Adiabatic process with $\Delta C_p = 0$ and with constant heat capacities

$$T = T_0 - \frac{\Delta H_{r298}(n'_{a0} - n'_a)}{\sum n_i C_{pi}} \tag{14}$$

This expression is substituted instead of Eq. (8) to find the volume with Eq. (9).

The density in terms of the molecular weight M is

$$\rho = \frac{M}{V} = \frac{PM_0 n_{t0}}{RT n_t}$$

Also, in terms of the tube length dL,

$$dV_r = 0.7854 D^2 dL$$

Combining,

$$-dP = \frac{0.046 W^{1.8}\mu^{0.2}RT[n_{t0} + \delta_a(n_{a0} - n_a)]}{gD^{6.8}M_0 n_{r0}P}\,dV_r \tag{7-88}$$

This is to be solved simultaneously with the flow reactor equation, Eq. (7-84). Alternatively, dV_r can be eliminated from Eq. (7-88) for a direct relation between P and n_a.

More accurate relations than Eqs. (7-86) and (7-87) are described in Sec. 11 of this Handbook.

RECYCLE AND SEPARATION MODES

All reactor modes can sometimes be advantageously operated with recycling of part of the product or intermediate streams. Heated or cooled recycle streams serve to moderate undesirable temperature travels, and they can be processed for changes in composition before being returned.

Say the recycle flow rate in a PFR is V'_r and the fresh feed rate is V'_0, with the ratio $R = V'_r/V'_0$. With a fresh feed concentration of C_0 and a product of C_2 the composite feed concentration is

$$C_1 = \frac{C_0 + RC_2}{1 + R} \tag{7-89}$$

The change in concentration across the reactor becomes

$$\Delta C = C_1 - C_2 = \frac{C_2 - C_0}{1 + R} \tag{7-90}$$

Accordingly, the change in concentration (or in temperature) across the reactor can be made as small as desired by upping the recycle ratio. Eventually, the reactor can become a differential unit with substantially constant temperature, while substantial differences will concurrently arise between the fresh feed inlet and the product withdrawal outlet. Such an operation is useful for obtaining experimental data for analysis of rate equations.

In the simplest case, where the product is recycled without change, the flow reactor equation at constant density with a power law is

$$-V'_e(1 + R)dC = kC^q dV_r$$

and

$$V_r = V'_0(1 + R)\int_{C_2}^{C_1} \frac{dC}{kC^q} \tag{7-91}$$

Recycling increases the size of the reactor and degrades the plug flow characteristics, so there must be practical compensation by adjustment of the temperature or composition.

Example 10: Reactor Size with Recycle For first-order reaction with $C_0/C_2 = 10$ and $R = 5$, $C_1/C_2 = 2.5$. The relative reactor sizes with recycle and without are

$$\text{Ratio} = \frac{(1 + 5)\ln(C_1/C_2)}{\ln(C_0/C_2)} = \frac{5.497}{4.605} = 1.193$$

With reversible reactions, recycling is warranted when improvement in conversion can be realized by removing some of the product in a separator and returning only unconverted material. In some CSTR operations, the product is removed continuously by extraction or azeotropic distillation. The gasoline addi-

TABLE 7-8   True Contact Time in A PFR

The ratio V_r/V_0' is of the volume of the reactor to the incoming volumetric rate and has the dimensions of time. It will be compared with the true residence time when the number of mols changes as reaction goes on, or P and T also change.

$$r_\alpha = k\left(\frac{n_\alpha}{V'}\right)^q = k\left(\frac{\pi}{RT}\right)^q\left(\frac{n_\alpha}{n_t}\right)^q$$

$$V' = \frac{n_t RT}{\pi}$$

The differential balance on the reactant is:

$$-dn_\alpha = r_\alpha dV_r$$

$$dV_r = -\frac{1}{k}\left(\frac{RT}{\pi}\right)^q\left(\frac{n_t}{n_\alpha}\right)^q dn_\alpha$$

$$\frac{dV_r}{V_0'} = -\frac{1}{kn_{t0}}\left(\frac{RT}{\pi}\right)^{q-1}\left(\frac{n_t}{n_\alpha}\right)^q dn_\alpha \qquad (1)$$

The definition of the rate of reaction and the law of mass action is:

$$r_\alpha = -\frac{1}{V'}\frac{dn_\alpha}{dt} = k\left(\frac{n_\alpha}{V'}\right)^q$$

Rearrange to

$$dt = -\frac{1}{V'}\frac{dn_\alpha}{r_\alpha} = -\frac{1}{kV'}\left(\frac{V'}{n_\alpha}\right)^q dn_\alpha = -\frac{1}{k}\left(\frac{RT}{\pi}\right)^{q-1}\left(\frac{n_t^{q-1}}{n_\alpha^q}\right)dn_\alpha \qquad (2)$$

Eqs. (1) and (2) are the desired comparison. Before integrating, substitute

$$n_t = n_{t0} + \delta_\alpha(n_{\alpha0} - n_\alpha)$$

Example: Take $q = 1$, $n_\alpha = n_{\alpha0}$.

$$\frac{V_r}{V_0'} = \int_{n_\alpha}^{n_{\alpha0}}\frac{(\delta_\alpha+1)n_{\alpha0} - \delta_\alpha n_\alpha}{n_\alpha}\,dn_\alpha = \frac{1}{k}\left[(\delta_\alpha+1)\ln\frac{1}{1-x} - \delta_\alpha x\right]$$

$$t = \frac{1}{k}\int_{n_\alpha}^{n_{\alpha0}}\frac{dn_\alpha}{n_\alpha} = \frac{1}{k}\ln\frac{1}{1-x}$$

The ratio

$$y = \frac{t}{V_r/V_0'} = \delta_\alpha + 1 - \frac{\delta_\alpha x}{\ln[1/(1-x)]}$$

$$\begin{array}{lll} >1 & \text{when} & \delta_\alpha < 0 \\ <1 & \text{when} & \delta_\alpha > 0 \end{array}$$

tive methyl-*tert*-butyl ether is made in a distillation column where reaction and simultaneous separation occur.

HEAT EFFECTS

The heat balance of a reactor is made up of three terms: Heat of reaction + Heat transfer = Gain of sensible and latent heats by the mixture. This establishes the temperature as a function of the composition

$$T = f(n_a)$$

which may be substituted into the equations of the specific rate and the equilibrium constant

$$k = \exp\left[A + \frac{B}{f(n_a)}\right]$$

$$K_e = \exp\left[C + \frac{D}{f(n_a)}\right]$$

With these substitutions the rate equation remains a function of the composition alone.

Heat balances of several kinds of reactors are summarized in Tables 7-5, 7-6, 7-7 and 7-10.

Enthalpy changes of processes depend only on the end states. Normally the enthalpy change of reaction is known at some standard tem-

TABLE 7-9   Integration of Rate Equations of a PFR at Constant Pressure

Tubular flow reactors usually operate at nearly constant pressure. For a reactant A, the differential material balance is:

$$-dn_\alpha = n_{\alpha0}dx = -V'dC_\alpha = V'C_{\alpha0}dx = r_\alpha dV_r$$

One form of the integration is:

$$\frac{V_r}{n_{\alpha0}} = \int_{x_0}^{x}\frac{dx}{r_\alpha}, \qquad \frac{\text{reactor volume}}{\text{molal input rate}}$$

$$n_t = n_{t0} + \delta_\alpha(n_{\alpha0} - n_\alpha) = n_{t0} + \delta_\alpha x$$

$$V' = n_t RT/\pi$$

$$C_\alpha = \frac{n_\alpha}{V'} = \left(\frac{\pi}{RT}\right)\left(\frac{n_\alpha}{n_t}\right) = \left(\frac{\pi}{RT}\right)\left(\frac{n_{\alpha0} - x}{n_{t0} + \delta_\alpha x}\right)$$

The rate equations will be stated in these terms for a number of reactions. In all these cases, the integrands are ratios of second-degree equations. The moderately complex integrations are accomplished with the aid of a table of integrals, or by MATHEMATICA, or numerically when the constants are known.

$$\begin{array}{ll} 2\,A \Rightarrow M & (1) \\ A + B \Rightarrow M & (2) \\ 2A \Leftrightarrow M & (3) \\ A + B \Leftrightarrow M & (4) \end{array}$$

Part (1):

$$\delta_\alpha = \frac{(1-2)}{2} = -0.5$$

$$V' = \left(\frac{RT}{\pi}\right)(n_{t0} - 0.5x)$$

$$\frac{V_r}{n_{\alpha0}} = \int\frac{dx}{r_\alpha} = \frac{1}{k}(\overline{RT})^2\int_{x_0}^{x}\left(\frac{n_{t0} - 0.5x}{n_{\alpha0} - x}\right)^2 dx$$

Part (2):

$$\delta_\alpha = \frac{(1-2)}{1} = -1$$

$$V' = \left(\frac{RT}{\pi}\right)(n_{t0} - x)$$

$$\frac{V_r}{n_{\alpha0}} = \frac{1}{k}\left(\frac{\pi}{RT}\right)^2\int_{x_0}^{x}\frac{(n_{\alpha0} - x)(n_{b0} - x)}{(n_{t0} - x)^2}\,dx$$

Part (3):

$$\delta_a = -0.5$$

$$V' = \left(\frac{RT}{\pi}\right)(n_{t0} - 0.5x)$$

$$r_\alpha = k_1\left(\frac{RT}{\pi}\right)^2\left(\frac{n_{\alpha0} - x}{n_{t0} - 0.5x}\right)^2 - k_2\left(\frac{RT}{\pi}\right)\left(\frac{n_{m0} + 0.5n_{\alpha0}x}{n_{t0} - 0.5x}\right)$$

Part (4):

$$\delta_\alpha = -0.5$$

$$r_\alpha = k_1\left(\frac{RT}{\pi}\right)^2\frac{(n_{\alpha0} - x)(n_{b0} - x)}{(n_{t0} - 0.5x)^2} - k_2\left(\frac{RT}{\pi}\right)\left(\frac{n_{m0} + 0.5n_{\alpha0}x}{n_{t0} - 0.5x}\right)$$

perature, $T_b = 298$ K (536 R), for instance. The simplest formulation of the heat balance, accordingly, is to consider the reaction to occur at this temperature, transfer whatever heat is required, and raise the enthalpy of the reaction products to their final values.

Batch Reactions   For a batch reaction, the heat balance is

$$-(\Delta H_r)_{T_b}(n_{a0} - n_a) + Q = \sum n_i(H_{iT} - H_{iT_b}) \qquad (7\text{-}92)$$

TABLE 7-10   Material and Energy Balances of a Packed Bed Reactor

Diffusivity and thermal conductivity are taken appreciable only in the radial direction.

Material balance equation:

$$\frac{\partial x}{\partial z} - \frac{D}{u}\left(\frac{\partial^2 x}{\partial r^2} + \frac{1}{r}\frac{\partial x}{\partial r}\right) - \frac{\rho}{u_0 C_0} r_c = 0 \qquad (1)$$

Energy balance equation:

$$\frac{\partial T}{\partial z} - \frac{k}{GC_p}\left(\frac{\partial^2 T}{\partial r^2} + \frac{1}{r}\frac{\partial T}{\partial r}\right) + \frac{\Delta H_r \rho}{GC_p} r_c = 0 \qquad (2)$$

At the inlet:

$$x(0, r) = x_0 \qquad (3)$$

$$T(0, r) = T_0 \qquad (4)$$

At the center:

$$r = 0, \qquad \frac{\partial x}{\partial r} = \frac{\partial T}{\partial r} = 0 \qquad (5)$$

At the wall:

$$r = R, \qquad \frac{\partial x}{\partial r} = 0 \qquad (6)$$

$$\frac{\partial T}{\partial r} = \frac{U}{k}(T' - T) \qquad (7)$$

When the temperature T' of the heat-transfer medium is not constant, another enthalpy balance must be formulated to relate T' with the process temperature T.

A numerical solution of these equations may be obtained in terms of finite difference equivalents, taking m radial increments and n axial ones. With the following equivalents for the derivatives, the solution may be carried out by direct iteration:

$$r = m(\Delta r)$$

$$z = n(\Delta z) \qquad (8)$$

$$\frac{\partial T}{\partial z} = \frac{T_{m,n+1} - T_{m,n}}{\Delta z} \qquad (9)$$

$$\frac{\partial T}{\partial r} = \frac{T_{m+1,n} - T_{m,n}}{\Delta r} \qquad (10)$$

$$\frac{\partial^2 T}{\partial r^2} = \frac{T_{m+1,n} - 2T_{m,n} + T_{m-1,n}}{(\Delta r)^2} \qquad (11)$$

Expressions for the x derivatives are of the same form:

r_c = rate of reaction, a function of s and T
G = mass flow rate, mass/(time)(superficial cross section)
u = linear velocity
D = diffusivity
k = thermal conductivity

SOURCE: Adapted from Walas, *Chemical Process Equipment Selection and Design*, Butterworth-Heinemann, 1990.

The solvent, as well as any other inerts, and the mass of the vessel are included in this summation. The heat exchange through a jacket or coils at temperature T_m is

$$Q = UA(T_m - T) \qquad (7-93)$$

When phase changes are absent,

$$-(\Delta H_r)_{T_b}(n_{a0} - n_a) + UA(T_m - T) = \sum n_i \int_{T_b}^{T} C_{pi}\, dT \qquad (7-94)$$

When the mixture can be characterized by an overall heat capacity,

$$-(\Delta H_r)_{T_b}(n_{a0} - n_a) + UA(T_m - T) = V_r \rho C_p(T - T_b) \qquad (7-95)$$

or

$$-(\Delta H_r)_{T_b}(C_{a0} - C_a) + \left(\frac{UA}{V_r}\right)(T_m - T) = \rho C_p(T - T_b) \qquad (7-96)$$

CSTR Reactions   For a CSTR reaction, the quantities n_i are molal flow rates. Per unit of time,

$$-(\Delta H_r)_{T_b} V_r r_a + Q = \sum n_i (H_{iT} - H_{iT_b}) \qquad (7-97)$$

$$\Rightarrow \sum n_i \int_{T_b}^{T} C_{pi}\, dT \qquad (7-98)$$

The last equation applies in the absence of phase change.

Plug Flow Reactions   The differential relations in a cylindrical vessel are

$$dA = \left(\frac{4}{D}\right) dV_r \qquad -\Delta H_{rT}\, dn_a + \frac{4UA}{D}\, dV_r = \sum n_i C_{pi}\, dT \qquad (7-99)$$

$$\Rightarrow n_t C_{pt}\, dT \qquad (7-100)$$

Note that the enthalpy change of reaction is a function of temperature, but a mean value often is adequate.

The various heat balances are to be solved simultaneously with the appropriate material balances, but when the temperatures can be solved for explicitly their equivalents are simply substituted into the equations for k and K_e and the material balance is solved alone.

Packed Bed Reactors   The commonest vessels are cylindrical. They will have gradients of composition and temperature in the radial and axial directions. The partial differential equations of the material and energy balances are summarized in Table 7-10. Example 4 of "Modeling of Chemical Reactions" in Sec. 23 is an application of such equations.

A variety of provisions for heat transfer are illustrated in Figs. 23-1 to 23-3 and elsewhere in Sec. 23.

UNSTEADY CONDITIONS WITH ACCUMULATION TERMS

Unsteady material and energy balances are formulated with the conservation law, Eq. (7-68). The sink term of a material balance is $V_r c_a$ and the accumulation term is the time derivative of the content of reactant in the vessel, or $\partial(V_r C_a)/\partial t$, where both V_r and C_a depend on the time. An unsteady condition in the sense used in this section always has an accumulation term. This sense of unsteadiness excludes the batch reactor where conditions do change with time but are taken account of in the sink term. Startup and shutdown periods of batch reactors, however, are classified as unsteady; their equations are developed in the "Batch Reactors" subsection. For a semibatch operation in which some of the reactants are preloaded and the others are fed in gradually, equations are developed in Example 11, following.

For a CSTR the unsteady material balance is

$$V'C_{a0} = V'C_a + V_r r_a + \frac{d(V_r C_a)}{dt} \qquad (7-101)$$

Enthalpy balances also will have accumulation terms.

Conditions that give rise to unsteadiness are changes in feed rate, composition, or temperature. In the case of Fig. 7-6, a sinusoidal input of feed rate is introduced. The output concentration also appears to vary sinusoidally. The amplitude of the response is lower as the specific rate is increased.

If a sinusoidal variation of the temperature of the heat transfer medium in the jacket or coil occurs, say

$$T_m = T_{m0}(1 + \alpha \sin \beta t) \qquad (7-102)$$

the balances will be

$$-\Delta H_r V_r r_a + UA(T_m - T) = V'\rho C_p(T - T_0) + \rho V_r C_p \frac{dT}{dt} \qquad (7-103)$$

$$V'C_{a0} = V'C_a + V_r r_a + V_r \frac{dC_a}{dt} \qquad (7-104)$$

Since each input of mass to a perfect plug flow unit is independent of what has been input previously, its condition as it moves along the

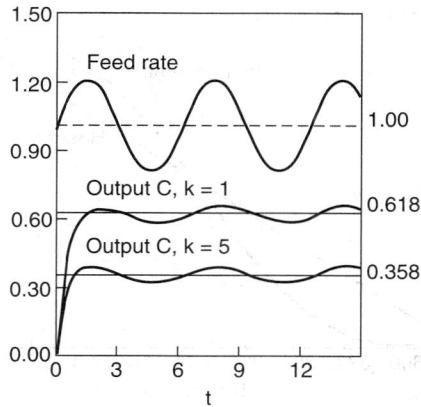

FIG. 7-6 Sinusoidal input of feed rate to a CSTR. Input: $F = 1 + 0.2 \sin(t)$; output: $dC/dt = (1 - C)[1 + 0.2 \sin(t)] - kc^2$. Straight lines are for constant feed rate.

reactor will be determined solely by its initial condition and its residence time, independently of what comes before and after. Practically, of course, some interaction will occur at the boundaries of successive inputs of different compositions or temperatures. This is governed by diffusional behaviors that are beyond the scope of the present work.

Example 11: Balances of a Semibatch Process   The reaction A + B ⇒ Products is carried out by first charging B into the vessel to a concentration C_{b0} and a volume V_{r0}, then feeding a solution of concentration C_{a0} at volumetric rate V' for a time t.

Volume of solution in the tank:

$$V_r = V_{r0} + V't \qquad (7\text{-}105)$$

Stoichiometric balance:

$$V'tC_{a0} - V_rC_a = V_{r0}C_{b0} - V_rC_b$$

$$C_b = C_a + \frac{V_{r0}C_{b0} - V'tC_{a0}}{V_{r0} + V't} \qquad (7\text{-}106)$$

Material balance on A:

$$\text{Input} = \text{Output} + \text{Sink} + \text{Accumulation}$$

$$V'C_{a0} = 0 + kV_rC_aC_b + \frac{d(C_aV_r)}{dt}$$

$$= kV_rC_aC_b + V_r\frac{dC_a}{dt} + C_aV'$$

$$\frac{dC_a}{dt} + kC_aC_b + \frac{V'C_a}{V_r} = \frac{V'C_{a0}}{V_r} \qquad (7\text{-}107)$$

Eqs. (7-105), (7-106), and (7-107) are combined into

$$\frac{dC_a}{dt} = \frac{V'}{V_{r0} + V't}(C_{a0} - C_a) - kC_a\left(C_a + \frac{V_{r0}C_{b0} - V'tC_{a0}}{V_{r0} + V't}\right) \qquad (7\text{-}108)$$

A numerical integration is required.

LARGE SCALE OPERATIONS

INTRODUCTION

In this category are included a number of topics that become especially significant on the industrial scale. Some of this material is covered at length in Sec. 23, so only an outline is provided here.

MULTIPLE STEADY STATES

Phenomena of multiple steady states and instabilities occur particularly with nonisothermal CSTRs. Some isothermal processes with hyperbolic rate equations and processes with porous catalysts also can have such behavior.

Mathematically, multiplicities become evident when heat and material balances are combined. Both are functions of temperature, the latter through the rate equation which depends on temperature by way of the Arrhenius law. The curves representing these balances may intersect in several points. For first order in a CSTR, the material balance in terms of the fraction converted can be written

$$x = \frac{k\bar{t}}{1 + k\bar{t}}, \qquad k = \exp\left(a - \frac{b}{T}\right) \qquad (7\text{-}109)$$

and the energy balance

Heat generation = Sensible heat gain

$$\frac{-\Delta H_r V_r C_f(1-x)}{\bar{t}} + UA(T_m - T) = \rho C_p V_r(T - T_f) \qquad (7\text{-}110)$$

These balances can be plotted two ways, as shown in Fig. 7-7:

1. x from both equations can be plotted against T, with the intersections at the steady state values of T and corresponding values of x.

2. The LHS (heat generation) and RHS (heat removal) of Eq. (7-110) are plotted against T after x has been eliminated between the two balances; the intersections identify the same steady state temperatures as the plot in Fig. 7-7a.

Conditions at which the slope of the heat generation line is greater than that of the heat removal line are unstable, and where it is less the condition is stable (see Fig. 7-7b). At an unstable point, any fluctua-

tion in conditions will move the temperature to a neighboring point. Control systems always produce small fluctuations of the process variables, as in the sinusoidal case of Fig. 7-6. If the fluctuations occur while the system is at an unstable point, the steadiness will disappear. In the case of Fig. 7-7c, as the unstable position is approached ($T = 280$, $C = 2.4$) the profiles of T and C become erratic and eventually degenerate to the condition at the stable point on the right (Figs. 7-7d and 7-7e).

Either of the two stable operating conditions can be selected by adjusting the positions of the curves so that only one intersection is obtained. In a plant, long-time unstable operation is unlikely because of imprecise temperature control.

Plug flow reactors with recycle exhibit some of the characteristics of CSTRs, including the possibility of multiple steady states. This topic is explored by Perlmutter (*Stability of Chemical Reactors*, Prentice-Hall, 1972).

Endothermic reactions possess only one steady state.

For complex reactions and with multistage CSTRs, more than three steady states can exist (as in Fig. 23-17c). Most of the work on multiplicities and instabilities has been done only on paper. No plant studies and a very few laboratory studies are mentioned in the comprehensive reviews of Razon and Schmitz (*Chem. Eng. Sci.*, **42**, 1,005–1,047 [1987]) and Morbidelli et al. (in Carberry and Varma, *Chemical Reaction and Reactor Engineering*, Dekker, 1987, pp. 973–1,054).

NONIDEAL BEHAVIOR

Reactors that are nominally CSTRs or PFRs may in practice deviate substantially from ideal mixing or nonmixing. This topic is developed at length in Sec. 23, so only a few summary statements are made here. More information about this topic also may be found in Nauman and Buffham (*Mixing in Continuous Flow Systems*, Wiley, 1983).

Laminar Flow   With highly viscous fluids the linear velocity along a streamline varies with the radial position. Laminar flow is characteristic of some polymeric systems. Figure 23-21 shows how the conversion is poorer in laminar flow than with uniform flow over the

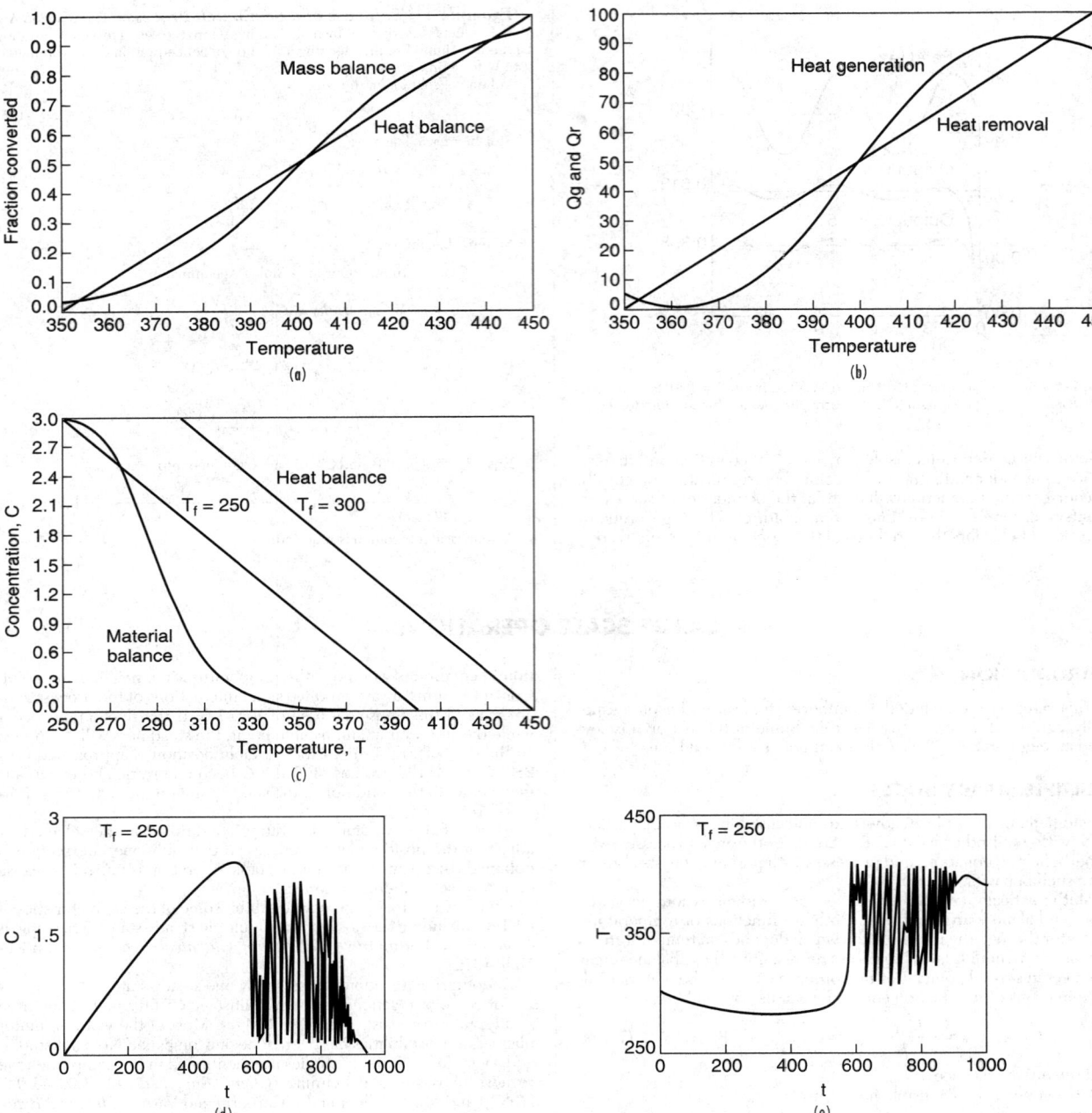

FIG. 7-7 Multiplicity and instability of first-order reactions in CSTRs. For (a) and (b), $k = \exp(25 - 10{,}000/T)$, $x = k/(1 + k)$, $200x + (350 - T) = T - 350$. For curves (c) to (e), $k = \exp(25 - 7{,}550/T)$, $C = 3/(1 + 300k)$, $C = 3 - 0.2(T - T_f)$. Unsteady state, $300\,dC/dt = 3 - (1 + 300k)C$, $C_0 = 0$; $300\,dT/dt = T_f - T + 15{,}000kC$, $T_0 = 300$. Temperature in °C.

cross section for first- and second-order reactions. Another adverse effect with viscous solutions is poor heat transfer. Accordingly, stirred tanks are often preferred to tubular units for such applications. The equations for radial and axial distributions of composition and temperature in laminar flow are studied by Nauman (*Chemical Reactor Design*, Wiley, 1987, pp. 165–203).

Residence Time Distribution (RTD) This is established by injecting a known amount of tracer into the feed stream and monitor-

ing its concentration in the effluent. At present there are no correlations of this kind of behavior that could be used for design of a new process, but such information about existing units is of value for diagnostic purposes.

The RTD is a distinctive characteristic of mixing behavior. In Fig. 7-3e, the CSTR has an RTD that varies as the negative exponential of the time and the PFR is represented by a vertical line at $t_r = 1$. Multistage units and many packed beds have bell-shaped RTDs, like that of the five-stage unit of Fig. 7-3e and the large-scale units of Figs. 23-10 and 23-11. An equation that represents such shapes is called the Erlang,

$$\text{RTD} = \frac{n^n}{\Gamma(n)} \, t_r^{n-1} \exp(-nt_r) \tag{7-111}$$

where n is interpreted as a number of CSTRs in series. When n is integral, $\Gamma(n)$ is replaced by $(n-1)!$. Plots of these curves are shown in Fig. 23-9.

When the RTD of a vessel is known, its performance as a reactor for a first-order reaction, and the range within which its performance will fall for other orders, can be predicted.

Segregated Flow   A real example is *bead polymerization* of styrene and some other materials. The reactant is in the form of individual small beads suspended in a fluid and retarded from agglomeration by colloids on their surfaces. Accordingly, they go through the reactor as independent bodies and attain conversions under batch conditions with their individual residence times. This is called *segregated flow*. With a particular RTD, conversion is a maximum with this flow pattern. The mean conversion of all the segregated elements then is given by

$$\frac{\overline{C}}{C_0} = \int_0^\infty (\text{RTD}) \left(\frac{C}{C_0} \right)_{\text{batch}} dt_r \tag{7-112}$$

For first order

$$\left(\frac{C}{C_0} \right)_{\text{batch}} = \exp(-k\bar{t}t_r) \tag{7-113}$$

For second order

$$\left(\frac{C}{C_0} \right)_{\text{batch}} = \frac{1}{(1 + kC_0\bar{t}t_r)} \tag{7-114}$$

Example 12: Segregated Flow   The pilot unit of Fig. 23-11 with $n = 9.3$ has

$$\text{RTD} = 13188 t_r^{8.3} \exp(-9.3t_r)$$

Some values of mean concentration ratio $\overline{C}/C_0$ of first- and second-order reactions obtained with Eq. (7-112) are:

$k\bar{t}$ or $kC_0\bar{t}$	1	2	5	10
First order $\overline{C}/C_0$	0.386	0.163	0.018	
Second order $\overline{C}/C_0$	0.513	0.349	0.180	0.100
Second order PFR	0.500	0.333	0.167	0.091

Maximum Mixedness   With a particular RTD, this pattern provides a lower limit to the attainable conversion. It is explained in Sec. 23. Some comparisons of conversions with different flow patterns are made in Fig. 23-14. Segregated conversion is easier to calculate and is often regarded as a somewhat plausible mechanism, so it is often the only one taken into account.

Dispersion   In tubes, and particularly in packed beds, the flow pattern is disturbed by eddies whose effect is taken into account by a dispersion coefficient in Fick's diffusion law. A PFR has a dispersion coefficient of 0 and a CSTR of ∞. Some rough correlations of the Peclet number uL/D in terms of Reynolds and Schmidt numbers are Eqs. (23-47) to (23-49). There is also a relation between the Peclet number and the value of n of the RTD equation, Eq. (7-111). The dispersion model is sometimes said to be an adequate representation of a reactor with a "small" deviation from plug flow, without specifying the magnitude of *small*. As a point of superiority to the RTD model, the dispersion model does have the empirical correlations that have been cited and can therefore be used for design purposes within the limits of those correlations.

OPTIMUM CONDITIONS

Optimization of a process is an activity whereby the *best* conditions are found for attainment of a maximum or minimum of some desired objective. In the broadest sense, an industrial process has maximum profit as its goal, but there are also problems with less-ambitious goals that do not involve money or the whole plant.

The best quality to be found may be a temperature, a temperature program, a concentration, a conversion, a yield of preferred product, a cycle period for a batch reaction, a daily production level, a kind of reactor, a size for a reactor, an arrangement of reactor elements, provisions for heat transfer, profit or cost, and so on—a maximum or minimum of some of these factors. Among the constraints that may be imposed on the process are temperature range, pressure range, corrosiveness, waste disposal, and others.

Once the objective and the constraints have been set, a mathematical model of the process can be subjected to a search strategy to find the optimum. Simple calculus is adequate for some problems, or Lagrange multipliers can be used for constrained extrema. When a full plant simulation can be made, various alternatives can be put through the computer. Such an operation is called *flowsheeting*. A chapter is devoted to this topic by Edgar and Himmelblau (*Optimization of Chemical Processes*, McGraw-Hill, 1988) where they list a number of commercially available software packages for this purpose, one of the first of which was Flowtran.

With many variables and constraints, linear and nonlinear programming may be applicable, as well as various numerical gradient search methods. Maximum principle and dynamic programming are laborious and have had only limited applications in this area. The various mathematical techniques are explained and illustrated, for instance, by Edgar and Himmelblau (*Optimization of Chemical Processes*, McGraw-Hill, 1988).

A few specific conclusions about optimum performance can be stated:

1. The minimum total volume of a CSTR battery for first-order reaction, and near-minimum for second-order, is obtained when all vessels are the same size.
2. An economical optimum number of CSTRs and their auxiliaries in series is 4 to 5.
3. In a sequence of PFR and CSTR, better performance is obtained with the PFR last. Performance of reversible reactions is improved with the CSTR at a higher temperature.
4. For the consecutive reactions $A \Rightarrow B \Rightarrow C$, a higher yield of intermediate B is obtained in batch reactors or PFRs than in CSTRs.
5. When the desirable product of a complex reaction is favored by a high concentration of some reactant, batch or semibatch reactors can be made superior to CSTRs.
6. Conversion by a reversible reaction is enhanced by starting out at high temperature and ending at low temperature if equilibrium conversion drops off at high temperature.
7. For a reversible reaction, the minimum size or maximum conversion is obtained when the rate of reaction is kept at a maximum at each conversion by adjustment of the temperature.

Variables   It is possible to identify a large number of variables that influence the design and performance of a chemical reactor with heat transfer, from the vessel size and type; catalyst distribution among the beds; catalyst type, size, and porosity; to the geometry of the heat-transfer surface, such as tube diameter, length, pitch, and so on. Experience has shown, however, that the reactor temperature, and often also the pressure, are the primary variables; feed compositions and velocities are of secondary importance; and the geometric characteristics of the catalyst and heat-exchange provisions are tertiary factors. Tertiary factors are usually set by standard plant practice. Many of the major optimization studies cited by Westerterp et al. (1984), for instance, are devoted to reactor temperature as a means of optimization.

The complexity of temperature regulation of three major commercial reversible processes are represented in Figs. 23-3a, 23-3e, and 23-3f. Presumably, these profiles have been established by fine-tuning the operations over a period of time.

Objective Function   This is the quantity for which a minimax is sought. For a complete manufacturing plant, it is related closely to the economy of the plant. Subsidiary problems may be to optimize conversion, production, selectivity, energy consumption, and so on in terms of temperature, pressure, catalyst, or other pertinent variables.

Case Studies   Several collections of more or less detailed solutions of optimization problems are cited, as follows.

1. Of the 23 studies listed under "Modeling of Chemical Reactors" in Sec. 23, a number are optimization oriented. Added to them

may be a detailed study of an existing sulfuric acid plant by Crowe et al. (*Chemical Plant Simulation*, Prentice-Hall, 1971).

2. Chen (*Process Reactor Design*, Allyn & Bacon, 1983) does the following examples mostly with simple calculus:

Batch reactors—optimum residence time for series and complex reactions, minimum cost, optimal operating temperature, and maximum rate of reaction

CSTRs—minimum volume of battery, maximum yield, optimal temperature for reversible reaction, minimum total cost, reactor volume with recycle, maximum profit for reversible reaction with recycle, and heat loss

Tubular flow reactors—minimum volume for second-order reversible reactions, maximum yield of consecutive reactions, minimum cost with and without recycle, and maximum profit with recycle

Packed bed reactor optimization

Size comparison for first- and second-order and reversible reactions

Selectivity of parallel and consecutive reactions and of reactions in a porous catalyst

3. Edgar and Himmelblau (*Optimization of Chemical Processes*, McGraw-Hill, 524–550, 1988) supply many references to other problems in the literature:

Optimal residence time for the reactions A $\Leftrightarrow$ B followed by B $\Rightarrow$ P or X

Optimal time for a biochemical CSTR

Selection of feedstock for thermal cracking to ethylene by linear programming

Maximum yield from a four-stage CSTR by nonlinear programming

Optimal design of ammonia synthesis by differential equation solution and a numerical gradient search

A C_4 alkylation process by sequential quadratic programming

4. Westerterp, van Swaaij, and Beenackers (*Chemical Reactor Design and Operation*, Wiley, 1984, pp. 674–746) also supply many references to other problems in the literature:

Optimized costs for several gas phase reactions: (1) A + B $\Rightarrow$ P; (2) A + B $\Leftrightarrow$ P; and (3) A + B $\Rightarrow$ P, A + 2B $\Rightarrow$ X, P + B $\Rightarrow$ X

Ammonia cold-shot converter

Maximum yield of first-order consecutive reactions in CSTR by application of Lagrange multipliers

Autothermal reactor for methanol synthesis using a numerical search technique

Minimum reactor volumes of isothermal and nonisothermal cascades by dynamic programming

Optimum temperature profiles of 2A $\Rightarrow$ B $\Rightarrow$ P by the maximum principle

Optimizing the temperature for A $\Rightarrow$ P and A $\Rightarrow$ X by the maximum principle

Westerterp et al. (1984; see Case Study 4, preceding) conclude, "Thanks to mathematical techniques and computing aids now available, any optimization problem can be solved, provided it is realistic and properly stated. The difficulties of optimization lie mainly in providing the pertinent data and in an adequate construction of the objective function."

HETEROGENEOUS REACTIONS

Heterogeneous reactions of industrial significance occur between all combinations of gas, liquid, and solid phases. The solids may be inert or reactive or catalysts in granular form. Some noncatalytic examples are listed in Table 7-11, and processes with solid catalysts are listed under "Catalysis" in Sec. 23. Equipment and operating conditions of heterogeneous processes are covered at some length in Sec. 23; only some highlights will be pointed out here.

Reactants migrate between phases in order to react: from gas phase to liquid, from fluid to solid, and between liquids when the reaction occurs in both phases. One of the liquids usually is aqueous. Resistance to mass transfer may have a strong effect on the overall rate of reaction. A principal factor is the *interfacial area*. Its magnitude is enhanced by agitation, spraying, sparging, use of trays or packing, and by size reduction or increase of the porosity of solids. These are the same operations that are used to effect physical mass transfer between

TABLE 7-11 Industrial Noncatalytic Heterogeneous Reaction

Gas/solid
  Action of chlorine on uranium oxide to recover volatile uranium chloride
  Removal of iron oxide impurity from titanium oxide by volatilization by action of chlorine
  Combustion and gasification of coal
  Manufacture of hydrogen by action of steam on iron
  Manufacture of blue gas by action of steam on carbon
  Calcium cyanamide by action of atmospheric nitrogen on calcium carbide
  Burning of iron sulfide ores with air
  Nitriding of steel

Liquid/solid
  Ion exchange
  Acetylene by action of water on calcium carbide
  Cyaniding of steel
  Hydration of lime
  Action of liquid sulfuric acid on solid sodium chloride or on phosphate rock or on sodium nitrate
  Leaching of uranium ores with sulfuric acid

Gas/liquid
  Sodium thiosulfate by action of sulfur dioxide on aqueous sodium carbonate and sodium sulfide
  Sodium nitrite by action of nitric oxide and oxygen on aqueous sodium carbonate
  Sodium hypochlorite by action of chlorine on aqueous sodium hydroxide
  Ammonium nitrate by action of ammonia on aqueous nitric acid
  Nitric acid by absorption of nitric oxide in water
  Recovery of iodine by action of sulfur dioxide on aqueous sodium iodate
  Hydrogenation of vegetable oils with gaseous hydrogen
  Desulfurization of gases by scrubbing with aqueous ethanolamines

Liquid/liquid
  Caustic soda by reaction of sodium amalgam and water
  Nitration of organic compounds with aqueous nitric acid
  Formation of soaps by action of aqueous alkalies on fats or fatty acids
  Sulfur removal from petroleum fractions by aqueous ethanolamines
  Treating of petroleum products with sulfuric acid

Solid/solid
  Manufacture of cement
  Boron carbide from boron oxide and carbon
  Calcium silicate from lime and silica
  Calcium carbide by reaction of lime and carbon
  Leblanc soda ash

Gas/liquid/solid
  Hydrogenation or liquefaction of coal in oil slurry

SOURCE: Adapted from Walas, *Reaction Kinetics for Chemical Engineers*, McGraw-Hill, 1959; Butterworths, 1989.

phases and the equipment can be similar, except that more heat transfer may be needed because of substantial heats of reaction.

Chemical reaction always enhances the rate of mass transfer between phases. The possible magnitudes of such enhancements are indicated in Tables 23-6 and 23-7. They are no more predictable than are specific rates of chemical reactions and must be found experimentally for each case, or in the relatively sparse literature on the subject.

Mechanisms The most widely investigated heterogeneous reactions have been gas/liquid and fluid/solid catalyst. The Hatta theory or Langmuir-Hinshelwood mechanisms can suggest the forms of rate equations but they always involve parameters to be found empirically. Because liquid diffusivities are low, most liquid/liquid reactions are believed to be mass-transfer controlled. In some cases the phase in which reaction occurs has been identified, but there are cases where both phases are active. Phase-transfer catalysts enhance the transfer of reactant from an aqueous to an organic phase and thus speed up reactions. Mass transfer responds less strongly to change of temperature than does chemical rate, so this feature can be used to discriminate between possible controlling mechanisms. A sensitivity to stirring rate or to a change in linear velocity also will indicate the presence of major resistance to mass transfer. No single pattern appears to hold for reactions of solids, but much is known about the behavior of important operations like cement manufacturing, ore roasting, and lime burning.

Reaction and Separation Some multiphase operations combine simultaneous reaction and separation. A few examples follow.

1. The yield of furfural from xylose is improved by countercurrent extraction with tetralin (Schoenemann, *Proc. 2d Europ. Symp. Chem. React. Eng.*, Pergamon, 1961, p. 30).

2. The reaction of vinyl acetate and stearic acid makes vinyl stearate and acetic acid but also some unwanted ethylidene acetate. A high selectivity is obtained by reaction in a distillation column with acetic acid overhead and vinyl stearate to the bottom (Geelen and Wiffels, *Proc. 3d Europ. Symp. Chem. React. Eng.*, Pergamon, 1964, p. 125).

3. The hydrolysis of fats is improved by running in a countercurrent extraction column (Donders et al., *Proc. 4th Europ. Symp. Chem. React. Eng.*, Pergamon, 1968, pp. 159–168).

4. In the production of KNO_3 from KCl and HNO_3, the product HCl is removed continuously from the aqueous phase by contact with amyl alcohol, thus forcing the reaction to completion (Baniel and Blumberg, *Chim. Ind.*, **4**, 27 [1957]).

5. Methyl-*tert*-butyl ether, a gasoline additive, is made from isobutene and methanol with distillation in a bed of acidic ion-exchange resin catalyst. The MTBE goes to the bottom with purity above 99 percent and unreacted materials overhead.

ACQUISITION OF DATA

INTRODUCTION

Kinetic data are acquired in the laboratory as a basis for design of large-scale equipment or for an understanding of its performance, or for the interpretation of possible reaction mechanisms. All levels of sophistication of equipment, statistical design of experiments, execution, and statistical analysis of the data are reported in the literature. Before serious work is undertaken, the appropriate literature should be consulted. The bibliography of Shah (*Gas-Liquid-Solid Reactor Design,* McGraw-Hill, 1979), for instance, has 145 items classified into 22 categories of reactor types.

The criteria for selection of laboratory reactors include equipment cost, ease of operation, ease of data analysis, accuracy, versatility, temperature uniformity, and controllability, suitability for mixed phases, and scale-up feasibility.

A number of factors limit the accuracy with which parameters for the design of commercial equipment can be determined. The parameters may depend on transport properties for heat and mass transfer that have been determined under nonreacting conditions. Inevitably, subtle differences exist between large and small scale. Experimental uncertainty is also a factor, so that under good conditions with modern equipment kinetic parameters can never be determined more precisely than ±5 to 10 percent (Hofmann, in de Lasa, *Chemical Reactor Design and Technology,* Martinus Nijhoff, 1986, p. 72).

Composition The law of mass action is expressed as a rate in terms of chemical compositions of the participants, so ultimately the variation of composition with time must be found. The composition is determined in terms of a property that is measured by some instrument and calibrated in terms of composition. Among the measures that have been used are titration, pressure, refractive index, density, chromatography, spectrometry, polarimetry, conductimetry, absorbance, and magnetic resonance. In some cases the composition may vary linearly with the observed property, but in every case a calibration is needed. Before kinetic analysis is undertaken, the data are converted to composition as a function of time (C, t), or to composition and temperature as functions of time (C, T, t). In a steady CSTR the rate is observed as a function of residence time.

When a reaction has many participants, which may be the case even of apparently simple processes like pyrolysis of ethane or synthesis of methanol, a factorial or other experimental design can be made and the data subjected to a *response surface analysis* (Davies, *Design and Analysis of Industrial Experiments,* Oliver & Boyd, 1954). A quadratic of this type for the variables x_1, x_2, and x_3 is

$$r = k_1 x_1 + k_2 x_2 + k_3 x_3 + k_{11} x_1^2 + k_{22} x_2^2 + k_{33} x_3^2 + k_{12} x_1 x_2 + k_{13} x_1 x_3 + k_{23} x_2 x_3$$

$$(7\text{-}115)$$

Analysis of such a correlation may reveal the significant variables and interactions, and may suggest some model, say of the L-H type, that could be analyzed in more detail by a regression process. The variables x_i could be various parameters of heterogeneous processes as well as concentrations. An application of this method to isomerization of *n*-pentane is given by Kittrel and Erjavec (*Ind. Eng. Chem. Proc. Des. Dev.,* **7**, 321 [1968]).

The constants of rate equations of single reactions often can be found by one of the linearization schemes of Fig. 7-1. Nonlinear regression methods can treat any kind of rate equation, even models made up of differential and algebraic equations together, for instance

$$\frac{dA}{dt} = -k_1 A$$

$$\frac{dB}{dt} = k_1 A - k_2 B^2 + k_3 C$$

$$C = A_0 + B_0 + C_0 - A - B$$

Software for these procedures is supplied, for example, by Constantinides (*Applied Numerical Methods with Personal Computers,* McGraw-Hill, 1987, pp. 577–614, with diskette) and by the commercial product *SimuSolv* (Mitchell and Gauthier Associates, 200 Baker Street, Concord, MA 01742). These do the integration, find the constants and their statistical criteria, and make the plots. *SimuSolv* is claimed "to provide maximum efficiency in problem solving with minimum involvement in computational procedures." Since the computer does the work, many possibilities may be considered. For the reaction cyclohexanol to cyclohexanone, 36 experiments at 6 temperature levels were made and more than 50 rate equations were tested (Hofmann, in de Lasa, *Chemical Reactor Design and Technology,* Martinus Nijhoff, 1986, p. 72). A rate equation for methanol from CO_2 and H_2 was selected from 44 possibilities by Beenackers and Graaf (in Cheremisinoff, *Handbook of Heat and Mass Transfer,* vol. 3, Gulf Publishing, 1989, pp. 671–699). They used a spinning basket reactor like the item shown in Fig. 23-29c.

EQUIPMENT

Many configurations of laboratory reactors have been employed. Rase (*Chemical Reactor Design for Process Plants,* Wiley, 1977) and Shah (*Gas-Liquid-Solid Reactor Design,* McGraw-Hill, 1979) each have about 25 sketches, and Shah's bibliography has 145 items classified into 22 categories of reactor types. Jankowski et al. (*Chemische Technik,* **30**, 441–446 [1978]) illustrate 25 different kinds of gradientless laboratory reactors for use with solid catalysts.

Laboratory reactors are of two main types:

1. Designed to obtain such fundamental data as chemical rates free of mass transfer resistances or other complications. Some of the heterogeneous reactors of Fig. 23-29, for instance, employ known interfacial areas, thus avoiding one uncertainty.

2. Simulations of the kinds of reactor intended for the pilot or plant scale. How to do the scale-up to the plant size, however, is a sizable problem in itself.

Batch Reactors In the simplest kind of investigation, reactants can be loaded into a number of ampules, kept in a thermostatic bath for various periods, and analyzed.

In terms of cost and versatility, the stirred batch reactor is the unit of choice for homogeneous or slurry reactions and even gas/liquid reactions when provision is made for recirculation of the gas. They are especially suited to reactions with half-lives in excess of 10 min. Samples are taken at intervals and the reaction is stopped by cooling, usually by at least 50°C (122°F), by dilution, or by destroying a residual reactant such as an acid or base; analysis can then be made at leisure. Analytic methods that do not necessitate termination of reaction include measurements of (1) the amount of gas produced, (2) the gas pressure in a constant volume vessel, (3) absorption of light, (4) electrical or thermal conductivity, (5) polarography, (6) viscosity of polymerization, and so on. The readings of any instrument should be calibrated to chemical composition or concentration. Operation may be isothermal, with the important effect of temperature determined from several isothermal runs, or the composition and temperature may be recorded simultaneously and the data regressed simultaneously. Finding the parameters of the nonisothermal equation $r = \exp (a + b/T) \, C^q$ is only a little more difficult than for $r = kC^q$. Rates, dC/dt, are found by numerical differentiation of (C, t) data.

On the laboratory scale, it is usually safe to assume that a batch reactor is stirred to uniform composition, but for critical cases such as high viscosities this could be checked with tracer tests.

CSTRs and other devices that require flow control are more expensive and difficult to operate. Particularly in steady operation, however, the great merit of CSTRs is their isothermicity and the fact that their mathematical representation is algebraic, involving no differential equations, thus making data analysis simpler.

For laboratory research purposes, CSTRs are considered feasible for holding times of 1 to 4,000 s, reactor volumes of 2 to 400 cm³ (0.122 to 24.4 in³) and flow rates of 0.1 to 2.0 cm³/s.

Flow Reactors Fast reactions and those in the gas phase are generally done in tubular flow reactors, just as they are often done on the commercial scale. Some heterogeneous reactors are shown in Fig. 23-29; the item in Fig. 23-29g is suited to liquid/liquid as well as gas/liquid. Stirred tanks, bubble and packed towers, and other commercial types are also used. The operation of such units can sometimes be predicted from independent data of chemical and mass transfer rates, correlations of interfacial areas, droplet sizes, and other data.

Usually it is not possible to measure compositions along a TFR, although temperatures can sometimes be measured. Mostly TFRs are kept at nearly constant temperatures. Small-diameter tubes immersed in a fluidized sand bed or molten lead or salt can hold quite constant temperatures of a few hundred degrees. A recycle unit like that shown in Fig. 23-29a can be operated as a differential reactor with arbitrarily small conversion and temperature change. This and the CSTR are the preferred laboratory devices nowadays, unless the budget allows for only a batch stirred flask. Test work in a tubular flow unit may be desirable if the commercial unit is to be of that type, although rate data from any kind of laboratory equipment are adaptable to the design of most kinds of large-scale equipment. Larger TFRs may be used in pilot plants to test predictions by data from gradientless reactors.

Multiple Phases Reactions between gas/liquid, liquid/liquid, and fluid/solid phases are often tested in CSTRs. Other laboratory types are suggested by the commercial units depicted in appropriate sketches in Sec. 23. Liquids can be reacted with gases of low solubilities in stirred vessels, with the liquid charged first and the gas fed continuously at the rate of reaction or dissolution, sometimes with recirculation in larger units. The reactors of Fig. 23-29 are designed to have known interfacial areas. Most equipment for gas absorption without reaction is adaptable to absorption with reaction. The many types of equipment for liquid/liquid extraction also are adaptable to reactions of immiscible phases.

Solid Catalysts Processes with solid catalysts are affected by diffusion of heat and mass (1) within the pores of the pellet, (2) between the fluid and the particle, and (3) axially and radially within the packed bed. Criteria in terms of various dimensionless groups have been developed to tell when these effects are appreciable. They are discussed by Mears (*Ind. Eng. Chem. Proc. Des. Devel.*, **10**, 541–547 [1971]; *Ind. Eng. Chem. Fund.*, **15**, 20–23 [1976]) and Satterfield (*Heterogeneous Catalysis in Practice*, McGraw-Hill, 1991, p. 491).

For catalytic investigations, the rotating basket or fixed basket with internal recirculation are the standard devices nowadays, usually more convenient and less expensive than equipment with external recirculation. In the fixed basket type, an internal recirculation rate of 10 to 15 or so times the feed rate effectively eliminates external diffusional resistance, and temperature gradients. A unit holding 50 cm³ (3.05 in³) of catalyst can operate up to 800 K (1440 R) and 50 bar (725 psi).

When deactivation occurs rapidly (in a few seconds during catalytic cracking, for instance), the fresh activity can be found with a transport reactor through which both reactants and fresh catalyst flow without slip and with short contact time. Since catalysts often are sensitive to traces of impurities, the time-deactivation of the catalyst usually can be evaluated only with commercial feedstock, preferably in a pilot plant.

Physical properties of catalysts also may need to be checked periodically, including pellet size, specific surface, porosity, pore size and size distribution, and effective diffusivity. The effectiveness of a porous catalyst is found by measuring conversions with successively smaller pellets until no further change occurs. These topics are touched on by Satterfield (*Heterogeneous Catalysis in Industrial Practice*, McGraw-Hill, 1991).

REFERENCES FOR LABORATORY REACTORS

Berty, "Laboratory reactors for catalytic studies", in Leach, ed., *Applied Industrial Catalysis*, vol. 1, Academic, 1983, pp. 41–57.

Danckwerts, *Gas-Liquid Reactions*, McGraw-Hill, 1970.

Hofmann, "Industrial process kinetics and parameter estimation", in *ACS Advances in Chemistry,* **109**, 519–534 (1972); "Kinetic data analysis and parameter estimation", in de Lasa, ed., *Chemical Reactor Design and Technology*, Martinus Nijhoff, 1986, pp. 69–105.

Horak and Pasek, *Design of Industrial Chemical Reactors from Laboratory Data*, Heyden, 1978.

Rase, *Chemical Reactor Design for Process Plants*, Wiley, 1977, pp. 195–259. 124 references.

Shah, *Gas-Liquid-Solid Reactor Design*, McGraw-Hill, 1979, pp. 149–179. 145 references.

SOLVED PROBLEMS

These numerical problems deal with ideal types of batch, continuously stirred, and plug flow reactors, for which the formulas are summarized in Tables 7-5 to 7-7. They find parameters of rate equations, conversions, vessel sizes, or operating conditions. Numerical methods are adopted for most integrations and differential equations. Several ODE softwares are readily available, including POLYMATH, which is obtainable through the AIChE. A larger and broader collection of solutions is provided by Walas (*Chemical Reaction Engineering Handbook of Solved Problems,* Gordon & Breach, 1995).

P1. EQUILIBRIUM OF FORMATION OF ETHYLBENZENE

Ethylbenzene is made from benzene and ethylene in the gas phase at 260°C and 40 atm.

$$C_6H_6 + C_2H_4 \Leftrightarrow C_6H_5C_2H_5$$

Equimolal proportions of the reactants are used. Thermodynamic data at 298 K are tabulated. The specific heats are averages. Find: (1) the enthalpy change of reaction at 298 and 573 K; (2) equilibrium constant at 298 and 573 K; (3) fractional conversion at 573 K.

	C_p	ΔH_f	ΔG_f
C_6H_6	28	19,820	30,989
C_2H_4	5	12,496	16,282
$C_6H_5C_2H_5$	38	7,120	31,208
Δ	−5	−25,196	−16,063

$$\Delta H_T = \Delta H_{298} + \int_{298}^{T} \Delta C_p \, dT = -25{,}196 - 5(T - 298)$$

$$= -26{,}576 \text{ at } 573 \text{ K} \tag{1}$$

$$\ln K_{298} = \frac{-\Delta G_{298}}{298R} = \frac{16{,}063}{(1.987)(298)} = 26.90$$

$$\frac{d \ln K}{dT} = \frac{\Delta H}{RT^2} = -\frac{23{,}706}{RT^2} - \frac{5}{RT}$$

$$\ln K_{298} = 26.9 - \int_{298}^{573} \left(\frac{11{,}854}{T^2} + \frac{2.50}{T} \right) dT = 6.17 \tag{2}$$

$$K = 485$$

$$= \frac{x(2 - x)}{40(1 - x)^2}$$

$$x = 0.9929, \text{ fraction converted} \tag{3}$$

P2. OPTIMUM CYCLE PERIOD WITH DOWNTIME

Find the optimum cycle period for a first-order batch reaction with a downtime of ϑ_d h per batch.

$$-\frac{dC}{d\vartheta} = kC$$

$$\vartheta = \frac{1}{k} \ln \left(\frac{C_0}{C} \right)$$

Number of daily batches:

$$n = \frac{24}{\vartheta + \vartheta_d}$$

Daily yield:

$$y = V_r(C_0 - C)_n = \frac{24V_r(C_0 - C)}{1/k \ln (C_0/C) + \vartheta_d} = \frac{24kV_r C_0(1 - C/C_0)}{\ln (C_0/C) + k\vartheta_d}$$

The ordinate of the plot is $y/24kV_rC_0$ which is proportional to the daily yield. The peaks in this curve are at these values of the parameters:

$k\vartheta_d$	0.01	0.10	1	5
C/C_0	0.87	0.65	0.32	0.12

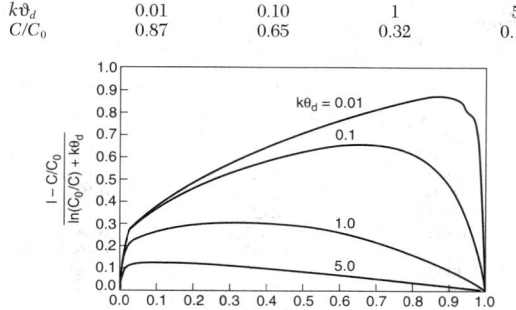

P3. PARALLEL REACTIONS OF BUTADIENE

Butadiene (B) reacts with acrolein (A) and also forms a dimer according to the reactions

$$C_4H_6 + C_3H_4O \overset{1}{\Rightarrow} C_7H_{10}O, \quad 2\,C_4H_6 \overset{2}{\Rightarrow} C_8H_{12}$$

The reaction is carried out in a closed vessel at 330°C, starting at 1 atm with equal concentrations of A and B, 0.010 g mol/L each. Specific rates are $k_1 = 5.900$ and $k_2 = 1.443$ L/(g mol·min). Find (1) B as a function of A; (2) A and B as functions of t.

$$-\frac{dA}{dt} = k_1 AB = 5.9AB \tag{1}$$

$$-\frac{dB}{dt} = k_1 AB + k_2 B^2 = 5.9AB + 1.443B^2 \tag{2}$$

Dividing these equations,

$$\frac{dB}{dA} - \frac{k_2 B}{k_1 A} = 1 \tag{3}$$

This is a linear equation whose solution is

$$B = \frac{k_1}{k_1 - k_2} A + IA^{k_2/k_1} = 1.32A - 0.010A^{0.245} \tag{4}$$

The integration constant was evaluated with $A_0 = B_0 = 0.010$.
  Substituting (4) into (1),

$$-\frac{dA}{dt} = 5.9A(1.32A - 0.010A^{0.245}) \tag{5}$$

The variables are separable, but an integration in closed form is not possible because of the odd exponent. Numerical integration followed by substitution into (4) will provide both A and B as functions of t. The plots, however, are of solutions of the original differential equations with ODE.

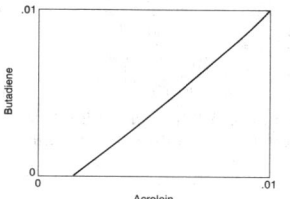
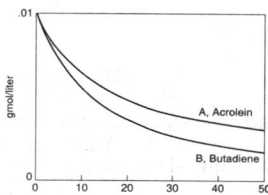

P4. BATCH REACTION WITH HEAT TRANSFER

A second-order reaction proceeds in a batch reactor provided with heat transfer. Initial conditions are $T_0 = 350$ and $C_0 = 1$. Other data are:

$$k = \exp \left(16 - \frac{5{,}000}{T} \right) \quad \text{ft}^3/(\text{lb mol·h}) \tag{1}$$

$$\Delta H_r = -(5000 + 5T) \quad \text{Btu/lb mol} \tag{2}$$

$$\rho C_p = 50$$

The rate of heat transfer is

$$Q = UA(300 - T) \quad \text{Btu/h} \tag{3}$$

The temperature T and the time t will be found in terms of fractional conversion x when $UA/V_r = 0$ or 150.
  The rate equation may be written:

$$\frac{dt}{dx} = \frac{1}{kC_0(1 - x)^2} \tag{4}$$

The differential heat balance is

$$\rho C_p V_r dT = Q dt - \Delta H_r V_r C_0 dx$$

Substituting for dt from Eq. (4) and rearranging,

$$\frac{dT}{dx} = \frac{1}{\rho C_p} \left[\frac{Q}{V_r kC_0(1 - x)^2} - \Delta H_r C_0 \right]$$

$$= 0.02 \left[\frac{UA(300 - T)}{V_r k(1 - x)^2} + 5000 + 5T \right] \tag{5}$$

Equations (1), (4), and (5) are solved simultaneously with $UA/V_r = 0$ or 150. In the adiabatic case, the temperature tends to run away.

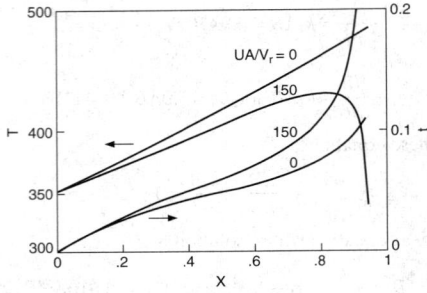

P5. A SEMIBATCH PROCESS

A tank is charged initially with $v_{r0} = 100$ L of a solution of concentration $C_{b0} = 2$ g mol/L. Another solution is then pumped in at $V' = 5$ L/min with concentration $C_{a0} = 0.8$ until a stoichiometric amount has been added. The rate equation is

$$r = 0.015 C_a C_b \text{ g mol/(L·min)}$$

Find the concentration during the filling period and for 50 min afterward.

$$V_r = 100 + 5t \tag{1}$$

$$C_b = C_a + \frac{100(2) - 5(0.8)t}{100 + 5t} = C_a + \frac{40 - 0.8t}{20 + t} \tag{2}$$

$$\frac{dC_a}{dt} = \frac{0.8 - C_a}{20 + t} - 0.015 C_a \left(C_a + \frac{40 - 0.8t}{20 + t} \right) \tag{3}$$

The input is continued until 200 lb mol of A have been added, which is for 50 min. Eq. (3) is integrated for this time interval. After input is discontinued the rate equation is

$$-\frac{dC_a}{dt} = k C_a^2 \tag{4}$$

At $t = 50$, $C_a = C_{a1} = 0.4467$.

$$C_a = \frac{1}{1/C_{a1} + k(t - 50)} = \frac{1}{2.2386 + 0.015(t - 50)} \tag{5}$$

Plots are shown for several specific rates, including $k = 0$ when no reaction takes place.

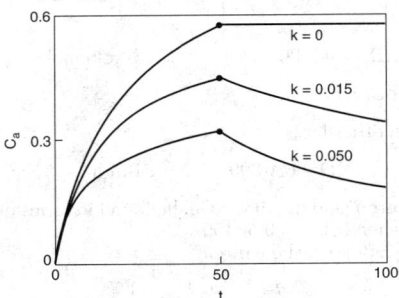

P6. OPTIMUM REACTION TEMPERATURE WITH DOWNTIME

A liquid phase reaction $2A \rightleftharpoons B + C$ has the rate equation

$$r_a = k\left(C_a^2 - \frac{C_b C_c}{K_e} \right) = k C_{a0}^2 \left[(1 - f)^2 - \frac{f^2}{K_e} \right], \quad \text{kg mol/(m}^3\text{·h)}$$

where f = fractional conversion
$C_{a0} = 1$

$$k = \exp\left(4.5 - \frac{2,500}{T} \right)$$

$$K_e = \exp\left(28.8 - 0.037T - \frac{5,178}{T} \right)$$

The downtime is 1 h per batch. Find the temperature at which the daily production is a maximum.

The reaction time of one batch is

$$t_b = \frac{1}{k} \int_0^f \frac{df}{(1 - f)^2 - f^2/K_e} \tag{1}$$

$$\text{Batches/day} = \frac{24}{t_b + 1}$$

$$\text{Daily production} = \frac{24}{t_b + 1} V_r C_{a0} f$$

Maximize $P = f/(t_b + 1)$ as a function of temperature. Eq. (1) is integrated with POLYMATH for several temperatures and the results are plotted. The tabulation gives the integration at 550 K. The peak value of $P = f/(t_b + 1) = 0.1941$ at 550 K, $t_b = 0.6$, and $f = 0.3105$.

$$\text{Maximum daily production} = 0.1941(24) V_r C_{a0}$$

$$= 4.66 V_r \text{ kg mol/d}$$

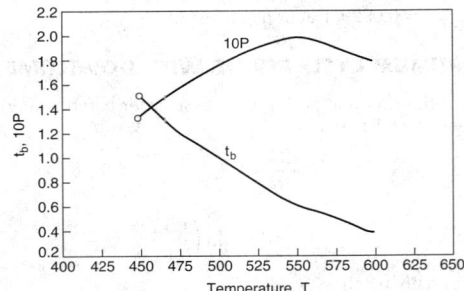

The equations:

$$\frac{d(f)}{d(t)} = k * \left((1 - f)**2 - f** \frac{2}{ke} \right)$$

$$x = 550$$

$$k = \exp\left(4.5 - \frac{2,500}{x} \right)$$

$$ke = \exp\left(28.8 - \frac{5,178}{x - .037 * x} \right)$$

$$p = f/(t + 1)$$

Initial values: $t_0 = 0.0$ $f_0 = 0.0$
Final value: $t_f = 2.0000$

t	f	p
0.0	0.0	0.0
0.2000	0.1562	0.1302
0.4000	0.2535	0.1811
0.6000	0.3105	0.1941
0.8000	0.3427	0.1904
1.0000	0.3605	0.1802
1.2000	0.3702	0.1683
1.4000	0.3755	0.1565
1.6000	0.3784	0.1455
1.8000	0.3799	0.1357
2.0000	0.3807	0.1269

P7. RATE EQUATIONS FROM CSTR DATA

For the consecutive reactions $2A \Rightarrow B$ and $2B \Rightarrow C$, concentrations were measured as functions of residence time in a CSTR. In all experiments, $C_{a0} = 1$ lb mol/ft^3. Volumetric flow rate was constant. The data are tabulated in the first three columns. Check the proposed rate equations,

$$r_a = k_1 C_a^\alpha$$

$$r_b = -0.5 k_1 C_a^\alpha + k_2 C_b^\beta$$

Write and rearrange the material balances on the CSTR.

$$C_{a0} = C_a + \bar{t}r_a$$

$$r_a = \frac{C_{a0} - C_a}{\bar{t}} = k_1 C_a^{\alpha} \tag{1}$$

$$r_b = \frac{C_{b0} - C_b}{\bar{t}} = -0.5k_1 C_a^{\alpha} + k_2 C_b^{\beta} = -0.5r_a + k_2 C_b^{\beta} \tag{2}$$

Numerical values of r_a, r_b, and $r_b + 0.5r_a$ are tabulated. The constants of the rate equations are evaluated from the plots of the linearized equations,

$$\ln r_a = \ln k_1 + \alpha \ln C_a = -2.30 + 2.001 \ln C_a$$

$$\ln (r_b + 0.5r_a) = \ln k_2 + \beta \ln C_b = -4.606 + 0.9979 \ln C_b$$

which make the rate equations

$$r_a = 0.1003 C_a^{2.00} \tag{3}$$

$$r_b = -0.0502 C_a^2 + 0.01 C_b^{0.998} \tag{4}$$

$\bar{t}$	C_a	C_b	r_a	$-r_b$	$r_b + 0.5r_a$
10	1.000	0.4545	0.100	0.04545	0.00455
20	0.780	0.5083	0.061	0.02542	0.00508
40	0.592	0.5028	0.0352	0.01257	0.00503
100	0.400	0.400	0.0160	0.0040	0.0040
450	0.200	0.1636	0.0040	0.000364	0.00164

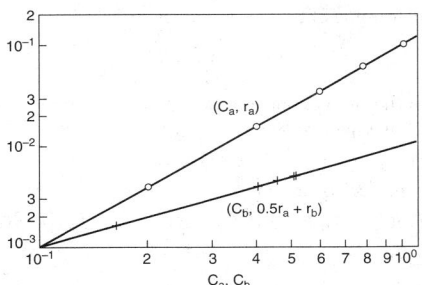

P8. COMPARISON OF BATCH AND CSTR OPERATIONS

A solution containing 0.5 lb mol/ft³ of reactive component is to be treated at 25 ft³/h. The rate equation is

$$r = \frac{-dC}{dt} = 2.33 C^{1.7} \text{ lb mol/(ft}^3\cdot\text{h)}$$

1. If the downtime is 45 min per batch, what size reactor is needed for 90% conversion?
2. What percentage conversion is attained with a two-stage CSTR, each vessel being 50 ft³?

Part 1: The integral of the rate equation is solved for the time,

$$t = \frac{1}{0.7k}(C^{-0.7} - C_0^{-0.7}) = \frac{1}{0.7(2.33)}\left(\frac{1}{0.05^{0.7}} - \frac{1}{0.5^{0.7}}\right)$$

$$= 4.00 \text{ h}$$

Number of batches $= \dfrac{24}{(4 + 0.75)} = 5.053/\text{d}$

Reactor volume $V_r = \dfrac{24(25)}{5.053} = 118.7 \text{ ft}^3$

Part 2:

$$\tau = \frac{50}{25} = 2$$

$$0.5 = C_1 + \tau r = C_1 + 2(2.33)C_1^{1.7}$$

$$C_1 = C_2 + 2(2.33)C_2^{1.7}$$

The solution is,

$$C_1 = 0.1994, \qquad = 60.1\% \text{ conversion}$$
$$C_2 = 0.1025, \qquad = 79.5\% \text{ conversion}$$

P9. INSTANTANEOUS AND GRADUAL FEED RATES

Initially a reactor contains 2 m³ of a solvent. A solution containing 2 kg mol/m³ of reactant A is pumped in at the rate of 0.06 m³/min until the volume becomes 4 m³. The rate equation is $r_a = 0.25 C_a$, 1/min. Compare the time-composition profile of this operation with charging all of the feed instantaneously.

During the filling period,

$$V_r = 2 + 0.06t$$

$$V' C_{a0} = kV_r C_a + \frac{d(V_r C_a)}{dt} = kV_r C_a + V_r \frac{dC_a}{dt} + C_a \frac{dV_r}{dt}$$

$$0.06(2) = 0.25(2 + 0.06t)C_a + (2 + 0.06t)\frac{dC_a}{dt} + 0.06C_a$$

$$\frac{dC_a}{dt} = \frac{0.12 - (0.56 + 0.015t)C_a}{2 + 0.06t}, \qquad C_{a0} = 0 \tag{1}$$

When all of A is charged at the beginning,

$$\frac{dC_a}{dt} = -0.25C_a, \qquad C_{a0} = 0.5 \tag{2}$$

The integrals of these two equations are plotted. A peak value, $C_a = 0.1695$, is reached in the first operation at $t = 10$.

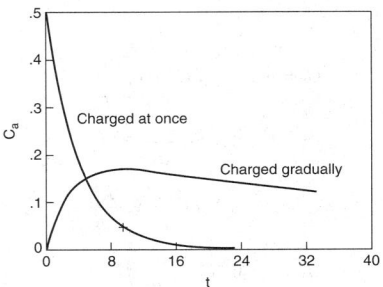

P10. FILLING AND UNSTEADY OPERATING PERIOD OF A CSTR

A stirred reactor is being charged at 5 ft³/min with a concentration of 2 mol/ft³. The reactor has a capacity of 150 ft³ but is initially empty. The rate of reaction is

$$r = 0.02C^2 \text{ lb mol/(ft}^3\cdot\text{min)}.$$

After the tank is filled, pumping is continued and overflow is permitted at the same flow rate. Find the concentration in the tank when it first becomes full, and find how long it takes for the effluent concentration to get within 95% of the steady state value.

Filling period:

$$V_r = V't$$

$$V'C_0 = kV_r C^2 + \frac{dV_r C}{dt} = kV_r C^2 + \left(C + t\frac{dC}{dt}\right)V'$$

$$\frac{dC}{dt} = \frac{C_0 - ktC^2 - C}{t} = \frac{2 - 0.02tC^2 - C}{t}, \qquad C = 2 \text{ when } t = 0$$

The numerical solution is $C = 1.3269$ when $t = 30$.

Unsteady period:

$$V'C_0 = V'C + kV_r C^2 + V_r \frac{dC}{dt}$$

$$\frac{dC}{dt} = \frac{C_0 - C - k\tau C^2}{\tau} = \frac{2 - C - 0.02(30)C^2}{30},$$

$$C = 1.3269 \text{ when } t = 30.$$

The variables are separable, but the plot is of a numerical solution. The steady state concentration is 1.1736.

At 95% approach to steady state from the condition at $t = 30$,

$$C = 0.05(1.3269) + 0.95(1.1736) = 1.1813$$

From a printout of the solution, $t = 67.4$ min at this value.

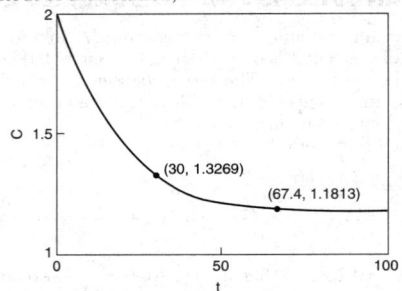

P11.   SECOND-ORDER REACTION IN TWO STAGES

A second order reaction is conducted in two equal CSTR stages. The residence time per stage is $\tau = 1$ and the specific rate is $kC_0 = 0.5$. Feed concentration is C_0. Two cases are to be examined: (1) with pure solvent initially in the tanks; and (2) with concentrations C_0 initially in both tanks, that is, with $C_{10} = C_{20} = C_0$.

The unsteady balances on the two reactors are

$$f_i = \frac{C_i}{C_0}$$

$$FC_0 = FC_1 + V_r kC_1^2 + V_r\frac{dC_1}{dt}$$

$$1 = f_1 + 0.5f_1^2 + \frac{df_1}{dt} \qquad (1)$$

$$FC_1 = FC_2 + V_r kC_2^2 + V_r\frac{dC_2}{dt}$$

$$f_1 = f_2 + 0.5f_2^2 + \frac{df_2}{dt} \qquad (2)$$

The steady state values are the same for both starting conditions, obtained by zeroing the derivatives in Eqs. (1) and (2). Then

$$f_1 = 0.5702, \qquad f_2 = 0.7321$$

The plots are of numerical solutions.

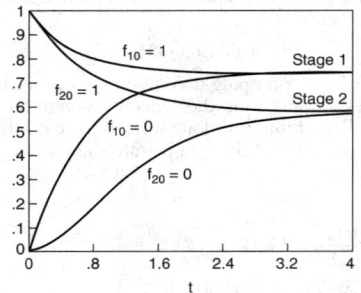

P12.   BUTADIENE DIMERIZATION IN A TFR

A mixture of 0.5 mol of steam per mol of butadiene is dimerized in a tubular reactor at 640°C and 1 atm. The forward specific rate is $k = 118$ g mol/(L·h·atm^2) and the equilibrium constant is 1.27. Find the length of 10-cm ID tube for 40% conversion when the total feed rate is 9 kg mol/h.

$$2A \Leftrightarrow B$$

$$n_{a0} = 6 \text{ kg mol/h}$$

$$n_t = n_s + n_a + n_b = 0.5n_{a0} + n_a + 0.5(n_{a0} - n_a) = n_{a0} + 0.5n_a$$

$$P_a = \frac{n_a}{n_{a0} + 0.5n_a} \qquad (1)$$

$$P_b = \frac{0.5(n_{a0} - n_a)}{n_{a0} + 0.5n_a}$$

$$r_a = k\left(P_a^2 - \frac{P_b}{K_e}\right) = \frac{118}{n_{a0} + 0.5n_a}\left(\frac{n_a^2}{n_{a0} + 0.5n_a} - \frac{n_{a0} - n_a}{2(1.27)}\right) \qquad (1)$$

Put $n_{a0} = 6$, substitute Eq. (1) into the flow reactor equation, and integrate numerically.

$$V_r = \int_{3.6}^{6} \frac{dn_a}{r_a} = 0.0905 \text{ m}^3$$

$$L = \frac{0.0905(10^6)}{78.5} = 1{,}153 \text{ cm}$$

P13.   AUTOCATALYTIC REACTION WITH RECYCLE

Part of the effluent from a PFR is returned to the inlet. The recycle ratio is R, fresh feed rate is F_0

$$R = \frac{F_r}{F_0}$$

$$F_t = F_r + F_0 = F_0(R + 1)$$

The concentration of the mixed feed is

$$C_{at} = \frac{C_{a0} + RC_{af}}{1 + R}$$

where C_{af} is the outlet concentration. For the autocatalytic reaction A $\Rightarrow$ B, the rate equation is

$$r_a = kC_aC_b = kC_a(C_{a0} - C_a)$$

The flow reactor equation is

$$-F_t dC_a = -F_0(R + 1)dC_a = r_a dV_r = kC_a(C_{a0} - C_a)dV_r$$

$$\frac{kV_r}{F_0} = (R + 1)\int_{C_{af}}^{C_{at}} \frac{dC_a}{C_a(C_{a0} - C_a)}$$

The plot is for $C_{a0} = 2$ and $C_{af} = 0.04$. The minimum reactor size is at a recycle ratio $R = 0.23$ and mixed feed $C_{at} = 1.57$.

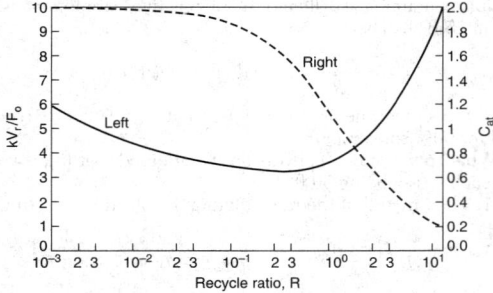

P14.   MINIMUM RESIDENCE TIME IN A PFR

A reversible reaction A $\Leftrightarrow$ B is conducted in a plug flow reactor. The rate equation is

$$r = kC_{a0}\left(1 - x - \frac{x}{K_e}\right)$$

where $C_{a0} = 4$

$$k = \exp\left(17.2 - \frac{5{,}800}{T}\right)$$

$$K_e = \exp\left(-24.7 + \frac{9{,}000}{T}\right)$$

Find the conditions for minimum V_r/V' when conversion is 80%.

The flow reactor equation is

$$-dn_a = V'C_{a0}dx = kC_{a0}\left(\frac{1-x-x}{K_e}\right)dV_r$$

$$\frac{V_r}{V'} = \frac{1}{k}\int_0^{0.8}\frac{dx}{1-x-x/K_e} = \frac{1}{k(1+1/K_e)}\ln\frac{1}{0.2-0.8/K_e} \quad (1)$$

The plot of this equation shows the minimum to be $V_r/V' = 2.04$ at $T = 340$ K.

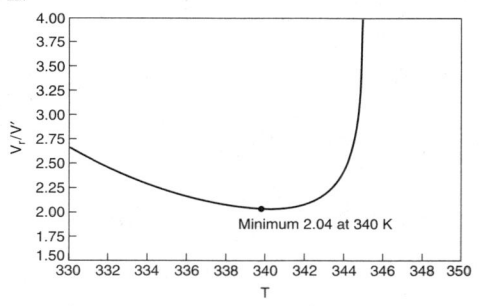

P15. HEAT TRANSFER IN A CYLINDRICAL REACTOR

A reaction $A \Rightarrow 2B$ runs in a tube provided with a cooling jacket that keeps the wall at 630 R. Inlet is pure A at 650 R and 50 atm. Other data are stated in the following. Find the profiles of temperature and conversion along the reactor, both with heat transfer and adiabatically.

Tube diameter $D = \dfrac{1}{6}$ ft

$C_{pa} = 20$, $C_{pb} = 15$ Btu/(lb mol R)

$\Delta H_r = -8,000$ Btu/(lb mol A)

$k = \exp\left(7.82 - \dfrac{3,000}{T}\right) \quad (1)$

Heat transfer coefficient $U = 5$ Btu/(ft$^3 \cdot$h R)

Heat transfer area $dA = (4/D)dV_r = 24dV_r$

Rate equation:

$$r_a = k\left(\frac{n_a}{V}\right) = k\left(\frac{\pi}{RT}\right)\left(\frac{n_a}{n_t}\right) = \frac{50k}{0.729T}\left(\frac{n_a}{2n_{a0}-n_a}\right)$$

$$= \frac{68.6k}{T}\left(\frac{1-x}{1+x}\right) \quad (2)$$

$$x = 1 - \frac{n_a}{n_{a0}}$$

Flow reactor:

$$-dn_a = n_{a0}dx = r_a dV_r$$

$$\frac{dx}{d(V_r/n_{a0})} = r_a \quad (3)$$

Heat balance over a differential volume dV_r:

$$\Delta H_r dn_a = -\Delta H_r r_a\, dV_r$$

$$= \sum n_i C_{pi}dT + U(T-T_w)dA$$

$$= n_{a0}[20(1-x)+15(2x)dT + 24U(T-T_w)dV_r$$

$$\frac{dT}{d(V_r/n_{a0})} = \frac{-\Delta H_r r_a - 24U(T-T_w)}{20-10x} = \frac{8,000r_a - 120(T-630)}{20-10x} \quad (4)$$

Differential Eqs. (3) and (4) are solved simultaneously with auxiliary Eqs. (1) and (2) by ODE. The solutions with $U = 5$ and $U = 0$ are shown.

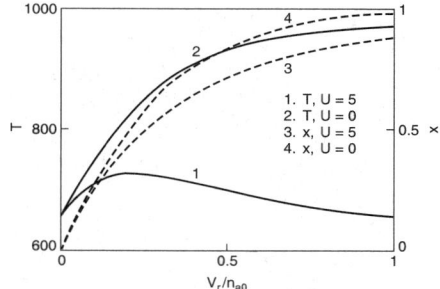

P16. PRESSURE DROP AND CONVERSION IN A PFR

A reaction $A \Rightarrow 3B$ takes place in a tubular flow reactor at constant temperature and an inlet pressure of 5 atm. The rate equation is

$$r_a = k\left(\frac{n_a}{V'}\right) = \frac{kP}{RT}\left(\frac{n_a}{3n_{a0}-n_a}\right) = \frac{kP}{RT}\left(\frac{1-x}{1+2x}\right)$$

When put into the plug flow equation,

$$n_{a0}dx = \frac{kP}{RT}\left(\frac{1-x}{1+2x}\right)AdL$$

or

$$\frac{dx}{dL} = 0.02\frac{P(1-x)}{1+2x} \quad (1)$$

where several factors have been combined into the numerical coefficient.

The pressure gradient due to friction is proportional to the flowing mole rate, $1 + 2x$, and inversely to the density or the pressure. Here again, several factors are incorporated into a numerical coefficient, making

$$-\frac{dP}{dL} = 0.6\frac{1+2x}{P} \quad (2)$$

The numbered equations are integrated and plotted. They show the typical fall in pressure as conversion with an increase in the number of moles proceeds at constant pressure, $x = 0.48$ when $L = 10$.

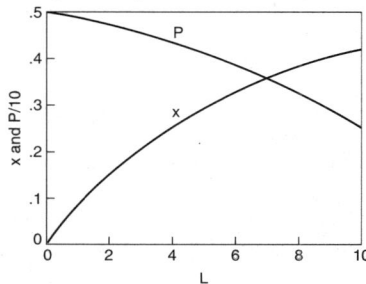

Process Control

Thomas F. Edgar, Ph.D., *Professor of Chemical Engineering, University of Texas, Austin, TX. (Advanced Control Systems, Process Measurements, Section Editor)*

Cecil L. Smith, Ph.D., *Principal, Cecil L. Smith Inc., Baton Rouge, LA. (Batch Process Control, Telemetering and Transmission, Digital Technology for Process Control, Process Control and Plant Safety)*

F. Greg Shinskey, B.S.Ch.E., *Consultant (retired from Foxboro Co.), North Sandwich, NH. (Fundamentals of Process Dynamics and Control, Unit Operations Control)*

George W. Gassman, B.S.M.E., *Senior Research Specialist, Final Control Systems, Fisher Controls International, Inc., Marshalltown, IA. (Controllers, Final Control Elements, and Regulators)*

Paul J. Schafbuch, Ph.D., *Senior Research Specialist, Final Control Systems, Fisher Controls International, Inc., Marshalltown, IA. (Controllers, Final Control Elements, and Regulators)*

Thomas J. McAvoy, Ph.D., *Professor of Chemical Engineering, University of Maryland, College Park, MD. (Fundamentals of Process Dynamics and Control)*

Dale E. Seborg, Ph.D., *Professor of Chemical Engineering, University of California, Santa Barbara, CA. (Advanced Control Systems)*

Nomenclature

Symbol	Definition	Symbol	Definition
A	Area	R, r	Set point
A_a	Actuator area	R_T	Resistance in temperature sensor
A_c	Amplitude of controlled variable	R_1	Valve resistance
A_m	Output amplitude limits	s	Laplace transform variable
A_v	Cross sectional area of valve	$\mathbf{s}$	Search direction
A_1	Cross sectional area of tank	S_i	Step response coefficient
b	Controller output bias	t	Time
B	Bottoms flow rate	T	Temperature
B_i°	Limit on control	T_b	Base temperature
c, C	Controlled variable	T_f	Exhaust temperature
c_A	Concentration of A	T_R	Reset time
C_d	Discharge coefficient	u	Controller output
C_i	Inlet concentration	U	Heat transfer coefficient
C_i°	Limit on control move	V	Volume
C_L	Specific heat of liquid	V_s	Product value
C_o	Integration constant	w	Mass flow rate
C_r	Heat capacity of reactants	w_i	Weighting factor
C_v	Valve flow coefficient	W	Steam flow rate
D	Distillate flow rate	x	Mass fraction
D_i°	Limit on output	x_i	Optimization variable
$D(s)$	Decoupler transfer function	x_T	Pressure drop ratio factor
e	Error	X	Transform of deviation variable
E	Economy of evaporator	y	Process output, controlled variable, valve travel
f	Function of time	Y	Controller tuning law, expansion factor
F, f	Feed flow rate	z	z-transform variable
F_L	Pressure recovery factor	z_i	Feed mole fraction (distillation)
g_c	Unit conversion constant	Z	Compressibility factor
g_i	Algebraic inequality constraint		
G	Transfer function		**Greek symbols**
G_c	Controller transfer function		
G_f	Feedforward controller transfer function	α	Digital filter coefficient
G_L	Load transfer function	α_T	Temperature coefficient of resistance
G_m	Sensor transfer function	β	Resistance thermometer parameter
G_p	Process transfer function	γ	Ratio of specific heats
G_t	Transmitter transfer function	δ	Move suppression factor
G_v	Valve transfer function	Δq	Load step change
h_i	Algebraic equality constraints	Δt	Time step
h_1	Liquid head in tank	ΔT	Temperature change
H	Latent heat of vaporization	Δu	Control move
i	Summation index	ε	Spectral emissivity, step size
I_i	Impulse response coefficient	ζ	Damping factor (second order system)
j	Time index	θ	Time delay
J	Objective function or performance index	λ	Relative gain array parameter, wavelength
k	Time index	Λ	Relative gain array
k_f	Flow coefficient	ξ	Deviation variable
k_r	Kinetic rate constant	ρ	Density
K	Gain	σ	Stefan-Boltzmann constant
K_c	Controller gain	Σ_τ	Total response time
K_L	Load transfer function gain	τ	Time constant
K_m	Measurement gain	τ_D	Derivative time (PID controller)
K_p	Process gain	τ_F	Filter time constant
K_u	Ultimate controller gain (stability)	τ_I	Integral time (PID controller)
L	Disturbance or load variable	τ_L	Load time constant
L_p	Sound pressure level	τ_n	Natural period of closed loop
m, M	Manipulated variable	τ_p	Process time constant
m_c	Number of constraints	τ_o	Period of oscillation
M_o	Mass flow	ϕ_{PI}	Phase lag
M_r	Mass of reactants		
M_w	Molecular weight		**Subscripts**
n	Number of data points, number of stages or effects		
N	Number of inputs/outputs, model horizon	A	Species A
p_1	Pressure	b	Best
p_a	Actuator pressure	c	Controller
p_v	Vapor pressure	eff	Effective
P	Proportional band (%)	F	Feedforward
P_u	Proportional band (ultimate)	i	Initial, inlet
q	Radiated energy flux	L	Load, disturbance
q_b	Energy flux to a black body	m	Measurement or sensor
Q_a	Flow rate	p	Process
r_c	Number of constraints	s	Steady state
R	Equal percentage valve characteristic	set	Set point value
		t	Transmitter
		u	Ultimate
		v	Valve

FUNDAMENTALS OF PROCESS DYNAMICS AND CONTROL

THE GENERAL CONTROL SYSTEM

A process is shown in Fig. 8-1 with a manipulated input M, a load input L, and a controlled output C, which could be flow, pressure, liquid level, temperature, composition, or any other inventory, environmental, or quality variable that is to be held at a desired value identified as the set point R. The load may be a single variable or aggregate of variables acting either independently or manipulated for other purposes, affecting the controlled variable much as the manipulated variable does. Changes in load may occur randomly as caused by changes in weather, diurnally with ambient temperature, manually when operators change production rate, stepwise when equipment is switched in or out of service, or cyclically as the result of oscillations in other control loops. Variations in load will drive the controlled variable away from set point, requiring a corresponding change in the manipulated variable to bring it back. The manipulated variable must also change to move the controlled variable from one set point to another.

An open-loop system positions the manipulated variable either manually or on a programmed basis, without using any process measurements. This operation is acceptable for well-defined processes without disturbances. An automanual transfer switch is provided to allow manual adjustment of the manipulated variable in case the process or the control system is not performing satisfactorily.

A closed-loop system uses the measurement of one or more process variables to move the manipulated variable to achieve control. Closed-loop systems may include feedforward, feedback, or both.

Feedback Control In a feedback control loop, the controlled variable is compared to the set point R, with the difference, deviation, or error e acted upon by the controller to move m in such a way as to minimize the error. This action is specifically negative feedback, in that an increase in deviation moves m so as to decrease the deviation. (Positive feedback would cause the deviation to expand rather than diminish and therefore does not regulate.) The action of the controller is selectable to allow use on process gains of both signs.

The controller has tuning parameters related to proportional, integral, derivative, lag, deadtime, and sampling functions. A negative feedback loop will oscillate if the controller gain is too high, but if it is too low, control will be ineffective. The controller parameters must be properly related to the process parameters to ensure closed-loop stability while still providing effective control. This is accomplished first by the proper selection of control modes to satisfy the requirements of the process, and second by the appropriate tuning of those modes.

Feedforward Control A feedforward system uses measurements of disturbance variables to position the manipulated variable in such a way as to minimize any resulting deviation. The disturbance variables could be either measured loads or the set point, the former being more common. The feedforward gain must be set precisely to offset the deviation of the controlled variable from the set point.

Feedforward control is usually combined with feedback control to eliminate any offset resulting from inaccurate measurements and calculations and unmeasured load components. The feedback controller can either bias or multiply the feedforward calculation.

Computer Control Computers have been used to replace analog PID controllers, either by setting set points of lower level controllers in supervisory control, or by driving valves directly in direct digital control. Single-station digital controllers perform PID control in one or two loops, including computing functions such as mathematical operations, characterization, lags, and deadtime, with digital logic and alarms. Distributed control systems provide all these functions, with the digital processor shared among many control loops; separate processors may be used for displays, communications, file servers, and the like. A host computer may be added to perform high-level operations such as scheduling, optimization, and multivariable control. More details on computer control are provided later in this section.

PROCESS DYNAMICS AND MATHEMATICAL MODELS

GENERAL REFERENCES: Seborg, Edgar, and Mellichamp, *Process Dynamics and Control*, Wiley, New York 1989; Marlin, *Process Control*, McGraw-Hill, New York, 1995; Ogunnaike and Ray, *Process Dynamics Modeling and Control*, Oxford University Press, New York, 1994; Smith and Corripio, *Principles and Practices of Automatic Process Control*, Wiley, New York, 1985

Open-Loop versus Closed-Loop Dynamics It is common in industry to manipulate coolant in a jacketed reactor in order to control conditions in the reactor itself. A simplified schematic diagram of such a reactor control system is shown in Fig. 8-2. Assume that the reactor temperature is adjusted by a controller that increases the coolant flow in proportion to the difference between the desired reactor temperature and the temperature that is measured. The proportionality constant is K_c. If a small change in the temperature of the inlet stream occurs, then depending on the value of K_c, one might observe the reactor temperature responses shown in Fig. 8-3. The top plot shows the case for no control ($K_c = 0$), which is called the open loop, or the normal dynamic response of the process by itself. As K_c increases, several effects can be noted. First, the reactor temperature responds faster and faster. Second, for the initial increases in K_c, the maximum deviation in the reactor temperature becomes smaller. Both of these effects are desirable so that disturbances from normal operation have

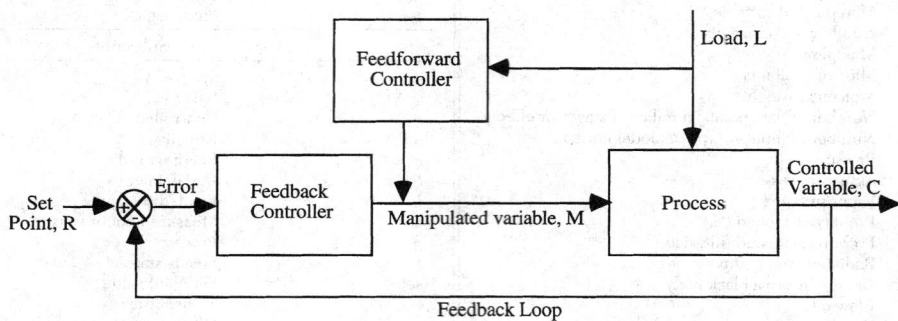

FIG. 8-1 Block diagram for feedforward and feedback control.

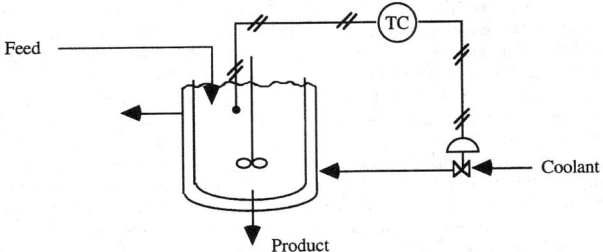

FIG. 8-2   Reactor control system.

as small an effect as possible on the process under study. As the gain is increased further, eventually a point is reached where the reactor temperature oscillates indefinitely, which is undesirable. This point is called the stability limit, where $K_c = K_u$, the ultimate controller gain. Increasing K_c further causes the magnitude of the oscillations to increase, with the result that the control valve will cycle between full open and closed.

The responses shown in Fig. 8-3 are typical of the vast majority of regulatory loops encountered in the process industries. Figure 8-3 shows that there is an optimal choice for K_c, somewhere between 0 (no control) and K_u (stability limit). If one has a dynamic model of a process, then this model can be used to calculate controller settings. In Fig. 8-3, no time scale is given, but rather the figure shows relative responses. A well-designed controller might be able to speed up the response of a process by a factor of roughly two to four. Exactly how fast the control system responds is determined by the dynamics of the process itself.

Physical Models versus Empirical Models   In developing a dynamic process model, there are two distinct approaches that can be taken. The first involves models based on first principles, called physical models, and the second involves empirical models. The conservation laws of mass, energy, and momentum form the basis for developing physical models. The resulting models typically involve sets of differential and algebraic equations that must be solved simultaneously. Empirical models, by contrast, involve postulating the form of a dynamic model, usually as a transfer function, which is discussed below. This transfer function contains a number of parameters that need to be estimated. For the development of both physical and empirical models, the most expensive step normally involves verification of their accuracy in predicting plant behavior.

To illustrate the development of a physical model, a simplified treatment of the reactor, shown in Fig. 8-2, is used. It is assumed that the reactor is operating isothermally and that the inlet and exit volumetric flows and densities are the same. There are two components, A and B, in the reactor, and a single first order reaction of $A \rightarrow B$ takes place. The inlet concentration of A, which we shall call c_i, varies with time. A dynamic mass balance for the concentration of A (c_A) can be written as follows:

$$V \frac{dc_A}{dt} = Fc_i - Fc_A - k_r V_c \tag{8-1}$$

In Eq. (8-1), the flow in of A is Fc_i, the flow out is Fc_A, and the loss via reaction is $k_r V c_A$, where V = reactor volume and k_r = kinetic rate constant. In this example, c_i is the input, or forcing variable, and c_A is the output variable. If V, F, and k_r are constant, Eq. (8-1) can be rearranged by dividing by $(F + k_r V)$ so that it only contains two groups of parameters. The result is:

$$\tau \frac{dc}{dt} = Kc_i - c_A \tag{8-2}$$

where $\tau = V/(F + k_r V)$ and $K = F/(F + k_r V)$. For this example, the resulting model is a first-order differential equation in which τ is called the **time constant** and K the **process gain**.

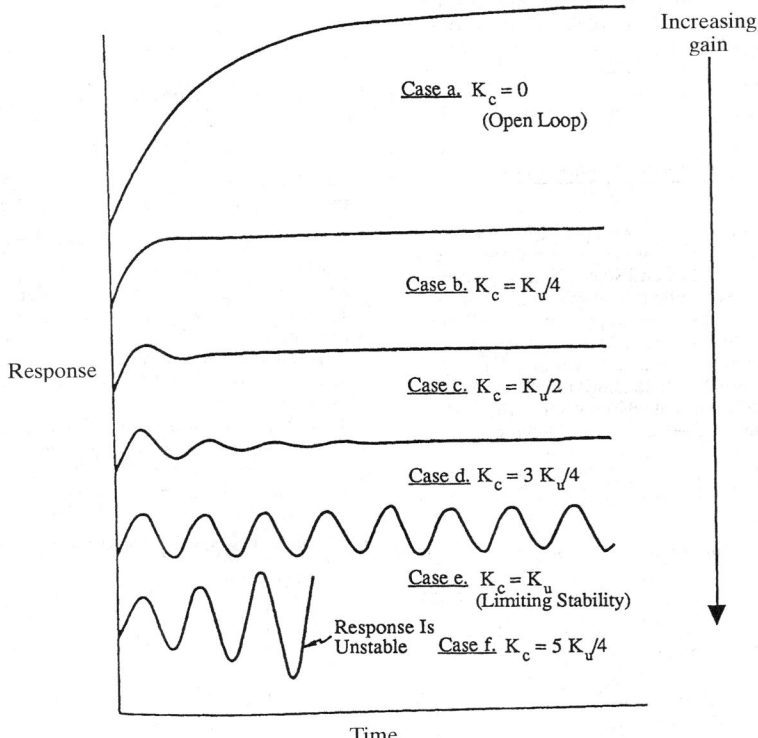

FIG. 8-3   Typical control system responses.

As an alternative to deriving Eq. (8-2) from a dynamic mass balance, one could simply postulate a first-order differential equation to be valid (empirical modeling). Then it would be necessary to estimate values for τ and K so that the postulated model described the reactor's dynamic response. The advantage of the physical model over the empirical model is that the physical model gives insight into how reactor parameters affect the values of τ, and K, which in turn affects the dynamic response of the reactor.

Nonlinear versus Linear Models If V, F, and k are constant, then Eq. (8-1) is an example of a linear differential equation model. In a linear equation, the output and input variables and their derivatives only appear to the first power. If the rate of reaction were second order, then the resulting dynamic mass balance would be:

$$V \frac{dc}{dt} = Fc_i - Fc_A - k_r Vc_A^2 \qquad (8\text{-}3)$$

Since c_A appears in this equation to the second power, the equation is nonlinear.

The difference between linear systems and nonlinear systems can be seen by considering the steady state behavior of Eq. (8-1) compared to Eq. (8-3) (the left-hand side is zero; i.e., $dc_A/dt = 0$). For a given change in c_i, Δc_i, the change in c_A calculated from Eq. (8-1), or Δc, is always proportional to Δc_i, and the proportionality constant is K [see Eq. (8-2)]. The change in the output of a system divided by a change in the input to the system is called the *process gain*. Linear systems have constant process gains for all changes in the input. By contrast, Eq. (8-3) gives a Δc that varies in proportion to Δc_i but with the proportionality factor being a function of the concentration levels in the reactor. Thus, depending on where the reactor operates, a change in c_i produces different changes in c_A. In this case, the process has a nonlinear gain. Systems with nonlinear gains are more difficult to control than linear systems that have constant gains.

Simulation of Dynamic Models Linear dynamic models are particularly useful for analyzing control-system behavior. The insight gained through linear analysis is invaluable. However, accurate dynamic process models can involve large sets of nonlinear equations. Analytical solution of these models is not possible. Thus, in these cases, one must turn to simulation approaches to study process dynamics and the effect of process control. Equation (8-3) will be used to illustrate the simulation of nonlinear processes. If dc_A/dt on the left-hand side of Eq. (8-3) is replaced with its finite difference approximation, one gets:

$$c_A(t + \Delta t) = \frac{c_A(t) + \Delta t \cdot [Fc_i(t) - Fc_A(t) - k_r Vc_A(t)^2]}{V} \qquad (8\text{-}4)$$

Starting with an initial value of c_A and knowing $c_i(t)$, Eq. (8-4) can be solved for $c_A(t + \Delta t)$. Once $c_A(t + \Delta t)$ is known, the process can be repeated to calculate $c_A(t + 2\Delta t)$, and so on. This approach is called the Euler integration method; while it is simple, it is not necessarily the best approach to numerically integrating nonlinear differential equations. To achieve accurate solutions with an Euler approach, one often needs to take small steps in time, Δt. A number of more sophisticated approaches are available that allow much larger step sizes to be taken but require additional calculations. One widely used approach is the fourth-order Runge Kutta method, which involves the following calculations:

define
$$f(c_A,t) = \frac{Fc_i(t) - Fc_A - k_r Vc_A^2}{V} \qquad (8\text{-}5)$$

then
$$c_A(t + \Delta t) = c_A(t) + \Delta t(m_1 + 2m_2 + 2m_3 + m_4) \qquad (8\text{-}6)$$

with
$$m_1 = f[c_A(t), t] \qquad (8\text{-}7)$$

$$m_2 = f\left[c_A(t) + \frac{m_1 \Delta t}{2}, t + \frac{\Delta t}{2}\right] \qquad (8\text{-}8)$$

$$m_3 = f\left[c_A(t) + \frac{m_2 \Delta t}{2}, t + \frac{\Delta t}{2}\right] \qquad (8\text{-}9)$$

$$m_4 = f[c_A(t) + m_3 \Delta t, t + \Delta t] \qquad (8\text{-}10)$$

In this method, the m_i's are calculated sequentially in order to take a step in time. Even though this method requires calculation of the four additional m_i values, for equivalent accuracy the fourth-order Runge Kutta method can result in a faster numerical solution, since a larger step, Δt, can be taken with it. Increasingly sophisticated simulation packages are being used to calculate the dynamic behavior of processes and test control system behavior. These packages have good user interfaces, and they can handle stiff systems where some variables respond on a time scale that is much much faster or slower than other variables. A simple Euler approach cannot effectively handle stiff systems, which frequently occur in chemical-process models.

Laplace Transforms When mathematical models are used to describe process dynamics in conjunction with control-system analysis, the models generally involve linear differential equations. Laplace transforms are very effective for solving linear differential equations. The key advantage of using Laplace transforms is that they convert differential equations into algebraic equations. The resulting algebraic equations are easier to solve than the original differential equations. When the Laplace transform is applied to a linear differential equation in time, the result is an algebraic equation in a new variable, s, called the Laplace variable. To get the solution to the original differential equation, one needs to invert the Laplace transform. Table 8-1 gives a number of useful Laplace transform pairs, and more extensive tables are available (Seborg, Edgar, and Mellichamp, *Process Dynamics and Control*, Wiley, New York, 1989).

To illustrate how Laplace transforms work, consider the problem of solving Eq. (8-2), subject to the initial condition that $c_A = 0$ at $t = 0$, and c_i is constant. If c_A were not initially zero, one would define a deviation variable between c_A and its initial value ($c_A - c_0$). Then the transfer function would be developed using this deviation variable. Taking the Laplace transform of both sides of Eq. (8-2) gives:

$$\mathcal{L}\left(\frac{\tau dc_A}{dt}\right) = \mathcal{L}(Kc_i) - \mathcal{L}(c_A) \qquad (8\text{-}11)$$

Denoting the $\mathcal{L}(c)$ as $C_A(s)$ and using the relationships in Table 8-1 gives:

$$\tau s C_A(s) = \frac{Kc_i}{s} - C_A(s) \qquad (8\text{-}12)$$

Equation (8-12) can be solved for C_A to give:

$$C_A(s) = \frac{Kc_i/s}{\tau s + 1} \qquad (8\text{-}13)$$

Using the entries in Table 8-1, Eq. (8-13) can be inverted to give the transient response of c_A as:

$$c_A(t) = (Kc_i)(1 - e^{-t/\tau}) \qquad (8\text{-}14)$$

Equation (8-14) shows that c_A starts from 0 and builds up exponentially to a final concentration of Kc_i. Note that to get Eq. (8-14), it was only necessary to solve the algebraic Eq. (8-12) and then find the inverse of $C_A(s)$ in Table 8-1. The original differential equation was not solved directly. In general, techniques such as partial fraction expansion must be used to solve higher order differential equations with Laplace transforms.

Transfer Functions and Block Diagrams A very convenient and compact method of representing the process dynamics of linear systems involves the use of transfer functions and block diagrams. A transfer function can be obtained by starting with a physical model as

TABLE 8-1 Frequently Used Laplace Transforms

Time function, $f(t)$	Transform, $F(s)$
A	A/s^2
At	A/s
Ae^{-at}	$A/(s + a)$
$A(1 - e^{-t/\tau})$	$A/[s(\tau s + 1)]$
$A \sin(\omega t)$	$A\omega/(s^2 + \omega^2)$
$f(t - \theta)$	$e^{-\theta s}F(s)$
df/dt	$sF(s) - f(0)$
$\int f(t)\, dt$	$F(s)/s$

discussed previously. If the physical model is nonlinear, then it first needs to be linearized around an operating point. The resulting linearized model is then approximately valid in a region around this operating point. To illustrate how transfer functions are developed, Eq. (8-2) will again be used. First, one defines deviation variables, which are the process variables minus their steady state values at the operating point. For Eq. (8-2), there would be deviation variables for both c_A and c_i, and these are defined as:

$$\xi = c_A - c_s \qquad (8\text{-}15)$$

$$\xi_i = c_i - c_{is} \qquad (8\text{-}16)$$

where the subscript s stands for steady state. Substitution of Eq. (8-15) and (8-16) into Eq. (8-2) gives:

$$\tau \frac{d\xi}{dt} = K\xi_i - \xi + (Kc_{is} - c_s) \qquad (8\text{-}17)$$

The term in parentheses in Eq. (8-17) is zero at steady state and thus it can be dropped. Next the Laplace transform is taken, and the resulting algebraic equation solved. Denoting $X(s)$ as the Laplace transform of ξ and $X_i(s)$ as the transform of ξ_i, the final transfer function can be written as:

$$\frac{X}{X_i} = \frac{K}{\tau s + 1} \qquad (8\text{-}18)$$

Equation (8-18) is an example of a first-order transfer function. As mentioned above, an alternative to formally deriving Eq. (8-18) involves simply postulating its form and then identifying its two parameters, the **process gain** K and **time constant** τ, to fit the process under study. In fitting the parameters, data can be generated by forcing the process. If step forcing is used, then the resulting response is called the **process reaction curve**. Often transfer functions are placed in block diagrams, as shown in Fig. 8-4. Block diagrams show how changes in an input variable affect an output variable. Block diagrams are a means of concisely representing the dynamics of a process under study. Since linearity is assumed in developing a block diagram, if more than one variable affects an output, the contributions from each can be added together.

Continuous versus Discrete Models   The preceding discussion has focused on systems where variables change continuously with time. Most real processes have variables that are continuous in nature, such as temperature, pressure, and flow. However, some processes involve discrete events, such as the starting or stopping of a pump. In addition, modern plants are controlled by digital computers, which are discrete by nature. In controlling a process, a digital system samples variables at a fixed rate, and the resulting system is a sampled data system. From one sampling instant until the next, variables are assumed to remain fixed at their sampled values. Similarly, in controlling a process, a digital computer sends out signals to control elements, usually valves, at discrete instants of time. These signals remain fixed until the next sampling instant.

Figure 8-5 illustrates the concept of sampling a continuous function. At integer values of the sampling rate, Δt, the value of the variable to be sampled is measured and held until the next sampling instant. To deal with sampled data systems, the z transform has been developed. The z transform of the function given in Fig. 8-5 is defined as

$$Z(f) = \sum_{n=0}^{\infty} f(n\,\Delta t) z^{-n} \qquad (8\text{-}19)$$

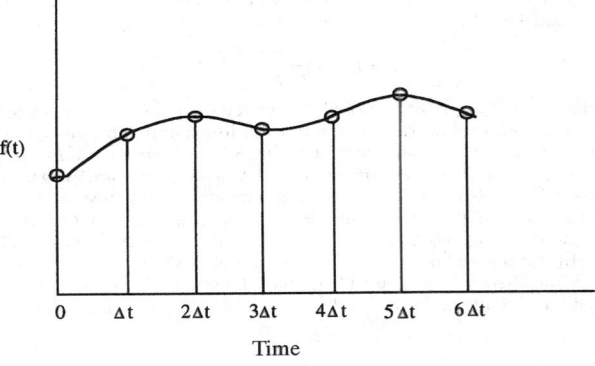

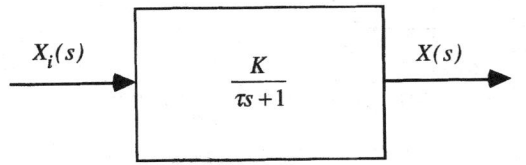

FIG. 8-4   First-order transfer function.

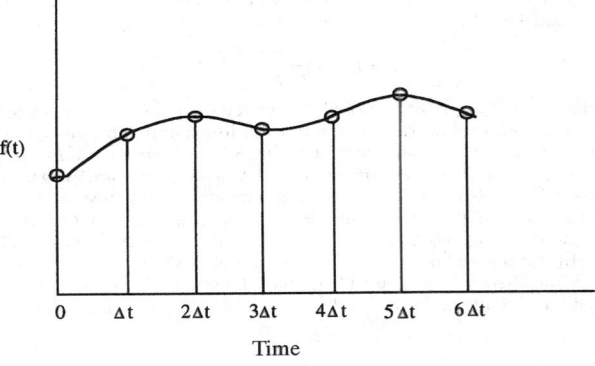

FIG. 8-5   Sampled data example.

In an analogous manner to Laplace transforms, one can develop transfer functions in the z domain as well as block diagrams. Tables of z transform pairs have been published (Seborg, Edgar, and Mellichamp, *Process Dynamics and Control*, Wiley, New York, 1989) so that the discrete transfer functions can be inverted back to the time domain. The inverse gives the value of the function at the discrete sampling instants. Sampling a continuous variable results in a loss of information. However, in practical applications, sampling is fast enough that the loss is typically insignificant and the difference between continuous and discrete modeling is small in terms of its effect on control. Increasingly, model predictive controllers that make use of discrete dynamic models are being used in the process industries. The purpose of these controllers is to guide a process to optimum operating points. These model predictive control algorithms are typically run at much slower sampling rates than are used for basic control loops such as flow control or pressure control. The discrete dynamic models used are normally developed from data generated from plant testing as discussed hereafter. For a detailed discussion of modeling sampled data systems, the interested reader is referred to textbooks on digital control (Astrom and Wittenmark, *Computer Controlled Systems*, Prentice Hall, Englewood Cliffs, NJ, 1984).

Process Characteristics in Transfer Functions   In many cases, process characteristics are expressed in the form of transfer functions. In the previous discussion, a reactor example was used to illustrate how a transfer function could be derived. Here, another system involving flow out of a tank, shown in Fig. 8-6, is considered.

Proportional Element   First, consider the outflow through the exit valve on the tank. If the flow through the line is turbulent, then Bernoulli's equation can be used to relate the flow rate through the valve to the pressure drop across the valve as:

$$f_1 = k_f A_v \sqrt{2g_c(h_1 - h_0)} \qquad (8\text{-}20)$$

where f_1 = flow rate, k_f = flow coefficient, A_v = cross sectional area of the restriction, g_c = constant, h_1 = liquid head in tank, and h_0 = atmo-

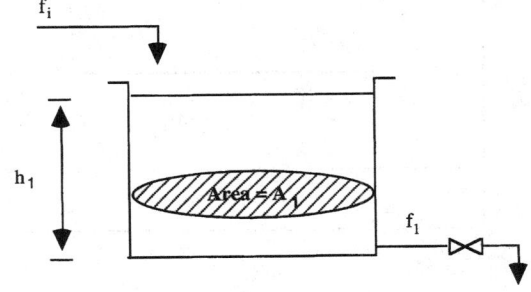

FIG. 8-6   Single tank with exit valve.

spheric pressure. This relationship between flow and head is nonlinear, and it can be linearized around a particular operating point to give:

$$f_1 - f_{1s} = \left(\frac{1}{R_1}\right)(h_1 - h_{1s}) \tag{8-21}$$

where $R_1 = f_{1s}/(g_c k_f^2 A^2)$ is called the resistance of the valve in analogy with an electrical resistance. The transfer function relating changes in flow to changes in head is shown in Fig. 8-7, and it is an example of a pure gain system with no dynamics. In this case, the process gain is $K = 1/R_1$. Such a system has an instantaneous dynamic response, and for a step change in head, there is an immediate step change in flow, as shown in Fig. 8-8. The exact magnitude of the step in flow depends on the operating flow, f_{1s}, as the definition of R_1 shows.

First-Order Lag (Time Constant Element)   Next consider the system to be the tank itself. A dynamic mass balance on the tank gives:

$$A_1 \frac{dh_1}{dt} = f_i - f_1 \tag{8-22}$$

where A_1 is the cross sectional area of the tank and f_i is the inlet flow. By substituting Eq. (8-21) into Eq. (8-22), and following the approach discussed above for deriving transfer functions, one can develop the transfer function relating changes in h_1 to changes in f_i. The resulting transfer function is another example of a first-order system, shown in Fig. 8-4, and it has a gain, $K = R_1$, and a time constant, $\tau_1 = R_1 A_1$. For a step change in f_i, h_1 follows a decaying exponential response from its initial value, h_{1s}, to a final value of $h_{1s} + R_1 \Delta f_i$ (Fig. 8-9). At a time equal to τ_1, the transient in h_1 is 63 percent finished; and at $3\tau_1$, the response is 95 percent finished. These percentages are the same for all first-order processes. Thus, knowledge of the time constant of a first-order process gives insight into how fast the process responds to sudden input changes.

Capacity Element   Now consider the case where the valve in Fig. 8-7 is replaced with a pump. In this case, it is reasonable to assume that the exit flow from the tank is independent of the level in the tank. For such a case, Eq. (8-22) still holds, except that f_1 no longer depends on h_1. For changes in f_i, the transfer function relating changes in h_1 to changes in f_i is shown in Fig. 8-10. This is an example of a pure capacity process, also called an integrating system. The cross sectional area of the tank is the chemical process equivalent of an electrical capacitor. If the inlet flow is step forced while the outlet is held

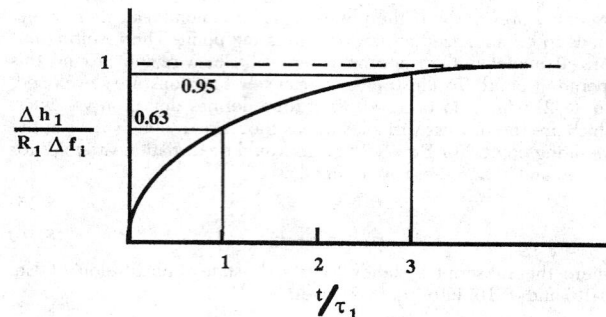

FIG. 8-9   Response of first-order system.

constant, then the level builds up linearly as shown in Fig. 8-11. Eventually the liquid would overflow the tank.

Second-Order Element   Because of their linear nature, transfer functions can be combined in a straightforward manner. Consider the two tank system shown in Fig. 8-12. For tank 1, the transfer function relating changes in f_1 to changes in f_i can be obtained by combining two first order transfer functions to give:

$$\frac{F_1(s)}{F_i(s)} = \frac{1}{R_1 A_1 s + 1} \tag{8-23}$$

Since f_1 is the inlet flow to tank 2, the transfer function relating changes in h_2 to changes in f_1 has the same form as that given in Fig. 8-4:

$$\frac{H_2(s)}{F_1(s)} = \frac{R_2}{A_2 R_2 s + 1} \tag{8-24}$$

Equations (8-23) and (8-24) can be multiplied together to give the final transfer function relating changes in h_2 to changes in f_i as shown in Fig. 8-13. This is an example of a second-order transfer function. This transfer function has a gain $R_1 R_2$ and two time constants, $R_1 A_1$ and $R_2 A_2$. For two equal tanks, a step change in f_i produces the S-shaped response in level in the second tank shown in Fig. 8-14.

General Second-Order Element   Figure 8-3 illustrates the fact that closed loop systems often exhibit oscillatory behavior. A general

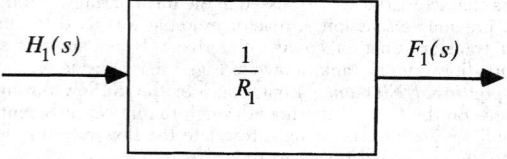

FIG. 8-7   Proportional element transfer function.

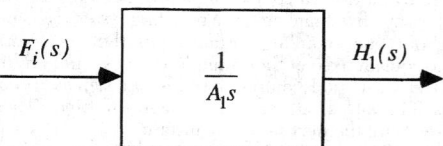

FIG. 8-10   Pure capacity transfer function.

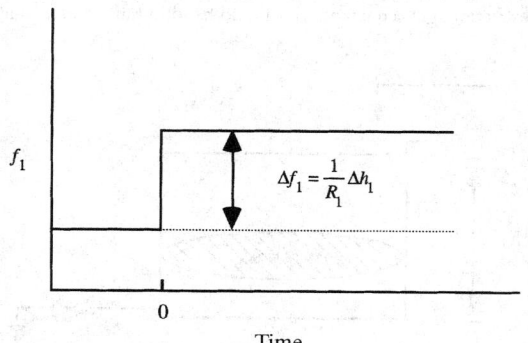

FIG. 8-8   Response of proportional element.

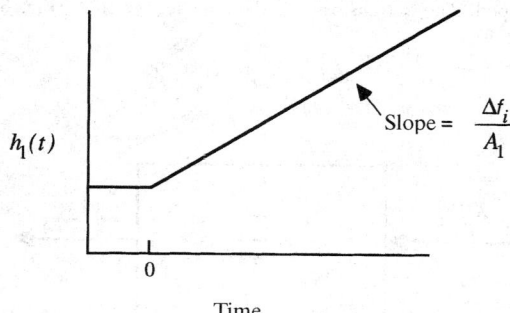

FIG. 8-11   Response of pure capacity system.

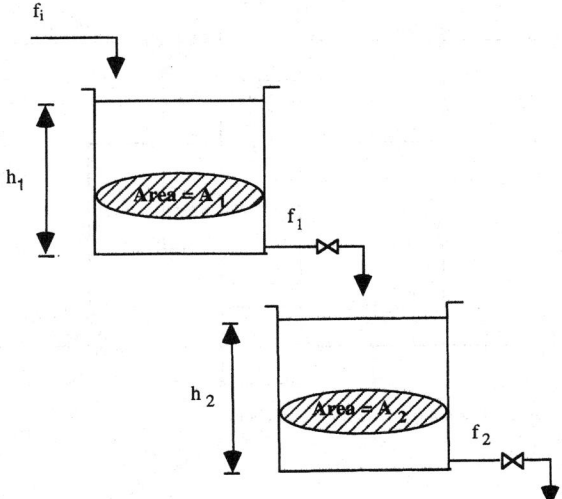

FIG. 8-12   Two tanks in series.

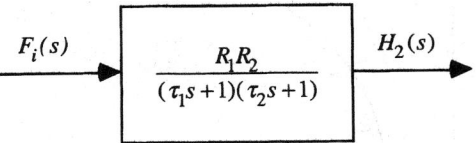

FIG. 8-13   Second-order transfer function.

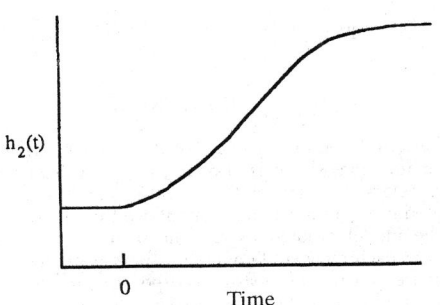

FIG. 8-14   Response of second-order system.

second-order transfer function that can exhibit oscillatory behavior is important for the study of automatic control systems. Such a transfer function is given in Fig. 8-15. For a step input, the transient responses shown in Fig. 8-16 result. As can be seen when $\zeta < 1$, the response oscillates and when $\zeta > 1$, the response is S-shaped. Few open-loop chemical processes exhibit an oscillating response; most exhibit an S-shaped step response.

Distance-Velocity Lag (Dead-Time Element)   The dead-time element, commonly called a distance-velocity lag, is often encountered in process systems. For example, if a temperature-measuring element is located downstream from a heat exchanger, a time delay occurs before the heated fluid leaving the exchanger arrives at the temperature measurement point. If some element of a system produces a dead-time of θ time units, then an input to that unit, $f(t)$, will be reproduced at the output as $f(t - \theta)$. The transfer function for a pure dead-time element is shown in Fig. 8-17, and the transient response of the element is shown in Fig. 8-18.

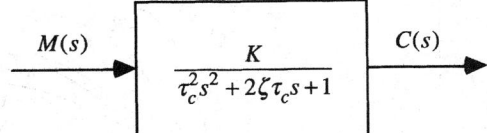

FIG. 8-15   General second-order transfer function.

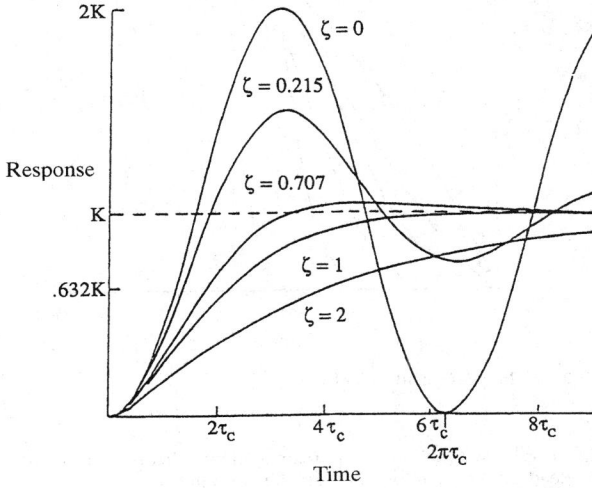

FIG. 8-16   Response of general second-order system.

Higher-Order Lags   If a process is described by a series of n first-order lags, the overall system response becomes proportionally slower with each lag added. The special case of a series of n first-order lags with equal time constants has a transfer function given by:

$$G(s) = \frac{K}{(\tau s + 1)^n} \qquad (8\text{-}25)$$

The step response of this transfer function is shown in Fig. 8-19. Note that all curves reach about 60 percent of their final value at $t = n\tau$.

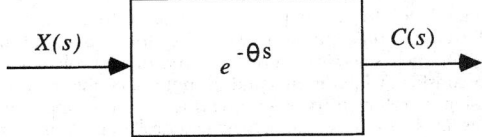

FIG. 8-17   Dead-time transfer function.

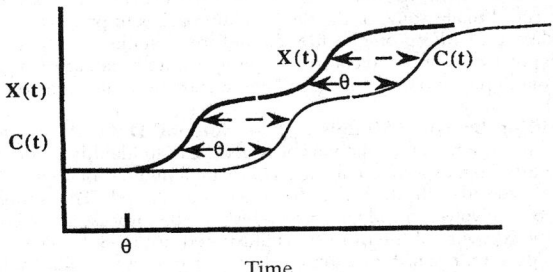

FIG. 8-18   Response of dead-time system.

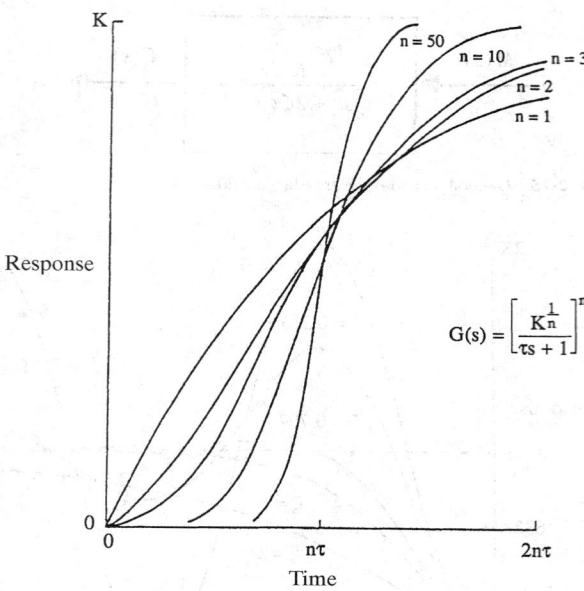

FIG. 8-19   Response of nth order lags.

$$G(s) = \left[\frac{K^{\frac{1}{n}}}{\tau s + 1}\right]^n$$

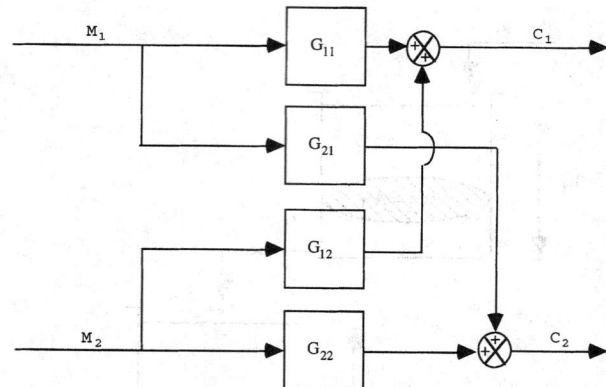

FIG. 8-20   Example of 2×2 transfer function.

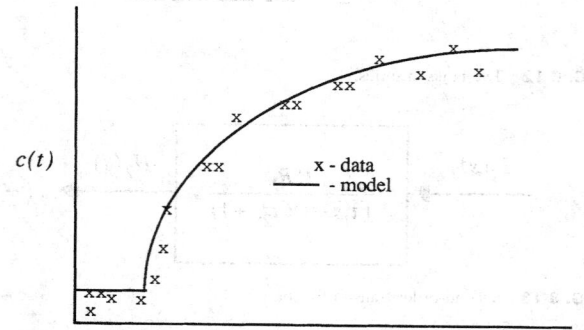

FIG. 8-21   Plot of experimental data.

Higher-order systems can be approximated by a first or second-order plus dead-time system for control system design.

Multiinput, Multioutput Systems   The dynamic systems considered up to this point have been examples of single-input, single-output (SISO) systems. In chemical processes, one often encounters systems where one input can affect more than one output. For example, assume that one is studying a distillation tower in which both reflux and boilup are manipulated for control purposes. If the output variables are the top and bottom product compositions, then each input affects both outputs. For this distillation example, the process is referred to as a 2×2 system to indicate the number of inputs and outputs. In general, multiinput, multioutput (MIMO) systems can have n inputs and m outputs with $n \neq m$, and they can be nonlinear. Such a system would be called an $n \times m$ system. An example of a transfer function for a 2×2 linear system is given in Fig. 8-20. Note that since linear systems are involved, the effects of the two inputs on each output are additive. In many process-control systems, one input is selected to control one output in a MIMO system. For m output there would be m such selections. For this type of control strategy, one needs to consider which inputs and outputs to couple together, and this problem is referred to as loop pairing. Another important issue that arises involves interaction between control loops. When one loop makes a change in its manipulated variable, the change affects the other loops in the system. These changes are the direct result of the multivariable nature of the process. In some cases, the interaction can be so severe that overall control-system performance is drastically reduced. Finally, some of the modern approaches to process control tackle the MIMO problem directly, and they simultaneously use all manipulated variables to control all output variables rather than pairing one input to one output (see later section on multivariable control).

Fitting Dynamic Models to Experimental Data   In developing empirical transfer functions, it is necessary to identify model parameters from experimental data. There are a number of approaches to process identification that have been published. The simplest approach involves introducing a step test into the process and recording the response of the process, as illustrated in Fig. 8-21. The x's in the figure represent the recorded data. For purposes of illustration, the process under study will be assumed to be first order with dead-time and have the transfer function:

$$G(s) = \frac{C(s)}{M(s)} = K \exp(-\theta s)/(\tau s + 1) \qquad (8\text{-}26)$$

The response produced by Eq. (8-26), $c(t)$, can be found by inverting the transfer function, and it is also shown in Fig. 8-21 for a set of model parameters, K, τ, and θ, fitted to the data. These parameters are calculated using optimization to minimize the squared difference between the model predictions and the data, i.e., a least squares approach. Let each measured data point be represented by c_j (measured response), t_j (time of measured response), $j = 1$ to n. Then the least squares problem can be formulated as:

$$\min_{\tau, \theta, K} \sum_{j=0}^{n} [c_j - \hat{c}(t_j)]^2 \qquad (8\text{-}27)$$

which can be solved to calculate the optimal values of K, τ, and θ. A number of software packages are available for minimizing Eq. (8-27).

One operational problem that step forcing causes is the fact that the process under study is moved away from its steady state operating point. Plant managers may be reluctant to allow large steady state changes, since normal production will be disturbed by the changes. As a result, alternative methods of forcing actual processes have been developed, and these included pulse testing and pseudo random binary signal (PRBS) forcing, both of which are illustrated in Fig. 8-22. With pulse forcing, one introduces a step, and then after a period of time the input is returned to its original value. The result is that the process dynamics are excited, but after the forcing, the process returns to its original steady state. PRBS forcing involves a series of pulses of fixed height and random duration, as shown in Fig. 8-22. The advantage of PRBS is that forcing can be concentrated on particular frequency ranges that are important for control-system design.

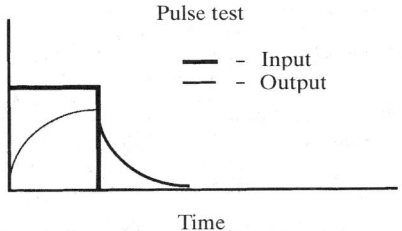

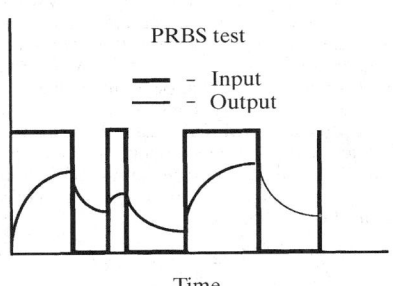

FIG. 8-22 Pulse and PRBS testing.

Transfer function models are linear in nature, but chemical processes are known to exhibit nonlinear behavior. One could use the same type of optimization objective as given in Eq. (8-26) to determine parameters in nonlinear first-principle models, such as Eq. (8-3) presented earlier. Also, nonlinear empirical models, such as neural network models, have recently been proposed for process applications. The key to the use of these nonlinear empirical models is having high-quality process data, which allows the important nonlinearities to be identified.

FEEDBACK CONTROL SYSTEM CHARACTERISTICS

GENERAL REFERENCES: Shinskey, *Feedback Controllers for the Process Industries*, McGraw-Hill, New York, 1994; Seborg, Edgar, and Mellichamp, *Process Dynamics and Control*, Wiley, New York, 1989.

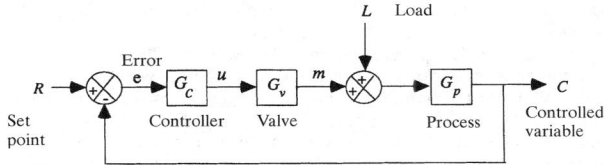

FIG. 8-23 Both load regulation and setpoint response require high gains for the feedback controller.

There are two objectives in applying feedback control: regulating the controlled variable at set point following changes in load, and responding to set-point changes; the latter called servo operation. In fluid processes, almost all control loops must contend with variations in load; therefore, regulation is of primary importance. While most loops will operate continuously at fixed set points, frequent changes in set points can occur in flow loops and in batch production. The most common mechanism for achieving both objectives is feedback control, because it is the simplest and most universally applicable approach to the problem.

Closing the Loop The simplest representation of the closed feedback loop is shown in Fig. 8-23. The load is shown entering the process at the same point as the manipulated variable because that is the most common point of entry, and also because, lacking better information, the transfer function gains in the path of the manipulated variable are the best estimates of those in the load path. In general, the load never impacts directly on the controlled variable without passing through the dominant lag in the process. Where the load is unmeasured, its current value can be observed to be the controller output required to keep the controlled variable C at set point R.

If the loop is opened, either by placing the controller in manual operation or by setting its gains to zero, the load will have complete influence over the controlled variable, and the set point will have none. Only by closing the loop with controller gains as high as possible will the influence of the load be minimized and that of the set point be maximized. There is a practical limit to the controller gains, however, at the point where the controlled variable develops a uniform oscillation (see Fig. 8-24). This is defined as the limit of stability, and it is reached when the product of gains in the loop $|G_cG_vG_p|$ for that frequency of oscillation is equal to 1.0. If a change in a parameter in the loop causes an increase from this condition, oscillations will expand, creating a dangerous situation where safe limits of operation could be exceeded. Consequently, control loops should be left in a condition

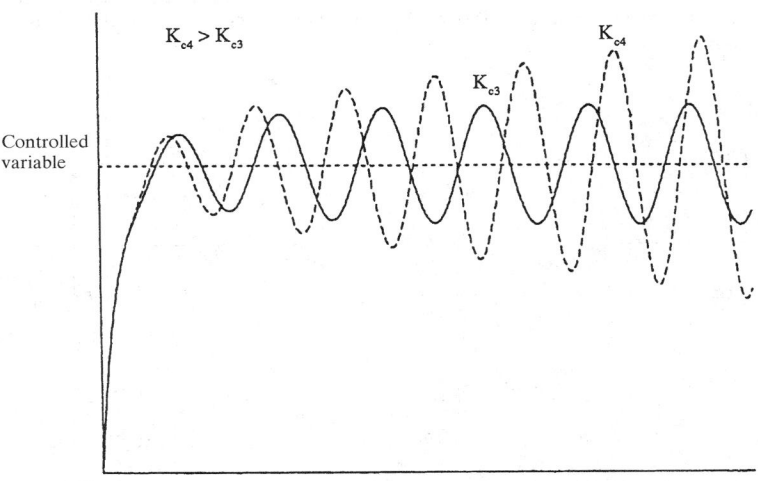

FIG. 8-24 Transition to instability as controller gain increases.

where the loop gain is less than 1.0 by a safe margin that allows for possible variations in process parameters.

In controller design, a choice must be made between performance and robustness. Performance is a measure of how well a given controller with certain parameter settings regulates a variable relative to the best loop performance with optimal controller settings. Robustness is a measure of how small a change in a process parameter is required to bring the loop from its current state to the limit of stability. Increasing controller performance by raising its gains can be expected to decrease robustness. Both performance and robustness are functions of the process being controlled, the selection of the controller, and the tuning of the controller parameters.

On/Off Control An on/off controller is used for manipulated variables having only two states. They commonly control temperatures in homes, electric water-heaters and refrigerators, and pressure and liquid level in pumped storage systems. On/off control is satisfactory where slow cycling is acceptable because it always leads to cycling when the load lies between the two states of the manipulated variable. The cycle will be positioned symmetrically about the set point only if the normal value of the load is equidistant between the two states of the manipulated variable. The period of the symmetrical cycle will be approximately 4θ, where θ is the deadtime in the loop. If the load is not centered between the states of the manipulated variable, the period will tend to increase, and the cycle follows a sawtooth pattern.

Every on/off controller has some degree of deadband, also known as lockup, or differential gap. Its function is to prevent erratic switching between states, thereby extending the life of contacts and motors. Instead of changing states precisely when the controlled variable crosses set point, the controller will change states at two different points for increasing and decreasing signals. The difference between these two switching points is the deadband (see Fig. 8-25); it increases the amplitude and period of the cycle, similar to the effect of dead time.

A three-state controller is used to drive either a pair of independent on/off actuators such as heating and cooling valves, or a bidirectional motorized actuator. The controller is actually two on/off controllers, each with deadband, separated by a dead zone. When the controlled variable lies within the dead zone, neither output is energized. This controller can drive a motorized valve to the point where the manipulated variable matches the load, thereby avoiding cycling.

Proportional Control A proportional controller moves its output proportional to the deviation in the controlled variable from set point:

$$u = K_c e + b = \frac{100}{P} e + b \qquad (8\text{-}28)$$

where $e = \pm(r - c)$, the sign selected to produce negative feedback. In some controllers, proportional gain K_c is expressed as a pure number; in others, it is set as $100/P$, where P is the proportional band in percent. The output bias b of the controller is also known as manual reset. The proportional controller is not a good regulator, because any change in output to a change in load results in a corresponding change in the controlled variable. To minimize the resulting offset, the bias should be set at the best estimate of the load and the proportional band set as low as possible. Processes requiring a proportional band of more than a few percent will control with unacceptable values of offset.

Proportional control is most often used for liquid level where variations in the controlled variable carry no economic penalty, and where other control modes can easily destabilize the loop. It is actually recommended for controlling the level in a surge tank when manipulating the flow of feed to a critical downstream process. By setting the proportional band just under 100 percent, the level is allowed to vary over the full range of the tank capacity as inflow fluctuates, thereby minimizing the resulting rate of change of manipulated outflow. This technique is called averaging level control.

Proportional-plus-Integral (PI) Control Integral action eliminates the offset described above by moving the controller output at a rate proportional to the deviation from set point. Although available alone in an integral controller, it is most often combined with proportional action in a PI controller:

$$u = \frac{100}{P} \left(e + \frac{1}{\tau_I} \int e \, dt \right) + C_0 \qquad (8\text{-}29)$$

where τ_I is the integral time constant in minutes; in some controllers, it is introduced as integral gain or reset rate $1/\tau_I$ in repeats per minute. The last term in the equation is the constant of integration, the value the controller output has when integration begins.

The PI controller is by far the most commonly used controller in the process industries. The summation of the deviation with its integral in the above equation can be interpreted in terms of frequency response of the controller (Seborg, Edgar, and Mellichamp, *Process Dynamics and Control*, Wiley, New York, 1989). The PI controller produces a phase lag between zero and 90 degrees:

$$\phi_{PI} = -\tan^{-1} \frac{\tau_0}{2\pi\tau_I} \qquad (8\text{-}30)$$

where τ_0 is the period of oscillation of the loop. The phase angle should be kept between 15 degrees for lag-dominant processes and 45 degrees for dead-time-dominant processes for optimum results.

Proportional-plus-Integral-plus-Derivative (PID) Control The derivative mode moves the controller output as a function of the rate-of-change of the controlled variable, which adds phase lead to the controller, increasing its speed of response. It is normally combined

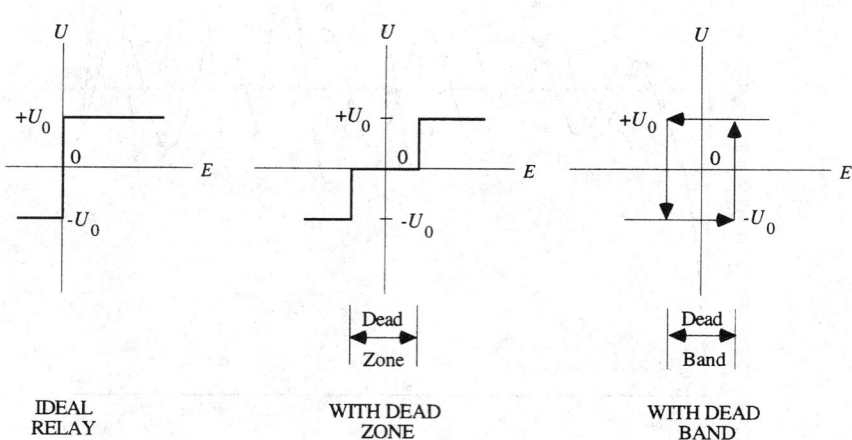

FIG. 8-25 On/off controller characteristics.

with proportional and integral modes. The noninteracting form of the PID controller appears functionally as:

$$u = \frac{100}{P}\left(e + \frac{1}{\tau_I}\int e\,dt + \tau_D\frac{dc}{dt}\right) + C_0 \qquad (8\text{-}31)$$

where τ_D is the derivative time constant. Note that derivative action is applied to the controlled variable rather than to the deviation, as it should not be applied to the set point; the selection of the sign for the derivative term must be consistent with the action of the controller. Figure 8-26 compares typical loop responses for P, PI, and PID controllers, along with the uncontrolled case.

In some analog PID controllers, the integral and derivative terms are combined serially rather than in parallel as done in the last equation. This results in interaction between these modes, such that the effective values of the controller parameters differ from their set values as follows:

$$\tau_{I_{\text{eff}}} = \tau_I + \tau_D$$

$$\tau_{D_{\text{eff}}} = \frac{1}{1/\tau_D + 1/\tau_I}$$

$$K_c = \frac{100}{P}\left(1 + \frac{\tau_D}{\tau_I}\right) \qquad (8\text{-}32)$$

The performance of the interacting controller is almost as good as the noninteracting controller on most processes, but the tuning rules differ because of the above relationships. With digital PID controllers, the noninteracting version is commonly used.

There is always a gain limit placed upon the derivative term—a value of 10 is typical. However, interaction decreases the derivative gain below this value by the factor $1 + \tau_D/\tau_I$, which is the reason for the decreased performance of the interacting PID controller. Sampling in a digital controller has a similar effect, limiting derivative gain to the ratio of derivative time to the sample interval of the controller. Noise on the controlled variable is amplified by derivative action, preventing its use in controlling flow and liquid level. Derivative action is recommended for control of temperature and composition, reducing the integrated error (IE) by a factor of two over PI control with no loss in robustness (Shinskey, *Feedback Controllers for the Process Industries*, McGraw-Hill, New York, 1994).

CONTROLLER TUNING

The performance of a controller depends as much on its tuning as its design. Tuning must be applied by the end user to fit the controller to the controlled process. There are many different approaches to controller tuning based on the particular performance criteria selected, whether load or set-point changes are most important, whether the process is lag- or deadtime-dominant, and the availability of information about the process dynamics. The earliest definitive work in this field was done at the Taylor Instrument Company by Ziegler and Nichols (Trans. ASME, 759, 1942), tuning PI and interacting PID controllers for optimum response to step load changes applied to lag-dominant processes. While these tuning rules are still in use, they are approximate and do not apply to set-point changes, dead-time-dominant processes, or noninteracting PID controllers (Seborg, Edgar, and Mellichamp, *Process Dynamics and Control*, Wiley, New York, 1989).

Controller Performance Criteria   The most useful measures of controller performance in an industrial setting are the maximum deviation in the controlled variable resulting from a disturbance and its integral. The disturbance could be to the set point or to the load, depending on the variable being controlled and its context in the process. The size of the deviation and its integral are proportional to the size of the disturbance (if the loop is linear at the operating point). While actual disturbances in a plant setting may appear to be random, the controller needs a reliable test to determine how well it is tuned. The disturbance of choice for test purposes is the step, because it can be applied manually, and by containing all frequencies including zero, it exercises all modes of the controller. When tuned optimally for step disturbances, the controller should be well-tuned for most other disturbances as well.

Figure 8-27 shows the optimum response of a controlled variable to a step change in load. A step change in load may be simulated by stepping the controller output while it is in the manual mode followed immediately by transfer to automatic. The maximum deviation is the most important criterion for variables that could exceed safe operating levels such as steam pressure, drum level, and steam temperature in a boiler. The same rule applies to product quality, which could violate specifications and therefore be rejected. If the product can be accumulated in a downstream storage tank, however, its average quality is more important, and this is a function of the deviation integrated over the residence time in the tank. Deviation in the other direction, where the product is better than specification, is safe, but it increases production costs in proportion to the integrated deviation because quality is given away.

For a PI or PID controller, the integrated deviation—better known as the integrated error IE—is related to the controller settings:

$$\text{IE} = \frac{\Delta u P \tau_I}{100} \qquad (8\text{-}33)$$

where Δu is the difference in controller output between two steady states, as required by a change in load or set point. The proportional band P and integral time τ_I are the indicated settings of the controller

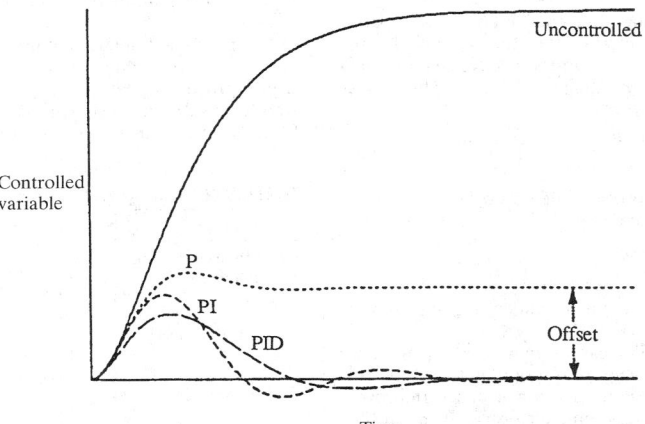

FIG. 8-26   Response for a step change in disturbance with tuned P, PI, and PID controllers and with no control.

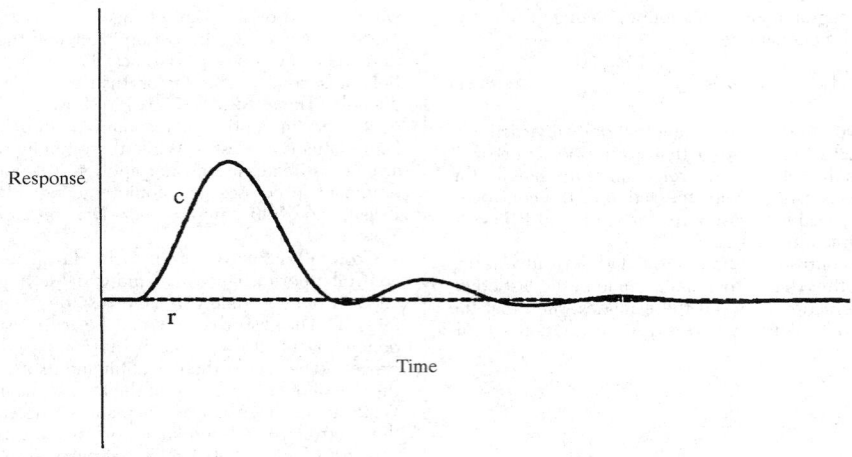

FIG. 8-27 The minimum-IAE response to a step load change has little overshoot and is well-damped.

for both interacting and noninteracting PID controllers. Although the derivative term does not appear in the relationship, its use typically allows a 50 percent reduction in integral time and therefore in IE. The integral time in the IE expression should be augmented by the sample interval if the controller is digital, the time constant of any filter used, and the value of any deadtime compensator.

It would appear from the above that minimizing IE is simply a matter of minimizing the P and τ_I settings of the controller. However, settings will be reached that produce uniform oscillations—an unacceptable situation. It is preferable, instead, to find a combination of controller settings that minimize integrated absolute error IAE, which for both load and set-point changes is a well-damped response with minimal overshoot. Figure 8-27 is an example of a minimum-IAE response to a step change in load for a lag-dominant process. Because of the very small overshoot, the IAE will be only slightly larger than the IE. Loops that are tuned to minimize IAE tend to be close to minimum IE and also minimum peak deviation.

The performance of a controller (and its tuning) must be based on what is achievable for a given process. The concept of best practical IE (IE$_b$) for a step change in load Δq can be estimated (Shinskey, *Feedback Controllers for the Process Industries,* McGraw-Hill, New York, 1994):

$$\text{IE}_b = \Delta q K_L \tau_L (1 - e^{-\theta/\tau_L}) \qquad (8\text{-}34)$$

where K_L is the gain and τ_L the primary time constant in the load path, and θ the dead time in the manipulated path to the controlled variable. If the load or its gain is unknown, Δu and $K(= K_v K_p)$ may be substituted. If the process is non-self-regulating (i.e., it is an integrator), the relationship is

$$\text{IE}_b = \frac{\Delta q \theta^2}{\tau_1} \qquad (8\text{-}35)$$

where τ_1 is the time constant of the process integrator. The peak deviation with the best practical response curve is:

$$e_b = \frac{\text{IE}_b}{\theta + \tau_2} \qquad (8\text{-}36)$$

where τ_2 is the time constant of a common secondary lag (e.g., in the measuring device).

The performance for any controller can be measured against this standard by comparing the IE it achieves in responding to a step load change with the best practical IE. Potential performance improvements by tuning PI controllers on lag-dominant processes lie in the 20–30 percent range, while for PID controllers they fall between 40–60 percent, varying with secondary lags.

Tuning Methods Based on Known Process Models The most accurate tuning rules for controllers have been based on simulation, where the process parameters can be specified and IAE and IE can be integrated during the simulation as an indication of performance. Controller settings are then iterated until a minimum IAE is reached for a given disturbance. These optimum settings are then related to the parameters of the simulated process in tables, graphs, or equations, as a guide to tuning controllers for processes whose parameters are known (Seborg, Edgar, and Mellichamp, *Process Dynamics and Control,* Wiley, New York, 1989). This is a multidimensional problem, however, in that the relationships change as a function of process type, controller type, and source of disturbance.

Table 8-2 summarizes these rules for minimum-IAE load response for the most common controllers. The process gain K and time constant τ_m are obtained from the product of G_v and G_p in Fig. 8-23. Derivative action is not effective for dead-time-dominant processes. For non-self-regulating processes, τ is the time constant of the integrator. The last category of distributed lag includes all heat-transfer processes, backmixed vessels, and processes having multiple interacting lags such as distillation columns; $\sum \tau$ represents the total response time of these processes (i.e., the time required for 63 percent complete response to a step input). Any secondary lag, sampling interval, or filter time constant should be added to dead-time θ.

The principal limitation to using these rules is that the true process parameters are often unknown. Steady-state gain K can be calculated from a process model or determined from the steady-state results of a step test as $\Delta c/\Delta u$, as shown in Fig. 8-28. The test will not be viable, however, if the time constant of the process τ_m is longer than a few

TABLE 8-2 Tuning Rules Using Known Process Parameters

Process	Controller	P	τ_I	τ_D
Dead-time-dominant	PI	250K	0.5 θ	
Lag-dominant	PI	106K θ/τ_m	4.0 θ	
	PID$_n$	77K θ/τ_m	1.8 θ	0.45 θ
	PID$_i$	106K θ/τ_m	1.5 θ	0.55 θ
Non-self-regulating	PI	106 θ/τ_1	4.0 θ	
	PID$_n$	78 θ/τ_1	1.9 θ	0.48 θ
	PID$_i$	108 θ/τ_1	1.6 θ	0.58 θ
Distributed lags	PI	20K	0.50 $\sum \tau$	
	PID$_n$	10K	0.30 $\sum \tau$	0.09 $\sum \tau$
	PID$_i$	15K	0.25 $\sum \tau$	0.10 $\sum \tau$

NOTE: n = noninteracting; i = interacting controller modes

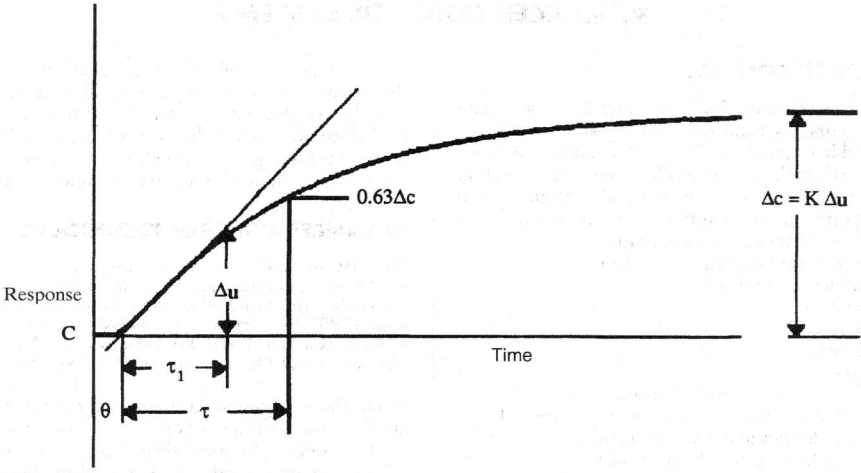

FIG. 8-28 If a steady state can be reached, gain K and time constant τ can be estimated from a step response; if not, use τ_1 instead.

minutes, since five time constants must elapse to approach a steady state within one percent, and unrequested disturbances may intervene. Estimated dead-time θ is the time from the step to the intercept of a straight line tangent to the steepest part of the response curve. The estimated time constant τ is the time from that point to 63 percent of the complete response. In the presence of a secondary lag, these results will not be completely accurate, however. The time for 63 percent response may be more accurately calculated as the residence time of the process: its volume divided by current volumetric flow rate.

Tuning Methods When Process Model Is Unknown Ziegler and Nichols developed two tuning methods for processes with unknown parameters. The open-loop method uses a step test without waiting for a steady state to be reached and is therefore applicable to very slow processes. Deadtime is estimated from the intercept of the tangent in Fig. 8-28, whose slope is also used. If the process is non-self-regulating, the controlled variable will continue to follow this slope, changing an amount equal to Δu in a time equal to its time constant. This time estimate τ_1 is used along with θ to tune controllers according to Table 8-3, applicable to lag-dominant processes.

A more recent tuning approach uses integral criteria such as the integral of the squared error (ISE), integral of the absolute error (IAE), and the time-weighted IAE (ITAE) of Seborg, Edgar, and Mellichamp (*Process Dynamics and Control*, Wiley, New York, 1989). The controller parameters are selected to minimize various integrals. Power-law correlations for PID controller settings have been tabulated for a range of first-order model parameters. The best tuning parameters have been fitted using a general equation, $Y = A(\theta/\tau)^B$, where Y depends on the particular controller mode to be evaluated (K_C, τ_I, τ_D).

There are several features of the correlations that should be noted:

1.   The controller gain is inversely proportional to the process gain for constant dead time and time constant.

2.   The allowable controller gain is higher when the ratio of dead time to time constant becomes smaller. This is because dead time has a destabilizing effect on the control system, limiting the controller gain, while a larger time constant generally demands a higher controller gain.

A recent addition to the model-based tuning correlations is Internal Model Control (Rivera, Morari, and Skogestad, "Internal Model Control 4: PID Controller Design," *IEC Proc. Des. Dev.*, **25**, 252, 1986), which offers some advantages over the other methods described here. However, the correlations are similar to the ones discussed above. Other plant testing and controller design approaches such as frequency response can be used for more complicated models.

The Ziegler and Nichols closed-loop method requires forcing the loop to cycle uniformly under proportional control. The natural period τ_n of the cycle—the proportional controller contributes no phase shift to alter it—is used to set the optimum integral and derivative time constants. The optimum proportional band is set relative to the undamped proportional band P_u, which produced the uniform oscillation. Table 8-4 lists the tuning rules for a lag-dominant process. A uniform cycle can also be forced using on/off control to cycle the manipulated variable between two limits. The period of the cycle will be close to τ_n if the cycle is symmetrical; the peak-to-peak amplitude of the controlled variable A_c divided by the output limits A_m is a measure of process gain at that period and is therefore related to P_u for the proportional cycle:

$$P_u = 100 \, \frac{\pi}{4} \frac{A_c}{A_m} \qquad (8\text{-}37)$$

The factor $\pi/4$ compensates for the square wave in the output. Tuning rules are given in Table 8-4.

TABLE 8-3 Tuning Rules Using Slope and Intercept

Controller	P	τ_I	τ_D
PI	$150\ \theta/\tau$	$3.5\ \theta$	—
PID_n	$75\ \theta/\tau$	$2.1\ \theta$	$0.63\ \theta$
PID_i	$113\ \theta/\tau$	$1.8\ \theta$	$0.70\ \theta$

NOTE: n = noninteracting, i = interacting controller modes

TABLE 8-4 Tuning Rules Using Proportional Cycle

Controller	P	τ_I	τ_D
PI	$1.70\ P_u$	$0.81\ \tau_n$	—
PID_n	$1.30\ P_u$	$0.48\ \tau_n$	$0.11\ \tau_n$
PID_i	$1.80\ P_u$	$0.39\ \tau_n$	$0.14\ \tau_n$

ADVANCED CONTROL SYSTEMS

BENEFITS OF ADVANCED CONTROL

The economics of most processes are determined by the steady-state operating conditions. Excursions from these steady-state conditions generally average out and have an insignificant effect on the economics of the process, except when the excursions lead to off-specification products. In order to enhance the economic performance of a process, the steady-state operating conditions must be altered in a manner that leads to more efficient process operation.

The following hierarchy is used for process control:
Level 0: Measurement devices and actuators
Level 1: Regulatory control
Level 2: Supervisory control
Level 3: Production control
Level 4: Information technology

Levels 2, 3, and 4 clearly affect the process economics, as all three levels are directed to optimizing the process in some manner. However, level 0 (measurement devices and actuators) and level 1 (regulatory control) would appear to have no effect on process economics. Their direct effect is indeed minimal, but indirectly, they have a major effect. Basically, these levels provide the foundation for all higher levels. A process cannot be optimized until it can be operated consistently at the prescribed targets. Thus, a high degree of regulatory control must be the first goal of any automation effort. In turn, the measurements and actuators provide the process interface for regulatory control.

For most processes, the optimum operating point is determined by a constraint. The constraint might be a product specification (a product stream can contain no more than 2 percent ethane); violation of this constraint causes off-specification product. The constraint might be an equipment limit (vessel pressure rating is 300 psig); violation of this constraint causes the equipment protection mechanism (pressure relief device) to activate. As the penalties are serious, violation of such constraints must be very infrequent.

If the regulatory control system were perfect, the target could be set exactly equal to the constraint (that is, the target for the pressure controller could be set at the vessel relief pressure). However, no regulatory control system is perfect. Therefore, the value specified for the target must be on the safe side of the constraint, thus giving the control system some "elbow room." How much depends on the following:

1. *The performance of the control system (i.e., how effectively it responds to disturbances).* The faster the control system reacts to a disturbance, the closer the process can be operated to the constraint.

2. *The magnitude of the disturbances to which the control system must respond.* If the magnitude of the major disturbances can be reduced, the process can be operated closer to the constraint.

One measure of the performance of a control system is the variance of the controlled variable from the target. Both improving the control system and reducing the disturbances will lead to a lower variance in the controlled variable.

In a few applications, improving the control system leads to a reduction in off-specification product and thus improved process economics. However, in most situations, the process is operated sufficiently far from the constraint that very little, if any, off-specification product results from control system deficiencies. Management often places considerable emphasis on avoiding off-spec production, so consequently the target is actually set far more conservatively than it should be.

In most applications, simply improving the control system does not directly lead to improved process economics. Instead, the control system improvement must be accompanied by shifting the target closer to the constraint. There is always a cost of operating a process in a conservative manner. The cost may be a lower production rate, a lower process efficiency, a product giveaway, or otherwise. When management places extreme emphasis on avoiding off-spec production, the natural reaction is to operate very conservatively, thus incurring other costs.

The immediate objective of an advanced control effort is to reduce the variance in an important controlled variable. However, this effort must be coupled with a commitment to adjust the target for this controlled variable so that the process is operated closer to the constraint. In large throughput (commodity) processes, very small shifts in operating targets can lead to large economic returns.

ADVANCED CONTROL TECHNIQUES

GENERAL REFERENCES: Seborg, Edgar, and Mellichamp, *Process Dynamics and Control,* John Wiley and Sons, New York, 1989. Stephanopoulos, *Chemical Process Control: An Introduction to Theory and Practice,* Prentice Hall, Englewood Cliffs, New Jersey, 1984. Shinskey, *Process Control Systems,* 3d ed., McGraw-Hill, New York, 1988. Ogunnaike and Ray, *Process Dynamics, Modeling, and Control,* Oxford University Press, New York, 1994.

While the single-loop PID controller is satisfactory in many process applications, it does not perform well for processes with slow dynamics, time delays, frequent disturbances, or multivariable interactions. We discuss several advanced control methods hereafter that can be implemented via computer control, namely feedforward control, cascade control, time-delay compensation, selective and override control, adaptive control, fuzzy logic control, and statistical process control.

Feedforward Control If the process exhibits slow dynamic response and disturbances are frequent, then the application of feedforward control may be advantageous. Feedforward (FF) control differs from feedback (FB) control in that the primary disturbance or load (L) is measured via a sensor and the manipulated variable (m) is adjusted so that deviations in the controlled variable from the set point are minimized or eliminated (see Fig. 8-29). By taking control action based on measured disturbances rather than controlled variable error, the controller can reject disturbances before they affect the controlled variable c. In order to determine the appropriate settings for the manipulated variable, one must develop mathematical models that relate:

1. The effect of the manipulated variable on the controlled variable

2. The effect of the disturbance on the controlled variable

These models can be based on steady-state or dynamic analysis. The performance of the feedforward controller depends on the accuracy of both models. If the models are exact, then feedforward control offers the potential of perfect control (i.e., holding the controlled variable precisely at the set point at all times because of the ability to predict the appropriate control action). However, since most mathematical models are only approximate and since not all disturbances are measurable, it is standard practice to utilize feedforward control in conjunction with feedback control. Table 8-5 lists the relative advantages and disadvantages of feedforward and feedback control. By combining the two control methods, the strengths of both schemes can be utilized.

FF control therefore attempts to eliminate the effects of measurable disturbances, while FB control would correct for unmeasurable

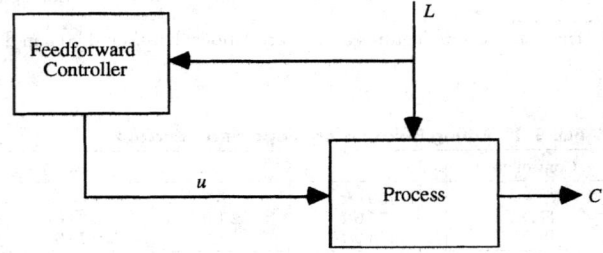

FIG. 8-29 Block diagram for feedforward control configuration.

TABLE 8-5 Relative Advantages and Disadvantages of Feedback and Feedforward

Advantages	Disadvantages
Feedforward	
• Acts before the effect of a disturbance has been felt by the system	• Requires measurement of all possible disturbances and their direct measurement
• Good for systems with large time constant or deadtime	• Cannot cope with unmeasured disturbances
• Does not introduce instability in the closed-loop response	• Sensitive to process/model error
Feedback	
• Does not require identification and measurement of any disturbance for corrective action	• Control action not taken until the effect of the disturbance has been felt by the system
• Does not require an explicit process model	• Unsatisfactory for processes with large time constants and frequent disturbances
• Controller can be robust to process/model errors	• May cause instability in the closed-loop response

disturbances and modeling errors. This is often referred to as feedback trim. These controllers have become widely accepted in the chemical process industries since the 1960s.

Design Based on Material and Energy Balances Consider a heat exchanger example (see Fig. 8-30) to illustrate the use of FF and FB control. The control objective is to maintain T_2, the exit liquid temperature, at the desired value (or set point) T_{set} despite variations in inlet liquid flow rate F and inlet liquid temperature T_1. This is done by manipulating W, the steam flow rate. A feedback control scheme would entail measuring T_2, comparing T_2 to T_{set}, and then adjusting W. A feedforward control scheme requires measuring F and T_1, and adjusting W (knowing T_{set}), in order to control exit temperature, T_2.

Figure 8-31 shows the control system diagrams for FB and FF control. A feedforward control algorithm can be designed for the heat exchanger in the following manner. Using a steady-state energy balance and assuming no heat loss from the heat exchanger,

$$WH = FC(T_2 - T_1) \tag{8-38}$$

where H = latent heat of vaporization
C_L = specific heat of liquid.

Rearranging Eq. (8-38),

$$W = \frac{C_L}{H} F(T_2 - T_1) \tag{8-39}$$

or

$$W = K_1 F(T_2 - T_1) \tag{8-40}$$

with

$$K_1 = \frac{C_L}{H} \tag{8-41}$$

Replace T_2 by T_{set}:

$$W = K_1 F(T_{\text{set}} - T_1) \tag{8-42}$$

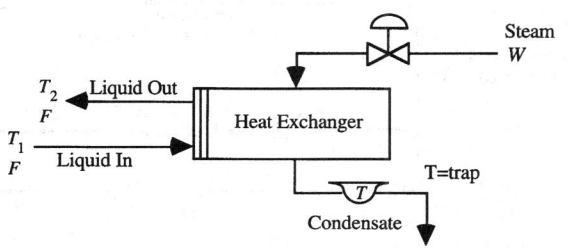

FIG. 8-30 A heat exchanger diagram.

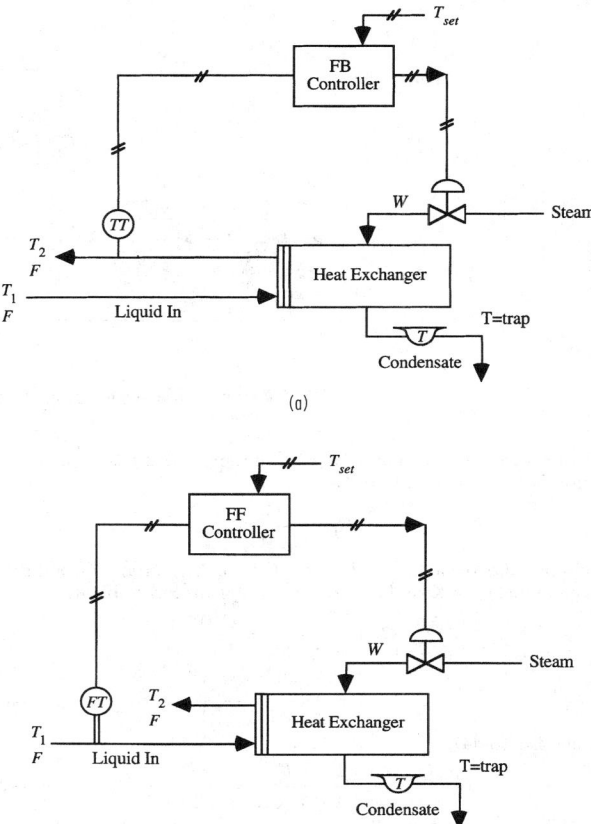

FIG. 8-31 (*a*) Feedback control of a heat exchanger. (*b*) Feedforward control of a heat exchanger.

Equation (8-42) can be used in the FF calculation, assuming one knows the physical properties C_L and H. Of course, it is probable that the model will contain errors (e.g., unmeasured heat losses, incorrect C_L or H). Therefore, K_1 can be designated as an adjustable parameter that can be tuned. The use of a physical model for FF control is desirable since it provides a physical basis for the control law and gives an a priori estimate of what the tuning parameters are. Note that such a model could be nonlinear [e.g., in Eq. (8-42), F and T_{set} are multiplied].

Block Diagram Analysis One shortcoming of this feedforward design procedure is that it is based on the steady-state characteristics of the process and as such, neglects process dynamics (i.e., how fast the controlled variable responds to changes in the load and manipulated variables). Thus, it is often necessary to include "dynamic compensation" in the feedforward controller. The most direct method of designing the FF dynamic compensator is to use a block diagram of a general process, as shown in Fig. 8-32. G_t represents the disturbance transmitter, G_f is the feedforward controller, G_L relates the load to the controlled variable, G_v is the valve, and G_p is the process. G_m is the output transmitter and G_c is the feedback controller. All blocks correspond to transfer functions (via Laplace transforms).

Using block diagram algebra and Laplace transform variables, the controlled variable $C(s)$ is given by

$$C(s) = \frac{G_t G_f L(s) + G_L L(s)}{1 + G_m G_c G_v G_p} \tag{8-43}$$

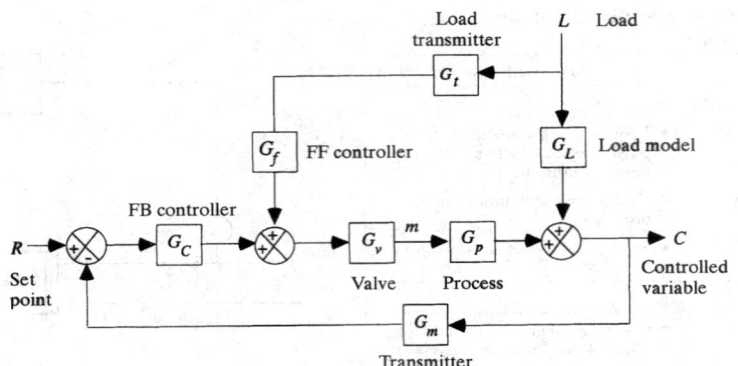

FIG. 8-32   Block diagram for feedback-feedforward control.

For disturbance rejection $[L(s) \neq 0]$ we require that $C(s) = 0$, or zero error. Solving Eq. (8-43) for G_f,

$$G_f = \frac{-G_L}{G_t G_v G_p} \qquad (8\text{-}44)$$

Suppose the dynamics of G_L and G_p are first order; in addition, assume that $G_v = K_v$ and $G_t = K_t$ (constant gains for simplicity).

$$G_L(s) = \frac{K_L}{\tau_L s + 1} = \frac{C(s)}{L(s)} \qquad (8\text{-}45)$$

$$G_p(s) = \frac{K_p}{\tau_p s + 1} = \frac{C(s)}{U(s)} \qquad (8\text{-}46)$$

Using Eq. (8-44),

$$G_f(s) = -\frac{\tau_p s + 1}{\tau_L s + 1} \cdot \frac{K_L}{K_p K_v K_t} = \frac{-K(\tau_p s + 1)}{\tau_L s + 1} \qquad (8\text{-}47)$$

The above FF controller can be implemented using analog elements or more commonly by a digital computer. Figure 8-33 compares typical responses for PID FB control, steady-state FF control ($s = 0$), dynamic FF control, and combined FF/FB control. In practice, the engineer can tune K, τ_p, and τ_L in the field to improve the performance of the FF controller. The feedforward controller can also be simplified to provide steady-state feedforward control. This is done by setting $s = 0$ in $G_f(s)$. This might be appropriate if there is uncertainty in the dynamic models for G_L and G_p.

Other Considerations in Feedforward Control   The tuning of feedforward and feedback control systems can be performed independently. In analyzing the block diagram in Fig. 8-32, note that G_f is chosen to cancel out the effects of the disturbance $L(s)$ as long as there are no model errors. For the feedback loop, therefore, the effects of $L(s)$ can also be ignored, which for the servo case is:

$$\frac{C(s)}{R(s)} = \frac{G_c G_v G_p K_m}{1 + G_c G_v G_p G_m} \qquad (8\text{-}48)$$

Note that the characteristic equation will be unchanged for the FF + FB system, hence system stability will be unaffected by the presence of the FF controller. In general, the tuning of the FB controller can be less conservative than for the case of FB alone, since smaller excursions from the set point will result. This in turn would make the dynamic model $G_p(s)$ more accurate.

The tuning of the controller in the feedback loop can be theoretically performed independent of the feedforward loop (i.e., the feedforward loop does not introduce instability in the closed-loop response). For more information on feedforward/feedback control applications and design of such controllers, refer to the general references.

Cascade Control   One of the disadvantages of using conventional feedback control for processes with large time lags or delays is that disturbances are not recognized until after the controlled variable deviates from its set point. In these processes, correction by feedback control is generally slow and results in long-term deviation from set point. One way to improve the dynamic response to load changes is by using a secondary measurement point and a secondary controller; the secondary measurement point is located so that it recognizes the upset condition before the primary controlled variable is affected.

One such approach is called cascade control, which is routinely used in most modern computer control systems. Consider a chemical reactor, where reactor temperature is to be controlled by coolant flow to the jacket of the reactor (Fig. 8-34). The reactor temperature can be influenced by changes in disturbance variables such as feed rate or feed temperature; a feedback controller could be employed to compensate for such disturbances by adjusting a valve on the coolant flow to the reactor jacket. However, suppose an increase occurs in the

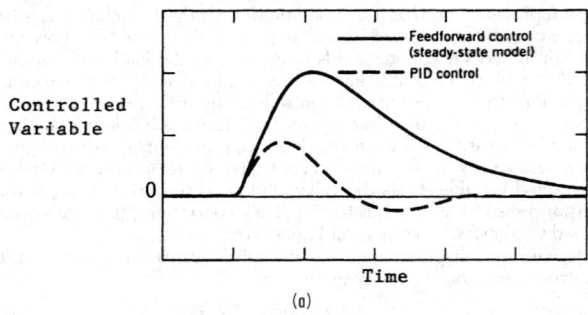

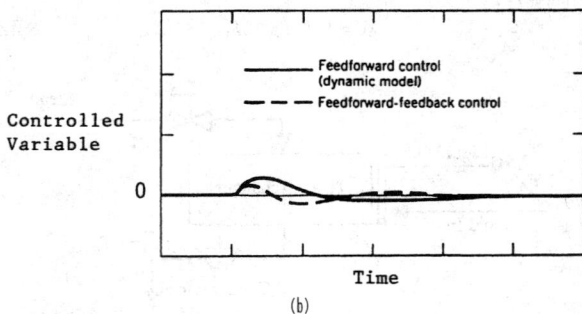

FIG. 8-33   (a) Comparison of FF (steady state model) and PID FB control for load change; (b) comparison of FF (dynamic model) and combined FF/FB control.

coolant temperature as a result of changes in the plant coolant system. This will cause a change in the reactor temperature measurement, although such a change will not occur quickly, and the corrective action taken by the controller will be delayed.

Cascade control is one solution to this problem (see Fig. 8-35). Here the jacket temperature is measured, and an error signal is sent from this point to the coolant control valve; this reduces coolant flow, maintaining the heat transfer rate to the reactor at a constant level and rejecting the disturbance. The cascade control configuration will also adjust the setting of the coolant control valve when an error occurs in reactor temperature. The cascade control scheme shown in Fig. 8-35 contains two controllers. The primary controller is the reactor temperature coolant temperature controller. It measures the reactor temperature, compares it to the set point, and computes an output, which is the set point for the coolant flow rate controller. This secondary controller compares the set point to the coolant temperature measurement and adjusts the valve. The principal advantage of cascade control is that the secondary measurement (jacket temperature) is located closer to a potential disturbance in order to improve the closed-loop response.

Figure 8-36 shows the block diagram for a general cascade control system. In tuning of a cascade control system, the secondary controller (in the inner loop) is tuned first with the primary controller in manual. Often only a proportional controller is needed for the secondary loop, since offset in the secondary loop can be treated by using proportional plus integral action in the primary loop. When the primary controller is transferred to automatic, it can be tuned using the techniques described earlier in this section. For more information on theoretical analysis of cascade control systems, see the general references for a discussion of applications of cascade control.

Time-Delay Compensation Time delays are a common occurrence in the process industries because of the presence of recycle loops, fluid-flow distance lags, and "dead time" in composition measurements resulting from use of chromatographic analysis. The presence of a time delay in a process severely limits the performance of a conventional PID control system, reducing the stability margin of the closed-loop control system. Consequently, the controller gain must be reduced below that which could be used for a process without delay. Thus, the response of the closed-loop system will be sluggish compared to that of the system with no time delay.

In order to improve the performance of time-delay systems, special control algorithms have been developed to provide time-delay compensation.

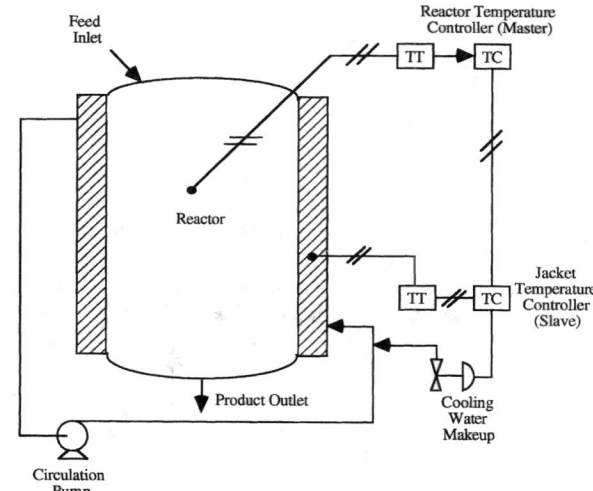

FIG. 8-35 Cascade control of an exothermic chemical reactor.

pensation. The Smith predictor technique is the best known algorithm; a related method is called the analytical predictor. Various investigators have found that based on integral squared error, the performance of the Smith predictor can be as much as 30 percent better than for a conventional controller.

The Smith predictor is a model-based control strategy that involves a more complicated block diagram than that for a conventional feedback controller, although a PID controller is still central to the control strategy (see Fig. 8-37). The key concept is based on better coordination of the timing of manipulated variable action. The loop configuration takes into account the fact that the current controlled variable measurement is not a result of the current manipulated variable action, but the value taken 0 time units earlier. Time-delay compensation can yield excellent performance; however, if the process model parameters change (especially the time delay), the Smith predictor performance will deteriorate and is not recommended unless other precautions are taken.

Selective and Override Control When there are more controlled variables than manipulated variables, a common solution to this problem is to use a selector to choose the appropriate process variable from among a number of available measurements. Selectors can be based on either multiple measurement points, multiple final control elements, or multiple controllers, as discussed below. Selectors are used to improve the control system performance as well as to protect equipment from unsafe operating conditions.

One type of selector device chooses as its output signal the highest (or lowest) of two or more input signals. This approach is often referred to as auctioneering. On instrumentation diagrams, the symbol HS denotes high selector and LS a low selector. For example, a high selector can be used to determine the hot-spot temperature in a fixed-bed chemical reactor. In this case, the output from the high selector is the input to the temperature controller. In an exothermic catalytic reaction, the process may run away due to disturbances or changes in the reactor. Immediate action should be taken to prevent a dangerous rise in temperature. Because a hot spot may potentially develop at one of several possible locations in the reactor, multiple (redundant) measurement points should be employed. This approach minimizes the time required to identify when a temperature has risen too high at some point in the bed.

The use of high or low limits for process variables is another type of selective control, called an override. The feature of anti-reset windup in feedback controllers is a type of override. Another example is a distillation column with lower and upper limits on the heat input to the column reboiler. The minimum level ensures that liquid will remain

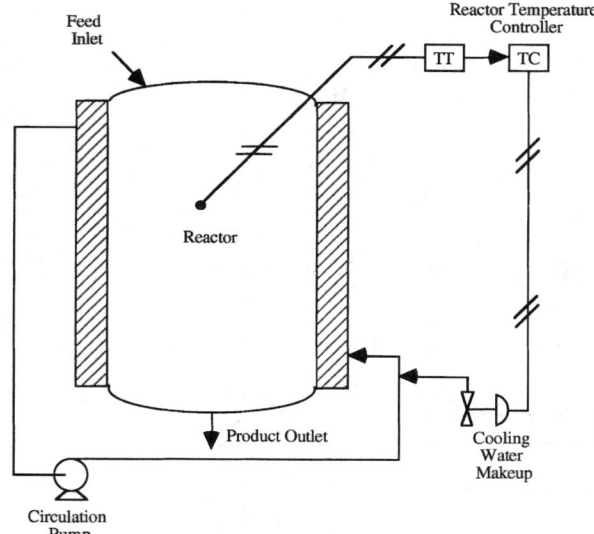

FIG. 8-34 Conventional control of an exothermic chemical reactor.

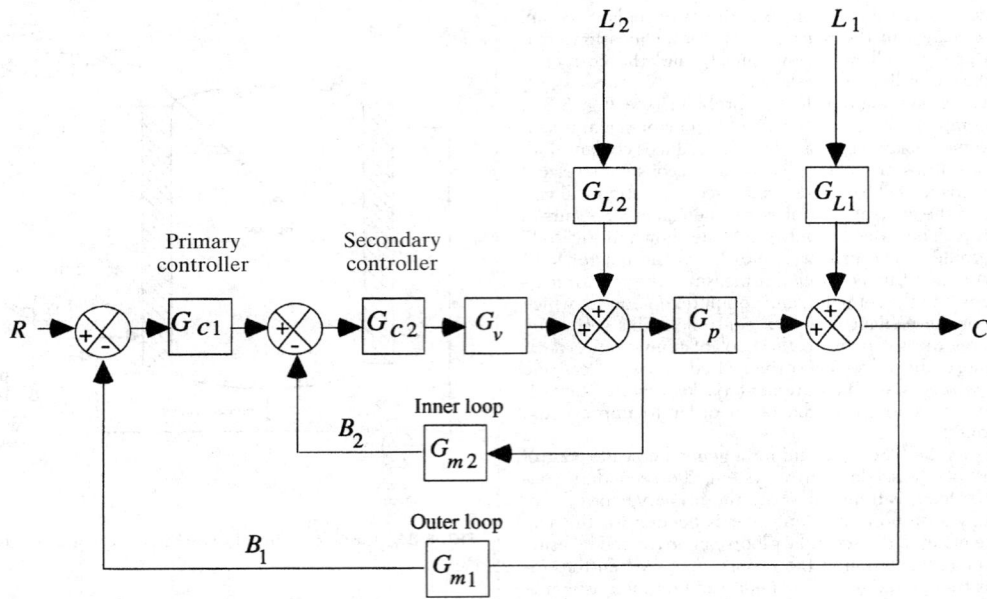

FIG. 8-36   Block diagram of the cascade control system. For a chemical reactor, L_1 would correspond to a feed temperature or composition disturbance, while L_2 would be a change in the cooling water temperature.

on the trays, while the upper limit is determined by the onset of flooding. Overrides are also used in forced-draft combustion-control systems to prevent an imbalance between air flow and fuel flow, which could result in unsafe operating conditions.

Other types of selective systems employ multiple final control elements or multiple controllers. In some applications, several manipulated variables are used to control a single process variable (also called split-range control). Typical examples include the adjustment of both inflow and outflow from a chemical reactor in order to control reactor pressure or the use of both acid and base to control pH in waste-water treatment. In this approach, the selector chooses from several controller outputs which final control element should be adjusted (Marlin, *Process Control*, McGraw-Hill, New York, 1995).

Adaptive Control   Process control problems inevitably require on-line tuning of the controller constants to achieve a satisfactory degree of control. If the process operating conditions or the environment changes significantly, the controller may have to be retuned. If these changes occur quite frequently, then adaptive control techniques should be considered. An adaptive control system is one in which the controller parameters are adjusted automatically to compensate for changing process conditions.

During the 1980s, several adaptive controllers were field-tested and commercialized in the U.S. and abroad, including products by ASEA (Sweden), Leeds and Northrup, Foxboro, and Sattcontrol. At the present time, some form of adaptive tuning is available on almost all PID controllers. The ASEA adaptive controller, Novatune, was

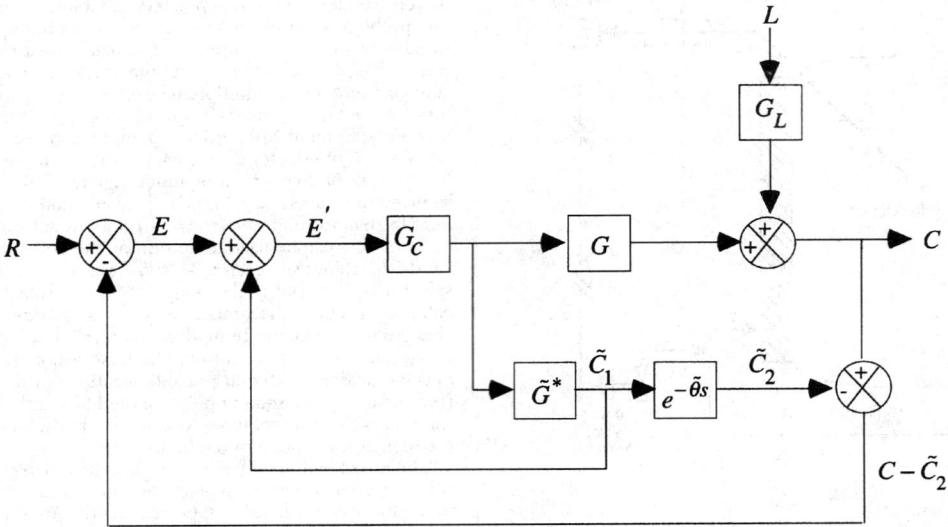

FIG. 8-37   Block diagram of the Smith predictor. The process model used in the controller is $\tilde{G} = \tilde{G}°e^{-\tilde{\theta}s}$ ($\tilde{G}°$ = model without delay; $e^{-\tilde{\theta}s}$ = time delay element).

announced in 1983 and is generally based on minimum-variance-control algorithms. Both feedforward and feedback control capabilities reside in the hardware. The unit has been tested successfully in reactor and paper machine control applications in Europe and in pH control of wastewater in the United States.

Foxboro developed a self-tuning PID controller that is based on a so-called "expert system" approach for adjustment of the controller parameters. The on-line tuning of K_c, τ_I, and τ_D is based on the closed-loop transient response to a step change in set point. By evaluating the salient characteristics of the response (e.g., the decay ratio, overshoot, and closed-loop period), the controller parameters can be updated without actually finding a new process model. The details of the algorithm, however, are proprietary.

The Sattcontroller (also marketed by Fisher-Rosemount) has an autotuning function that is based on placing the process in a controlled oscillation at very low amplitude, comparable with that of the noise level of the process. This is done via a relay-type step function with hysteresis. The autotuner identifies the dynamic parameters of the process (the ultimate gain and period) and automatically calculates K_c, τ_I, and τ_D using empirical tuning rules. Gain scheduling can also be implemented with this controller, using up to three sets of PID controller parameters.

The subject of adaptive control is one of current interest. New algorithms are presently under development, but these need to be field-tested before industrial acceptance can be expected. It is clear, however, that digital computers will be required for implementation of self-adaptive controllers due to their complexity. An adaptive controller is inherently nonlinear and therefore more complicated than the conventional PID controller.

Fuzzy Logic Control   The application of fuzzy logic to process control requires the concepts of fuzzy rules and fuzzy inference. A fuzzy rule, also known as a fuzzy IF-THEN statement, has the form:

$$\text{If } x \text{ then } y$$

where x specifies a vector of input variables and corresponding membership values and y specifies an output variable and its corresponding membership value. For example,

$$\text{if input1} = \text{high}$$
$$\text{and input2} = \text{low},$$
$$\text{then output} = \text{medium}.$$

Three functions are required to perform logical inferencing with the fuzzy rules. The fuzzy AND is the product of a rule's input membership values, generating a weight for the rule's output. The fuzzy OR is a normalized sum of the weights assigned to each rule that contributes to a particular decision. The third function used is defuzzification, which generates a crisp final output. In one approach, the crisp output is the weighted average of the peak element values: $\sum [w(i) \, p(i)] / \sum [w(i)]$.

With a single feedback control architecture, information that is readily available to the algorithm includes the error signal, the difference between the process variable and the set point variable, change in error from previous cycles to the current cycle, changes to the set point variable, change of the manipulated variable from cycle to cycle, and the change in the process variable from past to present. In addition, multiple combinations of the system response data are available. As long as the irregularity lies in that dimension wherein fuzzy decisions are being based or associated, the result should be enhanced performance. This enhanced performance should be demonstrated in both the transient and steady-state response. If the system tends to have changing dynamic characteristics or exhibits nonlinearities, fuzzy logic control should offer a better alternative to using constant PID settings. Most fuzzy logic software begins building its information base during the autotune function. In fact, the majority of the information used in the early stages of system startup comes from the autotune solutions.

In addition to single-loop process controllers, products that have benefited from the implementation of fuzzy logic are:
- Camcorders with automatic compensation for operator-injected noise such as shaking and moving
- Elevators with decreased wait time, making intelligent floor decisions and minimizing travel and power consumption

- Antilock braking systems with quickly reacting independent wheel decisions based on current and acquired knowledge
- Television with automatic color, brightness, and acoustic control based on signal and environmental conditions

Sometimes fuzzy logic controllers are combined with pattern recognition software such as artificial neural networks (Kosko, *Neural Networks and Fuzzy Systems*, Prentice Hall, Englewood Cliffs, New Jersey, 1992).

Statistical Process Control   Statistical process control (SPC), also called statistical quality control (SQC), involves the application of statistical concepts to determine whether a process is operating satisfactorily. The ideas involved in statistical quality control are over fifty years old, but only recently with the growing worldwide focus on increased productivity have applications of SPC become widespread. If a process is operating satisfactorily (or "in control"), then the variation of product quality falls within acceptable bounds, usually the minimum and maximum values of a specified composition or property (product specification).

Figure 8-38 illustrates the typical spread of values of the controlled variable that might be expected to occur under steady-state operating conditions. The mean and root mean square (RMS) deviation are identified in Fig. 8-38 and can be computed from a series of n observations $c_1, c_2, \ldots c_n$ as follows:

$$\text{mean: } \bar{c} = \frac{1}{n} \sum_{i=1}^{n} c_i \qquad (8\text{-}49)$$

RMS deviation:

$$\sigma = \left[\frac{1}{n} \sum_{i=1}^{n} (c_i - \bar{c})^2 \right]^{1/2} \qquad (8\text{-}50)$$

The RMS deviation is a measure of the spread of values for c around the mean. A large value of σ indicates that wide variations in c occur. The probability that the controlled variable lies between the values of c_1 and c_2 is given by the area under the distribution between c_1 and c_2 (histogram). If the histogram follows a normal probability distribution, then 99.7 percent of all observations should lie with $\pm 3\sigma$ of the mean (between the lower and upper control limits). These limits are used to determine the quality of control.

If all data from a process lie within the $\pm 3\sigma$ limits, then we conclude that nothing unusual has happened during the recorded time period. The process environment is relatively unchanged, and the product quality lies within specification. On the other hand, if repeated violations of the $\pm 3\sigma$ limits occur, then the process environment has changed and the process is out of control.

One way to codify abnormal behavior is the so-called Western Electric rules, which identify cases where a process is out of control:
1. One point that occurs outside the upper or lower control limits
2. Any seven consecutive points lying on the same side of the center line (mean)
3. Any seven consecutive points that increase or decrease
4. Any nonrandom pattern

In the above list, one assumes that sample values are independent (i.e., not correlated).

There are important economic consequences of a process being out of control; for example, product waste and customer dissatisfaction. Hence, statistical process control does provide a way to continuously monitor process performance and improve product quality. A typical process may go out of control due to several reasons, including
- Persistent disturbances from the weather
- An undetected grade change in raw materials
- A malfunctioning instrument or control system

Statistical quality control is a diagnostic tool—that is, an indicator of quality problems—but it does not identify the source of the problem or the corrective action to be taken. The Shewhart chart provides a way to analyze variability of a single measurement, as discussed in the following example. The data in Fig. 8-39 were obtained from the monitoring of pH in a yarn-soaking kettle used in textile manufacturing (Seborg, Edgar, and Mellichamp, *Process Dynamics and Control*, Wiley, New York, 1989). Because pH has a crucial influence on color and durability of the yarn, it is important to maintain pH within a range that gives the best results for both characteristics. The pH is considered to be in control between values of 4.25 and 4.64. At the

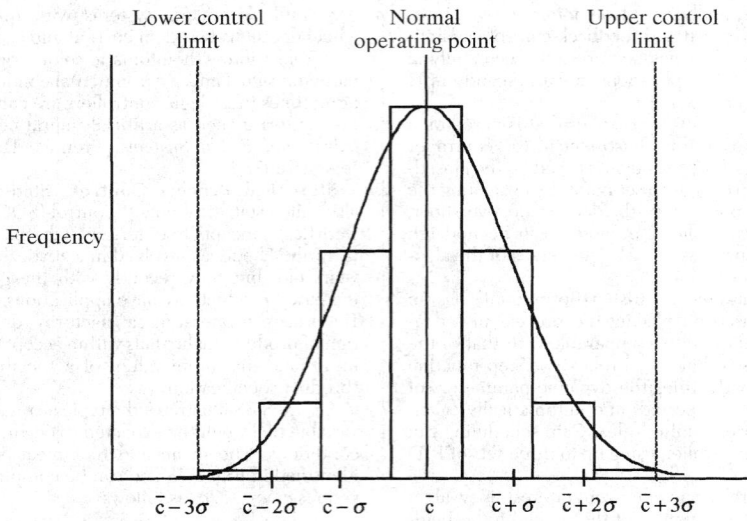

FIG. 8-38 Histogram plotting frequency of occurrence. c = mean, σ = rms deviation. Also shown is fit by normal probability distribution.

25th day, the data show that pH is out of control; this might imply that a property change in the raw material has occurred and must be corrected with the supplier. However, a real-time correction would be preferable. In Fig. 8-39, the pH was adjusted by slowly adding more acid to the vats until it came back into control (on day 29).

In continuous processes where automatic feedback control has been implemented, the feedback mechanism theoretically ensures that product quality is at or near the set point regardless of process disturbances. This, of course, requires that an appropriate manipulated variable has been identified for adjusting the product quality. However, even under feedback control, there may be daily variations of product quality because of disturbances or equipment or instrument malfunctions. These occurrences can be analyzed using the concepts of statistical quality control.

More details on statistical process control are available in several textbooks (Grant and Leavenworth, *Statistical Quality Control*, McGraw-Hill, New York, 1980; Montgomery, *Introduction to Statistical Quality Control*, Wiley, New York, 1985).

MULTIVARIABLE CONTROL PROBLEMS

GENERAL REFERENCES: Shinskey, F. G., *Process Control Systems*, 3d ed., McGraw-Hill, New York, 1988. Seborg, D. E., T. F. Edgar, and D. A. Mellichamp, *Process Dynamics and Control*, Wiley, New York, 1989. McAvoy, T. J., *Interaction Analysis*, ISA, Research Triangle Park, North Carolina, 1983.

Process control books and journal articles tend to emphasize problems with a single controlled variable. In contrast, most practical problems are *multivariable* control problems because many process variables must be controlled. In fact, for virtually any important industrial process, at least two variables must be controlled: product quality and throughput. In this section, strategies for multivariable control problems are considered.

Three examples of simple multivariable control problems are shown in Fig. 8-40. The in-line blending system blends pure components A and B to produce a product stream with flow rate w and mass fraction of A, x. Adjusting either inlet flow rate w_A or w_B affects *both* of the controlled variables w and x. For the pH neutralization process in Figure 8-40(b), liquid level h and the pH of the exit stream are to be controlled by adjusting the acid and base flow rates w_a and w_b. Each of the manipulated variables affects both of the controlled variables. Thus, both the blending system and the pH neutralization process are said to exhibit strong process interactions. In contrast, the process interactions for the gas-liquid separator in Fig. 8-40(c) are not as strong because one manipulated variable, liquid flow rate L, has only a small and indirect effect on one controlled variable, pressure P.

Strong process interactions can cause serious problems if a conventional multiloop feedback control scheme (e.g., PI or PID controllers) is employed. The process interactions can produce undesirable control loop interactions where the controllers fight each other. Also, it may be difficult to determine the best pairing of controlled and manipulated variables. For example, in the in-line blending process in Fig. 8-40(a), should w be controlled with w_A and x with w_B, or vice versa?

Control Strategies for Multivariable Control Problems If a conventional multiloop control strategy performs poorly due to control loop interactions, a number of solutions are available:

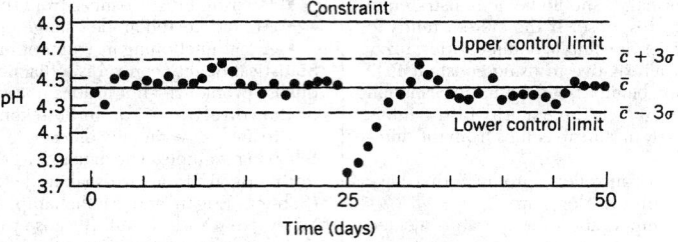

FIG. 8-39 Process control chart for the average daily pH readings.

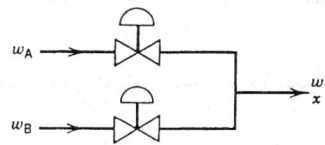

In-line blending system

(a)

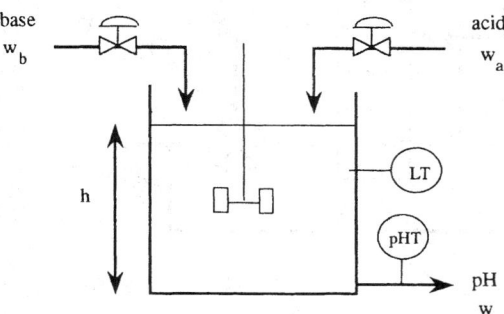

pH neutralization process

(b)

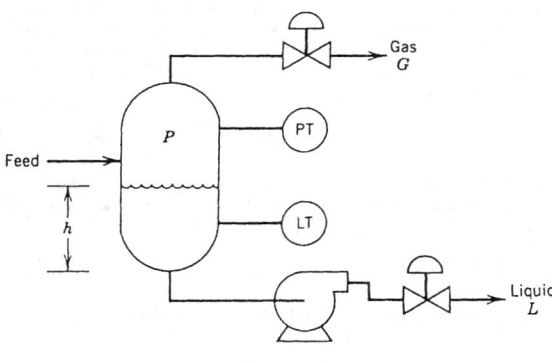

Gas liquid separator

(c)

FIG. 8-40   Physical examples of multivariable control problems.

a.   Detune one or more of the control loops
b.   Choose different controlled or manipulated variables (or pairings)
c.   Use a decoupling control system
d.   Use a multivariable control scheme (e.g., model predictive control)

Detuning a controller (e.g., using a smaller controller gain or a larger reset time) tends to reduce control loop interactions by sacrificing the performance for the detuned loops. This approach may be acceptable if some of the controlled variables are faster or less important than others.

The selection of controlled and manipulated variables is of crucial importance in designing a control system. In particular, a judicious choice may significantly reduce control loop interactions. For the blending process in Fig. 8-40(a), a straightforward control strategy would be to control x by adjusting w_A, and w by adjusting w_B. But

physical intuition suggests that it would be better to control x by adjusting the ratio $w_A/(w_A + w_B)$ and to control product flow rate w by the sum $w_A + w_B$. Thus, the new manipulated variables would be: $M_1 = w_A/(w_A + w_B)$ and $M_2 = w_A + w_B$. In this control scheme, M_1 only affects x and M_2 only affects w. Thus, the control loop interactions have been eliminated. Similarly, for the pH neutralization process in Fig. 8-40(b), the control loop interactions would be greatly reduced if pH is controlled by $M_1 = w_a/(w_a + w_b)$ and liquid level h is controlled by $M_2 = w_a + w_b$.

Decoupling Control Systems   Decoupling control systems provide an alternative approach for reducing control loop interactions. The basic idea is to use additional controllers called "decouplers" to compensate for undesirable process interactions.

As an illustrative example, consider the simplified block diagram for a representative decoupling control system shown in Fig. 8-41. The two controlled variables C_1 and C_2 and two manipulated variables M_1 and M_2 are related by four process transfer functions, G_{p11}, G_{p12}, and so on. For example, G_{p11} denotes the transfer function between M_1 and C_1:

$$\frac{C_1(s)}{M_1(s)} = G_{p11}(s) \qquad (8\text{-}51)$$

Figure 8-41 includes two conventional feedback controllers: G_{c1} controls C_1 by manipulating M_1, and G_{c2} controls C_2 by manipulating M_2. The output signals from the feedback controllers serve as input signals to the two decouplers D_{12} and D_{21}. The block diagram is in a simplified form because the load variables and transfer functions for the final control elements and sensors have been omitted.

The function of the decouplers is to compensate for the undesirable process interactions represented by G_{p12} and G_{p21}. Suppose that the process transfer functions are all known. Then the ideal design equations are:

$$D_{12}(s) = -\frac{G_{p12}(s)}{G_{p11}(s)} \qquad (8\text{-}52)$$

$$D_{21}(s) = -\frac{G_{p21}(s)}{G_{p22}(s)} \qquad (8\text{-}53)$$

These decoupler design equations are very similar to the ones for feedforward control in an earlier section. In fact, decoupling can be interpreted as a type of feedforward control where the input signal is the output of a feedback controller rather than a measured load variable.

In principle, ideal decoupling eliminates control loop interactions and allows the closed-loop system to behave as a set of independent control loops. But in practice, this ideal behavior is not attained for a variety of reasons, including imperfect process models and the presence of saturation constraints on controller outputs and manipulated variables. Furthermore, the ideal decoupler design equations in (8-52) and (8-53) may not be physically realizable and thus would have to be approximated.

In practice, other types of decouplers and decoupling control configurations have been employed. For example, in *partial decoupling*, only a single decoupler is employed (i.e., either D_{12} or D_{21} in Fig. 8-41 is set equal to zero). This approach tends to be more robust than complete decoupling and is preferred when one of the controlled variables is more important than the other. *Static decouplers* can be used to reduce the steady-state interactions between control loops. They can be designed by replacing the transfer functions in Eqs. (8-52) and (8-53) with the corresponding steady-state gains,

$$D_{12}(s) = -\frac{K_{p12}}{K_{p11}} \qquad (8\text{-}54)$$

$$D_{21}(s) = -\frac{K_{p21}}{K_{p22}} \qquad (8\text{-}55)$$

The advantage of static decoupling is that less process information is required: namely, only steady-state gains. *Nonlinear decouplers* can be used when the process behavior is nonlinear.

Pairing of Controlled and Manipulated Variables   A key decision in multiloop-control-system design is the pairing of manipu-

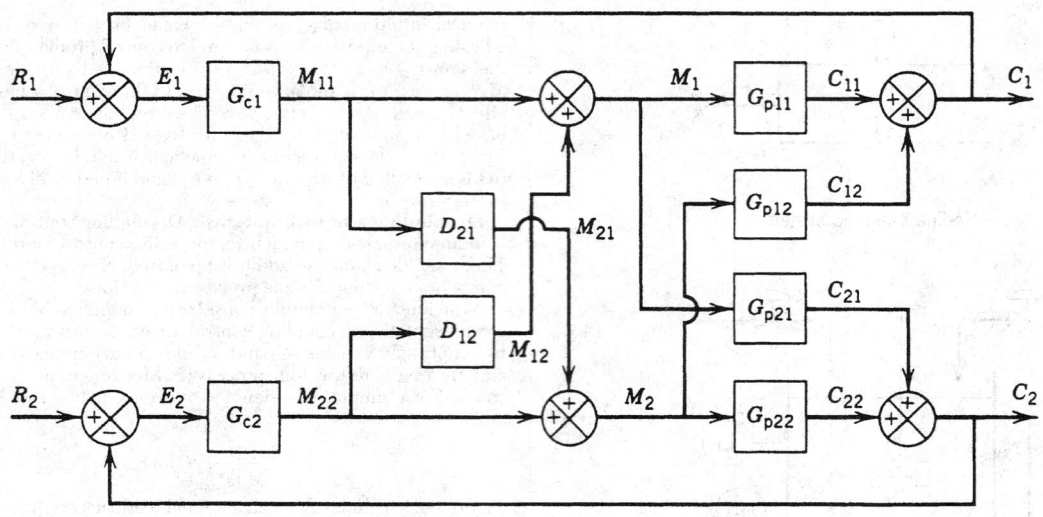

FIG. 8-41 Decoupling control system.

lated and controlled variables. This is referred to as the controller-pairing problem. Suppose there are N controlled variables and N manipulated variables. Then $N!$ distinct control configurations exist. For example, if $N = 5$, then there are 120 different multiloop control schemes. In practice, many of them would be rejected based on physical insight or previous experience. But a smaller number (e.g., 5–15) may appear to be feasible and further analysis would be warranted. Thus, it is very useful to have a simple method for choosing the most promising control configuration.

The most popular and widely used technique for determining the best controller pairing is the relative gain array (RGA) method (Bristol, "On a New Measure of Process Interaction," *IEEE Trans. Auto. Control*, AC-11, 133, 1966). The RGA method provides two important items of information:

1. A measure of the degree of process interactions between the manipulated and controlled variables
2. A recommended controller pairing

An important advantage of the RGA method is that it requires minimal process information: namely, steady-state gains. Another advantage is that the results are independent of both the physical units used and the scaling of the process variables. The chief disadvantage of the RGA method is that it neglects process dynamics, which can be an important factor in the pairing decision. Thus, the RGA analysis should be supplemented with an evaluation of process dynamics. Although extensions of the RGA method that incorporate process dynamics have been reported, these extensions have not been widely applied.

RGA Method for 2 × 2 Control Problems To illustrate the use of the RGA method, consider a control problem with two inputs and two outputs. The more general case of $N \times N$ control problems is considered elsewhere (McAvoy, *Interaction Analysis*, ISA, Research Triangle Park, North Carolina, 1983). As a starting point, it is assumed that a linear, steady-state process model is available,

$$C_1 = K_{11}M_1 + K_{12}M_2 \tag{8-56}$$

$$C_2 = K_{21}M_1 + K_{22}M_2 \tag{8-57}$$

where M_1 and M_2 are steady-state values of the manipulated inputs; C_1 and C_2 are steady-state values of the controlled outputs; and the values K are steady-state gains. The C and M variables are deviation variables from nominal steady-state values. This process model could be obtained in a variety of ways, such as by linearizing a theoretical model or by calculating steady-state gains from experimental data or a steady-state simulation.

By definition, the relative gain λ_{ij} between the ith manipulated variable and the jth controlled variable is defined as:

$$\lambda_{ij} = \frac{\text{open-loop gain between } C_i \text{ and } M_j}{\text{closed-loop gain between } C_i \text{ and } M_j} \tag{8-58}$$

where the open-loop gain is simply K_{ij} from Eqs. (8-56) and (8-57). The closed-loop gain is defined to be the steady-state gain between M_j and C_i when the other control loop is closed and no offset occurs due to the presence of integral control action. The RGA for the 2 × 2 process is denoted by

$$\Lambda = \begin{pmatrix} \lambda_{11} & \lambda_{12} \\ \lambda_{21} & \lambda_{22} \end{pmatrix} \tag{8-59}$$

The RGA has the important normalization property that the sum of the elements in each row and each column is exactly one. Consequently, the RGA in Eq. (8-59) can be written as

$$\Lambda = \begin{pmatrix} \lambda & 1-\lambda \\ 1-\lambda & \lambda \end{pmatrix} \tag{8-60}$$

where λ can be calculated from the following formula:

$$\lambda = \frac{1}{1 - \dfrac{K_{12} K_{21}}{K_{11} K_{22}}} \tag{8-61}$$

Ideally, the relative gains that correspond to the proposed controller pairing should have a value of one since Eq. (8-58) implies that the open and closed-loop gains are then identical. If a relative gain equals one, the steady-state operation of this loop will not be affected when the other control loop is changed from manual to automatic, or vice versa. Consequently, the recommendation for the best controller pairing is to pair the controlled and manipulated variables so that the corresponding relative gains are positive and close to one.

RGA Example In order to illustrate use of the RGA method, consider the following steady-state version of a transfer function model for a pilot-scale, methanol-water distillation column (Wood and Berry, "Terminal Composition Control of a Binary Distillation Column," *Chem. Eng. Sci.*, **28**, 1707, 1973): $K_{11} = 12.8$, $K_{12} = -18.9$, $K_{21} = 6.6$, and $K_{22} = -19.4$. It follows that $\lambda = 2$ and

$$\Lambda = \begin{pmatrix} 2 & -1 \\ -1 & 2 \end{pmatrix} \tag{8-62}$$

Thus it is concluded that the column is fairly interacting and the recommended controller pairing is to pair C_1 with M_1 and C_2 with M_2.

MODEL PREDICTIVE CONTROL

Introduction The model-based control strategy that has been most widely applied in the process industries is *model predictive control* (MPC). It is a general method that is especially well-suited for difficult multiinput, multioutput (MIMO) control problems where there are significant interactions between the manipulated inputs and the controlled outputs. Unlike other model-based control strategies, MPC can easily accommodate inequality constraints on input and output variables such as upper and lower limits or rate-of-change limits.

A key feature of MPC is that future process behavior is predicted using a dynamic model and available measurements. The controller outputs are calculated so as to minimize the difference between the predicted process response and the desired response. At each sampling instant, the control calculations are repeated and the predictions updated based on current measurements. In typical industrial applications, the set point and target values for the MPC calculations are updated using on-line optimization based on a steady-state model of the process. Constraints on the controlled and manipulated variables can be routinely included in both the MPC and optimization calculations. The extensive MPC literature includes survey articles (Garcia, Prett, and Morari, *Automatica*, **25**, 335, 1989; Richalet, *Automatica*, **29**, 1251, 1993) and books (Prett and Garcia, *Fundamental Process Control*, Butterworths, Stoneham, Massachusetts, 1988; Soeterboek, *Predictive Control—A Unified Approach*, Prentice Hall, Englewood Cliffs, New Jersey, 1991).

The current widespread interest in MPC techniques was initiated by pioneering research performed by two industrial groups in the 1970s. Shell Oil (Houston, TX) reported their Dynamic Matrix Control (DMC) approach in 1979, while a similar technique, marketed as IDCOM, was published by a small French company, ADERSA, in 1978. Since then, there have been over one thousand applications of these and related MPC techniques in oil refineries and petrochemical plants around the world. Thus, MPC has had a substantial impact and is currently the method of choice for difficult multivariable control problems in these industries. However, relatively few applications have been reported in other process industries, even though MPC is a very general approach that is not limited to a particular industry.

Advantages and Disadvantages of MPC Model Predictive Control offers a number of important advantages:

1. It is a general control strategy for MIMO processes with inequality constraints on input and output variables.

2. It can easily accommodate difficult or unusual dynamic behavior such as large time delays and inverse responses.

3. Since the control calculations are based on optimizing control system performance, MPC can be readily integrated with on-line optimization strategies to optimize plant performance.

4. The control strategy can be easily updated on-line to compensate for changes in process conditions, constraints, or performance criteria.

But current versions of MPC have significant disadvantages:

1. The MPC strategy is very different from conventional multi-loop control strategies and thus initially unfamiliar to plant personnel.

2. The MPC calculations can be relatively complicated (e.g., solving an LP or QP problem at each sampling instant) and thus require a significant amount of computer resources and effort.

3. The development of a dynamic model from plant data is time consuming, typically requiring one to three weeks of around-the-clock plant tests.

4. Since empirical models are generally used, they are only valid over the range of conditions that were considered during the plant tests.

5. Theoretical studies have demonstrated that MPC can perform poorly for some types of process disturbances, especially when output constraints are employed (Lundstrom, Lee, Morari, and Skogestad, *Computers Chem. Eng.*, **19**, 409, 1995).

Since MPC has been widely used and has had considerable impact, there is a broad consensus that its advantages far outweigh its disadvantages.

Economic Incentives for Automation Projects Industrial applications of advanced process control strategies such as MPC are motivated by the need for improvements regarding safety, product quality, environmental standards, and economic operation of the process. One view of the economics incentives for advanced automation techniques is illustrated in Fig. 8-42. Distributed control systems (DCS) are widely used for data acquisition and conventional single-loop (PID) control. Usually, they are the most expensive part of the entire control system. The addition of advanced regulatory control systems such as decouplers, selective controls, and time-delay compensation can provide additional benefits for a modest incremental cost. But experience has indicated that the major benefits can be obtained for relatively small incremental costs through a combination of MPC and on-line optimization. The results in Fig. 8-42 are shown qualitatively, rather than quantitatively, because the actual costs and benefits are application-dependent.

A key reason why MPC has become a major commercial and technical success is that there are numerous vendors who are licensed to market MPC products and install them on a turnkey basis. Consequently, even medium-sized companies are able to take advantage of this new technology. Payout times of 3–12 months have been reported.

Basic Features of MPC Model predictive control strategies have a number of distinguishing features:

1. A dynamic model of the process is used to predict the future outputs over a *prediction horizon* consisting of the next *p* sampling periods.

2. A reference trajectory is used to represent the desired output response over the prediction horizon.

3. Inequality constraints on the input and output variables can be included as an option.

4. At each sampling instant, a control policy consisting of the next *m* control moves is calculated. The control calculations are based on minimizing a quadratic or linear performance index over the prediction horizon while satisfying the constraints.

5. The performance index is expressed in terms of future control moves and the predicted deviations from the reference trajectory.

6. A *receding horizon approach* is employed. At each sampling instant, only the first control move (of the *m* moves that were calculated) is actually implemented. Then the predictions and control calculations are repeated at the next sampling instant.

These distinguishing features of MPC will now be described in more detail.

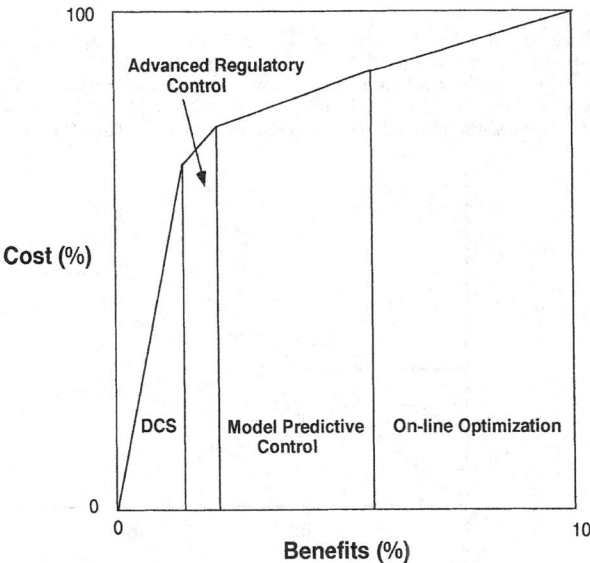

FIG. 8-42 Economic incentives for automation projects in the process industries.

A key feature of MPC is that a dynamic model of the process is used to predict future values of the controlled outputs. There is considerable flexibility concerning the choice of the dynamic model. For example, a physical model based on first principles (e.g., mass and energy balances) or an empirical model could be selected. Also, the empirical model could be a linear model (e.g., transfer function, step response model, or state space model) or a nonlinear model (e.g., neural net model). However, most industrial applications of MPC have relied on linear empirical models, which may include simple nonlinear transformations of process variables.

The original formulations of MPC (i.e., DMC and IDCOM) were based on empirical linear models expressed in either step-response or impulse-response form. For simplicity, we will consider only a single-input, single-output (SISO) model. However, the SISO model can be easily generalized to the MIMO models that are used in industrial applications. The step response model relating a single controlled variable y and a single manipulated variable u can be expressed as

$$\hat{y}(k) = \sum_{i=1}^{N} S_i \Delta u(k-i) + y(0) \qquad (8\text{-}63)$$

where $\hat{y}(k)$ is the predicted value of y at the k-sampling instant; $u(k)$ is the value of the manipulated input at time k; and the model parameters S_i are referred to as the *step-response coefficients*. The initial value $y(0)$ is assumed to be known. The change in the manipulated input from one sampling instant to the next is denoted by

$$\Delta u(k) \triangleq u(k) - u(k-1) \qquad (8\text{-}64)$$

The step-response model is also referred to as a finite impulse response (FIR) model or a discrete convolution model.

In principle, the step-response coefficients can be determined from the output response to a step change in the input. A typical response to a unit step change in input u is shown in Fig. 8-43. The step response coefficients S_i are simply the values of the output variable at the sampling instants, after the initial value $y(0)$ has been subtracted. Theoretically, they can be determined from a single-step response, but, in practice, a number of "bump tests" are required to compensate for unanticipated disturbances, process nonlinearities, and noisy measurements.

The step-response model in Eq. (8-63) is equivalent to the following impulse response model:

$$\hat{y}(k) = \sum_{i=1}^{N} I_i u(k-i) + y(0) \qquad (8\text{-}65)$$

where the impulse response coefficients I_i are related to the step-response coefficients by $I_i = S_i - S_{i-1}$. Step- and impulse-response models typically contain a large number of parameters because the model horizon N is usually quite large ($30 < N < 70$). In fact, these models are often referred to as nonparametric models. The DMC version of MPC is based on step-response models, while IDCOM utilizes impulse response models.

The receding horizon feature of MPC is shown in Fig. 8-44 with the current sampling instant denoted by k. Past input signals [$u(i)$ for $i < k$] are used to predict the output at the next p sampling instants [$\hat{y}(k+i)$ for $i = 1, 2, \ldots, p$]. The control calculations are performed to generate an m-step control policy [$u(k), u(k+1), \ldots, u(k+m)$], which optimizes the performance index. The first control action, $u(k)$, is implemented. Then at the next sampling instant $(k+1)$, the prediction and control calculations are repeated in order to determine $u(k+1)$. In Fig. 8-44, the reference trajectory (or target) is considered to be constant. Other possibilities include a gradual or step set point change that can be generated by on-line optimization.

The performance index for MPC applications is usually a linear or quadratic function of the predicted errors and calculated future control moves. For example, the following quadratic performance index has been widely used:

$$\min_{\Delta \boldsymbol{u}(k)} J = \sum_{i=1}^{p} w_i e^2(k+i) + \delta \sum_{i=1}^{m} \Delta u^2(k+i-1) \qquad (8\text{-}66)$$

The value $e(k+i)$ denotes the predicted error at time $(k+i)$,

$$e(k+i) \triangleq r(k+i) - \hat{y}(k+i) \qquad (8\text{-}67)$$

where $r(k+i)$ is the reference value at time $k+i$, and $\Delta \boldsymbol{u}(k)$ denotes the vector of current and future control moves over the next m sampling instants:

$$\Delta \boldsymbol{u}(k) = [\Delta u(k), \Delta u(k+1), \ldots, \Delta u(k+m-1)]^T \qquad (8\text{-}68)$$

Equation (8-66) contains two types of design parameters that can also be used for tuning purposes. The move suppression factor δ penalizes large control moves, while the weighting factors w_i allow the predicted errors to be weighed differently at each time step, if desired.

Inequality constraints on the future inputs or their rates of change are widely used in the MPC calculations. For example, if both upper and lower limits are required, the constraints could be expressed as:

$$B_{i_*} \leq u(k+i) \leq B_i^\circ \qquad \text{for } i = 1, 2, \ldots, m \qquad (8\text{-}69)$$

$$C_{i_*} \leq \Delta u(k+i) \leq C_i^\circ \qquad \text{for } i = 1, 2, \ldots, m \qquad (8\text{-}70)$$

where B_i and C_i are constants. Constraints on the predicted outputs are sometimes included as well:

$$D_{i_*} \leq \hat{y}(k+i) \leq D_i^\circ \qquad \text{for } i = 1, 2, \ldots, p \qquad (8\text{-}71)$$

The minimization of the quadratic performance index in Eq. (8-66), subject to the constraints in Eq. (8-69) to (8-71) and the step-response model in Eq. (8-63), can be formulated as a standard quadratic pro-

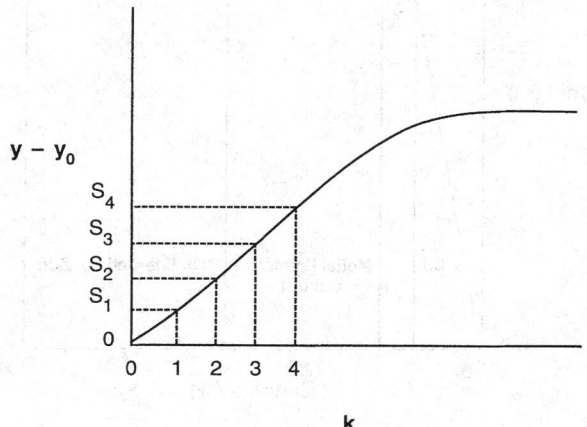

FIG. 8-43 Step response for u, a unit step change in the input.

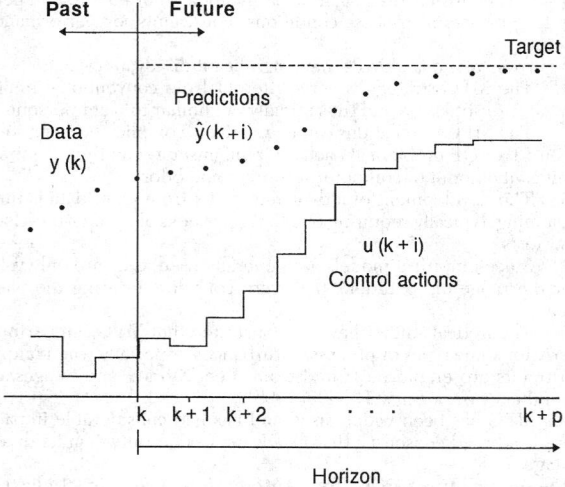

FIG. 8-44 The "moving horizon" approach of model predictive control.

gramming (QP) problem. Consequently, efficient QP solution techniques can be employed. When the inequality constraints in Eqs. (8-69) to (8-71) are omitted, the optimization problem has an analytical solution (Prett and Garcia, *Fundamental Process Control,* Butterworths, Stoneham, Massachusetts, 1988; Soeterboek, *Predictive Control—A Unified Approach,* Prentice Hall, Englewood Cliffs, New Jersey, 1991). If the quadratic terms in Eq. (8-66) are replaced by linear terms, a linear programming program (LP) problem results that can also be solved using standard methods.

This MPC formulation for SISO control problems can easily be extended to MIMO problems.

Implementation Issues   A critical factor in the successful application of any model-based technique is the availability of a suitable dynamic model. In typical MPC applications, an empirical model is identified from data acquired during extensive plant tests. The experiments generally consist of a series of bump tests in the manipulated variables. Typically, the manipulated variables are adjusted one at a time and the plant tests require a period of one to three weeks. The step or impulse response coefficients are then calculated using linear-regression techniques such as least-squares methods. However, details concerning the procedures utilized in the plant tests and subsequent model identification are considered to be proprietary information. The scaling and conditioning of plant data for use in model identification and control calculations can be key factors in the success of the application.

The MPC control problem illustrated in Eqs. (8-66) to (8-71) contains a variety of design parameters: model horizon N, prediction horizon p, control horizon m, weighting factors w_i, move suppression factor δ, the constraint limits B_i, C_i, and D_i, and the sampling period Δt. Some of these parameters can be used to tune the MPC strategy, notably the move suppression factor δ, but details remain largely proprietary. One commercial controller, Honeywell's RMPCT® (Robust Multivariable Predictive Control Technology), provides default tuning parameters based on the dynamic process model and the model uncertainty.

Integration of MPC and On-Line Optimization   As indicated in Fig. 8-42, significant potential benefits can be realized by using a combination of MPC and on-line optimization. At the present time, most commercial MPC packages integrate the two methodologies in a hierarchical configuration such as the one shown in Fig. 8-45. The MPC calculations are performed quite often (e.g., every 1–10 min) and implemented as set points for PID control loops at the DCS level. The targets and constraints for the MPC calculations are generated by solving a steady-state optimization problem (LP or QP) based on a linear process model. These calculations may be performed as often as the MPC calculations. As an option, the targets and constraints for the LP or QP optimization can be generated from a nonlinear process model using a nonlinear optimization technique. These calculations tend to be performed less frequently (e.g., every 1–24 hours) due to the complexity of the calculations and the process models.

The combination of MPC and frequent on-line optimization has been successfully applied in oil refineries and petrochemical plants around the world.

REAL-TIME PROCESS OPTIMIZATION

GENERAL REFERENCES: Biles and Swain, *Optimization and Industrial Experimentation,* Wiley—Interscience, New York, 1980. Dantzig, *Linear Programming and Extensions,* Princeton University Press, Princeton, New Jersey, 1963. Edgar and Himmelblau, *Optimization of Chemical Processes,* McGraw-Hill, New York, 1987. Fletcher, *Practical Methods of Optimization.* Wiley, New York, 1980. Gill, Murray, and Wright, *Practical Optimization,* Academic Press, New York, 1981. Murtagh, *Advanced Linear Programming,* McGraw-Hill, New York, 1983. Murty, *Linear Programming,* Wiley, New York, 1983. Reklaitis, Ravindran, and Ragsdell, *Engineering Optimization,* Wiley—Interscience, New York, 1984.

The chemical industry has undergone significant changes during the past 20 years due to the increased cost of energy and raw materials, more stringent environmental regulations, and intense worldwide competition. Modifications of both plant-design procedures and plant operating conditions have been implemented in order to reduce costs

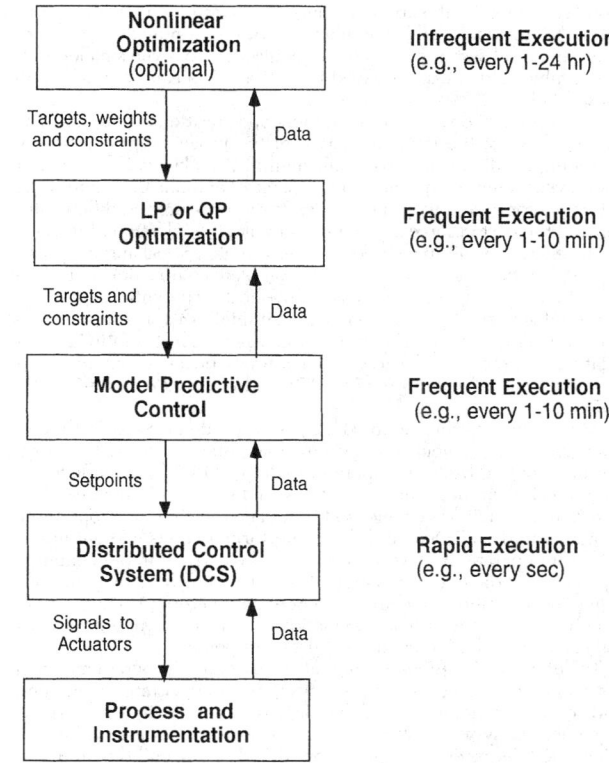

FIG. 8-45   Hierarchical control configuration for MPC and on-line optimization.

and meet constraints. One of the most important engineering tools that can be employed in such activities is optimization. As plant computers have become more powerful, the size and complexity of problems that can be solved by optimization techniques have correspondingly expanded. A wide variety of problems in the operation and analysis of chemical plants (as well as many other industrial processes) can be solved by optimization. Real-time optimization means that the process-operating conditions (set points) are evaluated on a regular basis and optimized. Sometimes this is called steady-state optimization or supervisory control. This section examines the basic characteristics of optimization problems and their solution techniques and describes some representative benefits and applications in the chemical and petroleum industries.

Typical problems in chemical engineering process design or plant operation have many possible solutions. Optimization is concerned with selecting the best among the entire set by efficient quantitative methods. Computers and associated software make the computations involved in the selection feasible and cost-effective. Engineers work to improve the initial design of equipment and strive for enhancements in the operation of the equipment once it is installed in order to realize the most production, the greatest profit, the maximum cost, the least energy usage, and so on. In plant operations, benefits arise from improved plant performance, such as improved yields of valuable products (or reduced yields of contaminants), reduced energy consumption, higher processing rates, and longer times between shutdowns. Optimization can also lead to reduced maintenance costs, less equipment wear, and better staff utilization. It is helpful to systematically identify the objective, constraints, and degrees of freedom in a process or a plant if such benefits as improved quality of designs, faster and more reliable troubleshooting, and faster decision making are to be achieved.

Optimization can take place at many levels in a company, ranging from a complex combination of plants and distribution facilities down through individual plants, combinations of units, individual pieces of equipment, subsystems in a piece of equipment, or even smaller entities. Problems that can be solved by optimization can be found at all these levels.

While process design and equipment specification are usually performed prior to the implementation of the process, optimization of operating conditions is carried out monthly, weekly, daily, hourly, or even every minute. Optimization of plant operations determines the set points for each unit at the temperatures, pressures, and flow rates that are the best in some sense. For example, the selection of the percentage of excess air in a process heater is quite critical and involves a balance on the fuel-air ratio to assure complete combustion and at the same time make the maximum use of the heating potential of the fuel. Typical day-to-day optimization in a plant minimizes steam consumption or cooling water consumption, optimizes the reflux ratio in a distillation column, or allocates raw materials on an economic basis [Latour, *Hydro Proc.*, **58**(6), 73, 1979, and *Hydro. Proc.*, **58**(7), 219, 1979].

A real-time optimization (RTO) system determines set point changes and implements them via the computer control system without intervention from unit operators. The RTO system completes all data transfer, optimization calculations, and set point implementation before unit conditions change and invalidate the computed optimum. In addition, the RTO system should perform all tasks without upsetting plant operations. Several steps are necessary for implementation of RTO, including determination of the plant steady state, data gathering and validation, updating of model parameters (if necessary) to match current operations, calculation of the new (optimized) set points, and the implementation of these set points.

To determine if a process unit is at steady state, a program monitors key plant measurements (e.g., compositions, product rates, feed rates, and so on) and determines if the plant is steady enough to start the sequence. Only when all of the key measurements are within the allowable tolerances is the plant considered steady and the optimization sequence started. Tolerances for each measurement can be tuned separately. Measured data are then collected by the optimization computer. The optimization system runs a program to screen the measurements for unreasonable data (gross error detection). This validity checking automatically modifies the model updating calculation to reflect any bad data or when equipment is taken out of service. Data validation and reconciliation (on-line or off-line) is an extremely critical part of any optimization system.

The optimization system then may run a parameter-fitting case that updates model parameters to match current plant operation. The integrated process model calculates such items as exchanger heat transfer coefficients, reactor performance parameters, furnace efficiencies, and heat and material balances for the entire plant. Parameter fitting allows for continual updating of the model to account for plant deviations and degradation of process equipment. After completion of the parameter fitting, the information regarding the current plant constraints, the control status data, and the economic values for feed products, utilities, and other operating costs are collected. The economic values are updated by the planning and scheduling department on a regular basis. The optimization system then calculates the optimized set points. The steady-state condition of the plant is re-checked after the optimization case is successfully completed. If the plant is still steady, then the values of the optimization targets are transferred to the process-control system for implementation. After a line-out period, the process-control computer resumes the steady-state detection calculations, restarting the cycle.

Essential Features of Optimization Problems   The solution of optimization problems involves the use of various tools of mathematics. Consequently, the formulation of an optimization problem requires the use of mathematical expressions. From a practical viewpoint, it is important to mesh properly the problem statement with the anticipated solution technique. Every optimization problem contains three essential categories:

1. An objective function to be optimized (revenue function, cost function, etc.)
2. Equality constraints (equations)

3. Inequality constraints (inequalities)

Categories 2 and 3 comprise the model of the process or equipment; category 1 is sometimes called the economic model.

No single method or algorithm of optimization exists that can be applied efficiently to all problems. The method chosen for any particular case will depend primarily on (1) the character of the objective function, (2) the nature of the constraints, and (3) the number of independent and dependent variables. Table 8-6 summarizes the six general steps for the analysis and solution of optimization problems (Edgar and Himmelblau, *Optimization of Chemical Processes*, McGraw-Hill, New York, 1988). You do not have to follow the cited order exactly, but you should cover all of the steps eventually. Shortcuts in the procedure are allowable, and the easy steps can be performed first. Steps 1, 2, and 3 deal with the mathematical definition of the problem: identification of variables and specification of the objective function and statement of the constraints. If the process to be optimized is very complex, it may be necessary to reformulate the problem so that it can be solved with reasonable effort. Later in this section, we discuss the development of mathematical models for the process and the objective function (the economic model).

Step 5 in Table 8-6 involves the computation of the optimum point. Quite a few techniques exist to obtain the optimal solution for a problem. We describe several classes of methods below; Fig. 8-46 is a diagram for selection of individual optimization techniques. In general, the solution of most optimization problems involves the use of a digital computer to obtain numerical answers. Over the past 15 years, substantial progress has been made in developing efficient and robust digital methods for optimization calculations. Much is known about which methods are most successful. Virtually all numerical optimization methods involve iteration, and the effectiveness of a given technique often depends on a good first guess for the values of the variables at the optimal solution. After the optimum is computed, a sensitivity analysis for the objective function value should be performed to determine the effects of errors or uncertainty in the objective function, mathematical model, or other constraints.

Development of Process (Mathematical) Models   Constraints in optimization problems arise from physical bounds on the variables, empirical relations, physical laws, and so on. The mathematical relations describing the process also comprise constraints. Two general categories of models exist:

1. Those based on physical theory
2. Those based on strictly empirical descriptions

Mathematical models based on physical and chemical laws (e.g., mass and energy balances, thermodynamics, chemical reaction kinetics) are frequently employed in optimization applications. These models are conceptually attractive because a general model for any system size can be developed before the system is constructed. On the other hand, an empirical model can be devised that simply correlates input-output data without any physiochemical analysis of the process. For

TABLE 8-6   The Six Steps Used to Solve Optimization Problems

1. Analyze the process itself so that the process variables and specific characteristics of interest are defined (i.e., make a list of all of the variables).

2. Determine the criterion for optimization and specify the objective function in terms of the above variables together with coefficients. This step provides the performance model (sometimes called the economic model when appropriate).

3. Develop via mathematical expressions a valid process or equipment model that relates the input-output variables of the process and associated coefficients. Include both equality and inequality constraints. Use well-known physical principles (mass balances, energy balances), empirical relations, implicit concepts, and external restrictions. Identify the independent and dependent variables (number of degrees of freedom).

4. If the problem formulation is too large in scope, (a) break it up into manageable parts and/or (b) simplify the objective function and model.

5. Apply a suitable optimization technique to the mathematical statement of the problem.

6. Check the answers and examine the sensitivity of the result to changes in the coefficients in the problem and the assumptions.

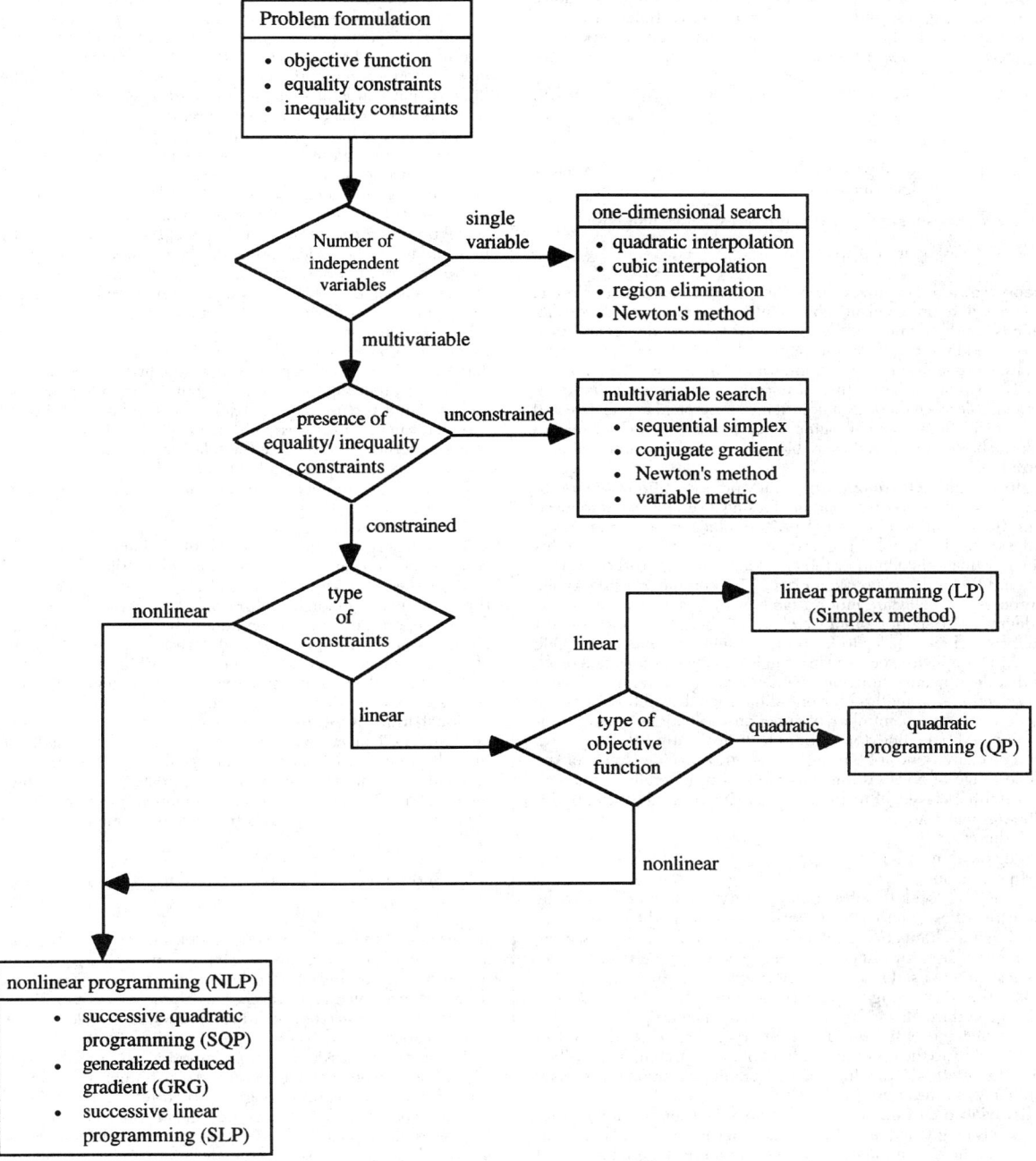

FIG. 8-46   Diagram for selection of optimization techniques with algebraic constraints and objective function.

these models, optimization is often used to fit a model to process data, using a procedure called parameter estimation. The well-known least squares curve-fitting procedure is based on optimization theory, assuming that the model parameters are contained linearly in the model. One example is the yield matrix, where the percentage yield of each product in a unit operation is estimated for each feed component using process data rather than employing a mechanistic set of chemical reactions.

Formulation of the Objective Function   The formulation of objective functions is one of the crucial steps in the application of optimization to a practical problem. You must be able to translate the desired objective into mathematical terms. In the chemical process industries, the objective function often is expressed in units of currency (e.g., U.S. dollars) because the normal industrial goal is to minimize costs or maximize profits subject to a variety of constraints.

A typical economic model involves the costs of raw materials, values of products, costs of production, as functions of operating conditions, projected sales figures, and the like. An objective function can be expressed in terms of these quantities; for example, annual operating profit ($/yr) might be expressed as:

$$J = \sum_s F_s V_s - \sum_r F_r C_r - OC \qquad (8\text{-}72)$$

where J = profit/time

$\displaystyle\sum_s F_s V_s$ = sum of product flow rates times respective product values (income)

$\displaystyle\sum_r F_r C_r$ = sum of feed flows times respective unit costs

OC = operating costs/time

Unconstrained Optimization   Unconstrained optimization refers to the case where no inequality constraints are present and all equality constraints can be eliminated by solving for selected dependent variables followed by substitution for them in the objective function. Very few realistic problems in process optimization are unconstrained. However, it is desirable to have efficient unconstrained optimization techniques available since these techniques must be applied in real time and iterative calculations cost computer time. The two classes of unconstrained techniques are single-variable optimization and multivariable optimization.

Single Variable Optimization   Many process optimization problems can be reduced to the variation of a single variable so as to maximize profit or some other overall process objective function. Some examples of single variable optimization include optimizing the reflux ratio in a distillation column or the air/fuel ratio in a furnace (Martin, Latour, and Richard, *Chem. Engr. Prog.*, **77**, September, 1981). While most processes actually are multivariable processes with several operating degrees of freedom, often we choose to optimize only the most important variable in order to keep the strategy uncomplicated. One characteristic implicitly required in a single variable optimization problem is that the objective function J be unimodal in the variable x.

The selection of a method for one-dimensional search is based on the trade-off between number of function evaluations versus computer time. We can find the optimum by evaluating the objective function at many values of x using a small grid spacing (Δx) over the allowable range of x values, but this method is generally inefficient. There are three classes of techniques that can be used efficiently for one-dimensional search:

1.   Indirect
2.   Region elimination
3.   Interpolation

Indirect methods seek to solve the necessary condition $dJ/dx = 0$ by iteration, but these methods are not as popular as the second two classes. Region elimination methods include equal interval search, dichotomous search (or bisecting), Fibonacci search, and golden section. These methods do not use information on the shape of the function (other than being unimodal) and thus tend to be rather conservative. The third class of techniques uses repeated polynomial fitting to predict the optimum. These interpolation methods tend to converge rapidly to the optimum without being very complicated. Two interpolation methods, quadratic and cubic interpolation, have been used in many optimization packages.

Multivariable Optimization   In multivariable optimization problems, there is no guarantee that the optimum can be reached in a reasonable amount of computer time. The numerical optimization of general nonlinear multivariable objective functions requires that efficient and robust techniques be employed. Efficiency is important since iteration is employed. For example, in multivariable "grid" search for a problem with four independent variables, an equally spaced grid for each variable is prescribed. For ten values of each of the four variables, there would be 10^4 total function evaluations required to find the best answer for the grid intersections. However, this computational effort still may not yield a result close enough to the true optimum, requiring further search. Therefore, grid search is a very inefficient method for most problems involving many variables.

In multivariable optimization, the difficulty of dealing with multi-variable functions is usually resolved by treating the problem as a series of one-dimensional searches. For a given starting point, a search direction $\mathbf{s}$ is specified, and the optimum is found by searching along that direction. The step size ε is the distance moved along $\mathbf{s}$. Then a new search direction is determined, followed by another one-dimensional search. The algorithm used to specify the search direction depends on the optimization method.

There are two basic types of unconstrained optimization algorithms: (1) those requiring function derivatives and (2) those that do not. The nonderivative methods are of interest in optimization applications because these methods can be readily adapted to the case in which experiments are carried out directly on the process. In such cases, an actual process measurement (such as yield) can be the objective function, and no mathematical model for the process is required. Methods that do not require derivatives are called direct methods and include sequential simplex (Nelder-Meade) and Powell's method. The sequential simplex method is quite satisfactory for optimization with two or three independent variables, is simple to understand, and is fairly easy to execute. Powell's method is more efficient than the simplex method and is based on the concept of conjugate search directions.

The second class of multivariable optimization techniques in principle requires the use of partial derivatives, although finite difference formulas can be substituted for derivatives; such techniques are called indirect methods and include the following classes:

1.   Steepest descent (gradient) method
2.   Conjugate gradient (Fletcher-Reeves) method
3.   Newton's method
4.   Quasi-Newton methods

The steepest descent method is quite old and utilizes the intuitive concept of moving in the direction where the objective function changes the most. However, it is clearly not as efficient as the other three. Conjugate gradient utilizes only first-derivative information, as does steepest descent, but generates improved search directions. Newton's method requires second derivative information but is very efficient, while quasi-Newton retains most of the benefits of Newton's method but utilizes only first derivative information. All of these techniques are also used with constrained optimization.

Constrained Optimization   When constraints exist and cannot be eliminated in an optimization problem, more general methods must be employed than those described above, since the unconstrained optimum may correspond to unrealistic values of the operating variables. The general form of a nonlinear programming problem allows for a nonlinear objective function and nonlinear constraints, or

Minimize $J(x_1, x_2, \ldots, x_n)$

Subject to $h_i(x_1, x_2, \ldots, x_n) = 0 \qquad (i = 1, r_c)$

$g_i(x_1, x_2, \ldots, x_n) \geq 0 \qquad (i = 1, m_c) \qquad (8\text{-}73)$

In this case, there are n process variables with r_c equality constraints and m_c inequality constraints. Such problems pose a serious challenge to performing optimization calculations in a reasonable amount of time. Typical constraints in chemical process optimization include operating conditions (temperatures, pressures, and flows have limits), storage capacities, and product purity specifications.

An important class of constrained optimization problems is one in which both the objective function and constraints are linear. The solution of these problems is highly structured and can be obtained rapidly. The accepted procedure, linear programming (LP), has become quite popular in the past twenty years, solving a wide range of industrial problems. It is increasingly being used for on-line optimization. For processing plants, there are several different kinds of linear constraints that may arise, making the LP method of great utility.

1.   Production limitation due to equipment throughput restrictions, storage limits, or market constraints.
2.   Raw material (feedstock) limitation.
3.   Safety restrictions on allowable operating temperatures and pressures.
4.   Physical property specifications placed on the composition of the final product. For blends of various products, we usually assume that a composite property can be calculated through the averaging of pure component physical properties.

5. Material and energy balances of the steady-state model.

The optimum in linear programming lies at the constraint intersections, which was generalized to any number of variables and constraints by George Dantzig. The Simplex algorithm is a matrix-based numerical procedure for which many digital computer codes exist, both for mainframe and microcomputers (Edgar and Himmelblau, *Optimization of Chemical Processes,* McGraw-Hill, New York, 1987; Schrage, *Linear, Integer, and Quadratic Programming with LINDO,* Scientific Press, Palo Alto, California, 1983). The algorithm can handle virtually any number of inequality constraints and any number of variables in the objective function and utilizes the observation that only the constraint boundaries need to be examined to find the optimum. In some instances, nonlinear optimization problems even with nonlinear constraints can be linearized so that the LP algorithm can be employed to solve them (called successive linear programming or SLP). In the process industries, the Simplex algorithm has been applied to a wide range of problems, including refinery scheduling, olefins production, the optimal allocation of boiler fuel, and the optimization of a total plant.

Nonlinear Programming   The most general case for optimization occurs when both the objective function and constraints are nonlinear, a case referred to as nonlinear programming. While the idea behind the search methods used for unconstrained multivariable problems are applicable, the presence of constraints complicates the solution procedure.

In practice, one of the best current general algorithms (best on the basis of many tests) using iterative linearization is the Generalized Reduced Gradient algorithm (GRG). The GRG algorithm employs linear or linearized constraints, defines new variables that are normal to the constraints, and expresses the gradient (or other search direction) in terms of this normal basis (Liebman, Lasdon, Schrage, and Waren, *GINO,* Scientific Press, Palo Alto, California, 1986). Other established types of constrained optimization methods include the following types of algorithms:

1. Penalty functions with augmented Lagrangian method (an enhancement of the classical Lagrange multiplier method)
2. Successive quadratic programming

All of these methods have been utilized to solve nonlinear programming problems in the field of chemical engineering design and operations (Lasdon and Waren, *Oper. Res.,* **5,** 34, 1980). Nonlinear programming is receiving increased usage in the area of real-time optimization.

One important class of nonlinear programming techniques is called quadratic programming (QP), where the objective function is quadratic and the constraints are linear. While the solution is iterative, it can be obtained quickly as in linear programming. This is the basis for the newest type of constrained multivariable control algorithms called model predictive control. The dominant method used in the refining industry utilizes the solution of a QP and is called dynamic matrix control or DMC. See the earlier subsection on model predictive control for more details.

EXPERT SYSTEMS

An expert system is a computer program that uses an expert's knowledge in a particular domain to solve a narrowly focused, complex problem. An off-line system uses information entered manually and produces results in visual form to guide the user in solving the problem at hand. An on-line system uses information taken directly from process measurements to perform tasks automatically or instruct or alert operating personnel to the status of the plant.

Each expert system has a rule base created by the expert to respond the way the expert would to sets of input information. Expert systems used for plant diagnostics and management usually have an open rule base, which can be changed and augmented as more experience accumulates and more tasks are to be automated. The system begins as an empty shell with an assortment of functions such as equation-solving, logic, and simulation, as well as input and display tools to allow an expert to construct a proprietary rule base. The "expert" in this case would be the person or persons having the deepest knowledge about the process, its problems, its symptoms, and remedies. Converting these inputs into meaningful outputs is the principal task in constructing a rule base. Skill at computer programming is especially helpful, although most shells allow rules to be entered in the vernacular. First-principles models (deep knowledge) produce the most accurate results, although heuristics are always required to establish limits.

A closed expert system is one designed by an expert to be sold in quantity for use by others (where open systems tend to be unique). It is closed to keep users from altering the rule base and thereby changing the product. Common examples in process control are autotuning and self-tuning controllers whose rule base is designed by one or more experts in that field. Once packaged and sold, its rule base cannot be changed in the field no matter how poorly it performs the task; revisions must be made by the manufacturer in later releases as for any software product.

The development vehicle used to create and test the rule base must be as flexible as possible, allowing easy alterations and expansion of the rule base with whatever displays can convey the most information. The delivery vehicle, however, should be virtually transparent to the user, conveying only as much information as needed to solve the problem at hand. Self-tuning controllers can perform their task without explicitly informing users, but their output and status is available on demand, and their operation may be easily limited or interrupted.

To be successful, the scope of an expert system must be limited to a narrow group of common problems that are readily solved by conventional means, and where the return on investment is greatest. Widening the scope usually requires more complex methods and treats less common problems having lower return.

UNIT OPERATIONS CONTROL

PROCESS AND INSTRUMENTATION DIAGRAMS

GENERAL REFERENCES: Shinskey, *Process Control Systems,* 3d ed., McGraw-Hill, New York, 1988. Luyben, *Practical Distillation Control,* Van Nostrand Reinhold, New York, 1992.

The process and instrumentation (P&I) diagram provides a graphical representation of the control configuration for the process. The P&I diagrams illustrate the measurement devices that provide inputs to the control strategy, the actuators that will implement the results of the control calculations, and the function blocks that provide the control logic.

The symbology for drawing P&I diagrams generally follows standards developed by one of the following organizations:

1. International Society for Measurement and Control (ISA). The chemicals, refining, and foods industries generally follow this standard.

2. Scientific Apparatus Manufacturers Association (SAMA). The fossil-fuel electric utility industry generally follows this standard.

Both organizations update their standards from time to time, primarily because the continuing evolutions in control-system hardware provide additional possibilities for implementing control schemes.

Although arguments can be made for the advantages of each symbology, the practices within an industry seem to be mainly the result of historical practice with no indication of any significant shift. Most companies adopt one of the standards but then tailor or extend the symbology to best suit their internal practices. Such companies maintain an internal document and/or drawing that specifies the symbology used on their P&I diagrams. Their internal personnel and all contractors are instructed to adhere to this symbology when developing P&I diagrams.

Figure 8-47 presents a P&I diagram for a simple temperature control loop that adheres to the ISA symbology. The measurement

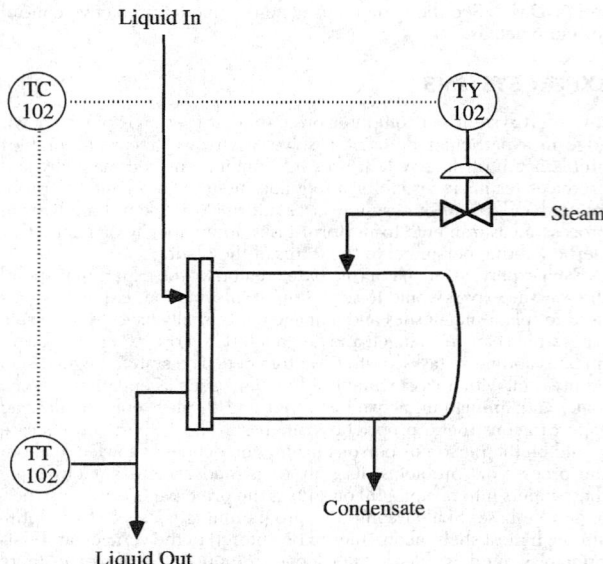

FIG. 8-47 Example of a process and instrument diagram.

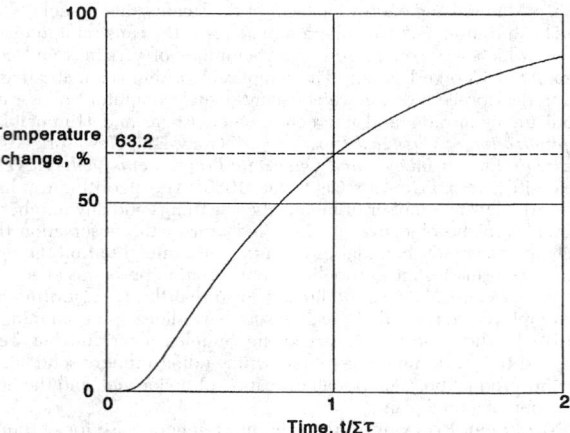

FIG. 8-48 Temperature leaving a heat exchanger responds as a distributed lag, the gain and time constant of which vary inversely with flow.

devices and most elements of the control logic are represented by circles. In Figure 8-47, circles are used to designate the following:

1. TT102 is the temperature measurement device.
2. TC102 is the temperature controller.
3. TY102 is the current-to-pneumatic (I/P) transducer.

The symbol for the control valve in Fig. 8-47 is for a pneumatic positioning valve without a valve positioner.

Electronic signals (that is, 4–20 milliamp current loops) are represented by dashed lines. In Fig. 8-47, these include the following:

1. The signal from the measurement device to the controller.
2. The signal from the controller to the I/P transducer.

Pneumatic signals are represented by solid lines that are cross-hatched intermittently. The signal from the I/P transducer to the pneumatic positioning valve is pneumatic.

The ISA symbology provides different symbols for different types of actuators. Furthermore, variations for the controller symbol distinguish control algorithms implemented in DCS technology from panel-mounted single-loop controllers.

CONTROL OF HEAT EXCHANGERS

Steam-Heated Exchangers Steam, the most common heating medium, transfers its latent heat in condensing, causing heat flow to be proportional to steam flow. Thus, a measurement of steam flow is essentially a measure of heat transfer. Consider raising a liquid from temperature T_1 to T_2 by condensing steam:

$$Q = WH = FC_L(T_2 - T_1) \qquad (8-74)$$

where W and H are the mass flow of steam and its latent heat, F and C_L are the mass flow and specific heat of the liquid, and Q is the rate of heat transfer. The response of controlled temperature to steam flow is linear:

$$\frac{dT_2}{dW} = \frac{H}{FC_L} \qquad (8-75)$$

However, the steady-state process gain described by this derivative varies inversely with liquid flow: Adding a given increment of heat flow to a smaller flow of liquid produces a greater temperature rise.

Dynamically, the response of liquid temperature to a step in steam flow is that of a distributed lag, shown in Fig. 8-48. The time required to reach 63 percent complete response, $\sum \tau$, is essentially the residence time of the fluid in the exchanger, which is its volume divided

by its flow. The residence time then varies inversely with flow. Table 8-2 gives optimum settings for PI and PID controllers for distributed lags, the proportional band varying directly with steady-state gain, and integral and derivative settings directly with $\sum \tau$. Since both these parameters vary inversely with liquid flow, fixed settings for the temperature controller are optimal at only one flow rate.

Undamped oscillations will be produced when the flow decreases by one-third from the value at which the controller was optimally tuned, whereas increasing flow rates produces an overdamped response. The stable operating range can be broadened to one-half the original flow by using an equal-percentage steam valve whose gain varies directly with flow. The best solution is to adapt the PID settings to change inversely with measured flow, thereby keeping the controller optimally tuned for all flow rates.

Feedforward control can also be applied by multiplying the liquid flow measurement—after dynamic compensation—by the output of the temperature controller, the result used to set steam flow in cascade. Feedforward is capable of a reduction in integrated error as much as a hundredfold but requires the use of a steam-flow loop and dynamic compensator to approach this.

Steam flow is sometimes controlled by manipulating a valve in the condensate line rather than the steam line, because it is smaller and hence less costly. Heat transfer, then, is changed by raising or lowering the level of condensate flooding the heat-transfer surface, an operation that is slower than manipulating a steam valve. Protection also needs to be provided against an open condensate valve blowing steam into the condensate system.

Exchange of Sensible Heat When there is no change in phase, heat transfer is no longer linear with flow of the manipulated stream, as illustrated by Fig. 8-49. Here again, an equal-percentage valve should be used on that stream to linearize the loop. The variable dynamics of the distributed lag apply, limiting the stable operating range in the same way as for the steam-heated exchanger. These heat exchangers are also sensitive to variations in the temperature of the manipulated stream, an increasingly common problem where heat is being recovered at variable temperatures for reuse in heat transfer.

Figure 8-50 shows a temperature controller (TC) setting a heat-flow controller (QC) in cascade. A measurement of the manipulated flow is multiplied by its temperature difference across the heat exchanger to calculate the current heat-transfer rate, using the right side of Eq. (8-74). Variations in supply temperature, then, appear as variations in calculated heat transfer, which the QC can quickly correct by adjusting the manipulated flow. An equal-percentage valve is still required to linearize the secondary loop, but the primary loop of temperature-setting heat flow is linear. Feedforward can be added by multiplying the dynamically compensated flow measurement of the other fluid by the output of the temperature controller.

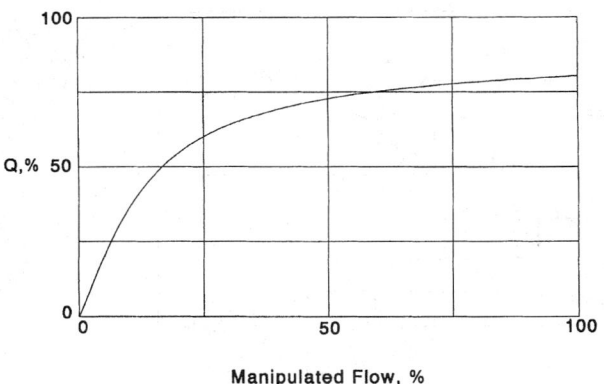

FIG. 8-49   Heat-transfer rate in sensible-heat exchange varies nonlinearly with flow of the manipulated fluid.

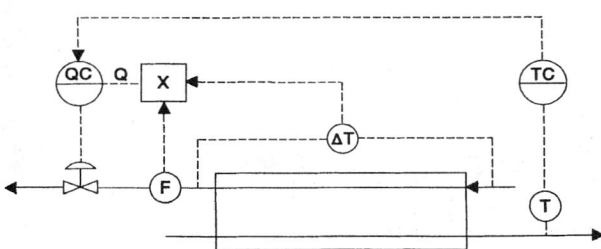

FIG. 8-50   Manipulating heat flow linearizes the loop and protects against variations in supply temperature.

When manipulating a stream whose flow is independently determined, such as flow of a product or a heat-transfer fluid from a fired heater, a three-way valve is used to divert the required flow to the heat exchanger. This does not alter the linearity of the process or its sensitivity to supply variations and even adds the possibility of independent flow variations. The three-way valve should have equal-percentage characteristics, and heat-flow control may be even more beneficial.

DISTILLATION COLUMN CONTROL

Distillation columns have four or more closed loops—increasing with the number of product streams and their specifications—all of which interact with each other to some extent. Because of this interaction, there are many possible ways to pair manipulated and controlled variables through controllers and other mathematical functions with widely differing degrees of effectiveness. Columns also differ from each other, so that no single rule of configuring control loops can be applied successfully to all. The following rules apply to the most common separations.

Controlling Quality of a Single Product   If one of the products of a column is far more valuable than the others, its quality should be controlled to satisfy given specifications, and its recovery should be maximized by minimizing losses of its principal component in other streams. This is achieved by maximizing reflux ratio consistent with flooding limits on trays, which means maximizing the flow of reflux or vapor, whichever is limiting. The same rule should be followed when heating and cooling have little value. A typical example is the separation of high-purity propylene from much lower-valued propane, usually achieved with waste heat from quench water from the cracking reactors.

The most important factor affecting product quality is the material balance. In separating a feed stream F into distillate D and bottom B products, an overall mole-flow balance must be maintained:

$$F = D + B \qquad (8\text{-}76)$$

as well as a balance on each component:

$$Fz_i = Dy_i + Bx_i \qquad (8\text{-}77)$$

where z, y, and x are mol fractions of component i in the respective streams. Combining these equations gives a relationship between the composition of the products and their relative portion of the feed:

$$\frac{D}{F} = 1 - \frac{B}{F} = \frac{z_i - x_i}{y_i - x_i} \qquad (8\text{-}78)$$

From the above, it can be seen that control of either x_i or y_i requires both product flow rates to change with feed rate and feed composition.

Figure 8-51 shows a propylene-propane fractionator controlled at maximum boilup by the differential pressure controller (DPC) across the trays. This loop is fast enough to reject upsets in the temperature of the quench water quite easily. Pressure is controlled by manipulating the heat-transfer surface in the condenser through flooding. If the condenser should become overloaded, pressure will rise above set point, but this has no significant effect on the other control loops. Temperature measurements on this column are not helpful, as the difference between the component boiling points is too small. Propane content in the propylene distillate is measured by a chromatographic analyzer sampling the overhead vapor for fast response and is controlled by the analyzer controller (AC) manipulating the ratio of distillate to feed rates. The feedforward signal from feed rate is dynamically compensated by $f(t)$ and nonlinearly characterized by $f(x)$ to account for variations in propylene recovery as feed rate changes. Distillate flow can be measured and controlled more accurately than reflux flow by a factor equal to the reflux ratio—in this column, typically between 10 and 20. Therefore, reflux flow is placed under accumulator level control (LC). Yet composition responds to the difference between boilup and reflux. To eliminate the lag inherent in the response of the level controller, reflux flow is driven by the subtractor in the direction opposite to distillate flow—this is essential to fast response of the composition loop.

Controlling Quality of Two Products   Where the two products have similar value, or where heating and cooling costs are comparable to product losses, the compositions of both products should be controlled. This introduces the possibility of strong interaction between the two composition loops, as they tend to have the same speed of response. To minimize interaction, most columns should have distillate composition controlled by reflux ratio and bottom composition by boilup or preferably boilup-to-bottom ratio. These loops are insensitive to variations in feed rate, eliminating the need for feedforward control, and they also reject heat-balance upsets quite effectively. Figure 8-52 shows a depropanizer controlled by reflux and boilup ratios. The actual mechanism through which these ratios are manipulated is as $D/(L + D)$ and $B/(V + B)$, where L is reflux flow and V is vapor boilup, which decouples the temperature loops from the liquid-level loops. Column pressure here is controlled by flooding both condenser and accumulator; however, there is no LC on the accumulator, so this arrangement will not function with an overloaded condenser. Temperatures are used as indications of composition in this column because of the substantial difference in boiling points between propane and butanes. However, off-key components such as ethane do effect the accuracy of the relationship so that an analyzer controller is used to set the top temperature controller (TC) in cascade.

If the products from a column are especially pure, even this configuration may produce excessive interaction between the composition loops. Then the composition of the less pure product should be controlled by manipulating its own flow; the composition of the remaining product should be controlled by manipulating reflux ratio if it is the distillate or boilup ratio if it is the bottom product.

CHEMICAL REACTORS

Composition Control   The first requirement for successful control of a chemical reactor is to establish the proper stoichiometry, that is, to control the flow rates of the reactants in the proportions needed

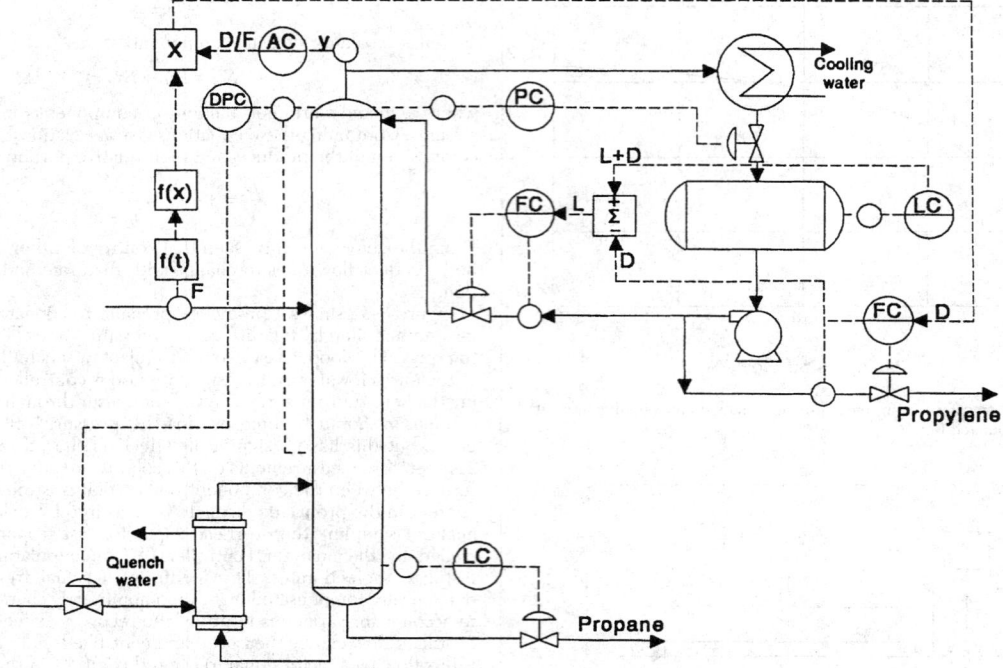

FIG. 8-51   The quality of high-purity propylene should be controlled by manipulating the material balance.

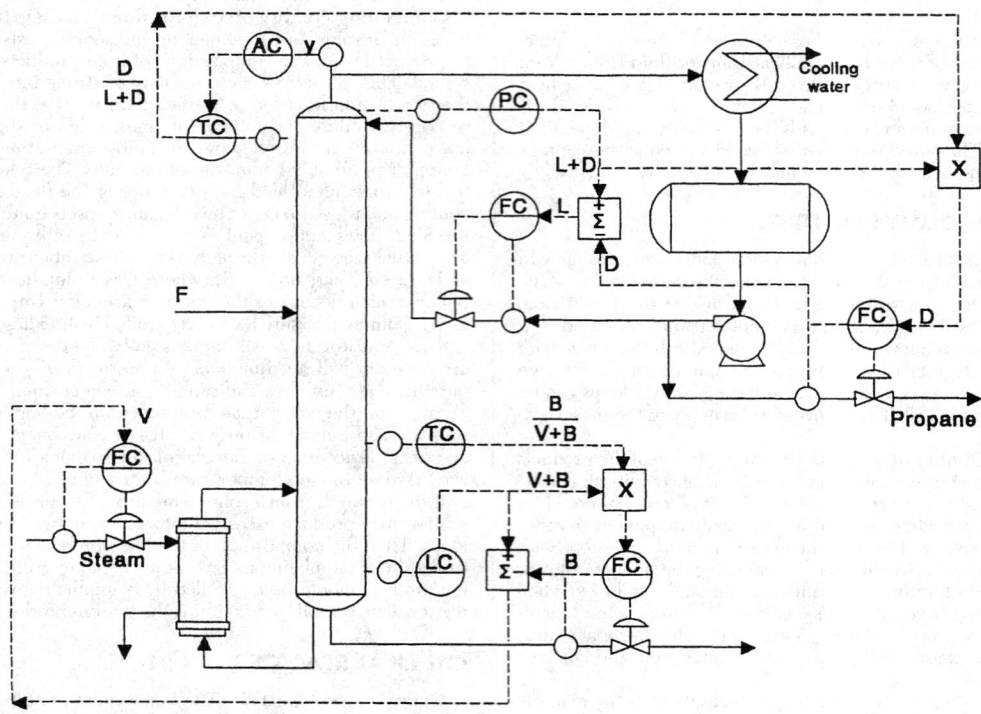

FIG. 8-52   Depropanizers require control of both products, here using reflux-ratio and boilup-ratio manipulation.

to satisfy the reaction chemistry. In a continuous reactor, this begins by setting ingredient flow rates in ratio to one another. However, because of variations in the purity of the feed streams and inaccuracy in flow metering, some indication of excess reactant such as pH or a composition measurement should be used to trim the ratios. Many reactions are incomplete, leaving one or more reactants unconverted. They are separated from the products of the reaction and recycled to the reactor, usually contaminated with inert components. While reactants can be recycled to complete conversion (extinction), inerts can accumulate to the point of impeding the reaction and must be purged from the system. Inerts include noncondensible gases that must be vented and nonvolatiles from which volatile products must be stripped.

If one of the reactants differs in phase from the others and the products, it may be manipulated to close the material balance on that phase. For example, a gas reacting with liquids to produce a liquid product may be added, as it is consumed to control reactor pressure; a gaseous purge would be necessary. Similarly, a liquid reacting with a gas to produce a gaseous product could be added, as it is consumed to control liquid level in the reactor; a liquid purge would be required. Where a large excess of one reactant A is used to minimize side reactions, the unreacted excess is typically sent to a storage tank for recycling. Its flow from the recycle storage tank is set in the desired ratio to the flow of reactant B, with the flow of fresh A manipulated to control recycle tank level if the feed is a liquid or tank pressure if it is a gas. Some catalysts travel with the reactants and must be recycled in the same way.

With batch reactors, it may be possible to add all reactants in their proper quantities initially if the reaction rate can be controlled by injection of initiator or adjustment of temperature. In semibatch operation, one key ingredient is flow-controlled into the batch at a rate that sets the production. This ingredient should not be manipulated for temperature control of an exothermic reactor, as the loop includes two dominant lags—concentration of the reactant and heat capacity of the reaction mass—and can easily go unstable.

Temperature Control   Reactor temperature should always be controlled by heat transfer. Endothermic reactions require heat and therefore are eminently self-regulating. Exothermic reactions produce heat, which tends to raise reaction temperature, thereby increasing reaction rate and producing more heat. This positive feedback is countered by negative feedback in the cooling system, which removes more heat as reactor temperature rises. Most continuous reactors have enough heat-transfer surface relative to reaction mass so that negative feedback dominates and they are self-regulating. But most batch reactors do not and are therefore steady-state unstable. Unstable reactors are controllable, but the temperature controller requires a high gain, and the cooling system must have enough margin to accommodate the largest expected disturbance in heat load.

Figure 8-53 shows the recommended system for controlling the temperature of an exothermic reactor, either continuous or batch. The circulating pump on the coolant loop is absolutely essential to effective temperature control in keeping dead time minimum and constant—without it, dead time varies inversely with cooling load, causing limit cycling at low loads. Heating is usually required to raise the temperature to reaction conditions, although it is often locked out in a batch reactor once initiator is introduced. The valves are operated in split range, the heating valve opening from 50–100 percent of controller output, and the cooling valve opening from 0–50 percent. The cascade system linearizes the reactor temperature loop, speeds its response, and protects it from disturbances in the cooling system. The flow of heat removed per unit of coolant flow is directly proportional to the temperature rise of the coolant, which varies with both the temperature of the reactor and the rate of heat transfer from it. Using an equal-percentage cooling valve helps compensate for this nonlinearity, although it is incomplete—a preferred arrangement would be to manipulate coolant flow using a heat-flow controller as described in Fig. 8-50.

The flow of heat across the heat-transfer surface is linear with both temperatures, leaving the primary loop with a constant gain. Using the coolant exit rather than inlet temperature as the secondary controlled variable moves the jacket dynamics from the primary to the secondary

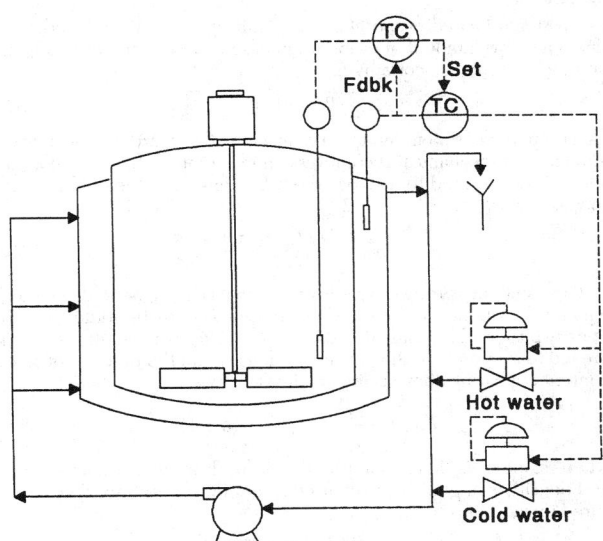

FIG. 8-53   The reactor temperature controller sets coolant outlet temperature in cascade, with primary integral feedback taken from the secondary temperature measurement.

loop, reducing the period of the primary loop. Performance and robustness are both improved by using the secondary temperature measurement as the feedback signal to the integral mode of the primary controller. (This feature may only be available with controllers that integrate by positive feedback.) This places the entire secondary loop in the integral path of the primary controller, effectively pacing its integral time to the rate at which the secondary temperature is able to respond. The primary controller may also be left in the automatic mode at all times without integral windup.

The primary time constant of the reactor is

$$\tau_1 = \frac{M_r C_r}{UA} \tag{8-79}$$

where M_r and C_r are the mass and heat capacity of the reactants, and U and A are the overall heat-transfer coefficient and area respectively. This system was tested on a pilot reactor where the heat-transfer area and mass could both be changed by a factor of two, changing τ_1 by a factor of four as confirmed by observations of rates of temperature rise. Yet the controllers configured as described in Fig. 8-53 did not require retuning as τ_1 varied. The primary controller should be PID, and the secondary controller at least PI in this system; if the secondary controller has no integral mode, the primary controller will control with offset. Set point overshoot in batch reactor control can be avoided by setting derivative time of the primary controller higher than its integral time, but this is only effective with interacting PID controllers.

CONTROLLING EVAPORATORS

The most important consideration in controlling the quality of concentrate from an evaporator is forcing the vapor rate to match the flow of excess solvent entering in the feed. The mass flow of solid material entering and leaving are equal in the steady state:

$$M_0 x_0 = M_n x_n \tag{8-80}$$

where M_0 and x_0 are the mass flow and solid fraction of the feed, and M_n and x_n are their values in the product after n effects of evaporation. The total solvent evaporated from all the effects must then be

$$\sum W = M_0 - M_n = M_0 \left(1 - \frac{x_0}{x_n} \right) \tag{8-81}$$

For a steam-heated evaporator, each unit of steam W_0 applied produces a known amount of evaporation based on the number of effects and their fractional economy E:

$$\sum W = nEW_0 \qquad (8\text{-}82)$$

(A comparable statement can be made with regard to the power applied to a mechanical recompression evaporator.) In summary, the steam flow required to increase the solid content of the feed from x_0 to x_n is

$$W_0 = \frac{M_0(1 - x_0/x_n)}{nE} \qquad (8\text{-}83)$$

The usual measuring device for feed flow is a magnetic flowmeter, which is a volumetric device whose output F must be multiplied by density ρ to produce mass flow M_0. For most aqueous solutions which are fed to evaporators, the product of density and the function of solid content appearing above is linear with density:

$$F\rho \left(1 - \frac{x_0}{x_n}\right) \approx F[1 - m(\rho - 1)] \qquad (8\text{-}84)$$

where slope m is determined by the desired product concentration, and density is in g/ml. The required steam flow in lb/h for feed measured in gal/min is then

$$W_0 = \frac{500F[1 - m(\rho - 1)]}{nE} \qquad (8\text{-}85)$$

where the factor of 500 converts gal/min of water to lb/h. The factor nE is about 1.74 for a double-effect evaporator and 2.74 for a triple-effect. Using a thermocompressor (ejector) driven with 150-lb/in² steam on a single-effect evaporator gives an nE of 2.05; it essentially adds the equivalent of one effect to the evaporator train.

A cocurrent evaporator train with its controls is illustrated in Fig. 8-54. The control system applies equally well to countercurrent or mixed-feed evaporators, the principal difference being the tuning of the dynamic compensator $f(t)$, which must be done in the field to minimize the short-term effects of changes in feed flow on product quality. Solid concentration in the product is usually measured as density; feedback trim is applied by the AC adjusting slope m of the density function, which is the only term related to x_n. This recalibrates the system whenever x_n must move to a new set point.

The accuracy of the system depends on controlling heat flow; therefore, if steam pressure varies, compensation must be applied to correct for both steam density and enthalpy as a function of pressure. Some evaporators must use unreliable sources of low-pressure steam. In this case, the measurement of pressure-compensated steam flow can be used to set feed flow by solving the last equation for F using W_0 as a variable. The steam-flow controller would be set for a given production rate, but the dynamically compensated steam-flow measurement would be the input signal to calculate the feed-flow set point. Both of these configurations are widely used in controlling corn-syrup concentrators.

DRYING OPERATIONS

Controlling dryers is much different than controlling evaporators because on-line measurements of feed rate and composition and product composition are rarely available. Most dryers transfer moisture from wet feed into hot dry air in a single pass. The process is generally very self-regulating, in that moisture becomes progressively harder to remove from the product as it dries: This is known as falling-rate drying. Controlling the temperature of the air leaving a cocurrent dryer tends to regulate the moisture in the product, as long as feed rate and the moisture in the feed and air are reasonably constant. At constant outlet air temperature, product moisture tends to rise with all three of these variables.

In the absence of moisture analyzers, regulation of product quality can be improved by raising the temperature of the exhaust air in proportion to the evaporative load. The evaporative load can be estimated by the loss in temperature of the air passing through the dryer in the steady state. Changes in load are first observed in upsets in exhaust temperature at a given inlet temperature; the controller then responds by returning the exhaust air to its original temperature by changing that of the inlet air.

Figure 8-55 illustrates the simplest application of this principal as the linear relationship

$$T_0 = T_b + K\Delta T \qquad (8\text{-}86)$$

where T_0 is the set point for exhaust temperature elevated above a base temperature T_b, corresponding to zero-load operation, and ΔT is the drop in air temperature from inlet to outlet. Coefficient K must be set to regulate product moisture over the expected range of evaporative load. If set too low, product moisture will increase with increasing load; if set too high, it will decrease with increasing load. While K can be estimated from the model of a dryer, it does depend on the rate-of-

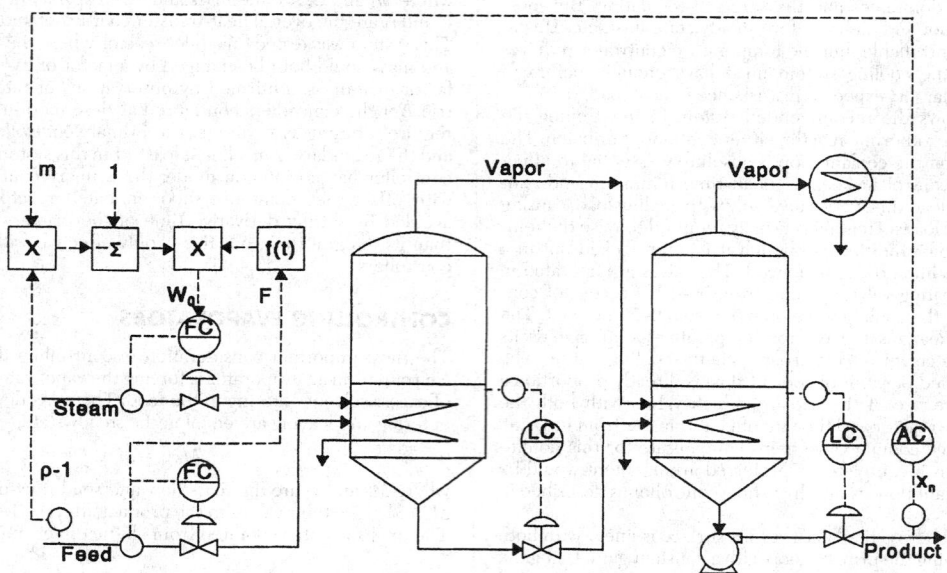

FIG. 8-54 Controlling evaporators requires matching steam flow and evaporative load, here using feedforward control.

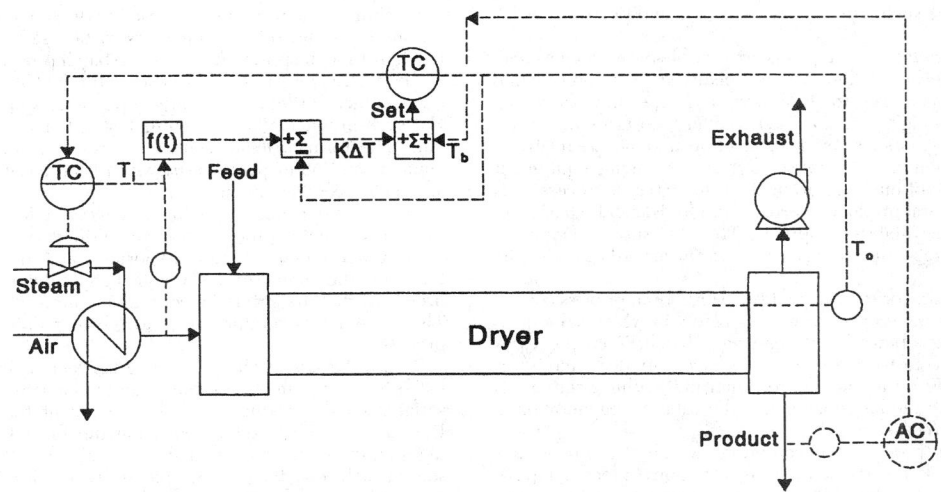

FIG. 8-55   Product moisture from a cocurrent dryer can be regulated through temperature control indexed to heat load.

drying curve for the product, its particle size, and whether the load variations are due primarily to changes in feed rate or feed moisture.

It is important to have the most accurate measurement of exhaust temperature attainable. Note that Fig. 8-55 shows the sensor inserted into the dryer upstream of the rotating seal, because leakage there could cause the temperature in the exhaust duct to read low—even lower than the wet-bulb temperature, an impossibility without leakage of either heat or outside air.

The calculation of exhaust-temperature set point forms a positive-feedback loop capable of destabilizing the dryer. For example, an increase in load causes the controller to raise inlet temperature, which will in turn raise the calculated set point calling for a further increase in inlet temperature. The gain in the set point loop, K, typically is well below the gain of the exhaust temperature measurement responding to the same change in inlet temperature. Negative feedback then dominates in the steady state, but the response of the exhaust temperature measurement is delayed by the dryer. A similar lag $f(t)$ is shown inserted in the set point loop to prevent positive feedback from dominating in the short term, which could cause cycling.

If product moisture is measured off-line, analytical results can be used to adjust K and T_b manually. If an on-line analyzer is used, the analyzer controller would be most effective in adjusting the bias T_b, as shown in the figure.

While the rotary dryer shown is commonly used for grains and minerals, this system has been successfully applied to fluid-bed drying of plastic pellets, air-lift drying of wood fibers, and spray drying of milk solids. The air may be steam-heated as shown or heated by direct combustion of fuel, provided that a representative measurement of inlet air temperature can be made. If it cannot, then evaporative load can be inferred from a measurement of fuel flow, replacing ΔT in the set point calculation.

If the feed flows countercurrent to the air, as is the case when drying granulated sugar, exhaust temperature does not respond to variations in product moisture. For these dryers, product moisture can better be regulated by controlling its temperature at the point of discharge. Conveyor-type dryers are usually divided into a number of zones, each separately heated with recirculation of air which raises its wet-bulb temperature. Only the last two zones may require indexing of exhaust-air temperature as a function of ΔT.

Batch drying, used on small lots like pharmaceuticals, begins operation by blowing air at constant inlet temperature through saturated product in constant-rate drying, where ΔT is constant at its maximum value ΔT_c. When product moisture reaches the point where falling-rate drying begins, the exhaust temperature begins to rise. The desired product moisture will be reached at a corresponding exhaust temperature T_f, which is related to the temperature T_c observed during constant-rate drying, as well as ΔT_c:

$$T_f = T_c + K\Delta T_c \qquad (8\text{-}87)$$

The control system requires the values of T_c and ΔT_c observed during the first minutes of operation to be stored as the basis for the above calculation of end point. When the exhaust temperature then reaches the value calculated, drying is terminated. Coefficient K can be estimated from models but requires adjustment on-line to reach product specifications repeatedly. Products having different moisture specifications or particle size will require different settings of K, but the system does compensate for variations in feed moisture, batch size, air moisture, and inlet temperature. Some exhaust air may be recirculated to control the dewpoint of the inlet air, thereby conserving energy toward the end of the batch and when the ambient air is especially dry.

BATCH PROCESS CONTROL

BATCH VERSUS CONTINUOUS PROCESSES

GENERAL REFERENCES: Fisher, *Batch Control Systems: Design, Application, and Implementation*, ISA, Research Triangle Park, North Carolina, 1990; Rosenof and Ghosh, *Batch Process Automation*, Van Nostrand Reinhold, New York, 1987.

When categorizing process plants, the following two extremes can be identified:

1. *Commodity plants.*   These plants are custom-designed to produce large amounts of a single product (or a primary product plus one or more secondary products). An example is a chlorine plant, where the primary product is chlorine and the secondary products are hydrogen and sodium hydroxide. Usually the margins (product value less manufacturing costs) for the products from commodity plants are small, so the plants must be designed and operated for best possible efficiencies. Although a few are batch, most commodity plants are

continuous. Factors such as energy costs are life-and-death issues for such plants.

2. *Specialty plants.*   These plants are capable of producing small amounts of a variety of products. Such plants are common in fine chemicals, pharmaceuticals, foods, and so on. In specialty plants, the margins are usually high, so factors such as energy costs are important but not life-and-death issues. As the production amounts are relatively small, it is not economically feasible to dedicate processing equipment to the manufacture of only one product. Instead, batch processing is utilized so that several products (perhaps hundreds) can be manufactured with the same process equipment. The key issue in such plants is to manufacture consistently each product in accordance with its specifications.

The above two categories represent the extremes in process configurations. The term *semibatch* designates plants in which some processing is continuous but other processing is batch. Even processes that are considered to be continuous can have a modest amount of batch processing. For example, the reformer unit within a refinery is thought of as a continuous process, but the catalyst regeneration is normally a batch process.

In a continuous process, the conditions within the process are largely the same from one day to the next. Variations in feed composition, plant utilities (e.g., cooling water temperature), catalyst activities, and other variables occur, but normally these changes are either about an average (e.g., feed compositions) or exhibit a gradual change over an extended period of time (e.g., catalyst activities). Summary data such as hourly averages, daily averages, and the like are meaningful in a continuous process.

In a batch process, the conditions within the process are continually changing. The technology for making a given product is contained in the product recipe that is specific to that product. Such recipes normally state the following:

1. *Raw material amounts.*   This is the stuff needed to make the product.

2. *Processing instructions.*   This is what must be done with the stuff in order to make the desired product.

This concept of a recipe is quite consistent with the recipes found in cookbooks.

Sometimes the term *recipe* is used to designate only the raw material amounts and other parameters to be used in manufacturing a batch. Although appropriate for some batch processes, this concept is far too restrictive for others. For some products, the differences from one product to the next are largely physical as opposed to chemical. For such products, the processing instructions are especially important. The term *formula* is more appropriate for the raw material amounts and other parameters, with *recipe* designating the formula and the processing instructions.

The above concept of a recipe permits the following three different categories of batch processes to be identified:

1. *Cyclical batch.*   Both the formula and the processing instructions are the same from batch to batch. Batch operations within processes that are primarily continuous often fall into this category. The catalyst regenerator within a reformer unit is a cyclical batch process.

2. *Multigrade.*   The processing instructions are the same from batch to batch, but the formula can be changed to produce modest variations in the product. In a batch PVC plant, the different grades of PVC are manufactured by changing the formula. In a batch pulp digester, the processing of each batch or *cook* is the same, but at the start of each cook, the process operator is permitted to change the formula values for chemical-to-wood ratios, cook time, cook temperature, and so on.

3. *Flexible batch.*   Both the formula and the processing instructions can change from batch to batch. Emulsion polymerization reactors are a good example of a flexible batch facility. The recipe for each product must detail both the raw materials required and how conditions within the reactor must be sequenced in order to make the desired product.

Of these, the flexible batch is by far the most difficult to automate and requires a far more sophisticated control system than either the cyclical batch or the multigrade batch facility.

Batches and Recipes   Each batch of product is manufactured in accordance with a product recipe, which contains all information (formula and processing instructions) required to make a batch of the product (see Fig. 8-56). For each batch of product, there will be one and only one product recipe. However, a given product recipe is nor-

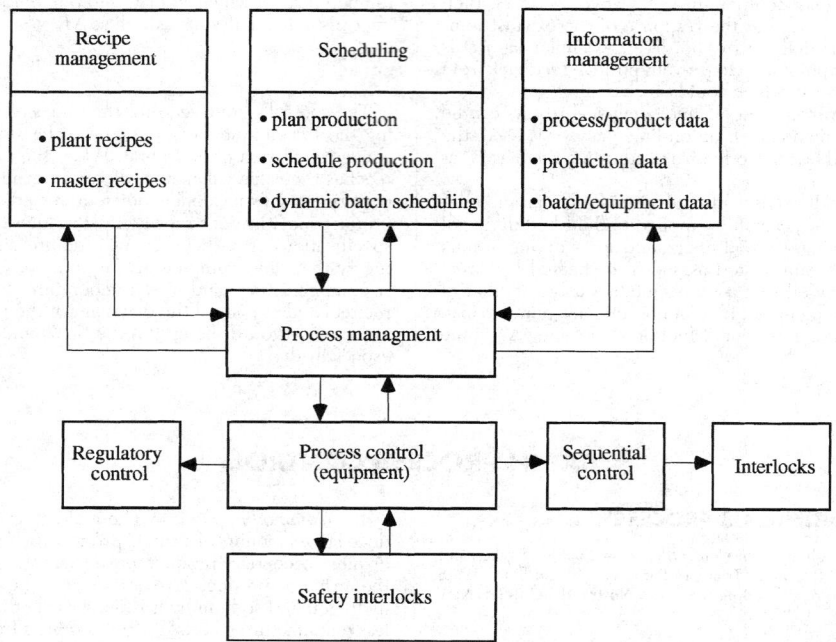

FIG. 8-56   Batch control overview.

mally used to make several batches of product. To uniquely identify a batch of product, each batch is assigned a unique identifier called the batch ID. Most companies adopt a convention for generating the batch ID, but this convention varies from one company to the next.

In most batch facilities, more than one batch of product will be in some stage of production at any given time. The batches in progress may or may not be using the same recipe. The maximum number of batches that can be in progress at any given time is a function of the equipment configuration for the plant.

The existence of multiple batches in progress at a given time presents numerous opportunities for the process operator to make errors, such as charging a material to the wrong batch. Charging a material to the wrong batch is almost always detrimental to the batch to which the material is incorrectly charged. Unless this error is recognized quickly so that the proper charge can be made, the error is also detrimental to the batch to which the charge was supposed to be made. Such errors usually lead to an off-specification batch, but the consequences could be more serious and result in a hazardous condition.

Recipe management refers to the assumption of such duties by the control system. Each batch of product is tracked throughout its production, which may involve multiple processing operations on various pieces of processing equipment. Recipe management assures that all actions specified in the product recipe are performed on each batch of product made in accordance with that recipe. As the batch proceeds from one piece of processing equipment to the next, recipe management is also responsible for assuring that the proper type of process equipment is used and that this processing equipment is not currently in use by another batch.

By assuming such responsibilities, the control system greatly reduces the incidences where operator error results in off-specification batches. Such a reduction in error is essential to implement just-in-time production practices, where each batch of product is manufactured at the last possible moment. When a batch (or batches) are made today for shipment by overnight truck, there is insufficient time for producing another batch to make up for an off-specification batch.

Routing and Production Monitoring   In some facilities, batches are individually scheduled. However, in most facilities, production is scheduled by product runs, where a run is the production of a stated quantity of a given product. From the stated quantity and the standard yield of each batch, the number of batches can be determined. As this is normally more than one batch of product, a production run is normally a sequence of some number of batches of the same product.

In executing a production run, the following issues must be addressed (see Fig. 8-56):

1. *Processing equipment must be dedicated to making the run.* More than one run is normally in progress at a given time. The maximum number of runs simultaneously in progress depends on the equipment configuration of the plant. Routing involves determining which processing equipment will be used for each production run.

2. *Raw material must be utilized.* When a production run is scheduled, the necessary raw materials must be allocated to the production run. As the individual batches proceed, the consumption of raw materials must be monitored for consistency with the allocation of raw materials to the production run.

3. *The production quantity for the run must be achieved by executing the appropriate number of batches.* The number of batches is determined from a standard yield for each batch. However, some batches may achieve yields higher than the standard yield, but other batches may achieve yields lower than the standard yield. The actual yields from each batch must be monitored and significant deviations from the expected yields must be communicated to those responsible for scheduling production.

The last two activities are key components of production monitoring, although production monitoring may also involve other activities such as tracking equipment utilization.

Production Scheduling   In this regard, it is important to distinguish between scheduling runs (sometimes called long-term scheduling) and assigning equipment to runs (sometimes called routing or short-term scheduling). As used herein, production scheduling refers to scheduling runs and is usually a corporate-level as opposed to a plant-level function. Short-term scheduling or routing was previously discussed and is implemented at the plant level.

The long-term scheduling is basically a material resources planning (MRP) activity involving the following:

1. *Forecasting.* Orders for long-delivery raw materials are issued at the corporate level based on the forecast for the demand for products. The current inventory of such raw materials is also maintained at the corporate level. This constitutes the resources from which products can be manufactured.

2. *The orders for products.* Orders are normally received at the corporate level and then assigned to individual plants for production and shipment. Although the scheduling of some products is based on required product inventory levels, scheduling based on orders and shipping directly to the customer (usually referred to as just-in-time) avoids the costs associated with maintaining product inventories.

3. *Plant locations and capacities.* While producing a product at the nearest plant usually lowers transportation costs, plant capacity limitations sometimes dictate otherwise.

Any company competing in the world economy needs the flexibility to accept orders on a worldwide basis and then assign them to individual plants to be filled. Such a function is logically implemented within the corporate-level information technology framework.

BATCH AUTOMATION FUNCTIONS

Automating a batch facility requires a spectrum of functions.

Interlocks   Some of these are provided for safety and are properly called safety interlocks. However, others are provided to avoid mistakes in processing the batch and are properly called process interlocks.

Discrete Device States   Discrete devices such as two-position valves can be driven to either of two possible states. Such devices can be optionally outfitted with limit switches that indicate the state of the device. For two-position valves, the following combinations are possible:

1. No limit switches
2. One limit switch on the closed position
3. One limit switch on the open position
4. Two limit switches

In process-control terminology, the discrete device driver is the software routine that generates the output to a discrete device such as a valve and also monitors the state feedback information to ascertain that the discrete device actually attains the desired state. Given the variety of discrete devices used in batch facilities, this logic must include a variety of capabilities. For example, valves do not instantly change states, but instead each valve exhibits a travel time for the change from one state to another. To accommodate this characteristic of the field device, the processing logic within the discrete device driver must provide for a user-specified transition time for each field device. When equipped with limit switches, the potential states for a valve are as follows:

1. *Open.* The valve has been commanded to open, and the limit switch inputs are consistent with the open state.

2. *Closed.* The valve has been commanded to close, and the limit switch inputs are consistent with the closed state.

3. *Transition.* This is a temporary state that is only possible after the valve has been commanded to change state. The limit switch inputs are not consistent with the commanded state, but the transition time has not expired.

4. *Invalid.* The transition time has expired, and the limit switch inputs are not consistent with the commanded state for the valve.

The invalid state is an abnormal condition that is generally handled in a manner similar to process alarms. The transition state is not considered to be an abnormal state but may be implemented in either of the following ways:

1. *Drive and wait.* Further actions are delayed until the device attains its commanded state.

2. *Drive and proceed.* Further actions are initiated while the device is in the transition state.

The latter is generally necessary for devices with long travel times, such as flush-fitting reactor discharge valves that are motor-driven.

Closing such valves is normally via drive and wait; however, drive and proceed is usually appropriate when opening the valve.

Although two-state devices are most common, the need occasionally arises for devices with three or more states. For example, an agitator may be on high speed, on slow speed, or off.

Process States   Batch processing usually involves imposing the proper sequence of states on the process. For example, a simple blending sequence might be as follows:

1. Transfer specified amount of material from tank *A* to tank *R*. The process state is "Transfer from *A*."
2. Transfer specified amount of material from tank *B* to tank *R*. The process state is "Transfer from *B*."
3. Agitate for specified period of time. The process state is "Agitate without cooling."
4. Cool (with agitation) to specified target temperature. The process state is "Agitate with cooling."

For each process state, the various discrete devices are expected to be in a specified device state. For process state "Transfer from *A*," the device states might be as follows:

1. Tank *A* discharge valve: open.
2. Tank *R* inlet valve: open.
3. Tank *A* transfer pump: running.
4. Tank *R* agitator: off.
5. Tank *R* cooling valve: closed.

For many batch processes, process state representations are a very convenient mechanism for representing the batch logic. A grid or table can be constructed, with the process states as rows and the discrete device states as columns (or vice versa). For each process state, the state of every discrete device is specified to be one of the following:

1. Device state 0, which may be valve closed, agitator off, and so on
2. Device state 1, which may be valve open, agitator on, and so on
3. No change or don't care

This representation is easily understandable by those knowledgeable about the process technology and is a convenient mechanism for conveying the process requirements to the control engineers responsible for implementing the batch logic.

Many batch software packages also recognize process states. A configuration tool is provided to define a process state. With such a mechanism, the batch logic does not need to drive individual devices but can simply command that the desired process state be achieved. The system software then drives the discrete devices to the device states required for the target process state. This normally includes the following:

1. Generating the necessary commands to drive each device to its proper state
2. Monitoring the transition status of each device to determine when all devices have attained their proper states
3. Continuing to monitor the state of each device to assure that the devices remain in their proper states

Should any discrete device not remain in its target state, failure logic must be initiated.

We will use the control of a simple mixing process (Fig. 8-57) to demonstrate various batch control strategies found in commercial systems. To start the operation sequence, a solenoid valve (VN7) is opened to introduce liquid *A*. When the liquid level in the tank reaches an intermediate level (LH2), flow *B* is started to turn on the mixer. When the liquid level is high (LXH2), flow *B* is stopped and the discharge valve is opened (VN9). The discharge valve is closed and the motor stopped when the tank level reaches the low limit (LL2). The operator may start another mixing cycle by depressing the start button again. It should be noted that this simplified control strategy does not deal with emergency process conditions. Timing of equipment sequencing, such as making sure valve 8 is closed before opening the discharge valve, is not considered. However, this example fully demonstrates the device interlocking and signal latching often encountered in sequential process control.

This process is event triggered and can be easily programmed using sequential logic [Figure 8-58*a*]. Many PLC implementations start the programming phase with sequential logic design. Gate 1 ensures that

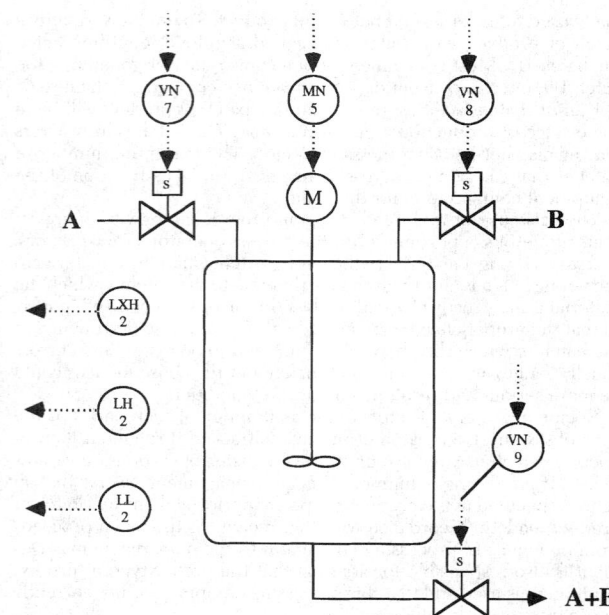

FIG. 8-57   Process schematics of a mixing tank.

the process will not start, when requested, if the tank level is not low. Gate 3 opens valve 7 for flow *A* only if valve 8 is not opened. Gate 2 latches the operator request once valve 7 is opened such that the operator may release the push button. Gate 4 starts flow *B* and the mixer motor when the intermediate level is reached. The start signal is fed into gate 3 to terminate flow *A*. At the high tank level, gate 6 opens the discharge valve. This signal is fed into gate 4 to stop flow *B* and the mixer motor. Gate 5 latches in the discharge signal until the tank is drained. Note that for a DCS, this sequential logic can be entered entirely as Boolean functional blocks.

Figure 8-58*b* is the ladder logic diagram for the same mixing process. It involves rungs of parallel circuits containing relays (the circles) and contacts. Parallel bars on the rungs represent contacts. A slashed pair of bars depict a normally closed contact. A normally open momentary contact is shown on rung 1 in Fig. 8-58*b*. The ladder logic and diagram builder in PLCs can be programmed easily because there are only a limited number of symbols required in ladder logic diagrams.

The translation from sequential logic to ladder logic is straightforward. In general, two or more contacts on the same rung forms an AND gate. Contacts on branches of a rung form an OR gate. For example, contact C1 on rung 1 is normally open, unless the tank level is low. Contact CR8 is normally closed unless relay CR8 on rung 2 is energized. An operator-actuated push button, HS4, and the contact C1 forms an AND gate equivalent to gate 1 in Fig. 8-58*a*. Therefore, when the operator depresses the push button when the tank level is low, relay CR7 is energized, which closes contact CR7 on branch rung 1A. Once contact CR7 is latched in, the operator may release the button. The junction connecting rungs 1 and 1A is equivalent to the output of the OR gate 2 in Fig. 8-58*a*.

Regulatory Control   For most batch processes, the discrete logic requirements overshadow the continuous control requirements. For many batch processes, the continuous control can be provided by simple loops for flow, pressure, level, and temperature. However, very sophisticated advanced control techniques are occasionally applied. As temperature control is especially critical in reactors, the simple feedback approach is replaced by model-based strategies that rival if not exceed the sophistication of advanced control loops in continuous plants.

In some installations, alternative approaches for regulatory control

Sequential logic diagram.

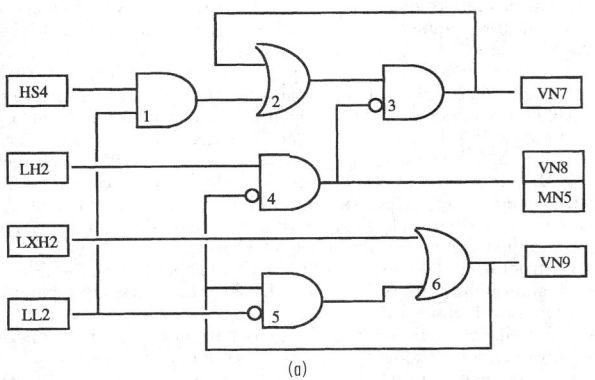

(a)

Ladder logic diagram.

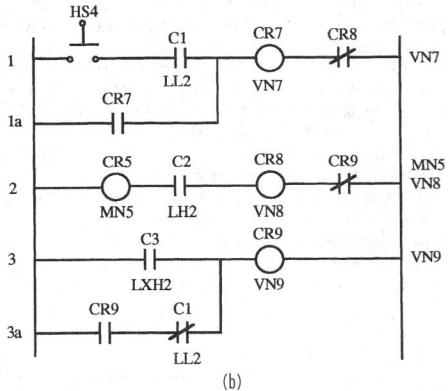

(b)

FIG. 8-58 Logic diagrams for the control of the mixing tank.

may be required. Where a variety of products are manufactured, the reactor may be equipped with alternative heat-removal capabilities, including the following:

1. Jacket filled with cooling water. Most such jackets are once-through, but some are recirculating.
2. Heat exchanger in a pump-around loop.
3. Reflux condenser.

The heat removal capability to be used usually depends on the product being manufactured. Therefore, regulatory loops must be configured for each possible option, and sometimes for certain combinations of the possible options. These loops are enabled and disabled depending on the product being manufactured.

The interface between continuous controls and discrete controls is also important. For example, a feed might be metered into a reactor at a variable rate, depending on another feed or possibly on reactor temperature. However, the product recipe calls for a specified quantity of this feed. The flow must be totalized (i.e., integrated), and when the flow total attains a specified value, the feed must be terminated.

The discrete logic must have access to operational parameters such as controller modes. That is, the discrete logic must be able to switch a controller to manual, auto, or cascade. Furthermore, the discrete logic must be able to force the controller output to a specified value.

Sequence Logic Sequence logic must not be confused with discrete logic. Discrete logic is especially suitable for interlocks or per-

missives, e.g., the reactor discharge valve must be closed in order for the feed valve to be opened. Sequence logic is used to force the process to attain the proper sequence of states. For example, a feed preparation might be to first charge A, then charge B, then mix, and finally cool. Although discrete logic can be used to implement sequence logic, other alternatives are often more attractive.

Sequence logic is often, but not necessarily, coupled with the concept of a process state. Basically, the sequence logic determines when the process should proceed from the current state to the next, and sometimes what the next state should be.

Sequence logic must encompass both normal and abnormal process operations. Thus, sequence logic is often viewed as consisting of two distinct but related parts:

1. *Normal logic.* This sequence logic provides for the normal or expected progression from one process state to another.
2. *Failure logic.* This logic provides for responding to abnormal conditions, such as equipment failures.

Of these, the failure logic can easily be the most demanding. The simplest approach is to stop or hold on any abnormal condition, and let the process operator sort things out. However, this is not always acceptable. Some failures lead to hazardous conditions that require immediate action; waiting for the operator to decide what to do is not acceptable. The appropriate response to such situations is best determined in conjunction with the process hazards analysis.

No single approach has evolved as the preferred way to implement sequence logic. The approaches utilized include the following:

1. *Discrete logic.* Sequence logic can be implemented via ladder logic, and this approach is common when sequence logic is implemented in programmable logic controllers (PLCs).
2. *Programming languages.* Traditional procedural languages do not provide the necessary constructs for implementing sequence logic. This necessitates one of the following:

a. Special languages. The necessary extensions for sequence logic are provided by extending the syntax of the programming language. This is the most common approach within distributed control systems (DCSs). The early implementations used BASIC as the starting point for the extensions; the later implementations used C as the starting point. A major problem with this approach is portability, especially from one manufacturer to the next but sometimes from one product version to the next within the same manufacturer's product line.

b. Subroutine or function libraries. The facilities for sequence logic are provided via subroutines or functions that can be referenced from programs written in FORTRAN or C. This requires a general-purpose program development environment and excellent facilities to trap the inevitable errors in such programs. Operating systems with such capabilities have long been available on the larger computers, but not for the microprocessors utilized within DCS systems. However, such operating systems are becoming more common within DCS systems.

3. *State machines.* This technology is commonly applied within the discrete manufacturing industries. However, its migration to process batch applications has been limited.
4. *Graphical implementations.* For sequence logic, the flowchart traditionally used to represent the logic of computer programs must be extended to provide parallel execution paths. Such extensions have been implemented in a graphical representation called Grafcet. As process engineers have demonstrated a strong dislike for ladder logic, PLC manufacturers are considering providing Grafcet either in addition to or as an alternative to ladder logic.

As none of the above have been able to dominate the industry, it is quite possible that future developments will provide a superior approach for implementing sequence logic.

BATCH PRODUCTION FACILITIES

Especially for flexible batch applications, the batch logic must be properly structured in order to be implemented and maintained in a reasonable manner. An underlying requirement is that the batch process equipment be properly structured. The following structure is appropriate for most batch production facilities.

Plant A plant is the collection of production facilities at a geographical site. The production facilities at a site normally share warehousing, utilities, and the like.

Equipment Suite An equipment suite is the collection of equipment available for producing a group of products. Normally, this group of products is similar in certain respects. For example, they might all be manufactured from the same major raw materials. Within the equipment suite, material transfer and metering capabilities are available for these raw materials. The equipment suite contains all of the necessary types of processing equipment (reactors, separators, and so on) required to convert the raw materials into salable products. A plant may consist of only one suite of equipment, but large plants usually contain multiple equipment suites.

Process Unit or Batch Unit A process unit is a collection of processing equipment that can, at least at certain times, be operated in a manner completely independent from the remainder of the plant. A process unit normally provides a specific function in the production of a batch of product. For example, a process unit might be a reactor complete with all associated equipment (jacket, recirculation pump, reflux condenser, and so on). However, each feed preparation tank is usually a separate process unit. With this separation, preparation of the feed for the next batch can be started as soon as the feed tank is emptied for the current batch.

All but the very simplest equipment suites contain multiple process units. The minimum number of process units is one for each type of processing equipment required to make a batch of product. However, many equipment suites contain multiple process units of each type. In such equipment suites, multiple batches and multiple production runs can be in progress at a given time.

Item of Equipment An item of equipment is a hardware item that performs a specific purpose. Examples are pumps, heat exchangers, agitators, and the like. A process unit could consist of a single item of equipment, but most process units consist of several items of equipment that must be operated in harmony in order to achieve the function expected of the process unit.

Device A device is the smallest element of interest to batch logic. Examples of devices include measurement devices and actuators.

STRUCTURED BATCH LOGIC

Flexible batch applications must be pursued using a structured approach to batch logic. In such applications, the same processing equipment is used to make a variety of products. In most facilities, little or no proprietary technology is associated with the equipment itself; the proprietary technology is how this equipment is used to produce each of the products.

The primary objective of the structured approach is to separate cleanly the following two aspects of the batch logic:

Product Technology Basically, this encompasses the product technology, such as how to mix certain molecules to make other molecules. This technology ultimately determines the chemical and physical properties of the final product. The product recipe is the principal source for the product technology.

Process Technology The process equipment permits certain processing operations (e.g., heat to a specified temperature) to be undertaken. Each processing operation will involve certain actions (e.g., opening appropriate valves).

The need to keep these two aspects separated is best illustrated by a situation where the same product is to be made at different plants. While it is possible that the processing equipment at the two plants is identical, this is rarely the case. Suppose one plant uses steam for heating its vessels, but the other uses a hot oil system as the source of heat. When a product recipe requires that material is to be heated to a specified temperature, each plant can accomplish this objective, but will go about it in quite different ways.

The ideal case for a product recipe is as follows:
1. Contains all of the product technology required to make a product
2. Contains no equipment-dependent information, that is, no process technology

In the previous example, such a recipe would simply state that the product must be heated to a specified temperature. Whether heating is undertaken with steam or hot oil is irrelevant to the product technology. By restricting the product recipe to a given product technology, the same product recipe can be used to make products at different sites. Timing diagrams (such as Fig. 8-59) are one way to represent a recipe.

At a given site, the specific approach to be used to heat a vessel is important. The traditional approach is for an engineer at each site to expand the product recipe into a document that explains in detail how the product is to be made at the site. This document goes by various names, although standard operating procedure or SOP is a common one. Depending on the level of detail to which it is written, the SOP could specify exactly which valves must be opened in order to heat the contents of a vessel. Thus, the SOP is site-dependent, and contains both product technology and process technology.

In structuring the logic for a flexible batch application, the following organization permits product technology to be cleanly separated from process technology:
- A recipe consists of a formula and one or more processing operations. Ideally, only product technology is contained in a recipe.
- A processing operation consists of one or more phases. Ideally, only product technology is contained in a processing operation.
- A phase consists of one or more actions. Ideally, only process technology is contained in a phase.

In this structure, the recipe and processing operations would be the same at each site that manufactures the product. However, the logic that comprises each phase would be specific to a given site. Using the heating example from above, each site would require a phase to heat the contents of the vessel. However, the logic within the phase at one site would accomplish the heating by opening the appropriate steam valves, while the logic at the other site would accomplish the heating by opening the appropriate hot oil valves.

Usually the critical part of structuring batch logic is the definition of the phases. There are two ways to approach this:
1. Examine the recipes for the current products for commonality, and structure the phases to reflect this commonality.
2. Examine the processing equipment to determine what processing capabilities are possible, and write phases to accomplish each possible processing capability.

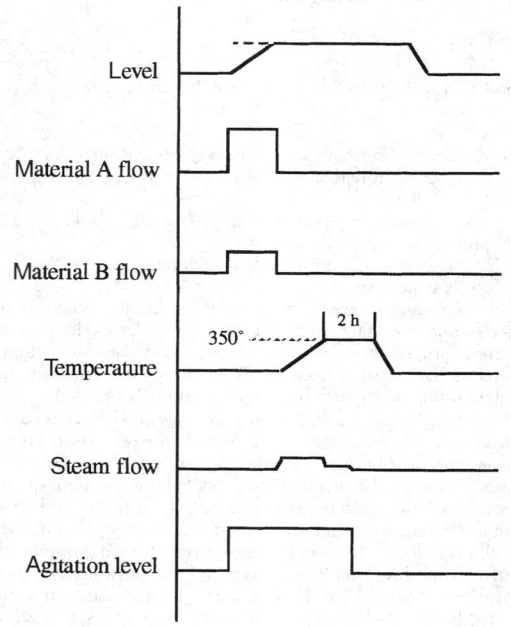

FIG. 8-59 Sample process timing diagram.

There is the additional philosophical issue of whether to have a large number of simple phases with few options each, or a small number of complex phases with numerous options. The issues are a little different from structuring a complex computer program into subprograms. Each possible alternative will have advantages and disadvantages.

As the phase contains no product technology, the implementation of a phase must be undertaken by those familiar with the process equipment. Furthermore, they should undertake this on the basis that the result will be used to make a variety of products, not just those that are initially contemplated. The development of the phase logic must also encompass all equipment-related safety issues. The phase should accomplish a clearly defined objective, so the implementers should be able to thoroughly consider all relevant issues in accomplishing this objective. The phase logic is defined in detail, implemented in the control system, and then thoroughly tested. Except when the processing equipment is modified, future modifications to the phase should be infrequent. The result should be a very dependable module that can serve as a building block for batch logic.

Even for flexible batch applications, a comprehensive menu of phases should permit most new products to be implemented using currently existing phases. By reusing exising phases, numerous advantages accrue:

1. The engineering effort to introduce a new recipe at a site is reduced.
2. The product is more likely to be on-spec the first time, thus avoiding the need to dispose of off-spec product.
3. The new product can be supplied to customers sooner, hopefully before competitors can supply the product.

There is also a distinct advantage in maintenance. When a problem with a phase is discovered and the phase logic is corrected, the correction is effectively implemented in all recipes that use the phase. If a change is implemented in the processing equipment, the affected phases must be modified accordingly and then thoroughly tested. These modifications are also effectively implemented in all recipes that use these phases.

PROCESS MEASUREMENTS

General References: Benedict, *Fundamentals of Temperature, Pressure, and Flow Measurements,* Wiley, New York, 1969. Considine, *Process Instruments and Control Handbook,* McGraw-Hill, New York, 1993. Considine and Ross, *Handbook of Applied Instrumentation,* McGraw-Hill, New York, 1964. Doebelin, *Measurement Systems: Application and Design,* 4th ed. McGraw-Hill, New York, 1990. Ginesi and Annarummo, "User Tips for Mass, Volume Flowmeters," *Tech,* 41, April, 1994. *ISA Transducer Compendium,* 2d ed., Plenum, New York, 1969. Liptak, *Instrument Engineers Handbook,* Chilton, Philadelphia, 1995. Michalski, Eckersdorf, and McGhee, *Temperature Measurement,* Wiley, Chichester, 1991. Nichols, G. D., *On-Line Process Analyzers,* Wiley, New York, 1988.

GENERAL CONSIDERATIONS

Process measurements encompass the application of the principles of metrology to the process in question. The objective is to obtain values for the current conditions within the process and make this information available in a form usable by either the control system, process operators, or any other entity that needs to know. The term "measured variable" or "process variable" designates the process condition that is being determined.

Process measurements fall into two categories:

1. *Continuous measurements.* An example of a continuous measurement is a level measurement device that determines the liquid level in a tank (in meters).
2. *Discrete measurements.* An example of a discrete measurement is a level switch that indicates the presence or absence of liquid at the location at which the level switch is installed.

In continuous processes, most process control applications rely on continuous measurements. In batch processes, many of the process control applications will utilize discrete as well as continuous measurements. In both types of processes, the safety interlocks and process interlocks rely largely on discrete measurements.

Continuous Measurements In most applications, continuous measurements are considerably more ambitious than discrete measurements. Basically, discrete measurements involve a yes/no decision, whereas continuous measurements may entail considerable signal processing.

The components of a typical continuous measurement device are as follows:

• *Sensor.* This component produces a signal that is related in a known manner to the process variable of interest. The sensors in use today are primarily of the electrical analog variety, and the signal is in the form of a voltage, a resistance, a capacitance, or some other directly measurable electrical quantity. Prior to the mid 1970s, instruments tended to use sensors whose signal was mechanical in nature, and thus compatible with pneumatic technology. Since that time the fraction of sensors that are digital in nature has grown considerably, often eliminating the need for analog-to-digital conversion.

• *Signal processing.* The signal from most sensors is related in a nonlinear fashion to the process variable of interest. In order for the output of the measurement device to be linear with respect to the process variable of interest, linearization is required. Furthermore, the signal from the sensor might be affected by variables other than the process variable. In this case, additional variables must be sensed and the signal from the sensor compensated to account for the other variables. For example, reference junction compensation is required for thermocouples (except when used for differential temperature measurements).

• *Transmitter.* The measurement device output must be a signal that can be transmitted over some distance. Where electronic analog transmission is used, the low range on the transmitter output is 4 milliamps, and the upper range is 20 milliamps. Microprocessor-based transmitters (often referred to as smart transmitters) are usually capable of transmitting the measured variable digitally in engineering units.

Accuracy and Repeatability Definitions of terminology pertaining to process measurements can be obtained from standard S51.1 from the International Society of Measurment and Control (ISA) and standard RC20-11 from the Scientific Apparatus Manufacturers Association (SAMA), both of which are updated periodically. An appreciation of accuracy and repeatability is especially important. Some applications depend on the accuracy of the instrument, but other applications depend on repeatability. Excellent accuracy implies excellent repeatability; however, an instrument can have poor accuracy but excellent repeatability. In some applications, this is acceptable, as discussed below.

Range and Span A continuous measurement device is expected to provide credible values of the measured value between a lower range and an upper range. The difference between the upper range and the lower range is the span of the measurement device. The maximum value for the upper range and the minimum value for the lower range depend on the principles on which the measurement device is based and on the design chosen by the manufacturer of the measurement device. If the measured variable is greater than the upper range or less than the lower range, the measured variable is said to be out-of-range or the measurement device is said to be overranged.

Accuracy Accuracy refers to the difference between the measured value and the true value of the measured variable. Unfortunately, the true value is never known, so in practice accuracy refers to the difference between the measured value and an accepted standard value for the measured variable.

Accuracy can be expressed in a number of ways:
1. As an absolute difference in the units of the measured variable
2. As a percent of the current reading
3. As a percent of the span of the measured variable
4. As a percent of the upper range of the span

For process measurements, accuracy as a percent of span is the most common.

Manufacturers of measurement devices always state the accuracy of the instrument. However, these statements always specify specific or reference conditions at which the measurement device will perform with the stated accuracy, with temperature and pressure most often appearing in the reference conditions. When the measurement device is applied at other conditions, the accuracy is affected. Manufacturers usually also provide some statements on how accuracy is affected when the conditions of use deviate from the referenced conditions in the statement of accuracy. Although appropriate calibration procedures can minimize some of these effects, rarely can they be totally eliminated. It is easily possible for such effects to cause a measurement device with a stated accuracy of 0.25 percent of span at reference conditions to ultimately provide measured values with accuracies of 1 percent or less. Microprocessor-based measurement devices usually provide better accuracy than the traditional electronic measurement devices.

In practice, most attention is given to accuracy when the measured variable is the basis for billing, such as in custody transfer applications. However, whenever a measurement device provides data to any type of optimization strategy, accuracy is very important.

Repeatability Repeatability refers to the difference between the measurements when the process conditions are the same. This can also be viewed from the opposite perspective. If the measured values are the same, repeatability refers to the difference between the process conditions.

For regulatory control, repeatability is of major interest. The basic objective of regulatory control is to maintain uniform process operation. Suppose that on two different occasions, it is desired that the temperature in a vessel be 80°C. The regulatory control system takes appropriate actions to bring the measured variable to 80°C. The difference between the process conditions at these two times is determined by the repeatability of the measurement device.

In the use of temperature measurement for control of the separation in a distillation column, repeatability is crucial but accuracy is not. Composition control for the overhead product would be based on a measurement of the temperature on one of the trays in the rectifying section. A target would be provided for this temperature. However, at periodic intervals, a sample of the overhead product is analyzed in the laboratory and the information provided to the process operator. Should this analysis be outside acceptable limits, the operator would adjust the set point for the temperature. This procedure effectively compensates for an inaccurate temperature measurement; however, the success of this approach requires good repeatability from the temperature measurement.

Dynamics of Process Measurements Especially where the measurement device is incorporated into a closed loop control configuration, dynamics are important. The dynamic characteristics depend on the nature of the measurement device, and also on the nature of components associated with the measurement device (for example, thermowells and sample conditioning equipment). The term *measurement system* designates the measurement device and its associated components.

The following dynamics are commonly exhibited by measurement systems:
• *Time constants.* Where there is a capacity and a throughput, the measurement device will exhibit a time constant. For example, any temperature measurement device has a thermal capacity (mass times heat capacity) and a heat flow term (heat transfer coefficient and area). Both the temperature measurement device and its associated thermowell will exhibit behavior typical of time constants.
• *Dead time.* Probably the best example of a measurement device that exhibits pure dead time is the chromatograph, because the analysis is not available for some time after a sample is injected. Additional dead time results from the transportation lag within the sample

system. Even continuous analyzer installations are plagued by dead time from the sample system.
• *Underdamped behavior.* Measurement devices with mechanical components often have a natural harmonic and can exhibit underdamped behavior. The displacer type of level measurement device is capable of such behavior.

While the manufacturers of measurement devices can supply some information on the dynamic characteristics of their devices, interpretation is often difficult. Measurement device dynamics are quoted on varying bases, such as rise time, time to 63 percent response, settling time, and so on. Even where the time to 63 percent response is quoted, it might not be safe to assume that the measurement device exhibits first-order behavior.

Where the manufacturer of the measurement device does not supply the associated equipment (thermowells, sample conditioning equipment, and the like), the user must incorporate the characteristics of these components to obtain the dynamics of the measurement system.

An additional complication is that most dynamic data are stated for configurations involving reference materials such as water, air, and so on. The nature of the process material will affect the dynamic characteristics. For example, a thermowell will exhibit different characteristics when immersed in a viscous organic emulsion than when immersed in water. It is often difficult to extrapolate the available data to process conditions of interest.

Similarly, it is often impossible, or at least very difficult, to experimentally determine the characteristics of a measurement system under the conditions where it is used. It is certainly possible to fill an emulsion polymerization reactor with water and determine the dynamic characteristics of the temperature measurement system. However, it is not possible to determine these characteristics when the reactor is filled with the emulsion under polymerization conditions.

The primary impact of unfavorable measurement dynamics is on the performance of closed loop control systems. This explains why most control engineers are very concerned with measurement dynamics. The goal to improve the dynamic characteristics of measurement devices is made difficult because the discussion regarding measurement dynamics is often subjective.

Selection Criteria The selection of a measurement device entails a number of considerations given below, some of which are almost entirely subjective.
1. *Measurement span.* The measurement span required for the measured variable must lie entirely within the instrument's envelope of performance.
2. *Performance.* Depending on the application, accuracy, repeatability, or perhaps some other measure of performance is appropriate. Where closed loop control is contemplated, speed of response must be included.
3. *Reliability.* Data available from the manufacturers can be expressed in various ways and at various reference conditions. Often, previous experience with the measurement device within the purchaser's organization is weighted most heavily.
4. *Materials of construction.* The instrument must withstand the process conditions to which it is exposed. This encompasses considerations such as operating temperatures, operating pressures, corrosion, and abrasion. For some applications, seals or purges may be necessary.
5. *Prior use.* For the first installation of a specific measurement device at a site, training of maintenance personnel and purchases of spare parts might be necessary.
6. *Potential for releasing process materials to the environment.* Fugitive emissions are receiving ever increasing attention. Exposure considerations, both immediate and long term, for maintenance personnel are especially important when the process fluid is either corrosive or toxic.
7. *Electrical classification.* Article 500 of the National Electric Code provides for the classification of the hazardous nature of the process area in which the measurement device will be installed. If the measurement device is not inherently compatible with this classification, suitable enclosures must be purchased and included in the installation costs.

8. *Physical access.* Subsequent to installation, maintenance personnel must have physical access to the measurement device for maintenance and calibration. If additional structural facilities are required, they must be included in the installation costs.

9. *Cost.* There are two aspects of the cost:

a. Initial purchase and installation (capital costs).

b. Recurring costs (operational expense). This encompasses instrument maintenance, instrument calibration, consumables (for example, titrating solutions must be purchased for automatic titrators), and any other costs entailed in keeping the measurement device in service.

Calibration Calibration entails the adjustment of a measurement device so that the value from the measurement device agrees with the value from a standard. The International Standards Organization (ISO) has developed a number of standards specifically directed to calibration of measurement devices. Furthermore, compliance with the ISO 9000 standards requires that the working standard used to calibrate a measurement device must be traceable to an internationally recognized standard such as those maintained by the National Institute of Standards and Technology (NIST).

Within most companies, the responsibility for calibrating measurement devices is delegated to a specific department. Often, this department may also be responsible for maintaining the measurement device. The specific calibration procedures depend on the type of measurement device. The frequency of calibration is normally predetermined, but earlier action may be dictated if the values from the measurement device become suspect.

Calibration of some measurement devices involves comparing the measured value with the value from the working standard. Pressure and differential pressure transmitters are calibrated in this manner. Calibration of analyzers normally involves using the measurement device to analyze a specially prepared sample whose composition is known. These and similar approaches can be applied to most measurement devices.

Flow is an important measurement whose calibration presents some challenges. When a flow measurement device is used in applications such as custody transfer, provision is made to pass a known flow through the meter. However, such a provision is costly and is not available for most in-process flowmeters. Without such a provision, a true calibration of the flow element itself is not possible. For orifice meters, calibration of the flowmeter normally involves calibration of the differential pressure transmitter, and the orifice plate is usually only inspected for deformation, abrasion, and so on. Similarly, calibration of a magnetic flowmeter normally involves calibration of the voltage measurement circuitry, which is analogous to calibration of the differential pressure transmitter for an orifice meter.

TEMPERATURE MEASUREMENTS

Measurement of the hotness or coldness of a body or fluid is commonplace in the process industries. Temperature-measuring devices utilize systems with properties that vary with temperature in a simple, reproducible manner and thus can be calibrated against known references (sometimes called *secondary thermometers*). The three dominant measurement devices used in automatic control are thermocouples, resistance thermometers, and pyrometers and are applicable over different temperature regimes.

Thermocouples Temperature measurements using thermocouples are based on the discovery by Seebeck in 1821 that an electric current flows in a continuous circuit of two different metallic wires if the two junctions are at different temperatures. The thermocouple may be represented diagrammatically as shown in Fig. 8-60. A and B are the two metals, and T_1 and T_2 are the temperatures of the junctions. Let T_1 and T_2 be the reference junction (cold junction) and the measuring junction, respectively. If the thermoelectric current i flows in the direction indicated in Fig. 8-60, metal A is customarily referred to as thermoelectrically positive to metal B. Metal pairs used for thermocouples include platinum-rhodium (the most popular and accurate), chromel-alumel, copper-constantan, and iron-constantan. The thermal emf is a measure of the difference in temperature between T_2 and T_1. In control systems the reference junction is usually located at

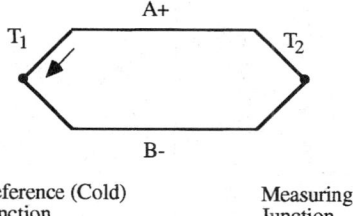

FIG. 8-60 Basic circuit of Seebeck effect.

the emf-measuring device. The reference junction may be held at constant temperature such as in an ice bath or a thermostated oven, or it may be at ambient temperature but electrically compensated (cold-junction-compensated circuit) so that it appears to be held at a constant temperature.

Resistance Thermometers The resistance thermometer depends upon the inherent characteristics of materials to change in electrical resistance when they undergo a change in temperature. Industrial resistance thermometers are usually constructed of platinum, copper, or nickel, and more recently semiconducting materials such as thermistors are being used. Basically, a resistance thermometer is an instrument for measuring electrical resistance that is calibrated in units of temperature instead of in units of resistance (typically ohms). Several common forms of bridge circuits are employed in industrial resistance thermometry, the most common being the Wheatstone bridge. A resistance thermometer detector (RTD) consists of a resistance conductor (metal), which generally shows an increase in resistance with temperature. The following equation represents the variation of resistance with temperature (°C):

$$R_T = R_0(1 + a_1T + a_2T^2 + \cdots + a_nT^n)$$

$$R_0 = \text{resistance at } 0°C \tag{8-88}$$

The temperature coefficient of resistance α_T is expressed as:

$$\alpha_T = \frac{1}{R_T}\frac{dR_T}{dT} \tag{8-89}$$

For most metals, α_T is positive. For many pure metals, the coefficient is essentially constant and stable over large portions of their useful range. Typical resistance versus temperature curves for platinum, copper, and nickel are given in Fig. 8-61, with platinum usually the metal of choice. Platinum has a useful range of –200°C to 800°C, while Nickel (–80°C to 320°C) and copper (–100°C to 100°C) are more limited. Detailed resistance versus temperature tables are available from the National Bureau of Standards and suppliers of resistance thermometers. Table 8-7 gives recommended temperature measurement ranges for thermocouples and RTDs. Resistance thermometers are receiving increased usage because they are about ten times more accurate than thermocouples.

Thermistors Thermistors are nonlinear temperature-dependent resistors, and normally only the materials with negative temperature

TABLE 8-7 Recommended Temperature Measurement Ranges for RTDs and Thermocouples

Resistance thermometer detectors (RTDs)	
100V Pt	–200°C–+850°C
120V Ni	–80°C–+320°C
Thermocouples	
Type B	700°C–+1820°C
Type E	–175°C–+1000°C
Type J	–185°C–+1200°C
Type K	–175°C–+1372°C
Type N	0°C–+1300°C
Type R	125°C–+1768°C
Type S	150°C–+1768°C
Type T	–170°C–+400°C

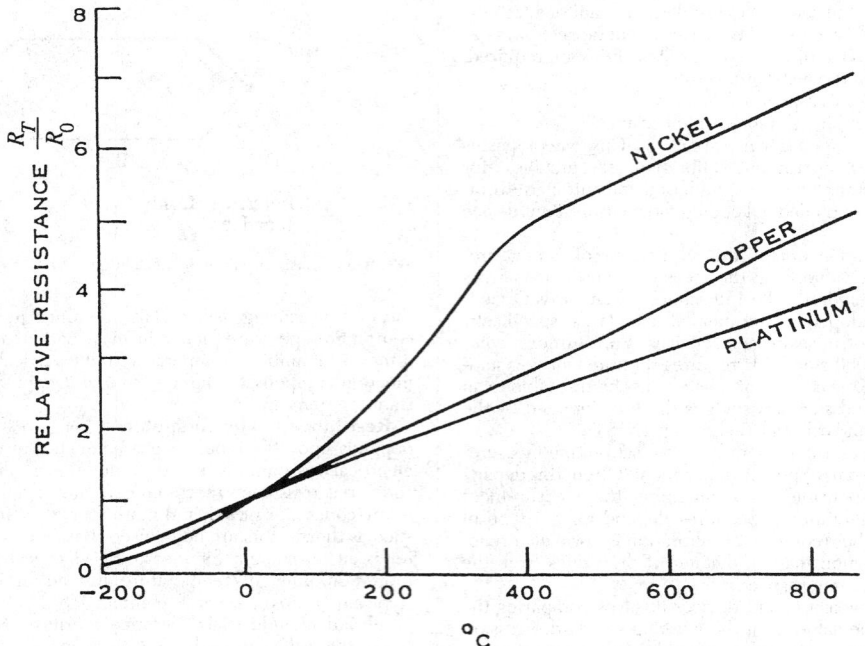

FIG. 8-61 Typical resistance-thermometer curves for platinum, copper, and nickel wire, where R_T = resistance at temperature T and R_0 = resistance at 0°C.

coefficient of resistance (NTC type) are used. The resistance is related to temperature as:

$$R_T = R_{T_r} \exp\left[\beta\left(\frac{1}{T} - \frac{1}{T_r}\right)\right] \qquad (8\text{-}90)$$

where T_r is a reference temperature, which is generally 298 K. Thus

$$\alpha_T = \frac{1}{R_T}\frac{dR_T}{dT} \qquad (8\text{-}91)$$

The value of β is of the order of 4000, so at room temperature (298 K), $\alpha_T = -0.045$ for thermistor and 0.0035 for 100 Ω Platinum RTD. Compared with RTDs, NTC type thermistors are advantageous in that the detector dimension can be made small, the resistance value is higher (less affected by the resistances of the connecting leads), the temperature sensitivity is higher, and the thermal inertia of the sensor is low. Disadvantages of thermistors to RTDs include nonlinear characteristics and low measuring temperature range.

Filled-System Thermometers The filled-system thermometer is designed to provide an indication of temperature some distance removed from the point of measurement. The measuring element (bulb) contains a gas or liquid that changes in volume, pressure, or vapor pressure with temperature. This change is communicated through a capillary tube to a Bourdon tube or other pressure- or volume-sensitive device. The Bourdon tube responds so as to provide a motion related to the bulb temperature. Those systems that respond to volume changes are completely filled with a liquid. Systems that respond to pressure changes either are filled with a gas or are partially filled with a volatile liquid. Changes in gas or vapor pressure with changes in bulb temperatures are carried through the capillary to the Bourdon. The latter bulbs are sometimes constructed so that the capillary is filled with a nonvolatile liquid.

Fluid-filled bulbs deliver enough power to drive controller mechanisms and even directly actuate control valves. These devices are characterized by large thermal capacity, which sometimes leads to slow response, particularly when they are enclosed in a thermal well for process measurements. Filled-system thermometers are used extensively in industrial processes for a number of reasons. The simplicity

of these devices allows rugged construction, minimizing the possibility of failure with a low level of maintenance, and inexpensive overall design of control equipment. In case of system failure, the entire unit must be replaced or repaired.

As normally used in the process industries, the sensitivity and percentage of span accuracy of these thermometers are generally the equal of those of other temperature-measuring instruments. Sensitivity and absolute accuracy are not the equal of those of short-span electrical instruments used in connection with resistance-thermometer bulbs. Also, the maximum temperature is somewhat limited.

Bimetal Thermometers Thermostatic bimetal can be defined as a composite material made up of strips of two or more metals fastened together. This composite, because of the different expansion rates of its components, tends to change curvature when subjected to a change in temperature. With one end of a straight strip fixed, the other end deflects in proportion to the temperature change, the square of the length, and inversely as the thickness, throughout the linear portion of the deflection characteristic curve. If a bimetallic strip is wound into a helix or a spiral and one end is fixed, the other end will rotate when heat is applied. For a thermometer with uniform scale divisions, a bimetal must be designed to have linear deflection over the desired temperature range. Bimetal thermometers are used at temperatures ranging from 580°C down to −180°C and lower. However, at the low temperatures the rate of deflection drops off quite rapidly. Bimetal thermometers do not have long-time stability at temperatures above 430°C.

Pyrometers Planck's distribution law gives the radiated energy flux $q_b(\lambda, T)d\lambda$ in the wavelength range λ to $\lambda + d\lambda$ from a black surface:

$$q_b(\lambda, T) = \frac{C_1}{\lambda^5}\frac{1}{e^{c_2/\lambda T} - 1} \qquad (8\text{-}92)$$

where $C_1 = 3.7418 \times 10^{10}$ μW μm^4 cm^{-2}, and $C_2 = 14{,}388$ μm K.

If the target object is a black body and if the pyrometer has a detector that measures the specific wavelength signal from the object, the temperature of the object can be exactly estimated from Eq. (8-92). While it is possible to construct a physical body that closely approximates black body behavior, most real-world objects are not black bodies. The deviation from a black body can be described by the spectral emissivity

$$\varepsilon_T = \frac{q(T)}{q_b(T)} \qquad (8\text{-}93)$$

where $q(\lambda, T)$ is the radiated energy flux from a real body in the wavelength range λ to $\lambda + d\lambda$ and $0 < \varepsilon_{\lambda,T} < 1$. Integrating Eq. (8-92) over all wavelengths gives the Stefan-Boltzmann equation

$$q_b(T) = \int_0^\infty q_b(\lambda, T)\, d\lambda$$
$$= \sigma T^4 \qquad (8\text{-}94)$$

where σ is the Stefan-Boltzmann constant. Similar to Eq. (8-93), the emissivity ε_T for the total radiation is

$$\varepsilon_T = \frac{q(T)}{q_b(T)} \qquad (8\text{-}95)$$

where $q(T)$ is the radiated energy flux from a real body with emissivity ε_T.

Total Radiation Pyrometers   In total radiation pyrometers, the thermal radiation is detected over a large range of wavelengths from the object at high temperature. The detector is normally a thermopile, which is built by connecting several thermocouples in series to increase the temperature measurement range. The pyrometer is calibrated for black bodies, so the indicated temperature T_p should be converted for non-black body temperature.

Photoelectric Pyrometers   Photoelectric pyrometers belong to the class of band radiation pyrometers. The thermal inertia of thermal radiation detectors does not permit the measurement of rapidly changing temperatures. For example, the smallest time constant of a thermal detector is about 1 msec, while the smallest time constant of a photoelectric detector can be about 1 or 2 sec. Photoelectric pyrometers may use photoconductors, photodiodes, photovoltaic cells, or vacuum photocells. Photoconductors are built from glass plates with thin film coatings of 1 μm thickness, using PbS, CdS, PbSe or PbTe. When the incident radiation has the same wavelength as the materials are able to absorb, the captured incident photons free photoelectrons, which form an electric current. Photodiodes in germanium or silicon are operated with a reverse bias voltage applied. Under the influence of the incident radiation their conductivity as well as their reverse saturation current is proportional to the intensity of the radiation within the spectral response band from 0.4 to 1.7 μm for Ge and 0.6 to 1.1 μm for Si. Because of the above characteristics, the operating range of a photoelectric pyrometer can be either spectral or in a specific band. Photoelectric pyrometers can be applied for a specific choice of the wavelength.

Disappearing Filament Pyrometers   Disappearing filament pyrometers can be classified as spectral pyrometers. The brightness of a lamp filament is changed by adjusting the lamp current until the filament disappears against the background of the target, at which point the temperature is measured. Since the detector is the human eye, it is difficult to calibrate for on-line measurements.

Ratio Pyrometers   The ratio pyrometer is also called the two-color pyrometer. Two different wavelengths are utilized for detecting the radiated signal. If one uses Wien's law for small values of λT, the detected signals from spectral radiant energy flux emitted at the wavelengths λ_1 and λ_2 with emissivities ε_{λ_1} and ε_{λ_2} are

$$S_{\lambda_1} = K C_1 \varepsilon_{\lambda_1} \lambda_1^{-5} \exp^{-C_2/\lambda_1 T} \qquad (8\text{-}96)$$

$$S_{\lambda_2} = K C_1 \varepsilon_{\lambda_2} \lambda_2^{-5} \exp^{-C_2/\lambda_2 T} \qquad (8\text{-}97)$$

The ratio of the signals S_{λ_1} and S_{λ_2} is

$$\frac{S_{\lambda_1}}{S_{\lambda_2}} = \frac{\varepsilon_{\lambda_1}}{\varepsilon_{\lambda_2}} \left(\frac{\lambda_2}{\lambda_1} \right)^5 \exp\left[\frac{C_2}{T} \left(\frac{1}{\lambda_2} - \frac{1}{\lambda_1} \right) \right] \qquad (8\text{-}98)$$

Nonblack or nongrey bodies are characterized by wavelength dependence of their spectral emissivity. Let T_c be defined as the temperature of the body corresponding to the temperature of a black body. If the ratio of its radiant intensities at the wavelengths λ_1, and λ_2 equals

the ratio of the radiant intensities of the nonblack body, whose temperature is to be measured at the same wavelength, then Wien's law gives

$$\frac{\varepsilon_{\lambda_1} \exp^{-C_2/\lambda_1 T}}{\varepsilon_{\lambda_2} \exp^{-C_2/\lambda_2 T}} = \frac{\exp^{-C_2/\lambda_1 T_c}}{\exp^{-C_2/\lambda_2 T_c}} \qquad (8\text{-}99)$$

where T is the true temperature of the body. Rearranging Eq. (8-99) gives

$$T = \left[\frac{\ln \varepsilon_{\lambda_1}/\varepsilon_{\lambda_2}}{C_2 \left(\dfrac{1}{\lambda_1} - \dfrac{1}{\lambda_2} \right)} + \frac{1}{T_c} \right]^{-1} \qquad (8\text{-}100)$$

For black or grey bodies, Eq. (8-98) reduces to

$$\frac{S_{\lambda_1}}{S_{\lambda_2}} = \left(\frac{\lambda_2}{\lambda_1} \right)^5 \exp\left[\frac{C_2}{T} \left(\frac{1}{\lambda_2} - \frac{1}{\lambda_1} \right) \right] \qquad (8\text{-}101)$$

Thus, by measuring S_{λ_1} and S_{λ_2}, the temperature T can be estimated.

Accuracy of Pyrometers   Most of the temperature estimation methods for pyrometers assume that the object is either a grey body or has known emissivity values. The emissivity of the nonblack body depends on the internal state or the surface geometry of the objects. Also, the medium through which the thermal radiation passes is not always transparent. These inherent uncertainties of the emissivity values make the accurate estimation of the temperature of the target objects difficult. Proper selection of the pyrometer and accurate emissivity values can provide a high level of accuracy.

PRESSURE MEASUREMENTS

Pressure defined as force per unit area is usually expressed in terms of familiar units of weight-force and area or the height of a column of liquid that produces a like pressure at its base. Process pressure-measuring devices may be divided into three groups: (1) those that are based on the measurement of the height of a liquid column, (2) those that are based on the measurement of the distortion of an elastic pressure chamber, and (3) electrical sensing devices.

Liquid-Column Methods   Liquid-column pressure-measuring devices are those in which the pressure being measured is balanced against the pressure exerted by a column of liquid. If the density of the liquid is known, the height of the liquid column is a measure of the pressure. Most forms of liquid-column pressure-measuring devices are commonly called manometers. When the height of the liquid is observed visually, the liquid columns are contained in glass or other transparent tubes. The height of the liquid column may be measured in length units or be calibrated in pressure units. Depending on the pressure range, water and mercury are the liquids most frequently used. Since the density of the liquid used varies with temperature, the temperature must be taken into account for accurate pressure measurements.

Elastic-Element Methods   Elastic-element pressure-measuring devices are those in which the measured pressure deforms some elastic material (usually metallic) within its elastic limit, the magnitude of the deformation being approximately proportional to the applied pressure. These devices may be loosely classified into three types: Bourdon tube, bellows, and diaphragm.

Bourdon-Tube Elements   Probably the most frequently used process pressure-indicating device is the C-spring Bourdon-tube pressure gauge. Gauges of this general type are available in a wide variety of pressure ranges and materials of construction. Materials are selected on the basis of pressure range, resistance to corrosion by the process materials, and effect of temperature on calibration. Gauges calibrated with pressure, vacuum, compound (combination pressure and vacuum), and suppressed-zero ranges are available.

Bellows Element   The bellows element is an axially elastic cylinder with deep folds or convolutions. The bellows may be used unopposed, or it may be restrained by an opposing spring. The pressure to be measured may be applied either to the inside or to the space outside the bellows, with the other side exposed to atmospheric pressure. For measurement of absolute pressure either the inside or the space outside of the bellows can be evacuated and sealed. Differential pressures may be measured by applying the pressures to opposite sides of a single bellows or to two opposing bellows.

Diaphragm Elements Diaphragm elements may be classified into two principal types: those that utilize the elastic characteristics of the diaphragm and those that are opposed by a spring or other separate elastic element. The first type usually consists of one or more capsules, each composed of two diaphragms bonded together by soldering, brazing, or welding. The diaphragms are flat or corrugated circular metallic disks. Metals commonly used in diaphragm elements include brass, phosphor bronze, beryllium copper, and stainless steel. Ranges are available from fractions of an inch of water to about 206.8 kPa gauge. The second type of diaphragm is used for containing the pressure and exerting a force on the opposing elastic element. The diaphragm is a flexible or slack diaphragm of rubber, leather, impregnated fabric, or plastic. Movement of the diaphragm is opposed by a spring that determines the deflection for a given pressure. This type of diaphragm is used for the measurement of extremely low pressure, vacuum, or differential pressure.

Electrical Methods

Strain Gauges When a wire or other electrical conductor is stretched elastically, its length is increased and its diameter is decreased. Both of these dimensional changes result in an increase in the electrical resistance of the conductor. Devices utilizing resistance-wire grids for measuring small distortions in elastically stressed materials are commonly called strain gauges. Pressure-measuring elements utilizing strain gauges are available in a wide variety of forms. They usually consist of one of the elastic elements described earlier to which one or more strain gauges have been attached to measure the deformation. There are two basic strain-gauge forms: bonded and unbonded. Bonded strain gauges are those which are bonded directly to the surface of the elastic element whose strain is to be measured. The unbonded-strain-gauge transducer consists of a fixed frame and an armature which moves with respect to the frame in response to the measured pressure. The strain-gauge wire filaments are stretched between the armature and frame. The strain gauges are usually connected electrically in a Wheatstone-bridge configuration.

Strain-gauge pressure transducers are manufactured in many forms for measuring gauge, absolute, and differential pressures and vacuum. Full-scale ranges from 25.4 mm of water to 10,134 MPa are available. Strain gauges bonded directly to a diaphragm pressure-sensitive element usually have an extremely fast response time and are suitable for high-frequency dynamic-pressure measurements.

Piezoresistive Transducers A variation of the conventional strain-gauge pressure transducer uses bonded single-crystal semiconductor wafers, usually silicon, whose resistance varies with strain or distortion. Transducer construction and electrical configurations are similar to those using conventional strain gauges. A permanent magnetic field is applied perpendicular to the resonating sensor. An AC current causes the resonator to vibrate, and the resonant frequency is a function of the pressure (tension) of the resonator. The principal advantages of piezoresistive transducers are a much higher bridge voltage output and smaller size. Full-scale output voltages of 50 to 100 mV/V of excitation are typical. Some newer devices provide digital rather than analog output.

Piezoelectric Transducers Certain crystals produce a potential difference between their surfaces when stressed in appropriate directions. Piezoelectric pressure transducers generate a potential difference proportional to a pressure-generated stress. Because of the extremely high electrical impedance of piezoelectric crystals at low frequency, these transducers are usually not suitable for measurement of static process pressures.

FLOW MEASUREMENTS

Flow, defined as volume per unit of time at specified temperature and pressure conditions, is generally measured by positive-displacement or rate meters. The term "positive-displacement meter" applies to a device in which the flow is divided into isolated measured volumes when the number of fillings of these volumes is counted in some man-

ner. The term "rate meter" applies to all types of flowmeters through which the material passes without being divided into isolated quantities. Movement of the material is usually sensed by a primary measuring element that activates a secondary device. The flow rate is then inferred from the response of the secondary device by means of known physical laws or from empirical relationships.

The principal classes of flow-measuring instruments used in the process industries are variable-head, variable-area, positive-displacement, and turbine instruments, mass flowmeters, vortex-shedding and ultrasonic flowmeters, magnetic flowmeters, and more recently, Coriolis mass flowmeters. Head meters are covered in more detail in Sec. 5.

Orifice Meter The most widely used flowmeter involves placing a fixed-area flow restriction (an orifice) in the pipe carrying the fluid. This flow restriction causes a pressure drop that can be related to flow rate. The sharp-edge orifice is popular because of its simplicity, low cost, and the large amount of research data on its behavior. For the orifice meter, the flow rate Q_a for a liquid is given by

$$Q_a = \frac{C_d A_2}{\sqrt{1 - (A_2/A_1)^2}} \cdot \sqrt{\frac{2(p_1 - p_2)}{\rho}} \qquad (8\text{-}102)$$

where $p_1 - p_2$ is the pressure drop, ρ is the density, A_1 is the pipe cross-sectional area, A_2 is the orifice cross-sectional area, and C_d is the discharge coefficient. The discharge coefficient C_d varies with the Reynolds number at the orifice and can be calibrated with a single fluid, such as water (typically $C_d \approx 0.6$). If the orifice and pressure taps are constructed according to certain standard dimensions, quite accurate (about 0.4 to 0.8 percent error) values of C_d may be obtained. It should also be noted that the standard calibration data assume no significant flow disturbances such as elbows, valves, and so on, for a certain minimum distance upstream of the orifice. The presence of such disturbances close to the orifice can cause errors of as much as 15 percent. Accuracy in measurements limits the meter to a range of 3:1. The orifice has a relatively large permanent pressure loss that must be made up by the pumping machinery.

Venturi Meter The venturi tube operates on exactly the same principle as the orifice [see Eq. (8-102)]. Discharge coefficients of venturis are larger than those for orifices and vary from about 0.94 to 0.99. A venturi gives a definite improvement in power losses over an orifice and is often indicated for measuring very large flow rates, where power losses can become economically significant. The initial higher cost of a venturi over an orifice may thus be offset by reduced operating costs.

Rotameter A rotameter consists of a vertical tube with a tapered bore in which a float changes position with the flow rate through the tube. For a given flow rate the float remains stationary since the vertical forces of differential pressure, gravity, viscosity, and buoyancy are balanced. The float position is the output of the meter and can be made essentially linear with flow rate by making the tube area vary linearly with the vertical distance.

Turbine Meter If a turbine wheel is placed in a pipe containing a flowing fluid, its rotary speed depends on the flow rate of the fluid. A turbine can be designed whose speed varies linearly with flow rate. The speed can be measured accurately by counting the rate at which turbine blades pass a given point, using magnetic pickup to produce voltage pulses. By feeding these pulses to an electronic pulse-rate meter, one can measure flow rate by summing the pulses during a timed interval. Turbine meters are available with full-scale flow rates ranging from about 0.1 to 30,000 gpm for liquids and 0.1 to 15,000 ft³/min for air. Nonlinearity can be less than 0.05 percent in the larger sizes. Pressure drop across the meter varies with the square of flow rate and is about 3 to 10 psi at full flow. Turbine meters can follow flow transients quite accurately since their fluid/mechanical time constant is of the order of 2 to 10 msec.

Vortex-Shedding Flowmeters These flowmeters take advantage of vortex shedding, which occurs when a fluid flows past a non-streamlined object (a blunt body). The flow cannot follow the shape of the object and separates from it, forming turbulent vortices or eddies at the object's side surfaces. As the vortices move downstream, they grow in size and are eventually shed or detached from the object.

Shedding takes place alternately at either side of the object, and the rate of vortex formation and shedding is directly proportional to the volumetric flow rate. The vortices are counted and used to develop a signal linearly proportional to the flow rate. The digital signals can easily be totaled over an interval of time to yield the flow rate. Accuracy can be maintained regardless of density, viscosity, temperature, or pressure when the Reynolds number is greater than 10,000. There is usually a low flow cutoff point below which the meter output is clamped at zero. This flowmeter is recommended for use with relatively clean, low viscosity liquids, gases, and vapors, and rangeability of 10:1 to 20:1 is typical. A sufficient length of straight-run pipe is necessary to prevent distortion in the fluid velocity profile.

Ultrasonic Flowmeters All ultrasonic flowmeters are based upon the variable time delays of received sound waves that arise when a flowing liquid's rate of flow is varied. Two fundamental measurement techniques, depending upon liquid cleanliness, are generally used. In the first technique, two opposing transducers are inserted in a pipe so that one transducer is downstream from the other. These transducers are then used to measure the difference between the velocity at which the sound travels with the direction of flow and velocity at which it travels against the direction of flow. The differential velocity is measured either by (1) direct time delays using sound wave burst or (2) frequency shifts derived from beat-together, continuous signals. The frequency-measurement technique is usually preferred because of its simplicity and independence of the liquid static velocity. A relatively clean liquid is required to preserve the uniqueness of the measurement path.

In the second technique, the flowing liquid must contain scatters in the form of particles or bubbles that will reflect the sound waves. These scatters should be traveling at the velocity of the liquid. A Doppler method is applied by transmitting sound waves along the flow path and measuring the frequency shift in the returned signal from the scatters in the process fluid. This frequency shift is proportional to liquid velocity.

Magnetic Flowmeters The principle behind these flowmeters is Faraday's law of electromagnetic inductance. The magnitude of the voltage induced in a conductive medium moving at right angles through a magnetic field is directly proportional to the product of the magnetic flux density, the velocity of the medium, and the path length between the probes. A minimum value of fluid conductivity is required to make this approach viable. The pressure of multiple phases or undissolved solids can affect the accuracy of the measurement if the velocities of the phases are different than that for straight-run pipe. Magmeters are very accurate over wide flow ranges and are especially accurate at low flow rates. Typical applications include metering viscous fluids, slurries, or highly corrosive chemicals. Because magmeters should be filled with fluid, the preferred installation is in vertical lines with flow going upwards. However, magmeters can be used in tight piping schemes where it is impractical to have long pipe runs, typically requiring lengths equivalent to five or more pipe diameters.

Coriolis Mass Flowmeters Coriolis mass flowmeters utilize a vibrating tube in which Coriolis acceleration of a fluid in a flow loop can be created and measured. They can be used with virtually any liquid and are extremely insensitive to operating conditions, with high pressure over ranges of 100:1. These meters are more expensive than volumetric meters and range in size from $\frac{1}{16}$ to 6 inches. Due to the circuitous path of flow through the meter, Coriolis flowmeters exhibit higher than average pressure changes. The meter should be installed so that it will remain full of fluid, with the best installation in a vertical pipe with flow going upward. There is no Reynolds number limitation with this meter, and it is quite insensitive to velocity profile distortions and swirl, hence there is no requirement for straight piping upstream.

LEVEL MEASUREMENTS

The measurement of level can be defined as the determination of the location of the interface between two fluids, separable by gravity, with respect to a fixed datum plane. The most common level measurement is that of the interface between a liquid and a gas. Other level mea-

surements frequently encountered are the interface between two liquids, between a granular or fluidized solid and a gas, and between a liquid and its vapor.

A commonly used basis for classification of level devices is as follows: float-actuated, displacer, and head devices, and a miscellaneous group that depends mainly on fluid characteristics.

Float-Actuated Devices Float-actuated devices are characterized by a buoyant member that floats at the interface between two fluids. Since a significant force is usually required to move the indicating mechanism, float-actuated devices are generally limited to liquid-gas interfaces. By properly weighting the float, they can be used to measure liquid-liquid interfaces. Float-actuated devices may be classified on the basis of the method used to couple the float motion to the indicating system as discussed below.

Chain or Tape Float Gauge In these types of gauges, the float is connected to the indicating mechanism by means of a flexible chain or tape. These gauges are commonly used in large atmospheric storage tanks. The gauge-board type is provided with a counterweight to keep the tape or chain taut. The tape is stored in the gauge head on a spring-loaded reel. The float is usually a pancake-shaped hollow metal float with guide wires from top to bottom of the tank to constrain it.

Lever and Shaft Mechanisms In pressurized vessels, float-actuated lever and shaft mechanisms are frequently used for level measurement. This type of mechanism consists of a hollow metal float and lever attached to a rotary shaft, which transmits the float motion to the outside of the vessel through a rotary seal.

Magnetically Coupled Devices A variety of float-actuated level devices that transmit the float motion by means of magnetic coupling have been developed. Typical of this class of devices are magnetically operated level switches and magnetic-bond float gauges. A typical magnetic-bond float gauge consists of a hollow magnet-carrying float that rides along a vertical nonmagnetic guide tube. The follower magnet is connected and drives an indicating dial similar to that on a conventional tape float gauge. The float and guide tube are in contact with the measured fluid and come in a variety of materials for resistance to corrosion and to withstand high pressures or vacuum. Weighted floats for liquid-liquid interfaces are available.

Head Devices A variety of devices utilize hydrostatic head as a measure of level. As in the case of displacer devices, accurate level measurement by hydrostatic head requires an accurate knowledge of the densities of both heavier-phase and lighter-phase fluids. The majority of this class of systems utilize standard-pressure and differential-pressure measuring devices.

Bubble-Tube Systems The commonly used bubble-tube system sharply reduces restrictions on the location of the measuring element. In order to eliminate or reduce variations in pressure drop due to the gas flow rate, a constant differential regulator is commonly employed to maintain a constant gas flow rate. Since the flow of gas through the bubble tube prevents entry of the process liquid into the measuring system, this technique is particularly useful with corrosive or viscous liquids, liquids subject to freezing, and liquids containing entrained solids.

Electrical Methods Two electrical characteristics of fluids—conductivity and dielectric constant—are frequently used to distinguish between two phases for level-measurement purposes. An application of electrical conductivity is the fixed-point level detection of a conductive liquid such as high and low water levels. A voltage is applied between two electrodes inserted into the vessel at different levels. When both electrodes are immersed in the liquid, a current flows. Capacitance-type level measurements are based on the fact that the electrical capacitance between two electrodes varies with the dielectric constant of the material between them. A typical continuous level-measurement system consists of a rod electrode positioned vertically in a vessel, the other electrode usually being the metallic vessel wall. The electrical capacitance between the electrodes is a measure of the height of the interface along the rod electrode. The rod is usually conductively insulated from process fluids by a coating of plastic. The dielectric constant of most liquids and solids is markedly higher than that of gases and vapors. The dielectric constant of water and other polar liquids is also higher than that of hydrocarbons and other nonpolar liquids.

Thermal Methods Level-measuring systems may be based on the difference in thermal characteristics between the fluids, such as temperature or thermal conductivity. A fixed-point level sensor based on the difference in thermal conductivity between two fluids consists of an electrically heated thermistor inserted into the vessel. The temperature of the thermistor and consequently its electrical resistance increase as the thermal conductivity of the fluid in which it is immersed decreases. Since the thermal conductivity of liquids is markedly higher than that of vapors, such a device can be used as a point level detector for liquid-vapor interface.

Sonic Methods A fixed-point level detector based on sonic-propagation characteristics is available for detection of a liquid-vapor interface. This device uses a piezoelectric transmitter and receiver, separated by a short gap. When the gap is filled with liquid, ultrasonic energy is transmitted across the gap, and the receiver actuates a relay. With a vapor filling the gap, the transmission of ultrasonic energy is insufficient to actuate the receiver.

PHYSICAL PROPERTY MEASUREMENTS

Physical-property measurements are sometimes equivalent to composition analyzers, because the composition can frequently be inferred from the measurement of a selected physical property.

Density and Specific Gravity For binary or pseudobinary mixtures of liquids or gases or a solution of a solid or gas in a solvent, the density is a function of the composition at a given temperature and pressure. Specific gravity is the ratio of the density of a noncompressible substance to the density of water at the same physical conditions. For nonideal solutions, empirical calibration will give the relationship between density and composition. Several types of measuring devices are described below.

Liquid Column Density may be determined by measuring the gauge pressure at the base of a fixed-height liquid column open to the atmosphere. If the process system is closed, then a differential pressure measurement is made between the bottom of the fixed height liquid column and the vapor over the column. If vapor space is not always present, the differential-pressure measurement is made between the bottom and top of a fixed-height column with the top measurement being made at a point below the liquid surface.

Displacement There are a variety of density-measurement devices based on displacement techniques. A hydrometer is a constant-weight, variable-immersion device. The degree of immersion, when the weight of the hydrometer equals the weight of the displaced liquid, is a measure of the density. The hydrometer is adaptable to manual or automatic usage. Another modification includes a magnetic float suspended below a solenoid, the varying magnetic field maintaining the float at a constant distance from the solenoid. Change in position of the float, resulting from a density change, excites an electrical system which increases or decreases the current through the solenoid.

Direct Mass Measurement One type of densitometer measures the natural vibration frequency and relates the amplitude to changes in density. The density sensor is a U-shaped tube held stationary at its node points and allowed to vibrate at its natural frequency. At the curved end of the U is an electrochemical device that periodically strikes the tube. At the other end of the U, the fluid is continuously passed through the tube. Between strikes, the tube vibrates at its natural frequency. The frequency changes directly in proportion to changes in density. A pickup device at the curved end of the U measures the frequency and electronically determines the fluid density. This technique is useful because it is not affected by the optical properties of the fluid. However, particulate matter in the process fluid can affect the accuracy.

Radiation-Density Gauges Gamma radiation may be used to measure the density of material inside a pipe or process vessel. The equipment is basically the same as for level measurement, except that here the pipe or vessel must be filled over the effective, irradiated sample volume. The source is mounted on one side of the pipe or vessel and the detector on the other side with appropriate safety radiation shielding surrounding the installation. Cesium 137 is used as the radiation source for path lengths under 610 mm (24 in) and cobalt 60 above 610 mm. The detector is usually an ionization gauge. The absorption of the gamma radiation is a function of density. Since the absorption path includes the pipe or vessel walls, an empirical calibration is used. Appropriate corrections must be made for the source intensity decay with time.

Viscosity Continuous viscometers generally measure either the resistance to flow or the drag or torque produced by movement of an element (moving surface) through the fluid. Each installation is normally applied over a narrow range of viscosities. Empirical calibration over this range allows use on both newtonian and nonnewtonian fluids. One such device uses a piston inside a cylinder. The hydrodynamic pressure of the process fluid raises the piston to a preset height. Then the inlet valve closes and the piston is allowed to free-fall, and the time of travel (typically a few seconds) is a measure of viscosity. Other geometries include the rotation of a spindle inside a sample chamber and a vibrating probe immersed in the fluid. Because viscosity depends on temperature, the viscosity measurement must be thermostated with a heater or cooler.

Refractive-Index When light travels from one medium (e.g., air or glass) into another (e.g., a liquid), it undergoes a change of velocity and, if the angle of incidence is not 90°, a change of direction. For a given interface, angle, temperature, and wavelength of light the amount of deviation or refraction will depend on the composition of the liquid. If the sample is transparent, the normal method is to measure the refraction of light transmitted through the glass-sample interface. If the sample is opaque, the reflectance near the critical angle at a glass-sample interface is measured. In an on-line refractometer, the process fluid is separated from the optics by a prism material. A beam of light is focused on a point in the fluid which creates a conic section of light at the prism, striking the fluid at different angles (greater than or less than the critical angle). The critical angle depends on the species concentrations; as the critical angle changes, the proportions of reflected and refracted light change. A photodetector produces a voltage signal proportional to the light refracted, when compared to a reference signal. Refractometers can be used with opaque fluids and in streams that contain particulates.

Dielectric Constant The dielectric constant of material represents its ability to reduce the electric force between two charges separated in space. This property is useful in process control for polymers, ceramic materials, and semiconductors. Dielectric constants are measured with respect to vacuum (1.0); typical values range from 2 (benzene) to 33 (methanol) to 80 (water). The value for water is higher than for most plastics. A measuring cell is made of glass or some other insulating material and is usually doughnut-shaped, with the cylinders coated with metal, which constitute the plates of the capacitor.

Thermal Conductivity All gases and vapor have the ability to conduct heat from a heat source. At a given temperature and physical environment, radiation, and convection heat losses will be stabilized and the temperature of the heat source will be mainly dependent on the thermal conductivity and thus the composition of the surrounding gases. Thermal-conductivity analyzers normally consist of a sample cell and a reference cell, each containing a combined heat source and detector. These cells are normally contained in a metal block with two small cavities in which the detectors are mounted. The sample flows through the sample-cell cavity past the detector. The reference cell is an identical cavity with a detector through which a known gas flows. The combined heat source and detectors are normally either wire filaments or thermistors heated by a constant current. Since their resistance is a function of temperature, the sample-detector resistance will vary with sample composition while the reference-detector resistance will remain constant. The output from the detector bridge will be a function of sample composition.

CHEMICAL COMPOSITION ANALYZERS

A number of composition analyzers used for process monitoring and control require chemical conversion of one or more sample components preceding quantitative measurement. These reactions include formation of suspended solids for turbidimetric measurement, formation of colored materials for colorimetric detection, selective oxidation or reduction for electrochemical measurement, and formation of electrolytes for measurement by electrical conductance. Some nonvolatile materials may be separated and measured by gas chromatography after conversion into volatile derivatives.

Chromatographic Analyzers Chromatographic analyzers are widely used for the separation and measurement of volatile compounds and of compounds that can be quantitatively converted into volatile derivatives. These materials are separated by placing a portion of the sample in a chromatographic column and carrying the compounds through the column with a gas stream. As a result of the different affinities of the sample components for the column packing, the compounds emerge successively as binary mixtures with the carrier gas. A detector at the column outlet measures some physical property which can be related to the concentrations of the compounds in the carrier gas. Both the concentration peak height and the peak height-time integral, (i.e., peak area) can be related to the concentration of the compound in the original sample. The two detectors most commonly used for process chromatographs are the thermal-conductivity detector and the hydrogen-flame ionization detector. Thermal-conductivity detectors, discussed earlier, require calibration for the thermal response of each compound. Hydrogen-flame ionization detectors are more complicated than thermal-conductivity detectors but are capable of 100 to 10,000 times greater sensitivity for hydrocarbons and organic compounds. For ultrasensitive detection of trace impurities, carrier gases must be specially purified.

Infrared Analyzers Many gaseous and liquid compounds absorb infrared radiation. The degree of absorption at specific wavelengths depends on molecular structure and concentration. There are two common detector types for nondispersive infrared analyzers. These analyzers normally have two beams of radiation, an analyzing and a reference beam. One type of detector consists of two gas-filled cells separated by a diaphragm. As the amount of infrared energy absorbed by the detector gas in one cell changes, the cell pressure changes. This causes movement in the diaphragm, which in turn causes a change in capacitance between the diaphragm and a reference electrode. This change in electrical capacitance is measured as the output. The second type of detector consists of two thermopiles or two bolometers, one in each of the two radiation beams. The infrared radiation absorbed by the detector is measured by a differential thermocouple output or a resistance-thermometer (bolometer) bridge circuit.

With gas-filled detectors, a chopped light system is normally used in which one side of the detector sees the source through the analyzing beam and the other side the reference beam, alternating at a frequency of a few hertz.

Ultraviolet and Visible-Radiation Analyzers Many gas and liquid compounds absorb radiation in the near-ultraviolet or visible region. For example, organic compounds containing aromatic and carbonyl structural groups are good absorbers in the ultraviolet region. Also many inorganic salts and gases absorb in the ultraviolet or visible region. In contrast, straight-chain and saturated hydrocarbons, inert gases, air, and water vapor are essentially transparent. Process analyzers are designed to measure the absorbance in a particular wavelength band. The desired band is normally isolated by means of optical filters. When the absorbance is in the visible region, the term "colorimetry" is used. A phototube is the normal detector. Appropriate optical filters are used to limit the energy reaching the detector to the desired level and the desired wavelength region. Since absorption by the sample is logarithmic if a sufficiently narrow wavelength region is used, an exponential amplifier is sometimes used to compensate and produce a linear output.

Paramagnetism A few gases including O_2, NO, and NO_2 exhibit paramagnetic properties as a result of unpaired electrons. In a nonuniform magnetic field, paramagnetic gases, because of their magnetic susceptibility, tend to move toward the strongest part of the field, thus displacing diamagnetic gases. Paramagnetic susceptibility of these gases decreases with temperature. These effects permit measurement of the concentration of the strongest paramagnetic gas, oxygen. This analyzer used a dumbbell suspended in the magnetic field which is repelled or attracted toward the magnetic field depending on the magnetic susceptibility of the gas.

ELECTROANALYTICAL INSTRUMENTS

Conductometric Analysis Solutions of electrolytes in ionizing solvents (e.g., water) conduct current when an electrical potential is applied across electrodes immersed in the solution. Conductance is a function of ion concentration, ionic charge, and ion mobility. Conductance measurements are ideally suited for measurement of the concentration of a single strong electrolyte in dilute solutions. At higher concentrations, conductance becomes a complex, nonlinear function of concentration requiring suitable calibration for quantitative measurements.

Measurement of pH The primary detecting element in pH measurement is the glass electrode. A potential is developed at the pH-sensitive glass membrane as a result of differences in hydrogen ion activity in the sample and a standard solution contained within the electrode. This potential measured relative to the potential of the reference electrode gives a voltage that is expressed as pH. Instrumentation for pH measurement is among the most widely used process-measurement devices. Rugged electrode systems and highly reliable electronic circuits have been developed for this use.

After installation, the majority of pH measurement problems are sensor-related, mostly on the reference side, including junction plugging, poisoning, and depletion of electrolyte. For the glass (measuring electrode), common difficulties are broken or cracked glass, coating, and etching or abrasion. Symptoms such as drift, sluggish response, unstable readings, and inability to calibrate are indications of measurement problems. On-line diagnostics such as impedance measurements, wiring checks, and electrode temperature are now available in most instruments. Other characteristics that can be measured off-line include efficiency or slope and asymmetry potential (offset), which indicate whether the unit should be cleaned or changed [Nichols, *Chem. Engr. Prog.,* **90**(12), 64, 1994; McMillan, *Chem. Engr. Prog.,* **87**(12), 30, 1991].

Specific-Ion Electrodes In addition to the pH glass electrode specific for hydrogen ions, a number of electrodes that are selective for the measurement of other ions have been developed. This selectivity is obtained through the composition of the electrode membrane (glass, polymer, or liquid-liquid) and the composition of the electrode. These electrodes are subject to interference from other ions, and the response is a function of the total ionic strength of the solution. However, electrodes have been designed to be highly selective for specific ions, and when properly used, these provide valuable process measurements.

MOISTURE MEASUREMENT

Moisture measurements are important in the process industries because moisture can foul products, poison reactions, damage equipment, or cause explosions. Moisture measurements include both absolute-moisture methods and relative-humidity methods. The absolute methods are those that provide a primary output that can be directly calibrated in terms of dew-point temperature, molar concentration, or weight concentration. Loss of weight on heating is the most familiar of these methods. The relative-humidity methods are those that provide a primary output that can be more directly calibrated in terms of percentage of saturation of moisture.

Dew-Point Method For many applications, the dew point is the desired moisture measurement. When concentration is desired, the relation between water content and dew point is well-known and available. The dew-point method requires an inert surface whose temperature can be adjusted and measured, a sample gas stream flowing past the surface, a manipulated variable for adjusting the surface temperature to the dew point, and a means of detecting the onset of condensation.

Although the presence of condensate can be detected electrically, the original and most often used method is the optical detection of change in light reflection from an inert metallic-surface mirror. Some instruments measure the attenuation of reflected light at the onset of condensation. Others measure the increase of light dispersed and scattered by the condensate instead of, or in addition to, the reflected-light measurement. Surface cooling is obtained with an expendable refrigerant liquid, conventional mechanical refrigeration, or thermoelectric cooling. Surface-temperature measurement is usually made with a thermocouple or a thermistor.

Piezoelectric Method   A piezoelectric crystal in a suitable oscillator circuit will oscillate at a frequency dependent on its mass. If the crystal has a stable hygroscopic film on its surface, the equivalent mass of the crystal varies with the mass of water sorbed in the film. Thus, the frequency of oscillation depends on the water in the film. The analyzer contains two such crystals in matched oscillator circuits. Typically, valves alternately direct the sample to one crystal and a dry gas to the other on a 30-s cycle. The oscillator frequencies of the two circuits are compared electronically, and the output is the difference between the two frequencies. This output is then representative of the moisture content of the sample. The output frequency is usually converted to a variable DC voltage for meter readout and recording. Multiple ranges are provided for measurement from about 1 ppm to near saturation. The dry reference gas is preferably the same as the sample except for the moisture content of the sample. Other reference gases which are adsorbed in a manner similar to the dried sample gas may be used. The dry gas is usually supplied by an automatic dryer. The method requires a vapor sample to the detector. Mist striking the detector destroys the accuracy of measurement until it vaporizes or is washed off the crystals. Water droplets or mist may destroy the hygroscopic film, thus requiring crystal replacement. Vaporization or gas-liquid strippers may sometimes be used for the analysis of moisture in liquids.

Capacitance Method   Several analyzers utilize the high dielectric constant of water for its detection in solutions. The alternating electric current through a capacitor containing all or part of the sample between the capacitor plates is measured. Selectivity and sensitivity are enhanced by increasing the concentration of moisture in the cell by filling the capacitor sample cell with a moisture-specific sorbent as part of the dielectric. This both increases the moisture content and reduces the amount of other interfering sample components. Granulated alumina is the most frequently used sorbent. These detectors may be cleaned and recharged easily and with satisfactory reproducibility if the sorbent itself is uniform.

Oxide Sensors   Aluminum oxide can be used as a sensor for moisture analysis. A conductivity cell has one electrode node of aluminum, which is anodized to form a thin film of aluminum oxide, followed by coating with a thin layer of gold (the opposite electrode). Moisture is selectively adsorbed through the gold layer and into the hygroscopic aluminum oxide layer, which in turn determines the electrical conductivity between gold and aluminum oxide. This value can be related to ppm water in the sample. This sensor can operate between near vacuum to several hundred atmospheres, and it is independent of flow rate (including static conditions). Temperature, however, must be carefully monitored. A similar device is based on phosphorous pentoxide. Moisture content influences the electrical current between two inert metal electrodes, which are fabricated as a helix on the inner wall of a tubular nonconductive sample cell. For a constant DC voltage applied to the electrodes, a current flows which is proportional to moisture. The moisture is absorbed into the hygroscopic phosphorous pentoxide, where the current electrolyzes the water molecules into hydrogen and oxygen. This sensor will handle moisture up to 1000 ppm and 6 atm pressure. Similar to the aluminum oxide ion, temperature control is very important.

Photometric Moisture Analysis   This analyzer requires a light source, a filter wheel rotated by a synchronous motor, a sample cell, a detector to measure the light transmitted, and associated electronics. Water has two absorption bands in the near infrared region at 1400 and 1900 nm. This analyzer can measure moisture in liquid or gaseous samples at levels from 5 ppm up to 100 percent, depending on other chemical species in the sample. Response time is less than 1 s, and samples can be run up to 300°C and 400 psig.

OTHER TRANSDUCERS

Gear Train   Rotary motion and angular position are easily transduced by various types of gear arrangements. A gear train in conjunction with a mechanical counter is a direct and effective way to obtain a digital readout of shaft rotations. The numbers on the counter can mean anything desired, depending on the gear ratio and the actuating device used to turn the shaft. A pointer attached to a gear train can be used to indicate a number of revolutions or a small fraction of a revolution for any specified pointer rotation.

Differential Transformer   These devices produce an AC electrical output from linear movement of an armature. They are very versatile in that they can be designed for a full range of output with any range of armature travel up to several inches. The transformers have one or two primaries and two secondaries connected to oppose each other. With an AC voltage applied to the primary, the output voltage depends on the position of the armature and the coupling. Such devices produce accuracies of 0.5 to 1.0 percent of full scale and are used to transmit forces, pressures, differential pressures, or weights up to 1500 m. They can also be designed to transmit rotary motion.

Hall-Effect Sensors   Some semiconductor materials exhibit a phenomenon in the presence of a magnetic field which is adaptable to sensing devices. When a current is passed through one pair of wires attached to a semiconductor, such as germanium, another pair of wires properly attached and oriented with respect to the semiconductor will develop a voltage proportional to the magnetic field present and the current in the other pair of wires. Holding the exciting current constant and moving a permanent magnet near the semiconductor produce a voltage output proportional to the movement of the magnet. The magnet may be attached to a process-variable measurement device which moves the magnet as the variable changes. Hall-effect devices provide high speed of response, excellent temperature stability, and no physical contact.

SAMPLING SYSTEMS FOR PROCESS ANALYZERS

The sampling system consists of all the equipment required to present a process analyzer with a clean representative sample of a process stream and to dispose of that sample. When the analyzer is part of an automatic control loop, the reliability of the sampling system is as important as the reliability of the analyzer or the control equipment. Sampling systems have several functions. The sample must be withdrawn from the process, transported, conditioned, introduced into the analyzer, and disposed. Probably the most common problem in sample-system design is the lack of realistic information concerning the properties of the process material at the sampling point. Another common problem is the lack of information regarding the conditioning required so that the analyzer may utilize the sample without malfunction for long periods of time. Some samples require enough conditioning and treating that the sampling systems become equivalent to miniature online processing plants. These systems possess many of the same fabrication, reliability, and operating problems as small-scale pilot plants except that the sampling system must generally operate reliably for much longer periods of time.

Selecting the Sampling Point   The selection of the sampling point is based primarily on supplying the analyzer with a sample whose composition or physical properties are pertinent to the control function to be performed. Other considerations include selecting locations that provide representative homogeneous samples with minimum transport delay, locations that collect a minimum of contaminating material, and locations that are accessible for test and maintenance procedures.

Sample Withdrawal from Process   A number of considerations are involved in the design of sample-withdrawal devices that will provide representative samples. For example, in a horizontal pipe that conveys process fluid, a sample point on the bottom of the pipe will collect a maximum amount of rust, scale, or other solid materials being carried along by the process fluid. In a gas stream, such a location will also collect a maximum amount of liquid contaminants. A sample point on the top side of a pipe will, for liquid streams, collect a maximum amount of vapor contaminants being carried along. Bends in the piping that produce swirls or cause centrifugal concentration of the denser phase may cause maximum contamination to be at unexpected locations. Two-phase process materials are difficult to sample for a total-composition representative sample.

A typical method for obtaining a sample of process fluid well away from vessel or pipe walls is an eduction tube inserted through a packing gland. This sampling method withdraws liquid sample and vaporizes it for transporting to the analyzer location. The transport lag time from the end of the probe to the vaporizer is minimized by using tubing having a small internal volume compared with pipe and valve volumes.

This sample probe may be removed for maintenance and reinstalled without shutting down the process. The eduction tube is made of material that will not corrode so that it will slide through the packing gland even after long periods of service. There may be a small amount of process-fluid leakage until the tubing is withdrawn sufficiently to close the gate valve. A swaged ferrule on the end of the tube prevents accidental ejection of the eduction tube prior to removal of the packing gland. The section of pipe surrounding the eduction tube and extending into the process vessel provides mechanical protection for the eduction tube.

Sample Transport   Transport time, the time elapsed between sample withdrawal from the process and its introduction into the analyzer, should be minimized, particularly if the analyzer is an automatic analyzer-controller. Any sample-transport time in the analyzer-controller loop must be treated as equivalent to process dead time in determining conventional feedback controller settings or in evaluating controller performance. Reduction in transport time usually means transporting the sample in the vapor state.

Design considerations for sample-lines are as follows:
1.   The structural strength or protection must be compatible with the area through which the sample line runs.
2.   Line size and length must be small enough to meet transport-time requirements without excessive pressure drop or excessive bypass of sample at the analyzer input.
3.   Line size and internal-surface quality must be adequate to prevent clogging by the contaminants in the sample.
4.   The prevention of a change of state of the sample may require insulation, refrigeration, or heating of the sample line.
5.   Sample-line material must be such as to minimize corrosion due to sample or environment.

Sample Conditioning   Sample conditioning usually involves the removal of contaminants or some deleterious component from the sample mixture and/or the adjustment of temperature, pressure, and flow rate of the sample to values acceptable to the analyzer. Some of the more common contaminants that must be removed are rust, scale, corrosion products, deposits due to chemical reactions, and tar. In sampling some process streams, the material to be removed may include the primary-process product such as polymer or the main constituent of the stream such as oil. In other cases, the material to be removed is present in trace quantities. For example, water in an online chromatograph sample can damage the chromatographic column packing. When contaminants or other materials that will hinder analysis represent a large percentage of the stream composition, their removal may significantly alter the integrity of the sample. In some cases, removal must be done as part of the analysis function so that removed material can be accounted for. In other cases, proper calibration of the analyzer output will suffice.

TELEMETERING AND TRANSMISSION

ANALOG SIGNAL TRANSMISSION

Modern control systems permit the measurement device, the control unit, and the final actuator to be physically separated by several hundred meters, if necessary. This requires the transmission of the measured variable from the measurement device to the control unit, and the transmission of the controller output from the control unit to the final actuator.

In each case, transmission of a single value in only one direction is required. Such requirements can be met by analog signal transmission. A span is defined for the value to be transmitted, and the value is basically transmitted as a percent of this span. For the measured variable, the logical span is the measurement span. For the controller output, the logical span is the range of the final actuator (e.g., valve fully closed to valve fully open).

For pneumatic transmission systems, the signal range used for the transmission is 3 to 15 psig. In each pneumatic transmission system, there can be only one transmitter, but there can be any number of receivers. When most measurement devices were pneumatic, pneumatic transmission was the logical choice. However, with the displacement of pneumatic measurement devices by electronic devices, pneumatic transmission is becoming less common but is unlikely to totally disappear.

In order for electronic transmission systems to be less susceptible to interference from magnetic fields, current is used for the transmission signal instead of voltage. The signal range is 4 to 20 milliamps. In each circuit or "current loop," there can be only one transmitter. There can be more than one receiver, but not an unlimited number. For each receiver, a 250 ohm "range resistor" is inserted into the current loop, which provides a 1- to 5-volt input to the receiving device. The number of receivers is limited by the power available from the transmitter.

Both pneumatic and electronic transmission use a "live zero." This enables the receiver to distinguish a transmitted value of zero percent of span from a transmitter or transmission system failure. Transmission of zero percent of span provides a signal of 4 milliamps in electronic transmission. Should the transmitter or the transmission system fail (i.e., an open circuit in a current loop), the signal level would be zero milliamps.

For most measurement variable transmissions, the lower range of the measurement span corresponds to 4 milliamps and the upper range of the measurement span corresponds to 20 milliamps. On an open circuit, the measured variable would fail to its lower range. In some applications, this is undesirable. For example, in a fired heater that is heating material to a target temperature, failure of the temperature measurement to its lower span value would drive the output of the combustion control logic to the maximum possible firing rate. In such applications, the analog transmission signal is normally inverted, with the upper range of the measurement span corresponding to 4 milliamps and the lower range of the measurement span corresponding to 20 milliamps. On an open circuit, the measured variable would fail to its upper range. For the fired heater, failure of the measured variable to its upper span would drive the output of the combustion control logic to the minimum firing rate.

DIGITAL SYSTEMS

With the advent of the microprocessor, digital technology began to be used for data collection, feedback control, and all other information processing requirements in production facilities. Such systems must acquire data from a variety of measurement devices, and control systems must drive final actuators.

Analog Input and Outputs   Analog inputs are generally divided into two categories:

1. *High level.* Where the source is a process transmitter, the range resistor in the current loop converts the 4–20 milliamp signal into a 1–5 volt signal. The conversion equipment can be unipolar (i.e., capable of processing only positive voltages).

2. *Low level.* The most common low level signals are inputs from thermocouples. These inputs rarely exceed 30 millivolts, and could be zero or even negative. The conversion equipment must be bipolar (i.e., capable of processing positive and negative voltages).

Ultimately, such signals are converted to digital values via an analog-to-digital (A/D) converter. However, the A/D converter is normally preceded by two other components:

1. *Multiplexer.* This permits one A/D converter to service multiple analog inputs. The number of inputs to a multiplexer is usually between 8 and 256.

2. *Amplifier.* As A/D converters require high level signals, a high gain amplifier is required to convert low-level signals into high-level signals.

One of the important parameters for the A/D converter is its resolution. The resolution is stated in terms of the number of significant binary digits (bits) in the digital value. As the repeatability of most process transmitters is around 0.1 percent, the minimum acceptable resolution for a bipolar A/D converter is 12 bits, which translates to 11 data bits plus one bit for the sign. With this resolution, the analog input values can be represented to 1 part in 2^{11}, or one part in 2048. Normally, a 5-volt input is converted to a digital value of 2000, which effectively gives a resolution of 1 part in 2000 or 0.05 percent. Very few process control systems utilize resolutions higher than 14 bits, which translates to a resolution of 1 part in 8000 or 0.0125 percent.

For 4–20 milliamp inputs, the resolution is not quite as good as stated above. For a 12-bit bipolar A/D converter, 1-volt converts to a digital value of 400. Thus, the range for the digital value is 400 to 2000, making the effective input resolution 1 part in 1600, or 0.0625 percent.

On the output side, dedicated digital-to-analog converters are provided for each analog output. Outputs are normally unipolar, and require a lower resolution than inputs. A 10-bit resolution is normally sufficient, giving a resolution of 1 part in 1000 or 0.1%.

Pulse Inputs Where the sensor within the measurement device is digital in nature, analog-to-digital conversion can be avoided. For rotational devices, the rotational element can be outfitted with a shaft encoder that generates a known number of pulses per revolution. The digital system can process such inputs in either of the following ways:

1. Count the number of pulses over a fixed interval of time.

2. Determine the time for a specified number of pulses.

3. Determine the duration of time between the leading (or trailing) edges of successive pulses.

Of these, the first option is the most commonly used in process applications.

Turbine flowmeters are probably the most common example where pulse inputs are used. Another example is a watt-hour meter. Basically any measurement device that involves a rotational element can be interfaced via pulses.

Occasionally, a nonrotational measurement device can generate pulse outputs. One example is the vortex shedding meter, where a pulse can be generated when each vortex passes over the detector.

Serial Interfaces Some very important measurement devices cannot be reasonably interfaced via either analog or pulse inputs. Two examples are the following:

1. Chromatographs can perform a total composition analysis for a sample. It is possible but inconvenient to provide an analog input for each component. Furthermore, it is often desirable to capture other information, such as the time that the analysis was made (normally the time the sample was injected).

2. Load cells are capable of resolutions of 1 part in 100,000. A/D converters for analog inputs cannot even approach such resolutions.

One approach to interfacing with such devices is serial interfaces. This involves two aspects:

1. *Hardware interface.* The RS-232 interface standard is the basis for most serial interfaces.

2. *Protocol.* This is interpreting the sequence of characters transmitted by the measurement device. There are no standards for protocols, which means that custom software is required.

One advantage of serial interfaces is that two-way communication is possible. For example, a "tare" command can be issued to a load cell.

Microprocessor-Based Transmitters The cost of microprocessor technology has declined to the point where it is economically feasible to incorporate a microprocessor into each transmitter. Such microprocessor-based transmitters are often referred to as "smart" transmitters. As opposed to conventional or "dumb" transmitters, the smart transmitters offer the following capabilities:

1. Checks on the internal electronics, such as verifying that the voltage levels of internal power supplies are within specifications.

2. Checks on environmental conditions within the instruments, such as verifying that the case temperature is within specifications.

3. Compensation of the measured value for conditions within the instrument, such as compensating the output of a pressure transmitter for the temperature within the transmitter. Smart transmitters are much less affected by temperature and pressure variations than conventional transmitters.

4. Compensation of the measured value for other process conditions, such as compensating the output of a capacitance level transmitter for variations in process temperature.

5. Linearizing the output of the transmitter. Functions such as square root extraction of the differential pressure for a head-type flowmeter can be done within the instrument instead of within the control system.

6. Configuring the transmitter from a remote location, such as changing the span of the transmitter output.

7. Automatic recalibration of the transmitter. Although this is highly desired by users, the capabilities, if any, in this respect depend on the type of measurement.

Due to these capabilities, smart transmitters offer improved performance over conventional transmitters.

Transmitter/Actuator Networks With the advent of smart transmitters and smart actuators, the limitations of the 4–20 milliamp analog signal transmission retard the full utilization of the capabilities of the smart devices. For smart transmitters, the following capabilities are required:

1. *Transmission of more than one value from a transmitter.* Information beyond the measured variable is available from the smart transmitter. For example, a smart pressure transmitter can also report the temperature within its housing. Knowing that this temperature is above normal values permits corrective action to be taken before the device fails. Such information is especially important during the initial commissioning of a plant.

2. *Bidirectional transmission.* Configuration parameters such as span, engineering units, resolution, and so on, must be communicated to the smart transmitter.

Similar capabilities are required for smart actuators.

In order to meet their initial requirements, several manufacturers have developed digital communications capabilities for communicating with smart transmitters. These can be used either in addition to or in lieu of the 4–20 milliamp signal. Although most manufacturers release enough information on their communications features to permit another manufacturer to provide compatible instruments (and in some cases provide an open communication standard), the communications capability provided by a manufacturer may be proprietary.

Users purchase their transmitters from a variety of manufacturers, so this situation limits the full utilization of the capabilities of smart transmitters and valves. Efforts to develop a standard for a communications network have not proceeded smoothly. The International Society for Measurement and Control (ISA) has attempted to develop a standard generally referred to as fieldbus. The standards effort attempted to develop a world standard, encompassing European, Japanese, and American products. This effort focused on developing a single standard with which all manufacturers would comply. Currently, efforts are mostly being directed to providing the capability for interoperability between the products of the manufacturers with competing communications networks. Meanwhile, users are reluctant to make major commitments, and are continuing to rely primarily on the traditional 4–20 milliamp transmission.

FILTERING AND SMOOTHING

A signal received from a process transmitter generally contains the following features:

1. Low-frequency process disturbances. The control system is expected to react to these disturbances.
2. High-frequency process disturbances. The frequency of these disturbances is beyond the capability of the control system to effectively react.
3. Measurement noise.
4. Stray electrical pickup, primarily 50- or 60-cycle AC. Frequencies are measured in Hertz (Hz), with 60-cycle AC being a 60-Hz frequency.

The objective of filtering and smoothing is to remove the last three components, leaving only the low frequency process disturbances.

Normally this has to be accomplished using the proper combination of analog and digital filters. Sampling a continuous signal results in a phenomenon often referred to as aliasing or foldover. In order to represent a sinusoidal signal, a minimum of four samples are required during each cycle. That is, the sampling interval must be at least 1/4th the period of the sinusoid. Consequently, when a signal is sampled at a frequency ω_s, all frequencies higher than $(\pi/2)\omega_s$ cannot be represented at their original frequency. Instead, they are present in the sampled signal with their original amplitude but at a lower frequency harmonic.

Because of the aliasing or foldover issues, a combination of analog and digital filtering is usually required. The sampler (i.e., the A/D converter) must be preceded by an analog filter that rejects those high-frequency components such as stray electrical pickup that would result in foldover when sampled. In commercial products, analog filters are normally incorporated into the input processing hardware by the manufacturer. The software then permits the user to specify digital filtering to remove any undesirable low-frequency components.

On the analog side, the filter is often the conventional resistor-capacitor or RC filter. However, other possibilities exist. For example, one type of A/D converter is called an "integrating A/D" because the converter basically integrates the input signal over a fixed interval of time. By making the interval 1/60th second, this approach provides excellent rejection of any 60-Hz electrical noise.

On the digital side, the input processing software generally provides for smoothing via the exponentially weighted moving average, which is the digital counterpart to the RC network analog filter. The smoothing equation is as follows:

$$y_i = \alpha x_i + (1 - \alpha)y_{i-1} \qquad (8\text{-}103)$$

where x_i = current value of input
y_i = current output from filter
y_{i-1} = previous output from filter
α = filter coefficient

The degree of smoothing is determined by the filter coefficient α, with $\alpha = 1$ being no smoothing and $\alpha = 0$ being infinite smoothing (no effect of new measurements). The filter coefficient α is related to the filter time constant τ_F and the sampling interval Δt by the following equation:

$$\alpha = 1 - \exp\left(\frac{-\Delta t}{\tau_F}\right) \qquad (8\text{-}104)$$

or by the approximation

$$\alpha = \frac{\Delta t}{\Delta t + \tau_F} \qquad (8\text{-}105)$$

Another approach to smoothing is to use the arithmetic moving average, which is represented by the following equation:

$$y_i = \frac{\left[\sum_{j=1}^{n} x_{i+1-j}\right]}{n} \qquad (8\text{-}106)$$

The term "moving" is applied because the filter software maintains a storage array with the previous n values of the input. When a new value is received, the oldest value in the storage array is replaced with the new value, and the arithmetic average recomputed. This permits the filtered value to be updated each time a new input value is received.

In process applications, determining τ_F (or α) for the exponential filter and n for the moving average filter is often done merely by observing the behavior of the filtered value. If the filtered value is "bouncing," the degree of smoothing (that is, τ_F or n) is increased. This can easily lead to an excessive degree of filtering, which will limit the performance of any control system that uses the filtered value. The degree of filtering is best determined from the frequency spectrum of the measured input, but such information is rarely available for process measurements.

ALARMS

The purpose of an alarm is to alert the process operator to a process condition that requires immediate attention. An alarm is said to occur whenever the abnormal condition is detected and the alert is issued. An alarm is said to return to normal when the abnormal condition no longer exists.

Analog alarms can be defined on measured variables, calculated variables, controller outputs, and the like. For analog alarms, the following possibilities exist:

1. *High/low alarms.*   A high alarm is generated when the value is greater than or equal to the value specified for the high-alarm limit. A low alarm is generated when the value is less than or equal to the value specified for the low-alarm limit.
2. *Deviation alarms.*   An alarm limit and a target are specified. A high deviation alarm is generated when the value is greater than or equal to the target plus the deviation alarm limit. A low deviation alarm is generated when the value is less than or equal to the target minus the deviation alarm limit.
3. *Trend or rate-of-change alarms.*   A limit is specified for the maximum rate of change, usually specified as a change in the measured value per minute. A high trend alarm is generated when the rate of change of the variable is greater than or equal to the value specified for the trend alarm limit. A low trend alarm is generated when the rate of change of the variable is less than or equal to the negative of the value specified for the trend alarm limit.

Most systems permit multiple alarms of a given type to be configured for a given value. For example, configuring three high alarms provides a high alarm, a high-high alarm, and a high-high-high alarm.

One operational problem with analog alarms is that noise in the variable can cause multiple alarms whenever its value approaches a limit. This can be avoided by defining a deadband on the alarm. For example, a high alarm would be processed as follows:

1. *Occurrence.*   The high alarm is generated when the value is greater than or equal to the value specified for the high-alarm limit.
2. *Return to normal.*   The high-alarm return to normal is generated when the value is less than or equal to the high alarm limit less the deadband.

As the degree of noise varies from one input to the next, the deadband must be individually configurable for each alarm.

Discrete alarms can be defined on discrete inputs, limit switch inputs from on/off actuators and so on. For discrete alarms, the following possibilities exist:

1. *Status alarms.*   An expected or normal state is specified for the discrete value. A status alarm is generated when the discrete value is other than its expected or normal state.
2. *Change-of-state alarm.*   A change-of-state alarm is generated on any change of the discrete value.

The expected sequence of events on an alarm is basically as follows:

1. The alarm occurs. This usually activates an audible annunciator.
2. The alarm occurrence is acknowledged by the process operator. When all alarms have been acknowledged, the audible annunciator is silenced.
3. Corrective action is initiated by the process operator.
4. The alarm condition returns to normal.

However, additional requirements are imposed at some plants. Sometimes the process operator must acknowledge the alarm's return to normal. Some plants require that the alarm occurrence be reissued if the alarm remains in the occurred state longer than a specified period of time. Consequently, some "personalization" of the alarming facilities is done.

When alarms were largely hardware-based (i.e., the panel alarm systems), the purchase and installation of the alarm hardware imposed a certain discipline on the configuration of alarms. With digital systems, the suppliers have made it extremely easy to configure alarms. In fact, it is sometimes easier to configure alarms on a measured value than not to configure the alarms. Furthermore, the engineer assigned the responsibility for defining alarms should ensure that an abnormal process condition will not go undetected because an alarm has not been configured. When alarms are defined on every measured and calculated variable, the result is an excessive number of alarms, most of which are duplicative and unnecessary.

The accident at the Three Mile Island nuclear plant clearly demonstrated that an alarm system can be counterproductive. An excessive number of alarms can distract the operator's attention from the real problem that needs to be addressed. Alarms that merely tell the operator something that is already known do the same. In fact, a very good definition of a nuisance alarm is one that informs the operator of a situation of which the operator is already aware. The only problem with applying this definition is determining what the operator already knows.

Unless some discipline is imposed, engineering personnel, especially where contractors are involved, will define far more alarms than plant operations require. This situation may be addressed by simply setting the alarm limits to values such that the alarms never occur. However, changes in alarms and alarm limits are changes from the perspective of the Process Safety Management regulations. It is prudent to impose the necessary discipline to avoid an excessive number of alarms. Potential guidelines are as follows:

1. For each alarm, a specific action is expected from the process operator. Operator actions such as "call maintenance" are inappropriate with modern systems. If maintenance needs to know, modern systems can inform maintenance directly.

2. Alarms should be restricted to abnormal situations for which the process operator is responsible. A high alarm on the temperature in one of the control system cabinets should not be issued to the process operator. Correcting this situation is the responsibility of maintenance, not the process operator.

3. Process operators are expected to be exercising normal surveillance of the process. Therefore, alarms are not appropriate for situations known to the operator either through previous alarms or through normal process surveillance. The "sleeping operator" problem can be addressed by far more effective means than the alarm system.

4. When the process is operating normally, no alarms should be triggered. Within the electric utility industry, this design objective is known as "darkboard." Application of darkboard is especially important in batch plants, where much of the process equipment is operated intermittently.

Ultimately, guidelines such as those above will be taken seriously only if production management carefully configures the alarms. The consequences of excessive and redundant alarms will be felt primarily by those responsible for production operations. Therefore, production management must make adequate resources available for reviewing and analyzing the proposed alarm configurations.

Another serious distraction to a process operator is the multiple alarm event, where a single event within the process results in multiple alarms. When the operator must individually acknowledge each alarm, considerable time can be lost in silencing the obnoxious annunciator before the real problem is addressed. Air-handling systems are especially vulnerable to this, where any fluctuation in pressure (for example, resulting from a blower trip) can cause a number of pressure alarms to occur. Point alarms (high alarms, low alarms, status alarms, etc.) are especially vulnerable to the multiple alarm event. This can be addressed in one of two ways:

1. *Ganging alarms.* Instead of individually issuing the point alarms, all alarms associated with a certain aspect of the process are simply wired to give a single trouble alarm. The responsibility rests entirely with the operator to determine the nature of the problem.

2. *Intelligent alarms.* Logic is incorporated into the alarm system to determine the nature of the problem and then issue a single alarm to the process operator. Sometimes this is called an expert system.

While the intelligent alarm approach is clearly preferable, substantial process analysis is required to support intelligent alarming. Meeting the following two objectives is quite challenging:

1. The alarm logic must consistently detect abnormal conditions within the process.

2. The alarm logic must not issue an alert to an abnormal condition when in fact none exists.

Often the latter case is more challenging than the former.

Logically, the intelligent alarm effort must be linked to the process hazards analysis. Developing an effective intelligent alarming system requires substantial commitments of effort, involving both process engineers, control systems engineers, and production personnel. Methodologies such as expert systems can facilitate the implementation of an intelligent alarming system, but they must still be based on a sound analysis of the potential process hazards.

DIGITAL TECHNOLOGY FOR PROCESS CONTROL

GENERAL REFERENCES: Fortier, *Design and Analysis of Distributed Real-Time Systems,* McGraw-Hill, New York, 1985; Hawryszkiewycs, *Database Analysis and Design,* Science Research Associates Inc., Chicago, 1984; Khambata, *Microprocessors/Microcomputers: Architecture, Software, and Systems,* 2d ed., Wiley, New York, 1987; Liptak, *Instrument Engineers Handbook,* Chilton Book Company, Philadelphia, 1995; Mellichamp (ed.), *Real-Time Computing with Applications to Data Acquisition and Control,* Van Nostrand Reinhold, New York, 1983.

Since the 1970s, process controls have evolved from pneumatic analog technology to electronic analog technology to microprocessor-based controls. Electronic analog technology has virtually disappeared from process controls. Pneumatic controls continue to be manufactured, but they are relegated to special situations where pneumatics can offer a unique advantage. Process controls are dominated by programmable electronic systems (PES), most of which are based on microprocessor technology.

HIERARCHY OF INFORMATION SYSTEMS

Coupling digital controls with networking technology permits information to be passed from level-to-level within a corporation at high rates of speed. This technology is capable of presenting the measured variable from a flow transmitter installed in a plant in a remote location anywhere in the world to the company headquarters in less than a second.

A hierarchical representation of the information flow within a company leads to a better understanding of how information is passed from one layer to the next. Such representations can be developed in varying degrees of detail, and most companies have developed one that describes their specific practices. The following hierarchy consists of five levels.

Measurement Devices and Actuators Often referred to as level 0, this layer couples the control and information systems to the process. The measurement devices provide information on the current conditions within the process. The actuators permit control decisions to be imposed on the process. Although traditionally analog, smart transmitters and smart valves based on microprocessor technology will eventually dominate this layer.

Regulatory Controls   The objective of this layer is to operate the process at or near the targets supplied by others, be it the process operator or a higher layer in the hierarchy. In order to achieve consistent process operations, a high degree of automatic control is required from the regulatory layer. The direct result is a reduction in variance in the key process variables. More uniform product quality is an obvious benefit. However, consistent process operation is a prerequisite for optimizing the process operations. To ensure success for the upper level functions, the first objective of any automation effort must be to achieve a high degree of regulatory control.

Supervisory Controls   The regulatory layer blindly attempts to operate the process at the specified targets, regardless of the appropriateness of these targets. Determining the most appropriate targets is the responsibility of the supervisory layer. Given the current production targets for a unit, supervisory control determines how the process can be best operated to meet the production targets. Usually this optimization has a limited scope, being confined to a single production unit or possibly even a single unit operation within a production unit. Supervisory control translates changes in factors such as current process efficiencies, current energy costs, cooling medium temperatures, and so on, to changes in process operating targets so as to optimize process operations.

Production Controls   The nature of the production control logic differs greatly between continuous and batch plants. A good example of production control in a continuous process is refinery optimization. From the assay of the incoming crude oil, the values of the various possible refined products, the contractual commitments to deliver certain products, the performance measures of the various units within a refinery, and the like, it is possible to determine the mix of products that optimizes the economic return from processing this crude. The solution of this problem involves many relationships and constraints and is solved with techniques such as linear programming.

In a batch plant, production control often takes the form of routing or short-term scheduling. For a multiproduct batch plant, determining the long term schedule is basically a manufacturing resource planning (MRP) problem, where the specific products to be manufactured and the amounts to be manufactured are determined from the outstanding orders, the raw materials available for production, the production capacities of the process equipment, and other factors. The goal of the MRP effort is the long-term schedule, which is a list of the products to be manufactured over a specified period of time (often one week). For each product on the list, a target amount is also specified. To manufacture this amount usually involves several batches. The term "production run" often refers to the sequence of batches required to make the target amount of product, so in effect the long term schedule is a list of production runs.

Most multiproduct batch plants have more than one piece of equipment of each type. Routing refers to determining the specific pieces of equipment that will be used to manufacture each run on the long term production schedule. For example, the plant might have five reactors, eight neutralization tanks, three grinders, and four packing machines. For a given run, a rather large number of possible routes are possible. Furthermore, rarely is only one run in progress at a given time. The objective of routing is to determine the specific pieces of production equipment to be used for each run on the long-term production schedule. Given the dynamic nature of the production process (equipment failures, insertion/deletion of runs into the long-term schedule, etc.), the solution of the routing problem continues to be quite challenging.

Corporate Information Systems   Terms such as *management information systems* (MIS) and *information technology* (IT) are frequently used to designate the upper levels of computer systems within a corporation. From a control perspective, the functions performed at this level are normally long-term and/or strategic. For example, in a processing plant, long-term contracts are required with the providers of the feedstocks. A forecast must be developed for the demand for possible products from the plant. This demand must be translated into needed raw materials, and then contracts executed with the suppliers to deliver these materials on a relatively uniform schedule.

While most companies within the process industries recognize the importance of information technology in managing their businesses, this technology has been a source of considerable frustration and disappointment. Schedule delays, cost overruns, and failure of the final product to perform as expected have often eroded the credibility of information technology. However, immense potential remains for the technology, and process companies have no choice but to seek continuous improvement.

DISTRIBUTED CONTROL SYSTEMS

Although digital control technology was first applied to process control in 1959, the total dependence of the early centralized architectures on a single computer for all control and operator interface functions resulted in complex systems with dubious reliability. Adding a second processor increased both the complexity and the cost. Consequently, many installations provided analog backup systems to protect against a computer malfunction.

Microprocessor technology permitted these technical issues to be addressed in a cost-effective manner. In the mid-1970s, a process control architecture referred to as a distributed control system (DCS) was introduced and almost instantly became a commercial success. A DCS consists of some number of microprocessor-based nodes that are interconnected by a digital communications network, often called a data highway. The key features of this architecture are as follows:

1. The process control functions and the operator interface, also referred to as man-machine interface (MMI) or human-machine interface (HMI), is provided by separate nodes. This approach is referred to as *split-architecture*, and it permits considerable flexibility in choosing a configuration that most appropriately meets the needs of the application.

2. The process control functions can be distributed functionally and/or geographically. Functional distribution permits related control functions to be grouped and implemented in a single node. Geographical distribution permits the process control nodes to be physically located near the equipment being controlled. As the digital communications network is based on local area network (LAN) technology, the nodes within the DCS can be physically separated by thousands of meters.

3. Redundancy can be provided where appropriate, the following being typical:

   a. Multiple operator interface nodes can be provided to reduce the impact of an operator interface node failure.

   b. The digital communications network is normally redundant to the extent that at least two independent paths are available between any two nodes of the DCS.

   c. Consisting basically of processor and memory, the process control nodes are highly reliable, with mean-times-between-failures approaching 100 years. Redundant configurations are available for especially critical applications.

4. As the data within the DCS are digital in nature, interfaces to upper level computers are technically easier to implement. Unfortunately, the proprietary nature of the communications networks within commercial DCS products complicate the implementation of such interfaces. Truly open DCS architectures, at least as the term "open" is used in the mainstream of computing, are not yet available.

Figure 8-62 depicts a hypothetical distributed control system. A number of different unit configurations are illustrated. This system consists of many commonly used DCS components, including multiplexers (MUXs), single/multiple-loop controllers, programmable logic controllers (PLCs), and smart devices. A typical system includes the following elements as well:

• *Host computers.*   These are the most powerful computers in the system, capable of performing functions not normally available in other units. They act as the arbitrator unit to route internodal communications. An operator interface is supported and various peripheral devices are coordinated. Computationally intensive tasks, such as optimization or advanced control strategies, are processed here.

• *Data highway.*   This is the communication link between com-

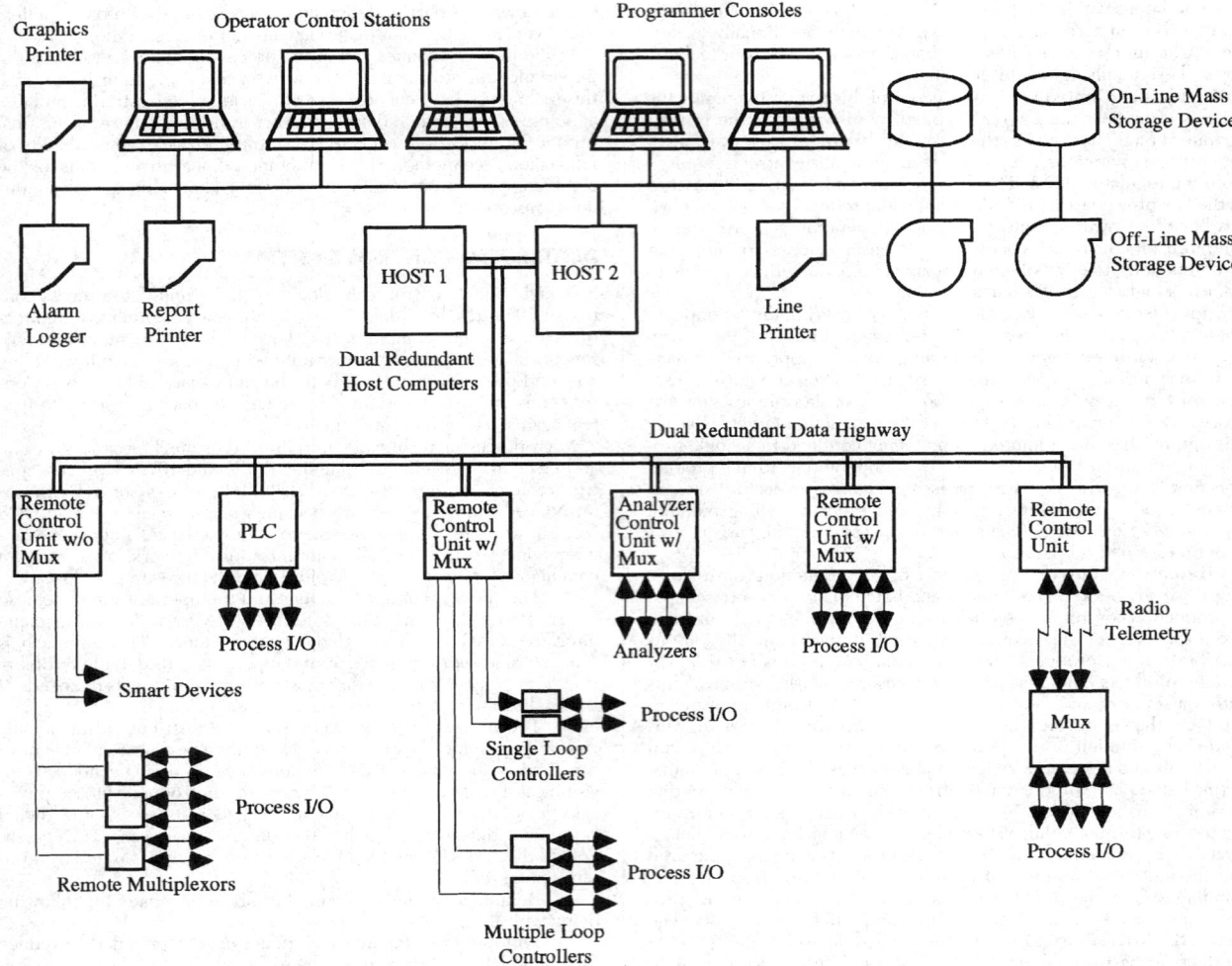

FIG. 8-62 A typical DDC system.

ponents of a network. Coaxial cable is often used. A redundant pair is normally supplied to reduce possibility of link failure.

• *Real-time clocks (RTCs).* Real-time systems are required to respond to events, as they occur, in a timely manner. This is especially crucial in process control systems where control actions applied at the wrong time may amplify process deviations or destabilize the processes. The nodes in the systems are interrupted periodically by the real-time clocks to maintain the actual elapsed times.

• *Operator control stations.* These typically consist of color graphics monitors with special keyboards, in addition to a conventional alphanumeric keyboard, containing keys to perform dedicated functions. Operators may supervise and control processes from these stations. A control station may contain a number of printers for alarm logging, report printing, or hard-copying process graphics.

• *Remote control units.* These units are used to control unit processes. Basic control functions such as the PID algorithm are implemented here. Depending on other hardware components used, data acquisition capability may be required to perform digital control. They may be configured to supply process set points to single-loop controllers. Radio telemetry may be installed to communicate with MUX units located at great distances.

• *Programmer consoles.* These are programming terminals. Developing system software on the host machines is a common prac-

tice by many system suppliers. This eliminates compatibility problems between development and target environments. Programming capability is normally retained when the system is delivered such that system users may develop their own application programs.

• *Mass storage device.* Typically, fixed-head hard disk drives are used to store active data, including on-line and historical databases and non-memory-resident programs. Memory-resident programs are stored to allow loading at system startups. The tape drives are used for archives and backups.

DISTRIBUTED DATABASE AND THE DATABASE MANAGER

A database is a centralized location for data storage. The use of databases enhances system performance by maintaining complex relations between data elements while reducing data redundancy. A database may be built based on the relational model, the entity relationship model, or some other model. The database manager is a system utility program or programs acting as the gatekeeper to the databases. All functions retrieving or modifying data must submit a request to the manager. Information required to access the database include the tag name of the database entity, often referred to as a point, the attributes to be accessed, and the values if modifying. The database manager maintains the integrity of the databases by executing a request only when not processing other conflicting requests. Although a number of functions may read the same data item at the same time, writing by a number of functions or simultaneous read and write of the same data item is not permitted.

To allow flexibility, the database manager must also perform point addition or deletion. However, the ability to create a point type or to add or delete attributes of a point type is not normally required because, unlike other data processing systems, a process control system normally involves a fixed number of point types and related attributes. For example, analog and binary input and output types are required for process I/O points. Related attributes for these point types include tag names, values, and hardware addresses. Different system manufacturers may define different point types using different data structures. We will discuss other commonly used point types and attributes as they appear.

Historical Database Subsystem We have discussed the use of on-line databases. An historical database is built similar to an on-line database. Unlike their on-line counterparts, the information stored in a historical database is not normally accessed directly by other subsystems for process control and monitoring. Periodic reports and long-term trends are generated based on the archived data. The reports are often used for long-term planning and system performance evaluations such as statistical process (quality) control. The trends may be used to detect process drifts or to compare process variations at different times.

The historical data is sampled at user-specified intervals. A typical process plant contains a large number of data points, but it is not feasible to store data for all points at all times. The user determines if a data point should be included in the list of archive points. Most systems provide archive-point menu displays. The operators are able to add or delete data points to the archive point lists. The sampling periods are normally some multiples of their base scan frequencies. However, some systems allow historical data sampling of arbitrary intervals. This is necessary when intermediate virtual data points that do not have the scan frequency attribute are involved. The archive point lists are continuously scanned by the historical database software. On-line databases are polled for data. The times of data retrieval are recorded with the data obtained. To conserve storage space, different data compression techniques are employed by various manufacturers.

The historical data may be used for long-term trending. The live trends data are displayed but not stored. Therefore, these trends cannot be recalled once cleared off the screens. The historical trend of any archive point may be displayed at any time because the values used are extracted from the archived data. Zooming, that is, axis scaling, is allowed by most systems. As a result, the displayed data point intervals may not be multiples of stored data intervals. Many systems provide data interpolation and smoothing functions to process the stored data when they are displayed. The live and historical trend displays are superior to strip charts in many ways. In addition to conventional trend recording, the zoom-in capability allows close examination of recorded data, whereas zoom-out compresses long-term data within a screen. Exact data sampled at any time point can be extracted by cursor positioning. Strip-chart recorders have disappeared from many modern plants.

Periodic reports, including shift, daily, weekly, monthly, and quarterly reports, are printed based on archived data. Some reports may contain simply the stored data in certain specific arrangements. More often, quantities such as mean values, standard deviations, or other calculated values are included. Instead of hard-coding reports to user specifications, many system suppliers provide report generation packages in the form of metalanguages. These packages allow users to configure report formats suitable for their particular requirements. The report generator interprets the configuration files prepared by the users to create reports. Due to the infrequent execution, the report generator is normally operated in the batch mode.

DCS manufacturers have devoted considerable efforts to make it easy to implement and enhance process control configurations within their products. Although programming in the traditional sense is possible within most products, the majority of the functions required for a process control application can be implemented by configuring as opposed to programming.

PROCESS CONTROL LANGUAGE

A digital control system involves software development. The introduction of high-level programming languages such as FORTRAN and BASIC in the 1960s was considered a major breakthrough. For process control applications, some companies have incorporated libraries of software routines for these languages, but others have developed speciality pseudolanguages. These implementations are characterized by their statement-oriented language structure. Although substantial savings in time and efforts can be realized, software development costs continue to be significant.

The most successful and user-friendly approach, which is now adopted by virtually all commercial systems, is the fill-in-the-forms or table-driven process control languages (PCLs). The core of these languages is a number of basic functional blocks or software modules. All modules are defined as database points. Using a module is analogous to calling a subroutine in conventional programs.

In general, each module contains some inputs and an output. The programming involves soft-wiring outputs of blocks to inputs of other blocks. Some modules may require additional parameters to direct module execution. The users are required to fill in the sources of input values, the destinations of output values, and the parameters in blanks of forms or tables prepared for the modules. The source and destination blanks may be filled with process I/Os when appropriate. To connect modules, some systems require filling the tag names of modules originating or receiving data. Additional programs are often required to resolve ambiguities when connecting multiple input-output modules. Another method involves the use of intermediate data points. The blanks in a pair of interconnecting modules are filled with the tag name of the same data point. Batch jobs and/or interactive data entry may be performed to fill the databases. A completed control strategy resembles a data flow diagram. The soft-wiring of modules is similar to hard-wiring analog-electronic circuits in analog computers.

Additional database space must be allocated when intermediate data points are used. A system can be designed to use process I/O points as intermediates. However, the data acquisition software must be programmed to bypass these points when scanned. All system builders provide virtual data point types if the intermediate data storage scheme is adopted. These points are not scanned by the data acquisition software. Memory space requirements are reduced by eliminating unnecessary attributes such as hardware addresses and scan frequencies. It should be noted that the fill-in-the-forms technique is applicable to all data point types.

All process control languages contain PID controller blocks. The digital PID controller is normally programmed to execute in velocity form. A pulse duration output may be used to receive the velocity output directly. Where positional signal is expected, an operating mode bit is used to enable an internal integrator. This flexibility is not normally available in analog controllers. Unlike an analog controller, the three modes in a digital PID controller do not interact. This simplifies the tuning effort. In addition to the tuning constants, a typical digital PID controller contains some entries not normally found in an analog controller:

• When a process error is below certain tolerable deadband, the controller ceases modifying output. This is referred to as gap action.

• The magnitude of change in a velocity output is limited by a change clamp.

• A pair of output clamps is used to restrict a positional output value from exceeding specified limits.

• The controller action can be disabled by triggering a binary deactivate input signal, during process startup, shutdown, or when some abnormal conditions exist.

Although modules are supplied and their internal configurations are different from system to system, their basic functionalities are the same.

SINGLE-LOOP CONTROLS

With the exception of pneumatic controllers for special applications, commercial single-loop controllers are almost entirely microprocessor-based. The most basic products provide only the PID control algorithm, but the more powerful versions provide a set of general-purpose algorithms comparable to those in a DCS. For applications such as cascade control and multivariable control, the manufacturers of single-loop controllers provide multiloop versions of their products. These multiloop controllers have much in common with the process control node in a DCS.

Single-loop controllers provide both the process control functions and the operator interface function. This makes them ideally suited to very small applications, where only two or three loops are required. However, it is possible to couple single-loop controllers to a personal computer (PC) to provide the operator interface function. Such installations are extremely cost effective, and with the keen competition in PC-based products, the capabilities are comparable and sometimes even better than that provided by a DCS. However, this approach makes sense only up to about 25 loops.

Initially, the microprocessor-based single-loop controllers made the power of digital control affordable to those with small processes. To compete with these products in small applications, the DCS suppliers have introduced micro-DCS versions of their products. As a PC-based operator interface is usually a component of the micro-DCS, there is sometimes little distinction between a micro-DCS and a system consisting of single-loop controllers coupled to a PC-based operator interface.

PROGRAMMABLE LOGIC CONTROLLERS

The programmable logic controller (PLC) was the first digital technology to successfully compete with conventional technology in industrial control applications. Initially developed in the early 1970s for applications within the manufacturing industries (principally automotive), the PLC proved to be superb for implementing discrete logic. The earliest PLCs were limited to discrete I/O, basic Boolean logic functions (AND, OR, NOT), timers, and counters. However, versions soon appeared with analog I/O, math functions, PID control algorithms, and other functions required for process control applications.

Developed to replace hard-wired relay logic, the early PLCs were "programmed" using the same ladder logic diagrams used to represent logic implemented with hard-wired relays. As the initial target market was electrical, programming in ladder logic was a definite

advantage, and some union contracts specifically required that such discrete logic be presented as ladder diagrams. However, ladder logic is not the programming medium preferred by instrument engineers, which hampers the acceptance of PLCs for process control. Alternatives to ladder logic are available for programming PLCs, but established perceptions are slow to change.

Developed specifically for implementing discrete logic, PLCs continue to provide the best route to implementing such logic. The manufacturers of PLCs provide robust, cost-effective discrete I/O modules. Regardless of its acceptability, ladder logic is the most efficient means for implementing discrete logic. Because PLCs scan the discrete logic very rapidly, a 100-millisecond scan rate is considered very slow for a PLC. The process control modules of a DCS often implement discrete logic using function blocks, which is less efficient than ladder logic and normally results in a slower scan rate. A few DCS process control modules have used ladder logic to implement discrete logic, but their discrete I/O capabilities and slow scan rates rarely match that of a PLC. Consequently, for applications heavy with discrete logic, most DCS suppliers will incorporate one or more PLCs into their system.

Being excellent at discrete logic, PLCs are a potential candidate for implementing interlocks. Process interlocks are clearly acceptable for implementation within a PLC. Implementation of safety interlocks in programmable electronic systems (such as a PLC) is not universally accepted. Many organizations continue to require that all safety interlocks be hard-wired, but implementing safety interlocks in a PLC that is dedicated to safety functions is accepted by some as being equivalent to the hard-wired approach.

INTERCOMPUTER COMMUNICATIONS

A group of computers becomes a network when intercomputer communication is established. Prior to the 1980s, all system suppliers used proprietary protocols to network their systems. Ad hoc approaches were sometimes used to connect third-party equipment, which was not cost-effective in system maintenance, upgrade, and expansion. The recent introduction of standardized protocols has led to a decrease in initial capital cost. Most current DCS network protocol designs are based on the ISO-OSI° seven-layer model.

The most notable effort in standardizing plant automation protocols

---

° Abbreviated from International Standards Organization-Open System Interconnection. They are the physical, data link, network, transports session, presentation, and application layers. Only the physical, data link, and application layers are present in the mini-MAP.

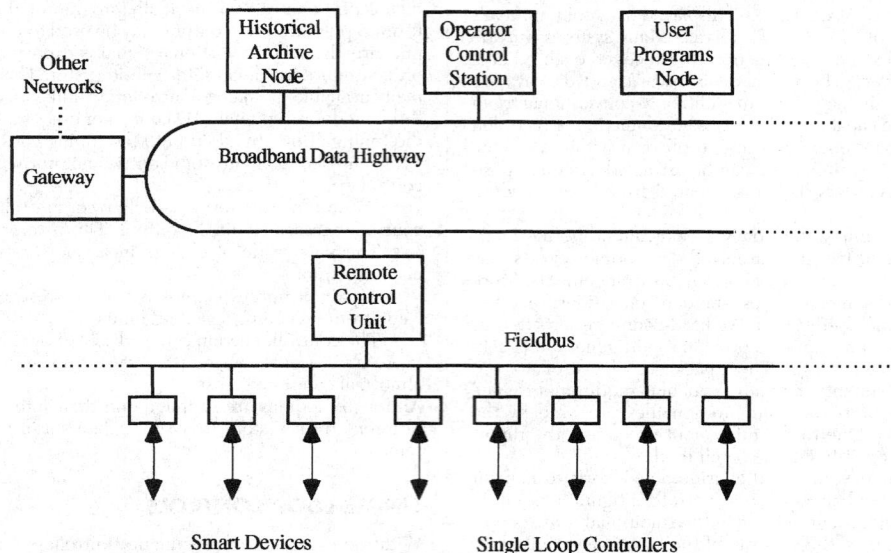

FIG. 8-63 A DCS using broadband data highway and fieldbus.

was initiated by General Motors in the early 1980s. This culminated in the Manufacturing Automation Protocol (MAP), which adopted the ISO-OSI standards as its basis. MAP specifies a broadband backbone local area network (LAN), which incorporates a selection of existing standard protocols suitable for manufacturing automation while defining additional protocols where no standard previously existed. Although intended for discrete component systems, MAP has evolved to address the integration of DCSs used in process control as well. Due to various technical reasons, MAP has gained limited acceptance by the process industries as of 1990. Engineering societies, including ISA and IEEE, and many operating companies are collaborating to refine MAP for wider support.

More microprocessor-based process equipment, such as smart instruments and single-loop controllers, with digital communications capability are now becoming available and are used extensively in process plants. A fieldbus, which is a low-cost protocol, is necessary to perform efficient communication between the DCS and these devices. So-called mini-MAP architecture was developed to satisfy process control and instrumentation requirements while incorporating existing ISA standards. It is intended to improve access time while allowing communications to a large number of microprocessor-based devices. The mini-MAP contains only three of the seven layers specified by the ISO-OSI model; therefore, a mini-MAP device cannot communicate with devices on the MAP bus directly. The development of MAP/EPA (Enhanced Performance Architecture) is parallel to that of the mini-MAP. This scheme adopts the full seven-layer model with a reduced set of MAP protocols. The MAP/EPA is compatible to both the complete MAP and the mini-MAP. Another benefit of standardizing the fieldbus is that it encourages third-party traditional equipment manufacturers to enter the smart equipment market, resulting in increased competition and improved equipment quality. Figure 8-63 illustrates a LAN-based DCS.

Irrespective of the protocol used, communication programs act as servers to the database manager. When some functions request data from a remote node, the database manager will transfer the request to the remote node database manager via the communication programs. The remote node communication programs will relay the request to the resident database manager and return the obtained data. The remote database access and the existence of communications equipment and software are transparent to plant operators.

CONTROLLERS, FINAL CONTROL ELEMENTS, AND REGULATORS

GENERAL REFERENCES: Baumann, *Control Valve Primer*, 2d ed., ISA, 1994; Considine, *Process/Industrial Instruments & Controls Handbook*, 4th ed., McGraw-Hill, 1993; Driskell, *Control-Valve Selection and Sizing*, ISA, 1983; Hammitt, *Cavitation and Multiphase Flow Phenomena*, McGraw-Hill, 1980; Norton, *Fundamentals of Noise and Vibration Analysis for Engineers*, Cambridge University Press, 1989; Ulanski, *Valve and Actuator Technology*, McGraw-Hill, 1991.

External control of the process is achieved by devices that are specially designed, selected and configured for the intended process-control application. The text below covers three very common function classifications of process-control devices: controllers, final control elements, and regulators.

The process controller is the "master" of the process-control system. It accepts a set point and other inputs and generates an output or outputs that it computes from a rule or set of rules that are part of its internal configuration. The controller output serves as an input to another controller or, more often, as an input to a final control element. The final control element is the device that affects the flow in the piping system of the process. The final control element serves as an interface between the process controller and the process. Control valves and adjustable speed pumps are the principal types discussed.

Regulators, though not controllers or final control elements, perform the combined function of these two devices (controller and final control element) along with the measurement function commonly associated with the process variable transmitter. The uniqueness, control performance, and widespread usage of the regulator make it deserving of a functional grouping of its own.

ELECTRONIC AND PNEUMATIC CONTROLLERS

Electronic Controllers   Almost all of the electronic process controllers used today are microprocessor-based devices. These processor-based controllers contain, or have access to, input/output (I/O) interface electronics that allow various types of signals to enter and leave the controller's processor. The controller, depending on its type, uses sufficient read-only-memory (ROM) and read/write-accessible-memory (RAM) to perform the controller function.

The resolution of the analog I/O channels of the controller vary somewhat, with 12-bit and 14-bit conversions quite common. Sample rates for the majority of the constant sample rate controllers range from 1 to 10 samples/second. Hard-wired single-pole, low-pass filters are installed on the analog inputs to the controller to protect the sampler from aliasing errors.

Distributed Control Systems   Some knowledge of the distributed control system (DCS) is useful in understanding electronic controllers. A DCS is a process control system with sufficient performance to support large-scale real-time process applications. The DCS has (1) an operations workstation with a cathode ray tube (CRT) for display; (2) a controller subsystem that supports various types of controllers and controller functions; (3) an I/O subsystem for transducing data; (4) a higher-level computing platform for performing process supervision, information processing, and analysis; and (5) communication networks to tie the DCS subsystems, plant areas, and other plant systems together.

The component controllers used in the controller subsystem portion of the DCS can be of various types and include multiloop controllers, programmable logic controllers, personal computer controllers, single-loop controllers, and fieldbus controllers. The type of electronic controller utilized depends on the size and functional characteristic of the process application being controlled. See the earlier section on distributed control systems.

Multiloop Controllers   The multiloop controller is a DCS network device that uses a single 32-bit microprocessor to provide control functions to many process loops. The controller operates independent of the other devices on the DCS network and can support from 20 to 500 loops. Data acquisition capability for up to 1000 analog and discrete I/O channels or more can also be provided by this controller.

The multiloop controller contains a variety of function blocks (for example, PID, totalizer, lead/lag compensator, ratio control, alarm, sequencer, and Boolean) that can be "soft-wired" together to form complex control strategies. The multiloop controller, as part of a DCS, communicates with other controllers and man/machine interface (MMI) devices also on the DCS network.

Programmable Logic Controllers   The programmable logic controller (PLC) originated as a solid-state replacement for the hard-wired relay control panel and was first used in the automotive industry for discrete manufacturing control. Today, PLCs are used to implement Boolean logic functions, timers, counters, and some math functions and PID control. PLCs are often used with on/off process control valves.

PLCs are classified by the number of the I/O functions supported. There are several sizes available, with the smallest PLCs supporting less than 128 I/O channels and the largest supporting over 1023 I/O channels. I/O modules are available that support high-current motor loads, general-purpose voltage and current loads, discrete inputs, analog I/O and special-purpose I/O for servomotors, stepping motors, high-speed pulse counting, resolvers, decoders, multiplexed displays, and keyboards. PLCs often come with lights or other discrete indicators to determine the status of key I/O channels.

When used as an alternative to a DCS, the PLC is programmed with a handheld or computer-based loader. The PLC is typically programmed with basic ladder logic or a high-level computer language such as BASIC, FORTRAN, or C. Programmable logic controllers use 16- or 32-bit microprocessors and offer some form of point-to-point communications such as RS-232C, RS-422, or RS-485.

Personal Computer Controller Because of its high performance at low cost and its unexcelled ease of use, application of the personal computer (PC) as a platform for process controllers is growing. When configured to perform scan, control, alarm, and data acquisition (SCADA) functions and combined with a spreadsheet or database management application, the PC controller can be a low-cost, basic alternative to the DCS or PLC.

Using the PC for control requires installation of a board into the expansion slot in the computer, or the PC can be connected to an external I/O module using a standard communication port on the PC (RS-232, RS-422, or IEEE-488). The controller card/module supports 16- or 32-bit microprocessors. Standardization and high volume in the PC market has produced a large selection of hardware and software tools for PC controllers.

Single-Loop Controller The single-loop controller (SLC) is a process controller that produces a single output. SLCs can be pneumatic, analog electronic, or microprocessor-based. Pneumatic SLCs are discussed in the pneumatic controller section, and analog electronic SLC is not discussed because it has been virtually replaced by the microprocessor-based design.

The microprocessor-based SLC uses an 8- or 16-bit microprocessor with a small number of digital and analog process input channels with control logic for the I/O incorporated within the controller. Analog inputs and outputs are available in the standard ranges (1–5 volts DC and 4–20 mA DC). Direct process inputs for temperature sensors (thermistor RTD and thermocouple types) are available. Binary outputs are also available. The face of the SLC has some form of visible display and pushbuttons that are used to view or adjust control values and configuration. SLCs are available for mounting in panel openings as small as 48 mm by 48 mm (1.9 by 1.9 inches).

The processor based SLC allows the operator to select a control strategy from a predefined set of control functions. Control functions include PID, on/off, lead/lag, adder/subtractor, multiply/divider, filter functions, signal selector, peak detector, and analog track. SLCs feature auto/manual transfer switching, multi-setpoint, self diagnostics, gain scheduling, and perhaps also time sequencing. Most processor-based SLCs have self-tuning PID control algorithms.

Sample times for the microprocessor-based SLCs vary from 0.1 to 0.4 seconds. Low-pass analog electronic filters are installed on the process inputs to stop aliasing errors caused by fast changes in the process signal. Input filter time constants are typically in the range from 0.1 to 1 s. Microprocessor-based SLCs may be made part of a DCS by using the communication port (RS-488 is common) on the controller or may be operated in a standalone mode independent of the DCS.

Fieldbus Controller The benefits of eliminating all analog communication links to and from the devices in the process loop (including final control elements and measurement transmitters) have stimulated considerable interest in standardizing a suitable digital fieldbus communication network. Although a universal network standard is not currently complete (see "Digital Field Communications" in this section), several manufacturers have made available field devices that feature basic process-controller functionality. These controllers, known as fieldbus controllers, reside in the final control element or measurement transmitter and are considered to be an option available with these control devices. A suitable communications modem is present in the device to interface with a proprietary, PC-based, or hybrid analog/digital bus network.

Presently, fieldbus controllers are single-loop controllers with 8- and 16-bit microprocessors and are options to digital field-control devices. These controllers support the basic PID control algorithm

and are projected to increase in functionality as the controller market develops. Parameters relating to intrinsic safety, communication type and bit rate, level of DCS support, and ultimate controller performance differentiate currently available fieldbus controllers.

Controller Reliability and Application Trends Critical process-control applications demand a high level of reliability from the electronic controller. Some methods that improve the reliability of electronic controllers include: (1) focusing on robust circuit design using quality components; (2) using redundant circuits, modules, or subsystems where necessary; (3) using small-sized backup systems when needed; (4) reducing repair time and using more powerful diagnostics; and (5) distributing functionality to more independent modules to limit the impact of a failed module.

Currently, the trend in process control is away from centralized process control and toward an increased number of small distributed-control or PLC systems. This trend will put emphasis on the evolution of the fieldbus controller and continued growth of the PC-based controller. Also, as hardware and software improves, the functionality of the controller will increase, and the supporting hardware will be physically smaller. Hence, the traditional lines between the DCS and the PLC will become less distinct as systems will be capable of supporting either function set.

Pneumatic Controllers The pneumatic controller is an automatic controller that uses pneumatic pressure as a power source and generates a single pneumatic output pressure. The pneumatic controller is used in single-loop control applications and is often installed on the control valve or on an adjacent pipestand or wall in close proximity to the control valve and/or measurement transmitter. Pneumatic controllers are used in areas where it would be hazardous to use electronic equipment, in locations without power, in situations where maintenance personnel are more familiar with pneumatic controllers, or in applications where replacement with modern electronic controls has not been justified.

Process-variable feedback for the controller is achieved by one of two methods. The process variable can (1) be measured and transmitted to the controller by using a separate measurement transmitter with a 0.2–1.0-bar (3–15-psig) pneumatic output, or (2) be sensed directly by the controller, which contains the measurement sensor within its enclosure. Controllers with integral sensing elements are available that sense pressure, differential pressure, temperature, and level. Some controller designs have the set point adjustment knob in the controller, making set point adjustment a local and manual operation. Other types receive a set point from a remotely located pneumatic source, such as a manual air set regulator or another controller, to achieve set point adjustment. There are versions of the pneumatic controller that support the useful one-, two-, and three-mode combinations of proportional, integral, and derivative actions. Other options include auto/manual transfer stations, antireset windup circuitry, on/off control, and process-variable and set point indicators.

Pneumatic controllers are made of Bourdon tubes, bellows, diaphragms, springs, levers, cams, and other fundamental transducers to accomplish the control function. If operated on clean, dry plant air, they offer good performance and are extremely reliable. Pneumatic controllers are available with one or two stages of pneumatic amplification, with the two-stage designs having faster dynamic response characteristics.

An example of a pneumatic PI controller is shown in Fig. 8-64a. This controller has two stages of pneumatic amplification and a Bourdon tube input element that measures process pressure. The Bourdon tube element is a flattened tube that has been formed into a curve so that changes in pressure inside the tube cause vertical motions to occur at the ungrounded end. This motion is transferred to the left end of the beam, as shown.

The resulting motion of the beam is detected by the pneumatic nozzle amplifier, which, by proper sizing of the nozzle and fixed orifice diameters, causes the pressure internal to the nozzle to rise and fall with vertical beam motion. The internal nozzle pressure is routed to the pneumatic relay. The relay, which is constructed like the booster relay described in the "Valve Control Devices" subsection, has a direct linear input-to-output pressure characteristic. The output of the relay is the controller's output and is piped away to the final control element.

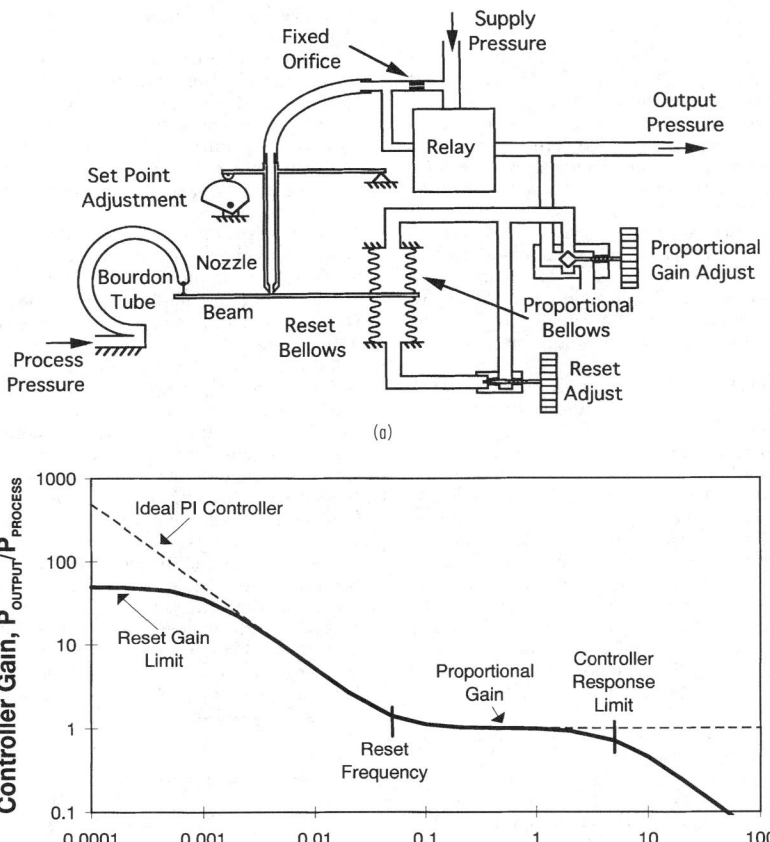

FIG. 8-64 Pneumatic controller: (*a*) example; (*b*) frequency response characteristic.

To generate the *P* and *I* control modes, a feedback circuit consisting of two bellows and two small metering valves has been added to the pneumatic amplifier system described above. The first valve is the proportional gain valve and is adjusted to provide an output pressure that is ratioed to the output pressure from the relay. The ratio used is set by manual adjustment of this valve. The output from the proportional gain valve is connected to a proportional feedback bellows, which provides negative feedback around the pneumatic amplifiers. The output pressure from the proportional gain valve is also connected to another valve, called the reset valve. This valve meters flow to and from a second bellows, known as the reset bellows. The resistance provided by the valve and the capacitance relating to the volume of the bellows forms a time constant that becomes the reset time constant for the controller.

The frequency response characteristics of a pneumatic PI controller and an ideal PI controller are shown in Fig. 8-64*b*. Notice that the gain of the pneumatic controller reaches a limit at the lowest frequencies. This limit is due to a less-than-infinite amount of forward amplifier gain. The manufacturer of the controller designs the forward gain term to be as high as possible to better approximate ideal reset action in the controller but never to reach the ideal. Reset gain for available pneumatic controllers runs between 20 and 100 times the gain implied by a proportional gain of unity. Unity proportional gain implies that a 100 percent change in process input (i.e., a full range change) will generate a 100 percent change in controller output.

The reset time, which is user-adjustable, can range from 0.05 seconds to 80 minutes or more, depending on controller design. The reset time constant, when converted to frequency $1/2(T_R)$ Hz (where T_R is the reset time in seconds), determines the frequency where the reset and proportional response characteristics of the controller merge (see Fig. 8-64*b*). Tuning the reset adjustment on the controller moves the reset frequency to the left or right along the frequency axis and thereby affects the reset action of the controller.

The response limit for the controller is a function of the design of the relay, the size of the load volume to which the controller is attached, the setting of the proportional band valve, and the forward gain designed into the pneumatic amplifiers. The frequency response limit (sometimes called the controller bandwidth) for a pneumatic controller into a small instrument load volume is in the 5- to 8-Hz range for two-stage pneumatic designs like the one shown. The dynamic response of the controller to set point changes is essentially the same as that indicated for process-variable changes. The set point adjustment mechanism affects the vertical motion of the nozzle over the beam and results in actions in the controller similar to those produced by changes in the process pressure.

The main shortcomings of the pneumatic controller is its lack of flexibility when compared to modern electronic controller designs. Increased range of adjustability, choice of alternate control algorithms, the communication link to the control system, and other features and services provided by the electronic controller make it a superior choice in most of today's applications.

CONTROL VALVES

A control valve consists of a valve, an actuator, and possibly one or more valve-control devices. The valves discussed in this section are applicable to throttling control (i.e., where flow through the valve is regulated to any desired amount between maximum and minimum limits). Other valves such as check, isolation, and relief valves are addressed in the next subsection. As defined, control valves are automatic control devices that modify the fluid flow rate as specified by the controller.

Valves   Valves are categorized according to their design style. These styles can be grouped into type of stem motion—linear or rotary. The valve stem is the rod, shaft, or spindle that connects the actuator with the closure member (i.e., a movable part of the valve that is positioned in the flow path to modify the rate of flow). Motion of either type is known as travel. The major categories are described briefly below.

Globe and Angle   The most common linear stem-motion control valve is the globe valve. The name comes from the globular shaped cavities around the port. In general, a port is any fluid passageway, but often the reference is to the passage that is blocked off by the closure member when the valve is closed. In globe valves, the closure member is called a plug. The plug in the valve shown in Fig. 8-65 is guided by a large-diameter port and moves within the port to provide the flow control orifice of the valve. A very popular alternate construction is a cage-guided plug as illustrated in Fig. 8-66. In many such designs, openings in the cage provide the flow control orifices. The valve seat is the zone of contact between the moving closure member and the stationary valve body, which shuts off the flow when the valve is closed. Often the seat in the body is on a replaceable part known as a seat ring. This stationary seat can also be designed as an integral part of the cage. Plugs may also be port-guided by wings or a skirt that fits snugly into the seat-ring bore.

One distinct advantage of cage guiding is the use of balanced plugs in single-port designs. The unbalanced plug depicted in Fig. 8-65 is subjected to a static pressure force equal to the port area times the valve pressure differential (plus the stem area times the downstream pressure) when the valve is closed. In the balanced design (Fig. 8-66),

note that both the top and bottom of the plug are subjected to the same downstream pressure when the valve is closed. Leakage via the plug-to-cage clearance is prevented by a plug seal. Both plug types are subjected to hydrostatic force due to internal pressure acting on the stem area and to dynamic flow forces when the valve is flowing.

The plug, cage, seat ring, and associated seals are known as the trim. A key feature of globe valves is that they allow maintenance of the trim via a removable bonnet without removing the valve body from the line. Bonnets are typically bolted on but may be threaded in smaller sizes.

Angle valves are an alternate form of the globe valve. They often share the same trim options and have the top-entry bonnet style. Angle valves can eliminate the need for an elbow but are especially useful when direct impingement of the process fluid on the body wall is to be avoided. Sometimes it is not practical to package a long trim within a globe body, so an angle body is used. Some angle bodies are self draining, which is an important feature for dangerous fluids.

Butterfly   The classic design of butterfly valves is shown in Fig. 8-67. Its chief advantage is high capacity in a small package and a very low initial cost. Much of the size and cost advantage is due to the wafer body design, which is clamped between two pipeline flanges. In the simplest design, there is no seal as such, merely a small clearance gap between the disc OD and the body ID. Often a true seal is provided by a resilient material in the body that is engaged via an interference fit with the disc. In a lined butterfly valve, this material covers the entire body ID and extends around the body ends to eliminate the need for pipeline joint gaskets. In a fully lined valve, the disc is also coated to minimize corrosion or erosion.

A high-performance butterfly valve has a disc that is offset from the shaft center line. This eccentricity causes the seating surface to move away from the seal once the disc is out of the closed position, reducing friction and seal wear. Also known as an eccentric disc valve, its advantage is improved shutoff while maintaining high ultimate capacity at a

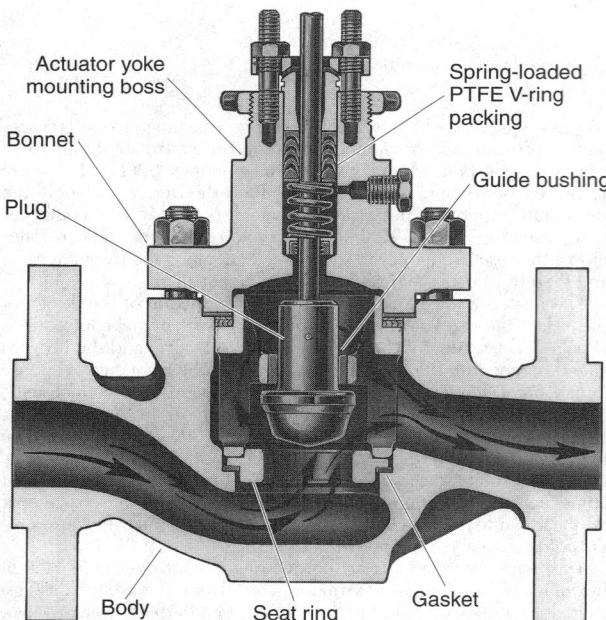

FIG. 8-65 Post-guided contour-plug globe valve with metal seat and raised-face flange end connections. (*Courtesy Fisher-Rosemount.*)

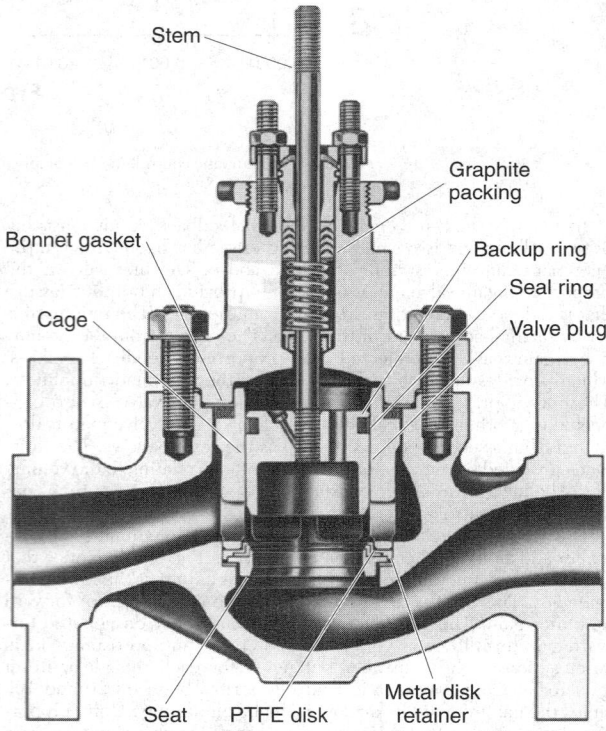

FIG. 8-66 Cage-guided balanced-plug globe valve with polymer seat and plug seal. (*Courtesy Fisher-Rosemount.*)

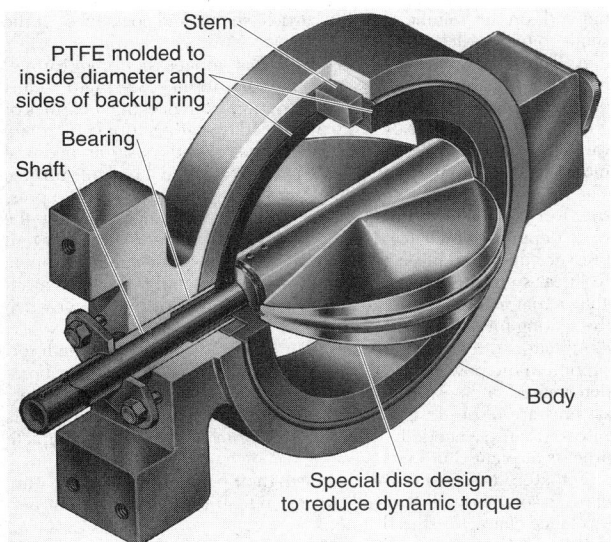

FIG. 8-67 Partial cutaway of wafer-style lined butterfly valve. (*Courtesy Fisher-Rosemount.*)

reasonable cost. This cost advantage relative to other design styles is particularly true in sizes above 6-inch nominal pipe size (NPS). Improved shutoff is due to advances in seal technologies, including polymer, flexing metal, combination metal with polymer inserts, and so on, many utilizing pressure assist.

Ball   Ball valves get their name from the shape of the closure member. One version uses a full spherical member with a cylindrical bore through it. The ball is rotated ¼ turn from the full-closed to the full-open position. If the bore is the same diameter as the mating-pipe fitting ID, the valve is referred to as full-bore. If the hole is undersized, the ball valve is considered to be a venturi style. A segmented ball is a portion of a hollow sphere—large enough to block the port when closed. Segmented balls often have a V-shaped contour along one edge, which provides a desirable flow characteristic (see Fig. 8-68). Both full-ball and segmented-ball valves are known for their low resistance to flow when full open. Shutoff leakage is minimized through the use of flexing or spring-loaded elastomeric or metal seals.

Bodies are usually in two or three pieces or have a removable retainer to facilitate installing seals. End connections are usually flanged or threaded in small sizes, although segmented-ball valves are offered in wafer style also.

Plug   There are two substantially different rotary-valve design categories referred to as plug valves. The first consists of a cylindrical or slightly conical plug with a port through it. The plug rotates to vary the flow much like a ball valve. The body is top-entry but is geometrically simpler than a globe valve and thus can be lined with fluorocarbon polymer to protect against corrosion. These plug valves have excellent shutoff but are generally not for modulating service due to high friction. A variation of the basic design (similar to the eccentric butterfly disc) only makes sealing contact in the closed position and is used for control.

The other rotary plug design is portrayed in Fig. 8-69. The seating surface is substantially offset from the shaft, producing a ball-valve-like motion with the additional cam action of the plug into the seat when closing. In reverse flow, high-velocity fluid motion is directed inward—impinging on itself and only contacting the plug and seat ring.

Multi-Port   This term refers to any valve or manifold of valves with more than one inlet or outlet. For throttling control, the three-way body is used for blending (two inlets, one outlet) or as a divertor (one inlet, two outlets). A three-way valve is most commonly a special globelike body with special trim that allows flow both over and under the plug. Two rotary valves and a pipe tee can also be used. Special three-, four-, and five-way ball-valve designs are used for switching applications.

Special Application Valves

Digital Valves   True digital valves consist of discrete solenoid-operated flow ports that are sized according to binary weighing. The valve can be designed with sharp-edged orifices or with streamlined nozzles that can be used for flow metering. Precise control of the throttling-control orifice is the strength of the digital valve. Digital valves are mechanically complicated and expensive, and they have considerably reduced maximum flow capacities over the globe and rotary valve styles.

Cryogenic Service   Valves designed to minimize heat absorption for throttling liquids and gases below 80 K are called cryogenic service valves. These valves are designed with small valve bodies to minimize heat absorption and long bonnets between the valve and actuator to allow for extra layers of insulation around the valve. For extreme cases, vacuum jacketing can be constructed around the entire valve to minimize heat influx.

High Pressure   Valves used for pressures nominally above 760

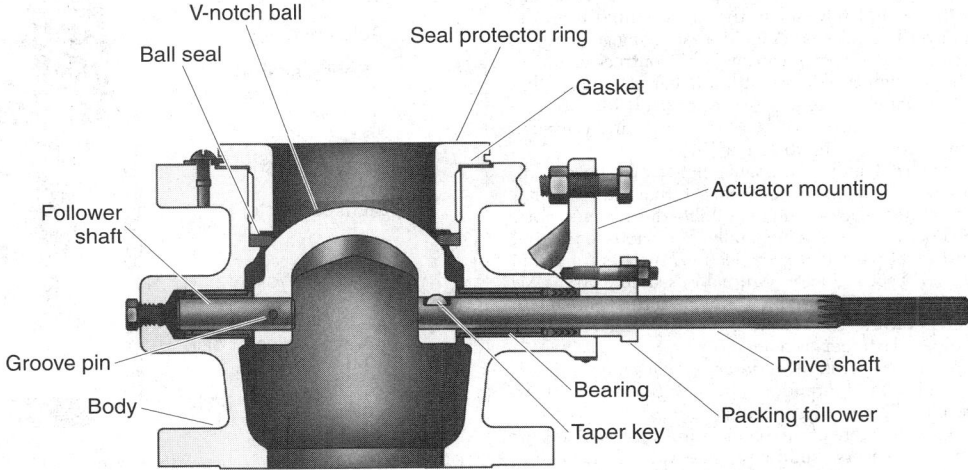

FIG. 8-68 Segmented ball valve. Partial view of actuator mounting shown 90° out of position. (*Courtesy Fisher-Rosemount.*)

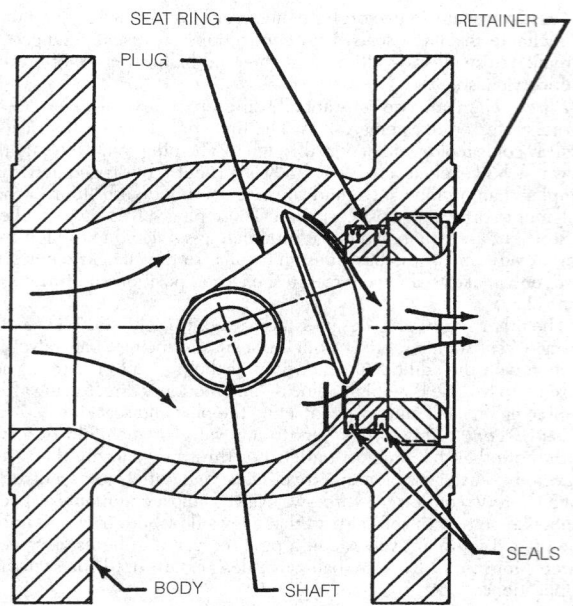

FIG. 8-69 Eccentric plug valve shown in erosion-resistant reverse flow direction. Shaded components can be made of hard metal or ceramic materials. (*Courtesy Fisher-Rosemount.*)

bar (11,000 psi, pressures above ANSI Class 4500) are often custom-designed for specific applications. Normally, these valves are of the plug type and use specially hardened plug and seat assemblies. Internal surfaces are polished, and internal corners and intersecting bores are smoothed to reduce high localized stresses in the valve body. Steam loops in the valve body are available to raise the body temperature to increase ductility and impact strength of the body material.

High-Viscous Process Used most extensively by the polymer industry, the valve for high-viscous fluids is designed with smooth finished internal passages to prevent stagnation and polymer degradation. These valves are available with integral body passages through which a heat-transfer fluid is pumped to keep the valve and process fluid heated.

Pinch The industrial equivalent of controlling flow by pinching a soda straw is the pinch valve. Valves of this type use fabric-reinforced elastomer sleeves that completely isolate the process fluid from the metal parts in the valve. The valve is actuated by applying air pressure directly to the outside of the sleeve, causing it to contract or pinch. Another method is to pinch the sleeve with a linear actuator with a specially attached foot. Pinch valves are used extensively for corrosive material service and erosive slurry service. This type of valve is used in applications with pressure drops up to 10 bar (145 psi).

Fire-Safe Valves that handle flammable fluids may have additional safety-related requirements for minimal external leakage, minimal internal (downstream) leakage, and operability during and after a fire. Being fire-safe does not mean being totally impervious to fire, but a sample valve must meet particular specifications such as American Petroleum Institute (API) 607, Factory Mutual Research Corp. (FM) 7440, or the British Standard 5146 under a simulated fire test. Due to very high flame temperature, metal seating (either primary or as a backup to a burned-out elastomer) is mandatory.

Solids Metering The control valves described earlier are primarily used for the control of fluid (liquid or gas) flow. Sometimes these valves, particularly the ball, butterfly, or sliding gate valves, are used to throttle dry or slurry solids. More often, special throttling mechanisms like venturi ejectors, conveyers, knife-type gate valves, or rotating vane valves are used. The particular solids-metering valve hardware

depends on the volume, density, particle shape, and coarseness of the solids to be handled.

Actuators An actuator is a device that applies the force (torque) necessary to cause a valve's closure member to move. Actuators must overcome pressure and flow forces; friction from packing, bearings or guide surfaces, and seals; and provide the seating force. In rotary valves, maximum friction occurs in the closed position and the moment necessary to overcome it is referred to as breakout torque. The rotary valve shaft torque generated by steady-state flow and pressure forces is called dynamic torque. It may tend to open or close the valve depending on valve design and travel. Dynamic torque per unit pressure differential is largest in butterfly valves at roughly 70° open. In linear stem-motion valves, the flow forces should not exceed available actuator force, but this is usually accounted for by default when the seating force is provided.

Actuators often provide a failsafe function. In the event of an interruption in the power source, the actuator will place the valve in a predetermined safe position, usually either full open or full closed. Safety systems are often designed to trigger local failsafe action at specific valves to cause a needed action to occur, which may not be a complete process or plant shutdown.

Actuators are classified according to their power source. The nature of these sources leads naturally to design features that make their performance characteristics distinct.

Pneumatic Despite the availability of more sophisticated alternatives, the pneumatically driven actuator is still by far the most popular type. Historically the most common has been the spring and diaphragm design (Fig. 8-70). The compressed air input signal fills a chamber sealed by an elastomeric diaphragm. The pressure force on the diaphragm plate causes a spring to be compressed and the actuator stem to move. This spring provides the failsafe function and contributes to the dynamic stiffness of the actuator. If the accompanying valve is "push-down-to-close," the actuator depicted in Fig. 8-70 would be described as "air-to-close" or synonymously as fail-open. A

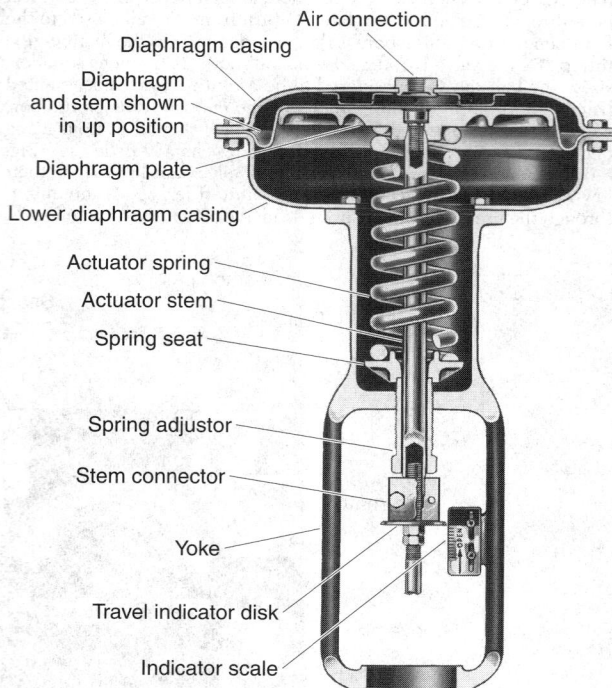

FIG. 8-70 Spring and diaphragm actuator with an "up" fail-safe mode. Spring adjuster allows slight alteration of bench set. (*Courtesy Fisher-Rosemount.*)

slightly different design yields "air-to-open" or fail-closed action. The spring is typically precompressed to provide a significant available force in the failed position (e.g., to provide seating load). The spring also provides a proportional relationship between the force generated by air pressure and stem position. The pressure range over which a spring and diaphragm actuator strokes in the absence of valve forces is known as the bench set. The chief advantages of spring and diaphragm actuators are their high reliability, low cost, adequate dynamic response, and failsafe action—all of which are inherent in their simple design.

Alternately, the pressurized chamber can be formed by a circular piston with a seal on its outer edge sliding within a cylindrical bore. Higher operating pressure (6 bar [~90 psig] is typical) and longer strokes are possible. Piston actuators can be spring-opposed but many times are in a dual-acting configuration (i.e., compressed air is applied to both sides of the piston with the net force determined from the pressure difference—see Fig. 8-71). Dynamic stiffness is usually higher with piston designs than with spring and diaphragm actuators; see "Positioner/Actuator Stiffness." Failsafe action, if necessary, is achieved without a spring through the use of additional solenoid valves, trip valves, or relays. See "Valve Control Devices."

Motion Conversion   Actuator power units with translational output can be adapted to rotary valves that generally need 90° or less rotation. A lever is attached to the rotating shaft and a link with pivoting means on the end connects to the linear output of the power unit, an arrangement similar to an internal combustion engine crankshaft, connecting rod, and piston. When the actuator piston, or more commonly the diaphragm plate, is designed to tilt, one pivot can be eliminated (see Fig. 8-71). Scotch yoke and rack and pinion arrangements are also commonly used, especially with piston power units. Friction and changing mechanical advantage of these motion-conversion mechanisms means the available torque may vary greatly with travel. One notable exception is vane-style rotary actuators whose offset "piston" pivots, giving direct rotary output.

Hydraulic   The design of typical hydraulic actuators is similar to double-acting piston pneumatic types. One key advantage is the high pressure (typically 35 to 70 bar [500 to 1000 psi]), which leads to high thrust in a smaller package. The incompressible nature of the

hydraulic oil means these actuators have very high dynamic stiffness. The incompressibility and small chamber size connote fast stroking speed and good frequency response. The disadvantages include high initial cost, especially when considering the hydraulic supply. Maintenance is much more difficult than with pneumatics, especially on the hydraulic positioner.

Electrohydraulic actuators have similar performance characteristics and cost/maintenance ramifications. The main difference is that they contain their own electric-powered hydraulic pump. The pump may run continuously or be switched on when a change in position is required. Their main application is remote sites without an air supply when a failsafe spring return is needed.

Electric   The most common electric actuators use a typical motor—three-phase AC induction, capacitor-start split-phase induction, or DC. Normally the motor output passes through a large gear reduction and, if linear motion output is required, a ball screw or thread. These devices can provide large thrust, especially given their size. Lost motion in the gearing system does create backlash, but if not operating across a thrust reversal, this type of actuator has very high stiffness. Usually the gearing system is self-locking, which means that forces on the closure member cannot move it by spinning a nonenergized motor. This behavior is called a lock-in-last-position failsafe mode. Some gear systems (e.g., low-reduction spur gears) can be backdriven. A solenoid-activated mechanical brake or locking current to motor field coils is added to provide lock-in-last-position fail mode. A battery backup system for a DC motor can guard against power failures. Otherwise, an electric actuator is not acceptable if fail-open/closed action is mandatory. Using electrical power requires environmental enclosures and explosion protection, especially in hydrocarbon-processing facilities; see the full discussion in "Valve Control Devices."

Unless sophisticated speed-control power electronics is used, position modulation is achieved via bang-bang control. Mechanical inertia causes overshoot, which is (1) minimized by braking and/or (2) hidden by adding dead band to the position control. Without these provisions, high starting currents would cause motors to overheat from constant "hunting" within the position loop. Travel is limited with power interruption switches or with force (torque) electromechanical cutouts when the closed position is against a mechanical stop (e.g., a globe valve). Electric actuators are often used for on/off service. Stepper motors can be used instead, and they, as their name implies, move in fixed incremental steps. Through gear reduction, the typical number of increments for 90° rotation range from 5000 to 10,000; hence positioning resolution at the actuator is excellent. Position overshoot is not an issue, and added dead band need only be a few steps away.

An electromagnetic solenoid can be used to directly actuate the plug on very small linear stem-motion valves. A solenoid is usually designed as a two-position device, so this valve control is on/off. Special solenoids with position feedback can provide proportional action for modulating control. Force requirements of medium-sized valves can be met with piloted plug designs, which use process pressure to assist the solenoid force. Piloted plugs are also used to minimize the size of common pneumatic actuators, especially when there is need for high seating load.

Manual   A manually positioned valve is by definition not an automatic control valve, but it may be involved with process control. For rotary valves, the manual operator can be as simple as a lever, but a wheel driving a gear reduction is necessary in larger-size valves. Linear motion is normally created with a wheel turning a screw-type device. A manual override is usually available as an option for the powered actuators listed above. For spring-opposed designs, an adjustable travel stop will work as a one-way manual input. In more complex designs the handwheel can provide loop control override via an engagement means. Some gear-reduction systems of electric actuators allow the manual positioning to be independent from the automatic positioning without declutching.

Valve-Control Devices   Devices mounted on the control valve that interface various forms of input signals, monitor and transmit valve position, or modify valve response are valve-control devices. In some applications, several auxiliary devices are used together on the

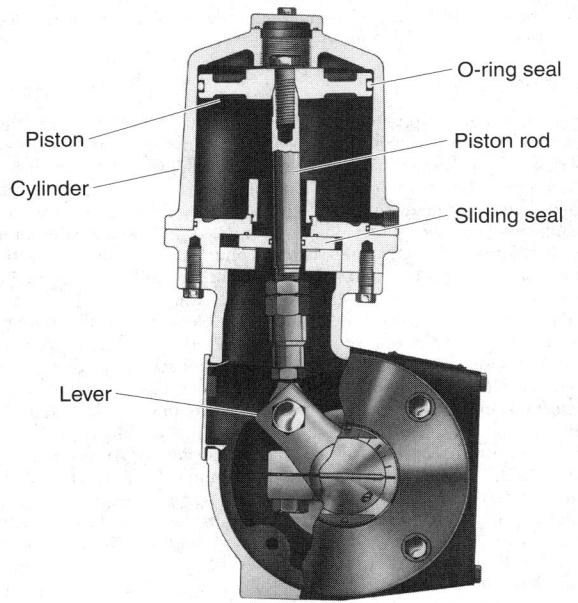

FIG. 8-71 Double-acting piston rotary actuator with lever and tilting piston for motion conversion. (*Courtesy Fisher-Rosemount.*)

O-ring seal

Piston

Piston rod

Cylinder

Sliding seal

Lever

same control valve. For example, mounted on the control valve, one may find a current-to-pressure transducer, a valve positioner, a volume booster relay, a solenoid valve, a trip valve, a limit switch, a process controller, and/or a stem-position transmitter. Figure 8-72 shows a valve positioner mounted on the yoke leg of a spring and diaphragm actuator.

As most throttling control valves are still operated by pneumatic actuators, the control-valve device descriptions that follow relate primarily to devices that are used with pneumatic actuators. The function of hydraulic and electrical counterparts are very similar. Specific details on a particular valve-control device are available from the vendor of the device.

Transducers   The current-to-pressure transducer (I/P transducer) is a conversion interface that accepts a standard 4–20 mA input current from the process controller and converts it to a pneumatic output in a standard pneumatic pressure range (normally 0.2–1.0 bar [3–15 psig] or, less frequently, 0.4–2.0 bar [6–30 psig]). The output pressure generated by the transducer is connected directly to the pressure connection on a spring-opposed diaphragm actuator or to the input of a pneumatic valve positioner.

Figure 8-73a is the schematic of a basic I/P transducer. The transducer shown is characterized by (1) an input conversion that generates an angular displacement of the beam proportional to the input current, (2) a pneumatic amplifier stage that converts the resulting angu-

FIG. 8-72   Valve and actuator with valve positioner attached. (*Courtesy Fisher-Rosemount.*)

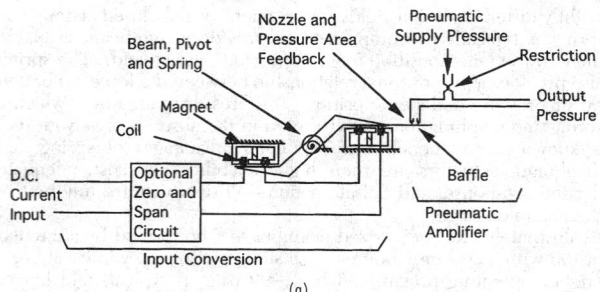

(a)

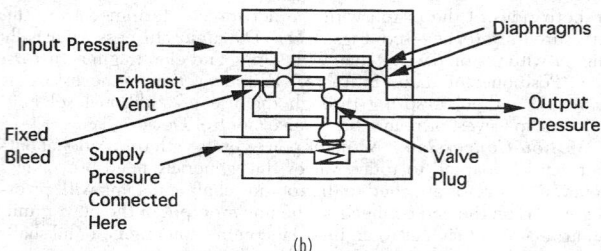

(b)

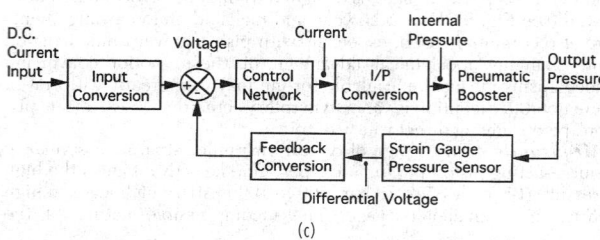

(c)

FIG. 8-73   Current to pressure transducer components parts: (*a*) direct current to pressure conversion; (*b*) pneumatic booster amplifier (relay); (*c*) block diagram of a modern I/P transducer.

lar displacement to pneumatic pressure, and (3) a pressure area that serves as a means to return the beam back to very near its original position when the new output pressure is achieved. The result is a device that generates a pressure output that tracks the input current signal. The transducer shown in Fig. 8-73a is used to provide pressure to small load volumes (normally 4.0 in³ or less), such as a positioner or booster input. With only one stage of pneumatic amplification, the flow capacity of this transducer is limited and not sufficient to provide responsive load pressure directly to a pneumatic actuator.

The flow capacity of the transducer can be increased by adding a booster relay like the one shown in Fig. 8-73b. The flow capacity of the booster relay is nominally fifty to one hundred times that of the nozzle amplifier shown in Fig. 8-73a and makes the combined transducer/booster suitably responsive to operate pneumatic actuators. This type of transducer is stable into all sizes of load volumes and produces measured accuracy (see Instrument Society of America [ISA]-S51.1-1979, "Process Instrumentation Terminology" for the definition of measured accuracy) of 0.5 percent to 1.0 percent of span.

Better measured accuracy results from the transducer design shown in Fig. 8-73c. In this design, pressure feedback is taken at the output of the booster-relay stage and fed back to the main summer. This allows the transducer to correct for errors generated in the pneumatic booster as well as errors in the I/P-conversion stage. Also, particularly with the new analog electric and digital versions of this design, PID control is used in the transducer-control network to give extremely good static accuracy, fast dynamic response, and reasonable stability into a wide range of load volumes (small instrument bellows to large actuators). Also, environmental factors such as temperature change, vibration, and supply pressure fluctuation affect this type of transducer the least. Even a perfectly accurate I/P transducer cannot compensate for stem-position errors generated by friction, backlash, and varying force loads coming from the actuator and valve. To do this compensation, a different control-valve device, known as a valve positioner, is required.

Valve Positioners The valve positioner, when combined with an appropriate actuator, forms a complete closed-loop valve-position control system. This system makes the valve stem conform to the input signal coming from the process controller in spite of force loads that the actuator may encounter while moving the control valve. Usually, the valve positioner is contained in its own enclosure and is mounted on the control valve.

The key parts of the positioner/actuator system, shown in Fig. 8-74a, are (1) an input-conversion network, (2) a stem-position feedback network, (3) a summing junction, (4) an amplifier network, and (5) an actuator.

The input-conversion network shown is the interface between the input signal and the summer. This block converts the input current or pressure (from an I/P transducer or a pneumatic process controller) to a voltage, an electric current, a force, torque, displacement or other particular variable that can be directly used by the summer. The input conversion usually contains a means to adjust the slope and offset of the block to provide for a means of spanning and zeroing the positioner during calibration. In addition, means for changing the sense (known as "action") of the input/output characteristic are oftentimes addressed in this block. Also, exponential, logarithmic or other predetermined characterization can be put in this block to provide a characteristic that is useful in offsetting or reinforcing a nonlinear valve or process characteristic.

The stem-position feedback network converts stem travel to a useful form for the summer. This block includes the feedback linkage, which varies with actuator type. Depending on positioner design, the stem-position feedback network can provide span and zero and characterization functions similar to that described for the input-conversion block.

The amplifier network provides signal conversion and suitable static and dynamic compensation for good positioner performance. Control from this block usually reduces down to a form of proportional or proportional plus derivative control. The output from this block in the case of a pneumatic positioner is a single connection to the spring and diaphragm actuator or two connections for push-pull operation of a springless piston actuator. The action of the amplifier network and the action of the stem-position feedback can be reversed together to provide for reversed positioner action.

By design, the gain of the amplifier network shown in Fig. 8-74a is made very large. Large gain in the amplifier network means that only a small proportional deviation will be required to position the actuator through its active range of travels. This means that the signals into the summer track very closely and that the gain of the input-conversion block and the stem-position feedback block determine the closed-loop relationship between the input signal and the stem travel.

Large amplifier gain also means that only a small amount of additional stem-travel deviation will result when large external force loads are applied to the actuator stem. For example, if the positioner's amplifier network has a gain of 50 and assuming that high packing-box friction loads require 25 percent of the actuator's range of thrust to move the actuator, then only 25 percent/50 or 0.5 percent deviation between input signal and output travel will result due to valve friction.

Figure 8-74b is an example of a pneumatic positioner/actuator. The input signal is a pneumatic pressure that (1) moves the summing beam, which (2) operates the spool valve amplifier, which (3) provides flow to and from the piston actuator, which (4) causes the actuator to move and continue moving until (5) the feedback force returns the beam to its original position and stops valve travel at a new position. Typical positioner operation is thereby achieved.

Static performance measurements related to positioner/actuator operation are: conformity, measured accuracy, hysteresis, dead band, repeatability, and locked stem-pressure gain. Definitions and standardized test procedures for determining these measurements can be found in ISA-S75.13-1989, "Method of Evaluating the Performance of Positioners with Analog Input Signals and Pneumatic Output".

Dynamics of Pneumatic Positioners Dynamically, the pneumatic positioner is characterized by the combined effects of gain and capacitance and nonlinear effects such as valve friction and flow-capacity saturation. Generally, there is a threshold level of input signal below which the positioner output will not respond at all. This band is on the order of 0.1 percent of input span for pneumatic positioners but can be larger if significant valve friction is present. Above this threshold level, but below the level that causes velocity saturation, the positioner is approximately linear and likened to a second-order low-pass filter (see Fig. 8-75a). Natural frequencies range from 0.3 to 3.0 Hz and damping ratios of 0.6 to 2.0 are common and dependent on positioner design and the physical size of the actuator volume. At higher drive levels, the flow capacity of the positioner is reached and attenuation of the resulting travel begins.

Positioner/Actuator Stiffness Minimizing the effect of dynamic loads on valve-stem travel is an important characteristic of the positioner/actuator. Stem position must be maintained in spite of changing reaction forces caused by valve throttling. These forces can be random in nature (buffeting force) or result from a negative sloped force/stem travel characteristic (negative gradient); either could result in valve-stem instability and loss of control. To reduce and eliminate the effect of these forces, the effective stiffness of the positioner/actuator must be made sufficiently high to maintain adequate stability of the valve stem.

The stiffness characteristic of the positioner/actuator varies with frequency. Figure 8-75b indicates the stiffness of the positioner/actuator is high at low frequencies and is directly related to the locked-stem pressure gain provided by the positioner. As frequency increases, a dip in the stiffness curve results from dynamic gain attenuation in the pneumatic amplifiers in the positioner. The value at the bottom of the dip is the sum of the mechanical stiffness of the spring in the actu-

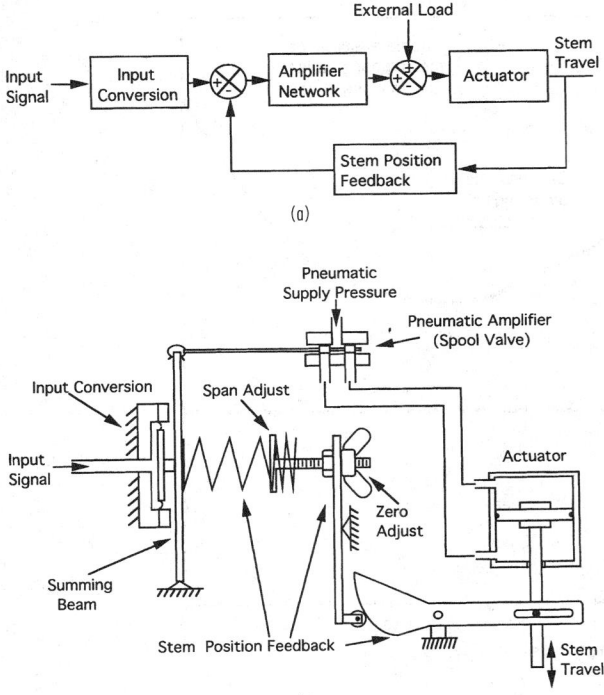

FIG. 8-74 Positioner/actuators: (*a*) generic block diagram; (*b*) an example of a pneumatic positioner/actuator.

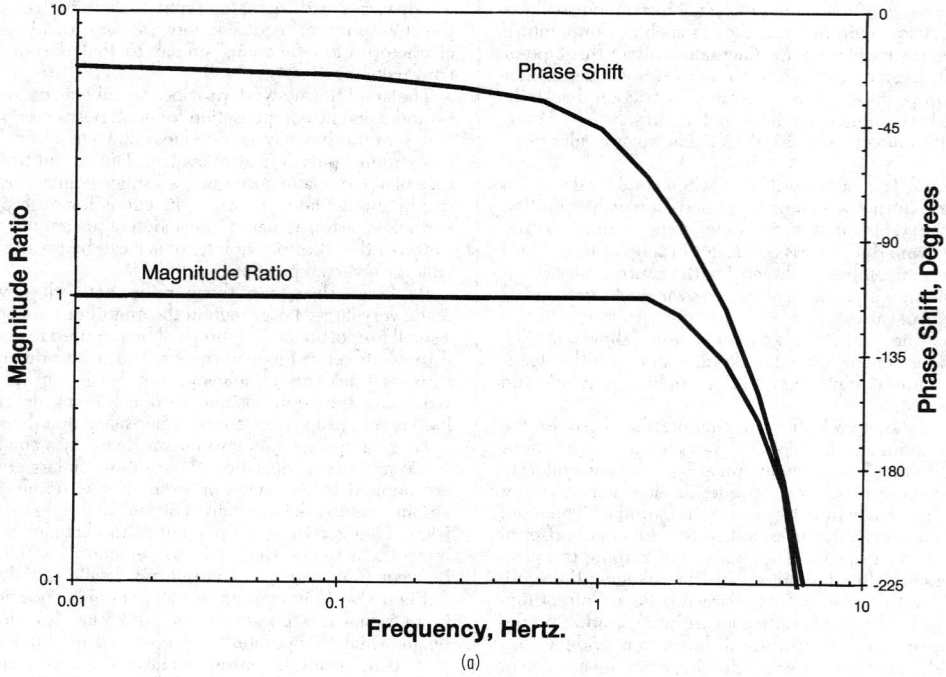

(a)

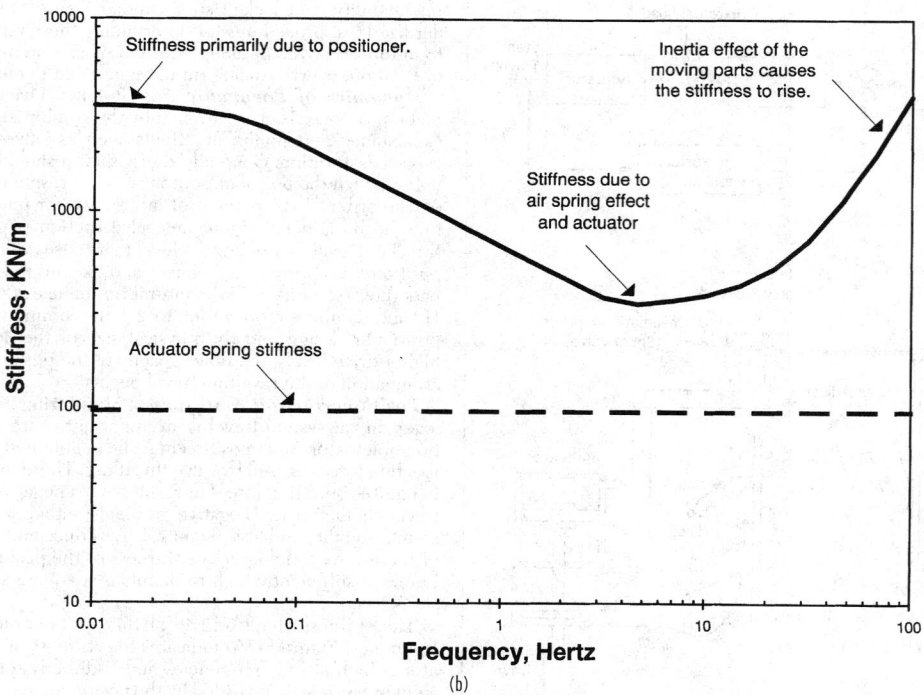

(b)

FIG. 8-75 Frequency response curves for a pneumatic positioner/actuator: (*a*) input signal to stem travel for a 69-inch^2 spring and diaphragm actuator with a 1.5-inch total travel and 3–15 psig input pressure; (*b*) dynamic stiffness for the same positioner/actuator.

ator and the air spring effect produced by air enclosed in the actuator casing.

The air spring effect results from adiabatic expansion and compression of the air in the actuator casing. Numerically, the small perturbation value for air spring stiffness in Newtons/meter is given by Eq. (8-107).

$$\text{Air spring rate} = \frac{\gamma P_x A_a^2 a}{V} \qquad (8\text{-}107)$$

where γ is ratio of specific heats (1.4 for air), P_a is the actuator pressure in Pascal absolute, A_a is the actuator pressure area in m², and V is the internal actuator volume in m³.

Notice in the figure that the minimum stiffness value (mechanical spring stiffness + air spring stiffness) is several times larger than the stiffness produced by the spring in the actuator (shown as a dotted line) by itself. This indicates that the air spring stiffness is quite significant and worth considering in actuator design and actuator sizing. To the right of the dip, the inertia effects of the mass of the moving parts of the valve and actuator cause the overall system stiffness to rise with increasing frequency.

Positioner Application   Positioners are widely used on pneumatic valve actuators. More often than not, they provide improved process-loop control because they reduce valve-related nonlinearity. Dynamically, positioners maintain their ability to improve control-valve performance for sinusoidal input frequencies up to about one half of the positioner bandwidth. At input frequencies greater than this, the attenuation in the positioner amplifier network gets large, and valve nonlinearity begins to affect final control-element performance more significantly. Because of this, the most successful use of the positioner occurs when the positioner-response bandwidth is greater than twice that of the most dominant time lag in the process loop.

Some typical examples of where the dynamics of the positioner are sufficiently fast to improve process control are the following:

1. *In a distributed control system (DCS) process loop with an electronic transmitter.*   The DCS controller and the electronic transmitter have time constants that are dominant over the positioner response. Positioner operation is therefore beneficial in reducing valve-related nonlinearity.

2. *In a process loop with a pneumatic controller and a large process time constant.*   Here the process time constant is dominant, and the positioner will improve the linearity of the final control element. Some common processes with large time constants that benefit from positioner application are liquid level, temperature, large volume gas pressure, and mixing.

3. Additional situations where valve positioners are used are as follows:

• On springless actuators where the actuator is not usable for throttling control without position feedback.

• When split ranging is required to control two or more valves sequentially. In the case of two valves, the smaller control valve is calibrated to open in the lower half of the input signal range and a larger valve is calibrated to open in the upper half of the input signal range. Calibrating the input command signal range in this way is known as split-range operation and increases the practical range of throttling process flows over that of a single valve.

• In open-loop control applications where best static accuracy is needed.

On occasion, positioner use can degrade process control. Such is the case when the process controller, the process, and the process transmitter have time constants that are similar or smaller than that of the positioner/actuator. This situation is characterized by low process-controller P gain (P gain < 0.5), and hunting or limit cycling of the process variable is observed. Improvements here can be made by doing one of the following:

• Install a dominant first-order low-pass filter in the loop ahead of the positioner and retune the process loop. This should allow increased proportional gain in the process loop and reduce hunting. Possible means for adding the filter include adding it to the firmware of the DCS controller, by adding an external RC network on the output of the process controller or by enabling the filter function in the

input of the positioner if it is available. Also, some transducers, when connected directly to the actuator, form a dominant first-order lag that can be used to stabilize the process loop.

• Select a positioner with a faster response characteristic.

Booster Relays   The booster relay is a single-stage power amplifier having a fixed gain relationship between the input and output pressures. The device is packaged as a complete standalone unit with pipe-thread connections for input, output, and supply pressure. The booster amplifier shown in Fig. 8-73*b* shows the basic construction of the booster relay. Enhanced versions are available that provide specific features such as: (1) variable gain to split the output range of a pneumatic controller to operate more than one valve or to provide additional actuator force; (2) low hysteresis for relaying measurement and control signals; (3) high flow capacity for increased actuator-stroking speed; and (4) arithmetic, logic, or other compensation functions for control system design.

A particular type of booster relay, called a dead-band booster, is shown in Fig. 8-76. This booster is designed to be used exclusively between the output of a valve positioner and the input to a pneumatic actuator. It is designed to provide extra flow capacity to stroke the actuator faster than with the positioner alone. The dead-band booster is designed intentionally with a large dead band (approximately 5 percent of the input span), elastomer seats for tight shutoff, and an adjustable bypass valve connected between the input and the output of the booster. The bypass valve is tuned to provide the best compromise between increased actuator stroking speed and positioner/actuator stability.

With the exception of the dead-band booster, the application of booster relays has diminished somewhat by the increased use of current-to-pressure transducers, electropneumatic positioners, and electronic control systems. Transducers and valve positioners serve much the same functionality as the booster relay in addition to interfacing with the electronic process controller.

Solenoid Valves   The electric solenoid valve has two output states. When sufficient electric current is supplied to the coil, an internal armature moves against a spring to an extreme position. This motion causes an attached pneumatic or hydraulic valve to operate. When current is removed, the spring returns the armature and the attached solenoid valve to the deenergized position. An intermediate pilot stage is sometimes used when additional force is required to operate the main solenoid valve. Generally, solenoid valves are used to pressurize or vent the actuator casing for on/off control-valve application and safety shutdown applications.

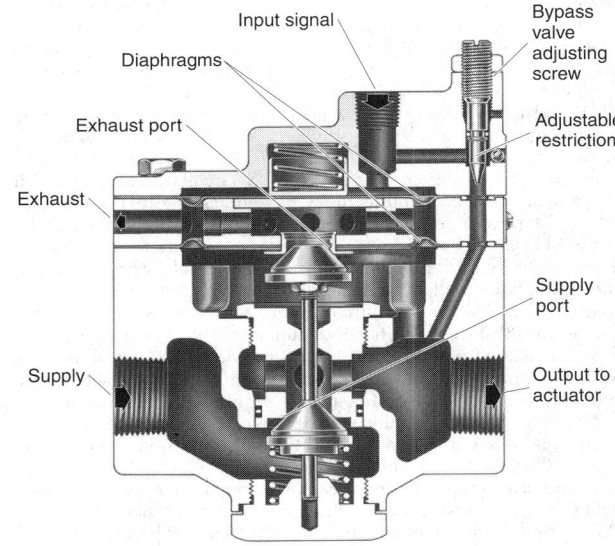

FIG. 8-76   Dead-band booster relay. (*Courtesy Fisher-Rosemount.*)

Trip Valves The trip valve is part of a system that is used where a specific valve action (i.e., fail up, fail down, or lock in last position) is required when pneumatic supply pressure to the control valve falls below a preset level. Trip systems are used primarily on springless piston actuators requiring fail-open or fail-closed action. An air storage or "volume" tank and a check valve are used with the trip valve to provide power to stroke the valve when supply pressure is lost. Trip valves are designed with hysteresis around the trip point to avoid instability when the trip pressure and the reset pressure settings are too close to the same value.

Limit Switches and Stem-Position Transmitters Travel-limit switches, position switches, and valve-position transmitters are devices that, when mounted on the valve, actuator, damper, louver, or other throttling element, detect the component's relative position. The switches are used to operate alarms, signal lights, relays, solenoid valves, or discrete inputs into the control system. The valve-position transmitter generates a 4–20-mA output that is proportional to the position of the valve.

Fire and Explosion Protection Electrical equipment can be a source of ignition in environments with combustible concentrations of gas, liquid, dust, fibers, or flyings. Most of the time it is possible to locate the electronic equipment away from these hazardous areas. However, where electric or electronic valve-mounted instruments must be used in areas where there is a hazard of fire or explosion, the equipment must be designed to meet requirements for safety. Articles 500 through 504 of the National Electrical Code address definitions of hazardous locations and the requirements for electrical devices in locations where fire or explosion hazard exists. NFPA (National Fire Protection Agency) 497M addresses the properties and group classification of gas, vapor, and dust for electrical devices in hazardous locations. With valve-mounted accessories, the approved protection concepts most often used for safety protection are explosion-proof, intrinsically safe, nonincendive, and dust-ignition-proof.

The explosion-proof enclosure is designed such that an explosion in the interior of the enclosure containing the electronic circuits will be contained. The enclosure will not allow sufficient flame to escape to the exterior to cause an ignition. Also, a surface temperature rating is given to the device. This rating must indicate a lower surface temperature than the ignition temperature of the gas in the hazardous area.

Explosion-proof enclosures are characterized by strong metal enclosures with special close-fitting access covers and breathers that contain an ignition to the inside of the enclosure. Field wiring in the hazardous environment is enclosed in a metal conduit of the mineral-insulated-cable type. All conduit and cable connections or cable terminations are threaded and explosion-proof. Conduit seals are put into the conduit or cable system at locations defined by the National Electric Code (Article 501) to prevent gas and vapor leakage and to prevent flames from passing from one part of the conduit system to the other.

The intrinsically safe (I.S.) control-valve device contains circuits that are incapable of releasing sufficient electrical or thermal energy to cause ignition of a specified hazardous mixture under normal or fault operating conditions of the circuit. I.S. circuits are designed with voltage- and current-limiting networks added where necessary to achieve approved levels of safety. I.S. field wiring need not be enclosed in metal conduit but must be kept separate from wiring for nonintrinsically safe circuits. Intrinsically safe field wiring must be energy limited, usually by a Zener diode barrier circuit located in the control room. The manufacture of the intrinsically safe control-valve devices must list the identification number of the control drawing on the nameplate attached to the approved device. The control drawing contains information showing approved combinations of accessories and other connected apparatus such as Zener diode energy barriers.

ANSI/ISA S12.12, "Nonincendive Electrical Equipment for Use in Class I and Class II, Division 2 and Class III, Division 1 and 2 Hazardous (Classified) Locations," addresses requirements for nonincendive electrical equipment and wiring. Nonincendive apparatus and/or field wiring refers to approved equipment or wiring that is incapable of imparting sufficient energy to ignite the specified hazardous atmosphere under normal circuit operating conditions. Nonincendive protection is considered for applications where hazardous concentrations

of flammable gas and vapors or combustible dusts are only present under abnormal operating conditions or in those applications where easily ignited fibers or flyings are present in sufficient quantities to cause ignitable mixtures. For applications where hazardous concentrations of flammable gas and vapors or combustible dust are present continuously, intermittently, or periodically, more stringent protection (see Division 1, National Electrical Code, Article 500) offered by explosion-proof or intrinsically safe concepts is required.

The dust-ignition-proof protection concept excludes dust from entering the device enclosure and will not permit arcs, sparks, or heat generated by the device to cause ignition of external suspensions or accumulations of the dust. Enclosure requirements can be found in ANSI/UL 1203-1994, "Explosion-Proof and Dust-Ignition-Proof Electrical Equipment for Use in Hazardous Locations."

Certified testing and approval for control-valve devices used in hazardous locations is normally procured by the manufacturer of the device. The manufacturer often goes to a third party laboratory for testing and certification. Applicable approval standards are available from CSA, CENELEC, FM, SAA, and UL.

Environmental Enclosures Enclosures for valve accessories are sometimes required to provide protection from specific environmental conditions. The National Electrical Manufacturers Association (NEMA) provides descriptions and test methods for equipment used in specific environmental conditions in NEMA 250. Protection against rain, windblown dust, hose-directed water, and external ice formation are examples of environmental conditions that are covered by NEMA standards.

Also, the electronic control-valve device's level of immunity to, and emission of, electromagnetic interference (EMI) can be an issue in the chemical-valve environment. EMI requirements for the control-valve devices are presently mandatory in the European Community but voluntary in the United States, Japan, and the rest of the world. International Electrotechnical Commission (IEC) 801, Parts 1 through 4, "Electromagnetic Compatibility for Industrial Process Measurement and Control Equipment," defines tests and requirements for control-device immunity. Immunity and emission standards are addressed in CENELEC (European Committee for Electrotechnical Standardization) EN 50 081-1:1992, EN 50 081-2:1993, EN 50 082-1:1992, and prEN 50 082-2:1994.

Digital Field Communications An increasing number of valve-mounted devices are available that support digital communications in addition to, or in place of, the traditional 4–20 mA current signal. These control-valve devices have increased functionality, resulting in reduced setup time, improved control, combined functionality of traditionally separate devices, and control-valve diagnostic capability. Digital communications also allow the control system to become completely distributed where, for example, the process PID controller could reside in the valve positioner or in the process transmitter.

The high-performance, all-digital, multidrop communication protocol for use in the process-control industry is known as fieldbus. Presently there are several regional and industry-based fieldbus standards including the French standard, FIP (NFC 4660x approved by UTE), the German standard, Profibus (DIN 19245 approved by DKE), and proprietary standards by DCS vendors. As of 1997, none of these fieldbus standards have been adopted by international standards organizations.

The International Electrotechnical Commission (IEC) Standards Committee 65C Working Group 6 (IEC SC65C WG6) and the ISA Standards and Practices Committee 50 (ISA SP50) are presently working on a fieldbus standard, but at the time of this writing, the standard is unfinished. One interim solution supported by some valve-device products is the hybrid communication method, where both analog and digital communication capabilities are present in the same device. This scheme has the advantage of allowing the communicating valve-control device to be retrofit into a traditional 4–20 mA current loop and still support digital communications between the final control element and the control room. Here the current signal is used to communicate the primary signal value, and the digital communication channel carries secondary variable information, configuration information, calibration information, and alert and diagnostic information.

An example of a hybrid protocol that is open (not proprietary) and

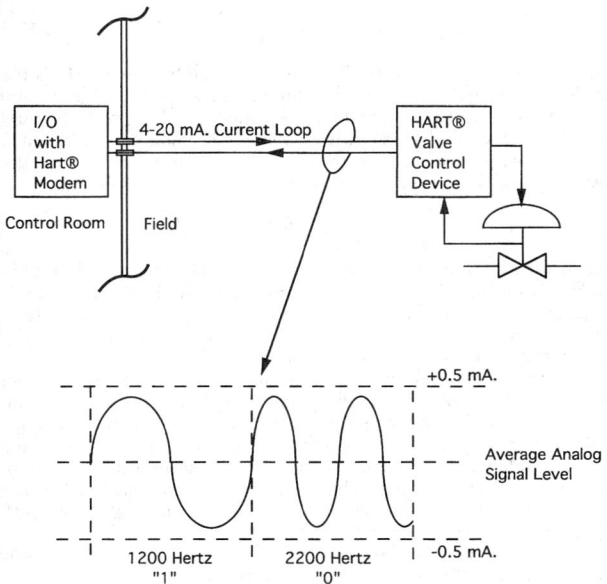

FIG. 8-77 Hybrid point-to-point communications between the control room and the control valve device.

in use by several manufacturers of control-valve devices is known as HART®° (Highway Addressable Remote Transducer) protocol (see Fig. 8-77). With this protocol, the digital communications occur over the same two wires that provide the 4–20 mA process control signal without disrupting the process signal. The protocol uses the frequency-shift keying (FSK) technique where two individual frequencies, one representing the mark and the other representing the space, are superimposed on the 4–20 mA current signal. As the average value of the signals used is zero, there is no DC offset value added to the 4–20 mA signal.

Valve Application Technology   Functional requirements and the properties of the controlled fluid determine which valve and actuator types are best for a specific application. If demands are modest and no unique valve features are required, the valve-design style selection may be determined solely by cost. If so, general-purpose globe or angle valves provide exceptional value, especially in sizes less than 3-inch NPS and hence are very popular. Beyond type selection, there are many other valve specifications that must be determined properly in order to ultimately yield-improved process control.

Materials and Pressure Ratings   Valves must be constructed from materials that are sufficiently immune to corrosive or erosive action by the process fluid. Common body materials are cast iron, steel, stainless steel, high-nickel alloys, and copper alloys such as bronze. Trim materials usually need a greater immunity due to the higher fluid velocity in the throttling region. High hardness is desirable in erosive and cavitating applications. Heat-treated and precipitation-hardened stainless steels are common. High hardness is also good for guiding, bearing, and seating surfaces; cobalt-chromium alloys are utilized in cast or wrought form and frequently as welded overlays called hard facing. In less stringent situations, chrome plating, heat-treated nickel coatings, and ion nitriding are used. Tungsten carbide and ceramic trim are warranted in extremely erosive services. See Sec. 28, "Materials of Construction," for specific material properties.

Since the valve body is a pressurized vessel, it is usually designed to comply with a standardized system of pressure ratings. Two common systems are described in the standards ANSI B16.34 and DIN 2401.

Internal pressure limits under these standards are divided into broad classes, with specific limits being a function of material and temperature. Manufacturers also assign their own pressure ratings based on internal design rules. A common insignia is "250 WOG," which means a pressure rating of 250 psig (~17 bar) in water, oil, or gas at ambient temperature. "Storage and Process Vessels" in Sec. 10 provides introductory information on compliance of pressure-vessel design to industry codes (e.g., ASME Boiler and Pressure Vessel Code—Section VIII, ASME B31.3 Chemical Plant and Petroleum Refinery Piping).

Valve bodies are also standardized to mate with common piping connections: flanged, butt-weld end, socket-weld end, and screwed end. Dimensional information for some of these joints and class pressure-temperature ratings are included in Sec. 10, "Process Plant Piping." Control valves have their own standardized face-to-face dimensions that are governed by ISA Standards S75.03, 04, 12, 14, 15, 16, 20, and 22. Butterfly valves are also governed by API 609 and Manufacturers Standardization Society (MSS) SP-67 and 68.

Sizing   Throttling control valves must be selected to pass the required flow rate given expected pressure conditions. Sizing is not merely matching the end connection size with surrounding piping; it is a key step in ensuring that the process can be properly controlled. Sizing methods range from simple models based on elementary fluid mechanics to very complex models when unusual thermodynamics or nonideal behaviors occur. Basic sizing practices have been standardized upon (e.g., ISA S75.01) and are implemented as PC-based programs by manufacturers. The following is a discussion of very basic sizing equations and the associated physics.

Regardless of the particular process variable being controlled (e.g., temperature, level, pH), the output of a control valve is flow rate. The throttling valve performs its function of manipulating flow rate by virtue of being an adjustable resistance to flow. Flow rate and pressure conditions are normally known when a process is designed and the valve resistance must be matched accordingly. In the tradition of orifice and nozzle discharge coefficients, this resistance is embodied in the valve flow coefficient C_v. The mass flow rate (w) in kg/h is given for a liquid by

$$w = 27.3 C_v \sqrt{\rho(p_1 - p_2)} \qquad (8-108)$$

where p_1 and p_2 are upstream and downstream static pressure in bar, respectively. The density of the fluid ρ is expressed in kg/m³. This equation is valid for nonvaporizing, turbulent-flow conditions for a valve with no attached fittings. The relationship can be derived from the principles of conservation of mass and energy. A more complete presentation of sizing relationships is given in ISA S75.01, including provisions for pipe reducers, vaporizing liquids, and Reynolds number effects.

While the above equation gives the relationship between pressure and flow from a macroscopic point of view, it does not explain what is going on inside the valve. Valves create a resistance to flow by restricting the cross sectional area of the flow passage and also by forcing the fluid to change direction as it passes through the body and trim. The conservation of mass principle dictates that, for steady flow, density × average velocity × cross sectional area equals a constant. The average velocity of the fluid stream at the minimum restriction in the valve is therefore much higher than at the inlet. Note that due to the abrupt nature of the flow contraction that forms the minimum passage, the main fluid stream may separate from the passage walls and form a jet that has an even smaller cross section, the so-called vena contracta. The ratio of minimum stream area to the corresponding passage area is called the contraction coefficient. As the fluid expands from the minimum cross sectional area to the full passage area in the downstream piping, large amounts of turbulence are generated. Direction changes can also induce significant amounts of turbulence.

Some of the potential energy that was stored in the fluid by pressurizing it (e.g., the work done by a pump) is first converted into the kinetic energy of the fast-moving fluid at the vena contracta. Some of that kinetic energy turns into the kinetic energy of turbulence. As the turbulent eddies break down into smaller and smaller structures, viscous effects ultimately convert all of the turbulent energy into heat. Therefore, a valve converts fluid energy from one form to another.

For many valve constructions, it is reasonable to approximate the fluid transition from the valve inlet to the minimum cross section of the flow stream as an isentropic or lossless process. This minimum pressure, p_{vc}, can be estimated from the Bernoulli relationship. See Sec. 6 ("Fluid and Particle Mechanics") for more background information. Downstream of the vena contracta, the flow is definitely not lossless due to all the turbulence that is generated. As the flow passage area increases and the fluid slows down, some of the kinetic energy of the fluid is converted back to potential energy as the pressure recovers. The remaining energy that is permanently lost via turbulence accounts for the permanent pressure or head loss of the valve. The relative amount of pressure that is recouped determines whether the valve is considered to be high or low recovery. See Fig. 8-78 for an illustration of how the mean pressure changes as fluid moves through a valve. The flow-passage geometry at and downstream of the vena contracta primarily determines the amount of recovery. The amount of recovery is quantified by the *liquid* pressure recovery factor F_L where

$$F_L = \sqrt{\frac{p_1 - p_2}{p_1 - p_{vc}}} \qquad (8\text{-}109)$$

A key limitation of sizing Eq. (8-109) is the limitation to incompressible fluids. For gases and vapors, density is dependent on pressure. For convenience, compressible fluids are often assumed to follow the ideal-gas-law model. Deviations from ideal behavior are corrected for, to first order, with nonunity values of compressibility factor Z. (See Sec. 2, "Physical and Chemical Data," for definitions and data for common fluids.) For compressible fluids

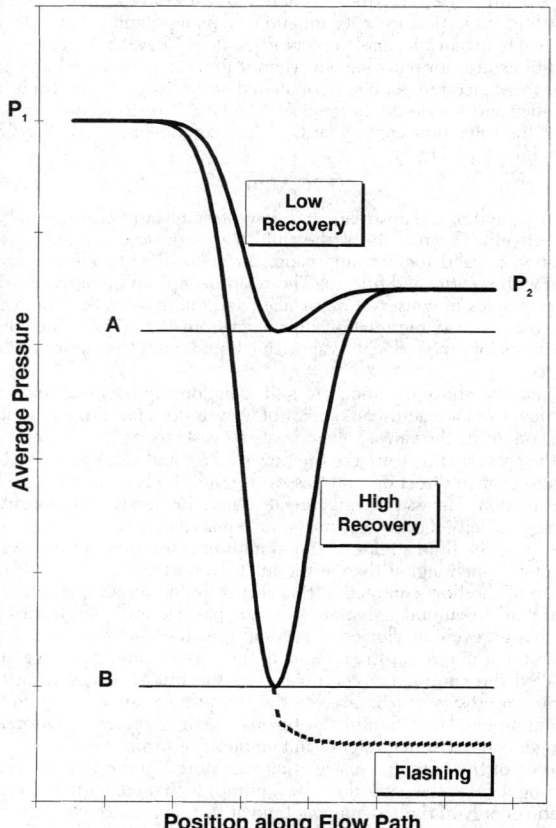

FIG. 8-78 Generic depictions of average pressure at subsequent cross sections throughout a control valve. F_Ls selected for illustration are 0.9 and 0.63 for low and high recovery, respectively. Internal pressure in the high-recovery valve is shown as a dashed line for flashing conditions ($p_2 < p_v$) with p_v = B.

$$w = 94.8 C_v P_1 Y \sqrt{\frac{x M_w}{T_1 Z}} \qquad (8\text{-}110)$$

where P_1 is in bar absolute, T_1 is inlet temperature in K, M_w is the molecular weight, and x is the dimensionless pressure-drop ratio $(p_1 - p_2)/p_1$. The expansion factor Y accounts for changes in the fluid density as the fluid passes through the valve and for variation in the contraction coefficient with pressure drop. For convenience the experimental data is approximated by a simple relationship:

$$Y = 1 - \frac{1.4x}{3x_T\gamma} \qquad \text{for} \qquad x \le \frac{x_T\gamma}{1.4} \qquad (8\text{-}111)$$

where γ is the ratio of specific heats and x_T is the pressure drop ratio factor. Even though a fluid may be compressible, if the value of x is small, the flow will behave as though it is incompressible. In the limit as x goes to zero, Eq. (8-110) reduces to the incompressible form Eq. (8-108) with ρ expressed via the ideal-gas equation of state.

Compressible fluids exhibit a phenomenon known as choking. Given a nozzle geometry with fixed inlet conditions, the mass flow rate will increase as P_2 is decreased up to a maximum amount at the critical pressure drop. The velocity at the vena contracta has reached sonic and a standing shock has formed. This shock causes a step change in pressure as flow passes through it, and further reduction in P_2 does not increase mass flow. x_T is a parameter of the flow model that relates to the critical pressure-drop ratio but also accounts for valve geometry effects. The value of x_T varies with flow-path geometry; a rough estimate for conventional valves is one-half. In the choked case,

$$x > \frac{x_T\gamma}{1.4} \qquad \text{and} \qquad Y = 0.67 \qquad (8\text{-}112)$$

Noise Control   Sound is a fluctuation of air pressure that can be detected by the human ear. Sound travels through any fluid (e.g., the air) as a compression/expansion wave. This wave travels radially outward in all directions from the sound source. The pressure wave induces an oscillating motion in the transmitting medium that is superimposed on any other net motion it may have. These waves are reflected, refracted, scattered, and absorbed as they encounter solid objects. Sound is transmitted through solids in a complex array of types of elastic waves. Sound is characterized by its amplitude, frequency, phase, and direction of propagation.

Sound strength is therefore location-dependent and is often quantified as a sound pressure level (L_p) in dB based on the root-mean-square (rms) sound-pressure (p_S) value, where

$$L_p = 10 \log_{10}\left(\frac{p_S}{p_{\text{reference}}}\right)^2 \qquad (8\text{-}113)$$

For airborne sound, the reference pressure is 2×10^{-5} Pa (29×10^{-10} psi), which is nominally the human threshold of hearing at 1000 Hz. The corresponding sound pressure level is 0 dB. Conversation is about 50 dB, and a jackhammer operator is subject to 100 dB. Extreme levels such as a jet engine at takeoff might produce 140 dB at a distance of 3 m, which is a pressure amplitude of 200 Pa (29×10^{-3} psi). These examples demonstrate both the sensitivity and wide dynamic range of the human ear.

Traveling sound waves carry energy. Sound intensity I is a measure of the power passing through a unit area in a specified direction and is related to p_S. Measuring sound intensity in a process plant gives clues as to the location of the source. As one moves away from the source, the fact that the energy is spread over a larger area requires that sound pressure level decrease. For example, doubling one's distance from a point source reduces the L_p by 6 dB. Viscous action from the induced fluid motion absorbs additional acoustic energy. However, in free air, this viscous damping is negligible over short distances (on the order of a meter).

Noise is a group of sounds with many nonharmonic frequency components of varying amplitudes and random phase. The turbulence generated by a throttling valve creates noise. As a valve converts potential energy to heat, some of it becomes acoustic energy as an intermediate step. Valves handling large amounts of compressible fluid through a large pressure change create the most noise because more total power is being transformed. Liquid flows are noisy only under special circumstances as will be seen in the next subsection. Due to the random nature of turbulence and the broad distribution of length and velocity scales of turbulent eddies, valve-generated sound is usually random, broad-spectrum noise. Total sound pressure level from two such statistically uncorrelated sources is (in dB):

$$L_p = 10 \log_{10} \left[\frac{(p_{S1})^2 + (p_{S2})^2}{(p_{\text{reference}})^2} \right]$$ (8-114)

For example, two sources of equal strength combine to create an L_p that is 3 dB higher.

While noise is annoying to listen to, the real reasons for being concerned about noise are its impact on people and equipment. Hearing loss can occur due to long-term exposure to moderately high or even short exposure to very high noise levels. The U.S. Occupational Safety and Health Act (OSHA) has specific guidelines for permissible levels and exposure times. The human ear has a frequency-dependent sensitivity to sound. When the effect on humans is the criteria, L_p measurements are weighted to account for the ear's response. This so-called A-weighted scale is defined in ANSI S1.4 and is commonly reported as L_{pA}. Figure 8-79 illustrates the difference between actual and perceived airborne sound pressure level. At sufficiently high levels, noise and the associated vibration can damage equipment.

There are two approaches to fluid-generated noise control—source or path treatment. Path treatment means absorbing or blocking the transmission of noise after it has been created. The pipe itself is a barrier. The sound pressure level inside a standard schedule pipe is roughly 40–60 dB higher than on the outside. Thicker walled pipe reduces levels somewhat more, and adding acoustical insulation on the outside of the pipe reduces ambient levels up to 10 dB per inch of thickness. Since noise propagates relatively unimpeded inside the pipe, barrier approaches require the entire downstream piping system to be treated in order to be totally effective. In-line silencers place absorbent material inside the flow stream, thus reducing the level of the internally propagating noise. Noise reductions up to 25 dB can be achieved economically with silencers.

The other approach to valve noise problems is the use of quiet trim. Two basic strategies are used to reduce the initial production of noise—dividing the flow stream into multiple paths and using several flow resistances in series. L_p is proportional to mass flow and is dependent on vena contracta velocity. If each path is an independent source, it is easy to show from Eq. (8-114) that p_s^2 is inversely proportional to the number of passages; additionally, smaller passage size shifts the predominate spectral content to higher frequencies, where structural resonance may be less of a problem. Series resistances or multiple stages can reduce maximum velocity and/or produce back pressure to keep jets issuing from multiple passages acting independently. While some of the basic principles are understood, predicting noise for a particle-flow passage requires some empirical data as a basis. Valve manufacturers have developed noise-prediction methods for the valves they build. ISA S75.17 is a public-domain methodology for standard (nonlow noise) valve types, although treatment of some multistage, multipath types is underway. Low-noise hardware consists of special cages in linear stem valves, perforated domes or plates and multichannel inserts in rotary valves, and separate devices that use multiple fixed restrictions.

Cavitation and Flashing From the discussion on pressure recovery it was seen that the pressure at the vena contracta can be much lower than the downstream pressure. If the pressure on a liquid falls below its vapor pressure (p_v), the liquid will vaporize. Due to the effect of surface tension, this vapor phase will first appear as bubbles. These bubbles are carried downstream with the flow, where they collapse if the pressure recovers to a value above p_v. This pressure-driven process of vapor-bubble formation and collapse is known as cavitation.

Cavitation has three negative side effects in valves—noise and vibration, material removal, and reduced flow. The bubble-collapse process is a violent asymmetrical implosion that forms a high-speed microjet and induces pressure waves in the fluid. This hydrodynamic noise and the mechanical vibration that it can produce are far stronger than other noise-generation sources in liquid flows. If implosions occur adjacent to a solid component, minute pieces of material can be removed, which, over time, will leave a rough, cinderlike surface.

The presence of vapor in the vena contracta region puts an upper limit on the amount of liquid that will pass through a valve. A mixture of vapor and liquid has a lower density than the liquid alone. While Eq. (8-108) is not applicable to two-phase flows because pressure changes are redistributed due to varying density and the two phases do not necessarily have the same average velocity, it does suggest that lower density reduces total mass flow rate. Figure 8-80 illustrates a typical flow-rate-to-pressure-drop relationship. As with compressible gas flow at a given p_1, flow increases as p_2 is decreased until the flow chokes (i.e., no additional fluid will pass). The transition between incompressible and choked flow is gradual because, within the convoluted flow passages of valves, the pressure is actually an uneven distribution at each cross section, and, consequently, vapor-formation zones increase gradually. In fact, isolated zones of bubble formation or incipient cavitation often occur at pressure drops well below that at which a reduction in flow is noticeable. The similarity between liquid and gas choking is not serendipitous; it is surmised that the two-phase fluid is traveling at the mixture's sonic velocity in the throat when choked. Complex fluids with components having varying vapor pressures and/or entrained noncondensable gases (e.g., crude oil) will exhibit soft vaporization/implosion transitions.

There are several methods to reduce cavitation or at least its negative side effects. Material damage is slowed by using harder materials and by directing the cavitating stream away from passage walls (e.g., with an angle body flowing down). Sometimes the system can be designed to place the valve in a higher p_2 location or add downstream resistance, which creates back pressure. A low recovery valve has a higher minimum pressure for a given p_2 and so is a means to eliminate the cavitation itself, not just its side effects. In Fig. 8-78, if $p_v <$ "B" neither valve will cavitate substantially. For $p_v >$ "B" but $<$ "A," the

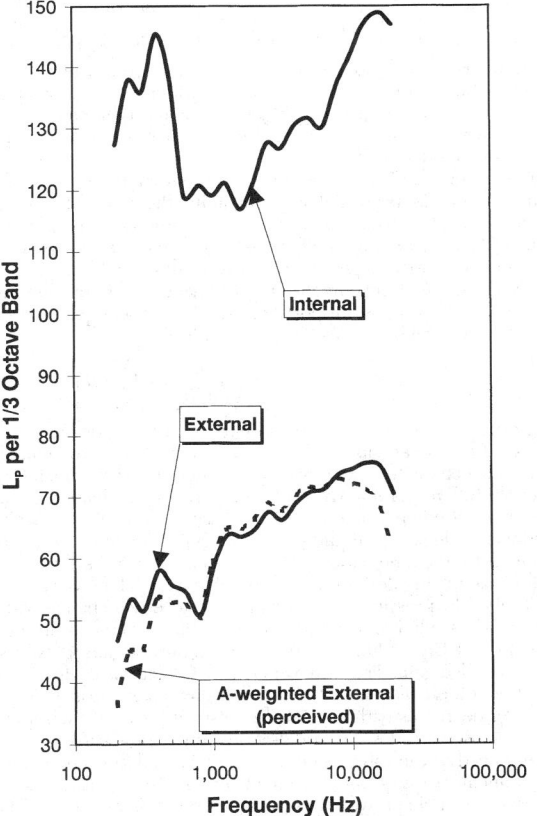

FIG. 8-79 Valve-generated sound pressure level spectrums.

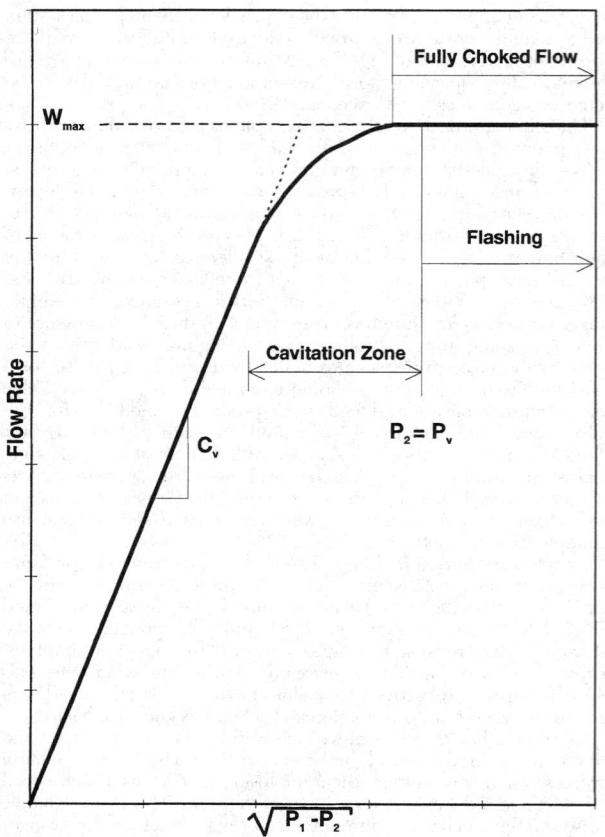

FIG. 8-80 Liquid flow rate versus pressure drop (assuming constant P_1 and P_v).

high recovery valve will cavitate substantially, but the low recovery valve will not. Special anticavitation trims are available for globe/angle valves and more recently for some rotary valves. These trims use multiple contraction/expansion stages or other distributed resistances to boost F_L to values sometimes near unity.

If p_2 is below p_v, the two-phase mixture will continue to vaporize in the body outlet and/or downstream pipe until all liquid phase is gone, a condition known as flashing. The resulting huge increase in specific volume leads to high velocities, and any remaining liquid droplets acquire much of the higher vapor-phase velocity. Impingement of these droplets can produce material damage, but it differs from cavitation damage because it exhibits a smooth surface. Hard materials and directing the two-phase jets away from solid surfaces are means to avoid this damage.

Seals, Bearings, and Packing Systems In addition to their control function, valves often need to provide shutoff. ANSI B16.104, FCI 70-2 (1991), and IEC 534-4 all recognize six standard classifications and define their as-shipped qualification tests. Class I is an amount agreed to by user and supplier with no test needed. Classes II, III, and IV are based on an air test with maximum leakage of 0.5 percent, 0.1 percent, and 0.01 percent of rated capacity, respectively. Class V restricts leakage to 5×10^{-6} ml of water per second per mm of port diameter per bar differential. Class VI allows 0.15 to 6.75 ml per minute of air to escape depending on port size; this class implies the need for interference-fit elastomeric seals. With the exception of Class V, all classes are based on standardized pressure conditions that may not represent actual conditions. Therefore, it is difficult to estimate leakage in service. Leakage normally increases over time as seals

and seating surfaces become nicked or worn. Leak passages across the seat-contact line, known as wire drawing, may form and become worse over time—even in hard metal seats under sufficiently high pressure differentials.

Polymers used for seat and plug seals and internal static seals include: PTFE (polytetrafluoroethylene) and other fluorocarbons, polyethylene, nylon, polyether-ether-ketone, and acetal. Fluorocarbons are often carbon or glass-filled to improve mechanical properties and heat resistance. Temperature and chemical compatibility with the process fluid are the key selection criteria. Polymer-lined bearings and guides are used to decrease friction, which lessens dead band and reduces actuator force requirements. See Sec. 28, "Materials of Construction," for properties.

Packing forms the pressure-tight seal, where the stem protrudes through the pressure boundary. Packing is typically made from PTFE or, for high temperature, a bonded graphite. If the process fluid is toxic, more sophisticated systems such as dual packing, live-loaded, or a flexible metal bellows may be warranted. Packing friction can significantly degrade control performance. Pipe, bonnet, and internal-trim joint gaskets are typically a flat sheet composite. Gaskets intended to absorb dimensional mismatch are typically made from filled spiral-wound flat stainless steel wire with PTFE or graphite filler. The use of asbestos in packing and gaskets has largely been eliminated.

Flow Characteristics The relationship between valve flow and valve travel is called the valve-flow characteristic. The purpose of flow characterization is to make loop dynamics independent of load, so that a single controller tuning remains optimal for all loads. Valve gain is one factor affecting loop dynamics. In general, gain is the ratio of change in output to change in input. The input of a valve is travel (y) and the output is flow (w). Since pressure conditions at the valve can depend on flow (hence travel), valve gain is

$$\frac{dw}{dy} = \frac{\partial w}{\partial C_v}\frac{dC_v}{dy} + \frac{\partial w}{\partial p_1}\frac{dp_1}{dy} + \frac{\partial w}{\partial p_2}\frac{dp_2}{dy} \qquad (8\text{-}115)$$

An inherent valve flow characteristic is defined as the relationship between flow rate and travel, under constant pressure conditions. Since the last two terms in Eq. (8-115) are zero in this case, the inherent characteristic is necessarily also the relationship between flow coefficient and travel.

Figure 8-81 shows three common inherent characteristics. A linear characteristic has a constant slope, meaning the inherent valve gain is a constant. The most popular characteristic is equal-percentage, which gets its name from the fact that equal changes in travel produce equal-percentage changes in the existing flow coefficient. In other words, the slope of the curve is proportional to C_v or equivalently that inherent valve gain is proportional to flow. The equal-percentage characteristic can be expressed mathematically by

$$C_v(y) = (\text{rated } C_v) \exp\left[\left(\frac{y}{\text{rated } y} - 1\right)\ln R\right] \qquad (8\text{-}116)$$

This expression represents a set of curves parameterized by R. Note that C_v ($y = 0$) equals (rated C_v)/R rather than zero; real equal-percentage characteristics deviate from theory at some small travel to meet shutoff requirements. An equal-percentage characteristic provides perfect compensation for a process that has gain inversely proportional to flow (e.g., liquid pressure). Quick opening does not have a standardized mathematical definition. Its shape arises naturally from high-capacity plug designs used in on/off service globe valves.

Frequently, pressure conditions at the valve will change with flow rate. This so-called process influence [the last two terms on the right hand side of Eq. (8-115)] combine with inherent gain to express the installed valve gain. The flow-versus-travel relationship for a specific set of conditions is called the installed flow characteristic. Typically, valve Δp decreases with load, since pressure losses in the piping system increase with flow. Figure 8-82 illustrates how allocation of total system head to the valve influences the installed flow characteristics. For a linear or quick-opening characteristic, this transition toward a concave down shape would be more extreme. This effect of typical process pressure variation, which causes equal-percentage character-

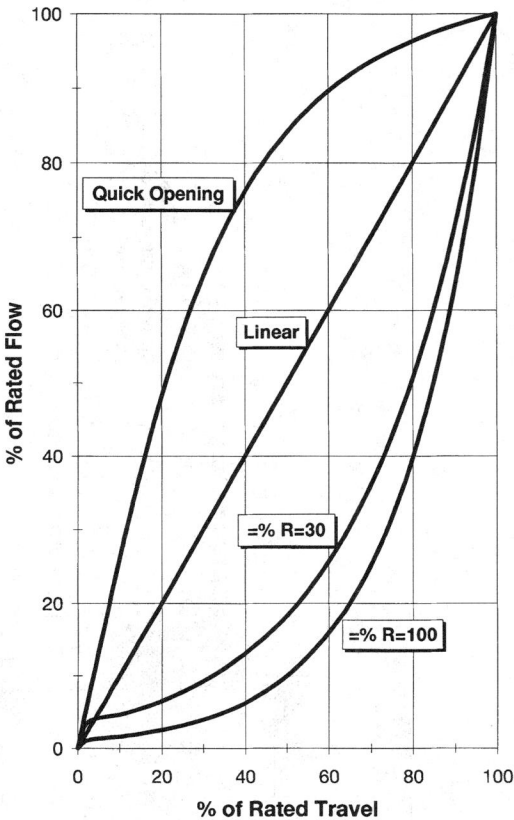

FIG. 8-81   Typical inherent flow characteristics.

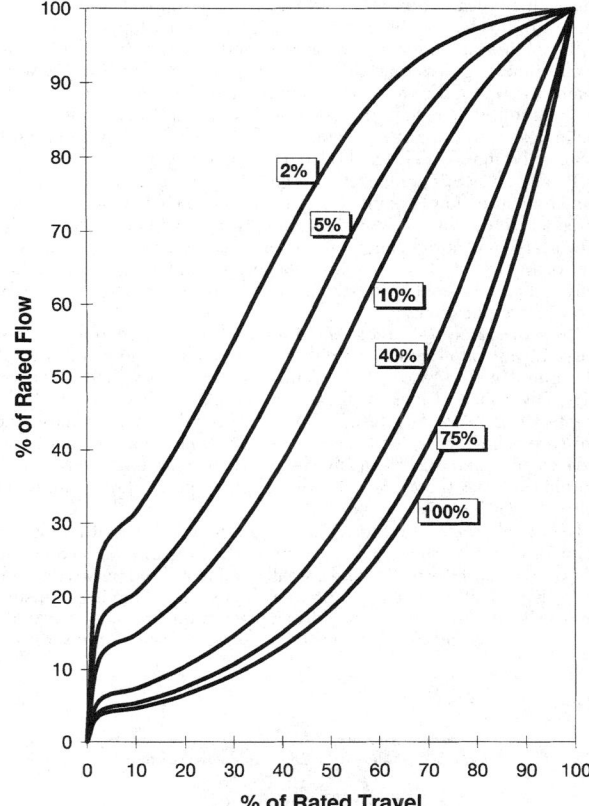

FIG. 8-82   Installed flow characteristic as a function of percent of total system head allocated to the control valve (assuming constant head pump, no elevation head loss, and an R equal 30 equal-percentage inherent characteristic).

istics to have fairly constant installed gain, is one reason the equal-percentage characteristic is the most popular.

Due to clearance flow, flow force gradients, seal friction, and the like, flow cannot be throttled to an arbitrarily small value. Installed rangeability is the ratio of maximum to minimum controllable flow. The actuator and positioner, as well as the valve, influence the installed rangeability. Inherent rangeability is defined as the ratio of the largest to the smallest C_v within which the characteristic meets specified criteria (see ISA S75.11). The R value in the equal-percentage definition is a theoretical rangeability only. While high installed rangeability is desirable, it is also important not to oversize a valve; otherwise, turndown (ratio of maximum normal to minimum controllable flow) will be limited.

Sliding stem valves are characterized by altering the contour of the plug when the port and plug determine the minimum (controlling) flow area. Passage area versus travel is also easily manipulated in characterized cage designs. Inherent rangeability varies widely, but typical values are 30 for contoured plugs and 20–50 for characterized cages. While these types of valves can be characterized, the degree to which manufacturers conform to the mathematical ideal is revealed by plotting measured C_v versus travel. Note that ideal equal percentage will plot as a straight line on a semilog graph. Custom characteristics that compensate for a specific process are possible.

Rotary stem-valve designs are normally offered only in their naturally occurring characteristic, since it is difficult to appreciably alter this. If additional characterization is required, the positioner or controller may be characterized. However, these approaches are less direct, since it is possible for device nonlinearity and dynamics to distort the compensation.

OTHER PROCESS VALVES

In addition to the throttling control valve, other types of process valves are used to manipulate the process.

Valves for On/Off Applications   Valves are often required for service that is primarily nonthrottling in nature. Valves in this category, depending on the service requirements, may be of the same design as the types used for throttling control or, as in the case of gate valves, different in design. Valves in this category usually have tight shutoff when they are closed and low pressure drops when they are wide open. The on/off valve can be operated manually, such as by handwheel or lever; or automatically, with pneumatic or electric actuators.

Batch   Batch process operation is an application requiring on/off valve service. Here the valve is opened and closed to provide reactant, catalyst, or product to and from the batch reactor. Like the throttling control valve, the valve used in this service must be designed to open and close thousands of times. For this reason, valves used in this application are often the same valves used in continuous throttling applications. Ball valves are especially useful in batch operations. The ball valve has a straight-through flow passage that reduces pressure drop in the wide-open state and provides tight shutoff capability when closed. In addition, the segmented ball valve provides for shearing action between the ball and the ball seat that promotes closure in slurry service.

Isolation   A means for pressure-isolating control valves, pumps, and other piping hardware for installation and maintenance is another common application for an on/off valve. In this application, the valve is required to have tight shutoff so that leakage is stopped when the piping system is under repair. As the need to cycle the valve in this application is far less than that of a throttling control valve, the wear characteristics of the valve are less important. Also, because there are many required in a plant, the isolation valve needs to be reliable, simple in design and simple in operation. The gate valve, shown in Figure 8-83, is the most widely used valve in this application.

The gate valve is composed of a gate-like disc that moves perpendicular to the flow stream. The disc is moved up and down by a threaded screw that is rotated to effect disc movement. Because the disc is large and at right angles to the process pressure, large seat loading for tight shutoff is possible. Wear produced by high seat loading during the movement of the disk prohibits the use of the gate valve for throttling applications.

Pressure Relief Valves   Definitions for pressure relief valves, relief valves, pilot-operated pressure relief valves and safety valves, are found in the ASME Boiler and Pressure Vessel Code, Section VIII, Division 1, "Rules for Construction of Pressure Vessels," Paragraphs UG-125 and UG-126. The pressure-relief valve is an automatic pressure relieving device designed to open when normal conditions are exceeded and to close again when normal conditions are restored. Within this class there are relief valves, pilot operated pressure relief valves, and safety valves.

Relief valves (see Fig. 8-84) have spring-loaded disks that close a main orifice against a pressure source. As pressure rises, the disk begins to rise off the orifice and a small amount of fluid passes through the valve. Continued rise in pressure above the opening pressure causes the disk to open the orifice in a proportional fashion. The main orifice reduces and closes when the pressure returns to the set pres-

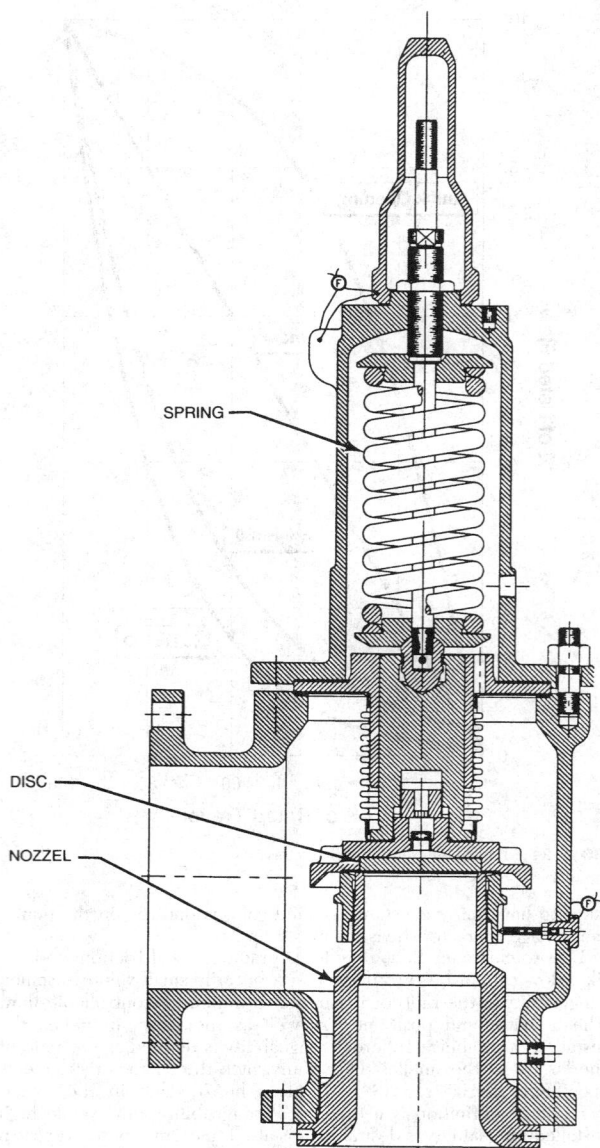

FIG. 8-84   Relief valve. (*Courtesy Teledyne Fluid Systems, Farris Engineering.*)

sure. Additional sensitivity to over-pressure conditions can be improved by adding an auxiliary pressure relief valve (pilot) to the basic pressure relief valve. This combination is known as a pilot-operated pressure relief valve.

The safety valve is similar to the relief valve except it is designed to open fully, or pop, with only a small amount of pressure over the rated limit. Conventional safety valves are sensitive to downstream pressure and may have unsatisfactory operating characteristics in variable back pressure applications. The balanced safety relief valve is available and minimizes the effect of downstream pressure on performance.

Check Valves   The purpose of a check valve is to allow relatively unimpeded flow in the desired direction but to prevent flow in the reverse direction. Two common designs are swing-type and lift-type check valves—the names of which denote the motion of the closure member. In the forward direction, flow forces overcome the weight of

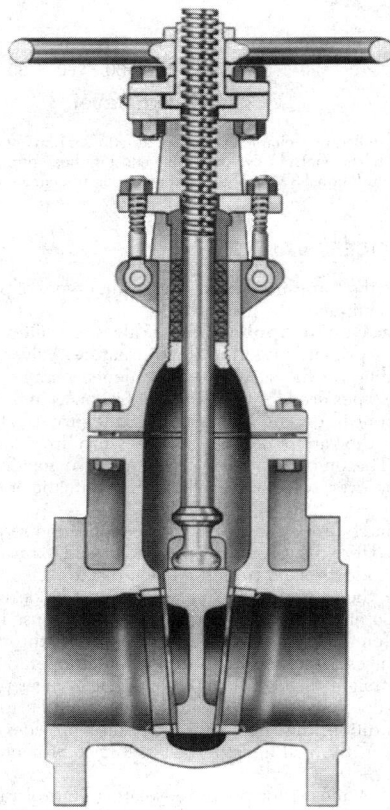

FIG. 8-83   Gate valve. (*Courtesy Crane Valves.*)

the member or a spring to open the flow passage. With reverse pressure conditions, flow forces drive the closure member into the valve seat, thus providing shutoff.

ADJUSTABLE SPEED PUMPS

An alternative to throttling a process with a process-control valve and a fixed speed pump is by adjusting the speed of the process pump and not using a throttling control valve at all. Pump speed can be varied by using variable-speed prime movers such as turbines, motors with magnetic or hydraulic couplings, and electric motors. Each of these methods of modulating pump speed has its own strengths and weaknesses but all offer energy savings and dynamic performance advantages over throttling with a control valve.

The centrifugal pump directly driven by a variable-speed electric motor is the most commonly used hardware combination for adjustable speed pumping. The motor is operated by an electronic-motor speed controller whose function is to generate the voltage or current waveform required by the motor to make the speed of the motor track the input command input signal from the process controller.

The most popular form of motor speed control for adjustable-speed pumping is the voltage-controlled pulse-width-modulated (PWM) frequency synthesizer and AC squirrel-cage induction motor combination. The flexibility of application of the PWM motor drive and its 90 percent+ electrical efficiency along with the proven ruggedness of the traditional AC induction motor makes this combination popular.

From an energy-consumption standpoint, the power required to maintain steady process flow with an adjustable-speed pump system (three-phase PWM drive and a squirrel-cage induction motor driving a centrifugal pump on water) is less than that required with a conventional control valve and a fixed speed pump. Figure 8-85 shows this to be the case for a system where 100 percent of the pressure loss is due to flow velocity losses. At 75 percent flow, Fig. 8-85 shows the constant speed-pump/control-valve use at a 10.1-kW rate where throttling with the adjustable speed pump and no control valve used at a 4.1-kW rate. This trend of reduced energy consumption is true for the entire range of flows, although amounts vary.

From a dynamic-response standpoint, the adjustable speed pump has a dynamic characteristic that is more suitable in process-control applications than those characteristics of control valves. The small amplitude response of an adjustable speed pump does not contain the dead band or the dead time commonly found in the small amplitude response of the control valve. Nonlinearities associated with frictions in the valve and discontinuities in the pneumatic portion of the control-valve instrumentation are not present with electronic variable-speed drive technology. As a result, process control with the adjustable speed pump does not exhibit limit cycles, problems related to low controller gain and generally degraded process loop performance caused by control valve nonlinearities.

Unlike the control valve, the centrifugal pump has poor or nonexistent shutoff capability. A flow check valve or an automated on/off valve may be required to achieve shutoff requirements. This requirement may be met by automating an existing isolation valve in retrofit applications.

REGULATORS

A regulator is a compact device that maintains the process variable at a specific value in spite of disturbances in load flow. It combines the functions of the measurement sensor, controller, and final control element into one self-contained device. Regulators are available to control pressure, differential pressure, temperature, flow, liquid level, and other basic process variables. They are used to control the differential across a filter press, heat exchanger, or orifice plate. Regulators are used for monitoring pressure variables for redundancy, flow check, and liquid surge relief.

Regulators may be used in gas blanketing systems to maintain a protective environment above any liquid stored in a tank or vessel as the liquid is pumped out. When the temperature of the vessel is suddenly cooled, the regulator maintains the tank pressure and protects the walls of the tank from possible collapse. Regulators are known for their fast dynamic response. The absence of time delay that often comes with more sophisticated control systems makes the regulator useful in applications requiring fast corrective action.

Regulators are designed to operate on the process pressures in the pipeline without any other sources of energy. Upstream and downstream pressures are used to supply and exhaust the regulator. Exhausting is back to the downstream piping so that no contamination or leakage to the external environment occurs. This makes regulators useful in remote locations where power is not available or where external venting is not allowed.

The regulator is limited to operating on processes with clean, nonslurry process fluids. The small orifice and valve assemblies contained in the regulator can plug and malfunction if the process fluid that operates the regulator is not sufficiently clean.

Regulators are normally not suited to systems that require constant set point adjustment. Although regulators are available with capability to respond to remote set point adjustment, this feature adds complexity to the regulator and may be better addressed by a control-valve-based system. In the simplest of regulators, tuning of the regulator for best control is accomplished by changing a spring, an orifice, or a nozzle.

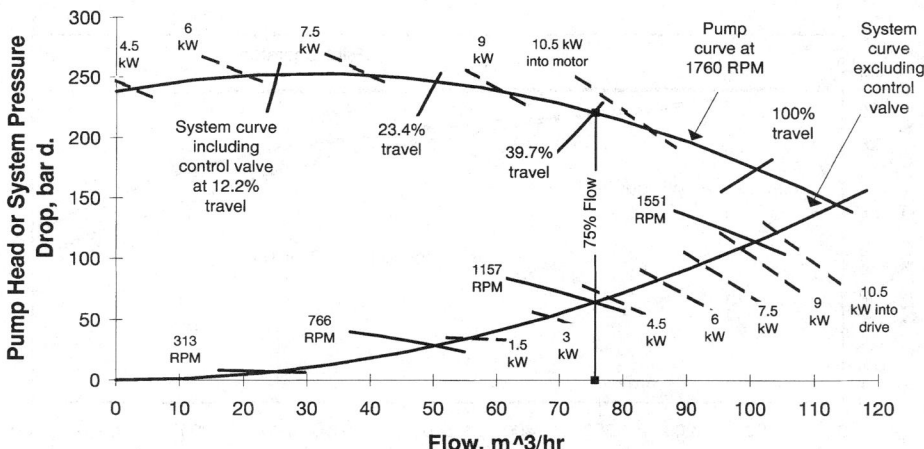

FIG. 8-85   Pressure, flow, and power for a throttling process using (1) a control valve and a constant speed pump and (2) an adjustable speed pump.

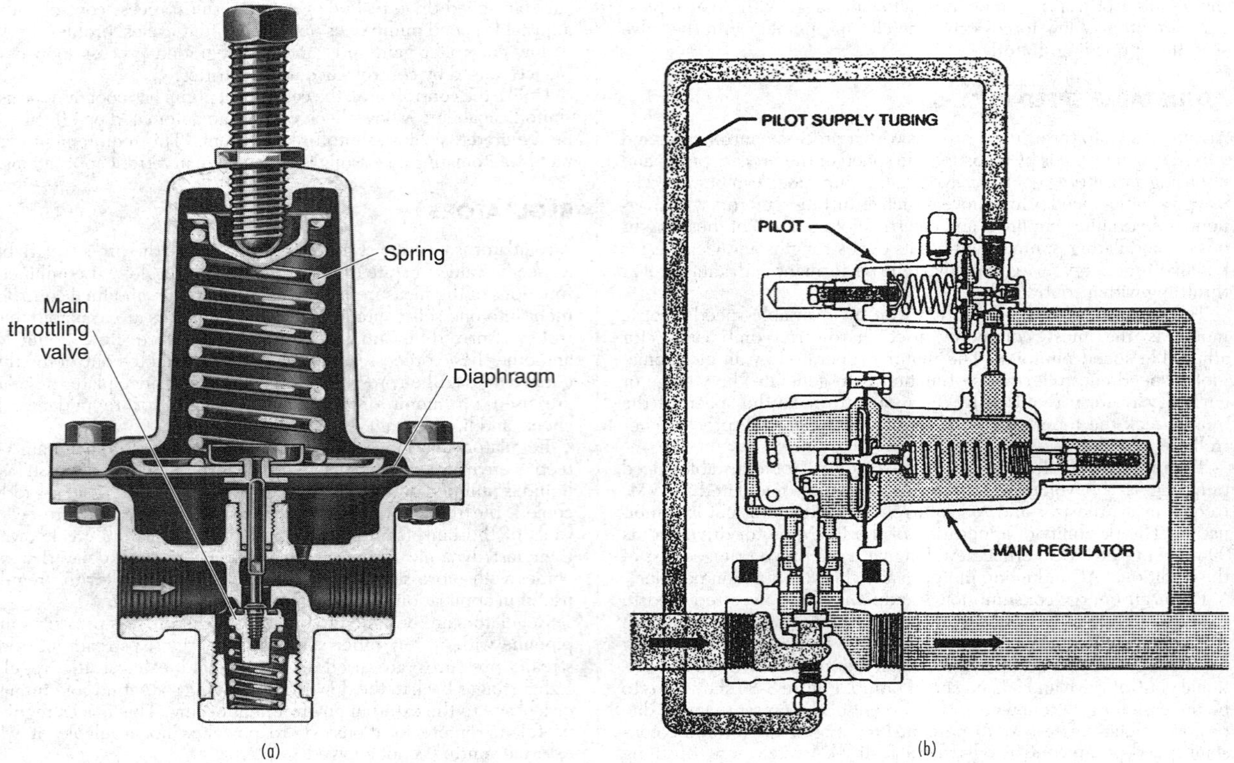

FIG. 8-86 Regulators: (a) self-operated; (b) pilot-operated. (*Courtesy Fisher-Rosemount.*)

Self-Operated Regulators Self-operated regulators are the simplest form of regulator. This regulator (see Fig. 8-86a) is composed of a main throttling valve, a diaphragm or piston to sense pressure, and a spring. The self-contained regulator is completely operated by the process fluid, and no outside control lines or pilot stage is used. In general, self-operated regulators are simple in construction, easy to operate and maintain, and are usually stable devices. Except for some of the pitot tube types, self-operated regulators have very good dynamic response characteristics. This is because any change in the controlled variable registers directly and immediately upon the main diaphragm to produce a quick response to the disturbance.

The disadvantage of the self-operated regulator is that it is not generally capable of maintaining a set point as load flow is increased. Because of the proportional nature of the spring and diaphragm-throttling effect, offset from set point occurs in the controlled variable as flow increases. Figure 8-87 shows a typical regulation curve for the self-contained regulator.

Reduced set point offset with increasing load flow can be achieved

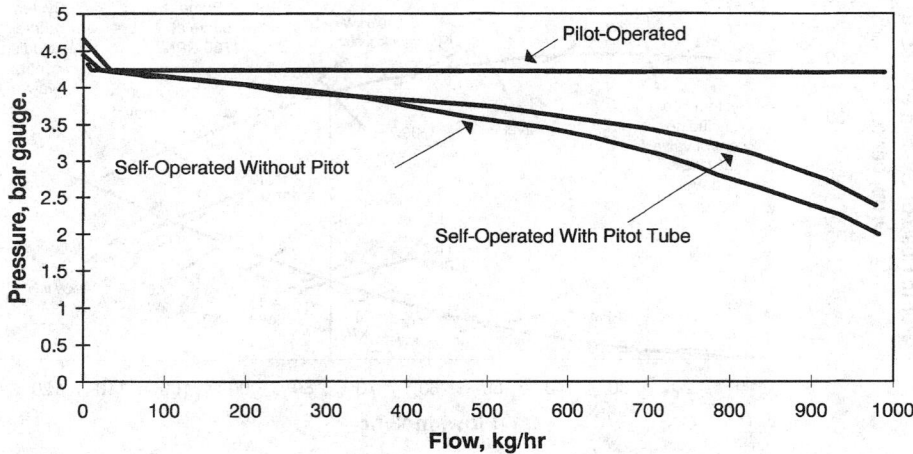

FIG. 8-87 Pressure regulation curves for three regulator types.

by adding a pitot tube to the self-operated regulator. The tube is positioned somewhere near the vena contracta of the main regulator valve. As flow though the valve increases, the measured feedback pressure from the pitot tube drops below the control pressure. This causes the main valve to open or boost more than it would if the static value of control pressure were acting on the diaphragm. The resultant effect keeps the control pressure closer to set point and thus prevents a large drop in process pressure during high-load-flow conditions. Figure 8-87 shows the improvement that the pitot-tube regulator provides over the regulator without the tube. A side effect of adding a pitot-tube method is that the response of the regulator can be slowed due to the restriction provided by the pitot tube.

Pilot-Operated Regulators Another category of regulators uses a pilot stage to provide the load pressure on the main diaphragm. This pilot is a regulator itself that has the ability to multiply a small change in downstream pressure into a large change in pressure applied to the regulator diaphragm. Due to this high-gain feature, pilot-operated regulators can achieve a dramatic improvement in steady-state accuracy over that achieved with a self-operated regulator. Figure 8-87 shows for regulation at high flows the pilot-operated regulator is best of the three regulators shown.

The main limitation of the pilot-operated regulator is stability. When the gain in the pilot amplifier is raised too much, the loop can become unstable and oscillate or hunt. The two-path pilot regulator (see *b*) is also available. This regulator combines the effects of self-operated and the pilot-operated styles and mathematically produces the equivalent of proportional plus reset control of the process pressure.

Over-Pressure Protection Figure 8-87 shows a characteristic rise in control pressure that occurs at low or zero flow. This lockup tail is due to the effects of imperfect plug and seat alignment and the elastomeric effects of the main throttle valve. If for some reason the main throttle valve fails to completely shut off, or if the valve shuts off but the control pressure continues to rise for other reasons, the lockup tail could get very large, and the control pressure could rise to extremely high valves. Damage to the regulator or the downstream pressure volume could occur.

To avoid this situation, some regulators are designed with a built-in over-pressure relief mechanism. Over-pressure relief circuits usually are composed of a spring-opposed diaphragm and valve assembly that vents the downstream piping when the control pressure rises above the set point pressure.

PROCESS CONTROL AND PLANT SAFETY

GENERAL REFERENCES: *Guidelines for Safe Automation of Chemical Processes*, AIChE Center for Chemical Process Safety, New York, 1993.

Accidents in chemical plants make headline news, especially when there is loss of life or the general public is affected in even the slightest way. This increases the public's concern and may lead to government action. The terms *hazard* and *risk* are defined as follows:

• *Hazard.* A potential source of harm to people, property, or the environment

• *Risk.* Possibility of injury, loss, or an environmental accident created by a hazard

Safety is the freedom from hazards and thus the absence of any associated risks. Unfortunately, absolute safety cannot be realized.

The design and implementation of safety systems must be undertaken with a view of two issues:

• *Regulatory.* The safety system must be consistent with all applicable codes and standards as well as "generally accepted good engineering practices."

• *Technical.* Just meeting all applicable regulations and "following the crowd" does not relieve a company of its responsibilities. The safety system must work.

The regulatory environment will continue to change. As of this writing, the key regulatory instrument is OSHA 29 CFR 1910.119 that pertains to process safety management within plants in which certain chemicals are present.

In addition to government regulation, industry groups and professional societies are producing documents ranging from standards to guidelines. Instrument Society of America Standard S84.01, "Application of Safety Instrumented Systems for the Process Industries," is in draft form at the date of this writing. The *Guidelines for Safe Automation of Chemical Processes* from the American Institute of Chemical Engineers' Center for Chemical Process Safety (1993) provides a comprehensive coverage of the various aspects of safety, and, although short on specifics, it is very useful to operating companies developing their own specific safety practices (that is, it does not tell you what to do, but it helps you decide what is proper for your plant).

The ultimate responsibility for safety rests with the operating company; OSHA 1910.119 is clear on this. Each company is expected to develop (and enforce) its own practices in the design, installation, testing, and maintenance of safety systems. Fortunately, some companies make these documents public. Monsanto's *Safety System Design Practices* was published in its entirety in the proceedings of the International Symposium and Workshop on Safe Chemical Process Automation, Houston, Texas, September 27–29, 1994 (available from the American Institute of Chemical Engineers' Center for Chemical Process Safety).

ROLE OF AUTOMATION IN PLANT SAFETY

As microprocessor-based controls displaced hardwired electronic and pneumatic controls, the impact on plant safety has definitely been positive. When automated procedures replace manual procedures for routine operations, the probability of human errors leading to hazardous situations is lowered. The enhanced capability for presenting information to the process operators in a timely manner and in the most meaningful form increases the operator's awareness of the current conditions in the process. Process operators are expected to exercise due diligence in the supervision of the process, and timely recognition of an abnormal situation reduces the likelihood that the situation will progress to the hazardous state. Figure 8-88 depicts the layers of safety protection in a typical chemical plant.

Although microprocessor-based process controls enhance plant safety, their primary objective is efficient process operation. Manual

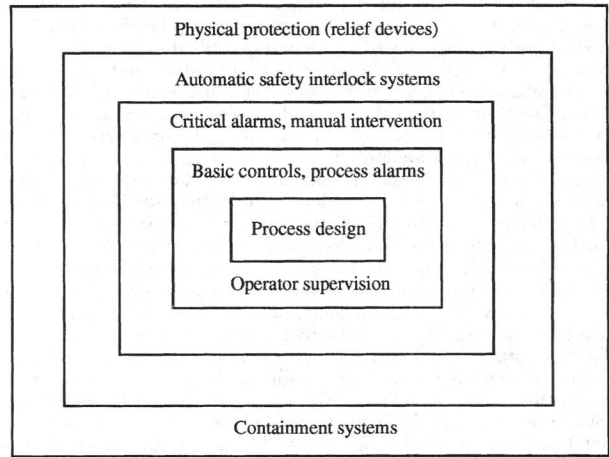

FIG. 8-88 Layers of safety protection in chemical plants.

operations are automated to reduce variability, to minimize the time required, to increase productivity, and so on. Remaining competitive in the world market demands that the plant be operated in the best manner possible, and microprocessor-based process controls provide numerous functions that make this possible. Safety is never compromised in the effort to increase competitiveness, but enhanced safety is a by-product of the process-control function and is not a primary objective.

By attempting to maintain process conditions at or near their design values, the process controls also attempt to prevent abnormal conditions from developing within the process. Although process controls can be viewed as a protective layer, this is really a by-product and not the primary function. Where the objective of a function is specifically to reduce risk, the implementation is normally not within the process controls. Instead, the implementation is within a separate system specifically provided to reduce risk. This system is generally referred to as the safety interlock system.

As safety begins with the process design, an inherently safe process is the objective of modern plant designs. When this cannot be achieved, process hazards of varying severity will exist. Where these hazards put plant workers and/or the general public at risk, some form of protective system is required. Process safety management addresses the various issues, ranging from assessment of the process hazard to assuring the integrity of the protective equipment installed to cope with the hazard. When the protective system is an automatic action, it is incorporated into the safety interlock system, not within the process controls.

INTEGRITY OF PROCESS CONTROL SYSTEMS

Ensuring the integrity of process controls involves both hardware issues, software issues, and human issues. Of these, the hardware issues are usually the easiest to assess and the software issues the most difficult.

The hardware issues are addressed by providing various degrees of redundancy, by providing multiple sources of power and/or an uninterruptible power supply, and the like. The manufacturers of process controls provide a variety of configuration options. Where the process is inherently safe and infrequent shutdowns can be tolerated, nonredundant configurations are acceptable. For more demanding situations, an appropriate requirement might be that no single component failure can render the process-control system inoperable. For the very critical situations, triple-redundant controls with voting logic might be appropriate. The difficulty is assessing what is required for a given process.

Another difficulty is assessing the potential for human errors. If redundancy is accompanied with increased complexity, the resulting increased potential for human errors must be taken into consideration. Redundant systems require maintenance procedures that can correct problems in one part of the system while the remainder of the system is in full operation. When conducting maintenance in such situations, the consequences of human errors can be rather unpleasant.

The use of programmable systems for process control present some possibilities for failures that do not exist in hard-wired electromechanical implementations. Probably the one of most concern is latent defects or "bugs" in the software, either the software provided by the supplier or the software developed by the user. The source of this problem is very simple. There is no methodology available that can be applied to obtain absolute assurance that a given set of software is completely free of defects. Increased confidence in a set of software is achieved via extensive testing, but no amount of testing results in absolute assurance that there are no defects. This is especially true of real-time systems, where the software can easily be exposed to a sequence of events that was not anticipated. Just because the software performs correctly for each event individually does not mean that it will perform correctly when two (or more) events occur at nearly the same time. This is further complicated by the fact that the defect may not be in the programming; it may be in how the software was designed to respond to the events.

The testing of any collection of software is made more difficult as the complexity of the software increases. Software for process control has progressively become more complex, mainly because the requirements have progressively become more demanding. To remain competitive in the world market, processes must be operated at higher production rates, within narrower operating ranges, closer to equipment limits, and so on. Demanding applications require sophisticated control strategies, which translate into more complex software. Even with the best efforts of both supplier and user, complex software systems are unlikely to be completely free of defects.

CONSIDERATIONS IN IMPLEMENTATION OF SAFETY INTERLOCK SYSTEMS

Where hazardous conditions can develop within a process, a protective system of some type must be provided. Sometimes these are in the form of process hardware such as pressure relief devices. However, sometimes logic must be provided for the specific purpose of taking the process to a state where the hazardous condition cannot exist. The term *safety interlock system* is normally used to designate such logic.

The purpose of the logic within the safety interlock system is very different from the logic within the process controls. Fortunately, the logic within the safety interlock system is normally much simpler than the logic within the process controls. This simplicity means that a hardwired implementation of the safety interlock system is usually an option. Should a programmable implementation be chosen, this simplicity means that latent defects in the software are less likely to be present. Most safety systems only have to do simple things, but they must do them very, very well.

The difference in the nature of process controls and safety interlock systems leads to the conclusion that these two should be physically separated (see Fig. 8-89). That is, safety interlocks should not be piggy-backed onto a process-control system. Instead, the safety interlocks should be provided by equipment, either hard-wired or programmable, that is dedicated to the safety functions. As the process controls become more complex, faults are more likely. Separation means that faults within the process controls have no consequences in the safety interlock system.

Modifications to the process controls are more frequent than modifications to the safety interlock system. Therefore, physically separating the safety interlock system from the process controls provides the following benefits:

1. The possibility of a change to the process controls leading to an unintentional change to the safety interlock system is eliminated.

2. The possibility of a human error in the maintenance of the process controls having consequences for the safety interlock system is eliminated.

3. Management of change is simplified.

4. Administrative procedures for software-version control are more manageable.

Separation also applies to the measurement devices and actuators.

Although the traditional point of reference for safety interlock systems is a hard-wired implementation, a programmed implementation is an alternative. The potential for latent defects in software implementation is a definite concern. Another concern is that solid-state components are not guaranteed to fail to the safe state. The former is addressed by extensive testing; the latter is addressed by manufacturer-supplied and/or user-supplied diagnostics that are routinely executed by the processor within the safety interlock system. Although issues must be addressed in programmable implementations, the hard-wired implementations are not perfect either.

Where a programmed implementation is deemed to be acceptable, the choice is usually a programmable logic controller (PLC) that is dedicated to the safety function. PLCs are programmed using the traditional relay ladder diagrams used for hard-wired implementations. The facilities for developing, testing, and troubleshooting PLCs are excellent. However, for PLCs used in safety interlock systems, administrative procedures must be developed and implemented to address the following issues:

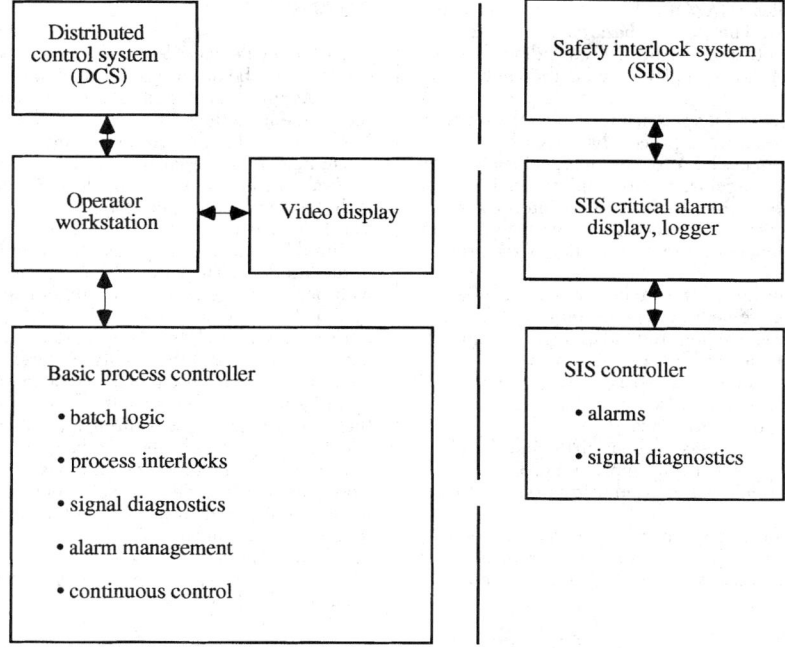

FIG. 8-89 Total control system with parallel tasks.

1. Version controls for the PLC program must be implemented and rigidly enforced. Revisions to the program must be reviewed in detail and thoroughly tested before implementing in the PLC. The various versions must be clearly identified so that there can be no doubt as to what logic is provided by each version of the program.

2. The version of the program that is currently being executed by the PLC must be known with absolute certainty. It must simply not be possible for a revised version of the program undergoing testing to be downloaded to the PLC.

Constant vigilance is required to prevent lapses in such administrative procedures.

INTERLOCKS

An interlock is a protective response initiated on the detection of a process hazard. The interlock system consists of the measurement devices, logic solvers, and final control elements that recognize the hazard and initiate an appropriate response. Most interlocks consist of one or more logic conditions that detect out-of-limit process conditions and respond by driving the final control elements to the safe states. For example, one must specify that a valve fails open or fails closed.

The potential that the logic within the interlock could contain a defect or bug is a strong incentive to keep it simple. Within process plants, most interlocks are implemented with discrete logic, which means either hard-wired electromechanical devices or programmable logic controllers.

Interlocks within process plants can be broadly classified as follows:

1. *Safety interlocks.* These are designed to protect the public, the plant personnel, and possibly the plant equipment from process hazards.

2. *Process interlocks.* These are designed to prevent process conditions that would unduly stress equipment (perhaps leading to minor damage), lead to off-specification product, and so on.

Basically, the process interlocks address hazards whose consequences essentially lead to a monetary loss, possibly even a short plant shut-down. The more serious hazards are addressed by the safety interlocks.

Implementation of process interlocks within process control systems is perfectly acceptable. Furthermore, it is also permissible (and probably advisable) that responsible operations personnel be authorized to bypass or ignore a process. Safety interlocks must be implemented within the separate safety interlock system. Bypassing or ignoring safety interlocks by operations personnel is simply not permitted. When this is necessary for actions such as verifying that the interlock continues to be functional, such situations must be infrequent and incorporated into the design of the interlock.

Safety interlocks are assigned to categories that reflect the severity of the consequences should the interlock fail to perform as intended. The specific categories used within a company is completely at the discretion of the company. However, most companies use categories that distinguish among the following:

1. *Hazards that pose a risk to the public.* Complete redundancy is normally required.

2. *Hazards that could lead to injury of company personnel.* Partial redundancy is often required (for example, redundant measurements but not redundant logic).

3. *Hazards that could result in major equipment damage and consequently lengthy plant downtime.* No redundancy is normally required for these, although redundancy is always an option.

Situations that result in minor equipment damage that can be quickly repaired do not generally require a safety interlock; however, a process interlock might be appropriate.

A process hazards analysis is intended to identify the safety interlocks required for a process and to provide the following for each:

1. The hazard that is to be addressed by the safety interlock.

2. The classification of the safety interlock.

3. The logic for the safety interlock, including inputs from measurement devices and outputs to actuators.

The process hazards analysis is conducted by an experienced, multidisciplinary team that examines the process design, the plant equipment, operating procedures, and so on, using techniques such as hazard and operability studies (HAZOP), failure mode and effect analysis (FEMA), and others. The process hazards analysis recommends appropriate measures to reduce the risk, including (but not limited to) the safety interlocks to be implemented in the safety interlock system.

Diversity is recognized as a useful approach to reduce the number of defects. The team that conducts the process hazards analysis does not implement the safety interlocks but provides the specifications for the safety interlocks to another organization for implementation. This organization reviews the specifications for each safety interlock, seeking clarifications as necessary from the process hazards analysis team and bringing any perceived deficiencies to the attention of the process hazards analysis team.

Diversity can be used to further advantage in redundant configurations. Where redundant measurement devices are required, different technology can be used for each. Where redundant logic is required, one can be programmed and one hard-wired.

Reliability of the interlock systems has two aspects:
1. It must react should the hazard arise.
2. It must not react when there is no hazard.

Emergency shutdowns often pose risks in themselves, and therefore they should be undertaken only when truly appropriate. The need to avoid extraneous shutdowns is not just to avoid disruption in production operations.

Although safety interlocks can inappropriately initiate shutdowns, the process interlocks are usually the major source of problems. It is possible to configure so many process interlocks that it is not possible to operate the plant.

TESTING

As part of the detailed design of each safety interlock, written test procedures must be developed for the following purposes:
1. Assure that the initial implementation complies with the requirements defined by the process hazards analysis team.
2. Assure that the interlock (hardware, software, and I/O) continues to function as designed. The design must also determine the time interval on which this must be done. Often these tests must be done with the plant in full operation.

The former is the responsibility of the implementation team and is required for the initial implementation and following any modification to the interlock. The latter is the responsibility of plant maintenance, with plant management responsible for seeing that it is done on the specified interval of time.

Execution of each test must be documented, showing when it was done, by whom, and the results. Failures must be analyzed for possible changes in the design or implementation of the interlock.

These tests must encompass the complete interlock system, from the measurement devices through the final control elements. Merely simulating inputs and checking the outputs is not sufficient. The tests must duplicate the process conditions and operating environments as closely as possible. The measurement devices and final control elements are exposed to process and ambient conditions and thus are usually the most likely to fail. Valves that remain in the same position for extended periods of time may stick in that position and not operate when needed. The easiest component to test is the logic; however, this is the least likely to fail.

Process Economics*

F. A. Holland, D.Sc., Ph.D., *Consultant in Heat Energy Recycling; Research Professor, University of Salford, England; Fellow, Institution of Chemical Engineers, London. (Section Editor)*

J. K. Wilkinson, M.Sc., *Consultant Chemical Engineer; Fellow, Institution of Chemical Engineers, London.*

° The contribution of the late Mr. F. A. Watson, who was an author for the Sixth edition, is acknowledged.

9-2 PROCESS ECONOMICS

Nomenclature and Units

Symbol	Definition	Units	Symbol	Definition	Units
a	Empirical constant in general equations	Various	f_d	Discount factor, $(1 + i)^{-n}$	Dimensionless
A	Annual income or expenditure particularized by the subscript	\$/year	f_i	Compound-interest factor, $(1 + i)^n$	Dimensionless
A_A	Annual allowances against tax other than for depreciation of fixed assets	\$/year	f_k	Capitalized-cost factor, f_{AP}/i	Dimensionless
A_D	Annual writing down (depreciation) of fixed assets, allowable against tax	\$/year	f_p	Piping-cost factor defined by Eq. (9-249)	Dimensionless
(ATR)	Asset-turnover ratio defined by Eq. (9-131)	Dimensionless	$f(x)$	Distribution function of x variously defined	Dimensionless
b	Empirical constant in general equations	Various	F	Future value of a sum of money	\$
b_c	Deviation from budgeted capacity	Dimensionless	F_n	Sum of f_d for Years 1 to n	Dimensionless
B	Parametric constant in Eq. (9-204)	Dimensionless	i	Interest rate per period, usually annual, often the cost of capital	Dimensionless
c	Empirical constant in general equations	Various	i_e	Effective interest rate defined by Eq. (9-111)	Dimensionless
c	Cost (or income) per unit of sales or production particularized by the subscript	\$/unit	i_m	Minimum acceptable interest rate defined by Eq. (9-107)	Dimensionless
c_B	Cost of base heat supply	\$/unit	i_r	Entrepreneurial-risk interest rate	Dimensionless
c_D	Cost of heat energy delivered by a heat pump defined by Eq. (9-240)	\$/GJ	i'	Nominal annual interest rate	Dimensionless
c_I	Cost of high-grade energy supplied to the compressor of a vapor compression heat pump	\$/GJ	I	Value of inventory particularized by the subscript	\$
c_L	Cost of labor per unit of production	\$/hour	k_n	Constants in Eq. (9-81)	Various
c°	Standard cost particularized by the subscript	\$/hour	K	Effective value of the first unit of production	\$/unit, time/unit, etc.
C	Cost particularized by the subscript	\$	$\ln(a)$	Logarithm to the base e of a	Dimensionless
C_{CT}	Installed cost of a cooling tower	\$	$\log(a)$	Logarithm to the base 10 of a	Dimensionless
C_{DS}	Installed cost of a demineralized-water system	\$	m	Number of interest periods due per year	Dimensionless
$(C_{EQ})_{DEL}$	Delivered-equipment cost	\$	m	Number of units removed from inventory	Dimensionless
C_K	Capitalized cost of a fixed asset defined by Eq. (9-47)	\$	(MSF)	Measured-survival function defined by Eq. (9-106)	Dimensionless
C_L	Cost of land and other nondepreciable assets	\$	n	Number of years, units, etc.	Dimensionless
C_{RS}	Installed cost of a refrigeration system	\$	N	Slope of the learning curve defined by Eq. (9-64)	Dimensionless
C_{RW}	Installed cost of a river-water supply system	\$	N	Number of inventory orders per year	Dimensionless
C_{WS}	Installed cost of a water-softening system	\$	(NPV)	Net present value	\$
(CI)	Cost index as used in Eq. (9-246)	Dimensionless	$p(x)$	Probability of the variable having the value x	Dimensionless
$(COP)_A$	Actual coefficient of performance of a heat pump	Dimensionless	P	Present value of a sum of money	\$
(CR)	Capital ratio defined by Eq. (9-134)	Year	P_a	Production time worked	Hour
(CRR)	Capital-rate-of-return ratio defined by Eq. (9-56)	Year	P_b	Budgeted production	Standard hour
(CSR)	Contribution-sales ratio defined by Eq. (9-236)	Dimensionless	P_e	Production efficiency defined by Eq. (9-216)	Dimensionless
d	Empirical constant in general equations	Various	P_1	Level of productive activity defined by Eq. (9-217)	Dimensionless
d	Symbol indicating differentiation	Dimensionless	P_s	Actual production rate	Standard hour
(DR)	Debt ratio defined by Eq. (9-139)	Dimensionless	P_s'	Book value of asset at the end of year s'	\$
(DCFRR)	Discounted-cash-flow rate of return	Year^{-1}	P_w	Budgeted working time	Hour
e	Empirical constant in general equations	Various	(PBP)	Payback period defined by Eq. (9-30)	Year
e	Base of natural logarithms, 2.71828	Dimensionless	(PM)	Profit margin defined by Eq. (9-127)	Dimensionless
$\exp(a)$	Exponential function of a, e^a	Dimensionless	(PSR)	Profit-sales ratio defined by Eq. (9-235)	Dimensionless
(EMIP)	Equivalent maximum investment period defined by Eq. (9-55)	Year	q	Quantity defining the scale of operation	Various
f_{AF}	Annuity future-worth factor, $i[(1+i)^n - 1]^{-1}$	Dimensionless	Q_D	Process-heat-rate requirement	GJ/hour
			r	Fraction of range of the independent variable	Dimensionless
f_{AP}	Annuity present-worth factor, $f_{AF}(1+i)^n$	Dimensionless	R	Production rate	Units/year
			R°	Standard production rate	Units/year
			R_B	Breakeven production rate	Units/year

Nomenclature and Units (*Concluded*)

Symbol	Definition	Units
R_0	Scheduled production rate	Units/year
R_S	Sales rate	Units/year
(ROA)	Return on assets defined by Eq. (9-129)	Dimensionless
(ROE)	Return on equity defined by Eq. Eq. (9-130)	Dimensionless
(ROI)	Return on investment defined by Eq. (9-128)	Dimensionless
s	Scheduled number of productive years	
s'	Number of productive years to date	
$s°$	Sample standard deviation	Various
S	Scrap value of a depreciable asset	$
t	Fractional tax rate payable on adjusted income	Dimensionless
t_C	Time taken to construct plant	Years
t_{SU}	Time taken to start up plant	Years
T	Auxiliary variable defined by Eq. (9-92)	Various
U	Size of inventory order	Units
V	Variable cost of inventory order	$/unit
W	Power supplied at shaft of a heat pump	GJ/hour
x	General variable	
$\bar{x}$	Mean value of x	Various
X	Cumulative production from startup	Units
y	Cumulative probability	Dimensionless
y	Operating time of a heat pump	Hours/year
Y	Cumulative average cost, production time, etc.	$/unit, hour/unit, etc.
Y	Operating-labor rate in Eq. (9-204)	labor-hour/ton
$\bar{Y}$	Cumulative-average batch cost, etc.	$/unit, etc.
z	Standard score defined by Eq. (9-73)	Dimensionless
	Greek symbols	
α	Proportionality factor in Eq. (9-168)	Dimensionless
β	Proportionality factor in Eq. (9-171)	Dimensionless
β	Exponent in Eqs. (9-106) and (9-117)	Dimensionless
δ	Symbol indicating partial differentiation	Dimensionless
Δ	Symbol indicating a difference of like quantities	Dimensionless
η	Contribution efficiency defined by Eq. (9-119)	Dimensionless
η	Margin of safety defined by Eq. (9-229)	Dimensionless
θ	Time taken to produce a given amount of product	Hour
σ	Population standard deviation	Various
Σ	Symbol indicating a sum of like quantities	Dimensionless
ϕ	Fractional increase in production rate	Dimensionless
ϕ_P	Parameter defined with Eq. (9-254)	Dimensionless
ψ	Parameter defined with Eq. (9-241)	Dimensionless
χ	Plant capacity in Eq. (9-204)	Tons/day
χ	Weight of product per unit of raw material	Dimensionless
	Subscripts	
A	Allowance against tax other than for capital depreciation	
BD	Depreciation allowance shown in company balance sheet	
BL	Within project boundary limits	

Symbol	Definition
BOH	Budgeted overhead
CF	Cash flow after payment of tax and expenses
CI	Cash income after payment of expenses
DCF	Discounted cash flow
DME	Direct manufacturing expense
FC	Fixed capital
FE	Fixed expense
FGE	Fixed general expense
FIFO	On a first-in–first-out basis
FIN	Financial-resources inventory
FME	Fixed manufacturing expense
FOH	Fixed overhead
GE	General expense
GP	Gross profit
IME	Indirect manufacturing expense
INV	Inventory
IO	Inventory-orders cost
IT	Income tax payable
IW	Inventory working cost
L	Labor-earnings index
L	Lower-quartile value of the variable
LIFO	Last-in–first-out basis
max	Maximum value
M	Median value of the variable
ME	Manufacturing expense
N	At agreed normal production rate
NCI	Net cash income after payment of tax
NOH	Overhead cost at agreed normal production rate
NNP	Net profit after payment of tax
NP	Net profit before payment of tax
OH	Overhead cost
P	Profit
RM	Raw material
s'	In the s'th productive year
S	From sales and other income
SAV	On a simple-average basis
ST	Steel-price index
SVOH	Semivariable overhead
TC	Total capital
TE	Total expense
TFE	Total fixed expense
TVE	Total variable expense
U	Utilities
U	Upper-quartile value of the variable
VE	Variable expense
VGE	Variable general expense
VME	Variable manufacturing expense
VOH	Variable overhead expense
W	Weighted value
WC	Working capital
WAV	On a weighted-average basis
1, 2, j, n	1st, 2d, jth, nth item, year, etc.

GENERAL REFERENCES: Allen, D. H., *Economic Evaluation of Projects,* 3d ed., Institution of Chemical Engineers, Rugby, England, 1991. Aries, R. S. and R. D. Newton, *Chemical Engineering Cost Estimation,* McGraw-Hill, New York, 1955. Baasel, W. D., *Preliminary Chemical Engineering Plant Design,* 2d ed., Van Nostrand Reinold, New York, 1989. Barish, N. N. and S. Kaplan, *Economic Analysis for Engineering and Management Decision Making,* 2d ed., McGraw-Hill, New York, 1978. Bierman, H., Jr. and S. Smidt, *The Capital Budgeting Decision, Economic Analysis and Financing of Investment Projects,* 7th ed., Macmillan, London, 1988. Canada, J. R. and J. A. White, *Capital Investment Decision: Analysis for Management and Engineering,* 2d ed., Prentice Hall, Englewood Cliffs, NJ, 1980. Carsberg, B. and A. Hope, *Business Investment Decisions under Inflation,* Macdonald & Evans, London, 1976. Chemical Engineering (ed.), *Modern Cost Engineering,* McGraw-Hill, New York, 1979. Garvin, W. W., *Introduction to Linear Programming,* McGraw-Hill, New York, 1960. Gass, S. I., *Linear Programming,* McGraw-Hill, New York, 1985. Granger, C. W. J., *Forecasting in Business and Economics,* Academic Press, New York, 1980. Hackney, J. W. and K. K. Humphreys (ed.), *Control and Management of Capital Projects,* 2d ed., McGraw-Hill, New York, 1991. Happel, J., W. H. Kapfer, B. J. Blewitt, P. T. Shannon, and D. G. Jordan, *Process Economics,* American Institute of Chemical Engineers, New York, 1974. Hill, D. A. and L. E. Rockley, *Secrets of Successful Financial Management,* Heinemann, London, 1990. Holland, F. A., F. A. Watson, and J. K. Wilkinson, *Introduction to Process Economics,* 2d ed., Wiley, London, 1983. Humphreys, K. K. (ed.), *Jelen's Cost and Optimization Engineering,* 3d ed., McGraw-Hill, New York, 1991. Institution of Chemical Engineers (ed.), *A Guide to Capital Cost Estimation,* Institution of Chemical Engineers, Rugby, England, 1988. Jordan, R. B., *How to Use the Learning Curve,* Materials Management Institute, Boston, 1965. Kharbanda, O. P. and E. A. Stallworthy, *Capital Cost Estimating in the Process Industries,* 2d ed., Butterworth-Heinemann, London, 1988. Kirkman, P. R., *Accounting under Inflationary Conditions,* 2d ed., Routledge, Chapman & Hall, London, 1978. Liddle, C. J. and A. M. Gerrard, *The Application of Computers to Capital Cost Estimation,* Institution of Chemical Engineers, Rugby, England, 1975. Loomba, N. P., *Linear Programming,* McGraw-Hill, New York, 1964. Merrett, A. J. and A. Sykes, *The Finance and Analysis of Capital Projects,* Longman, London, 1963. Merrett, A. J. and A. Sykes, *Capital Budgeting and Company Finance,* Longman, London, 1966. Ostwald, P. F., *Engineering Cost Estimating,* 3d ed., Prentice Hall, Englewood Cliffs, NJ, 1991. Park, W. R. and D. E. Jackson, *Cost Engineering Analysis,* 2d ed., Wiley, New York, 1984. Peters, M. S. and K. D. Timmerhaus, *Plant Design and Economics for Chemical Engineers,* 4th ed., McGraw-Hill, New York, 1991. Pilcher, R., *Principles of Construction Management,* 3d ed., McGraw-Hill, New York, 1992. Popper, H. (ed.), *Modern Cost Estimating Techniques,* McGraw-Hill, New York, 1970. Raiffa, H. and R. Schlaifer, *Applied Statistical Decision Theory,* Harper & Row (Harvard Business), New York, 1984. Ridge, W. J., *Value Analysis for Better Management,* American Management Association, New York, 1969. Rose, L. M., *Engineering Investment Decisions: Planning under Uncertainty,* Elsevier, Amsterdam, 1976. Rudd, D. F. and C. C. Watson, *The Strategy of Process Engineering,* Wiley, New York, 1968. Thorne, H. C. and J. B. Weaver (ed.), *Investment Appraisal for Chemical Engineers,* American Institute of Chemical Engineers, New York, 1991. Weaver, J. B., 'Project Selection in the 1980's', *Chem. Eng. News* **37–46** (Nov. 2, 1981). Wells, G. L., *Process Engineering with Economic Objectives,* Wiley, New York, 1973. Wilkes, F. M., *Capital Budgeting Techniques,* 2d ed., Wiley, London, 1983. Wood, E. G., *Costing Matters for Managers,* Beekman Publications, London, 1977. Woods, D. R., *Process Design and Engineering,* Prentice Hall, Englewood Cliffs, NJ, 1993. Wright, M. G., *Financial Management,* McGraw-Hill, London, 1970.

NOMENCLATURE

An attempt has been made to bring together most of the methods currently available for project evaluation and to present them in such a way as to make the methods amenable to modern computational techniques. To this end the practices of accountants and others have been reduced, where possible, to mathematical equations which are usually solvable with an electronic hand calculator equipped with scientific function keys. To make the equations suitable for use on high-speed computers an attempt has been made to devise a nomenclature which is suitable for machines using ALGOL, COBOL, or FORTRAN compilers. The number of letters and numbers used to define a variable has usually been limited to five. The letters are mnemonic in English wherever possible and are derived in two ways. First, when a standard accountancy phrase exists for a term, this has been abbreviated in capital letters and enclosed in parentheses, e.g., (ATR), for assets-to-turnover ratio; (DCFRR), for discounted-cash-flow rate of return. Clearly, the parentheses are omitted when the letter group is used to define the variable name for the computer. Second, a general symbol is defined for a type of variable and is modified by a mnemonic subscript, e.g., an annual cash quantity A_{TC}, annual total capital outlay, \$/year. Clearly, the symbols are written on one line when the letter group is used to define a variable name for the computer. In other cases, when well-known standard symbols exist, they have been adopted, e.g., z for the standard score as used in the normal distribution. Also, a, b, c, d, and e have been used to denote empirical constants and x and y to denote general variables where their use does not clash with other meanings of the same symbols.

The coverage in this section is so wide that nomenclature has sometimes proved a problem which has required the use of primes, asterisks, and other symbols not universally acceptable in the naming of computer variables. However, it is realized that each individual will program only his or her preferred methods, which will release some symbols for other uses. Also, it is not difficult to replace a forbidden symbol by an acceptable one; e.g., c_{RM} might be rendered CARM and P'_S as PSP by using A for asterisk and P for prime. For compilers which recognize only one alphabetical case, an extra prefix can be used to distinguish between uppercase and lowercase letters, for which purpose the letters U and L have been used only in a restricted way in the nomenclature.

It is, of course, impossible to allow for all possible variations of equation requirements and machine capability, but it is hoped that the nomenclature in the table presented at the beginning of the section will prove adequate for most purposes and will be capable of logical extension to other more specialized requirements.

INVESTMENT AND PROFITABILITY

In order to assess the profitability of projects and processes it is necessary to define precisely the various parameters.

Annual Costs, Profits, and Cash Flows To a large extent, accountancy is concerned with annual costs. To avoid confusion with other costs, annual costs will be referred to by the letter A.

The revenue from the annual sales of product A_S, minus the total annual cost or expense required to produce and sell the product A_{TE}, excluding any annual provision for plant depreciation, is the annual cash income A_{CI}:

$$A_{CI} = A_S - A_{TE} \qquad (9\text{-}1)$$

Net annual cash income A_{NCI} is the annual cash income A_{CI}, minus the annual amount of tax A_{IT}:

$$A_{NCI} = A_{CI} - A_{IT} \tag{9-2}$$

Taxable income is $(A_{CI} - A_D - A_A)$, where A_D is the annual writing-down allowance and A_A is the annual amount of any other allowances. A distinction is made between the writing-down allowance permissible for the computation of tax due, the actual depreciation in value of an asset, and the book depreciation in value of that asset as shown in the company position statement. There is no necessary connection between these values unless specified by law, although the first two or all three are often assigned the same value in practice. Some governments give cash incentives to encourage companies to build plants in otherwise unattractive areas. Neither A_D nor A_A involves any expenditure of cash, since they are merely book transactions. The annual amount of tax A_{IT} is given by

$$A_{IT} = (A_{CI} - A_D - A_A)t \tag{9-3}$$

where t is the fractional tax rate. The value of t is determined by the appropriate tax authority and is subject to change. For most developed countries the value of t is about 0.35 or 35 percent.

The annual amount of tax A_{IT} included in Eq. (9-2) does not necessarily correspond to the annual cash income A_{CI} in the same year. The tax payments in Eq. (9-2) should be those actually paid in that year. In the United States, companies pay about 80 percent of the tax on estimated current-year earnings in the same year. In the United Kingdom, companies do not pay tax until at least 9 months after the end of the accounting period, which, for the most part, amounts to paying tax on the previous year's earnings. When assessing projects for different countries, engineers should acquaint themselves with the tax situation in those countries.

In modern methods of profitability assessment, cash flows are more meaningful than profits, which tend to be rather loosely defined. The net annual cash flow after tax is given by

$$A_{CF} = A_{NCI} - A_{TC} \tag{9-4}$$

where A_{TC} is the annual expenditure of capital, which is not necessarily zero after the plant has been built. For example, working capital, plant additions, or modifications may be required in future years.

The total annual expense A_{TE} required to produce and sell a product can be written as the sum of the annual general expense A_{GE} and the annual manufacturing cost or expense A_{ME}:

$$A_{TE} = A_{GE} + A_{ME} \tag{9-5}$$

Annual general expense A_{GE} arises from the following items: adminis-

tration, sales, shipping of product, advertising and marketing, technical service, research and development, and finance.

The terms gross annual profit A_{GP} and net annual profit A_{NP} are commonly used by accountants and misused by others. Normally, both A_{GP} and A_{NP} are calculated before tax is deducted. Gross annual profit A_{GP} is given by

$$A_{GP} = A_S - A_{ME} - A_{BD} \tag{9-6}$$

where A_{BD} is the balance-sheet annual depreciation charge, which is not necessarily the same as A_D used in Eq. (9-3) for tax purposes. Net annual profit A_{NP} is simply

$$A_{NP} = A_{GP} - A_{GE} \tag{9-7}$$

Equation (9-7) can also be written as

$$A_{NP} = A_{CI} - A_{BD} \tag{9-8}$$

Net annual profit after tax A_{NNP} can be written as

$$A_{NNP} = A_{NCI} - A_{BD} \tag{9-9}$$

The relationships among the various annual costs given by Eqs. (9-1) through (9-9) are illustrated diagrammatically in Fig. 9-1. The top half of the diagram shows the tools of the accountant; the bottom half, those of the engineer. The net annual cash flow A_{CF}, which excludes any provision for balance-sheet depreciation A_{BD}, is used in two of the more modern methods of profitability assessment: the net-present-value (NPV) method and the discounted-cash-flow-rate-of-return (DCFRR) method. In both methods, depreciation is inherently taken care of by calculations which include capital recovery.

Annual general expense A_{GE} can be written as the sum of the fixed and variable general expenses:

$$A_{GE} = A_{FGE} + A_{VGE} \tag{9-10}$$

Similarly, annual manufacturing expense A_{ME} can be written as the sum of the fixed and variable manufacturing expenses:

$$A_{ME} = A_{FME} + A_{VME} \tag{9-11}$$

A variable expense is considered to be one which is directly proportional to the rate of production R_P or of sales R_S as is most appropriate to the case under consideration. Unless the variation in finished-product inventory is large when compared with the total production over the period in question, it is usually sufficiently accurate to consider R_P and R_S to be represented by the same-numerical-value R units of sale or production per year. A fixed expense is then considered to be one which is not directly proportional to R, such as overhead charges. Fixed expenses are not necessarily constant but may be sub-

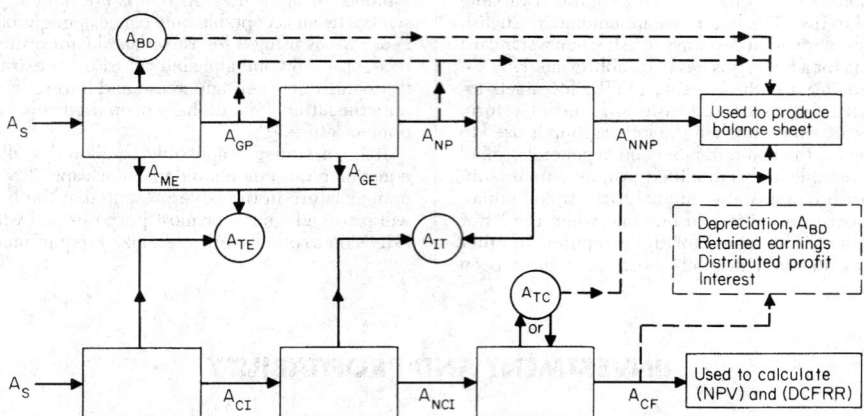

FIG. 9-1 Relationship between annual costs, annual profits, and cash flows for a project. A_{BD} = annual depreciation allowance; A_{CF} = annual net cash flow after tax; A_{CI} = annual cash income; A_{GE} = annual general expense; A_{GP} = annual gross profit; A_{IT} = annual tax; A_{ME} = annual manufacturing cost; A_{NCI} = annual net cash income; A_{NNP} = annual net profit after taxes; A_{NP} = annual net profit; A_S = annual sales; A_{TC} = annual total cost; (DCFRR) = discounted-cash-flow rate of return; (NPV) = net present value.

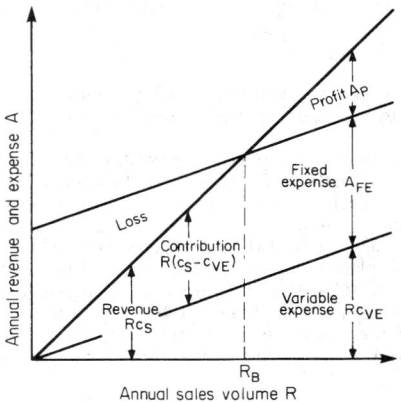

FIG. 9-2 Conventional breakeven chart.

ject to stepwise variation at different levels of production. Some authors consider such steps as included in a semivariable expense, which is less amenable to mathematical analysis than the above division of expenses.

Contribution and Breakeven Charts These can be used to give valuable preliminary information prior to the use of the more sophisticated and time-consuming methods based on discounted cash flow. If the sales price per unit of sales is c_S and the variable expense is c_{VE} per unit of production, Eq. (9-7) can be rewritten as

$$A_{NP} = R(c_S - c_{VE}) - A_{FE} \qquad (9\text{-}12)$$

where $R(c_S - c_{VE})$ is known as the annual contribution. The net annual profit is zero at an annual production rate

$$R_B = A_{FE}/(c_S - c_{VE}) \qquad (9\text{-}13)$$

where R_B is the breakeven production rate.

Breakeven charts can be plotted in any of the three forms shown in Figs. 9-2, 9-3, and 9-4. The abscissa shown as annual sales volume R is also frequently plotted as a percentage of the designed production or sales capacity R_0. In the case of ships, aircraft, etc., it is then called the percentage utilization. The percentage margin of safety is defined as $100(R_0 - R_B)/R_0$.

A decrease in selling price c_S will decrease the slope of the lines in Figs. 9-2, 9-3, and 9-4 and increase the required breakeven value R_B for a given level of fixed expense A_{FE}.

Capital Costs The total capital cost C_{TC} of a project consists of the fixed-capital cost C_{FC} plus the working-capital cost C_{WC}, plus the cost of land and other nondepreciable costs C_L:

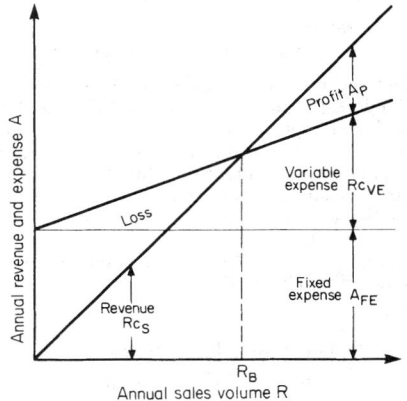

FIG. 9-3 Breakeven chart showing fixed expense as a burden cost.

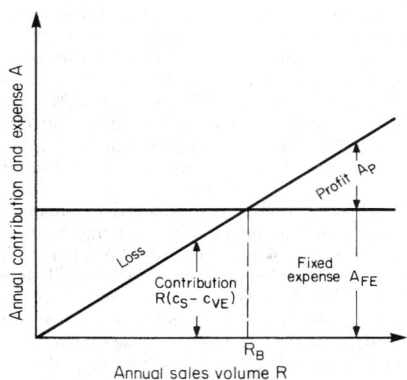

FIG. 9-4 Breakeven chart showing relationship between contribution and fixed expense.

$$C_{TC} = C_{FC} + C_{WC} + C_L \qquad (9\text{-}14)$$

The project may be a complete plant, an addition to an existing plant, or a plant modification.

The working-capital cost of a process or a business normally includes the items shown in Table 9-1. Since working capital is completely recoverable at any time, in theory if not in practice, no tax allowance is made for its depreciation. Changes in working capital arising from varying trade credits or payroll or inventory levels are usually treated as a necessary business expense except when they exceed the tax debt due. If the annual income is negative, additional working capital must be provided and included in the A_{TC} for that year. The value of land and other nondepreciables often increases over the working life of the project. These are therefore not treated in the same way as other capital investments but are shown to have made a (taxable) profit or loss only when the capital is finally recovered.

Working capital may vary from a very small fraction of the total capital cost to almost the whole of the invested capital, depending on the process and the industry. For example, in jewelry-store operations, the fixed capital is very small in comparison with the working capital. On the other hand, in the chemical-process industries, the working capital is likely to be in the region of 10 to 20 percent of the value of the fixed-capital investment.

Depreciation The term "depreciation" is used in a number of different contexts. The most common are:

1. A tax allowance
2. A cost of operation
3. A means of building up a fund to finance plant replacement
4. A measure of falling value

In the first case, the annual taxable income is reduced by an annual depreciation charge or allowance which has the effect of reducing the annual amount of tax payable. The annual depreciation charge is merely a book transaction and does not involve any expenditure of cash. The method of determining the annual depreciation charge must be agreed to by the appropriate tax authority.

In the second case, depreciation is considered to be a manufacturing cost in the same way as labor cost or raw-materials cost. However,

TABLE 9-1 Working-Capital Costs

Raw materials for plant startup
Raw-materials, intermediate, and finished-product inventories
Cost of handling and transportation of materials to and from stores
Cost of inventory control, warehouse, associated insurance, security
   arrangements, etc.
Money to carry accounts receivable (i.e., credit extended to customers) less
   accounts payable (i.e., credit extended by suppliers)
Money to meet payrolls when starting up
Readily available cash for emergencies
Any additional cash required to operate the process or business

it is more difficult to estimate a depreciation cost per unit of product than it is to do so for labor or raw-materials costs. In the net-present-value (NPV) and discounted-cash-flow-rate-of-return (DCFRR) methods of measuring profitability, depreciation, as a cost of operation, is implicitly accounted for. (NPV) and (DCFRR) give measures of return after a project has generated sufficient income to repay, among other things, the original investment and any interest charges that the invested money would otherwise have brought into the company.

In the third case, depreciation is considered as a means of providing for plant replacement. In the rapidly changing modern chemical-process industries, many plants will never be replaced because the processes or products have become obsolete during their working life. Management should be free to invest in the most profitable projects available, and the creation of special-purpose funds may hinder this. However, it is desirable to designate a proportion of the retained income as a fund from which to finance new capital projects. These are likely to differ substantially from the projects that originally generated the income.

In the fourth case, a plant or a piece of equipment has a limited useful life. The primary reason for the decrease in value is the decrease in future life and the consequent decrease in the number of years for which income will be earned. At the end of its life, the equipment may be worth nothing, or it may have a salvage or scrap value S. Thus a fixed-capital cost C_{FC} depreciates in value during its useful life of s years by an amount that is equal to $(C_{FC} - S)$. The useful life is taken from the startup of the plant.

On the basis of straight-line depreciation, the average annual amount of depreciation A_D over a service life of s years is given by

$$A_D = (C_{FC} - S)/s \qquad (9-15)$$

The book value after the first year P_1 is given by

$$P_1 = C_{FC} - A_D \qquad (9-16)$$

The book value at the end of a specified number of years s' is given by

$$P_{s'} = C_{FC} - s'A_D \qquad (9-17)$$

The principal use of a particular depreciation rate is for tax purposes. The permitted annual depreciation is subtracted from the annual income before the latter is taxed. The basis for depreciation in a particular case is a matter of agreement between the taxation authority and the company, in conformity with tax laws.

Other commonly used methods of computing depreciation are the declining-balance method (also known as the fixed-percentage method) and the sum-of-years-digits method.

On the basis of declining-balance (fixed-percentage) depreciation, the book value at the end of the first year is given by

$$P_1 = C_{FC}(1 - r) \qquad (9-18)$$

where r is a fraction to be agreed with the taxation authority.

The book value at the end of specified number of years s' is given by

$$P_{s'} = C_{FC}(1 - r)^{s'} \qquad (9-19)$$

When the fraction r is chosen to be $2/s$, i.e., twice the reciprocal of the service life s, the method is called the double-declining-balance method.

The declining-balance method of depreciation allows equipment or plant to be depreciated by a greater amount during the earlier years than during the later years. This method does not allow equipment or plant to be depreciated to a zero value at the end of the service life.

On the basis of sum-of-years-digits depreciation, the annual amount of depreciation for a specified number of years s' for a plant of fixed-capital cost C_{FC}, scrap value S, and service life s is given by

$$A_{Ds'} = \left(\frac{s - s' + 1}{1 + 2 + 3 + \cdots + s}\right)(C_{FC} - S) \qquad (9-20)$$

Equation (9-20) can also be rewritten in the form

$$A_{Ds'} = \left[\frac{2(s - s' + 1)}{s(s + 1)}\right](C_{FC} - S) \qquad (9-21)$$

It can be shown that the book value at the end of a particular year s' is

$$P_{s'} = 2\left[\frac{1 + 2 + \cdots + (s - s')}{s(s - 1)}\right](C_{FC} - S) + S \qquad (9-22)$$

The sum-of-years-digits depreciation allows equipment or plant to be depreciated by a greater amount during the early years than during the later years.

A fourth method of computing depreciation (now seldom used) is the sinking-fund method. In this method, the annual depreciation A_D is the same for each year of the life of the equipment or plant. The series of equal amounts of depreciation A_D, invested at a fractional interest rate i and made at the end of each year over the life of the equipment or plant of s years, is used to build up a future sum of money equal to $(C_{FC} - S)$. This last is the fixed-capital cost of the equipment or plant minus its salvage or scrap value and is the total amount of depreciation during its useful life. The equation relating $(C_{FC} - S)$ and A_D is simply the annual cost or payment equation, written either as

$$C_{FC} - S = A_D\left[\frac{(1 + i)^s - 1}{i}\right] \qquad (9-23)$$

or

$$C_{FC} - S = \frac{A_D}{f_{AF}} \qquad (9-24)$$

where f_{AF} is the annuity future-worth factor given by

$$f_{AF} = i/[(1 + i)^s - 1]$$

In the sinking-fund method of depreciation, the effect of interest is to make the annual decrease of the book value of the equipment or plant less in the early than in the later years with consequent higher tax due in the earlier years when recovery of the capital is most important.

It is preferable not to think of annual depreciation as a contribution to a fund to replace equipment at the end of its life but as part of the difference between the revenue and the expenditure, which difference is tax-free.

Some of the preceding methods of computing depreciation are not allowed by taxation authorities in certain countries. When calculating depreciation, it is necessary to obtain details of the methods and rates permitted by the appropriate authority and to use the information provided.

Figure 9-5 shows the fall in book value with time for a piece of equipment having a fixed-capital cost of $120,000, a useful life of 10 years, and a scrap value of $20,000. This fall in value is calculated by using (1) straight-line depreciation, (2) double-declining depreciation, and (3) sum-of-years-digits depreciation.

Traditional Measures of Profitability

Rate-of-Return Methods Although traditional rate-of-return methods have the advantage of simplicity, they can yield very misleading results. They are based on the relation

Percent rate of return
$$= [(\text{annual profit})/(\text{invested capital})]100 \qquad (9-25)$$

Since different meanings are ascribed to both annual profit and invested capital in Eq. (9-25), it is important to define the terms precisely. The invested capital may refer to the original total capital investment, the depreciated investment, the average investment, the current value of the investment, or something else. The annual profit may refer to the net annual profit before tax A_{NP}, the net annual profit after tax A_{NNP}, the annual cash income before tax A_{CI}, or the annual cash income after tax A_{NCI}.

The fractional interest rate of return based on the net annual profit after tax and the original investment is

$$i = A_{NNP}/C_{TC} \qquad (9-26)$$

which can be written in terms of Eq. (9-9) as

$$i = (A_{NCI}/C_{TC}) - (A_{BD}/C_{TC}) \qquad (9-27)$$

where A_{BD} is the balance-sheet annual depreciation. The main disadvantage of using Eq. (9-27) is that the fractional depreciation rate

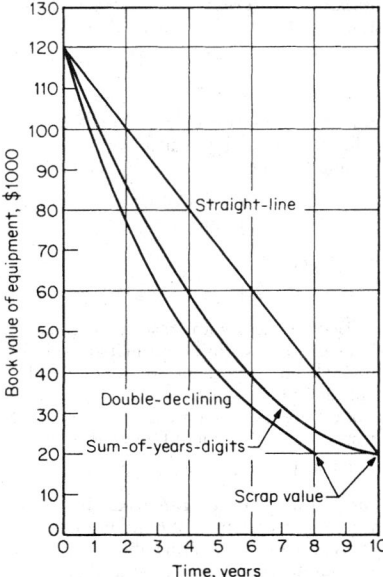

FIG. 9-5   Book value against time for various depreciation methods.

A_{BD}/C_{TC} is arbitrarily assessed. Its value will affect the fractional rate of return considerably and may lead to erroneous conclusions when making comparisons between different companies. This is particularly true when making international comparisons.

Figures 9-6, 9-7, and 9-8 show the effect of the depreciation method on profit for a project described by the following data:

Net annual cash income after tax A_{NCI} = \$25,500 in each of 10 years

Fixed-capital cost C_{FC} = \$120,000

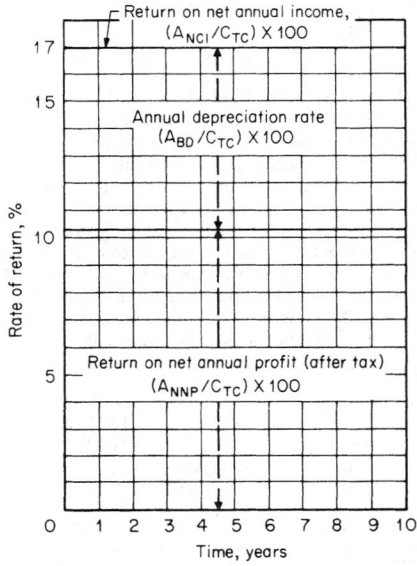

FIG. 9-6   Effect of straight-line depreciation on rate of return for a project. A_{BD} = annual depreciation allowance; A_{NCI} = annual net cash income after tax; A_{NNP} = annual net profit after payment of tax; C_{TC} = total capital cost.

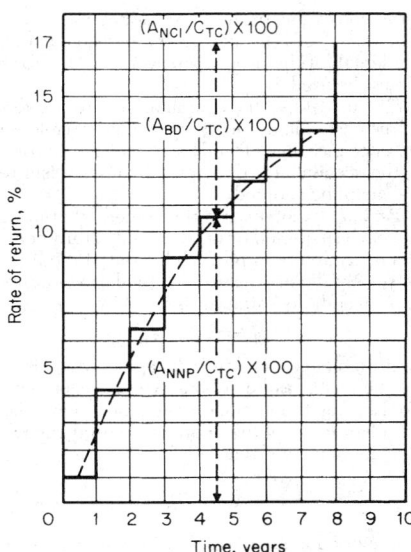

FIG. 9-7   Effect of double-declining depreciation on rate of return for a project.

Estimated salvage value of plant items S = \$20,000

Working capital C_{WC} = \$10,000

Cost of land C_L = \$20,000

In Eq. (9-27), i can be taken either on the basis of the net annual cash income for a particular year or on the basis of an average net annual cash income over the length of the life of the project. The equations corresponding to Eq. (9-26) based on depreciated and average investment are given respectively as follows:

$$i = A_{NNP}/(P_{s'} + C_{WC} + C_L) \qquad (9\text{-}28)$$

and

$$i = 2A_{NNP}/(C_{FC} + S + 2C_{WC} + 2C_L) \qquad (9\text{-}29)$$

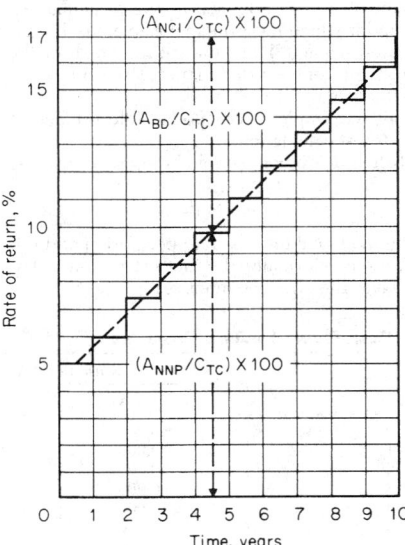

FIG. 9-8   Effect of sum-of-years-digits depreciation on rate of return for a project.

where $P_{s'}$ is the book value of the fixed-capital investment at the end of a particular year s'. If i is taken on the basis of average values for A_{NNP} over the length of the project, an average value for the working capital C_{WC} must be used.

In Eqs. (9-28) and (9-29), the computations are based on unchanging values of the cost of land and other nondepreciable costs C_L. This is unrealistic, since the value of land has a tendency to rise. In such circumstances, the accountancy principle of conservatism requires that the lowest valuation be adopted.

Payback Period Another traditional method of measuring profitability is the payback period or fixed-capital-return period. Actually, this is really a measure not of profitability but of the time it takes for cash flows to recoup the original fixed-capital expenditure.

The net annual cash flow after tax is given by

$$A_{CF} = A_{NCI} - A_{TC} \qquad (9\text{-}4)$$

where A_{TC} is the annual expenditure of capital, which is not necessarily zero after the plant has been built. The payback period (PBP) is the time required for the cumulative net cash flow taken from the startup of the plant to equal the depreciable fixed-capital investment $(C_{FC} - S)$. It is the value of s' that satisfies

$$\sum_{s'=0}^{s'=(\text{PBP})} A_{CF} = C_{FC} - S \qquad (9\text{-}30)$$

The payback-period method takes no account of cash flows or profits received after the breakeven point has been reached. The method is based on the premise that the earlier the fixed capital is recovered, the better the project. However, this approach can be misleading.

Let us consider projects A and B, having net annual cash flows as listed in Table 9-2. Both projects have initial fixed-capital expenditures of $100,000. On the basis of payback period, project A is the more desirable since the fixed-capital expenditure is recovered in 3 years, compared with 5 years for project B. However, project B runs for 7 years with a cumulative net cash flow of $110,000. This is obviously more profitable than project A, which runs for only 4 years with a cumulative net cash flow of only $10,000.

Time Value of Money A large part of business activity is based on money that can be loaned or borrowed. When money is loaned, there is always a risk that it may not be returned. A sum of money called interest is the inducement offered to make the risk acceptable. When money is borrowed, interest is paid for the use of the money over a period of time. Conversely, when money is loaned, interest is received.

The amount of a loan is known as the principal. The longer the period of time for which the principal is loaned, the greater the total amount of interest paid. Thus, the future worth of the money F is greater than its present worth P. The relationship between F and P depends on the type of interest used.

Table 9-3 gives examples of compound-interest factors and example compound-interest calculations.

Simple Interest When simple interest is used, F and P are related by

$$F = P(1 + ni) \qquad (9\text{-}31)$$

where i is the fractional interest rate per period and n is the number of interest periods. Normally, the interest period is 1 year, in which case i is known as the effective interest rate.

Annual Compound Interest It is more common to use compound interest, in which F and P are related by

$$F = P(1 + i)^n \qquad (9\text{-}32)$$

or

$$F = Pf_i \qquad (9\text{-}33)$$

where the compound-interest factor $f_i = (1 + i)^n$. Values for compound-interest factors are readily available in tables.

The present value P of a future sum of money F is

$$P = F/(1 + i)^n \qquad (9\text{-}34)$$

or

$$F = P/f_d \qquad (9\text{-}35)$$

where the discount factor f_d is

$$f_d = 1/f_i = 1/[(1 + i)^n]$$

Values for the discount factors are readily available in tables which show that it will take 7.3 years for the principal to double in amount if compounded annually at 10 percent per year and 14.2 years if compounded annually at 5 percent per year.

For the case of different annual fractional interest rates $(i_1, i_2, \ldots, i_n$ in successive years), Eq. (9-32) should be written in the form

$$F = P(1 + i_1)(1 + i_2)(1 + i_3) \cdots (1 + i_n) \qquad (9\text{-}36)$$

Short-Interval Compound Interest If interest payments become due m times per year at compound interest, mn payments are required in n years. The nominal annual interest rate i' is divided by m to give the effective interest rate per period. Hence,

$$F = P[1 + (i'/m)]^{mn} \qquad (9\text{-}37)$$

It follows that the effective annual interest i is given by

$$i = [1 + (i'/m)]^m - 1 \qquad (9\text{-}38)$$

The annual interest rate equivalent to a compound-interest rate of 5 percent per month (i.e., $i'/m = 0.05$) is calculated from Eq. (9-38) to be

$$i = (1 + 0.05)^{12} - 1 = 0.796, \text{ or } 79.6 \text{ percent/year}$$

Continuous Compound Interest As m approaches infinity, the time interval between payments becomes infinitesimally small, and in the limit Eq. (9-37) reduces to

$$F = P \exp(i'n) \qquad (9\text{-}39)$$

A comparison of Eqs. (9-32) and (9-39) shows that the nominal interest rate i' on a continuous basis is related to the effective interest rate i on an annual basis by

$$\exp(i'n) = (1 + i)^n \qquad (9\text{-}40)$$

Numerically, the difference between continuous and annual compounding is small. In practice, it is probably far smaller than the errors in the estimated cash-flow data. Annual compound interest conforms more closely to current acceptable accounting practice. However, the small difference between continuous and annual compounding may be significant when applied to very large sums of money.

Let us suppose that $100 is invested at a nominal interest rate of 5 percent. We then compute the future worth of the investment after 2 years and also compute the effective annual interest rate for the following kinds of interest: (1) simple, (2) annual compound, (3) monthly compound, (4) daily compound, and (5) continuous compound. The following tabulation shows the results of the calculations, along with the appropriate equation to be used:

Interest type	Equation	Future worth F	Effective rate i, %	Equation
1	(9-31)	$110.000	5	(9-31)
2	(9-32)	$110.250	5	(9-38)
3	(9-37)	$110.495	5.117	(9-38)
4	(9-37)	$110.516	5.1267	(9-38)
5	(9-39)	$110.517	5.1271	(9-38)

When computing the effective annual rate for continuous compounding, the first term of Eq. (9-38), $[1 + (i'/m)]^m$, approaches $e^{i'}$ as m approaches infinity.

TABLE 9-2 Cash Flows for Two Projects

Year	Cash flows A_{CF}	
	Project A	Project B
0	$100,000	$100,000
1	50,000	0
2	30,000	10,000
3	20,000	20,000
4	10,000	30,000
5	0	40,000
6	0	50,000
7	0	60,000
$\sum A_{CF}$	$10,000	$110,000
Payback period (PBP)	3 years	5 years

TABLE 9-3 Compound Interest Factors*

(For examples demonstrating use see end of table.)

5% Compound Interest Factors

n	Single payment: Compound-amount factor, Given P, to find F, $(1+i)^n$	Single payment: Present-worth factor, Given F, to find P, $\frac{1}{(1+i)^n}$	Uniform annual series: Sinking-fund factor, Given F, to find A, $\frac{i}{(1+i)^n-1}$	Uniform annual series: Capital-recovery factor, Given P, to find A, $\frac{i(1+i)^n}{(1+i)^n-1}$	Uniform annual series: Compound-amount factor, Given A, to find F, $\frac{(1+i)^n-1}{i}$	Uniform annual series: Present-worth factor, Given A, to find P, $\frac{(1+i)^n-1}{i(1+i)^n}$
1	1.050	0.9524	1.00000	1.05000	1.000	0.952
2	1.103	.9070	.48780	.53780	2.050	1.859
3	1.158	.8638	.31721	.36721	3.153	2.723
4	1.216	.8227	.23201	.28201	4.310	3.546
5	1.276	.7835	.18097	.23097	5.526	4.329
6	1.340	.7462	.14702	.19702	6.802	5.076
7	1.407	.7107	.12282	.17282	8.142	5.786
8	1.477	.6768	.10472	.15472	9.549	6.463
9	1.551	.6446	.09069	.14069	11.027	7.108
10	1.629	.6139	.07950	.12950	12.578	7.722
11	1.710	.5847	.07039	.12039	14.207	8.306
12	1.796	.5568	.06283	.11283	15.917	8.863
13	1.886	.5303	.05646	.10646	17.713	9.394
14	1.980	.5051	.05102	.10102	19.599	9.899
15	2.079	.4810	.04634	.09634	21.579	10.380
16	2.183	.4581	.04227	.09227	23.657	10.838
17	2.292	.4363	.03870	.08870	25.840	11.274
18	2.407	.4155	.03555	.08555	28.132	11.690
19	2.527	.3957	.03275	.08275	30.539	12.085
20	2.653	.3769	.03024	.08024	33.066	12.462
21	2.786	.3589	.02800	.07800	35.719	12.821
22	2.925	.3418	.02597	.07597	38.505	13.163
23	3.072	.3256	.02414	.07414	41.430	13.489
24	3.225	.3101	.02247	.07247	44.502	13.799
25	3.386	.2953	.02095	.07095	47.727	14.094
26	3.556	.2812	.01956	.06956	51.113	14.375
27	3.733	.2678	.01829	.06829	54.669	14.643
28	3.920	.2551	.01712	.06712	58.403	14.898
29	4.116	.2429	.01605	.06605	62.323	15.141
30	4.322	.2314	.01505	.06505	66.439	15.372
31	4.538	.2204	.01413	.06413	70.761	15.593
32	4.765	.2099	.01328	.06328	75.299	15.803
33	5.003	.1999	.01249	.06249	80.064	16.003
34	5.253	.1904	.01176	.06176	85.067	16.193
35	5.516	.1813	.01107	.06107	90.320	16.374
40	7.040	.1420	.00828	.05828	120.800	17.159
45	8.985	.1113	.00626	.05626	159.700	17.774
50	11.467	.0872	.00478	.05478	209.348	18.256
55	14.636	.0683	.00367	.05367	272.713	18.633
60	18.679	.0535	.00283	.05283	353.584	18.929
65	23.840	.0419	.00219	.05219	456.798	19.161
70	30.426	.0329	.00170	.05170	588.529	19.343
75	38.833	.0258	.00132	.05132	756.654	19.485
80	49.561	.0202	.00103	.05103	971.229	19.596
85	63.254	.0158	.00080	.05080	1,245.087	19.684
90	80.730	.0124	.00063	.05063	1,594.607	19.752
95	103.035	.0097	.00049	.05049	2,040.694	19.806
100	131.501	.0076	.00038	.05038	2,610.025	19.848

6% Compound Interest Factors

n	Single payment: Compound-amount factor, Given P, to find F, $(1+i)^n$	Single payment: Present-worth factor, Given F, to find P, $\frac{1}{(1+i)^n}$	Uniform annual series: Sinking-fund factor, Given F, to find A, $\frac{i}{(1+i)^n-1}$	Uniform annual series: Capital-recovery factor, Given P, to find A, $\frac{i(1+i)^n}{(1+i)^n-1}$	Uniform annual series: Compound-amount factor, Given A, to find F, $\frac{(1+i)^n-1}{i}$	Uniform annual series: Present-worth factor, Given A, to find P, $\frac{(1+i)^n-1}{i(1+i)^n}$
1	1.060	0.9434	1.00000	1.06000	1.000	0.943
2	1.124	.8900	.48544	.54544	2.060	1.833
3	1.191	.8396	.31411	.37411	3.184	2.673
4	1.262	.7921	.22859	.28859	4.375	3.465
5	1.338	.7473	.17740	.23740	5.637	4.212
6	1.419	.7050	.14336	.20336	6.975	4.917
7	1.504	.6651	.11914	.17914	8.394	5.582
8	1.594	.6274	.10104	.16104	9.897	6.210
9	1.689	.5919	.08702	.14702	11.491	6.802
10	1.791	.5584	.07587	.13587	13.181	7.360
11	1.898	.5268	.06679	.12679	14.972	7.887
12	2.012	.4970	.05928	.11928	16.870	8.384
13	2.133	.4688	.05296	.11296	18.882	8.853
14	2.261	.4423	.04758	.10758	21.015	9.295
15	2.397	.4173	.04296	.10296	23.276	9.712
16	2.540	.3936	.03895	.09895	25.673	10.106
17	2.693	.3714	.03544	.09544	28.213	10.477
18	2.854	.3503	.03236	.09236	30.906	10.828
19	3.026	.3305	.02962	.08962	33.760	11.158
20	3.207	.3118	.02718	.08718	36.786	11.470
21	3.400	.2942	.02500	.08500	39.993	11.764
22	3.604	.2775	.02305	.08305	43.392	12.042
23	3.820	.2618	.02128	.08128	46.996	12.303
24	4.049	.2470	.01968	.07968	50.816	12.550
25	4.292	.2330	.01823	.07823	54.865	12.783
26	4.549	.2198	.01690	.07690	59.156	13.003
27	4.822	.2074	.01570	.07570	63.706	13.211
28	5.112	.1956	.01459	.07459	68.528	13.406
29	5.418	.1846	.01358	.07358	73.640	13.591
30	5.743	.1741	.01265	.07265	79.058	13.765
31	6.088	.1643	.01179	.07179	84.802	13.929
32	6.453	.1550	.01100	.07100	90.890	14.084
33	6.841	.1462	.01027	.07027	97.343	14.230
34	7.251	.1379	.00960	.06960	104.184	14.368
35	7.686	.1301	.00897	.06897	111.435	14.498
40	10.286	.0972	.00646	.06646	154.762	15.046
45	13.765	.0727	.00470	.06470	212.744	15.456
50	18.420	.0543	.00344	.06344	290.336	15.762
55	24.650	.0406	.00254	.06254	394.172	15.991
60	32.988	.0303	.00188	.06188	533.128	16.161
65	44.145	.0227	.00139	.06139	719.083	16.289
70	59.076	.0169	.00103	.06103	967.932	16.385
75	79.057	.0126	.00077	.06077	1,300.949	16.456
80	105.796	.0095	.00057	.06057	1,746.600	16.509
85	141.579	.0071	.00043	.06043	2,342.982	16.549
90	189.465	.0053	.00032	.06032	3,141.075	16.579
95	253.546	.0039	.00024	.06024	4,209.104	16.601
100	339.302	.0029	.00018	.06018	5,638.368	16.618

TABLE 9-3   Compound Interest Factors (*Concluded*)

Examples of Use of Table and Factors

Given: $2500 is invested now at 5 percent.
Required: Accumulated value in 10 years (i.e., the amount of a given principal).

Solution:
$$F = P(1 + i)^n = \$2500 \times 1.05^{10}$$
$$\text{Compound-amount factor} = (1 + i)^n = 1.05^{10} = 1.629$$
$$F = \$2500 \times 1.629 = \$4062.50$$

Given: $19,500 will be required in 5 years to replace equipment now in use.
Required: With interest available at 3 percent, what sum must be deposited in the bank at present to provide the required capital (i.e., the principal which will amount to a given sum)?

Solution:
$$P = F \frac{1}{(1 + i)^n} = \$19,500 \frac{1}{1.03^5}$$
$$\text{Present-worth factor} = 1/(1 + i)^n = 1/1.03^5 = 0.8626$$
$$P = \$19,500 \times 0.8626 = \$16,821$$

Given: $50,000 will be required in 10 years to purchase equipment.
Required: With interest available at 4 percent, what sum must be deposited each year to provide the required capital (i.e., the annuity which will amount to a given fund)?

Solution:
$$A = F \frac{i}{(1 + i)^n - 1} = \$50,000 \frac{0.04}{1.04^{10} - 1}$$
$$\text{Sinking-fund factor} = \frac{i}{(1 + i)^n - 1} = \frac{0.04}{1.04^{10} - 1} = 0.08329$$
$$A = \$50,000 \times 0.08329 = \$4,164$$

Given: $20,000 is invested at 10 percent interest.
Required: Annual sum that can be withdrawn over a 20-year period (i.e., the annuity provided by a given capital).

Solution:
$$A = P \frac{i(1 + i)^n}{(1 + i)^n - 1} = \$20,000 \frac{0.10 \times 1.10^{20}}{1.10^{20} - 1}$$
$$\text{Capital-recovery factor} = \frac{i(1 + i)^n}{(1 + i)^n - 1} = \frac{0.10 \times 1.10^{20}}{1.10^{20} - 1} = 0.11746$$
$$A = \$20,000 \times 0.11746 = \$2349.20$$

Given: $500 is invested each year at 8 percent interest.
Required: Accumulated value in 15 years (i.e., amount of an annuity).

Solution:
$$F = A \frac{(1 + i)^n - 1}{i} = \$500 \frac{1.08^{15} - 1}{0.08}$$
$$\text{Compound-amount factor} = \frac{(1 - i)^n - 1}{i} = \frac{1.08^{15} - 1}{0.08} = 27.152$$
$$F = \$500 \times 27.152 = \$13,576$$

Given: $8000 is required annually for 25 years.
Required: Sum that must be deposited now at 6 percent interest.

Solution:
$$P = A \frac{(1 + i)^n - 1}{i(1 + i)^n} = \$8000 \frac{1.06^{25} - 1}{0.06 \times 1.06^{25}}$$
$$\text{Present-worth factor} = \frac{(1 + i)^n - 1}{i(1 + i)^n} = \frac{1.06^{25} - 1}{0.06 \times 1.06^{25}} = 12.783$$
$$P = \$8000 \times 12.783 = \$102,264$$

*Factors presented for two interest rates only. By using the appropriate formulas, values for other interest rates may be calculated.

Annual Cost or Payment A series of equal annual payments A invested at a fractional interest rate i at the end of each year over a period of n years may be used to build up a future sum of money F. These relations are given by

$$F = A \left[\frac{(1+i)^n - 1}{i} \right] \qquad (9\text{-}41)$$

or

$$F = A/f_{AF} \qquad (9\text{-}42)$$

where the annuity future-worth factor is

$$f_{AF} = i/[(1+i)^n - 1]$$

Values for f_{AF} are readily available in tables.

Equation (9-41) can be combined with Eq. (9-34) to yield

$$P = A \left[\frac{(1+i)^n - 1}{i(1+i)^n} \right] \qquad (9\text{-}43)$$

$$P = A/f_{AP} \qquad (9\text{-}44)$$

where P is the present worth of the series of future equal annual payments A and the annuity present-worth factor is

$$f_{AP} = [i(1+i)^n]/[(1+i)^n - 1]$$

Values for f_{AP} are also available in tables.

Alternatively, the annual payment A required to build up a future sum of money F with a present value of P is given by

$$A = F f_{AF} \qquad (9\text{-}45)$$

$$A = P f_{AP} \qquad (9\text{-}46)$$

Equation (9-41) represents the future sum of a series of uniform annual payments that are invested at a stated interest rate over a period of years. This procedure defines an ordinary annuity. Other forms of annuities include the annuity due, in which payments are made at the beginning of the year instead of at the end; and the deferred annuity, in which the first payment is deferred for a definite number of years.

Capitalized Cost A piece of equipment of fixed-capital cost C_{FC} will have a finite life of n years. The capitalized cost of the equipment C_K is defined by

$$(C_K - C_{FC})(1+i)^n = C_K - S \qquad (9\text{-}47)$$

C_K is in excess of C_{FC} by an amount which, when compounded at an annual interest rate i for n years, will have a future worth of C_K less the salvage or scrap value S. If the renewal cost of the equipment remains constant at $(C_{FC} - S)$ and the interest rate remains constant at i, then C_K is the amount of capital required to replace the equipment in perpetuity.

Equation (9-47) may be rewritten as

$$C_K = \left[C_{FC} - \frac{S}{(1+i)^n} \right] \left[\frac{(1+i)^n}{(1+i)^n - 1} \right] \qquad (9\text{-}48)$$

or

$$C_K = (C_{FC} - S f_d) f_k \qquad (9\text{-}49)$$

where f_d is the discount factor and f_k, the capitalized-cost factor, is

$$f_k = [(1+i)^n]/[(1+i)^n - 1]$$

Values for each factor are available in tables.

Example 1: Capitalized Cost of Equipment A piece of equipment has been installed at a cost of $100,000 and is expected to have a working life of 10 years with a scrap value of $20,000. Let us calculate the capitalized cost of the equipment based on an annual compound-interest rate of 5 percent.

Therefore, we substitute values into Eq. (9-48) to give

$$C_K = \left[\$100{,}000 - \frac{\$20{,}000}{(1+0.05)^{10}} \right] \left[\frac{(1+0.05)^{10}}{(1+0.05)^{10} - 1} \right]$$

$$C_K = [\$100{,}000 - (\$20{,}000/1.62889)](2.59009)$$

$$C_K = \$227{,}207$$

Modern Measures of Profitability An investment in a manufacturing process must earn more than the cost of capital for it to be worthwhile. The larger the additional earnings, the more profitable the venture and the greater the justification for putting the capital at risk. A profitability estimate is an attempt to quantify the desirability of taking this risk.

The ways of assessing profitability to be considered in this section are (1) discounted-cash-flow rate of return (DCFRR), (2) net present value (NPV) based on a particular discount rate, (3) equivalent maximum investment period (EMIP), (4) interest-recovery period (IRP), and (5) discounted breakeven point (DBEP).

Cash Flow Let us consider a project in which $C_{FC} = \$1{,}000{,}000$, $C_{WC} = \$90{,}000$, and $C_L = \$10{,}000$. Hence, $C_{TC} = \$1{,}100{,}000$ from Eq. (9-14). If all this capital expenditure occurs in Year 0 of the project, then $A_{TC} = \$1{,}100{,}000$ in Year 0 and $-A_{TC} = -\$1{,}100{,}000$. From Eq. (9-4), it is seen that any capital expenditure makes a negative contribution to the net annual cash flow A_{CF}.

Let us consider another project in which the fixed-capital expenditure is spread over 2 years, according to the following pattern:

$$C_{FC} = C_{FC0} + C_{FC1}$$

Year 0		Year 1	
$C_{FC0} =$	$400,000	$C_{FC1} =$	$600,000
$C_L =$	10,000	$C_{WC} =$	90,000
$A_{TC} =$	410,000	$A_{TC} =$	690,000

In the final year of the project, the working capital and the land are recovered, which in this case cost a total of $100,000. Thus, in the final year of the project, $A_{TC} = -\$100{,}000$ and $-A_{TC} = +\$100{,}000$. From Eq. (9-4), it is seen that any capital recovery makes a positive contribution to the net annual cash flow.

During the development and construction stages of a project, A_{CI} and A_{IT} are both zero in Eqs. (9-2) and (9-4). For this period, the cash flow for the project is negative and is given by

$$A_{CF} = -A_{TC} \qquad (9\text{-}50)$$

Figure 9-9 shows the cash-flow stages in a project. The expenditure during the research and development stage is normally relatively small. It will usually include some preliminary process design and a market survey. Once the decision to go ahead with the project has been taken, detailed process-engineering design will commence, and the rate of expenditure starts to increase. The rate is increased still further when equipment is purchased and construction gets under way. There is no return on the investment until the plant is started up. Even during startup, there is some additional expenditure. Once the plant is operating smoothly, an inflow of cash is established. During the early stages of a project, there may be a tax credit because of the existence of expenses without corresponding income.

Discounted Cash Flow The present value P of a future sum of money F is given by

$$P = F f_d \qquad (9\text{-}51)$$

where $f_d = 1/(1+i)^n$, the discount factor. Values for this factor are readily available in tables. For example, $90,909 invested at an annual interest rate of 10 percent becomes $100,000 after 1 year. Similarly, $38,554 invested at 10 percent becomes $100,000 after 10 years.

Thus, cash flow in the early years of a project has a greater value than the same amount in the later years of a project. Therefore, it pays to receive money as soon as possible and to delay paying out money for as long as possible.

Time is taken into account by using the annual discounted cash flow A_{DCF}, which is related to the annual cash flow A_{CF} and the discount factor f_d by

$$A_{DCF} = A_{CF} f_d \qquad (9\text{-}52)$$

Thus, at the end of any year n,

$$(A_{DCF})_n = (A_{CF})_n/(1+i)^n$$

The sum of the annual discounted cash flows over n years, $\sum A_{DCF}$, is known as the net present value (NPV) of the project:

$$(\text{NPV}) = \sum_0^n (A_{DCF})_n \qquad (9\text{-}53)$$

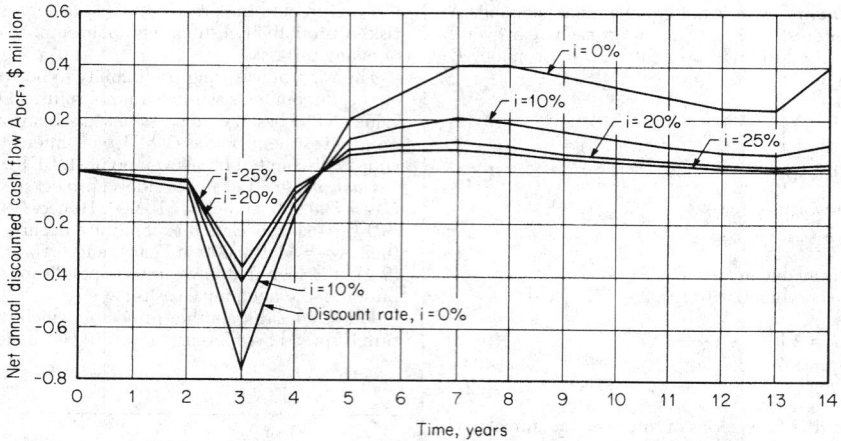

FIG. 9-9   Effect of discount rate on cash flows.

The value of (NPV) is directly dependent on the choice of the fractional interest rate i. An interest rate can be selected to make (NPV) = 0 after a chosen number of years. This value of i is found from

$$\sum_0^n (A_{DCF})_n = \frac{(A_{CF})_0}{(1+i)^0} + \frac{(A_{CF})_1}{(1+i)^1} + \cdots + \frac{(A_{CF})_n}{(1+i)^n} = 0 \qquad (9\text{-}54)$$

Equation (9-54) may be solved for i either graphically or by an iterative trial-and-error procedure. The value of i given by Eq. (9-54) is known as the discounted-cash-flow rate of return (DCFRR). It is also known as the profitability index, true rate of return, investor's rate of return, and interest rate of return.

Cash-Flow Curves   Figure 9-9 shows the cash-flow stages in a project together with their discounted-cash-flow values for the data given in Table 9-4. In addition to cash-flow and discounted-cash-flow curves, it is also instructive to plot cumulative-cash-flow and cumulative-discounted-cash-flow curves. These are shown in Fig. 9-10 for the data in Table 9-4.

The cost of capital may also be considered as the interest rate at which money can be invested instead of putting it at risk in a manufacturing process. Let us consider the process data listed in Table 9-4 and plotted in Fig. 9-10. If the cost of capital is 10 percent, then the appropriate discounted-cash-flow curve in Fig. 9-10 is *abcdef*. Up to point e, or 8.49 years, the capital is at risk. Point e is the discounted breakeven point (DBEP). At this point, the manufacturing process

has paid back its capital and produced the same return as an equivalent amount of capital invested at a compound-interest rate of 10 percent. Beyond the breakeven point, the capital is no longer at risk and any cash flow above the horizontal baseline, $\sum A_{DCF} = 0$, is in excess of the return on an equivalent amount of capital invested at a compound-interest rate of 10 percent. Thus, the greater the area above the baseline, the more profitable the process.

When (NPV) and (DCFRR) are computed, depreciation is not considered as a separate expense. It is simply used as a permitted writing-down allowance to reduce the annual amount of tax in accordance with the rules applying in the country of earning. The tax payable is deducted in accordance with Eq. (9-2) in the year in which it is paid, which may differ from the year in which the corresponding income was earned.

A (DCFRR) of, say, 15 percent implies that 15 percent per year will be earned on the investment, in addition to which the project generates sufficient money to repay the original investment plus any interest payable on borrowed capital plus all taxes and expenses.

It is not normally possible to make a comprehensive assessment of profitability with a single number. The shape of the cumulative-cash-flow and cumulative-discounted-cash-flow curves both before and after the breakeven point is an important factor.

D. H. Allen [*Chem. Eng.*, **74**, 75–78 (July 3, 1967)] accounted for the shape of the cumulative-undiscounted-cash-flow curve up to the

TABLE 9-4   Annual Cash Flows and Discounted Cash Flows for a Project

Year	A_{CF}, $	$\sum A_{CF}$, $	Discounted at 10%			Discounted at 20%			Discounted at 25%		
			f_d	A_{DCF}, $	$\sum A_{DCF}$, $	f_d	A_{DCF}, $	$\sum A_{DCF}$, $	f_d	A_{DCF}, $	$\sum A_{DCF}$, $
0	−10,000	−10,000	1.00000	−10,000	−10,000	1.00000	−10,000	−10,000	1.00000	−10,000	−10,000
1	−30,000	−40,000	0.90909	−27,273	−37,273	0.83333	−25,000	−35,000	0.80000	−24,000	−34,000
2	−60,000	−100,000	0.82645	−49,587	−86,860	0.69444	−41,666	−76,666	0.64000	−38,400	−72,400
3	−750,000	−850,000	0.75131	−563,483	−650,343	0.57870	−434,025	−510,691	0.51200	−384,000	−456,400
4	−150,000	−1,000,000	0.68301	−102,452	−752,795	0.48225	−72,338	−583,029	0.40960	−61,440	−517,840
5	+200,000	−800,000	0.62092	+124,184	−628,611	0.40188	+80,376	−502,653	0.32768	+65,536	−452,304
6	+300,000	−500,000	0.56447	+169,341	−459,270	0.33490	+100,470	−402,183	0.26214	+78,642	−373,662
7	+400,000	−100,000	0.51316	+205,264	−254,006	0.27908	+111,632	−290,551	0.20972	+83,888	−289,774
8	+400,000	+300,000	0.46651	+186,604	−67,402	0.23257	+93,028	−197,523	0.16777	+67,108	−222,666
9	+360,000	+660,000	0.42410	+152,676	+85,274	0.19381	+69,772	−127,751	0.13422	+48,319	−174,347
10	+320,000	+980,000	0.38554	+123,373	+208,647	0.16151	+51,683	−76,068	0.10737	+34,358	−139,989
11	+280,000	+1,260,000	0.35049	+98,137	+306,784	0.13459	+37,685	−38,383	0.08590	+24,052	−115,937
12	+240,000	+1,500,000	0.31863	+76,471	+383,255	0.11216	+26,918	−11,465	0.06872	+16,493	−99,444
13	+240,000	+1,740,000	0.28966	+69,518	+452,773	0.09346	+22,430	+10,965	0.05498	+13,195	−86,249
14	+400,000	+2,140,000	0.26333	+105,332	+558,105	0.07789	+31,156	+42,121	0.04398	+17,592	−68,657

NOTE: A_{CF} is net annual cash flow, A_{DCF} is net annual discounted cash flow, f_d is discount factor at stated interest, $\sum A_{CF}$ is cumulative cash flow, and $\sum A_{DCF}$ is cumulative discounted cash flow.

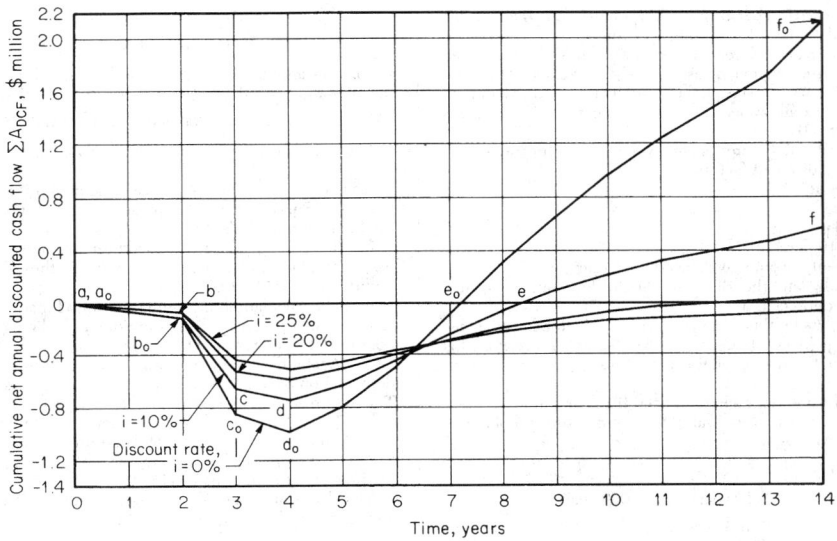

FIG. 9-10   Effect of discount rate on cumulative cash flows.

breakeven point e_0 in Fig. 9-10 by using a parameter known as the equivalent maximum investment period (EMIP), which is defined as

$$(\text{EMIP}) = \frac{\text{area } (a_0 \text{ to } e_0)}{(\sum A_{CF})_{\max}} \quad \text{for} \quad A_{CF} \leq 0 \quad (9\text{-}55)$$

where the area $(a_0 \text{ to } e_0)$ refers to the area below the horizontal baseline $(\sum A_{CF} = 0)$ on the cumulative-cash-flow curve in Fig. 9-10. The sum $(\sum A_{CF})_{\max}$ is the maximum cumulative expenditure on the project, which is given by point d_0 in Fig. 9-10. (EMIP) is a time in years. It is the equivalent period during which the total project debt would be outstanding if it were all incurred at one instant and all repaid at one instant. Clearly, the shorter the (EMIP), the more attractive the project.

Allen accounted for the shape of the cumulative-cash-flow curve

beyond the breakeven point by using a parameter known as the interest-recovery period (IRP). This is the time period (illustrated in Fig. 9-11) that makes the area $(e_0 \text{ to } f_0)$ above the horizontal baseline equal to the area $(a_0 \text{ to } e_0)$ below the horizontal baseline on the cumulative-cash-flow curve.

C. G. Sinclair [*Chem. Process. Eng.*, **47**, 147 (1966)] has considered similar parameters to the (EMIP) and (IRP) based on a cumulative-discounted-cash-flow curve.

Consideration of the cash-flow stages in Fig. 9-10 shows the factors that can affect the (EMIP) and (IRP). If the required capital investment is increased, it is necessary to increase the rate of income after startup for the (EMIP) to remain the same. In order to have the (EMIP) small, it is necessary to keep the research and development, design, and construction stages short.

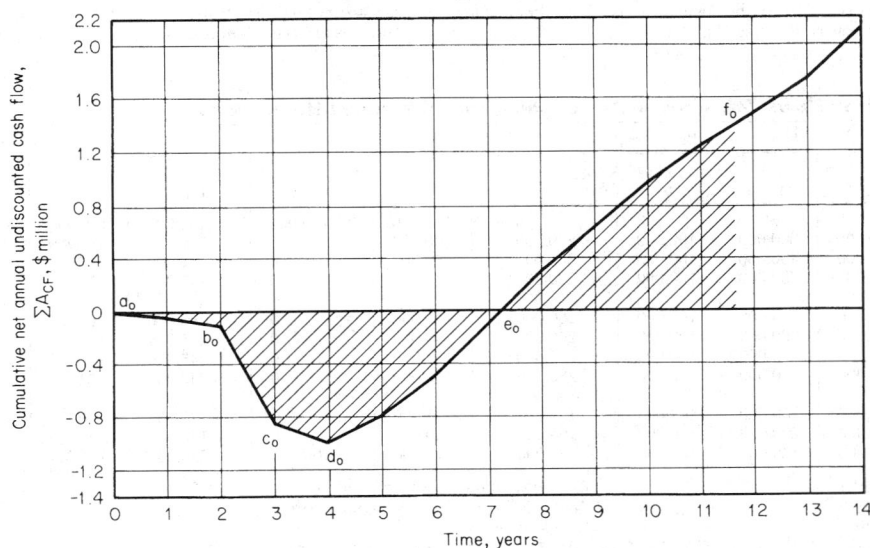

FIG. 9-11   Cumulative cash flow against time, showing interest recovery period.

Example 2: Net Present Value for Different Depreciation Methods The following data describe a project. Revenue from annual sales and the total annual expense over a 10-year period are given in the first three columns of Table 9-5. The fixed-capital investment C_{FC} is $1,000,000. Plant items have a zero salvage value. Working capital C_{WC} is $90,000, and cost of land C_L is $10,000. There are no tax allowances other than depreciation; i.e., A_A is zero. The fractional tax rate t is 0.50.

We shall calculate for these data the net present value (NPV) for the following depreciation methods and discount factors:

a. Straight-line, 10 percent
b. Straight-line, 20 percent
c. Double-declining, 10 percent
d. Sum-of-years-digits, 10 percent
e. Straight-line, 10 percent; income tax delayed for 1 year

In addition, we shall calculate the discounted-cash-flow rate of return (DCFRR) with straight-line depreciation.

a. We begin the calculations for this example by finding the total capital cost C_{TC} for the project from Eq. (9-14). Here, C_{TC} = $1,100,000. In Year 0, this amount is the same as the net annual capital expenditure A_{TC} and is listed in Table 9-5.

The annual rate of straight-line depreciation of the fixed-capital investment C_{FC}, from $1,000,000 at startup to a salvage value S, of zero at the end of a productive life s of 10 years, is given by

$$A_D = (C_{FC} - S)/s$$

$$A_D = (\$1,000,000 - \$0)/10 \text{ years} = \$100,000/\text{year}$$

The annual cash income A_{CI} for Year 1, when A_s = $400,000 per year and A_{TE} = $100,000 per year, is, from Eq. (9-1), $300,000 per year. Values for subsequent years are calculated in the same way and listed in Table 9-4.

Annual amount of tax A_{IT} for Year 1, when A_{CI} = $300,000 per year, A_D = $100,000 per year, A_A = $0 per year, and t = 0.5, is found from Eq. (9-3) to be

$$A_{IT} = [(\$300,000 - \$100,000 - \$0)/\text{year}](0.5)$$

$$= \$100,000/\text{year}$$

Values for subsequent years are calculated in the same way and listed in Table 9-4.

Net annual cash flow (after tax) A_{CF} for Year 0, when A_{CI} = $0 per year, A_{IT} = $0 per year, and A_{TC} = $1,100,000 per year, is found from Eq. (9-4) to be

$$A_{CF} = \$0/\text{year} - \$1,100,000/\text{year} = -\$1,100,000/\text{year}$$

Net annual cash flow (after tax) A_{CF} for Year 1, when A_{CI} = $300,000 per year, A_{IT} = $100,000 per year, and A_{TC} = $0 per year, is found from Eqs. (9-2) and (9-4) to be

$$A_{CF} = \$200,000/\text{year} - \$0/\text{year} = \$200,000/\text{year}$$

Values for the years up to and including Year 9 are calculated in the same way and listed in Table 9-5.

At the end of Year 10, the working capital (C_{WC} = $90,000) and the cost of land (C_L = $10,000) are recovered, so that the annual expenditure of capital A_{TC} in Year 10 is -$100,000 per year. Hence, the net annual cash flow (after tax) for Year 10 must reflect this recovery. By using Eq. (9-4),

$$A_{CF} = \$110,000/\text{year} - (-\$100,000/\text{year})$$

$$= \$210,000/\text{year}$$

The net annual discounted cash flow A_{DCF} for Year 1, when A_{CF} = $200,000 per year and f_d = 0.90909 (for i = 10 percent), is found from Eq. (9-52) to be

$$A_{DCF} = (\$200,000/\text{year})(0.90909) = \$181,820/\text{year}$$

Values for subsequent years are calculated in the same way and listed in Table 9-5.

The net present value (NPV) is found by summing the values of A_{DCF} for each year, as in Eq. (9-53). The net present value is found to be $276,210, as given by the final entry in Table 9-5.

b. The same procedure is used for i = 20 percent. The discount factors to be used in a table similar to Table 9-5 must be those for 20 percent. The (NPV) is found to be -$151,020.

c. The calculations are similar to those for subexample *a* except that depreciation is computed by using the double-declining method of Eq. (9-19). The net present value is found to be $288,530.

d. Again, the calculations are similar to those for subexample *a* except that depreciation is computed by using the sum-of-years-digits method of Eq. (9-20). The net present value is found to be $316,610.

e. The calculations follow the same procedure as for subexample *a*, but the annual amount of tax A_{IT} is calculated for a particular year and then deducted from the annual cash income A_{CI} for the following year. The net present value for Year 11 is found to be $341,980.

The discounted-cash-flow rate of return (DCFRR) can readily be obtained approximately by interpolation of the (NPV) for i = 10 percent and i = 20 percent:

$$(\text{DCFRR}) = 0.100 + [(\$276,210)(0.20 - 0.10)]/[\$276,210 - (-\$151,020)]$$

$$(\text{DCFRR}) = 0.164, \text{ or } 16.4 \text{ percent}$$

The calculation of (DCFRR) usually requires a trial-and-error solution of Eq. (9-57), but rapidly convergent methods are available [N. H. Wild, *Chem. Eng.*, **83**, 153–154 (Apr. 12, 1976)]. For simplicity linear interpolation is often used.

A comparison of the (NPV) values for a 10 percent discount factor shows clearly that double-declining depreciation is more advantageous than straight-line depreciation and that sum-of-years-digits depreciation is more advantageous than the double-declining method. However, a significant advantage is obtained by delaying the payment of tax for 1 year even with straight-line depreciation.

This example is a simplified one. The cost of the working capital is assumed to be paid for in Year 0 and returned in Year 10. In practice, working capital increases with the production rate. Thus there may be an annual expenditure on working capital in a number of years subsequent to Year 0. Except in loss-making years, this is usually treated as an expense of the process. In loss-making years the cash injection for working capital is included in the A_{TC} for that year.

Analysis of Techniques Both the (NPV) and the (DCFRR) methods are based on discounted cash flows and in that sense are vari-

TABLE 9-5 Annual Cash Flows, Straight-Line Depreciation, and 10 Percent Discount Factor

Year					Before tax							After tax	
	A_S, $	A_{TE}, $	A_{CI}, $	$A_D + A_A$, $	$A_{CI} - A_D - A_A$, $	A_{IT}, $	A_{TC}, $	A_{CF}, $	f_d	A_{DCF}, $	(NPV), $
0	0	0	0	0	0	0	+1,100,000	−1,100,000	1.0000	−1,100,000	−1,100,000
1	400,000	100,000	300,000	100,000	200,000	100,000	0	200,000	0.90909	181,820	−918,180
2	500,000	100,000	400,000	100,000	300,000	150,000	0	250,000	0.82645	206,610	−711,570
3	500,000	110,000	390,000	100,000	290,000	145,000	0	245,000	0.75131	184,070	−527,500
4	500,000	120,000	380,000	100,000	280,000	140,000	0	240,000	0.68301	163,920	−363,580
5	520,000	130,000	390,000	100,000	290,000	145,000	0	245,000	0.62092	152,120	−211,460
6	520,000	130,000	390,000	100,000	290,000	145,000	0	245,000	0.56447	138,300	−73,160
7	520,000	140,000	380,000	100,000	280,000	140,000	0	240,000	0.51316	123,160	+50,000
8	390,000	140,000	250,000	100,000	150,000	75,000	0	175,000	0.46651	81,640	+131,640
9	350,000	150,000	200,000	100,000	100,000	50,000	0	150,000	0.42410	63,610	+195,250
10	280,000	160,000	120,000	100,000	20,000	10,000	−100,000	210,000	0.38554	80,960	+276,210

A_S = revenue from annual sales.
A_{TE} = total annual expense.
A_{CI} = annual cash income.
$A_D + A_A$ = annual depreciation and other tax allowances.
$A_{CI} - A_D - A_A$ = taxable income.
$A_{IT} = (A_{CI} - A_D - A_A)t$ = amount of tax at t = 0.5.

A_{TC} = total annual capital expenditure.
$A_{CF} = A_{CI} - A_{IT} - A_{TC}$ = net annual cash flow.
f_d = discount factor at 10%.
A_{DCF} = net annual discounted cash flow.
(NPV) = $\sum A_{DCF}$ = net present value.

ations of the same basic method. However, when ranking different projects on the basis of profitability, they can produce different results.

Discounted-cash-flow rate of return (DCFRR) has the advantage of being unique and readily understood. However, when used alone, it gives no indication of the scale of the operation. The (NPV) indicates the monetary return, but unlike that of the (DCFRR) its value depends on the base year chosen for the calculation. Additional information is needed before its significance can be appreciated. However, when a company is considering investment in a portfolio of projects, individual (NPV)s have the advantage of being additive. This is not true of (DCFRR)s.

Increasing use is being made of the capital-rate-of-return ratio (CRR), which is the net present value (NPV) divided by the maximum cumulative expenditure or maximum net outlay, $-(\sum A_{CF})_{max}$.

$$(CRR) = (NPV)/(\sum A_{CF})_{max} \quad \text{for} \quad A_{CF} \leq 0 \quad (9\text{-}56)$$

The maximum net outlay is very important, since no matter how profitable a project is, the matter is academic if the company is unable to raise the money to undertake the project.

An (NPV) or (DCFRR) estimation will be no better than the accuracy of the projected cash flows over the life of the project. Clearly, one is likely to predict cash flows more accurately for 2 or 3 years ahead than, say, for 9 or 10 years ahead. However, since the cash flows for the later years are discounted to a greater extent than the cash flows for the earlier years, the latter have less effect on the overall estimation. Nevertheless, the difficulty of predicting cash flows in later years and the inherent lack of confidence in these predictions are serious disadvantages of the (DCFRR) method. In this respect (NPV)s are more useful since they are calculated for each year of a project. Thus, a project with a favorable (NPV) in the early years is a promising one.

One way of overcoming these disadvantages of the (DCFRR) method is to make estimates of the times required to reach certain values of (DCFRR). For example, how many years will it take to reach (DCFRR)s of 10 percent, 15 percent, 20 percent per year, etc.? Although (DCFRR) trial-and-error calculations and (NPV) calculations are tedious if done manually, computer programs which are suitable for programmable pocket calculators can readily be written to make calculations easier.

It is possible for some projects to reach a stage at which repairs, replacements, etc., can exceed net earnings in a particular year. In this case the cumulative-discounted-cash-flow or net-present-value curve plotted against time has a genuine maximum.

It is important when appraising by (NPV) and (DCFRR) not to consider the past in profitability estimations. Good money should never follow bad. It is unwise to continue to put money into a project if a more profitable project exists, even though this course may involve scrapping an expensive plant. Other considerations may, however, outweigh purely financial criteria in a particular case.

No single value for a profitability estimate should be accepted without further consideration. An intelligent consideration of the cumulative-cash-flow and cumulative-discounted-cash-flow curves such as those shown in Fig. 9-10, together with experience and good judgment, is the best way of assessing the financial merit of a project.

When considering future projects, top management will most likely require the discounted-cash-flow rate of return and the payback period. However, the estimators should also supply management with the following:

Cumulative discounted-cash-flow or (NPV) curve for a discount rate of 10 percent per year or other agreed aftertax cost of capital

Maximum net outlay, $(\sum A_{CF})_{max}$, for $A_{CF} \leq 0$

Discounted breakeven point (DBEP)

Plot of capital-return ratio (CRR) against time over the life of the project for a discount rate at the cost of capital

Number of years to reach discounted-cash-flow rates of return of, say, 15 and 25 percent per year respectively

Comparisons on the basis of time can be summarized by the following:

Duration of the project

Breakeven point (BEP)

Discounted breakeven point (DBEP)

Equivalent maximum investment period (EMIP)

Interest-recovery period (IRP)

Payback period (PBP)

Comparisons on the basis of cash can be summarized by the following:

Maximum cumulative expenditure on the project, $(\sum A_{CF})_{max}$, for $A_{CF} \leq 0$

Maximum discounted cumulative expenditure on the project

Cumulative net annual cash flow $\sum A_{CF}$

Cumulative net annual discounted cash flow $\sum A_{DCF}$ or net present value (NPV)

Capitalized cost C_K

Comparisons on the basis of interest can be summarized as (1) the net present value (NPV) and (2) the discounted-cash-flow rate of return (DCFRR), which from Eqs. (9-53) and (9-54) is given formally as the fractional interest rate i which satisfies the relationship

$$(NPV) = \sum_{0}^{n} (A_{DCF})_n = 0 \quad (9\text{-}57)$$

When comparing project profitability, the ranking on the basis of net present value (NPV) may differ from that on the basis of discounted-cash-flow rate of return (DCFRR). Let us consider the data for two projects:

Cost of capital	Project C	Project D
i, %	(NPV), $	(NPV), $
4	+100,000	+62,000
8	+41,000	+28,000
12	−2,000	+10,000
16	−32,000	−4,000

These (NPV) data are plotted against the cost of capital, as shown in Fig. 9-12. The discounted-cash-flow rate of return is the value of i that satisfies Eq. (9-5). From Fig. 9-12, (NPV) = 0 at a (DCFRR) of 11.8 percent for project C and 14.7 percent for project D. Thus, on the basis of (DCFRR), project D is more profitable than project C.

The (NPV) of project C is equal to that of project D at a cost of capital $i = 9.8$ percent. If the cost of capital is greater than 9.8 percent, project D has the higher (NPV) and is, therefore, the more profitable. If the cost of capital is less than 9.8 percent, project C has the higher (NPV) and is the more profitable.

Benefit of Early Cash Flows It pays to receive cash inflows as early as possible and to delay cash outflows as long as possible.

Let us consider the net annual cash flows (after tax) A_{CF} for projects E, F, and G, listed in Table 9-6. The cumulative annual cash flows $\sum A_{CF}$ and cumulative discounted annual cash flows $\sum A_{DCF}$, using a discount of 10 percent for these projects, are also listed in Table 9-6. We notice that the cumulative annual cash flow for each project is +$1000.

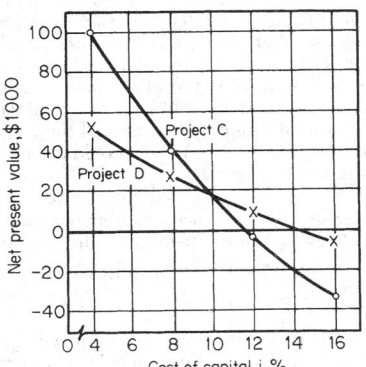

FIG. 9-12 Effect of cost of capital on net present value.

TABLE 9-6 Cash-Flow Data for Projects E, F, and G

Year	A_{CF}, $	$\sum A_{CF}$, $	f_d	A_{DCF}, $	$\sum A_{DCF} =$ (NPV), $
			Discounted at 10%		
			Project E		
0	−5000	−5000	1.0000	−5000	−5000
1	+3000	−2000	0.90909	+2727	−2273
2	+2000	0	0.82645	+1653	−620
3	+1000	+1000	0.75131	+751	+131
			Project F		
0	−5000	−5000	1.0000	−5000	−5000
1	+1000	−4000	0.90909	+909	−4091
2	+2000	−2000	0.82645	+1653	−2438
3	+3000	+1000	0.75131	+2254	−184
			Project G		
0	−5000	−5000	1.0000	−5000	−5000
1	+2000	−3000	0.90909	+1818	−3182
2	+2000	−1000	0.82645	+1653	−1529
3	+2000	+1000	0.75131	+1503	−26

The (DCFRR) is the discount rate that satisfies Eq. (9-57) in the final year of the project. We can approximate the (DCFRR) for each project as follows:

For project E,

$$\sum A_{CF} = +\$1000 \text{ in Year 3 for } i = 0 \text{ percent}$$

$$\sum A_{DCF} = +\$131 \text{ in Year 3 for } i = 10 \text{ percent}$$

$$\sum A_{DCF} = \$0 \text{ in Year 3 for } i = \text{(DCFRR)}$$

Therefore,

$$1000/(1000 − 131) \cong \text{(DCFRR)}/10$$

$$\text{(DCFRR)} \cong 11.5 \text{ percent}$$

Similarly for project F,

$$1000/(1000 + 184) \cong \text{(DCFRR)}/10$$

$$\text{(DCFRR)} \cong 8.4 \text{ percent}$$

Similarly for project G,

$$1000/(1000 + 26) \cong \text{(DCFRR)}/10$$

$$\text{(DCFRR)} \cong 9.7 \text{ percent}$$

In terms of net present value (NPV), the projects in order of merit are E, G, and F, with (NPV)s of +$131, −$26, and −$184 respectively. In terms of (DCFRR), the projects in order of merit are also E, G, and F, with (DCFRR) values of 11.5 percent, 9.7 percent, and 8.4 percent respectively.

When to Scrap an Existing Process Let us suppose that a company invests $50,000 in a manufacturing process that has positive net annual flows (after tax) A_{CF} of $10,000 in each year. During the third year of operation, an alternative process becomes available. The new process would require an investment of $40,000 but would have positive net annual cash flows (after tax) of $20,000 in each year. The cost of capital is 10 percent, and it is estimated that a market will exist for the product for at least 6 more years. Should the company continue with the existing process (project H), or should it scrap project H and adopt the new process (project I)?

The net annual cash flows A_{CF} and cumulative discounted annual cash flow $\sum A_{DCF}$ for a discount factor of 10 percent are listed in Table 9-7 for the two projects. At the end of Year 9, the net present values are

$$\text{(NPV)} = +\$35,390 \text{ for project I}$$

$$\text{(NPV)} = +\$7591 \text{ for project H}$$

The difference is +$27,779, which is numerically greater than the money lost by the end of Year 3 for project H. Thus project H should be scrapped, and the new project I adopted if only economic reasons need to be considered. Recovery of working capital and the cost of

TABLE 9-7 Cash-Flow Data for Projects H and I

Year	A_{CF}, $	f_d	A_{DCF}, $	$\sum A_{DCF} =$ (NPV), $
		Discounted at 10%		
		Project H		
0	−50,000	1.0000	−50,000	−50,000
1	+10,000	0.90909	9,091	−40,909
2	+10,000	0.82645	8,265	−32,644
3	+10,000	0.75131	7,513	−25,131
4	+10,000	0.68301	6,830	−18,301
5	+10,000	0.62092	6,209	−12,092
6	+10,000	0.56447	5,645	−6,447
7	+10,000	0.51316	5,132	−1,315
8	+10,000	0.46651	4,665	+3,350
9	+10,000	0.42410	4,241	+7,591
		Project I		
3	−40,000	0.75131	−30,052	−30,052
4	+20,000	0.68301	+13,660	−16,392
5	+20,000	0.62092	+12,418	−3,974
6	+20,000	0.56447	+11,289	+7,315
7	+20,000	0.51316	+10,263	+17,578
8	+20,000	0.46651	+9,330	+26,908
9	+20,000	0.42410	+8,482	+35,390

land have been neglected since the latter is the same for each project and the former would also favor project I.

Incremental Comparisons A company may have the choice of, say, investing $10,000 in project J, which will give a (DCFRR) of 16 percent, or $7000 in project K, which will give a (DCFRR) of 18 percent. Should it spend $10,000 on project J or spend only $7000 on project K and invest the difference of $3000 elsewhere?

Both projects have lives of 10 years and constant positive net annual cash flows A_{CF} of $2069 and $1558 for projects J and K respectively. The corresponding (NPV)s at a discount factor of 10 percent are +$2710 and +$2560 respectively. These data are summarized as follows:

	Project J	Project K	Project (J − K)
A_{CF}, $, in Year 0	−10,000	−7,000	−3,000
A_{CF}, $, in each of Years 1–10	+2,069	+1,558	+511
(NPV), i = 10 percent, $	+2,710	+2,560	+150
(DCFRR), percent	16	18	12.4

From the difference in cash flows between the projects, the discounted-cash-flow rate of return (DCFRR) for project (J-K) can be shown as 12.4 percent. This is significantly lower than for either project J or project K. Thus, if the $3000 can be invested to give a return greater than 12.4 percent, project K should be chosen in preference to project J.

Comparisons on the Basis of Capitalized Cost A machine in a process generates a positive net cash flow of $1000. Two alternatives are available: machine L, costing $2000, requires replacement every 4 years, and machine M, costing $3000, requires replacement every 6 years. Neither machine has any scrap value. The cost of capital is 10 percent. Which machine is the more profitable to operate?

In this case, the lives of the machines are unequal, and the comparison is conveniently made on the basis of capitalized cost. This puts lives on the same basis, which is an infinite number of years. The net annual cash flows generated by each machine are equal.

The capitalized cost C_K of a piece of fixed-capital cost C_{FC} is the amount of capital required to ensure that the equipment may be renewed in perpetuity. For a piece of equipment with no scrap value, C_K is given by

$$C_K = C_{FC} \left[\frac{(1 + i)^n}{(1 + i)^n - 1} \right] \tag{9-58}$$

For machine L,

$$C_K = (\$2000)(3.15471) = \$6309.42$$

For machine M,

$$C_K = (\$3000)(2.29607) = \$6888.21$$

Thus, machine L with the lower capitalized cost is the more profitable to operate.

Relationship between (PBP) and (DCFRR)   For the case of a single lump-sum capital expenditure C_{FC} which generates a constant annual cash flow A_{CF} in each subsequent year, the payback period is given by the equation

$$(PBP) = C_{FC}/A_{CF} \qquad (9\text{-}59)$$

if the scrap value of the capital outlay may be taken as zero.

For this simplified case the net present value (NPV) after n years with money invested at a required aftertax compound annual fractional interest rate i is given by the equation

$$(NPV) = C_{FC} - A_{CF}F_n \qquad (9\text{-}60)$$

where

$$F_n = \sum_1^n \frac{1}{(1+i)^n}$$

When (NPV) = 0, the value of i given by Eq. (9-60) is the discounted-cash-flow rate of return (DCFRR), and in this case Eqs. (9-59) and (9-60) can be combined to give:

$$(PBP) = F_n \qquad (9\text{-}61)$$

Figure 9-13 is a plot of Eq. (9-61) in the form of the number of years n required to reach a certain discounted-cash-flow rate of return (DCFRR) for a given payback period (PBP). The figure is a modification of plots previously published by A. G. Bates [*Hydrocarbon Process.*, **45**, 181–186 (March 1966)], C. Estrup [*Br. Chem. Eng.*, **16**, 171 (February–March 1971)], and F. A. Holland and F. A. Watson [*Process Eng. Econ.*, **1**, 293–299 (December 1976)].

In the limiting case when n approaches infinity, Eq. (9-61) can be written as

$$(DCFRR)_{max} = 1/(PBP) \qquad (9\text{-}62)$$

which means, for example, that if the payback period is 4 years, the maximum possible discounted-cash-flow rate of return which can be reached is 25 percent. The corresponding (DCFRR) for (PBP) = 10 years is 10 percent.

Equations (9-59), (9-60), (9-61), and (9-62) may be used as they stand to assess expenditure on energy-conservation measures since a constant amount of energy is saved in each year subsequent to the capital outlay. However, the annual cash flows A_{CF} corresponding to the energy savings remain constant only if there is no inflation or if the money values are corrected to their purchasing power at the time of the capital expenditure.

Sensitivity Analysis   An economic study should pinpoint the areas most susceptible to change. It is easier to predict expenses than either sales or profits. Fairly accurate estimates of capital costs and processing costs can be made. However, for the most part, errors in these estimates have a correspondingly smaller effect than changes in sales price, sales volume, and the costs of raw materials and distribution.

Sales and raw-materials prices may be affected by any of the following: discounts and allowances, availability of substitutes, contract pricing, government regulations, quality and form of the materials, and competition. Sales volume may be affected by any of the following: new uses for the product, new markets, advertising, quality, overcapacity, replacement by another product, competition, and timing of entry into the market.

Distribution costs depend on plant location, physical state of the material (whether liquid, gas, or solid), nature of the material (whether corrosive, explosive, flammable, perishable, or toxic), freight rates, and labor costs. Distribution costs may be affected by any of the following: new methods of materials handling, safety regulations, productivity agreements, wage rates, transportation systems, storage systems, quality, losses, and seasonal effects.

It is worthwhile to make tables or plot curves that show the effect of variations in costs and prices on profitability. This procedure is called **sensitivity analysis.** Its purpose is to determine to which factors the profitability of a project is most sensitive. Sensitivity analysis should always be carried out to observe the effect of departures from expected values.

For many years, companies and countries have lived with the problem of inflation, or the falling value of money. Costs—in particular, labor costs—tend to rise each year. Failure to account for this trend in predicting future cash flows can lead to serious errors and misleading profitability estimates.

Another important factor is the tendency of product prices to fall as the total national or international volume of production increases. Sales prices may fall by 20 percent for a doubling in volume or production.

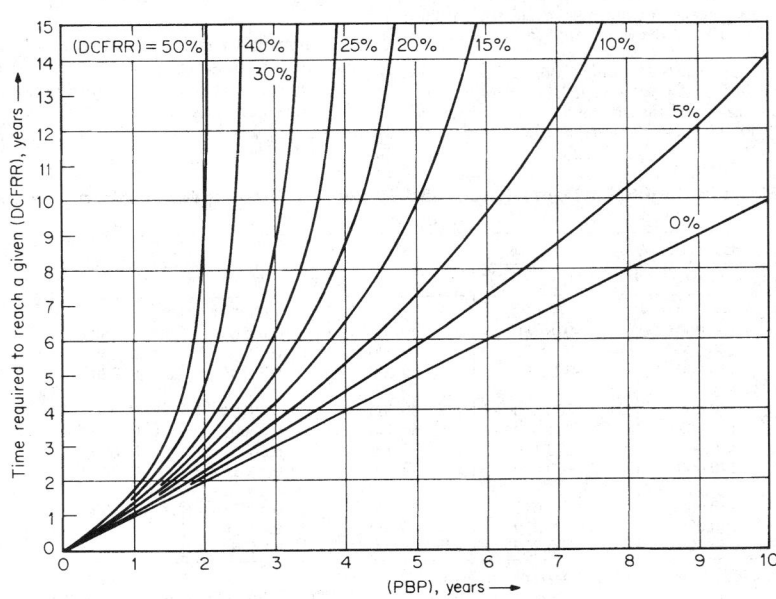

FIG. 9-13   Relationship between payback period and discounted-cash-flow rate of return.

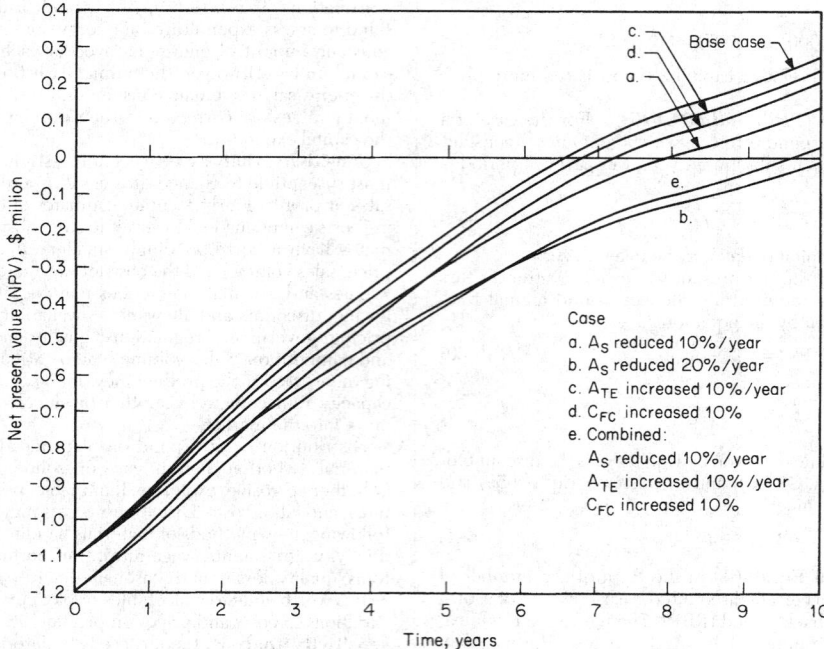

FIG. 9-14   Net present value against time, showing effect of adverse changes in cash flows.

No profitability estimate is better than the inherent accuracy of the data.

Example 3: Sensitivity Analysis   The following data describe a project. Revenue from annual sales and total annual expense over a 10-year period are given in the first three columns of Table 9-5. The fixed-capital investment C_{FC} is $1 million. Plant items have a zero salvage value. Working capital C_{WC} is $90,000, and the cost of land C_L is $10,000. There are no tax allowances other than depreciation; i.e., A_A is zero. The fractional tax rate t is 0.50. For this project, the net present value for a 10 percent discount factor and straight-line depreciation was shown to be $276,210 and the discounted-cash-flow rate of return to be 16.4 percent per year.

We shall use these data and the accompanying information of Table 9-5 as the base case and calculate for straight-line depreciation the net present value (NPV) with a 10 percent discount factor and the discounted-cash-flow rate of return (DCFRR) for the project with the following situations.

Case	Modification
a	Revenue A_S reduced by 10 percent per year
b	Revenue A_S reduced by 20 percent per year
c	Total expense A_{TE} increased by 10 percent per year
d	Fixed-capital investment increased by 10 percent
e	A_S reduced by 10 percent per year, A_{TE} increased by 10 percent per year, and C_{FC} increased by 10 percent

The results are shown in Figs. 9-14 and 9-15 and Tables 9-8 and 9-9.

Learning Curves   It is usual to learn from experience. Consequently, the time taken to produce an article, the number of spoiled batches, the cost per unit of production, etc., tend to decrease with the number of units produced. The relationships are expressed for the ideal case by

TABLE 9-8   Annual Cash Flows, Straight-Line Depreciation, and 10 Percent Discount Factor When Revenue Is Reduced by 10 Percent per Year

Year	Base case A_S, $	Base case A_{TE}, $	(NPV), $	ΔA_S, $	f_d	ΔA_{DCF}, $	Δ(NPV), $	Reduced (NPV), $
0	0	0	−1,100,000	0	1.0000	0	0	−1,100,000
1	400,000	100,000	−918,180	40,000	0.90909	18,180	18,180	−936,360
2	500,000	100,000	−711,570	50,000	0.82645	20,660	38,840	−750,410
3	500,000	110,000	−527,500	50,000	0.75131	18,780	57,620	−585,120
4	500,000	120,000	−363,580	50,000	0.68301	17,070	74,690	−438,270
5	520,000	130,000	−211,460	52,000	0.62092	16,140	90,830	−302,290
6	520,000	130,000	−73,160	52,000	0.56447	14,680	105,510	−178,670
7	520,000	140,000	+50,000	52,000	0.51316	13,340	118,850	−68,850
8	390,000	140,000	+131,640	39,000	0.46651	9,100	127,950	+3,690
9	350,000	150,000	+195,250	35,000	0.42410	7,420	135,370	+59,880
10	280,000	160,000	+276,210	28,000	0.38554	5,390	140,760	+135,450

A_S = base revenue from annual sales before tax.
A_{TE} = base total annual expense before tax.
(NPV) = base net present value after tax.
ΔA_S = decrease in annual revenue.

f_d = discount factor at 10%.
ΔA_{DCF} = decrease in net discounted cash flow at income tax rate = 0.5.
Δ(NPV) = $\Sigma \Delta A_{DCF}$ = decrease in net present value.
Reduced (NPV) = ΣA_{DCF} = reduced net present value after tax.

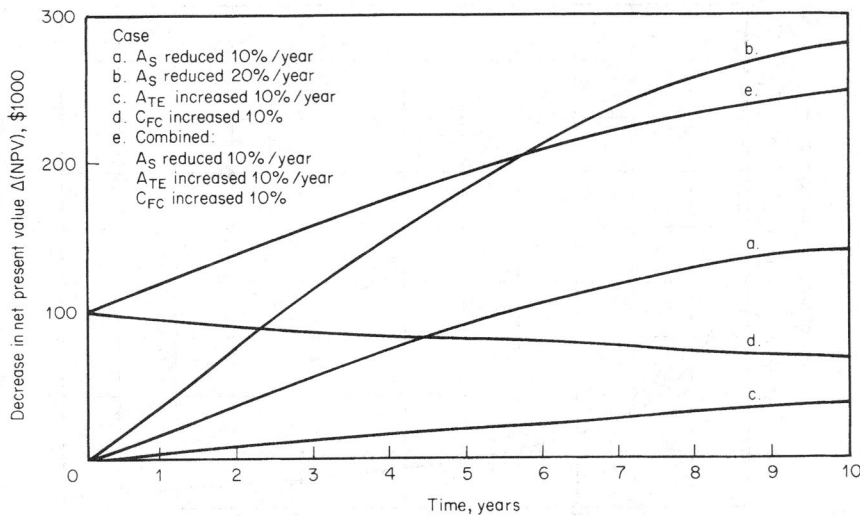

FIG. 9-15 Decrease in net present value against time resulting from adverse changes in cash flows.

$$Y = KX^N \qquad (9\text{-}63)$$

where Y = cumulative-average cost, production time, etc., per unit
X = cumulative production, units
K = effective value of first unit produced
N = slope of straight-line plot of Y versus X on log-log paper

The particular learning curve is usually characterized by the percentage reduction in the cumulative average value Y when the number of units X is doubled. From this definition it follows that

$$N = \log\,(\text{characteristic}/100)/\log 2 \qquad (9\text{-}64)$$

The cost c_{ME} of the last unit of a block bringing the cumulative production to X units is, from Eq. (9-63),

$$c_{ME} = K[X^{N+1} - (X-1)^{N+1}] \qquad (9\text{-}65)$$

These unit costs, or the time taken to produce the last unit, etc., may be plotted on cartesian coordinates against the number of units produced to provide a standard against which the performance of a new employee, a new machine, etc., can be judged. Figure 9-16 shows such a plot for the subsequent example.

In general, cost data will be available for multiple units. Typically, the cost of production for 1 week or of a specific order is computed and an average cost per unit obtained. This average value $\overline{Y}$ for the batch should be plotted against the corresponding learning-curve value $\overline{X}$ calculated by Eq. (9-66):

$$\overline{X}^N = (X_2^{N+1} - X_1^{N+1})/(X_2 - X_1) \qquad (9\text{-}66)$$

where X_1 and X_2 are the cumulative production before and after the batch. This form of the equation is useful when only the previous production history of the process is known, from the serial numbers or otherwise.

TABLE 9-9 Summary of Results of Sensitivity Analysis

Case	(NPV), $ $i = 10\%$	(DCFRR), %
Base case	276,210	16.4
A_S reduced 10% per year	135,450	12.9
A_S reduced 20% per year	−5,330	9.8
A_{TE} increased 10% per year	238,430	15.0
C_{FC} increased 10%	206,890	14.0
Combined: A_S reduced 10% per year A_{TE} increased 10% per year C_{FC} increased 10%	28,420	10.6

A straight line may be fitted to the (X,Y) or $(\overline{X},\overline{Y})$ pairs of data when plotted on log-log graph paper from which the slope N and the intercept $\log K$ with $\overline{X} = 1$ may be read. Alternatively, the method of least squares may be used to estimate the values of K and N, giving the best fit to the available data.

It will be noted that a value of $N = 0$, corresponding to a characteristic of 100 percent for the learning curve, implies that the value of Y is independent of X. This would imply that learning by experience was not possible and thus corresponds to an optimally designed process or one for which the costs are determined by external factors. Similarly, a value of $N = -1$, corresponding to the 50 percent learning curve, implies that the cost of production is inversely proportional to the number produced, which is absurd. Projects having characteristics less than 70 percent are impractical. Low characteristics are typical of hasty entry into a market in an attempt to preempt it. Characteristics tend to increase with experience, so that established and mature projects are likely to have characteristics around 95 percent. Characteristics close to 100 percent are unlikely to be achieved because of random factors such as changes in personnel, accidents, supply delays, etc. Figure 9-17 represents a typical practical case, from which it can be seen that the curve has a point of inflexion but eventually settles down to an approximately straight line of lower slope than that of the conventionally defined learning curve. At some point it is useful to change to the equation of this mature project line.

Significant changes in working, such as the introduction of new equipment, the influx of a large number of inexperienced workers, or a temporary reduction in skills after a long shutdown, may produce a sudden increase in all the cumulative-average curves. The simplest way to handle this, when the next accurate costing is available, is to deduct the value of X obtained from the curve from that actually achieved and to use this value as a constant correction to X until the next break in the curve is reached. If the causes of such steps recur, the size of the step can often be related to a particular cause. In such cases the estimated step change can be used for predictions until the next accurately determined values are obtained.

Applications for the learning curve are already extensive, and new uses can often be found. Care is needed in applying the techniques to ensure that it is possible for learning to take place. In projecting prices, etc., unusual items, such as the cost of the special setting up of tools or factory rearrangements, should be excluded from the production costs used to establish the learning curve. In times of inflation, costs should be corrected for the effects of inflation in the manner to be shown subsequently. Production times or spoilage rates are not affected by cost allocations or inflation and may prove to be better

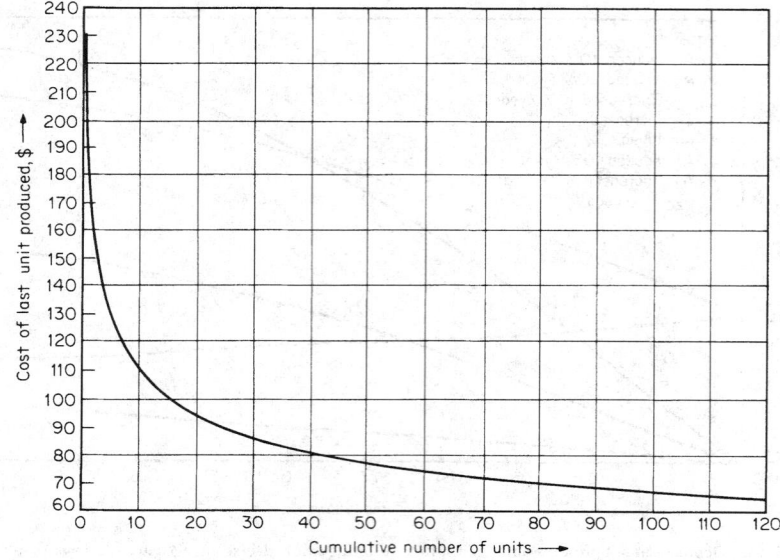

FIG. 9-16 Cartesian plot of learning curve.

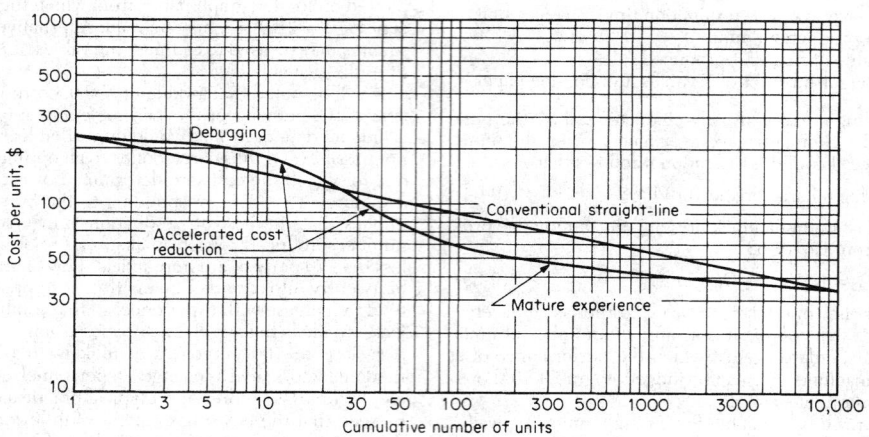

FIG. 9-17 Logarithmic plot of learning curve.

standards of performance where appropriate. However, the learning curve is often required when preparing quotations for batch production runs, particularly when competition is likely to be keen. In such cases the average cost of the production run $\overline{Y}$ between cumulative production totals of X_1 and X_2 may be estimated by Eq. (9-67) when the previous cumulative-average cost Y_1 is known:

$$\overline{Y} = Y_1[(X_2/X_1)^{N+1} - 1]/[(X_2/X_1) - 1] \qquad (9\text{-}67)$$

In process engineering, fractional units can often be produced so that the learning curve can be treated as being continuous. When only discrete numbers of units can be produced, the learning curve is strictly a histogram. In order to allow for this it is sufficient to increase the value of X by half a unit before applying the above equations. The difference is significant at small values of X, such as may be used for the initial estimates of K and N. As the project matures, it is better to use the equations as presented, as the cost of the first unit K is an entirely notional one. Major technological changes should, of course, be treated as the start of a new project.

R. B. Jordan (*How to Use the Learning Curve,* Materials Management Institute, Boston, 1965) discusses the uses of the learning curve

extensively and provides many tables of factors. The uses considered include estimating starting costs, determining labor requirements, establishing factory cost targets, checking employee-training progress, the make-or-buy decision, aid in purchasing negotiations, and aid in establishing a selling price.

Example 4: Estimation of Average Cost of Incremental Units
The cost of an initial batch of 21 units, exclusive of special tools and setting-up costs, averaged $120 per unit. The average cost of the next batch of 80 units was $75.81. Let us establish the learning curve implied by these data and hence estimate the probable average cost of the next 50 units. We shall establish also the unit-cost curve to be used as a control during follow-up orders.

If the batch units are capable of continuous subdivision, we proceed as follows. We substitute the given values of the cumulative-average cost Y and cumulative production X for the first batch into Eq. (9-63) to give, by taking logarithms of each side,

$$\log 120 = \log K + N \log 21$$

The cost of the first batch is $120 \times 21 = \$2520$, and that of the second batch is $75.81 \times 80 = \$6065$. The total cost of the first 101 units is therefore $8585, with a cumulative-average unit cost of $85. We substitute as before to give

$$\log 85 = \log K + N \log 101$$

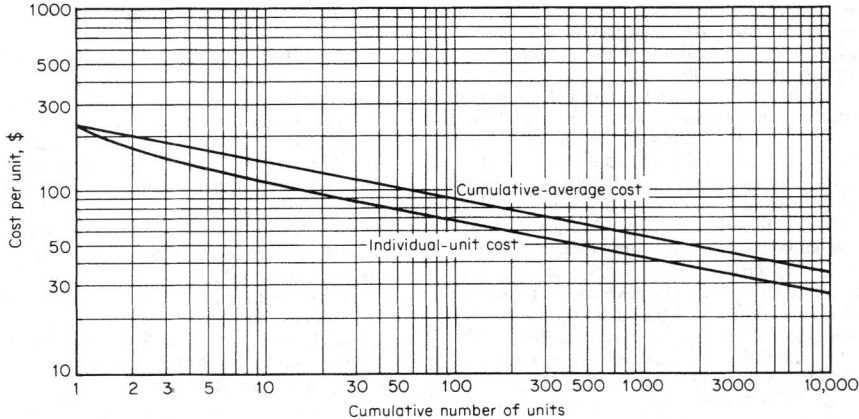

FIG. 9-18 Effect of learning on the average cost of a product.

From these equations it follows that $K = \$234.15$ and $N = -0.2196$. This line is plotted in Fig. 9-18.

From Eq. (9-64) it follows that the value of the characteristic of this learning curve = 100 antilog $(-0.2196 \log 2) = 85.0$ percent. From Eq. (9-65) the production cost of the third unit is

$$c_{ME} = (\$234.15)(3^{0.7804} - 2^{0.7804}) = \$149.70$$

Values calculated in this way are plotted in Fig. 9-16 and also in Fig. 9-18. It will be noted that after about 10 units this latter curve becomes parallel to the cumulative-average-cost curve and that the Y values are $(N + 1)$ times those obtained from the latter curve.

Since the cumulative-average cost Y_2 of the first 101 units was $85, it follows from Eq. (9-67) that the average cost of the third batch of 50 units, bringing the cumulative total to 151, is given by

$$\overline{Y}_3 = (\$85)[(151/101)^{0.7804} - 1]/[(151/101) - 1]$$

$$= \$63.30 \text{ per unit}$$

This may be used as a cost guide when quoting the order.

If the units of production may not be subdivided, the procedure is similar except that all X values are increased by 0.5 unit in establishing the curves. The results are not sufficiently different to be significant for estimation purposes.

To the above costs must be added back any unit costs omitted from those to which learning might bring improvement. These will normally include overheads and specific charges on the project such as the unit cost of special tools, jigs, etc.

Risk and Uncertainty Discounted-cash-flow rates of return (DCFRR) and net present values (NPV) for future projects can never be predicted absolutely because the cash-flow data for such projects are subject to uncertainty. Therefore, when stating predicted values of (DCFRR) and (NPV) for projects, it is also desirable to give a measure of confidence in the predictions.

For example, for a particular project it may be estimated that there is a 90 percent chance of the (DCFRR) being greater than 10 percent, a 50 percent chance of its being greater than 16 percent, and only a 10 percent chance of its being greater than 20 percent. Management retains the power of decision to proceed with the project or not, but the probability data provide desirable information for the decision.

The estimation of probabilities requires the use of statistics. Thus statistical methods play an increasing role in decision making.

Predictions from Limited Data Predictions of future sales price, sales volume, etc., are normally based on a very limited amount of data about past events. Furthermore, it would not be convenient to use the entire population of past events even if it were available. A statistic is a measure, based on limited information from a sample, that allows the corresponding parameter of the population to be estimated.

The mean value $\overline{x}$ of a property x is a statistic based on a sample of n items defined by

$$\overline{x} = (x_1 + x_2 + x_3 + \cdots + x_n)/n \tag{9-68}$$

The mean $\overline{x}$ is the statistic corresponding to the population parameters μ, which is the arithmetic average of all the items in the population. In many cases, not all the x values will be different. In such circumstances, Eq. (9-68) can be written as

$$\overline{x} = [\sum x_i f(x_i)]/[\sum f(x_i)] \tag{9-69}$$

where $f(x_i)$ is the frequency with which a particular value x_i occurs. It is often convenient to divide the frequency of occurrence by the total number of items. In this case, $f(x_i)$ becomes the relative frequency of occurrence of the value x_i, and $\sum f(x_i) = 1$.

The values of x may be either discrete or continuous. The number of sales of, say, automobiles in any one day must be an integer. If a business sells 4 automobiles, this represents all possible values of x in the range of 3.5 to 4.5.

When x represents a continuous variable quantity, it is sometimes convenient to take the total or relative frequency of occurrences within a given range of x values. These frequencies can then be plotted against the midvalues of x to form a histogram. In this case, the ordinate should be the frequency per unit of width x. This makes the area under any bar proportional to the probability that the value of x will lie in the given range. If the relative frequency is plotted as ordinate, the sum of the areas under the bars is unity.

If x is a continuous variable and the interval ranges are made smaller and smaller, a smooth curve will eventually result. The area under such a curve between x_1 and x_2 represents the probability that a randomly selected item will have a value of x lying in the range x_1 to x_2. This is the information that is desired.

Data available from past experience can be used to generate frequency distribution curves. It is essential for a company to have an efficient commercial-intelligence system to assess market conditions.

Accuracy of sales forecasting can also be increased by a careful study of past sales records, price trends, etc. However, the uncertainty of an estimate increases the farther into the future that the estimate is projected.

Estimates of sales income and other types of forecasts are usually based on the opinions of experts. Experts should be able to estimate maximum, minimum, and most likely, or modal, values for a quantity. The modal value is not necessarily midway between the minimum and maximum values, since many distributions are skewed. An expert may be asked to estimate the probability of the occurrence of certain values on each side of the mode. When experts are questioned separately, the procedure is known as the **Delphic method.** Strictly speaking, this method requires that the opinion of each expert be assessed by a coordinator, who then feeds the results back to see if the opinions of one expert are modified by those of others. The process is repeated until agreement is reached. In practice, the procedure is too tedious to be repeated more than once.

It is useful to compare the past predictions of each expert with the results obtained in practice. This information enables the opinions to be weighted by the coordinator. When the experts work in close collaboration, it is not possible to avoid some collusion. In this case, it is better to arrive at a single consensus opinion by a free and open discussion. This is the think-tank method. Its main disadvantage is that rank or aggressiveness might unduly weight one or more opinions.

The opinions of the experts, however obtained, provide a basis for plotting a frequency or probability distribution curve. If the relative frequency is plotted as ordinate, the total area under the curve is unity. The area under the curve between two values of the quantity is the probability that a randomly selected value will fall in the range between the two values of the quantity. These probabilities are mere estimates, and their reliability depends on the skill of the forecasters.

The estimated (DCFRR) and the estimated (NPV) are both functions of the estimated cumulative revenue from annual sales $\sum A_S$, the estimated cumulative total annual cost or expense $\sum A_{TE}$, and the estimated fixed capital cost C_{FC} of the plant. The revenue from annual sales for each year is in turn the product of the sales price and sales volume. Initially it is desirable to select those values from the distribution curves of $\sum A_S$, $\sum A_{TE}$, and C_{FC} which enable the maximum and minimum (DCFRR) and (NPV) to be calculated.

If the maximum values of (DCFRR) and (NPV) are not acceptable to the company, the project should promptly be rejected. If the minimum values of (DCFRR) and (NPV) are acceptable, a detailed assessment should be made. If the maximum values of (DCFRR) and (NPV) are acceptable but the minimum values are not, the feasibility study should be continued.

Mathematical Models for Distribution Curves Mathematical models have been developed to fit the various distribution curves. It is most unlikely that any frequency distribution curve obtained in practice will exactly fit a curve plotted from any of these mathematical models. Nevertheless, the approximations are extremely useful, particularly in view of the inherent inaccuracies of practical data. The most common are the binomial, Poisson, and normal, or gaussian, distributions.

A normal distribution curve is bell-shaped (see Sec. 3). The curve obeys the relationship

$$f(x) = \frac{\exp - [(x - \mu)^2 / 2\sigma^2]}{\sigma (2\pi)^{0.5}} \tag{9-70}$$

where σ is known as the true standard deviation. The standard deviation $s°$ from a sample is given by

$$s° = \left[\frac{\sum (x_i - \bar{x})^2 f(x_i)}{\sum f(x_i) - 1} \right]^{0.5} \tag{9-71}$$

The standard deviation $s°$ for the sample corresponds to the true standard deviation σ for the whole population in the same way that the mean $\bar{x}$ of the sample corresponds to the arithmetic average μ for the whole population. Equation (9-70) can be written more compactly as

$$f(z) = [\exp (-z^2/2)] / [(2\pi)^{0.5}] \tag{9-72}$$

where the standard score z is

$$z = (x - \mu)/\sigma \tag{9-73}$$

The area under the curve of $f(z)$ is unity if the abscissa extends from minus infinity to plus infinity. The area under the curve between z_1 and z_2 is the probability that a randomly selected value of x will lie in the range z_1 and z_2, since this is the relative frequency with which that range of values would be represented in an infinite number of trials.

An event that will definitely occur has a probability of unity. An event that will definitely not occur has a probability of zero.

Equation (9-72) can be integrated between limits to determine the probability that a random value lies between the selected limits. Extensive tables of $f(z)$ and the associated integral are available (see Sec. 3).

A frequency distribution curve can be used to plot a cumulative-frequency curve. This is the curve of most importance in business decisions and can be plotted from a normal frequency distribution curve (see Sec. 3). The cumulative curve represents the probability of a random value z having a value of, say, z_1 or less.

If a property or variable c is a function of several other variables x_1, x_2, etc., it can be written in the form

$$c = \phi(x_1, x_2, \ldots x_n) \tag{9-74}$$

If each x is a normally distributed independent variable, then

$$\sigma_c^2 = \left(\frac{\partial c}{\partial x_1} \right)^2 \sigma_1^2 + \left(\frac{\partial c}{\partial x_2} \right)^2 \sigma_2^2 + \cdots + \left(\frac{\partial c}{\partial x_n} \right)^2 \sigma_n^2 \tag{9-75}$$

where σ_c is the standard deviation of the variable c and σ_1, σ_2, etc., are the standard deviations of the variables x_1, x_2, etc.

Many distributions occurring in business situations are not symmetrical but skewed, and the normal distribution curve is not a good fit. However, when data are based on estimates of future trends, the accuracy of the normal approximation is usually acceptable. This is particularly the case as the number of component variables x_1, x_2, etc., in Eq. (9-74) increases. Although distributions of the individual variables (x_1, x_2, etc.) may be skewed, the distribution of the property or variable c tends to approach the normal distribution.

Let us consider an event that must have one of two outcomes. It must either occur with probability p_1 or fail to occur with probability p_2. Since these are exclusive events and the probability that something will happen is unity, it follows that

$$p_1 + p_2 = 1 \tag{9-76}$$

Provided that no learning process is involved (so that the value of p_1 is not influenced by previous results), the probability of x successes in n trials is given by the term containing p_1^x in the expansion of the binomial:

$$(p_1 + p_2)^n = p_1^n + \cdots + \frac{n!}{x!(n-x)!} p_1^x p_2^{(n-x)} \cdots + p_2^n \tag{9-77}$$

where x and n are integers and $x!$ (read as x factorial) is the product of all integers from unity to x.

Example 5: Probability Calculation If a six-sided die marked with the numbers 1, 2, 3, 4, 5, and 6 is thrown, the probability that any given number will be uppermost is 1/6. If the die is thrown twice in succession, then the probability of a given sequence of numbers occurring, say, 5 followed by 6, is $(1/6)(1/6) = 1/36$. The chance of any particular number occurring 0, 1, 2, 3, or 4 times in four throws of the die (or in a simultaneous throw of four dice) is given by the successive terms of Eq. (9-77), expanded as

$(\tfrac{5}{6} + \tfrac{1}{6})^4 = (1)(\tfrac{5}{6})^4(\tfrac{1}{6})^0 + (4)(\tfrac{5}{6})^3(\tfrac{1}{6})^1 + (6)(\tfrac{5}{6})^2(\tfrac{1}{6})^2 + (4)(\tfrac{5}{6})^1(\tfrac{1}{6})^3$

$+ (1)(\tfrac{5}{6})^0(\tfrac{1}{6})^4 = 0.4823 + 0.3858 + 0.1157 + 0.0154 + 0.0008 = 1$

The distribution of the number of successes is skewed toward the low numbers. In particular, there is only a slightly better than even-money chance that any given number will occur even once in four throws. Such highly unsymmetrical distributions cannot be approximated by the normal distribution curve.

However, an increasing number of throws will result in totals that are close to the normal distribution. This fact can be used to approximate such a distribution without the enormous labor of the calculations required by the use of Eq. (9-77).

Possible values of the total of four throws of a die are integers from 4 to 24 and hence represent values in the range from 3.5 to 24.5. The mean value $\bar{x}$ of this range is given by Eq. (9-68) as $\bar{x} = (3.5 + 24.5)/2 = 14.0$.

The cumulative probability of a normally distributed variable lying within 4 standard deviations of the mean is 0.49997. Therefore, it is more than 99.99 percent (0.49997/0.50000) certain that a random value will be within $\pm 4\sigma$ from the mean. For practical purposes, σ may be taken as one-eighth of the range of certainty, and the standard deviation can be obtained:

$$s° \cong (24.5 - 3.5)/8 = 2.625$$

From Eq. (9-73) the standard score becomes

$$z = (x - \mu)/\sigma \cong (x - \bar{x})/s°$$

For a total score of 4 (i.e., $x = 4$), the standard score is approximately $z = (4 - 14)/2.625 = -3.81$. Since the normal curve is symmetrical about $z = 0$, the height of the ordinate at $z = -3.81$ is the same as that at $z = +3.81$. From tables of values of cumulative probabilities of the normal distribution, the height of the ordinate is 0.0003 in units of $1/\sigma$. The relative frequency of 4 occurring is thus approximately $0.0003/2.625 = 0.0001$.

This concept can be used to translate Delphic or other opinions into probability distributions and hence into useful decision-making tools.

Example 6: Calculation of Probability of Meeting a Sales Demand A store that is open 5 days a week is to promote a new product. The manager believes that not more than 5 units will be sold in any one day, but he cannot be more precise about the probable sales pattern. Stocks are delivered once per week. What size should the first order be to give a 95 percent certainty of meeting demand?

Since the product is sold in units, the possible range of weekly sales is from -0.5 to $+25.5$ units. Therefore, the mean of the sales distribution will be

$$\bar{x} \cong [25.5 - (-0.5)]/2 = 13$$

The standard deviation for this example will be

$$s° \cong [25.5 - (-0.5)]/8 = 3.25$$

From this, the approximate frequency distribution of daily sales can be derived by using Eqs. (9-70), (9-72), or (9-73). The desired area to the right of $z = 0$ for the normal probability distribution curve is $0.95 - 0.50 = 0.45$. For this value the standard score $z = 1.645$.

$$x \cong 13 + (1.645)(3.25) = 18.35$$

Hence, to be 95 percent certain of meeting demand, 19 units should be purchased.

If the value of n in Eq. (9-77) is large and neither p_1 nor p_2 is too close to zero, the binomial distribution can be approximated by

$$z = \frac{x - np_1}{\sqrt{np_1p_2}} \qquad (9\text{-}78)$$

The approximation of Eq. (9-78) is good enough for most purposes if np_1 and np_2 are each greater than 5.

Example 7: Calculation of Probability of Sales   The records of a business show that never more than 1 item is sold in a day and that 2 sales per week can be expected. What is the probability of selling between 90 and 120 items in a 300-day year?

In a year consisting of 50 weeks of 6 days, the mean or expected value of the distribution is 100 items. The probability of a sale of an item on a given day is $p_1 = 100/300 = 1/3$, and of no sale is $p_2 = 2/3$. From Eq. (9-78),

$$z = \frac{x - (300)(1/3)}{\sqrt{(300)(1/3)(2/3)}} = \frac{x - 100}{8.165}$$

The integral range of 90 to 120 items contains all possible values of x from 89.5 to 120.5. For $x = 89.5$, $z = -1.286$; and for $x = 120.5$, $z = 2.511$.

The cumulative probability of a standard score of 1.286 is 0.11, while that of a standard score of 2.511 is 0.99. Therefore, the probability of annual sales in the range of 90 to 120 items is $(0.99 - 0.11) = 0.88$, or 88 percent.

There are times when the frequency measurement is an integral number of events in a given segment of a continuum, for example, the number of automobiles passing a given point in 1 h or the number of leaks in a given length of hosepipe. In such cases, the correct frequency distribution is the Poisson distribution, in which the probability of x events per unit of a continuum occurring is given by

$$f(x) = \lambda^x e^{-\lambda}/x! \qquad (9\text{-}79)$$

where x is an integer, e is the base of natural logarithms, and λ is a parameter of the system $\lambda = \mu = \sigma^2$.

As λ increases, the Poisson distribution approaches the normal distribution, with the relationship

$$z = (x - \lambda)/\sqrt{\lambda} \qquad (9\text{-}80)$$

When the value of p_1 is very close to zero in Eq. (9-77), so that the occurrence of the event is rare, the binomial distribution can be approximated by the Poisson distribution with $\lambda = np_1$ when $n > 50$ while $np_1 < 5$.

Example 8: Calculation of Probability of Equipment Breakdown   The daily chance of a breakdown in a production line operated continuously for 300 days per year is estimated at 1 percent from past performance. Let us estimate the probability of 4 or more breakdowns in the coming year.

For $n = 300$ and $p_1 = 0.01$, $\lambda = np_1 = 3 < 5$, the probability of no breakdown is found from Eq. (9-79) to be

$$f(0) = (3)^0 e^{-3}/0! = 1/e^3 = 0.0498$$

Similarly,

Breakdowns	Probability
1	$f(1) = (3)^1 e^{-3}/1! = 0.1496$
2	$f(2) = (3)^2 e^{-3}/2! = 0.2240$
3	$f(3) = (3)^3 e^{-3}/3! = 0.2240$

Since something must happen, the probability of 4 or more breakdowns is

$$1 - 0.0498 - 0.1496 - 0.2240 - 0.2240 = 0.3526$$

A simple trial will show how much more easily the preceding calculation is carried out than direct use of Eq. (9-77).

The necessary value of λ may often be established as in the following example.

Example 9: Calculation of Probability of Machine Failures
In a production period of 100 days, 0, 1, 2, 3, and 4 machine failures occurred in a single day on 41, 37, 15, 6, and 1 occasions respectively. Let us fit a Poisson distribution to the data and estimate the maximum number of machine failures likely to occur in 1 day of a 300-day year.

The mean number of failures is found from Eq. (9-69) to be

$$\bar{x} = \frac{0(41) + 1(37) + 2(15) + 3(6) + 4(1)}{41 + 37 + 15 + 6 + 1} = 0.89$$

The standard deviation is found from Eq. (9-71):

$$s° = \left[\frac{\sum (x_i - \bar{x}) f(x_i)}{\sum f(x_i) - 1} \right]^{0.5}$$

The steps for calculating the numerator and denominator for this equation are tabulated as follows:

$(x_i - \bar{x})^2$	$f(x_i)$	$(x_i - \bar{x})^2 f(x_i)$
$(0 - 0.89)^2$	41	32.48
$(1 - 0.89)^2$	37	0.45
$(2 - 0.89)^2$	15	18.48
$(3 - 0.89)^2$	6	26.71
$(4 - 0.89)^2$	1	9.67
	$\sum (x_i - \bar{x})^{-2} f(x_i)$	$= 87.79$
	$\sum f(x_i) - 1$	$= 99$

Therefore, $s° = (87.79/99)^{0.5} = 0.9417$.

The Poisson distribution is a good fit since

$$\lambda = \mu \cong \bar{x} = 0.8900$$

and

$$\lambda = \sigma^2 \cong (s°)^2 \cong 0.8868$$

The Poisson distribution is found from Eq. (9-79) to be

$$f(x) = [(0.89)^x e^{-0.89}]/x!$$

By substituting the appropriate values of x for this example into the preceding equation, we find $f(0) = 0.4107$, $f(1) = 0.3655$, $f(2) = 0.1627$, $f(3) = 0.0483$, $f(4) = 0.0107$, $f(5) = 0.0019$, and $f(6) = 0.0003$. Hence, in 300 days the expected maximum number of breakdowns in 1 day is 5 since $(300)f(6) = 0.09$ occurrence.

In many business applications, Eq. (9-74) can be reduced to the linear relationship

$$c = k_1 x_1 + k_2 x_2 + \cdots + k_n x_n \qquad (9\text{-}81)$$

where the k's are constants. Equation (9-75) then becomes

$$\sigma_c^2 = k_1^2 \sigma_1^2 + k_2^2 \sigma_2^2 + \cdots + k_n^2 \sigma_n^2 \qquad (9\text{-}82)$$

On the other hand, for a product function such as

$$c = x_1 x_2 \qquad (9\text{-}83)$$

Eq. (9-75) can be written in the form

$$\sigma_c^2/c^2 = \sigma_1^2/x_1^2 + \sigma_2^2/x_1^2 \qquad (9\text{-}84)$$

The discounted-cash-flow rate of return (DCFRR) and net present value (NPV) are functions of the cumulative revenue from annual sales $\sum A_{TE}$ and the fixed-capital cost of the plant C_{FC}, among other factors.

Equation (9-75) can be written for (DCFRR) and for (NPV) as

$$\sigma_{(DCFRR)}^2 = \left[\frac{\partial (DCFRR)}{\partial \sum A_S} \right]^2 \sigma_{\sum A_S}^2 + \left[\frac{\partial (DCFRR)}{\partial \sum A_{TE}} \right]^2 \sigma_{\sum A_{TE}}^2$$

$$+ \left[\frac{\partial (DCFRR)}{\partial C_{FC}} \right]^2 \sigma_{C_{FC}}^2 \qquad (9\text{-}85)$$

$$\sigma^2_{(NPV)} = \frac{\partial (NPV)^2}{\partial \sum A_S} \sigma^2_{\sum A_S} + \frac{\partial (NPV)^2}{\partial \sum A_{TE}} \sigma^2_{\sum A_{TE}} + \frac{\partial (NPV)^2}{\partial C_{FC}} \sigma^2_{C_{FC}} \quad (9\text{-}86)$$

The revenue from annual sales A_S of a product at an annual production rate R and sales price of c_s per unit of production is

$$A_S = Rc_S \quad (9\text{-}87)$$

Equation (9-84) can be written as:

$$\sigma^2_{A_S} = (A_S/R)^2 \sigma^2_R + (A_S/c_s)^2 \sigma^2_{c_s} \quad (9\text{-}88)$$

An extensive example illustrating the use of Eqs. (9-81) through (9-86) in establishing the probability of attaining a given value of the net present value or less in a particular year of a project was presented by Holland et al. [F. A. Holland, F. A. Watson, and J. K. Wilkinson, *Chem. Eng.*, **81,** 105–110 (Jan. 7, 1974)]. The result is shown in Fig. 9-19.

Decision makers often prefer to have graphs showing the probability of attaining a value greater than a given value. Such curves are easily obtained by subtracting the probability of achieving a given value or less from 100 percent. Figure 9-20 was obtained in this way and shows the probability of attaining a (DCFRR) greater than a given value.

Monte Carlo Method   The Monte Carlo method makes use of random numbers. A digital computer can be used to generate pseudo-random numbers in the range from 0 to 1. To describe the use of random numbers, let us consider the frequency distribution curve of a particular factor, e.g., sales volume. Each value of the sales volume has a certain probability of occurrence. The cumulative probability of that value (or less) being realized is a number in the range from 0 to 1. Thus, a random number in the same range can be used to select a random value of the sales volume.

In the same way, random values of the other factors can be obtained. These can then be combined to give random values of (DCFRR) and (NPV) and, in turn, used to plot cumulative-probability curves for (DCFRR) and (NPV). The computer may be required to perform some 10,000 to 50,000 calculations.

The use of the Monte Carlo method in project appraisal was illustrated by Holland et al. [F. A. Holland, F. A. Watson, and J. K. Wilkinson, *Chem. Eng.*, **81,** 76–79 (Feb. 4, 1974)]. The cumulative-probability curves of (DCFRR) and (NPV) can never be more accurate than the opinions on which they are based, and comparable accuracy can be obtained by the use of S-shaped curves with relatively small computational effort.

S-Shaped Curves   K. D. Tocher (*The Art of Simulation,* rev. ed., English Universities Press, London, 1967) presented a comprehen-

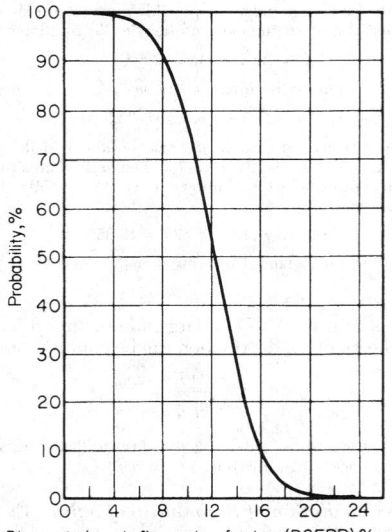

FIG. 9-20   Probability of a given discounted-cash-flow rate of return or more for a project.

sive treatment of the generation of random and pseudorandom numbers and their use in a wide range of simulated processes. He also considered sampling techniques from the various statistical distributions and the design of simulated processes. It will be noted that the cumulative distribution curves are S-shaped, and Tocher (op. cit., p. 16) recommended as a general equation for such curves

$$x = a + by + cy^2 + d(1 - y)^2 \ln y + ey^2 \ln (1 - y) \quad (9\text{-}89)$$

in which x varies from $-\infty$ to $+\infty$ as y varies from 0 to 1. The underlying frequency curve corresponding to Eq. (9-89) is

$$\frac{1}{p(x)} = \frac{dx}{dy} = b + 2cy + d(1 - y) \left(\frac{1 - y}{y} - 2 \ln y \right)$$
$$+ ey \left[2 \ln (1 - y) - \frac{y}{1 - y} \right] \quad (9\text{-}90)$$

If necessary, the fit can be improved by increasing the order of the polynomial part of Eq. (9-89), so that this approach provides a very flexible method of simulation of a cumulative-frequency distribution. The method can even be extended to J-shaped curves, which are characterized by a maximum frequency at $x = 0$ and decreasing frequency for increasing values of x, by considering the reflexion of the curve in the y axis to exist. The resulting single maximum curve can then be sampled correctly by Monte Carlo methods if the vertical scale is halved and only absolute values of x are considered.

When the data do not warrant the accuracy of Eq. (9-89) or Eq. (9-90), simpler curves will usually suffice if the frequency distribution may be assumed to have a single maximum value.

Let us consider a product which is sold entirely on the basis of personal recommendation. The rate of sale will depend on the number of people who have already bought the product. Thus initially sales will increase exponentially. Eventually the market will be saturated, and only replacement purchases will be made. If the frequency curve may be assumed to be symmetrical about a single maximum value, the cumulative distribution curve is known as the logistics curve and is defined by Eq. (9-91):

$$y = c/[1 + a \exp (-bx)] \quad (9\text{-}91)$$

where y varies between zero and c as x ranges from $-\infty$ to $+\infty$.

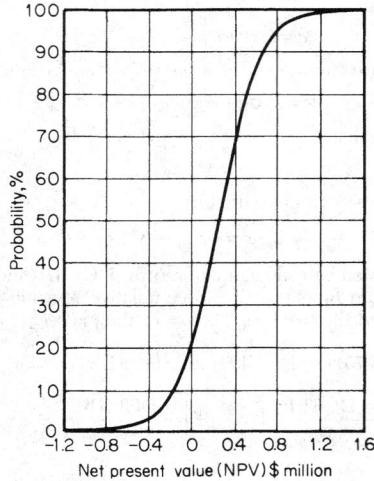

FIG. 9-19   Probability of a given net present value or less for a project.

Although only three constants appear explicitly in Eq. (9-91), two further constants are implied by the choice of zero as the lower bound of y and the point of inflexion at $y = c/2$. The usual use of Eq. (9-91) is in sales forecasting, in which case y is sales demand and x is time. If such a curve already exists, the value of c can be read as the upper asymptote and a and b obtained by the use of an auxiliary variable T where

$$T = x_2 \text{ (at } y = r_2 c) - x_1 \text{ (at } y = r_1 c) \tag{9-92}$$

$$b = [\ln (1/r_1 - 1) - \ln (1/r_2 - 1)]/T \tag{9-93}$$

$$a = (1/r_1 - 1) \exp (bx_1) \tag{9-94}$$

or

$$a = (1/r_2 - 1) \exp (bx_2)$$

If the values of a obtained from Eq. (9-94) differ significantly, the logistics curve is not a suitable representation of the data.

Example 10: Logistics Curve We shall derive the logistics curve representing the cumulative-frequency distributions of the normal distribution curve defined by Eqs. (9-72) and (9-73). In this case, y varies between a cumulative probability of zero and unity as z varies from $-\infty$ to $+\infty$. Since the upper bound is unity, $c = 1$. From Table 9-10 the area under the right-hand side of the curve between $z = 0$ and $z = z$ may be read. Since the frequency curve is symmetrical about the mean, this is also the area between $z = 0$ and $z = z$. Hence, the area under the frequency curve, which represents the cumulative probability, is 0.50000 at $z = 0$ and the 80 percentile, for which the area is 0.80000, corresponds to the value $z = 0.842$. We substitute these values into Eqs. (9-92) through (9-94) to give

$$T = 0.842 - 0.000 = 0.842$$

$$b = [\ln (1/0.50 - 1) - \ln (1/0.80 - 1)]/0.842 = 1.6464$$

$$a = 1.0000 \text{ or } 1.00000$$

From Eq. (9-91) the corresponding logistic curve is

$$y = [1 + \exp (-1.6464z)]^{-1}$$

The cumulative-frequency function calculated from this simple expression is compared with the precise value in Table 9-10.

TABLE 9-10 Data for Normal Distribution Curve

Standard score, z	Ordinate of normal distribution curve, $p(z)$		Area under normal curve, cumulative probability, $y = \int_0^z p(z)\, dz$	
	Precise	Estimated	Precise	Estimated
0.000	0.3989	0.4116	0.0000	0.00000
0.100	0.3970	0.4088	0.03983	0.04017
0.200	0.3910	0.4006	0.07926	0.08158
0.253	0.3864	0.3943	0.10000	0.10265
0.300	0.3814	0.3875	0.11791	0.12103
0.400	0.3683	0.3700	0.15542	0.15894
0.500	0.3521	0.3491	0.19146	0.19492
0.524	0.3478	0.3436	0.20000	0.20323
0.600	0.3332	0.3255	0.22575	0.22866
0.700	0.3123	0.3003	0.25804	0.25996
0.800	0.2897	0.2744	0.28814	0.28870
0.842	0.2798	0.2634	0.30000	0.30000
0.900	0.2661	0.2484	0.31594	0.31484
1.000	0.2420	0.2231	0.34134	0.33840
1.200	0.1942	0.1761	0.38493	0.37822
1.282	0.1953	0.1587	0.40000	0.39194
1.400	0.1497	0.1358	0.41234	0.40929
1.600	0.1109	0.1029	0.44520	0.43303
1.645	0.1032	0.0964	0.45000	0.43752
1.800	0.0790	0.0769	0.46407	0.45092
2.000	0.0540	0.0569	0.47725	0.46418
2.500	0.0175	0.0260	0.49379	0.48395
3.000	0.0044	0.0116	0.49865	0.49289
4.000	0.0001	0.0023	0.49997	0.49862
∞	0.0000	0.0000	0.50000	0.50000

When a cumulative-frequency curve can be satisfactorily represented by a logistics curve, the underlying frequency curve can be obtained by differentiation of Eq. (9-91) as

$$p(x) = \frac{dy}{dx} = \frac{abc \exp (-bx)}{[1 + a \exp (-bx)]^2} \tag{9-95}$$

The probability-density function for the normal distribution curve calculated from Eq. (9-95) by using the values of a, b, and c obtained in Example 10 is also compared with precise values in Table 9-10. In such symmetrical cases the best fit is to be expected when the median or 50 percentile x_M is used in conjunction with the lower quartile or 25 percentile x_L or with the upper quartile or 75 percentile x_U. These statistics are frequently quoted, and determination of values of a, b, and c by using x_M with x_L and with x_U is an indication of the symmetry of the curve. When the agreement is reasonable, the mean values of b so determined should be used to calculate the corresponding value of a.

In practice most distribution curves are not symmetrical about the median but are inherently skewed. The effect of an advertising campaign is usually to increase the rate of sales in the early years. It may also increase the level of mature demand for the product, but this mature demand must be asymptotic to a finite upper limit of sales c. Such a curve is positively skewed since $(x_M - x_L) < (x_U - x_M)$. This situation can often be approximated by the Gompertz curve defined by Eq. (9-96):

$$\ln y = \ln c - a \exp (-bx) \tag{9-96}$$

which has its point of inflexion at $0.3679 c$. In terms of the upper and lower quartiles and the median,

$$b = 0.8794/(x_U - x_M) \tag{9-97}$$

$$b = 0.6931/(x_M - x_L) \tag{9-98}$$

$$a = 0.6931 \exp (bx_M) \tag{9-99}$$

The suitability of the Gompertz fit to the curve can be assessed by comparing the values of b calculated from Eqs. (9-97) and (9-98), and, if suitable, the average value of b may be used in Eq. (9-99) to calculate the corresponding value of a to ensure a fit at the median and reasonable accuracy over the more important practical range within a couple of standard deviations on either side of the median.

The underlying frequency distribution curve of the Gompertz curve may be obtained by differentiation of Eq. (9-96) to give

$$p(x) = dy/dx = yab \exp (-bx) \tag{9-100}$$

The logistic and Gompertz curves are of the general shape illustrated by Fig. 9-19. They may be adapted to fit curves of the general shape illustrated by Fig. 9-20 by a little mathematical manipulation. As an example, let us consider the current ratio, the ratio of current assets to current debts, as is quoted in Dun & Bradstreet statistics. A typical value for United States industrial chemical companies might be listed as $x_L = 1.82$, $x_M = 2.59$, and $x_U = 3.25$. First, we notice that $(x_M - x_L) > (x_U - x_M)$. This curve is, therefore, negatively skewed, or reversed S-shaped, and the logistics curve is not suitable. Nor can the Gompertz equation be used directly. However, it is clear that if the curve is drawn upside down and backward, the transformed curve will be positively skewed. Mathematically, this is equivalent to interchanging the upper and lower bounds and considering the dependent variable to be $(c - y)$. In the present case the quoted values represent the cumulative probabilities that the current ratio will be less than the quoted value and hence the value of y ranges between zero and unity. Hence, $c = 1$. In the transformed curve $x_L = 3.25$, $x_M = 2.59$, and $x_U = 1.82$. Hence, from Eq. (9-97)

$$b = 0.8794/(1.82 - 2.59) = -1.1421$$

and from Eq. (9-98)

$$b = 0.6931/(2.59 - 3.25) = -1.0502$$

The variation is within 5 percent of the mean value of $b = -1.0961$, and the transformed curve should be sufficiently accurate for many purposes. From Eq. (9-99)

$$a = 0.6931 \exp [(-1.0961)(2.59)] = 0.04054$$

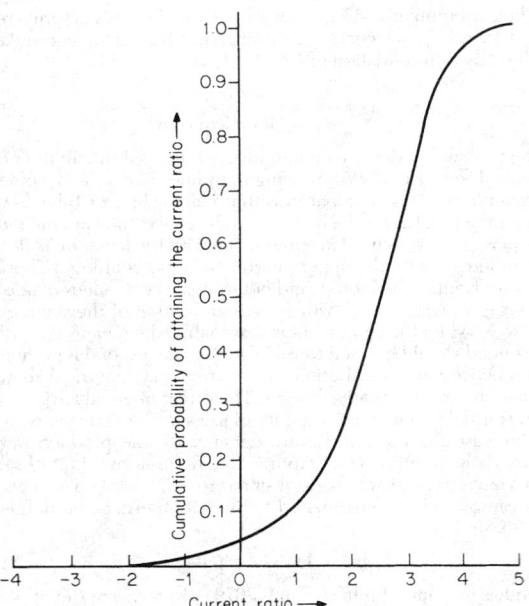

FIG. 9-21 Cumulative probability of a given current ratio.

Hence, from Eq. (9-96) the equation of the transformed curve is

$$\ln(1 - y) = -0.04054 \exp(1.0961x)$$

Since $d(1 - y) = -dy$, the corresponding underlying frequency distribution curve is from Eq. (9-100):

$$p(x) = +0.0444(1 - y)\exp(1.0961x)$$

Values of y and $p(x)$ calculated from last two equations are plotted in Figs. 9-21 and 9-22 respectively.

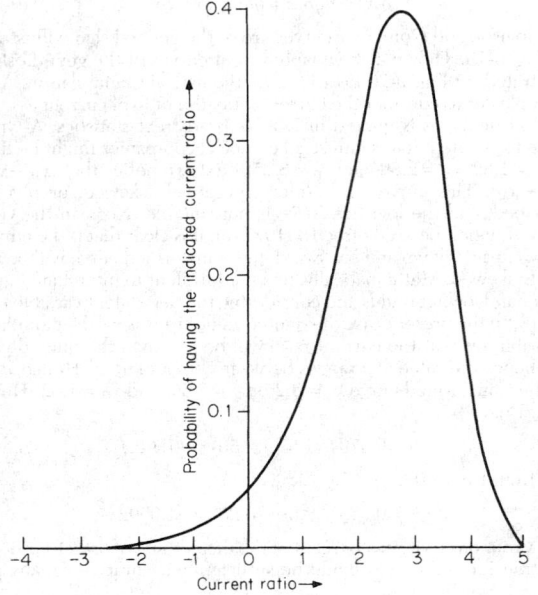

FIG. 9-22 Probability of a given current ratio.

In all such S-shaped curves the range of x is from $-\infty$ to $+\infty$, so that there is always a finite possibility of negative values of x occurring. In the present case the definition of the current ratio makes values of x below zero meaningless. The error of some 4 percent in the cumulative-probability curve implied by this factor may be tolerable in a given case.

It can be shown [J. J. Molder and E. G. Rogers, *Manage. Sci.*, **15**, B-76 (1968)] that for continuous events it is possible to estimate the mean and standard deviation of a skewed distribution from estimates of a low value, a most likely or modal value, and a high value. It is suggested, since it is difficult to make very fine subjective judgments as to probabilities, that the range most likely to be accurate is that for which there is a 10 percent chance of a value less than the low value and a 10 percent chance of a value greater than the high value. These values will usually imply a skewed distribution. For the suggested 80 percent confidence level the best available estimates are

$$\bar{x} = (\text{low value}) + [(2)(\text{modal value})] + (\text{high value}) \quad (9\text{-}101)$$

$$s° = [(\text{high value}) - (\text{low value})]/2.65 \quad (9\text{-}102)$$

On this basis an alternative approach to risk analysis is the parameter method [D. O. Cooper and L. B. Davidson, *Chem. Eng. Prog.*, **72**, 73–78 (November 1976)].

Example 11: Parameter Method of Risk Analysis Let us consider the project outlined in Table 9-5. It is estimated that the basic data represent the most likely values and that there is a 10 percent chance that A_S will be reduced by more than 20 percent or will be increased by more than 5 percent. In the same way the low and high levels at 10 percent probability for A_{TE} are considered to be 5 percent below and 25 percent above the base figures respectively. The low and high values for C_{FC} are considered to be 5 percent below and 30 percent above the base figure, while changes in other parameters are considered to be immaterial.

With a cost of capital i of 10 percent the various cash flows can be discounted and summed. Thus for the base cases $\sum A_s f_d = \$2,815,600$, $\sum A_{TE}f_d = \$754,716$, $\sum A_D f_d = \$614,457$, and $\sum C_{WC}f_d = \$61,446$. With corporate taxes payable at 50 percent the aftertax cash flows of the first three items are $(1 - 0.50)$ of the sums calculated above. The discounted working capital and the fixed-capital outlay are not subject to tax. These most probable values are listed and summed in Table 9-11 and, after adjustment for tax, give the modal value of the (NPV) as $\$276,224$.

A reduction of 20 percent in A_s for each year will result in a 20 percent reduction in $\sum A_s f_d$ below the modal value, i.e., a reduction of $(0.20)(\$2,815,600) = \$563,120$. The aftertax effect of this reduction on the contribution to (NPV) is $(1 - 0.50)(\$563,120) = \$281,560$, making the low value $\$1,407,800 - \$281,560 = \$1,126,240$ or, more directly, $(0.8)(\$1,407,800)$. Other values in Table 9-11 are calculated in a similar manner.

The mean value of each of the distributions is obtained from these high, modal, and low values by the use of Eq. (9-101). If the distribution is skewed, the mean and the mode will not coincide. However, the mean values may be summed to give the mean value of the (NPV) as $\$161,266$. The standard deviation of each of the distributions is calculated by the use of Eq. (9-75). The fact that the (NPV) of the mean or the mode is the sum of the individual mean or modal values implies that Eq. (9-81) is appropriate with all the k's equal to unity. Hence, by Eq. (9-81) the standard deviation of the (NPV) is the root mean square of the individual standard deviations. In the present case $s° = \$166,840$ for the (NPV).

If the resulting distribution is assumed to be normal, then the cumulative distribution curve can immediately be generated. From Table 9-10, a standard score of 4 corresponds to a probability of $0.5 + 0.49997 = 0.99997$ and one of -4 to a probability of $0.5 - 0.49997 = 0.00003$, virtually unity and zero respectively. From Eq. (9-73) a standard score of 4 corresponds to an (NPV) of $\$161,266 + (4)(\$166,840) = \$828,626$ and one of -4 to an (NPV) of $\$506,094$. Values of (NPV) corresponding to other confidence limits may be calculated in the same way and plotted to give the curve of Fig. 9-23.

TABLE 9-11 Data for Risk Analysis

Parameter	(NPV), $/year			Mean	Standard deviation
	Low	Modal	High		
A_S	1,126,240	1,407,800	1,478,190	1,355,008	132,811
A_{TE}	−471,698	−377,358	−358,490	−396,226	−42,720
C_{FC}	−900,603	−692,772	−658,133	−736,070	−91,498
C_{WC}	−61,446	−61,446	−61,446	−61,446	0

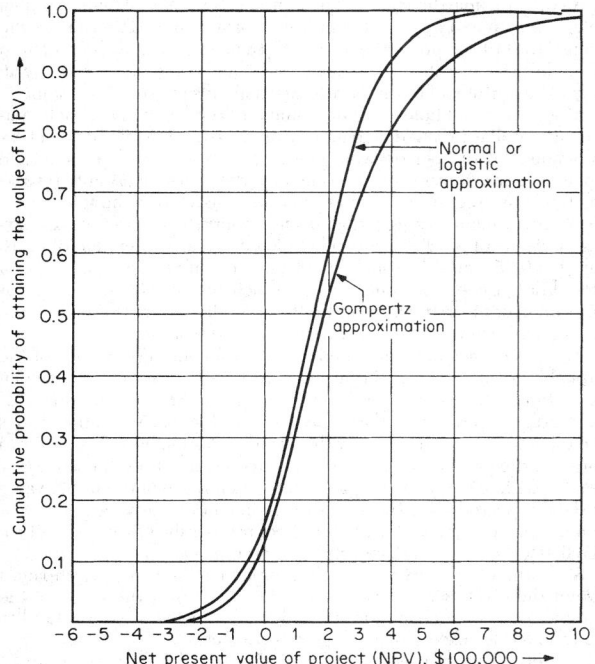

FIG. 9-23 Cumulative probability of a given net present value or less for a project showing normal and Gompertz approximations.

As has been stated, with the uncertainties attached to many business assessments of the range of various factors, the central-limit theorem implies that the assumption of a normal distribution of the main variable is sufficiently accurate provided that there are several factors contributing to that main variable. The results are as informative as most Monte Carlo estimates and have the advantage that they can be rapidly obtained without recourse to a digital computer, although a good desk calculator speeds the work. Strictly, the variables should be independent and additive. Thus it is better, for example, to treat $(A_S - A_{TVE})$ as a single variable since both sales income and total variable expense are related to the annual rate of sales R. In such cases the standard deviation of A_{TVE} would be added to or subtracted from that of A_S before squaring to obtain the variance according as the uncertainty of the group was greater or less than that of the individual factors. Also, when a product such as $A_S = Rc_S$ is involved, Eq. (9-84) should be used to estimate the variance rather than Eq. (9-82). When the predominant uncertainties are multiplied together, a log-normal distribution may provide a better final distribution. A similar technique may be applied to the (DCFRR) provided that Eq. (9-85) is

used in place of Eq. (9-86) to estimate the overall variance of the main variable.

When the estimates are well founded, the skewness may be preserved by using a distribution such as the Gompertz. The median of that curve occurs as $y = 0.5\, c$, while the point of inflexion corresponds to the mode at $y = c/\exp(1) = 0.3679\, c$. The statistician Karl Pearson suggested as a simple measure of skewness

$$\text{Skewness} = 3\,(\text{mean} - \text{median})/\sigma \qquad (9\text{-}103)$$

with an empirical approximation in terms of the mode given by

$$(\text{Mean} - \text{mode}) = 3\,(\text{mean} - \text{median}) \qquad (9\text{-}104)$$

Applying these equations to the present problem,

$$(\text{Mean} - \text{mode}) = \$161{,}266 - \$276{,}224 = -\$114{,}958$$

$$\text{Skewness} = -\$114{,}958/\$166{,}840 = -0.6890$$

For symmetrical distributions, such as the logistic or normal, the skewness should be zero.

The Gompertz distribution requires the distribution to be positively skewed, which can be achieved by treating $-(\text{NPV})$ as the independent variable and $(c - y)$ as the dependent variable. From Eq. (9-104) the median of the distribution is given approximately as

$$\text{Median} = [\$161{,}266 - (-\$114{,}958)]/3 = \$199{,}585$$

Substituting values into Eq. (9-96) with $-(\text{NPV})$ as the independent variable to give, since the range of y is zero to unity,

$$\ln(1 - 0.5) = \ln(1) - a\exp[(-b)(-\$199{,}585)]$$

$$\ln(1 - 1/e) = \ln(1) - a\exp[(-b)(-\$276{,}224)]$$

whence $\quad b = \dfrac{-\ln[\ln(0.5)/\ln(1 - 1/e)]}{[(-\$199{,}585) - (-\$276{,}224)]} = -5.388 \times 10^{-6}/\$$

$$a = \frac{-\ln 0.5}{\exp[(5.388 \times 10^{-6}/\$)(-\$199{,}585)]} = 2.0315$$

The Gompertz curve of the distribution is then, in terms of (NPV),

$$\ln y = -2.0315\exp[-5.388 \times 10^{-6}(\text{NPV})]$$

For the same degree of certainty as before, the minimum value of the (NPV) is likely to be

$$\ln\left[\frac{\ln(0.00003)}{-2.0315}\right] - 5.388 \times 10^{-6} = -\$303{,}365$$

and the maximum of $\$2{,}064{,}569$ calculated in the same way for $y = 0.99997$. Other values are calculated in the same way and are plotted as in Fig. 9-23.

Decision Trees   In a typical decision tree, illustrated in a very simplified form by Fig. 9-24, each node represents a decision point (DP) at which one or more alternatives are available. Some quantifiable result of each alternative is chosen as a basis for comparison: for example, the net present value (NPV). A value is assigned to the probability of attaining each result, either cumulative or not as required. These may be obtained by the methods just described or otherwise. The estimates are subject to the restriction that the sum of the proba-

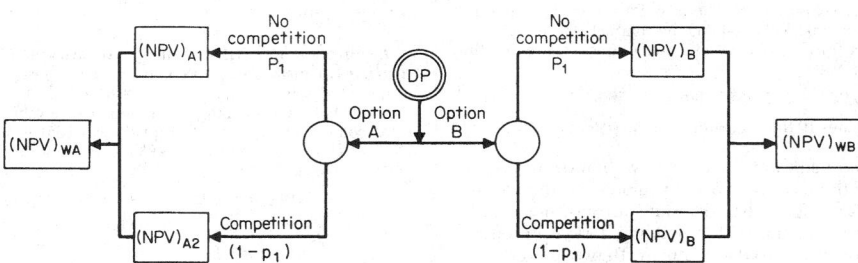

FIG. 9-24   Effect of decision-tree options on net present value.

bilities for all branches leaving each node shall be unity since some decision must be taken there.

In considering two investments, we shall let option B be a safe investment having a base net present value $(NPV)_B$ that is independent of any competition. We shall let option A yield a net present value $(NPV)_{A1}$ if no competition exists and $(NPV)_{A2}$ if competition exists. We shall then let the probabilities of no competition and competition be p_1 and p_2 respectively. Then p_2 must equal $(1 - p_1)$.

The expected (NPV) for option A can be written, from Eq. (9-105), which follows, as

$$(NPV)_{WA} = p_1(NPV)_{A1} + (1 - p_1)(NPV)_{A2}$$

where $(NPV)_{WA}$ is the weighted net present value for option A based on the probabilities of encountering no competition p_1 and of encountering competition $(1 - p_1)$.

In the same way the expected (NPV) for option B is given by

$$(NPV)_{WB} = p_1(NPV)_B + (1 - p_1)(NPV)_B = (NPV)_B$$

The gain in the expected value of option A over option B is thus

$$\Delta(NPV)_W = (NPV)_{WA} - (NPV)_{WB}$$

Let us suppose that the options represented in Fig. 9-24 were such that $(NPV)_B = (0.5)(NPV)_{A1} = 2(NPV)_{A2}$. Then substitution leads to

$$\Delta(NPV)_W = [2p_1 + 0.5(1 - p_1) - 1](NPV)_B$$
$$= (1.5p_1 - 0.5)(NPV)_B$$

The choice is immaterial when $\Delta(NPV)_W = 0$, i.e., when $p_1 = 1/3$. If the probability of no competition is greater than 1/3, option A should be chosen; otherwise option B should be chosen.

The technique is based on the methods of linear algebra and the theory of games. When the problem contains many multibranched decision points, a computer may be needed to follow all possible paths and list them in order of desirability in terms of the quantitative criterion chosen. The decision maker may then concentrate on the routes at the top of the list and choose from among them by using other, possibly subjective criteria. The technique has many uses which are well covered in an extensive literature and will not be further considered here.

Numerical Measures of Risk Without risk and the reward for successfully accepting risk, there would be no business activity. In estimating the probabilities of attaining various levels of net present value (NPV) and discounted-cash-flow rate of return (DCFRR), there was a spread in the possible values of (NPV) and (DCFRR). A number of methods have been suggested for assessing risks and rewards to be expected from projects.

Let us consider a proposed project in which there is a probability p_1 that a net present value $(NPV)_1$ will result, a probability p_2 that $(NPV)_2$ will result, etc. A weighted average $(NPV)_w$, known as the expected value, can then be calculated from

$$(NPV)_w = p_1(NPV)_1 + p_2(NPV)_2 + \cdots \qquad (9\text{-}105)$$

where $p_1 + p_2 + \cdots = 1.0$.

Analogous equations may be written for other additive measures of profitability such as net profit.

Example 12: Expected Value of Net Profit Let us consider a contractor who stands to make a net profit of $100,000 on a contract. The cost of preparing the bid on the contract is $10,000. There are four competing contractors, each with a probability $p_1 = 0.25$ of obtaining the contract. Thus, each contractor has a probability $p_2 = 0.75$ of not obtaining the contract. Therefore, the expected value of the project is

$$0.25(\$100,000) + 0.75(-\$10,000) = \$17,500$$

In this case, the potential gain is 10 times greater than the potential loss.

If the potential loss can bankrupt the company, then decisions are not necessarily made on the basis of expected value even though the potential gain may be very high. Also, decisions are not necessarily made on the basis of expected value if the potential loss represents a relatively small amount of money to the company. Between these two extremes, expected value can be a very useful criterion, particularly for a company with a large number of projects.

A company may be considering a project with a very high potential rate of return and a low risk, but it may prove impossible to raise the money to start the project. Conversely, the company may be prepared to undertake an extremely risky project if the investment is trivial. Thus, the attitude of a company to risk depends on the circumstances.

Money does not hold the same value for each company or each individual. A dollar may keep a pauper from starvation while being a trivial amount to the person who gave it. Attempts have been made to quantify a company's attitude to money, risk, and uncertainty by asking business executives a number of questions such as the following:

"Your company has signed a business contract with potential after-tax proceeds of P. The probability of achieving the net gain of P is, say, $p_1 = 0.75$, and the probability of a net loss of P is $p_2 = 0.25$. If you would rather keep the contract, how much cash would you accept for your interest in it? If you would rather be released from the contract, how much cash would you pay to be released from it?"

The same questions may then be asked for different values of the probabilities p_1 and p_2. The answers to these questions can give an indication of the importance to the company of P at various levels of risk and are used to plot the utility curve in Fig. 9-25. Positive values are the amounts of money that the company would accept in order to forgo participation. Negative values are the amounts the company would pay in order to avoid participation. Only when the utility value and the expected value (i.e., the straight line in Fig. 9-25) are the same can net present value (NPV) and discounted-cash-flow rate of return (DCFRR) be justified as investment criteria.

Since the utility curve has such a subjective basis, most companies prefer the objectivity of (NPV) and (DCFRR) over the range of the normal income and expenditure budget. Subjective methods tend to be reserved for exceptionally high risk projects.

A utility curve such as that in Fig. 9-25 is specific to a certain sum of money. The curve is likely to be different for, say, $P = \$10,000$. Figure 9-25 can only be used to consider projects that fall within the range of −$100,000 to +$100,000. Other utility curves must be used to cover projects that lie outside this range.

R. O. Swalm ["Utility Theory—Insight into Risk Taking," *Harv. Bus. Rev.*, **44**, 123–136 (November–December 1966)] found that many business executives had difficulty in appreciating fine shades of odds and confined his considerations to even-money bets. He asked various executives to state what guaranteed sum of money they considered equivalent to a gamble related to the toss of a coin. If the coin fell on one side, they would win a given sum of money; if the coin fell on the other side, they would get nothing.

Swalm started by considering a sum of money equivalent to twice the maximum expenditure that the executive could authorize in 1 year. This was used to obtain a further utility. In this way, a utility curve could be sketched. Swalm chose an arbitrary utility scale based on a range of −120 utiles to +120 utiles. (NOTE: It is as incorrect to compare utiles by ratio as it is to imply that an object at 30°C is twice as hot as an object at 15°C.)

Swalm found that most executives are conservative in their expenditure and that the patterns of utility curves are very similar if plotted with an ordinate range of ±1 unit. The unit, in this case, is the maximum authorized annual expenditure of the executive. Such curves may appear to differ quite widely when plotted in terms of absolute money values. The curves also show that executives tend to be more conservative when considering a loss than they do when considering a reduced gain.

Example 13: Evaluation of Investment Priorities Using Probability Calculations A company is considering investment in one or more of three projects, A, B, and C. We wish to evaluate the investment priorities if the probabilities of attaining various net present values (NPV) are as listed in the third column of Table 9-11. Equation (9-105) gives the expected value for $(NPV)_w$. Hence for project A, $(NPV)_w$ is computed from the data in Table 9-12 and found to be

$$(NPV)_w = 0.1(\$95,000) + 0.8(\$45,000) + 0.1(-\$75,000)$$
$$(NPV)_w = \$9,500 + \$36,000 - \$7,500$$
$$(NPV)_w = \$38,000$$

Corresponding values for projects B and C are calculated in the same way and are listed in Table 9-12.

TABLE 9-12 Comparison of Projects in Terms of Expected Value and Expected Utility

Project	(NPV), $	Probability, p	Equivalent probability of winning $100,000	Probable utility	Expected value, $	Expected utility, $
A	95,000	0.1	0.80	0.080	9,500	8,000
	45,000	0.8	0.34	0.272	36,000	27,200
	−75,000	0.1	0.03	0.003	−7,500	300
		1.0		0.355	38,000	35,500
B	50,000	0.2	0.37	0.074	10,000	7,400
	20,000	0.6	0.23	0.138	12,000	13,800
	−60,000	0.2	0.04	0.008	−12,000	800
		1.0		0.220	10,000	22,000
C	45,000	0.1	0.35	0.035	4,500	3,500
	10,000	0.6	0.20	0.120	6,000	12,000
	−60,000	0.3	0.04	0.012	−18,000	1,200
		1.0		0.167	−7,500	16,700

In project A, the probability $p = 0.1$ for (NPV) = $95,000. Figure 9-25 shows that $95,000 is the amount of money that this company would pay for a 0.8 probability of gaining $100,000. There is, therefore, a 0.2 probability of losing $100,000. In this case, a probability of $p = 0.1$ of attaining $95,000 is equivalent to a probable utility of $(0.1)(0.8) = 0.08$ of gaining $100,000.

Equation (9-105) can be used to calculate the expected utility if the probabilities p_1, p_2, etc., are replaced by the probable utilities and if the net present values $(NPV)_1$, $(NPV)_2$, etc., are each replaced by $100,000. For project A, the expected utility U_w is

$$U_w = [0.1(0.8) + 0.8(0.34) + 0.1(0.03)]\$100,000$$

$$U_w = \$8,000 + \$27,200 + \$300$$

$$U_w = \$35,500$$

Corresponding values for projects B and C are calculated in the same way and are listed in Table 9-12.

The straight line in Fig. 9-25 represents the situation in which the expected value and the expected utility are equal over the range of −$100,000 to

+$100,000. In this case, decisions can be taken on the basis of the highest expected value as a routine matter. In other cases, decisions should be made on the basis of the highest expected utility. The utility curve in Fig. 9-25 represents the present attitude of management to $100,000. This curve should be updated as the company's business position changes. In this example, the utility curve is above the straight line. This represents a tendency on the part of the company's decision makers to gamble. When it is below the straight line, the utility curve implies conservatism. The investment priorities should be to implement project A and then, if finance is available, project B.

It might appear that project C should also be considered in view of the expected utility of $16,700. However, it is better to do nothing than to implement project C. The utility of doing nothing, which is equivalent to paying $0, is read from Fig. 9-25 to be 0.17. This gives a corresponding probable utility of $(1.0)(0.17)(\$100,000)$, or $17,000. This is a better result than investing in project C.

In this example, the order of priorities based on expected utilities is the same as that based on expected values. However, the order of priorities is clear-cut on the basis of expected value but much less so on the basis of expected utility.

Capital is at risk until the breakeven point has been reached. It is common practice to give consideration to the discounted breakeven point (DBEP), the time at which the (NPV) is zero when discounting at the cost of capital. At any time after the (DBEP), the project will have recovered its cost and provided a greater return on the capital than the cost of capital. It is customary for management to spread risk by diversifying the activities of a company among a portfolio of projects.

R. L. Reul [*Chem. Eng.* (*London*), **238,** CE 120–125 (May 1970)] has defined a parameter, which he calls the measured-survival function (MSF), given by

$$(MSF) = 1 - (1 - p)^\beta \qquad (9\text{-}106)$$

where (MSF) is the probability that a portfolio of bets with a similar strategy will at least break even, and β is the amount of one win divided by the amount of each bet. Reul has applied Eq. (9-106) to the research and development activities of a company. Equation (9-106) is based on the simplified assumption that a project either succeeds with probability p and achieves the expected reward or fails completely with probability $(1 - p)$. Therefore, (MSF) is the probability of at least

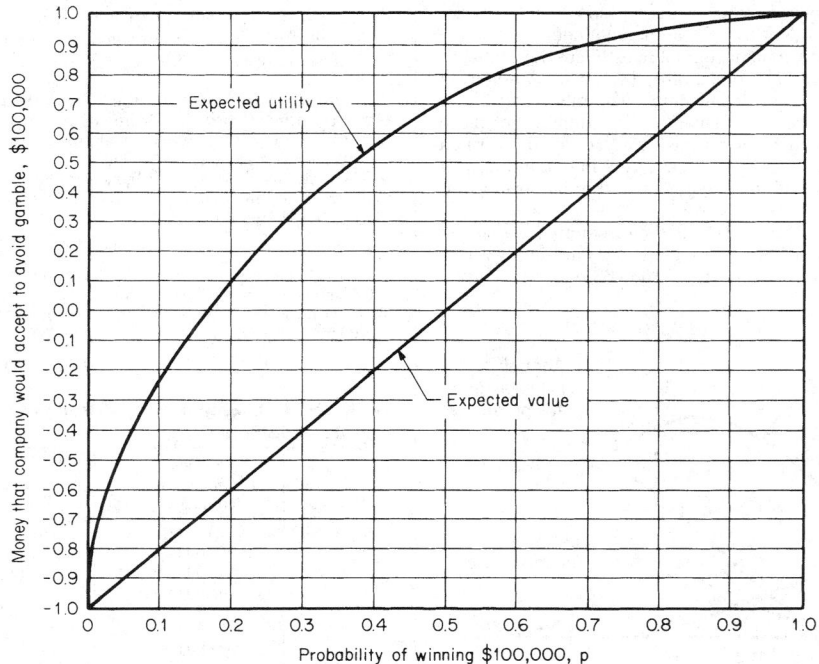

FIG. 9-25 Utility-function plot for $100,000.

one success when β similar projects are undertaken and represents a conservative measure of risk. It follows that $\beta > 1$ and hence that $(MSF) > p$. Many projects may result in greater returns or have an increased probability of attaining a given return if more money is spent. Each alternative derivable result from a given project is treated as a separate risk in the portfolio.

Research and development activities do not, in themselves, produce a salable product. Thus, they cannot directly generate a return on capital outlay. A successful research and development project is one that results in an activity that earns revenue for the company. The life cycle of the revenue from an individual product may be as shown in Fig. 9-26.

This revenue has to pay not only for the successful project but for all the unsuccessful research and development activities. It is common practice to consider all R&D as a portfolio. Disbursements for R&D are relatively flexible and can be switched from less favorable to more favorable projects at short notice.

When considering individual projects, β should be taken as the lesser of

$$\beta = \frac{\text{expected proceeds if project is successful}}{\text{disbursement on project}}$$

or

$$\beta = \frac{\text{total expenditure on all projects over budget period}}{\text{expenditure on project over budget period}}$$

Because the projects in a portfolio will usually have different probabilities of success and different rewards for success, β and p in Eq. (9-106) are conservatively estimated as follows:

$$\beta = \frac{\text{total annual proceeds if all projects are successful}}{\text{total annual disbursements on all projects}}$$

$$p = \frac{\text{total expected value of all projects}}{\text{total proceeds if all projects are successful}}$$

The expected value can be calculated from Eq. (9-105).

The relationship between (MSF), p, and $1/\beta$ in Eq. (9-106) is shown graphically in Fig. 9-27. It is the responsibility of management to decide on an acceptable value of the (MSF) for its company. The value chosen will depend on the company's attitude to risk that can be quantified in the form of a utility curve such as the one shown in Fig. 9-25, from which a value of equivalent (MSF) can be obtained.

It is also the responsibility of management to estimate the probabilities for the success of individual projects after due consideration of all the data provided by the various departments. The rate of return on investment that is acceptable to management is a function of these responsibilities. Each industry has a reasonably well defined return on investment that reflects the degree of risk inherent in that industry. If management decisions are faulty, the company either will overspend or will miss opportunities.

With a disbursement of $1000 in Year 0, the discounted breakeven point (DBEP) will be reached in 3 years at a compound-interest rate of 30 percent if the annual net profit $A_{NP} = \$550.63$ per year. Thus, a

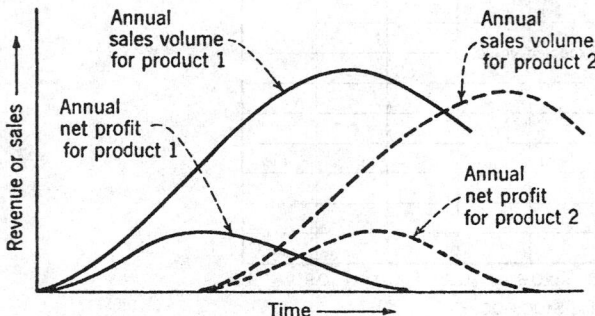

FIG. 9-26 Life cycle of products.

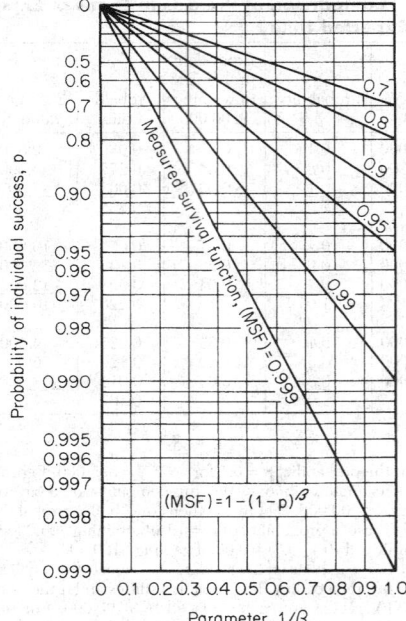

FIG. 9-27 Measured-survival-function plot.

discounted-cash-flow rate of return (DCFRR) of 30 percent corresponds to

$$\frac{1}{\beta} = \frac{\$1000}{(3 \text{ years})(\$550.63/\text{year})} = 0.61$$

For $1/\beta = 0.61$ and an (MSF) = 0.999, the probability of individual success is read from Fig. 9-27 to be $p = 0.985$. Similarly, it can be deduced that if (MSF) = 0.999 and $p = 0.95$, a (DCFRR) of 45 percent is required; if breakeven in 20 years is acceptable, then a (DCFRR) of only 10 percent is needed.

Example 14: Estimation of Probability of a Research and Development Program Breaking Even Details of the estimates for the current research and development program of a company are given in Table 9-13. We shall estimate the probability that this portfolio will at least break even.

The total annual proposed disbursement for R&D is $500,000. The effective total annual income if all projects reached their anticipated income would be $1,300,000. Therefore,

$$\beta = \$1,300,000/\$500,000 = 2.600$$

Project A has an expected value of $(0.95)(\$500,000/\text{year}) = \$475,000/\text{year}$; project B has an expected value of $(0.90)(\$400,000/\text{year}) = \$360,000/\text{year}$; and so on. We sum these values to obtain the total expected value of the portfolio as $1,109,500 per year. Hence,

$$p = (\$1,109,500/\text{year})/(\$1,300,000/\text{year}) = 0.8535$$

TABLE 9-13 Example of a Portfolio of Projects for a Research and Development Program

Project	Proposed disbursement for coming year	Annual aftertax income if successful	Probability of success
A	$125,000	$ 500,000	0.95
B	100,000	400,000	0.90
C	100,000	125,000	0.80
D	80,000	100,000	0.75
E	50,000	60,000	0.70
F	20,000	30,000	0.65
G	20,000	70,000	0.50
H	5,000	15,000	0.20
Totals	$500,000	$1,400,000	

From Fig. 9-27, for a probability of success $p = 0.8535$ and a value $1/\beta = 1/2.600 = 0.3846$, the (MSF) is 99.3 percent. This is the probability that this portfolio will at least break even.

Alternatively, we can substitute the values for p and β into Eq. (9-106) to get

$$(MSF) = 1 - (1 - 0.8535)^{2.6} = 0.9932, \text{ or } 99.32 \text{ percent}$$

The (MSF) and utility curves can be related.

Example 15: Utility-Function Curve Let us sketch a utility-function curve that is equivalent to the following pattern of measured-survival functions (MSF), which expresses the observed strategy of a particular manager when spending an authorized annual budget of $1,000,000:

Case	Potential proceeds annually, $	(MSF), %
a	Above 600,000	99.9
b	300,000–600,000	95.0
c	0–300,000	65.0
d	Losses	75.0

We shall plot the resultant curve on a utility scale of ± 120 utiles against a potential gain of $\pm \$1,000,000$.

The required axes range from $-\$1,000,000$ per year to $+\$1,000,000$ per year, and from -120 utiles to $+120$ utiles. Utiles can be compared by ratio on an absolute scale only. Hence, for purposes of calculation the axes are moved to provide a working range of $0 per year to $2,000,000 per year and 0 to 240 utiles as in Fig. 9-28. On these axes, a potential gain of $600,000 per year corresponds to an absolute amount of $(600,000 + 1,000,000) = \$1,600,000$ per year, and a potential loss of $200,000 per year to an absolute amount of $(-200,000 + 1,000,000) = \$800,000$ per year.

a. For annual proceeds above $600,000 per year, (MSF) is 99.9 percent. If the certainty of an annual gain of $600,000 has to be abandoned in an effort to obtain an annual gain of $1,000,000, then on an absolute scale

$$\beta = (\$2,000,000/\text{year})/(\$1,600,000) = 1.2500$$

With $1/\beta = 0.8000$ and (MSF) $= 99.9$ percent, we find the required probability of success by solving Eq. (9-106) for p:

$$p = 1 - (1 - 0.999)^{0.800} = 0.996$$

The utility of an amount of money is its utility when it is certain to be obtained, multiplied by its probability of being attained. On a scale in which an absolute annual income of $2,000,000 per year has a utility of 240 utiles, the utility of $1,600,000 is $(0.996)(240)$, or 239 utiles.

b. For annual proceeds between $300,000 and $600,000, (MSF) $= 95$ percent. If the certainty of an annual gain of $300,000 has to be abandoned to obtain an annual gain of $600,000, then, as before,

$$1/\beta = \$1,300,000/\text{year}/\$1,600,000/\text{year} = 0.8125$$

$$p = 1 - (1 - 0.95)^{0.8125} = 0.912$$

Since, to this manager, the utility of an absolute income of $1,600,000 is 239 utiles, the value of $1,300,000 is $(0.912)(239) = 218$ utiles. On the original scales, potential annual proceeds of $300,000 have a utility of $(218 - 120)$, or 98 utiles.

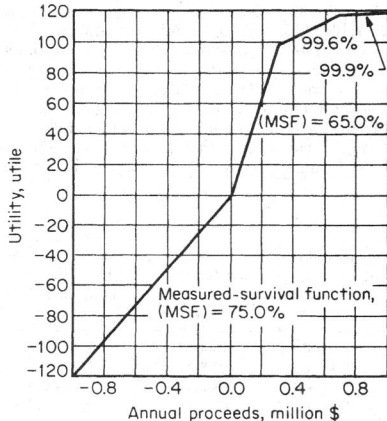

FIG. 9-28 Utility-function plot illustrating managerial strategy.

c and *d.* Values of utility at other potential annual gains are calculated in the same way and shown graphically in Fig. 9-28.

This strategy is extremely conservative when high gains are possible but becomes less so for smaller potential gains. If potential losses are involved, the strategy is a fair one for which (NPV) would be an accurate guide for choosing alternatives.

Insurance and Risk In the venture-premium method of assessment, risky investments are required to yield a rate of return that adds a premium to the cost of finance. D. F. Rudd and C. C. Watson (*The Strategy of Process Engineering*, Wiley, New York, 1968, p. 91) consider this relationship:

$$i_m = i + i_r \tag{9-107}$$

where i is the cost of capital, i_m is the minimum acceptable interest rate of return on the investment, and i_r is known as the risk rate.

They suggested that each project should pay an insurance premium i_r to guarantee the expected profits. The magnitude of i_r is proportional to the amount of capital to be risked. It is also a function of the degree of risk involved. Working capital and capital for auxiliary facilities are assumed to be risk-free. Thus, the risk rate is applied only to the fraction of the capital investment likely to be lost if the project is unexpectedly terminated.

The main objection to the venture-premium method is that the assessment of the riskiness of a project may be too subjective. This could lead to the rejection of potentially attractive proposals and the acceptance of projects that merely appear to be risk-free.

Insurance is protection against risk. Commercial insurance companies minimize their own risks by covering a large number of individuals against a given risk and also by offering coverage on a wide variety of different types of risk. It is frequently quite difficult to assess the probability of success of a particular research and development project. It is much easier for an insurance company to assess its probabilities from its casualty tables.

Businesses tend to provide their own insurance cover when individual claims are likely to be a small fraction of the available capital. The cost of commercial insurance is about 30 percent higher than would be necessary to cover the same risk in one's own company. However, for low-probability, high-cost risks, most businesses prefer to insure with a commercial insurance company. Such risks include loss of plant or buildings due to fire and losses of revenue due to delays in startup or strikes.

It is also becoming necessary to insure against factors not normally considered until recently. These include possible lawsuits for polluting the environment. The cost of insurance increases the annual total expense A_{TE}. Thus, overinsurance can lead to an unnecessary decrease in profitability. The management of a company must ultimately judge its own risks.

As an example, let us calculate the required risk rate for a project that is described by the following: (1) risk strategy is equivalent to an (MSF) of 99 percent, (2) payback of risk capital is 3 years, (3) cost of capital i is 10 percent, and (4) probability of complete success of the project is estimated as 95 percent.

First, we calculate the value of β in Eq. (9-106). For this project, with (MSF) $= 0.99$ and $p = 0.95$,

$$\beta = \frac{\log[1 - (MSF)]}{\log(1 - p)} = \frac{\log(0.01)}{\log(0.05)} = 1.537$$

To recover this amount of capital and interest in 3 years, the average net annual cash flow A_{CF} required is

$$A_{CF} = 1.537/3 = 0.5124 \ (\$/\text{year})/(\$ \text{ invested})$$

In effect, in computing the average net annual cash flow per dollar invested, the value of f_{AP} of Eq. (9-46) has been obtained for this example. From tables of the annuity present-worth factor f_{AP} the value of the interest rate is found to be $i_m = 0.25$ when $f_{AP} = 0.5124$ with $n = 3$ years.

Hence, by substituting appropriate values into Eq. (9-107) and solving for the required risk rate,

$$i_r = i_m - i$$
$$= 25 - 10 = 15 \text{ percent}$$

based on the payback period of the risk capital. All capital C_{WC} is completely recoverable without risk and requires interest only at 10 percent. The unrecovered part of the risk capital C_{FC} attracts the additional risk interest rate of 15 percent, which should be reduced as the risk capital is written down.

A different view of risk is expressed in Eq. (9-108):

$$[1 + (\text{DCFRR})] = (1 + i)(1 + i'_r) \qquad (9\text{-}108)$$

The (DCFRR) represents the return on all capital invested after such capital has been paid back, together with any interest incurred by borrowing it, and after payment of all expenses, including taxes, associated with the project. It thus represents the entrepreneurial return to the company for managing the total capital employed. If the cost of capital i is set at the best risk-free use of that capital, such as the interest rate on a bank deposit or on government bonds, etc., i'_r represents the increased entrepreneurial return on the capital for taking the risks involved. This is a useful concept since the probability of achieving a given (DCFRR), and hence of a particular value of i'_r, may be estimated by the methods detailed previously. We notice that i, as so defined, implies that all taxes and interest have been paid. Thus, $100 deposited in a bank at a rate of 10 percent with half of the money borrowed at 15 percent and corporation tax at 40 percent would result in a risk-free income after tax of

$$[(\$100)(0.10/\text{year}) - (0.5)(\$100)(0.15/\text{year})](1 - 0.40) = \$1.5/\text{year}$$

The same money invested in a project with a (DCFRR) of 10 percent would, by Eq. (9-108), obtain an entrepreneurial return $i'_r = 8.37$ percent on the whole investment, i.e., $8.37/$100. Investment of the entrepreneur's own money would only achieve an aftertax return of $(0.1)(1 - 0.40) = 6$ percent on $50, or $3/$100 of total investment. The incentive to the entrepreneur to manage the project thus corresponds to a tax-free income of $5.37/$100 of total investment. In practice, money is borrowed from more than one source at different interest rates and at different tax liabilities. The effective cost of capital in such cases can be obtained by an extension of the above reasoning and is treated in detail by A. J. Merrett and A. Sykes (*Capital Budgeting and Company Finance*, Longmans, London, 1966, pp. 30–48).

Inflation It is currently necessary to evaluate the profitability of proposed investments whose future earnings are virtually certain to be eroded by inflation. It has been common practice to ignore the effects of inflation. This is done on the reasonable grounds that predicting the market rate of interest, and thus the appropriate discount rate for future cash flows, is difficult enough without having to worry about inflation as well. But failure at least to try to predict inflation rates and take them into account can greatly distort a project's economics, especially at the double-digit rates that have been found throughout the world. It is the common experience that a given amount of money buys less and less of goods and services as time goes by. The problem is to express this experience quantitatively.

Published figures for inflation rates are based on some particular mixture of goods and services that is chosen to represent the material wants of the average citizen. If a given quantity of this specific mixture cost $100 last year and now costs $120, then the mix has suffered a 20 percent rate of inflation. The purchasing power of the currency (i.e., of the $120) in respect of these goods and services has consequently fallen by a factor of ($120 − 100)/$120, or 16.7 percent.

Two kinds of inflation can be considered: general, or open, inflation and repressed, or differential, inflation. In the first case, all costs and

prices increase at a uniform rate. Thus, the same rate of inflation will be calculated regardless of the particular mixture of goods and services chosen. In the second case, the rate of inflation will depend on the spending spectrum of the individual or company. For instance, a given company's labor costs and material costs may inflate at different rates. To quite a large extent, inflation becomes repressed, or differential, in such fields as taxation, import control, and price restriction.

The effect of inflation on the real value of future earnings from a project should not be confused with the effect of the market rates of interest on those earnings. Strictly speaking, the market interest rate and the inflation rate are not fully independent, at least according to some economic theorists. However, they are here treated as being separate. Because of each effect, a dollar of project income next year has a smaller true value than does a dollar in hand today. The interest-rate effect could be offset because a dollar could be financially invested at the prevailing interest rate and the dollar plus interest earnings recouped in a year. By contrast, the inflation effect comes about simply because a dollar can buy more now than a year hence because of an irreversible rise in prices. The distinction is clarified in the following subsections.

Effect of Inflation on (NPV) When computing the (NPV) for a proposed project, error arises if the actual cash flows are simply added together instead of adjusting all the values to their purchasing power in a particular year. The reason lies in the basis of (NPV) calculations. We shall rewrite Eq. (9-57) to give

$$(\text{NPV}) = A_{CF_0} + \sum_1^n \frac{A_{CF_n}}{(1 + i)^n} \qquad (9\text{-}109)$$

Equation (9-109) is valid for the case of no inflation. In the case of general inflation at a fractional rate i_i, this equation can be written in the modified form

$$(\text{NPV}) = A_{CF_0} + \sum_1^n \frac{A_{CF_0}}{(1 + i)^n(1 + i_i)^n} \qquad (9\text{-}110)$$

Equation (9-110) enables all the net annual cash flows to be corrected to their purchasing power in Year 0. If the inflation rate is zero, Eq. (9-110) becomes identical with Eq. (9-109).

The following example illustrates the effect of inflation on (NPV) as well as on the taxes the company pays.

Example 16: Effect of Inflation on Net Present Value Let us consider a simplified project in which $1,100,000 of capital is spent in Year 0, $1,000,000 for fixed-capital items and $100,000 for working capital. The fixed capital is depreciated on a straight-line basis to a book value of zero at the end of Year 5. The annual sales revenue in Years 1 through 5 is $500,000. There is no inflation. The $100,000 of working capital is recovered at the end of Year 5. The taxation rate is 50 percent, and the market interest rate is 10 percent. Table 9-14 lists the cash-flow data for this project, showing that the (NPV) at the end of Year 5 is $99,326 by using Eq. (9-109).

Let us modify this example by assuming that there is a general inflation rate of 20 percent per year and that the project analyst ignores the inflation and (inappropriately) applies Eq. (9-109). The revenue and expense data for this case are shown in Table 9-15, yielding an (NPV) of $431,269. When Eq. (9-109) is (inappropriately) used for the same example with various other rates of inflation, the resulting (NPV)s can be plotted as the upper line in Fig. 9-29.

If the inflation is correctly taken into account by applying Eq. (9-110), the results are strikingly different. By further discounting the discounted cash flows A_{DCF} of Table 9-15 by the f_d factors corresponding to an inflation rate of 20 percent before summing, it can be seen that the project actually incurs a negative (NPV) of $208,733 in uninflated-money terms. The lower line in Fig. 9-29

TABLE 9-14 (NPV) Calculations with No Inflation

Year, n	Net capital expenditure, A_{TC}	Revenue from sales, A_S	Total expenses, A_{TE}	Cash income, $A_{CI} (= A_S - A_{TE})$	Depreciation charge, A_D	Taxable income, $(A_{CI} - A_D)$	Amount of tax at $t = 0.5$, A_{IT} $[= (A_{CI} - A_D)t]$	Net cash flow after tax, A_{CF} $(= A_{CI} - A_{IT} - A_{TC})$	Discount factor at $i = 10\%$, f_d $\left[= \dfrac{1}{(1 + 0.1)^n}\right]$	Discounted net cash flow, A_{DCF} $[= A_{CF}(f_d)]$	Net present value (NPV), $\left(= \sum_0^n A_{DCF}\right)$
0	$1,100,000	0	0	0	0	0	0	−$1,100,000	1.00000	−$1,100,000	−$1,100,000
1	0	$500,000	$100,000	$400,000	$200,000	$200,000	$100,000	300,000	0.90909	272,727	−827,273
2	0	500,000	100,000	400,000	200,000	200,000	100,000	300,000	0.82645	247,935	−579,338
3	0	500,000	100,000	400,000	200,000	200,000	100,000	300,000	0.75131	225,393	−353,945
4	0	500,000	100,000	400,000	200,000	200,000	100,000	300,000	0.68301	204,903	−149,042
5	−100,000	500,000	100,000	400,000	200,000	200,000	100,000	400,000	0.62092	248,368	+99,326

TABLE 9-15   (NPV) Calculations with Inflation Present But Not Allowed For

Year, n	Net capital expenditure, A_{TC}	Revenue from sales, A_S	Total expenses, A_{TE}	Cash income, A_{CI}	Depreciation charge, A_D	Taxable income, $(A_{CI} - A_D)$	Amount of tax at $t =$ 0.5, A_{IT}	Net cash flow, A_{CF}	Discount factor at $i = 10\%$, f_d	Discounted net cash flow, A_{DCF}	Net present value (NPV)
0	$1,100,000	0	0	0	0	0	0	−$1,100,000	1.00000	−$1,100,000	−$1,100,000
1	0	$500,000	$100,000	$400,000	$200,000	$200,000	$100,000	300,000	0.90909	272,727	−827,273
2	0	600,000	120,000	480,000	200,000	280,000	140,000	340,000	0.82645	280,993	−546,280
3	0	720,000	144,000	576,000	200,000	376,000	188,000	388,000	0.75131	291,508	−254,772
4	0	864,000	172,800	691,200	200,000	491,200	245,600	445,600	0.68301	304,349	+49,577
5	−100,000	1,036,800	207,360	829,440	200,000	629,440	314,720	614,720	0.62092	381,692	+431,269

extends the example by assuming other rates of inflation. Figure 9-29 shows that the effect of inflation, if not taken into account, is to make a project seem more profitable than it actually is.

Table 9-15 shows that the total amount of tax actually paid over the 5-year period was $988,320. This becomes $534,272 in uninflated-money terms when the tax for each year is corrected to its purchasing power in Year 0, using f_d factors for the 20 percent inflation rate employed for the example. Calculations for other rates of inflation can also be made, and the results plotted as in Fig. 9-30.

This confirms that although the tax paid will increase with inflation, the gain to the government is more apparent than real. It is interesting to note that although the tax paid corrected to its purchasing power in Year 0 is almost constant irrespective of the inflation rate, it does go through a maximum at an inflation rate of about 17 percent in this example.

Effect of Inflation on (DCFRR)   A net annual cash flow A_{CF} will have a cash value of $A_{CF}(1 + i)$ 1 year later if invested at a fractional interest rate i. If there is inflation at an annual rate i_i, then an effective rate of return or interest rate i_e can be defined by the equation

$$A_{CF}(1 + i_e) = [A_{CF}(1 + i)]/(1 + i_i) \qquad (9\text{-}111)$$

which can be simplified and rewritten to give

$$i_e = i - i_i - i_e i_i \qquad (9\text{-}112)$$

In the context of the discounted-cash-flow rate of return, Eq. (9-112) becomes

$$i_e = (DCFRR) - i_i - i_e i_i \qquad (9\text{-}113)$$

In this equation, (DCFRR) can be viewed as the nominal discounted-cash-flow rate of return uncorrected for inflation and i_e can be thought of as the true or real discounted-cash-flow rate of return.

Instead of using Eq. (9-113), it is unfortunately common practice to try to obtain the true or effective rate of return by calculating the nominal (DCFRR), based on actual net annual cash flows uncorrected for inflation, and then subtracting the inflation rate from it as if

$$i_e = (DCFRR) - i_i \qquad (9\text{-}114)$$

Equation (9-113) shows that Eq. (9-114) is only approximately true and should be used, if at all, solely for low interest rates. Let us consider the case of a nominal (DCFRR) of 5 percent and an inflation rate of 3 percent. Equation (9-14) yields an approximate effective return rate of 2 percent, compared with the real effective rate of 1.94 percent given by Eq. (9-113); i.e., there is an error of 3.1 percent. Now let us consider the case of a nominal (DCFRR) of 25 percent and an inflation rate of 23 percent. Equation (9-114) yields an approximate effective return rate of 2 percent, compared with 1.63 percent from Eq. (9-113); in this case, the error that results is 22.7 percent.

Inflation, (DCFRR), and Payback Period   More insight into the effect of inflation on (DCFRR) calculations can be gained by considering the payback period (PBP), which is defined as the elapsed time necessary for the positive aftertax cash flows from the project to

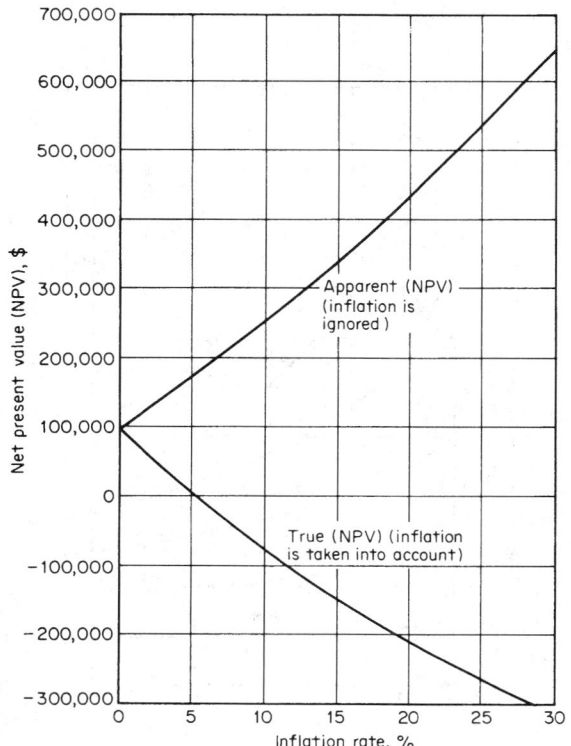

FIG. 9-29   Effect of inflation rate on net present value for a project.

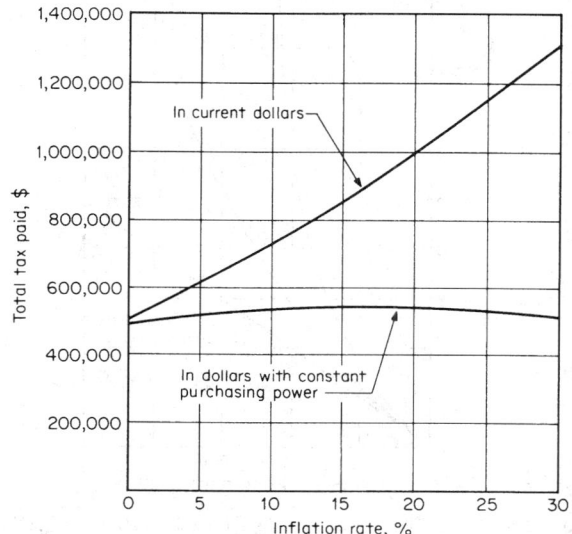

FIG. 9-30   Effect of inflation rate on taxes paid for a project.

recoup the original fixed-capital expenditure. In this definition, the cash flows are not discounted to allow for the market rate of interest or for the inflation rate, so that a project with a given (PBP) could show various values for its (DCFRR) and a given (DCFRR) could pertain to projects with various payback periods.

We shall consider the simple case of (1) a single capital expenditure made immediately before the start of production and (2) equal positive net annual cash flows A_{CF} in all the productive years of the project. For this case, Eq. (9-109) can be rewritten in terms of the payback period and the (DCFRR) as follows:

$$(PBP) = \sum_{1}^{n} \frac{1}{[(1 + (DCFRR)]^n} \qquad (9\text{-}115)$$

The relationship set out in Eq. (9-115) can also be viewed via a different chain of causality with (DCFRR) as a given parameter, (PBP) as the independent variable, and n as the variable whose value is being sought. Such an approach is the basis for the lines in Fig. 9-31, each of which shows the number of years of project life required to achieve an effective interest rate or a (DCFRR) of 20 percent by projects having various payback periods. The three lines differ from each other with respect to the matter of inflation.

If there is no inflation, then the middle line pertains. Because there is no inflation, the nominal (DCFRR) is equal to or identical with i_e, the real discounted-cash-flow rate of return, as can be seen from the relationship expressed in Eq. (9-113).

When inflation does exist, the relevant parameter is i_e, which is different from the nominal (DCFRR). Equation (9-113), manipulated into equivalent form,

$$(DCFRR) = (1 + i_e)(1 + i_i) - 1$$

shows that in order to achieve an i_e of 20 percent when the general inflation rate is likewise 20 percent, a project must generate a nominal (DCFRR) of 44 percent. This is the basis for the uppermost line in Fig. 9-31. Other lines pertaining to other rates of inflation could be plotted in the same way.

Let us assume that 20 percent inflation prevails but that the analyst ignores it and mistakenly takes a (DCFRR) of the project at its nominal value instead of converting it to an i_e. Equation (9-115) rearranged into the form

$$i_e = [1 + (DCFRR)]/(1 + i_i) - 1 \qquad (9\text{-}116)$$

shows that with a nominal (DCFRR) of 20 percent and a general inflation rate of 20 percent, the true or effective rate of interest is zero. This is the basis for the lowest line in Fig. 9-31. Points for lines corresponding to other rates of inflation could be plotted onto that figure. Plots similar to Fig. 9-31 can be drawn for other (DCFRR) values.

Figure 9-31 shows that the elapsed time necessary to reach a nominal (DCFRR) for a given project decreases sharply with inflation. This figure, like Fig. 9-29, shows that the effect of inflation is to make a project seem more profitable than it actually is.

The magnitude of the effect comes through even more clearly in Fig. 9-32, a plot of the time needed to reach a nominal (DCFRR) of 20 percent against the inflation rate for various values of (PBP). This plot also shows that the longer the payback period, the greater the increase in apparent profitability of the project.

The true rates of return i_e can be calculated from Eq. (9-116) to be 20, 9.09, 0, and −7.69 percent respectively for general inflation rates of 0, 10, 20, and 30 percent. Thus, although the time required for a project with a payback period of 4 years to reach a nominal (DCFRR) of 20 percent is reduced from almost 9 years under conditions of no inflation to less than 3½ years for 30 percent inflation, the true rate of return that prevails for the latter condition is −7.69 percent, implying that the project loses money in real terms.

It is interesting to note that, in order to reach a real (DCFRR) or i_e of 20 percent within a reasonable project lifetime when the general inflation rate is 20 percent, it follows from Fig. 9-31 that the payback period for the project must not be much in excess of 2 years.

Although it is difficult to carry out economic-feasibility studies on projects in a time of high inflation, it is important to try to predict inflation rates and allow for them in such studies.

When different people talk about inflation, they often adopt different concepts without realizing it. The area of conceptual uncertainty

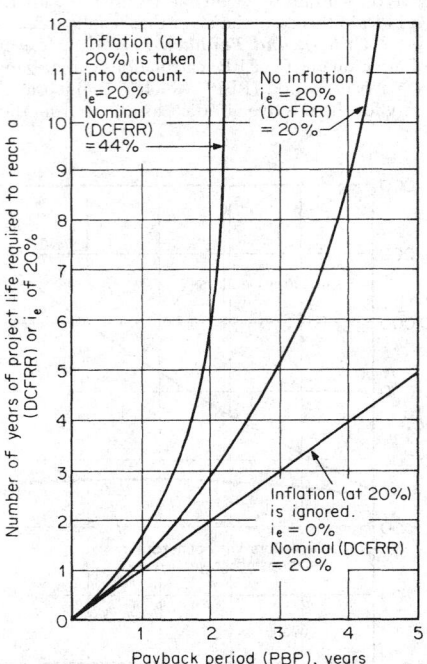

FIG. 9-31 Effect of inflation rate on the relationship between the payback period and the discounted-cash-flow rate of return.

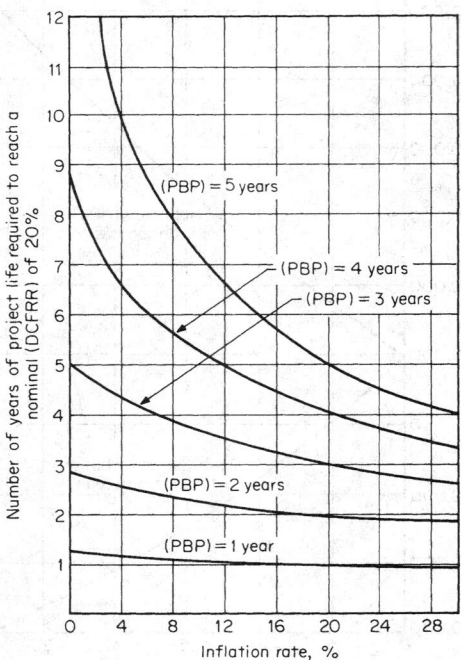

FIG. 9-32 Adverse effect of inflation for higher payback periods.

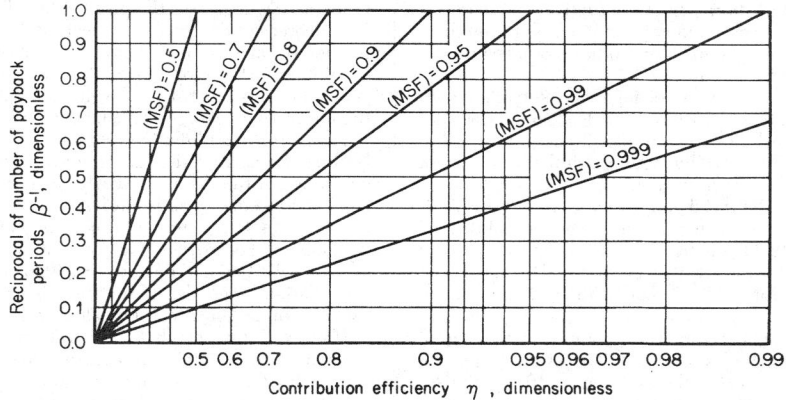

FIG. 9-33 Relationship between measured-survival function, number of payback periods, and contribution efficiency.

can be said to lie somewhere between the upper and lower lines shown on Fig. 9-31 in most cases.

Inflation and the (MSF) By applying the measured-survival-function concept to manufacturing projects rather than to research and development, we can define a modified (MSF) for a given project as

$$(MSF) = 1 - (1 - \eta)^\beta \qquad (9\text{-}117)$$

Here, β is the number of payback periods that have elapsed since the project started to generate positive net annual cash flows A_{CF} up to any given year n since project startup. It is given by

$$\beta = \frac{\sum_{n=0}^{n=n} (A_{CF})_n}{C_{FC} - S} \qquad (9\text{-}118)$$

If all the net annual cash flows in Eq. (9-118) are based on their purchasing power in Year 0, then β is independent of inflation.

As for the contribution efficiency η, it is the ratio of (1) the annual profit that can actually be achieved in a given year for a given sales volume to (2) the profit that could be obtained if no repayment of capital or interest were required and all fixed-expense items were credited free to the project. It is defined by

$$\eta = [R(c_s - c_{VE}) - A_{FE}]/[R(c_s - c_{VE})] \qquad (9\text{-}119)$$

where R is the annual production rate or sales volume in physical units, c_s is the sales price per unit, c_{VE} is the variable production and selling cost per unit, and A_{FE} is the annual fixed cost.

If the project gets a "free ride," i.e., if A_{FE} is zero, then η takes on its maximum possible value of unity. Conversely, if the project and its production rate are only at the breakeven point, then η becomes zero. Therefore, contribution efficiency can be regarded as a measure of the probability of success for the project.

The relationship between the number of payback periods, the contribution efficiency, and the measured-survival function as set out in Eq. (9-117) is plotted in Fig. 9-33.

The contribution efficiency defined by Eq. (9-119) may vary from year to year. In that case, Eq. (9-117) can be written in the modified form

$$(MSF) = 1 - [(1 - \eta_1)(1 - \eta_2) \cdots (1 - \eta_n)]^{\beta/n} \qquad (9\text{-}120)$$

where $\eta_1, \eta_2, \ldots \eta_n$ are the contribution efficiencies in Years 1, 2, ... n respectively.

As in the case of the (MSF) defined by Reul for research and development projects, it is the responsibility of management in a particular manufacturing company to decide on an acceptable level of (MSF) for

manufacturing projects. That decision reflects and helps quantify the company's attitude toward risk.

Thus, (MSF) should in practice be regarded as a given or predetermined variable, and Eq. (9-117) accordingly becomes more useful if it is rearranged. For instance, the values of contribution efficiency for a given value of (MSF) are related to the number of elapsed payback periods by

$$\eta = 1 - [1 - (MSF)]^{1/\beta} \qquad (9\text{-}121)$$

If the acceptable (MSF) is 0.9, this can be satisfied by a project having $\eta = 0.9$ and $\beta = 1$, or a project having $\eta = 0.684$ and $\beta = 2$, and so on. Once Eq. (9-121) has been used to calculate a required contribution efficiency [given the (MSF) and the expected number of payback periods of project life], Eq. (9-119) can be applied to determine the necessary selling price if R, c_{VE}, and A_{FE} are known. Similarly, Eq. (9-119) can be used to find the required production rate if c_S is known.

It is also possible to combine (MSF) considerations with evaluation of the true discounted-cash-flow rate of return (DCFRR) by using Eq. (9-62). The relationship of Eq. (9-59) is independent of inflation if all money values are based on those prevailing in the startup year. For this case, Fig. 9-34 shows the true (DCFRR) reached in a given time, expressed as the number of elapsed payback periods β for various values of the payback period.

Let us consider a project having a contribution efficiency of 0.684 and a payback period of 3 years. Figure 9-33 shows that when two payback periods have elapsed, a measured-survival function of 0.9 has

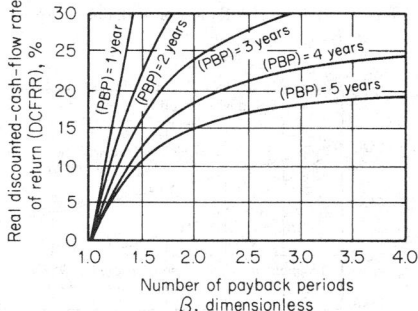

FIG. 9-34 Real discounted-cash-flow rate of return against number of payback periods for various payback periods.

been attained. In addition, Fig. 9-34 shows that the discounted-cash-flow rate of return reached at that time is 24 percent.

Effects of Differential Inflation   Inflation can be general or differential. In the first case, all costs and prices increase at a uniform rate. In the second, government controls and other factors cause the various costs and prices to inflate at different rates.

The onset of general inflation does not change the value of the contribution efficiency η, as can be seen from Eq. (9-119), and it does not affect the value of β if the cash flows in Eq. (9-118) are converted to their purchasing power in Year 0. Thus, general inflation does not cause the measured-survival function to change.

Differential inflation, on the other hand, can affect the measured-survival function. We shall assume, for instance, that the sales price per unit product c_s in Eq. (9-119) is frozen at a constant level while some or all of the production costs are allowed to rise. This causes the value of η to decrease; therefore, (MSF) likewise decreases, as can be seen from Eq. (9-117).

Let us consider the effect of differential inflation on the overall profitability of the project of the last example. The effect of general inflation on this project showed that the apparent profitability rises sharply, to an (NPV) of $431,269 at a general inflation rate of 20 percent. However, when the cash flows of the (NPV) are properly corrected to their purchasing power in Year 0, the (NPV) instead becomes $208,733.

The effect of differential inflation on this project emerges in Fig. 9-35, with all (NPV)s corrected to their purchasing power in Year 0. The top line shows (NPV) for various rates of general inflation. The bottom line shows (NPV) for the differential-inflation case in which only the costs are allowed to increase while product selling price and thus cash income remain constant from year to year. The middle line shows the effect of general inflation when the price rises are delayed by 1 year. The figure confirms that both of these situations take away from the attractiveness of the project.

The effect upon total taxes paid, when they are corrected to their purchasing power in Year 0, is shown in Fig. 9-36. Differential inflation not only decreases the profitability of the project to its owner but also decreases the revenue received by the taxing authority. The method of calculation is identical to that of the earlier example.

Another instance of differential inflation occurs when the prices of goods and services rise uniformly but the cost of borrowing money, the interest rate charged on a loan, does not rise.

If the fractional inflation rate is i_i, a fractional interest rate i_L on a loan can be corrected to an effective rate of interest by Eq. (9-116) with i_L substituted for (DCFRR). The effect of various amounts of loan, borrowed at various interest rates i_L, on the net present value of a particular, fairly simple project is shown in Fig. 9-37. Thus, if $25,000 were borrowed at an interest rate of 15 percent for the project, the (NPV) would be about $43,000 at a zero inflation rate. But if the inflation for goods and services i_i is 10 percent, the effective interest rate for that loan can be calculated from Eq. (9-116) to be only 4.55 percent. It is seen from Fig. 9-37 that this increases the (NPV) of the project to $48,000. This confirms the economic advantage of borrowing at a fixed interest rate in a time of general inflation.

A topical aspect of differential inflation is the question of energy costs. Will the cost of a particular fuel rise or fall in relation to prices in general, and if so, what effect will this have on the economics of a project?

Example 17: Effect of Fuel Cost on Project Economics   A process unit is heated by gas. We assume that $100 spent on energy-conservation measures for this particular unit at the end of 1980 would save 200 therms (21.1 GJ) of gas energy in each subsequent year. If the cost of gas in 1980 is $x per therm, the annual dollar savings at 1980 prices are $200x. The (NPV) at the end of year n for this project is

$$(NPV) = -100 + \sum_{1}^{n} \frac{(200)x}{(1+i)^n}$$

if the appropriate discount factor is i.

This is independent of inflation provided that the cost of gas rises in line with any general rate of inflation. However, if the real cost of gas rises at a fractional annual rate r over and above the general inflation rate, it should be modified into the form

$$(NPV) = -100 + \sum_{1}^{n} \frac{(200x)(1+r)^n}{(1+i)^n}$$

This equation confirms that as the gas price rises because of inflation, the attractiveness of the conservation project also rises.

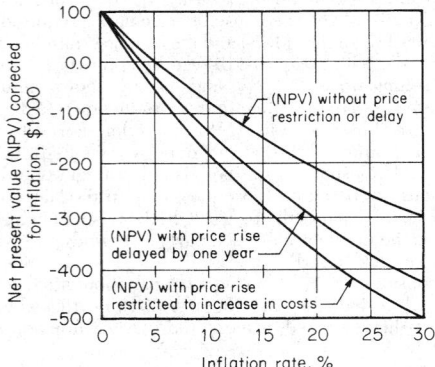

FIG. 9-35   Effect of differential inflation on inflation-corrected net present value.

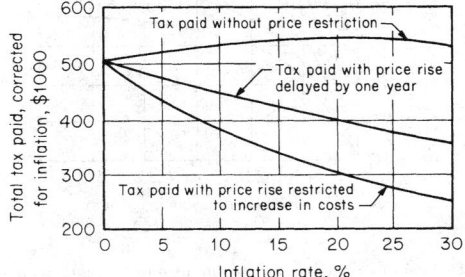

FIG. 9-36   Effect of differential inflation on inflation-corrected tax revenue.

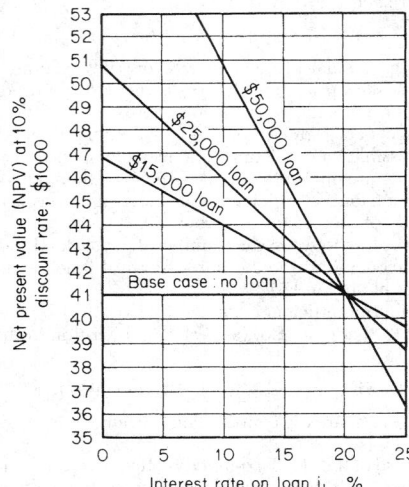

FIG. 9-37   Effect of loan interest rate on the net present value of a project.

ACCOUNTING AND COST CONTROL

Principles of Accounting Accounting is the art of recording business transactions in a systematic manner. Financial statements are both the basis for and the result of management decisions. Such statements can tell managers or engineers a great deal about their company, provided that they can interpret the information correctly.

Since a fair allocation of costs requires considerable technical knowledge of operations in the chemical-process industries, a close liaison between the senior process engineers and the accountants in a company is desirable. Indeed, the success of a company depends on a combination of financial, technical, and managerial skills.

Accounting is also the language of business, and the different departments of management use it to communicate within a broad context of financial and cost terms. Engineers involved in feasibility studies and detailed process evaluations are dependent for financial information on the company accountants, especially for information on the way in which the company intends to allocate its overhead costs. It is vital that engineers correctly interpret such information and that they can, if necessary, make the accountants understand the effect of the chosen method of allocation.

The method of allocating overheads can seriously affect the assigned costs of a project and hence the apparent cash flows for that project. Since these cash flows are used to assess profitability by the net-present-value (NPV) and discounted-cash-flow-rate-of-return (DCFRR) methods, unfair allocation of overhead costs can result in a wrong choice between alternative projects.

In addition to understanding the principles of accountancy and obtaining a working knowledge of its practical techniques, engineers should be aware of possible inaccuracies of accounting information in the same way that they allow for errors in any technical data.

At first acquaintance, the language of accountancy appears illogical to most engineers. Although accountants normally express themselves in tabular form, the basis of all their practice can be simply expressed by

$$\text{Capital} = \text{assets} - \text{liabilities} \qquad (9\text{-}122)$$

Equation (9-122) can alternatively be written as

$$\text{Assets} = \text{capital} + \text{liabilities} \qquad (9\text{-}123)$$

Capital, often referred to as net worth, is the money value of the business, since assets are the money values of things the business owns while liabilities are the money values of the things the business owes.

Most engineers have great difficulty in thinking of capital (also known as ownership) as a liability. This is easily overcome once it is realized that a business is a legal entity in its own right, owing money to the individuals who own it. This realization is absolutely essential when considering large companies with stockholders and is used for consistency even for sole ownerships and partnerships. If an individual puts up $10,000 capital to start a business, then that business has a liability to repay $10,000 to the individual.

It is even more difficult to think of profit as being a liability. Profit is the increase in money value available for distribution to the owners and effectively represents the interest obtained on the capital. If the profit is not distributed, it represents an increase in capital by the normal concept of compound interest. Thus, if the individual's business makes a profit of $5000, the liability to the individual is increased to $15,000. With this concept in mind, Eq. (9-123) can be expanded to

$$\text{Assets} = \text{capital} + \text{liabilities} + \text{profit} \qquad (9\text{-}124)$$

where the capital is considered as the cash investment in the business and is distinguished from the resultant profit in the same way that principal and interest are separated.

Profit (as referred to above) is the difference between the total cash revenue from sales and the total of all costs and other expenses incurred in making those sales. With this definition, Eq. (9-124) can be further expanded to

$$\text{Assets} + \text{expenses} = \text{capital} + \text{liabilities} + \text{revenue from sales} \qquad (9\text{-}125)$$

Engineers usually have the greatest difficulty in regarding an expense as being equivalent to an asset, as is implied by Eq. (9-125). Let us consider a one-person business. We assume for a given period

a profit of $5000 and total expenses excluding the individual's earnings of $8000. Also we assume that the individual's labor to the business in this period is worth $12,000. The revenue required from sales would be $25,000. Effectively, the individual has made a personal income of $17,000 in the period but has apportioned it to the business as $12,000 expense for the individual's labor and $5000 return on capital. In larger businesses, there will also be those who receive salaries but do not hold stock and, therefore, receive no profits and stockholders who receive profits but no salaries. Thus, the difference between expenses and profits is very practical.

The period covered by the published accounts of a company is usually 1 year, but the details from which these accounts are compiled are entered daily in a journal. The journal is a chronological listing of every transaction of the business, with details of the corresponding income or expenditure. For the smallest businesses, this may provide sufficient documentation, but in most cases the unsystematic nature of the journal can lead to computational errors. Therefore, the usual practice is to keep accounts that are listings of transactions related to a specific topic such as "purchase-of-oil account." This account would list the cost of each purchase of oil, together with the date of purchase, as extracted from the journal.

Principles of Double-Entry Accounting Many of the accounts involve both income and expenditure. The general practice is to keep accounts by the double-entry system, which may be summarized by

$$\text{Debits} = \text{credits} \qquad (9\text{-}126)$$

The principle of double entry dates from the fifteenth century and is based on the premise that every transaction involves a giver and a receiver of value. Double entry requires that each transaction be entered into two accounts, the convention being that the account of the giver is credited and the account of the receiver is debited with the same amount of money, as noted in the journal. For convenience, each account is divided centrally, and the debit items are entered on the right-hand side. It is also usual to provide a cross-reference to the journal entry so that errors and omissions can be checked.

Let us consider the purchase of $50,000 worth of plant equipment by company A, paid for by check. The accounting entries are: debit the plant-equipment account $50,000, and credit the bank account $50,000. The plant-equipment account is then said to have a debit balance of $50,000, and the bank account a credit balance of $50,000, if these happen to be the only entries.

If company A then sells $100,000 worth of product that is paid for by check, the accounting entries are: credit the sales account $100,000, and debit the bank account $100,000. The bank account will now have a debit balance of ($100,000 − $50,000) = $50,000, and the sales account a credit balance of $100,000, if this happens to be the only sale to date in the accounting period.

In principle, the debiting and crediting of accounts are relatively straightforward. However, a great deal of practice is essential in order to achieve proficiency. Although it is not at all necessary for engineers to compete with professional accountants in this field, engineers should appreciate what accountants do and why they do it.

Of the accounts considered in the preceding illustrations, the plant-equipment and bank accounts are asset accounts, and the sales account is a liability account. To increase an asset, debit the asset account; to increase a liability, credit the liability account. Conversely, to decrease an asset, credit the asset account; to decrease a liability, debit the liability account.

Closing the Books At the end of the accounting period, the individual accounts are closed by balancing each in accordance with Eq. (9-126). The balances are transferred either to the balance sheet in the case of capital expenditure or to the income statement in the case of revenue expenditure. An alternative name for the balance sheet is the position statement; the income statement is also called the trading and profit-and-loss account.

The purpose of capital expenditure, such as the purchase of a piece of plant equipment for $50,000, is to earn future revenue. In contrast, the purpose of revenue expenditure is to maintain existing business.

TABLE 9-16   Income Statement for ABC Company

Revenue		
Sales revenue	$1,900,000	
Other revenue	100,000	
		$2,000,000
Expenses		
Raw materials	953,000	
Wages	185,000	
Utilities	44,000	
Depreciation	68,000	
Other expenses	376,000	
Income taxes	194,000	
		1,820,000
Net profit (after tax)		$ 180,000

Revenue expenditure includes the direct material costs and direct labor costs incurred in the manufacture of a product, together with the associated overheads that include maintenance of the plant. Since these expenses are debits, the debit balance for a given accounting period is obtained by adding up the debit balances from each individual expenditure account. Similarly, since revenues from sales and other income are credits, the credit balance for a given accounting period is obtained by adding up the credit balances from each individual income or revenue account.

To ascertain profit or loss (calculated as income minus expenditure for a given accounting period), income and expenditure must be matched. For example, any rent paid in advance beyond the current accounting period should not be included in the profit or loss calculation. Similarly, goods sold but not yet paid for in a given accounting period should not be included in the revenue total for that period.

An income statement such as the one shown in Table 9-16 is used to obtain the profit or loss for a given period. The debit and credit balances of all the accounts that do not represent expenditure or income for a given accounting period are entered as assets and liabilities in a balance sheet such as that shown in Table 9-17.

There is no rigid format for either the income statement or the balance sheet. Tables 9-16 and 9-17 show common layouts for the income statement and balance sheet respectively, but these are not the only forms. For example, vertical balance sheets, with the assets listed above the liabilities and equity, are also popular.

Some expenditures are partly capital and partly revenue. For example, repair and improvement work may be done on a plant simultaneously. In this case, the repair work should be classified as revenue expenditure and the plant-improvement work as capital expenditure.

Accounting Concepts and Conventions   Accounting is based on the following concepts: (1) money measurement, (2) business entity, (3) going concern, (4) cost, and (5) matching.

Concept 1.   "Money measurement" means that only those facts that can be represented in monetary terms are recorded. The balance sheet and income statement for a company give no indication as to what might happen in the future. The company may be about to be successfully sued for a large sum of money, or a competitor may be launching a new product that will seriously reduce future sales of the company's products.

Concept 2.   "Business entity" means that accounts are kept for the company quite independently of the people who may own the company. For example, if an individual puts an additional $10,000 into a one-person business, the accounts show that the business is $10,000 richer. They do not show that the individual's personal wealth has been depleted by $10,000.

Concept 3.   "Going concern" means that the accounting is based on the premise that the business will continue indefinitely. It is most unlikely that the values of the assets shown in the balance sheet are what the assets would realize if sold. No attempt is made in normal accounting to measure the value of the business to a potential buyer.

Concept 4.   "Cost" means that the assets are normally shown in the balance sheet at cost price together with their subsequent depreciation. Some assets such as land may be considerably more valuable than when originally purchased, but no indication of this is given in the balance sheet. However, some governments now require a note giving the current estimated value of the land.

Concept 5.   "Matching" means that the revenue in a given accounting period should correspond to the expenses for that accounting period.

Accounting is also based on the following conventions: (1) materiality, (2) conservatism, or prudence, and (3) consistency.

Materiality deals with determining whether certain expenditures will have a significant effect on a company's accounting procedures. This is a matter of judgment that is to be made by each company. Obviously, the purchase of a vehicle is a material item, but writing paper or tools for maintenance are less obvious. Although such items may last well beyond the current accounting period, it may not be worth the accounting effort to treat them as material items. Some companies will treat a particular item as capital; other companies, as expenditure. Clearly, the purchase of a piece of equipment costing, say, $1000, will be regarded as less material by a giant company than by a small one.

Conservatism, or prudence, means monetary values that tend to understate rather than overstate the profit are taken.

Consistency means that accounting items are normally treated in

TABLE 9-17   Balance Sheet for XYZ Company

Assets (thousands of dollars)		Liabilities and stockholders' equity (thousands of dollars)	
Current assets		**Current liabilities**	
Cash	$ 38,893	Notes payable	$ 34,507
Notes and accounts receivable	110,740	Accounts payable and accrued liabilities	106,433
Inventories:		Accrued taxes	7,264
Finished products	17,396	Total current liabilities	148,204
Work in process	56,690		
Raw materials and supplies (at cost)	35,790	**Long-term liabilities**	67,677
Total inventories	109,876	Deferred income taxes	13,225
Total current assets	259,509	Other deferred credits	2,307
Investments and long-term receivables (at cost)	94,009	**Stockholders' equity**	
		Common stock, $20 par value	
Property, plant, and equipment (at cost)		Shares authorized, 7,750,000	
Land	6,110	Shares issued, 4,794,450	95,889
Buildings	63,848	Capital in excess of par value of common stock	31,798
Machinery and equipment	106,185	Retained earnings	101,492
	176,143	Total stockholders' equity	229,179
Less accumulated depreciation	75,163		
Net property, plant and equipment	100,980		
Prepaid and deferred, charges	6,094		
Total assets	$460,592	Total liabilities and stockholders' equity	$460,592

the same way over an indefinite number of years. For example, an individual item would not be treated as an expenditure during one year and as a capital item during the next year without good reason being given.

Balance Sheet   The balance sheet, also called the position statement, presents an accounting view of the financial status of a company at a particular point in time. A typical balance sheet is shown in Table 9-17. Although a balance sheet has two sides that balance, it is not part of the double-entry system. In fact, it is not an account but rather a statement listing all the assets of a company and the various claims against these assets on the last day of the accounting period. The assets must be equal to the claims against them at all times. Those who have claims against the assets are the owners (stockholders in a business corporation) and the people to whom the company owes money. In the case of the latter, the company is said to have liabilities to its creditors. The total claim against the assets is often labeled "liabilities and owners' equity."

Assets are classified as current or fixed, and liabilities as current or long-term. Fixed assets are material items that have a relatively long life and normally include land, buildings, plant, vehicles, etc. They are held for the specific purpose of earning revenue and are not for sale in the normal course of business. Current assets include cash and those items that can be fairly easily converted into cash, such as raw-materials inventories, etc. In contrast to fixed assets, current assets are acquired for the specific purpose of conversion into cash in the normal course of business. However, what is regarded as a fixed asset by one type of company might be regarded as a current asset by another. For example, a chemical company would normally classify its vehicles as a fixed asset. However, a company whose primary business was to sell vehicles would classify them as a current asset.

Similarly, the distinction between current and long-term liabilities is also not clear-cut. Current liabilities include accounts payable (money owed to creditors), taxes payable, dividends payable, etc., if due within a year. Long-term liabilities include deferred income taxes, bonds, notes, etc., that do not have to be paid within a year. The owners' equity includes the par, or face, value of the capital received from stockholders and any retained earnings. The balance sheet shows only the nominal value and not the current or real value of this capital.

A balance sheet includes items that are not regarded as assets or liabilities in normal language, such as expenditures carried forward and accumulated profits.

Accountants regard assets as resources that have not yet been used up. Assets are normally shown on the balance sheet at cost minus accumulated depreciation. In this sense, the depreciation charge for an accounting period is the means of converting a part of an asset into a current expenditure that is then listed as an expense in the income statement.

Let us consider plant equipment costing $1 million and purchased on Jan. 1, 1988. Table 9-18 shows the provision for the depreciation account for 1988, 1989, and 1990 for straight-line depreciation, assuming a service life of 10 years and zero scrap value. The credit entries of $100,000 for the depreciation in each year are balanced by the depreciation charge of $100,000 debited to the income statement (or trading and profit-and-loss account) in each year. Table 9-19 shows the corre-

TABLE 9-18   Provision for Depreciation of Plant-Equipment Account

1988			
		Jan. 1 Balance brought down	0
Dec. 31 Balance carried down	$100,000	Dec. 31 Debited to income statement	$100,000
1989			
		Jan. 1 Balance brought down	$100,000
Dec. 31 Balance carried down	$200,000	Dec. 31 Debited to income statement	100,000
	$200,000		$200,000
1990			
		Jan. 1 Balance brought down	$200,000
Dec. 31 Balance carried down	$300,000	Dec. 31 Debited to income statement	100,000
	$300,000		$300,000

TABLE 9-19   Balance-Sheet Entries

As of Dec. 31, 1988		
Plant equipment at cost	$1,000,000	
Less depreciation to date	100,000	
		$900,000
As of Dec. 31, 1989		
Plant equipment at cost	$1,000,000	
Less depreciation to date	200,000	
		$800,000
As of Dec. 31, 1990		
Plant equipment at cost	$1,000,000	
Less depreciation to date	300,000	
		$700,000

sponding entries in the balance sheets for the years 1988, 1989, and 1990. Entries for subsequent years are made in the same way.

A balance sheet is true only for one particular point in time; it tells nothing about the trends in a company. However, by comparing balance sheets for successive years, management can follow changes in the various items. If the observed trend is undesirable, management can take corrective action. Since the accounting period of 1 year is long for most businesses, it is usual to draw up balance sheets at more frequent intervals for control purposes. These may be less formal than those issued annually to the stockholders. In general, balance sheets are less useful to management than are income statements.

Income Statement   Income statements range from the very simple presentation shown in Table 9-16 to the more informative and more complex presentation shown in Table 9-20. The income statement shows the revenue and the corresponding expenses that were incurred to earn that revenue over a period of time. It is the most obvious measure of the efficiency of a business. Although published income statements are normally for 1-year periods, many companies use monthly income statements for internal purposes.

Income statements are very useful tools to assist management in controlling a business and planning for the future. Since management needs to follow the trends of the normal expenses, extraordinary expenses such as those incurred as a result of a major fire or flood should be shown separately.

If revenue and expenses are not properly matched, an understatement or an overstatement of profit may occur. If raw materials were previously purchased at a lower cost than their current cost, profit will be overstated. Any overstatement of profit will mean that more tax will be paid.

One of the most important items in an income statement is depreciation expense. Although depreciation should not be thought of as a means to build up a fund to replace plant, it nevertheless does enable money to be retained in the business by reducing the profit available for distribution to stockholders. It is of course a duty of both accountants and management to see that sufficient money is retained in the business to replace assets and to invest such money in other processes or outside investment.

A further duty of accountants and management is to ensure that the company always has sufficient working capital to enable it to carry on its business.

Types of Accountancy   The traditional work of accountants has been to prepare balance sheets and income statements. Nowadays, accountants are becoming increasingly concerned with forward planning. Modern accountancy can roughly be divided into two branches, financial accountancy and management or cost accountancy.

Financial accountancy is concerned with stewardship. This involves the preparation of balance sheets and income statements that represent the interest of stockholders and are consistent with existing legal requirements. Taxation is an important element of financial accounting.

Management accounting is concerned with decision making and control. This is the branch of accountancy closest to the interest of most process engineers. Management accounting is concerned with standard costing, budgetary control, and investment decisions.

Accounting statements present only facts that can be expressed in financial terms. They do not indicate whether a company is developing new products that will ensure a sound business future. A company

TABLE 9-20 Income Statement for a Mature Year for a New Chemical Product, Produced at 10 Million lb/Year

			Unit values, cents/lb	% sales revenue, %
Revenue from annual sales A_S		$2,000,000	20.00	100.0
Direct manufacturing expense A_{DME}				
Raw materials	$ 884,000		8.84	44.2
Catalysts and solvents	69,000		0.69	3.4
Operating labor	102,000		1.02	5.1
Operating supervision	20,000		0.20	1.0
Utilities	22,000		0.22	1.1
Operating maintenance	21,000		0.21	1.1
Operating supplies	4,000		0.04	0.2
Royalties and patents	10,000		0.10	0.5
Total A_{DME}	$1,132,000	$1,132,000	11.32	56.6
Indirect manufacturing expense A_{IME}				
Payroll overhead	28,000		0.28	1.4
Central laboratory	10,000		0.10	0.5
General plant overhead	52,000		0.52	2.6
Packaging and storage	22,000		0.22	1.1
Property taxes	14,000		0.14	0.7
Insurance	6,000		0.06	0.3
Total A_{IME}	$132,000	$132,000	1.32	6.6
Total manufacturing expense (excluding depreciation) A_{ME}		$1,264,000	12.64	63.2
Depreciation A_{BD}		68,000	0.68	3.4
Other expenses				
Administration	74,000		0.74	3.7
Sales and shipping	124,000		1.24	6.2
Advertising and marketing	40,000		0.40	2.0
Technical service	10,000		0.10	0.5
Research and development	60,000		0.60	3.0
Total other expenses	308,000	308,000	3.08	15.4
Total expense A_{TE}		$1,640,000	16.40	82.0
Net annual profit A_{NP}		$360,000	3.60	18.0

may have impressive current financial statements and yet be heading for bankruptcy in a few years' time if provision is not being made for the introduction of sufficient new products or services.

Financing Assets by Equity and Debt

Financial Ratios Probably the most commonly mentioned ratio is the profit margin (PM), defined as

$$(PM) = \frac{\text{net annual profit}}{\text{revenue from annual sales}} 100 \qquad (9\text{-}127)$$

Another common ratio is the return on investment (ROI), defined as

$$(ROI) = \frac{\text{net annual profit}}{\text{investment}} 100 \qquad (9\text{-}128)$$

In both Eq. (9-127) and Eq. (9-128), the net annual profit can be either before or after tax. It can also include interest and dividends receivable, etc.

Obviously, the net annual profit must be clearly defined before comparisons are made with other companies. Similarly the term "investment" in Eq. (9-128) can have a variety of meanings. The two most common ones (used when assessing the profitability of companies as opposed to projects) are total assets and owners' equity or capital employed. In the first case, Eq. (9-128) can be written as

$$(ROA) = \frac{\text{net annual profit}}{\text{total assets}} 100 \qquad (9\text{-}129)$$

where (ROA) is called the return on assets. In the second case, Eq. (9-128) can be written as

$$(ROE) = \frac{\text{net annual profit}}{\text{stockholders' equity}} 100 \qquad (9\text{-}130)$$

where (ROE) is the return on equity.

Asset-turnover ratio (ATR) is a commonly used measure of company performance, defined as

$$(ATR) = \frac{\text{revenue from annual sales}}{\text{total assets}} 100 \qquad (9\text{-}131)$$

A comparison between Eqs. (9-127), (9-129), and (9-131) shows that

$$(ROA) = (ATR)(PM) \qquad (9\text{-}132)$$

Thus (ROA) can be improved by increasing either (ATR) or (PM). A variation of Eq. (9-131) is the fixed-asset turnover ratio (FATR), defined as

$$(FATR) = \frac{\text{revenue from annual sales}}{\text{fixed assets}} 100 \qquad (9\text{-}133)$$

Clearly, (FATR) is of less value than (ATR) when applied to companies that use relatively large amounts of working capital. The (FATR) is the inverse of the capital ratio (CR) for single projects. (CR) is defined as

$$(CR) = C_{FC}/A_S \qquad (9\text{-}134)$$

where C_{FC} is the fixed-capital cost for a green-fields (grass-roots) site and A_S is the revenue from annual sales.

The fixed assets in Eq. (9-133) and those included in the total assets in Eqs. (9-129) and (9-133) are usually taken at their written-down, or book, value, which may differ significantly from their market value. This is one disadvantage in using Eqs. (9-129), (9-131), and (9-133).

The revenue from annual sales referred to in Eqs. (9-127), (9-131), and (9-132) is normally taken to be the gross turnover, which includes intergroup sales. However, intergroup sales are eliminated in consolidated or group accounts. Again, revenue from annual sales must be clearly defined before comparisons are made with other companies.

Let us consider the simplified balance-sheet or position statement shown in Table 9-21. Essentially, total assets are related to liabilities and stockholders' equity by

$$\text{Total assets} = \text{stockholders' equity} + \text{total debt} \qquad (9\text{-}135)$$

Equation (9-135) can also be written as

$$\text{Stockholders' equity} = \text{total assets} - \text{total debt} \qquad (9\text{-}136)$$

Equations (9-130) and (9-136) can be combined to give

$$(ROE) = \frac{\text{net annual profit}}{\text{total assets} - \text{total debt}} 100 \qquad (9\text{-}137)$$

Equation (9-137) can also be written to include a quantity called the debt ratio (DR), which gives

$$(ROE) = \frac{\text{net annual profit}}{\text{total assets}} \left[\frac{100}{1 - (DR)} \right] \qquad (9\text{-}138)$$

TABLE 9-21 Simplified Balance Sheets for Companies X and Y

X Company balance sheet

	Total debt	0
	Stockholders' equity	$100,000
Total assets $100,000	Total liabilities and stockholders' equity	$100,000

Y Company balance sheet

	Total debt	$ 50,000
	Stockholders' equity	$ 50,000
Total assets $100,000	Total liabilities and stockholders' equity	$100,000

where (DR) is the debt ratio as given by

$$(DR) = \frac{total\ debt}{total\ assets} \qquad (9\text{-}139)$$

Return on assets (ROA) can be related to the return on equity (ROE) by combining Eqs. (9-129) and (9-138):

$$(ROA) = (ROE)/[1 - (DR)] \qquad (9\text{-}140)$$

(ROE) can also be related to the asset-turnover ratio (ATR) and the profit margin (PM) by combining Eqs. (9-132) and (9-140):

$$(ROE) = [(ATR)(PM)]/[1 - (DR)] \qquad (9\text{-}141)$$

Financing by Debt, or Leverage   The debt ratio (DR) is also known as the leverage, or gearing, ratio. Highly levered companies have a high proportion of debt to total assets. At first glance, it may appear that the use of leverage is a simple way of increasing the return on equity (ROE). However, interest charges have to be paid on the debt. Whether leverage is a good thing or not will depend on exactly what the interest charges are in relation to the return on assets and the return on equity.

Let us consider the simplified balance sheets of two companies, X and Y, shown in Table 9-21. Companies X and Y have a debt, or leverage, ratio of zero and 0.5 respectively. Let us assume that the debt is of the debenture type for tax purposes and that the interest rate is 10 percent per annum. The return on equity (ROE) after tax is given in Table 9-22 for companies X and Y for various values of net annual profit A_{NP} before tax. A_{NNP} is the net annual profit after tax. The data for Table 9-21 are plotted in Fig. 9-38. This figure shows that leverage has no effect on the (ROE) when the interest rate charged for the borrowed money is equal to the return on assets (ROA) before tax. Leverage provides increased (ROE) values when the (ROA) is greater than the interest rate charged for the borrowed money and decreased (ROE) values when it is less.

The greater the debt, or leverage, ratio (DR), the more sensitive the (ROE) is to a change in (ROA) and the steeper the slope of the line in Fig. 9-38. Dividends to stockholders are paid out of the net annual profit after tax A_{NNP}, from which the (ROE) after tax in Fig. 9-38 is calculated. Thus, the higher the leverage, the greater the financial risk to the stockholder. This risk is not the same as the business risk of the company, which is a function of its overall prospects in its particular industry. Leverage increases the return to the stockholders when the (ROA) is higher than the interest rate on debt and decreases the return when the (ROA) is lower than the interest rate.

Whether the assets of a company are financed largely by stockholders' equity (also called net worth), or largely by debt, or by some combination of the two depends on a number of factors. If sales do not fluctuate, a company is in a good position to pay the fixed interest charges on debt. This is also the case if the revenue from sales is steadily increasing. In this case, any new common stock issued by the company is likely to command a good price, and it also increases the attractiveness of equity financing.

The attitude of management is also an important factor in deter-

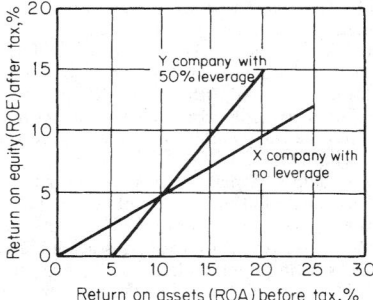

FIG. 9-38   Effect of leverage on the return on equity.

mining how much debt financing is used. In a small firm in which management owns most of the equity, management may be very reluctant to issue further amounts of common stock that would lead to a dilution of its control. Furthermore, if management has great confidence in future prospects, it will wish to ensure the maximum return for itself. In contrast, the equity in a large company is widely distributed, and the issue of further amounts of common stock has little effect on the control of the company.

The difference between equity financing and debt financing is not always clear-cut. For example, preferred stock can be classified as stockholders' equity or debt, depending on who is doing the financial analysis.

Equity Financing   Typically, the company balance sheet will show the stockholders' equity and list the preferred stock, common stock, and retained earnings as in Table 9-23.

The issue of common stock is the basic method of financing a company. Common stockholders take the ultimate risk in a business because they have no right to a return on their investment. However, they have the right to elect the directors of the company, who in turn are responsible for the management of the business. Stockholders are likely to vote the board of directors out if adequate dividends are not paid. Usually the liability of stockholders is limited to the nominal, or par, value of their stock, and hence they can lose only what they have already paid for the stock. If the liability is not limited by law, the personal assets of the stockholders are at risk in the event of company bankruptcy, in proportion to the amount of stock held.

Preferred stock is often used as an alternative to debt when companies do not wish to issue additional common stock or to incur the fixed interest charges required to finance debt. Preferred stockholders are not normally allowed to vote for the board of directors. They have the right to receive fixed amounts of dividends before common stockholders are paid any dividends. However, a company does not have to pay dividends. The board of directors may decide to pay small or no dividends in a particular year. Holders of cumulative preferred stock

TABLE 9-22   Return on Equity after Tax for Companies X and Y

(ROA) before tax	5%	10%	15%	20%
		X company		
A_{NP}	$5,000	$10,000	$15,000	$20,000
Less tax at 50%	($2,500)	($ 5,000)	($ 7,500)	($10,000)
A_{NNP}	$2,500	$ 5,000	$ 7,500	$10,000
(ROE) after tax	2½%	5%	7½%	10%
		Y company		
A_{NP} before interest	$5,000	$10,000	$15,000	$20,000
Less interest	($5,000)	($ 5,000)	($ 5,000)	($ 5,000)
A_{NP} after interest	0	$ 5,000	$10,000	$15,000
Less tax at 50%	0	($ 2,500)	($ 5,000)	($ 7,500)
A_{NNP}	0	$ 2,500	$ 5,000	$ 7,500
(ROE) after tax	0	5%	10%	15%

TABLE 9-23   Stockholders' Equity as Shown in Section of a Company's Balance Sheet

Preferred stock, par value $100 per share Authorized 2000 issued and outstanding	1,500	$ 150,000	
Less discount on preferred stock		(10,000)	
Preferred-stock equity			$ 140,000
Common stock, par value $10 per share Authorized and issued 100,000 shares	$1,000,000		
Amount paid in excess of par	100,000		
		$1,100,000	
Retained earnings		100,000	
Common-stock equity			$1,200,000
Total stockholders' equity			$1,340,000

are entitled to receive compensation for the previous underpayment of dividends when the company again pays dividends.

Common stockholders have a right to the residual assets of a company in the event of dissolution or liquidation but only after all the creditors and then any liabilities to the preferred stockholders have been paid. The larger the proportion of debt financing in a company, the smaller the amount the common stockholders are likely to receive if the company is liquidated.

Common stockholders normally have a preemptive right to the first option to purchase any additional issues of common stock. This prevents management from using an additional issue of common stock to override the control exercised by existing stockholders. Preemptive rights also protect existing stockholders from having the value of their shares decreased by such dilution, since the same net earnings would be spread over more units of stock.

Let us consider the very simplified case of a company with 100,000 shares of common stock, each with a market value of $10, giving a total market value of $1,000,000. If a further 50,000 shares are sold at $4 each, the total market value of the 150,000 shares is $1,200,000, or $8 each. This means that the new stockholders have gained at the expense of the original ones. The preemptive right is designed to prevent this. In practice, the situation is rather more complex than is indicated here.

Both common and preferred stocks normally have a par, or nominal, value. In the case of common stock, the market value at the time of issue usually differs from the par value. Stock can be issued either at a premium or at a discount, depending on prevailing economic conditions and the strength of the company. The difference between the actual amount paid and the par value is listed in the stockholders'-equity section of the balance sheet, as shown in Table 9-23. The issuance of stock at a premium or a discount is done to protect existing stockholders.

In the case of preferred stock, the par value has more meaning than with common stock, since it is the amount due preferred stockholders if the company goes into liquidation, provided that this is a condition of issue.

The advantage of using common stock to finance assets is that it does not incur fixed interest charges. Furthermore, there is no maturity date, as there is with all loans and most preference issues. Common stock can often be issued more easily than debt can be financed. However, the flotation costs of common stock can be quite high, especially when stock values are depressed, so that large discounts for the stock are needed to induce purchase.

Stockholders' equity in a company is made up of the capital contributed by the stockholders and the capital generated from retained earnings. The presence of retained earnings on a balance sheet, as shown in Table 9-23, does not necessarily mean that they are matched by an equal amount of cash. In fact, there may be little or no cash available. The retained earnings shown on a balance sheet may be largely fictitious. For example, the assets on a balance sheet may be worth less than shown by at least the value of the retained earnings.

Purchase and Sale of Equities   Stockholders usually require an adequate return on their investment, and the quoted price of the stock reflects the consensus opinion of investors as to the current health of the company. Purchases or sales are normally made through stockbrokers.

Most stock transactions are completed through organized security exchanges on which the stock is listed. Such exchanges have physical existence in the form of buildings located in different regions of the country. Each exchange has members who are often the nominated representatives of large brokerage firms having offices in various cities. These offices are in constant telephone and telegraph communication with the members at the exchange, passing on requests to buy or sell specified stocks. Since brokers live by commissions and charges on transactions, they attempt to match such requests either directly or by dealings with other brokers. In the United Kingdom, brokers must deal through an independent "jobber," similar in function to a specialist broker, who quotes a low price for sales and a higher one for purchases before the jobber knows whether the broker is buying or selling. The difference represents the jobber's margin, or "turn." If requests to buy exceed offers for sale, the price of the stock rises until someone is tempted to sell. Conversely, if an excess of stock is offered for sale, the price is likely to fall.

It is an advantage to a company to be listed on a stock exchange since its investors can more easily sell their stock if they decide to do so. This increased liquidity makes investors more willing to accept a lower rate of return, which effectively lowers the cost of capital to the company.

Because dealings in the stock of a listed company are published, a healthy company engenders confidence that makes it easier to obtain other forms of finance. In the absence of a regular market, stock transactions are necessarily infrequent, and prices are liable to wide fluctuation, which may make creditors wary and possibly lead to bankruptcy proceedings. Such dealings are usually referred to as "over-the-counter" and are confined to the relatively few specialist brokers who hold inventories of such stock and are prepared to "make a market" in them or are limited to private transactions.

Retained Earnings   Much confusion is caused by the practice of dividing retained earnings under various headings such as reserve for replacement of plant, reserve for contingencies, etc. This procedure also restricts the flexibility of management in expenditure decisions.

The amount of retained earnings shown on a balance sheet should not be taken as a measure of the amount of future dividends that the company is likely to pay. A contract may exist that specifies a minimum balance of retained earnings, which is then not available for dividends until bonds issued by the company have been retired.

Dividends can be paid either as cash or in the form of an additional issue of stock. A stock split is really a stock dividend, and both are used to reduce the price of stock when management considers that it is too high. A stock dividend is essentially a transfer of retained earnings to the common-stock account and makes the amount transferred unavailable for future dividends. A stock dividend may be used in place of a cash dividend when a company is short of cash.

Debt Financing   In practice, debt financing covers a variety of fixed-income securities, both long-term and short-term. The most common forms of long-term debt are bonds, mortgages, and debentures.

A bond is simply a long-term promissory note. It is a contract established between borrower and lender in a document called an indenture. A bond indenture includes a detailed description of assets that are pledged, together with any protective clauses and provisions for redemption. A trustee is appointed to look after the interest of the bondholders. The trustee is normally a commercial bank. Bonds may be issued with a call provision that enables a company to redeem its bonds at any date earlier than scheduled. Obviously, this would be an advantage to a company in times of falling interest rates. However, a company has to pay more than the par value of the bond for this privilege. The additional amount is called the bond premium.

Sometimes a company uses a sinking fund to retire a bond. A series of equal annual payments A, invested at a fractional interest rate i and made at the end of each year over a period of n years, is equivalent to a sum of money of present value P, given by

$$A = Pf_{AP} \qquad (9\text{-}46)$$

where f_{AP} is the annuity present-worth factor, which is

$$f_{AP} = [i(1+i)^n]/[(1+i)^n - 1]$$

A company may use a sinking fund in a variety of ways, but the simplest is to pay a fixed amount A at the end of each year to buy and retire bonds until after n years all the bonds have been retired. This annual payment may prove a significant strain on the resources of a company. Failure to make the payment could result in bankruptcy. In the case of income bonds, a company is required only to pay interest when it earns it.

A mortgage is a bond in which specific real assets are pledged as security. A senior mortgage has a prior claim on assets. A junior mortgage is normally a second mortgage on the residual value of the assets. A blanket mortgage is a pledge on all real property owned by a company.

A debenture is an unsecured bond. Strong companies are in a better position to issue debentures than weak companies since they have less need to pledge specific assets. Debenture holders are really general creditors. Subordinated debenture holders have claims on assets only after the claims of certain other claimants have been met. The issue of subordinated debentures provides a tax advantage for a com-

TABLE 9-24 Comparative Ratios for Selected United States Industry Groups for 1992*

Industry	Agricultural chemicals	Paints and allied products	Petroleum refining	Plastic materials and resins	Soap and other detergents
Number of companies	2391	238	98	234	66
Ratio					
$\dfrac{\text{Current assets}}{\text{Current debt}}$	2.0	2.8	1.3	1.8	2.2
$\dfrac{\text{Net profit} \times 100, \%}{\text{net sales}}$	1.9	1.8	1.5	1.5	3.8
$\dfrac{\text{Net profit} \times 100, \%}{\text{net worth}}$	9.3	6.8	6.7	14.3	16.3
$\dfrac{\text{Net sales}}{\text{Net working capital}}$	9.3	5.2	15.8	9.9	5.9
Collection period, days	27.8	39.6	34.8	39.3	44.7
$\dfrac{\text{Net sales}}{\text{Inventory}}$	11.4	8.1	14.6	11.9	8.9
$\dfrac{\text{Fixed assets} \times 100, \%}{\text{Net worth}}$	43.0	30.8	131.5	20.0	42.3
$\dfrac{\text{Current debt} \times 100, \%}{\text{Net worth}}$	52.1	42.0	68.4	96.8	46.9
$\dfrac{\text{Total debt} \times 100, \%}{\text{Net worth}}$	71.9	66.2	147.7	111.0	61.4
$\dfrac{\text{Current debt} \times 100, \%}{\text{Inventory}}$	116.1	89.0	193.7	143.6	119.4

*Reprinted with the special permission of Dun & Bradstreet International.
Numbers above are median values.

pany compared with the issue of preferred stock because the interest payable is a tax-deductible expense.

A financial analyst looking at a company from a potential common stockholder's point of view is likely to classify preferred stock as debt. In contrast, bondholders and general creditors are likely to regard preferred stock as additional equity. Since preferred stock is a hybrid type of security, it may be issued by a company whose management is divided over the question of whether to use equity or debt to finance additional assets. However, preferred stock does have the disadvantage that the dividends are not allowed as a tax-deductible expense.

Comparative Company Data Table 9-24 gives comparative company data that have been compiled by Dun & Bradstreet for various types of processing industries. The median value for each ratio is given.

Row 1 in Table 9-24 is the

$$\text{Current ratio} = \frac{\text{current assets}}{\text{current liabilities}} \qquad (9\text{-}142)$$

Compare

$$\text{Quick ratio} = \frac{\text{liquid assets}}{\text{current liabilities}} \qquad (9\text{-}143)$$

Row 2 in Table 9-24 is the profit margin (PM) of Eq. (9-127). In this case, the net profit referred to is the net annual profit after tax and depreciation A_{NNP}. The net sales is the revenue from annual sales A_S after deductions for returns, allowances, and discounts for gross sales.

Row 3 in Table 9-24 is the return on equity (ROE) of Eq. (9-130). In this case, the net worth is the tangible net worth representing the sum of the preferred and common stocks and the surplus and undistributed profits or retained earnings, less any intangible items such as goodwill, etc.

Row 7 in Table (9-24) is the

Average collection period

$$= \frac{\text{average value of accounts receivable}}{\text{revenue from sales per day}} \qquad (9\text{-}144)$$

The funded debt (referred to in row 14) consists of mortgages,

bonds, debentures, serial notes, or other obligations with maturity of more than 1 year from the statement date.

Robert Morris Associates also compiles extensive comparative company data for various industries. In addition to ratios similar to the Dun & Bradstreet ratios shown in Table 9-24, Robert Morris Associates gives very useful breakdowns of assets and liabilities for various industries. Table 9-25 shows a breakdown of assets and liabilities for United States manufacturers of industrial inorganic chemicals.

Application of Overall Company Ratios The various ratios for a hypothetical company are listed in Table 9-26. The balance sheet shown in Table 9-27 has been built up from the ratios in Table 9-26 in terms of the revenue from net annual sales A_S.

Let us calculate the following values for the right-hand side of the balance sheet as follows:

From ratio 5

$$\text{Net worth} = A_S/2.50 = 0.4\, A_S$$

From ratio 11

$$\text{Total debt} = (0.4\, A_S)(0.65) = 0.26\, A_S$$

From ratio 10

$$\text{Current debt} = (0.4\, A_S)(0.35) = 0.14\, A_S$$
$$\text{Long-term debt} = \text{total debt} - \text{current debt}$$
$$\text{Long-term debt} = 0.26\, A_S - 0.14\, A_S = 0.12\, A_S$$

We calculate the following values for the left-hand side of the balance sheet:

From ratio 99-45

$$\text{Fixed assets} = (0.4\, A_S)(0.74) = 0.29\, A_S$$

From ratio 1

$$\text{Current assets} = (0.14\, A_S)(2.60) = 0.36\, A_S$$

From ratio 8

$$\text{Inventory} = A_S/7.14 = 0.14\, A_S$$

From ratio 7

$$\text{Accounts receivable} = (A_S/365)(61) = 0.167\, A_S$$

TABLE 9-25   Typical Balance Sheet for a United States Manufacturer of Industrial Inorganic Chemicals

Assets	%	Liabilities	%
Cash	4.3	Short-term due to banks	5.6
Marketable securities	2.2	Due to trade	14.5
Net receivables	26.2	Income taxes	3.1
Net inventory	24.6	Current maturities long-term debt	2.1
All other current assets	0.9	All other current liabilities	5.0
Total current assets	58.3	Total current debt	30.3
Fixed assets	36.3		
All other noncurrent assets	5.4	Noncurrent debt unsubordinated	18.3
		Total unsubordinated debt	48.6
		Subordinated debt	3.1
		Tangible net worth	48.3
Total assets	100.0	Total liabilities and stockholders' equity	100.0

Abridged from *Annual Statement Studies*, 1973 ed., copyright 1973 by Robert Morris Associates, Philadelphia.

Cash and short-term investments = total current assets
$$\text{Cash and short-term investments} = \text{total current assets} - (\text{inventory} + \text{accounts receivable})$$
$$\text{Cash and short-term investments} = 0.364\,A_S - 0.307\,A_S = 0.057\,A_S$$

In addition to the data for the balance sheet, we calculate the net annual profit (after tax), i.e., ratio 2, to be $A_{NNP} = 0.04\,A_S$.

TABLE 9-26   Ratios for a Typical Industrial Chemical Company

No.	Ratio
1	$\dfrac{\text{Current assets}}{\text{Current debt}} = 2.60$
2	$\left(\dfrac{\text{Net profit}}{\text{Net sales}}\right)100 = 4.00$
3	$\left(\dfrac{\text{Net profit}}{\text{Net worth}}\right)100 = 10.0$
4	$\left(\dfrac{\text{Net profit}}{\text{Net working capital}}\right)100 = 18.18$
5	$\dfrac{\text{Net sales}}{\text{Net worth}} = 2.50$
6	$\dfrac{\text{Net sales}}{\text{Net working capital}} = 4.50$
7	$\text{Collection period} = \dfrac{\text{accounts receivable}}{\text{sales per day}} = 61 \text{ days}$
8	$\dfrac{\text{Net sales}}{\text{Inventory}} = 7.14$
9	$\left(\dfrac{\text{Fixed assets}}{\text{Net worth}}\right)100 = 74.00$
10	$\left(\dfrac{\text{Current debt}}{\text{Net worth}}\right)100 = 35.00$
11	$\left(\dfrac{\text{Total debt}}{\text{Net worth}}\right)100 = 65.00$
12	$\left(\dfrac{\text{Inventory}}{\text{Net working capital}}\right)100 = 63.00$
13	$\left(\dfrac{\text{Current debt}}{\text{Inventory}}\right)100 = 100.00$
14	$\left(\dfrac{\text{Funded debt}}{\text{Net working capital}}\right)100 = 76.50$

In practice, the ratios are obtained from the information published in the balance sheet. The advantage of the above presentation is that it relates everything to the revenue from net annual sales and hence underlines the importance of sales.

Careful study of the ratios can produce many inferences as to the health of the company. For example, the leverage, or debt, ratio (DR) for this example is

$$(DR) = \frac{\text{total debt}}{\text{total assets}} = \frac{0.260\,A_S}{0.660\,A_S} = 0.40$$

This value is quite low and does not present any problems of control by debtors, such as can arise when (DR) is greater than 1.

From Table 9-27 we calculate the ratio for

$$\frac{\text{Current debt}}{\text{Cash + short-term investments}} = \frac{0.140\,A_S}{0.057\,A_S} = 2.45$$

Therefore, requests for early repayment by more than 40 percent of the debtors could be met. Hence, no liquidity problems are likely to arise, and advantage can be taken of discounts for early payment. Also, the current debt could be met by sale of the inventory, which takes $(0.140\,A_S/A_S)(365)$, or 51 days. The quick ratio is $1/2.45 = 0.407$.

If it is assumed that current debtors are due for payment within 61 days, the same time as that allowed to creditors, no bankruptcy petitions are likely.

The profit of 10 percent, indicated by ratio 3 in Table 9-26, will be reduced by any dividend due to preferred stockholders, because such payments are not part of fixed-debt expenses; the residue is shared among the ordinary stockholders. If all the long-term debts were in redeemable 6 percent preferred shares, then (from ratio 3) the net annual profit (after tax) is $A_{NNP} = 0.10(0.40\,A_S)$, or $0.04\,A_S$. Interest due on preferred shares is $0.06(0.12\,A_S)$, or $0.0072\,A_S$. Therefore, the earnings for the ordinary shares are

$$(0.04\,A_S - 0.0072\,A_S)/(0.4\,A_S - 0.12\,A_S) = 0.1171$$

This value corresponds to 11.71 cents per dollar of common stockholders' equity.

If it is assumed that available interest rates offered by banks, government, etc., for no-risk investment of capital are 10 percent, then the maximum economic market price of $100 stock units in this hypothetical company is about $117. If all the debt is in bonds, etc., earnings on ordinary stock would be 10 cents per dollar of net worth, and the maximum economic price of the stock would be about $100 unless stock prices were expected to rise.

Other ratios can easily be deduced from those listed. For example, the return on assets (ROA) and the asset-turnover ratio (ATR) are

$$(ROA) = \frac{\text{net annual profit}}{\text{total assets}}100 = \frac{0.04\,A_S}{0.66\,A_S} = 6.06$$

$$(ATR) = \frac{\text{revenue from annual sales}}{\text{total assets}}100$$

$$(ATR) = (A_S/0.66\,A_S)100 = 151.5$$

TABLE 9-27 Balance Sheet for a Typical Industrial Chemical Company, Dec. 31, 1991

Assets		Liabilities and stockholders' equity	
Current assets		Liabilities	
Cash	$0.057 A_S$	Current debt	$0.140 A_S$
Accounts receivable	$0.167 A_S$	Long-term debt	$0.120 A_S$
Inventory	$0.140 A_S$		
Total current assets	$0.364 A_S$	Total debt	$0.260 A_S$
Fixed assets	$0.296 A_S$	Net worth	$0.400 A_S$
Total assets	$0.660 A_S$	Total liabilities and stockholders' equity	$0.660 A_S$

Cost of Capital   The value of the interest rate of return used in calculating the net present value (NPV) of a project is usually referred to as the cost of capital. It is not a constant value since it depends on the financial structure of the company, the policy of the company toward a particular project, the local method of assessing taxation, and, in some cases, the measure of risk associated with the particular project. The last-named factor is best dealt with by calculating the entrepreneur's risk allowance inherent in the project i'_r from Eq. (9-108), written in the form

$$i'_r = [1 + (\text{DCFRR})]/(1 + i) - 1$$

where i is the cost of capital exclusive of the risk allowance. The value of i'_r should be compared with the probability of exceeding or of failing to achieve an (NPV) of zero when using that value of i. The decision to proceed can then be made with a full knowledge of the odds against success. The decision can be related to the company attitude to budgets of the relevant size by the use of probable utilities, as has already been discussed. Cash flows used in calculating (NPV) and (DCFRR) should, of course, be corrected for the anticipated rates of inflation, preferably to the time when the utility curve was obtained. This is important since inflation is likely to have a distorting effect on utility curves obtained at different times. This may be due to an unconscious wish to protect against inflation by achieving higher rewards while assigning less importance to any losses incurred, thus tending toward a gambling outlook.

In the absence of a risk allowance the cost of capital becomes a technical financial computation based on sources of funds and company policy. As such it will usually be presented as a figure specified for use in a particular appraisal and is therefore of little concern to the project assessor. However, the following résumé indicates the kinds of factors to be considered.

In most companies the objective of company policy is to maximize the financial return to the equity stockholders. This is not invariably the case, since a young company will often plow back an unusually large proportion of its profits to encourage growth. Also, it is increasingly the case that projects are undertaken to restore or preserve an environmental amenity or to bring work into a particular locality. In such circumstances a low value of the cost of capital might be assigned to the project. In many government projects a limited loss is acceptable, in which case the value of i would be negative.

When the objective is to maximize the aftertax return to the stockholders, a balance must be struck between the proportion of aftertax company profits which are retained to permit growth of the company assets and the proportion which are distributed to provide an income for the stockholders in the form of dividends. The latter will usually be subject to personal income taxes, sometimes at higher-than-normal rates. The growth potential should be reflected in an increased value of the stock as quoted on the stock exchange. Such growth may result in the imposition of capital gains or inheritance taxes. The selection of the right proportion of earnings to be retained is crucial since this affects the appeal of the company to investors and hence its credit worthiness in the eyes of creditors. The optimum split is influenced by the type of investor since institutional tax rates and exemptions often differ from those applied to private investors. It is for this reason that the optimum split is sensitive to local taxation policy.

Most companies can maintain a given level of business only by continuous reinvestment in plant and equipment. If company growth is required, additional investment is essential. In general, a company has only three sources of new money, namely, cash received from the sale of newly issued shares, retained earnings, and debt capital of all kinds including deferred taxes. In certain circumstances cash grants may be forthcoming from government sources. Each of these sources has its own effective rate of interest, and it is the weighted average of these rates which constitutes the cost of capital exclusive of risk allowance.

There is no interest payable directly on equity stocks, but there is a concealed rate expected by investors. Without the expectation of a certain return on their investment they would not invest in a new issue, nor would they retain existing holdings of stock. The sale of stocks on the stock exchange does not affect the cash holding of the company, but new issues must be at prices lower than existing values quoted on the exchange unless great confidence exists that the new money will produce an increased income greatly in excess of the reduction in earnings per share caused by the new issue. Stock carrying a fixed interest normally has the interest treated as an allowable expense before tax in the same way as a bank overdraft, which is a relatively short-term source of debt. Deferred taxes carry an interest rate which, like an overdraft, is normally compounded daily at a nominal annual rate but naturally is not an allowable expense for tax purposes. Cash owing on outstanding bills carries a notional rate of interest since in many cases prompt settlement of bills would attract a cash discount.

Example 18: Risk-Free Cost of Capital   A company requires an investment of $100,000 in new plant to maintain its present sales. Let us determine the current cost of capital to the company and the risk-free cost of capital that it should assign to the plant-replacement project, given the following data.

Company assets: from stock sales	$	300,000
from retained earnings		200,000
as bills due		100,000
as deferred taxes		200,000
as bank overdraft		200,000
Total assets		$1,000,000
Current annual income	$	200,000

Bills are due on monthly account with a 2 percent discount for cash. Overdraft and deferred-tax interest are compounded daily at nominal annual interest rates of 15 and 9 percent respectively. Corporation tax, capital gains tax, and personal income tax rates are 50, 40, and 30 percent respectively. The current rate of inflation is at 8 percent per year. The traditional return expected by investors is 7 percent per year net of all taxes in real terms.

The interest-rate equivalent of the cash discounts is 2 percent per month, since this discount could be obtained every month if payment were to be made at the beginning of the month rather than, as at present, at its end. Since the bills are settled monthly, the notional interest is paid monthly and should not be compounded. The discount is equivalent to 12 monthly simple-interest payments per year. Hence, from Eq. (9-31) the effective annual interest rate on discounts = $(12)(0.02) = 0.24 = 24$ percent. It would, therefore, be a good use of surplus cash to reduce this debt as quickly as possible. This would require cash equivalent to one-sixth of the annual bills due, or $16,700, to be available. It can, therefore, be assumed that this level of liquidity is not available for capital projects, either as working capital to reduce the debt or for fixed-capital projects. Further, since the new project will not increase sales, it cannot generate further debt of this kind. Hence, this source is not available to capitalize the new project.

Since the overdraft is payable daily at a nominal annual interest rate of 15 percent, it follows from Eq. (9-38) that the effective annual interest rate on overdraft = $(1 + 0.15/365)^{365} - 1 = 16.18$ percent. Similarly, the effective annual interest rate on deferred tax = $(1 + 0.09/365)^{365} - 1 = 9.42$ percent.

The new plant will not increase sales and will therefore not increase the tax debt, so that this source is not available to capitalize the project. An increase in overdraft may be available, subject to a maximum imposed by the acceptable gearing of the company.

Since the liquidity of the company is so low, it is possible that it is already extended to its maximum debt, in which case the gearing

$$\frac{\text{Total equity}}{\text{Total debt}} = \frac{\$300,000 + \$200,000}{\$100,000 + \$200,000 + \$200,000} = 1.00$$

Since neither increased bills due nor increased tax debt is available to finance the new project, this implies that the required $100,000 of new capital will be available as $50,000 from increased overdraft and $50,000 from increased equity. The effective interest rate on the equity involved must therefore be calculated.

Equity is available from two sources. First, the company can sell new stock which, if in the form of ordinary shares, carries no interest payment. Although this course appears cheap, its use for projects which do not increase earnings, at least to a compensatory level, is usually inadvisable. This leaves retained earnings as the most likely source of equity for the present project.

Equity holders require a real return on their outlay, which they assume to be at the stock-market price if this differs from the face value of the stock, of 7 percent net of all taxes. Retained earnings attract a 40 percent capital gains tax; hence the actual interest rate required on distribution forgone is $7/(1 - 0.40) = 11.67$ percent. This is in real terms and at a time of 8 percent inflation rate must be increased in cash terms to $(1 + 0.1167)(1.08) - 1 = 20.60$ percent.

In the same way the effective interest rate on distributed earnings, on which an income tax of 30 percent is payable, may be calculated to be $[0.07/(1 - 0.30) + 1](1.08) - 1 = 18.80$ percent. If the shares are currently valued at par by the stock exchange, this would require distribution, on the $300,000 of issued equity, of $(\$300,000)(0.188) = \$56,400$. This amount is required after corporation tax has been paid on the earnings of $200,000. Thus the earnings which can be retained are $(\$200,000)(1 - 0.50) - \$56,400 = \$43,600$. This is close to the $50,000 required. If the company were well regarded on the stock exchange, a slight reduction in distributed dividends might not reduce share values since the purpose is to maintain future earnings. However, it would be possible for share values to be reduced by up to $(\$56,400 - \$50,000)/\$56,400 = 0.1135 = 11.35$ percent. If this happened, the effective interest rate on the retained earnings should be increased to $20.60/(1 - 0.1135) = 23.24$ percent. These rates are net of corporation tax, which is the correct form for use in (NPV) calculations.

The interest rate due on deferred tax is also net of corporation tax at 9.42 percent. The interest payable on overdrafts is an expense fully allowable against tax, so that the effective aftertax rate is reduced to $(16.18)(1 - 0.50) = 8.09$ percent. Similarly, as the advantage forgone on the discounts would tend to increase company profits and hence tax due, the effective aftertax gain is reduced to $(24)(1 - 0.50) = 12.00$ percent.

The present cost of capital to the company is the weighted-average interest payable on the various sources of funds on an after-corporation-tax basis. This is readily calculable to be at least $[(\$300,000)(0.00) + (\$200,000)(20.60) + (\$100,000)(12.00) + (\$200,000)(9.42) + (\$200,000)(8.09)]/(\$1,000,000) = 8.82$ percent. The cost of capital to the new project, with only two sources, should be $[(\$50,000)(20.60) + (\$50,000)(8.09)]/(\$100,000) = 14.35$ percent.

Since this project is essential if current production is to be maintained, many companies would assess the cost of capital at somewhere near the lower value. Values of cost of capital in the region of 10 percent are to be expected in developed countries at the present time.

We notice in particular that inflation does not affect quoted interest rates when assessing present values of cost of capital. It must, however, be taken into account in assessing the interest rate on the dividend which will be expected by investors.

As has been stated, it is alternatively possible to assign to the cost of capital the best risk-free return available on the money. The assessment then proceeds as discussed in connection with Eq. (9-108).

Management and Cost Accounting In any given time period, cost may be divided into expired and unexpired cost. An expired cost is an expense; an unexpired cost is an asset. This division is the basis for income statements and balance sheets.

Cost accounting is the name traditionally given to accounting for manufacturing costs. The manufacturing cost of a product is traditionally taken as the sum of the costs for (1) direct materials, (2) direct labor, (3) manufacturing overheads, and (4) administration, selling, and finance.

Two methods are in general use in accounting for manufacturing costs, absorption costing and marginal costing. In absorption costing, which is the traditional method, all manufacturing overhead costs are included in the cost of sales. In marginal costing only variable manufacturing overhead costs are included in the cost of sales. Marginal costing is more valuable than absorption costing in decision making.

However, it is sometimes quite difficult to separate costs and particularly manufacturing overhead costs into fixed and variable components. In the long term virtually all costs are variable. The difference between the two methods assumes great importance in inventory evaluation. In cost accounting, costs are identified with cost centers. These are accounting devices which may or may not have a physical existence. In the simplest case of a plant manufacturing a single product, the entire plant may be the cost center.

In practice there are two major classifications of cost accounting systems, job costing and process costing. In the former, costs are collected for each job or batch irrespective of the accounting period. This system is normally used in construction work. Process costing is normally used in continuous and semicontinuous processes. Costs are collected for a specific accounting period.

Allocation of Overheads How overheads are allocated can affect the total cost of a product and, hence, the estimated future cash flows for a project. Since these cash flows are used in the net-present-value (NPV) and discounted-cash-flow-rate-of-return (DCFRR) methods for estimating profitability, erroneous allocations could result in the wrong choice of project.

The modern trend is for overhead costs to become an increasing proportion of total product costs. This results from the ever-greater sophistication of process plants. Therefore, it is highly desirable that chemical engineers should have some say in the allocation of overheads and that this should not be left entirely to accountants.

Direct costs are those that can be directly charged to a single product. The most obvious direct cost is for raw materials, of which the quantity consumed is directly proportional to the amount of product manufactured. Direct process labor is also considered to be a direct cost.

However, many costs cannot be directly charged to an individual product. These so-called indirect, burden, or overhead costs range from the lighting and heating required for the plant and offices to the cafeteria and medical facilities provided. When several products are made in a plant, it becomes increasingly difficult to allocate overheads correctly among the various products.

A number of different methods are commonly used to estimate the amount of overhead to be allocated to an individual product. These methods are necessary because accountancy costs become prohibitive for charging all costs directly to an individual product. Unfortunately, there is always an arbitrary element inherent in the process of allocation.

Overheads in the chemical-process industries are commonly calculated as a percentage of (1) direct materials cost, (2) direct labor cost, or (3) prime or direct costs. Other methods of allocating overheads are on the basis of (1) plant area, (2) number of employees, (3) capital value, and (4) electric power.

These listings do not include all the methods in use. The validity of a particular method depends on the process and the industry. An inappropriate method can lead to misleading and even absurd results.

Let us consider the manufacture of metal ornaments. The processing cost, exclusive of material, may vary very little for a wide range of materials. However, the direct materials cost will be much greater for precious than for base metals. In this case, an overhead allocation on the basis of direct material costs could be very misleading, while one based on direct labor cost could be quite accurate.

Problems can also arise when allocating overheads on the basis of direct labor cost. Let us consider a company that evaluates overheads at 125 percent of direct labor cost. A process plant employs seven operators, each with a direct cost of $10,000 per budget period. As a result of a works-study exercise, it is found that the plant can operate satisfactorily with six operators. The actual cost saving is likely to be far nearer to the direct labor savings of $10,000 per period than to the calculated saving of $10,000 + \$10,000(125/100) = \$22,500$ per period. The $22,500 calculated saving is the direct labor cost plus overheads taken as 125 percent of the direct labor cost.

A thorough analysis should be made before production is stopped on a product that is losing money. Although direct costs of the discontinued product will be saved, overheads are not eliminated, as might be inferred from taking overheads as a percentage of direct material, direct labor, or prime costs. The plant is still there, together with its associated costs for interest charges, insurance, painting, some maintenance, etc. Continued production can still make a useful contribution to such overheads. This contribution is lost if production is stopped and must then be borne by other products.

Problems can arise with each of the methods used for allocating overheads. Two process plants may occupy similar areas yet have vastly different material or labor costs. Problems can also arise with an individual plant that can be used to make different products, as Example 19 will show.

Example 19: Overhead in Two Different Projects   Let us consider a plant that can make either product A or product B. At normal capacity, the overhead cost is known to be $2.50 per unit. Product A has a direct materials cost of $8 per unit and a direct labor cost of $2 per unit. For simplicity, the prime cost is here taken as the sum of these two costs, i.e., $10 per unit.

In Table 9-28, a correct overhead cost of $2.50 per unit at normal capacity is calculated by taking either 31.25 percent of the direct materials cost, 125 percent of the direct labor cost, or 25 percent of the prime cost. All these methods give a total cost of $12.50 per unit and a profit of $1.50 per unit for a selling price of $14 per unit.

The alternative, product B, has a direct materials cost of $6 per unit and a direct labor cost of $4 per unit. In Table 9-28 overhead costs of $1.875 per unit, $5 per unit, and $2.50 per unit are calculated by taking 31.25 percent of direct materials cost, 125 percent of direct labor cost, and 25 percent of prime cost respectively. Total costs are $11.875 per unit, $15 per unit, and $12.50 per unit respectively and profits of $2.125 per unit, −$1 per unit, and $1.50 per unit respectively, for a selling price of $14 per unit.

An alternative to allocating overheads by using a single method is to classify the various overheads into groups and to use the most appropriate allocation for each group. For example, depreciation would be allocated on the basis of capital cost, while indirect labor might be allocated either on the basis of direct labor cost or on the number of employees. Clearly, this alternative method is more complex, increases the associated accountancy costs, and is prone to misinterpretation and possibly abuse.

Inventory Evaluation and Control

Inventory Effect on Cash Income and Profit   When the annual production rate is equal to the annual sales volume, the revenue from annual sales A_S is

$$A_S = Rc_s \tag{9-87}$$

where R is the production rate, units per year, and c_s is the sales price per unit. In this case, the annual cash income A_{CI} and the net annual profit A_{NP} before tax are given respectively from Eqs. (9-1), (9-8), and (9-12) by

$$A_{CI} = R(c_s - c_{TVE}) - A_{TFE} \tag{9-145}$$

$$A_{NP} = R(c_s - c_{TVE}) - A_{TFE} - A_{BD} \tag{9-146}$$

where c_{TVE} is the total variable expense per unit of production, A_{TFE} is the total annual fixed expense required to produce and sell a product but excluding any annual provision for plant depreciation, and A_{BD} is the balance-sheet annual depreciation charge.

However, in a given accounting period the sales volume may differ from the volume of production. In this case, the inventory of finished product I_1 at the beginning of the accounting period will differ from

that at the end, I_2, and Eqs. (9-145) and (9-146) need to be written in modified form as

$$A_{CI} = R(c_s - c_{TVE}) - A_{TFE} + (I_1 - I_2) \tag{9-147}$$

$$A_{NP} = R(c_s - c_{TVE}) - (A_{TFE} + A_{BD}) + (I_1 - I_2) \tag{9-148}$$

If the annual sales volume exceeds the annual production rate R by an amount ΔR, then Eqs. (9-147) and (9-148) can be written as

$$A_{CI} = R(c_s - c_{TVE}) - A_{TFE} + \Delta R(c_s - c_{INV}) \tag{9-149}$$

$$A_{NP} = R(c_s - c_{TVE}) - (A_{TFE} + A_{BD}) + \Delta R(c_s - c_{INV}) \tag{9-150}$$

where c_{INV} is the value per unit of inventory of the finished product.

Clearly, the value of c_{INV} affects both the annual cash income and the net annual profit. Since annual cash incomes are the basic data for (NPV) and (DCFRR) methods of estimating profitability, the actual value per unit of inventory is of direct importance for chemical engineers engaged in economic assessments.

Let us divide Eq. (9-150) by Eq. (9-87) to give

$$\frac{A_{NP}}{A_S} = \left(1 - \frac{c_{TVE}}{c_S}\right) - \frac{(A_{TFE} + A_{BD})}{A_S} + \left(\frac{\Delta R}{R}\right)\left(1 - \frac{c_{INV}}{c_S}\right) \tag{9-151}$$

$$\frac{A_{NP}}{A_S} = (CSR) - \frac{(A_{TFE} + A_{BD})}{A_S} + \left(\frac{\Delta R}{R}\right)\left(1 - \frac{c_{INV}}{c_S}\right) \tag{9-152}$$

where $(CSR) = 1 - (c_{TVE}/c_S)$ is the contribution to the sales-price ratio. When the ratio of net annual profit to revenue from annual sales A_{NP}/A_S is expressed as a percentage it is known as the profit margin.

The terms (CSR) and $(A_{TFE} + A_{BD})/A_S$ have a similar order of magnitude in chemical processing. For example, (CSR) values of 0.1 to 0.4 are typical, and these are quite close to a value of 0.2 that is common in general chemical processing for $(A_{TFE} + A_{BD})/A_S$. Thus, the profit margin is very sensitive to the value c_{INV} per unit of inventory.

For example, let us consider that both (CSR) and $(A_{TFE} + A_{BD})/A_S$ are equal to 0.3. In this case, Eq. (9-152) can be written as

$$A_{NP}/A_S = (\Delta R/R)(1 - c_{INV}/c_S) \tag{9-153}$$

If the sales volume exceeds the annual production rate by 10 percent and the inventory is valued at the sales price, then Eq. (9-153) shows that the profit margin is $(A_{NP}/A_S)100 = 0$ percent. If the inventory is valued at the total variable cost, then the profit margin $(A_{NP}/A_S)100 = (0.1)(1 - 0.7)(100) = 3$ percent. Hence, the value of the inventory is of vital importance.

Unfortunately, there is no universally accepted method for valuing inventory. The value of c_{INV} per unit can be taken on any of the following bases:

1. Direct material plus direct labor cost.
2. Direct material plus direct labor plus other direct production expenses. (This is the total variable production cost.)
3. Total variable production cost plus fixed production overhead cost.
4. Total variable cost. (This includes both production and general expenses.)
5. Total cost. (This includes variable and fixed production and general expenses.)

Methods 1, 2, and 4 are termed direct costing, variable costing, and marginal costing respectively. Although direct costing is being increas-

TABLE 9-28   Profits of Products with Different Methods of Overhead Allocation

	Product A—basis of overhead allocation			Product B—basis of overhead allocation		
	31.25% of direct materials cost, $/unit	125% of direct labor cost, $/unit	25% of prime cost, $/unit	31.25% of direct materials cost, $/unit	125% of direct labor cost, $/unit	25% of prime cost, $/unit
Direct materials cost	8.00	8.00	8.00	6.000	6.000	6.000
Direct labor cost	2.00	2.00	2.00	4.000	4.000	4.000
Prime or direct cost	10.00	10.00	10.00	10.000	10.000	10.000
Overhead cost	2.50	2.50	2.50	1.875	5.000	2.500
Total cost	12.50	12.50	12.50	11.875	15.000	12.500
Selling price	14.00	14.00	14.00	14.000	14.000	14.000
Profit/(loss)	1.50	1.50	1.50	2.125	(1.000)	1.500

ingly used for internal accounting and control purposes, it is not acceptable to tax authorities as a basis for calculating profit. Tax authorities and most accountants favor method 3, which is known as absorption costing. We have already seen that this method has the disadvantage that the fixed-overhead cost per unit is determined for a particular normal production rate. If the production rate exceeds the normal, there is overabsorption of fixed overheads. Conversely, if the production rate falls below the normal, there is underabsorption of fixed overheads. Method 5 is known as full-absorption costing.

Most people agree that general expenses incurred in administration, selling, distribution, etc., should not be included in the cost of inventory. In fact, many feel that no costs should be absorbed before they have been incurred. In general, method 2 is favored by engineers and method 3 by most accountants. However, the accountancy convention is to value at either cost or market value, whichever is lower. In the methods considered, either actual or standard costs can be used. Note that method 3 shows a higher profit than method 2 when sales volume exceeds the production rate and a lower profit when the production rate exceeds sales volume.

The total profits (before tax) over the life of a project are independent of the method used to value inventory. Over a project life of n years, Eq. (9-148) can be written as

$$\sum_0^n A_{NP} = \sum_0^n R(c_S - c_{TVE}) - \sum_0^n (A_{TFE} + A_{BD}) + \sum_0^n (I_m - I_{m+1}) \quad (9\text{-}154)$$

where the last term in Eq. (9-154) becomes $\sum_0^n (I_m - I_{m+1}) = I_0 - I_{n+1}$. Since there will be no material in inventory in the year before the project starts or in the year after it terminates, $I_0 = I_{n+1} = 0$. Hence, total profits do not depend on individual values for I.

However, the annual profit A_{NP} (before tax) does depend on the value of the inventory. Since the tax payable in any individual year is based on A_{NP}, the net annual profit A_{NNP} (after tax) is also dependent on the method chosen for valuing inventory. Frequently, a particular method for valuing inventory is chosen to delay payment of tax as long as it is legally possible to do so.

So far, only the inventory of finished product has been considered. There are also inventories of raw materials and work in process, i.e., partially processed materials or intermediate products, to be considered. It is necessary to modify Eqs. (9-147) through (9-154) accordingly to take these inventories into account.

Effect of Raw-Materials Prices   Raw materials for the chemical-process industries are subject to relatively wide variations in price. These effects on profits will now be considered.

When the price of raw materials varies from week to week, not all the units in storage will have been purchased at the same price. Let us consider χ units in storage at the start of the inventory period, purchased at a price c_1 per unit. Additional quantities χ_2, χ_3, etc., are purchased at prices c_2, c_3, etc., per unit respectively until finally χ_n units are purchased at the latest price of c_n per unit at the end of the inventory period. The total value of the inventory C_{INV} at the end of the inventory period (in the absence of any withdrawal) is given by

$$C_{INV} = \sum_1^n \chi_j c_j \quad (9\text{-}155)$$

The value of the inventory I at any given time depends on the values ascribed to the units withdrawn from inventory. There are five methods for valuing inventory: (1) FIFO (first-in–first-out), (2) LIFO (last-in–first-out), (3) average cost, (4) standard cost, and (5) market value.

In the FIFO method, the units taken out of storage are valued at their purchase price beginning with the earliest item purchased. If a number of units m are removed from inventory during the period, the total cost of these items on a FIFO basis is given by:

$$C_{FIFO} = \sum_1^p \chi_j c_j + \left(m - \sum_1^p \chi_j\right) c_{p+1} \quad (9\text{-}156)$$

providing that the value of m satisfies

$$m < \sum_0^n \chi_j$$

and where p is the largest integer, such that

$$\sum_1^p \chi < m$$

Hence, the value of the inventory, I_{FIFO} at any given time is

$$I_{FIFO} = C_I - C_{FIFO} \quad (9\text{-}157)$$

In terms of Eqs. (9-155) and (9-156), Eq. (9-157) can be written as

$$I_{FIFO} = \sum_1^n \chi_j c_j - \sum_1^p \chi_j c_j - \left(m - \sum_1^p \chi_j\right) c_{p+1} \quad (9\text{-}158)$$

In the LIFO method, the m units taken out of storage are valued at their purchase price, beginning with the latest item purchased. In a similar manner, the value of the material C_{LIFO} taken out of inventory is given by

$$C_{LIFO} = \sum_p^n \chi_j c_j + \left(m - \sum_p^n \chi_j\right) c_{p-1} \quad (9\text{-}159)$$

where p is the smallest integer, such that

$$\sum_p^n \chi_j \leq m$$

Hence, the value of the inventory I_{LIFO} at any given time is

$$I_{LIFO} = C_I - C_{LIFO} \quad (9\text{-}160)$$

In terms of Eqs. (9-155) and (9-159), Eq. (9-160) can be written

$$I_{LIFO} = \sum_1^n \chi_j c_j - \sum_p^n \chi_j c_j - \left(m - \sum_p^n \chi_j\right) c_{p-1} \quad (9\text{-}161)$$

Example 20: Inventory Computation   Let us consider 10 successive batches of raw materials, of 1000 units, purchased in a time of rising prices in which $c_1 = \$0.10$ per unit, $c_2 = \$0.11$ per unit, etc., as listed in Table 9-29. The total cost of the purchases in the inventory is found from Eq. (9-155) to be $\$1450$.

Let us calculate the value of the raw-materials inventory after, say, 5500 units have been withdrawn from inventory, first by using FIFO and then by using LIFO.

We substitute the appropriate quantities into Eq. (9-158) for FIFO, keeping in mind that $p = 5$, $n = 10$, and $m = 5500$, to get

$$I_{FIFO} = \$1450 - \$600 - (5500 - 5000)(\$0.15)$$
$$I_{FIFO} = \$775$$

In a similar manner, we substitute values into Eq. (9-161) for LIFO (except that p is now 6) to get

$$I_{LIFO} = \$1450 - \$850 - (5500 - 5000)(\$0.14)$$
$$I_{LIFO} = \$530$$

Thus, in a time of rising raw-materials prices, the FIFO method gives a higher value for the remaining inventory than will LIFO. In a time of falling prices, the FIFO method will give a lower value for the remaining inventory than will LIFO.

Average-Cost Basis for Inventory   Either a simple average or a weighted average can be used to value inventory cost.

Using the simple-average method, the value C_{SAV} of the material taken out of inventory is given by

$$C_{SAV} = m(c_1 + c_n)/2 \quad (9\text{-}162)$$

TABLE 9-29   Costs of Inventory with Rising Prices

p	x_j, units	c_j, \$/unit	$x_j c_j$, \$	$\sum_1^p x_j$, units	$\sum_1^p x_j c_j$, \$	$\sum_p^n x_j$, units	$\sum_p^n x_j c_j$, \$
1	1,000	0.10	100	1,000	100	10,000	1,450
2	1,000	0.11	110	2,000	210	9,000	1,350
3	1,000	0.12	120	3,000	330	8,000	1,240
4	1,000	0.13	130	4,000	460	7,000	1,120
5	1,000	0.14	140	5,000	600	6,000	990
6	1,000	0.15	150	6,000	750	5,000	850
7	1,000	0.16	160	7,000	910	4,000	700
8	1,000	0.17	170	8,000	1,080	3,000	540
9	1,000	0.18	180	9,000	1,260	2,000	370
10	1,000	0.19	190	10,000	1,450	1,000	190

The value of the inventory I_{SAV} at any given time is

$$I_{SAV} = C_I - C_{SAV} \qquad (9\text{-}163)$$

In terms of Eqs. (9-155) and (9-162), Eq. (9-163) can be written as

$$I_{SAV} = \sum_1^n \chi_j c_j - [m(c_1 + c_n)/2] \qquad (9\text{-}164)$$

We shall consider the value of the inventory after 5500 units have been withdrawn, using the data listed in Table 9-29. On the basis of a simple average, the materials withdrawn are priced at ($0.10 + $0.19)/2 = $0.145 per unit. Since the total cost of the purchases in the raw-materials inventory is found from Eq. (9-155) to be $1450, the value of the inventory after 5500 units have been withdrawn is calculated from Eq. (9-164) to be

$$I_{SAV} = \$1450 - (5500)(\$0.145) = \$652.50$$

For the weighted-average method, the value of the material C_{WAV} taken out of inventory is given by

$$C_{WAV} = m \sum_1^n \chi_j c_j / \sum_1^n \chi_j \qquad (9\text{-}165)$$

$$I_{WAV} = C_I - C_{WAV} \qquad (9\text{-}166)$$

In terms of Eqs. (9-153) and (9-164), Eq. (9-166) can be written

$$I_{WAV} = \sum_1^n \chi_j c_j \left[1 - \left(m / \sum_1^n \chi_j \right) \right] \qquad (9\text{-}167)$$

Let us use Eq. (9-167) to value the inventory, after 5500 units have been withdrawn, by employing the data of Table 9-29.

$$I_{WAV} = \$1450 \, [1 - (5500/10,000)] = \$652.50$$

For this example, the values of I_{SAV} and I_{WAV} are the same because the batches purchased are of equal size and the prices are linearly progressive. This is a combination rarely found in practice.

A more realistic example, in which the buyer seeks to purchase at the lowest price, is provided by the data of Table 9-30. The quantities bought vary according to price, but some may have been made at high prices to maintain production.

On the basis of Table 9-30, when 5500 items have been removed from inventory, the value of the inventory by using the FIFO, LIFO, simple-average, and weighted average methods respectively is

$$I_{\text{FIFO}} = \$1175 - \$560 - (5500 - 5200)(\$0.10)$$
$$I_{\text{FIFO}} = \$573.00$$
$$I_{\text{LIFO}} = \$1175 - \$645 - (5500 - 5000)(\$0.10)$$
$$I_{\text{LIFO}} = \$480.00$$
$$I_{SAV} = \$1175 - 5500 \, [(\$0.10 + \$0.15)/2]$$
$$I_{SAV} = \$487.50$$
$$I_{WAV} = \$1175[1 - (5000/10,000)]$$
$$I_{WAV} = \$528.75$$

Pros and Cons of Inventory Valuation In the standard-price method of inventory valuation, all materials are taken out at the same price. In addition to simplicity, the method has the advantage that the efficiency of raw-materials purchase is constantly checked.

In both the average-cost and the standard-cost methods of valuing inventory, materials are not charged out at actual cost. Thus, the amount of profit or loss for the period may be varied by the method chosen to value the inventory. For this reason, accountants usually insist that the method of inventory valuation be consistent from period to period. This causes inertia but does not prevent a change of method when it can be justified. In such cases, it is usual to inform stockholders of the change because the influence on declared profits can be large.

Unfortunately, there is no right or wrong way to value inventory, although certain methods are not allowed in certain countries for tax-assessment purposes. For example, LIFO is not allowed in the United Kingdom. As a general rule, the method used should be the one that gives the lowest tax liability. However, it is generally accepted that consistency is also a virtue in inventory valuation.

It is important to realize that the method used to value inventory for cost accounting purposes is not necessarily the one used to draw up the balance sheet and financial accounts. In this case, inventory is valued either at the cost given by another method or at the market value, whichever is lower.

Inventory Control The optimum size of inventory depends on the type of industry and on the skills available to the individual company. Inventories are high in the tobacco industry and low in perishable-foods businesses. The larger the inventory, the larger the warehousing and associated costs. These costs include insurance, taxes, depreciation, handling and security charges, etc., and can be taken as roughly proportional to the value of the inventory I. The annual cost A_{IW} of maintaining an inventory of value I is given by

$$A_{IW} = \alpha I \qquad (9\text{-}168)$$

where α is the proportionality factor which is of the order of 0.25 for many industries.

In contrast, some costs can be reduced as a result of larger inventories. For example, larger discounts can be obtained on bulk purchases and deliveries. In addition, larger inventories reduce the risk of losing sales and goodwill through interruptions to production and consequently running out of stocks of finished goods.

The cost of placing an order for materials is partly fixed and partly variable. The annual cost of ordering A_{IO} is given by

$$A_{IO} = FN + VR \qquad (9\text{-}169)$$

where F is the fixed cost per order, N is the number of orders per year, V is the variable cost of ordering per unit of production, and R is the annual production rate. Although the administrative cost of placing an order will be more or less fixed, the shipping costs are proportional to the size of the order and, hence, for the year are proportional to the annual production rate. The total annual cost of inventory A_I is the sum of Eqs. (9-168) and (9-169):

$$A_I = \alpha I + FN + VR \qquad (9\text{-}170)$$

For a given number of orders per year N, the value of the inventory is proportional to the magnitude of the average individual order U. Hence, Eq. (9-170) can be written as

$$A_I = \beta U + (FR/U) + VR \qquad (9\text{-}171)$$

where β is a proportionality factor and the number of orders per year N has been written as R/U.

By differentiating Eq. (9-171) with respect to U, the optimum size of order can be estimated. The differential is:

$$dA_I/dU = \beta - (FR/U^2) \qquad (9\text{-}172)$$

Setting the right-hand side of Eq. (9-172) equal to zero yields

$$U = \sqrt{FR/\beta} \qquad (9\text{-}173)$$

where U is now the optimum size of order.

A number of models have been developed to enable managers to handle inventories in the most profitable manner. These models can be applied to other elements of working capital, such as cash.

TABLE 9-30 Costs of Inventory with Fluctuating Prices

p	x_j, units	c_j, \$/unit	$x_j c_j$, \$	$\sum_1^p x_j$ units	$\sum_1^p x_j c_j$, \$	$\sum_p^n x_j$ units	$\sum_p^n x_j c_j$, \$
1	1,000	0.10	100	1,000	100	10,000	1,175
2	1,500	0.11	165	2,500	265	9,000	1,075
3	500	0.13	65	3,000	330	7,500	910
4	2,000	0.10	200	5,000	530	7,000	845
5	200	0.15	30	5,200	560	5,000	645
6	1,000	0.14	140	6,200	700	4,800	615
7	2,500	0.11	275	8,700	975	3,800	475
8	300	0.15	45	9,000	1,020	1,300	200
9	100	0.20	20	9,100	1,040	1,000	155
10	900	0.15	135	10,000	1,175	900	135

Working Capital   The amount and disposition of working capital and the efficiency of its use determine the immediate prospects for future growth in a company. The bulk of managerial effort in a company is directly or indirectly concerned with the manipulation of working capital. Insufficient or misused working capital is the commonest cause of business failure.

Engineers concerned with cost estimations tend to make estimates of the fixed-capital cost of a project, leaving considerations of working capital to the accountants. Although the estimation of fixed-capital cost is more straightforward from an engineering point of view, the estimation of working capital is of vital importance both for an individual project and for the company as a whole.

Working capital can range from about 10 percent to almost 100 percent of the invested capital, depending on the industry, and is an important factor in the profitability index of a business. For this reason, it is best to compare the performance of an individual company with that of others that are as similar as possible.

Gross working capital is normally defined as total current assets, while net working capital is current assets minus current liabilities. Current assets normally amount to more than half of the total assets of a company. When accountants wish to emphasize working capital in a company, they present the balance sheet in the vertical form, as shown in Table 9-31. In this table, the stockholders' equity of $91,650 funds the sum of $38,650 (net working capital) and the sum of $53,000 (fixed assets less the long-term loan). The flow of working capital is diagrammatically illustrated in Fig. 9-39.

The necessary working capital varies with sales volume or production rate. For example, sales and hence accounts receivable will double for a doubling in sales volume. In addition, an increase in sales volume normally requires increased inventories of raw materials, work in progress, and finished goods, all of which tie up capital. An increase in sales volume may lead to a relative shortage of working capital. In turn, this may mean that accounts payable cannot be paid in time and that valuable cash discounts may be lost or interest and penalty charges incurred. Creditors may take legal action to obtain payments and thereby put an additional strain on the current assets of the company to provide legal fees. Often, such action leads other creditors to take similar steps, which may lead a fundamentally sound company into bankruptcy.

A shortage of cash may prevent a company from taking advantage of large discounts available for bulk purchase of raw materials. The importance of the availability of adequate cash or near cash can be seen by considering an account payable within 28 days, with a 2 percent discount allowed if paid within 7 days. If cash is not available to pay the account within 7 days, this is then equivalent to paying 2 percent interest on the money for the remaining 21-day period, or an annual compound-interest rate of more than 41 percent.

Adequate cash and a history of prompt payment of accounts strengthen the credit standing of a company and make it easier to obtain bank loans, etc. A company needs additional cash as a contingency against fires, floods, strikes, etc., as well as for additional advertising required to counteract the activities of competitors. This additional money is normally held as interest-bearing investments that can be turned into cash on short notice.

TABLE 9-31   Balance Sheet for BCD Company, Dec. 31, 1991

Current assets		
Cash	$14,575	
Accounts receivable	35,575	
Inventories		
Raw materials	6,000	
Work in process	3,750	
Finished product	10,000	
Gross working capital		$69,900
Current liabilities		
Accounts payable	25,000	
Notes payable	3,000	
Bank loans	3,000	
Accruals payable	250	
		31,250
Net working capital		$38,650
Fixed assets	75,000	
Long-term loan	22,000	
		$53,000
Stockholders' equity		$91,650

Working-Capital Ratios   Financial analysts make extensive use of ratios in assessing the economic health of a company. For evaluating the ability of a company to successfully maintain and develop its immediate business activities, analysts apply a current ratio and a quick (or acid-test) ratio, as given by

$$\text{Current ratio} = \frac{\text{current assets}}{\text{current liabilities}} \qquad (9\text{-}142)$$

$$\text{Quick ratio} = \frac{\text{liquid assets}}{\text{current liabilities}} \qquad (9\text{-}143)$$

Liquid assets are those that can be realized almost immediately, such as cash, accounts receivable, and marketable securities. Although inventories are current assets, they must not be regarded as liquid assets because they cannot usually be converted into cash without winding up the business.

Although a high current ratio is desirable, this may be achieved by having unnecessarily high inventories that bring no profit except when commodity prices are rising rapidly. The quick ratio is less misleading in this respect.

Good management practice will hold inventories at the lowest possible levels consistent with customer satisfaction and efficient plant operation. Excessive inventories are unproductive and are an investment having little or no rate of return. Excessive inventories should be maintained only when supplies are erratic or rising in price. Management should normally aim for a high inventory-turnover ratio, as given by:

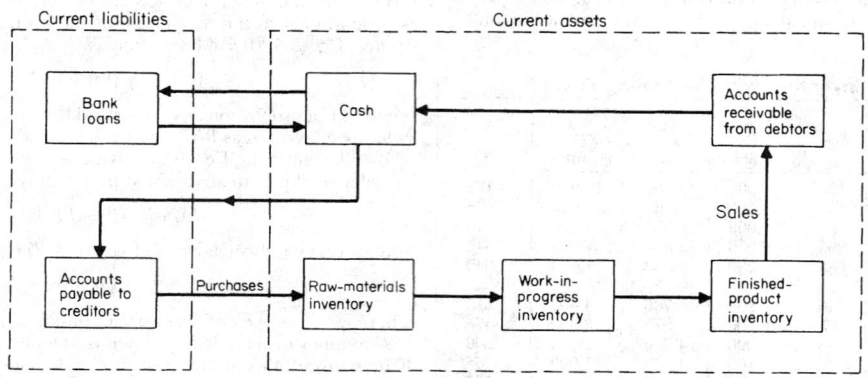

FIG. 9-39   Flow of working capital showing relationship between current liabilities and current assets.

$$\text{Inventory-turnover ratio} = \frac{\text{revenue from annual sales}}{\text{average value of inventory}} \quad (9\text{-}174)$$

Similarly, good management practice is to hold accounts receivable at a low level and to have a high accounts-receivable-turnover ratio, as given by

Accounts-receivable-turnover ratio

$$= \frac{\text{revenue from annual sales}}{\text{average value of accounts receivable}} \quad (9\text{-}175)$$

Alternatively, it is good practice to have a low average collection period, as given by

Average collection period

$$= \frac{\text{average value of accounts receivable}}{\text{revenue from sales per day}} \quad (9\text{-}144)$$

Equation (9-144) gives the number of days during which sales are tied up in receivables.

An accounts-receivable-turnover ratio of 12 is considered fairly good for a manufacturing company. This implies an average collection period of about 1 month. The price obtained for the goods should include an allowance for interest (otherwise obtainable) on the money tied up for such a period.

Since ratios, like balance sheets, refer to a particular point in time, they have a limited use unless they are compared with previous values. A study of ratio trends indicates whether or not a company is approaching a working-capital or a liquidity crisis and may enable management to compare the performance of the company with that of competitors.

Funds Statement   A typical funds statement is shown in Table 9-32. It displays the change in net working capital and can be obtained from a statement of changes in working capital, such as the one shown in Table 9-33. A funds statement shows where the cash came from and how it was used.

Changes in working capital ΔC_{WC} in an annual accounting period can be represented by

$$\Delta C_{WC} = A_S - A_{TE} - \sum \Delta C_{FC} - \sum \Delta C_L + \sum \Delta C_{FIN} \quad (9\text{-}176)$$

where A_S is the revenue from annual sales of a product, A_{TE} is the total cost or expense required to produce and sell the product but excluding any annual provision for plant depreciation, $\sum \Delta C_{FC}$ is the sum of the changes in depreciable fixed assets, $\sum \Delta C_L$ is the sum of the changes in nondepreciable fixed assets, and $\sum \Delta C_{FIN}$ is the sum of the

changes in financial resources such as loans, bonds, preferred stock, common stock, etc.

Equation (9-176) can also be written as

$$\Delta C_{WC} = A_{CI} - \sum \Delta C_{FC} - \sum \Delta C_L + \sum \Delta C_{FIN} \quad (9\text{-}177)$$

where A_{CI} is the annual cash income, which is the main source of funds for most companies. In this case, the annual cash income excludes all noncash expenses such as the balance-sheet annual depreciation charge A_{BD}, which is purely a book transaction.

A positive value of any term in Eq. (9-177) implies an increase in working capital, and a negative value a decrease. For example, the sale of fixed assets such as plant, buildings, land, etc., is a source of cash, and the purchase of fixed assets uses up cash. Similarly, an increase in financial resources in the form of loans and stock and bond issues is a source of cash, and a decrease in financial resources in the form of repayment of loans, retirement of stocks and bonds, and the payment of cash dividends uses up cash. (Note that a stock dividend as opposed to a cash dividend does not use up cash.)

The relation between (1) net annual cash flow A_{CF} after tax for individual projects, (2) the annual amount of tax A_{IT}, and (3) the annual expenditure of capital A_{YC} is

$$A_{CF} = A_{CI} - A_{IT} - A_{YC} \quad (9\text{-}178)$$

Equation (9-178) for a single project is really analogous to Eq. (9-177) for a company.

An income statement or profit-and-loss account gives the net annual profit A_{NP} before tax. In order to assess the annual cash income A_{CI} as a source of funds from the value of the net annual profit A_{NP} given in the income statement, it is necessary to add back all noncash expenses such as the balance-sheet annual depreciation charge A_{BD}. This practice sometimes erroneously suggests that depreciation is a source of funds, whereas cash income is the only source of funds.

Although A_{BD} does not affect working capital in any way, the annual depreciation charge A_D does affect the annual amount of tax A_{IT} given by

$$A_{IT} = (A_{CI} - A_D - A_A)t \quad (9\text{-}3)$$

where t is the fractional tax rate and A_A is the annual amount of any other allowances. Thus, the net annual cash income is affected by depreciation allowances, as follows:

$$A_{NCI} = A_{CI} - A_{IT} \quad (9\text{-}2)$$

In this sense, depreciation makes working capital available by reducing the cash outflow for taxes.

When a fixed asset is sold at a price that differs from its book value, an accounting gain or loss is recorded. This gain or loss does not affect working capital, which has simply been increased by the amount of cash received from the sale. For example, if an item of plant is sold for $40,000 (whatever its book value), the increase in working capital resulting from the sale is $40,000. If the book value of the plant had been $50,000, the accounting loss of $10,000 on the sale would be included as an expense when calculating the cash income. This $10,000 must be added back to the cash income in order to get the cash income excluding noncash expenses as required for A_{CI}. Conversely, any accounting gain on the sale of a fixed asset must be subtracted if it has been included in the cash income.

Book values of fixed assets are determined by the balance-sheet annual depreciation charges A_{BD}, which do not affect working capital. Although the accounting gain or loss on the sale of a fixed asset is based on its book value, working capital is not affected by depreciation assessments.

Transactions that change the character of the net working capital but do not affect its value occur in a company. For example, a cash payment of $10,000 for accounts payable reduces both the current asset of cash by $10,000 and the current liability of accounts payable by $10,000, leaving the net working capital unchanged. However, this transaction affects both the current and the quick ratios.

After the balance sheet and the income statement or profit-and-loss account, the funds statement is generally regarded as the most important financial document. However, many financial managers regard a statement showing changes in cash as being of equal importance.

As with balance sheets and income statements, there is no rigid format for funds statements. These vary in the amount of detail given and

TABLE 9-32   Funds Statement for Year Ending Dec. 31, 1991

Sources of funds		
Cash income from operations		$15,000
New finance		10,000
Sale of plant equipment		2,000
Total sources of funds		$27,000
Applications of funds		
Purchase of land	$3,000	
Cash dividend on stock	9,000	
Total applications of funds		$12,000
Increase in net working capital		$15,000

TABLE 9-33   Statement of Changes in Working Capital

	Jan. 1, 1991	Jan. 1, 1992	Change
Cash	$ 80,000	$ 94,000	+$14,000
Accounts receivable	20,000	35,000	+$15,000
Inventories	30,000	25,000	−5,000
Total current assets	$130,000	$154,000	+$24,000
Current liabilities	($ 60,000)	($ 69,000)	(+$ 9,000)
Net working capital	$ 70,000	$ 85,000	+$15,000
Current ratio	2.17	2.23	
Quick ratio	1.67	1.87	

TABLE 9-34   Statement of Changes in Working Capital and Sources and Applications of Funds for BCD Company

	Balance sheets		Change in funds	
	Dec. 31, 1991	Dec. 31, 1992	Applications	Sources
Cash	$ 14,575	$ 13,000		$ 1,575
Accounts receivable	35,575	36,000	$ 425	
Inventories	19,750	20,750	1,000	
Total current assets	$ 69,900	$ 69,750		
Accounts payable	$ 25,000	$ 27,000		2,000
Notes payable	3,000	2,500	500	
Bank loans	3,000	2,000	1,000	
Accruals payable	250	300		50
Total current liabilities	$ 31,250	$ 31,800		
Net working capital	$ 38,650	$ 37,950		
Fixed assets	75,000	85,000	10,000	
Net assets	$113,650	$122,950		
Long-term loan	$ 22,000	$ 22,000		
Stockholders' equity	$ 91,650	$100,950		9,300
			$12,925	$12,925

also in their layout. Funds statements help management in planning for the future, for example, in timing and making financial provision for future expenditures.

A typical management accountant's statement for changes in working capital and sources and applications of funds is shown in Table 9-34. This is based on the following relation: an increase in application of funds equals an increase in sources of funds. The relation can also be expressed as follows: an increase in assets plus a decrease in liabilities equals an increase in liabilities plus a decrease in assets.

In Table 9-34, the cash income for the year has been absorbed in the increase in stockholders' equity, which consists of issued stock plus retained profits or income.

General Considerations   Many people regard the management of working capital as essentially a cash-flow problem. Certainly, expansion involves increased investment in fixed assets and in the various items that comprise the current assets, all of which require cash. If this leads to a shortage of cash so that a company cannot pay its liabilities, then a situation called overtrading results, and the company may ultimately be forced into liquidation. Therefore, it is essential that a company have access to readily available cash or sources of short-term financing, preferably without having to specify a particular asset as collateral, since such collateral can only be pledged to an agreed realizable value that is much less than its true value.

Accounts payable, also called trade credit, are the major source of short-term financing. Accounts payable normally amount to about 40 percent of the current liabilities of a manufacturing company. Such short-term financing is relatively expensive when available discounts are lost.

The second most important source of short-term financing is notes payable from commercial banks. Banks normally require a borrower to maintain a compensating balance. For example, if a company requires a loan of $100,000, it must borrow more than this, say, $120,000 (on which it pays interest), in order to maintain a minimum checking-account balance of $20,000. Commercial banks also provide a wide variety of other services that can be of great help to companies in temporary financial difficulties.

The third most important source of short-term financing is the commercial paper, or promissory notes, of large companies. This is the cheapest form of finance. However, the amount of money available is a function of the excess liquidity of the large companies at a given time, and the money may have to be repaid at relatively short notice.

Sometimes, fixed assets are purchased via short-term loans, which can lead to liquidity problems. For the most part, fixed assets should be financed from long-term or permanent capital such as stocks or bonds. The proven ability of management to handle working capital efficiently will put a company in a better position to obtain such long-term capital when required, because the confidence of bankers and stockholders will have been obtained.

Budgets and Cost Control   R. J. Bull (*Accounting in Business*, 2d ed., Butterworth, London, 1972, p. 163) defined a budget as a comprehensive and coordinated plan, expressed in monetary terms, directing and controlling the resources and trading activities of an enterprise for some specified period in the future. A budget is not a forecast. A forecast is an estimate of the future which may or may not be attained. A budget is an overall objective based on a forecast. In addition, a budget defines the detailed objectives to be achieved by various levels of management in an organization. Since the achievement of these objectives requires the complete cooperation of management, the targets set must be realistic in terms of resources and past performance.

A comparison of actual with budgeted results can be used as the basis for control at the company, departmental, plant, or project level. In addition, a continuing record of performance should be maintained to provide the data for the preparation of further budgets.

Since company accounts are normally published annually, 1 year is commonly taken as a budget period. However, budget periods can vary widely depending on the nature of the operation. For example, a sales budget may be for a period of, say, 3 or 6 months, while the budget period for the installation of, say, a nuclear power station would extend over many years.

The basic objectives of budgets are planning and control. The first step is to determine the limiting factor. For example, budgeted sales cannot exceed the maximum productive capacity of the available plant. Since all the activities in the plan are interrelated, the extent of the plan is determined by the limiting factor.

After the plan is put into operation, actual progress is monitored against established standards. These data may subsequently lead to the plan's being modified in order to achieve the objectives more effectively.

R. Pilcher (*Appraisal and Control of Project Costs*, McGraw-Hill, Maidenhead, England, 1973, p. 233) stated the main purposes of a cost control system to be:

1. To provide immediate warning of uneconomic operations in both the long and the short terms
2. To provide the relevant feedback, carefully qualified in detail by all the conditions under which the work has been carried out, to the estimator who is responsible for establishing the standards in the past and in the future
3. To provide data to assist in the valuation of those variations that will arise during the course of the work
4. To promote cost consciousness
5. To summarize progress

Budgeted income statements are identical in form to actual income statements. However, the budgeted numbers are objectives rather than achievements. Budgetary models based on mathematical equations are increasingly being used. These may be used to determine rapidly the effect of changes in variables. Variance analysis is discussed in the treatment of manufacturing-cost estimation.

MANUFACTURING-COST ESTIMATION

The annual manufacturing cost or expense A_{ME} can be written as the sum of the direct manufacturing or prime cost A_{DME} and the indirect manufacturing or overhead cost A_{IME}:

$$A_{ME} = A_{DME} + A_{IME} \qquad (9\text{-}179)$$

The determination of direct or prime costs is more straightforward than the determination of indirect or overhead costs. When more than one product is involved, the question arises as to the correct distribution of overhead costs between the various products.

In addition to fixed and variable costs there are mixed or semivariable costs. These have a fixed and a variable element. The variable element can vary with production linearly, stepwise, or in a curvilinear manner. However, it is a convenient simplification to divide all manufacturing costs into fixed and linearly variable costs.

General Considerations   Manufacturing costs are best considered in the context of the manufacturing, trading, and profit-and-loss accounts. Typical examples of these are shown in Tables 9-35, 9-36, and 9-37, respectively. These are based on the conventional accountancy period of 1 year.

The gross annual profit A_{GP} in Table 9-36 is dependent on the balance-sheet annual depreciation charge A_{BD}, which is not necessarily the same as the depreciation allowance used for tax purposes. Since A_{BD} is arbitrarily chosen, it can be used to make the gross annual profit A_{GP} high or low according to the company policy.

The gross annual profit A_{GP} is also dependent on the method of valuing the inventory. For example, raw materials may have been purchased at the beginning of the accounting year at, say, 9 cents per kilogram. The purchase price may have risen to, say, 12 cents per kilogram at the end of the accounting year. If valuation of the inventory is made at the higher purchase price, the production cost is lower and the gross profit higher than if the valuation is made at the lower purchase price. Although in this case the profit looks better, a higher tax is payable.

TABLE 9-35   Manufacturing Account

Inventory of raw materials, Jan. 1, 1991	$ 35,000	
Add purchases	870,000	
Add carriage inward	25,000	
	$ 930,000	
Less inventory of raw materials Dec. 31, 1991	51,000	
Cost of raw materials consumed	$ 879,000	
Direct wages	122,000	
Direct utilities	22,000	
Other direct expenses	104,000	
Prime cost or direct manufacturing	$1,127,000	$1,127,000
expense, A_{DME}		
Payroll overhead	28,000	
General plant overhead	52,000	
Other indirect expenses	52,000	
Indirect manufacturing expense, A_{IME}	$ 132,000	
Depreciation, A_{BD}	68,000	
	$ 200,000	200,000
		$1,327,000
Add work in progress, Jan. 1, 1991		30,000
		$1,357,000
Less work in progress, Dec. 31, 1991		35,000
Production cost of products		$1,322,000

TABLE 9-36   Trading Account

Inventory of finished products, Jan. 1, 1991		$ 200,000
Add production cost of products		1,322,000
		$1,522,000
Less inventory of finished products, Dec. 31, 1991		190,000
		$1,332,000
Gross profit		668,000
Sales		$2,000,000

TABLE 9-37   Profit-and-Loss Account

Administration	$ 74,000	
Sales and shipping	124,000	
Advertising and marketing	40,000	
Technical service	10,000	
Research and development	60,000	
	$308,000	$308,000
Net profit before taxes		360,000
Gross profit		$668,000

The profit is also dependent on the method of valuing the work in progress. We shall consider the manufacture of 100,000 kg of product with a prime or direct manufacturing cost of 10 cents per kilogram and an additional indirect manufacturing expense of 5 cents per kilogram. We assume that 90,000 kg of the product is sold and 10,000 kg is stored in inventory. The value of the inventory is $1000 on the basis of prime or direct cost and $1500 when the indirect manufacturing expense is included. It is still a controversial question as to whether manufacturing overheads should be absorbed in the cost. The latter is known as absorption costing and is the traditional accounting method. Direct costing is being increasingly used and is particularly favored by engineers.

The annual cash income A_{CI} before tax, in terms of the revenue from annual sales A_S of a product and the components of the total annual cost or expense required to produce and sell the product (excluding any allowance for plant depreciation), is expressed by

$$A_{CI} = [A_S - (A_{VGE} + A_{VME})] - (A_{FGE} + A_{FME}) \qquad (9\text{-}180)$$

where A_{VGE} and A_{VME} are the annual variable general and variable manufacturing expenses respectively and A_{FGE} and A_{FME} are the annual fixed general and fixed manufacturing expenses respectively.

Revenue from annual sales is

$$A_S = Rc_S \qquad (9\text{-}181)$$

where R is the production rate and c_S the sales price per unit of production.

Similarly, A_{VGE} can be taken as proportional to the annual production rate $\dot R$ and the variable general expense per unit of production c_{VGE}:

$$A_{VGE} = Rc_{VGE} \qquad (9\text{-}182)$$

Likewise, for A_{VME},

$$A_{VME} = Rc_{VME} \qquad (9\text{-}183)$$

In practice, annual direct variable costs such as raw materials, utilities, etc., are not always proportional to the production rate.

Substituting Eqs. (9-182) and (9-183) into the first term on the right of Eq. (9-180) yields

$$A_S - (A_{VGE} + A_{VME}) = R[c_S - (c_{VGE} + c_{VME})] \qquad (9\text{-}184)$$

In Eq. (9-184), the term $[c_S - (c_{VGE} + c_{VME})]$ is the contribution to cash income.

For simultaneous production of more than one product, Eq. (9-184) can be written as

$$A_S - (A_{VGE} + A_{VME}) = R_1 \{(c_S)_1 - [(c_{VME})_1 + (c_{VME})_1]\}$$
$$+ R_2 \{(c_S)_2 - [(c_{VGE})_2 + (c_{VME})_2]\} + \dots \qquad (9\text{-}185)$$

where the subscripts 1, 2, etc., refer to the various coproducts.

The contribution to cash income made by a particular product depends on the method of accounting. Widely different values for the contribution can be calculated by using different methods of assigning manufacturing expenses.

The cost of each item in a cost estimate should be presented in such a way that the estimate can be modified and updated at any time in the future when revised data become available.

Published data and shortcut estimating methods can be used to calculate the approximate manufacturing cost of a new product. However, most companies have extensive data on various items of cost such as overheads, property taxes, etc. These data should be used whenever possible to give the estimate that is most valid for a particular company.

For a new product, the ratio of manufacturing expense to sales price c_{ME}/c_S should be compared with the ratio of total manufacturing cost or expense to sales revenue for the company as a whole. If the ratio c_{ME}/c_S is less than or equal to the ratio for the company, then the proposed sales price appears to be reasonable and the product is probably commercially viable. This comparison is, of course, used only as an approximate guide in preliminary assessments.

One Main Product Plus By-Products   We shall let one unit of raw material yield χ_1, χ_2, etc., weights of products 1, 2, etc., respectively. The variable general expense per unit of raw material will be

c_{VGE} and the variable manufacturing expense per unit of raw material will be c_{VME}. The unit of raw material may be either a single material or a mixture of several components. I. Leibson and G. A. Trischman [*Chem. Eng.*, **78**, 69–74 (May 31, 1971)] showed the effects on manufacturing expenses of alternate feedstocks having different compositions and costs for producing the same products.

If product 1 is the main product, then this carries all the variable expenses, so that

$$(c_{VGE})_1 = c_{VGE}/\chi_1 \qquad (9\text{-}186)$$

$$(c_{VME})_1 = c_{VME}/\chi_1 \qquad (9\text{-}187)$$

where $(c_{VGE})_1$ and $(c_{VME})_1$ are the variable general and variable manufacturing expenses respectively per unit of product 1.

For by-product 2, etc., $(c_{VGE})_2 = 0$ and $(c_{VME})_2 = 0$, etc. For this case, Eq. (9-184) becomes

$$A_S - (A_{VGE} + A_{VME})$$
$$= R_1 [(c_S)_1 - (c_{VGE} + c_{VME})/\chi_1] + R_2(c_S)_2 + \dots \qquad (9\text{-}188)$$

If R is the annual rate at which the raw material is consumed, then

$$R_1 = \chi_1 R \qquad (9\text{-}189)$$

$$R_2 = \chi_2 R \qquad (9\text{-}190)$$

In terms of one unit of raw material, Eq. (9-185) can be combined with Eq. (9-189) and (9-190) and written as

$$[A_S - (A_{VGE} + A_{VME})]/R$$
$$= \chi_1(c_S)_1 - (c_{VGE} + c_{VME}) + \chi_2(c_S)_2 + \dots \qquad (9\text{-}191)$$

Equation (9-191) gives the contribution of products 1, 2, etc., per unit of raw material. In this particular case, the contribution of the main product per unit weight of raw material is $\chi_1(c_S)_1 - (c_{VGE} + c_{VME})$. The contribution of product 2 is $\chi_2(c_S)_2$, i.e., its selling price per unit weight of raw material.

Two Main Products

By Weight We shall let one unit of raw material yield χ_1 and χ_2 weights of products 1 and 2 respectively. The variable general expense per unit of raw material will be c_{VGE} and the variable manufacturing expense per unit of raw material c_{VME}. In practice, it is rare for $\chi_1 + \chi_2$ to be exactly unity.

If the variable expenses are shared by weight, then

$$(c_{VGE})_1 = \chi_1 c_{VGE}/[(\chi_1 + \chi_2)\chi_1] \qquad (9\text{-}192)$$

$$(c_{VME})_1 = \chi_1 c_{VME}/[(\chi_1 + \chi_2)\chi_1] \qquad (9\text{-}193)$$

$$(c_{VGE})_2 = \chi_2 c_{VGE}/[(\chi_1 + \chi_2)\chi_2] \qquad (9\text{-}194)$$

$$(c_{VME})_2 = \chi_2 c_{VME}/[(\chi_1 + \chi_2)\chi_2] \qquad (9\text{-}195)$$

where $(c_{VGE})_1$ and $(c_{VME})_1$ are the variable general and variable manufacturing expenses per unit of product 1 respectively and $(c_{VGE})_2$ and $(c_{VME})_2$ are the comparable expenses for product 2.

For this example, Eq. (9-185) becomes

$$A_S - (A_{VGE} + A_{VME}) = R_1\left[(c_S)_1 - \frac{(c_{VGE} + c_{VME})}{(\chi_1 + \chi_2)}\right]$$
$$+ R_2\left[(c_S)_2 - \frac{(c_{VGE} + c_{VME})}{(\chi_1 + \chi_2)}\right] \qquad (9\text{-}196)$$

In Eqs. (9-192) through (9-196) the sum of χ_1 and χ_2 may be equal to or less than 1. The above analysis can be extended for any number of coproducts.

In terms of one unit of raw material, Eq. (9-196) can be combined with Eqs. (9-189) and (9-190) and written as

$$\frac{A_S - (A_{VGE} + A_{VME})}{R} = \chi_1\left[(c_S)_1 - \frac{(c_{VGE} + c_{VME})}{(\chi_1 + \chi_2)}\right]$$
$$+ \chi_2\left[(c_S)_2 - \frac{(c_{VGE} + c_{VME})}{(\chi_1 + \chi_2)}\right] \qquad (9\text{-}197)$$

Equation (9-197) gives the contribution of products 1 and 2 per unit weight of raw material.

By Value We shall let one unit of raw material yield χ_1 and χ_2 weights of products 1 and 2 respectively, with values of $\chi_1(c_S)_1$ and $\chi_2(c_S)_2$ respectively; $(c_S)_1$ and $(c_S)_2$ are the sales prices of products 1 and 2 per unit of production.

We shall let the variable general expense per unit of raw material be c_{VGE} and the variable manufacturing expense per unit of raw material be c_{VME}. If the variable expenses are shared by value then

$$(c_{VGE})_1 = \frac{\chi_1(c_S)_1 c_{VGE}}{[\chi_1(c_S)_1 + \chi_2(c_S)_2]\chi_1} \qquad (9\text{-}198)$$

$$(c_{VME})_1 = \frac{\chi_1(c_S)_1 c_{VME}}{[\chi_1(c_S)_1 + \chi_2(c_S)_2]\chi_1} \qquad (9\text{-}199)$$

$$(c_{VGE})_2 = \frac{\chi_2(c_S)_2 c_{VGE}}{[\chi_1(c_S)_1 + \chi_2(c_S)_2]\chi_2} \qquad (9\text{-}200)$$

$$(c_{VME})_2 = \frac{\chi_2(c_S)_2 c_{VME}}{[\chi_1(c_S)_1 + \chi_2(c_S)_2]\chi_2} \qquad (9\text{-}201)$$

where $(c_{VGE})_1$ and $(c_{VME})_1$ are the variable general and variable manufacturing expenses per unit of product 1 respectively and $(c_{VGE})_2$ and $(c_{VME})_2$ are the comparable expenses for product 2.

For this case, Eq. (9-185) becomes

$$A_S - (A_{VGE} + A_{VME}) = R_1\left\{(c_S)_1 - \frac{(c_S)_1(c_{VGE} + c_{VME})}{[\chi_1(c_S)_1 + \chi_2(c_S)_2]}\right\}$$
$$+ R_2\left\{(c_S)_2 - \frac{(c_S)_2(c_{VGE} + c_{VME})}{[\chi_1(c_S)_1 + \chi_2(c_S)_2]}\right\} \qquad (9\text{-}202)$$

This analysis can be extended for any number of coproducts.

In terms of one unit of raw material, Eq. (9-202) can be combined with Eqs. (9-189) and (9-190) and written as

$$\frac{A_S - (A_{VGE} + A_{VME})}{R} = \chi_1\left\{(c_S)_1 - \frac{(c_S)_1(c_{VGE} + c_{VME})}{[\chi_1(c_S)_1 + \chi_2(c_S)_2]}\right\}$$
$$+ \chi_2\left\{(c_S)_2 - \frac{(c_S)_2(c_{VGE} + c_{VME})}{[\chi_1(c_S)_1 + \chi_2(c_S)_2]}\right\} \qquad (9\text{-}203)$$

Equation (9-203) gives the contribution of products 1 and 2 per unit weight of raw material.

Example 21: Calculation of Contributions to Income for Multiple Products One kilogram of raw material is used to manufacture $\chi_1 = 0.32$ kg of product 1 and $\chi_2 = 0.64$ kg of product 2. The balance of the raw material goes to waste. Product 1 sells at $(c_S)_1 = 40$ cents per kilogram, and product 2 sells at $(c_S)_2 = 12$ cents per kilogram. The variable general and variable manufacturing expenses, including raw materials, $c_{VGE} + c_{VME}$, total 10 cents per kilogram.

Let us calculate (*a*) the total contribution to cash income per kilogram of raw material and (*b*) the individual contribution of each product to cash income per kilogram of raw material for the following:

Case	Condition
1	Product 1 as the main product charged for all the variable expenses
2	Product 2 as the main product charged for all the variable expenses
3	Products 1 and 2 sharing the variable expenses on the basis of weight
4	Products 1 and 2 sharing the variable expenses on the basis of value

Case 1. The total contribution per kilogram of raw material is obtained by substituting the appropriate values into Eq. (9-191). For these conditions,

$$\chi_1(c_S)_1 + \chi_2(c_S)_2 - (c_{VGE} + c_{VME}) = 0.32(40 \text{ cents/kg})$$
$$+ 0.64 (12 \text{ cents/kg}) - 10 \text{ cents/kg}$$
$$= 12.8 \text{ cents/kg} + 7.68 \text{ cents/kg}$$
$$- 10 \text{ cents/kg}$$
$$= 10.48 \text{ cents/kg}$$

The individual contribution of product 1 as the main product is 12.8 cents − 10 cents = 2.8 cents per kilogram of raw material. The individual contribution of product 2 as the by-product is 7.68 cents per kilogram of raw material.

Case 2. The total contribution when product 2 is the main product is also obtained from Eq. (9-191) and as in Case 1 is found to be 10.48 cents per kilogram.

The individual contribution of product 2 as the main product is 7.68 cents − 10 cents = −2.32 cents per kilogram of raw material. The individual contribution of product 1 as the by-product is 12.8 cents per kilogram of raw material.

Case 3. When products 1 and 2 share the variable expenses on the basis of weight, the total contribution per kilogram of raw material is found by substituting the unit costs into Eq. (9-197). Values for each term are

$$\chi_1(c_S)_1 = 12.8 \text{ cents/kg} \qquad \text{(same as Case 1)}$$

$$\chi_2(c_S)_2 = 7.68 \text{ cents/kg} \qquad \text{(same as Case 1)}$$

$$[\chi_1/(\chi_1 + \chi_2)](c_{VGE} + c_{VME}) = (0.32/0.96)(10 \text{ cents/kg}) = 3.34 \text{ cents/kg}$$

$$[\chi_2/(\chi_1 + \chi_2)](c_{VGE} + c_{VME}) = (0.64/0.96)(10 \text{ cents/kg}) = 6.66 \text{ cents/kg}$$

Therefore, the total contribution becomes 12.8 + 7.68 − 3.34 − 6.66 = 10.48 cents per kilogram.

The individual contribution of product 1 is 12.8 cents − 3.34 cents = 9.46 cents per kilogram of raw material. The individual contribution of product 2 is 7.68 cents − 6.66 cents = 1.02 cents per kilogram of raw material.

Case 4. When products 1 and 2 share the variable expenses on the basis of value, the total contribution per kilogram of raw material is found by substituting the unit costs into Eq. (9-203). Values of each term are

$$\chi_1(c_s)_1 = 12.8 \text{ cents/kg} \qquad \text{(same as Case 1)}$$

$$\chi_2(c_s)_2 = 7.68 \text{ cents/kg} \qquad \text{(same as Case 1)}$$

$$\frac{\chi_1(c_s)_1(c_{VGE} + c_{VME})}{\chi_1(c_s)_1 + \chi_2(c_s)_2} = \frac{(12.8)(10)}{12.8 + 7.68} = 6.24 \text{ cents/kg}$$

$$\frac{\chi_2(c_s)_2(c_{VGE} + c_{VME})}{\chi_1(c_s)_1 + \chi_2(c_s)_2} = \frac{(7.68)(10)}{12.8 + 7.68} = 3.76 \text{ cents/kg}$$

Therefore, the total contribution becomes 12.8 + 7.68 − 6.24 − 3.76 = 10.48 cents per kilogram.

The individual contribution of product 1 is found as 12.8 cents − 6.24 cents = 6.56 cents per kilogram of raw material. The individual contribution of product 2 then becomes 7.68 cents − 3.76 cents = 3.92 cents per kilogram of raw material.

Direct Manufacturing Costs Direct manufacturing costs include raw materials, operating labor, utilities, and some miscellaneous items. A summary of the characteristics of each follows.

Raw Materials The cost of raw materials is normally the largest item of expense in the manufacturing cost of a product. The quantities of raw materials consumed can be calculated from material balances.

Material costs are conveniently presented in tables that give the following: name of material, form and grade, method of delivery, unit of measure, cost per unit, source of cost, annual consumption, annual cost, fractional consumption per unit of production, and cost per unit of production.

Net consumption of materials should be used for catalysts, solvents, filter aids, etc., that may have a recovery value. Current prices of chemicals are published in various trade journals. However, quotations from suppliers should be used whenever possible.

It may be possible for a company to negotiate the purchase of a material at a cost per unit that is significantly lower than the current published price. This is particularly true if large quantities are involved. Thus, estimates should be presented for both minimum and maximum costs. Price trends, availability, and quality are other factors that should be considered. A knowledge of price trends is particularly important for a product that a company may not manufacture for several years.

The yield in a chemical reaction determines the quantities of materials in the material balance. Assumed yields are used to obtain approximate exploratory estimates. In this case, possible ranges should be given. Firmer estimates require yields based on laboratory or, preferably, pilot-plant work.

Operating Labor The cost of operating labor is the second largest item of expense in the manufacturing cost. Labor requirements for a process can be estimated from an intelligent study of the equipment flow sheet, paying careful attention to the various primary process steps such as fractionation, filtration, etc. The hourly wage rate should be that currently paid in the company. Once the number of persons required per shift has been estimated for a particular production rate, the annual labor cost and the labor cost per unit of production can be estimated.

H. E. Wessel [*Chem. Eng.*, **59**, 209–210 (July 1952)] made a study of the operating-labor requirements in the United States chemical industry and presented the data as a plot of labor-hours per ton per processing step versus plant capacity in tons per day. These data can be represented by:

$$\log_{10} Y = 0.783 \log_{10} \chi + 1.252 + B \qquad (9\text{-}204)$$

where Y is the operating-labor-hours per ton per processing step; χ is plant capacity, tons per day; and B is a constant having values of 0.132 when multiple units are used to increase capacity or when the process is completely batch, of 0 for the average chemical-processing plant, and of −0.167 for large, highly automated plants or plants concerned with fluid processing.

Wessel's data for the United States chemical industry refer to the short ton equal to 2000 lb, or 907.2 kg. Labor requirements are higher in countries with lower productivities.

The approximate cost of supervision for operating labor is equivalent to 10 percent of the labor cost for simple operations and 25 percent for complex operations.

Utilities These include steam, cooling water, process water, electricity, fuel, compressed air, and refrigeration. The consumption of utilities can be estimated from the material and energy balances for the process, together with the equipment flow sheet.

Let us consider a cooler in the equipment flow sheet. The required rate of heat removal is known from the balances, and the rate of cooling water can be calculated once the inlet and outlet temperatures of the water have been specified. The calculation of the consumption of other utilities is also straightforward. Allowances should be made for wastage.

The current cost per unit for each utility is usually well known in a company. Thus, the annual cost for utilities and the utilities cost per unit of production can be estimated. The latter is normally much smaller than the raw-materials and labor costs. However, a great deal more work is involved in calculating the utilities cost than for any other item in the manufacturing cost.

Unfortunately, there are no satisfactory shortcut methods for doing this. When the utilities cost is relatively small, it may be possible to make an intelligent guess on the basis of known costs for similar processes in the company. Alternatively, published data for the consumption of utilities per unit of production for various processes may be used.

Miscellaneous Direct Costs Estimates for the cost of maintenance and repairs, operating supplies, royalties, and patents are best based on company records for similar processes. A rough average value for the annual cost of maintenance is 6 percent of the capital cost of the plant. This percentage can vary from 2 to 10 percent, depending on the severity of plant operation. Approximately half of the maintenance costs are for materials and half for labor. Royalty and patents costs are in the order of 1 to 5 percent of the sales price of the product.

Indirect Manufacturing Costs Estimates for the cost of payroll overhead, control laboratory, general plant overhead, packaging, and storage facilities are best based on company records for similar processes.

Payroll overhead includes the cost of pensions, holidays, sick pay, etc., and is normally between 15 and 20 percent of the operating-labor cost. Laboratory work is required for product quality control, and its cost is approximately 10 to 20 percent of the operating-labor cost.

Plant overhead includes the cost of medical, safety, recreational, effluent-disposal, and warehousing facilities, etc. In general, the larger the plant, the lower the overhead per unit of production. Plant-overhead costs can vary between 15 and 150 percent of the operating-labor cost. Packaging costs depend on the physical and chemical nature of the product as well as on its use and value. The cost of packaging is as high as one-third of the selling price for soaps and pharmaceuticals.

Rapid Manufacturing-Cost Estimates Fixed manufacturing costs are a function of the fixed-capital investment and are independent of the production rate of the plant. Property taxes or rates depend on location. They may be taken as 2 percent of the fixed-capital cost of the plant in the absence of specific data. The cost of insurance depends on both location and the hazardous nature of the materials handled. This cost is normally of the order of 1 percent of the fixed-capital cost of the plant.

The manufacturing cost of a product is the sum of the processing or conversion cost and the cost of raw materials. The processing cost can be roughly broken down into three parts: investment-related cost, labor-related cost, and utility cost.

Companies usually include in the charge for overhead the following items: operating supplies, supervision, indirect payroll expenses, plant protection, plant office, general plant overhead, and control laboratory. This overhead charge is frequently taken as an equivalent percentage of the direct labor cost.

The percentage is best obtained from company records. Although it can vary over a wide range, a reasonable value is 125 percent. In this case, the labor-related cost would be 2.25 times the direct labor cost. For a 6000-h year and N persons per shift earning $c_L\$$ per hour, the annual labor-related cost would be 13,500 $(c_L N)$.

Let us consider a plant of fixed-capital cost C_{FC}. If the annual property taxes are taken as 0.02 C_{FC}, insurance as 0.01 C_{FC}, and maintenance as 0.06 C_{FC}, the annual investment-related cost would be 0.09 C_{FC}. Annual utilities cost is A_U. The annual processing cost A_p can be represented by

$$A_p = \eta_1 C_{FC} + \eta_2 c_L N + A_U \qquad (9\text{-}205)$$

where the factors η_1 and η_2 can be obtained from the data available in a particular company and have the dimensions of year^{-1} and hours per year respectively.

Substituting the information previously given into Eq. (9-205) yields the relationship

$$A_p = 0.09 C_{FC} + 13{,}500 c_L N + A_U \qquad (9\text{-}206)$$

Equation (9-206) represents very closely the manufacturing costs of a particular company and is typical of the coefficients to be expected.

The annual processing cost A_{P2} for a similar plant of a different size designed for an annual production rate R_2 can be approximately calculated from an equation of the form

$$A_{P2} = \eta_1 C_{FC1}(R_2/R_1)^{0.7} + \eta_2 c_L N_1 (R_2/R_1)^{0.25} + A_{U1}(R_2/R_1) \qquad (9\text{-}207)$$

(F. A. Holland, F. A. Watson, and J. K. Wilkinson, *Introduction to Process Economics*, 2d ed. Wiley, London and New York, 1983, p. 158).

The processing cost per unit of production for a plant with an annual production R_2 can be approximately calculated from

$$\frac{A_{P2}}{R_2} = \frac{100}{R_2}\left[\eta_1 C_{FC}\left(\frac{R_2}{R_1}\right)^{0.7} + \eta_2 c_L N_1\left(\frac{R_2}{R_1}\right)^{0.25} + A_{U1}\left(\frac{R_2}{R_1}\right)\right] \qquad (9\text{-}208)$$

where A_{P2}/R_2 is in cents per kilogram.

Equation (9-208) can be used to compute data for plots such as Fig. 9-40, which shows the decrease in processing cost per unit of production A_{P2}/R_2 with increasing plant size.

Manufacturing Cost as a Basis for Product Pricing   Pricing on the basis of cost plus a fair profit has the disadvantage of ignoring demand. The modern approach is to price on the basis of market research. However, the classic cost-plus-fair-profit approach can still give useful complementary information. This can be done by any of the following three methods:

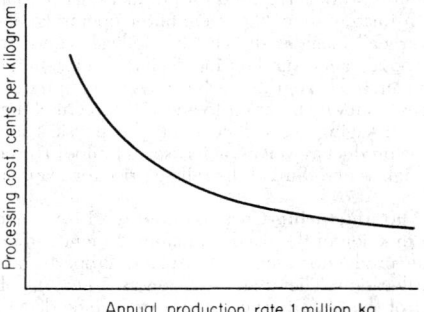

FIG. 9-40   Decrease in processing cost against increase in plant size.

1. Absorption pricing
2. Rate-of-return pricing
3. Marginal pricing

The gross annual profit A_{GP} for a product is given by

$$A_{GP} = A_S - A_{ME} - A_{BD} \qquad (9\text{-}6)$$

where A_S is the revenue from annual sales, A_{ME} the annual manufacturing cost, and A_{BD} the balance-sheet annual depreciation charge.

Equation (9-6) can also be rewritten in the form

$$A_S = A_{GP} + A_{VME} + A_{FME} + A_{BD} \qquad (9\text{-}209)$$

where A_{VME} and A_{FME} are the annual variable and fixed manufacturing costs or expenses respectively.

Equation (9-209) can also be rewritten as

$$c_S = c_{VME} + (A_{GP}/R) + (A_{FME} + A_{BD})/R \qquad (9\text{-}210)$$

where R is the annual sales volume taken as equal to the annual production rate, c_S is the sales price per unit of production, and c_{VME} is the variable manufacturing expense per unit of production.

Absorption pricing is based on a normal annual production rate R. The gross profit per unit A_{GP}/R is taken as a fixed percentage χ of the fixed plus variable manufacturing costs given by the equation

$$\frac{A_{GP}}{R} = \left(\frac{\chi}{100}\right)\left[c_{VME} + \left(\frac{A_{FME} + A_{BD}}{R}\right)\right] \qquad (9\text{-}211)$$

We combine Eqs. (9-210) and (9-211) to give

$$c_S = \left(\frac{100 + \chi}{100}\right)\left[c_{VME} + \left(\frac{A_{FME} + A_{BD}}{R}\right)\right] \qquad (9\text{-}212)$$

Equations (9-210), (9-211), and (9-212) are based on a fixed normal annual production rate R.

Let us consider a change in annual production rate to $R + \Delta R$. In order to maintain the gross profit per unit as A_{GP}/R, the sales price per unit of production would need to be $c_S - \Delta c_S$. For this case Eq. (9-210) can be written in the modified form

$$c_S - \Delta c_S = c_{VME} + (A_{GP}/R) + [(A_{FME} + A_{BD})/(R + \Delta R)] \qquad (9\text{-}213)$$

We subtract Eq. (9-213) from Eq. (9-212) to give

$$\Delta c_S = \frac{\Delta R}{R}\left(\frac{A_{FME} + A_{BD}}{R + \Delta R}\right) \qquad (9\text{-}214)$$

Equation (9-214) gives the overpricing Δc_S per unit of production for an increase in annual production rate ΔR. Equation (9-214) also gives the underpricing Δc_S per unit of production for a decrease in annual production rate ΔR. In the first case the fixed costs or overheads are said to be overabsorbed and in the second case underabsorbed.

Absorption pricing is rigid and arbitrary and may result in business being turned away if the fixed sales price c_S cannot be obtained even though the business may give a useful contribution to fixed costs.

Rate-of-return pricing is a modified form of absorption pricing. It is based on the equation

$$\frac{A_{GP}}{A_{ME}} = \left(\frac{C_{TC}}{A_{ME}}\right)\left(\frac{A_{GP}}{C_{TC}}\right) \qquad (9\text{-}215)$$

where C_{TC} is the total capital employed. Equation (9-215) can also be written as

Percentage markup on cost
    = (capital-turnover ratio)(projected rate of return on capital)

The percentage markup on cost is calculated for a known capital-turnover ratio and a desired rate of return on capital. As with absorption pricing, the percentage markup on manufacturing cost per unit of production is calculated for a normal annual production rate. If this production rate is exceeded, the rate of return on capital will be higher than projected because of the decrease in unit cost. Conversely, if the production rate is lower than normal, the rate of return on capital will be lower than projected because of the increase in unit cost. For production rates both higher and lower than the normal production rate, the percentage markup is based on the normal unit cost. Thus the method is strictly valid only for the normal production rate.

With marginal pricing a company chooses its selling prices so as to maximize the total contribution $\sum R(c_S - c_{VME})$ from its various products. The method is particularly useful for large multiproduct companies with extensive existing facilities since it is the marginal cost that must be considered as the base case when entering into competition with another company. The lowest acceptable price for a product is that which gives the lowest worthwhile contribution to fixed costs. Marginal pricing enables a company to develop a more aggressive pricing policy than when using absorption or rate-of-return pricing.

Absorption, rate-of-return, and marginal pricing have been considered here on the basis of manufacturing cost. Total cost, which is the sum of manufacturing and general costs, can also be considered as the basis. In this case the appropriate profit to consider is the net annual profit rather than the gross annual profit.

Standard Costs for Budgetary Control For convenience and simplicity, we shall consider the total cost of a manufactured product to be the sum of the material, labor, and overhead costs. Standard costs are those that have been predetermined and budgeted for the manufacture of a given amount of product in a given time. The deviation of the actual cost from the standard cost is called the variance. It is far easier to make comparisons between periods by using variances than by using actual production data. The different variances for material, labor, and overhead costs are listed in Table 9-38.

Standard costing is extensively used in budgetary-control systems. Criteria for the establishment of standards range from the maximum possible under ideal conditions to those expected under normal conditions. Past or historical costs are not always the best basis for setting up standards because past performance may have been unnecessarily inefficient.

Static and Flexible Budgets Overhead cost can significantly affect the profitability of a project and is the only cost outside the control of the project manager. The project is expected to contribute a definite amount toward the expenses of the company and will be charged this amount even if the production rate is zero. This is the fixed component of the overhead cost and will include directly allocable costs such as depreciation and a proportion of general costs such as office salaries and heating.

Other nonproduction costs such as indirect labor may vary linearly with the production rate and represent the variable component of the overhead. Costs that are neither fixed nor variable but occur in dis-

TABLE 9-38 Variances

Revenue from sales $Rc_S - R°c_S°$	=	Sales price $R(c_S - c_S°)$ +	Sales volume $c_S°(R - R°)$	
Profit $Rc_{NP} - R°c_{NP}°$	=	Unit profit $R(c_{NP} - c_{NP}°)$ +	Quantity $c_{NP}°(R - R°)$	
Total cost $Rc_{TE} - R°c_{TE}°$	=	Unit cost $R(c_{TE} - c_{TE}°)$ +	Production rate $c_{TE}°(R - R°)$	
Direct material cost $Rc_{RM} - R°c_{RM}°$	=	Material price $R(c_{RM} - c_{RM}°)$ +	Material usage $c_{RM}°(R - R°)$	
Direct labor cost $\theta c_L - \theta°c_L°$	=	Wage rate $\theta(c_L - c_L°)$ +	Labor efficiency $c_L°(\theta - \theta°)$	
Overhead cost $C_{OH} - \theta°c_L°$	=	Budgeted cost $(C_{OH} - \theta c_{BOH})$ +	Volume $\theta(c_{BOH} - c_{NOH}°)$ +	Efficiency $c_{NOH}°(\theta - \theta°)$

c_{BOH} = Flexible budgeted overhead cost, \$/h.
c_L = Actual labor cost, \$/h.
$c_{NOH}°$ = Standard overhead cost based on normal production rate, \$/h.
c_{NP} = Actual net profit before tax, \$/unit.
C_{OH} = Actual overhead cost, \$/period.
c_{RM} = Actual raw-materials cost, \$/unit.
c_S = Actual selling price, \$/unit.
c_{TE} = Actual total cost, \$/unit.
R = Actual quantity, units/period.
$R°$ = Standard quantity, units/period.
θ = Actual time to produce a given quantity, h.
$\theta°$ = Standard time to produce a given quantity, h.
$c_{NP} = c_S - c_{TE}$.
NOTE: The asterisk on all items not otherwise defined indicates the standard cost for that item.

TABLE 9-39 Flexible Budget for Overhead Costs

Overhead cost, \$/month	Production, 1 million lb/month				
	8	9	10	11	12
Fixed	40,000	40,000	40,000	40,000	40,000
Variable	40,000	45,000	50,000	55,000	60,000
Semivariable	40,000	40,000	60,000	60,000	90,000
Total	120,000	125,000	150,000	155,000	190,000

crete steps at various production levels (such as supervisory labor) are the semivariable component of the overhead cost. It is an easy matter to determine these various components for various production rates and list them as shown in Table 9-39.

Two types of overhead budget are currently in use. The static (often referred to as the fixed) budgeted overhead cost is related to the standard budgeted production rate. The flexible budgeted overhead cost is that shown as the total cost in Table 9-39. Values for intermediate production rates are often obtained by interpolation. This is justifiable only when semivariable costs are a negligible part of overhead costs.

Flexible budgeting is more widely used than static budgeting despite certain logical difficulties. This is so because production in many cases is seasonal and the use of a static production norm might distort evaluation of performance. Variances are the difference between the actual costs expended and the budgeted costs expected. Variances are unfavorable if positive and favorable if negative. Any variance should be explained and, if necessary, controlled; the largest variance should be considered first.

Let us consider the overhead-cost data for Table 9-39 with 10 million kg per month as the standard production rate. The static budgeted overhead is then \$150,000 per month, or 1.5 cents per kilogram. We assume that the actual overhead is \$186,000 for a month in which 12 million kg was produced. Then, the static budgeted overhead cost would be 12 million(1.5), or \$180,000 per month. Therefore, the variance is \$186,000 − \$180,000 = +\$6000, which is unfavorable because \$6000 more was spent than was anticipated.

From Table 9-39 we find that the flexible budgeted overhead cost for a production rate of 12 million kg per month is \$190,000. The corresponding variance is \$186,000 minus \$190,000, or −\$4,000, which is favorable because \$4,000 less was spent than was anticipated. Thus, the use of flexible budgeting makes this particular performance look better without changing either the production rate or a single cost of the planned budget.

The Standard Hour The standard hour can be defined as the number of units of output expected to be produced in 1 h. It is often used as a measure of output rate by cost accountants.

Let us consider a batch processing unit that can produce either 1000 kg of product A in a cycle time of 5 h or 900 kg of product B in a cycle time of 3 h. Thus, for this processing unit a standard hour is 200 kg of product A or 300 kg of product B. In a budget period of, say, 1000 h, it is possible to produce 200,000 kg of product A, or 300,000 kg of product B, or any appropriate combination of the two products.

For example, let us assume that production requirements are twice as great for product A as for product B, i.e., a ratio of 600 kg, or 3 standard hours, of product A to 300 kg, or 1 standard hour, of product B. On this basis, for a budget period of 1000 h, 750 standard hours [(¾)1000] would be used to produce 750(200) = 150,000 kg of product A, and 250 standard hours [(1/4)1,000] would be used to produce 250(300) = 75,000 kg of product B.

Production efficiency P_e can be calculated from

$$P_e = (P_s/P_a)100 \qquad (9\text{-}216)$$

where P_s is the actual production rate in standard hours and P_a is the actual hours worked.

The level of production activity P_ℓ can be calculated from

$$P_\ell = (P_s/P_b)100 \qquad (9\text{-}217)$$

where P_b is the budgeted production in standard hours.

The deviation from budgeted capacity b_c can be calculated from

$$b_c = (P_a/P_w)100 \qquad (9\text{-}218)$$

where P_w is the budgeted number of working hours.

TABLE 9-40   Sales, Profits, and Manufacturing Costs

Component	Actual, $/period	Standard, $/period	Variance, $/period
Direct materials cost	325,080	350,000	−24,920
Direct labor cost	55,001	51,600	+3,401
Overhead cost	75,000	68,370	+6,630
Manufacturing cost	455,081	469,970	−14,889
Revenue from sales	588,240	570,000	+18,240
Gross profit	133,159	100,030	+33,129

Actual Costs versus Standard Costs   Let us consider the sales, profits, and manufacturing-cost data in Table 9-40. The gross profit is $33,129 per period better than expected. Clearly, there is less incentive to investigate overall costs when the profit variance is favorable than if the profit were less than expected. However, standard costing enables an objective analysis of the data, whether good or bad, to be made.

The individual variances in Table 9-40 show that the increased profit is due to reduced material costs, which affect manufacturing costs to a greater extent than increased labor and overhead costs. The individual variances also show that to an even greater extent the increased profit is due to the increase in sales revenue.

Clearly, management will wish to investigate both labor and overhead costs for any inefficiencies and to ascertain the reasons for the improved sales revenue. If necessary, the standard values can be revised.

We notice that profit is obtained as the difference between two large cash sums and that variances of some 3 percent in manufacturing costs and sales revenue have resulted in a variance of some 33 percent in gross profit.

Table 9-40 is a very simplified presentation. In a full standard costing system, the direct material, direct labor, and overhead variances are broken down into component parts to enable an even closer look at the operation. Standard costing is an invaluable aid to management for controlling a business.

Variances: Direct Material Cost   Since the variance in direct material cost $\Delta\,(Rc_{RM})$ is the difference between actual cost and standard cost,

$$\Delta\,(Rc_{RM}) = Rc_{RM} - R^\circ c_{RM}^\circ \qquad (9\text{-}219)$$

where R is the actual quantity and R° is the standard quantity, units per period, c_{RM} is the actual price, and c_{RM}° is the standard price, $ per unit.

Equation (9-219) can be written in an expanded form:

$$\Delta\,(Rc_{RM}) = R(c_{RM} - c_{RM}^\circ) + c_{RM}^\circ(R - R^\circ) \qquad (9\text{-}220)$$

where $R(c_{RM} - c_{RM}^\circ)$, known as the direct-material-price variance, is the actual quantity multiplied by the deviation in unit price, and $c_{RM}^\circ(R - R^\circ)$, known as the direct-material-usage variance, is the standard unit price multiplied by the deviation in quantity.

By using the data of Table 9-41, let us calculate the direct materials cost and the standard direct materials cost as

$$Rc_{RM} = 1,806,000(0.18) = \$325,080/\text{period}$$

$$R^\circ c_{RM}^\circ = 1,750,000(0.20) = \$350,000/\text{period}$$

From Eq. (9-219) we calculate the direct-material-cost variance as −$24,920 per period. This variance is favorable. However, by using the relations of Eq. (9-220) we calculate a direct-material-price variance

TABLE 9-41   Cost Data for Problems

Factor	Actual value	Standard value
Raw-materials cost, $/unit	$c_{RM} = 0.18$	$c_{RM}^\circ = 0.20$
Direct labor cost, $/unit	$c_L = 8.45$	$c_L^\circ = 8.00$
Production rate, units/period	$R = 1,806,000$	$R^\circ = 1,750,000$
Production time h/period	$\theta = 6509$	$\theta_N^\circ = 6250$
Fixed and semivariable overhead cost, $/period		$C_{FOH}^\circ = 60,000$
Variable overhead cost, $/h		$C_{VOH}^\circ = 1.00$
Overhead cost, $/period	$C_{OH} = 75,000$	

of $(1,806,000)(0.18 - 0.20) = -\$36,120$ per period. This variance is favorable. Likewise, we calculate a direct-material-usage variance of $(0.20)(1,806,000 - 1,750,000) = \$11,200$ per period. This variance is unfavorable.

In this case, the favorable direct-material-cost variance was achieved because of a lower unit price despite an inefficient material usage, which needs to be investigated. (There is no room for complacency since the lower unit price may well be temporary.)

In the case of mixtures of raw materials, the direct-material-usage variance can be further subdivided into (1) a direct-material-mixture variance and (2) a direct-material-yield variance. The former is due to the difference between the actual and standard mixture compositions, and the latter to the difference between the actual and standard yields. Here, the standard yield is the output expected from the standard input of material. The yield variance denotes the extent of loss of material. The direct-material-mixture variance can be illustrated by Example 22.

Example 22: Direct-Material-Mixture Variance   A standard mixture of 100 units of material contains 70 percent of material A at $0.08 per unit and 30 percent of material B at $0.12 per unit. The standard mixture cost is $70(0.08) + 30(0.12) = \$9.20$.

Now let us consider a mixture of 100 units containing 75 percent of material A and 25 percent of material B. The cost of this mixture at standard prices is $75(0.08) + 25(0.12) = \$9.00$. The direct-material-mixture variance is $9.00 − $9.20 = −$0.20 and is favorable. The favorable variance has been brought about by using more of the lower-priced material A and less of the higher-priced material B.

The direct-material-yield variance is illustrated as follows. Let us assume that the standard mixture (cost $9.20 for 100 units) has a standard loss of 20 percent, making the cost $9.20 for 80 units, or $0.115 per unit of output. Now let us consider the actual loss to be 30 percent, leaving 70 units of output for each 100 units of input. The direct-material-yield variance is $0.115(80 − 70) = $1.15 and is unfavorable.

Variances: Direct Labor Cost   Since the variance in direct labor cost $\Delta\,(\theta c_L)$ is the difference between actual cost and standard cost,

$$\Delta\,(\theta c_L) = \theta c_L - \theta^\circ c_L^\circ \qquad (9\text{-}221)$$

where c_L is the actual pay or wage rate, $ per hour; c_L° is the standard pay or wage rate, $ per hour; θ is the actual time taken to produce a given quantity of product in a given period, hours; and θ° is the standard time taken to produce a given quantity of product in a given period, hours.

Equation (9-221) can also be written in expanded form:

$$\Delta\,(\theta c_L) = \theta(c_L - c_L^\circ) + c_L^\circ(\theta - \theta^\circ) \qquad (9\text{-}222)$$

where $\theta(c_L - c_L^\circ)$, known as the direct pay, or wage-rate, variance, is the actual time taken to produce a given output multiplied by the deviation in wage rate, and $c_L^\circ(\theta - \theta^\circ)$, known as the direct-labor-efficiency variance, is the standard wage rate multiplied by the deviation in time taken to produce a given output.

By using the data of Table 9-41, we calculate that 1 standard hour corresponds to

$$(1,750,000/6250) = 280 \text{ units/standard hour}$$

Standard time to actual production is

$$\theta^\circ = (1,806,000/280) = 6450 \text{ standard hours/period}$$

Direct labor cost is

$$\theta c_L = 6509(8.45) = \$55,001/\text{period}$$

Standard direct labor cost is

$$\theta^\circ c_L^\circ = 6450(8.00) = \$51,600/\text{period}$$

From Eq. (9-219) we calculate the direct-labor-cost variance as $3401 per period. This variance is unfavorable. However, by using the relations of Eq. (9-220), we calculate a direct pay, or wage-rate, variance of $\theta(c_L - c_L^\circ) = 6509(8.45 - 8.00) = \$2929$ per period. This direct pay variance is unfavorable. Likewise, we calculate the direct-labor-efficiency variance as $c_L^\circ(\theta - \theta^\circ) = 8.00(6509 - 6450) = \$472$ per period. This variance is also unfavorable.

For this example, the adverse direct-labor cost variance of $3401 is due to both a higher wage rate per hour and a higher number of labor-hours.

The direct-labor-cost variance can, if necessary, be broken down into a direct-labor-idle-time variance in addition to the direct-wage-rate and direct-labor-efficiency variances. The direct-labor-idle-time variance is simply the number of idle labor-hours in the period multiplied by the standard wage rate. This is rarely relevant to the conditions existing in process plants except when maintenance is involved.

Variances: Overhead Cost The variance in overhead cost ΔC_{OH} is the difference between actual overhead cost and static standard overhead cost,

$$\Delta C_{OH} = C_{OH} - \theta^{\circ} c_{NOH}^{\circ} \qquad (9\text{-}223)$$

where C_{OH} is the actual overhead cost incurred in a given period, $ per period; c_{NOH}° is the static standard overhead cost based on the normal production rate, $ per hour; θ is the actual time taken to produce a given quantity of product in a given period; and θ° is the standard time taken to produce a given quantity of product in a given period, hours.

Equation (9-223) can also be written in expanded form

$$\Delta C_{OH} = (C_{OH} - \theta c_{BOH}) + \theta(c_{BOH} - c_{NOH}^{\circ}) + c_{NOH}^{\circ} (\theta - \theta^{\circ}) \qquad (9\text{-}224)$$

where c_{BOH} is the flexible budgeted overhead cost at the actual production rate or operating capacity.

In Eq. (9-223), $(C_{OH} - \theta c_{BOH})$ is known as the budgeted overhead-cost variance, $\theta(c_{BOH} - c_{NOH}^{\circ})$ as the overhead-volume variance, and $c_{NOH}^{\circ}(\theta - \theta^{\circ})$ as the overhead-efficiency variance. The last is analogous to the labor-efficiency variance and is the standard overhead rate multiplied by the deviation in time taken to produce a given output.

Also in Eq. (9-224), c_{BOH} is simply the flexible budgeted overhead cost in dollars per hour for the actual production rate, and the overhead-volume variance $\theta(c_{BOH} - c_{NOH}^{\circ})$ is the actual time taken to produce a given output multiplied by the difference between the flexible budgeted overhead cost and the standard overhead cost in dollars per hour. The budgeted overhead-cost variance $(C_{OH} - \theta c_{BOH})$ is the difference between the actual overhead cost and the actual time (in hours) required to produce the given output multiplied by the flexible budgeted overhead cost (in dollars per hour).

We shall write the fixed overhead cost for the budget period as C°_{FOH}, the semivariable overhead cost as C°_{SVOH}, and the standard hours to produce the agreed normal production as θ°_N. The standard overhead cost at the agreed normal production rate can then be calculated from

$$c_{NOH}^{\circ} = [(C_{FOH}^{\circ} + C_{SVOH}^{\circ})/\theta_N^{\circ}] + c_{VOH}^{\circ} \qquad (9\text{-}225)$$

where c_{VOH}° is the standard variable overhead cost, $ per hour.

For production rates that differ from the agreed normal rate, the flexible budgeted overhead cost is given by

$$c_{BOH} = [(C_{FOH}^{\circ} + C_{SVOH}^{\circ})/\theta] + c_{VOH}^{\circ} \qquad (9\text{-}226)$$

where θ is the actual hours taken to produce a given amount of product.

For production rates lower than normal, the fixed overheads are underused, and the flexible budgeted overhead cost c_{BOH} is greater than the standard overhead cost c_{NOH}°. For production rates higher than normal, c_{BOH} is less than c_{NOH}°.

It is common practice in cost accountancy to treat the standard semivariable cost C_{SVOH}° at the normal production rate as part of the standard fixed cost. In this case, Eqs. (9-225) and (9-226) can be written respectively as

$$c_{NOH}^{\circ} = (C_{FOH}^{\circ}/\theta_N^{\circ}) + c_{VOH}^{\circ} \qquad (9\text{-}227)$$

$$c_{BOH} = (C_{FOH}^{\circ}/\theta) + c_{VOH}^{\circ} \qquad (9\text{-}228)$$

By using the data of Table 9-40 in Eq. (9-227), we calculate the standard overhead cost to be

$$c_{NOH}^{\circ} = (60,000/6250) + 1.00 = \$10.60/h$$

From Eq. (9-228) and the data in Table 9-40 we calculate the flexible budgeted overhead cost to be

$$c_{BOH} = (60,000/6509) + 1.00 = \$10.22/h$$

By substituting into Eq. (9-223), we calculate the overhead cost variance:

$$\Delta c_{OH} = 75,000 - 6450(10.60) = \$6630$$

This variance is unfavorable. (Note that standard time for actual production was previously calculated to be 6450 h per period.)

The overhead cost variance comprises (1) budgeted overhead-cost variance, (2) overhead-volume variance, and (3) overhead-efficiency variance. The calculations for each follow:

$$c_{OH} - \theta c_{BOH} = 75,000 - 6509(10.22) = \$8478$$

Budgeted overhead-cost variance is positive and, therefore, unfavorable.

$$\theta(c_{BOH} - c_{NOH}^{\circ}) = 6509(10.22 - 10.60) = -\$2473$$

Overhead volume variance is negative and favorable.

$$c_{NOH}^{\circ}(\theta - \theta^{\circ}) = 10.60(6509 - 6450) = \$625$$

Overhead efficiency variance is positive and unfavorable.

The total variances in each category are listed in Table 9-40.

Chemical engineers usually make detailed evaluations of costs rather than evaluations for profits or sales. However, the latter can be analyzed in a similar manner to costs by using the equations shown in Table 9-38. For this purpose, the sign convention will be reversed because an increase in sales or profits would be considered favorable, whereas an increase in cost would be considered unfavorable. The equations can be applied to both batch and continuous processes.

Budgets can be used for both forward planning and control. Variances show managers what their costs should have been and how near they came to meeting budgeted values. Managers will be able to assess, over a number of budget periods, the rate of improvement in performance in their areas of responsibility. A good budgetary system not only should provide detailed information and an appraisal of performance but also should motivate people to improve performance.

Contribution Analysis Contribution analysis can be used to make rapid assessments of the effect of changes in manufacturing costs on profitability. A dimensionless contribution efficiency η can be defined by rewriting Eq. (9-12) in the form

$$\eta = \frac{R(c_S - c_{VE}) - A_{FE}}{R(c_S - c_{VE})} \qquad (9\text{-}229)$$

[F. A. Holland and F. A. Watson, *Eng. Process Econ.*, **1**, 135–143 (1976)].

This represents the ratio of the net annual profit A_{NP} actually achieved divided by the profit which could be obtained if no repayment of capital or interest were required and all fixed-expense items were credited free to the project. The contribution efficiency η is also the profit per unit of contribution. A value for η of unity would be obtained for a very high production rate R whether c_S is greater or less than c_{VE}. For the unusual case of c_S being equal to c_{VE}, the value of η would become negatively infinite for a finite annual fixed expense A_{FE} or positively infinite if A_{FE} became negative because of excessive subsidy of expenses. However, for most projects which are intended to pay their own expenses and taxes, A_{NP} must be positive, and hence c_S is usually greater than c_{VE}, so that η will normally have values in the range of zero to unity. For projects which are not intended to make a profit but are provided for their social or amenity value, the aim should be to bring the value of η as near to zero as possible.

The breakeven production rate R_B is defined by Eq. (9-13) as the production rate at which the project makes neither a profit nor a loss. Equation (9-13) and (9-229) can be combined to give

$$\eta = (R - R_B)/R \qquad (9\text{-}230)$$

which shows that the contribution efficiency η is a function of the production rate R and that η has the value zero when the production rate is the breakeven production rate R_B. For all real projects R_B will be positively finite while R cannot be less than zero, and hence the practical range of η is from negative infinity to unity.

At first glance it might appear that it is desirable to have a value of η as near to unity as possible. However, this is not necessarily so. Reference to Eq. (9-229) will show that if the unit contribution $(c_S - c_{VE})$

is positive, as it must be if the project is ever to make a profit, a value of unity for the contribution efficiency η implies a negligible value of the annual fixed expense A_{FE} when compared with the annual contribution $R(c_S - c_{VE})$. In such cases either $R(c_S - c_{VE})$ is very large, thus attracting competition in spite of high capital charges, or the fixed expenses are very low, so that it is easy for many small competitors to enter the market. The result of such competition often leads to a rapid reduction in sales price c_S. Ball-point pens and electronic calculators were both drastically reduced in price as a result of competitors entering the market.

Since the variable expense per unit of production c_{VE} is by definition independent of production rate, the unit contribution $(c_S - c_{VE})$ and hence the value of η will be reduced. On the other hand, a large company may be in a stronger and more stable position with a modest contribution efficiency and relatively high fixed costs which will deter competitors from entering the market and thereby depressing the sales price. In this argument it is implicit that c_S is also independent of output from the project under consideration. In many cases this will be the case since if many small buyers can choose from many alternative producers, the individual producer cannot adjust the price to suit its output, while at the opposite extreme a group of producers that are in a position to make such adjustments are also likely to attract the attention of antitrust legislation.

It is of interest to be able to examine the effect of changes in productivity on the profitability of projects. Historically, labor costs have been regarded as variable costs, implying that if workers doubled their output their net wages also doubled. This may have been the case for some piecework rates, but it is generally not true today. It is not normally possible to reduce the work force in step with falling demand or to recruit and train labor in step with increasing demand. In general it is better to consider labor as a fixed cost, with any part of a production bonus which is truly proportional to output included in the variable expense c_{VE}. If the annual fixed expense A_{FE} varies significantly with production rate R owing to this factor, then the breakeven chart will consist of curves, the simplicity of the method is lost, and it will be assumed that a particular change in productivity agreements implies a step change in A_{FE} and/or in c_{VE}. Let us consider an increase in productivity for the same fixed labor cost, with other fixed costs remaining the same. We shall let the original production rate R be increased by an increase in productivity by a fraction ϕ. Therefore,

$$\Delta R = R\phi \tag{9-231}$$

where ΔR is the increase in sales volume or production rate. By substitution into Eq. (9-229), the resulting increase in profit ΔA_{NP} is given by

$$A_{NP} + \Delta A_{NP} = A_S + \Delta A_S - (A_{VE} + \Delta A_{VE}) - (A_{FE} + 0) \tag{9-232}$$

We shall subtract A_{NP} from Eq. (9-232) to give

$$\Delta A_{NP} = \Delta A_S - \Delta A_{VE} = \Delta R(c_S - c_{VE}) \tag{9-233}$$

It follows from Eqs. (9-12), (9-229), and (9-233) that

$$\Delta A_{NP}/A_{NP} = \phi/\eta \tag{9-234}$$

Thus a change of productivity ϕ of 10 percent will result in a 10 percent increase in profit when $\eta = 1$ and a very large increase in profit when η is close to zero. If η is negative, increased productivity reduces the profit (or increases the loss).

Equation (9-234) illustrates the enormous influence that go-slow tactics can have on the profitability of companies and processes which have low contribution efficiencies, since a slowdown has little effect on A_{FE}. It is sometimes the case that in different countries productivity per worker varies considerably in similar industries. When poor productivity is not the result of technical or capital inadequacy, it should be possible to increase profitability without a proportionate increase in A_{FE}.

Breakeven charts present a snapshot of the present situation by means of graphs which are generally drawn in the manner shown in Figs. 9-2, 9-3, and 9-4. Since the lines are straight, this implies that c_S, c_{VE}, and A_{FE} will remain constant over the range of variation of R, which is of interest. The values would be based on the production rate currently achieved (or scheduled), since all the data are available from the financial analysis of current production, so that c_S would be $(A_S)_0/R_0$, and so on.

Two well-known ratios used by financial analysts are the profit-to-sales ratio (PSR) and the contribution-to-sales ratio (CSR). The profit-to-sales ratio is simply Eq. (9-127) for profit margin (PM) rewritten as the ratio

$$(PSR) = A_{NP}/Rc_S \tag{9-235}$$

The contribution-to-sales ratio

$$(CSR) = (A_S - A_{VE})/A_S = (c_S - c_{VE})/c_S \tag{9-236}$$

We substitute these values into Eq. (9-229) to give

$$\eta = (PSR)/(CSR) \tag{9-237}$$

The contribution efficiency at the scheduled output η_0 is given by substituting the value of the scheduled output into Eq. (9-229) to give

$$\eta_0 = \frac{R_0(c_S - c_{VE}) - A_{FE}}{R_0(c_S - c_{VE})}$$

The characteristic shape of a given breakeven chart of the type represented by Figs. (9-2), (9-3), and (9-4) can be defined by the two ratios (CSR) and η_0, while the scale of the project can be defined by a single annual cost such as A_{FE}. This information may be used for the rapid investigation of the likely effect on current profits obtainable by changes in various factors such as prices, expenses, and throughput. It should be noted that this technique is not intended to replace discounted methods of investment appraisal but to provide a rapid assessment of the probable effect of changes in current conditions. If the current profitability is always maximized, then the discounted-cash-flow present value will always be made as great as conditions in a changing world will permit.

Valuation of Recycled Heat Energy   The rising cost of energy is having an inflationary effect on manufacturing costs. One obvious way to reduce energy costs is to recycle heat energy whenever possible [S. A. K. El-Meniawy, F. A. Watson, and F. A. Holland, *Indian Chem. Eng.,* **22** (July–September 1980)].

Heat pumps are particularly suitable for recycling heat energy in the chemical-process industries. For the outlay of an additional fixed-capital expenditure C_{FC} on a heat-pump system, a considerable reduction in the annual heating cost can be effected.

Let us consider a process unit requiring heat at the rate of Q_D GJ/h operating for y h in a year. We shall let the unit cost of this base heating requirement be c_B \$ per gigajoule. Therefore the annual heating cost for this unit is $Q_D y c_B$ \$ per year.

We then consider the use of a heat pump to supply this heat so that

$$Q_D = W(COP)_A \tag{9-238}$$

where W is the rate of energy input to the compressor in gigajoules per hour and $(COP)_A$ is the actual coefficient of performance of the heat pump.

When interest charges are involved, the fixed-capital expenditure on a heat-pump system C_{FC} can be related to an annual cost A_{FC} for the estimated life of the heat pump in years by the equation

$$A_{FC} = C_{FC} f_{AP} \tag{9-239}$$

where $f_{AP} = [i(1 + i)^n]/[(1 + i)^n - 1]$, the annuity present-worth factor, and i is the fractional interest rate per year payable on the borrowed money.

A given value for A_{FC} enables a cost in, say, dollars per gigajoule, to be assigned to the heat energy made available by a heat pump.

We shall consider a heat-pump system which operates for y h/year and consumes W GJ/h of high-grade energy to drive the compressor. We shall let the unit cost of the input energy to the compressor be c_I \$ per gigajoule. The annual amount of heat delivered by the heat pump is $Q_D y$, which in terms of Eq. (9-238) can be written as $W(COP)_A y$ in gigajoules per year. The annual cost of this delivered heat, neglecting any maintenance cost, is $(W c_I y + A_{FC})$ in dollars per year. Therefore the unit cost c_D of the heat energy delivered by a compressor-driven pump is

$$c_D = (W c_I y + A_{FC})/[W(COP)_A y] \text{ \$/GJ} \tag{9-240}$$

The ratio of the unit costs of the delivered heat energy and the input energy can be obtained by combining Eqs. (9-239) and (9-240) to give

$$(c_D/c_I) = [(1 + \psi)/(\text{COP})_A] \qquad (9\text{-}241)$$

where $\psi = (C_{FC}/W)(f_{AP}/yc_I)$. ψ is a dimensionless parameter which contains cost and usage data for a particular heat pump; it should have as low a value as possible in order to minimize the unit cost of the delivered heat c_D. The cost C_{FC}/W is the fixed-capital cost per unit of input energy in \$ $(\text{GJh}^{-1})^{-1}$, and this should also be as low as possible consistent with a long life and good reliability for the heat pump. Clearly y, the number of operating hours per year, should approach as closely as possible to the maximum value of 8760 h for a 365-day year.

Since all costs refer to a given year, Eqs. (9-239), (9-240), and (9-241) are independent of inflation.

The annual cost of heat delivered by the heat pump is $Q_D y c_D$, where the unit cost c_D is given by Eq. (9-240). Therefore, the annual saving on heating costs in dollars per year is $Q_D y(c_B - c_D)$, which can also be written in terms of Eq. (9-238) as $W(\text{COP})_A y(c_B - c_D)$.

The payback period in years for a heat-pump system is the additional fixed-capital cost C_{FC} divided by the annual saving on heating costs. This can be written as

$$(\text{PBP}) = \left(\frac{C_{FC}}{W}\right)\left[\frac{1}{(\text{COP})_A y(c_B - c_D)}\right] \qquad (9\text{-}242)$$

which can also be written in terms of Eq. (9-241) as

$$(\text{PBP}) = \left(\frac{C_{FC}}{W}\right)\left\{\frac{1}{y[(\text{COP})_A c_B - c_I(1 + \psi)]}\right\} \qquad (9\text{-}243)$$

For the special case of the unit cost of the input energy to the compressor being the same as the unit cost of the base heat supply, i.e., $c_B = c_I$, Eq. (9-243) simplifies to

$$(\text{PBP}) = \left(\frac{C_{FC}}{W}\right)\frac{1}{yc_I[(\text{COP})_A - (1 + \psi)]} \qquad (9\text{-}244)$$

Equation (9-244) can be used to calculate the payback period when electricity, oil, or gas, etc., is used to drive the compressor and also to provide the base heating.

Equation (9-244) shows that, to have a low payback period (PBP), C_{FC}/W and ψ should be small and y, $(\text{COP})_A$, and c_I large. Clearly as the unit cost of input energy c_I increases, the economics of heat pumps becomes more favorable.

The value of ψ will for most cases be less than 0.2 and with the right application may well be less than 0.1. Values for the annuity present-worth factor will in most cases be less than 0.15.

Since 1 bbl (0.159 m^3) of oil is normally quoted as having a thermal-energy value of 6.12 GJ, a world oil price of, say, US\$40 per barrel is equivalent to US\$6.54 per gigajoule.

For simplicity, we substitute $c_I = $ US\$6⅔ per gigajoule and a conservative value of $[(\text{COP})_A - (1 + \psi)] = 3$ into Eq. (9-244) to give

$$(\text{PBP}) = (C_{FC}/W)(1/20y) \qquad (9\text{-}245)$$

where (PBP) is the payback period in years, y is the operating hours per year, and (C_{FC}/W) is the fixed-capital cost in US\$ $(\text{GJh}^{-1})^{-1}$ of primary-energy input.

Equation (9-245) shows that in this particular case the fixed-capital cost per unit of input energy (C_{FC}/W) must not exceed \$160,000 $(\text{GJh}^{-1})^{-1}$, or \$576 per kilowatt, to have a 1-year payback period if the heat pump is operational for 8000 h/year. For this case the corresponding value of ψ is about 0.12 for a heat pump with an operating life of 10 years purchased with money borrowed at a 10 percent rate of interest.

Equation (9-245) also shows that the fixed-capital cost per unit of energy input (C_{FC}/W) must not exceed \$40,000 $(\text{GJh}^{-1})^{-1}$, or \$144 per kilowatt, to have a 1-year payback period if the heat pump is operational for only 2000 h/year. For this case the corresponding value of ψ is also about 0.12 for a heat pump with an operating life of 10 years purchased with money borrowed at a 10 percent rate of interest.

FIXED-CAPITAL-COST ESTIMATION

Total Capital Cost The installed cost of the fixed-capital investment C_{FC} is obviously an essential item which must be forecast before an investment decision can be made. It forms part of the total capital investment C_{TC}, defined by Eq. (9-14). The fixed-capital investment is usually regarded as the capital needed to provide all the depreciable facilities. It is sometimes divided into two classes by defining battery limits and auxiliary facilities for the project. The boundary for battery limits includes all manufacturing equipment but excludes administrative offices, storage areas, utilities, and other essential and nonessential auxiliary facilities.

Cost Indices The value of money will change because of inflation and deflation. Hence cost data can be accurate only at the time when they are obtained and soon go out of date. Data from cost records of equipment and projects purchased in the past may be converted to present-day values by means of a cost index. The present cost of the item is found by multiplying the historical cost by the ratio of the present cost index divided by the index applicable at the previous date. Ideally each cost item affected by inflation should be forecast separately. Labor costs, construction costs, raw-materials and energy prices, and product prices all change at different rates. Composite indices are derived by adding weighted fractions of the component indices. Most cost indices represent national averages, and local values may differ considerably.

Table 9-42 presents information on some cost indices for the United States. *Engineering News-Record* updates its construction-cost index in March, June, September, and December. The *Oil and Gas Journal* gives the Nelson-Farrar refinery indices in the first issue of each quarter. The *Chemical Engineering* plant-cost index and Marshall and Swift equipment-cost index are given in each issue of the publication *Chemical Engineering*. Derivation of the base values is referred to in the respective publications.

Table 9-43 is based on the method suggested by J. Cran [*Eng. Process Econ.*, **2**, 89–90 (1977)]. He showed that reasonably accurate plant-cost indices for various countries could be derived by using two component indices in the equation

$$(\text{CI})_P = 0.327(\text{CI})_{ST} + 0.673(\text{CI})_L \qquad (9\text{-}246)$$

where $(\text{CI})_{ST}$ is the steel-price index and $(\text{CI})_L$ the earnings index for labor in the particular country. Most of the data required can be obtained from the *United Nations Monthly Bulletin of Statistics* or the Organization for Economic Cooperation and Development (OECD) annual review of the iron and steel industry. In Table 9-43, the plant-cost indices have been brought to a common base of 1980 = 100. The values given do not relate costs in one country to those in another country, as this involves many complex and difficult problems. However, the table indicates the inflationary trends in plant costs since 1980 for each of the countries listed.

Types and Accuracy of Estimates Capital-cost estimates may be required for a variety of reasons, among others to enable feasibility studies to be carried out, to enable a manufacturing company to select from alternative investments, to assist in selection from alternative designs, to provide information for planning the appropriation of capital, and to enable a contractor to bid on a new project. It is therefore essential to achieve the greatest accuracy of estimation with a minimum expenditure of time and money.

Two simple rules are invaluable in aiding the production of consistently accurate estimates:

1. Check the completeness of the project scope.
2. Reduce the effect of bias by using statistically proven methods of estimation based on experience.

Estimates which are lower than actual project costs are often the result of sizable omissions of equipment, services, or auxiliary facilities

TABLE 9-42   Cost Indices for the United States

| Year | Price indices* | | | Construction cost indices | | | |
	Consumer	Producer	Industrial chemicals	Chemical equipment†	Process plants‡	Petroleum refinery§	General construction¶
1980	100	100	100	660	261	286	303
1981	110	108	113	721	297	315	330
1982	117	113	106	746	314	340	357
1983	121	114	103	761	317	357	380
1984	126	115	102	780	323	370	387
1985	131	115	100	790	325	374	392
1986	133	114	97	798	318	380	401
1987	138	117	102	814	324	391	412
1988	144	122	106	852	343	406	422
1989	151	127	108	895	355	417	429
1990	159	132	111	915	358	427	441
1991	165	132	111	931	361	436	452
1992	170	133	110	943	358	445	470
1993	175	136	111	967	359	453	489

*Bureau of Labor Statistics, 1980 = 100.
†Marshall & Swift, *Chem. Eng.*, 1926 = 100.
‡*Chem. Eng.*, plant index, 1957–1959 = 100.
§Nelson-Farrar refinery index, *Oil Gas J.*, 1967 = 100.
¶*Eng. News Rec.*, 1967 = 100.

rather than of errors in pricing or estimation methods. To avoid this, the use of a checklist of items involved in a new project as given in Table 9-44 can be invaluable.

The first stage toward producing an accurate estimate is to use a standard cost code for all construction projects. Table 9-45 shows a suitable numerical cost code, and Table 9-46 shows a typical alphabetical-numerical code. The cost-code system can be used throughout the estimating and construction stages for the collection of cost data by manual or computer methods. There are numerous types of fixed-capital-cost estimates, but in 1958 the American Association of Cost Engineers defined five types as follows:

1. *Order-of-magnitude estimate (ratio estimate).* Rule-of-thumb method based on cost data for previous similar types of plant; probable error within 10 to 50 percent.
2. *Study estimate (factored estimate).* Better than order-of-magnitude; requires knowledge of major items of equipment; used for feasibility surveys; probable error up to 30 percent.
3. *Preliminary estimate (budget-authorization estimate).* Requires more detailed information than study estimate; probable error up to 20 percent.
4. *Definitive estimate (project-control estimate).* Based on considerable data prior to preparation of completed drawings and specifications; probable error within 10 percent.
5. *Detailed estimate (firm or contractor's estimate).* Requires completed drawings, specifications, and site surveys; probable error within 5 percent.

Greater accuracy of estimation may be achieved, within limits, by the expenditure of more time and money. The greater the accuracy required, the greater the time and effort needed to obtain the design and cost data prior to making the estimate.

W. R. Park investigated the cost and accuracy of estimates for a project with a total cost of $1 million as shown in Fig. 9-41 (*Cost Engi-*

neering Analysis, Wiley, New York, 1973, p. 133). Table 9-47 shows typical average costs for producing estimates [adapted from A. Pikulik and H. E. Diaz, *Chem. Eng.,* **84,** 106–122 (Oct. 10, 1977)].

Rapid Estimations

Ratio Methods   J. E. Haselbarth [*Chem. Eng.,* **74,** 214–215 (Dec. 4, 1967)] published data giving the total capital investment per unit of annual production capacity C_{TC}/R. Table 9-48 lists data for many processes involving production units constructed on a previously developed site. Plants built on a green-field site would cost about 30 to 40 percent more, but enlargements of an existing plant would cost about 20 to 30 percent less than the values given in Table 9-48. Total fixed-capital investments for installations within the battery limits are given in Table 9-48. These refer to North American values corresponding to a Marshall and Swift index of 1000 and are adapted from the data of D. R. Woods (*Financial Decision Making in the Process Industry,* Prentice Hall, Englewood Cliffs, NJ, 1975, pp. 288–290). L. Lynn and R. F. Howland [*Chem. Eng.,* **67,** 131–136 (Feb. 8, 1960)] studied the capital ratios for 17 process industries and summarized data for more than 1000 processes. The capital ratio (CR) for a plant erected on a green-field site is defined as the ratio of the fixed-capital investment C_{FC} to the annual sales revenue A_S:

$$(CR) = C_{FC}/A_S \qquad (9\text{-}247)$$

However, Lynn and Howland included in the fixed-capital cost not only money invested in production and storage facilities but also that invested in land, research and development costs, and any auxiliary facilities necessary to support the process. Typical values of capital ratios for the year 1958 are listed in Table 9-49.

Both of the preceding methods are relatively inaccurate and can be used only for rough screening. They have the advantage that an esti-

TABLE 9-43   International Plant-Cost Indices*

Year	Australia	Belgium	Canada	Denmark	France	Germany	Italy	Japan	Netherlands	Spain	Sweden	United Kingdom	United States
1985	151	134	137	143	171	121	201	112	119	190	151	148	124
1986	160	135	142	148	178	124	206	111	121	207	162	156	122
1987	170	138	146	159	186	128	217	110	123	222	171	166	123
1988	180	141	153	163	191	133	231	115	124	233	183	182	131
1989	193	147	160	177	200	137	253	122	126	249	199	200	137
1990	203	152	166	184	204	142	260	126	130	267	215	216	136
1991	208	161	173	193	210	150	269	127	133	283	227	232	138
1992	215	167	174	199	217	160	286	127	140	303	232	241	136
1993	218	171	180	198	223	162	292	129	142	315	244	240	136

*From *Process Eng.,* London, U.K. (monthly). Annual averages for each year (1980 = 100).

TABLE 9-44   Checklist of Items for Fixed-Capital-Cost Estimates

Land:
  Surveys
  Fees
  Property cost
Site development:
  Site clearing
  Grading
  Roads, access and on-site
  Walkways
  Railroads
  Fence
  Parking areas
  Other paved areas
  Wharves and piers
  Recreational facilities
  Landscaping
Process buildings:
  (List as required) Include in each as required substructure, superstructures, platforms, supports, stairways, ladders, access ways, cranes, monorails, hoists, elevators
Auxiliary buildings:
  Administration and office
  Medical or dispensary
  Cafeteria
  Garage
  Product warehouse(s)
  Parts or stores warehouse
  Maintenance shops—electric, piping, sheet metal, machine, welding, carpenters, instrument
  Guard and safety
  Hose houses
  Change houses
  Smoking stations (in hazardous plants)
  Personnel building
  Shipping office and platforms
  Research laboratory
  Control laboratories
Building services:
  Plumbing
  Heating
  Ventilation
  Dust collection
  Air conditioning
  Sprinkler systems
  Elevators, escalators
  Building lighting
  Telephones
  Fire alarm
  Paging
  Intercommunication systems
  Painting
Process equipment:
  (List carefully from checked flow sheets)
Nonprocess equipment:
  Office furniture and equipment
  Cafeteria equipment
  Safety and medical equipment
  Shop equipment
  Automotive heavy maintenance and yard material-handling equipment
  Laboratory equipment
  Lockers and locker-room benches
  Garage equipment
  Shelves, bins, pallets, hand trucks
  Housekeeping equipment
  Fire extinguishers, hoses, fire engines
Process appurtenances:
  Piping—carbon steel, alloy, cast iron, lead-lined, aluminum, copper, asbestos-cement, ceramic, plastic, rubber, reinforced concrete
  Pipe hangers, fittings, valves
  Insulation—piping, equipment
  Instruments
  Instrument panels
  Electrical—panels, switches, motors, conduit, wire, fittings, feeders, grounding, instrument and control wiring

Utilities:
  Boiler plant
  Incinerator
  Ash disposal
  Boiler feed-water treatment
  Electric generation
  Electrical substations
  Refrigeration plant
  Air plant
  Wells
  River intake
  Primary water treatment—filtration, coagulation, aeration
  Secondary water treatment—deionization, demineralization, pH and hardness control
  Cooling towers
  Water storage
  Effluent outfall
  Process-waste sewers
  Process-waste pumping stations
  Sanitary-waste sewers
  Sanitary-waste pumping stations
  Impounders, collection basins
  Waste treatment, including gases
  Storm sewers
Yard distribution and facilities (outside battery limits):
  Process pipe lines—steam, condensate, water, gas, fuel oil, air, fire, instrument, and electric lines
  Raw-material and finished-product handling equipment—elevators, hoists, conveyors, airveyors, cranes
  Raw-material and finished-product storage—tanks, spheres, drums, bins, silos
  Fuel receiving, blending, and storage
  Product loading stations
  Track and truck scales
Miscellaneous:
  Demolition and alteration work
  Catalysts
  Chemicals (initial charge only)
  Spare parts and noninstalled equipment spares
  Surplus equipment, supplies and equipment allowance
  Equipment rentals (for construction)
  Premium time (for construction)
  Inflation cost allowance
  Freight charges
  Taxes and insurance
  Duties
  Allowance for modifications and extra construction work during startup
Engineering costs:
  Administrative
  Process, project, and general engineering
  Drafting
  Cost engineering
  Procurement, expediting, and inspection
  Travel and living expense
  Reproductions
  Communications
  Scale model
  Outside architect and engineering fees
Construction expense:
  Construction, operation, and maintenance of temporary sheds, offices, roads, parking lots, railroads, electrical, piping, communication, and fencing
  Construction tools and equipment
  Warehouse personnel and expense
  Construction supervision
  Accounting and timekeeping
  Purchasing, expediting, and traffic
  Safety and medical
  Guards and watchmen
  Travel and transportation allowance for craft labor
  Fringe benefits
  Housekeeping
  Weather protection
  Permits, special licenses, field tests
  Rental of off-site space
  Contractor's home office expense and fees
  Taxes and insurance, interest

TABLE 9-45   Standard Cost Code (Numerical) Summary

Code Group	Subdivision
Direct costs:	
0–349	Process section (process equipment listed in flow-sheet sequence 1, 2, 3, 4, 5, . . . , N, preceded by process identification number)
350–399	Site development
400–449	Process buildings
450–489	Auxiliary buildings
490–499	Nonprocess equipment
500–599	Process appurtenances (piping, insulation, instrumentation, electrical, etc.)
600–699	Utilities and yard services (boiler plant, refrigeration, compressed air, water supply and treatment, effluents, fire protection, yard piping, yard electrical, yard materials handling, raw and finished-product storage)
700–729	Substructures
730–749	Superstructures
750	Painting
760–769	Building services
770–799	Demolition and alteration to existing structure
800–819	Surplus equipment, supplies and materials, royalty payments
820–830	Design modifications, construction modifications, and extra work during startup
Indirect costs:	
850–874	Home office engineering
875–889	Architect's and engineer's charges and fees
900–969	Construction expenses
970–994	Taxes and insurance
995	Contractor's home office expenses
996	Contractor's fees

mate can be made in a few minutes, and they do not require design work or process flow sheets.

Step Count Methods These methods, used for order-of-magnitude estimates, are based on the definition of the functional units required to carry out the process. A functional unit is a signifi-

TABLE 9-46   Standard Cost Code (Alphabetical-Numerical)

A	Site work and foundations
B	Buildings (less foundations)
C	Steel structures and platforms (other than buildings)
D	Heat exchangers
E	Fractionating towers
F	Tanks and drums
G	Pumps and pump drivers
H	Compressors and blowers
J	Reactors and converters
K	Grinding, crushing, and classifying equipment
L	Materials-handling equipment
M	Fired heaters
N	Catalysts and chemicals
O	Laboratory equipment
P	Piping
Q	Instruments and controls
R	Electrical
S	Insulation and painting
T	Utility equipment (boilers, generators, refrigeration)
U	Plant and building accessories (railroads, fence, etc.)
V	Laboratory equipment
W	Safety equipment
X	Warehouse spares
100	Equipment and materials delivered and installed (A through X)
200	Sales and use taxes
300	Temporary facilities
320	Construction equipment, tools, and supplies
340	Construction supervision
360	Field office expense
380	Warehousing expense
400	Payroll taxes and insurance
500	Home office engineering costs
600	Procurement costs
700	Resident engineering
800	Royalty payments
900	Engineering administrative overhead and profit
950	Constructor's administrative overhead and profit

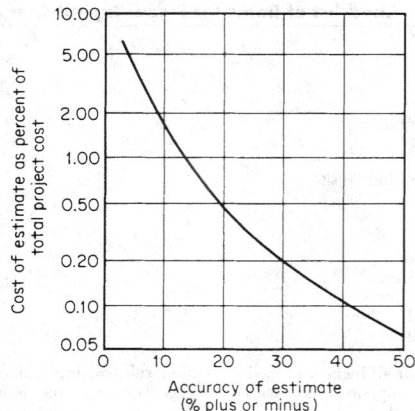

FIG. 9-41 Relationship between cost and accuracy of cost estimations.

cant process step, including all the equipment and ancillaries to operate. The sum of the costs of the functional units gives an estimate of the total capital cost of the plant. Usually a functional unit is a unit operation, a unit process, or a separation involving energy transfer, moving parts, and possibly a high level of internals.

Pumping and heat exchange form part of a functional unit. "In-process" storage is ignored, but large storage requirements for raw materials, intermediates, or products are usually estimated separately. Sometimes there are difficulties in the identification and definition of the functional units involved in a process.

For a particular process, the capital cost per functional unit is given by

$$C_F = f(q, T, p, M_c, CI) \qquad (9\text{-}248)$$

where q is capacity, T is operating temperature, p is pressure, M_c is a materials of construction factor, and CI is a relevant cost index.

Typical methods are those of F. C. Zevnik and R. L. Buchanan [*Chem. Eng. Progr.*, **59**, 70–77 (Feb. 1963)] and J. H. Taylor (*Eng. & Proc. Econ.*, **2**, 259–267, 1977). The former is mainly a graphical method of estimating the cost per functional unit (C_F) based on the capacity, the maximum pressure, the maximum temperature, and the materials of construction. The Taylor method requires the determination of the **costliness index,** which is dependent on the complexity of the process. A simpler method was suggested by S. R. Timms (M.Phil. thesis, Aston University, England, 1980) to give the battery limits cost for gas phase processes only in U.S. dollars with a Marshall and Swift index of 1000. The simple equation is

$$C_{BL} = 11{,}800Nq^{0.615} \qquad (9\text{-}249a)$$

Taking into account materials of construction, temperature, and pressure effects, this becomes

$$C_{BL} = 5500Nq^{0.639}M_c(T_{max})^{0.066}(p_{max})^{-0.016} \qquad (9\text{-}249b)$$

where N is the number of functional units; q is the capacity (tons/y); M_c is the materials of construction factor: 1.0 for carbon steel, 1.15 for

TABLE 9-47   Typical Average Costs for Making Estimates (1990)*

	Cost of project		
Type of estimate	Less than $2 million	$2 to $10 million	$10 to $100 million
Order of magnitude estimate	$ 3,000	$ 6,000	$ 13,000
Study estimate	20,000	40,000	80,000
Preliminary estimate	50,000	80,000	130,000
Definitive estimate	80,000	160,000	320,000
Detailed estimate	200,000	520,000	1,000,000

*Adapted from A. Pikulik and H. E. Diaz, "Cost Estimating for Major Process Equipment." *Chem. Eng.*, **84**, 106–122 (Oct. 10, 1977).

TABLE 9-48   Capital-Cost Data for Processing Plants*

Product	Process route	Size, 1000 metric tons/year	Approximate cost† $ \times 10^6$	Total capital cost ($) metric ton/yr of product	Size range 1000 metric tons/year	Exponent n‡
			Chemical plants			
Acetaldehyde	Ethylene	50	13.7	274	20–150	0.70
Acetic acid	Methanol/CO	10	6.6	660	3–30	0.68
Acetone	Propylene	100	35.0	350	30–300	0.45
Ammonia	Steam reforming	100	26.0	260	30–300	0.70
Ammonium nitrate	Ammonia/nitric acid	100	6.0	60	30–300	0.65
Butanol	Propylene/CO/H_2O	50	44.0	880	20–150	0.40
Chlorine	Electrolysis of NaCl	50	31.0	620	20–150	0.45
Ethylene	Refinery gases	50	14.4	288	20–150	0.83
Ethylene oxide	Ethylene/O_2	50	55.0	1100	20–150	0.78
Formaldehyde (37%)	Methanol	10	17.7	1770	3–30	0.55
Glycol	Ethylene/Cl_2	5	16.6	3320	2–20	0.75
Hydrofluoric acid	Hydrogen fluoride/H_2O	10	8.8	880	3–30	0.68
Methanol	CO_2/natural gas/steam	60	14.4	240	20–200	0.60
Nitric acid (conc.)	Ammonia oxidation	100	6.6	66	30–300	0.60
Phosphoric acid	Calcium phosphate/H_2SO_4	5	3.3	660	2–20	0.60
Polyethylene (high density)	Ethylene	5	17.7	3540	2–20	0.65
Propylene	Refinery gases	10	3.3	330	3–30	0.70
Sulfuric acid	Sulfur	100	3.3	33	30–300	0.65
Urea	Ammonia/CO_2	60	8.8	147	20–200	0.70
			Refinery units			
Alkylation (H_2SO_4)	Catalytic	10	21.0	2100	3–30	0.60
Coking (delayed)	Thermal	10	28.8	2880	3–30	0.38
Coking (fluid)	Thermal	10	17.7	1770	3–30	0.42
Cracking (fluid)	Catalytic	10	17.7	1770	3–30	0.70
Cracking	Thermal	10	5.5	550	3–30	0.70
Distillation (atm)	65% vaporized	100	35.4	354	30–300	0.90
Distillation (vac.)	65% vaporized	100	21.0	210	30–300	0.70
Hydrotreating	Catalytic desulfurization	10	3.3	330	3–30	0.65
Reforming	Catalytic	10	32.0	3200	3–30	0.60
Polymerization	Catalytic	10	5.5	550	3–30	0.58

*Adapted from M. S. Peters and K. D. Timmerhaus, "Plant Design and Economics for Chemical Engineers", 4th ed., McGraw-Hill, New York, 1991.
†All costs are approximate U.S.A. values with M & S = 1000, assuming 330 operating days per year.
‡Exponents apply roughly for threefold capacity ratio extending either way from the plant size given.

low-grade stainless steel, 1.2 for medium-grade stainless steel, 1.3 for high-grade stainless steel; T_{max} is the maximum process temperature (K); and p_{max} is the maximum process pressure (bars).

For liquid and/or solid handling, A. V. Bridgwater and F. D. G. Bossom (*Proceedings of the 6th International Cost Engineering Conference,* Mexico, October, 1980) suggested the following equations:

$$C_{BL} = 2160N(q/s)^{0.675} \qquad (9\text{-}250)$$

for plant capacities (q) above 60,000 tons/y, where C_{BL} is the capital cost for battery limits in U.S. dollars (M & S = 1,000); N is the number of functional units; and s is the reactor conversion (weight of desired

TABLE 9-49   Capital Ratios for Process Industries

Industry	Capital ratio,* 1958
Chemicals, general	2.02
Carbon black	3.98
Explosives	1.64
Glass	1.46
Fibers, synthetic	3.44
Foodstuffs, processed	0.66
Inorganics, heavy	2.24
Nonferrous metals	3.31
Petroleum	3.08
Pharmaceuticals	0.92
Pigments, paints, and inks	1.04
Pulp and paper	2.01
Resins and plastics	1.90
Rubber	1.04
Soap and detergents	0.69
Steel	2.78
Sulfur	1.97
Average	2.01

*Capital ratio = (fixed-capital investment)/(annual sales revenue).

reactor product/weight of reactor input). Thus, q/s represents the feed throughput in tons per year. For plant capacities below 60,000 tons per year, the equation becomes:

$$C_{BL} = 189,300N(q/s)^{0.30} \qquad (9\text{-}251)$$

In general, the step count method of estimation can be applied to any special situation to derive a model equation for that particular industry or group of processes.

Exponential Methods   Rapid capital-cost estimates can be made by using capacity-ratio exponents based on existing cost data of a company or drawn from published correlations.

If the cost of a piece of equipment or plant of size or capacity q_1 is C_1, then the cost of a similar piece of equipment or plant of size or capacity q_2 can be calculated from

$$C_2 = C_1(q_2/q_1)^n \qquad (9\text{-}252)$$

where the value of the exponent n depends on the type of equipment or plant. Cost indices should be used to bring the cost data to a common year. Table 9-48 gives typical values of n for various processes along with the cost of a plant of given capacity at a particular time and the capacity range of applicability. For process plants, capacity is expressed in terms of annual production capacity in metric tons per year.

Exponential cost correlations have been developed for individual items of equipment. Care must be taken in determining whether the cost of the equipment has been expressed as free on board (FOB), delivered (DEL), or installed (INST), as this is not always clearly stated. In many cases the cost must be correlated in terms of parameters related to capacity such as surface area for heat exchangers or power for grinding equipment. There are four main sources of error in such cost correlations:

1. Oversimplification by correlating the cost of equipment in terms of a single variable

2. Representation of data by using a simple exponential relationship
3. Failure to include the effects of technological improvements
4. Errors incurred because of special circumstances

Table 9-50 gives typical values of the exponent n for many types of equipment. Prices are North American with a Marshall and Swift index of 1000, mainly for carbon steel equipment.

Factor Estimations Most factor methods for estimating the total installed cost of a process plant are based on a combination of materials, labor, and overhead cost components. These can be conveniently grouped as

1. Cost of major items of equipment
2. Cost of complete installation of equipment
3. Auxiliary equipment to make the process work
4. Engineering and field expenses
5. Contractor's fees and contingencies

A great variety of factors are in use, depending on the time available and the accuracy expected. Normally the input information required is the base cost. Determination of this cost usually requires a knowledge of equipment sizes, probably using mass and energy balances for the proposed process.

Equipment or Base Cost The total cost of the main-plant items is generally used as the base cost. Again, care must be taken with equipment costs which may be quoted as installed (INST), delivered to site (DEL), or free on board the delivery vehicle at the place of manufacture or other specified location (FOB).

Base equipment includes all equipment within the battery limits whose cost is as significant as the cost of a pump. For example, storage tanks, knockout drums, accumulators, heat exchangers, and pumps are classed as main-plant items (MPI). Early in the development of the process-flow diagram, it is advisable to increase the estimated (MPI) cost by 10 to 20 percent to allow for later additions. When the scope of the process has been well defined, (MPI) costs should be increased by 1 to 10 percent.

For order-of-magnitude estimates the cost of equipment delivered $(C_{EQ})_{DEL}$ varies approximately from 1.1 to 1.25 times the FOB cost $(C_{EQ})_{FOB}$. The factor would be at the lower end of the range for domestic purchases and at the higher end for imports. Installation costs include unpacking, mounting, and connecting up to existing auxiliaries or utilities. The cost of equipment installed $(C_{EQ})_{INST}$ varies with type and size but generally ranges from 1.4 to 2.2 times the delivered-equipment cost $(C_{EQ})_{DEL}$.

Single-factor methods collect the various items of expenditure into one factor, which is usually used to multiply the total cost of delivered equipment $\sum (C_{EQ})_{DEL}$ to give the fixed-capital cost for plant within the battery limits:

$$(C_{FC})_{BL} = f \sum (C_{EQ})_{DEL} \qquad (9\text{-}253)$$

Typical values for single factors f for battery-limit-plant costs (for a carbon steel plant including auxiliaries but not land) are as follows:

Solids processing (S)	3.8
Solids-fluid processing (S-F)	4.1
Fluid processing (F)	4.8

Thus the factors vary with the type of processing, although the boundaries between the classifications are not clear-cut and considerable judgment is required in selection of the correct factor.

Multiple-factor methods include the cost contributions for each given activity, which can be added together to give an overall factor. This factor can be used to multiply the total cost of delivered equipment $\sum (C_{EQ})_{DEL}$ to produce an estimate of the total fixed-capital investment either for grass-roots or for battery-limit plants. The costs may be divided into four groups:

1. Cost of plant within battery limits
2. Cost of auxiliaries
3. Cost of engineering and field expenses
4. Cost of contractor's fees plus contingency allowance

Table 9-51 gives typical values of such factors for carbon steel installations taken from the data of D. R. Woods (*Financial Decision Making in the Process Industry*, Prentice Hall, Englewood Cliffs, NJ, 1975, p. 184). Auxiliaries and site preparation are given as factors of the delivered-equipment cost in Table 9-51, whereas C. A. Miller [*Chem.*

Eng., **72**, 226–236 (Sept. 13. 1965)] expresses auxiliary costs as factors of the battery-limit (BL) cost. Table 9-52 gives the factors from the breakdown of Miller, which is more detailed than that of Woods.

Example 23: Estimation of Total Installed Cost of a Plant Let us estimate the total installed cost for a grass-roots plant producing an organic chemical (S-F process) on a continuous basis. We assume that the total cost of delivered equipment $\sum (C_{EQ})_{DEL}$ is $1 million and use suitable factors from Table 9-51.

The estimated values for the various contributions are given in Table 9-53, resulting in an estimate of $4,280,000 for the total fixed-capital investment, including a contingency factor.

A multiple-factor method for predesign cost estimating has been put forward by D. H. Allen and R. C. Page [*Chem. Eng.*, **82**, 142–150 (Mar. 3, 1975)] for fluid-type plants (F) that include some vapor processing. The method requires the following input information:

1. Plant flow sheet giving main-plant items and process streams
2. Total process-stream input per year
3. Extreme temperature and pressure conditions, if any
4. Materials of construction for main-plant items
5. Operating phases for each main-plant item
6. Expectation of any unusually high or low, direct or indirect initial costs

By means of 12 procedural steps involving the input information, several equations, graphs, and tables, the total cost of delivered equipment $\sum (C_{EQ})_{DEL}$ is estimated. This is then converted into a grass-roots investment estimate by dividing $\sum (C_{EQ})_{DEL}$ by a single factor ranging from 15 to 30 percent (average value, 21 percent). The method is rapid and is claimed to be accurate within −20 to +25 percent, but it has only been tested by using data published in the literature for eight plants.

Multiple-Factor Methods That Separate Materials and Labor These methods have become increasingly popular. While they are similar to the preceding methods, labor and materials costs are considered separately. Hence it is possible to allow for variations in efficiency and labor costs in different localities or countries. H. C. Bauman (*Fundamentals of Cost Engineering in the Chemical Industry*, Van Nostrand Reinhold, New York, 1964, p. 295) divides most of the components of Table 9-51 into material and labor components, quoting the data as ranges and medians of the percentage of the total fixed-capital investment. In Table 9-54, Bauman's data have been converted to factors of the delivered-equipment cost for a grass-roots installation.

A study has been made by A. V. Bridgwater [*Chem. Eng.*, **86**, 119–121 (Nov. 5, 1979)] of the geographical variations in capital costs. He concluded that because of trade and competition basic equipment costs do not vary significantly in the industrialized countries of the western world. The main differences in construction costs at various international locations are due to variations in labor costs and productivity, the use of specialized equipment, and sundry local factors. Table 9-55 gives location factors for the construction of chemical plants of similar function in various countries (1993 values). The factors have been corrected by Bridgwater for location variations in labor costs and efficiency and converted at the average value of the exchange rate.

Factor Methods Using the Modular Approach These are methods used for estimating the cost of major-equipment units and have been proposed by several authors. Perhaps the most comprehensive is the method suggested by K. M. Guthrie [*Chem. Eng.*, **76**, 114–142 (Mar. 24, 1969)]. Table 9-56 gives average factors for major-equipment items based on a $(C_{EQ})_{FOB}$ cost for carbon steel units. To the FOB cost of the item is added, by means of factors, the total materials cost to complete the module M. Erection and setting costs L are added as a factor or calculated from the L/M cost ratio to give $M + L = X$, the direct module cost. Indirect costs, such as freight, taxes, insurance, engineering, and field expense, are added to $(M + L)$ to give the total module cost. This excludes contingency allowances, contractor's fees, auxiliaries, site development, land, and industrial buildings, which may have to be added when applicable. The factors in Table 9-56 were based on mid-1968 prices for a United States Gulf Coast location.

TABLE 9-50 Typical Exponents for Equipment Cost versus Capacity

Equipment	Size	Unit	Approximate cost, $000	Size range	Exponent
Agitator, turbine, top entry, open, FOB	10 (7.5)	hp (kW)	7.0	2–30 (1.5–22.4)	0.45
Agitator, turbine, top entry, closed, FOB	10 (7.5)	hp (kW)	10.7	2–200 (1.5–150)	0.56
Blower, centrifugal, 4 lbf/in² (27.6 kN/m²), DEL, excluding motor	10 (4.72)	10^3 sft³/min (sm³/s)	67	0.5–150 (0.24–71)	0.60
Cone crusher, FOB, crusher only	100 (74.6)	hp (kW)	130	30–300 (22.4–224)	0.92
Jaw crusher, FOB, excluding motor	10 (7.5)	hp (kW)	34	1–60 (0.75–44.7)	0.65
Jaw crusher, FOB, excluding motor	100 (74.6)	hp (kW)	284	60–400 (44.7–300)	0.81
Centrifugal pump, C/S, FOB, excluding motor	10 (7.5)	hp (kW)	1.6	0.5–40 (0.37–30)	0.30
Centrifugal pump, C/S, FOB, excluding motor	100 (74.6)	hp (kW)	4.4	40–400 (30–300)	0.67
Conveyor, belt, C/S, FOB, excluding motor	100 (9.3)	ft² (m²)	6.7	60–200 (5.6–18.6)	0.50
Conveyor, screw, C/S, DEL, excluding motor	70 (540)	ft × m diameter (m × mm diameter)	10	50–100 (390–780)	0.46
Centrifuge, automatic batch, horizontal, C/S, FOB	20 (1.86)	Filter area, ft² (m²)	100	7–80 (0.65–7.43)	0.65
Compressor, reciprocating, <1000 lbf/in², FOB, including motor	300 (224)	hp (kW)	133	1–20000 (0.75–1490)	0.84
Crystallizer, forced circulation, C/S, FOB	100 (91)	ton/day (Mg/day)	283	10–1000 (9.1–970)	0.59
Dryer, drum, C/S, FOB, excluding motor	100 (9.3)	ft² (m²)	73	10–400 (0.9–37)	0.52
Dryer, vacuum, shelf, C/S, FOB, excluding trays, vacuum equipment	100 (9.3)	ft² (m²)	17	15–1000 (1.4–93)	0.56
Dust collector, cloth, shaker type, FOB, including motors	10^4 (4.7)	sft³/min (m³/s)	17	10^3–5 × 10^4 (0.47–23.6)	0.79
Dust collector, multicyclones, FOB	10^4 (4.7)	sft³/min (m³/s)	7	10^3–1.5 × 10^5 (0.47–70.8)	0.66
	10^4 (4.7)	ft³/min at 40°C (m³/s)	77	10^3–8 × 10^4 (0.47–73.8)	0.39
Electrostatic precipitator, FOB	2 × 10^5 (94)	(m³/s)	383	8 × 10^4–10^6 (37.8–472)	0.81
Ejector, single-stage, 100 psig, steam, FOB	3 (10^{-2})	lb/h (air/mmHg absolute)	2.7	0.2–30 (6.8 × 10^{-4}–0.1)	0.50
Ejector, two-stage, FOB, including condenser, piping	1 (3.4 × 10^{-3})	[kg/h/(N/m²)]	6.3	0.2–10 (6.8 × 10^{-4}–3.4 × 10^{-2})	0.43
Ejector, multistage, FOB, including condenser, piping	10 (3.4 × 10^{-2})	[kg/h/(N/m²)]	16.7	0.2–100 (6.8 × 10^{-4}–0.34)	0.26
Filter, vertical-pressure leaf, C/S, DEL	100 (9.3)	ft² (m²)	17	30–1500 (2.8–140)	0.57
Filter, plate and frame, C/S, DEL	100 (9.3)	ft² (m²)	5.7	10–1000 (0.9–93)	0.55
Filter, vacuum rotary drum, C/S, FOB, including motor	100 (9.3)	ft² (m²)	63.3	10–1500 (0.9–140)	0.48
Heat exchanger, shell-tube, floating head, C/S, DEL; fixed tube × 0.85; U tube × 0.87; kettle × 1.35	1000 (93)	ft² (m²)	21.7	20–20000 (1.9–1860)	0.59
Heat exchanger, thermal screw, C/S, FOB, excluding motor	100 (9.3)	ft² (m²)	33	10–400 (0.9–37)	0.78
Kettle, jacketed, glass-lined, FOB	100 (0.38)	U.S. gal (m³)	53	50–1000 (0.2–3.8)	0.48
Motors, ac induction, wound rotor, TEFC, FOB	10 (7.5)	hp (kW)	12.3	10–25 (7.5–18.6)	0.56
Motors, ac induction, wound rotor, TEFC, FOB	70 (52)	hp (kW)	19.3	25–200 (18.6–149)	0.77
Piping, typical straight run, C/S, FOB, $/ft					
Installed: $/ft × 6 to 7	6 (152)	Nominal diameter in (mm)	0.0093	1–24 (25–610)	1.33
Complex network: FOB $/ft × 2 Installed: $/ft × 13					
Pressure vessel horizontal drum (150 psig), C/S	1000 (3.8)	U.S. gal (m³)	6.3	100–80000 (0.4–302)	0.62
Jacketed reactors, including mixer, FOB	100 (0.38)	U.S. gal (m³)	9.3	10–4000 (0.04–15.1)	0.53
Refrigeration, packaged mechanical, INST	100 (351.7)	U.S. tons (kW)	133	10–1000 (35.2–3520)	0.73
Screen, vibrating, single-deck, DEL, including motor	500 (46)	ft² (m²)	10	150–700 (14–65)	0.62
Stack, carbon steel		ft (m)	—	20–150 (6.1–45.7)	1.00
Tanks: atm, horizontal cylinder, C/S, FOB	1000 (3.8)	U.S. gal (m³)	4.7	100–40000 (0.4–151)	0.57
Vertical cylinder, C/S, FOB	1000 (3.8)	U.S. gal (m³)	3.3	100–20000 (0.4–76)	0.30
Vertical jacketed, C/S, FOB	1000 (3.8)	U.S. gal (m³)	15	70–1500 (0.26–5.7)	0.57
Vertical agitated, C/S, FOB, including motor	1000 (3.8)	U.S. gal (m³)	12.3	100–20000 (0.4–76)	0.50
Towers, distillation including internals, INST	4000 (trays)	$\left(\dfrac{\text{feed, lb/year}}{10^6}\right)^{0.65}$	3300	300–30000	1.00

NOTE: All costs are North American values with M & S = 1000.

TABLE 9-51 Factors to Convert Delivered-Equipment Costs into Fixed-Capital Investment

Details	Grass-roots plants			Battery-limit installations		
	Solids processing	Solids-fluid processing	Fluid processing	Solids processing	Solids-fluid processing	Fluid processing
Equipment, delivered	1.00	1.00	1.00	1.00	1.00	1.00
Installed	0.19–0.23	0.39–0.43	0.76	0.45	0.39	0.27–0.47
Piping	0.07–0.23	0.30–0.39	0.33	0.16	0.31	0.66–1.20
Structural steel foundations, reinforced concrete			0.28			0–0.13
Electrical	0.13–0.25	0.08–0.17	0.09	0.10	0.10	0.09–0.11
Instruments	0.03–0.12	0.13	0.13	0.09	0.13	
Battery-limits building and service	0.33–0.50	0.26–0.35	0.45	0.25	0.39	0.18–0.34
Excavation and site preparation	0.03–0.18	0.08–0.22		0.13	0.10	0.10
Auxiliaries	0.14–0.30	0.48–0.55	Included above	0.40	0.55	0.70
Total physical plant	2.37	2.97	3.04	2.58	2.97	3.50
Field expense	0.10–0.12	0.35–0.43		0.39	0.34	0.41
Engineering		0.35–0.43	0.41	0.33	0.32	0.33
Direct plant costs	2.48	3.73	3.45	3.30	3.63	4.24
Contractor's fees, overhead, profit	0.30–0.33	0.09–0.17	0.17	0.17	0.18	0.21
Contingency	0.26	0.39	0.36	0.34	0.36	0.42
C_{FC}: total fixed-capital investment	3.06	4.27	3.98	3.81	4.17	4.87

TABLE 9-52 Factor Method of Miller (Based on Delivered-Equipment Costs = 100)*

		Battery-limit costs (range of factors in percent of basic equipment); average unit cost of main-plant item (MPI)						
		Under $9000	$9000 to $15,000	$15,000 to $21,000	$21,000 to $30,000	$30,000 to $39,000	$39,000 to $51,000	$51,000
Field erection of basic equipment	High percentage of equipment involving high field labor	23/18	21/17	19.5/16	18.5/15	17.5/14.2	16.5/13.5	15.5/13
	Average (mild steel) equipment	18/12.5	17/11.5	16/10.8	15/10	14.2/9.2	13.5/8.5	13/8
	High percentage of corrosion materials and other high-unit-cost equipment involving little field erection	12.5/7.5	11.5/6.7	10.8/6	10/5.5	9.2/5.2	8.5/5	8/4.8
Equipment foundations and structural supports	High: predominance of compressors or mild steel equipment requiring heavy foundations			17/12	15/10	14/9	12/8	10.5/6
	Average: for mild steel fabricated-equipment solids			12.5/7	11/6	9.5/5	8/4	7/3
	Average: for predominance of alloy and other high-unit-price fabricated equipment	7/3	8/3	8.5/3	7.5/3	6.5/2.5	5.5/2	4.5/1.5
	Low: equipment more or less sitting on floor	5/0	4/0	3/0	2.5/0	2/0	1.5/0	1/0
	Piling or rock excavation	Increase above values by 25 to 100%						
Piping, including ductwork but excluding insulation	High: gases and liquids, petrochemicals, plants with substantial ductwork	105/65	90/58	80/48	70/40	58/34	50/30	42/25
	Average for chemical plants: liquids, electrolytic plants	65/33	58/27	48/22	40/16	34/12	30/10	25/9
	Liquids and solids	33/13	27/10	22/8	16/6	12/5	10/4	9/3
	Low: solids	13/5	10/4	8/3	6/2	5/1	4/0	3/0
Insulation of equipment only	Very high: substantial mild steel equipment requiring lagging and very low temperatures	13/10	11.5/8.5	10/7.4	9/6.2	7.8/5.3	6.8/4.5	5.8/3.5
	High: substantial equipment requiring lagging and high temperatures (petrochemicals)	10.3/7.5	9/6.3	7.8/5.2	6.7/4.2	5.7/3.4	4.7/3.8	4.8/2.5
	Average for chemical plants	7.8/3.4	6.5/2.6	5.5/2.1	4.5/1.7	3.6/1.4	2.9/1.1	2.2/.8
	Low	3.5/0	2.7/0	2.2/0	1.8/0	1.5/0	1.2/0	1/0
Insulation of piping only	Very high: substantial mild steel piping requiring lagging and very low temperatures	22/16	19/13	16/11	14/9	12/7	9/5	6/3.5
	High: substantial piping requiring lagging and high temperatures (petrochemicals)	18/14	15/12	13/10	11/8	9/6	7/4	4.5/2.5
	Average for chemical plants	16/12	14/10	12/8	10/6	8/4	6/2	4/2
	Low	14/8	12/6	10/5	8/4	6/3	4/2	2/1

TABLE 9-52   Factor Method of Miller (Based on Delivered-Equipment Costs = 100) *(Concluded)*

		Battery-limit costs (range of factors in percent of basic equipment); average unit cost of main-plant item (MPI)						
		Under $9000	$9000 to $15,000	$15,000 to $21,000	$21,000 to $30,000	$30,000 to $39,000	$39,000 tc $51,000	$51,000
All electrical except building, lighting, and instrumentation	Electrolytic plants, including rectification equipment		55/42	50/38	45/33	40/30	35/26	
	Plants with mild steel equipment, heavy drives, solids	26/17	22.5/15	19.5/12.5	17/10	14/8.5	12/7	10/6
	Plants with alloy or high-unit-cost equipment, chemical and petrochemical plants	18/9.5	15.5/8.5	13/6.5	11/5.5	9/4.5	7.3/3.5	6/2.5
Instrumentation°	Substantial instrumentation, central control panels, petrochemicals		58/31	46/24	37/18	29/13	23/10	18/7
	Miscellaneous chemical plants		32/13	26/10	20/7	15/5	11/3	8/2
	Little instrumentation, solids		21/9	17/7	13/5	10/3	7/2	5/1
Miscellaneous, including site preparation, painting, and other items not accounted for above	Top of range—large complicated processes; bottom of range—smaller, simple processes	Range for all values of basic equipment is 6 to 1%						

Buildings—architectural and structural, excludes building services†	Building evaluation when most process units are located inside buildings				
		High, brick and steel	Medium	Economical	Evaluation
	Quality of construction	+4	+2	0	
		Very high unit cost equipment	Mostly alloy steel	Mixed materials	Costly carbon steel
	Type of equipment	−3	−2	−1	0
		Very high	Intermediate	Atmosphere	
	Operations pressures	−2	−1	0	
cost equipment, chemical	Building class = algebraic sum =				

		Average unit cost of MPI						
	Building class	Under $9000	$9000 to $15,000	$15,000 to $21,000	$21,000 to $30,000	$30,000 to $39,000	$39,000 to $51,000	$51,000
Cost of process Units inside buildings	+2	92/68	82/61	74/56	67/49	59/44	52/39	46/33
	+1 to −1	72/49	62/43	56/38	51/33	45/29	41/26	36/21
	−2	50/37	44/33	40/29	35/25	30/21	27/18	23/15
Open-air plants with minor buildings		37/16	32/13	28/11	24/8	20/6	17/4	14/2

Building services‡	High	Normal	Low
Compressed air for general service only	4	1½	0.5
Electric lighting	18	9	5
Sprinklers	10	6	3
Plumbing	20	12	3
Heating	25	16	8
Ventilation:			
Without air conditioning	18	8	0
With air conditioning	45	35	25
Total overall average§	85	55	20

The above factors apply to those items normally classified as building services. They do not include (1) services located outside the building such as substations, outside sewers, and outside water lines, all of which are considered to be outside the battery limit as well as outside the building; and (2) process services.

°Courtesy C. A. Miller of Canadian Industries Ltd. and the American Association of Cost Engineers.

NOTE: The average unit cost of the main-plant items is the total cost of the MPI divided by the total number of items. Figures include up to 3 percent for BL outside lighting, which is not covered in building services.

°Total instrumentation cost does not vary a great deal with size and hence is not readily calculated as a percentage of basic equipment. This is particularly true for distillation systems. If in doubt, detailed estimates should be made.

†When building specifications and dimensions are known, a high-speed building-cost estimate is recommended, especially if buildings are a significant item of cost. If a separate estimate is not possible, evaluate the buildings as shown before selecting the factors.

‡The following factors are for battery-limit (process) buildings only and are expressed in percentage of the building architectural and structural cost. They are not related to the basic equipment cost.

§The totals provide the ranges for the type of building involved and are useful when individual service requirements are not known. Note that the overall averages are not the sum of the individual columns.

TABLE 9-53 Estimate Using Factors from Table 9-51

Details (solids-fluid, grass-roots plant)	Factor assumed	Cost, $	Percentage of total
Equipment, delivered	1.00	1,000,000	23.4
Installed	0.41	410,000	9.6
Piping	0.34	340,000	8.0
Electrical	0.13	130,000	3.0
Instruments	0.13	130,000	3.0
Battery-limit building and service	0.30	300,000	7.0
Excavation and site preparation	0.15	150,000	3.5
Auxiliaries	0.52	520,000	12.2
Total physical plant	2.98	2,980,000	69.7
Field expense	0.39	390,000	9.1
Engineering	0.39	390,000	9.1
Direct plant costs	3.76	3,760,000	87.9
Contractor's fees, overhead, profit	0.13	130,000	3.0
Contingency	0.39	390,000	9.1
Total fixed-capital investment	4.28	4,280,000	100.0

TABLE 9-54 Typical Factors with Separation of Materials and Labor*

	Total factor	Materials factor	Labor factor
Equipment delivered	1.00		
Installation	0.09		0.09
Instruments installed	0.13	0.09	0.04
Piping	0.29	0.155	0.135
Foundations and steel	0.18	0.08	0.10
Insulation painting	0.11	0.025	0.085
Electrical	0.18	0.06	0.12
Battery-limit building	0.21	0.13	0.08
Site preparation	0.08		
Auxiliaries	0.55		
Physical-plant cost	2.82		
Engineering and home office	0.31	0.01	0.30
Field expense	0.43	0.30	0.13
Direct plant cost	3.56		
Contractor's fees	0.17		
Contingency	0.39		
Fixed-capital cost	4.12		

*Based on the data of H. C. Bauman, *Fundamentals of Cost Engineering in the Chemical Industry*, Van Nostrand Reinhold, New York, 1964, p. 295, for essentially carbon steel equipment.

A. Pikulik and H. E. Diaz [*Chem. Eng.*, **84**, 106–122 (Oct. 10, 1977)] presented a graphical method for estimating the fabricated cost of distillation columns and pressure vessels, storage tanks, fired heaters, pumps and drivers, compressors and drivers, and vacuum equipment.

TABLE 9-55 Location Factors for Chemical Plants of Similar Functions (1993 Values)

Location	Factor (United States = 1.0)
Australia	1.04
Austria	0.85
Belgium	0.70
Canada	1.14
Central Africa	1.51
Central America	1.20
Denmark	0.85
Finland	0.88
France	0.73
Germany	0.76
Greece	0.80
India	
imported element	0.80
indigenous element	0.25
Ireland	0.70
Italy	0.79
Japan	1.46
Malaysia	0.42
Middle East	0.84
New Zealand	1.27
North Africa	
imported element	0.65
indigenous element	0.44
Norway	0.92
Portugal	1.00
South Africa	0.90
South America	1.36
Spain	0.83
Sweden	0.75
Switzerland	0.94
Turkey	0.80
United Kingdom	0.76
United States	1.00

NOTE:

1. Increase a factor by 10 percent for each 1000 mi or part of 1000 mi that the new plant is distant from a major manufacturing or import center, or both.

2. When materials or labor, or both, are obtained from more than a single source, prorate the appropriate factors.

3. Investment incentives have been ignored.

Equipment Costs The cost of delivered equipment forms the basis of most methods of estimating the fixed-capital cost. The equipment required can usually be divided into (1) processing equipment, (2) equipment for handling and storage of raw materials, and (3) finished products handling and storage equipment.

Quotations for equipment costs from fabricators or suppliers are the most accurate. Therefore most companies base their costs on (1) quotations from fabricators, (2) past purchase records updated with

TABLE 9-56 Factors for Individual Items*

	Exchangers			Vessels				
Details	Furnaces	Shell and tube	Air-cooled	Vertical	Horizontal	Pump and driver	Compressor and driver	Tanks
FOB equipment	1.00	1.00	1.00	1.00	1.00	1.00	1.00	1.00
Piping	0.18	0.46	0.18	0.61	0.42	0.30	0.21	
Concrete	0.10	0.05	0.02	0.10	0.06	0.04	0.12	
Steel		0.03		0.08				
Instruments	0.04	0.10	0.05	0.12	0.06	0.03	0.08	
Electrical	0.02	0.02	0.12	0.05	0.05	0.31	0.16	
Insulation		0.05		0.08	0.05	0.03	0.03	
Paint				0.01	0.01	0.01	0.01	
Total materials = M	1.34	1.71	1.38	2.05	1.65	1.72	1.61	1.20
Erection and setting (L)	0.30	0.63	0.38	0.95	0.59	0.70	0.58	0.13
X, excluding site preparation and auxiliaries ($M + L$)	1.64	2.34	1.76	3.00	2.24	2.42	2.19	1.33
Freight, insurance, taxes, engineering, home office, construction		0.08		0.08	0.08	0.08	0.08	0.08
Overhead or field expense	0.60	0.95	0.70	1.12	0.92	0.97	0.97	
Total module factor	2.24	3.37	2.46	4.20	3.24	3.47	3.24	1.41

*From K. M. Guthrie, *Chem. Eng.*, **76**, 114–142 (Mar. 24, 1969). Based on FOB equipment cost = 100 (carbon steel).

appropriate cost indices, and (3) exponential methods of adjusting prices for capacity changes.

A large number of graphs for estimating the costs of various types of process equipment and auxiliaries are presented in the excellent text by M. S. Peters and K. D. Timmerhaus (*Plant Design and Economics for Chemical Engineers,* 4th ed., McGraw-Hill, New York, 1991, Chaps. 14–16) at January 1990 prices. Information on process equipment costs also appears, from time to time, in various journals (see Table 9-50). Although these published cost data are extremely useful for making rapid estimates, no published data can compete with the detailed and usually confidential cost records of large companies.

Piping Estimation The cost of fabrication and installation of process-plant piping appears to range from 18 to 61 percent of the FOB equipment cost as indicated in Table 9-56. This would normally represent about 7 to 15 percent of the installed plant cost and is obviously a significant item. The various available piping-estimation methods are as follows:

1. Detailed pricing from piping drawings
2. Guthrie method
3. Dickson N method
4. Pricing by weight of specific types of pipe
5. Price estimation by cost per joint
6. Pricing as a factor of equipment cost
7. Pricing as a factor of total plant installed cost

The first five methods are applicable only after rigorous circuit analysis and when piping layouts and isometric drawings or scale models are available for quantity takeoff (e.g., pipe size, length, and specification, flanges and valve count, etc.).

Guthrie's method [K. M. Guthrie, *Chem. Eng.*, **76**, 201–216 (Apr. 14, 1969)] is mainly graphical, using average mid-1968 costs for a United States Gulf Coast location.

The Dickson N method [R. A. Dickson, *Chem. Eng.*, **54**, 121–123 (November 1947)] is a variation of the detailed price takeoff. Various circuits for each type of pipe are completely priced for a base size. Another chart gives an N factor for all other pipe sizes. Multiplying the cost of the circuit for the base size by the appropriate N factor yields the estimated cost of the new circuit of the desired pipe size. The method depends for its accuracy on periodic repricing of the base-size circuits in order to keep the base charts up to date.

Estimating by weight requires virtually complete takeoff, including weight calculations and a full record of past costs on this basis. Its only advantage lies in the time saved in the detailed estimates of the cost of piping components.

Estimating by cost per joint depends on the accumulation of past data, analyzed and conveniently correlated for use. The main advantage of the method lies in the fact that good engineering flow sheets can be used for the estimation.

Figure 9-42 is a plot of the number of labor-hours of field erection time per joint against the nominal pipe size of shop-fabricated carbon steel and low-alloy pipe. The unit of work measurement used in this method is the pipe joint, requiring two joints for couplings and valves, three for tees, etc., as most of the labor-hours involved in pipe erection are expended in making connections. The additional costs of handling, suspending, and placing lengths of pipe in position are included in the chart.

It should be noted that Fig. 9-42 gives labor-hours only. Material costs must be obtained by price takeoff from drawings on which all valves and instrument connections are shown. Pipe lengths and fittings are taken off by referring to the equipment-layout plan and elevation drawing. The graph of Fig. 9-42 can be updated by using actual costs for a specific job, in which case the labor cost per joint represents a total labor cost including all the factors applicable to labor shown in Table 9-57. It should be possible to analyze statistically uniform data from a number of complete jobs to determine the value of each factor for various project locations.

Methods 6 and 7 are simpler procedures, using factors for estimating piping costs when neither flow sheets nor detailed piping drawings are available. Tables 9-51, 9-52, 9-54, and 9-56 include typical values of piping factors based on total equipment cost, delivered or FOB, as indicated in the particular table. These methods require some degree of judgment in selecting the appropriate factor, based on experience gained by comparing piping costs for similar previously installed process plants.

A rough method of estimating the piping factor as a percentage of the total delivered cost of major process equipment (excluding instruments and electrical items) was presented by E. S. Sokullu in the form

$$f_P = 11\phi_P^{1.6} \qquad (9\text{-}254)$$

where ϕ_P = (number of actual pipes on flow diagram)/(number of major process equipment units) [*Chem. Eng.*, **76**, 148–150 (Feb. 10, 1969)].

The equipment-unit method would appear to give more accuracy than the preceding methods, particularly for unfamiliar process arrangements. It requires the accumulation of piping costs for various sizes of main-plant items such as pumps, heat exchangers, evaporators, tanks, and columns. Basically it is assumed that piping designs for specific items are similar for most projects. Statistical analysis of such data shows good agreement with the more detailed takeoff pricing methods. Since for most processes the length of pipe used is a small proportion of the total piping cost, the assumption of an average length of piping per main-plant item, based on actual costs for several previous jobs, should give sufficient accuracy. Correction for escalation of costs can be carried out by using a single cost index, unlike methods 1 to 5.

Most of the factorial methods of estimation given previously, with the exception of the method of Allen and Page (loc. cit.), tend to estimate costs which are based on carbon steel equipment or installations. Table 9-58 gives typical multiplying factors for converting carbon steel costs to equivalent-alloy costs for a few items of equipment. (Adapted from *A Guide to Capital Cost Estimating*, 3d ed., Institution of Chemical Engineers, Rugby, England, 1988, p. 70.)

Electrical and Instrumentation Estimation These costs usually range from 4 to 10 percent of the total installed plant cost, with a median value of about 7.5 percent. As with piping estimation, the process design must be almost completed before detailed drawings and specifications can be prepared for estimating purposes. However, actual electrical costs can be up to 100 percent higher than estimated costs, and so it is important to attempt to maintain the accuracy range within reasonable limits.

TABLE 9-57 Components of Total Installed Piping Cost

Material	Pipe, valves, fittings, nuts, bolts, gaskets, and hangers
Labor	Cut, erect, align, fit, bolt, thread or weld, and test
Indirect costs	Handle and haul, store, scaffold, lost time, tools and rentals, contractor's overhead and profit
Factors applicable to labor-hours	Craft rate, productivity, height, and complexity
Crafts involved in piping erection	Pipefitters, laborers, carpenters, warehouse workers, teamsters, and operating engineers

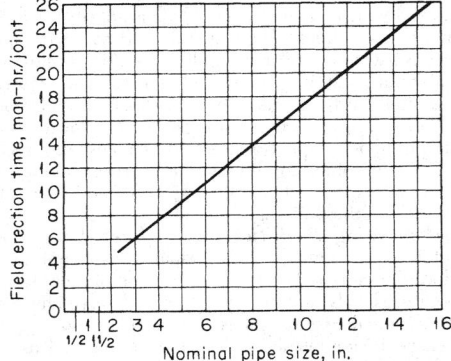

FIG. 9-42 Labor-hours required to erect large quantities of shop-fabricated steel and low-alloy piping.

TABLE 9-58 Typical Factors for Converting Carbon Steel Cost to Equivalent-Alloy Costs

Material	Pumps, etc.	Other equipment
All carbon steel	1.00	1.00
Stainless steel, Type 410	1.43	2.00
Stainless steel, Type 304	1.70	2.80
Stainless steel, Type 316	1.80	2.90
Stainless steel, Type 310	2.00	3.33
Rubber-lined steel	1.43	1.25
Bronze	1.54	
Monel	3.33	

Material	Heat exchangers
Carbon steel shell and tubes	1.00
Carbon steel shell, aluminum tubes	1.25
Carbon steel shell, monel tubes	2.08
Carbon steel shell, 304 stainless tubes	1.67
304 stainless steel shell and tubes	2.86

TABLE 9-59 Electrical and Instrumentation as Percentage of Total Installed Plant Cost

Type of plant	Electrical (process and service)		Instrumentation	
	Range, %	Median, %	Range, %	Median, %
Solids plants	3.7–10.7	5.4 %	0.3–6.0	0.8
Grass-roots process	4.0–7.9	5.9 %	1.9–4.3	3.2
Battery-limit process	4.3–10.1	7.5 %	0.1–7.9	3.7

TABLE 9-60 Component Electrical Costs as Percentage of Total Electrical Cost

Component	Range, %	Median, %
Power wiring	25–50	40
Lighting	7–25	12
Transformation and service	9–65	40
Instrument-control wiring	3–8	5

During the design stages, frequent changes in the type and sizes of equipment lead to delays in establishing electrical requirements. Hence it is very difficult to obtain a detailed estimate of the cost of the electrical part of the project. For order-of-magnitude and study estimates, an appropriate factor in the range 4 to 10 percent of the total installed plant cost can be used. However, for budget-authorization or preliminary estimates requiring an accuracy within 5 percent more accurate methods are necessary.

The methods available for electrical estimates are as follows:
1. Detailed takeoff
2. Factored electrical cost as a percentage of total installed plant cost for specific types of plant
3. Unit pricing

The detailed-takeoff method can rarely if ever be used. When detailed drawings are available, costs may be estimated by pricing materials and components from suppliers' catalogs or, for special items, from quotations. Handbooks are available which give typical values of the labor-hours required to perform units of installation work, such as installation of switches, starters, motors, conduit wiring, and push buttons of various sizes, for both hazardous and nonhazardous areas. Labor rates can be obtained from various government statistical sources or elsewhere. For the United States the National Electrical Contractors Association publishes an excellent manual of electrical costs. From the complete plans and specifications, the estimator can take off materials, estimate the labor cost, apply appropriate factors for labor efficiency, productivity, and local conditions, and achieve good results.

The factor estimate, if based on tested actual data, gives good results in the study estimate and often proves adequate at the preliminary estimate stage. It is essential to accumulate from past experience data showing actual electrical costs (1) as a percentage of total installed plant cost and (2) as a percentage of installed equipment costs. Studies of electrical installations for more than 100 plants (H. C. Bauman, *Fundamentals of Cost Engineering in the Chemical Industry*, Van Nostrand Reinhold, New York, 1964, p. 134) showed electrical costs ranging from about 4 to 11 percent of total plant cost, with a median for battery-limit process plants of 7.5 percent. The corresponding range based on installed equipment costs was 15 to 40 percent, with a median of 26 percent. Thus, it appears that there is a better correlation between electrical costs and total installed plant cost than with installed equipment costs. Table 9-59, taken from Bauman's data, gives typical values of electrical costs as a percentage of total installed plant cost. Cost ranges for installed instrumentation costs are also included in Table 9-59, as these would form part of electrical costs. The ranges of values are rather wide, depending on the degree of automatic control required.

Electrical costs involve four main components: (1) power wiring, (2) lighting, (3) transformation and service, and (4) instrument and control wiring. A breakdown of these component costs as a percentage of total electrical cost is given in Table 9-60.

The unit-cost method can give a quick and accurate estimation, provided it is based on accumulated data from many jobs on various types of plant. The actual data are analyzed to provide unit-cost information for electrical components as follows:
1. Total installed cost per motor
2. Total installed cost per lighting outlet by type
3. Total installed cost of receptacles by type (incandescent, fluorescent, etc.)
4. Total installed cost for each wired instrumentation point
5. Total installed cost for each unit of transformation
6. Total installed cost per lineal foot of distribution by type (overhead bare and insulated, underground)
7. Total installed cost of each interlock point

Each unit cost contains all the costs involved in the installation of that unit. For motors installed costs include the starter, conduit, wire, and a proportionate share of the service panelboard and busbars. The motor cost is not included since this will be part of the equipment cost. In the case of lighting, the installed cost includes the lighting fixtures, the conduit and wire, and a proportional share of the lighting panelboard and service switching costs.

Auxiliaries Estimation Chemical-plant auxiliaries normally include all structures, equipment, and services which are not directly involved in the process. Within this broad range there are two major classifications, utilities and service facilities.

The typical cost range for auxiliaries is from 20 to 40 percent of the total installed plant cost. For a small continuous-process plant making a single product, the cost of auxiliaries would lie in the lower part of the range, while for large multiprocess grass-roots plants the factor would tend to be near the upper limit of the range.

Auxiliary Buildings Typical variations in the cost of auxiliaries for a variety of process plants are given in Table 9-61. The widest variation is shown for auxiliary buildings, which is not surprising in view of the many types and quality of materials and the wide variation in methods of construction. For example, amenities buildings such as offices, cafeteria, first-aid rooms, gatehouses, and control rooms would necessitate fairly expensive brick and plaster-wall construction. On the other hand, services buildings such as substations, switch rooms, and pump or compressor houses would cost about 5 to 10 percent less. Provision of air conditioning, furniture, and equipment for cafeteria, laboratory, and office buildings would add about 50 percent to the basic cost of the building.

Steam-Generating Facilities These form the second largest investment item for chemical-plant auxiliary equipment. Variations in capacity, location indoors or outdoors, the type of fuel used, pressure and temperature levels, and the type of process served have an important effect on actual cost as well as on cost relative to other auxiliary items. Package boiler installations can be purchased as shop-built units which are assembled, piped, and wired ready to be erected on the owner's foundations. They are available in units up to about 136,000 kg/h (300,000 lb/h), although units larger than about 45,360

TABLE 9-61   Typical Ranges of Auxiliary Facilities as Percentage of Total Installed Plant Cost

Grass roots and large additions

	Range, %	Median, %
Auxiliary buildings	3–9	5.0
Steam generation	2.6–6	3.0
Refrigeration, including distribution	1–3	2.0
Water supply, cooling, and pumping	0.4–3.7	1.8
Finished-product storage	0.7–2.4	1.5
Process-waste systems	0.4–1.8	1.1
Raw-materials storage	0.3–3.2	1.1
Steam distribution	0.2–2	1.0
Electrical distribution	0.4–2.1	1.0
Air compressor and distribution	0.2–3.0	1.0
Water distribution	0.1–2	0.9
Fire protection system	0.3–1.0	0.7
Water treatment	0.2–1.1	0.6
Railroads	0.3–0.9	0.6
Roads and walks	0.2–1.2	0.6
Gas supply and distribution	0.2–0.4	0.3
Sanitary-waste disposal	0.1–0.4	0.3
Communications	0.1–0.3	0.2
Yard and fence lighting	0.1–0.3	0.2

kg/h (100,000 lb/h) may be available only on a semierected basis. It is usually necessary to obtain firm price quotations that take into account all the factors involved. Housing the boiler installations in buildings will generally increase the cost by about 7 to 9 percent per kilogram-hour (15 to 20 percent per pound-hour) of steam-generating capacity over the cost for outdoor installations or for installations in existing buildings.

For most chemical plants, process steam is used at pressures of 1.825 MN/m² (250 psig), saturated or lower. When combined heat and power generation is economically justified, the steam may be generated at about 5.96 MN/m² (850 psig) appropriately superheated and used to drive back-pressure steam turbines passing out process steam at the required pressure level.

Electricity   A reliable and adequate electricity supply is usually available through government or private enterprises. Owing to the increasing cost of purchased electricity, many companies have installed combined heating and power (CHP) generation systems. A cogeneration plant may (1) be owned and operated by the industrial user or the utility, (2) serve or be isolated from one or more industrial users, or (3) form an integral part of the local utility grid. Typical costs for generating steam range from $7.90 to $9.50 per 1000 kg ($3.60 to $4.30 per 1000 lb) at 3550 kPa (500 psig); $3.70 to $7.70 per 1000 kg ($1.70 to $3.50 per 1000 lb) at 790 kPa (100 psig); and $2.00 to $3.70 per 1000 kg ($0.90 to $1.70 per 1000 lb) for exhaust steam. The above costs apply for a Marshall and Swift index of 1000.

For most plants, electric distribution systems start at the power company's service point on the plant's property. The choice of an electrical system will depend on many factors relating to the particular project. A wide range of items, from switchgear, transformers, and motor control centers to cabling, earthing, and lighting, are required. Normally, competitive quotations would be obtained to take an estimate beyond the study stage.

Various types of overhead and underground distribution systems may be used depending on local conditions. Generally, an overhead system will incur only about 30 percent of the cost of an underground distribution system.

Water Systems   These systems usually form the third highest cost item in chemical-plant auxiliaries, with cooling towers representing the largest part of the investment. Although the installed cost increases with the terminal temperature range, an approximate cost correlation is given by

$$C_{CT} = 100q^{0.87} \qquad (9\text{-}255)$$

where C_{CT} is the installed cost of the cooling tower in United States dollars for a Marshall and Swift (M & S) index of 1000 and q is the capacity in United States gallons per minute over the range from $(1)(10^3)$ to $(1)(10^5)$ U.S. gal/min.

River-water pumping and filtering installations can be approximately correlated by

$$C_{RW} = 0.65q^{0.81} \qquad (9\text{-}256)$$

where C_{RW} is the installed cost of the river-water system in United States dollars for an M & S index of 1000 and q is the capacity over the range from $(4)(10^5)$ to $(1)(10^7)$ U.S. gal/day.

Similarly, installed costs of water-softening systems can be correlated in United States dollars (M & S = 1000) as follows:

$$C_{WS} = 1380q^{0.44} \qquad (9\text{-}257)$$

over the range of capacity from $(3)(10^7)$ to $(1)(10^9)$ U.S. gal/day and of demineralizing systems by

$$C_{DS} = 0.17q^{1.9} \qquad (9\text{-}258)$$

over the range of q values from $(1)(10^4)$ to $(4)(10^5)$ U.S. gal/day. Actual water-treatment costs may vary widely from the above, depending on the quality of the water, the percentage of dissolved solids, and the total hardness.

Refrigeration Systems   These systems are being used increasingly in chemical processing. Installed costs of packaged mechanical units in United States dollars (M & S = 1000) can be approximately correlated by

$$C_{RS} = 4630q^{0.73} \qquad (9\text{-}259)$$

where q is the capacity in tons of refrigeration over the range from 10 to 1000 tons. One ton of refrigeration is equivalent to a rate of heat removal of 3.517 kW (12,000 Btu/h).

Roads and Walkways   The cost of roads and walkways in chemical plants is difficult to estimate, since these vary with type of construction and thickness of applied cover. Some typical unit costs for roads are as follows: For 305-mm (12-in) gravel base covered with 76-mm (3-in) asphalt, the cost is $17.10 per square meter ($14.30 per square yard); for a reinforced concrete slab with a 152-mm (6-in) subbase, the cost is from $28.40 to $35.10 per square meter ($23.80 to $29.30 per square yard), depending on the thickness of concrete (for M & S = 1000).

Installed costs for railroads, including switches and frogs, can be roughly estimated as follows (for M & S = 1000):

Linear meters	Linear feet	$/meter	$/foot
152–305	500–1000	230.00	70.00
305–915	1000–3000	210.00	64.00
915–3050	3000–10000	200.00	61.00
Above 3050	Above 10000	187.00	57.00

Usually the cost of roads and walkways amounts to 0.2 to 1.2 percent of the fixed capital cost with a typical value of 0.6 percent. Similarly, railroads cost 0.3 to 0.9 percent of the fixed capital cost, having an average value of 0.6 percent.

Use of Computers in Cost Estimation   A large part of estimation consists of the collection and storage of data obtained from records of actual plant costs. The data then must be correlated and updated and the required information rapidly retrieved for use in further cost estimations. A comprehensive survey (C. J. Liddle and A. M. Gerrard, *The Application of Computers to Capital Cost Estimation*, Institution of Chemical Engineers, London, 1975, pp. 6–17) suggests that large chemical manufacturers, equipment vendors, and some contractors are using the computer increasingly for data retrieval, followed by simple correlation and the application of factorial methods to cost estimation.

In the case of equipment vendors, the computer's contribution appears to be particularly worthwhile owing to the elimination of estimating errors in producing price quotations. Several companies have developed an automated quotation system to overcome delay and inaccuracy in estimating and bidding. Such systems appear to have been developed by firms already possessing significant computing facilities, since the cost of computer time is small compared with the cost of the computer. Qualitatively, operating costs for an automated quotation system appear to be about half of those of a manual system of price quotation. The methods of estimation used are based on the manual methods described previously.

Several of the larger chemical manufacturers, particularly those in the petrochemicals field, have developed computer packages based on manual methods of factorial estimating. Usually the input data consists of the cost of each main-plant item (MPI) obtained from quotations or historical records. The program then estimates the costs of erection, piping, instrumentation, electricals, civil engineering, and lagging for each (MPI) in turn by adding a series of factors. These account for the complexity of the process and the constructional difficulties for each (MPI) to produce an estimate of the overall plant cost. It is obviously necessary to introduce appropriate inflation indices to bring the estimated costs up to date.

For process plants, it is often possible to use these cost-estimation programs with a design or flow-sheet program to optimize on a particular component or even over the whole plant (*A Guide to Capital Cost Estimating*, 3d ed. Institution of Chemical Engineers, Rugby, England, 1988, pp. 48–49). However, it must be remembered that optimization is expensive on computer time, although there appear to be no data available on the cost effectiveness of the computer in this area. It is also possible to incorporate the capital-cost estimate in an investment evaluation involving forecasts of expenses and revenue from sales. Thus, by means of the computer design and costing can be brought together. There is an immediate feedback of information, resulting in improved design and lower costs. In some types of plants, costing data can be fed in as a subroutine to the design programs. All these possibilities assume that the total cost of using the computer is not unreasonable.

Startup Costs Startup problems can reduce aftertax earnings during the early years, the most serious effect being to delay the startup of production, causing a loss of earnings. An accurate estimate of startup time and cost can help in (1) predicting the availability of new products, (2) planning market entry, and (3) estimating the overall profitability because of more accurate cash-flow forecasts and (NPV) calculations [R. P. Feldman, *Chem. Eng.*, **76**, 87–90 (Nov. 3, 1969)].

Startup costs are defined as the total of those costs directly related to bringing a new production facility into operation. They should not include the costs of entering or expanding a business. Hence startup costs include the following:

1. All expenses due to changes in process and equipment after completion of construction but excluding those due to changes in project scope
2. All labor costs after completing construction, especially those incurred in checking the functioning of equipment
3. All costs incurred during the startup period but excluding normal operating expenses
4. Expenses for training plant personnel even if incurred before startup has officially begun
5. All research and development costs incurred during startup

The following expenses should *not* be included:
1. Marketing costs
2. Expense of training sales representatives
3. Penalties for shipping outside optimum freight areas
4. Costs associated with starting a new company
5. Lost sales unless there is a contract with a penalty
6. Profit lost due to timing

Startup time may be defined as the time span between the end of construction and the beginning of normal operation. Hence it should start when the contractor finishes the whole plant or a specified section of it to enable comparisons to be made with other startup times. It is usual to define "normal" operation as (1) operations at a certain percentage of design capacity, (2) a specified number of days of continuous operation, or (3) the capability of making products of a specified purity.

It is essential for project and production management to agree beforehand on the definition to be applied. Obtaining agreement on the definition of "normal" operation is important since (1) it sets a target for field personnel, (2) it ensures that everyone is striving for the same target, (3) it permits comparisons with other plants, and (4) it determines a cutoff point for completion of startup. It may be necessary to wait until the plant is running well to obtain the actual total cost of startup.

For control purposes it is advisable to estimate startup cost and time beforehand and then try to stay within the estimates. The general parameters which can be used to estimate startup cost C_{SU}, which are usually between 2 and 20 percent of the battery-limit fixed-capital cost, are as follows:

1. Direct fixed-capital cost for plant (battery-limit capital), $(C_{FC})_{BL}$
2. Newness of process and technology, b
3. Newness of type and size of equipment, c
4. Labor quality and quantity, d
5. Interplant dependency, e

Hence startup cost may be expressed as

$$C_{SU} = (C_{FC})_{BL}[0.10 + b + c + d + ne] \qquad (9\text{-}260)$$

When applied to large air-separation and ammonia plants (1000 to 1400 metric tons/day), the following values for the parameters can be used:

$b = 0.05$ for a radically new process
$\quad = 0.02$ for a relatively new process
$\quad = -0.02$ for an old process

$c = 0.07$ if radically new
$\quad = 0.04$ if very new
$\quad = 0.02$ if relatively new
$\quad = -0.03$ if old

$d = 0.04$ if labor is in very short supply
$\quad = 0.02$ if labor is in short supply
$\quad = -0.1$ if labor is in surplus supply

$e = 0.04$ if plant is very dependent on another
$\quad = 0.02$ if moderately dependent on another
$\quad = -0.02$ if independent

and

n = number of plants or sections making up the process chain

Startup time t_{SU} for these plants may be estimated from construction time t_C by developing an equation similar to Eq. (9-260):

$$t_{SU} = t_C(0.15 + b + c + d + ne) \qquad (9\text{-}261)$$

For the same type of plant the values of the parameters are

$b = 0.15$ for a radically new process
$\quad = 0.05$ for a relatively new process
$\quad = -0.01$ for an old process

$c = 0.15$ if radically new
$\quad = 0.08$ if very new
$\quad = 0.05$ if relatively new
$\quad = -0.01$ if old

$d = 0.15$ if labor is in very short supply
$\quad = 0.05$ if labor is in short supply
$\quad = -0.01$ if labor is in surplus supply

$e = 0.25$ if plant is very dependent on another
$\quad = 0.10$ if moderately dependent on another
$\quad = -0.02$ if independent

and

n = number of plants or sections making up the process chain

It should be noted that these values are based on previous experience with certain types of plants, but appropriate values which apply to other processes and locations could be selected.

Construction Time The duration of construction is difficult to estimate owing to the large number of variables involved. In general, estimates of construction time tend to be overoptimistic, especially for larger projects. Usually projects costing less than $5 million at 1993 prices can be completed in 10 to 18 months, while those costing more than $10 million may take from 18 to 42 months to complete. Delays of up to 12 months behind schedule are quite possible, particularly when there are labor problems. As mentioned previously, such delays will usually result in increased construction costs. Often, a more serious effect is loss of earnings resulting from a delayed startup. Both of these factors increase the payback period and reduce the attainable net present value and discounted-cash-flow rate of return of the project.

Project Control   Having made a good estimate of the capital cost and the expected construction time, it is essential to introduce an effective system for controlling expenditure of time and money during construction. Good capital-cost control can cut down expenditures even when the definitive estimate is not very accurate. It is most important for management to receive early warnings if overruns in expenditure or time are likely to occur.

Effective cost control should start from the beginning of the project at the research and development stage and continue through the design and estimating stages to initial operation of the plant [J. W. Hackney and K. K. Humphreys (eds.), *Control and Management of Capital Projects*, 2d ed., McGraw-Hill, New York, 1991]. The stages discussed here are the later steps after authorization of funds and during project construction. After the purchasing department has placed the orders for equipment and materials, speed and efficiency during the construction stage is most important in ensuring the financial success of the project. Field expenditure during construction can amount to 30 to 60 percent of the fixed-capital cost and includes the costs of all labor, installed equipment, and materials together with associated process piping, electrical instrumentation, and insulation. Construction therefore requires efficient execution and prompt feedback of progress information, necessitating a good cost-control system.

Figure 9-43 shows the flow of information needed for cost control. The chart assumes a definitive estimate which has been linked to a standard code of accounts. As construction proceeds, up-to-date cost-control reports are supplied to the field cost engineer. From the home office the engineer receives monthly reports of engineering and drafting labor-hours used and money expended, together with a list of drawings and specifications completed up to that time. Monthly expenditures and current commitments come in coded detail from the job ledgers of the accountants. Timekeepers' records give details of craft and nonmanual labor-hour expenditures. Quantities of equipment and material held on site are reported daily by quantity surveyors to the construction superintendent. All purchase orders are posted in the ledgers as current commitments, whether they are placed at the home office or in the field, and an up-to-date warehouse inventory is maintained.

At the end of each month, the field cost engineer collects all current information on a detailed cost report form. As these are actual costs, they can be used to estimate future job costs to completion. Daily reports of unit-cost progress for concrete, excavation, masonry, steel, piping, and electrical work, etc., are then used to predict possible overruns or underruns for the various items. Analysis and comparison with the original estimate point out trouble spots for early attention. If an item is running into difficulty, it is red-flagged to the resident and project engineers for remedial action.

In practice, the existence of a tight cost-control system tends to spread a cost consciousness among the personnel involved in the project. Such an awareness, even in construction-equipment maintenance and job housekeeping, can lead to efficient cost control throughout.

Cost reports should be brief but informative, preferably in summary form. They should report expenditures and commitments, estimated costs to complete, and expected overruns or underruns of the authorized budget for each important item of cost. Brief notes should emphasize significant deviations from predicted cost. Any large, persistent overrun should have already been investigated and reported to the project and construction managers for immediate attention. If an expected overrun cannot be avoided, the current summary cost report should serve as justification for a request for additional funds.

When organized efficiently, the cost-control system should require no more paperwork than for the normal construction procedure. The cost of cost control appears to vary between 0.2 and 0.5 percent of the total project value. Proper use of the normal records available for craft-labor time, warehouse-inventory control, and the usual accounting purposes should be adequate. The savings achieved by good cost control should far exceed the additional costs of operating the system. Additional details on the technique are given by H. C. Bauman (*Fundamentals of Cost Engineering in the Chemical Industry*, Van Nostrand Reinhold, New York, 1964, pp. 190–196).

Scheduling construction to ensure that the project is completed in the shortest possible time is an essential part of project control. The project-control estimate defines to a large extent the construction-time schedule. It is then possible to prepare a master schedule from the control estimate by carefully sequencing and synchronizing the installation work according to past experience. Drawings are usually completed in predictable order owing to the dependence of certain designs on preceding work. The normal order of completion of drawings and specifications is (1) site work, (2) substructures, (3) equipment and building superstructures, (4) equipment layouts, (5) piping, (6) insulation, (7) instrumentation, and (8) electrical work.

Detailed planning and scheduling then involve establishing the items of work required and determining the correct sequence of work and the number of persons required to perform each item of work.

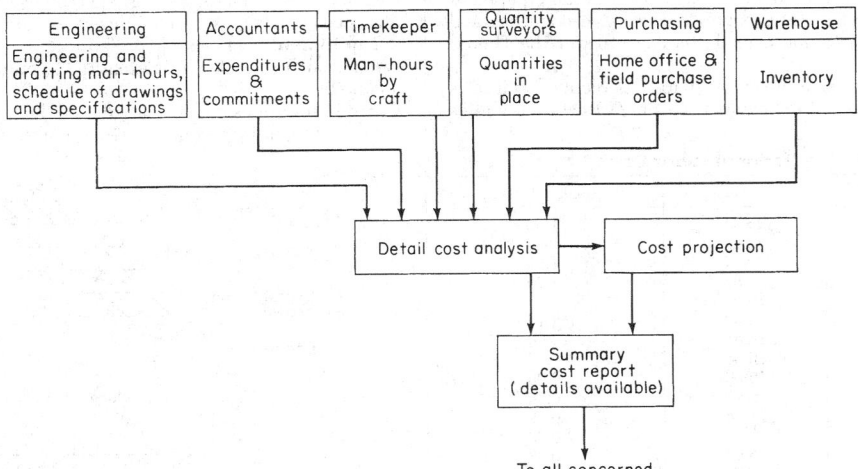

FIG. 9-43   Information flow for cost control.

From this information it is possible to prepare bar charts by using a 4-week month and noting on the chart the interdependence of the various functions. The starting time for each class of work is fixed on the chart, and the duration is calculated from the labor-hours allocated to that work from the control estimate. Work should progress smoothly as time elapses, but the operations must be linked by the order of necessary precedence. Starting times for the various items of work will be staggered as drawings are released and also to smooth out labor requirements.

Sketching the bar chart is commenced by inserting the arrival dates of key items, observing any necessary precedence. Estimates of the duration of erection time can be made to obtain the starting date for process piping. Since the supporting structure must be in place when the key item arrives, it is possible to work back along the bar chart to the preparation of the foundations. From the complete bar chart, built up in a similar manner, a tentative startup date can be set after allowing a few weeks for tidying up bits and pieces. Some activities can be speeded up, but it is necessary to estimate the increased cost of so doing.

Actual progress made with construction work can be indicated on the bar chart by filling in the open bars according to the percentage toward completion. Comparison of the actual progress bar for the whole project with the cumulative labor-hour curve indicates whether the job is ahead of schedule or not. If corrective action is required, effort should be concentrated on the key or critical items.

Large projects will usually require network analysis using the critical-path method (CPM) or program evaluation and review technique (PERT) in the planning, scheduling, and progress-control stages. Examples of bar charts and a fuller description of network analysis are given by J. W. Hackney and K. K. Humphreys (eds.), *Control and Management of Capital Projects*, 2d ed., McGraw-Hill, New York, 1991. A detailed treatment of the use of PERT and CPM techniques as applied to contract bidding strategy and to project control is given by L. A. Swanson and H. L. Pazer (*Pertsim: Text and Simulation*, International Textbook, Scranton, Pa., 1969), who present a hand simulation technique based on probabilistic methods.

Overseas Construction Costs   Although Table 9-55 gives location factors for the construction of chemical plants of similar function in various countries at 1993 values, these may vary differentially over a period of time owing to local changes in labor costs and productivity. Hence, it is often necessary to estimate the various components of overseas construction costs separately. Equipment and material prices will depend on local labor costs and the availability of raw materials. If the basic materials have to be imported, costs in the source area become important and import duties and freight charges must be added.

Equipment and material normally amount to about 40 to 45 percent of the costs of a typical chemical plant. In general, equipment and material costs are slightly cheaper in European countries and Japan, whereas in Mexico and Canada they are nearer the United States average.

Construction labor makes up about 20 to 35 percent of total costs for a chemical plant. Table 9-62 compares average 1994 hourly rates for various types of construction labor in several countries with those for the United States.

Fringe benefits are known in countries other than the United States as "social charges"; they vary considerably in degree of coverage from country to country. Typical allowances in these countries include family benefits based on number of children, health service, maternity benefits, disability allowances, grants for funeral expenses, old-age and war pensions, unemployment benefits, and pension schemes. Additional fringe benefits may include paid holidays, starting allowances for new workers, relocation grants, severance pay, profit sharing, production bonuses, special gratuities, and sometimes housing allowances. It is essential to investigate the local situation thoroughly to determine the benefits payable and the additional cost on the basic hourly wage rate.

Labor productivity is very much dependent on the health and well-being of the workers and also on the availability of laborsaving tools and construction equipment. The frequency of strikes, holidays, slowdowns, and political unrest will also depress productivity. Closed-shop practices or demarcation disputes will also affect the productivity of labor. The use of standard equipment, parts, and methods tends to improve productivity.

In a particular country, productivity will depend largely on the number of hours worked per week. Production will increase with the number of hours worked during the week, but as more overtime is worked, fatigue will produce a falloff in productivity.

In the United States, construction craft labor usually work a normal 40 h/week. The United Kingdom operates a 40-h schedule, although there is strong pressure to reduce this to 38 h/week. European countries tend to work a normal 40 h/week, and some Far Eastern countries may work up to 45 h/week.

Productivity of local craft labor also depends on the use and availability of modern mechanical tools and construction equipment. Normally, the low cost of labor in certain countries tends to cut out the purchase or hire of sophisticated laborsaving equipment and to encourage the employment of large pools of labor, particularly in developing countries such as India, Pakistan, southeast Asian countries, and many African countries. In turn, this usually leads to higher construction costs. The use of laborsaving equipment is prevalent in Canada, western Europe, Japan, and, to an increasing extent, the Middle East.

Complete Plant Costs   It is difficult to compare costs of domestic and overseas plants owing to the wide variation in types of plants and sizes and the rapid changes in technology. Useful data are scarce, and the following comparisons must be used with caution and then only for order-of-magnitude estimates of fixed-capital costs.

The method uses a breakdown of costs for a typical chemical plant installed in the United States, as shown in Fig. 9-44. Costs of equipment, appurtenances, construction, and engineering with material and labor separate are given as a percentage of total installed United States costs. The four components of cost are defined as follows:

Equipment includes all prefabricated machines, appliances, or systems such as tanks, heat exchangers, pumps, motors, switchgear, and boilers.

Appurtenances are auxiliary items which cover materials, such as pipes, valves, fittings, conduit, wire, tubing, and insulation.

TABLE 9-62 Comparative National Labor Costs

Hourly labor costs	$	Relative to U.S.
Germany	25.35	1.53
Switzerland	22.49	1.36
Belgium	20.84	1.25
Netherlands	19.82	1.20
Japan	19.05	1.15
Austria	19.02	1.15
United States	16.58	1.00
France	16.05	0.97
Italy	15.69	0.95
Britain	12.90	0.78
Ireland	11.94	0.72
Spain	11.36	0.69
Greece	6.29	0.38
Portugal	4.70	0.28

NOTE: Figures include fringe benefits.
SOURCE: Adapted from Economist Intelligence Unit January 1994.

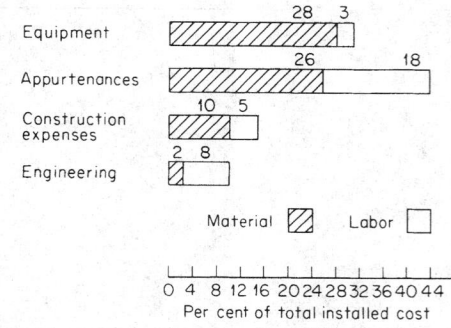

FIG. 9-44 Typical breakdown of chemical-plant costs by major component.

Construction expense includes the cost of construction equipment, tools, sheds, railroad trackage, road materials, welding machines, scaffolding, and timber, which are all used in construction but do not form a permanent part of the plant.

Engineering is mainly labor but has a small component cost which can be classified with equipment and materials, such as tools, paper, pencils, and reproduction costs.

In total, labor amounts to 34 percent and material to 66 percent of total installed costs.

Table 9-63 uses the data of Fig. 9-44 to compare the relative fixed-capital costs for plant construction in other countries with those for the United States. The relative cost ratios were developed from data similar to those in Table 9-62. Labor ratios were corrected for the different local rates and hours per working week, job duration, and degree of mechanization available in other countries. Some of these factors are difficult to estimate, and the final "total" ratios give a reasonable order-of-magnitude value for relative construction costs for equivalent plants in the countries indicated.

TABLE 9-63 Relative Plant Construction Costs in Various Countries Compared with the United States

Country	Equipment	Material	Labor	Engineering	Total
United States	0.28	0.38	0.26	0.08	1.00
England	0.26	0.41	0.18	0.05	0.90
Italy	0.22	0.32	0.31	0.05	0.90
Mexico	0.26	0.35	0.35	0.04	1.00
Australia	0.38	0.54	0.29	0.09	1.30
Canada	0.32	0.44	0.31	0.08	1.15
France	0.27	0.40	0.22	0.06	0.95
Germany	0.26	0.36	0.32	0.06	1.00
Japan	0.29	0.35	0.22	0.04	0.90

The choice of an overseas manufacturing site involves the consideration of many political and economic factors in addition to costs. Table 9-64 gives a list of 92 items which should be taken into account when choosing a plant location for manufacturing abroad.

TABLE 9-64 Factors in Choosing a Foreign Manufacturing Site

Economic factors
  Size of GNP and rate of growth
  Is there a working development plan?
  Resistance to recession
  Relative dependence on imports and exports
  Foreign-exchange position
  Balance-of-payments outlook
  Stability of currency; convertibility
  Remittance and repatriation regulations
  Balance of economy (industry-agriculture-trade)
  Size of market for your products; rate of growth
  Size of population; rate of growth
  Per capita income; rate of growth
  Income distribution
  Current or prospective membership in a customs union
  Price levels; rate of inflation
Political factors
  Stability of government; its form
  Presence or absence of class antagonism
  Special political, ethnic, and social problems
  Attitude toward private and foreign investment
  Acceptability of United States investment by government
  Acceptability of United States investment by customers and competitors
  Presence or absence of nationalization threat
  Presence or absence of state industries
  Do state industries receive favored treatment?
  Concentration of influence in small groups
  Treaty of friendship or establishment with United States?
Government factors
  Are fiscal and monetary policies sound?
  Freedom from bureaucratic red tape
  Fairness and honesty of administrative procedures
  Degree of antiforeign or anti–United States discrimination
  Fairness of courts
  Clear and modern corporate investment laws
  Patentability of your products
  Presence or absence of price controls
  Restrictions on 100 percent United States or foreign ownership
Geographic factors
  Efficiency of transport (railways, waterways, highways)
  Port facilities
  Free ports, free zones, bonded warehouses
  Proximity of site to export markets
  Proximity of site to suppliers, customers
  Proximity to raw-material sources
  Existing supporting industry
  Availability of local raw materials
  Availability of power, water, gas
  Reliability of utilities
  Waste-disposal facilities
  Can exports be easily made?
  Can imports be easily made?

Geographic factors (*Continued*)
  Are plant sites readily available?
  Cost of suitable land
Labor factors
  Availability of English-speaking managerial, technical, office personnel
  Availability of skilled labor
  Availability of semiskilled and unskilled labor
  Level of worker productivity
  Training facilities
  Outlook for increase in labor supply
  Degree of skill and discipline at all levels
  Tranquillity of labor relations
  Presence or absence of militant or Communist-dominated unions
  Degree of labor voice in management
  Freedom to hire and fire
  Compulsory and voluntary fringe benefits
  Social security taxes
  Total cost, including fringes, compared with alternative sites
  Compulsory or customary profit sharing
Tax factors
  Tax rates (corporate and personal income, capital, withholding, turnover, excise, payroll, capital gains, customs, other indirect and local taxes)
  General tax morality
  Fairness and incorruptibility of tax authorities
  Long-term trend for taxes
  Taxation of export income and income earned abroad
  Tax incentives for new businesses
  Depreciation rates
  Tax-loss carry-forward and carry-back
  Joint tax treaties
  Duty and tax drawbacks when imported goods are exported
  Availability of tariff protection
Capital-sources factors
  Availability of local capital
  Costs of local borrowing
  Normal terms for local borrowing
  Availability of convertible currencies locally
  Modern banking system
  Government credit aids to new businesses
  Availability and cost of export financing, insurance
  Do United States or European capital sources favor loans here?
Business factors
  Availability of United States government investment insurance
  General business morality
  State of marketing and distribution system
  Are administrative procedures simple and effective?
  Normal profit margins in general, in your industry
  Competitive situation in your industry; is it cartelized?
  What are antitrust and restrictive practices laws, and do they conflict with United States laws?
  Availability of amenities for United States expatriate executives and families

Transport and Storage of Fluids

Meherwan P. Boyce, Ph.D., *President & CEO, Boyce Engineering International, Inc.; Fellow, American Society of Mechanical Engineers; Registered Professional Engineer.*

Nomenclature and Units

In this listing, symbols used in the section are defined in a general way and appropriate SI and U.S. customary units are given. Specific definitions, as denoted by subscripts, are stated at the place of application in the section. Some specialized symbols used in the section are defined only at the place of application.

Symbol	Definition	SI units	U.S. customary units
A	Area	m²	ft²
A	Factor for determining minimum value of R_1		
A_∞	Free-stream speed of sound		
a	Area	m²	ft²
a	Duct or channel width	m	ft
a	Coefficient, general		
B	Height	m	ft
b	Duct or channel height	m	ft
b	Coefficient, general		
C	Coefficient, general		
C	Conductance	m³/s	ft³/s
C	Sum of mechanical allowances (thread or groove depth) plus corrosion or erosion allowances	mm	in
C	Cold-spring factor		
C	Constant		
C_a	Capillary number	Dimensionless	Dimensionless
C_1	Estimated self-spring or relaxation factor		
c_p	Constant-pressure specific heat	J/(kg·K)	Btu/(lb·°R)
c_v	Constant-volume specific heat	J/(kg·K)	Btu/(lb·°R)
D	Diameter	m	ft
D, D_0	Outside diameter of pipe	mm	in
d	Diameter	m	ft
E	Modulus of elasticity	N/m²	lbf/ft²
E	Quality factor		
E_a	As-installed Young's modulus	MPa	kip/in² (ksi)
E_c	Casting quality factor		
E_j	Joint quality factor		
E_m	Minimum value of Young's modulus	MPa	kip/in² (ksi)
F	Force	N	lbf
F	Friction loss	(N·m)/kg	(ft·lbf)/lb
F	Correction factor	Dimensionless	Dimensionless
f	Frequency	Hz	1/s
f	Friction factor	Dimensionless	Dimensionless
f	Stress-range reduction factor		
G	Mass velocity	kg/(s·m²)	lb/(s·ft²)
g	Local acceleration due to gravity	m/s²	ft/s²
g_c	Dimensional constant	1.0 (kg·m)/(N·s²)	32.2 (lb·ft)/(lbf·s²)
H	Depth of liquid	m	ft
H, h	Head of fluid, height	m	ft
H_{ad}	Adiabatic head	N·m/kg	lbf·ft/lbm
h	Flexibility characteristic		
h	Height of truncated cone; depth of head	m	in
i	Specific enthalpy	J/kg	Btu/lb
i	Stress-intensification factor		
i_i	In-plane stress-intensification factor		
i_o	Out-plane stress intensification factor		
I	Electric current	A	A
J	Mechanical equivalent of heat	1.0 (N·m)/J	778 (ft·lbf)/Btu
K	Index, constant or flow parameter		

Symbol	Definition	SI units	U.S. customary units
K	Fluid bulk modulus of elasticity	N/m²	lbf/ft²
K_1	Constant in empirical flexibility equation		
k	Ratio of specific heats	Dimensionless	Dimensionless
k	Flexibility factor		
k	Adiabatic exponent c_p/c_v		
L	Length	m	ft
L	Developed length of piping between anchors	m	ft
L	Dish radius	m	in
M	Molecular weight	kg/mol	lb/mol
M_i, m_i	In-plane bending moment	N·mm	in·lbf
M_o	Out-plane bending moment	N·mm	in·lbf
M_t	Torsional moment	N·mm	in·lbf
M_∞	Free stream Mach number		
m	Mass	kg	lb
m	Thickness	m	ft
N	Number of data points or items	Dimensionless	Dimensionless
N	Frictional resistance	Dimensionless	Dimensionless
N	Equivalent full temperature cycles		
N_S	Strouhal number	Dimensionless	Dimensionless
N_{De}	Dean number	Dimensionless	Dimensionless
N_{Fr}	Froude number	Dimensionless	Dimensionless
N_{Re}	Reynolds number	Dimensionless	Dimensionless
N_{We}	Weber number	Dimensionless	Dimensionless
NPSH	Net positive suction head	m	ft
n	Polytropic exponent		
n	Pulsation frequency	Hz	1/s
n	Constant, general		
n	Number of items	Dimensionless	Dimensionless
P	Design gauge pressure	kPa	lbf/in²
P_{ad}	Adiabatic power	kW	hp
p	Pressure	Pa	lbf/ft²
p	Power	kW	hp
Q	Heat	J	Btu
Q	Volume	m³	ft³
Q	Volume rate of flow (liquids)	m³/h	gal/min
Q	Volume rate of flow (gases)	m³/h	ft³/min (cfm)
q	Volume flow rate	m³/s	ft³/s
R	Gas constant	8314 J/(K·mol)	1545 (ft·lbf)/(mol·°R)
R	Radius	m	ft
R	Electrical resistance	Ω	Ω
R	Head reading	m	ft
R	Range of reaction forces or moments in flexibility analysis	N or N·mm	lbf or in·lbf
R	Cylinder radius	m	ft
R	Universal gas constant	J/(kg·K)	(ft·lbf)/(lbm·°R)
R_a	Estimated instantaneous reaction force or moment at installation temperature	N or N·mm	lbf or in·lbf
R_m	Estimated instantaneous maximum reaction force or moment at maximum or minimum metal temperature	N or N·mm	lbf or in·lbf
R_1	Effective radius of miter bend	mm	in

Nomenclature and Units (*Concluded*)

Symbol	Definition	SI units	U.S. customary units
r	Radius	m	ft
r	Pressure ratio	Dimensionless	Dimensionless
r_c	Critical pressure ratio		
r_k	Knuckle radius	m	in
r_2	Mean radius of pipe using nominal wall thickness $\bar{T}$	mm	in
S	Specific surface area	m^2/m^3	ft^2/ft^3
S	Fluid head loss	Dimensionless	Dimensionless
S	Specific energy loss	m/s^2	lbf/lb
S	Speed	m^3/s	ft^3/s
S	Basic allowable stress for metals, excluding factor E, or bolt design stress	MPa	kip/in² (ksi)
S_A	Allowable stress range for displacement stress	MPa	kip/in² (ksi)
S_E	Computed displacement-stress range	MPa	kip/in² (ksi)
S_L	Sum of longitudinal stresses	MPa	kip/in² (ksi)
S_T	Allowable stress at test temperature	MPa	kip/in² (ksi)
S_b	Resultant bending stress	MPa	kip/in² (ksi)
S_c	Basic allowable stress at minimum metal temperature expected	MPa	kip/in² (ksi)
S_h	Basic allowable stress at maximum metal temperature expected	MPa	kip/in² (ksi)
S_t	Torsional stress	MPa	kip/in² (ksi)
s	Specific gravity		
s	Specific entropy	J/(kg·K)	Btu/(lb·°R)
T	Temperature	K (°C)	°R (°F)
T_s	Effective branch-wall thickness	mm	in
$\bar{T}$	Nominal wall thickness of pipe	mm	in
$\bar{T}_b$	Nominal branch-pipe wall thickness	mm	in
$\bar{T}_h$	Nominal header-pipe wall thickness	mm	in
t	Head or shell radius	mm	in
t	Pressure design thickness	mm	in
t	Time	s	s
t_m	Minimum required thickness, including mechanical, corrosion, and erosion allowances	mm	in
t_r	Pad or saddle thickness	mm	in
U	Straight-line distance between anchors	m	ft
u	Specific internal energy	J/kg	Btu/lb
u	Velocity	m/s	ft/s
V	Velocity	m/s	ft/s
V	Volume	m^3	ft^3

Symbol	Definition	SI units	U.S. customary units
v	Specific volume	m^3/kg	ft^3/lb
W	Work	N·m	lbf·ft
W	Weight	kg	lb
w	Weight flow rate	kg/s	lb/s
x	Weight fraction	Dimensionless	Dimensionless
x	Distance or length	m	ft
x	Value of expression $[(p_2/p_1)^{(k-1/k)} - 1]$		
Y	Expansion factor	Dimensionless	Dimensionless
y	Distance or length	m	ft
y	Resultant of total displacement strains	mm	in
Z	Section modulus of pipe	mm^3	in^3
Z	Vertical distance	m	ft
Z_e	Effective section modulus for branch	mm^3	in^3
z	Gas-compressibility factor	Dimensionless	Dimensionless
z	Vertical distance	m	ft

	Greek symbols		
α	Viscous-resistance coefficient	$1/m^2$	$1/ft^2$
α	Angle	°	°
σ	Half-included angle	°	°
α, β, θ	Angles	°	°
β	Inertial-resistance coefficient	1/m	1/ft
β	Ratio of diameters	Dimensionless	Dimensionless
Γ	Liquid loading	kg/(s·m)	lb/(s·ft)
Γ	Pulsation intensity	Dimensionless	Dimensionless
δ	Thickness	m	ft
ε	Wall roughness	m	ft
ε	Voidage—fractional free volume	Dimensionless	Dimensionless
η	Viscosity, nonnewtonian fluids	Pa·s	lb/(ft·s)
η_{ad}	Adiabatic efficiency		
η_p	Polytropic efficiency		
θ	Angle	°	°
λ	Molecular mean free-path length	m	ft
μ	Viscosity	Pa·s	lb/(ft·s)
ν	Kinematic viscosity	m^2/s	ft^2/s
ρ	Density	kg/m^3	lb/ft^3
σ	Surface tension	N/m	lbf/ft
σ_c	Cavitation number	Dimensionless	Dimensionless
τ	Shear stress	N/m^2	lbf/ft^2
ϕ	Shape factor	Dimensionless	Dimensionless
ϕ	Angle	°	°
ψ	Flow coefficient		
ψ	Pressure coefficient		
ψ	Sphericity	Dimensionless	Dimensionless

INTRODUCTION

Transportation and the storage of fluids (gases and liquids) involves the understanding of the properties and behavior of fluids. The study of fluid dynamics is the study of fluids and their motion in a force field.

Flows can be classified into two major categories: (a) incompressible and (b) compressible flow. Most liquids fall into the incompressible-flow category, while most gases are compressible in nature. A perfect fluid can be defined as a fluid that is nonviscous and nonconducting. Fluid flow, compressible or incompressible, can be classified by the ratio of the inertial forces to the viscous forces. This ratio is represented by the Reynolds number (N_{Re}). At a low Reynolds number, the flow is considered to be laminar, and at high Reynolds numbers, the flow is considered to be turbulent. The limiting types of flow are the inertialess flow, sometimes called Stokes flow, and the inviscid flow that occurs at an infinitely large Reynolds number. Reynolds numbers (dimensionless) for flow in a pipe is given as:

$$N_{Re} = \frac{\rho V D}{\mu} \qquad (10\text{-}1)$$

where ρ is the density of the fluid, V the velocity, D the diameter, and μ the viscosity of the fluid. In fluid motion where the frictional forces interact with the inertia forces, it is important to consider the ratio of the viscosity μ to the density ρ. This ratio is known as the kinematic viscosity (ν). Tables 10-1 and 10-2 give the kinematic viscosity for several fluids. A flow is considered to be *adiabatic* when there is no transfer of heat between the fluid and its surroundings. An isentropic flow is one in which the entropy of each fluid element remains constant.

To fully understand the mechanics of flow, the following definitions explain the behavior of various types of fluids in both their static and flowing states.

A perfect fluid is a nonviscous, nonconducting fluid. An example of this type of fluid would be a fluid that has a very small viscosity and conductivity and is at a high Reynolds number. An ideal gas is one that obeys the equation of state:

$$\frac{P}{\rho} = RT \qquad (10\text{-}2)$$

where P = pressure, ρ = density, R is the gas constant per unit mass, and T = temperature.

A flowing fluid is acted upon by many forces that result in changes in pressure, temperature, stress, and strain. A fluid is said to be isotropic when the relations between the components of stress and those of the rate of strain are the same in all directions. The fluid is said to be Newtonian when this relationship is linear. These pressures and temperatures must be fully understood so that the entire flow picture can be described.

The *static pressure* in a fluid has the same value in all directions and can be considered as a scalar point function. It is the pressure of a flowing fluid. It is normal to the surface on which it acts and at any given point has the same magnitude irrespective of the orientation of the surface. The static pressure arises because of the random motion in the fluid of the molecules that make up the fluid. In a diffuser or nozzle, there is an increase or decrease in the static pressure due to the change in velocity of the moving fluid.

Total Pressure is the pressure that would occur if the fluid were brought to rest in a reversible adiabatic process. Many texts and engineers use the words *total* and *stagnation* to describe the flow characteristics interchangeably. To be accurate, the stagnation pressure is the pressure that would occur if the fluid were brought to rest adiabatically or diabatically.

Total pressure will only change in a fluid if shaft work or work of extraneous forces are introduced. Therefore, total pressure would increase in the impeller of a compressor or pump; it would remain constant in the diffuser. Similarly, total pressure would decrease in the turbine impeller but would remain constant in the nozzles.

Static temperature is the temperature of the flowing fluid. Like static pressure, it arises because of the random motion of the fluid molecules. Static temperature is in most practical installations impossible to measure since it can be measured only by a thermometer or thermocouple at rest relative to the flowing fluid that is moving with the fluid. Static temperature will increase in a diffuser and decrease in a nozzle.

Total temperature is the temperature that would occur when the fluid is brought to rest in a reversible adiabatic manner. Just like its counterpart *total pressure, total* and *stagnation temperatures* are used interchangeably by many test engineers.

Dynamic temperature and pressure are the difference between the total and static conditions.

$$P_d = P_T - P_s \qquad (10\text{-}3)$$
$$T_d = T_T - T_s \qquad (10\text{-}4)$$

where subscript d refers to dynamic, T to total, and s to static.

Another helpful formula is:

$$P_K = \frac{1}{2} \rho V^2 \qquad (10\text{-}5)$$

For incompressible fluids, $P_K = P_d$.

TABLE 10.2 Kinematic Viscosity

Liquid	Temperature		$\nu \times 10^6$ (ft²/sec)
	°C	°F	
Glycerine	20	68	7319
Mercury	0	32	1.35
Mercury	100	212	0.980
Lubricating oil	20	68	4306
Lubricating oil	40	104	1076
Lubricating oil	60	140	323

TABLE 10.1 Density, Viscosity, and Kinematic Viscosity of Water and Air in Terms of Temperature

Temperature		Water			Air at a pressure of 760 mm Hg (14.696 lbf/in²)		
(°C)	(°F)	Density ρ (lbf sec²/ft⁴)	Viscosity $\mu \times 10^6$ (lbf sec/ft²)	Kinematic viscosity $\nu \times 10^6$ (ft²/sec)	Density ρ (lbf sec²/ft⁴)	Viscosity $\mu \times 10^6$ (lbf sec/ft²)	Kinematic viscosity $\nu \times 10^6$ (ft²/sec)
−20	−4	—	—	—	0.00270	0.326	122
−10	14	—	—	—	0.00261	0.338	130
0	32	1.939	37.5	19.4	0.00251	0.350	140
10	50	1.939	27.2	14.0	0.00242	0.362	150
20	68	1.935	21.1	10.9	0.00234	0.375	160
40	104	1.924	13.68	7.11	0.00217	0.399	183
60	140	1.907	9.89	5.19	0.00205	0.424	207
80	176	1.886	7.45	3.96	0.00192	0.449	234
100	212	1.861	5.92	3.19	0.00183	0.477	264

Conversion factors: 1 kp sec²/m⁴ = 0·01903 lbf sec²/ft⁴ (= slug/ft³)
1 lbf sec²/ft⁴ = 32.1719 lb/ft³   (lb = lb mass; lbf = lb force)
1 kp sec²/m⁴ = 9.80665 kg/m³   (kg = kg mass; kp = kg force)
1 kg/m³ = 16.02 lb/ft³

MEASUREMENT OF FLOW

This subsection deals with the techniques of measuring pressures, temperatures, velocities, and flow rates of flowing fluids.

STATIC PRESSURE

Local Static Pressure   In a moving fluid, the local static pressure is equal to the pressure on a surface which moves with the fluid or to the normal pressure (for newtonian fluids) on a stationary surface which parallels the flow. The pressure on such a surface is measured by making a small hole perpendicular to the surface and connecting the opening to a pressure-sensing element (Fig. 10-1a). The hole is known as a piezometer opening or pressure tap.

Measurement of local static pressure is frequently difficult or impractical. If the channel is so small that introduction of any solid object disturbs the flow pattern and increases the velocity, there will be a reduction and redistribution of the static pressure. If the flow is in straight parallel lines, aside from the fluctuations of normal turbulence, the flat disk (Fig. 10-1b) and the bent tube (Fig. 10-1c) give satisfactory results when properly aligned with the stream. Slight misalignments can cause serious errors. Diameter of the disk should be 20 times its thickness and 40 times the static opening; the face must be flat and smooth, with the knife edges made by beveling the underside. The piezometer tube, such as that in Fig. 10-1c, should have openings with size and spacing as specified for a pitot-static tube (Fig. 10-6).

Readings given by open straight tubes (Fig. 10-1d and 1e are too low due to flow separation. Readings of closed tubes oriented perpendicularly to the axis of the stream and provided with side openings (Fig. 10-1e) may be low by as much as two velocity heads.

Average Static Pressure   In most cases, the object of a static-pressure measurement is to obtain a suitable average value for substitution in Bernoulli's theorem or in an equivalent flow formula. This can be done simply only when the flow is in straight lines parallel to the confining walls, such as in straight ducts at sufficient distance downstream from bends (2 diameters) or other disturbances. For such streams, the sum of static head and gravitational potential head is the same at all points in a cross section taken perpendicularly to the axis of flow. Thus the exact location of a piezometer opening about the periphery of such a cross section is immaterial provided its elevation is known. However, in stating the static pressure, the custom is to give the value at the elevation corresponding to the centerline of the stream.

With flow in curved passages or with swirling flow, determination of a true average static pressure is, in general, impractical. In metering, straightening vanes are often placed upstream of the pressure tap to eliminate swirl. Fig. 10-2 shows various flow equalizers and straighteners.

Specifications for Piezometer Taps   The size of a static opening should be small compared with the diameter of the pipe and yet large compared with the scale of surface irregularities. For reliable results, it is essential that (1) the surface in which the hole is made be substantially smooth and parallel to the flow for some distance on either side of the opening, and (2) the opening be flush with the surface and possess no "burr" or other irregularity around its edge.

Rounding of the edge is often employed to ensure absence of a burr. Pressure readings will be high if the tap is inclined upstream, is rounded excessively on the upstream side, has a burr on the downstream side, or has an excessive countersink or recess. Pressure readings will be low if the tap is inclined downstream, is rounded excessively on the downstream side, has a burr on the upstream side, or protrudes into the flow stream. Errors resulting from these faults can be large.

Recommendations for **pressure-tap dimensions** are summarized in Table 10-3. Data from several references were used in arriving at these composite values. The length of a pressure-tap opening prior to

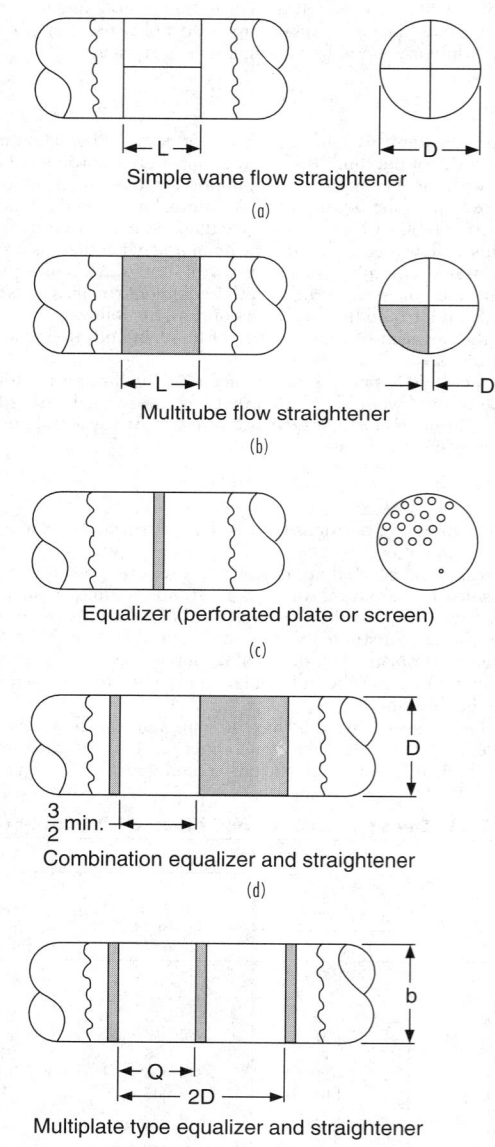

Simple vane flow straightener

(a)

Multitube flow straightener

(b)

Equalizer (perforated plate or screen)

(c)

Combination equalizer and straightener

(d)

Multiplate type equalizer and straightener

(e)

FIG. 10-2   Flow equalizers and straighteners [*Power Test Code 10, Compressors and Exhausters, Amer. Soc. of Mechanical Engineers, 1965*].

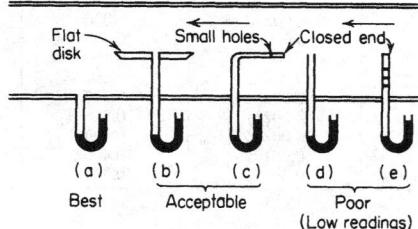

| (a) | (b) | (c) | (d) | (e) |
| Best | Acceptable | | Poor (Low readings) | |

FIG. 10-1   Measurement of static pressure.

TABLE 10-3 Pressure-Tap Holes

Nominal inside pipe diameter, in	Maximum diameter of pressure tap, mm (in)	Radius of hole-edge rounding, mm (in)
1	3.18 (⅛)	<0.40 (1/64)
2	6.35 (¼)	0.40 (1/64)
3	9.53 (⅜)	0.40–0.79 (1/64–1/32)
4	12.7 (½)	0.79 (1/32)
8	12.7 (½)	0.79–1.59 (1/32–1/16)
16	19.1 (¾)	0.79–1.59 (1/32–1/16)

any enlargement in the tap channel should be at least two tap diameters, preferably three or more.

A **piezometer ring** is a toroidal manifold into which are connected several sidewall static taps located around the perimeter of a common cross section. Its intent is to give an average pressure if differences in pressure other than those due to static head exist around the perimeter. However, there is generally no assurance that a true average is provided thereby. The principal advantage of the ring is that use of several holes in place of a single hole reduces the possibility of completely plugging the static openings.

For information on prediction of static-hole error, see Shaw, *J. Fluid Mech.,* **7,** 550–564 (1960); Livesey, Jackson, and Southern, *Aircr. Eng.,* **34,** 43–47 (February 1962).

For nonnewtonian fluids, pressure readings with taps may also be low because of fluid-elasticity effects. This error can be largely eliminated by using flush-mounted diaphragms.

For information on the pressure-hole error for nonnewtonian fluids, see Han and Kim, *Trans. Soc. Rheol.,* **17,** 151–174 (1973); Novotny and Eckert, *Trans. Soc. Rheol.,* **17,** 227–241 (1973); and Higashitani and Lodge, *Trans. Soc. Rheol.,* **19,** 307–336 (1975).

Dynamic pressure may be measured by use of a pitot tube that is a simple impact tube. These tubes measure the pressure at a point where the velocity of the fluid is brought to zero. Pitot tubes must be parallel to the flow. The pitot tube is sensitive to yaw or angle attack. In general angles of attack over 10° should be avoided. In cases where the flow direction is unknown, it is recommended to use a Kiel probe. Figure 10-3 shows a Kiel probe. This probe will read accurately to an angle of about 22° with the flow.

Special Tubes A variety of special forms of the pitot tube have been evolved. Folsom (loc. cit.) gives a description of many of these special types together with a comprehensive bibliography. Included

are the impact tube for **boundary-layer** measurements and **shielded total-pressure tubes.** The latter are insensitive to angle of attack up to 40°.

Chue [*Prog. Aerosp. Sci.,* **16,** 147–223 (1975)] reviews the use of the pitot tube and allied pressure probes for impact pressure, static pressure, dynamic pressure, flow direction and local velocity, skin friction, and flow measurements.

A reversed pitot tube, also known as a **pitometer,** has one pressure opening facing upstream and the other facing downstream. Coefficient *C* for this type is on the order of 0.85. This gives about a 40 percent increase in pressure differential as compared with standard pitot tubes and is an advantage at low velocities. There are commercially available very compact types of pitometers which require relatively small openings for their insertion into a duct.

The **pitot-venturi** flow element is capable of developing a pressure differential 5 to 10 times that of a standard pitot tube. This is accomplished by employing a pair of concentric venturi elements in place of the pitot probe. The low-pressure tap is connected to the throat of the inner venturi, which in turn discharges into the throat of the outer venturi. For a discussion of performance and application of this flow element, see Stoll, *Trans. Am. Soc. Mech. Eng.,* **73,** 963–969 (1951).

TOTAL TEMPERATURE

For most points requiring temperature monitoring, either thermocouples or resistive thermal detectors (RTD's) can be used. Each type of temperature transducer has its own advantages and disadvantages, and both should be considered when temperature is to be measured. Since there is considerable confusion in this area, a short discussion of the two types of transducers is necessary.

Thermocouples The various types of thermocouples provide transducers suitable for measuring temperatures from −330 to 5000°F (−201 to 2760°C). The useful ranges for the various types are shown in Fig. 10-4. Thermocouples function by producing a voltage proportional to the temperature differences between two junctions of dissimilar metals. By measuring this voltage, the temperature difference can be determined. It is assumed that the temperature is known at one of the junctions; therefore, the temperature at the other junction can be determined. Since the thermocouples produce a voltage, no external power supply is required to the test junction; however, for accurate measurement, a reference junction is required. For a temperature monitoring system, reference junctions must be placed at

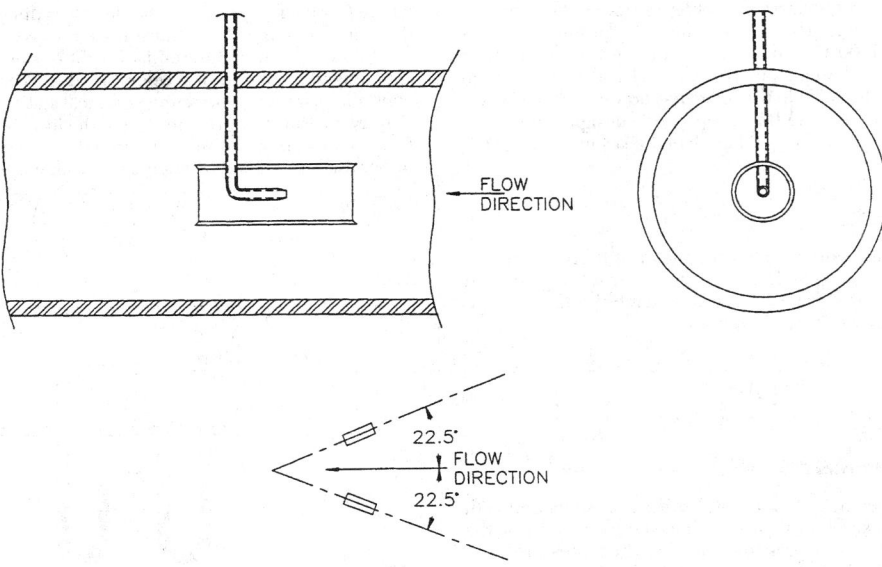

FIG. 10-3 Kiel probe. Accurate measurements can be made at angles up to 22.5° with the flow stream.

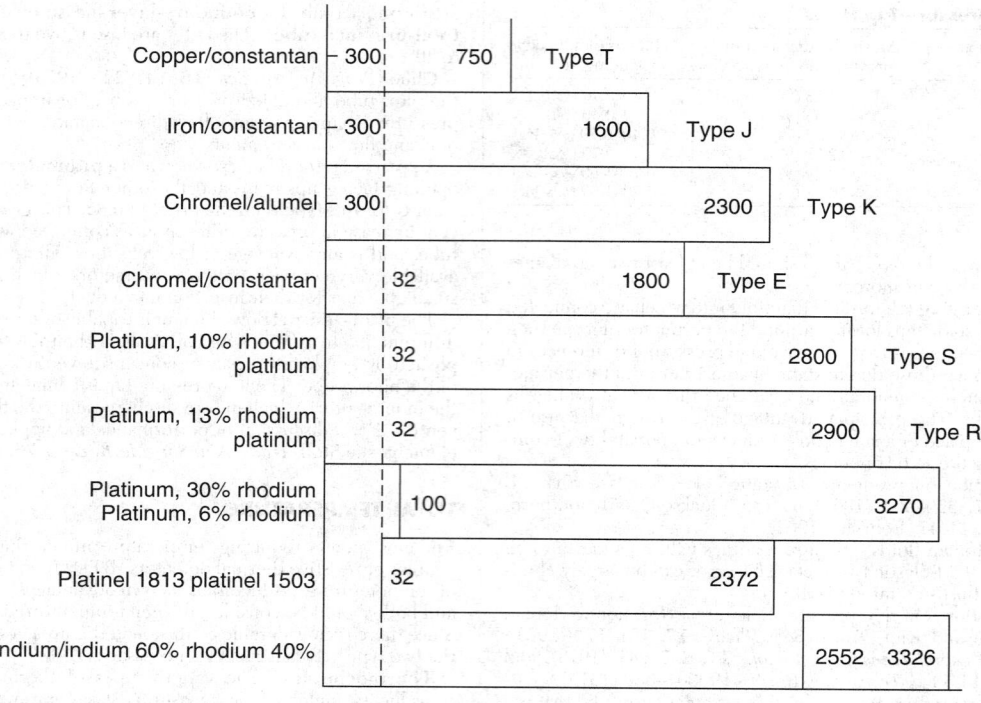

FIG. 10-4   Ranges of various thermocouples.

each thermocouple or similar thermocouple wire installed from the thermocouple to the monitor where there is a reference junction. Properly designed thermocouple systems can be accurate to approximately ±2°F (±1°C).

Resistive Thermal Detectors (RTD)   RTDs determine temperature by measuring the change in resistance of an element due to temperature. Platinum is generally utilized in RTDs because it remains mechanically and electrically stable, resists contaminations, and can be highly refined. The useful range of platinum RTDs is –454–1832°F (–270–1000°C). Since the temperature is determined by the resistance in the element, any type of electrical conductor can be utilized to connect the RTD to the indicator; however, an electrical current must be provided to the RTD. A properly designed temperature monitoring system utilizing RTDs can be accurate ±0.02°F (±0.01°C).

STATIC TEMPERATURE

Since this temperature requires the thermometer or thermocouple to be at rest relative to the flowing fluid, it is impractical to measure. It can be, however, calculated from the measurement of total temperature and total and static pressure.

$$T_S = \frac{T_O}{\left(\dfrac{P_O}{P_S}\right)^{(k-1)/k}} \tag{10-6}$$

VELOCITY MEASUREMENTS

Pitot Tubes   The combination of pitot tubes in conjunction with sidewall static taps measures local or point velocities by measuring the difference between the total pressure and the static pressure. The pitot tube shown in Fig. 10-5 consists of an impact tube whose opening faces directly into the stream to measure impact pressure, plus one

or more sidewall taps to measure local static pressure. The combined pitot-static tube shown in Fig. 10-6 consists of a jacketed impact tube with one or more rows of holes, 0.51 to 1.02 mm (0.02 to 0.04 in) in diameter, in the jacket to measure the static pressure. Velocity V_0 m/s (ft/s) at the point where the tip is located is given by

$$V_0 = C \sqrt{2g_c \, \Delta h} = C \sqrt{2g_c \, (P_T - P_S)/P_S} \tag{10-7}$$

where C = coefficient, dimensionless; g_c = dimensional constant; Δh = dynamic pressure ($\Delta h_s g/g_c$), expressed in (N·m)/kg [(ft·lbf)/lb or ft of fluid flowing]; Δh_s = differential height of static liquid column corresponding to Δh; g = local acceleration due to gravity; g_c = dimensional constant; p_i = impact pressure; p_0 = local static pressure; and ρ_0 = fluid density measured at pressure p_0 and the local temperature. With gases at velocities above 60 m/s (about 200 ft/s), compressibility becomes important, and the following equation should be used:

$$V_0 = C \sqrt{\frac{2g_c k}{k-1} \left(\frac{p_0}{\rho_0}\right)\left[\left(\frac{p_i}{p_0}\right)^{(k-1)/k} - 1\right]} \tag{10-8}$$

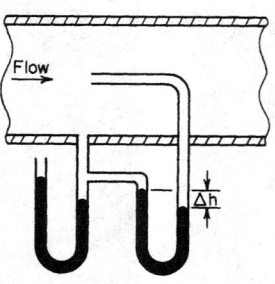

FIG. 10-5   Pitot tube with sidewall static tap.

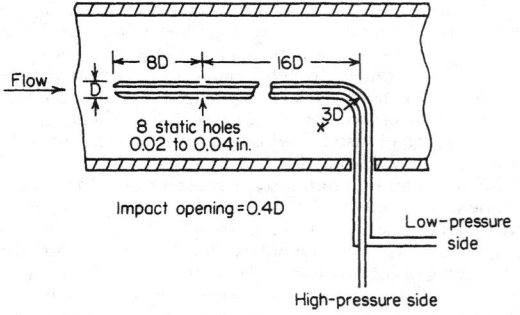

FIG. 10-6 Pitot-static tube.

where k is the ratio of specific heat at constant pressure to that at constant volume. (See ASME Research Committee on Fluid Meters Report, op. cit., p. 105.) Coefficient C is usually close to 1.00 (± 0.01) for simple pitot tubes (Fig. 10-5) and generally ranges between 0.98 and 1.00 for pitot-static tubes (Fig. 10-6).

There are certain limitations on the range of usefulness of pitot tubes. With gases, the differential is very small at low velocities; e.g., at 4.6 m/s (15.1 ft/s) the differential is only about 1.30 mm (0.051 in) of water (20°C) for air at 1 atm (20°C), which represents a lower limit for 1 percent error even when one uses a micromanometer with a precision of 0.0254 mm (0.001 in) of water. Equation does not apply for Mach numbers greater than 0.7 because of the interference of shock waves. For supersonic flow, local Mach numbers can be calculated from a knowledge of the dynamic and true static pressures. The free stream Mach number (M_∞) is defined as the ratio of the speed of the stream (V_∞) to the speed of sound in the free stream:

$$A_\infty = \sqrt{\left(\frac{\partial P}{\partial \rho}\right)_{s=c}} \qquad (10\text{-}9)$$

$$M_\infty = \frac{V_\infty}{\sqrt{\left(\frac{\partial P}{\partial \rho}\right)_{s=c}}} \qquad (10\text{-}10)$$

where S is the entropy. For isentropic flow, this relationship and pressure can be written as:

$$M_\infty = \frac{V_\infty}{\sqrt{kRT_s}} \qquad (10\text{-}11)$$

The relationships between total and static temperature and pressure are given by the following relationship:

$$\frac{T_T}{T_S} = 1 + \frac{k-1}{2} M^2 \qquad (10\text{-}12)$$

$$\frac{P_T}{P_S} = \left(1 + \frac{k-1}{2} M^2\right)^{(k-1)/k} \qquad (10\text{-}13)$$

With **liquids** at low velocities, the effect of the Reynolds number upon the coefficient is important. The coefficients are appreciably less than unity for Reynolds numbers less than 500 for pitot tubes and for Reynolds numbers less than 2300 for pitot-static tubes [see Folsom, *Trans. Am. Soc. Mech. Eng.*, **78**, 1447–1460 (1956)]. Reynolds numbers here are based on the probe outside diameter. Operation at low Reynolds numbers requires prior calibration of the probe.

The pitot-static tube is also sensitive to **yaw** or **angle of attack** than is the simple pitot tube because of the sensitivity of the static taps to orientation. The error involved is strongly dependent upon the exact probe dimensions. In general, angles greater than 10° should be avoided if the velocity error is to be 1 percent or less.

Disturbances upstream of the probe can cause large errors, in part because of the turbulence generated and its effect on the static-pressure measurement. A calming section of at least 50 pipe diameters is desirable. If this is not possible, the use of straightening vanes or a honeycomb is advisable.

The effect of **pulsating flow** on pitot-tube accuracy is treated by Ower et al., op. cit., pp. 310–312. For sinusoidal velocity fluctuations, the ratio of indicated velocity to actual mean velocity is given by the factor $\sqrt{1 + \lambda^2/2}$, where λ is the velocity excursion as a fraction of the mean velocity. Thus, the indicated velocity would be about 6 percent high for velocity fluctuations of ± 50 percent, and pulsations greater than ± 20 percent should be damped to avoid errors greater than 1 percent. The error increases as the frequency of flow oscillations approaches the natural frequency of the pitot tube and the density of the measuring fluid approaches the density of the process fluid [see Horlock and Daneshyar, *J. Mech. Eng. Sci.*, **15**, 144–152 (1973)].

Pressures substantially lower than true impact pressures are obtained with pitot tubes in turbulent flow of dilute polymer solutions [see Halliwell and Lewkowicz, *Phys. Fluids*, **18**, 1617–1625 (1975)].

Traversing for Mean Velocity   Mean velocity in a duct can be obtained by dividing the cross section into a number of equal areas, finding the local velocity at a representative point in each, and averaging the results. In the case of **rectangular passages,** the cross section is usually divided into small squares or rectangles and the velocity is found at the center of each. In circular pipes, the cross section is divided into several equal annular areas as shown in Fig. 10-7. Read-

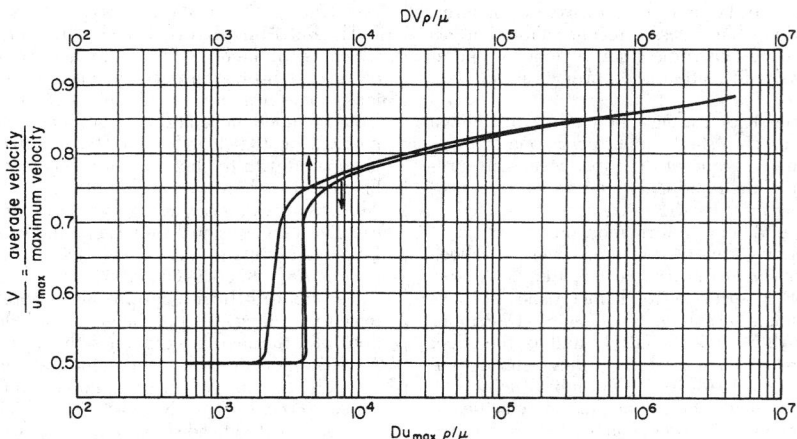

FIG. 10-7 Velocity ration versus Reynolds number for smooth circular pipes. [*Based on data from Rothfus, Archer, Klimas, and Sikchi*, Am. Inst. Chem. Eng. J., 3,208 (1957).]

ings of velocity are made at the intersections of a diameter and the set of circles which bisect the annuli and the central circle.

For an N-point traverse on a circular cross section, make readings on each side of the cross section at

$$100 \times \sqrt{(2n-1)/N} \text{ percent} \quad (n = 1, 2, 3 \text{ to } N/2)$$

of the pipe radius from the center. Traversing several diameters spaced at equal angles about the pipe is required if the velocity distribution is unsymmetrical. With a normal velocity distribution in a circular pipe, a 10-point traverse theoretically gives a mean velocity 0.3 percent high; a 20-point traverse, 0.1 percent high.

For normal velocity distribution in straight circular pipes at locations preceded by runs of at least 50 diameters without pipe fittings or other obstructions, the graph in Fig. 10-7 shows the ratio of mean velocity V to velocity at the center u_{max} plotted against the Reynolds number, where D = inside pipe diameter, ρ = fluid density, and μ = fluid viscosity, all in consistent units. Mean velocity is readily determined from this graph and a pitot reading at the center of the pipe if the quantity $D u_{max} \rho/\mu$ is less than 2000 or greater than 5000. The method is unreliable at intermediate values of the Reynolds number.

Methods for determining mean flow rate from probe measurements under nonideal conditions are described by Mandersloot, Hicks, and Langejan [*Chem. Eng. (London)*, no. 232, CE370-CE380 (1969)].

Anemometers   An anemometer may be any instrument for measurement of gas velocity, e.g., a pitot tube, but usually the term refers to one of the following types.

The vane **anemometer** is a delicate revolution counter with jeweled bearings, actuated by a small windmill, usually 75 to 100 mm (about 3 to 4 in) in diameter, constructed of flat or slightly curved radially disposed vanes. Gas velocity is determined by using a stopwatch to find the time interval required to pass a given number of meters (feet) of gas as indicated by the counter. The velocity so obtained is inversely proportional to gas density. If the original calibration was carried out in a gas of density ρ_0 and the density of the gas stream being metered is ρ_1, the true gas velocity can be found as follows: From the calibration curve for the instrument, find $V_{t,0}$ corresponding to the quantity $V_m \sqrt{\rho_1/\rho_0}$, where V_m = measured velocity. Then the actual velocity $V_{t,1}$ is equal to $V_{t,0} \sqrt{\rho_0/\rho_1}$. In general, when working with air, the effects of atmospheric-density changes can be neglected for all velocities above 1.5 m/s (about 5 ft/s). In all cases, care must be taken to hold the anemometer well away from one's body or from any object not normally present in the stream.

Vane anemometers can be used for gas-velocity measurements in the range of 0.3 to 45 m/s (about 1 to 150 ft/s), although a given instrument generally has about a twentyfold velocity range. Bearing friction has to be minimized in instruments designed for accuracy at the low end of the range, while ample rotor and vane rigidity must be provided for measurements at the higher velocities. Vane anemometers are sensitive to shock and cannot be used in corrosive atmospheres. Therefore, accuracy is questionable unless a recent calibration has been made and the history of the instrument subsequent to calibration is known. For additional information, see Ower et al., op. cit., chap. VIII.

A **turbine flowmeter** consists of a straight flow tube containing a turbine which is free to rotate on a shaft supported by one or more bearings and located on the centerline of the tube. Means are provided for magnetic detection of the rotational speed, which is proportional to the volumetric flow rate. Its use is generally restricted to clean, noncorrosive fluids. Additional information on construction, operation, range, and accuracy can be obtained from Holzbock (*Instruments for Measurement and Control*, 2d ed., Reinhold, New York, 1962, pp. 155–162). For performance characteristics of these meters with liquids, see Shafer, *J. Basic Eng.*, **84**, 471–485 (December 1962); or May, *Chem. Eng.*, **78**(5), 105–108 (1971); and for the effect of density and Reynolds number when used in gas flowmetering, see Lee and Evans, *J. Basic Eng.*, **82**, 1043–1057 (December 1965).

The **current meter** is generally used for measuring velocities in open channels such as rivers and irrigation channels. There are two types, the cup meter and the propeller meter. The former is more widely used. It consists of six conical cups mounted on a vertical axis pivoted at the ends and free to rotate between the rigid arms of a

U-shaped clevis to which a vaned tailpiece is attached. The wheel rotates because of the difference in drag for the two sides of the cup, and a signal proportional to the revolutions of the wheel is generated. The velocity is determined from the count over a period of time. The current meter is generally useful in the range of 0.15 to 4.5 m/s (about 0.5 to 15 ft/s) with an accuracy of ± 2 percent. For additional information see Creager and Justin, *Hydroelectric Handbook*, 2d ed., Wiley, New York, 1950, pp. 42–46.

The **hot-wire anemometer** consists essentially of an electrically heated fine wire (generally platinum) exposed to the gas stream whose velocity is being measured. An increase in fluid velocity, other things being equal, increases the rate of heat flow from the wire to the gas, thereby tending to cool the wire and alter its electrical resistance. In a constant-current anemometer, gas velocity is determined by measuring the resulting wire resistance; in the constant-resistance type, gas velocity is determined from the current required to maintain the wire temperature, and thus the resistance, constant. The difference in the two types is primarily in the electric circuits and instruments employed.

The hot-wire anemometer can, with suitable calibration, accurately measure velocities from about 0.15 m/s (0.5 ft/s) to supersonic velocities and detect velocity fluctuations with frequencies up to 200,000 Hz. Fairly rugged, inexpensive units can be built for the measurement of mean velocities in the range of 0.15 to 30 m/s (about 0.5 to 100 ft/s). More elaborate, compensated units are commercially available for use in unsteady flow and turbulence measurements. In calibrating a hot-wire anemometer, it is preferable to use the same gas, temperature, and pressure as will be encountered in the intended application. In this case the quantity $I^2 R_w/\Delta t$ can be plotted against $\sqrt{V}$, where I = hot-wire current, R_w = hot-wire resistance, Δt = difference between the wire temperature and the gas bulk temperature, and V = mean local velocity. A procedure is given by Wasan and Baid [*Am. Inst. Chem. Eng. J.*, **17**, 729–731 (1971)] for use when it is impractical to calibrate with the same gas composition or conditions of temperature and pressure. Andrews, Bradley, and Hundy [*Int. J. Heat Mass Transfer*, **15**, 1765–1786 (1972)] give a calibration correlation for measurement of small gas velocities. The hot-wire anemometer is treated in considerable detail in Dean, op. cit., chap. VI; in Ladenburg et al., op. cit., art. F-2; by Grant and Kronauer, *Symposium on Measurement in Unsteady Flow*, American Society of Mechanical Engineers, New York, 1962, pp. 44–53; ASME Research Committee on Fluid Meters Report, op. cit., pp. 105–107; and by Compte-Bellot, *Ann. Rev. Fluid Mech.*, **8**, pp. 209–231 (1976).

The hot-wire anemometer can be modified for liquid measurements, although difficulties are encountered because of bubbles and dirt adhering to the wire. See Stevens, Borden, and Strausser, David Taylor Model Basin Rep. 953, December 1956; Middlebrook and Piret, *Ind. Eng. Chem.*, **42**, 1511–1513 (1950); and Piret et al., *Ind. Eng. Chem.*, **39**, 1098–1103 (1947).

The **hot-film anemometer** has been developed for applications in which use of the hot-wire anemometer presents problems. It consists of a platinum-film sensing element deposited on a glass substrate. Various geometries can be used. The most common involves a wedge with a $30°$ included angle at the end of a tapered rod. The wedge is commonly 1 mm (0.039 in) long and 0.2 mm (0.0079 in) wide on each face. Compared with the hot wire, it is less susceptible to fouling by bubbles or dirt when used in liquids, has greater mechanical strength when used with gases at high velocities and high temperatures, and can give a higher signal-to-noise ratio. For additional information see Ling and Hubbard, *J. Aeronaut. Sci.*, **23**, 890–891 (1956); and Ling, *J. Basic Eng.*, **82**, 629–634 (1960).

The **heated-thermocouple anemometer** measures gas velocity from the cooling effect of the gas stream flowing across the hot junctions of a thermopile supplied with constant electrical power input. Alternate junctions are maintained at ambient temperature, thus compensating for the effect of ambient temperature. For details see Bunker, *Proc. Instrum. Soc. Am.*, **9**, pap. 54-43-2 (1954).

A glass-coated bead **thermistor anemometer** can be used for the measurement of low fluid velocities, down to 0.001 m/s (0.003 ft/s) in air and 0.0002 m/s (0.0007 ft/s) in water [see Murphy and Sparks, *Ind. Eng. Chem. Fundam.*, **7**, 642–645 (1968)].

The **laser-Doppler anemometer** measures local fluid velocity from the change in frequency of radiation, between a stationary source and a receiver, due to scattering by particles along the wave path. A laser is commonly used as the source of incident illumination. The measurements are essentially independent of local temperature and pressure. This technique can be used in many different flow systems with transparent fluids containing particles whose velocity is actually measured. For a brief review of the laser-Doppler technique see Goldstein, *Appl. Mech. Rev.*, **27**, 753–760 (1974). For additional details see Durst, Melling, and Whitelaw, *Principles and Practice of Laser-Doppler Anemometry,* Academic, New York, 1976.

Flow Visualization A great many techniques have been developed for the visualization of velocity patterns, particularly for use in water-tunnel and wind-tunnel studies. In the case of liquids, the more common methods of revealing flow lines involve the use of dye traces, the addition of aluminum flake, plastic particles, globules of equal density liquid (dibutyl phatalate and kerosene) and glass spheres, and the use of polarized light with a doubly refractive liquid or suspension. For the last-named techniques, called *flow birefringence,* see Prados and Peebles, *Am. Inst. Chem. Eng. J.,* **5**, 225–234 (1959). The velocity pattern for laminar flow in a two-dimensional system can be quantitatively mapped by using an electrolytic-tank analog or a conductive-paper analog with a suitable combination of resistances, sources, and sinks. The hydrogen-bubble technique has been proposed for flow visualization and velocity field mapping in liquids. A fine wire, usually of the order of 0.013 to 0.05 mm (0.0005 to 0.002 in) in diameter, is employed as the negative electrode of a direct-current circuit in a water channel. Hydrogen bubbles, formed at the wire by periodic electrical pulses, are swept off by hydrodynamic forces and follow the flow. The bubbles are made visible by lighting at an oblique angle to the direction of view. For details see Schraub, Kline, Henry, Runstadler, and Little, *J. Basic. Eng.,* **87**, 429–444 (1965); or Davis and Fox, *J. Basic Eng.,* **89**, 771–781 (1967).

Thomas and Rice [*J. Appl. Mech.,* **40**, 321–325 (1973)] applied the hydrogen-bubble technique for velocity measurements in thin liquid films. Durelli and Norgard [*Exp. Mech.,* **12**, 169–177 (1972)] compare the flow birefringence and hydrogen-bubble techniques.

In the case of **gases,** flow lines can be revealed through the use of smoke traces or the addition of a lightweight powder such as balsa dust to the stream. One of the best smoke generators is the reaction of titanium tetrachloride with moisture in the air. A woodsmoke-generation system is described by Yu, Sparrow, and Eckert [*Int. J. Heat Mass Transfer,* **15**, 557–558 (1972)]. Tufts of wool or nylon attached at one end to a solid surface can be used to reveal flow phenomena in the vicinity of the surface. Optical methods commonly employed depend upon changes in the refractive index resulting from the presence of heated wires or secondary streams in the flow field or upon changes in density in the primary gas as a result of compressibility effects. The three common techniques are the shadowgraph, the schlieren, and the interferometer. All three theoretically can give quantitative information on the velocity profiles in a two-dimensional system, but in practice only the interferometer is commonly so used. The optical methods are described by Ladenburg et al. (op. cit., pp. 3–108). For additional information on other methods, see Goldstein, *Modern Developments in Fluid Dynamics,* vol. I, London, 1938, pp. 280–296.

The **water table** is frequently used to simulate two-dimensional compressible-flow phenomena in gases. It provides an effective, low-cost means for velocity and pressure-distribution studies or for flow visualization using either shadowgraph or schlieren techniques. In the water table, the wave velocity corresponds to the velocity of sound in the gas, streaming water flow corresponds to subsonic flow, shooting water flow corresponds to supersonic flow, and a hydraulic jump corresponds to a shock wave. From precise measurements of water depth, it is possible to calculate corresponding gas temperatures, pressures, and densities. For information on water-table design and operation see Orlin, Lindner, and Bitterly, *Application of the Analogy between Water Flow with a Free Surface and Two-Dimensional Compressible Gas Flow,* NACA Rep. 875, 1947, or Mathews, *The Design, Operation, and Uses of the Water Channel as an Instrument for the Investigation of Compressible-Flow Phenomena,* NACA Tech. Note

2008, 1950. Additional theoretical background can be obtained from Preiswerk, *Application of the Methods of Gas Dynamics to Water Flows with Free Surface,* part I: *Flows with No Energy Dissipation,* NACA Tech. Mem. 934, 1940; part II: *Flows with Momentum Discontinuities (Hydraulic Jumps),* NACA Tech. Mem. 935, 1940.

HEAD METERS

General Principles If a constriction is placed in a closed channel carrying a stream of fluid, there will be an increase in velocity, and hence an increase in kinetic energy, at the point of constriction. From an energy balance, as given by Bernoulli's theorem (see subsection "Energy Balance"), there must be a corresponding reduction in pressure. Rate of discharge from the constriction can be calculated by knowing this pressure reduction, the area available for flow at the constriction, the density of the fluid, and the coefficient of discharge C. The last-named is defined as the ratio of actual flow to the theoretical flow and makes allowance for stream contraction and frictional effects. The metering characteristics of commonly used head meters are reviewed and grouped by Halmi [*J. Fluids Eng.,* **95**, 127–141 (1973)].

The term **static head** generally denotes the pressure in a fluid due to the head of fluid above the point in question. Its magnitude is given by the application of Newton's law (force = mass × acceleration). In the case of **liquids** (constant density), the static head p_h Pa (lbf/ft^2) is given by

$$p_h = h\rho g/g_c \qquad (10\text{-}14a)$$

where h = head of liquid above the point, m (ft); ρ = liquid density; g = local acceleration due to gravity; and g_c = dimensional constant.

The head developed in a compressor or pump is the energy force per unit mass. In the measuring systems it is often misnamed as (ft) while the units are really ft-lb/lbm or kilojoules.

For a compressor or turbine, it is represented by the following relationship:

$$E = U_1 V_{\theta 1} - U_2 V_{\theta 2} \qquad (10\text{-}14b)$$

where U is the blade speed and V_θ is the tangential velocity component of absolute velocity. This equation is known as the Euler equation.

Liquid-Column Manometers The **height,** or **head** [Eq. (10-14a)], to which a fluid rises in an open vertical tube attached to an apparatus containing a liquid is a direct measure of the pressure at the point of attachment and is frequently used to show the level of liquids in tanks and vessels. This same principle can be applied with U-tube gauges (Fig. 10-8a) and equivalent devices (such as that shown in Fig. 10-8b) to measure pressure in terms of the head of a fluid other than the one under test. Most of these gauges may be used either as **open** or as **differential manometers.** The manometric fluid that constitutes the measured liquid column of these gauges may be any liquid immiscible with the fluid under pressure. For high vacuums or for high pressures and large pressure differences, the gauge liquid is a high-density liquid, generally mercury; for low pressures and small pressure differences, a low-density liquid (e.g., alcohol, water, or carbon tetrachloride) is used.

The **open U tube** (Fig. 10-8a) and the **open gauge** (Fig. 10-8b) each show a reading h_M m (ft) of manometric fluid. If the interface of the manometric fluid and the fluid of which the pressure is wanted is K m (ft) below the point of attachment, A, ρ_A is the density of the lat-

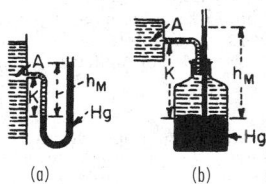

FIG. 10-8 Open manometers.

ter fluid at A, and ρ_M is that of the manometric fluid, then gauge pressure p_A Pa (lbf/ft²) at A is

$$p_A = (h_M\rho_M - K\rho_A)(g/g_c) \qquad (10\text{-}15)°$$

where g = local acceleration due to gravity and g_c = dimensional constant. The head H_A at A as meters (feet) of the fluid at that point is

$$h_A = h_M(\rho_M/\rho_A) - K \qquad (10\text{-}16)°$$

When a gas pressure is measured, unless it is very high, ρ_A is so much smaller than ρ_M that the terms involving K in these formulas are negligible.

The **differential U tube** (Fig. 10-9) shows the pressure difference between taps A and B to be

$$p_A - p_B = [h_M(\rho_M - \rho_A) + K_A\rho_A - K_B\rho_B](g/g_c) \qquad (10\text{-}17)°$$

where h_M is the difference in height of the manometric fluid in the U tube; K_A and K_B are the vertical distances of the upper surface of the manometric fluid above A and B respectively; ρ_A and ρ_B are the densities of the fluids at A and B respectively; and ρ_M is the density of the manometric fluid. If either pressure tap is above the higher level of manometric fluid, the corresponding K is taken to be negative. Valve D, which is kept closed when the gauge is in use, is used to vent off gas which may accumulate at these high points.

The **inverted differential U tube,** in which the manometric fluid may be a gas or a light liquid, can be used to measure liquid pressure differentials, especially for the flow of slurries where solids tend to settle out. Additional details on the use of this manometer can be obtained from Doolittle (op. cit., p. 18).

Closed U tubes (Fig. 10-10) using mercury as the manometric fluid serve to measure directly the absolute pressure p of a fluid, provided that the space between the closed end and the mercury is substantially a perfect vacuum.

The **mercury barometer** (Fig. 10-11) indicates directly the absolute pressure of the atmosphere in terms of height of the mercury column. Normal (standard) barometric pressure is 101.325 kPa by definition. Equivalents of this pressure in other units are 760 mm mercury (at 0°C), 29.921 inHg (at 0°C), 14.696 lbf/in², and 1 atm. For cases in which barometer readings, when expressed by the height of a mercury column, must be corrected to standard temperature (usually 0°C), appropriate temperature correction factors are given in ASME PTC, op. cit., pp. 23–26; and Weast, *Handbook of Chemistry and Physics*, 59th ed., Chemical Rubber, Cleveland, 1978–1979, pp. E39–E41.

Tube Size for Manometers To avoid capillary error, tube diameter should be sufficiently large and the manometric fluids of such densities that the effect of capillarity is negligible in comparison with the gauge reading. The effect of capillarity is practically negligible for tubes with inside diameters 12.7 mm (½ in) or larger (see ASME PTC, op. cit., p. 15). Small diameters are generally permissible for U tubes because the capillary displacement in one leg tends to cancel that in the other.

The capillary rise in a small vertical open tube of circular cross section dipping into a pool of liquid is given by

$$h = \frac{4\sigma g_c \cos\theta}{gD(\rho_1 - \rho_2)} \qquad (10\text{-}18)$$

Here σ = surface tension, D = inside diameter, ρ_1 and ρ_2 are the densities of the liquid and gas (or light liquid) respectively, g = local acceleration due to gravity, g_c = dimensional constant, and θ is the contact angle subtended by the heavier fluid. For most organic liquids and water, the contact angle θ is zero against glass, provided the glass is wet with a film of the liquid; for mercury against glass, θ = 140° (*International Critical Tables*, vol. IV, McGraw-Hill, New York, 1928, pp. 434–435). For further discussion of capillarity, see Schwartz, *Ind. Eng. Chem.*, **61**(1), 10–21 (1969).

° The line leading from the pressure tap to the gauge is assumed to be filled with fluid of the same density as that in the apparatus at the location of the pressure tap; if this is not the case, ρ_A is the density of the fluid actually filling the gauge line, and the value given for h_A must be multiplied by ρ_A/ρ, where ρ is the density of the fluid whose head is being measured.

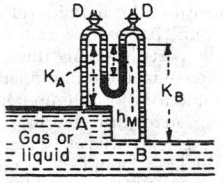

FIG. 10-9 Differential U tube.

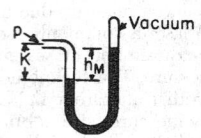

FIG. 10-10 Closed U tube.

Multiplying Gauges To attain the requisite precision in measurement of small pressure differences by liquid-column manometers, means must often be devised to magnify the readings. Of the schemes that follow, the second and third may give tenfold multiplication; the fourth, as much as thirtyfold. In general, the greater the multiplication, the more elaborate must be the precautions in the use of the gauge if the gain in precision is not to be illusory.

1. *Change of manometric fluid.* In open manometers, choose a fluid of lower density. In differential manometers, choose a fluid such that the difference between its density and that of the fluid being measured is as small as possible.

2. *Inclined U tube* (Fig. 10-12). If the reading R m (ft) is taken as shown and R_0 m (ft) is the zero reading, by making the substitution $h_M = (R - R_0)\sin\theta$, the formulas of preceding paragraphs give ($p_A - p_B$) when the corresponding upright U tube is replaced by one inclined. For precise work, the gauge should be calibrated because of possible variations in tube diameter and slope.

3. *The draft gauge* (Fig. 10-13). Commonly used for low gas heads, this gauge has for one leg of the U a reservoir of much larger bore than the tubing that forms the inclined leg. Hence variations of level in the inclined tube produce little change in level in the reservoir. Although h_M may be readily computed in terms of reading R and the dimensions of the tube, calibration of the gauge is preferable; often the changes of level in the reservoir are not negligible, and also variations in tube diameter may introduce serious error into the computation. Commercial gauges are often provided with a scale giving h_M directly in height of water column, provided a particular liquid (often not water) fills the tube; failure to appreciate that the scale is incorrect unless the gauge is filled with the specified liquid is a frequent source of error. If the scale reads correctly when the density of the gauge liquid is ρ_0, then the reading must be multiplied by ρ/ρ_0 if the density of the fluid actually in use is ρ.

4. *Two-fluid U tube* (Fig. 10-14). This is a highly sensitive device for measuring small gas heads. Let A be the cross-sectional area of each of the reservoirs and a that of the tube forming the U; let ρ_1 be

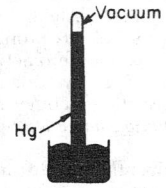

FIG. 10-11 Mercury barometer.

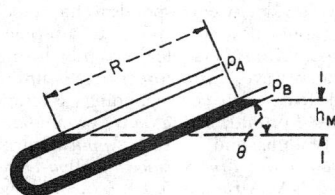

FIG. 10-12 Inclined U tube.

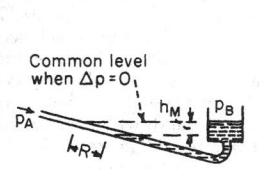

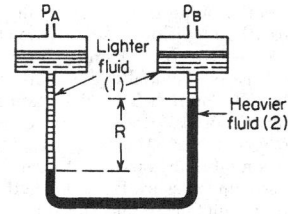

FIG. 10-13 Draft gauge.   **FIG. 10-14** Two-fluid U tube.

the density of the lighter fluid and ρ_2 that of the heavier fluid; and if R is the reading and R_0 its value with zero pressure difference, then the pressure difference is

$$p_A - p_B = (R - R_0)\left(\rho_2 - \rho_1 + \frac{a}{A}\rho_1\right)\frac{g}{g_c} \qquad (10\text{-}19)$$

where g = local acceleration due to gravity and g_c = dimensional constant.

When A/a is sufficiently large, the term $(a/A)\rho_1$ in Eq. (10-19) becomes negligible in comparison with the difference $(\rho_2 - \rho_1)$. However, this term should not be omitted without due consideration. In applying Eq. (10-19), the densities of the gauge liquids may not be taken from tables without the possibility of introducing serious error, for each liquid may dissolve appreciable quantities of the other. Before the gauge is filled, the liquids should be shaken together, and the actual densities of the two layers should be measured for the temperature at which the gauge is to be used. When high magnification is being sought, the U tube may have to be enclosed in a constant-temperature bath so that $(\rho_2 - \rho_1)$ may be accurately known. In general, if highest accuracy is desired, the gauge should be calibrated.

Several **micromanometers,** based on the liquid-column principle and possessing extreme precision and sensitivity, have been developed for measuring minute gas-pressure differences and for calibrating low-range gauges. Some of these micromanometers are available commercially. These micromanometers are free from errors due to capillarity and, aside from checking the micrometer scale, require no calibration. See Doolittle, op. cit., p. 21.

Mechanical Pressure Gauges The **Bourdon-tube gauge** indicates pressure by the amount of flection under internal pressure of an oval tube bent in an arc of a circle and closed at one end. These gauges are commercially available for all pressures below atmospheric and for pressures up to 700 MPa (about 100,000 lbf/in²) above atmospheric. Details on Bourdon-type gauges are given by Harland [*Mach. Des.,* **40**(22), 69–74 (Sept. 19, 1968)].

A **diaphragm gauge** depends for its indication on the deflection of a diaphragm, usually metallic, when subjected to a difference of pressure between the two faces. These gauges are available for the same general purposes as Bourdon gauges but are not usually employed for high pressures. The aneroid barometer is a type of diaphragm gauge.

Small **pressure transducers with flush-mounted diaphragms** are commercially available for the measurement of either steady or fluctuating pressures up to 100 MPa (about 15,000 lbf/in²). The metallic diaphragms are as small as 4.8 mm (³⁄₁₆ in) in diameter. The transducer is mounted on the apparatus containing the fluid whose pressure is to be measured so that the diaphragm is flush with the inner surface of the apparatus. Deflection of the diaphragm is measured by unbonded strain gauges and recorded electrically.

With nonnewtonian fluids the pressure measured at the wall with non-flush-mounted pressure gauges may be in error (see subsection "Static Pressure").

Bourdon and diaphragm gauges that show both pressure and vacuum indications on the same dial are called **compound gauges.**

Conditions of Use Bourdon tubes should not be exposed to temperatures over about 65°C (about 150°F) unless the tubes are specifically designed for such operation. When the pressure of a hotter fluid is to be measured, some type of liquid seal should be used to keep the hot fluid from the tube. In using either a Bourdon or a diaphragm gauge to measure gas pressure, if the gauge is below the

pressure tap of the apparatus so that liquid can collect in the lead, the gauge reading will be too high by an amount equal to the hydrostatic head of the accumulated liquid.

For measuring pressures of corrosive fluids, slurries, and similar process fluids which may foul Bourdon tubes, a **chemical gauge,** consisting of a Bourdon gauge equipped with an appropriate flexible diaphragm to seal off the process fluid, may be used. The combined volume of the tube and the connection between the diaphragm and the tube is filled with an inert liquid. These gauges are available commercially.

Further details on pressure-measuring devices are found in Sec. 22.

Calibration of Gauges Simple **liquid-column manometers** do not require calibration if they are so constructed as to minimize errors due to capillarity (see subsection "Liquid-Column Manometers"). If the scales used to measure the readings have been checked against a standard, the accuracy of the gauges depends solely upon the precision of determining the position of the liquid surfaces. Hence liquid-column manometers are primary standards used to calibrate other gauges.

For **high pressures** and, with commercial mechanical gauges, even for quite moderate pressures, a deadweight gauge (see ASME PTC, op. cit., pp. 36–41; Doolittle, op. cit., p. 33; Jones, op. cit., p. 43; Sweeney, op. cit., p. 104; and Tongue, op. cit., p. 29) is commonly used as the primary standard because it is safer and more convenient than use of manometers. When manometers are used as high-pressure standards, an extremely high mercury column may be avoided by connecting a number of the usual U tubes in series. Multiplying gauges are standardized by comparing them with a micromanometer. Procedure in the calibration of a gauge consists merely of connecting it, in parallel with a standard gauge, to a reservoir wherein constant pressure may be maintained. Readings of the unknown gauge are then made for various reservoir pressures as determined by the standard.

Calibration of **high-vacuum gauges** is described by Sellenger [*Vacuum,* **18**(12), 645–650 (1968)].

Venturi Meters The standard Herschel-type venturi meter consists of a short length of straight tubing connected at either end to the pipe line by conical sections (see Fig. 10-15). Recommended proportions (ASME PTC, op. cit., p. 17) are entrance cone angle $\alpha_1 = 21 \pm 2°$, exit cone angle $\alpha_2 = 5$ to 15°, throat length = one throat diameter, and upstream tap located 0.25 to 0.5 pipe diameter upstream of the entrance cone. The straight and conical sections should be joined by smooth curved surfaces for best results.

The practical working equation for weight rate of discharge, adopted by the ASME Research Committee on Fluid Meters for use with either gases or liquids, is

$$w = q_1\rho_1 = CYA_2\sqrt{\frac{2g_c(p_1 - p_2)\rho_1}{1 - \beta^4}}$$
$$= KYA_2\sqrt{2g_c(p_1 - p_2)\rho_1} \qquad (10\text{-}20)$$

where A_2 = cross-sectional area of throat; C = coefficient of discharge, dimensionless; g_c = dimensional constant; $K = C/\sqrt{1 - \beta^4}$, dimensionless; p_1, p_2 = pressure at upstream and downstream static pressure taps respectively; q_1 = volumetric rate of discharge measured at upstream

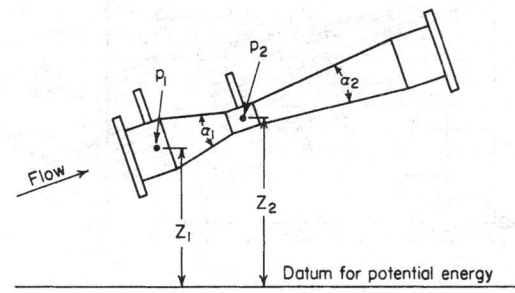

FIG. 10-15 Herschel-type venturi tube.

pressure and temperature; w = weight rate of discharge; Y = expansion factor, dimensionless; β = ratio of throat diameter to pipe diameter, dimensionless; and ρ_1 = density at upstream pressure and temperature.

For the flow of **gases**, expansion factor Y, which allows for the change in gas density as it expands adiabatically from p_1 to p_2, is given by

$$Y = \sqrt{r^{2/k}\left(\frac{k}{k-1}\right)\left(\frac{1-r^{(k-1)/k}}{1-r}\right)\left(\frac{1-\beta^4}{1-\beta^4 r^{2/k}}\right)} \qquad (10\text{-}21)$$

for venturi meters and flow nozzles, where $r = p_2/p_1$ and k = specific heat ratio c_p/c_v. Values of Y computed from Eq. (10-21) are given in Fig. 10-16 as a function of r, k, and β.

For the flow of **liquids**, expansion factor Y is unity. The change in potential energy in the case of an inclined or vertical venturi meter must be allowed for. Equation (10-20) is accordingly modified to give

$$w = q_1\rho = CA_2\sqrt{\frac{[2g_c(p_1 - p_2) + 2g\rho(Z_1 - Z_2)]\rho}{1 - \beta^4}} \qquad (10\text{-}22)$$

where g = local acceleration due to gravity and Z_1, Z_2 = vertical heights above an arbitrary datum plane corresponding to the centerline pressure-reading locations for p_1 and p_2 respectively.

Value of the **discharge coefficient** C for a **Herschel-type venturi meter** depends upon the Reynolds number and to a minor extent upon the size of the venturi, increasing with diameter. A plot of C versus pipe Reynolds number is given in *ASME PTC*, op. cit., p. 19. A value of 0.984 can be used for pipe Reynolds numbers larger than 200,000.

Permanent pressure loss for a Herschel-type venturi tube depends upon diameter ratio β and discharge cone angle α_2. It ranges from 10 to 15 percent of the pressure differential $(p_1 - p_2)$ for small angles (5 to 7°) and from 10 to 30 percent for large angles (15°), with the larger losses occurring at low values of β (see *ASME PTC*, op. cit., p. 12). See Benedict, *J. Fluids Eng.*, **99**, 245–248 (1977), for a general equation for pressure loss for venturis installed in pipes or with plenum inlets.

For flow measurement of **steam and water mixtures** with a Herschel-type venturi in 2½-in- and 3-in-diameter pipes, see Collins and Gacesa, *J. Basic Eng.*, **93**, 11–21 (1971).

A variety of **short-tube** venturi meters are available commercially. They require less space for installation and are generally (although not always) characterized by a greater pressure loss than the corresponding Herschel-type venturi meter. Discharge coefficients vary widely for different types, and individual calibration is recommended if the manufacturer's calibration is not available. Results of tests on the Dall flow tube are given by Miner [*Trans. Am. Soc. Mech. Eng.*, **78**, 475–479 (1956)] and Dowdell [*Instrum. Control Syst.*, **33**, 1006–1009 (1960)]; and on the Gentile flow tube (also called Beth flow tube or Foster flow tube) by Hooper [*Trans. Am. Soc. Mech. Eng.*, **72**, 1099–1110 (1950)].

The use of a **multiventuri system** (in which an inner venturi discharges into the throat of an outer venturi) to increase both the dif-

ferential pressure for a given flow rate and the signal-to-loss ratio is described by Klomp and Sovran [*J. Basic Eng.*, **94**, 39–45 (1972)].

Flow Nozzles A simple form of flow nozzle is shown in Fig. 10-17. It consists essentially of a short cylinder with a flared approach section. The approach cross section is preferably elliptical in shape but may be conical. Recommended contours for long-radius flow nozzles are given in *ASME PTC*, op. cit., p. 13. In general, the length of the straight portion of the throat is about one-half throat diameter, the upstream pressure tap is located about one pipe diameter from the nozzle inlet face, and the downstream pressure tap about one-half pipe diameter from the inlet face. For subsonic flow, the pressures at points 2 and 3 will be practically identical. If a conical inlet is preferred, the inlet and throat geometry specified for a Herschel-type venturi meter can be used, omitting the expansion section.

Rate of discharge through a flow nozzle for subcritical flow can be determined by the equations given for venturi meters, Eq. (10-20) for gases and Eq. (10-22) for liquids. The expansion factor Y for nozzles is the same as that for venturi meters [Eq. (10-21), Fig. 10-16]. The value of the discharge coefficient C depends primarily upon the pipe Reynolds number and to a lesser extent upon the diameter ratio β. Curves of recommended coefficients for long-radius flow nozzles with pressure taps located one pipe diameter upstream and one-half pipe diameter downstream of the inlet face of the nozzle are given in *ASME PTC*, op. cit., p. 15. In general, coefficients range from 0.95 at a pipe Reynolds number of 10,000 to 0.99 at 1,000,000.

The performance characteristics of pipe-wall-tap nozzles (Fig. 10-17) and throat-tap nozzles are reviewed by Wyler and Benedict [*J. Eng. Power*, **97**, 569–575 (1975)].

Permanent pressure loss across a subsonic flow nozzle is approximated by

$$p_1 - p_4 = \frac{1 - \beta^2}{1 + \beta^2}(p_1 - p_2) \qquad (10\text{-}23)$$

where p_1, p_2, p_4 = static pressures measured at the locations shown in Fig. 10-17; and β = ratio of nozzle throat diameter to pipe diameter, dimensionless. Equation (10-23) is based on a momentum balance assuming constant fluid density (see Lapple et al., *Fluid and Particle Mechanics*, University of Delaware, Newark, 1951, p. 13).

See Benedict, loc. cit., for a general equation for pressure loss for nozzles installed in pipes or with plenum inlets. Nozzles show higher loss than venturis. Permanent pressure loss for laminar flow depends on the Reynolds number in addition to β. For details, see Alvi, Sridharan, and Lakshamana Rao, *J. Fluids Eng.*, **100**, 299–307 (1978).

Critical Flow Nozzle For a given set of upstream conditions, the rate of discharge of a gas from a nozzle will increase for a decrease in the absolute pressure ratio p_2/p_1 until the linear velocity in the throat reaches that of sound in the gas at that location. The value of p_2/p_1 for which the acoustic velocity is just attained is called the critical pressure ratio r_c. The actual pressure in the throat will not fall below $p_1 r_c$ even if a much lower pressure exists downstream.

The **critical pressure ratio** r_c can be obtained from the following theoretical equation, which assumes a perfect gas and a frictionless nozzle:

$$r_c^{(1-k)/k} + \left(\frac{k-1}{2}\right)\beta^4 r_c^{2/k} = \frac{k+1}{2} \qquad (10\text{-}24)$$

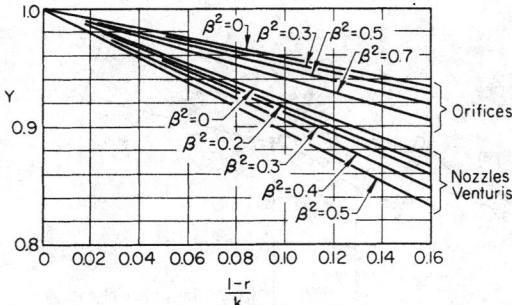

FIG. 10-16 Values of expansion factor Y for orifices, nozzles, and venturis.

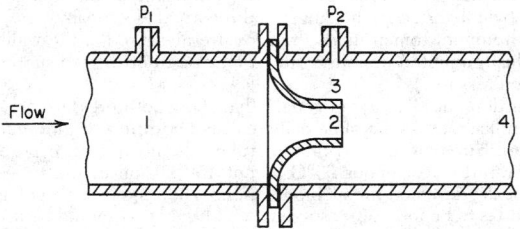

FIG. 10-17 Flow-nozzle assembly.

This reduces, for $\beta \le 0.2$, to

$$r_c = \left(\frac{2}{k+1}\right)^{k/(k-1)} \tag{10-25}$$

where k = ratio of specific heats c_p/c_v and β = diameter ratio. A table of values of r_c as a function of k and β is given in the ASME Research Committee on Fluid Meters Report, op. cit., p. 68. For small values of β, $r_c = 0.487$ for $k = 1.667$, 0.528 for $k = 1.40$, 0.546 for $k = 1.30$, and 0.574 for $k = 1.15$.

Under **critical flow conditions,** only the upstream conditions p_1, v_1, and T_1 need be known to determine flow rate, which, for $\beta \le 0.2$, is given by

$$w_{max} = CA_2 \sqrt{g_c k \left(\frac{p_1}{v_1}\right)\left(\frac{2}{k+1}\right)^{(k+1)/(k-1)}} \tag{10-26}$$

For a **perfect gas,** this corresponds to

$$w_{max} = CA_2 p_1 \sqrt{g_c k \left(\frac{M}{RT_1}\right)\left(\frac{2}{k+1}\right)^{(k+1)/(k-1)}} \tag{10-27}$$

For air, Eq. (10-26) reduces to

$$w_{max} = C_1 CA_2 p_1 / \sqrt{T_1} \tag{10-28}$$

where A_2 = cross-sectional area of throat; C = coefficient of discharge, dimensionless; g_c = dimensional constant; k = ratio of specific heats, c_p/c_v; M = molecular weight; p_1 = pressure on upstream side of nozzle; R = gas constant; T_1 = absolute temperature on upstream side of nozzle; v_1 = specific volume on upstream side of nozzle; C_1 = dimensional constant, 0.0405 SI units (0.533 U.S. customary units); and w_{max} = maximum-weight flow rate.

Discharge coefficients for critical flow nozzles are, in general, the same as those for subsonic nozzles. See Grace and Lapple, *Trans. Am. Soc. Mech. Eng.*, **73**, 639–647 (1951); and Szaniszlo, *J. Eng. Power,* **97**, 521–526 (1975). Arnberg, Britton, and Seidl [*J. Fluids Eng., 96,* 111–123 (1974)] present discharge-coefficient correlations for circular-arc venturi meters at critical flow. For the calculation of the flow of natural gas through nozzles under critical-flow conditions, see Johnson, *J. Basic Eng., 92,* 580–589 (1970).

Orifice Meters A **square-edged** or **sharp-edged** orifice, as shown in Fig. 10-18, is a clean-cut square-edged hole with straight walls perpendicular to the flat upstream face of a thin plate placed crosswise of the channel. The stream issuing from such an orifice

attains its minimum cross section (vena contracta) at a distance downstream of the orifice which varies with the ratio β of orifice to pipe diameter (see Fig. 10-19).

For a centered circular orifice in a pipe, the pressure differential is customarily measured between one of the following pressure-tap pairs. Except in the case of flange taps, all measurements of distance from the orifice are made from the upstream face of the plate.

1. *Corner taps.* Static holes drilled one in the upstream and one in the downstream flange, with the openings as close as possible to the orifice plate.
2. *Radius taps.* Static holes located one pipe diameter upstream and one-half pipe diameter downstream from the plate.
3. *Pipe taps.* Static holes located 2½ pipe diameters upstream and eight pipe diameters downstream from the plate.
4. *Flange taps.* Static holes located 25.4 mm (1 in) upstream and 25.4 mm (1 in) downstream from the plate.
5. *Vena-contracta taps.* The upstream static hole is one-half to two pipe diameters from the plate. The downstream tap is located at the position of minimum pressure (see Fig. 10-19).

Radius taps are best from a practical standpoint; the downstream pressure tap is located at about the mean position of the vena contracta, and the upstream tap is sufficiently far upstream to be unaffected by distortion of the flow in the immediate vicinity of the orifice (in practice, the upstream tap can be as much as two pipe diameters from the plate without affecting the results). Vena-contracta taps give the largest differential head for a given rate of flow but are inconvenient if the orifice size is changed from time to time. Corner taps offer the sometimes great advantage that the pressure taps can be built into the plate carrying the orifice. Thus the entire apparatus can be quickly inserted in a pipe line at any convenient flanged joint without having to drill holes in the pipe. Flange taps are similarly convenient, since by merely replacing standard flanges with special orifice flanges, suitable pressure taps are made available. Pipe taps give the lowest differential pressure, the value obtained being close to the permanent pressure loss.

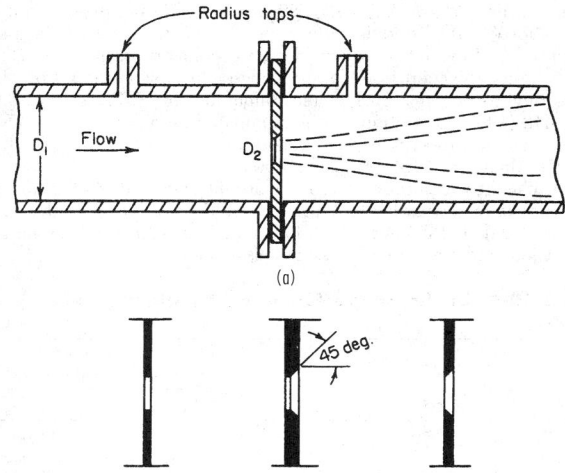

FIG. 10-18 Square-edged or sharp-edged orifices. The plate at the orifice opening must not be thicker than one-thirtieth of the pipe diameter, one-eighth of the orifice diameter, or one-fourth of the distance from the pipe wall to the edge of the opening. (*a*) Pipe-line orifice. (*b*) Types of plates.

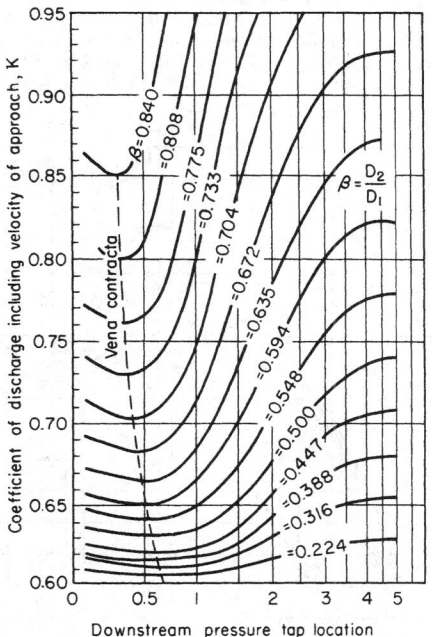

FIG. 10-19 Coefficient of discharge for square-edged circular orifices for $N_{Re} > 30,000$ with the upstream tap located between one and two pipe diameters from the orifice plate [*Spitzglass*, Trans. Am. Soc. Mech. Eng., 44, 919 (1922).]

Rate of discharge through an orifice meter is given by Eq. (10-8) for either liquids or gases. For the case of subsonic flow of a gas ($r_c < r < 1.0$), the expansion factor Y for orifices is approximated by

$$Y = 1 - [(1 - r)/k](0.41 + 0.35\beta^4) \qquad (10\text{-}29)$$

where r = ratio of downstream to upstream static pressure (p_2/p_1), k = ratio of specific heats (c_p/c_v), and β = diameter ratio. (See Fig. 10-16.) Values of Y for supercritical flow of a gas ($r < r_c$) through orifices are given by Benedict [*J. Basic Eng.*, **93**, 121–137 (1971)]. For the case of **liquids**, expansion factor Y is unity, and Eq. (10-22) should be used, since it allows for any difference in elevation between the upstream and downstream taps.

Coefficient of discharge C for a given orifice type is a function of the Reynolds number N_{Re} (based on orifice diameter and velocity) and diameter ratio β. At Reynolds numbers greater than about 30,000, the coefficients are substantially constant. For square-edged or sharp-edged concentric circular orifices, the value will fall between 0.595 and 0.620 for vena-contracta or radius taps for β up to 0.8 and for flange taps for β up to 0.5. Figure 10-19 gives the coefficient of discharge K, including the velocity-of-approach factor ($1/\sqrt{1 - \beta^4}$), as a function of β and the location of the downstream tap. Precise values of K are given in *ASME PTC*, op. cit., pp. 20–39, for flange taps, radius taps, vena-contracta taps, and corner taps. Precise values of C are given in the ASME Research Committee on Fluid Meters Report, op. cit., pp. 202–207, for the first three types of taps.

The discharge coefficient of sharp-edged orifices was shown by Benedict, Wyler, and Brandt [*J. Eng. Power*, **97**, 576–582 (1975)] to increase with edge roundness. Typical as-purchased orifice plates may exhibit deviations on the order of 1 to 2 percent from ASME values of the discharge coefficient.

In the transition region (N_{Re} between 50 and 30,000), the coefficients are generally higher than the above values. Although calibration is generally advisable in this region, the curves given in Fig. 10-20 for corner and vena-contracta taps can be used as a guide. In the laminar-flow region ($N_{Re} < 50$), the coefficient C is proportional to $\sqrt{N_{Re}}$. For $1 < N_{Re} < 100$, Johansen [*Proc. R. Soc.* (*London*), **A121**, 231–245 (1930)] presents discharge-coefficient data for sharp-edged orifices with corner taps. For $N_{Re} < 1$, Miller and Nemecek [ASME Paper 58-A-106 (1958)] present correlations giving coefficients for sharp-edged orifices and short-pipe orifices (L/D from 2 to 10). For short-pipe orifices (L/D from 1 to 4), Dickerson and Rice [*J. Basic Eng.*, **91**,

546–548 (1969)] give coefficients for the intermediate range ($27 < N_{Re} < 7000$). See also subsection "Contraction and Entrance Losses."

Permanent pressure loss across a concentric circular orifice with radius or vena-contracta taps can be approximated for turbulent flow by

$$(p_1 - p_4)/(p_1 - p_2) = 1 - \beta^2 \qquad (10\text{-}30)$$

where p_1, p_2 = upstream and downstream pressure-tap readings respectively, p_4 = fully recovered pressure (four to eight pipe diameters downstream of the orifice), and β = diameter ratio. See *ASME PTC*, op. cit., Fig. 5.

See Benedict, *J. Fluids Eng.*, **99**, 245–248 (1977), for a general equation for pressure loss for orifices installed in pipes or with plenum inlets. Orifices show higher loss than nozzles or venturis. Permanent pressure loss for laminar flow depends on the Reynolds number in addition to β. See Alvi, Sridharan, and Lakshmana Rao, loc. cit., for details.

For the case of **critical flow** through a square- or sharp-edged concentric circular orifice (where $r \leq r_c$, as discussed earlier in this subsection), use Eqs. (10-26), (10-27), and (10-28) as given for critical-flow nozzles. However, unlike nozzles, the flow through a sharp-edged orifice continues to increase as the downstream pressure drops below that corresponding to the critical pressure ratio r_c. This is due to an increase in the cross section of the vena contracta as the downstream pressure is reduced, giving a corresponding increase in the coefficient of discharge. At $r = r_c$, C is about 0.75, while at $r \cong 0$, C has increased to about 0.84. See Grace and Lapple, loc. cit.; and Benedict, *J. Basic Eng.*, **93**, 99–120 (1971).

Measurements by Harris and Magnall [*Trans. Inst. Chem. Eng.* (*London*), **50**, 61–68 (1972)] with a venturi ($\beta = 0.62$) and orifices with radius taps ($\beta = 0.60–0.75$) indicate that the discharge coefficient for **nonnewtonian fluids**, in the range N'_{Re} (generalized Reynolds number) 3500 to 100,000, is approximately the same as for newtonian fluids at the same Reynolds number.

Quadrant-edge orifices have holes with rounded edges on the upstream side of the plate. The quadrant-edge radius is equal to the thickness of the plate at the orifice location. The advantages claimed for this type versus the square- or sharp-edged orifice are constant-discharge coefficients extending to lower Reynolds numbers and less possibility of significant changes in coefficient because of erosion or other damage to the inlet shape.

Values of discharge coefficient C and Reynolds numbers limit for constant C are presented in Table 10-4, based on Ramamoorthy and Seetharamiah [*J. Basic Eng.*, **88**, 9–13 (1966)] and Bogema and Monkmeyer (*J. Basic Eng.*, **82**, 729–734 (1960)). At Reynolds numbers above those listed for the upper limits, the coefficients rise abruptly. As Reynolds numbers decrease below those listed for the lower limits, the coefficients pass through a hump and then drop off. According to Bogema, Spring, and Ramamoorthy [*J. Basic Eng.*, **84**, 415–418 (1962)], the hump can be eliminated by placing a fine-mesh screen about three pipe diameters upstream of the orifice. This reduces the lower N_{Re} limit to about 500.

Permanent pressure loss across quadrant-edge orifices for turbulent flow is somewhat lower than given by Eq. (10-30). See Alvi, Sridharan, and Lakshmana Rao, loc. cit., for values of discharge coefficient and permanent pressure loss in laminar flow.

TABLE 10-4 Discharge Coefficients for Quadrant-Edge Orifices

β	C‡	K‡	Limiting N_{Re}^* for constant coefficient†	
			Lower	Upper
0.225	0.770	0.771	5,000	60,000
0.400	0.780	0.790	5,000	150,000
0.500	0.824	0.851	4,000	200,000
0.600	0.856	0.918	3,000	120,000
0.630	0.885	0.964	3,000	105,000

*Based on pipe diameter and velocity.
†For a precision of about ±0.5 percent.
‡Can be used with corner taps, flange taps, or radius taps.

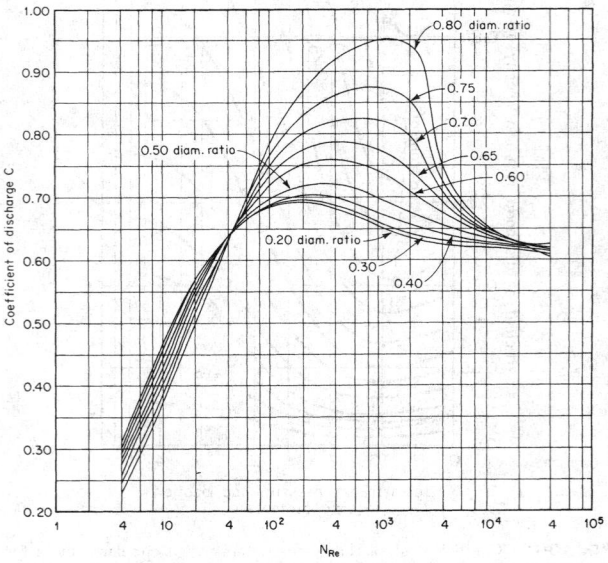

FIG. 10-20 Coefficient of discharge for square-edged circular orifices with corner taps. [*Tuve and Sprenkle*, Instruments, **6**, 201 (1933).]

Segmental and **eccentric orifices** are frequently used for gas metering when there is a possibility that entrained liquids or solids would otherwise accumulate in front of a concentric circular orifice. This can be avoided if the opening is placed on the lower side of the pipe. For liquid flow with entrained gas, the opening is placed on the upper side. The pressure taps should be located on the opposite side of the pipe from the opening.

Coefficient C for a square-edged eccentric circular orifice (with opening tangent to pipe wall) varies from about 0.61 to 0.63 for β's from 0.3 to 0.5, respectively, and pipe Reynolds numbers > 10,000 for either vena-contracta or flange taps (where β = diameter ratio). For square-edged segmental orifices, the coefficient C falls generally between 0.63 and 0.64 for $0.3 \leq \beta \leq 0.5$ and pipe Reynolds numbers > 10,000, for vena-contracta or flange taps, where β = diameter ratio for an equivalent circular orifice = $\sqrt{\alpha}$ (α = ratio of orifice to pipe cross-sectional areas). Values of expansion factor Y are slightly higher than for concentric circular orifices, and the location of the vena contracta is moved farther downstream as compared with concentric circular orifices. For further details, see ASME Research Committee on Fluid Meters Report, op. cit., pp. 210–213.

For permanent pressure loss with segmental and eccentric orifices with laminar pipe flow see Lakshmana Rao and Sridharan, *Proc. Am. Soc. Civ. Eng., J. Hydraul. Div.,* **98** (HY 11), 2015–2034 (1972).

Annular orifices can also be used to advantage for gas metering when there is a possibility of entrained liquids or solids and for liquid metering with entrained gas present in small concentrations. Coefficient K was found by Bell and Bergelin [*Trans. Am. Soc. Mech. Eng.,* **79**, 593–601 (1957)] to range from about 0.63 to 0.67 for annulus Reynolds numbers in the range of 100 to 20,000 respectively for values of $2L/(D - d)$ less than 1 where L = thickness of orifice at outer edge, D = inside pipe diameter, and d = diameter of orifice disk. The annulus Reynolds number is defined as

$$N_{Re} = (D - d)(G/\mu) \qquad (10\text{-}31)$$

where G = mass velocity pV through orifice opening and μ = fluid viscosity. The above coefficients were determined for β's (= d/D) in the range of 0.95 to 0.996 and with pressure taps located 19 mm (¾ in) upstream of the disk and 230 mm (9 in) downstream in a 5.25-in-diameter pipe.

Elbow Meters   A pipe elbow can be used as a flowmeter for liquids if the differential centrifugal head generated between the inner and outer radii of the bend is measured by means of pressure taps located midway around the bend. Equation (10-22) can be used, except that the pressure-difference term $(p_1 - p_2)$ is now taken to be the differential centrifugal pressure and β is taken as zero if one assumes no change in cross section between the pipe and the bend. The discharge coefficient should preferably be determined by calibration, but as a guide it can be estimated within ± 6 percent for circular pipe for Reynolds numbers greater than 10^5 from $C = 0.98 \sqrt{R_c/2D}$, where R_c = radius of curvature of the centerline and D = inside pipe diameter in consistent units. See Murdock, Foltz, and Gregory, *J. Basic Eng.,* **86**, 498–506 (1964); or the ASME Research Committee on Fluid Meters Report, op. cit., pp. 75–77.

Accuracy   Square-edged orifices and venturi tubes have been so extensively studied and standardized that reproducibilities within 1 to 2 percent can be expected between standard meters when new and clean. This is therefore the order of reliability to be had, if one assumes (1) accurate measurement of meter differential, (2) selection of the coefficient of discharge from recommended published literature, (3) accurate knowledge of fluid density, (4) accurate measurement of critical meter dimensions, (5) smooth upstream face of orifice, and (6) proper location of the meter with respect to other flow-disturbing elements in the system. Care must also be taken to avoid even slight corrosion or fouling during use.

Presence of **swirling flow** or an **abnormal velocity distribution** upstream of the metering element can cause serious metering error unless calibration in place is employed or sufficient straight pipe is inserted between the meter and the source of disturbance. Table 10-5 gives the minimum lengths of straight pipe required to avoid appreciable error due to the presence of certain fittings and valves either upstream or downstream of an orifice or nozzle. These values were

extracted from plots presented by Sprenkle [*Trans. Am. Soc. Mech. Eng.,* **67**, 345–360 (1945)]. Table 10-5 also shows the reduction in spacing made possible by the use of straightening vanes between the fittings and the meter. Entirely adequate straightening vanes can be provided by fitting a bundle of thin-wall tubes within the pipe. The center-to-center distance between tubes should not exceed one-fourth of the pipe diameter, and the bundle length should be at least 8 times this distance.

The distances specified in Table 10-5 will be conservative if applied to venturi meters. For specific information on requirements for venturi meters, see a discussion by Pardoe appended to Sprenkle (op. cit.). Extensive data on the effect of installation on the coefficients of venturi meters are given elsewhere by Pardoe [*Trans. Am. Soc. Mech. Eng.,* **65**, 337–349 (1943)].

In the presence of **flow pulsations,** the indications of head meters such as orifices, nozzles, and venturis will often be undependable for several reasons. First, the measured pressure differential will tend to be high, since the pressure differential is proportional to the square of flow rate for a head meter, and the square root of the mean differential pressure is always greater than the mean of the square roots of the differential pressures. Second, there is a phase shift as the wave passes through the metering restriction which can affect the differential. Third, pulsations can be set up in the manometer leads themselves. Frequency of the pulsation also plays a part. At low frequencies, the meter reading can generally faithfully follow the flow pulsations, but at high frequencies it cannot. This is due to inertia of the fluid in the manometer leads or of the manometric fluid, whereupon the meter would give a reading intermediate between the maximum and minimum flows but having no readily predictable relation to the mean flow. Pressure transducers with flush-mounted diaphragms can be used together with high-speed recording equipment to provide accurate records of the pressure profiles at the upstream and downstream pressure taps, which can then be analyzed and translated into a mean flow rate.

TABLE 10-5   Locations of Orifices and Nozzles Relative to Pipe Fittings

Type of fitting upstream	$\dfrac{D_2}{D_1}$	Distance, upstream fitting to orifice		Distance, vanes to orifice	Distance, nearest downstream fitting from orifice
		Without straight-ening vanes	With straight-ening vanes		
Single 90° ell, tee, or cross used as ell	0.2	6			2
	0.4	6			
	0.6	9	9		
	0.8	20	12	8	4
2 short-radius 90° ells in form of S	0.2	7			2
	0.4	8	8		
	0.6	13	10	6	
	0.8	25	15	11	4
2 long- or short-radius 90° ells in perpendicular planes	0.2	15	9	5	2
	0.4	18	10	6	
	0.6	25	11	7	
	0.8	40	13	9	4
Contraction or enlargement	0.2	8	Vanes		2
	0.4	9	have no		
	0.6	10	advantage		
	0.8	15			4
Globe valve or stop check	0.2	9	9	5	2
	0.4	10	10	6	
	0.6	13	10	6	
	0.8	21	13	9	4
Gate valve, wide open, or plug cocks	0.2	6	Same as globe valve		2
	0.4	6			
	0.6	8			
	0.8	14			4

Distances in pipe diameters, D_1

The rather general practice of producing a steady differential reading by placing restrictions in the manometer leads can result in a reading which, under a fixed set of conditions, may be useful in control of an operation but which has no readily predictable relation to the actual average flow. If calibration is employed to compensate for the presence of pulsations, complete reproduction of operating conditions, including source of pulsations and waveform, is necessary to ensure reasonable accuracy.

According to Head [*Trans. Am. Soc. Mech. Eng.,* **78,** 1471–1479 (1956)], a pulsation-intensity limit of $\Gamma = 0.1$ is recommended as a practical pulsation threshold below which the performance of all types of flowmeters will differ negligibly from steady-flow performance (an error of less than 1 percent in flow due to pulsation). Γ is the peak-to-trough flow variation expressed as a fraction of the average flow rate. According to the ASME Research Committee on Fluid Meters Report (op. cit., pp. 34–35), the fractional metering error E for **liquid flow** through a head meter is given by

$$(1 + E)^2 = 1 + \Gamma^2/8 \qquad (10\text{-}32)$$

When the pulsation amplitude is such as to result in a greater-than-permissible metering error, consideration should be given to installation of a pulsation damper between the source of pulsations and the flowmeter. References to methods of pulsation-damper design are given in the subsection "Unsteady-State Behavior."

Pulsations are most likely to be encountered in discharge lines from reciprocating pumps or compressors and in lines supplying steam to reciprocating machinery. For **gas flow,** a combination involving a surge chamber and a constriction in the line can be used to damp out the pulsations to an acceptable level. The surge chamber is generally located as close to the pulsation source as possible, with the constriction between the surge chamber and the metering element. This arrangement can be used for either a suction or a discharge line. For such an arrangement, the metering error has been found to be a function of the Hodgson number N_H, which is defined as

$$N_H = Qn\,\Delta p_s/qp_s \qquad (10\text{-}33)$$

where Q = volume of surge chamber and pipe between metering element and pulsation source; n = pulsation frequency; Δp_s = permanent pressure drop between metering element and surge chamber; q = average volume flow rate, based on gas density in the surge chamber; and p_s = pressure in surge chamber.

Herning and Schmid [*Z. Ver. Dtsch. Ing.,* **82,** 1107–1114 (1938)] presented charts for a simplex double-acting compressor for the prediction of metering error as a function of the Hodgson number and s, the ratio of piston discharge time to total time per stroke. Table 10-6a gives the minimum Hodgson numbers required to reduce the metering error to 1 percent as given by the charts (for specific heat ratios between 1.28 and 1.37). Schmid [*Z. Ver. Dtsch. Ing.,* **84,** 596–598 (1940)] presented similar charts for a duplex double-acting compressor and a triplex double-acting compressor for a specific heat ratio of 1.37. Table 10-6b gives the minimum Hodgson numbers correspond-

ing to a 1 percent metering error for these cases. The value of $Q\,\Delta p_s$ can be calculated from the appropriate Hodgson number, and appropriate values of Q and Δp_s selected so as to satisfy this minimum requirement.

AREA METERS

General Principles   The underlying principle of an ideal area meter is the same as that of a head meter of the orifice type (see subsection "Orifice Meters"). The stream to be measured is throttled by a constriction, but instead of observing the variation with flow of the differential head across an orifice of fixed size, the constriction of an area meter is so arranged that its size is varied to accommodate the flow while the differential head is held constant.

A simple example of an area meter is a gate valve of the rising-stem type provided with static-pressure taps before and after the gate and a means for measuring the stem position. In most common types of area meters, the variation of the opening is automatically brought about by the motion of a weighted piston or float supported by the fluid. Two different cylinder- and piston-type area meters are described in the ASME Research Committee on Fluid Meters Report, op. cit., pp. 82–83.

Rotameters   The rotameter, an example of which is shown in Fig. 10-21, has become one of the most popular flowmeters in the chemical-process industries. It consists essentially of a plummet, or "float," which is free to move up or down in a vertical, slightly tapered tube having its small end down. The fluid enters the lower end of the tube and causes the float to rise until the annular area between the float and the wall of the tube is such that the pressure drop across this constriction is just sufficient to support the float. Typically, the tapered tube is of glass and carries etched upon it a nearly linear scale on which the position of the float may be visually noted as an indication of the flow.

Interchangeable precision-bore glass tubes and metal metering tubes are available. Rotameters have proved satisfactory both for gases and for liquids at high and at low pressures. A single instrument can readily cover a tenfold range of flow, and by providing floats of different densities a two-hundredfold range is practicable. Rotameters are available with pneumatic, electric, and electronic transmitters for actuating remote recorders, integrators, and automatic flow con-

TABLE 10-6a Minimum Hodgson Numbers
Simplex double-acting compressor

s	N_H	s	N_H
0.167	1.31	0.667	0.60
0.333	1.00	0.833	0.43
0.50	0.80	1.00	0.34

TABLE 10-6b Minimum Hodgson Numbers

Duplex double-acting compressor		Triplex double-acting compressor	
s	N_H	s	N_H
0.167	1.00	0.167	0.85
0.333	0.70	0.333	0.30
0.50	0.30	0.50	0.15
0.667	0.10	0.667	0.06
0.833	0.05	0.833	0.00
1.00	0.00	1.00	0.00

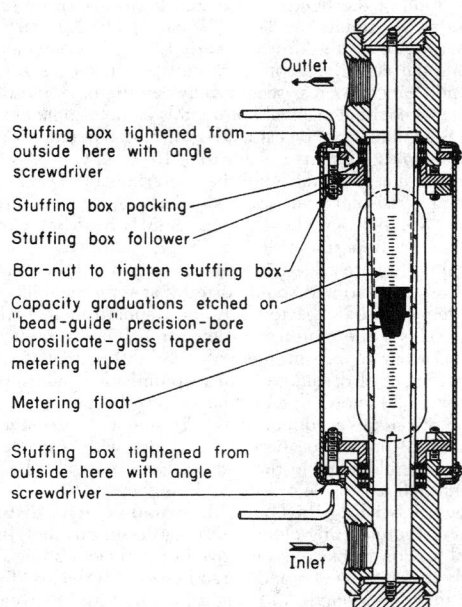

Outlet

Stuffing box tightened from outside here with angle screwdriver

Stuffing box packing

Stuffing box follower

Bar-nut to tighten stuffing box

Capacity graduations etched on "bead-guide" precision-bore borosilicate-glass tapered metering tube

Metering float

Stuffing box tightened from outside here with angle screwdriver

Inlet

FIG. 10-21   Rotameter.

trollers (see Considine, op. cit., pp. 4-35–4-36, and Sec. 22 of this *Handbook*).

Rotameters require no straight runs of pipe before or after the point of installation. Pressure losses are substantially constant over the whole flow range. In experimental work, for greatest precision, a rotameter should be calibrated with the fluid which is to be metered. However, most modern rotameters are precision-made so that their performance closely corresponds to a master calibration plot for the type in question. Such a plot is supplied with the meter upon purchase.

According to Head [*Trans. Am. Soc. Mech. Eng.*, **76**, 851–862 (1954)], flow rate through a rotameter can be obtained from

$$w = q\rho = KD_f \sqrt{\frac{W_f(\rho_f - \rho)\rho}{\rho_f}} \qquad (10\text{-}34)$$

and

$$K = \phi \left[\frac{D_t}{D_f}, \frac{\mu}{\sqrt{\dfrac{W_f(\rho_f - \rho)\rho}{\rho_f}}} \right] \qquad (10\text{-}35)$$

where w = weight flow rate; q = volume flow rate; ρ = fluid density; K = flow parameter, $\text{m}^{1/2}/\text{s}$ ($\text{ft}^{1/2}/\text{s}$); D_f = float diameter at constriction; W_f = float weight; ρ_f = float density; D_t = tube diameter at point of constriction; and μ = fluid viscosity. The appropriate value of K is obtained from a composite correlation of K versus the parameters shown in Eq. (10-35) corresponding to the float shape being used. The relation of D_t to the rotameter reading is also required for the tube taper and size being used.

The ratio of flow rates for two different fluids A and B at the same rotameter reading is given by

$$\frac{w_A}{w_B} = \frac{K_A}{K_B} \sqrt{\frac{(\rho_f - \rho_A)\rho_A}{(\rho_f - \rho_B)\rho_B}} \qquad (10\text{-}36)$$

A measure of self-compensation, with respect to weight rate of flow, for fluid-density changes can be introduced through the use of a float with a density twice that of the fluid being metered, in which case an increase of 10 percent in ρ will produce a decrease of only 0.5 percent in w for the same reading. The extent of immunity to changes in fluid viscosity depends upon the shape of the float.

According to Baird and Cheema [*Can. J. Chem. Eng.*, **47**, 226–232 (1969)], the presence of square-wave pulsations can cause a rotameter to overread by as much as 100 percent. The higher the pulsation frequency, the less the float oscillation, although the error can still be appreciable even when the frequency is high enough so that the float is virtually stationary. Use of a damping chamber between the pulsation source and the rotameter will reduce the error.

Additional information on rotameter theory is presented by Fischer [*Chem. Eng.*, **59**(6), 180–184 (1952)], Coleman [*Trans. Inst. Chem. Eng.*, **34**, 339–350 (1956)], and McCabe and Smith (*Unit Operations of Chemical Engineering*, 3d ed., McGraw-Hill, New York, 1976, pp. 215–218).

MASS FLOWMETERS

General Principles   There are two main types of mass flowmeters: (1) the so-called true mass flowmeter, which responds directly to mass flow rate, and (2) the inferential mass flowmeter, which commonly measures volume flow rate and fluid density separately. A variety of types of true mass flowmeters have been developed, including the following: (*a*) the Magnus-effect mass flowmeter, (*b*) the axial-flow, transverse-momentum mass flowmeter, (*c*) the radial-flow, transverse-momentum mass flowmeter, (*d*) the gyroscopic transverse-momentum mass flowmeter, and (*e*) the thermal mass flowmeter. Type *b* is the basis for several commercial mass flowmeters, one version of which is briefly described here.

Axial-Flow Transverse-Momentum Mass Flowmeter   This type is also referred to as an angular-momentum mass flowmeter. One embodiment of its principle involves the use of axial flow through a driven impeller and a turbine in series. The impeller imparts angular momentum to the fluid, which in turn causes a torque to be imparted

to the turbine, which is restrained from rotating by a spring. The torque, which can be measured, is proportional to the rotational speed of the impeller and the mass flow rate.

Inferential Mass Flowmeter   There are several types in this category, including the following:

1. *Head meters with density compensation.*   Head meters such as orifices, venturis, or nozzles can be used with one of a variety of densitometers [e.g., based on (*a*) buoyant force on a float, (*b*) hydraulic coupling, (*c*) voltage output from a piezoelectric crystal, or (*d*) radiation absorption]. The signal from the head meter, which is proportional to ρV^2 (where ρ = fluid density and V = fluid velocity), is multiplied by ρ given by the densitometer. The square root of the product is proportional to the mass flow rate.

2. *Head meters with velocity compensation.*   The signal from the head meter, which is proportional to ρV^2, is divided by the signal from a velocity meter to give a signal proportional to the mass flow rate.

3. *Velocity meters with density compensation.*   The signal from the velocity meter (e.g., turbine meter, electromagnetic meter, or sonic velocity meter) is multiplied by the signal from a densitometer to give a signal proportional to the mass flow rate.

Additional information on mass-flowmeter principles can be obtained from Yeaple (*Hydraulic and Pneumatic Power and Control*, McGraw-Hill, New York, 1966, pp. 125–128), Halsell [*Instrum. Soc. Am. J.*, **7**, 49–62 (June 1960)], and Flanagan and Colman [*Control*, **7**, 242–245 (1963)]. Information on commercially available mass flowmeters is given in the latter two references.

WEIRS

Liquid flow in an open channel may be metered by means of a weir, which consists of a dam over which, or through a notch in which, the liquid flows. The terms "rectangular weir," "triangular weir," etc., generally refer to the shape of the notch in a notched weir. All weirs considered here have flat upstream faces that are perpendicular to the bed and walls of the channel.

Sharp-edged weirs have edges like those of square or sharp-edged orifices (see subsection "Orifice Meters"). Notched weirs are ordinarily sharp-edged. Weirs not in the sharp-edged class are, for the most part, those described as **broad-crested weirs.**

The head h_0 on a weir is the liquid-level height above the crest or base of the notch. The head must be measured sufficiently far upstream to avoid the drop in level occasioned by the overfall which begins at a distance about $2h_0$ upstream from the weir. Surface-level measurements should be made a distance of $3h_0$ or more upstream, preferably by using a stilling box equipped with a high-precision level gauge, e.g., a hook gauge or float gauge.

With sharp-edged weirs, the sheet of discharging liquid, called the "nappe," contracts as it leaves the opening and free discharge occurs. Rounding the upstream edge will reduce the contraction and increase the flow rate for a given head. A clinging nappe may result if the head is very small, if the edge is well rounded, or if air cannot flow in beneath the nappe. This, in turn, results in an increase in the discharge rate for a given head as compared with that for a free nappe. For further information on the effect of the nappe, see Gibson, *Hydraulics and Its Applications*, 5th ed., Constable, London, 1952; and Chow, *Open-Channel Hydraulics*, McGraw-Hill, New York, 1959.

Flow through a **rectangular weir** (Fig. 10-22) is given by

$$q = 0.415(L - 0.2h_0)h_0^{1.5}\sqrt{2g} \qquad (10\text{-}37)$$

where q = volume flow rate, L = crest length, h_0 = weir head, and g =

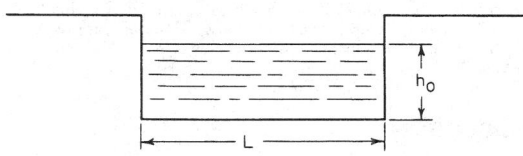

FIG. 10-22   Rectangular weir.

local acceleration due to gravity. This is known as the modified Francis formula for a rectangular sharp-edged weir with two end corrections; it applies when the velocity-of-approach correction is small. The Francis formula agrees with experiments within 3 percent if (1) L is greater than $2h_0$, (2) velocity of approach is 0.6 m/s (2 ft/s) or less, (3) height of crest above bottom of channel is at least $3h_0$, and (4) h_0 is not less than 0.09 m (0.3 ft).

Narrow rectangular notches ($h_0 > L$) have been found to give about 93 percent of the discharge predicted by the Francis formula. Thus

$$q = 0.386 L h_0^{1.5} \sqrt{2g} \tag{10-38}$$

In this case, no end corrections are applied even though the formula applies only for sharp-edged weirs. See Schoder and Dawson, *Hydraulics*, McGraw-Hill, New York, 1934, p. 175, for further details.

The **triangular-notch weir** has the advantage that a single notch can accommodate a wide range of flow rates, although this in turn reduces its accuracy. The discharge for sharp- or square-edged weirs is given by

$$q = (0.31 h_0^{2.5} \sqrt{2g})/\tan \phi \tag{10-39}$$

See Eq. (10-37) for nomenclature. Angle ϕ is illustrated in Fig. 10-23. Equations (10-37), (10-38), and (10-39) are applicable only to the flow of water. However, for the case of triangular-notch weirs Lenz [*Trans. Am. Soc. Civ. Eng.*, **108**, 759–802 (1943)] has presented correlations predicting the effect of viscosity over the range of 0.001 to 0.15 Pa·s (1 to 150 cP) and surface tension over the range of 0.03 to 0.07 N/m (30 to 70 dyn/cm). His equation predicts about an 8 percent increase in flow for a liquid of 0.1-Pa·s (100-cP) viscosity compared with water at 0.001 Pa·s (1 cP) and about a 1 percent increase for a liquid with one-half the surface tension of water. For fluids of moderate viscosity, Ranga Raju and Asawa [*Proc. Am. Soc. Civ. Eng., J. Hydraul. Div.*, **103** (HY 10), 1227–1231 (1977)] find that the effect of viscosity and surface tension on the discharge flow rate for rectangular and triangular-notch ($\phi = 45°$) weirs can be neglected when

$$(N_{Re})^{0.2}(N_{We})^{0.6} > 900 \tag{10-40}$$

where N_{Re} (Reynolds number) $= \sqrt{gh_0^3}/\nu$, g = local acceleration due to gravity, h_0 = weir head, ν = kinematic viscosity; N_{We} (Weber number) $= \rho g h_0^2/g_c \sigma$, ρ = density, g_c = dimensional constant, and σ = surface tension.

For the flow of high-viscosity liquids over rectangular weirs, see Slocum, *Can. J. Chem. Eng.*, **42**, 196–200 (1964). His correlation is based on data for liquids with viscosities in the range of 2.5 to 500 Pa·s (25 to 5000 cP), in which range the discharge decreases markedly for a given head as viscosity is increased.

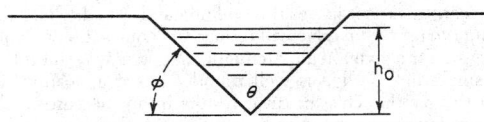

FIG. 10-23 Triangular weir.

Information on other types of weirs can be obtained from Addison, op. cit.; Gibson, *Hydraulics and Its Applications*, 5th ed., Constable, London, 1952; Henderson, *Open Channel Flow*, Macmillan, New York, 1966; Linford, *Flow Measurement and Meters*, Spon, London, 1949; Lakshmana Rao, "Theory of Weirs," in *Advances in Hydroscience*, vol. 10, Academic, New York, 1975; and Urquhart, *Civil Engineering Handbook*, 4th ed., McGraw-Hill, New York, 1959.

TWO-PHASE SYSTEMS

It is generally preferable to meter each of the individual components of a two-phase mixture separately prior to mixing, since it is difficult to meter such mixtures accurately. Problems arise because of fluctuations in composition with time and variations in composition over the cross section of the channel. Information on metering of such mixtures can be obtained from the following sources.

Gas-Solid Mixtures Carlson, Frazier, and Engdahl [*Trans. Am. Soc. Mech. Eng.*, **70**, 65–79 (1948)] describe the use of a **flow nozzle** and a **square-edged orifice** in series for the measurement of both the gas rate and the solids rate in the flow of a finely divided solid-in-gas mixture. The nozzle differential is sensitive to the flow of both phases, whereas the orifice differential is not influenced by the solids flow.

Farbar [*Trans. Am. Soc. Mech. Eng.*, **75**, 943–951 (1953)] describes how a **venturi meter** can be used to measure solids flow rate in a gas-solids mixture when the gas rate is held constant. Separate calibration curves (solids flow versus differential) are required for each gas rate of interest.

Cheng, Tung, and Soo [*J. Eng. Power*, **92**, 135–149 (1970)] describe the use of an **electrostatic probe** for measurement of solids flow in a gas-solids mixture.

Goldberg and Boothroyd [*Br. Chem. Eng.*, **14**, 1705–1708 (1969)] describe several types of solids-in-gas flowmeters and give an extensive bibliography.

Gas-Liquid Mixtures An empirical equation was developed by Murdock [*J. Basic Eng.*, **84**, 419–433 (1962)] for the measurement of gas-liquid mixtures using **sharp-edged orifice** plates with either radius, flange, or pipe taps.

An equation for use with **venturi meters** was given by Chisholm [*Br. Chem. Eng.*, **12**, 454–457 (1967)]. A procedure for determining steam quality via pressure-drop measurement with upflow through either venturi meters or sharp-edged orifice plates was given by Collins and Gacesa [*J. Basic Eng.*, **93**, 11–21 (1971)].

Liquid-Solid Mixtures Liptak [*Chem. Eng.*, **74**(4), 151–158 (1967)] discusses a variety of techniques that can be used for the measurement of solids-in-liquid suspensions or slurries. These include metering pumps, weigh tanks, magnetic flowmeter, ultrasonic flowmeter, gyroscope flowmeter, etc.

Shirato, Gotoh, Osasa, and Usami [*J. Chem. Eng. Japan*, **1**, 164–167 (January 1968)] present a method for determining the mass flow rate of suspended solids in a liquid stream wherein the liquid velocity is measured by an electromagnetic flowmeter and the flow of solids is calculated from the pressure drops across each of two vertical sections of pipe of different diameter through which the suspension flows in series.

PUMPING OF LIQUIDS AND GASES

GENERAL REFERENCES: Paul N. Garay, P. E., *Pump Application Desk Book*, Fairmont Press, 1993. John W. Dufor and William E. Nelson, *Centrifugal Pump Sourcebook*, McGraw-Hill, 1992. *Process Pumps*, ITT Fluid Technology Corporation, 1992. James Corley, "The Vibration Analysis of Pumps: A Tutorial," Fourth International Pump Symposium, Texas A & M University, Houston, Texas, May 1987.

INTRODUCTION

A pump is a physical contrivance that is used to deliver fluids from one location to another through conduits. Over the years, numerous pump

designs have evolved to meet differing requirements. The basic requirements to define the application are suction and delivery pressures, pressure loss in transmission, and the flow rate. Special requirements may exist in food, pharmaceutical, nuclear, and other industries that impose material selection requirements of the pump. The primary means of transfer of energy to the fluid that causes flow are gravity, displacement, centrifugal force, electromagnetic force, transfer of momentum, mechanical impulse, and a combination of these energy-transfer mechanisms. Gravity and centrifugal force are the most common energy-transfer mechanisms in use.

Pump designs have largely been standardized. Based on application experience, numerous standards have come into existence. As special projects and new application situations for pumps develop, these standards will be updated and revised. Common pump standards are:

1. American Petroleum Institute (API) Standard 610, Centrifugal Pumps for Refinery Service
2. American Waterworks Association (AWWA) E101, Deep Well Vertical Turbine Pumps
3. Underwriters Laboratories (UL) UL 51, UL343, UL1081, UL448, UL1247
4. National Fire Protection Agency (NFPA) NFPA-20 Centrifugal Fire Pumps
5. American Society of Mechanical Engineers (ASME)
6. American National Standards Institute
7. Hydraulic Institute Standards (Application)

These standards specify design, construction, and testing details such as material selection, shop inspection and tests, drawings and other uses required, clearances, construction procedures, and so on.

There are four (4) major types of pumps: (1) positive displacement, (2) dynamic (kinetic), (3) lift, and (4) electromagnetic. Piston pumps are positive displacement pumps. The most common centrifugal pumps are of dynamic type; ancient bucket-type pumps are lift pumps; and electromagnetic pumps use electromagnetic force and are common in modern reactors. Canned pumps are also becoming popular in the petrochemical industry because of the drive to minimize fugitive emissions. Figure 10-24 shows pump classification:

TERMINOLOGY

Displacement   Discharge of a fluid from a vessel by partially or completely displacing its internal volume with a second fluid or by mechanical means is the principle upon which a great many fluid-transport devices operate. Included in this group are reciprocating-piston and diaphragm machines, rotary-vane and gear types, fluid piston compressors, acid eggs, and air lifts.

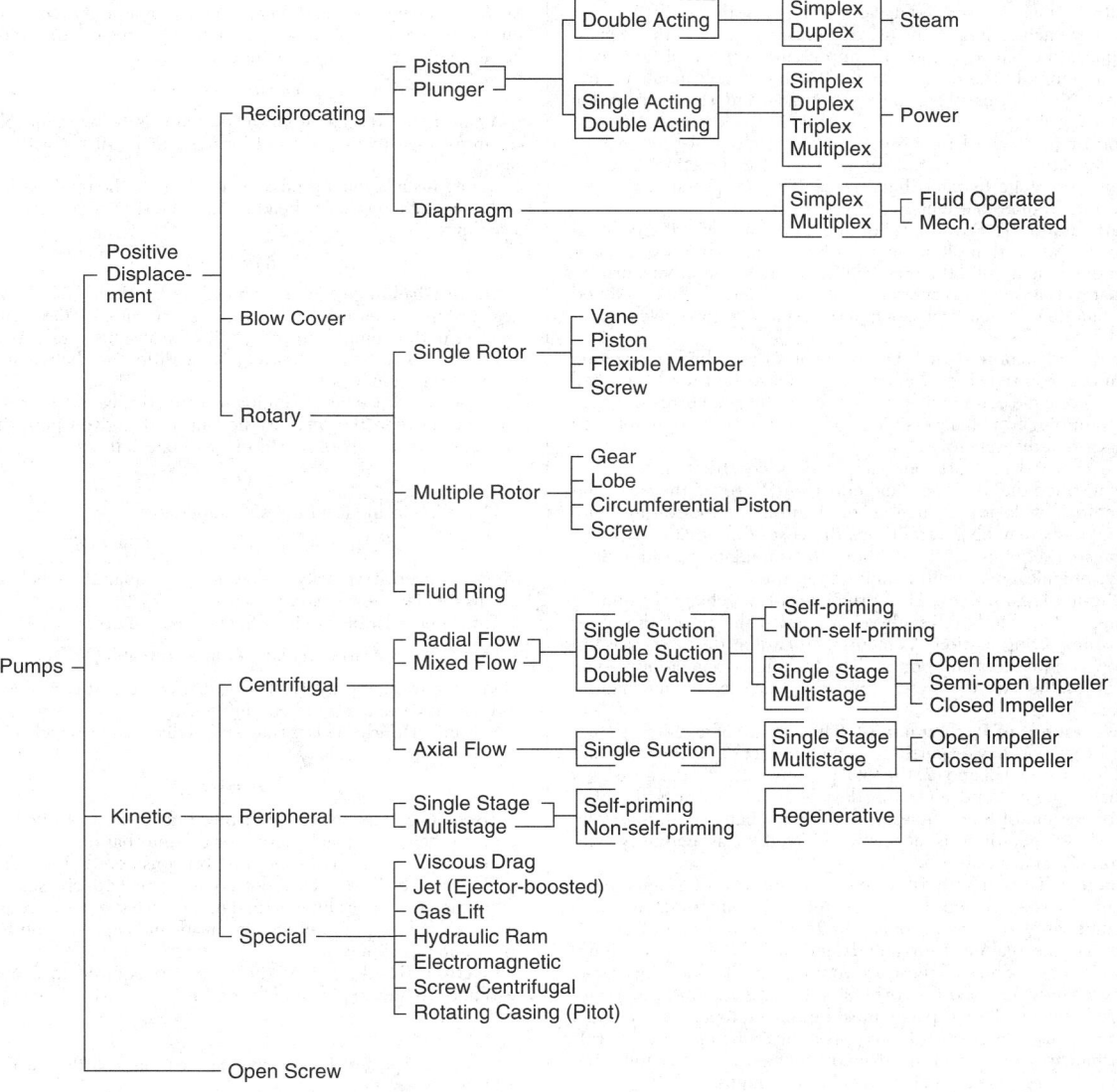

FIG. 10-24   Classification of pumps (*Courtesy of Hydraulic Institute*).

The large variety of displacement-type fluid-transport devices makes it difficult to list characteristics common to each. However, for most types it is correct to state that (1) they are adaptable to high-pressure operation, (2) the flow rate through the pump is variable (auxiliary damping systems may be employed to reduce the magnitude of pressure pulsation and flow variation), (3) mechanical considerations limit maximum throughputs, and (4) the devices are capable of efficient performance at extremely low-volume throughput rates.

Centrifugal Force   Centrifugal force is applied by means of the centrifugal pump or compressor. Though the physical appearance of the many types of centrifugal pumps and compressors varies greatly, the basic function of each is the same, i.e., to produce kinetic energy by the action of centrifugal force and then to convert this energy into pressure by efficiently reducing the velocity of the flowing fluid.

In general, centrifugal fluid-transport devices have these characteristics: (1) discharge is relatively free of pulsation; (2) mechanical design lends itself to high throughputs, capacity limitations are rarely a problem; (3) the devices are capable of efficient performance over a wide range of pressures and capacities even at constant-speed operation; (4) discharge pressure is a function of fluid density; and (5) these are relatively small high-speed devices and less costly.

A device which combines the use of centrifugal force with mechanical impulse to produce an increase in pressure is the axial-flow compressor or pump. In this device the fluid travels roughly parallel to the shaft through a series of alternately rotating and stationary radial blades having airfoil cross sections. The fluid is accelerated in the axial direction by mechanical impulses from the rotating blades; concurrently, a positive-pressure gradient in the radial direction is established in each stage by centrifugal force. The net pressure rise per stage results from both effects.

Electromagnetic Force   When the fluid is an electrical conductor, as is the case with molten metals, it is possible to impress an electromagnetic field around the fluid conduit in such a way that a driving force that will cause flow is created. Such pumps have been developed for the handling of heat-transfer liquids, especially for nuclear reactors.

Transfer of Momentum   Deceleration of one fluid (motivating fluid) in order to transfer its momentum to a second fluid (pumped fluid) is a principle commonly used in the handling of corrosive materials, in pumping from inaccessible depths, or for evacuation. Jets and eductors are in this category.

Absence of moving parts and simplicity of construction have frequently justified the use of jets and eductors. However, they are relatively inefficient devices. When air or steam is the motivating fluid, operating costs may be several times the cost of alternative types of fluid-transport equipment. In addition, environmental considerations in today's chemical plants often inhibit their use.

Mechanical Impulse   The principle of mechanical impulse when applied to fluids is usually combined with one of the other means of imparting motion. As mentioned earlier, this is the case in axial-flow compressors and pumps. The turbine or regenerative-type pump is another device which functions partially by mechanical impulse.

Measurement of Performance   The amount of useful work that any fluid-transport device performs is the product of (1) the mass rate of fluid flow through it and (2) the total pressure differential measured immediately before and after the device, usually expressed in the height of column of fluid equivalent under adiabatic conditions. The first of these quantities is normally referred to as **capacity,** and the second is known as **head.**

Capacity   This quantity is expressed in the following units. In SI units capacity is expressed in cubic meters per hour (m³/h) for both liquids and gases. In U.S. customary units it is expressed in U.S. gallons per minute (gal/min) for liquids and in cubic feet per minute (ft³/min) for gases. Since all these are volume units, the density or specific gravity must be used for conversion to mass rate of flow. When gases are being handled, capacity must be related to a pressure and a temperature, usually the conditions prevailing at the machine inlet. It is important to note that all heads and other terms in the following equations are expressed in height of column of liquid.

Total Dynamic Head   The total dynamic head H of a pump is the total discharge head h_d minus the total suction head h_s.

Total Suction Head   This is the reading h_{gs} of a gauge at the suction flange of a pump (corrected to the pump centerline°), plus the barometer reading and the velocity head h_{vs} at the point of gauge attachment:

$$h_s = h_{gs} + atm + h_{vs} \qquad (10\text{-}41)$$

If the gauge pressure at the suction flange is less than atmospheric, requiring use of a vacuum gauge, this reading is used for h_{gs} in Eq. (10-41) with a negative sign.

Before installation it is possible to estimate the total suction head as follows:

$$h_s = h_{ss} - h_{fs} \qquad (10\text{-}42)$$

where h_{ss} = static suction head and h_{fs} = suction friction head.

Static Suction Head   The static suction head h_{ss} is the vertical distance measured from the free surface of the liquid source to the pump centerline plus the absolute pressure at the liquid surface.

Total Discharge Head   The total discharge head h_d is the reading h_{gd} of a gauge at the discharge flange of a pump (corrected to the pump centerline°), plus the barometer reading and the velocity head h_{vd} at the point of gauge attachment:

$$h_d = h_{gd} + atm + h_{vd} \qquad (10\text{-}43)$$

Again, if the discharge gauge pressure is below atmospheric, the vacuum-gauge reading is used for h_{gd} in Eq. (10-43) with a negative sign.

Before installation it is possible to estimate the total discharge head from the static discharge head h_{sd} and the discharge friction head h_{fd} as follows:

$$h_d = h_{sd} + h_{fd} \qquad (10\text{-}44)$$

Static Discharge Head   The static discharge head h_{sd} is the vertical distance measured from the free surface of the liquid in the receiver to the pump centerline,° plus the absolute pressure at the liquid surface. **Total static head** h_{ts} is the difference between discharge and suction static heads.

Velocity   Since most liquids are practically incompressible, the relation between the quantity flowing past a given point in a given time and the velocity of flow is expressed as follows:

$$Q = Av \qquad (10\text{-}45)$$

This relationship in SI units is as follows:

$$v \text{ (for circular conduits)} = 3.54 \, Q/d^2 \qquad (10\text{-}46)$$

where v = average velocity of flow, m/s; Q = quantity of flow, m³/h; and d = inside diameter of conduit, cm.

This same relationship in U.S. customary units is

$$v \text{ (for circular conduits)} = 0.409 \, Q/d^2 \qquad (10\text{-}47)$$

where v = average velocity of flow, ft/s; Q = quantity of flow, gal/min; and d = inside diameter of conduit, in.

Velocity Head   This is the vertical distance by which a body must fall to acquire the velocity v.

$$h_v = v^2/2g \qquad (10\text{-}48)$$

Viscosity   (See Sec. 5 for further information.) In flowing liquids the existence of internal friction or the internal resistance to relative motion of the fluid particles must be considered. This resistance is called viscosity. The viscosity of liquids usually decreases with rising temperature. Viscous liquids tend to increase the power required by a pump, to reduce pump efficiency, head, and capacity, and to increase friction in pipe lines.

Friction Head   This is the pressure required to overcome the resistance to flow in pipe and fittings. It is dealt with in detail in Sec. 5.

° On vertical pumps, the correction should be made to the eye of the suction impeller.

Work Performed in Pumping   To cause liquid to flow, work must be expended. A pump may raise the liquid to a higher elevation, force it into a vessel at higher pressure, provide the head to overcome pipe friction, or perform any combination of these. Regardless of the service required of a pump, all energy imparted to the liquid in performing this service must be accounted for; consistent units for all quantities must be employed in arriving at the work or power performed.

When arriving at the performance of a pump, it is customary to calculate its **power output,** which is the product of (1) the total dynamic head and (2) the mass of liquid pumped in a given time. In SI units power is expressed in kilowatts; horsepower is the conventional unit used in the United States.

In SI units,

$$kW = HQ\rho/3.670 \times 10^5 \qquad (10\text{-}49)$$

where kW is the pump power output, kW; H = total dynamic head, N·m/kg (column of liquid); Q = capacity, m^3/h; and ρ = liquid density, kg/m^3.

When the total dynamic head H is expressed in pascals, then

$$kW = HQ/3.599 \times 10^6 \qquad (10\text{-}50)$$

In U.S. customary units,

$$hp = HQs/3.960 \times 10^3 \qquad (10\text{-}51)$$

where hp is the pump-power output, hp; H = total dynamic head, lbf·ft/lbm (column of liquid); Q = capacity, U.S. gal/min; and s = liquid specific gravity.

When the total dynamic head H is expressed in pounds-force per square inch, then

$$hp = HQ/1.714 \times 10^3 \qquad (10\text{-}52)$$

The **power input** to a pump is greater than the **power output** because of internal losses resulting from friction, leakage, etc. The efficiency of a pump is therefore defined as

$$Pump\ efficiency = (power\ output)/(power\ input) \qquad (10\text{-}53)$$

Suction Limitations of a Pump   Whenever the pressure in a liquid drops below the vapor pressure corresponding to its temperature, the liquid will vaporize. When this happens within an operating pump, the vapor bubbles will be carried along to a point of higher pressure, where they suddenly collapse. This phenomenon is known as **cavitation.** Cavitation in a pump should be avoided, as it is accompanied by metal removal, vibration, reduced flow, loss in efficiency, and noise. When the absolute suction pressure is low, cavitation may occur in the pump inlet and damage result in the pump suction and on the impeller vanes near the inlet edges. To avoid this phenomenon, it is necessary to maintain a **required net positive suction head** $(NPSH)_R$, which is the equivalent total head of liquid at the pump centerline less the vapor pressure p. Each pump manufacturer publishes curves relating $(NPSH)_R$ to capacity and speed for each pump.

When a pump installation is being designed, the **available net positive suction head** $(NPSH)_A$ must be equal to or greater than the $(NPSH)_R$ for the desired capacity. The $(NPSH)_A$ can be calculated as follows:

$$(NPSH)_A = h_{ss} - h_{fs} - p \qquad (10\text{-}54)$$

If $(NPSH)_A$ is to be checked on an existing installation, it can be determined as follows:

$$(NPSH)_A = atm + h_{gs} - p + h_{vs} \qquad (10\text{-}55)$$

Practically, the NPSH required for operation without cavitation and vibration in the pump is somewhat greater than the theoretical. The actual $(NPSH)_R$ depends on the characteristics of the liquid, the total head, the pump speed, the capacity, and impeller design. Any suction condition which reduces $(NPSH)_A$ below that required to prevent cavitation at the desired capacity will produce an unsatisfactory installation and can lead to mechanical difficulty.

NPSH Requirements for Other Liquids   NPSH values depend on the fluid being pumped. Since water is considered a standard fluid

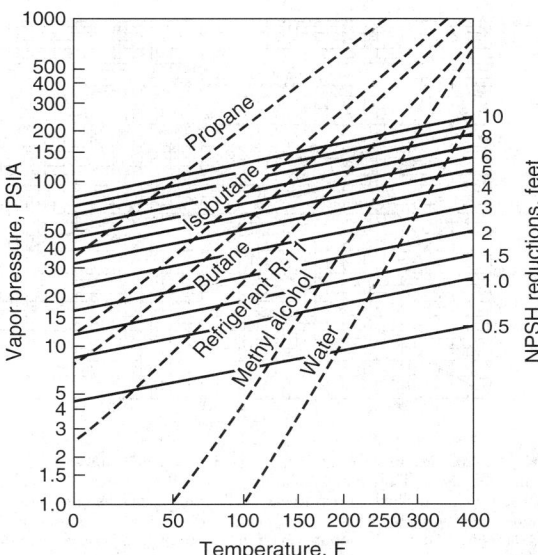

FIG. 10-25   NPSH reductions for pumps handling hydrocarbon liquids and high-temperature water. This chart has been constructed from test data obtained using the liquids shown (*Hydraulic Institute Standards*).

for pumping, various correction methods have been developed to evaluate NPSH when pumping other fluids. The most recent of these corrective methods has been developed by Hydraulic Institute and is shown in Fig. 10-25.

The chart shown in Fig. 10-25 is for pure liquids. Extrapolation of data beyond the ranges indicated in the graph may not produce accurate results. Figure 10-25 shows the variation of vapor pressure and NPSH reductions for various hydrocarbons and hot water as a function of temperature. Certain rules apply while using this chart. When using the chart for hot water, if the NPSH reduction is greater than one-half of the NPSH required for cold water, deduct one-half of cold water NPSH to obtain the corrected NPSH required. On the other hand, if the value read on the chart is less than one-half of cold water NPSH, deduct this chart value from the cold water NPSH to obtain the corrected NPSH.

Example 1: NPSH Calculation   Suppose a selected pump requires a minimum NPSH of 16 ft (4.9 m) when pumping cold water; What will be the NPSH limitation to pump propane at 55°F (12.8°C) with a vapor pressure of 100 psi? Using the chart in Fig. 10-25, NPSH reduction for propane gives 9.5 ft (2.9 m). This is greater than one-half of cold water NPSH of 16 ft (4.9 m). The corrected NPSH is therefore 8 ft (2.2 m) or one-half of cold water NPSH.

PUMP SELECTION

When selecting pumps for any service, it is necessary to know the liquid to be handled, the total dynamic head, the suction and discharge heads, and, in most cases, the temperature, viscosity, vapor pressure, and specific gravity. In the chemical industry, the task of pump selection is frequently further complicated by the presence of solids in the liquid and liquid corrosion characteristics requiring special materials of construction. Solids may accelerate erosion and corrosion, have a tendency to agglomerate, or require delicate handling to prevent undesirable degradation.

Range of Operation   Because of the wide variety of pump types and the number of factors which determine the selection of any one type for a specific installation, the designer must first eliminate all but those types of reasonable possibility. Since range of operation is always an important consideration, Fig. 10-26 should be of assistance. The boundaries shown for each pump type are at best approximate, as unusual applications for which the best selection contradicts the chart

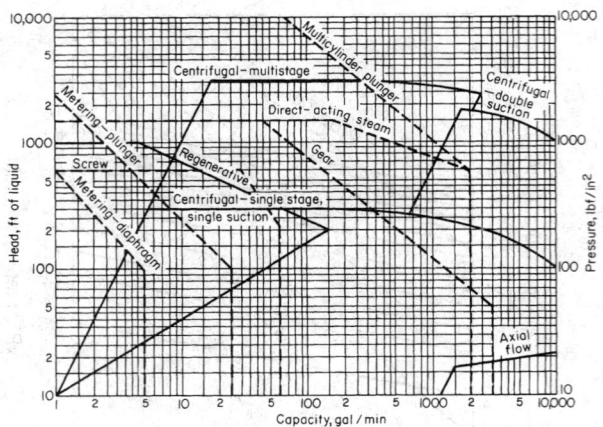

FIG. 10-26 Pump coverage chart based on normal ranges of operation of commercially available types. Solid lines: use left ordinate, head scale. Broken lines: use right ordinate, pressure scale. To convert gallons per minute to cubic meters per hour, multiply by 0.2271; to convert feet to meters, multiply by 0.3048; and to convert pounds-force per square inch to kilopascals, multiply by 6.895.

will arise. In most cases, however, Fig. 10-26 will prove useful in limiting consideration to two or three types of pumps.

Pump Materials of Construction In the chemical industry, the selection of pump materials of construction is dictated by considerations of corrosion, erosion, personnel safety, and liquid contamination. The experience of pump manufacturers is often valuable in selecting materials. See section on materials.

Presence of Solids When a pump is required to pump a liquid containing suspended solids, there are unique requirements which must be considered. Adequate clear-liquid hydraulic performance and the use of carefully selected materials of construction may not be all that is required for satisfactory pump selection. Dimensions of all internal passages are critical. Pockets and dead spots, areas where solids can accumulate, must be avoided. Close internal clearances are undesirable because of abrasion. Flushing connections for continuous or intermittent use should be provided.

For installations in which suspended solids must be handled with a minimum of solids breakage or degradation, such as pumps feeding filter presses, special attention is required; either a low-shear positive-displacement pump or a recessed-impeller centrifugal pump may be called for.

Ease of maintenance is of increasing importance in today's economy. Chemical pump installations that require annual maintenance costing 2 or 3 times the original investment are not uncommon. In most cases this expense is the result of improper selection.

CENTRIFUGAL PUMPS

The centrifugal pump is the type most widely used in the chemical industry for transferring liquids of all types—raw materials, materials in manufacture, and finished products—as well as for general services of water supply, boiler feed, condenser circulation, condensate return, etc. These pumps are available through a vast range of sizes, in capacities from 0.5 m³/h to 2×10^4 m³/h (2 gal/min to 10^5 gal/min), and for discharge heads (pressures) from a few meters to approximately 48 MPa (7000 lbf/in²). The size and type best suited to a particular application can be determined only by an engineering study of the problem.

The primary advantages of a centrifugal pump are simplicity, low first cost, uniform (nonpulsating) flow, small floor space, low maintenance expense, quiet operation, and adaptability for use with a motor or a turbine drive.

A centrifugal pump, in its simplest form, consists of an impeller rotating within a casing. The **impeller** consists of a number of blades,

either open or shrouded, mounted on a shaft that projects outside the casing. Its axis of rotation may be either horizontal or vertical, to suit the work to be done. **Closed-type,** or **shrouded,** impellers are generally the most efficient. **Open-** or **semiopen-type** impellers are used for viscous liquids or for liquids containing solid materials and on many small pumps for general service. Impellers may be of the **single-suction** or the **double-suction** type—single if the liquid enters from one side, double if it enters from both sides.

Casings There are three general types of casings, but each consists of a chamber in which the impeller rotates, provided with inlet and exit for the liquid being pumped. The simplest form is the **circular casing,** consisting of an annular chamber around the impeller; no attempt is made to overcome the losses that will arise from eddies and shock when the liquid leaving the impeller at relatively high velocities enters this chamber. Such casings are seldom used.

Volute casings take the form of a spiral increasing uniformly in cross-sectional area as the outlet is approached. The volute efficiently converts the velocity energy imparted to the liquid by the impeller into pressure energy.

A third type of casing is used in **diffuser-type** or turbine pumps. In this type, **guide vanes** or **diffusers** are interposed between the impeller discharge and the casing chamber. Losses are kept to a minimum in a well-designed pump of this type, and improved efficiency is obtained over a wider range of capacities. This construction is often used in multistage high-head pumps.

Action of a Centrifugal Pump Briefly, the action of a centrifugal pump may be shown by Fig. 10-27. Power from an outside source is applied to shaft A, rotating the impeller B within the stationary casing C. The blades of the impeller in revolving produce a reduction in pressure at the entrance or eye of the impeller. This causes liquid to flow into the impeller from the suction pipe D. This liquid is forced outward along the blades at increasing tangential velocity. The velocity head it has acquired when it leaves the blade tips is changed to pressure head as the liquid passes into the volute chamber and thence out the discharge E.

Centrifugal-Pump Characteristics Figure 10-28 shows a typical characteristic curve of a centrifugal pump. It is important to note that at any fixed speed the pump will operate along this curve and at no other points. For instance, on the curve shown, at 45.5 m³/h (200 gal/min) the pump will generate 26.5-m (87-ft) head. If the head is increased to 30.48 m (100 ft), 27.25 m³/h (120 gal/min) will be delivered. It is not possible to reduce the capacity to 27.25 m³/h (120 gal/min) at 26.5-m (87-ft) head unless the discharge is throttled so that 30.48 m (100 ft) is actually generated within the pump. On pumps with variable-speed drivers such as steam turbines, it is possible to change the characteristic curve, as shown by Fig. 10-29.

As shown in Eq. (10-48), the head depends upon the velocity of the fluid, which in turn depends upon the capability of the impeller to transfer energy to the fluid. This is a function of the fluid viscosity and the impeller design. It is important to remember that the head produced will be the same for any liquid of the same viscosity. The pressure rise, however, will vary in proportion to the specific gravity.

For quick pump selection, manufacturers often give the most essential performance details for a whole range of pump sizes. Figure 10-30 shows typical performance data for a range of process pumps based on suction and discharge pipes and impeller diameters. The performance data consists of pump flow rate and head. Once a pump

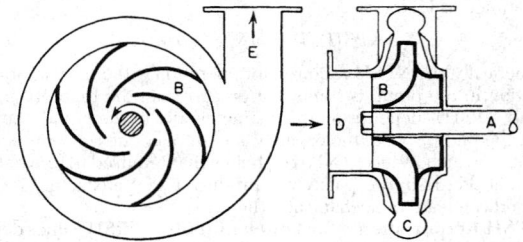

FIG. 10-27 A simple centrifugal pump.

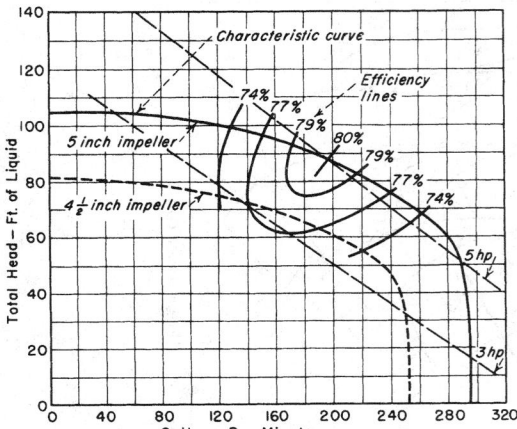

FIG. 10-28 Characteristic curve of a centrifugal pump operating at a constant speed of 3450 r/min. To convert gallons per minute to cubic meters per hour, multiply by 0.2271; to convert feet to meters, multiply by 0.3048; to convert horsepower to kilowatts, multiply by 0.746; and to convert inches to centimeters, multiply by 2.54.

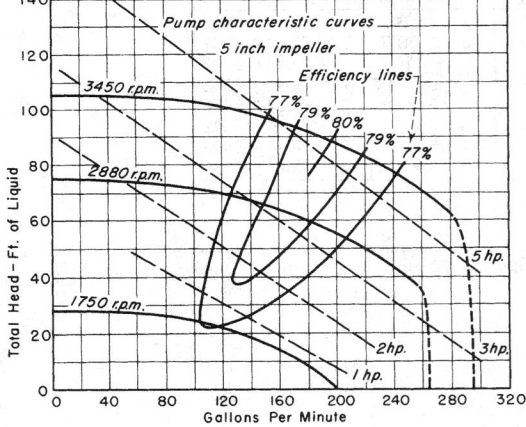

FIG. 10-29 Characteristic curve of a centrifugal pump at various speeds. To convert gallons per minute to cubic meters per hour, multiply by 0.2271; to convert feet to meters, multiply by 0.3048; to convert horsepower to kilowatts, multiply by 0.746; and to convert inches to centimeters, multiply by 2.54.

meets a required specification, then a more detailed performance data for the particular pump can be easily found based on the curve reference number. Figure 10-31 shows a more detailed pump performance curve that includes, in addition to pump head and flow, the break horsepower required, NPSH required, number of vanes, and pump efficiency for a range of impeller diameters.

If detailed manufacturer-specified performance curves are not available for a different size of the pump or operating condition, a best estimate of the off-design performance of pumps can be obtained through similarity relationship or the affinity laws. These are:

1. Capacity (Q) is proportional to impeller rotational speed (N).
2. Head (h) varies as square of the impeller rotational speed.
3. Break horsepower (BHP) varies as the cube of the impeller rotational speed.

These equations can be expressed mathematically and appear in Table 10-7.

TABLE 10-7   The Affinity Laws

	Constant impeller diameter	Constant impeller speed
Capacity	$\dfrac{Q_1}{Q_2}=\dfrac{N_1}{N_2}$	$\dfrac{Q_1}{Q_2}=\dfrac{D_1}{D_2}$
Head	$\dfrac{H_1}{H_2}=\dfrac{(N_1)^2}{(N_2)^2}$	$\dfrac{h_1}{h_2}=\dfrac{(D_1)^2}{(D_2)^2}$
Break horsepower	$\dfrac{BHP_1}{BHP_2}=\dfrac{(N_1)^3}{(N_2)^3}$	$\dfrac{BHP_1}{BHP_2}=\dfrac{(P_1)^3}{(P_2)^3}$

System Curves   In addition to the pump design, the operational performance of a pump depends upon factors such as the downstream load characteristics, pipe friction, and valve performance. Typically, head and flow follow the following relationship:

$$\frac{(Q_2)^2}{(Q_1)^2}=\frac{h_2}{h_1} \qquad (10\text{-}56)$$

where the subscript 1 refers to the design condition and 2 to the actual conditions. The above equation indicates that head will change as a square of the water flow rate.

Figure 10-32 shows the schematic of a pump, moving a fluid from tank A to tank B, both of which are at the same level. The only force that the pump has to overcome in this case is the pipe function, variation of which with fluid flow rate is also shown in the figure. On the other for the use shown in Figure 10-33, the pump in addition to pipe friction should overcome head due to difference in elevation between tanks A and B. In this case, elevation head is constant, whereas the head required to overcome friction depends on the flow rate. Figure 10-34 shows the pump performance requirement of a valve opening and closing.

Pump Selection   One of the parameters that is extremely useful in selecting a pump for a particular application is specific speed N_s. Specific speed of a pump can be evaluated based on its design speed, flow, and head.

$$N_s=\frac{NQ^{0.5}}{H^{0.75}} \qquad \text{or} \qquad N_s=\frac{NQ^{1/2}}{H^{3/4}} \qquad (10\text{-}57)$$

where N = rpm, Q is flow rate in gpm, and H is head in ft·lb$_f$/lbm.

Specific speed is a parameter that defines the speed at which impellers of geometrically similar design have to be run to discharge one gallon per minute against a one-foot head. In general, pumps with a low specific speed have a low capacity and high specific speed, high capacity. Specific speeds of different types of pumps are shown in Table 10-8 for comparison.

Another parameter that helps in evaluating the pump suction limitations, such as cavitation, is suction specific speed.

$$S=\frac{NQ^{1/2}}{(NPSH)^{3/4}} \qquad (10\text{-}58)$$

Typically, for single-suction pumps, suction-specific speed above 11,000 is considered excellent. Below 7000 is poor and 7000–9000 is of an average design. Similarly, for double-suction pumps, suction-specific speed above 14,000 is considered excellent, below 7000 is poor, and 9000–11,000 is average.

Figure 10-35 shows the schematic of specific-speed variation for different types of pumps. The figure clearly indicates that, as the specific speed increases, the ratio of the impeller outer diameter D_1 to inlet or eye diameter D_2 decreases, tending to become unity for pumps of axial-flow type.

TABLE 10-8   Specific Speeds of Different Types of Pumps

Pump type	Specific speed range
Below 2,000	Process pumps and feed pumps
2,000–5,000	Turbine pumps
4,000–10,000	Mixed-flow pumps
9,000–15,000	Axial-flow pumps

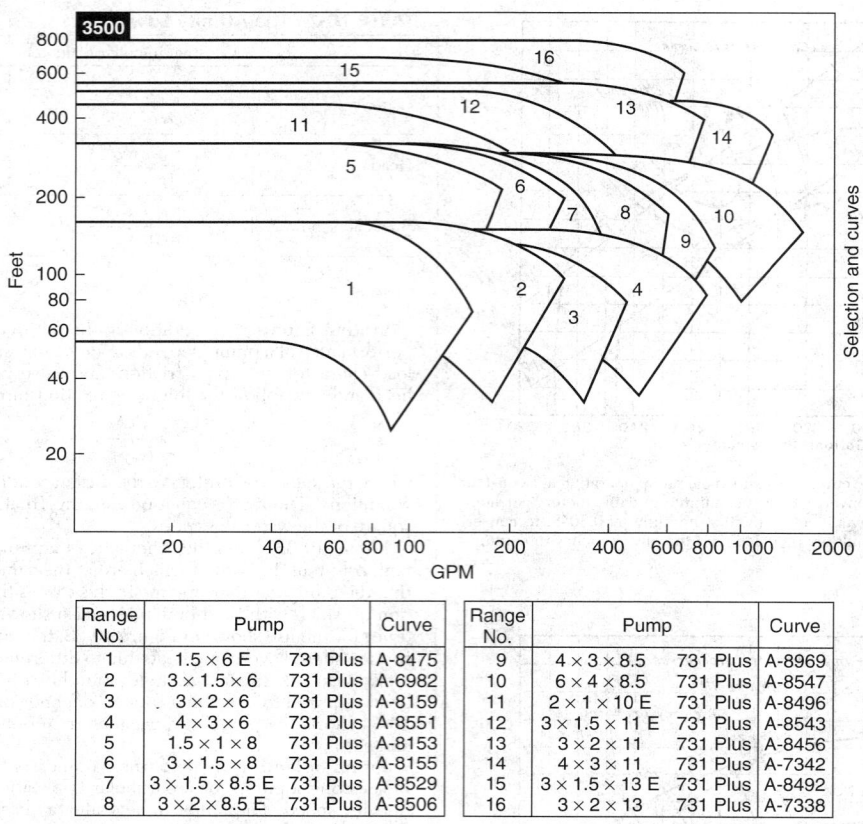

Range No.	Pump		Curve	Range No.	Pump		Curve
1	1.5 × 6 E	731 Plus	A-8475	9	4 × 3 × 8.5	731 Plus	A-8969
2	3 × 1.5 × 6	731 Plus	A-6982	10	6 × 4 × 8.5	731 Plus	A-8547
3	3 × 2 × 6	731 Plus	A-8159	11	2 × 1 × 10 E	731 Plus	A-8496
4	4 × 3 × 6	731 Plus	A-8551	12	3 × 1.5 × 11 E	731 Plus	A-8543
5	1.5 × 1 × 8	731 Plus	A-8153	13	3 × 2 × 11	731 Plus	A-8456
6	3 × 1.5 × 8	731 Plus	A-8155	14	4 × 3 × 11	731 Plus	A-7342
7	3 × 1.5 × 8.5 E	731 Plus	A-8529	15	3 × 1.5 × 13 E	731 Plus	A-8492
8	3 × 2 × 8.5 E	731 Plus	A-8506	16	3 × 2 × 13	731 Plus	A-7338

FIG. 10-30   Performance curves for a range of open impeller pumps.

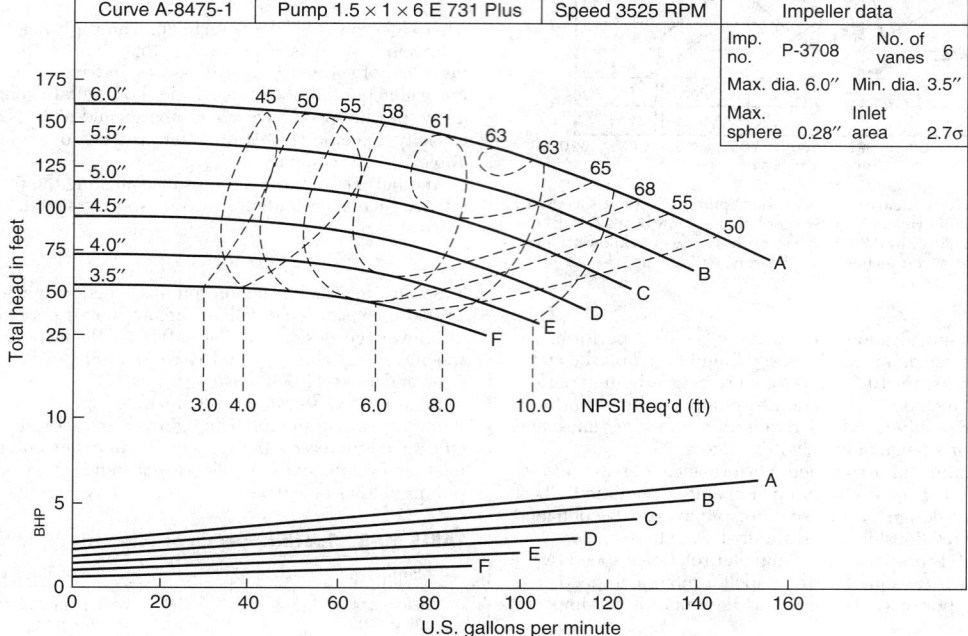

FIG. 10-31   Typical pump performance curve. The curve is shown for water at 85°F. If the specific gravity of the fluid is other than unity, BHP must be corrected.

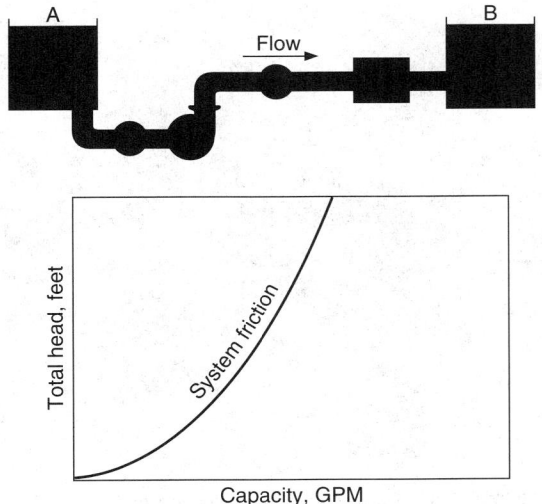

FIG. 10-32 Variation of total head versus flow rate to overcome friction.

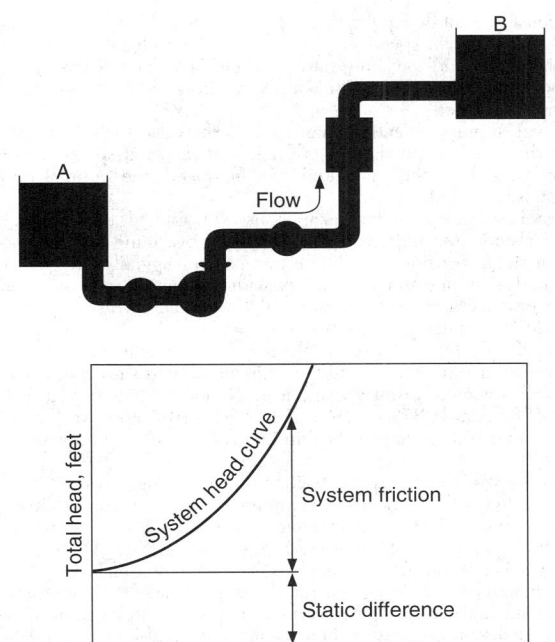

FIG. 10-33 Variation of total head as a function of flow rate to overcome both friction and static head.

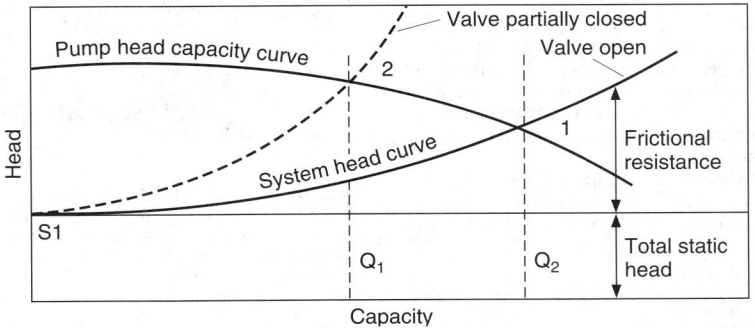

FIG. 10-34 Typical steady-state response of a pump system with a valve fully and partially open.

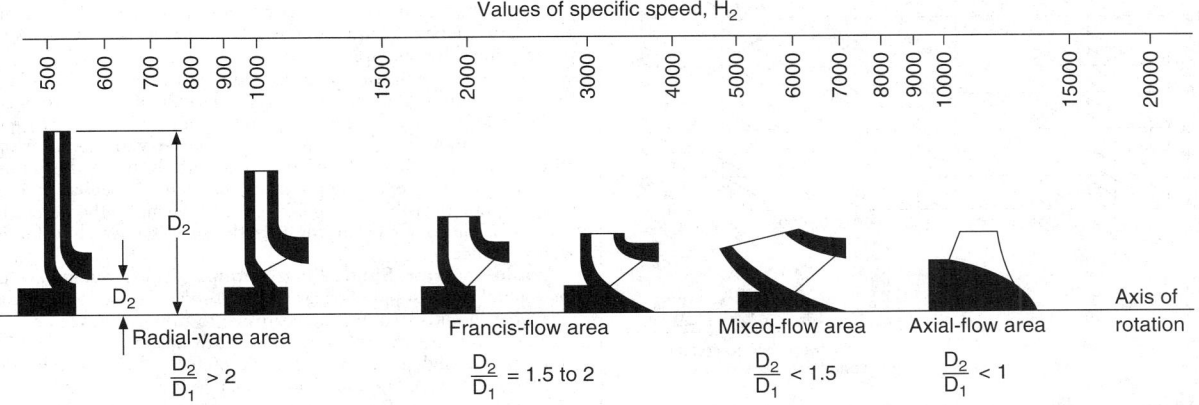

FIG. 10-35 Specific speed variations of different types of pump.

Typically, axial flow pumps are of high flow and low head type and have a high specific speed. On the other hand, purely radial pumps are of high head and low flow rate capability and have a low specific speed. Obviously, a pump with a moderate flow and head has an average specific speed.

A typical pump selection chart such as shown in Fig. 10-36 calculates the specific speed for a given flow, head, and speed requirements. Based on the calculated specific speed, the optimal pump design is indicated.

Process Pumps This term is usually applied to single-stage pedestal-mounted units with single-suction overhung impellers and with a single packing box. These pumps are ruggedly designed for ease in dismantling and accessibility, with mechanical seals or packing arrangements, and are built especially to handle corrosive or otherwise difficult-to-handle liquids.

Specifically but not exclusively for the chemical industry, most pump manufacturers now build to national standards **horizontal and vertical process pumps.** American National Standards Institute (ANSI) Standards B73.1—1977 and B73.2—1975 apply to the horizontal (Fig. 10-37a) and vertical in-line (Fig. 10-37b) pumps respectively.

The horizontal pumps are available for capacities up to 900 m³/h (4000 gal/min); the vertical in-line pumps, for capacities up to 320 m³/h (1400 gal/min). Both horizontal and vertical in-line pumps are available for heads up to 120 m (400 ft). The intent of each ANSI specification is that pumps from all vendors for a given nominal capacity and total dynamic head at a given rotative speed shall be dimensionally interchangeable with respect to mounting, size, and location of suction and discharge nozzles, input shaft, base plate, and foundation bolts.

The vertical in-line pumps, although relatively new additions, are finding considerable use in chemical and petrochemical plants in the United States. An inspection of the two designs will make clear the relative advantages and disadvantages of each.

Chemical pumps are available in a variety of materials. Metal pumps are the most widely used. Although they may be obtained in iron, bronze, and iron with bronze fittings, an increasing number of pumps of ductile-iron, steel, and nickel alloys are being used. Pumps are also available in glass, glass-lined iron, carbon, rubber, rubber-lined metal, ceramics, and a variety of plastics, such units usually being employed for special purposes.

Sealing the Centrifugal Chemical Pump Although detailed treatment of **shaft seals** is presented in the subsection "Sealing of Rotating Shafts," it is appropriate to mention here the special problems

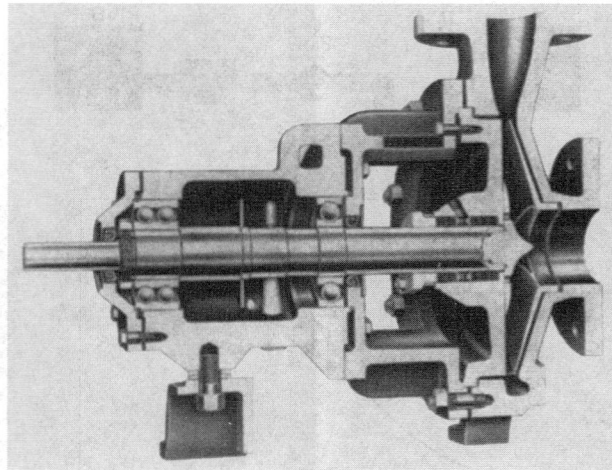

FIG. 10-37a Horizontal process pump conforming to American National Standards Institute (ANSI) Standard B73.1-1977.

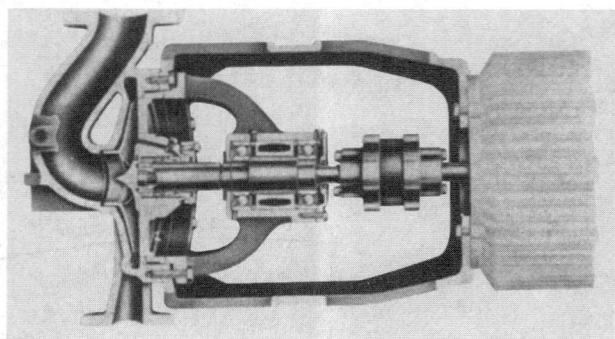

FIG. 10-37b Vertical in-line process pump conforming to ANSI Standard B73.2-1975. The pump shown is driven by a motor through flexible coupling. Not shown but also conforming to ANSI Standard B73.2 are vertical in-line pumps with rigid couplings and with no coupling (impeller-mounted on an extended motor shaft).

of sealing centrifugal chemical pumps. Current practice demands that packing boxes be designed to accommodate both packing and mechanical seals. With either type of seal, one consideration is of paramount importance in chemical service: the liquid present at the sealing surfaces must be free of solids. Consequently, it is necessary to provide a secondary compatible liquid to flush the seal or packing whenever the process liquid is not absolutely clean.

The use of **packing** requires the continuous escape of liquid past the seal to minimize and to carry away the frictional heat developed. If the effluent is toxic or corrosive, quench glands or catch pans are usually employed. Although packing can be adjusted with the pump operating, leaking mechanical seals require shutting down the pump to correct the leak. Properly applied and maintained **mechanical seals** usually show no visible leakage. In general, owing to the more effective performance of mechanical seals, they have gained almost universal acceptance.

Double-Suction Single-Stage Pumps These pumps are used for general water-supply and circulating service and for chemical service when liquids that are noncorrosive to iron or bronze are being handled. They are available for capacities from about 5.7 m³/h (25 gal/min) up to as high as 1.136×10^4 m³/h (50,000 gal/min) and heads up to 304 m (1000 ft). Such units are available in iron, bronze, and iron with bronze fittings. Other materials increase the cost; when they are required, a standard chemical pump is usually more economical.

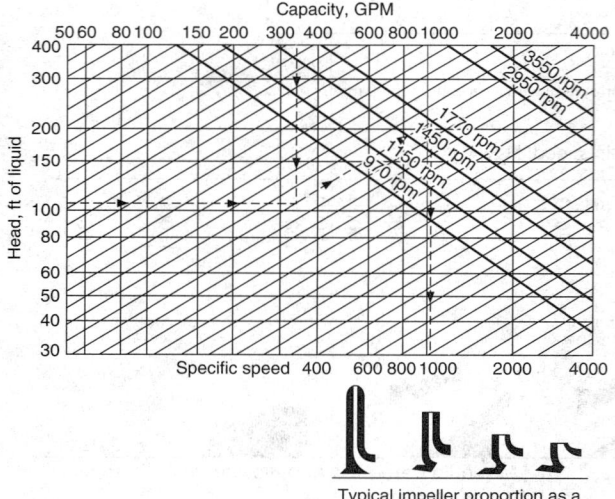

FIG. 10-36 Relationships between specific speed, rotative speed, and impeller proportions (*Worthington Pump Inc., Pump World, vol. 4, no. 2, 1978*).

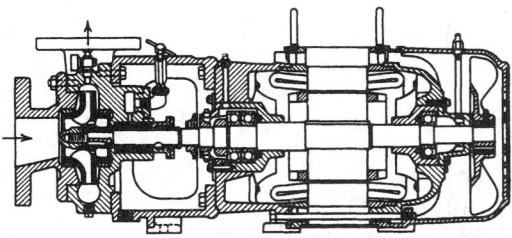

FIG. 10-38   Close-coupled pump.

Close-Coupled Pumps   (Fig. 10-38) Pumps equipped with a built-in electric motor or sometimes steam-turbine-driven (i.e., with pump impeller and driver on the same shaft) are known as close-coupled pumps. Such units are extremely compact and are suitable for a variety of services for which standard iron and bronze materials are satisfactory. They are available in capacities up to about 450 m³/h (2000 gal/min) for heads up to about 73 m (240 ft). Two-stage units in the smaller sizes are available for heads to around 150 m (500 ft).

Canned-Motor Pumps   (Fig. 10-39) These pumps command considerable attention in the chemical industry. They are close-coupled units in which the cavity housing the motor rotor and the pump casing are interconnected. As a result, the motor bearings run in the process liquid and all seals are eliminated. Because the process liquid is the bearing lubricant, abrasive solids cannot be tolerated. Standard single-stage canned-motor pumps are available for flows up to 160 m³/h (700 gal/min) and heads up to 76 m (250 ft). Two-stage units are available for heads up to 183 m (600 ft). Canned-motor pumps are being widely used for handling organic solvents, organic heat-transfer liquids, and light oils as well as many clean toxic or hazardous liquids or for installations in which leakage is an economic problem.

Vertical Pumps   In the chemical industry, the term **vertical process pump** (Fig. 10-40) generally applies to a pump with a vertical shaft having a length from drive end to impeller of approximately 1 m (3.1 ft) minimum to 20 m (66 ft) or more. Vertical pumps are used as either **wet-pit pumps** (immersed) or **dry-pit pumps** (externally mounted) in conjunction with stationary or mobile tanks containing difficult-to-handle liquids. They have the following advantages: the liquid level is above the impeller, and the pump is thus self-priming; and the shaft seal is above the liquid level and is not wetted by the pumped liquid, which simplifies the sealing task. When no bottom connections are permitted on the tank (a safety consideration for highly corrosive or toxic liquid), the vertical wet-pit pump may be the only logical choice.

These pumps have the following disadvantages: intermediate or line bearings are generally required when the shaft length exceeds about 3 m (10 ft) in order to avoid shaft resonance problems; these

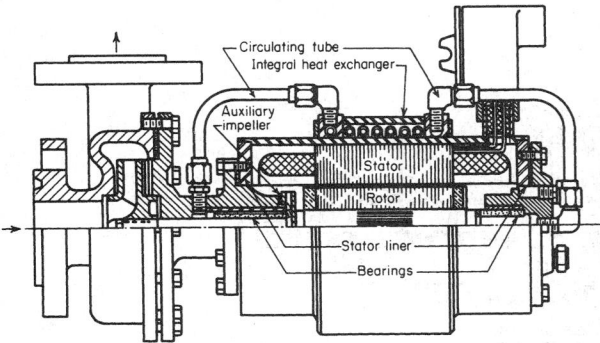

FIG. 10-39   Canned-motor pump (*Courtesy of Chempump Division, Crane Co.*)

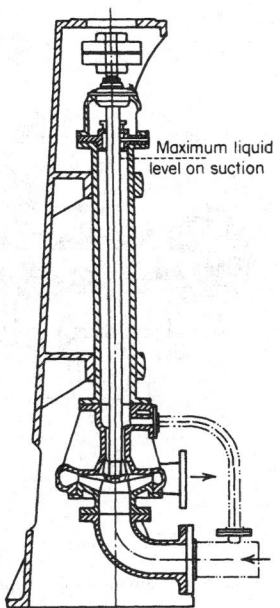

FIG. 10-40   Vertical process pump for dry-pit mounting. (*Courtesy of Lawrence Pumps, Inc.*)

bearings must be lubricated whenever the shaft is rotating. Since all wetted parts must be corrosion-resistant, low-cost materials may not be suitable for the shaft, column, etc. Maintenance is more costly since the pumps are larger and more difficult to handle.

For abrasive service, vertical cantilever designs requiring no line or foot bearings are available. Generally, these pumps are limited to about a 1-m (3.1-ft) maximum shaft length. Vertical pumps are also used to pump waters to reservoirs. One such application in the Los Angeles water basin has fourteen 4-stage pumps, each pump requiring 80,000 Hp to drive them.

Sump Pumps   These are small single-stage vertical pumps used to drain shallow pits or sumps. They are of the same general construction as vertical process pumps but are not designed for severe operating conditions.

Multistage Centrifugal Pumps   These pumps are used for services requiring heads (pressures) higher than can be generated by a single impeller. All impellers are in series, the liquid passing from one impeller to the next and finally to the pump discharge. The total head then is the summation of the heads of the individual impellers. Deep-well pumps, high-pressure water-supply pumps, boiler-feed pumps, fire pumps, and charge pumps for refinery processes are examples of multistage pumps required for various services.

Multistage pumps may be of the **volute type** (Fig. 10-41), with single- or double-suction impellers (Fig. 10-42), or of the **diffuser type** (Fig. 10-43). They may have horizontally split casings or, for extremely high pressures, 20 to 40 MPa (3000 to 6000 lbf/in²), vertically split barrel-type exterior casings with inner casings containing diffusers, interstage passages, etc.

PROPELLER AND TURBINE PUMPS

Axial-Flow (Propeller) Pumps   (Fig. 10-44) These pumps are essentially very-high-capacity low-head units. Normally they are designed for flows in excess of 450 m³/h (2000 gal/min) against heads of 15 m (50 ft) or less. They are used to great advantage in closed-loop circulation systems in which the pump casing becomes merely an elbow in the line. A common installation is for calandria circulation. A characteristic curve of an axial-flow pump is given in Fig. 10-45.

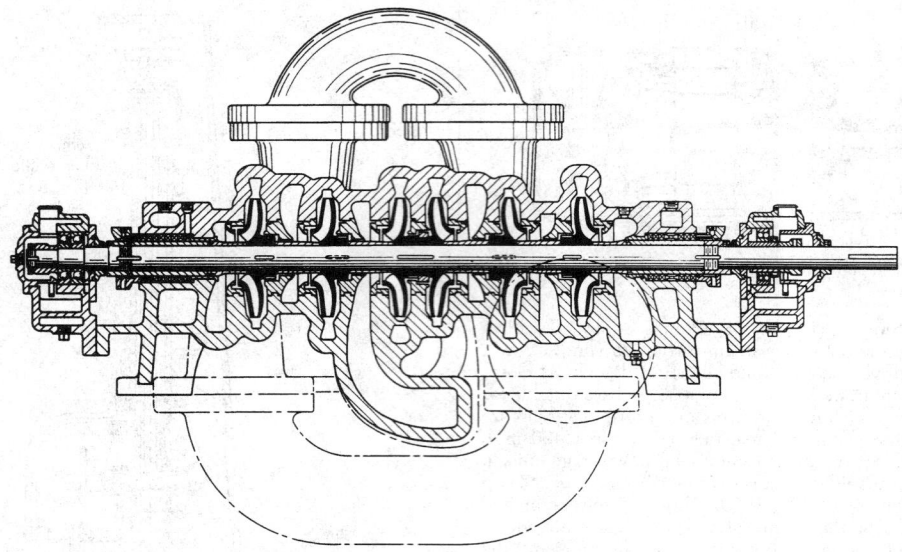

FIG. 10-41 Six-stage volute-type pump.

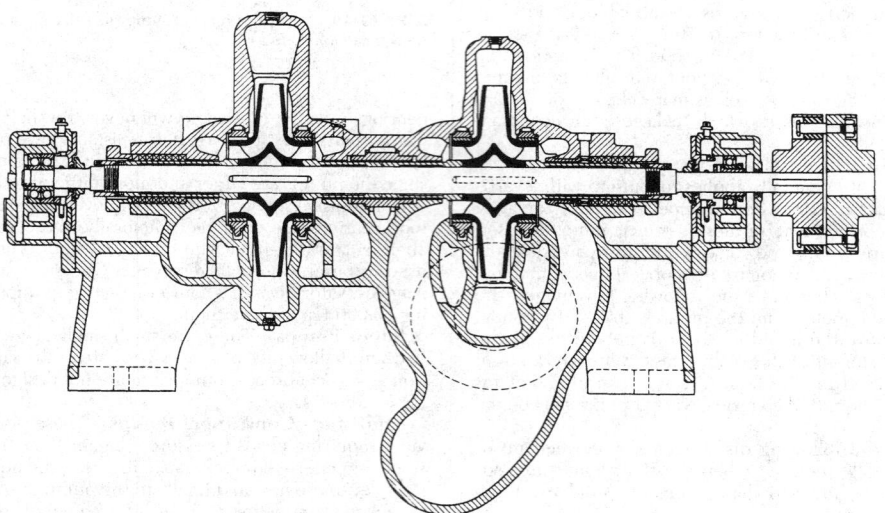

FIG. 10-42 Two-stage pump having double-suction impellers.

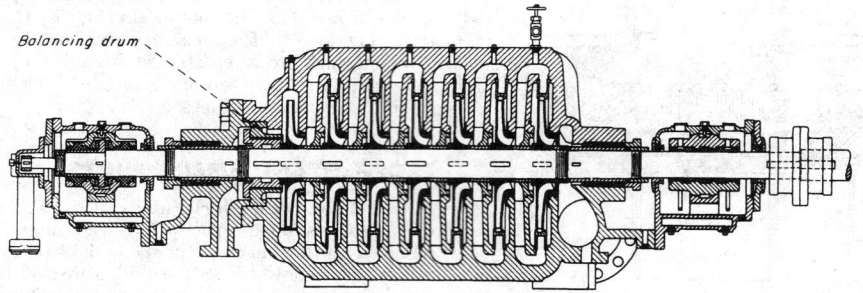

FIG. 10-43 Seven-stage diffuser-type pump.

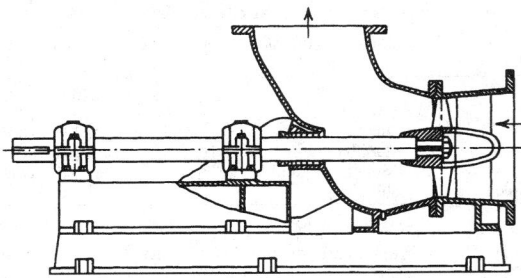

FIG. 10-44 Axial-flow elbow-type propeller pump. (*Courtesy of Lawrence Pumps, Inc.*)

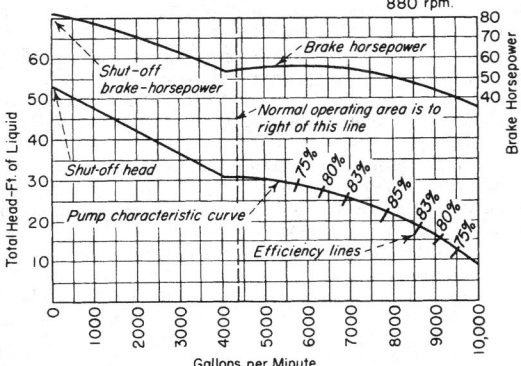

FIG. 10-45 Characteristic curve of an axial-flow pump. To convert gallons per minute to cubic meters per hour, multiply by 0.2271; to convert feet to meters, multiply by 0.3048; and to convert horsepower to kilowatts, multiply by 0.746.

Turbine Pumps The term "turbine pump" is applied to units with mixed-flow (part axial and part centrifugal) impellers. Such units are available in capacities from 20 m³/h (100 gal/min) upward for heads up to about 30 m (100 ft) per stage. Turbine pumps are usually vertical.

A common form of turbine pump is the vertical pump, which has the pump element mounted at the bottom of a column that serves as the discharge pipe (see Fig. 10-46). Such units are immersed in the liquid to be pumped and are commonly used for wells, condenser circulating water, large-volume drainage, etc. Another form of the pump has a shell surrounding the pumping element which is connected to the intake pipe. In this form, the pump is used on condensate service in power plants and for process work in oil refineries.

Regenerative Pumps Also referred to as turbine pumps because of the shape of the impeller, regenerative pumps employ a combination of mechanical impulse and centrifugal force to produce heads of several hundred meters (feet) at low volumes, usually less than 20 m³/h (100 gal/min). The impeller, which rotates at high speed with small clearances, has many short radial passages milled on each side at the periphery. Similar channels are milled in the mating surfaces of the casing. Upon entering, the liquid is directed into the impeller passages and proceeds in a spiral pattern around the periphery, passing alternately from the impeller to the casing and receiving successive impulses as it does so. Figure 10-47 illustrates a typical performance-characteristic curve.

These pumps are particularly useful when low volumes of low-viscosity liquids must be handled at higher pressures than are normally available with centrifugal pumps. Close clearances limit their use to clean liquids. For very high heads, multistage units are available.

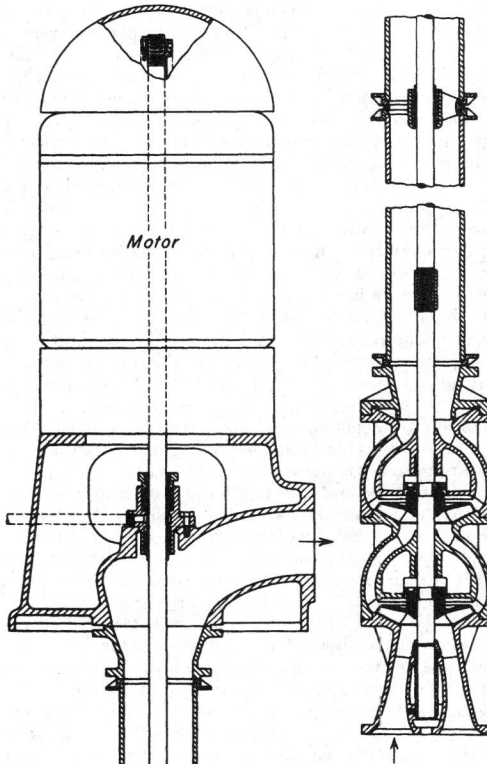

FIG. 10-46 Vertical multistage turbine, or mixed-flow, pump.

POSITIVE-DISPLACEMENT PUMPS

Whereas the total dynamic head developed by a centrifugal, mixed-flow, or axial-flow pump is uniquely determined for any given flow by the speed at which it rotates, **positive-displacement pumps** and those which approach positive displacement will ideally produce whatever head is impressed upon them by the system restrictions to flow. Actually, with slippage neglected, the maximum head attainable is determined by the power available in the drive and the strength of the pump parts. An automatic relief valve set to open at a safe pressure

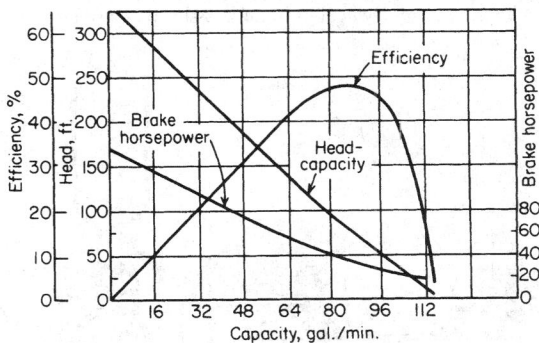

FIG. 10-47 Characteristic curves of a regenerative pump. To convert gallons per minute to cubic meters per hour, multiply by 0.2271; to convert feet to meters, multiply by 0.3048; and to convert horsepower to kilowatts, multiply by 0.746.

higher than the normal or maximum discharge pressure is generally required on the discharge side of all positive-displacement pumps.

In general, overall efficiencies of positive-displacement pumps are higher than those of centrifugal equipment because internal losses are minimized. On the other hand, the flexibility of each piece of equipment in handling a wide range of capacities is somewhat limited.

Positive-displacement pumps may be of either the **reciprocating** or the **rotary** type. In all positive-displacement pumps, a cavity or cavities are alternately filled and emptied of the pumped fluid by the action of the pump.

Reciprocating Pumps   There are three classes of reciprocating pumps: **piston pumps, plunger pumps,** and **diaphragm pumps.** Basically, the action of the liquid-transferring parts of these pumps is the same, a cylindrical piston, plunger, or bucket or a round diaphragm being caused to pass or flex back and forth in a chamber. The device is equipped with valves for the inlet and discharge of the liquid being pumped, and the operation of these valves is related in a definite manner to the motions of the piston. In all modern-design reciprocating pumps, the suction and discharge valves are operated by pressure difference. That is, when the pump is on its suction stroke and the pump cavity is increasing in volume, the pressure is lowered within the pump cavity, permitting the higher suction pressure to open the suction valve and allowing liquid to flow into the pump. At the same time, the higher discharge-line pressure holds the discharge valve closed. Likewise on the discharge stroke, as the pump cavity is decreasing in volume, the higher pressure developed in the pump cavity holds the suction valve closed and opens the discharge valve to expel liquid from the pump into the discharge line.

The *overall efficiency* of these pumps varies from about 50 percent for the small pumps to about 90 percent or more for the larger sizes.

As shown in Fig. 10-48, reciprocating pumps, except when used for metering service, are frequently provided on the discharge side with gas-charged chambers, the purpose of which is to limit pressure pulsation and to provide a more uniform flow in the discharge line. In many installations, surge chambers are required on the suction side as well. Piping layouts should be studied to determine the most effective size and location. If surge chambers are used, provision should be made to keep the chamber charged with gas. A surge chamber filled with liquid is of no value. A liquid-level gauge is desirable to permit checking the amount of gas in the chamber.

Reciprocating pumps may be of **single-cylinder** or **multicylinder** design. Multicylinder pumps have all cylinders in parallel for increased capacity. Piston-type pumps may be single-acting or double-acting; i.e., pumping may be accomplished from one or both ends of the piston. Plunger pumps are always single-acting. The following tabulation (Table 10-9) provides data on the flow variation of reciprocating pumps of various designs.

Piston Pumps   There are two ordinary types of piston pumps, simplex double-acting pumps and duplex double-acting pumps.

Simplex Double-Acting Pumps   These pumps may be direct-acting (i.e., direct-connected to a steam cylinder) or power-driven (through a crank and flywheel from the crosshead of a steam engine).

TABLE 10-9   Flow Variation of Reciprocating Pumps

Number of cylinders	Single- or double-acting	Flow variation per stroke from mean, percent
Single	Single	+220 to −100
Single	Double	+60 to −100
Duplex	Single	+24.1 to −100
Duplex	Double	+6.1 to −21.5
Triplex	Single and double	+1.8 to −16.9
Quintuplex	Single	+1.8 to −5.2

Figure 10-48 is a direct-acting pump, designed for use at pressures up to 0.690 MPa (100 lbf/in²). In this figure, the piston consists of disks A and B, with packing rings C between them. A bronze liner for the water cylinder is shown at D. Suction valves are E_1 and E_2. Discharge valves are F_1 and F_2.

Duplex Double-Acting Pumps   These pumps differ primarily from those of the simplex type in having two cylinders whose operation is coordinated. They may be direct-acting, steam-driven, or power-driven with crank and flywheel.

A duplex outside-end-packed **plunger pump** with pot valves, of the type used with hydraulic presses and for similar service, is shown in Fig. 10-49. In this drawing, plunger A is direct-connected to rod B, while plunger C is operated from the rod by means of yoke D and tie rods.

Plunger pumps differ from piston pumps in that they have one or more constant-diameter plungers reciprocating through packing glands and displacing liquid from cylinders in which there is considerable radial clearance. They are always single-acting, in the sense that only one end of the plunger is used in pumping the liquid.

Plunger pumps are available with one, two, three, four, five, or even more cylinders. Simplex and duplex units are often built in a horizontal design. Those with three or more cylinders are usually of vertical

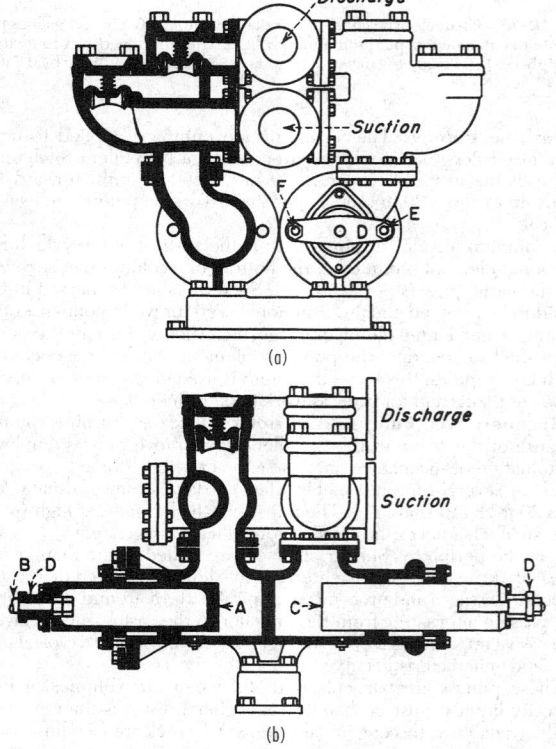

(a)

(b)

FIG. 10-49   Duplex single-acting plunger pump.

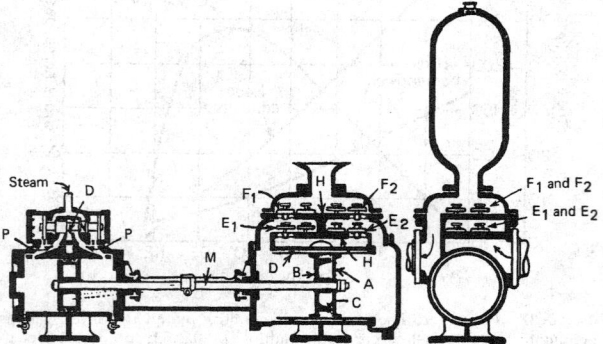

FIG. 10-48   Double-acting steam-driven reciprocating pump.

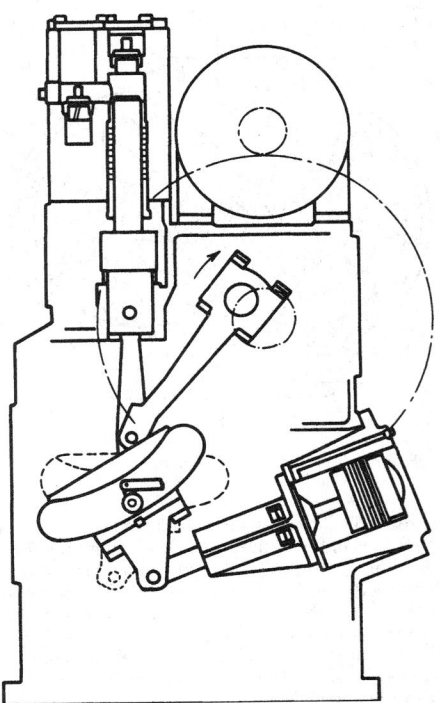

FIG. 10-50 Adrich-Groff variable-stroke power pump. (*Courtesy of Ingersoll-Rand.*)

design. The driver may be an electric motor, a steam or gas engine, or a steam turbine. This is the common type of **power pump.** An example, arranged for belt drive, is shown in Fig. 10-50 from which the action may be readily traced.

Occasionally plunger pumps are constructed with opposed cylinders and plungers connected by yokes and tie rods; this arrangement, in effect, constitutes a double-acting unit.

Simplex plunger pumps mounted singly or in gangs with a common drive are quite commonly used as **metering** or **proportioning pumps** (Fig. 10-51). Frequently a variable-speed drive or a stroke-adjusting mechanism is provided to vary the flow as desired. These pumps are designed to measure or control the flow of liquid within a deviation of ±2 percent with capacities up to 11.35 m³/h (50 gal/min) and pressures as high as 68.9 MPa (10,000 lbf/in²).

Diaphragm Pumps   These pumps perform similarly to piston and plunger pumps, but the reciprocating driving member is a flexible diaphragm fabricated of metal, rubber, or plastic. The chief advantage of this arrangement is the elimination of all packing and seals exposed to the liquid being pumped. This, of course, is an important asset for equipment required to handle hazardous or toxic liquids.

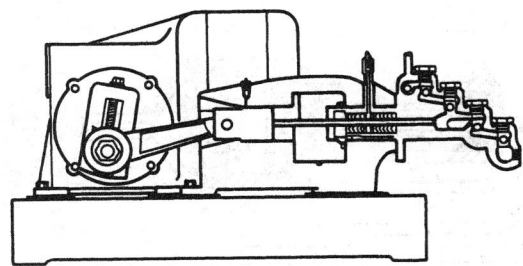

FIG. 10-51 Plunger-type metering pump. (*Courtesy of Milton Roy Co.*)

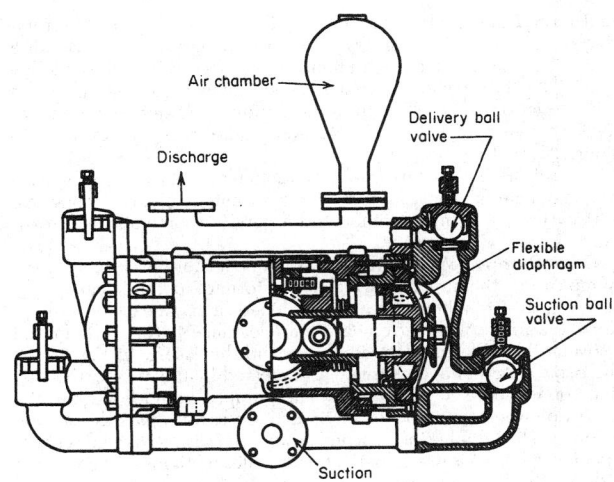

FIG. 10-52 Mechanically actuated diaphragm pump.

A common type of low-capacity diaphragm pump designed for metering service employs a plunger working in oil to actuate a metallic or plastic diaphragm. Built for pressures in excess of 6.895 MPa (1000 lbf/in²) with flow rates up to about 1.135 m³/h (5 gal/min) per cylinder, such pumps possess all the characteristics of plunger-type metering pumps with the added advantage that the pumping head can be mounted in a remote (even a submerged) location entirely separate from the drive.

Figure 10-52 shows a high-capacity 22.7-m³/h (100-gal/min) pump with actuation provided by a mechanical linkage.

Pneumatically Actuated Diaphragm Pumps   (Fig. 10-53) These pumps require no power source other than plant compressed air. They must have a flooded suction, and the pressure is, of course, limited to the available air pressure. Because of their slow speed and large valves, they are well suited to the gentle handling of liquids for which degradation of suspended solids should be avoided.

A major consideration in the application of diaphragm pumps is the realization that diaphragm failure will probably occur eventually. The consequences of such failure should be realistically appraised before selection, and maintenance procedures should be established accordingly.

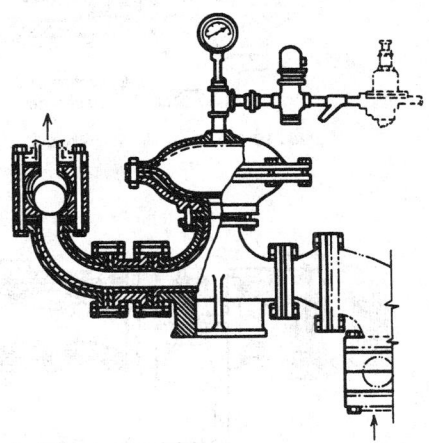

FIG. 10-53 Pneumatically actuated diaphragm pump for slurry service. (*Courtesy of Dorr-Olvier Inc.*)

Rotary Pumps   In rotary pumps the liquid is displaced by rotation of one or more members within a stationary housing. Because internal clearances, although minute, are a necessity in all but a few special types, capacity decreases somewhat with increasing pump differential pressure. Therefore, these pumps are not truly positive-displacement pumps. However, for many other reasons they are considered as such.

The selection of materials of construction for rotary pumps is critical. The materials must be corrosion-resistant, compatible when one part is running against another, and capable of some abrasion resistance.

Gear Pumps   When two or more impellers are used in a rotary-pump casing, the impellers will take the form of toothed-gear wheels as in Fig. 10-54, of helical gears, or of lobed cams. In each case, these impellers rotate with extremely small clearance between them and between the surfaces of the impellers and the casing. In Fig. 10-54, the two toothed impellers rotate as indicated by the arrows; the suction connection is at the bottom. The pumped liquid flows into the spaces between the impeller teeth as these cavities pass the suction opening. The liquid is then carried around the casing to the discharge opening, where it is forced out of the impeller teeth mesh. The arrows indicate this flow of liquid.

Rotary pumps are available in two general classes, interior-bearing and exterior-bearing. The **interior-bearing type** is used for handling liquids of a lubricating nature, and the **exterior-bearing type** is used with nonlubricating liquids. The interior-bearing pump is lubricated by the liquid being pumped, and the exterior-bearing type is oil-lubricated.

The use of spur gears in gear pumps will produce in the discharge pulsations having a frequency equivalent to the number of teeth on both gears multiplied by the speed of rotation. The amplitude of these disturbances is a function of tooth design. The pulsations can be reduced markedly by the use of rotors with helical teeth. This in turn introduces end thrust, which can be eliminated by the use of double-helical or herringbone teeth.

Screw Pumps   A modification of the helical gear pump is the screw pump. Both gear and screw pumps are positive displacement

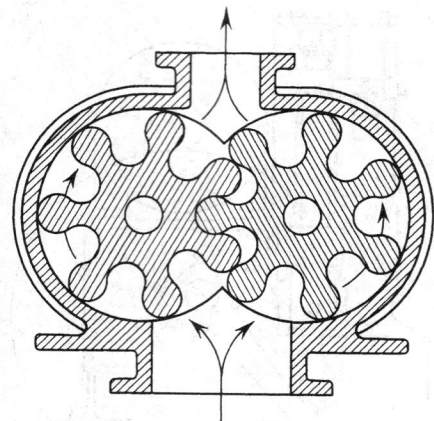

FIG. 10-54   Positive-displacement gear-type rotary pump.

pumps. Figure 10-55 illustrates a two-rotor version in which the liquid is fed to either the center or the ends, depending upon the direction of rotation, and progresses axially in the cavities formed by the meshing threads or teeth. In three-rotor versions, the center rotor is the driving member while the other two are driven. Figure 10-56 shows still another arrangement, in which a metal rotor of unique design rotates without clearance in an elastomeric stationary sleeve.

Screw pumps, because of multiple dams that reduce slip, are well adapted for producing higher pressure rises, for example, 6.895 MPa (1000 lbf/in^2), especially when handling viscous liquids such as heavy oils. The all-metal pumps are generally subject to the same limitations on handling abrasive solids as conventional gear pumps. In addition, the wide bearing spans usually demand that the liquid have considerable lubricity to prevent metal-to-metal contact.

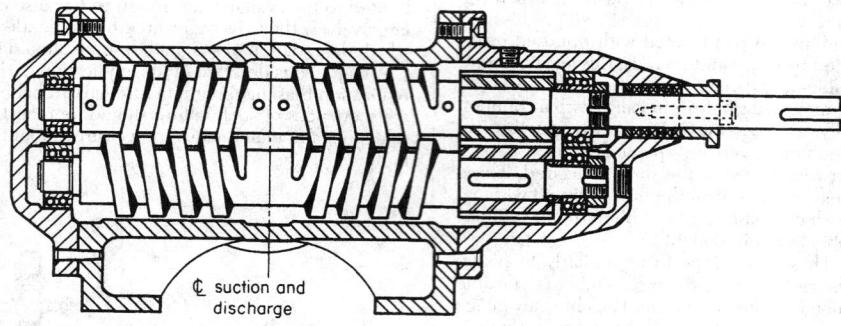

FIG. 10-55   Two-rotor screw pump. (*Courtesy of Warren Quimby Pump Co.*)

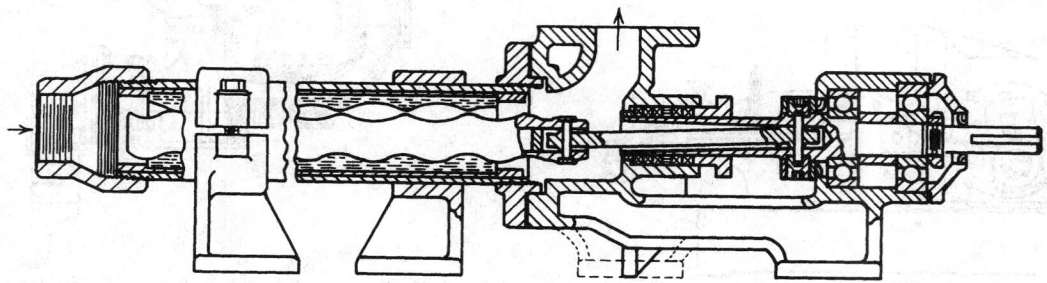

FIG. 10-56   Single-rotor screw pump with an elastomeric lining. (*Courtesy of Moyno Pump Division, Robbins & Myers, Inc.*)

Among the liquids handled by rotary pumps are mineral oils, vegetable oils, animal oils, greases, glucose, viscose, molasses, paints, varnish, shellac, lacquers, alcohols, catsup, brine, mayonnaise, sizing, soap, tanning liquors, vinegar, and ink. Some screw-type units are specially designed for the gentle handling of large solids suspended in the liquid.

Fluid-Displacement Pumps In addition to pumps that depend on the mechanical action of pistons, plungers, or impellers to move the liquid, other devices for this purpose employ displacement by a secondary fluid. This group includes air lifts and acid eggs.

The **air lift** is a device for raising liquid by means of compressed air. In the past it was widely used for pumping wells, but it has been less widely used since the development of efficient centrifugal pumps. It operates by introducing compressed air into the liquid near the bottom of the well. The air-and-liquid mixture, being lighter than liquid alone, rises in the well casing. The advantage of this system of pumping lies in the fact that there are no moving parts in the well. The pumping equipment is an air compressor, which can be located on the surface.

A simplified sketch of an air lift for this purpose is shown in Fig. 10-57. Ingersoll-Rand has developed empirical information on air-lift performance which is available upon request.

An important application of the gas-lift principle involves the extraction of oil from wells. There are several references to both practical and theoretical work involving gas lift performance and related problems. Recommended sources are American Petroleum Institute, *Drilling and Production Practices*, 1952, pp. 257–317, and 1939, p. 266; *Trans. Am. Soc. Mining Metall. Eng.*, **92**, 296–313 (1931), **103**, 170–186 (1933), **118**, 56–70 (1936), **192**, 317–326 (1951), **189**, 73–82 (1950), and **198**, 271–278 (1953); *Trans Am. Soc. Mining Metall.*, and *Pet. Eng.*, **213** (1958), and **207**, 17–24 (1956); and *Univ. Wisconsin Bull., Eng. Ser.*, **6**, no. 7 (1911, reprinted 1914).

An **acid egg**, or **blowcase**, consists of an egg-shaped container which can be filled with a charge of liquid that is to be pumped. This container is fitted with an inlet pipe for the charge, an outlet pipe for the discharge, and a pipe for the admission of compressed air or gas, as illustrated in Fig. 10-58. Pressure of air or gas on the surface of the liquid forces it out of the discharge pipe. Such pumps can be hand-operated or arranged for semiautomatic or automatic operation.

JET PUMPS

Jet pumps are a class of liquid-handling device that makes use of the momentum of one fluid to move another.

Ejectors and **injectors** are the two types of jet pumps of interest to chemical engineers. The ejector, also called the siphon, exhauster, or eductor, is designed for use in operations in which the head pumped against is low and is less than the head of the fluid used for pumping.

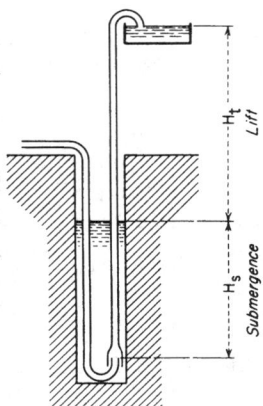

FIG. 10-57 Simplified sketch of an air lift, showing submergence and total head.

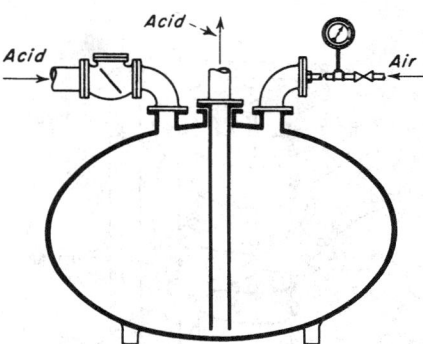

FIG. 10-58 A form of acid egg. External controls required for automatic operation are not shown.

The injector is a special type of jet pump, operated by steam and used for boiler feed and similar services, in which the fluid being pumped is discharged into a space under the same pressure as that of the steam being used to operate the injector.

Figure 10-59 shows a simple design for a jet pump of the ejector type. The pumping fluid enters through the nozzle at the left and passes through the venturi nozzle at the center and out of the discharge opening at the right. As it passes into the venturi nozzle, it develops a suction that causes some of the fluid in the suction chamber to be entrained with the stream and delivered through this discharge.

The efficiency of an ejector or jet pump is low, being only a few percent. The head developed by the ejector is also low except in special types. The device has the disadvantage of diluting the fluid pumped by mixing it with the pumping fluid. In steam injectors for boiler feed and similar services in which the heat of the steam is recovered, efficiency is close to 100 percent.

The simple ejector or siphon is widely used, in spite of its low efficiency, for transferring liquids from one tank to another, for lifting acids, alkalies, or solid-containing liquids of an abrasive nature, and for emptying sumps.

ELECTROMAGNETIC PUMPS

The necessity of circulating liquid-metal heat-transfer media in nuclear-reactor systems has led to development of electromagnetic pumps. All electromagnetic pumps utilize the motor principle: a conductor in a magnetic field, carrying a current which flows at right angles to the direction of the field, has a force exerted on it, the force being mutually perpendicular to both the field and the current. In all electromagnetic pumps, the fluid is the conductor. This force, suitably directed in the fluid, manifests itself as a pressure if the fluid is suitably contained. The field and current can be produced in a number of different ways and the force utilized variously.

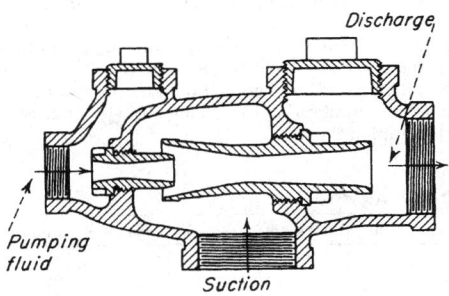

FIG. 10-59 Simple ejector using a liquid-motivating fluid.

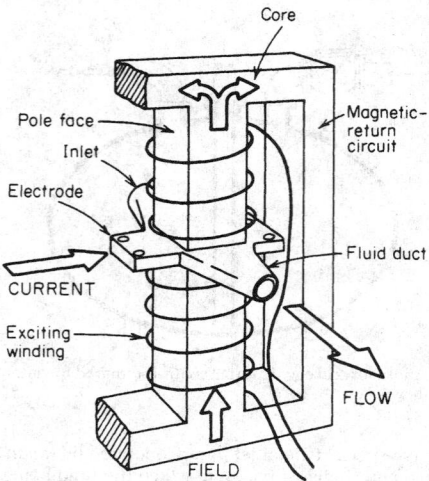

FIG. 10-60 Simplified diagram of a direct-current-operated electromagnetic pump.

Both alternating- and direct-current units are available. While dc pumps (Fig. 10-60) are simpler, their high-current requirement is a definite limitation; ac pumps can readily obtain high currents by making use of transformers. Multipole induction ac pumps have been built in helical and linear configurations. Helical units are effective for relatively high heads and low flows, while linear induction pumps are best suited for large flows at moderate heads. Electromagnetic pumps are available for flow rates up to 2.271×10^3 m³/h (10,000 gal/min), and pressures up to 2 MPa (300 lbf/in²) are practical. Performance characteristics resemble those of centrifugal pumps.

VIBRATION MONITORING

One of the major factors that causes pump failure is vibration, which usually causes seal damage and oil leakage. Vibration in pumps is caused by numerous factors such as cavitation, impeller unbalance, loose bearings, and pipe pulsations. Typically, large-amplitude vibration occurs when the frequency of vibration coincides with that of the natural frequency of the pump system. This results in a catastrophic operating condition that should be avoided. If the natural frequency is close to the upper end of the operating speed range, then the pump system should be stiffened to reduce vibration. On the other hand, if the natural frequency is close to the lower end of the operating range, the unit should be made more flexible. During startup, the pump system may go through its system natural frequency, and vibration can occur. Continuous operation at this operating point should be avoided.

ASME recommends periodic monitoring of all pumps. Pump vibration level should fall within the prescribed limits. The reference vibration level is measured during acceptance testing. This level is specified by the manufacturer.

During periodic maintenance, the vibration level should not exceed alert level (see Table 10-10). If the measured level exceeds the alert level then preventive maintenance should be performed, by diagnosing the cause of vibration and reducing the vibration level prior to continued operation.

TABLE 10-10   Alert Levels

Reference value mils.	Alert mils., microns	Action required mils., microns
$V_r < 0.5$	1.0	1.5
$0.5 < V_r < 2.0$	$2V_r$	$3V_r$
$2.0 < V_r < 5.0$	$2+V_r$	$4+V_r$
$5.0 < V_r$	$1.4V_r$	$1.8V_r$

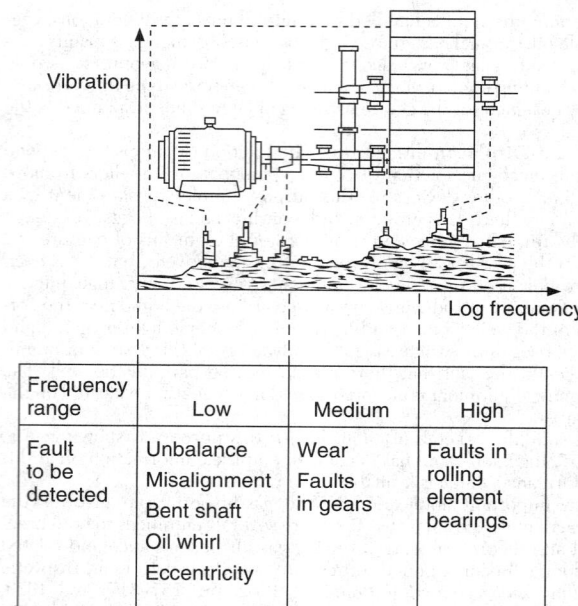

FIG. 10-61 Frequency range of typical machinery faults.

Typical problems and their vibration frequency ranges are shown in Fig. 10-61.

Collection and analysis of vibration signatures is a complex procedure. By looking at a vibration spectrum, one can identify which components of the pump system are responsible for a particular frequency component. Comparison of vibration signatures at periodic intervals reveals if a particular component is deteriorating. The following example illustrates evaluation of the frequency composition of an electric motor gear pump system.

Example 2: Vibration   Consider an electric motor rotating at 1800 rpm driving an 8-vane centrifugal pump rotating at 600 rpm. For this 3:1 speed reduction, assume a gear box having two gears of 100 and 300 tooth. Since 60 Hz is 1 rpm,

Motor frequency = 1800/60 = 30 Hz
Pump frequency = 600/60 = 10 Hz
Gear mesh frequency = 300 teeth × 600 rpm = 3000 Hz
Vane frequency = 8 × 600 rpm = 80 Hz

An ideal vibration spectra for this motor-gear pump assembly would appear as shown in Fig. 10-62.

Figure 10-63 shows an actual pump vibration spectra. In the figure, several amplitude peaks occur at several frequencies.

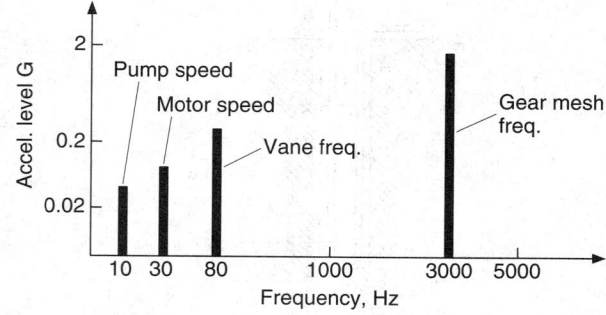

FIG. 10-62 An ideal vibration spectra from an electric motor pump assembly.

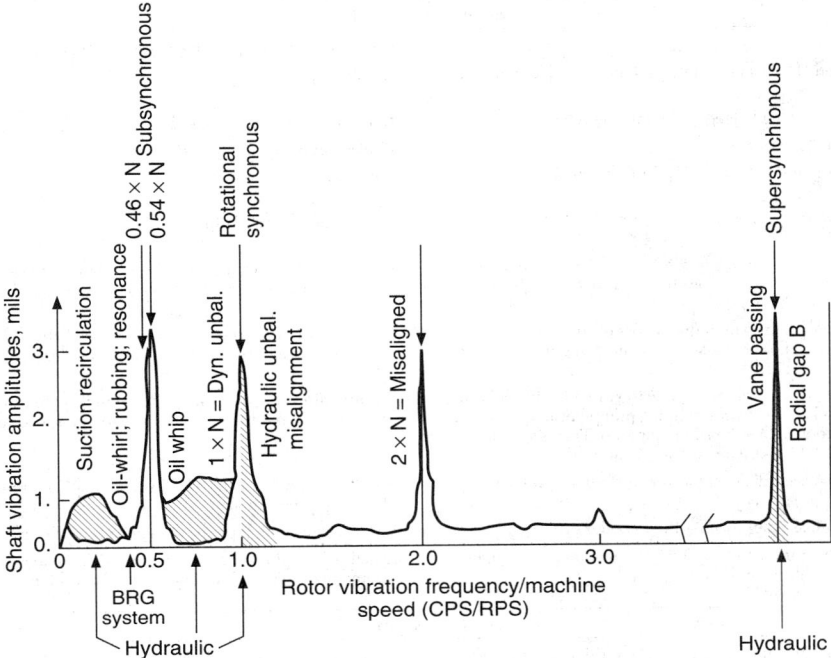

FIG. 10-63 An actual pump vibration spectra.

PUMP DIAGNOSTICS

As the mechanical integrity of the pump system changes, the amplitude of vibration levels change. In some cases, in order to identify the source of vibration, pump speed may have to be varied, as these problems are frequency- or resonance-dependent. Pump impeller imbalance and cavitation are related to this category. Table 10-11 classifies different types of pump-related problems, their possible causes and corrective actions.

Typical pump-related problems are classified under
1. Cavitation-type problems
2. Capacity-type problems
3. Motor overload problems

It is advisable in most of these cases to use accelerometers. Displacement probes will not give the high-frequency signals and velocity probes because their mechanical design is very directional and prone to deterioration. Figure 10-64 shows the signal from the various types of probes.

PUMP SPECIFICATIONS

Pump specifications depend upon numerous factors but mostly on application. Typically, the following factors should be considered while preparing a specification.
1. Application, scope and type
2. Service conditions
3. Operating conditions
4. Consternation-application-specific details and special considerations
   - Casing connection
   - Impeller details
   - Shaft
   - Shifting box details—lubrications, sealing, etc.
   - Bearing frame and bearings
   - Base plate and couplings
   - Materials
   - Special operating conditions and miscellaneous items

Table 10-12 is based on the API and ASME codes.

COMPRESSION OF GASES

Theory of Compression In any continuous compression process the relation of absolute pressure p to volume V is expressed by the formula

$$pV^n = C = \text{constant} \tag{10-59}$$

The plot of pressure versus volume for each value of exponent n is known as the **polytropic** curve. Since the work W performed in proceeding from p_1 to p_2 along any polytropic curve (Fig. 10-65) is

$$W = \int_1^2 p \, dV \tag{10-60}$$

it follows that the amount of work required is dependent upon the polytropic curve involved and increases with increasing values of n. The path requiring the least amount of input work is $n = 1$, which is equivalent to **isothermal** compression. For **adiabatic** compression (i.e., no heat is being added or taken away during the process), $n = k = $ ratio of specific heat at constant pressure to that at constant volume.

Since most compressors operate along a polytropic path approaching the adiabatic, compressor calculations are generally based on the adiabatic curve.

Some formulas based upon the adiabatic equation and useful in compressor work are as follows:

Pressure, volume, and **temperature** relations for perfect gases:

$$p_2/p_1 = (V_1/V_2)^k \tag{10-61}$$

$$T_2/T_1 = (V_1/V_2)^{k-1} \tag{10-62}$$

$$p_2/p_1 = (T_2/T_1)^{k/(k-1)} \tag{10-63}$$

Adiabatic Calculations Adiabatic head is expressed as follows: In SI units,

$$H_{ad} = \frac{k \times RT_1}{k-1}\left[\left(\frac{p_2}{p_1}\right)^{(k-1)/k} - 1\right] \tag{10-64a}$$

where H_{ad} = adiabatic head, N·m/kg; R = gas constant, J/(kg·K) = 8314/ molecular weight; T_1 = inlet gas temperature, K; p_1 = absolute inlet pressure, kPa; and p_2 = absolute discharge pressure, kPa.

TABLE 10-11 Pump Problems

Possible causes	Corrective action
Cavitating-type problems	
Plugged suction screen.	Check for indications of the presence of screen. Remove and clean screen.
Piping gaskets with undersized IDs installed, a very common problem in small pumps.	Install proper-sized gaskets.
Column tray parts or ceramic packing lodged in the impeller eye.	Remove suction piping and debris.
Deteriorated impeller eye due to corrosion.	Replace impeller and overhaul pump.
Flow rate is high enough above design that NPSH for flow rate has increased above NPSH.	Reduce flow rate to that of design.
Lined pipe collapsed at gasket area or ID due to buildup of corrosion products between liner and carbon-steel pipe.	Replace deteriorated piping.
Poor suction piping layout, too many elbows in too many planes, a tee branch almost directly feeding the suction of the other pump, or not enough straight run before the suction flange of the pump.	Redesign piping layout, using fewer elbows and laterals for tees, and have five or more straight pipe diameters before suction flange.
Vertical pumps experience a vortex formation due to loss of submergence required by the pump. Observe the suction surface while the pump is in operation, if possible.	Review causes of vortexing. Consider installation of a vortex breaker such as a bell mouth umbrella or changes to sump design.
Spare pump begins to cavitate when attempt is made to switch it with the running pump. The spare is "backed off" by the running pump because its shutoff head is less than the head produced by the running pump. This is a frequent problem when one pump is turbine-driven and one is motor-driven.	Throttle discharge of running pump until spare can get in system. Slow down running pump if it is a turbine or variable-speed motor.
Suction piping configuration causes adverse fluid rotation when approaching impeller.	Install sufficient straight run of suction piping, or install vanes in piping to break up prerotation.
Velocity of the liquid is too high as it approaches the impeller eye.	Install larger suction piping or reduce flow through pump.
Pump is operating at a low-flow-producing suction recirculation in the impeller eye. This results in a cavitationlike sound.	Install bypass piping back to suction vessel to increase flow through pump. Remember bypass flow may have to be as high as 50 percent of design flow.
Capacity-type problems	
Check the discharge block valve opening first. It may be partially closed and thus the problem.	Open block valve completely.
Wear-ring clearances are excessive (closed impeller design).	Overhaul pump. Renew wear rings if clearance is about twice design value for energy and performance reasons.
Impeller-to-case or head clearances are excessive (open impeller design).	Reposition impeller to obtain correct clearance.
Air leaks into the system if the pump suction is below atmospheric pressure.	Take actions as needed to eliminate air leaks.
Increase in piping friction to the discharge vessel due to the following: 1. Gate has fallen off the discharge valve stem. 2. Spring is broken in the spring-type check valve. 3. Check valve flapper pin is worn, and flapper will not swing open. 4. Lined pipe collapsing. 5. Control valve stroke improperly set, causing too much pressure drop.	Take the following actions: 1. Repair or replace gate valve. 2. Repair valve by replacing spring. 3. Overhaul check valve; restore proper clearance to pin and flapper bore. 4. Replace damaged pipe. 5. Adjust control valve stroke as necessary.
Suction and/or discharge vessel levels are not correct, a problem mostly seen in lower-speed pumps.	Calibrate level controllers as necessary.
Motor running backward or impeller of double suction design is mounted backward. Discharge pressure developed in both cases is about one-half design value.	Check for proper rotation and mounting of impeller. Reverse motor leads if necessary.
Entrained gas from the process lowering NPSH available.	Reduce entrained gas in liquid by process changes as needed.
Polymer or scale buildup in discharge nozzle areas.	Shut down pump and remove scale or deposits.
Mechanical seal in suction system under vacuum is leaking air into system, causing pump curve to drop.	Change percentage balance of seal faces or increase spring tension.
The pump may have formed a vortex at high flow rates or low liquid level. Does the vessel have a vortex breaker? Does the incoming flow cause the surface to swirl or be agitated?	Reduce flow to design rates. Raise liquid level in suction vessel. Install vortex breaker in suction vessel.
Variable-speed motor running too slowly.	Adjust motor speed as needed.
Bypassing is occurring between volute channels in a double-volute pump casing due to a casting defect or extreme erosion.	Overhaul pump; repair eroded area.
The positions of impellers are not centered with diffuser vanes. Several impellers will cause vibration and lower head output.	Overhaul pump; reposition individual impellers as needed. Reposition whole rotor by changing thrust collar locator spacer.
When the suction system is under vacuum, the spare pump has difficulty getting into system.	Install a positive-pressure steam (from running pump) to fill the suction line from the block valve through the check valve.
Certain pump designs use an internal bypass orifice port to alter head-flow curve. High liquid velocities often erode the orifice, causing the pump to go farther out on the pump curve. The system head curve increase corrects the flow back up the curve.	Overhaul pump, restore orifice to correct size.
Replacement impeller is not correct casting pattern; therefore NPSH required is different.	Overhaul pump, replace impeller with correct pattern.
Volute and cutwater area of casing is severely eroded.	Overhaul pump; replace casing or repair by welding. Stress-relieve after welding as needed.

TABLE 10-11 Pump Problems (*Concluded*)

Possible causes	Corrective action
Overload problems	
Polymer buildup between wear surfaces (rings or vanes).	Remove buildup to restore clearances.
Excessive wear ring (closed impeller) or cover-case clearance (open impeller).	Replace wear rings or adjust axial clearance of open impeller. In severe cases, cover or case must be replaced.
Pump circulating excessive liquid back to suction through a breakdown bushing or a diffuser gasket area.	Overhaul pump, replacing parts as needed.
Minimum-flow loop left open at normal rates, or bypass around control valve is open.	Close minimum-flow loop or control valve bypass valve.
Discharge piping leaking under liquid level in sump-type design.	Inspect piping for leakage. Replace as needed.
Electrical switch gear problems cause one phase to have low amperage.	Check out switch gear and repair as necessary.
Specific gravity is higher than design specification.	Change process to adjust specific gravity to design value, or throttle pump to reduce horsepower requirements. This will not correct problem with some vertical turbine pumps that have a flat horsepower-required curve.
Pump motor not sized for end of curve operation.	Replace motor with one of larger size, or reduce flow rate.
Open impeller has slight rub on casing. Most often occurs in operations from 250 to 400°F due to piping strain and differential growth in the pump.	Increase clearance of impeller to casing.
A replacement impeller was not trimmed to the correct diameter.	Remove impeller from pump and turn to correct diameter.

In U.S. customary units,

$$H_{ad} = \frac{k}{k-1} RT_1 \left[\left(\frac{p_2}{p_1} \right)^{(k-1)/k} - 1 \right] \qquad (10\text{-}64b)$$

where H_{ad} = adiabatic head, ft·lbf/lbm; R = gas constant, (ft·lbf)/(lbm·°R) = 1545/molecular weight; T_1 = inlet gas temperature, °R; p_1 = absolute inlet pressure, lbf/in²; and p_2 = absolute discharge pressure, lbf/in².

The **work** expended on the gas during compression is equal to the product of the adiabatic head and the mass flow of gas handled. Therefore, the adiabatic power is as follows:

In SI units,

$$kW_{ad} = \frac{WH_{ad}}{10^3} = \frac{k \times WRT_1}{k-1} \left[\left(\frac{p_2}{p_1} \right)^{(k-1)/k} - 1 \right] \qquad (10\text{-}65a)$$

or

$$kW_{ad} = 2.78 \times 10^{-4} \frac{k}{k-1} Q_1 p_1 \left[\left(\frac{p_2}{p_1} \right)^{(k-1)/k} - 1 \right] \qquad (10\text{-}66a)$$

where kW_{ad} = power, kW; W = mass flow, kg/s × 9.806 N/kg; and Q_1 = volume rate of gas flow, m³/h, at compressor inlet conditions.

In U.S. customary units,

$$hp_{ad} = \frac{WH_{ad}}{550} = \frac{k}{k-1} \frac{WRT_1}{550} \left[\left(\frac{p_2}{p_1} \right)^{(k-1)/k} - 1 \right] \qquad (10\text{-}65b)$$

or

$$hp_{ad} = 4.36 \times 10^{-3} \frac{k}{k-1} Q_1 p_1 \left[\left(\frac{p_2}{p_1} \right)^{(k-1)/k} - 1 \right] \qquad (10\text{-}66b)$$

where hp_{ad} = power, hp; W = mass flow, lb/s; and Q_1 = volume rate of gas flow, ft³/min.

Adiabatic discharge temperature is

$$T_2 = T_1 (p_2/p_1)^{(k-1)/k} \qquad (10\text{-}67)$$

The work in a compressor under ideal conditions as previously shown occurs at constant entropy. The actual process is a polytropic process as shown in Fig. 10-65 and given by the equation of state Pv^n = constant.

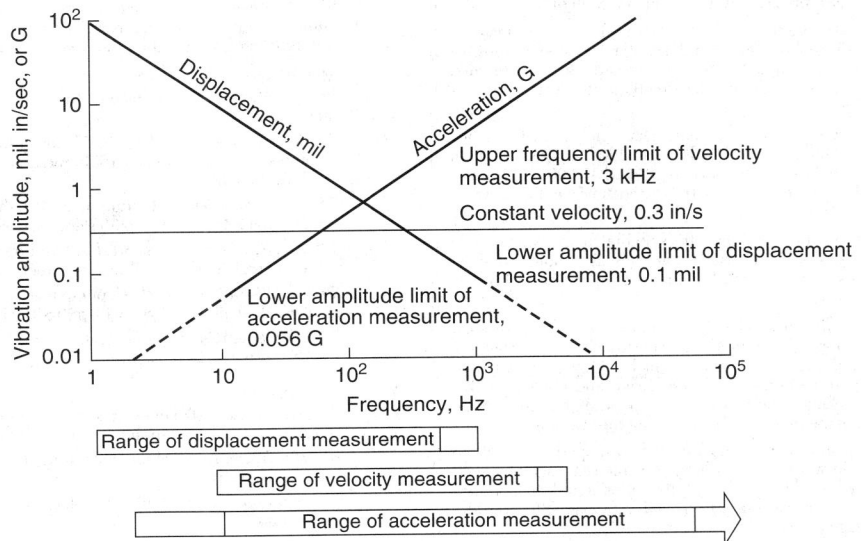

FIG. 10-64 Limitations on machinery vibrations analysis systems and transducers.

TABLE 10-12 API and ASME Codes

Specification	Description
1.0	**Scope:** This specification covers horizontal, end suction, vertically split, single-stage centrifugal pumps with top centerline discharge and "back pullout" feature.
2.0	**Service Conditions:** Pump shall be designed to operate satisfactorily with a reasonable service life when operated either intermittently or continuously in typical process applications.
3.0	**Operating Conditions:**

3.0 Operating Conditions:

Capacity _____ U.S. gallons per minute _____
Head (_____ ft total head) (_____ psig). Speed _____ rpm
Suction Pressure (_____ ft head) (positive) (lift) (_____ psig)
Liquid to be handled _____
Specific gravity _____ Viscosity (_____)
Temperature of liquid at inlet _____°F
Solids content _____% _____ Max. size

4.0 Pump Construction:

4.1 Casing. Casing shall be vertically split with self-venting top centerline discharge, with an integral foot located directly under the casing for added support. All casings shall be of the "back pullout" design with suction and discharge nozzles cast integrally. Casings shall be provided with bosses in suction and discharge nozzles, and in bottom of casing for gauge taps and drain tap. (Threaded taps with plugs shall be provided for these features.)

4.2 Casing Connections. Connections shall be A.N.S.I. flat-faced flanges. [Cast iron (125) (250) psig rated] [Duron metal, steel, alloy steel (150) (300) psig rated]

4.3 Casing Joint Gasket. A confined-type nonasbestos gasket suitable for corrosive service shall be provided at the casing joint.

4.4 Impeller. Fully-open impeller with front edge having contoured vanes curving into the suction for minimum NPSH requirements and maximum efficiency shall be provided. A hex head shall be cast in the eye of the impeller to facilitate removal, and eliminate need for special impeller removing tool. All impellers shall have radial "pump-out" vanes on the back side to reduce stuffing box pressure and aid in eliminating collection of solids at stuffing box throat. Impellers shall be balanced within A.N.S.I. guidelines to ISO tolerances.

  4.4.1 Impeller Clearance Adjustment. All pumps shall have provisions for adjustment of axial clearance between the leading edge of the impeller and casing. This adjustment shall be made by a precision microdial adjustment at the outboard bearing housing, which moves the impeller forward toward the suction wall of the casing.

4.5 Shafts. Shafts shall be suitable for hook-type sleeve. Shaft material shall be (SAE 1045 steel on Duron and 316 stainless steel pumps) or (AISI 316 stainless steel on CD-4MCu pumps and #20 stainless steel pumps). Shaft deflection shall not exceed .005 at the vertical centerline of the impeller.

4.6 Shaft Sleeve. Renewable hook-type shaft sleeve that extends through the stuffing box and gland shall be provided. Shaft sleeve shall be (316 stainless steel), (#20 stainless steel) or (XH-800 Ni-chrome-boron coated 316 stainless steel with coated surface hardness of approximately 800 Brinnell).

4.7 Stuffing Box. Stuffing box shall be suitable for packing, single (inside or outside) or double-inside mechanical seal without modifications. Stuffing box shall be accurately centered by machined rabbit fits on case and frame adapter.

  4.7.1 Packed Stuffing Box. The standard packed stuffing box shall consist of five rings of graphited nonasbestos packing; a stainless steel packing base ring in the bottom of the box to prevent extrusion of the packing past the throat; a teflon seal cage, and a two-piece 316 stainless steel packing gland to insure even pressure on the packing. Ample space shall be provided for repacking the stuffing box.

    4.7.1.1 Lubrication-Packed Stuffing Box. A tapped hole shall be provided in the stuffing box directly over the seal cage for lubrication and cooling of the packing. Lubrication liquid shall be supplied (from an external source) (through a by-pass line from the pump discharge nozzle).

  4.7.2 Stuffing Box with Mechanical Seal. Mechanical seal shall be of the (single inside) (single outside) (double inside) (cartridge) type and (balanced) (unbalanced).

  Stuffing box is to be (standard) (oversize) (oversize tapered).

Suitable space shall be provided in the standard and oversized stuffing box for supplying a (throttle bushing) (dilution control bushing) with single seals. Throttle bushings and dilution control bushings shall be made of (glass-filled teflon) (a suitable metal material).

  4.7.2.1 Lubrication—Stuffing Box with Mechanical Seals. Suitable tapped connections shall be provided to effectively lubricate, cool, flush, quench, etc., as required by the application or recommendations of the mechanical seal manufacturer.

4.8 Bearing Frame and Bearings:

4.8.1 Bearing Frame. Frames shall be equipped with axial radiating fins extending the length of the frame to aid in heat dissipation. Frame shall be provided with ductile iron outboard bearing housing. Both ends of the frame shall be provided with lip-type oil seals and labyrinth-type deflectors of metallic reinforced synthetic rubber to prevent the entrance of contaminants.

4.8.2 Bearings. Pump bearings shall be heavy-duty, antifriction ball-type on both ends. The single row inboard bearing, nearest the impeller, shall be free to float within the frame and shall carry only radial load. The double row outboard bearing (F4-G1 and F4-I1) or duplex angular contact bearing (F4-H1), coupling end, shall be locked in place to carry radial and axial thrust loads. Bearings shall be designed for a minimum life of 20,000 hours in any normal pump operating range.

4.9 *Bearing Lubrication.* Ball bearings shall be oil-mist—lubricated by means of a slinger. The oil slinger shall be mounted on the shaft between the bearings to provide equal lubrication to both bearings. Bulls-eye oil-sight glasses shall be provided on both sides of the frame to provide a positive means of checking the proper oil level from either side of the pump. A tapped and plugged hole shall also be provided in both sides of the frame to mount bottle-type constant-level oilers where desired. A tapped and plugged hole shall be provided on both sides for optional straight-through oil cooling device.

5.0 Baseplate and Coupling:

5.1 Baseplate. Baseplates shall be rigid and suitable for mounting pump and motor. Baseplates shall be of channel steel construction.

5.2 Coupling. Coupling shall be flexible-spacer type. Coupling shall have at least three-and-one-half-inch spacer length for ease of rotating element removal. Both coupling hubs shall be provided with flats 180° apart to facilitate removal of impeller. Coupling shall not require lubrication.°

6.0 Mechanical Modifications Required for High Temperature:

6.1 Modifications Required, Temperature Range 250–350°F. Pumps for operation in this range shall be provided with a water-jacketed stuffing box.

6.2 Modifications Required, Temperature Range 351–550°F (Maximum). Pumps for operation in this range shall be provided with a water-jacketed stuffing box and a water-cooled bearing frame.

7.0 Materials: Pump materials shall be selected to suit the particular service requirements.

7.1 Cast Iron—316 SS Fitted. 15″ only; pump shall have cast iron casing and stuffing box cover. 316 SS metal impeller; shaft shall be 1045 steel with 316 SS sleeve.

7.2 All Duron Metal. All pump materials shall be Duron metal. Shaft shall be 1045 steel, with 316 SS sleeve. 316 SS metal impeller optional.

7.3 All AISI 316 Stainless Steel. All pump materials shall be AISI 316 stainless steel. Shaft should be 1045 steel, with 316 SS sleeve.

7.4 All #20 Stainless Steel. All pump materials shall be #20 SS stainless steel. Shaft shall be 316 SS, with #20 SS sleeve.

7.5 All CD-4MCu. All pump materials shall be CD-4MCu. Shaft shall be 316 SS, with #20 SS sleeve.

8.0 Miscellaneous:

8.1 Nameplates. All nameplates and other data plates shall be stainless steel, suitably secured to the pump.

8.2 Hardware. All machine bolts, stud nuts, and capscrews shall be of the hex-head type.

8.3 Rotation. Pump shall have clockwise rotation viewed from its driven end.

8.4 Parts Numbering. Parts shall be completely identified with a numerical system (no alphabetical letters) to facilitate parts inventory control and stocking. Each part shall be properly identified by a separate number, and those parts that are identical shall have the same number to effect minimum spare parts inventory.

°Omit if not applicable.

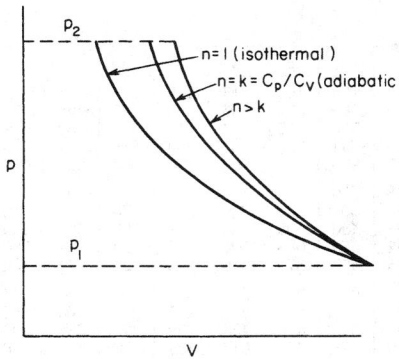

FIG. 10-65 Polytropic compression curves.

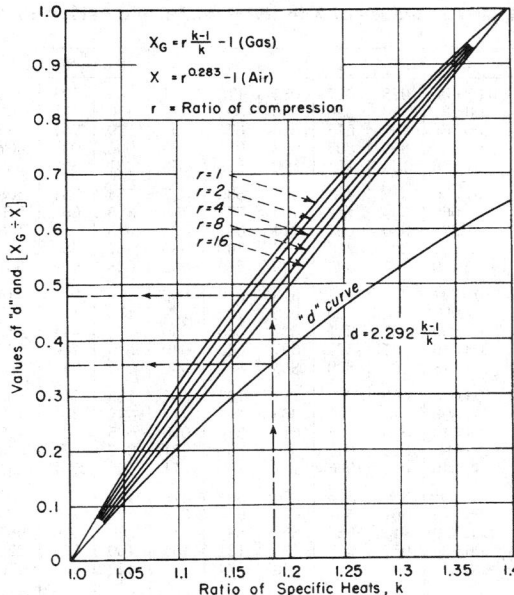

FIG. 10-66 Factors for use in adiabatic formula. Values of X to be used in finding X_G may be obtained from Table 10-3. (*By permission of Compressed Air Data.*)

Adiabatic efficiency is given by the following relationship

$$\eta_{ad} = \frac{\text{Ideal work}}{\text{Actual work}} \qquad (10\text{-}68)$$

In terms of the change total temperatures the relationship can be written as:

$$\eta_{ad} = \frac{T_2 - T_1}{T_{2a} - T_1} \qquad (10\text{-}69)$$

where T_{2a} is the total actual discharge temperature of the gas. The adiabatic efficiency can be represented in terms of total pressure change:

$$\eta_{ad} = \frac{\left(\dfrac{P_2}{P_1}\right)^{(k-1)/k} - 1}{\left(\dfrac{P_2}{P_1}\right)^{(n-1)/n} - 1} \qquad (10\text{-}70)$$

Polytropic head can be expressed by the following relationship.

$$H_{ad} = \frac{n}{n-1} ZRT_1 \left\{ \left[\left(\frac{P_2}{P_1} \right)^{(n-1)/n} - 1 \right] \right\} \qquad (10\text{-}71)$$

Likewise for polytropic efficiency, which is often considered as the small stage efficiency, or the hydraulic efficiency:

$$\eta_{pc} = \frac{(k-1)/k}{(n-1)/n} \qquad (10\text{-}72)$$

Polytropic efficiency is the limited value of the isentropic efficiency as the pressure ratio approaches 1.0, and the value of the polytropic efficiency is higher than the corresponding adiabatic efficiency as seen in Fig. 10-66.

A characteristic of polytropic efficiency is that the polytropic efficiency of a multistage unit is equal to the stage efficiency if each stage has the same efficiency.

Air and a number of other gases have a value of $k = 1.39$ to 1.41. To simplify calculations for these gases, tables have been made of the bracketed expression $[(p_2/p_1)^{(k-1)/k} - 1]$ in these equations for a value of $k = 1.395$. These are known as X factors, and they are given in Table 10-13. By using X factors, the adiabatic formulas for $k = 1.395$ read as follows:

Adiabatic temperature, pressure, and volume relations:

$$V_1/V_2 = p_2/[(X+1)p_1] \qquad (10\text{-}73)$$

$$T_2/T_1 = X + 1 \qquad (10\text{-}74)$$

$$T_2 - T_1 = T_1 X = T_2[X/(X+1)] \qquad (10\text{-}75)$$

Adiabatic power:
In SI units,

$$kW_{ad} = 9.81 \times 10^{-4} Q_1 p_1 X \qquad (10\text{-}76a)$$

In U.S. customary units,

$$hp_{ad} = 0.0154 Q_1 p_1 X \qquad (10\text{-}76b)$$

Adiabatic discharge temperature:

$$T_2 = T_1(X + 1) \qquad (10\text{-}77)$$

To find the X factor X_G for a gas of any k value refer to Fig. 6-34. This figure gives values of X_G/X for gases having specific-heat ratios between 1.0 and 1.4. The factor X_G is then the product of X_G/X from Fig. 10-66 and the value X from Table 10-13 for desired compression ratio.

Adiabatic power for gases other than air:
In SI units,

$$kW_{ad} = 6.37 \times 10^{-4} Q_1 p_1 X/d \qquad (10\text{-}78a)$$

In U.S. customary units,

$$hp_{ad} = 1 \times 10^{-2} Q_1 p_1 X/d \qquad (10\text{-}78b)$$

where $d = 2.922 (k-1)/k$.

If the compression cycle approaches the isothermal condition, $pV =$ constant, as is the case when several stages with intercoolers are used, a simple approximation of the power is obtained from the following formula:
In SI units,

$$kW = 2.78 \times 10^{-4} Q_1 p_1 \ln p_2/p_1 \qquad (10\text{-}79a)$$

In U.S. customary units,

$$hp = 4.4 \times 10^{-3} Q_1 p_1 \ln p_2/p_1 \qquad (10\text{-}79b)$$

For multistage compressors of N_s number of stages with adiabatic compression in each stage, equal division of work between stages, and intercooling to the intake temperature, the following formulas are helpful:
In SI units,

$$kW_{ad} = \frac{6.37 \times 10^{-4} N_s Q_1 p_1}{d} \left(\sqrt[N_s]{X_G + 1} - 1 \right) \qquad (10\text{-}80a)$$

TABLE 10-13   Values of X for Normal Air and Perfect Diatomic Gases

$$X = r^{0.283} - 1$$

r	0	1	2	3	4	5	6	7	8	9	r	0	1	2	3	4	5	6	7	8	9
1.00	0.00 000	028	057	085	113	141	169	198	226	254	1.75	0.17 160	179	198	217	236	255	274	292	311	330
1.01	.00 282	310	338	366	394	422	450	478	506	534	1.76	.17 349	368	387	406	425	443	462	481	500	519
1.02	.00 562	590	618	646	673	701	729	757	785	812	1.77	.17 538	556	575	594	613	631	650	669	688	706
1.03	.00 840	868	895	923	951	978	006	034	061	089	1.78	.17 725	744	762	781	800	818	837	856	874	893
1.04	.01 116	144	171	199	226	253	281	308	336	363	1.79	.17 912	930	949	968	986	005	023	042	061	079
1.05	.01 390	418	445	472	500	527	554	581	608	636	1.80	.18 098	116	135	153	172	191	209	228	246	265
1.06	.01 663	690	717	744	771	798	825	852	879	906	1.81	.18 283	302	320	339	357	376	394	412	431	449
1.07	.01 933	960	987	014	041	068	095	122	148	175	1.82	.18 468	486	505	523	541	560	578	596	615	633
1.08	.02 202	229	255	282	309	336	362	389	416	442	1.83	.18 652	670	688	707	725	743	762	780	798	816
1.09	.02 469	495	522	549	575	602	628	655	681	708	1.84	.18 835	853	871	890	908	926	944	962	981	999
1.10	.02 734	760	787	813	840	866	892	919	945	971	1.85	.19 017	035	054	072	090	108	126	144	163	181
1.11	.02 997	024	050	076	102	129	155	181	207	233	1.86	.19 199	217	235	253	271	289	308	326	344	362
1.12	.03 259	285	311	337	363	389	415	441	467	493	1.87	.19 380	398	416	434	452	470	488	506	524	542
1.13	.03 519	545	571	597	623	649	675	700	726	752	1.88	.19 560	578	596	614	632	650	668	686	704	722
1.14	.03 778	804	829	855	881	906	932	958	983	009	1.89	.19 740	758	776	794	811	829	847	865	883	901
1.15	.04 035	060	086	111	137	162	188	213	239	264	1.90	.19 919	937	954	972	990	008	026	044	061	079
1.16	.04 290	315	341	366	391	417	442	467	493	518	1.91	.20 097	115	133	150	168	186	204	221	239	257
1.17	.04 543	569	594	619	644	670	695	720	745	770	1.92	.20 275	292	310	328	345	363	381	399	416	434
1.18	.04 796	821	846	871	896	921	946	971	996	021	1.93	.20 452	469	487	504	522	540	557	575	593	610
1.19	.05 046	071	096	121	146	171	196	221	245	270	1.94	.20 628	645	663	681	698	716	733	751	768	786
1.20	.05 295	320	345	370	394	419	444	469	493	518	1.95	.20 804	821	839	856	874	891	909	926	944	961
1.21	.05 543	567	592	617	641	666	691	715	740	764	1.96	.20 979	996	013	031	048	066	083	101	118	135
1.22	.05 789	813	838	862	887	911	936	960	985	009	1.97	.21 153	170	188	205	222	240	257	275	292	309
1.23	.06 034	058	082	107	131	155	180	204	228	253	1.98	.21 327	344	361	379	396	413	431	448	465	482
1.24	.06 277	301	325	350	374	398	422	446	470	495	1.99	.21 500	517	534	552	569	586	603	620	638	655
1.25	.06 519	543	567	591	615	639	663	687	711	735	2.00	.21 672	689	707	724	741	758	775	792	810	827
1.26	.06 759	783	807	831	855	879	903	927	951	974	2.01	.21 844	861	878	895	913	930	947	964	981	998
1.27	.06 998	022	046	070	094	117	141	165	189	212	2.02	.22 015	032	049	066	084	101	118	135	152	169
1.28	.07 236	260	283	307	331	354	378	402	425	449	2.03	.22 186	203	220	237	254	271	288	305	322	339
1.29	.07 472	496	520	543	567	590	614	637	661	684	2.04	.22 356	373	390	407	424	441	458	474	491	508
1.30	.07 708	731	754	778	801	825	848	871	895	918	2.05	.22 525	542	559	576	593	610	627	644	660	677
1.31	.07 941	965	988	011	035	058	081	104	128	151	2.06	.22 694	711	728	745	762	778	795	812	829	846
1.32	.08 174	197	220	243	267	290	313	336	359	382	2.07	.22 863	879	896	913	930	946	963	980	997	013
1.33	.08 405	428	451	474	497	520	543	566	589	612	2.08	.23 030	047	064	080	097	114	130	147	164	181
1.34	.08 635	658	681	704	727	750	773	795	818	841	2.09	.23 197	214	231	247	264	281	297	314	331	347
1.35	.08 864	887	910	932	955	978	001	023	046	069	2.10	.23 364	380	397	414	430	447	463	480	497	513
1.36	.09 092	114	137	160	182	205	228	250	273	295	2.11	.23 530	546	563	579	596	613	629	646	662	679
1.37	.09 318	341	363	386	408	431	453	476	498	521	2.12	.23 695	712	728	745	761	778	794	811	827	844
1.38	.09 543	566	588	611	633	655	678	700	723	745	2.13	.23 860	877	893	909	926	942	959	975	992	008
1.39	.09 767	790	812	834	857	879	901	923	946	968	2.14	.24 024	041	057	074	090	106	123	139	155	172
1.40	.09 990	012	035	057	079	101	123	145	168	190	2.15	.24 188	204	221	237	253	270	286	302	319	335
1.41	.10 212	234	256	278	300	322	344	366	389	411	2.16	.24 351	368	384	400	416	433	449	465	481	498
1.42	.10 433	455	477	499	521	542	564	586	608	630	2.17	.24 514	530	546	563	579	595	611	627	644	660
1.43	.10 652	674	696	718	740	761	783	805	827	849	2.18	.24 676	692	708	724	741	757	773	789	805	821
1.44	.10 871	892	914	936	958	979	001	023	045	066	2.19	.24 838	854	870	886	902	918	934	950	966	983
1.45	.11 088	110	131	153	175	196	218	239	261	283	2.20	.24 999	015	031	047	063	079	095	111	127	143
1.46	.11 304	326	347	369	390	412	433	455	476	498	2.21	.25 159	175	191	207	223	239	255	271	287	303
1.47	.11 520	541	562	584	605	627	648	669	691	712	2.22	.25 319	335	351	367	383	399	415	431	447	463
1.48	.11 734	755	776	798	819	840	862	883	904	925	2.23	.25 479	495	511	526	542	558	574	590	606	622
1.49	.11 947	968	989	010	032	053	074	095	116	138	2.24	.25 638	654	669	685	701	717	733	749	765	780
1.50	.12 159	180	201	222	243	264	286	307	328	349	2.25	.25 796	812	828	844	859	875	891	907	923	938
1.51	.12 370	391	412	433	454	475	496	517	538	559	2.26	.25 954	970	986	001	017	033	049	064	080	096
1.52	.12 580	601	622	643	664	685	706	726	747	768	2.27	.26 112	127	143	159	175	190	206	222	237	253
1.53	.12 789	810	831	852	872	893	914	935	956	977	2.28	.26 269	284	300	316	331	347	363	378	394	409
1.54	.12 997	018	039	060	080	101	122	142	163	184	2.29	.26 425	441	456	472	488	503	519	534	550	566
1.55	.13 205	225	246	266	287	308	328	349	370	390	2.30	.26 581	597	612	628	643	659	675	690	706	721
1.56	.13 411	431	452	472	493	513	534	554	575	595	2.31	.26 737	752	768	783	799	814	830	845	861	876
1.57	.13 616	636	657	677	698	718	739	759	780	800	2.32	.26 892	907	923	938	954	969	984	000	015	031
1.58	.13 820	841	861	881	902	922	942	963	983	003	2.33	.27 046	062	077	092	108	123	139	154	169	185
1.59	.14 024	044	064	085	105	125	145	165	186	206	2.34	.27 200	216	231	246	262	277	292	308	323	338
1.60	.14 226	246	267	287	307	327	347	367	387	408	2.35	.27 354	369	384	400	415	430	446	461	476	492
1.61	.14 428	448	468	488	508	528	548	568	588	608	2.36	.27 507	522	538	553	568	583	599	614	629	644
1.62	.14 628	648	668	688	708	728	748	768	788	808	2.37	.27 660	675	690	705	721	736	751	766	781	797
1.63	.14 828	848	868	888	908	928	948	968	988	007	2.38	.27 812	827	842	857	873	888	903	918	933	948
1.64	.15 027	047	067	087	107	126	146	166	186	206	2.39	.27 964	979	994	009	024	039	054	070	085	100
1.65	.15 225	245	265	284	304	324	344	363	383	403	2.40	.28 115	130	145	160	175	190	205	220	236	251
1.66	.15 423	442	462	481	501	521	540	560	580	599	2.41	.28 266	281	296	311	326	341	356	371	386	401
1.67	.15 619	638	658	678	697	717	736	756	775	795	2.42	.28 416	431	446	461	476	491	506	521	536	551
1.68	.15 814	834	853	873	892	912	931	951	970	990	2.43	.28 566	581	596	611	626	641	656	671	686	701
1.69	.16 009	028	048	067	087	106	125	145	164	184	2.44	.28 716	730	745	760	775	790	805	820	835	850
1.70	.16 203	222	242	261	280	299	319	338	357	377	2.45	.28 865	879	894	909	924	939	954	969	984	998
1.71	.16 396	415	434	454	473	492	511	531	550	569	2.46	.29 013	028	043	058	073	087	102	117	132	147
1.72	.16 588	607	626	646	665	684	703	722	741	760	2.47	.29 162	176	191	206	221	235	250	265	280	295
1.73	.16 780	799	818	837	856	875	894	913	932	951	2.48	.29 309	324	339	353	368	383	398	412	427	442
1.74	.16 970	989	008	027	046	065	084	103	122	141	2.49	.29 457	471	486	501	515	530	545	559	574	589

TABLE 10-13 Values of X for Normal Air and Perfect Diatomic Gases (*Continued*)

r	0	1	2	3	4	5	6	7	8	9	r	0	1	2	3	4	5	6	7	8	9
2.50	0.29 604	618	633	647	662	677	691	706	721	735	2.75	0.33 147	161	174	188	202	215	229	243	256	270
2.51	.29 750	765	779	794	808	823	838	852	867	881	2.76	.33 284	297	311	325	338	352	366	379	393	407
2.52	.29 896	911	925	940	954	969	984	998	013	027	2.77	.33 420	434	448	461	475	488	502	516	529	543
2.53	.30 042	056	071	085	100	114	129	144	158	173	2.78	.33 556	570	584	597	611	624	638	651	665	679
2.54	.30 187	202	216	231	245	260	274	289	303	318	2.79	.33 692	706	719	733	746	760	773	787	801	814
2.55	.30 332	346	361	375	390	404	419	433	448	462	2.80	.33 828	841	855	868	882	895	909	922	936	949
2.56	.30 476	491	505	520	534	548	563	577	592	606	2.81	.33 963	976	990	003	017	030	044	057	070	084
2.57	.30 620	635	649	663	678	692	707	721	735	750	2.82	.34 097	111	124	138	151	165	178	191	205	218
2.58	.30 764	778	793	807	821	836	850	864	879	893	2.83	.34 232	245	259	272	285	299	312	326	339	352
2.59	.30 907	921	936	950	964	979	993	007	021	036	2.84	.34 366	379	393	406	419	433	446	459	473	486
2.60	.31 050	064	079	093	107	121	136	150	164	178	2.85	.34 500	513	526	540	553	566	580	593	606	620
2.61	.31 193	207	221	235	249	264	278	292	306	320	2.86	.34 633	646	660	673	686	700	713	726	739	753
2.62	.31 335	349	363	377	391	405	420	434	448	462	2.87	.34 766	779	793	806	819	832	846	859	872	886
2.63	.31 476	490	505	519	533	547	561	575	589	603	2.88	.34 899	912	925	939	952	965	978	991	005	018
2.64	.31 618	632	646	660	674	688	702	716	730	744	2.89	.35 031	044	058	071	084	097	110	124	137	150
2.65	.31 759	773	787	801	815	829	843	857	871	885	2.90	.35 163	176	190	203	216	229	242	255	269	282
2.66	.31 899	913	927	941	955	969	983	997	011	025	2.91	.35 295	308	321	334	347	361	374	387	400	413
2.67	.32 039	053	067	081	095	109	123	137	151	165	2.92	.35 426	439	452	466	479	492	505	518	531	544
2.68	.32 179	193	207	221	235	249	262	276	290	304	2.93	.35 557	570	584	597	610	623	636	649	662	675
2.69	.32 318	332	346	360	374	388	402	416	429	443	2.94	.35 688	701	714	727	740	753	767	780	793	806
2.70	.32 457	471	485	499	513	527	540	554	568	582	2.95	.35 819	832	845	858	871	884	897	910	923	936
2.71	.32 596	610	624	637	651	665	679	693	707	720	2.96	.35 949	962	975	988	001	014	027	040	053	066
2.72	.32 734	748	762	776	789	803	817	831	845	858	2.97	.36 079	092	105	118	131	144	157	169	182	195
2.73	.32 872	886	900	913	927	941	955	968	982	996	2.98	.36 208	221	234	247	260	273	286	299	312	324
2.74	.33 010	023	037	051	065	078	092	106	119	133	2.99	.36 337	350	363	376	389	402	415	428	440	453

r	0	1	2	3	4	5	6	7	8	9
3.0	0.3647	0.3659	0.3672	0.3685	0.3698	0.3711	0.3723	0.3736	0.3749	0.3761
3.1	.3774	.3786	.3799	.3811	.3824	.3836	.3849	.3861	.3874	.3886
3.2	.3898	.3911	.3923	.3935	.3947	.3959	.3971	.3984	.3996	.4008
3.3	.4020	.4032	.4044	.4056	.4068	.4080	.4091	.4103	.4115	.4127
3.4	.4139	.4150	.4162	.4174	.4186	.4197	.4209	.4220	.4232	.4244
3.5	.4255	.4267	.4278	.4290	.4301	.4313	.4324	.4335	.4347	.4358
3.6	.4369	.4380	.4392	.4403	.4414	.4425	.4437	.4448	.4459	.4470
3.7	.4481	.4492	.4503	.4514	.4525	.4536	.4547	.4558	.4569	.4580
3.8	.4591	.4602	.4612	.4623	.4634	.4645	.4656	.4666	.4677	.4688
3.9	.4698	.4709	.4720	.4730	.4741	.4752	.4762	.4773	.4783	.4794
4.0	.4804	.4815	.4825	.4835	.4846	.4856	.4867	.4877	.4887	.4898
4.1	.4908	.4918	.4928	.4939	.4949	.4959	.4970	.4980	.4990	.5000
4.2	.5010	.5020	.5030	.5040	.5050	.5060	.5070	.5080	.5090	.5100
4.3	.5110	.5120	.5130	.5140	.5150	.5160	.5170	.5179	.5189	.5199
4.4	.5209	.5219	.5228	.5238	.5248	.5258	.5267	.5277	.5287	.5296
4.5	.5306	.5316	.5325	.5335	.5344	.5354	.5363	.5373	.5382	.5392
4.6	.5401	.5411	.5420	.5430	.5439	.5449	.5458	.5467	.5477	.5486
4.7	.5495	.5505	.5514	.5523	.5533	.5542	.5551	.5560	.5570	.5579
4.8	.5588	.5597	.5606	.5616	.5625	.5634	.5643	.5652	.5661	.5670
4.9	.5679	.5688	.5697	.5706	.5715	.5724	.5733	.5742	.5751	.5760
5.0	.5769	.5778	.5787	.5796	.5805	.5814	.5822	.5831	.5840	.5849
5.1	.5858	.5867	.5875	.5884	.5893	.5902	.5910	.5919	.5928	.5936
5.2	.5945	.5954	.5962	.5971	.5980	.5988	.5997	.6006	.6014	.6023
5.3	.6031	.6040	.6048	.6057	.6065	.6074	.6082	.6091	.6099	.6108
5.4	.6116	.6125	.6133	.6142	.6150	.6159	.6167	.6175	.6184	.6192
5.5	.6200	.6209	.6217	.6225	.6234	.6242	.6250	.6258	.6267	.6275
5.6	.6283	.6291	.6300	.6308	.6316	.6324	.6332	.6340	.6349	.6357
5.7	.6365	.6373	.6381	.6389	.6397	.6405	.6413	.6421	.6430	.6438
5.8	.6446	.6454	.6462	.6470	.6478	.6486	.6494	.6502	.6509	.6517
5.9	.6525	.6533	.6541	.6549	.6557	.6565	.6573	.6581	.6588	.6596
6.0	.6604	.6612	.6620	.6628	.6635	.6643	.6651	.6659	.6666	.6674
6.1	.6682	.6690	.6697	.6705	.6713	.6721	.6729	.6736	.6744	.6752
6.2	.6759	.6767	.6774	.6782	.6789	.6797	.6805	.6812	.6820	.6827
6.3	.6835	.6843	.6850	.6858	.6865	.6873	.6888	.6895	.6903
6.4	.6910	.6918	.6925	.6933	.6940	.6948	.6955	.6963	.6970	.6978
6.5	.6985	.6992	.7000	.7007	.7014	.7021	.7028	.7036	.7043	.7050
6.6	.7058	.7065	.7073	.7080	.7087	.7095	.7102	.7110	.7117	.7124
6.7	.7131	.7138	.7145	.7153	.7160	.7167	.7174	.7181	.7189	.7196
6.8	.7203	.7210	.7217	.7224	.7232	.7239	.7246	.7253	.7260	.7267
6.9	.7274	.7281	.7288	.7295	.7302	.7309	.7316	.7323	.7330	.7338
7.0	.7345	.7352	.7359	.7366	.7373	.7380	.7386	.7393	.7400	.7407
7.1	.7414	.7421	.7428	.7435	.7442	.7449	.7456	.7463	.7470	.7477
7.2	.7483	.7490	.7497	.7504	.7511	.7518	.7524	.7531	.7538	.7545
7.3	.7552	.7559	.7565	.7572	.7579	.7586	.7592	.7599	.7606	.7613
7.4	.7620	.7626	.7633	.7640	.7646	.7653	.7660	.7666	.7673	.7680

TABLE 10-13 Values of X for Normal Air and Perfect Diatomic Gases (Concluded)

r	0	1	2	3	4	5	6	7	8	9
7.5	0.7687	0.7693	0.7700	0.7706	0.7713	0.7720	0.7726	0.7733	0.7740	0.7746
7.6	.7753	.7760	.7766	.7773	.7779	.7786	.7792	.7799	.7806	.7813
7.7	.7819	.7825	.7832	.7838	.7845	.7851	.7858	.7864	.7871	.7877
7.8	.7884	.7890	.7897	.7903	.7910	.7916	.7923	.7929	.7936	.7942
7.9	.7949	.7955	.7961	.7968	.7974	.7981	.7987	.7993	.8000	.8006
8.0	.8013	.8019	.8025	.8032	.8038	.8044	.8051	.8057	.8063	.8070
8.1	.8076	.8082	.8089	.8095	.8101	.8108	.8114	.8120	.8126	.8133
8.2	.8139	.8145	.8151	.8158	.8164	.8170	.8176	.8183	.8189	.8195
8.3	.8201	.8207	.8214	.8220	.8226	.8232	.8238	.8245	.8251	.8257
8.4	.8263	.8269	.8275	.8281	.8288	.8294	.8300	.8306	.8312	.8318
8.5	.8324	.8330	.8336	.8343	.8349	.8355	.8361	.8367	.8373	.8379
8.6	.8385	.8391	.8397	.8403	.8409	.8415	.8421	.8427	.8433	.8439
8.7	.8445	.8451	.8457	.8463	.8469	.8475	.8481	.8487	.8493	.8499
8.8	.8505	.8511	.8517	.8523	.8529	.8535	.8541	.8547	.8552	.8558
8.9	.8564	.8570	.8576	.8582	.8588	.8594	.8600	.8605	.8611	.8617
9.0	.8623	.8629	.8635	.8641	.8646	.8652	.8658	.8664	.8670	.8676
9.1	.8681	.8687	.8693	.8699	.8705	.8710	.8716	.8722	.8728	.8734
9.2	.8739	.8745	.8751	.8757	.8762	.8768	.8774	.8779	.8785	.8791
9.3	.8797	.8802	.8808	.8814	.8819	.8825	.8831	.8837	.8842	.8848
9.4	.8854	.8859	.8865	.8871	.8876	.8882	.8888	.8893	.8899	.8905
9.5	.8910	.8916	.8921	.8927	.8933	.8938	.8944	.8949	.8955	.8961
9.6	.8966	.8972	.8977	.8983	.8989	.8994	.9000	.9005	.9011	.9016
9.7	.9022	.9028	.9033	.9039	.9044	.9050	.9055	.9061	.9066	.9072
9.8	.9077	.9083	.9088	.9094	.9099	.9105	.9110	.9116	.9121	.9127
9.9	.9132	.9138	.9143	.9149	.9154	.9159	.9165	.9170	.9176	.9181
10.0	.9187	.9192	.9198	.9203	.9208	.9214	.9219	.9225	.9230	.9235
10.1	.9241	.9246	.9252	.9257	.9262	.9268	.9273	.9278	.9284	.9289
10.2	.9295	.9300	.9305	.9311	.9316	.9321	.9327	.9332	.9337	.9343
10.3	.9348	.9353	.9358	.9364	.9369	.9374	.9380	.9385	.9390	.9396
10.4	.9401	.9406	.9411	.9417	.9422	.9427	.9432	.9438	.9443	.9448
10.5	.9453	.9459	.9464	.9469	.9474	.9480	.9485	.9490	.9495	.9500
10.6	.9506	.9511	.9516	.9521	.9526	.9532	.9537	.9542	.9547	.9552
10.7	.9558	.9563	.9568	.9573	.9578	.9583	.9589	.9594	.9599	.9604
10.8	.9609	.9614	.9619	.9625	.9630	.9635	.9640	.9645	.9650	.9655
10.9	.9660	.9665	.9671	.9676	.9681	.9686	.9691	.9696	.9701	.9706
11.0	.9711	.9716	.9721	.9726	.9732	.9737	.9742	.9747	.9752	.9757
11.1	.9762	.9767	.9772	.9777	.9782	.9787	.9792	.9797	.9802	.9807
11.2	.9812	.9817	.9822	.9827	.9832	.9837	.9842	.9847	.9852	.9857
11.3	.9862	.9867	.9872	.9877	.9882	.9887	.9892	.9897	.9902	.9907
11.4	.9912	.9916	.9921	.9926	.9931	.9936	.9941	.9946	.9951	.9956
11.5	.9961	.9966	.9971	.9975	.9980	.9985	.9990	.9995	1.0000	1.0005
11.6	1.0010	1.0015	1.0019	1.0024	1.0029	1.0034	1.0039	1.0044	1.0049	1.0054
11.7	1.0058	1.0063	1.0068	1.0073	1.0078	1.0083	1.0087	1.0092	1.0097	1.0102
11.8	1.0107	1.0112	1.0116	1.0121	1.0126	1.0131	1.0136	1.0140	1.0145	1.0150
11.9	1.0155	1.0160	1.0164	1.0169	1.0174	1.0179	1.0184	1.0188	1.0193	1.0198
12.0	1.0203	1.0207	1.0212	1.0217	1.0222	1.0226	1.0231	1.0236	1.0241	1.0245

Printed by permission of Compressed Air Data.

Taken from Moss and Smith, Engineering Computations for Air and Gases, *Trans. Am. Soc. Mech. Engrs.*, vol. 52, 1930, paper APM-52-8. For nozzles $r = p_1/p_2$. For compressors and exhausters $r = p_2/p_1$.

r	X	r	X	r	X	r	X	r	X	r	X	r	X	r	X	r	X
12.5	1.0428	15.0	1.1520	17.5	1.2479	20.0	1.3345	22.5	1.4136	25.0	1.4867	27.5	1.5546	30.0	1.6183	32.5	1.6783
13.0	1.0666	15.5	1.1720	18.0	1.2659	20.5	1.3509	23.0	1.4287	25.5	1.5006	28.0	1.5678	30.5	1.6306	33.0	1.6899
13.5	1.0887	16.0	1.1916	18.5	1.2835	21.0	1.3669	23.5	1.4435	26.0	1.5144	28.5	1.5794	31.0	1.6434	33.5	1.7014
14.0	1.1103	16.5	1.2108	19.0	1.3008	21.5	1.3828	24.0	1.4581	26.5	1.5280	29.0	1.5933	31.5	1.6547	34.0	1.7127
14.5	1.1314	17.0	1.2295	19.5	1.3189	22.0	1.3983	24.5	1.4725	27.0	1.5414	29.5	1.6059	32.0	1.6666	34.5	1.7240

Values of X from 12.5 to 34.5 calculated by Ingersoll-Rand Co.

In U.S. customary units,

$$\text{hp}_{ad} = \frac{1 \times 10^{-2} N_s Q_1 p_1}{d} (\sqrt[N_y]{X_G + 1} - 1) \qquad (10\text{-}80b)$$

$$T_2 = T_1 \sqrt[N_y]{X_G + 1} \qquad (10\text{-}81)$$

To be able to decide which type of compressor would best fit the job, we should first divide the compressors into three main categories: positive displacement, centrifugal, and axial flow. In general terms, positive displacement compressors are used for high pressure and low flow characteristics; centrifugal compressors are used for medium to high pressure delivery and medium flow; and axial flow compressors are low pressure and high flow.

Compressor Selection To select the most satisfactory compression equipment, engineers must consider a wide variety of types, each of which offers peculiar advantages for particular applications. Among the major factors to be considered are flow rate, head or pressure, temperature limitations, method of sealing, method of lubrication, power consumption, serviceability, and cost.

To be able to decide which compressor best fits the job, the engineer must analyze the flow characteristics of the units. The following dimensionless numbers describe the flow characteristics.

Reynolds number is the ratio of the inertia forces to the viscous forces

$$N_{\text{Re}} = \frac{\rho V D}{\mu} \qquad (10\text{-}82)$$

where ρ is the density of the gas, V is the velocity of the gas, D is the diameter of the impeller, and μ is the viscosity of the gas.

The specific speed compares the adiabatic head and flow rate in geometrically similar machines at various speeds.

$$N_s = \frac{N \sqrt{Q}}{H_{ad}^{3/4}} \qquad (10\text{-}83)$$

where N is the speed of rotation of the compressor, Q is the volume flow rate, and H is the adiabatic head.

The specific diameter compares head and flow rates in geometrically similar machines at various diameters

$$D_S = \frac{D H^{1/4}}{\sqrt{Q}} \qquad (10\text{-}84)$$

The flow coefficient is the capacity of the flow rate of the machine

$$\phi = \frac{Q^1}{ND^3} \qquad (10\text{-}85)$$

The pressure coefficient is the pressure or the pressure rise of the machine

$$\Psi = \frac{H}{N^2 D^2} \qquad (10\text{-}86)$$

In selecting the machines of choice, the use of specific speed and diameter best describe the flow. Figure 10-67 shows the characteristics of the three types of compressors. Other considerations in chemical plant service such as problems with gases which may be corrosive or have abrasive solids in suspension must be dealt with. Gases at elevated temperatures may create a potential explosion hazard, while air at the same temperatures may be handled quite normally; minute amounts of lubricating oil or water may contaminate the process gas and so may not be permissible, and for continuous-process use, a high degree of equipment reliability is required, since frequent shutdowns for inspection or maintenance cannot be tolerated.

FANS AND BLOWERS

Fans are used for low pressures where generally the delivery pressure is less than 3.447 kPa (0.5 lb/in²), and blowers are used for higher pressures. However, they are usually below delivery pressures of 10.32 kPa (1.5 lbf/in²). These units can either be centrifugal or the axial-flow type.

Fans and blowers are used for many types of ventilating work such as air-conditioning systems. In large buildings, blowers are often used due to the high delivery pressures needed to overcome the pressure drop in the ventilation system. Most of these blowers are of the centrifugal type. Blowers are also used to supply draft air to boilers and furnaces. Fans are used to move large volumes of air or gas through ducts, supplying air for drying, conveying material suspended in the gas stream, removing fumes, condensing towers and other high-flow, low-pressure applications.

Axial-flow fans are designed to handle very high flow rates and low pressure. The disc-type fans are similar to those of a household fan. They are usually for general circulation or exhaust work without ducts.

The so-called propeller-type fans with blades that are aerodynamically designed (as seen in Fig. 10-68) can consist of two or more stages. The air in these fans enters in an axial direction and leaves in an axial direction. The fans usually have inlet guide vanes followed by a rotating blade, followed by a stationary (stator) blade.

Centrifugal Blowers These blowers have air or gases entering in the axial direction and being discharged in the radial direction. These blowers have 3 types of blades, radial or straight blades, forward curved blades, and backward curved blades (Figs. 10-69–10-71).

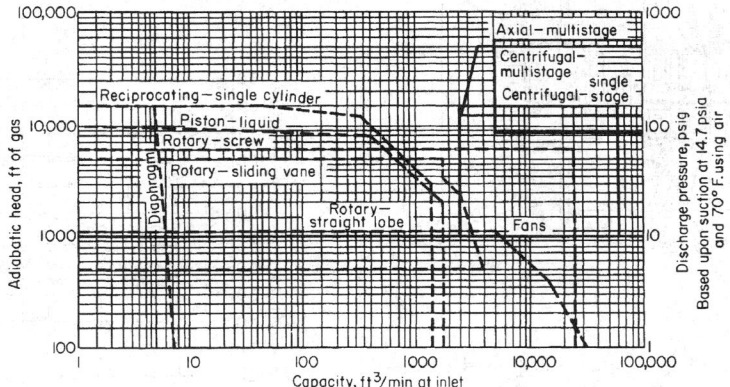

FIG. 10-67 Compressor coverage chart based on the normal range of operation of commercially available types shown. Solid lines: use left ordinate, head. Broken lines: use right ordinate, pressure. To convert cubic feet per minute to cubic meters per hour, multiply by 1.699; to convert feet to meters, multiply by 0.3048; and to convert pounds-force per square inch to kilopascals, multiply by 6.895; (°F − 32)% = °C.

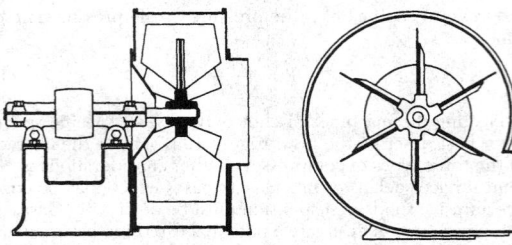

FIG. 10-68 Straight-blade, or steel-plate, fan.

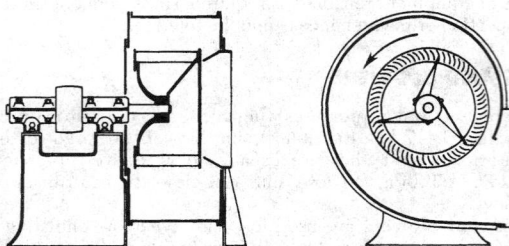

FIG. 10-69 Forward-curved blade, or "scirocco"-type, fan.

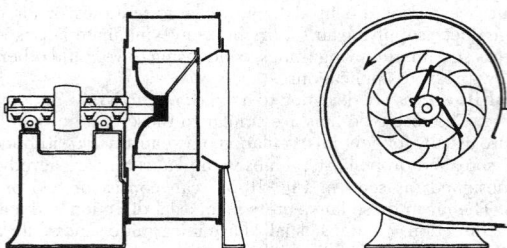

FIG. 10-70 Backward-curved-blade fan.

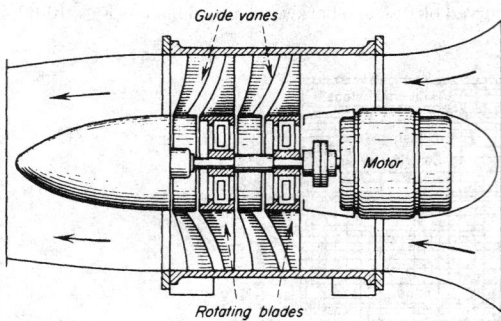

FIG. 10-71 Two-stage axial-flow fan.

Radial blade blowers as seen in Fig. 10-68 are usually used in large-diameter or high-temperature applications. The blades being radial in direction have very low stresses as compared to the backward or forward curve blades. The rotors have anywhere between 4 to 12 blades and usually operate at low speeds. These fans are used in exhaust work especially for gases at high temperature and with suspensions in the flow stream.

Forward-Curved Blade Blowers These blowers discharge the gas at a very high velocity. The pressure supplied by this blower is lower than that produced in the other two blade characteristics. The number of blades in such a rotor can be large—up to 50 blades—and the speed is high—usually 3600–1800 rpm in 60-cycle countries and 3000–1500 rpm in 50-cycle countries.

Backward-Curved Blade Blowers These blowers are used when a higher discharge pressure is needed. It is used over a wide range of applications. Both the forward and backward curved blades do have much higher stresses than the radial bladed blower.

The centrifugal blower produces energy in the air stream by the centrifugal force and imparts a velocity to the gas by the blades. Forward curved blades impart the most velocity to the gas. The scroll-shaped volute diffuses the air and creates an increase in the static pressure by reducing the gas velocity. The change in total pressure occurs in the impeller—this is usually a small change. The static pressure is increased both in the impeller and the diffuses section. Operating efficiencies of the fan range from 40–80 percent. The discharge total pressure is the summation of the static pressure and the velocity head.

The power needed to drive the fan can be computed as follows.

$$\text{Power (kw)} = 2.72 \times 10^{-5}QP \qquad (10\text{-}87)$$

where Q is the fan volume (m³/hr) and P is the total discharge pressure in cm of water column.

In U.S. customary units,

$$\text{hp} = 1.57 \times 10^{-4}Qp \qquad (10\text{-}88)$$

where hp is the fan power output, hp; Q is the fan volume, ft³/min; and p is the fan-operating pressure, inches water column.

$$\text{Efficiency} = \frac{\text{air power output}}{\text{shaft power input}} \qquad (10\text{-}89)$$

Fan Performance The performance of a centrifugal fan varies with changes in conditions such as temperature, speed, and density of the gas being handled. It is important to keep this in mind in using the catalog data of various fan manufacturers, since such data are usually based on stated standard conditions. Corrections must be made for variations from these standards. The usual variations are as follows:

When speed varies, (1) capacity varies directly as the speed ratio, (2) pressure varies as the square of the speed ratio, and (3) horsepower varies as the cube of the speed ratio.

When the temperature of air or gas varies, horsepower and pressure vary inversely as the absolute temperature, speed and capacity being constant. See Fig. 10-72.

When the density of air or gas varies, horsepower and pressure vary directly as the density, speed and capacity being constant.

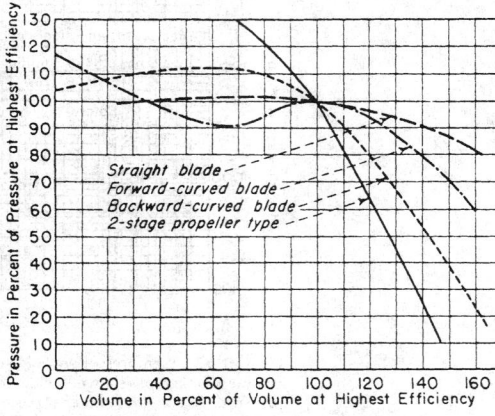

FIG. 10-72 Approximate characteristic curves of various types of fans.

COMPRESSORS

Compressors are used to handle large volumes of gas at pressure increases from 10.32 kPa ($1.5\,lbg/in^2$) to several hundred kPa (lbg/in^2). We can divide compressors into two major categories:

1. Continuous-flow compressors.
   a. Centrifugal compressors
   b. Axial flow compressors
2. Positive displacement compressors
   a. Rotary compressors
   b. Reciprocating compressors

Continuous-Flow Compressors   Continuous-flow compressors are machines where the flow is continuous, unlike positive displacement machines where the flow is fluctuating. Continuous-flow compressors are also classified as turbomachines. These types of machines are widely used in the chemical and petroleum industry for many services. They are also used extensively in many other industries such as the iron and steel industry, pipeline boosters, and on offshore platforms for reinjection compressors. Continuous-flow machines are usually much smaller in size and produce much less vibration than their counterpart, positive displacement units.

Centrifugal Compressors   The flow in a centrifugal compressor enters the impeller in an axial direction and exits in a radial direction.

In a typical centrifugal compressor, the fluid is forced through the impeller by rapidly rotating impeller blades. The velocity of the fluid is converted to pressure, partially in the impeller and partially in the stationary diffusers. Most of the velocity leaving the impeller is converted into pressure energy in the diffuser as shown in Fig. 10-73. It is normal practice to design the compressor so that half the pressure rise takes place in the impeller and the other half in the diffuser. The diffuser consists of a vaneless space, a vane that is tangential to the impeller, or a combination of both. These vane passages diverge to convert the velocity head into pressure energy.

Centrifugal compressors in general are used for higher pressure ratios and lower flow rates compared to lower-stage pressure ratios and higher flow rates in axial compressors. The pressure ratio in a single-stage centrifugal compressor varies depending on the industry and application. In the petrochemical industry the single stage pressure ratio is about 1.2:1. Centrifugal compressors used in the aerospace

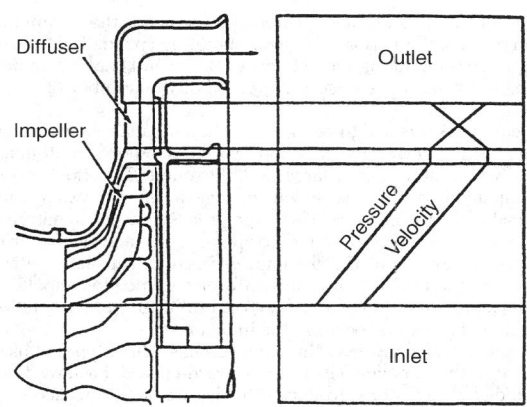

FIG. 10-73   Pressure and velocity through a centrifugal compressor.

industry, usually as a compressor of a gas turbine, have pressure ratios between 3:1 to as high as 9:1 per stage.

In the petrochemical industry, the centrifugal compressors consist mainly of casings with multiple stages. In many instances, multiple casings are also used, and to reduce the power required to drive these multiple casings, there are intercoolers between them. Each casing can have up to 9 stages. In some cases, intercoolers are also used between single stages of compressor to reduce the power required for compression. These compressors are usually driven by gas turbines, steam turbines, and electric motors. Speed-increasing gears may be used in conjunction with these drivers to obtain the high speeds at which many of these units operate. Rotative speeds as high as 50,000 rpm are not uncommon. Most of the petrochemical units run between 9,000–15,000 rpm.

The compressor's operating range is between two major regions as seen in Fig. 10-74, which is a performance map of a centrifugal compressor. These two regions are *surge,* which is the lower flow limit of

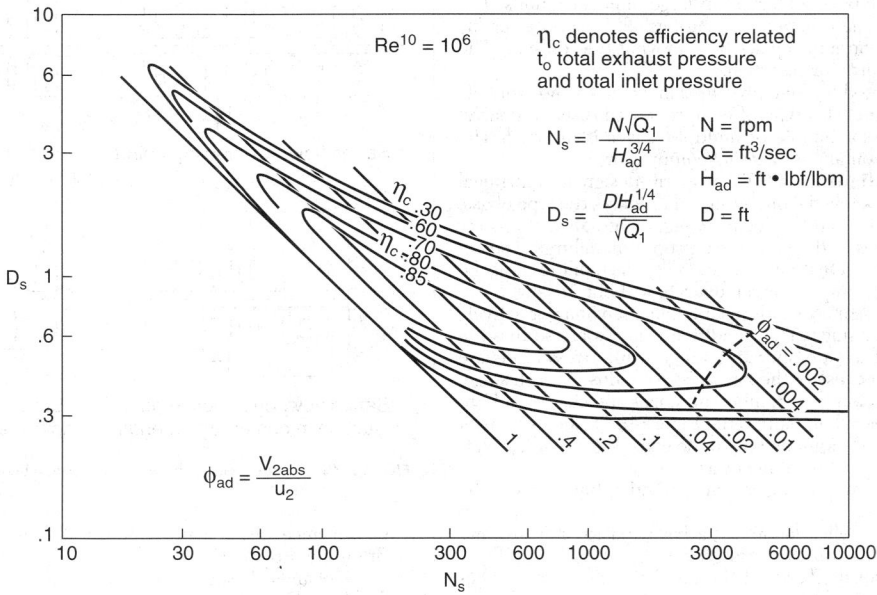

FIG. 10-74   Centrifugal compressor map. (*Balje, O. E., "A Study of Reynolds Number Effects in Turbomachinery,"* Journal of Engineering for Power, *ASME Trans., vol. 86, series A, p. 227*).

stable operation, and *choke* or *stonewall*, which is the maximum flow through the compressor at a given operating speed. The centrifugal compressor's operating range between surge and choke is reduced as the pressure ratio per stage is increased or the number of stages are added.

A compressor is said to be in surge when the main flow through the compressor reverses its direction. Surge is often symptomized by excessive vibration and a large audible sound. This flow reversal is accompanied with a very violent change in energy, which causes a reversal of the thrust force. The surge process is cyclic in nature and if allowed to cycle for some time, irreparable damage can occur to the compressor. In a centrifugal compressor, surge is usually initiated at the exit of the impeller or at the diffuser entrance for impellers producing a pressure ratio of less than 3:1. For higher pressure ratios, the initiation of surge can occur in the inducer.

A centrifugal compressor impeller can have three types of blades at the exit of the impeller. These are forward-curved, backward-curved, and radial blades. Forward-curved blades are not often used in a centrifugal compressor's impeller because of the very high-velocity discharge at the compressor that would require conversion of the high velocity to a pressure head in the diffuser, which is accompanied by high losses. Radial blades are used in impellers of high pressure ratio since the stress levels are minimal. Backward-curved blades give the highest efficiency and the largest operating margin of any of the various types of blades in an impeller. Most centrifugal compressors in the petrochemical industry use backward-curved impellers because of the higher efficiency and larger operating range.

Process compressors have impellers with very low pressure ratio impellers and thus large surge-to-choke margins. The common method of classifying process-type centrifugal compressors is based on the number of impellers and the casing design. Sectionalized casing types have impellers that are usually mounted on the extended motor shaft, and similar sections are bolted together to obtain the desired number of stages. Casing material is either steel or cast iron. These machines require minimum supervision and maintenance and are quite economic in their operating range. The sectionalized casing design is used extensively in supplying air for combustion in ovens and furnaces.

The horizontally split type have casings split horizontally at the midsection and the top. The bottom halves are bolted and doweled together. This design type is preferred for large multistage units. The internal parts such as shaft, impellers, bearings, and seals are readily accessible for inspection and repairs by removing the top half. The casing material is cast iron or cast steel.

Barrel casings are used for high pressures in which the horizontally split joint is inadequate. This type of compressor consists of a barrel into which a compressor bundle of multiple stages is inserted. The bundle is itself a horizontally split casing compressor.

Compressor Configuration   To properly design a centrifugal compressor, one must know the operating conditions—the type of gas, its pressure, temperature, and molecular weight. One must also know the corrosive properties of the gas so that proper metallurgical selection can be made. Gas fluctuations due to process instabilities must be pinpointed so that the compressor can operate without surging.

Centrifugal compressors for industrial applications have relatively low pressure ratios per stage. This condition is necessary so that the compressors can have a wide operating range while stress levels are kept at a minimum. Because of the low pressure ratios for each stage, a single machine may have a number of stages in one "barrel" to achieve the desired overall pressure ratio. Figure 10-75 shows some of the many configurations. Some factors to be considered when selecting a configuration to meet plant needs are:

1.   Intercooling between stages can considerably reduce the power consumed.

2.   Back-to-back impellers allow for a balanced rotor thrust and minimize overloading the thrust bearings.

3.   Cold inlet or hot discharge at the middle of the case reduces oil-seal and lubrication problems.

4.   Single inlet or single discharge reduces external piping problems.

5.   Balance planes that are easily accessible in the field can appreciably reduce field-balancing times.

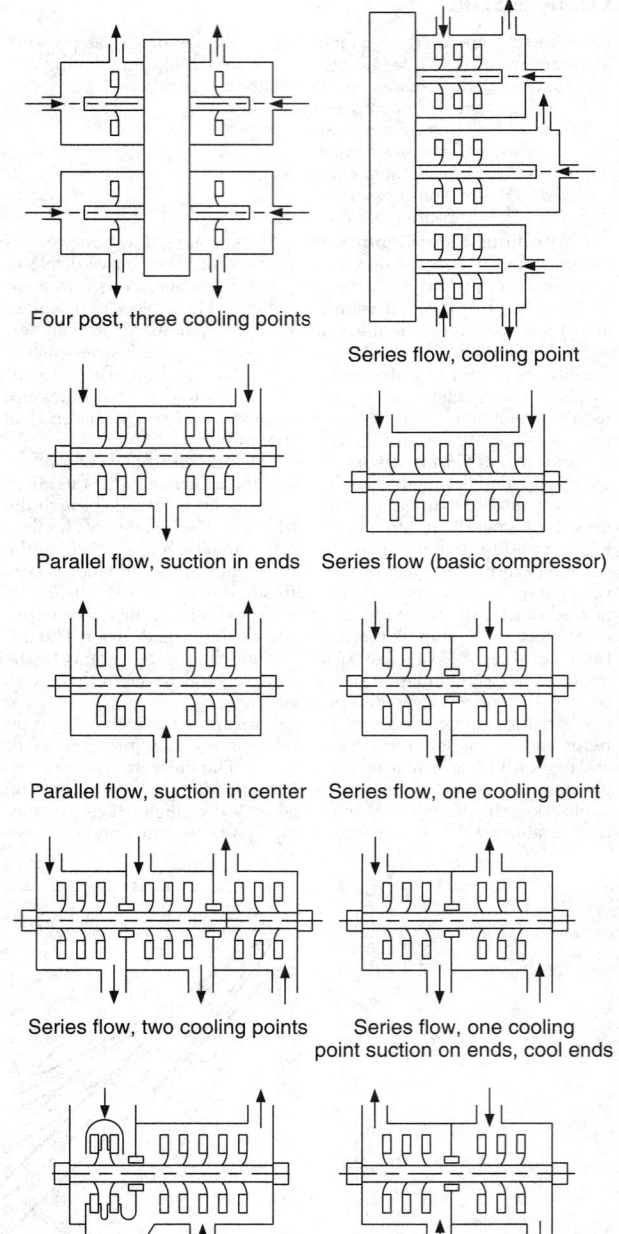

Four post, three cooling points

Series flow, cooling point

Parallel flow, suction in ends   Series flow (basic compressor)

Parallel flow, suction in center   Series flow, one cooling point

Series flow, two cooling points   Series flow, one cooling point suction on ends, cool ends

Series flow, one cooling point, suction in center, warm ends   Series flow, with double flow inlet and side stream

FIG. 10-75   Various configurations of centrifugal compressors.

6.   Balance piston with no external leakage will greatly reduce wear on the thrust bearings.

7.   Hot and cold sections of the case that are adjacent to each other will reduce thermal gradients and thus reduce case distortion.

8.   Horizontally split casings are easier to open for inspection than vertically split ones, reducing maintenance time.

9.   Overhung rotors present an easier alignment problem because

shaft-end alignment is necessary only at the coupling between the compressor and driver.

10.   Smaller, high-pressure compressors that do the same job will reduce foundation problems but will have greatly reduced operational range.

Impeller Fabrication   Centrifugal-compressor impellers are either shrouded or unshrouded. Open, shrouded impellers that are mainly used in single-stage applications are made by investment-casting techniques or by three-dimensional milling. Such impellers are used, in most cases, for the high-pressure-ratio stages. The shrouded impeller is commonly used in the process compressor because of its low pressure ratio stages. The low tip stresses in this application make it a feasible design. Figure 10-76 shows several fabrication techniques. The most common type of construction is seen in A and B where the blades are fillet-welded to the hub and shroud. In B, the welds are full penetration. The disadvantage in this type of construction is the obstruction of the aerodynamic passage. In C, the blades are partially machined with the covers and then butt-welded down the middle. For backward lean-angled blades, this technique has not been very successful, and there has been difficulty in achieving a smooth contour around the leading edge.

D illustrates a slot-welding technique and is used where blade-passage height is too small (or the backward lean-angle too high) to permit conventional fillet welding. In E, an electron-beam technique is shown. Its major disadvantage is that electron-beam welds should preferably be stressed in tension but, for the configuration of E, they are in shear. The configurations of G through J use rivets. Where the rivet heads protrude into the passage aerodynamic performance is reduced. Riveted impellers were used in the 1960s—they are very rarely used now. Elongation of these rivets occurs at certain critical surge conditions and can lead to major failures.

Materials for fabricating these impellers are usually low-alloy steels, such as AISI 4140 or AISI 4340. AISI 4140 is satisfactory for most applications; AISI 4340 is used for large impellers requiring higher strengths. For corrosive gases, AISI 410 stainless steel (about 12 percent chromium) is used. Monel K-500 is employed in halogen gas atmospheres and oxygen compressors because of its resistance to sparking. Titanium impellers have been applied to chlorine service. Aluminum-alloy impellers have been used in great numbers, especially at lower temperatures (below 300°F). With new developments in aluminum alloys, this range is increasing. Aluminum and titanium are sometimes selected because of their low density. This low density can cause a shift in the critical speed of the rotor, which may be advantageous.

Axial Flow Compressors   Axial flow compressors are used mainly as compressors for gas turbines. They are also used in the steel industry as blast furnace blowers and in the chemical industry for large nitric acid plants. They are mainly used for applications where the head required is low and the flow large.

Figure 10-77 shows a typical axial-flow compressor. The rotating element consists of a single drum to which are attached several rows of decreasing-height blades having airfoil cross sections. Between each rotating blade row is a stationary blade row. All blade angles and areas are designed precisely for a given performance and high efficiency. The use of multiple stages permits overall pressure increases up to 30:1. The efficiency in an axial flow compressor is higher than the centrifugal compressor.

Pressure ratio per casing can be comparable with those of centrifugal equipment, although flow rates are considerably higher for a given casing diameter because of the greater area of the flow path. The pressure ratio per stage is less than in a centrifugal compressor. The pressure ratio per stage in industrial compressors is between 1:05–1:15 per stage and for aeroturbines 1.1–1.2 per stage.

The axial flow compressors used in gas turbines vary depending on the type of turbines. The industrial-type gas turbine has an axial flow compressor of a rugged construction. These units have blades that have low aspect ratio (R = blade height/blade chord) with minimum streamline curvation, and the shafts are support on sleeve-type bearings. The industrial gas turbine compressor has also a lower pressure ratio per stage (stage = rotor + stationary blade), giving a low blade loading. This also gives a larger operating range than its counterpart

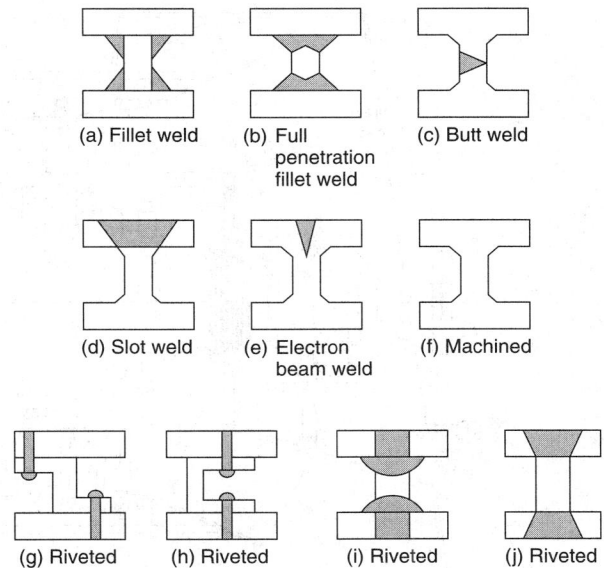

FIG. 10-76   Several fabrication techniques for centrifugal impellers.

the aero axial gas turbine compressor but considerably less than the centrifugal compressor.

The axial flow compressors in aero gas turbines are heavily loaded. The aspect ratio of the blades, especially the first few stages, can be as high as 4.0, and the effect of streamline curvature is substantial. The streamline configuration is a function of the annular passage area, the camber and thickness distribution of the blade, and the flow angles at the inlet and outlet of the blades. The shafts on these units are supported on antifriction bearings (roller or ball bearings).

The operation of the axial flow compressor is a function of the rotational speed of the blades and the turning of the flow in the rotor. The stationary blades (stator) are used to diffuse the flow and convert the velocity increased in the rotor to a pressure increase. One rotor and one stator make up a stage in a compressor. One additional row of fixed blades (inlet guide vanes) is frequently used at the compressor inlet to ensure that air enters the first stage rotors at the desired angle. In addition to the stators, another diffuser at the exit of the compressor further diffuses the gas and, in the case of gas turbines, controls its velocity entering the combustor. The axial flow compressor has a much smaller operating range "Surge to Choke" than its counterpart in the centrifugal compressor. Because of the steep characteristics of the head/flow capacity curve, the surge point is usually within 10 percent of the design point.

The axial flow compressor has three distinct stall phenomena. Rotating stall and individual blade stall are aerodynamic phenomena. Stall flutter is an aeroelastic phenomenon. Rotating stall (propagating stall) consists of large stall zones covering several blade passages and propagates in the direction of the rotor and at some fraction of rotor speed. The number of stall zones and the propagating rates vary considerably. Rotating stall is the most prevalent type of stall phenomenon. Individual blade stall occurs when all the blades around the compressor annulus stall simultaneously without the occurrence of the stall propagation mechanism. The phenomena of stall flutter is caused by self-excitation of the blade and is aeroelastic. It must be distinguished from classic flutter, since classic flutter is a coupled torsional-flexural vibration that occurs when the freestream velocity over an airfoil section reaches a certain critical velocity. Stall flutter, on the other hand, is a phenomenon that occurs due to the stalling of the flow around a blade. Blade stall causes Karman vortices in the airfoil wake. Whenever the frequency of the vortices coincides with the natural frequency of airfoil, flutter will occur. Stall flutter is a major cause of compressor-blade failure.

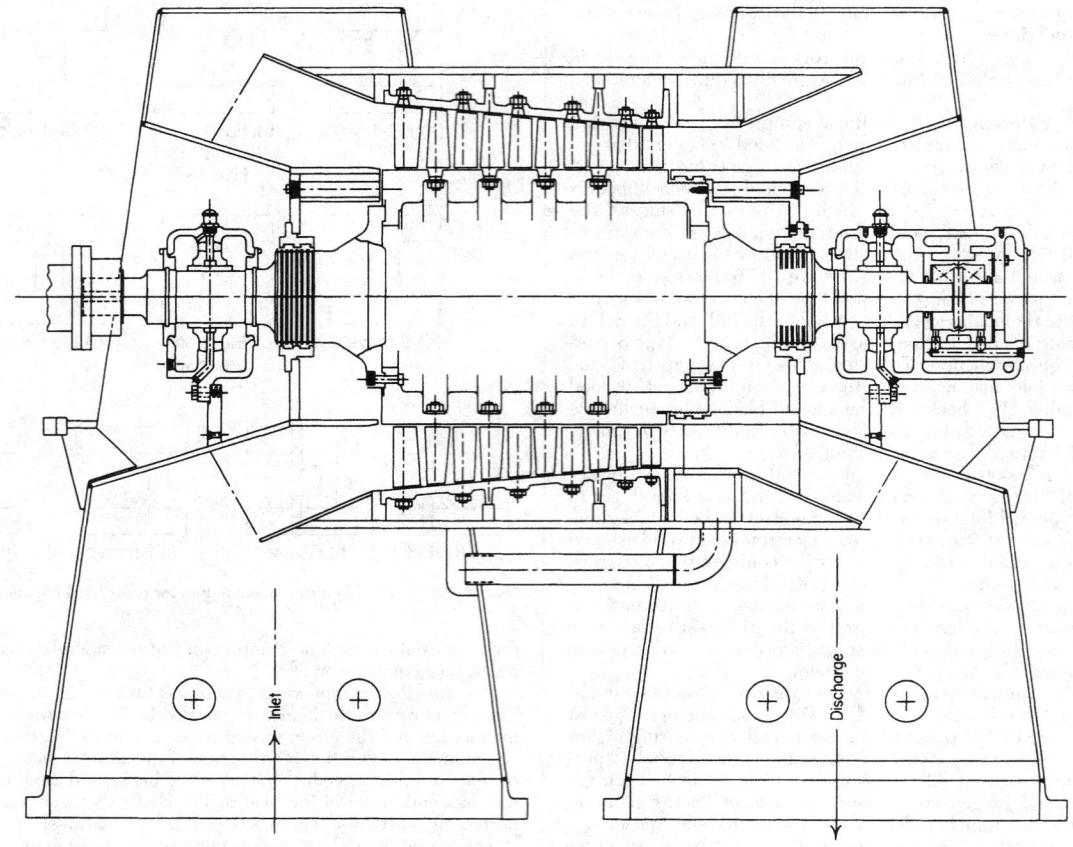

FIG. 10-77   Axial-flow compressor. (*Courtesy of Allis-Chalmers Corporation.*)

Positive Displacement Compressors   Positive displacement compressors are machines that are essentially constant volume machines with variable discharge pressures. These machines can be divided into two types:

1.   Rotary compressors
2.   Reciprocating compressors

Many users consider rotary compressors, such as the "Rootes"-type blower, as turbomachines because their behavior in terms of the rotor dynamics is very close to centrifugal and axial flow machinery. Unlike the reciprocating machines, the rotary machines do not have a very high vibration problem but, like the reciprocating machines, they are positive displacement machines.

Rotary Compressors   Rotary compressors are machines of the positive-displacement type. Such units are essentially constant-volume machines with variable discharge pressure. The volume can be varied only by changing the speed or by bypassing or wasting some of the capacity of the machine. The discharge pressure will vary with the resistance on the discharge side of the system. A characteristic curve typical of the form produced by these rotary units is shown in Fig. 10-78. Rotary compressors are generally classified as of the straight-lobe type, screw type, sliding-vane type, and liquid-piston type.

Straight-Lobe Type   This type is illustrated in Fig. 10-79. Such units are available for pressure differentials up to about 83 kPa (12 lbf/in²) and capacities up to 2.549 × 10⁴ m³/h (15,000 ft³/min). Sometimes multiple units are operated in series to produce higher pressures; individual-stage pressure differentials are limited by the shaft deflection, which must necessarily be kept small to maintain rotor and casing clearance.

Screw-Type   This type of rotary compressor, as shown in Fig. 10-80, is capable of handling capacities up to about 4.248 × 10⁴ m³/h (25,000 ft³/min) at pressure ratios of 4:1 and higher. Relatively small-diameter rotors allow rotative speeds of several thousand rev/min. Unlike the straight-lobe rotary machine, it has male and female rotors whose rotation causes the axial progression of successive sealed cavities. These machines are staged with intercoolers when such an

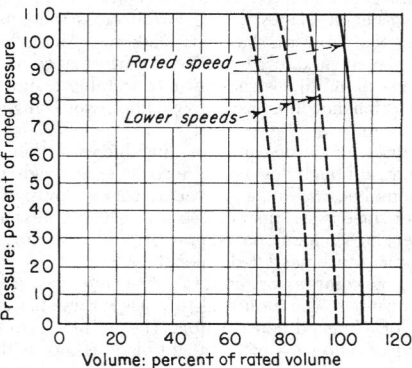

FIG.   10-78   Approximate   performance   curves   for   a   rotary   positive-displacement compressor. The safety valve in discharge line or bypass must be set to operate at a safe value determined by construction.

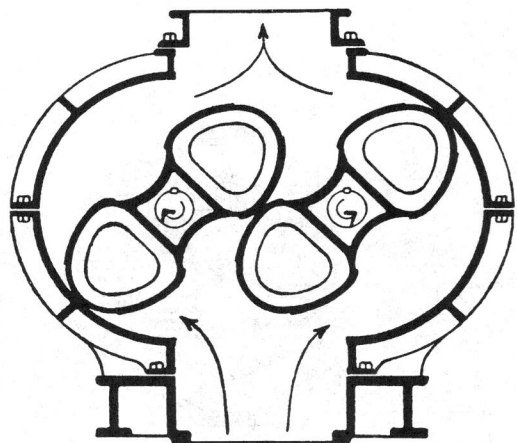

FIG. 10-79   Two-impeller type of rotary positive-displacement blower.

arrangement is advisable. Their high-speed operation usually necessitates the use of suction- and discharge-noise suppressors. The bearings used are sleeve-type bearings. Due to the side pressures experienced, tilting pad bearings are highly recommended.

Sliding-Vane Type   This type is illustrated in Fig. 10-81. These units are offered for operating pressures up to 0.86 MPa (125 lbf/in²) and in capacities up to 3.4×10^3 m³/h (2000 ft³/min). Generally, pressure ratios per stage are limited to 4:1. Lubrication of the vanes is required, and the air or gas stream therefore contains lubricating oil.

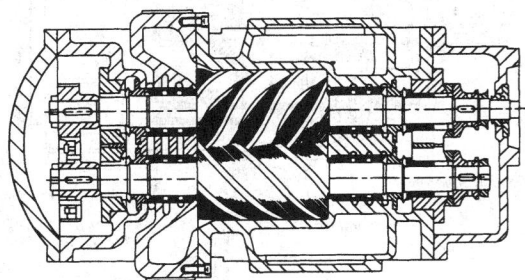

FIG. 10-80   Screw-type rotary compressor.

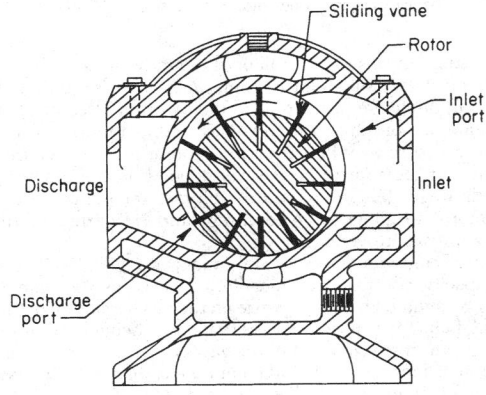

FIG. 10-81   Sliding-vane type of rotary compressor.

Liquid-Piston Type   This type is illustrated in Fig. 10-82. These compressors are offered as single-stage units for pressure differentials up to about 0.52 MPa (75 lbf/in²) in the smaller sizes and capacities up to 6.8×10^3 m³/h (4000 ft³/min) when used with a lower pressure differential. Staging is employed for higher pressure differentials. These units have found wide application as vacuum pumps on wet-vacuum service. Inlet and discharge ports are located in the impeller hub. As the vaned impeller rotates, centrifugal force drives the sealing liquid against the walls of the elliptical housing, causing the air to be successively drawn into the vane cavities and expelled against discharge pressure. The sealing liquid must be externally cooled unless it is used in a once-through system. A separator is usually employed in the discharge line to minimize carryover of entrained liquid. Compressor capacity can be considerably reduced if the gas is highly soluble in the sealing liquid.

The liquid-piston type of compressor has been of particular advantage when hazardous gases are being handled. Because of the gas-liquid contact and because of the much greater liquid specific heat, the gas-temperature rise is very small.

Reciprocating Compressors   Reciprocating compressors are used mainly when high-pressure head is required at a low flow. Reciprocating compressors are furnished in either single-stage or multistage types. The number of stages is determined by the required compressor ratio p_2/p_1. The compression ratio per stage is generally limited to 4, although low-capacity units are furnished with compression ratios of 8 and even higher. Generally, the maximum compression ratio is determined by the maximum allowable discharge-gas temperature.

Single-acting air-cooled and water-cooled air compressors are available in sizes up to about 75 kW (100 hp). Such units are available in one, two, three, or four stages for pressure as high as 24 MPa (3500 lbf/in²). These machines are seldom used for gas compression because of the difficulty of preventing gas leakage and contamination of the lubricating oil.

The compressors most commonly used for compressing gases have a crosshead to which the connecting rod and piston rod are connected. This provides a straight-line motion for the piston rod and permits simple packing to be used. Figure 10-83 illustrates a simple single-stage machine of this type having a double-acting piston. Either single-acting (Fig. 10-84) or double-acting pistons (Fig. 10-85) may be used, depending on the size of the machine and the number of stages. In some machines double-acting pistons are used in the first stages and single-acting in the later stages.

On multistage machines, intercoolers are provided between stages. These heat exchangers remove the heat of compression from the gas and reduce its temperature to approximately the temperature existing at the compressor intake. Such cooling reduces the volume of gas going to the high-pressure cylinders, reduces the power required for compression, and keeps the temperature within safe operating limits.

Figure 10-86 illustrates a two-stage compressor end such as might be used on the compressor illustrated in Fig. 10-83.

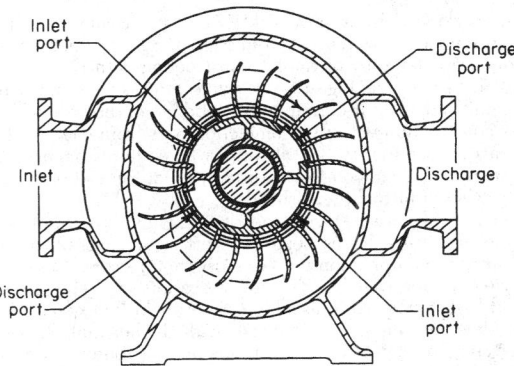

FIG. 10-82   Liquid-piston type of rotary compressor.

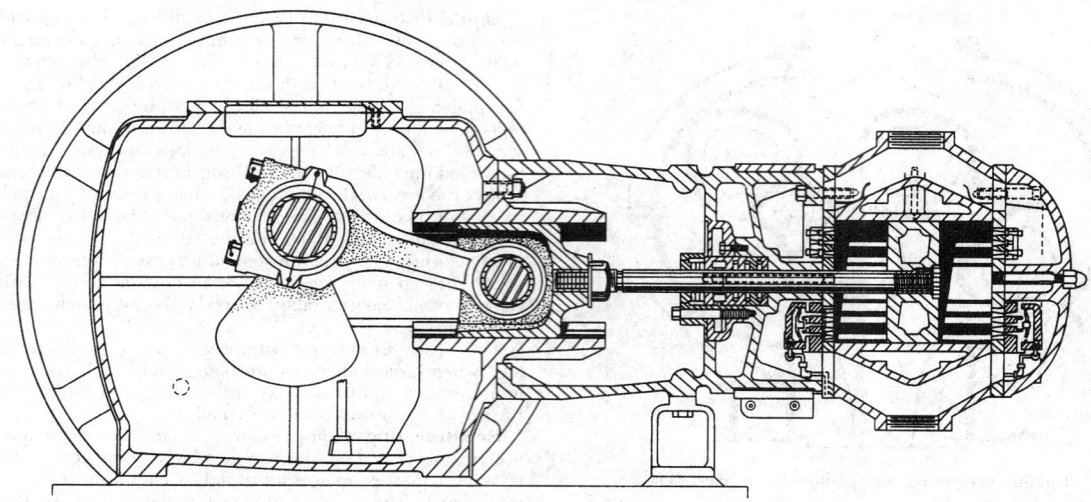

FIG. 10-83 Typical single-stage, double-acting water-cooled compressor.

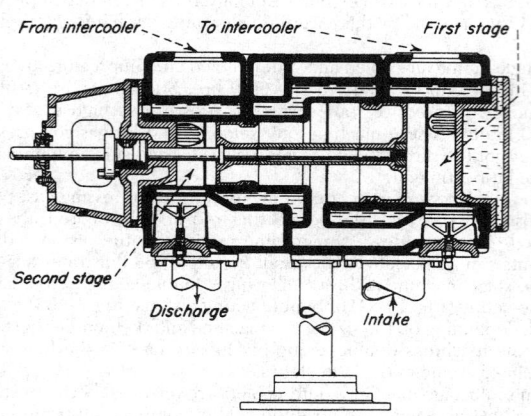

FIG. 10-84 Two-stage single-acting opposed piston in a single step-type cylinder.

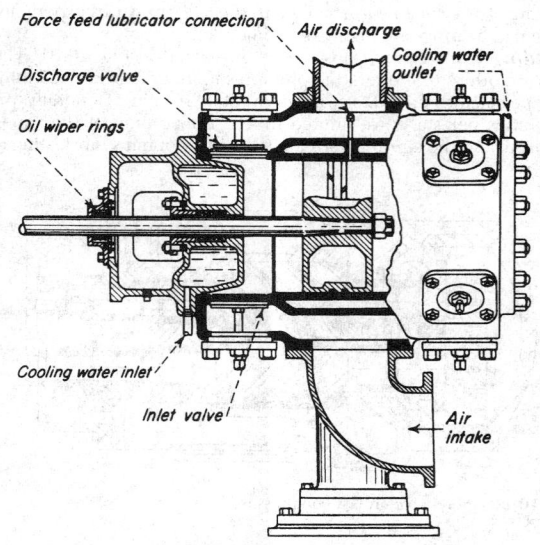

FIG. 10-85 Typical double-acting compressor piston and cylinder.

Compressors with horizontal cylinders such as illustrated in Figs. 10-83 to 10-86 are most commonly used because of their accessibility. However, machines are also built with vertical cylinders and other arrangements such as right-angle (one horizontal and one vertical cylinder) and V-angle.

Compressors up to around 75 kW (100 hp) usually have a single center-throw crank, as illustrated in Fig. 10-83. In larger sizes compressors are commonly of duplex construction with cranks on each end of the shaft (see Fig. 10-87). Some large synchronous motor-driven units are of four-corner construction; i.e., they are of double-duplex construction with two connecting rods from each of the two crank throws (see Fig. 10-88). Steam-driven compressors have one or more steam cylinders connected directly by piston rod or tie rods to the gas-cylinder piston or crosshead.

Valve Losses   Above piston speeds of 2.5 m/s (500 ft/min), suction and discharge valve losses begin to exert significant effects on the actual internal compression ratio of most compressors, depending on the valve port area available. The obvious results are high temperature rise and higher power requirements than might be expected. These effects become more pronounced with higher-molecular-weight gases. Valve problems can be a very major contributor to down time experienced by these machines.

Control Devices   In many installations the use of gas is intermit-tent, and some means of controlling the output of the compressor is therefore necessary. In other cases constant output is required despite variations in discharge pressure, and the control device must operate to maintain a constant compressor speed. Compressor capacity, speed, or pressure may be varied in accordance with requirements. The nature of the control device will depend on the function to be regulated. Regulation of pressure, volume, temperature, or some other factor determines the type of regulation required and the type of the compressor driver.

The most common control requirement is regulation of capacity. Many capacity controls, or unloading devices, as they are usually termed, are actuated by the pressure on the discharge side of the compressor. A falling pressure indicates that gas is being used faster than it is being compressed and that more gas is required. A rising pressure indicates that more gas is being compressed than is being used and that less gas is required.

An obvious method of controlling the capacity of a compressor is to

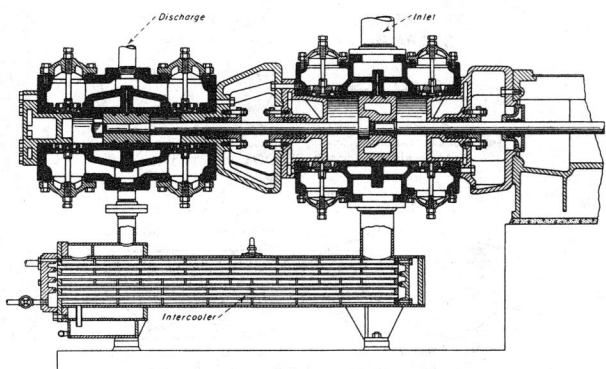

FIG. 10-86 Two-stage double-acting compressor cylinders with intercooler.

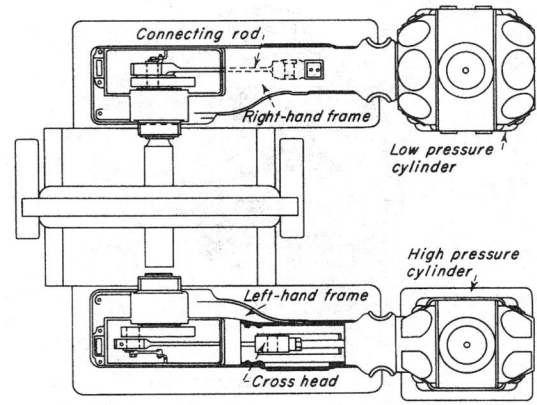

FIG. 10-87 Duplex two-stage compressor (plan view).

vary the speed. This method is applicable to units driven by variable-speed drivers such as steam pistons, steam turbines, gas engines, diesel engines, etc. In these cases the regulator actuates the steam-admission or fuel-admission valve on the compressor driver and thus controls the speed.

Motor-driven compressors usually operate at constant speed, and other methods of controlling the capacity are necessary. On reciprocating compressors discharging into receivers, up to about 75 kW (100 hp), two types of control are usually available. These are automatic-start-and-stop control and constant-speed control.

Automatic-start-and-stop control, as its name implies, stops or starts the compressor by means of a pressure-actuated switch as the gas demand varies. It should be used only when the demand for gas will be intermittent.

Constant-speed control should be used when gas demand is fairly constant. With this type of control, the compressor runs continuously but compresses only when gas is needed. Three methods of unloading the compressor with this type of control are in common use: (1) **closed suction unloaders,** (2) **open inlet-valve unloaders,** and (3) **clearance unloaders.** The closed suction unloader consists of a pressure-actuated valve which shuts off the compressor intake. Open inlet-valve unloaders (see Fig. 10-89) operate to hold the compressor inlet valves open and thereby prevent compression. Clearance unloaders (see Fig. 10-90) consist of pockets or small reservoirs which are opened when unloading is desired. The gas is compressed into

them on the compression stroke and reexpands into the cylinder on the return stroke, thus preventing the compression of additional gas.

It is sometimes desirable to have a compressor equipped with both constant-speed and automatic-start-and-stop control. When this is done, a switch allows immediate selection of either type.

Motor-driven reciprocating compressors above about 75 kW (100 hp) in size are usually equipped with a step control. This is in reality a variation of constant-speed control in which unloading is accomplished in a series of steps, varying from full load down to no load. **Three-step control** (full load, one-half load, and no load) is usually accomplished with inlet-valve unloaders. **Five-step control** (full load, three-fourths load, one-half load, one-fourth load, and no load) is accomplished by means of clearance pockets (see Fig. 10-91). On some machines, inlet-valve and clearance-control unloading are used in combination.

Although such control devices are usually automatically operated, manual operation is satisfactory for some services. When manual operation is provided, it often consists of a valve or valves to open and close clearance pockets. In some cases, a movable cylinder head is provided for variable clearance in the cylinder (see Fig. 10-92).

When no capacity control or unloading device is provided, it is necessary to provide bypasses between the inlet and discharge in order that the compressor can be started against no load (see Fig. 10-93).

Nonlubricated Cylinders Most compressors use oil to lubricate

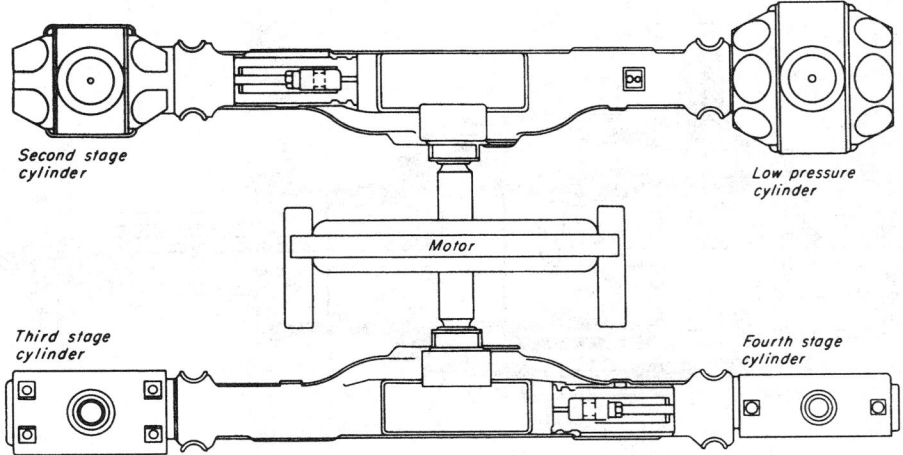

FIG. 10-88 Four-corner four-stage compressor (plan view).

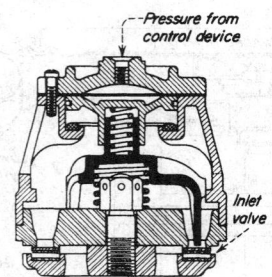

FIG. 10-89 Inlet-valve unloader.

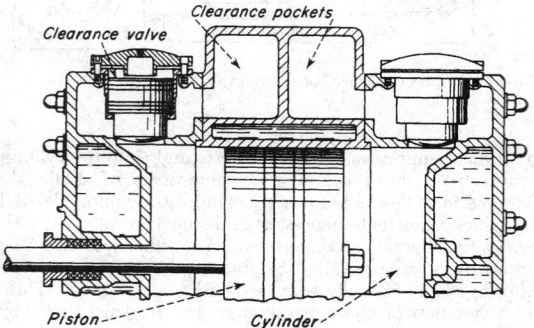

FIG. 10-90 Clearance-control cylinder. *(Courtesy of Ingersoll-Rand.)*

the cylinder. In some processes, however, the slightest oil contamination is objectionable. For such cases a number of manufacturers furnish a "nonlubricated" cylinder (see Fig. 10-94). The piston on these cylinders is equipped with piston rings of graphitic carbon or Teflon° as well as pads or rings of the same material to maintain proper clearance between the piston and the cylinder. Plastic packing of a type that requires no lubricant is used on the stuffing box. Although oil-wiper rings are used on the piston rod where it leaves the compressor frame, minute quantities of oil might conceivably enter the cylinder on the rod. If even such small amounts of oil are objectionable, an extended cylinder connecting piece can be furnished. This simply

° ®Du Pont tetrafluoroethylene fluorocarbon resin.

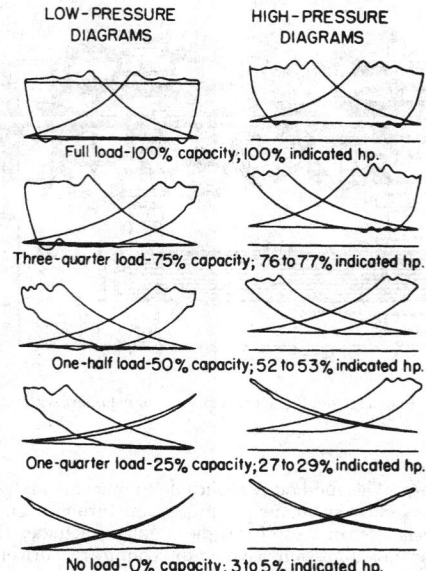

FIG. 10-91 Actual indicator diagram of a two-stage compressor showing the operation of clearance control at five load points.

lengthens the piston rod enough so that no portion of the rod can alternately enter the frame and the cylinder.

In many cases, a small amount of gas leaking through the packing is objectionable. Special connecting pieces are furnished between the cylinder and the frame, which may be either single-compartment or double-compartment. These may be furnished gastight and vented back to the suction or filled with a sealing gas or fluid and held under a slight pressure.

High-Pressure Compressors There is a definite trend in the chemical industry toward the use of high-pressure compressors with discharge pressures of from 34.5 to 172 MPa (5000 to 25,000 lbf/in²) and with capacities from 8.5×10^3 to 42.5×10^3 m³/h (5000 to 25,000 ft³/min). These require special design, and a complete knowledge of the characteristics of the gas is necessary. In most cases, these types of applications use the barrel-type centrifugal compressor.

The gas usually deviates considerably from the perfect-gas laws, and in many cases temperature or other limitations necessitate a thor-

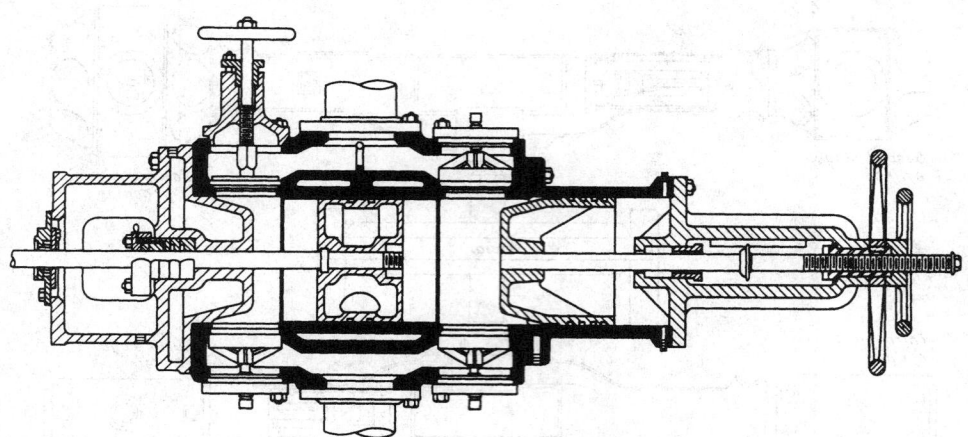

FIG. 10-92 Sectional view of a cylinder equipped with a hand-operated valve lifter on one end and a variable-volume clearance pocket at other end.

Starting compressor	Stopping compressor
Start with A and D open	Close - - - - C
Close - - - - - - D	Close - - - B
Close - - - - - - - A	Open - - - - A and D
Open - - - - - - - B	
Slowly open - - - C	

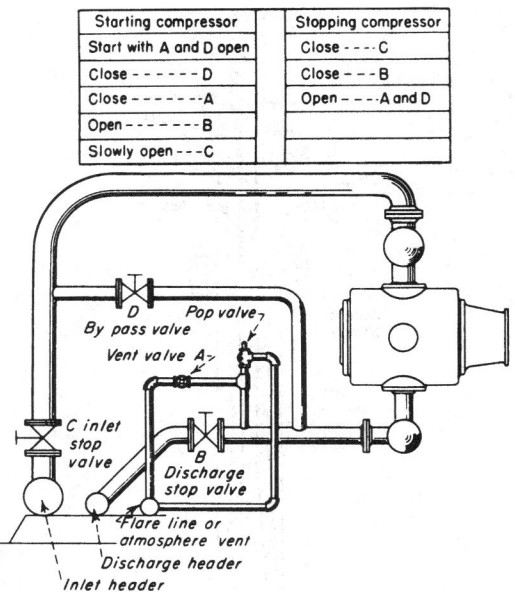

FIG. 10-93 Bypass arrangement for a single-stage compressor. On multistage machines, each stage is bypassed in a similar manner. Such an arrangement is necessary for no-load starting.

ough engineering study of the problem. These compressors usually have five, six, seven, or eight stages, and the cylinders must be properly proportioned to meet the various limitations involved and also to balance the load among the various stages. In many cases, scrubbing or other processing is carried on between stages. High-pressure cylinders are steel forgings with single-acting plungers (see Fig. 10-95). The compressors are usually designed so that the pressure load against the plunger is opposed by one or more single-acting pistons of the lower pressure stages. Piston-rod packing is usually of the segmental-ring metallic type. Accurate fitting and correct lubrication are very important. High-pressure compressor valves are designed for the

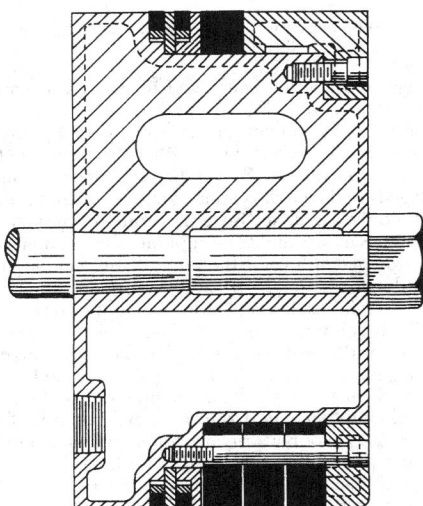

FIG. 10-94 Piston equipped with carbon piston and wearing rings for a nonlubricated cylinder.

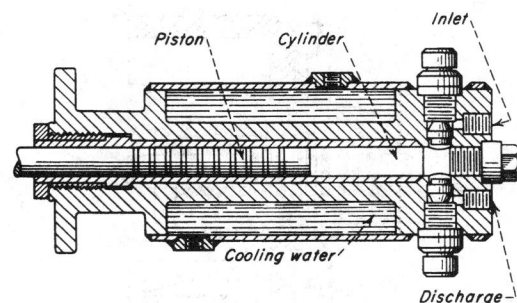

FIG. 10-95 Forged-steel single-acting high-pressure cylinder.

conditions involved. Extremely high-grade engineering and skill are necessary.

Piston-Rod Packing Proper piston-rod packing is important. Many types are available, and the most suitable is determined by the gas handled and the operating conditions for a particular unit.

There are many types and compositions of soft packing, semimetallic packing, and metallic packing. In many cases, metallic packing is to be recommended. A typical low-pressure packing arrangement is shown in Fig. 10-96. A high-pressure packing arrangement is shown in Fig. 10-97.

When wet, volatile, or hazardous gases are handled or when the service is intermittent, an auxiliary packing gland and soft packing are usually employed (see Fig. 10-98).

Metallic Diaphragm Compressors (Fig. 10-99) These are available for small quantities [up to about 17 m³/h (10 ft³/min)] for compression ratios as high as 10:1 per stage. Temperature rise is not a

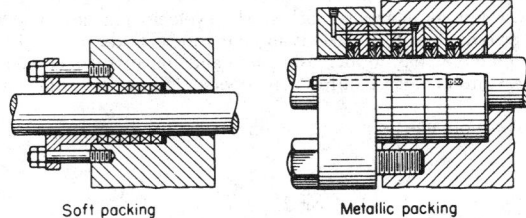

Soft packing   Metallic packing

FIG. 10-96 Typical packing arrangements for low-pressure cylinders.

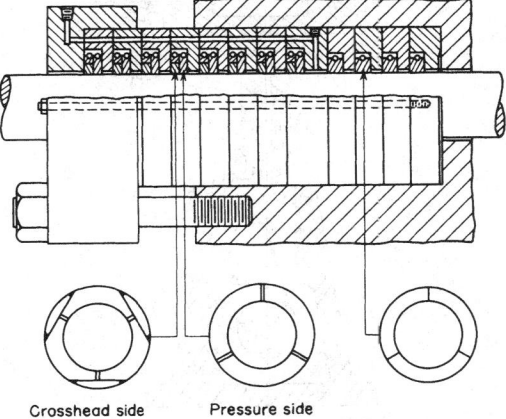

Crosshead side   Pressure side

FIG. 10-97 Typical packing arrangement, using metallic packing, for high-pressure cylinders.

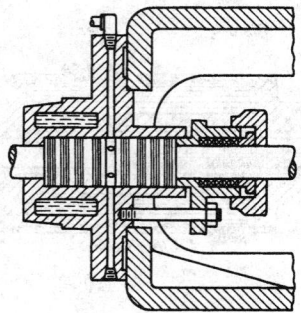

FIG. 10-98   Soft packing in an auxiliary stuffing box for handling gases.

serious problem, as the large wall area relative to the gas volume permits sufficient heat transfer to approach isothermal compression. These compressors possess the advantage of having no seals for the process gas. The diaphragm is actuated hydraulically by a plunger pump.

EJECTORS

An ejector is a simplified type of vacuum pump or compressor which has no pistons, valves, rotors, or other moving parts. Figure 10-100 illustrates a steam-jet ejector. It consists essentially of a nozzle which discharges a high-velocity jet across a suction chamber that is connected to the equipment to be evacuated. The gas is entrained by the steam and carried into a venturi-shaped diffuser which converts the velocity energy into pressure energy. Figure 10-101 shows a large-sized ejector, sometimes called a booster ejector, with multiple nozzles. Nozzles are devices in subsonic flow that have a decreasing area and accelerate the flow. They convert pressure energy to velocity energy. A minimum area is reached when velocity reaches sonic flow. In supersonic flow, the nozzle is an increasing area device. A diffuser in subsonic flow has an increasing area and converts velocity energy into pressure energy. A diffuser in supersonic flow has a decreasing area.

Two or more ejectors may be connected in series or stages. Also, a

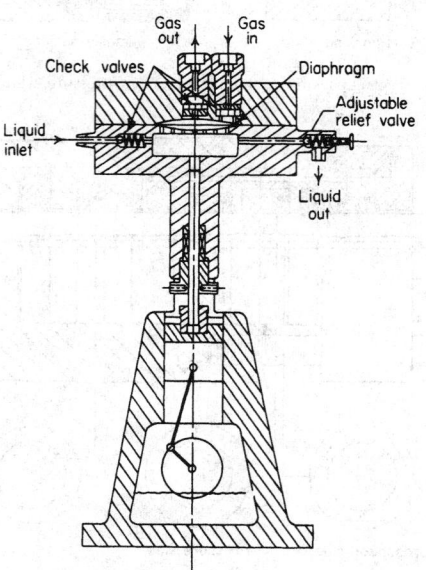

FIG. 10-99   High-pressure, low-capacity compressor having a hydraulically actuated diaphragm. (*Pressure Products Industries.*)

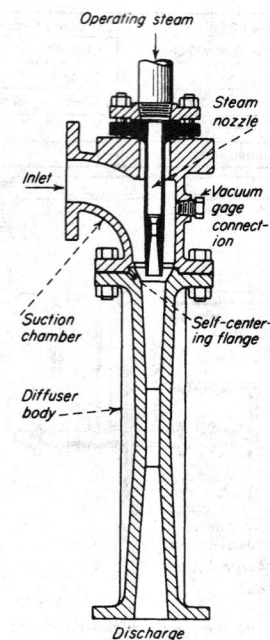

FIG. 10-100   Typical steam-jet ejector.

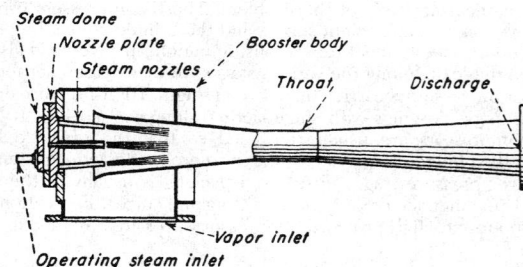

FIG. 10-101   Booster ejector with multiple steam nozzles.

number of ejectors may be connected in parallel to handle larger quantities of gas or vapor.

Liquid- or air-cooled condensers are usually used between stages. Liquid-cooled condensers may be of either the direct-contact (barometric) or the surface type. By condensing vapor the load on the following stage is reduced, thus minimizing its size and reducing consumption of motive gas. Likewise, a precondenser installed ahead of an ejector reduces its size and consumption if the suction gas contains vapors that are condensable at the temperature condition available. An **aftercondenser** is frequently used to condense vapors from the final stage, although this does not affect ejector performance.

Ejector Performance   The performance of any ejector is a function of the area of the motive-gas nozzle and venturi throat, pressure of the motive gas, suction and discharge pressures, and ratios of specific heats, molecular weights, and temperatures. Figure 10-102, based on the assumption of **constant-area mixing,** is useful in evaluating single-stage-ejector performance for compression ratios up to 10 and area ratios up to 100 (see Fig. 10-103 for notation).

For example,[*] assume that it is desired to evacuate air at 2.94 lbf/in² with a steam ejector discharging to 14.7 lbf/in² with available steam

---

[*] All data are given in U.S. customary units since the charts are in these units. Conversion factors to SI units are given on the charts.

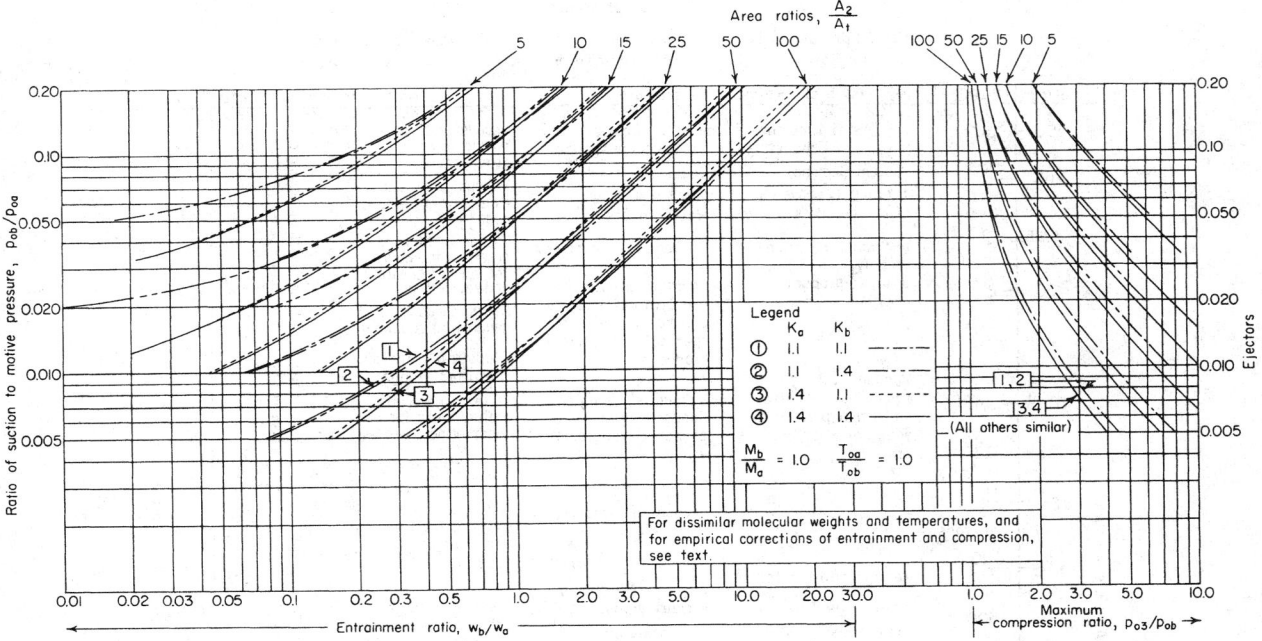

FIG. 10-102 Design curves for optimum single-stage ejectors. [*DeFrate and Hoerl*, Chem. Eng. Prog., **55**, *Symp. Ser. 21*, 46 (1959).]

pressure of 100 lbf/in². Entering the chart at $p_{03}/p_{0b} = 5.0$, at $p_{0b}/p_{0a} = 2.94/100 = 0.0294$ the optimum area ratio is 12. Proceeding horizontally to the left, w_b/w_a is approximately 0.15 lb of air per 1 lb of steam. This value must be corrected for the temperature and molecular-weight differences of the two fluids by Eq. (10-90).

$$w/w_a = w_b/w_a \sqrt{T_{0a}M_b/T_{0b}M_a} \qquad (10\text{-}90)$$

In addition, there are empirical correction factors which should be applied. Laboratory tests show that for ejectors with constant-area mixing the actual entrainment and compression ratios will be approximately 90 percent of the calculated values and even less at very small values of p_{0b}/p_{0a}. This compensates for ignoring wall friction in the mixing section and irreversibilities in the nozzle and diffuser. In theory, each point on a given design curve of Fig. 10-102 is associated with an optimum ejector for prevailing operating conditions. Adjacent points on the same curve represent theoretically different ejectors for the new conditions, the difference being that for each ratio of p_{0b}/p_{0a} there is an optimum area for the exit of the motive-gas nozzle. In practice, however, a segment of a given curve for constant A_2/A_t represents the performance of a single ejector satisfactorily for estimating purposes, provided that the suction pressure lies within 20 to 130 percent of the design suction pressure and the motive pressure within 80 to 120 percent of design motive pressure. Thus the curves can be used to

select an optimum ejector for the design point and to estimate its performance at off-design conditions within the limits noted. Final ejector selection should, of course, be made with the assistance of a manufacturer of such equipment.

Uses of Ejectors   For the operating range of steam-jet ejectors in vacuum applications, see the subsection "Vacuum Systems."

The choice of the most suitable type of ejector for a given application depends upon the following factors:

1. *Steam pressure.*   Ejector selection should be based upon the minimum pressure in the supply line selected to serve the unit.

2. *Water temperature.*   Selection is based on the maximum water temperature.

3. *Suction pressure and temperature.*   Overall process requirements should be considered. Selection is usually governed by the minimum suction pressure required (the highest vacuum).

4. *Capacity required.*   Again overall process requirements should be considered, but selection is usually governed by the capacity required at the minimum process pressure.

Ejectors are easy to operate and require little maintenance. Installation costs are low. Since they have no moving parts, they have long life, sustained efficiency, and low maintenance cost. Ejectors are suitable for handling practically any type of gas or vapor. They are also suitable for handling wet or dry mixtures or gases containing sticky or solid matter such as chaff or dust.

Ejectors are available in many materials of construction to suit process requirements. If the gases or vapors are not corrosive, the diffuser is usually constructed of cast iron and the steam nozzle of stainless steel. For more corrosive gases and vapors, many combinations of materials such as bronze, various stainless-steel alloys, and other corrosion-resistant metals, carbon, and glass can be used.

VACUUM SYSTEMS

Figure 10-104 illustrates the level of vacuum normally required to perform many of the common manufacturing processes. The attainment of various levels is related to available equipment in Fig. 10-105.

Vacuum Equipment   The equipment shown in Fig. 10-105 has been discussed elsewhere in this section with the exception of the **dif-**

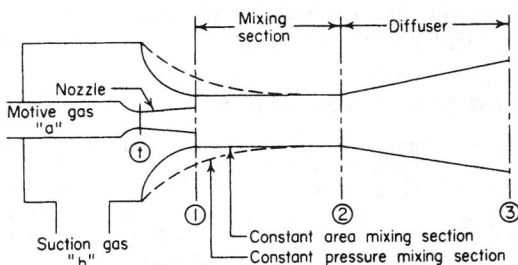

FIG. 10-103   Notation for Fig. 10-102.

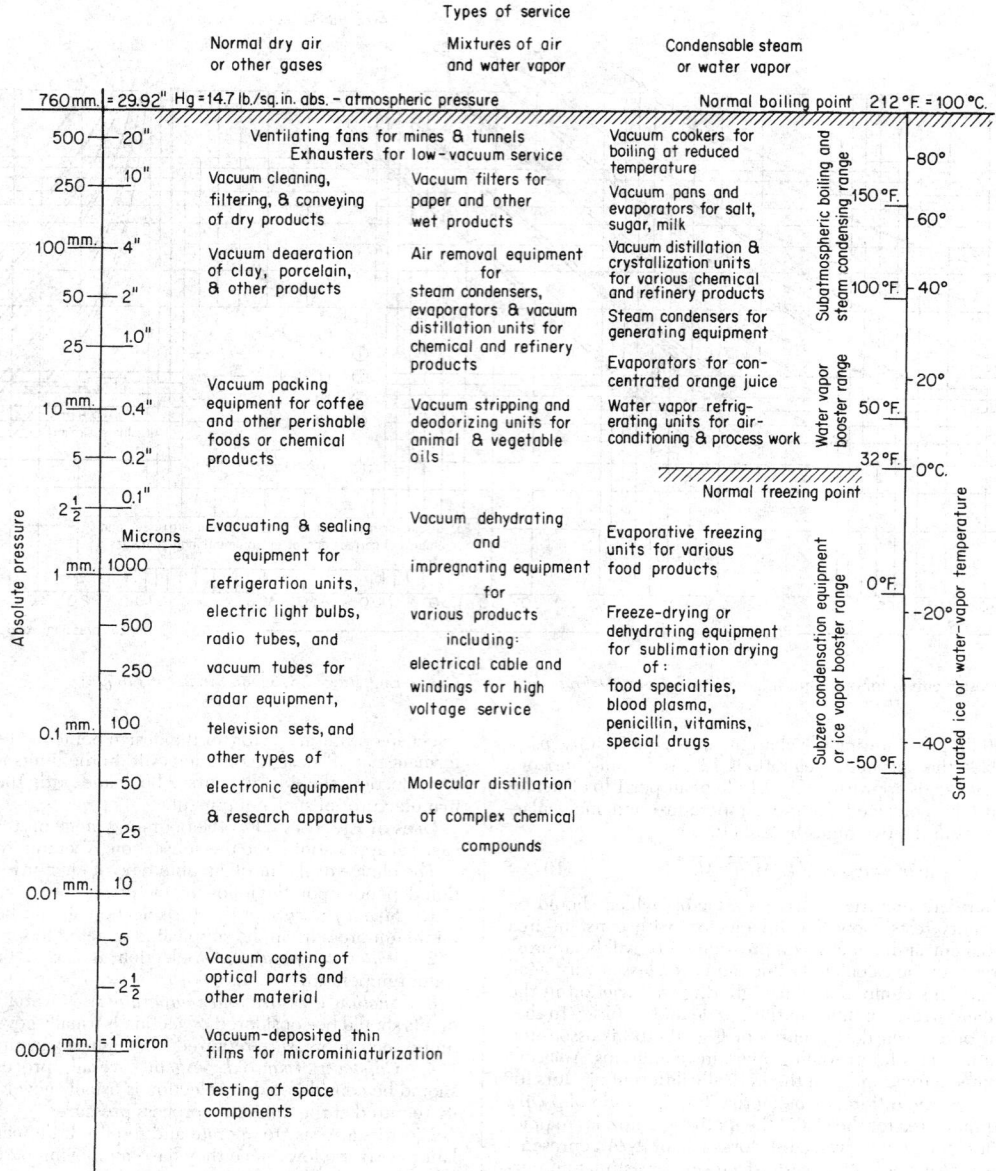

FIG. 10-104   Vacuum levels normally required to perform common manufacturing processes. (*Courtesy of* Compressed Air *magazine.*)

fusion pump. Figure 10-106 depicts a typical design. A liquid of low absolute vapor pressure is boiled in the reservoir. The vapor is ejected at high velocity in a downward direction through multiple jets and is condensed on the walls, which are cooled by the surrounding coils. Molecules of the gas being pumped enter the vapor stream and are driven downward by collisions with the vapor molecules. The gas molecules are removed through the discharge line by a backing pump such as a rotary oil-sealed unit.

Diffusion pumps operate at very low pressures. The ultimate vacuum attainable depends somewhat upon the vapor pressure of the pump liquid at the temperature of the condensing surfaces. By providing a cold trap between the diffusion pump and the region being evacuated, pressures as low as 10^{-7} mmHg absolute are achieved in

this manner. Liquids used for diffusion pumps are mercury and oils of low vapor pressure. Silicone oils have excellent characteristics for this service.

SEALING OF ROTATING SHAFTS

Seals are very important and often critical components in large rotating machinery especially on high-pressure and high-speed equipment. The principal sealing systems used between the rotor and stationary elements fall into two main categories: (1) noncontacting seals and (2) face seals. These seals are an integral part of the rotating system, they affect the dynamic operating characteristics of the machine. The stiffness and damping factors will be changed by the seal geometry and

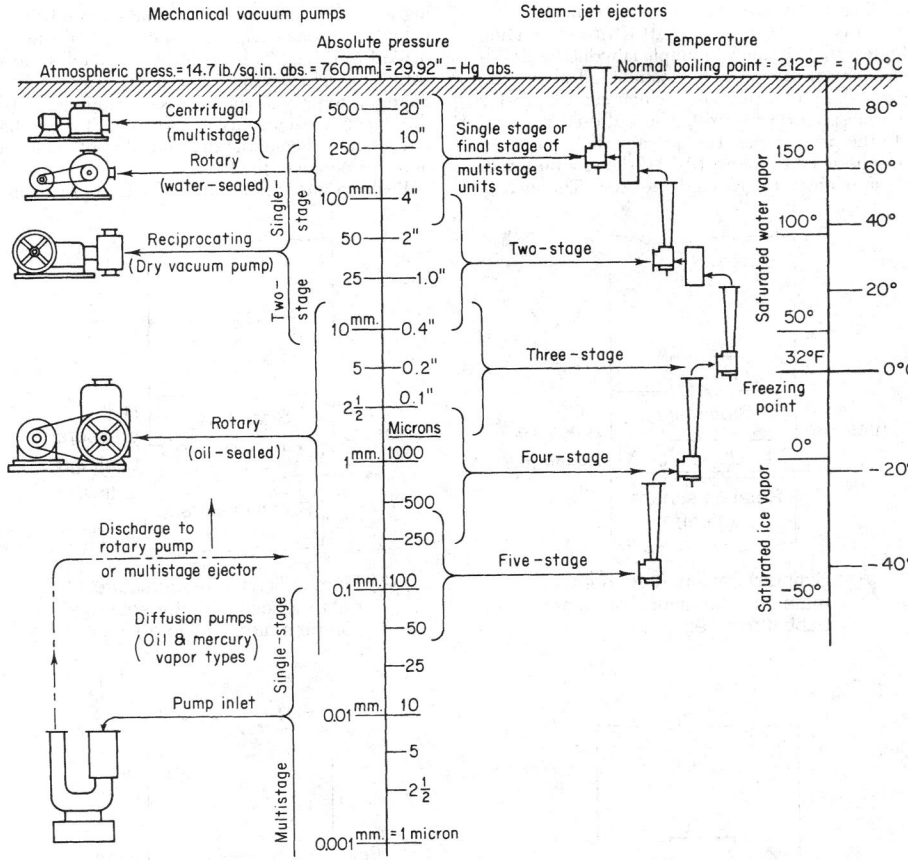

FIG. 10-105   Vacuum levels attainable with various types of equipment. (*Courtesy of* Compressed Air *magazine.*)

pressures. In operation the rotating shafts have both radial and axial movement. Therefore any seal must be flexible and compact to ensure maximum sealing minimum effect on rotor dynamics.

Noncontact Seals   Noncontact seals are used extensively in gas service in high speed rotating equipment. These seals have good mechanical reliability and minimum impact on the rotor dynamics of the system. They are not positive sealing. There are two types of non-contact seals: (1) labyrinth seals and (2) ring seals.

Labyrinth Seals   The labyrinth is one of the simplest of the many sealing devices. It consists of a series of circumferential strips of metal extending from the shaft or from the bore of the shaft housing to form a cascade of annular orifices. Labyrinth seal leakage is greater than that of clearance bushings, contact seals, or filmriding seals.

The major advantages of labyrinth seals are their simplicity, reliability, tolerance to dirt, system adaptability, very low shaft power consumption, material selection flexibility, minimal effect on rotor dynamics, back diffusion reduction, integration of pressure, lack of pressure limitations, and tolerance to gross thermal variations. The major disadvantages are the high leakage, loss of machine efficiency, increased buffering costs, tolerance to ingestion of particulates with resulting damage to other critical items such as bearings, the possibility of the cavity clogging due to low gas velocities or back diffusion, and the inability to provide a simple seal system that meets OSHA or EPA standards. Because of some of the foregoing disadvantages, many machines are being converted to other types of seals.

Labyrinth seals are simple to manufacture and can be made from conventional materials. Early designs of labyrinth seals used knife-edge seals and relatively large chambers or pockets between the knives. These relatively long knives are easily subject to damage. The modern, more functional, and more reliable labyrinth seals consist of sturdy, closely spaced lands. Some labyrinth seals are shown in Fig. 10-107. Figure 10-107a is the simplest form of the seal. Figure

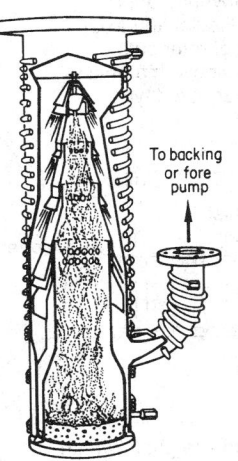

To backing or fore pump

FIG. 10-106   Typical diffusion pump. (*Courtesy of* Compressed Air *magazine.*)

10-107*b* shows a grooved seal; it is more difficult to manufacture but produces a tighter seal. Figures 10-107*c* and 10-107*d* are rotating labyrinth-type seals. Figure 10-107*e* shows a simple labyrinth seal with a buffered gas for which pressure must be maintained above the process gas pressure and the outlet pressure (which can be greater than or less than the atmospheric pressure). The buffered gas produces a fluid barrier to the process gas. The eductor sucks gas from the vent near the atmospheric end. Figure 10-107*f* shows a buffered, stepped labyrinth. The step labyrinth gives a tighter seal. The match-

ing stationary seal is usually manufactured from soft materials such as babbitt or bronze, while the stationary or rotating labyrinth lands are made from steel. This composition enables the seal to be assembled with minimal clearance. The lands can therefore cut into the softer materials to provide the necessary running clearances for adjusting to the dynamic excursions of the rotor. To maintain maximum sealing efficiency, it is essential that the labyrinth lands maintain sharp edges in the direction of the flow.

Leakage past these labyrinths is approximately inversely propor-

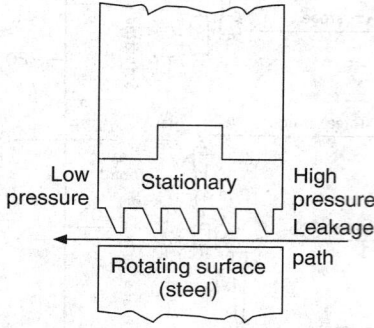

(a) Simplest design. (Labyrinth materials: aluminum, bronze, babbitt or steel)

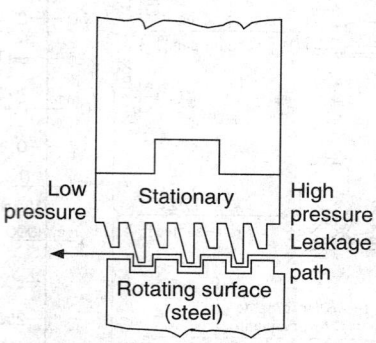

(b) More difficult to manufacture but produces a tighter seal. (Same material as in a.)

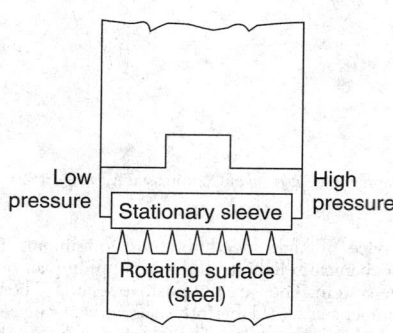

(c) Rotating labyrinth type, before operation. (Sleeve material: babbitt, aluminum, nonmetallic or other soft materials)

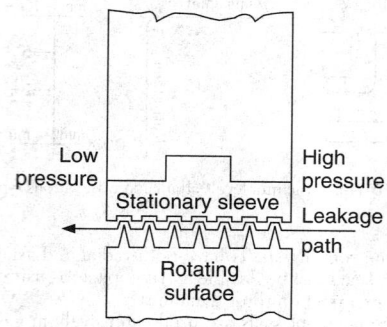

(d) Rotating labyrinth, after operation. Radial and axial movement of rotor cuts grooves in sleeve material to simulate staggered type shown in b.

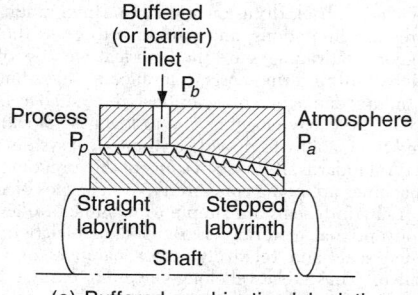

(e) Buffered combination labyrinth

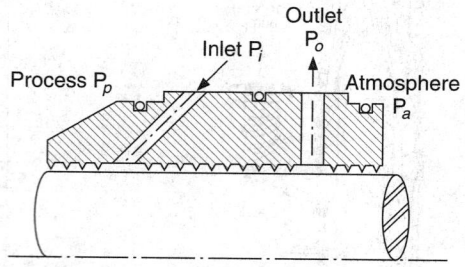

(f) Buffered-vented straight labyrinth

FIG. 10-107   Various configurations of labyrinth seals.

tional to the square root of the number of labyrinth lands. This translates into the following relationship if leakage is to be cut in half in a four labyrinth seal: The number of labyrinths would have to be increased to 16. The Elgi leakage formula can be modified and written as:

$$m_\ell = AK \left[\frac{(g/V_o)(P_o - P_n)}{n + \ln (P_n/P_o)} \right]^{1/2} \qquad (10\text{-}91)$$

where
m_ℓ = leakage
A = leakage Area of single throttling
K = labyrinth constant ($K = .9$ for straight labyrinths, $K = .75$ for staggered labyrinths)
P_o = absolute pressure before the labyrinth
P_n = absolute pressure after the last labyrinth
V_o = specific volume before the labyrinth
n = number of lands

The leakage of a labyrinth seal can be kept to a minimum by providing: (1) minimum clearance between the seal lands and the seal sleeve, (2) sharp edges on the lands to reduce the flow discharge coefficient, and (3) grooves or steps in the flow path for reducing dynamic head carryover from stage to stage.

The labyrinth sleeve can be flexibly mounted to permit radial motion for self-aligning effects. In practice, a radial clearance of under 0.008 is difficult to achieve.

Ring Seals The restrictive ring seal is essentially a series of sleeves in which the bores form a small clearance around the shaft. Thus, the leakage is limited by the flow resistance in the restricted area and controlled by the laminar or turbulent friction. There are two types of ring seals: (1) fixed seal rings and (2) floating seal rings. The floating rings permit a much smaller leakage; they can be either the segmented type as shown in Fig. 10-108a or the rigid type as shown in Fig. 10-108b.

Fixed Seal Rings The fixed-seal ring consists of a long sleeve affixed to a housing in which the shaft rotates with small clearances. Long assemblies must be used to keep leakage within a reasonable limit. Long seal assemblies aggravate alignment and rubbing problems, thus requiring shafts to operate below their capacity. The fixed bushing seal operates with appreciable eccentricity and, combined with large clearances, produces large leakages, thus making this kind of seal impractical where leakage is undesirable.

Floating Seal Rings Clearance seals that are free to move in a radial direction are known as floating seals. The floating characteristics permit them to move freely, thus avoiding severe rubs. Due to differential thermal expansion between the shaft and bushing, the bushings should be made of material with a higher coefficient of thermal expansion. This is achieved by shrinking the carbon into a metallic retaining ring with a coefficient of expansion that equals or exceeds that of the shaft material. It is advisable in high shearing applications to lock the bushings against rotation.

Buildup of dirt and other foreign material lodged between the seal ring and seat will create an excessive spin and damage on the floating seal ring unit. It is therefore improper to use soft material such as babbitt and silver as seal rings.

Packing Seal A common type of rotating shaft seal consists of packing composed of fibers which are first woven, twisted, or braided into strands and then formed into coils, spirals, or rings. To ensure initial lubrication and to facilitate installation, the basic materials are often impregnated. Common materials are asbestos fabric, braided and twisted asbestos, rubber and duck, flax, jute, and metallic braids. The so-called plastic packings can be made up with varying amounts of fiber combined with a binder and lubricant for high-speed applications. Maximum temperatures that base materials of packings withstand and still give good service are as follows:

	°C	°F
Flax	38	100
Cotton	93	200
Duck and rubber	149	300
Rubber	177	350
Metallic (lead-based)	218	425
Asbestos 1003	260	500
Asbestos 204	371	700
Metallic (aluminum-based)	552	1025
Metallic (copper-based)	829	1525

Packing may not provide a completely leak-free seal. With shaft surface speeds less than approximately 2.5 m/s (500 ft/min), the packing may be adjusted to seal completely. However, for higher speeds some leakage is required for lubrication, friction reduction, and cooling.

Application of Packing Coils and spirals are cut to form closed or nearly closed rings in the stuffing box. Clearance between ends should be sufficient to allow for fitting and possible expansion due to increased temperature or liquid absorption of the packing while in operation.

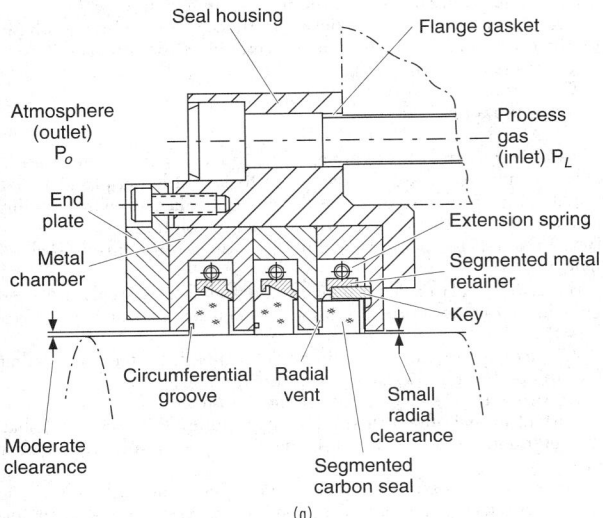

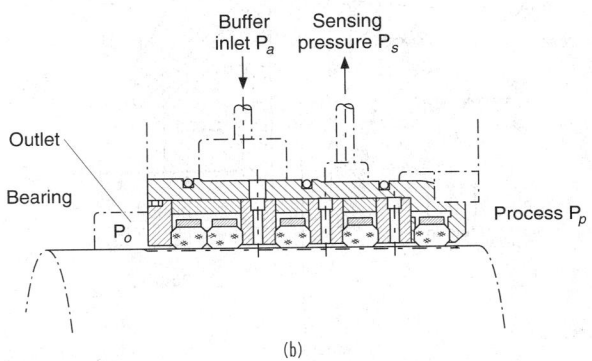

FIG. 10-108 Floating-type restrictive ring seal.

The correct form of the ring joint depends on materials and service requirements. Braided and flexible metallic packings usually have butt or square joints (Fig. 10-109a). With other packing material, service experience indicates that rings cut with bevel or skive joints (Fig. 10-109b) are more satisfactory. A slight advantage of the bevel joint over the butt joint is that the bevel permits a certain amount of sliding action, thus absorbing a portion of ring expansion.

In the manufacture of packings, the proper grade and type of **lubricant** is usually impregnated for each service for which the packing is recommended. However, it may be desirable to replenish the lubricant during the normal life of the packing. Lack of lubrication causes packing to become hard and lose its resiliency, thus increasing friction, shortening packing life, and increasing operating costs.

An effective auxiliary device frequently used with packing and rotary shafts is the **seal cage** (or **lantern ring**), shown in Fig. 10-110. The seal cage provides an annulus around the shaft for the introduction of a lubricant, oil, grease, etc. The seal cage is also used to introduce liquid for cooling, to prevent the entrance of atmospheric air, or to prevent the infiltration of abrasives from the process liquid.

The chief advantage of packing over other types of seals is the ease with which it can be adjusted or replaced. Most equipment is designed so that disassembly of major components is not required to remove or add packing rings. The major disadvantages of a packing-type seal are (1) short life, (2) requirement for frequent adjustment, and (3) need for some leakage to provide lubrication and cooling.

Mechanical Face Seals   This type of seal forms a running seal between flat precision-finished surfaces. It is an excellent seal against leakages. The sealing surfaces are planes perpendicular to the rotating shaft, and the forces that hold the contact faces are parallel to the shaft axis. For a seal to function properly, there are four sealing points:
1. Stuffing box face
2. Leakage down the shaft
3. Mating ring in the gland plate
4. Dynamic faces

Mechanical Seal Selection   There are many factors that govern the selection of seals. These factors apply to any type of seal:
1. Product
2. Seal environment
3. Seal arrangement
4. Equipment
5. Secondary packing
6. Seal face combinations
7. Seal gland plate
8. Main seal body

Product   Physical and chemical properties of the liquid or gas being sealed places constraints on the type of material, design, and arrangement of the seal.

Pressure.   Pressure affects the choice of material and whether balanced or unbalanced seal design can be used. Most unbalanced seals are good up to 100 psig stuffing box pressure. Over 100 psig, balanced seals should be used.

Temperature.   The temperature of the liquid being pumped is important because it affects the seal face material selection as well as the wear life of the seal face.

Lubricity.   In any mechanical seal design, there is rubbing motion between the dynamic seal faces. This rubbing motion is often lubricated by the fluid being pumped. Most seal manufacturers limit the speed of their seals to 90 ft/sec (30 m/sec). This is primarily due to centrifugal forces acting on the seal, which tends to restrict the seal's axial flexibility.

Abrasion.   If there are entrained solids in the liquid, it is desirable to have a flushed single inside type with a face combination of very hard material.

Corrosion.   This affects the type of seal body: what spring material, what face material, and what type of elastomer or gasket material. The corrosion rate will affect the decision of whether to use a single or multiple spring design because the spring can usually tolerate a greater amount of corrosion without weakening it appreciably.

Seal Environment   The design of the seal environment is based on the product and the four general parameters that regulate it:
1. Pressure control
2. Temperature control
3. Fluid replacement
4. Atmospheric air elimination

Seal Arrangement   There are four types of seal arrangements:
1. Double seals are standard with toxic and lethal products, but maintenance problems and seal design contribute to poor reliability. The double face-to-face seal may be a better solution.
2. Do not use a double seal in dirty service—the inside seal will hang up.
3. API standards for balanced and unbalanced seals are good guidelines; too low a pressure for a balanced seal may encourage face lift-off.
4. Arrangement of the seal will determine its success more than the vendor. Over 100 arrangements are available.

Equipment   The geometry of the pump or compressor is very important in seal effectiveness. Different pumps with the same shaft diameter and the total differential head can present different sealing problems.

Secondary Packing   Much more emphasis should be placed on secondary packing especially if Teflon is used. A wide variation in performance is seen between various seal vendors, depending on seal arrangement there can be difference in mating ring packing.

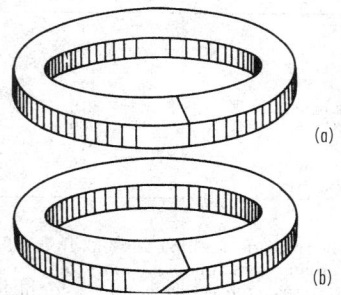

FIG. 10-109   Butt (a) and skive (b) joints for compression packing rings.

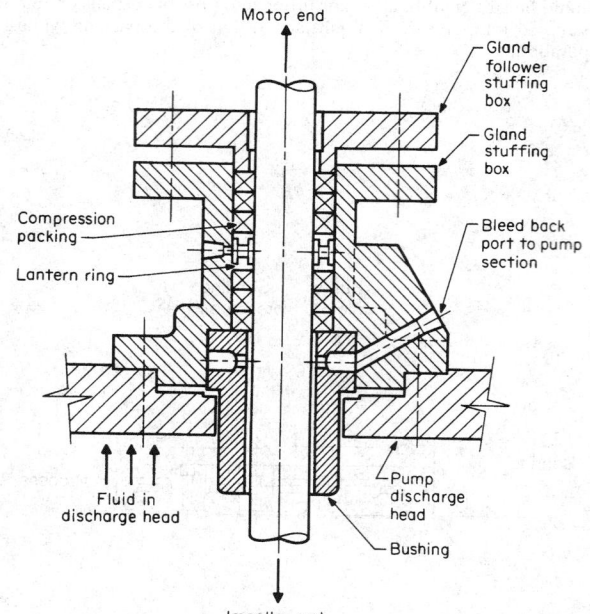

FIG. 10-110   Seal cage or lantern ring. (*Courtesy of Crane Packing Co.*)

Seal Face Combinations   The dynamic of seal faces is better understood today. Seal-face combinations have come a long way in the past 8–10 years. Stellite is being phased out of the petroleum and petrochemical applications. Better grades of ceramic are available, cost of tungsten has come down, and relapping of tungsten are available near most industrial areas. Silicon carbide is being used in abrasive service.

Seal Gland Plate   The seal gland plate is caught in between the pump vendor and the seal vendor. Special glands should be furnished by seal vendors, especially if they require heating, quenching, and drain with a floating-throat bushing. Gland designs are complex and may have to be revisited, especially if seals are changed.

Main Seal Body   The term *seal body* makes reference to all rotating parts on a pusher seal, excluding shaft packing and seal ring. In many cases it is the chief reason to avoid a particular design for a particular service.

Basically, most mechanical seals have the following components as seen in Fig. 10-111.

1. Rotating seal ring
2. Stationary seal ring
3. Spring devices to provide pressure
4. Static seals

A loading device such as a spring is needed to ensure that in the event of loss or hydraulic pressure the sealing surfaces are kept closed. The amount of the load on the sealing area is determined by the degree of "seal balance." Figure 10-113 shows what seal balance means. A completely balanced seal is when the only force exerted on the sealing surfaces is the spring force; i.e., hydraulic pressure does not act on the sealing surface. The type of spring depends on the space available, loading characteristics, and the seal environment. Based on these considerations, either a single or multiple spring can be used. In small axial space, belleville springs, finger washers, or curved washers can be used.

Shaft-sealing elements can be split up into two groups. The first type may be called pusher-type seals and includes the O-ring, V-ring, U-cup, and wedge configurations. Figure 10-116 shows some typical pusher-type seals. The second type is the bellow-type seals, which differ from the pusher-type seals in that they form a static seal between themselves and the shaft.

Internal and External Seals   Mechanical seals are classified broadly as internal or external. **Internal seals** (Fig. 10-112) are installed with all seal components exposed to the fluid sealed. The advantages of this arrangement are (1) the ability to seal against high pressure, since the hydrostatic force is normally in the same direction as the spring force; (2) protection of seal parts from external mechanical damage; and (3) reduction in the shaft length required.

For high-pressure installations, it is possible to balance partially or fully the hydrostatic force on the rotating member of an internal seal by using a stepped shaft or shaft sleeve (Fig. 10-113). This method of

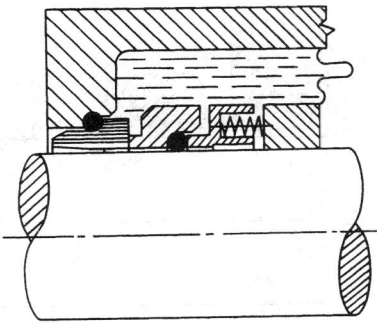

FIG. 10-112   Internal mechanical seal.

relieving face pressure is an effective way of decreasing power consumption and extending seal life.

When abrasive solids are present and it is not permissible to introduce appreciable quantities of a secondary flushing fluid into the process, double internal seals are sometimes used (Fig. 10-114). Both sealing faces are protected by the flushing fluid injected between them even though the inward flow is negligible.

External seals (Fig. 10-115) are installed with all seal components protected from the process fluid. The advantages of this arrangement are that (1) fewer critical materials of construction are required, (2) installation and setting are somewhat simpler because of the exposed position of the parts, and (3) stuffing-box size is not a limiting factor. Hydraulic balancing is accomplished by proper proportioning of the seal face and secondary seal diameters.

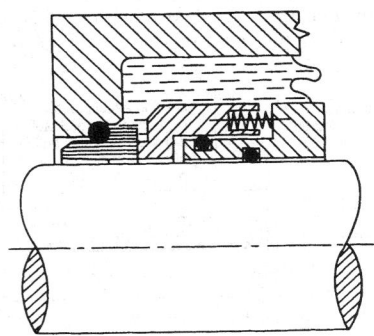

FIG. 10-113   Balanced internal mechanical seal.

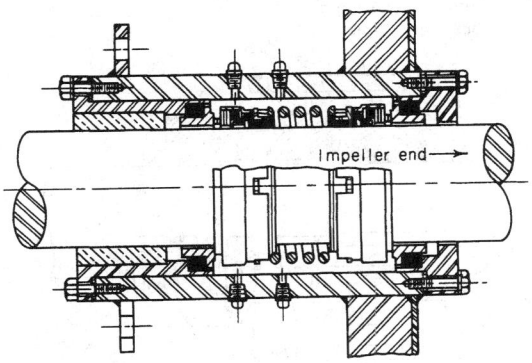

FIG. 10-114   Internal bellows-type double mechanical seal.

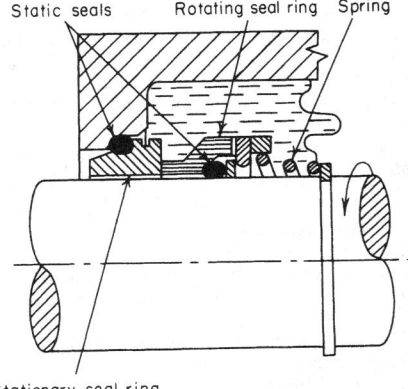

FIG. 10-111   Mechanical-seal components.

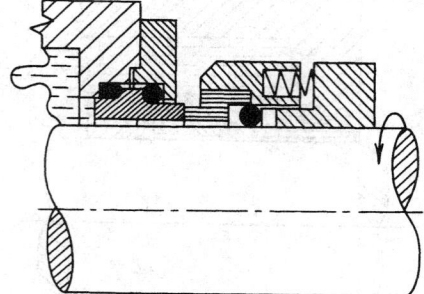

FIG. 10-115   External mechanical seal.

Throttle bushings (Fig. 10-117) are commonly used with single internal or external seals when solids are present in the fluid and the inflow of a flushing fluid is not objectionable. These close-clearance bushings are intended to serve as flow restrictions through which the maintenance of a small inward flow of flushing fluid prevents the entrance of a process fluid into the stuffing box.

A typical complex seal utilizes both the noncontact and mechanical aspects of sealing. Figure 10-118 shows such a seal with its two major elements. This type of seal will normally have buffering via a labyrinth seal and a positive shutdown device. For shutdown, the carbon ring is tightly sandwiched between the rotating seal ring and the stationary sleeve with gas pressure to prevent gas from leaking out when no oil pressure is available.

In operation seal oil pressure is about 30–50 psi over the process gas pressure. The high-pressure oil enters the top and completely fills the seal cavity. A small percentage is forced across the carbon ring seal faces. The rotative speed of the carbon ring can be anywhere between zero and full rotational speed. Oil crossing the seal faces contacts the process gas and therefore "contaminated oil." The contaminated oil

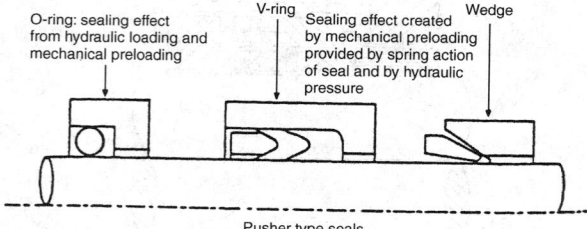

FIG. 10-116   Various types of shaft-sealing elements.

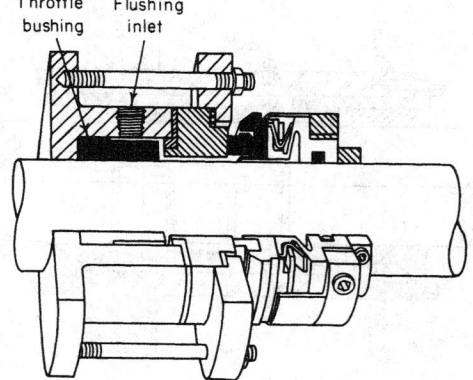

FIG. 10-117   External mechanical seal and throttle bushing.

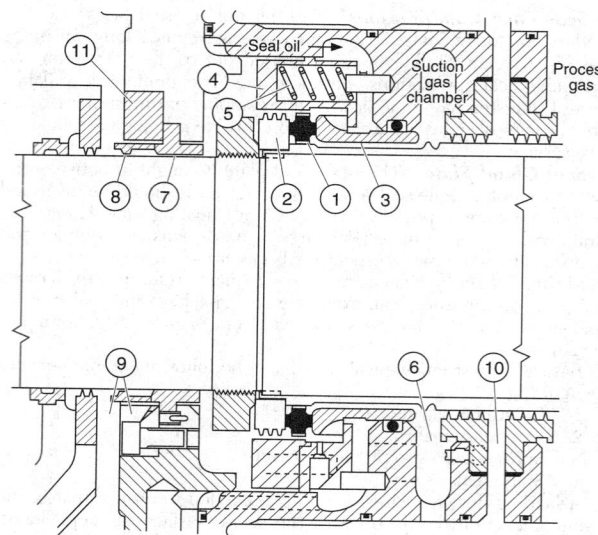

1. Rotating carbon ring	7. Floating babbitt-faced steel ring
2. Rotating seal ring	8. Seal wiper ring
3. Stationary sleeve	9. Seal oil drain line
4. Spring retainer	10. Buffer gas injection port
5. Spring	11. Bypass orifice
6. Gas and contaminated oil drain	

FIG. 10-118   Mechanical contact shaft seal.

leaves through the contaminated oil drain to a degassifier for purification. The majority of the oil flows through the uncontaminated seal oil drain.

Materials   Springs and other metallic components are available in a wide variety of alloys and are usually selected on the basis of temperature and corrosion conditions. The use of a particular mechanical seal is frequently restricted by the temperature limitations of the organic materials used in the static seals. Most elastomers are limited to about 121°C (250°F). Teflon will withstand temperatures of 260°C (500°F) but softens appreciably above 204°C (400°F). Glass-filled Teflon is dimensionally stable up to 232 to 260°C (450 to 500°F).

One of the most common elements used for seal faces is carbon. Although compatible with most process media, carbon is affected by strong oxidizing agents, including fuming nitric acid, hydrogen chloride, and high-temperature air [above 316°C (600°F)]. Normal mating-face materials for carbon are tungsten or chromium carbide, hard steel, stainless steel, or one of the cast irons.

Other sealing-face combinations that have been satisfactory in corrosive service are carbide against carbide, ceramic against ceramic, ceramic against carbon, and carbon against glass. The ceramics have also been mated with the various hard-facing alloys. When selecting seal materials the possibility of galvanic corrosion must also be considered.

BEARINGS

Many factors enter into the selection of the proper design for bearings. Some of these factors are:

1. Shaft speed range.
2. Maximum shaft misalignment that can be tolerated.
3. Critical speed analysis and the influence of bearing stiffness on this analysis
4. Loading of the compressor impellers
5. Oil temperatures and viscosity
6. Foundation stiffness
7. Axial movement that can be tolerated
8. Type of lubrication system and its contamination
9. Maximum vibration levels that can be tolerated

Types of Bearings Figure 10-119 shows a number of different types of journal bearings. A description of a few of the pertinent types of journal bearings is given here:

1. *Plain journal.* Bearing is bored with equal amounts of clearance (on the order of one and one-half to two thousands of an inch per inch of journal diameter) between the journal and bearing.

2. *Circumferential grooved bearing.* Normally the oil groove is half the bearing length. This configuration provides better coolings but reduces load capacity by dividing the bearing into two parts.

3. *Cylindrical bore bearings.* Another common bearing type used in turbines. It has a split construction with two axial oil-feed grooves at the split.

4. *Pressure or pressure dam.* Used in many places where bearing stability is required, this bearing is a plain journal bearing with a pressure pocket cut in the unloaded half. This pocket is approximately $\frac{1}{32}$ of an inch deep with a width 50 percent of the bearing length. This groove or channel covers an arc of 135° and terminates abruptly in a sharp-edge-edge dam. The direction or rotation is such that the oil is pumped down the channel toward the sharp edge. Pressure dam bearings are for one direction of rotation. They can be used in conjunction with cylindrical bore bearings as shown in Fig. 10-119.

5. *Lemon bore or elliptical.* This bearing is bored with shims split line, which are removed before installation. The resulting shape approximates an ellipse with the major axis clearance approximately twice the minor axis clearance. Elliptical bearings are for both directions of rotation.

6. *Three-lobe bearing.* The three-lobe bearing is not commonly used in turbomachines. It has a moderate load-carrying capacity and can be operated in both directions.

7. *Offset halves.* In principle, this bearing acts very similar to a pressure dam bearing. Its load-carrying capacity is good. It is restricted to one direction of rotation.

8. *Tilt-pad bearings.* This bearing is the most common bearing type in today's machines. It consists of several bearing pads posed around the circumference of the shaft. Each pad is able to tilt to assume the most effective working position. This bearing also offers the greatest increase in fatigue life because of the following advantages:

• Thermal conductive backing material to dissipate heat developed in oil film.

• A thin babbitt layer can be centrifugally cast with a uniform thickness of about 0.005 inch. Thick babbitts greatly reduce bearing life. Babbitt thickness in the neighborhood of .01 reduce the bearing life by more than half.

• Oil film thickness is critical in bearing stiffness calculations. In a tilting-pad bearing, one can change this thickness in a number of ways: (a) change the number of pads; (b) direct the load on or in between the pads; (c) change the axial length of pad.

Bearing type	Load capacity	Suitable direction of rotation	Resistance to half-speed whirl	Stiffness and damping
Cylindrical bore	Good		Worst	Moderate
Cylindrical bore with dammed groove	Good			Moderate
Lemon bore	Good			Moderate
Three lobe	Moderate		Increasing	Good
Offset halves	Good			Excellent
Tilting pad	Moderate		Best	Good

FIG. 10-119 Comparison of general bearing types.

Bearing type		Load capacity	Suitable direction of rotation	Tolerance of changing load/speed	Tolerance of misalignment	Space requirement
Plain washer 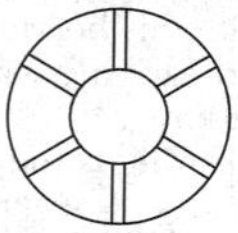		Poor	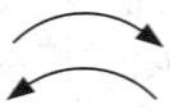	Good	Moderate	Compact
Taper land 	Bidirectional	Moderate	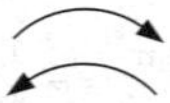	Poor	Poor	Compact
	Unidirectional	Good		Poor	Poor	Compact
Tilting pad 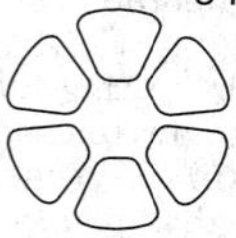	Bidirectional	Good		Good	Good	Greater
	Unidirectional	Good		Good	Good	Greater

FIG. 10-120 Comparison of thrust-bearing types.

The previous list contains some of the most common types of journal bearings. They are listed in the order of growing stability. All of the bearings designed for increased stability are obtained at higher manufacturing costs and reduced efficiency. The antiwhirl bearings all impose a parasitic load on the journal, which causes higher-power losses to the bearings and in turn requires higher oil flow to cool the bearing.

Thrust Bearings The most important function of a thrust bearing is to resist the unbalanced force in a machine's working fluid and to maintain the rotor in its position (within prescribed limits). A complete analysis of the thrust load must be conducted. As mentioned earlier, compressors with back-to-back rotors reduce this load greatly on thrust bearings. Figure 10-120 shows a number of thrust-bearing types. Plain, grooved thrust washers are rarely used with any continuous load, and their use tends to be confined to cases where the thrust load is very short duration or possibly occurs at standstill or low speed only. Occasionally, this type of bearing is used for light loads (less than 50 lb/in²), and in these circumstances the operation is probably hydrodynamic due to small distortions present in the nominally flat bearing surface.

When significant continuous loads have to be taken on a thrust washer, it is necessary to machine into the bearing surface a profile to generate a fluid film. This profile can be either a tapered wedge or occasionally a small step.

The tapered-land thrust bearing, when properly designed, can take and support a load equal to a tilting-pad thrust bearing. With perfect alignment, it can match the load of even a self-equalizing tilting-pad thrust bearing that pivots on the back of the pad along a radial line. For variable-speed operation, tilting-pad thrust bearings as shown in Fig. 10-121 are advantageous when compared to conventional taper-land bearings. The pads are free to pivot to form a proper angle for lubrication over a wide speed range. The self-leveling feature equalizes individual pad loadings and reduces the sensitivity to shaft misalignments that may occur during service. The major drawback of this bearing type is that standard designs require more axial space than a nonequalizing thrust bearing.

The thrust-carrying capacity can be greatly improved by maintaining pad flatness and removing heat from the loaded zone. By the use of high thermal conductivity backing materials with proper thickness

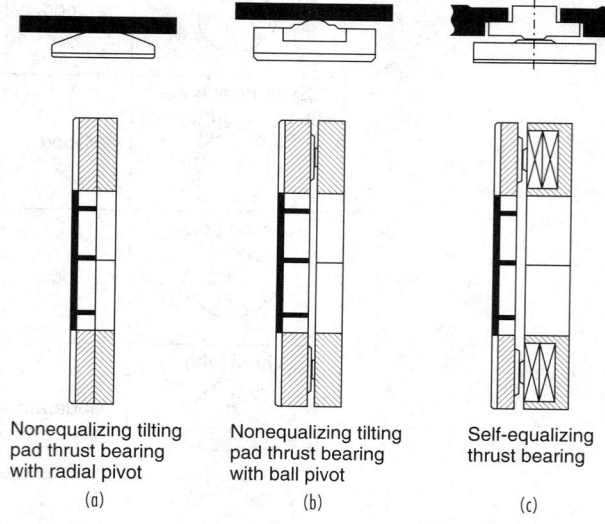

Nonequalizing tilting pad thrust bearing with radial pivot

(a)

Nonequalizing tilting pad thrust bearing with ball pivot

(b)

Self-equalizing thrust bearing

(c)

FIG. 10-121 Various types of thrust bearings.

and proper support, the maximum continuous thrust limit can be increased to 1000 psi or more. This new limit can be used to increase either the factor of safety and improve the surge capacity of a given size bearing or reduce the thrust bearing size and consequently the losses generated for a given load.

Since the higher thermal conductivity material (copper or bronze) is a much better bearing material than the conventional steel backing, it is possible to reduce the babbitt thickness to .010–.030 inch. Embedded thermocouples and RTDs will signal distress in the bearing if properly positioned. Temperature-monitoring systems have been found to be more accurate than axial-position indicators, which tend to have linearity problems at high temperatures.

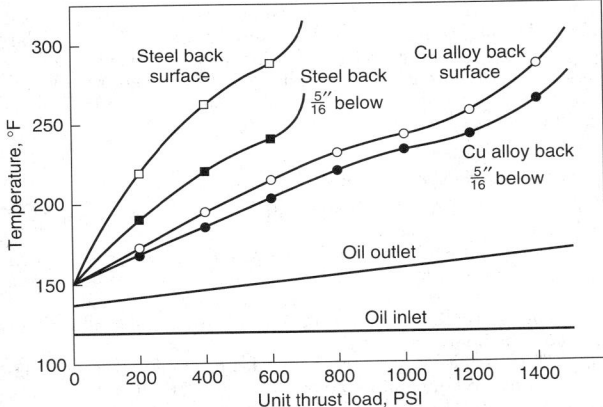

FIG. 10-122 Thrust-bearing temperature characteristics.

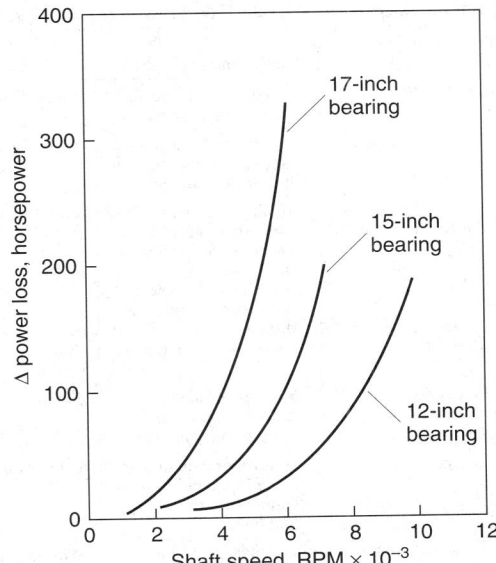

FIG. 10-123 Difference in total-power-loss data test minus catalog frictional losses versus shaft speed for 6 × 6 pad double-element thrust bearings.

In a change from steel-backing to copper-backing, a different set of temperature limiting criteria should be used. Figure 10-122 shows a typical set of curves for the two backing materials. This chart also shows that drain oil temperature is a poor indicator of bearing operating conditions because there is very little change in drain oil temperature from low load to failure load.

Thrust-Bearing Power Loss   The power consumed by various thrust bearing types is an important consideration in any system. Power losses must be accurately predicted so that turbine efficiency can be computed and the oil supply system properly designed. Figure 10-123 shows a typical power consumption in thrust bearings as a function of unit speed. The total power loss is usually about 0.8–10 percent of the total rate power of the unit. New vector lube bearings reduce the horsepower loss by as much as 30 percent. In large vertical pumps, thrust bearings take not only the load caused by

the fluid but also the load caused by the weight of the entire assembly (shaft and impellers). In some large pumps these could be about 60 ft (20 m) high and weigh 16 tons. The thrust bearing for such a pump is over 5 ft (1.7 meters) in diameter with each thrust pad weighing over 110 lb or (50 kg). In such cases, the entire pump assembly is first floated before the unit is started.

PROCESS-PLANT PIPING

CODES AND STANDARDS

Units: Pipe and Tubing Sizes and Ratings   In this subsection pipe and tubing sizes are generally quoted in units of inches. To convert inches to millimeters, multiply by 25.4. Ratings are given in pounds. To convert pounds to kilograms, multiply by 0.454.

Pressure-Piping Code   The code for pressure piping (ANSI B31) consists of a number of sections which collectively constitute the code. Table 10-14 shows the status of the B31 code as of December 1980. The sections are published as separate documents for simplicity and convenience. The sections differ extensively.

The Chemical Plant and Petroleum Refinery Piping Code (ANSI B31.3) is a section of ANSI B31. It was derived from a merging of the code groups for chemical-plant (B31.6) and petroleum-refinery (B31.3) piping into a single committee. Some of the significant requirements of ANSI B31.3, Petroleum Refinery Piping (1980 edition), are summarized in the following presentation, which is aimed primarily at welded and seamless construction.

Where the word "code" is used in this subsection of the *Handbook* without other identification, it refers to the B31.3 section of ANSI B31. The code has been extensively quoted in this subsection of the *Handbook* with the permission of the publisher. The code is published by and copies are available from the American Society of Mechanical Engineers (ASME), 345 East 47th Street, New York, New York 10017. References to the ASME code are to the ASME Boiler and Pressure Vessel Code, also published by the American Society of Mechanical Engineers.

National Standards   The American National Standards Institute (ANSI) and the American Petroleum Institute (API) have established dimensional standards for the most widely used piping components. Lists of these standards as well as specifications for pipe and fitting materials and testing methods of the American Society for Testing and Materials (ASTM), American Welding Society (AWS) specifications, and standards of the Manufacturers Standardization Society of the Valve and Fittings Industry (MSS) can be found in the ANSI B31 code sections. Many of these standards contain pressure-temperature ratings which will be of assistance to engineers in their design function. The use of published standards does not eliminate the need for engineering judgment. For example, although the code calculation formulas recognize the need to provide an allowance for corrosion, the standard rating tables for valves, flanges, fittings, etc., do not incorporate a corresponding allowance.

The introduction to the code sets forth engineering requirements deemed necessary for the safe design and construction of piping systems. While safety is the basic consideration of the code, this factor alone will not necessarily govern final specifications for any pressure piping system.

Designers are cautioned that the code is not a design handbook and does not do away with the need for competent engineering judgment.

Governmental Regulations: OSHA   Sections of the ANSI B31 code have been adopted with certain reservations or revisions by some state and local authorities as local codes.

The specific requirements for piping systems in certain services have been promulgated as Occupational Safety and Health Act

TABLE 10-14 Status of ANSI B31 Code for Pressure Piping

Standard number and designation	Scope and application	Remarks*
B31.1.0 Power Piping	For all piping in steam-generating stations	Latest issue: 1980
B31.2 Fuel Gas Piping	For fuel gas for steam-generating stations and industrial buildings	Latest issue: 1968
B31.3 Chemical Plant and Petroleum Refinery Piping	For all piping within the property limits of facilities engaged in the processing or handling of chemical, petroleum, or related products unless specifically excluded by the code	Latest issue: 1980
B31.4 Liquid Petroleum Transportation Piping Systems	For liquid crude or refined products in cross-country pipe lines	Latest issue: 1979
B31.5 Refrigeration Piping	For refrigeration piping in packaged units and commercial or public buildings	Latest issue: 1974
B31.7 Nuclear Power Piping	For fluids whose loss from the system could cause radiation hazard to plant personnel or the general public	Withdrawn; see ASME Boiler and Pressure Vessel Code, Sec. 3
B31.8 Gas Transmission and Distribution Systems	For gases in cross-country pipe lines as well as for city distribution lines	Latest issue: 1975

*Addenda are issued at intervals between publication of complete editions. Information on the latest issues can be obtained from the American Society of Mechanical Engineers, 345 East 47th Street, New York, N.Y. 10017.

(OSHA) regulations. These rules and regulations will presumably be revised and supplemented from time to time and may include specific requirements not contemplated in Sec. B31.3.

CODE CONTENTS AND SCOPE

The code prescribes minimum requirements for the materials, design, fabrication, assembly, support, erection, examination, inspection, and testing of piping systems subject to pressure or vacuum. The scope of the piping covered by B31.3 is illustrated in Fig. 10-124. It applies to all fluids including fluidized solids and to all services except as noted in the figure.

Some of the more significant requirements of ANSI B31.3 (1980 edition) have been summarized and incorporated in this section of the *Handbook*. For a more comprehensive treatment of code requirements engineers are referred to the B31.3 code and the standards referenced therein.

PIPE-SYSTEM MATERIALS

The selection of material to resist deterioration in service is outside the scope of the B31.3 code (see Sec. 23). Experience has, however, resulted in the following material considerations extracted from the code with the permission of the publisher, the American Society of Mechanical Engineers, New York.

General Considerations Considerations to be evaluated when selecting piping materials are (1) possible exposure to fire with respect to the loss of strength, degradation temperature, melting point, or combustibility of the pipe or support material; (2) ability of thermal insulation to protect the pipe from fire; (3) susceptibility of the pipe to brittle failure, possibly resulting in fragmentation hazards, or failure from thermal shock when exposed to fire or fire-fighting measures; (4) susceptibility of the piping material to crevice corrosion in stag-

nant confined areas (screwed joints) or adverse electrolytic effects if the metal is subject to contact with a dissimilar metal; (5) the suitability of packing, seals, gaskets, and lubricants or sealants used on threads as well as compatibility with the fluid handled; and (6) the refrigerating effect of a sudden loss of pressure on volatile fluids in determining the lowest expected service temperature.

Specific Material Precautions

Metals The following characteristics are to be evaluated when applying certain metals in piping:

1. *Irons: cast, malleable, and high silicon (14.5 percent).* Their lack of ductility and their sensitivity to thermal and mechanical shock.

2. *Carbon steel and low- and intermediate-alloy steels*
   a. The possibility of embrittlement when handling alkaline or strong caustic fluids.
   b. The possible conversion of carbides to graphite during long-time exposure to temperature above 427°C (800°F) of carbon steels, plain nickel steel, carbon-manganese steel, manganese-vanadium steel, and carbon-silicon steel.
   c. The possible conversion of carbides to graphite during long-time exposure to temperatures above 468°C (875°F) of carbon-molybdenum steel, manganese-molybdenum-vanadium steel, and chromium-vanadium steel.
   d. The advantages of silicon-killed carbon steel (0.1 percent silicon minimum) for temperatures above 480°C (900°F).
   e. The possibility of hydrogen damage when piping material is exposed to hydrogen or to aqueous acid solutions under certain temperature-pressure conditions.
   f. The possibility of deterioration when piping material is exposed to hydrogen sulfide.

3. *High-alloy (stainless) steels*
   a. The possibility of stress-corrosion cracking of austenitic stainless steels exposed to media such aa chlorides and other halides either internally or externally. The latter can result from improper selection or application of thermal insulation.
   b. The susceptibility to intergranular corrosion of austenitic stainless steels after sufficient exposure to temperatures between 427 and 871°C (800 and 1600°F) unless stabilized or low-carbon grades are used.
   c. The susceptibility to intercrystalline attack of austenitic stainless steels on contact with zinc or lead above their melting points or with many lead and zinc compounds at similarly elevated temperatures.
   d. The brittleness of ferritic stainless steels at room temperature after service at temperatures above 370°C (700°F).

4. *Nickel and nickel-base alloys*
   a. The susceptibility to grain boundary attack of nickel and nickel-base alloys not containing chromium when exposed to small quantities of sulfur at temperatures above 315°C (600°F).
   b. The susceptibility to grain boundary attack of nickel-base alloys containing chromium at temperatures above 595°C (1100°F) under reducing conditions and above 760°C (1400°F) under oxidizing conditions.
   c. The possibility of stress-corrosion cracking of nickel-copper alloy (70 Ni-30 Cu) in hydrofluoric acid vapor if the alloy is highly stressed or contains residual stresses from forming or welding.

5. *Aluminum and aluminum alloys*
   a. The compatibility with aluminum of thread compounds used in aluminum threaded joints to prevent seizing and galling.
   b. The possibility of corrosion from concrete, mortar, lime, plaster, or other alkaline materials used in buildings or other structures.
   c. The susceptibility of alloys 5154, 5087, 5083, and 5456 to exfoliation or intergranular attack; and the upper temperature limit of 65°C (150°F) to avoid such deterioration.

6. *Copper and copper alloys*
   a. The possibility of dezincification of brass alloys.
   b. The susceptibility to stress-corrosion cracking of copper-based alloys.
   c. The possibility of unstable acetylide formation when exposed to acetylene.

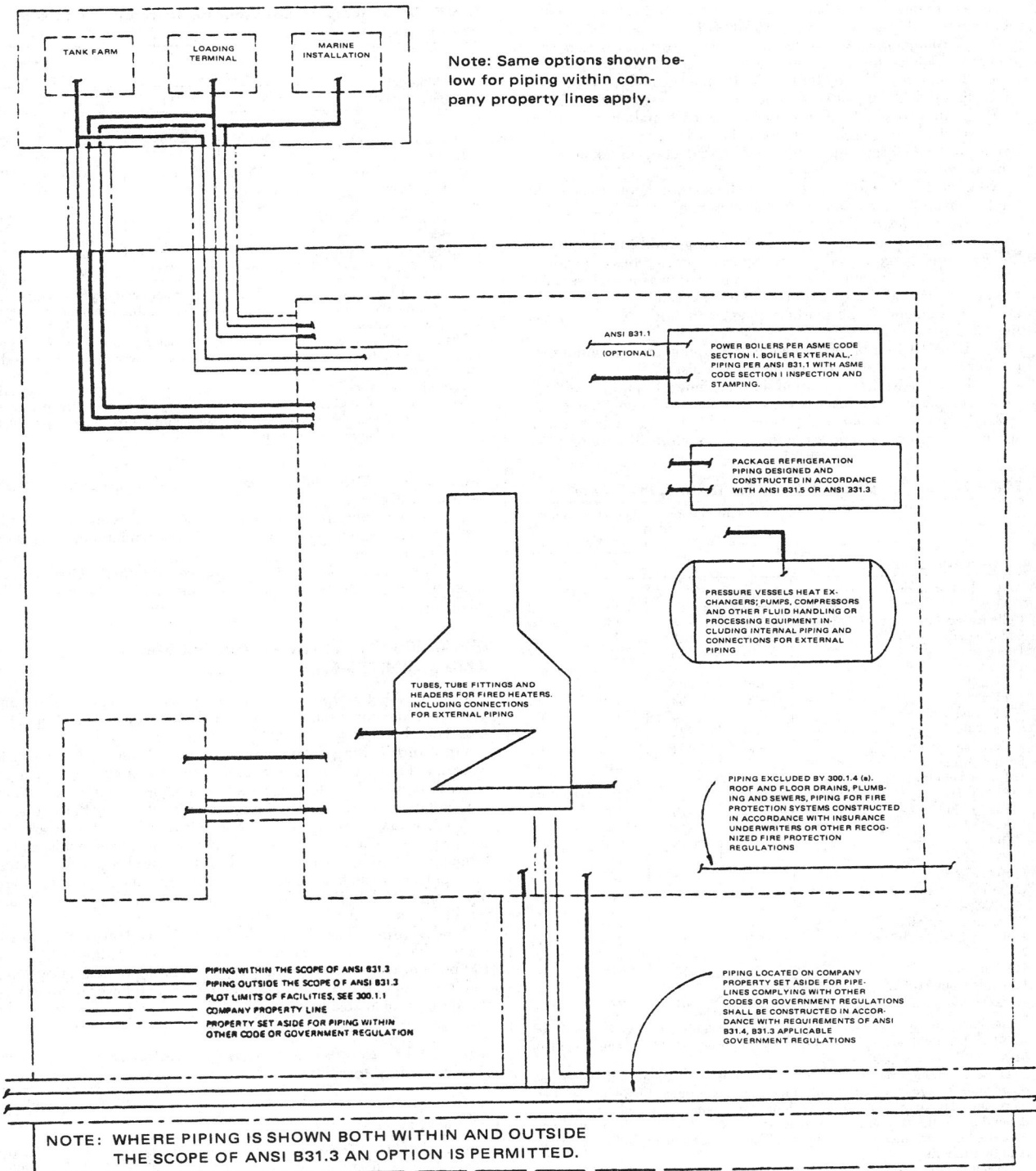

FIG. 10-124   Scope of piping covered by the Chemical Plant and Petroleum Refinery Piping Code, ANSI B31.3 (*From ASME Chemical Plant and Petroleum Refinery Piping Code, ANSI B31.3—1980; reproduced with permission of the publisher, the American Society of Mechanical Engineers, New York.*)

7. *Titanium and titanium alloys.* The possibility of deterioration of titanium and its alloys above 315°C (600°F).

8. *Zirconium and zirconium alloys.* The possibility of deterioration of zirconium and zirconium alloys above 315°C (600°F).

9. *Tantalum.* Above 300°C (570°F), the possibility of reactivity of tantalum with all gases except the inert gases. Below 300°C (570°F), the possibility of embrittlement of tantalum by nascent (monatomic) hydrogen (but not molecular hydrogen). Nascent hydrogen is produced by galvanic action or as a product of corrosion by certain chemicals.

Nonmetals   The following are specific considerations to be evaluated when applying certain nonmetals in piping:

1. *Thermoplastics*

a. If thermoplastic piping is used above ground for compressed air or other compressed gases, special precautions should be observed. In determining the needed safeguarding for such services, the energetics and the specific failure mechanism need to be evaluated. Encasement of the plastic piping in shatter-resistant material may be considered.

b. Table 10-15 lists recommended minimum and maximum temperature limits for thermoplastic pipe materials.

c. Table 10-16 lists minimum and maximum temperature limits for thermoplastic materials used as nonpressure retaining linings.

2. *Reinforced thermosetting resins.* Table 10-17 lists the normally accepted maximum temperature limits for reinforced-thermosetting-

TABLE 10-15   Temperature Limits for Thermoplastic Pipe*

| Material (generic type) | Recommended temperature limits | | | |
| | Minimum | | Maximum | |
	°F	°C	°F	°C
Acrylonitrile-butadiene-styrene (ABS)	−30	−34	180	82
Cellulose acetate butyrate (CAB)	0	−18	140	60
Chlorinated polyether	0	−18	210	99
Polyacetal	0	−18	170	77
Polyethylene				
PE 1404	−30	−34	100	38
PE 2305	−30	−34	120	49
PE 2306	−30	−34	140	60
PE 3306	−30	−34	160	71
PE 3406	−30	−34	180	82
Polypropylene	30	−01	210	99
Poly (vinyl chloride)				
PVC 1120	0	−18	150	66
PVC 1220	0	−18	150	66
PVC 2110	0	−18	130	54
PVC 2112	0	−18	130	54
PVC 2116	0	−18	150	66
PVC 2120	0	−18	150	66
Chlorinated poly (vinyl chloride) (CPVC, 4120)	0	−18	210	99
Poly (vinylidene chloride)	40	4	160	71
Poly (vinylidene fluoride)	0	−18	275	135
Nylon	−30	−34	180	82
Polybutylene	0	−18	210	99
Poly (phenylene oxide) (POP 2125)	30	−01	210	99

*Extracted from the Chemical Plant and Petroleum Refinery Piping Code, ANSI B31.3–1980, with permission of the publisher, the American Society of Mechanical Engineers, New York.

These recommendations are for low-pressure applications with water and other fluids that do not significantly affect the properties of the particular thermoplastic. The upper temperature limits are reduced at higher pressures, depending on the combination of fluid and expected service life. Lower temperature limits are affected more by installation, environment, and safeguarding than by strength.

Because of low thermal conductivity, temperature gradients through the pipe wall may be substantial. Tabulated limits apply where more than half the wall thickness is at or above the stated temperature.

These recommendations apply only to products covered by ASTM standards listed in Appendix A, Table 3, of the code. Manufacturers should be consulted for temperature limits on the specific types and kinds of plastic not covered by those ASTM standards.

TABLE 10-16   Temperature Limits for Thermoplastics Used as Linings*

| Material (generic type) | Minimum temperature | | Maximum temperature | |
	°F	°C	°F	°C
Poly (tetrafluoroethylene)	−325	−198	500	260
Poly (fluorinated ethylene propylene)	−325	−198	400	204
Poly (vinylidene chloride)	0	−18	175	79
Poly (vinylidene fluoride)	0	−18	275	135
Polypropylene	0	−18	225	107
Poly (perfluoroalkoxy)	−325	−198	500	260

*Extracted from the Chemical Plant and Petroleum Refinery Piping Code, ANSI B31.3–1980, with permission of the publisher, the American Society of Mechanical Engineers.

Listed temperature limits apply to lining material only. Rules for establishing temperature limits for components being listed are covered elsewhere in this code.

These temperature limits are based on material tests and do not necessarily reflect evidence of successful use as piping-component linings at these temperatures. The designer should contact the manufacturer for specific applications, particularly as temperature limits are approached.

resin materials. The minimum recommended temperature is −29°C (−20°F) in all cases.

3. *Asbestos cement.* The normally accepted temperature limits for asbestos cement piping are −18°C (0°F) minimum and 93°C (200°F) maximum.

4. *Borosilicate glass and impregnated graphite.* Their lack of ductility and sensitivity to thermal and mechanical shock should be taken into account.

METALLIC PIPE SYSTEMS: CARBON STEEL AND STAINLESS STEEL

The ferrous-metal piping systems comprising wrought carbon and alloy steels including stainless steels are the most widely used and the most completely covered by national standards.

Pipe and Tubing   Pipe and tubing are divided into two main classes, seamless and welded. Seamless pipe, as a trade designation, refers to pipe made by forging a solid round, piercing it by simultaneously rotating and forcing it over a piercer point and further reducing it by rolling and drawing. However, seamless pipe and tubing are also produced by extrusion, casting into static or centrifugal molds, and by forging and boring. Seamless pipe has the same kilopascal (pounds-force per square inch) strength throughout the wall. Pierced seamless pipe frequently has the inside surface eccentric to the outside surface, resulting in nonuniform wall thickness.

Welded pipe is made from rolled strips formed into cylinders and seam-welded by various methods. The welds are credited with 60 to 100 percent of the strength of the pipe wall depending on welding and inspection procedures. Larger diameters and lower ratios of wall thickness to diameter can be obtained in welded pipe than can be

TABLE 10-17   Temperature Limits for Reinforced Thermosetting Resins*†

| Material (generic type) | Maximum temperature | |
	°C	°F
Epoxy, glass-fiber-reinforced	149	300
Polyester, glass-fiber-reinforced	93	200
Furan, glass-fiber-reinforced	93	200
Furan, carbon-reinforced	93	200

*Extracted from the Chemical Plant and Petroleum Refinery Piping Code, ANSI B31.3–1980, with permission of the publisher, the American Society of Mechanical Engineers, New York.

†Minimum recommended temperature of all materials is −29°C (−20°F).

obtained in seamless pipe (other than cast pipe). Uniform wall thickness is obtained. Hydrostatic testing does not reveal very short lengths of partially completed weld. This presents a possibility that small leaks may develop prematurely when corrosive fluids are being handled or the pipe is exposed to external corrosion. The weld must be taken into account in developing procedures for bending, flaring, and expanding the welded pipe.

Additional thickness and additional size and wall-thickness combinations are available as tubing. Two common classifications of tubing are "pressure" and "mechanical." Wall thickness (gauge) is specified as either "average wall" or "minimum wall." Minimum wall is more costly than average wall, and because of closer wall-thickness and diametral tolerance, both gauge systems make pressure tubing more costly than pipe. However, average-wall carbon steel electric-resistance-welded tubing, sizes $2\frac{3}{8}$, $2\frac{7}{8}$, $3\frac{1}{2}$, and $4\frac{1}{2}$ in outside diameter, produced from coiled strip on progressive forming rolls and electromagnetically rather than pressure-tested, competes vigorously with pipe.

Table 10-18 gives standard size and wall-thickness combinations together with capacity and weight.

Joints   Pipe must be joined to pipe and to other components. Optimum design requires a minimum of assembly labor and provides the same resistance possessed by the pipe to (1) internal pressure as regards both rupture and leakage, (2) bending moments arising from spanning long distances between supports or from thermal expansion in piping containing offsets, (3) axial strain arising from internal pressure acting on changes in direction, blanks or closed valves, or thermal contraction in straight runs, and (4) rupture or leakage in event of fire.

However, joints in pipe buried in the soil, where the position of each length and component is fixed, need provide the same resistance as the pipe to internal pressure only; in event of earth settlement, the joints may be required to yield to resulting bending moments without leakage. Also, in piping subject to thermal expansion and contraction, some joints may be required to yield to resulting bending moments and axial strains without leakage.

The ideal pipe joint is free from changes in any dimension of the flow passage or direction of flow which would increase pressure drop or prevent complete drainage. It is free from crevices in which corrosion might be accelerated. It would require a minimum of labor to disassemble. Required frequency for disassembling the joint must be considered in making the selection. Generally speaking, joints which are easy to disassemble are deficient in one or more of the other requirements of the ideal joint.

Most joints involve modifications of the components being joined; those with the desired modifications can usually be purchased.

Welded Joints   The most widely used joint in piping systems is the **butt-weld joint** (Fig. 10-125). In all ductile pipe metals which can be welded, pipe, elbows, tees, laterals, reducers, caps, valves, flanges, and V-clamp joints are available in all sizes and wall thicknesses with ends prepared for butt welding. Joint strength equal to the original pipe (except for work-hardened pipes which are annealed by the welding), unimpaired flow pattern, and generally unimpaired corrosion resistance more than compensate for the necessary careful alignment, skilled labor, and equipment required.

Plain-end pipe used for socket-weld joints (Fig. 10-126) is available in all sizes, but fittings and valves with socket-weld ends are limited to sizes 3 in and smaller, for which the extra cost of the socket is outweighed by much easier alignment and less skill needed in welding. The joint is not so resistant to bending stress as the butt-welded joint but is otherwise equal, except that for some fluids the crevice between the pipe and the socket may promote corrosion. ANSI B16.11—1973,

Forged Steel Fittings, Socket-Welding and Threaded, requires that the wall thickness of the socket must be equal to or greater than 1.25 times the minimum pipe wall.

Branch Welds   These welds eliminate the purchase of tees and require no more weld metal than tees (Fig. 10-127). If the branch approaches the size of the run, careful end preparation of the branch pipe is required and the run pipe is weakened by the branch weld. See subsection "Pressure Design of Metallic Components: Wall Thickness" for rules for reinforcement. Reinforcing pads and fittings are commercially available. Use of the fittings facilitates visual inspection of the branch weld. See subsection "Welding, Brazing, or Soldering" for rules for welded joints.

Threaded Joints   Pipe with **taper-pipe-thread** ends (Fig. 10-128), per ANSI B2.1, is available 12 in and smaller, subject to minimum-wall limitations. Fittings and valves with taper-pipe-thread ends are available in most pipe metals.

Principal use of threaded joints is in sizes 2 in and smaller, in metals for which the most economically produced walls are thick enough to withstand considerable pressure and corrosion after reduction in thickness due to threading. For threaded joints over 2 in, assembly labor size and cost of tools increase rapidly. Careful alignment, required at the start of assembly and during rotation of the components, as well as variation in length produced by diametral tolerances in the threads, severely limits preassembly of the components. Threading is not a precise machining operation, and filler materials known as "pipe dope" are necessary to block the spiral leakage path.

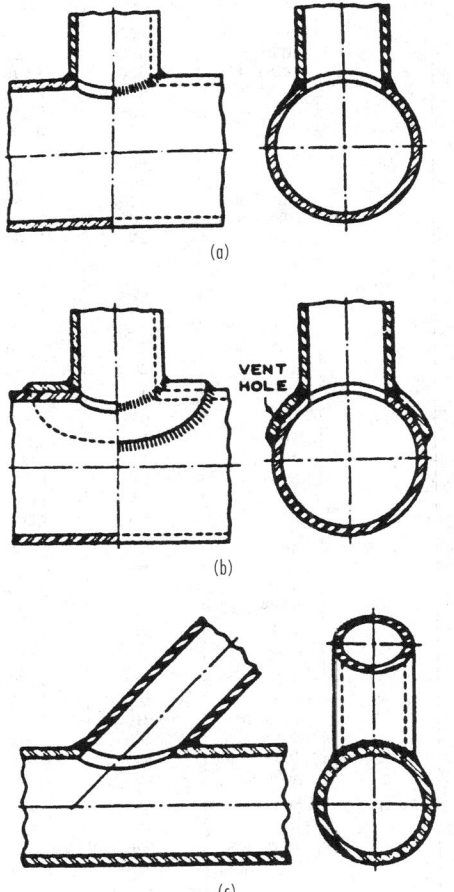

FIG. 10-127   Branch welds. (a) Without added reinforcement. (b) With added reinforcement. (c) Angular branch.

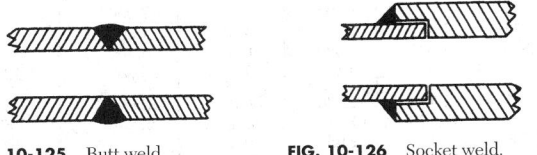

FIG. 10-125   Butt weld.   **FIG. 10-126**   Socket weld.

TABLE 10-18 Properties of Steel Pipe

Nominal pipe size, in	Outside diameter, in	Schedule no.	Wall thickness, in	Inside diameter, in	Cross-sectional area		Circumference, ft, or surface, ft²/ft of length		Capacity at 1-ft/s velocity		Weight of plain-end pipe, lb/ft
					Metal, in²	Flow, ft²	Outside	Inside	U.S. gal/min	lb/h water	
⅛	0.405	10S	0.049	0.307	0.055	0.00051	0.106	0.0804	0.231	115.5	0.19
		40ST, 40S	.068	.269	.072	.00040	.106	.0705	.179	89.5	.24
		80XS, 80S	.095	.215	.093	.00025	.106	.0563	.113	56.5	.31
¼	0.540	10S	.065	.410	.097	.00092	.141	.107	.412	206.5	.33
		40ST, 40S	.088	.364	.125	.00072	.141	.095	.323	161.5	.42
		80XS, 80S	.119	.302	.157	.00050	.141	.079	.224	112.0	.54
⅜	0.675	10S	.065	.545	.125	.00162	.177	.143	.727	363.5	.42
		40ST, 40S	.091	.493	.167	.00133	.177	.129	.596	298.0	.57
		80XS, 80S	.126	.423	.217	.00098	.177	.111	.440	220.0	.74
½	0.840	5S	.065	.710	.158	.00275	.220	.186	1.234	617.0	.54
		10S	.083	.674	.197	.00248	.220	.176	1.112	556.0	.67
		40ST, 40S	.109	.622	.250	.00211	.220	.163	0.945	472.0	.85
		80XS, 80S	.147	.546	.320	.00163	.220	.143	0.730	365.0	1.09
		160	.188	.464	.385	.00117	.220	.122	0.527	263.5	1.31
		XX	.294	.252	.504	.00035	.220	.066	0.155	77.5	1.71
¾	1.050	5S	.065	.920	.201	.00461	.275	.241	2.072	1036.0	0.69
		10S	.083	.884	.252	.00426	.275	.231	1.903	951.5	0.86
		40ST, 40S	.113	.824	.333	.00371	.275	.216	1.665	832.5	1.13
		80XS, 80S	.154	.742	.433	.00300	.275	.194	1.345	672.5	1.47
		160	.219	.612	.572	.00204	.275	.160	0.917	458.5	1.94
		XX	.308	.434	.718	.00103	.275	.114	0.461	230.5	2.44
1	1.315	5S	.065	1.185	.255	.00768	.344	.310	3.449	1725	0.87
		10S	.109	1.097	.413	.00656	.344	.287	2.946	1473	1.40
		40ST, 40S	.133	1.049	.494	.00600	.344	.275	2.690	1345	1.68
		80XS, 80S	.179	0.957	.639	.00499	.344	.250	2.240	1120	2.17
		160	.250	0.815	.836	.00362	.344	.213	1.625	812.5	2.84
		XX	.358	0.599	1.076	.00196	.344	.157	0.878	439.0	3.66
1¼	1.660	5S	.065	1.530	0.326	.01277	.435	.401	5.73	2865	1.11
		10S	.109	1.442	0.531	.01134	.435	.378	5.09	2545	1.81
		40ST, 40S	.140	1.380	0.668	.01040	.435	.361	4.57	2285	2.27
		80XS, 80S	.191	1.278	0.881	.00891	.435	.335	3.99	1995	3.00
		160	.250	1.160	1.107	.00734	.435	.304	3.29	1645	3.76
		XX	.382	0.896	1.534	.00438	.435	.235	1.97	985	5.21
1½	1.900	5S	.065	1.770	0.375	.01709	.497	.463	7.67	3835	1.28
		10S	.109	1.682	0.614	.01543	.497	.440	6.94	3465	2.09
		40ST, 40S	.145	1.610	0.800	.01414	.497	.421	6.34	3170	2.72
		80XS, 80S	.200	1.500	1.069	.01225	.497	.393	5.49	2745	3.63
		160	.281	1.338	1.429	.00976	.497	.350	4.38	2190	4.86
		XX	.400	1.100	1.885	.00660	.497	.288	2.96	1480	6.41
2	2.375	5S	.065	2.245	0.472	.02749	.622	.588	12.34	6170	1.61
		10S	.109	2.157	0.776	.02538	.622	.565	11.39	5695	2.64
		40ST, 40S	.154	2.067	1.075	.02330	.622	.541	10.45	5225	3.65
		80ST, 80S	.218	1.939	1.477	.02050	.622	.508	9.20	4600	5.02
		160	.344	1.687	2.195	.01552	.622	.436	6.97	3485	7.46
		XX	.436	1.503	2.656	.01232	.622	.393	5.53	2765	9.03
2½	2.875	5S	.083	2.709	0.728	.04003	.753	.709	17.97	8985	2.48
		10S	.120	2.635	1.039	.03787	.753	.690	17.00	8500	3.53
		40ST, 40S	.203	2.469	1.704	.03322	.753	.647	14.92	7460	5.79
		80XS, 80S	.276	2.323	2.254	.02942	.753	.608	13.20	6600	7.66
		160	.375	2.125	2.945	.02463	.753	.556	11.07	5535	10.01
		XX	.552	1.771	4.028	.01711	.753	.464	7.68	3840	13.69
3	3.500	5S	.083	3.334	0.891	.06063	.916	.873	27.21	13,605	3.03
		10S	.120	3.260	1.274	.05796	.916	.853	26.02	13,010	4.33
		40ST, 40S	.216	3.068	2.228	.05130	.916	.803	23.00	11,500	7.58
		80XS, 80S	.300	2.900	3.016	.04587	.916	.759	20.55	10,275	10.25
		160	.438	2.624	4.213	.03755	.916	.687	16.86	8430	14.32
		XX	.600	2.300	5.466	.02885	.916	.602	12.95	6475	18.58
3⅓	4.0	5S	.083	3.834	1.021	.08017	1.047	1.004	35.98	17,990	3.48
		10S	.120	3.760	1.463	.07711	1.047	.984	34.61	17,305	4.97
		40ST, 40S	.226	3.548	2.680	.06870	1.047	.929	30.80	15,400	9.11
		80XS, 80S	.318	3.364	3.678	.06170	1.047	.881	27.70	13,850	12.50
4	4.5	5S	.083	4.334	1.152	.10245	1.178	1.135	46.0	23,000	3.92
		10S	.120	4.260	1.651	.09898	1.178	1.115	44.4	22,200	5.61
		40ST, 40S	.237	4.026	3.17	.08840	1.178	1.054	39.6	19,800	10.79
		80XS, 80S	.337	3.826	4.41	.07986	1.178	1.002	35.8	17,900	14.98

TABLE 10-18   Properties of Steel Pipe (*Continued*)

Nominal pipe size, in	Outside diameter, in	Schedule no.	Wall thickness, in	Inside diameter, in	Cross-sectional area		Circumference, ft, or surface, ft²/ft of length		Capacity at 1-ft/s velocity		Weight of plain-end pipe, lb/ft
					Metal, in²	Flow, ft²	Outside	Inside	U.S. gal/min	lb/h water	
		120	0.438	3.624	5.58	0.07170	1.178	0.949	32.2	16,100	19.00
		160	.531	3.438	6.62	.06647	1.178	0.900	28.9	14,450	22.51
		XX	.674	3.152	8.10	.05419	1.178	0.825	24.3	12,150	27.54
5	5.563	5S	.109	5.345	1.87	.1558	1.456	1.399	69.9	34,950	6.36
		10S	.134	5.295	2.29	.1529	1.456	1.386	68.6	34,300	7.77
		40ST, 40S	.258	5.047	4.30	.1390	1.456	1.321	62.3	31,150	14.62
		80XS, 80S	.375	4.813	6.11	.1263	1.456	1.260	57.7	28,850	20.78
		120	.500	4.563	7.95	.1136	1.456	1.195	51.0	25,500	27.04
		160	.625	4.313	9.70	.1015	1.456	1.129	45.5	22,750	32.96
		XX	.750	4.063	11.34	.0900	1.456	1.064	40.4	20,200	38.55
6	6.625	5S	.109	6.407	2.23	.2239	1.734	1.677	100.5	50,250	7.60
		10S	.134	6.357	2.73	.2204	1.734	1.664	98.9	49,450	9.29
		40ST, 40S	.280	6.065	5.58	.2006	1.734	1.588	90.0	45,000	18.97
		80XS, 80S	.432	5.761	8.40	.1810	1.734	1.508	81.1	40,550	28.57
		120	.562	5.501	10.70	.1650	1.734	1.440	73.9	36,950	36.39
		160	.719	5.187	13.34	.1467	1.734	1.358	65.9	32,950	45.34
		XX	.864	4.897	15.64	.1308	1.734	1.282	58.7	29,350	53.16
8	8.625	5S	.109	8.407	2.915	.3855	2.258	2.201	173.0	86,500	9.93
		10S	.148	8.329	3.941	.3784	2.258	2.180	169.8	84,900	13.40
		20	.250	8.125	6.578	.3601	2.258	2.127	161.5	80,750	22.36
		30	.277	8.071	7.265	.3553	2.258	2.113	159.4	79,700	24.70
		40ST, 40S	.322	7.981	8.399	.3474	2.258	2.089	155.7	77,850	28.55
		60	.406	7.813	10.48	.3329	2.258	2.045	149.4	74,700	35.64
		80XS, 80S	.500	7.625	12.76	.3171	2.258	1.996	142.3	71,150	43.39
		100	.594	7.437	14.99	.3017	2.258	1.947	135.4	67,700	50.95
		120	.719	7.187	17.86	.2817	2.258	1.882	126.4	63,200	60.71
		140	.812	7.001	19.93	.2673	2.258	1.833	120.0	60,000	67.76
		XX	.875	6.875	21.30	.2578	2.258	1.800	115.7	57,850	72.42
		160	.906	6.813	21.97	.2532	2.258	1.784	113.5	56,750	74.69
10	10.75	5S	.134	10.482	4.47	.5993	2.814	2.744	269.0	134,500	15.19
		10S	.165	10.420	5.49	.5922	2.814	2.728	265.8	132,900	18.65
		20	.250	10.250	8.25	.5731	2.814	2.685	257.0	128,500	28.04
		30	.307	10.136	10.07	.5603	2.814	2.655	252.0	126,000	34.24
		40ST, 40S	.365	10.020	11.91	.5475	2.814	2.620	246.0	123,000	40.48
		80S, 60XS	.500	9.750	16.10	.5185	2.814	2.550	233.0	116,500	54.74
		80	.594	9.562	18.95	.4987	2.814	2.503	223.4	111,700	64.43
		100	.719	9.312	22.66	.4729	2.814	2.438	212.3	106,150	77.03
		120	.844	9.062	26.27	.4479	2.814	2.372	201.0	100,500	89.29
		140, XX	1.000	8.750	30.63	.4176	2.814	2.291	188.0	94,000	104.13
		160	1.125	8.500	34.02	.3941	2.814	2.225	177.0	88,500	115.64
12	12.75	5S	0.156	12.438	6.17	.8438	3.338	3.26	378.7	189,350	20.98
		10S	0.180	12.390	7.11	.8373	3.338	3.24	375.8	187,900	24.17
		20	0.250	12.250	9.82	.8185	3.338	3.21	367.0	183,500	33.38
		30	0.330	12.090	12.88	.7972	3.338	3.17	358.0	179,000	43.77
		ST, 40S	0.375	12.000	14.58	.7854	3.338	3.14	352.5	176,250	49.56
		40	0.406	11.938	15.74	.7773	3.338	3.13	349.0	174,500	53.52
		XS, 80S	0.500	11.750	19.24	.7530	3.338	3.08	338.0	169,000	65.42
		60	0.562	11.626	21.52	.7372	3.338	3.04	331.0	165,500	73.15
		80	0.688	11.374	26.07	.7056	3.338	2.98	316.7	158,350	88.63
		100	0.844	11.062	31.57	.6674	3.338	2.90	299.6	149,800	107.32
		120, XX	1.000	10.750	36.91	.6303	3.338	2.81	283.0	141,500	125.49
		140	1.125	10.500	41.09	.6013	3.338	2.75	270.0	135,000	139.67
		160	1.312	10.126	47.14	.5592	3.338	2.65	251.0	125,500	160.27
14	14	5S	0.156	13.688	6.78	1.0219	3.665	3.58	459	229,500	23.07
		10S	0.188	13.624	8.16	1.0125	3.665	3.57	454	227,000	27.73
		10	0.250	13.500	10.80	0.9940	3.665	3.53	446	223,000	36.71
		20	0.312	13.376	13.42	0.9750	3.665	3.50	438	219,000	45.61
		30, ST	0.375	13.250	16.05	0.9575	3.665	3.47	430	215,000	54.57
		40	0.438	13.124	18.66	0.9397	3.665	3.44	422	211,000	63.44
		XS	0.500	13.000	21.21	0.9218	3.665	3.40	414	207,000	72.09
		60	0.594	12.812	25.02	0.8957	3.665	3.35	402	201,000	85.05
		80	0.750	12.500	31.22	0.8522	3.665	3.27	382	191,000	106.13
		100	0.938	12.124	38.49	0.8017	3.665	3.17	360	180,000	130.85
		120	1.094	11.812	44.36	0.7610	3.665	3.09	342	171,000	150.79
		140	1.250	11.500	50.07	0.7213	3.665	3.01	324	162,000	170.21
		160	1.406	11.188	55.63	0.6827	3.665	2.93	306	153,000	189.11
16	16	5S	0.165	15.670	8.21	1.3393	4.189	4.10	601	300,500	27.90
		10S	0.188	15.624	9.34	1.3314	4.189	4.09	598	299,000	31.75
		10	0.250	15.500	12.37	1.3104	4.189	4.06	587	293,500	42.05

TABLE 10-18 Properties of Steel Pipe (Concluded)

Nominal pipe size, in	Outside diameter, in	Schedule no.	Wall thickness, in	Inside diameter, in	Cross-sectional area		Circumference, ft, or surface, ft²/ft of length		Capacity at 1-ft/s velocity		Weight of plain-end pipe, lb/ft
					Metal, in²	Flow, ft²	Outside	Inside	U.S. gal/min	lb/h water	
		20	0.312	15.376	15.38	1.2985	4.189	4.03	578	289,000	52.27
		30, ST	0.375	15.250	18.41	1.2680	4.189	3.99	568	284,000	62.58
		40, XS	0.500	15.000	24.35	1.2272	4.189	3.93	550	275,000	82.77
		60	0.656	14.688	31.62	1.1766	4.189	3.85	528	264,000	107.50
		80	0.844	14.312	40.19	1.1171	4.189	3.75	501	250,500	136.61
		100	1.031	13.938	48.48	1.0596	4.189	3.65	474	237,000	164.82
		120	1.219	13.562	56.61	1.0032	4.189	3.55	450	225,000	192.43
		140	1.438	13.124	65.79	0.9394	4.189	3.44	422	211,000	223.64
		160	1.594	12.812	72.14	0.8953	4.189	3.35	402	201,000	245.25
18	18	5S	0.165	17.670	9.25	1.7029	4.712	4.63	764	382,000	31.43
		10S	0.188	17.624	10.52	1.6941	4.712	4.61	760	379,400	35.76
		10	0.250	17.500	13.94	1.6703	4.712	4.58	750	375,000	47.39
		20	0.312	17.376	17.34	1.6468	4.712	4.55	739	369,500	58.94
		ST	0.375	17.250	20.76	1.6230	4.712	4.52	728	364,000	70.59
		30	0.438	17.124	24.16	1.5993	4.712	4.48	718	359,000	82.15
		XS	0.500	17.000	27.49	1.5763	4.712	4.45	707	353,500	93.45
		40	0.562	16.876	30.79	1.5533	4.712	4.42	697	348,500	104.67
		60	0.750	16.500	40.64	1.4849	4.712	4.32	666	333,000	138.17
		80	0.938	16.124	50.28	1.4180	4.712	4.22	636	318,000	170.92
		100	1.156	15.688	61.17	1.3423	4.712	4.11	602	301,000	207.96
		120	1.375	15.250	71.82	1.2684	4.712	3.99	569	284,500	244.14
		140	1.562	14.876	80.66	1.2070	4.712	3.89	540	270,000	274.22
		160	1.781	14.438	90.75	1.1370	4.712	3.78	510	255,000	308.50
20	20	5S	0.188	19.624	11.70	2.1004	5.236	5.14	943	471,500	39.78
		10S	0.218	19.564	13.55	2.0878	5.236	5.12	937	467,500	46.06
		10	0.250	19.500	15.51	2.0740	5.236	5.11	930	465,000	52.73
		20, ST	0.375	19.250	23.12	2.0211	5.236	5.04	902	451,000	78.60
		30, XS	0.500	19.000	30.63	1.9689	5.236	4.97	883	441,500	104.13
		40	0.594	18.812	36.21	1.9302	5.236	4.92	866	433,000	123.11
		60	0.812	18.376	48.95	1.8417	5.236	4.81	826	413,000	166.40
		80	1.031	17.938	61.44	1.7550	5.236	4.70	787	393,500	208.87
		100	1.281	17.438	75.33	1.6585	5.236	4.57	744	372,000	256.10
		120	1.500	17.000	87.18	1.5763	5.236	4.45	707	353,500	296.37
		140	1.750	16.500	100.3	1.4849	5.236	4.32	665	332,500	341.09
		160	1.969	16.062	111.5	1.4071	5.236	4.21	632	316,000	397.17
24	24	5S	0.218	23.564	16.29	3.0285	6.283	6.17	1359	679,500	55.37
		10, 10S	0.250	23.500	18.65	3.012	6.283	6.15	1350	675,000	63.41
		20, ST	0.375	23.250	27.83	2.948	6.283	6.09	1325	662,500	94.62
		XS	0.500	23.000	36.90	2.885	6.283	6.02	1295	642,500	125.49
		30	0.562	22.876	41.39	2.854	6.283	5.99	1281	640,500	140.68
		40	0.688	22.624	50.39	2.792	6.283	5.92	1253	626,500	171.29
		60	0.969	22.062	70.11	2.655	6.283	5.78	1192	596,000	238.35
		80	1.219	21.562	87.24	2.536	6.283	5.64	1138	569,000	296.58
		100	1.531	20.938	108.1	2.391	6.283	5.48	1073	536,500	367.39
		120	1.812	20.376	126.3	2.264	6.283	5.33	1016	508,000	429.39
		140	2.062	19.876	142.1	2.155	6.283	5.20	965	482,500	483.12
		160	2.344	19.312	159.5	2.034	6.283	5.06	913	456,500	542.13
30	30	5S	0.250	29.500	23.37	4.746	7.854	7.72	2130	1,065,000	79.43
		10, 10S	0.312	29.376	29.10	4.707	7.854	7.69	2110	1,055,000	98.93
		ST	0.375	29.250	34.90	4.666	7.854	7.66	2094	1,048,000	118.65
		20, XS	0.500	29.000	46.34	4.587	7.854	7.59	2055	1,027,500	157.53
		30	0.625	28.750	57.68	4.508	7.854	7.53	2020	1,010,000	196.08

5S, 10S, and 40S are extracted from Stainless Steel Pipe, ANSI B36.19—1976, with permission of the publisher, the American Society of Mechanical Engineers, New York. ST = standard wall, XS = extra strong wall, XX = double extra strong wall, and Schedules 10 through 160 are extracted from Wrought-Steel and Wrought-Iron Pipe, ANSI B36.10—1975, with permission of the same publisher. Decimal thicknesses for respective pipe sizes represent their nominal or average wall dimensions. Mill tolerances as high as ±12½ percent are permitted.

Plain-end pipe is produced by a square cut. Pipe is also shipped from the mills threaded, with a threaded coupling on one end, or with the ends beveled for welding, or grooved or sized for patented couplings. Weights per foot for threaded and coupled pipe are slightly greater because of the weight of the coupling, but it is not available larger than 12 in or lighter than Schedule 30 sizes 8 through 12 in, or Schedule 40 6 in and smaller.

To convert inches to millimeters, multiply by 25.4; to convert square inches to square millimeters, multiply by 645; to convert feet to meters, multiply by 0.3048; to convert square feet to square meters, multiply by 0.0929; to convert pounds per foot to kilograms per meter, multiply by 1.49; to convert gallons to cubic meters, multiply by 3.7854 × 10⁻³; and to convert pounds to kilograms, multiply by 0.4536.

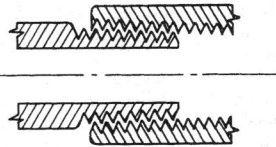

FIG. 10-128   Taper pipe thread.

Threads notch the pipe and cause loss of strength and fatigue resistance. Enlargement and contraction of the flow passage at threaded joints creates turbulence; sometimes corrosion and erosion are aggravated at the point where the pipe has already been thinned by threading. The tendency of pipe wrenches to crush pipe and fittings limits the torque available for tightening threaded joints. For low-pressure systems, a slight rotation in the joint may be used to impart flexibility to the system, but this same rotation, unwanted, may cause leaks to develop in higher-pressure systems. In some metals, galling occurs when threaded joints are disassembled.

Straight Pipe Threads   These are confined to light-weight couplings in sizes 2 in and smaller (Fig. 10-129). Manufacturers of threaded pipe ship it with such couplings installed on one end of each pipe. The joint obtained is inferior to that obtained with taper threads. The code limits the joint shown in Fig. 10-129 to 1.0 MPa (150 lbf/in²) gauge maximum, 182°C (360°F) maximum, and to nonflammable, nontoxic fluids.

When both components of a threaded joint are of weldable metal, the joint may be **seal-welded** as shown in Fig. 10-130. Seal welds may be used only to prevent leakage of threaded joints. They are not considered as contributing any strength to the joint. This type of joint is limited to new construction and is not suitable as a repair procedure, since pipe dope in the threads would interfere with welding. This method provides tight joints with a minimum of welding labor. When threaded joints used to join materials with widely different coefficients of thermal expansion are subject to temperature cycling, seal welding may be needed to prevent leakage.

To assist in assembly and disassembly of both threaded and welded systems, **union joints** (Fig. 10-131) are used. They comprise metal-to-metal seats drawn together by a shouldered straight thread nut and are available both in couplings for joining two lengths of pipe and on the ends of some fittings. On threaded piping systems in which disassembly is not contemplated, union joints installed at intervals permit future further tightening of threaded joints. Tightening of heavy unions yields tight joints even if the pipe is slightly misaligned at the start of tightening.

Flanged Joints   For sizes larger than 2 in when disassembly is contemplated, the flanged joint (Fig. 10-132) is most widely used.

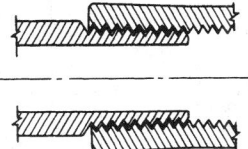

FIG. 10-129   Taper pipe to straight coupling thread.

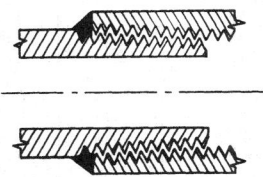

FIG. 10-130   Taper pipe thread seal-welded.

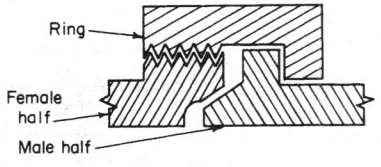

FIG. 10-131   Union.

Figures 10-133 and 10-134 illustrate the wide variety of types and facings available. Though flanged joints consume a large volume of metal, precise machining is required only on the facing. Flanged joints do not impose severe diametral tolerances on the pipe. Careful alignment prior to assembly of flat-face and raised-face flanges is not required, and the necessary wrenches are far smaller than those for screwed assembly for the same size of pipe.

Manufacturers offer **flanged-end pipe** in only a few metals. Otherwise, flanges are attached to pipe by various types of joints (Fig. 10-133). The lap joint involves a modification of the pipe which may be formed from the pipe itself or by welding a ring or a lap-joint stub end to it. **Flanged-end fittings** and valves are available in all sizes of most pipe metals.

Welding-neck flanges provide joints as strong as the pipe under all types of static and cycling loading. Slip-on, socket-weld, and lap-joint flanges provide joints as strong as the pipe under static loading but have lower resistance to cyclic stresses (see Table 10-54). Lap-joint flanges avoid the necessity of orienting flanges so that vertical and horizontal centerlines are halfway between bolt holes and permit orientation of the stems of flanged valves at any angle needed to provide clearance. The tolerance is ⅛ in in the bolt holes; the necessity of making sure that the gasket does not protrude into the flow channel results in some disturbance of the flow pattern when flat-face and raised-face flanges are used. This can be eliminated by using welding-neck or socket-weld flanges with male-and-female or tongue-and-groove facings.

Dimensions of alloy and carbon steel and cast-iron pipe flanges with flat and raised faces are given in Tables 10-19 to 10-25 (see Fig. 10-134). The dimensions were extracted from Cast-Iron Pipe Flanges and Flanged Fittings, ANSI B16.1—1975, and Steel Pipe Flanges and Flanged Fittings, ANSI 16.5—1977, with the permission of the publisher, the American Society of Mechanical Engineers, New York. Against cast-iron flanged fittings or valves, steel pipe flanges are often preferred to cast-iron flanges because they permit welded rather than

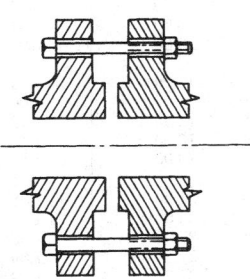

FIG. 10-132   Flanged joint.

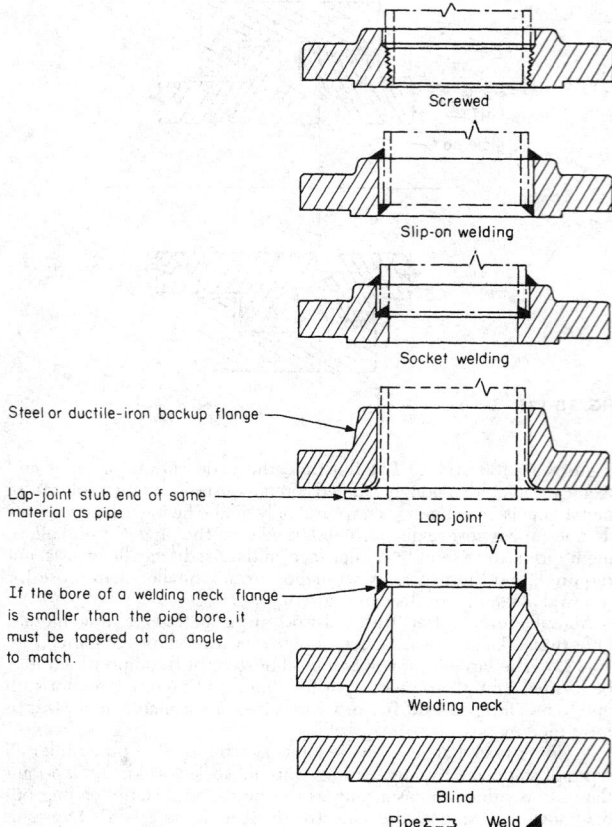

Steel or ductile-iron backup flange

Lap-joint stub end of same material as pipe

If the bore of a welding neck flange is smaller than the pipe bore, it must be tapered at an angle to match.

FIG. 10-133   Types of carbon and alloy steel flanges.

screwed assembly to the pipe and because cast-iron pipe flanges, not being reinforced by the pipe, are not so resistant to abuse as flanges cast integrally on cast-iron fittings.

Facing of flanges for alloy and carbon steel pipe and fittings is shown in Fig. 10-134; 125-lb cast-iron pipe and fitting flanges have flat faces, which with full-face gaskets minimize bending stresses; 250-lb cast-iron pipe and fitting flanges have 1.5-mm (1/16-in) raised faces (wider than on steel flanges) for the same purpose. Carbon steel and ductile- (nodular-) iron lap-joint flanges are widely used as backup flanges with stub ends in piping systems of austenitic stainless steel and other expensive materials to reduce costs (see Fig. 10-133). The code prohibits the use of ductile-iron flanges at temperatures above 343°C (650°F). When the type of facing affects the length through the hub dimension of flanges, correct dimensions for commonly used facings can be determined from the dimensional data in Fig. 10-134.

Gaskets   Gaskets must resist corrosion by the fluids handled. The more expensive male-and-female or tongue-and-groove facings may be required to seat hard gaskets adequately. With these facings the gasket generally cannot blow out. Flanged joints, by placing the gasket material under heavy compression and permitting only edge attack by the fluid handled, can use gasket materials which in other joints might not satisfactorily resist the fluid handled.

The finish of flange facings varies with the manufacturer. For raised-face or male mating surfaces the finish usually consists of a continuous spiral groove formed by a round-nosed tool or a V tool (serrated finish). Female surfaces are smooth-finished (i.e., without definite tool markings). Other finishes are concentric-grooved, lapped, or mirror (cold-water). The latter two are usually for application without gaskets.

In general, for 300-lb ANSI and lower-rated flanges compressed asbestos-sheet gaskets are used [400°C (750°F) maximum]. The metal-asbestos spiral-wound type is used for higher pressure and temperature services [593°C (1100°F) maximum], including services involving cyclic or difficultly contained fluids. The development of substitutes for asbestos in gaskets is being actively pursued because of the health hazard associated with asbestos. Metal-TFE and metal-graphite spiral-wound gaskets are available and may seal better than metal-asbestos gaskets. Spiral-wound gaskets are also used widely in high-pressure steam services. Spiral-wound gaskets should preferably be used with a smooth-flange finish.

TABLE 10-19   Dimensions of Class 150-lb Flanges for Use with Steel Pipe*

All dimensions in inches

Nominal pipe size	Outside diameter of flange	Thickness of flange, minimum	Diameter of bolt circle	Diameter of bolts	No. of bolts	Length through hub			
						Threaded slip-on socket welding	Lap joint	Welding neck	ANSI B16.1, screwed (125-lb)
1/2	3.50	0.44	2.38	1/2	4	0.62	0.62	1.88	
3/4	3.88	0.50	2.75	1/2	4	0.62	0.62	2.06	
1	4.25	0.56	3.12	1/2	4	0.69	0.69	2.19	0.69
1 1/4	4.62	0.62	3.50	1/2	4	0.81	0.81	2.25	0.81
1 1/2	5.00	0.69	3.88	1/2	4	0.88	0.88	2.44	0.88
2	6.00	0.75	4.75	5/8	4	1.00	1.00	2.50	1.00
2 1/2	7.00	0.88	5.50	5/8	4	1.12	1.12	2.75	1.12
3	7.50	0.94	6.00	5/8	4	1.19	1.19	2.75	1.19
3 1/2	8.50	0.94	7.00	5/8	8	1.25	1.25	2.81	1.25
4	9.00	0.94	7.50	5/8	8	1.31	1.31	3.00	1.31
5	10.00	0.94	8.50	3/4	8	1.44	1.44	3.50	1.44
6	11.00	1.00	9.50	3/4	8	1.56	1.56	3.50	1.56
8	13.50	1.12	11.75	3/4	8	1.75	1.75	4.00	1.75
10	16.00	1.19	14.25	7/8	12	1.94	1.94	4.00	1.94
12	19.00	1.25	17.00	7/8	12	2.19	2.19	4.50	2.19
14	21.00	1.38	18.75	1	12	2.25	3.12	5.00	2.25
16	23.50	1.44	21.25	1	16	2.50	3.44	5.00	2.50
18	25.00	1.56	22.75	1 1/8	16	2.69	3.81	5.50	2.69
20	27.50	1.69	25.00	1 1/8	20	2.88	4.06	5.69	2.88
24	32.00	1.88	29.50	1 1/4	20	3.25	4.38	6.00	3.25

*Dimensions from ANSI B16.5—1977, unless otherwise noted. To convert inches to millimeters, multiply by 25.4.

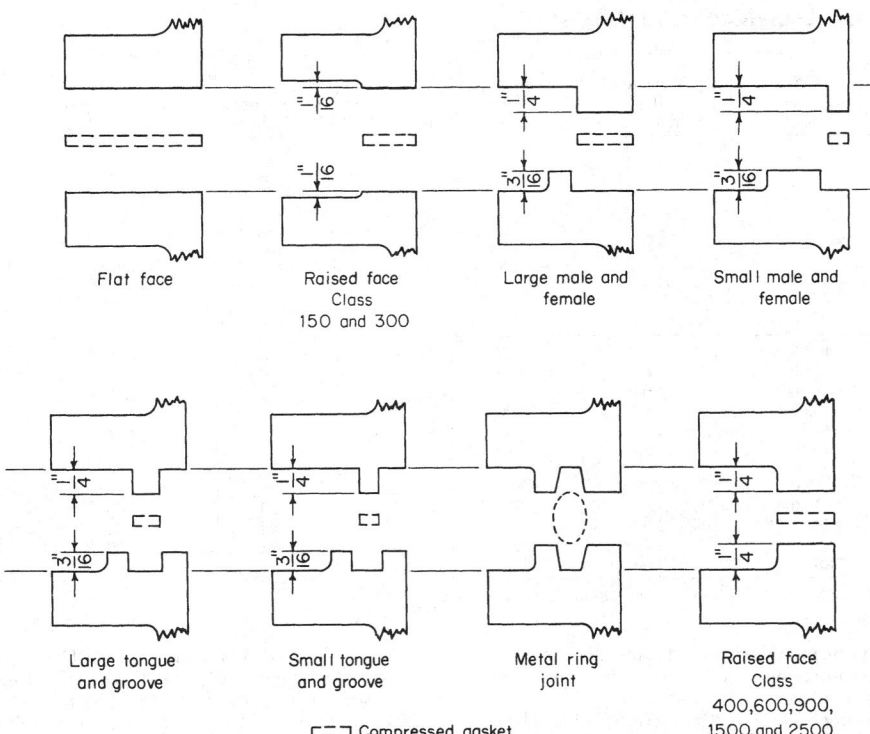

FIG. 10-134 Flange facings, illustrated on welding-neck flanges. (On small male-and-female facings the outside diameter of the male face is less than the outside diameter of the pipe, so this facing does not apply to screwed or slip-on flanges. A similar joint can be made with screwed flanges and threaded pipe by projecting the pipe through one flange and recessing it in the other. However, pipe thicker than Schedule 40 is required to avoid crushing gaskets.) To convert inches to millimeters, multiply by 25.4.

TABLE 10-20 Dimensions of Class 300 Flanges for Use with Steel Pipe*

All dimensions in inches

Nominal pipe size	Outside diameter of flange	Thickness of flange, minimum	Diameter of bolt circle	Diameter of bolts	No. of bolts	Length through hub			ANSI B16.1, screwed (Class 250)
						Threaded slip-on socket welding	Lap joint	Welding neck	
½	3.75	0.56	2.62	½	4	0.88	0.88	2.06	
¾	4.62	0.62	3.25	⅝	4	1.00	1.00	2.25	
1	4.88	0.69	3.50	⅝	4	1.06	1.06	2.44	0.88
1¼	5.25	0.75	3.88	⅝	4	1.06	1.06	2.56	1.00
1½	6.12	0.81	4.50	¾	4	1.19	1.19	2.69	1.12
2	6.50	0.88	5.00	⅝	8	1.31	1.31	2.75	1.25
2½	7.50	1.00	5.88	¾	8	1.50	1.50	3.00	1.43
3	8.25	1.12	6.62	¾	8	1.69	1.69	3.12	1.56
3½	9.00	1.19	7.25	¾	8	1.75	1.75	3.19	1.62
4	10.00	1.25	7.88	¾	8	1.88	1.88	3.38	1.75
5	11.00	1.38	9.25	¾	8	2.00	2.00	3.88	1.88
6	12.50	1.44	10.62	¾	12	2.06	2.06	3.88	1.94
8	15.00	1.62	13.00	⅞	12	2.44	2.44	4.38	2.19
10	17.50	1.88	15.25	1	16	2.62	3.75	4.62	2.38
12	20.50	2.00	17.75	1⅛	16	2.88	4.00	5.12	2.56
14	23.00	2.12	20.25	1⅛	20	3.00	4.38	5.62	2.69
16	25.50	2.25	22.50	1¼	20	3.25	4.75	5.75	2.88
18	28.00	2.38	24.75	1¼	24	3.50	5.12	6.25	
20	30.50	2.50	27.00	1¼	24	3.75	5.50	6.38	
24	36.00	2.75	32.00	1½	24	4.19	6.00	6.62	

*Dimensions from ANSI B16.5—1977, unless otherwise noted. To convert inches to millimeters, multiply by 25.4.

TABLE 10-21 Dimensions of Class 400 Steel Flanges*

All dimensions in inches

Nominal pipe size	Outside diameter of flange	Thickness of flange, minimum	Diameter of bolt circle	Diameter of bolts	No. of bolts	Length through hub		
						Threaded slip-on socket welding	Lap joint	Welding neck
½								
¾								
1								
1¼								
1½			Use Class 600 dimensions in these sizes.					
2								
2½								
3								
3½								
4	10.00	1.38	7.88	⅞	8	2.00	2.00	3.50
5	11.00	1.50	9.25	⅞	8	2.12	2.12	4.00
6	12.50	1.62	10.62	⅞	12	2.25	2.25	4.06
8	15.00	1.88	13.00	1	12	2.69	2.69	4.62
10	17.50	2.12	15.25	1¼	16	2.88	4.00	4.88
12	20.50	2.25	17.75	1¼	16	3.12	4.25	5.38
14	23.00	2.38	20.25	1¼	20	3.31	4.62	5.88
16	25.50	2.50	22.50	1⅜	20	3.69	5.00	6.00
18	28.00	2.62	24.75	1⅜	24	3.88	5.38	6.50
20	30.50	2.75	27.00	1½	24	4.00	5.75	6.62
24	36.00	3.00	32.00	1¾	24	4.50	6.25	6.88

*Dimensions from ANSI B16.5—1977. To convert inches to millimeters, multiply by 25.4.

The spiral-wound type furnished with a solid metallic ring on the outside to limit gasket compression provides protection against blow-out when used with raised facing.

Metal-Ring-Joint Facing This is the most costly facing. The ring must be softer than the flange and is usually a softer grade of the same metal as the flange. It is used where other gasket materials are destroyed by the fluid being handled. In event of fire, it does not leak. Because the surfaces that the gasket contacts are below the flange face, it is the least likely facing to be damaged in handling. Compared with raised or smooth faces, it is more difficult to disassemble because the flanges can be separated only in the axial direction.

Bolting Bolting requirements for ANSI flanged joints are pre-sented in the code. For joining two steel flanges, by reference to ANSI B16.5, Steel Pipe Flanges and Flanged Fittings, the code requires alloy steel bolting, except that bolting for 150- and 300-lb flanges at 204°C (400°F) and lower may be made of ASTM A307 Grade B low-carbon externally threaded fasteners. The code limits this exception to −29°C (−20°F) minimum.

Steel 150-lb flanges may be bolted to cast-iron valves, fittings, or other cast-iron piping components having either Class 125 cast integral or screwed flanges. If such construction is used, it is preferred that the 1.5-mm (¹⁄₁₆-in) raised face on steel flanges be removed. If the raised face is removed and a flat-ring gasket extending to the inner edge of the bolt holes is used, the bolting shall not be stronger than

TABLE 10-22 Dimensions of Class 600 Steel Flanges*

All dimensions in inches

Nominal pipe size	Outside diameter of flange	Thickness of flange, minimum	Diameter of bolt circle	Diameter of bolts	No. of bolts	Length through hub		
						Threaded slip-on socket welding	Lap joint	Welding neck
½	3.75	0.56	2.62	½	4	0.88	0.88	2.06
¾	4.62	0.62	3.25	⅝	4	1.00	1.00	2.25
1	4.88	0.69	3.50	⅝	4	1.06	1.06	2.44
1¼	5.25	0.81	3.88	⅝	4	1.12	1.12	2.62
1½	6.12	0.88	4.50	¾	4	1.25	1.25	2.75
2	6.50	1.00	5.00	⅝	8	1.44	1.44	2.88
2½	7.50	1.12	5.88	¾	8	1.62	1.62	3.12
3	8.25	1.25	6.62	¾	8	1.81	1.81	3.25
3½	9.00	1.38	7.25	⅞	8	1.94	1.94	3.38
4	10.75	1.50	8.50	⅞	8	2.12	2.12	4.00
5	13.00	1.75	10.50	1	8	2.38	2.38	4.50
6	14.00	1.88	11.50	1	12	2.62	2.62	4.62
8	16.50	2.19	13.75	1⅛	12	3.00	3.00	5.25
10	20.00	2.50	17.00	1¼	16	3.38	4.38	6.00
12	22.00	2.62	19.25	1¼	20	3.62	4.62	6.12
14	23.75	2.75	20.75	1¾	20	3.69	5.00	6.50
16	27.00	3.00	23.75	1½	20	4.19	5.50	7.00
18	29.25	3.25	25.75	1⅝	20	4.62	6.00	7.25
20	32.00	3.50	28.50	1⅝	24	5.00	6.50	7.50
24	37.00	4.00	33.00	1⅞	24	5.50	7.25	8.00

*Dimensions from ANSI B16.5—1977. To convert inches to millimeters, multiply by 25.4.

TABLE 10-23 Dimensions of Class 900 Steel Flanges*

All dimensions in inches

Nominal pipe size	Outside diameter of flange	Thickness of flange, minimum	Diameter of bolt circle	Diameter of bolts	No. of bolts	Length through hub Threaded slip-on socket welding	Lap joint	Welding neck
½								
¾								
1								
1¼			Use Class 1500 dimensions in these sizes.					
1½								
2								
2½								
3	9.50	1.50	7.50	⅞	8	2.12	2.12	4.00
4	11.50	1.75	9.25	1⅛	8	2.75	2.75	4.50
5	13.75	2.00	11.00	1¼	8	3.12	3.12	5.00
6	15.00	2.19	12.50	1⅛	12	3.38	3.38	5.50
8	18.50	2.50	15.50	1⅜	12	4.00	4.50	6.38
10	21.50	2.75	18.50	1⅜	16	4.25	5.00	7.25
12	24.00	3.12	21.00	1⅜	20	4.62	5.62	7.88
14	25.25	3.38	22.00	1½	20	5.12	6.12	8.38
16	27.75	3.50	24.25	1⅝	20	5.25	6.50	8.50
18	31.00	4.00	27.00	1⅞	20	6.00	7.50	9.00
20	33.75	4.25	29.50	2	20	6.25	8.25	9.75
24	41.00	5.50	35.50	2½	20	8.00	10.50	11.50

*Dimensions from ANSI B16.5—1977. To convert inches to millimeters, multiply by 25.4.

carbon steel per ASTM A307 Grade B; if a full-face gasket is used, the bolting may be heat-treated carbon steel or alloy steel (ASTM A193). If the raised face of the steel flange is not removed, the bolting shall not be stronger than carbon steel ASTM A307 Grade B.

Steel 300-lb flanges may be bolted to cast-iron valves, fittings, or other cast-iron piping components having either Class 250 cast-iron integral or screwed flanges, without any change in the raised face on either flange. If such construction is used, the bolting shall not be stronger than carbon steel, ASTM A307 Grade B.

Cast-iron 25-lb and Class 125 integral or screwed companion flanges may be used with a full-face gasket or with a flat-ring gasket extending to the inner edge of the bolts. When a full-face gasket is used, the bolting may be of heat-treated carbon steel or alloy steel (ASTM A193). When a flat-ring gasket is used, the bolting shall not be stronger than carbon steel, per ASTM A307 Grade B.

When two Class 250 cast-iron integral or screwed companion flanges having 1.5-mm (¹⁄₁₆-in) raised faces are bolted together, the bolting shall not be stronger than carbon steel, per ASTM A307 Grade B.

Other Types of Piping Joints Packed-gland joints (Fig. 10-135) require no special end preparation of pipe but do require careful control of the diameter of the pipe. Thus the supplier of the pipe should be notified when packed-gland joints are to be used. Cast- and ductile-iron pipe, fittings, and valves are available with the bell cast on one or

TABLE 10-24 Dimensions of Class 1500 Steel Flanges*

All dimensions in inches

Nominal pipe size	Outside diameter of flange	Thickness of flange, minimum	Diameter of bolt circle	Diameter of bolts	No. of bolts	Length through hub Threaded slip-on socket welding	Lap joint	Welding neck
½	4.75	0.88	3.25	¾	4	1.25	1.25	2.38
¾	5.12	1.00	3.50	¾	4	1.38	1.38	2.75
1	5.88	1.12	4.00	⅞	4	1.62	1.62	2.88
1¼	6.25	1.12	4.38	⅞	4	1.62	1.62	2.88
1½	7.00	1.25	4.88	1	4	1.75	1.75	3.25
2	8.50	1.50	6.50	⅞	8	2.25	2.25	4.00
2½	9.62	1.62	7.50	1	8	2.50	2.50	4.12
3	10.50	1.88	8.00	1⅛	8	2.88	2.88	4.62
4	12.25	2.12	9.50	1¼	8	3.56	3.56	4.88
5	14.75	2.88	11.50	1½	8	4.12	4.12	6.12
6	15.50	3.25	12.50	1⅜	12	4.69	4.69	6.75
8	19.00	3.62	15.50	1⅝	12	5.62	5.62	8.38
10	23.00	4.25	19.00	1⅞	12	6.25	7.00	10.00
12	26.50	4.88	22.50	2	16	7.12	8.62	11.12
14	29.50	5.25	25.00	2¼	16		9.50	11.75
16	32.50	5.75	27.75	2½	16		10.25	12.25
18	36.00	6.38	30.50	2¾	16		10.88	12.88
20	38.75	7.00	32.75	3	16		11.50	14.00
24	46.00	8.00	39.00	3½	16		13.00	16.00

*Dimensions from ANSI B15.5—1977. To convert inches to millimeters, multiply by 25.4.

TABLE 10-25 Dimensions of Class 2500 Steel Flanges*

All dimensions in inches

Nominal pipe size	Outside diameter of flange	Thickness of flange, minimum	Diameter of bolt circle	Diameter of bolts	No. of bolts	Length through hub		
						Threaded	Lap joint	Welding neck
½	5.25	1.19	3.50	¾	4	1.56	1.56	2.88
¾	5.50	1.25	3.75	¾	4	1.69	1.69	3.12
1	6.25	1.38	4.25	⅞	4	1.88	1.88	3.50
1¼	7.25	1.50	5.12	1	4	2.06	2.06	3.75
1½	8.00	1.75	5.75	1⅛	4	2.38	2.38	4.38
2	9.25	2.00	6.75	1	8	2.75	2.75	5.00
2½	10.50	2.25	7.75	1⅛	8	3.12	3.12	5.62
3	12.00	2.62	9.00	1¼	8	3.62	3.62	6.62
4	14.00	3.00	10.75	1½	8	4.25	4.25	7.50
5	16.50	3.62	12.75	1¾	8	5.12	5.12	9.00
6	19.00	4.25	14.50	2	8	6.00	6.00	10.75
8	21.75	5.00	17.25	2	12	7.00	7.00	12.50
10	26.50	6.50	21.25	2½	12	9.00	9.00	16.50
12	30.00	7.25	24.38	2¾	12	10.00	10.00	18.25

*Dimensions from ANSI B16.5—1977. To convert inches to millimeters, multiply by 25.4.

more ends. Glands, bolts, and gaskets are shipped with the pipe. Couplings equipped with packed glands at each end, known as Dresser couplings, are available in several metals. The joints can be assembled with small wrenches and unskilled labor, in limited space, and if necessary, under water.

Packed-gland joints are designed to take the same hoop stress as the pipe. They do not resist bending moments or axial forces tending to separate the joints but yield to them to an extent indicated by the vendor's allowable-angular-deflection and end movement specifications. Further angular or end movement produces leakage, but end movement can be limited by harnessing or bridling with a combination of rods and welded clips or clamps, or by anchoring to existing or new structures. The crevice between the bell and the spigot may promote corrosion. The joints are widely used in underground lines. They are not affected by limited earth settlement, and friction of the earth prevents end separation. When disassembly by moving pipe axially is not practical, packed-joint couplings which can be slid entirely onto one of the two lengths joined are available. However, the tendency of the packing to adhere to the pipe makes this difficult.

Poured joints (Fig. 10-136) require no special end preparation of the pipe or diametral control. They are used for brittle materials. Pipe, fittings, and valves are furnished with the bells cast on one or more ends. The pouring compound may be molten, or chemical-setting, or merely compacted; these choices are listed in descending order of ability to hold pressure. These joints cannot absorb angular or axial movement without leaking. Disassembly for maintenance is accomplished by cutting the pipe and reassembly by the use of a coupling with a bell at each end.

Push-on joints (Fig. 10-137) require diametral control of the end of the pipe. They are used for brittle materials. Pipe, fittings, and valves are furnished with the bells cast on one or more ends. Considerable force is required to push the spigot through the O ring; this is reduced by the extension on the O ring, which causes the friction of the pipe to elongate the cross section of the main portion of the O ring.

Push-on joints do not resist bending moments or axial forces tending to separate the joints but yield to them to an extent limited by the vendor's allowable-angular-deflection and end-movement specifications. End movement can be limited by harnessing or bridling with a combination of rods and clamps, or by anchoring to existing or new structures. The joints are widely used on underground lines. They are not affected by limited earth settlement, and friction of the earth prevents end separation. A lubricant is used on the O ring during assembly. After this disappears, the O ring bonds somewhat to the spigot and disassembly is very difficult. Disassembly for maintenance is accomplished by cutting the pipe and reassembly by use of a coupling with a packed-gland joint on each end.

Expanded joints (Fig. 10-138) are confined to the smaller pipe sizes of ductile metals. A smooth finish is required on the outside of the pipe and on the faces of the ridges inside the bore. Pipe and bore must have the same coefficient of thermal expansion. Furthermore, it is essential that the pipe metal have a lower yield point than the metal

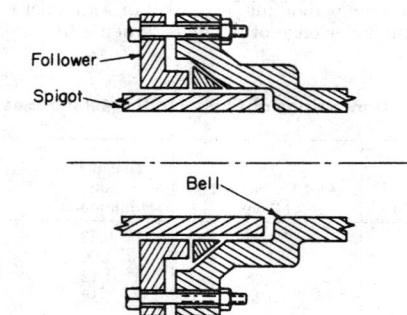

FIG. 10-136 Packed-gland joint.

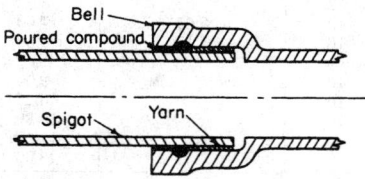

FIG. 10-135 Poured joint.

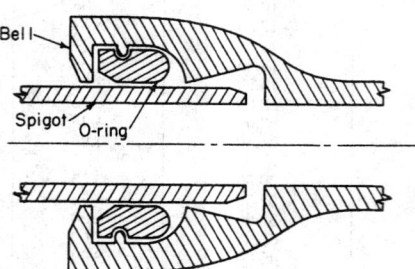

FIG. 10-137 Push-on joint.

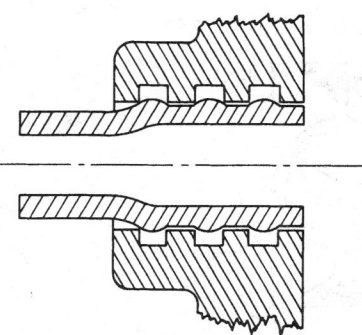

FIG. 10-138 Expanded joint.

compared with cut grooves, they are easier to form and reduce the metal wall less. However, they slightly reduce the flow area. They are limited to thin walls of ductile material, while cut grooves, because of their reduction of the pipe wall, are limited to thick walls. In the larger pipe sizes, some commonly used wall thicknesses are too thick for rolled grooves but too thin for cut grooves. The thinning of the walls impairs resistance to corrosion and erosion but not to internal pressure, because the thinned area is reinforced by the coupling.

Control of outside diameter is important. Permissible minus tolerance is limited, since it impairs the grip of the couplings. Plus tolerance makes it necessary to cut the cut grooves more deeply, increasing the thinning of the wall. Plus tolerance is not a problem with rolled grooves, since they are confined to walls thin enough so that the couplings can compress the pipe. Pipe is available from vendors already grooved and also with heavier-wall grooved ends welded on.

Grooved joints resist axial forces tending to separate the joints. Angular deflection, up to the limit specified by the vendor, may be used to absorb thermal expansion and to permit the piping to be laid on uneven ground. Compared with flanged joints, grooved joints will not pull misaligned pipe into alignment, and thus they require more support, but otherwise they require less labor for handling, assembly, and disassembly.

Gaskets are self-sealing against both internal and external pressure and are available in a wide variety of elastomers. However, successful performance of an elastomer as a flange gasket does not necessarily mean equally satisfactory performance in a grooved joint, since exposure to the fluid in the latter is much greater and hardening has a greater unfavorable effect. It is customary to use couplings which are resistant to corrosion by the fluid in the pipe, but couplings which would contaminate the fluid may be used.

V-clamp joints (Fig. 10-140) are attached to the pipe by butt-weld or expanded joints. Theoretically, there is only one relative position of the parts in which the conical surfaces of the clamp are completely in contact with the conical surfaces of the stub ends. In actual practice, there is considerable flexing of the stub ends and the clamp; also complete contact is not required. This permits use of elastomeric gaskets

into which it is expanded, except in cases in which the metal into which it is expanded is a thin cylinder temporarily backed by clamped heavy semicylindrical metal shells with a high yield point. An expanding tool is required, one for each size of pipe.

After completion of the joint, it is difficult to determine whether the increase in the inside diameter of the pipe represents permanent stretch of the bore of the mating part or flow of metal into the grooves of the bore. An excess of the latter results in excessive thinning of the tube, while an insufficiency of the latter may cause the pipe to pull out of the bore under axial loading. In a variation, the expanded joint is combined with a flared joint to increase resistance to axial load. These joints are used to attach unions and Lovekin flanges to pipe. For alloy piping, composite Lovekin flanges in which the bore and raised face portion are made of the alloy, retained in the steel balance of the flange by an offset, are available.

Grooved joints (Fig. 10-139) are divided into two classes, cut grooves and rolled grooves. Rolled grooves are preferred because,

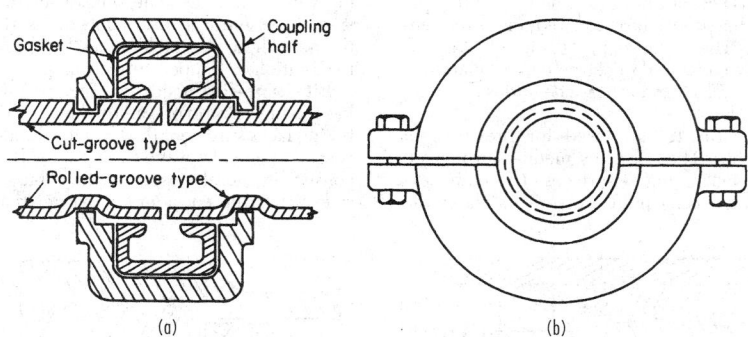

FIG. 10-139 Grooved joint. (*a*) Section. (*b*) End view.

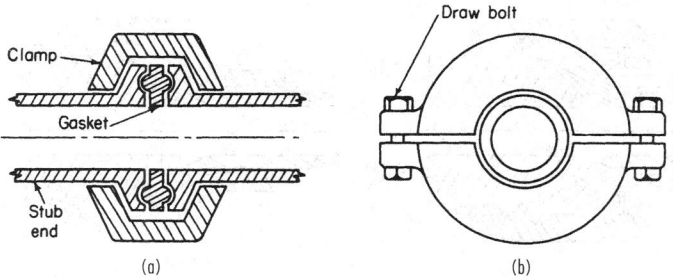

FIG. 10-140 V-clamp joint. (*a*) Section. (*b*) End view.

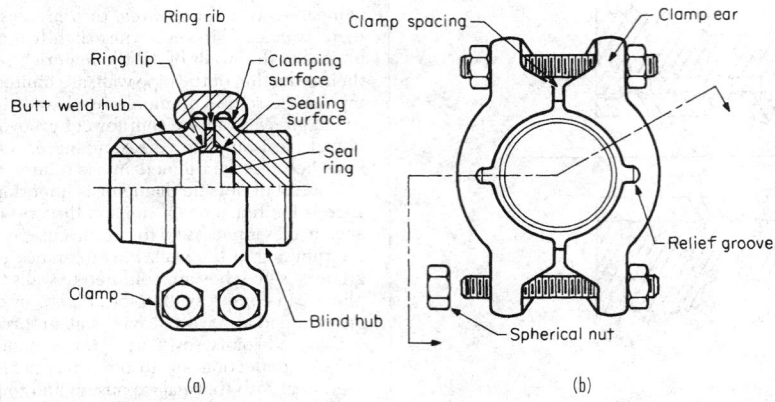

FIG. 10-141   Seal-ring joint. (*Courtesy of Gray Tool Co.*)

as well as metal gaskets. Fittings are also available with integral conical shouldered ends.

Conical ends vary from machined forgings to roll-formed tubing, and clamps vary from machined forgings to bands to which several roll-formed channels are attached at their centers by spot welding. A hinge may be inserted in the band as a substitute for one of the draw bolts. Latches may also be substituted for draw bolts.

Compared with flanges, V-clamp joints use less metal, require less labor for assembly, and are less likely to leak under wide-range rapid temperature cycling. However, they are more susceptible to failure or damage from overtightening. They are widely used for high-alloy piping subject to periodic cleaning or relocation. Manufactured as forgings, they are used in carbon steel with metal gaskets for very high pressures. They resist both axial strain and bending moments. Each size of each type of joint is customarily rated by the vendor for both internal pressure and bending moment.

Seal-ring joints (Fig. 10-141) consist of hubs attached to the pipe, generally by welding. The joint is proprietary and sold under the registered trade name of Grayloc. The metal seal ring is in effect a self-energizing gasket. This joint is widely used in petrochemical plants for service at the higher pressures. Valves and other accessories are manufactured with Grayloc hub ends.

Pressure-seal joints (Fig. 10-142) are used for pressures of 4.4 MPa (600 lbf/in²) and higher. They use less metal than flanged joints but require much more machining of surfaces. There are several designs, in all of which increasing fluid pressure increases the

force holding the sealing surfaces against each other. These joints are widely used as bonnet joints in carbon and alloy steel valves.

Tubing Joints   **Flared-fitting joints** (see Fig. 10-143) are used for ductile tubing when the ratio of wall thickness to the diameter is small enough to permit flaring without cracking the inside surface. The tubing must have a smooth interior surface. The three-piece type avoids torsional strain on the tubing and minimizes vibration fatigue on the flared portion of the tubing. More labor is required for assembly, but the fitting is more resistant to temperature cycling than other tubing fittings and is unlikely to be damaged by overtightening, and its efficiency is not impaired by repeated assembly and disassembly. Size is limited because of the large number of machined surfaces. The nut and, in the three-piece type, the sleeve need not be of the same material as the tubing. For these fittings, less control of tubing diameter is required.

Compression-fitting joints (Fig. 10-144) are used for ductile tubing with thin walls. The outside of the tubing must be clean and smooth. Assembly consists only of inserting the tubing and tightening the nut. These are the least costly tubing fittings but are not resistant to vibration or temperature cycling.

Bite-type-fitting joints (Fig. 10-145) are used when the tubing has too high a ratio of wall thickness to diameter for flaring, when the tubing lacks sufficient ductility for flaring, and for low assembly-labor cost. The outside of the tubing must be clean and smooth. Assembly consists in merely inserting the tubing and tightening the nut. The sleeve must be considerably harder than the tubing yet still ductile

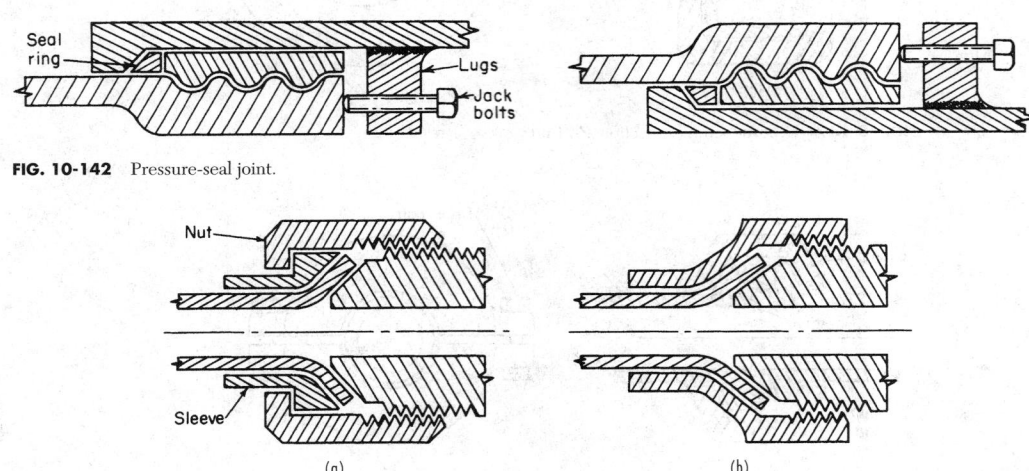

FIG. 10-142   Pressure-seal joint.

FIG. 10-143   Flared-fitting joint. (*a*) Three-piece. (*b*) Two-piece.

enough to be diametrally compressed and must be as resistant to corrosion by the fluid handled as the tubing. The fittings are resistant to vibration but not to wide-range rapid temperature cycling. Compared with flared fittings, they are less suited for repeated assembly and disassembly, require closer diametral control of the tubing, and are more susceptible to damage from overtightening. They are widely used for oil-filled hydraulic systems at all pressures.

O-ring seal joints (Fig. 10-146) are also used for applications requiring heavy-wall tubing. The outside of the tubing must be clean and smooth. The joint may be assembled repeatedly, and as long as the tubing is not damaged, leaks can usually be corrected by replacing the O ring and the antiextrusion washer. This joint is used extensively in oil-filled hydraulic systems.

Soldered Joints   (Fig. 10-147) These joints require precise control of the diameter of the pipe or tubing and of the cup in the fitting in order to cause the solder to draw into the clearance between the cup and the tubing by capillary action (Fig. 10-147). Extrusion provides this diametral control, and the joints are most widely used in copper. A 50 percent lead, 50 percent tin solder is used for tempera-

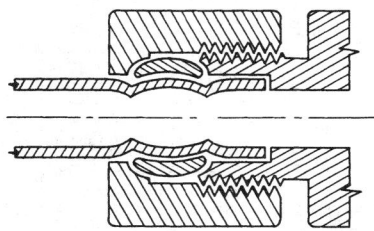

FIG. 10-144   Compression-fitting joint.

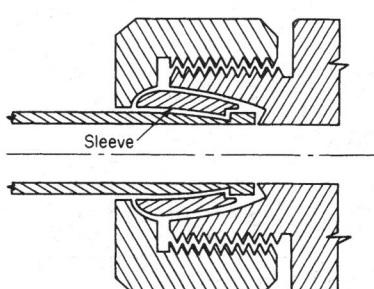

Sleeve

FIG. 10-145   Bite-type-fitting joint.

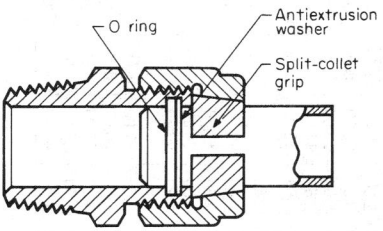

O ring

Antiextrusion washer

Split-collet grip

FIG. 10-146   O-ring seal joint. *(Courtesy of the Lenz Co.)*

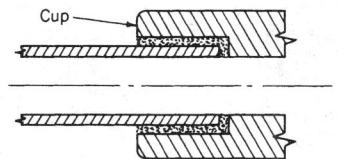

Cup

FIG. 10-147   Soldered, brazed, or cemented joint.

tures up to 93°C (200°F). Careful cleaning of the outside of the tubing and inside of the cup is required.

Heat for soldering is usually obtained from torches. The high conductivity of copper makes it necessary to use large flames for the larger sizes, and for this reason the location in which the joint will be made must be carefully considered. Soldered joints are most widely used in sizes 2 in and smaller for which heat requirements are less burdensome. Soldered joints should not be used in areas where plant fires are likely because exposure to fires results in rapid and complete failure of the joints. Properly made, the joints are completely impervious. The code permits the use of soldered joints only for Category D fluid service and then only if the system is not subject to severe cyclic condions.

Silver Brazed Joints   These are similar to soldered joints except that a temperature of about 600°C (1100°F) is required. A 15 percent silver, 80 percent copper, 5 percent phosphorus solder is used for copper and copper alloys, while 45 percent silver, 15 percent copper, 16 percent zinc, 24 percent cadmium solders are used for copper, copper alloys, carbon steel, and alloy steel. Silver-brazed joints are used for temperatures up to 200°C (400°F). Cast-bronze fittings and valves with preinserted rings of 15 percent silver, 80 percent copper, 5 percent phosphorus brazing alloy are available.

Silver-brazed joints are used when temperature or the combination of temperature and pressure is beyond the range of soldered joints. They are also more reliable in the event of plant fires and are more resistant to vibration. If they are used for fluids that are flammable, toxic, or damaging to human tissue, appropriate safeguarding is required by the code. There are OSHA regulations governing the use of silver brazing alloys containing cadmium and other toxic materials.

Bends and Fittings   Directional changes in piping systems require bends and elbow fittings. Bends may be made cold or hot. The outside wall is thinned by an amount that varies with the procedure used. Subsequent annealing is required for some materials. To prevent wrinkling and excessive flattening, sand packing is required for hot bending, and sand packing or flexible mandrels may be necessary for cold bending, depending on the ratios of the outside diameter of the pipe to the centerline radius of the bend and to the wall thickness of the pipe. For bends with a centerline radius of five nominal pipe diameters, internal support is not required when the wall thickness is at least 6 percent of the outside diameter of the pipe. Wrinkled bends are made by progressively heating the pipe only on the side which will be the inside of the bend.

Elbow fittings may be cast, forged, or hot- or cold-formed from short pieces of pipe or made by welding together pieces of miter-cut pipe. The thinning of pipe during the forming of elbows is compensated for by starting with heavier walls.

Flow in bends and elbow fittings is more turbulent than in straight pipe, thus increasing corrosion and erosion. This can be countered by selecting a component with greater radius of curvature, thicker wall, or smoother interior contour, but this is seldom economical in miter elbows.

Compared with elbow fittings, bends with a centerline radius of three or five nominal pipe diameters save the cost of joints and reduce pressure drop. Such bends are not suited for installation in a bank of pipes of unequal size when the bends are in the same plane as the bank.

Flanged fittings are used when pipe is likely to be dismantled for frequent cleaning or extensive revision, for lined piping systems, or for seasonal insertion of blanks as a substitute for valves. They are also used in areas where welding is not permitted. Cast fittings are usually flanged. Table 10-26 gives dimensions for flanged fittings. Dimensions of carbon and alloy steel **butt-welding fittings** are shown in Table 10-27. Butt-welding fittings are available in the wall thicknesses shown in Table 10-18. Butt-welding elbows with short, straight pipe extensions at the ends are also available for insertion in slip-on flanges. Schedule 5 and Schedule 10 stainless-steel butt-welding fittings are also available with such extensions for expanding into stainless-steel hubs mechanically locked in carbon steel ANSI B16.5 dimension flanges. The use of expanded joints (Fig. 10-138) is restricted by the code.

Forged fittings made by boring out solid forgings are available with socket-weld (Fig. 10-126) or with screwed ends in sizes through 4 in, but 2 in is the usual upper size limit for use. ANSI B16.11—1973 gives minimum dimensions for socket-weld 3000- and 6000-lb classes and

TABLE 10-26 Dimensions of Flanged Fittings*

All dimensions in inches

Elbow   Long-Radius Elbow   45° Elbow   Tee   Cross   45° Lateral   Reducer (—— Concentric, - - - Eccentric)

Nominal pipe size	ANSI B16.5, Class 150 / ANSI B16.1, Class 125						ANSI B16.5, Class 300 / ANSI B16.1, Class 250						ANSI B16.5, Class 400					ANSI B16.5, Class 600				
	AA	BB	CC	EE	FF	GG	AA	BB	CC	EE	FF	GG	AA	CC	EE	FF	GG	AA	CC	EE	FF	GG
½													Use Class 600 dimensions in these sizes					3.25	2.00	5.75	1.75	5.00
¾																		3.75	2.50	6.75	2.00	5.00
1	3.50	5.00	1.75	5.75	1.75	4.50	4.00	5.00	2.25	6.50	2.00	4.50						4.25	2.50	7.25	2.25	5.00
1¼	3.75	5.50	2.00	6.25	1.75	4.50	4.25	5.50	2.50	7.25	2.25	4.50						4.50	2.75	8.00	2.50	5.00
1½	4.00	6.00	2.25	7.00	2.00	4.50	4.50	6.00	2.75	8.50	2.50	4.50						4.75	3.00	9.00	2.75	5.00
2	4.50	6.50	2.50	8.00	2.50	5.00	5.00	6.50	3.00	9.00	2.50	5.00						5.75	4.25	10.25	3.50	6.00
2½	5.00	7.00	3.00	9.50	2.50	5.50	5.50	7.00	3.50	10.50	2.50	5.50						6.50	4.50	11.50	3.50	6.75
3	5.50	7.75	3.00	10.00	3.00	6.00	6.00	7.75	3.50	11.00	3.00	6.00						7.00	5.00	12.75	4.00	7.25
3½	6.00	8.50	3.50	11.50	3.00	6.50	6.50	8.50	4.00	12.50	3.00	6.50						7.50	5.50	14.00	4.50	7.75
4	6.50	9.00	4.00	12.00	3.00	7.00	7.00	9.00	4.50	13.50	3.00	7.00	8.00	5.50	16.00	4.50	8.25	8.50	6.00	16.50	4.50	8.75
5	7.50	10.25	4.50	13.50	3.50	8.00	8.00	10.25	5.00	15.00	3.50	8.00	9.00	6.00	16.75	5.00	9.25	10.00	7.00	19.50	6.00	10.25
6	8.00	11.50	5.00	14.50	3.50	9.00	8.50	11.50	5.50	17.50	4.00	9.00	9.75	6.25	18.75	5.25	10.00	11.00	7.50	21.00	6.50	11.25
8	9.00	14.00	5.50	17.50	4.50	11.00	10.00	14.00	6.00	20.50	5.00	11.00	11.75	6.75	22.25	5.75	12.00	13.00	8.50	24.50	7.00	13.25
10	11.00	16.50	6.50	20.50	5.00	12.00	11.50	16.50	7.00	24.00	5.50	12.00	13.25	7.75	25.75	6.25	13.50	15.50	9.50	29.50	8.00	15.75
12	12.00	19.00	7.50	24.50	5.50	14.00	13.00	19.00	8.00	27.50	6.00	14.00	15.00	8.75	29.75	6.50	15.25	16.50	10.00	31.50	8.50	16.75
14	14.00	21.50	7.50	27.00	6.00	16.00	15.00	21.50	8.50	31.00	6.50	16.00	16.25	9.25	32.75	7.00	16.50	17.50	10.75	34.25	9.00	17.75
16	15.00	24.00	8.00	30.00	6.50	18.00	16.50	24.00	9.50	34.50	7.50	18.00	17.75	10.25	36.25	8.00	18.50	19.50	11.75	38.50	10.00	19.75
18	16.50	26.50	8.50	32.00	7.00	19.00	18.00	26.50	10.00	37.50	8.00	19.00	19.25	10.75	39.25	8.50	19.50	21.50	12.25	42.00	10.50	21.75
20	18.00	29.00	9.50	35.00	8.00	20.00	19.50	29.00	10.50	40.50	8.50	20.00	20.75	11.25	42.75	9.00	21.00	23.50	13.00	45.50	11.00	23.75
24	22.00	34.00	11.00	40.50	9.00	24.00	22.50	34.00	12.00	47.50	10.00	24.00	24.25	12.75	50.25	10.50	24.50	27.50	14.75	53.00	13.00	27.75

Nominal pipe size	ANSI B16.5, Class 900					ANSI B16.5, Class 1500					ANSI B16.5, Class 2500				
	AA	CC	EE	FF	GG	AA	CC	EE	FF	GG	AA	CC	EE	FF	GG
½						4.25	3.00				5.19				
¾						4.50	3.25				5.37				
1						5.00	3.50	9.00	2.50	5.00	6.06	4.00			
1¼						5.50	4.00	10.00	3.00	5.75	6.87	4.25			
1½						6.00	4.25	11.00	3.50	6.25	7.56	4.75			
2	Use Class 1500 dimensions in these sizes					7.25	4.75	13.25	4.00	7.25	8.87	5.75	15.25	5.25	9.50
2½						8.25	5.25	15.25	4.50	8.25	10.00	6.25	17.25	5.75	10.50
3	7.50	5.50	14.50	4.50	7.75	9.25	5.75	17.25	5.00	9.25	11.37	7.25	19.75	6.75	11.75
4	9.00	6.50	17.50	5.50	9.25	10.75	7.25	19.25	6.00	10.75	13.25	8.50	23.00	7.75	13.50
5	11.00	7.50	21.00	6.50	11.25	13.25	8.75	23.25	7.50	13.75	15.62	10.00	27.25	9.25	15.75
6	12.00	8.00	22.50	6.50	12.25	13.88	9.38	24.88	8.12	14.50	18.00	11.50	31.25	10.50	18.00
8	14.50	9.00	27.50	7.50	14.75	16.38	10.88	29.88	9.12	17.00	20.12	12.75	35.25	11.75	20.50
10	16.50	10.00	31.50	8.50	16.75	19.50	12.00	36.00	10.25	20.25	25.00	16.00	43.25	14.75	25.50
12	19.00	11.00	34.50	9.00	17.75	22.25	13.25	40.75	12.00	23.00	28.00	17.75	49.25	16.25	29.00
14	20.25	11.50	36.50	9.50	19.00	24.75	14.25	44.00	12.50	25.75					
16	22.25	12.50	40.75	10.50	21.00	27.25	16.25	48.25	14.75	28.25					
18	24.00	13.25	45.50	12.00	24.50	30.25	17.75	53.25	16.50	31.50					
20	26.00	14.50	50.25	13.00	26.50	32.75	18.75	57.75	17.75	34.00					
24	30.50	18.00	60.00	15.50	30.50	38.25	20.75	67.25	20.50	39.75					

*Outline drawings show ¼-in (6.5-mm) raised face machined onto flange, as for ANSI B16.5 400-lb and higher. ANSI B16.1 250-lb and ANSI B16.5 150- and 300-lb have 1/16-in (1.5-mm) raised face; ANSI B16.1 125-lb has no raised face. See Tables 10-19 through 10-25 for flange drillings. Dimensions for 400- and 600-lb fittings are identical for sizes ½ to 3½ in inclusive. Dimensions for 900- and 1500-lb fittings are identical for sizes ½ to 2½ in inclusive. To convert inches to millimeters, multiply by 25.4. The dimensions were extracted from Cast-Iron Pipe Flanges and Flanged Fittings, ANSI B16.1—1975, and Steel Pipe Flanges and Flanged Fittings, ANSI B16.5—1977, with permission of the publisher, the American Society of Mechanical Engineers, New York.

TABLE 10-27 Butt-Welding Fittings*

All dimensions in inches

  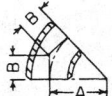 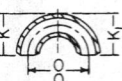 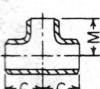   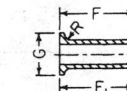

| 90° elbows
A for long
radius
A₁ for short
radius | 90° elbows
long radius
reducing | 45° elbows
long radius | 180° bends
O for long radius
O₁ for long radius
K for long radius
K₁ for short radius | Tee
straight
and
reducing
(M is for
straight
tees only) | Reducers
Concentric

Eccentric | Caps | Stub ends
F for A.N.S.I. B16.9
F₁ for MSS-SP-43 |

Pipe size	A	K	A1	K1	B	O	O1	M, C	H	E†	G	F	F1	R‡
½	1.50	1.88			0.62	3.00		1.00		1.00	1.38	3.00	2.00	0.12
¾ (1)	1.12	1.69			0.44	2.25		1.12	1.50	1.00	1.69	3.00	2.00	0.12
1	1.50	2.19	1.00	1.62	0.88	3.00	2.00	1.50	2.00	1.50	2.00	4.00	2.00	0.12
1¼	1.88	2.75	1.25	2.06	1.00	3.75	2.50	1.88	2.00	1.50	2.50	4.00	2.00	0.19
1½	2.25	3.25	1.50	2.44	1.12	4.50	3.00	2.25	2.50	1.50	2.88	4.00	2.00	0.25
2	3.00	4.19	2.00	3.19	1.38	6.00	4.00	2.50	3.00	1.50	3.62	6.00	2.50	0.31
2½	3.75	5.19	2.50	3.94	1.75	7.50	5.00	3.00	3.50	1.50	4.12	6.00	2.50	0.31
3	4.50	6.25	3.00	4.75	2.00	9.00	6.00	3.38	3.50	2.00	5.00	6.00	2.50	0.38
3½	5.25	7.25	3.50	5.50	2.25	10.50	7.00	3.75	4.00	2.50	5.50	6.00	3.00	0.38
4	6.00	8.25	4.00	6.25	2.50	12.00	8.00	4.12	4.00	2.50	6.19	6.00	3.00	0.44
5	7.50	10.31	5.00	7.75	3.12	15.00	10.00	4.88	5.00	3.00	7.31	8.00	3.00	0.44
6	9.00	12.31	6.00	9.31	3.75	18.00	12.00	5.62	5.50	3.50	8.50	8.00	3.50	0.50
8	12.00	16.31	8.00	12.31	5.00	24.00	16.00	7.00	6.00	4.00	10.62	8.00	4.00	0.50
10	15.00	20.38	10.00	15.38	6.25	30.00	20.00	8.50	7.00	5.00	12.75	10.00	5.00	0.50
12	18.00	24.38	12.00	18.38	7.50	36.00	24.00	10.00	8.00	6.00	15.00	10.00	6.00	0.50
14	21.00	28.00	14.00	21.00	8.75	42.00	28.00	11.00	13.00	6.50	16.25	12.00	6.00	0.50
16	24.00	32.00	16.00	24.00	10.00	48.00	32.00	12.00	14.00	7.00	18.50	12.00	6.00	0.50
18	27.00	36.00	18.00	27.00	11.25	54.00	36.00	13.50	15.00	8.00	21.00	12.00	6.00	0.50
20	30.00	40.00	20.00	30.00	12.50	60.00	40.00	15.00	20.00	9.00	23.00	12.00	6.00	0.50
24	36.00	48.00	24.00	36.00	15.00	72.00	48.00	17.00	20.00	10.50	27.25	12.00	6.00	0.50

*Extracted from Wrought-Steel Butt-Welding Fittings, ANSI B16.9—1978, and from Wrought-Steel Butt-Welding Short-Radius Elbows and Returns, ANSI B16.28—1978, with permission of the publisher, the American Society of Mechanical Engineers, New York. A and B dimensions of 1.50 and 0.75 in respectively may be furnished for NPS ¾ at the manufacturer's option. O and K dimensions may likewise be furnished in 2.00 in and 3.00 in respectively.

†For wall thicknesses greater than extra heavy, E is greater than shown here for sizes 2 in and larger.

‡For MSS SP-43 type B stub ends, which are designed to be backed up by slip-on flanges, R = ¹⁄₃₂ in for 4 in and smaller and ¹⁄₁₆ in for 6 through 12 in. To convert inches to millimeters multiply by 25.4.

for screwed 2000-, 3000-, and 6000-lb classes. It also contains pressure-temperature ratings for the classes in various ferrous alloys. The use of socket-weld and screwed fittings is restricted by the code.

Steel forged fittings with screwed ends may be installed without pipe dope in the threads and seal-welded (Fig. 10-130) to secure bubble-tight joints with a minimum of welders' labor. They are not subject to deformation by pipe wrenches, and such couplings, bushings, and plugs are often used with the screwed fittings below.

ANSI B16.3—1977 gives dimensions of 150-lb **malleable-iron screwed fittings** through the 6-in size for 1.0 MPa (150 lbf/in²) saturated steam and 2.1 MPa (300 lbf/in²) at room temperature and for 300-lb malleable-iron screwed fittings through the 3-in size for 2.1 MPa (300 lbf/in²) steam at 290°C (550°F) and 7.0 MPa (1000 lbf/in²) at room temperature. These fittings are available with male threads or unions on one end for installation in confined spaces. Major use is in 150-lb elbows, tees, and reducers in sizes 2 in and smaller. They are less costly than forged fittings but cannot be seal-welded. The code does not permit the use of malleable iron in toxic service or in flammable service above either 150°C (300°F) or 2.76 MPa (400 lbf/in²) gauge.

ANSI B16.4—1977 gives dimensions of 125-lb **cast-iron screwed fittings** through the 12-in size for 0.86 MPa (125 lb/in²) saturated steam and 1.2 MPa (175 lbf/in²) at 66°C (150°F) and of 250-lb cast-iron screwed fittings through the 12-in size for 1.72 MPa (250 lbf/in²) saturated steam and for 2.76 MPa (400 lbf/in²) at 66°C (150°F). The 125-lb fittings are made in regular 90° and 45° elbows, reducing elbows, regular and reducing tees, and crosses. The 250-lb fittings are made only

in straight sizes. Major use is in 125-lb elbows, tees, and reducers in low-pressure noncritical service. The code does not permit the use of cast iron in toxic service or aboveground within process unit limits for flammable-fluid service above 150°C (300°F) or 1.0 MPa (150 lbf/in²).

Tees Tees may be cast, forged, or hot- or cold-formed from short pieces of pipe. Though it is impossible to have the same flow simultaneously through all three end connections, it is not economical to produce or stock the great variety of tees which accurate sizing of end connections requires. It is customary to stock only tees with the two end (run) connections of the same size and the branch connection either of the same size as the run connections or one, two, or three sizes smaller. Adjacent reducers or reducing elbow fittings are used for other size reductions. Branch connections (see subsection "Joints") are often more economical than tees, particularly when the ratio of branch to run is small.

Reducers Reducers may be cast, forged, or hot- or cold-formed from short pieces of pipe. End connections may be concentric or eccentric, that is, tangent to the same plane at one point on their circumference. For pipe supported by hangers, concentric reducers permit maintenance of the same hanger length; for pipe laid on structural steel, eccentric reducers permit maintaining the same elevation of top of steel. Eccentric reducers with the common tangent plane below permit complete drainage of branched horizontal piping systems through branches smaller than the main. With the common tangent plane above, they permit liquid flow in horizontal lines to sweep the line free of gas or vapor.

Reducing elbow fittings permit change of direction and concentric size reduction in the same fitting.

Valves Valve bodies may be cast, forged, machined from bar stock, or fabricated from welded plate. Steel valves are available with screwed or socket-weld ends in the smaller sizes. Bronze and brass screwed-end valves are widely used for low-pressure service in steel systems. Table 10-28 gives contact-surface-of-face to contact-surface-of-face dimensions for flanged ferrous valves and end-to-end dimensions for butt-welding ferrous valves. Drilling of end flanges is shown in Tables 10-19 to 10-25. Bolt holes are located so that the stem is equidistant from the centerline of two bolt holes. Even if removal for maintenance is not anticipated, flanged valves are frequently used instead of butt-welding-end valves because they permit insertion of blanks for isolating sections of a loop piping system.

Ferrous valves are also available in nodular (ductile) iron, which has tensile strength and yield point approximately equal to cast carbon steel at temperatures of 343°C (650°F) and below and only slightly less elongation.

Valves serve not only to regulate the flow of fluids but also to isolate piping or equipment for maintenance without interrupting other connected units. Valve design should keep pressure, temperature changes, and strain from connected piping from distorting or misaligning the sealing surfaces. The sealing surfaces should be of such material and design that the valve will remain tight over a reasonable service period. The principal types are named, described, compared, and illustrated with line diagrams in subsequent subsections. In the line diagrams, the operating stem is shown in solid black, direction of flow by arrows on a thin solid line, and motion of valve parts by arrows on a dotted line. Moving parts are drawn with solid lines in the nearly closed position and with dotted lines in the fully open position. Packing is represented by an X in a square.

Gate Valves These valves are designed in two types (Fig. 10-148). The wedge-shaped-gate, **inclined-seat** type is most commonly used. The wedge gate is usually solid but may be flexible (partly cut into halves by a plane at right angles to the pipe) or split (completely cleft by such a plane). Flexible and split wedges minimize galling of the sealing surfaces by distorting more easily to match angularly misaligned seats. In the double-disk **parallel-seat** type, an inclined-plane device mounted between the disks converts stem force to axial force, pressing the disks against the seats after the disks have been positioned for closing. This gate assembly distorts automatically to match both angular misalignment of the seats and longitudinal shrinkage of the valve body on cooling.

When shearing high-velocity flow of dense fluids, the gate assemblies shake violently, and for this service solid-wedge or flexible-

wedge valves are preferred. When valve operation is manual, small bypass valves installed in parallel with the main valve may be used to eliminate the shake problem and to minimize manual effort in opening and closing the valves. Double-disk parallel-seat valves should be installed with the stem essentially vertical. All wedge gate valves are equipped with tongue-and-groove guides to keep the gate sealing surfaces from clattering on the seats and marring them during opening and closing. Depending on the velocity and density of the fluid stream being sheared, these guiding surfaces may be as cast, machined, or hard-surfaced and ground.

Gate valves may have nonrising stems, inside-screw rising stems, or outside-screw rising stems, listed in order of decreasing exposure of the stem threads to the fluid handled. Rising-stem valves require more space, but the position of the stem visually indicates the position of the gate. Indication is clearest on the outside-screw rising-stem valves, and on these the stem threads and thrust collars may be lubricated, reducing operating effort. The stem connection to the gate assembly prevents the stem from rotating.

Gate valves are used to minimize pressure drop in the open position and to stop the flow of fluid rather than to regulate it. The problem, when the valve is closed, of pressure buildup in the bonnet from cold liquids expanding or chemical action between fluid and bonnet should be solved by a relief valve or by notching the upstream seat ring.

Globe Valves (Fig. 10-149) These are designed as either inside-screw rising-stem or outside-screw rising-stem. Small valves generally are of the inside-screw type, while in larger sizes the outside-screw type is preferred. In most designs the disks are free to rotate on the stems; this prevents galling between the disk and the seat.

In the larger sizes, with conical seats, this swivel may permit enough misalignment to prevent proper sealing between the disk and the seat. When the valve is close to an elbow on the upstream side, the swivel also permits uneven distribution of the fluid to spin the disk on the stem. Guides above the disk, below the disk, or both are used to prevent misalignment and spinning. Misalignment can also be prevented by the use of spherical seats and designing the disk so that the pressure point of the stem on the disk is at the center of the sphere. In some designs, spinning and misalignment are prevented by rigidly attaching the disk to the stem, preventing rotation of the stem by lugs which ride along the yoke, and using a yoke bushing as in outside-screw-and-yoke gate valves.

Large globe valves should be installed with stems vertical. Globe valves are preferably installed with the higher-pressure side connected to the top of the disk. Exceptions occur (1) when blocked flow caused by separation of the disk from the stem would damage equipment or (2) when the valve is installed in seldom-used vertical drain lines in which accumulation of rust, scale, or sludge might prevent opening the valve.

Pressure drop through globe valves is much greater than that for gate valves. In Y-type globe valves, the stem and seat are at about 45° to the pipe instead of 90°. This reduces pressure drop but impairs alignment of seat and disk.

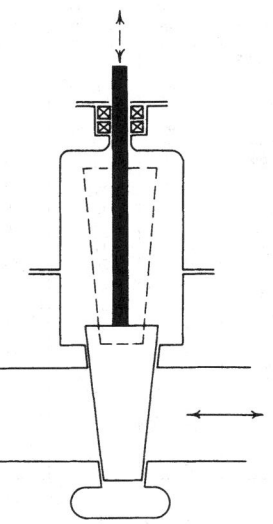

FIG. 10-148 Gate valve.

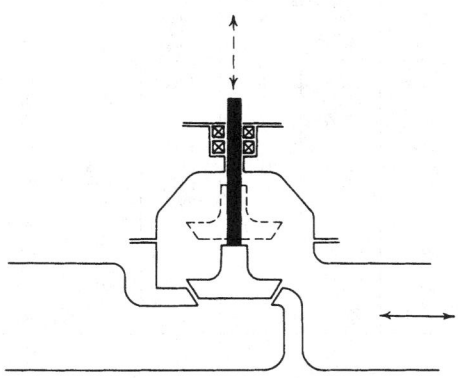

FIG. 10-149 Globe valve.

TABLE 10-28 Dimensions of Valves*

All dimensions in inches

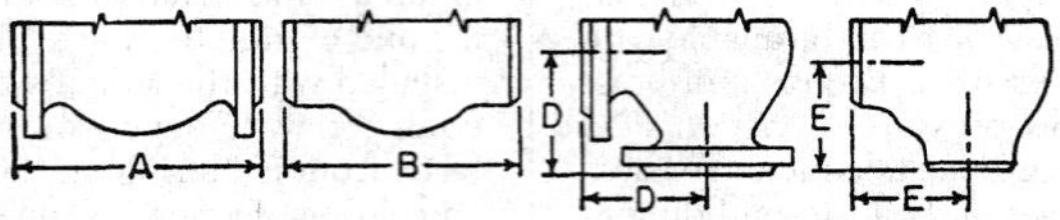

Nominal valve size	Class 125 cast iron					Class 150 steel, MSS-SP-42 through 12-in size					Class 250 cast iron		
	Flanged end					Flanged end	Welding end	Flanged end and welding end			Flanged end		
	Gate					Gate	Gate				Gate		
	Solid wedge *A*	Double disk *A*	Globe and lift check *A*	Angle and lift check *D*	Swing check *A*	Solid wedge and double disk *A*	Solid wedge and double disk *B*	Globe and lift check *A* and *B*	Angle and lift check *D* and *E*	Swing check *A* and *B*	Solid wedge and double disk *A*	Globe, lift check, and swing check *A*	Angle and lift check *D*
¼						4	4	4	2	4			
⅜						4	4	4	2	4			
½						4¼	4¼	4¼	2¼	4¼			
¾						4⅝	4⅝	4⅝	2½	4⅝			
1						5	5	5	2¾	5			
1¼						5½	5½	5½	3	5½			
1½						6½	6½	6½	3¼	6½			
2	7	7	8	4	8	7	8½	8	4	8	8½	10½	5¼
2½	7½	7½	8½	4¼	8½	7½	9½	8½	4¼	8½	9½	11½	5¾
3	8	8	9½	4¾	9½	8	11⅛	9½	4¾	9½	11⅛	12½	6¼
3½	8½	8½				†					12	14	7
4	9	9	11½	5¾	11½	9	12	11½	5¾	11½	15	15¾	7⅞
5	10	10	13	6½	13	10	15	14	7	13	15⅞	17½	8¾
6	10½	10½	14	7	14	10½	15⅞	16	8	14	16½	21	10½
8	11½	11½	19½	9¾	19½	11½	16½	19½	9¾	19½	18	24½	12¼
10	13	13	24½	12¼	24½	13	18	24½	12¼	24½	19¾	28	14
12	14	14	27½	13¾	27½	14	19¾	27½	13¾	27½	22½	†	
14	15	†	31	15½	31	15	22½	31	15½	31	24	†	
16	16	†	36	18	†	16	24	36	18	†	26		
18	17	†			†	17	26			†	28		
20	18	†			†	18	28			†			
24	20	†			†	20	32			†	31		

Nominal valve size	Class 300 steel				Class 400 steel				Class 600 steel						
	Flanged end and welding end				Flanged end and welding end				Flanged end and welding end						
	Gate				Gate				Gate					Angle and lift check	
	Solid wedge and double disk *A* and *B*	Globe and lift check *A* and *B*	Angle and lift check *D* and *E*	Swing check *A* and *B*	Solid wedge *A* and *B*	Double disk *A* and *B*	Globe, lift check, and swing check *A* and *B*	Angle and lift check *D* and *E*	Solid wedge *A* and *B*	Double disk *A* and *B*	Short pattern‡ *B*	Regular globe, regular lift check, swing check *A* and *B*	Short pattern‡ globe, short pattern lift check *B*	Regular *D* and *E*	Short pattern *E*
½	5½	6	3		6½		6½	3¼	6½			6½		3¼	
¾	6	7	3½		7½		7½	3¾	7½			7½		3¾	
1	6½§	8	4	8½	8½	8½	8½	4¼	8½	8½	5¼	8½	5¼	4¼	
1¼	7§	8½	4¼	9	9	9	9	4½	9	9	5¾	9	5¾	4½	
1½	7½	9	4½	9½	9½	9½	9½	4¾	9½	9½	6	9½	6	4¾	
2	8½	10½	5¼	10½	11½	11½	11½	5¾	11½	11½	7	11½	7	5¾	4¼
2½	9½	11½	5¾	11½	13	13	13	6½	13	13	8½	13	8½	6½	5
3	11⅛	12½	6¼	12½	14	14	14	7	14	14	10	14	10	7	6
4	12	14	7	14	16	16	16	8	17	17	12	17	12	8½	7
5	15	15¾	7⅞	15¾	18	18	18	9	20	20	15	20	15	10	8½
6	15⅞	17½	8¾	17½	19½	19½	19½	9¾	22	22	18	22	18	11	10
8	16½	22	11	21	23½	23½	23½	11¾	26	26	23	26	23	13	
10	18	24½	12¼	24½	26½	26½	26½	13¼	31	31	28	31	28	15½	
12	19¾	28	14	28	30	30	30	15	33	33	32	33	32	16½	
14	30			†	32½	30½	†		35	35	35	†			
16	33			†	35½	35½	†		39	39	39	†			
18	36			†	38½	38½	†		43	43	43	†			
20	39			†	41½	41½	†		47	47	47	†			
22	43			†	45	45	†		51	51		†			
24	45			†	48½	48½	†		55	55	55	†			

TABLE 10-28 Dimensions of Valves *(Concluded)*

	Class 900 steel							Class 1500 steel				
	Flanged end and welding end							Flanged end and welding end				
	Gate			Regular globe regular lift check, swing check	Short pattern‡ globe, short pattern lift check	Angle and lift check		Gate			Globe, lift check, swing check	Angle and lift check
Nominal valve size	Solid wedge *A* and *B*	Double disk *A* and *B*	Short pattern‡ *B*	*A* and *B*	*B*	Regular *D* and *E*	Short pattern *E*	Solid wedge *A* and *B*	Double disk *A* and *B*	Short pattern‡ *B*	*A* and *B*	*D* and *E*
¾				9		4½					9	4½
1	10		5½	10		5		10		5½	10	5
1¼	11		6½	11		5½		11		6½	11	5½
1½	12		7	12		6		12		7	12	6
2	14½	14½	8½	14½		7¼		14½	14½	8½	14½	7¼
2½	16½	16½	10	16½		8¼		16½	16½	10	16½	8¼
3	15	15	12	15	12	7½	6	18½	18½	12	18½	9¼
4	18	18	14	18	14	9	7	21½	21½	16	21½	10¾
5	22	22	17	22	17	11	8½	26½	26½	19	26½	13¼
6	24	24	20	24	20	12	10	27¾	27¾	22	27¾	13⅞
8	29	29	26	29	26	14½	13	32¾	32¾	28	32¾	16⅜
10	33	33	31	33	31	16½	15½	39	39	34	39	19½
12	38	38	36	38	36	19	18	44½	44½	39	44½	22¼
14	40½	40½	39	40½	39	20¼	19½	49½	49½	42	49½	24¾
16	44½	44½	43					54½	54½	47		
18	48	48	†					60½	60½	53		
20	52	52	†					65½	65½	58		
24	61	61	†					76½	76½			

	Class 2500 steel				
	Flanged end and welding end				
	Gate			Globe, lift check, swing check	Angle and lift check
Nominal valve size	Solid wedge *A* and *B*	Double disk *A* and *B*	Short pattern‡ *B*	*A* and *B*	*B*
½	10⅜			10⅜	5 3/16
¾	10¾			10¾	5⅜
1	12⅛		7 5/16	12⅛	6 1/16
1¼	13¾		9⅛	13¾	6⅞
1½	15⅛		9⅛	15⅛	7 9/16
2	17¾	17¾	11	17¾	8⅞
2½	20	20	13	20	10
3	22¾	22¾	14½	22¾	11⅜
4	26½	26½	18	26½	13¼
5	31¼	31¼	21	31¼	15⅝
6	36	36	24	36	18
8	40¼	40¼	30	40¼	20⅛
10	50	50	36	50	25
12	56	56	41	56	28
14			44		
16			49		
18			55		

NOTE: Outline drawings for flanged valves shown ¼-in raised face machined onto flange, as for 400-lb cast-steel valves; 150- and 300-lb cast-steel valves and 250-lb cast-iron valves have 1/16-in raised faces; 125-lb cast-iron and 150-lb corrosion-resistant valves covered by MSS-SP-42 have no raised faces.

*Extracted from Face-to-Face and End-to-End Dimensions of Ferrous Valves, ANSI B16.10—1973, with permission of the publisher, the American Society of Mechanical Engineers, New York. To convert inches to millimeters, multiply by 25.4.

†Not shown in ANSI B16.10 but commercially available.

‡These dimensions apply to pressure-seal or flangeless bonnet valves only.

§Solid wedge only.

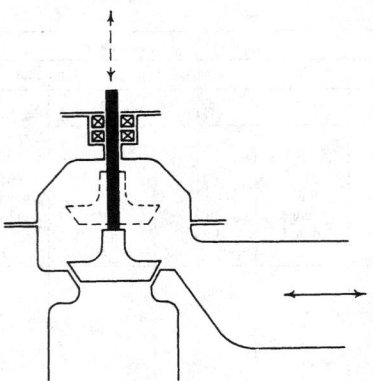

FIG. 10-150   Angle valve.

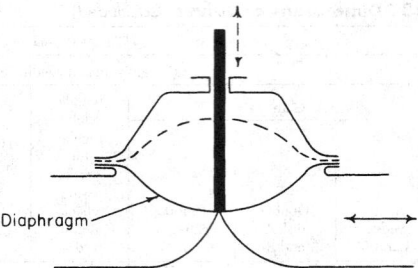

FIG. 10-151   Diaphragm valve.

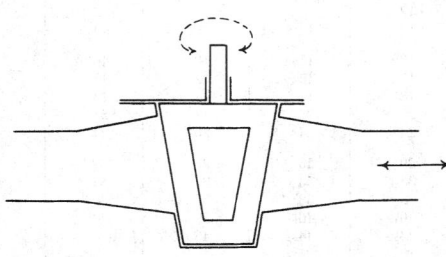

FIG. 10-152   Plug cock.

Globe valves in horizontal lines prevent complete drainage. Seat-wiper valves in which the disk may be rotated by a separate stem inside and concentric with the main stem are used to clear the seats of solid deposits.

Angle Valves   These valves are similar to globe valves; the same bonnet, stem, and disk are used for both (Fig. 10-150). They combine an elbow fitting and a globe valve into one component with a substantial saving in pressure drop. Flanged angle valves are easier to remove and replace than flanged globe valves.

Diaphragm Valves   These valves are limited to pressures of approximately 50 lbf/in² (Fig. 10-151). The fabric-reinforced diaphragms may be made from natural rubber, from a synthetic rubber, or from natural or synthetic rubbers faced with Teflon° fluorocarbon resin. The simple shape of the body makes lining it economical. Elastomers have shorter lives as diaphragms than as linings because of flexing but still provide satisfactory service. Plastic bodies, which have low moduli of elasticity compared with metals, are practical in diaphragm valves since alignment and distortion are minor problems.

These valves are excellent for fluids containing suspended solids and can be installed in any position. Models in which the dam is very low, reducing pressure drop to a negligible quantity and permitting complete drainage in horizontal lines, are available. However, drainage can be obtained with any model simply by installing it with the stem horizontal. The only maintenance required is replacement of the diaphragm, which can be done very quickly without removing the valve from the line.

Plug Cocks   These valves (Fig. 10-152) are limited to temperatures below 260°C (500°F) since differential expansion between the plug and the body results in seizure. The size and shape of the port divide these valves into different types. In order of increasing cost they are short venturi, reduced rectangular port; long venturi, reduced rectangular port; full rectangular port; and full round port.

In lever-sealed plug cocks, tapered plugs are used. The plugs are raised by turning one lever, rotated by another lever, and reseated by the first lever. **Lubricated** plug cocks may use straight or tapered plugs. The tapered plugs may be raised slightly, to reduce turning effort, by injection of the lubricant, which also acts as a seal. Plastic is used in nonlubricated plug cocks as a body liner, a plug coating, or port seals in the body or on the plug.

In plug cocks other than lever-sealed plug cocks, the contact area between plug and body is large, and gearing is usually used in sizes 6 in and larger to minimize operating effort. There are several lever-sealed plug cocks incorporating mechanisms which convert the rotary motion of a handwheel into sequenced motion of the two levers.

For lubricated plug cocks, the lubricant must have limited viscosity change over the range of operating temperature, must have low solubility in the fluid handled, and must be applied regularly. There must be no chemical reaction between the lubricant and the fluid which

would harden or soften the lubricant or contaminate the fluid. For these reasons, lubricated plug cocks are most often used when there are a large number handling the same or closely related fluids at approximately the same temperature.

Lever-sealed plug cocks are used for throttling service. Because of the large contact area between plug and body, if a plug cock is operable, there is little likelihood of leakage when closed, and the handle position is a clearly visible indication of the valve position.

Ball Valves   (Figs. 10-153 and 10-154) These valves are limited to temperatures that have little effect on their plastic seats. Since the sealing element is a ball, its alignment with the axis of the stem is not essential to tight shutoff. In free-ball valves the ball is free to move axially. Pressure differential across the valve forces the ball in the closed position against the downstream seat and the latter against the body. In fixed-ball valves, the ball rotates on stem extensions, with the bearings sealed with O rings. Plastic seats may be compressed or spring-loaded against the ball and the body by the assembly of the valves, or they may be forced against the ball by pressure across the valve acting against O rings which seal between the seat and the body.

Ball valves in which the ball and seats are inserted from above are known as top-entry ball valves. Replacement of seats is easiest in this

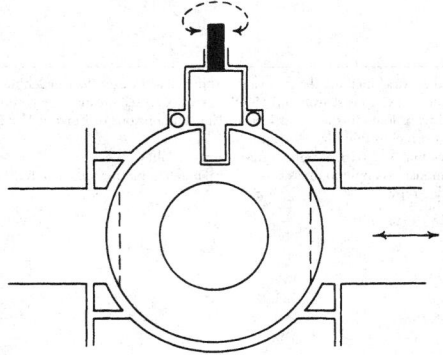

FIG. 10-153   Ball valve; free ball.

° Du Pont TFE fluorocarbon resin.

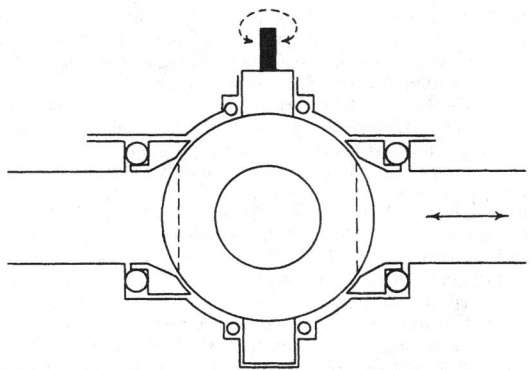

FIG. 10-154   Ball valve; fixed ball.

type. The others are known as split-body valves. Some of these incorporate bolted assembly which permits their use as joints for assembly of the piping. Replacement of seats in this type is easiest when the body consists of three pieces with the ball and the seats contained in the middle piece.

For the larger sizes in high-pressure service, the fixed-ball type with O-ring seat seals requires less operating effort. However, these require two different plastic materials with resistance to the fluid and its temperature. Like plug cocks, ball valves may be either restricted-port or full-port, but the ports are always round and pressure drop is low.

Butterfly Valves   These valves (Fig. 10-155) occupy less space in the line than any other valves. Relatively tight sealing without excessive operating torque and seat wear is accomplished by a variety of methods, such as resilient seats, piston rings on the disk, and inclining the stem to limit contact between the portions of disk closest to the stem and the body seat to a few degrees of curvature.

Fluid-pressure distribution tends to close the valve. For this reason, the smaller manually operated valves have a latching device on the handle, and the larger manually operated valves use worm gearing on the stem. This hydraulic unbalance is proportional to the pressure drop and, with line velocities exceeding 7.6 m/s (25 ft/s), is the principal component in the torque required to operate the valves. Compared with other valves for low-pressure drops, these valves can be operated by smaller hydraulic cylinders. In this service butterfly valves with insert bodies for bolting between existing flanges with bolts that pass by the body are the lowest-first-cost valve in pipe sizes 10 in and larger. Pressure drop is quite high compared with that of gate valves.

Swing Check Valves   These valves (Fig. 10-156) are used to prevent reversal of flow. Normal design is for use only in horizontal lines, where the force of gravity on the disk is at a maximum at the start of closing and at a minimum at the end of closing. Unlike most other valves, check valves are more likely to leak at low pressure than at high pressure, since fluid pressure alone forces the disk to conform to the seat. For this reason elastomers are often mounted on the disk. Swing check valves are available with low-cost insert bodies.

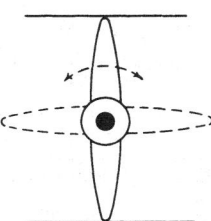

FIG. 10-155   Butterfly valve.

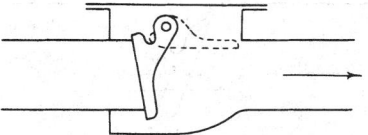

FIG. 10-156   Swing check valve.

Lift Check Valves   These valves (Figs. 10-157 to 10-159) are made in three styles. Vertical lift check valves are for installation in vertical lines, where the flow is normally upward; globe check valves are for use in horizontal lines; angle check valves are for installation where a vertical line with upward flow turns horizontal. Globe and angle check valves normally incorporate an integral dashpot above the disk to slow the motion of the disk and reduce wear. In vertical lift check valves, this feature is found only in the larger sizes. Springs may be incorporated in the dashpots to speed closing, but this increases the pressure drop. Lift checks should not be used when the fluid contains suspended solids.

Tilting-Disk Check Valves   These valves (Fig. 10-160) may be installed in a horizontal line or in lines in which the flow is vertically upward. The pivot point is located so that the distribution of pressure in the fluid handled speeds the closing but arrests slamming. Compared with swing check valves of the same size, pressure drop is less at low velocities but greater at high velocities.

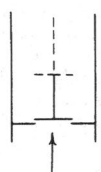

FIG. 10-157   Lift check valve, vertical.

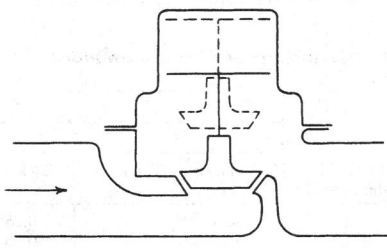

FIG. 10-158   Lift check valve, globe.

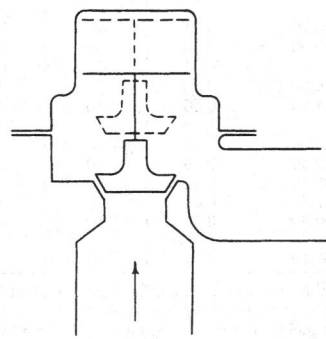

FIG. 10-159   Lift check valve, angle.

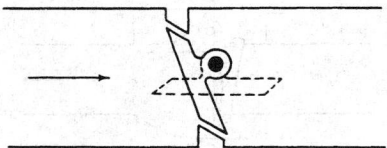

FIG. 10-160 Tilting-disk check valve.

Closure at the instant of reversal of flow is most nearly attained in these valves. This timing of closure is not the whole solution to noise and shock at check valves. For example, if cessation of pressure at the inlet of a valve produces flashing of the decelerating stream downstream from the valve or if stoppage of flow is caused by a sudden closure of a valve some distance downstream from the check valve and the stoppage is followed by returning water hammer, slower closure may be necessary. For these applications, tilting-disk check valves are equipped with external dashpots. They are also available with low-cost insert bodies.

Valve Trim Various alloys are available for valve parts such as seats, disks, and stems which must retain smooth finish for successful operation. The problem in seat materials is fivefold: (1) resistance to corrosion by the fluid handled and to oxidation at high temperatures, (2) resistance to erosion by suspended solids in the fluid, (3) prevention of galling (seizure at point of contact) by differences in material or hardness or both, (4) maintenance of high strength at high temperature, and (5) avoidance of distortion.

All valve trim materials have coefficients of thermal expansion which exceed those of cast or forged carbon steel by 24 to 45 percent and tend to cause distortion of seats and disks. To some extent leakage from this cause is prevented by closing the valve more tightly. Inserting a ring of high-temperature elastomer or plastic, either in or alongside the trim metal in the seat or disk, prevents leakage from this cause.

CAST IRON, DUCTILE IRON, AND HIGH-SILICON IRON

Cast Iron and Ductile Iron Cast iron and ductile iron provide more metal for less cost than steel in piping systems and are widely used in low-pressure services in which internal and external corrosion may cause a considerable loss of metal. They are widely used for underground water distribution. Cement lining is available at a nominal cost for handling water causing tuberculation.

Ductile iron has an elongation of 10 percent or more compared with essentially nil elongation for cast iron and has for all practical purposes supplanted cast iron as a cast piping material. It is usually centrifugally cast in rapidly revolving molds. This manufacturing method improves tensile strength and reduces porosity. Ductile-iron pipe is manufactured to ANSI A21.51—1976 and is available in nominal sizes from 3 through 54 in. Wall thicknesses are specified by seven standard thickness classes. Table 10-29 gives the outside diameter and standard thickness for various rated water working pressures for centrifugally cast ductile-iron pipe. The required wall thickness for underground installations increases with internal pressure, depth of laying, and weight of vehicles operating over the pipe. It is reduced by the degree to which the soil surrounding the pipe provides uniform support along the pipe and around the lower 180°. Tables are provided in ANSI A21.51 for determining wall-thickness-class recommendations for various installation conditions. The poured joint (Fig. 10-135) has been almost entirely superseded by the mechanical joint (Fig. 10-136) and the push-on joint (Fig. 10-137), which are better suited to wet trenches, bad weather, and unskilled labor and minimize strain on the pipe from ground settlement. Lengths vary between 5 and 6 m (between 18 and 20 ft), depending on the supplier. Stock fittings are designed for 1.72-MPa (250-lbf/in^2) cast iron or 2.41-MPa (350-lbf/in^2) ductile iron in sizes through 12 in and for 1.0- and 1.72-MPa (150- and 250-lbf/in^2) cast iron or 2.41-MPa (350-lbf/in^2) ductile iron in sizes 14 in and larger. Stock fittings include 22½° and 11¼° bends. Ductile-iron pipe is also supplied with flanges that match the dimensions of Class 125 flanges shown in ANSI B16.1 (see Table 10-19). These flanges are assembled to the pipe barrel by threaded joints.

High-Silicon Iron Duriron is a high-silicon iron containing approximately 14.5 percent silicon and 0.85 percent carbon. Durichlor is a special high-silicon iron containing appreciable amounts of molybdenum.

These alloys are available in the cast form only. Pipe and fittings are cast with the upset ends being joined by split flanges. Integrally cast flanged pipe is also available. Allowable working pressures cannot be

TABLE 10-29 Dimensions of Ductile-Iron Pipe*

Standard thickness for internal pressure†

Pipe size, in.	Outside diameter, in	Rated water working pressure, lbf/in²‡									
		150		200		250		300		350	
		Thickness, in	Thickness class	Thickness, in	Thickness class	Thickness, in	Thickness class	Thickness, in	Thickness class	Thickness, in	Thickness class
3	3.96	0.25	51	0.25	51	0.25	51	0.25	51	0.25	51
4	4.80	0.26	51	0.26	51	0.26	51	0.26	51	0.26	51
6	6.90	0.25	50	0.25	50	0.25	50	0.25	50	0.25	50
8	9.05	0.27	50	0.27	50	0.27	50	0.27	50	0.27	50
10	11.10	0.29	50	0.29	50	0.29	50	0.29	50	0.29	50
12	13.20	0.31	50	0.31	50	0.31	50	0.31	50	0.31	50
14	15.30	0.33	50	0.33	50	0.33	50	0.33	50	0.33	50
16	17.40	0.34	50	0.34	50	0.34	50	0.34	50	0.34	50
18	19.50	0.35	50	0.35	50	0.35	50	0.35	50	0.35	50
20	21.60	0.36	50	0.36	50	0.36	50	0.36	50	0.39	51
24	25.80	0.38	50	0.38	50	0.38	50	0.41	51	0.44	52
30	32.00	0.39	50	0.39	50	0.43	51	0.47	52	0.51	53
36	38.30	0.43	50	0.43	50	0.48	51	0.53	52	0.58	53
42	44.50	0.47	50	0.47	50	0.53	51	0.59	52	0.65	53
48	50.80	0.51	50	0.51	50	0.58	51	0.65	52	0.72	53
54	57.10	0.57	50	0.57	50	0.65	51	0.73	52	0.81	53

*Extracted from the American National Standard for Ductile-Iron Pipe, Centrifugally Cast in Metal Molds or Sand-Lined Molds, for Water or Other Liquids, ANSI A21.51—1976, with permission of the publisher, the American Society of Mechanical Engineers, New York.
†To convert from inches to millimeters, multiply by 25.4; to convert pounds-force per square inch to megapascals, multiply by 0.006895.
‡These pipe walls are adequate for the rated working pressure plus a surge allowance of 100 lbf/in². For the effect of laying conditions and depth of bury, see ANSI A21.51.

TABLE 10-30 High-Silicon Iron Pipe*

	Split flanged ends				Bell-and-spigot ends			
Size, inside diam., in	Outside diam., in	Wall thickness, in	Standard† length, ft	Weight per piece, lb	Outside diam., in	Wall thickness, in	Standard† length, ft	Weight per piece, lb
1	1¾	⅜	3	15				
1½	2¼	⅜	3	18	2⅛	5⁄16	3	20
2	2¾	⅜	4	32	2⅝	5⁄16	4	30
2½	3¼	⅜	5	45				
3	3⅞	7⁄16	5	62	311⁄16	11⁄32	5	68
4	4⅞	7⁄16	5	100	4⅝	5⁄16	5	89
6	7	½	5	180	611⁄16	11⁄32	5	133
8	9¼	⅝	6	265	9	½	5	232
10	11½	¾	6	433	11¼	⅝	5	341
12	14	1	6	694	13¼	⅝	5	463
15					16¾	⅞	5	680

*The Duriron Co.

†Laying lengths; lengths less than standard are available.

NOTE: To convert inches to millimeters, multiply by 25.4; to convert feet to meters, multiply by 0.3048; and to convert pounds to kilograms, multiply by 0.4536.

stated in the manner customary for other types of pipe because of such variables as thermal shock, pulsating pressures, and the corrosive fluids being handled. Although rupture does not occur below 2.76-MPa (400-lbf/in^2) pressure in sizes up to and including 6 in, 0.3 MPa (50 lbf/in^2) is a normal recommendation, even though the pipe has been used for pressure considerably in excess of that figure.

Table 10-30 lists sizes 1 to 12 in, and larger sizes can be obtained. Bell-and-spigot pipe is produced in the weights and dimensions shown in Table 10-30; fittings are available. New hubless pipe utilizing TFE gaskets and stainless steel clamps to make a mechanical joint are now available (the trade name is Duriron MJ).

The coefficient of linear expansion of these alloys in the temperature range of 21 to 100°C (70 to 212°F) is 12.2×10^{-6}/°C (6.8×10^{-6}/°F), which is slightly above that of cast iron (National Bureau of Standards). Since these alloys have practically no elasticity, it is necessary to use expansion joints in relatively short pipe lines. Connections for flanged pipe, fittings, valves, and pumps are made to 125-lb American Standard drilling.

The use of high-silicon iron in flammable-fluid service or in Category M fluid service is prohibited by the code.

NONFERROUS-METAL PIPING SYSTEMS

Aluminum Seamless aluminum pipe and tube are produced by extrusion in essentially pure aluminum and in several alloys; 6-, 9-, and 12-m (20-, 30-, and 40-ft) lengths are available. Alloying and mill treatment improve physical properties, but welding reduces them. Essentially pure aluminum has an ultimate tensile strength of 65.5 MPa (9500 lbf/in^2) subject to a slight increase by mill treatment which is lost during welding. Alloy 6061, which contains 0.25 percent copper, 0.6 percent silicon, 1 percent magnesium, and 0.25 percent chromium, has an ultimate tensile strength of 124 MPa (18,000 lbf/in^2) in the annealed condition, 262 MPa (38,000 lbf/in^2), mill-treated as 6061-T6, and 165 MPa (24,000 lbf/in^2) at welded joints. Extensive use is made of alloy 1060, which is 99.6 percent pure aluminum, for hydrogen peroxide; of alloy 3003, which contains 1.2 percent manganese, for high-purity chemicals; and of alloys 6063 and 6061 for many other services. Alloy 6063 is the same as 6061 minus the chromium and has slightly lower mechanical properties.

Aluminum is not embrittled by low temperatures and is not subject to external corrosion when exposed to normal atmospheres. At 200°C (400°F) its strength is less than half that at room temperature. It is attacked by alkalies, by traces of copper, nickel, mercury, and other heavy-metal ions, and by prolonged contact with wet insulation. It suffers from galvanic corrosion when coupled to copper, nickel, or lead-base alloys but not when coupled to galvanized iron or austenitic stainless steel.

Aluminum pipe is stocked in 3003, 6061, and 6063 Schedule 40 through 10 in, Schedule 30, 8 through 10 in, and standard-weight 12-in size. It is also stocked in 6063 as Schedule 5 through 6 in and Schedule 10 through 8 in (see Table 10-18).

Threaded **aluminum fittings** are seldom recommended for process piping. Wrought fittings with welding ends (see Table 10-27 for dimensions) and with grooved joint ends are available. Wrought 6061-T6 flanges with dimensions per Table 10-19 are also available. Cast flanges and flanged fittings, sand-cast as alloy B214, 3.8 percent magnesium alloy with 90-MPA (13,000-lbf/in^2) yield strength, or permanent mold cast as alloy 356-T6, 7 percent silicon, 0.3 percent magnesium alloy with 185-MPa (27,000-lbf/in^2) yield strength are available, but consideration must be given to the fact that the modulus of elasticity of aluminum is only slightly more than one-third that of ferrous alloys. See Table 10-26 for dimensions.

Aluminum-body diaphragm and ball valves are used extensively.

Copper and Copper Alloys Seamless copper, bronze, brass, copper-nickel-alloy, and copper-silicon-alloy pipe and tubing are produced by extrusion. Tubing is available in outside-diameter sizes from 1⁄16 to 16 in and in a range of wall thicknesses varying from 0.005 in for the smallest tubing to 0.75 in for the 16-in size. Tubing is usually specified by outside diameter and wall thickness.

Seamless copper tubing is sold in water-tubing sizes (ASTM B88 and B306). These sizes are identified by a "standard" size designation dimensionally ⅛ in less than the nominal outside diameter. The tubing is also sold as outside-diameter copper tubing (ASTM B280).

Copper tubing is widely used in offices and laboratories for water, steam tracing, pneumatic control systems, compressed air, refrigeration, and inert-gas piping. Connections are made with flared-fitting joints (Fig. 10-143), compression-fitting joints (Fig. 10-144), bite-type-fitting joints (Fig. 10-145), and soldered or brazed joints (Fig. 10-147). Figure 10-147 is most economical for ¾-in size and larger. Ease of handling and bending favors the use of copper; it will usually survive a freeze-up without failure.

Copper water tubing ASTM B88 with dimensions and tolerances as given in Table 10-31 is available drawn or annealed in straight lengths of 6.1 m (20 ft) in types K, L, and M through 8-in size. Type K is available in 5.5-m (18-ft) lengths in 10-in size and 3.6-m (12-ft) lengths in 12-in size. Type L is available in 6.1-m (20-ft) lengths in 10-in size and 5.5-m (18-ft) lengths in 12-in size. Type M is available in 6.1-m (20-ft) lengths through 12-in size. All three types are available in 18.3-m (60-ft) or 30-m (100-ft) coils in sizes up to 1 in, in 18.3-m (60-ft) coils in 1¼- and 1½-in sizes, and in 12.2- or 13.7-m (40- or 45-ft) coils in 2-in size.

DWV tubing, ASTM B280, is available in 6-m (20-ft) straight lengths in the following size-wall combinations: 1¼ in, 0.040-in wall; 1½ in, 0.042-in wall; 2 in, 0.042-in wall; 3 in, 0.045-in wall; 4 in, 0.058-in wall; 5 in, 0.072-in wall; and 6 in, 0.083-in wall. DWV is available only in drawn temper. Outside-diameter copper tubing B280 is available in annealed or drawn temper, depending on size; it is used for refrigeration field service, automotive applications, and general service. Dimensions and tolerances are shown in Table 10-32. Drawn temper is available in 6.1-m (20-ft) straight lengths; annealed temper, in 15.2-m (50-ft) coils.

Too high a temperature or too long a heating period when silver-brazing ruins red-brass solder-joint fittings more quickly than wrought-copper fittings. The former are available in larger sizes. Yellow brass fails from dezincification in some waters.

Red-brass and bronze valves are available with female solder-joint ends for soldered copper-tubing piping systems.

Copper pipe is available per ASTM B42 with dimensions as in Table 10-33. Butt-welding fittings (Table 10-27) are available to fit copper pipe, as are screwed fittings per ANSI B16.15, but solder-end fittings of approximately the same dimensions as the screwed fittings and silver-brazing alloy comprise the usual method of assembly. Red-brass or bronze valves with ends identical to the fittings are available. Flanges and flanged fittings are seldom used, since soldered or silver-brazed joints can be melted apart and reassembled.

TABLE 10-31 Copper Water Tubing—Types K, L, M (ASTM B88)*

Standard size, in	Nominal outside diameter, in	Average outside diameter tolerance, in†		Nominal wall thickness and tolerances, in						Theoretical weight, lb/ft		
				Type K		Type L		Type M				
		Annealed	Drawn	Wall thickness	Tolerance‡	Wall thickness	Tolerance‡	Wall thickness	Tolerance‡	Type K	Type L	Type M
¼	0.375	0.002	0.001	0.035	0.004	0.030	0.0035	§	§	0.145	0.126	§
⅜	0.500	0.0025	0.001	0.049	0.004	0.035	0.0035	0.025	0.0025	0.269	0.198	0.145
½	0.625	0.0025	0.001	0.049	0.004	0.040	0.0035	0.028	0.0025	0.344	0.285	0.204
⅝	0.750	0.0025	0.001	0.049	0.004	0.042	0.0035	§	§	0.418	0.362	§
¾	0.875	0.003	0.001	0.065	0.0045	0.045	0.004	0.032	0.003	0.641	0.455	0.328
1	1.125	0.0035	0.0015	0.065	0.0045	0.050	0.004	0.035	0.0035	0.839	0.655	0.465
1¼	1.375	0.004	0.0015	0.065	0.0045	0.055	0.0045	0.042	0.0035	1.04	0.884	0.682
1½	1.625	0.0045	0.002	0.072	0.005	0.060	0.0045	0.049	0.004	1.36	1.14	0.940
2	2.125	0.005	0.002	0.083	0.007	0.070	0.006	0.058	0.006	2.06	1.75	1.46
2½	2.625	0.005	0.002	0.095	0.007	0.080	0.006	0.065	0.006	2.93	2.48	2.03
3	3.125	0.005	0.002	0.109	0.007	0.090	0.007	0.072	0.006	4.00	3.33	2.68
3½	3.625	0.005	0.002	0.120	0.008	0.100	0.007	0.083	0.007	5.12	4.29	3.58
4	4.125	0.005	0.002	0.134	0.010	0.110	0.009	0.095	0.009	6.51	5.38	4.66
5	5.125	0.005	0.002	0.160	0.010	0.125	0.010	0.109	0.009	9.67	7.61	6.66
6	6.125	0.005	0.002	0.192	0.012	0.140	0.011	0.122	0.010	13.9	10.2	8.92
8	8.125	0.006	+0.002 −0.004	0.271	0.016	0.200	0.014	0.170	0.014	25.9	19.3	16.5
10	10.125	0.008	+0.002 −0.006	0.338	0.018	0.250	0.016	0.212	0.015	40.3	30.1	25.6
12	12.125	0.008	+0.002 −0.006	0.405	0.020	0.280	0.018	0.254	0.016	57.8	40.4	36.7

*Copyright American Society for Testing and Materials, 1916 Race Street, Philadelphia, Pa. 19103; reprinted/adapted with permission. To convert inches to millimeters, multiply by 25.4; to convert pounds per foot to kilograms per meter, multiply by 1.49.
†The average outside diameter of a tube is the average of the maximum and minimum outside diameter, as determined at any one cross section of the tube.
‡Maximum deviation at any one point.
§Indicates that the material is not generally available or that no tolerance has been established.

TABLE 10-32 Copper Outside-Diameter Tubing for Refrigeration Field Service and Automotive and General Service (ASTM B280)*

For mechanical or soldered fittings

Standard size, in	Outside diameter, in (mm)	Wall thickness, in (mm)	Weight, lb/ft (kg/m)	Tolerances†	
				Average outside diameter, plus and minus, in (mm)‡	Wall thickness, plus and minus, in (mm)
			For coil		
⅛	0.125 (3.18)	0.030 (0.762)	0.0347 (0.0516)	0.002 (0.051)	0.003 (0.076)
3/16	0.187 (4.75)	0.030 (0.762)	0.0575 (0.0856)	0.002 (0.051)	0.0025 (0.064)
¼	0.250 (6.35)	0.030 (0.762)	0.0804 (0.120)	0.002 (0.051)	0.0025 (0.064)
5/16	0.312 (7.92)	0.032 (0.813)	0.109 (0.162)	0.002 (0.051)	0.0025 (0.064)
⅜	0.375 (9.52)	0.032 (0.813)	0.134 (0.199)	0.002 (0.051)	0.0025 (0.064)
½	0.500 (12.7)	0.032 (0.813)	0.182 (0.271)	0.002 (0.051)	0.0025 (0.064)
⅝	0.625 (15.9)	0.035 (0.889)	0.251 (0.373)	0.002 (0.051)	0.0030 (0.076)
¾	0.750 (19.1)	0.035 (0.889)	0.305 (0.454)	0.0025 (0.064)	0.0035 (0.089)
¾	0.750 (19.1)	0.042 (1.07)	0.362 (0.539)	0.0025 (0.064)	0.0035 (0.089)
⅞	0.875 (22.3)	0.045 (1.14)	0.455 (0.677)	0.003 (0.076)	0.004 (0.10)
1⅛	1.125 (28.6)	0.050 (1.27)	0.665 (0.975)	0.0035 (0.089)	0.004 (0.10)
1⅜	1.375 (34.9)	0.055 (1.40)	0.884 (1.32)	0.004 (0.10)	0.0045 (0.11)
1⅝	1.625 (41.3)	0.060 (1.52)	1.14 (1.70)	0.0045 (0.11)	0.0045 (0.11)
			For straight lengths (applicable to drawn-temper tube only)		
⅜	0.375 (9.52)	0.030 (0.762)	0.126 (0.187)	0.001 (0.025)	0.0035 (0.089)
½	0.500 (12.7)	0.035 (0.889)	0.198 (0.146)	0.001 (0.025)	0.0035 (0.089)
⅝	0.625 (15.9)	0.040 (1.02)	0.285 (0.424)	0.001 (0.025)	0.0035 (0.089)
¾	0.750 (19.1)	0.042 (1.07)	0.362 (0.539)	0.001 (0.025)	0.0035 (0.089)
⅞	0.875 (22.3)	0.045 (1.14)	0.455 (0.677)	0.001 (0.025)	0.004 (0.10)
1⅛	1.125 (28.6)	0.050 (1.27)	0.655 (0.975)	0.0015 (0.038)	0.004 (0.10)
1⅜	1.375 (34.9)	0.055 (1.40)	0.884 (1.32)	0.0015 (0.038)	0.0045 (0.11)
1⅝	1.625 (41.3)	0.060 (1.52)	1.14 (1.70)	0.002 (0.051)	0.0045 (0.11)
2⅛	2.125 (54.0)	0.070 (1.78)	1.75 (2.60)	0.002 (0.051)	0.006 (0.15)
2⅝	2.625 (66.7)	0.080 (2.03)	2.48 (3.69)	0.002 (0.051)	0.006 (0.15)
3⅛	3.125 (79.4)	0.090 (2.29)	3.33 (4.96)	0.002 (0.051)	0.007 (0.18)
3⅝	3.625 (92.1)	0.100 (2.54)	4.29 (6.38)	0.002 (0.051)	0.007 (0.18)
4⅛	4.125 (105)	0.110 (2.79)	5.38 (8.01)	0.002 (0.051)	0.009 (0.23)

*Copyright American Society for Testing and Materials, 1916 Race Street, Philadelphia, Pa. 19103; reprinted/adapted with permission.
†The tolerances listed represent the maximum deviation at any point.
‡The average outside diameter of a tube is the average of the maximum and minimum outside diameters as determined at any one cross section of the tube.

TABLE 10-33   Copper and Red-Brass Pipe (ASTM B42 and B43)*: Standard Dimensions, Weights, and Tolerances

Standard pipe size, in	Nominal outside diameter, in (mm)	Average outside diameter tolerances, in (mm), all minus†	Nominal wall thickness, in (mm)	Tolerance, in (mm)‡	Theoretical weight, lb/ft (kg/m) Red brass	Theoretical weight, lb/ft (kg/m) Copper
			Regular pipe			
⅛	0.405 (10.3)	0.004 (0.10)	0.062 (1.57)	0.004 (0.10)	0.253 (0.376)	0.259 (0.385)
¼	0.540 (13.7)	0.004 (0.10)	0.082 (2.08)	0.005 (0.13)	0.447 (0.665)	0.457 (0.680)
⅜	0.675 (17.1)	0.005 (0.13)	0.090 (2.29)	0.005 (0.13)	0.627 (0.933)	0.641 (0.954)
½	0.840 (21.3)	0.005 (0.13)	0.107 (2.72)	0.006 (0.15)	0.934 (1.39)	0.955 (1.42)
¾	1.050 (26.7)	0.006 (0.15)	0.114 (2.90)	0.006 (0.15)	1.27 (1.89)	1.30 (1.93)
1	1.315 (33.4)	0.006 (0.15)	0.126 (3.20)	0.007 (0.18)	1.78 (2.65)	1.82 (2.71)
1¼	1.660 (42.2)	0.006 (0.15)	0.146 (3.71)	0.008 (0.20)	2.63 (3.91)	2.69 (4.00)
1½	1.900 (48.3)	0.006 (0.15)	0.150 (3.81)	0.008 (0.20)	3.13 (4.66)	3.20 (4.76)
2	2.375 (60.3)	0.008 (0.20)	0.156 (3.96)	0.009 (0.23)	4.12 (6.13)	4.22 (6.28)
2½	2.875 (73.0)	0.008 (0.20)	0.187 (4.75)	0.010 (0.25)	5.99 (8.91)	6.12 (9.11)
3	3.500 (88.9)	0.010 (0.25)	0.219 (5.56)	0.012 (0.30)	8.56 (12.7)	8.76 (13.0)
3½	4.000 (102)	0.010 (0.25)	0.250 (6.35)	0.013 (0.33)	11.2 (16.7)	11.4 (17.0)
4	4.500 (114)	0.012 (0.30)	0.250 (6.35)	0.014 (0.36)	12.7 (18.9)	12.9 (19.2)
5	5.562 (141)	0.014 (0.36)	0.250 (6.35)	0.014 (0.36)	15.8 (23.5)	16.2 (24.1)
6	6.625 (168)	0.016 (0.41)	0.250 (6.35)	0.014 (0.36)	19.0 (28.3)	19.4 (28.9)
8	8.625 (219)	0.020 (0.51)	0.312 (7.92)	0.022 (0.56)	30.9 (46.0)	31.6 (47.0)
10	10.750 (273)	0.022 (0.56)	0.365 (9.27)	0.030 (0.76)	45.2 (67.3)	46.2 (68.7)
12	12.750 (324)	0.024 (0.61)	0.375 (9.52)	0.030 (0.76)	55.3 (82.3)	56.5 (84.1)
			Extra strong pipe			
⅛	0.405 (10.3)	0.004 (0.10)	0.100 (2.54)	0.006 (0.15)	0.363 (0.540)	0.371 (0.552)
¼	0.540 (13.7)	0.004 (0.10)	0.123 (3.12)	0.007 (0.18)	0.611 (0.909)	0.625 (0.930)
⅜	0.675 (17.1)	0.005 (0.13)	0.127 (3.23)	0.007 (0.18)	0.829 (1.23)	0.847 (1.26)
½	0.840 (21.3)	0.005 (0.13)	0.149 (3.78)	0.008 (0.20)	1.23 (1.83)	1.25 (1.86)
¾	1.050 (26.7)	0.006 (0.15)	0.157 (3.99)	0.009 (0.23)	1.67 (2.48)	1.71 (2.54)
1	1.315 (33.4)	0.006 (0.15)	0.182 (4.62)	0.010 (0.25)	2.46 (3.66)	2.51 (3.73)
1¼	1.660 (42.2)	0.006 (0.15)	0.194 (4.93)	0.010 (0.25)	3.39 (5.04)	3.46 (5.15)
1½	1.900 (48.3)	0.006 (0.15)	0.203 (5.16)	0.011 (0.28)	4.10 (6.10)	4.19 (6.23)
2	2.375 (60.3)	0.008 (0.20)	0.221 (5.61)	0.012 (0.30)	5.67 (8.44)	5.80 (8.63)
2½	2.875 (73.0)	0.008 (0.20)	0.280 (7.11)	0.015 (0.38)	8.66 (12.9)	8.85 (13.2)
3	3.500 (88.9)	0.010 (0.25)	0.304 (7.72)	0.016 (0.41)	11.6 (17.3)	11.8 (17.6)
3½	4.000 (102)	0.010 (0.25)	0.321 (8.15)	0.017 (0.43)	14.1 (21.0)	14.4 (21.4)
4	4.500 (114)	0.012 (0.30)	0.341 (8.66)	0.018 (0.46)	16.9 (25.1)	17.3 (25.7)
5	5.562 (141)	0.014 (0.36)	0.375 (9.52)	0.019 (0.48)	23.2 (34.5)	23.7 (35.3)
6	6.625 (168)	0.016 (0.41)	0.437 (11.1)	0.027 (0.69)	32.2 (47.9)	32.9 (49.0)
8	8.625 (219)	0.020 (0.51)	0.500 (12.7)	0.035 (0.89)	48.4 (72.0)	49.5 (73.7)
10	10.750 (273)	0.022 (0.56)	0.500 (12.7)	0.040 (1.0)	61.1 (90.9)	62.4 (92.9)

*Copyright American Society for Testing and Materials, 1916 Race Street, Philadelphia, Pa. 19103; reprinted/adapted with permission. All tolerances are plus and minus except as otherwise indicated.

†The average outside diameter of a tube is the average of the maximum and minimum outside diameters as determined at any one cross section of the tube.

‡Maximum deviation at any one point.

Threadless copper pipe, thinner than ASTM B42, is available with dimensions as in Table 10-34. Solder-end fittings similar to ANSI B16.15 screwed fittings and solder-end valves are used with this pipe.

Copper pipe is attacked by water originating in granite substrata, and for this reason red-brass pipe per ASTM B43 with red-brass screwed or solder-end fittings is sometimes used in its place.

70 percent copper, 30 percent nickel and 90 percent copper, 10 percent nickel ASTM B466 are available as seamless pipe and welding fittings for handling brackish water in Schedule 10 and regular copper pipe thicknesses.

Copper-silicon alloy (96 percent cooper, 3 percent silicon, 1 percent manganese), per ASTM B315, is furnished as seamless pipe and welding fittings in regular and extra-strong copper pipe thicknesses. It is easier to weld than copper.

Lead and Lead-Lined Steel Pipe   Lead and lead-lined steel pipe have been essentially eliminated as piping materials owing to health hazards in fabrication and installation and to environmental objections. Lead has been replaced by suitable plastic, reinforced plastic, plastic-lined steel, or high-alloy materials.

Magnesium   Extruded magnesium tubing is available per ASTM B217–58 alloyed with aluminum, manganese, or zinc. Ultimate and yield strengths at 204°C (400°F) are about one-half those at room temperature. Outside-diameter range is ¼ through 8 in. Wall thickness ranges from a minimum of 0.028 in to a maximum of 0.031 in for the ¼-in diameter and from a minimum of 0.250 in to a maximum of 1.0 in for the 8-in diameter.

Nickel and Nickel Alloys   A wide range of ferrous and nonferrous nickel and nickel-bearing alloys are available. They are usually selected because of their improved resistance to chemical attack or their superior resistance to the effects of high temperature. In general terms their cost and corrosion resistance are somewhat a function of their nickel content. The 300 Series stainless steels are the most generally used. Some other frequently used alloys are listed in Table 10-35 together with their nominal compositions. For metallurgical and corrosion resistance data, see Sec. 28.

Titanium   Pipe per ASTM B337 is available welded or seamless via one of the following processes: extrusion, centrifugal casting, machining of bar stock, or powder compaction; Schedule 5S, 10S, 40S, and 80S, ⅛- through 24-in size. Extruded and drawn tubing per ASTM B338 is available from ¼-in outside diameter, 0.020- through 0.083-in wall, up through 3-in outside diameter. Cast welding fittings, flanges, and valves are also available. Titanium is used at temperatures

TABLE 10-34   Hard-Drawn Copper Threadless Pipe (ASTM B302)*

Standard pipe size, in	Nominal dimensions, in (mm)			Cross-sectional area of bore, in² (cm²)	Nominal weight, lb/ft (kg/m)	Tolerances, in (mm)	
	Outside diameter	Inside diameter	Wall thickness			Average outside diameter, all minus†	Wall thickness, plus and minus
¼	0.540 (13.7)	0.410 (10.4)	0.065 (1.65)	0.132 (0.852)	0.376 (0.559)	0.004 (0.10)	0.0035 (0.089)
⅜	0.675 (17.1)	0.545 (13.8)	0.065 (1.65)	0.233 (1.50)	0.483 (0.719)	0.004 (0.10)	0.004 (0.10)
½	0.840 (21.3)	0.710 (18.0)	0.065 (1.65)	0.396 (2.55)	0.613 (0.912)	0.005 (0.13)	0.004 (0.10)
¾	1.050 (26.7)	0.920 (23.4)	0.065 (1.65)	0.665 (4.29)	0.780 (1.16)	0.005 (0.13)	0.004 (0.10)
1	1.315 (33.4)	1.185 (30.1)	0.065 (1.65)	1.10 (7.10)	0.989 (1.47)	0.005 (0.13)	0.004 (0.10)
1¼	1.660 (42.2)	1.530 (38.9)	0.065 (1.65)	1.84 (11.9)	1.26 (1.87)	0.006 (0.15)	0.004 (0.10)
1½	1.900 (48.3)	1.770 (45.0)	0.065 (1.65)	2.46 (15.9)	1.45 (2.16)	0.006 (0.15)	0.004 (0.10)
2	2.375 (60.3)	2.245 (57.0)	0.065 (1.65)	3.96 (25.5)	1.83 (272)	0.007 (0.18)	0.006 (0.15)
2½	2.875 (73.0)	2.745 (69.7)	0.065 (1.65)	5.92 (38.2)	2.22 (3.30)	0.007 (0.18)	0.006 (0.15)
3	3.500 (88.9)	3.334 (84.7)	0.083 (2.11)	8.73 (56.3)	3.45 (5.13)	0.008 (0.20)	0.007 (0.18)
3½	4.000 (102)	3.810 (96.8)	0.095 (2.41)	11.4 (73.5)	4.52 (6.73)	0.008 (0.20)	0.007 (0.18)
4	4.500 (114)	4.286 (109)	0.107 (2.72)	14.4 (92.9)	5.72 (8.51)	0.010 (0.25)	0.009 (0.23)
5	5.562 (141)	5.298 (135)	0.132 (3.40)	22.0 (142)	8.73 (13.0)	0.012 (0.30)	0.010 (0.25)
6	6.625 (168)	6.309 (160)	0.158 (4.01)	31.3 (202)	12.4 (18.5)	0.014 (0.36)	0.010 (0.25)
8	8.625 (219)	8.215 (209)	0.205 (5.21)	53.0 (342)	21.0 (31.2)	0.018 (0.46)	0.014 (0.36)
10	10.750 (273)	10.238 (260)	0.256 (6.50)	82.3 (531)	32.7 (48.7)	0.018 (0.46)	0.016 (0.41)
12	12.750 (324)	12.124 (308)	0.313 (7.95)	115 (742)	47.4 (70.5)	0.018 (0.46)	0.020 (0.51)

*Copyright American Society for Testing and Materials, 1916 Race Street, Philadelphia, Pa. 19103; reprinted/adapted with permission.
†The average outside diameter of a tube is the average of the maximum and minimum outside diameters, as determined at any one cross section of the tube.

TABLE 10-35   Common Nickel and Nickel-Bearing Alloys

Common trade name or registered trademark	Code designation	Alloy no.	ASTM specification (pipe)	Nominal composition, %										
				Ni	Cr	Mo	Fe	C[a]	Si[a]	Mn	Cu	Cb	Co	W
Type 304 stainless steel		S30400	A312	9	19		70	0.08		2.0				
Type 316 stainless steel		S31600	A312	11	18	2.5	66.5	0.08		2.0				
Carpenter 20cb[b]	Ni-Cr-Fe-Mo-Cu-Cb stabilized	N08020	B464	33	20	2.5	38.5	0.06		2.0	3	1		
Incoloy 800[c]	Ni-Fe-Cr	N08800	B407	32.5	21		46	0.05	0.5	0.8	0.4			
Incoloy 825[c]	Ni-Fe-Cr-Mo-Cu	N08825	B423	42	21.5	3	30	0.03	0.2	0.5	2.2			
Hastelloy C-276[d]	Ni-Mo-Cr low carbon	N10276	B575[e]	54	15	16	5	0.02	0.08	1			2.5	4
Hastelloy B-2[d]	Ni-Mo	N10001	B333[e]	64	1	28	2	0.02	0.1	1				
Inconel 625[c]	Ni-Cr-Mo-Cb	N06625	B444	61	21.5	9	2.5	0.05	0.2	0.2		4		
Inconel 600[c]	Ni-Cr-Fe	N06600	B167	76	15.5		8	0.08	0.2	0.5	0.2			
Monel 400[c]	Ni-Cu	N04400	B165	66			1.2	0.20	0.2	1	31.5			
Nickel 200[c]	Ni	N02200	B161	99+			0.2	0.08						
Hastelloy G[d]	Ni-Cr-Fe-Mo-Cu	N06007	B622	42	22.2	6.5	19.5	0.05	1	1.5	2	2.2[f]	2.5[a]	1[a]

[a] Maximum.
[b] Registered trademark, Carpenter Technology Corp.
[c] Registered trademark, Huntington Alloys, Inc.
[d] Registered trademark, Cabot Corp.
[e] Plate.
[f] Cb + Ta.

up to 315°C (600°F). It is extremely notch-sensitive. Titanium alloys such as 6 Al-4V, with higher tensile strengths than straight titanium, are available. Unfortunately, they lack the corrosion resistance and weldability of the unalloyed material.

Zirconium (Tin 1.2 to 1.7 Percent)   Tubing is available seamless ranging from ½- outside diameter by 0.030-in wall to 8-in outside diameter by 0.4-in wall, and welded up through 30-in outside diameter by ⅛-in wall. Cast valves and fittings are also available.

Flexible Metal Hose   Deeply corrugated thin brass, bronze, Monel, aluminum, and steel tubes are covered with flexible braided-wire jackets to form flexible metal hose. Both tube and braid are brazed or welded to pipe-thread, union, or flanged ends. Failures are often the result of corrosion of the braided-wire jacket or of a poor jacket-to-fitting weld. Inside diameters range from ⅛ to 12 in. Maximum recommended temperature for bronze hose is approximately 230°C (450°F). Metal thickness is much less than for straight tube for the same pressure-temperature conditions; so accurate data on corrosion and erosion are required to make proper selection.

NONMETALLIC PIPE AND LINED PIPE SYSTEMS

Asbestos Cement   Asbestos-cement pipe is seamless pipe made of silica and portland cement, compacted under heavy pressure, uniformly reinforced with asbestos fiber, and thoroughly cured. The interior surface is smooth, does not corrode, and does not tuberculate. Under normal conditions of operation, asbestos cement will handle solutions within a pH range of 4.5 to 14. It is a brittle material and undergoes expansion on wetting. There are stringent OSHA regulations pertaining to the fabrication and use of asbestos-containing materials. The most widely used joints are push-on joints. This pipe is used extensively for underground water systems, for paper-mill slurries and wastes, and for mine water. The push-on joints limit the temperature to 65°C (150°F). The light weight of the pipe minimizes handling labor, but careful handling is required to avoid damage. This pipe is available with an epoxy lining which increases its corrosion resistance.

Asbestos-cement fittings and valves are not available, but flanged fabricated-steel fittings lined with segments of asbestos-cement pipe and cement-lined cast-iron fittings with end bells for push-on joint to asbestos-cement pipe can be obtained. Adapters to regular cast-iron fittings are also available. When the pipe is installed aboveground, two guided supports per length of pipe are recommended, and when push-on joints are used, internal pressure thrusts at changes in direction, at reducers, at dead ends, and at valves must be resisted by braces. When poured flanges are used, expansion joints must be used also with braces to resist corresponding pressure thrust.

Pressure Pipe   This pipe is made in three classes corresponding to working pressures of 0.7, 1.0, and 1.4 MPa (100, 150, and 200 lbf/in²) (Table 10-36).

Gravity Sewer Pipe   This pipe is made in five classes for varying depths of bury, trench dimension, soil, and vehicular loading (Table 10-37).

Impervious Graphite   Impervious-graphite pipe, fittings, and valve bodies are made of electric-furnace graphite which, after extruding or molding, is rendered impervious by impregnation with synthetic resins. When impregnated with phenolic resin, it is resistant to most acids (including hydrofluoric), salts, and organic compounds. When impregnated with modified phenolic resin, it is resistant to strong alkalies and highly oxidizing materials. Ultimate tensile strength is low, 17.2 MPa (2500 lbf/in²), and the material is brittle, but the modulus of elasticity is only 15,168 MPa (2.2 × 10⁶ lbf/in²). The material is highly resistant to thermal shock and is available with glass-cloth and resin armor for protection against physical abuse. Maximum continuous operating temperature is 170°C (340°F). Components are designed for operating pressure which increases from 0.3 MPa (50 lbf/in²) at 170°C (340°F) to 0.5 MPa (75 lbf/in²) at 21°C (70°F).

Table 10-38 lists standard sizes of pipe; ½-, ¾-, and ⅞-in sizes are heat-exchanger tubing, and standard fittings are not available for these sizes. Pipe is shipped threaded on request. National Form straight threads are used. Fittings made from the same material with the same thread form are available and include laps which can be screwed on the ends of pipe and stub ends which can be screwed into the fittings, both for the purpose of making flanged lap joints. All threaded joints are permanently bonded by special cements. Flanged joints use split cast-iron backup flanges which have 150-lb ANSI B16.5 bolting in sizes 6 in and smaller and 300-lb ANSI B16.5 bolting in sizes 8 in and larger. Asbestos sheet packing is used between the flange and the back of the lap to equalize bearing. Pipe can be sawed

TABLE 10-36   Asbestos-Cement Pressure Pipe*

Nominal size	Length, ft	Class 100†			Class 150†			Class 200†		
		Inside diam., in	Wall, in‡	Wt., lb/ft§	Inside diam., in	Wall, in‡	Wt., lb/ft§	Inside diam., in	Wall, in‡	Wt., lb/ft§
4	13	3.95	0.35	6.3	3.95	0.43	7.6	3.95	0.43	9.3
6	13	5.85	.42	10.6	5.85	.53	13.0	5.70	.60	15.4
8	13	7.85	.47	15.8	7.85	.63	19.9	7.60	.75	23.9
10	13	9.85	.52	21.8	10.00	.83	32.0	9.63	1.01	37.2
12	13	11.70	.64	29.7	12.00	.96	43.8	11.56	1.18	51.7
14	13	13.59	.74	38.9	14.00	1.11	58.5	13.59	1.31	69.0
16	13	15.50	.83	48.8	16.00	1.23	73.0	15.50	1.48	89.2

*Johns-Manville Co.
†Equivalent to working pressure, lb/sq in.
‡Minimum thickness of machined end; balance of pipe is thicker.
§Pipe plus push-on joint coupling.
NOTE: To convert inches to millimeters, multiply by 25.4; to convert pounds per foot to kilograms per meter, multiply by 1.49; to convert pounds-force per square inch to megapascals, multiply by 0.00689; and to convert feet to meters, multiply by 0.3048.

TABLE 10-37   Asbestos-Cement Gravity Sewer Pipe*

Nominal size	Inside diam., in	Class 1500†		Class 2400†		Class 3300†		Class 4000†		Class 5000†	
		Wall, in‡	Wt., lb/ft	Wall, in‡	Wt., lb/ft	Wall, in‡	Wt., lb/ft	Wall, in‡	Wt., lb/ft	Wall, in‡	Wt., lb/ft
6	6.00	0.46	8.5	0.49	9.5	0.57	11.1				
8	8.00	.51	12.6	.52	13.3	.61	15.6				
10	10.00	.56	17.6	.58	18.9	.68	22.0	0.75	24.3	0.85	27.6
12	12.05	.61	22.8	.63	24.3	.75	28.8	.82	31.5	0.93	35.8
14	14.05			.68	30.3	.81	35.8	.89	39.3	1.00	44.3
16	16.05			.73	37.0	.86	43.1	.95	47.6	1.07	53.7
18	18.05			.77	43.6	.91	50.9	1.01	56.5	1.13	63.3
20	20.05			.81	50.7	.96	59.2	1.06	65.4	1.19	73.6
24	24.05			.89	66.4	1.05	77.2	1.16	85.3	1.30	95.8
30	30.05					1.17	106.8	1.30	118.8	1.45	132.7
36	36.05							1.42	155.0	1.59	173.8

Standard pipe length is 13 ft except 6 in Class 1500 is 10 ft and 8 in Class 1500 may also be 10 ft.
*Johns-Manville Co.
†Crushing strength per A.S.T.M. three-edge bearing method.
‡Thickness of wall of pipe excluding machined ends. Same coupling is used for all classes; it protects the machined ends from crushing loads.
NOTE: To convert inches to millimeters, multiply by 25.4; to convert pounds per foot to kilograms per meter, multiply by 1.49; and to convert feet to meters, multiply by 0.3048.

TABLE 10-38   Standard Sizes of Impervious Graphite Pipe*

Nominal pipe size, in	Inside diameter, in	Outside diameter, in	Wall thickness, in	Maximum length, ft	Average weight, lb/ft	Inside cross-sectional area, ft²	Circumference, ft, or surface, ft²/ft of length	
							Inside	Outside
1	1	1½	¼	9	0.74	.00545	0.262	0.393
1½	1½	2	¼	9	1.1	.01227	.393	.524
2	2	2¾	⅜	9	1.7	.0218	.524	.687
2½	2⅜	3	5⁄16	9	2.0	.0308	.622	.785
3	3	4	½	9	5.4	.0491	.785	1.047
4	4	5¼	⅝	9	8.1	.0873	1.047	1.374
6	6	7½	¾	9	15.6	.1965	1.571	1.964
8	8⅛	9 11⁄16	25⁄32	6	23.2	.360	2.127	2.536
10	10⅛	12 21⁄32	1 17⁄64	6	44.2	.559	2.650	3.313

*Courtesy Union Carbide Corporation, Carbon Products Division.
NOTE: To convert inches to millimeters, multiply by 25.4; to convert feet to meters, multiply by 0.3048; to convert pounds per foot to kilograms per meter, multiply by 1.49; and to convert square feet to square meters, multiply by 0.0929.

to length in the field and threaded with special tools. Synthetic elastomeric and Teflon gaskets are available. Diaphragm valves with impervious graphite bodies are available in sizes from 1 through 6 in. Maximum recommended support spacing is 2.7 m (9 ft), and valves should be supported independently.

Cement-Lined Steel   Cement-lined steel pipe is made by lining steel pipe with special cement. Its use prevents pickup of iron by the fluid handled, corrosion of the metal by brackish water, and growth of tuberculation. Threaded pipe in sizes from ¾ to 4 in is stocked; however, cement-lined pipe in sizes smaller than 1½ is not considered practical for common use.

The coefficients of expansion of iron and cement are nearly alike. Table 10-39 gives dimensions of cement-lined pipe.

Cement-lined carbon steel pipe larger than 4 in is shipped with flanged or welding ends. Welding does not damage the lining, which forms a slag protecting the weld. Shop cement lining of carbon steel pipe is covered by AWWA C205. Cement-lined carbon steel butt-welding fittings and flanged cast-iron fittings are available. AWWA C602 includes cement lining of both cast-iron and carbon steel water lines in place.

Chemical Ware   Acidproof chemical-stoneware pipe and fittings withstand most acid, alkali, or other corrosives, the main exception being hydrofluoric acid. The range of sizes made with the bell-and-spigot joint and with plain butt ends is shown in Table 10-40.

Plain butt-end pipe is furnished with cemented-on flanges with ANSI B16.1 drilling or (for use in ventilating work in which the space is too limited for bell-and-spigot pipe) with a ring for joining with a steel band. Medium-pressure chemical-stoneware pipe armored with glass fiber reinforced with furan resin can be obtained. Flanges with ANSI B16.1 drilling bear against hubs formed from the armor.

Fittings and plug valves with ends to match the various types of pipe are available.

Vitrified-Clay Sewer Pipe   This pipe is resistant to very dilute chemicals except hydrofluoric acid and is produced as standard-strength and extra-strength (ASTM C700). It is used for sewage, industrial waste, and storm water at atmospheric pressure. Elbows, Y branches, tees, reducers, and increasers are available. Assembly is by poured joints which allow for ample angular deflection. Joint com-

pounds are of the hot-pour type or the cold mastic type; both adhere tightly to the scored clay surfaces but remain flexible enough to prevent leakage in the event of earth settlement. Pipe is also available with bituminous or plastic material die-cast on the outside of the spigot and the inside of the bell. The interfaces are a snug fit cemented by applying a solvent to them at the time of assembly. Dimensions of pipe are given in Table 10-41. Choice between standard and extra strength is based on earth and vehicular loading.

Concrete   Unreinforced-concrete sewer pipe is made with poured joint ends in sizes from 4 to 24 in conforming to ASTM C14. Reinforced-concrete culvert, storm-drain, and sewer pipe is made with poured joint or push-on joint ends conforming to ASTM C76 in five classes of reinforcement area and wall thickness in sizes from 12 through 108 in. Essentially the same pipe, except that it has push-on joint ends only, is available for water pressures up to 0.31 MPa (45 lbf/in²) in sizes 12 through 96 in and lengths up through 5.6 m (16 ft) conforming to AWWA C302.

For higher water pressures, a steel cylinder approximately 1.6 mm (1⁄16 in) thick is embedded in the wall of the pipe, which prevents leakage through cracks, and to this there may be added prestressed circumferential reinforcing wire applied after the cylinder has been stiffened by cement lining. Such pipe is available in accordance with AWWA C300, sizes 20 through 96 in, for pressures 0.27 through 1.8 MPa (40 through 260 lbf/in²), and in accordance with AWWA C301, sizes 16 through 96 in. Push-on joints are used. Pipe is also available with steel lugs welded to the reinforcing cages and projecting through the outside surface of the pipe for "bridling." This is known as "subaqueous pipe." Concrete fittings are also available. Con-

TABLE 10-39   Cement-Lined Carbon-steel Pipe*

Standard pipe size, in	Inside diam. after lining, in	Thickness of lining, in	Weight, per ft, lb	Standard pipe size, in	Inside diam. after lining, in	Thickness of lining, in	Weight per ft, lb
¾	0.70	0.06	1.3	3	2.70	0.13	8.3
1	.90	.07	1.9	4	3.60	.16	12.0
1¼	1.20	.08	2.5	6	5.40	.25	24.0
1½	1.40	.09	3.0	8	7.40	.25	32.0
2	1.80	.10	4.1	10	9.40	.30	43.0
2½	2.20	.10	6.6	12	11.40	.30	55.0

*To convert inches to millimeters, multiply by 25.4; to convert pounds per foot to kilograms per meter, multiply by 1.49.

TABLE 10-40   Chemical Stoneware: Bell-and-Spigot and Plain Butt-End Pipe*

Inside diam., in	Outside diam., in	Wall thickness, in
1½	2¼	⅜
2	2¾	⅜
3	4	½
4	5	½
5	6	½
6	7¼	⅝
8	9½	¾
10	11¾	⅞
12	13¾	⅞
14	15¾	⅞
15	17	1
16	18	1
18	20	1
20	22	1

Standard lengths up to 5 ft.
*Maurice A. Knight Co.
NOTE: To convert inches to millimeters, multiply by 25.4; to convert feet to meters, multiply by 0.3048.

TABLE 10-41 Vitrified-Clay Sewer Pipe*

Nominal size	Min. laying length, ft	Min. outside diam. of barrel, in	Min. wall thickness Standard strength, in	Min. wall thickness Extra strength, in
4	2	4⅞	⁷⁄₁₆	
6	2	7¹⁄₁₆	½	⁹⁄₁₆
8	2	9¼	⁹⁄₁₆	¾
10	2	11½	¹¹⁄₁₆	⅞
12	2	13¾	¹³⁄₁₆	1¹⁄₁₆
16	3	17³⁄₁₆	¹⁵⁄₁₆	1⅜
18	3	20⅝	1⅛	1¾
21	3	24⅛	1⁵⁄₁₆	2
24	3	27½	1½	2¼
27	3	31	1¹¹⁄₁₆	2½
30	3	34⅜	1⅞	2¾
33	3	37⅝	2	3
36	3	40¾	2¹⁄₁₆	3¼

*To convert inches to millimeters, multiply by 25.4.

crete piping systems can be lined with special salt-glazed vitrified-clay liner plates, joined with a die-cast asphalt joint. Concrete pressure pipe is competitive with cement-lined ductile iron for underground plant water systems.

Glass Pipe and Tubing These are made from heat- and chemical-resistant borosilicate glass (e.g., Corning Glass Works No. 7740) ASTM C599. This glass is highly stable in acids and resists attack by alkalies in solutions in which pH is 8 or less. It is attacked by hydrofluoric acid and glacial phosphoric acid. Some important physical properties are:

Modulus of elasticity	9,750,000 lb/in² (67,224 MPa)
Specific gravity	2.23
Specific heat	0.20
Thermal conductivity at 75°F	8.1 Btu/(h·ft²)(°F/in)[1.168 W/(m·K)]

Conical flanged glass pipe (Fig. 10-161) is made in the sizes shown in Table 10-42 and in lengths from 0.15 to 3 m (6 in to 10 ft). Maximum recommended working pressure is 0.3 MPa (50 lbf/in²) through 3-in size, 0.24 MPa (35 lbf/in²) for 4-in size, and 0.14 MPa (20 lbf/in²) for 6-in size. Maximum sudden temperature differential is 93°C (200°F) through 3-in size, 80°C (175°F) for 4-in size, and 71°C (160°F) for 6-in size. Maximum operating temperature is 232°C (450°F). A complete line of fittings is available, and special parts are made to order. Thermal-expansion stresses should be completely relieved by tied Teflon corrugated expansion joints and offsets. Temperature rating may be limited by joint design and materials. Hangers should be padded to avoid scratching pipe, should fit loosely, and should be located 0.3 m (1 ft) from each end of each 3-m (10-ft) length.

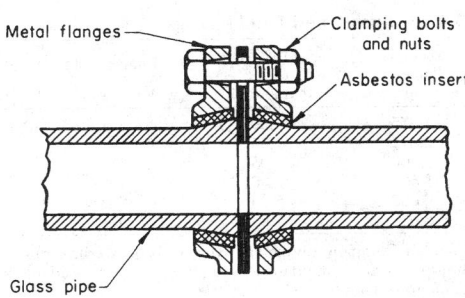

FIG. 10-161 Conical flanged joint.

Glass pipe can be furnished with an epoxy-resin coating reinforced with woven glass fiber to protect it from abuse. Equipped with special ball couplings, this may be used for 1-MPa (150-lbf/in²) pressure.

For very low pressures, beaded-end pipe equipped with single-bolt band-type couplings is available.

Glass-Lined Steel Pipe This pipe is fully resistant to all acids except hydrofluoric and concentrated phosphoric acids at temperatures up to 121°C (250°F). It is also resistant to alkaline solutions at moderate temperatures. Glass-lined steel pipe can be used at temperatures up to 232°C (450°F) under some exposure conditions provided there are no excessive sudden temperature changes. The operating pressure rating of commonly available systems is 1 MPa (150 lbf/in²). The glass lining is approximately 1.6 mm (¹⁄₁₆ in) thick. It is made by lining Schedule 40 steel pipe. Fittings are available in glass-lined cast iron, ductile iron, and steel. The fitting rating and recommended applications for fittings depend on the substrate material. Standard pipe sizes available are 1½ through 8 in. Larger-diameter pipe up to 48 in is available on a custom-order basis. A range of standard lengths is generally available from stock in 1½- through 4-in sizes. See Table 10-43 for dimensional data. Special Pfaudler-design steel split flanges drilled to ANSI Class 150 dimensions are used for assembly of the system.

Chemical-Porcelain Pipe Made of dense, nonporous material and fired at 1230°C (2250°F), chemical-porcelain pipe, fittings, and valves are inert to all acids except hydrofluoric but are not usually recommended for alkalies. Surfaces, except when ground for gasketing, are usually glazed for easy cleaning. Working pressures of 0.3 to 0.7 MPa (50 to 100 lbf/in²) are recommended for valves and piping. Temperatures of 200°C (400°F) or more can be used, but sudden thermal shocks must be avoided.

Cast-iron flanges (ANSI B16.1, 125-lb bolt spacing) are permanently attached to the porcelain with high strength acid-resistant cement. Flanged chemical-porcelain 90° and 45° elbows, tees, crosses, reducers, caps, and globe valves of the Y pattern are available. Armored chemical porcelain is furnished with 1.5- to 2.4-mm- (¹⁄₁₆- to

TABLE 10-42 Glass Pipe and Tubing: Conical Flanged Joint*

Pipe size, in (mm)	Pipe outside diameter, in (mm)	Cone outside diameter, in (mm)	Wall thickness, in (mm)	Cone angle, °	Approximate weight per foot, lb (kg)
1 (25)	1⁵⁄₁₆ ± 0.016 (33 ± 0.4)	1⁹⁄₁₆ ± 0.016 (40 ± 0.4)	⁵⁄₃₂ ± 0.016 (4.0 ± 0.4)	12	0.6 (0.27)
1½ (38)	1²⁷⁄₃₂ ± 0.020 (47 ± 0.5)	2⅛ ± 0.016 (54 ± 0.4)	¹¹⁄₆₄ ± 0.016 (4.4 ± 0.4)	12	1.0 (0.45)
2 (51)	2¹¹⁄₃₂ ± 0.040 (60 ± 1.0)	2⅝ ± 0.020 (67 ± 0.5)	¹¹⁄₆₄ ± 0.020 (4.4 ± 0.5)	12	1.13 (0.51)
3 (76)	3¹³⁄₃₂ ± 0.56 (87 ± 1.4)	3²⁵⁄₃₂ ± 0.031 (96 ± 0.8)	¹³⁄₆₄ ± 0.021 (5.2 ± 0.5)	12	2.0 (0.91)
4 (102)	4¹⁷⁄₃₂ ± 0.068 (115 ± 1.7)	5²³⁄₆₄ ± 0.016 (136 ± 0.4)	¹⁷⁄₆₄ ± 0.025 (6.7 ± 0.6)	21	3.4 (1.5)
6 (152)	6²¹⁄₃₂ ± 0.075 (169 ± 1.9)	7.553 ± 0.016 (192 ± 0.4)	⁵⁄₁₆ ± 0.040 (7.9 ± 1.0)	21	6.3 (2.9)

*From Corning Glass Works. See Fig. 10-161.
NOTE: To convert feet to meters, multiply by 0.3048.

TABLE 10-43   Glass-Lined Steel Pipe*

Size, in	Outside diameter, in	Approximate inside diameter, in	Range of standard lengths, in	
			Minimum†	Maximum
1½	1.875	1.50	3½	120
2	2.375	1.95	4	120
3	3.500	2.95	4½	120
4	4.500	3.90	4½	120
6‡	6.625	5.95	5	120
8‡	8.625	7.85	5½	120

*From Pfaudler Company, division of Sybron Corp. To convert inches to millimeters, multiply by 25.4. Standard-length pipe spools are available in the following increments of length:

For lengths, in	Standard lengths available in length increments, in
3½–6	½
6–8	2
8–10	1
10–12	2
12–120	6

†Spacers are available in ½-in increments for making up lengths of less than the minimum spool length shown.

‡Spool lengths less than 120 in are available but are not standard.

³⁄₃₂-in-) thick woven glass cloth impregnated with and bonded to the porcelain by plastic cement. The armor is continuous end to end and runs under the flanges. It prevents abuse from cracking the porcelain and, if the porcelain is cracked, prevents rupture.

Fused Silica or Fused Quartz   Containing 99.8 percent silicon dioxide, fused silica and fused quartz can be obtained as opaque or transparent pipe and tubing. The melting point is 1710°C (3100°F). Tensile strength is approximately 48 MPa (7000 lbf/in²); specific gravity is about 2.2. The pipe and tubing can be used continuously at temperatures up to 1000°C (1830°F) and intermittently up to 1500°C (2730°F). The material's chief assets are noncontamination of most chemicals in high-temperature service, thermal-shock resistance, and high-temperature electrical insulating characteristics.

Transparent tubing is available in inside diameters from 1 to 125 mm in a range of wall thicknesses. Satin-surface tubing is available in inside diameters from ¹⁄₁₆ to 2 in, and sand-surface pipe and tubing are available in ½- to 24-in inside diameters and lengths up to 6 m (20 ft). Sand-surface pipe and tubing are obtainable in wall thicknesses varying from ⅛ to 1 in. Pipe and tubing sections in both opaque and transparent fused silica or fused quartz can be readily machine-ground to special tolerances for pressure joints or other purposes. Also, fused-silica piping and tubing can be reprocessed to meet special-design requirements. Manufacturers should be consulted for specific details.

Wood and Wood-Lined Steel Pipe   Douglas fir, white pine, redwood, and cypress are the most common woods used for wood pipe. Wood-lined steel pipe is suitable for temperatures up to 82°C (180°F) and for pressures from 1.4 MPa (200 lbf/in²) for the 4-in size, through 0.86 MPa (125 lbf/in²) for the 10-in size, to 0.7 MPA (100 lbf/in²) for sizes larger than 10 in. For fume stacks and similar uses, wood-stave pipe with rods on 0.3-m (1-ft) centers is most satisfactory because it permits periodic tightening. In recent years reinforced plastics have supplanted wood pipe in most applications.

Plastic-Lined and Rubber-Lined Steel Pipe   Use of a variety of polymeric materials as liners for steel pipe rather than as piping systems solves problems which the relatively low tensile strength of the polymer at elevated temperature and high thermal expansion, compared with steel, would produce. The steel outer shell permits much wider spacing of supports, reliable flanged joints, and higher pressure and temperature in the piping. The size range is 1 through 12 in. The systems are flanged with 125-lb cast-iron, 150-lb ductile-iron, and 150- and 300-lb steel flanges. The linings are factory-installed in both pipe and fittings. Lengths are available up to 6 m (20 ft). Lined ball, diaphragm, and check valves and plug cocks are available.

One method of manufacture consists of inserting the liner into an oversize, approximately Schedule 40 steel tube and swaging the assembly to produce iron-pipe-size outside diameter, firmly engaging the liner which projects from both ends of the pipe. Flanges are then screwed onto the pipe, and the projecting liner is hot-flared over the flange faces nearly to the bolt holes. In another method, the liner is pushed into steel pipe having cold-flared laps backed up by flanges at the ends and then hot-flared over the faces of the laps. Pipe lengths made by either method may be shortened in the field and reflared with special procedures and tools. Square and tapered spacers are furnished to adjust for small discrepancies in assembly.

Saran Liners   Saran (Dow Chemical Co.) polyvinylidene chloride liners have excellent resistance to hydrochloric acid. Maximum temperature is 80°C (175°F).

Polypropylene Liners   Polypropylene liners (Hercules Incorporated) are used in sulfuric acid service. At 10 to 30 percent concentration the upper temperature limit is 93°C (200°F). In the range of 50 to 93 percent concentration, this drops from 66 to 24°C (from 150 to 75°F).

Kynar Liners   Kynar (Pennwalt Chemicals Corp.) vinylidene fluoride liners are used for many chemicals, including bromine and 50 percent hydrochloric acid.

TFE-, PFA-, and FEP-Lined Steel Pipe   These are available in sizes from 1 through 12 in and in lengths through 6 m (20 ft). The liners are not affected by any concentration of acids, alkalies, or solvents, but vent holes or internal grooving is required in the steel pipe to release gases which permeate through the liners. Manufacturers should be consulted before use in vacuum service. Experience has determined that practical upper temperature limits are 204°C (400°F) for TFE (polytetrafluoroethylene) and PFA (perfluoroalkoxy) and 149°C (300°F) for FEP (fluoroethylene polymer); 150-lb and 300-lb ductile-iron or steel flanged lined fittings and valves are used. The nonadhesive properties of the liner make it ideal for handling sticky or viscous substances. Thickness of the lining varies from 1.5 to 3.8 mm (60 to 150 mil), depending on pipe size. Only flanged joints are used.

Rubber-Lined Pipe   This pipe is made in lengths up to 6 m (20 ft) with seamless, straight seam-welded and some types of spiral-welded pipe using various types of natural and synthetic adhering rubber. The type of rubber is selected to provide the most suitable lining for the specific service. In general, soft rubber is used for abrasion resistance, semihard for general service, and hard for the more severe service conditions. Multiple-ply lining and combinations of hard and soft rubber are available. The thickness of lining ranges from 3.2 to 6.4 mm (⅛ to ¼ in) depending on the service, the type of rubber, and the method of lining. Cast-steel, ductile-iron, and cast-iron flanged fittings are available rubber-lined. The fittings are usually purchased by the vendor since absence of porosity on the inner surface is essential. Pipe is flanged before rubber lining, and welding elbows and tees may be incorporated at one end of the length of pipe, subject to the conditions that the size of the pipe and the location of the fittings are such that the operator doing the lining can place a hand on any point on the interior surface of the fitting. Welds must be ground smooth on the inside, and a radius is required at the inner edge of the flange face.

The rubber lining is extended out over the face of flanges. With hard-rubber lining, a gasket is required. With soft-rubber lining, coating or a polyethylene sheet is required in place of a gasket to avoid bonding of the lining of one flange to the lining on the other and to permit disassembly of the flanged joint. Also, for pressures over 0.86 MPa (125 lbf/in²), the tendency of soft-rubber linings to extrude out between the flanges may be prevented by terminating the lining inside the bolt holes and filling the balance of the space between the flange faces with a Masonite spacer of the proper thickness. Hard-rubber-lined gate, diaphragm, and swing check valves are available. In the gate valves, stem, wedge assembly, and seat rings, and in the check valves, hinge pin, flapper arm, disk, and seat ring must be made of metal resistant to the solution handled.

Plastic Pipe   In contrast to other piping materials, plastic pipe is free from internal and external corrosion, is easily cut and joined, and does not cause galvanic corrosion when coupled to other materials. Allowable stresses and upper temperature limits are low. Normal operation is in the creep range. Fluids for which a plastic is not suited penetrate and soften it rather than dissolve surface layers. Coefficients of thermal expansion are high. The use of thermoplastic pipe in flammable service aboveground is prohibited by the code.

Support spacing must be much closer than for carbon steel. As temperature increases, the allowable stress for many plastic pipes decreases very rapidly, and heat from sunlight or adjacent hot uninsulated equipment has a marked effect. Successful economical underground use of plastic pipe does not necessarily indicate similar economies outdoors aboveground.

Plastic tubing is widely used for instrument air-signal connections.

Methods of joining include threaded joints with IPS dimensions, solvent-welded joints, heat-fused joints, and insert fittings. Schedules 40 and 80 (see Table 10-18) have been used as a source for standardized dimensions at joints. Some plastics are available in several grades with allowable stresses varying by a factor of 2 to 1. For the same plastic, ½-in Schedule 40 pipe of the strongest grade may have 4 times the allowable internal pressure of the weakest grade of a 2-in Schedule 40 pipe. For this reason, the plastic-pipe industry is shifting to standard dimension ratios (approximately the same ratio of diameter to wall thickness over a wide range of pipe sizes).

ASTM and the Plastics Pipe Institute, a division of the Society of the Plastics Industry, have established identifications for plastic pipe in which the first group of letters identifies the plastic, the two following numbers identify the grade of that plastic, and the last two numbers represent the design stress in the nearest lower (0.7-MPa (100-lbf/in²) unit at 23°C (73.4°F).

Polyethylene Polyethylene (PE) pipe and tubing are available in sizes 42 in and smaller. They have excellent resistance at room temperature to salts, sodium and ammonium hydroxides, and sulfuric, nitric, and hydrochloric acids. Pipe and tubing are produced by extrusion from resins whose density varies with the manufacturing process. Physical properties and therefore wall thickness depend on the particular resin used. About 3 percent carbon black is added to provide resistance to ultraviolet light. Use of higher-density resin reduces splitting and pinholing in service and increases the strength of the material and the maximum service temperature.

ASTM D2104 covers PE pipe in sizes ½ through 6 in, with IPS Schedule 40 outside and inside diameters for insert-fitting joints. ASTM D2239 covers five standard dimension ratios of pipe diameter to wall thickness in sizes ½ through 6 in, with IPS Schedule 40 inside diameter for insert-fitting joints. ASTM D2447 covers sizes ½ through 12 in, with IPS Schedule 40 and 80 outside and inside diameters for use with heat-fusion socket-type and butt-type fittings. ASTM D3035 covers standard dimension ratios of pipe sizes from ½ through 6 in with IPS outside diameters. All these specifications cover five PE materials (see Table 10-15). Hydrostatic design stresses within the recommended temperature limits are given in Appendix A, Table 3, of the code. The hydrostatic design stress is the maximum tensile hoop stress due to internal hydrostatic water pressure that can be applied continuously with a high degree of certainty that failure of the pipe will not occur. Biaxially oriented polyethylene (PEO) pipe (ASTM D3287) has a higher hydrostatic design stress than PE pipe.

Polyethylene water piping is not damaged by freezing. Pipe and tubing 2 in and smaller are shipped in coils several hundred feet in length.

Clamped-insert joints (Fig. 10-162) are used for flexible plastic pipe up through the 2-in size. Friction between the pipe and the spud is developed both by forcing the spud into the pipe and by tightening the clamp. For the larger sizes, which have thicker walls, these meth-

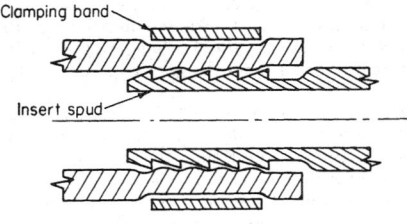

FIG. 10-162 Clamped-insert joint.

ods cannot develop adequate friction. The joints also have high pressure drop. Stainless-steel bands are available. Inserts are available in nylon, polypropylene, and a variety of metals. A significant use for PE and PP pipe is the technique of rehabilitating deteriorated pipe lines by lining them with plastic pipe. Lining an existing pipe with plastic pipe has a large cost advantage over replacing the line, particularly if replacement of the old line would require excavation.

Polyvinyl chloride Polyvinyl chloride (PVC) and chlorinated polyvinyl chloride (CPVC) pipe and tubing are available in sizes 12 in and smaller for PVC and 4 in and smaller for CPVC. They have excellent resistance at room temperature to salts, ammonium hydroxide, and sulfuric, nitric, acetic, and hydrochloric acid but may be damaged by ketones, aromatics, and some chlorinated hydrocarbons.

Five PVC pipe materials having characteristic chemical resistance, impact strength, and hydrostatic design stresses are included in the group of ASTM pipe specifications pertaining to PVC. While all these materials have a −18°C (0°F) minimum-recommended-temperature limit (see Table 10-15), Code PVC-1120 and Code PVC-1220 materials become brittle at and below 4°C (40°F). On the other hand, Code PVC-2110, Code PVC-2112, and Code PVC-2216 materials have higher impact resistance but a lower hydrostatic design stress at elevated temperatures. Code PVC-2120 has the best combination of both properties. Allowable hydrostatic design stresses are given in Appendix A, Table 3, of the code, although no stresses are provided for temperatures above 38°C (100°F). The hydrostatic design stresses at 23°C (73.4°F) are 13.8 MPA (2000 lbf/in²) for PVC-1120, PVC-1220, and PVC-2120, 11.0 MPa (1600 lbf/in²) for PVC-2116 and CPVC-4116, 8.6 MPa (1250 lbf/in²) for PVC-2116, and 6.9 MPa (1000 lbf/in²) for PVC-2110. ASTM D1785 covers sizes ⅛ through 12 in of PVC pipe in IPS Schedules 40, 80, and 120, except that Schedule 120 starts at ½ in and is not IPS for sizes from ½ through 3 in. ASTM D2241 covers the same size range but with IPS outside diameter and seven standard dimension ratios: 13.5, 17, 21, 26, 32.5, 41, and 64.

ASTM D2513 covers pipe in sizes from ½ through 12 in in both IPS outside diameter and plastic-tubing diameters from ¼ through 1¾ in with standard-dimension-ratio wall thicknesses. This product is intended for gas service. ASTM D2672 covers bell-end pipe in sizes from ⅛ through 8 in in IPS Schedule 40 and in IPS outside diameter and the same standard dimension ratios for wall thicknesses as in D2241. The pipe is intended to be joined by cementing. ASTM D2740 covers PVC-tubing diameters from ½ through 1¼ in with standard-dimension-ratio wall thicknesses.

Solvent-cemented joints (Fig. 10-147) are standard, but screwed joints are sometimes used with Schedule 80 pipe. Cemented joints must not be disturbed for 5 min and achieve full strength in 1 day. Because of the difference in thermal expansion, joints between PVC pipe and metal pipe should be flanged, using a PVC flange on the PVC pipe and a full-face gasket. Flanges are available with ANSI B16.5 150-lb drilling. Ball valves, Y-type globe valves, and diaphragm valves are available in PVC.

Polypropylene Polypropylene (PP) pipe and fittings have excellent resistance to most common organic and mineral acids and their salts, strong and weak alkalies, and many organic chemicals. They are available in sizes ½ through 6 in, in Schedules 40 and 80, but are not covered as such by ASTM specifications.

Reinforced-Thermosetting-Resin (RTR) Pipe Glass-reinforced epoxy resin has good resistance to nonoxidizing acids, alkalies, salt water, and corrosive gases. The glass reinforcement is many times stronger at room temperature than plastics, does not lose strength with increasing temperature, and reinforces the resin effectively up to 149°C (300°F). (See Table 10-17 for temperature limits.) The glass reinforcement is located near the outside wall, protected from the contents by a thick wall of resin and protected from the atmosphere by a thin wall of resin. Stock sizes are 2 through 12 in.

Pipe is supplied in 6- and 12-m (20- and 40-ft) lengths. It is more economical for long, straight runs than for systems containing numerous fittings. When the pipe is sawed to nonfactory lengths, it must be sawed very carefully to avoid cracking the interior plastic zone. A two-component cement may be used to bond lengths into socket couplings or flanges or cemented-joint fittings. Curing of the cement is temperature-sensitive; it sets to full strength in 45 min at 93°C (200°F), in

TABLE 10-44   Typical Hanger-Spacing Ranges Recommended for Reinforced-Thermosetting-Resin Pipe

Nominal pipe size, in	2	3	4	6	8	10	12
Hanger-spacing range, ft	5–8	6–9	6–10	8–11	9–13	10–14	11–15

NOTE: Consult pipe manufacturer for recommended hanger spacing for the specific RTR pipe being used. Tabulated values are based on a specific gravity of 1.25 for the contents of the pipe. To convert feet to meters, multiply by 0.3048.

12 h at 38°C (100°F), and in 24 h at 10°C (50°F). Extensive use is made of shop-fabricated flanged preassemblies. Only flanged joints are used to bond to metallic piping systems. Compared with that of other plastics, the ratio of fitting cost to pipe cost is high. Cemented-joint fittings and flanged fittings are available. Flanged lined metallic valves are used.

RTR is more flexible than metallic pipe and consequently requires closer support spacing. While the recommended spacing varies among manufacturers and with the type of product, Table 10-44 gives typical hanger-spacing ranges. The pipe fabricator should be consulted for recommended hanger spacing on the specific pipe-wall construction being used.

Epoxy resin has a higher strength at elevated temperatures than polyester resins but is not as resistant to attack by some fluids. Some glass-reinforced epoxy-resin pipe is made with a polyester-resin liner. The coefficient of thermal expansion of glass-reinforced resin pipe is higher than that for carbon steel but much less than that for plastics.

Glass-reinforced polyester is the most widely used reinforced-resin system. A wide choice of polyester resins is available. The bisphenol resins resist strong acids as well as alkaline solutions. The size range is 2 through 12 in; the temperature range is shown in Table 10-17. Diameters are not standardized. Adhesive-cemented socket joints and hand-lay-up reinforced butt joints are used. For the latter, reinforcement consists of layers of glass cloth saturated with adhesive cement.

Haveg 41NA   This is a proprietary thermoset plastic consisting of a phenol-formaldehyde resin and nonasbestos silicate fillers. It is furnished as pipe and fittings with several types of joints and is resistant to most acidic chemicals, especially hydrochloric acid. The standard joint uses split cast-iron flanges set in tapered grooves machined in the outside of the pipe. A facing and grooving tool is available. Standard lengths are 1.2 m (4 ft) in the ½- and ¾-in sizes and 3 m (10 ft) in all other sizes.

Flanges are drilled per ANSI B16.5, except that the bolt holes are smaller. Figure 10-163 shows pressure-temperature ratings for standard-wall pipe with standard joints. Pipe and fittings with cemented sleeve joints are also available for use when external corrosion might destroy cast-iron flanges. Y-type globe valves, diaphragm valves, and foot and check valves are available.

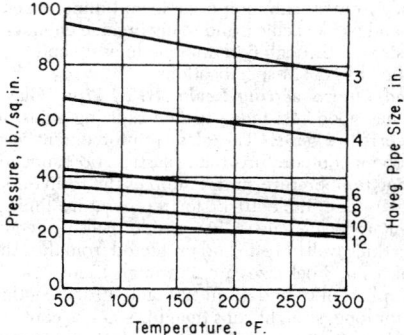

FIG. 10-163   Operating pressure-temperature ratings for Haveg 41NA and 61NA pipe and fittings. (°F − 32)% = °C; to convert pounds-force per square inch to kilopascals, multiply by 6.895; to convert inches to millimeters, multiply by 25.4.

Haveg 61NA   A proprietary nonasbestos silicate-filled furfuryl alcohol-formaldehyde resin pipe, Haveg 61NA is highly resistant to most acids and, with some reservations, to sodium hydroxide. It is also resistant to many hydrocarbons, halogenated organic compounds, and organic acids. Its pressure-temperature ratings are shown in Figure 10-163.

PIPING-SYSTEM DESIGN

Safeguarding   Safeguarding may be defined as the provision of protective measures as required to ensure the safe operation of a proposed piping system. General considerations to be evaluated should include (1) the hazardous properties of the fluid, (2) the quantity of fluid which could be released by a piping failure, (3) the effect of a failure (such as possible loss of cooling water) on overall plant safety, (4) evaluation of effects on a reaction with the environment (i.e., possibility of a nearby source of ignition), (5) the probable extent of exposure of operating or maintenance personnel, and (6) the relative inherent safety of the piping by virtue of materials of construction, methods of joining, and history of service reliability.

Evaluation of safeguarding requirements might include engineered protection against possible failures such as thermal insulation, armor, guards, barricades, and damping for protection against severe vibration, water hammer, or cyclic operating conditions. Simple means to protect people and property such as shields for valve bonnets, flanged joints, and sight glasses should not be overlooked. The necessity for means to shut off or control flow in the event of a piping failure such as block valves or excess-flow valves should be examined.

Classification of Fluid Services   The code applies to piping systems as illustrated in Fig. 10-124, but two categories of fluid services are segregated for special consideration as follows:

Category D   fluid service is defined as "a fluid service to which all the following apply: (1) the fluid handled is nonflammable and nontoxic; (2) the design gage pressure does not exceed 150 psi (1.0 MPa); and (3) the design temperature is between −20°F. (−29°C.) and 360°F. (182°C.)."

Category M   fluid service is defined as "a fluid service in which a single exposure to a very small quantity of a toxic fluid, caused by leakage, can produce serious irreversible harm to persons on breathing or bodily contact, even when prompt restorative measures are taken."

The code assigns to the owner the responsibility for identifying those fluid services which are in Categories D and M. The design and fabrication requirements for Class M toxic-service piping are beyond the scope of this *Handbook*. See ANSI B31.3—1976, chap. VIII.

Design Conditions   Definitions of the temperatures, pressures, and various forces applicable to the design of piping systems are as follows:

Design Pressure   The design pressure of a piping system shall not be less than the pressure at the most severe condition of coincident pressure and temperature resulting in the greatest required component thickness or rating.

Design Temperature   The design temperature is the material temperature representing the most severe condition of coincident pressure and temperature. For uninsulated metallic pipe with fluid below 38°C (100°F), the metal temperature is taken as the fluid temperature.

With fluid at or above 38°C (100°F) and without external insulation, the metal temperature is taken as a percentage of the fluid temperature unless a lower temperature is determined by test or calculation. For pipe, threaded and welding-end valves, fittings, and other components with a wall thickness comparable with that of the pipe, the percentage is 95 percent; for flanges and flanged valves and fittings, 90 percent; for lap-joint flanges, 85 percent; and for bolting, 80 percent.

With external insulation, the metal temperature is taken as the fluid temperature unless service data, tests, or calculations justify lower values. For internally insulated pipe, the design metal temperature shall be calculated or obtained from tests.

Ambient Influences   If cooling results in a vacuum, the design must provide for external pressure or a vacuum breaker installed; also provision must be made for thermal expansion of contents trapped between or in closed valves. Nonmetallic or nonmetallic-lined pipe may require protection when ambient temperature exceeds design temperature.

Occasional variations of pressure or temperature, or both, above operating levels are characteristic of certain services. If the following criteria are met, such variations need not be considered in determining pressure-temperature design conditions. Otherwise, the most severe conditions of coincident pressure and temperature during the variation shall be used to determine design conditions. (Application of pressures exceeding pressure-temperature ratings of valves may under certain conditions cause loss of seat tightness or difficulty of operation. Such an application is the owner's responsibility.)

All the following criteria must be met:

1. The piping system shall have no pressure-containing components of cast iron or other nonductile metal.

2. Nominal pressure stresses shall not exceed the yield strength at temperature (see Table 10-49 and S_y data in ASME Code, Sec. VIII, Division 2).

3. Combined longitudinal stresses S_L shall not exceed the limits established in the code (see pressure design of piping components for S_L limitations).

4. The number of cycles (or variations) shall not exceed 7000 during the life of the piping system.

5. Occasional variations above design conditions shall remain within one of the following limits for pressure design:

• When the variation lasts no more than 10 h at any one time and no more than 100 h per year, it is permissible to exceed the pressure rating or the allowable stress for pressure design at the temperature of the increased condition by not more than 33 percent.

• When the variation lasts no more than 50 h at any one time and not more than 500 h per year, it is permissible to exceed the pressure rating or the allowable stress for pressure design at the temperature of the increased condition by not more than 20 percent.

Dynamic Effects   Design must provide for impact (hydraulic shock, etc.), wind (exposed piping), earthquake (see ANSI A58.1), discharge reactions, and vibrations (of piping arrangement and support).

Weight considerations include (1) live loads (contents, ice, and snow), (2) dead loads (pipe, valves, insulation, etc.), and (3) test loads (test fluid).

Thermal-expansion and -contraction loads occur when a piping system is prevented from free thermal expansion or contraction as a result of anchors and restraints or undergoes large, rapid temperature changes or unequal temperature distribution because of an injection of cold liquid striking the wall of a pipe carrying hot gas.

Design Criteria: Metallic Pipe   The code uses three different approaches to design, as follows:

1. It provides for the use of dimensionally standardized components at their published pressure-temperature ratings.

2. It provides design formulas and maximum stresses.

3. It prohibits the use of materials, components, or assembly methods in certain conditions.

Components Having Specific Ratings   These are listed in ANSI, API, and industry standards. These ratings are acceptable for design pressures and temperatures unless limited in the code. A list of component standards is given in Appendix E of the code. The following rating tables covering commonly used components have been extracted from the original document with permission of the publisher, the American Society of Mechanical Engineers, New York: Table 10-45 lists pressure-temperature ratings for flanges, flanged fittings, and flanged valves; and Table 10-46 lists hydrostatic-shell test pressures for flanges, flanged fittings, and flanged valves. Flanged joints, flanged valves in the open position, and flanged fittings may be subjected to system hydrostatic tests at a pressure not to exceed the hydrostatic-shell test pressure. Flanged valves in the closed position may be subjected to a system hydrostatic test at a pressure not to exceed 110 percent of the 100°F rating of the valve unless otherwise limited by the manufacturer.

Pressure-temperature ratings for soldered and brazed copper-tubing joints are given in Tables 10-47 and 10-48 respectively.

Components without Specific Ratings   Components such as pipe and butt-welding fittings are generally furnished in nominal thicknesses. Fittings are rated for the same allowable pressures as pipe of the same nominal thickness and, along with pipe, are rated by the rules for pressure design and other provisions of the code.

Pressure Design of Metallic Components: Wall Thickness   External-pressure stress evaluation of piping is the same as for pressure vessels. But an important difference exists when one is establishing design pressure and wall thickness for internal pressure as a result of the ASME Boiler and Pressure Vessel Code's requirement that the relief-valve setting be not higher than the design pressure. For vessels this means that the design is for a pressure 10 percent more or less above the intended maximum operating pressure to avoid popping or leakage from the valve during normal operation. However, on piping the design pressure and temperature are taken as the maximum intended operating pressure and coincident temperature combination which results in the maximum thickness. The temporary increased operating conditions listed under "Design Criteria" cover temporary operation at pressures that cause relief valves to leak or open fully. Allowable stresses for nearly 1000 materials are contained in the code. For convenience, the allowable stresses for commonly used materials have been extracted from the code and listed in Table 10-49.

For **straight metal pipe under internal pressure** the formula for minimum required wall thickness t_m is applicable for D_o/t ratios greater than 6. The more conservative Barlow and Lamé equations may also be used. Equation (10-92) includes a factor Y varying with material and temperature to account for the redistribution of circumferential stress which occurs under steady-state creep at high temperature and permits slightly lesser thickness at this range.

$$t_m = \frac{PD_o}{2(SE + PY)} + C \qquad (10\text{-}92)$$

where (in consistent units)

$\quad P$ = design pressure
$\quad D_o$ = outside diameter of pipe
$\quad C$ = sum of allowances for corrosion, erosion, and any thread or groove depth. For threaded components the depth is h of ANSI B2.1, and for grooved components the depth is the depth removed (plus 1/64 in when no tolerance is specified).
$\quad SE$ = allowable stress (see Table 10-49)
$\quad S$ = basic allowable stress for materials, excluding casting, joint, or structural-grade quality factors
$\quad E$ = quality factor. The quality factor E is one or the product of more than one of the following quality factors: casting quality factor E_c, joint quality factor E_j (see Fig. 10-164), and structural-grade quality factor E_s of 0.92.
$\quad Y$ = coefficient having value in Table 10-50 for ductile ferrous materials, 0.4 for ductile nonferrous materials, and zero for brittle materials such as cast iron.
$\quad t_m$ = minimum required thickness, in, to which manufacturing tolerance must be added when specifying pipe thickness on purchase orders. [Most ASTM specifications to which mill pipe is normally obtained permit minimum wall to be 12½ percent less than nominal. ASTM A155 for fusion-welded pipe permits minimum wall 0.25 mm (0.01 in) less than nominal plate thickness.] Pipe with t equal to or greater than $D/6$ or P/SE greater than 0.385 requires special consideration.

In addition to establishing the wall thickness for internal pressure, the stress values in Table 10-49 control other portions of the design. The **sum of the longitudinal stresses** S_L (in the corroded condition) due to internal pressure, weight of pipe and contents between supports, and other sustained loadings such as friction between a laid (not hung) long length of straight cold pipe and its supports when it is placed in service, shall not exceed the value of S_h. In this determination, for pipe with welded longitudinal seams, the longitudinal weld joint factor is disregarded. Also, when **thermal-expansion or contraction strains** are taken up primarily by bending or torsion, the local stresses so produced are limited to the following range designated as S_A:

$$S_A = f(1.25S_c + 0.25S_h) \qquad (10\text{-}93)$$

TABLE 10-45   Pressure-Temperature Ratings for Flanges, Flanged Fittings, and Flanged Valves of Typical Materials,[4] lbf/in²

Material group	1.1	1.5	1.9	1.10	1.13	2.1	2.2	2.3	2.6	2.7
Materials temperature, °F	Carbon — Normal	C, ½Mo	1, 1¼Cr, ½Mo	2¼Cr, 1Mo	5Cr, ½Mo	Type 304	Type 316	Type 304L Type 316L	Type 309	Type 310
150-lb class										
−20 to 100	285	265	290			275	275	230	260	
200	260		260			235	240	195	230	
300	230		230			205	215	175	220	
400			200			180	195	160	200	
500			170				170	145	170	
600			140				140	140	140	
650			125				125	125	125	
700			110				110	110	110	
750			95				95	95	95	
800			80				80	80	80	
850			65				65	65	65	
900			50				50		50	
950			35				35		35	
1000			20				20		20	
300-lb class										
−20 to 100	740	695	750	750	750	720	720	600	670	
200	675	680	710	715	750	600	620	505	605	
300	655	655	675	675	730	530	560	455	570	
400	635	640	660	650	705	470	515	415	535	
500	600	620		640	665	435	480	380	505	
600	550			605		415	450	360	480	
650	535			590		410	445	350	465	
700	535			570		405	430	345	455	
750	505			530		400	425	335	445	
800	410			510	500	395	415	330	435	
850	270			485	440	390	405	320	425	
900	170			450	355	385	395		415	
950	105	280		380	260	375	385		385	
1000	50	165	225	270	190	325	365		335	350
1050			140	200	140	310	360		290	335
1100			95	115	105	260	325		225	290
1150			50	105	70	195	275		170	245
1200			35	55	45	155	205		130	205
1250						110	180		100	160
1300						85	140		80	120
1350						60	105		60	80
1400						50	75		45	55
1450						35	60		30	40
1500						25	40		25	25
400-lb class°										
−20 to 100	990	925	1000	1000	1000	960	960	800	895	
200	900	905	950	955	1000	800	825	675	805	
300	875	870	895	905	970	705	745	605	760	
400	845	855	880	865	940	630	685	550	710	
500	800	830		855	885	585	635	510	670	
600	730			805		555	600	480	635	
650	715			785		545	590	470	620	
700	710			755		540	575	460	610	
750	670			710		530	565	450	595	
800	550			675	665	525	555	440	580	
850	355			650	585	520	540	430	565	
900	230			600	470	510	525		555	
950	140	375		505	350	500	515		515	
1000	70	220	300	355	255	430	485		450	465
1050			185	265	190	410	480		390	445
1100			130	150	140	345	430		300	390
1150			70	140	90	260	365		230	330
1200			45	75	60	205	275		175	275
1250						145	245		135	215
1300						110	185		105	160
1350						85	140		80	105
1400						65	100		60	75
1450						45	80		40	50
1500						30	55		30	30

TABLE 10-45   Pressure-Temperature Ratings for Flanges, Flanged Fittings, and Flanged Valves of Typical Materials,[4] lbf/in² (Continued)

Materials temperature, °F	Carbon steel Normal (1.1)	Carbon steel C, ½Mo (1.5)	1, 1¼Cr, ½Mo (1.9)	2¼Cr, 1Mo (1.10)	5Cr, ½Mo (1.13)	Type 304 (2.1)	Type 316 (2.2)	Type 304L / Type 316L (2.3)	Type 309 (2.6)	Type 310 (2.7)
				600-lb class						
−20 to 100	1480	1390	1500	1500	1500	1440	1440	1200	1345	
200	1350	1360	1425	1430	1500	1200	1240	1015	1210	
300	1315	1305	1345	1355	1455	1055	1120	910	1140	
400	1270	1280	1315	1295	1410	940	1030	825	1065	
500	1200	1245	1285	1280	1330	875	955	765	1010	
600	1095		1210			830	905	720	955	
650	1075		1175			815	890	700	930	
700	1065		1135			805	865	685	910	
750	1010		1065			795	845	670	895	
800	825		1015		995	790	830	660	870	
850	535		975		880	780	810	645	850	
900	345		900		705	770	790		830	
950	205	560	755		520	750	775		775	
1000	105	330	445	535	385	645	725		670	700
1050			275	400	280	620	720		585	665
1100			190	225	205	515	645		445	585
1150			105	205	140	390	550		345	495
1200			70	110	90	310	410		260	410
1250						220	365		200	325
1300						165	275		160	240
1350						125	205		115	160
1400						90	150		90	110
1450						70	115		60	75
1500						50	85		50	50
				900-lb class						
−20 to 100	2220	2085	2250	2250	2250	2160	2160	1800	2015	
200	2025	2035	2135	2150	2250	1800	1860	1520	1815	
300	1970	1955	2020	2030	2185	1585	1680	1360	1705	
400	1900	1920	1975	1945	2115	1410	1540	1240	1600	
500	1795	1865	1925	1920	1995	1310	1435	1145	1510	
600	1640		1815			1245	1355	1080	1435	
650	1610		1765			1225	1330	1050	1395	
700	1600		1705			1210	1295	1030	1370	
750	1510		1595			1195	1270	1010	1340	
800	1235		1525		1490	1180	1245	985	1305	
850	805		1460		1315	1165	1215	965	1275	
900	515		1350		1060	1150	1180		1245	
950	310	845	1130		780	1125	1160		1160	
1000	155	495	670	805	575	965	1090		1010	1050
1050			410	595	420	925	1080		875	1000
1100			290	340	310	770	965		670	875
1150			155	310	205	585	825		515	740
1200			105	165	135	465	620		390	620
1250						330	545		300	485
1300						245	410		235	360
1350						185	310		175	235
1400						145	225		135	165
1450						105	175		95	115
1500						70	125		70	70
				1500-lb class						
−20 to 100	3705	3470	3750	3750	3750	3600	3600	3600	3360	
200	3375	3395	3560	3580	3750	3000	3095	2530	3025	
300	3280	3260	3365	3385	3640	2640	2795	2270	2845	
400	3170	3200	3290	3240	3530	2350	2570	2065	2665	
500	2995	3105	3210	3200	3325	2185	2390	1910	2520	
600	2735		3025			2075	2255	1800	2390	
650	2685		2940			2040	2220	1750	2330	
700	2665		2840			2015	2160	1715	2280	
750	2520		2660			1990	2110	1680	2230	
800	2060		2540		2485	1970	2075	1645	2170	
850	1340		2435		2195	1945	2030	1610	2125	
900	860		2245		1765	1920	1970		2075	

TABLE 10-45   Pressure-Temperature Ratings for Flanges, Flanged Fittings, and Flanged Valves of Typical Materials,[4] lbf/in² (Continued)

Material group	1.1	1.5	1.9	1.10	1.13	2.1	2.2	2.3	2.6	2.7
Materials temperature, °F	Carbon steel (Normal)	Carbon steel (C, ½Mo)	1, 1¼Cr, ½Mo	2¼Cr, 1Mo	5Cr, ½Mo	Type 304	Type 316	Type 304L / Type 316L	Type 309	Type 310
1500-lb class (Cont.)										
950	515	1405	1885		1305	1870	1930		1930	
1000	260	825	1115	1340	960	1610	1820		1680	1750
1050			685	995	705	1545	1800		1460	1665
1100			480	565	515	1285	1610		1115	1460
1150			260	515	345	980	1370		860	1235
1200			170	275	225	770	1030		650	1030
1250						550	910		495	805
1300						410	685		395	600
1350						310	515		290	395
1400						240	380		225	275
1450						170	290		155	190
1500						120	205		120	120
2500-lb class										
−20 to 100	6170	5785	6250	6250	6250	6000	6000	5000	5600	
200	5625	5660	5930	5965	6250	5000	5160	4220	5040	
300	5470	5435	5605	5640	6070	4400	4660	3780	4740	
400	5280	5330	5485	5400	5880	3920	4280	3440	4440	
500	4990	5180	5350	5330	5540	3640	3980	3180	4200	
600	4560		5040			3460	3760	3000	3980	
650	4475		4905			3400	3700	2920	3880	
700	4440		4730			3360	3600	2860	3800	
750	4200		4430			3320	3520	2800	3720	
800	3430		4230		4145	3280	3460	2740	3620	
850	2230		4060		3660	3240	3380	2680	3540	
900	1430		3745		2945	3200	3280		3460	
950	860	2345	3145		2170	3120	3220		3220	
1000	430	1370	1860	2230	1600	2685	3030		2800	2915
1050			1145	1660	1170	2570	3000		2430	2770
1100			800	945	860	2145	2685		1860	2430
1150			430	860	570	1630	2285		1430	2060
1200			285	460	370	1285	1715		1085	1715
1250						915	1515		830	1345
1300						685	1145		660	1000
1350						515	860		485	660
1400						400	630		370	460
1450						285	485		260	315
1500						200	345		200	200

°For Group 1.1, do not use ASTM A181 Grade I or II materials.

NOTES:
1. Ratings shown apply to other material groups when column-dividing lines have been omitted.
2. Temperature notes for all material groups in Table 10-45.

Material group	Materials[3] (specification—grade)	See Notes
1.1	A105, A181-II, A216-WCB, A515-70	a, h
	A516-70	a, g
	A350-LF2, A537-C1.1	d
1.5	A182-F1, A204-A, A204-B, A217-WC1	b, h
	A352-LC1	d, m
1.9	A182-F11, A182-F12, A387-11, C1.2	m, c
	A217-WC6	j, m
1.10	A182-F22, A387-22, C1.2	c
	A217-WC9	m, j
1.13	A182-F5a, A217-C5	m
2.1	A182-F304, A182-F304H	n
	A240-304, A351-CF8	n, o
	A351-CF3	f
2.2	A182-F316, A182-F316H, A240-316	n, o
	A240-317, A351-CF8M	n, o
	A351-CF3M	g
2.3	A182-F304L, A240-304L	f
	A182-F316L, A240-316L	g
2.6	A240-309S, A351-CH8, A351-CH20	n, o
2.7	A182-F310, A240-310S	k, n
	A351-CK20	n

TABLE 10-45 Pressure-Temperature Ratings for Flanges, Flanged Fittings, and Flanged Valves of Typical Materials,[4] lbf/in² (Concluded)

a. Permissible but not recommended for prolonged use above about 800°F.
b. Permissible but not recommended for prolonged use above about 850°F.
c. Permissible but not recommended for prolonged use above about 1100°F.
d. Not to be used over 650°F.
f. Not to be used over 800°F.
g. Not to be used over 850°F.
h. Not to be used over 1000°F.
i. Not to be used over 1050°F.
j. Not to be used over 1100°F.
k. For service temperatures 1050°F and above, assurance must be provided that grain size is not finer than ASTM No. 6.
l. Only killed steel shall be used above 850°F.
m. Use normalized and tempered material only.
n. At temperatures over 1000°F, use only when the carbon content is 0.04 percent or higher.
o. See ANSI B16.5 for heat treatment for service temperatures over 1000°F.
p. The ratings at −20 to 100°F given for the materials covered shall also apply at lower temperatures. The ratings for low-temperature service of the cast and forged materials listed in ASTM A352 and A350 shall be taken the same as the −20 to 100°F ratings for carbon steel.
q. Some of the materials listed in the rating tables undergo a decrease in impact resistance at temperatures lower than −20°F to such an extent as to be unable to resist safely shock loadings, sudden changes of stress, or high stress concentration.
   3. See ANSI B16.5, Table 1A, for additional information and notes relating to specific materials.
   4. Extracted from Steel Pipe Flanges and Flanged Fittings, ANSI B16.5—1977 and B16.34—1977, with permission of the publisher, the American Society of Mechanical Engineers, New York.
   5. A product used under the jurisdiction of the ASME Boiler and Pressure Vessel Code and the ANSI Code for Pressure Piping B31.1 is subject to any limitation of those codes. This includes any maximum-temperature limitation for a material or a code rule governing the use of a material at a low temperature.
   6. (°F − 32)⅝ = °C; to convert pounds-force per square inch to megapascals, multiply by 0.006895.

where S_c = S from Table 10-49 at a minimum (cold) metal tempera-
          ture normally expected during operation or shutdown
          (See Note 13, Table 10-49)
        S_h = S from Table 10-49 at maximum (hot) metal tempera-
          ture normally expected during operation or shutdown
          (See Note 13, Table 10-49)
        f = stress-range reduction factor for total number of full
          temperature cycles over expected life (See Table 10-51)

When the anticipated number of cycles is substantially less than 7000, useful information can be obtained from ASME Boiler and Pressure Vessel Code, Sec. III, "Nuclear Vessels."

However, if the sum of longitudinal stresses S_L enumerated is less than their stated limit S_h, the difference may be added to the term $0.25S_h$ in the equation limiting the stress range:

$$S_A = f[1.25(S_c + S_h) - S_L] \qquad (10\text{-}94)$$

For flanges of nonstandard dimensions or for sizes beyond the scope of the approved standards, design shall be in accordance with the requirements of the ASME Boiler and Pressure Vessel Code, Sec. VIII, except that requirements for fabrication, assembly, inspection testing, and the pressure and temperature limits for materials of the Piping Code are to prevail. Countermoment flanges of flat face or otherwise providing a reaction outside the bolt circle are permitted if

TABLE 10-46 Hydrostatic-Shell Test Pressures for Flanges, Flanged Fittings, and Flanged Valves of Typical Materials*

Material group no.	Shell test pressures by class, lbf/in² gauge						
	150	300	400	600	900	1500	2500
1.1	450	1125	1500	2225	3350	5575	9275
1.5	400	1050	1400	2100	3150	5225	8700
1.9	450	1125	1500	2250	3375	5625	9375
1.10	450	1125	1500	2250	3375	5625	9375
1.13	450	1125	1500	2250	3375	5625	9375
2.1	425	1100	1450	2175	3250	5400	9000
2.2	425	1100	1450	2175	3250	5400	9000
2.3	350	900	1200	1800	2700	4500	7500
2.6	400	1025	1350	2025	3025	5050	8400
2.7	400	1025	1350	2025	3025	5050	8400

*Extracted from Steel Pipe Flanges and Flanged Fittings, ANSI B16.5—1977, with permission of the publisher, the American Society of Mechanical Engineers, New York. Test temperature not to exceed 125°F. (°F − 32)⅝ = °C; to convert pounds-force per square inch to megapascals, multiply by 0.006895.

designed or tested in accordance with code requirements under pressure-containing components "not covered by standards and for which design formulas or procedures are not given."

In accordance with listed standards, **blind flanges** may be used at their pressure-temperature ratings. The minimum thickness of nonstandard blind flanges shall be the same as for a bolted flat cover, in accordance with the rules of the ASME Boiler and Pressure Vessel Code, Sec. VIII.

Operational blanks shall be of the same thickness as blind flanges or may be calculated by the following formula (use consistent units):

$$t = d \sqrt{3P/16S} \qquad (10\text{-}95)$$

where d = inside diameter of gasket for raised- or flat (plain)-face
          flanges, or the gasket pitch diameter for retained
          gasketed flanges
        P = internal design pressure or external design pressure
        S = applicable allowable stress

Valves must comply with the applicable standards listed in Appendix E of the code and with the allowable pressure-temperature limits established thereby but not beyond the code-established service or materials limitations. Special valves must meet the same requirements as for countermoment flanges.

The code contains no specific rules for the design of **fittings** other than as branch openings. Ratings established by recognized standards are acceptable, however. ANSI Standard B16.5 for steel-flanged fittings incorporates a 1.5 shape factor and thus requires the entire fitting to be 50 percent heavier than a simple cylinder in order to provide reinforcement for openings and/or general shape. ANSI B16.9 for butt-welding fittings, on the other hand, requires only that the fittings be able to withstand the calculated bursting strength of the straight pipe with which they are to be used.

The thickness of **pipe bends** shall be determined as for straight pipe, provided the bending operation does not result in a difference between maximum and minimum diameters greater than 8 and 3 percent of the nominal outside diameter of the pipe for internal and external pressure respectively.

The maximum allowable internal pressure for multiple miter bends shall be the lesser value calculated from Eqs. (10-96) and (10-97). These equations are not applicable when θ exceeds 22.5°.

$$P = \frac{SEt}{r_2}\left(\frac{t}{t + 0.643 \tan\theta \sqrt{r_2 t}}\right) \qquad (10\text{-}96)$$

$$P = \frac{SEt}{r_2}\left(\frac{R_1 - r_2}{R_1 - 0.5r_2}\right) \qquad (10\text{-}97)$$

TABLE 10-47 Strength of Solder Joints*

Maximum recommended pressure-temperature ratings for solder joints made with copper tubing and wrought-copper and -bronze or cast-bronze solder-joint pressure fittings and using representative commercial solders

Joining material used in joints	Working temperatures, °F	Maximum working pressure, lbf/in²			
		$\frac{1}{8}$ to 1 in, inclusive†	$1\frac{1}{4}$ to 2 in, inclusive†	$2\frac{1}{2}$ to 4 in, inclusive†	5 to 8 in, inclusive†
50-50 tin-lead solder‡	100	200	175	150	135
	150	150	125	100	90
	200	100	90	75	70
	250	85	75	50	45
95-5 tin-antimony solder	100	500	400	300	270
	150	400	350	275	250
	200	300	250	200	180
	250	200	175	150	135

NOTE: For extremely low working temperatures (in the 0 to −200°F range) it is recommended that a joining material melting at or above 1100°F be used. (Joining materials with melting points in excess of 800°F are defined as "brazing" alloys by the American Welding Society.) See Table 10-48.

*Extracted from ANSI B16.22—1973 with permission of the publisher, the American Society of Mechanical Engineers, New York. (°F − 32)$\frac{5}{9}$ = °C; to convert inches to millimeters, multiply by 25.4; to convert pounds-force per square inch to megapascals, multiply by 0.006895.

†Standard water-tubing sizes.

‡ASTM B32.66T Alloy Grade 50A.

where nomenclature is the same as for straight pipe except as follows (see Fig. 10-165):

t = pressure design thickness
r_2 = mean radius of pipe
R_1 = effective radius of miter bend, defined as the shortest distance from the pipe centerline to the intersection of the planes of adjacent miter joints
θ = angle of miter cut, °
α = angle of change in direction at miter joint
  = 2θ, °

TABLE 10-48 Strength of Silver-Brazed Joints*

Maximum recommended pressure-temperature ratings for brazed joints made with copper tubing and copper or copper-alloy fittings and using representative commercial brazing alloys

Outside-diameter size, in	lbf/in²			
	150°F (S = 5100 lbf/in²)	250°F (S = 4700 lbf/in²)	350°F (S = 4000 lbf/in²)	400°F (S = 3000 lbf/in²)
$\frac{1}{8}$	1790	1650	1400	1050
$\frac{3}{16}$	1190	1100	940	700
$\frac{1}{4}$	890	825	700	525
$\frac{5}{16}$	840	780	660	500
$\frac{3}{8}$	780	720	615	460
$\frac{1}{2}$	680	625	530	400
$\frac{5}{8}$	615	565	480	360
$\frac{3}{4}$	535	495	420	315
$\frac{7}{8}$	490	450	385	290
$1\frac{1}{8}$	420	390	330	250
$1\frac{3}{8}$	380	350	295	220
$1\frac{5}{8}$	350	320	275	205
$2\frac{1}{8}$	310	285	245	180
$2\frac{5}{8}$	286	265	225	170
$3\frac{1}{8}$	270	250	190	140
$3\frac{5}{8}$	260	240	200	150
$4\frac{1}{8}$	250	230	195	145
$5\frac{1}{8}$	225	210	180	135
$6\frac{1}{8}$	215	195	165	125

*Extracted from ANSI B16.41—January 1977 draft, with permission of the publisher, the American Society of Mechanical Engineers, New York. (°F − 32)$\frac{5}{9}$ = °C; to convert inches to millimeters, multiply by 25.4; to convert pounds-force per square inch to megapascals, multiply by 0.006895.

For compliance with the code, the value of R_1 shall not be less than that given by Eq. (10-98):

$$R_1 = A/\tan\theta + D/2 \qquad (10\text{-}98)$$

where A has the following empirical values (**not valid in SI units**):

t, in	A
≤0.5	1.0
0.5 < t < 0.88	2t
≥0.88	(2t/3) + 1.17

Piping branch connections involve the same considerations as pressure-vessel nozzles. However, outlet size in proportion to piping header size is unavoidably much greater for piping. The current Piping Code rules for calculation of branch-connection reinforcement are similar to those of the ASME Boiler and Pressure Vessel Code, Sec. VIII, Division I (1980 edition) for a branch with axis at right angles to the header axis. If the branch connection makes an angle β with the header axis from 45 to 90°, the Piping Code requires that the area to be replaced be increased by dividing it by sin β. In such cases the half width of the reinforcing zone measured along the header axis is similarly increased, except that it may not exceed the outside diameter of the header. Some details of commonly used reinforced branch connections are given in Fig. 10-166.

The rules provide that a branch connection has adequate strength for pressure if a fitting (tee, lateral, or cross) is in accordance with an approved standard and is used within the pressure-temperature limitations or if the connection is made by welding a coupling or half coupling (wall thickness not less than the branch anywhere in reinforcement zone or less than extra heavy or 3000 lb) to the run and provided the ratio of branch to run diameters is not greater than one-fourth and that the branch is not greater than 2 in nominal diameter.

Dimensions of extra-heavy couplings are given in the *Steel Products Manual* published by the American Iron and Steel Institute. In ANSI B16.11—1966, 2000-lb couplings were superseded by 3000-lb couplings.

ANSI B31.3 states that the reinforcement area for resistance to external pressure is to be at least one-half of that required to resist internal pressure.

The code provides no guidance for analysis but requires that external and internal **attachments** be designed to avoid flattening of the pipe, excessive localized bending stresses, or harmful thermal gradients, with further emphasis on minimizing stress concentrations in cyclic service.

No.	Type of joint		Type of seam	Examination	Factor, E_j
1	Furnace butt weld, continuous		Straight	As required by listed specifications	0.60
2	Electric resistance weld		Straight or spiral	As required by listed specifications	0.85
3	Electric fusion weld				
	a Single butt weld (with or without filler metal)		Straight or spiral	As required by listed specifications or this code	0.80
				Additionally spot-radiographed per ANSI B31.3, par. 336.6.1	0.90
				Additionally 100 percent radiographed per ANSI B31.3, par. 336.4.5	1.00
	b Double butt weld (with or without filler metal)		Straight or spiral (except as provided in 4*b*)	As required by listed specification or this code	0.85
				Additionally spot-radiographed per ANSI B31.3, par. 336.6.1	0.90
				Additionally 100 percent radiographed per ANSI B31.3, par. 336.4.5	1.00
4	Per specific specifications				
	a ASTM A211	As permitted in specifications	Spiral	As required by specifications	0.75
	b Double submerged arc-welded pipe per API 5L or 5LX		Straight with one or two seams	As required by specifications, additionally examined by radiography for lengths of 200 mm (8 in) at each end	0.95

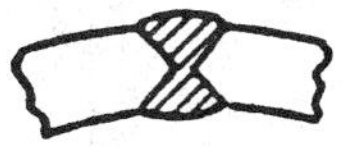

FIG. 10-164 Longitudinal and spiral-weld joint factor E_j. NOTE: It is not permitted to increase the joint quality factor by additional examination for joints 1, 2, and 4a. (*Extracted from ANSI B31.3—1980, with permission of the publisher, the American Society of Mechanical Engineers, New York.*)

The code provides design requirements for **closures** which are flat, ellipsoidal, spherically dished, hemispherical, conical (without transition knuckles), conical convex to pressure, toriconical concave to pressure, and toriconical convex to pressure.

Openings in closures over 50 percent in diameter are designed as flanges in flat closures and as reducers in other closures. Openings of not over one-half of the diameter are to be reinforced as branch connections.

Thermal Expansion and Flexibility: Metallic Piping ANSI B31.3 requires that piping systems have sufficient flexibility to prevent thermal expansion or contraction or the movement of piping supports or terminals from causing (1) failure of piping supports from overstress or fatigue; (2) leakage at joints; or (3) detrimental stresses or distortions in piping or in connected equipment (pumps, turbines, or valves, for example), resulting from excessive thrusts or movements in the piping.

To assure that a system meets these requirements, the computed displacement–stress range S_E shall not exceed the allowable stress range S_A [Eqs. (10-93) and (10-94)], the reaction forces R_m [Eq. (10-105)] shall not be detrimental to supports or connected equipment, and movement of the piping shall be within any prescribed limits.

Displacement Strains Strains result from piping being displaced from its unrestrained position:

1. *Thermal displacements.* A piping system will undergo dimensional changes with any change in temperature. If it is constrained from free movement by terminals, guides, and anchors, it will be displaced from its unrestrained position.

2. *Reaction displacements.* If the restraints are not considered rigid and there is a predictable movement of the restraint under load, this may be treated as a compensating displacement.

3. *Externally imposed displacements.* Externally caused movement of restraints will impose displacements on the piping in addition to those related to thermal effects. Such movements may result from causes such as wind sway or temperature changes in connected equipment.

Total Displacement Strains Thermal displacements, reaction displacements, and externally imposed displacements all have equivalent effects on the piping system and must be considered together in determining total displacement strains in a piping system.

Expansion strains may be taken up in three ways: by bending, by torsion, or by axial compression. In the first two cases maximum stress occurs at the extreme fibers of the cross section at the critical location. In the third case the entire cross-sectional area over the entire length is for practical purposes equally stressed.

Bending or torsional flexibility may be provided by bends, loops, or offsets; by corrugated pipe or expansion joints of the bellows type; or by other devices permitting rotational movement. These devices must be anchored or otherwise suitably connected to resist end forces from fluid pressure, frictional resistance to pipe movement, and other causes.

Axial flexibility may be provided by expansion joints of the slipjoint or bellows types, suitably anchored and guided to resist end forces from fluid pressure, frictional resistance to movement, and other causes.

Displacement Stresses Stresses may be considered proportional to the total displacement strain only if the strains are well distributed and not excessive at any point. The methods outlined here and in the code are applicable only to such a system. Poor distribution of strains (unbalanced systems) may result from:

1. Highly stressed small-size pipe runs in series with large and relatively stiff pipe runs

TABLE 10-49 Allowable Stresses in Tension for Materials (4, 13, 28)*

Specifications are ASTM unless otherwise indicated. Numbers in parentheses refer to notes at end of table.

Material	Specification	P no. (23)	Grade	Class	Factor, E	Minimum tensile strength, kip/in²	Minimum yield strength, kip/in²	Notes	Minimum temperature (18)	Minimum temperature to 100	200	300	400	500	600
Iron															
Centrifugally cast pipe															
	FS-WW-P421c							8, 10, 17	−20	6.0	6.0	6.0	6.0		
	AWWA C106							8, 10, 17	−20	6.0	6.0	6.0	6.0		
	AWWA C108							8, 10, 17	−20	6.0	6.0	6.0	6.0		
Carbon steel															
Seamless pipe and tubing															
	A53	1	A	Type S		48.0	30.0	1, 2	−20	16.0	16.0	16.0	16.0	16.0	14.8
	A53	1	B	Type S		60.0	35.0	1, 2	−20	20.0	20.0	20.0	20.0	18.9	17.3
	A106	1	A			48.0	30.0	2	−20	16.0	16.0	16.0	16.0	16.0	14.8
	A106	1	B			60.0	35.0	2	−20	20.0	20.0	20.0	20.0	18.9	17.3
	A106	1	C			70.0	40.0	2	−20	23.3	23.3	23.3	22.9	21.6	19.7
	A120	1						21	−20	12.0	11.4				
	A333	1	1			55.0	30.0	1, 2	−50	18.3	18.3	17.7	17.2	16.2	14.8
	A333	1	6			60.0	35.0	2	−50	20.0	20.0	20.0	20.0	18.9	17.3
	API 5L	1	A			48.0	30.0	1, 2	−20	16.0	16.0	16.0	16.0	16.0	14.8
	API 5L	1	B			60.0	35.0	1, 2	−20	20.0	20.0	20.0	20.0	18.9	17.3
	API 5LX	SP2	X42			60.0	42.0	37, 38	−20	20.0	20.0	20.0	20.0		
	API 5LX	SP3	X46			63.0	46.0	37, 38	−20	21.0	21.0	21.0	21.0		
	API 5LX	SP3	X52			66.0	52.0	37, 38	−20	22.0	22.0	22.0	22.0		
	API 5LX	SP3	X52			72.0	52.0	37, 38	−20	24.0	24.0	24.0	24.0		
Electric-resistance-welded pipe															
	A53	1	A	Type E	0.85	48.0	30.0	1, 2	−20	13.6	13.6	13.6	13.6	13.6	12.6
	A53	1	B	Type E	0.85	60.0	35.0	1, 2	−20	17.0	17.0	17.0	17.0	16.1	14.7
	A120	1			0.85			21	−20	10.2	9.7				
	A135	1	A		0.85	48.0	30.0	1, 2	−20	13.6	13.6	13.6	13.6	13.6	12.6
	A135	1	B		0.85	60.0	35.0	1, 2	−20	17.0	17.0	17.0	17.0	16.1	14.7
	A333	1	1		0.85	55.0	30.0	1, 2	−50	15.6	15.6	15.0	14.6	13.8	12.6
	A333	1	6		0.85	60.0	35.0	2	−50	17.0	17.0	17.0	17.0	16.1	14.7
	A587	1			0.85	48.0	30.0	1, 2	−20	13.6	13.6	13.6	13.6	13.6	12.6
	API 5L	1	A25	I and II	0.85	45.0	25.0	1, 2	−20	12.8	12.8	12.3	11.8		
	API 5L	1	A		0.85	48.0	30.0	1, 2	−20	13.6	13.6	13.6	13.6	13.6	12.6
	API 5L	1	B		0.85	60.0	35.0	1, 2	−20	17.0	17.0	17.0	17.0	16.1	14.7
	API 5L	SP2	X42		0.85	60.0	42.0	37, 38	−20	17.0	17.0	17.0	17.0		
	API 5LX	SP3	X46		0.85	63.0	46.0	37, 38	−20	17.9	17.9	17.9	17.9		
	API 5LX	SP3	X52		0.85	66.0	52.0	37, 38	−20	18.7	18.7	18.7	18.7		
	API 5LX	SP3	X52		0.85	72.0	52.0	37, 38	−20	20.4	20.4	20.4	20.4		
Electric-fusion-welded pipe (straight seam)															
A570 GR A	A134	1			0.74	45.0	25.0	5, 21	−20	11.1	10.5	10.0			
A570 GR B	A134	1			0.74	49.0	30.0	5, 21	−20	12.1	11.4	10.9			
A570 GR C	A134	1			0.74	52.0	33.0	5, 21	−20	12.8	12.1	11.6			
A570 GR D	A134	1			0.74	55.0	40.0	5, 21	−20	13.6	12.8	12.2			
A570 GR E	A134	1			0.74	58.0	42.0	5, 21	−20	14.3	13.5	12.9			
Low- and intermediate-alloy steel															
Seamless pipe															
3½ Ni	A333	9B	3			65.0	35.0		−150	21.7	19.6	19.6	18.7	17.8	16.8
¾ Cr, ¾ Ni, Cu, Al	A333	4	4			60.0	35.0		−150	20.0	19.1	18.2	17.3	16.4	15.5
2¼ Ni	A333	9A	7			65.0	35.0		−100	21.7	19.6	19.6	18.7	17.6	16.8
9 Ni	A333	11A-SG1	8			100.0	75.0	40	−320	31.7	31.7				
C, ½ Mo	A335	3	P1			55.0	30.0	3	−20	18.3	18.3	17.5	16.9	16.3	15.7
5 Cr, ½ Mo	A335	5	P5			60.0	30.0		−20	20.0	18.1	17.4	17.2	17.1	16.8
1¼ Cr, ½ Mo	A335	4	P11			60.0	30.0		−20	20.0	18.7	18.0	17.5	17.2	16.7
2¼ Cr, 1 Mo	A335	5	P22			60.0	30.0		−20	20.0	18.5	18.0	17.9	17.9	16.8
Stainless steel															
Seamless pipe and tubing															
18Cr, 8Ni pipe	A312	8	TP304			75.0	30.0	7, 14, 16, 20	−425	20.0	20.0	20.0	18.7	17.5	16.4
18Cr, 8Ni pipe	A312	8	TP304H			75.0	30.0	16	−325	20.0	20.0	20.0	18.7	17.5	16.4
18Cr, 8Ni pipe	A312	8	TP304L			70.0	25.0		−425	16.7	16.7	16.7	15.8	14.8	14.0
25Cr, 20Ni pipe	A312	8	TP310			75.0	30.0	19, 24, 32	−325	20.0	20.0	20.0	20.0	20.0	19.2
25Cr, 20Ni pipe	A312	8	TP310			75.0	30.0	6, 19, 24, 32	−325	20.0	20.0	20.0	20.0	20.0	19.2
16Cr, 12Ni, 2Mo	A312	8	TP316			75.0	30.0	14, 16	−325	20.0	20.0	20.0	19.3	17.9	17.0
16Cr, 12Ni, 2Mo pipe	A312	8	TP316H			75.0	30.0	16	−325	20.0	20.0	20.0	19.3	17.9	17.0
16Cr, 12Ni, 2Mo pipe	A312	8	TP316L			70.0	25.0		−325	16.7	16.7	16.7	15.5	14.4	13.5
18Cr, 10Ni, Cb pipe	A312	8	TP347			75.0	30.0	7, 14	−425	20.0	20.0	20.0	20.0	19.9	19.3
18Cr, 10Ni, Cb pipe	A312	8	TP347H			75.0	30.0		−325	20.0	20.0	20.0	20.0	19.9	19.3
Centrifugally cast pipe															
18Cr, 8Ni	A451	8	CPF8		0.90	70.0	30.0	14, 15, 16	−425	18.0	18.0	17.8	15.8	14.8	14.1
18Cr, 10Ni, 2Mo	A451	8	CPF8M		0.90	70.0	30.0	14, 15, 16	−425	18.0	18.0	18.0	17.5	16.3	15.4
18Cr, 10Ni, Cb	A451	8	CPF8C		0.90	70.0	30.0	7, 14, 15	−325	18.0	18.0	18.0	18.0	17.4	16.5
15Cr, 13Ni, 2Mo, Cb	A451	8	CPF10MC		0.90	70.0	30.0	7, 11, 14, 15	−325	18.0					
23Cr, 13Ni	A451	8	CPH8		0.90	65.0	28.0	11, 14, 15, 19	−325	16.8	16.8	16.8	16.8	16.8	16.2
23Cr, 13Ni	A451	8	CPH10 or CPH 20		0.90	70.0	30.0	9, 11, 14, 15, 19, 24	−325	18.0	18.0	18.0	18.0	18.0	17.3
25Cr, 20Ni	A451	8	CPK20		0.90	65.0	28.0	14, 15, 19, 24	−325	16.8	16.8	16.8	16.8	16.8	16.2
18Cr, 8Ni	A452	8	TP304H		0.85	75.0	30.0	15, 16	−325	17.0	17.0	17.0	15.9	14.8	14.0
16Cr, 12Ni, 2Mo	A452	8	TP316H		0.85	75.0	30.0	15, 16	−325	17.0	17.0	17.0	16.4	15.2	14.4
18Cr, 10Ni, Cb	A452	8	TP347H		0.85	75.0	30.0	15	−325	17.0	17.0	17.0	17.0	16.9	16.2

650	700	750	800	850	900	950	1000	1050	1100	1150	1200	1250	1300	1350	1400	1450	1500
14.5	14.4	10.7	9.3	7.9	6.5	4.5	2.5	1.6	1.0								
17.0	16.8	13.0	10.8	8.7	6.5	4.5	2.5	1.6	1.0								
14.5	14.4	10.7	9.3	7.9	6.5	4.5	2.5	1.6	1.0								
17.0	16.8	13.0	10.8	8.7	6.5	4.5	2.5	1.6	1.0								
19.4	19.2	14.8	12.0														
14.5	14.4	12.0	10.2	8.3	6.5	4.5	2.5	1.6	1.0								
17.0	16.8	13.0	10.8	8.7	6.5	4.5	2.5	1.6	1.0								
14.5	14.4	10.7	9.3	7.9	6.5	4.5	2.5	1.6	1.0								
17.0	16.8	13.0	10.8	8.7	6.5	4.5	2.5	1.6	1.0								
12.3	12.2	9.1	7.9	6.7	5.5	3.8	2.1	1.4	0.9								
14.5	14.0	11.0	9.2	7.4	5.5	3.8	2.1	1.4	0.9								
12.3	12.2	9.1	7.9	6.7	5.5	3.8	2.1	1.4	0.9								
14.5	14.0	11.0	9.2	7.4	5.5	3.8	2.1										
12.3	12.2	10.2	8.7	7.1	5.5	3.8	2.1	1.4	0.9								
14.5	14.0	11.0	9.2	7.4	5.5	3.8	2.1	1.4	0.9								
12.3																	
12.3	12.2	9.1	7.9	6.7	5.5	3.8	2.1	1.4	0.9								
14.5	14.0	11.0	9.2	7.4	5.5	3.8	2.1	1.4	0.9								
16.3	15.5	13.9	11.4	9.0	6.5	4.5	2.5	1.6	1.0								
15.0																	
16.3	15.5	13.9	11.4	9.0	6.5	4.5	2.5	1.6	1.0								
15.4	15.1	13.8	13.5	13.1	12.7	8.2	4.8										
16.6	16.3	13.2	12.8	12.1	10.9	8.0	5.8	4.2	2.9	2.0	1.3						
16.2	15.6	15.0	15.0	14.4	13.1	11.0	7.8	5.5	4.0	2.5	1.2						
17.9	17.9	17.9	15.2	14.5	12.8	11.0	7.8	5.8	4.2	3.0	2.0						
16.2	16.0	15.6	15.2	14.9	14.6	14.4	13.8	12.2	9.7	7.7	6.0	4.7	3.7	2.9	2.3	1.8	1.4
16.2	16.0	15.6	15.2	14.9	14.6	14.4	13.8	12.2	9.7	7.7	6.0	4.7	3.7	2.9	2.3	1.8	1.4
13.7	13.5	13.3	13.0	12.8	11.9	9.9	7.8	6.3	5.1	4.0	3.2	2.6	2.1	1.7	1.1	1.0	0.9
18.8	18.3	18.0	17.5	14.6	13.9	12.5	11.0	7.1	5.0	3.6	2.5	1.5	0.8	0.5	0.4	0.3	0.2
18.8	18.3	18.0	17.5	14.6	13.9	12.5	11.0	9.8	8.5	7.3	6.0	4.8	3.5	2.3	1.6	1.1	0.8
16.7	16.3	16.1	15.9	15.7	15.5	15.4	15.3	14.5	12.4	9.8	7.4	5.5	4.1	3.1	2.3	1.7	1.3
16.7	16.3	16.1	15.9	15.7	15.5	15.4	15.3	14.5	12.4	9.8	7.4	5.5	4.1	3.1	2.3	1.7	1.3
13.2	12.9	12.6	12.4	12.1	11.8	11.5	11.2	10.8	10.2	8.8	6.4	4.7	3.5	2.5	1.8	1.3	1.0
19.0	18.6	18.5	18.3	15.4	14.9	14.8	14.0	12.1	9.1	6.1	4.4	3.3	2.2	1.5	1.2	0.9	0.8
19.0	18.6	18.5	18.3	18.2	18.1	18.1	18.0	17.1	14.2	10.5	7.9	5.9	4.4	3.2	2.5	1.8	1.3
13.8	13.6	13.4	13.3	11.6	11.4	11.1	9.7	8.6	6.7	5.2	4.0	2.9	2.2	1.6	1.2	0.9	0.7
15.0	14.6	14.2	14.0	13.2	13.0	12.6	11.8	10.3	8.4	7.2	6.1	4.8	3.6	2.7	2.1	1.7	1.3
16.2	15.8	15.5	15.4	12.6	12.5	12.3	12.1	11.7	9.7	7.2	4.5	3.2	2.4	1.8	1.4	1.0	0.9
15.7	15.4	15.1	14.7	11.5	11.2	10.6	9.4	7.6	5.8	4.5	3.3	2.6	2.1	1.5	1.2	0.8	0.7
16.9	16.5	16.2	15.7	12.2	12.0	11.2	9.5	7.6	5.8	4.5	3.3	2.6	2.1	1.5	1.2	0.8	0.7
15.7	15.4	15.1	14.7	11.5	11.2	10.7	9.9	8.8	7.6	6.5	5.4	4.3	3.1	2.1	1.4	1.0	0.7
13.8	13.5	13.2	12.8	12.7	12.4	12.2	11.7	10.3	8.3	6.5	5.1	4.0	3.1	2.4	1.9	1.5	1.2
14.2	13.8	13.6	13.5	13.3	13.2	13.1	13.0	12.3	10.5	8.3	6.3	4.6	3.5	2.6	1.9	1.4	1.0
16.1	15.8	15.7	15.5	15.4	15.4	15.4	15.3	14.5	12.1	8.9	6.7	5.0	3.7	2.7	2.1	1.5	1.1

TABLE 10-49 Allowable Stresses in Tension for Materials (4, 13, 28) (Continued)

Specifications are ASTM unless otherwise indicated. Numbers in parentheses refer to notes at end of table.

Material	Specification	P no. (23)	Grade	Class	Factor, E	Minimum tensile strength, kip/in²	Minimum yield strength, kip/in²	Notes	Minimum temperature (18)	Minimum temperature to 100	200	300	400	500	600
Electric-fusion-welded pipe and tubing															
18Cr, 8Ni pipe	A312	8	TP304		0.85	75.0	30.0	14, 16	−425	17.0	17.0	17.0	15.9	14.8	14.0
18Cr, 8Ni pipe	A312	8	TP304H		0.85	75.0	30.0	16	−325	17.0	17.0	17.0	15.9	14.8	14.0
18Cr, 8Ni pipe	A312	8	TP304L		0.85	70.0	25.0		−425	14.2	14.2	14.2	13.4	12.5	11.9
23Cr, 12Ni pipe	A312	8	TP309		0.85	75.0	30.0	19, 24, 32	−325	17.0	17.0	17.0	17.0	17.0	16.3
25Cr, 20Ni pipe	A312	8	TP310		0.85	75.0	30.0	19, 24, 32	−325	17.0	17.0	17.0	17.0	17.0	16.3
25Cr, 20Ni pipe	A312	8	TP310		0.85	75.0	30.0	6, 19, 24, 32	−325	17.0	17.0	17.0	17.0	17.0	16.3
16Cr, 12Ni, 2Mo pipe	A312	8	TP316		0.85	75.0	30.0	14, 16	−325	17.0	17.0	17.0	16.4	15.2	14.4
16Cr, 12Ni, 2Mo pipe	A312	8	TP316H		0.85	75.0	30.0	16	−325	17.0	17.0	17.0	16.4	15.2	14.4
16Cr, 12Ni, 2Mo pipe	A312	8	TP316L		0.85	70.0	25.0		−325	14.2	14.2	14.2	13.2	12.2	11.5
18Cr, 13Ni, 3Mo pipe	A312	8	TP317		0.85	75.0	30.0	14, 16	−325	17.0	17.0	17.0	16.4	15.2	14.4
18Cr, 10Ni, Ti pipe	A312	8	TP321		0.85	75.0	30.0	7, 14	−325	17.0	17.0	17.0	15.8	14.7	13.9
18Cr, 10Ni, Ti pipe	A312	8	TP321H		0.85	75.0	30.0		−325	17.0	17.0	17.0	15.8	14.7	13.9
18Cr, 10Ni, Cb pipe	A312	8	TP347		0.85	75.0	30.0	7, 14	−425	17.0	17.0	17.0	17.0	16.9	16.4
18Cr, 10Ni, Cb pipe	A312	8	TP347H		0.85	75.0	30.0		−325	17.0	17.0	17.0	17.0	16.9	16.4

Material	Specification	P no. (23), (30)	Temper	Class	Size range, in	Factor, E	Minimum tensile strength, kip/in²	Minimum yield strength, kip/in²	Notes	Minimum temperature (18)
Copper and copper alloy Seamless pipe and tubing										
Copper pipe	B42	31	Drawn	102, 120, 122	⅛–2, inclusive		45.0	40.0	9, 27	−325
Copper tubing	B88	31	Annealed	C10200, C12000, C12200			30.0	9.0	9, 29	−325
Copper tubing	B88	31	Drawn	C10200, C12000, C12200			36.0	30.0	9, 27, 29	−325
Cu, Ni 90/10	B466	34	Annealed	C70600			38.0	13.0	9	−325
Cu, Ni 70/30	B466	34	Annealed	C71500			50.0	18.0	9	−325

Material	Specification	P no. (23)	Grade	Class	Size range, in	Factor, E	Minimum tensile strength, kip/in²	Minimum yield strength, kip/in²	Notes	Minimum temperature	Minimum temperature to 100	200	300	400	500	600
Nickel and nickel alloy Seamless pipe and tubing																
Nickel	B161	41	200 (N02200)	Annealed	5 OD and under		55.0	15.0		−325	10.0	10.0	10.0	10.0	10.0	10.0
Nickel	B161	41	200 (N02200)	Annealed	Over 5 OD		55.0	12.0		−325	8.0	8.0	8.0	8.0	8.0	8.0
Low-C Ni	B161	41	201 (N02201)	Annealed	5 OD and under		50.0	12.0		−325	8.0	7.7	7.5	7.5	7.5	7.5
Low-C Ni	B161	41	201 (N02201)	Annealed	Over 5 OD		50.0	10.0		−325	6.7	6.4	6.3	6.2	6.2	6.2
Ni, Cu	B165	42	400 (N04400)	Annealed	5 OD and under		70.0	28.0		−325	18.7	16.4	15.4	14.8	14.8	14.8
Ni, Cu	B165	42	400 (N04400)	Annealed	Over 5 OD		70.0	25.0		−325	16.7	14.7	13.7	13.2	13.2	13.2
Ni, Cr, Fe	B167	43	600 (N06600)	Hot-finished or hot-finished annealed	5 OD and under		80.0	30.0		−325	20.0	20.0	20.0	20.0	20.0	20.0
Ni, Cr, Fe	B167	43	600 (N06600)	Hot-finished or hot-finished annealed	Over 5 OD		75.0	25.0		−325	16.7	16.7	16.7	16.7	16.7	16.7
Ni, Fe, Cr	B407	45	800 H (N08800)	Cold-drawn solution annealed or hot-finished			65.0	25.0	39	−325	16.7	16.7	16.7	16.7	16.7	16.5
Ni, Cr, Mo, Cb	B444	43	625 (N06625)	Annealed			120.0	60.0	42	−325	30.0	30.0	30.0	28.2	27.0	26.4
Welded pipe																
Ni, Mo	B619		B (N10001)	Solution-annealed		0.85	100.0	45.0		−325	25.5	25.5	25.5	25.5	25.5	25.5
Ni, Mo	B619	44	B-2 (N10665)	Solution-annealed		0.85	110.0	51.0		−325	23.4	23.4	23.4	23.4	23.4	23.1
Ni, Mo, Cr	B619		C-4 (N06455)	Solution-annealed		0.85	100.0	40.0		−325	21.2	21.2	21.2	21.2	21.0	20.7
Ni, Mo, Cr	B619	44	C276 (N10276)	Solution-annealed		0.85	100.0	41.0		−325	23.2	23.2	23.2	23.2	22.9	21.6
Ni, Cr, Fe, Mo, Cu	B619	45	G1 (N06007)	Solution-annealed		0.85	90.0	35.0		−325	19.1	19.1	19.1	18.6	18.3	17.9
Ni, Cr, Mo, Fe	B619	45	X (N06002)	Solution-annealed		0.85	100.0	40.0		−325	22.6	20.5	19.8	19.5	18.9	17.9
Ni, Fe, Cr, Mo	B619	45	20-MOD (N08320)	Solution annealed		0.85	75.0	28.0		−325	15.9	15.9	15.8	15.2	15.0	14.9

Specification	P no.	Grade	Temper	Size range, in	Minimum tensile strength, kip/in²	Minimum yield strength, kip/in²	Notes	Minimum temperature, (18)	Metal temperature, °F (22)						
									Minimum temperature, to 100	150	200	250	300	350	400
Aluminum alloy Seamless pipe and tubing															
B210	21	1060	0	0.018–0.500	8.5	2.5	26	−452	1.7	1.7	1.6	1.5	1.3	1.1	0.8
B210	21	3003	0	0.010–0.500	14.0	5.0	26	−452	3.3	3.3	3.3	3.1	2.4	1.8	1.4
B210	23	6061	T4	0.025–0.500	30.0	16.0	12, 26	−452	10.0	10.0	10.0	9.8	9.2	7.9	5.6
B210	23	6061	T4, T6 welded		24.0		35	−452	8.0	8.0	8.0	7.9	7.4	6.1	4.3

650	700	750	800	850	900	950	1000	1050	1100	1150	1200	1250	1300	1350	1400	1450	1500
13.7	13.6	13.2	12.9	12.7	12.5	12.2	11.7	10.3	8.3	6.5	5.1	4.0	3.1	2.5	2.0	1.5	1.2
13.7	13.6	13.2	12.9	12.7	12.5	12.2	11.7	10.3	8.3	6.5	5.1	4.0	3.1	2.5	2.0	1.5	1.2
11.6	11.4	11.3	11.0	10.9	10.1	8.4	6.6	5.4	4.3	3.4	2.8	2.2	1.8	1.4	0.9	0.8	0.7
16.0	15.6	15.3	14.9	12.4	11.8	10.6	8.9	7.2	5.5	4.2	3.2	2.5	2.0	1.5	1.1	0.8	0.6
16.0	15.6	15.3	14.9	12.4	11.8	10.6	9.3	6.0	4.2	3.1	2.1	1.2	0.6	0.4	0.3	0.2	0.2
16.0	15.6	15.3	14.9	12.4	11.8	10.6	9.3	8.3	7.2	6.2	5.1	4.0	3.0	2.0	1.4	0.9	0.6
14.2	13.8	13.6	13.5	13.3	13.2	13.1	13.0	12.3	10.5	8.3	6.3	4.6	3.5	2.6	1.9	1.4	1.1
14.2	13.8	13.6	13.5	13.3	13.2	13.1	13.0	12.3	10.5	8.3	6.3	4.6	3.5	2.6	1.9	1.4	1.1
11.2	10.9	10.7	10.5	10.3	10.0	9.8	9.5	9.2	8.7	7.4	5.4	4.0	3.0	2.1	1.6	1.1	0.9
14.2	13.9	13.6	13.5	13.3	13.2	13.1	13.0	12.3	10.5	8.3	6.3	4.6	3.5	2.6	1.9	1.4	1.1
13.6	13.4	13.3	13.1	13.0	13.0	12.9	11.7	8.2	5.8	4.2	3.1	2.2	1.4	0.9	0.6	0.4	0.3
13.6	13.4	13.3	13.1	13.0	13.0	12.9	11.9	9.9	7.7	5.9	4.5	3.5	2.7	2.1	1.6	1.2	0.9
16.1	15.8	15.7	15.5	13.1	12.7	12.3	11.9	10.3	7.7	5.2	3.7	2.8	1.9	1.3	1.0	0.8	0.6
16.1	15.8	15.7	15.6	15.5	15.4	15.4	15.3	14.5	12.1	8.9	6.7	5.0	3.7	2.7	2.1	1.5	1.1

Metal temperature, °F (22)

Minimum temperature to 100	150	200	250	300	350	400	450	500	550	600	650	700
15.0	11.2	11.2	11.2	11.0	10.3	4.2						
6.0	6.0	5.9	5.8	5.0	3.8	2.5	1.5	0.8				
12.0	9.0	8.7	8.3	8.0	5.0	2.5	1.5	0.8				
8.7	8.3	8.1	8.0	7.8	7.7	7.5	7.3	7.2	7.0	6.0		
12.0	11.6	11.3	11.0	10.8	10.6	10.3	10.1	9.9	9.8	9.6	9.5	9.4

Metal temperature, °F (22)

650	700	750	800	850	900	950	1000	1050	1100	1150	1200	1250	1300	1350	1400	1450	1500
7.5	7.4	7.3	7.2	5.8	4.5	3.7	3.0	2.4	2.0	1.5	1.2						
6.2	6.2	6.1	5.9	5.8	4.5	3.7	3.0	2.4	2.0	1.5	1.2						
14.8	14.8	14.6	14.2														
13.2	13.2	13.0	12.7														
20.0	20.0	20.0	20.0	19.6	16.0	10.6	7.0	4.5	3.0	2.2	2.0						
16.7	16.7	16.7	16.7	16.5	15.9	10.6	7.0	4.5	3.0	2.2	2.0						
16.0	15.7	15.4	15.3	15.1	14.8	14.6	14.4	13.7	13.5	11.2	8.4	6.9	5.4	4.5	3.6	3.0	2.5
26.0	26.0	26.0	26.0	26.0	26.0	26.0	26.0	26.0	26.0	21.0	13.2						
25.0	25.5	24.5	23.5														
23.1	23.0	22.9	22.8														
20.5	20.4	20.0	19.5														
21.0	20.4	20.0	19.5	19.2	18.9	18.7	18.5										
17.8	17.8	17.6	17.4	17.2	17.0	16.6	16.1										
17.6	17.3	17.0	16.8	16.7	16.7	16.3	15.8	15.3	14.9	12.3	9.6	8.0	6.5				
14.9	14.9	14.8	14.6														

Design stresses for bolting materials

Material	Specification	Grade	Size range, in	Minimum tensile strength, kip/in²	Minimum yield strength, kip/in²	Notes	Minimum temperature (18)	Minimum temperature, to 100	200	300	400	500	600	650
Carbon steel														
	A307	B		60.0		22	−20	13.7	13.7	13.7	13.7∣	13.7∥		
	A325			105.0			−20	19.3	19.3	19.3	19.3∣	19.3∥	19.3	19.3
	A194	1, 2				25	−20							
	A194	2H				25	−50							
Alloy steel														
Cr, Mo	A193	B7	2½ and under	125.0	105.0	33	−20	25.0	25.0	25.0	25.0	25.0	25.0	25.0
Cr, 0.2Mo	A193	B7M	2½ and under	100.0	80.0		−50	20.0	20.0	20.0	20.0	20.0	20.0	20.0
Cr, Mo, V	A193	B16	2½ and under	125.0	105.0		−20	25.0	25.0	25.0	25.0	25.0	25.0	25.0
C, Mo	A194	4				25								
Cr, Mo	A320	L7, L7A, L7B, L7C	2½ and under	125.0	105.0	31	−150	25.0	25.0	25.0	25.0	20.0	20.0	20.0
Stainless steel														
12 Cr	A193	B6	4 and under	110.0	85.0	19, 31	−20	21.2	21.2	21.2	21.2	21.2	21.2	21.2
304 solution-treated	A193	B8, Cl. 1		75.0	30.0	31, 32, 41	−325	18.8	15.6	14.0	12.9	12.1	11.4	11.2
316 solution treated	A193	B8M, Cl. 1		75.0	30.0	31, 32, 41	−325	18.8	16.1	14.6	13.3	12.5	11.8	11.5
304 strain-hardened	A193	B8, Cl. 2	Up to ¾	125.0	100.0	31, 32, 41	−325	25.0						
			¾ to 1	115.0	80.0	31, 32, 41	−325	20.0						
			Over 1 to 1¼	105.0	65.0	31, 32, 41	−325	16.2						
			Over 1¼ to 1½	100.0	50.0	31, 32, 41	−325	12.5						
316 strain-hardened	A193	B8M, Cl. 2	Up to ¾	110.0	95.0	31, 32, 41	−325	22.0	22.0	22.0	22.0	22.0	22.0	22.0
			¾ to 1	100.0	80.0	31, 32, 41	−325	20.0	20.0	20.0	20.0	20.0	20.0	20.0
			Over 1 to 1¼	95.0	65.0	31, 32, 41	−325	16.2	16.2	16.2	16.2	16.2	16.2	16.2
			Over 1¼ to 1½	90.0	50.0	31, 32, 41	−325	12.5	12.5	12.5	12.5	12.5	12.5	12.5
14 Cr, 24 Ni	A453	660A/B		130.0	85.0	19, 31	−20	21.3	20.7	20.5	20.4	20.3	20.2	20.2

Material	Specification	Grade	Temper	Size range, in	Minimum strength, kip/in²	Minimum yield strength, kip/in²	Notes	Minimum temperature (18)
Aluminum and aluminum-base alloy								
	B211	2024	T4	0.500–4.500	62.0	42.0	34, 35	−325
	B211	6061	T6, T651	0.125–8.000	42.0	35.0	34, 35	−325
Copper and copper-base alloy								
Cu, Si	B98	C65500, C66100	Soft		52.0	15.0	43	−325
Cu, Si	B98	C65100	Bolt	Over ½ to 1	75.0	45.0		−325
Al, Bronze	B150	C64200		Over ½ to 1	85.0	45.0		−325
Al, Bronze	B150	C63000		½ to 1	100.0	50.0		−325
Al, Bronze	B150	C61400		Over ½ to 1	75.0	35.0		−325
Nickel and nickel-base alloy								
Nickel	B160	200 (N02200)	Cold-drawn		65.0	40.0		−325
Low C, Ni	B160	201 (N02100)	Annealed hot-finished		50.0	10.0		−325
Ni, Cu	B164	400 (N04400)	Hot-finished	All except hexagonal over 2½	80.0	40.0		−325
Ni, Cu	B164	400 (N04400)	Cold-drawn stress-relieved		84.0	50.0	36	−325
Ni, Cr, Fe	B166	600 (N06600)	Annealed		80.0	35.0		−325

NOTES:

Special note for the sixth edition: At this time, metric equivalents have not been provided for the allowable-stress tables of the piping code B31.3. They may be computed by the following relationships; (°F − 32) × ⁵⁄₉ = °C; lbf/in² (stress) × 6.895 × 10⁻³ = MPa.

1. For temperatures above 480°C (900°F) consider the advantages of killed steel.

2. Conversion of carbides to graphite may occur after prolonged exposure to temperatures over 425°C (800°F).

3. Conversion of carbides to graphite may occur after prolonged exposure to temperatures over 468°C (875°F).

4. In shaded areas, allowable-stress values which are printed in *italics* exceed two-thirds of the expected yield strength at temperature. All other allowable-stress values in shaded areas are equal to 90 percent of expected yield strength at temperature. See ANSI B31.3.

5. A quality factor of 92 percent is included for structural grade.

6. The higher stress values at 566°C (1050°F) and above for this material shall be used only when the steel has an austenitic micrograin size No. 6 or less (coarser grain) as defined in ASTM E112. Otherwise the lower stress values shall be used.

7. For temperatures above 538°C (1000°F), these stress values may be used only if the material has been heat-treated at a temperature of 1090°C (2000°F) minimum.

8. There are restrictions in the code on the use of this material.

9. For use in code piping at the stated allowable stresses, the tensile and yield strengths listed in these tables must be verified by tensile tests at the mill; such tests shall be specified in the purchase order.

10. Pressure-temperature ratings of cast and forged parts as published in standards referenced in this code section may be used for parts meeting requirements of these standards. Allowable stresses for castings and forgings, where listed, are for use in the design of special components not furnished in accordance with such standards.

11. Certain forms of this material, as stated in Table 10-57, must be impact-tested to qualify for service below −29°C (−20°F). Alternatively, if provisions for impact testing are included in the material specification as supplementary requirements and are invoked, the material may be used down to the temperature at which the test was conducted in accordance with the specification.

12. For welded construction with work-hardened grades, use the stresses for annealed material; for welded construction with precipitation-hardened grades, use the special allowable stresses for welded construction given in the tables.

13. *SE* values shown in this table for welded pipe include the joint quality factor E_j for the longitudinal weld as required by Fig. 10-164 and, when applicable, the structural-grade quality factor E_S of 0.92. For some code computations, particularly with regard to expansion, flexibility, structural attachments, supports, and restraints, the longitudinal-joint quality factor E_j need not be considered. To determine the allowable stress *S* for use in code computations not utilizing the joint quality factor E_j divide the value *SE* shown in this table by the longitudinal-joint quality factor E_j tabulated in Fig. 10-164.

14. For temperatures above 38°C (100°F) these stress values apply only when the carbon content is 0.04 percent or higher.

15. Stress values shown include the casting quality factor shown in this table. Higher stress values can be used if special inspection is accomplished.

16. These unstabilized grades of stainless steel have an increasing tendency to intergranular carbide precipitation as the carbon content increases above 0.03 percent.

17. The allowable stress to be used for this gray-cast-iron material at its upper temperature limit of 232°C (450°F) is the same as that shown in the 204°C (400°F) column.

Metal temperature, °F (22)

700	750	800	850	900	950	1000	1050	1100	1150	1200	1250	1300	1350	1400	1450	1500
25.0	23.6	21.0	17.0	12.5	8.5	4.5										
20.0	20.0	18.5	16.2	12.5	8.5	4.5										
25.0	25.0	25.0	23.5	20.5	16.0	11.0	6.3	2.8								
20.0	20.0	20.0	16.2	12.5	8.5	4.5										
21.2	21.2 ǀ	19.6	15.6	12.0												
11.0	10.8	10.5	10.3	10.1	9.9	9.7	9.5	8.8	7.7	6.0	4.7	3.7	2.9	2.3	1.8	1.4
11.3	11.0	10.9	10.8	10.7	10.7	10.6	10.5	10.3	9.3	7.4	5.4	4.1	3.0	2.2	1.7	1.2
22.0	22.0	22.0														
20.0	20.0	20.0														
16.2	16.2	16.2														
12.5	12.5	12.5														
20.1	20.0	19.9	19.9	19.9	19.8	19.8										

Metal temperature, °F (22)

Minimum temperature, to 100	200	300	400	500	600	650	700	750	800	850	900	950	1000	1050	1100	1150	1200
10.5	10.5	10.4	4.5														
8.4	8.4	8.4	4.4														
10.0	10.0	10.0															
11.3	11.3	11.3															
21.3	21.3	21.3	20.8	12.6	9.9												
25.0	25.0	25.0	25.0	20.7	12.0	8.5	6.0										
18.8	18.8	18.8	18.4	16.1													
10.0	10.0	10.0	10.0	10.0	10.0												
6.7	6.4	6.3	6.2	6.2	6.2	6.2	6.2	6.0	5.9	5.8	4.8	3.7	3.0	2.4	2.0	1.5	1.2
20.0	20.0	20.0	20.0	20.0	20.0	20.0	19.2	18.5	14.5	8.5	4.0						
12.5	12.5	12.5	12.5	12.5													
20.0	20.0	20.0	20.0	20.0	20.0	19.8	19.6	19.4	19.1	18.7	16.0	10.6	7.0	4.5	3.0	2.2	2.0

18. The minimum temperature shown is that design minimum temperature for which the material is normally suitable without impact testing other than that required by the material specification. However, the use of a material at a design minimum temperature below −29°C (−20°F) is established by rules elsewhere in the code, including any necessary impact-test requirements.

19. These steels are intended for use at high temperatures; however, they may have low ductility and/or low impact properties at room temperature after being used above the temperature indicated by the single bar (ǀ).

20. For pipe sizes NPS 8 and larger and for wall thicknesses of Schedule 140 or heavier, the minimum specification tensile strength is 483 MPa (70.0 kip/in²).

21. There are restrictions on the use of this material in the text of the code.

22. A single bar (ǀ) in these stress tables indicates that there are conditions other than stress which affect usage above or below the temperature as described in other referenced notes. A double bar (ǁ) after a tabled stress indicates that use of the material is prohibited above that temperature.

23. See ANSI B31.3 for a description of P-number groupings.

24. This material when used below −29°C (−20°F) requires impact testing if the carbon content is above 0.10 percent.

25. This is a product specification. No design stresses are necessary. Limitations on metal temperature for materials covered by this specification are:

	°C	°F
Grades 1 and 2	−29 to 480	−20 to 900
Grade 2H	−45 to 595	−50 to 1100
Grade 3	−29 to 595	−20 to 1100
Grade 4	−100 to 595	−150 to 1100
Grade 6	−29 to 425	−20 to 800
Grade 8FA (see Note 24)	−29 to 425	−20 to 800
Grades 8MA and 8TA	−198 to 815	−325 to 1500
Grades 8A and 8CA	−254 to 815	−425 to 1500

26. For use in code piping at the stated allowable stresses, the required minimum tensile and yield properties must be verified by tensile test at the mill. If such tests are not mandatory in the ASTM specification, they shall be specified in the purchase order.

27. After use above the temperature indicated by a single bar (ǀ), use at a lower temperature shall be based on the stress values allowed for the annealed condition of the material.

TABLE 10-49 Allowable Stresses in Tension for Materials (4, 13, 28) (Concluded)

28. The *SE* values in Table 10-49 are equal to the basic allowable stresses in tension *S* multiplied by a quality factor *E* (see subsection "Pressure Design of Metallic Components: Wall Thickness"). The design stress values for bolting materials are equal to the basic allowable stresses *S*. The stress values in shear shall be 0.80 times the allowable stresses in tension derived from tabulated values in Table 10-49 adjusted when applicable in accordance with Note 13. Stress values in bearing shall be twice those in shear.

29. Yield strengths listed are not included in ASTM specifications. The value shown is based on yield strengths of materials with similar characteristics.

30. The letter *a* indicates alloys which are not recommended for welding and which, if welded, must be individually qualified. The letter *b* indicates copper-base alloys which must be individually qualified.

31. These stress values are established from a consideration of strength only and will be satisfactory for average service. For bolted joints when freedom from leakage over a long period of time without retightening is required, lower stress values may be necessary as determined from the flexibility of the flange and bolts and corresponding relaxation properties.

32. For temperatures above 538°C (1000°F), these stress values apply only when the carbon content is 0.04 percent or higher.

33. For use at temperatures below −29 through −45°C (−20 through −50°F) this material must be quenched and tempered.

34. The stress values given for this material are not applicable when either welding or thermal cutting is employed.

35. For stress-relieved tempers (T351, T3510, T3511, T451, T4510, T4511, T651, T6510, T6511) stress values for material in the listed temper shall be used.

36. The maximum operating temperature is arbitrarily set at 260°C (500°F) because harder temper adversely affects design stress in the creep-rupture-temperature ranges.

37. Pipe produced to this specification is not intended for high-temperature service. The stress values apply to either nonexpanded or cold-expanded material in the as-rolled, normalized, or normalized and tempered condition.

38. Special P numbers SP-1, SP-2, and SP-3 of carbon steels are not included in P No. 1 because of a possible high-carbon-high-manganese combination which would require special consideration in qualification. Qualification of any high-carbon-high-manganese grade may be extended to other grades in its group.

39. Annealed at approximately 1150°C (2100°F).

40. If no welding is employed in the fabrication of piping from these materials, the allowable stress values may be increased to 230 MPa (33.3 kip/in²).

41. For all design temperatures, the maximum hardness shall be Rockwell C35 immediately under the thread roots. The hardness shall be taken on a flat area at least 3 mm (⅛ in) across, prepared by removing threads. No more material than necessary shall be removed to prepare the area. Hardness determination shall be made at the same frequency as tensile tests.

42. The minimum tensile strength of the reduced section tensile specimen in accordance with QW-462.1 of ASME Code Sec. IX shall not be less than 758 MPa (110.0 kip/in²).

43. Copper-silicon alloys are not always suitable when exposed to certain media and high temperature, particularly above 100°C (212°F). Users should satisfy themselves that the alloy selected is satisfactory for the service for which it is to be used.

*Table 10-49 and notes have been extracted from the Chemical Plant and Petroleum Refinery Piping Code, ANSI B31.3–1980, with permission of the publisher, the American Society of Mechanical Engineers, New York.

TABLE 10-50 Values of Coefficient *Y* When *t* Is Less Than *D/6**

Materials	Temperature, °C (°F)					
	485 (900) and lower	510 (950)	540 (1000)	560 (1050)	595 (1100)	620 (1150) and higher
Ferritic steels	0.4	0.5	0.7	0.7	0.7	0.7
Austenitic steels	0.4	0.4	0.4	0.4	0.5	0.7
Other ductile metals	0.4	0.4	0.4	0.4	0.4	0.4
Cast iron	0.0					

*Extracted from ANSI B31.3—1980, with permission of the publisher, the American Society of Mechanical Engineers, New York.

TABLE 10-51 Stress-Range Reduction Factors *f**

Cycles, number	Factor, *f*
7000 and less	1.0
7000–14,000	0.9
14,000–22,000	0.8
22,000–45,000	0.7
45,000–100,000	0.6
Over 100,000	0.5

*Extracted from ANSI B31.3—1980, with permission of the publisher, the American Society of Mechanical Engineers, New York.

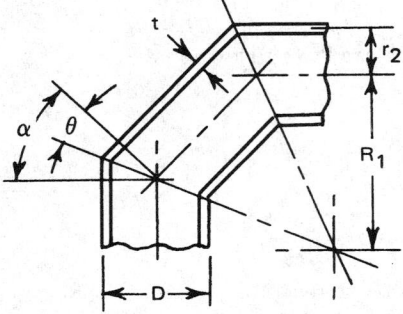

FIG. 10-165 Nomenclature for miter bends. (*Extracted from the Chemical Plant and Petroleum Refinery Code, ANSI B31.3—1976, with permission of the publisher, the American Society of Mechanical Engineers, New York.*)

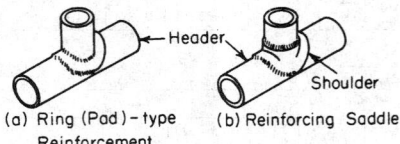
(a) Ring (Pad)-type Reinforcement   (b) Reinforcing Saddle

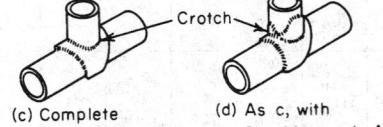
(c) Complete Encirclement Pad   (d) As c, with Shoulder-pads Added

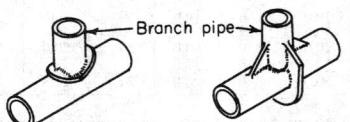
(f) Reinforcing Collar   (e) Horseshoe- and-Gusset type Reinforcement

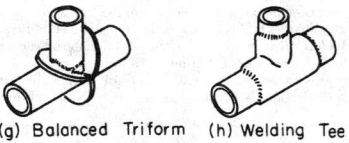
(g) Balanced Triform   (h) Welding Tee

FIG. 10-166 Types of reinforcement for branch connections. (*From Kellogg, Design of Piping Systems, Wiley, New York, 1965.*)

2. Local reduction in size or wall thickness or local use of a material having reduced yield strength (for example, girth welds of substantially lower strength than the base metal)

3. A line configuration in a system of uniform size in which expansion or contraction must be absorbed largely in a short offset from the major portion of the run

If unbalanced layouts cannot be avoided, appropriate analytical methods must be applied to assure adequate flexibility. If the designer determines that a piping system does not have adequate inherent flexibility, additional flexibility may be provided by adding bends, loops, offsets, swivel joints, corrugated pipe, expansion joints of the bellows or slip-joint type, or other devices. Suitable anchoring must be provided.

As contrasted with stress from sustained loads such as internal pressure or weight, displacement stresses may be permitted to cause limited overstrain in various portions of a piping system. When the system is operated initially at its greatest displacement condition, any yielding reduces stress. When the system is returned to its original condition, there occurs a redistribution of stresses which is referred to as self-springing. It is similar to cold springing in its effects.

While stresses resulting from thermal strain tend to diminish with time, the algebraic difference in displacement condition and in either the original (as-installed) condition or any anticipated condition with a greater opposite effect than the extreme displacement condition remains substantially constant during any one cycle of operation. This difference is defined as the displacement-stress range, and it is a determining factor in the design of piping for flexibility. See Eqs. (10-93) and (10-94) for the allowable stress range S_A and Eq. (10-100) for the computed stress range S_E.

Cold Spring   Cold spring is the intentional deformation of piping during assembly to produce a desired initial displacement and stress. For pipe operating at a temperature higher than that at which it was installed, cold spring is accomplished by fabricating it slightly shorter than design length. Cold spring is beneficial in that it serves to balance the magnitude of stress under initial and extreme displacement conditions. When cold spring is properly applied, there is less likelihood of overstrain during initial operation; hence, it is recommended especially for piping materials of limited ductility. There is also less deviation from as-installed dimensions during initial operation, so that hangers will not be displaced as far from their original settings.

Inasmuch as the service life of a system is affected more by the range of stress variation than by the magnitude of stress at a given time, no credit for cold spring is permitted in stress-range calculations. However, in calculating the thrusts and moments when actual reactions as well as their range of variations are significant, credit is given for cold spring.

Values of thermal-expansion coefficients to be used in determining total displacement strains for computing the stress range are determined from Table 10-52 as the algebraic difference between the value at design maximum temperature and that at the design minimum temperature for the thermal cycle under analysis.

Values for Reactions   Values of thermal displacements to be used in determining total displacement strains for the computation of reactions on supports and connected equipment shall be determined as the algebraic difference between the value at design maximum (or minimum) temperature for the thermal cycle under analysis and the value at the temperature expected during installation.

The as-installed and maximum or minimum moduli of elasticity, E_a and E_m respectively, shall be taken as the values shown in Table 10-53. Poisson's ratio may be taken as 0.3 at all temperatures for all metals.

The allowable stress range for displacement stresses S_A and permissible additive stresses shall be as specified in Eqs. (10-93) and (10-94) for systems primarily stressed in bending and/or torsion. For pipe or piping components containing longitudinal welds the basic allowable stress S may be used to determine S_A. (See Table 10-49, Note 13.)

Nominal thicknesses and outside diameters of pipe and fittings shall be used in flexibility calculations.

In the absence of more directly applicable data, the flexibility factor k and stress-intensification factor i shown in Table 10-54 may be used in flexibility calculations in Eq. (10-101). For piping components or attachments (such as valves, strainers, anchor rings, and bands) not covered in the table, suitable stress-intensification factors may be assumed by comparison of their significant geometry with that of the components shown.

Requirements for Analysis   No formal analysis of adequate flexibility is required in systems which (1) are duplicates of successfully operating installations or replacements without significant change of systems with a satisfactory service record; (2) can readily be judged

adequate by comparison with previously analyzed systems; or (3) are of uniform size, have no more than two points of fixation, have no intermediate restraints, and fall within the limitations of empirical Eq. (10-99):[*]

$$\frac{Dy}{(L-U)^2} \le K_1 \qquad (10\text{-}99)$$

where D = outside diameter of pipe, in (mm)
   y = resultant of total displacement strains, in (mm), to be absorbed by the piping system
   L = developed length of piping between anchors, ft (m)
   U = anchor distance, straight line between anchors, ft (m)
   K_1 = 0.03 for U.S. customary units listed
     = 208.3 for SI units listed in parentheses

1.   All systems not meeting these criteria shall be analyzed by simplified, approximate, or comprehensive methods of analysis appropriate for the specific case.

2.   Approximate or simplified methods may be applied only if they are used in the range of configurations for which their adequacy has been demonstrated.

3.   Acceptable comprehensive methods of analysis include analytical and chart methods which provide an evaluation of the forces, moments, and stresses caused by displacement strains.

4.   Comprehensive analysis shall take into account stress-intensification factors for any component other than straight pipe. Credit may be taken for the extra flexibility of such a component.

In calculating the flexibility of a piping system between anchor points, the system shall be treated as a whole. The significance of all parts of the line and of all restraints introduced for the purpose of reducing moments and forces on equipment or small branch lines and also the restraint introduced by support friction shall be recognized. Consider all displacements over the temperature range defined by operating and shutdown conditions.

Flexibility Stresses   Bending and torsional stresses shall be computed using the as-installed modulus of elasticity E_a and then combined in accordance with Eq. (10-100) to determine the computed displacement stress range S_E, which shall not exceed the allowable stress range S_A [Eqs. (10-93) and (10-94).]

$$S_E = \sqrt{S_b^2 + 4S_t^2} \qquad (10\text{-}100)$$

where S_b = resultant bending stress, lbf/in² (MPa)
   $S_t = M_t/2Z$ = torsional stress, lbf/in² (MPa)
   M_t = torsional moment, in·lbf (N·mm)
   Z = section modulus of pipe, in³ (mm³)

The resultant bending stresses S_b to be used in Eq. (10-100) for elbows and miter bends shall be calculated in accordance with Eq. (10-101), with moments as shown in Fig. (10-167):

$$S_b = \frac{\sqrt{(i_i M_i)^2 + (i_o M_o)^2}}{Z} \qquad (10\text{-}101)$$

where S_b = resultant bending stress, lbf/in² (MPa)
   i_i = in-plane stress-intensification factor from Table 10-54
   i_o = out-plane stress-intensification factor from Table 10-54
   M_i = in-plane bending moment, in·lbf (N·mm)
   M_o = out-plane bending moment, in·lbf (N·mm)
   Z = section modulus of pipe, in³ (mm³)

The resultant bending stresses S_b to be used in Eq. (10-100) for branch connections shall be calculated in accordance with Eqs. (10-102) and (10-103), with moments as shown in Fig. 10-168.

[*] WARNING: No general proof can be offered that this equation will yield accurate or consistently conservative results. It is not applicable to systems used under severe cyclic conditions. It should be used with caution in configurations such as unequal leg U bends ($L/U > 2.5$) or near-straight sawtooth runs, or for large thin-wall pipe ($i \ge 5$), or when extraneous displacements (not in the direction connecting anchor points) constitute a large part of the total displacement. There is no assurance that terminal reactions will be acceptably low even if a piping system falls within the limitations of Eq. (10-99).

TABLE 10-52 Thermal-Expansion Coefficients, U.S. Customary Units, for Metals*

Mean coefficient of linear thermal expansion between 70°F and indicated temperature, μin/(in·°F)

Temperature, °F	Carbon steel, carbon-molybdenum low-chromium (through 3 Cr Mo)	5 Cr Mo through 9 Cr Mo	Austenitic stainless steels, 18 Cr, 8 Ni	12 Cr 17 Cr 27 Cr	25 Cr, 20 Ni	Monel 67 Ni, 30 Cu	3½ Nickel	Aluminum	Gray cast iron	Bronze	Brass	70 Cu, 30 Ni	Ni-Fe-Cr	Ni-Cr-Fe	Ductile iron
-325	5.00	4.70	8.15	4.30		5.55	4.76	9.90		8.40	8.20	6.65			
-300	5.07	4.77	8.21	4.36		5.72	4.90	10.04		8.45	8.24	6.76			
-275	5.14	4.84	8.28	4.41		5.89	5.01	10.18		8.50	8.29	6.86			
-250	5.21	4.91	8.34	4.47		6.06	5.15	10.33		8.55	8.33	6.97			
-225	5.28	4.98	8.41	4.53		6.23	5.30	10.47		8.60	8.37	7.08			
-200	5.35	5.05	8.47	4.59		6.40	5.45	10.61		8.65	8.41	7.19			4.65
-175	5.42	5.12	8.54	4.64		6.57	5.52	10.76		8.70	8.46	7.29			4.76
-150	5.50	5.20	8.60	4.70		6.75	5.59	10.90		8.75	8.50	7.40			4.87
-125	5.57	5.26	8.66	4.78		6.85	5.67	11.08		8.85	8.61	7.50			4.98
-100	5.65	5.32	8.75	4.85		6.95	5.78	11.25		8.95	8.73	7.60			5.10
-75	5.72	5.38	8.83	4.93		7.05	5.83	11.43		9.05	8.84	7.70			5.20
-50	5.80	5.45	8.90	5.00		7.15	5.88	11.60		9.15	8.95	7.80			5.30
-25	5.85	5.51	8.94	5.05		7.22	5.94	11.73		9.23	9.03	7.87			5.40
0	5.90	5.56	8.98	5.10		7.28	6.00	11.86		9.32	9.11	7.94			5.50
25	5.96	5.62	9.03	5.14		7.35	6.08	11.99		9.40	9.18	8.02			5.58
50	6.01	5.67	9.07	5.19		7.41	6.16	12.12		9.49	9.26	8.09			5.66
70	6.07	5.73	9.11	5.24		7.48	6.25	12.25		9.57	9.34	8.16		7.13	5.74
100	6.13	5.79	9.16	5.29		7.55	6.33	12.39		9.66	9.42	8.24		7.20	5.82
125	6.19	5.85	9.20	5.34		7.62	6.36	12.53		9.75	9.51	8.31		7.25	5.87
150	6.25	5.92	9.25	5.40		7.70	6.39	12.67		9.85	9.59	8.39		7.30	5.92
175	6.31	5.98	9.29	5.45		7.77	6.42	12.81		9.93	9.68	8.46		7.35	5.97
200	6.38	6.04	9.34	5.50	8.79	7.84	6.45	12.95	5.75	10.03	9.76	8.54	7.90	7.40	6.02
225	6.43	6.08	9.37	5.54	8.81	7.89	6.50	13.03	5.80	10.05	9.82	8.58	8.01	7.44	6.08
250	6.49	6.12	9.41	5.58	8.83	7.93	6.55	13.12	5.84	10.08	9.88	8.63	8.12	7.48	6.14
275	6.54	6.15	9.44	5.62	8.85	7.98	6.60	13.20	5.89	10.10	9.94	8.67	8.24	7.52	6.20
300	6.60	6.19	9.47	5.66	8.87	8.02	6.65	13.28	5.93	10.12	10.00	8.71	8.35	7.56	6.25
325	6.65	6.23	9.50	5.70	8.89	8.07	6.69	13.36	5.97	10.15	10.06	8.76	8.46	7.60	6.31
350	6.71	6.27	9.53	5.74	8.90	8.11	6.73	13.44	6.02	10.18	10.11	8.81	8.57	7.63	6.37
375	6.76	6.30	9.56	5.77	8.91	8.16	6.77	13.52	6.06	10.20	10.17	8.85	8.69	7.67	6.43
400	6.82	6.34	9.59	5.81	8.92	8.20	6.80	13.60	6.10	10.23	10.23	8.90	8.80	7.70	6.48
425	6.87	6.38	9.62	5.85	8.92	8.25	6.83	13.68	6.15	10.25	10.29		8.82	7.72	6.57
450	6.92	6.42	9.65	5.89	8.92	8.30	6.86	13.75	6.19	10.28	10.35		8.85	7.75	6.66
475	6.97	6.46	9.67	5.92	8.92	8.35	6.89	13.83	6.24	10.30	10.41		8.87	7.77	6.75
500	7.02	6.50	9.70	5.96	8.93	8.40	6.93	13.90	6.28	10.32	10.47		8.90	7.80	6.85
525	7.07	6.54	9.73	6.00	8.93	8.45	6.97	13.98	6.33	10.35	10.53		8.92	7.82	6.88
550	7.12	6.58	9.76	6.05	8.93	8.49	7.01	14.05	6.38	10.38	10.58		8.95	7.85	6.92
575	7.17	6.62	9.79	6.09	8.93	8.54	7.04	14.13	6.42	10.41	10.64		8.97	7.88	6.95
600	7.23	6.66	9.82	6.13	8.94	8.58	7.08	14.20	6.47	10.44	10.69		9.00	7.90	6.98
625	7.28	6.70	9.85	6.17	8.94	8.63	7.12		6.52	10.46	10.75		9.02	7.92	7.02
650	7.33	6.73	9.87	6.20	8.95	8.68	7.16		6.56	10.48	10.81		9.05	7.95	7.04

Temp													
675	7.38	6.77	9.90	6.23	8.95	8.73	7.19	6.61	10.50	10.86	9.07	7.98	7.08
700	7.44	6.80	9.92	6.26	8.96	8.78	7.22	6.65	10.52	10.92	9.10	8.00	7.11
725	7.49	6.84	9.95	6.29	8.96	8.83	7.25	6.70	10.55	10.98	9.12	8.02	7.14
750	7.54	6.88	9.99	6.33	8.96	8.87	7.29	6.74	10.57	11.04	9.15	8.05	7.18
775	7.59	6.92	10.02	6.36	8.96	8.92	7.31	6.79	10.60	11.10	9.17	8.08	7.22
800	7.65	6.96	10.05	6.39	8.97	8.96	7.34	6.83	10.62	11.16	9.20	8.10	7.25
825	7.70	7.00	10.08	6.42	8.97	9.01	7.37	6.87	10.65	11.22	9.22		7.27
850	7.75	7.03	10.11	6.46	8.98	9.06	7.40	6.92	10.67	11.28	9.25		7.31
875	7.79	7.07	10.13	6.49	8.99	9.11	7.43	6.96	10.70	11.34	9.27		7.34
900	7.84	7.10	10.16	6.52	9.00	9.16	7.45	7.00	10.72	11.40	9.30		7.37
925	7.87	7.13	10.19	6.55	9.05	9.21	7.47	7.05	10.74	11.46	9.32		7.41
950	7.91	7.16	10.23	6.58	9.10	9.25	7.49	7.10	10.76	11.52	9.35		7.44
975	7.94	7.19	10.26	6.60	9.15	9.30	7.52	7.14	10.78	11.57	9.37		7.47
1000	7.97	7.22	10.29	6.63	9.18	9.34	7.55	7.19	10.80	11.63	9.40		7.50
1025	8.01	7.25	10.32	6.65	9.20	9.39			10.83	11.69	9.42		
1050	8.05	7.27	10.34	6.68	9.22	9.43			10.85	11.74	9.45		
1075	8.08	7.30	10.37	6.70	9.24	9.48			10.88	11.80	9.47		
1100	8.12	7.32	10.39	6.72	9.25	9.52			10.90	11.85	9.50		
1125	8.14	7.34	10.41	6.74	9.29	9.57			10.93	11.91	9.52		
1150	8.16	7.37	10.44	6.75	9.33	9.61			10.95	11.97	9.55		
1175	8.17	7.39	10.46	6.77	9.36	9.66			10.98	12.03	9.57		
1200	8.19	7.41	10.48	6.78	9.39	9.70			11.00	12.09	9.60		
1225	8.21	7.43	10.50	6.80	9.43	9.75					9.64		
1250	8.24	7.45	10.51	6.82	9.47	9.79					9.68		
1275	8.26	7.47	10.53	6.83	9.50	9.84					9.71		
1300	8.28	7.49	10.54	6.85	9.53	9.88					9.75		
1325	8.30	7.51	10.56	6.86	9.53	9.92					9.79		
1350	8.32	7.52	10.57	6.88	9.54	9.96					9.83		
1375	8.34	7.54	10.59	6.89	9.55	10.00					9.86		
1400	8.36	7.55	10.60	6.90	9.56	10.04					9.90		
1425			10.64								9.94		
1450			10.68								9.98		
1475			10.72								10.01		
1500			10.77								10.05		

*Extracted from the Chemical Plant and Petroleum Refinery Piping Code, ANSI B31.3—1980, with permission of the publisher, the American Society of Mechanical Engineers, New York. These data are for information, and it is not implied that materials are suitable for all the temperatures shown. (°F − 32)5/9 = °C; to convert microinches per inch-degree Fahrenheit to meters per meter-degree Kelvin, multiply by 1.8.

TABLE 10-53 Modulus of Elasticity, U.S. Customary Units, for Metals*

Material	E = Modulus of elasticity, lbf/in² (multiply tabulated values by 10⁶)																	
	Temperature, °F																	
	−325	−200	−100	70	200	300	400	500	600	700	800	900	1000	1100	1200	1300	1400	1500
Modulus of elasticity: ferrous materials																		
Carbon steels with carbon content 0.30 percent or less, 3½ Ni	30.0	29.5	29.0	27.9	27.7	27.4	27.0	26.4	25.7	24.8	23.4	18.5	15.4	13.0				
Carbon steels with carbon content above 0.30 percent	31.0	30.6	30.4	29.9	29.5	29.0	28.3	27.4	26.7	25.4	23.8	21.5	18.8	15.0	11.2			
Carbon-molydenum steels, low-chromium steels through 3 Cr Mo	31.0	30.6	30.4	29.9	29.5	29.0	28.6	28.0	27.4	26.6	25.7	24.5	23.0	20.4	15.6			
Intermediate-chromium steels (5 Cr Mo through 9 Cr Mo)	29.4	28.5	28.1	27.4	27.1	26.8	26.4	26.0	25.4	24.9	24.2	23.5	22.8	21.9	20.8	19.5	18.1	
Austenitic steels (TP304, 310, 316, 321, 347)	30.4	29.9	29.4	28.3	27.7	27.1	26.6	26.1	25.4	24.8	24.1	23.4	22.7	22.0	21.3	20.7	19.3	17.9
Straight chromium steels (12 Cr, 17 Cr, 27 Cr)	30.8	30.3	29.8	29.2	28.7	28.3	27.7	27.0	26.0	24.8	23.1	21.1	18.6	15.6	12.2			
Gray cast iron				13.4	13.2	12.9	12.6	12.2	11.7	11.0	10.2							

Material	E = Modulus of elasticity, lbf/in² (multiply tabulated values by 10⁶)															
	Temperature, °F															
	−325	−200	−100	70	100	200	300	400	500	600	700	800	900	1000	1100	1200
Modulus of elasticity: nonferrous materials																
Monel (67 Ni, 30 Cu) and (66 Ni, 29 Cu—Al)	26.8	26.6	26.4	26.0	26.0	26.0	25.8	25.6	25.4	24.7	23.1	21.0	18.6	16.0	14.3	13.0
Copper-nickel (70 Cu, 30 Ni)				21.6	21.5	21.2	20.9	20.6	20.3	20.0	19.7	19.4				
Aluminum alloys	11.3	10.9	10.6	10.1	10.0	9.8	9.5	8.7	7.7							
Copper (99.98 percent Cu)	17.0	16.7	16.5	16.0	15.8	15.6	15.4	15.1	14.7	14.2	13.7					
Commercial brass (66 Cu, 34 Zn)	15.0	14.7	14.5	14.0	13.9	13.7	13.5	13.0	12.7	12.2	11.8					
Leaded tin bronze (88 Cu, 6 Sn, 1.5 Pb, 4.5 Zn)	14.2	13.8	13.5	13.0	12.9	12.7	12.4	12.0	11.7	11.3	10.9					

*Extracted from the Chemical Plant and Petroleum Refinery Piping Code, ANSI B31.3—1980, with permission of the publisher, the American Society of Mechanical Engineers, New York. These data are for information, and it is not implied that materials are suitable for all the temperatures shown. (°F − 32)⅝ = °C; to convert pounds-force per square inch to megapascals, multiply by 0.006895.

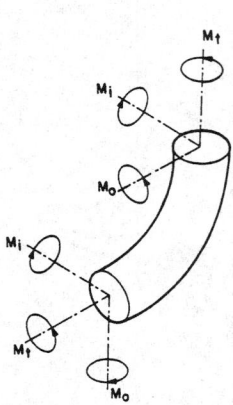

FIG. 10-167 Moments in bends. (*Extracted from the Chemical Plant and Petroleum Refinery Piping Code, ANSI B31.3—1976, with permission of the publisher, the American Society of Mechanical Engineers, New York.*)

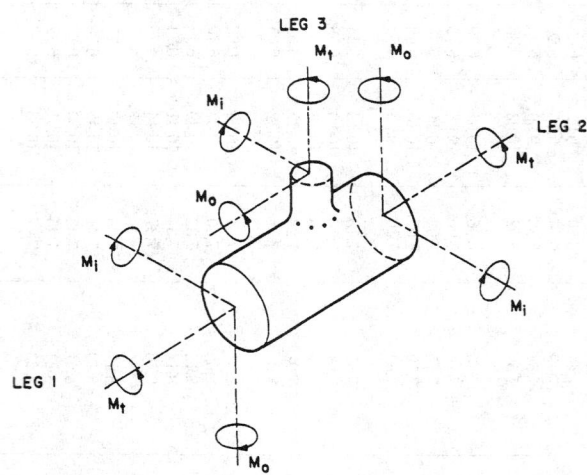

FIG. 10-168 Moments in branch connections. (*Extracted from the Chemical Plant and Petroleum Refiner Piping Code, ANSI B31.3—1976, with permission of the publisher, the American Society of Mechanical Engineers, New York.*)

TABLE 10-54   Flexibility Factor _k_ and Stress-Intensification Factor _i_ *

| Description | Stress intensification factor[a,h] | | | Flexibility characteristic h | Sketch |
	Flexibility factor k	Out-plane, i_o	In-plane, i_i		
Welding elbow[a,b,c,f,i] or pipe bend	$\dfrac{1.65}{h}$	$\dfrac{0.75}{h^{2/3}}$	$\dfrac{0.9}{h^{2/3}}$	$\dfrac{\overline{T}R_1}{(r_2)^2}$	
Closely spaced miter bend[a,b,c] $s < r_2\,(1+\tan\theta)$	$\dfrac{1.52}{h^{5/6}}$	$\dfrac{0.9}{h^{2/3}}$	$\dfrac{0.9}{h^{2/3}}$	$\dfrac{\cot\theta}{2}\dfrac{\overline{T}_s}{(r_2)^2}$	
Single miter bend[a,b] or widely spaced miter bend $s \geq r_2\,(1+\tan\theta)$	$\dfrac{1.52}{h^{5/6}}$	$\dfrac{0.9}{h^{2/3}}$	$\dfrac{0.9}{h^{2/3}}$	$\dfrac{1+\cot\theta}{2}\dfrac{\overline{T}}{r_2}$	
Welding tee[a,b,f] per ANSI B16.9 with $r_x > \tfrac{1}{8} D_b$ $T_c \geq 1.5\,\overline{T}$	1	$\dfrac{0.9}{h^{2/3}}$	$\tfrac{3}{4}\,i_o + \tfrac{1}{4}$	$4.4\,\dfrac{\overline{T}}{r_2}$	
Reinforced fabricated[a,b,e] tee with pad or saddle	1	$\dfrac{0.9}{h^{2.3}}$	$\tfrac{3}{4}\,i_o + \tfrac{1}{4}$	$\dfrac{(\overline{T} + 1/2\,t_r)^{5/2}}{\overline{T}^{3/2}\,r_2}$	
Unreinforced[a,b] fabricated tee	1	$\dfrac{0.9}{h^{2/3}}$	$\tfrac{3}{4}\,i_o + \tfrac{1}{4}$	$\dfrac{\overline{T}}{r_2}$	
Extruded[a,b] welding tee $T_c < 1.5\,\overline{T}$	1	$\dfrac{0.9}{h^{2/3}}$	$\tfrac{3}{4}\,i_o + \tfrac{1}{4}$	$\left(1 + \dfrac{r_x}{r_2}\right)\dfrac{\overline{T}}{r_2}$	
Welded-in[a,b] contour insert $r_x \geq \tfrac{1}{8} D_b$ $T_c \geq 1.5\,\overline{T}$	1	$\dfrac{0.9}{h^{2/3}}$	$\tfrac{3}{4}\,i_o + \tfrac{1}{4}$	$4.4\,\dfrac{\overline{T}}{r_2}$	
Branch[a,b,g] welded-on fitting (integrally reinforced)	1	$\dfrac{0.9}{h^{2/3}}$	$\dfrac{0.9}{h^{2/3}}$	$3.3\,\dfrac{\overline{T}}{r_2}$	

TABLE 10-54   Flexibility Factor k and Stress-Intensification Factor i (*Concluded*)

Description	Flexibility factor k	Stress-intensification factor i
Butt-welded joint, reducer, or weld-neck flange	1	1.0
Double-welded slip-on flange	1	1.2
Fillet welded joint or pocket-weld flange	1	1.3
Lap-joint flange (with ANSI B16.9 lap-joint stub)	1	1.6
Screwed pipe joint or screwed flange	1	2.3
Corrugated straight pipe or corrugated or creased bend[d]	5	2.5

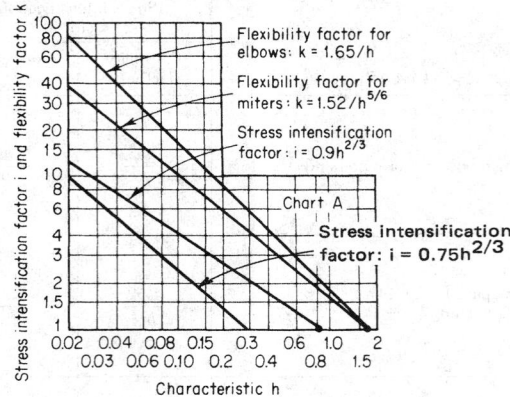

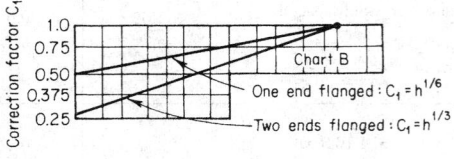

[a]The flexibility factor k applies to bending in any plane. The flexibility factors k and stress intensification factors i shall not be less than unity; factors for torsion equal unity. Both factors apply over the effective arc length (shown by heavy centerlines in the sketches) for curved and miter bends and to the intersection point for tees.

[b]The values of k and i can be read directly from Chart A by entering with the characteristic h computed from the formulas given above. Nomenclature is as follows:

T = for elbows and miter bends, the nominal wall thickness of the fitting, in (mm)

= for tees, the nominal wall thickness of the matching pipe, in (mm)

$\overline{T}_c$ = the crotch thickness of tees, in (mm)

t_r = pad or saddle thickness, in (mm)

θ = one-half angle between adjacent miter axes, °

r_2 = mean radius of matching pipe, in (mm)

R_1 = bend radius of welding elbow or pipe bend, in (mm)

r_x = radius of curvature of external contoured portion of outlet, in, measured in the plane containing the axes of the run and branch.

s = miter spacing at centerline, in (mm)

D_b = outside diameter of branch, in (mm)

[c]When flanges are attached to one or both ends, the values of k and i shall be corrected by the factors C_1, which can be read directly from Chart B, entering with the computed h.

[d]Factors shown apply to bending. Flexibility factor for torsion equals 0.9.

[e]When t_r is $> 1\frac{1}{2}\,T$, use $h = 4(\overline{T}/r_2)$.

[f]Designers are cautioned that cast butt-welded fittings may have considerably heavier walls than that of the pipe with which they are used. Large errors may be introduced unless the effect of these greater thicknesses is considered.

[g]Designers must assure themselves that this fabrication has a pressure rating equivalent to that of straight pipe.

[h]A single intensification factor equal to $0.9/h^{2/3}$ may be used for both i_i and i_o if desired.

[i]In large-diameter thin-wall elbows and bends, pressure can significantly affect the magnitudes of k and i. To correct values from the table,

$$\text{Divide } k \text{ by } \left[1 + 6\left(\frac{P}{E_c}\right)\left(\frac{r_2}{t}\right)^{7/3}\left(\frac{R_1}{r_2}\right)^{1/3}\right] \qquad \text{Divide } i \text{ by } \left[1 + 3.25\left(\frac{P}{E_c}\right)\left(\frac{r_2}{t}\right)^{5/2}\left(\frac{R_1}{r_2}\right)^{2/3}\right] \qquad (10\text{-}107)$$

*Extracted from the Chemical Plant and Petroleum Refinery Piping Code, ANSI B31.3—1980, with permission of the publisher, the American Society of Mechanical Engineers, New York.

For header (legs 1 and 2):

$$S_b = \frac{\sqrt{(i_i m_i)^2 + (i_o M_o)^2}}{Z} \qquad (10\text{-}102)$$

For branch (leg 3):

$$S_b = \frac{\sqrt{(i_i m_i)^2 + (i_o M_o)^2}}{Z_e} \qquad (10\text{-}103)$$

where S_b = resultant bending stress, lbf/in^2 (MPa)

Z_e = effective section modulus for branch, in^3 (mm^3)

$$Z_e = \pi\, r_2^2 T_s \qquad (10\text{-}104)$$

r_2 = mean branch cross-sectional radius, in (mm)

T_s = effective branch wall thickness, in (mm) [lesser of $\overline{T}_h$ and $(i_o)(\overline{T}_b)$]

$\overline{T}_h$ = thickness of pipe matching run of tee or header exclusive of reinforcing elements, in (mm)

$\overline{T}_b$ = thickness of pipe matching branch, in (mm)

i_o = out-plane stress-intensification factor (Table 10-54)

i_i = in-plane stress-intensification factor (Table 10-54)

Allowable stress range S_A and permissible additive stresses shall be computed in accordance with Eqs. (10-93) and (10-94).

Required Weld Quality Assurance   Any weld at which S_E exceeds $0.8\,S_A$ for any portion of a piping system and the equivalent number of cycles N exceeds 7000 shall be fully examined in accordance with the requirements for severe cyclic service (presented later in this section).

Reactions: Metallic Piping   Reaction forces and moments to be used in the design of restraints and supports and in evaluating the effects of piping displacements on connected equipment shall be based on the reaction range R for the extreme displacement conditions, considering the range previously defined for reactions and using E_a. The designer shall consider instantaneous maximum values of forces and moments in the original and extreme displacement conditions as well as the reaction range in making these evaluations.

Maximum Reactions for Simple Systems   For two-anchor systems without intermediate restraints, the maximum instantaneous values of reaction forces and moments may be estimated from Eqs. (10-105) and (10-106).

1. *For extreme displacement conditions, R_m.* The temperature for this computation is the design maximum or design minimum temperature as previously defined for reactions, whichever produces the larger reaction:

$$R_m = R\left(1 - \frac{2C}{3}\right)\frac{E_m}{E_a} \qquad (10\text{-}105)$$

where C = cold-spring factor varying from zero for no cold spring to 1.0 for 100 percent cold spring. (The factor $\frac{2}{3}$ is based on experience, which shows that specified cold spring cannot be fully assured even with elaborate precautions.)

E_a = modulus of elasticity at installation temperature, lbf/in² (MPa)

E_m = modulus of elasticity at design maximum or design minimum temperature, lbf/in² (MPa)

R = range of reaction forces or moments (derived from flexibility analysis) corresponding to the full displacement-stress range and based on E_a, lbf or in·lbf (N or N·mm)

R_m = estimated instantaneous maximum reaction force or moment at design maximum or design minimum temperature, lbf or in·lbf (N or N·mm)

2. *For original condition, R_a.* The temperature for this computation is the expected temperature at which the piping is to be assembled.

$$R_a = CR \quad \text{or} \quad C_1 R, \text{ whichever is greater} \qquad (10\text{-}106)$$

where nomenclature is as for Eq. (10-105) and

$C_1 = 1 - (S_h E_a / S_E E_m)$

= estimated self-spring or relaxation factor (use zero if value of C_1 is negative)

R_a = estimated instantaneous reaction force or moment at installation temperature, lbf or in·lbf (N or N·mm)

S_E = computed displacement-stress range, lbf/in² (MPa). See Eq. (10-100).

S_h = See Eq. (10-93).

Maximum Reactions for Complex Systems   For multianchor systems and for two-anchor systems with intermediate restraints, Eqs. (10-105) and (10-106) are not applicable. Each case must be studied to estimate the location, nature, and extent of local overstrain and its effect on stress distribution and reactions.

Acceptable comprehensive methods of analysis are analytical, model-test, and chart methods, which evaluate for the entire piping system under consideration the forces, moments, and stresses caused by bending and torsion from a simultaneous consideration of terminal and intermediate restraints to thermal expansion and include all external movements transmitted under thermal change to the piping by its terminal and intermediate attachments. Correction factors, as provided by the details of these rules, must be applied for the stress intensification of curved pipe and branch connections and may be applied for the increased flexibility of such component parts.

Brock [in Crocker (ed.), *Piping Handbook,* 5th ed., McGraw-Hill, New York, 1967, sec. 4] provides further data on methods of analysis.

Expansion Joints   All the foregoing applies to "stiff piping systems," i.e., systems without expansion joints (see detail 1 of Fig. 10-169). When space limitations, process requirements, or other considerations result in configurations of insufficient flexibility, capacity for deflection within allowable stress range limits may be increased successively by the use of one or more hinged bellows expansion joints, viz., semirigid (detail 2) and nonrigid (detail 3) systems, and expansion effects essentially eliminated by a free-movement joint (detail 4) system. Expansion joints for semirigid and nonrigid systems are restrained against longitudinal and lateral movement by the hinges with the expansion element under bending movement only and are known as "rotation" or "hinged" joints (see Fig. 10-170). Semirigid systems are limited to one plane; nonrigid systems require a minimum of three joints for two-dimensional and five joints for three-dimensional expansion movement.

Joints similar to that shown in Fig. 10-170, except with two pairs of hinge pins equally spaced around a gimbal ring, achieve similar results with a lesser number of joints.

Expansion joints for free-movement systems can be designed for axial or offset movement alone, or for combined axial and offset movements (see Fig. 10-171). For offset movement alone, the end load due to pressure and weight can be transferred across the joint by tie rods or structural members (see Fig. 10-172). For axial or combined movements, anchors must be provided to absorb the unbalanced pressure load and force bellows to deflect.

Commercial bellows elements are usually light-gauge [of the order of (0.05 to 0.10 in) thick] and are available in stainless and other alloy steels, copper, and other nonferrous materials. Multi-ply bellows, bellows with external reinforcing rings, and toroidal contour bellows are available for higher pressures. Since bellows elements are ordinarily rated for strain ranges which involve repetitive yielding, predictable

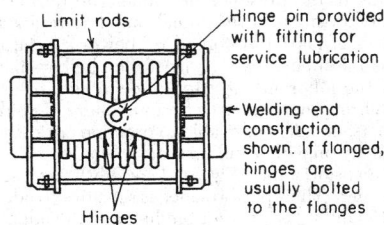

FIG. 10-170   Hinged expansion joint. (*From Kellogg,* Design of Piping Systems, *Wiley, New York, 1965.*)

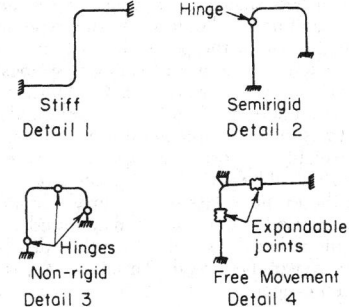

FIG. 10-169   Flexibility classification for piping systems. (*From Kellogg,* Design of Piping Systems, *Wiley, New York, 1965.*)

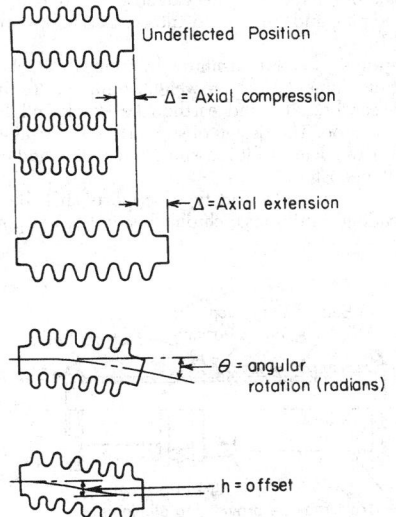

FIG. 10-171   Action of expansion bellows under various movements. (*From Kellogg,* Design of Piping Systems, *Wiley, New York, 1965.*)

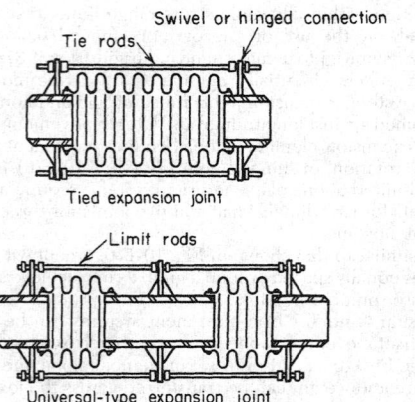

FIG. 10-172 Constrained-bellows expansion joints. (*From Kellogg*, Design of Piping Systems, *Wiley, New York, 1965.*)

performance is assured only by adequate fabrication controls and knowledge of the potential fatigue performance of each design. The attendant cold work can affect corrosion resistance and promote susceptibility to corrosion fatigue or stress corrosion; joints in a horizontal position cannot be drained and have frequently undergone pitting or cracking due to the presence of condensate during operation or offstream. For low-pressure essentially nonhazardous service, non-metallic bellows of fabric-reinforced rubber or special materials are sometimes used. For corrosive service Teflon bellows may be used.

Because of the inherently greater susceptibility of expansion bellows to failure from unexpected corrosion, failure of guides to control joint movements, etc., it is advisable to examine critically their design choice in comparison with a stiff system.

Slip-type expansion joints (Fig. 10-173) substitute packing (ring or plastic) for bellows. Their performance is sensitive to adequate design with respect to guiding to prevent binding and the adequacy of stuffing boxes and attendant packing, sealant, and lubrication. Anchors must be provided for the unbalanced pressure force and for the friction forces to move the joint. The latter can be much higher than the elastic force required to deflect a bellows joint. Rotary packed joints, ball joints, and other special joints can absorb end load.

Corrugated pipe and corrugated and creased bends are also used to decrease stiffness.

Pipe Supports Loads transmitted by piping to attached equipment and supporting elements include weight, temperature- and pressure-induced effects, vibration, wind, earthquake, and shock, and thermal expansion and contraction. The design of supports and restraints is based on concurrently acting loads (if it is assumed that wind and earthquake do not act simultaneously).

Resilient and constant-effort-type supports shall be designed for maximum loading conditions including test unless temporary supports are provided.

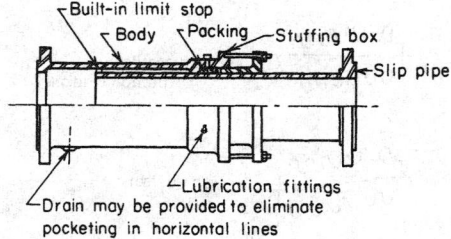

FIG. 10-173 Slip-type expansion joint. (*From Kellogg*, Design of Piping Systems, *Wiley, New York, 1965.*)

Though not specified in the code, supports for discharge piping from relief valves must be adequate to withstand the jet reaction produced by their discharge.

The code states further that pipe-supporting elements shall (1) avoid excessive interference with thermal expansion and contraction of pipe which is otherwise adequately flexible; (2) be such that they do not contribute to leakage at joints or excessive sag in piping requiring drainage; (3) be designed to prevent overstress, resonance, or disengagement due to variation of load with temperature; also, so that combined longitudinal stresses in the piping shall not exceed the code allowable limits; (4) be such that a complete release of the piping load will be prevented in the event of spring failure or misalignment, weight transfer, or added load due to test during erection; (5) be of steel or wrought iron; (6) be of alloy steel or protected from temperature when the temperature limit for carbon steel may be exceeded; (7) not be cast iron except for roller bases, rollers, anchor bases, etc., under mainly compression loading; (8) not be malleable or nodular iron except for pipe clamps, beam clamps, hanger flanges, clips, bases, and swivel rings; (9) not be wood except for supports mainly in compression when the pipe temperature is at or below ambient; and (10) have threads for screw adjustment which shall conform to ANSI B1.1.

A supporting element used as an anchor shall be designed to maintain an essentially fixed position.

To protect terminal equipment or other (weaker) portions of the system, restraints (such as anchors and guides) shall be provided where necessary to control movement or to direct expansion into those portions of the system that are adequate to absorb them. The design, arrangement, and location of restraints shall ensure that expansion-joint movements occur in the directions for which the joint is designed. In addition to the other thermal forces and moments, the effects of friction in other supports of the system shall be considered in the design of such anchors and guides.

Anchors for Expansion Joints Anchors (such as those of the corrugated, omega, disk, or slip type) shall be designed to withstand the algebraic sum of the forces at the maximum pressure and temperature at which the joint is to be used. These forces are:

1. Pressure thrust, which is the product of the effective thrust area times the maximum pressure to which the joint will be subjected during normal operation. (For slip joints the effective thrust area shall be computed by using the outside diameter of the pipe. For corrugated, omega, or disk-type joints, the effective thrust area shall be that area recommended by the joint manufacturer. If this information is unobtainable, the effective area shall be computed by using the maximum inside diameter of the expansion-joint bellows.)

2. The force required to compress or extend the joint in an amount equal to the calculated expansion movement.

3. The force required to overcome the static friction of the pipe in expanding or contracting on its supports, from installed to operating position. The length of pipe considered should be that located between the anchor and the expansion joint.

Support Fixtures Hanger rods may be pipe straps, chains, bars, or threaded rods which permit free movement for thermal expansion or contraction. Sliding supports shall be designed for friction and bearing loads. Brackets shall be designed to withstand movements due to friction in addition to other loads. Spring-type supports shall be designed for weight load at the point of attachment and to prevent misalignment, buckling, or eccentric loading of springs, and provided with stops to prevent spring overtravel. Compensating-type spring hangers are recommended for high-temperature and critical-service piping to make the supporting force uniform with appreciable movement. Counterweight supports shall have stops to limit travel. Hydraulic supports shall be provided with safety devices and stops to support load in the event of loss of pressure. Vibration dampers or sway braces may be used to limit vibration amplitude.

The code requires that the safe load for threaded hanger rods be based on the root area of the threads. This, however, assumes concentric loading. When hanger rods move to a nonvertical position so that the load is transferred from the rod to the supporting structure via the edge of one flat of the nut on the rod, it is necessary to consider the root area to be reduced by one-third. If a clamp is connected to a vertical line to support its weight, it is recommended that shear lugs be welded to the pipe, or that the clamp be located below a fitting or flange, to prevent slippage. Consideration shall be given to the localized stresses induced in the piping by the integral attachment. Typical pipe supports are shown in Fig. 10-174.

Much piping is supported from structures installed for other pur-

poses. It is common practice to use beam formulas for tubular sections to determine stress, maximum deflection, and maximum slope of piping in **spans between supports.** When piping is supported from structures installed for that sole purpose and those structures rest on driven piles, detailed calculations are usually made to determine max-

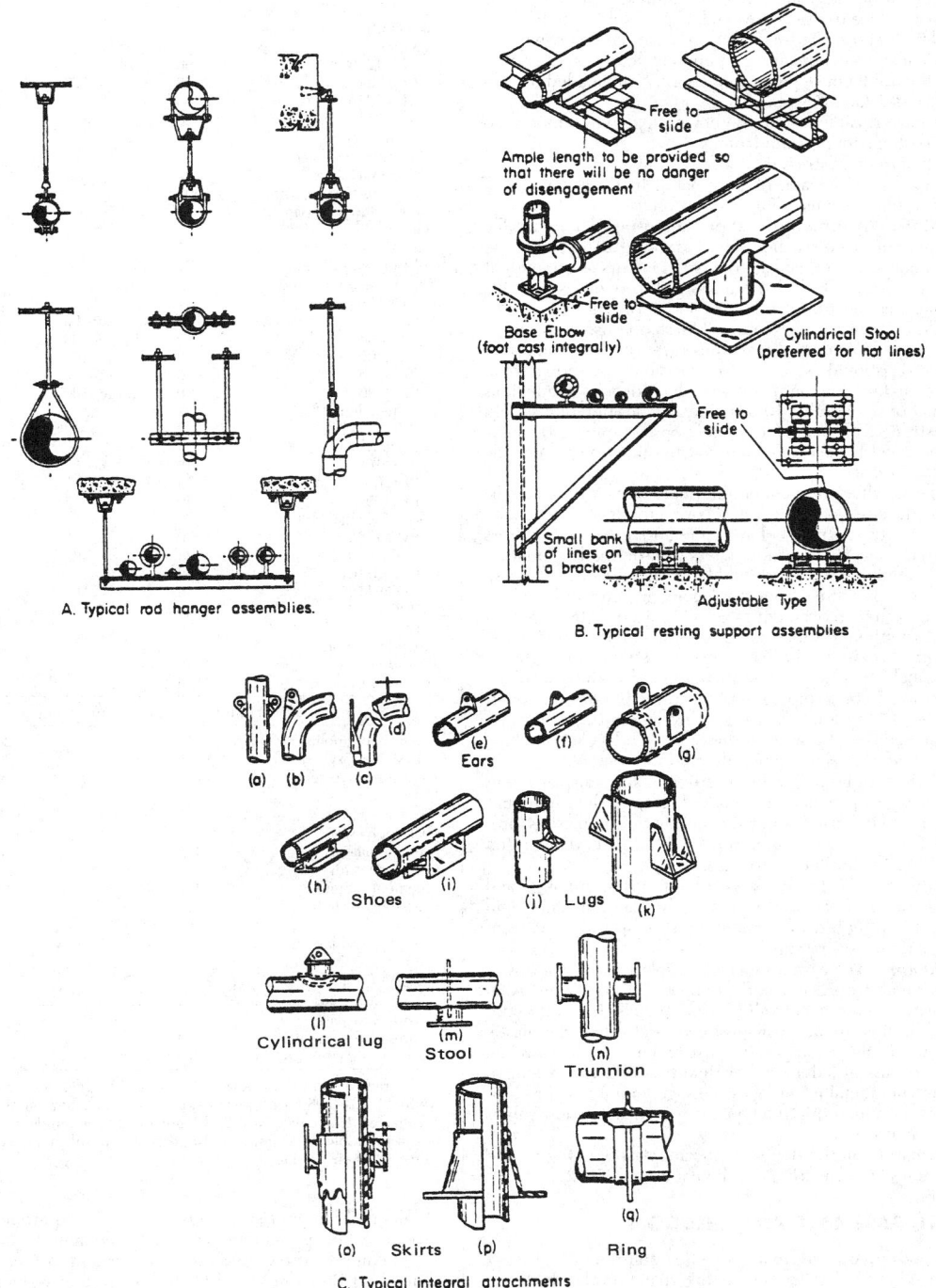

A. Typical rod hanger assemblies.

B. Typical resting support assemblies

C. Typical integral attachments

FIG. 10-174 Typical pipe supports and attachments. (*From Kellogg,* Design of Piping Systems, *Wiley, New York, 1965.*)

imum permissible spans. Limits imposed on maximum slope to make the contents of the line drain to the lower end require calculations made on the weight per foot of the empty line. To avoid interference with other components, maximum deflection should be limited to 25.4 mm (1 in).

Pipe hangers are essentially frictionless but require taller pipe-support structures which cost more than structures on which pipe is laid. Devices that reduce friction between laid pipe subject to thermal movement and its supports are used to accomplish the following:

1. Reduce loads on anchors or on equipment acting as anchors.
2. Reduce the tendency of pipe acting as a column loaded by friction at supports to buckle sideways off supports.
3. Reduce nonvertical loads imposed by piping on its supports so as to minimize cost of support foundations.
4. Reduce longitudinal stress in pipe.

Linear bearing surfaces made of fluorinated hydrocarbons or of graphite and also rollers are used for this purpose.

Design Criteria: Nonmetallic Pipe   In using a nonmetallic material, designers must satisfy themselves as to the adequacy of the material and its manufacture, considering such factors as strength at design temperature, impact- and thermal-shock properties, toxicity, methods of making connections, and possible deterioration in service. Rating information, based usually on ASTM standards or specifications, is generally available from the manufacturers of these materials. Particular attention should be given to provisions for the thermal expansion of nonmetallic piping materials, which may be as much as 5 to 10 times that of steel (Table 10-55). Special consideration should be given to the strength of small pipe connections to piping and equipment and to the need for extra flexibility at the junction of metallic and nonmetallic systems.

Table 10-56 gives values for the modulus of elasticity for nonmetals; however, no specific stress-limiting criteria or methods of stress analysis are presented. Stress-strain behavior of most nonmetals differs considerably from that of metals and is less well-defined for mathematic analysis. The piping system should be designed and laid out so that flexural stresses resulting from displacement due to expansion, contraction, and other movement are minimized. This concept requires special attention to supports, terminals, and other restraints.

Displacement Strains   The concepts of strain imposed by restraint of thermal expansion or contraction and by external movement described for metallic piping apply in principle to nonmetals. Nevertheless, the assumption that stresses throughout the piping system can be predicted from these strains because of fully elastic behavior of the piping materials is not generally valid for nonmetals.

In thermoplastics and some thermosetting resins, displacement strains are not likely to produce immediate failure of the piping but may result in detrimental distortion. Especially in thermoplastics, progressive deformation may occur upon repeated thermal cycling or on prolonged exposure to elevated temperature.

In brittle nonmetallics (such as porcelain, glass, impregnated graphite, etc.) and some thermosetting resins, the materials show rigid behavior and develop high displacement stresses up to the point of sudden breakage due to overstrain.

Elastic Behavior   The assumption that displacement strains will produce proportional stress over a sufficiently wide range to justify an elastic-stress analysis often is not valid for nonmetals. In brittle nonmetallic piping, strains initially will produce relatively large elastic stresses. The total displacement strain must be kept small, however, since overstrain results in failure rather than plastic deformation. In plastic and resin nonmetallic piping strains generally will produce stresses of the overstrained (plastic) type even at relatively low values of total displacement strain.

Further information on the design of thermoplastic piping can be found in the Plastics Pipe Institute's Technical Report TR-21.

FABRICATION, ASSEMBLY, AND ERECTION

Welding, Brazing, or Soldering   Code requirements dealing with fabrication are more detailed for welding than for other methods of joining, since welding is used not only to join two pipes end to end but also to fabricate fittings which replace seamless fittings such as

TABLE 10-55   Thermal Expansion Coefficients: Nonmetals*

	Mean coefficients (divide table values by 10^6)			
Material description	in/(in·°F)	Range, °F	mm/mm, °C	Range, °C
Thermoplastics				
Acetal AP2012	2		4	
Acrylonitrile-butadiene-styrene				
ABS 1208	60		108	
ABS 1210	55	45–55	99	8–12
ABS 1316	40		72	
ABS 2112	40		72	
Cellulose acetate butyrate				
CAB MH08	80		144	
CAB 5004	95		171	
Chlorinated poly (vinyl chloride)				
CPVC 4120	35		63	
Polybutylene PB 2110	72		130	
Polyether, chlorinated	45		81	
Polyethylene				
PE 1404	100	46–100	180	8–38
PE 2305	90	46–100	162	8–38
PE 2306	80	46–100	144	8–38
PE 3306	70	46–100	126	8–38
PE 3406	60	46–100	108	8–38
Polyphenylene POP 2125	30		54	
Polypropylene				
PP1110	48	33–67	86	0–20
PP1208	43		77	
PP2105	40		72	
Poly(vinyl chloride)				
PVC 1120	30	23–37	54	−5–+3
PVC 1220	35	34–40	63	1–4
PVC 2110	50		90	
PVC 2112	45		81	
PVC 2116	40	37–45	72	3–8
PVC 2120	30		54	
Vinylidine fluoride	85		153	
Vinylidine/vinyl chloride	100		180	
Reinforced thermosetting resins				
Asbestos-phenolic	11–30		20–54	
Asbestos-epoxy	11–30		20–54	
Asbestos-polyester	11–30		20–54	
Glass-epoxy, centrifugal-cast	9–13		16–23	
Glass-polyester, centrifugal-cast	9–15		16–27	
Glass-polyester, filament-wound	9–11		16–20	
Glass-polyester, hand lay-up	12–15		22–27	
Glass-epoxy, filament-wound	9–13		16–23	
Other nonmetallic materials				
Borosilicate glass	1.8		3	
Impregnated graphite	2.4		4	
Hard rubber (Buna N)	40		72	

*Extracted from the Chemical Plant and Petroleum Refinery Piping Code, ANSI B31.3—1980, with permission of the publisher, the American Society of Mechanical Engineers, New York. Individual compounds may vary from the values in the table by as much as 10 percent. Consult manufacturers for specific values for their products.

elbows and lap-joint stub ends. The code requirements for welding processes and operators are essentially the same as covered in the subsection on pressure vessels (i.e., qualification to Sec. IX of the ASME Boiler and Pressure Vessel Code) except that welding processes are not restricted, the material grouping (P number) must be in accordance with Appendix A, and welding positions are related to pipe posi-

TABLE 10-56   Modulus of Elasticity: Nonmetals*

Material description	E, kip/in² (73.4°F)	E, MPa (23°C)
Thermoplastics		
Acetal	410	2830
ABS, type 1210	250	1725
ABS, type 1316	340	2345
CAB	120	830
PVC, type 1120	420	2895
PVC, type 1220	410	2830
PVC, type 2110	340	2345
PVC, type 2116	380	2620
Chlorinated PVC	420	2895
Chlorinated polyether	160	1105
PE, type 2306	90	620
PE, type 3306	130	895
PE, type 3406	150	1035
Polypropylene	120	825
Vinylidene/vinyl chloride	100	690
Vinylidene fluoride	120	825
Thermosetting resins, axially reinforced		
Epoxy-asbestos	1200	8280
Phenolic-asbestos	1200	8280
Epoxy-glass, centrifugally cast	1200–1900	8280–13100
Epoxy-glass, filament-wound	1100–2000	7580–13800
Polyester-glass, centrifugally cast	1200–1900	8280–13100
Polyester-glass, hand lay-up	800–1000	5510–6900
Other		
Borosilicate glass	9800	67,600
Impregnated graphite	2300	15,900
Hard rubber (Buna N)	300	2070

*Extracted from the Chemical Plant and Petroleum Refinery Piping Code, ANSI B31.3—1980, with permission of the publisher, the American Society of Mechanical Engineers, New York.

tion. The code also permits one fabricator to accept welders or welding operators qualified by another employer without requalification when welding pipe by the same or equivalent procedure. Procedure qualification may include a requirement for low-temperature toughness tests. See Table 10-57.

Filler metal is required to conform with the requirements of Sec. IX. Backing rings (of ferrous material), when used, shall be of weldable quality with sulfur limited to 0.05 percent. Backing rings of nonferrous and nonmetallic materials may be used provided they are proved satisfactory by procedure-qualification tests and provided their use has been approved by the designer.

The code requires internal alignment within the dimensional limits specified in the welding procedure and the engineering design without specific dimensional limitations. Internal trimming is permitted for correcting internal misalignment provided such trimming does not result in a finished wall thickness before welding of less than required minimum wall thickness t_m. When necessary, weld metal may be deposited on the inside or outside of the component to provide alignment or sufficient material for trimming.

Table 10-58 is a digest of code requirements for the quality of welds. The defects referred to are illustrated in Fig. 10-175.

The qualification of brazing procedures, brazers, or brazing operations is required in accordance with the requirements of Part QB, Sec. IX, ASME Code, except that for Category D fluid service at design temperatures not over 93°C (200°F). Such qualification is at the owner's option. The clearance between surfaces to be joined by brazing or soldering shall be no larger than is necessary to allow complete capillary distribution of the filler metal.

The only requirement for solderers is that they follow the procedure in the *Copper Tube Handbook* of the Copper Development Association.

Bending and Forming Pipe may be bent to any radius for which the bend-arc surface will be free of cracks and substantially free of buckles. The use of bends which are creased or corrugated is permitted. Bending may be by any hot or cold method permissible within the radii and material characteristics of the pipe being bent.

Postbend heat treatment may be required for bends in some mate-

rials; its necessity depends on the severity of the bend. The details of these requirements are spelled out in the code. Piping components may be formed by any suitable hot or cold pressing, rolling, forging, hammering, spinning, drawing, or other method. Thickness after forming shall be not less than required by design. Special rules cover the forming and pressure design verification of flared laps. Hot bending and hot forming shall be done within a temperature range consistent with material characteristics, end use, or postoperation heat treatment.

The development of fabrication facilities for bending pipe to the radius of commercial butt-welding long-radius elbows and forming flared metallic (Van Stone) laps on pipe are important techniques in reducing welded-piping costs. These techniques save both the cost of the ell or stub end and the welding operation required to attach the fitting to the pipe.

Preheating and Heat Treatment Preheating and postoperation heat treatment are used to avert or relieve the detrimental effects of the high temperature and severe thermal gradients inherent in the welding of metals. In addition, heat treatment may be needed to relieve residual stresses created during the bending or forming of metals. The code provisions shown in Tables 10-59 and 10-60 represent basic practices which are acceptable for most applications of welding, bending, and forming, but they are not necessarily suitable for all service conditions. The specification of more or less stringent preheating and heat-treating requirements is a function of those responsible for the engineering design.

Joining Nonmetallic Pipe Thermoplastic piping may be joined by a qualified hot-gas welding procedure, a qualified solvent-cement procedure, or by a qualified heat-fusion procedure. The general welding and heat-fusion procedures are described in ASTM D-2657 and solvent-cement procedures in ASTM D-2855. Two other techniques, for flared joints and elastomeric-sealed joints, are described in ASTM D-3140 and D-3139, respectively.

In joining reinforced thermosetting pipe it is particularly important that the pipe be cut without chipping or cracking it. It is also important to sand, file, or grind any mold-release agent from the surfaces to be cemented. Joints are built up layer by layer of adhesive-saturated reinforcement by following the manufacturer's recommended procedure. Application of adhesive to the surfaces to be joined and assembly of these surfaces shall produce a continuous bond and provide an adhesive seal to protect the reinforcement from attack by the contents of the pipe. Unfilled or unbonded areas of the joint are considered defects and must be repaired.

Assembly and Erection Flanged-joint faces shall be aligned to the design plane to within 1⁄16 in/ft (0.5 percent) maximum measured across any diameter, and flange bolt holes shall be aligned to within 3.2-mm (1⁄8-in) maximum offset. Flanged joints involving flanges with widely differing mechanical properties shall be assembled with extra care, and tightening to a predetermined torque is recommended.

The use of flat washers under bolt heads and nuts is a code requirement when assembling nonmetallic flanges. It is preferred that the bolts extend completely through their nuts; however, a lack of complete thread engagement not exceeding one thread is permitted by the code. In assembling nonmetallic lined joints consideration must be given to the need and means for maintaining electrical continuity when static sparking could occur. The assembly of cast-iron bell-and-spigot piping is covered in AWWA Standard C600.

Screwed joints which are intended to be seal-welded shall be made up without any thread compound.

EXAMINATION, INSPECTION, AND TESTING

Examination and Inspection The code differentiates between examination and inspection. "Examination" applies to quality-control functions performed by personnel of the piping manufacturer, fabricator, or erector. "Inspection" applies to functions performed for the owner by the authorized inspector.

The authorized inspector shall be designated by the owner and shall be the owner, an employee of the owner, an employee of an engineering or scientific organization, or an employee of a recognized insurance or inspection company acting as the owner's agent. The inspector

TABLE 10-57 Requirements for Low-Temperature Toughness Tests*

Type of Material		Column A		Column B
		At or above minimum temperature listed in Table 10-49 or Table 10-15		Below minimum temperatures listed in Table 10-49 or Table 10-15
Listed metallic materials	Ductile iron, malleable iron, Carbon steel, ASTM A36, ASTM A283	1. No additional requirements.		1. Shall not be used.
	All other carbon steel, low-intermediate, and high-alloy steels, ferritic steels	Base metal	Deposited weld metal and heat-affected zone (See Note 1)	2. Except when conditions conform to Note 2, the material shall be heat-treated to control its microstructure by a method appropriate to the material as outlined in the specification applicable to the product form and then impact-tested. (See Note 1.) Deposited weld metal and heat-affected zone shall be impact-tested.
		2a. No additional requirements.	2b. When materials are fabricated or assembled by welding, the deposited weld metal and heat-affected zone shall be impact-tested if the design temperature is below −29°C (−20°F) unless conditions conform to Note 2.	
	Austenitic stainless steel	3a. If (1) the carbon content by analysis is greater than 0.10 percent or (2) the material is not in the solution-heat-treated condition, then impact testing is required for design temperatures below −29°C (−20°F). See Note 2.	3b. When materials are fabricated or assembled by welding, the deposited weld metal shall be impact-tested for design temperature below −29°C (−20°F) unless conditions conform to Note 2.	3. The material shall be impact-tested. See Note 2.
	Austenitic ductile iron, ASTM A571	4a. No additional requirements.	4b. Welding not permitted.	4. The material shall be impact-tested. This material shall not be used at design minimum temperatures lower than −196°C (−320°F). Welding is not permitted.
	Aluminum alloy, copper, copper alloy, nickel, nickel alloy, unalloyed titanium	5a. No additional requirements.	5b. No additional requirements except that when the composition of the filler metal is outside the range of composition for the base metal, testing shall be in accordance with column B, item 5.	5. Low-temperature tests such as tensile elongation and sharp-notch tensile strength (compared with unnotched tensile strength) shall have been conducted to provide assurance to the designer that the material and the deposited weld metal are suitable at the design minimum temperatures.
Listed nonmetallic materials		6. No additional requirements.		6. Below the recommended minimum temperatures, the designer shall have test results at or below the lowest expected service temperature which assure that the materials will have adequate toughness and are suitable at the design minimum temperatures.
Unlisted materials		Unlisted materials which conform to a published specification and are of composition, heat treatment, and product form comparable with those of listed materials shall be subject to the same requirements as the listed materials. All other unlisted materials conforming to a published specification shall be qualified as required by the applicable item in col. B.		

*Extracted from the Chemical Plant and Petroleum Refinery Piping Code, ANSI B31.3—1980, with permission of the publisher, the American Society of Mechanical Engineers, New York.

NOTE: These toughness-test requirements are in addition to tests required by the material specification.

1. Any tests and associated acceptance criteria which are part of the welding-procedure qualification for filler materials and heat-affected zone need not be repeated.

2. Impact testing is not required if the design temperature is below −29°C (−20°F) but at or above −46°C (−50°F) and the maximum operating pressure of the fabricated or assembled components will not exceed 25 percent of the maximum allowable design pressure at ambient temperature and the combined longitudinal stress due to pressure, dead weight, and displacement strain (see Par. 319.2.1) does not exceed 41 MPa (6000 lbf/in²).

shall not represent or be an employee of the piping erector, the manufacturer, or the fabricator unless the owner is also the erector, the manufacturer, or the fabricator.

The authorized inspector shall have a minimum of 10 years' experience in the design, fabrication, or inspection of industrial pressure piping. Each 20 percent of satisfactory work toward an engineering degree accredited by the Engineers' Council for Professional Development shall be considered equivalent to 1 year's experience, up to 5 years total.

It is the owner's responsibility, exercised through the authorized inspector, to verify that all required examinations and testing have been completed and to inspect the piping to the extent necessary to be

TABLE 10-58   Limitations on Imperfections in Welds*

Imperfection†	When required examination is	Girth and miter-joint butt welds	Longitudinal butt welds‡	Fillet, socket, seal, and reinforcement attachment welds	Welded branch connections and fabricated laps
Cracks or lack of fusion	Any	None permitted	None permitted	None permitted	None permitted
Incomplete penetration	100% radiography	None permitted	None permitted	NA	None permitted
	Visual or random or spot radiography	A	None permitted	NA	A and H
Internal porosity	100% radiography	B	B	NA	B and H
	Random or spot radiography	C	C	NA	C and H
Slag inclusions or elongated defects	100% radiography	D	D	NA	D and H
	Random or spot radiography	E	E	NA	B and H
Undercutting	Any	Lesser of 1/32 in or $\overline{T}w/4$	None permitted	Lesser of 1/32 in or $\overline{T}w/4$	Lesser of 1/32 in or $\overline{T}w/4$
Surface porosity and exposed slag inclusion (3/16-in nominal wall thickness and less)		None permitted	None permitted	None permitted	None permitted
Concave root surface (suck-up)		F	F	NA	F and H
Weld reinforcement		G	G	G	G and H

NOTES:

NA: Not applicable.

A: The lesser of 1/32 in or 0.2 $\overline{T}w$ deep. The total length of such imperfections shall not exceed 1.5 in (38 mm) in any 6 in (150 mm) of weld length. (See Fig. 10-175).

B: An individual pocket of porosity shall not exceed the lesser of $\overline{T}w/3$ or 1/8 in in its greatest dimension. The total area of porosity projected radially through the weld shall not exceed an area equivalent to 3 times the area of a single maximum pocket allowable in any square inch (645 mm²) of projected weld area.

C: An individual pocket of porosity shall not exceed the lesser of $\overline{T}w/2$ or 1/8 in in its greatest dimension. The total area of porosity projected radially through the weld shall not exceed an area equivalent to 3 times the area of a single maximum pocket allowable in any square inch (645 mm²) of projected weld area.

D: The developed length of any single slag inclusion or elongated defect shall not exceed $\overline{T}w/3$. The total cumulative developed length of slag inclusions and/or elongated defects shall not exceed $\overline{T}w$ in any 12 $\overline{T}w$ length of weld. The width of a slag inclusion shall not exceed the lesser of 1/16 in or $\overline{T}w/3$.

E: The developed length of any single slag inclusion or elongated defect shall not exceed 2 $\overline{T}w$. The total cumulative developed length of slag inclusions and/or elongated defects shall not exceed 4 $\overline{T}w$ in any 6-in length of weld. The width of a slag inclusion shall not exceed the lesser of 1/8 in or $\overline{T}w/2$.

F: For single-sided welded joints, concavity of the root surface shall not reduce the total thickness of the joint, including reinforcement, to less than the thickness of the thinner of the components being jointed.

G: External weld reinforcement and internal weld protrusion shall be fused with and shall merge smoothly into the component surface. The thickness of external weld reinforcement and internal weld protrusion (when no backing ring is used) shall not exceed the following:

Wall thickness $\overline{T}w$, in	Weld reinforcement or protrusion, in, maximum
1/4 and under	1/16
over 1/4 through 1/2	1/8
over 1/2 through 1	5/32
over 1 (25.4 mm)	3/16

H: These requirements apply only to butt welds.

*Extracted from the Chemical Plant and Petroleum Refinery Piping Code, ANSI B31.3—1980, with permission of the publisher, the American Society of Mechanical Engineers, New York. To convert inches to millimeters, multiply by 25.4.

†See Fig. 10-175 for illustration of the defects.

‡This column applies to welds not made in accordance with a standard listed in Appendix A or Appendix E of the code.

satisfied that it conforms to all applicable requirements of the code and the engineering design. This verification may include certifications and records pertaining to materials, components, heat treatment, examination and testing, and qualifications of operators and procedures. The authorized inspector may delegate the performance of inspection to a qualified person.

Inspection does not relieve the manufacturer, the fabricator, or the erector of responsibility for providing materials, components, and skill in accordance with requirements of the code and the engineering design, performing all required examinations, and preparing records of examinations and tests for the inspector's use.

Examination Methods   The code establishes the types of examinations for evaluating various types of imperfections (see Table 10-61).

Personnel performing examinations other than visual shall be qualified in accordance with applicable portions of SNT TC-1A, *Recommended Practice for Nondestructive Testing Personnel Qualification and Certification.* Procedures shall be qualified as required in Par. T-150, Art. 1, Sec. V of the ASME Code. Limitations on imperfections shall be in accordance with the engineering design but shall at least meet the requirements of the code (see Tables 10-58 and 10-59) for the specific type of examination. Repairs shall be made as applicable.

Visual Examination   This consists of observation of the portion of components, joints, and other piping elements that are or can be exposed to view before, during, or after manufacture, fabrication, assembly, erection, inspection, or testing. The examination includes verification of code and engineering design requirements for materials and components, dimensions, joint preparation, alignment, welding or joining, supports, assembly, and erection.

Visual examination shall be performed in accordance with Art. 9, Sec. V of the ASME Code.

Magnetic-Particle Examination   This examination shall be performed in accordance with Art. 7, Sec. V of the ASME Code.

Liquid-Penetrant Examination   This examination shall be performed in accordance with Art. 6, Sec. V of the ASME Code.

Radiographic Examination   The following definitions apply to radiography required by the code or by the engineering design:

1. "Random radiography" applies only to girth butt welds. It is radiographic examination of the complete circumference of a specified percentage of the girth butt welds in a designated lot of piping.

2. "100 percent radiography" applies only to girth butt welds unless otherwise specified in the engineering design. It is defined as radiographic examination of the complete circumference of all the girth butt welds in a designated lot of piping. If the engineering design specifies that 100 percent radiography shall include welds other than girth butt welds, the examination shall include the full length of all such welds.

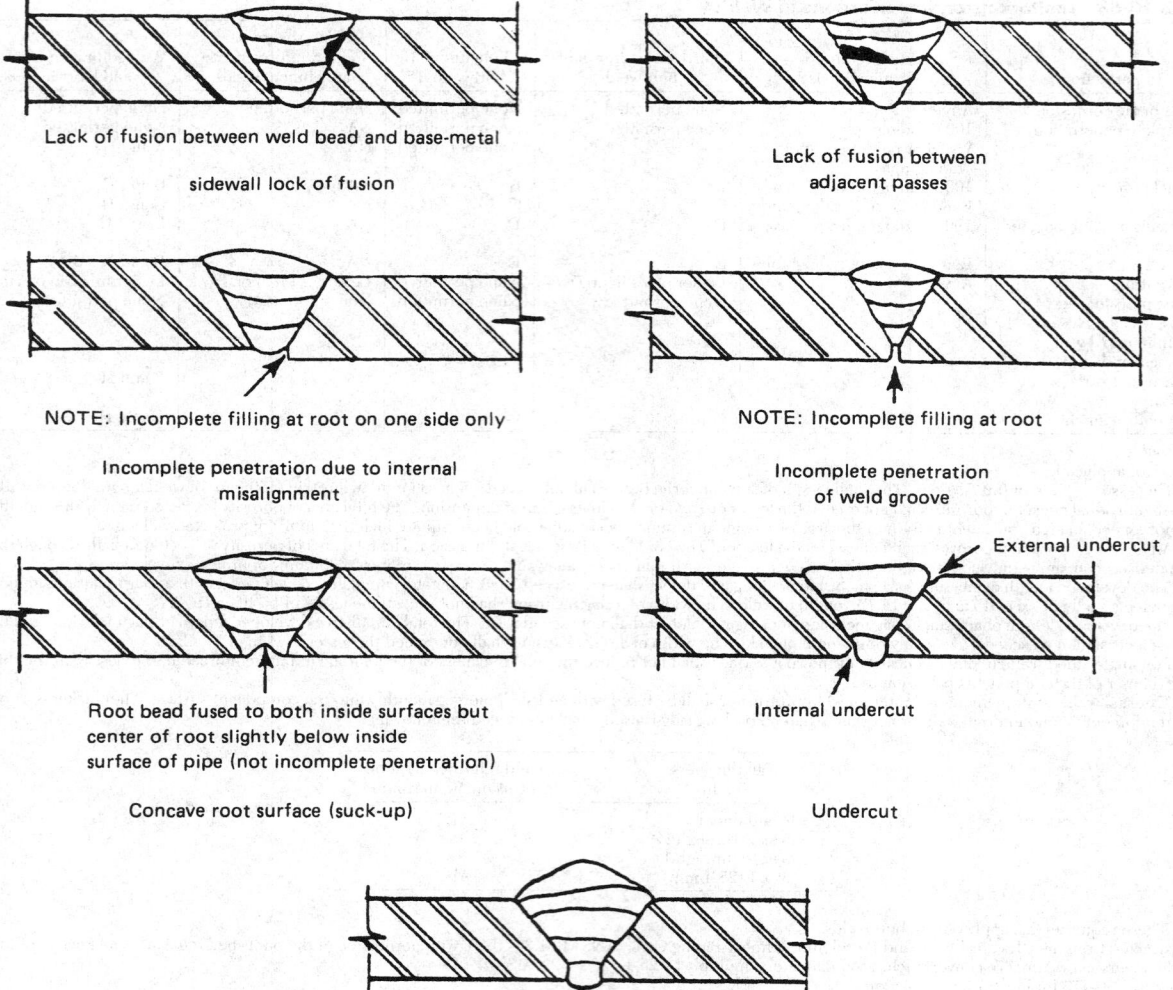

Lack of fusion between weld bead and base-metal

sidewall lock of fusion

Lack of fusion between
adjacent passes

NOTE: Incomplete filling at root on one side only

Incomplete penetration due to internal
misalignment

NOTE: Incomplete filling at root

Incomplete penetration
of weld groove

Root bead fused to both inside surfaces but
center of root slightly below inside
surface of pipe (not incomplete penetration)

Concave root surface (suck-up)

External undercut

Internal undercut

Undercut

Excess external reinforcement

FIG. 10-175   Typical weld imperfections. (*Extracted from the Chemical Plant and Petroleum Refinery Piping Code, ANSI B31.3—1976, with permission of the publisher, the American Society of Mechanical Engineers, New York.*)

3.   "Spot radiography" is the practice of making a single-exposure radiograph at a point within a specified extent of welding. Required coverage for a single spot radiograph is as follows:

• For longitudinal welds, at least 150 mm (6 in) of weld length.

• For girth, miter, and branch welds in piping 2½ in NPS and smaller, a single elliptical exposure which encompasses the entire weld circumference, and in piping larger than 2½ in NPS, at least 25 percent of the inside circumference or 150 mm (6 in), whichever is less.

Radiography of components other than castings and of welds shall be in accordance with Art. 2, Sec. V of the ASME Code. Limitations on imperfections in components other than castings and welds shall be as stated in Table 10-58 for the degree of radiography involved.

Ultrasonic Examination   Ultrasonic examination of welds shall be in accordance with Art. 5, Sec. V of the ASME Code, except that the modifications stated in Par. 336.4.6 of the code shall be substituted for T-535.1(*d*)(2).

Type and Extent of Required Examination   The intent of examinations is to provide the examiner and the inspector with reasonable assurance that the requirements of the code and the engineering design have been met. For P-number 3, 4, and 5 materials,

examination shall be performed after any heat treatment has been completed.

Examination Normally Required   Piping not covered by Category D fluid service or severe cyclic conditions shall be examined as follows or to any greater extent specified in the engineering design.

1.   *Visual examination*

a.   Sufficient materials and components, selected at random, to satisfy the examiner that they conform to specifications and are free from damage.

b.   At least 5 percent of fabrication. For welds, each welder's or welding operator's work shall be represented, though not necessarily each type of weld for each welder or welding operator. Limitations on imperfections shall be as stated in Table 10-58.

c.   100 percent of fabrication for longitudinal welds other than those in components made to material specifications recognized in the code. Limitations on imperfections are those of Table 10-58.

d.   Random examination of the assembly of threaded, bolted, and other joints to satisfy the examiner that they conform to requirements.

e.   Random examination during erection of piping, including checking of alignments, supports, and cold spring.

TABLE 10-59 Preheat Temperatures*

Base-metal P number[†]	Weld-metal analysis A number[‡]	Base-material group	Nominal wall thickness mm	in	Minimum specified tensile strength, base metal MPa	kip/in²	Minimum temperature Required °C	°F	Recommended °C	°F
1	1	Carbon steel	<25.4	<1	≤490	≤71			10	50
			≥25.4	≥1	All	All			80	175
			All	All	>490	>71			80	175
3	2, 11	Alloy steels Cr ½% maximum	<12.7	<½	≤490	≤71			10	50
			≥12.7	≥½	All	All			80	175
			All	All	≥490	>71			80	175
4	3	Alloy steels Cr > ½% to 2%	All	All	All	All	150	300		
5	4, 5	Alloy steels Cr 2¼% to 10%	All	All	All	All	175	350		
6	6	High-alloy steels: martensitic	All	All	All	All			150§	300§
7	7	High-alloy steels: ferritic	All	All	All	All			10	50
8	8, 9	High-alloy steels: austenitic	All	All	All	All			10	50
9A, 9B, 9C	10	Nickel alloy steels	All	All	All	All			95	200
10A		Mn-V steel	All	All	All	All			80	175
10B		Cr-V steel	All	All	All	All			150	300
11A Group 1		9% Ni steel	All	All	All	All			10	50
P21-P52			All	All	All	All			10	50

*Extracted from the Chemical Plant and Petroleum Refinery Piping Code, ANSI B31.3—1980, with permission of the publisher, the American Society Mechanical Engineers, New York.
†P number from ASME Code, Sec. IX, Table QW-422.
‡A number from ASME Code, Sec. IX, Table QW-442.
§Maximum interpass temperature 315°C (600°F).

TABLE 10-60 Requirements for Heat Treatment*

Base-metal P number[†]	Weld-metal analysis A number[‡]	Material group	Nominal wall thickness mm	in	Minimum specified tensile strength, base metal MPa	kip/in²	Metal temperature range °C	°F	h/in, nominal wall§	Minimum time, h	Brinell hardness, maximum
1	1	Carbon steel	≤19	≤¾	All	All	None	None	1	1	
			>19	>¾			595–650	1100–1200			
3	2, 11	Alloy steels Cr ½% max	≤19	≤¾	≤490	≤71	None	None			
			>19	>¾	All	All	595–720	1100–1325	1	1	225
			All	All	>490	>71	595–720	1100–1325	1	1	225
4	3	Alloy steels Cr > ½% to 2%	≤12.7	≤½	≤490	≤71	None	None			
			>12.7	>½	All	All	705–745	1300–1375	1	2	225
			All	All	>490	>71	705–745	1300–1375	1	2	225
5	4, 5	Alloy steels Cr 2¼% to 10%	≤½ and ≤3% Cr and ≤0.15% C		All	All	None	None			
			>½ or >3% Cr or >0.15% C		All	All	705–760	1300–1400	1	2	241
6	6	High-alloy steels: martensitic	All	All	All	All	730–790	1350–1450	1	2	241
		A240, Gr 429	All	All	All	All	620–660	1150–1225	1	2	241
7	7	High-alloy steels: ferritic	All	All	All	All	None	None			
8	8, 9	High-alloy steels: austenitic	All	All	All	All	None	None			
9A	10	Nickel alloy steels	≤19	≤¾	All	All	None	None			
9B			>19	>¾	All	All	595–635	1100–1175	½	1	
10A		Mn-V steel	≤19	≤¾	All	All	None	None			
			>19	>¾	All	All	595–705	1100–1300	1	1	225
			All	All	>490	>71	595–705	1100–1300	1	1	225
10B		Cr-V steel	≤12.7	≤½	≤490	≤71	None	None			
			>12.7	>½	All	All	595–730	1100–1350	1	1	225
			All	All	>490	>71	595–730	1100–1350	1	1	225
11A, Group 1		9% Ni steel	≤51	≤2	All	All	None	None			
			>51	>2	All	All	550–585	1025–1085	1	1	
							[Cooling rate > 150°C (300°F)/h to 315°C (600°F)]				

*Extracted from the Chemical Plant and Petroleum Refinery Piping Code, ANSI B31.3—1980, with permission of the publisher, the American Society of Mechanical Engineers, New York.
†P number from ASME Code, Sec. IX, Table QW-422, Special P numbers (SP-1, SP-2, SP-3) require special consideration in procedure qualification. The required thermal treatment shall be established by the engineering design and demonstrated by the procedure qualification.
‡A number from ASME Code, Sec. IX, Table QW-422.
§For SI equivalent, h/mm, divide h/in by 25.

TABLE 10-61   Types of Examination for Evaluating Imperfections*

| | Type of examination | | | |
| | | | Ultrasonic or radiographic | |
Type of imperfection	Visual	Liquid-penetrant or magnetic-particle	Random	100%
Crack	X	X	X	X
Incomplete penetration	X		X	X
Lack of fusion	X		X	X
Weld undercutting	X			
Weld reinforcement	X			
Internal porosity			X	X
External porosity	X			
Internal slag inclusions			X	X
External slag inclusions	X			
Concave root surface	X		X	X

*Extracted from the Chemical Plant and Petroleum Refinery Piping Code, ANSI B31.3—1980, with permission of the publisher, the American Society of Mechanical Engineers, New York. For limitations on imperfections in welds see Table 10-58.

f.   Examination of erected piping for evidence of damage that would require repair or replacement and for other evident deviations from the intent of the design.
    2.   *Other examination.*   When piping is intended for service at temperatures above 186°C (366°F) or gauge pressures above 1.0 MPa (150 lbf/in²) as designated in the engineering design, a minimum of 5 percent of circumferential butt welds shall be examined fully by random radiography or ultrasonic examination. The welds to be examined shall be selected to ensure that the work product of each individual welder or welding operator doing the production welding is included. They shall also be selected to maximize coverage of intersections with longitudinal joints. A minimum of 38 mm (1½ in) of the longitudinal welds shall be examined. In-process examination may be substituted for all or part of the radiographic or ultrasonic examination on a weld-for-weld basis if specified in the engineering design.
    3.   *In-process examination.*   In-process examination comprises visual examination of the following as applicable:
    a.   Joint preparation and cleanliness
    b.   Preheating
    c.   Fit-up and internal alignment prior to welding
    d.   Weld position, electrode, and other variables specified by the welding procedure
    e.   Condition of the root pass after cleaning (external and, where accessible, internal), aided by liquid-penetrant or magnetic-particle examination when specified in the engineering design
    f.   Slag removal and weld condition between passes
    g.   Appearance of the finished weld
    4.   *Certification and records for components and materials.*   The examiner shall be assured, by examination of certification, records, or other evidence, that the materials and components are of the specified grades and that they have received required heat treatment, examination, and testing. The examiner shall provide the inspector with a certification that all quality-control requirements of the code and of the engineering design have been met.
    Category D Fluid-Service Piping   This piping, as designated in the engineering design, shall be visually examined to the extent necessary to satisfy the examiner that components, materials, and work conform to the requirements of the code and the engineering design.
    Piping Subject to Severe Cyclic Conditions   Piping for other than Category D fluids to be used under severe cyclic conditions shall be examined as follows or to any greater extent specified in the engineering design.
    1.   *Visual examination*
    a.   All fabrication threaded, bolted, and other joints shall be examined.
    b.   All piping erection shall be examined to verify dimensions and alignment. Supports, guides, and points of cold spring shall be checked to assure that movement of the piping under all conditions of

start-up, operation, and shutdown will be accommodated without binding or constraint.
    2.   *Other examination.*   All circumferential butt welds and all fabricated branch connection welds comparable to Fig. 10-127 shall be examined by 100 percent radiography or (if specified in the engineering design) by ultrasonic examination. Limitations on imperfections are as specified in Table 10-58. The code also requires that a welding procedure which promotes a smooth, fully penetrated internal surface be employed and that the external surface of the completed weld be free of undercutting and finished to within 500 AARH. Socket welds and nonradiographed branch-connection welds shall be examined by magnetic-particle or liquid-penetrant methods.
    Impact Testing   Materials conforming to ASTM specifications listed in the code may generally be used at temperatures down to the lowest temperature listed for that material in the stress table without additional testing. When welding or other operations are performed on these materials, additional low-temperature toughness tests may be required. The code requirements are listed in Table 10-57.
    Pressure Testing   Prior to initial operation, installed piping shall be pressure-tested to assure tightness except as permitted for Category D fluid service described later. The pressure test shall be maintained for a sufficient time to determine the presence of any leaks but not less than 10 min.
    If repairs or additions are made following the pressure tests, the affected piping shall be retested except that, in the case of minor repairs or additions, the owner may waive retest requirements when precautionary measures are taken to assure sound construction.
    When pressure tests are conducted at metal temperatures near the ductile-to-brittle transition temperature of the material, the possibility of brittle fracture shall be considered.
    The test shall be hydrostatic, using water, with the following exceptions. If there is a possibility of damage due to freezing or if the operating fluid or piping material would be adversely affected by water, any other suitable liquid may be used. If a flammable liquid is used, its flash point shall not be less than 50°C (120°F), and consideration shall be given to the test environment.
    The hydrostatic-test pressure at any point in the system shall be as follows:
    1.   Not less than 1½ times the design pressure.
    2.   For a design temperature above the test temperature, the minimum test pressure shall be as calculated by the following formula:

$$P_T = 1.5 \, PS_T/S \qquad (10\text{-}108)$$

where P_T = test hydrostatic gauge pressure, MPa (lbf/in²)
            P = internal design pressure, MPa (lbf/in²)
            S_T = allowable stress at test temperature, MPa (lbf/in²)
            S = allowable stress at design temperature, MPa (lbf/in²)

    If the test pressure as so defined would produce a stress in excess of the yield strength at test temperature, the test pressure may be reduced to the maximum pressure that will not exceed the yield strength at test temperature.
    A preliminary air test at not more than 0.17-MPa (25-lbf/in²) gauge pressure may be made prior to hydrostatic test in order to locate major leaks.
    If hydrostatic testing is not considered practicable by the owner, a pneumatic test in accordance with the following procedure may be substituted, using air or another nonflammable gas.
    If the piping is tested pneumatically, the test pressure shall be 110 percent of the design pressure. Pneumatic testing involves a hazard owing to the possible release of energy stored in compressed gas. Therefore, particular care must be taken to minimize the chance of brittle failure of metals and thermoplastics. The test temperature is important in this regard and must be considered when material is chosen in the original design. Any pneumatic test shall include a preliminary check at not more than 0.17-MPa (25-lbf/in²) gauge pressure. The pressure shall be increased gradually in steps providing sufficient time to allow the piping to equalize strains during test and to check for leaks. If the test liquid in the system is subject to thermal expansion, precautions shall be taken to avoid excessive pressure.
    At the owner's option, a piping system used only for Category D

fluid service as defined in the subsection "Classification of Fluid Service" may be tested at the normal operating conditions of the system during or prior to initial operation by examining for leaks at every joint not previously tested. A preliminary check shall be made at not more than 0.17-MPa (25-lbf/in²) gauge pressure when the contained fluid is a gas or a vapor. The pressure shall be increased gradually in steps providing sufficient time to allow the piping to equalize strains during testing and to check for leaks.

Tests alternative to those required by these provisions may be applied under certain conditions described in the code.

Piping required to have a sensitive leak test shall be tested by the gas- and bubble-formation testing method specified in Art. 10, Sec. V of the ASME Code or by another method demonstrated to have equal or greater sensitivity. The sensitivity of the test shall be at least (100 Pa·mL)/s [(10³ atm·mL)/s] under test conditions. If a hydrostatic pressure test is used, it shall be carried out after the sensitive leak test.

Records shall be kept of each piping installation during the testing.

COMPARISON OF PIPING-SYSTEM COSTS

Piping may represent as much as 25 percent of the cost of a chemical-process plant. The installed cost of piping systems varies widely with the materials of construction and the complexity of the system. A study of piping costs shows that the most economical choice of material for a simple straight piping run may not be the most economical for a complex installation made up of many short runs involving numerous fittings and valves. The economics also depends heavily on the pipe size and fabrication techniques employed. Fabrication methods such as bending to standard long-radius-elbow dimensions and machine-flaring lap joints have a large effect on the cost of fabricating pipe from ductile materials suited to these techniques. Cost reductions of as high as 35 percent are quoted by some custom fabricators utilizing advanced techniques.

Figure 10-176 is based on data extracted from a comparison of the installed cost of piping systems of various materials published by the Dow Chemical Co. The chart shows the relative cost ratios for systems of various materials based on two installations, one consisting of 152 m (500 ft) of 2-in pipe in a complex piping arrangement and the other of 305 m (1000 ft) of 2-in pipe in a straight-run piping arrangement. Figure 10-176 is based on field-fabrication construction techniques using welding stubs, the method commonly used by contractors. A considerably different ranking would result from using other construction methods such as machine-formed lap joints and bends in place of welding elbows. Piping-cost experience shows that it is difficult to generalize and reflect accurate piping-cost comparisons. For an accurate comparison the cost for each type of material must be estimated individually on the basis of the actual fabrication and installation methods that will be used and the conditions anticipated for the proposed installation.

FORCES OF PIPING ON PROCESS MACHINERY AND PIPING VIBRATION

The reliability of process rotating machinery is affected by the quality of the process piping installation. Excessive external forces and moments upset casing alignment and can reduce clearance between motor and casing. Further, the bearings, seals, and coupling can be adversely affected, resulting in repeated failures that may be correctly diagnosed as misalignment, and may have excessive piping forces as the root causes. Most turbine and compressor manufacturers have prescribed specification or will follow NEMA standards for allowable nozzle loading.

Prior to any machinery alignment procedure, it is imperative to check for machine pipe strain. This is accomplished by the placement of dial indicators on the shaft and then loosening the hold-down bolts. Movements of greater than 1 mil are considered indication of a pipe strain condition.

This is an important practical problem area, as piping vibration can cause considerable downtime or even pipe failure.

Pipe vibration is caused by:
1. Internal flow (pulsation)

2. Plant machinery (such as compressors, pumps)

Pulsation can be problematic and difficult to predict. Pulsations are also dependent on acoustic resonance characteristics.

When a pulsation frequency coincides with a mechanical or acoustic resonance, severe vibration can result. A common cause for pulsation is the presence of flow control valves or pressure regulators. These often operate with high pressure drops (i.e., high flow velocities), which can result in the generation of severe pulsation. Flashing and cavitation can also contribute.

Modern-day piping design codes can model the vibration situation, and problems can thus be resolved in the design phases.

HEAT TRACING OF PIPING SYSTEMS

Heat tracing is used to maintain pipes and the material that pipes contain at temperatures above the ambient temperature. Two common uses of heat tracing are preventing water pipes from freezing and maintaining fuel oil pipes at high enough temperatures such that the viscosity of the fuel oil will allow easy pumping. Heat tracing is also used to prevent the condensation of a liquid from a gas and to prevent the solidification of a liquid metal.

A heat-tracing system is often more expensive on an installed cost basis than the piping system it is protecting, and it will also have significant operating costs. A recent study on heat-tracing costs by a major chemical company showed installed costs of $31/ft to $142/ft and yearly operating costs of $1.40/ft to $16.66/ft. In addition to being a major cost, the heat-tracing system is an important component of the reliability of a piping system. A failure in the heat-tracing system will often render the piping system inoperable. For example, with a water freeze protection system, the piping system may be destroyed by the expansion of water as it freezes if the heat-tracing system fails.

The vast majority of heat-traced pipes are insulated to minimize heat loss to the environment. A heat input of 2 to 10 watts per foot is generally required to prevent an insulated pipe from freezing. With high wind speeds, an uninsulated pipe could require well over 100 watts per foot to prevent freezing. Such a high heat input would be very expensive.

Heat tracing for insulated pipes is generally only required for the period when the material in the pipe is not flowing. The heat loss of an insulated pipe is very small compared to the heat capacity of a flowing fluid. Unless the pipe is extremely long (several thousands of feet), the temperature drop of a flowing fluid will not be significant.

The three major methods of avoiding heat tracing are:
1. Changing the ambient temperature around the pipe to a temperature that will avoid low-temperature problems. Burying water pipes below the frost line or running them through a heated building are the two most common examples of this method.
2. Emptying a pipe after it is used. Arranging the piping such that it drains itself when not in use, can be an effective method of avoiding the need for heat tracing. Some infrequently used lines can be pigged or blown out with compressed air. This technique is not recommended for commonly used lines due to the high labor requirement.
3. Arranging a process such that some lines have continuous flow can eliminate the need for tracing these lines. This technique is generally not recommended because a failure that causes a flow stoppage can lead to blocked or broken pipes.

Some combination of these techniques may be used to minimize the quantity of traced pipes. However, the majority of pipes containing fluids that must be kept above the minimum ambient temperature are generally going to require heat tracing.

Types of Heat-Tracing Systems Industrial heat-tracing systems are generally fluid systems or electrical systems. In fluid systems, a pipe or tube called the *tracer* is attached to the pipe being traced, and a warm fluid is put through it. The tracer is placed under the insulation. Steam is by far the most common fluid used in the tracer, although ethylene glycol and more exotic heat-transfer fluids are used. In electrical systems, an electrical heating cable is placed against the pipe under the insulation.

Fluid Tracing Systems Steam tracing is the most common type of industrial pipe tracing. In 1960, over 95 percent of industrial tracing systems were steam traced. By 1995, improvements in electric

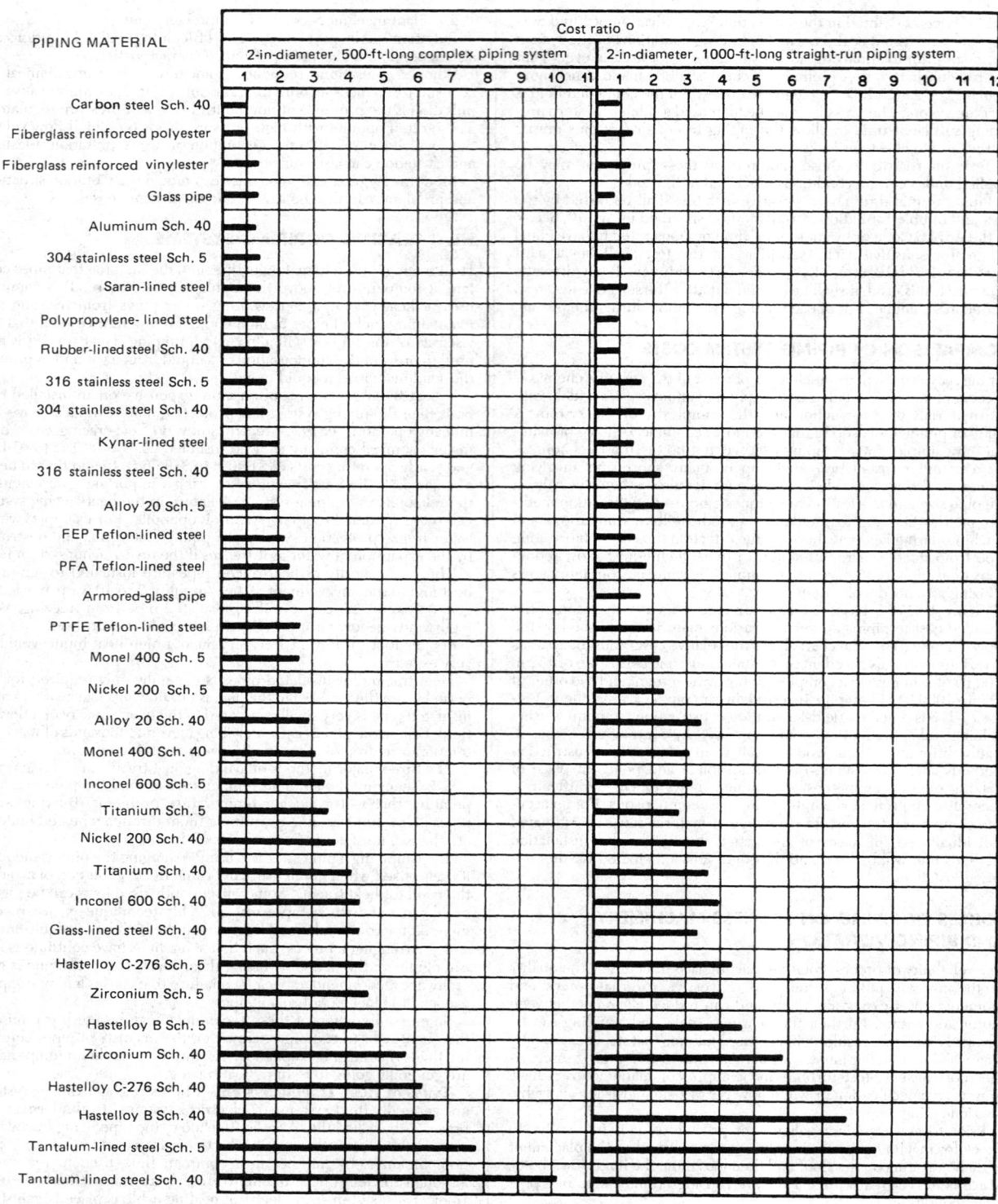

FIG. 10-176 Cost rankings and cost ratios for various piping materials. This figure is based on field-fabrication construction techniques using welding stubs, as this is the method most often employed by contractors. A considerably different ranking would result from using other construction methods, such as machined-formed lap joints, for the alloy pipe. °Cost ratio = (cost of listed item)/(cost of Schedule 40 carbon steel piping system, field-fabricated by using welding stubs). (*Extracted with permission from* Installed Cost of Corrosion Resistant Piping, *copyright 1977, Dow Chemical Co.*)

heating technology increased the electric share to 30 to 40 percent, but steam tracing is still the most common system. Fluid systems other than steam are rather uncommon and account for less than 5 percent of tracing systems.

Half-inch copper tubing is commonly used for steam tracing. Three-eighths-inch tubing is also used, but the effective circuit length is then decreased from 150 feet to about 60 feet. In some corrosive environments, stainless steel tubing is used, and occasionally standard carbon steel pipe (one half inch to one inch) is used as the tracer.

In addition to the tracer, a steam tracing system (Fig. 10-177) consists of steam supply lines to transport steam from the existing steam lines to the traced pipe, a steam trap to remove the condensate and hold back the steam, and in most cases a condensate return system to return the condensate to the existing condensate return system. In the past, a significant percentage of condensate from steam tracing was simply dumped to drains, but increased energy costs and environmental rules have caused almost all condensate from new steam tracing systems to be returned. This has significantly increased the initial cost of steam tracing systems.

Applications requiring accurate temperature control are generally limited to electric tracing. For example chocolate lines cannot be exposed to steam temperatures or the product will degrade and if caustic soda is heated above 150°F it becomes extremely corrosive to carbon steel pipes.

For some applications, either steam or electricity is simply not available and this makes the decision. It is rarely economic to install a steam boiler just for tracing. Steam tracing is generally considered only when a boiler already exists or is going to be installed for some other primary purpose. Additional electric capacity can be provided in most situations for reasonable costs. It is considerably more expensive to supply steam from a long distance than it is to provide electricity. Unless steam is available close to the pipes being traced, the automatic choice is usually electric tracing.

For most applications, particularly in processing plants, either steam tracing or electric tracing could be used, and the correct choice is dependent on the installed costs and the operating costs of the competing systems.

TABLE 10-62   Steam versus Electric Tracing*

Temperature maintained	TIC			TOC		
	Steam	Electric	Ratio S/E	Steam	Electric	Ratio S/E
50°F	22,265	7,733	2.88	1,671	334	5.00
150°F	22,265	13,113	1.70	4,356	1,892	2.30
250°F	22,807	17,624	1.29	5,348	2,114	2.53
400°F	26,924	14,056	1.92	6,724	3,942	1.71

*Specifications: 400 feet of four-inch pipe, $25/hr labor, $0.07/kWh, $4.00/1,000# steam, 100-foot supply lines. TIC = total installed cost; TOC = total operating costs.

Economics of Steam Tracing versus Electric Tracing   The question of the economics of various tracing systems has been examined thoroughly. All of these papers have concluded that electric tracing is generally less expensive to install and significantly less expensive to operate. Electric tracing has significant cost advantages in terms of installation because less labor is required than steam tracing. However, it is clear that there are some special cases where steam tracing is more economical.

The two key variables in the decision to use steam tracing or electric tracing are the temperature at which the pipe must be maintained and the distance to the supply of steam and a source of electric power.

Table 10-62 shows the installed costs and operating costs for 400 feet of four-inch pipe, maintained at four different temperatures, with supply lengths of 100 ft. for both electricity and steam and $25/hr labor.

The major advantages of a steam tracing system are:

1. *High heat output.*   Due to its high temperature, a steam tracing system provides a large amount of heat to the pipe. There is a very high heat transfer rate between the metallic tracer and a metallic pipe. Even with damage to the insulation system, there is very little chance of a low temperature failure with a steam-tracing system.

2. *High reliability.*   Many things can go wrong with a steam tracing system but, very few of the potential problems lead to a heat tracing failure. Steam traps fail, but they usually fail in the open position,

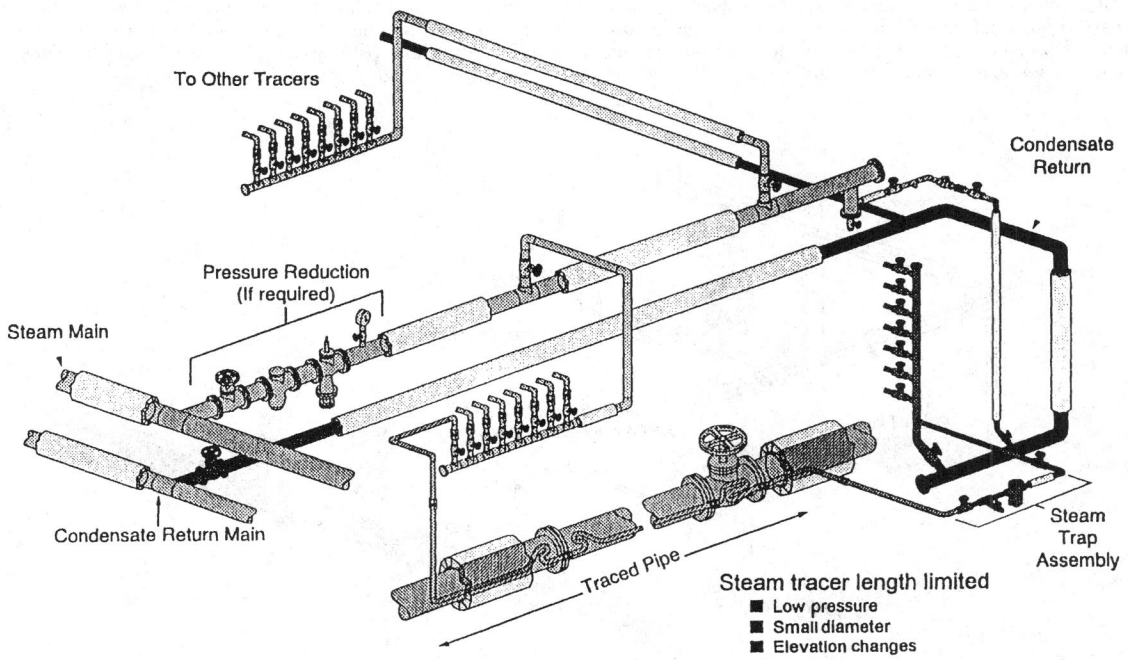

FIG. 10-177   Steam tracing system.

allowing for a continuous flow of steam to the tracer. Other problems such as steam leaks that can cause wet insulation are generally prevented from becoming heat-tracing failures by the extremely high heat output of a steam tracer. Also, a tracing tube is capable of withstanding a large amount of mechanical abuse without failure.

3. *Safety.* While steam burns are fairly common, there are generally fewer safety concerns than with electric tracing.

4. *Common usage.* Steam tracing has been around for many years and many operators are familiar with the system. Because of this familiarity, failures due to operator error are not very common.

The weaknesses of a steam-tracing system are:

1. *High installed costs.* The incremental piping required for the steam supply system and the condensate return system must be installed, insulated, and, in the case of the supply system, additional steam traps are often required. The tracer itself is not expensive, but the labor required for installation is relatively high. Studies have shown that steam tracing systems typically cost from 50 to 150 percent more than a comparable electric tracing system.

2. *Energy inefficiency.* A steam tracing system's total energy use is often more than twenty times the actual energy requirement to keep the pipe at the desired temperature. The steam tracer itself puts out significantly more energy than required. The steam traps use energy even when they are properly operating and waste large amounts of energy when they fail in the open position, which is the common failure mode. Steam leaks waste large amounts of energy, and both the steam supply system and the condensate return system use significant amounts of energy.

3. *Poor temperature control.* A steam tracing system offers very little temperature control capability. The steam is at a constant temperature (50 psig steam is 300°F) usually well above that desired for the pipe. The pipe will reach an equilibrium temperature somewhere between the steam temperature and the ambient temperature. However, the section of pipe against the steam tracer will effectively be at the steam temperature. This is a serious problem for temperature-sensitive fluids such as food products. It also represents a problem with fluids such as bases and acids, which are not damaged by high temperatures but often become extremely corrosive to piping systems at higher temperatures.

4. *High maintenance costs.* Leaks must be repaired and steam traps must be checked and replaced if they have failed. Numerous studies have shown that, due to the energy lost through leaks and failed steam traps, an extensive maintenance program is an excellent investment. Steam maintenance costs are so high that for low-

temperature maintenance applications, total steam operating costs are sometimes greater than electric operating costs, even if no value is placed on the steam.

Electric Tracing An electric tracing system (see Fig. 10-178) consists of an electric heater placed against the pipe under the thermal insulation, the supply of electricity to the tracer, and any control or monitoring system that may be used (optional). The supply of electricity to the tracer usually consists of an electrical panel and electrical conduit or cable trays. Depending on the size of the tracing system and the capacity of the existing electrical system, an additional transformer may be required.

Advantages of Electric Tracing

1. *Lower installed and operating costs.* Most studies have shown that electric tracing is less expensive to install and less expensive to operate. This is true for most applications. However, for some applications, the installed costs of steam tracing are equal to or less than electric tracing.

2. *Reliability.* In the past, electric heat tracing had a well-deserved reputation for poor reliability. However, since the introduction of self-regulating heaters in 1971, the reliability of electric heat tracing has improved dramatically. Self-regulating heaters cannot destroy themselves with their own heat output. This eliminates the most common failure mode of polymer-insulated constant wattage heaters. Also, the technology used to manufacture mineral-insulated cables, high-temperature electric heat tracing, has improved significantly, and this has improved their reliability.

3. *Temperature control.* Even without a thermostat or any control system, an electric tracing system usually provides better temperature control than a steam tracing system. With thermostatic or electronic control, very accurate temperature control can be achieved.

4. *Safety.* The use of self-regulating heaters and ground leakage circuit breakers has answered the safety concerns of most engineers considering electric tracing. Self-regulating heaters eliminate the problems from high-temperature failures, and ground leakage circuit breakers minimize the danger of an electrical fault to ground, causing injury or death.

5. *Monitoring capability.* One question often asked about any heat-tracing system is, "How do I know it is working?" Electric tracing now has available almost any level of monitoring desired. The temperature at any point can be monitored with both high and low alarm capability. This capability has allowed many users to switch to electric tracing with a high degree of confidence.

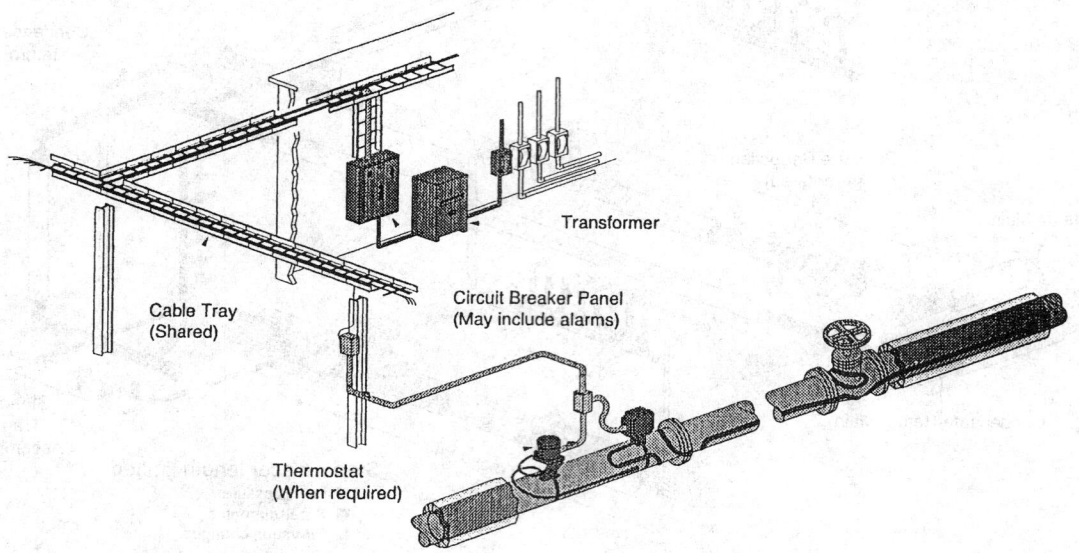

Cable Tray
(Shared)

Transformer

Circuit Breaker Panel
(May include alarms)

Thermostat
(When required)

FIG. 10-178 Electrical heat tracing system.

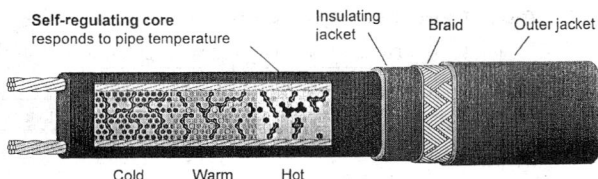

FIG. 10-179   Self-regulating heating cable.

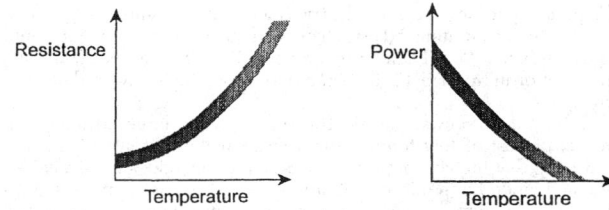

FIG. 10-180   Self regulation.

6. *Energy efficiency.*   Electric heat tracing can accurately provide the energy required for each application without the large additional energy use of a steam system. Unlike steam tracing systems, other parts of the system do not use significant amounts of energy.

Disadvantages of Electric Tracing

1. *Poor reputation.*   In the past, electric tracing has been less than reliable. Due to past failures, some operating personnel are unwilling to take a chance on any electric tracing.

2. *Design requirements.*   A slightly higher level of design expertise is required for electric tracing than for steam tracing.

3. *Lower power output.*   Since electric tracing does not provide a large multiple of the required energy, it is less forgiving to problems such as damaged insulation or below design ambient temperatures. Most designers include a 10 to 20 percent safety factor in the heat loss calculation to cover these potential problems. Also, a somewhat higher than required design temperature is often specified to provide an additional safety margin. For example, many water systems are designed to maintain 50°F to prevent freezing.

Types of Electric Tracing   Self-regulating electric tracing (see Fig. 10-179) is by far the most popular type of electric tracing. The heating element in a self-regulating heater is a conductive polymer between the bus wires. This conductive polymer increases its resistance as its temperature increases. The increase in resistance with temperature causes the heater to lower its heat output at any point where its temperature increases (Fig. 10-180). This self-regulating effect eliminates the most common failure mode of constant wattage electric heaters, which is destruction of the heater by its own heat output.

Because self-regulating heaters are parallel heaters, they may be cut to length at any point without changing their power output per unit of length. This makes them much easier to deal with in the field. They may be terminated, teed, or spliced in the field with hazardous-area-approved components.

MI Cables (mineral insulated cables, Fig. 10-181) are the electric heat tracers of choice for high-temperature applications. High-temperature applications are generally considered to maintain temperatures above 250°F or exposure temperatures above 420°F where self-regulating heaters cannot be used. MI cable consists of one or two heating wires, magnesium oxide insulation (from whence it gets its name), and an outer metal sheath. Today the metal sheath is generally inconel. This eliminates both the corrosion problems with copper sheaths and the stress cracking problems with stainless steel.

MI cables can maintain temperatures up to 1200°F and withstand exposure to up to 1500°F. The major disadvantage of MI cable is that it must be factory-fabricated to length. It is very difficult to terminate or splice the heater in the field. This means pipe measurements are necessary before the heaters are ordered. Also, any damage to an MI cable generally requires a complete new heater. It's not as easy to splice in a good section as with self-regulating heaters.

Polymer-insulated constant wattage electric heaters are slightly cheaper than self-regulating heaters, but they are generally being replaced with self-regulating heaters due to inferior reliability. These heaters tend to destroy themselves with their own heat output when they are overlapped at valves or flanges. Since overlapping self-regulating heaters is the standard installation technique, it is difficult to prevent this technique from being used on the similar-looking constant-wattage heaters.

SECT (skin-effect current tracing) is a special type of electric tracing employing a tracing pipe, usually welded to the pipe being traced, that is used for extremely long lines. With SECT tracing circuits, up to 10 miles can be powered from one power point. All SECT systems are specially designed by heat-tracing vendors.

Impedance tracing uses the pipe being traced to carry the current and generate the heat. Less than 1 percent of electric heat-tracing systems use this method. Low voltages and special electrical isolation techniques are used. Impedance heating is useful when extremely high heat densities are required, like when a pipe containing aluminum metal must be melted from room temperature on a regular basis. Most impedance systems are specially designed by heat tracing vendors.

Choosing the Best Tracing System   Some applications require either steam tracing or electric tracing regardless of the relative economics. For example, a large line that is regularly allowed to cool and needs to be quickly heated would require steam tracing because of its much higher heat output capability. In most heat-up applications, steam tracing is used with heat-transfer cement, and the heat output is increased by a factor of up to 10. This is much more heat than would

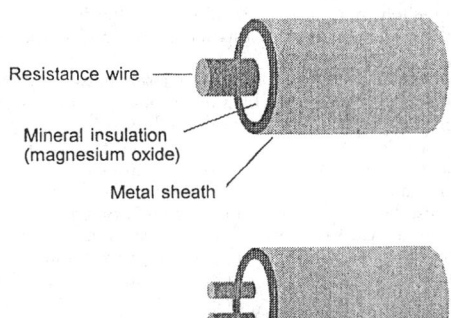

Resistance wire

Mineral insulation (magnesium oxide)

Metal sheath

FIG. 10-181   Mineral insulated cable (MI cable).

Series resistance circuit

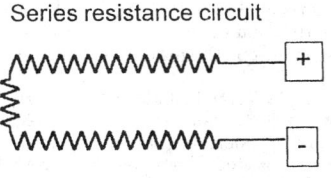

be practical to provide with electric tracing. For example, a half-inch copper tube containing 50 psig steam with heat transfer cement would provide over 1100 BTU/hr/ft to a pipe at 50°F. This is over 300 watts per foot or more than 15 times the output of a high-powered electric tracer.

Table 10-62 shows that electric tracing has a large advantage in terms of cost at low temperatures and smaller but still significant advantages at higher temperatures. Steam tracing does relatively better at higher temperatures because steam tracing supplies significantly more power than is necessary to maintain a pipe at low temperatures. Table 10-62 indicates that there is very little difference between the steam tracing system at 50°F and the system at 250°F. However, the electric system more than doubles in cost between these two temperatures because more heaters, higher powered heaters, and higher temperature heaters are required.

The effect of supply lengths on a 150°F system can be seen from Table 10-63. Steam supply pipe is much more expensive to run than

TABLE 10-63 Effect of Supply Lengths

Ratio of Steam TIC to Electric TIC Maintained at 150°F

Steam supply length	Electric supply length		
	40 feet	100 feet	300 feet
40 feet	1.1	1.0	0.7
100 feet	1.9	1.7	1.1
300 feet	4.9	4.2	2.9

electrical conduit. With each system having relatively short supply lines (40 feet each) the electric system has only a small cost advantage (10 percent, or a ratio of 1.1). This ratio is 2.1 at 50°F and 0.8 at 250°F. However, as the supply lengths increase, electric tracing has a large cost advantage.

STORAGE AND PROCESS VESSELS

STORAGE OF LIQUIDS

Atmospheric Tanks The term *atmospheric tank* as used here applies to any tank that is designed to be used within plus or minus several hundred pascals (a few pounds per square foot) of atmospheric pressure. It may be either open to the atmosphere or enclosed. Minimum cost is usually obtained with a vertical cylindrical shape and a relatively flat bottom at ground level.

American Petroleum Institute (API) The institute has developed a series of atmospheric tank standards and specifications. Some of these are:

API Specification 12B, Bolted Production Tanks
API Specification 12D, Large Welded Production Tanks
API Specification 12F, Small Welded Production Tanks
API Standard 650, Steel Tanks for Oil Storage

American Water Works Association (AWWA) The association has many standards dealing with water handling and storage. A list of its publications is given in the *AWWA Handbook* (annually). AWWA D100, Standard for Steel Tanks—Standpipes, Reservoirs, and Elevated Tanks for Water Storage, contains rules for design and fabrication.

Although AWWA tanks are intended for water, they could be used for the storage of other liquids.

Underwriters Laboratories Inc. has published the following tank standards:

UL 58, Steel Underground Tanks for Flammable and Combustible Liquids
UL 142, Steel Aboveground Tanks for Flammable and Combustible Liquids

UL 58 covers horizontal steel tanks up to 190 m³ (50,000 gal), with a maximum diameter of 3.66 m (12 ft), and a maximum length of six diameters. Thickness and a number of design and fabrication details are given. UL 142 covers horizontal steel tanks up to 190 m³ (50,000 gal) (like UL 58), and vertical tanks up to 10.7-m (35-ft) height. Thickness and other details are given. The maximum diameter for a vertical tank is not specified.

The Underwriters Standards overlap API, but include tanks that are too small for API Standards. Underwriters Standards are, however, not as detailed as API and therefore put more responsibility on the designer. They do not specify grades of steel other than requiring weldability. Designers should also place their own limits on the diameter (or thickness) of vertical tanks. They can obtain guidance from API.

Posttensioned Concrete This material is frequently used for tanks to about 57,000 m³ (15 × 10⁶ gal), usually containing water. Their design is treated in detail by Creasy (*Prestressed Concrete Cylindrical Tanks*, Wiley, New York, 1961). For the most economical design of

large open tanks at ground levels, he recommends limiting vertical height to 6 m (20 ft). Seepage can be a problem if unlined concrete is used with some liquids (e.g., gasoline).

Elevated Tanks These can supply a large flow when required, but pump capacities need be only for average flow. Thus, they may save on pump and piping investment. They also provide flow after pump failure, an important consideration for fire systems.

Open Tanks These may be used to store materials that will not be harmed by water, weather, or atmospheric pollution. Otherwise, a roof, either fixed or floating, is required. **Fixed roofs** are usually either domed or coned. Large tanks have coned roofs with intermediate supports. Since negligible pressure is involved, snow and wind are the principal design loads. Local building codes often give required values.

Fixed-roof atmospheric tanks require **vents** to prevent pressure changes which would otherwise result from temperature changes and withdrawal or addition of liquid. API Standard 2000, Venting Atmospheric and Low Pressure Storage Tanks, gives practical rules for vent design. The principles of this standard can be applied to fluids other than petroleum products. Excessive losses of volatile liquids, particularly those with flash points below 38°C (100°F), may result from the use of open vents on fixed-roof tanks. Sometimes vents are manifolded and led to a vent tank, or the vapor may be extracted by a recovery system.

An effective way of preventing vent loss is to use one of the many types of **variable-volume tanks.** These are built under API Standard 650. They may have floating roofs of the double-deck or the single-deck type. There are lifter-roof types in which the roof either has a skirt moving up and down in an annular liquid seal or is connected to the tank shell by a flexible membrane. A fabric expansion chamber housed in a compartment on top of the tank roof also permits variation in volume.

Floating Roofs These must have a seal between the roof and the tank shell. If not protected by a fixed roof, they must have drains for the removal of water, and the tank shell must have a "wind girder" to avoid distortion. An industry has developed to retrofit existing tanks with floating roofs. Much detail on the various types of tank roofs is given in manufacturers' literature. Figure 10-182 shows types. These roofs cause less condensation buildup and are highly recommended.

Pressure Tanks Vertical cylindrical tanks constructed with domed or coned roofs, which operate at pressures above several hundred pascals (a few pounds per square foot) but which are still relatively close to atmospheric pressure, can be built according to API Standard 650. The pressure force acting against the roof is transmitted to the shell, which may have sufficient weight to resist it. If not, the uplift will act on the tank bottom. The strength of the bottom, however, is limited, and if it is not sufficient, an anchor ring or a heavy

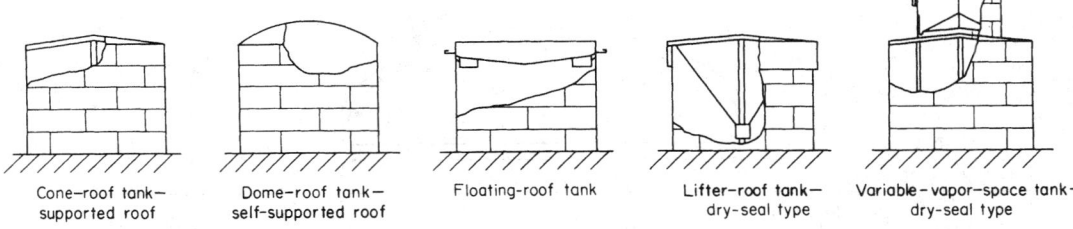

FIG. 10-182 Some types of atmospheric storage tanks.

foundation must be used. In the larger sizes uplift forces limit this style of tank to very low pressures.

As the size or the pressure goes up, curvature on all surfaces becomes necessary. Tanks in this category, up to and including a pressure of 103.4 kPa (15 lbf/in²), can be built according to API Standard 620. Shapes used are spheres, ellipsoids, toroidal structures, and circular cylinders with torispherical, ellipsoidal, or hemispherical heads. The ASME Pressure Vessel Code (Sec. VIII of the ASME Boiler and Pressure Vessel Code), although not required below 103.4 kPa (15 lbf/in²), is also useful for designing such tanks.

Tanks that could be subjected to vacuum should be provided with vacuum-breaking valves or be designed for vacuum (external pressure). The ASME Pressure Vessel Code contains design procedures.

Calculation of Tank Volume A tank may be a single geometrical element, such as a cylinder, a sphere, or an ellipsoid. It may also have a compound form, such as a cylinder with hemispherical ends or a combination of a toroid and a sphere. To determine the volume, each geometrical element usually must be calculated separately. Calculations for a full tank are usually simple, but calculations for partially filled tanks may be complicated.

To calculate the volume of a **partially filled horizontal cylinder** refer to Fig. 10-183. Calculate the angle α in degrees. Any units of length can be used, but they must be the same for H, R, and L. The liquid volume

$$V = LR^2\left(\frac{\alpha}{57.30} - \sin\alpha\cos\alpha\right) \qquad (10\text{-}109)$$

This formula may be used for any depth of liquid between zero and the full tank, provided the algebraic signs are observed. If H is greater than R, $\sin\alpha\cos\alpha$ will be negative and thus will add numerically to $\alpha/57.30$. Table 10-64 gives liquid volume, for a partially filled horizontal cylinder, as a fraction of the total volume, for the dimensionless ratio H/D or $H/2R$.

The **volumes of heads** must be calculated separately and added to the volume of the cylindrical portion of the tank. The four types of heads most frequently used are the standard dished head,° torispherical or ASME head, ellipsoidal head, and hemispherical head. Dimensions and volumes for all four of these types are given in *Lukens Spun Heads*, Lukens Inc., Coatesville, Pennsylvania. Approximate volumes can also be calculated by the formulas in Table 10-65. Consistent units must be used in these formulas.

A partially filled horizontal tank requires the determination of the partial volume of the heads. The Lukens catalog gives approximate volumes for partially filled (axis horizontal) standard ASME and ellipsoidal heads. A formula for **partially filled heads**, by Doolittle [*Ind. Eng. Chem.*, **21**, 322–323 (1928)], is

$$V = 0.215\,H^2\,(3R - H) \qquad (10\text{-}110)$$

where in consistent units V = volume, R = radius, and H = depth of liquid. Doolittle made some simplifying assumptions which affect the volume given by the equation, but the equation is satisfactory for

° The standard dished head does not comply with the ASME Pressure Vessel Code.

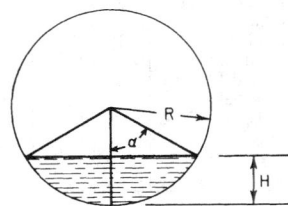

FIG. 10-183 Calculation of partially filled horizontal tanks. H = depth of liquid; R = radius; D = diameter; L = length; α = half of the included angle; and $\cos\alpha = 1 - H/R = 1 - 2H/D$.

determining the volume as a fraction of the entire head. This fraction, calculated by Doolittle's formula, is given in Table 10-66 as a function of H/D_i (H is the depth of liquid, and D_i is the inside diameter). Table 10-66 can be used for standard dished, torispherical, ellipsoidal, and hemispherical heads with an error of less than 2 percent of the volume of the entire head. The error is zero when $H/D_i = 0$, 0.5, and 1.0. Table 10-66 cannot be used for conical heads.

When a tank volume cannot be calculated or when greater precision is required, **calibration** may be necessary. This is done by draining (or filling) the tank and measuring the volume of liquid. The

TABLE 10-64 Volume of Partially Filled Horizontal Cylinders

H/D	Fraction of volume	H/D	Fraction of volume	H/D	Fraction of volume	H/D	Fraction of volume
0.01	0.00169	0.26	0.20660	0.51	0.51273	0.76	0.81545
.02	.00477	.27	.21784	.52	.52546	.77	.82625
.03	.00874	.28	.22921	.53	.53818	.78	.83688
.04	.01342	.29	.24070	.54	.55088	.79	.84734
.05	.01869	.30	.25231	.55	.56356	.80	.85762
.06	.02450	.31	.26348	.56	.57621	.81	.86771
.07	.03077	.32	.27587	.57	.58884	.82	.87760
.08	.03748	.33	.28779	.58	.60142	.83	.88727
.09	.04458	.34	.29981	.59	.61397	.84	.89673
.10	.05204	.35	.31192	.60	.62647	.85	.90594
.11	.05985	.36	.32410	.61	.63892	.86	.91491
.12	.06797	.37	.33636	.62	.65131	.87	.92361
.13	.07639	.38	.34869	.63	.66364	.88	.93203
.14	.08509	.39	.36108	.64	.67590	.89	.94015
.15	.09406	.40	.37353	.65	.68808	.90	.94796
.16	.10327	.41	.38603	.66	.70019	.91	.95542
.17	.11273	.42	.39858	.67	.71221	.92	.96252
.18	.12240	.43	.41116	.68	.72413	.93	.96923
.19	.13229	.44	.42379	.69	.73652	.94	.97550
.20	.14238	.45	.43644	.70	.74769	.95	.98131
.21	.15266	.46	.44912	.71	.75930	.96	.98658
.22	.16312	.47	.46182	.72	.77079	.97	.99126
.23	.17375	.48	.47454	.73	.78216	.98	.99523
.24	.18455	.49	.48727	.74	.79340	.99	.99831
.25	.19550	.50	.50000	.75	.80450	1.00	1.00000

TABLE 10-65   Volumes of Heads*

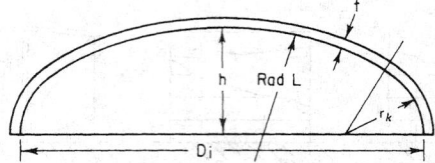

Type of head	Knuckle radius, r_k	h	L	Volume	% Error	Remarks
Standard dished	Approx. $3t$		Approx. D_i	Approx. $0.050D_i^3 + 1.65tD_i^2$	± 10	h varies with t
Torispherical or A.S.M.E.	$0.06L$		D_i	$0.0809D_i^3$	± 0.1	r_k must be the larger of $0.06L$ and $3t$
Torispherical or A.S.M.E.	$3t$		D_i	Approx. $0.513hD_i^2$	± 8	
Ellipsoidal				$\pi D_i^2 h/6$	0	
Ellipsoidal		$D_i/4$		$\pi D_i^3/24$	0	Standard proportions
Hemispherical		$D_i/2$	$D_i/2$	$\pi D_i^3/12$	0	
Conical				$\pi h(D_i^2 + D_i d + d^2)/12$	0	Truncated cone h = height d = diameter at small end

*Use consistent units.

measurement may be made by weighing, by a calibrated fluid meter, or by repeatedly filling small measuring tanks which have been calibrated by weight.

Container Materials and Safety   Storage tanks are made of almost any structural material. Steel and reinforced concrete are most widely used. Plastics and glass-reinforced plastics are used for tanks up to about 230 m³ (60,000 gal). Resistance to corrosion, light weight, and lower cost are their advantages. Plastic and glass coatings are also applied to steel tanks. Aluminum and other nonferrous metals are used when their special properties are required. When expensive metals such as tantalum are required, they may be applied as tank linings or as clad metals.

Some grades of steel listed by API and AWWA Standards are of lower quality than is customarily used for pressure vessels. The stresses allowed by these standards are also higher than those allowed by the ASME Pressure Vessel Code. Small tanks containing nontoxic substances are not particularly hazardous and can tolerate a reduced factor of safety. Tanks containing highly toxic substances and very large tanks containing any substance can be hazardous. The designer must consider the magnitude of the hazard. The possibility of brittle behavior of ferrous metal should be taken into account in specifying materials (see subsection "Safety in Design").

TABLE 10-66   Volume of Partially Filled Heads on Horizontal Tanks*

H/D_i	Fraction of volume	H/D_i	Fraction of volume	H/D_i	Fraction of volume	H/D_i	Fraction of volume
0.02	0.0012	0.28	0.1913	0.52	0.530	0.78	0.8761
.04	.0047	.30	.216	.54	.560	.80	.8960
.06	.0104	.32	.242	.56	.590	.82	.9145
.08	.0182	.34	.268	.58	.619	.84	.9314
.10	.0280	.36	.295	.60	.648	.86	.9467
.12	.0397	.38	.323	.62	.677	.88	.9603
.14	.0533	.40	.352	.64	.705	.90	.9720
.16	.0686	.42	.381	.66	.732	.92	.9818
.18	.0855	.44	.410	.68	.758	.94	.9896
.20	.1040	.46	.440	.70	.784	.96	.9953
.22	.1239	.48	.470	.72	.8087	.98	.9988
.24	.1451	.50	.500	.74	.8324	1.00	1.0000
.26	.1676			.76	.8549		

*Based on Eq. (10-110).

Volume 1 of National Fire Codes (National Fire Protection Association, Quincy, Massachusetts) contains recommendations (Code 30) for venting, drainage, and dike construction of tanks for **flammable liquids.**

Container Insulation   Tanks containing materials above atmospheric temperature may require insulation to reduce loss of heat. Almost any of the commonly used insulating materials can be employed. Calcium silicate, glass fiber, mineral wool, cellular glass, and plastic foams are among those used. Tanks exposed to weather must have jackets or protective coatings, usually asphalt, to keep water out of the insulation.

Tanks with contents at lower than atmospheric temperature may require insulation to minimize heat absorption. The insulation must have a vapor barrier at the outside to prevent condensation of atmospheric moisture from reducing its effectiveness. An insulation not damaged by moisture is preferable. The insulation techniques presently used for refrigerated systems can be applied (see subsection "Low-Temperature and Cryogenic Storage").

Tank Supports   Large vertical atmospheric steel tanks may be built on a base of about 150 cm (6 in) of sand, gravel, or crushed stone if the subsoil has adequate bearing strength. It can be level or slightly coned, depending on the shape of the tank bottom. The porous base provides drainage in case of leaks. A few feet beyond the tank perimeter the surface should drop about 1 m (3 ft) to assure proper drainage of the subsoil. API Standard 650, Appendix B, and API Standard 620, Appendix C, give recommendations for tank foundations.

The bearing pressure of the tank and contents must not exceed the **bearing strength** of the soil. Local building codes usually specify allowable soil loading. Some approximate bearing values are:

	kPa	Tons/ft²
Soft clay (can be crumbled between fingers)	100	1
Dry fine sand	200	2
Dry fine sand with clay	300	3
Coarse sand	300	3
Dry hard clay (requires a pick to dig it)	350	3.5
Gravel	400	4
Rock	1000–4000	10–40

For high, heavy tanks, a foundation ring may be needed. Prestressed concrete tanks are sufficiently heavy to require foundation rings. Foundations must extend below the frost line. Some tanks that are not flat-bottomed may also be supported by soil if it is suitably graded and drained. When soil does not have adequate bearing strength, it may be excavated and backfilled with a suitable soil, or piles capped with a concrete mat may be required.

Spheres, spheroids, and toroids use steel or concrete saddles or are supported by columns. Some may rest directly on soil. Horizontal cylindrical tanks should have two rather than multiple saddles to avoid indeterminate load distribution. Small horizontal tanks are sometimes supported by legs. Most tanks must be designed to resist the reactions of the saddles or legs, and they may require reinforcing. Neglect of this can cause collapse. Tanks without stiffeners usually need to make contact with the saddles on at least 2.1 rad (120°) of their circumference. An elevated steel tank may have either a circle of steel columns or a large central steel standpipe. Concrete tanks usually have concrete columns. Tanks are often supported by buildings.

Pond and Underground Storage Low-cost liquid materials, if they will not be damaged by rain or atmospheric pollution, may be stored in **ponds.** A pond may be excavated or formed by damming a ravine. To prevent loss by seepage, the soil which will be submerged may require treatment to make it sufficiently impervious. This can also be accomplished by lining the pond with concrete, plastic film, or some other barrier. Prevention of seepage is especially necessary if the pond contains material that could contaminate present or future water supplies.

Underground Storage Investment in both storage facilities and land can often be reduced by underground storage. Porous media between impervious rocks are also used. Cavities can be formed in salt domes and beds by dissolving the salt and pumping it out. Geological formations suitable for some of these methods can be found in numerous locations. The most extensive application has been the storage of petroleum products, both liquid and gaseous, in the southwestern part of the United States. Chemicals have been handled in this way. Information on some installations is given in articles by Billue, Haight and Bernard, and Nixon [*Pet. Refiner*, **33**, 108–116 (1954)]. Another useful reference is *Relationships between Selected Physical Parameters and Cost Responses for the Deep-Well Disposal of Aqueous Industrial Wastes*, Technical Report to the U.S. Public Health Service, EHE 07-6801, CRWR28, by the Center for Research in Water Resources, University of Texas, Austin, August 1968. It contains an extensive bibliography.

Water is also stored underground when suitable formations are available. When an excess of surface water is available part of the time, the excess is treated, if required, and pumped into the ground to be retrieved when needed. Sometimes pumping is unnecessary, and it will seep into the ground.

Underground chambers are also constructed in frozen earth (see subsection "Low-Temperature and Cryogenic Storage"). Underground tunnel or tank storage is often the most practical way of storing hazardous or radioactive materials. A cover of 30 m (100 ft) of rock or dense earth can exert a pressure of about 690 kPa (100 lbf/in^2).

STORAGE OF GASES

Gas Holders Gas is sometimes stored in expandable gas holders of either the liquid-seal or dry-seal type. The liquid-seal holder is a familiar sight. It has a cylindrical container, closed at the top, and varies its volume by moving it up and down in an annular water-filled seal tank. The seal tank may be staged in several lifts (as many as five). Seal tanks have been built in sizes up to 280,000 m^3 (10×10^6 ft^3). The dry-seal holder has a rigid top attached to the sidewalls by a flexible fabric diaphragm which permits it to move up and down. It does not involve the weight and foundation costs of the liquid-seal holder. Additional information on gas holders can be found in *Gas Engineers Handbook,* Industrial Press, New York, 1966.

Solution of Gases in Liquids Certain gases will dissolve readily in liquids. In some cases in which the quantities are not large, this may be a practical storage procedure. Examples of gases that can be handled in this way are ammonia in water, acetylene in acetone, and hydrogen chloride in water. Whether or not this method is used depends mainly on whether the end use requires the anhydrous or the liquid state. Pressure may be either atmospheric or elevated. The

solution of acetylene in acetone is also a safety feature because of the instability of acetylene.

Storage in Pressure Vessels, Bottles, and Pipe Lines The distinction between pressure vessels, bottles, and pipes is arbitrary. They can all be used for storing gases under pressure. A storage pressure vessel is usually a permanent installation. Storing a gas under pressure not only reduces its volume but also in many cases liquefies it at ambient temperature. Some gases in this category are carbon dioxide, several petroleum gases, chlorine, ammonia, sulfur dioxide, and some types of Freon. Pressure tanks are frequently installed underground.

Liquefied petroleum gas (LPG) is the subject of API Standard 2510, The Design and Construction of Liquefied Petroleum Gas Installations at Marine and Pipeline Terminals, Natural Gas Processing Plants, Refineries, and Tank Farms. This standard in turn refers to:

1. National Fire Protection Association (NFPA) Standard 58, Standard for the Storage and Handling of Liquefied Petroleum Gases
2. NFPA Standard 59, Standard for the Storage and Handling of Liquefied Petroleum Gases at Utility Gas Plants
3. NFPA Standard 59A, Standard for the Production, Storage, and Handling of Liquefied Natural Gas (LNG)

The API Standard gives considerable information on the construction and safety features of such installations. It also recommends minimum distances from property lines. The user may wish to obtain added safety by increasing these distances.

The term **bottle** is usually applied to a pressure vessel that is small enough to be conveniently portable. Bottles range from about 57 L (2 ft^3) down to CO$_2$ capsules of about 16.4 mL (1 in^3). Bottles are convenient for small quantities of many gases, including air, hydrogen, nitrogen, oxygen, argon, acetylene, Freon, and petroleum gas. Some are one-time-use disposable containers.

Pipe Lines A pipe line is not ordinarily a storage device. Pipes, however, have been buried in a series of connected parallel lines and used for storage. This avoids the necessity of providing foundations, and the earth protects the pipe from extremes of temperature. The economics of such an installation would be doubtful if it were designed to the same stresses as a pressure vessel. Storage is also obtained by increasing the pressure in operating pipe lines and thus using the pipe volume as a tank.

Low-Temperature and Cryogenic Storage This type is used for gases that liquefy under pressure at atmospheric temperature. In cryogenic storage the gas is at, or near to, atmospheric pressure and remains liquid because of low temperature. A system may also operate with a combination of pressure and reduced temperature. The term "cryogenic" usually refers to temperatures below −101°C (−150°F). Some gases, however, liquefy between −101°C and ambient temperatures. The principle is the same, but cryogenic temperatures create different problems with insulation and construction materials.

The liquefied gas must be maintained at or below its boiling point. Refrigeration can be used, but the usual practice is to cool by evaporation. The quantity of liquid evaporated is minimized by insulation. The vapor may be vented to the atmosphere (wasteful), it may be compressed and reliquefied, or it may be used.

At very low temperatures with liquid air and similar substances, the tank may have double walls with the interspace evacuated. The well-known Dewar flask is an example. Large tanks and even pipe lines are now built this way. An alternative is to use double walls without vacuum but with an insulating material in the interspace. Perlite and plastic foams are two insulating materials employed in this way. Sometimes both insulation and vacuum are used.

Materials Materials for liquefied-gas containers must be suitable for the temperatures, and they must not be brittle. Some carbon steels can be used down to −59°C (−75°F), and low-alloy steels to −101°C (−150°F) and sometimes −129°C (−200°F). Below these temperatures austenitic stainless steel (AISI 300 series) and aluminum are the principal materials.

Low temperatures involve problems of **differential thermal expansion.** With the outer wall at ambient temperature and the inner wall at the liquid boiling point, relative movement must be accommodated. Some systems for accomplishing this are patented. The Gaz

Transport of France reduces dimensional change by using a thin inner liner of Invar. Another patented French system accommodates this change by means of the flexibility of thin metal which is creased. The creases run in two directions, and the form of the crossings of the creases is a feature of the system.

Low-temperature tanks may be installed underground to take advantage of the insulating value of the earth. Frozen-earth storage is also used. The frozen earth forms the tank. Some installations using this technique have been unsuccessful because of excessive heat absorption.

COST OF STORAGE FACILITIES

Contractors' bids offer the most reliable information on cost. Order-of-magnitude costs, however, may be required for preliminary studies. One way of estimating them is to obtain cost information from similar facilities and scale it to the proposed installation. Costs of steel storage tanks and vessels have been found to vary approximately as the 0.6 to 0.7 power of their weight [see Happel, *Chemical Process Economics,* Wiley, 1958, p. 267; also Williams, *Chem. Eng.,* **54**(12), 124 (1947)]. All estimates based on the costs of existing equipment must be corrected for changes in the price index from the date when the equipment was built. Considerable uncertainty is involved in adjusting data more than a few years old.

Based on a survey in 1994 for storage tanks, the prices for field-erected tanks are for multiple-tank installations erected by the contractor on foundations provided by the owner. Some cost information on tanks is given in various references cited in Sec. 25. Cost data vary considerably from one reference to another.

Prestressed (posttensioned) concrete tanks cost about 20 percent more than steel tanks of the same capacity. Once installed, however, concrete tanks require very little maintenance. A true comparison with steel would, therefore, require evaluating the maintenance cost of both types.

BULK TRANSPORT OF FLUIDS

Transportation is often an important part of product cost. Bulk transportation may provide significant savings. When there is a choice between two or more forms of transportation, the competition may result in rate reduction. Transportation is subject to considerable regulation, which will be discussed in some detail under specific headings.

Pipe Lines For quantities of fluid which an economic investigation indicates are sufficiently large and continuous to justify the investment, pipe lines are one of the lowest-cost means of transportation. They have been built up to 1.22 m (48 in) or more in diameter and about 3200 km (2000 mi) in length for oil, gas, and other products. Water is usually not transported more than 160 to 320 km (100 to 200 miles), but the conduits may be much greater than 1.22 m (48 in) in diameter. Open canals are also used for water transportation.

Petroleum pipe lines before 1969 were built to ASA (now ANSI) Standard B31.4 for liquids and Standard B31.8 for gas. These standards were seldom mandatory because few states adopted them. The U.S. Department of Transportation (DOT), which now has responsibility for pipe-line regulation, issued Title 49, Part 192—Transportation of Natural Gas and Other Gas by Pipeline: Minimum Safety Standards, and Part 195—Transportation of Liquids by Pipeline. These contain considerable material from B31.4 and B31.8. They allow generally higher stresses than the ASME Pressure Vessel Code would allow for steels of comparable strength. The enforcement of their regulations is presently left to the states and is therefore somewhat uncertain.

Pipe-line pumping stations usually range from 16 to 160 km (10 to 100 miles) apart, with maximum pressures up to 6900 kPa (1000 lbf/in^2) and velocities up to 3 m/s (10 ft/s) for liquid. Gas pipe lines have higher velocities and may have greater spacing of stations.

Tanks Tank cars (single and multiple tank), tank trucks, portable tanks, drums, barrels, carboys, and cans are used to transport fluids (see Figs. 10-184–10-186). Interstate transportation is regulated by the DOT. There are other regulating agencies—state, local, and private. Railroads make rules determining what they will accept, some states require compliance with DOT specifications on intrastate movements, and tunnel authorities as well as fire chiefs apply restrictions. Water shipments involve regulations of the U.S. Coast Guard. The American Bureau of Shipping sets rules for design and construction which are recognized by insurance underwriters.

The most pertinent **DOT regulations** (*Code of Federal Regulations,* Title 18, Parts 171–179 and 397) were published by R. M. Graziano (then agent and attorney for carriers and freight forwarders) in his tariff titled *Hazardous Materials Regulations of the Department of Transportation* (1978). New tariffs identified by number are issued at intervals, and interim revisions are sent out. Agents change at intervals.

Graziano's tariff lists many regulated (dangerous) commodities (Part 172, DOT regulations) for transportation. This includes those that are poisonous, flammable, oxidizing, corrosive, explosive, radioactive, and compressed gases. Part 178 covers specifications for all types of containers from carboys to large portable tanks and tank trucks. Part 179 deals with tank-car construction.

An Association of American Railroads (AAR) publication, *Specifications for Tank Cars,* covers many requirements beyond the DOT regulations.

Some additional details are given later. Because of frequent changes, it is always necessary to check the latest rules. The **shipper,** not the carrier, has the ultimate responsibility for shipping in the correct container.

Tank Cars These range in size from about 7.6 to 182 m^3 (2000 to 48,000 gal), and a car may be single or multiunit. The DOT now limits them to 130 m^3 (34,500 gal) and 120,000 kg (263,000 lb) gross mass. Large cars usually result in lower investment per cubic meter and take lower shipping rates. Cars may be insulated to reduce heating or cooling of the contents. Certain liquefied gases may be carried in insulated cars; temperatures are maintained by evaporation (see subsection "Low-Temperature and Cryogenic Storage"). Cars may be heated by steam coils or by electricity. Some products are loaded hot, solidify in transport, and are melted for removal. Some low-temperature cargoes must be unloaded within a given time (usually 30 days) to prevent pressure buildup.

Tank cars are classified as pressure or general-purpose. Pressure cars have relief-valve settings of 517 kPa (75 lbf/in^2) and above. Those designated as general-purpose cars are, nevertheless, pressure vessels and may have relief valves or rupture disks. The DOT specification code number indicates the type of car. For instance, 105A500W indicates a pressure car with a test pressure of 3447 kPa (500 lbf/in^2) and a relief-valve setting of 2585 kPa (375 lbf/in^2). In most cases, loading and unloading valves, safety valves, and vent valves must be in a dome or an enclosure.

Companies shipping dangerous materials sometimes build tank cars with metal thicker than required by the specifications in order to reduce the possibility of leakage during a wreck or fire. The punching of couplers or rail ends into heads of tanks is a hazard.

Older tank cars have a center sill or beam running the entire length of the car. Most modern cars have no continuous sill, only short stub sills at each end. Cars with full sills have tanks anchored longitudinally at the center of the sill. The anchor is designed to be weaker than either the tank shell or the doubler plate between anchor and shell. Cars with stub sills have similar safeguards. Anchors and other parts are designed to meet AAR requirements.

The impact forces on car couplers put high stresses in sills, anchors, and doublers. This may start fatigue cracks in the shell, particularly at the corners of welded doubler plates. With brittle steel in cold weather, such cracks sometimes cause complete rupture of the tank. Large end radii on the doublers and tougher steels will reduce this hazard. Inspection of older cars can reveal cracks before failure.

A difference between tank cars and most pressure vessels is that tank cars are designed in terms of the theoretical ultimate or bursting strength of the tank. The test pressure is usually 40 percent of the bursting pressure (sometimes less). The safety valves are set at 75 percent of the test pressure. Thus, the maximum operating pressure is usually 30 percent of the bursting pressure. This gives a nominal factor of safety of 3.3, compared with 4.0 for Division 1 of the ASME Pressure Vessel Code.

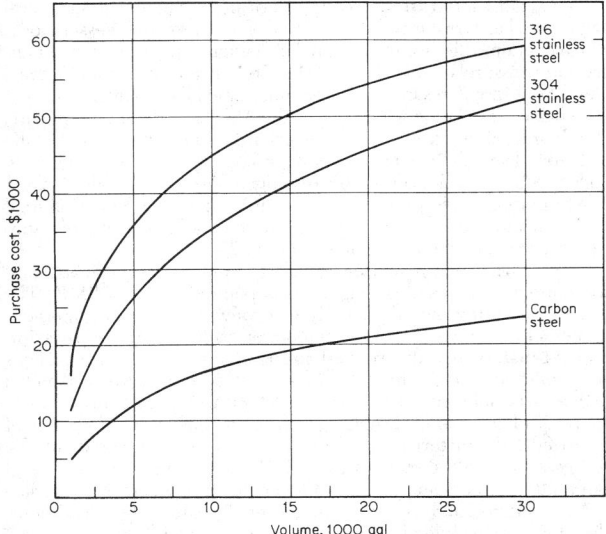

FIG. 10-184 Cost of shop-fabricated tanks in mid-1980 with ¼-in walls. Multiplying factors on carbon steel costs for other materials are: carbon steel, 1.0; rubber-lined carbon steel, 1.5; aluminum, 1.6; glass-lined carbon steel, 4.5; and fiber-reinforced plastic, 0.75 to 1.5. Multiplying factors on type 316 stainless-steel costs for other materials are: 316 stainless steel, 1.0; Monel, 2.0; Inconel, 2.0; nickel, 2.0; titanium, 3.2; and Hastelloy C, 3.8. Multiplying factors for wall thicknesses different from ¼ in are:

Thickness, in	Carbon steel	304 stainless steel	316 stainless steel
½	1.4	1.8	1.8
¾	2.1	2.5	2.6
1	2.7	3.3	3.5

To convert gallons to cubic meters, multiply by 3.785×10^{-3}.

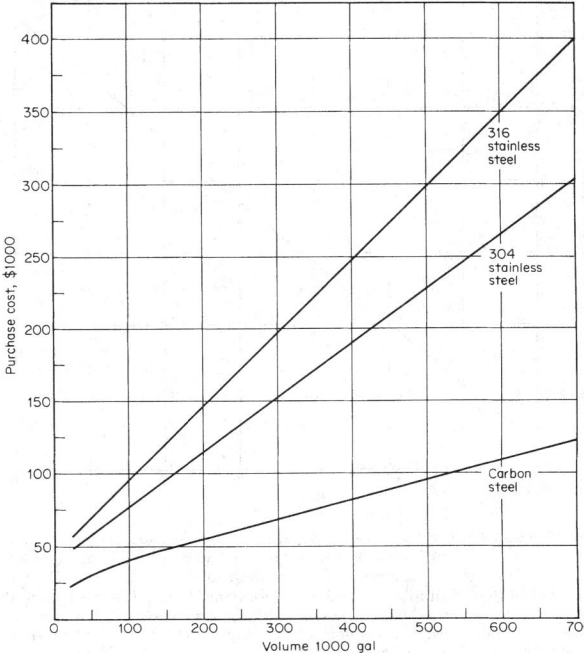

FIG. 10-185 Cost of small field-erected tanks in mid-1980, including stairs, platforms, and a normal complement of nozzles. The carbon steel curve is for API Standard 650 tanks, and the others are for stainless-steel tanks for atmospheric pressure with flat bottoms. The curves are for tanks purchased in quantities of three or more at a Gulf Coast site. Multiplying factors for other materials are: 316 stainless steel, 1.5; Monel, 2.0; Inconel, 2.0; nickel, 2.0; titanium, 3.2; and Hastelloy C, 3.8. Allowances should be added to the factored costs as follows: 10 percent for stiffener rings, 20 percent for API Standard 620, 15 percent for quantity of one tank, 10 percent for quantity of two tanks, 15 percent of steel cost for congested working area, 50 percent of steel cost for an integral steel dike. To convert gallons to cubic meters, multiply by 3.785×10^{-3}.

The DOT rules require that pressure cars have relief valves designed to limit pressure to 82.5 percent (with certain exceptions) of test pressure (110 percent of maximum operating pressure) when exposed to fire. Appendix A of AAR Specifications deals with the flow capacity of relief devices. The formulas apply to cars in the upright position with the device discharging vapor. They may not protect the car adequately when it is overturned and the device is discharging liquid.

Appendix B of AAR Specifications deals with the certification of facilities. Fabrication, repairing, testing, and specialty work on tank cars must be done in certified facilities. The AAR certifies shops to build cars of certain materials, to do test work on cars, or to make certain repairs and alterations.

Tank Trucks These trucks may have single, compartmented, or multiple tanks. Many of their requirements are similar to those for tank cars, except that thinner shells are permitted in most cases. Trucks for nonhazardous materials are subject to few regulations other than the normal highway laws governing all motor vehicles. But trucks carrying hazardous materials must comply with DOT regulations, Parts 173, 177, 178, and 397. Maximum weight, axle loading, and length are governed by state highway regulations. Many states have limits in the vicinity of 31,750 kg (70,000 lb) total mass, 14,500 kg (32,000 lb) for tandem axles, and 18.3 m (60 ft) or less overall length. Some allow tandem trailers.

Truck cargo tanks (for dangerous materials) are built under Part 173 and Subpart J of Part 178, DOT regulations. This includes Specifications MC-306, MC-307, MC-312, and MC-331. MC-331 is required for compressed gas. Subpart J requires tanks for pressures above 345 kPa (50 lbf/in²) in one case and 103 kPa (15 lbf/in²) in another to be built according to the ASME Pressure Vessel Code. A particular issue of the code is specified.

Because of the demands of highway service, the DOT specifications have a number of requirements in addition to the ASME Code. These include design for impact forces and rollover protection for fittings.

Portable tanks, drums, or bottles are shipped by rail, ship, air, or truck. Portable tanks containing hazardous materials must conform to DOT regulations, Parts 173 and 178, Subpart H.

Some tanks are designed to be shipped by trailer and transferred to railcars or ships (see following discussion).

Marine Transportation Seagoing **tankers** are for high tonnage. The traditional tanker uses the ship structure as a tank. It is subdivided into a number of tanks by means of transverse bulkheads and a centerline bulkhead. More than one product can be carried. An elaborate piping system connects the tanks to a pumping plant which can discharge or transfer the cargo. Harbor and docking facilities appear to be the only limit to tanker size. The largest tanker size to date is about 500,000 deadweight tons. In the United States, tankers are built to specifications of the American Bureau of Shipping and the U.S. Coast Guard.

Low-temperature liquefied gases are shipped in special ships with insulation between the hull and an inner tank. Poisonous materials are shipped in separate tanks built into the ship. This prevents tank leakage from contaminating harbors. Separate tanks are also used to transport pressurized gases.

Barges are used on inland waterways. Popular sizes are up to 16 m (52½) wide by 76 m (250 ft) long, with 2.6 m (8½ ft) to 4.3 m (14 ft) draft. Cargo requirements and waterway limitations determine design. Use of barges of uniform size facilitates rafting them together.

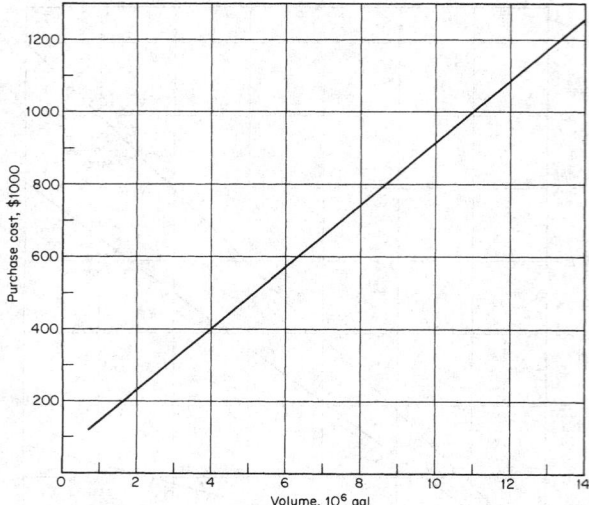

FIG. 10-186 Cost of large field-erected tanks in mid-1990, including stairs, platforms, and a normal complement of nozzles; curve for carbon steel API Standard 650 tanks in quantities of three or more at a Gulf Coast site. For type 304 stainless steel, multiply cost by 2.5; and for type 316 stainless steel, multiply cost by 3.5. Allowances should be added to the factored costs as follows: 10 percent for stiffener rings, 20 percent for API Standard 620, 15 percent for quantity of one tank, 10 percent for quantity of two tanks, and 15 percent of carbon steel cost for a congested working area. To convert gallons to cubic meters, multiply by 3.785×10^{-3}.

Portable tanks may be stowed in the holds of conventional cargo ships or special container ships, or they may be fastened on deck.

Container ships have guides in the hold and on deck which hold boxlike containers or tanks. The tank is latched to a trailer chassis and hauled to shipside. A movable gantry, sometimes permanently installed on the ship, hoists the tank from the trailer and lowers it into the guides on the ship. This system achieves large savings in labor, but its application is sometimes limited by lack of agreement between ship operators and unions.

Portable tanks for regulated commodities in marine transportation must be designed and built under Coast Guard regulations (see discussion under "Pressure Vessels").

Materials of Construction for Bulk Transport Because of the more severe service, construction materials for transportation usually are more restricted than for storage. Most large pipe lines are constructed of steel conforming to API Specification 5L or 5LX. Most tanks (cars, etc.) are built of pressure-vessel steels or AAR specification steels, with a few of aluminum or stainless steel. Carbon steel tanks may be lined with rubber, plastic, nickel, glass, or other materials. In many cases this is practical and cheaper than using a stainless-steel tank. Other materials for tank construction may be proposed and used if approved by the appropriate authorities (AAR and DOT).

PRESSURE VESSELS

This discussion of pressure vessels is intended as an overview of the codes most frequently used for the design and construction of pressure vessels. Chemical engineers who design or specify pressure vessels should determine the federal and local laws relevant to the problem and then refer to the most recent issue of the pertinent code or standard before proceeding. Laws, codes, and standards are frequently changed.

A pressure vessel is a closed container of limited length (in contrast to the indefinite length of piping). Its smallest dimension is considerably larger than the connecting piping, and it is subject to pressures above 7 or 14 kPa (1 or 2 lbf/in²). It is distinguished from a boiler, which in most cases is used to generate steam for use external to itself.

Code Administration The American Society of Mechanical Engineers has written the ASME Boiler and Pressure Vessel Code, which contains rules for the design, fabrication, and inspection of boilers and pressure vessels. The ASME Code is an American National Standard. Most states in the United States and all Canadian provinces have passed legislation which makes the ASME Code or certain parts of it their legal requirement. Only a few jurisdictions have adopted the code for all vessels. The others apply it to certain types of vessels or to boilers. States employ inspectors (usually under a chief boiler inspector) to enforce code provisions. The authorities also depend a great deal on insurance company inspectors to see that boilers and pressure vessels are maintained in a safe condition.

The ASME Code is written by a large committee and many subcommittees, composed of engineers appointed by the ASME. The Code Committee meets regularly to review the code and consider requests for its revision, interpretation, or extension. **Interpretation and extension** are accomplished through "code cases." The decisions are published in *Mechanical Engineering*. Code cases are also mailed to those who subscribe to the service. A typical code case might be the approval of the use of a metal which is not presently on the list of approved code materials. Inquiries relative to code cases should be addressed to the secretary of the ASME Boiler and Pressure Vessel Committee, American Society of Mechanical Engineers, New York.

A new edition of the code is issued every 3 years. Between editions, alterations are handled by issuing semiannual addenda, which may be purchased by subscription. The ASME considers any issue of the code to be adequate and safe, but some government authorities specify certain issues of the code as their legal requirement.

Inspection Authority The National Board of Boiler and Pressure Vessel Inspectors is composed of the chief inspectors of states and municipalities in the United States and Canadian provinces which have made any part of the Boiler and Pressure Vessel Code a legal requirement. This board promotes uniform enforcement of boiler and pressure-vessel rules. One of the board's important activities is providing examinations for, and commissioning of, inspectors. Inspectors so qualified and employed by an insurance company, state, municipality, or Canadian province may inspect a pressure vessel and permit it to be stamped ASME—NB (National Board). An inspector employed by a vessel user may authorize the use of only the ASME stamp. The ASME Code Committee authorizes fabricators to use the various ASME stamps. The stamps, however, may be applied to a vessel only with the approval of the inspector.

The ASME Boiler and Pressure Vessel Code consists of eleven sections as follows:

    I.   Power Boilers
   II.   Materials
       a.   Ferrous
       b.   Nonferrous
       c.   Welding rods, electrodes, and filler metals
       d.   Properties
  III.   Rules for Construction of Nuclear Power Plant Components
  IV.   Heating Boilers
   V.   Nondestructive Examination
  VI.   Rules for Care and Operation of Heating Boilers
 VII.   Guidelines for the Care of Power Boilers
VIII.   Pressure Vessels
  IX.   Welding and Brazing Qualifications
   X.   Fiber-Reinforced Plastic Pressure Vessels
  XI.   Rules for Inservice Inspection of Nuclear Power Plant Components

Pressure vessels (as distinguished from boilers) are involved with Secs. II, III, V, VIII, IX, X, and XI. Section VIII, Division I, is the Pressure Vessel Code as it existed in the past (and will continue). Division 2 was brought out as a means of permitting higher design stresses while ensuring at least as great a degree of safety as in Division 1. These two divisions plus Secs. III and X will be discussed briefly here. They refer to Secs. II and IX.

ASME Code Section VIII, Division 1 Most pressure vessels used in the process industry in the United States are designed and constructed in accordance with Sec. VIII, Division 1 (see Fig. 10-187). This division is divided into three subsections followed by appendixes.

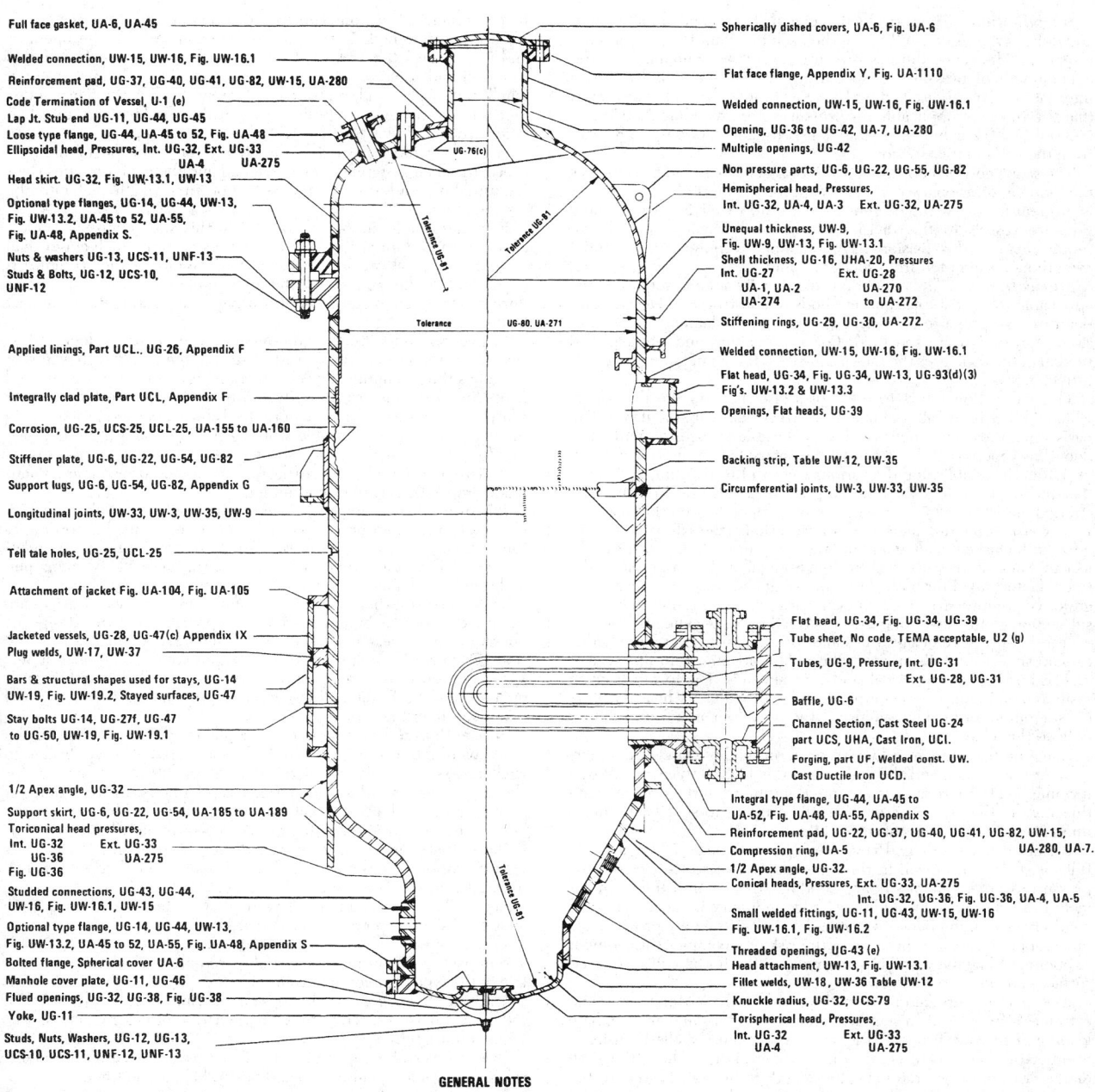

Full face gasket, UA-6, UA-45

Welded connection, UW-15, UW-16, Fig. UW-16.1

Reinforcement pad, UG-37, UG-40, UG-41, UG-82, UW-15, UA-280

Code Termination of Vessel, U-1 (e)

Lap Jt. Stub end UG-11, UG-44, UG-45

Loose type flange, UG-44, UA-45 to 52, Fig. UA-48

Ellipsoidal head, Pressures, Int. UG-32, Ext. UG-33
UA-4   UA-275

Head skirt, UG-32, Fig. UW-13.1, UW-13

Optional type flanges, UG-14, UG-44, UW-13,
Fig. UW-13.2, UA-45 to 52, UA-55,
Fig. UA-48, Appendix S.

Nuts & washers UG-13, UCS-11, UNF-13

Studs & Bolts, UG-12, UCS-10,
UNF-12

Applied linings, Part UCL., UG-26, Appendix F

Integrally clad plate, Part UCL, Appendix F

Corrosion, UG-25, UCS-25, UCL-25, UA-155 to UA-160

Stiffener plate, UG-6, UG-22, UG-54, UG-82

Support lugs, UG-6, UG-54, UG-82, Appendix G

Longitudinal joints, UW-33, UW-3, UW-35, UW-9

Tell tale holes, UG-25, UCL-25

Attachment of jacket Fig. UA-104, Fig. UA-105

Jacketed vessels, UG-28, UG-47(c) Appendix IX

Plug welds, UW-17, UW-37

Bars & structural shapes used for stays, UG-14
UW-19, Fig. UW-19.2, Stayed surfaces, UG-47

Stay bolts UG-14, UG-27f, UG-47
to UG-50, UW-19, Fig. UW-19.1

1/2 Apex angle, UG-32

Support skirt, UG-6, UG-22, UG-54, UA-185 to UA-189

Toriconical head pressures,
Int. UG-32   Ext. UG-33
UG-36   UA-275
Fig. UG-36

Studded connections, UG-43, UG-44,
UW-16, Fig. UW-16.1, UW-15

Optional type flange, UG-14, UG-44, UW-13,
Fig. UW-13.2, UA-45 to 52, Fig. UA-48, Appendix S

Bolted flange, Spherical cover UA-6

Manhole cover plate, UG-11, UG-46

Flued openings, UG-32, UG-38, Fig. UG-38

Yoke, UG-11

Studs, Nuts, Washers, UG-12, UG-13,
UCS-10, UCS-11, UNF-12, UNF-13

Spherically dished covers, UA-6, Fig. UA-6

Flat face flange, Appendix Y, Fig. UA-1110

Welded connection, UW-15, UW-16, Fig. UW-16.1

Opening, UG-36 to UG-42, UA-7, UA-280

Multiple openings, UG-42

Non pressure parts, UG-6, UG-22, UG-55, UG-82

Hemispherical head, Pressures,
Int. UG-32, UA-4, UA-3 Ext. UG-32, UA-275

Unequal thickness, UW-9,
Fig. UW-9, UW-13, Fig. UW-13.1

Shell thickness, UG-16, UHA-20, Pressures
Int. UG-27   Ext. UG-28
UA-1, UA-2   UA-270
UA-274   to UA-272

Stiffening rings, UG-29, UG-30, UA-272.

Welded connection, UW-15, UW-16, Fig. UW-16.1

Flat head, UG-34, Fig. UG-34, UW-13, UW-93(d)(3)
Fig's. UW-13.2 & UW-13.3

Openings, Flat heads, UG-39

Backing strip, Table UW-12, UW-35

Circumferential joints, UW-3, UW-33, UW-35

Flat head, UG-34, Fig. UG-34, UG-39

Tube sheet, No code, TEMA acceptable, U2 (g)

Tubes, UG-9, Pressure, Int. UG-31
Ext. UG-28, UG-31

Baffle, UG-6

Channel Section, Cast Steel UG-24
part UCS, UHA, Cast Iron, UCI.

Forging, part UF, Welded const. UW.

Cast Ductile Iron UCD.

Integral type flange, UG-44, UA-45 to
UA-52, Fig. UA-48, UA-55, Appendix S

Reinforcement pad, UG-22, UG-37, UG-40, UG-41, UG-82, UW-15,
UA-280, UA-7.

Compression ring, UA-5

1/2 Apex angle, UG-32.

Conical heads, Pressures, Ext. UG-33, UA-275
Int. UG-32, UG-36, Fig. UG-36, UA-4, UA-5

Small welded fittings, UG-11, UG-43, UW-15, UW-16
Fig. UW-16.1, Fig. UW-16.2

Threaded openings, UG-43 (e)

Head attachment, UW-13, Fig. UW-13.1

Fillet welds, UW-18, UW-36 Table UW-12

Knuckle radius, UG-32, UCS-79

Torispherical head, Pressures,
Int. UG-32   Ext. UG-33
UA-4   UA-275

GENERAL NOTES

HEAT TREATMENT UG-85, UW-10, UW-40,
UCS-56, TABLE UCS-56, UCS-79(d)
UCS-85, UNF-56, UHA-32, UHA-105,
& UCL-34

INSPECTION UG-90 THRU UG-97, U-1 (j)

JOINT EFFICIENCY UW-12, & TABLE UW-12

LETHAL SERVICE UW-2(a),UCD-2, & UCI-2

LOADINGS UG-22

LOW TEMPERATURE UG-84, UW-2(b),
UCS-65, UCS-66, UCS-67,
UNF-65, & UCL-27

MATERIALS UG-5 THRU UG-15, UG-18, UG-77,
UCL-11 & UW-5
TABLES NF-1 & NF-2

PRESSURE, DESIGN UG-19, & UG-21
MAX. ALLOWABLE WORKING UG-98

TEMPERATURE, DESIGN UG-19, UG-20

**PRESSURE VESSELS SUBJECT TO
DIRECT FIRING** UW-2(d), U-1(h)

RADIOGRAPHIC EXAM, UW-11, UW-51, UW-52,
UCS-57, UNF-57, UHA-33, & UCL-35
SPOT EXAM OF WELDED JOINT UW-52
NO RADIOGRAPH "W-11(c)

RELIEF DEVICES UG-125 THROUGH UG-136, APP. XI

REPAIRS UG-78, UW-38, UW-40(d)

STRESS MAX. ALLOW., VALUE UG-23,
UW-12(c), UNF-23, UHA-23, UCL-23

TEST, HYDROSTATIC, UG-99, UCI-99, UCL-52,
& UA-60
PNEUMATIC UW-50 & UG-100
PROOF, UG-101
NON-DESTRUCTIVE, UG-103, UNF-58,
& UHA-34
MAG. PART. UA-70 THRU UA-73
LIQ. PENE. UA-91 THRU UA-95
ULTRASONIC, UA-901 THRU
UA-904
IMPACT, UG-84, UCS-66, UHA-51, NF-6

STAMPING & DATA, UG-115 THRU UG-120

UNFIRED STEAM BOILERS, UW-2(c), U-1(g)

FIG. 10-187 ASME Code Sec. VIII, Division 1: applicable paragraphs for design and construction details. (*Courtesy of Missouri Boiler and Tank Co.*).

Introduction   The Introduction contains the scope of the division and defines the responsibilities of the user, the manufacturer, and the inspector. The scope defines pressure vessels as containers for the containment of pressure. It specifically excludes vessels having an internal pressure not exceeding 103 kPa (15 lbf/in²) and further states that the rules are applicable for pressures not exceeding 20,670 kPa (3000 lbf/in²). For higher pressures it is usually necessary to deviate from the rules in this division.

The scope covers many other less basic exclusions, and inasmuch as the scope is occasionally revised, except for the most obvious cases, it is prudent to review the current issue before specifying or designing pressure vessels to this division. Any vessel which meets all the requirements of this division may be stamped with the code *U* symbol even though exempted from such stamping.

Subsection A   This subsection contains the general requirements applicable to all materials and methods of construction. Design temperature and pressure are defined here, and the loadings to be considered in design are specified. For stress failure and yielding, this section of the code uses the maximum-stress theory of failure as its criterion.

This subsection refers to the tables elsewhere in the division in which the maximum allowable tensile-stress values are tabulated. The basis for the establishment of these allowable stresses is defined in detail in Appendix P; however, as the safety factors used were very important in establishing the various rules of this division, it is noted that the safety factors for internal-pressure loads are 4 on ultimate strength and 1.6 or 1.5 on yield strength, depending on the material. For external-pressure loads on cylindrical shells, the safety factors are 3 for both elastic buckling and plastic collapse. For other shapes subject to external pressure and for longitudinal shell compression, the safety factors are 4 for both elastic buckling and plastic collapse. Longitudinal compressive stress in cylindrical elements is limited in this subsection by the lower of either stress failure or buckling failure.

Internal-pressure design rules and formulas are given for cylindrical and spherical shells and for ellipsoidal, torispherical (often called ASME heads), hemispherical, and conical heads. The formulas given assume membrane-stress failure, although the rules for heads include consideration for buckling failure in the transition area from cylinder to head (knuckle area).

Longitudinal joints in cylinders are more highly stressed than circumferential joints, and the code takes this fact into account. When forming heads, there is usually some thinning from the original plate thickness in the knuckle area, and it is prudent to specify the minimum allowable thickness at this point.

Unstayed flat heads and covers can be designed by very specific rules and formulas given in this subsection. The stresses caused by pressure on these members are bending stresses, and the formulas include an allowance for additional edge moments induced when the head, cover, or blind flange is attached by bolts. Rules are provided for quick-opening closures because of the risk of incomplete attachment or opening while the vessel is pressurized. Rules for braced and stayed surfaces are also provided.

External-pressure failure of shells can result from overstress at one extreme or from elastic instability at the other or at some intermediate loading. The code provides the solution for most shells by using a number of charts. One chart is used for cylinders where the shell diameter-to-thickness ratio and the length-to-diameter ratio are the variables. The rest of the charts depict curves relating the geometry of cylinders and spheres to allowable stress by curves which are determined from the modulus of elasticity, tangent modulus, and yield strength at temperatures for various materials or classes of materials. The text of this subsection explains how the allowable stress is determined from the charts for cylinders, spheres, and hemispherical, ellipsoidal, torispherical, and conical heads.

Frequently cost savings for cylindrical shells can result from reducing the effective length-to-diameter ratio and thereby reducing shell thickness. This can be accomplished by adding circumferential stiffeners to the shell. Rules are included for designing and locating the stiffeners.

Openings are always required in pressure-vessel shells and heads. Stress intensification is created by the existence of a hole in an other-

wise symmetrical section. The code compensates for this by an area-replacement method. It takes a cross section through the opening, and it measures the area of the metal of the required shell that is removed and replaces it in the cross section by additional material (shell wall, nozzle wall, reinforcing plate, or weld) within certain distances of the opening centerline. These rules and formulas for calculation are included in Subsec. A.

When a cylindrical shell is drilled for the insertion of multiple tubes, the shell is significantly weakened and the code provides rules for tube-hole patterns and the reduction in strength that must be accommodated.

Fabrication tolerances are covered in this subsection. The tolerances permitted for shells for external pressure are much closer than those for internal pressure because the stability of the structure is dependent on the symmetry. Other paragraphs cover repair of defects during fabrication, material identification, heat treatment, and impact testing.

Inspection and testing requirements are covered in detail. Most vessels are required to be hydrostatic-tested (generally with water) at 1½ times the maximum allowable working pressure. Some enameled (glass-lined) vessels are permitted to be hydrostatic-tested at lower pressures. Pneumatic tests are permitted and are carried to at least 1¼ times the maximum allowable working pressure, and there is provision for proof testing when the strength of the vessel or any of its parts cannot be computed with satisfactory assurance of accuracy. Pneumatic or proof tests are rarely conducted.

Pressure-relief-device requirements are defined in Subsec. A. Set point and maximum pressure during relief are defined according to the service, the cause of overpressure, and the number of relief devices. Safety, safety relief, relief valves, rupture disk, breaking pin, and rules on tolerances for the relieving point are given.

Testing, certification, and installation rules for relieving devices are extensive. Every chemical engineer responsible for the design or operation of process units should become very familiar with these rules. The pressure-relief-device paragraphs are the only parts of Sec. VIII, Division 1, that are concerned with the installation and ongoing operation of the facility; all other rules apply only to the design and manufacture of the vessel.

Subsection B   This subsection contains rules pertaining to the methods of fabrication of pressure vessels. Part UW is applicable to welded vessels. Service restrictions are defined. Lethal service is for "lethal substances," which are defined as poisonous gases or liquids of such a nature that a very small amount of the gas or the vapor of the liquid mixed or unmixed with air is dangerous to life when inhaled. It is stated that it is the user's responsibility to advise the designer or manufacturer if the service is lethal. All vessels in lethal service shall have all butt-welded joints fully radiographed, and when practical, joints shall be butt-welded. All vessels fabricated of carbon or low-alloy steel shall be postweld-heat-treated.

Low-temperature service is defined as being below −29°C (−20°F), and impact testing of many materials is required. The code is restrictive in the type of welding permitted.

Unfired steam boilers with design pressures exceeding 345 kPa (50 lbf/in²) have restrictive rules on welded-joint design, and all butt joints require full radiography.

Pressure vessels subject to direct firing have special requirements relative to welded-joint design and postweld heat treatment.

This subsection includes rules governing welded-joint designs and the degree of radiography, with efficiencies for welded joints specified as functions of the quality of joint. These efficiencies are used in the formulas in Subsec. A for determining vessel thicknesses.

Details are provided for head-to-shell welds, tube sheet-to-shell welds, and nozzle-to-shell welds. Acceptable forms of welded stay-bolts and plug and slot welds for staying plates are given here.

Rules for the welded fabrication of pressure vessels cover welding processes, manufacturer's record keeping on welding procedures, welder qualification, cleaning, fit-up alignment tolerances, and repair of weld defects. Procedures for postweld heat treatment are detailed. Checking the procedures and welders and radiographic and ultrasonic examination of welded joints are covered.

Requirements for vessels fabricated by forging in Part UF include unique design requirements with particular concern for stress risers, fabrication, heat treatment, repair of defects, and inspection. Vessels fabricated by brazing are covered in Part UB. Brazed vessels cannot be used in lethal service, for unfired steam boilers, or for direct firing. Permitted brazing processes as well as testing of brazed joints for strength are covered. Fabrication and inspection rules are also included.

Subsection C   This subsection contains requirements pertaining to classes of materials. Carbon and low-alloy steels are governed by Part UCS, nonferrous materials by Part UNF, high-alloy steels by Part UHA, and steels with tensile properties enhanced by heat treatment by Part UHT. Each of these parts includes tables of maximum allowable stress values for all code materials for a range of metal temperatures. These stress values include appropriate safety factors. Rules governing the application, fabrication, and heat treatment of the vessels are included in each part.

Part UHT also contains more stringent details for nozzle welding that are required for some of these high-strength materials. Part UCI has rules for cast-iron construction, Part UCL has rules for welded vessels of clad plate as lined vessels, and Part UCD has rules for ductile-iron pressure vessels.

A relatively recent addition to the code is Part ULW, which contains requirements for vessels fabricated by layered construction. This type of construction is most frequently used for high pressures, usually in excess of 13,800 kPa (2000 lbf/in^2).

There are several methods of layering in common use: (1) thick layers shrunk together; (2) thin layers, each wrapped over the other and the longitudinal seam welded by using the prior layer as backup; and (3) thin layers spirally wrapped. The code rules are written for either thick or thin layers. Rules and details are provided for all the usual welded joints and nozzle reinforcement. Supports for layered vessels require special consideration, in that only the outer layer could contribute to the support. For lethal service only the inner shell and inner heads need comply with the requirements in Subsec. B. Inasmuch as radiography would not be practical for inspection of many of the welds, extensive use is made of magnetic-particle and ultrasonic inspection. When radiography is required, the code warns the inspector that indications sufficient for rejection in single-wall vessels may be acceptable. Vent holes are specified through each layer down to the inner shell to prevent buildup of pressure between layers in the event of leakage at the inner shell.

Mandatory Appendixes   These include a section on supplementary design formulas for shells not covered in Subsec. A. Formulas are given for thick shells, heads, and dished covers. Another appendix gives very specific rules, formulas, and charts for the design of bolted-flange connections. The nature of these rules is such that they are readily programmable for a digital computer, and most flanges now are designed by using computers. One appendix includes only the charts used for calculating shells for external pressure discussed previously. Jacketed vessels are covered in a separate appendix in which very specific rules are given, particularly for the attachment of the jacket to the inner shell. Other appendixes cover inspection and quality control.

Nonmandatory Appendixes   These cover a number of subjects, primarily suggested good practices and other aids in understanding the code and in designing with the code. Several current nonmandatory appendixes will probably become mandatory.

Figure 10-188 illustrates a pressure vessel with the applicable code paragraphs noted for the various elements. Additional important paragraphs are referenced at the bottom of the figure.

ASME Code Section VIII, Division 2   Paragraph A-100e of Division 2 states: "In relation to the rules of Division 1 of Section VIII, these rules of Division 2 are more restrictive in the choice of materials which may be used but permit higher design stress intensity values to be employed in the range of temperatures over which the design stress intensity value is controlled by the ultimate strength or the yield strength; more precise design procedures are required and some common design details are prohibited; permissible fabrication procedures are specifically delineated and more complete testing and inspection are required." Most Division 2 vessels fabricated to date have been large or intended for high pressure and, therefore, expensive when

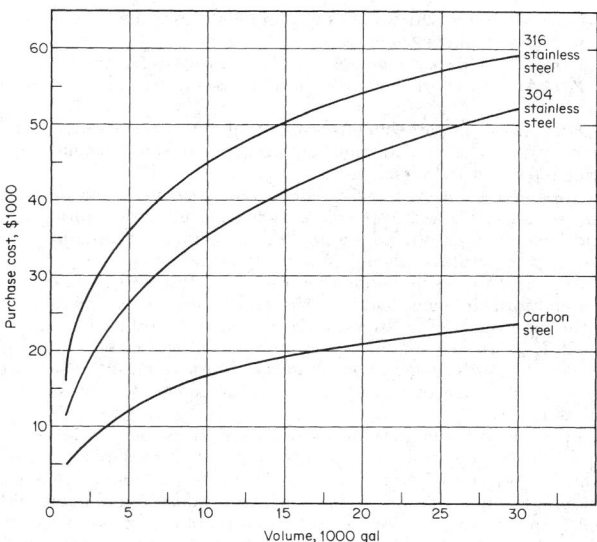

FIG. 10-188   ASME Code Sec. VIII, Division 1: applicable paragraphs for design and construction details. (*Courtesy of Missouri Boiler and Tank Co.*)

the material and labor savings resulting from smaller safety factors have been greater than the additional engineering, administrative, and inspection costs.

The organization of Division 2 differs from that of Division 1.

Part A   This part gives the scope of the division, establishes its jurisdiction, and sets forth the responsibilities of the user and the manufacturer. Of particular importance is the fact that no upper limitation in pressure is specified and that a user's design specification is required. The user or the user's agent shall provide requirements for intended operating conditions in such detail as to constitute an adequate basis for selecting materials and designing, fabricating, and inspecting the vessel. The user's design specification shall include the method of supporting the vessel and any requirement for a fatigue analysis. If a fatigue analysis is required, the user must provide information in sufficient detail so that an analysis for cyclic operation can be made.

Part AM   This part lists permitted individual construction materials, applicable specifications, special requirements, design stress-intensity values, and other property information. Of particular importance are the ultrasonic-test and toughness requirements. Among the properties for which data are included are thermal conductivity and diffusivity, coefficient of thermal expansion, modulus of elasticity, and yield strength. The design stress-intensity values include a safety factor of 3 on ultimate strength at temperature or 1.5 on yield strength at temperature.

Part AD   This part contains requirements for the design of vessels. The rules of Division 2 are based on the maximum-shear theory of failure for stress failure and yielding. Higher stresses are permitted when wind or earthquake loads are considered. Any rules for determining the need for fatigue analysis are given here.

Rules for the design of shells of revolution under internal pressure differ from the Division 1 rules, particularly the rules for formed heads when plastic deformation in the knuckle area is the failure criterion. Shells of revolution for external pressure are determined on the same criterion, including safety factors, as in Division 1. Reinforcement for openings uses the same area-replacement method as Division 1; however, in many cases the reinforcement metal must be closer to the opening centerline.

The rest of the rules in Part AD for flat heads, bolted and studded connections, quick-actuating closures, and layered vessels essentially duplicate Division 1. The rules for support skirts are more definitive in Division 2.

Part AF This part contains requirements governing the fabrication of vessels and vessel parts.

Part AR This part contains rules for pressure-relieving devices.

Part AI This part contains requirements controlling inspection of vessel.

Part AT This part contains testing requirements and procedures.

Part AS This part contains requirements for stamping and certifying the vessel and vessel parts.

Appendixes Appendix 1 defines the basis used for defining stress-intensity values. Appendix 2 contains external-pressure charts, and Appendix 3 has the rules for bolted-flange connections; these two are exact duplicates of the equivalent appendixes in Division 1.

Appendix 4 gives definitions and rules for stress analysis for shells, flat and formed heads, and tube sheets, layered vessels, and nozzles including discontinuity stresses. Of particular importance are Table 4-120.1, "Classification of Stresses for Some Typical Cases," and Fig. 4-130.1, "Stress Categories and Limits of Stress Intensity." These are very useful in that they clarify a number of paragraphs and simplify stress analysis.

Appendix 5 contains rules and data for stress analysis for cyclic operation. Except in short-cycle batch processes, pressure vessels are usually subject to few cycles in their projected lifetime, and the endurance-limit data used in the machinery industries are not applicable. Curves are given for a broad spectrum of materials, covering a range from 10 to 1 million cycles with allowable stress values as high as 650,000 lbf/in². This low-cycle fatigue has been developed from strain-fatigue work in which stress values are obtained by multiplying the strains by the modulus of elasticity. Stresses of this magnitude cannot occur, but strains do. The curves given have a factor of safety of 2 on stress or 20 on cycles.

Appendix 6 contains requirements of experimental stress analysis, Appendix 8 has acceptance standards for radiographic examination, Appendix 9 covers nondestructive examination, Appendix 10 gives rules for capacity conversions for safety valves, and Appendix 18 details quality-control-system requirements.

The remaining appendixes are nonmandatory but useful to engineers working with the code.

General Considerations Most pressure vessels for the chemical-process industry will continue to be designed and built to the rules of Sec. VIII, Division 1. While the rules of Sec. VIII, Division 2, will frequently provide thinner elements, the cost of the engineering analysis, stress analysis and higher-quality construction, material control, and inspection required by these rules frequently exceeds the savings from the use of thinner walls.

Additional ASME Code Considerations

ASME Code Sec. III: Nuclear Power Plant Components This section of the code includes vessels, storage tanks, and concrete containment vessels as well as other nonvessel items.

ASME Code Sec. X: Fiberglass–Reinforced-Plastic Pressure Vessels This section is limited to four types of vessels: bag-molded and centrifugally cast, each limited to 1,000 kPa (150 lbf/in²); filament-wound with cut filaments limited to 10,000 kPa (1500 lbf/in²); and filament-wound with uncut filaments limited to 21,000 kPa (3000 lbf/in²). Operating temperatures are limited to the range from +66°C (150°F) to −54°C (−65°F). Low modulus of elasticity and other property differences between metal and plastic required that many of the procedures in Sec. X be different from those in the sections governing metal vessels. The requirement that at least one vessel of a particular design and fabrication shall be tested to destruction has prevented this section from being widely used. The results from the combined fatigue and burst test must give the design pressure a safety factor of 6 to the burst pressure.

Safety in Design Designing a pressure vessel in accordance with the code will, under most circumstances, provide adequate safety. In the code's own words, however, the rules "cover minimum construction requirements for the design, fabrication, inspection, and certification of pressure vessels." The significant word is "minimum." The **ultimate responsibility** for safety rests with the user and the designer. They must decide whether anything beyond code requirements is necessary. The code cannot foresee and provide for all the unusual conditions to which a pressure vessel might be exposed. If it tried to do so, the majority of pressure vessels would be unnecessarily restricted. Some of the conditions that a vessel might encounter are unusually low temperatures, unusual thermal stresses, stress ratcheting caused by thermal cycling, vibration of tall vessels excited by von Karman vortices caused by wind, very high pressures, runaway chemical reactions, repeated local overheating, explosions, exposure to fire, exposure to materials that rapidly attack the metal, containment of extremely toxic materials, and very large sizes of vessels. Large vessels, although they may contain nonhazardous materials, could, by their very size, create a serious hazard if they burst. The failure of the Boston molasses tank in 1919 killed 12 people. For pressure vessels which are outside code jurisdiction, there are sometimes special hazards in very-high-strength materials and plastics. There may be many others which the designers should recognize if they encounter them.

Metal fatigue, when it is present, is a serious hazard. Section VIII, Division 1, mentions rapidly fluctuating pressures. Division 2 and Sec. III do require a fatigue analysis. In extreme cases vessel contents may affect the fatigue strength (endurance limit) of the material. This is corrosion fatigue. Although most ASME Code materials are not particularly sensitive to corrosion fatigue, even they may suffer an endurance limit loss of 50 percent in some environments. High-strength heat-treated steels, on the other hand, are very sensitive to corrosion fatigue. It is not unusual to find some of these which lose 75 percent of their endurance in corrosive environments. In fact, in corrosion fatigue many steels do not have an endurance limit. The curve of stress versus cycles to failure (S/N curve) continues to slope downward regardless of the number of cycles.

Brittle fracture is probably the most insidious type of pressure-vessel failure. Without brittle fracture, a pressure vessel could be pressurized approximately to its ultimate strength before failure. With brittle behavior some vessels have failed well below their design pressures (which are about 25 percent of the theoretical bursting pressures). In order to reduce the possibility of brittle behavior, Division 2 and Sec. III require impact tests.

The subject of brittle fracture has been understood only since about 1950, and knowledge of some of its aspects is still inadequate. A notched or cracked plate of pressure-vessel steel, stressed at 66°C (150°F), would elongate and absorb considerable energy before breaking. It would have a ductile or plastic fracture. As the temperature is lowered, a point is reached at which the plate would fail in a brittle manner with a flat fracture surface and almost no elongation. The transition from ductile to brittle fracture actually takes place over a temperature range, but a point in this range is selected as the **transition temperature.** One of the ways of determining this temperature is the Charpy impact test (see ASTM Specification E-23). After the transition temperature has been determined by laboratory impact tests, it must be correlated with service experience on full-size plates. The literature on brittle fracture contains information on the relation of impact tests to service experience on some carbon steels.

A more precise but more elaborate method of dealing with the ductile-brittle transition is the **fracture-analysis diagram.** This uses a transition known as the **nil-ductility temperature** (NDT), which is determined by the drop-weight test (ASTM Standard E208) or the drop-weight tear test (ASTM Standard E436). The application of this diagram is explained in two papers by Pellini and Puzak (*Trans. Am. Soc. Mech. Eng.,* 429 (October 1964); Welding Res. Counc. Bull. 88, 1963.

Section VIII, Division 1, is rather lax with respect to brittle fracture. It allows the use of many steels down to −29°C (−20°F) without a check on toughness. Occasional brittle failures show that some vessels are operating below the nil-ductility temperature, i.e., the lower limit of ductility. Division 2 has resolved this problem by requiring impact tests in certain cases. Tougher grades of steel, such as the SA516 steels (in preference to SA515 steel), are available for a small price premium. Stress relief, steel made to fine-grain practice, and normalizing all reduce the hazard of brittle fracture.

Nondestructive testing of both the plate and the finished vessel is important to safety. In the analysis of fracture hazards, it is important to know the size of the flaws that may be present in the completed vessel. The four most widely used methods of examination are radiographic, magnetic-particle, liquid-penetrant, and ultrasonic.

Radiographic examination is either by **x-rays** or by **gamma radiation.** The former has greater penetrating power, but the latter is more portable. Few x-ray machines can penetrate beyond 300-mm (12-in) thickness.

Ultrasonic techniques use vibrations with a frequency between 0.5 and 20 MHz transmitted to the metal by a transducer. The instrument sends out a series of pulses. These show on a cathode-ray screen as they are sent out and again when they return after being reflected from the opposite side of the member. If there is a crack or an inclusion along the way, it will reflect part of the beam. The initial pulse and its reflection from the back of the member are separated on the screen by a distance which represents the thickness. The reflection from a flaw will fall between these signals and indicate its magnitude and position. Ultrasonic examination can be used for almost any thickness of material from a fraction of an inch to several feet. Its use is dependent upon the shape of the body because irregular surfaces may give confusing reflections. Ultrasonic transducers can transmit pulses normal to the surface or at an angle. Transducers transmitting pulses that are oblique to the surface can solve a number of special inspection problems.

Magnetic-particle examination is used only on magnetic materials. Magnetic flux is passed through the part in a path parallel to the surface. Fine magnetic particles, when dusted over the surface, will concentrate near the edges of a crack. The sensitivity of magnetic-particle examination is proportional to the sine of the angle between the direction of the magnetic flux and the direction of the crack. To be sure of picking up all cracks, it is necessary to probe the area in two directions.

Liquid-penetrant examination involves wetting the surface with a fluid which penetrates open cracks. After the excess liquid has been wiped off, the surface is coated with a material which will reveal any liquid that has penetrated the cracks. In some systems a colored dye will seep out of cracks and stain whitewash. Another system uses a penetrant that becomes fluorescent under ultraviolet light.

Each of these four popular methods has its advantages. Frequently, best results are obtained by using more than one method. Magnetic particles or liquid penetrants are effective on surface cracks. Radiography and ultrasonics are necessary for subsurface flaws. *No known method of nondestructive testing can guarantee the absence of flaws.* There are other less widely used methods of examination. Among these are eddy-current, electrical-resistance, acoustics, and thermal testing. *Nondestructive Testing Handbook* [Robert C. McMaster (ed.), Ronald, New York, 1959] gives information on many testing techniques.

The **eddy-current technique** involves an alternating-current coil along and close to the surface being examined. The electrical impedance of the coil is affected by flaws in the structure or changes in composition. Commercially, the principal use of eddy-current testing is for the examination of tubing. It could, however, be used for testing other things.

The **electrical-resistance method** involves passing an electric current through the structure and exploring the surface with voltage probes. Flaws, cracks, or inclusions will cause a disturbance in the voltage gradient on the surface. Railroads have used this method for many years to locate transverse cracks in rails.

The **hydrostatic test** is, in one sense, a method of examination of a vessel. It can reveal gross flaws, inadequate design, and flange leaks. Many believe that a hydrostatic test guarantees the safety of a vessel. This is not necessarily so. A vessel that has passed a hydrostatic test is probably safer than one that has not been tested. It can, however, still fail in service, even on the next application of pressure. Care in material selection, examination, and fabrication do more to guarantee vessel integrity than the hydrostatic test.

The ASME Codes recommend that hydrostatic tests be run at a temperature that is usually above the nil-ductility temperature of the material. This is, in effect, a pressure-temperature treatment of the vessel. When tested in the relatively ductile condition above the nil-ductility temperature, the material will yield at the tips of cracks and flaws and at points of high residual weld stress. This procedure will actually reduce the residual stresses and cause a redistribution at crack tips. The vessel will then be in a safer condition for subsequent operation. This procedure is sometimes referred to as **notch nullification.**

It is possible to design a hydrostatic test in such a way that it probably will be a proof test of the vessel. This usually requires, among other things, that the test be run at a temperature as low as and preferably lower than the minimum operating temperature of the vessel. Proof tests of this type are run on vessels built of ultrahigh-strength steel to operate at cryogenic temperatures.

Other Regulations and Standards   Pressure vessels may come under many types of regulation, depending on where they are and what they contain. Although many states have adopted the ASME Boiler and Pressure Vessel Code, either in total or in part, any state or municipality may enact its own requirements. The federal government regulates some pressure vessels through the Department of Transportation, which includes the Coast Guard. If pressure vessels are shipped into foreign countries, they may face additional regulations.

Pressure vessels carried aboard United States-registered ships must conform to rules of the **U.S. Coast Guard.** Subchapter F of Title 46, *Code of Federal Regulations,* covers marine engineering. Of this, Parts 50 through 61 and 98 include pressure vessels. Many of the rules are similar to those in the ASME Code, but there are differences.

The **American Bureau of Shipping** (ABS) has rules that insurance underwriters require for the design and construction of pressure vessels which are a permanent part of a ship. Pressure cargo tanks may be permanently attached and come under these rules. Such tanks supported at several points are independent of the ship's structure and are distinguished from "integral cargo tanks" such as those in a tanker. ABS has pressure vessel rules in two of its publications. Most of them are in *Rules for Building and Classing Steel Vessels.*

Standards of Tubular Exchanger Manufacturers Association (TEMA) give recommendations for the construction of tubular heat exchangers. Although TEMA is not a regulatory body and there is no legal requirement for the use of its standards, they are widely accepted as a good basis for design. By specifying TEMA standards, one can obtain adequate equipment without having to write detailed specifications for each piece. TEMA gives formulas for the thickness of tube sheets. Such formulas are not in ASME Codes. (See further discussion of TEMA in Sec. 11.)

Vessels with Unusual Construction   High pressures create design problems. The ASME Code Sec. VIII, Division 1, applies to vessels rated for pressures up to 20,670 kPa (3000 lbf/in²). Division 2 is unlimited. At high pressures, special designs not necessarily in accordance with the code are sometimes used. At such pressures, a vessel designed for ordinary low-carbon-steel plate, particularly in large diameters, would become too thick for practical fabrication by ordinary methods. The alternatives are to make the vessel of high-strength plate, use a solid forging, or use multilayer construction.

High-strength steels with tensile strengths over 1380 MPa (200,000 lbf/in²) are limited largely to applications for which weight is very important. Welding procedures are carefully controlled, and preheat is used. These materials are brittle at almost any temperature, and vessels must be designed to prevent brittle fracture. Flat spots and variations in curvature are avoided. Openings and changes in shape require appropriate design. The maximum permissible size of flaws is determined by fracture mechanics, and the method of examination must assure as much as possible that larger flaws are not present. All methods of nondestructive testing may be used. Such vessels require the most sophisticated techniques in design, fabrication, and operation.

Solid forgings are frequently used in construction for pressure vessels above 20,670 kPa (3000 lbf/in²) and even lower. Almost any shell thickness can be obtained, but most of them range between 50 and 300 mm (2 and 12 in). The ASME Code lists forging materials with tensile strengths from 414 to 930 MPa (from 60,000 to 135,000 lbf/in²). Brittle fracture is a possibility, and the hazard increases with thickness. Furthermore, some forging alloys have nil-ductility temperatures as high as 121°C (250°F). A forged vessel should have an NDT at least 17°C (30°F) below the design temperature. In operation, it should be slowly and uniformly heated at least to NDT before it is subjected to pressure. During construction, nondestructive testing should be used to detect dangerous cracks or flaws. Section VIII of the ASME Code, particularly Division 2, gives design and testing techniques.

As the size of a forged vessel increases, the sizes of ingot and handling equipment become larger. The cost may increase faster than the weight. The problems of getting sound material and avoiding brittle fracture also become more difficult. Some of these problems are avoided by use of **multilayer construction.** In this type of vessel, the heads and flanges are made of forgings, and the cylindrical portion is built up by a series of layers of thin material. The thickness of these layers may be between 3 and 50 mm (⅛ and 2 in), depending on the type of construction. There is an inner lining which may be different from the outer layers.

Although there are multilayer vessels as small as 380-mm (15-in) inside diameter and 2400 mm (8 ft) long, their principal advantage applies to the larger sizes. When properly made, a multilayer vessel is probably safer than a vessel with a solid wall. The layers of thin material are tougher and less susceptible to brittle fracture, have less probability of defects, and have the statistical advantage of a number of small elements instead of a single large one. The heads, flanges, and welds, of course, have the same hazards as other thick members. Proper attention is necessary to avoid cracks in these members.

There are several assembly techniques. One frequently used is to form successive layers in half cylinders and butt-weld them over the previous layers. In doing this, the welds are staggered so that they do not fall together. This type of construction usually uses plates from 6 to 12 mm (¼ to ½ in) thick. Another method is to weld each layer separately to form a cylinder and then shrink it over the previous layers. Layers up to about 50-mm (2-in) thickness are assembled in this way. A third method of fabrication is to wind the layers as a continuous sheet. This technique is used in Japan. The Wickel construction, fabricated in Germany, uses helical winding of interlocking metal strip. Each method has its advantages and disadvantages, and choice will depend upon circumstances.

Because of the possibility of voids between layers, it is preferable not to use multilayer vessels in applications where they will be subjected to fatigue. Inward thermal gradients (inside temperature lower than outside temperature) are also undesirable.

Articles on these vessels have been written by Fratcher [*Pet. Refiner,* **34**(11), 137 (1954)] and by Strelzoff, Pan, and Miller [*Chem. Eng.,* **75**(21), 143–150 (1968)].

Vessels for high-temperature service may be beyond the temperature limits of the stress tables in the ASME Codes. Section VIII, Division 1, makes provision for construction of pressure vessels up to 650°C (1200°F) for carbon and low-alloy steel and up to 815°C (1500°F) for stainless steels (300 series). If a vessel is required for temperatures above these values and above 103 kPa (15 lbf/in²), it would be necessary, in a code state, to get permission from the state authorities to build it as a special project. Above 815°C (1500°F), even the 300 series stainless steels are weak, and creep rates increase rapidly. If the metal which resists the pressure operates at these temperatures, the vessel pressure and size will be limited. The vessel must also be expendable because its life will be short. Long exposure to high temperature may cause the metal to deteriorate and become brittle. Sometimes, however, economics favor this type of operation.

One way to circumvent the problem of low metal strength is to use a metal inner liner surrounded by insulating material, which in turn is confined by a pressure vessel. The liner, in some cases, may have perforations which will allow pressure to pass through the insulation and act on the outer shell, which is kept cool to obtain normal strength. The liner has no pressure differential acting on it and, therefore, does not need much strength. Ceramic linings are also useful for high-temperature work.

Lined vessels are used for many applications. Any type of lining can be used in an ASME Code vessel, provided it is compatible with the metal of the vessel and the contents. Glass, rubber, plastics, rare metals, and ceramics are a few types. The lining may be installed separately, or if a metal is used, it may be in the form of clad plate. The cladding on plate can sometimes be considered as a stress-carrying part of the vessel.

A **ceramic lining** when used with high temperature acts as an insulator so that the steel outer shell is at a moderate temperature while the temperature at the inside of the lining may be very high. Ceramic linings may be of unstressed brick, or prestressed brick, or cast in place. Cast ceramic linings or unstressed brick may develop cracks and are used when the contents of the vessel will not damage the outer shell. They are usually designed so that the high temperature at the inside will expand them sufficiently to make them tight in the outer (and cooler) shell. This, however, is not usually sufficient to prevent some penetration by the product.

Prestressed-brick linings can be used to protect the outer shell. In this case, the bricks are installed with a special thermosetting-resin mortar. After lining, the vessel is subjected to internal pressure and heat. This expands the steel vessel shell, and the mortar expands to take up the space. The pressure and temperature must be at least as high as the maximum that will be encountered in service. After the mortar has set, reduction of pressure and temperature will allow the vessel to contract, putting the brick in compression. The upper temperature limit for this construction is about 190°C (375°F). The installation of such linings is highly specialized work done by a few companies. Great care is usually exercised in operation to protect the vessel from exposure to unsymmetrical temperature gradients. Side nozzles and other unsymmetrical designs are avoided insofar as possible.

Concrete pressure vessels may be used in applications that require large sizes. Such vessels, if made of steel, would be too large and heavy to ship. Through the use of posttensioned (prestressed) concrete, the vessel is fabricated on the site. In this construction, the reinforcing steel is placed in tubes or plastic covers, which are cast into the concrete. Tension is applied to the steel after the concrete has acquired most of its strength.

Concrete nuclear reactor vessels, of the order of magnitude of 15-m (50-ft) inside diameter and length, have inner linings of steel which confine the pressure. After fabrication of the liner, the tubes for the cables or wires are put in place and the concrete is poured. High-strength reinforcing steel is used. Because there are thousands of reinforcing tendons in the concrete vessel, there is a statistical factor of safety. The failure of 1 or even 10 tendons would have little effect on the overall structure.

Plastic pressure vessels have the *advantages of chemical resistance* and light weight. Above 103 kPa (15 lbf/in²), with certain exceptions, they must be designed according to the ASME Code section (see "Storage of Gases") and are confined to the three types of approved code construction. Below 103 kPa (15 lbf/in²), any construction may be used. Even in this pressure range, however, the code should be used for guidance. Solid plastics, because of low strength and creep, can be used only for the lowest pressures and sizes. A stress of a few hundred pounds-force per square inch is the maximum for most plastics. To obtain higher strength, the filled plastics or filament-wound vessels, specified by the code, must be used. Solid-plastic parts, however, are often employed inside a steel shell, particularly for heat exchangers.

Graphite and ceramic vessels are used fully armored; that is, they are enclosed within metal pressure vessels. These materials are also used for boxlike vessels with backing plates on the sides. The plates are drawn together by tie bolts, thus putting the material in compression so that it can withstand low pressure.

Vessel Codes Other Than ASME   Different design and construction rules are used in other countries. Chemical engineers concerned with pressure vessels outside the United States must become familiar with local pressure-vessel laws and regulations. *Boilers and Pressure Vessels,* an international survey of design and approval requirements published by the British Standards Institution, Maylands Avenue, Hemel Hempstead, Hertfordshire, England, in 1975, gives pertinent information for 76 political jurisdictions.

The British Code (British Standards) and the West German Code (*A. D. Merkblätter*) in addition to the ASME Code are most commonly permitted, although Netherlands, Sweden, and France also have codes. The major difference between the codes lies in factors of safety and in whether or not ultimate strength is considered. ASME Code, Sec. VIII, Division 1, vessels are generally heavier than vessels built to the other codes; however, the differences in allowable stress for a given material are less in the higher temperature (creep) range.

Engineers and metallurgists have developed alloys to comply economically with individual codes. In West Germany, where design stress is determined from yield strength and creep-rupture strength and no allowance is made for ultimate strength, steels which have a very high yield-strength-to-ultimate-strength ratio are used.

Other differences between codes include different bases for the design of reinforcement for openings and the design of flanges and heads. Some codes include rules for the design of heat-exchanger tube sheets, while others (ASME Code) do not. The Dutch Code (*Grondslagen*) includes very specific rules for calculation of wind loads, while the ASME Code leaves this entirely to the designer.

There are also significant differences in construction and inspection rules. Unless engineers make a detailed study of the individual codes and keep current, they will be well advised to make use of responsible experts for any of the codes.

Vessel Design and Construction   The ASME Code lists a number of loads that must be considered in designing a pressure vessel. Among them are impact, weight of the vessel under operating and test conditions, superimposed loads from other equipment and piping, wind and earthquake loads, temperature-gradient stresses, and localized loadings from internal and external supports. In general, the code gives no values for these loads or methods for determining them, and no formulas are given for determining the stresses from these loads. Engineers must be knowledgeable in mechanics and strength of materials to solve these problems.

Some of the problems are treated by Brownell and Young, *Process Equipment Design*, Wiley, New York, 1959. ASME papers treat others, and a number of books published by the ASME are collections of papers on pressure-vessel design: *Pressure Vessels and Piping Design: Collected Papers, 1927–1959; Pressure Vessels and Piping Design and Analysis*, four volumes; and *International Conference: Pressure Vessel Technology*, published annually.

Throughout the year the Welding Research Council publishes bulletins which are final reports from projects sponsored by the council, important papers presented before engineering societies, and other reports of current interest which are not published in *Welding Research*. A large number of the published bulletins are pertinent for vessel designers.

Care of Pressure Vessels   Protection against **excessive pressure** is largely taken care of by code requirements for relief devices. Exposure to fire is also covered by the code. The code, however, does not provide for the possibility of local overheating and weakening of a vessel in a fire. Insulation reduces the required relieving capacity and also reduces the possibility of local overheating.

A pressure-reducing valve in a line leading to a pressure vessel is not adequate protection against overpressure. Its failure will subject the vessel to full line pressure.

Vessels that have an operating cycle which involves the solidification and remelting of solids can develop excessive pressures. A solid plug of material may seal off one end of the vessel. If heat is applied at that end to cause melting, the expansion of the liquid can build up a high pressure and possibly result in yielding or rupture. Solidification in connecting piping can create similar problems.

Some vessels may be exposed to a runaway chemical reaction or even an explosion. This requires relief valves, rupture disks, or, in extreme cases, a barricade (the vessel is expendable). A vessel with a large rupture disk needs anchors designed for the jet thrust when the disk blows.

Vacuum must be considered. It is nearly always possible that the contents of a vessel might contract or condense sufficiently to subject it to an internal vacuum. If the vessel cannot withstand the vacuum, it must have vacuum-breaking valves.

Improper operation of a process may result in the vessel's **exceeding design temperature.** Proper control is the only solution to this problem. Maintenance procedures can also cause excessive temperatures. Sometimes the contents of a vessel may be burned out with torches. If the flame impinges on the vessel shell, overheating and damage may occur.

Excessively low temperature may involve the hazard of brittle fracture. A vessel that is out of use in cold weather could be at a subzero temperature and well below its nil-ductility temperature. In startup, the vessel should be warmed slowly and uniformly until it is above the NDT. A safe value is 38°C (100°F) for plate if the NDT is unknown. The vessel should not be pressurized until this temperature is exceeded. Even after the NDT has been passed, excessively rapid heating or cooling can cause high thermal stresses.

Corrosion is probably the greatest threat to vessel life. Partially filled vessels frequently have severe pitting at the liquid-vapor interface. Vessels usually do not have a corrosion allowance on the outside. Lack of protection against the weather or against the drip of corrosive chemicals can reduce vessel life. Insulation may contain damaging substances. Chlorides in insulating materials can cause cracking of stainless steels.

There are many ways in which a pressure vessel can suffer **mechanical damage.** The shells can be dented or even punctured, they can be dropped or have hoisting cables improperly attached, bolts can be broken, flanges are bent by excessive bolt tightening, gasket contact faces can be scratched and dented, rotating paddles can drag against the shell and cause wear, and a flange can be bolted up with a gasket half in the groove and half out. Most of these forms of damage can be prevented by care and common sense. If damage is repaired by straightening, as with a dented shell, it may be necessary to stress-relieve the repaired area. Some steels are susceptible to embrittlement by aging after severe straining. A safer procedure is to cut out the damaged area and replace it.

The National Board Inspection Code, published by the National Board of Boiler and Pressure Vessel Inspectors, Columbus, Ohio, is helpful. Any repair, however, is acceptable if it is made in accordance with the rules of the Pressure Vessel Code.

Pressure vessels should be **inspected periodically.** No rule can be given for the frequency of these inspections. Frequency depends on operating conditions. If the early inspections of a vessel indicate a low corrosion rate, intervals between inspections may be lengthened. Some vessels are inspected at 5-year intervals; others, as frequently as once a year. Measurement of corrosion is an important inspection item. One of the most convenient ways of measuring thickness (and corrosion) is to use an ultrasonic gauge. The location of the corrosion and whether it is uniform or localized in deep pits should be observed and reported. Cracks, any type of distortion, and leaks should be observed. Cracks are particularly dangerous because they can lead to sudden failure. Insulation is usually left in place during inspection of insulated vessels. If, however, severe external corrosion is suspected, the insulation should be removed. All forms of nondestructive testing are useful for examinations.

Care in **reassembling** the vessel is particularly important. Gaskets should be properly located, particularly if they are in grooves. Bolts should be tightened in proper sequence. In some critical cases and with large bolts, it is necessary to control bolt tightening by torque wrenches, micrometers, patented bolt-tightening devices, or heating bolts. After assembly, vessels are sometimes given a hydrostatic test.

Pressure-Vessel Cost and Weight   The curves of Fig. 10-188 can be used for estimating cost (freight allowed) when a weight estimate is not available. The cost is based on some 1990 pressure-vessel costs. The prices are plotted as a function of vessel volume for average vessels 6.35 mm (¼ in) thick which are not of unusual design. Correction factors for other thicknesses are given. Complicated vessels could cost considerably more. Guthrie [*Chem. Eng.*, **76**(6), 114–142 (1969)] also gives pressure-vessel cost data.

When vessels have complicated construction (large, heavy bolted connections, support skirts, etc.), it is preferable to estimate their weight and apply a unit cost in dollars per pound. Some data for vessels purchased in 1968 are plotted in Fig. 10-189. There is a variation of about 2 to 1 between the lowest and the highest costs. The unit FOB cost of carbon steel and type 304 stainless steel was found to vary as the −0.34 power of the weight. Stainless-steel vessels frequently include considerable carbon steel in the form of support skirts, brackets, legs, lap-joint flanges, bolts, etc. In calculating the equivalent weight of a stainless-steel vessel, each pound of carbon should be considered equivalent to 0.4 lb of stainless.

Pressure-vessel weights are obtained by calculating the cylindrical

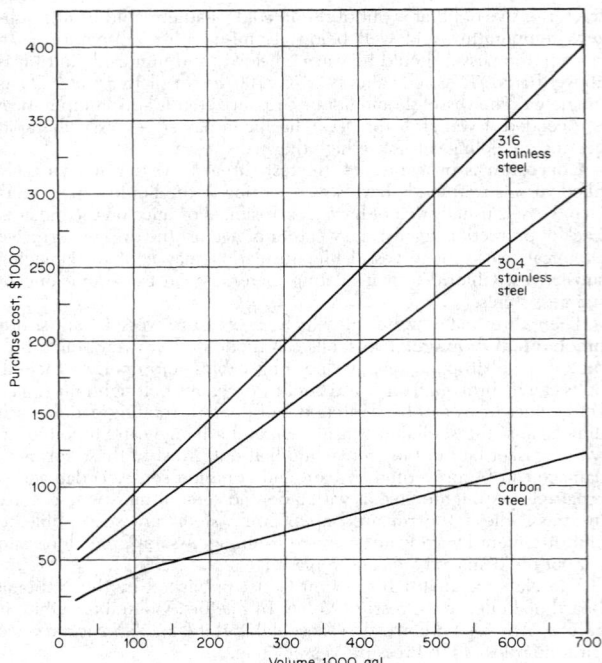

FIG. 10-189 Cost per pound of pressure vessels (1968). For carbon steel, $C = 9.05\ W^{-0.34}$; for type 304 stainless steel, $C = 25.6\ W^{-0.34}$; and for type 316 stainless steel, $C = 34.2\ W^{-0.34}$; where C = FOB cost in dollars per pound and W = weight in pounds. To convert pounds to kilograms, multiply by 0.454.

shell and heads separately and then adding the weights of nozzles and attachments. Steel weighs 0.283 lb/in³ and 40.7 lb/ft³ for 1-in plate. Metal in heads can be approximated by calculating the area of the blank (disk) used for forming the head. The required diameter of blank can be calculated by multiplying the head outside diameter by

TABLE 10-67 Factors for Estimating Diameters of Blanks for Formed Heads

	Ratio d/t	Blank diameter factor
A.S.M.E. head	Over 50	1.09
	30–50	1.11
	20–30	1.15
Ellipsoidal head	Over 20	1.24
	10–20	1.30
Hemispherical head	Over 30	1.60
	18–30	1.65
	10–18	1.70

d = head diameter
t = nominal minimum head thickness

TABLE 10-68 Extra Thickness Allowances for Formed Heads*

	Extra thickness, in		
	A.S.M.E. and Ellipsoidal		
Minimum head thickness, in	Head o.d. up to 150 in incl.	Head o.d. over 150 in	Hemispherical
Up to 0.99	1/16	1/8	3/16
1 to 1.99	1/8	1/8	3/8
2 to 2.99	1/4	1/4	5/8

*Lukens, Inc.

the approximate factors given in Table 10-67. These factors make no allowance for the straight flange which is a cylindrical extension that is formed on the head. The blank diameter obtained from these factors must be increased by twice the length of straight flange, which is usually 1½ to 2 in but can be up to several inches in length. Manufacturers' catalogs give weights of heads.

Forming a head thins it in certain areas. To obtain the required minimum thickness of a head, it is necessary to use a plate that is initially thicker. Table 10-68 gives allowances for additional thickness.

Nozzles and flanges may add considerably to the weight of a vessel. Their weights can be obtained from manufacturers' catalogs (Taylor Forge Division of Gulf & Western Industries, Inc., Tube Turns Inc., Ladish Co., Lenape Forge, and others). Other parts such as skirts, legs, support brackets, and other details must be calculated.

Heat-Transfer Equipment

Richard L. Shilling, P.E., B.S.M., B.E.M.E., *Manager of Engineering Development, Brown Fintube Company—a Koch Engineering Company; Member, American Society of Mechanical Engineers. (Shell-and-Tube Heat Exchangers, Hairpin/Double Pipe Heat Exchangers, Air Cooled Heat Exchangers, Heating and Cooling of Tanks, Fouling and Scaling, Heat Exchangers for Solids—significant contribution by Arthur D. Holt, Thermal Insulation—significant contribution by Herbert A. Moak) (Section Editor)*

Kenneth J. Bell, Ph.D., P.E., *Regents Professor Emeritus, School of Chemical Engineering, Oklahoma State University; Member, American Institute of Chemical Engineers. (Thermal Design of Heat Exchangers, Condenser, Reboilers)*

Patrick M. Bernhagen, P.E., B.S.M.E., *Senior Mechanical Engineer, Foster Wheeler USA Corporation, American Society of Mechanical Engineers. (Compact and Non-Tubular Heat Exchangers)*

Thomas M. Flynn, Ph.D., P.E., *Cryogenic Engineer, President CRYOCO, Louisville, Colorado; Member, American Institute of Chemical Engineers. (Cryogenic Processes)*

Victor M. Goldschmidt, Ph.D., P.E., *Professor of Mechanical Engineering, Purdue University, West Lafayette, Indiana. (Air Conditioning)*

Predrag S. Hrnjak, Ph.D., V.Res., *Assistant Professor, University of Illinois at Ubana Champaign and Principal Investigator—U of I Air Conditioning and Refrigeration Center, Assistant Professor, University of Belgrade; Member, International Institute of Refrigeration, American Society of Heating, Refrigeration and Air Conditioning. (Refrigeration)*

F. C. Standiford, M.S., P.E., *Member, American Institute of Chemical Engineers, American Chemical Society. (Thermal Design of Evaporators, Evaporators)*

Klaus D. Timmerhaus, Ph.D., P.E., *Professor and President's Teaching Scholar, University of Colorado, Boulder, Colorado; Fellow, American Institute of Chemical Engineers, American Society for Engineering Education, American Association for the Advancement of Science, Member, American Astronautical Society, National Academy of Engineering, Austrian Academy of Science, International Institute of Refrigeration, American Society of Heating, Refrigerating and Air Conditioning Engineers, American Society of Environmental Engineers, Engineering Society for Advancing Mobility on Land, Sea, Air, and Space, Sigma Xi, The Research Society. (Cryogenic Processes)*

THERMAL DESIGN OF HEAT-TRANSFER EQUIPMENT

INTRODUCTION TO THERMAL DESIGN

Design methods for several important classes of process heat-transfer equipment are presented in the following portions of Sec. 11. Mechanical descriptions and specifications of equipment are given in this section and should be read in conjunction with the use of this material. It is impossible to present here a comprehensive treatment of heat-exchanger selection, design, and application. The best general references in this field are Hewitt, Shires, and Bott, *Process Heat Transfer,* CRC Press, Boca Raton, FL, 1994; and Schlünder (ed.), *Heat Exchanger Design Handbook,* Begell House, New York, 1983.

Approach to Heat-Exchanger Design The proper use of basic heat-transfer knowledge in the design of practical heat-transfer equipment is an art. Designers must be constantly aware of the differences between the idealized conditions for and under which the basic knowledge was obtained and the real conditions of the mechanical expression of their design and its environment. The result must satisfy process and operational requirements (such as availability, flexibility, and maintainability) and do so economically. An important part of any design process is to consider and offset the consequences of error in the basic knowledge, in its subsequent incorporation into a design method, in the translation of design into equipment, or in the operation of the equipment and the process. Heat-exchanger design is not a highly accurate art under the best of conditions.

The design of a process heat exchanger usually proceeds through the following steps:

1. Process conditions (stream compositions, flow rates, temperatures, pressures) must be specified.
2. Required physical properties over the temperature and pressure ranges of interest must be obtained.
3. The type of heat exchanger to be employed is chosen.
4. A preliminary estimate of the size of the exchanger is made, using a heat-transfer coefficient appropriate to the fluids, the process, and the equipment.
5. A first design is chosen, complete in all details necessary to carry out the design calculations.
6. The design chosen in step 5 is evaluated, or *rated,* as to its ability to meet the process specifications with respect to both heat transfer and pressure drop.
7. On the basis of the result of step 6, a new configuration is chosen if necessary and step 6 is repeated. If the first design was inadequate to meet the required heat load, it is usually necessary to increase the size of the exchanger while still remaining within specified or feasible limits of pressure drop, tube length, shell diameter, etc. This will sometimes mean going to multiple-exchanger configurations. If the first design more than meets heat-load requirements or does not use all the allowable pressure drop, a less expensive exchanger can usually be designed to fulfill process requirements.
8. The final design should meet process requirements (within reasonable expectations of error) at lowest cost. The lowest cost should include operation and maintenance costs and credit for ability to meet long-term process changes, as well as installed (capital) cost. Exchangers should not be selected entirely on a lowest-first-cost basis, which frequently results in future penalties.

Overall Heat-Transfer Coefficient The basic design equation for a heat exchanger is

$$dA = dQ/U\,\Delta T \tag{11-1}$$

where dA is the element of surface area required to transfer an amount of heat dQ at a point in the exchanger where the overall heat-transfer coefficient is U and where the overall bulk temperature difference between the two streams is ΔT. The overall heat-transfer coefficient is related to the individual film heat-transfer coefficients and fouling and wall resistances by Eq. (11-2). Basing U_o on the outside surface area A_o results in

$$U_o = \frac{1}{1/h_o + R_{do} + xA_o/k_w A_{wm} + (1/h_i + R_{di})A_o/A_i} \tag{11-2}$$

Equation (11-1) can be formally integrated to give the outside area required to transfer the total heat load Q_T:

$$A_o = \int_0^{Q_T} \frac{dQ}{U_o\,\Delta T} \tag{11-3}$$

To integrate Eq. (11-3), U_o and ΔT must be known as functions of Q. For some problems, U_o varies strongly and nonlinearly throughout the exchanger. In these cases, it is necessary to evaluate U_o and ΔT at several intermediate values and numerically or graphically integrate. For many practical cases, it is possible to calculate a constant mean overall coefficient U_{om} from Eq. (11-2) and define a corresponding mean value of ΔT_m, such that

$$A_o = Q_T/U_{om}\,\Delta T_m \tag{11-4}$$

Care must be taken that U_o does not vary too strongly, that the proper equations and conditions are chosen for calculating the individual coefficients, and that the mean temperature difference is the correct one for the specified exchanger configuration.

Mean Temperature Difference The temperature difference between the two fluids in the heat exchanger will, in general, vary from point to point. The mean temperature difference (ΔT_m or MTD) can be calculated from the terminal temperatures of the two streams if the following assumptions are valid:

1. All elements of a given fluid stream have the same thermal history in passing through the exchanger.[*]
2. The exchanger operates at steady state.
3. The specific heat is constant for each stream (or if either stream undergoes an isothermal phase transition).
4. The overall heat-transfer coefficient is constant.
5. Heat losses are negligible.

Countercurrent or Cocurrent Flow If the flow of the streams is either *completely* countercurrent or completely cocurrent or if one or both streams are isothermal (condensing or vaporizing a pure component with negligible pressure change), the correct MTD is the logarithmic-mean temperature difference (LMTD), defined as

$$\text{LMTD} = \Delta T_{lm} = \frac{(t_1' - t_2'') - (t_2' - t_1'')}{\ln\left(\dfrac{t_1' - t_2''}{t_2' - t_1''}\right)} \tag{11-5a}$$

for *countercurrent flow* (Fig. 11-1a) and

$$\text{LMTD} = \Delta T_{lm} = \frac{(t_1' - t_1'') - (t_2' - t_2'')}{\ln\left(\dfrac{t_1' - t_1''}{t_2' - t_2''}\right)} \tag{11-5b}$$

for *cocurrent flow* (Fig. 11-1b)

If U is not constant but a linear function of ΔT, the correct value of

---

[*] This assumption is vital but is usually omitted or less satisfactorily stated as "each stream is well mixed at each point." In a heat exchanger with substantial bypassing of the heat-transfer surface, e.g., a typical baffled shell-and-tube exchanger, this condition is not satisfied. However, the error is in some degree offset if the same MTD formulation used in reducing experimental heat-transfer data to obtain the basic correlation is used in applying the correlation to design a heat exchanger. The compensation is not in general exact, and insight and judgment are required in the use of the MTD formulations. Particularly, in the design of an exchanger with a very close temperature approach, bypassing may result in an exchanger that is inefficient and even thermodynamically incapable of meeting specified outlet temperatures.

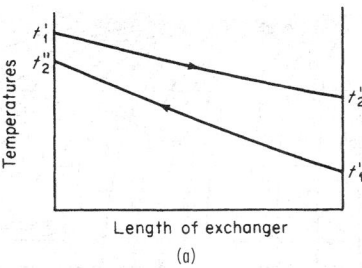

(a)

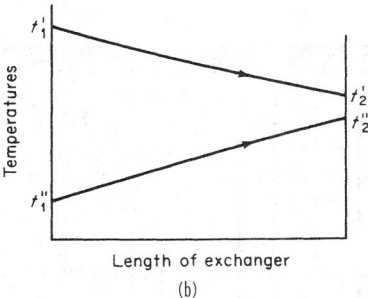

(b)

FIG. 11-1 Temperature profiles in heat exchangers. (*a*) Countercurrent. (*b*) Cocurrent.

$U_{om}\Delta T_m$ to use in Eq. (11-4) is [Colburn, *Ind. Eng. Chem.*, **25**, 873 (1933)]

$$U_{om}\Delta T_m = \frac{U_o''(t_1' - t_2'') - U_o'(t_2' - t_1'')}{\ln\left(\dfrac{U_o''(t_1' - t_2'')}{U_o'(t_2' - t_1'')}\right)} \qquad (11\text{-}6a)$$

for *countercurrent flow*, where U_o'' is the overall coefficient evaluated when the stream temperatures are t_1' and t_2'' and U_o' is evaluated at t_2' and t_1''. The corresponding equation for *cocurrent flow* is

$$U_{om}\Delta T_m = \frac{U_o''(t_1' - t_1'') - U_o'(t_2' - t_2'')}{\ln\left(\dfrac{U_o''(t_1' - t_1'')}{U_o'(t_2' - t_2'')}\right)} \qquad (11\text{-}6b)$$

where U_o' is evaluated at t_2' and t_2'' and U_o'' is evaluated at t_1' and t_1''. To use these equations, it is necessary to calculate two values of U_o.°

The use of Eq. (11-6) will frequently give satisfactory results even if U_o is not strictly linear with temperature difference.

Reversed, Mixed, or Cross-Flow If the flow pattern in the exchanger is not completely countercurrent or cocurrent, it is necessary to apply a **correction factor** F_T by which the LMTD is multiplied to obtain the appropriate MTD. These corrections have been mathematically derived for flow patterns of interest, still by making assumptions 1 to 5 [see Bowman, Mueller, and Nagle, *Trans. Am. Soc. Mech. Eng.*, **62**, 283 (1940) or Hewitt, et al. op. cit.]. For a common flow pattern, the 1-2 exchanger (Fig. 11-2), the correction factor F_T is given in Fig. 11-4a, which is also valid for finding F_T for a 1-2 exchanger in which the shell-side flow direction is reversed from that shown in Fig. 11-2. Figure 11-4a is also applicable with negligible error to exchangers with one shell pass and any number of tube passes. Values of F_T less than 0.8 (0.75 at the very lowest) are generally unacceptable because the exchanger configuration chosen is inefficient; the chart is difficult to read accurately; and even a small violation of the first assumption underlying the MTD will invalidate the mathematical derivation and lead to a thermodynamically inoperable exchanger.

Correction-factor charts are also available for exchangers with more

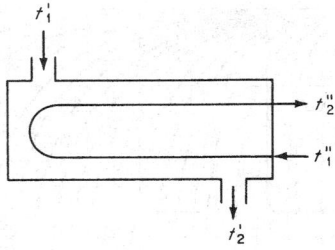

FIG. 11-2 Diagram of a 1-2 exchanger (one well-baffled shell pass and two tube passes with an equal number of tubes in each pass).

than one shell pass provided by a longitudinal shell-side baffle. However, these exchangers are seldom used in practice because of mechanical complications in their construction. Also thermal and physical leakages across the longitudinal baffle further reduce the mean temperature difference and are not properly incorporated into the correction-factor charts. Such charts are useful, however, when it is necessary to construct a multiple-shell exchanger train such as that shown in Fig. 11-3 and are included here for two, three, four, and six *separate identical* shells and two or more tube passes per shell in Fig. 11-4b, c, d, and e. If only one tube pass per shell is required, the piping can and should be arranged to provide pure countercurrent flow, in which case the LMTD is used with no correction.

Cross-flow exchangers of various kinds are also important and require correction to be applied to the LMTD calculated by assuming countercurrent flow. Several cases are given in Fig. 11-4f, g, h, i, and j.

Many other MTD correction-factor charts have been prepared for various configurations. The F_T charts are often employed to make approximate corrections for configurations even in cases for which they are not completely valid.

THERMAL DESIGN FOR SINGLE-PHASE HEAT TRANSFER

Double-Pipe Heat Exchangers The design of double-pipe heat exchangers is straightforward. It is generally conservative to neglect natural-convection and entrance effects in turbulent flow. In laminar flow, natural convection effects can increase the theoretical Graetz prediction by a factor of 3 or 4 for fully developed flows. Pressure drop is calculated by using the correlations given in Sec. 6.

If the inner tube is longitudinally finned on the outside surface, the equivalent diameter is used as the characteristic length in both the Reynolds-number and the heat-transfer correlations. The fin effi-

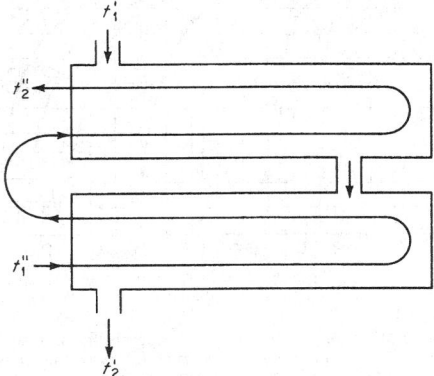

FIG. 11-3 Diagram of a 2-4 exchanger (two separate identical well-baffled shells and four or more tube passes).

° This task can be avoided if a hydrocarbon stream is the limiting resistance by the use of the caloric temperature charts developed by Colburn [*Ind. Eng. Chem.*, **25**, 873 (1933)].

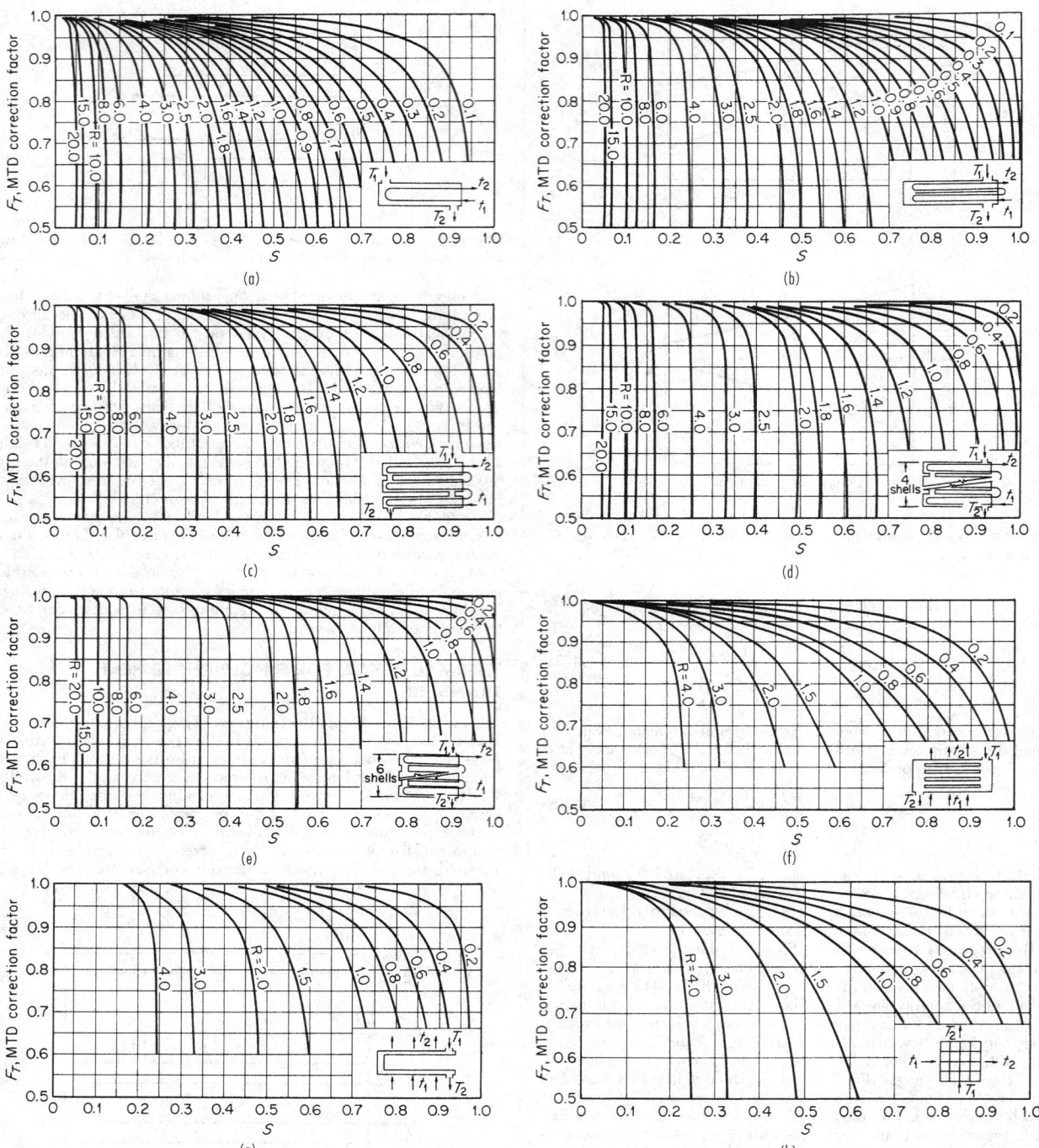

FIG. 11-4 LMTD correction factors for heat exchangers. In all charts, $R = (T_1 - T_2)/(t_2 - t_1)$ and $S = (t_2 - t_1)/(T_1 - t_1)$. (a) One shell pass, two or more tube passes. (b) Two shell passes, four or more tube passes. (c) Three shell passes, six or more tube passes. (d) Four shell passes, eight or more tube passes. (e) Six shell passes. twelve or more tube passes. (f) Cross-flow, one shell pass, one or more parallel rows of tubes. (g) Cross-flow, two passes, two rows of tubes; for more than two passes, use $F_T = 1.0$. (h) Cross-flow, one shell pass, one tube pass, both fluids unmixed

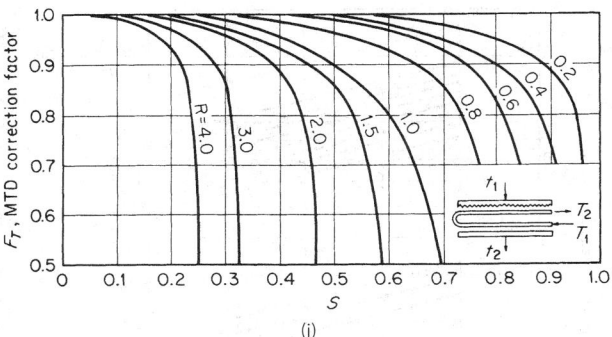

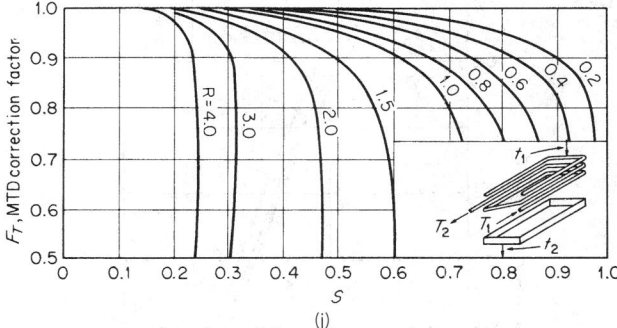

FIG. 11-4   (*Continued*) LMTD correction factors for heat exchangers. In all charts, $R = (T_1 - T_2)/(t_2 - t_1)$ and $S = (t_2 - t_1)/(T_1 - t_1)$. (*i*) Cross-flow (drip type), two horizontal passes with U-bend connections (trombone type). (*j*) Cross-flow (drip type), helical coils with two turns.

ciency must also be known to calculate an *effective* outside area to use in Eq. (11-2).

Fittings contribute strongly to the pressure drop on the annulus side. General methods for predicting this are not reliable, and manufacturer's data should be used when available.

Double-pipe exchangers are often piped in complex series-parallel arrangements on both sides. The MTD to be used has been derived for some of these arrangements and is reported in Kern (*Process Heat Transfer*, McGraw-Hill, New York, 1950). More complex cases may require trial-and-error balancing of the heat loads and rate equations for subsections or even for individual exchangers in the bank.

Baffled Shell-and-Tube Exchangers   The method given here is based on the research summarized in Final Report, Cooperative Research Program on Shell and Tube Heat Exchangers, Univ. Del. Eng. Exp. Sta. Bull. 5 (June 1963). The method assumes that the shell-side heat transfer and pressure-drop characteristics are equal to those of the ideal tube bank corresponding to the cross-flow sections of the exchanger, modified for the distortion of flow pattern introduced by the baffles and the presence of leakage and bypass flow through the various clearances required by mechanical construction.

It is assumed that process conditions and physical properties are known and the following are known or specified: tube outside diameter D_o, tube geometrical arrangement (unit cell), shell inside diameter D_s, shell outer tube limit D_{otl}, baffle cut l_c, baffle spacing l_s, and number of sealing strips N_{ss}. The effective tube length between tube sheets L may be either specified or calculated after the heat-transfer coefficient has been determined. If additional specific information (e.g., tube-baffle clearance) is available, the exact values (instead of estimates) of certain parameters may be used in the calculation with some improvement in accuracy. To complete the rating, it is necessary to know also the tube material and wall thickness or inside diameter.

This rating method, though apparently generally the best in the open literature, is not extremely accurate. An exhaustive study by Palen and Taborek [*Chem. Eng. Prog. Symp. Ser.* 92, **65,** 53 (1969)] showed that this method predicted shell-side coefficients from about 50 percent low to 100 percent high, while the pressure-drop range was from about 50 percent low to 200 percent high. The mean error for heat transfer was about 15 percent low (safe) for all Reynolds numbers, while the mean error for pressure drop was from about 5 percent low (unsafe) at Reynolds numbers above 1000 to about 100 percent high at Reynolds numbers below 10.

Calculation of Shell-Side Geometrical Parameters

1. *Total number of tubes in exchanger N_t.* If not known by direct count, estimate using Eq. (11-84) or (11-85).

2. *Tube pitch parallel to flow p_p and normal to flow p_n.* These quantities are needed only for estimating other parameters. If a detailed drawing of the exchanger is available, it is better to obtain these other parameters by direct count or calculation. The pitches are described by Fig. 11-5 and read therefrom for common tube layouts.

3. *Number of tube rows crossed in one cross-flow section N_c.* Count from exchanger drawing or estimate from

$$N_c = \frac{D_s[1 - 2(l_c/D_s)]}{p_p} \tag{11-7}$$

4. *Fraction of total tubes in cross-flow F_c*

$$F_c = \frac{1}{\pi} \left[\pi + 2 \frac{D_s - 2l_c}{D_{otl}} \sin\left(\cos^{-1}\frac{D_s - 2l_c}{D_{otl}}\right) - 2\cos^{-1}\frac{D_s - 2l_c}{D_{otl}} \right] \tag{11-8}$$

F_c is plotted in Fig. 11-6. This figure is strictly applicable only to splitting, floating-head construction but may be used for other situations with minor error.

5. *Number of effective cross-flow rows in each window N_{cw}*

$$N_{cw} = \frac{0.8l_c}{p_p} \tag{11-9}$$

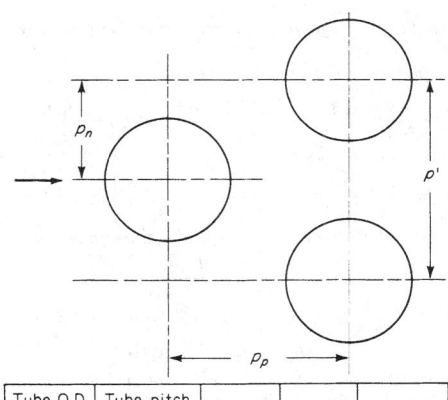

Tube O.D. D_o, in.	Tube pitch p', in.	Layout	p_p, in.	p_n, in.
0.625	0.812		0.704	0.406
0.750	0.938		0.814	0.469
0.750	1		1.000	1.000
0.750	1		0.707	0.707
0.750	1		0.866	0.500
1.000	1.250		1.250	1.250
1.000	1.250		0.884	0.884
1.000	1.250		1.082	0.625

FIG. 11-5   Values of tube pitch for common tube layouts. To convert inches to meters, multiply by 0.0254. Not that D_o, p', p_p, and p_n have units of inches.

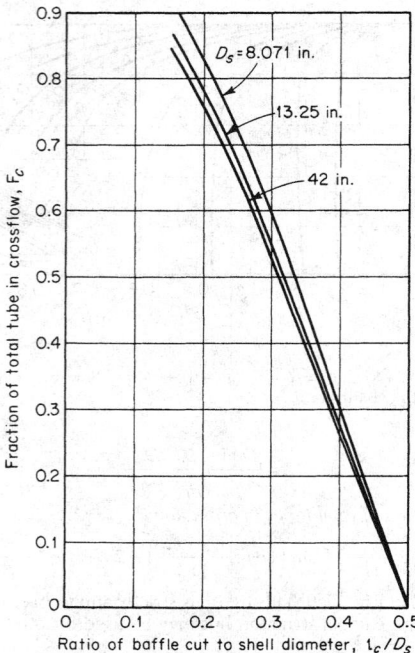

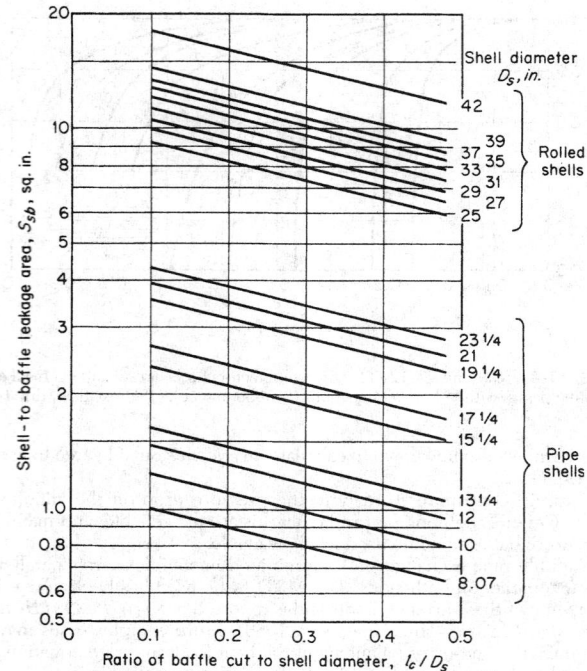

FIG. 11-6 Estimation of fraction of tubes in cross-flow F_c [Eq. (11-8)]. To convert inches to meters, multiply by 0.0254. Note that l_c and D_s have units of inches.

FIG. 11-7 Estimation of shell-to-baffle leakage area [Eq. (11-13)]. To convert inches to meters, multiply by 0.0254; to convert square inches to square meters, multiply by $(6.45)(10^{-4})$. Note that l_c and D_s have units of inches.

6. *Cross-flow area at or near centerline for one cross-flow section S_m*
    a.   For rotated and in-line square layouts:

$$S_m = l_s \left[D_s - D_{otl} + \frac{D_{otl} - D_o}{p_n} (p' - D_o) \right] \quad \text{m}^2 \text{ (ft}^2\text{)} \qquad (11\text{-}10a)$$

    b.   For triangular layouts:

$$S_m = l_s \left[D_s - D_{otl} + \frac{D_{otl} - D_o}{p'} (p' - D_o) \right] \quad \text{m}^2 \text{ (ft}^2\text{)} \qquad (11\text{-}10b)$$

7. *Fraction of cross-flow area available for bypass flow F_{bp}*

$$F_{bp} = \frac{(D_s - D_{otl}) l_s}{S_m} \qquad (11\text{-}11)$$

8. *Tube-to-baffle leakage area for one baffle S_{tb}.* Estimate from

$$S_{tb} = b D_o N_T (1 + F_c) \quad \text{m}^2 \text{ (ft}^2\text{)} \qquad (11\text{-}12)$$

where $b = (6.223)(10^{-4})$ (SI) or $(1.701)(10^{-4})$ (U.S. customary). These values are based on Tubular Exchanger Manufacturers Association (TEMA) Class R construction which specifies 1/32-in diametral clearance between tube and baffle. Values should be modified if extra tight or loose construction is specified or if clogging by dirt is anticipated.

9. *Shell-to-baffle leakage area for one baffle S_{sb}.* If diametral shell-baffle clearance δ_{sb} is known, S_{sb} can be calculated from

$$S_{sb} = \frac{D_s \, \delta_{sb}}{2} \left[\pi - \cos^{-1}\left(1 - \frac{2l_c}{D_s} \right) \right] \quad \text{m}^2 \text{ (ft}^2\text{)} \qquad (11\text{-}13)$$

where the value of the term $\cos^{-1}(1 - 2l_c/D_s)$ is in radians and is between 0 and $\pi/2$. S_{sb} is plotted in Fig. 11-7, based on TEMA Class R standards. Since pipe shells are generally limited to diameters below 24 in, the larger sizes are shown by using the rolled-shell specification. Allowance should be made for especially tight or loose construction.

10. *Area for flow through window S_w.* This area is obtained as the difference between the gross window area S_{wg} and the window area occupied by tubes S_{wt}:

$$S_w = S_{wg} - S_{wt} \qquad (11\text{-}14)$$

$$S_{wg} = \frac{D_s^2}{4} \left[\cos^{-1}\left(1 - 2\frac{l_c}{D_s} \right) \right.$$
$$\left. - \left(1 - 2\frac{l_c}{D_s} \right) \sqrt{1 - \left(1 - 2\frac{l_c}{D_s} \right)^2} \right] \quad \text{m}^2 \text{ (ft}^2\text{)} \quad (11\text{-}15)$$

S_{wg} is plotted in Fig. 11-8. S_{wt} can be calculated from

$$S_{wt} = (N_T/8)(1 - F_c)\pi D_o^2 \qquad \text{m}^2 \text{ (ft}^2\text{)} \qquad (11\text{-}16)$$

11. *Equivalent diameter of window D_w [required only if laminar flow, defined as $(N_{Re})_s \leq 100$, exists]*

$$D_w = \frac{4S_w}{(\pi/2)N_T(1 - F_c)\,D_o + D_s \theta_b} \qquad \text{m (ft)} \qquad (11\text{-}17)$$

where θ_b is the baffle-cut angle given by

$$\theta_b = 2 \cos^{-1}\left(1 - \frac{2l_c}{D_s} \right) \qquad \text{rad} \qquad (11\text{-}18)$$

12. *Number of baffles N_b*

$$N_b = \frac{L - 2le}{l_s} + 1 \qquad (11\text{-}19)$$

where *le* is the entrance/exit baffle spacing, often different from the central baffle spacing. The effective tube length L must be known to calculate N_b, which is needed to calculate shell-side pressure drop. In designing an exchanger, the shell-side coefficient may be calculated and the required exchanger length for heat transfer obtained before N_b is calculated.

Shell-Side Heat-Transfer Coefficient Calculation
1. Calculate the *shell-side Reynolds number* $(N_{Re})_s$.

$$(N_{Re})_s = D_o W / \mu_b S_m \qquad (11\text{-}20)$$

where W = mass flow rate and μ_b = viscosity at bulk temperature. The arithmetic mean bulk shell-side fluid temperature is usually adequate

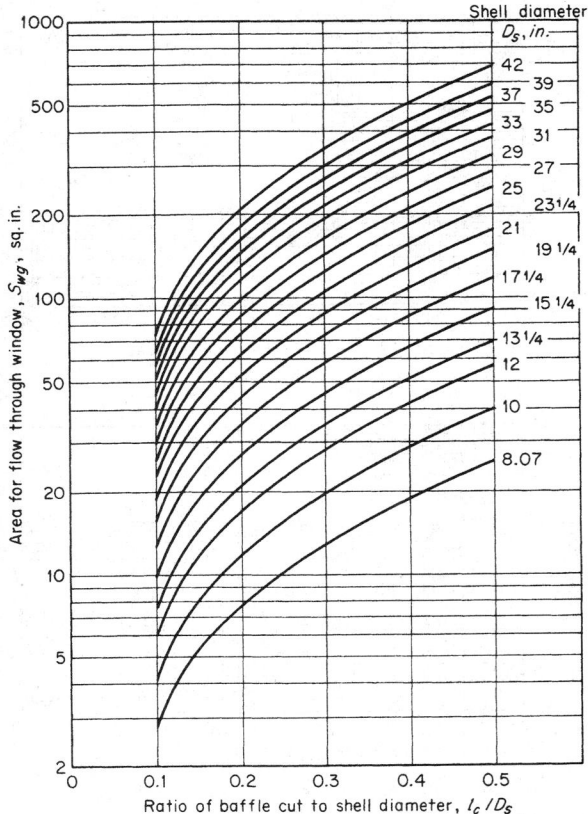

FIG. 11-8 Estimation of window cross-flow area [Eq. (11-15)]. To convert inches to meters, multiply by 0.0254. Note that l_c and D_s have units of inches.

to evaluate all bulk properties of the shell-side fluid. For large temperature ranges or for viscosity that is very sensitive to temperature change, special care must be taken, such as using Eq. (11-6).

2. Find j_k from the ideal-tube bank curve for a given tube layout at the calculated value of $(N_{Re})_s$, using Fig. 11-9, which is adapted from ideal-tube-bank data obtained at Delaware by Bergelin et al. [*Trans. Am. Soc. Mech. Eng.*, **74**, 953 (1952) and the Grimison correlation [*Trans. Am. Soc. Mech. Eng.*, **59**, 583 (1937)].

3. Calculate the shell-side *heat-transfer coefficient for an ideal tube bank* h_k.

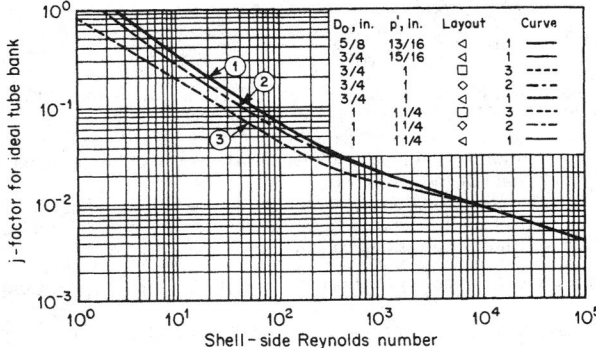

FIG. 11-9 Correlation of j factor for ideal tube bank. To convert inches to meters, multiply by 0.0254. Note that p' and D_o have units of inches.

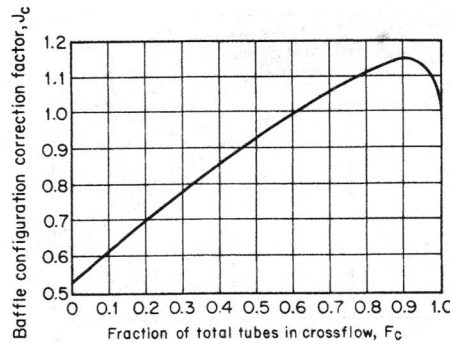

FIG. 11-10 Correction factor for baffle-configuration effects.

$$h_k = j_k c \frac{W}{S_m} \left(\frac{k}{c\mu}\right)^{2/3} \left(\frac{\mu_b}{\mu_w}\right)^{0.14} \qquad (11\text{-}21)$$

where c is the specific heat, k is the thermal conductivity, and μ_w is the viscosity evaluated at the mean surface temperature.

4. Find the correction factor for baffle-configuration effects J_c from Fig. 11-10.

5. Find the correction factor for baffle-leakage effects J_l from Fig. 11-11.

6. Find the correction factor for bundle-bypassing effects J_b from Fig. 11-12.

7. Find the correction factor for adverse temperature-gradient buildup at low Reynolds number J_r:

a. If $(N_{Re})_s < 100$, find J_r° from Fig. 11-13, knowing N_b and $(N_c + N_{cw})$.

b. If $(N_{Re})_s \leq 20$, $J_r = J_r^{\circ}$.

c. If $20 < (N_{Re})_s < 100$, find J_r from Fig. 11-14, knowing J_r° and $(N_{Re})_s$.

8. Calculate the *shell-side heat-transfer coefficient for the exchanger* h_s from

$$h_s = h_k J_c J_l J_b J_r \qquad (11\text{-}22)$$

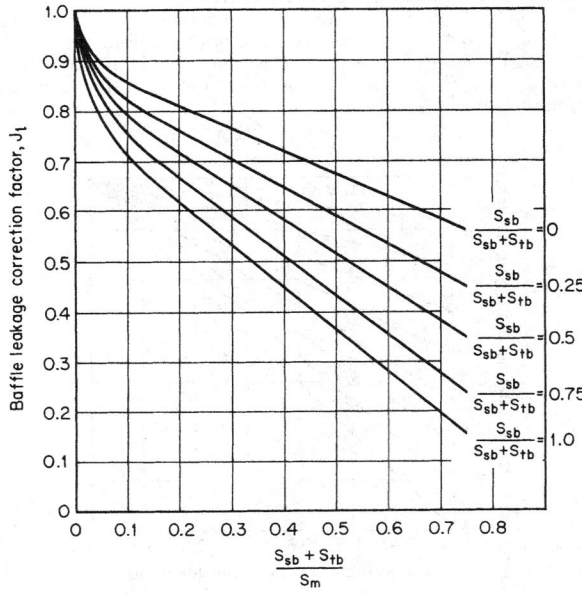

FIG. 11-11 Correction factor for baffle-leakage effects.

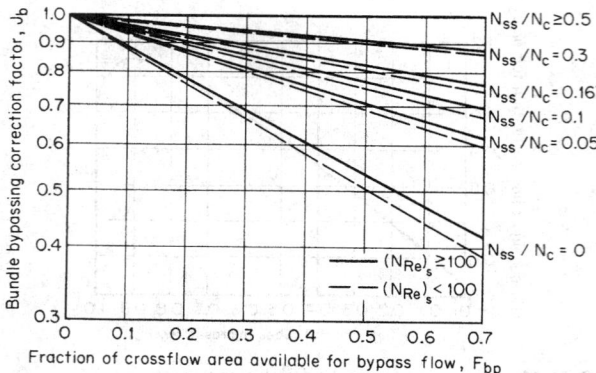

FIG. 11-12   Correction factor for bypass flow.

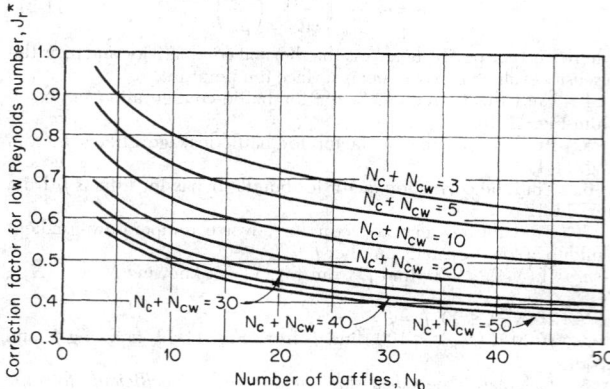

FIG. 11-13   Basic correction factor for adverse temperature gradient at low Reynolds numbers.

Shell-Side Pressure-Drop Calculation

1.   Find f_k from the ideal-tube-bank friction-factor curve for the given tube layout at the calculated value of $(N_{Re})_s$, using Fig. 11-15a for triangular and rotated square arrays and Fig. 11-15b for in-line square arrays. These curves are adapted from Bergelin et al. and Grimison (loc. cit.).

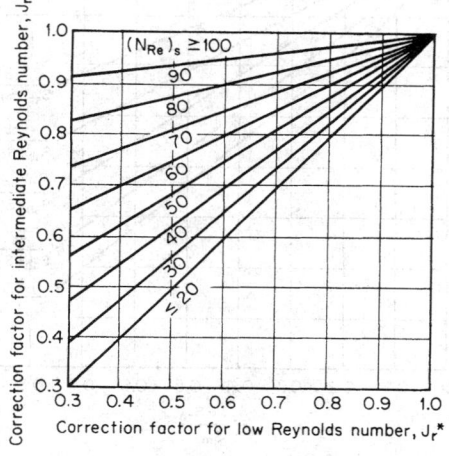

FIG. 11-14   Correction factor for adverse temperature gradient at intermediate Reynolds numbers.

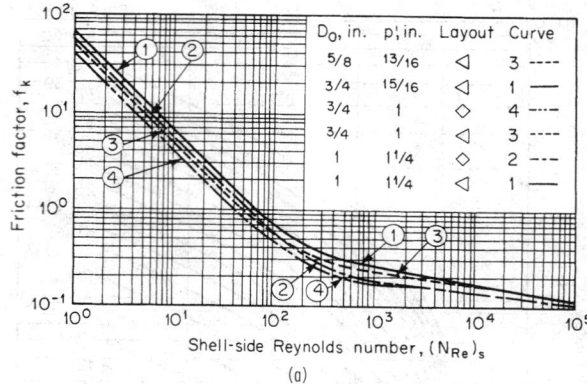

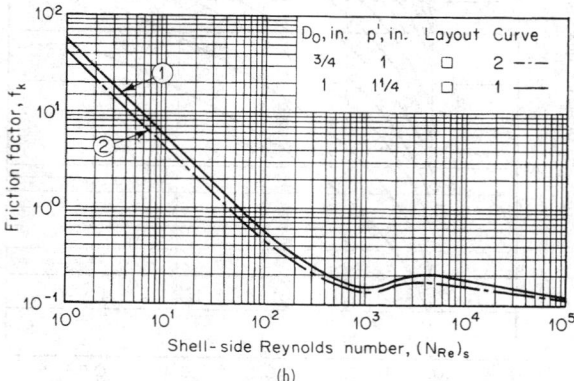

FIG. 11-15   Correction of friction factors for ideal tube banks. (a) Triangular and rotated square arrays. (b) In-line square arrays.

2.   Calculate the *pressure drop for an ideal cross-flow section.*

$$\Delta P_{bk} = b \frac{f_k W^2 N_c}{\rho S_m^2} \left(\frac{\mu_w}{\mu_b} \right)^{0.14} \tag{11-23}$$

where $b = (2.0)(10^{-3})$ (SI) or $(9.9)(10^{-5})$ (U.S. customary).

3.   Calculate the *pressure drop for an ideal window section.* If $(N_{Re})_s \geq 100$,

$$\Delta P_{wk} = b \frac{W^2 (2 + 0.6 N_{cw})}{S_m S_w \rho} \tag{11-24a}$$

where $b = (5)(10^{-4})$ (SI) or $(2.49)(10^{-5})$ (U.S. customary).
   If $(N_{Re})_s < 100$,

$$\Delta P_{wk} = b_1 \frac{\mu_b W}{S_m S_w \rho} \left(\frac{N_{cw}}{p' - D_o} + \frac{l_s}{D_w^2} \right) + b_2 \frac{W^2}{S_m S_w \rho} \tag{11-24b}$$

where $b_1 = (1.681)(10^{-5})$ (SI) or $(1.08)(10^{-4})$ (U.S. customary), and $b_2 = (9.99)(10^{-4})$ (SI) or $(4.97)(10^{-5})$ (U.S. customary).

4.   Find the correction factor for the effect of baffle leakage on pressure drop R_l from Fig. 11-16. Curves shown are not to be extrapolated beyond the points shown.

5.   Find the correction factor for bundle bypass R_b from Fig. 11-17.

6.   Calculate the *pressure drop across the shell side* (excluding nozzles). Units for pressure drop are lbf/ft².

$$\Delta P_s = [(N_b - 1)(\Delta P_{bk})R_b + N_b \Delta P_{wk}]R_l + 2 \Delta P_{bk} R_b \left(1 + \frac{N_{cw}}{N_c} \right) \tag{11-25}$$

The values of h_s and ΔP_s calculated by this procedure are for clean exchangers and are intended to be as accurate as possible, not conservative. A fouled exchanger will generally give lower heat-transfer rates, as reflected by the dirt resistances incorporated into Eq. (11-2), and higher pressure drops. Some estimate of **fouling effects** on pres-

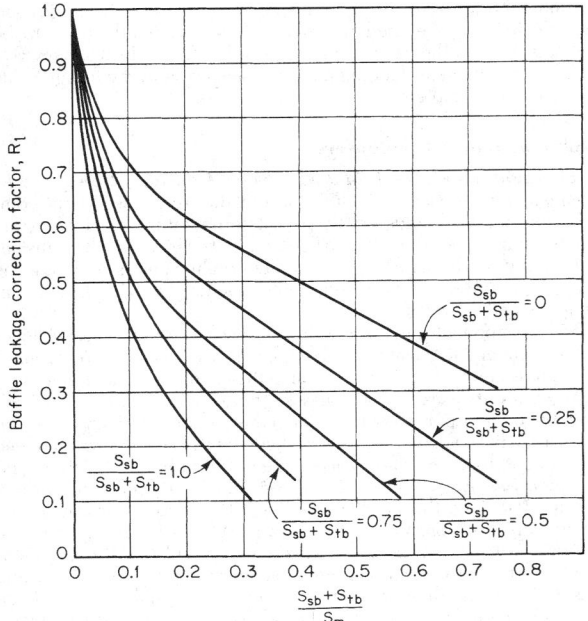

FIG. 11-16   Correction factor for baffle-leakage effect on pressure drop.

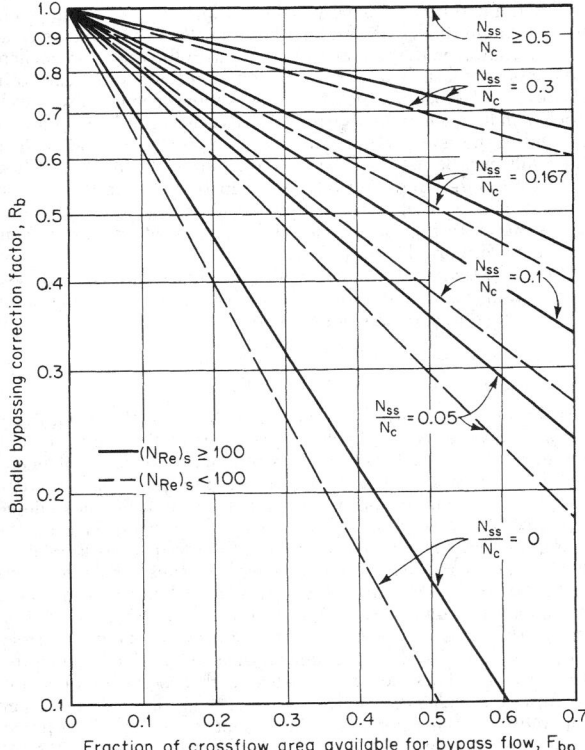

FIG. 11-17   Correction factor on pressure drop for bypass flow.

sure drop may be made by using the methods just given by assuming that the fouling deposit blocks the leakage and possibly the bypass areas. The fouling may also decrease the clearance between tubes and significantly increase the pressure drop in cross-flow.

THERMAL DESIGN OF CONDENSERS

Single-Component Condensers

Mean Temperature Difference   In condensing a single component at its saturation temperature, the entire resistance to heat transfer on the condensing side is generally assumed to be in the layer of condensate. A mean condensing coefficient is calculated from the appropriate correlation and combined with the other resistances in Eq. (11-2). The overall coefficient is then used with the LMTD (no F_T correction is necessary for isothermal condensation) to give the required area, even though the condensing coefficient and hence U are not constant throughout the condenser.

If the vapor is **superheated** at the inlet, the vapor may first be desuperheated by sensible heat transfer from the vapor. This occurs if the surface temperature is above the saturation temperature, and a single-phase heat-transfer correlation is used. If the surface is below the saturation temperature, condensation will occur directly from the superheated vapor, and the effective coefficient is determined from the appropriate condensation correlation, using the saturation temperature in the LMTD. To determine whether or not condensation will occur directly from the superheated vapor, calculate the surface temperature by assuming single-phase heat transfer.

$$T_{\text{surface}} = T_{\text{vapor}} - \frac{U}{h}\,(T_{\text{vapor}} - T_{\text{coolant}}) \qquad (11\text{-}26)$$

where h is the sensible heat-transfer coefficient for the vapor, U is calculated by using h, and both are on the same area basis. If $T_{\text{surface}} > T_{\text{saturation}}$, no condensation occurs at that point and the heat flux is actually higher than if $T_{\text{surface}} \leq T_{\text{saturation}}$ and condensation did occur. It is generally conservative to design a pure-component desuperheater-condenser as if the entire heat load were transferred by condensation, using the saturation temperature in the LMTD.

The design of an integral **condensate subcooling section** is more difficult, especially if close temperature approach is required. The condensate layer on the surface is on the average subcooled by one-third to one-half of the temperature drop across the film, and this is often sufficient if the condensate is not reheated by raining through the vapor. If the condensing-subcooling process is carried out inside tubes or in the shell of a vertical condenser, the single-phase subcooling section can be treated separately, giving an area that is added onto that needed for condensation. If the subcooling is achieved on the shell side of a horizontal condenser by flooding some of the bottom tubes with a weir or level controller, the rate and heat-balance equations must be solved for each section to obtain the area required.

Pressure drop on the condensing side reduces the final condensing temperature and the MTD and should always be checked. In designs requiring close approach between inlet coolant and exit condensate (subcooled or not), underestimation of pressure drop on the condensing side can lead to an exchanger that cannot meet specified terminal temperatures. Since pressure-drop calculations in two-phase flows such as condensation are relatively inaccurate, designers must consider carefully the consequences of a larger-than-calculated pressure drop.

Horizontal In-Shell Condensers   The mean **condensing coefficient** for the outside of a bank of horizontal tubes is calculated from Eq. (5-93) for a single tube, corrected for the number of tubes in a vertical row. For undisturbed laminar flow over all the tubes, Eq. (5-97) is, for realistic condenser sizes, overly conservative because of rippling, splashing, and turbulent flow (*Process Heat Transfer*, McGraw-Hill, New York, 1950). Kern proposed an exponent of −⅙ on the basis of experience, while Freon-11 data of Short and Brown (*General Discussion on Heat Transfer*, Institute of Mechanical Engineers, London, 1951) indicate independence of the number of tube rows. It seems reasonable to use no correction for inviscid liquids and Kern's correction for viscous condensates. For a cylindrical tube bundle, where N varies, it is customary to take N equal to two-thirds of the maximum or centerline value.

Baffles in a horizontal in-shell condenser are oriented with the cuts vertical to facilitate drainage and eliminate the possibility of flooding in the upward cross-flow sections. **Pressure drop** on the vapor side can be estimated by the data and method of Diehl and Unruh [*Pet. Refiner*, **36**(10), 147 (1957); **37**(10), 124 (1958)].

High vapor velocities across the tubes enhance the condensing coefficient. There is no correlation in the open literature to permit designers to take advantage of this. Since the vapor flow rate varies along the length, an incremental calculation procedure would be required in any case. In general, the pressure drops required to gain significant benefit are above those allowed in most process applications.

Vertical In-Shell Condensers Condensers are often designed so that condensation occurs on the outside of vertical tubes. Equation (5-88) is valid as long as the condensate film is laminar. When it becomes turbulent, Fig. 5-10 or Colburn's equation [*Trans. Am. Inst. Chem. Eng.*, **30**, 187 (1933–1934) may be used.

Some judgment is required in the use of these correlations because of construction features of the condenser. The tubes must be supported by baffles, usually with maximum cut (45 percent of the shell diameter) and maximum spacing to minimize pressure drop. The flow of the condensate is interrupted by the baffles, which may draw off or redistribute the liquid and which will also cause some splashing of free-falling drops onto the tubes.

For **subcooling,** a liquid inventory may be maintained in the bottom end of the shell by means of a weir or a liquid-level-controller. The subcooling heat-transfer coefficient is given by the correlations for natural convection on a vertical surface [Eqs. (5-33a), (5-33b)], with the pool assumed to be well mixed (isothermal) at the subcooled condensate exit temperature. Pressure drop may be estimated by the shell-side procedure.

Horizontal In-Tube Condensers Condensation of a vapor inside horizontal tubes occurs in kettle and horizontal thermosiphon reboilers and in air-cooled condensers. In-tube condensation also offers certain advantages for condensation of multicomponent mixtures, discussed in the subsection "Multicomponent Condensers." The various in-tube correlations are closely connected to the **two-phase flow pattern** in the tube [*Chem. Eng. Prog. Symp. Ser.*, **66**(102), 150 (1970)]. At low flow rates, when gravity dominates the flow pattern, Eq. (5-101) may be used. At high flow rates, the flow and heat transfer are governed by vapor shear on the condensate film, and Eq. (5-100a) is valid. A simple and generally conservative procedure is to calculate the coefficient for a given case by both correlations and use the *larger* one.

Pressure drop during condensation inside horizontal tubes can be computed by using the correlations for two-phase flow given in Sec. 6 and neglecting the pressure recovery due to deceleration of the flow.

Vertical In-Tube Condensation Vertical-tube condensers are generally designed so that vapor and liquid flow cocurrently downward; if pressure drop is not a limiting consideration, this configuration can result in higher heat-transfer coefficients than shell-side condensation and has particular advantages for multicomponent condensation. If gravity controls, the mean heat-transfer coefficient for condensation is given by Figs. 5-9 and 5-10. If vapor shear controls, Eq. (5-99a) is applicable. It is generally conservative to calculate the coefficients by both methods and choose the *higher* value. The pressure drop can be calculated by using the Lockhart-Martinelli method [*Chem. Eng. Prog.*, **45**, 39 (1945)] for friction loss, neglecting momentum and hydrostatic effects.

Vertical in-tube condensers are often designed for **reflux or knock-back application** in reactors or distillation columns. In this case, vapor flow is upward, countercurrent to the liquid flow on the tube wall; the vapor shear acts to thicken and retard the drainage of the condensate film, reducing the coefficient. Neither the fluid dynamics nor the heat transfer is well understood in this case, but Soliman, Schuster, and Berenson [*J. Heat Transfer*, **90**, 267–276 (1968)] discuss the problem and suggest a computational method. The Diehl-Koppany correlation [*Chem. Eng. Prog. Symp. Ser.* 92, **65** (1969)] may be used to estimate the maximum allowable vapor velocity at the tube inlet. If the vapor velocity is great enough, the liquid film will be carried upward; this design has been employed in a few cases in which only part of the stream is to be condensed. This velocity cannot be accurately computed, and a very conservative (high) outlet velocity must be used if unstable flow and flooding are to be avoided; 3 times the vapor velocity given by the Diehl-Koppany correlation for incipient flooding has been suggested as the design value for completely stable operation.

Multicomponent Condensers

Thermodynamic and Mass-Transfer Considerations Multicomponent vapor mixture includes several different cases: all the components may be liquids at the lowest temperature reached in the condensing side, or there may be components which dissolve substantially in the condensate even though their boiling points are below the exit temperature, or one or more components may be both noncondensable and nearly insoluble.

Multicomponent condensation always involves sensible-heat changes in the vapor and liquid along with the latent-heat load. Compositions of both phases in general change through the condenser, and **concentration gradients** exist in both phases. Temperature and concentration profiles and transport rates at a point in the condenser usually cannot be calculated, but the binary cases have been treated: condensation of one component in the presence of a completely insoluble gas [Colburn and Hougen, *Ind. Eng. Chem.*, **26**, 1178–1182 (1934); and Colburn and Edison, *Ind. Eng. Chem.*, **33**, 457–458 (1941)] and condensation of a binary vapor [Colburn and Drew, *Trans. Am. Inst. Chem. Eng.*, **33**, 196–215 (1937)]. It is necessary to know or calculate diffusion coefficients for the system, and a reasonable approximate method to avoid this difficulty and the reiterative calculations is desirable. To integrate the point conditions over the total condensation requires the temperature, composition enthalpy, and flow-rate profiles as functions of the heat removed. These are calculated from component thermodynamic data if the vapor and liquid are assumed to be in equilibrium at the local vapor temperature. This assumption is not exactly true, since the condensate and the liquid-vapor interface (where equilibrium does exist) are intermediate in temperature between the coolant and the vapor.

In calculating the condensing curve, it is generally assumed that the vapor and liquid flow collinearly and in intimate contact so that composition equilibrium is maintained between the total streams at all points. If, however, the condensate drops out of the vapor (as can happen in horizontal shell-side condensation), and flows to the exit without further interaction, the remaining vapor becomes excessively enriched in light components with a decrease in condensing temperature and in the temperature difference between vapor and coolant. The result may be not only a small reduction in the amount of heat transferred in the condenser but also an inability to condense totally the light ends even at reduced throughput or with the addition of more surface. To prevent the liquid from segregating, in-tube condensation is preferred in critical cases.

Thermal Design If the controlling resistance for heat and mass transfer in the vapor is sensible-heat removal from the cooling vapor, the following design equation is obtained:

$$A = \int_0^{Q_T} \frac{1 + U'Z_H/h_{sv}}{U'(T_v - T_c)} \, dQ \qquad (11\text{-}27)$$

U' is the overall heat-transfer coefficient between the vapor-liquid interface and the coolant, including condensate film, dirt and wall resistances, and coolant. The condensate film coefficient is calculated from the appropriate equation or correlation for pure vapor condensation for the geometry and flow regime involved, using mean liquid properties. Z_H is the ratio of the sensible heat removed from the vapor-gas stream to the total heat transferred; this quantity is obtained from thermodynamic calculations and may vary substantially from one end of the condenser to the other, especially when removing vapor from a noncondensable gas. The sensible-heat-transfer coefficient for the vapor-gas stream h_{sv} is calculated by using the appropriate correlation or design method for the geometry involved, neglecting the presence of the liquid. As the vapor condenses, this coefficient decreases and must be calculated at several points in the process. T_v and T_c are temperatures of the vapor and of the coolant respectively. This procedure is similar in principle to that of Ward [*Petro/Chem. Eng.*, **32**(11), 42–48 (1960)]. It may be nonconservative for condensing steam and other high-latent-heat substances, in which case it may be necessary to increase the calculated area by 25 to 50 percent.

Pressure drop on the condensing side may be estimated by judicious application of the methods suggested for pure-component condensation, taking into account the generally nonlinear decrease of vapor-gas flow rate with heat removal.

THERMAL DESIGN OF REBOILERS

For a **single-component reboiler design,** attention is focused upon the mechanism of heat and momentum transfer at the hot surface. In *multicomponent systems,* the light components are preferentially vaporized at the surface, and the process becomes limited by their rate of diffusion. The net effect is to decrease the effective temperature difference between the hot surface and the bulk of the boiling liquid. If one attempts to vaporize too high a fraction of the feed liquid to the reboiler, the temperature difference between surface and liquid is reduced to the point that nucleation and vapor generation on the surface are suppressed and heat transfer to the liquid proceeds at the lower rate associated with single-phase natural convection. The only safe procedure in design for wide-boiling-range mixtures is to vaporize such a limited fraction of the feed that the boiling point of the remaining liquid mixture is still at least 5.5°C (10°F) below the surface temperature. Positive flow of the unvaporized liquid through and out of the reboiler should be provided.

Kettle Reboilers It has been generally assumed that kettle reboilers operate in the pool boiling mode, but with a lower peak heat flux because of vapor binding and blanketing of the upper tubes in the bundle. There is some evidence that vapor generation in the bundle causes a high circulation rate through the bundle. The result is that, at the lower heat fluxes, the kettle reboiler actually gives higher heat-transfer coefficients than a single tube. Present understanding of the recirculation phenomenon is insufficient to take advantage of this in design. Available nucleate pool boiling correlations are only very approximate, failing to account for differences in the nucleation characteristics of different surfaces. The Mostinski correlation [Eq. (5-102)] and the McNelly correlation [Eq. (5-103)] are generally the best for single components or narrow-boiling-range mixtures at low fluxes, though they may give errors of 40 to 50 percent. Experimental heat-transfer coefficients for pool boiling of a given liquid on a given surface should be used if available. The bundle **peak heat flux** is a function of tube-bundle geometry, especially of tube-packing density; in the absence of better information, the Palen-Small modification [Eq. (5-108)] of the Zuber maximum-heat-flux correlation is recommended.

A general method for analyzing kettle reboiler performance is by Fair and Klip, *Chem. Eng. Prog.* **79**(3), 86 (1983). It is effectively limited to computer application.

Kettle reboilers are generally assumed to require negligible pressure drop. It is important to provide good longitudinal liquid flow paths within the shell so that the liquid is uniformly distributed along the entire length of the tubes and excessive local vaporization and vapor binding are avoided.

This method may also be used for the thermal design of **horizontal thermosiphon reboilers.** The recirculation rate and pressure profile of the thermosiphon loop can be calculated by the methods of Fair [*Pet. Refiner*, **39**(2), 105–123 (1960)].

Vertical Thermosiphon Reboilers Vertical thermosiphon reboilers operate by natural circulation of the liquid from the still through the downcomer to the reboiler and of the two-phase mixture from the reboiler through the return piping. The flow is induced by the hydrostatic pressure imbalance between the liquid in the downcomer and the two-phase mixture in the reboiler tubes. Thermosiphons do not require any pump for recirculation and are generally regarded as less likely to foul in service because of the relatively high two-phase velocities obtained in the tubes. Heavy components are not likely to accumulate in the thermosiphon, but they are more difficult to design satisfactorily than kettle reboilers, especially in vacuum operation. Several shortcut methods have been suggested for thermosiphon design, but they must generally be used with caution. The method due to Fair (loc. cit.), based upon two-phase flow correlations, is the most complete in the open literature but requires a computer for practical use. Fair also suggests a shortcut method that is satisfactory for preliminary design and can be reasonably done by hand.

Forced-Recirculation Reboilers In forced-recirculation reboilers, a pump is used to ensure circulation of the liquid past the heat transfer surface. Force-recirculation reboilers may be designed so that boiling occurs inside vertical tubes, inside horizontal tubes, or on the shell side. For forced boiling inside vertical tubes, Fair's method (loc. cit.) may be employed, making only the minor modification that the recirculation rate is fixed and does not need to be balanced against the pressure available in the downcomer. Excess pressure required to circulate the two-phase fluid through the tubes and back into the column is supplied by the pump, which must develop a positive pressure increase in the liquid.

Fair's method may also be modified to design forced-recirculation reboilers with horizontal tubes. In this case the hydrostatic-head-pressure effect through the tubes is zero but must be considered in the two-phase return lines to the column.

The same procedure may be applied in principle to design of forced-recirculation reboilers with shell-side vapor generation. Little is known about two-phase flow on the shell side, but a reasonable estimate of the friction pressure drop can be made from the data of Diehl and Unruh [*Pet. Refiner*, **36**(10), 147 (1957); **37**(10), 124 (1958)]. No void-fraction data are available to permit accurate estimation of the hydrostatic or acceleration terms. These may be roughly estimated by assuming homogeneous flow.

THERMAL DESIGN OF EVAPORATORS

Heat duties of evaporator heating surfaces are usually determined by conventional heat and material balance calculations. Heating surface areas are normally, but not always taken as those in contact with the material being evaporated. It is the heat transfer ΔT that presents the most difficulty in deriving or applying heat-transfer coefficients. The total ΔT between heat source and heat sink is never all available for heat transfer. Since energy usually is carried to and from an evaporator body or effect by condensible vapors, loss in pressure represents a loss in ΔT. Such losses include pressure drop through entrainment separators, friction in vapor piping, and acceleration losses into and out of the piping. The latter loss has often been overlooked, even though it can be many times greater than the friction loss. Similarly, friction and acceleration losses past the heating surface, such as in a falling film evaporator, cause a loss of ΔT that may or may not have been included in the heat transfer ΔT when reporting experimental results. Boiling-point rise, the difference between the boiling point of the solution and the condensing point of the solvent at the same pressure, is another loss. Experimental data are almost always corrected for boiling-point rise, but plant data are suspect when based on temperature measurements because vapor at the point of measurement may still contain some superheat, which represents but a very small fraction of the heat given up when the vapor condenses but may represent a substantial fraction of the actual net ΔT available for heat transfer. A ΔT loss that must be considered in forced-circulation evaporators is that due to temperature rise through the heater, a consequence of the heat being absorbed there as sensible heat. A further loss may occur when the heater effluent flashes as it enters the vapor-liquid separator. Some of the liquid may not reach the surface and flash to equilibrium with the vapor pressure in the separator, instead of recirculating to the heater, raising the average temperature at which heat is absorbed and further reducing the net ΔT. Whether or not these ΔT losses are allowed for in the heat-transfer coefficients reported depends on the method of measurement. Simply basing the liquid temperature on the measured vapor head pressure may ignore both—or only the latter if temperature rise through the heater is estimated separately from known heat input and circulation rate. In general, when calculating overall heat-transfer coefficients from individual-film coefficients, all of these losses must be allowed for, while when using reported overall coefficients care must be exercised to determine which losses may already have been included in the heat transfer ΔT.

Forced-Circulation Evaporators In evaporators of this type in which hydrostatic head prevents boiling at the heating surface, **heat-**

transfer coefficients can be predicted from the usual correlations for condensing steam (Fig. 5-10) and forced-convection sensible heating [Eq. (5-50)]. The liquid film coefficient is improved if boiling is not completely suppressed. When only the film next to the wall is above the boiling point, Boarts, Badger, and Meisenberg [*Ind. Eng. Chem.*, **29**, 912 (1937)] found that results could be correlated by Eq. (5-50) by using a constant of 0.0278 instead of 0.023. In such cases, the course of the liquid temperature can still be calculated from known circulation rate and heat input.

When the bulk of the liquid is boiling in part of the tube length, the film coefficient is even higher. However, the liquid temperature starts dropping as soon as full boiling develops, and it is difficult to estimate the course of the temperature curve. It is certainly safe to estimate heat transfer on the basis that no bulk boiling occurs. Fragen and Badger [*Ind. Eng. Chem.*, **28**, 534 (1936)] obtained an **empirical correlation** of overall heat-transfer coefficients in this type of evaporator, based on the ΔT at the heater inlet:

In U.S. customary units

$$U = 2020D^{0.57}(V_s)^{3.6/L}/\mu^{0.25}\,\Delta T^{0.1} \tag{11-28}$$

where D = mean tube diameter, V_s = inlet velocity, L = tube length, and μ = liquid viscosity. This equation is based primarily on experiments with copper tubes of 0.022 m (8/8 in) outside diameter, 0.00165 m (16 gauge), 2.44 m (8 ft) long, but it includes some work with 0.0127-m (½-in) tubes 2.44 m (8 ft) long and 0.0254-m (1-in) tubes 3.66 m (12 ft) long.

Long-Tube Vertical Evaporators   In the rising-film version of this type of evaporator, there is usually a nonboiling zone in the bottom section and a boiling zone in the top section. The length of the nonboiling zone depends on heat-transfer characteristics in the two zones and on pressure drop during two-phase flow in the boiling zone. The work of Martinelli and coworkers [Lockhart and Martinelli, *Chem. Eng. Prog.*, **45**, 39–48 (January 1949); and Martinelli and Nelson, *Trans. Am. Soc. Mech. Eng.*, **70**, 695–702 (August 1948)] permits a prediction of pressure drop, and a number of correlations are available for estimating film coefficients of heat transfer in the two zones. In estimating pressure drop, integrated curves similar to those presented by Martinelli and Nelson are the easiest to use. The curves for pure water are shown in Figs. 11-18 and 11-19, based on the assumption that the flow of both vapor and liquid would be turbulent if each were flowing alone in the tube. Similar curves can be prepared if one or both flows are laminar or if the properties of the liquid differ appreciably from the properties of pure water. The **acceleration pressure drop** ΔP_a is calculated from the equation

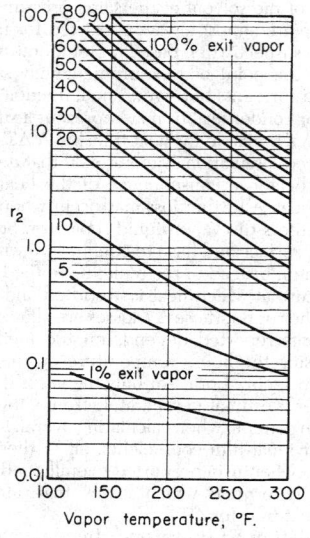

FIG. 11-18   Acceleration losses in boiling flow. °C = (°F – 32)/1.8.

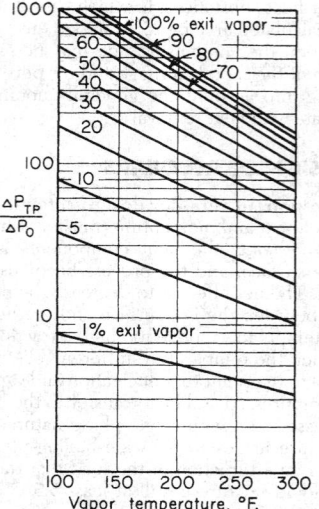

FIG. 11-19   Friction pressure drop in boiling flow. °C = (°F – 32)/1.8.

$$\Delta P_a = br_2 G^2/32.2 \tag{11-29}$$

where $b = (2.6)(10^7)$(SI) and 1.0 (U.S. customary) and using r_2 from Fig. 11-18. The frictional pressure drop is derived from Fig. 11-19, which shows the ratio of two-phase pressure drop to that of the entering liquid flowing alone.

Pressure drop due to hydrostatic head can be calculated from liquid holdup R_1. For nonfoaming dilute aqueous solutions, R_1 can be estimated from $R_1 = 1/[1 + 2.5(V/L)(\rho_l/\rho_v)^{1/2}]$. Liquid holdup, which represents the ratio of liquid-only velocity to actual liquid velocity, also appears to be the principal determinant of the convective coefficient in the boiling zone (Dengler, Sc.D. thesis, MIT, 1952). In other words, the convective coefficient is that calculated from Eq. (5-50) by using the liquid-only velocity divided by R_1 in the Reynolds number. Nucleate boiling augments convective heat transfer, primarily when ΔT's are high and the convective coefficient is low [Chen, *Ind. Eng. Chem. Process Des. Dev.*, **5**, 322 (1966)].

Film coefficients for the **boiling of liquids other than water** have been investigated. Coulson and McNelly [*Trans. Inst. Chem. Eng.*, **34**, 247 (1956)] derived the following relation, which also correlated the data of Badger and coworkers [*Chem. Metall. Eng.*, **46**, 640 (1939); *Chem. Eng.*, **61**(2), 183 (1954); and *Trans. Am. Inst. Chem. Eng.*, **33**, 392 (1937); **35**, 17 (1939); **36**, 759 (1940)] on water:

$$N_{\mathrm{Nu}} = (1.3 + b\,D)(N_{\mathrm{Pr}})_l^{0.9}(N_{\mathrm{Re}})_l^{0.23}(N_{\mathrm{Re}})_g^{0.34}\left(\frac{\rho_l}{\rho_g}\right)^{0.25}\left(\frac{\mu_g}{\mu_l}\right) \tag{11-30}$$

where $b = 128$ (SI) or 39 (U.S. customary), N_{Nu} = Nusselt number based on liquid thermal conductivity, D = tube diameter, and the remaining terms are dimensionless groupings of liquid Prandtl number, liquid Reynolds number, vapor Reynolds number, and ratios of densities and viscosities. The Reynolds numbers are calculated on the basis of each fluid flowing by itself in the tube.

Additional corrections must be applied when the fraction of vapor is so high that the remaining liquid does not wet the tube wall or when the velocity of the mixture at the tube exits approaches sonic velocity. McAdams, Woods, and Bryan (*Trans. Am. Soc. Mech. Eng.*, 1940), Dengler and Addoms (loc. cit.), and Stroebe, Baker, and Badger [*Ind. Eng. Chem.*, **31**, 200 (1939)] encountered dry-wall conditions and reduced coefficients when the weight fraction of vapor exceeded about 80 percent. Schweppe and Foust [*Chem. Eng. Prog.*, **49**, *Symp. Ser.* 5, 77 (1953)] and Harvey and Foust (ibid., p. 91) found that "sonic choking" occurred at surprisingly low flow rates.

The simplified method of calculation outlined includes no allowance for the **effect of surface tension.** Stroebe, Baker, and Badger (loc. cit.) found that by adding a small amount of surface-active agent the boiling-film coefficient varied inversely as the square of the surface tension. Coulson and Mehta [*Trans. Inst. Chem. Eng.*, **31**, 208 (1953)] found the exponent to be −1.4. The higher coefficients at low surface tension are offset to some extent by a higher pressure drop, probably because the more intimate mixture existing at low surface tension causes the liquid fraction to be accelerated to a velocity closer to that of the vapor. The pressure drop due to acceleration ΔP_a derived from Fig. 11-18 allows for some slippage. In the limiting case, such as might be approached at low surface tension, the acceleration pressure drop in which "fog" flow is assumed (no slippage) can be determined from the equation

$$\Delta P_a' = \frac{y(V_g - V_l)G^2}{g_c} \qquad (11\text{-}31)$$

where
y = fraction vapor by weight
V_g, V_l = specific volume gas, liquid
G = mass velocity

While the foregoing methods are valuable for detailed evaporator design or for evaluating the effect of changes in conditions on performance, they are cumbersome to use when making preliminary designs or cost estimates. Figure 11-20 gives the general range of **overall long-tube vertical- (LTV) evaporator heat-transfer coefficients** usually encountered in commercial practice. The higher coefficients are encountered when evaporating dilute solutions and the lower range when evaporating viscous liquids. The dashed curve represents the approximate lower limit, for liquids with viscosities of about 0.1 Pa·s (100 cP). The LTV evaporator does not work well at low temperature differences, as indicated by the results shown in Fig. 11-21 for seawater in 0.051-m (2-in), 0.0028-m (12-gauge) brass tubes 7.32 m (24 ft) long (W. L. Badger Associates, Inc., U.S. Department of the Interior, Office of Saline Water Rep. 26, December 1959, OTS Publ. PB 161290). The feed was at its boiling point at the vapor-head pressure, and feed rates varied from 0.025 to 0.050 kg/(s·tube) [200 to 400 lb/(h·tube)] at the higher temperature to 0.038 to 0.125 kg/(s·tube) [300 to 1000 lb/(h·tube)] at the lowest temperature.

Falling film evaporators find their widest use at low temperature differences—also at low temperatures. Under most operating conditions encountered, heat transfer is almost all by pure convection, with a negligible contribution from nucleate boiling. Film coefficients on the condensing side may be estimated from Dukler's correlation, [*Chem. Eng. Prog.* **55**, 62 1950]. The same Dukler correlation presents curves covering falling film heat transfer to nonboiling liquids that are equally applicable to the falling film evaporator [Sinek and Young, *Chem. Eng. Prog.* **58**, No. 12, 74 (1962)]. Kunz and Yerazunis [*J. Heat Transfer* **8**, 413 (1969)] have

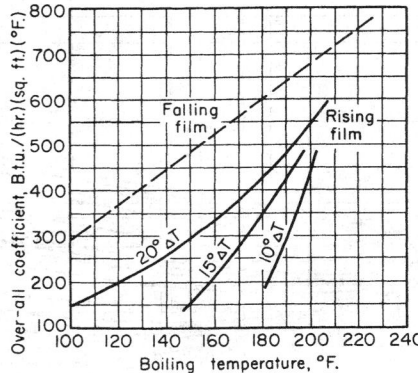

FIG. 11-21 Heat-transfer coefficients in LTV seawater evaporators. °C = (°F − 32)/1.8; to convert British thermal units per hour-square foot-degrees Fahrenheit to joules per square meter-second-kelvins, multiply by 5.6783.

since extended the range of physical properties covered, as shown in Fig. 11-22. The boiling point in the tubes of such an evaporator is higher than in the vapor head because of both frictional-pressure drop and the head needed to accelerate the vapor to the tube-exit velocity. These factors, which can easily be predicted, make the overall apparent coefficients somewhat lower than those for nonboiling conditions. Figure 11-21 shows overall apparent heat-transfer coefficients determined in a falling-film seawater evaporator using the same tubes and flow rates as for the rising-film tests (W. L. Badger Associates, Inc., loc. cit.).

Short-Tube Vertical Evaporators Coefficients can be estimated by the same detailed method described for recirculating LTV evaporators. Performance is primarily a function of temperature level, temperature difference, and viscosity. While liquid level can also have an important influence, this is usually encountered only at levels lower than considered safe in commercial operation. **Overall heat-transfer coefficients** are shown in Fig. 11-23 for a basket-type evaporator (one with an annular downtake) when boiling water with 0.051-m (2-in) outside-diameter 0.0028-m-wall (12-gauge), 1.22-m-(4-ft) long steel tubes [Badger and Shepard, *Chem. Metall. Eng.*, **23**, 281 (1920)]. Liquid level was maintained at the top tube sheet. Foust, Baker, and Badger [*Ind. Eng. Chem.*, **31**, 206 (1939)] measured recirculating velocities and heat-transfer coefficients in the same evaporator except with 0.064-m (2.5-in) 0.0034-m-wall (10-gauge), 1.22-m- (4-ft-) long tubes and temperature differences from 7 to 26°C (12 to 46°F). In the normal range of liquid levels, their results can be expressed as

$$U_c = \frac{b(\Delta T_c)^{0.22} N_{Pr}^{0.4}}{(V_g - V_l)^{0.37}} \qquad (11\text{-}32)$$

where b = 153 (SI) or 375 (U.S. customary) and the subscript c refers to true liquid temperature, which under these conditions was about 0.56°C (1°F) above the vapor-head temperature. This work was done with water.

No detailed tests have been reported for the performance of propeller calandrias. Not enough is known regarding the performance of the propellers themselves under the cavitating conditions usually encountered to permit predicting circulation rates. In many cases, it appears that the propeller does no good in accelerating heat transfer over the transfer for natural circulation (Fig. 11-23).

Miscellaneous Evaporator Types **Horizontal-tube evaporators** operating with partially or fully submerged heating surfaces behave in much the same way as short-tube verticals, and heat-transfer coefficients are of the same order of magnitude. Some test results for water were published by Badger [*Trans. Am. Inst. Chem. Eng.*, **13**, 139 (1921)]. When operating unsubmerged, their heat transfer performance is roughly comparable to the falling-film vertical tube evaporator. Condensing coefficients inside the tubes can be derived from Nusselt's theory which, based on a constant-heat flux rather than a constant film ΔT, gives:

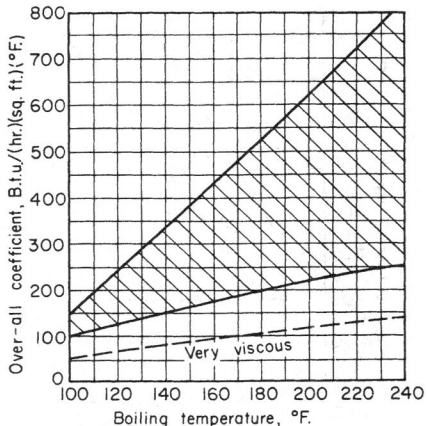

FIG. 11-20 General range of long-tube vertical- (LTV) evaporator coefficients. °C = (°F − 32)/1.8; to convert British thermal units per hour-square foot-degrees Fahrenheit to joules per square meter-second-kelvins, multiply by 5.6783.

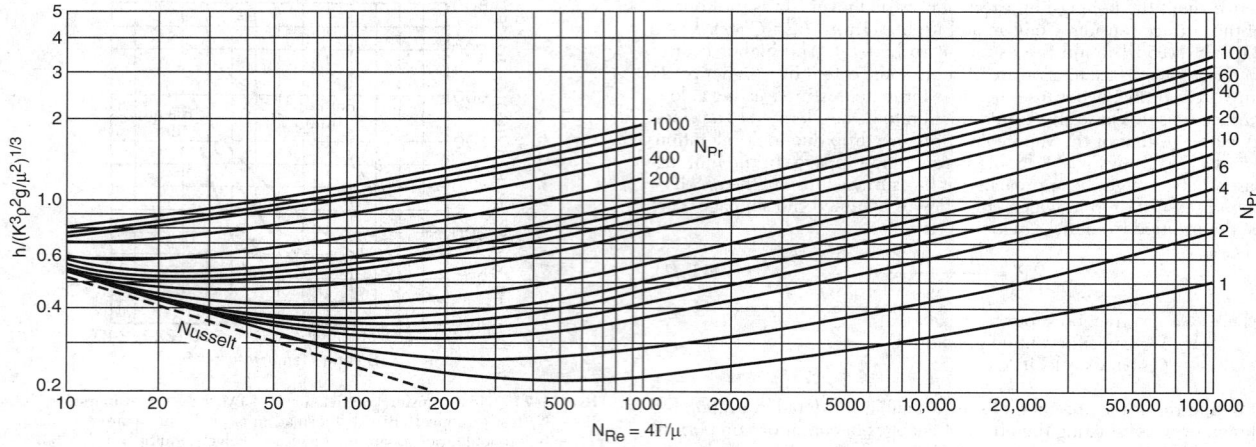

FIG. 11-22   Kunz and Yerazunis Correlation for falling-film heat transfer.

$$\frac{h}{(k^3\rho^2g/\mu^2)^{1/3}} = 1.59(4\Gamma/\mu)^{-1/3} \qquad (11\text{-}33a)$$

For the boiling side, a correlation based on seawater tests gave:

$$\frac{h}{(k^3\rho^2g/\mu^2)^{1/3}} = 0.0147(4\Gamma/\mu)^{1/3}(D)^{-1/3} \qquad (11\text{-}33b)$$

where Γ is based on feed-rate per unit length of the top tube in each vertical row of tubes and D is in meters.

Heat-transfer coefficients in clean coiled-tube evaporators for seawater are shown in Fig. 11-24 [Hillier, *Proc. Inst. Mech. Eng. (London)*, **1B**(7), 295 (1953)]. The tubes were of copper.

Heat-transfer coefficients in **agitated-film evaporators** depend primarily on liquid viscosity. This type is usually justifiable only for very viscous materials. Figure 11-25 shows general ranges of overall coefficients [Hauschild, *Chem. Ing. Tech.*, **25**, 573 (1953); Lindsey, *Chem. Eng.*, **60**(4), 227 (1953); and Leniger and Veldstra, *Chem. Ing. Tech.*, **31**, 493 (1959)]. When used with nonviscous fluids, a wiped-film evaporator having fluted external surfaces can exhibit very high coefficients (Lustenader et al., *Trans. Am. Soc. Mech. Eng.*, Paper 59-SA-30, 1959), although at a probably unwarranted first cost.

Heat Transfer from Various Metal Surfaces   In an early work, Pridgeon and Badger [*Ind. Eng. Chem.*, **16**, 474 (1924)] published test results on copper and iron tubes in a horizontal-tube evaporator that indicated an extreme **effect of surface cleanliness** on heat-transfer coefficients. However, the high degree of cleanliness needed for high coefficients was difficult to achieve, and the tube layout and

liquid level were changed during the course of the tests so as to make direct comparison of results difficult. Other workers have found little or no effect of conditions of surface or tube material on boiling-film coefficients in the range of commercial operating conditions [Averin, *Izv. Akad. Nauk SSSR Otd. Tekh. Nauk*, no. 3, p. 116, 1954; and Coulson and McNelly, *Trans. Inst. Chem. Eng.*, **34**, 247 (1956)].

Work in connection with desalination of seawater has shown that **specially modified surfaces** can have a profound effect on heat-transfer coefficients in evaporators. Figure 11-26 (Alexander and Hoffman, Oak Ridge National Laboratory TM-2203) compares overall coefficients for some of these surfaces when boiling fresh water in 0.051-m (2-in) tubes 2.44-m (8-ft) long at atmospheric pressure in both upflow and downflow. The area basis used was the nominal outside area. Tube 20 was a smooth 0.0016-m- (0.062-in-) wall aluminum brass tube that had accumulated about 6 years of fouling in seawater service and exhibited a fouling resistance of about $(2.6)(10^{-5})$ $(m^2 \cdot s \cdot K)/$ J [0.00015 (ft$^2 \cdot$h$\cdot$°F)/Btu]. Tube 23 was a clean aluminum tube with 20 spiral corrugations of 0.0032-m (⅛-in) radius on a 0.254-m (10-in) pitch indented into the tube. Tube 48 was a clean copper tube that had 50 longitudinal flutes pressed into the wall (General Electric double-flute profile, Diedrich, U.S. Patent 3,244,601, Apr. 5, 1966). Tubes 47 and 39 had a specially patterned porous sintered-metal deposit on the boiling side to promote nucleate boiling (Minton, U.S.

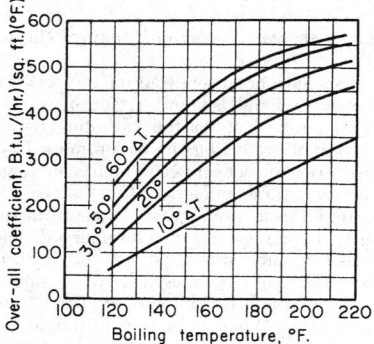

FIG. 11-23   Heat-transfer coefficients for water in short-tube evaporators. °C = (°F − 32)/1.8; to convert British thermal units per hour-square foot-degrees Fahrenheit to joules per square meter-second-kelvins, multiply by 5.6783.

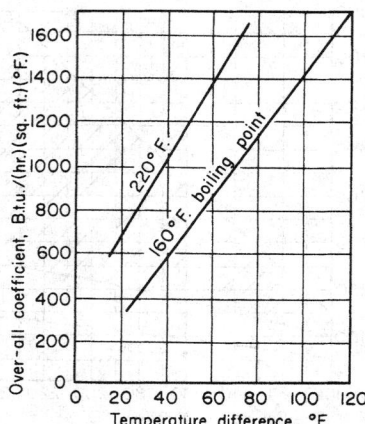

FIG. 11-24   Heat-transfer coefficients for seawater in coil-tube evaporators. °C = (°F − 32)/1.8; to convert British thermal units per hour-square foot-degrees Fahrenheit to joules per square meter-second-kelvins, multiply by 5.6783.

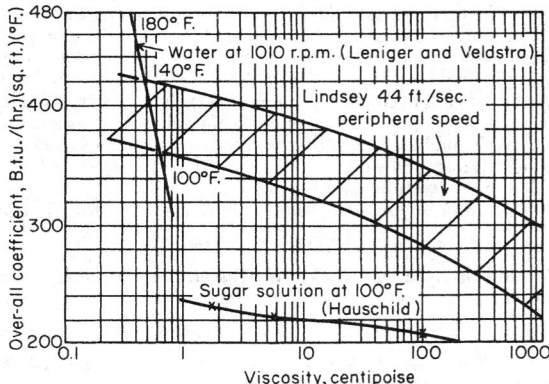

FIG. 11-25 Overall heat-transfer coefficients in agitated-film evaporators. °C = (°F − 32)/1.8; to convert British thermal units per hour-square foot-degrees Fahrenheit to joules per square meter-second-kelvins, multiply by 5.6783; to convert centipoises to pascal-seconds, multiply by 10^{-3}.

Patent 3,384,154, May 21, 1968). Both of these tubes also had steam-side coatings to promote dropwise condensation—parylene for tube 47 and gold plating for tube 39.

Of these special surfaces, only the **double-fluted tube** has seen extended services. Most of the gain in heat-transfer coefficient is due to the condensing side; the flutes tend to collect the condensate and leave the lands bare [Carnavos, *Proc. First Int. Symp. Water Desalination,* **2,** 205 (1965)]. The condensing-film coefficient (based on the actual outside area, which is 28 percent greater than the nominal area) may be approximated from the equation

$$h = b\left(\frac{k^3\rho^2 g}{\mu^2}\right)^{1/3}\left(\frac{\mu\lambda}{L}\right)^{1/3}\left(\frac{q}{A}\right)^{-0.833} \qquad (11\text{-}34a)$$

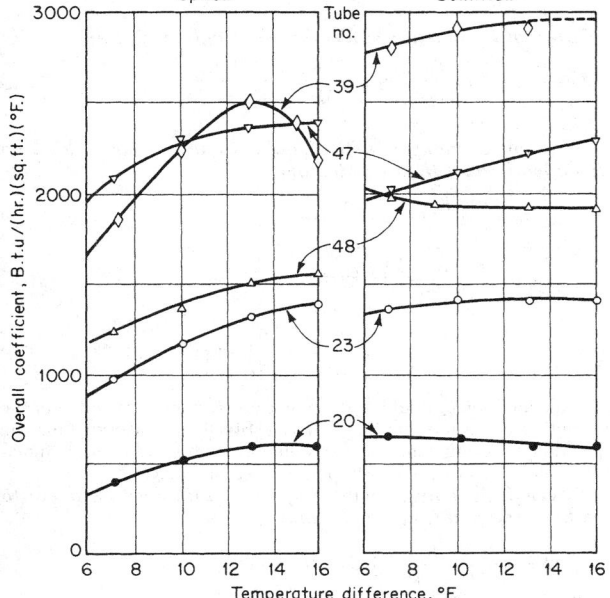

FIG. 11-26 Heat-transfer coefficients for enhanced surfaces. °C = (°F − 32)/1.8; to convert British thermal units per hour-square foot-degrees Fahrenheit to joules per square meter-second-kelvins, multiply by 5.6783. (*By permission from Oak Ridge National Laboratory TM-2203.*)

where $b = 2100$ (SI) or 1180 (U.S. customary). The boiling-side coefficient (based on actual inside area) for salt water in downflow may be approximated from the equation

$$h = 0.035(k^3\rho^2 g/\mu^2)^{1/3}(4\Gamma/\mu)^{1/3} \qquad (11\text{-}34b)$$

The boiling-film coefficient is about 30 percent lower for pure water than it is for salt water or seawater. There is as yet no accepted explanation for the superior performance in salt water. This phenomenon is also seen in evaporation from smooth tubes.

Effect of Fluid Properties on Heat Transfer   Most of the heat-transfer data reported in the preceding paragraphs were obtained with water or with dilute solutions having properties close to those of water. Heat transfer with other materials will depend on the type of evaporator used. For forced-circulation evaporators, methods have been presented to calculate the effect of changes in fluid properties. For natural-circulation evaporators, **viscosity** is the most important variable as far as aqueous solutions are concerned. Badger (*Heat Transfer and Evaporation,* Chemical Catalog, New York, 1926, pp. 133–134) found that, as a rough rule, overall heat-transfer coefficients varied in inverse proportion to viscosity if the boiling film was the main resistance to heat transfer. When handling molasses solutions in a forced-circulation evaporator in which boiling was allowed to occur in the tubes, Coates and Badger [*Trans. Am. Inst. Chem. Eng.,* **32,** 49 (1936)] found that from 0.005 to 0.03 Pa·s (5 to 30 cP) the overall heat-transfer coefficient could be represented by $U = b/\mu_f^{1.24}$, where $b = 2.55$ (SI) or 7043 (U.S. customary). Fragen and Badger [*Ind. Eng. Chem.,* **28,** 534 (1936)] correlated overall coefficients on sugar and sulfite liquor in the same evaporator for viscosities to 0.242 Pa·s (242 cP) and found a relationship that included the viscosity raised only to the 0.25 power.

Little work has been published on the effect of viscosity on heat transfer in the long-tube vertical evaporator. Cessna, Leintz, and Badger [*Trans. Am. Inst. Chem. Eng.,* **36,** 759 (1940)] found that the overall coefficient in the nonboiling zone varied inversely as the 0.7 power of viscosity (with sugar solutions). Coulson and Mehta [*Trans. Inst. Chem. Eng.,* **31,** 208 (1953)] found the exponent to be −0.44, and Stroebe, Baker, and Badger (loc. cit.) arrived at an exponent of −0.3 for the effect of viscosity on the film coefficient in the boiling zone.

Kerr (Louisiana Agr. Exp. Sta. Bull. 149) obtained plant data shown in Fig. 11-27 on various types of full-sized evaporators for cane sugar. These are invariably forward-feed evaporators concentrating to about 50° Brix, corresponding to a viscosity on the order of 0.005 Pa·s (5 cP) in the last effect. In Fig. 11-27 curve *A* is for short-tube verticals with central downtake, *B* is for standard horizontal tube evaporators, *C* is for Lillie evaporators (which were horizontal-tube machines with no liquor level but having recirculating liquor showered over the tubes), and *D* is for long-tube vertical evaporators. These curves show apparent coefficients, but sugar solutions have boiling-point rises low enough not to affect the results noticeably. Kerr also obtained the data shown in Fig.

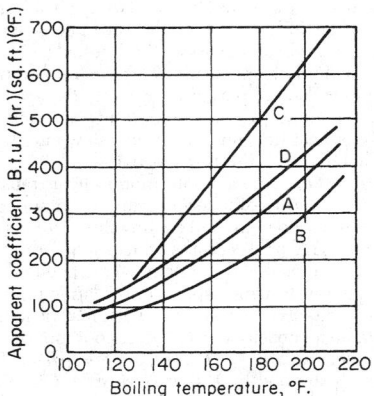

FIG. 11-27 Kerr's tests with full-sized sugar evaporators. °C = (°F − 32)/1.8; to convert British thermal units per hour-square foot-degrees Fahrenheit to joules per square meter-second-kelvins, multiply by 5.6783.

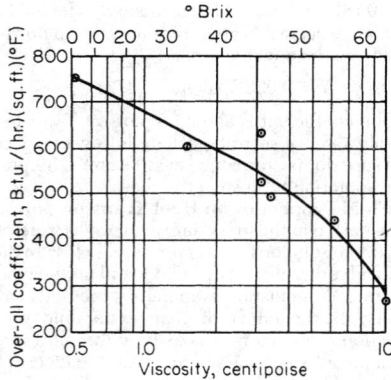

FIG. 11-28 Effect of viscosity on heat transfer in short-tube vertical evaporator. To convert centipoises to pascal-seconds, multiply by 10^{-3}; to convert British thermal units per hour-square foot-degrees Fahrenheit to joules per square meter-second-kelvins, multiply by 5.6783.

11-28 on a laboratory short-tube vertical evaporator with 0.44- by 0.61-m (1¾- by 24-in) tubes. This work was done with sugar juices boiling at 57°C (135°F) and an 11°C (20°F) temperature difference.

Effect of Noncondensables on Heat Transfer Most of the heat transfer in evaporators does not occur from pure steam but from vapor evolved in a preceding effect. This vapor usually contains inert gases—from air leakage if the preceding effect was under vacuum, from air entrained or dissolved in the feed, or from gases liberated by decomposition reactions. To prevent these inerts from seriously impeding heat transfer, the gases must be channeled past the heating surface and vented from the system while the gas concentration is still quite low. The influence of inert gases on heat transfer is due partially to the effect on ΔT of lowering the partial pressure and hence condensing temperature of the steam. The primary effect, however, results from the formation at the heating surface of an insulating blanket of gas through which the steam must diffuse before it can condense. The latter effect can be treated as an added resistance or fouling factor equal to 6.5×10^{-5} times the local mole percent inert gas (in $J^{-1} \cdot s \cdot m^2 \cdot K$) [Standiford, *Chem. Eng. Prog.*, **75**, 59–62 (July 1979)]. The effect on ΔT is readily calculated from Dalton's law. Inert-gas concentrations may vary by a factor of 100 or more between vapor inlet and vent outlet, so these relationships should be integrated through the tube bundle.

BATCH OPERATIONS: HEATING AND COOLING OF VESSELS

Nomenclature (Use consistent units.) A = heat-transfer surface; C, c = specific heats of hot and cold fluids respectively; L_0 = flow rate of liquid added to tank; M = mass of fluid in tank; T, t = temperature of hot and cold fluids respectively; T_1, t_1 = temperatures at beginning of heating or cooling period or at inlet; T_2, t_2 = temperature at end of period or at outlet; T_0, t_0 = temperature of liquid added to tank; U = coefficient of heat transfer; and W, w = flow rate through external exchanger of hot and cold fluids respectively.

Applications One typical application in heat transfer with batch operations is the heating of a reactor mix, maintaining temperature during a reaction period, and then cooling the products after the reaction is complete. This subsection is concerned with the heating and cooling of such systems in either unknown or specified periods.

The technique for deriving expressions relating time for heating or cooling agitated batches to coil or jacket area, heat-transfer coefficients, and the heat capacity of the vessel contents was developed by Bowman, Mueller, and Nagle [*Trans. Am. Soc. Mech. Eng.*, **62**, 283–294 (1940)] and extended by Fisher [*Ind. Eng. Chem.*, **36**, 939–942 (1944)] and Chaddock and Sanders [*Trans. Am. Inst. Chem. Eng.*, **40**, 203–210 (1944)] to external heat exchangers. Kern (*Process Heat Transfer*, McGraw-Hill, New York, 1950, Chap. 18) collected and published the results of these investigators.

The assumptions made were that (1) U is constant for the process

and over the entire surface, (2) liquid flow rates are constant, (3) specific heats are constant for the process, (4) the heating or cooling medium has a constant inlet temperature, (5) agitation produces a uniform batch fluid temperature, (6) no partial phase changes occur, and (7) heat losses are negligible. The developed equations are as follows. If any of the assumptions do not apply to a system being designed, new equations should be developed or appropriate corrections made. Heat exchangers are counterflow except for the 1-2 exchangers, which are one-shell-pass, two-tube-pass, parallel-flow counterflow.

Coil-in-Tank or Jacketed Vessel: Isothermal Heating Medium

$$\ln (T_1 - t_1)/(T_1 - t_2) = UA\theta/Mc \qquad (11\text{-}35)$$

Cooling-in-Tank or Jacketed Vessel: Isothermal Cooling Medium

$$\ln (T_1 - t_1)/(T_2 - t_1) = UA\theta/MC \qquad (11\text{-}35a)$$

Coil-in-Tank or Jacketed Vessel: Nonisothermal Heating Medium

$$\ln \frac{T_1 - t_1}{T_1 - t_2} = \frac{WC}{Mc}\left(\frac{K_1 - 1}{K_1}\right)\theta \qquad (11\text{-}35b)$$

where $K_1 = e^{UA/WC}$

Coil-in-Tank: Nonisothermal Cooling Medium

$$\ln \frac{T_1 - t_1}{T_2 - t_1} = \frac{wc}{MC}\left(\frac{K_2 - 1}{K_2}\right)\theta \qquad (11\text{-}35c)$$

where $K_2 = e^{UA/wc}$

External Heat Exchanger: Isothermal Heating Medium

$$\ln \frac{T_1 - t_1}{T_1 - t_2} = \frac{wc}{Mc}\left(\frac{K_2 - 1}{K_2}\right)\theta \qquad (11\text{-}35d)$$

External Exchanger: Isothermal Cooling Medium

$$\ln \frac{T_1 - t_1}{T_2 - t_1} = \frac{WC}{MC}\left(\frac{K_1 - 1}{K_1}\right)\theta \qquad (11\text{-}35e)$$

External Exchanger: Nonisothermal Heating Medium

$$\ln \frac{T_1 - t_1}{T_1 - t_2} = \left(\frac{K_3 - 1}{M}\right)\left(\frac{wWC}{K_3 wc - WC}\right)\theta \qquad (11\text{-}35f)$$

where $K_3 = e^{UA(1/WC - 1/wc)}$

External Exchanger: Nonisothermal Cooling Medium

$$\ln \frac{T_1 - t_1}{T_2 - t_1} = \left(\frac{K_4 - 1}{M}\right)\left(\frac{Wwc}{K_4 wc - WC}\right)\theta \qquad (11\text{-}35g)$$

where $K_4 = e^{UA(1/WC - 1/wc)}$

External Exchanger with Liquid Continuously Added to Tank: Isothermal Heating Medium

$$\ln \frac{t_1 - t_0 - \dfrac{w}{L_0}\left(\dfrac{K_2 - 1}{K_2}\right)(T_1 - t_1)}{t_2 - t_0 - \dfrac{w}{L_0}\left(\dfrac{K_2 - 1}{K_2}\right)(T_1 - t_2)}$$

$$= \left[\frac{w}{L_0}\left(\frac{K_2 - 1}{K_2}\right) + 1\right]\ln \frac{M + L_0\theta}{M} \qquad (11\text{-}35h)$$

If the addition of liquid to the tank causes an average endothermic or exothermic heat of solution, $\pm q_s$ J/kg (Btu/lb) of makeup, it may be included by adding $\pm q_s/c_0$ to both the numerator and the denominator of the left side. The subscript 0 refers to the makeup.

External Exchanger with Liquid Continuously Added to Tank: Isothermal Cooling Medium

$$\ln \frac{T_0 - T_1 - \dfrac{W}{L_0}\left(\dfrac{K_1 - 1}{K_1}\right)(T_1 - t_1)}{T_0 - T_2 - \dfrac{W}{L_0}\left(\dfrac{K_1 - 1}{K_1}\right)(T_2 - t_1)}$$

$$= \left[1 - \frac{W}{L_0}\left(\frac{K_1 - 1}{K_1}\right)\right]\ln \frac{M + L_0\theta}{M} \qquad (11\text{-}35i)$$

The heat-of-solution effects can be included by adding $\pm q_s/C_0$ to both the numerator and the denominator of the left side.

External Exchanger with Liquid Continuously Added to Tank: Nonisothermal Heating Medium

$$\ln \frac{t_0 - t_1 + \dfrac{wWC(K_5-1)(T_1-t_1)}{L_0(K_5WC-wc)}}{t_0 - t_2 + \dfrac{wWC(K_5-1)(T_1-t_2)}{L_0(K_5WC-wc)}}$$

$$= \left[\frac{wWC(K_5-1)}{L_0(K_5Wc-wc)} + 1\right] \ln \frac{M + L_0\theta}{M} \quad (11\text{-}35j)$$

where $K_5 = e^{(UA/wc)(1-wc/WC)}$

The heat-of-solution effects can be included by adding $\pm q_s/c_0$ to both the numerator and the denominator of the left side.

External Exchanger with Liquid Continuously Added to Tank: Nonisothermal Cooling Medium

$$\ln \frac{T_0 - T_1 - \dfrac{Wwc(K_6-1)(T_1-t_1)}{L_0(K_6wc-WC)}}{T_0 - T_2 - \dfrac{Wwc(K_6-1)(T_2-t_1)}{L_0(K_6wc-WC)}}$$

$$= \left[\frac{Wwc(K_6-1)}{L_0(K_6wc-WC)} + 1\right] \ln \frac{M + L_0\theta}{M} \quad (11\text{-}35k)$$

where $K_6 = e^{(UA/WC)(1-WC/wc)}$

The heat-of-solution effects can be included by adding $\pm q_s/C_0$ to both the numerator and the denominator of the left side.

Heating and Cooling Agitated Batches: 1-2 Parallel Flow-Counterflow

$$\frac{UA}{wc} = \frac{1}{\sqrt{R^2+1}} \ln \frac{2 - S(R+1-\sqrt{R^2+1})}{2 - S(R+1+\sqrt{R^2+1})} \quad (11\text{-}35l)$$

$$R = \frac{T_1 - T_2}{t' - t} = \frac{wc}{WC} \quad \text{and} \quad S = \frac{t' - t}{T_1 - t}$$

$$\frac{2 - S(R+1-\sqrt{R^2+1})}{2 - S(R+1+\sqrt{R^2+1})} = e^{(UA/wc)\sqrt{R^2+1}} = K_7 \quad (11\text{-}35m)$$

$$S = \frac{2(K_7-1)}{K_7(R+1+R^2+1) - (R+1-R^2+1)}$$

External 1-2 Exchanger: Heating

$$\ln(T_1-t_1)/(T_1-t_2) = (Sw/M)\theta \quad (11\text{-}35n)$$

External 1-2 Exchanger: Cooling

$$\ln(T_1-t_1)/(T_2-t_1) = S(wc/MC)\theta \quad (11\text{-}35o)$$

The cases of multipass exchangers with liquid continuously added to the tank are covered by Kern, as cited earlier. An alternative method for all multipass-exchanger gases, including those presented as well as cases with two or more shells in series, is as follows:

1. Determine UA for using the applicable equations for counterflow heat exchangers.
2. Use the initial batch temperature T_1 or t_1.
3. Calculate the outlet temperature from the exchanger of each fluid. (This will require trial-and-error methods.)
4. Note the F_T correction factor for the corrected mean temperature difference. (See Fig. 11-4.)
5. Repeat steps 2, 3, and 4 by using the final batch temperature T_2 and t_2.
6. Use the average of the two values for F, then increase the required multipass UA as follows:

$$UA(\text{multipass}) = UA(\text{counterflow})/F_T$$

In general, values of F_T below 0.8 are uneconomical and should be avoided. F_T can be raised by increasing the flow rate of either or both

of the flow streams. Increasing flow rates to give values well above 0.8 is a matter of economic justification.

If F_T varies widely from one end of the range to the other, F_T should be determined for one or more intermediate points. The average should then be determined for each step which has been established and the average of these taken for use in step 6.

Effect of External Heat Loss or Gain If heat loss or gain through the vessel walls cannot be neglected, equations which include this heat transfer can be developed by using energy balances similar to those used for the derivations of equations given previously. Basically, these equations must be modified by adding a heat-loss or heat-gain term.

A simpler procedure, which is probably acceptable for most practical cases, is to ratio UA or θ either up or down in accordance with the required modification in total heat load over time θ.

Another procedure, which is more accurate for the external-heat-exchanger cases, is to use an equivalent value for MC (for a vessel being heated) derived from the following energy balance:

$$Q = (Mc)_e(t_2-t_1) = Mc(t_2-t_1) + U'A'(MTD')\theta \quad (11\text{-}35p)$$

where Q is the total heat transferred over time θ, $U'A'$ is the heat-transfer coefficient for heat loss times the area for heat loss, and MTD' is the mean temperature difference for the heat loss.

A similar energy balance would apply to a vessel being cooled.

Internal Coil or Jacket Plus External Heat Exchanger This case can be most simply handled by treating it as two separate problems. M is divided into two separate masses M_1 and $(M-M_1)$, and the appropriate equations given earlier are applied to each part of the system. Time θ, of course, must be the same for both parts.

Equivalent-Area Concept The preceding equations for batch operations, particularly Eq. 11-35 can be applied for the calculation of heat loss from tanks which are allowed to cool over an extended period of time. However, different surfaces of a tank, such as the top (which would not be in contact with the tank contents) and the bottom, may have coefficients of heat transfer which are different from those of the vertical tank walls. The simplest way to resolve this difficulty is to use an equivalent area A_e in the appropriate equations where

$$A_e = A_bU_b/U_s + A_tU_t/U_s + A_s \quad (11\text{-}35q)$$

and the subscripts b, s, and t refer to the bottom, sides, and top respectively. U is usually taken as U_s. Table 11-1 lists typical values for U_s and expressions for A_e for various tank configurations.

Nonagitated Batches Cases in which vessel contents are vertically stratified, rather than uniform, in temperature, have been treated by Kern (op. cit.). These are of little practical importance except for tall, slender vessels heated or cooled with external exchangers. The result is that a smaller exchanger is required than for an equivalent agitated batch system that is uniform.

Storage Tanks The equations for batch operations with agitation may be applied to storage tanks even though the tanks are not agitated. This approach gives conservative results. The important cases (nonsteady state) are:

1. *Tanks cool; contents remain liquid.* This case is relatively simple and can easily be handled by the equations given earlier.
2. *Tanks cool, contents partially freeze, and solids drop to bottom or rise to top.* This case requires a two-step calculation. The first step is handled as in case 1. The second step is calculated by assuming an isothermal system at the freezing point. It is possible, given time and a sufficiently low ambient temperature, for tank contents to freeze solid.
3. *Tanks cool and partially freeze; solids form a layer of self-insulation.* This complex case, which has been known to occur with heavy hydrocarbons and mixtures of hydrocarbons, has been discussed by Stuhlbarg [*Pet. Refiner*, **38**, 143 (Apr. 1, 1959)]. The contents in the center of such tanks have been known to remain warm and liquid even after several years of cooling.

It is very important that a melt-out riser be installed whenever tank contents are expected to freeze on prolonged shutdown. The purpose is to provide a molten chimney through the crust for relief of thermal expansion or cavitation if fluids are to be pumped out or recirculated through an external exchanger. An external heat tracer, properly

TABLE 11-1 Typical Values for Use with Eqs. (11-36) to (11-44)*

Application	Fluid	U_s	A_s
Tanks on legs, outdoors, not insulated	Oil	3.7	$0.22\,A_t + A_b + A_s$
	Water at 150°F.	5.1	$0.16\,A_t + A_b + A_s$
Tanks on legs, outdoors, insulated 1 in.	Oil	0.45	$0.7\,A_t + A_b + A_s$
	Water	0.43	$0.67\,A_t + A_b + A_s$
Tanks on legs, indoors, not insulated	Oil	1.5	$0.53\,A_t + A_b + A_s$
	Water	1.8	$0.35\,A_t + A_b + A_s$
Tanks on legs, indoors, insulated 1 in.	Oil	0.36	$0.8\,A_t + A_b + A_s$
	Water	0.37	$0.73\,A_t + A_b + A_s$
Flat-bottom tanks,† outdoors, not insulated	Oil	3.7	$0.22\,A_t + A_s + 0.43\,D_t$
	Water	5.1	$0.16\,A_t + A_s + 0.31\,D_t$
Flat-bottom tanks,† outdoors, insulated 1 in.	Oil	0.36	$0.7\,A_t + A_s + 3.9\,D_t$
	Water	0.37	$0.16\,A_t + A_s + 3.7\,D_t$
Flat-bottom tanks, indoors, not insulated	Oil	1.5	$0.53\,A_t + A_s + 1.1\,D_t$
	Water	1.8	$0.35\,A_t + A_s + 0.9\,D_t$
Flat-bottom tanks, indoors, insulated 1 in.	Oil	0.36	$0.8\,A_t + A_s + 4.4\,D_t$
	Water	0.37	$0.73\,A_t + A_s + 4.5\,D_t$

*Based on typical coefficients.
†The ratio $(t - t_g)(t - t')$ assumed at 0.85 for outdoor tanks. °C = (°F − 32)/1.8; to convert British thermal units per hour-square foot-degrees Fahrenheit to joules per square meter-second-kelvins, multiply by 5.6783.

located, will serve the same purpose but may require more remelt time before pumping can be started.

THERMAL DESIGN OF TANK COILS

The thermal design of tank coils involves the determination of the area of heat-transfer surface required to maintain the contents of the tank at a constant temperature or to raise or lower the temperature of the contents by a specified magnitude over a fixed time.

Nomenclature A = area; A_b = area of tank bottom; A_c = area of coil; A_e = equivalent area; A_s = area of sides; A_t = area of top; A_1 = equivalent area receiving heat from external coils; A_2 = equivalent area not covered with external coils; D_t = diameter of tank; F = design (safety) factor; h = film coefficient; h_a = coefficient of ambient air; h_c = coefficient of coil; h_h = coefficient of heating medium; h_i = coefficient of liquid phase of tank contents or tube-side coefficient referred to outside of coil; h_z = coefficient of insulation; k = thermal conductivity; k_g = thermal conductivity of ground below tank; M = mass of tank contents when full; t = temperature; t_a = temperature of ambient air; t_d = temperature of dead-air space; t_f = temperature of contents at end of heating; t_g = temperature of ground below tank; t_h = temperature of heating medium; t_0 = temperature of contents at beginning of heating; U = overall coefficient; U_b = coefficient at tank bottom; U_c = coefficient of coil; U_d = coefficient of dead air to the tank contents; U_i = coefficient through insulation; U_s = coefficient at sides; U_t = coefficient at top; and U_2 = coefficient at area A_2.

Typical coil coefficients are listed in Table 11-2. More exact values can be calculated by using the methods for natural convection or forced convection given elsewhere in this section.

Maintenance of Temperature Tanks are often maintained at temperature with internal coils if the following equations are assumed to be applicable:

$$q = U_s A_e (T - t') \qquad (11\text{-}36)$$

and

$$A_c = q/U_c(MTD) \qquad (11\text{-}36a)$$

These make no allowance for unexpected shutdowns. One method of allowing for shutdown is to add a safety factor to Eq. 11-36a.

In the case of a tank maintained at temperature with internal coils, the coils are usually designed to cover only a portion of the tank. The temperature t_d of the dead-air space between the coils and the tank is obtained from

$$U_d A_1(t_d - t) = U_2 A_2(t - t') \qquad (11\text{-}37)$$

The heat load is

$$q = U_d A_1(t_d - t) + A_1 U_i(t_d - t') \qquad (11\text{-}38)$$

The coil area is

$$A_c = \frac{qF}{U_c(t_h - t_d)_m} \qquad (11\text{-}39)$$

where F is a safety factor.

Heating

Heating with Internal Coil from Initial Temperature for Specified Time

$$Q = Wc(t_f - t_o) \qquad (11\text{-}40)$$

$$A_c = \left[\frac{Q}{\theta_h} + U_s A_e\left(\frac{t_f + t_o}{2} - t'\right)\right]\left[\frac{1}{U_c[t_h - (t_f + t_o)/2]}\right](F) \qquad (11\text{-}41)$$

where θ_h is the length of heating period. This equation may also be used when the tank contents have cooled from t_f to t_o and must be reheated to t_f. If the contents cool during a time θ_c, the temperature at the end of this cooling period is obtained from

$$\ln\left(\frac{t_f - t'}{t_o - t'}\right) = \frac{U_s A_e \theta_c}{Wc} \qquad (11\text{-}42)$$

Heating with External Coil from Initial Temperature for Specified Time The temperature of the dead-air space is obtained from

$$U_d A_1[t_d - 0.5(t_f - t_o)] = U_2 A_2[0.5(t_f - t_o) - t'] + Q/\theta_h \qquad (11\text{-}43)$$

The heat load is

$$q = U_i A_1(t_d - t') + U_2 A_2[0.5(t_f - t_o) - t'] + Q/\theta_h \qquad (11\text{-}44)$$

The coil area is obtained from Eq. 11-39.

The safety factor used in the calculations is a matter of judgment based on confidence in the design. A value of 1.10 is normally not considered excessive. Typical design parameters are shown in Tables 11-1 and 11-2.

HEATING AND COOLING OF TANKS

Tank Coils Pipe tank coils are made in a wide variety of configurations, depending upon the application and shape of the vessel. **Helical** and **spiral** coils are most commonly shop-fabricated, while the **hairpin** pattern is generally field-fabricated. The helical coils are used principally in process tanks and pressure vessels when large areas for rapid heating or cooling are required. In general, heating coils are placed low in the tank, and cooling coils are placed high or distributed uniformly through the vertical height.

Stocks which tend to solidify on cooling require uniform coverage of the bottom or agitation. A maximum spacing of 0.6 m (2 ft) between turns of 50.8-mm (2-in) and larger pipe and a close approach to the tank wall are recommended. For smaller pipe or for low-temperature heating media, closer spacing should be used. In the case of the com-

TABLE 11-2 Overall Heat-Transfer Coefficients for Coils Immersed in Liquids

U Expressed as Btu/(h · ft² · °F)

Substance inside coil	Substance outside coil	Coil material	Agitation	U
Steam	Water	Lead	Agitated	70
Steam	Sugar and molasses solutions	Copper	None	50–240
Steam	Boiling aqueous solution			600
Cold water	Dilute organic dye intermediate	Lead	Turboagitator at 95 r.p.m.	300
Cold water	Warm water	Wrought iron	Air bubbled into water surrounding coil	150–300
Cold water	Hot water	Lead	0.40 r.p.m. paddle stirrer	90–360
Brine	Amino acids		30 r.p.m.	100
Cold water	25% oleum at 60°C.	Wrought iron	Agitated	20
Water	Aqueous solution	Lead	500 r.p.m. sleeve propeller	250
Water	8% NaOH		22 r.p.m.	155
Steam	Fatty acid	Copper (pancake)	None	96–100
Milk	Water		Agitation	300
Cold water	Hot water	Copper	None	105–180
60°F. water	50% aqueous sugar solution	Lead	Mild	50–60
Steam and hydrogen at 1500 lb./sq. in.	60°F. water	Steel		100–165
Steam 110–146 lb./sq. in. gage	Vegetable oil	Steel	None	23–29
Steam	Vegetable oil	Steel	Various	39–72
Cold water	Vegetable oil	Steel	Various	29–72

NOTES: Chilton, Drew, and Jebens [*Ind. Eng. Chem.*, **36**, 510 (1944)] give film coefficients for heating and cooling agitated fluids using a coil in a jacketed vessel.
Because of the many factors affecting heat transfer, such as viscosity, temperature difference, and coil size, the values in this table should be used primarily for preliminary design estimates and checking calculated coefficients.
°C = (°F − 32)/1.8; to convert British thermal units per hour-square foot-degrees Fahrenheit to joules per square meter-second-kelvins, multiply by 5.6783.

mon hairpin coils in vertical cylindrical tanks, this means adding an encircling ring within 152 mm (6 in) of the tank wall (see Fig. 11-29a for this and other typical coil layouts). The coils should be set directly on the bottom or raised not more than 50.8 to 152 mm (2 to 6 in), depending upon the difficulty of remelting the solids, in order to permit free movement of product within the vessel. The coil inlet should be above the liquid level (or an internal melt-out riser installed) to provide a molten path for liquid expansion or venting of vapors.

Coils may be sloped to facilitate drainage. When it is impossible to do so and remain close enough to the bottom to get proper remelting, the coils should be blown out after usage in cold weather to avoid damage by freezing.

Most coils are firmly clamped (but not welded) to supports. **Supports** should allow expansion but be rigid enough to prevent uncontrolled motion (see Fig. 11-29b). Nuts and bolts should be securely fastened. Reinforcement of the inlet and outlet connections through the tank wall is recommended, since bending stresses due to thermal expansion are usually high at such points.

In general, 50.8- and 63.4-mm (2- and 2½-in) coils are the most economical for shop fabrication and 38.1- and 50.8-mm (1½- and 2-in) for field fabrication. The tube-side heat-transfer coefficient, high-pressure, or layout problems may lead to the use of smaller-size pipe.

The wall thickness selected varies with the service and material. Carbon steel coils are often made from schedule 80 or heavier pipe to allow for corrosion. When stainless-steel or other high-alloy coils are not subject to corrosion or excessive pressure, they may be of schedule 5 or 10 pipe to keep costs at a minimum, although high-quality welding is required for these thin walls to assure trouble-free service.

Methods for calculating heat loss from tanks and the sizing of tank coils have been published by Stuhlbarg [*Pet. Refiner*, **38**, 143 (April 1959)].

Fin-tube coils are used for fluids which have poor heat-transfer characteristics to provide more surface for the same configuration at reduced cost or when temperature-driven fouling is to be minimized. Fin tubing is not generally used when bottom coverage is important. **Fin-tube tank heaters** are compact prefabricated bundles which can be brought into tanks through manholes. These are normally installed vertically with longitudinal fins to produce good convection currents. To keep the heaters low in the tank, they can be installed horizontally with helical fins or with perforated longitudinal fins to prevent entrapment. Fin tubing is often used for heat-sensitive material because of the lower surface temperature for the same heating medium, resulting in a lesser tendency to foul.

Plate or panel coils made from two metal sheets with one or both embossed to form passages for a heating or cooling medium can be used in lieu of pipe coils. Panel coils are relatively light in weight, easy to install, and easily removed for cleaning. They are available in a range of standard sizes and in both flat and curved patterns. Process tanks have been built by using panel coils for the sides or bottom. A serpentine construction is generally utilized when liquid flows through the unit. Header-type construction is used with steam or other condensing media.

Standard **glass coils** with 0.18 to 11.1 m² (2 to 120 ft²) of heat-transfer surface are available. Also available are plate-type units made of **impervious graphite.**

Teflon Immersion Coils   Immersion coils made of Teflon fluorocarbon resin are available with 2.5-mm (0.10-in) ID tubes to

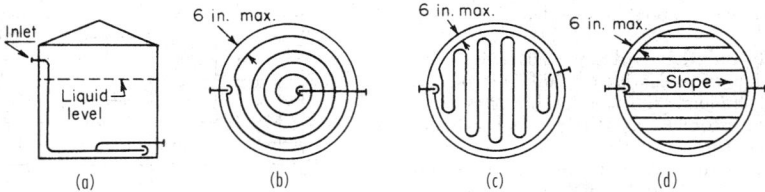

FIG. 11-29a Typical coil designs for good bottom coverage. (*a*) Elevated inlet on spiral coil. (*b*) Spiral with recircling ring. (*c*) Hairpin with encircling ring. (*d*) Ring header type.

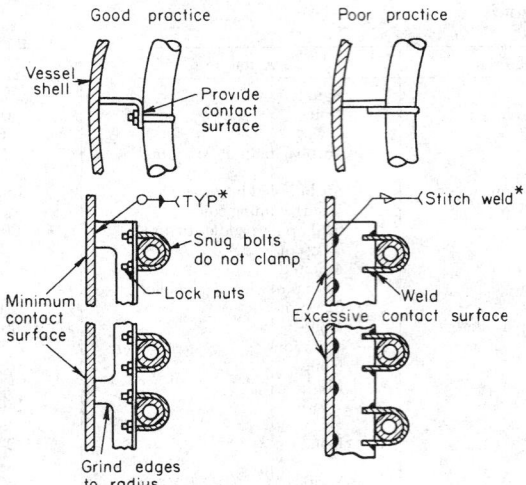

FIG. 11-29b Right and wrong ways to support coils. [*Chem. Eng.*, 172 (May 16, 1960).]

increase overall heat-transfer efficiency. The flexible bundles are available with 100, 160, 280, 500, and 650 tubes with standard lengths varying in 0.6-m (2-ft) increments between 1.2 and 4.8 m (4 and 16 ft). These coils are most commonly used in metal-finishing baths and are adaptable to service in reaction vessels, crystallizers, and tanks where corrosive fluids are used.

Bayonet Heaters   A bayonet-tube element consists of an outer and an inner tube. These elements are inserted into tanks and process vessels for heating and cooling purposes. Often the outer tube is of expensive alloy or nonmetallic (e.g., glass, impervious graphite), while the inner tube is of carbon steel. In glass construction, elements with 50.8- or 76.2-mm (2- or 3-in) glass pipe [with lengths to 2.7 m (9 ft)] are in contact with the external fluid, with an inner tube of metal.

External Coils and Tracers   Tanks, vessels, and pipe lines can be equipped for heating or cooling purposes with external coils. These are generally 9.8 to 19 mm (⅜ to ¾ in) so as to provide good distribution over the surface and are often of soft copper or aluminum, which can be bent by hand to the contour of the tank or line. When necessary to avoid "hot spots," the tracer is so mounted that it does not touch the tank.

External coils spaced away from the tank wall exhibit a coefficient of around 5.7 W/(m²·°C) [1 Btu/(h·ft² of coil surface·°F)]. Direct contact with the tank wall produces higher coefficients, but these are difficult to predict since they are strongly dependent upon the degree of contact. The use of **heat-transfer cements** does improve performance. These puttylike materials of high thermal conductivity are troweled or caulked into the space between the coil and the tank or pipe surface.

Costs of the cements (in 1960) varied from 37 to 63 cents per pound, with requirements running from about 0.27 lb/ft of ⅜-in outside-diameter tubing to 1.48 lb/ft of 1-in pipe. Panel coils require ½ to 1 lb/ft². A rule of thumb for preliminary estimating is that the per-foot installed cost of tracer with cement is about double that of the tracer alone.

Jacketed Vessels   Jacketing is often used for vessels needing frequent cleaning and for glass-lined vessels which are difficult to equip with internal coils. The jacket eliminates the need for the coil yet gives a better overall coefficient than external coils. However, only a limited heat-transfer area is available. The conventional jacket is of simple construction and is frequently used. It is most effective with a condensing vapor. A liquid heat-transfer fluid does not maintain uniform flow characteristics in such a jacket. Nozzles, which set up a swirling motion in the jacket, are effective in improving heat transfer. Wall thicknesses are often high unless reinforcement rings are installed.

Spiral baffles, which are sometimes installed for liquid services to improve heat transfer and prevent channeling, can be designed to serve as reinforcements. A spiral-wound channel welded to the vessel wall is an alternative to the spiral baffle which is more predictable in performance, since cross-baffle leakage is eliminated, and is reportedly lower in cost [Feichtinger, *Chem. Eng.*, **67**, 197 (Sept. 5, 1960)].

The half-pipe jacket is used when high jacket pressures are required. The flow pattern of a liquid heat-transfer fluid can be controlled and designed for effective heat transfer. The dimple jacket offers structural advantages and is the most economical for high jacket pressures. The low volumetric capacity produces a fast response to temperature changes.

EXTENDED OR FINNED SURFACES

Finned-Surface Application   Extended or finned surfaces are often used when one film coefficient is substantially lower than the other, the goal being to make $h_o A_{oe} \approx h_i A_i$. A few typical fin configurations are shown in Fig. 11-30a. Longitudinal fins are used in double-pipe exchangers. Transverse fins are used in cross-flow and shell-and-tube configurations. High transverse fins are used mainly with low-pressure gases; low fins are used for boiling and condensation of nonaqueous streams as well as for sensible-heat transfer. Finned surfaces have been proven to be a successful means of controlling temperature driven fouling such as coking and scaling. Fin spacing should be great enough to avoid entrapment of particulate matter in the fluid stream (5mm Minimum Spacing).

The area added by the fin is not as efficient for heat transfer as bare tube surface owing to resistance to conduction through the fin. The effective heat-transfer area is

$$A_{oe} = A_{uf} + A_f \Omega \qquad (11\text{-}45)$$

The fin efficiency is found from mathematically derived relations, in which the film heat-transfer coefficient is assumed to be constant over the entire fin and temperature gradients across the thickness of the fin have been neglected (see Kraus, *Extended Surfaces*, Spartan Books, Baltimore, 1963). The efficiency curves for some common fin configurations are given in Figs. 11-30a and 11-30b.

High Fins   To calculate heat-transfer coefficients for cross-flow to a transversely finned surface, it is best to use a correlation based on experimental data for that surface. Such data are not often available, and a more general correlation must be used, making allowance for the possible error. Probably the best general correlation for bundles of finned tubes is given by Schmidt [*Kaltetechnik*, **15**, 98–102, 370–378 (1963)]:

$$h D_r / k = K (D_r \rho V'_{\max} / \mu)^{0.625} R_f^{-0.375} N_{Pr}^{1/3} \qquad (11\text{-}46)$$

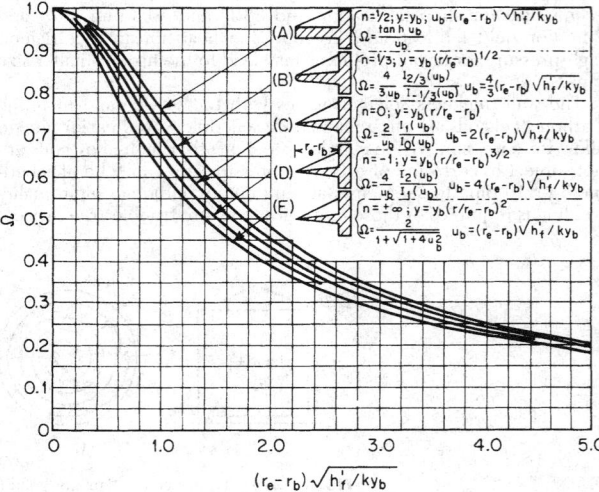

FIG. 11-30a Efficiencies for several longitudinal fin configurations.

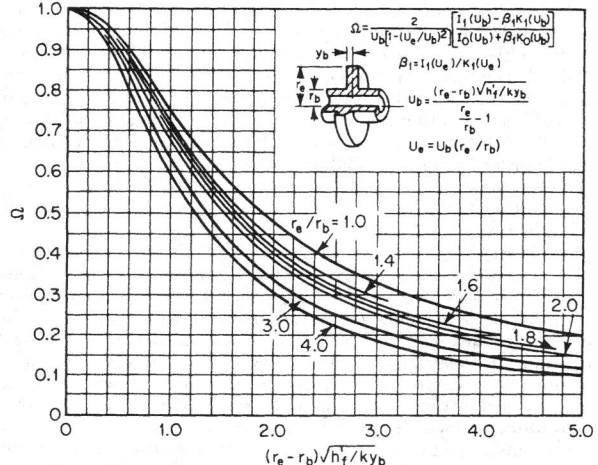

$$\Omega = \frac{2}{U_b[1-(U_e/U_b)^2]}\left[\frac{I_1(U_b)-\beta_1K_1(U_b)}{I_0(U_b)+\beta_1K_0(U_b)}\right]$$

$$\beta_1 = I_1(U_e)/K_1(U_e)$$

$$U_b = \frac{(r_e-r_b)\sqrt{h_f'/ky_b}}{\frac{r_e}{r_b}-1}$$

$$U_e = U_b(r_e/r_b)$$

FIG. 11-30b Efficiencies for annular fins of constant thickness.

where $K = 0.45$ for staggered tube arrays and 0.30 for in-line tube arrays: D_r is the root or base diameter of the tube; V'_{max} is the maximum velocity through the tube bank, i.e., the velocity through the minimum flow area between adjacent tubes; and R_f is the ratio of the total outside surface area of the tube (including fins) to the surface of a tube having the same root diameter but without fins.

Pressure drop is particularly sensitive to geometrical parameters, and available correlations should be extrapolated to geometries different from those on which the correlation is based only with great caution and conservatism. The best correlation is that of Robinson and Briggs [*Chem. Eng. Prog.*, **62**, *Symp. Ser.* 64, 177–184 (1966)].

Low Fins Low-finned tubing is generally used in shell-and-tube configurations. For sensible-heat transfer, only minor modifications are needed to permit the shell-side method given earlier to be used for both heat transfer and pressure [see Briggs, Katz, and Young, *Chem. Eng. Prog.*, **59**(11), 49–59 (1963)]. For condensing on low-finned tubes in horizontal bundles, the Nusselt correlation is generally satisfactory for low-surface-tension [$\sigma < (3)(10^{-6})$N/m (30 dyn/cm)] condensates fins of finned surfaces should not be closely spaced for high-surface-tension condensates (notably water), which do not drain easily.

The modified Palen-Small method can be employed for reboiler design using finned tubes, but the maximum flux is calculated from A_o, the total outside heat-transfer area including fins. The resulting value of q_{max} refers to A_o.

FOULING AND SCALING

Fouling refers to any change in the solid boundary separating two heat transfer fluids, whether by dirt accumulation or other means, which results in a decrease in the rate of heat transfer occurring across that boundary. Fouling may be classified by mechanism into six basic categories:

1. *Corrosion fouling.* The heat transfer surface reacts chemically with elements of the fluid stream producing a less conductive, corrosion layer on all or part of the surface.

2. *Biofouling.* Organisms present in the fluid stream are attracted to the warm heat-transfer surface where they attach, grow, and reproduce. The two subgroups are microbiofoulants such as slime and algae and macrobiofoulants such as snails and barnacles.

3. *Particulate fouling.* Particles held in suspension in the flow stream will deposit out on the heat-transfer surface in areas of sufficiently lower velocity.

4. *Chemical reaction fouling* (ex.—Coking). Chemical reaction of the fluid takes place on the heat-transfer surface producing an adhering solid product of reaction.

5. *Precipitation fouling* (ex.—Scaling). A fluid containing some dissolved material becomes supersaturated with respect to this mate-

rial at the temperatures seen at the heat-transfer surface. This results in a crystallization of the material which "plates out" on the warmer surface.

6. *Freezing fouling.* Overcooling of a fluid below the fluid's freezing point at the heat-transfer surface causes solidification and coating of the heat-transfer surface.

Control of Fouling Once the combination of mechanisms contributing to a particular fouling problem are recognized, methods to substantially reduce the fouling rate may be implemented. For the case of **corrosion fouling**, the common solution is to choose a less corrosive material of construction balancing material cost with equipment life. In cases of **biofouling**, the use of copper alloys and/or chemical treatment of the fluid stream to control organism growth and reproduction are the most common solutions.

In the case of **particulate fouling**, one of the more common types, insuring a sufficient flow velocity and minimizing areas of lower velocities and stagnant flows to help keep particles in suspension is the most common means of dealing with the problem. For water, the recommended tubeside minimum velocity is about 0.9 to 1.0 m/s. This may not always be possible for moderate to high-viscosity fluids where the resulting pressure drop can be prohibitive.

Special care should be taken in the application of any velocity requirement to the shellside of segmental-baffled bundles due to the many different flow streams and velocities present during operation, the unavoidable existence of high-fouling areas of flow stagnation, and the danger of flow-induced tube vibration. In general, shellside-particulate fouling will be greatest for segmentally baffled bundles in the regions of low velocity and the TEMA-fouling factors (which are based upon the use of this bundle type) should be used. However, since the 1940's, there have been a host of successful, low-fouling exchangers developed, some tubular and some not, which have in common the elimination of the cross-flow plate baffle and provide practically no regions of flow stagnation at the heat-transfer surface. Some examples are the plate and frame exchanger, the spiral plate exchanger, and the twisted tube exchanger, all of which have dispensed with baffles altogether and use the heat-transfer surface itself for bundle support. The general rule for these designs is to provide between 25 and 30 percent excess surface to compensate for potential fouling, although this can vary in special applications.

For the remaining classifications—**polymerization, precipitation,** and **freezing**—fouling is the direct result of temperature extremes at the heat-transfer surface and is reduced by reducing the temperature difference between the heat-transfer surface and the bulk-fluid stream. Conventional wisdom says to increase velocity, thus increasing the local heat-transfer coefficient to bring the heat-transfer surface temperature closer to the bulk-fluid temperature. However, due to a practical limit on the amount of heat-transfer coefficient increase available by increasing velocity, this approach, although better than nothing, is often not satisfactory by itself.

A more effective means of reducing the temperature difference is by using, in concert with adequate velocities, some form of extended surface. As discussed by Shilling (*Proceedings of the 10th International Heat Transfer Conference*, Brighton, U.K., **4**, p. 423), this will tend to reduce the temperature extremes between fluid and heat transfer surface and not only reduce the rate of fouling but make the heat exchanger generally less sensitive to the effects of any fouling that does occur. In cases where unfinned tubing in a triangular tube layout would not be acceptable because fouling buildup and eventual mechanical cleaning are inevitable, extended surface should be used only when the exchanger construction allows access for cleaning.

Fouling Transients and Operating Periods Three common behaviors are noted in the development of a fouling film over a period of time. One is the so-called *asymptotic fouling* in which the speed of fouling resistance increase decreases over time as it approaches some asymptotic value beyond which no further fouling can occur. This is commonly found in temperature-driven fouling. A second is *linear fouling* in which the increase in fouling resistance follows a straight line over the time of operation. This could be experienced in a case of severe particulate fouling where the accumulation of dirt during the time of operation did not appreciably increase velocities to mitigate the problem. The third, *falling rate fouling,* is neither linear nor asymptotic but instead lies somewhere between these two extremes.

The rate of fouling decreases with time but does not appear to approach an asymptotic maximum during the time of operation. This is the most common type of fouling in the process industry and is usually the result of a combination of different fouling mechanisms occurring together.

The optimum operating period between cleanings depends upon the rate and type of fouling, the heat exchanger used (i.e. baffle type, use of extended surface, and velocity and pressure drop design constraints), and the ease with which the heat exchanger may be removed from service for cleaning. As noted above, care must be taken in the use of fouling factors for exchanger design, especially if the exchanger configuration has been selected specifically to minimize fouling accumulation. An oversurfaced heat exchanger which will not foul enough to operate properly can be almost as much a problem as an undersized exchanger. This is especially true in steam-heated exchangers where the ratio of design MTD to minimum achievable MTD is less than U_clean divided by U_fouled.

Removal of Fouling Deposits Chemical removal of fouling can be achieved in some cases by weak acid, special solvents, and so on. Other deposits adhere weakly and can be washed off by periodic operation at very high velocities or by flushing with a high-velocity steam or water jet or using a sand-water slurry. These methods may be applied to both the shell side and tube side without pulling the bundle. Many fouling deposits, however, must be removed by positive mechanical action such as rodding, turbining, or scraping the surface. These techniques may be applied inside of tubes without pulling the bundle but can be applied on the shellside only after bundle removal. Even then there is limited access because of the tube pitch and rotated square or large triangular layouts are recommended. In many cases, it has been found that designs developed to minimize fouling often develop a fouling layer which is more easily removed.

Fouling Resistances There are no published methods for predicting fouling resistances a priori. The accumulated experience of exchanger designers and users was assembled more than 40 years ago based primarily upon segmental-baffled exchanger bundles and may be found in the *Standards of Tubular Exchanger Manufacturers Association* (TEMA). In the absence of other information, the fouling resistances contained therein may be used.

TYPICAL HEAT-TRANSFER COEFFICIENTS

Typical overall heat-transfer coefficients are given in Tables 11-3 through 11-8. Values from these tables may be used for preliminary estimating purposes. They should not be used in place of the design methods described elsewhere in this section, although they may serve as a useful check on the results obtained by those design methods.

THERMAL DESIGN FOR SOLIDS PROCESSING

Solids in divided form, such as powders, pellets, and lumps, are heated and/or cooled in chemical processing for a variety of objectives such as solidification or fusing (Sec. 11), drying and water removal (Sec. 20), solvent recovery (Secs. 13 and 20), sublimation (Sec. 17), chemical reactions (Sec. 20), and oxidation. For process and mechanical-design considerations, see the referenced sections.

Thermal design concerns itself with sizing the equipment to effect the heat transfer necessary to carry on the process. The design equation is the familiar one basic to all modes of heat transfer, namely,

$$A = Q/U \Delta t \tag{11-47}$$

where A = effective heat-transfer surface, Q = quantity of heat required to be transferred, Δt = temperature difference of the process, and U = overall heat-transfer coefficient. It is helpful to define the modes of heat transfer and the corresponding overall coefficient as U_{co} = overall heat-transfer coefficient for (indirect through-a-wall) *conduction*, U_{cv} = overall heat-transfer coefficient for the little-used *convection* mechanism, U_{ct} = heat-transfer coefficient for the *contactive* mechanism in which the gaseous-phase heat carrier passes directly through the solids bed, and U_{ra} = heat-transfer coefficient for *radiation*.

There are two general methods for determining numerical values

for U_{co}, U_{cv}, U_{ct}, and U_{ra}. One is by analysis of actual operating data. Values so obtained are used on geometrically similar systems of a size not too different from the equipment from which the data were obtained. The second method is predictive and is based on the material properties and certain operating parameters. Relative values of the coefficients for the various modes of heat transfer at temperatures up to 980°C (1800°F) are as follows (Holt, Paper 11, Fourth National Heat Transfer Conference, Buffalo, 1960):

Convective	1
Radiant	2
Conductive	20
Contactive	200

Because heat-transfer equipment for solids is generally an adaptation of a primarily material-handling device, the area of heat transfer is often small in relation to the overall size of the equipment. Also peculiar to solids heat transfer is that the Δt varies for the different heat-transfer mechanisms. With a knowledge of these mechanisms, the Δt term generally is readily estimated from temperature limitations imposed by the burden characteristics and/or the construction.

Conductive Heat Transfer Heat-transfer equipment in which heat is transferred by conduction is so constructed that the solids load (burden) is separated from the heating medium by a wall.

For a high proportion of applications, Δt is the log-mean temperature difference. Values of U_{co} are reported in Secs. 11, 15, 17, and 19. A *predictive* equation for U_{co} is

$$U_{co} = \left(\frac{h}{h - 2ca/d_m} \right)\left(\frac{2ca}{d_m} \right) \tag{11-48}$$

where h = wall film coefficient, c = volumetric heat capacity, d_m = depth of the burden, and α = thermal diffusivity. Relevant thermal properties of various materials are given in Table 11-9. For details of terminology, equation development, numerical values of terms in typical equipment and use, see Holt [*Chem. Eng.*, **69**, 107 (Jan. 8, 1962)].

Equation (11-48) is applicable to burdens in the solid, liquid, or gaseous phase, either static or in laminar motion; it is applicable to solidification equipment and to divided-solids equipment such as metal belts, moving trays, stationary vertical tubes, and stationary-shell fluidizers.

Fixed (or packed) bed operation occurs when the fluid velocity is low or the particle size is large so that fluidization does not occur. For such operation, Jakob (*Heat Transfer*, vol. 2, Wiley, New York, 1957) gives

$$hD_t/k = b_1 b D_t^{0.17}(D_p G/\mu)^{0.83}(c\mu/k) \tag{11-49a}$$

where $b_1 = 1.22$ (SI) or 1.0 (U.S. customary), $h = U_{co}$ = overall coefficient between the inner container surface and the fluid stream,

$$b = .2366 + .0092 \left(\frac{D_p}{D_t} \right) - 4.0672 \left(\frac{D_p}{D_t} \right)^2$$
$$+ 18.229 \left(\frac{D_p}{D_t} \right)^3 - 11.837 \left(\frac{D_p}{D_t} \right)^4 \tag{11-49b}$$

D_p = particle diameter, D_t = vessel diameter, (note that D_p/D_t has units of foot per foot in the equation), G = superficial mass velocity, k = fluid thermal conductivity, μ = fluid viscosity, and c = fluid specific heat. Other correlations are those of Leva [*Ind. Eng. Chem.*, **42**, 2498 (1950)]:

$$h = 0.813 \frac{k}{D_t} e^{-6D_p/D_t} \left(\frac{D_p G}{\mu} \right)^{0.90} \qquad \text{for } \frac{D_p}{D_t} < 0.35 \tag{11-50a}$$

$$h = 0.125 \frac{k}{D_t} \left(\frac{D_p G}{\mu} \right)^{0.75} \qquad \text{for } 0.35 < \frac{D_p}{D_t} < 0.60 \tag{11-50b}$$

and Calderbank and Pogerski [*Trans. Inst. Chem. Eng.* (London), **35**, 195 (1957)]:

$$hD_p/k = 3.6(D_p G/\mu\epsilon_v)^{0.365} \tag{11-51}$$

where ϵ_v = fraction voids in the bed.

A technique for calculating radial temperature gradients in a packed bed is given by Smith (*Chemical Engineering Kinetics*, McGraw-Hill, New York, 1956).

TABLE 11-3 Typical Overall Heat-Transfer Coefficients in Tubular Heat Exchangers

$U = \text{Btu}/(°F \cdot ft^2 \cdot h)$

Shell side	Tube side	Design U	Includes total dirt	Shell side	Tube side	Design U	Includes total dirt
Liquid-liquid media				Dowtherm vapor	Dowtherm liquid	80–120	.0015
Aroclor 1248	Jet fuels	100–150	0.0015	Gas-plant tar	Steam	40–50	.0055
Cutback asphalt	Water	10–20	.01	High-boiling hydrocarbons V	Water	20–50	.003
Demineralized water	Water	300–500	.001	Low-boiling hydrocarbons A	Water	80–200	.003
Ethanol amine (MEA or DEA) 10–25% solutions	Water or DEA, or MEA solutions	140–200	.003	Hydrocarbon vapors (partial condenser)	Oil	25–40	.004
Fuel oil	Water	15–25	.007	Organic solvents A	Water	100–200	.003
Fuel oil	Oil	10–15	.008	Organic solvents high NC, A	Water or brine	20–60	.003
Gasoline	Water	60–100	.003	Organic solvents low NC, V	Water or brine	50–120	.003
Heavy oils	Heavy oils	10–40	.004	Kerosene	Water	30–65	.004
Heavy oils	Water	15–50	.005	Kerosene	Oil	20–30	.005
Hydrogen-rich reformer stream	Hydrogen-rich reformer stream	90–120	.002	Naphtha	Water	50–75	.005
				Naphtha	Oil	20–30	.005
Kerosene or gas oil	Water	25–50	.005	Stabilizer reflux vapors	Water	80–120	.003
Kerosene or gas oil	Oil	20–35	.005	Steam	Feed water	400–1000	.0005
Kerosene or jet fuels	Trichlorethylene	40–50	.0015	Steam	No. 6 fuel oil	15–25	.0055
Jacket water	Water	230–300	.002	Steam	No. 2 fuel oil	60–90	.0025
Lube oil (low viscosity)	Water	25–50	.002	Sulfur dioxide	Water	150–200	.003
Lube oil (high viscosity)	Water	40–80	.003	Tall-oil derivatives, vegetable oils (vapor)	Water	20–50	.004
Lube oil	Oil	11–20	.006	Water	Aromatic vapor-stream azeotrope	40–80	.005
Naphtha	Water	50–70	.005				
Naphtha	Oil	25–35	.005				
Organic solvents	Water	50–150	.003	Gas-liquid media			
Organic solvents	Brine	35–90	.003	Air, N₂, etc. (compressed)	Water or brine	40–80	.005
Organic solvents	Organic solvents	20–60	.002	Air, N₂, etc., A	Water or brine	10–50	.005
Tall oil derivatives, vegetable oil, etc.	Water	20–50	.004	Water or brine	Air, N₂ (compressed)	20–40	.005
Water	Caustic soda solutions (10–30%)	100–250	.003	Water or brine	Air, N₂, etc., A	5–20	.005
Water	Water	200–250	.003	Water	Hydrogen containing natural-gas mixtures	80–125	.003
Wax distillate	Water	15–25	.005				
Wax distillate	Oil	13–23	.005	Vaporizers			
Condensing vapor-liquid media				Anhydrous ammonia	Steam condensing	150–300	.0015
Alcohol vapor	Water	100–200	.002	Chlorine	Steam condensing	150–300	.0015
Asphalt (450°F.)	Dowtherm vapor	40–60	.006	Chlorine	Light heat-transfer oil	40–60	.0015
Dowtherm vapor	Tall oil and derivatives	60–80	.004	Propane, butane, etc.	Steam condensing	200–300	.0015
				Water	Steam condensing	250–400	.0015

NC = noncondensable gas present.
V = vacuum.
A = atmospheric pressure.
Dirt (or fouling factor) units are (h · ft² · °F)/Btu.
To convert British thermal units per hour-square foot-degrees Fahrenheit to joules per square meter-second-kelvins, multiply by 5.6783; to convert hours per square foot-degree Fahrenheit-British thermal units to square meters per second-kelvin-joules, multiply by 0.1761.

TABLE 11-4 Typical Overall Heat-Transfer Coefficients in Refinery Service

Btu/(°F · ft² · h)

	Fluid	API gravity	Fouling factor (one stream)	Reboiler, steam-heated	Condenser, water-cooled*	Exchangers, liquid to liquid (tube-side fluid designation appears below)			Reboiler (heating liquid designated below)			Condenser (cooling liquid designated below)			
						C	G	H	C	G†	K	D	F	G	J
A	Propane		0.001	160	95	85	85	80	110	95	35				
B	Butane		.001	155	90	80	75	75	105	90	35	80	55	40	30
C	400°F. end-point gasoline	50	.001	120	80	70	65	60	65	50	30				
D	Virgin light naphtha	70	.001	140	85	70	55	55	75	60	35	75			
E	Virgin heavy naphtha	45	.001	95	75	65	55	50	55	45	30	70	50	35	30
F	Kerosene	40	.001	85	60	60	55	50		45	25		50	35	30
G	Light gas oil	30	.002	70	50	60	50	50		40	25	70	45	30	30
H	Heavy gas oil	22	.003	60	45	55	50	45	50	40	20	70	40	30	20
J	Reduced crude	17	.005			55	45	40							
K	Heavy fuel oil (tar)	10	.005			50	40	35							

Fouling factor, water side 0.0002; heating or cooling streams are shown at top of columns as C, D, F, G, etc.; to convert British thermal units per hour-square foot-degrees Fahrenheit to joules per square meter-second-kelvins, multiply by 5.6783; to convert hours per square foot-degree Fahrenheit-British thermal units to square meters per second-kelvin-joules, multiply by 0.1761.
*Cooler, water-cooled, rates are about 5 percent lower.
†With heavy gas oil (H) as heating medium, rates are about 5 percent lower.

TABLE 11-5 Overall Coefficients for Air-Cooled Exchangers on Bare-Tube Basis

Btu/(°F · ft² · h)

Condensing	Coefficient	Liquid cooling	Coefficient
Ammonia	110	Engine-jacket water	125
Freon-12	70	Fuel oil	25
Gasoline	80	Light gas oil	65
Light hydrocarbons	90	Light hydrocarbons	85
Light naphtha	75	Light naphtha	70
Heavy naphtha	65	Reformer liquid	
Reformer reactor		streams	70
effluent	70	Residuum	15
Low-pressure steam	135	Tar	7
Overhead vapors	65		

Gas cooling	Operating pressure, lb./sq. in. gage	Pressure drop, lb./sq. in.	Coefficient
Air or flue gas	50	0.1 to 0.5	10
	100	2	20
	100	5	30
Hydrocarbon gas	35	1	35
	125	3	55
	1000	5	80
Ammonia reactor stream			85

Bare-tube external surface is 0.262 ft²/ft.

Fin-tube surface/bare-tube surface ratio is 16.9.

To convert British thermal units per hour-square foot-degrees Fahrenheit to joules per square meter-second-kelvins, multiply by 5.6783; to convert pounds-force per square inch to kilopascals, multiply by 6.895.

Fluidization occurs when the fluid flow rate is great enough so that the pressure drop across the bed equals the weight of the bed. As stated previously, the solids film thickness adjacent to the wall d_m is difficult to measure and/or predict. Wen and Fau [*Chem. Eng.*, **64**(7), 254 (1957)] give for *external walls*:

$$h = bk(c_s\rho_s)^{0.4}(G\eta/\mu N_f)^{0.36} \quad (11\text{-}51a)$$

where $b = 0.29$ (SI) or 11.6 (U.S. customary), c_s = heat capacity of solid, ρ_s = particle density, η = fluidization efficiency (Fig. 11-31) and N_f = bed expansion ratio (Fig. 11-32). *For internal walls,* Wen and Fau give

$$h_i = bhG^{-0.37} \quad (11\text{-}51b)$$

where $b = 0.78$ (SI) or 9 (U.S. customary), h_i is the coefficient for internal walls, and h is calculated from Eq. (11-51a). G_{mf}, the minimum fluidizing velocity, is defined by

$$G_{mf} = \frac{b\rho_g^{1.1}(\rho_s - \rho_g)^{0.9}D_p^2}{\mu} \quad (11\text{-}51c)$$

where $b = (1.23)(10^{-2})$ (SI) or $(5.23)(10^5)$ (U.S. customary).

Wender and Cooper [*Am. Inst. Chem. Eng. J.*, **4**, 15 (1958)] developed an empirical correlation for *internal walls*:

$$\frac{hD_p/k}{1-\epsilon_v}\left(\frac{k}{c\rho}\right)^{0.43} = bC_R\left(\frac{D_pG}{\mu}\right)^{0.23}\left(\frac{c_s}{c_g}\right)^{0.80}\left(\frac{\rho_s}{\rho_g}\right)^{0.66} \quad (11\text{-}52a)$$

where $b = (3.51)(10^{-4})$ (SI) or 0.033 (U.S. customary) and C_R = correction for displacement of the immersed tube from the axis of the vessel (see the reference). For external walls:

$$\frac{hD_p}{k_g(1-\epsilon_v)(c_s\rho_s/c_g\rho_g)} = f(1 + 7.5e^{-x}) \quad (11\text{-}52b)$$

where $x = 0.44L_Hc_s/D_tc_g$ and f is given by Fig. 11-33. An important feature of this equation is inclusion of the ratio of bed depth to vessel diameter L_H/D_t.

For **dilute fluidized beds** on the shell side of an unbaffled tubular bundle Genetti and Knudsen [*Inst. Chem. Eng. (London) Symp. Ser.* 3,172 (1968)] obtained the relation:

$$\frac{hD_p}{k} = \frac{5\phi(1-\epsilon_v)}{\left[1 + \dfrac{580}{N_{Re}}\left(\dfrac{k_s}{D_p^{1.5}c_s\rho_g g^{0.5}}\right)\left(\dfrac{\rho_s}{\rho_g}\right)^{1.1}\left(\dfrac{G_{mf}}{G}\right)^{7/3}\right]^2} \quad (11\text{-}53a)$$

where ϕ = particle surface area per area of sphere of same diameter. When particle transport occurred through the bundle, the heat-transfer coefficients could be predicted by

TABLE 11-6 Panel Coils Immersed in Liquid: Overall Average Heat-Transfer Coefficients*

U expressed in Btu/(h · ft² · °F)

Hot side	Cold side	Clean-surface coefficients		Design coefficients, considering usual fouling in this service	
		Natural convection	Forced convection	Natural convection	Forced convection
Heating applications:					
Steam	Watery solution	250–500	300–550	100–200	150–275
Steam	Light oils	50–70	110–140	40–45	60–110
Steam	Medium lube oil	40–60	100–130	35–40	50–100
Steam	Bunker C or No. 6 fuel oil	20–40	70–90	15–30	60–80
Steam	Tar or asphalt	15–35	50–70	15–25	40–60
Steam	Molten sulfur	35–45	45–55	20–35	35–45
Steam	Molten paraffin	35–45	45–55	25–35	40–50
Steam	Air or gases	2–4	5–10	1–3	4–8
Steam	Molasses or corn sirup	20–40	70–90	15–30	60–80
High temperature hot water	Watery solutions	115–140	200–250	70–100	110–160
High temperature heat-transfer oil	Tar or asphalt	12–30	45–65	10–20	30–50
Dowtherm or Aroclor	Tar or asphalt	15–30	50–60	12–20	30–50
Cooling applications:					
Water	Watery solution	110–135	195–245	65–95	105–155
Water	Quench oil	10–15	25–45	7–10	15–25
Water	Medium lube oil	8–12	20–30	5–8	10–20
Water	Molasses or corn sirup	7–10	18–26	4–7	8–15
Water	Air or gases	2–4	5–10	1–3	4–8
Freon or ammonia	Watery solution	35–45	60–90	20–35	40–60
Calcium or sodium brine	Watery solution	100–120	175–200	50–75	80–125

*Tranter Manufacturing, Inc.

NOTE: To convert British thermal units per hour-square foot-degrees Fahrenheit to joules per square meter-second-kelvins, multiply by 5.6783.

TABLE 11-7 Jacketed Vessels: Overall Coefficients

Jacket fluid	Fluid in vessel	Wall material	Overall $U°$ Btu/(h · ft² · °F)	Overall $U°$ J/(m² · s · K)
Steam	Water	Stainless steel	150–300	850–1700
Steam	Aqueous solution	Stainless steel	80–200	450–1140
Steam	Organics	Stainless steel	50–150	285–850
Steam	Light oil	Stainless steel	60–160	340–910
Steam	Heavy oil	Stainless steel	10–50	57–285
Brine	Water	Stainless steel	40–180	230–1625
Brine	Aqueous solution	Stainless steel	35–150	200–850
Brine	Organics	Stainless steel	30–120	170–680
Brine	Light oil	Stainless steel	35–130	200–740
Brine	Heavy oil	Stainless steel	10–30	57–170
Heat-transfer oil	Water	Stainless steel	50–200	285–1140
Heat-transfer oil	Aqueous solution	Stainless steel	40–170	230–965
Heat-transfer oil	Organics	Stainless steel	30–120	170–680
Heat-transfer oil	Light oil	Stainless steel	35–130	200–740
Heat-transfer oil	Heavy oil	Stainless steel	10–40	57–230
Steam	Water	Glass-lined CS	70–100	400–570
Steam	Aqueous solution	Glass-lined CS	50–85	285–480
Steam	Organics	Glass-lined CS	30–70	170–400
Steam	Light oil	Glass-lined CS	40–75	230–425
Steam	Heavy oil	Glass-lined CS	10–40	57–230
Brine	Water	Glass-lined CS	30–80	170–450
Brine	Aqueous solution	Glass-lined CS	25–70	140–400
Brine	Organics	Glass-lined CS	20–60	115–340
Brine	Light oil	Glass-lined CS	25–65	140–370
Brine	Heavy oil	Glass-lined CS	10–30	57–170
Heat-transfer oil	Water	Glass-lined CS	30–80	170–450
Heat-transfer oil	Aqueous solution	Glass-lined CS	25–70	140–400
Heat-transfer oil	Organics	Glass-lined CS	25–65	140–370
Heat-transfer oil	Light oil	Glass-lined CS	20–70	115–400
Heat-transfer oil	Heavy oil	Glass-lined CS	10–35	57–200

°Values listed are for moderate nonproximity agitation. CS = carbon steel.

$$j_H = 0.14(N_{Re}/\phi)^{-0.68} \qquad (11\text{-}53b)$$

In Eqs. (11-53a) and (11-53b), N_{Re} is based on particle diameter and superficial fluid velocity.

Zenz and Othmer (see "Introduction: General References") give an excellent summary of fluidized bed-to-wall heat-transfer investigations.

Solidification involves heavy heat loads transferred essentially at a steady temperature difference. It also involves the varying values of liquid- and solid-phase thickness and thermal diffusivity. When these are substantial and/or in the case of a liquid flowing over a changing solid

layer interface, Siegel and Savino (ASME Paper 67-WA/Ht-34, November 1967) offer equations and charts for prediction of the layer-growth time. For solidification (or melting) of a slab or a semi-infinite bar, initially at its transition temperature, the position of the interface is given by the one-dimensional Newmann's solution given in Carslaw and Jaeger (*Conduction of Heat in Solids*, Clarendon Press, Oxford, 1959).

Later work by Hashem and Sliepcevich [*Chem. Eng. Prog., 63, Symp. Ser.* 79, 35, 42 (1967)] offers more accurate second-order finite-difference equations.

The heat-transfer rate is found to be substantially higher under conditions of **agitation.** The heat transfer is usually said to occur by com-

TABLE 11-8 External Coils; Typical Overall Coefficients*

U expressed in Btu/(h · ft² · °F)

Type of coil	Coil spacing, in.†	Fluid in coil	Fluid in vessel	Temp. range, °F.	U‡ without cement	U with heat-transfer cement
⅜ in. o.d. copper tubing attached with bands at 24-in. spacing	2	5 to 50 lb./sq. in. gage steam	Water under light agitation	158–210	1–5	42–46
	3⅛			158–210	1–5	50–53
	6¼			158–210	1–5	60–64
	12½ or greater			158–210	1–5	69–72
⅜ in. o.d. copper tubing attached with bands at 24-in. spacing	2	50 lb./sq. in. gage steam	No. 6 fuel oil under light agitation	158–258	1–5	20–30
	3⅛			158–258	1–5	25–38
	6¼			158–240	1–5	30–40
	12½ or greater			158–238	1–5	35–46
Panel coils		50 lb./sq. in. gage steam	Boiling water	212	29	48–54
		Water	Water	158–212	8–30	19–48
		Water	No. 6 fuel oil	228–278	6–15	24–56
			Water	130–150	7	15
			No. 6 fuel oil	130–150	4	9–19

°Data courtesy of Thermon Manufacturing Co.

†External surface of tubing or side of panel coil facing tank.

‡For tubing, the coefficients are more dependent upon tightness of the coil against the tank than upon either fluid. The low end of the range is recommended.

NOTE: To convert British thermal units per hour-square foot-degrees Fahrenheit to joules per square meter-second-kelvins, multiply by 5 6783; to convert inches to meters, multiply by 0.0254; and to convert pounds-force per square inch to kilopascals, multiply by 6.895.

TABLE 11-9 Thermal Properties of Various Materials as Affecting Conductive Heat Transfer

Material	Thermal conductivity, B.t.u./(hr.)(sq. ft.)(°F./ft.)	Volume specific heat, B.t.u./(cu. ft.)(°F.)	Thermal diffusivity, sq. ft./hr.
Air	0.0183	0.016	1.143
Water	0.3766	62.5	0.0755
Double steel plate, sand divider	0.207	19.1	0.0108
Sand	0.207	19.1	0.0108
Powdered iron	0.0533	12.1	0.0044
Magnetite iron ore	0.212	63	0.0033
Aerocat catalysts	0.163	20	0.0062
Table salt	0.168	12.6	0.0133
Bone char	0.0877	16.9	0.0051
Pitch coke	0.333	16.2	0.0198
Phenolformaldehyde resin granules	0.0416	10.5	0.0042
Phenolformaldehyde resin powder	0.070	10	0.0070
Powdered coal	0.070	15	0.0047

To convert British thermal units per hour-square foot-degrees Fahrenheit to joules per meter-second-kelvins, multiply by 1.7307; to convert British thermal units per cubic foot-degrees Fahrenheit to joules per cubic meter-kelvins, multiply by $(6.707)(10^4)$; and to convert square feet per hour to square meters per second, multiply by $(2.581)(10^{-5})$.

bined conductive and convective modes. A discussion and explanation are given by Holt [*Chem. Eng.,* **69**(1), 110 (1962)]. Prediction of U_{co} by Eq. (11-48) can be accomplished by replacing α by α_e, the effective thermal diffusivity of the bed. To date so little work has been performed in evaluating the effect of mixing parameters that few predictions can be made. However, for agitated liquid-phase devices Eq. (18-19) is applicable. Holt (loc. cit.) shows that this equation can be converted for solids heat transfer to yield

$$U_{co} = a'c_s D_t^{-0.3} N^{0.7}(\cos \omega)^{0.2} \qquad (11\text{-}54)$$

where D_t = agitator or vessel diameter; N = turning speed, r/min; ω = effective angle of repose of the burden; and a' is a proportionality constant. This is applicable for such devices as agitated pans, agitated kettles, spiral conveyors, and rotating shells.

The solids passage time through **rotary devices** is given by Saemann [*Chem. Eng. Prog.,* **47**, 508, (1951)]:

$$\theta = 0.318L \sin \omega/S_r ND_t \qquad (11\text{-}55a)$$

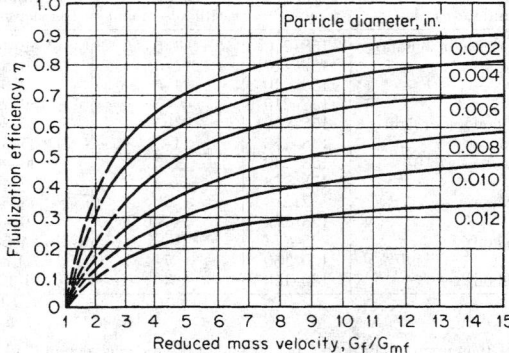

FIG. 11-31 Fluidization efficiency.

and by Marshall and Friedman [*Chem. Eng. Prog.,* **45**, 482–493, 573–588 (1949)]:

$$\theta = (0.23L/S_r N^{0.9}D_t) \pm (0.6BLG/F_a) \qquad (11\text{-}55b)$$

where the second term of Eq. (11-55b) is positive for counterflow of air, negative for concurrent flow, and zero for indirect rotary shells. From these equations a predictive equation is developed for rotary-shell devices, which is analogous to Eq. (11-54):

$$U_{co} = \frac{b'c_s D_t N^{0.9}Y}{(\Delta t)L \sin \omega} \qquad (11\text{-}56)$$

where θ = solids-bed passage time through the shell, min; S_r = shell slope; L = shell length; Y = percent fill; and b' is a proportionality constant.

Vibratory devices which constantly agitate the solids bed maintain a relatively constant value for U_{co} such that

$$U_{co} = a'c_s \alpha_e \qquad (11\text{-}57)$$

with U_{co} having a nominal value of 114 J/(m²·s·K) [20 Btu/(h·ft²·°F)].

Contactive (Direct) Heat Transfer Contactive heat-transfer equipment is so constructed that the particulate burden in solid phase is directly exposed to and permeated by the heating or cooling medium (Sec. 20). The carrier may either heat or cool the solids. A large amount of the industrial heat processing of solids is effected by this mechanism. Physically, these can be classified into packed beds and various degrees of agitated beds from dilute to dense fluidized beds.

The temperature difference for heat transfer is the log-mean temperature difference when the particles are large and/or the beds packed, or the difference between the inlet fluid temperature t_3 and average exhausting fluid temperature t_4, expressed $\Delta_3 t_4$, for small particles. The use of the log mean for packed beds has been confirmed by Thodos and Wilkins (Second American Institute of Chemical Engineers-IIQPR Meeting, Paper 30D, Tampa, May 1968). When fluid and solid flow directions are axially concurrent and particle size is small, as in a vertical-shell fluid bed, the temperature of the exiting solids t_2 (which is also that of exiting gas t_4) is used as $\Delta_3 t_2$, as shown by Levenspiel, Olson, and Walton [*Ind. Eng. Chem.,* **44**, 1478 (1952)], Marshall [*Chem. Eng. Prog.,* **50**, Monogr. Ser. 2, 77 (1954)], Leva (*Fluidization,* McGraw-Hill, New York, 1959), and Holt (Fourth Int. Heat Transfer Conf. Paper 11, American Institute of Chemical Engineers-American Society of Mechanical Engineers, Buffalo, 1960). This temperature difference is also applicable for well-fluidized beds of small particles in cross-flow as in various vibratory carriers.

The **packed-bed-to-fluid heat-transfer coefficient** has been investigated by Baumeister and Bennett [*Am. Inst. Chem. Eng. J.,* **4**, 69 (1958)], who proposed the equation

$$j_H = (h/cG)(c\mu/k)^{2/3} = aN_{Re}^m \qquad (11\text{-}58)$$

where N_{Re} is based on particle diameter and superficial fluid velocity. Values of a and m are as follows:

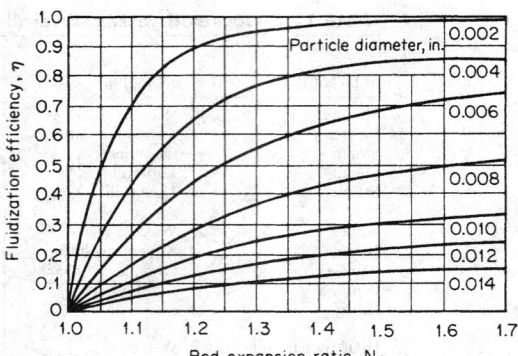

FIG. 11-32 Bed expansion ratio.

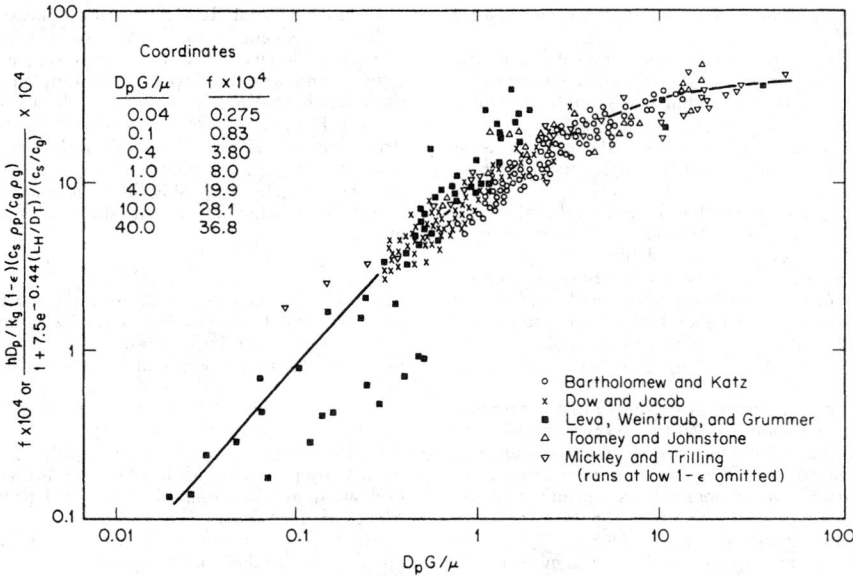

FIG. 11-33 f factor for Eq. (11-52b).

D_t/D_p (dimensionless)	a	m
10.7	1.58	−0.40
16.0	0.95	−0.30
25.7	0.92	−0.28
>30	0.90	−0.28

Glaser and Thodos [*Am. Inst. Chem. Eng. J.,* **4,** 63 (1958)] give a correlation involving individual particle shape and bed porosity. Kunii and Suzuki [*Int. J. Heat Mass Transfer,* **10,** 845 (1967)] discuss heat and mass transfer in packed beds of fine particles.

Particle-to-fluid heat-transfer coefficients in gas **fluidized beds** are predicted by the relation (Zenz and Othmer, op. cit.)

$$\frac{hD_p}{k} = 0.017(D_pG_{mf}/\mu)^{1.21} \tag{11-59a}$$

where G_{mf} is the superficial mass velocity at incipient fluidization.

A more general equation is given by Frantz [*Chem. Eng.,* **69**(20), 89 (1962)]:

$$hD_p/k = 0.015(D_pG/\mu)^{1.6}(c\mu/k)^{0.67} \tag{11-59b}$$

where h is based on true gas temperature.

Bed-to-wall coefficients in dilute-phase transport generally can be predicted by an equation of the form of Eq. (5-50). For example, Bonilla et al. (American Institute of Chemical Engineers Heat Transfer Symp., Atlantic City, N.J., December 1951) found for 1- to 2-μm chalk particles in water up to 8 percent by volume that the coefficient on Eq. (5-50) is 0.029 where k, ρ, and c were arithmetic weighted averages and the viscosity was taken equal to the coefficient of rigidity. Farber and Morley [*Ind. Eng. Chem.,* **49,** 1143 (1957)] found the coefficient on Eq. (5-50) to be 0.025 for the upward flow of air transporting silica-alumina catalyst particles at rates less than 2 kg solids kg air (2 lb solids/lb air). Physical properties used were those of the transporting gas. See Zenz and Othmer (op. cit.) for additional details covering wider porosity ranges.

The thermal performance of **cylindrical rotating shell** units is based upon a volumetric heat-transfer coefficient

$$U_{ct} = \frac{Q}{V_r(\Delta t)} \tag{11-60a}$$

where V_r = volume. This term indirectly includes an area factor so that thermal performance is governed by a cross-sectional area rather than by a heated area. Use of the heated area is possible, however:

$$U_{ct} = \frac{Q}{(\Delta_3 t_2)A} \quad \text{or} \quad \frac{Q}{(\Delta_3 t_4)A} \tag{11-60b}$$

For **heat transfer directly to solids,** predictive equations give directly the volume V or the heat-transfer area A, as determined by heat balance and airflow rate. For devices with gas flow normal to a fluidized-solids bed,

$$A = \frac{Q}{\Delta t_p(c\rho_g)(F_g)} \tag{11-61}$$

where $\Delta t_p = \Delta_3 t_4$ as explained above, $c\rho$ = volumetric specific heat, and F_g = gas flow rate. For air, $c\rho$ at normal temperature and pressure is about 1100 J/(m³·K) [0.0167 Btu/(ft³·°F)]; so

$$A = \frac{bQ}{(\Delta_3 t_4)F_g} \tag{11-62}$$

where $b = 0.0009$ (SI) or 60 (U.S. customary). Another such equation, for stationary vertical-shell and some horizontal rotary-shell and pneumatic-transport devices in which the gas flow is parallel with and directionally concurrent with the fluidized bed, is the same as Eq. (11-62) with $\Delta_3 t_4$ replaced by $\Delta_3 t_2$. If the operation involves drying or chemical reaction, the heat load Q is much greater than for sensible-heat transfer only. Also, the gas flow rate to provide moisture carry-off and stoichiometric requirements must be considered and simultaneously provided. A good treatise on the latter is given by Pinkey and Plint (*Miner. Process.,* June 1968, p. 17).

Evaporative cooling is a special patented technique that often can be advantageously employed in cooling solids by contactive heat transfer. The drying operation is terminated *before* the desired final moisture content is reached, and solids temperature is at a moderate value. The cooling operation involves contacting the burden (preferably fluidized) with air at normal temperature and pressure. The air adiabatically absorbs and carries off a large part of the moisture and, in doing so, picks up heat from the warm (or hot) solids particles to supply the latent heat demand of evaporation. For entering solids at temperatures of 180°C (350°F) and less with normal heat-capacity values of 0.85 to 1.0 kJ/(kg·K) [0.2 to 0.25 Btu/(lb·°F)], the effect can be calculated by:

1. Using 285 m³ (1000 ft³) of airflow at normal temperature and pressure at 40 percent relative humidity to carry off 0.45 kg (1 lb) of

water [latent heat 2326 kJ/kg (1000 Btu/lb)] and to lower temperature by 22 to 28°C (40 to 50°F).

2.   Using the lowered solids temperature as t_3 and calculating the remainder of the heat to be removed in the regular manner by Eq. (11-62). The required air quantity for (2) must be equal to or greater than that for (1).

When the solids heat capacity is higher (as is the case for most organic materials), the temperature reduction is inversely proportional to the heat capacity.

A nominal result of this technique is that the required airflow rate and equipment size is about two-thirds of that when evaporative cooling is not used. See Sec. 20 for equipment available.

Convective Heat Transfer   Equipment using the true convective mechanism when the heated particles are mixed with (and remain with) the cold particles is used so infrequently that performance and sizing equations are not available. Such a device is the pebble heater as described by Norton (*Chem. Metall. Eng.*, July 1946). For operation data, see Sec. 9.

Convective heat transfer is often used as an adjunct to other modes, particularly to the conductive mode. It is often more convenient to consider the agitative effect a performance-improvement influence on the thermal diffusivity factor α, modifying it to α_e, the effective value.

A *pseudo-convective heat-transfer* operation is one in which the heating gas (generally air) is passed over a bed of solids. Its use is almost exclusively limited to drying operations (see Sec. 12, tray and shelf dryers). The operation, sometimes termed direct, is more akin to the conductive mechanism. For this operation, Tsao and Wheelock [*Chem. Eng.*, **74**(13), 201 (1967)] predict the heat-transfer coefficient when radiative and conductive effects are absent by

$$h = bG^{0.8} \tag{11-63}$$

where $b = 14.31$ (SI) or 0.0128 (U.S. customary), h = convective heat transfer, and G = gas flow rate.

The **drying rate** is given by

$$K_{cv} = \frac{h(T_d - T_w)}{\lambda} \tag{11-64}$$

where K_{cv} = drying rate, for constant-rate period, kg/(m²·s) [lb/(h·ft²)]; T_d and T_w = respective dry-bulb and wet-bulb temperatures of the air; and λ = latent heat of evaporation at temperature T_w. Note here that the temperature-difference determination of the operation is a simple linear one and of a steady-state nature. Also note that the operation is a function of the airflow rate. Further, the solids are granular with a fairly uniform size, have reasonable capillary voids, are of a firm texture, and have the particle surface wetted.

The coefficient h is also used to predict (in the constant-rate period) the total overall air-to-solids heat-transfer coefficient U_{cv} by

$$1/U_{cv} = 1/h + x/k \tag{11-65}$$

where k = solids thermal conductivity and x is evaluated from

$$x = \frac{z(X_c - X_o)}{X_c - X_e} \tag{11-65a}$$

where z = bed (or slab) thickness and is the total thickness when drying and/or heat transfer is from one side only but is one-half of the thickness when drying and/or heat transfer is simultaneously from both sides; X_o, X_c, and X_e are respectively the initial (or feed-stock), critical, and equilibrium (with the drying air) moisture contents of the solids, all in kg H₂O/kg dry solids (lb H₂O/lb dry solids). This coefficient is used to predict the *instantaneous* drying rate

$$-\frac{W}{A}\frac{dX}{d\theta} = \frac{U_{cv}(T_d - T_w)}{\lambda} \tag{11-66}$$

By rearrangement, this can be made into a design equation as follows:

$$A = -\frac{W\lambda(dX/d\theta)}{U_{cv}(T_d - T_w)} \tag{11-67}$$

where W = weight of dry solids in the equipment, λ = latent heat of evaporation, and θ = drying time. The reader should refer to the full reference article by Tsao and Wheelock (loc. cit.) for other solids conditions qualifying the use of these equations.

Radiative Heat Transfer   Heat-transfer equipment using the radiative mechanism for divided solids is constructed as a "table" which is stationary, as with trays, or moving, as with a belt, and/or agitated, as with a vibrated pan, to distribute and expose the burden in a plane parallel to (but not in contact with) the plane of the radiant-heat sources. Presence of air is not necessary (see Sec. 12 for vacuum-shelf dryers and Sec. 22 for resublimation). In fact, if air in the intervening space has a high humidity or CO₂ content, it acts as an energy absorber, thereby depressing the performance.

For the radiative mechanism, the temperature difference is evaluated as

$$\Delta t = T_e^4 - T_r^4 \tag{11-68}$$

where T_e = absolute temperature of the radiant-heat source, K (°R); and T_r = absolute temperature of the bed of divided solids, K (°R).

Numerical values for U_{ra} for use in the general design equation may be calculated from experimental data by

$$U_{ra} = \frac{Q}{A(T_e^4 - T_r^4)} \tag{11-69}$$

The literature to date offers practically no such values. However, enough proprietary work has been performed to present a reliable evaluation for the comparison of mechanisms (see "Introduction: Modes of Heat Transfer").

For the radiative mechanism of heat transfer to solids, the rate equation for parallel-surface operations is

$$q_{ra} = b(T_e^4 - T_r^4)i_f \tag{11-70}$$

where $b = (5.67)(10^{-8})$(SI) or $(0.172)(10^{-8})$(U.S. customary), q_{ra} = radiative heat flux, and i_f = an interchange factor which is evaluated from

$$1/i_f = 1/e_s + 1/e_r - 1 \tag{11-70a}$$

where e_s = coefficient of emissivity of the source and e_r = "emissivity" (or "absorptivity") of the receiver, which is the divided-solids bed. For the emissivity values, particularly of the heat source e_s, an important consideration is the wavelength at which the radiant source emits as well as the flux density of the emission. Data for these values are available from Polentz [*Chem. Eng.*, **65**(7), 137; (8), 151 (1958)] and Adlam (*Radiant Heating*, Industrial Press, New York, p. 40). Both give radiated flux density versus wavelength at varying temperatures. Often, the seemingly cooler but longer wavelength source is the better selection.

Emitting sources are (1) pipes, tubes, and platters carrying steam, 2100 kPa (300 lbf/in²); (2) electrical-conducting glass plates, 150 to 315°C (300 to 600°F) range; (3) light-bulb type (tungsten-filament resistance heater); (4) modules of refractory brick for gas burning at high temperatures and high fluxes; and (5) modules of quartz tubes, also operable at high temperatures and fluxes. For some emissivity values see Table 11-10.

For *predictive work*, where U_{ra} is desired for sizing, this can be obtained by dividing the flux rate q_{ra} by Δt:

$$U_{ra} = q_{ra}/(T_e^4 - T_r^4) = i_f b \tag{11-71}$$

where $b = (5.67)(10^{-8})$ (SI) or $(0.172)(10^{-8})$ (U.S. customary). Hence:

$$A = \frac{Q}{U_{ra}(T_e^4 - T_r^4)} \tag{11-72}$$

where A = bed area of solids in the equipment.

Important considerations in the application of the foregoing equations are:

1.   Since the temperature of the emitter is generally known (preselected or readily determined in an actual operation), the absorptivity value e_r is the unknown. This absorptivity is partly a measure of the ability of radiant heat to penetrate the body of a solid particle (or a moisture film) instantly, as compared with diffusional heat transfer by conduction. Such instant penetration greatly reduces processing time and case-hardening effects. Moisture release and other mass transfer, however, still progress by diffusional means.

2.   In one of the major applications of radiative devices (drying),

TABLE 11-10 Normal Total Emissivity of Various Surfaces

A. Metals and Their Oxides					
Surface	t, °F.°	Emissivity°	Surface	t, °F.°	Emissivity°
Aluminum			Sheet steel, strong rough oxide layer	75	0.80
Highly polished plate, 98.3% pure	440–1070	0.039–0.057	Dense shiny oxide layer	75	0.82
Polished plate	73	0.040	Cast plate:		
Rough plate	78	0.055	Smooth	73	0.80
Oxidized at 1110°F	390–1110	0.11–0.19	Rough	73	0.82
Aluminum-surfaced roofing	100	0.216	Cast iron, rough, strongly oxidized	100–480	0.95
Calorized surfaces, heated at 1110°F.			Wrought iron, dull oxidized	70–680	0.94
Copper	390–1110	0.18–0.19	Steel plate, rough	100–700	0.94–0.97
Steel	390–1110	0.52–0.57	High temperature alloy steels (see Nickel		
Brass			Alloys)		
Highly polished:			Molten metal		
73.2% Cu, 26.7% Zn	476–674	0.028–0.031	Cast iron	2370–2550	0.29
62.4% Cu, 36.8% Zn, 0.4% Pb, 0.3% Al	494–710	0.033–0.037	Mild steel	2910–3270	0.28
82.9% Cu, 17.0% Zn	530	0.030	Lead		
Hard rolled, polished:			Pure (99.96%), unoxidized	260–440	0.057–0.075
But direction of polishing visible	70	0.038	Gray oxidized	75	0.281
But somewhat attacked	73	0.043	Oxidized at 390°F	390	0.63
But traces of stearin from polish left on	75	0.053	Mercury	32–212	0.09–0.12
Polished	100–600	0.096	Molybdenum filament	1340–4700	0.096–0.292
Rolled plate, natural surface	72	0.06	Monel metal, oxidized at 1110°F	390–1110	0.41–0.46
Rubbed with coarse emery	72	0.20	Nickel		
Dull plate	120–660	0.22	Electroplated on polished iron, then		
Oxidized by heating at 1110°F	390–1110	0.61–0.59	polished	74	0.045
Chromium; see Nickel Alloys for Ni-Cr steels	100–1000	0.08–0.26	Technically pure (98.9% Ni, + Mn),		
Copper			polished	440–710	0.07–0.087
Carefully polished electrolytic copper	176	0.018	Electropolated on pickled iron, not		
Commercial, emeried, polished, but pits			polished	68	0.11
remaining	66	0.030	Wire	368–1844	0.096–0.186
Commercial, scraped shiny but not			Plate, oxidized by heating at 1110°F	390–1110	0.37–0.48
mirror-like	72	0.072	Nickel oxide	1200–2290	0.59–0.86
Polished	242	0.023	Nickel alloys		
Plate, heated long time, covered with			Chromnickel	125–1894	0.64–0.76
thick oxide layer	77	0.78	Nickelin (18–32 Ni; 55–68 Cu; 20 Zn), gray		
Plate heated at 1110°F	390–1110	0.57	oxidized	70	0.262
Cuprous oxide	1470–2010	0.66–0.54	KA-2S alloy steel (8% Ni; 18% Cr), light		
Molten copper	1970–2330	0.16–0.13	silvery, rough, brown, after heating	420–914	0.44–0.36
Gold			After 42 hr. heating at 980°F	420–980	0.62–0.73
Pure, highly polished	440–1160	0.018–0.035	NCT-3 alloy (20% Ni; 25% Cr.), brown,		
Iron and steel			splotched, oxidized from service	420–980	0.90–0.97
Metallic surfaces (or very thin oxide			NCT-6 alloy (60% Ni; 12% Cr), smooth,		
layer):			black, firm adhesive oxide coat from		
Electrolytic iron, highly polished	350–440	0.052–0.064	service	520–1045	0.89–0.82
Polished iron	800–1880	0.144–0.377	Platinum		
Iron freshly emeried	68	0.242	Pure, polished plate	440–1160	0.054–0.104
Cast iron, polished	392	0.21	Strip	1700–2960	0.12–0.17
Wrought iron, highly polished	100–480	0.28	Filament	80–2240	0.036–0.192
Cast iron, newly turned	72	0.435	Wire	440–2510	0.073–0.182
Polished steel casting	1420–1900	0.52–0.56	Silver		
Ground sheet steel	1720–2010	0.55–0.61	Polished, pure	440–1160	0.0198–0.0324
Smooth sheet iron	1650–1900	0.55–0.60	Polished	100–700	0.0221–0.0312
Cast iron, turned on lathe	1620–1810	0.60–0.70	Steel, see Iron		
Oxidized surfaces:			Tantalum filament	2420–5430	0.194–0.31
Iron plate, pickled, then rusted red	68	0.612	Tin—bright tinned iron sheet	76	0.043 and 0.064
Completely rusted	67	0.685	Tungsten		
Rolled sheet steel	70	0.657	Filament, aged	80–6000	0.032–0.35
Oxidized iron	212	0.736	Filament	6000	0.39
Cast iron, oxidized at 1100°F	390–1110	0.64–0.78	Zinc		
Steel, oxidized at 1100°F	390–1110	0.79	Commercial, 99.1% pure, polished	440–620	0.045–0.053
Smooth oxidized electrolytic iron	260–980	0.78–0.82	Oxidized by heating at 750°F.	750	0.11
Iron oxide	930–2190	0.85–0.89	Galvanized sheet iron, fairly bright	82	0.228
Rough ingot iron	1700–2040	0.87–0.95	Galvanized sheet iron, gray oxidized	75	0.276

B. Refractories, Building Materials, Paints, and Miscellaneous					
Asbestos			Carbon		
Board	74	0.96	T-carbon (Gebr. Siemens) 0.9% ash	260–1160	0.81–0.79
Paper	100–700	0.93–0.945	(this started with emissivity at 260°F.		
Brick			of 0.72, but on heating changed to		
Red, rough, but no gross irregularities	70	0.93	values given)		
Silica, unglazed, rough	1832	0.80	Carbon filament	1900–2560	0.526
Silica, glazed, rough	2012	0.85	Candle soot	206–520	0.952
Grog brick, glazed	2012	0.75	Lampblack-waterglass coating	209–362	0.959–0.947
See Refractory Materials below.					

TABLE 11-10 Normal Total Emissivity of Various Surfaces (*Concluded*)

B. Refractories, Building Materials, Paints, and Miscellaneous

Surface	t, °F.°	Emissivity°	Surface	t, °F.°	Emissivity°
Same	260–440	0.957–0.952	Oil paints, sixteen different, all colors	212	0.92–0.96
Thin layer on iron plate	69	0.927	Aluminum paints and lacquers		
Thick coat	68	0.967	10% Al, 22% lacquer body, on rough or		
Lampblack, 0.003 in. or thicker	100–700	0.945	smooth surface	212	0.52
Enamel, white fused, on iron	66	0.897	26% Al, 27% lacquer body, on rough or		
Glass, smooth	72	0.937	smooth surface	212	0.3
Gypsum, 0.02 in. thick on smooth or			Other Al paints, varying age and Al		
blackened plate	70	0.903	content	212	0.27–0.67
Marble, light gray, polished	72	0.931	Al lacquer, varnish binder, on rough plate	70	0.39
Oak, planed	70	0.895	Al paint, after heating to 620°F	300–600	0.35
Oil layers on polished nickel (lube oil)	68		Paper, thin		
Polished surface, alone		0.045	Pasted on tinned iron plate	66	0.924
+0.001-in. oil		0.27	On rough iron plate	66	0.929
+0.002-in. oil		0.46	On black lacquered plate	66	0.944
+0.005-in. oil		0.72	Plaster, rough lime	50–190	0.91
Infinitely thick oil layer		0.82	Porcelain, glazed	72	0.924
Oil layers on aluminum foil (linseed oil)			Quartz, rough, fused	70	0.932
Al foil	212	0.087†	Refractory materials, 40 different	1110–1830	
+1 coat oil	212	0.561	poor radiators		$\begin{bmatrix}0.65\\0.70\end{bmatrix} - 0.75$
+2 coats oil	212	0.574			
Paints, lacquers, varnishes			good radiators		$\begin{bmatrix}0.80\\0.85\end{bmatrix} - \begin{bmatrix}0.85\\0.90\end{bmatrix}$
Snowhite enamel varnish or rough iron					
plate	73	0.906	Roofing paper	69	0.91
Black shiny lacquer, sprayed on iron	76	0.875	Rubber		
Black shiny shellac on tinned iron sheet	70	0.821	Hard, glossy plate	74	0.945
Black matte shellac	170–295	0.91	Soft, gray, rough (reclaimed)	76	0.859
Black lacquer	100–200	0.80–0.95	Serpentine, polished	74	0.900
Flat black lacquer	100–200	0.96–0.98	Water	32–212	0.95–0.963
White lacquer	100–200	0.80–0.95			

°When two temperatures and two emissivities are given, they correspond, first to first and second to second, and linear interpolation is permissible. °C = (°F − 32)/1.8.
†Although this value is probably high, it is given for comparison with the data by the same investigator to show the effect of oil layers. See Aluminum, Part A of this table.

the surface-held moisture is a good heat absorber in the 2- to 7-μm wavelength range. Therefore, the absorptivity, color, and nature of the solids are of little importance.

3. For drying, it is important to provide a small amount of venting air to carry away the water vapor. This is needed for two reasons. First, water vapor is a good absorber of 2- to 7-μm energy. Second, water-vapor accumulation depresses further vapor release by the solids. If the air over the solids is kept fairly dry by venting, very little heat is carried off, because dry air does not absorb radiant heat.

4. For some of the devices, when the overall conversion efficiency has been determined, the application is primarily a matter of computing the required heat load. It should be kept in mind, however, that there are two conversion efficiencies that must be differentiated. One measure of efficiency is that with which the source converts input energy to output radiated energy. The other is the overall efficiency that measures the proportion of input energy that is actually absorbed by the solids. This latter is, of course, the one that really matters.

Other applications of radiant-heat processing of solids are the toasting, puffing, and baking of foods and the low-temperature roasting and preheating of plastic powder or pellets. Since the determination of heat loads for these operations is not well established, bench and pilot tests are generally necessary. Such processes require a fast input of heat and higher heat fluxes than can generally be provided by indirect equipment. Because of this, infrared-equipment size and space requirements are often much lower.

Although direct contactive heat transfer can provide high temperatures and heat concentrations and at the same time be small in size, its use may not always be preferable because of undesired side effects such as drying, contamination, case hardening, shrinkage, off color, and dusting.

When radiating and receiving surfaces are not in parallel, as in rotary-kiln devices, and the solids burden bed may be only intermittently exposed and/or agitated, the calculation and procedures become very complex, with photometric methods of optics requiring consideration. The following equation for heat transfer, which allows for convective effects, is commonly used by designers of **high-temperature furnaces:**

$$q_{ra} = Q/A = b\sigma\,[(T_g/100)^4 - (T_s/100)^4] \qquad (11\text{-}73)$$

where $b = 5.67$ (SI) or 0.172 (U.S. customary); Q = total furnace heat transfer; σ = an emissivity factor with recommended values of 0.74 for gas, 0.75 for oil, and 0.81 for coal; A = effective area for absorbing heat (here the solids burden exposed area); T_g = exiting-combustion-gas absolute temperature; and T_s = absorbing surface temperature.

In rotary devices, reradiation from the exposed shell surface to the solids bed is a major design consideration. A treatise on furnaces, including radiative heat-transfer effects, is given by Ellwood and Danatos [*Chem. Eng.*, **73**(8), 174 (1966)]. For discussion of radiation heat-transfer computational methods, heat fluxes obtainable, and emissivity values, see Schornshort and Viskanta (ASME Paper 68-H 7-32), Sherman (ASME Paper 56-A-111), and the following subsection.

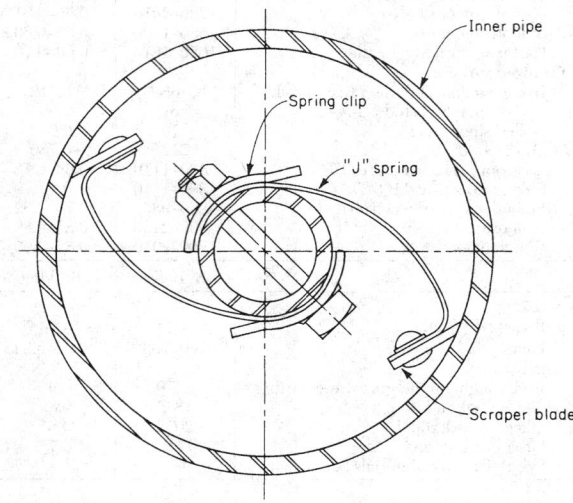

FIG. 11-34 Scraper blade of scraped-surface exchanger. (*Henry Vogt Machine Co., Inc.*)

SCRAPED-SURFACE EXCHANGERS

Scraped-surface exchangers have a rotating element with spring-loaded scraper blades to scrape the inside surface (Fig. 11-34). Generally a double-pipe construction is used; the scraping mechanism is in the inner pipe, where the process fluid flows; and the cooling or heating medium is in the outer pipe. The most common size has 6-in inside and 8-in outside pipes. Also available are 3- by 4-in, 8- by 10-in, and 12- by 14-in sizes (in × 25.4 = mm). These double-pipe units are commonly connected in series and arranged in double stands.

For **chilling** and **crystallizing** with an evaporating refrigerant, a 27-in shell with seven 6-in pipes is available (Henry Vogt Machine Co.). In direct contact with the scraped surface is the process fluid which may deposit crystals upon chilling or be extremely fouling or of very high viscosity. Motors, chain drives, appropriate guards, and so on are required for the rotating element. For chilling service with a refrigerant in the outer shell, an accumulator drum is mounted on top of the unit.

Scraped-surface exchangers are particularly suitable for heat transfer with crystallization, heat transfer with severe fouling of surfaces, heat transfer with solvent extraction, and heat transfer of high-viscosity fluids. They are extensively used in paraffin-wax plants and in petrochemical plants for crystallization.

TEMA-STYLE SHELL-AND-TUBE HEAT EXCHANGERS

TYPES AND DEFINITIONS

TEMA-style shell-and-tube-type exchangers constitute the bulk of the unfired heat-transfer equipment in chemical-process plants, although increasing emphasis has been developing in other designs. These exchangers are illustrated in Fig. 11-35, and their features are summarized in Table 11-11.

TEMA Numbering and Type Designation Recommended practice for the designation of TEMA-style shell-and-tube heat exchangers by numbers and letters has been established by the Tubular Exchanger Manufacturers Association (TEMA). This information from the sixth edition of the TEMA Standards is reproduced in the following paragraphs.

It is recommended that heat-exchanger size and type be designated by numbers and letters.

1. *Size.* Sizes of shells (and tube bundles) shall be designated by numbers describing shell (and tube-bundle) diameters and tube lengths as follows:

2. *Diameter.* The nominal diameter shall be the inside diameter of the shell in inches, rounded off to the nearest integer. For kettle reboilers the nominal diameter shall be the port diameter followed by the shell diameter, each rounded off to the nearest integer.

3. *Length.* The nominal length shall be the tube length in inches. Tube length for straight tubes shall be taken as the actual overall length. For U tubes the length shall be taken as the straight length from end of tube to bend tangent.

4. *Type.* Type designation shall be by letters describing stationary head, shell (omitted for bundles only), and rear head, in that order, as indicated in Fig. 11-1.

Typical Examples (A) Split-ring floating-heat exchanger with removable channel and cover, single-pass shell, 591-mm (23¼-in) inside diameter with tubes 4.9 m (16 ft) long. SIZE 23–192 TYPE AES.

(B) U-tube exchanger with bonnet-type stationary head, split-flow shell, 483-mm (19-in) inside diameter with tubes 21-m (7-ft) straight length. SIZE 19–84 TYPE GBU.

(C) Pull-through floating-heat-kettle-type reboiler having stationary head integral with tube sheet, 584-mm (23-in) port diameter and 940-mm (37-in) inside shell diameter with tubes 4.9-m (16-ft) long. SIZE 23/37–192 TYPE CKT.

(D) Fixed-tube sheet exchanger with removable channel and cover, bonnet-type rear head, two-pass shell, 841-mm (33⅓-in) diameter with tubes 2.4 m (8-ft) long. SIZE 33–96 TYPE AFM.

(E) Fixed-tube sheet exchanger having stationary and rear heads integral with tube sheets, single-pass shell, 432-mm (17-in) inside diameter with tubes 4.9-m (16-ft) long. SIZE 17–192 TYPE CEN.

Functional Definitions Heat-transfer equipment can be designated by type (e.g., fixed tube sheet, outside packed head, etc.) or by

TABLE 11-11 Features of TEMA Shell-and-Tube-Type Exchangers*

Type of design	Fixed tube sheet	U-tube	Packed lantern-ring floating head	Internal floating head (split backing ring)	Outside-packed floating head	Pull-through floating head
T.E.M.A. rear-head type	L or M or N	U	W	S	P	T
Relative cost increases from A (least expensive) through E (most expensive)	B	A	C	E	D	E
Provision for differential expansion	Expansion joint in shell	Individual tubes free to expand	Floating head	Floating head	Floating head	Floating head
Removable bundle	No	Yes	Yes	Yes	Yes	Yes
Replacement bundle possible	No	Yes	Yes	Yes	Yes	Yes
Individual tubes replaceable	Yes	Only those in outside row†	Yes	Yes	Yes	Yes
Tube cleaning by chemicals inside and outside	Yes	Yes	Yes	Yes	Yes	Yes
Interior tube cleaning mechanically	Yes	Special tools required	Yes	Yes	Yes	Yes
Exterior tube cleaning mechanically:						
Triangular pitch	No	No‡	No‡	No‡	No‡	No‡
Square pitch	No	Yes	Yes	Yes	Yes	Yes
Hydraulic-jet cleaning:						
Tube interior	Yes	Special tools required	Yes	Yes	Yes	Yes
Tube exterior	No	Yes	Yes	Yes	Yes	Yes
Double tube sheet feasible	Yes	Yes	No	No	Yes	No
Number of tube passes	No practical limitations	Any even number possible	Limited to one or two passes	No practical limitations§	No practical limitations	No practical limitations§
Internal gaskets eliminated	Yes	Yes	Yes	No	Yes	No

NOTE: Relative costs A and B are not significantly different and interchange for long lengths of tubing.
*Modified from page a-8 of the Patterson-Kelley Co. Manual No. 700A, Heat Exchangers.
†U-tube bundles have been built with tube supports which permit the U-bends to be spread apart and tubes inside of the bundle replaced.
‡Normal triangular pitch does not permit mechanical cleaning. With a wide triangular pitch, which is equal to 2 (tube diameter plus cleaning lane)/√3, mechanical cleaning is possible on removable bundles. This wide spacing is infrequently used.
§For odd number of tube side passes, floating head requires packed joint or expansion joint.

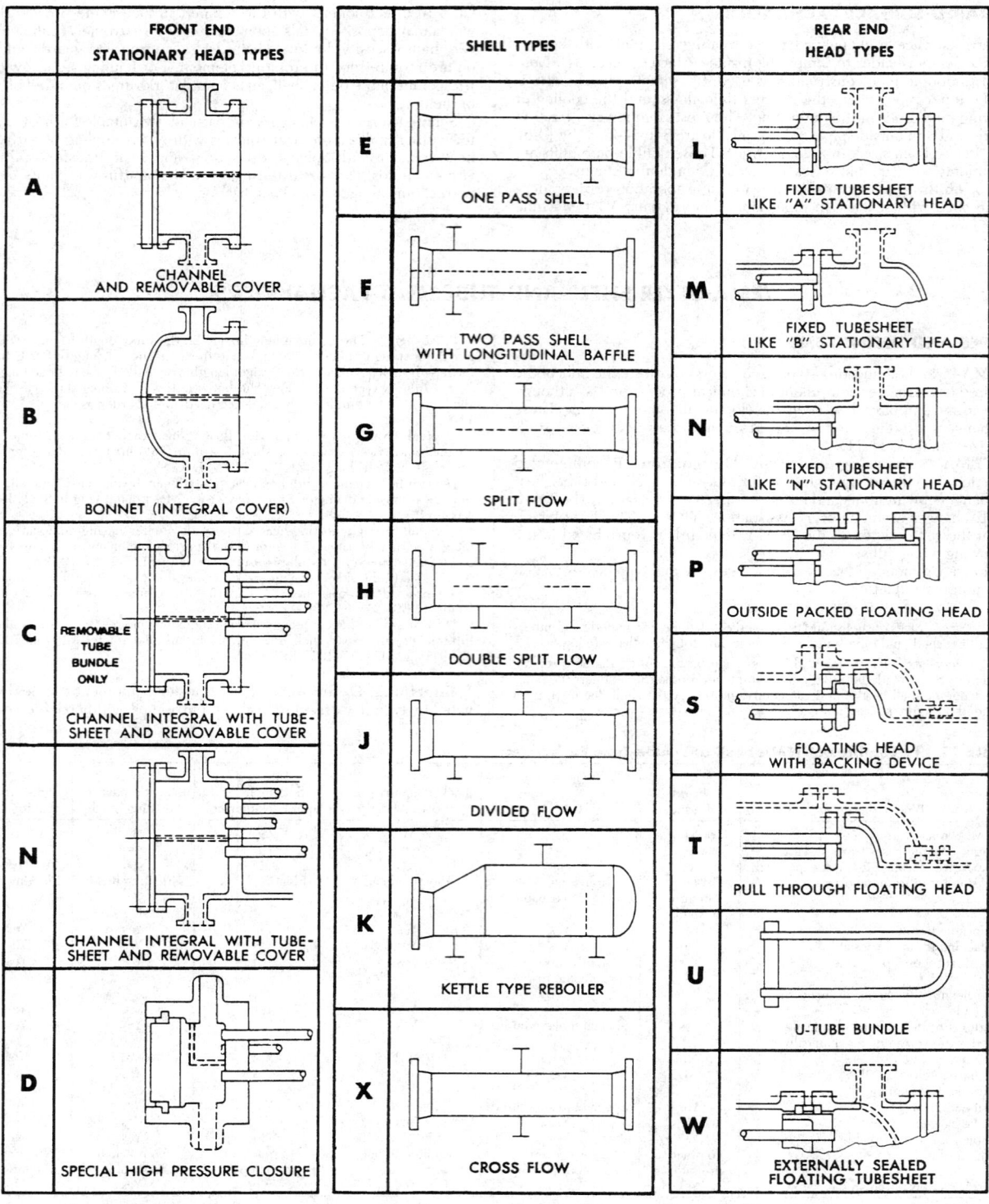

FIG. 11-35 TEMA-type designations for shell-and-tube heat exchangers. (Standards of Tubular Exchanger Manufacturers Association, 6th ed., 1978.)

function (chiller, condenser, cooler, etc.). Almost any type of unit can be used to perform any or all of the listed functions. Many of these terms have been defined by Donahue [*Pet. Process.*, 103 (March 1956)].

Equipment	Function
Chiller	Cools a fluid to a temperature below that obtainable if water only were used as a coolant. It uses a refrigerant such as ammonia or Freon.
Condenser	Condenses a vapor or mixture of vapors, either alone or in the presence of a noncondensable gas.
Partial condenser	Condenses vapors at a point high enough to provide a temperature difference sufficient to preheat a cold stream of process fluid. This saves heat and eliminates the need for providing a separate preheater (using flame or steam).
Final condenser	Condenses the vapors to a final storage temperature of approximately 37.8°C (100°F). It uses water cooling, which means that the transferred heat is lost to the process.
Cooler	Cools liquids or gases by means of water.
Exchanger	Performs a double function: (1) heats a cold fluid by (2) using a hot fluid which it cools. None of the transferred heat is lost.
Heater	Imparts sensible heat to a liquid or a gas by means of condensing steam or Dowtherm.
Reboiler	Connected to the bottom of a fractionating tower, it provides the reboil heat necessary for distillation. The heating medium may be either steam or a hot-process fluid.
Thermosiphon reboiler	Natural circulation of the boiling medium is obtained by maintaining sufficient liquid head to provide for circulation.
Forced-circulation reboiler	A pump is used to force liquid through the reboiler.
Steam generator	Generates steam for use elsewhere in the plant by using the available high-level heat in tar or a heavy oil.
Superheater	Heats a vapor above the saturation temperature.
Vaporizer	A heater which vaporizes part of the liquid.
Waste-heat boiler	Produces steam; similar to steam generator, except that the heating medium is a hot gas or liquid produced in a chemical reaction.

GENERAL DESIGN CONSIDERATIONS

Selection of Flow Path In selecting the flow path for two fluids through an exchanger, several general approaches are used. The tube-side fluid is more corrosive or dirtier or at a higher pressure. The shell-side fluid is a liquid of high viscosity or a gas.

When alloy construction for one of the two fluids is required, a carbon steel shell combined with alloy tube-side parts is less expensive than alloy in contact with the shell-side fluid combined with carbon steel headers.

Cleaning of the inside of tubes is more readily done than cleaning of exterior surfaces.

For gauge pressures in excess of 2068 kPa (300 lbf/in²) for one of the fluids, the less expensive construction has the high-pressure fluid in the tubes.

For a given pressure drop, higher heat-transfer coefficients are obtained on the shell side than on the tube side.

Heat-exchanger shutdowns are most often caused by fouling, corrosion, and erosion.

Construction Codes "Rules for Construction of Pressure Vessels, Division 1," which is part of Section VIII of the ASME Boiler and Pressure Vessel Code (American Society of Mechanical Engineers), serves as a construction code by providing minimum standards. New editions of the code are usually issued every 3 years. Interim revisions are made semiannually in the form of addenda. Compliance with ASME Code requirements is mandatory in much of the United States and Canada. Originally these rules were not prepared for heat exchangers. However, the welded joint between tube sheet and shell of the fixed-tube-sheet heat exchanger is now included. A nonmandatory

appendix on tube-to-tube-sheet joints is also included. Additional rules for heat exchangers are being developed.

Standards of Tubular Exchanger Manufacturers Association, 6th ed., 1978 (commonly referred to as the TEMA Standards), serve to supplement and define the ASME Code for all shell-and-tube-type heat-exchanger applications (other than double-pipe construction). TEMA Class R design is "for the generally severe requirements of petroleum and related processing applications. Equipment fabricated in accordance with these standards is designed for safety and durability under the rigorous service and maintenance conditions in such applications." TEMA Class C design is "for the generally moderate requirements of commercial and general process applications," while TEMA Class B is "for chemical process service."

The mechanical-design requirements are identical for all three classes of construction. The differences between the TEMA classes are minor and were listed by Rubin [*Hydrocarbon Process.*, **59**, 92 (June 1980)].

Among the topics of the TEMA Standards are nomenclature, fabrication tolerances, inspection, guarantees, tubes, shells, baffles and support plates, floating heads, gaskets, tube sheets, channels, nozzles, end flanges and bolting, material specifications, and fouling resistances.

Shell and Tube Heat Exchangers for General Refinery Services, API Standard 660, 4th ed., 1982, is published by the American Petroleum Institute to supplement both the TEMA Standards and the ASME Code. Many companies in the chemical and petroleum processing fields have their own standards to supplement these various requirements. *The Interrelationships between Codes, Standards, and Customer Specifications for Process Heat Transfer Equipment* is a symposium volume which was edited by F. L. Rubin and published by ASME in December 1979. (See discussion of pressure-vessel codes in Sec. 6.)

Design pressures and temperatures for exchangers usually are specified with a margin of safety beyond the conditions expected in service. Design pressure is generally about 172 kPa (25 lbf/in²) greater than the maximum expected during operation or at pump shutoff. Design temperature is commonly 14°C (25°F) greater than the maximum temperature in service.

Tube Bundle Vibration Damage from tube vibration has become an increasing problem as plate baffled heat exchangers are designed for higher flow rates and pressure drops. The most effective method of dealing with this problem is the avoidance of cross flow by use of tube support baffles which promote only longitudinal flow. However, even then, strict attention must be given the bundle area under the shell inlet nozzle where flow is introduced through the side of the shell. TEMA has devoted an entire section in its standards to this topic. In general, the mechanisms of tube vibration are as follows:

Vortex Shedding The vortex-shedding frequency of the fluid in cross-flow over the tubes may coincide with a natural frequency of the tubes and excite large resonant vibration amplitudes.

Fluid-Elastic Coupling Fluid flowing over tubes causes them to vibrate with a whirling motion. The mechanism of fluid-elastic coupling occurs when a "critical" velocity is exceeded and the vibration then becomes self-excited and grows in amplitude. This mechanism frequently occurs in process heat exchangers which suffer vibration damage.

Pressure Fluctuation Turbulent pressure fluctuations which develop in the wake of a cylinder or are carried to the cylinder from upstream may provide a potential mechanism for tube vibration. The tubes respond to the portion of the energy spectrum that is close to their natural frequency.

Acoustic Coupling When the shell-side fluid is a low-density gas, acoustic resonance or coupling develops when the standing waves in the shell are in phase with vortex shedding from the tubes. The standing waves are perpendicular to the axis of the tubes and to the direction of cross-flow. Damage to the tubes is rare. However, the noise can be extremely painful.

Testing Upon completion of shop fabrication and also during maintenance operations it is desirable hydrostatically to test the shell side of tubular exchangers so that visual examination of tube ends can be made. Leaking tubes can be readily located and serviced. When leaks are determined without access to the tube ends, it is necessary to reroll or reweld all the tube-to-tube-sheet joints with possible damage to the satisfactory joints.

Testing for leaks in heat exchangers was discussed by Rubin [*Chem. Eng.*, **68**, 160–166 (July 24, 1961)].

Performance testing of heat exchangers is described in the American Institute of Chemical Engineers' *Standard Testing Procedure for Heat Exchangers*, Sec. 1. "Sensible Heat Transfer in Shell-and-Tube-Type Equipment."

PRINCIPAL TYPES OF CONSTRUCTION

Figure 11-36 shows details of the construction of the TEMA types of shell-and-tube heat exchangers. These and other types are discussed in the following paragraphs.

Fixed-Tube-Sheet Heat Exchangers Fixed-tube-sheet exchangers (Fig. 11-36b) are used more often than any other type, and the frequency of use has been increasing in recent years. The tube sheets are welded to the shell and serve as flanges to which the tube-side headers are bolted. This construction requires that the shell and tube-sheet materials be weldable to each other.

When such welding is not possible, a "blind"-gasket type of construction is utilized. The blind gasket is not accessible for maintenance or replacement once the unit has been constructed. This construction is used for steam surface condensers, which operate under vacuum.

The tube-side header (or channel) may be welded to the tube sheet, as shown in Fig. 11-35 for type C and N heads. This type of construc-tion is less costly than types B and M or A and L and still offers the advantage that tubes may be examined and replaced without disturbing the tube-side piping connections.

There is no limitation on the number of tube-side passes. Shell-side passes can be one or more, although shells with more than two shell-side passes are rarely used.

Tubes can completely fill the heat-exchanger shell. Clearance between the outermost tubes and the shell is only the minimum necessary for fabrication. Between the inside of the shell and the baffles some clearance must be provided so that baffles can slide into the shell. Fabrication tolerances then require some additional clearance between the outside of the baffles and the outermost tubes. The edge distance between the outer tube limit (OTL) and the baffle diameter must be sufficient to prevent vibration of the tubes from breaking through the baffle holes. The outermost tube must be contained within the OTL. Clearances between the inside shell diameter and OTL are 13 mm (½ in) for 635-mm-(25-in-) inside-diameter shells and up, 11 mm (⁷⁄₁₆ in) for 254- through 610-mm (10- through 24-in) pipe shells, and slightly less for smaller-diameter pipe shells.

Tubes can be replaced. Tube-side headers, channel covers, gaskets, etc., are accessible for maintenance and replacement. Neither the shell-side baffle structure nor the blind gasket is accessible. During tube removal, a tube may break within the shell. When this occurs, it is most difficult to remove or to replace the tube. The usual procedure is to plug the appropriate holes in the tube sheets.

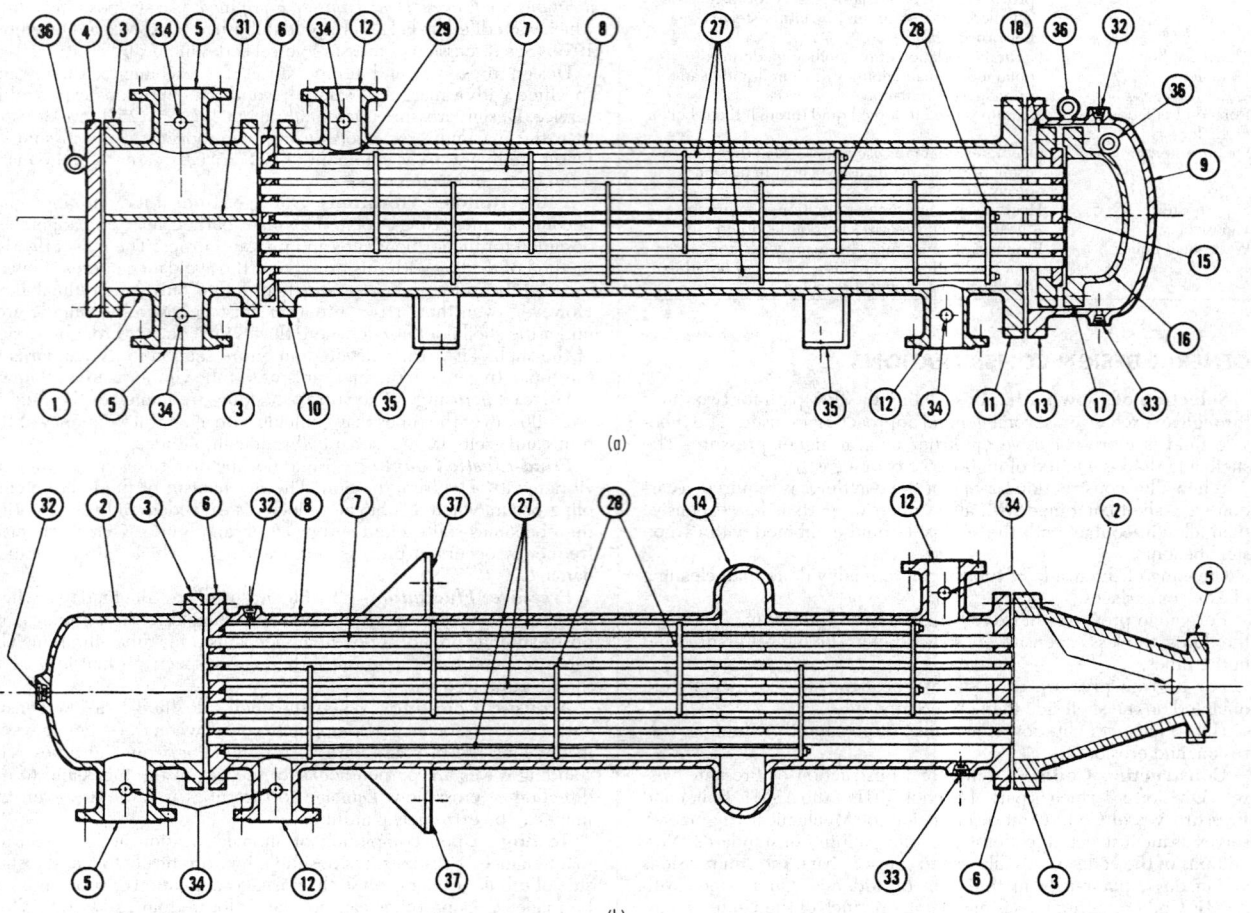

FIG. 11-36 Heat-exchanger-component nomenclature. (*a*) Internal-floating-head exchanger (with floating-head backing device). Type AES. (*b*) Fixed-tube-sheet exchanger. Type BEM. (*Standard of Tubular Exchanger Manufacturers Association, 6th ed., 1978.*)

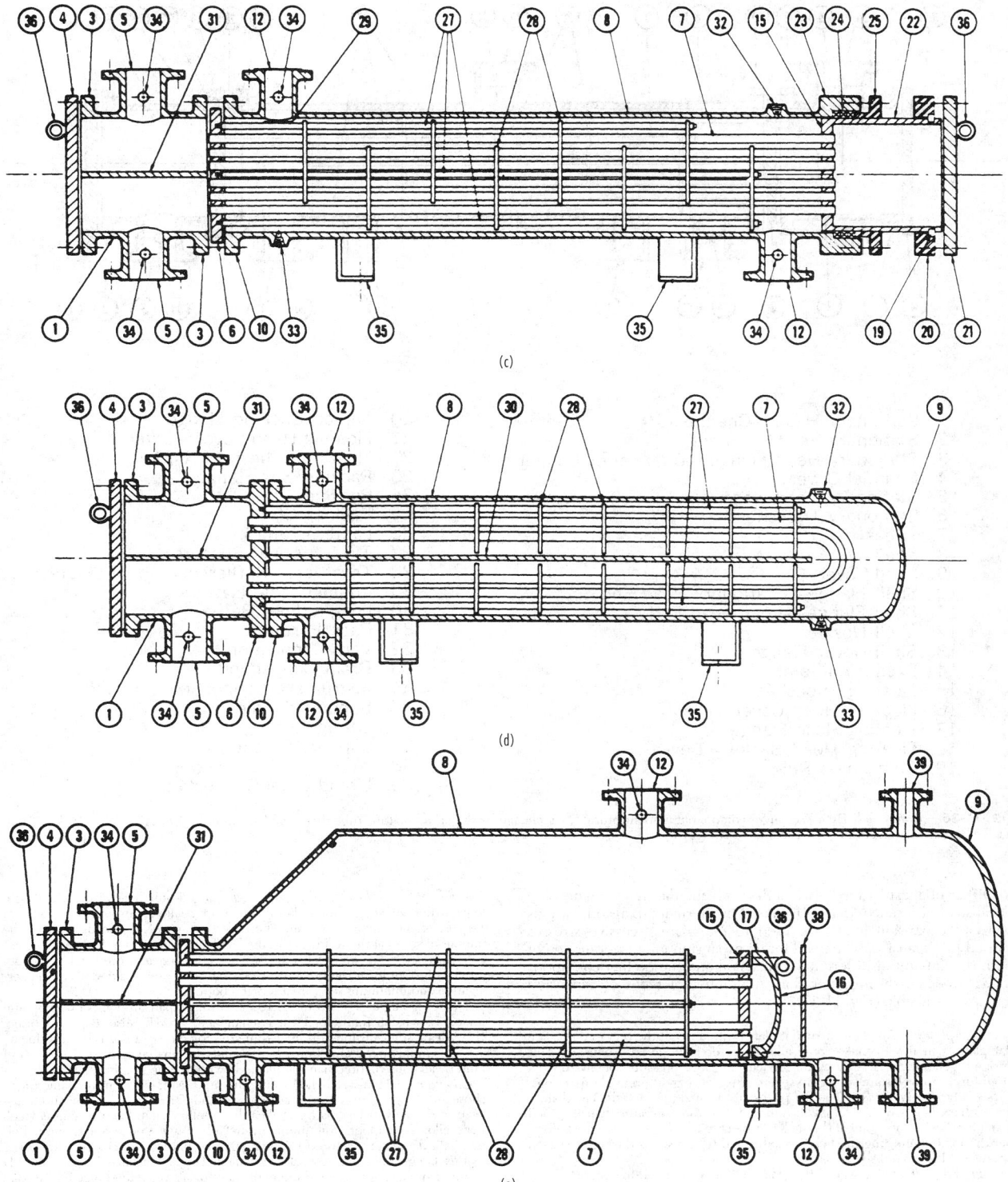

FIG. 11-36 (*Continued*) Heat-exchanger-component nomenclature. (*c*) Outside-packed floating-head exchanger. Type AEP. (*d*) U-tube heat exchanger. Type CFU. (*e*) Kettle-type floating-head reboiler. Type AKT. (*Standard of Tubular Exchanger Manufacturers Association, 6th ed., 1978.*)

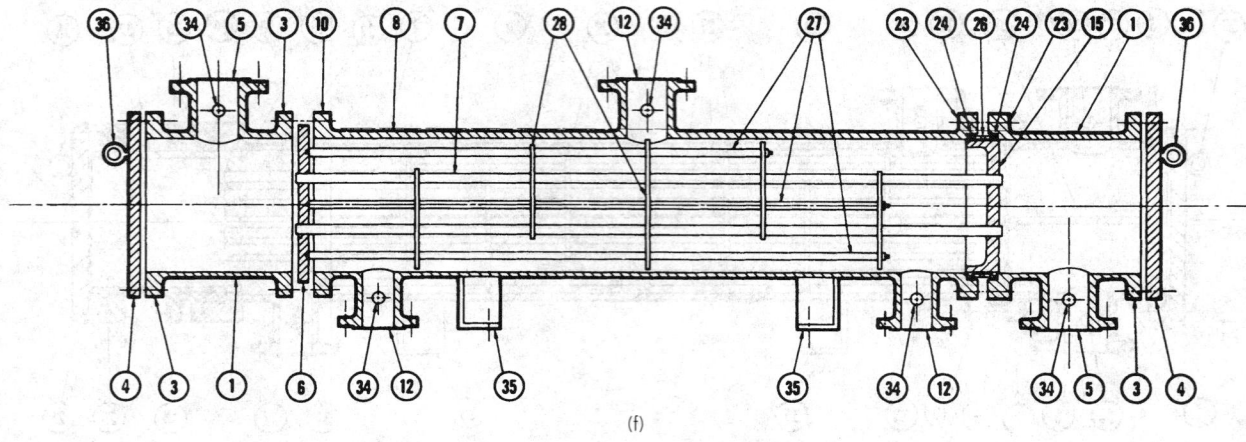

(f)

1. Stationary Head—Channel
2. Stationary Head—Bonnet
3. Stationary Head Flange—Channel or Bonnet
4. Channel Cover
5. Stationary Head Nozzle
6. Stationary Tubesheet
7. Tubes
8. Shell
9. Shell Cover
10. Shell Flange—Stationary Head End
11. Shell Flange—Rear Head End
12. Shell Nozzle
13. Shell Cover Flange
14. Expansion Joint
15. Floating Tubesheet
16. Floating Head Cover
17. Floating Head Flange
18. Floating Head Backing Device
19. Split Shear Ring

20. Slip-on Backing Flange
21. Floating Head Cover—External
22. Floating Tubesheet Skirt
23. Packing Box Flange
24. Packing
25. Packing Gland
26. Lantern Ring
27. Tie Rods and Spacers
28. Transverse Baffles or Support Plates
29. Impingement Plate
30. Longitudinal Baffle
31. Pass Partition
32. Vent Connection
33. Drain Connection
34. Instrument Connection
35. Support Saddle
36. Lifting Lug
37. Support Bracket
38. Weir
39. Liquid Level Connection

FIG. 11-36   (*Continued*) Heat-exchanger-component nomenclature. (*f*) Exchanger with packed floating tube sheet and lantern ring. Type AJW. (Standard of Tubular Exchanger Manufacturers Association, *6th ed., 1978.*)

Differential expansion between the shell and the tubes can develop because of differences in length caused by thermal expansion. Various types of expansion joints are used to eliminate excessive stresses caused by expansion. The need for an expansion joint is a function of both the amount of differential expansion and the cycling conditions to be expected during operation. A number of types of expansion joints are available (Fig. 11-37).

a. Flat plates.   Two concentric flat plates with a bar at the outer edges. The flat plates can flex to make some allowance for differential expansion. This design is generally used for vacuum service and gauge pressures below 103 kPa (15 lbf/in^2). All welds are subject to severe stress during differential expansion.

b. Flanged-only heads.   The flat plates are flanged (or curved). The diameter of these heads is generally 203 mm (8 in) or more greater than the shell diameter. The welded joint at the shell is subject to the stress referred to before, but the joint connecting the heads is subjected to less stress during expansion because of the curved shape.

c. Flared shell or pipe segments.   The shell may be flared to connect with a pipe section, or a pipe may be halved and quartered to produce a ring.

d. Formed heads.   A pair of dished-only or elliptical or flanged and dished heads can be used. These are welded together or connected by a ring. This type of joint is similar to the flanged-only-head type but apparently is subject to less stress.

e. Flanged and flued heads.   A pair of flanged-only heads is provided with concentric reverse flue holes. These heads are relatively expensive because of the cost of the fluing operation. The curved shape of the heads reduces the amount of stress at the welds to the shell and also connecting the heads.

f. Toroidal.   The toroidal joint has a mathematically predictable smooth stress pattern of low magnitude, with maximum stresses at sidewalls of the corrugation and minimum stresses at top and bottom.

The foregoing designs were discussed as ring expansion joints by Kopp and Sayre, "Expansion Joints for Heat Exchangers" (ASME Misc. Pap., vol. 6, no. 211). All are statically indeterminate but are subjected to analysis by introducing various simplifying assumptions. Some joints in current industrial use are of lighter wall construction than is indicated by the method of this paper.

g. Bellows.   Thin-wall bellows joints are produced by various manufacturers. These are designed for differential expansion and are tested for axial and transverse movement as well as for cyclical life. Bellows may be of stainless steel, nickel alloys, or copper. (Aluminum, Monel, phosphor bronze, and titanium bellows have been manufactured.) Welding nipples of the same composition as the heat-exchanger shell are generally furnished. The bellows may be hydraulically formed from a single piece of metal or may consist of welded pieces. External insulation covers of carbon steel are often provided to protect the light-gauge bellows from damage. The cover also prevents insulation from interfering with movement of the bellows (see *h*).

h. Toroidal bellows.   For high-pressure service the bellows type of joint has been modified so that movement is taken up by thin-wall small-diameter bel-

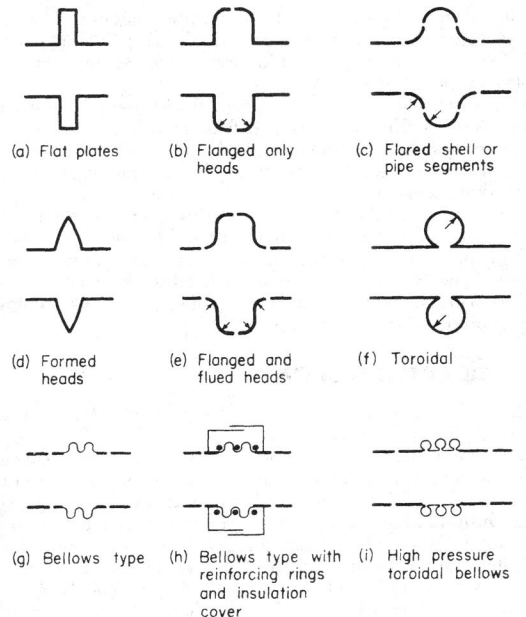

(a) Flat plates   (b) Flanged only heads   (c) Flared shell or pipe segments

(d) Formed heads   (e) Flanged and flued heads   (f) Toroidal

(g) Bellows type   (h) Bellows type with reinforcing rings and insulation cover   (i) High pressure toroidal bellows

FIG. 11-37   Expansion joints.

lows of a toroidal shape. Thickness of parts under high pressure is reduced considerably (see *f*).

Improper handling during manufacture, transit, installation, or maintenance of the heat exchanger equipped with the thin-wall-bellows type or toroidal type of expansion joint can damage the joint. In larger units these light-wall joints are particularly susceptible to damage, and some designers prefer the use of the heavier walls of formed heads.

Chemical-plant exchangers requiring expansion joints most commonly have used the flanged-and-flued-head type. There is a trend toward more common use of the light-wall-bellows type.

U-Tube Heat Exchanger (Fig. 11-36*d*)   The tube bundle consists of a stationary tube sheet, U tubes (or hairpin tubes), baffles or support plates, and appropriate tie rods and spacers. The tube bundle can be removed from the heat-exchanger shell. A tube-side header (stationary head) and a shell with integral shell cover, which is welded to the shell, are provided. Each tube is free to expand or contract without any limitation being placed upon it by the other tubes.

The U-tube bundle has the advantage of providing minimum clearance between the outer tube limit and the inside of the shell for any of the removable-tube-bundle constructions. Clearances are of the same magnitude as for fixed-tube-sheet heat exchangers.

The number of tube holes in a given shell is less than that for a fixed-tube-sheet exchanger because of limitations on bending tubes of a very short radius.

The U-tube design offers the advantage of reducing the number of joints. In high-pressure construction this feature becomes of considerable importance in reducing both initial and maintenance costs. The use of U-tube construction has increased significantly with the development of hydraulic tube cleaners, which can remove fouling residues from both the straight and the U-bend portions of the tubes.

Mechanical cleaning of the inside of the tubes was described by John [*Chem. Eng., 66,* 187–192 (Dec. 14, 1959)]. Rods and conventional mechanical tube cleaners cannot pass from one end of the U tube to the other. Power-driven tube cleaners, which can clean both the straight legs of the tubes and the bends, are available.

Hydraulic jetting with water forced through spray nozzles at high pressure for cleaning tube interiors and exteriors of removal bundles is reported by Canaday ("Hydraulic Jetting Tools for Cleaning Heat Exchangers," ASME Pap. 58-A-217, unpublished).

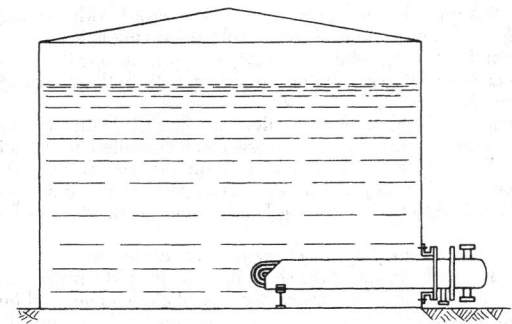

FIG. 11-38   Tank suction heater.

The tank suction heater, as illustrated in Fig. 11-38, contains a U-tube bundle. This design is often used with outdoor storage tanks for heavy fuel oils, tar, molasses, and similar fluids whose viscosity must be lowered to permit easy pumping. Uusally the tube-side heating medium is steam. One end of the heater shell is open, and the liquid being heated passes across the outside of the tubes. Pumping costs can be reduced without heating the entire contents of the tank. Bare tube and integral low-fin tubes are provided with baffles. Longitudinal fin-tube heaters are not baffled. Fins are most often used to minimize the fouling potential in these fluids.

Kettle-type reboilers, evaporators, etc., are often U-tube exchangers with enlarged shell sections for vapor-liquid separation. The U-tube bundle replaces the floating-heat bundle of Fig. 11-36*e*.

The U-tube exchanger with copper tubes, cast-iron header, and other parts of carbon steel is used for water and steam services in office buildings, schools, hospitals, hotels, etc. Nonferrous tube sheets and admiralty or 90-10 copper-nickel tubes are the most frequently used substitute materials. These standard exchangers are available from a number of manufacturers at costs far below those of custom-built process-industry equipment.

Packed-Lantern-Ring Exchanger (Fig. 11-36*f*)   This construction is the least costly of the straight-tube removable bundle types. The shell- and tube-side fluids are each contained by separate rings of packing separated by a lantern ring and are installed at the floating tube sheet. The lantern ring is provided with weep holes. Any leakage passing the packing goes through the weep holes and then drops to the ground. Leakage at the packing will not result in mixing within the exchanger of the two fluids.

The width of the floating tube sheet must be great enough to allow for the packings, the lantern ring, and differential expansion. Sometimes a small skirt is attached to a thin tube sheet to provide the required bearing surface for packings and lantern ring.

The clearance between the outer tube limit and the inside of the shell is slightly larger than that for fixed-tube-sheet and U-tube exchangers. The use of a floating-tube-sheet skirt increases this clearance. Without the skirt the clearance must make allowance for tube-hole distortion during tube rolling near the outside edge of the tube sheet or for tube-end welding at the floating tube sheet.

The packed-lantern-ring construction is generally limited to design temperatures below 191°C (375°F) and to the mild services of water, steam, air, lubricating oil, etc. Design gauge pressure does not exceed 2068 kPa (300 lbf/in²) for pipe shell exchangers and is limited to 1034 kPa (150 lbf/in²) for 610- to 1067-mm- (24- to 42-in-) diameter shells.

Outside-Packed Floating-Head Exchanger (Fig. 11-36*c*)   The shell-side fluid is contained by rings of packing, which are compressed within a stuffing box by a packing follower ring. This construction was frequently used in the chemical industry, but in recent years usage has decreased. The removable-bundle construction accommodates differential expansion between shell and tubes and is used for shell-side service up to 4137 kPa gauge pressure (600 lbf/in²) at 316°C (600°F). There are no limitations upon the number of tube-side passes or upon the tube-side design pressure and temperature. The outside-packed floating-head exchanger was the most commonly used type of removable-bundle construction in chemical-plant service.

The floating-tube-sheet skirt, where in contact with the rings of packing, has fine machine finish. A split shear ring is inserted into a groove in the floating-tube-sheet skirt. A slip-on backing flange, which in service is held in place by the shear ring, bolts to the external floating-head cover.

The floating-head cover is usually a circular disk. With an odd number of tube-side passes, an axial nozzle can be installed in such a floating-head cover. If a side nozzle is required, the circular disk is replaced by either a dished head or a channel barrel (similar to Fig. 11-36*f*) bolted between floating-head cover and floating-tube-sheet skirt.

The outer tube limit approaches the inside of the skirt but is farther removed from the inside of the shell than for any of the previously discussed constructions. Clearances between shell diameter and bundle OTL are 22 mm (⅞ in) for small-diameter pipe shells, 44 mm (1¾ in) for large-diameter pipe shells, and 58 mm (2 1/16 in) for moderate-diameter plate shells.

Internal Floating-Head Exchanger (Fig. 11-36*a*) The internal floating-head design is used extensively in petroleum-refinery service, but in recent years there has been a decline in usage.

The tube bundle is removable, and the floating tube sheet moves (or floats) to accommodate differential expansion between shell and tubes. The outer tube limit approaches the inside diameter of the gasket at the floating tube sheet. Clearances (between shell and OTL) are 29 mm (1⅛ in) for pipe shells and 37 mm (1 7/16 in) for moderate-diameter plate shells.

A split backing ring and bolting usually hold the floating-head cover at the floating tube sheet. These are located beyond the end of the shell and within the larger-diameter shell cover. Shell cover, split backing ring, and floating-head cover must be removed before the tube bundle can pass through the exchanger shell.

With an even number of tube-side passes the floating-head cover serves as return cover for the tube-side fluid. With an odd number of passes a nozzle pipe must extend from the floating-head cover through the shell cover. Provision for both differential expansion and tube-bundle removal must be made.

Pull-Through Floating-Head Exchanger (Fig. 11-36*e*) Construction is similar to that of the internal-floating-head split-backing-ring exchanger except that the floating-head cover bolts directly to the floating tube sheet. The tube bundle can be withdrawn from the shell without removing either shell cover or floating-head cover. This feature reduces maintenance time during inspection and repair.

The large clearance between the tubes and the shell must provide for both the gasket and the bolting at the floating-head cover. This clearance is about 2 to 2½ times that required by the split-ring design. Sealing strips or dummy tubes are often installed to reduce bypassing of the tube bundle.

Falling-Film Exchangers Falling-film shell-and-tube heat exchangers have been developed for a wide variety of services and are described by Sack [*Chem. Eng. Prog.,* **63**, 55 (July 1967)]. The fluid enters at the top of the vertical tubes. Distributors or slotted tubes put the liquid in film flow in the inside surface of the tubes, and the film adheres to the tube surface while falling to the bottom of the tubes. The film can be cooled, heated, evaporated, or frozen by means of the proper heat-transfer medium outside the tubes. Tube distributors have been developed for a wide range of applications. Fixed tube sheets, with or without expansion joints, and outside-packed-head designs are used.

Principal advantages are high rate of heat transfer, no internal pressure drop, short time of contact (very important for heat-sensitive materials), easy accessibility to tubes for cleaning, and, in some cases, prevention of leakage from one side to another.

These falling-film exchangers are used in various services as described in the following paragraphs.

Liquid Coolers and Condensers Dirty water can be used as the cooling medium. The top of the cooler is open to the atmosphere for access to tubes. These can be cleaned without shutting down the cooler by removing the distributors one at a time and scrubbing the tubes.

Evaporators These are used extensively for the concentration of ammonium nitrate, urea, and other chemicals sensitive to heat when minimum contact time is desirable. Air is sometimes introduced in the tubes to lower the partial pressure of liquids whose boiling points are high. These evaporators are built for pressure or vacuum and with top or bottom vapor removal.

Absorbers These have a two-phase flow system. The absorbing medium is put in film flow during its fall downward on the tubes as it is cooled by a cooling medium outside the tubes. The film absorbs the gas which is introduced into the tubes. This operation can be cocurrent or countercurrent.

Freezers By cooling the falling film to its freezing point, these exchangers convert a variety of chemicals to the solid phase. The most common application is the production of sized ice and paradichlorobenzene. Selective freezing is used for isolating isomers. By melting the solid material and refreezing in several stages, a higher degree of purity of product can be obtained.

TUBE-SIDE CONSTRUCTION

Tube-Side Header The tube-side header (or stationary head) contains one or more flow nozzles.

The **bonnet** (Fig. 11-35B) bolts to the shell. It is necessary to remove the bonnet in order to examine the tube ends. The fixed-tube-sheet exchanger of Fig. 11-36*b* has bonnets at both ends of the shell.

The **channel** (Fig. 11-35A) has a removable channel cover. The tube ends can be examined by removing this cover without disturbing the piping connections to the channel nozzles. The channel can bolt to the shell as shown in Fig. 11-36*a* and *c*. The Type C and Type N channels of Fig. 11-35 are welded to the tube sheet. This design is comparable in cost with the bonnet but has the advantages of permitting access to the tubes without disturbing the piping connections and of eliminating a gasketed joint.

Special High-Pressure Closures (Fig. 11-35D) The channel barrel and the tube sheet are generally forged. The removable channel cover is seated in place by hydrostatic pressure, while a shear ring subjected to shearing stress absorbs the end force. For pressures above 6205 kPa (900 lbf/in²) these designs are generally more economical than bolted constructions, which require larger flanges and bolting as pressure increases in order to contain the end force with bolts in tension. Relatively light-gauge internal pass partitions are provided to direct the flow of tube-side fluids but are designed only for the differential pressure across the tube bundle.

Tube-Side Passes Most exchangers have an even number of tube-side passes. The fixed-tube-sheet exchanger (which has no shell cover) usually has a return cover without any flow nozzles as shown in Fig. 11-35M; Types L and N are also used. All removable-bundle designs (except for the U tube) have a floating-head cover directing the flow of tube-side fluid at the floating tube sheet.

Tubes Standard heat-exchanger tubing is ¼, ⅜, ½, ⅝, ¾, 1, 1¼, and 1½ in in outside diameter (in × 25.4 = mm). Wall thickness is measured in Birmingham wire gauge (BWG) units. (A comprehensive list of tubing characteristics and sizes is given in section 9, table D-7 of TEMA.) The most commonly used tubes in chemical plants and petroleum refineries are 19- and 25-mm (¾- and 1-in) outside diameter. Standard tube lengths are 8, 10, 12, 16, and 20 ft, with 20 ft now the most common (ft × 0.3048 = m).

Manufacturing tolerances for steel, stainless-steel, and nickel-alloy tubes are such that the tubing is produced to either average or minimum wall thickness. Seamless carbon steel tube of minimum wall thickness may vary from 0 to 20 percent above the nominal wall thickness. Average-wall seamless tubing has an allowable variation of plus or minus 10 percent. Welded carbon steel tube is produced to closer tolerances (0 to plus 18 percent on minimum wall; plus or minus 9 percent on average wall). Tubing of aluminum, copper, and their alloys can be drawn easily and usually is made to minimum wall specifications.

Common practice is to specify **exchanger surface** in terms of total external square feet of tubing. The effective outside heat-transfer surface is based on the length of tubes measured between the inner faces of tube sheets. In most heat exchangers there is little difference between the total and the effective surface. Significant differences are usually found in high-pressure and double-tube-sheet designs.

Integrally finned tube, which is available in a variety of alloys and sizes, is being used in shell-and-tube heat exchangers. The fins are radially extruded from thick-walled tube to a height of 1.6 mm (⅟16 in) spaced at 1.33 mm (19 fins per inch) or to a height of 3.2 mm (⅛ in) spaced at 2.3 mm (11 fins per inch). External surface is approximately 2½ times the outside surface of a bare tube with the same outside diameter. Also available are 0.93-mm- (0.037-in-) high fins spaced 0.91 mm (28 fins per inch) with an external surface about 3.5 times the surface of the bare tube. Bare ends of nominal tube diameter are provided, while the fin height is slightly less than this diameter. The tube can be inserted into a conventional tube bundle and rolled or welded to the tube sheet by the same means, used for bare tubes. An integrally finned tube rolled into a tube sheet with double serrations and flared at the inlet is shown in Fig. 11-39. Internally finned tubes have been manufactured but have limited application.

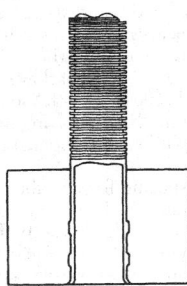

FIG. 11-39 Integrally finned tube rolled into tube sheet with double serrations and flared inlet. (*Woverine Division, UOP, Inc.*)

Longitudinal fins are commonly used in double-pipe exchangers upon the outside of the inner tube. U-tube and conventional removable tube bundles are also made from such tubing. The ratio of external to internal surface generally is about 10 or 15:1.

Transverse fins upon tubes are used in low-pressure gas services. The primary application is in air-cooled heat exchangers (as discussed under that heading), but shell-and-tube exchangers with these tubes are in service.

Rolled Tube Joints Expanded tube-to-tube-sheet joints are standard. Properly rolled joints have uniform tightness to minimize tube fractures, stress corrosion, tube-sheet ligament pushover and enlargement, and dishing of the tube sheet. Tubes are expanded into the tube sheet for a length of two tube diameters, or 50 mm (2 in), or tube-sheet thickness minus 3 mm (⅛ in). Generally tubes are rolled for the last of these alternatives. The expanded portion should never extend beyond the shell-side face of the tube sheet, since removing such a tube is extremely difficult. Methods and tools for tube removal and tube rolling were discussed by John [*Chem. Eng.,* **66,** 77–80 (Dec. 28, 1959)], and rolling techniques by Bach [*Pet. Refiner,* **39,** 8, 104 (1960)].

Tube ends may be projecting, flush, flared, or beaded (listed in order of usage). The flare or bell-mouth tube end is usually restricted to water service in condensers and serves to reduce erosion near the tube inlet.

For moderate general process requirements at gauge pressures less than 2058 kPa (300 lbf/in²) and less than 177°C (350°F), tube-sheet holes without grooves are standard. For all other services with expanded tubes at least two grooves in each tube hole are common. The number of grooves is sometimes changed to one or three in proportion to tube-sheet thickness.

Expanding the tube into the **grooved tube holes** provides a stronger joint but results in greater difficulties during tube removal.

Welded Tube Joints When suitable materials of construction are used, the tube ends may be welded to the tube sheets. Welded joints may be seal-welded "for additional tightness beyond that of tube rolling" or may be strength-welded. Strength-welded joints have been found satisfactory in very severe services. Welded joints may or may not be rolled before or after welding.

The variables in tube-end welding were discussed in two unpublished papers (Emhardt, "Heat Exchanger Tube-to-Tubesheet Joints," ASME Pap. 69-WA/HT-47; and Reynolds, "Tube Welding for Conventional and Nuclear Power Plant Heat Exchangers," ASME Pap. 69-WA/HT-24), which were presented at the November 1969 meeting of the American Society of Mechanical Engineers.

Tube-end rolling before welding may leave lubricant from the tube expander in the tube hole. Fouling during normal operation followed by maintenance operations will leave various impurities in and near the tube ends. Satisfactory welds are rarely possible under such conditions, since tube-end welding requires extreme cleanliness in the area to be welded.

Tube **expansion after welding** has been found useful for low and moderate pressures. In high-pressure service tube rolling has not been able to prevent leakage after weld failure.

Double-Tube-Sheet Joints This design prevents the passage of either fluid into the other because of leakage at the tube-to-tube-sheet joints, which are generally the weakest points in heat exchangers. Any leakage at these joints admits the fluid to the gap between the tube sheets. Mechanical design, fabrication, and maintenance of double-tube-sheet designs require special consideration.

SHELL-SIDE CONSTRUCTION

Shell Sizes Heat-exchanger shells are generally made from standard-wall steel pipe in sizes up to 305-mm (12-in) diameter; from 9.5-mm (⅜-in) wall pipe in sizes from 356 to 610 mm (14 to 24 in); and from steel plate rolled at discrete intervals in larger sizes. Clearances between the outer tube limit and the shell are discussed elsewhere in connection with the different types of construction.

The following formulae may be used to estimate tube counts for various bundle sizes and tube passes. The estimated values include the removal of tubes to provide an entrance area for shell nozzle sizes of one-fifth the shell diameter. Due to the large effect from other parameters such as design pressure/corrosion allowance, baffle cuts, seal strips, and so on, these are to be used as estimates only. Exact tube counts are part of the design package of most reputable exchanger design software and are normally used for the final design.

Triangular tube layouts with pitch equal to 1.25 times the tube outside diameter:

$$C = 0.75 \, (D/d) - 36; \text{ where } D = \text{Bundle O.D.} \, d = \text{Tube O.D.}$$

Range of accuracy: $-24 \le C \le 24$.

$$1 \text{ Tube Pass: } N_t = 1298. + 74.86C + 1.283C^2 - .0078C^3 - .0006C^4 \quad (11\text{-}74a)$$
$$2 \text{ Tube Pass: } N_t = 1266. + 73.58C + 1.234C^2 - .0071C^3 - .0005C^4 \quad (11\text{-}74b)$$
$$4 \text{ Tube Pass: } N_t = 1196. + 70.79C + 1.180C^2 - .0059C^3 - .0004C^4 \quad (11\text{-}74c)$$
$$6 \text{ Tube Pass: } N_t = 1166. + 70.72C + 1.269C^2 - .0074C^3 - .0006C^4 \quad (11\text{-}74d)$$

Square tube layouts with pitch equal to 1.25 times the tube outside diameter:

$$C = (D/d) - 36.; \text{ where } D = \text{Bundle O.D.} \, d = \text{Tube O.D.}$$

Range of accuracy: $-24 \le C \le 24$.

$$1 \text{ Tube Pass: } N_t = 593.6 + 33.52C + .3782C^2 - .0012C^3 + .0001C^4 \quad (11\text{-}75a)$$
$$2 \text{ Tube Pass: } N_t = 578.8 + 33.36C + .3847C^2 - .0013C^3 + .0001C^4 \quad (11\text{-}75b)$$
$$4 \text{ Tube Pass: } N_t = 562.0 + 33.04C + .3661C^2 - .0016C^3 + .0002C^4 \quad (11\text{-}75c)$$
$$6 \text{ Tube Pass: } N_t = 550.4 + 32.49C + .3873C^2 - .0013C^3 + .0001C^4 \quad (11\text{-}75d)$$

Shell-Side Arrangements The **one-pass shell** (Fig. 11-35E) is the most commonly used arrangement. Condensers from single component vapors often have the nozzles moved to the center of the shell for vacuum and steam services.

A solid longitudinal baffle is provided to form a two-pass shell (Fig. 11-35F). It may be insulated to improve thermal efficiency. (See further discussion on baffles). A two-pass shell can improve thermal effectiveness at a cost lower than for two shells in series.

For **split flow** (Fig. 11-35G), the longitudinal baffle may be solid or perforated. The latter feature is used with condensing vapors.

A **double-split-flow** design is shown in Fig. 11-35H. The longitudinal baffles may be solid or perforated.

The **divided flow** design (Fig. 11-35J), mechanically is like the one-pass shell except for the addition of a nozzle. Divided flow is used to meet low-pressure-drop requirements.

The **kettle reboiler** is shown in Fig. 11-35K. When nucleate boiling is to be done on the shell-side, this common design provides adequate dome space for separation of vapor and liquid above the tube bundle and surge capacity beyond the weir near the shell cover.

BAFFLES AND TUBE BUNDLES

The **tube bundle** is the most important part of a tubular heat exchanger. The tubes generally constitute the most expensive component of the exchanger and are the one most likely to corrode. Tube sheets, baffles, or support plates, tie rods, and usually spacers complete the bundle.

Minimum **baffle spacing** is generally one-fifth of the shell diameter and not less than 50.8 mm (2 in). Maximum baffle spacing is limited by the requirement to provide adequate support for the tubes. The maximum unsupported tube span in inches equals $74\,d^{0.75}$ (where d is the outside tube diameter in inches). The unsupported tube span is reduced by about 12 percent for aluminum, copper, and their alloys.

Baffles are provided for heat-transfer purposes. When shell-side baffles are not required for heat-transfer purposes, as may be the case in condensers or reboilers, tube supports are installed.

Segmental Baffles   Segmental or cross-flow baffles are standard. Single, double, and triple segmental baffles are used. Baffle cuts are illustrated in Fig. 11-40. The double segmental baffle reduces cross-flow velocity for a given baffle spacing. The triple segmental baffle reduces both cross-flow and long-flow velocities and has been identified as the "window-cut" baffle.

Baffle cuts are expressed as the ratio of segment opening height to shell inside diameter. Cross-flow baffles with horizontal cut are shown in Fig. 11-36a, c, and f. This arrangement is not satisfactory for horizontal condensers, since the condensate can be trapped between baffles, or for dirty fluids in which the dirt might settle out. Vertical-cut baffles are used for side-to-side flow in horizontal exchangers with condensing fluids or dirty fluids. Baffles are notched to assure complete drainage when the units are taken out of service. (These notches permit some bypassing of the tube bundle during normal operation.)

Tubes are most commonly arranged on an equilateral triangular pitch. Tubes are arranged on a square pitch primarily for mechanical cleaning purposes in removable-bundle exchangers.

Maximum baffle cut is limited to about 45 percent for single segmental baffles so that every pair of baffles will support each tube. Tube bundles are generally provided with baffles cut so that at least one row of tubes passes through all the baffles or support plates. These tubes hold the entire bundle together. In pipe-shell exchangers with a horizontal baffle cut and a horizontal pass rib for directing tube-side flow in the channel, the maximum baffle cut, which permits a minimum of one row of tubes to pass through all baffles, is approximately 33 percent in small shells and 40 percent in larger pipe shells.

Maximum shell-side heat-transfer rates in forced convection are apparently obtained by cross-flow of the fluid at right angles to the tubes. In order to maximize this type of flow some heat exchangers are built with segmental-cut baffles and with "no tubes in the window" (or the baffle cutout). Maximum baffle spacing may thus equal maximum unsupported-tube span, while conventional baffle spacing is limited to one-half of this span.

The maximum baffle spacing for no tubes in the window of single segmental baffles is unlimited when intermediate supports are provided. These are cut on both sides of the baffle and therefore do not affect the flow of the shell-side fluid. Each support engages all the tubes; the supports are spaced to provide adequate support for the tubes.

Rod Baffles   Rod or bar baffles have either rods or bars extending through the lanes between rows of tubes. A baffle set can consist of a baffle with rods in all the vertical lanes and another baffle with

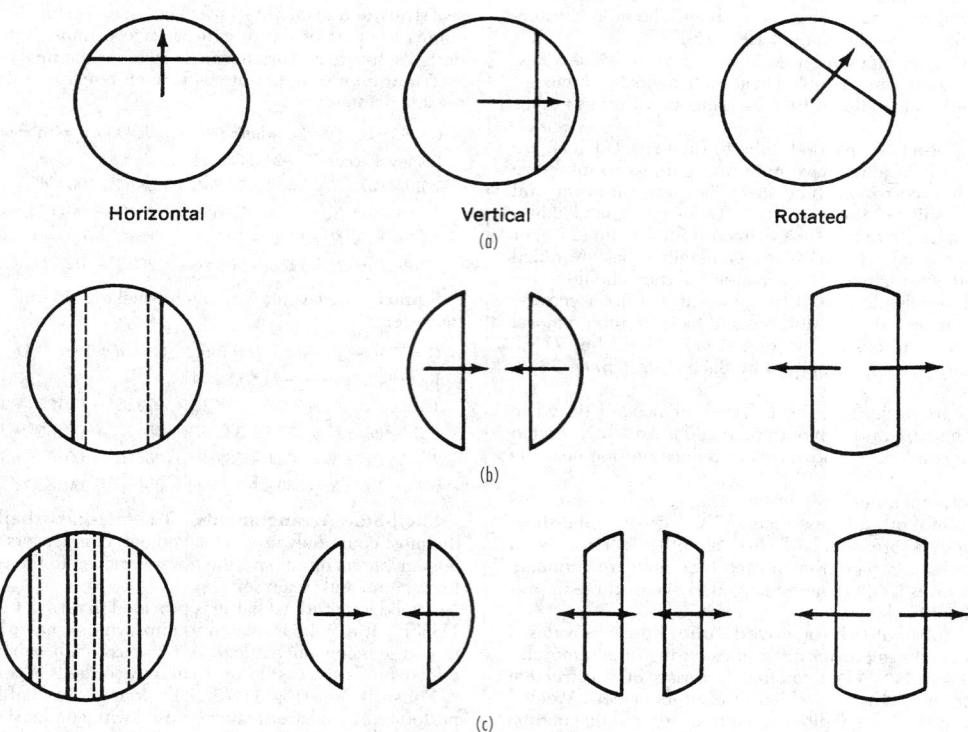

Horizontal   Vertical   Rotated
                                      (a)

(b)

(c)

FIG. 11-40   Baffle cuts. (*a*) Baffle cuts for single segmental baffles. (*b*) Baffle cuts for double segmental baffles. (*c*) Baffle cuts for triple segmental baffles.

rods in all the horizontal lanes between the tubes. The shell-side flow is uniform and parallel to the tubes. Stagnant areas do not exist.

One device uses four baffles in a baffle set. Only half of either the vertical or the horizontal tube lanes in a baffle have rods. The new design apparently provides a maximum shell-side heat-transfer coefficient for a given pressure drop.

Tie Rods and Spacers   Tie rods are used to hold the baffles in place with spacers, which are pieces of tubing or pipe placed on the rods to locate the baffles. Occasionally baffles are welded to the tie rods, and spacers are eliminated. Properly located tie rods and spacers serve both to hold the bundle together and to reduce bypassing of the tubes.

In very large fixed-tube-sheet units, in which concentricity of shells decreases, baffles are occasionally welded to the shell to eliminate bypassing between the baffle and the shell.

Metal baffles are standard. Occasionally plastic baffles are used either to reduce corrosion or in vibratory service, in which metal baffles may cut the tubes.

Impingement Baffle   The tube bundle is customarily protected against impingement by the incoming fluid at the shell inlet nozzle when the shell-side fluid is at a high velocity, is condensing, or is a two-phase fluid. Minimum entrance area about the nozzle is generally equal to the inlet nozzle area. Exit nozzles also require adequate area between the tubes and the nozzles. A full bundle without any provision for shell inlet nozzle area can increase the velocity of the inlet fluid by as much as 300 percent with a consequent loss in pressure.

Impingement baffles are generally made of rectangular plate, although circular plates are more desirable. Rods and other devices are sometimes used to protect the tubes from impingement. In order to maintain a maximum tube count the impingement plate is often placed in a conical nozzle opening or in a dome cap above the shell.

Impingement baffles or flow-distribution devices are recommended for axial tube-side nozzles when entrance velocity is high.

Vapor Distribution   Relatively large shell inlet nozzles, which may be used in condensers under low pressure or vacuum, require provision for uniform vapor distribution.

Tube-Bundle Bypassing   Shell-side heat-transfer rates are maximized when bypassing of the tube bundle is at a minimum. The most significant bypass stream is generally between the outer tube limit and the inside of the shell. The clearance between tubes and shell is at a minimum for fixed-tube-sheet construction and is greatest for straight-tube removable bundles.

Arrangements to reduce tube-bundle bypassing include:

Dummy tubes.   These tubes do not pass through the tube sheets and can be located close to the inside of the shell.

Tie rods with spacers.   These hold the baffles in place but can be located to prevent bypassing.

Sealing strips.   These longitudinal strips either extend from baffle to baffle or may be inserted in slots cut into the baffles.

Dummy tubes or tie rods with spacers may be located within the pass partition lanes (and between the baffle cuts) in order to ensure maximum bundle penetration by the shell-side fluid.

When tubes are omitted from the tube layout to provide entrance area about an impingement plate, the need for sealing strips or other devices to cause proper bundle penetration by the shell-side fluid is increased.

Longitudinal Flow Baffles   In fixed-tube-sheet construction with multipass shells, the baffle is usually welded to the shell and positive assurance against bypassing results. Removable tube bundles have a sealing device between the shell and the longitudinal baffle. Flexible light-gauge sealing strips and various packing devices have been used. Removable U-tube bundles with four tube-side passes and two shell-side passes can be installed in shells with the longitudinal baffle welded in place.

In split-flow shells the longitudinal baffle may be installed without a positive seal at the edges if design conditions are not seriously affected by a limited amount of bypassing.

Fouling in petroleum-refinery service has necessitated rough treatment of tube bundles during cleaning operations. Many refineries avoid the use of longitudinal baffles, since the sealing devices are subject to damage during cleaning and maintenance operations.

CORROSION IN HEAT EXCHANGERS

Some of the special considerations in regard to heat-exchanger corrosion are discussed in this subsection. A more extended presentation in Sec. 23 covers corrosion and its various forms as well as materials of construction.

Materials of Construction   The most common material of construction for heat exchangers is carbon steel. Stainless-steel construction throughout is sometimes used in chemical-plant service and on rare occasions in petroleum refining. Many exchangers are constructed from dissimilar metals. Such combinations are functioning satisfactorily in certain services. Extreme care in their selection is required since electrolytic attack can develop.

Carbon steel and alloy combinations appear in Table 11-12 "Alloys" in chemical- and petrochemical-plant service in approximate order of use are stainless-steel series 300, nickel, Monel, copper alloy, aluminum, Inconel, stainless-steel series 400, and other alloys. In petroleum-refinery service the frequency order shifts, with copper alloy (for water-cooled units) in first place and low-alloy steel in second place. In some segments of the petroleum industry copper alloy, stainless series 400, low-alloy steel, and aluminum are becoming the most commonly used alloys.

Copper-alloy tubing, particularly inhibited admiralty, is generally used with cooling water. Copper-alloy tube sheets and baffles are generally of naval brass.

Aluminum alloy (and in particular alclad aluminum) tubing is sometimes used in water service. The alclad alloy has a sacrificial aluminum-alloy layer metallurgically bonded to a core alloy.

Tube-side headers for water service are made in a wide variety of materials: carbon steel, copper alloy, cast iron, and lead-lined or plastic-lined or specially painted carbon steel.

Bimetallic Tubes   When corrosive requirements or temperature conditions do not permit the use of a single alloy for the tubes, bimetallic (or duplex) tubes may be used. These can be made from almost any possible combination of metals. Tube sizes and gauges can be varied. For thin gauges the wall thickness is generally divided equally between the two components. In heavier gauges the more expensive component may comprise from a fifth to a third of the total thickness.

The component materials comply with applicable ASTM specifications, but after manufacture the outer component may increase in hardness beyond specification limits, and special care is required during the tube-rolling operation. When the harder material is on the outside, precautions must be exercised to expand the tube properly. When the inner material is considerably softer, rolling may not be practical unless ferrules of the soft material are used.

In order to eliminate galvanic action the outer tube material may be stripped from the tube ends and replaced with ferrules of the inner tube material. When the end of a tube with a ferrule is expanded or welded to a tube sheet, the tube-side fluid can contact only the inner tube material, while the outer material is exposed to the shell-side fluid.

Bimetallic tubes are available from a small number of tube mills and are manufactured only on special order and in large quantities.

TABLE 11-12   Dissimilar Materials in Heat-Exchanger Construction

Part	Relative use	1	2	3	4	5	6
	Relative cost	A	B	C	D	C	E
Tubes		●	●	●	●	●	●
Tube sheets			●		●	●	●
Tube-side headers				●	●		●
Baffles					●	●	●
Shell							●

Carbon steel replaced by an alloy when ● appears.
Relative use: from 1 (most popular) through 6 (least popular) combinations.
Relative cost: from A (least expensive) to E (most expensive).

Clad Tube Sheets Usually tube sheets and other exchanger parts are of a solid metal. Clad or bimetallic tube sheets are used to reduce costs or because no single metal is satisfactory for the corrosive conditions. The alloy material (e.g., stainless steel, Monel) is generally bonded or clad to a carbon steel backing material. In fixed-tube-sheet construction a copper-alloy-clad tube sheet can be welded to a steel shell, while most copper-alloy tube sheets cannot be welded to steel in a manner acceptable to ASME Code authorities.

Clad tube sheets in service with carbon steel backer material include stainless-steel types 304, 304L, 316, 316L, and 317, Monel, Inconel, nickel, naval rolled brass, copper, admiralty, silicon bronze, and titanium. Naval rolled brass and Monel clad on stainless steel are also in service.

Ferrous-alloy-clad tube sheets are generally prepared by a weld overlay process in which the alloy material is deposited by welding upon the face of the tube sheet. Precautions are required to produce a weld deposit free of defects, since these may permit the process fluid to attack the base metal below the alloy. Copper-alloy-clad tube sheets are prepared by brazing the alloy to the carbon steel backing material.

Clad materials can be prepared by bonding techniques, which involve rolling, heat treatment, explosive bonding, etc. When properly manufactured, the two metals do not separate because of thermal-expansion differences encountered in service. Applied tube-sheet facings prepared by tack welding at the outer edges of alloy and base metal or by bolting together the two metals are in limited use.

Nonmetallic Construction Shell-and-tube exchangers with glass tubes 14 mm (0.551 in) in diameter and 1 mm (0.039 in) thick with tube lengths from 2.015 m (79.3 in) to 4.015 m (158 in) are available. Steel shell exchangers have a maximum design pressure of 517 kPa (75 lbf/in²). Glass shell exchangers have a maximum design gauge pressure of 103 kPa (15 lbf/in²). Shell diameters are 229 mm (9 in), 305 mm (12 in), and 457 mm (18 in). Heat-transfer surface ranges from 3.16 to 51 m² (34 to 550 ft²). Each tube is free to expand, since a Teflon sealer sheet is used at the tube-to-tube-sheet joint.

Impervious graphite heat-exchanger equipment is made in a variety of forms, including outside-packed-head shell-and-tube exchangers. They are fabricated with impervious graphite tubes and tube-side headers and metallic shells. Single units containing up to 1300 m² (14,000 ft²) of heat-transfer surface are available.

Teflon heat exchangers of special construction are described later in this section.

Fabrication Expanding the tube into the tube sheet reduces the tube wall thickness and work-hardens the metal. The induced stresses can lead to **stress corrosion.** Differential expansion between tubes and shell in fixed-tube-sheet exchangers can develop stresses, which lead to stress corrosion.

When austenitic stainless-steel tubes are used for corrosion resistance, a close fit between the tube and the tube hole is recommended in order to minimize work hardening and the resulting loss of corrosion resistance.

In order to facilitate removal and replacement of tubes it is customary to roller-expand the tubes to within 3 mm (⅛ in) of the shell-side face of the tube sheet. A 3-mm- (⅛-in-) long gap is thus created between the tube and the tube hole at this tube-sheet face. In some services this gap has been found to be a focal point for corrosion.

It is standard practice to provide a chamfer at the inside edges of tube holes in tube sheets to prevent cutting of the tubes and to remove burrs produced by drilling or reaming the tube sheet. In the lower tube sheet of vertical units this chamfer serves as a pocket to collect material, dirt, etc., and to serve as a corrosion center.

Adequate venting of exchangers is required both for proper operation and to reduce corrosion. Improper venting of the water side of exchangers can cause alternate wetting and drying and accompanying chloride concentration, which is particularly destructive to the series 300 stainless steels.

Certain corrosive conditions require that special consideration be given to complete drainage when the unit is taken out of service. Particular consideration is required for the upper surfaces of tube sheets in vertical heat exchangers, for sagging tubes, and for shell-side baffles in horizontal units.

SHELL-AND-TUBE EXCHANGER COSTS

Basic costs of shell-and-tube heat exchangers made in the United States of carbon steel construction in 1958 are shown in Fig. 11-41.

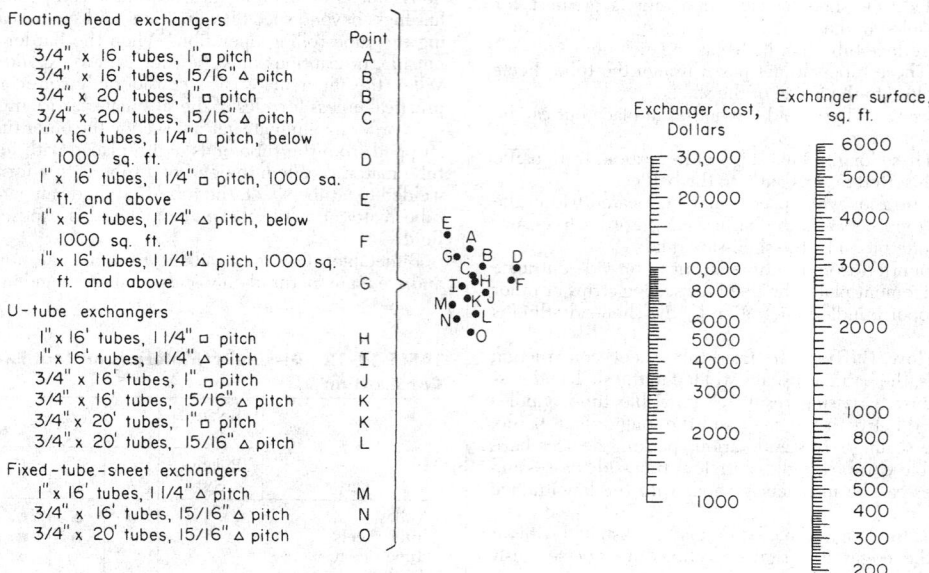

FIG. 11-41 Costs of basic exchangers—all steel, TEMA Class R, 150 lbf/in², 1958. To convert pounds-force per square inch to kilopascals, multiply by 6.895; to convert square feet to square meters, multiply by 0.0929; to convert inches to millimeters, multiply by 25.4; and to convert feet to meters, multiply by 0.3048.

Cost data for shell-and-tube exchangers from 15 sources were correlated and found to be consistent when scaled by the Marshall and Swift index [Woods et al., *Can. J. Chem. Eng.*, **54**, 469–489 (December 1976)].

Costs of shell-and-tube heat exchangers can be estimated from Fig. 11-41 and Tables 11-13 and 11-14. These 1960 costs should be updated by use of the Marshall and Swift Index, which appears in each issue of *Chemical Engineering*. Note that during periods of high and low demand for heat exchangers the prices in the marketplace may vary significantly from those determined by this method.

Small heat exchangers and exchangers bought in small quantities are likely to be more costly than indicated.

Standard heat exchangers (which are in some instances off-the-shelf items) are available in sizes ranging from 1.9 to 37 m² (20 to 400 ft²) at costs lower than for custom-built units. Steel costs are approximately one-half, admiralty tube-side costs are two-thirds, and stainless costs are three-fourths of those for equivalent custom-built exchangers.

Kettle-type-reboiler costs are 15 to 25 percent greater than for equivalent internal-floating-head or U-tube exchangers. The higher extra is applicable with relatively large kettle-to-port-diameter ratios and with increased internals (e.g., vapor-liquid separators, foam breakers, sight glasses).

To estimate exchanger costs for varying construction details and alloys, first determine the base cost of a similar heat exchanger of basic construction (carbon steel, Class R, 150 lbf/in²) from Fig. 11-41. From Table 11-13, select appropriate extras for higher pressure rating and for alloy construction of tube sheets and baffles, shell and shell cover, and channel and floating-head cover. Compute these extras in accordance with the notes below the table. For tubes other than welded carbon steel, compute the extra by multiplying the exchanger surface by the appropriate cost per square foot from Table 11-14.

When points for 20-ft-long tubes do not appear in Fig. 11-41, use 0.95 times the cost of the equivalent 16-ft-long exchanger. Length variation of steel heat exchangers affects costs by approximately $1 per square foot. Shell diameters for a given surface are approximately equal for U-tube and floating-head construction.

Low-fin tubes (1/16-in-high fins) provide 2.5 times the surface per lineal foot. Surface required should be divided by 2.5; then use Fig. 11-41 to determine basic cost of the heat exchanger. Actual surface times extra costs (from Table 11-14) should then be added to determine cost of fin-tube exchanger.

TABLE 11-13   Extras for Pressure and Alloy Construction and Surface and Weights*

Percent of steel base price, 1500-lbf/in² working pressure

	Shell diameters, in												
	12	14	16	18	20	22	24	27	30	33	36	39	42
Pressure†													
300 lbf/in²	7	7	8	8	9	9	10	11	11	12	13	14	15
450 lbf/in²	18	19	20	21	22	23	24	27	29	31	32	33	35
600 lbf/in²	28	29	31	33	35	37	39	40	41	32	44	45	50
Alloy													
All-steel heat exchanger	100	100	100	100	100	100	100	100	100	100	100	100	100
Tube sheets and baffles													
Naval rolled brass	14	17	19	21	22	22	22	22	22	23	24	24	25
Monel	24	31	35	37	39	39	40	40	41	41	41	41	42
1¼ Cr, ½ Mo	6	7	7	7	8	8	8	8	8	9	10	10	11
4–6 Cr, ½ Mo	19	22	24	25	26	26	26	25	25	25	26	26	26
11–13 Cr (stainless 410)	21	24	26	27	27	27	27	27	27	27	27	27	28
Stainless 304	22	27	29	30	31	31	31	31	30	30	30	31	31
Shell and shell cover													
Monel	45	48	51	52	53	52	52	51	49	47	45	44	44
1¼ Cr, ½ Mo	20	22	24	25	25	25	24	22	20	19	18	17	17
4–6 Cr, ½ Mo	28	31	33	35	35	35	34	32	30	28	27	26	26
11–13 Cr (stainless 410)	29	33	35	36	36	36	35	34	32	30	29	27	27
Stainless 304	32	34	36	37	38	37	37	35	33	31	30	29	28
Channel and floating-head cover													
Monel	40	42	42	43	42	41	40	37	34	32	31	40	30
1¼ Cr, ½ Mo	23	24	24	25	24	24	23	22	21	21	21	20	20
4–6 Cr, ½ Mo	36	37	38	38	37	36	34	31	29	27	26	25	24
11–13 Cr (stainless 410)	37	38	39	39	38	37	35	32	30	28	27	26	25
Stainless 304	37	39	39	39	38	37	36	33	31	29	28	26	26
Surface													
Surface, ft², internal floating head, ¾-in OD by 1-in square pitch, 16 ft 0 in, tube‡	251	302	438	565	726	890	1040	1470	1820	2270	2740	3220	3700
1-in OD by 1¼-in square pitch, 16-ft 0-in tube§	218	252	352	470	620	755	876	1260	1560	1860	2360	2770	3200
Weight, lb, internal floating head, 1-in OD, 14 BWG tube	2750	3150	4200	5300	6600	7800	9400	11,500	14,300	17,600	20,500	24,000	29,000

*Modified from E. N. Sieder and G. H. Elliot, *Pet. Refiner*, **39**(5), 223 (1960).
†Total extra is 0.7 × pressure extra on shell side plus 0.3 × pressure extra on tube side.
‡Fixed-tube-sheet construction with ¾-in OD tube on 15/16-in triangular pitch provides 36 percent more surface.
§Fixed-tube-sheet construction with 1-in OD tube on 1¼-in triangular pitch provides 18 percent more surface.
For an all-steel heat exchanger with mixed design pressures the total extra for pressure is 0.7 × pressure extra on shell side plus 0.3 × pressure extra tube side.
For an exchanger with alloy parts and a design pressure of 150 lbf/in², the alloy extras are added. For shell and shell cover the combined alloy-pressure extra is the alloy extra times the shell-side pressure extra/100. For channel and floating-head cover the combined alloy-pressure extra is the alloy extra times the tube-side pressure extra/100. For tube sheets and baffles the combined alloy-pressure extra is the alloy extra times the higher-pressure extra times 0.9/100. (The 0.9 factor is included since baffle thickness does not increase because of pressure.)
NOTE: To convert pounds-force per square inch to kilopascals, multiply by 6.895; to convert square feet to square meters, multiply by 0.0929; and to convert inches to millimeters, multiply by 25.4.

TABLE 11-14   Base Quantity Extra Cost for Tube Gauge and Alloy

Dollars per square foot

	¾-in OD tubes			1-in OD tubes		
	16 BWG	14 BWG	12 BWG	16 BWG	14 BWG	12 BWG
Carbon steel	0	0.02	0.06	0	0.01	0.07
Admiralty	0.78	1.20	1.81	0.94	1.39	2.03
(T-11) 1¼ Cr, ½ Mo	1.01	1.04	1.11	0.79	0.82	0.95
(T-5) 4–6 Cr	1.61	1.65	1.74	1.28	1.32	1.48
Stainless 410 welded	2.62	3.16	4.12	2.40	2.89	3.96
Stainless 410 seamless	3.10	3.58	4.63	2.84	3.31	4.47
Stainless 304 welded	2.50	3.05	3.99	2.32	2.83	3.88
Stainless 304 seamless	3.86	4.43	5.69	3.53	4.08	5.46
Stainless 316 welded	3.40	4.17	5.41	3.25	3.99	5.36
Stainless 316 seamless	7.02	7.95	10.01	6.37	7.27	9.53
90-10 cupronickel	1.33	1.89	2.67	1.50	2.09	2.90
Monel	4.25	5.22	6.68	4.01	4.97	6.47
Low fin						
Carbon steel	0.22	0.23		0.18	0.19	
Admiralty	0.58	0.75		0.70	0.87	
90-10 cupronickel	0.72	0.96		0.86	1.06	

NOTE: To convert inches to millimeters, multiply by 25.4.

HAIRPIN/DOUBLE-PIPE HEAT EXCHANGERS

PRINCIPLES OF CONSTRUCTION

Hairpin heat exchangers (often also referred to as "double pipes") are characterized by a construction form which imparts a U-shaped appearance to the heat exchanger. In its classical sense, the term *double pipe* refers to a heat exchanger consisting of a pipe within a pipe, usually of a straight-leg construction with no bends. However, due to the need for removable bundle construction and the ability to handle differential thermal expansion while avoiding the use of expansion joints (often the weak point of the exchanger), the current U-shaped configuration has become the standard in the industry (Fig. 11-42). A further departure from the classical definition comes when more than one pipe or tube is used to make a tube bundle, complete with tubesheets and tube supports similar to the TEMA style exchanger.

Hairpin heat exchangers consist of two shell assemblies housing a common set of tubes and interconnected by a return-bend cover referred to as the *bonnet*. The shell is supported by means of bracket assemblies designed to cradle both shells simultaneously. These brackets are configured to permit the modular assembly of many hairpin sections into an exchanger bank for inexpensive future-expansion capability and for providing the very long thermal lengths demanded by special process applications.

The bracket construction permits support of the exchanger without fixing the supports to the shell. This provides for thermal movement of the shells within the brackets and prevents the transfer of thermal stresses into the process piping. In special cases the brackets may be welded to the shell. However, this is usually avoided due to the resulting loss of flexibility in field installation and equipment reuse at other sites and an increase in piping stresses.

The hairpin heat exchanger, unlike the removable bundle TEMA styles, is designed for bundle insertion and removal from the return end rather than the tubesheet end. This is accomplished by means of removable split rings which slide into grooves machined around the outside of each tubesheet and lock the tubesheets to the external closure flanges. This provides a distinct advantage in maintenance since bundle removal

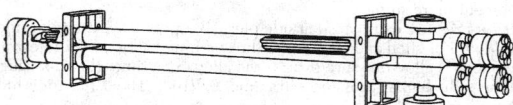

FIG. 11-42 Double-pipe-exchanger section with longitudinal fins. (*Brown Fin-tube Co.*)

takes place at the exchanger end furthest from the plant process piping without disturbing any gasketed joints of this piping.

FINNED DOUBLE PIPES

The design of the classical single-tube double-pipe heat exchanger is an exercise in pure longitudinal flow with the shellside and tubeside coefficients differing primarily due to variations in flow areas. Adding longitudinal fins gives the more common double-pipe configuration (Table 11-15). Increasing the number of tubes yields the *multitube* hairpin.

MULTITUBE HAIRPINS

For years, the slightly higher mechanical-design complexity of the hairpin heat exchanger relegated it to only the smallest process requirements with shell sizes not exceeding 100 mm. In the early 1970s the maximum available sizes were increased to between 300 and 400 mm depending upon the manufacturer. At the present time, due to recent advances in design technology, hairpin exchangers are routinely produced in shell sizes between 50 (2 in) and 800 mm (30 in) for a wide range of pressures and temperatures and have been made in larger sizes as well. Table 11-16 gives common hairpin tube counts and areas for 19 mm (¾ in) O.D. tubes arranged on a 24 mm (15/16 in) triangular tube layout.

The hairpin width and the centerline distance of the two legs (shells) of the hairpin heat exchanger are limited by the outside diameter of the closure flanges at the tubesheets. This diameter, in turn, is a function of the design pressures. As a general rule, for low-to-moderate design pressures (less than 15 bar), the center-to-center distance is approximately 1.5 to 1.8 times the shell outside diameter, with this ratio decreasing slightly for the larger sizes.

TABLE 11-15   Double-Pipe Hairpin Section Data

Shell pipe O.D.		Inner pipe O.D.		Fin height		Fin count	Surface-area-per-unit length	
mm	in	mm	in	mm	in	(max)	sq m/m	sq ft/ft
60.33	2.375	25.4	1.000	12.7	0.50	24	0.692	2.27
88.9	3.500	48.26	1.900	12.7	0.50	36	1.07	3.51
114.3	4.500	48.26	1.900	25.4	1.00	36	1.98	6.51
114.3	4.500	60.33	2.375	19.05	0.75	40	1.72	5.63
114.3	4.500	73.03	2.875	12.70	0.50	48	1.45	4.76
141.3	5.563	88.9	3.500	17.46	0.6875	56	2.24	7.34
168.3	6.625	114.3	4.500	17.46	0.6875	72	2.88	9.44

TABLE 11-16 Multitube Hairpin Section Data

Size	Shell O.D. mm	Shell O.D. in	Shell thickness mm	Shell thickness in	Tube count 19 mm	Surface area for 6.1 m (20 ft.) nominal length sq m	sq ft
03-MT	88.9	3.500	5.49	0.216	5	3.75	40.4
04-MT	114.3	4.500	6.02	0.237	9	6.73	72.4
05-MT	141.3	5.563	6.55	0.258	14	10.5	113.2
06-MT	168.3	6.625	7.11	0.280	22	16.7	179.6
08-MT	219.1	8.625	8.18	0.322	42	32.0	344.3
10-MT	273.1	10.75	9.27	0.365	68	52.5	564.7
12-MT	323.9	12.75	9.53	0.375	109	84.7	912.1
14-MT	355.6	14.00	9.53	0.375	136	107.	1159.
16-MT	406.4	16.00	9.53	0.375	187	148.	1594.
18-MT	457.2	18.00	9.53	0.375	241	191.	2054.
20-MT	508.0	20.00	9.53	0.375	304	244.	2622.
22-MT	558.8	22.00	9.53	0.375	380	307.	3307.
24-MT	609.6	24.00	9.53	0.375	463	378.	4065.
26-MT	660.4	26.00	9.53	0.375	559	453.	4879.
28-MT	711.2	28.00	9.53	0.375	649	529.	5698.
30-MT	762.0	30.00	11.11	0.4375	752	630.	6776.

One interesting consequence of this fact is the inability to construct a hairpin tube bundle having the smallest radius bends common to a conventional U-tube, TEMA shell, and tube bundle. In fact, in the larger hairpin sizes the tubes might be better described as curved rather than bent. The smallest U-bend diameters are greater than the outside diameter of shells less than 300 mm in size. The U-bend diameters are greater than 300 mm in larger shells. As a general rule, mechanical tube cleaning around the radius of a U-bend may be accomplished with a flexible shaft-cleaning tool for bend diameters greater than ten times the tube's inside diameter. This permits the tool to pass around the curve of the tube bend without binding.

In all of these configurations, maintaining longitudinal flow on both the shellside and tubeside allows the decision for placement of a fluid stream on either one side or the other to be based upon design efficiency (mass flow rates, fluid properties, pressure drops, and veloci-

ties) and not because there is any greater tendency to foul on one side than the other. Experience has shown that, in cases where fouling is influenced by flow velocity, overall fouling in tube bundles is less in properly designed longitudinal flow bundles where areas of low velocity can be avoided without flow-induced tube vibration.

This same freedom of stream choice is not as readily applied when a segmental baffle is used. In those designs, the baffle's creation of low velocities and stagnant flow areas on the outside of the bundle can result in increased shellside fouling at various locations of the bundle. The basis for choosing the stream in those cases will be similar to the common shell and tube heat exchanger. At times a specific selection of stream side must be made regardless of tube-support mechanism in expectation of an unresolvable fouling problem. However, this is often the exception rather than the rule.

DESIGN APPLICATIONS

One benefit of the hairpin exchanger is its ability to handle high tubeside pressures at a lower cost than other removable-bundle exchangers. This is due in part to the lack of pass partitions at the tubesheets which complicate the gasketing design process. Present mechanical design technology has allowed the building of dependable, removable-bundle, hairpin multitubes at tubeside pressures of 825 bar (12,000 psi).

The best known use of the hairpin is its operation in true countercurrent flow which yields the most efficient design for processes that have a close temperature approach or temperature cross. However, maintaining countercurrent flow in a tubular heat exchanger usually implies one tube pass for each shell pass. As recently as 30 years ago, the lack of inexpensive, multiple-tube pass capability often diluted the advantages gained from countercurrent flow.

The early attempts to solve this problem led to investigations into the area of heat transfer augmentation. This familiarity with augmentation techniques inevitably led to improvements in the efficiency and capacity of the small heat exchangers. The result has been the application of the hairpin heat exchanger to the solution of unique process problems, such as dependable, once-through, convective boilers offering high-exit qualities, especially in cases of process-temperature crosses.

AIR-COOLED HEAT EXCHANGERS

AIR-COOLED HEAT EXCHANGERS

Atmospheric air has been used for many years to cool and condense fluids in areas of water scarcity. During the 1960s the use of air-cooled heat exchangers grew rapidly in the United States and elsewhere. In Europe, where seasonal variations in ambient temperatures are relatively small, air-cooled exchangers are used for the greater part of process cooling. In some new plants all cooling is done with air. Increased use of air-cooled heat exchangers has resulted from lack of available water, significant increases in water costs, and concern for water pollution.

Air-cooled heat exchangers include a tube bundle, which generally has spiral-wound fins upon the tubes, and a fan, which moves air across the tubes and is provided with a driver. Electric motors are the most commonly used drivers; typical drive arrangements require a V belt or a direct right-angle gear. A plenum and structural supports are basic components. Louvers are often used:

A bay generally has two tube bundles installed in parallel. These may be in the same or different services. Each bay is usually served by two (or more) fans and is furnished with a structure, a plenum, and other attendant equipment.

The location of air-cooled heat exchangers must consider the large space requirements and the possible recirculation of heated air because of the effect of prevailing winds upon buildings, fired heaters, towers, various items of equipment, and other air-cooled exchangers. Inlet air temperature at the exchanger can be significantly higher than

the ambient air temperature at a nearby weather station. See *Air-Cooled Heat Exchangers for General Refinery Services*, API Standard 661, 2d ed., January 1978, for information on refinery-process air-cooled heat exchangers.

Forced and Induced Draft The forced-draft unit, which is illustrated in Fig. 11-43 pushes air across the finned tube surface. The fans are located below the tube bundles. The induced-draft design has the fan above the bundle, and the air is pulled across the finned tube surface. In theory, a primary advantage of the forced-draft unit is that less power is required. This is true when the air-temperature rise exceeds 30°C (54°F).

Air-cooled heat exchangers are generally arranged in banks with several exchangers installed side by side. The height of the bundle aboveground must be one-half of the tube length to produce an inlet velocity equal to the face velocity. This requirement applies both to ground-mounted exchangers and to those pipe-rack-installed exchangers which have a fire deck above the pipe rack.

The forced-draft design offers better accessibility to the fan for on-stream maintenance and fan-blade adjustment. The design also provides a fan and V-belt assembly, which are not exposed to the hot-air stream that exits from the unit. Structural costs are less, and mechanical life is longer.

Induced-draft design provides more even distribution of air across the bundle, since air velocity approaching the bundle is relatively low. This design is better suited for exchangers designed for a close approach of product outlet temperature to ambient-air temperature.

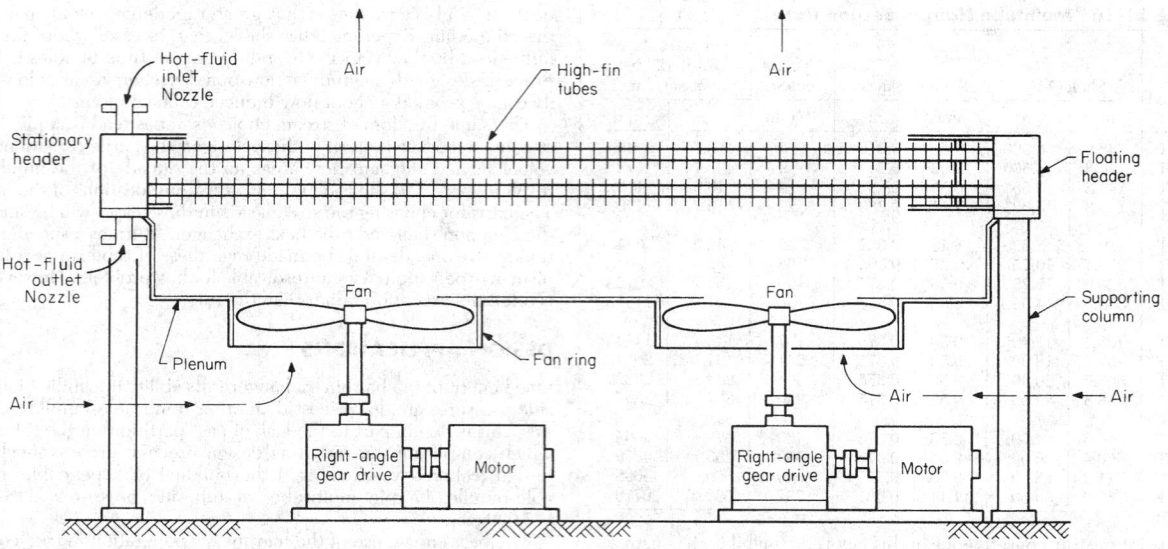

FIG. 11-43   Forced-draft air-cooled heat exchanger. [*Chem. Eng.*, **114** (Mar. 27, 1978).]

Induced-draft units are less likely to recirculate the hot exhaust air, since the exit air velocity is several times that of the forced-draft unit. Induced-draft design more readily permits the installation of the air-cooled equipment above other mechanical equipment such as pipe racks or shell-and-tube exchangers.

In a service in which sudden temperature change would cause upset and loss of product, the induced-draft unit gives more protection in that only a fraction of the surface (as compared with the forced-draft unit) is exposed to rainfall, sleet, or snow.

Tube Bundle   The principal parts of the tube bundle are the finned tubes and the header. Most commonly used is the plug header, which is a welded box that is illustrated in Fig. 11-44. The finned tubes are described in a subsequent paragraph. The components of a tube bundle are identified in the figure.

The second most commonly used header is a cover-plate header. The cover plate is bolted to the top, bottom, and end plates of the header. Removing the cover plate provides direct access to the tubes without the necessity of removing individual threaded plugs.

Other types of headers include the bonnet-type header, which is constructed similarly to the bonnet construction of shell-and-tube heat exchangers; manifold-type headers, which are made from pipe and have tubes welded into the manifold; and billet-type headers, made from a solid piece of material with machined channels for distributing the fluid. Serpentine-type tube bundles are sometimes used for very viscous fluids. A single continuous flow path through pipe is provided.

Tube bundles are designed to be rigid and self-contained and are mounted so that they expand independently of the supporting structure.

The face area of the tube bundle is its length times width. The net free area for air flow through the bundle is about 50 percent of the face area of the bundle.

The standard air face velocity (FV) is the velocity of standard air passing through the tube bundle and generally ranges from 1.5 to 3.6 m/s (300 to 700 ft/min).

Tubing   The 25.4-mm (1-in) outside-diameter tube is most commonly used. Fin heights vary from 12.7 to 15.9 mm (0.5 to 0.625 in), fin spacing from 3.6 to 2.3 mm (7 to 11 per linear inch), and tube triangular pitch from 50.8 to 63.5 mm (2.0 to 2.5 in). Ratio of extended surface to bare-tube outside surface varies from about 7 to 20. The

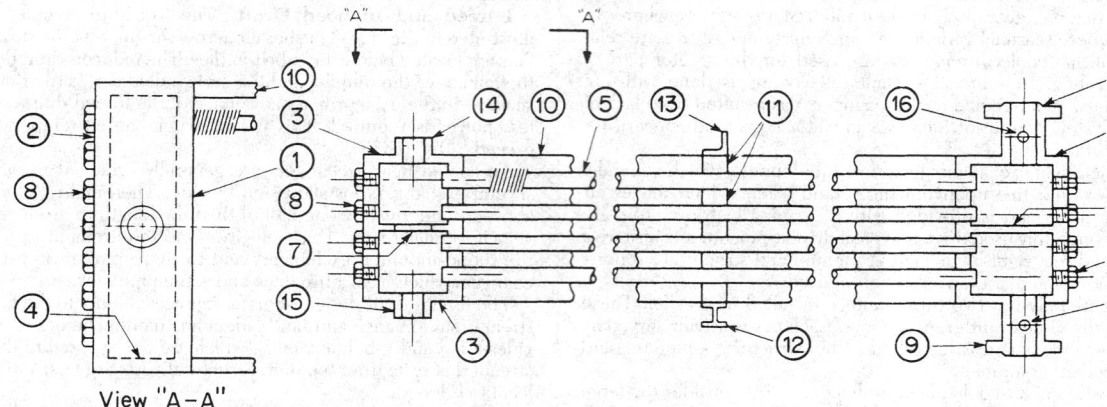

FIG. 11-44   Typical construction of a tube bundle with plug headers: (1) tube sheet; (2) plug sheet; (3) top and bottom plates; (4) end plate; (5) tube; (6) pass partition; (7) stiffener; (8) plug; (9) nozzle; (10) side frame; (11) tube spacer; (12) tube-support cross member; (13) tube keeper; (14) vent; (15) drain; (16) instrument connection. (API Standard 661.)

38-mm (1½-in) tube has been used for flue-gas and viscous-oil service. Tube size, fin heights, and fin spacing can be further varied.

Tube lengths vary and may be as great as 18.3 m (60 ft). When tube length exceeds 12.2 m (40 ft), three fans are generally installed in each bay. Frequently used tube lengths vary from 6.1 to 12.2 m (20 to 40 ft).

Finned-Tube Construction The following are descriptions of commonly used finned-tube constructions (Fig. 11-45).

1. *Embedded.* Rectangular-cross-section aluminum fin which is wrapped under tension and mechanically embedded in a groove 0.25 ± 0.05 mm (0.010 ± 0.002 in) deep, spirally cut into the outside surface of a tube.

2. *Integral (or extruded).* An aluminum outer tube from which fins have been formed by extrusion, mechanically bonded to an inner tube or liner.

3. *Overlapped footed.* L-shaped aluminum fin wrapped under tension over the outside surface of a tube, with the tube fully covered by the overlapped feet under and between the fins.

4. *Footed.* L-shaped aluminum fin wrapped under tension over the outside surface of a tube with the tube fully covered by the feet between the fins.

5. *Bonded.* Tubes on which fins are bonded to the outside surface by hot-dip galvanizing, brazing, or welding.

Typical metal design temperatures for these finned-tube constructions are 399°C (750°F) embedded, 288°C (550°F) integral, 232°C (450°F) overlapped footed, and 177°C (350°F) footed.

Tube ends are left bare to permit insertion of the tubes into appropriate holes in the headers or tube sheets. Tube ends are usually roller-expanded into these tube holes.

Fans Axial-flow fans are large-volume, low-pressure devices. Fan diameters are selected to give velocity pressures of approximately 2.5 mm (0.1 in) of water. Total fan efficiency (fan, driver, and transmission device) is about 75 percent, and fan drives usually have a minimum of 95 percent mechanical efficiency.

Usually fans are provided with four or six blades. Larger fans may have more blades. Fan diameter is generally slightly less than the width of the bay.

At the fan-tip speeds required for economical performance, a large amount of noise is produced. The predominant source of noise is vortex shedding at the trailing edge of the fan blade. Noise control of air-cooled exchangers is required by the Occupational Safety and Health Act (OSHA). API Standard 661 (*Air-Cooled Heat Exchangers for General Refinery Services,* 2d ed., January 1978) has the purchaser specifying sound-pressure-level (SPL) values per fan at a location designated by the purchaser and also specifying sound-power-level (PWL) values per fan. These are designated at the following octave-band-center frequencies: 63, 125, 250, 1000, 2000, 4000, 8000, and also the dBa value (the dBa is a weighted single-value sound-pressure level).

Reducing the fan-tip speed results in a straight-line reduction in air flow while the noise level decreases. The API Standard limits fan-tip speed to 61 m/s (12,000 ft/min) for typical constructions. Fan-design changes which reduce noise include increasing the number of fan blades, increasing the width of the fan blades, and reducing the clearance between fan tip and fan ring.

Both the quantity of air and the developed static pressure of fans in air-cooled heat exchangers are lower than indicated by fan manufacturers' test data, which are applicable to testing-facility tolerances and not to heat-exchanger constructions.

The axial-flow fan is inherently a device for moving a consistent volume of air when blade setting and speed of rotation are constant. Variation in the amount of air flow can be obtained by adjusting the blade angle of the fan and the speed of rotation. The blade angle can be either (1) permanently fixed, (2) hand-adjustable, or (3) automatically adjusted. Air delivery and power are a direct function of blade pitch angle.

Fan mounting should provide a minimum of one-half to three-fourths diameter between fan and ground on a forced-draft heat exchanger and one-half diameter between tubes and fan on an induced-draft cooler.

Fan blades can be made of aluminum, molded plastic, laminated plastic, carbon steel, stainless steel, and Monel.

Fan Drivers Electric motors or steam turbines are most commonly used. These connect with gears or V belts. (Gas engines connected through gears and hydraulic motors either direct-connected or connected through gears are in use. Fans may be driven by a prime mover such as a compressor with a V-belt takeoff from the flywheel to a jack shaft and then through a gear or V belt to the fan. Direct motor drive is generally limited to small-diameter fans.

V-belt drive assemblies are generally used with fans 3 m (10 ft) and less in diameter and motors of 22.4 kW (30 hp) and less.

Right-angle gear drive is preferred for fans over 3 m (10 ft) in diameter, for electric motors over 22.4 kW (30 hp), and with steam-turbine drives.

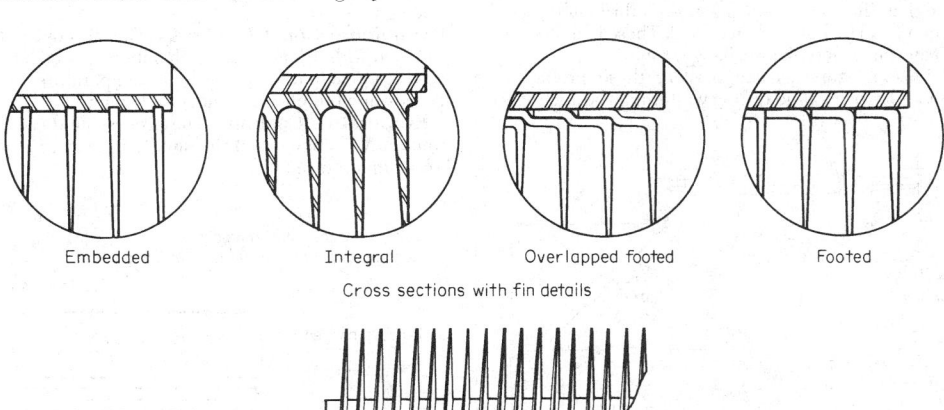

Embedded   Integral   Overlapped footed   Footed

Cross sections with fin details

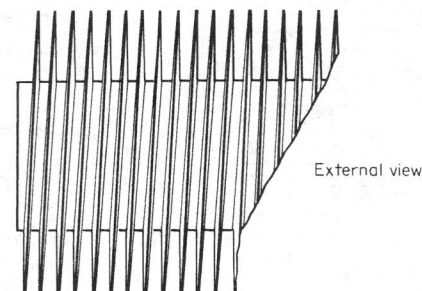

External view

FIG. 11-45 Finned-tube construction.

Fan Ring and Plenum Chambers The air must be distributed from the circular fan to the rectangular face of the tube bundle. The air velocity at the fan is between 3.8 and 10.2 m/s (750 and 2000 ft/in). The plenum-chamber depth (from fan to tube bundle) is dependent upon the fan dispersion angle (Fig. 11-46), which should have a maximum value of 45°.

The fan ring is made to commercial tolerances for the relatively large diameter fan. These tolerances are greater than those upon closely machined fan rings used for small-diameter laboratory-performance testing. Fan performance is directly affected by this increased clearance between the blade tip and the ring, and adequate provision in design must be made for the reduction in air flow. API Standard 661 requires that fan-tip clearance be a maximum of 0.5 percent of the fan diameter for diameters between 1.9 and 3.8 m (6.25 and 12.5 ft). Maximum clearance is 9.5 mm (⅜ in) for smaller fans and 19 mm (¾ in) for larger fans.

The depth of the fan ring is critical. Worsham (ASME Pap. 59-PET-27, Petroleum Mechanical Engineering Conference, Houston, 1959) reports an increase in flow varying from 5 to 15 percent with the same power consumption when the depth of a fan ring was doubled. The percentage increase was proportional to the volume of air and static pressure against which the fan was operating.

When making a selection, the stall-out condition, which develops when the fan cannot produce any more air regardless of power input, should be considered.

Air-Flow Control Process operating requirements and weather conditions are considered in determining the method of controlling air flow. The most common methods include simple on-off control, on-off step control (in the case of multiple-driver units), two-speed-motor control, variable-speed drivers, controllable fan pitch, manually or automatically adjustable louvers, and air recirculation.

Winterization is the provision of design features, procedures, or systems for air-cooled heat exchangers to avoid process-fluid operating problems resulting from low-temperature inlet air. These include fluid freezing, pour point, wax formation, hydrate formation, laminar flow, and condensation at the dew point (which may initiate corrosion). Freezing points for some commonly encountered fluids in refinery service include: benzene, 5.6°C (42°F); *p*-xylene 15.5°C (55.9°F); cyclohexane, 6.6°C (43.8°F); phenol, 40.9°C (105.6°F); monoethanolamine, 10.3°C (50.5°F); and diethanolamine, 25.1°C (77.2°F). Water solutions of these organic compounds are likely to freeze in air-cooled exchangers during winter service. Paraffinic and olefinic gases (C_1 through C_4) saturated with water vapor form hydrates when cooled. These hydrates are solid crystals which can collect and plug exchanger tubes.

Air-flow control in some services can prevent these problems. Cocurrent flow of air and process fluid during winter may be adequate

to prevent problems. (Normal design has countercurrent flow of air and process fluid.) In some services when the hottest process fluid is in the bottom tubes, which are exposed to the lowest-temperature air, winterization problems may be eliminated.

Following are references which deal with problems in low-temperature environments: Brown and Benkley, "Heat Exchangers in Cold Service—A Contractor's View," *Chem. Eng. Prog.,* **70,** 59–62 (July 1974); Franklin and Munn, "Problems with Heat Exchangers in Low Temperature Environments," *Chem. Eng. Prog.,* **70,** 63–67 (July 1974); Newell, "Air-Cooled Heat Exchangers in Low Temperature Environments: A Critique," *Chem. Eng. Prog.,* **70,** 86–91 (October 1974); Rubin, "Winterizing Air Cooled Heat Exchangers," *Hydrocarbon Process.,* **59,** 147–149 (October 1980); Shipes, "Air-Cooled Heat Exchangers in Cold Climates," *Chem. Eng. Prog.,* **70,** 53–58 (July 1974).

Air Recirculation Recirculation of air which has been heated as it crosses the tube bundle provides the best means of preventing operating problems due to low-temperature inlet air. Internal recirculation is the movement of air within a bay so that the heated air which has crossed the bundle is directed by a fan with reverse flow across another part of the bundle. Wind skirts and louvers are generally provided to minimize the entry of low-temperature air from the surroundings. Contained internal recirculation uses louvers within the bay to control the flow of warm air in the bay as illustrated in Fig. 11-47. Note that low-temperature inlet air has access to the tube bundle.

External recirculation is the movement of the heated air within the bay to an external duct, where this air mixes with inlet air, and the mixture serves as the cooling fluid within the bay. Inlet air does not have direct access to the tube bundle; an adequate mixing chamber is essential. Recirculation over the end of the exchanger is illustrated in Fig. 11-48. Over-the-side recirculation also is used. External recirculation systems maintain the desired low temperature of the air crossing the tube bundle.

Trim Coolers Conventional air-cooled heat exchangers can cool the process fluid to within 8.3°C (15°F) of the design dry-bulb temperature. When a lower process outlet temperature is required, a trim cooler is installed in series with the air-cooled heat exchanger. The water-cooled trim cooler can be designed for a 5.6 to 11.1°C (10 to 20°F) approach to the wet-bulb temperature (which in the United States is about 8.3°C (15°F) less than the dry-bulb temperature). In arid areas the difference between dry- and wet-bulb temperatures is much greater.

Humidification Chambers The air-cooled heat exchanger is provided with humidification chambers in which the air is cooled to a close approach to the wet-bulb temperature before entering the finned-tube bundle of the heat exchanger.

Evaporative Cooling The process fluid can be cooled by using evaporative cooling with the sink temperature approaching the wet-bulb temperature.

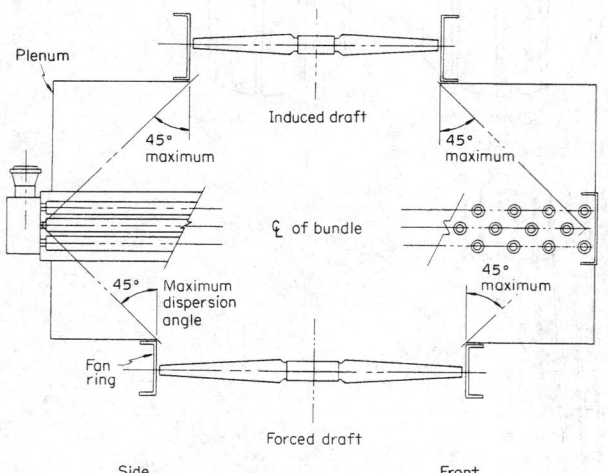

FIG. 11-46 Fan dispersion angle. (API Standard 661.)

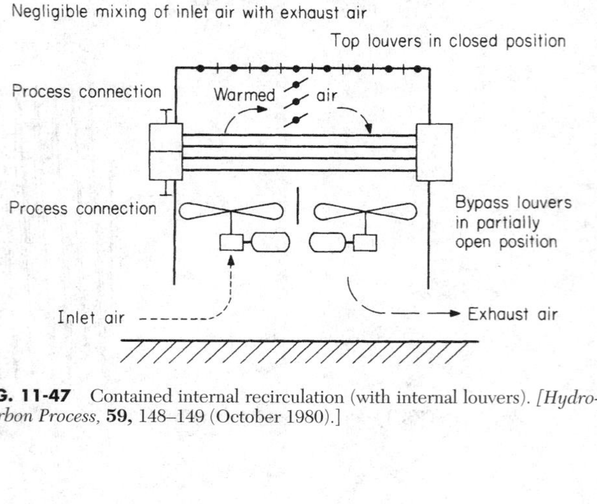

FIG. 11-47 Contained internal recirculation (with internal louvers). [*Hydrocarbon Process,* **59,** 148–149 (October 1980).]

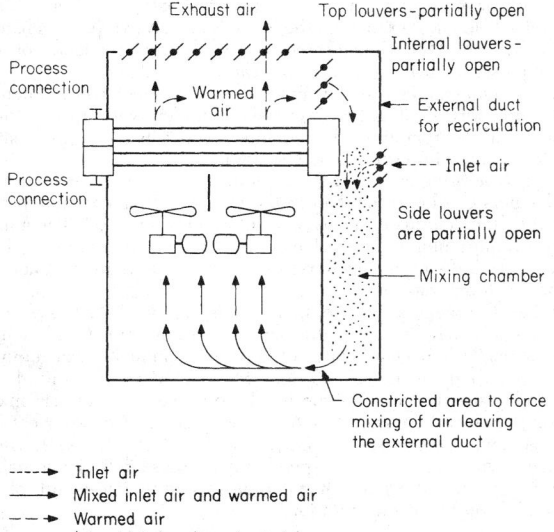

Exhaust air   Top louvers - partially open

Process
connection   Internal louvers -
partially open

Warmed
air   External duct
for recirculation

Inlet air

Side louvers
are partially open

Process
connection   Mixing chamber

Constricted area to force
mixing of air leaving
the external duct

- - - - →   Inlet air

⟶   Mixed inlet air and warmed air

- - →   Warmed air
(air which has been heated by
crossing the tube bundle)

FIG. 11-48   External recirculation with adequate mixing chamber. [*Hydrocarbon Process*, **59**, 148–149 (October 1980).]

Steam Condensers   Air-cooled steam condensers have been fabricated with a single tube-side pass and several rows of tubes. The bottom row has a higher temperature difference than the top row, since the air has been heated as it crosses the rows of tubes. The bottom row condenses all the entering steam before the steam has traversed the length of the tube. The top row, with a lower temperature driving force, does not condense all the entering steam. At the exit header, uncondensed steam flows from the top row into the bottom row. Since noncondensable gases are always present in steam, these accumulate within the bottom row because steam is entering from both ends of the tube. Performance suffers.

Various solutions have been used. These include orifices to regulate the flow into each tube, a "blow-through steam" technique with a vent condenser, complete separation of each row of tubes, and inclined tubes.

Air-Cooled Overhead Condensers   Air-cooled overhead condensers (AOC) have been designed and installed above distillation columns as integral parts of distillation systems. The condensers generally have inclined tubes, with air flow over the finned surfaces induced by a fan. Prevailing wind affects both structural design and performance.

AOC provide the additional advantages of reducing ground-space requirements and piping and pumping requirements and of providing smoother column operation.

The downflow condenser is used mainly for nonisothermal condensation. Vapors enter through a header at the top and flow downward. The reflux condenser is used for isothermal and small-temperature-change conditions. Vapors enter at the bottom of the tubes.

AOC usage first developed in Europe but became more prevalent in the United States during the 1960s. A state-of-the-art article was published by Dehne [*Chem. Eng. Prog.*, **64**, 51 (July 1969)].

Air-Cooled Heat-Exchanger Costs   The cost data that appear in Table 11-17 are unchanged from those published in the 1963 edition of this *Handbook*. In 1969 Guthrie [*Chem. Eng.*, **75**, 114 (Mar. 24, 1969)] presented cost data for field-erected air-cooled exchangers. These costs are only 25 percent greater than those of Table 11-17 and include the costs of steel stairways, indirect subcontractor charges, and field-erection charges. Since minimal field costs would be this high (i.e., 25 percent of purchase price), the basic costs appear to be unchanged. (Guthrie indicated a cost band of plus or minus 25 percent.) Preliminary design and cost estimating of air-

TABLE 11-17   Air-Cooled Heat-Exchanger Costs (1970)

Surface (bare tube), sq. ft.	500	1000	2000	3000	5000
Cost for 12-row-deep bundle, dollars/square foot	9.0	7.6	6.8	5.7	5.3
Factor for bundle depth:					
6 rows	1.07	1.07	1.07	1.12	1.12
4 rows	1.2	1.2	1.2	1.3	1.3
3 rows	1.25	1.25	1.25	1.5	1.5

Base: Bare-tube external surface 1 in. o.d. by 12 B.W.G. by 24 ft. 0 in. steel tube with 8 aluminum fins per inch ⅝-in. high. Steel headers. 150 lb./sq. in. design pressure. V-belt drive and explosion-proof motor. Bare-tube surface 0.262 sq. ft./ft. Fin-tube surface/bare-tube surface ratio is 16.9.

Factors: 20 ft. tube length	1.05	
30 ft. tube length	0.95	
18 B.W.G. admiralty tube	1.04	
16 B.W.G. admiralty tube	1.12	

NOTE: To convert feet to meters, multiply by 0.3048; to convert square feet to square meters, multiply by 0.0929; and to convert inches to millimeters, multiply by 25.4.

cooled heat exchangers have been discussed by J. E. Lerner ["Simplified Air Cooler Estimating," *Hydrocarbon Process.*, **52**, 93–100 (February 1972)].

Design Considerations

1. *Design dry-bulb temperature.*   The typically selected value is the temperature which is equaled or exceeded 2½ percent of the time during the warmest consecutive 4 months. Since air temperatures at industrial sites are frequently higher than those used for these weather-data reports, it is good practice to add 1 to 3°C (2 to 6°F) to the tabulated value.

2. *Air recirculation.*   Prevailing winds and the locations and elevations of buildings, equipment, fired heaters, etc., require consideration. All air-cooled heat exchangers in a bank are of one type, i.e., all forced-draft or all induced-draft. Banks of air-cooled exchangers must be placed far enough apart to minimize air recirculation.

3. *Wintertime operations.*   In addition to the previously discussed problems of winterization, provision must be made for heavy rain, strong winds, freezing of moisture upon the fins, etc.

4. *Noise.*   Two identical fans have a noise level 3 dBa higher than one fan, while eight identical fans have a noise level 9 dBa higher than a single fan. Noise level at the plant site is affected by the exchanger position, the reflective surfaces near the fan, the hardness of these surfaces, and noise from adjacent equipment. The extensive use of air-cooled heat exchangers contributes significantly to plant noise level.

5. *Ground area and space requirements.*   Comparisons of the overall space requirements for plants using air cooling versus water cooling are not consistent. Some air-cooled units are installed above other equipment—pipe racks, shell-and-tube exchangers, etc. Some plants avoid such installations because of safety considerations, as discussed later.

6. *Safety.*   Leaks in air-cooled units are directly to the atmosphere and can cause fire hazards or toxic-fume hazards. However, the large air flow through an air-cooled exchanger greatly reduces any concentration of toxic fluids. Segal [*Pet. Refiner*, **38**, 106 (April 1959)] reports that air-fin coolers "are not located over pumps, compressors, electrical switchgear, control houses and, in general, the amount of equipment such as drums and shell-and-tube exchangers located beneath them are minimized."

Pipe-rack-mounted air-cooled heat exchangers with flammable fluids generally have concrete fire decks which isolate the exchangers from the piping.

7. *Atmospheric corrosion.*   Air-cooled heat exchangers should not be located where corrosive vapors and fumes from vent stacks will pass through them.

8. *Air-side fouling.*   Air-side fouling is generally negligible.

9. *Process-side cleaning.*   Either chemical or mechanical cleaning on the inside of the tubes can readily be accomplished.

10. *Process-side design pressure.*   The high-pressure process fluid is always in the tubes. Tube-side headers are relatively small as

compared with water-cooled units when the high pressure is generally on the shell side. High-pressure design of rectangular headers is complicated. The plug-type header is normally used for design gauge pressures to 13,790 kPa (2000 lbf/in²) and has been used to 62,000 kPa (9000 lbf/in²). The use of threaded plugs at these pressures creates problems. Removable cover plate headers are generally limited to gauge pressures of 2068 kPa (300 lbf/in²). The expensive billet-type header is used for high-pressure service.

11. *Bond resistance.* Vibration and thermal cycling affect the bond resistance of the various types of tubes in different manners and thus affect the amount of heat transfer through the fin tube.

12. *Approach temperature.* The approach temperature, which is the difference between the process-fluid outlet temperature and the design dry-bulb air temperature, has a practical minimum of 8 to 14°C (15 to 25°F). When a lower process-fluid outlet temperature is required, an air-humidification chamber can be provided to reduce the inlet air temperature toward the wet-bulb temperature. A 5.6°C (10°F) approach is feasible. Since typical summer wet-bulb design temperatures in the United States are 8.3°C (15°F) lower than dry-bulb temperatures, the outlet process-fluid temperature can be 3°C (5°F) below the dry-bulb temperature.

13. *Mean-temperature-difference (MTD) correction factor.* When the outlet temperatures of both fluids are identical, the MTD correction factor for a 1:2 shell-and-tube exchanger (one pass shell side, two or more passes tube side) is approximately 0.8. For a single-pass air-cooled heat exchanger the factor is 0.91. A two-pass exchanger has a factor of 0.96, while a three-pass exchanger has a factor of 0.99 when passes are arranged for counterflow.

14. *Maintenance cost.* Maintenance for air-cooled equipment as compared with shell-and-tube coolers (complete with cooling-tower costs) indicates that air-cooling maintenance costs are approximately 0.3 to 0.5 those for water-cooled equipment.

15. *Operating costs.* Power requirements for air-cooled heat exchangers can be lower than at the summer design condition provided that an adequate means of air-flow control is used. The annual power requirement for an exchanger is a function of the means of air-flow control, the exchanger service, the air-temperature rise, and the approach temperature.

When the mean annual temperature is 16.7°C (30°F) lower than the design dry-bulb temperature and when both fans in a bay have automatically controllable pitch of fan blades, annual power required has been found to be 22, 36, and 54 percent respectively of that needed at the design condition for three process services [Frank L. Rubin, "Power Requirements Are Lower for Air-Cooled Heat Exchangers with AV Fans," *Oil Gas J.*, 165–167 (Oct. 11, 1982)]. Alternatively, when fans have two-speed motors, these deliver one-half of the design flow of air at half speed and use only one-eighth of the power of the full-speed condition.

COMPACT AND NONTUBULAR HEAT EXCHANGERS

COMPACT HEAT EXCHANGERS

With equipment costs rising and limited available plot space, compact heat exchangers are gaining a larger portion of the heat exchange market. Numerous types use special enhancement techniques to achieve the required heat transfer in smaller plot areas and, in many cases, less initial investment. As with all items that afford a benefit there is a series of restrictions that limit the effectiveness or application of these special heat exchanger products. In most products discussed some of these considerations are presented, but a thorough review with reputable suppliers of these products is the only positive way to select a compact heat exchanger. The following guidelines will assist in prequalifying one of these.

PLATE-AND-FRAME EXCHANGERS

There are two major types gasketed and welded-plate heat exchangers. Each shall be discussed individually.

GASKETED-PLATE EXCHANGERS (G. PHE)

Description This type is the fastest growing of the compact exchangers and the most recognized (see Fig. 11-49). A series of corrugated alloy material channel plates, bounded by elastomeric gaskets are hung off and guided by longitudinal carrying bars, then compressed by large-diameter tightening bolts between two pressure retaining frame plates (cover plates). The frame and channel plates have portholes which allow the process fluids to enter alternating flow passages (the space between two adjacent-channel plates). Gaskets around the periphery of the channel plate prevent leakage to the atmosphere and also prevent process fluids from coming in contact with the frame plates. No interfluid leakage is possible in the port area due to a dual-gasket seal.

The frame plates are typically epoxy-painted carbon-steel material and can be designed per most pressure vessel codes. Design limitations are in the Table 11-18. The channel plates are always an alloy material with 304SS as a minimum (see Table 11-18 for other materials).

Channel plates are typically 0.4 to 0.8 mm thick and have corrugation depths of 2 to 10 mm. Special Wide Gap (WG PHE) plates are available, in limited sizes, for slurry applications with depths of approximately 16 mm. The channel plates are compressed to achieve metal-to-metal contact for pressure-retaining integrity. These narrow gaps and high number of contact points which change fluid flow direction, combine to create a very high turbulence between the plates. This means high individual-heat-transfer coefficients (up to 14200 W/m² °C), but also very high pressure drops per length as well. To compensate, the channel plate lengths are usually short, most under 2 and few over 3 meters in length. In general, the same pressure drops as conventional exchangers are used without loss of the enhanced heat transfer.

Expansion of the initial unit is easily performed in the field without special considerations. The original frame length typically has an additional capacity of 15–20 percent more channel plates (i.e., surface area). In fact, if a known future capacity is available during fabrication stages, a longer carrying bar could be installed, and later, increasing the surface area would be easily handled. When the expansion is needed, simply untighten the carrying bolts, pull back the frame plate, add the additional channel plates, and tighten the frame plate.

Applications Most PHE applications are liquid-liquid services but there are numerous steam heater and evaporator uses from their heritage in the food industry. Industrial users typically have chevron style channel plates while some food applications are washboard style.

Fine particulate slurries in concentrations up to 70 percent by weight are possible with standard channel spacings. Wide-gap units are used with larger particle sizes. Typical particle size should not exceed 75 percent of the *single* plate (not total channel) gap.

Close temperature approaches and tight temperature control possible with PHE's and the ability to sanitize the entire heat transfer surface easily were a major benefit in the food industry.

Multiple services in a single frame are possible.

Gasket selection is one of the most critical and limiting factors in PHE usage. Table 11-19 gives some guidelines for fluid compatibility. Even trace fluid components need to be considered. The higher the operating temperature and pressure, the shorter the anticipated gasket life. *Always* consult the supplier on gasket selection and obtain an estimated or guaranteed lifetime.

The major applications are, but not limited to, as follows:

Temperature cross applications   (lean/rich solvent)
Close approaches   (fresh water/seawater)

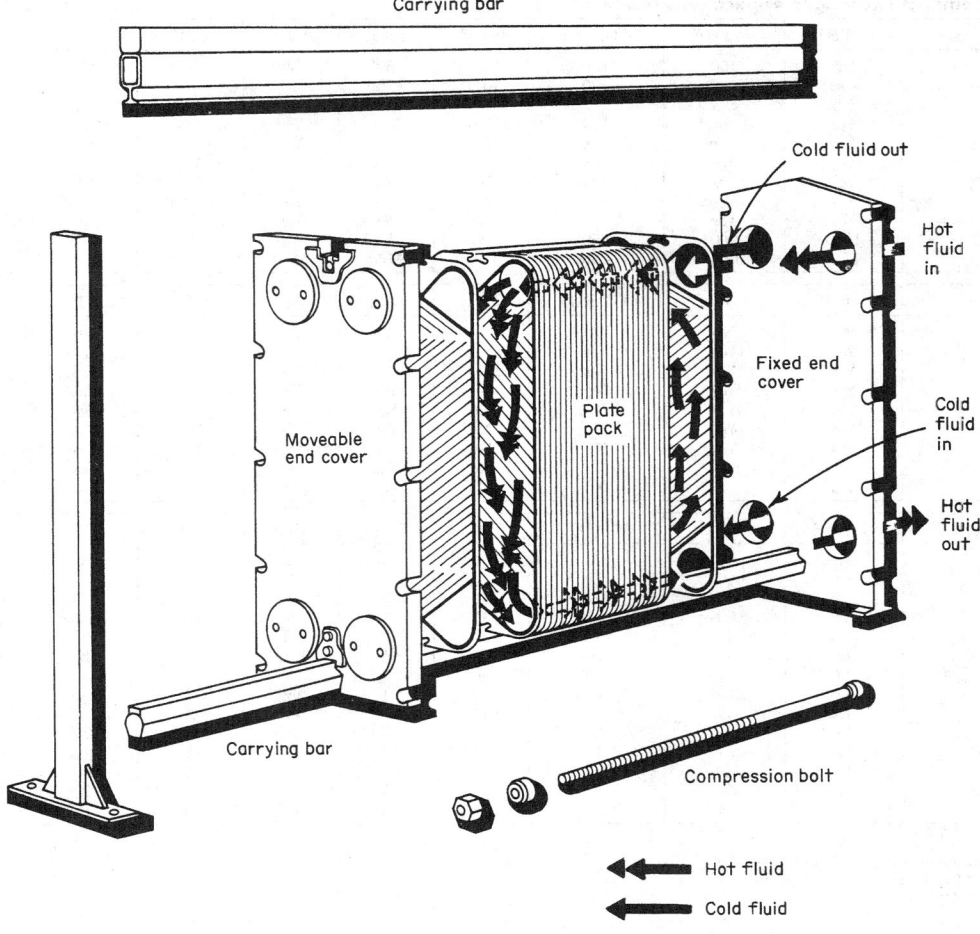

FIG. 11-49   Plate-and-frame heat exchanger. Hot fluid flows down between alternate plates, and cold fluid flows up between alternate plates. (*Thermal Division, Alfa-Laval, Inc.*)

Viscous fluids   (emulsions)
Sterilized surface required   (food, pharmaceutical)
Polished surface required   (latex, pharmaceutical)
Future expansion required
Space restrictions
Barrier coolant services   (closed-loop coolers)
Slurry applications   (TiO$_2$, Kaolin, precipitated
                                                  calcium carbonate, and beet
                                                  sugar raw juice)

Design   Standard channel-plate designs, unique to each manufacturer, are developed with limited modifications of each plates' corrugation depths and included angles. Manufacturers combine their different style plates to custom-fit each service. Due to the possible combinations, it is impossible to present a way to exactly size PHEs. However, it is possible to estimate areas for new units and to predict performance of existing units with different conditions (chevron-type channel plates are presented).

The fixed length and limited corrugation included angles on channel plates makes the NTU method of sizing practical. (Waterlike fluids are assumed for the following examples).

$$\text{NTU} = \frac{\Delta t \text{ of either side}}{\text{LMTD}} \qquad (11\text{-}76)$$

Most plates have NTU values of 0.5 to 4.0, with 2.0 to 3.0 as the most

common, (multipass shell and tube exchangers are typically less than 0.75). The more closely the fluid profile matches that of the channel plate, the smaller the required surface area. Attempting to increase the service NTU beyond the plate's NTU capability causes oversurfacing (inefficiency).

True sizing from scratch is impractical since a pressure balance on a channel-to-channel basis, from channel closest to inlet to furthest, must be achieved and when mixed plate angles are used; this is quite a challenge. Computer sizing is not just a benefit, it is a necessity for supplier's selection. Averaging methods are recommended to perform any sizing calculations.

From the APV heat-transfer handbook—*Design & Application of Paraflow-Plate Heat Exchangers* and J. Marriott's article, "Where and How To Use Plate Heat Exchangers," Chemical Engineering, April 5, 1971, there are the following equations for plate heat transfer.

$$\text{Nu} = \frac{h\text{De}}{k} = 0.28 * (\text{Re})^{0.65} * (\text{Pr})^{0.4} \qquad (11\text{-}77)$$

where De = 2 × depth of single-plate corrugation

$$G = \frac{W}{\text{Np} * w * \text{De}} \qquad (11\text{-}78)$$

Width of the plate (w) is measured from inside to inside of the channel gasket. If not available, use the tear-sheet drawing width and sub-

TABLE 11-18 Compact Exchanger Applications Guide

Design conditions	G. PHE	W. PHE	WG. PHE	BHE	DBL	MLT	STE	CP	SHE	THE
Design temperature °C	165	150	150	185	+500	+500	+500	450	+400	+500
Minimum metal temp °C	−30	−30	−30	−160	−160	−160	−160	−160	−160	−160
Design pressure MPa	2.5	2.5	0.7	3.1	+20	+20	+20	3.1	2.0	+20
Inspect for leakage	Yes	Partial	Yes	No	Yes	Yes	Yes	Partial	Yes	Yes
Mechanical cleaning	Yes	Yes/no	Yes	No	Yes	Yes	Yes/no	Yes	Yes	Yes
Chemical cleaning	Yes	Yes	Yes	Yes	Yes	Yes	Yes	Yes	Yes	Yes
Expansion capability	Yes	Yes	Yes	No	No	No	No	No	No	No
Repair	Yes	Yes/no	Yes	No	Yes	Yes	Partial	Partial	Partial	Yes
Temperature cross	Yes	Yes	Yes	Yes	Yes	Yes	Yes	Yes	Yes	No°
Surface area/unit m²	1850	900	250	50	10	150	60	275	450	High
Holdup volume	Low	Low	Low	Low	Med	Med	Low	Low	Med	High

Materials	G. PHE	W. PHE	WG. PHE	BHE	DBL	MLT	STE	CP	SHE	THE
Mild steel	No	No	No	No	Yes	Yes	Yes	Yes	Yes	Yes
Stainless	Yes	Yes	Yes	Yes	Yes	Yes	Yes	Yes	Yes	Yes
Titanium	Yes	Yes	Yes	No	Yes	Yes	Yes	Yes	Yes	Yes
Hastalloy	Yes	Yes	No	No	Yes	Yes	Yes	Yes	Yes	Yes
Nickel	Yes	Yes	No	No	Yes	Yes	Yes	Yes	Yes	Yes
Alloy 20	Yes	Yes	No	No	Yes	Yes	Yes	Yes	Yes	Yes
Incoloy 825	Yes	Yes	No	No	Yes	Yes	Yes	Yes	Yes	Yes
Monel	Yes	Yes	No	No	Yes	Yes	Yes	Yes	Yes	Yes
Impervious graphite	Yes	No	No	No	No	No	No	No	No	Yes

Service	G. PHE	W. PHE	WG. PHE	BHE	DBL	MLT	STE	CP	SHE	THE
Clean fluids	A	A	A	A	A	A	A	A	A	A
Gasket incompatibility	D	A/D	D	A	A	A	A	A	A	A
Medium viscosity	A/B	A/B	A/B	B	A	A	A/B	A/B	A	A
High viscosity	A/B	A/B	A/B	D	A	A	A/B	A/B	A	A
Slurries & pulp (fine)	B/D	D	A/B	C	A	A/B	C	B	A	A/D
Slurries & pulp (coarse)	D	D	B	D	A	B/C	D	B	A	A/D
Refrigerants	D	A	D	A	A	A	B/C	A	A	A
Thermal fluids	D	A/B	A	A/B	A	A	C	A	A	A
Vent condensers	D	D	D	D	A/D	A	A	B/C	A	A
Process condenser	D	C	D	D	A/D	A	A	B/C	B	A
Vacuum reboil/cond	D	D	D	D	A/D	B	A	B/C	C	A/C
Evaporator	D	C	C	A	B	B	A	B/C	C	A
Tight temp control	A	A	A	A	A	A	A	B	A	C
High scaling	B	B	A	D	A	A/B	B/C	B	B	A/D

°Multipass
Adapted from Alfa-Laval and Vicarb literature
A—Very good C—Fair
B—Good D—Poor

TABLE 11-19 Elastomer Selection Guide

	Uses	Avoid
Nitrile (NBR)	Oil resistant Fat resistant Food stuffs Mineral oil Water	Oxidants Acids Aromatics Alkalies Alcohols
Resin cured butyl (IIR)	Acids Lyes Strong alkalies Strong phosphoric acid Dilute mineral acids Ketones Amines Water	Fats and fatty acids Petroleum oils Chlorinated hydrocarbons Liquids with dissolved chlorine Mineral oil Oxygen rich demin. water Strong oxidants
Ethylene-propylene (EPDM)	Oxidizing agents Dilute acids Amines Water (Mostly any IIR fluid)	Oils Hot & conc. acids Very strong oxidants Fats & fatty acids Chlorinated hydrocarbons
Viton (FKM, FPM)	Water Petroleum oils Many inorganic acids (Most all NBR fluids)	Amines Ketones Esters Organic acids Liquid ammonia

tract two times the bolt diameter and subtract another 50 mm. For depth of corrugation ask supplier, or take the compressed plate pack dimension, divide by the number of plates and subtract the plate thickness from the result. The number of passages (Np) is the number of plates minus 1 then divided by 2.

Typical overall coefficients to start a rough sizing are as below. Use these in conjunction with the NTU calculated for the process. The closer the NTU matches the plate (say between 2.0 and 3.0), the higher the range of listed coefficients can be used. The narrower (smaller) the depth of corrugation, the higher the coefficient (and pressure drop), but also the lower the ability to carry through any particulate.

Water-water	5700–7400 W/(m² °C)
Steam-water	5700–7400 W/(m² °C)
Glycol/Glycol	2300–4000 W/(m² °C)
Amine/Amine	3400–5000 W/(m² °C)
Crude/Emulsion	400–1700 W/(m² °C)

Pressure drops typically can match conventional tubular exchangers. Again from the APV handbook an average correlation is as follows:

$$\Delta P = \frac{2fG^2L}{g\rho De} \qquad (11\text{-}79)$$

where $f = 2.5 \, (GDe/\mu)^{-0.3}$
g = gravitational constant

Fouling factors are typically $\frac{1}{10}$ of TEMA values or a percent oversurfacing of 10–20 percent is used. ("Sizing Plate Exchangers" Jeff Kerner, *Chemical Engineering*, November 1993).

LMTD is calculated like a 1 pass-1 pass shell and tube with no F correction factor required in most cases.

Overall coefficients are determined like shell and tube exchangers; that is, sum all the resistances, then invert. The resistances include the hot-side coefficient, the cold-side coefficient, the fouling factor (usually only a total value not individual values per fluid side) and the wall resistance.

WELDED- AND BRAZED-PLATE EXCHANGERS (W. PHE & BHE)

The title of this group of plate exchangers has been used for a great variety of designs for various applications from normal gasketed-plate exchanger services to air-preheater services on fired heaters or boilers. The intent here is to discuss more traditional heat-exchanger designs, not the heat-recovery designs on fired equipment flue-gas streams. Many similarities exist between these products but the manufacturing techniques are quite different due to the normal operating conditions these units experience.

To overcome the gasket limitations, PHE manufacturers have developed welded-plate exchangers. There are numerous approaches to this solution: weld plate pairs together with the other fluid-side conventionally gasketed, weld up both sides but use a horizontal stacking of plates method of assembly, entirely braze the plates together with copper or nickel brazing, diffusion bond then pressure form plates and bond etched, passage plates.

Most methods of welded-plate manufacturing do not allow for inspection of the heat-transfer surface, mechanical cleaning of that surface, and have limited ability to repair or plug off damage channels. Consider these limitations when the fluid is heavily fouling, has solids, or in general the repair or plugging ability for severe services.

One of the previous types has an additional consideration of the brazing material to consider for fluid compatibility. The brazing compound entirely coats both fluid's heat-transfer surfaces.

The second type, a Compabloc (CP) from Vicarb, has the advantage of removable cover plates, similar to air-cooled exchanger headers, to observe both fluids surface area. The fluids flow at 90° angles to each other on a horizontal plane. LMTD correction factors approach 1.0 for Compabloc just like the other welded and gasketed PHEs. Hydroblasting of Compabloc surfaces is also possible. The Compabloc has higher operating conditions than PHE's or W-PHE.

The performances and estimating methods of welded PHEs match those of gasketed PHEs in most cases, but normally the Compabloc, with larger depth of corrugations, can be lower in overall coefficient. Some extensions of the design operating conditions are possible with welded PHEs, most notably is that cryogenic applications are possible. Pressure vessel code acceptance is available on most units.

SPIRAL-PLATE EXCHANGERS (SHE)

Description   The *spiral-plate heat exchanger* (SHE) may be one exchanger selected primarily on its virtues and not on its initial cost. SHEs offer high reliability and on-line performance in many severely fouling services such as slurries.

The SHE is formed by rolling two strips of plate, with welded-on spacer studs, upon each other into clock-spring shape. This forms two passages. Passages are sealed off on one end of the SHE by welding a bar to the plates; hot and cold fluid passages are sealed off on opposite ends of the SHE. A single rectangular flow passage is now formed for each fluid, producing very high shear rates compared to tubular designs. Removable covers are provided on each end to access and clean the entire heat transfer surface. Pure countercurrent flow is achieved and LMTD correction factor is essentially = 1.0.

Since there are no dead spaces in a SHE, the helical flow pattern combines to entrain any solids and create high turbulence creating a self-cleaning flow passage.

There are no thermal-expansion problems in spirals. Since the center of the unit is not fixed, it can torque to relieve stress.

The SHE can be expensive when only one fluid requires a high-alloy material. Since the heat-transfer plate contacts both fluids, it is required to be fabricated out of the higher alloy. SHEs can be fabricated out of any material that can be cold-worked and welded.

The channel spacings can be different on each side to match the flow rates and pressure drops of the process design. The spacer studs are also adjusted in their pitch to match the fluid characteristics.

As the coiled plate spirals outward, the plate thickness increases from a minimum of 2 mm to a maximum (as required by pressure) up to 10 mm. This means relatively thick material separates the two fluids compared to tubing of conventional exchangers. Pressure vessel code conformance is a common request.

Applications   The most common applications that fit SHE are slurries. The rectangular channel provides high shear and turbulence to sweep the surface clear of blockage and causes no distribution problems associated with other exchanger types. A localized restriction causes an increase in local velocity which aids in keeping the unit free flowing. Only fibers that are long and stringy cause SHE to have a blockage it cannot clear itself.

As an additional antifoulant measure, SHEs have been coated with a phenolic lining. This provides some degree of corrosion protection as well, but this is not guaranteed due to pinholes in the lining process.

There are three types of SHE to fit different applications:

Type I is the spiral-spiral flow pattern. It is used for all heating and cooling services and can accommodate temperature crosses such as lean/rich services in one unit. The removable covers on each end allow access to one side at a time to perform maintenance on that fluid side. Never remove a cover with one side under pressure as the unit will telescope out like a collapsible cup.

Type II units are the condenser and reboiler designs. One side is spiral flow and the other side is in cross flow. These SHEs provide very stable designs for vacuum condensing and reboiling services. A SHE can be fitted with special mounting connections for reflux-type vent-condenser applications. The vertically mounted SHE directly attaches on the column or tank.

Type III units are a combination of the Type I and Type II where part is in spiral flow and part is in cross flow. This SHE can condense and subcool in a single unit.

The unique channel arrangement has been used to provide on-line cleaning, by switching fluid sides to clean the fouling (caused by the fluid that previously flowed there) off the surface. Phosphoric acid coolers use pond water for cooling and both sides foul; water, as you expect, and phosphoric acid deposit crystals. By reversing the flow sides, the water dissolves the acid crystals and the acid clears up the organic fouling. SHEs are also used as oleum coolers, sludge coolers/heaters, slop oil heaters, and in other services where multiple-flow-passage designs have not performed well.

Design   A thorough article by P.E. Minton of Union Carbide called "Designing Spiral-Plate Heat Exchangers," appeared in *Chemical Engineering*, May 4, 1970. It covers the design in detail. Also an article in *Chemical Engineering Progress* titled "Applications of Spiral Plate Heat Exchangers" by A. Hargis, A. Beckman, and J. Loicano appeared in July 1967, provides formulae for heat-transfer and pressure-drop calculations.

Spacings are from 6.35 to 31.75 mm (in 6.35 mm increments) with 9.5 mm the most common. Stud densities are 60×60 to 110×110 mm, the former the most common. The width (measured to the spiral flow passage), is from 150 to 2500 mm (in 150 mm increments). By varying the spacing and the width, separately for each fluid, velocities can be maintained at optimum rates to avoid fouling tendencies or utilize the allowable pressure drop most effectively. Diameters can reach 1500 mm. The total surface areas exceed 465 sqm. Materials that work harder are not suitable for spirals since hot-forming is not possible and heat treatment after forming is impractical.

$$\text{Nu} = \frac{H\text{De}}{k} = 0.0315 \, (\text{Re})^{0.8} \, (\text{Pr})^{0.25} \, (\mu/\mu_w)^{0.17} \qquad (11\text{-}80)$$

where
$$\text{De} = 2 \times \text{spacing}$$
$$\text{Flow area} = \text{width} \times \text{spacing}$$

$$\Delta P = \frac{LV^2\rho}{1.705\text{E}{-}03} * 1.45 \quad (1.45 \text{ for } 60 \times 60 \text{ mm studs}) \qquad (11\text{-}81)$$

LMTD and overall coefficient are calculated like in PHE section above.

BRAZED-PLATE-FIN HEAT EXCHANGER

Brazed-aluminum-plate-fin heat exchangers (or core exchangers or cold boxes) as they are sometimes called, were first manufactured for the aircraft industry during World War II. In 1950, the first tonnage air-separation plant with these compact, lightweight, reversing heat exchangers began producing oxygen for a steel mill. Aluminum-plate-fin exchangers are used in the process and gas-separation industries, particularly for services below −45°C.

Core exchangers are made up of a stack of rectangular sheets of aluminum separated by a wavy, usually perforated, aluminum fin. Two ends are sealed off to form a passage (see Fig. 11-50). The layers have the wavy fins and sealed ends alternating at 90° to each. Aluminum half-pipe-type headers are attached to the open ends to route the fluids into the alternating passages. Fluids usually flow at this same 90° angle to each other. Variations in the fin height, number of passages, and the length and width of the prime sheet allow for the core exchanger to match the needs of the intended service.

Design conditions range in pressures from full vacuum to 96.5 bar g and in temperatures from −269°C to 200°C. This is accomplished meeting the quality standards of most pressure vessel codes.

Applications are varied for this highly efficient, compact exchanger. Mainly it is seen in the cryogenic fluid services of air-separation plants, refrigeration trains like in ethylene plants, and in natural-gas processing plants. Fluids can be all vapor, liquid, condensing, or vaporizing. Multifluid exchangers and multiservice cores, that is one exchanger with up to 10 different fluids, are common for this type of product. Cold boxes are a group of cores assembled into a single structure or module, prepiped for minimum field connections. (Data obtained from ALTEC INTERNATIONAL. For detailed information refer to *GPSA Engineering Handbook* Section 9.)

PLATE-FIN TUBULAR EXCHANGERS (PFE)

Description These shell and tube exchangers are designed to use a group of tightly spaced plate fins to increase the shellside heat transfer performance as fins do on double-pipe exchangers. In this design, a series of very thin plates (fins), usually of copper or aluminum material, are punched to the same pattern as the tube layout, spaced very close together, and mechanically bonded to the tube. Fin spacing is 315–785 FPM (Fins Per Meter) with 550 FPM most common. The fin thicknesses are 0.24 mm for aluminum and 0.19 mm for copper. Surface-area ratios over bare prime-tube units can be 20:1 to 30:1. The cost of the additional plate-fin material, without a reduction in shell diameter in many cases, and increased fabrication has to be offset by the total reduction of plot space and prime tube-surface area. The more costly the prime tube or plot space cost, the better the payout for this design. A rectangular tube layout is normally used, *no tubes in the window* (NTIW). The window area (where no tubes are) of the plate-fins are cut out. This causes a larger shell diameter for a given tube count compared to conventional tubular units. A dome area on top and bottom of the inside of the shell has been created for the fluid to flow along the tube length. In order to exit the unit the fluid must flow across the plate-finned tube bundle with extremely low pressure loss. The units from the outside and from the tubeside appear like any conventional shell and tube exchanger.

Applications Two principal applications are rotating equipment oil coolers and compressor inter- and after-coolers. Although seemingly different applications, both rely on the shellside finning to enhance the heat transfer of low heat-transfer characteristic fluids,

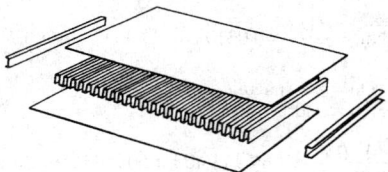

FIG. 11-50 Exploded view of a typical plate-fin arrangement. (*Trane Co.*)

viscous oils, and gases. By nature of the fluids and their applications, both are clean servicing. The tightly spaced fins would be a maintenance problem otherwise.

Design The economics usually work out in the favor of gas coolers when the centrifugal machine's flow rate reaches about 5000 scfm. The pressure loss can be kept to 7.0 kPa in most cases. When the ratio of $A_i h_i$ to $A_s h_s$ is 20:1, is another point to consider these plate-fin designs. Vibration is practically impossible with this design, and uses in reciprocating compressors are possible due to this.

Marine and hydraulic-oil coolers use these characteristics to enhance the coefficient of otherwise poorly performing fluids. The higher metallurgies in marine applications like 90/10 Cu-Ni afford the higher cost of plate-fin design to be offset by the less amount of alloy material being used. On small hydraulic coolers, these fins usually allow one to two size smaller coolers for the package and save skid space and initial cost.

Always check on metallurgy compatibility and cleanliness of the shellside fluid! (Data provided by Bos-Hatten and ITT-Standard.)

SPIRAL-TUBE EXCHANGERS (STE)

Description These exchangers are typically a series of stacked helical-coiled tubes connected to manifolds, then inserted into a casing or shell. They have many advantages like spiral-plate designs, such as avoiding differential expansion problems, acceleration effects of the helical flow increasing the heat transfer coefficient, and compactness of plot area. They are typically selected because of their economical design.

The most common form has both sides in helical flow patterns, pure countercurrent flow is followed and the LMTD correction factor approaches 1.0. Temperature crosses are possible in single units. Like the spiral-plate unit, different configurations are possible for special applications.

Tube material includes any that can be formed into a coil, but usually copper, copper alloys, and stainless steel are most common. The casing or shell material can be cast iron, cast steel, cast bronze, fabricated steel, stainless, and other high-alloy materials. Units are available with pressure vessel code conformance.

The data provided herein has been supplied by Graham Mfg. for their units called Heliflow.

Applications The common Heliflow applications are tank-vent condensers, sample coolers, pump-seal coolers, and steam-jet vacuum condensers. Instant water heaters, glycol/water services, and cryogenic vaporizers use the spiral tube's ability to reduce thermally induced stresses caused in these applications.

Many other applications are well suited for spiral tube units but many believe only small surface areas are possible with these units. Graham Mfg. states units are available to 60 m². Their ability to polish the surfaces, double-wall the coil, use finned coil, and insert static mixers, among others configurations in design, make them quite flexible. Tubeside design pressures can be up to 69000 kPa. A cross-flow design on the external surface of the coil is particularly useful in steam-jet ejector condensing service. These Heliflow units, can be made very cost-effective, especially in small units. The main differences, compared to spiral plate, is that the tubeside cannot be cleaned except chemically and that multiple flow passages make tubeside slurry applications (or fouling) impractical.

Design The fluid flow is similar to the spiral-plate exchangers, but through parallel tube passages. Graham Mfg. has a liquid-liquid sizing pamphlet available from their local distributor. An article by M.A. Noble, J.S. Kamlani, and J.J. McKetta "Heat Transfer in Spiral Coils", was published in *Petroleum Engineer*, April 1952 p. 723, discussing sizing techniques.

The tubeside fluid must be clean or at least chemically cleanable. With a large number of tubes in the coil, cleaning of inside surfaces is not totally reliable. Fluids that attack stressed materials such as chlorides should be reviewed as to proper coil-material selection. Fluids that contain solids can be a problem due to erosion of relatively thin coil materials unlike the thick plates in spiral-plate units and multiple, parallel, fluid passages compared to a single passage in spiral-plate units.

GRAPHITE HEAT EXCHANGERS

Impervious graphite exchangers now come in a variety of geometries to suit the particular requirements of the service. They include cubic block form, drilled cylinder block, shell and tube, and plate and frame.

Description   Graphite is one of three crystalline forms of carbon. The other two are diamond and charcoal. Graphite has a hexagonal crystal structure, diamond is cubic, and charcoal is amorphous. Graphite is inert to most chemicals and resists corrosion attack. It is however porous and to be used, it must be impregnated with a resin sealer. Two main resins used are phenolic and PTFE with furan (one currently being phased out of production). Selection of resins include chemical compatibility, operating temperatures, and type of unit to be used. For proper selection, consult with a graphite supplier.

Shell-and-tube units in graphite were started by Karbate in 1939. The European market started using block design in the 1940s. Both technologies utilize the high thermal conductivity of the graphite material to compensate for the poor mechanical strength. The thicker materials needed to sustain pressure do not adversely impede the heat transfer. Maximum design pressures range from 0.35 to 1.0 kPa depending on type and size of exchanger. Design temperature is dependent on the fluids and resin selection, the maximum is 230 °C.

In all situations, the graphite heat transfer surface is contained within a metal structure or a shell (graphite lined on process side) to maintain the design pressure. For shell and tube units, the design is a packed floating tubesheet at both ends within a shell and channel. For stacked block design, the standardize blocks are glued together with special adhesives and compressed within a framework that includes manifold connections for each fluid. The cylindrical block unit is a combination of the above two with blocks glued together and surrounded by a pressure retaining shell. Pressure vessel code conformance of the units is possible due to the metallic components of these designs. Since welding of graphite is not possible, the selection and application of the adhesives used are critical to the proper operating of these units. Tube to tubesheet joints are glued since rolling of tubes into tubesheet is not possible. The packed channels and gasketed manifold connections are two areas of additional concern when selecting sealants for these units.

Applications and Design   The major applications for these units are in the acid-related industries. Sulfuric, phosphoric, and hydrochloric acids require either very costly metals or impervious graphite. Usually graphite is the more cost-effective material to be used. Applications are increasing in the herbicide and pharmaceutical industries as new products with chlorine and fluorine compounds expand. Services are coolers, condensers, and evaporators, basically all services requiring this material. Types of units are shell-and-tube, block-type (circular and rectangular), and plate-and-frame-type exchangers. The design of the shell-and-tube units are the same as any but the design characteristics of tubes, spacing, and thickness are unique to the graphite design. The block and plate and frame also can be evaluated using techniques previously addressed but again, the unique characteristics of the graphite materials require input from a reputable supplier. Most designs will need the supplier to provide the most cost-effective design for the immediate and future operation of the exchangers. Also, consider the entire system design as some condensers and/or evaporators can be integral with their associated column.

CASCADE COOLERS

Cascade coolers are a series of standard pipes, usually manifolded in parallel, and connected in series by vertically or horizontally oriented U-bends. Process fluid flows inside the pipe entering at the bottom and water trickles from the top downward over the external pipe surface. The water is collected from a trough under the pipe sections, cooled, and recirculated over the pipe sections. The pipe material can be any of the metallic and also glass, impervious graphite, and ceramics. The tubeside coefficient and pressure drop is as in any circular duct. The water coefficient (with Re number less than 2100) is calculated from the following equation by W.H. McAdams, T.B. Drew, and G.S. Bays Jr., from the ASME trans. **62**, 627–631 (1940).

$h = 218 * (G'/D_o)^{1/3}$ (W/m² °C) (11-82)
$G' = m/(2L)$
m = water rate (kg/hr)
L = length of each pipe section (meter)
D_o = outside diameter of pipe (meter)

LMTD corrections are per Fig. 11-4 i or j depending on U-bend orientation.

BAYONET-TUBE EXCHANGERS

This type of exchanger gets its name from its design which is similar to a bayonet sword and its associated scabbard or sheath. The bayonet tube is a smaller-diameter tube inserted into a larger-diameter tube that has been capped at one end. The fluid flow is typically entering the inner tube, exiting, hitting the cap of the larger tube, and returning the opposite direction in the annular area. The design eliminates any thermal expansion problems. It also creates a unique nonfreeze-type tube-side for steam heating of cryogenic fluids, the inner tube steam keeps the annulus condensate from freezing against the cold shellside fluid. This design can be expensive on a surface-area basis due to the need of a double channel design and only the outer tube surface is used to transfer heat. LMTD calculations for nonisothermal fluid are quite extensive and those applications are far too few to attempt to define it. The heat transfer is like the annular calculation of a double-pipe unit. The shellside is a conventional-baffled shell-and-tube design.

ATMOSPHERIC SECTIONS

These consist of a rectangular bundle of tubes in similar fashion to air cooler bundles, placed just under the cooled water distribution section of a cooling tower. It, in essence, combines the exchanger and cooling tower into a single piece of equipment. This design is only practical for single-service cooler/condenser applications, and expansion capabilities are not provided. The process fluid flows inside the tubes and the cooling tower provides cool water that flows over the outside of the tube bundle. Water quality is critical for these applications to prevent fouling or corrosive attack on the outside of the tube surfaces and to prevent blockage of the spray nozzles. The initial and operating costs are lower than separate cooling tower and exchanger. Principal applications now are in the HVAC, Refrigeration and Industrial systems. Sometimes these are called "Wet Surface Air Coolers"

$$h = 1729 \ [(\text{m}^2/\text{hr})/\text{face area m}^2]^{1/3} \qquad (11\text{-}83)$$

NONMETALLIC HEAT EXCHANGERS

Another growing field is that of nonmetallic heat exchanger designs which typically are of the shell and tube or coiled-tubing type. The graphite units were previously discussed but numerous other materials are available. The materials include Teflon, PVDF, glass, ceramic, and others as the need arises.

When using these types of products, consider the following topics and discuss the application openly with experienced suppliers.

1.   The tube-to-tubesheet joint, how is it made? Many use "O" rings to add another material to the selection process. Preference should be given to a fusing technique of similar material.

2.   What size tube or flow passage is available? Small tubes plug unless filtration is installed. Size of filtering is needed from the supplier.

3.   These materials are very sensitive to temperature and pressure. Thermal or pressure shocks must be avoided.

4.   Thermal conductivity of these materials is very low and affects the overall coefficient. When several materials are compatible, explore all of them, as final cost is not always the same as raw material costs.

PVDF HEAT EXCHANGERS

These shell-and-tube-type exchangers are similar to the Teflon designs but have some mechanical advantages over Teflon units. First the tubes are available in 9.5 mm sizes which reduces the chances of plugging that are found in Teflon units with unfiltered fluids. Second, the material has higher strength even at lower temperatures almost double. Larger units are possible with PVDF materials.

Tube to tubesheet joints, a weakness of most nonmetallic units, are fused by special techniques that do not severely affect the chemical suitability of the unit. Some nonmetallics use Teflon or "O" rings that add an extra consideration to material selection.

The shell is usually a steel design and, like the graphite units before, can obtain pressure-vessel certification.

CERAMIC HEAT EXCHANGERS

These include glass, silicon carbide, and similar variations. Even larger tubes are available in these materials, up to 19-mm diameter. They have high thermal conductivities and are usually very smooth surfaces to resist fouling. Very high material/fluid compatibility is seen for these products, not many fluids are excluded. Brittleness is a consideration of these materials and a complete discussion of the service with an experienced supplier is warranted. The major selection criteria to explore is the use of "O" rings and other associated joints at tubesheet. The shell is steel in most cases.

TEFLON HEAT EXCHANGERS

Teflon tube shell-and-tube heat exchangers (Ametek) made with tubes of chemically inert Teflon fluorocarbon resin are available. The tubes are 0.25-in OD by 0.20-in ID, 0.175-in OD by 0.160-in ID, or 0.125-in OD by 0.100-in ID (in × 25.4 equal mm). The larger tubes are primarily used when pressure-drop limitations or particles reduce the effectiveness of smaller tubes. These heat exchangers generally operate at higher pressure drops than conventional units and are best suited for relatively clean fluids. Being chemically inert, the tubing has many applications in which other materials corrode. Fouling is negligible because of the antistick properties of Teflon.

The heat exchangers are of single-pass, countercurrent-flow design with removable tube bundles. Tube bundles are made of straight flexible tubes of Teflon joined together in integral honeycomb tube sheets. Baffles and O-ring gaskets are made of Teflon. Standard shell diameters are 102, 204, and 254 mm (4, 8, and 10 in). Tube counts range from 105 to 2000. Surface varies from 1.9 to 87 m^2 (20 to 940 ft^2). Tube lengths vary from 0.9 to 4.9 m (3 to 16 ft). At 37.8°C (100°F) maximum operating gauge pressures are 690 kPa (100 lbf/in^2) internal and 379 kPa (55 lbf/in^2) external. At 149°C (300°F) the maximum pressures are 207 kPa (30 lbf/in^2) internal and 124 kPa (18 lbf/in^2) external.

HEAT EXCHANGERS FOR SOLIDS

This section describes equipment for heat transfer to or from solids by the indirect mode. Such equipment is so constructed that the solids load (burden) is separated from the heat-carrier medium by a wall; the two phases are never in direct contact. Heat transfer is by conduction based on diffusion laws. Equipment in which the phases are in direct contact is covered in other sections of this *Handbook,* principally in Sec. 20.

Some of the devices covered here handle the solids burden in a static or laminar-flowing bed. Other devices can be considered as continuously agitated kettles in their heat-transfer aspect. For the latter, unit-area performance rates are higher.

Computational and graphical methods for predicting performance are given for both major heat-transfer aspects in Sec. 10. In solids heat processing with indirect equipment, the engineer should remember that the heat-transfer capability of the wall is many times that of the solids burden. Hence the solids properties and bed geometry govern the rate of heat transfer. This is more fully explained earlier in this section. Only limited resultant (not predictive) and "experience" data are given here.

EQUIPMENT FOR SOLIDIFICATION

A frequent operation in the chemical field is the removal of heat from a material in a molten state to effect its conversion to the solid state. When the operation is carried on batchwise, it is termed casting, but when done continuously, it is termed flaking. Because of rapid heat transfer and temperature variations, jacketed types are limited to an initial melt temperature of 232°C (450°F). Higher temperatures [to 316°C (600°F)] require extreme care in jacket design and cooling-liquid flow pattern. Best performance and greatest capacity are obtained by (1) holding precooling to the minimum and (2) optimizing the cake thickness. The latter cannot always be done from the heat-transfer standpoint, as size specifications for the end product may dictate thickness.

Table Type   This is a simple flat metal sheet with slightly upturned edges and jacketed on the underside for coolant flow. For many years this was the mainstay of food processors. Table types are still widely used when production is in small batches, when considerable batch-to-batch variation occurs, for pilot investigation, and when the cost of continuous devices is unjustifiable. Slab thicknesses are usually in the range of 13 to 25 mm (½ to 1 in). These units are homemade, with no standards available. Initial cost is low, but operating labor is high.

Agitated-Pan Type   A natural evolution from the table type is a circular flat surface with jacketing on the underside for coolant flow and the added feature of a stirring means to sweep over the heat-transfer surface. This device is the agitated-pan type (Fig. 11-51). It is a batch-operation device. Because of its age and versatility it still serves a variety of heat-transfer operations for the chemical-process industries. While the most prevalent designation is agitated-pan dryer (in this mode, the burden is heated rather than cooled), considerable use is made of it for solidification applications. In this field, it is particularly suitable for processing burdens that change phase (1) slowly, by "thickening," (2) over a wide temperature range, (3) to an amorphous solid form, or (4) to a soft semigummy form (versus the usual hard crystalline structure).

The stirring produces the end product in the desired divided-solids form. Hence, it is frequently termed a "granulator" or a "crystallizer." A variety of factory-made sizes in various materials of construction are available. Initial cost is modest, while operating cost is rather high (as is true of all batch devices), but the ability to process "gummy" burdens and/or simultaneously effect two unit operations often yields an economical application.

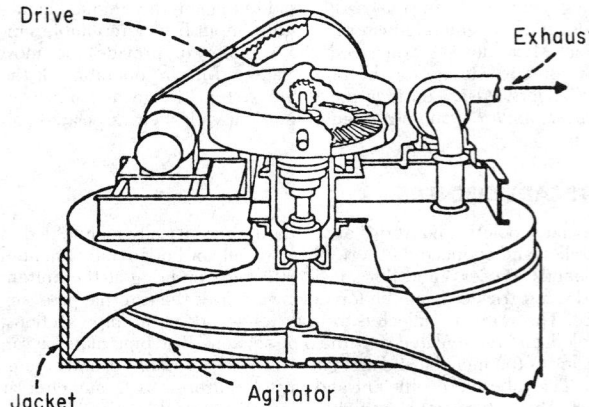

FIG. 11-51   Heat-transfer equipment for solidification (with agitation); agitated-pan type.

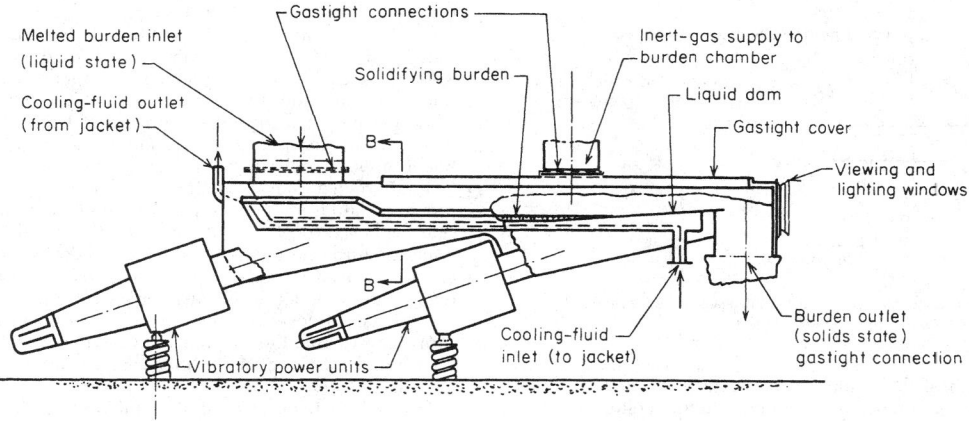

FIG. 11-52 Heat-transfer equipment for batch solidification; vibrating-conveyor type. (*Courtesy of Jeffrey Mfg. Co.*)

Vibratory Type This construction (Fig. 11-52) takes advantage of the burden's special needs and the characteristic of vibratory actuation. A flammable burden requires the use of an inert atmosphere over it and a suitable nonhazardous fluid in the jacket. The vibratory action permits construction of rigid self-cleaning chambers with simple flexible connections. When solidification has been completed and vibrators started, the intense vibratory motion of the whole deck structure (as a rigid unit) breaks free the friable cake [up to 76 mm (3 in) thick], shatters it into lumps, and conveys it up over the dam to discharge. Heat-transfer performance is good, with overall coefficient U of about 68 W/(m²·°C) [12 Btu/(h·ft²·°F)] and values of heat flux q in the order of 11,670 W/m² [3700 Btu/(h·ft²)]. Application of timing-cycle controls and a surge hopper for the discharge solids facilitates automatic operation of the caster and continuous operation of subsequent equipment.

Belt Types The patented metal-belt type (Fig. 11-53a), termed the "water-bed" conveyor, features a thin wall, a well-agitated fluid side for a thin water film (there are no rigid welded jackets to fail), a stainless-steel or Swedish-iron conveyor belt "floated" on the water

with the aid of guides, no removal knife, and cleanability. It is mostly used for cake thicknesses of 3.2 to 15.9 mm (⅛ to ⅝ in) at speeds up to 15 m/min (50 ft/min), with 45.7-m (150-ft) pulley centers common. For 25- to 32-mm (1- to 1¼-in) cake, another belt on top to give two-sided cooling is frequently used. Applications are in food operations for cooling to harden candies, cheeses, gelatins, margarines, gums, etc.; and in chemical operations for solidification of sulfur, greases, resins, soaps, waxes, chloride salts, and some insecticides. Heat transfer is good, with sulfur solidification showing values of $q = 5800$ W/m² [1850 Btu/(h·ft²)] and $U = 96$ W/(m²·°C) [17 Btu/(h·ft²·°F)] for a 7.9-mm (⁵⁄₁₆-in) cake.

The submerged metal belt (Fig. 11-53b) is a special version of the metal belt to meet the peculiar handling properties of pitch in its solidification process. Although adhesive to a dry metal wall, pitch will not stick to the submerged wetted belt or rubber edge strips. Submergence helps to offset the very poor thermal conductivity through two-sided heat transfer.

A fairly recent application of the water-cooled metal belt to solidification duty is shown in Fig. 11-54. The operation is termed pastilliz-

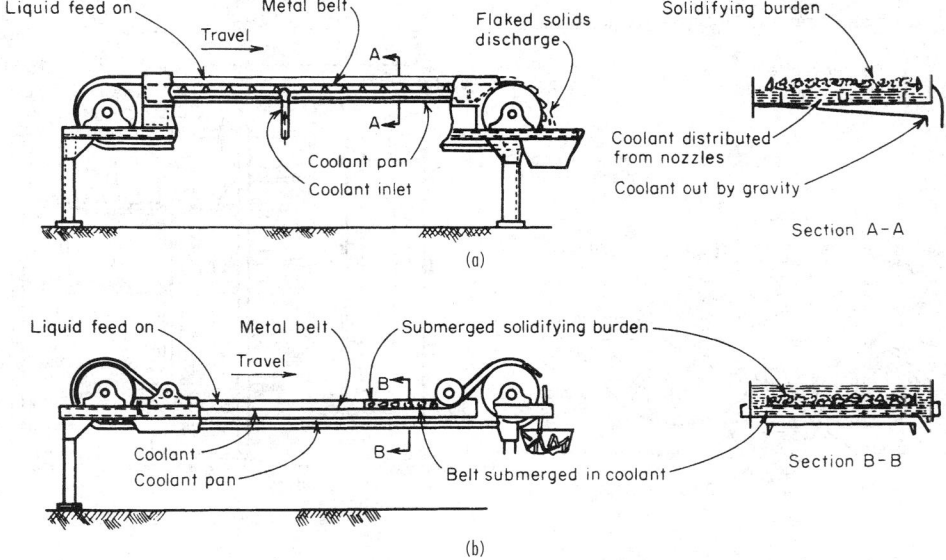

FIG. 11-53 Heat-transfer equipment for continuous solidification. (*a*) Cooled metal belt. (*Courtesy of Sandvik, Inc.*) (*b*) Submerged metal belt. (*Courtesy of Sandvik, Inc.*)

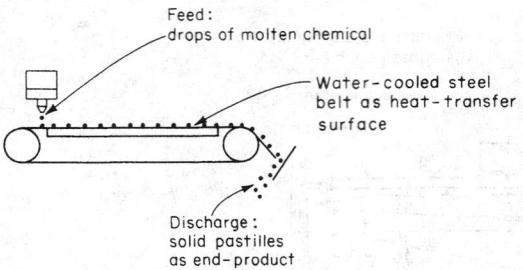

FIG. 11-54   Heat-transfer equipment for solidification; belt type for the operation of pastillization. (*Courtesy of Sandvik, Inc.*)

ing from the form of the solidified end product, termed "pastilles." The novel feature is a one-step operation from the molten liquid to a fairly uniformly sized and shaped product without intermediate operations on the solid phase.

Another development features a nonmetallic belt [*Plast. Des. Process.,* **13** (July 1968)]. When rapid heat transfer is the objective, a glass-fiber, Teflon-coated construction in a thickness as little as 0.08 mm (0.003 in) is selected for use. No performance data are available, but presumably the thin belt permits rapid heat transfer while taking advantage of the nonsticking property of Teflon. Another development [*Food Process. Mark.,* **69** (March 1969)] is extending the capability of belt solidification by providing use of subzero temperatures.

Rotating-Drum Type   This type (Fig. 11-55 a and b) is not an adaptation of a material-handling device (though volumetric material throughput is a first consideration) but is designed specifically for

heat-transfer service. It is well engineered, established, and widely used. The twin-drum type (Fig. 11-55b) is best suited to thin [0.4- to 6-mm (¹⁄₆₄ to ¼-in)] cake production. For temperatures to 149°C (300°F) the coolant water is piped in and siphoned out. Spray application of coolant water to the inside is employed for high-temperature work, permitting feed temperatures to at least 538°C (1000°F), or double those for jacketed equipment. Vaporizing refrigerants are readily applicable for very low temperature work.

The burden must have a definite solidification temperature to assure proper pickup from the feed pan. This limitation can be overcome by side feeding through an auxiliary rotating spreader roll. Application limits are further extended by special feed devices for burdens having oxidation-sensitive and/or supercooling characteristics. The standard double-drum model turns downward, with adjustable roll spacing to control sheet thickness. The newer twin-drum model (Fig. 11-55b) turns upward and, though subject to variable cake thickness, handles viscous and indefinite solidification-temperature-point burden materials well.

Drums have been successfully applied to a wide range of chemical products, both inorganic and organic, pharmaceutical compounds, waxes, soaps, insecticides, food products to a limited extent (including lard cooling), and even flake-ice production. A novel application is that of using a water-cooled roll to pick up from a molten-lead bath and turn out a 1.2-m- (4-ft-) wide continuous sheet, weighing 4.9 kg/m² (1 lb/ft²), which is ideal for a sound barrier. This technique is more economical than other sheeting methods [*Mech. Eng.,* **631** (March 1968)].

Heat-transfer performance of drums, in terms of reported heat flux is: for an 80°C (176°F) melting-point wax, 7880 W/m² [2500 Btu/(h·ft²)]; for a 130°C (266°F) melting-point organic chemical, 20,000 W/m² [6500 Btu/(h·ft²)]; and for high- [318°C (604°F)] melting-point caustic soda (water-sprayed in drum), 95,000 to 125,000 W/m² [30,000

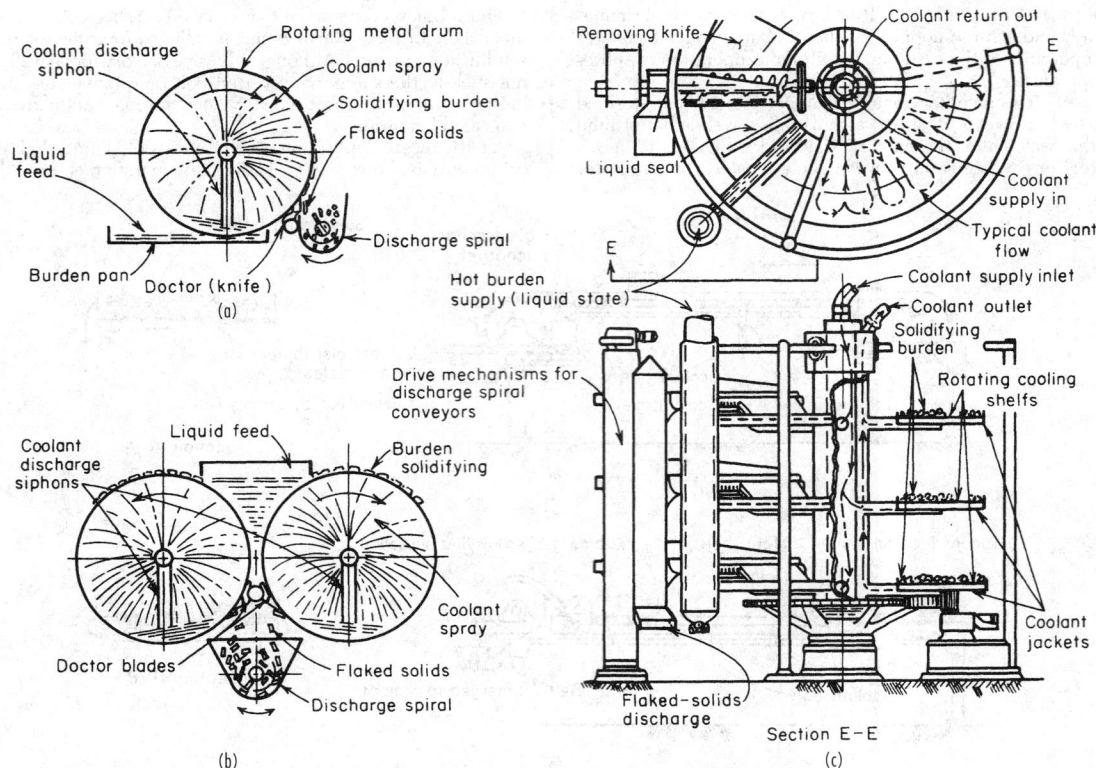

FIG. 11-55   Heat-transfer equipment for continuous solidification. (*a*) Single drum. (*b*) Twin drum. (*c*) Roto-shelf. (*Courtesy of Buflovak Division, Blaw-Knox Food & Chemical Equipment, Inc.*)

to 40,000 Btu/(h · ft²)], with overall coefficients of 340 to 450 W/(m²·°C) [60 to 80 Btu/(h·ft²·°F)]. An innovation that is claimed often to increase these performance values by as much as 300 percent is the addition of hoods to apply impinging streams of heated air to the solidifying and drying solids surface as the drums carry it upward [*Chem. Eng.*, **74**, 152 (June 19, 1967)]. Similar rotating-drum indirect heat-transfer equipment is also extensively used for drying duty on liquids and thick slurries of solids (see Sec. 20).

Rotating-Shelf Type   The patented **Roto-shelf** type (Fig. 11-55c) features (1) a large heat-transfer surface provided over a small floor space and in a small building volume, (2) easy floor cleaning, (3) non-hazardous machinery, (4) stainless-steel surfaces, (5) good control range, and (6) substantial capacity by providing as needed 1 to 10 shelves operated in parallel. It is best suited for thick-cake production and burden materials having an indefinite solidification temperature. Solidification of liquid sulfur into 13- to 19-mm- (½- to ¾-in-) thick lumps is a successful application. Heat transfer, by liquid-coolant circulation through jackets, limits feed temperatures to 204°C (400°F). Heat-transfer rate, controlled by the thick cake rather than by equipment construction, should be equivalent to the belt type. Thermal performance is aided by applying water sprayed directly to the burden top to obtain two-sided cooling.

EQUIPMENT FOR FUSION OF SOLIDS

The thermal duty here is the opposite of solidification operations. The indirect heat-transfer equipment suitable for one operation is not suitable for the other because of the material-handling rather than the thermal aspects. Whether the temperature of transformation is a definite or a ranging one is of little importance in the selection of equipment for fusion. The burden is much agitated, but the beds are deep.

Only fair overall coefficient values may be expected, although heat-flux values are good.

Horizontal-Tank Type   This type (Fig. 11-56a) is used to transfer heat for melting or cooking dry powdered solids, rendering lard from meat-scrap solids, and drying divided solids. Heat-transfer coefficients are 17 to 85 W/(m²·°C) [3 to 15 Btu/(h·ft²·°F)] for drying and 28 to 140 W/(m²·°C) [5 to 25 Btu/(h·ft²·°F)] for vacuum and/or solvent recovery.

Vertical Agitated-Kettle Type   Shown in Fig. 11-56b this type is used to cook, melt to the liquid state, and provide or remove reaction heat for solids that vary greatly in "body" during the process so that material handling is a real problem. The virtues are simplicity and 100 percent cleanability. These often outweigh the poor heat-transfer aspect. These devices are available from the small jacketed type illustrated to huge cast-iron direct-underfired bowls for calcining gypsum. Temperature limits vary with construction; the simpler jackets allow temperatures to 371°C (700°F) (as with Dowtherm), which is not true of all jacketed equipment.

Mill Type   Figure 11-56c shows one model of roll construction used. Note the ruggedness, as it is a *power device* as well as one for *indirect* heat transfer, employed to knead and heat a mixture of dry powdered-solid ingredients with the objective of reacting and reforming via fusion to a consolidated product. In this compounding operation, frictional heat generated by the kneading may require heat-flow reversal (by cooling). Heat-flow control and temperature-level considerations often predominate over heat-transfer performance. Power and mixing considerations, rather than heat transfer, govern. The two-roll mill shown is employed in compounding raw plastic, rubber, and rubberlike elastomer stocks. Multiple-roll mills less knives (termed calenders) are used for continuous sheet or film production in widths up to 2.3 m (7.7 ft). Similar equipment is employed in the chemical compounding of inks, dyes, paint pigments, and the like.

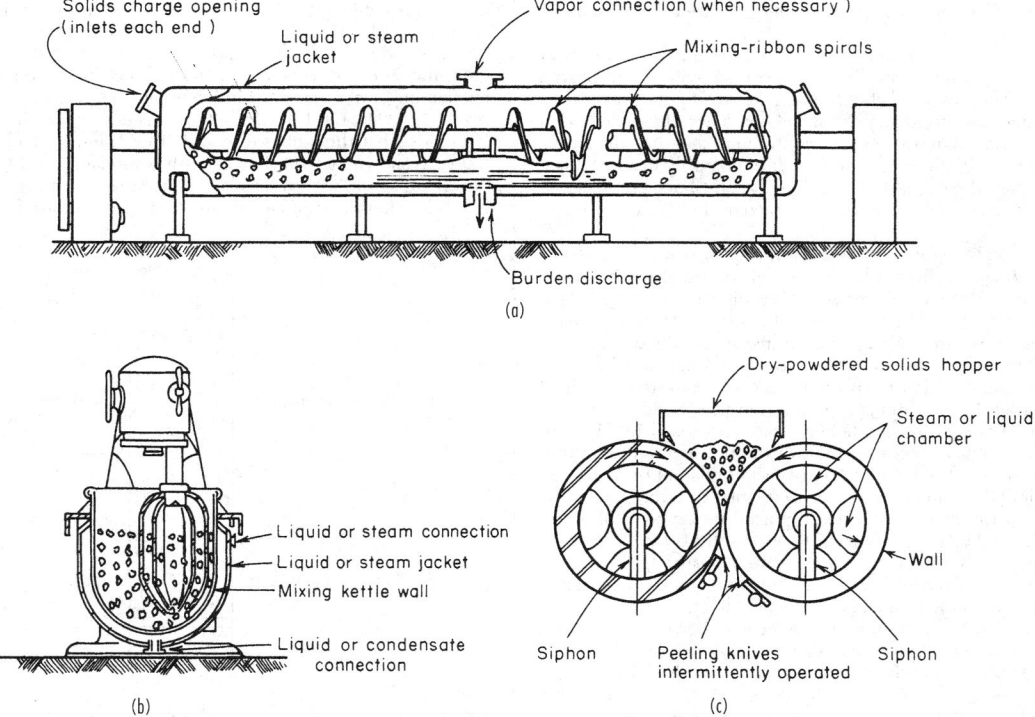

FIG. 11-56   Heat-transfer equipment for fusion of solids. (*a*) Horizontal-tank type. (*Courtesy of Struthers Wells Corp.*) (*b*) Agitated kettle. (*Courtesy of Read-Standard Division, Capital Products Co.*) (*c*) Double-drum mill. (*Courtesy of Farrel-Birmingham Co.*)

HEAT-TRANSFER EQUIPMENT FOR SHEETED SOLIDS

Cylinder Heat-Transfer Units   Sometimes called "can" dryers or drying rolls, these devices are differentiated from drum dryers in that they are used for solids in flexible continuous-sheet form, whereas drum dryers are used for liquid or paste forms. The construction of the individual cylinders, or drums, is similar in most respects to that of drum dryers. Special designs are used to obtain uniform distribution of steam within large drums when uniform heating across the drum surface is critical.

A cylinder dryer may consist of one large cylindrical drum, such as the so-called Yankee dryer, but more often it comprises a number of drums arranged so that a continuous sheet of material may pass over them in series. Typical of this arrangement are Fourdrinier-paper-machine dryers, cellophane dryers, slashers for textile piece goods and fibers, etc. The multiple cylinders are arranged in various ways. Generally they are staggered in two horizontal rows. In any one row, the cylinders are placed close together. The sheet material contacts the undersurface of the lower rolls and passes over the upper rolls, contacting 60 to 70 percent of the cylinder surface. The cylinders may also be arranged in a single horizontal row, in more than two horizontal rows, or in one or more vertical rows. When it is desired to contact only one side of the sheet with the cylinder surface, unheated guide rolls are used to conduct the sheeting from one cylinder to the next. For sheet materials that shrink on processing, it is frequently necessary to drive the cylinders at progressively slower speeds through the dryer. This requires elaborate individual electric drives on each cylinder.

Cylinder dryers usually operate at atmospheric pressure. However, the Minton paper dryer is designed for operation under vacuum. The drying cylinders are usually heated by steam, but occasionally single cylinders may be gas-heated, as in the case of the Pease blueprinting machine. Upon contacting the cylinder surface, wet sheet material is first heated to an equilibrium temperature somewhere between the wet-bulb temperature of the surrounding air and the boiling point of the liquid under the prevailing total pressure. The heat-transfer resistance of the vapor layer between the sheet and the cylinder surface may be significant.

These cylinder units are applicable to almost any form of sheet material that is not injuriously affected by contact with steam-heated metal surfaces. They are used chiefly when the sheet possesses certain properties such as a tendency to shrink or lacks the mechanical strength necessary for most types of continuous-sheeting air dryers. Applications are to dry films of various sorts, paper pulp in sheet form, paper sheets, paperboard, textile piece goods and fibers, etc. In some cases, imparting a special finish to the surface of the sheet may be an objective.

The **heat-transfer performance capacity** of cylinder dryers is not easy to estimate without a knowledge of the sheet temperature, which, in turn, is difficult to predict. According to published data, steam temperature is the largest single factor affecting capacity. Overall evaporation rates based on the total surface area of the dryers cover a range of 3.4 to 23 kg water/(h·m²) [0.7 to 4.8 lb water/(h·ft²)].

The value of the **coefficient of heat transfer** from steam to sheet is determined by the conditions prevailing on the inside and on the surface of the dryers. Low coefficients may be caused by (1) poor removal of air or other noncondensables from the steam in the cylinders, (2) poor removal of condensate, (3) accumulation of oil or rust on the interior of the drums, and (4) accumulation of a fiber lint on the outer surface of the drums. In a test reported by Lewis et al. [*Pulp Pap. Mag. Can.*, **22** (February 1927)] on a sulfite-paper dryer, in which the actual sheet temperatures were measured, a value of 187 W/(m²·°C) [33 Btu/(h·ft²·°F)] was obtained for the coefficient of heat flow between the steam and the paper sheet.

Operating-cost data for these units are meager. Power costs may be estimated by assuming 1 hp per cylinder for diameters of 1.2 to 1.8 m (4 to 6 ft). Data on labor and maintenance costs are also lacking.

The size of commercial cylinder dryers covers a wide range. The individual rolls may vary in diameter from 0.6 to 1.8 m (2 to 6 ft) and up to 8.5 m (28 ft) in width. In some cases, the width of rolls decreases throughout the dryer in order to conform to the shrinkage of the sheet. A single-cylinder dryer, such as the Yankee dryer, generally has a diameter between 2.7 and 4.6 m (9 and 15 ft).

HEAT-TRANSFER EQUIPMENT FOR DIVIDED SOLIDS

Most equipment for this service is some adaptation of a *material-handling* device whether or not the transport ability is desired. The old vertical tube and the vertical shell (fluidizer) are exceptions. Material-handling problems, plant transport needs, power, and maintenance are prime considerations in equipment selection and frequently overshadow heat-transfer and capital-cost considerations. Material handling is generally the most important aspect. Material-handling characteristics of the divided solids may vary during heat processing. The body changes are usually important in drying, occasionally significant for heating, and only on occasion important for cooling. The ability to minimize effects of changes is a major consideration in equipment selection. Dehydration operations are better performed on contactive apparatus (see Sec. 12) that provides air to carry off released water vapor before a semiliquid form develops.

Some types of equipment are convertible from heat removal to heat supply by simply changing the temperature level of the fluid or air. Other types require an auxiliary change. Others require constructional changes. Temperature limits for the equipment generally vary with the thermal operation. The kind of thermal operation has a major effect on heat-transfer values. For drying, overall coefficients are substantially higher in the presence of substantial moisture for the constant-rate period than in finishing. However, a stiff "body" occurrence due to moisture can prevent a normal "mixing" with an adverse effect on the coefficient.

Fluidized-Bed Type   Known as the cylindrical fluidizer, this operates with a bed of **fluidized solids** (Fig. 11-57). It is an indirect heat-transfer version of the contactive type in Sec. 17. An application disadvantage is the need for batch operation unless some short circuiting can be tolerated. Solids-cooling applications are few, as they can be more effectively accomplished by the fluidizing gas via the contactive mechanism that is referred to in Sec. 11. Heating applications are many and varied. These are subject to one shortcoming, which is the dissipation of the heat input by carry-off in the fluidizing gas. Heat-transfer performance for the indirect mode to solids has been outstanding, with overall coefficients in the range of 570 to 850 W/(m²·°C) [100 to 150 Btu/(h·ft²·°F)]. This device with its thin film does for solids what the falling-film and other thin-film techniques do for fluids, as shown by Holt (Pap. 11, 4th National Heat-Transfer Conference, August 1960). In a design innovation with high heat-transfer capability, heat is supplied indirectly to the fluidized solids through

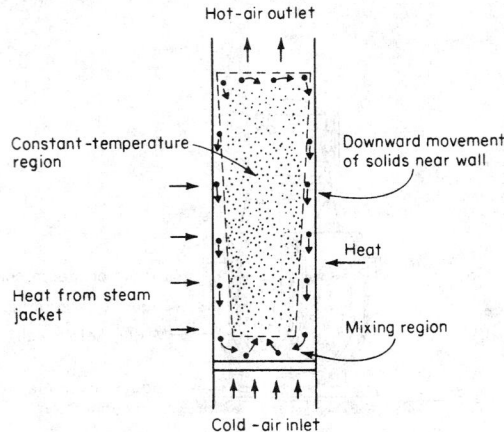

FIG. 11-57   Heat-transfer equipment for divided solids; stationary vertical-shell type. The indirect fluidizer.

the walls of in-bed, horizontally placed, finned tubes [Petrie, Freeby, and Buckham, *Chem. Eng. Prog.*, **64**(7), 45 (1968)].

Moving-Bed Type This concept uses a single-pass tube bundle in a vertical shell with the divided solids flowing by gravity in the tubes. It is little used for solids. A major difficulty in divided-solids applications is the problem of charging and discharging with uniformity. A second is poor heat-transfer rates. Because of these limitations, this tube-bundle type is not the workhorse for solids that it is for liquid and gas-phase heat exchange.

However, there are applications in which the nature of a specific chemical reactor system requires indirect heating or cooling of a moving bed of divided solids. One of these is the segregation process which through a gaseous reaction frees chemically combined copper in an ore to a free copper form which permits easy, efficient subsequent recovery [Pinkey and Plint, *Miner. Process.*, 17–30 (June 1968)]. The apparatus construction and principle of operation are shown in Fig. 11-58. The functioning is abetted by a novel heat-exchange provision of a fluidized sand bed in the jacket. This provides a much higher unit heat-input rate (coefficient value) than would the usual low-density hot-combustion-gas flow.

Agitated-Pan Type This device (Fig. 11-52) is not an adaptation of a material-handling device but was developed many years ago primarily for heat-transfer purposes. As such, it has found wide application. In spite of its batch operation with high attendant labor costs, it is still used for processing divided solids when no phase change is occurring. Simplicity and easy cleanout make the unit a wise selection for handling small, experimental, and even some production runs when quite a variety of burden materials are heat-processed. Both heating and cooling are feasible with it, but greatest use has been for drying [see Sec. 12 and Uhl and Root, *Chem. Eng. Prog.*, **63**(7), 8 (1967)]. This device, because it can be readily covered (as shown in the illustration) and a vacuum drawn or special atmosphere provided, features versatility to widen its use. For drying granular solids, the heat-transfer rate ranges from 28 to 227 W/(m²·°C) [5 to 40 Btu/

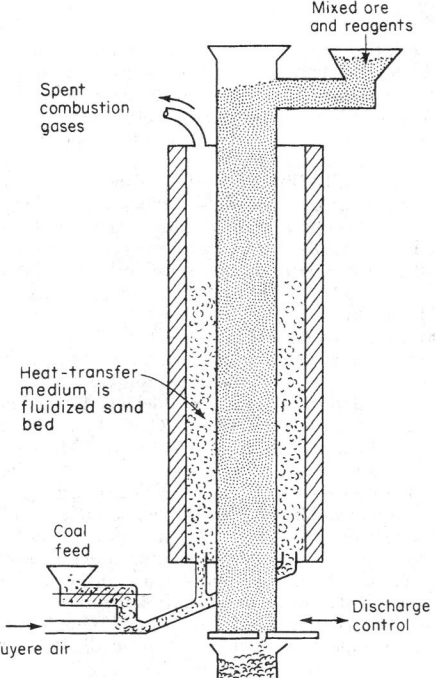

FIG. 11-58 Stationary vertical-tube type of indirect heat-transfer equipment with divided solids inside tubes, laminar solids flow and steady-state heat conditions.

(h·ft²·°F)]. For atmospheric applications, thermal efficiency ranges from 65 to 75 percent. For vacuum applications, it is about 70 to 80 percent. These devices are available from several sources, fabricated of various metals used in chemical processes.

Kneading Devices These are closely related to the agitated pan but differ as being primarily mixing devices with heat transfer a secondary consideration. Heat transfer is provided by jacketed construction of the main body and is effected by a coolant, hot water, or steam. These devices are applicable for the compounding of divided solids by mechanical rather than chemical action. Application is largely in the pharmaceutical and food-processing industries. For a more complete description, illustrations, performance, and power requirements, refer to Sec. 19.

Shelf Devices Equipment having heated and/or cooled shelves is available but is little used for divided-solids heat processing. Most extensive use of stationary shelves is freezing of packaged solids for food industries and for freeze drying by sublimation (see Sec. 22).

Rotating-Shell Devices These (see Fig. 11-59) are installed horizontally, whereas stationary-shell installations are vertical. Material-handling aspects are of greater importance than thermal performance. Thermal results are customarily given in terms of overall coefficient on the basis of the total area provided, which varies greatly with the design. The effective use, chiefly percent fill factor, varies widely, affecting the reliability of stated coefficient values. For performance calculations see Sec. 10 on heat-processing theory for solids. These devices are variously used for cooling, heating, and drying and are the workhorses for heat-processing divided solids in the large-capacity range. Different modifications are used for each of the three operations.

The **plain** type (Fig. 11-59a) features simplicity and yet versatility through various end-construction modifications enabling wide and varied applications. Thermal performance is strongly affected by the "body" characteristics of the burden because of its dependency for material handling on frictional contact. Hence, performance ranges from well-agitated beds with good thin-film heat-transfer rates to poorly agitated beds with poor thick-film heat-transfer rates. Temperature limits in application are (1) low-range cooling with shell dipped in water, 400°C (750°F) and less; (2) intermediate cooling with forced circulation of tank water, to 760°C (1400°F); (3) primary cooling, above 760°C (1400°F), water copiously sprayed and loading kept light; (4) low-range heating, below steam temperature, hot-water dip; and (5) high-range heating by tempered combustion gases or ribbon radiant-gas burners.

The **flighted** type (Fig. 11-59b) is a first-step modification of the plain type. The simple flight addition improves heat-transfer performance. This type is most effective on semifluid burdens which slide readily. Flighted models are restricted from applications in which soft-cake sticking occurs, breakage must be minimized, and abrasion is severe. A special flighting is one having the cross section compartmented into four lesser areas with ducts between. Hot gases are drawn through the ducts en route from the outer oven to the stack to provide about 75 percent more heating surface, improving efficiency and capacity with a modest cost increase. Another similar unit has the flights made in a triangular-duct cross section with hot gases drawn through.

The **tubed-shell** type (Fig. 11-59c) is basically the same device more commonly known as a "steam-tube rotary dryer" (see Sec. 20). The rotation, combined with slight inclination from the horizontal, moves the shell-side solids through it continuously. This type features good mixing with the objective of increased heat-transfer performance. Tube-side fluid may be water, steam, or combustion gas. Bottom discharge slots in the shell are used so that heat-transfer-medium supply and removal can be made through the ends; these restrict wide-range loading and make the tubed type inapplicable for floody materials. These units are seldom applicable for sticky, soft-caking, scaling, or heat-sensitive burdens. They are not recommended for abrasive materials. This type has high thermal efficiency because heat loss is minimized. **Heat-transfer coefficient** values are: water, 34 W/(m²·°C) [6 Btu/(h·ft²·°F)]; steam, same, with heat flux reliably constant at 3800 W/m² [1200 Btu/(h·ft²)]; and gas, 17 W/(m²·°C) [3 Btu/(h·ft²·°F)], with a high temperature difference. Although from the

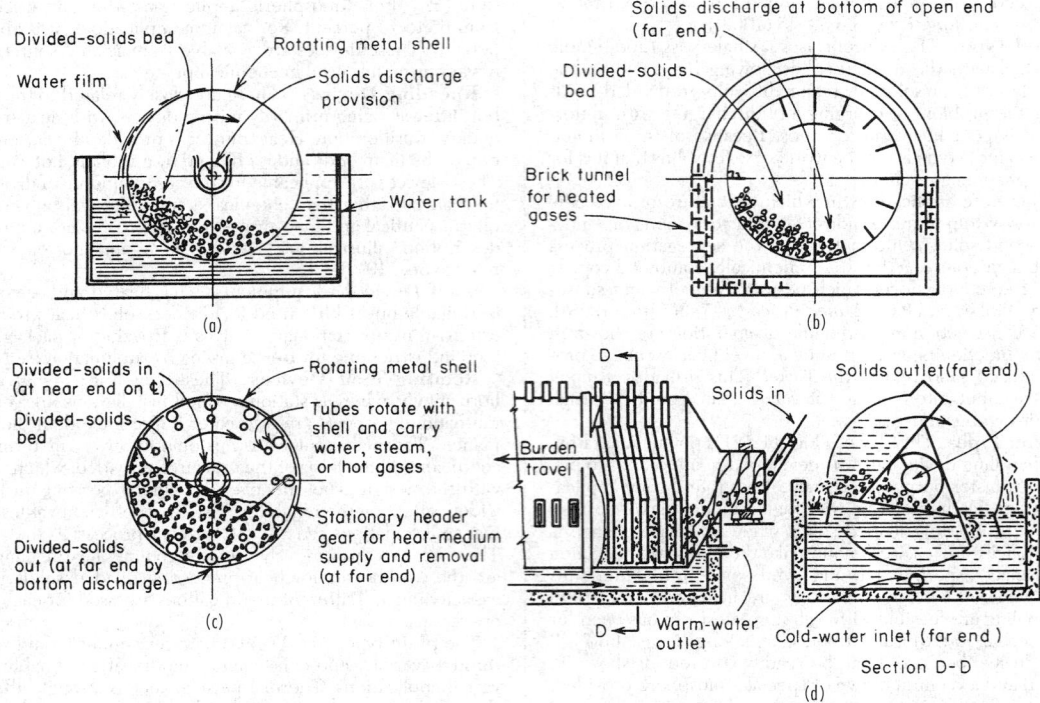

FIG. 11-59 Rotating shells as indirect heat-transfer equipment. (*a*) Plain. (*Courtesy of BSP Corp.*) (*b*) Flighted. (*Courtesy of BSP Corp.*) (*c*) Tubed. (*d*) Deep-finned type. (*Courtesy of Link-Belt Co.*)

preceding discussion the device may seem rather limited, it is nevertheless widely used for drying, with condensing steam predominating as the heat-carrying fluid. But with water or refrigerants flowing in the tubes, it is also effective for cooling operations. The units are custom-built by several manufacturers in a wide range of sizes and materials. A few fabricators that specialize in this type of equipment have accumulated a vast store of data for determining application sizing.

The patented **deep-finned** type in Fig. 11-59d is named the "Rotofin cooler." It features loading with a small layer thickness, excellent mixing to give a good effective diffusivity value, and a thin fluid-side film. Unlike other rotating-shell types, it is installed horizontally, and the burden is moved positively by the fins acting as an Archimedes spiral. Rotational speed and spiral pitch determine travel time. For cooling, this type is applicable to both secondary and intermediate cooling duties. Applications include solids in small lumps [9 mm (¾ in)] and granular size [6 mm and less (¼ to 0 in)] with no larger pieces to plug the fins, solids that have a free-flowing body characteristic with no sticking or caking tendencies, and drying of solids that have a low moisture and powder content unless special modifications are made for substantial vapor and dust handling. Thermal performance is very good, with overall coefficients to 110 W/(m²·°C) [20 Btu/(h·ft²·°F)], with one-half of these coefficients nominal for cooling based on the total area provided (nearly double those reported for other indirect rotaries).

Conveyor-Belt Devices   The metal-belt type (Fig. 11-55) is the only device in this classification of material-handling equipment that has had serious effort expended on it to adapt it to indirect heat-transfer service with divided solids. It features a lightweight construction of a large area with a thin metal wall. Indirect-cooling applications have been made with poor thermal performance, as could be expected with a static layer. Auxiliary plowlike mixing devices, which are considered an absolute necessity to secure any worthwhile results for this service, restrict applications.

Spiral-Conveyor Devices   Figure 11-60 illustrates the major

adaptations of this widely used class of material-handling equipment to indirect heat-transfer purposes. These conveyors can be considered for heat-transfer purposes as continuously agitated kettles. The adaptation of Fig. 11-60d offers a batch-operated version for evaporation duty. For this service, all are package-priced and package-shipped items requiring few, if any, auxiliaries.

The **jacketed solid-flight** type (Fig. 11-60a) is the standard low-cost (parts-basis-priced) material-handling device, with a simple jacket added and employed for secondary-range heat transfer of an incidental nature. Heat-transfer coefficients are as low as 11 to 34 W/(m²·°C) [2 to 6 Btu/(h·ft²·°F)] on sensible heat transfer and 11 to 68 W/(m²·°C) [2 to 12 Btu/(h·ft²·°F)] on drying because of substantial static solids-side film.

The **small-spiral–large-shaft** type (Fig. 11-60b) is inserted in a solids-product line as pipe banks are in a fluid line, solely as a heat-transfer device. It features a thin burden ring carried at a high rotative speed and subjected to two-sided conductance to yield an estimated heat-transfer coefficient of 285 W/(m²·°C) [50 Btu/(h·ft²·°F)], thereby ranking thermally next to the shell-fluidizer type. This device for powdered solids is comparable with the Votator of the fluid field.

Figure 11-60c shows a fairly new spiral device with a medium-heavy annular solids bed and having the combination of a jacketed, stationary outer shell with moving paddles that carry the heat-transfer fluid. A unique feature of this device to increase volumetric throughput, by providing an overall greater temperature drop, is that the heat medium is supplied to and withdrawn from the rotor paddles by a parallel piping arrangement in the rotor shaft. This is a unique flow arrangement compared with the usual series flow. In addition, the rotor carries burden-agitating spikes which give it the trade name of Porcupine Heat-Processor (*Chem. Equip. News,* April 1966; and Uhl and Root, AIChE Prepr. **21,** 11th National Heat-Transfer Conference, August 1967).

The **large-spiral hollow-flight** type (Fig. 11-60d) is an adaptation, with external bearings, full fill, and salient construction points as

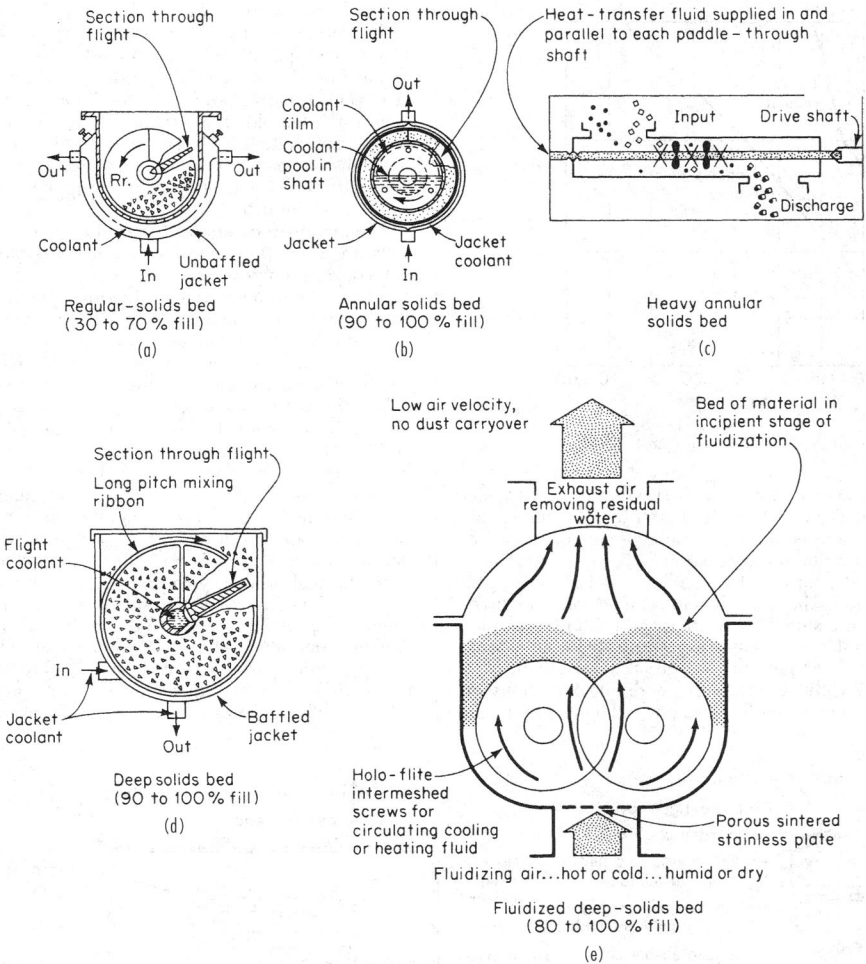

FIG. 11-60   Spiral-conveyor adaptations as heat-transfer equipment. (*a*) Standard jacketed solid flight. (*Courtesy of Jeffrey Mfg. Co.*) (*b*) Small spiral, large shaft. (*Courtesy of Fuller Co.*) (*c*) "Porcupine" medium shaft. (*Courtesy of Bethlehem Corp.*) (*d*) Large spiral, hollow flight. (*Courtesy of Rietz Mfg. Co.*) (*e*) Fluidized-bed large spiral, helical flight. (*Courtesy of Western Precipitation Division, Joy Mfg. Co.*)

shown, that is highly versatile in application. Heat-transfer coefficients are 34 to 57 W/(m²·°C) [6 to 10 Btu/(h·ft²·°F)] for poor, 45 to 85 W/(m²·°C) [8 to 15 Btu/(h·ft²·°F)] for fair, and 57 to 114 W/(m²·°C) [10 to 20 Btu/(h·ft²·°F)] for wet conductors. A popular version of this employs two such spirals in one material-handling chamber for a pugmill agitation of the deep solids bed. The spirals are seldom heated. The shaft and shell are heated.

Another deep-bed spiral-activated solids-transport device is shown by Fig. 11-60*e*. The flights carry a heat-transfer medium as well as the jacket. A unique feature of this device which is purported to increase heat-transfer capability in a given equipment space and cost is the dense-phase fluidization of the deep bed that promotes agitation and moisture removal on drying operations.

Double-Cone Blending Devices   The original purpose of these devices was mixing (see Sec. 19). Adaptations have been made; so many models now are primarily for indirect heat-transfer processing. A jacket on the shell carries the heat-transfer medium. The mixing action, which breaks up agglomerates (but also causes some degradation), provides very effective burden exposure to the heat-transfer surface. On drying operations, the vapor release (which in a static bed is a slow diffusional process) takes place relatively quickly. To provide vapor removal from

the burden chamber, a hollow shaft is used. Many of these devices carry the hollow-shaft feature a step further by adding a rotating seal and drawing a vacuum. This increases thermal performance notably and makes the device a natural for solvent-recovery operations.

These devices are replacing the older tank and spiral-conveyor devices. Better provisions for speed and ease of fill and discharge (without powered rotation) minimize downtime to make this batch-operated device attractive. Heat-transfer coefficients ranging from 28 to 200 W/(m²·°C) [5 to 35 Btu/(h·ft²·°F)] are obtained. However, if caking on the heat-transfer walls is serious, then values may drop to 5.5 or 11 W/(m²·°C) [1 or 2 Btu/(h·ft²·°F)], constituting a misapplication. The double cone is available in a fairly wide range of sizes and construction materials. The users are the fine-chemical, pharmaceutical, and biological-preparation industries.

A novel variation is a cylindrical model equipped with a tube bundle to resemble a shell-and-tube heat exchanger with a bloated shell [*Chem. Process.*, **20** (Nov. 15, 1968)]. Conical ends provide for redistribution of burden between passes. The improved heat-transfer performance is shown by Fig. 11-61.

Vibratory-Conveyor Devices   Figure 11-62 shows the various adaptations of vibratory material-handling equipment for indirect

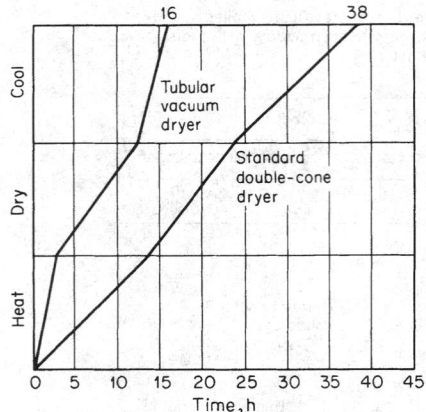

FIG. 11-61   Performance of tubed, blender heat-transfer device.

heat-transfer service on divided solids. The basic vibratory-equipment data are given in Sec. 21. These indirect heat-transfer adaptations feature simplicity, nonhazardous construction, nondegradation, nondusting, no wear, ready conveying-rate variation [1.5 to 4.5 m/min (5 to 15 ft/min)], and good heat-transfer coefficient—115 W/(m²·°C) [20 Btu/(h·ft²·°F)] for sand. They usually require feed-rate and distribution auxiliaries. They are suited for heating and cooling of divided solids in powdered, granular, or moist forms but no sticky, liquefying, or floody ones. Terminal-temperature differences less than 11°C (20°F) on cooling and 17°C (30°F) on heating or drying operations are seldom practical. These devices are for medium and light capacities.

The **heavy-duty jacketed** type (Fig. 11-62a) is a special custom-built adaptation of a heavy-duty vibratory conveyor shown in Fig. 11-60. Its application is continuously to cool the crushed material [from about 177°C (350°F)] produced by the vibratory-type "caster" of Fig. 11-53. It does not have the liquid dam and is made in longer lengths that employ L, switchback, and S arrangements on one floor. The capacity rate is 27,200 to 31,700 kg/h (30 to 35 tons/h) with heat-transfer coefficients in the order of 142 to 170 W/(m²·°C) [25 to 30 Btu/(h·ft²·°F)]. For heating or drying applications, it employs steam to 414 kPa (60 lbf/in²).

The **jacketed or coolant-spraying** type (Fig. 11-62b) is designed to assure a very thin, highly agitated liquid-side film and the same initial coolant temperature over the entire length. It is frequently employed for transporting substantial quantities of hot solids, with cooling as an incidental consideration. For heating or drying applications, hot water or steam at a gauge pressure of 7 kPa (1 lbf/in²) may be employed. This type is widely used because of its versatility, simplicity, cleanability, and good thermal performance.

The **light-duty jacketed** type (Fig. 11-62c) is designed for use of air as a heat carrier. The flow through the jacket is highly turbulent and is usually counterflow. On long installations, the air flow is parallel to every two sections for more heat-carrying capacity and a fairly uniform surface temperature. The outstanding feature is that a wide range of temperature control is obtained by merely changing the heat-carrier temperature level from as low as atmospheric moisture condensation will allow to 204°C (400°F). On heating operations, a very good thermal efficiency can be obtained by insulating the machine and recycling the air. While heat-transfer rating is good, the heat-removal capacity is limited. Cooler units are often used in series with like units operated as dryers or when clean water is unavailable. Drying applications are for heat-sensitive [49 to 132°C (120° to 270°F)] products; when temperatures higher than steam at a gauge pressure of 7 kPa (1 lbf/in²) can provide are wanted but heavy-duty equipment is

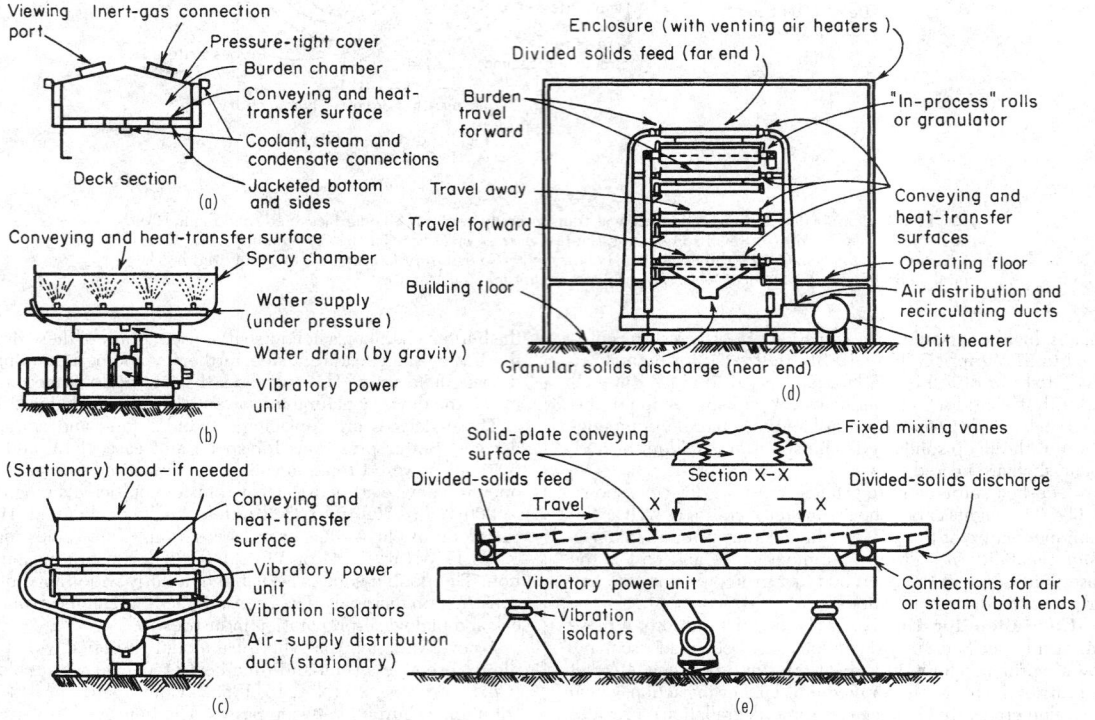

FIG. 11-62   Vibratory-conveyor adaptations as indirect heat-transfer equipment. (*a*) Heavy-duty jacketed for liquid coolant or high-pressure steam. (*b*) Jacketed for coolant spraying. (*c*) Light-duty jacketed construction. (*d*) Jacketed for air or steam in tiered arrangement. (*e*) Jacketed for air or steam with Mix-R-Step surface. (*Courtesy of Jeffrey Mfg. Co.*)

too costly; when the jacket-corrosion hazard of steam is unwanted; when headroom space is at a premium; and for highly abrasive burden materials such as fritted or crushed glasses and porcelains.

The **tiered arrangement** (Fig. 11-62d) employs the units of Fig. 11-62 with either air or steam at a gauge pressure of 7 kPa (1 lbf/in²) as a heat medium. These are custom-designed and built to provide a large amount of heat-transfer surface in a small space with the minimum of transport and to provide a complete processing system. These receive a damp material, resize while in process by granulators or rolls, finish dry, cool, and deliver to packaging or tableting. The applications are primarily in the fine-chemical, food, and pharmaceutical manufacturing fields.

The **Mix-R-Step** type in Fig. 11-62e is an adaptation of a vibratory conveyor. It features better heat-transfer rates, practically doubling the coefficient values of the standard flat surface and trebling heat-flux values, as the layer depth can be increased from the normal 13 to 25 and 32 mm (½ to 1 and 1¼ in). It may be provided on decks jacketed for air, steam, or water spray. It is also often applicable when an infrared heat source is mounted overhead to supplement the indirect or as the sole heat source.

Elevator Devices   The **vibratory elevating-spiral** type (Fig. 11-63) adapts divided-solids-elevating material-handling equipment to heat-transfer service. It features a large heat-transfer area over a small floor space and employs a reciprocating shaker motion to effect transport. Applications, layer depth, and capacities are restricted, as burdens must be of such "body" character as to convey uphill by the microhopping-transport principle. The type lacks self-emptying ability. Complete washdown and cleaning is a feature not inherent in any other elevating device. A typical application is the cooling of a low-density plastic powder at the rate of 544 kg/h (1200 lb/h).

Another elevator adaptation is that for a **spiral-type elevating** device developed for ground cement and thus limited to fine powdery burdens. The spiral operates inside a cylindrical shell, which is externally cooled by a falling film of water. The spiral not only elevates the material in a thin layer against the wall but keeps it agitated to achieve high heat-transfer rates. Specific operating data are not available [*Chem. Eng. Prog.*, **68**(7), 113 (1968)]. The falling-water film, besides being ideal thermally, by virtue of no jacket pressure very greatly reduces the hazard that the cooling water may contact the water-sensitive burden in process. Surfaces wet by water are accessible for cleaning. A fair range of sizes is available, with material-handling capacities to 60 tons/h.

Pneumatic-Conveying Devices   See Sec. 21 for descriptions, ratings, and design factors on these devices. Use is primarily for transport purposes, and heat transfer is a very secondary consideration.

Applications have largely been for plastics in powder and pellet forms. By modifications, needed cooling operations have been simultaneously effected with transport to stock storage [*Plast. Des. Process.*, **28** (December 1968)].

Heat-transfer aspects and performance were studied and reported on by Depew and Farbar (ASME Pap. 62-HT-14, September 1962). Heat-transfer coefficient characteristics are similar to those shown in Sec. 11 for the indirectly heated fluid bed. Another frequent application on plastics is a small, rather incidental but necessary amount of drying required for plastic pellets and powders on receipt when shipped in bulk to the users. Pneumatic conveyers modified for heat transfer can handle this readily.

A pneumatic-transport device designed primarily for heat-sensitive products is shown in Fig. 11-64. This was introduced into the United States after 5 years' use in Europe [*Chem. Eng.*, **76**, 54 (June 16, 1969)].

Both the shell and the rotor carry steam as a heating medium to effect indirect transfer as the burden briefly contacts those surfaces rather than from the transport air, as is normally the case. The rotor turns slowly (1 to 10 r/min) to control, by deflectors, product distribution and prevent caking on walls. The carrier gas can be inert, as nitrogen, and also recycled through appropriate auxiliaries for solvent recovery. Application is limited to burdens that (1) are fine and uniformly grained for the pneumatic transport, (2) dry very fast, and (3) have very little, if any, sticking or decomposition characteristics. Feeds can carry 5 to 100 percent moisture (dry basis) and discharge at 0.1 to 2 percent. Wall temperatures range from 100 to 170°C (212 to 340°F) for steam and lower for a hot-water-heat source. Pressure drops are in order of 500 to 1500 mmH₂O (20 to 60 inH₂O). Steam consumption approaches that of a contractive-mechanism dryer down to a low value of 2.9 kg steam/kg water (2.9 lb steam/lb water). Available burden capacities are 91 to 5900 kg/h (200 to 13,000 lb/h).

Vacuum-Shelf Types   These are very old devices, being a version of the table type. Early-day use was for drying (see Sec. 12). Heat transfer is slow even when supplemented by vacuum, which is 90 percent or more of present-day use. The newer vacuum blender and cone devices are taking over many applications. The slow heat-transfer rate is quite satisfactory in a major application, freeze drying, which is a sublimation operation (see Sec. 22 for description) in which the water must be retained in the solid state during its removal. Then slow diffusional processes govern. Another extensive application is freezing packaged foods for preservation purposes.

Available sizes range from shelf areas of 0.4 to 67 m² (4 to 726 ft²). These are available in several manufacturers' standards, either as system components or with auxiliary gear as packaged systems.

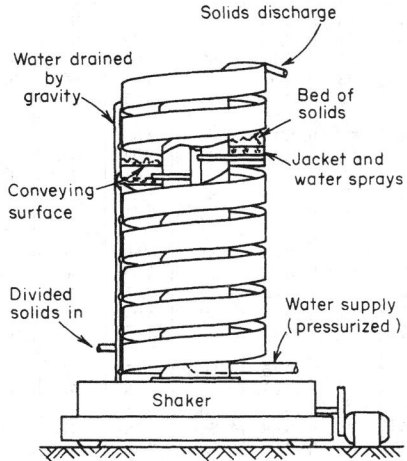

FIG. 11-63   Elevator type as heat-transfer equipment. (*Courtesy of Carrier Conveyor Corp.*)

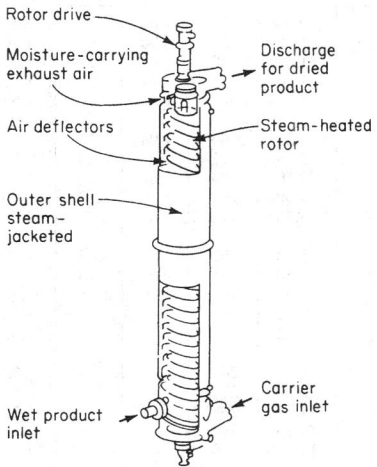

FIG. 11-64   A pneumatic-transport adaptation for heat-transfer duty. (*Courtesy of Werner & Pfleiderer Corp.*)

THERMAL INSULATION

Materials or combinations of materials which have air- or gas-filled pockets or void spaces that retard the transfer of heat with reasonable effectiveness are thermal insulators. Such materials may be particulate and/or fibrous, with or without binders, or may be assembled, such as multiple heat-reflecting surfaces that incorporate air- or gas-filled void spaces.

The ability of a material to retard the flow of heat is expressed by its thermal **conductivity** (for unit thickness) or **conductance** (for a specific thickness). Low values for thermal conductivity or conductance (or high thermal resistivity or resistance value) are characteristics of thermal insulation.

Heat is transferred by radiation, conduction, and convection. Radiation is the primary mode and can occur even in a vacuum. The amount of heat transferred for a given area is relative to the temperature differential and emissivity from the radiating to the absorbing surface. Conduction is due to molecular motion and occurs within gases, liquids, and solids. The tighter the molecular structure, the higher the rate of transfer. As an example, steel conducts heat at a rate approximately 600 times that of typical thermal-insulation materials. Convection is due to mass motion and occurs only in fluids. The prime purpose of a thermal-insulation system is to minimize the amount of heat transferred.

INSULATION MATERIALS

Materials   Thermal insulations are produced from many materials or combinations of materials in various forms, sizes, shapes, and thickness. The most commonly available materials fall within the following categories:

Fibrous or cellular—mineral. Alumina, asbestos, glass, perlite, rock, silica, slag, or vermiculite.

Fibrous or cellular—organic. Cane, cotton, wood, and wood bark (cork).

Cellular organic plastics. Elastomer, polystyrene, polyisocyanate, polyisocyanurate, and polyvinyl acetate.

Cements. Insulating and/or finishing.

Heat-reflecting metals (reflective). Aluminum, nickel, stainless steel.

Available forms. Blanket (felt and batt), block, cements, loose fill, foil and sheet, formed or foamed in place, flexible, rigid, and semirigid.

The actual thicknesses of piping insulation differ from the nominal values. Dimensional data of ASTM Standard C585 appear in Table 11-20.

Thermal Conductivity (K Factor)   Depending on the type of insulation, the thermal conductivity (K factor) can vary with age, manufacturer, moisture content, and temperature. Typical published values are shown in Fig. 11-65. Mean temperature is equal to the arithmetic average of the temperatures on both sides of the insulating material.

Actual system heat loss (or gain) will normally exceed calculated values because of projections, axial and longitudinal seams, expansion-contraction openings, moisture, workers' skill, and physical abuse.

Finishes   Thermal insulations require an external covering (finish) to provide protection against entry of water or process fluids, mechanical damage, and ultraviolet degradation of foamed materials. In some cases the finish can reduce the flame-spread rating and/or provide fire protection.

The finish may be a coating (paint, asphaltic, resinous, or polymeric), a membrane (coated felt or paper, metal foil, or laminate of plastic, paper, foil or coatings), or sheet material (fabric, metal, or plastic).

Finishes for systems operating below 2°C (35°F) must be sealed and retard vapor transmission. Those from 2°C (35°F) through 27°C (80°F) should retard vapor transmission (to prevent surface condensation), and those above 27°C (80°F) should prevent water entry and allow moisture to escape.

Metal finishes are more durable, require less maintenance, reduce heat loss, and, if uncoated, increase the surface temperature on hot systems.

TABLE 11-20   Thicknesses of Piping Insulation

in mm	Outer diameter		Insulation, nominal thickness													
			1 25		1½ 38		2 51		2½ 64		3 76		3½ 89		4 102	
Nominal iron-pipe size, in			Approximate wall thickness													
	in	mm	in	mm	in	mm	in	mm	in	mm	in	mm	in	mm	in	mm
½	0.84	21	1.01	26	1.57	40	2.07	53	2.88	73	3.38	86	3.88	99	4.38	111
¾	1.05	27	0.90	23	1.46	37	1.96	50	2.78	71	3.28	83	3.78	96	4.28	109
1	1.32	33	1.08	27	1.58	40	2.12	54	2.64	67	3.14	80	3.64	92	4.14	105
1¼	1.66	42	0.91	23	1.66	42	1.94	49	2.47	63	2.97	75	3.47	88	3.97	101
1½	1.90	48	1.04	26	1.54	39	2.35	60	2.85	72	3.35	85	3.85	98	4.42	112
2	2.38	60	1.04	26	1.58	40	2.10	53	2.60	66	3.10	79	3.60	91	4.17	106
2½	2.88	73	1.04	26	1.86	47	2.36	60	2.86	73	3.36	85	3.92	100	4.42	112
3	3.50	89	1.02	26	1.54	39	2.04	52	2.54	65	3.04	77	3.61	92	4.11	104
3½	4.00	102	1.30	33	1.80	46	2.30	58	2.80	71	3.36	85	3.86	98	4.36	111
4	4.50	114	1.04	26	1.54	39	2.04	52	2.54	65	3.11	79	3.61	92	4.11	104
4½	5.00	127	1.30	33	1.80	46	2.30	58	2.86	73	3.36	85	3.86	98	4.48	114
5	5.56	141	0.99	25	1.49	38	1.99	51	2.56	65	3.06	78	3.56	90	4.18	106
6	6.62	168	0.96	24	1.46	37	2.02	51	2.52	64	3.02	77	3.65	93	4.15	105
7	7.62	194			1.52	39	2.02	51	2.52	64	3.15	80	3.65	93	4.15	105
8	8.62	219			1.52	39	2.02	51	2.65	67	3.15	80	3.65	93	4.15	105
9	9.62	244			1.52	39	2.15	55	2.65	67	3.15	80	3.65	93	4.15	105
10	10.75	273			1.58	40	2.08	53	2.58	66	3.08	78	3.58	91	4.08	104
11	11.75	298			1.58	40	2.08	53	2.58	66	3.08	78	3.58	91	4.08	104
12	12.75	324			1.58	40	2.08	53	2.58	66	3.08	78	3.58	91	4.08	104
14	14.00	356			1.46	37	1.96	50	2.46	62	2.96	75	3.46	88	3.96	101
Over 14, up to and including 36					1.46	37	1.96	50	2.46	62	2.96	75	3.46	88	3.96	101

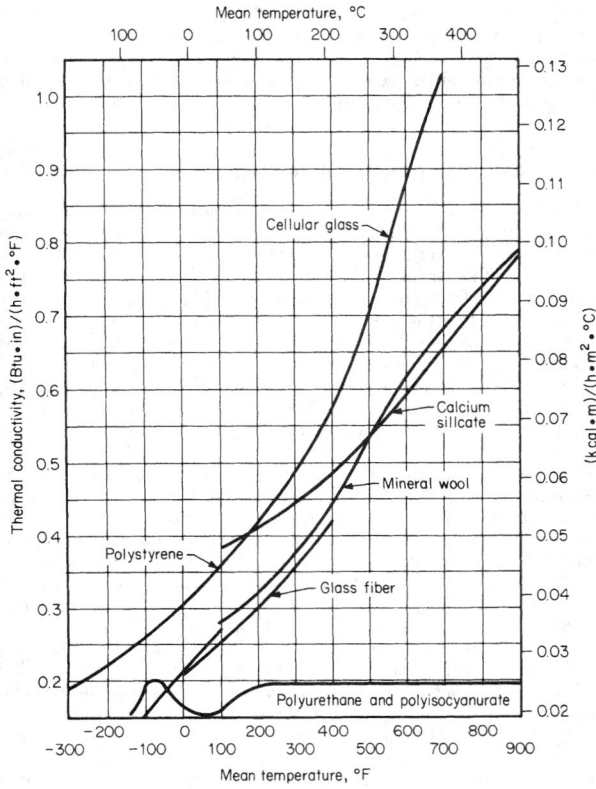

FIG. 11-65 Thermal conductivity of insulating materials.

SYSTEM SELECTION

A combination of insulation and finish produces the thermal-insulation system. Selection of these components depends on the purpose for which the system is to be used. No single system performs satisfactorily from the cryogenic through the elevated-temperature range. Systems operating below freezing have a low vapor pressure, and atmospheric moisture is pushed into the insulation system, while the reverse is true for hot systems. Some general guidelines for system selection follow.

Cryogenic [−273 to −101°C (−459 to −150°F)] High Vacuum This technique is based on the Dewar flask, which is a double-walled vessel with reflective surfaces on the evacuated side to reduce radiation losses. Figure 11-66 shows a typical laboratory-size Dewar. Figure 11-67 shows a semiportable type. Radiation losses can be further reduced by filling the cavity with powders such as perlite or silica prior to pulling the vacuum.

Multilayer Multilayer systems consist of series of radiation-reflective shields of low emittance separated by fillers or spacers of very low conductance and exposed to a high vacuum.

Foamed or Cellular Cellular plastics such as polyurethane and polystyrene do not hold up or perform well in the cryogenic temperature range because of permeation of the cell structure by water vapor, which in turn increases the heat-transfer rate. Cellular glass holds up better and is less permeable.

Low Temperature [−101 to −1°C (−150 to +30°F)] Cellular glass, glass fiber, polyurethane foam, and polystyrene foam are frequently used for this service range. A vapor-retarder finish with a perm rating less than 0.02 is required. In addition, it is good practice to coat all contact surfaces of the insulation with a vapor-retardant mastic to prevent moisture migration when the finish is damaged or is not properly maintained. Closed-cell insulation should not be relied

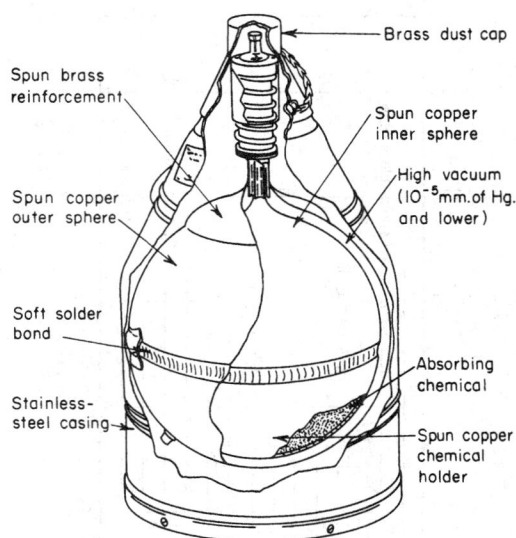

FIG. 11-66 Dewar flask.

on as the vapor retarder. Hairline cracks can develop, cells can break down, glass-fiber binders are absorbent, and moisture can enter at joints between all materials.

Moderate and High Temperature [over 2°C (36°F)] Cellular or fibrous materials are normally used. See Fig. 11-68 for nominal

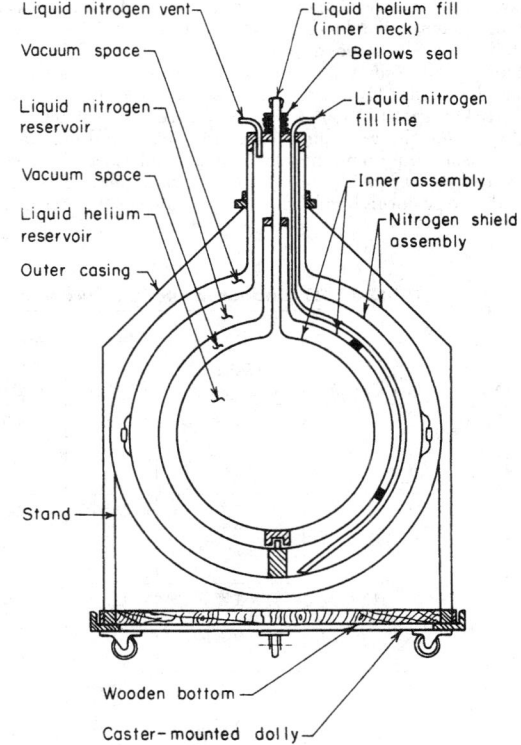

FIG. 11-67 Hydrogen bottle.

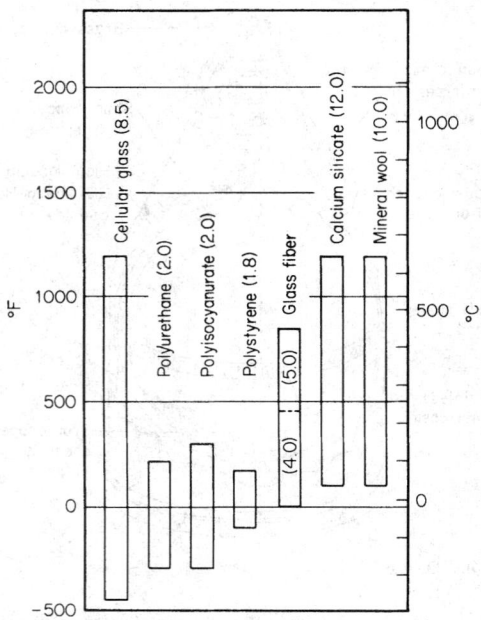

FIG. 11-68 Insulating materials and applicable temperature ranges.

temperature range. Nonwicking insulation is desirable for systems operating below 100°C (212°F).

Other Considerations **Autoignition** can occur if combustible fluids are absorbed by wicking-type insulations. **Chloride stress corrosion** of austenitic stainless steel can occur when chlorides are concentrated on metal surfaces at or above approximately 60°C (140°F). The chlorides can come from sources other than the insulation. Some calcium silicates are formulated to exceed the requirements of the MIL-I-24244A specification. **Fire resistance** of insulations varies widely. Calcium silicate, cellular glass, glass fiber, and mineral wool are fire-resistant but do not perform equally under actual fire conditions. A steel jacket provides protection, but aluminum does not.

Traced pipe performs better with a nonwicking insulation which has

low thermal conductivity. Underground systems are very difficult to keep dry permanently. Methods of insulation include factory-preinsulated pouring types and conventionally applied types. Corrosion can occur under wet insulation. A protective coating, applied directly to the metal surface, may be required.

ECONOMIC THICKNESS OF INSULATION

Optimal economic insulation thickness may be determined by various methods. Two of these are the minimum-total-cost method and the incremental-cost method (or marginal-cost method). The minimum-total-cost method involves the actual calculations of lost energy and insulation costs for each insulation thickness. The thickness producing the lowest total cost is the optimal economic solution. The optimum thickness is determined to be the point where the last dollar invested in insulation results in exactly $1 in energy-cost savings ("ETI—Economic Thickness for Industrial Insulation," Conservation Pap. **46,** Federal Energy Administration, August 1976). The incremental-cost method provides a simplified and direct solution for the least-cost thickness.

The total-cost method does *not* in general provide a satisfactory means for making most insulation investment decisions, since an economic return on investment is required by investors and the method does not properly consider this factor. Return on investment is considered by Rubin ("Piping Insulation—Economics and Profits," in *Practical Considerations in Piping Analysis,* ASME Symposium, vol. **69,** 1982, pp. 27–46). The incremental method used in this reference requires that each incremental ½ in of insulation provide the predetermined return on investment. The minimum thickness of installed insulation is used as a base for calculations. The incremental installed capital cost for each additional ½ in of insulation is determined. The energy saved for each increment is then determined. The value of this energy varies directly with the temperature level [e.g., steam at 538°C (1000°F) has a greater value than condensate at 100°C (212°F)]. The final increment selected for use is required either to provide a satisfactory return on investment or to have a suitable payback period.

Recommended Thickness of Insulation Indoor insulation thickness appears in Table 11-21, and outdoor thickness appears in Table 11-22. These selections were based upon calcium silicate insulation with a suitable aluminum jacket. However, the variation in thickness for fiberglass, cellular glass, and rock wool is minimal. Fiberglass is available for maximum temperatures of 260, 343, and 454°C (500, 650, and 850°F). Rock wool, cellular glass, and calcium silicate are used up to 649°C (1200°F).

TABLE 11-21 Indoor Insulation Thickness, 80°F Still Ambient Air*

		Minimum pipe temperature, °F							
		Energy cost, $/million Btu							
Pipe size, in	Insulation thickness, in	1	2	3	4	5	6	7	8
¾	1½	950	600	550	400	350	300	250	250
	2				1100	1000	900	800	750
	2½				1750	1050	950	850	800
	3								1200
1	1½	1200	800	600	500	450	400	350	300
	2			1200	1000	900	800	700	700
	2½					1200	1050	1000	900
	3						1100	1150	950
1½	1½	1100	750	550	450	400	400	350	300
	2				1000	850	700	650	600	500
	2½					1050	900	800	750	650
	3							1150	1100	1000
2	1½	1050	700	500	450	400	350	300	300
	2			1050	850	750	700	600	600
	2½			1100	950	1000	750	700	650
	3				1200	1050	950	850	800

TABLE 11-21   Indoor Insulation Thickness, 80°F Still Ambient Air* (*Concluded*)

Pipe size, in	Insulation thickness, in	Minimum pipe temperature, °F							
		Energy cost, $/million Btu							
		1	2	3	4	5	6	7	8
3	1½	950	650	500	400	350	300	300	250
	2		1100	900	700	600	550	500	450
	2½			1050	850	750	650	500	500
	3				1050	950	800	750	700
4	1½	950	600	500	400	350	300	300	250
	2		1100	850	700	600	550	500	450
	2½			1200	1000	850	750	700	650
	3				1050	900	800	750	700
	3½							1150	1050
6	1½	600	350	300	250	250	200	200	200
	2		1100	850	700	600	550	500	500
	2½			900	800	650	600	550	550
	3			1150	1000	850	750	700	600
	3½						1100	1000	900
	4								1200
8	2		1000	800	650	550	500	450	400
	2½		1050	850	700	600	550	500	450
	3				1050	900	800	750	700
	3½					1200	1100	1000	900
	4							1150	1100
10	2		1100	850	700	650	550	500	450
	2½		1200	900	750	700	600	550	500
	3			1050	900	750	700	600	550
	3½					1200	1050	950	900
	4								1200
12	2	1150	750	600	500	400	400	350	300
	2½		1000	800	650	550	500	450	400
	3			1200	1000	900	800	700	650
	3½					1200	1100	1000	900
	4						1150	1050	950
	4½						1200	1100	1000
14	2	1050	650	550	450	400	350	300	300
	2½		1000	800	650	550	500	450	400
	3			1100	950	800	700	650	600
	3½					1150	1000	950	850
	4					1200	1050	1000	900
	4½						1200	1100	1000
16	2	950	650	500	400	350	300	300	300
	2½		1000	800	700	600	550	500	450
	3		1200	950	800	700	600	550	500
	3½					1150	1050	950	850
	4					1200	1100	1000	900
	4½						1150	1050	950
18	2	1000	650	500	400	350	350	300	300
	2½		950	750	600	550	500	450	400
	3		1150	900	750	650	550	500	500
	3½					1200	1100	1000	900
	4						1150	1050	950
	4½						1200	1100	1000
20	2	1050	700	550	450	400	350	350	300
	2½		1000	800	600	550	500	450	400
	3		1150	900	750	650	550	500	500
	3½						1100	1000	950
	4						1150	1050	1000
	4½							1200	1100
24	2	950	600	500	400	350	300	300	250
	2½		1150	900	750	650	550	500	450
	3			1050	900	750	700	600	550
	3½					1100	1000	900	800
	4					1150	1050	950	850
	4½						1150	1050	950

*Aluminum-jacketed calcium silicate insulation with an emissivity factor of 0.05. To convert inches to millimeters, multiply by 25.4, to convert dollars per 1 million British thermal units to dollars per 1 million kilojoules, multiply by 0.948, °C = 5/9 (°F − 32).

TABLE 11-22 Outdoor Insulation Thickness, 7.5-mi/h Wind, 60°F Air*

Pipe size, in	Thickness, in	Minimum pipe temperature, °F							
		Energy cost, $/million Btu							
		1	2	3	4	5	6	7	8
¾	1	450	300	250	250	200	200	150	150
	1½	800	500	400	300	250	250	200	200
	2			1150	950	850	750	700	650
	2½			1100	1000	900	800	750	700
1	1	400	300	250	200	200	150	150	150
	1½	1000	650	500	400	350	300	300	250
	2			1100	900	800	700	600	600
	2½				1200	1050	950	850	800
	3					1100	1000	900	850
1½	1	350	250	200	200	150	150	150	150
	1½	900	600	450	350	300	300	250	250
	2		100	850	700	600	550	500	450
	2½			1150	950	800	750	700	600
	3					1200	1050	1000	900
2	1	350	250	200	150	150	150	150	150
	1½	900	550	450	400	300	300	250	250
	2		1150	900	750	650	600	550	500
	2½			1000	850	750	650	600	550
	3				1050	950	850	750	700
3	1	300	200	150	150	150	150	150	150
	1½	750	500	400	300	250	250	250	200
	2		950	750	600	500	450	400	350
	2½		1150	950	750	650	600	500	500
	3			1150	1000	850	750	650	600
	3½								1150
4	1	250	200	150	150	150	150	150	150
	1½	750	500	350	300	250	250	200	200
	2		950	750	600	500	450	400	350
	2½			1050	900	700	650	600	550
	3			1100	950	750	700	650	600
	3½						1200	1100	1000
6	1	250	150	150	150	150	150	150	150
	1½	450	300	200	200	150	150	150	150
	2		900	700	600	500	450	400	350
	2½		1050	800	650	600	500	450	400
	3			1050	900	750	700	600	550
	3½					1150	1050	950	850
	4							1200	1150
	4½								1200
8	1	250	200	150	150	150	150	150	150
	2		850	650	550	450	400	350	350
	2½		900	700	600	500	450	400	400
	3			1100	950	800	750	700	600
	3½					1150	1000	950	850
	4							1050	1000
10	2	200	150	150	150	150	150	150	150
	2½		1000	800	650	550	500	450	400
	3		1200	950	800	700	600	550	500
	3½					1100		900	800
	4							1150	1050
	4½							1200	1100
12	1½	250	150	150	150	150	150	150	150
	2	950	600	500	400	350	300	250	250
	2½		900	700	550	500	400	400	350
	3			1100	900	800	700	650	550
	3½					1100	1000	900	850
	4					1150	1050	950	900
	4½					1200	1100	1000	950
14	1½	250	150	150	150	150	150	150	150
	2	850	550	400	350	300	250	250	250
	2½		850	650	550	500	400	400	400
	3		1000	850	750	700	650	550	500
	3½				1200	1000	950	850	800
	4					1050	1000	900	850
	4½					1100	1000	950	

TABLE 11-22 Outdoor Insulation Thickness, 7.5-mi/h Wind, 60°F Air* (Concluded)

Pipe size, in	Thickness, in	Minimum pipe temperature, °F							
		Energy cost, $/million Btu							
		1	2	3	4	5	6	7	8
16	1½	250	150	150	150	150	150	150	150
	2	800	500	350	300	300	250	250	200
	2½		900	700	550	500	450	400	350
	3		1000	850	700	600	500	450	400
	3½				1200	1000	950	850	800
	4					1100	1000	900	850
	4½					1150	1000	950	900
18	1½	250	150	150	150	150	150	150	150
	2	850	550	400	350	300	250	250	200
	2½		800	650	500	450	400	350	350
	3		1000	800	650	550	500	450	400
	3½					1100	1000	900	850
20	1½	150	150	150	150	150	150	150	150
	2	900	550	450	350	300	300	250	250
	2½		850	650	550	450	400	350	350
	3		1000	800	650	550	500	450	400
	3½					1150	1050	950	900
	4					1200	1100	1000	950
	4½						1200	1100	1050
24	1½	150	150	150	150	150	150	150	150
	2	800	500	400	300	250	250	200	200
	2½		950	750	650	550	500	450	400
	3		1150	950	750	650	600	550	500
	3½				1150	1000	900	800	750
	4				1200	1050	950	850	800
	4½						1050	950	850
	5								

*Aluminum-jacketed calcium silicate insulation with an emissivity factor of 0.05. To convert inches to millimeters, multiply by 25.4; to convert miles per hour to kilometers per hour, multiply by 1.609; and to convert dollars per 1 million British thermal units to dollars per 1 million kilojoules, multiply by 0.948; °C = 5/9 (°F − 32).

The tables were based upon the cost of energy at the end of the first year, a 10 percent inflation rate on energy costs, a 15 percent interest cost, and a present-worth pretax profit of 40 percent per annum on the last increment of insulation thickness. Dual-layer insulation was used for 3½-in and greater thicknesses. The tables and a full explanation of their derivation appear in a paper by F. L. Rubin (op. cit.). Alternatively, the selected thicknesses have a payback period on the last nominal ½-in increment of 1.44 years as presented in a later paper by Rubin ["Can You Justify More Piping Insulation?" *Hydrocarbon Process.*, 152–155 (July 1982)].

Example 1 For 24-in pipe at 371°C (700°F) with an energy cost of $4/million Btu, select 2-in thickness for indoor and 2½-in for outdoor locations. [A 2½-in thickness would be chosen at 399°C (750°F) indoors and 3½-in outdoors.]

Example 2 For 16-in pipe at 343°C (650°F) with energy valued at $5/million Btu, select 2½-in insulation indoors [use 3-in thickness at 371°C (700°F)]. Outdoors choose 3-in insulation [use 3½-in dual-layer insulation at 538°C (1000°F)].

Example 3 For 12-in pipe at 593°C (1100°F) with an energy cost of $6/million Btu, select 3½-in thickness for an indoor installation and 4½-in thickness for an outdoor installation.

INSTALLATION PRACTICE

Pipe Depending on diameter, pipe is insulated with cylindrical half, third, or quarter sections or with flat segmental insulation. Fittings and valves are insulated with preformed insulation covers or with individual pieces cut from sectional straight pipe insulation.

Method of Securing Insulation with factory-applied jacketing may be secured with adhesive on the overlap, staples, tape, or wire, depending on the type of jacket and the outside diameter. Insulation which has a separate jacket is wired or banded in place before the jacket (finish) is applied.

Double Layer Pipe expansion is a significant factor at temperatures above 600°F (316°C). Above this temperature, insulation should be applied in a double layer with all joints staggered to prevent excessive heat loss and high surface temperature at joints opened by pipe expansion. This procedure also minimizes thermal stresses in the insulation.

Finish Covering for cylindrical surfaces ranges from asphalt-saturated or saturated and coated organic and asbestos paper, through laminates of such papers and plastic films or aluminum foil, to medium-gauge aluminum, galvanized steel, or stainless steel. Fittings and irregular surfaces may be covered with fabric-reinforced mastics or preformed metal or plastic covers. Finish selection depends on function and location. Vapor-barrier finishes may be in sheet form or a mastic, which may or may not require reinforcing, depending on the method of application, and additional protection may be required to prevent mechanical abuse and/or provide fire resistance. Criteria for selecting other finishes should include protection of insulation against water entry, mechanical abuse, or chemical attack. Appearance, life-cycle cost, and fire resistance may also be determining factors. Finish may be secured with tape, adhesive, bands, or screws. Fasteners which will penetrate vapor-retarder finishes should not be used.

Tanks, Vessels, and Equipment Flat, curved, and irregular surfaces such as tanks, vessels, boilers, and breechings are normally insulated with flat blocks, beveled lags, curved segments, blankets, or spray-applied insulation. Since no general procedure can apply to all materials and conditions, it is important that manufacturers' specifications and instructions be followed for specific insulation applications.

Method of Securing On small-diameter cylindrical vessels, the insulation may be secured by banding around the circumference. On larger cylindrical vessels, banding may be supplemented with angle-iron ledges to support the insulation and prevent slipping. On large flat and cylindrical surfaces, banding or wiring may be supplemented with various types of welded studs or pins. Breather springs may be required with bands to accommodate expansion and contraction.

Finish The materials are the same as for pipe and should satisfy the same criteria. Breather springs may be required with bands.

ADDITIONAL REFERENCES: *ASHRAE Handbook and Product Directory: Fundamentals,* American Society of Heating, Refrigerating and Air Condition-ing Engineers, Atlanta, 1981. Turner and Malloy, *Handbook of Thermal Insulation Design Economics for Pipes and Equipment,* Krieger, New York, 1980. Turner and Malloy, *Thermal Insulation Handbook,* McGraw-Hill, New York, 1981.

AIR CONDITIONING

INTRODUCTION

Air Conditioning is the process of treating air so as to control simultaneously its temperature, humidity, cleanliness, and distribution to meet the requirements of the conditioned space. The portions relating only to temperature and humidity control will be discussed here. For detailed discussions of air cleanliness and distribution, refer, for example, to the current edition of the *HVAC Applications* volume of the *A.S.H.R.A.E. Handbooks* (ASHRAE, 1791 Tullie Circle, N.E., Atlanta, Ga.). Applications of air conditioning include the promotion of human comfort and the maintenance of proper conditions for the manufacture, processing, and preserving of material and equipment. Also, in industrial environments where, for economical or other reasons, conditions cannot be made entirely comfortable, air conditioning may be used for maintaining the efficiency and health of workers at safe tolerance limits.

COMFORT AIR CONDITIONING

Comfort is influenced by temperature, humidity, air velocity, radiant heat, clothing, and work intensity. Psychological factors may also influence comfort, but their discussion is beyond the scope of this handbook. The reader is referred to Chap. 42 of the *HVAC Applications* volume of the *A.S.H.R.A.E. Handbooks* for a full discussion of the control of noise, which must also be considered in air-conditioning design. Figure 5 in Chap. 8 of the *HVAC Fundamentals* volume of the *A.S.H.R.A.E. Handbooks* relates the variables of ambient temperature, dew point temperature (or humidity ratio) to comfort under clothing and activity conditions typical for office space occupancy. It also shows boundary values for $ET°$, the effective temperature index. This index combines temperature and moisture conditions into a single index representing the temperature of an environment at 50 percent relative humidity resulting in the same heat transfer from the skin as for the actual case. Hence, the $ET°$ for 50 percent relative humidity is equal in value to the ambient dry-bulb temperature.

INDUSTRIAL AIR CONDITIONING

Industrial buildings have to be designed according to their intended use. For instance, the manufacture of hygroscopic materials (paper, textiles, foods, etc.) will require relatively tight controls of relative humidity. On the other hand, the storage of furs will demand relatively low temperatures, while the ambient in a facility manufacturing refractories might be acceptable at notably higher temperatures. Chapter 12 of the *HVAC Applications* volume of the *A.S.H.R.A.E. Handbooks* provides extensive tables of suggested temperatures and humidities for industrial air conditioning.

VENTILATION

In the design of comfort air-conditioning systems, odors arising from occupants, cooking, or other sources must be controlled. this is accomplished by introducing fresh air or purified recirculated air in sufficient quantities to reduce odor concentrations to an acceptable level by dilution. Recommended fresh-air requirements for different types of buildings are called for in A.S.H.R.A.E. standard 62-1989 "Ventilation for Acceptable Indoor Air Quality." These values range in the order of 15 to 30 cfm per person, according to application.

In industrial air-conditioning systems, harmful environmental gases, vapors, dusts, and fumes are often encountered. These contaminants can be controlled by exhaust systems at the source, by dilution ventilation, or by a combination of the two methods. When exhaust systems are used, it is necessary to introduce sufficient fresh air into the air-conditioned area to make up for that exhausted. Generally, low exhaust systems are used where the contaminant sources are concentrated and/or where the contaminant may be highly toxic. Where the contaminant comes from widely dispersed points, however, dilution ventilation is usually employed. Combinations of the two systems sometimes provide the least expensive installation. dilution ventilation is not appropriate for cases where large volumes of contaminant are released and cases where the employees must work near the contaminant source. The selection of dilution ventilation for cases with potential fire or smoke should be accompanied by careful study. Details for design of dilution ventilation systems are given in Chap. 25 of the *HVAC Applications* volume of the *A.S.H.R.A.E. Handbooks.* Chapter 27 of the same volume discusses industrial exhaust systems. Exhaust stacks should be high enough to adequately dispense the contaminated air and to prevent recirculation into fresh air intakes (Chap. 14 of the 1993 *HVAC Fundamentals* volume of the *A.S.H.R.A.E. Handbooks*). Depending on the contaminant and air pollution legislation, it may be necessary to reduce the contaminant emission rate by such methods as filtering, scrubbing, catalytic oxidation, or incineration.

AIR-CONDITIONING EQUIPMENT

Basically, an air-conditioning system consists of a fan unit which forces a mixture of fresh outdoor air and room air through a series of devices which act upon the air to clean it, to increase or decrease its temperature, and to increase or decrease its water-vapor content or humidity.

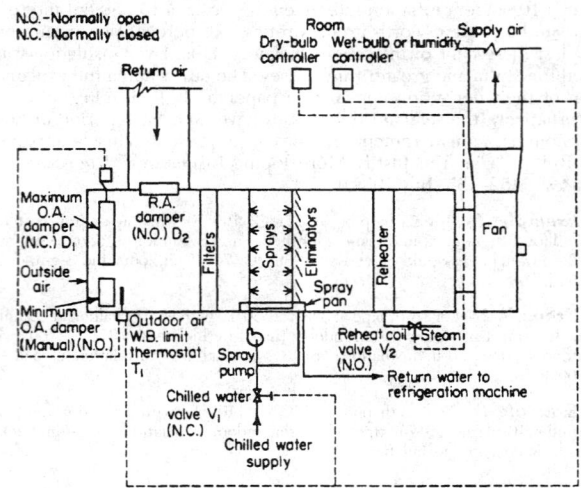

FIG. 11-69 Typical central-station air-conditioning unit and control system. On a rising room wet-bulb temperature, the wet-bulb branch-line air pressure increases through the reverse-acting outdoor-air wet-bulb temperature-limit thermostat T_1 to open gradually the maximum outdoor-air damper D_1 and simultaneously closr return-air damper D_2, then gradually open chilled-water valve V_1. On a rising room dry-bulb temperature, the dry-bulb branch-line air pressure gradually increases to close reheat steam valve V_2. When outdoor wet-bulb temperature exceeds the set point of the outdoor-air wet-bulb-limit thermostat T_1, which is set at the return-air wet-bulb temperature, this thermostat decreases branch-line pressure to close gradually maximum outdoor damper D_1 and simultaneously open return-air damper D_2. The reverse sequences are followed during the heating season.

In general, air conditioning equipment can be classified into two broad types: central (sometimes called field erected) and unitary (or packaged).

CENTRAL SYSTEMS

Figure 11-69 describes a typical central system. Either water or direct-expansion refrigerant coils or air washers may be used for cooling. Steam or hot-water coils are available for heating. Humidification may be provided by target-type water nozzles, pan humidifiers, air washers, or sprayed coils. Air cleaning is usually provided by cleanable or throwaway filters. Central-station air-conditioning units in capacities up to about 50,000 cu ft/min are available in prefabricated units.

The principle types of refrigeration equipment used in large central systems are: Reciprocating (up to 300 hp); helical rotary (up to 750 tons); absorption (up to 2000 tons); and centrifugal (up to 10,000 tons). The drives for the reciprocating, rotary, and centrifugal compressors may be electric motors, gas or steam turbines, or gas or diesel engines. The heat rejected from the condensors usually calls for cooling towers or air-cooled condensers; in some cases evaporative cooling might be practical.

UNITARY REFRIGERANT-BASED AIR-CONDITIONING SYSTEMS

These systems include window-mounted air conditioners and heat pumps, outdoor unitary equipment, indoor unitary equipment, unitary self-contained systems, and commercial self-contained systems. These are described in detail in the *HVAC Systems and Equipment* volume of the *A.S.H.R.A.E. Handbooks*. A detailed analysis of the proposed installation is usually necessary to select the air conditioning equipment which is best in overall performance. Each type of air conditioner has its own particular advantages and disadvantages. Important factors to be considered in the selection of air conditioning equipment are degree of temperature and humidity control required, investment, owning, and operating costs, and space requirements. Another important factor is the building itself, that is, whether it is new or existing construction. For example, for existing buildings where it may be inadvisable to install air-supply ducts, the self-contained or unit-type air conditioner may offer the greatest advantages in reduced installation costs. For large industrial processes where close temperature and humidity control are required, a central station system is usually employed.

TABLE 11-23   Outdoor Design Temperatures*

City-state	Winter Dry bulb, °F.	Summer Dry bulb, °F.	Summer Wet bulb, °F.	City-state	Winter Dry bulb, °F.	Summer Dry bulb, °F.	Summer Wet bulb, °F.
Akron, Ohio	−5	95	75	Milwaukee, Wis.	−15	95	75
Albany, N.Y.	−10	93	75	Minneapolis, Minn.	−20	95	75
Albuquerque, N.M.	0	95	70	Nashville, Tenn.	0	95	78
Atlanta, Ga.	10	95	76	New Haven, Conn.	0	95	75
Baltimore, Md.	0	95	78	New Orleans, La.	20	95	80
Billings, Mont.	−25	90	66	New York, N.Y.	0	95	75
Birmingham, Ala.	10	95	78	Newark, N.J.	0	95	75
Bloomfield, N.J.	0	95	75	Norfolk, Va.	15	95	78
Boise, Idaho	−10	95	65	Oakland, Calif.	30	85	65
Boston, Mass.	0	92	75	Oklahoma City, Okla.	0	101	77
Bridgeport, Conn.	0	95	75	Omaha, Nebr.	−10	95	78
Buffalo, N.Y.	−5	93	73	Peoria, Ill.	−10	96	76
Charleston, S.C.	15	95	78	Philadelphia, Pa.	0	95	78
Chattanooga, Tenn.	10	95	76	Phoenix, Ariz.	0	105	76
Chicago, Ill.	−10	95	75	Pittsburgh, Pa.	−5	95	75
Cincinnati, Ohio	0	95	78	Portland, Me.	−5	90	73
Cleveland, Ohio	0	95	75	Portland, Ore.	10	90	68
Columbus, Ohio	−10	95	76	Providence, R.I.	0	93	75
Dallas, Tex.	0	100	78	Reno, Nev.	−5	95	65
Dayton, Ohio	0	95	78	Richmond, Va.	15	95	78
Denver, Colo.	−10	95	64	Roanoke, Va.	0	95	76
Des Moines, Iowa	−15	95	78	Rochester, N.Y.	−5	95	75
Detroit, Mich.	−10	95	75	St. Louis, Mo.	0	95	78
Duluth, Minn.	−25	93	73	St. Paul, Minn.	−20	95	75
East Orange, N.J.	0	95	75	Salt Lake City, Utah	−10	95	65
El Paso, Tex.	10	100	69	San Antonio, Tex.	20	100	78
Erie, Pa.	−5	93	75	San Francisco, Calif.	35	85	65
Fitchburg, Mass.	−10	93	75	Schenectady, N.Y.	−10	93	75
Flint, Mich.	−10	95	75	Scranton, Pa.	−5	95	75
Fort Wayne, Ind.	−10	95	75	Seattle, Wash.	15	85	65
Fort Worth, Tex.	10	100	78	Shreveport, La.	20	100	78
Grand Rapids, Mich.	−10	95	75	Sioux City, Iowa	−20	95	78
Hartford, Conn.	0	93	75	Spokane, Wash.	−15	93	65
Houston, Tex.	20	95	80	Springfield, Mass.	−10	93	75
Indianapolis, Ind.	−10	95	76	Syracuse, N.Y.	−10	93	75
Jacksonville, Fla.	25	95	78	Tampa, Fla.	30	95	78
Jersey City, N.J.	0	95	75	Toledo, Ohio	−10	95	75
Kansas City, Mo.	−10	100	76	Tucson, Ariz.	25	105	72
Lincoln, Nebr.	−10	95	78	Tulsa, Okla.	0	101	77
Little Rock, Ark.	5	95	78	Washington, D.C.	0	95	78
Long Beach, Calif.	35	90	70	Wichita, Kans.	−10	100	75
Los Angeles, Calif.	35	90	70	Wilmington, Del.	0	95	78
Louisville, Ky.	0	95	78	Worcester, Mass.	0	93	75
Memphis, Tenn.	0	95	78	Youngstown, Ohio	−5	95	75
Miami, Fla.	35	91	79				

*Carrier, Cherne, Grant, and Roberts, *Modern Air Conditioning, Heating, and Ventilating*, 3d ed., p. 531, Pitman, New York, 1959.

LOAD CALCULATION

First step in the solution of an air-conditioning problem is to determine the proper design temperature conditions. Since both outdoor and indoor temperatures greatly influence the size of the equipment, the designer must exercise good judgment in selecting the proper conditions for his/her particular case. Table 11-23 lists winter and summer outdoor temperature conditions in common use for comfort applications for various United States cities. For critical-process air conditioning, it may be desirable to use a different set of outdoor temperature conditions. However, it is seldom good practice to design for the extreme maximum or minimum outside conditions. (See the 1993 *HVAC Fundamentals* volume of the *A.S.H.R.A.E. Handbooks*).

After the proper inside and summer outside temperature conditions for comfort and temperature conditions for process air conditioning have been selected, the next step is to calculate the space cooling load, which is made up of sensible heat and latent heat loads. The sensible heat load consists of (1) transmission through walls, roofs, floors, ceilings, and window glass, (2) solar and sky radiation, (3) heat gains from infiltration of outside air, (4) heat gains from people, lights, appliances, and power equipment (including the supply-air fan motor), and (5) heat to be removed from materials or products brought in at higher than room temperature. The latent heat load includes loads due to moisture (1) given off from people, appliances, and products and (2) from infiltration of outside air. The space total heat load is the sum of the sensible heat load and latent heat load of the space. The total refrigeration load consists of the total space load plus the sensible and latent heat loads from the outside air introduced at the conditioning unit.

The procedure for load calculation in nonresidential buildings should account for thermal mass (storage) effects as well as occupancy and other uses affecting the load. The load can in turn be strongly dependent on the nature of the building utilization; as an example, lightning might be a major component in the thermal load for a high-rise office building causing a need for cooling even in winter days. There are various approaches to load calculation, some requiring elaborate computer models. Chapter 26 of the 1993 *HVAC Fundamentals* volume of the *A.S.H.R.A.E. Handbooks* presents a step-by-step outline of the current methods in practice for load calculation.

REFRIGERATION

INTRODUCTION

Refrigeration is a process where heat is transferred from a lower- to a higher-temperature level by doing work on a system. In some systems heat transfer is used to provide the energy to drive the refrigeration cycle. All refrigeration systems are heat pumps ("pumps energy from a lower to a higher potential"). The term heat pump is mostly used to describe refrigeration system applications where heat rejected to the condenser is of primary interest.

There are many means to obtain refrigerating effect, but here three will be discussed: mechanical vapor refrigeration cycles, absorption and steam jet cycles due to their significance for industry.

Basic Principles   Since refrigeration is the practical application of the thermodynamics, comprehending the basic principles of thermodynamics is crucial for full understanding of refrigeration. Section 4 includes a through approach to the theory of thermodynamics. Since our goal is to understand refrigeration processes, cycles are of the crucial interest.

The Carnot refrigeration cycle is reversible and consists of adiabatic (isentropic due to reversible character) compression (1-2), isothermal rejection of heat (2-3), adiabatic expansion (3-4) and isothermal addition of heat (4-1). The temperature-entropy diagram is shown in Fig. 11-70. The Carnot cycle is an unattainable ideal which serves as a standard of comparison and it provides a convenient guide to the temperatures that should be maintained to achieve maximum effectiveness.

The measure of the system performance is *coefficient of performance* (COP). For refrigeration applications COP is the ratio of heat removed from the low-temperature level (Q_{low}) to the energy input (W):

$$COP_R = \frac{Q_{low}}{W} \tag{11-84}$$

For the heat pump (HP) operation, heat rejected at the high temperature (Q_{high}) is the objective, thus:

$$COP_{HP} = \frac{Q_{high}}{W} = \frac{Q + W}{W} = COP_R + 1 \tag{11-85}$$

For a Carnot cycle (where $\Delta Q = T\Delta s$), the COP for the refrigeration application becomes (note than T is absolute temperature [K]):

$$COP_R = \frac{T_{low}}{T_{high} - T_{low}} \tag{11-86}$$

and for heat pump application:

$$COP_{HP} = \frac{T_{high}}{T_{high} - T_{low}} \tag{11-87}$$

The COP in real refrigeration cycles is always less than for the ideal (Carnot) cycle and there is constant effort to achieve this ideal value.

Basic Refrigeration Methods   Three basic methods of refrigeration (mentioned above) use similar processes for obtaining refrigeration effect: evaporation in the evaporator, condensation in the condenser where heat is rejected to the environment, and expansion in a flow restrictor. The main difference is in the way compression is being done (Fig. 11-71): using mechanical work (in compressor), thermal energy (for absorption and desorption), or pressure difference (in ejector).

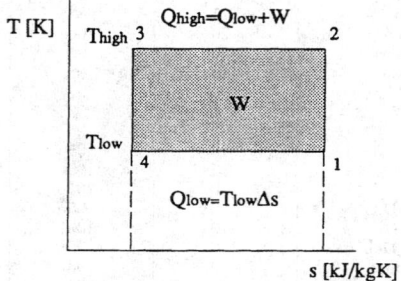

FIG. 11-70   Temperature-entropy diagram of the Carnot cycle.

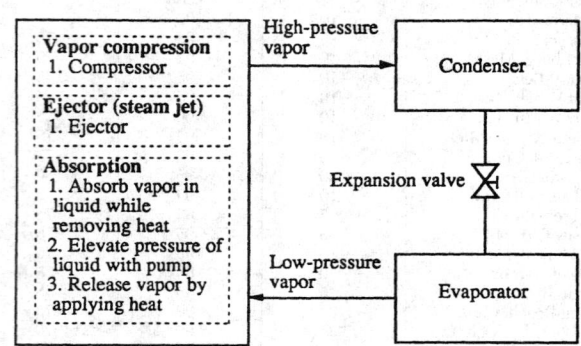

FIG. 11-71   Methods of transforming low-pressure vapor into high-pressure vapor in refrigeration systems (Stoecker, Refrigeration and Air Conditioning).

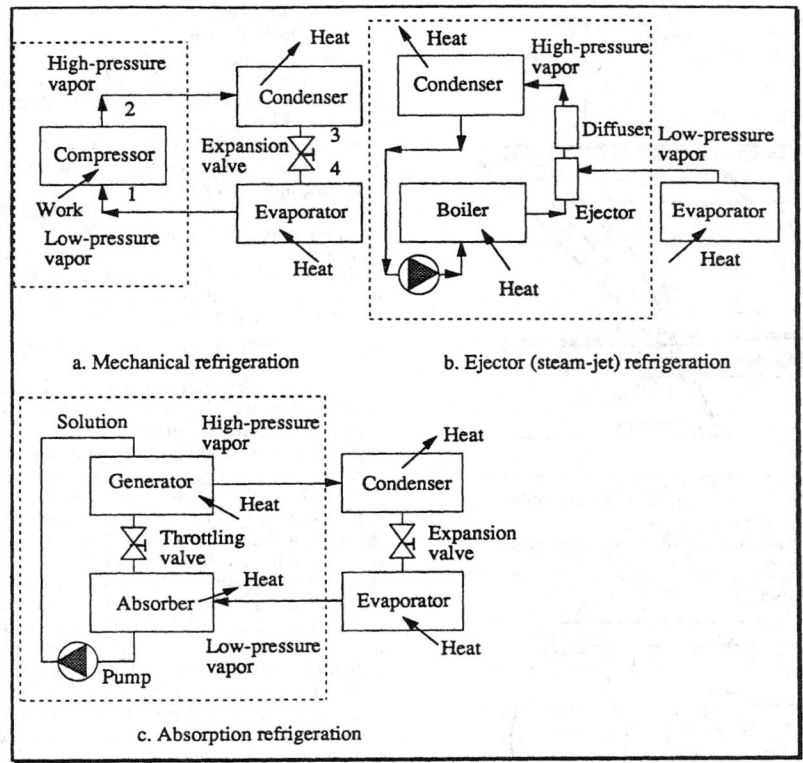

FIG. 11-72   Basic refrigeration systems.

In the next figure (Fig. 11-72) basic refrigeration systems are displayed more detailed. More elaborated approach is presented in the text.

MECHANICAL REFRIGERATION (VAPOR-COMPRESSION SYSTEMS)

Vapor-Compression Cycles   The most widely used refrigeration principle is vapor compression. Isothermal processes are realized through isobaric evaporation and condensation in the tubes. Standard vapor compression refrigeration cycle (counterclockwise Rankine cycle) is marked in Fig. 11-72a) by 1, 2, 3, 4.

Work that could be obtained in turbine is small, and iturbine is substituted for an expansion valve. For the reasons of proper compressor function, wet compression is substituted for an compression of dry vapor.

Although the *T-s* diagram is very useful for thermodynamic analysis, the pressure enthalpy diagram is used much more in refrigeration practice due to the fact that both evaporation and condensation are isobaric processes so that heat exchanged is equal to enthalpy difference $\Delta Q = \Delta h$. For the ideal, isentropic compression, the work could be also presented as enthalpy difference $\Delta W = \Delta h$. The vapor compression cycle (Rankine) is presented in Fig. 11-73 in *p-h* coordinates.

Figure 11-74 presents actual versus standard vapor-compression cycle. In reality, flow through the condenser and evaporator must be accompanied by pressure drop. There is always some subcooling in the condenser and superheating of the vapor entering the compressor-suction line, both due to continuing process in the heat exchangers and the influence of the environment. Subcooling and superheating are usually desirable to ensure only liquid enters the expansion device. Superheating is recommended as a precaution against droplets of liquid being carried over into the compressor.

There are many ways to increase cycle efficiency (COP). Some of them are better suited to one, but not for the other refrigerant. Sometimes, for the same refrigerant, the impact on COP could be different for various temperatures. One typical example is the use of a liquid-to-suction heat exchanger (Fig. 11-75).

The suction vapor coming from the evaporator could be used to

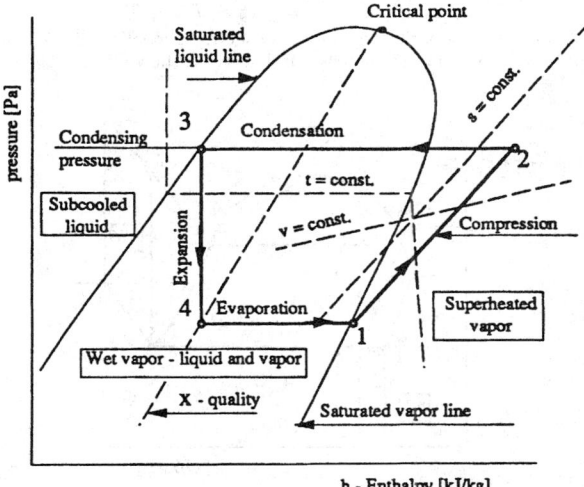

FIG. 11-73   *p-h* diagram for vapor-compression cycle.

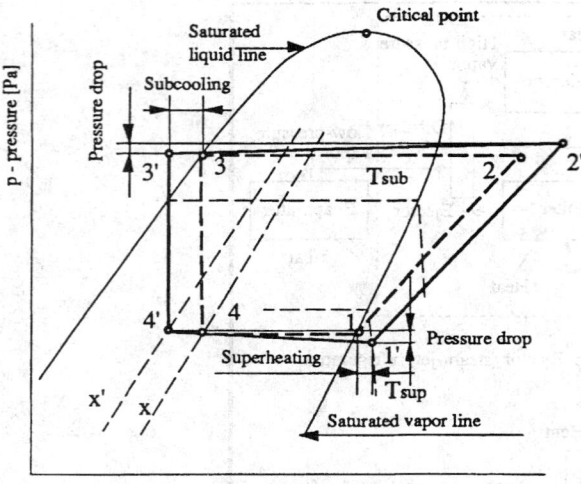

FIG. 11-74 Actual vapor-compression cycle compared with standard cycle.

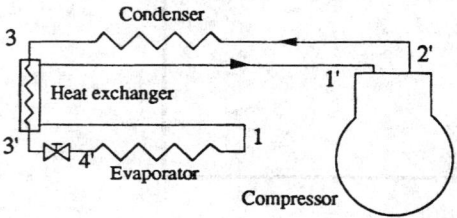

FIG. 11-75 Refrigeration system with a heat exchanger to subcool the liquid from the condenser.

subcool the liquid from the condenser. Graphic interpretation in T-s diagram for such a process is shown in Fig. 11-76. The result of the use of suction line heat exchanger is to increase the refrigeration effect ΔQ and to increase the work by ΔW. The change in COP is then:

$$\Delta COP = COP' - COP = \frac{Q + \Delta Q}{(P + \Delta P) - Q/P} \qquad (11\text{-}88)$$

When dry, or superheated, vapor is used to subcool the liquid, the COP in R12 systems will increase, and decrease the COP in NH_3 sys-

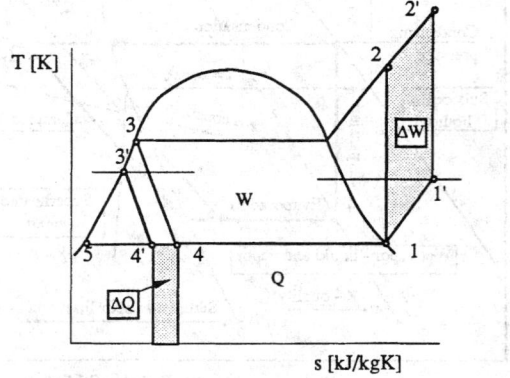

FIG. 11-76 Refrigeration system with a heat exchanger to subcool the liquid from the condenser.

tems. For R22 systems it could have both effects, depending on the operating regime. Generally, this measure is advantageous (COP is improved) for fluids with high, specific heat of liquid (less-inclined saturated-liquid line on the p-h diagram), small heat of evaporation h_{fg}, when vapor-specific heat is low (isobars in superheated regions are steep), and when the difference between evaporation and condensation temperature is high. Measures to increase COP should be studied for every refrigerant. Sometimes the purpose of the suction-line heat exchanger is not only to improve the COP, but to ensure that only the vapor reaches the compressor, particularly in the case of a malfunctioning expansion valve.

The system shown in Fig. 11-75 is direct expansion where dry or slightly superheated vapor leaves the evaporator. Such systems are predominantly used in small applications because of their simplicity and light weight. For the systems where efficiency is crucial (large industrial systems), recirculating systems (Fig. 11-77) are more appropriate.

Ammonia refrigeration plants are almost exclusively built as recirculating systems. The main advantage of recirculating versus direct expansion systems is better utilization of evaporator surface area. The diagram showing influence of quality on the local heat-transfer coefficients is shown in figure 11-90. It is clear that heat-transfer characteristics will be better if the outlet quality is lower than 1. Circulation could be achieved either by pumping (mechanical or gas) or using gravity (thermosyphon effect: density of pure liquid at the evaporator entrance is higher than density of the vapor-liquid mixture leaving the evaporator). The circulation ratio (ratio of actual mass flow rate to the evaporated mass flow rate) is higher than 1 and up to 5. Higher values are not recommended due to a small increase in heat-transfer rate for a significant increase in pumping costs.

Multistage Systems When the evaporation and condensing pressure (or temperature) difference is large, it is prudent to separate compression in two stages. The use of multistage systems opens up the opportunity to use flash-gas removal and intercooling as measures to improve performance of the system. One typical two-stage system with two evaporating temperatures and both flash-gas removal and intercooling is shown in figure 11-78. The purpose of the flash-tank intercooler is to: (1) separate vapor created in the expansion process, (2) cool superheated vapor from compressor discharge, and (3) to eventually separate existing droplets at the exit of the medium-temperature evaporator. The first measure will decrease the size of the low-stage compressor because it will not wastefully compress the portion of the flow which cannot perform the refrigeration and second will decrease the size of the high-stage compressor due to lowering the specific volume of the vapor from the low-stage compressor discharge, positively affecting operating temperatures of the high-stage compressor due to cooling effect.

If the refrigerating requirement at a low-evaporating temperature is Q_l and at the medium level is Q_m, then mass flow rates (m_1 and m_m respectively) needed are:

$$m_1 = \frac{Q_l}{h_1 - h_8} = \frac{Q_l}{h_1 - h_7} \qquad (11\text{-}89)$$

$$m_m = \frac{Q_m}{h_3 - h_6} \qquad (11\text{-}90)$$

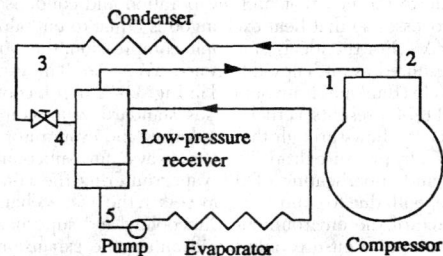

FIG. 11-77 Recirculation system.

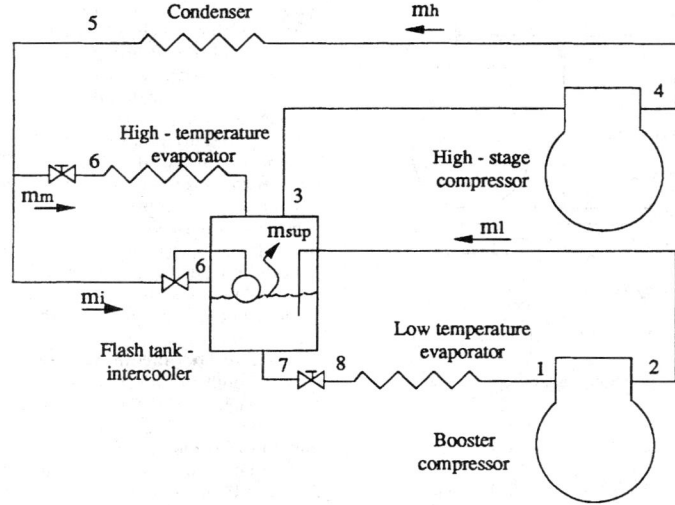

FIG. 11-78 Typical two-stage system with two evaporating temperatures, flash-gas removal, and intercooling.

The mass flow rate at the flash-tank inlet m_i consists of three components ($m_i = m_1 + m_{sup} + m_{flash}$):

m_1 = liquid at p_m feeding low temperature evaporator,

m_{sup} = liquid at p_m to evaporate in flash tank to cool superheated discharge,

m_{flash} = flashed refrigerant, used to cool remaining liquid.

Vapor component is:

$$m_{flash} = x_m * m_i \qquad (11\text{-}91)$$

and liquid component is:

$$(1 - x_m) * m_i = m_1 + m_{sup} \qquad (11\text{-}92)$$

Liquid part of flow to cool superheated compressor discharge is determined by:

$$m_{sup} = \frac{Q_l}{h_1 - h_8} * \frac{h_2 - h_3}{h_3 - h_7} = m_1 * \frac{h_2 - h_3}{h_{fgm}} \qquad (11\text{-}93)$$

Since the quality x_m is:

$$x_m = \frac{h_6 - h_7}{h_3 - h_7} \qquad (11\text{-}94)$$

mass flow rate through condenser and high-stage compressor m_h is finally:

$$m_h = m_m + m_i \qquad (11\text{-}95)$$

The optimum intermediate pressure for the two-stage refrigeration cycles is determined as the geometric mean between evaporation pressure (p_l) and condensing pressure (p_h, Fig. 11-79):

$$p_m = (\text{sqrt}) \left(\frac{p_h}{p_l} \right)$$

based on equal pressure ratios for low- and high-stage compressors. Optimum interstage pressure is slightly higher than the geometric mean of the suction and the discharge pressures, but, due to very flat optimum of power versus interstage pressure relation geometric mean, it is widely accepted for determining the intermediate pressure. Required pressure of intermediate-level evaporator may dictate interstage pressure other than determined as optimal.

Two-stage systems should be seriously considered when the evaporating temperature is below −20°C. Such designs will save on power and reduce compressor discharge temperatures, but will increase initial cost.

Cascade System   This is a reasonable choice in cases when the evaporating temperature is very low (below −60°C). When condensing pressures are to be in the rational limits, the same refrigerant has a high, specific volume at very low temperatures, requiring a large compressor. The evaporating pressure may be below atmospheric, which could cause moisture and air infiltration into the system if there is a leak. In other words, when the temperature difference between the medium that must be cooled and the environment is too high to be served with one refrigerant, it is wise to use different refrigerants in the high and low stages. Figure 11-80 shows a cascade system schematic diagram. There are basically two independent systems linked via a heat exchanger: the evaporator of the high-stage and the condenser of the low-stage system.

EQUIPMENT

Compressors   These could be classified by one criteria (the way the increase in pressure is obtained) as positive-displacement and

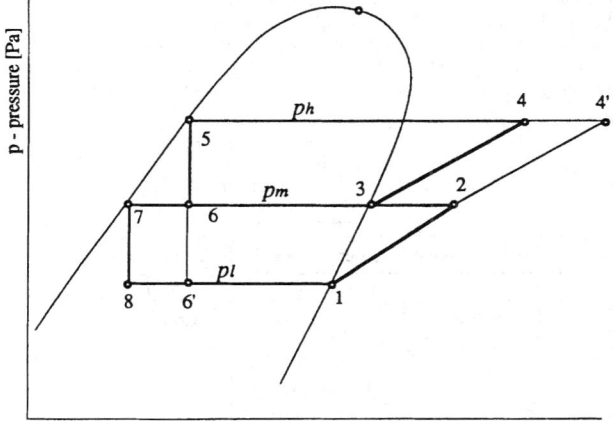

FIG. 11-79 Pressure enthalpy diagram for typical two-stage system with two evaporating temperatures, flash-gas removal, and intercooling.

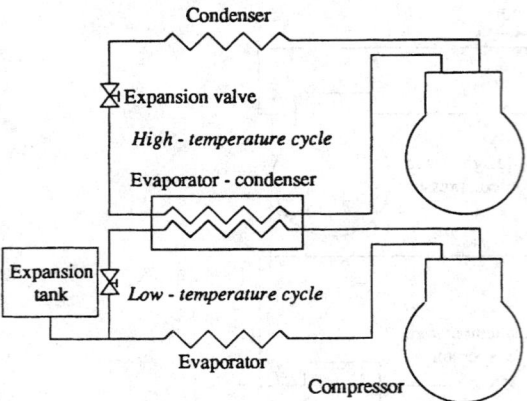

FIG. 11-80   Cascade system.

dynamic types as shown in Fig. 11-81 (see Sec. 10 for drawings and mechanical description of the various types of compressors). *Positive-displacement compressors* (PDC) are the machines that increase the pressure of the vapor by reducing the volume of the chamber. Typical PDC are reciprocating (in a variety of types) or rotary as screw (with one and two rotors), vane, scroll, and so on. Centrifugal or turbocompressors are machines where the pressure is raised converting some of kinetic energy obtained by a rotating mechanical element which continuously adds angular momentum to a steadily flowing fluid, similar to a fan or pump.

Generally, reciprocating compressors dominate in the range up to 300 kW refrigeration capacity. Centrifugal compressors are more accepted for the range over 500 kW, while screw compressors are in between with a tendency to go toward smaller capacities. The vane and the scroll compressors are finding their places primarily in very low capacity range (domestic refrigerators and the air conditioners), although vane compressors could be found in industrial compressors. Frequently, screw compressors operate as boosters, for the base load, while reciprocating compressors accommodate the variation of capacity, in the high stage. The major reason is for such design is the advantageous operation of screw compressors near full load and in design conditions, while reciprocating compressors seem to have better efficiencies at part-load operation than screw.

Using other criteria, compressors are classified as *open, semihermetic (accessible),* or *hermetic.* Open type is characterized by shaft extension out of compressor where it is coupled to the driving motor. When the electric motor is in the same housing with the compressor mechanism, it could be either hermetic or accessible (semihermetic). Hermetic compressors have welded enclosures, not designed to be repaired, and are generally manufactured for smaller capacities (sel-

dom over 30 kW), while semihermetic or an accessible type is located in the housing which is tightened by screws. Semihermetic compressors have all the advantages of hermetic (no sealing of moving parts, e.g., no refrigerant leakage at the seal shaft, no external motor mounting, no coupling alignment) and could be serviced, but it is more expensive.

Compared to other applications, refrigeration capacities in the chemical industry are usually high. That leads to wide usage of either centrifugal, screw, or high-capacity rotary compressors. Most centrifugal and screw compressors use economizers to minimize power and suction volume requirements. Generally, there is far greater use of open-drive type compressors in the chemical plants than in air-conditioning, commercial, or food refrigeration. Very frequently, compressor lube oil systems are provided with auxiliary oil pumps, filters, coolers, and other equipment to permit maintenance and repair without shut down.

Positive-Displacement Compressors   *Reciprocating compressors* are built in different sizes (up to about one megawatt refrigeration capacity per unit). Modern compressors are high-speed, mostly direct-coupled, single-acting, from one to mostly eight, and occasionally up to sixteen cylinders.

Two characteristics of compressors for refrigeration are the most important: refrigerating capacity and power. Typical characteristics are as presented in the Fig. 11-82.

Refrigerating capacity Q_e is the product of mass flow rate of refrigerant m and refrigerating effect R which is (for isobaric evaporation) $R = h_{\text{evaporator outlet}} - h_{\text{evaporator inlet}}$. Power P required for the compression, necessary for the motor selection, is the product of mass flow rate m and work of compression W. The latter is, for the isentropic compression, $W = h_{\text{discharge}} - h_{\text{suction}}$. Both of these characteristics could be calculated for the ideal (without losses) and for the actual compressor. Ideally, the mass flow rate is equal to the product of the compressor displacement V_i per unit time and the gas density ρ: $m = V_i * \rho$. The compressor displacement rate is volume swept through by the pistons (product of the cylinder number n, and volume of cylinder $V = $ stroke $* d^2\pi/4$) per second. In reality, the actual compressor delivers less refrigerant.

Ratio of the actual flow rate (entering compressor) to the displacement rate is the volumetric efficiency η_{va}. The volumetric efficiency is

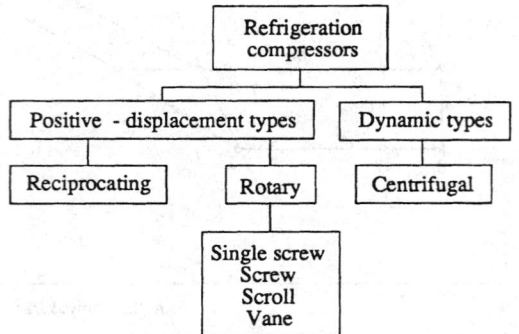

FIG. 11-81   Types of refrigeration compressors.

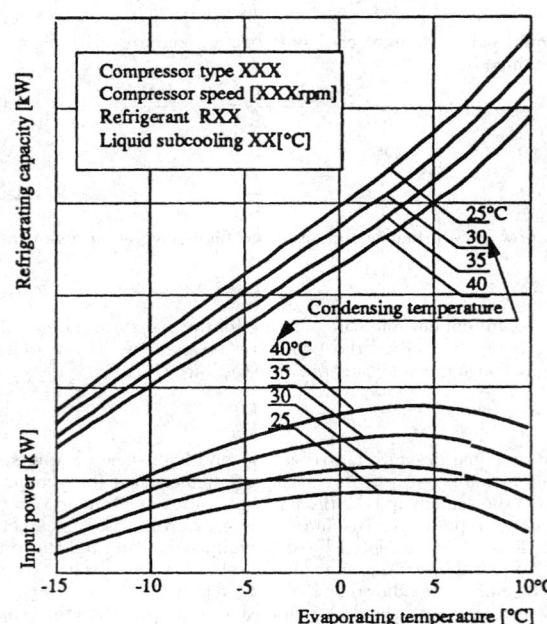

FIG. 11-82   Typical capacity and power-input curves for reciprocating compressor.

less then unity due to: reexpansion of the compressed vapor in clearance volume, pressure drop (through suction and discharge valves, strainers, manifolds, etc.), internal gas leakage (through the clearance between piston rings and cylinder walls, etc.), valve inefficiencies, and due to expansion of the vapor in the suction cycle caused by the heat exchanged (hot cylinder walls, oil, motor, etc.).

Similar to volumetric efficiency, isentropic (adiabatic) efficiency η_a is the ratio of the work required for isentropic compression of the gas to work input to the compressor shaft. The adiabatic efficiency is less than one mainly due to pressure drop through the valve ports and other restricted passages and the heating of the gas during compression.

Figure 11-83 presents the compression on a pressure-volume diagram for an ideal compressor with clearance volume (thin lines) and actual (thick lines). Compression in an ideal compressor without clearance is extended using dashed lines to the points I_d (end of discharge), line $I_d - I_s$ (suction), and I_s (beginning of suction). The area surrounded by the lines of compression, discharge, reexpansion and intake presents the work needed for compression. Actual compressor only appears to demand less work for compression due to smaller area in the p-V diagram. Mass flow rate for an ideal compressor is higher, which cannot be seen in the diagram. In reality, an actual compressor will have diabatic compression and reexpansion and higher-discharge and lower-suction pressures due to pressure drops in valves and lines. The slight increase in the pressure at the beginning of the discharge and suction is due to forces needed to initially open valves.

When the suction pressure is lowered, the influence of the clearance will increase, causing in the extreme cases the entire volume to be used for reexpansion, which drives the volumetric efficiency to zero.

There are various options for capacity control of reciprocating refrigeration compressors:

1. Opening the suction valves by some external force (oil from the lubricating system, discharge gas, electromagnets . . .).
2. Gas bypassing—returning discharge gas to suction (within the compressor or outside the compressor).
3. Controlling suction pressure by throttling in the suction line.
4. Controlling discharge pressure.
5. Adding reexpansion volume.
6. Changing the stroke.
7. Changing the compressor speed.

The first method is used most frequently. The next preference is for the last method, mostly used in small compressors due to problems with speed control of electrical motors. Other means of capacity control are very seldom utilized due to thermodynamic inefficiencies and design difficulties. Energy losses in a compressor, when capacity regulation is provided by lifting the suction valves, are due to friction of gas flowing in and out the unloaded cylinder. This is shown in Fig. 11-84 where the comparison is made for ideal partial load operation, reciprocating, and screw compressors.

Rotary compressors are also PDC types, but where refrigerant flow rotates during compression. Unlike the reciprocating type, rotary compressors have a built-in volume ratio which is defined as volume in cavity when the suction port is closed ($V_s = m * v_s$) over the volume in the cavity when the discharge port is uncovered ($V_d = m * v_d$). Built-in volume ratio determines for a given refrigerant and conditions the pressure ratio which is:

$$\frac{p_d}{p_s} = \left(\frac{v_s}{v_d}\right)^n \qquad (11\text{-}96)$$

where n represents the politropic exponent of compression.

In other words, in a reciprocating compressor the discharge valve opens when the pressure in the cylinder is slightly higher than the pressure in the high-pressure side of the system, while in rotary compressors the discharge pressure will be established only by inlet conditions and built-in volume ratio regardless of the system discharge pressure. Very seldom are the discharge and system (condensing) pressure equal, causing the situation shown in Fig. 11-85. When condensing pressure (p) is lower than discharge (p_2), shown as case (a), "over compression" will cause energy losses presented by the horn on the diagram. If the condensing pressure is higher, in the moment when the discharge port uncovers there will be flow of refrigerant

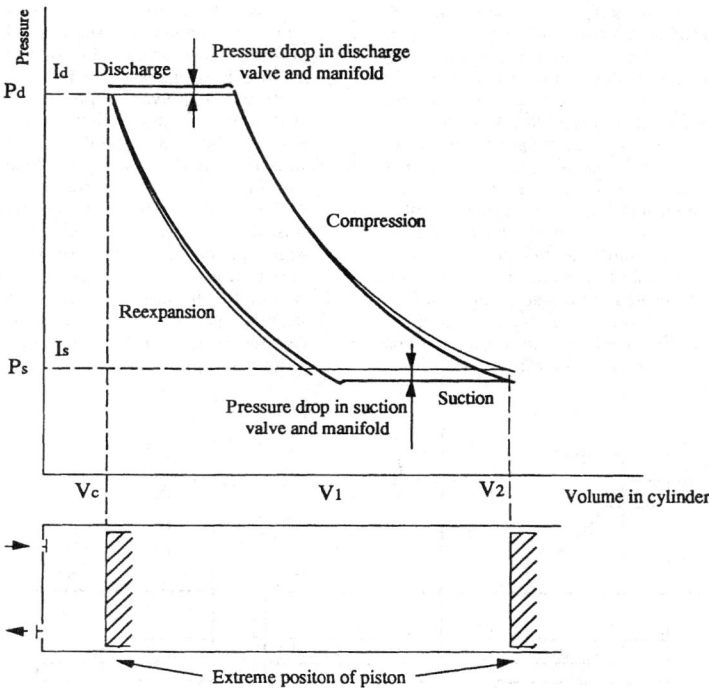

FIG. 11-83 Pressure-volume diagram of an ideal (thin line) and actual (thick line) reciprocating compressor.

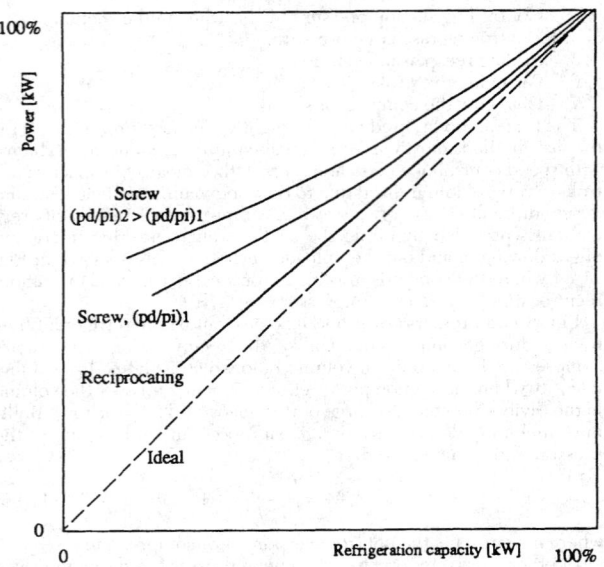

FIG. 11-84 Typical power-refrigeration capacity data for different types of compressors during partial, unloaded operation.

screw compressors.

The slide is located in the compressor casting below the rotors, allowing internal gas recirculation without compression. Slide valve is operated by a piston located in a hydraulic cylinder and actuated by high-pressure oil from both sides. When the compressor is started, the slide valve is fully open and the compressor is unloaded. To increase capacity, a solenoid valve on the hydraulic line opens, moving the piston in the direction of increasing capacity. In order to increase part-load efficiency, the slide valve is designed to consist of two parts, one traditional slide valve for capacity regulation and other for built-in volume adjustment.

Single screw compressors are a newer design (early 1960s) compared to twin screw compressors, and are manufactured in the range of capacity from 100 kW to 4 MW. The compressor screw is cylindrical with helical grooves mated with two star wheels (gaterotors) rotating in opposite direction from one another. Each tooth acts as the piston in the rotating "cylinder" formed by screw flute and cylindrical main-rotor casting.

As compression occurs concurrently in both halves of the compressor, radial forces are oppositely directed, resulting in negligible net-radial loads on the rotor bearings (unlike twin screw compressors), but there are some loads on the star wheel shafts.

Scroll compressors are currently used in relatively small-sized installations, predominantly for residential air-conditioning (up to 50 kW). They are recognized for low-noise operation. Two scrolls (free-standing, involute spirals bounded on one side by a flat plate) facing each other form a closed volume while one moves in a controlled orbit around a fixed point on the other, fixed scroll.

The suction gas which enters from the periphery is trapped by the scrolls. The closed volumes move radially inward until the discharge port is reached, when vapor is pressed out. The orbiting scroll is driven by a short-throw crank mechanism. Similar to screw compressors, internal leakage should be kept low, and is occurring in gaps between cylindrical surfaces and between the tips of the involute and the opposing scroll base plate.

Similar to the screw compressor, the scroll compressor is a constant-volume-ratio machine. Losses occur when operating conditions of the compressor do not match the built-in volume ratio (see Fig. 11-85).

Vane compressors are used in small, hermetic units, but sometimes as booster compressors in industrial applications. Two basic types are *fixed* (roller) or single-vane type and the *rotating* or multiple-vane type. In the single-vane type the rotor (called roller) is eccentrically placed in the cylinder so these two are always in contact. The contact line make the first separation between the suction and discharge chambers while the vane (spring-loaded divider) makes the second. In the multiple-vane compressors the rotor and the cylinder are not in the contact. The rotor has two or more sliding vanes which are held against the cylinder by centrifugal force. In the vane rotary compressors, no suction valves are needed. Since the gas enters the compressor continuously, gas pulsations are at minimum. Vane compressors have a high volumetric efficiency because of the small clearance volume and consequent low reexpansion losses. Rotary vane compressors have a low weight-to-displacement ratio, which makes them suitable for transport applications.

backwards into the compressor, causing losses shown in Fig. 11-85b and the last stage will be only discharge without compression. The case when the compressor discharge pressure is equal to the condensing pressure is shown in the Fig. 11-85c.

Double helical rotary (twin) screw compressors consist of two mating helically grooved rotors (male and female) with asymmetric profile, in a housing formed by two overlapped cylinders, with inlet and outlet ports. Developed relatively recently (in 1930s) the first twin screw compressors were used for air, and later (1950s) became popular for refrigeration. Screw compressors have some advantages over reciprocating compressors (fewer moving parts and more compact) but also some drawbacks (lower efficiency at off-design conditions, as discussed above, higher manufacturing cost due to complicated screw geometry, large separators and coolers for oil which is important as a sealant). Figure 11-86 shows the oil circuit of a screw compressor. Oil cooling could be provided by water, glycol, or refrigerant either in the heat-exchanger-utilizing-thermosyphon effect or the using-direct-expansion concept.

In order to overcome some inherent disadvantages, screw compressors have been initially used predominantly as booster (low-stage) compressors, and following development in capacity control and decreasing prices, they are widely used for high-stage applications. There are several methods for capacity regulation of screw compressors. One is variable speed drive, but a more economical first-cost concept is a slide valve that is used in some form by practically all

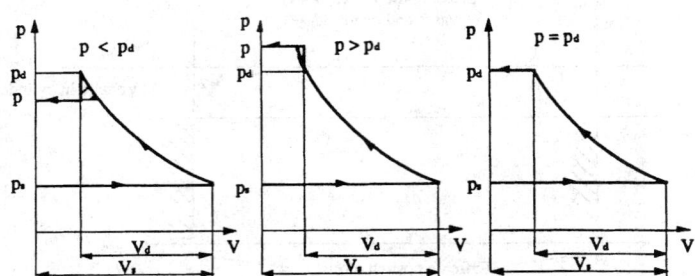

FIG. 11-85 Matching compressor built-in pressure ratio with actual pressure difference.

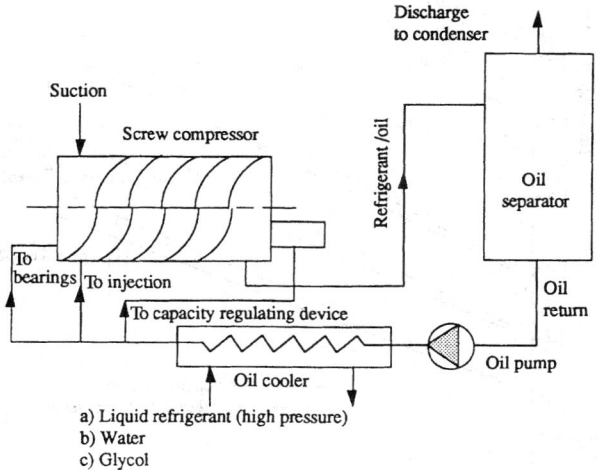

a) Liquid refrigerant (high pressure)
b) Water
c) Glycol

FIG. 11-86   Oil cooling in a screw compressor.

Centrifugal Compressors   These are sometimes called turbo-compressors and mostly serve refrigeration systems in the capacity range 200 to 10,000 kW. The main component is a spinning impeller wheel, backwards curved, which imparts energy to the gas being compressed. Some of the kinetic energy converts into pressure in a volute. Refrigerating centrifugal compressors are predominantly multistage, compared to other turbocompressors, that produce high-pressure ratios.

The torque T (Nm) the impeller ideally imparts to the gas is:

$$T = m\,(u_{tang.out}\,r_{out} - u_{tang.in}\,r_{in}) \qquad (11\text{-}97)$$

where: m (kg/s) = mass flow rate
r_{out} (m) = radius of exit of impeller
r_{in} (m) = radius of exit of impeller
$u_{tang.out}$ (m/s) = tangential velocity of refrigerant leaving impeller
$u_{tang.in}$ (m/s) = tangential velocity of refrigerant entering impeller

When refrigerant enters essentially radially, $u_{tang.in} = 0$ and torque becomes:

$$T = m * u_{tang.out} * r_{out} \qquad (11\text{-}98)$$

The power P (W), is the product of torque and rotative speed ω [l/s] so is

$$P = T * \omega = m * u_{tang.out} * r_{out} * \omega \qquad (11\text{-}99)$$

which for $u_{tang.out} = r_{out} * \omega$ becomes

$$P = m * u_{tang.out}^2 \qquad (11\text{-}100)$$

or for isentropic compression

$$P = m * \Delta h \qquad (11\text{-}101)$$

The performance of a centrifugal compressor (discharge to suction-pressure ratio vs. the flow rate) for different speeds is shown in Fig. 11-87. Lines of constant efficiencies show the maximum efficiency. Unstable operation sequence, called surging, occurs when compressors fails to operate in the range left of the surge envelope. It is characterized by noise and wide fluctuations of load on the compressor and the motor. The period of the cycle is usually 2 to 5 s, depending upon the size of the installation.

The capacity could be controlled by: (1) adjusting the prerotation vanes at the impeller inlet, (2) varying the speed, (3) varying the condenser pressure, and (4) bypassing discharge gas. The first two methods are predominantly used.

Condensers   These are heat exchangers that convert refrigerant

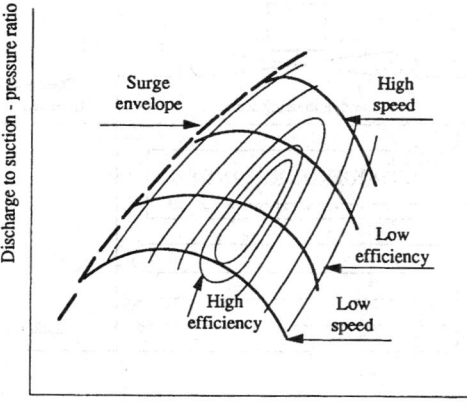

FIG. 11-87   Performance of the centrifugal compressor.

vapor to a liquid. Heat is transferred in three main phases: (1) desuperheating, (2) condensing, and (3) subcooling. In reality condensation occurs even in the superheated region and subcooling occurs in the condensation region. Three main types of refrigeration condensers are: air cooled, water cooled, and evaporative.

Air-cooled condensers are used mostly in air-conditioning and for smaller-refrigeration capacities. The main advantage is availability of cooling medium (air) but heat-transfer rates for the air side are far below values when water is used as a cooling medium. Condensation always occurs inside tubes, while the air side uses extended surface (fins).

The most common types of water-cooled refrigerant condensers are: (1) shell-and-tube, (2) shell-and-coil, (3) tube-in-tube, and (4) brazed-plate. *Shell-and-tube* condensers are built up to 30 MW capacity. Cooling water flows through the tubes in a single or multipass circuit. Fixed-tube sheet and straight-tube construction are common. Horizontal layout is typical, but sometimes vertical is used. Heat-transfer coefficients for the vertical types are lower due to poor condensate drainage, but less water of lower purity can be utilized. Condensation always occurs on the tubes, and often the lower portion of the shell is used as a receiver. In *shell-and-coil* condensers water circulates through one or more continuous or assembled coils contained within the shell while refrigerant condenses outside. The tubes cannot be mechanically cleaned nor replaced. *Tube-in-tube* condensers could be found in versions where condensation occurs either in the inner tube or in the annulus. Condensing coefficients are more difficult to predict, especially in the cases where tubes are formed in spiral. Mechanical cleaning is more complicated, sometimes impossible, and tubes are not replaceable. *Brazed-plate* condensers are constructed of plates brazed together to make up an assembly of separate channels. The plates are typically stainless steel, wave-style corrugated, enabling high heat-transfer rates. Performance calculation is difficult, with very few correlations available. The main advantage is the highest performance/volume (mass) ratio and the lowest refrigerant charge. The last mentioned advantage seems to be the most important feature for many applications where minimization of charge inventory is crucial.

Evaporative condensers (Fig. 11-88) are widely used due to lower condensing temperatures than in the air-cooled condensers and also lower than the water-cooled condenser combined with the cooling tower. Water demands are far lower than for water-cooled condensers. The chemical industry uses shell-and-tube condensers widely, although the use of air-cooled condensing equipment and evaporative condensers is on the increase.

Generally, cooling water is of a lower quality then normal, having also higher mud and silt content. Sometimes even replaceable copper tubes in shell-and-tube heat exchangers are required. It is advisable to use cupronickel instead of copper tubes (when water is high in chlorides) and to use conservative water side velocities (less then 2 m/s for copper tubes).

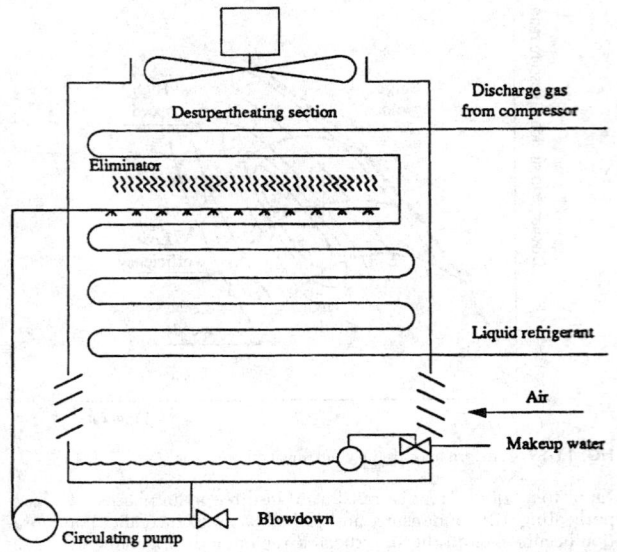

FIG. 11-88 Evaporative condenser with desuperheating coil.

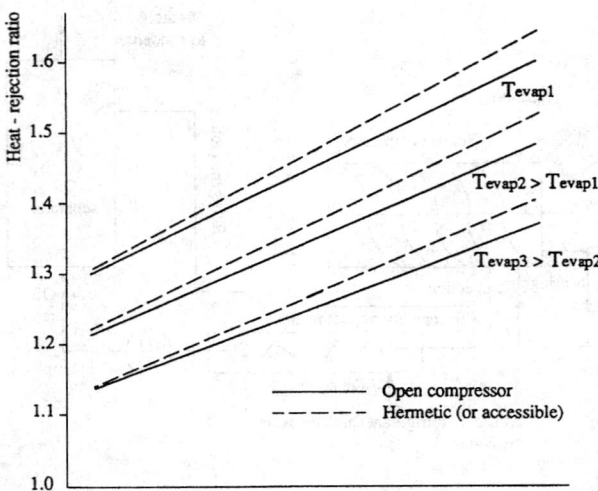

FIG. 11-89 Typical values of the heat-rejection ratio of the heat rejected at the condenser to the refrigerating capacity.

Evaporative condensers are used quite extensively. In most cases commercial evaporative condensers are not totally suitable for chemical plants due to the hostile atmosphere which usually abounds in vapor and dusts which can cause either chemical (corrosion) or mechanical problems (plugging of spray nozzles).

Air-cooled condensers are similar to evaporative in that the service dictates either the use of more expensive alloys in the tube construction or conventional materials of greater wall thickness.

Heat rejected in the condenser Q_{Cd} consists of heat absorbed in the evaporator Q_{Eevap} and energy W supplied by the compressor:

$$Q_{Cd} = Q_{Eevap} + W \qquad (11\text{-}102)$$

For the actual systems, compressor work will be higher than for ideal for the isentropic efficiency and other losses. In the case of hermetic or accessible compressors where an electrical motor is cooled by the refrigerant, condenser capacity should be:

$$Q_{Cd} = Q_{Eevap} + P_{EM} \qquad (11\text{-}103)$$

It is common that compressor manufacturers provide data for the ratio of the heat rejected at the condenser to the refrigeration capacity as shown in Fig. 11-89. The solid line represents data for the open compressors while the dotted line represents the hermetic and accessible compresors. The difference between solid and dotted line is due to all losses (mechanical and electrical in the electrical motor). Condenser design is based on the value:

$$Q_{Cd} = Q_{Eevap} * \text{heat-rejection ratio} \qquad (11\text{-}104)$$

Thermal and mechanical design of heat exchangers (condensers and evaporators) is presented earlier in this section.

Evaporators These are heat exchangers where refrigerant is evaporated while cooling the product, fluid, or body. Refrigerant could be in direct contact with the body that is being cooled, or some other medium could be used as secondary fluid. Mostly air, water, or antifreeze are fluids that are cooled. Design is strongly influenced by the application. Evaporators for air cooling will have in-tube evaporation of the refrigerant, while liquid chillers could have refrigerant evaporation inside or outside the tube. The heat-transfer coefficient for evaporation inside the tube (vs. length or quality) is shown in the Fig. 11-90. Fundamentals of the heat transfer in evaporators, as well as design aspects, are presented in Sec. 11. We will point out only some specific aspects of refrigeration applications.

Refrigeration evaporators could be classified according to the method of feed as either direct (dry) expansion or flooded (liquid overfeed). In dry-expansion the evaporator's outlet is dry or slightly superheated vapor. This limits the liquid feed to the amount that can be completely vaporized by the time it reaches the end of evaporator. In the liquid overfeed evaporator, the amount of liquid refrigerant circulating exceeds the amount evaporated by the circulation number. Decision on the type of the system to be used is one of the first in the design process. Direct-expansion evaporator is generally applied in smaller systems where compact design and the low first costs are crucial. Control of the refrigerant mass flow is then obtained by either a thermoexpansion valve or a capillary tube. Figure 11-90 suggests that the evaporator surface is the most effective in the regions with quality which is neither low nor high. In dry-expansion evaporators, inlet qualities are 10–20 percent, but when controlled by the thermoexpansion valve, vapor at the outlet is not only dry, but even superheated.

In recirculating systems saturated liquid ($x = 0$) is entering the evaporator. Either the pump or gravity will deliver more refrigerant liquid then will evaporate, so outlet quality could be lower than one. The ratio of refrigerant flow rate supplied to the evaporator overflow rate of refrigerant vaporized is the circulation ratio, n. When n increases, the coefficient of heat transfer will increase due to the wetted outlet of the evaporator and the increased velocity at the inlet (Fig. 11-91). In the range of $n = 2$ to 4, the overall U value for air cooler increases roughly by 20 to 30 percent compared to the direct-expansion case. Circulation rates higher than four are not efficient.

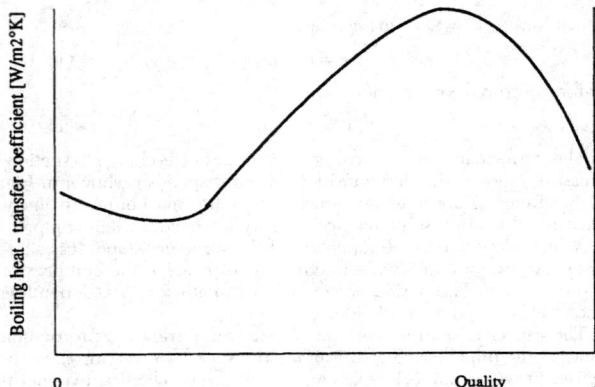

FIG. 11-90 Heat-transfer coefficient for boiling inside the tube.

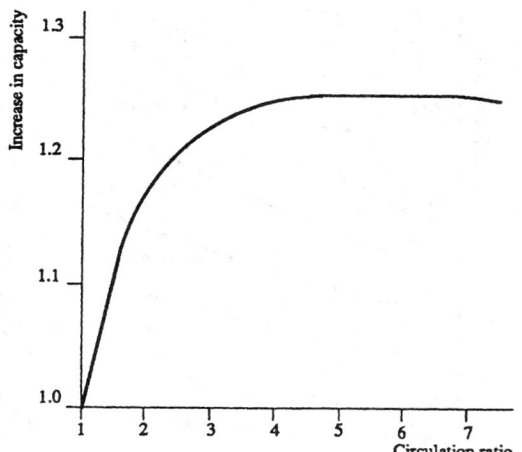

FIG. 11-91 Effect of circulation ratio on the overall heat-transfer coefficient of an air-cooling coil.

The price for an increase in heat-transfer characteristics is a more complex system with more auxiliary equipment: low-pressure receivers, refrigerant pumps, valves, and controls. Liquid refrigerant is predominantly pumped by mechanical pumps, however, sometimes gas at condensing pressure is used for pumping, in the variety of concepts.

The important characteristics of the refrigeration evaporators is the presence of the oil. The system is contaminated with oil in the compressor, in spite of reasonably efficient oil separators. Some systems will recirculate oil, when miscible with refrigerant, returning it to the compressor crankcase. This is found mostly in the systems using halocarbon refrigerants. Although oils that are miscible with ammonia exist, immiscibles are predominantly used. This inhibits the ammonia systems from recirculating the oil. In systems with oil recirculation when halocarbons are used special consideration should be given to proper sizing and layout of the pipes. Proper pipeline configuration, slopes, and velocities (to ensure oil circulation under all operating loads) are essential for good system operation. When refrigerant is lighter than the oil in systems with no oil recirculation, oil will be at the bottom of every volume with a top outlet. Then oil must be drained periodically to avoid decreasing the performance of the equipment.

It is essential for proper design to have the data for refrigerant-oil miscibility under all operating conditions. Some refrigerant-oil combinations will always be miscible, some always immiscible, but some will have both characteristics, depending on temperatures and pressures applied. Defrosting is the important issue for evaporators which are cooling air below freezing. Defrosting is done periodically, actuated predominantly by time relays, but other frost indicators are used (temperature, visual, or pressure-drop sensors). Defrost technique is determined mostly by fluids available and tolerable complexity of the system. Defrosting is done by the following mechanisms when the system is off:

• Hot (or cool) refrigerant gas (the predominant method in industrial applications)
• Water (defrosting from the outside, unlike hot gas defrost)
• Air (only when room temperature is above freezing)
• Electricity (for small systems where hot-gas defrost will be to complex and water is not available)
• Combinations of above.

System Analysis Design calculations are made on the basis of the close to the highest refrigeration load, however the system operates at the design conditions very seldom. The purpose of regulating devices is to adjust the system performance to cooling demands by decreasing the effect or performance of some component. Refrigeration systems have inherent self-regulating control which the engineer could rely on to a certain extent. When the refrigeration load starts to decrease, less refrigerant will evaporate. This causes a drop in evaporation temperature (as long as compressor capacity is unchanged) due to the imbalance in vapor being taken by the compressor and produced by evaporation in evaporator. With a drop in evaporation pressure, the compressor capacity will decrease due to: (1) lower vapor density (lower mass flow for the same volumetric flow rate) and (2) decrease in volumetric efficiency. On the other hand, when the evaporation temperature drops, for the unchanged temperature of the medium being cooled, the evaporator capacity will increase due to increase in the mean-temperature difference between refrigerant and cooled medium, causing a positive effect (increase) on the cooling load. With a decrease in the evaporation temperature the heat-rejection factor will increase causing an increase of heat rejected to the condenser, but refrigerant mass flow rate will decrease due to compressor characteristics. These will have an opposite effect on condenser load. Even a simplified analysis demonstrates the necessity for better understanding of system performance under different operating conditions. Two methods could be used for more accurate analysis. The traditional method of refrigeration-system analysis is through determination of balance points, while in recent years, system analysis is performed by system simulation or mathematical modeling, using mathematical (equation solving) rather than graphical (intersection of two curves) procedures. Systems with a small number of components such as the vapor-compression refrigeration system could be analyzed both ways. Graphical presentation, better suited for understanding trends is not appropriate for more complex systems, more detailed component description, and frequent change of parameters. There is a variety of different mathematical models tailored to fit specific systems, refrigerants, resources available, demands, and complexity. Although limited in its applications, graphical representation is valuable as the starting tool and for clear understanding of the system performance.

Refrigeration capacity q_e and power P curves for the reciprocating compressor are shown in Fig. 11-92. They are functions of temperatures of evaporation and condensation:

$$q_e = q_e(t_{evap}, t_{cd}) \qquad (11\text{-}105a)$$

and
$$P = P(t_{evap}, t_{cd}) \qquad (11\text{-}105b)$$

where q_e (kW) = refrigerating capacity
  P (kW) = power required by the compressor
  t_{evap} (°C) = evaporating temperature
  t_{cd} (°C) = condensing temperature.

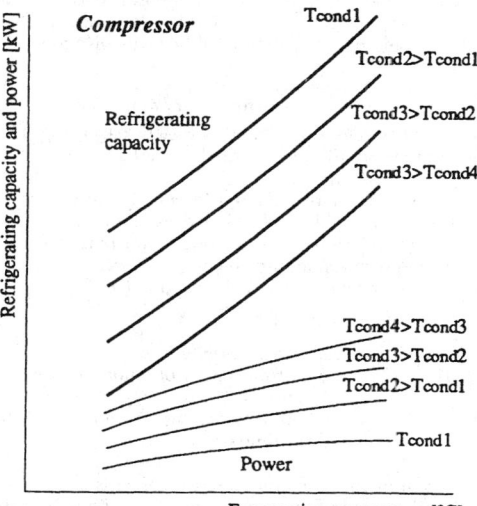

FIG. 11-92 Refrigerating capacity and power requirement for the reciprocating compressor.

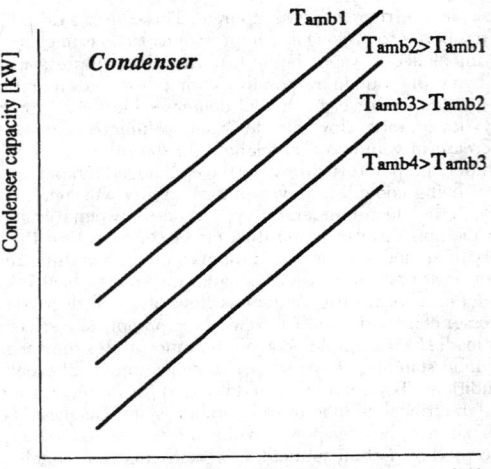

FIG. 11-93   Condenser performance.

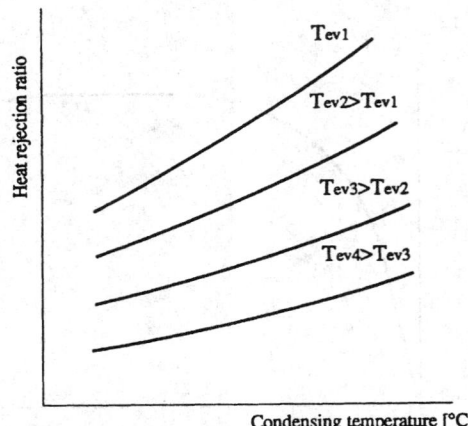

FIG. 11-94   Heat-rejection ratio.

A more detailed description of compressor performance is shown in the section on the refrigeration compressors.

Condenser performance, shown in figure 11-93, could be simplified as:

$$q_{cd} = F(t_{cd} - t_{amb}) \qquad (11\text{-}105c)$$

where q_{cd} (kW) = capacity of condenser;
   F (kW/°C) = capacity of condenser per unit inlet temperature difference ($F = U * A$);
   t_{amb} (°C) = ambient temperature (or temperature of condenser cooling medium).

In this analysis F will be constant but it could be described more accurately as a function of parameters influencing heat transfer in the condenser (temperature, pressure, flow rate, fluid thermodynamical, and thermophysical characteristics . . .).

Condenser performance should be expressed as "evaporating effect" to enable matching with compressor and evaporator performance. Condenser "evaporating effect" is the refrigeration capacity of an evaporator served by a particular condenser. It is the function of the cycle, evaporating temperature, and the compressor. The "evaporating effect" could be calculated from the heat-rejection ratio q_{cd}/q_e:

$$q_e = \frac{q_{cd}}{\text{heat-rejection ratio}} \qquad (11\text{-}105d)$$

The heat-rejection rate is presented in Fig. 11-94 (or Fig. 11-89).

Finally, the evaporating effect of the condenser is shown in Fig. 11-95.

The performance of the condensing-unit (compressor and condenser) subsystem could be developed as shown in Fig. 11-96 by superimposing two graphs, one for compressor performance and the other for condenser evaporating effect.

Evaporator performance could be simplified as:

$$q_e = F_{evap}(t_{amb} - t_{evap}) \qquad (11\text{-}106)$$

where q_e(kW) = evaporator capacity
   F_e (kW/°C) = evaporator capacity per unit inlet temperature difference
   t_{amb} (°C) = ambient temperature (or temperature of cooled body or fluid).

The diagram of the evaporator performance is shown in the Fig. 11-97. The character of the curvature of the lines (variable heat-transfer rate) indicates that the evaporator is cooling air. Influences of the flow rate of cooled fluid are also shown in this diagram; i.e., higher flow rate will increase heat transfer. The same effect could be shown

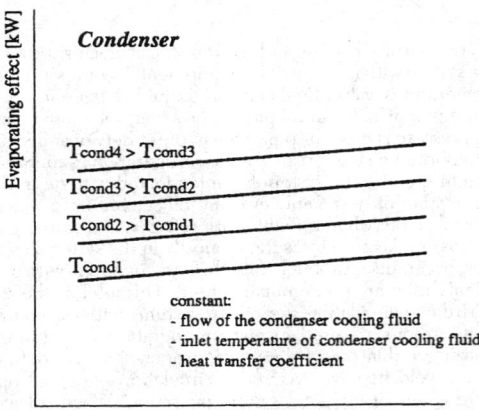

FIG. 11-95   Condenser evaporating effect.

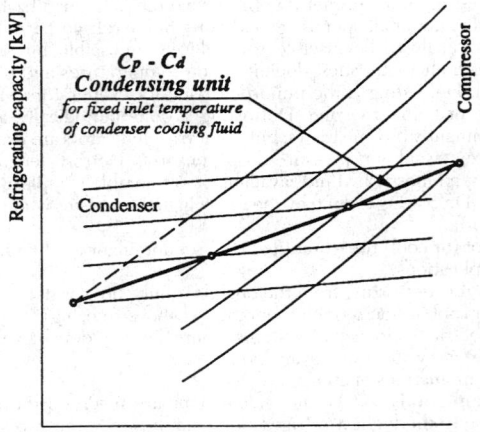

FIG. 11-96   Balance points of compressor and condenser determines performance of condensing unit for fixed temperature of condenser cooling fluid (flow rate and heat-transfer coefficient are constant).

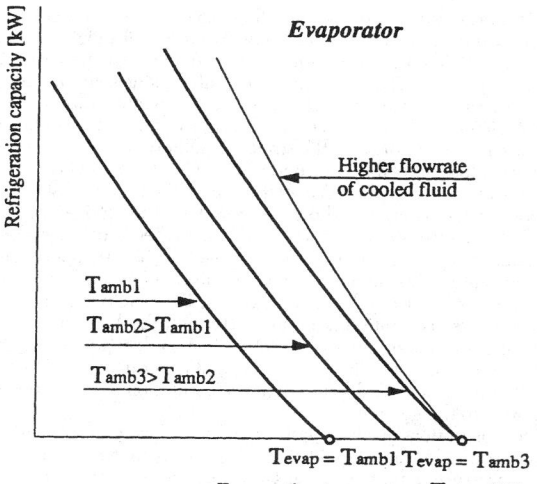

FIG. 11-97   Refrigerating capacity of evaporator.

in the condenser-performance curve. It is omitted only for the reasons of simplicity.

The performance of the complete system could be predicted by superimposing the diagrams for the condensing unit and the evaporator, as shown in Fig. 11-98. Point 1 reveals a balance for constant flows and inlet temperatures of chilled fluid and fluid for condenser cooling. When this point is transferred in the diagram for the condensing unit in the Figs. 11-95 or 11-96, the condensing temperature could be determined. When the temperature of entering fluid in the evaporator t_{amb1} is lowered to t_{amb2} the new operating conditions will be determined by the state at point 2. Evaporation temperature drops from t_{evap1} to t_{evap2}. If the evaporation temperature should be unchanged, the same reduction of inlet temperature could be achieved by reducing the capacity of the condensing unit from Cp to Cp*. The new operating point 3 shows reduction in capacity for Δ due to the reduction in the compressor or the condenser capacity.

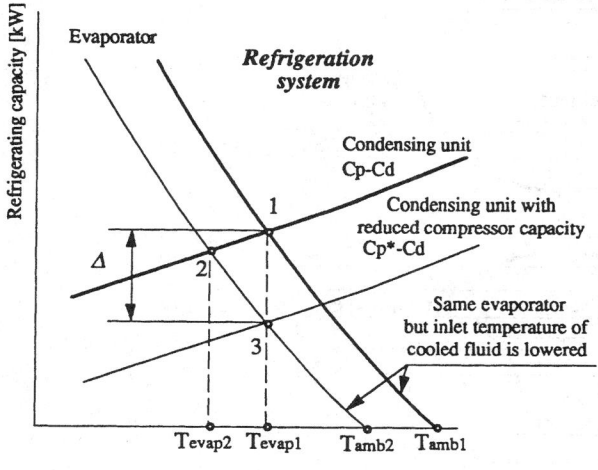

FIG. 11-98   Performance of complete refrigeration system (1), when there is reduction in heat load (2), and when for the same ambient (or inlet in evaporator) evaporation temperature is maintained constant by reducing capacity of compressor/condenser part (3).

Mathematical modeling is essentially the same process, but the description of the component performance is generally much more complex and detailed. This approach enables a user to vary more parameters easier, look into various possibilities for intervention, and predict the response of the system on different influences. Equation-solving does not necessarily have to be done by successive substitution or iteration as this procedure could suggest.

System, Equipment, and Refrigerant Selection   There is no universal rule which can be used to decide which system, equipment type, or refrigerant is the most appropriate for a given application. A number of variables influence the final-design decision:

- Refrigeration load
- Type of installation
- Temperature level of medium to be cooled
- Condensing media characteristics: type (water, air, . . .), temperature level, available quantities
- Energy source for driving the refrigeration unit (electricity, natural gas, steam, waste heat)
- Location and space available (urban areas, sensitive equipment around, limited space . . .)
- Funds available (i.e. initial vs. run-cost ratio)
- Safety requirements (explosive environment, aggressive fluids, . . .)
- Other demands (compatibility with existing systems, type of load, compactness, level of automatization, operating life, possibility to use process fluid as refrigerant)

Generally, vapor *compression systems* are considered first. They can be used for almost every task. Whenever it is possible, prefabricated elements or complete units are recommended. Reciprocating compressors are widely used for lower rates, more uneven heat loads (when frequent and wider range of capacity reduction is required). They ask for more space and have higher maintenance costs then centrifugal compressors, but are often the most economical in first costs. Centrifugal compressors are considered for huge capacities, when the evaporating temperature is not too low. Screw compressors are considered first when space in the machine room is limited, when system is operating long hours, and when periods between service should be longer.

Direct-expansions are more appropriate for smaller systems which should be compact, and where there are just one or few evaporators. Overfeed (recirculation) systems should be considered for all applications where first cost for additional equipment (surge drums, low-pressure receivers, refrigerant pumps, and accessories) is lower than the savings for the evaporator surface.

Choice of refrigerant is complex and not straightforward. For industrial applications, advantages of ammonia (thermodynamical and economical) overcome drawbacks which are mostly related to possible low-toxic and panics created by accidental leaks when used in urban areas. Halocarbons have many advantages (not toxic, not explosive, odorless . . .), but environmental issues and slightly inferior thermodynamical and thermophysical properties compared to ammonia or hydrocarbons as well as rising prices are giving the chance to other options. When this text was written the ozone-depletion issue was not resolved, R22 was still used but facing phaseout, and R134a was considered to be the best alternative for CFCs and HCFCs, having similar characteristics to the already banned R12. Very often, fluid to be cooled is used as a refrigerant in the chemical industry. Use of secondary refrigerants in combination with the ammonia central-refrigeration unit is becoming a viable alternative in many applications.

Absorption systems will be considered when there is low-cost low-pressure steam or waste heat available and evaporation temperature and refrigeration load are relatively high. Typical application range is for water chilling at 7–10°C, and capacities from 300 kW to 5 MW in a single unit. The main drawback is the difficulty in maintaining a tight system with the highly corrosive lithium bromide, and an operating pressure in the evaporator and the absorber below atmospheric.

Ejector (steam-jet) refrigeration systems are used for similar applications, when chilled water-outlet temperature is relatively high, when relatively cool condensing water and cheap steam at 7 bar are available, and for similar high duties (0.3–5 MW). Even though these systems usually have low first and maintenance costs, there are not many steam-jet systems running.

OTHER REFRIGERATION SYSTEMS APPLIED IN THE INDUSTRY

Absorption Refrigeration Systems Two main absorption systems are used in industrial application: lithium bromide-water and ammonia-water. Lithium bromide-water systems are limited to evaporation temperatures above freezing because water is used as the refrigerant, while the refrigerant in an ammonia-water system is ammonia and consequently it can be applied for the lower-temperature requirements.

Single-effect indirect-fired lithium bromide cycle is shown in Fig. 11-99. The machine consists of five major components:

Evaporator is the heat exchanger where refrigerant (water) evaporates (being sprayed over the tubes) due to low pressure in the vessel. Evaporation chills water flow inside the tubes that bring heat from the external system to be cooled.

Absorber is a component where strong absorber solution is used to absorb the water vapor flashed in the evaporator. A solution pump sprays the lithium bromide over the absorber tube section. Cool water is passing through the tubes taking refrigeration load, heat of dilution, heat to cool condensed water, and sensible heat for solution cooling.

Heat exchanger is used to improve efficiency of the cycle, reducing consumption of steam and condenser water.

Generator is a component where heat brought to a system in a tube section is used to restore the solution concentration by boiling off the water vapor absorbed in the absorber.

Condenser is an element where water vapor, boiled in the generator, is condensed, preparing pure water (refrigerant) for discharge to an evaporator.

Heat supplied to the generator is boiling weak (dilute) absorbent solution on the outside of the tubes. Evaporated water is condensed on the outside of the condenser tubes. Water utilized to cool the condenser is usually cooled in the cooling tower. Both condenser and generator are located in the same vessel, being at the absolute pressure of about 6 kPa. The water condensate passes through a liquid trap and enters the evaporator. Refrigerant (water) boils on the evaporator tubes and cools the water flow that brings the refrigeration load.

Refrigerant that is not evaporated flows to the recirculation pump to be sprayed over the evaporator tubes. Solution with high water concentration that enters the generator increases in concentration as water evaporates. The resulting strong, absorbent solution (solution with low water concentration) leaves the generator on its way to the heat exchanger. There the stream of high water concentration that flows to the generator cools the stream of solution with low water concentration that flows to the second vessel. The solution with low water concentration is distributed over the absorber tubes. Absorber and evaporator are located in the same vessel, so the refrigerant evaporated on the evaporator tubes is readily absorbed into the absorbent solution. The pressure in the second vessel during the operation is 7 kPa (absolute). Heat of absorption and dilution are removed by cooling water (usually from the cooling tower). The resulting solution with high water concentration is pumped through the heat exchanger to the generator, completing the cycle. Heat exchanger increases the efficiency of the system by preheating, that is, reducing the amount of heat that must be added to the high water solution before it begins to evaporate in the generator.

The absorption machine operation is analyzed by the use of a lithium bromide-water equilibrium diagram, as shown in Fig. 11-100. Vapor pressure is plotted against the mass concentration of lithium bromide in the solution. The corresponding saturation temperature for a given vapor pressure is shown on the left-hand side of the diagram. The line in the lower right corner of the diagram is the crystallization line. It indicates the point at which the solution will begin to change from liquid to solid, and this is the limit of the cycle. If the solution becomes overconcentrated, the absorption cycle will be interrupted owing to solidification, and capacity will not be restored until the unit is desolidified. This normally requires the addition of heat to the outside of the solution heat exchanger and the solution pump.

The diagram in Fig. 11-101 presents enthalpy data for LiBr-water solutions. It is needed for the thermal calculation of the cycle. Enthalpies for water and water vapor can be determined from the table of properties of water. The data in Fig. 11-101 are applicable to saturated or subcooled solutions and are based on a zero enthalpy of liquid water at 0°C and a zero enthalpy of solid LiBr at 25°C. Since

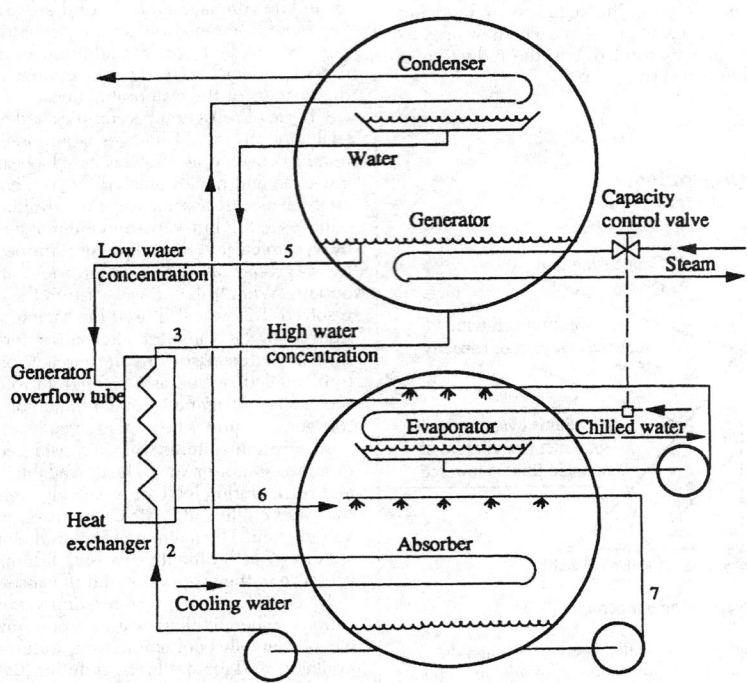

FIG. 11-99 Two-shell lithium bromide-water cycle chiller.

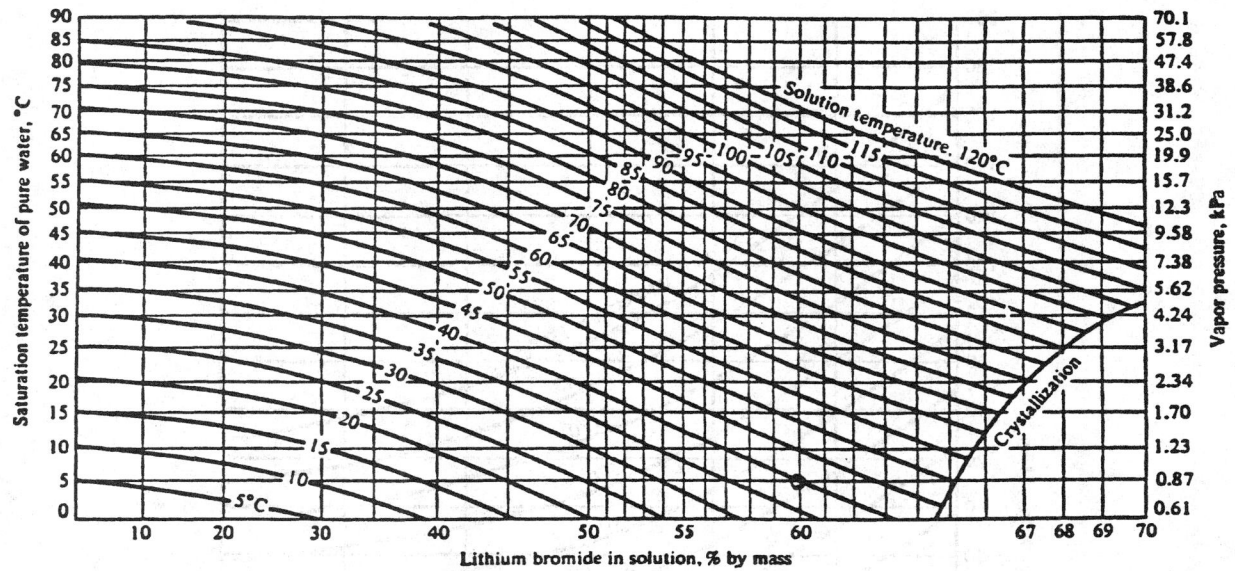

FIG. 11-100 Temperature-pressure-concentration diagram of saturated LiBr-water solutions (*W. F. Stoecker and J. W. Jones: Refrigeration and Air-Conditioning*)

the zero enthalpy for the water in the solution is the same as that in conventional tables of properties of water, the water property tables can be used in conjunction with diagram in Fig. 11-100.

Coefficient of performance of the absorption cycle is defined on the same principle as for the mechanical refrigeration:

$$\text{COP}_{\text{abs}} = \frac{\text{useful effect}}{\text{heat input}} = \frac{\text{refrigeration rate}}{\text{heat input at generator}}$$

but it should be noted that here denominator for the COP_{abs} is heat while for the mechanical refrigeration cycle it is work. Since these two forms of energy are not equal, COP_{abs} is not as low (0.6–0.8) as it appears compared to COP for mechanical system (2.5–3.5).

The double-effect absorption unit is shown in Fig. 11-102. All major components and operation of the double-effect absorption machine is similar to that for the single-effect machine. The primary generator, located in the vessel 1, is using an external heat source to evaporate water from dilute-absorbent (high water concentration) solution. Water vapor readily flows to the generator II where it is condensed on the tubes. The absorbent (LiBr) intermediate solution from generator I will pass through the heat exchanger on the way to the generator II where it is heated by the condensing water vapor. The throttling valve reduces pressure from vessel 1 (about 103 kPa absolute) to that of vessel 2. Following the reduction of pressure some water in the solution flashes to vapor, which is liquefied at the condenser. In the high temperature heat exchanger intermediate solution heats the weak (high water concentration) solution stream coming from the low temperature heat exchanger. In the low temperature heat exchanger strong solution is being cooled before entering the absorber. The absorber is on the same pressure as the evaporator. The double-effect absorption units achieve higher COPs than the single stage.

The ammonia-water absorption system was extensively used until the fifties when the LiBr-water combination became popular. Figure 11-103 shows a simplified ammonia-water absorption cycle. The refrigerant is ammonia, and the absorbent is dilute aqueous solution of ammonia. Ammonia-water systems differ from water-lithium bromide equipment to accommodate major differences: Water (here absorbent) is also volatile, so the regeneration of weak water solution to strong water solution is a fractional distillation. Different refrigerant (ammonia) causes different, much higher pressures: about 1100–2100 kPa absolute in condenser.

Ammonia vapor from the evaporator and the weak water solution from the generator are producing strong water solution in the absorber. Strong water solution is then separated in the rectifier producing (1) ammonia with some water vapor content and (2) very strong water solution at the bottom, in the generator. Heat in the generator vaporizes ammonia and the weak solution returns to absorber. On its way to the absorber the weak solution stream passes through the heat exchanger, heating strong solution from the absorber on the way to the rectifier. The other stream, mostly ammonia vapor but with some water-vapor content flows to the condenser. To remove water as much as possible, the vapor from the rectifier passes through the analyzer where it is additionally cooled. The remaining water escaped from the analyzer pass as liquid through the condenser and the evaporator to the absorber.

Ammonia-water units can be arranged for single-stage or cascaded two-stage operation. The advantage of two-staging is it creates the possibility of utilizing only part of the heat on the higher and the rest on the lower temperature level but the price is increase in first cost and heat required.

Ammonia-water and lithium bromide-water systems operate under comparable COP. The ammonia-water system is capable achieving evaporating temperatures below 0°C because the refrigerant is ammonia. Water as the refrigerant limits evaporating temperatures to no lower than freezing, better to 3°C. Advantage of the lithium bromide-water system is that it requires less equipment and operates at lower pressures. But this is also a drawback, because pressures are below atmospheric, causing air infiltration in the system which must be purged periodically. Due to corrosion problems, special inhibitors must be used in the lithium-bromide-water system. The infiltration of air in the ammonia-water system is also possible, but when evaporating temperature is below −33°C. This can result in formation of corrosive ammonium carbonate.

Further readings: *ASHRAE Handbook* 1994 Refrigeration Systems and Applications; Bogart, M., 1981: *Ammonia Absorption Refrigeration in Industrial Processes*, Gulf Publishing Co. Houston; Stoecker W. F. and Jones J. W. 1982: *Refrigeration and Air-Conditioning*, McGraw-Hill Book Company, New York.

Steam-Jet (Ejector) Systems These systems substitute an ejector for a mechanical compressor in a vapor compression system. Since refrigerant is water, maintaining temperatures lower than the environment requires that the pressure of water in the evaporator must be

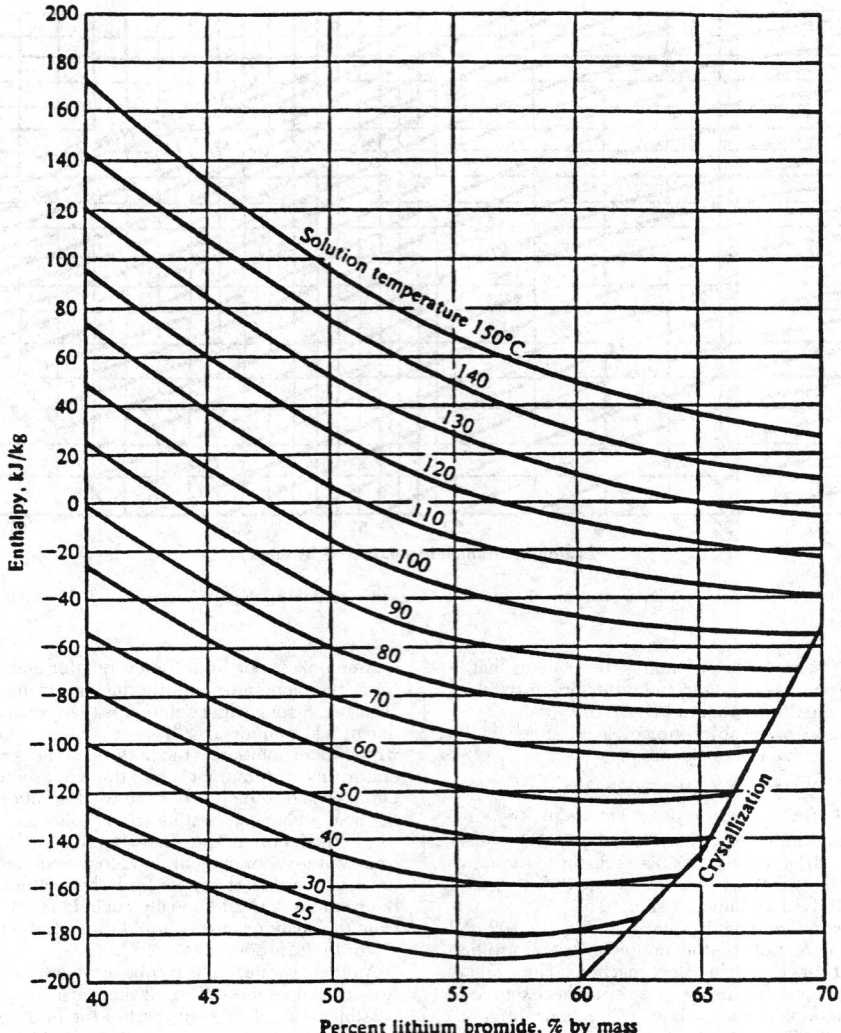

FIG. 11-101 Enthalpy of LiBr-water solutions (*W. F. Stoecker and J. W. Jones:* Refrigeration and Air-Conditioning).

below atmospheric. A typical arrangement for the steam-jet refrigeration cycle is shown in Fig. 11-104.

Main Components The main components of steam-jet refrigeration systems are:

1. *Primary steam ejector.* A kinetic device that utilizes the momentum of a high-velocity jet to entertain and accelerate a slower-moving medium into which it is directed. High-pressure steam is delivered to the nozzle of the ejector. The steam expands while flowing through the nozzle where the velocity increases rapidly. The velocity of steam leaving the nozzle is around 1200 m/s. Because of this high velocity, flash vapor from the tank is continually aspired into the moving steam. The mixture of steam and flash vapor then enters the diffuser section where the velocity is gradually reduced because of increasing cross-sectional area. The energy of the high-velocity steam compresses the vapor during its passage through the diffuser, raising it's temperature above the temperature of the condenser cooling water.

2. *Condenser.* The component of the system where the vapor condenses and where the heat is rejected. The rate of heat rejected is:

$$Q_{cond} = (W_s + W_w) \, hfg \qquad (11\text{-}107)$$

where: Q_{cond} = heat rejection (kW)
  W_s = primary booster steam rate (kg/s)
  W_w = flash vapor rate (kg/s)
  hfg = latent heat (kJ/kg)

The condenser design, surface area, and condenser cooling water quantity should be based on the highest cooling water temperature likely to be encountered. If the inlet cooling water temperature becomes hotter then the design, the primary booster (ejector) may cease functioning because of the increase in condenser pressure.

Two types of condensers could be used: the surface condenser (shown in Fig. 11-104) and the barometric or jet condenser (Fig. 11-105). The surface condenser is of shell-and-tube design with water flowing through the tubes and steam condensed on the outside surface. In the jet condenser, condensing water and the steam being condensed are mixed directly, and no tubes are provided. The jet condenser can be barometric or a low-level type. The barometric condenser requires a height of ~10 m above the level of the water in the hot well. A tailpipe of this length is needed so that condenser water and condensate can drain by gravity. In the low-level jet type, the tailpipe is eliminated, and it becomes necessary to remove the con-

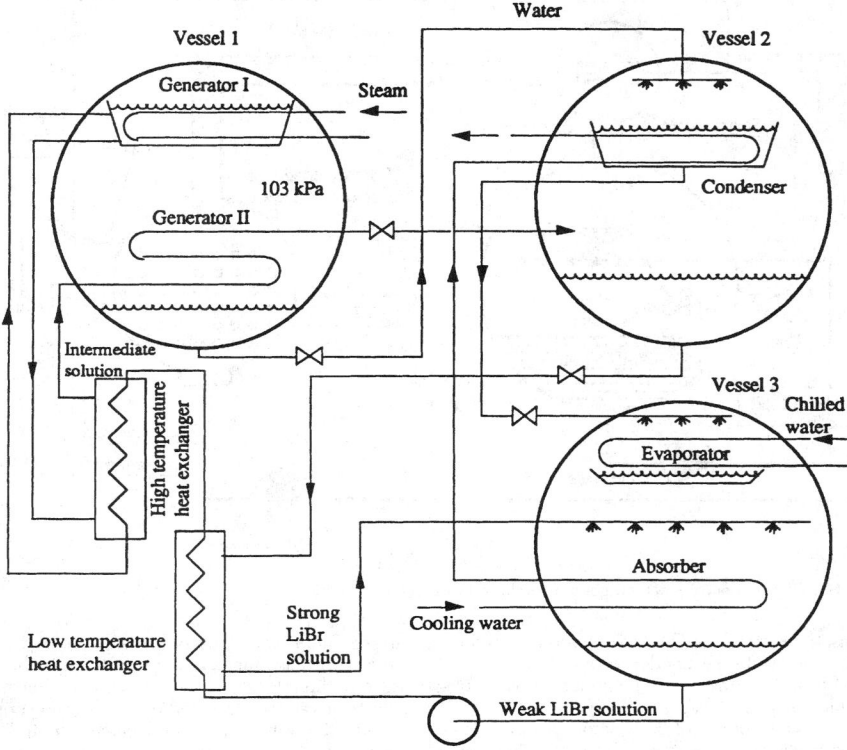

FIG. 11-102   Double-effect absorption unit.

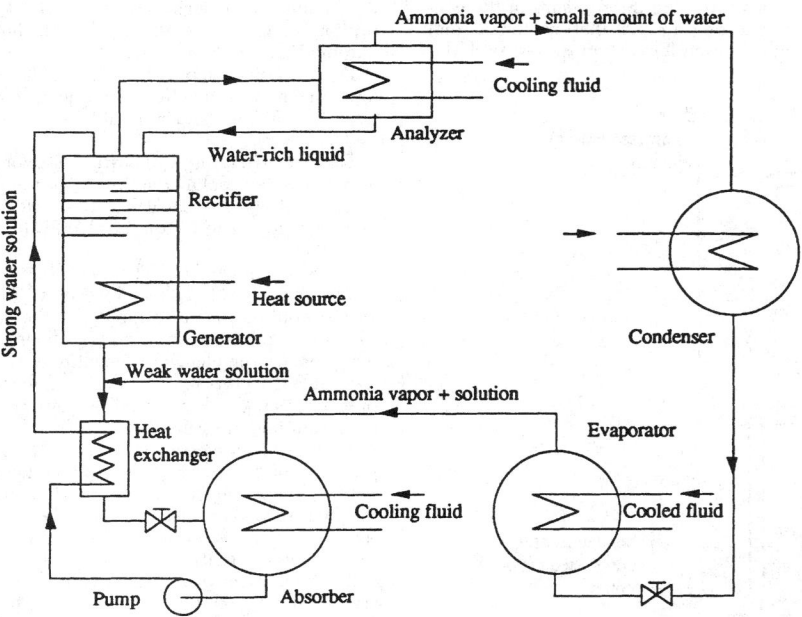

FIG. 11-103   Simplified ammonia-water absorption cycle.

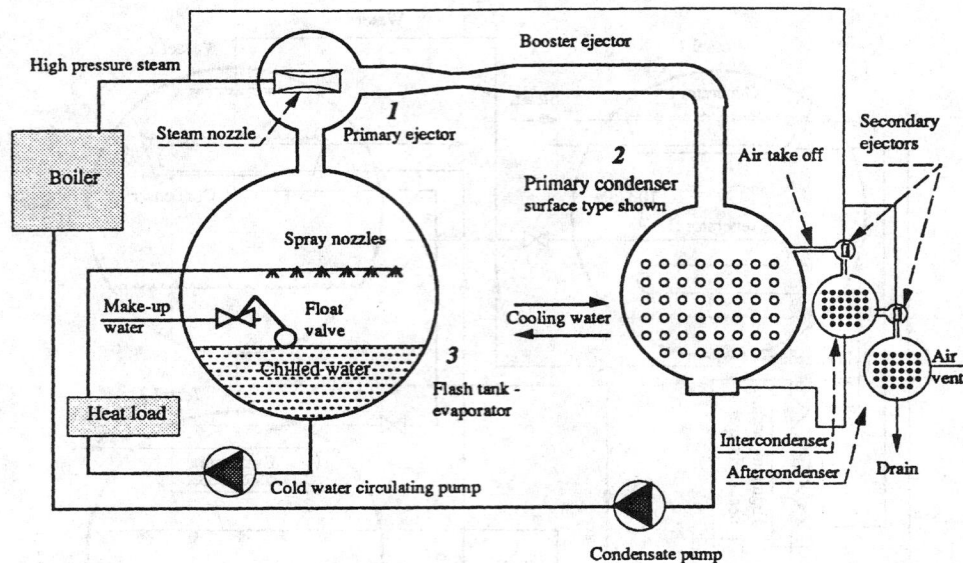

FIG. 11-104   Ejector (steam-jet) refrigeration cycle (with surface-type condenser).

denser water and condensate by pumping from the condenser to the hot well. The main advantages of the jet condenser are low maintenance with the absence of tubes and the fact that condenser water of varying degrees of cleanliness may be used.

3.   *Flash tank.*   This is the evaporator of the ejector system and is usually a large-volume vessel where large water-surface area is needed for efficient evaporative cooling action. Warm water returning from the process is sprayed into the flash chamber through nozzles (sometimes cascades are used for maximizing the contact surface, being less susceptible to carryover problems) and the cooled effluent is pumped from the bottom of the flash tank.

When the steam supply to one ejector of a group is closed, some means must be provided to for preventing the pressure in the condenser and flash tank from equalizing through that ejector. A compartmental flash tank is frequently used for such purposes. With this

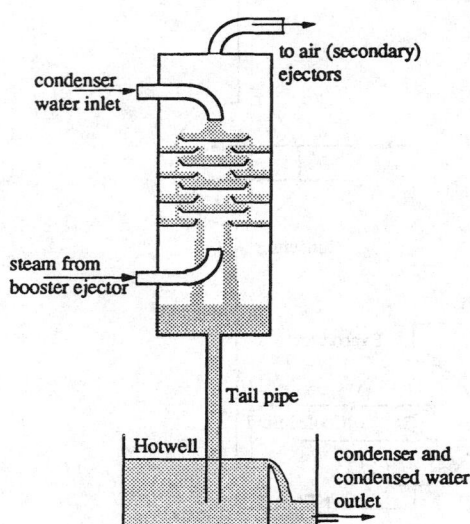

FIG. 11-105   Barometric condenser for steam-jet system.

arrangement, partitions are provided so that each booster is operating on its own flash tank. When the steam is shut off to any booster, the valve to the inlet spray water to that compartment also is closed.

A float valve is provided to control the supply of makeup water to replace the water vapor that has flashed off. The flash tank should be insulated.

Applications   The steam-jet refrigeration is suited for:

1.   Processes where direct vaporization is used for concentration or drying of heat-sensitive foods and chemicals, where, besides elimination of the heat exchanger, preservation of the product quality is an important advantage.

2.   Enabling the use of hard or even sea water for heat rejection e.g. for absorption of gases (CO_2, SO_2, ClO_2 . . .) in chilled water (desorption is provided simultaniously with chilling) when a direct contact barometric condenser is used.

Despite being simple, rugged, reliable, requiring low maintenance, low cost, and vibration free, steam-jet systems are not widely accepted in water chilling for air-conditioning due to characteristics of the cycle.

Factors Affecting Capacity   Ejector (steam-jet) units become attractive when cooling relatively high-temperature chilled water with a source of about 7 bar gauge waste steam and relatively cool condensing water. The factors involved with steam-jet capacity include the following:

1.   *Steam pressure.*   The main boosters can operate on steam pressures from as low as 0.15 bar up to 7 bar gauge. The quantity of steam required increases rapidly as the steam pressure drops (Fig. 11-106). The best steam rates are obtained with about 7 bar. Above this pressure the change in quantity of steam required is practically negligible. Ejectors must be designed for the highest available steam pressure, to take advantage of the lower steam consumption for various steam-inlet pressures.

The secondary ejector systems used for removing air require steam pressures of 2.5 bar or greater. When the available steam pressure is lower than this, an electrically driven vacuum pump is used for either the final secondary ejector or for the entire secondary group. The secondary ejectors normally require 0.2–0.3 kg/h of steam per kW of refrigeration capacity.

2.   *Condenser water temperature.*   In comparison with other vapor-compression systems, steam-jet machines require relatively large water quantities for condensation. The higher the inlet-water temperature, the higher are the water requirements (Fig. 11-107).

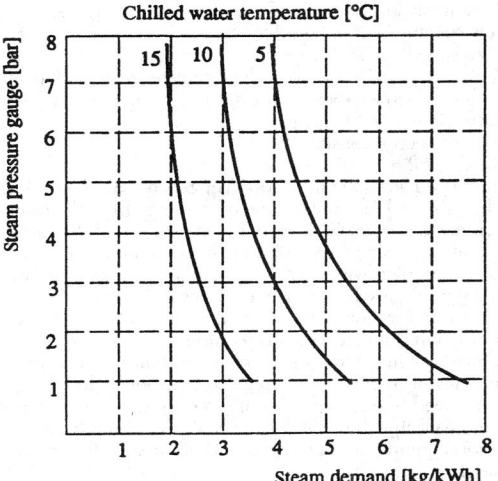

FIG. 11-106   Effect of steam pressure on steam demand at 38°C condenser temperature (ASHRAE 1983 Equipment Handbook).

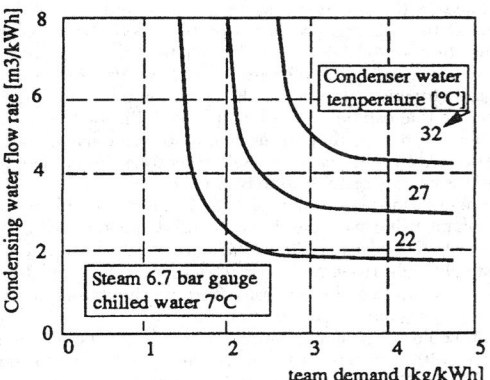

FIG. 11-107   Steam demand versus condenser-water flow rate.

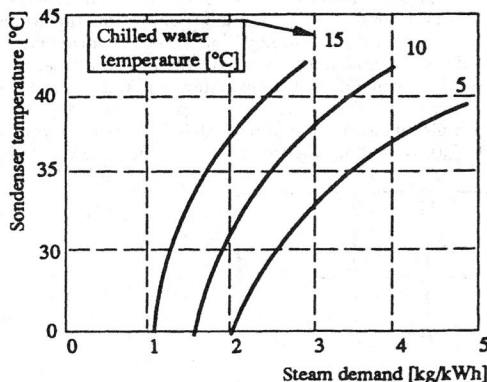

FIG. 11-108   Steam demand versus chilled-water temperature for typical steam-jet system (ASHRAE 1983 Equipment Handbook).

The condensing water temperature has an important effect on steam rate per refrigeration effect, rapidly decreasing with colder condenser cooling water. Figure 11-108 presents data on steam rate versus condenser water inlet for given chilled-water outlet temperatures and steam pressure.

3. *Chilled-water temperature.*   As the chilled-water outlet temperature decreases, the ratio of steam/refrigeration effect decreases, thus increasing condensing temperatures and/or increasing the condensing-water requirements.

Unlike other refrigeration systems, the chilled-water flow rate is of no particular importance in steam-jet system design, because there is, due to direct heat exchange, no influence of evaporator tube velocities and related temperature differences on heat-transfer rates. Widely varying return chilled-water temperatures have little effect on steam-jet equipment.

Multistage Systems   The majority of steam-jet systems being currently installed are multistage. Up to five stage systems are in commercial operation.

Capacity Control   The simplest way to regulate the capacity of most steam vacuum refrigeration systems is to furnish several primary boosters in parallel and operate only those required to handle the heat load. It is not uncommon to have as many as four main boosters on larger units for capacity variation. A simple automatic on-off type of control may be used for this purpose. By sensing the chilled-water temperature leaving the flash tank, a controller can turn steam on and off to each ejector as required.

Additionally, two other control systems which will regulate steam flow or condenser-water flow to the machine are available. As the condenser-water temperature decreases during various periods of the year, the absolute condenser pressure will decrease. This will permit the ejectors to operate on less steam because of the reduced discharge pressure. Either the steam flow or the condenser water quantities can be reduced in order to lower operating costs at other then design periods. The arrangement selected depends on cost considerations between the two flow quantities. Some systems have been arranged for a combination of the two, automatically reducing steam flow down to a point, followed by a reduction in condenser-water flow. For maximum operating efficiency, automatic control systems are usually justifiable in keeping operating cost to a minimum without excessive operator attention. In general, steam savings of about 10 percent of rated booster flow are realized for each 2.5°C reduction in condensing-water temperature below the design point.

In some cases, with relatively cold inlet condenser water it has been possible to adjust automatically the steam inlet pressure in response to chilled-water outlet temperatures. In general, however, this type of control is not possible because of the differences in temperature between the flash tank and the condenser. Under usual conditions of warm condenser-water temperatures, the main ejectors must compress water vapor over a relatively high ratio, requiring an ejector with entirely different operating characteristics. In most cases, when the ejector steam pressure is throttled, the capacity of the jet remains almost constant until the steam pressure is reduced to a point at which there is a sharp capacity decrease. At this point, the ejectors are unstable, and the capacity is severely curtailed. With a sufficient increase in steam pressure, the ejectors will once again become stable and operate at their deign capacity. In effect, steam jets have a vapor-handling capacity fixed by the pressure at the suction inlet. In order for the ejector to operate along its characteristic pumping curve, it requires a certain minimum steam flow rate which is fixed for any particular pressure in the condenser. (For further information on the design of ejectors, see Sec. 6)

Further reading and reference: *ASHRAE 1983 Equipment Handbook;* Spencer, E., 1961, *New Development in Steam Vacuum Refrigeration,* ASHRAE Transactions Vol. 67, p. 339

Refrigerants   A refrigerant is any body or substance which acts as a cooling agent by absorbing heat from another body or substance which has to be cooled. Primary refrigerants are those that are used in the refrigeration systems, where they alternately vaporize and condense as they absorb or and give off heat respectively. Secondary refrigerants are heat transfer fluids or heat carriers. Refrigerant pairs in absorption systems are ammonia-water and lithium bromide-water, while steam (water) is used as a refrigerant in ejector systems. Refrigerants used in the mechanical refrigeration systems are far more numerous.

A list of the most significant refrigerants is presented in the *ASHRAE Handbook* Fundamentals. More data are shown in the Chap. 3 of this handbook—"Physical and Chemical Data." Due to rapid changes in refrigerant, issue readers are advised to consult the most recent data and publications at the time of application.

The first refrigerants were natural: air, ammonia, CO_2, SO_2, and so on. Fast expansion of refrigeration in the second and third quarters of the 20th century is marked by the new refrigerants, chlorofluorocarbons (CFC) and hydrochlorofluorocarbons (HCFC). They are halocarbons which contain one or more of three halogens chlorine, fluorine, and bromine (Fig. 11-109). These refrigerants introduced many advantageous qualities compared to most of the existing refrigerants: odorless, nonflamable, nonexplosive, compatible with the most engineering materials, reasonably high COP, and nontoxic.

In the last decade, the refrigerant issue is extensively discussed due to the accepted hypothesis that the chlorine and bromine atoms from halocarbons released to the environment were using up ozone in the stratosphere, depleting it specially above the polar regions. Montreal Protocol and later agreements ban use of certain CFCs and halon compounds. It seems that all CFCs and most of the HCFCs will be out of production by the time this text will be published.

Chemical companies are trying to develop safe and efficient refrigerant for the refrigeration industry and application, but uncertainty in CFC and HCFC substitutes is still high. When this text was written HFCs were a promising solution. That is true especially for the R134a which seems to be the best alternative for R12. Substitutes for R22 and R502 are still under debate. Numerous ecologist and chemists are for an extended ban on HFCs as well, mostly due to significant use of CFCs in production of HFCs. Extensive research is ongoing to find new refrigerants. Many projects are aimed to design and study refrigerant mixtures, both azeotropic (mixture which behaves physically as a single, pure compound) and zeotropic having desirable qualities for the processes with temperature glides in the evaporator and the condenser.

Ammonia (R717) is the single natural refrigerant being used extensively (beside halocarbons). It is significant in industrial applications for its excellent thermodynamic and thermophysical characteristics. Many engineers are considering ammonia as a CFC substitute for various applications. Significant work is being done on reducing the refrigerant inventory and consequently problems related to leaks of this fluid with strong odor. There is growing interest in hydrocarbons in some contries, particularly in Europe. Indirect cooling (secondary refrigeration) is under reconsideration for many applications.

Due to the vibrant refrigerant issue it will be a challenge for every engineer to find the best solution for the particular application, but basic principles are the same. Good refrigerant should be:
- Safe: nontoxic, nonflamable, and nonexplosive
- Environmentally friendly
- Compatible with materials normally used in refrigeration: oils, metals, elastomers, etc.
- Desirable thermodynamic and thermophysical characteristics:
  High latent heat
  Low specific volume of vapor
  Low compression ratio
  Low viscosity

Reasonably low pressures for operating temperatures
Low specific heat of liquid
High specific heat of vapor
High conductivity and other heat transfer related characteristics
Reasonably low compressor discharge temperatures
Easily detected if leaking
High dielectric constant
Good stability

Secondary Refrigerants (Antifreezes or Brines) These are mostly liquids used for transporting heat energy from the remote heat source (process heat exchanger) to the evaporator of the refrigeration system. Antifreezes or brines do not change state in the process, but there are examples where some secondary refrigerants are either changing state themselves, or just particles which are carried in them.

Indirect refrigeration systems are more prevalent in the chemical industry than in the food industry, commercial refrigeration, or comfort air-conditioning. This is even more evident in the cases where a large amount of heat is to be removed or where a low temperature level is involved. Advantage of an indirect system is centralization of refrigeration equipment, which is specially important for relocation of refrigeration equipment in a nonhazardous area, both for people and equipment.

Salt Brines The typical curve of freezing point is shown in Fig. 11-110. Brine of concentration x (water concentration is 1-x) will not solidify at 0°C (freezing temperature for water, point A). When the temperature drops to B, the first crystal of ice is formed. As the temperature decreases to C, ice crystals continue to form and their mixture with the brine solution forms the slush. At the point C there will be part ice in the mixture $l_2/(l_1 + l_2)$, and liquid (brine) $l_1/(l_1 + l_2)$. At point D there is mixture of m_1 parts eutectic brine solution D_1 [concentration $m_1/(m_1 + m_2)$], and m_2 parts of ice [concentration $m_2/(m_1 + m_2)$]. Cooling the mixture below D solidifies the entire solution at the eutectic temperature. Eutectic temperature is the lowest temperature that can be reached with no solidification.

It is obvious that further strengthening of brine has no effect, and can cause a different reaction—salt sometimes freezes out in the installations where concentration is too high.

Sodium chloride, an ordinary salt (NaCl), is the least expensive per volume of any brine available. It can be used in contact with food and in open systems because of its low toxicity. Heat transfer coefficients are relatively high. However, its drawbacks are it has a relatively high freezing point and is highly corrosive (requires inhibitors thus must be checked on a regular schedule).

Calcium chloride ($CaCl_2$) is similar to NaCl. It is the second lowest-cost brine, with a somewhat lower freezing point (used for temperatures as low as −37°C). Highly corrosive and not appropriate for direct contact with food. Heat transfer coefficients are rapidly reduced at temperatures below −20°C. The presence of magnesium salts in either sodium or calcium chloride is undesirable because they tend to form sludge. Air and carbon dioxide are contaminants and excessive aeration of the brine should be prevented by use of close systems. Oxygen, required for corrosion, normally comes from the atmosphere and dissolves in the brine solution. Dilute brines dissolve oxygen more readily and are generally more corrosive than concentrated brines. It is believed that even a closed brine system will not prevent the infiltration of oxygen.

To adjust alkaline condition to pH 7.0–8.5 use caustic soda (to correct up to 7.0) or sodium dichromate (to reduce excessive alkalinity

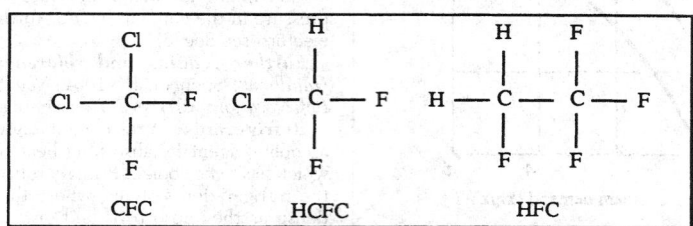

FIG. 11-109 Halocarbon refrigerants.

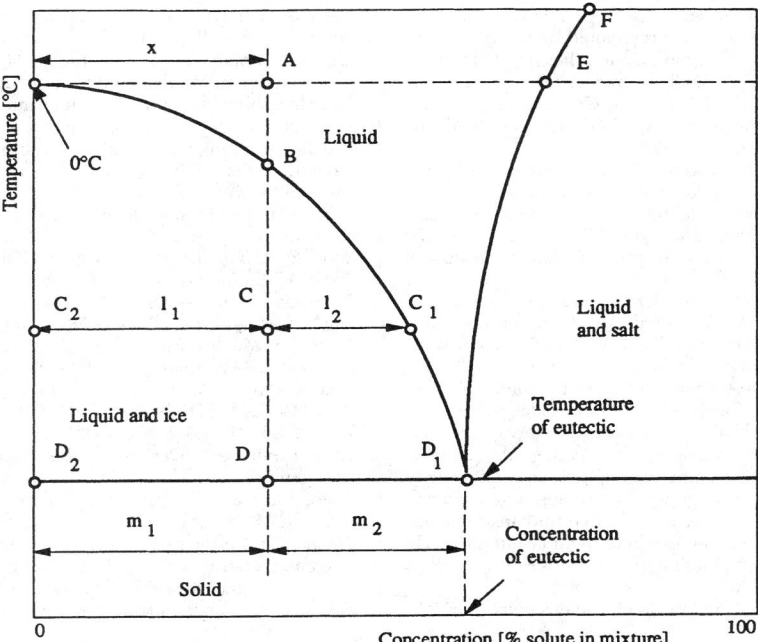

FIG. 11-110   Phase diagram of the brine.

below pH 8.5). Such slightly alkaline brines are generally less corrosive than neutral or acid ones, although with high alkalinity the activity may increase.

If the untreated brine has the proper pH value, the acidifying effect of the dichromate may be neutralized by adding commercial flake caustic soda (76 percent pure) in quantity that corresponds to 27 percent of sodium dichromate used. Caustic soda must be thoroughly dissolved in warm water before it is added to the brine.

Recommended inhibitor (sodium dichromate) concentrations are 2 kg/m³ of CaCl₂ and 3.2 kg/m³ of NaCl brine. Sodium dichromate when dissolved in water or brine makes the solution acid. Steel, iron, copper, or red brass can be used with brine circulating systems. Calcium chloride systems are generally equipped with all-iron-and-steel pumps and valves to prevent electrolysis in event of acidity. Copper and red brass tubing are used for calcium chloride evaporators. Sodium chloride systems are using all-iron or all-bronze pumps.

Organic Compounds (Inhibited Glycols)   *Ethylene glycol* is colorless and practically odorless and completely miscible with water. Advantages are low volatility and relatively low corrosivity when properly inhibited. Main drawbacks are relatively low heat-transfer coefficients at lower temperatures due to high viscosities (even higher than for propylene glycol). It is somewhat toxic, but less harmful than methanol water solutions. It is not appropriate for food industry and should not stand in open containers. Preferably waters that are classified as soft and are low in chloride and sulfate ions should be used for preparation of ethylene glycol solution.

Pure ethylene glycol freezes at −12.7°C. Exact composition and temperature for eutectic point are unknown, since solutions in this region turn to viscous, glassy mass that makes it difficult to determine the true freezing point. For the concentrations lower than eutectic, ice forms on freezing, while on the concentrated, solid glycol separates from the solution.

Ethylene glycol normally has pH of 8.8 to 9.2 and should not be used below 7.5. Addition of more inhibitor can not restore the solution to original condition. Once inhibitor has been depleted, it is recommended that the old glycol be removed from the system and the new charge be installed.

Propylene glycol is very similar to ethylene glycol, but it is not toxic

and is used in direct contact with food. It is more expensive and, having higher viscosity, shows lower heat transfer coefficients.

Methanol water is an alcohol-base compound. It is less expensive than other organic compounds and, due to lower viscosity, has better heat transfer and pressure drop characteristics. It is used up to −35°C. Disadvantages are (1) considered more toxic than ethylene glycol and thus more suitable for outdoor applications (2) flammable and could be assumed to be a potential fire hazard.

For ethylene glycol systems copper tubing is often used (up to 3 in), while pumps, cooler tubes, or coils are made of iron, steel, brass, copper, or aluminum. Galvanized tubes should not be used in ethylene glycol systems because of reaction of the inhibitor with the zinc.

Methanol water solutions are compatible with most materials but in sufficient concentration will badly corrode aluminum.

Ethanol water is a solution of denatured grain alcohol. Its main advantage is that it is nontoxic and thus is widely used in the food and chemical industry. By using corrosion inhibitors it could be made noncorrosive for brine service. It is more expensive than methanol water and has somewhat lower heat transfer coefficients. As an alcohol derivate it is flammable.

Secondary refrigerants shown below, listed under their generic names, are sold under different trade names. Some other secondary refrigerants appropriate for various refrigeration application will be listed under their trade names. More data could be obtained from the manufacurer.

Syltherm XLT (Dow Corning Corporation). A silicone polymer (Dimethyl Polysiloxane); recommended temperature range −70°C to 250°C; odorless; low in acute oral toxicity; noncorrosive toward metals and alloys commonly found in heat transfer systems.

Syltherm 800 (Dow Corning Corporation). A silicone polymer (Dimethyl Polysiloxane); recommended temperature range −40°C to 400°C; similar to Syltherm XLT, more appropriate for somewhat higher temperatures; flash point is 160°C.

D-limonene (Florida Chemicals). A compound of optically active terpene (C₁₀H₁₆) derived as an extract from orange and lemon oils; limited data shows very low viscosity at low temperatures—only one centipoise at −50°C; natural substance having questionable stability.

Therminol D-12 (Monsanto). A synthetic hydrocarbon; clear liquid;

recommended range −40°C to 250°C; not appropriate for contact with food; precautions against ignitions and fires should be taken with this product; could be found under trade names Santotherm or Gilotherm.

Therminol LT (Monsanto). Akylbenzene, synthetic aromatic ($C_{10}H_{14}$); recommended range −70°C to −180°C; not appropriate for contact with food; precautions against ignitions and fire should be taken dealing with this product.

Dowtherm J (Dow Corning Corporation). A mixture of isomers of an alkylated aromatic; recommended temperature range −70°C to 300°C; noncorrosive toward steel, common metals and alloys; combustible material; flash point 58°C; low toxic; prolonged and repeated exposure to vapors should be limited 10 ppm for daily exposures of eight hours.

Dowtherm Q (Dow Corning Corporation). A mixture of dyphenyle-hane and alkylated aromatics; recommended temperature range −30°C to 330°C; combustible material; flash point 120°C; considered low toxic, similar to Dowtherm J.

Safety in Refrigeration Systems This is of paramount importance and should be considered at every stage of installation.

The design engineer should have safety as the primary concern by choosing suitable system and refrigerant: selecting components, choosing materials and thicknesses of vessels, pipes, and relief valves of pressure vessels, proper venting of machine rooms, and arranging the equipment for convenient access for service and maintenance (piping arrangements, valve location, machine room layout, etc.). He or she should conform to the stipulation of the safety codes, which is also important for the purpose of professional liability.

During construction and installation, the installer's good decisions and judgments are crucial for safety, because design documentation never specifies all details. This is especially important when there is reconstruction or repair while the facility has been charged.

During operation the plant is in the hands of the operating personnel. They should be properly trained and familiar with the installation. Very often, accidents are caused by an improper practice, such as making an attempt to repair when proper preparation is not made. Operators should be trained in first-aid procedures and how to respond to emergencies.

Most frequently needed standards and codes are listed below, and the reader can find comments in: W. F. Stoecker: *Industrial Refrigeration,* Vol. 2. Ch. 12, Business News Publishing Co., Troy, MI. 1995; *ASHRAE Handbook* Refrigeration System and Applications, 1994, Ch. 51. Some important standards and codes on safety that a refrigeration engineer should consult are: ANSI/ASHRAE Standard 15-92—Safety Code for Mechanical Refrigeration, ASHRAE, Atlanta GA, 1992; ANSI/ASHRAE Standard 34-92—Number Designation of Refrigerants, ASHRAE, Atlanta GA, 1992; ANSI/ASME Boiler and Pressure Vessel Code, ASME, New York, 1989; ANSI/ASME Code for Pressure Piping, B31, B31.5–1987, ASME, New York, 1987; ANSI/IIAR 2—1984, Equipment, Design and Installation of Ammonia Mechanical Refrigeration Systems, IIAR, Chicago, 1984; IIAR Minimum Safety Criteria for a Safe Ammonia Refrigeration Systems, Bulletin 109; IIAR, IIAR Start-up, Inspection, and Maintenance of Ammonia Mechanical Refrigeration Systems, Bulletin 110, Chicago, 1988; IIAR Recommended Procedures in Event of Ammonia Spills, Bulletin No. 106, IIAR, Chicago, 1977; A Guide to Good Practices for the Operation of an Ammonia Refrigeration System, IIAR Bulletin R1, 1983.

CRYOGENIC PROCESSES

INTRODUCTION

Cryogenics, the production of low temperatures, is a major business in the United States with an annual market in excess of 12 billion dollars. It is a very diverse supporting technology, a means to an end and not an end in itself. For example, the combined production of oxygen and nitrogen, obtained by the cryogenic separation of air, accounts for 15 percent of the total annual U.S. production of 2.77×10^{11} kg (611 billion pounds) of organics and inorganics (1993 *C & EN Annual Report*). Liquid hydrogen production, in the last four decades, has risen from laboratory quantities to a level of over 2.1 kg/s, first spurred by nuclear weapons development and later by the United States space program. Similarly, the space age increased the need for liquid helium by more than a factor of ten, requiring the construction of large plants to separate helium from natural gas by cryogenic means. The demands for energy have likewise accelerated the construction of large base-load liquefied natural gas (LNG) plants around the world and have been responsible for the associated domestic LNG industry of today with its use of peak shaving plants.

Freezing as a means of preserving food dates back to 1840. However, today the food industry uses large quantities of liquid nitrogen for this purpose and as a refrigerant in frozen-food transport systems. In biological applications liquid-nitrogen cooled containers are routinely used to preserve whole blood, tissue, bone marrow, and animal semen for extended periods of time. Cryogenic surgery has become accepted in curing such involuntary disorders as Parkinson's disease. Medical analysis of patients has increased in sophistication with the use of *magnetic resonance imaging* (MRI) which utilizes cryogenically cooled superconducting magnets. Finally, one must recognize the role that cryogenics plays in the chemical-processing industry with the recovery of valuable feedstocks from natural gas streams, upgrading the heat content of fuel gas, purification of various process and waste streams, the production of ethylene, and so on.

PROPERTIES OF CRYOGENIC FLUIDS

There are presently several database programs of thermodynamic properties data developed specifically for fluids commonly associated with low temperature processing including helium, hydrogen, neon, nitrogen, oxygen, argon, and methane. For example, the NIST Standard Reference Database 12, Version 3.0 includes a total of 34 thermophysical properties for seventeen fluids in the database. Cryodata Inc. provides a similar computer version for 28 pure fluids as well as for mixtures incorporating many of these fluids. However, a few peculiarities associated with the fluids of helium, hydrogen, oxygen, and air need to be noted below.

Liquid helium-4 can exist in two different liquid phases: liquid helium I, the normal liquid, and liquid helium II, the superfluid, since under certain conditions the latter fluid acts as if it had no viscosity. The phase transition between the two liquid phases is identified as the lambda line and where this transition intersects the vapor-pressure curve is designated as the lambda point. Thus, there is no triple point for this fluid as for other fluids. In fact, solid helium can only exist under a pressure of 2.5 MPa or more.

A unique property of hydrogen is that it can exist in two different molecular forms: orthohydrogen and parahydrogen. (This is also true for deuterium, an isotope of hydrogen with an atomic mass of 2.) The thermodynamic equilibrium composition of the ortho- and para-varieties is temperature dependent. The equilibrium mixture of 75 percent orthohydrogen and 25 percent parahydrogen at ambient temperatures is recognized as normal hydrogen.

In contrast to other cryogenic fluids, liquid oxygen is slightly magnetic. It is also chemically reactive, particularly with hydrocarbon materials. Oxygen thus presents a safety problem and requires extra precautions in handling.

Since air is a mixture of predominantly nitrogen, oxygen, and a host of lesser impurities, there has been less interest in developing precise thermodynamic properties. The only recent correlation of thermodynamic properties is that published by Vasserman, et al. (Barouch, Israel Program for Scientific Translations, Jerusalem, 1970), and is based on the principle of corresponding states because of the scarcity of experimental data.

PROPERTIES OF SOLIDS

A knowledge of the properties and behavior of materials used in any cryogenic system is essential for proper design considerations. Often the choice of materials for the construction of cryogenic equipment will be dictated by other considerations besides mechanical properties as, for example, thermal conductivity (heat transfer along a structural member), thermal expansivity (expansion and contraction during cycling between ambient and low temperatures), and density (mass of the system). Since properties at low temperatures are often significantly different from those at ambient temperature, there is no substitute for test data on a truly representative sample specimen when designing for the limit of effectiveness of a cryogenic material or structure. For example, some metals including elements, intermetallic compounds and alloys exhibit the phenomenon of superconductivity at very low temperatures. The properties that are affected when a material becomes superconducting include specific heat, thermal conductivity, electrical resistance, magnetic permeability, and thermoelectric effect. As a result, the use of superconducting metals in the construction of equipment for temperatures lower than 10 K needs to be evaluated carefully. (High temperature superconductors because of their brittle ceramic structure are presently not considered as construction materials.)

Structural Properties at Low Temperatures   It is most convenient to classify metals by their lattice symmetry for low temperature mechanical properties considerations. The *face-centered-cubic* (fcc) metals and their alloys are most often used in the construction of cryogenic equipment. Al, Cu Ni, their alloys, and the austenitic stainless steels of the 18-8 type are fcc and do not exhibit an impact ductile-to-brittle transition at low temperatures. As a general rule, the mechanical properties of these metals with the exception of 2024-T4 aluminum, improve as the temperature is reduced. Since annealing of these metals and alloys can affect both the ultimate and yield strengths, care must be exercised under these conditions.

The *body-centered-cubic* (bcc) metals and alloys are normally classified as undesirable for low temperature construction. This class includes Fe, the martensitic steels (low carbon and the 400-series stainless steels), Mo, and Nb. If not brittle at room temperature, these materials exhibit a ductile-to-brittle transition at low temperatures. Cold working of some steels, in particular, can induce the austenite-to-martensite transition.

The *hexagonal-close-packed* (hcp) metals generally exhibit mechanical properties intermediate between those of the fcc and bcc metals. For example Zn encounters a ductile-to-brittle transition whereas Zr and pure Ti do not. The latter and their alloys with a hcp structure remain reasonably ductile at low temperatures and have been used for many applications where weight reduction and reduced heat leakage through the material have been important. However, small impurities of O, N, H, and C can have a detrimental effect on the low temperature ductility properties of Ti and its alloys.

Plastics increase in strength as the temperature is decreased, but this is also accompanied by a rapid decrease in elongation in a tensile test and a decrease in impact resistance. Teflon and glass-reinforced plastics retain appreciable impact resistance as the temperature is lowered. The glass-reinforced plastics also have high strength-to-weight and strength-to-thermal conductivity ratios. All elastomers, on the other hand, become brittle at low temperatures. Nevertheless, many of these materials including rubber, Mylar, and nylon can be used for static seal gaskets provided they are highly compressed at room temperature prior to cooling.

The strength of glass under constant loading also increases with decrease in temperature. Since failure occurs at a lower stress when the glass surface contains surface defects, the strength can be improved by tempering the surface.

Thermal Properties at Low Temperatures   For solids, the Debye model developed with the aid of statistical mechanics and quantum theory gives a satisfactory representation of the specific heat with temperature. Procedures for calculating values of Θ_D, the Debye characteristic temperature, using either elastic constants, the compressibility, the melting point, or the temperature dependence of the expansion coefficient are outlined by Barron (*Cryogenic Systems, 2d ed.*, Oxford University Press, 1985, pp 24–29).

Adequate prediction of the thermal conductivity for pure metals can be made by means of the Wiedeman-Franz law which states that the ratio of the thermal conductivity to the product of the electrical conductivity and the absolute temperature is a constant. This ratio for high-conductivity metals extrapolates essentially to the Sommerfeld value of 2.449×10^{-8} W Ω/K^2 at 0 K, but falls considerably below it at higher temperatures. High-purity aluminum and copper exhibit peaks in thermal conductivity between 20 to 50 K, but these peaks are rapidly suppressed with increased impurity levels and cold work of the metal. Some metals including Monel, Inconel, stainless steel, and structural and aluminum alloys show a steady decrease in thermal conductivity with a decrease in temperature.

All cryogenic liquids except hydrogen and helium have thermal conductivities that increase as the temperature is decreased. For these two exceptions, the thermal conductivity decreases with a decrease in temperature. The kinetic theory of gases correctly predicts the decrease in thermal conductivity of all gases when the temperature is lowered.

The expansion coefficient of a solid can be estimated with the aid of an approximate thermodynamic equation of state for solids which equates the thermal expansion coefficient β with the quantity $\gamma C_v \rho/B$ where γ is the Grüneisen dimensionless ratio, C_v is the specific heat of the solid, ρ is the density of the material, and B is the bulk modulus. For fcc metals the average value of the Grüneisen constant is near 2.3. However, there is a tendency for this constant to increase with atomic number.

Electrical Properties at Low Temperatures   The electrical resistivity of most pure metallic elements at ambient and moderately low temperatures is approximately proportional to the absolute temperature. At very low temperatures, however, the resistivity (with the exception of superconductors) approaches a residual value almost independent of temperature. Alloys, on the other hand, have resistivities much higher than those of their constituent elements and resistance-temperature coefficients that are quite low. The electrical resistivity of alloys as a consequence is largely independent of temperature and may often be of the same magnitude as the room temperature value.

Superconductivity   The physical state in which all resistance to the flow of direct-current electricity disappears is defined as superconductivity. The Bardeen-Cooper-Schriefer (BCS) theory has been reasonably successful in accounting for most of the basic features observed of the superconducting state for *low-temperature superconductors* (LTS) operating below 23 K. The advent of the ceramic *high-temperature superconductors* (HTS) by Bednorz and Miller (*Z. Phys.* B64, 189, 1989) has called for modifications to existing theories which have not been finalized to date. The massive interest in the new superconductors that can be cooled with liquid nitrogen is just now beginning to make its way into new applications.

Three important characteristics of the superconducting state are the critical temperature, the critical magnetic field, and the critical current. These parameters can be varied by using different materials or giving them special metallurgical treatments.

The alloy *niobium titanium* (NbTi) and the intermetallic compound of *niobium and tin* (Nb$_3$Sn) are the most technologically advanced LTS materials presently available. Even though NbTi has a lower critical field and critical current density, it is often selected because its metallurgical properties favor convenient wire fabrication. In contrast, Nb$_3$Sn is a very brittle material and requires wire fabrication under very well-defined temperature conditions.

There are presently four families of high-temperature superconductors under investigation for practical magnet applications. Table 11-25 shows that all HTS are copper oxide ceramics even though the oxygen content may vary. However, this variation generally has little effect on the physical properties of importance to superconductivity.

The most widely used development in HTS wire production is the powder-in-tube procedure with BSCCO ceramic materials. In this procedure very fine HTS powder, placed inside of a hollow silver tube, is fused as the tube length is mechanically increased to form a wire. Very high magnetic fields with this wire have been reported at 4 K; however, the performance degrades substantially above 20 to 30 K.

HTS materials, because of their ceramic nature, are quite brittle. This has introduced problems relative to the winding of superconducting magnets. One solution is to first wind the magnet with the powder-in-tube wire before the ceramic powder has been bonded and then heat treat the desired configuration to form the final product. Another solution is to form the superconductor into such fine fila-

TABLE 11-25   Composition and Critical Temperature, T_c for HTS Materials

Formula*	Accepted notation	Forms reported		Critical T_c, K
Y-Ba-Cu-O†	YBCO	123	124	80–92
Bi-Sr-Ca-Cu-O	BSCCO	2212	2223	80–110
Tl-Ba-Ca-Cu-O	TBCCO	Several		to 125
Hg-Ba-Ca-Cu-O	HBCCO	1201	1223	95–155‡

*Subscripts for compounds are not listed since there are generally several forms that can be produced (see column 3).

†Other rare earths may be substituted for Y (yttrium) providing new compounds with somewhat different properties.

‡Highest T_c obtained while subjecting sample to external pressure.

ments that they remain sufficiently flexible even after the powder has been heat treated.

REFRIGERATION AND LIQUEFACTION

A process for producing refrigeration or liquefaction at cryogenic temperatures usually involves ambient compression of a process fluid with heat rejection to a coolant. During the compression process, the enthalpy and entropy of the fluid are decreased. At the cryogenic temperature where heat is absorbed, the enthalpy and entropy are increased. The reduction in temperature of the process fluid is usually accomplished by heat exchange with a colder fluid and then followed by an expansion. This expansion may take place using either a throttling device (isenthalpic expansion) with only a reduction in temperature or a work-producing device (isentropic expansion) in which both temperature and enthalpy are decreased. Because of liquid withdrawal, a liquefaction system experiences an unbalanced flow in the heat exchanger while a refrigeration system with no liquid withdrawal system usually operates with a balanced flow in the heat exchanger, except where a portion of the flow is diverted through the work-producing expander.

Principles   The performance of a real refrigerator is measured by the coefficient of performance, COP, defined as

$$COP = \frac{Q}{W} = \frac{\text{heat removed at low temperature}}{\text{net work input}} \quad (11\text{-}108)$$

Another means of comparing the performance of a practical refrigerator is by the use of the figure of merit, FOM, defined as

$$FOM = \frac{COP}{COP_i} \quad (11\text{-}109)$$

where COP is the coefficient of performance of the actual refrigerator system and COP_i is the coefficient of performance for the thermodynamically ideal system. For a liquefier, the FOM is generally specified as

$$FOM = \frac{W_i/\dot{m}_f}{W/\dot{m}_f} \quad (11\text{-}110)$$

where W_i is the work of compression for the ideal cycle, W is the work of compression for the actual cycle, and $\dot{m}_f$ is the mass rate liquefied in the ideal or actual cycle.

The methods of refrigeration and/or liquefaction generally used include (1) vaporization of a liquid, (2) application of the Joule-Thomson effect in a gas, and (3) expansion of a gas in a work-producing engine. Normal commercial refrigeration generally is accomplished in a vapor-compression process. Temperatures to about 200 K can be obtained by cascading vapor-compression processes in which refrigeration is accomplished by liquid evaporation. Below this temperature, isenthalpic or isentropic expansions are generally used either singly or in combination. With few exceptions, refrigerators using these methods also absorb heat by vaporization of the liquid.

If refrigeration is to be accomplished at a temperature range where no suitable liquid exists to absorb heat by evaporation, then a cold gas must be available to absorb the heat. This is generally accomplished by using a work-producing expansion engine.

Expansion Types of Refrigerators   A thermodynamic process

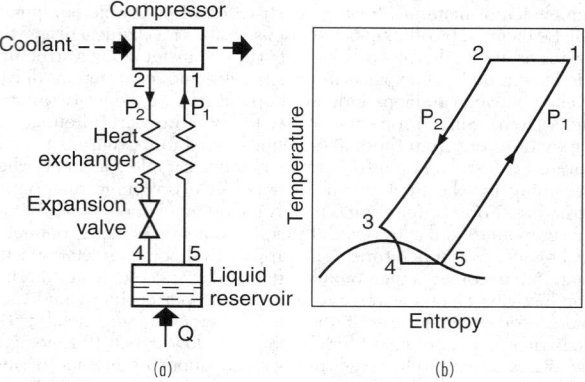

FIG. 11-111   Refrigerator using simple J-T cycle.

utilizing isenthalpic expansion to obtain cryogenic temperatures, and commonly referred to as the simple Linde or J-T cycle, is shown schematically with its corresponding temperature-entropy diagram in Fig. 11-111. The gaseous refrigerant is compressed at ambient temperature while essentially rejecting heat isothermally to a coolant. The compressed refrigerant is cooled countercurrently in a heat exchanger by the cold gas stream leaving the liquid reservoir before it enters the throttling valve. Upon expansion, Joule-Thomson cooling further reduces the temperature until, at steady state, a portion of the refrigerant is liquefied. For a refrigerator, the unliquefied fraction and the vapor formed by liquid evaporation from the absorbed heat Q are warmed in the heat exchanger before returning to the intake of the compressor. Assuming no heat inleaks, as well as negligible kinetic and potential energy changes in the fluid, the refrigeration duty Q is equivalent to $\dot{m}(h_1 - h_2)$, where the subscripts refer to the locations shown on Fig. 11-111. Applying Eq. 11-108, the coefficient of performance for the ideal J-T refrigerator is given by

$$COP = \frac{h_1 - h_2}{T_1[s_1 - s_2 - (h_1 - h_2)]} \quad (11\text{-}111)$$

For a simple J-T liquefier, the liquefied portion is continuously withdrawn from the reservoir and only the unliquefied portion of the fluid is warmed in the countercurrent heat exchanger and returned to the compressor. The fraction y that is liquefied is obtained by applying the first law to the heat exchanger, J-T valve, and liquid reservoir. This results in

$$y = \frac{h_1 - h_2}{h_1 - h_f} \quad (11\text{-}112)$$

where h_f is the specific enthalpy of the liquid being withdrawn. Maximum liquefaction occurs when the difference between h_1 and h_2 is maximized. To account for heat inleak, q_L, the relation needs to be modified to

$$y = \frac{h_1 - h_2 - q_L}{h_1 - h_f} \quad (11\text{-}113)$$

with a resultant decrease in the fraction liquefied.

Refrigerants used in this process have a critical temperature well below ambient; consequently liquefaction by direct compression is not possible. In addition, the inversion temperature of the refrigerant must be above ambient temperature to provide initial cooling by the J-T process. Auxiliary refrigeration is required if the simple J-T cycle is to be used to liquefy neon, hydrogen, or helium whose inversion temperatures are below ambient. Liquid nitrogen is the optimum refrigerant for hydrogen and neon liquefaction systems, while liquid hydrogen is the normal refrigerant for helium liquefaction systems.

To reduce the work of compression in this cycle a two-stage or dual-pressure process may be used whereby the pressure is reduced by two successive isenthalpic expansions. Since the isothermal work of compression is approximately proportional to the logarithm of the pressure ratio, and the Joule-Thomson cooling is roughly proportional to

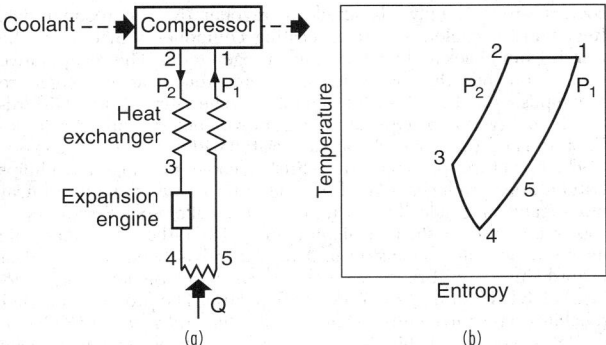

FIG. 11-112 Cold-gas refrigerator.

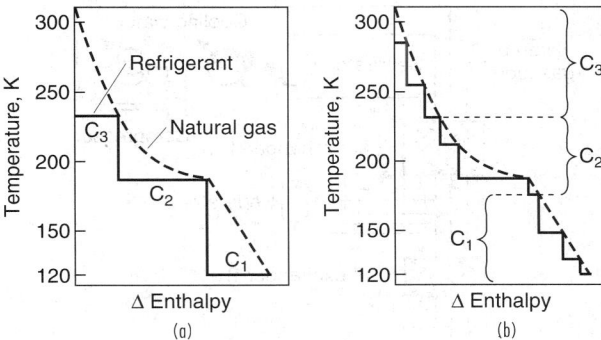

FIG. 11-114 Three- and nine-level cascade cycle cooling curve for natural gas.

the pressure difference, there is a greater reduction in compressor work than in refrigerating performance for this dual-pressure process.

In a work-producing expansion, the temperature of the process fluid is always reduced; hence, cooling does not depend on being below the inversion temperature prior to expansion. Additionally, the work-producing expansion results in a larger amount of cooling than in an isenthalpic expansion over the same pressure difference.

In large systems utilizing expanders, the work produced during expansion is conserved. In small refrigerators, the energy from the expansion is usually expended in a gas or hydraulic pump, or other suitable device. A schematic of a simple cold-gas refrigerator using this expansion principle and the corresponding temperature-entropy diagram is shown in Fig. 11-112. Gas compressed isothermally at ambient temperature is cooled in a heat exchanger by gas being warmed on its return to the compressor intake. Further cooling takes place during the engine expansion. In practice this expansion is never truly isentropic, and is reflected by path 3-4 on the temperature-entropy diagram. This specific refrigerator produces a cold gas which absorbs heat from 4-5 and provides a method of refrigeration that can be used to obtain temperatures between those of the boiling points of the lower-boiling cryogens.

It is not uncommon to utilize both the isentropic and isenthalpic expansions in a cycle. This is done to avoid the technical difficulties associated with the formation of liquid in the expander. The Claude or expansion engine cycle is an example of a combination of these meth-

ods and is shown in Fig. 11-113 along with the corresponding temperature-entropy diagram.

The mixed refrigerant cycle was developed to meet the need for liquefying large quantities of natural gas to minimize transportation costs of this fuel. This cycle resembles the classic cascade cycle in principle and may best be understood by referring to that cycle. In the latter, the natural gas stream after purification is cooled successively by vaporization of propane, ethylene, and methane. Each refrigerant may be vaporized at two or three pressure levels to increase the natural gas cooling efficiency, but at a cost of considerable increased process complexity.

Cooling curves for natural gas liquefaction by the cascade process are shown in Fig. 11-114. It is evident that the cascade cycle efficiency can be improved by increasing the number of refrigerants employed. For the same refrigeration capacity, the actual work required for the nine-level cascade cycle depicted is approximately 80 percent of that required by the three-level cascade cycle. This increase in efficiency is achieved by minimizing the temperature difference between the refrigerant and the natural gas stream throughout each increment of the cooling curve.

The mixed refrigerant cycle is a variation of the cascade cycle and involves the circulation of a single multicomponent refrigerant stream. The simplification of the compression and heat exchange services in such a cycle can offer potential for reduced capital expenditure over the conventional cascade cycle.

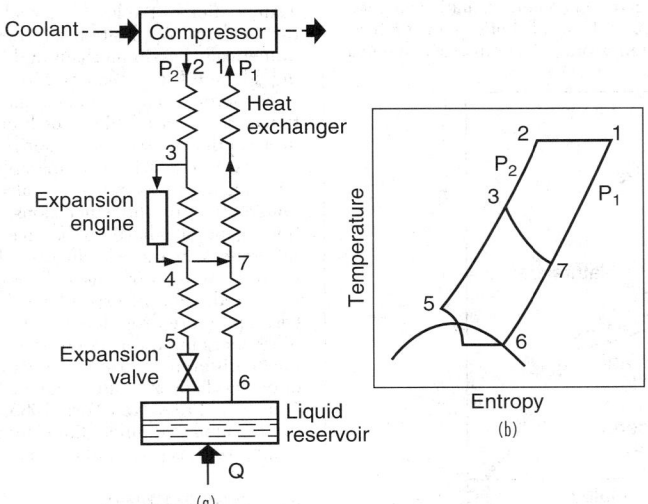

FIG. 11-113 Claude cycle refrigerator utilizing both expansion processes.

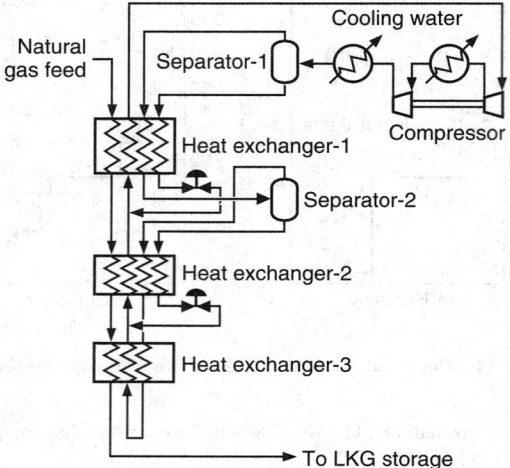

FIG. 11-115 Mixed-refrigerant cycle.

Figure 11-115 shows the basic concepts for a mixed refrigerant cycle (Gaumer, *Advances in Cryogenic Engineering* Vol. 31, Plenum, New York, 1986, p. 1095). Variations of the cycle are proprietary with those cryogenic engineering firms that have developed the technology. However, all of the mixed refrigerant processes use a carefully prepared refrigerant mix which is repeatedly condensed, vaporized, separated, and expanded. Thus, these processes require more complete knowledge of the thermodynamic properties of gaseous mixtures than those required in the expander or classical cascade cycles. This is particularly evident when cooling curves similar to the one shown in Fig. 11-116 are desired. An inspection of the mixed refrigerant cycle also shows that these processes must routinely handle two-phase flows in the heat exchangers.

Miniature Refrigerators   Expanded space and science projects have provided a need for miniature cryogenic coolers designated as cryocoolers. Such coolers provide useful cooling from a fraction of a watt to several watts at temperature levels from 1 to 90 K. These coolers are used to increase the sensitivity and signal-to-noise ratio of detectors by providing the required cryogenic operating temperatures as well as cooling the optical components to decrease the detector background radiation. The types of coolers developed to meet various specific requirements include solid cryogen coolers, radiative coolers, mechanical coolers, [3]He adsorption coolers, adiabatic demagnetization refrigerators, and liquid helium storage systems. Mechanical

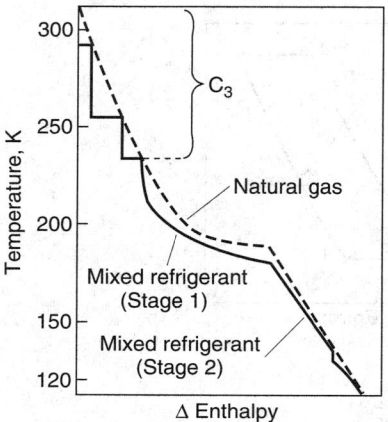

FIG. 11-116 Propane precooled mixed-refrigerant cycle cooling curve for natural gas.

coolers are generally classified as regenerative or recuperative. Regenerative coolers use reciprocating components that move the working fluid back and forth through a regenerator. The recuperative coolers, on the other hand, use countercurrent heat exchangers to accomplish the heat transfer operation. The Stirling and Gifford-McMahon cycles are typically regenerative coolers while the Joule-Thomson and Brayton cycles are associated with recuperative coolers.

The miniature split single-stage Stirling cooler developed by Philips Laboratories produces 5 W of cooling at 65 K with the aid of linear motors and magnetically moving parts. A smaller split Stirling cycle cooler that uses a stacked diaphragm spring rather than magnetic means to levitate the piston and displacer has been developed at Oxford University (Bradshaw, et al., *Advances in Cryogenic Engineering*, Vol. **31**, Plenum, New York, 1986, p. 801). The promise of higher reliability has spurred interest in the *pulse-tube refrigerator* (PTR). In the latest version of this device, an orifice and reservoir have been added to the warm end of the pulse tube (OPTR) to permit control of the phase shift required for optimum resonance in the system.

The Joule-Thomson cycle has also benefited from creative thinking. For example, Little (*6th International Cryocooler Conference*, Naval Postgraduate School, Monterey, CA, 1989, p. 3) has introduced a new method of fabricating J-T refrigerators using a photolithographic manufacturing process in which gas channels for the heat exchangers, expansion capillary, and liquid reservoir are etched on planar, glass substrates that are fused together to form the sealed refrigerator. These microminiature refrigerators have been made in a wide range of sizes and capacities.

Mixtures of highly polar gases are receiving considerable attention for J-T cycles since the magnitude of the Joule-Thomson coefficient increases with nonideality of the gas. Little (op.cit) and Longsworth (*8th International Cryocooler Conference*, JPL, Vail, CO, 1994, in press) have shown considerable ingenuity with gas mixtures for J-T cycles, developing small, lightweight, low-cost, but reliable cryocoolers for a number of applications.

Thermodynamic Analyses of Cycles   The thermodynamic quality measure of either a piece of equipment or an entire process is its reversibility. The second law, or more precisely the entropy increase, is an effective guide to this degree of irreversibility. However, to obtain a clearer picture of what these entropy increases mean, it has become convenient to relate such an analysis to the additional work that is required to overcome these irreversibilities. The fundamental equation for such an analysis is

$$W = W_{\text{rev}} + T_o \sum \dot{m}\Delta s \qquad (11\text{-}114)$$

where the total work, W, is the sum of the reversible work, W_{rev}, plus a summation of the losses in availability for various unit operations in the analysis. Application of this method has been demonstrated numerically by Timmerhaus and Flynn (*Cryogenic Process Engineering*, Plenum Press, 1989, p. 175).

Numerous analyses and comparisons of refrigeration and liquefaction cycles are available in the literature. Great care must be exercised in accepting these comparisons since it is quite difficult to put all processes on a strictly comparable basis. Many assumptions need to be made in the course of the calculations, and these can have considerable effect on the conclusions. Major factors upon which assumptions generally have to be made include heat leak, temperature differences in the exchangers, efficiencies of compressors and expanders, number of stages of compression, fraction of expander work recovered, state of expander exhaust, purity and condition of inlet gases, pressure drop due to fluid flow, and so on. In view of this fact, differences in power requirements of 10 to 20 percent can readily be due to differences in assumed variables and can negate the advantage of one cycle over another. Barron (*Cryogenic Systems*, 2d ed., Oxford University Press, New York, 1985, p. 94] has made an analysis of some of the more common liquefaction systems described earlier that emphasize this point rather well.

PROCESS EQUIPMENT

The equipment normally associated with cryogenic systems includes heat exchangers, compressors, expanders, throttling valves, and storage containers. Since the reciprocal or centrifugal compressors used generally operate at ambient temperatures, their operating principles are not covered here but in Sec. 10. Storage containers, are discussed later in Sec. 11.

Heat Exchangers Since most cryogens, with the exception of helium II behave as classical fluids, well-established principles of mechanics and thermodynamics at ambient temperature also apply for cryogens. Thus, similar conventional heat transfer correlations have been formulated for simple low-temperature heat exchangers. These correlations are described in terms of well-known dimensionless quantities such as the Nusselt, Reynolds, Prandtl, and Grashof numbers.

Because of the need to operate more efficiently at low temperatures, the simple heat exchangers have generally been replaced with more sophisticated types. Guidance for the development of such units for low-temperature service include the following factors:

1. Small temperature differences between inlet and exit streams to enhance efficiency.
2. Large surface area-to-volume ratio to minimize heat leak.
3. High heat transfer to reduce surface area.
4. Low mass to minimize start-up time.
5. Multichannel capability to minimize the number of exchangers.
6. High-pressure capability to provide design flexibility.
7. Low or reasonable pressure drop to minimize compression requirements.
8. High reliability with minimal maintenance to reduce shutdowns.

Problems sometimes occur in trying to minimize the temperature difference at the cold end of the heat exchanger, particularly if the specific heat of the warm fluid decreases with decreasing temperature as is the case with gaseous hydrogen.

The selection of an exchanger for low-temperature operation is usually determined by the process-design requirements, mechanical-design limitations, and economic considerations. Laboratory needs are generally met by concentric tube and extended surface exchangers, while industrial needs are most often met by the coiled-tube, plate-fin, reversing, and regenerator types of exchangers.

The coiled-tube heat exchanger offers unique advantages, especially when dealing with low-temperature design conditions where (1) simultaneous heat transfer between more than two streams is desired, (2) a large number of heat transfer units is required, and (3) high operating pressures are involved. Heat transfer for single-phase flow of either gas or liquid on the tubeside is generally well represented by either the Colburn correlation or modified forms of the Dittus-Boelter relationship.

The shape of the cooling and warming curves in coiled-tube heat exchangers is affected by the pressure drop in both the tube and shell-sides of the heat exchanger. This is particularly important for two-phase flows of multicomponent systems. For example, an increase in pressure drop on the shellside causes boiling to occur at a higher temperature, while an increase in pressure drop on the tubeside will cause condensation to occur at a lower temperature. The net result is both a decrease in the effective temperature difference between the two streams and a requirement for additional heat transfer area to compensate for these losses.

Plate-fin heat exchangers are about nine times as compact as conventional shell-and-tube heat exchangers with the same amount of surface area, weigh less than conventional heat exchangers, and withstand design pressures up to 6 MPa for temperatures between 4 and 340 K. Flow instability frequently becomes a limiting design parameter for plate-fin heat exchangers handling either boiling or condensing two-phase flows. This results in lower optimum economic mass flow velocities for plate-fin heat exchangers when compared with coiled-tube heat exchangers. The use of fins or extended surfaces in plate-fin or similar exchangers greatly increases the heat transfer area. Calculations using finned surfaces are outlined earlier in Sec. 11.

There are two basic approaches to heat-exchanger design for low temperatures: (1) the effectiveness-NTU approach and (2) the *log-mean-temperature-difference* (LMTD) approach. The LMTD approach is used most frequently when all the required mass flows are known and the area of the exchanger is to be determined. The effec-

tiveness-NTU approach is used more often when the inlet temperatures and the flow rates are specified for an exchanger with fixed area and the outlet temperatures are to be determined. Both methods are described earlier in Sec. 11.

System performance in cryogenic liquefiers and refrigerators is directly related to the effectiveness of the heat exchangers used in the system. For example, the liquid yield for a simple J-T cycle as given by Eq. 11-112 needs to be modified to

$$y = \frac{(h_1 - h_2) - (1 - \varepsilon)(h_1 - h_g)}{(h_1 - h_f) - (1 - \varepsilon)(h_1 - h_g)} \tag{11-115}$$

if the heat exchanger is less than 100 percent effective. Likewise, the heat exchanger ineffectiveness increases the work required for the system by an amount of

$$\Delta \dot{W} = \dot{m}(h_1 - h_g)(1 - \varepsilon) \tag{11-116}$$

Uninterrupted operation of heat exchangers at low temperatures requires removal of essentially all impurities present in the streams that are to be cooled. Equipment is readily available for the satisfactory removal of these impurities by both chemical and physical methods, but at increased operating expense. Another effective method for also accomplishing this impurity removal utilizes reversing heat exchangers. Proper functioning of the reversing heat exchanger is dependent upon the relationship between the pressures and temperatures of the two streams. Since the pressures are generally fixed by other factors, the purification function of the heat exchanger is normally controlled by selecting the right temperature difference throughout the heat exchanger. To assure that reevaporation takes place, these differences must be such that the vapor pressure of the impurity is greater than the partial pressure of the impurity in the purging stream.

Another type of reversing heat exchanger is the regenerator. As with all reversing heat exchangers, regenerators provide the simultaneous cooling and purification of gases in low-temperature processes. As noted earlier, reversing heat exchangers usually operate continuously. Regenerators do not operate continuously; instead, they operate by periodically storing heat in a packing during the first half of the cycle and then giving up this stored heat to the fluid during the second half of the cycle. Typically, a regenerator consists of two identical columns, which are packed with a porous solid material with a good heat capacity such as metal ribbon, through which the gases flow.

The low cost of the packing material, its large surface-area-per-unit volume, and the low-pressure drops encountered provide compelling arguments for utilizing regenerators. However, the intercontamination of fluid streams, caused by the periodic flow reversals, and the problems associated with designing regenerators to handle three or more fluids, has restricted their use to simple fluids, and favored adoption of plate-fin reversing heat exchangers.

Expanders The primary function of cryogenic expansion equipment is the reduction of the temperature of the gas being expanded to provide needed refrigeration. The expansion of a fluid to produce refrigeration may be carried out in two distinct ways: (1) in an expander where mechanical work is produced, and (2) in a Joule-Thomson valve where no work is produced.

Mechanical Expanders Reciprocating expanders are very similar in concept and design to reciprocating compressors. Generally these units are used with inlet pressures of 4 to 20 MPa. These machines operate at speeds up to 500 rpm. The thermal efficiencies (actual enthalpy difference/maximum possible enthalpy difference) range from about 75 percent for small units to 85 percent for large machines.

Turboexpanders have replaced reciprocating expanders in high-power installations as well as in small helium liquefiers. Sizes range from 0.75 to 7500 kW with flow rates up to 28 million m³/day. Today's large-tonnage air-separation plants are a reality due to the development of highly reliable and efficient turboexpanders. These expanders are being selected over other cryogenic equipment because of their ability to condense ethane and heavier hydrocarbons. This type of expander usually weighs and costs less and requires less space and operating personnel.

Turboexpanders can be classified as either axial or radial. Axial flow expanders have either impulse or reaction type blades and are suitable for multistage expanders because they permit a much easier flow path from one stage to the next. However, radial turboexpanders have lower stresses at a given tip speed, which permits them to run at higher speeds. This results in higher efficiencies with correspondingly lower energy requirements. As a consequence, most turboexpanders built today are of the radial type.

Joule-Thomson Valves   The principal function of a J-T valve is to obtain isenthalpic cooling of the gas flowing through the valve. These valves generally are needle-type valves modified for cryogenic operation. They are an important component in most refrigeration systems, particularly in the last stage of the liquefaction process. Joule-Thomson valves also offer an attractive alternative to turboexpanders for small-scale gas-recovery applications.

SEPARATION AND PURIFICATION SYSTEMS

The energy required to reversibly separate gas mixtures is the same as that necessary to isothermally compress each component in the mixture from the partial pressure of the gas in the mixture to the final pressure of the mixture. This reversible isothermal work is given by the familiar relation

$$\frac{-W_i}{\dot{m}} = T(s_1 - s_2) - (h_1 - h_2) \qquad (11\text{-}117)$$

where s_1 and h_1 refer to conditions before the separation and s_2 and h_2 refer to conditions after the separation. For a binary system, and assuming a perfect gas for both components, Eq. (11-117) simplifies to

$$\frac{-W_i}{\dot{m}} = RT\left(n_A \ln \frac{P_T}{p_A} + n_B \ln \frac{P_T}{p_B}\right) \qquad (11\text{-}118)$$

where n_A and n_B are the moles of A and B in the mixture, p_A and p_B are the partial pressures of these two components in the mixture, and P_T is the total pressure of the mixture. The figure of merit for a separation system is defined similar to that for a liquefaction system; see Eq. (11-110).

If the mixture to be separated is essentially a binary, both the McCabe-Thiele and the Ponchon-Savarit methods outlined in Sec. 13, with the appropriate cryogenic properties, can be used to obtain the ideal number of stages required. It should be noted, however, that it is not satisfactory to treat it as a binary mixture of oxygen and nitrogen if high purity (99 percent or better) oxygen is desired. The separation of oxygen from argon is a more difficult separation than oxygen from nitrogen and would require correspondingly many more plates. In fact, if the argon is not extracted from air, only 95 percent oxygen would be produced. The other rare gas constituents of air-helium, neon, krypton, and xenon are present in such small quantities and have boiling points so far removed from those of oxygen and nitrogen that they introduce no important complications.

Air-Separation Systems   Of the various separation schemes available today, the simplest is known as the Linde single-column system shown in Fig. 11-117 and first introduced in 1902. In it, purified compressed air passes through a precooling heat exchanger (if oxygen gas is the desired product, a three-channel exchanger for air, waste nitrogen, and oxygen gas is used; if liquid oxygen is to be recovered from the bottom of the column, a two-channel exchanger for air and waste nitrogen is employed), then through a coil in the boiler of the rectifying column where it is further cooled (acting at the same time as the boiler heat source); following this, it expands essentially to atmospheric pressure through a J-T valve and reenters the top of the column with the liquid providing the required reflux. Rectification occurs as liquid and vapor on each plate establish equilibrium. If oxygen gas is to be the product, purified air need only be compressed to a pressure of 3 to 6 MPa; if the product is to be liquid oxygen, a compressor outlet pressure of 20 MPa is necessary. Note that the Linde single-column separation system is simply a J-T liquefaction system with a substitution of a rectification column for the liquid reservoir. However, any of the other liquefaction systems discussed earlier could have been used to furnish the liquid for the column.

In a simple single-column process, although the oxygen purity is high, the nitrogen effluent stream is impure. The equilibrium vapor

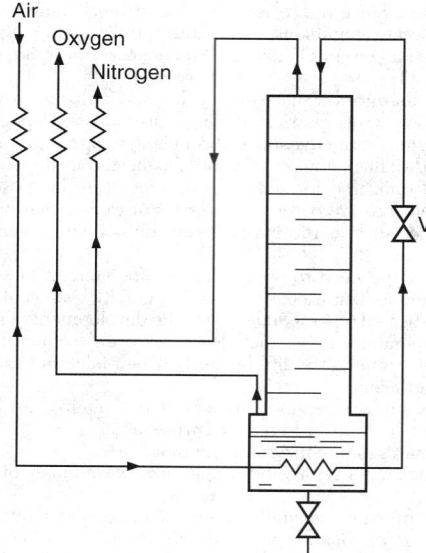

FIG. 11-117   Linde single-column air separator.

concentration of the overhead nitrogen effluent for an initial liquid mixture of 21 percent oxygen/79 percent nitrogen at 100 kPa is about 6 to 7 percent oxygen. Thus, the nitrogen waste gas stream with such an impurity may only be usable as a purge gas for certain conditions.

The impurity problem noted in the previous paragraph was solved by the introduction of the Linde double-column system shown in Fig. 11-118. Two rectification columns are placed one on top of the other (hence the name double-column system).

In this system the liquid air is introduced at an intermediate point B into the lower column, and a condenser-evaporator at the top of the

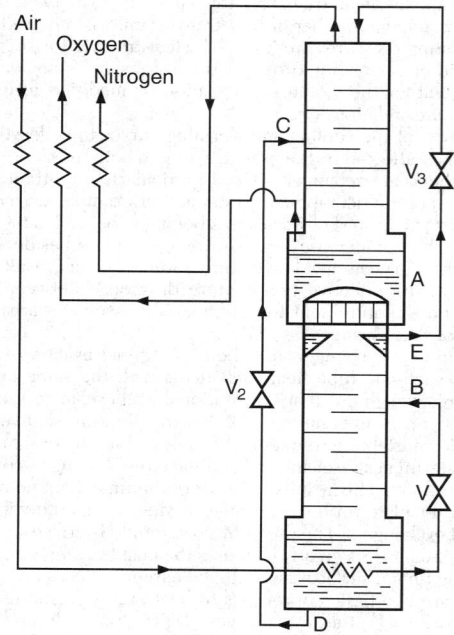

FIG. 11-118   Linde double-column air separator.

lower column makes the arrangement a complete reflux distillation column which delivers almost pure nitrogen at E. In order for the column simultaneously to deliver pure oxygen, the oxygen-rich liquid (about 45 percent O_2) from the bottom boiler is introduced at an intermediate level C in the upper column. The reflux and rectification in the upper column produce pure oxygen at the bottom and pure nitrogen at the top provided all major impurities are first removed from the column. More than enough liquid nitrogen is produced in the lower column for the needed reflux in both columns. Since the condenser must condense nitrogen vapor by evaporating liquid oxygen, it is necessary to operate the lower column at a higher pressure, about 500 kPa, while the upper column is operated at approximately 110 kPa. This requires a reduction in the pressure of the fluids from the lower column as they are admitted to the upper column.

In the cycle shown, gaseous oxygen and nitrogen are withdrawn at room temperature. Liquid oxygen could be withdrawn from point A and liquid nitrogen from point E, but in this case more refrigeration would be needed.

Even the best modern low-temperature air separation plant has an efficiency only a small fraction of the theoretical optimum, that is, about 15 to 20 percent. The principal sources of inefficiency are threefold: (1) the nonideality of the refrigerating process, (2) the imperfection of the heat exchangers, and (3) losses of refrigeration through heat leak.

Helium and Natural-Gas Systems Separation   Helium is produced primarily by separation of helium-rich natural gas. The helium content of the natural gas from plants operated by the U.S. Bureau of Mines normally has varied from 1 to 2 percent while the nitrogen content of the natural gas has varied from 12 to 80 percent. The remainder of the natural gas is methane, ethane, and heavier hydrocarbons.

A Bureau of Mines system for the separation of helium from natural gas is shown in Fig. 11-119. Since the major constituents of natural gas have boiling points very much different from that of helium, a distillation column is not necessary and the separation can be accomplished with condenser-evaporators.

The need to obtain greater recoveries of the C_2, C_3, and C_4's in natural gas has resulted in the expanded use of low-temperature processing of these streams. The majority of the natural gas processing at low temperatures to recover light hydrocarbons is now accomplished using the turboexpander cycle. Feed gas is normally available from 1 to 10 MPa. The gas is first dehydrated to a dew point of 200 K and lower. After dehydration the feed is cooled with cold residue gas. Liquid produced at this point is separated before entering the expander and sent to the condensate stabilizer. The gas from the separator is

expanded in a turboexpander where the exit stream can contain as much as 20 wt % liquid. This two-phase mixture is sent to the top section of the stabilizer which separates the two phases. The liquid is used as reflux in this unit while the cold gas exchanges heat with the fresh feed and is recompressed by the expander-driven compressor. Many variations to this cycle are possible and have been used in actual plants.

Gas Purification   The nature and concentration of impurities to be removed depends on the type of process involved. For example, in the production of large quantities of oxygen, various impurities must be removed to avoid plugging of the cold process lines or to avoid buildup of hazardous contaminants. The impurities in air that would contribute most to plugging would be water and carbon dioxide. Helium, hydrogen, and neon, on the other hand, will accumulate on the condensing side of the condenser-reboiler located between the two separation columns and will reduce the rate of heat transfer unless removed by intermittent purging. The buildup of acetylene, however, can prove to be dangerous even though the feed concentration in the air is no greater than 0.04 ppm.

Refrigeration purification is a relatively simple method for removing water, carbon dioxide, and certain other contaminants from a process stream by condensation or freezing. (Either regenerators or reversing heat exchangers may be used for this purpose since a flow reversal is periodically necessary to reevaporate and remove the solid deposits.) The effectiveness of this method depends upon the vapor pressure of the impurities relative to that of the major components of the process stream at the refrigeration temperature. Thus, assuming ideal gas behavior, the maximum impurity content in a gas stream after refrigeration would be inversely proportional to its vapor pressure. However, due to the departure from ideality at higher pressures, the impurity content will be considerably higher than predicted for the ideal situation. For example, the actual water vapor content in air will be over four times that predicted by ideal gas behavior at a temperature of 228 K and a pressure of 20 MPa.

Purification by a solid adsorbent is one of the most common low-temperature methods for removing impurities. Materials such as silica gel, carbon, and synthetic zeolites (molecular sieves) are widely used as adsorbents because of their extremely large effective surface areas. Most of the gels and carbon have pores of varying sizes in a given sample, but the synthetic zeolites are manufactured with closely controlled pore-size openings ranging from about 0.4 to 1.3 nm. This makes them even more selective than other adsorbents since it permits separation of gases on the basis of molecular size.

Information needed in the design of low-temperature adsorbers

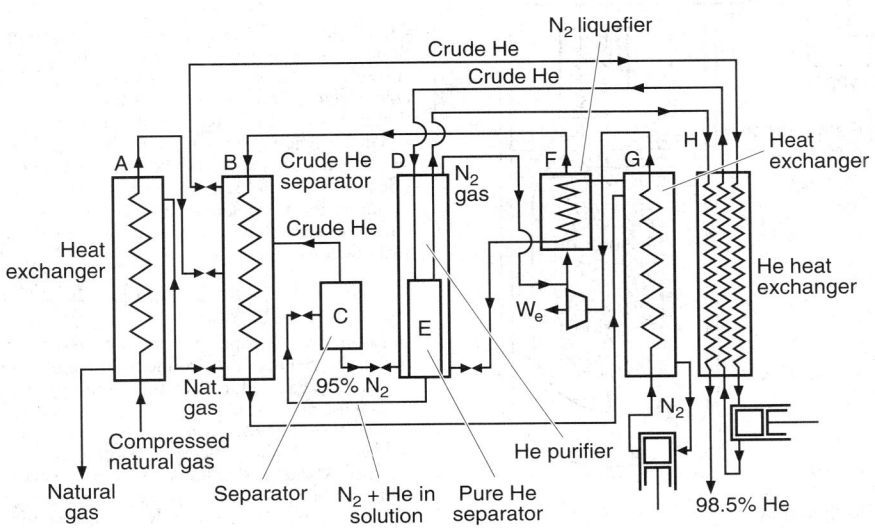

FIG. 11-119   Typical helium-separation plant as operated by the U.S. Bureau of Mines.

includes the equilibrium between the solid and gas and the rate of adsorption. Equilibrium data for the common systems generally are available from the suppliers of such material. The rate of adsorption is usually very rapid and the adsorption is essentially complete in a relatively narrow zone of the adsorber. If the concentration of the adsorbed gas is more than a trace, then heat of adsorption may also be a factor of importance in the design. (The heat of adsorption is usually of the same order or larger than the normal heat associated with the phase change.) Under such situations it is generally advisable to design the purification in two steps, that is, first removing a significant portion of the impurity either by condensation or chemical reaction and then completing the purification with a low-temperature adsorption system. A scheme combining the condensation and adsorption is shown in Fig. 11-120.

In normal plant operation at least two adsorption purifiers are employed—one in service while the other being desorbed of its impurities. In some cases there is an advantage in using an additional purifier by placing this unit in series with the adsorption unit to provide a backup if impurities are not trapped by the first unit. The cooling of the purifier must be effected with some of the purified gas to avoid adsorption during this period.

Experience in air separation plant operations and other cryogenic processing plants has shown that local freeze-out of impurities such as carbon dioxide can occur at concentrations well below the solubility limit. For this reason, the carbon dioxide content of the feed gas subject to the minimum operating temperature is usually kept below 50 ppm. The amine process and the molecular sieve adsorption process are the most widely used methods for carbon dioxide removal. The amine process involves adsorption of the impurity by a lean aqueous organic amine solution. With sufficient amine recirculation rate, the carbon dioxide in the treated gas can be reduced to less than 25 ppm. Oxygen is removed by a catalytic reaction with hydrogen to form water.

STORAGE AND TRANSFER SYSTEMS

Storage vessels range in type from low-performance containers, insulated by rigid foam or fibrous insulation where the liquid in the container boils away in a few hours, up to high-performance containers,

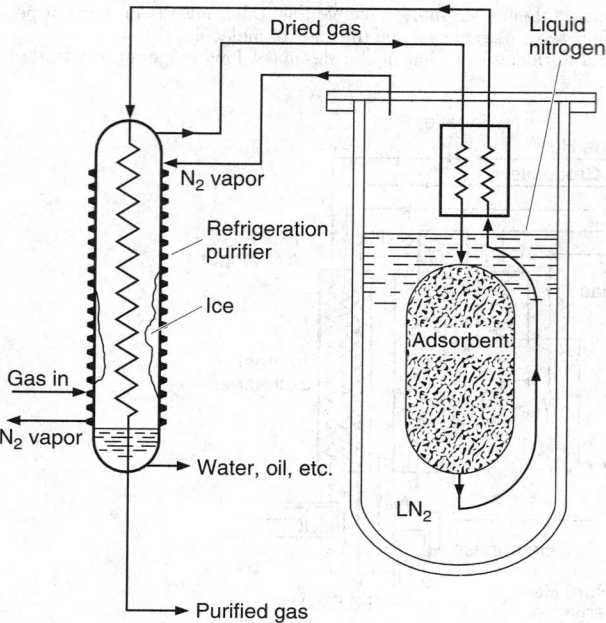

FIG. 11-120 Purifier using refrigeration and adsorption schemes in series.

insulated with multilayer insulations where less than 0.1 percent of the fluid contents is evaporated per day. In the more effective units, the storage container consists of an inner vessel which encloses the cryogenic fluid to be stored and an outer vessel or vacuum jacket. The latter maintains the vacuum necessary to make the insulation effective and at the same time serves as a vapor barrier to the migration of water and other condensibles to the cold surface of the inner vessel. Improvements have been made in the insulation used in these containers, but the vacuum-insulated double-walled Dewar is still the basic idea for high-performance cryogenic-fluid container designer.

Insulation Principles   The effectiveness of a liquefier or refrigerator is highly dependent upon the heat leak entering such a system. Since heat removal becomes more costly with a lowering in temperature as demonstrated by the Carnot limitation, most cryogenic systems employ some form of insulation to minimize the effect. The insulation strategy is to minimize radiative heat transfer, minimize convective heat transfer, and use only a minimum of solid conductance media. Factors considered in the selection of the most suitable insulation include its ruggedness, convenience, volume, weight, ease of fabrication and handling, and thermal effectiveness and cost. It is common practice to use an experimentally obtained apparent thermal conductivity to characterize the thermal effectiveness of various insulations. Typical k_a values for insulations used in cryogenic service are listed in Table 11-26.

TABLE 11-26   Representative Apparent Thermal Conductivity Values

Type of insulation	k_a, mW/m·K (77–300K)
Pure vacuum, 1.3×10^{-4} Pa	5
Foam insulation	26–35
Nonevacuated powders (perlite, silica aerogel)	19–44
Evacuated powders and fibers (1.3×10^{-1} Pa)	1–2
Opacified powdered insulations (1.3×10^{-1} Pa)	3.5×10^{-1}
Multilayer insulations (1.3×10^{-4} Pa)	$1.7–4 \times 10^{-2}$

Types of Insulation   Cryogenic insulations have generally been divided into five general categories: high vacuum, multilayer insulation, powder, foam, and special insulations. Each is discussed in turn in the following sections.

Vacuum Insulation   Heat transport across an evacuated space (1.3×10^{-4} Pa or lower), is by radiation and by conduction through the residual gas. The heat transfer by radiation generally is predominant and can be approximated by

$$\frac{\dot{Q}_r}{A_1} = \sigma \, (T_2^4 - T_1^4) \left[\frac{1}{e_1} + \frac{A_1}{A_2} \left(\frac{1}{e_2} - 1 \right) \right]^{-1} \quad (11\text{-}119)$$

where $\dot{Q}_r/A_1$ is the heat transfer by radiation-per-unit area, σ is the Stefan-Boltzmann constant, and e is the emissivity of the surfaces. The subscript 2 refers to the hot surface and the subscript 1 refers to the cold surface. The bracketed term on the right-hand side of this relation is designated as the overall emissivity factor, F_e.

The insertion of low-emissivity floating shields within the evacuated space can effectively reduce the heat transport by radiation. The effect of the shields is to greatly reduce the emissivity factor. For example, for N shields or ($N + 2$) surfaces, an emissivity of the outer and inner surface of e_o, and an emissivity of the shields of e_s, the emissivity factor reduces to

$$\left[2 \left(\frac{1}{e_o} + \frac{1}{e_s} - 1 \right) + \frac{(N-1)(2 - e_s)}{e_s} \right]^{-1} \quad (11\text{-}120)$$

In essence, one properly located low-emissivity shield can reduce the radiant heat transfer to around one-half of the rate without the shield, two shields can reduce this to around one-fourth of the rate without the shield, and so on.

Multilayer Insulation   Multilayer insulation consists of alternating layers of highly reflecting material, such as aluminum foil or aluminized Mylar, and a low-conductivity spacer material or insulator, such as fiberglass mat or paper, glass fabric, or nylon net, all under high vacuum. When properly applied at the optimum density, this type of insulation can have an apparent thermal conductivity as low as 10 to 50 µW/m·K between 20 and 300 K.

For a highly evacuated (on the order of 1.3×10^{-4} Pa) multilayer insulation, heat is transferred primarily by radiation and solid conduction through the spacer material. The apparent thermal conductivity of the insulation material under these conditions may be determined from

$$k_a = \frac{1}{N/\Delta x}\left\{h_s + \frac{\sigma e T_2^3}{2-e}\left[1+\left(\frac{T_1}{T_2}\right)^2\right]\left(1+\frac{T_1}{T_2}\right)\right\} \quad (11\text{-}121)$$

where $N/\Delta x$ is the number of complete layers (reflecting shield plus spacer) of the insulation-per-unit thickness, h_s is the solid conductance of the spacer material, σ is the Stefan-Boltzmann constant, e is the effective emissivity of the reflecting shield, and T_2 and T_1 are the temperatures of the warm and cold sides of the insulation, respectively. It is evident that the apparent thermal conductivity can be reduced by increasing the layer density up to a certain point. It is not obvious from the above relation that a compressive load affects the apparent thermal conductivity and thus the performance of a multilayer insulation. However, under a compressive load the solid conductance increases much more rapidly than $N/\Delta x$ resulting in an overall increase in k_a. Plots of heat flux versus compressive load on a logarithmic scale result in straight lines with slopes between 0.5 and 0.67.

The effective thermal conductivity values generally obtained in practice are at least a factor of two greater than the one-dimensional thermal conductivity values measured in the laboratory with carefully controlled techniques. This degradation in insulation thermal performance is caused by the combined presence of edge exposure to isothermal boundaries, gaps, joints, or penetrations in the insulation blanket required for structural supports, fill and vent lines, and high lateral thermal conductivity of these insulation systems.

Powder Insulation   A method of realizing some of the benefits of multiple floating shields without incurring the difficulties of awkward structural complexities is to use evacuated powder insulation. The penalty incurred in the use of this type of insulation, however, is a tenfold reduction in the overall thermal effectiveness of the insulation system over that obtained for multilayer insulation. In applications where this is not a serious factor, such as LNG storage facilities, and investment cost is of major concern, even unevacuated powder-insulation systems have found useful applications. The variation in apparent mean thermal conductivity of several powders as a function of interstitial gas pressure is shown in the familiar S-shaped curves of Fig. 11-121.

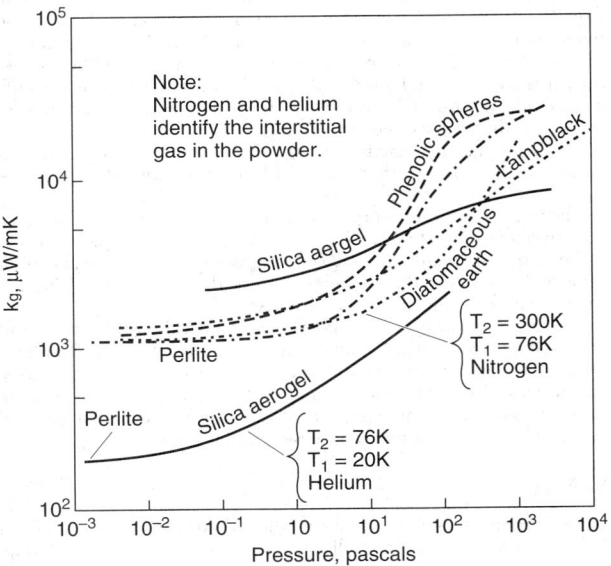

The apparent thermal conductivity of powder insulation at cryogenic temperatures is generally obtained from

$$k_a = \frac{k_g}{1 - V_r(1 - k_g/k_s)} \quad (11\text{-}122)$$

where k_g is the thermal conductivity of the gas within the insulation, k_s is the thermal conductivity of the powder, and V_r is the ratio of solid volume to the total volume. The amount of heat transport due to radiation through the powders can be reduced by the addition of metallic powders. A mixture containing approximately 40 to 50 wt % of a metallic powder gives the optimum performance.

Foam Insulation   Since foams are not homogeneous materials, their apparent thermal conductivity is dependent upon the bulk density of the insulation, the gas used to foam the insulation, and the mean temperature of the insulation. Heat conduction through a foam is determined by convection and radiation within the cells and by conduction in the solid structure. Evacuation of a foam is effective in reducing its thermal conductivity, indicating a partially open cellular structure, but the resulting values are still considerably higher than either multilayer or evacuated powder insulations.

Data on the thermal conductivity for a variety of foams used at cryogenic temperatures have been presented by Kropschot (*Cryogenic Technology*, R. W. Vance, ed., Wiley, New York, 1963, p. 239). Of all the foams, polyurethane and polystyrene have received the widest use at low temperatures. The major disadvantage of foams is that they tend to crack upon repeated thermal cycling and lose their insulation value.

Storage and Transfer Systems   In general, heat leak into a storage or transfer system for a cryogen is by (1) radiation and conduction through the insulation, and (2) conduction through any inner shell or transfer-line supports, piping leads, and access ports. Conduction losses are reduced by introducing long heat-leak paths, by making the cross sections for heat flow small, and by using materials with low thermal conductivity. Radiation losses, a major factor in the heat leak through insulations, are reduced with the use of radiation shields, such as multilayer insulation, boil-off vapor-cooled shields, and opacifiers in powder insulation.

Several considerations must be met when designing the inner vessel. The material of construction selected must be compatible with the stored cryogen. Nine percent nickel steels are acceptable for the higher-boiling cryogens ($T > 75$ K) while many aluminum alloys and austenitic steels are usually structurally acceptable throughout the entire temperature range. Because of its high thermal conductivity, aluminum is not a recommended material for piping and supports that must cross the insulation space. A change to a material of lower thermal conductivity for this purpose introduces a transition joint of a dissimilar material. Since such transition joints are generally mechanical in nature, leaks into the vacuum space develop upon repeated temperature cycling. In addition, the larger thermal coefficient of expansion of aluminum can pose still further support and cooldown problems.

Economic and cooldown considerations dictate that the shell of the storage container be as thin as possible. As a consequence, the inner container is designed to withstand only the internal pressure and bending forces while stiffening rings are used to support the weight of the fluid. The minimum thickness of the inner shell for a cylindrical vessel under such a design arrangement is given by Sec. VIII of the *ASME Boiler and Pressure Vessel Code*.

Since the outer shell of the storage container is subjected to atmospheric pressure on one side and evacuated conditions going down to 1.3×10^{-4} Pa on the other, consideration must be given to provide ample thickness of the material to withstand collapsing or buckling. Failure by elastic instability is covered by the ASME Code, in which design charts are available for the design of cylinders and spheres subjected to external pressure. Stiffening rings are also used on the outer shell to support the weight of the inner container and its contents as well as maintaining the sphericity of the shell.

The outer shell is normally constructed of carbon steel for economic reasons, unless aluminum is required to reduce the weight. Stainless-steel standoffs must be provided on the carbon steel outer shell for all piping penetrations to avoid direct contact with these penetrations when they are cold.

FIG. 11-121   Apparent mean thermal conductivities of several powder insulations as a function of interstitial gas pressure.

There are a variety of methods for supporting the inner shell within the outer shell and the cold transfer line within the outer line. Materials that have a high strength to thermal conductivity ratio are selected for these supports. Design of these supports for the inner shell must allow for shipping loads which may be several orders higher than in-service loads. Compression supports such as legs or pads may be used, but tension supports are more common. These may take the form of cables, welded straps, threaded bars, or a combination of these to provide restraint of the inner shell in several directions.

Most storage containers for cryogens are designed for a 10 percent ullage volume. The latter permits reasonable vaporization of the contents due to heat leak without incurring too rapid a buildup of the pressure in the container. This, in turn, permits closure of the container for short periods of time to either avoid partial loss of the contents or to transport flammable or hazardous cryogens safely from one location to another.

CRYOGENIC INSTRUMENTATION

Even though the combined production of cryogenic nitrogen and oxygen exceeds the production of any other chemical in the United States, the cryogenic industry does not appear to warrant a separate product line of instruments for diagnostic and control purposes. Low-temperature thermometry is the one exception. The general approach generally is that instruments developed for the usual CPI needs must be modified or accepted as is for cryogenic use.

Quite often problems arise when instruments for normal service are subjected to low temperature use. Since some metals become brittle at low temperatures, the instrument literally falls apart. Elastomeric gaskets and seals contract faster with decreasing temperatures than the surrounding metal parts, and the seal often is lost. Even hermetically sealed instruments can develop pin holes or small cracks to permit cryogenic liquids to enter these cases with time. Warming the instrument causes the trapped liquid to vaporize, sometimes generating excessive gas pressure and failure of the case.

Therefore, the first task in adapting normal instruments to cryogenic service is simply to give them a severe thermal shock by immersing them in liquid nitrogen repeatedly, and checking for mechanical integrity. This is the general issue; specific issues according to each type of measurement are discussed below.

Pressure This parameter is usually measured by the flush-mounted pressure transducer which consists of a force-summing device (bellow, diaphragm, bourdon tube, etc.) that translates the pressure into a displacement. The latter is then measured by an analog device (strain gage, piezoelectric crystal, variable distance between capacitor plates, and the like). Since these elements are likely to be made of different materials (bronze diaphragm, stainless-steel case, semiconductor strain gage), each will react to the temperature change in a different way. This is especially serious during cooldown, when the transient nature of material and construction prohibits all of the pressure-gage elements from being at the same temperature at the same time. Under steady-state conditions it is often possible to provide some temperature compensation through the well-known instrument technique of common-mode-rejection. Such compensation is generally not successful during transient temperature fluctuations. Only two courses of action are open: (1) hand-check each type of pressure transducer for thermal noise by thermally shocking it with immersion in liquid nitrogen; and (2) simplify the pressure-transducer construction to eliminate differences between materials. Some success has been observed in the latter area by manufacturers who make very small pressure sensing elements from a single semiconductor chip. The miniature size of these devices helps to reduce or eliminate temperature gradients across the device. The single-element nature of the pressure-gage assembly reduces differences in materials of construction.

Liquid Level The measurements for dense fluids such as liquid oxygen and liquid nitrogen are made in the conventional CPI approach using floats. Sight glasses cannot be used since radiation and thermal conduction would cause the cryogenic fluid within the sight glass to boil. The very light cryogens, liquid helium and liquid hydrogen, cannot sustain a float. Liquid hydrogen has the density of Styrofoam,™ about 70 g/l, making floating devices impractical. Some electrical analog is used for hydrogen and helium, most frequently a linear concentric-tube electrical capacitor. The dielectric constant of cryogens is related to their density by the Clausius-Mosotti relation. As the liquid level rises, the greater dielectric constant of the liquid between the tubes causes the overall capacitance to vary in a linear fashion. For best accuracy, these capacitance liquid-level measuring devices should be calibrated in place.

Flow The measurement of cryogenic fluids is most troublesome. Flow rate is not a natural physical parameter, like temperature, but is a derived quantity. A measurement of mass (or volume) must be made over a time interval to derive the flow rate. Because of this, any flow meter is only as good as its calibration. At this time, there is no national capability for calibrating cryogenic flowmeters. From data developed early in the nation's space program, considerable confidence has been developed in turbine-type flowmeters and in pressure-drop-type flowmeters. If all the usual ASTM guidelines are followed for meter installation, and if adequate temperature corrections are applied to changes in dimensions, then such meters can have an accuracy of ± 1 percent of their water calibrations. For very small flow applications, the Coriolis meters are promising. Vortex shedding flow meters appear useful for very large flow rates. Nonetheless, an actual calibration on the cryogen of interest is the only proof of accuracy.

Temperature The level of the temperature measurement (4 K, 20 K, 77 K, or higher) is the first issue to be considered. The second issue is the range needed (e.g., a few degrees around 90 K or 1 to 400 K). If the temperature level is that of air separation or liquefacting of natural gas (LNG), then the favorite choice is the platinum resistance thermometer (PRT). Platinum, as with all pure metals, has an electrical resistance that goes to zero as the absolute temperature decreases to zero. Accordingly, the lower useful limit of platinum is about 20 K, or liquid hydrogen temperatures. Below 20 K, semiconductor thermometers (germanium-, carbon-, or silicon-based) are preferred. Semiconductors have just the opposite resistance-temperature dependence of metals—their resistance increases as the temperature is lowered, as fewer valence electrons can be promoted into the conduction band at lower temperatures. Thus, semiconductors are usually chosen for temperatures from about 1 to 20 K.

If the temperature range of interest is large, say 1 to 400 K, then diode thermometers are recommended. Diodes have other advantages compared to resistance thermometers. By contrast, diode thermometers are very much smaller and faster. By selection of diodes all from the same melt, they may be made interchangeable. That is, one diode has the same calibration curve as another, which is not always the case with either semiconductor or metallic-resistance thermometers. It is well known, however, that diode thermometers may rectify an ac field, and thus may impose a dc noise on the diode output. Adequate shielding is required.

Special applications, such as in high-magnetic fields, require special thermometers. The carbon-glass and strontium-titinate resistance thermometers have the least magnetoresistance effects.

Thermocouples are unsurpassed for making temperature-difference measurements. The thermoelectric power of thermocouple materials makes them adequate for use at liquid-air temperatures and above. At 20 K and below, the thermoelectric power drops to a few μV/K, and their use in this range is as much art as science.

A descriptive flowchart has been prepared by Sparks (*Materials at Low Temperatures*, ASM, Metal Park, OH, 1983) to show the temperature range of cryogenic thermometers in general use today. Parese and Molinar (*Modern Gas-Based Temperature and Pressure Measurements*, Plenum, New York, 1992) provide details on gas- and vapor-pressure thermometry at these temperatures.

SAFETY

Past experience has shown that cryogenic fluids can be used safely in industrial environments as well as in typical laboratories provided all facilities are properly designed and maintained, and personnel handling these fluids are adequately trained and supervised. There are many hazards associated with cryogenic fluids. However, the principal ones are those associated with the response of the human body and the surroundings to the fluids and their vapors, and those associated with reactions between the fluids and their surroundings.

Physiological Hazards   Severe cold "burns" may be inflicted if the human body comes in contact with cryogenic fluids or with surfaces cooled by cryogenic fluids. Damage to the skin or tissue is similar to an ordinary burn. Because the body is composed mainly of water, the low temperature effectively freezes the tissue—damaging or destroying it. The severity of the burn depends upon the contact area and the contact time with prolonged contact resulting in deeper burns. Cold burns are accompanied by stinging sensations and pain similar to those of ordinary burns. The ordinary reaction is to withdraw that portion of the body that is in contact with the cold surface. Severe burns are seldom sustained if withdrawal is possible. Cold gases may not be damaging if the turbulence in the gas is low, particularly since the body can normally adjust for a heat loss of 95 J/m^2s for an area of limited exposure. If the heat loss becomes much greater than this, the skin temperature drops and freezing of the affected area may ensue. Freezing of facial tissue will occur in about 100 s if the heat loss is 2,300 J/m^2s.

Materials and Construction Hazards   Construction materials for noncryogenic service usually are chosen on the basis of tensile strength, fatigue life, weight, cost, ease of fabrication, corrosion resistance, and so on. When working with low temperatures the designer must consider the ductility of the material since low temperatures, as noted earlier, have the effect of making some construction materials brittle or less ductile. Some materials become brittle at low temperatures but still can absorb considerable impact, while others become brittle and lose their impact strength.

Flammability and Explosion Hazards   In order to have a fire or an explosion requires the combination of an oxidant, a fuel, and an ignition source. Generally the oxidizer will be oxygen. The latter may be available from a variety of sources including leakage or spillage, condensation of air on cryogenically cooled surfaces below 90 K, and buildup, as a solid impurity in liquid hydrogen. The fuel may be almost any noncompatible material or flammable gas; compatible materials can also act as fuels in the presence of extreme heat (strong

ignition sources). The ignition source may be a mechanical or electrostatic spark, flame, impact, heat by kinetic effects, friction, chemical reaction, and so on. Certain combinations of oxygen, fuel, and ignition sources will always result in fire or explosion. The order of magnitude of flammability and detonability limits for fuel-oxidant gaseous mixtures of two widely used cryogens is shown in Table 11-27.

TABLE 11-27   Flammability and Detonability Limits of Hydrogen and Methane Gas

Mixture	Flammability Limits (mol %)	Detonability Limits (mol %)
H_2-air	4–75	20–65
H_2-O_2	4–95	15–90
CH_4-air	5–15	6–14
CH_4-O_2	5–61	10–50

High-Pressure Gas Hazards   Potential hazards also exist in highly compressed gases because of the stored energy. In cryogenic systems such high pressures are obtained by gas compression during liquefaction or refrigeration, by pumping of liquids to high pressure followed by evaporation, and by confinement of cryogenic liquids with subsequent evaporation. If this confined gas is suddenly released through a rupture or break in a line, a significant thrust may be experienced. For example, the force generated by rupturing a 2.5-cm diameter valve located on a 13.9-MPa pressurized gas cylinder would be over 6670 N.

SUMMARY

It is obvious that the best designed facility is no better than the attention that is paid to safety. The latter is not considered once and forgotten. Rather, it is an ongoing activity that requires constant attention to every conceivable hazard that might be encountered. Because of its importance, safety, particularly at low temperatures, has received a large focus in the literature with its own safety manual prepared by NIST as well as by the British Cryogenics Council.

EVAPORATORS

GENERAL REFERENCES: Badger and Banchero, *Introduction to Chemical Engineering,* McGraw-Hill, New York, 1955. Standiford, *Chem. Eng., 70,* 158–176 (Dec. 9, 1963). *Testing Procedure for Evaporators,* American Institute of Chemical Engineers, 1979. *Upgrading Evaporators to Reduce Energy Consumption,* ERDA Technical Information Center, Oak Ridge, Tenn., 1977.

PRIMARY DESIGN PROBLEMS

Heat Transfer   This is the most important single factor in evaporator design, since the heating surface represents the largest part of evaporator cost. Other things being equal, the type of evaporator selected is the one having the highest heat-transfer cost coefficient under desired operating conditions in terms of $J/s \cdot K$ (British thermal units per hour per degree Fahrenheit) per dollar of installed cost. When power is required to induce circulation past the heating surface, the coefficient must be even higher to offset the cost of power for circulation.

Vapor-Liquid Separation   This design problem may be important for a number of reasons. The most important is usually prevention of entrainment because of value of product lost, pollution, contamination of the condensed vapor, or fouling or corrosion of the surfaces on which the vapor is condensed. Vapor-liquid separation in the vapor head may also be important when spray forms deposits on the walls, when vortices increase head requirements of circulating pumps, and when short circuiting allows vapor or unflashed liquid to be carried back to the circulating pump and heating element.

Evaporator performance is rated on the basis of **steam economy—** kilograms of solvent evaporated per kilogram of steam used. Heat is

required (1) to raise the feed from its initial temperature to the boiling temperature, (2) to provide the minimum thermodynamic energy to separate liquid solvent from the feed, and (3) to vaporize the solvent. The first of these can be changed appreciably by reducing the boiling temperature or by heat interchange between the feed and the residual product and/or condensate. The greatest increase in steam economy is achieved by reusing the vaporized solvent. This is done in a **multiple-effect evaporator** by using the vapor from one effect as the heating medium for another effect in which boiling takes place at a lower temperature and pressure. Another method of increasing the utilization of energy is to employ a **thermocompression** evaporator, in which the vapor is compressed so that it will condense at a temperature high enough to permit its use as the heating medium in the same evaporator.

Selection Problems   Aside from heat-transfer considerations, the selection of type of evaporator best suited for a particular service is governed by the characteristics of the feed and product. Points that must be considered are crystallization, salting and scaling, product quality, corrosion, and foaming. In the case of a **crystallizing evaporator,** the desirability of producing crystals of a definite uniform size usually limits the choice to evaporators having a positive means of circulation. **Salting,** which is the growth on body and heating-surface walls of a material having a solubility that increases with increase in temperature, is frequently encountered in crystallizing evaporators. It can be reduced or eliminated by keeping the evaporating liquid in close or frequent contact with a large surface area of crystallized solid. **Scaling** is the deposition and growth on body walls, and especially on heating surfaces, of a material undergoing an irreversible chemical reaction in the evaporator or having a solubility that decreases with an increase in temperature. Scaling can be reduced or eliminated in the same general manner as salting. Both salting and scaling liquids are usually best handled in evaporators that do not depend on boiling to induce circulation. **Fouling** is the formation of deposits other than salt or scale and may be due to corrosion, solid matter entering with the feed, or deposits formed by the condensing vapor.

Product Quality   Considerations of product quality may require low holdup time and low-temperature operation to avoid thermal degradation. The low holdup time eliminates some types of evaporators, and some types are also eliminated because of poor heat-transfer characteristics at low temperature. Product quality may also dictate special materials of construction to avoid metallic contamination or a catalytic effect on decomposition of the product. **Corrosion** may also influence evaporator selection, since the advantages of evaporators having high heat-transfer coefficients are more apparent when expensive materials of construction are indicated. Corrosion and erosion are frequently more severe in evaporators than in other types of equipment because of the high liquid and vapor velocities used, the frequent presence of solids in suspension, and the necessary concentration differences.

EVAPORATOR TYPES AND APPLICATIONS

Evaporators may be classified as follows:
1.   Heating medium separated from evaporating liquid by tubular heating surfaces.
2.   Heating medium confined by coils, jackets, double walls, flat plates, etc.
3.   Heating medium brought into direct contact with evaporating liquid.
4.   Heating by solar radiation.

By far the largest number of industrial evaporators employ tubular heating surfaces. Circulation of liquid past the heating surface may be induced by boiling or by mechanical means. In the latter case, boiling may or may not occur at the heating surface.

Forced-Circulation Evaporators (Fig. 11-122 _a,b,c_)   Although it may not be the most economical for many uses, the forced-circulation (FC) evaporator is suitable for the widest variety of evaporator applications. The use of a pump to ensure circulation past the heating surface makes possible separating the functions of heat transfer, vapor-liquid separation, and crystallization. The pump withdraws liquor from the flash chamber and forces it through the heating element back to the flash chamber. Circulation is maintained regardless of the evaporation rate; so this type of evaporator is well suited to **crystallizing operation,** in which solids must be maintained in suspension at all times. The liquid velocity past the heating surface is limited only by the pumping power needed or available and by accelerated corrosion and erosion at the higher velocities. **Tube velocities** normally range from a minimum of about 1.2 m/s (4 ft/s) in salt evaporators with copper or brass tubes and liquid containing 5 percent or more solids up to about 3 m/s (10 ft/s) in caustic evaporators having nickel tubes and liquid containing only a small amount of solids. Even higher velocities can be used when corrosion is not accelerated by erosion.

Highest heat-transfer coefficients are obtained in FC evaporators when the liquid is allowed to boil in the tubes, as in the type shown in Fig. 11-122_a_. The heating element projects into the vapor head, and the liquid level is maintained near and usually slightly below the top tube sheet. This type of FC evaporator is not well suited to salting solutions because boiling in the tubes increases the chances of salt deposit on the walls and the sudden flashing at the tube exits promotes excessive nucleation and production of fine crystals. Consequently, this type of evaporator is seldom used except when there are headroom limitations or when the liquid forms neither salt nor scale.

By far the largest number of forced-circulation evaporators are of the submerged-tube type, as shown in Fig. 11-122_b_. The heating element is placed far enough below the liquid level or return line to the flash chamber to prevent boiling in the tubes. Preferably, the hydrostatic head should be sufficient to prevent boiling even in a tube that is plugged (and hence at steam temperature), since this prevents salting of the entire tube. Evaporators of this type sometimes have hori-

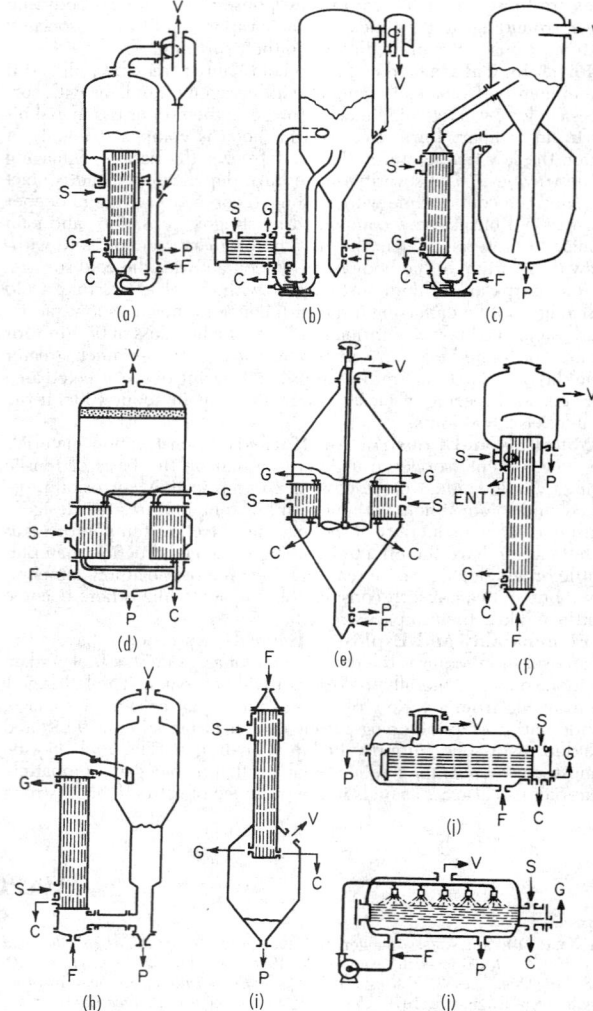

FIG. 11-122   Evaporator types. (_a_) Forced circulation. (_b_) Submerged-tube forced circulation. (_c_) Oslo-type crystallizer. (_d_) Short-tube vertical. (_e_) Propeller calandria. (_f_) Long-tube vertical. (_g_) Recirculating long-tube vertical. (_h_) Falling film. (_i,j_) Horizontal-tube evaporators. C = condensate; F = feed; G = vent; P = product; S = steam; V = vapor; ENT'T = separated entrainment outlet.

zontal heating elements (usually two-pass), but the vertical single-pass heating element is used whenever sufficient headroom is available. The vertical element usually has a lower friction loss and is easier to clean or retube than a horizontal heater. The submerged-tube forced-circulation evaporator is relatively immune to salting in the tubes, since no supersaturation is generated by evaporation in the tubes. The tendency toward scale formation is also reduced, since supersaturation in the heating element is generated only by a controlled amount of heating and not by both heating and evaporation.

The type of **vapor** head used with the FC evaporator is chosen to suit the product characteristics and may range from a simple centrifugal separator to the crystallizing chambers shown in Fig. 11-122_b_ and _c_. Figure 11-122_b_ shows a type frequently used for common salt. It is designed to circulate a slurry of crystals throughout the system. Figure 11-122_c_ shows a submerged-tube FC evaporator in which heating, flashing, and crystallization are completely separated. The crystallizing solids are maintained as a fluidized bed in the chamber below the vapor head and little or no solids circulate through the heater and flash chamber. This type is well adapted to growing coarse crystals, but the crystals usually approach a spherical shape, and careful design is required to avoid production of tines in the flash chamber.

In a submerged-tube FC evaporator, all heat is imparted as sensible heat, resulting in a temperature rise of the circulating liquor that reduces the overall temperature difference available for heat transfer. Temperature rise, tube proportions, tube velocity, and head requirements on the circulating pump all influence the selection of circulation rate. Head requirements are frequently difficult to estimate since they consist not only of the usual friction, entrance and contraction, and elevation losses when the return to the flash chamber is above the liquid level but also of increased friction losses due to flashing in the return line and vortex losses in the flash chamber. Circulation is sometimes limited by vapor in the pump suction line. This may be drawn in as a result of inadequate vapor-liquid separation or may come from vortices near the pump suction connection to the body or may be formed in the line itself by short circuiting from heater outlet to pump inlet of liquor that has not flashed completely to equilibrium at the pressure in the vapor head.

Advantages of forced-circulation evaporators:
1. High heat-transfer coefficients
2. Positive circulation
3. Relative freedom from salting, scaling, and fouling

Disadvantages of forced-circulation evaporators:
1. High cost
2. Power required for circulating pump
3. Relatively high holdup or residence time

Best applications of forced-circulation evaporators:
1. Crystalline product
2. Corrosive solutions
3. Viscous solutions

Frequent difficulties with forced-circulation evaporators:
1. Plugging of tube inlets by salt deposits detached from walls of equipment
2. Poor circulation due to higher than expected head losses
3. Salting due to boiling in tubes
4. Corrosion-erosion

Short-Tube Vertical Evaporators (Fig. 11-122d)   This is one of the earliest types still in widespread commercial use. Its principal use at present is in the evaporation of cane-sugar juice. Circulation past the heating surface is induced by boiling in the tubes, which are usually 50.8 to 76.2 mm (2 to 3 in) in diameter by 1.2 to 1.8 m (4 to 6 ft) long. The body is a vertical cylinder, usually of cast iron, and the tubes are expanded into horizontal tube sheets that span the body diameter. The circulation rate through the tubes is many times the feed rate; so there must be a return passage from above the top tube sheet to below the bottom tube sheet. Most commonly used is a central well or **downtake** as shown in Fig. 11-122d. So that friction losses through the downtake do not appreciably impede circulation up through the tubes, the area of the downtake should be of the same order of magnitude as the combined cross-sectional area of the tubes. This results in a downtake almost half of the diameter of the tube sheet.

Circulation and heat transfer in this type of evaporator are strongly affected by the liquid "level." Highest heat-transfer coefficients are achieved when the level, as indicated by an external gauge glass, is only about halfway up the tubes. Slight reductions in level below the optimum result in incomplete wetting of the tube walls with a consequent increased tendency to foul and a rapid reduction in capacity. When this type of evaporator is used with a liquid that can deposit salt or scale, it is customary to operate with the liquid level appreciably higher than the optimum and usually appreciably above the top tube sheet.

Circulation in the standard short-tube vertical evaporator is dependent entirely on boiling, and when boiling stops, any solids present settle out of suspension. Consequently, this type is seldom used as a crystallizing evaporator. By installing a propeller in the downtake, this objection can be overcome. Such an evaporator, usually called a **pro-**peller calandria,** is illustrated in Fig. 11-122e. The propeller is usually placed as low as possible to reduce cavitation and is shrouded by an extension of the downtake well. The use of the propeller can sometimes double the capacity of a short-tube vertical evaporator. The evaporator shown in Fig. 11-122e includes an elutriation leg for salt manufacture similar to that used on the FC evaporator of Fig. 11-122b. The shape of the bottom will, of course, depend on the particular application and on whether the propeller is driven from above or below. To avoid salting when the evaporator is used for crystallizing solutions, the liquid level must be kept appreciably above the top tube sheet.

Advantages of short-tube vertical evaporators:
1. High heat-transfer coefficients at high temperature differences
2. Low headroom
3. Easy mechanical descaling
4. Relatively inexpensive

Disadvantages of short-tube vertical evaporators:
1. Poor heat transfer at low temperature differences and low temperature
2. High floor space and weight
3. Relatively high holdup
4. Poor heat transfer with viscous liquids

Best applications of short-tube vertical evaporators:
1. Clear liquids
2. Crystalline product if propeller is used
3. Relatively noncorrosive liquids, since body is large and expensive if built of materials other than mild steel or cast iron
4. Mild scaling solutions requiring mechanical cleaning, since tubes are short and large in diameter

Long-Tube Vertical Evaporators (Fig. 11-122f, g, h)   More total evaporation is accomplished in this type than in all others combined because it is normally the **cheapest per unit of capacity.** The long-tube vertical (LTV) evaporator consists of a simple one-pass vertical shell-and-tube heat exchanger discharging into a relatively small vapor head. Normally, no liquid level is maintained in the vapor head, and the residence time of liquor is only a few seconds. The tubes are usually about 50.8 mm (2 in) in diameter but may be smaller than 25.4 mm (1 in). Tube length may vary from less than 6 to 10.7 m (20 to 35 ft) in the rising film version and to as great as 20 m (65 ft) in the falling film version. The evaporator is usually operated single-pass, concentrating from the feed to discharge density in just the time that it takes the liquid and evolved vapor to pass through a tube. An extreme case is the caustic high concentrator, producing a substantially anhydrous product at 370°C (700°F) from an inlet feed of 50 percent NaOH at 149°C (300°F) in one pass up 22-mm- (8/8-in-) outside-diameter nickel tubes 6 m (20 ft) long. The largest use of LTV evaporators is for concentrating black liquor in the pulp and paper industry. Because of the long tubes and relatively high heat-transfer coefficients, it is possible to achieve higher single-unit capacities in this type of evaporator than in any other.

The LTV evaporator shown in Fig. 11-122f is typical of those commonly used, especially for black liquor. Feed enters at the bottom of the tube and starts to boil partway up the tube, and the mixture of liquid and vapor leaving at the top at high velocity impinges against a deflector placed above the tube sheet. This deflector is effective both as a primary separator and as a foam breaker.

In many cases, as when the ratio of feed to evaporation or the ratio of feed to heating surface is low, it is desirable to provide for **recirculation of product** through the evaporator. This can be done in the type shown in Fig. 11-122f by adding a pipe connection between the product line and the feed line. Higher recirculation rates can be achieved in the type shown in Fig. 11-122g, which is used widely for condensed milk. By extending the enlarged portion of the vapor head still lower to provide storage space for liquor, this type can be used as a batch evaporator.

Liquid temperatures in the tubes of an LTV evaporator are far from uniform and are difficult to predict. At the lower end, the liquid is usually not boiling, and the liquor picks up heat as sensible heat. Since entering liquid velocities are usually very low, true heat-transfer coef-

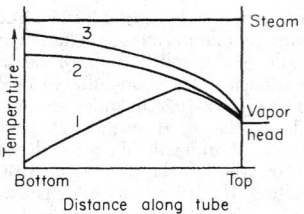

FIG. 11-123 Temperature variations in a long-tube vertical evaporator.

ficients are low in this nonboiling zone. At some point up the tube, the liquid starts to boil, and from that point on the liquid temperature decreases because of the reduction in static, friction, and acceleration heads until the vapor-liquid mixture reaches the top of the tubes at substantially vapor-head temperature. Thus the true temperature difference in the boiling zone is always less than the total temperature difference as measured from steam and vapor-head temperatures.

Although the true heat-transfer coefficients in the boiling zone are quite high, they are partially offset by the reduced temperature difference. The point in the tubes at which boiling starts and at which the maximum temperature is reached is sensitive to operating conditions, such as feed properties, feed temperature, feed rate, and heat flux. Figure 11-123 shows typical variations in liquid temperature in tubes of an LTV evaporator operating at a constant terminal temperature difference. Curve 1 shows the normal case in which the feed is not boiling at the tube inlet. Curve 2 gives an indication of the temperature difference lost when the feed enters at the boiling point. Curve 3 is for exactly the same conditions as curve 2 except that the feed contained 0.01 percent Teepol to reduce surface tension [Coulson and Mehta, *Trans. Inst. Chem. Eng.*, **31**, 208 (1953)]. The surface-active agent yields a more intimate mixture of vapor and liquid, with the result that liquid is accelerated to a velocity more nearly approaching the vapor velocity, thereby increasing the pressure drop in the tube. Although the surface-active agent caused an increase of more than 100 percent in the true heat-transfer coefficient, this was more than offset by the reduced temperature difference so that the net result was a reduction in evaporator capacity. This sensitivity of the LTV evaporator to changes in operating conditions is less pronounced at high than at low temperature differences and temperature levels.

The **falling-film** version of the LTV evaporator (Fig. 11-122*h*) eliminates these problems of hydrostatic head. Liquid is fed to the tops of the tubes and flows down the walls as a film. Vapor-liquid separation usually takes place at the bottom, although some evaporators of this type are arranged for vapor to rise through the tube countercurrently to the liquid. The pressure drop through the tubes is usually very small, and the boiling-liquid temperature is substantially the same as the vapor-head temperature. The falling-film evaporator is widely used for concentrating **heat-sensitive materials,** such as fruit juices, because the holdup time is very small, the liquid is not overheated during passage through the evaporator, and heat-transfer coefficients are high even at low boiling temperatures.

The principal problem with the falling-film LTV evaporator is that of **feed distribution** to the tubes. It is essential that all tube surfaces be wetted continually. This usually requires recirculation of the liquid unless the ratio of feed to evaporation is quite high. An alternative to the simple recirculation system of Fig. 11-122*h* is sometimes used when the feed undergoes an appreciable concentration change and the product is viscous and/or has a high boiling point rise. The feed chamber and vapor head are divided into a number of liquor compartments, and separate pumps are used to pass the liquor through the various banks of tubes in series, all in parallel as to steam and vapor pressures. The actual distribution of feed to the individual tubes of a falling-film evaporator may be accomplished by orifices at the inlet to each tube, by a perforated plate above the tube sheet, or by one or more spray nozzles.

Both rising- and falling-film LTV evaporators are generally unsuited to salting or severely scaling liquids. However, both are widely

used for black liquor, which presents a mild scaling problem, and also are used to carry solutions beyond saturation with respect to a crystallizing salt. In the latter case, deposits can usually be removed quickly by increasing the feed rate or reducing the steam rate in order to make the product unsaturated for a short time. The falling-film evaporator is not generally suited to liquids containing solids because of difficulty in plugging the feed distributors. However, it has been applied to the evaporation of saline waters saturated with $CaSO_4$ and containing 5 to 10 percent $CaSO_4$ seeds in suspension for scale prevention (Anderson, ASME Pap. 76-WA/Pwr-5, 1976).

Because of their simplicity of construction, compactness, and generally high heat-transfer coefficients, LTV evaporators are well suited to service with corrosive liquids. An example is the reconcentration of rayon spin-bath liquor, which is highly acid. These evaporators employ impervious graphite tubes, lead, rubber-covered or impervious graphite tube sheets, and rubber-lined vapor heads. Polished stainless-steel LTV evaporators are widely used for food products. The latter evaporators are usually similar to that shown in Fig. 11-122*g*, in which the heating element is at one side of the vapor head to permit easy access to the tubes for cleaning.

Advantages of long-tube vertical evaporators:
1. Low cost
2. Large heating surface in one body
3. Low holdup
4. Small floor space
5. Good heat-transfer coefficients at reasonable temperature differences (rising film)
6. Good heat-transfer coefficients at all temperature differences (falling film)

Disadvantages of long-tube vertical evaporators:
1. High headroom
2. Generally unsuitable for salting and severely scaling liquids
3. Poor heat-transfer coefficients of rising-film version at low temperature differences
4. Recirculation usually required for falling-film version

Best applications of long-tube vertical evaporators:
1. Clear liquids
2. Foaming liquids
3. Corrosive solutions
4. Large evaporation loads
5. High temperature differences—rising film, low temperature differences—falling film
6. Low-temperature operation—falling film
7. Vapor compression operation—falling film

Frequent difficulties with long-tube vertical evaporators:
1. Sensitivity of rising-film units to changes in operating conditions
2. Poor feed distribution to falling-film units

Horizontal-Tube Evaporators (Fig. 11-122*i*) In these types the steam is inside and the liquor outside the tubes. The submerged-tube version of Fig. 11-122*i* is seldom used except for the preparation of boiler feedwater. Low entrainment loss is the primary aim: the horizontal cylindrical shell yields a large disengagement area per unit of vessel volume. Special versions use deformed tubes between restrained tube sheets that crack off much of a scale deposit when sprayed with cold water. By showering liquor over the tubes in the version of Fig. 11-122*f* hydrostatic head losses are eliminated and heat-transfer performance is improved to that of the falling-film tubular type of Fig. 11-122*h*. Originally called the Lillie, this evaporator is now also called the spray-film or simply the horizontal-tube evaporator. Liquid distribution over the tubes is accomplished by sprays or perforated plates above the topmost tubes. Maintaining this distribution through the bundle to avoid overconcentrating the liquor is a problem unique to this type of evaporator. It is now used primarily for seawater evaporation.

Advantages of horizontal-tube evaporators:
1. Very low headroom
2. Large vapor-liquid disengaging area—submerged-tube type

3. Relatively low cost in small-capacity straight-tube type
4. Good heat-transfer coefficients
5. Easy semiautomatic descaling—bent-tube type

Disadvantages of horizontal-tube evaporators:
1. Unsuitable for salting liquids
2. Unsuitable for scaling liquids—straight-tube type
3. High cost—bent-tube type
4. Maintaining liquid distribution—film type

Best applications of horizontal-tube evaporators:
1. Limited headroom
2. Small capacity
3. Nonscaling nonsalting liquids—straight-tube type
4. Severely scaling liquids—bent-tube type

Miscellaneous Forms of Heating Surface   Special evaporator designs are sometimes indicated when heat loads are small, special product characteristics are desired, or the product is especially difficult to handle. **Jacketed kettles,** frequently with agitators, are used when the product is very viscous, batches are small, intimate mixing is required, and/or ease of cleaning is an important factor. Evaporators with steam in coiled **tubes** may be used for small capacities with scaling liquids in designs that permit "cold shocking," or complete withdrawal of the coil from the shell for manual scale removal. Other designs for scaling liquids employ flat-plate heat exchangers, since in general a scale deposit can be removed more easily from a flat plate than from a curved surface. One such design, the **channel-switching evaporator,** alternates the duty of either side of the heating surface periodically from boiling liquid to condensing vapor so that scale formed when the surface is in contact with boiling liquid is dissolved when the surface is next in contact with condensing vapor.

Agitated thin-film evaporators employ a heating surface consisting of one large-diameter tube that may be either straight or tapered, horizontal or vertical. Liquid is spread on the tube wall by a rotating assembly of blades that either maintain a close clearance from the wall or actually ride on the film of liquid on the wall. The expensive construction limits application to the most difficult materials. High agitation [on the order of 12 m/s (40 ft/s) rotor-tip speed] and power intensities of 2 to 20 kW m² (0.25 to 2.5 hp/ft²) permit handling extremely viscous materials. Residence times of only a few seconds permit concentration of heat-sensitive materials at temperatures and temperature differences higher than in other types [Mutzenberg, Parker, and Fischer. *Chem. Eng.,* **72,** 175–190 (Sept. 13, 1965)]. High feed-to-product ratios can be handled without recirculation.

Economic and process considerations usually dictate that agitated thin-film evaporators be operated in single-effect mode. Very high temperature differences can then be used: many are heated with Dowtherm or other high-temperature media. This permits achieving reasonable capacities in spite of the relatively low heat-transfer coefficients and the small surface that can be provided in a single tube [to about 20 m² (200 ft²)]. The structural need for wall thicknesses of 6 to 13 mm (¼ to ½ in) is a major reason for the relatively low heat-transfer coefficients when evaporating water-like materials.

Evaporators without Heating Surfaces   The **submerged-combustion** evaporator makes use of combustion gases bubbling through the liquid as the means of heat transfer. It consists simply of a tank to hold the liquid, a burner and gas distributor that can be lowered into the liquid, and a combustion-control system. Since there are no heating surfaces on which scale can deposit, this evaporator is well suited to use with severely scaling liquids. The ease of constructing the tank and burner of special alloys or nonmetallic materials makes practical the handling of highly corrosive solutions. However, since the vapor is mixed with large quantities of noncondensable gases, it is impossible to reuse the heat in this vapor, and installations are usually limited to areas of low fuel cost. One difficulty frequently encountered in the use of submerged-combustion evaporators is a high entrainment loss. Also, these evaporators cannot be used when control of crystal size is important.

Disk or **cascade evaporators** are used in the pulp and paper industry to recover heat and entrained chemicals from boiler stack gases and to effect a final concentration of the black liquor before it is burned in the boiler. These evaporators consist of a horizontal shaft on which are mounted disks perpendicular to the shaft or bars parallel to the shaft. The assembly is partially immersed in the thick black liquor so that films of liquor are carried into the hot-gas stream as the shaft rotates.

Some forms of **flash evaporators** require no heating surface. An example is a recrystallizing process for separating salts having normal solubility curves from salts having inverse solubility curves, as in separating sodium chloride from calcium sulfate [Richards, *Chem. Eng.,* **59**(3), 140 (1952)]. A suspension of raw solid feed in a recirculating brine stream is heated by direct steam injection. The increased temperature and dilution by the steam dissolve the salt having the normal solubility curve. The other salt remains undissolved and is separated from the hot solution before it is flashed to a lower temperature. The cooling and loss of water on flashing cause recrystallization of the salt having the normal solubility curve, which is separated from the brine before the brine is mixed with more solid feed for recycling to the heater. This system can be operated as a multiple effect by flashing down to the lower temperature in stages and using flash vapor from all but the last stage to heat the recycle brine by direct injection. In this process no net evaporation occurs from the total system, and the process cannot be used to concentrate solutions unless heating surfaces are added.

UTILIZATION OF TEMPERATURE DIFFERENCE

Temperature difference is the driving force for evaporator operation and usually is limited, as by compression ratio in vapor-compression evaporators and by available steam-pressure and heat-sink temperature in single- and multiple-effect evaporators. A fundamental objective of evaporator design is to make as much of this total temperature difference available for heat transfer as is economically justifiable. Some losses in temperature difference, such as those due to *boiling point rise* (BPR), are unavoidable. However, even these can be minimized, as by passing the liquor through effects or through different sections of a single effect in series so that only a portion of the heating surface is in contact with the strongest liquor.

Figure 11-124 shows approximate BPR losses for a number of process liquids. A correlation for concentrated solutions of many inorganic salts at the atmospheric pressure boiling point [Meranda and Furter, *J. Ch. and E. Data* **22**, 315-7 (1977)] is

$$BPR = 104.9 N_2^{1.14} \qquad (11\text{-}123)$$

where N_2 is the mole fraction of salts in solution. Correction to other pressures, when heats of solution are small, can be based on a constant ratio of vapor pressure of the solution to that of water at the same temperature.

The principal reducible loss in ΔT is that due to friction and to entrance and exit losses in vapor piping and entrainment separators. Pressure-drop losses here correspond to a reduction in condensing temperature of the vapor and hence a loss in available ΔT. These losses become most critical at the low-temperature end of the evaporator, both because of the increasing specific volume of the vapor and because of the reduced slope of the vapor-pressure curve. Sizing of vapor lines is part of the economic optimization of the evaporator, extra costs of larger vapor lines being balanced against savings in ΔT, which correspond to savings in heating-surface requirements. It should be noted that entrance and exit losses in vapor lines usually exceed by severalfold the straight-pipe friction losses, so they cannot be ignored.

VAPOR-LIQUID SEPARATION

Product losses in evaporator vapor may result from foaming, splashing, or entrainment. Primary separation of liquid from vapor is accomplished in the vapor head by making the horizontal plan area large enough so that most of the entrained droplets can settle out against the rising flow of vapor. Allowable velocities are governed by the Souders-Brown equation: $V = k\sqrt{(\rho_1 - \rho_v)/\rho_v}$, in which k depends on the size distribution of droplets and the decontamination factor F desired. For most evaporators and for F between 100 and 10,000, $k \cong$

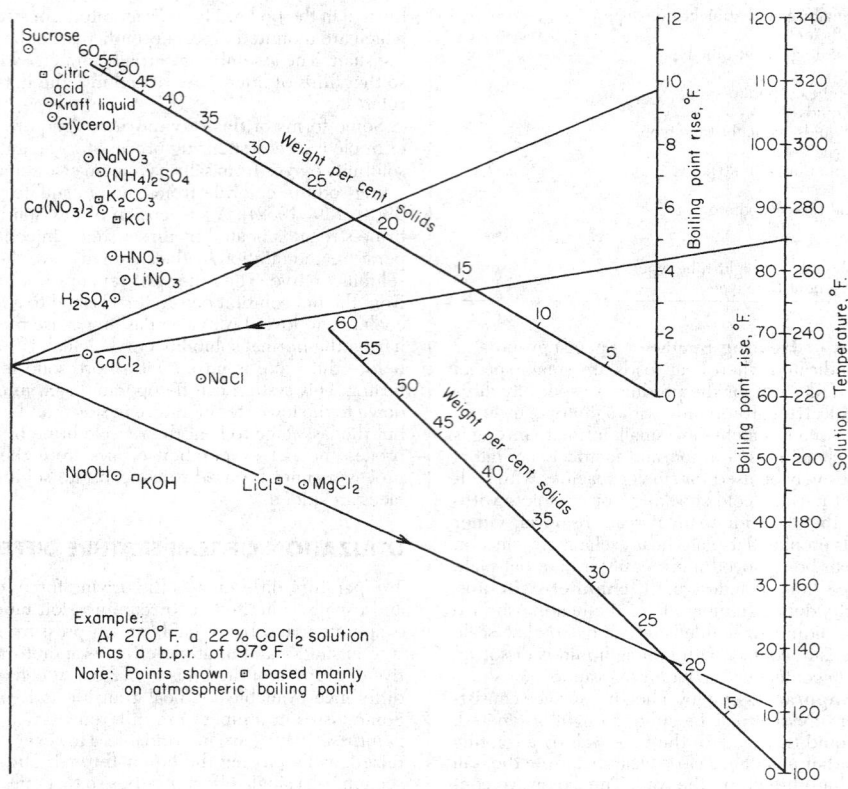

FIG. 11-124   Boiling-point rise of aqueous solutions. °C = 5/9 (°F − 32).

$0.245/(F − 50)^{0.4}$ (Standiford, *Chemical Engineers' Handbook*, 4th ed., McGraw-Hill, New York, 1963, p. 11–35). Higher values of k (to about 0.15) can be tolerated in the falling-film evaporator, where most of the entrainment separation occurs in the tubes, the vapor is scrubbed by liquor leaving the tubes, and the vapor must reverse direction to reach the outlet.

Foaming losses usually result from the presence in the evaporating liquid of colloids or of surface-tension depressants and finely divided solids. Antifoam agents are often effective. Other means of combating foam include the use of steam jets impinging on the foam surface, the removal of product at the surface layer, where the foaming agents seem to concentrate, and operation at a very low liquid level so that hot surfaces can break the foam. Impingement at high velocity against a baffle tends to break the foam mechanically, and this is the reason that the long-tube vertical, forced-circulation, and agitated-film evaporators are particularly effective with foaming liquids. Operating at lower temperatures and/or higher-dissolved solids concentrations may also reduce foaming tendencies.

Splashing losses are usually insignificant if a reasonable height has been provided between the liquid level and the top of the vapor head. The height required depends on the violence of boiling. Heights of 2.4 to 3.6 m (8 to 12 ft) or more are provided in short-tube vertical evaporators, in which the liquid and vapor leaving the tubes are projected upward. Less height is required in forced-circulation evaporators, in which the liquid is given a centrifugal motion or is projected downward as by a baffle. The same is true of long-tube vertical evaporators, in which the rising vapor-liquid mixture is projected against a baffle.

Entrainment losses by flashing are frequently encountered in an evaporator. If the feed is above the boiling point and is introduced above or only a short distance below the liquid level, entrainment losses may be excessive. This can occur in a short-tube-type evaporator if the feed is introduced at only one point below the lower tube

sheet (Kerr, Louisiana Agric. Expt. Stn. Bull. 149, 1915). The same difficulty may be encountered in forced-circulation evaporators having too high a temperature rise through the heating element and thus too wide a flashing range as the circulating liquid enters the body. Poor vacuum control, especially during startup, can cause the generation of far more vapor than the evaporator was designed to handle, with a consequent increase in entrainment.

Entrainment separators are frequently used to reduce product losses. There are a number of specialized designs available, practically all of which rely on a change in direction of the vapor flow when the vapor is traveling at high velocity. Typical separators are shown in Fig. 11-122, although not necessarily with the type of evaporator with which they may be used. The most common separator is the cyclone, which may have either a top or a bottom outlet as shown in Fig. 11-122a and b or may even be wrapped around the heating element of the next effect as shown in Fig. 11-122f. The separation efficiency of a cyclone increases with an increase in inlet velocity, although at the cost of some pressure drop, which means a loss in available temperature difference. Pressure drop in a cyclone is from 10 to 16 velocity heads [Lawrence, *Chem. Eng. Prog.*, **48**, 241 (1952)], based on the velocity in the inlet pipe. Such cyclones can be sized in the same manner as a cyclone dust collector (using velocities of about 30 m/s (100 ft/s) at atmospheric pressure) although sizes may be increased somewhat in order to reduce losses in available temperature difference.

Knitted wire mesh serves as an effective entrainment separator when it cannot easily be fouled by solids in the liquor. The mesh is available in woven metal wire of most alloys and is installed as a blanket across the top of the evaporator (Fig. 11-122d) or in a monitor of reduced diameter atop the vapor head. These separators have low-pressure drops, usually on the order of 13 mm (½ in) of water, and collection efficiency is above 99.8 percent in the range of vapor velocities from 2.5 to 6 m/s (8 to 20 ft/s) [Carpenter and Othmer, *Am. Inst. Chem.*

Eng. J., **1,** 549 (1955)]. Chevron (hook-and-vane) type separators are also used because of their higher-allowable velocities or because of their reduced tendency to foul with solids suspended in the entrained liquid.

EVAPORATOR ARRANGEMENT

Single-Effect Evaporators Single-effect evaporators are used when the required capacity is small, steam is cheap, the material is so corrosive that very expensive materials of construction are required, or the vapor is so contaminated that it cannot be reused. Single-effect evaporators may be operated in batch, semibatch, or continuous-batch modes or continuously. Strictly speaking, **batch evaporators** are ones in which filling, evaporating, and emptying are consecutive steps. This method of operation is rarely used since it requires that the body be large enough to hold the entire charge of feed and the heating element be placed low enough so as not to be uncovered when the volume is reduced to that of the product. The more usual method of operation is **semibatch,** in which feed is continually added to maintain a constant level until the entire charge reaches final density. **Continuous-batch** evaporators usually have a continuous feed and, over at least part of the cycle, a continuous discharge. One method of operation is to circulate from a storage tank to the evaporator and back until the entire tank is up to desired concentration and then finish in batches. **Continuous evaporators** have essentially continuous feed and discharge, and concentrations of both feed and product remain substantially constant.

Thermocompression The simplest means of reducing the energy requirements of evaporation is to compress the vapor from a single-effect evaporator so that the vapor can be used as the heating medium in the same evaporator. The compression may be accomplished by mechanical means or by a steam jet. In order to keep the compressor cost and power requirements within reason, the evaporator must work with a fairly narrow temperature difference, usually from about 5.5 to 11°C (10° to 20°F). This means that a large evaporator heating surface is needed, which usually makes the vapor-compression evaporator more expensive in first cost than a multiple-effect evaporator. However, total installation costs may be reduced when purchased power is the energy source, since the need for boiler and heat sink is eliminated. Substantial savings in operating cost are realized when electrical or mechanical power is available at a low cost relative to low-pressure steam, when only high-pressure steam is available to operate the evaporator, or when the cost of providing cooling water or other heat sink for a multiple-effect evaporator is high.

Mechanical thermocompression may employ reciprocating, rotary positive-displacement, centrifugal, or axial-flow compressors. Positive-displacement compressors are impractical for all but the smallest capacities, such as portable seawater evaporators. Axial-flow compressors can be built for capacities of more than 472 m³/s (1 × 10⁶ ft³/min). Centrifugal compressors are usually cheapest for the intermediate-capacity ranges that are normally encountered. In all cases, great care must be taken to keep entrainment at a minimum, since the vapor becomes superheated on compression and any liquid present will evaporate, leaving the dissolved solids behind. In some cases a vapor-scrubbing tower may be installed to protect the compressor. A mechanical recompression evaporator usually requires more heat than is available from the compressed vapor. Some of this extra heat can be obtained by preheating the feed with the condensate and, if possible, with the product. Rather extensive heat-exchange systems with close approach temperatures are usually justified, especially if the evaporator is operated at high temperature to reduce the volume of vapor to be compressed. When the product is a solid, an elutriation leg such as that shown in Fig. 11-122*b* is advantageous, since it cools the product almost to feed temperature. The remaining heat needed to maintain the evaporator in operation must be obtained from outside sources.

While theoretical compressor power requirements are reduced slightly by going to lower evaporating temperatures, the volume of vapor to be compressed and hence compressor size and cost increase so rapidly that low-temperature operation is more expensive than high-temperature operation. The requirement of low temperature for fruit-juice concentration has led to the development of an evaporator employing a **secondary fluid,** usually Freon or ammonia. In this evaporator, the vapor is condensed in an exchanger cooled by boiling Freon. The Freon, at a much higher vapor density than the water vapor, is then compressed to serve as the heating medium for the evaporator. This system requires that the latent heat be transferred through two surfaces instead of one, but the savings in compressor size and cost are enough to justify the extra cost of heating surface or the cost of compressing through a wider temperature range.

Steam-jet thermocompression is advantageous when steam is available at a pressure appreciably higher than can be used in the evaporator. The steam jet then serves as a reducing valve while doing some useful work. The efficiency of a steam jet is quite low and falls off rapidly when the jet is not used at the vapor-flow rate and terminal pressure conditions for which it was designed. Consequently multiple jets are used when wide variations in evaporation rate are expected. Because of the low first cost and the ability to handle large volumes of vapor, steam-jet thermocompressors are used to increase the economy of evaporators that must operate at low temperatures and hence cannot be operated in multiple effect. The steam-jet thermocompression evaporator has a heat input larger than that needed to balance the system, and some heat must be rejected. This is usually done by venting some of the vapor at the suction of the compressor.

Multiple-Effect Evaporation Multiple-effect evaporation is the principal means in use for economizing on energy consumption. Most such evaporators operate on a continuous basis, although for a few difficult materials a continuous-batch cycle may be employed. In a multiple-effect evaporator, steam from an outside source is condensed in the heating element of the first effect. If the feed to the effect is at a temperature near the boiling point in the first effect, 1 kg of steam will evaporate almost 1 kg of water. The first effect operates at (but is not controlled at) a boiling temperature high enough so that the evaporated water can serve as the heating medium of the second effect. Here almost another kilogram of water is evaporated, and this may go to a condenser if the evaporator is a double-effect or may be used as the heating medium of the third effect. This method may be repeated for any number of effects. Large evaporators having six and seven effects are common in the pulp and paper industry, and evaporators having as many as 17 effects have been built. As a first approximation, the **steam economy** of a multiple-effect evaporator will increase in proportion to the number of effects and usually will be somewhat less numerically than the number of effects.

The increased steam economy of a multiple-effect evaporator is gained at the expense of evaporator first cost. The total heat-transfer surface will increase substantially in proportion to the number of effects in the evaporator. This is only an approximation since going from one to two effects means that about half of the heat transfer is at a higher temperature level, where heat-transfer coefficients are generally higher. On the other hand, operating at lower temperature differences reduces the heat-transfer coefficient for many types of evaporator. If the material has an appreciable boiling-point elevation, this will also lower the available temperature difference. The only accurate means of predicting the changes in steam economy and surface requirements with changes in the number of effects is by detailed heat and material balances together with an analysis of the effect of changes in operating conditions on heat-transfer performance.

The approximate **temperature distribution** in a multiple-effect evaporator is under the control of the designer, but once built, the evaporator establishes its own equilibrium. Basically, the effects are a number of series resistances to heat transfer, each resistance being approximately proportional to $1/U_n A_n$. The total available temperature drop is divided between the effects in proportion to their resistances. If one effect starts to scale, its temperature drop will increase at the expense of the temperature drops across the other effects. This provides a convenient means of detecting a drop in heat-transfer coefficient in an effect of an operating evaporator. If the steam pressure and final vacuum do not change, the temperature in the effect that is scaling will decrease and the temperature in the preceding effect will increase.

The feed to a multiple-effect evaporator is usually transferred from one effect to another in series so that the ultimate product concentration is reached only in one effect of the evaporator. In **backward-feed** operation, the raw feed enters the last (coldest) effect, the discharge from this effect becomes the feed to the next-to-the-last effect, and so on until product is discharged from the first effect. This method of operation is advantageous when the feed is cold, since much less liquid must be heated to the higher temperature existing in the early effects. It is also used when the product is so viscous that high temperatures are needed to keep the viscosity low enough to give reasonable heat-transfer coefficients. When product viscosity is high but a hot product is not needed, the liquid from the first effect is sometimes flashed to a lower temperature in one or more stages and the flash vapor added to the vapor from one or more later effects of the evaporator.

In **forward-feed** operation, raw feed is introduced in the first effect and passed from effect to effect parallel to the steam flow. Product is withdrawn from the last effect. This method of operation is advantageous when the feed is hot or when the concentrated product would be damaged or would deposit scale at high temperature. Forward feed simplifies operation when liquor can be transferred by pressure difference alone, thus eliminating all intermediate liquor pumps. When the feed is cold, forward feed gives a low steam economy since an appreciable part of the prime steam is needed to heat the feed to the boiling point and thus accomplishes no evaporation. If forward feed is necessary and feed is cold, steam economy can be improved markedly by preheating the feed in stages with vapor bled from intermediate effects of the evaporator. This usually represents little increase in total heating surface or cost since the feed must be heated in any event and shell-and-tube heat exchangers are generally less expensive per unit of surface area than evaporator heating surface.

Mixed-feed operation is used only for special applications, as when liquor at an intermediate concentration and a certain temperature is desired for additional processing.

Parallel feed involves the introduction of raw feed and the withdrawal of product at each effect of the evaporator. It is used primarily when the feed is substantially saturated and the product is a solid. An example is the evaporation of brine to make common salt. Evaporators of the types shown in Fig. 11-122*b* or *e* are used, and the product is withdrawn as a slurry. In this case, parallel feed is desirable because the feed washes impurities from the salt leaving the body.

Heat-recovery systems are frequently incorporated in an evaporator to increase the steam economy. Ideally, product and evaporator condensate should leave the system at a temperature as low as possible. Also, heat should be recovered from these streams by exchange with feed or evaporating liquid at the highest possible temperature. This would normally require separate liquid-liquid heat exchangers, which add greatly to the complexity of the evaporator and are justifiable only in large plants. Normally, the loss in thermodynamic availability due to flashing is tolerated since the flash vapor can then be used directly in the evaporator effects. The most commonly used is a **condensate flash** system in which the condensate from each effect but the first (which normally must be returned to the boiler) is flashed in successive stages to the pressure in the heating element of each succeeding effect of the evaporator. Product flash tanks may also be used in a backward- or mixed-feed evaporator. In a forward-feed evaporator, the principal means of heat recovery may be by use of **feed preheaters** heated by vapor bled from each effect of the evaporator. In this case, condensate may be either flashed as before or used in a separate set of exchangers to accomplish some of the feed preheating. A feed preheated by last-effect vapor may also materially reduce condenser water requirements.

Seawater Evaporators The production of potable water from saline waters represents a large and growing field of application for evaporators. Extensive work done in this field to 1972 was summarized in the annual *Saline Water Conversion Reports* of the Office of Saline Water, U.S. Department of the Interior. **Steam economies** on the order of 10 kg evaporation/kg steam are usually justified because (1) unit production capacities are high, (2) fixed charges are low on capital used for public works (i.e., they use long amortization periods and have low interest rates, with no other return on investment considered), (3) heat-transfer performance is comparable with that of pure water, and (4) properly treated seawater causes little deterioration due to scaling or fouling.

Figure 11-125*a* shows a **multiple-effect** (falling-film) flow sheet as used for seawater. Twelve effects are needed for a steam economy of 10. Seawater is used to condense last-effect vapor, and a portion is then treated to prevent scaling and corrosion. Treatment usually consists of acidification to break down bicarbonates, followed by deaeration, which also removes the carbon dioxide generated. The treated seawater is then heated to successively higher temperatures by a portion of the vapor from each effect and finally is fed to the evaporating surface of the first effect. The vapor generated therein and the partially concentrated liquid are passed to the second effect, and so on until the last effect. The feed rate is adjusted relative to the steam rate so that the residual liquid from the last effect can carry away all the salts in solution, in a volume about one-third of that of the feed. Condensate formed in each effect but the first is flashed down to the following effects in sequence and constitutes the product of the evaporator.

As the feed-to-steam ratio is increased in the flow sheet of Fig. 11-125*a*, a point is reached where all the vapor is needed to preheat the feed and none is available for the evaporator tubes. This limiting case is the **multistage flash evaporator,** shown in its simplest form in Fig. 11-125*b*. Seawater is treated as before and then pumped through a number of feed heaters in series. It is given a final boost in temperature with prime steam in a **brine heater** before it is flashed down in series to provide the vapor needed by the feed heaters. The amount of steam required depends on the approach-temperature difference in the feed heaters and the flash range per stage. Condensate from the feed heaters is flashed down in the same manner as the brine.

Since the flow being heated is identical to the total flow being flashed, the temperature rise in each heater is equal to the flash range in each flasher. This temperature difference represents a loss from the temperature difference available for heat transfer. There are thus two ways of increasing the steam economy of such plants: increasing the heating surface and increasing the number of stages. Whereas the number of effects in a multiple-effect plant will be about 20 percent greater than the steam economy, the number of stages in a flash plant will be 3 to 4 times the steam economy. However, a large number of stages can be provided in a single vessel by means of internal bulkheads. The heat-exchanger tubing is placed in the same vessel, and the tubes usually are continuous through a number of stages. This requires ferrules or special close tube-hole clearances where the tubes pass through the internal bulkheads. In a plant for a steam economy of 10, the ratio of flow rate to heating surface is usually such that the seawater must pass through about 152 m of 19-mm (500 ft of ¾-in) tubing before it reaches the brine heater. This places a limitation on the physical arrangement of the vessels.

Inasmuch as it requires a flash range of about 61°C (110°F) to produce 1 kg of flash vapor for every 10 kg of seawater, the multistage flash evaporator requires handling a large volume of seawater relative to the product. In the flow sheet of Fig. 11-125*b* all this seawater must be deaerated and treated for scale prevention. In addition, the last-stage vacuum varies with the ambient seawater temperature, and ejector equipment must be sized for the worst condition. These difficulties can be eliminated by using the **recirculating multistage flash** flow sheet of Fig. 11-125*c*. The last few stages, called the **reject stages,** are cooled by a flow of seawater that can be varied to maintain a reasonable last-stage vacuum. A small portion of the last-stage brine is blown down to carry away the dissolved salts, and the balance is recirculated to the **heat-recovery stages.** This arrangement requires a much smaller makeup of fresh seawater and hence a lower treatment cost.

The multistage flash evaporator is similar to a multiple-effect forced-circulation evaporator, but with all the forced-circulation heaters in series. This has the advantage of requiring only one large-volume forced-circulation pump, but the sensible heating and short-circuiting losses in available temperature differences remain. A disadvantage of the flash evaporator is that the liquid throughout the system is at almost the discharge concentration. This has limited its industrial use to solutions in which no great concentration differences are required between feed and product and to where the liquid can be heated through wide temperature ranges without scaling. A partial

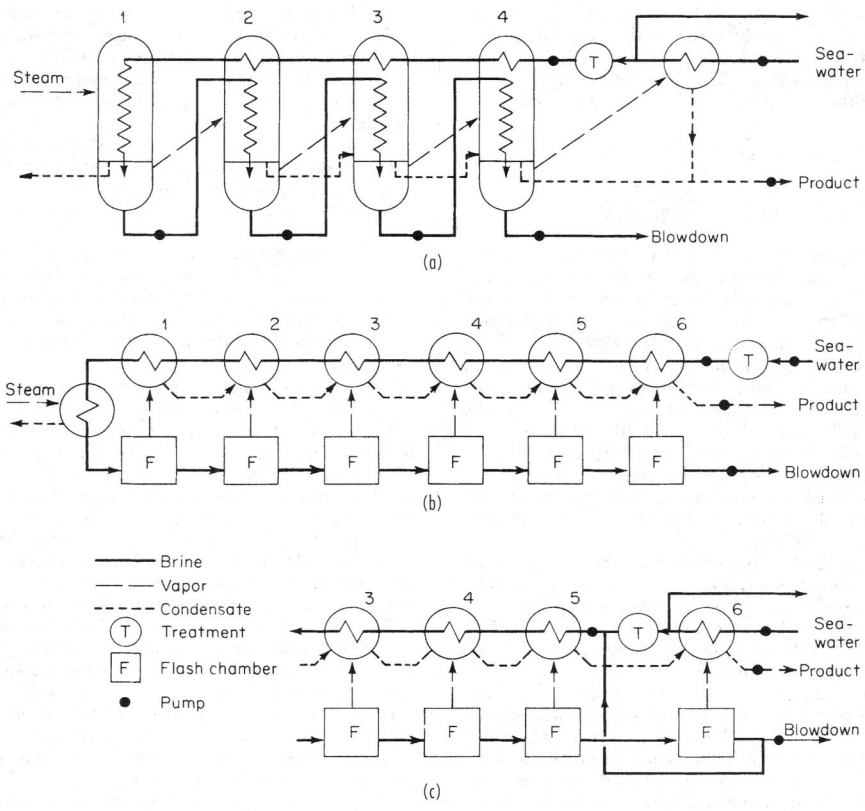

FIG. 11-125   Flow sheets for seawater evaporators. (*a*) Multiple effect (falling film). (*b*) Multistage flash (once-through). (*c*) Multistage flash (recirculating).

remedy is to arrange several multistage flash evaporators in series, the heat-rejection section of one being the brine heater of the next. This permits independent control of concentration but eliminates the principal advantage of the flash evaporator, which is the small number of pumps and vessels required. An unusual feature of the flash evaporator is that fouling of the heating surfaces reduces primarily the steam economy rather than the capacity of the evaporator. Capacity is not affected until the heat-rejection stages can no longer handle the increased flashing resulting from the increased heat input.

EVAPORATOR CALCULATIONS

Single-Effect Evaporators   The heat requirements of a single-effect continuous evaporator can be calculated by the usual methods of stoichiometry. If enthalpy data or specific heat and heat-of-solution data are not available, the heat requirement can be estimated as the sum of the heat needed to raise the feed from feed to product temperature and the heat required to evaporate the water. The latent heat of water is taken at the vapor-head pressure instead of at the product temperature in order to compensate partially for any heat of solution. If sufficient vapor-pressure data are available for the solution, methods are available to calculate the true latent heat from the slope of the Dühring line [Othmer, *Ind. Eng. Chem.,* **32,** 841 (1940)].

The heat requirements in batch evaporation are the same as those in continuous evaporation except that the temperature (and sometimes pressure) of the vapor changes during the course of the cycle. Since the enthalpy of water vapor changes but little relative to temperature, the difference between continuous and batch heat requirements is almost always negligible. More important usually is the effect of variation of fluid properties, such as viscosity and boiling-point rise, on heat transfer. These can only be estimated by a step-by-step calculation.

In selecting the **boiling temperature,** consideration must be given to the effect of temperature on heat-transfer characteristics of the type of evaporator to be used. Some evaporators show a marked drop in coefficient at low temperature—more than enough to offset any gain in available temperature difference. The condenser cooling-water temperature and cost must also be considered.

Thermocompression Evaporators   Thermocompression-evaporator calculations [Pridgeon, *Chem. Metall. Eng.,* **28,** 1109 (1923); Peter, *Chimia (Switzerland),* **3,** 114 (1949); Petzold, *Chem. Ing. Tech.,* **22,** 147 (1950); and Weimer, Dolf, and Austin, *Chem. Eng. Prog.,* **76**(11), 78 (1980)] are much the same as single-effect calculations with the added complication that the heat supplied to the evaporator from compressed vapor and other sources must exactly balance the heat requirements. Some knowledge of compressor efficiency is also required. Large axial-flow machines on the order of 236-m³/s (500,000-ft³/min) capacity may have efficiencies of 80 to 85 percent. Efficiency drops to about 75 percent for a 14-m³/s (30,000-ft³/min) centrifugal compressor. Steam-jet compressors have thermodynamic efficiencies on the order of only 25 to 30 percent.

Flash Evaporators   The calculation of a heat and material balance on a flash evaporator is relatively easy once it is understood that the temperature rise in each heater and temperature drop in each flasher must all be substantially equal. The steam economy E, kg evaporation/kg of 1055-kJ steam (lb/lb of 1000-Btu steam) may be approximated from

$$E = \left(1 - \frac{\Delta T}{1250}\right) \frac{\Delta T}{Y + R + \Delta T/N} \qquad (11\text{-}124)$$

where ΔT is the total temperature drop between feed to the first flasher and discharge from the last flasher. °C; N is the number of flash stages; Y is the approach between vapor temperature from the first flasher and liquid leaving the heater in which this vapor is condensed. °C (the approach is usually substantially constant for all stages); and R. °C. is the sum of the boiling-point rise and the short-circuiting loss in the first flash stage. The expression for the mean effective temperature difference Δt available for heat transfer then becomes

$$\Delta t = \frac{\Delta T}{N \ln \dfrac{1 - \Delta T/1250 - RE/\Delta T}{1 - \Delta T\,1250 - RE/\Delta T - E/N}} \qquad (11\text{-}125)$$

Multiple-Effect Evaporators A number of approximate methods have been published for estimating performance and heating-surface requirements of a multiple-effect evaporator [Coates and Pressburg, *Chem. Eng.*, **67**(6), 157 (1960); Coates, *Chem. Eng. Prog.*, **45**, 25 (1949); and Ray and Carnahan, *Trans. Am. Inst. Chem. Eng.*, **41**, 253 (1945)]. However, because of the wide variety of methods of feeding and the added complication of feed heaters and condensate flash systems, the only certain way of determining performance is by detailed heat and material balances. Algebraic solutions may be used, but if more than a few effects are involved, trial-and-error methods are usually quicker. These frequently involve trial-and-error within trial-and-error solutions. Usually, if condensate flash systems or feed heaters are involved, it is best to start at the first effect. The basic steps in the calculation are then as follows:

1. Estimate temperature distribution in the evaporator, taking into account boiling-point elevations. If all heating surfaces are to be equal, the temperature drop across each effect will be approximately inversely proportional to the heat-transfer coefficient in that effect.

2. Determine total evaporation required, and estimate steam consumption for the number of effects chosen.

3. From assumed feed temperature (forward feed) or feed flow (backward feed) to the first effect and assumed steam flow, calculate evaporation in the first effect. Repeat for each succeeding effect, checking intermediate assumptions as the calculation proceeds. Heat input from condensate flash can be incorporated easily since the condensate flow from the preceding effects will have already been determined.

4. The result of the calculation will be a feed to or a product discharge from the last effect that may not agree with actual requirements. The calculation must then be repeated with a new assumption of steam flow to the first effect.

5. These calculations should yield liquor concentrations in each effect that make possible a revised estimate of boiling-point rises. They also give the quantity of heat that must be transferred in each effect. From the heat loads, assumed temperature differences, and heat-transfer coefficients, heating-surface requirements can be determined. If the distribution of heating surface is not as desired, the entire calculation may need to be repeated with revised estimates of the temperature in each effect.

6. If sufficient data are available, heat-transfer coefficients under the proposed operating conditions can be calculated in greater detail and surface requirements readjusted.

Such calculations require considerable judgment to avoid repetitive trials but are usually well worth the effort. Sample calculations are given in the American Institute of Chemical Engineers *Testing Procedure for Evaporators* and by Badger and Banchero, *Introduction to Chemical Engineering*, McGraw-Hill, New York, 1955. These balances may be done by computer but programming time frequently exceeds the time needed to do them manually, especially when variations in flow sheet are to be investigated. The MASSBAL program of SACDA, London, Ont., provides a considerable degree of flexibility in this regard. Another program, not specific to evaporators, is ASPEN PLUS by Aspen Tech., Cambridge, MA. Many such programs include simplifying assumptions and approximations that are not explicitly stated and can lead to erroneous results.

Optimization The primary purpose of evaporator design is to enable production of the necessary amount of satisfactory product at the lowest total cost. This requires economic-balance calculations that may include a great number of variables. Among the possible variables are the following:

1. Initial steam pressure versus cost or availability.
2. Final vacuum versus water temperature, water cost, heat-transfer performance, and product quality.
3. Number of effects versus steam, water, and pump power cost.
4. Distribution of heating surface between effects versus evaporator cost.
5. Type of evaporator versus cost and continuity of operation.
6. Materials of construction versus product quality, tube life, evaporator life, and evaporator cost.
7. Corrosion, erosion, and power consumption versus tube velocity.
8. Downtime for retubing and repairs.
9. Operating-labor and maintenance requirements.
10. Method of feeding and use of heat-recovery systems.
11. Size of recovery heat exchangers.
12. Possible withdrawal of steam from an intermediate effect for use elsewhere.
13. Entrainment separation requirements.

The type of evaporator to be used and the materials of construction are generally selected on the basis of past experience with the material to be concentrated. The method of feeding can usually be decided on the basis of known feed temperature and the properties of feed and product. However, few of the listed variables are completely independent. For instance, if a large number of effects is to be used, with a consequent low temperature drop per effect, it is impractical to use a natural-circulation evaporator. If expensive materials of construction are desirable, it may be found that the forced-circulation evaporator is the cheapest and that only a few effects are justifiable.

The variable having the greatest influence on total cost is the number of effects in the evaporator. An economic balance can establish the optimum number where the number is not limited by such factors as viscosity, corrosiveness, freezing point, boiling-point rise, or thermal sensitivity. Under present United States conditions, savings in steam and water costs justify the extra capital, maintenance, and power costs of about seven effects in large commercial installations when the properties of the fluid are favorable, as in black-liquor evaporation. Under governmental financing conditions, as for plants to supply fresh water from seawater, evaporators containing from 12 to 30 or more effects can be justified.

As a general rule, the optimum number of effects increases with an increase in steam cost or plant size. Larger plants favor more effects, partly because they make it easier to install heat-recovery systems that increase the steam economy attainable with a given number of effects. Such recovery systems usually do not increase the total surface needed but do require that the heating surface be distributed between a greater number of pieces of equipment.

The most common evaporator design is based on the use of the same heating surface in each effect. This is by no means essential since few evaporators are "standard" or involve the use of the same patterns. In fact, there is no reason why all effects in an evaporator must be of the same type. For instance, the cheapest salt evaporator might use propeller calandrias for the early effects and forced-circulation effects at the low-temperature end, where their higher cost per unit area is more than offset by higher heat-transfer coefficients.

Bonilla [*Trans. Am. Inst. Chem. Eng.*, **41**, 529 (1945)] developed a simplified method for distributing the heating surface in a multiple-effect evaporator to achieve minimum cost. If the cost of the evaporator per unit area of heating surface is constant throughout, then minimum cost and area will be achieved if the ratio of area to temperature difference $A/\Delta T$ is the same for all effects. If the cost per unit area z varies, as when different tube materials or evaporator types are used, then $zA/\Delta T$ should be the same for all effects.

EVAPORATOR ACCESSORIES

Condensers The vapor from the last effect of an evaporator is usually removed by a condenser. **Surface condensers** are employed when mixing of condensate with condenser cooling water is not desired. They are for the most part shell-and-tube condensers with vapor on the shell side and a multipass flow of cooling water on the tube side. Heat loads, temperature differences, sizes, and costs are usually of the same order of magnitude as for another effect of the evaporator. Surface condensers use more cooling water and are so much more expensive that they are never used when a direct-contact condenser is suitable.

The most common type of direct-contact condenser is the counter-current **barometric condenser,** in which vapor is condensed by rising against a rain of cooling water. The condenser is set high enough so that water can discharge by gravity from the vacuum in the condenser. Such condensers are inexpensive and are economical on water consumption. They can usually be relied on to maintain a vacuum corresponding to a saturated-vapor temperature within 2.8°C (5°F) of the water temperature leaving the condenser [How, *Chem. Eng.,* **63**(2), 174 (1956)]. The ratio of water consumption to vapor condensed can be determined from the following equation:

$$\frac{\text{Water flow}}{\text{Vapor flow}} = \frac{H_v - h_2}{h_2 - h_1} \qquad (11\text{-}126)$$

where H_v = vapor enthalpy and h_1 and h_2 = water enthalpies entering and leaving the condenser. Another type of direct-contact condenser is the **jet** or **wet condenser,** which makes use of high-velocity jets of water both to condense the vapor and to force noncondensable gases out the tailpipe. This type of condenser is frequently placed below barometric height and requires a pump to remove the mixture of water and gases. Jet condensers usually require more water than the more common barometric-type condensers and cannot be throttled easily to conserve water when operating at low evaporation rates.

Vent Systems   Noncondensable gases may be present in the evaporator vapor as a result of leakage, air dissolved in the feed, or decomposition reactions in the feed. When the vapor is condensed in the succeeding effect, the noncondensables increase in concentration and impede heat transfer. This occurs partially because of the reduced partial pressure of vapor in the mixture but mainly because the vapor flow toward the heating surface creates a film of poorly conducting gas at the interface. (See page 11-14 for means of estimating the effect of noncondensable gases on the steam-film coefficient.) The most important means of reducing the influence of noncondensables on heat transfer is by properly channeling them past the heating surface. A positive vapor-flow path from inlet to vent outlet should be provided, and the path should preferably be tapered to avoid pockets of low velocity where noncondensables can be trapped. Excessive clearances and low-resistance channels that could bypass vapor directly from the inlet to the vent should be avoided [Standiford, *Chem. Eng. Prog.,* **75**, 59–62 (July 1979)].

In any event, noncondensable gases should be vented well before their concentration reaches 10 percent. Since gas concentrations are difficult to measure, the usual practice is to overvent. This means that an appreciable amount of vapor can be lost.

To help conserve steam economy, venting is usually done from the steam chest of one effect to the steam chest of the next. In this way, excess vapor in one vent does useful evaporation at a steam economy only about one less than the overall steam economy. Only when there are large amounts of noncondensable gases present, as in beet-sugar evaporation, is it desirable to pass the vents directly to the condenser to avoid serious losses in heat-transfer rates. In such cases, it can be worthwhile to recover heat from the vents in separate heat exchangers, which preheat the entering feed.

The noncondensable gases eventually reach the condenser (unless vented from an effect above atmospheric pressure to the atmosphere or to auxiliary vent condensers). These gases will be supplemented by air dissolved in the condenser water and by carbon dioxide given off on decomposition of bicarbonates in the water if a barometric condenser is used. These gases may be removed by the use of a water-jet-type condenser but are usually removed by a separate vacuum pump.

The vacuum pump is usually of the steam-jet type if high-pressure steam is available. If high-pressure steam is not available, more expensive mechanical pumps may be used. These may be either a water-ring (Hytor) type or a reciprocating pump.

The primary source of noncondensable gases usually is air dissolved in the condenser water. Figure 11-126 shows the dissolved-gas content of fresh water and seawater, calculated as equivalent air. The

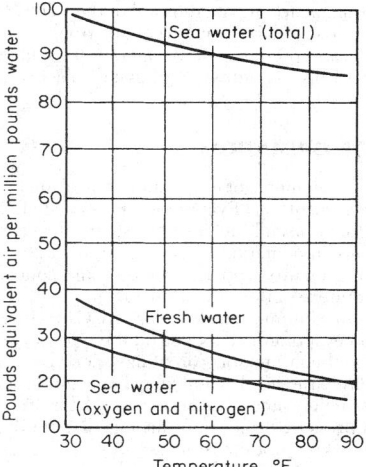

FIG. 11-126   Gas content of water saturated at atmospheric pressure. °C = 5/9 (°F − 32).

lower curve for seawater includes only dissolved oxygen and nitrogen. The upper curve includes carbon dioxide that can be evolved by complete breakdown of bicarbonate in seawater. Breakdown of bicarbonates is usually not appreciable in a condenser but may go almost to completion in a seawater evaporator. The large increase in gas volume as a result of possible bicarbonate breakdown is illustrative of the uncertainties involved in sizing vacuum systems.

By far the largest load on the vacuum pump is water vapor carried with the noncondensable gases. Standard power-plant practice assumes that the mixture leaving a surface condenser will have been cooled 4.2°C (7.5°F) below the saturation temperature of the vapor. This usually corresponds to about 2.5 kg of water vapor/kg of air. One advantage of the countercurrent barometric condenser is that it can cool the gases almost to the temperature of the incoming water and thus reduce the amount of water vapor carried with the air.

In some cases, as with pulp-mill liquors, the evaporator vapors contain constituents more volatile than water, such as methanol and sulfur compounds. Special precautions may be necessary to minimize the effects of these compounds on heat transfer, corrosion, and condensate quality. They can include removing most of the condensate countercurrent to the vapor entering an evaporator-heating element, channeling vapor and condensate flow to concentrate most of the "foul" constituents into the last fraction of vapor condensed (and keeping this condensate separate from the rest of the condensate), and flashing the warm evaporator feed to a lower pressure to remove much of the foul constituents in only a small amount of flash vapor. In all such cases, special care is needed to properly channel vapor flow past the heating surfaces so there is a positive flow from steam inlet to vent outlet with no pockets, where foul constituents or noncondensibles can accumulate.

Salt Removal   When an evaporator is used to make a crystalline product, a number of means are available for concentrating and removing the salt from the system. The simplest is to provide settling space in the evaporator itself. This is done in the types shown in Fig. 11-122*b, c,* and *e* by providing a relatively quiescent zone in which the salt can settle. Sufficiently high slurry densities can usually be achieved in this manner to reach the limit of pumpability. The evaporators are usually placed above barometric height so that the slurry can be discharged intermittently on a short time cycle. This permits the use of high velocities in large lines that have little tendency to plug.

If the amount of salts crystallized is on the order of a ton an hour or less, a salt trap may be used. This is simply a receiver that is connected to the bottom of the evaporator and is closed off from the evaporator periodically for emptying. Such traps are useful when insufficient headroom is available for gravity removal of the solids. However, traps require a great deal of labor, give frequent trouble with the shutoff valves, and also can upset evaporator operation completely if a trap is reconnected to the evaporator without first displacing all air with feed liquor.

EVAPORATOR OPERATION

The two principal elements of evaporator control are *evaporation rate* and *product concentration*. Evaporation rate in single- and multiple-effect evaporators is usually achieved by steam-flow control. Conventional-control instrumentation is used (see Sec. 22), with the added precaution that pressure drop across meter and control valve, which reduces temperature difference available for heat transfer, not be excessive when maximum capacity is desired. Capacity control of thermocompression evaporators depends on the type of compressor; positive-displacement compressors can utilize speed control or variations in operating pressure level. Centrifugal machines normally utilize adjustable inlet-guide vanes. Steam jets may have an adjustable spindle in the high-pressure orifice or be arranged as multiple jets that can individually be cut out of the system.

Product concentration can be controlled by any property of the solution that can be measured with the requisite accuracy and reliability. The preferred method is to impose control on rate of product withdrawal. Feed rates to the evaporator effects are then controlled by their levels. When level control is impossible, as with the rising-film LTV, product concentration is used to control the feed rate—frequently by rationing of feed to steam with the ration reset by product concentration, sometimes also by feed concentration. Other controls that may be needed include vacuum control of the last effect (usually by air bleed to the condenser) and temperature-level control of thermocompression evaporators (usually by adding makeup heat or by venting excess vapor, or both as feed or weather conditions vary). For more control detail, see *Measurement and Control in Water Desalination*, N. Lior, ed., pp. 241–305, Elsevier Science Publ. Co., NY, 1986.

Control of an evaporator requires more than proper instrumentation. Operator logs should reflect changes in basic characteristics, as by use of **pseudo heat-transfer** coefficients, which can detect obstructions to heat flow, hence to capacity. These are merely the ratio of any convenient measure of heat flow to the temperature drop across each effect. **Dilution** by wash and seal water should be monitored since it absorbs evaporative capacity. Detailed tests, routine measurements, and operating problems are covered more fully in *Testing Procedure for Evaporators* (loc. cit.) and by Standiford [*Chem. Eng. Prog.*, **58**(11), 80 (1962)].

By far the best application of computers to evaporators is for working up operators' data into the basic performance parameters such as heat-transfer coefficients, steam economy, and dilution.

Psychrometry, Evaporative Cooling, and Solids Drying

Charles G. Moyers, Ph D., P.E., *Principal Engineer, Union Carbide Corporation, Fellow, American Institute of Chemical Engineers. (Section Editor, Evaporative Cooling and Solids Drying)*

Glenn W. Baldwin, M.S., P.E., *Staff Engineer, Union Carbide Corporation, Member, American Institute of Chemical Engineers. (Psychrometry)*

Nomenclature and Units

In this table, symbols used in this section are defined in a general way. SI and customary U.S. units are listed. Specialized symbols are either defined at the point of application or in a separate table of nomenclature.

Symbol	Definition	SI units	U.S. customary units
A	Area	m^2	ft^2
a	Area	m^2	ft^2
C	Heat capacity	J/kg	Btu/lbm
D	Diameter	m	ft
D	Diffusivity	m^2/s	ft^2/s
d	Thickness	m	ft
E	Potential difference	V	V
F	Void fraction	Dimensionless	Dimensionless
G	Mass velocity	$k/(m^2 \cdot s)$	$lbm/(ft^2 \cdot s)$
H	Humidity	kg/kg	lbm/lbm
h	Heat-transfer coefficient	$J/(m^2 \cdot s \cdot K)$	$Btu/(ft^2 \cdot h \cdot °F)$
k	Mass-transfer coefficient	$kg/(m^2 \cdot s \cdot atm)$	$lbm/(h \cdot ft^2 \cdot atm)$
K	Thermal conductivity	$W/(m \cdot K)$	$Btu/(ft \cdot h \cdot °F)$
L	Length	m	ft
M	Molecular weight	kg/mol	lbm/mol
N	Rotational speed	l/s	l/min
P	Pressure	Pa	lbf/in^2
P	Pressure drop	Pa	lbf/ft^2
Q	Flow rate	m^3/s	ft^3/s
Q	Total heat flow	J/s	Btu/h
S	Slope	m/m	ft/ft
t	Time	s	s, h
t	Temperature	$°C$	$°F$
T	Absolute temperature	K	$°R$
U	Heat-transfer coefficient	$J/(m^2 \cdot s \cdot K)$	$Btu/(ft^2 \cdot h \cdot °F)$
u	Velocity	m/s	ft/s
V	Velocity	m/s	ft/s
V	Volume	m^3	ft^3
W	Mass	kg	lbm
Greek symbols			
ϵ	Void fraction	Dimensionless	Dimensionless
θ	Time	s	s, h
λ	Latent heat	J/kg	Btu/lbm
μ	Viscosity	$Pa \cdot s$	$lbm/(s \cdot ft)$
ρ	Density	kg/m^3	lbm/ft^3
σ	Surface tension	N/m	dyn/cm

PSYCHROMETRY*

GENERAL REFERENCES: *Handbook of Fundamentals,* American Society of Heating, Refrigerating and Air-Conditioning Engineers, New York, 1967. Quinn, "Humidity: The Neglected Parameter," *Test Eng.* (July 1968). Treybal, *Mass-Transfer Operations,* 3d ed. McGraw-Hill, New York, 1980. Wexler, *Humidity and Moisture,* vol. I, Reinhold, New York, 1965. Zimmerman and Lavine, Psychrometric Charts and Tables, Industrial Research Service, 2d ed., Dover, N.H., 1964.

Psychrometry is concerned with determination of the properties of gas-vapor mixtures. The air-water vapor system is by far the system most commonly encountered.

Principles involved in determining the properties of other systems are the same as with air-water vapor, with one major exception. Whereas the psychrometric ratio (ratio of heat-transfer coefficient to product of mass-transfer coefficient and humid heat, terms defined in the following subsection) for the air-water system can be taken as 1, the ratio for other systems in general does not equal 1. This has the effect of making the adiabatic-saturation temperature different from the wet-bulb temperature. Thus, for systems other than air-water vapor, calculation of psychrometric and drying problems is complicated by the necessity for point-to-point calculation of the temperature of the evaporating surface. For example, for the air-water system the temperature of the evaporating surface will be constant during the constant-rate drying period even though temperature and humidity of the gas stream change. For other systems, the temperature of the evaporating surface would change.

TERMINOLOGY

Terminology and relationships pertinent to psychrometry are:

Absolute humidity H equals the pounds of water vapor carried by 1 lb of dry air. If ideal-gas behavior is assumed, $H = M_w p/[M_a(P - p)]$, where M_w = molecular weight of water; M_a = molecular weight of air; p = partial pressure of water vapor, atm; and P = total pressure, atm.

When the partial pressure p of water vapor in the air at a given temperature equals the vapor pressure of water p_s at the same temperature, the air is saturated and the absolute humidity is designated the **saturation humidity** H_s.

Percentage absolute humidity (percentage saturation) is defined as the ratio of absolute humidity to saturation humidity and is given by $100 \, H/H_s = 100p(P - p_s)/[p_s(P - p)]$.

Percentage relative humidity is defined as the partial pressure of water vapor in air divided by the vapor pressure of water at the given temperature. Thus RH = $100p/p_s$.

Dew point, or **saturation temperature,** is the temperature at which a given mixture of water vapor and air is saturated, for example, the temperature at which water exerts a vapor pressure equal to the partial pressure of water vapor in the given mixture.

Humid heat c_s is the heat capacity of 1 lb of dry air and the moisture it contains. For most engineering calculations, $c_s = 0.24 + 0.45H$, where 0.24 and 0.45 are the heat capacities of dry air and water vapor, respectively, and both are assumed constant.

Humid volume is the volume in cubic feet of 1 lb of dry air and the water vapor it contains.

Saturated volume is the humid volume when the air is saturated.

Wet-bulb temperature is the dynamic equilibrium temperature attained by a water surface when the rate of heat transfer to the surface by convection equals the rate of mass transfer away from the surface. At equilibrium, if negligible change in the dry-bulb temperature is assumed, a heat balance on the surface is

$$k_g \lambda(p_s - p) = h_c(t - t_w) \qquad (12\text{-}1a)$$

where k_g = mass-transfer coefficient, lb/(h·ft^2·atm); λ = latent heat of vaporization, Btu/lb; p_s = vapor pressure of water at wet-bulb temperature, atm; p = partial pressure of water vapor in the environment, atm; h_c = heat-transfer coefficient, Btu/(h·ft^2·°F); t = temperature of air-water vapor mixture (dry-bulb temperature), °F; and t_w = wet-bulb temperature, °F. Under ordinary conditions the partial pressure and vapor pressure are small relative to the total pressure, and the wet-bulb equation can be written in terms of humidity differences as

$$H_s - H = (h_c/\lambda k')(t - t_w) \qquad (12\text{-}1b)$$

where k' = lb/(h·ft^2) (unit humidity difference) = $(M_a/M_w)k_g = 1.6k_g$.

Adiabatic-Saturation Temperature, or Constant-Enthalpy Lines If a stream of air is intimately mixed with a quantity of water at a temperature t_s in an adiabatic system, the temperature of the air will drop and its humidity will increase. If t_s is such that the air leaving the system is in equilibrium with the water, t_s will be the adiabatic-saturation temperature, and the line relating the temperature and humidity of the air is the adiabatic-saturation line. The equation for the adiabatic-saturation line is

$$H_s - H = (c_s/\lambda)(t - t_s) \qquad (12\text{-}2)$$

RELATION BETWEEN WET-BULB AND ADIABATIC-SATURATION TEMPERATURES

Experimentally it has been shown that for air-water systems the value of $h_c/k'c_s$, the **psychrometric ratio,** is approximately equal to 1. Under these conditions the wet-bulb temperatures and adiabatic-saturation temperatures are substantially equal and can be used interchangeably. The difference between adiabatic-saturation temperature and wet-bulb temperature increases with increasing humidity, but this effect is unimportant for most engineering calculations. An empirical formula for wet-bulb temperature determination of moist air at atmospheric pressure is presented by Liley [*Int. J. of Mechanical Engineering Education,* vol. 21, No. 2 (1993)].

For systems other than air-water vapor, the value of $h_c/k'c_s$ may differ appreciably from unity, and the wet-bulb and adiabatic-saturation temperatures are no longer equal. For these systems the psychrometric ratio may be obtained by determining h_c/k' from heat- and mass-transfer analogies such as the Chilton-Colburn analogy [*Ind. Eng. Chem.,* **26,** 1183 (1934)]. For low humidities this analogy gives

$$\frac{h_c}{k'} = c_s \left(\frac{\mu/\rho D_v}{c_s \mu/k}\right)^{2/3} \qquad (12\text{-}3)$$

where c_s = humid heat, Btu/(lb·°F); μ = viscosity, lb/(ft·h); ρ = density, lb/ft^3; D_v = diffusivity, ft^2/h; and k = thermal conductivity, Btu/(h·ft^2) (°F/ft). All properties should be evaluated for the gas mixture.

For the case of **flow past cylinders,** such as a wet-bulb thermometer, Bedingfield and Drew [*Ind. Eng. Chem.,* **42,** 1164 (1950)] obtained a correlation for their data on sublimation of cylinders into air and for the data of others on wet-bulb thermometers in air they give

$$h_c/k' = 0.294(\mu/\rho D_v)^{0.56} \qquad (12\text{-}4)$$

where the nomenclature is identical to that in Eq. (12-3). For evaporation into gases other than air, Eq. (12-3) with an exponent of 0.56 would apply.

Application of these equations is illustrated in Example 1.

° The contribution of Eno Bagnoli, E. I. du Pont de Nemours & Co., to material that was used from the sixth edition is acknowledged. (Psychrometry)

Example 1: Compare Wet-Bulb and Adiabatic-Saturation Temperatures   For the air-water system at atmospheric pressure, the measured values of dry-bulb and wet-bulb temperatures are 85 and 72°F respectively. Determine the absolute humidity and compare the wet-bulb temperature and adiabatic-saturation temperature. Assume that h_c/k' is given by Eq. (12-4).

Solution.   For relatively dry air the Schmidt number $\mu/\rho D_v$ is 0.60, and from Eq. (12-4) $h_c/k' = 0.294(0.60)^{0.56} = 0.221$. At 72°F the vapor pressure of water is 20.07 mmHg, and the latent heat of vaporization is 1051.6 Btu/lb. From Eq. (12-1b), $[20.07/(760 - 20.07)](18/29) - H = (0.221/1051.6)(85 - 72)$, or $H = 0.0140$ lb water/lb dry air. The humid heat is calculated as $c_s = 0.24 + 0.45(0.0140) = 0.246$. The adiabatic-saturation temperature is obtained from Eq. (12-2) as

$$H_s - 0.0140 = (0.246/1051.6)(85 - t_s)$$

Values of H_s and t_s are given by the saturation curve of the psychrometric chart, such as Fig. 12-2. By trial and error, $t_s = 72.1°F$, or the adiabatic-saturation temperature is 0.1°F higher than the wet-bulb temperature.

USE OF PSYCHROMETRIC CHARTS

Three **charts for the air-water vapor system** are given as Figs. 12-1 to 12-3 for low-, medium-, and high-temperature ranges. Figure 12-4 shows a modified Grosvenor chart, which is more familiar to the chemical engineer. These charts are for an absolute pressure of 1 atm. The corrections required at pressures different from atmospheric are given in Table 12-2. Figure 12-5 shows a psychrometric chart for combustion products in air. The thermodynamic properties of moist air are given in Table 12-1.

Examples Illustrating Use of Psychrometric Charts   In these examples the following nomenclature is used:

t = dry-bulb temperatures, °F
t_w = wet-bulb temperature, °F
t_d = dew-point temperature, °F
H = moisture content, lb water/lb dry air

ΔH = moisture added to or rejected from the air stream, lb water/lb dry air
h' = enthalpy at saturation, Btu/lb dry air
D = enthalpy deviation, Btu/lb dry air
$h = h' + D$ = true enthalpy, Btu/lb dry air
h_w = enthalpy of water added to or rejected from the system, Btu/lb dry air
q_a = heat added to the system, Btu/lb dry air
q_r = heat removed from system, Btu/lb dry air

Subscripts 1, 2, 3, etc., indicate entering and subsequent states.

Example 2: Determination of Moist Air Properties   Find the properties of moist air when the dry-bulb temperature is 80°F and the wet-bulb temperature is 67°F.

Solution.   Read directly from Fig. 12-2 (Fig. 12-6 shows the solution diagrammatically).

Moisture content H = 78 gr/lb dry air
= 0.011 lb water/lb dry air
Enthalpy at saturation h' = 31.6 Btu/lb dry air
Enthalpy deviation D = −0.1 Btu/lb dry air
True enthalpy h = 31.5 Btu/lb dry air
Specific volume v = 13.8 ft³/lb dry air
Relative humidity = 51 percent
Dew point t_d = 60.3°F

Example 3: Air Heating   Air is heated by a steam coil from 30°F dry-bulb temperature and 80 percent relative humidity to 75°F dry-bulb temperature. Find the relative humidity, wet-bulb temperature, and dew point of the heated air. Determine the quantity of heat added per pound of dry air.

Solution.   Reading directly from the psychrometric chart (Fig. 12-2),

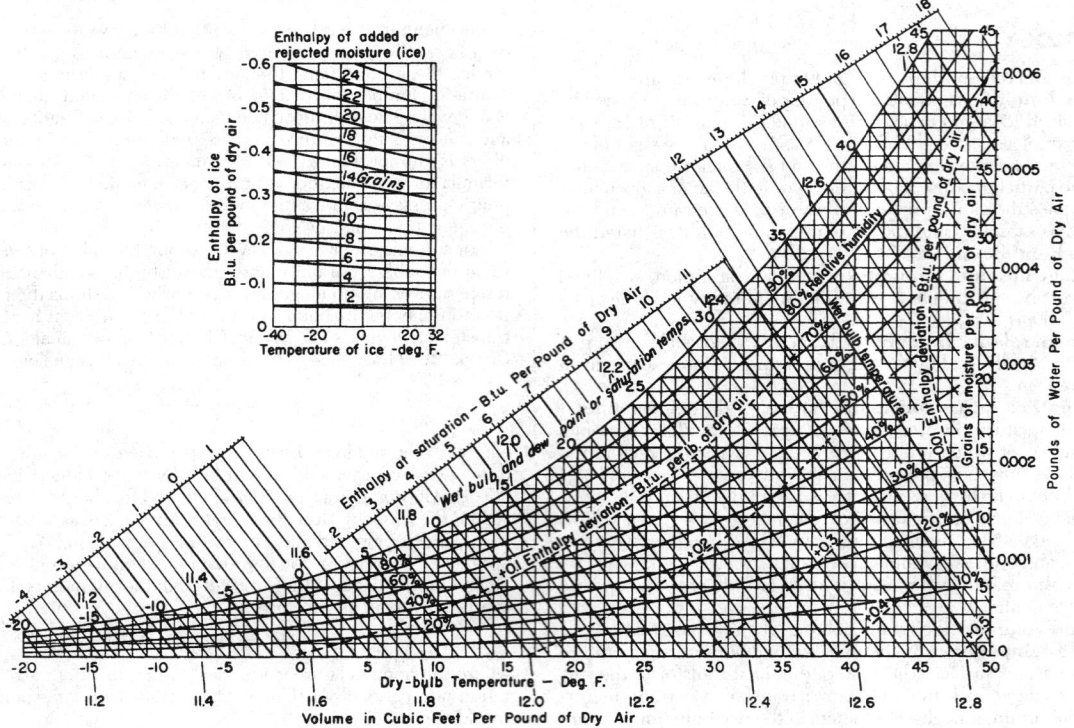

FIG. 12-1   Psychrometric chart—low temperatures. Barometric pressure, 29.92 inHg. To convert British thermal units per pound to joules per kilogram, multiply by 2326; to convert British thermal units per pound dry air-degree Fahrenheit to joules per kilogram-kelvin, multiply by 4186.8; and to convert cubic feet per pound to cubic meters per kilogram, multiply by 0.0624.

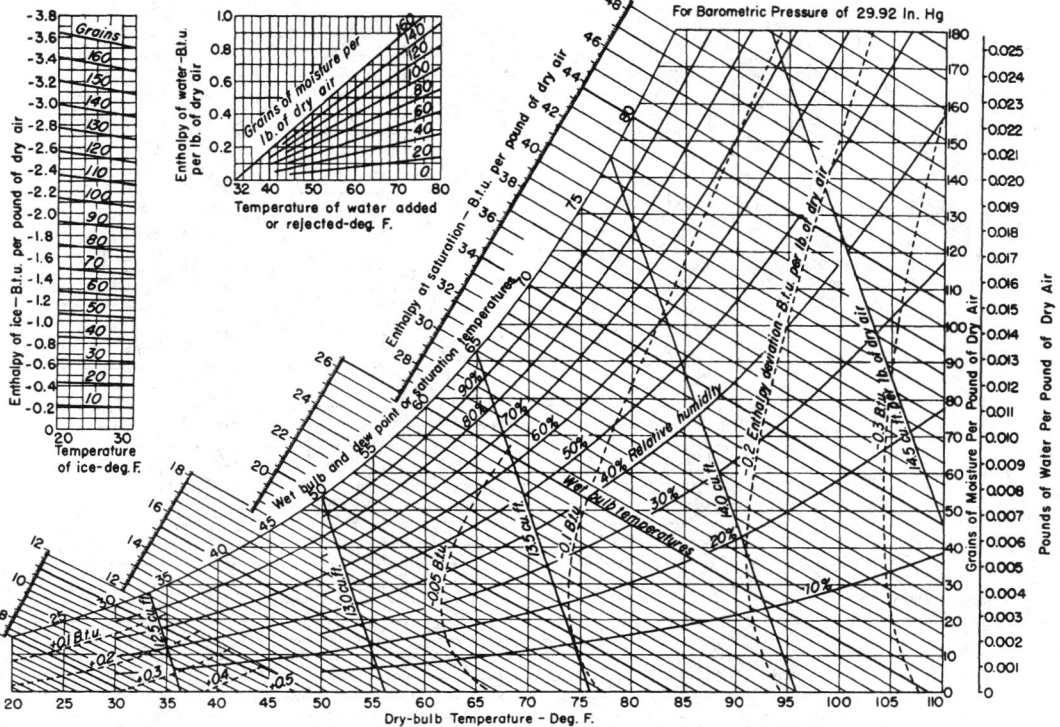

FIG. 12-2 Psychrometric chart—medium temperatures. Barometric pressure, 29.92 inHg. To convert British thermal units per pound dry air-degree Fahrenheit to joules per kilogram-kelvin, multiply by 4186.8; and to convert cubic feet per pound to cubic meters per kilogram, multiply by 0.0624.

Relative humidity = 15 percent

Wet-bulb temperature = 51.5°F

Dew point = 25.2°F

The enthalpy of the inlet air is obtained from Fig. 12-2 as $h_1 = h'_1 + D_1 = 10.1 + 0.06 = 10.16$ Btu/lb dry air; at the exit, $h_2 = h'_2 + D_2 = 21.1 - 0.1 = 21$ Btu/lb dry air. The heat added equals the enthalpy difference, or

$$q_a = \Delta h = h_2 - h_1 = 21 - 10.16 = 10.84 \text{ Btu/lb dry air}$$

If the enthalpy deviation is ignored, the heat added q_a is $\Delta h = 21.1 - 10.1 = 11$ Btu/lb dry air, or the result is 1.5 percent high. Figure 12-7 shows the heating path on the psychrometric chart.

Example 4: Evaporative Cooling Air at 95°F dry-bulb temperature and 70°F wet-bulb temperature contacts a water spray, where its relative humidity is increased to 90 percent. The spray water is recirculated; makeup water enters at 70°F. Determine exit dry-bulb temperature, wet-bulb temperature, change in enthalpy of the air, and quantity of moisture added per pound of dry air.

Solution. Figure 12-8 shows the path on a psychrometric chart. The leaving dry-bulb temperature is obtained directly from Fig. 12-2 as 72.2°F. Since the spray water enters at the wet-bulb temperature of 70°F and there is no heat added to or removed from it, this is by definition an adiabatic process and there will be no change in wet-bulb temperature. The only change in enthalpy is that from the heat content of the makeup water. This can be demonstrated as follows:

Inlet moisture $H_1 = 70$ gr/lb dry air

Exit moisture $H_2 = 107$ gr/lb dry air

$\Delta H = 37$ gr/lb dry air

Inlet enthalpy $h_1 = h'_1 + D_1 = 34.1 - 0.22$

$= 33.88$ Btu/lb dry air

Exit enthalpy $h_2 = h'_2 + D_2 = 34.1 - 0.02$

$= 34.08$ Btu/lb dry air

Enthalpy of added water $h_w = 0.2$ Btu/lb dry air (from small diagram, 37 gr at 70°F)

Then

$$q_a = h_2 - h_1 + h_w$$

$$= 34.08 - 33.88 + 0.2 = 0$$

Example 5: Cooling and Dehumidification Find the cooling load per pound of dry air resulting from infiltration of room air at 80°F dry-bulb temperature and 67°F wet-bulb temperature into a cooler maintained at 30°F dry-bulb and 28°F wet-bulb temperature, where moisture freezes on the coil, which is maintained at 20°F.

Solution. The path followed on a psychrometric chart is shown in Fig. 12-9.

Inlet enthalpy $h_1 = h'_1 + D_1 = 31.62 - 0.1$

$= 31.52$ Btu/lb dry air

Exit enthalpy $h_2 = h'_2 + D_2 = 10.1 + 0.06$

$= 10.16$ Btu/lb dry air

Inlet moisture $H_1 = 78$ gr/lb dry air

Exit moisture $H_2 = 19$ gr/lb dry air

Moisture rejected $\Delta H = 59$ gr/lb dry air

Enthalpy of rejected moisture $= -1.26$ Btu/lb dry air (from small diagram of Fig. 12-2)

Cooling load $q_r = 31.52 - 10.16 + 1.26$

$= 22.62$ Btu/lb dry air

Note that if the enthalpy deviations were ignored, the calculated cooling load would be about 5 percent low.

Example 6: Cooling Tower Determine water consumption and amount of heat dissipated per 1000 ft³/min of entering air at 90°F dry-bulb temperature and 70°F wet-bulb temperature when the air leaves saturated at 110°F and the makeup water is at 75°F.

Solution. The path followed is shown in Fig. 12-10.

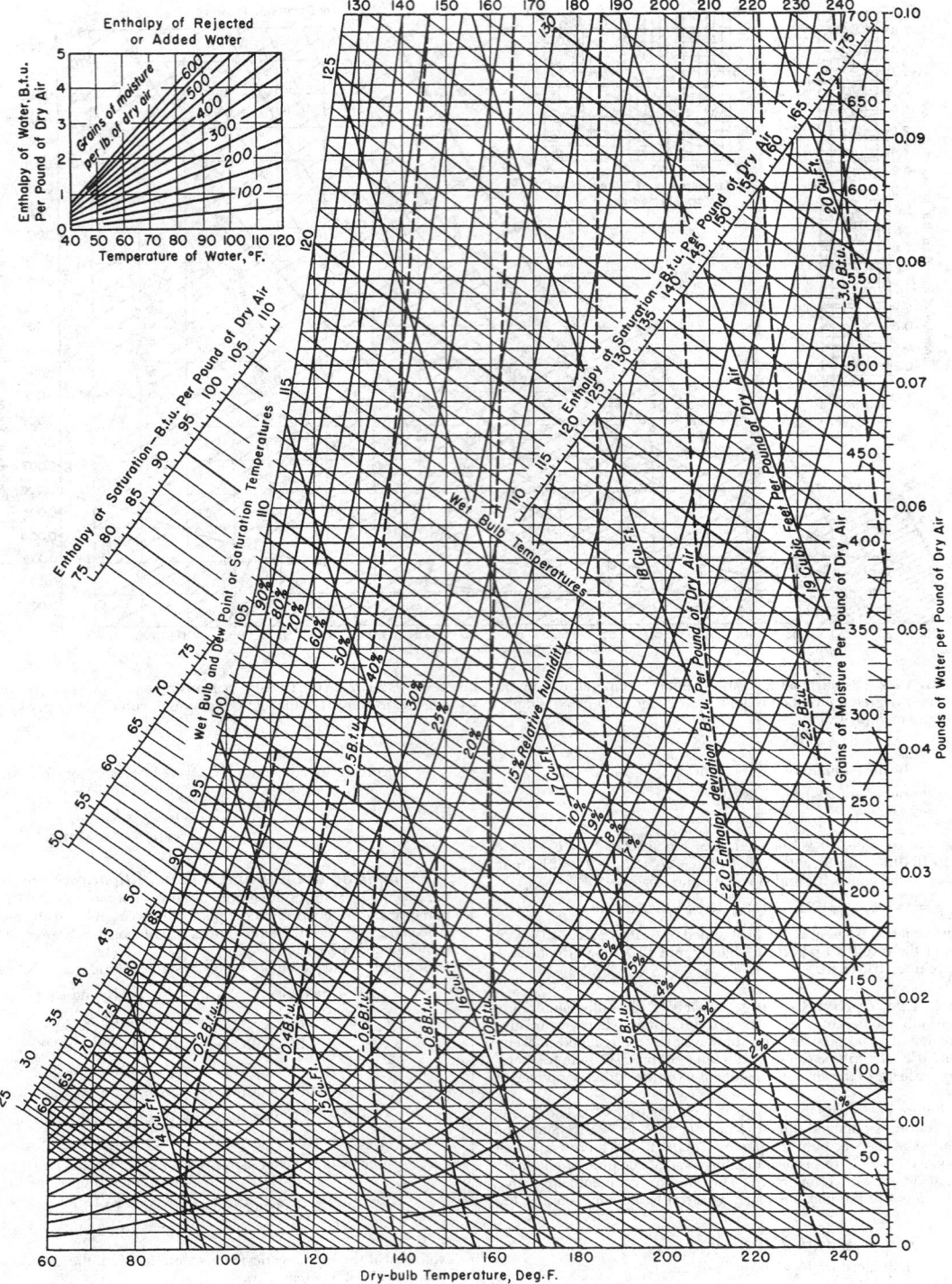

FIG. 12-3 Psychrometric chart—high temperatures. Barometric pressure, 29.92 inHg. To convert British thermal units per pound to joules per kilogram, multiply by 2326; to convert British thermal units per pound dry air-degree Fahrenheit to joules per kilogram-kelvin, multiply by 4186.8; and to convert cubic feet per pound to cubic meters per kilogram, multiply by 0.0624.

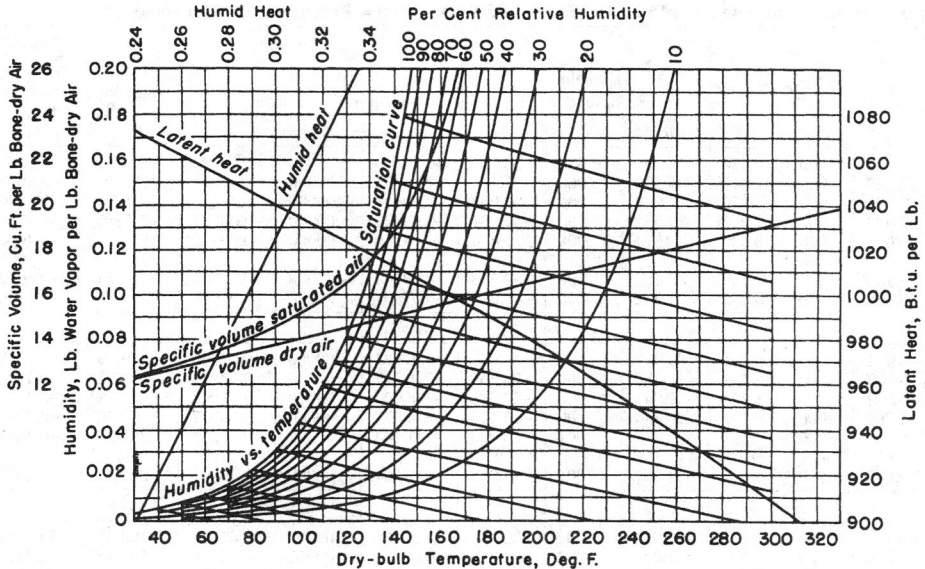

FIG. 12-4 Humidity chart for air-water vapor mixtures. To convert British thermal units per pound to joules per kilogram, multiply by 2326; to convert British thermal units per pound dry air-degree Fahrenheit to joules per kilogram-kelvin, multiply by 4186.8; and to convert cubic feet per pound to cubic meters per kilogram, multiply by 0.0624.

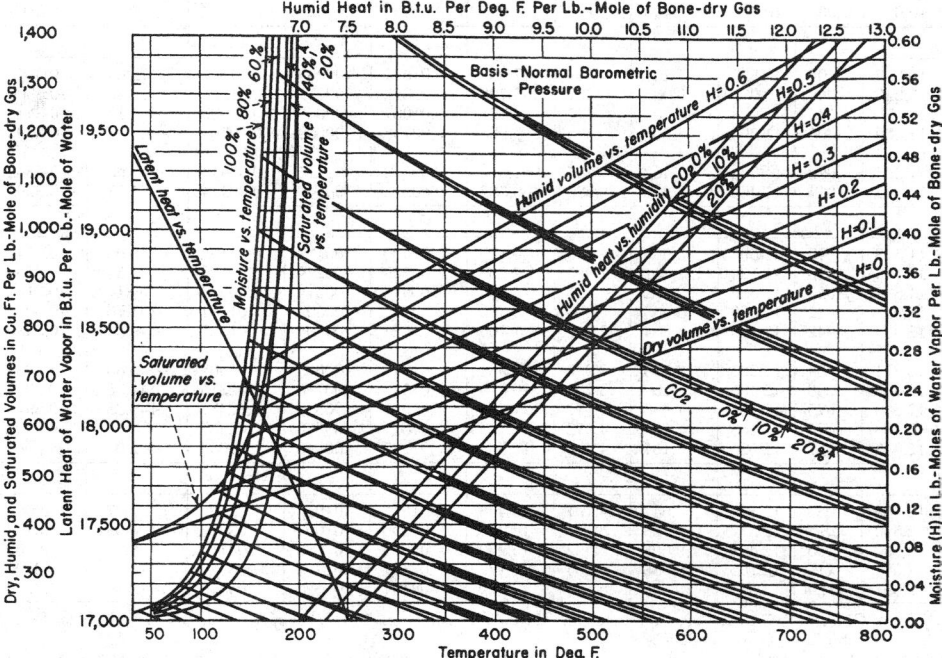

FIG. 12-5 Revised form of high-temperature psychrometric chart for air and combustion products, based on pound-moles of water vapor and dry gases. [*Hatta, Chem. Metall. Eng.,* **37,** *64 (1930).*]

TABLE 12-1 Thermodynamic Properties of Moist Air (Standard Atmospheric Pressure, 29.921 inHg)

Temp. t, °F	Saturation humidity $H_s \times 10^5$	Volume, cu. ft./lb. dry air			Enthalpy, B.t.u./lb. dry air			Entropy, B.t.u./(°F.)(lb. dry air)			Condensed water			Temp. t, °F
											Enthalpy, B.t.u./lb. h_w	Entropy, B.t.u./(lb.)(°F.) s_w	Vapor press., in. Hg $p_s \times 10^6$	
		v_a	v_{as}	v_s	h_a	h_{as}	h_s	s_a	s_{as}	s_s				
−160	0.2120	7.520	0.000	7.520	−38.504	0.000	−38.504	−0.10300	0.00000	−0.10300	−222.00	−0.4907	0.1009	−160
−155	.3869	7.647	.000	7.647	−37.296	.000	−37.296	−0.09901	.00000	−0.09901	−220.40	−0.4853	.1842	−155
−150	.6932	7.775	.000	7.775	−36.088	.000	−36.088	−0.09508	.00000	−0.09508	−218.77	−0.4800	.3301	−150
−145	1.219	7.902	.000	7.902	−34.881	.000	−34.881	−0.09121	.00000	−0.09121	−217.12	−0.4747	.5807	−145
−140	2.109	8.029	.000	8.029	−33.674	.000	−33.674	−0.08740	.00000	−0.08740	−215.44	−0.4695	1.004	−140
−135	3.586	8.156	.000	8.156	−32.468	.000	−32.468	−0.08365	.00000	−0.08365	−213.75	−0.4642	1.707	−135
−130	6.000	8.283	.000	8.283	−31.262	.000	−31.262	−0.07997	.00000	−0.07997	−212.03	−0.4590	2.858	−130
	$H_s \times 10^7$												$p_s \times 10^5$	
−125	0.9887	8.411	.000	8.411	−30.057	.000	−30.057	−0.07634	.00000	−0.07634	−210.28	−0.4538	0.4710	−125
−120	1.606	8.537	.000	8.537	−28.852	.000	−28.852	−0.07277	.00000	−0.07277	−208.52	−0.4485	.7653	−120
−115	2.571	8.664	.000	8.664	−27.648	.000	−27.648	−0.06924	.00000	−0.06924	−206.73	−0.4433	1.226	−115
−110	4.063	8.792	.000	8.792	−26.444	.000	−26.444	−0.06577	.00000	−0.06577	−204.92	−0.4381	1.939	−110
−105	6.340	8.919	.000	8.919	−25.240	.001	−25.239	−0.06234	.00000	−0.06234	−203.09	−0.4329	3.026	−105
−100	9.772	9.046	.000	9.046	−24.037	.001	−24.036	−0.05897	.00000	−0.05897	−201.23	−0.4277	4.666	−100
	$H_s \times 10^6$												$p_s \times 10^4$	
−95	1.489	9.173	.000	9.173	−22.835	.002	−22.833	−0.05565	.00000	−0.05565	−199.35	−0.4225	0.7111	−95
−90	2.242	9.300	.000	9.300	−21.631	.002	−21.629	−0.05237	.00001	−0.05236	−197.44	−0.4173	1.071	−90
−85	3.342	9.426	.000	9.426	−20.428	.003	−20.425	−0.04913	.00001	−0.04912	−195.51	−0.4121	1.597	−85
−80	4.930	9.553	.000	9.553	−19.225	.005	−19.220	−0.04595	.00001	−0.04594	−193.55	−0.4069	2.356	−80
−75	7.196	9.680	.000	9.680	−18.022	.007	−18.015	−0.04280	.00002	−0.04278	−191.57	−0.4017	3.441	−75
−70	10.40	9.806	.000	9.806	−16.820	.011	−16.809	−0.03969	.00003	−0.03966	−189.56	−0.3965	4.976	−70
−65	14.91	9.932	.000	9.932	−15.617	.015	−15.602	−0.03663	.00005	−0.03658	−187.53	−0.3913	7.130	−65
	$H_s \times 10^5$												$p_s \times 10^3$	
−60	2.118	10.059	.000	10.059	−14.416	.022	−14.394	−0.03360	.00006	−0.03354	−185.47	−0.3861	1.0127	−60
−55	2.982	10.186	.000	10.186	−13.214	.031	−13.183	−0.03061	.00009	−0.03052	−183.39	−0.3810	1.4258	−55
−50	4.163	10.313	.001	10.314	−12.012	.043	−11.969	−0.02766	.00012	−0.02754	−181.29	−0.3758	1.9910	−50
−45	5.766	10.440	.001	10.441	−10.811	.060	−10.751	−0.02474	.00015	−0.02459	−179.16	−0.3707	2.7578	−45
−40	7.925	10.566	.001	10.567	−9.609	.083	−9.526	−0.02186	.00021	−0.02165	−177.01	−0.3655	3.7906	−40
−35	10.81	10.693	.002	10.695	−8.408	.113	−8.295	−0.01902	.00028	−0.01874	−174.84	−0.3604	5.1713	−35
	$H_s \times 10^4$												$p_s \times 10^2$	
−30	1.464	10.820	.002	10.822	−7.207	.154	−7.053	−0.01621	.00038	−0.01583	−172.64	−0.3552	0.70046	−30
−25	1.969	10.946	.004	10.950	−6.005	.207	−5.798	−0.01342	.00051	−0.01291	−170.42	−0.3500	.94212	−25
−20	2.630	11.073	.005	11.078	−4.804	.277	−4.527	−0.01067	.00068	−0.00999	−168.17	−0.3449	1.2587	−20
−15	3.491	11.200	.006	11.206	−3.603	.368	−3.235	−0.00796	.00089	−0.00707	−165.90	−0.3398	1.6706	−15
−10	4.606	11.326	.008	11.334	−2.402	.487	−1.915	−0.00529	.00115	−0.00414	−163.60	−0.3346	2.2035	−10
−5	6.040	11.452	.011	11.463	−1.201	.639	−0.562	−0.00263	.00149	−0.00114	−161.28	−0.3295	2.8886	−5
	$H_s \times 10^3$													
0	0.7872	11.578	.015	11.593	0.000	.835	0.835	0.00000	.00192	0.00192	−158.93	−0.3244	3.7645	0
5	1.020	11.705	.019	11.724	1.201	1.085	2.286	.00260	.00246	.00506	−156.57	−0.3193	4.8779	5
10	1.315	11.831	.025	11.856	2.402	1.401	3.803	.00518	.00314	.00832	−154.17	−0.3141	6.2858	10
15	1.687	11.958	.032	11.990	3.603	1.800	5.403	.00772	.00399	.01171	−151.76	−0.3090	8.0565	15
20	2.152	12.084	.042	12.126	4.804	2.302	7.106	.01023	.00504	.01527	−149.31	−0.3039	10.272	20
25	2.733	12.211	.054	12.265	6.005	2.929	8.934	.01273	.00635	.01908	−146.85	−0.2988	13.032	25
30	3.454	12.338	.068	12.406	7.206	3.709	10.915	.01519	.00796	.02315	−144.36	−0.2936	16.452	30
32	3.788	12.388	.075	12.463	7.686	4.072	11.758	.01617	.00870	.02487	−143.36	−0.2916	18.035	32
32°	3.788	12.388	.075	12.463	7.686	4.072	11.758	.01617	.00870	.02487	0.04	0.0000	18.037	32°
34	4.107	12.438	.082	12.520	8.167	4.418	12.585	.01715	.00940	.02655	2.06	.0041	19.546	34
													p_s	
36	4.450	12.489	.089	12.578	8.647	4.791	13.438	.01812	.01016	.02828	4.07	.0081	0.21166	36
38	4.818	12.540	.097	12.637	9.128	5.191	14.319	.01909	.01097	.03006	6.08	.0122	.22904	38
40	5.213	12.590	.105	12.695	9.608	5.622	15.230	.02005	.01183	.03188	8.09	.0162	.24767	40
42	5.638	12.641	.114	12.755	10.088	6.084	16.172	.02101	.01275	.03376	10.09	.0202	.26763	42
44	6.091	12.691	.124	12.815	10.569	6.580	17.149	.02197	.01373	.03570	12.10	.0242	.28899	44

NOTE: Compiled by John A. Goff and S. Gratch. See also Keenan and Kaye. *Thermodynamic Properties of Air*, Wiley, New York, 1945. Enthalpy of dry air taken as zero at 0°F. Enthalpy of liquid water taken as zero at 32°F.

To convert British thermal units per pound to joules per kilogram, multiply by 2326; to convert British thermal units per pound dry air-degree Fahrenheit to joules per kilogram-kelvin, multiply by 4186.8; and to convert cubic feet per pound to cubic meters per kilogram, multiply by 0.0624.

°Entrapolated to represent metastable equilibrium with undercooled liquid.

TABLE 12-1 Thermodynamic Properties of Moist Air (Standard Atmospheric Pressure, 29.921 inHg) (Continued)

Temp. t, °F.	Saturation humidity $H_s \times 10^6$	Volume, cu. ft./lb. dry air			Enthalpy, B.t.u./lb. dry air			Entropy, B.t.u./(°F.)(lb. dry air)			Condensed water			Temp. t, °F.
		v_a	v_{as}	v_s	h_a	h_{as}	h_s	s_a	s_{as}	s_s	Enthalpy, B.t.u./lb. h_w	Entropy, B.t.u./(lb.)(°F.) s_w	Vapor press., in. Hg $p_s \times 10^6$	
46	6.578	12.742	.134	12.876	11.049	7.112	18.161	.02293	.01478	.03771	14.10	.0282	.31185	46
48	7.100	12.792	.146	12.938	11.530	7.681	19.211	.02387	.01591	.03978	16.11	.0321	.33629	48
50	7.658	12.843	.158	13.001	12.010	8.291	20.301	.02481	.01711	.04192	18.11	.0361	.36240	50
52	8.256	12.894	.170	13.064	12.491	8.945	21.436	.02575	.01839	.04414	20.11	.0400	.39028	52
54	8.894	12.944	.185	13.129	12.971	9.644	22.615	.02669	.01976	.04645	22.12	.0439	.42004	54
56	9.575	12.995	.200	13.195	13.452	10.39	23.84	.02762	.02121	.04883	24.12	.0478	.45176	56
58	10.30	13.045	.216	13.261	13.932	11.19	25.12	.02855	.02276	.05131	26.12	.0517	.48558	58
60	11.08	13.096	.233	13.329	14.413	12.05	26.46	.02948	.02441	.05389	28.12	.0555	.52159	60
62	11.91	13.147	.251	13.398	14.893	12.96	27.85	.03040	.02616	.05656	30.12	.0594	.55994	62
64	12.80	13.197	.271	13.468	15.374	13.94	29.31	.03132	.02803	.05935	32.12	.0632	.60073	64
66	13.74	13.247	.292	13.539	15.855	14.98	30.83	.03223	.03002	.06225	34.11	.0670	.64411	66
68	14.75	13.298	.315	13.613	16.335	16.09	32.42	.03314	.03213	.06527	36.11	.0708	.69019	68
	$H_s \times 10^2$													
70	1.582	13.348	.339	13.687	16.816	17.27	34.09	.03405	.03437	.06842	38.11	.0746	.73915	70
72	1.697	13.398	.364	13.762	17.297	18.53	35.83	.03495	.03675	.07170	40.11	.0784	.79112	72
74	1.819	13.449	.392	13.841	17.778	19.88	37.66	.03585	.03928	.07513	42.10	.0821	.84624	74
76	1.948	13.499	.422	13.921	18.259	21.31	39.57	.03675	.04197	.07872	44.10	.0859	.90470	76
78	2.086	13.550	.453	14.003	18.740	22.84	41.58	.03765	.04482	.08247	46.10	.0896	.96665	78
80	2.233	13.601	0.486	14.087	19.221	24.47	43.69	0.03854	0.04784	0.08638	48.10	0.0933	1.0323	80
82	2.389	13.651	.523	14.174	19.702	26.20	45.90	.03943	.05105	.09048	50.09	.0970	1.1017	82
84	2.555	13.702	.560	14.262	20.183	28.04	48.22	.04031	.05446	.09477	52.09	.1007	1.1752	84
86	2.731	13.752	.602	14.354	20.663	30.00	50.66	.04119	.05807	.09926	54.08	.1043	1.2529	86
88	2.919	13.803	.645	14.448	21.144	32.09	53.23	.04207	.06189	.10396	56.08	.1080	1.3351	88
90	3.118	13.853	.692	14.545	21.625	34.31	55.93	.04295	.06596	.10890	58.08	.1116	1.4219	90
92	3.330	13.904	.741	14.645	22.106	36.67	58.78	.04382	.07025	.11407	60.07	.1153	1.5135	92
94	3.556	13.954	.795	14.749	22.587	39.18	61.77	.04469	.07480	.11949	62.07	.1188	1.6102	94
96	3.795	14.005	.851	14.856	23.068	41.85	64.92	.04556	.07963	.12519	64.06	.1224	1.7123	96
98	4.049	14.056	.911	14.967	23.548	44.68	68.23	.04643	.08474	.13117	66.06	.1260	1.8199	98
100	4.319	14.106	.975	15.081	24.029	47.70	71.73	.04729	.09016	.13745	68.06	.1296	1.9333	100
102	4.606	14.157	1.043	15.200	24.510	50.91	75.42	.04815	.09591	.14406	70.05	.1332	2.0528	102
104	4.911	14.207	1.117	15.324	24.991	54.32	79.31	.04900	.1020	.1510	72.05	.1367	2.1786	104
	$H_s \times 10$													
106	0.5234	14.258	1.194	15.452	25.472	57.95	83.42	.04985	.1085	.1584	74.04	.1403	2.3109	106
108	.5578	14.308	1.278	15.586	25.953	61.80	87.76	.05070	.1153	.1660	76.04	.1438	2.4502	108
110	.5944	14.359	1.365	15.724	26.434	65.91	92.34	.05155	.1226	.1742	78.03	.1472	2.5966	110
112	.6333	14.409	1.460	15.869	26.915	70.27	97.18	.05239	.1302	.1826	80.03	.1508	2.7505	112
114	.6746	14.460	1.560	16.020	27.397	74.91	102.31	.05323	.1384	.1916	82.03	.1543	2.9123	114
116	.7185	14.510	1.668	16.178	27.878	79.85	107.73	.05407	.1470	.2011	84.02	.1577	3.0821	116
118	.7652	14.561	1.782	16.343	28.359	85.10	113.46	.05490	.1562	.2111	86.02	.1612	3.2603	118
120	.8149	14.611	1.905	16.516	28.841	90.70	119.54	.05573	.1659	.2216	88.01	.1646	3.4474	120
122	.8678	14.662	2.034	16.696	29.322	96.66	125.98	.05656	.1763	.2329	90.01	.1681	3.6436	122
124	.9242	14.712	2.174	16.886	29.804	103.0	132.8	.05739	.1872	.2446	92.01	.1715	3.8493	124
126	.9841	14.763	2.323	17.086	30.285	109.8	140.1	.05821	.1989	.2571	94.01	.1749	4.0649	126
128	1.048	14.813	2.482	17.295	30.766	117.0	147.8	.05903	.2113	.2703	96.00	.1783	4.2907	128
130	1.116	14.864	2.652	17.516	31.248	124.7	155.9	.05985	.2245	.2844	98.00	.1817	4.5272	130
132	1.189	14.915	2.834	17.749	31.729	133.0	164.7	.06067	.2386	.2993	100.00	.1851	4.7747	132
134	1.267	14.965	3.029	17.994	32.211	141.8	174.0	.06148	.2536	.3151	102.00	.1885	5.0337	134
136	1.350	15.016	3.237	18.253	32.692	151.2	183.9	.06229	.2695	.3318	104.00	.1918	5.3046	136
138	1.439	15.066	3.462	18.528	33.174	161.2	194.4	.06310	.2865	.3496	106.00	.1952	5.5878	138
	H_s													
140	0.1534	15.117	3.702	18.819	33.655	172.0	205.7	.06390	.3047	.3686	107.99	.1985	5.8838	140
142	.1636	15.167	3.961	19.128	34.136	183.6	217.7	.06470	.3241	.3888	109.99	.2018	6.1930	142
144	.1745	15.218	4.239	19.457	34.618	196.0	230.6	.06549	.3449	.4104	111.99	.2051	6.5160	144
146	.1862	15.268	4.539	19.807	35.099	209.3	244.4	.06629	.3672	.4335	113.99	.2084	6.8532	146
148	.1989	15.319	4.862	20.181	35.581	223.7	259.3	.06708	.3912	.4583	115.99	.2117	7.2051	148
150	.2125	15.369	5.211	20.580	36.063	239.2	275.3	.06787	.4169	.4848	117.99	.2150	7.5722	150
152	.2271	15.420	5.587	21.007	36.545	255.9	292.4	.06866	.4445	.5132	119.99	.2183	7.9550	152
154	.2430	15.470	5.996	21.466	37.026	273.9	310.9	.06945	.4743	.5438	121.99	.2216	8.3541	154
156	.2602	15.521	6.439	21.960	37.508	293.5	331.0	.07023	.5066	.5768	123.99	.2248	8.7701	156
158	.2788	15.571	6.922	22.493	37.990	314.7	352.7	.07101	.5415	.6125	125.99	.2281	9.2036	158

TABLE 12-1 Thermodynamic Properties of Moist Air (Standard Atmospheric Pressure, 29.921 inHg) (Concluded)

Temp. t, °F.	Saturation humidity $H_s \times 10^5$	Volume, cu. ft./lb. dry air			Enthalpy, B.t.u./lb. dry air			Entropy, B.t.u./(°F.)(lb. dry air)			Condensed water			Temp. t, °F.
											Enthalpy, B.t.u./lb. h_w	Entropy, B.t.u./(lb.)(°F.) s_w	Vapor press., in. Hg $p_s \times 10^6$	
		v_a	v_{as}	v_s	h_a	h_{as}	h_s	s_a	s_{as}	s_s				
160	.2990	15.622	7.446	23.068	38.472	337.8	376.3	.07179	.5793	.6511	128.00	.2313	9.6556	160
162	.3211	15.672	8.020	23.692	38.954	363.0	402.0	.07257	.6204	.6930	130.00	.2345	10.125	162
164	.3452	15.723	8.648	24.371	39.436	390.5	429.9	.07334	.6652	.7385	132.00	.2377	10.614	164
166	.3716	15.773	9.339	25.112	39.918	420.8	460.7	.07411	.7142	.7883	134.00	.2409	11.123	166
168	.4007	15.824	10.098	25.922	40.400	454.0	494.4	.07488	.7680	.8429	136.01	.2441	11.652	168
170	.4327	15.874	10.938	26.812	40.882	490.6	531.5	.07565	.8273	.9030	138.01	.2473	12.203	170
172	.4682	15.925	11.870	27.795	41.364	531.3	572.7	.07641	.8927	.9691	140.01	.2505	12.775	172
174	.5078	15.975	12.911	28.886	41.846	576.5	618.3	.07718	.9654	1.0426	142.02	.2537	13.369	174
176	.5519	16.026	14.074	30.100	42.328	627.1	669.4	.07794	1.047	1.125	144.02	.2568	13.987	176
178	.6016	16.076	15.386	31.462	42.810	684.1	726.9	.07870	1.137	1.216	146.03	.2600	14.628	178
180	.6578	16.127	16.870	32.997	43.292	748.5	791.8	.07946	1.240	1.319	148.03	.2631	15.294	180
182	.7218	16.177	18.565	34.742	43.775	821.9	865.7	.08021	1.357	1.437	150.04	.2662	15.985	182
184	.7953	16.228	20.513	36.741	44.257	906.2	950.5	.08096	1.490	1.571	152.04	.2693	16.702	184
186	.8805	16.278	22.775	39.053	44.740	1004	1049	.08171	1.645	1.727	154.05	.2724	17.446	186
188	.9802	16.329	25.427	41.756	45.222	1119	1164	.08245	1.825	1.907	156.06	.2755	18.217	188
190	1.099	16.379	28.580	44.959	45.704	1255	1301	.08320	2.039	2.122	158.07	.2786	19.017	190
192	1.241	16.430	32.375	48.805	46.187	1418	1464	.08394	2.296	2.380	160.07	.2817	19.845	192
194	1.416	16.480	37.036	53.516	46.670	1619	1666	.08468	2.609	2.694	162.08	.2848	20.704	194
196	1.635	16.531	42.885	59.416	47.153	1871	1918	.08542	3.002	3.087	164.09	.2879	21.594	196
198	1.917	16.581	50.426	67.007	47.636	2195	2243	.08616	3.507	3.593	166.10	.2910	22.514	198
200	2.295	16.632	60.510	77.142	48.119	2629	2677	.08689	4.179	4.266	168.11	.2940	23.468	200

TABLE 12-2 Additive Corrections for H, h, and v When Barometric Pressure Differs from Standard Barometer

Approximate altitude in feet

Wet-bulb temp. t_w	Sat. vapor press., in. Hg	−900 $\Delta p = +1$		900 $\Delta p = -1$		1800 $\Delta p = -2$		2700 $\Delta p = -3$		3700 $\Delta p = -4$		4800 $\Delta p = -5$		5900 $\Delta p = -6$	
		ΔH_s	Δh	ΔH_s	Δh	ΔH_s	Δh	ΔH_s	Δh	ΔH_s	Δh	ΔH_s	Δh	ΔH_s	Δh
−10	0.022	−0.10	−0.02	0.11	0.02	0.23	0.03	0.36	0.05	0.50	0.07	0.64	0.10	0.81	0.12
−8	.025	−0.12	−0.02	.12	.02	.26	.04	.40	.06	.55	.08	.72	.11	.90	.13
−6	.027	−0.13	−0.02	.14	.02	.29	.04	.44	.07	.62	.09	.80	.12	1.00	.15
−4	.030	−0.14	0.02	.15	.02	.32	.05	.50	.07	.69	.10	.89	.13	1.12	.17
−2	.034	−0.16	−0.02	.17	.02	.35	.05	.55	.08	.76	.11	.99	.15	1.24	.19
0	.038	−0.18	−0.03	.19	.03	.39	.06	.61	.09	.85	.13	1.10	.17	1.38	.21
2	.042	−0.20	−0.03	.21	.03	.44	.07	.68	.10	.94	.14	1.22	.19	1.53	.23
4	.046	−0.22	−0.03	.23	.03	.48	.07	.75	.11	1.05	.16	1.36	.21	1.70	.26
6	.051	−0.24	−0.04	.26	.04	.54	.08	.83	.13	1.16	.18	1.51	.23	1.89	.29
8	.057	−0.27	−0.04	.29	.04	.59	.09	.93	.14	1.28	.19	1.67	.25	2.09	.32
10	.063	−0.30	−0.04	.32	.05	.66	.10	1.03	.16	1.42	.22	1.85	.28	2.31	.35
12	.069	−0.33	−0.05	.35	.05	.73	.11	1.13	.17	1.57	.24	2.04	.31	2.56	.39
14	.077	−0.36	−0.05	.39	.06	.81	.12	1.25	.19	1.74	.26	2.26	.34	2.82	.43
16	.085	−0.40	−0.06	.43	.06	.89	.14	1.38	.21	1.92	.29	2.49	.38	3.12	.48
18	.093	−0.44	−0.07	.47	.07	.98	.15	1.53	.23	2.12	.32	2.75	.42	3.44	.53
20	.103	−0.49	−0.08	.52	.08	1.08	.17	1.68	.26	2.33	.36	3.03	.46	3.79	.58
22	.113	−0.5	−0.08	.6	.09	1.2	.18	1.9	.29	2.6	.40	3.4	.52	4.2	.64
24	.124	−0.6	−0.09	.6	.10	1.3	.20	2.1	.32	2.8	.43	3.7	.57	4.6	.71
26	.137	−0.7	−0.10	.7	.11	1.4	.22	2.3	.35	3.1	.48	4.1	.63	5.1	.78
28	.150	−0.7	−0.11	.8	.12	1.6	.24	2.5	.38	3.4	.52	4.5	.69	5.6	.86
30	.165	−0.8	−0.12	.8	.13	1.7	.27	2.7	.42	3.8	.58	4.9	.75	6.1	.92
32	.180	−0.9	−0.13	.9	.14	1.9	.29	3.0	.45	4.1	.63	5.3	.82	6.6	1.01
34	.197	−0.9	−0.14	1.0	.15	2.1	.32	3.2	.49	4.4	.68	5.7	.88	7.2	1.11
36	.212	−1.0	−0.15	1.1	.17	2.2	.35	3.5	.53	4.8	.74	6.2	.96	7.8	1.20
38	.229	−1.1	−0.17	1.2	.18	2.4	.37	3.8	.58	5.2	.80	6.8	1.05	8.4	1.30
40	.248	−1.2	−0.18	1.3	.20	2.6	.41	4.1	.63	5.7	.88	7.4	1.14	9.2	1.42
42	.268	−1.3	−0.20	1.4	.21	2.8	.44	4.4	.69	6.1	.94	8.0	1.23	10.0	1.54
44	.289	−1.4	−0.22	1.5	.23	3.1	.47	4.8	.74	6.7	1.04	8.7	1.34	10.8	1.67
46	.312	−1.5	−0.23	1.6	.25	3.3	.51	5.2	.80	7.2	1.11	9.4	1.45	11.7	1.81
48	.336	−1.6	−0.25	1.8	.27	3.6	.56	5.6	.87	7.8	1.21	10.2	1.58	12.6	1.95
50	.3624	−1.7	−0.27	1.9	.29	3.9	.60	6.1	.94	8.4	1.30	10.9	1.69	13.6	2.11
52	.3903	−1.9	−0.29	2.0	.32	4.2	.65	6.5	1.01	9.0	1.40	11.8	1.83	14.7	2.28
54	.4200	−2.0	−0.31	2.2	.34	4.5	.70	7.0	1.09	9.7	1.50	12.7	1.97	15.8	2.45
56	.4518	−2.2	−0.34	2.4	.37	4.9	.76	7.6	1.18	10.5	1.63	13.7	2.13	17.1	2.66
58	.4856	−2.3	−0.37	2.5	.39	5.3	.82	8.2	1.27	11.3	1.76	14.7	2.28	18.4	2.86

TABLE 12-2   Additive Corrections for H, h, and v When Barometric Pressure Differs from Standard Barometer (Concluded)

Approximate altitude in feet

Wet-bulb temp. t_w	Sat. vapor press., in. Hg	−900 $\Delta p = +1$		900 $\Delta p = -1$		1800 $\Delta p = -2$		2700 $\Delta p = -3$		3700 $\Delta p = -4$		4800 $\Delta p = -5$		5900 $\Delta p = -6$	
		ΔH_s	Δh	ΔH_s	Δh	ΔH_s	Δh	ΔH_s	Δh	ΔH_s	Δh	ΔH_s	Δh	ΔH_s	Δh
60	.522	−2.5	−0.40	2.7	.42	5.7	.88	8.8	1.37	12.2	1.90	15.9	2.47	19.9	3.09
62	.560	−2.7	−0.43	2.9	.46	6.1	.95	9.5	1.48	13.2	2.05	17.1	2.66	21.4	3.33
64	.601	−2.9	−0.46	3.2	.49	6.5	1.02	10.2	1.59	14.2	2.21	18.4	2.87	23.1	3.60
66	.644	−3.2	−0.50	3.4	.53	7.1	1.10	11.0	1.72	15.3	2.38	19.8	3.09	24.8	3.87
68	.690	−3.4	−0.53	3.7	.57	7.6	1.18	11.8	1.84	16.4	2.56	21.3	3.32	26.7	4.16
70	.739	−3.7	−0.57	3.9	.61	8.1	1.27	12.7	1.98	17.6	2.75	22.9	3.58	28.7	4.48
72	.791	−3.9	−0.61	4.2	.66	8.7	1.36	13.6	2.13	18.8	2.94	24.6	3.84	30.9	4.82
74	.846	−4.2	−0.66	4.6	.71	9.4	1.46	14.6	2.28	20.2	3.16	26.4	4.14	33.1	5.18
76	.905	−4.5	−0.71	4.9	.77	10.0	1.57	15.7	2.46	21.7	3.39	28.3	4.42	35.5	5.56
78	.967	−4.9	−0.76	5.2	.82	10.8	1.69	16.9	2.65	23.3	3.65	30.5	4.77	38.2	5.98
80	1.032	−5.2	−0.82	5.6	.88	11.6	1.82	18.1	2.84	25.1	3.93	32.7	5.13	41.0	6.43
82	1.102	−5.6	−0.88	6.0	.94	12.5	1.96	19.5	3.06	27.0	4.24	35.1	5.51	44.0	6.90
84	1.175	−6.0	−0.94	6.4	1.00	13.3	2.10	20.9	3.28	28.9	4.54	37.7	5.92	47.2	7.41
86	1.253	−6.4	−1.00	6.9	1.08	14.3	2.24	22.3	3.50	30.9	4.85	40.4	6.34	50.6	7.94
88	1.335	−6.9	−1.08	7.4	1.16	15.3	2.40	23.9	3.75	33.1	5.20	43.2	6.79	54.2	8.51
90	1.422	−7.4	−1.16	7.9	1.24	16.5	2.59	25.7	4.04	35.6	5.60	46.4	7.29	58.2	9.15
92	1.514	−7.9	−1.24	8.5	1.34	17.6	2.77	27.5	4.33	38.2	6.01	49.8	7.83	62.5	9.83
94	1.610	−8.5	−1.34	9.1	1.43	18.9	2.98	29.5	4.64	41.0	6.46	53.4	8.41	67.0	10.55
96	1.712	−9.1	−1.43	9.8	1.54	20.2	3.18	31.5	4.96	43.8	6.90	57.2	9.01	71.7	11.30
98	1.820	−9.7	−1.53	10.4	1.64	21.7	3.42	33.8	5.33	47.0	7.41	61.3	9.67	76.8	12.11
100	1.933	−10.4	−1.64	11.2	1.77	23.2	3.66	36.3	5.73	50.4	7.95	65.7	10.37	82.5	13.02
102	2.053	−11.1	−1.75	12.0	1.90	24.8	3.92	38.9	6.14	54.1	8.54	70.5	11.13	88.5	13.98
104	2.179	−11.9	−1.88	12.8	2.02	26.6	4.20	41.6	6.58	57.9	9.15	75.5	11.93	94.8	14.98
106	2.311	−12.8	−2.02	13.7	2.17	28.6	4.52	44.6	7.06	62.1	9.82	81.1	12.83	101.7	16.09
108	2.450	−13.7	−2.17	14.7	2.33	30.6	4.84	47.7	7.55	66.5	10.53	87.0	13.77	109.1	17.27
110	2.597	−14.7	−2.33	15.8	2.50	32.8	5.20	51.3	8.13	71.3	11.30	93.1	14.75	117.0	18.54
112	2.751	−15.7	−2.49	16.9	2.68	35.2	5.58	55.0	8.72	76.4	12.11	99.9	15.84	125.9	19.96
114	2.913	−16.9	−2.68	18.1	2.87	37.7	5.98	58.9	9.50	82.0	13.01	107.3	17.03	135.0	21.42
116	3.082	−18.0	−2.86	19.4	3.08	40.4	6.42	63.2	10.03	88.0	13.97	115.1	18.28	144.7	22.98
118	3.260	−19.3	−3.07	20.8	3.31	43.3	6.88	67.8	10.77	94.4	15.00	123.5	19.63	155.4	24.73
120	3.448	−20.7	−3.29	22.4	3.56	46.6	7.41	72.8	11.58	101.4	16.13	132.7	21.10	167.1	26.58
122	3.644	−22.2	−3.53	24.0	3.82	50.0	7.96	78.2	12.45	109.0	17.35	142.6	22.70	179.6	28.58
124	3.850	−23.8	−3.79	25.8	4.11	53.7	8.55	84.0	13.38	117.1	18.65	153.3	24.42	193.2	30.77
126	4.065	−25.6	−4.08	27.6	4.40	57.7	9.20	90.3	14.39	125.9	20.07	165.0	26.30	208.0	33.15
128	4.291	−27.5	−4.39	29.7	4.74	62.0	9.89	97.1	15.49	135.5	21.61	177.6	28.33	224.0	35.73
130	4.527	−29.5	−4.71	32.0	5.11	66.7	10.64	104.5	16.68	145.9	23.29	191.4	30.55	241.5	38.55
132	4.775	−31.8	−5.08	34.4	5.50	71.8	11.47	112.6	17.99	157.2	25.11	206.3	32.96	260.6	41.63
134	5.034	−34.2	−5.47	37.1	5.93	77.4	12.37	121.4	19.41	169.6	27.12	222.7	35.60	281.4	44.99
136	5.305	−36.8	−5.89	40.0	6.40	83.4	13.34	130.9	20.94	183.1	29.30	240.5	38.48	304.2	48.67
138	5.588	−39.7	−6.36	43.2	6.92	90.0	14.41	141.4	22.64	197.8	31.67	260.1	41.65	329.3	52.73
140	5.884	−42.8	−6.86	46.5	8.45	97.3	15.59	152.8	24.48	214.0	34.29	281.6	45.12	356.8	57.17

t = dry-bulb temperature, °F
t_w = wet-bulb temperature, °F
p = barometric pressure, inHg
Δp = pressure difference from standard barometer (inHg)
H = moisture content of air, gr/lb dry air
H_s = moisture content of air saturated at wet-bulb temperature (t_w), gr/lb dry air
ΔH = moisture-content correction of air when barometric pressure differs from standard barometer, gr/lb dry air
ΔH_s = moisture-content correction of air saturated at wet-bulb temperature when barometric pressure differs from standard barometer, gr/lb dry air
NOTE: To obtain ΔH reduce value of ΔH_s by 1 percent where $t - t_w = 24°F$ and correct proportionally when $t - t_w$ is not 24°F.
h = enthalpy of moist air, Btu/lb dry air
Δh = enthalpy correction when barometric pressure differs from standard barometer, for saturated or unsaturated air, Btu/lb dry air
v = volume of moist air, ft³/lb dry air

$$= \frac{0.754(t + 459.8)}{p}\left(1 + \frac{H}{4360}\right)$$

Example   At a barometric pressure of 25.92 with 220°F dry-bulb and 100°F wet-bulb temperatures, determine H, h, and v. $\Delta p = -4$, and from table $\Delta H_s = 50.4$. From note,

$$\Delta H = \Delta H_s - \left(\frac{120}{24} \times 0.01 \times 50.4\right) = 50.4 - 2.5 = 47.9$$

Therefore $H = 102$ (from chart) $+ 47.9 = 149.9$ gr/lb. dry air. From table $\Delta h = 7.95$. Therefore, $h =$ saturation enthalpy from chart + deviation + 7.95 = 71.7 − 2.0 + 7.95 = 77.65 Btu/lb dry air. From previous equation

$$v = \frac{0.754(220 + 459.7)}{25.92}\left(1 + \frac{149.9}{4360}\right) = 20.43 \text{ ft}^3/\text{lb dry air}$$

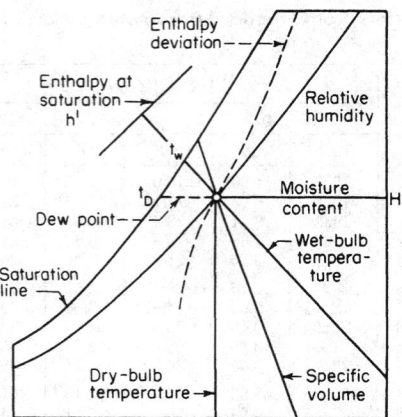

FIG. 12-6   Diagram of psychrometric chart showing the properties of moist air.

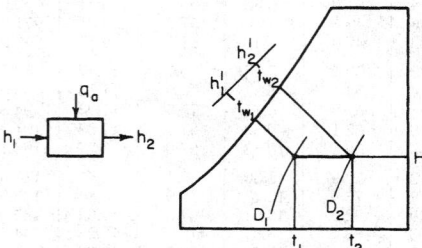

FIG. 12-7   Heating process.

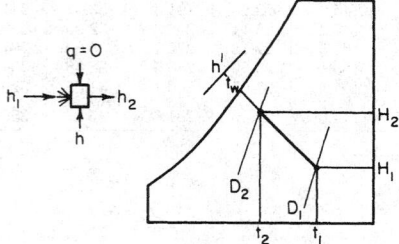

FIG. 12-8   Spray or evaporative cooling.

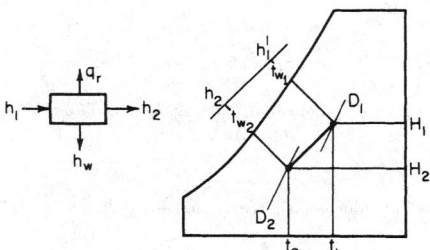

FIG. 12-9   Cooling and dehumidifying process.

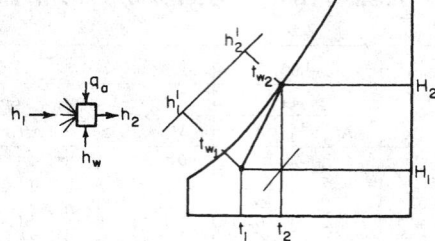

FIG. 12-10   Cooling tower.

Example 6: Cooling Tower (Continued)

$$\text{Exit moisture } H_2 = 416 \text{ gr/lb dry air}$$
$$\text{Inlet moisture } H_1 = 78 \text{ gr/lb dry air}$$
$$\text{Moisture added } \Delta H = 338 \text{ gr/lb dry air}$$
$$\text{Enthalpy of added moisture } h_w = 2.1 \text{ Btu/lb dry air (from small diagram}$$
$$\text{of Fig. 12-3)}$$

If greater precision is desired, h_w can be calculated as

$$h_w = (338/7000)(1)(75 - 32)$$
$$= 2.08 \text{ Btu/lb dry air}$$
$$\text{Enthalpy of inlet air } h_1 = h_1' + D_1 = 34.1 - 0.18$$
$$= 33.92 \text{ Btu/lb dry air}$$
$$\text{Enthalpy of exit air } h_2 = h_2' + D_2 = 92.34 + 0$$
$$= 92.34 \text{ Btu/lb dry air}$$
$$\text{Heat dissipated} = h_2 - h_1 - h_w$$
$$= 92.34 - 33.92 - 2.08$$
$$= 56.34 \text{ Btu/lb dry air}$$
$$\text{Specific volume of inlet air} = 14.1 \text{ ft}^3\text{/lb dry air}$$
$$\text{Total heat dissipated} = \frac{(1000)(56.34)}{14.1} = 3990 \text{ Btu/min}$$

Example 7: Recirculating Dryer

A dryer is removing 100 lb water/h from the material being dried. The air entering the dryer has a dry-bulb temperature of 180°F and a wet-bulb temperature of 110°F. The air leaves the dryer at 140°F. A portion of the air is recirculated after mixing with room air having a dry-bulb temperature of 75°F and a relative humidity of 60 percent. Determine quantity of air required, recirculation rate, and load on the preheater if it is assumed that the system is adiabatic. Neglect heatup of the feed and of the conveying equipment.

Solution.   The path followed is shown in Fig. 12-11.

$$\text{Humidity of room air } H_1 = 0.0113 \text{ lb/lb dry air}$$
$$\text{Humidity of air entering dryer } H_3 = 0.0418 \text{ lb/lb dry air}$$
$$\text{Humidity of air leaving dryer } H_4 = 0.0518 \text{ lb/lb dry air}$$
$$\text{Enthalpy of room air } h_1 = 30.2 - 0.3$$
$$= 29.9 \text{ Btu/lb dry air}$$
$$\text{Enthalpy of entering air } h_3 = 92.5 - 1.3$$
$$= 91.2 \text{ Btu/lb dry air}$$

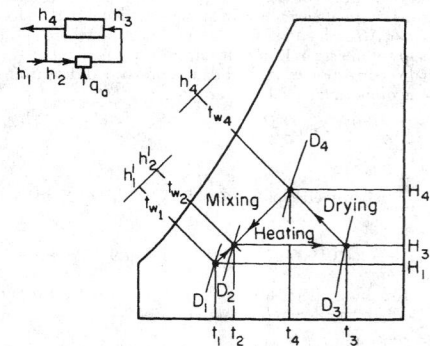

FIG. 12-11   Drying process with recirculation.

Enthalpy of leaving air $h_4 = 92.5 - 0.55$

$$= 91.95 \text{ Btu/lb dry air}$$

Quantity of air required is $100/(0.0518 - 0.0418) = 10,000$ lb dry air/h. At the dryer inlet the specific volume is 17.1 ft³/lb dry air. Air volume is $(10,000)(17.1)/60 = 2850$ ft³/min. Fraction exhausted is

$$\frac{X}{W_a} = \frac{0.0518 - 0.0418}{0.0518 - 0.0113} = 0.247$$

where X = quantity of fresh air and W_a = total air flow. Thus 75.3 percent of the air is recirculated. Load on the preheater is obtained from an enthalpy balance

$$q_a = 10,000(91.2) - 2470(29.9) - 7530(91.95)$$

$$= 146,000 \text{ Btu/h}$$

Use of Psychrometric Charts at Pressures Other Than Atmospheric   The psychrometric charts shown as Figs. 12-1 through 12-4 and the data of Table 12-1 are based on a system pressure of 1 atm (29.92 inHg). For other system pressures, these data must be corrected for the effect of pressure. Additive corrections to be applied to the atmospheric values of absolute humidity and enthalpy are given in Table 12-2.

The **specific volume** of moist air in cubic feet per pound of dry air can be determined for other pressures, if ideal-gas behavior is assumed, by the following equation:

$$v = \frac{0.754(t + 460)}{P}\left(1 + \frac{HM_a}{M_w}\right) \qquad (12\text{-}5)$$

where v = specific volume, ft³/lb dry air; t = dry-bulb temperature, °F; P = pressure, inHg; H = absolute humidity, lb water/lb dry air; M_a = molecular weight of air, lb/(lb·mol); and M_w = molecular weight of water vapor, lb/(lb·mol).

Relative humidity and **dew point** can be determined for other than atmospheric pressure from the partial pressure of water in the mixture and from the vapor pressure of water vapor. The partial pressure of water is calculated, if ideal-gas behavior is assumed, as

$$p = \frac{HP}{(M_w/M_a) + H} \qquad (12\text{-}6)$$

where p = partial pressure of water vapor, inHg; P = total pressure, inHg; H = absolute humidity, lb water/lb dry air, corrected to the actual pressure; M_a = molecular weight of air, lb/(lb·mol); and M_w = molecular weight of water vapor, lb/(lb·mol). The dew point of the mixture is then read directly from a table of vapor pressures as the temperature corresponding to the calculated partial pressure.

The relative humidity is obtained by dividing the calculated partial pressure by the vapor pressure of water at the dry-bulb temperature. Thus:

$$\text{Relative humidity} = 100 p/p_s \qquad (12\text{-}7)$$

where p = calculated partial pressure, inHg; and p_s = vapor pressure at dry-bulb temperature, inHg.

The preceding equations, which have assumed that both the air and the water vapor behave as ideal gases, are sufficiently accurate for most engineering calculations. If it is desired to remove the restriction that water vapor behave as an ideal gas, the actual density ratio should be used in place of the molecular-weight ratio in Eqs. (12-5) and (12-6).

Since the Schmidt number, Prandtl number, latent heat of vaporization, and humid heat are all essentially independent of pressure, the adiabatic-saturation-temperature and wet-bulb-temperature lines will be substantially equal at pressures different from atmospheric.

Example 8: Determination of Air Properties   For a barometric pressure of 25.92 inHg ($\Delta p = -4$), a dry-bulb temperature of 90°F, and a wet-bulb temperature of 70°F determine the following: absolute humidity, enthalpy, dew point, relative humidity, and specific volume.

Solution.   From Fig. 12-2, the moisture content is 78 gr/lb dry air = 0.0114 lb/lb dry air. From Table 12-2 at $t_w = 70$°F and $\Delta p = -4$ read $\Delta H_s = 17.6$ gr/lb dry air (additive correction for air saturated at the wet-bulb temperature).

$\Delta H = 17.6[1 - (20/24)(0.01)] = 17.4$, or actual humidity is $78 + 17.4 = 95.4$ gr/lb dry air, or 0.01362 lb/lb dry air. (See footnotes for Table 12-2.)

The enthalpy is obtained from Fig. 12-2 as $h = h' + D = 34.1 - 0.18 = 33.92$.

To this must be added the correction of 2.75 read from Table 12-2 for $\Delta p = -4$ and $t_w = 70$°F, giving the true enthalpy as $33.92 + 2.75 = 36.67$ Btu/lb dry air.

The partial pressure of water vapor is calculated from Eq. (12-6) as

$$p = \frac{HP}{(M_w/M_a) + H} = \frac{0.01362 \times 25.92}{0.622 + 0.01362} = 0.556 \text{ inHg}$$

From a table of vapor pressure, this corresponds to a dew point of 61.8°F.

Relative humidity is obtained from Eq. (12-7) as $100 \ p/p_s = (100 \times 0.556)/1.422 = 39.1$ percent.

The specific volume in cubic feet per pound of dry air is obtained from Eq. (12-5):

$$v = \frac{0.754(t + 460)}{25.92}\left(1 + \frac{HM_a}{M_w}\right)$$

$$= \frac{0.754(90 + 460)}{25.92}\left(1 + \frac{0.01362}{0.622}\right)$$

$$= 16.35 \text{ ft}^3/\text{lb dry air}$$

MEASUREMENT OF HUMIDITY

Dew-Point Method   The dew point of wet air is measured directly by observing the temperature at which moisture begins to form on an artificially cooled polished surface. The polished surface is usually cooled by evaporation of a low-boiling solvent such as ether, by vaporization of a condensed permanent gas such as carbon dioxide or liquid air, or by a temperature-regulated stream of water.

Although the dew-point method may be considered a fundamental technique for determining humidity, several uncertainties occur in its use. It is not always possible to measure precisely the temperature of the polished surface or to eliminate gradients across the surface. It is also difficult to detect the appearance or disappearance of fog; the usual practice is to take the dew point as the average of the temperatures when fog first appears on cooling and disappears on heating.

Wet-Bulb Method   Probably the most commonly used method for determining the humidity of a gas stream is the measurement of wet- and dry-bulb temperatures. The wet-bulb temperature is measured by contacting the air with a thermometer whose bulb is covered by a wick saturated with water. If the process is adiabatic, the thermometer bulb attains the wet-bulb temperature. When the wet- and dry-bulb temperatures are known, the humidity is readily obtained from charts such as Figs. 12-1 through 12-4. In order to obtain reliable information, care must be exercised to ensure that the wet-bulb thermometer remains wet and that radiation to the bulb is minimized. The latter is accomplished by making the relative velocity between wick and gas stream high [a velocity of 4.6 m/s (15 ft/s) is usually adequate for commonly used thermometers] or by the use of radiation shielding. Making sure that the wick remains wet is a mechanical problem, and the method used depends to a large extent on the particular arrangement. Again, as with the dew-point method, errors associated with the measurement of temperature can cause difficulty.

For measurement of atmospheric humidities the **sling psychrometer** is widely used. This is composed of a wet- and dry-bulb thermometer mounted in a sling which is whirled manually to give the desired gas velocity across the bulb. In the **Assmann psychrometer** the air is drawn past the bulbs by a motor-driven fan.

In addition to the mercury-in-glass thermometer, other temperature-sensing elements may be used for psychrometers. These include resistance thermometers, thermocouples, bimetal thermometers, and thermistors.

Mechanical Hygrometers   Materials such as human hair, wood fiber, and plastics have been used to measure humidity. These methods rely on a change in dimension with humidity.

Electric hygrometers measure the electrical resistance of a film of moisture-absorbing materials exposed to the gas. A wide variety of sensing elements have been used.

The **gravimetric method** is accepted as the most accurate humidity-measuring technique. In this method a known quantity of gas is passed over a moisture-absorbing chemical such as phosphorus pentoxide, and the increase in weight is determined.

EVAPORATIVE COOLING*

GENERAL REFERENCES: *ASHRAE Handbook and Product Directory: Equipment,* American Society of Heating, Refrigerating and Air-Conditioning Engineers, Atlanta, 1983. *Counterflow Cooling Tower Performance,* Pritchard Corporation, Kansas City, Mo., 1957. Hensley, "Cooling Tower Energy," *Heat. Piping Air Cond.* (October 1981). Kelley and Swenson, *Chem. Eng. Prog.,* **52,** 263 (1956). McAdams, *Heat Transmission,* 3d ed., McGraw-Hill, New York, 1954, pp. 356–365. Merkel, Z. *Ver. Dtsch. Ing. Forsch.,* no. 275 (1925). *The Parallel Path Wet-Dry Cooling Tower,* Marley Co., Mission Woods, Kan., 1972. *Performance Curves,* Cooling Tower Institute, Houston, 1967. *Plume Abatement and Water Conservation with Wet-Dry Cooling Tower,* Marley Co., Mission Woods, Kan., 1973. Tech. Bull. R-54-P-5, R-58-P-5, Marley Co., Mission Woods, Kan., 1957. Wood and Betts, *Engineer,* **189** (4912), 377, (4913), 349 (1950). Zivi and Brand, *Refrig. Eng.,* **64,** 8, 31–34, 90 (1956).

PRINCIPLES

The processes of cooling water are among the oldest known. Usually water is cooled by exposing its surface to air. Some of the processes are slow, such as the cooling of water on the surface of a pond; others are comparatively fast, such as the spraying of water into air. These processes all involve the exposure of water surface to air in varying degrees.

The heat-transfer process involves (1) latent heat transfer owing to vaporization of a small portion of the water and (2) sensible heat transfer owing to the difference in temperature of water and air. Approximately 80 percent of this heat transfer is due to latent heat and 20 percent to sensible heat.

Theoretical possible heat removal per pound of air circulated in a cooling tower depends on the temperature and moisture content of air. An indication of the moisture content of the air is its wet-bulb temperature. Ideally, then, the wet-bulb temperature is the lowest theoretical temperature to which the water can be cooled. Practically, the cold-water temperature approaches but does not equal the air wet-bulb temperature in a cooling tower; this is so because it is impossible to contact all the water with fresh air as the water drops through the wetted fill surface to the basin. The magnitude of approach to the wet-bulb temperature is dependent on tower design. Important factors are air-to-water contact time, amount of fill surface, and breakup of water into droplets. In actual practice, cooling towers are seldom designed for approaches closer than 2.8°C (5°F).

COOLING-TOWER THEORY

The most generally accepted theory of the cooling-tower heat-transfer process is that developed by Merkel (op. cit.). This analysis is based upon **enthalpy potential difference** as the driving force.

Each particle of water is assumed to be surrounded by a film of air, and the enthalpy difference between the film and surrounding air provides the driving force for the cooling process. In the integrated form the Merkel equation is

$$\frac{KaV}{L} = \int_{T_2}^{T_1} \frac{dT}{h' - h} \qquad (12\text{-}8)$$

where K = mass-transfer coefficient, lb water/(h·ft²); a = contact area, ft²/ft³ tower volume; V = active cooling volume, ft³/ft² of plan area; L = water rate, lb/(h·ft²); h' = enthalpy of saturated air at water temperature, Btu/lb; h = enthalpy of air stream, Btu/lb; and T_1 and T_2 = entering and leaving water temperatures, °F. The right-hand side of Eq. (12-8) is entirely in terms of air and water properties and is independent of tower dimensions.

Figure 12-12 illustrates water and air relationships and the driving potential which exist in a counterflow tower, where air flows parallel but opposite in direction to water flow. An understanding of this diagram is important in visualizing the cooling-tower process.

The water operating line is shown by line AB and is fixed by the inlet and outlet tower water temperatures. The air operating line begins at C, vertically below B and at a point having an enthalpy corresponding to that of the entering wet-bulb temperature. Line BC represents the initial driving force $(h' - h)$. In cooling water 1°F, the enthalpy per pound of air is increased 1 Btu multiplied by the ratio of pounds of water per pound of air. The liquid-gas ratio L/G is the slope of the operating line. The air leaving the tower is represented by point D. The cooling range is the projected length of line CD on the temperature scale. The cooling-tower approach is shown on the diagram as the difference between the cold-water temperature leaving the tower and the ambient wet-bulb temperature.

The coordinates refer directly to the temperature and enthalpy of any point on the water operating line but refer directly only to the enthalpy of a point on the air operating line. The corresponding wet-bulb temperature of any point on CD is found by projecting the point horizontally to the saturation curve, then vertically to the temperature coordinate. The integral [Eq. (12-8)] is represented by the area $ABCD$ in the diagram. This value is known as the **tower characteristic,** varying with the L/G ratio.

For example, an increase in entering wet-bulb temperature moves the origin C upward, and the line CD shifts to the right to maintain a constant KaV/L. If the cooling range increases, line CD lengthens. At a constant wet-bulb temperature, equilibrium is established by moving the line to the right to maintain a constant KaV/L. On the other hand, a change in L/G ratio changes the slope of CD, and the tower comes to equilibrium with a new KaV/L.

In order to predict tower performance it is necessary to know the required tower characteristics for fixed ambient and water conditions.

The tower characteristic KaV/L can be determined by integration. Normally used is the Chebyshev method for numerically evaluating the integral, whereby

$$\frac{KaV}{L} = \int_{T_2}^{T_1} \frac{dT}{h_w - h_a} \cong \frac{T_1 - T_2}{4} \left(\frac{1}{\Delta h_1} + \frac{1}{\Delta h_2} + \frac{1}{\Delta h_3} + \frac{1}{\Delta h_4} \right)$$

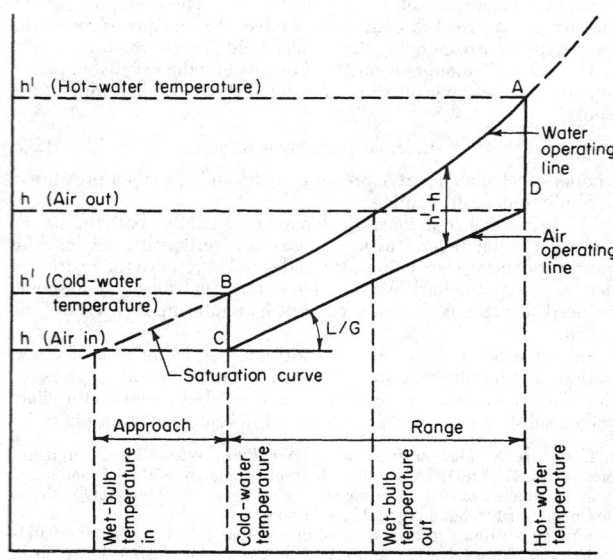

FIG. 12-12 Cooling-tower process heat balance. (*Marley Co.*)

* The contribution of Robert W. Norris, Robert W. Norris and Associates, Inc., to material that was used from the sixth edition is acknowledged. (Evaporative Cooling)

where h_w = enthalpy of air-water vapor mixture at bulk water temperature, Btu/lb dry air

h_a = enthalpy of air-water vapor mixture at wet-bulb temperature, Btu/lb dry air

Δh_1 = value of $(h_w - h_a)$ at $T_2 + 0.1(T_1 - T_2)$

Δh_2 = value of $(h_w - h_a)$ at $T_2 + 0.4(T_1 - T_2)$

Δh_3 = value of $(h_w - h_a)$ at $T_1 - 0.4(T_1 - T_2)$

Δh_4 = value of $(h_w - h_a)$ at $T_1 - 0.1(T_1 - T_2)$

Example 9: Calculation of Mass-Transfer Coefficient Group

Determine the theoretically required KaV/L value for a cooling duty from 105°F inlet water, 85°F outlet water, 78°F ambient wet-bulb temperature, and an L/G ratio of 0.97.

From air-water vapor-mixture tables, the enthalpy h_1 of the ambient air at 78°F wet-bulb temperature is 41.58 Btu/lb.

$$h_2 \text{ (leaving air)} = 41.58 + 0.97(105 - 85) = 60.98 \text{ Btu/lb}$$

T, °F.	h_{water}	h_{air}	$h_w - h_a$	$1/\Delta h$
$T_2 = 85$	49.43	$h_1 = 41.58$		
$T_2 + 0.1(20) = 87$	51.93	$h_1 + 0.1L/G(20) = 43.52$	$\Delta h_1 = 8.41$	0.119
$T_2 + 0.4(20) = 93$	60.25	$h_1 + 0.4L/G(20) = 49.34$	$\Delta h_2 = 10.91$	0.092
$T_1 - 0.4(20) = 97$	66.55	$h_2 - 0.4L/G(20) = 53.22$	$\Delta h_3 = 13.33$	0.075
$T_1 - 0.1(20) = 103$	77.34	$h_2 - 0.1L/G(20) = 59.04$	$\Delta h_4 = 18.30$	0.055
$T_1 = 105$	81.34	$h_2 = 60.98$		0.341

$$\frac{KaV}{L} = \frac{105 - 85}{4}(0.341) = 1.71$$

A quicker but less accurate method is by the use of a nomograph (Fig. 12-13) prepared by Wood and Betts (op. cit.).

Mechanical-draft cooling towers normally are designed for L/G ratios ranging from 0.75 to 1.50; accordingly, the values of KaV/L vary from 0.50 to 2.50. With these ranges in mind, an example of the use of the nomograph will readily explain the effect of changing variables.

Example 10: Application of Nomograph for Cooling-Tower Characteristics

If a given tower is operating with 20°F range, a cold-water temperature of 80°F, and a wet-bulb temperature of 70°F, a straight line may be drawn on the nomograph. If the L/G ratio is calculated to 1.0, then KaV/L may be established by a line drawn through $L/G = 1.0$ and parallel to the original line. The tower characteristic KaV/L is thus established at 1.42. If the wet-bulb temperature were to drop to 50°F, KaV/K and L/G ratios may be assumed to remain constant. A new line parallel to the original will then show that for the same range the cold-water temperature will be 70°F.

The nomograph provides an approximate solution; degree of accuracy will vary with changes in cooling as well as from tower to tower. Once the theoretical cooling-tower characteristic has been determined by numerical integration or from the nomograph for a given cooling duty, it is necessary to design the cooling-tower fill and air distribution to meet the theoretical tower characteristic. The Pritchard Corporation (op. cit.) has developed performance data on various tower-fill designs. These data are too extensive to include here, and those interested should consult this reference. See also Baker and Mart (Marley Co., Tech. Bull. R-52-P-10, Mission Woods, Kan.) and Zivi and Brand (loc. cit.).

MECHANICAL-DRAFT TOWERS

Two types of mechanical-draft towers are in use today: the forced-draft and the induced-draft. In the **forced-draft tower** the fan is mounted at the base, and air is forced in at the bottom and discharged at low velocity through the top. This arrangement has the advantage of locating the fan and drive outside the tower, where it is convenient for inspection, maintenance, and repairs. Since the equipment is out of the hot, humid top area of the tower, the fan is not subjected to corrosive conditions. However, because of the low exit-air velocity, the forced-draft tower is subjected to excessive recirculation of the humid

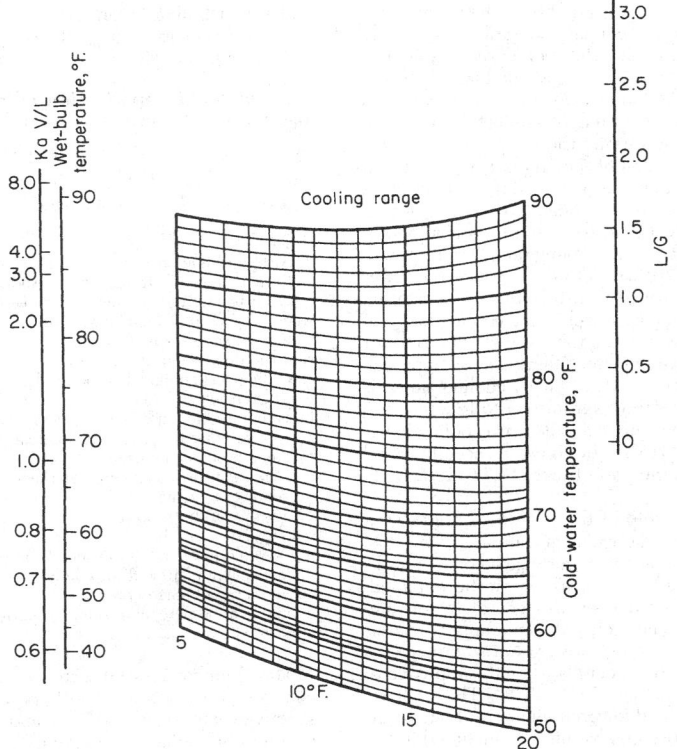

FIG. 12-13 Nomograph of cooling-tower characteristics. [*Wood and Betts*, Engineer, **189**(4912), 337 (1950).]

exhaust vapors back into the air intakes. Since the wet-bulb temperature of the exhaust air is considerably above the wet-bulb temperature of the ambient air, there is a decrease in performance evidenced by an increase in cold (leaving)-water temperature.

The **induced-draft tower** is the most common type used in the United States. It is further classified into counterflow and cross-flow design, depending on the relative flow directions of water and air. Thermodynamically, the **counterflow arrangement** is more efficient, since the coldest water contacts the coldest air, thus obtaining maximum enthalpy potential. The greater the cooling ranges and the more difficult the approaches, the more distinct are the advantages of the counterflow type. For example, with an L/G ratio of 1, an ambient wet-bulb temperature of 25.5°C (78°F), and an inlet water temperature of 35°C (95°F), the counterflow tower requires a KaV/L characteristic of 1.75 for a 2.8°C (5°F) approach, while a cross-flow tower requires a characteristic of 2.25 for the same approach. However, if the approach is increased to 3.9°C (7°F), both types of tower have approximately the same required KaV/L (within 1 percent).

The **cross-flow-tower** manufacturer may effectively reduce the tower characteristic at very low approaches by increasing the air quantity to give a lower L/G ratio. The increase in air flow is not necessarily achieved by increasing the air velocity but primarily by lengthening the tower to increase the air-flow cross-sectional area. It appears then that the cross-flow fill can be made progressively longer in the direction perpendicular to the air flow and shorter in the direction of the air flow until it almost loses its inherent potential-difference disadvantage. However, as this is done, fan power consumption increases.

Ultimately, the economic choice between counterflow and cross-flow is determined by the effectiveness of the fill, design conditions, and the costs of tower manufacture.

Performance of a given type of cooling tower is governed by the ratio of the weights of air to water and the time of contact between water and air. In commercial practice, the variation in the ratio of air to water is first obtained by keeping the air velocity constant at about 350 ft/(min·ft² of active tower area) and varying the water concentration, gal/(min·ft² of tower area). As a secondary operation, air velocity is varied to make the tower accommodate the cooling requirement.

Time of contact between water and air is governed largely by the time required for the water to discharge from the nozzles and fall through the tower to the basin. The time of contact is therefore obtained in a given type of unit by varying the height of the tower. Should the time of contact be insufficient, no amount of increase in the ratio of air to water will produce the desired cooling. It is therefore necessary to maintain a certain minimum height of cooling tower. When a wide approach of 8 to 11°C (15 to 20°F) to the wet-bulb temperature and a 13.9 to 19.4°C (25 to 35°F) cooling range are required, a relatively low cooling tower will suffice. A tower in which the water travels 4.6 to 6.1 m (15 to 20 ft) from the distributing system to the basin is sufficient. When a moderate approach and a cooling range of 13.9 to 19.4°C (25 to 35°F) are required, a tower in which the water travels 7.6 to 9.1 m (25 to 30 ft) is adequate. Where a close approach of 4.4°C (8°F) with a 13.9 to 19.4°C (25 to 35°F) cooling range is required, a tower in which the water travels from 10.7 to 12.2 m (35 to 40 ft) is required. It is usually not economical to design a cooling tower with an approach of less than 2.8°C (5°F), but it can be accomplished satisfactorily with a tower in which the water travels 10.7 to 12.2 m (35 to 40 ft).

Figure 12-14 shows the relationship of the hot-water, cold-water, and wet-bulb temperatures to the water concentration.° From this, the **minimum area** required for a given performance of a well-designed counterflow induced-draft cooling tower can be obtained. Figure 12-15 gives the horsepower per square foot of tower area required for a given performance. These curves do not apply to parallel or cross-flow cooling, since these processes are not so efficient as the counterflow process. Also, they do not apply when the approach to the cold-water temperature is less than 2.8°C (5°F). These charts should be considered approximate and for preliminary estimates only. Since many factors not shown in the graphs must be included in the

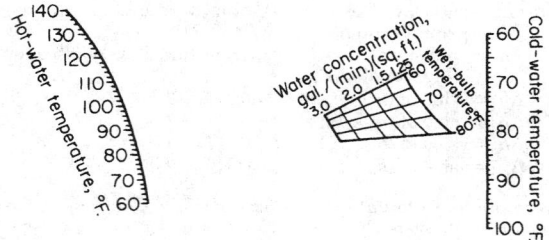

FIG. 12-14 Sizing chart for a counterflow induced-draft cooling tower, for induced-draft towers with (1) an upspray distributing system with 24 ft of fill or (2) a flume-type distributing system and 32 ft of fill. The chart will give approximations for towers of any height. (*Ecodyne Corp.*)

computation, the manufacturer should be consulted for final design recommendations.

The cooling performance of any tower containing a given depth of filling varies with the **water concentration.** It has been found that maximum contact and performance are obtained with a tower having a water concentration of 2 to 5 gal/(min·ft² of ground area). Thus the problem of calculating the size of a cooling tower becomes one of determining the proper concentration of water required to obtain the desired results. Once the necessary water concentration has been established, tower area can be calculated by dividing the gallons per minute circulated by the water concentration in gallons per minute-square foot. The required tower size then is a function of the following:
1. Cooling range (hot-water temperature minus cold-water temperature)
2. Approach to wet-bulb temperature (cold-water temperature minus wet-bulb temperature)
3. Quantity of water to be cooled
4. Wet-bulb temperature
5. Air velocity through the cell
6. Tower height

Example 11: Application of Sizing and Horsepower Charts
To illustrate the use of the charts, assume the following conditions:

$$\text{Hot-water temperature } T_1, °F = 102$$
$$\text{Cold-water temperature } T_2, °F = 78$$
$$\text{Wet-bulb temperature } t_w, °F = 70$$
$$\text{Water rate, gal/min} = 2000$$

A straight line on Fig. 12-14, connecting the points representing the design water and wet-bulb temperatures, shows that a water concentration of 2 gal/(min·ft²) is required. The area of the tower is calculated as 1000 ft² (quantity of water circulated divided by water concentration).

Fan horsepower is obtained from Fig. 12-15. Connecting the point representing 100 percent of standard tower performance with the turning point and extending this straight line to the horsepower scale show that it will require 0.041 hp/ft² of actual effective tower area. For a tower area of 1000 ft² 41.0 fan hp is required to perform the necessary cooling.

Suppose that the actual commercial tower size has an area of only 910 ft². Within reasonable limits, the shortage of actual area can be compensated for by an increase in air velocity through the tower. However, this requires boosting fan horsepower to achieve 110 percent of standard tower performance. From Fig. 12-15, the fan horsepower is found to be 0.057 hp/ft² of actual tower area, or 0.057 × 910 = 51.9 hp.

On the other hand, if the actual commercial tower area is 1110 ft², the cooling equivalent to 1000 ft² of standard tower area can be accomplished with less air and less fan horsepower. From Fig. 12-15, the fan horsepower for a tower operating at 90 percent of standard performance is 0.031 hp/ft² of actual tower area, or 34.5 hp.

This example illustrates the sensitivity of fan horsepower to small changes in tower area. The importance of designing a tower that is slightly oversize in ground area and of providing plenty of fan capacity becomes immediately apparent.

° See also London, Mason, and Boelter, loc. cit.; Lichtenstein, loc. cit.; Simpson and Sherwood, *J. Am. Soc. Refrig. Eng.*, **52**, 535, 574 (1946); Simons, *Chem. Metall. Eng.*, **49**(5), 138; (6), 83 (1942); **46**, 208 (1939); and Hutchinson and Spivey, *Trans. Inst. Chem. Eng.*, **20**, 14 (1942).

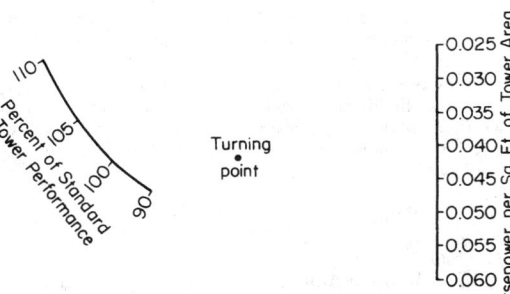

FIG. 12-15 Horsepower chart for a counterflow induced-draft cooling tower. [*Fluor Corp. (now Ecodyne Corp.)*]

Example 12: Application of Sizing Chart Assume the same cooling range and approach as used in Example 11 except that the wet-bulb temperature is lower. Design conditions would then be

$$
\begin{aligned}
\text{Water rate, gal/min} &= 2000 \\
\text{Temperature range } (T_1 - T_2), ^\circ\text{F} &= 24 \\
\text{Temperature approach } (T_2 - t_w), ^\circ\text{F} &= 8 \\
\text{Hot-water temperature } T_1, ^\circ\text{F} &= 92 \\
\text{Cold-water temperature } T_2, ^\circ\text{F} &= 68 \\
\text{Wet-bulb temperature } t_w, ^\circ\text{F} &= 60
\end{aligned}
$$

From Fig. 12-14, the water concentration required to perform the cooling is 1.75 gal/(min·ft²), giving a tower area of 1145 ft² versus 1000 ft² for a 70°F wet-bulb temperature. This shows that the lower the wet-bulb temperature for the same cooling range and approach, the larger is the area of the tower required and therefore the more difficult is the cooling job.

Figure 12-16 illustrates the type of performance curve furnished by the cooling-tower manufacturer. This shows the variation in performance with changes in wet-bulb and hot-water temperatures while the water quantity is maintained constant.

COOLING-TOWER OPERATION

Water Makeup Makeup requirements for a cooling tower consist of the summation of evaporation loss, drift loss, and blowdown. Therefore,

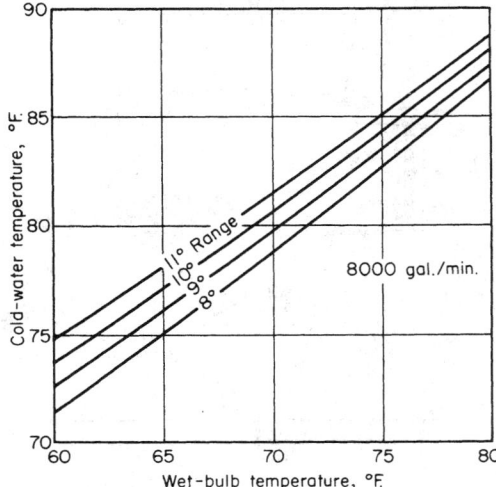

FIG. 12-16 Typical cooling-tower performance curve.

$$W_m = W_e + W_d + W_b \tag{12-9}$$

where W_m = makeup water, W_d = drift loss, and W_b = blowdown [consistent units, m³/(h·gal·min)].

Evaporation loss can be estimated by the following equation:

$$W_e = 0.00085 \, W_c (T_1 - T_2) \tag{12-10}$$

where W_c = circulating-water flow, gal/min at tower inlet
$T_1 - T_2$ = inlet-water temperature minus outlet-water temperature, °F

Drift is entrained water in the tower discharge vapors. Drift loss is a function of the drift-eliminator design, which typically varies between 0.1 and 0.2 percent of the water supplied to the tower. New developments in eliminator design make it possible to reduce drift loss well below 0.1 percent.

Blowdown discards a portion of the concentrated circulating water due to the evaporation process in order to lower the system solids concentration. The amount of blowdown can be calculated according to the number of cycles of concentration required to limit scale formation. Cycles of concentration are the ratio of dissolved solids in the recirculating water to dissolved solids in the makeup water. Since chlorides remain soluble on concentration, cycles of concentration are best expressed as the ratio of the chloride content of the circulating and makeup waters. Thus, the blowdown quantities required are determined from

$$\text{Cycles of concentration} = (W_e + W_b)/W_b \tag{12-11}$$

or $$W_b = W_e/(\text{cycles} - 1) \tag{12-12}$$

Cycles of concentration involved with cooling-tower operation normally range from three to five cycles. Below three cycles of concentration, excessive blowdown quantities are required and the addition of acid to limit scale formation should be considered.

Example 13: Calculation of Makeup Water Determine the amount of makeup required for a cooling tower with the following conditions:

Inlet water flow, m³/h (gal/min)	2270	(10,000)
Inlet water temperature, °C (°F)	37.77	(100)
Outlet water temperature, °C (°F)	29.44	(85)
Drift loss, percent	0.2	
Concentration, cycles	5	

Evaporation loss

$$
\begin{aligned}
W_e, \text{ m}^3/\text{h} &= 0.00085 \times 2270 \times (37.77 - 29.44) \times 1.8 \\
&= 28.9 \\
W_e, \text{ gal/min} &= 127.5
\end{aligned}
$$

Drift loss

$$
\begin{aligned}
W_d, \text{ m}^3/\text{h} &= 2270 \times 0.002 = 4.54 \\
W_d, \text{ gal/min} &= 20
\end{aligned}
$$

Blowdown

$$
\begin{aligned}
W_b, \text{ m}^3/\text{h} &= \frac{W_e}{(S-1)} = \frac{28.9}{4} = 7.24 \\
W_b, \text{ gal/min} &= 31.9
\end{aligned}
$$

Makeup

$$
\begin{aligned}
W_m, \text{ m}^3/\text{h} &= 28.9 + 4.54 + 7.24 = 40.7 \\
W_m, \text{ gal/min} &= 179.4
\end{aligned}
$$

Fan Horsepower In evaluating cooling-tower owning and operating costs, fan-horsepower requirements can be a significant factor. Large air quantities are circulated through cooling towers at exit velocities of about 10.2 m/s (2000 ft/min) maximum for induced-draft towers. Fan air-flow quantities depend upon tower-design factors, including such items as type of fill, tower configuration, and thermal-performance conditions.

The effective output of the fan is static air horsepower (SAHP), which is obtained by the following equation:

$$\text{SAHP} = \frac{Q(h_s)(d)}{33,000(12)} \qquad (12\text{-}13)$$

where Q = air volume, ft³/min; h_s = static head, in of water; and d = density of water at ambient temperature, lb/ft³.

Cooling-tower fan horsepower can be reduced substantially as the ambient wet-bulb temperature decreases if two-speed fan motors are used. Theoretically, operating at half speed will reduce air flow by 50 percent while decreasing horsepower to one-eighth of full-speed operation. However, actual half-speed operation will require about 17 percent of the horsepower at full speed as a result of the inherent motor losses at lighter loads.

Figure 12-17 shows a typical plot of outlet-water temperatures when a cooling tower is operated (1) in the fan-off position, (2) with the fan at half speed, and (3) with the fan at full speed. Note that at decreasing wet-bulb temperatures the water leaving the tower during half-speed operation could meet design water-temperature requirements of, say, 85°F. For example, for a 60°F wet-bulb, 20°F range, a leaving-water temperature slightly below 85°F is obtained with design water flow over the tower. If the fan had a 100-hp motor, 83 hp would be saved when operating it at half speed. In calculating savings, one should not overlook the advantage of having colder tower water available for the overall water-circulating system.

Recent developments in cooling-tower fan energy management also include automatic variable-pitch propeller-type fans and inverter-type devices to permit variable fan speeds. These schemes involve tracking the load at a *constant* leaving-water temperature.

The variable-pitch arrangement at constant motor speed changes the pitch of the blades through a pneumatic signal from the leaving-water temperature. As the thermal load and/or the ambient wet-bulb temperature decreases, the blade pitch reduces air flow and less fan energy is required.

Inverters make it possible to control a variable-speed fan by changing the frequency modulation. Standard alternating-current fan motors may be speed-regulated between 0 and 60 Hz. In using inverters for this application, it is important to avoid frequencies that would result in fan critical speeds.

Even though tower-fan energy savings can result from these arrangements, they may not constitute the best system approach. Power-plant steam condensers and refrigeration units, for example, can take advantage of colder tower water to reduce power consumption. Invariably, these system savings are much larger than cooling-tower fan savings with constant leaving-water temperatures. A refrigeration-unit condenser can utilize inlet-water temperatures down to 12.8°C (55°F) to reduce compressor energy consumption by 25 to 30 percent.

Pumping Horsepower   Another important factor in analyzing cooling-tower selections, especially in medium to large sizes, is the portion of pump horsepower directly attributed to the cooling tower. A counterflow type of tower with spray nozzles will have a pumping head equal to static lift plus nozzle pressure loss. A cross-flow type of tower with gravity flow enables a pumping head to equal static lift. A reduction in tower height therefore reduces static lift, thus reducing pump horsepower:

$$\text{Pump bhp} = \frac{\text{gal/min}(h_t)}{3960 \,(\text{pump efficiency})} \qquad (12\text{-}14)$$

where h_t = total head, ft.

Fogging and Plume Abatement   A phenomenon that occurs in cooling-tower operation is fogging, which produces a highly visible plume and possible icing hazards. Fogging results from mixing warm, highly saturated tower discharge air with cooler ambient air that lacks the capacity to absorb all the moisture as vapor. While in the past visible plumes have not been considered undesirable, properly locating towers to minimize possible sources of complaints has now received the necessary attention. In some instances, guyed high fan stacks have been used to reduce ground fog. Although tall stacks minimize the ground effects of plumes, they can do nothing about water-vapor saturation or visibility. The persistence of plumes is much greater in periods of low ambient temperatures.

More recently, environmental aspects have caused public awareness and concern over any visible plume, although many lay persons misconstrue cooling-tower discharge as harmful. This has resulted in a new development for plume abatement known as a wet-dry cooling-tower configuration. Reducing the relative humidity or moisture content of the tower discharge stream will reduce the frequency of plume formation. Figure 12-18 shows a "parallel path" arrangement that has been demonstrated to be technically sound but at substantially increased tower investment. Ambient air travels in parallel streams through the top dry-surface section and the evaporative section. Both sections benefit thermally by receiving cooler ambient air with the wet and dry air streams mixing after leaving their respective sections. Water flow is arranged in series, first flowing to the dry-coil section and then to the evaporation-fill section. A "series path" air-flow arrangement, in which dry coil sections can be located before or after the air traverses the evaporative section, also can be used. However, series-path air flow has the disadvantage of water impingement, which could result in coil scaling and restricted air flow.

Wet-dry cooling towers incorporating these designs are being used for large-tower industrial applications. At present they are not available for commercial applications.

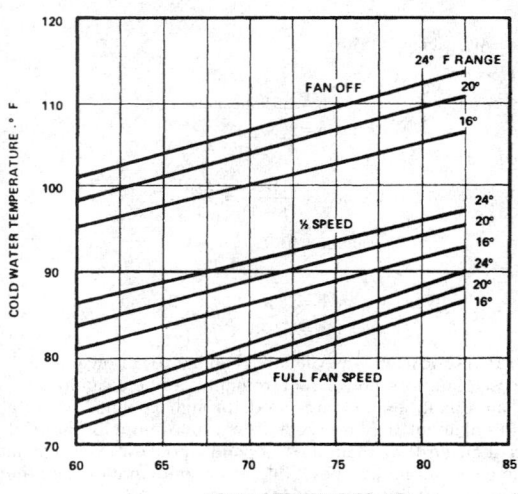

FIG. 12-17   Typical plot of cooling-tower performance at varying fan speeds.

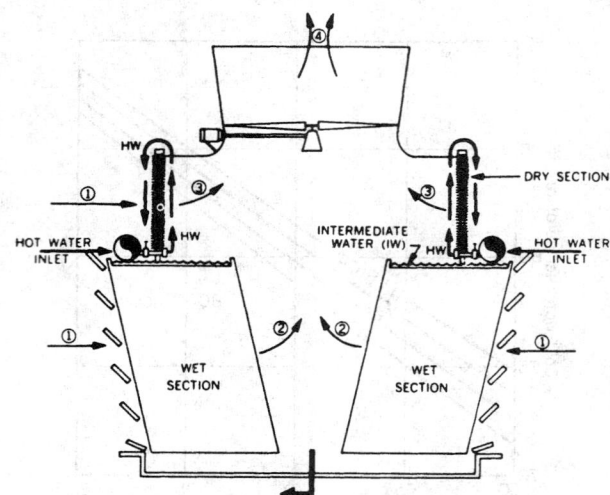

FIG. 12-18   Parallel-path cooling-tower arrangement for plume abatement. (*Marley Co.*)

Energy Management With today's emphasis on energy management, cooling towers have not been overlooked. During periods below 50°F ambient wet-bulb temperatures, cooling towers have the temperature capability to furnish chilled water directly to air-conditioning systems. For existing refrigeration–cooling-tower systems, piping can be installed to bypass the chiller to allow tower effluent to flow directly to cooling coils. After heat has been removed from the air stream, water returns directly to the cooling tower. Water temperature leaving the cooling tower is controlled between 8.9 and 12.2°C (48 and 54°F), usually by cycling cooling-tower fans. Depending upon the cleanliness of the cooling-tower water, it may be necessary to install a side-stream or full-flow filter to minimize contamination of the normally closed chilled-water circuit. Figure 12-19 shows the general arrangement of this system. Substantial savings can be realized during colder months by eliminating refrigeration-compressor energy.

Several other methods involving cooling towers have been used to reduce refrigeration energy consumption. These systems, as applied to centrifugal and absorption refrigeration machines, are known as thermocycle or free cooling systems. When water leaving the cooling tower is available below 10°C (50°F), the thermocycle system permits shutting down the compressor prime mover or reducing steam flow to an absorption unit. Figure 12-20 shows the arrangement for a centrifugal refrigeration unit.

The thermocycle system can be operated only when condensing water is available at a temperature lower than the required chilled-water-supply temperature. Modifications for a centrifugal refrigeration unit include the installation of a small liquid-refrigerant pump, cooler spray header nozzles, and a vapor bypass line between the cooler and the condenser. Without the compressor operating, a thermocycle capacity up to 35 percent of the refrigeration-unit rating can be produced.

The cooling-tower fan is operated at full speed to produce the coldest water temperature possible for a given ambient wet-bulb temperature. The large vapor bypass between the cooler and the condenser is opened along with the compressor suction damper or prerotational vanes. The heat removed from the chilled-water stream boils off refrigerant vapor from the cooler. This vapor flows mainly through the bypass line to the condenser, where it is condensed to a liquid. (Units having hot-gas bypass lines cannot be used for this purpose, as the pipe size is too small.) The liquid then returns to the cooler as in the normal refrigeration cycle. If the refrigeration unit contains internal float valves, they are bypassed manually or held open by an adjusting stem.

Thermocycle capacity is a function of the temperature difference between the chilled-water outlet temperature leaving the cooler and the inlet condenser water. The cycle finally stops when these two temperatures approach each other and there is not sufficient vapor pressure difference to permit flow between the heat exchangers.

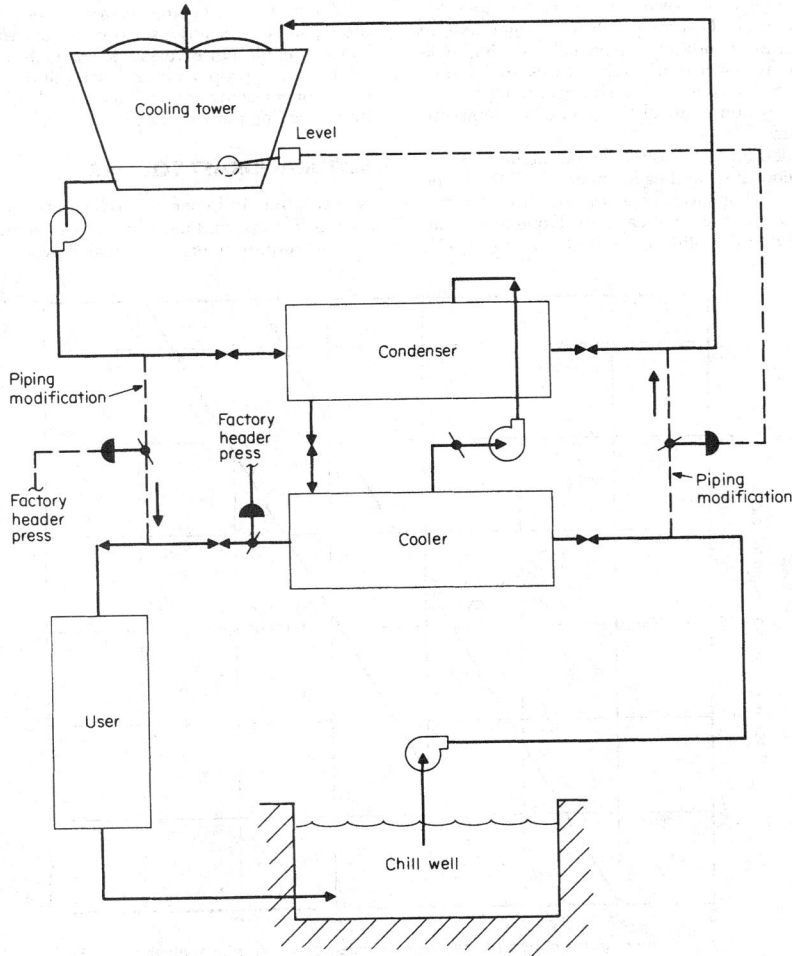

FIG. 12-19   Cooling-tower water for direct cooling during winter months.

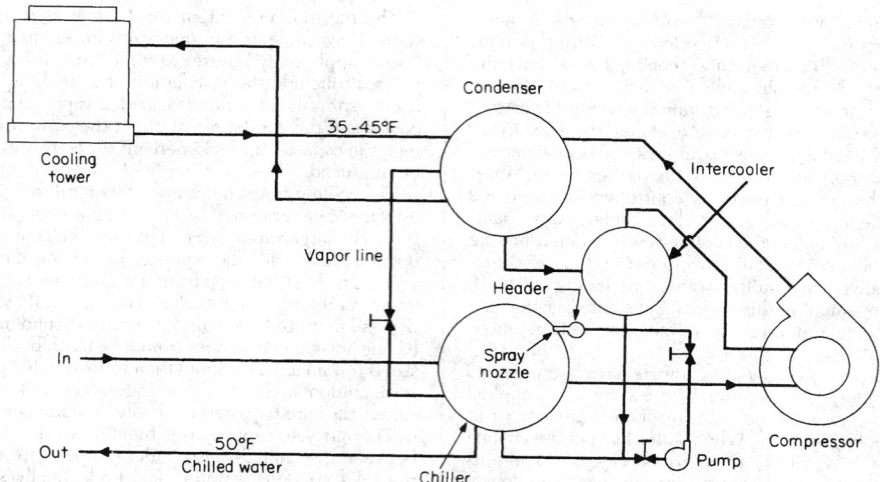

FIG. 12-20   Cooling-tower use on a thermocycle system during winter months.

Precise control of the outlet chilled-water temperature does not occur with thermocycle operation. This temperature is dependent on ambient wet-bulb-temperature conditions. Normally, during cold winter days little change occurs in wet-bulb temperatures, so that only slight water-temperature variations may occur. This would not be true of many spring and fall days, when relatively large climatic temperature swings can and do occur.

Refrigeration units modified for free cooling do not include the liquid-refrigerant pump and cooler spray header nozzles. Without the cooler refrigerant agitation for improved heat transfer, this arrangement allows up to about 20 percent of rated capacity. Expected capacities for both thermocycle and free cooling are indicated in Fig. 12-21.

In operating a cooling tower in the thermocycle or free-cooling mode, some precautions are necessary to minimize icing problems. These include fan reversals to circulate air down through the tower inlet louvers, proper water distribution, constant water flow over the tower, heat tracing of lines such as makeup lines as required, and maximum loading per tower cell.

NATURAL-DRAFT TOWERS

Natural-draft, or hyperbolic-type, towers have been in use since about 1916 in Europe and have become standard equipment for the water-cooling requirements of British power stations. They are primarily

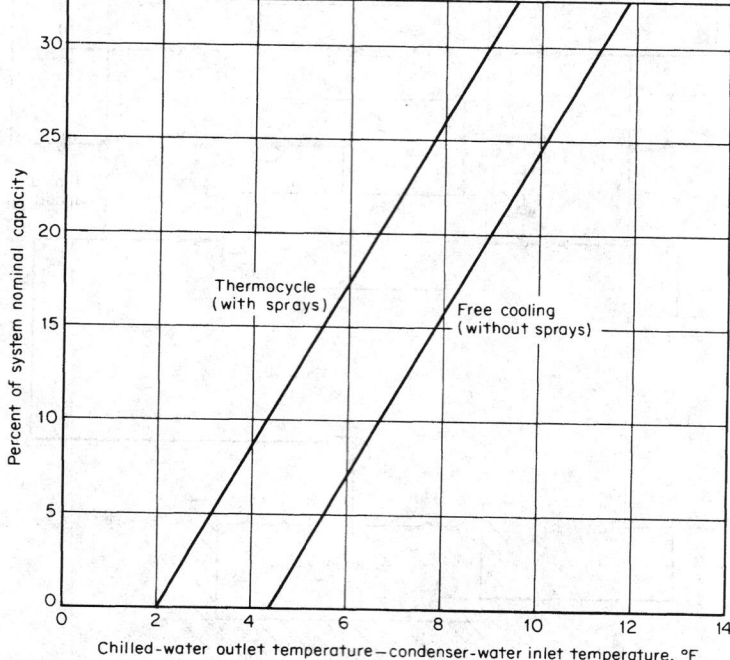

FIG. 12-21   Thermocycle and free-cooling-system capacities.

suited to very large cooling-water quantities, and the reinforced-concrete structures used are as large as diameters of 80.7 m (265 ft) and heights of 103.6 m (340 ft).

The design convenience obtained from the steady air flow of mechanical-draft towers is not realized in natural-draft-tower design. Air flow through a natural-draft tower is due largely to the difference in density between the cool inlet air and the warm exit air. The air leaving the stack is lighter than the ambient air, and a draft is created by chimney effect, thus eliminating the need for mechanical fans. McKelvey and Brooke (*The Industrial Cooling Tower*, Elsevier, New York, 1959, p. 108) note that natural-draft towers commonly operate at air-pressure differences in the region of 0.2-in water gauge when under full load. The mean velocity of the air above the tower packing is generally about 1.2 to 1.8 m/s (4 to 6 ft/s).

The performance of the natural-draft tower differs from that of the mechanical-draft tower in that the cooling is dependent upon the relative humidity as well as on the wet-bulb temperature. The draft will increase through the tower at high-humidity conditions because of the increase in available static pressure difference to promote air flow against internal resistances. Thus the higher the humidity at a given wet bulb, the colder the outlet water will be for a given set of conditions. This fundamental relationship has been used to advantage in Great Britain, where relative humidities are commonly 75 to 80 percent. Therefore, it is important in the design stages to determine correctly and specify the density of the entering and effluent air in addition to the usual tower-design conditions of range, approach, and water quantity. The performance relationship to humidity conditions makes exact control of outlet-water temperature difficult to achieve with the natural-draft tower.

Data for determining the size of natural-draft towers have been presented by Chilton [*Proc. Inst. Elec. Eng.*, **99**, 440 (1952)] and Rish and Steel (ASCE Symposium on Thermal Power Plants, October 1958). Chilton showed that the duty coefficient D_t of a tower is approximately constant over its normal range of operation and is related to tower size by an efficiency factor or performance coefficient C_t as follows:

$$D_t = (A \sqrt{Z_t})/(C_t \sqrt{C_t}) \qquad (12\text{-}15)$$

where A = base area of tower, ft^2, measured at pond sill level; and Z_t = height of tower, ft, measured above sill level. The duty coefficient may be determined from the formula

$$(W_L/D_t) = 90.59(\Delta h/\Delta T) \sqrt{\Delta t + 0.3124\Delta h} \qquad (12\text{-}16)$$

where Δh = change in total heat of the air passing through the tower, Btu/lb; ΔT = change of water temperature passing through tower, °F; Δt = difference between air temperature leaving the packing and inlet dry-bulb temperature, °F; and W_L = water load in the tower, lb/h. The air leaving the packing inside the tower is assumed to be saturated at a temperature halfway between the inlet- and outlet-water temperatures. A divergence between theory and practice of a few degrees in this latter assumption does not significantly affect the results, as the draft component depends on the ratio of the change of density to change of total heat and not on change of temperature alone.

Example 14: Duty Coefficient for a Hyperbolic Tower Determine the duty coefficient for a hyperbolic tower operating with

Temperature of water to tower, °F = 82
Leaving (recooled) water temperature, °F = 70
Temperature range ΔT, °F = 12
Dry-bulb air temperature t_2, °F = 57
Aspirated (ambient) wet-bulb air temperature t_{w2}, °F = 51.7
Water loading to tower W_L, lb/h = 38,200,000

$t_1 = (82° + 70°)/2 = 76°$ $h_1 = 39.8$ (from Fig. 12-2)

$t_2 = \underline{57°}$ $h_2 = \underline{21.3}$

$\Delta t = 19°$ $\Delta h = 18.5$ = 18.5

$W_L/D_t = 90.59(18.5/12) \sqrt{19 + 0.3124(18.5)} = 696$

$D_t = 38,200,000/696 = 55,000$

The performance coefficients usually attained have been about 5.2 for water loadings in excess of 750 lb/(h·ft^2), though new types of packing are improving

(lowering) it. By taking a C_t value of 5.0 and a tower height of 320 ft, the base area of the tower will be $(55,000)(5 \sqrt{5})/\sqrt{320} = 34,600$ ft^2, or the internal base diameter at sill level will be 210 ft. A ratio of height to base diameter of 3:2 is normally employed.

To determine how a natural-draft tower of any given duty coefficient will perform under varying conditions, Rish and Steel plotted the nomograph in Fig. 12-22. The straight line shown on the nomograph illustrates the conditions of Example 14.

SPRAY PONDS

Spray ponds provide an arrangement for lowering the temperature of water by evaporative cooling and, in so doing, greatly reduce the cooling area required in comparison with a cooling pond. A spray pond uses a number of nozzles which spray water into contact with the surrounding air. A well-designed spray nozzle should provide fine water drops but should not produce a mist which would be carried off as excessive drift loss.

Table 12-3 provides design data which will assist in the layout of a spray pond. The pond should be placed with its long axis at right angles to the prevailing summer wind. A long, narrow pond is more effective than a square one, so that decreasing pond width and increasing pond length will improve performance. Performance can also be increased by decreasing the amount of water sprayed per unit of pond area, increasing the height and fineness of spray drops, and increasing nozzle height above the basin sides.

Sufficient distance should be provided from the outer nozzles to keep spray from being carried over the sides of the basin. If it is not possible to provide 7.6 to 10.7 m (25 to 35 ft) of space, the pond should be enclosed with a louver fence, equal in height to the maximum height of the spray, to minimize drift loss. Also, during cold-weather periods, fogging can occur from the spray pond, so that consideration should be given to possible hazards to roadways or buildings in the immediate vicinity.

The physical designs and operating conditions of spray-pond installations vary greatly, and it is difficult to develop exact rating data that can be used for determining cooling performance in all cases. However, Fig. 12-23 shows the performance that can be obtained with a well-designed spray pond, based on a 21.1°C (70°F) wet-bulb temperature and a 2.2-m/s (5-mi/h) wind. This curve shows that a 3.3°C (6°F) approach to the wet bulb is possible at a 2.2°C (4°F) range, but at higher ranges the obtainable approach increases. If it is necessary to cool water through a large temperature range to a reasonably close approach, the spray pond could be staged. With this method, the water is initially sprayed, collected, and then resprayed in another part of a sectionalized pond basin.

Figure 12-24 shows performance curves for a spray pond used in steam-condensing service at varying wet-bulb and range conditions. Spray-pond performance can be calculated within reasonable accuracy on the basis of the leaving wet-bulb temperature of the air passing through the spray-filled volume. The air temperature leaving cannot exceed the warm water to the pond, and the closeness of approach will depend on the pond layout. In calculating the cooling obtained, the spray-filled volume is figured from a height equal to the elevation of the nozzles above the pond surface plus 0.3 m (1 ft) for each 7-kPa (1bf/in^2) nozzle pressure and a plan area extending 3 m (10 ft) beyond the outer nozzles. The air area involved is the projected area of a vertical plane through the filled volume and broadside to the direction of air movement. The horizontal distance that the air moves through the filled volume is considered the length of air travel.

Example 15: Cooling Capacity of a Spray Pond Determine the cooling capacity of a spray pond operating at the following conditions:

Water flow, gal/min	46,000
Spray-nozzle pressure, lb/in^2	7
Water flow per nozzle, gal/min	42.5
Effective area, length × width, ft^2	434 × 100
Effective height, ft	7 + 7 spray height
Wind velocity, ft/min	440

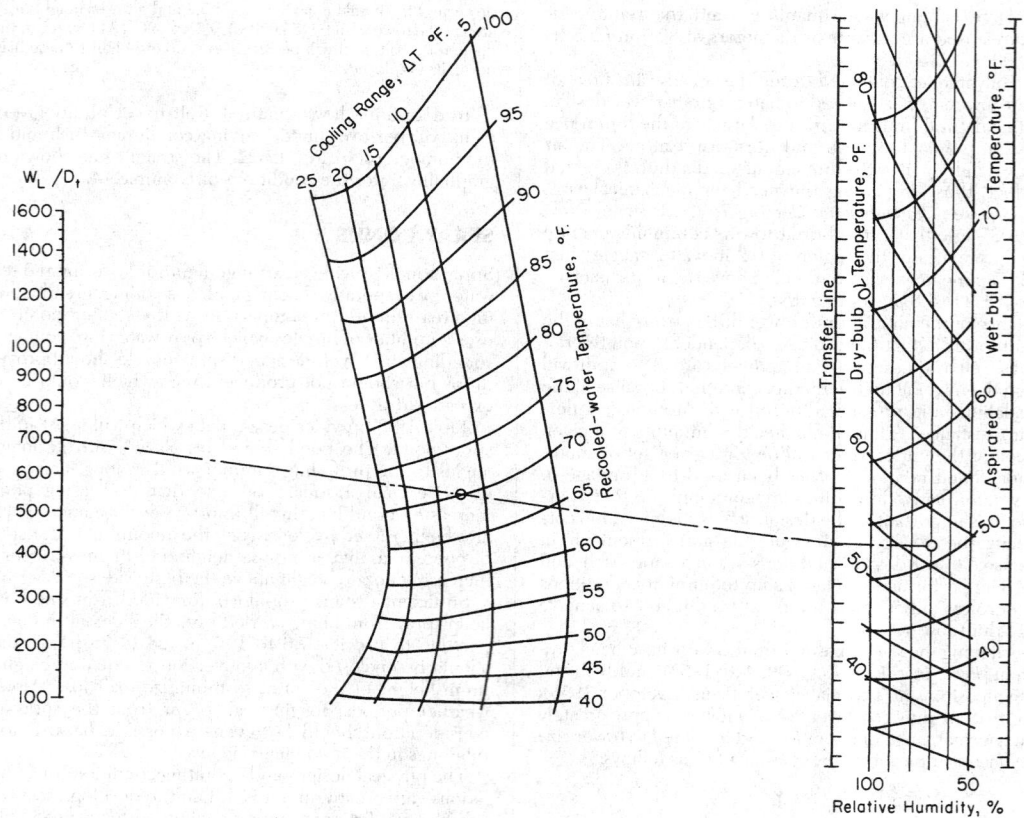

FIG. 12-22   Universal performance chart for natural-draft cooling towers. (*Rish and Steel,* ASCE Symposium on Thermal Power Plants, *October 1958.*)

Prevailing wind	Broadside to pond
Ambient wet-bulb temperature, °F	78
Water temperature in, °F	102

Effective air area = 434 × 14 = 6080 ft²

Air flow = 440 × 6080 = 2,680,000 ft³/min

$L = 46,000 \times 8.33 = 384,000$ lb water/min

$G = 2,680,000/14.3 = 187,500$ lb air/min

$L/G = 384,000/187,500 = 2.05$

h' at 78°F wet-bulb temperature = 41.58 Btu/lb (from Table 12-1). Assume water temperature out = 92°F.

TABLE 12-3   Spray-Pond Engineering Data and Design*

Recommendations	Usual	Minimum	Maximum
Nozzle capacity, gal./min. each	35–50	10	60
Nozzles per 12-ft. length of pipe	5–6	4	8
Height of nozzles above sides of basin, ft.	7–8	2	10
Nozzle pressure, lb./sq. in.	5–7	4	10
Size of nozzles and nozzle arms, in.	2	1¼	2½
Distance between spray lateral piping, ft.	25	13	38
Distance of nozzles from side of pond, unfenced, ft.	25–35	20	50
Distance of nozzles from side of pond, fenced, ft.	12–18	10	25
Height of louver fence, ft.	12	6	18
Depth of pond basin, ft.	4–5	2	7
Friction loss per 100 ft. pipe, in. of water	1–3	—	6
Design wind velocity, m.p.h.	5	3	10

*From Spray Pond Bull. SP-51, Marley Co., Mission Woods, Kan., p. 3.

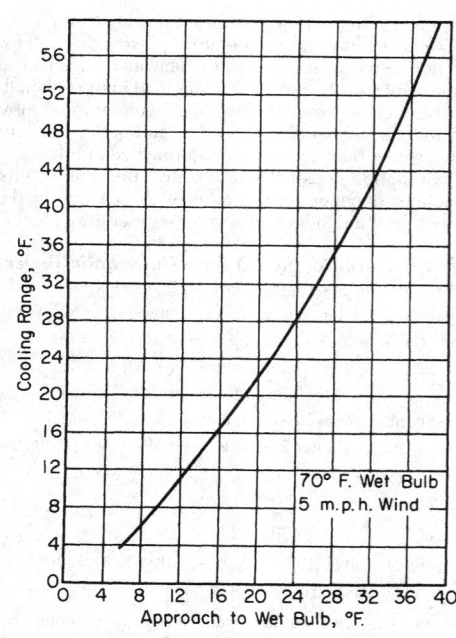

FIG. 12-23   Spray-pond performance curve.

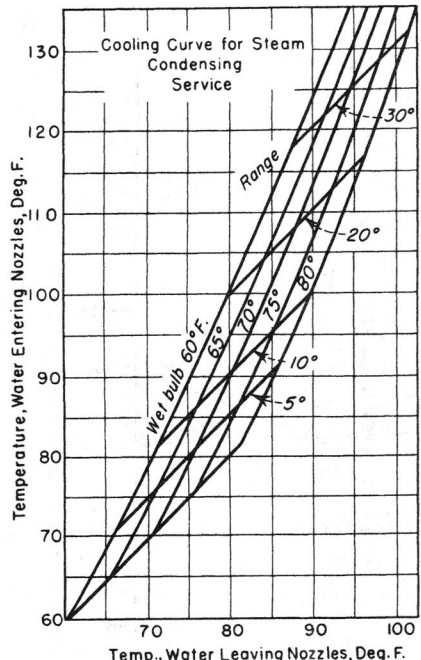

FIG. 12-24 Spray ponds: cooling curves for steam-condensing service.

$$\frac{L}{G} = \frac{(h'_2) \text{ air out} - (h'_1) \text{ air in}}{\text{water temperature in} - \text{water temperature out}} = 2.05 = \frac{h'_2 - 41.58}{10}$$

$h'_2 = 61.63$ Btu/lb

Corresponding wet-bulb temperature = 94°F air leaving pond.
Approach possible to air leaving (from Table 12-4) = −2°F.
Water temperature leaving spray pond = 94°F − 2°F = 92°F.

Since leaving-water temperature checks assumption, spray pond is capable of cooling 46,000 gal/min from 102 to 92°F with 78°F wet-bulb temperature and 5-mi/h wind. Total of 1080 spray nozzles required at 42.5 gal/min each, nozzles at 7-lbf/in² pressure.

COOLING PONDS

When large ground areas are available, cooling ponds offer a satisfactory method of removing heat from water. A pond may be constructed

at a relatively small investment by pushing up an earth dike 1.8 to 3.1 m (6 to 10 ft) high. For a successful pond installation, the soil must be reasonably impervious, and location in a flat area is desirable. Four principal heat-transfer processes are involved in obtaining cooling from an open pond. Heat is lost through evaporation, convection, and radiation and is gained through solar radiation. The required pond area depends on the number of degrees of cooling required and the net heat loss from each square foot of pond surface.

Langhaar [*Chem. Eng.*, **60**(8), 194 (1953)] states that under given atmospheric conditions a body of water would eventually come to a temperature at which heat loss would equal heat gain. This temperature is referred to as the equilibrium temperature and is designated as E in Fig. 12-25, a nomograph of cooling-pond performance. The equilibrium temperature is greatly affected by the amount of solar radiation, which is usually not known very accurately and which varies throughout the day. If a pond has at least a 24-h holdup, then daily average weather conditions may be used. For practical purposes, it is recommended that the equilibrium temperature be taken as equal to normal river-water or lake temperature for the specified weather conditions.

In order to cool to the equilibrium temperature, a pond of infinite size would be required for warm water. An approach of 1.7 to 2.2°C (3 to 4°F) is the lowest practicable in a pond of reasonable size. For a pond having more than a 24-h holdup, the leaving-water temperature will vary from the average by plus or minus 1.1°C (2°F) for a 0.9-m (5-ft) depth and 1.7°C (3°F) variation for a 0.9-m (3-ft) depth.

The area of pond required for a given cooling load is almost independent of pond depth. A depth of at least 0.9 m (3 ft) appears advisable to prevent excessive channeling of flow with ponds having irregular bottoms and to avoid large day-to-night changes in outlet temperature.

Factors considered to affect pond performance are air temperature, relative humidity, wind speed, and solar radiation. Items appearing to have only a minor effect include heat transfer between the earth and the pond, changing temperature and humidity of the air as it traverses the water, and rain.

Figure 12-25 provides a rapid method of determining the pond-area requirements for a given cooling duty. D_1 and D_2 are the approaches to equilibrium for the entering and leaving water, °F; V_w is the wind velocity, mi/h; product PQ represents the area of the pond surface, ft²/(gal·min) of flow to the pond. The P factor assumes a pond with uniform flow, without turbulence, and with the water warmer than the air.

Example 16: Calculation of Cooling-Pond Size Determine the required size of a cooling pond operating at the following conditions:

Relative humidity, percent = 50
Wind velocity, mi/h = 5
Dry-bulb air temperature, °F = 68
Solar heat gain, Btu/(h·ft²) = 100
Water quantity, gal/min = 10,000

TABLE 12-4 Degree Adjustment to Be Applied to Leaving-Air Wet-Bulb Temperature to Find Cooled-Water Temperatures of Spray Ponds*

Cooling range, °F.	Entering wet-bulb temp., † °F.	Adjustment, °F. Length of air travel, ft.‡		
		100	50	25
10	80	−3	+2	+4
	70	−2	+3	+5
	60	−1.5	+3.5	+5.5
15	80	−5.0	+1	+5
	70	−4.0	+2	+5.5
	60	−3.5	+2.5	+6
20	80	−7	0	+6
	70	−6	+1	+7
	60	−5.5	+1.5	+7.5

Cooled-water temperature = wet-bulb temperature of leaving air plus values shown.
*From "Heating, Ventilating, Air Conditioning Guide," 38th ed., p. 598 American Society of Heating, Refrigerating and Air Conditioning Engineers 1960.
†Wet-bulb temperature of air entering spray-filled volume.
‡Length of air travel through spray-filled volume.

TABLE 12-5 Maximum Expected Solar Radiation at Various North Latitudes*

B.t.u./(hr.)(sq. ft.)

	24-hr. avg. at north latitude				Noon value at north latitude			
	30°	35°	40°	45°	30°	35°	40°	45°
Jan. 1	65	50	40	30	240	205	170	135
Feb. 1	75	65	55	45	270	240	210	175
Mar. 1	90	80	75	65	305	285	255	230
Apr. 1	110	105	95	90	340	320	300	280
May 1	120	120	120	115	360	350	335	320
June 1	130	130	130	130	365	360	345	335
July 1	130	130	130	130	365	360	350	340
Aug. 1	125	125	125	120	360	350	340	325
Sept. 1	115	110	105	100	350	335	315	300
Oct. 1	100	90	80	75	315	295	270	245
Nov. 1	80	70	60	50	270	245	215	185
Dec. 1	65	55	45	35	240	210	175	140

*Langhaar, *Chem. Eng.*, **60**(8), 194 (1953).

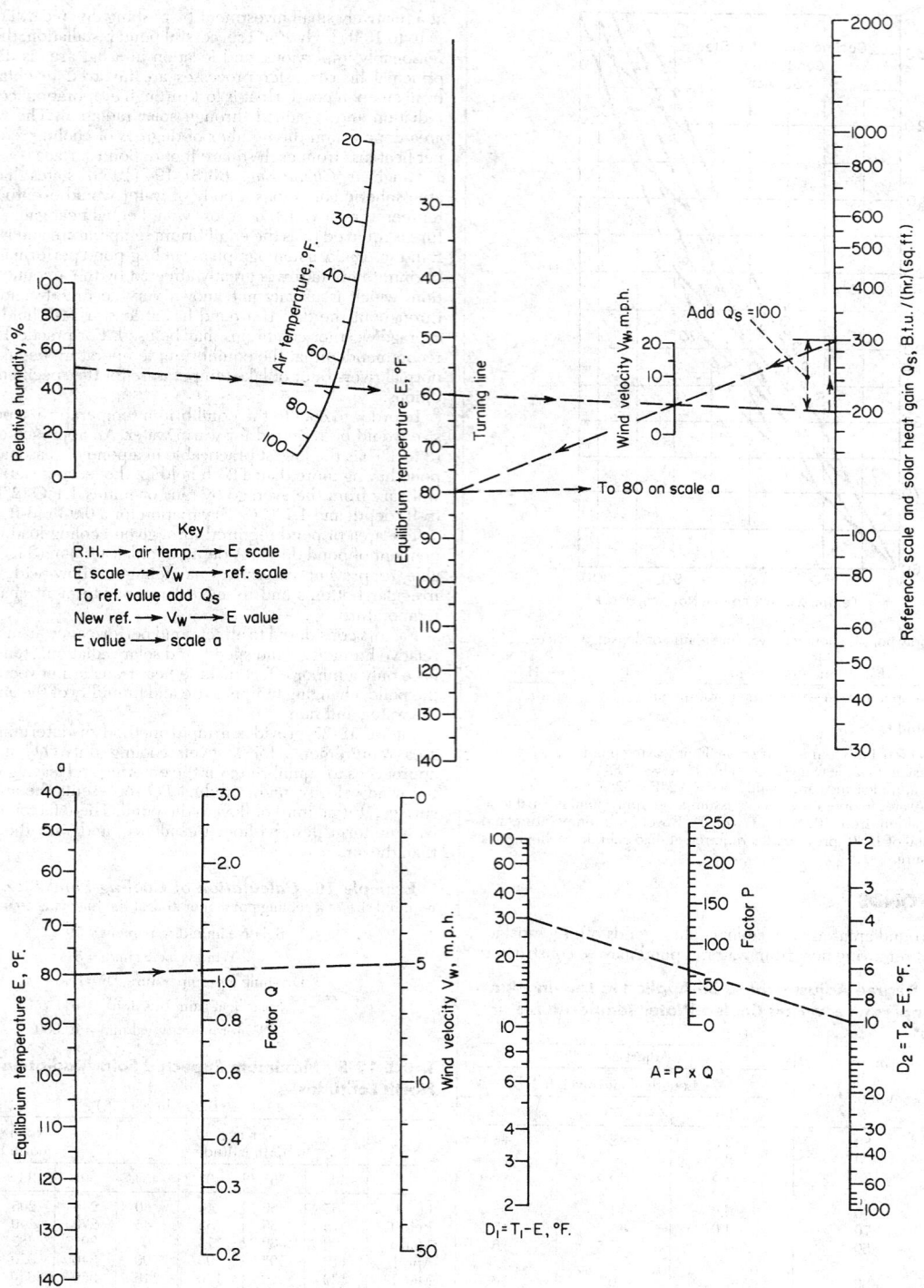

FIG. 12-25 Nomograph for determining cooling-pond performance and size. [*Langhar*, Chem. Eng., **60**(8), 194 (1953).]

Water inlet, °F = 110

Water outlet, °F = 90

From the nomograph, $P = 68$ and $Q = 1.07$, giving an area required of 73 ft²/(gal·min). Area for 10,000 gal/min is thus 730,000 ft², or 17 acres.

With a depth of 5 ft, total volume of the pond would amount to a 45.5-h holdup, which is more than the 24-h holdup required to maintain a fairly constant discharge temperature throughout the day.

Table 12-5 presents typical values of solar radiation on a horizontal surface, Btu/(h·ft²), based on analysis of Weather Bureau records for a number of stations throughout the United States. These are clear-day values, rarely exceeded even in the high arid regions. The normal or actual average monthly values are only 50 to 60 percent of the tabulated figures for most of the eastern United States and 80 to 90 percent in the arid southwest. Also, the solar radiation should be multiplied by the absorption coefficient for the pond, which appears to exceed 95 percent.

SOLIDS-DRYING FUNDAMENTALS*

GENERAL REFERENCES: Cook and DuMont, *Process Drying Practice*, McGraw-Hill, 1991. *Drying Technology—An International Journal*, Marcel Dekker, N.Y., 1982. Hall, *Dictionary of Drying*, Marcel Dekker, N.Y., 1979. Keey, *Introduction to Industrial Drying Operations*, Pergamon, N.Y., 1978. Masters, *Spray Drying Handbook*, Wiley, N.Y., 1990. Mujumdar, *Handbook of Industrial Drying*, Marcel Dekker, N.Y., 1987. Nonhebel and Moss, *Drying of Solids in the Chemical Industry*, CRC Press, Ohio, 1971. van't Land, *Industrial Drying Equipment*, Marcel Dekker, N.Y., 1991.

INTRODUCTION

The material on solids drying is divided into two subsections, "Solids-Drying Fundamentals," and "Solids-Drying Equipment." In this introductory part some elementary definitions are given. In solids-gas contacting equipment, the solids bed can exist in any of the following four conditions.

Static   This is a dense bed of solids in which each particle rests upon another at essentially the settled bulk density of the solids phase. Specifically, *there is no relative motion among solids particles* (Fig. 12-26).

Moving   This is a slightly expanded bed of solids in which the particles are separated only enough to flow one over another. Usually the flow is downward under the force of gravity, but upward motion by mechanical lifting or agitation may also occur within the process vessel. In some cases, lifting of the solids is accomplished in separate equipment, and solids flow in the presence of the gas phase is downward only. The latter is a moving bed as usually defined in the petroleum industry. In this definition, *solids motion is achieved by either mechanical agitation or gravity force* (Fig. 12-27).

Fluidized   This is an expanded condition in which the solids particles are supported by drag forces caused by the gas phase passing through the interstices among the particles at some critical velocity. It is an unstable condition in that the superficial gas velocity upward is less than the terminal setting velocity of the solids particles; the gas velocity is not sufficient to entrain and convey continuously all the solids. At the same time, there exist, within the stream of gas, eddies traveling at high enough velocities to lift the particles temporarily. Particle motion is continually upward and falling back. Specifically, the solids phase and the gas phase are intermixed and *together behave like a boiling fluid* (Fig. 12-28).

Dilute   This is a fully expanded condition in which the solids particles are so widely separated that they exert essentially no influence upon each other. Specifically, the solids phase is so fully dispersed in the gas that *the density of the suspension is essentially that of the gas phase alone* (Fig. 12-29). Commonly, this situation exists when the gas velocity at all points in the system exceeds the terminal settling velocity of the solids and the particles can be lifted and continuously conveyed by the gas; however, this is not always true. Gravity settling chambers such as prilling towers and countercurrent-flow spray dryers are two exceptions in which gas velocity is insufficient to entrain the solids completely.

Gas-Solids Contacting   Terms used in this section to describe the method by which gas may contact a bed of solids are the following:

1. *Parallel flow.*   The direction of gas flow is parallel to the surface of the solids phase. Contacting is primarily at the interface between phases, with possibly some penetration of gas into the voids among the solids near the surface. The solids bed is usually in a static condition (Fig. 12-30).

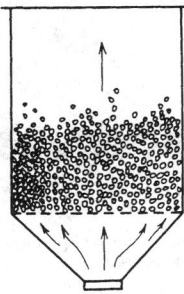

FIG. 12-28   Fluidized solids bed.

FIG. 12-26   Solids bed in static condition (tray dryer).

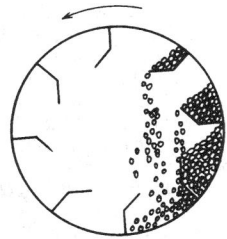

FIG. 12-27   Moving solids bed in a rotary dryer with lifters.

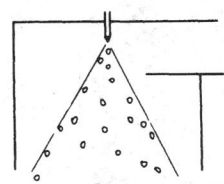

FIG. 12-29   Solids in a dilute condition near the top of a spray dryer.

° The contribution of George A. Schurr, E. I. du Pont de Nemours & Co., to material that was used from the sixth edition is acknowledged. (Solids Drying)

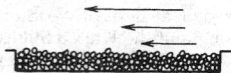

FIG. 12-30 Parallel gas flow over a static bed of solids.

2. *Perpendicular flow.* The direction of gas flow is normal to the phase interface. The gas impinges on the solids' surfaces. Again the solids bed is usually in a static condition (Fig. 12-31).

3. *Through circulation.* The gas penetrates and flows through interstices among the solids, circulating more or less freely around the individual particles (Fig. 12-32). This may occur when solids are in static, moving, fluidized, or dilute conditions.

Three additional terms require definition.

1. *Cocurrent gas flow.* The gas phase and solids particles both flow in the same direction (Fig. 12-33).

2. *Countercurrent gas flow.* The direction of gas flow is exactly opposite to the direction of solids movement.

3. *Cross-flow of gas.* The direction of gas flow is at a right angle to that of solids movement, across the solids bed (Fig. 12-34).

Because in a gas-solids-contacting operation heat transfer and mass transfer take place at the solids' surfaces, maximum process efficiency can be expected with a maximum exposure of solids surface to the gas phase, together with thorough mixing of gas and solids. Both are important. Within any arrangement of particulate solids, gas is present in the voids among the particles and contacts all surfaces except at the points of particle contact. When the solids bed is in a static or slightly moving condition, however, gas within the voids is cut off from the main body of the gas phase. Some transfer of energy and mass may occur by diffusion, but it is usually insignificant.

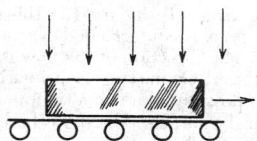

FIG. 12-31 Circulating gas impinging on a large solid object in perpendicular flow, in a roller-conveyor furnace.

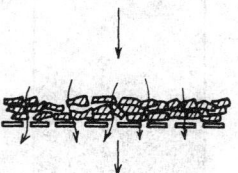

FIG. 12-32 Gas passing through a bed of preformed solids, in through circulation on a perforated-apron conveyor.

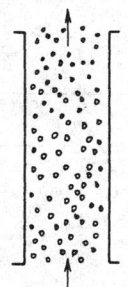

FIG. 12-33 Cocurrent gas-solids flow in a vertical-lift dilute-phase pneumatic conveyor.

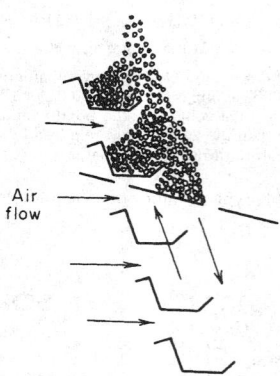

FIG. 12-34 Cross-flow of gas and solids in a cascade-type gravity dryer. (*Link-Belt Co., Multi-Louvre principle.*)

Equipment design and selection are governed by two factors:
1. Mechanical considerations
2. Solids flow and surface characteristics.

The former usually involves process temperature or isolation. Solids surface characteristics are important in that they control the extent to which an operation is diffusion-limited, i.e., diffusion into and out of the pores of a given solids particle, not through the voids among separate particles. The size of the solids particles, the surface-to-mass ratio, is also important in the evaluation of surface characteristics and the diffusion problem.

Gas-Solids Separations After the solids and gas have been brought together and mixed in a gas-solids contactor, it becomes necessary to separate the two phases. If the solids are sufficiently coarse and the gas velocity sufficiently low, it is possible to effect a complete gravitational separation in the primary contactor. Applications of this type are rare, however, and supplementary dust-collection equipment is commonly required. The recovery step may even dictate the type of primary contacting device selected. For example, when treating an extremely friable solid material, a deep fluidized-solids contactor might overload the collection system with fines, whereas the more gentle contacting of a traveling-screen contactor would be expected to produce a minimum of fines by attrition. Therefore, although gas-solids separation is usually considered as separate and distinct from the primary contacting operation, it is usually desirable to evaluate the separation problem at the same time that contacting methods are evaluated.

Definitions Drying generally refers to the removal of a liquid from a solid by **evaporation.** Mechanical methods for separating a liquid from a solid are not generally considered drying, although they often precede a drying operation, since it is less expensive and frequently easier to use mechanical methods than to use thermal methods.

This subsection presents the theory and fundamental concepts of the drying of solids.

Equipment commonly employed for the drying of solids is described both in this subsection in Sec. 12, where indirect heat transfer devices are discussed, and in Sec. 17 where fluidized beds are covered. Dryer control is discussed in Sec. 8. Excluding fluid beds this subsection contains mainly descriptions of direct-heat-transfer equipment. It also includes some indirect units; e.g., vacuum dryers, furnaces, steam-tube dryers, and rotary calciners.

Generally accepted terminology and definitions are given alphabetically in the following paragraphs.

Bound moisture in a solid is that liquid which exerts a vapor pressure less than that of the pure liquid at the given temperature. Liquid may become bound by retention in small capillaries, by solution in cell or fiber walls, by homogeneous solution throughout the solid, and by chemical or physical adsorption on solid surfaces.

Capillary flow is the flow of liquid through the interstices and over the surface of a solid, caused by liquid-solid molecular attraction.

Constant-rate period is that drying period during which the rate of water removal per unit of drying surface is constant.

Critical moisture content is the average moisture content when the constant-rate period ends.

Dry-weight basis expresses the moisture content of wet solid as kilograms of water per kilogram of bone-dry solid.

Equilibrium moisture content is the limiting moisture to which a given material can be dried under specific conditions of air temperature and humidity.

Falling-rate period is a drying period during which the instantaneous drying rate continually decreases.

Fiber-saturation point is the moisture content of cellular materials (e.g., wood) at which the cell walls are completely saturated while the cavities are liquid-free. It may be defined as the equilibrium moisture content as the humidity of the surrounding atmosphere approaches saturation.

Free-moisture content is that liquid which is removable at a given temperature and humidity. It may include bound and unbound moisture.

Funicular state is that condition in drying a porous body when capillary suction results in air being sucked into the pores.

Hygroscopic material is material that may contain bound moisture.

Initial moisture distribution refers to the moisture distribution throughout a solid at the start of drying.

Internal diffusion may be defined as the movement of liquid or vapor through a solid as the result of a concentration difference.

Moisture content of a solid is usually expressed as moisture quantity per unit weight of the dry or wet solid.

Moisture gradient refers to the distribution of water in a solid at a given moment in the drying process.

Nonhygroscopic material is material that can contain no bound moisture.

Pendular state is that state of a liquid in a porous solid when a continuous film of liquid no longer exists around and between discrete particles so that flow by capillary cannot occur. This state succeeds the funicular state.

Unaccomplished moisture change is the ratio of the free moisture present at any time to that initially present.

Unbound moisture in a hygroscopic material is that moisture in excess of the equilibrium moisture content corresponding to saturation humidity. All water in a nonhygroscopic material is unbound water.

Wet-weight basis expresses the moisture in a material as a percentage of the weight of the wet solid. Use of a dry-weight basis is recommended since the percentage change of moisture is constant for all moisture levels. When the wet-weight basis is used to express moisture content, a 2 or 3 percent change at high moisture contents (above 70 percent) actually represents a 15 to 20 percent change in evaporative load. See Fig. 12-35 for the relationship between the dry- and wet-weight bases.

APPLICATION OF PSYCHROMETRY TO DRYING

In any drying process, if an adequate supply of heat is assumed, the temperature and rate at which liquid vaporization occurs will depend on the vapor concentration in the surrounding atmosphere.

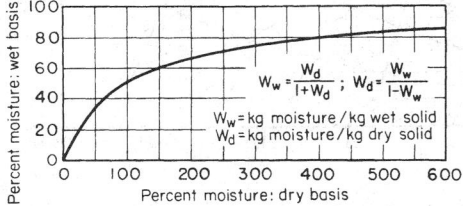

FIG. 12-35 Relationship between wet-weight and dry-weight bases.

In vacuum drying or other processes containing atmospheres of 100 percent vapor, the temperature of liquid vaporization will equal or exceed the saturation temperature of the liquid at the system pressure. (When a free liquid or wetted surface is present, drying will occur at the saturation temperature, just as free water at 101.325 kPa vaporizes in a 100 percent steam atmosphere at 100°C.)

On the other hand, when evolved vapor is purged from the dryer environment by using a second (inert) gas, the temperature at which vaporization occurs will depend on the concentration of vapor in the surrounding gas. In effect, the liquid must be heated to a temperature at which its vapor pressure equals or exceeds the partial pressure of vapor in the purge gas. In the reverse situation, condensation will occur.

In most drying operations, water is the liquid evaporated and air is the normally employed purge gas. For drying purposes, a psychrometric chart found very useful is that reproduced in Fig. 12-36.

1. The **wet-bulb or saturation temperature line** gives the maximum weight of water vapor that 1 kg of dry air can carry at the intersecting **dry-bulb temperature** shown on the abscissa at saturation humidity. The partial pressure of water in air equals the water-vapor pressure at that temperature. The saturation humidity is defined by

$$H_s = p_s/(P - p_s)18/28.9 \qquad (12\text{-}17)$$

where H_s = saturation humidity (kg/kg dry air), p_s = vapor pressure of water at temperature t_s, P = absolute pressure, and $18/28.9$ = ratio of molecular weights of water (18) and air (28.9). Similarly, the humidity at any condition less than saturation is given by

$$H = p/(P - p)18/28.9 \qquad (12\text{-}18)$$

2. The percent relative humidity is defined by

$$H_R = 100(p/p_s) \qquad (12\text{-}19)$$

where p = partial pressure of water vapor in the air, p_s = vapor pressure of water at the same temperature, and H_R = percent relative humidity.

3. Humid volumes are given by the curves entitled "Volume m³/kg dry air." The volumes are plotted as functions of absolute humidity and temperature. The difference between dry-air specific volume and humid-air volume at a given temperature is the volume of water vapor.

4. Enthalpy data are given on the basis of kilojoules per kilogram of dry air. **Enthalpy-at-saturation data** are accurate only at the saturation temperature and humidity. Enthalpy deviation curves permit enthalpy corrections for humidities less than saturation and show how the wet-bulb-temperature lines do not precisely coincide with constant-enthalpy, adiabatic cooling lines.

5. There are no lines for humid heats on Fig. 12-36. These may be calculated by

$$C_s = 1.0 + 1.87H \qquad (12\text{-}20)$$

where C_s = humid heat of moist air, kJ/(kg·K); 1.0 = specific heat of dry air, kJ/(kg·K); 1.87 = specific heat of water vapor, kJ/(kg·K); and H = absolute humidity, kg/kg dry air.

6. The wet-bulb-temperature lines represent also the adiabatic-saturation lines for air and water vapor only. These are based on the relationship

$$H_s - H = (C_s/\lambda)(t - t_s) \qquad (12\text{-}21)$$

where H_s and t_s = adiabatic saturation humidity and temperature respectively, corresponding to the air conditions represented by H and t, and C_s = humid heat for humidity H. The slope of the adiabatic-saturation curve is C_s/λ, where λ = latent heat of evaporation at t_s. These lines show the relationship between the temperature and humidity of air passing through a continuous dryer operating adiabatically.

The wet-bulb temperature is established by a dynamic equilibrium between heat and mass transfer when liquid evaporates from a small mass, such as the wet bulb of a thermometer, into a very large mass of gas such that the latter undergoes no temperature or humidity change. It is expressed by the relationship

$$h_c(t - t_w) = k'_g \lambda (H_w - H_a) \qquad (12\text{-}22)$$

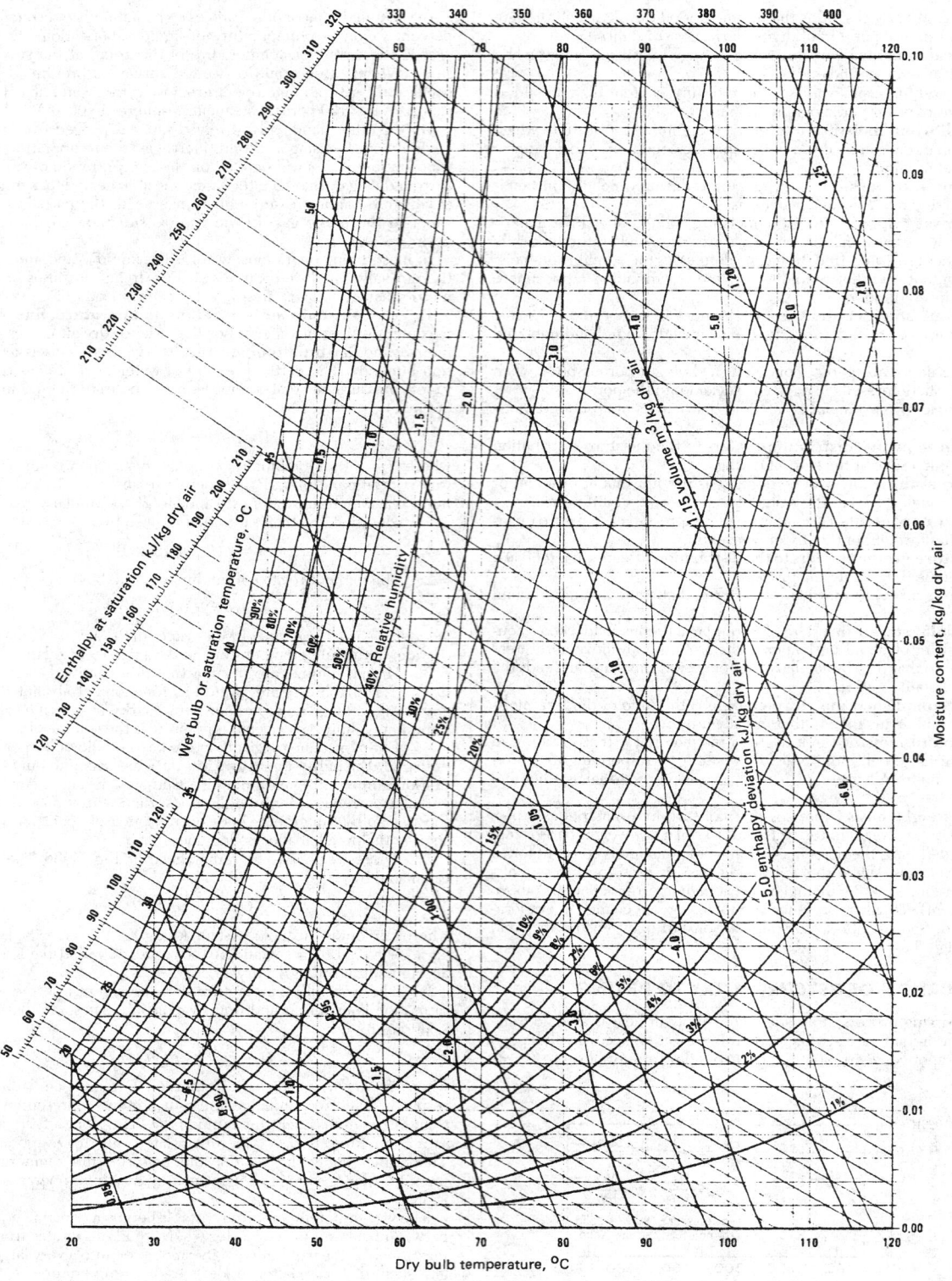

FIG. 12-36 Psychrometric chart: properties of air and water-vapor mixtures from 20 to 120°C. (*Carrier Corp.*)

where h_c = heat-transfer coefficient by convection, $J/(m^2 \cdot s \cdot K)$ [Btu/$(h \cdot ft^2 \cdot {}^\circ F)$]; t = air temperature, K; t_w = wet-bulb temperature of air, K; k'_g = mass-transfer coefficient, $kg/(s \cdot m^2)$ (kg/kg) [lb/$(h \cdot ft^2)$(lb/lb)]; λ = latent heat of evaporation at t_w, J/kg (Btu/lb); H_w = saturated humidity at t_w = kg/kg of dry air; and H_a = humidity of the surrounding air, kg/kg of dry air.

For air–water-vapor mixtures, it so happens that $h_c/k'_g = C_s$ approximately, although there is no theoretical reason for this. Hence, since the ratio $(H_w - H_a)/(t_w - t)$ equals $h_c/k'_g\lambda$, which represents the slope of the wet-bulb-temperature lines, it is also equal to C_s/λ, the slope of the adiabatic-saturation lines as shown previously.

A given humidity chart is precise only at the pressure for which it is evaluated. Most air–water-vapor charts are based on a pressure of 1 atm. Humidities read from these charts for given values of wet- and dry-bulb temperature apply only at an atmospheric pressure of 760 mmHg. If the total pressure is different from 760 mmHg, the humidity at a given wet-bulb and dry-bulb temperature must be corrected according to the following relationship.

$$H_a = H_o + 0.622p_w\left(\frac{1}{P - p_w} - \frac{1}{760 - p_w}\right) \quad (12\text{-}23)$$

where H_a = humidity of air at pressure P, kg/kg of dry air; H_o = humidity of air as read from a humidity chart based on 760-mm pressure at the observed wet- and dry-bulb temperatures, kg/kg dry air, p_w = vapor pressure of water at the observed wet-bulb temperature, mmHg; and P = the pressure at which the wet- and dry-bulb readings were taken. Similar corrections can be derived to correct specific volume, the saturation-humidity curve, and the relative-humidity curves.

HUMIDITY CHARTS FOR SOLVENT VAPORS

Humidity charts for other solvent vapors may be prepared in an analogous manner. There is one important difference involved, however, in that the wet-bulb temperature differs considerably from the adiabatic-saturation temperatures for vapors other than water.

Figures 12-37 to 12-39 show humidity charts for carbon tetrachloride, benzene, and toluene. The lines on these charts have been calculated in the manner outlined for air–water vapor except for the wet-bulb-temperature lines. The determination of these lines depends on data for the psychrometric ratio h_c/k'_g, as indicated by Eq. (12-22). For the charts shown, the wet-bulb-temperature lines are based on the following equation:

$$H_w - H = (\alpha h_c/\lambda_w k'_g)(t - t_w) \quad (12\text{-}24)$$

where α = radiation correction factor, a value of 1.06 having been used for these charts. Values of h_c/k'_g, obtained from values of $h_c/k'_g C_s$ as presented by Walker, Lewis, McAdams, and Gilliland (*Principles of Chemical Engineering*, 3d ed., McGraw-Hill, New York, 1937), where C_s = humid heat of air with respect to the vapor involved, are as follows:

Material	Carbon tetrachloride	Benzene	Toluene
$h_c/k'_g C_s$	0.51	0.54	0.47

A discussion of the theory of the relationship between h_c and k'_g may be found in the psychrometry part of this section. Because both theoretical and experimental values of h_c/k'_g apply only to dilute gas mixtures, the wet-bulb lines at high concentrations have been omitted. For a discussion of the precautions to be taken in making psychrometric determinations of solvent vapors at low solvent wet-bulb temperatures in the presence of water vapor, see the paper by Sherwood and Comings [*Trans. Am. Inst. Chem. Eng.*, **28**, 88 (1932)].

GENERAL CONDITIONS FOR DRYING

Solids drying encompasses two fundamental and simultaneous processes: (1) heat is transferred to evaporate liquid, and (2) mass is

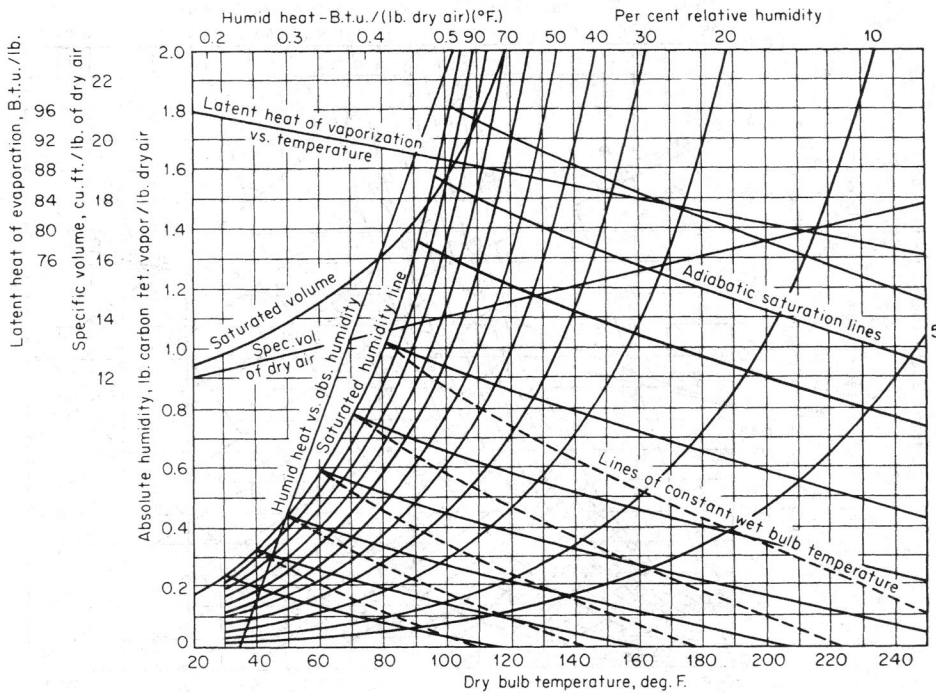

FIG. 12-37 Humidity chart for air-carbon tetrachloride vapor mixture. To convert British thermal units per pound to joules per kilogram, multiply by 2326; to convert British thermal units per pound dry air-degree Fahrenheit to joules per kilogram-kelvin, multiply by 4186.8; and to convert cubic feet per pound to cubic meters per kilogram, multiply by 0.0624.

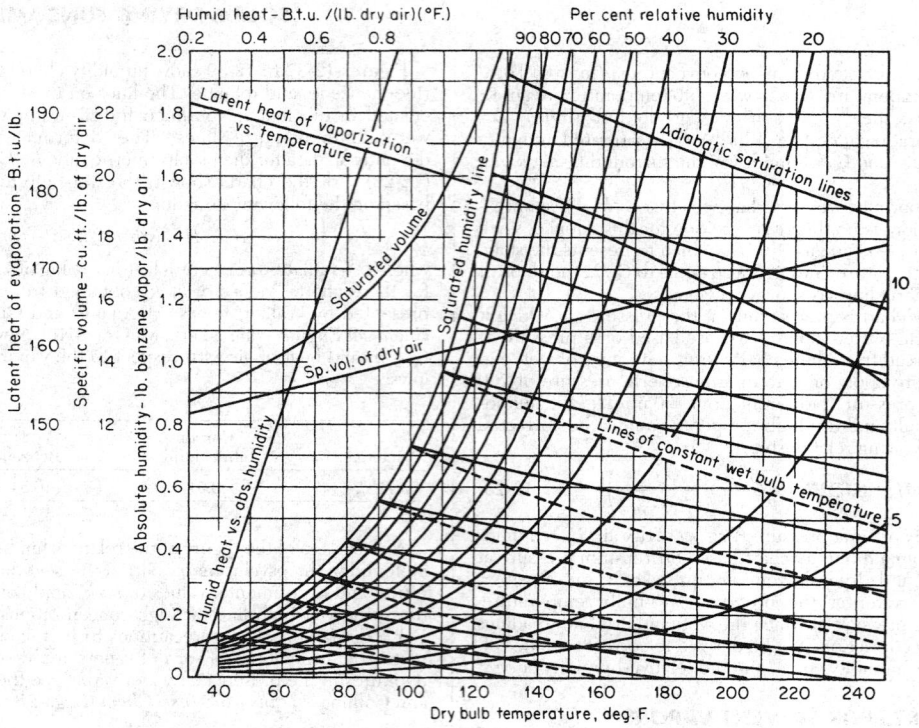

FIG. 12-38 Humidity chart for air-benzene-vapor mixture. To convert British thermal units per pound to joules per kilogram, multiply by 2326; to convert British thermal units per pound dry air-degree Fahrenheit to joules per kilogram-kelvin, multiply by 4186.8; and to convert cubic feet per pound to cubic meters per kilogram, multiply by 0.0624.

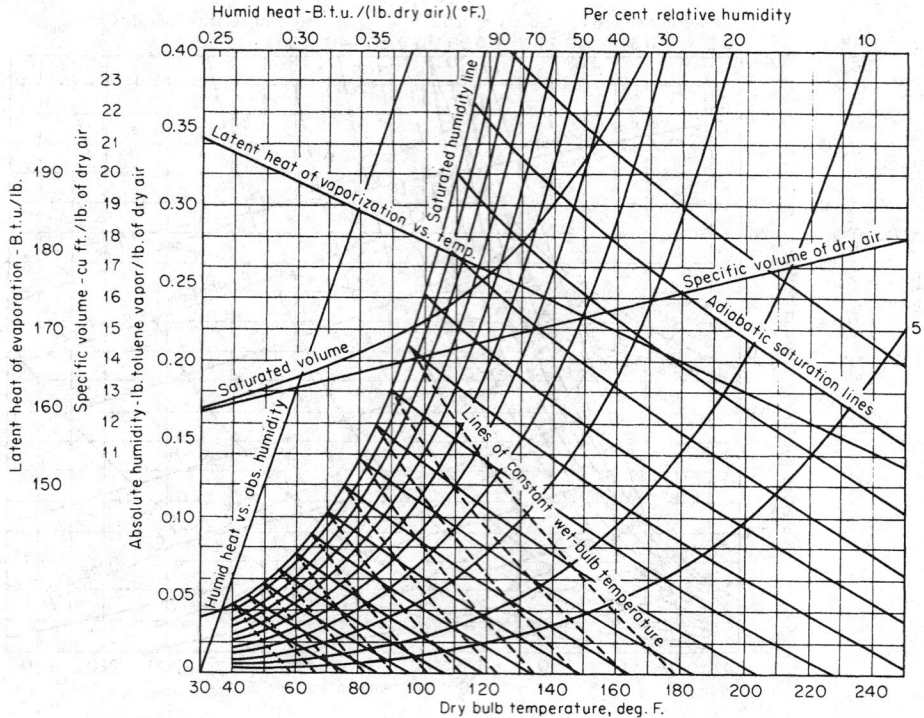

FIG. 12-39 Humidity chart for air-toluene-vapor mixture. To convert British thermal units per pound to joules per kilogram, multiply by 2326; to convert British thermal units per pound dry air-degree Fahrenheit to joules per kilogram-kelvin, multiply by 4186.8; and to convert cubic feet per pound to cubic meters per kilogram, multiply by 0.0624.

transferred as a liquid or vapor within the solid and as a vapor from the surface. The factors governing the rates of these processes determine the drying rate.

Commercial dryers differ fundamentally by the methods of heat transfer employed (see classification of dryers, Fig. 12-45). These industrial-dryer operations may utilize heat transfer by convection, conduction, radiation, or a combination of these. In each case, however, heat must flow to the outer surface and then into the interior of the solid. The single exception is dielectric and microwave drying, in which high-frequency electricity generates heat internally and produces a high temperature within the material and on its surface.

Mass is transferred in drying as a liquid and vapor within the solid and as vapor from the exposed surfaces. Movement within the solid results from a concentration gradient which is dependent on the characteristics of the solid. A solid to be dried may be porous or nonporous. It can also be hygroscopic or nonhygroscopic. Many solids fall intermediately between these two extremes, but it is generally convenient to consider the solid to be one or the other.

A study of how a solid dries may be based on the **internal** mechanism of liquid flow or on the effect of the **external** conditions of temperature, humidity, air flow, state of subdivision, etc., on the drying rate of the solids. The former procedure generally requires a fundamental study of the internal condition. The latter procedure, although less fundamental, is more generally used because the results have greater immediate application in equipment design and evaluation.

INTERNAL MECHANISM OF LIQUID FLOW

The structure of the solid determines the mechanism for which internal liquid flow may occur. These mechanisms can include (1) **diffusion** in continuous, homogeneous solids, (2) **capillary flow** in granular and porous solids, (3) flow caused by **shrinkage** and **pressure** gradients, (4) flow caused by **gravity,** and (5) flow caused by a **vaporization-condensation** sequence.

In general, one mechanism predominates at any given time in a solid during drying, but it is not uncommon to find different mechanisms predominating at different times during the drying cycle.

The study of internal moisture gradients establishes the particular mechanism which controls during the drying of a solid. The experimental determination of reliable moisture gradients is extremely difficult.

Hougen, McCauley, and Marshall [*Trans. Am. Inst. Chem. Eng.,* **36,** 183 (1940)] discussed the conditions under which capillary and diffusional flow may be expected in a drying solid and analyzed the published experimental moisture-gradient data for the two cases. Their curves indicate that capillary flow is typified by a moisture gradient involving a double curvature and point of inflection (Fig. 12-40a) while diffusional flow is a smooth curve, concave downward (Fig. 12-40b), as would be predicted from the diffusion equations. They also showed that the liquid-diffusion coefficient is usually a function

of moisture content which decreases with decreasing moisture. The effect of variable diffusivity is illustrated in Fig. 12-40b, where the dashed line is calculated for constant diffusivity and the solid line is experimental for the case in which the diffusion coefficient is moisture-dependent. Thus, the integrated diffusion equations assuming constant diffusivity only approximate the actual behavior.

These authors classified solids on the basis of capillary and diffusional flow:

Capillary Flow Moisture which is held in the interstices of solids, as liquid on the surface, or as free moisture in cell cavities, moves by gravity and capillary, provided that passageways for continuous flow are present. In drying, liquid flow resulting from capillarity applies to liquids not held in solution and to all moisture above the fiber-saturation point, as in textiles, paper, and leather, and to all moisture above the equilibrium moisture content at atmospheric saturations, as in fine powders and granular solids, such as paint pigments, minerals, clays, soil, and sand.

Vapor Diffusion Moisture may move by vapor diffusion through the solid, provided that a temperature gradient is established by heating, thus creating a vapor-pressure gradient. Vaporization and vapor diffusion may occur in any solid in which heating takes place at one surface and drying from the other and in which liquid is isolated between granules of solid.

Liquid Diffusion The movement of liquids by diffusion in solids is restricted to the equilibrium moisture content below the point of atmospheric saturation and to systems in which moisture and solid are mutually soluble. The first class applies to the last stages in the drying of clays, starches, flour, textiles, paper, and wood; the second class includes the drying of soaps, glues, gelatins, and pastes.

External Conditions The principal external variables involved in any drying study are temperature, humidity, air flow, state of subdivision of the solid, agitation of the solid, method of supporting the solid, and contact between hot surfaces and wet solid. All these variables will not necessarily occur in one problem.

PERIODS OF DRYING

When a solid is dried experimentally, data relating moisture content to time are usually obtained. These data are then plotted as moisture content (dry basis) W versus time θ, as shown in Fig. 12-41a. This curve represents the general case when a wet solid loses moisture first by evaporation from a saturated surface on the solid, followed in turn by a period of evaporation from a saturated surface of gradually decreasing area, and, finally, when the latter evaporates in the interior of the solid.

Figure 12-41a indicates that the drying rate is subject to variation with time or moisture content. This variation is better illustrated by graphically or numerically differentiating the curve and plotting $dW/d\theta$ versus W, as shown in Fig. 12-41b, or as $dW/d\theta$ versus θ, as shown in Fig. 12-41c. These **rate curves** illustrate that the drying process is not a smooth, continuous one in which a single mechanism controls throughout. Figure 12-41c has the advantage of showing how long each drying period lasts.

The section AB on each curve represents a **warming-up** period of the solids. Section BC on each curve represents the **constant-rate** period. Point C, where the constant rate ends and the drying rate begins falling, is termed the **critical-moisture** content. The curved portion CD on Fig. 12-41a is termed the **falling-rate** period and, as shown in Fig. 12-41b and c, is typified by a continuously changing rate throughout the remainder of the drying cycle. Point E (Fig. 12-41b) represents the point at which all the exposed surface becomes completely unsaturated and marks the start of that portion of the drying cycle during which the rate of internal moisture movement controls the drying rate. Portion CE in Fig. 12-41b is usually defined as the first falling-rate drying period; portion DE, as the second falling-rate period.

Constant-rate Period In the constant-rate period moisture movement within the solid is rapid enough to maintain a saturated condition at the surface, and the rate of drying is controlled by the rate of heat transferred to the evaporating surface. Drying proceeds by diffusion of vapor from the saturated surface of the material across a

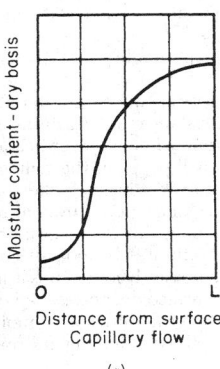

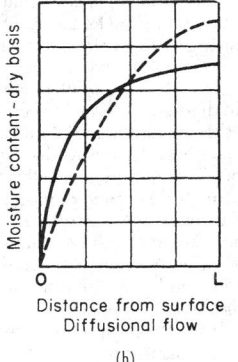

Capillary flow

Diffusional flow

(a)

(b)

FIG. 12-40 Two types of internal moisture gradients obtained in drying solids.

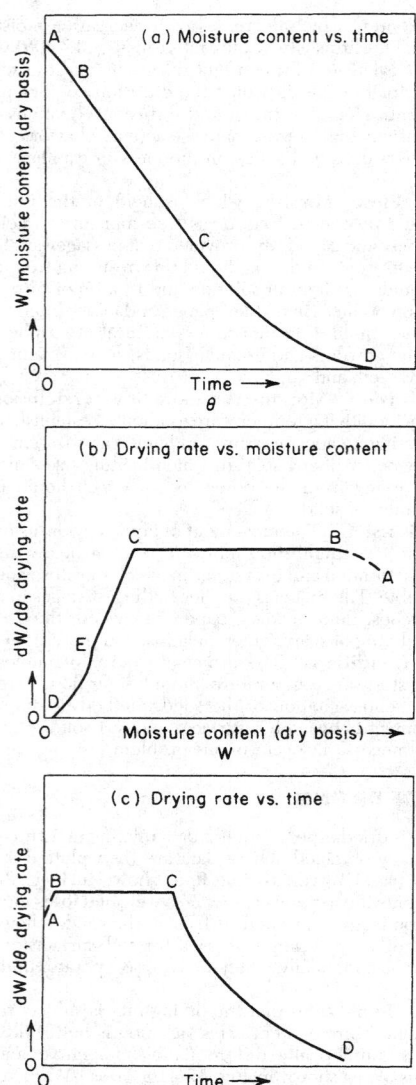

FIG. 12-41 The periods of drying.

stagnant air film into the environment. The rate of mass transfer balances the rate of heat transfer, and the temperature of the saturated surface remains constant. The mechanism of moisture removal is equivalent to evaporation from a body of water° and is essentially independent of the nature of the solids.

If heat is transferred solely by convection and in the absence of other heat effects, the surface temperature approaches the wet-bulb temperature. However, when heat is transferred by radiation, convection, or a combination of these and convection, the temperature at the saturated surface is between the wet-bulb temperature and the boiling point of water. Under these conditions, the rate of heat transfer is increased and a higher drying rate results.

When heat is transferred to a wet solid by convection to hot surfaces and heat transfer by convection is negligible, the solids approach the boiling-point temperature rather than the wet-bulb temperature. This method of heat transfer is utilized in indirect dryers (see classification

° The term "water" is used for convenience; this discussion applies equally to other liquids.

of dryers in Fig. 12-45). Radiation is also effective in increasing the constant rate by augmenting the convection heat transfer and raising the surface temperature above the wet-bulb temperature.

When the heat for evaporation in the constant-rate period is supplied by a hot gas, a dynamic equilibrium establishes the rate of heat transfer to the material and the rate of vapor removal from the surface:

$$dw/d\theta = h_t A \, \Delta t/\lambda = k_g A \, \Delta p \qquad (12\text{-}25)$$

where $dw/d\theta$ = drying rate, kg water/s; h_t = total heat-transfer coefficient, J/(m²·s·K) [Btu/(h·ft²·°F)]; A = area for heat transfer and evaporation, m²; λ = latent heat of evaporation at t'_s, J/kg (Btu/lb); k_g = mass-transfer coefficient, kg/(s·m²·atm) [lb/(h·ft²·atm)]; $\Delta t = t - t'_s$, where t = gas (dry-bulb) temperature, K; $p = p_s - p$, where p_s = vapor pressure of water at surface temperature t'_s, atm; and p = partial pressure of water vapor in the gas, atm.

The magnitude of the constant rate depends upon three factors:
1. The heat- or mass-transfer coefficient
2. The area exposed to the drying medium
3. The difference in temperature or humidity between the gas stream and the wet surface of the solid
All these factors are the external variables. The internal mechanism of liquid flow does not affect the constant rate.

For drying calculations, it is convenient to express Eq. (12-25) in terms of the decrease in moisture content rather than in the quantity of water evaporated. For evaporation from a tray of wet material, if no change in volume during drying is assumed, Eq. (12-25) becomes

$$dw/d\theta = (h/\rho_s \, d\lambda)(t - t'_s) \qquad (12\text{-}26)$$

where $dw/d\theta$ = drying rate, kg water/(s·kg dry solids); h_t = total heat-transfer coefficient, J/(m²·s·K) [Btu/(h·ft²·°F)]; ρ_s = bulk density dry material, kg/m³; d = thickness of bed, m; λ = latent heat of vaporization, J/kg (Btu/lb); t = air temperature, K; and t'_s = evaporating surface temperature, K. (Note that $dw/d\theta$ is inherently negative.)

A similar equation can be written for the through-circulation case:

$$dW/d\theta = (h_t a/\rho_s \lambda)(t - t'_s) \qquad (12\text{-}27)$$

where a = m² of heat-transfer area/m³ of bed, 1/m; and other symbols are as in Eq. (12-26).

The values of ρ_s and/or a must be known in order to use Eqs. (12-26) and (12-27). The value of a is difficult to estimate without experimental data. When the void fraction is known, a can sometimes be estimated from the following relationships:

For spherical particles,

$$a = \frac{6(1 - F)}{(D_p)_m} \qquad (12\text{-}28)$$

For uniform cylindrical particles,

$$a = \frac{4(0.5D_0 + Z)(1 - F)}{D_0 Z} \qquad (12\text{-}29)$$

where F = void fraction; $(D_p)_m$ = harmonic mean diameter of spherical particles, m; D_0 = diameter of cylinder, m; and Z = height of cylinder, m. For cylindrical particles that are long relative to their diameter, the term $0.5D_0$ in Eq. (12-29) can be neglected.

Falling-rate Period The falling-rate period begins at the critical moisture content when the constant-rate period ends. When the falling moisture content is above the critical moisture content, the whole drying process will occur under constant-rate conditions. If, on the other hand, the initial moisture content is below the critical moisture content, the entire drying process will occur in the falling-rate period. This period is usually divided into two zones: (1) the zone of **unsaturated surface drying** and (2) the zone where **internal moisture movement** controls. In the first zone, the entire evaporating surface can no longer be maintained and saturated by moisture movement within the solid. The drying rate decreases from the unsaturated portion, and hence the rate for the total surface decreases. Generally, the drying rate depends on factors affecting the diffusion of moisture away from the evaporating surface and those affecting the rate of internal moisture movement.

As drying proceeds, the point is reached where the evaporating surface is unsaturated. The point of evaporation moves into the solid, and the dry process enters the second falling-rate period. The drying rate is now governed by the rate of internal moisture movement; the influence of external variables diminishes. This period usually predominates in determining the overall drying time to lower moisture content.

LIQUID DIFFUSION

Diffusion-controlled mass transfer is assumed when the vapor or liquid flow conforms to Fick's second law of diffusion. This is stated in the unsteady-state-diffusion equation using mass-transfer notation as

$$\delta c/\delta\theta = D_{AB}(\delta^2 C/dx) \qquad (12\text{-}30)$$

where c = concentration of one component in a two-component phase of A and B, θ = diffusion time, x = distance in the direction of diffusion, and D_{AB} = binary diffusivity of the phase AB. This equation applies to diffusion in solids, stationary liquids, and stagnant gases.

The diffusion equation for the falling-rate drying period for a slab can be derived from the diffusion equation if one assumes that the surface is dry or at an equilibrium moisture content and that the initial moisture distribution is uniform. For these conditions, the following equation is obtained:

$$\frac{W - W_e}{W_c - W_e} = \frac{8}{\pi^2} \left[\sum_{n=0}^{n=\infty} \frac{1}{(2n+1)} e^{-(2n+1)^2 D_\ell \theta(\pi/2d)^2} \right] \qquad (12\text{-}31)$$

where W, W_e, and W_c = average moisture content (dry basis) at any time, θ, at the start of the falling-rate period and in equilibrium with the environment respectively, kg/kg; D_ℓ = liquid diffusivity, m^2/s; θ = time from start of falling-rate period, s; and d = one-half of the thickness of the solid layer through which diffusion occurs, m. When evaporation occurs from only one face, d = total thickness, m.

Equation (12-31) assumes that D_ℓ is constant; however, D_ℓ is rarely constant but varies with moisture content, temperature, and humidity. For long drying times, Eq. (12-31) simplifies to a limiting form of the diffusion equation as

$$\frac{W - W_e}{W_c - W_e} = \frac{8}{\pi^2} \left[e^{-D_\ell \theta(\pi/2d)^2} \right] \qquad (12\text{-}32)$$

Equation (12-32) may be differentiated to give the drying rate as

$$-\frac{dW}{d\theta} = \frac{\pi^2 D_\ell}{4d^2} (W - W_e) \qquad (12\text{-}33)$$

where $dW/d\theta$ = drying rate, kg/s.

When Eq. (12-31) is plotted on semilogarithmic graph paper, a straight line is obtained for values of $(W - W_e)/(W_c - W_e) < 0.6$. It is in the straight-line portion that the approximate form [Eq. (12-33)] applies.

Equations (12-31), (12-32), and (12-33) hold only for a slab-sheet solid whose thickness is small relative to the other two dimensions. For other shapes, reference should be made to Crank (*The Mathematics of Diffusion*, Oxford, London, 1956).

An approximate equation for the falling-rate period may be obtained by integration of Eq. (12-33). This gives an equation for materials in which moisture movement is controlled by diffusion:

$$\theta_f = \frac{4d^2}{D_\ell \pi^2} \ln \frac{W_c - W_e}{W - W_e} \qquad (12\text{-}34)$$

where θ_f = drying time in the falling-rate period.

Diffusion equations may also be used to study vapor diffusion in porous materials. It should be clear that all estimates based on relationships that assume constant diffusivity are approximations. Liquid diffusivity in solids usually decreases with moisture concentration. Liquid and vapor diffusivity also change, and material shrinks during drying.

CAPILLARY THEORY

If the porous size of a granular material is suitable, moisture may move from a region of high to one of low concentration as the result of capil-

lary action rather than by diffusion. The capillary theory assumes that a bed of nonporous spheres is composed of particles surrounding a space called a pore. These pores are connected by passages of various sizes. As water is progressively removed from the bed, the curvature of the water surface in the interstices of the top layer of spheres increases and a suction pressure resulting from curvature is set up. As the removal of water continues, the suction pressure attains a value in which air is drawn into the pore spaces between successive layers of spheres.

This entry suction or suction potential is a measure of the resultant forces tending to draw water from the interior of the bed to the surface. For a pore formed by regularly packed nonporous spheres, the suction potential is given by

$$P_s = x\sigma/r\rho g \qquad (12\text{-}35)$$

where P_s = suction potential, m of water; σ = surface tension; dyn/m; ρ = density of water, kg/m^3; $g = 9.8$ m/s^2; r = sphere radius, m; and x is a packing factor equal to 12.9 for rhombohedral and 4.8 for cubical packing.

As drying proceeds, the surface moisture evaporates, causing retreat of the surface menisci until the suction potential reaches a value given by Eq. (12-35). At this point, the pores of the surface will open, air will enter, and the moisture will redistribute itself with a slight lowering of the suction potential. As evaporation proceeds, the suction potential again increases until a slightly higher entry value is reached, when a further redistribution occurs.

The drying rate curve (Fig. 12-41) can be analyzed in terms of capillary theory. In region BC, there is a loss of moisture with a gradual increase in suction and emptying of the bulk of the larger pores in the solid. In region CE, there is an increase in suction as the moisture content decreases and finer pores are opened. Section ED represents a condition in which moisture is being removed by vapor diffusion from the interior of the body, although there is still sufficient water in the bed to give rise to capillary forces.

An approximate equation for use for materials in which moisture movement is controlled by capillary flow is given as

$$\theta_f = \frac{\rho_s \, d\lambda(W_c - W_e)}{h_t(t - t_s')} \ln \frac{W_c - W_e}{W - W_e} \qquad (12\text{-}36)$$

where θ_f = drying time in the falling-rate period.

Table 12-6 gives an approximate classification of materials that obey Eqs. (12-34) and (12-36).

CRITICAL MOISTURE CONTENT

To use the preceding equations for estimating drying times in the falling-rate period, it is necessary to know values of critical moisture content W_c. Such values are difficult to obtain without making actual drying tests, which in themselves would give the required drying time and thereby obviate solving the equations. However, when drying tests are not feasible, some estimate of critical moisture content must be made.

Values of critical moisture contents for some representative materials are given in Table 12-7 for drying by cross circulation and in Table 12-15 for drying by through circulation. The tabulated values are only approximate, since critical moisture content depends on the drying history. It appears that the constant-rate period ends when the moisture content at the surface reaches a specific value. Since the critical

TABLE 12-6 Materials Obeying Eqs. (12-34) and (12-38)

Materials obeying Eq. (12-34)	Materials obeying (Eq. 12-38)
1. Single-phase solid systems such as soap, gelatin, and glue	1. Coarse granular solids such as sand, paint pigments, and minerals
2. Wood and similar solids below the fiber-saturation point	2. Materials in which moisture flow occurs at concentrations above the equilibrium moisture content at atmospheric saturation or above the fiber-saturation point
3. Last stages of drying starches, textiles, paper, clay, hydrophilic solids, and other materials when bound water is being removed	

TABLE 12-7 Approximate Critical Moisture Contents Obtained on the Air Drying of Various Materials, Expressed as Percentage Water on the Dry Basis

Material	Thickness, in.	Critical moisture, % water, dry basis
Barium nitrate crystals, on trays	1.0	7
Beaverboard	0.17	Above 120
Brick clay	.62	14
Carbon pigment	1	40
Celotex	0.44	160
Chrome leather	.04	125
Copper carbonate (on trays)	1–1.5	60
English china clay	1	16
Flint clay refractory brick mix	2.0	13
Gelatin, initially 400 % water	0.1–0.2 (wet)	300
Iron blue pigment (on trays)	0.25–0.75	110
Kaolin		14
Lithol red	1	50
Lithopone press cake (in trays)	0.25	6.4
	.50	8.0
	.75	12.0
	1.0	16.0
Niter cake fines, on trays		Above 16
Paper, white eggshell	0.0075	41
Fine book	.005	33
Coated	.004	34
Newsprint		60–70
Plastic clay brick mix	2.0	19
Poplar wood	0.165	120
Prussian blue		40
Pulp lead, initially 140% water		Below 15
Rock salt (in trays)	1.0	7
Sand, 50–150 mesh	2.0	5
Sand, 200–325 mesh	2.0	10
Sand, through 325 mesh	2.0	21
Sea sand (on trays)	0.25	3
	.5	4.7
	.75	5.5
	1.0	5.9
	2.0	6.0
Silica brick mix	2.0	8
Sole leather	0.25	Above 90
Stannic tetrachloride sludge	1	180
Subsoil, clay fraction 55.4%		21
Subsoil, much higher clay content		35
Sulfite pulp	0.25–0.75	60–80
Sulfite pulp (pulp lap)	0.039	110
White lead		11
Whiting	0.25–1.5	6.9
Wool fabric, worsted		31
Wool, undyed serge		8

moisture content is the average moisture through the material, its value depends on the rate of drying, the thickness of the material, and the factors influencing moisture movement and resulting gradients within the solid. As a result, the critical moisture content increases with increased drying rate and with increased thickness of the mass of material being dried.

EQUILIBRIUM MOISTURE CONTENT

In drying solids it is important to distinguish between hygroscopic and nonhygroscopic materials. If a hygroscopic material is maintained in contact with air at constant temperature and humidity until equilibrium is reached, the material will attain a definite moisture content. This moisture is termed the equilibrium moisture content for the specified conditions. Equilibrium moisture may be adsorbed as a surface film or condensed in the fine capillaries of the solid at reduced pressure, and its concentration will vary with the temperature and humidity of the surrounding air. However, at low temperatures, e.g., 15 to 50°C, a plot of equilibrium moisture content versus percent relative humidity is essentially independent of temperature. At zero humidity the equilibrium moisture content of all materials is zero.

Equilibrium moisture content depends greatly on the nature of the solid. For nonporous, i.e., nonhygroscopic, materials, the equilibrium moisture content is essentially zero at all temperatures and humidities. For organic materials such as wood, paper, and soap, equilibrium moisture contents vary regularly over wide ranges as temperature and humidity change. In the special case of the dehydration of hydrated inorganic salts such as copper sulfate, sodium sulfate, or barium chloride, temperature and humidity control is very important in obtaining the desired degree of moisture removal, and the proper conditions must be determined from data on the water of hydration or crystallization as a function of air temperature and humidity.

Equilibrium moisture content of a solid is particularly important in drying because it represents the limiting moisture content for given conditions of humidity and temperature. If the material is dried to a moisture content less than it normally possesses in equilibrium with atmospheric air, it will return to its equilibrium value on storage unless special precautions are taken.

Equilibrium moisture content of a hygroscopic material may be determined in a number of ways, the only requirement being a source of constant-temperature and constant-humidity air. Determination may be made under static or dynamic conditions, although the latter case is preferred. A simple static procedure is to place a number of samples in ordinary laboratory desiccators containing sulfuric acid solutions of known concentrations which produce atmospheres of known relative humidity. The sample in each desiccator is weighed periodically until a constant weight is obtained. Moisture content at this final weight represents the equilibrium moisture content for the particular conditions.

The value of equilibrium moisture content, for many materials, depends on the direction in which equilibrium is approached. A different value is reached when a wet material loses moisture by desorption, as in drying, from that obtained when a dry material gains it by adsorption. For drying calculations the desorption values are preferred. In the general case, the equilibrum moisture content reached by losing moisture is higher than that reached by adsorbing it.

Equilibrium moisture content can be measured dynamically by placing the sample in a U tube through which is drawn a continuous flow of controlled-humidity air. Again the sample is weighed periodically until a constant weight is reached. Properly humidified air for such a procedure can be obtained by bubbling dry air through a large volume of a saturated salt solution which produces a definite degree of saturation of the air. Care must be taken to ensure that the air and salt solution reach equilibrium. Values of the humidity over various salt solutions may be found in Table 12-8. Several manufacturers supply chambers with dial-in temperature and humidity conditions for air-water systems.

ESTIMATIONS FOR TOTAL DRYING TIME

Estimates of both the constant-rate and the falling-rate periods are needed to estimate the total drying time for a given drying operation. If estimates for these periods are available, the total drying time is estimated by summing as

$$\theta_t = \theta_c + \theta_f \qquad (12\text{-}37)$$

where θ_t = total drying time, h; θ_c = drying time for constant-rate period, h; and θ_f = drying time for falling-rate period, h. The difficulty in estimating critical moisture content greatly reduces the number of drying cases in which calculation of a good estimate is possible.

ANALYSIS OF DRYING DATA TESTS ON PLANT DRYERS

When experiments are carried out to select a suitable dryer and to obtain design data, the effect of changes in various external variables is studied. These experiments should be conducted in an experimental unit that simulates the large-scale dryer from both the thermal and the material-handling aspects, and only material which is truly representative of full-scale production should be used.

Data expressing moisture content in terms of elapsed time should be obtained and the results plotted as shown in Fig. 12-42. For purposes of analysis, the moisture-time curve must be differentiated

TABLE 12-8 Maintenance of Constant Humidity

Solid phase	Max. temp., °C.	% humidity
$H_3PO_4 \cdot \frac{1}{2}H_2O$	24.5	9
$ZnCl_2 \cdot \frac{1}{2}H_2O$	20	10
$KC_2H_3O_2$	168	13
$LiCl \cdot H_2O$	20	15
$KC_2H_3O_2$	20	20
KF	100	22.9
$NaBr$	100	22.9
$CaCl_2 \cdot 6H_2O$	24.5	31
$CaCl_2 \cdot 6H_2O$	20	32.3
$CaCl_2 \cdot 6H_2O$	18.5	35
CrO_3	20	35
$CaCl_2 \cdot 6H_2O$	10	38
$CaCl_2 \cdot 6H_2O$	5	39.8
$K_2CO_3 \cdot 2H_2O$	24.5	43
$K_2CO_3 \cdot 2H_2O$	18.5	44
$Ca(NO_3)_2 \cdot 4H_2O$	24.5	51
$NaHSO_4 \cdot H_2O$	20	52
$Mg(NO_3)_2 \cdot 6H_2O$	24.5	52
$NaClO_3$	100	54
$Ca(NO_3)_2 \cdot 4H_2O$	18.5	56
$Mg(NO_3)_2 \cdot 6H_2O$	18.5	56
$NaBr \cdot 2H_2O$	20	58
$Mg(C_2H_3O_2)_2 \cdot 4H_2O$	20	65
$NaNO_2$	20	66
$(NH_4)_2SO_4$	108.2	75
$(NH_4)_2SO_4$	20	81
$NaC_2H_3O_2 \cdot 3H_2O$	20	76
$Na_2S_2O_3 \cdot 5H_2O$	20	78
NH_4Cl	20	79.5
NH_4Cl	25	79.3
NH_4Cl	30	77.5
KBr	20	84
Tl_2SO_4	104.7	84.8
$KHSO_4$	20	86
$Na_2CO_3 \cdot 10H_2O$	24.5	87
K_2CrO_4	20	88
$NaBrO_3$	20	92
$Na_2CO_3 \cdot 10H_2O$	18.5	92
$Na_2SO_4 \cdot 10H_2O$	20	93
$Na_2HPO_4 \cdot 12H_2O$	20	95
NaF	100	96.6
$Pb(NO_3)_2$	20	98
$TlNO_3$	100.3	98.7
$TlCl$	100.1	99.7

For a more complete list of salts, and for references to the literature see "International Critical Tables," vol. 1, p. 68.

graphically or numerically and the drying rates so obtained plotted to determine the nature and extent of the drying periods in the cycle. It is customary to plot drying rate versus moisture content as in Fig. 12-41b. Although instructive, this type of plot gives no information on duration of the drying periods. These are better shown by plots similar to Fig. 12-41c, in which drying rate is plotted as a function of time on either arithmetic or logarithmic coordinates. Logarithmic plots permit easy reading at low moisture contents or long times.

In order to determine whether a simple relationship exists in the falling-rate period, the unaccomplished moisture change, defined as ratio of free moisture in the solid at time θ to total free moisture present at start of the falling-rate period $(W - W_e)/(W_c - W_e)$, is plotted as a function of time on semilogarithmic paper. If a straight line is obtained such as curve B of Fig. 12-43 by using the upper scale of abscissa, either Eq. (12-34), for materials in which the moisture moves by diffusion, or Eq. (12-37), for materials in which the moisture movement is by capillary flow, may be applicable. If Eq. (12-37) applies, K_1, the slope of the falling-rate drying curve, is related to the constant drying rate. The latter is calculated from Eq. (12-38) and can be compared with the measured value. If the slopes agree, the moisture movement is by capillary flow. If the slopes do not agree, the moisture movement is by diffusion and the slope of the line should equal $\pi^2 D_\ell / 4d^2$.

The dependency of drying rate on material thickness must be established experimentally. With the effect of material thickness estab-

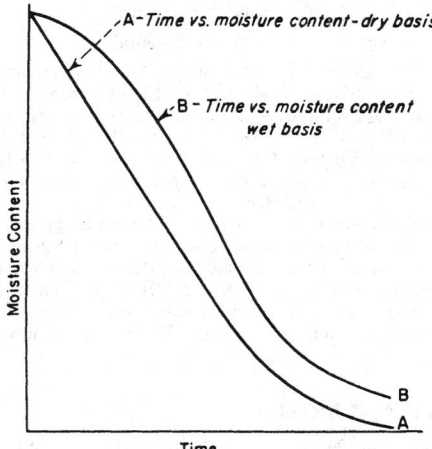

FIG. 12-42 Drying-time curves.

lished, liquid diffusivity can be calculated as indicated here. For this calculation, the theoretical values for an infinite slab are required. These are:

$\dfrac{D_\ell \theta}{d^2}$	0.02	0.05	0.10	0.15	0.20	0.30	0.50	1.0
$\dfrac{W - W_e}{W_c - W_e}$	0.84	0.75	0.642	0.563	0.496	0.387	0.238	0.069

A plot of these values is shown as curve A in Fig. 12-43.

If a straight line such as curve B of Fig. 12-43 represents the experimental data and if it has been established that the drying time varies inversely as the square of the thickness, the average liquid diffusivity can be obtained as follows. At a given value of $(W - W_e)/(W_c - W_e)$, read the corresponding value of $D_\ell \theta / d^2$ from curve A, Eq. (12-32), Fig. 12-43. At the same value of $(W - W_e)/(W_c - W_e)$, read the corresponding experimental value of θ from curve B (upper scale). Then

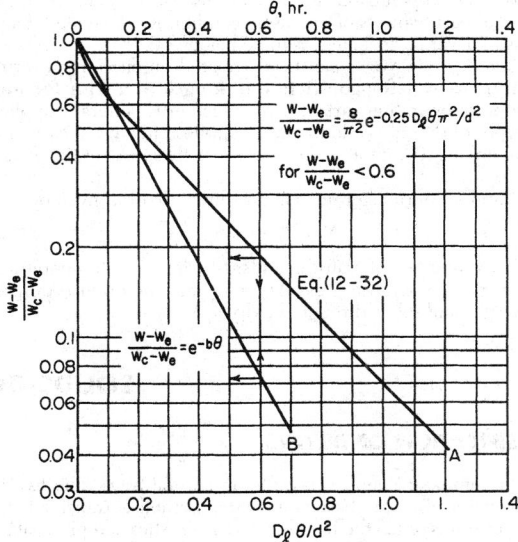

FIG. 12-43 Analysis of drying data.

$$D_{\ell_{avg}} = \frac{(D_\ell \theta/d^2) \text{ theoretical}}{(\theta/d^2) \text{ experimental}} \qquad (12\text{-}38)$$

where $D_{\ell_{avg}}$ = average experimental value of liquid diffusivity, m^2/h.

The value of diffusivity calculated from Eq. (12-38) must be recognized as an average value over the entire range of moisture change from $(W - W_e)/(W_c - W_e) = 1$ to the value $(W - W_c)/(W_c - W_e)$ at which θ/d^2 was evaluated. Further, Eq. (12-38) assumes that the theoretical curve is a straight line for all values of time. This is not true for values $(W - W_e)/(W_c - W_e)$ less than 0.6.

A more accurate value of $D_{\ell_{avg}}$ can be obtained by taking a ratio of slopes of the curves of Fig. 12-43. Thus the ratio of the slope of the experimental curve of unaccomplished moisture change versus drying time on a semilogarithmic plot [Eq. (12-32)] to the slope of the theoretical curve at the same unaccomplished moisture change, again on a semilogaritmic plot, equals the quantity D_ℓ/d^2. If d is known, D_ℓ can be evaluated.

TESTS ON PLANT DRYERS

Tests on plant-scale dryers are usually carried out to obtain design data for a specific material, to select a suitable dryer type, or to check present performance of an existing dryer with the objective of determining its capacity potential. In these tests overall performance data are obtained and the results used to make heat and material balances and to estimate overall drying rates or heat-transfer coefficients.

Generally, the minimum data to be taken in order to calculate the performance of a dryer are:
1. Inlet and outlet moisture contents
2. Inlet and outlet gas temperatures
3. Inlet and outlet material temperatures
4. Feed rate
5. Gas rate
6. Inlet and outlet humidities
7. Retention time or time of passage through the dryer
8. Fuel consumption

Whenever possible, moisture contents and temperatures should be measured at various points within the dryer.

Typical experimental and calculated results of a drying test for a continuous adiabatic convection dryer are shown in Fig. 12-44. Test data as elaborate as those shown are not usually justified economically except when basic studies, aimed at clarifying the effect of operating variables, are being carried out in order to arrive at a reliable design procedure. The completeness of the information which is sought in any given test depends on the ultimate use of the data. In any case data for at least two sets of operating conditions are needed if a good analysis of dryer performance is to be made.

Results of drying tests can be correlated empirically in terms of **overall heat-transfer coefficient** or **length of a transfer unit** as a function of operating variables. The former is generally applicable to all types of dryers, while the latter applies only in the case of continuous dryers. The relationship between these quantities is as follows.

The number of transfer units in any direct dryer is given by

$$N_t = (t_1 - t_2)/\Delta t_m \qquad (12\text{-}39)$$

where N_t = number of transfer units; t_1 = inlet gas temperature, K; t_2 = exit gas temperature; and Δt_m = mean temperature difference between gas and solids through the dryer, K.

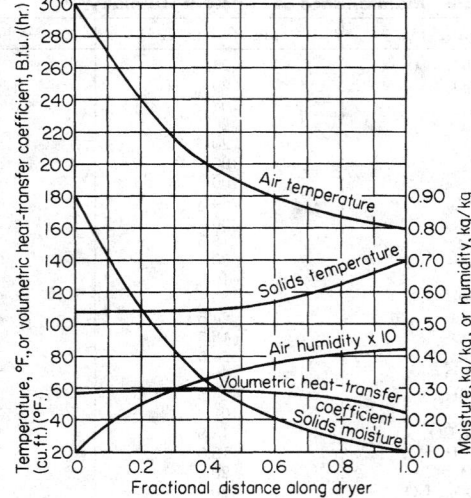

FIG. 12-44   Typical results of dryer-performance tests. To convert British thermal units per hour-cubic foot-degree Fahrenheit to joules per cubic meter-second-kelvin, multiply by 1.73.

The volumetric heat-transfer coefficient is given by

$$U_v = \frac{q_d}{V_d \, \Delta t_m} = \frac{w c_s (t_2 - t_1)}{A_d L_d \, \Delta t_m} \qquad (12\text{-}40)$$

where U_v = volumetric heat-transfer coefficient, $J/(m^3 \cdot s \cdot K)$ [$Btu/(h \cdot ft^3 \cdot {}^\circ F)$]; q = cross-sectional area of dryer, m^2; and L_d = dryer length, m.

The volumetric heat-transfer coefficients along the dryer are lower at the discharge end (Fig. 12-44) because of the internal resistance to moisture movement in the later stages of drying. When drying data are expressed in terms of overall performance, care and judgment should be exercised in extrapolating the results to other conditions, particularly conditions of different feed and product moisture. If, for example, the overall heat-transfer coefficients, from the data of Fig. 12-44, were used to predict a dryer design for reducing the product moisture below 10 percent, the design would be in error. Obviously, this problem can be circumvented by making sure that the final moisture in the experiments is below that desired in the product.

In any capacity test to determine the potential of a plant dryer, the effects of the following variables should be studied:
1. **Effect of increased temperature.**   This is often the simplest way to achieve increased capacity.
2. **Effect of increased final moisture.**   Because of the marked increase in drying time required to dry to low moisture contents, the permissible maximum final moisture should always be established.
3. **Effect of increasing air velocity** should be determined. Frequently, higher air rates are necessary to provide the required additional heat at higher capacities.
4. **Uniformity of air flow** should be established. Air-flow maldistribution can seriously reduce dryer capacity and efficiency.
5. Possible benefits from **air recirculation** should be considered.

SOLIDS-DRYING EQUIPMENT

CLASSIFICATION OF DRYERS

Drying equipment may be classified in several ways. The two most useful classifications are based on (1) the method of transferring heat to the wet solids or (2) the handling characteristics and physical properties of the wet material. The first method of classification reveals differences in dryer design and operation, while the second method is most useful in the selection of a group of dryers for preliminary consideration in a given drying problem.

A classification chart of drying equipment on the basis of heat transfer is shown in Fig. 12-45. This chart classifies dryers as direct or indirect, with subclasses of continuous or batchwise operation.

Direct Dryers   The general operating characteristics of direct dryers are these:

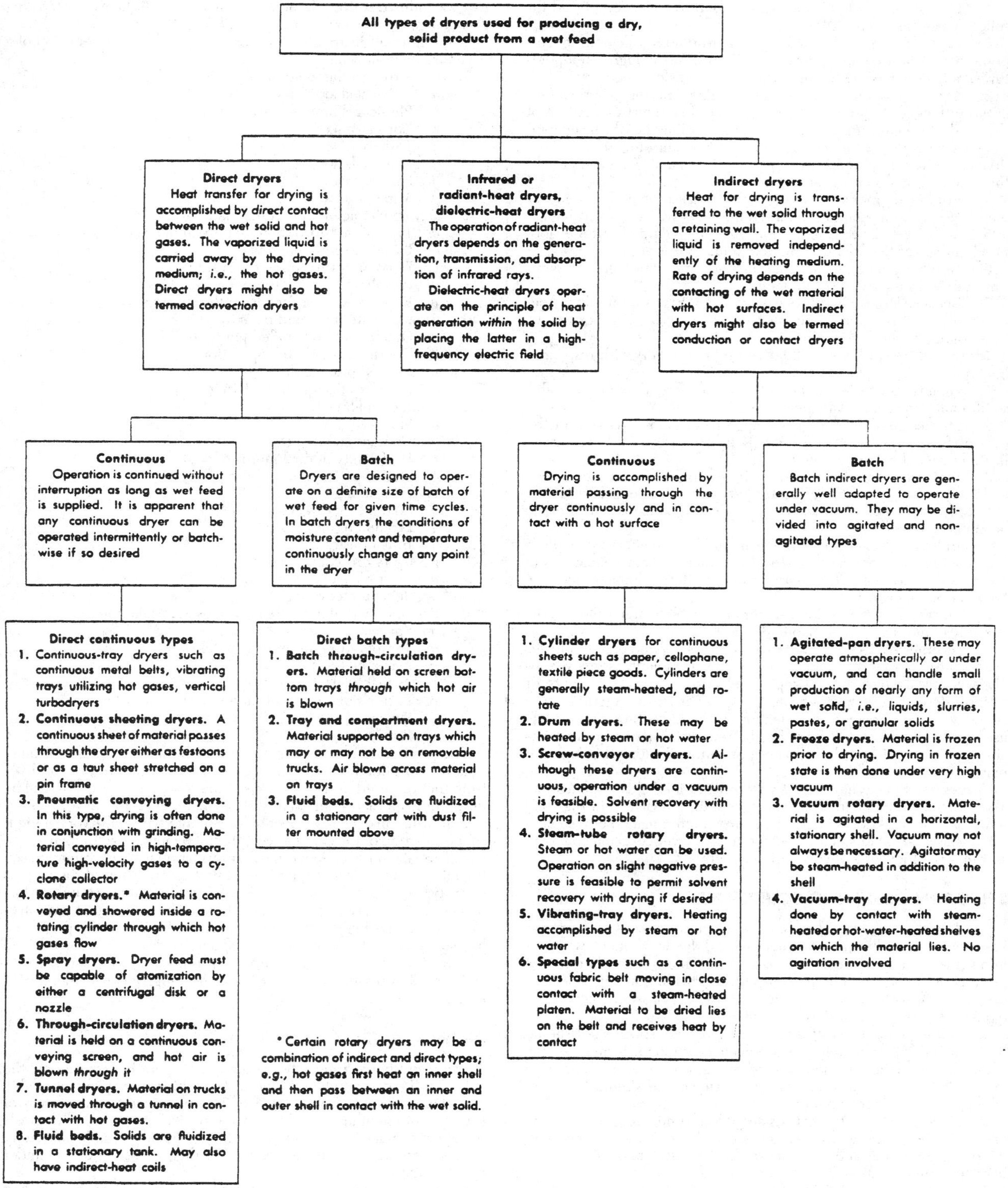

All types of dryers used for producing a dry, solid product from a wet feed

Direct dryers
Heat transfer for drying is accomplished by *direct* contact between the wet solid and hot gases. The vaporized liquid is carried away by the drying medium; *i.e.*, the hot gases. Direct dryers might also be termed convection dryers

Infrared or radiant-heat dryers, dielectric-heat dryers
The operation of radiant-heat dryers depends on the generation, transmission, and absorption of infrared rays.
Dielectric-heat dryers operate on the principle of heat generation *within* the solid by placing the latter in a high-frequency electric field

Indirect dryers
Heat for drying is transferred to the wet solid through a retaining wall. The vaporized liquid is removed independently of the heating medium. Rate of drying depends on the contacting of the wet material with hot surfaces. Indirect dryers might also be termed conduction or contact dryers

Continuous
Operation is continued without interruption as long as wet feed is supplied. It is apparent that any continuous dryer can be operated intermittently or batch-wise if so desired

Batch
Dryers are designed to operate on a definite size of batch of wet feed for given time cycles. In batch dryers the conditions of moisture content and temperature continuously change at any point in the dryer

Continuous
Drying is accomplished by material passing through the dryer continuously and in contact with a hot surface

Batch
Batch indirect dryers are generally well adapted to operate under vacuum. They may be divided into agitated and non-agitated types

Direct continuous types
1. **Continuous-tray dryers** such as continuous metal belts, vibrating trays utilizing hot gases, vertical turbodryers
2. **Continuous sheeting dryers.** A continuous sheet of material passes through the dryer either as festoons or as a taut sheet stretched on a pin frame
3. **Pneumatic conveying dryers.** In this type, drying is often done in conjunction with grinding. Material conveyed in high-temperature high-velocity gases to a cyclone collector
4. **Rotary dryers.*** Material is conveyed and showered inside a rotating cylinder through which hot gases flow
5. **Spray dryers.** Dryer feed must be capable of atomization by either a centrifugal disk or a nozzle
6. **Through-circulation dryers.** Material is held on a continuous conveying screen, and hot air is blown *through* it
7. **Tunnel dryers.** Material on trucks is moved through a tunnel in contact with hot gases
8. **Fluid beds.** Solids are fluidized in a stationary tank. May also have indirect-heat coils

Direct batch types
1. **Batch through-circulation dryers.** Material held on screen bottom trays *through* which hot air is blown
2. **Tray and compartment dryers.** Material supported on trays which may or may not be on removable trucks. Air blown across material on trays
3. **Fluid beds.** Solids are fluidized in a stationary cart with dust filter mounted above

1. **Cylinder dryers** for continuous sheets such as paper, cellophane, textile piece goods. Cylinders are generally steam-heated, and rotate
2. **Drum dryers.** These may be heated by steam or hot water
3. **Screw-conveyor dryers.** Although these dryers are continuous, operation under a vacuum is feasible. Solvent recovery with drying is possible
4. **Steam-tube rotary dryers.** Steam or hot water can be used. Operation on slight negative pressure is feasible to permit solvent recovery with drying if desired
5. **Vibrating-tray dryers.** Heating accomplished by steam or hot water
6. **Special types** such as a continuous fabric belt moving in close contact with a steam-heated platen. Material to be dried lies on the belt and receives heat by contact

1. **Agitated-pan dryers.** These may operate atmospherically or under vacuum, and can handle small production of nearly any form of wet solid, *i.e.*, liquids, slurries, pastes, or granular solids
2. **Freeze dryers.** Material is frozen prior to drying. Drying in frozen state is then done under very high vacuum
3. **Vacuum rotary dryers.** Material is agitated in a horizontal, stationary shell. Vacuum may not always be necessary. Agitator may be steam-heated in addition to the shell
4. **Vacuum-tray dryers.** Heating done by contact with steam-heated or hot-water-heated shelves on which the material lies. No agitation involved

* Certain rotary dryers may be a combination of indirect and direct types; e.g., hot gases first heat an inner shell and then pass between an inner and outer shell in contact with the wet solid.

FIG. 12-45   Classification of dryers, based on method of heat transfer. [*Revised from Marshall*, Heat, Piping Air Cond., **18**, 71 (1946).]

1. Direct contacting of hot gases with the solids is employed for solids heating and vapor removal.

2. Drying temperatures may range up to 1000 K, the limiting temperature for most common structural metals. At the higher temperatures, radiation becomes an important heat-transfer mechanism.

3. At gas temperatures below the boiling point, the vapor content of gas influences the rate of drying and the final moisture content of the solid. With gas temperatures above the boiling point throughout, the vapor content of the gas has only a slight retarding effect on the drying rate and final moisture content. Thus, superheated vapors of the liquid being removed can be used for drying.

4. For low-temperature drying, dehumidification of the drying air may be required when atmospheric humidities are excessively high.

5. A direct dryer consumes more fuel per pound of water evaporated, the lower the final moisture content.

6. Efficiency increases with an increase in the inlet-gas temperature for a constant exhaust temperature.

7. Because large amounts of gas are required to supply all the heat for drying, dust-recovery equipment may be very large and expensive when drying very small particles.

Indirect Dryers   These differ from direct dryers with respect to heat transfer and vapor removal:

1. Heat is transferred to the wet material by conduction through a solid retaining wall, usually metallic.

2. Surface temperatures may range from below freezing in the case of freeze dryers to above 800 K in the case of indirect dryers heated by combustion products.

3. Indirect dryers are suited to drying under reduced pressures and inert atmospheres to permit the recovery of solvents and to prevent the occurrence of explosive mixtures or the oxidation of easily decomposed materials.

4. Indirect dryers using condensing fluids as the heating medium are generally economical from the standpoint of heat consumption, since they furnish heat only in accordance with the demand made by the material being dried.

5. Dust recovery and dusty materials can be handled more satisfactorily in indirect dryers than in direct dryers.

Miscellaneous Dryers   Infrared dryers depend on the transfer of radiant energy to evaporate moisture. The radiant energy is supplied electrically by infrared lamps, by electric resistance elements, or by incandescent refractories heated by gas. The latter method has the added advantage of convection heating. Infrared heating is not widely used in the chemical industries for the removal of moisture. Its principal use is in baking or drying paint films and in heating thin layers of materials.

Dielectric dryers have not as yet found a wide field of application. Their fundamental characteristic of generating heat within the solid indicates potentialities for drying massive geometrical objects such as wood, sponge-rubber shapes, and ceramics. Power costs may range to 10 times the fuel costs of conventional methods.

SELECTION OF DRYING EQUIPMENT

1. *Initial selection of dryers.*   Select those dryers which appear best suited to handling the wet material and the dry product, which fit into the continuity of the process as a whole, and which will produce a product of the desired physical properties. This preliminary selection can be made with the aid of Table 12-9, which classifies the various types of dryers on the basis of the materials handled.

2. *Initial comparison of dryers.*   The dryers so selected should be evaluated approximately from available cost and performance data. From this evaluation, those dryers which appear to be uneconomical or unsuitable from the standpoint of performance should be eliminated from further consideration.

3. *Drying tests.*   Drying tests should be conducted in those dryers still under consideration. These tests will determine the optimum operating conditions and the product characteristics and will form the basis for firm quotations from equipment vendors.

4. *Final selection of dryer.*   From the results of the drying tests and quotations, the final selection of the most suitable dryer can be made. A recent article describing dryer scale-up methodology in the process industries was recently published by L. R. Genskow [*Drying Technology—An International Journal,* **12**(1&2), 47 (1994)].

The important factors to consider in the preliminary selection of a dryer are the following:

1. Properties of the material being handled
   a. Physical characteristics when wet
   b. Physical characteristics when dry
   c. Corrosiveness
   d. Toxicity
   e. Flammability
   f. Particle size
   g. Abrasiveness
2. Drying characteristics of the material
   a. Type of moisture (bound, unbound, or both)
   b. Initial moisture content
   c. Final moisture content (maximum)
   d. Permissible drying temperature
   e. Probable drying time for different dryers
3. Flow of material to and from the dryer
   a. Quantity to be handled per hour
   b. Continuous or batch operation
   c. Process prior to drying
   d. Process subsequent to drying
4. Product qualities
   a. Shrinkage
   b. Contamination
   c. Uniformity of final moisture content
   d. Decomposition of product
   e. Overdrying
   f. State of subdivision
   g. Product temperature
   h. Bulk density
5. Recovery problems
   a. Dust recovery
   b. Solvent recovery
6. Facilities available at site of proposed installation
   a. Space
   b. Temperature, humidity, and cleanliness of air
   c. Available fuels
   d. Available electric power
   e. Permissible noise, vibration, dust, or heat losses
   f. Source of wet feed
   g. Exhaust-gas outlets

The physical nature of the material to be handled is the primary item for consideration. A slurry will demand a different type of dryer from that required by a coarse crystalline solid, which, in turn, will be different from that required by a sheet material (Table 12-9).

Following preliminary selection of suitable types of dryers, a highspot evaluation of the size and cost should be made to eliminate those which are obviously uneconomical. Information for this evaluation can be obtained from material presented under discussion of the various dryer types. When data are inadequate, preliminary cost and performance data can usually be obtained from the equipment manufacturer. In comparing dryer performance, the factors in the preceding list which affect dryer performance should be properly weighed. The possibility of eliminating or simplifying processing steps which precede or follow drying, such as filtration, grinding, or conveying, should be carefully considered.

DRYING TESTS

These tests should establish the optimum operating conditions, the ability of the dryer to handle the material physically, product quality and characteristics, and dryer size. The principal manufacturers of drying equipment are usually prepared to perform the required tests on dryers simulating their equipment. Occasionally, simple laboratory experiments can serve to reduce further the number of dryers under consideration.

Once a given type and size of dryer has been installed, the product characteristics and drying capacity can be changed only within relatively narrow limits. Thus it is more economical and far more satisfac-

TABLE 12-9 Classification of Commercial Dryers Based on Materials Handled

Type of dryer	Liquids	Slurries	Pastes and sludges	Free-flowing powders	Granular, crystalline, or fibrous solids	Large solids, special forms and shapes	Continuous sheets	Discontinuous sheets
	True and colloidal solutions; emulsions. Examples: inorganic salt solutions, extracts, milk, blood, waste liquors, rubber latex, etc.	Pumpable suspensions. Examples: pigment slurries, soap and detergents, calcium carbonate, bentonite, clay slip, lead concentrates, etc.	Examples: filter-press cakes, sedimentation sludges, centrifuged solids, starch, etc.	100 mesh or less. Relatively free flowing in wet state. Dusty when dry. Examples: centrifuged precipitates, pigments, clay, cement.	Larger than 100 mesh. Examples: rayon staple, salt crystals, sand, ores, potato strips, synthetic rubber.	Examples: pottery, brick, rayon cakes, shotgun shells, hats, painted objects, rayon skeins, lumber.	Examples: paper, impregnated fabrics, cloth, cellophane, plastic sheets.	Examples: veneer, wallboard, photograph prints, leather, foam rubber sheets.
Vacuum freeze. Indirect type, batch or continuous operation	Usually used only for pharmaceuticals such as penicillin and blood plasma. Expensive. Used on heat-sensitive and readily oxidized materials	See comments under Liquids	See comments under Liquids	See comments under Liquids	Expensive. Usually used on pharmaceuticals and related products which cannot be dried successfully by other means. Applicable to fine chemicals	See comments under Granular solids	Applicable in special cases such as emulsion-coated films	See comments under Granular solids
Pan. Indirect type, batch operation	Atmospheric or vacuum. Suitable for small batches. Easily cleaned. Solvents can be recovered. Material agitated while dried	See comments under Liquids	See comments under Liquids	See comments under Liquids	Suitable for small batches. Easily cleaned. Material is agitated during drying, causing some degradation	Not applicable	Not applicable	Not applicable
Vacuum rotary. Indirect type, batch operation	Not applicable, except when pumping slowly on dry "heel"	May have application in special cases when pumping onto dry "heel"	Use is questionable. Material usually cakes to dryer walls and agitator. Solvents can be recovered	Suitable for nonsticking materials. Useful for large batches of heat-sensitive materials and for solvent recovery	Useful for large batches of heat-sensitive materials or where solvent is to be recovered. Product will suffer some grinding action. Dust collectors may be required.	Not applicable	Not applicable	Not applicable
Screw conveyor and indirect rotary. Indirect type, continuous operation	Applicable with dry-product recirculation	Applicable with dry-product recirculation	Generally requires recirculation of dry product. Little dusting occurs	Chief advantage is low dust loss. Well suited to most materials and capacities, particularly those requiring drying at steam temperature	Low dust loss. Material must not stick or be temperature-sensitive	Not applicable	Not applicable	Not applicable
Fluid beds. Batch, continuous, direct, and indirect	Applicable only with inert bed or dry-solids recirculator	See comments under Liquids	See comments under Liquids	Suitable, if not too dusty	Suitable for crystals, granules, and short fibers	Not applicable	Use hot inert particles for contacting	Use hot inert particles for contacting
Vibrating tray. Indirect type, continuous operation	Not applicable	Not applicable	Not applicable	Suitable for free-flowing materials	Suitable for free-flowing materials that can be conveyed on a vibrating tray	Not applicable	Not applicable	Not applicable

TABLE 12-9 Classification of Commercial Dryers Based on Materials Handled (*Concluded*)

	Liquids	Slurries	Pastes and sludges	Free-flowing powders	Granular, crystalline, or fibrous solids	Large solids, special forms and shapes	Continuous sheets	Discontinuous sheets
Drum. Indirect type, continuous operation	Single, double or twin. Atm. or vacuum operation. Product flaky and usually dusty. Maintenance costs may be high	See comments under Liquids. Twin-drum dryers are widely used	Can be used only when paste or sludge can be made to flow. See comments under Liquids	Not applicable	Not applicable	Not applicable	Not applicable	Not applicable
Cylinder. Indirect type, continuous operation	Not applicable	Not applicable	Not applicable	Not applicable	Not applicable	Not applicable	Suitable for thin or mechanically weak sheets which can be dried in contact with a heated surface. Special surface effects obtainable	Suitable for materials which need not be dried flat and which will not be injured by contact with hot drum
Infrared. Batch or continuous operation	Only for thin films	See comments under Liquids	See comments under Liquids (only for thin layers)	Only for thin layers	Primarily suited to drying surface moisture. Not suited for thick layers	Specially suited for drying and baking paint and enamels	Usually used in conjunction with other methods. Useful when there are space limitations	Useful for laboratory work or in conjunction with other methods
Dielectric. Batch or continuous operation	Very expensive	See comments under Liquids	See comments under Liquids	Very expensive	Very expensive	Rapid drying of large objects suited to this method	Applications for final stages of paper dryers	Successful on foam rubber. Not fully developed on other materials
Tray and compartment. Direct type, batch operation	Not applicable	For very small batch production. Laboratory drying	Suited to batch operation. At large capacities, investment and operating costs are high. Long drying times	Dusting may be a problem. See comments under Pastes and Sludges	Suited to batch operation. At large capacities, investment and operating costs are high. Long drying times	See comments under Granular solids	Not applicable	See comments under Granular solids
Batch through-circulation. Direct type, batch operation	Not applicable	Not applicable	Suitable only if material can be preformed. Suited to batch operation. Shorter drying time than tray dryers	Not applicable	Usually not suited for materials smaller than 30 mesh. Suited to small capacities and batch operation	Primarily useful for small objects	Not applicable	Not applicable
Tunnel. Continuous Tray. Direct type, continuous operation	Not applicable	Not applicable	Suitable for small and large-scale production.	See comments under Pastes and Sludges. Vertical-turbo applicable	Essentially large-scale, semicontinuous tray drying.	Suited to a wide variety of shapes and forms. Operation can be made continuous. Widely used	Not applicable	Suited for leather, wallboard, veneer.
Continuous through-circulation. Direct type, continuous operation	Not applicable	Only crystal filter dryer may be suited	Suitable for materials that can be preformed. Will handle large capacities. Roto-louvre requires dry-product recirculation	Not generally applicable, except Roto-louvre in certain cases	Usually not suited for materials smaller than 30 mesh. Material does not tumble, except in Roto-louvre dryer. Latter operates at higher temperatures	Suited to smaller objects that can be loaded on each other. Can be used to convey materials through heated zones. Roto-louvre not suited.	Not applicable	Special designs are required. Suited to veneers. Roto-louvre not applicable
Direct rotary. Direct type, continuous operation	Not applicable	Applicable with dry-product recirculation	Suitable only if product does not stick to walls and	Suitable for most materials and capacities, pro-	Suitable for most materials at most capacities. Dust-	Not applicable	Not applicable	Not applicable

Type of dryer	True and colloidal solutions; emulsions. Examples: inorganic salt solutions, extracts, milk, blood, waste liquors, rubber latex, etc. Applicable with dry-product recirculation	Pumpable suspensions. Examples: pigment slurries, soap and detergents, calcium carbonate, bentonite, clay slip, lead concentrates, etc.	Examples: filter-press cakes, sedimentation sludges, centrifuged solids, starch, etc. does not dust. Recirculation of product may prevent sticking	100 mesh or less. Relatively free flowing in wet state. Dusty when dry. Examples: centrifuged precipitates, pigments, clay, cement. vided that dusting is not too severe	Larger than 100 mesh. Examples: rayon staple, salt crystals, sand, ores, potato strips, synthetic rubber. ing or crystal abrasion will limit its use	Examples: pottery, brick, rayon cakes, shotgun shells, hats, painted objects, rayon skeins, lumber.	Examples: paper, impregnated fabrics, cloth, cellophane, plastic sheets.	Examples: veneer, wallboard, photograph prints, leather, foam rubber sheets.
Pneumatic conveying. Direct type, continuous operation	See comments under Slurries	Can be used only if product is recirculated to make feed suitable for handling	Usually requires recirculation of dry product to make suitable feed. Well suited to high capacities. Disintegration usually required	Suitable for materials that are easily suspended in a gas stream and lose moisture readily. Well suited to high capacities	Suitable for materials that are easily suspended in a gas stream. Well suited to high capacities. Product may suffer physical degradation	Not applicable	Not applicable	Not applicable
Spray. Direct type, continuous operation	Suited for large capacities. Product is usually powdery, spherical, and free-flowing. High temperatures can be used with heat-sensitive materials. Product may have low bulk density	See comments under Liquids. Pressure-nozzle atomizers subject to erosion	Requires special pumping equipment to feed the atomizer. See comments under Liquids	Not applicable	Not applicable	Not applicable	Not applicable	Not applicable
Continuous sheeting. Direct type, continuous operation	Not applicable	Not applicable	Not applicable	Not applicable	Not applicable	Not applicable	Different types are available for different requirements. Suitable for drying without contacting hot surfaces	Not applicable
Vacuum shelf. Indirect type, batch operation	Not applicable	Applicable for small-batch production	Suitable for batch operation, small capacities. Useful for heat-sensitive or readily oxidizable materials. Solvents can be recovered	See comments under Pastes and Sludges	Suitable for batch operation, small capacities. Useful for heat-sensitive or readily oxidizable materials. Solvents can be recovered.	See comments under Granular solids	Not applicable	See comments under Granular solids

tory to experiment in small-scale units than on the dryer that is finally installed.

On the basis of the results of the drying tests that establish size and operating characteristics, formal quotations and guarantees should be obtained from dryer manufacturers. Initial costs, installation costs, operating costs, product quality, dryer operability, and dryer flexibility can then be given proper weight in final evaluation and selection.

BATCH TRAY AND DRYERS

Description   A tray or compartment dryer is an enclosed, insulated housing in which solids are placed upon tiers of trays in the case of particulate solids or stacked in piles or upon shelves in the case of large objects. Heat transfer may be *direct* from gas to solids by circulation of large volumes of hot gas or *indirect* by use of heated shelves, radiator coils, or refractory walls inside the housing. In indirect-heat units, excepting vacuum-shelf equipment, circulation of a small quantity of gas is usually necessary to sweep moisture vapor from the compartment and prevent gas saturation and condensation. Compartment units are employed for the heating and drying of lumber, ceramics, sheet materials (supported on poles), painted and metal objects, and all forms of particulate solids.

Field of Application   Because of the high labor requirements usually associated with loading or unloading the compartments, batch compartment equipment is rarely economical except in the following situations:

1.   A long heating cycle is necessary because the size of the solid objects or permissible heating temperature requires a long holdup for internal diffusion of heat or moisture. This case may apply when the cycle will exceed 12 to 24 h.

2.   The production of several different products requires strict batch identity and thorough cleaning of equipment between batches. This is a situation existing in many small color-pigment-drying plants.

3.   The quantity of material to be processed does not justify investment in more expensive, continuous equipment. This case would apply in many pharmaceutical-drying operations.

Further, because of the nature of solids-gas contacting, which is usually by parallel flow and rarely by through circulation, heat transfer and mass transfer are comparatively inefficient. For this reason, use of tray and compartment equipment is restricted primarily to ordinary drying and heat-treating operations. Despite these harsh limitations, when the listed situations do exist, economical alternatives are difficult to develop.

Auxiliary Equipment   If noxious gases, fumes, or dust are given off during the operation, dust- or fume-recovery equipment will be necessary in the exhaust-gas system. Wet scrubbers are employed for the recovery of valuable solvents from dryers. In order to minimize heat losses, thorough insulation of the compartment with brick, asbestos, or other insulating compounds is necessary. Modern fabricated dryer-compartment panels usually have 7.5 to 15 cm of blanket insulation placed between the internal and external sheet-metal walls. Doors and other access openings should be gasketed and tight. In the case of tray and truck equipment, it is usually desirable to have available extra trays and trucks so that they can be preloaded for rapid emptying and loading of the compartment between cycles. Air filters and gas dryers are occasionally employed on the inlet-air system for direct-heat units.

Vacuum-shelf dryers require auxiliary stream jets or other vacuum-producing devices, interconducers for vapor removal, and occasionally wet scrubbers or (heated) bag-type dust collectors.

Uniform depth of loading in dryers and furnaces handling particulate solids is essential to consistent operation, minimum heating cycles, or control of final moisture. After a tray has been loaded, the bed should be leveled to a uniform depth. Special preform devices, noodle extruders, pelletizers, etc., are employed occasionally for preparing pastes and filter cakes so that screen bottom trays can be used and the advantages of through circulation approached.

Control of tray and compartment equipment is usually maintained by control of the circulating-air temperature (and humidity) and rarely by solids temperature. On vacuum units, control of the absolute pressure and heating-medium temperature is utilized. In direct dryers, cycle controllers are frequently employed to vary the air temper-

ature or velocity across the solids during the cycle, e.g., high air temperatures may be employed during a constant-rate drying period while the solids surface remains close to the air wet-bulb temperature. During the falling-rate periods, this temperature may be reduced to prevent casehardening or other degrading effects caused by overheating the solids surfaces. In addition, higher air velocities may be employed during early drying stages to improve heat transfer; however, after surface drying has been completed, this velocity may need to be reduced to prevent dusting. Two-speed circulating fans are employed commonly for this purpose.

Direct-Heat Tray Dryers   Satisfactory operation of tray-type dryers depends on maintaining a constant temperature and a uniform air velocity over all the material being dried.

Circulation of air at velocities of 1 to 10 m/s is desirable to improve the surface heat-transfer coefficient and to eliminate stagnant air pockets. Proper air flow in tray dryers depends on sufficient fan capacity, on the design of ductwork to modify sudden changes in direction, and on properly placed baffles. *Nonuniform air flow is one of the most serious problems in the operation of tray dryers.*

Tray dryers may be of the tray-truck or the stationary-tray type. In the former, the trays are loaded on trucks which are pushed into the dryer; in the latter, the trays are loaded directly into stationary racks within the dryer. Trucks may be fitted with flanged wheels to run on tracks or with flat swivel wheels. They may also be suspended from and moved on monorails. Trucks usually contain two tiers of trays, with 18 to 48 trays per tier, depending upon the tray dimensions.

Trays may be square or rectangular, with 0.5 to 1 m² per tray, and may be fabricated from any material compatible with corrosion and temperature conditions. When the trays are stacked in the truck, there should be a clearance of not less than 4 cm between the material in one tray and the bottom of the tray immediately above. When material characteristics and handling permit, the trays should have screen bottoms for additional drying area. Metal trays are preferable to nonmetallic trays, since they conduct heat more readily. Tray loadings range usually from 1 to 10 cm deep.

Steam is the usual heating medium, and a standard heater arrangement consists of a main heater before the circulating fan. When steam is not available or the drying load is small, electrical heat can be used. For temperatures above 450 K, products of combustion can be used, or indirect-fired air heaters.

Air is circulated by propeller or centrifugal fans; the fan is usually mounted within or directly above the dryer. Above 450 K, external or water-cooled bearings become necessary. Total pressure drop through the trays, heaters, and ductwork is usually in the range of 2.5 to 5 cm of water. Air recirculation is generally in the order of 80 to 95 percent except during the initial drying stage of rapid evaporation. Fresh air is drawn in by the circulating fan, frequently through dust filters. In most installations, air is exhausted by a separate small exhaust fan with a damper to control air-recirculation rates.

Prediction of Heat- and Mass-Transfer Coefficients in Direct-Heat Tray Dryers   In convection phenomena, heat-transfer coefficients depend on the geometry of the system, the gas velocity past the evaporating surface, and the physical properties of the drying gas. In estimating drying rates, the use of heat-transfer coefficients is preferred because they are usually more reliable than mass-transfer coefficients. In calculating mass-transfer coefficients from drying experiments; the partial pressure at the surface is usually inferred from the measured or calculated temperature of the evaporating surface. Small errors in temperature have negligible effect on the heat-transfer coefficient but introduce relatively large errors in the partial pressure and hence in the mass-transfer coefficient.

For many cases in drying, the heat-transfer coefficient can be expressed as

$$h_c = \alpha G^n / D_c p \qquad (12\text{-}41)$$

where h_c = heat-transfer coefficient, J/(m²·s·K)[Btu/(h·ft²·°F)]; G = mass velocity of drying gas, kg/(s·m²)[lb/(h·ft²)]; D_c = characteristic dimension of the system, m; and α, n, and p are empirical constants. When radiation and conduction effects are negligible, the constant rate of drying from a surface is thus given by the following heat-transfer expression derived from Eqs. (12-26) and (12-41):

$$dw/d\theta = (\alpha G^n A/\lambda D_c^m)(t - t_s') \tag{12-42}$$

When the liquid is water and the drying gas is air, t_s' is the wet-bulb temperature.

In order to estimate drying rates from Eq. (12-42) values of the empirical constants are required for the particular geometry under consideration. For flow parallel to plane plates, exponent n has been reported to range from 0.35 to 0.8 [Chu, Lane, and Conklin, *Ind. Eng. Chem.*, **45**, 1856 (1953); Wenzel and White, *Ind. Eng. Chem.*, **51**, 275 (1958)]. The differences in exponent have been attributed to differences in flow pattern in the space above the evaporating surface. In the absence of applicable specific data, the heat-transfer coefficient for the parallel-flow case can be taken, for estimating purposes, as

$$h = 8.8 G^{0.8}/D_c^{0.2} \tag{12-43}$$

where the experimental data have been weighted in favor of an exponent of 0.8 in conformity with the usual Colburn j factor and average values of the properties of air at 370 K have been incorporated.

Experimental data for drying from flat surfaces have been correlated by using the equivalent diameter of the flow channel or the length of the evaporating surface as the characteristic length dimension in the Reynolds number. However, the validity of one versus the other has not been established. The proper equivalent diameter probably depends at least on the geometry of the system, the roughness of the surface, and the flow conditions upstream of the evaporating surface. For most tray-drying calculations, the equivalent diameter (4 times the cross-sectional area divided by the perimeter of the flow channel) should be used.

For air flow impinging normally to the surface from slots, nozzles, or perforated plates, the heat-transfer coefficient can be obtained from the data of Friedman and Mueller (*Proceedings of the General Discussion on Heat Transfer*, Institution of Mechanical Engineers, London, and American Society of Mechanical Engineers, New York, 1951, pp. 138–142). These investigators give

$$h_c = \alpha G^{0.78} \tag{12-44}$$

where the gas mass velocity G is based on the total heat-transfer area and α is dependent on the plate open area, hole or slot size, and spacing between the plate, nozzle, or slot and the heat-transfer surface.

Most efficient performance is obtained with plates having open areas equal to 2 to 3 percent of the total heat-transfer area. The plate should be located at a distance equal to four to six hole (or equivalent) diameters from the heat-transfer surface.

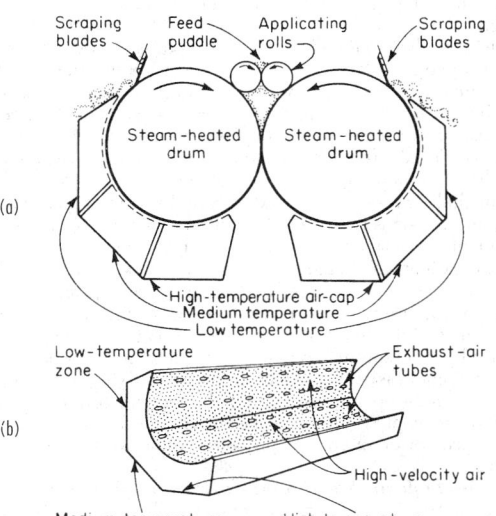

FIG. 12-46 Example of the use of air impingement in drying as a secondary heat source on a double-drum dryer. [*Chem. Eng.*, **197**, (*June 19, 1967*).]

Data from tests employing multiple slots, with a correction calculated for slot width, were reported by Korger and Kizek [*Int. J. Heat Mass Transfer*, London, **9**, 337 (1966)].

Air impingement is commonly employed for drying sheets, film, thin slabs, and coatings. Another application in which it is used is as a secondary heat source on drum and can dryers (see Fig. 12-46).

Determination of the Temperature of the Evaporating Surface in Direct-Heat Tray Dryers When radiation and conduction are negligible, the temperature of the evaporating surface approaches the wet-bulb temperature and is readily obtained from the humidity and dry-bulb temperature. Frequently, however, radiation and conduction cause the temperature of the evaporating surface to exceed the wet-bulb temperature. When this occurs, the true surface temperature must be estimated.

Under steady-state conditions the temperature of the evaporating surface increases until the rate of sensible heat transfer to the surface equals the rate of heat removed by evaporation from the surface. To calculate this temperature, it is convenient to modify Eq. (12-26) in terms of humidity rather than partial-pressure difference, as follows:

$$k_g(p_s - p) = k'(H_s - H) \tag{12-45}$$

where k' = mass-transfer coefficient, kg/(s·m²)(unit humidity difference), and $k' = pk_g(M_a/M_w)$ is a suitable approximation at low humidities; k_g = mass-transfer coefficient, kg/(s·m²·atm); M_a = molecular weight of air; M_w = molecular weight of the diffusing vapor; p_s = vapor pressure of the liquid at the temperature of the evaporating surface, atm; p = partial pressure of vapor in air, atm; H_s = saturation humidity of the air at the temperature of the drying surface, kg/kg dry air; H = humidity of the drying air, kg/kg dry air; and p = total pressure, atm. For air-water mixtures k' is approximately 1.6 k_g at atmospheric pressure.

A rate balance between evaporation and heat transfer when radiation occurs may be modified by means of the psychrometric ratio for air-water vapor mixtures to give:

$$\frac{\lambda}{C_s}(H_s - H) = (t - t_s') + \frac{h_t \epsilon}{h_c}(t_r - t_s') \tag{12-46}$$

where λ = latent heat of evaporation, J/kg at t_s'; $C_s = h_c/k'$, where h_c = convection heat-transfer coefficient, J/(s·m²·K), and C_s = heat capacity of humid air, J/(kg dry air·K), as defined in the subsection "Application of Psychrometry to Drying"; t = temperature of drying gases, K; t_s' = temperature of the wet surface, K; h_c = radiation heat-transfer coefficient, J/(s·m²·K); h_c = convection heat-transfer coefficient, J/(s·m²·K); t_r = temperature of source radiating heat to the wet surface, K; and ϵ = emissivity of surface receiving radiation.

Equation (12-46) may be solved by trial and error or graphically to estimate the true values of H_s and t_s' and, hence, the actual drying rate. The values of λ and h_r depend on the value of t_s' but can generally be considered constant over the range of temperatures usually encountered in air drying.

Frequently, particularly in tray drying, heat arrives at the evaporating surface from the tray walls by conduction through the wet material. For this case, in which both radiation and conduction are significant, the total heat-transfer coefficient is given by Shepherd, Brewer, and Hadlock [*Ind. Eng. Chem.*, **30**, 388 (1938)] as

$$h_t = (h_c + h_r)\left[1 + \frac{A_u}{1 + d(h_c + h_r)/k}\right] \tag{12-47}$$

where h_t = total heat-transfer coefficient, J/(s·m²·K); A_u = ratio of outside unwetted surface to evaporating surface area; d = depth of material in tray, m; and k = thermal conductivity of the wet material, J/[(s·m²)(K/m)]. Note that h_c must be corrected for emissivity of the surface. For insulated trays, the arithmetic average of inside and outside unwetted area should be used.

Equation (12-47) assumes that all heat sources are at the same temperature and that the convection coefficients to the evaporating surface and to the unwetted portions of the tray are equal. When radiation occurs from a source at a different temperature, the radiation coefficient can be corrected to the same basis by multiplying by the ratio $(t - t_s')/(t_r - t_s')$, where t, t_s', and t_r are the drying-gas, evaporating-surface, and radiator temperatures respectively.

A relationship for estimating the surface temperature t'_s, based on the use of Eq. (12-47) to determine h_t, is as follows:

$$(H_s - H) = (h_t C_s / \lambda h_c)(t - t'_s) \qquad (12\text{-}48)$$

Equation (12-48) can be solved numerically or graphically. Figure 12-47 indicates how H_s and t'_s may be determined graphically on a humidity chart by the point of intersection on the saturation-humidity curve of a straight line of slope $h_t C_s / \lambda h_c$ passing through point (H_s, t_s).

Performance Data for Direct-Heat Tray Dryers A standard two-truck dryer is illustrated in Fig. 12-48. Adjustable baffles or a perforated distribution plate is normally employed to develop 0.3 to 1.3 cm of water-pressure drop at the wall through which air enters the truck enclosure. This will enhance the uniformity of air distribution, from top to bottom, among the trays. In three (or more) truck ovens, air-reheat coils may be placed between trucks if the evaporative load is high. Means for reversing air-flow direction may also be provided in multiple-truck units.

Performance data on some typical tray and compartment dryers are tabulated in Table 12-10. These indicate that an overall rate of evaporation of 0.0025 to 0.025 kg water/(s·m²) of tray area may be expected from tray and tray-truck dryers. The thermal efficiency of this type of dryer will vary from 20 to 50 percent, depending on the drying temperature used and the humidity of the exhaust air. In drying to very low moisture contents under temperature restrictions, the thermal efficiency may be in the order of 10 percent. The major operating cost for a tray dryer is the labor involved in loading and unloading the trays. About two labor-hours are required to load and unload a standard two-truck tray dryer. In addition, about one-third to one-fifth of a

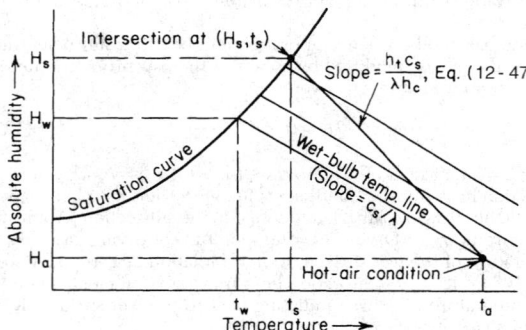

FIG. 12-47 Graphical estimation of surface temperature during constant-rate period.

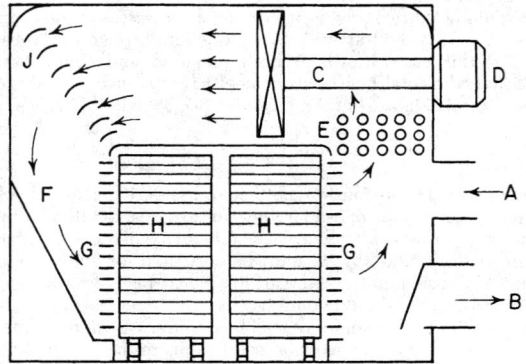

FIG. 12-48 Double-truck dryer. (*A*) Air-inlet duct. (*B*) Air-exhaust duct with damper. (*C*) Adjustable-pitch fan 1 to 15 hp. (*D*) Fan motor. (*E*) Fin heaters. (*F*) Plenum chamber. (*G*) Adjustable air-blast nozzles. (*H*) Trucks and trays. (*J*) Turning vanes.

worker's time is required to supervise the dryer during the drying period. Power for tray and compartment dryers will be approximately 1.1 kW per truck in the dryer. Maintenance will run from 3 to 5 percent of the installed cost per year.

BATCH THROUGH-CIRCULATION DRYERS

In one type of batch through-circulation dryer, heated air passes through a stationary permeable bed of the wet material placed on removable screen-bottom trays suitably supported in the dryer. This type is similar to a standard tray dryer except that hot air passes through the wet solid instead of across it. The pressure drop through the bed of material does not usually exceed about 2 cm of water. In another type, deep perforated-bottom trays are placed on top of plenum chambers in a closed-circuit hot-air-circulating system. In some food-drying plants, the material is placed in finishing bins with perforated bottoms; heated air passes up through the material and is removed from the top of the bin, reheated, and recirculated. The latter types involve a pressure drop through the bed of material of 1 to 8 cm of water at relatively low air rates. Table 12-11 gives performance data on three applications of batch through-circulation dryers. Batch through-circulation dryers are restricted in application to granular materials that permit free flow-through circulation of air. Drying times are usually much shorter than in parallel-flow tray dryers. Design methods are included in the subsection "Continuous Through-Circulation Dryers."

Vacuum-Shelf Dryers Vacuum-shelf dryers are indirect-heated batch dryers consisting of a vacuumtight chamber usually constructed of cast iron or steel plate, heated, supporting shelves within the chamber, a vacuum source, and usually a condenser. One or two doors are provided, depending on the size of the chamber. The doors are sealed with resilient gaskets of rubber or similar material (Fig. 12-49).

Hollow shelves of flat steel plate are fastened permanently inside the vacuum chamber and are connected in parallel to inlet and outlet headers. The heating medium, entering through one header and passing through the hollow shelves to the exit header, is generally steam, ranging in pressure from 700 kPa gauge to subatmospheric pressure for low-temperature operations. Low temperatures can be provided by circulating hot water, and high temperatures can be obtained by circulating hot oil or Dowtherm. Some small dryers employ electrically heated shelves. The material to be dried is placed in pans or trays on the heated shelves. The trays are generally of metal to ensure good heat transfer between the shelf and the tray.

Vacuum-shelf dryers may vary in size from 1 to 24 shelves, the largest chambers having overall dimensions of 6 m wide, 3 m long, and 2.5 m high.

Vacuum is applied to the chamber and vapor is removed through a large pipe which is connected to the chamber in a manner such that, if the vacuum is broken suddenly, the inrushing air will not greatly disturb the bed of material being dried. This line leads to a condenser where moisture or solvent that has been vaporized is condensed. The noncondensable exhaust gas goes to the vacuum source, which may be a wet or dry vacuum pump or a steam-jet ejector.

Vacuum-shelf dryers are used extensively for drying pharmaceuticals, temperature-sensitive or easily oxidizable materials, and materials so valuable that labor cost is insignificant. They are particularly useful for handling small batches of materials wet with toxic or valuable solvents. Recovery of the solvent is easily accomplished without danger of passing through an explosive range. Dusty materials may be dried with negligible dust loss. Hygroscopic materials may be completely dried at temperatures below that required in atmospheric dryers. The equipment is employed also for freeze-drying processes, for metallizing-furnace operations, and for the manufacture of semiconductor parts in controlled atmospheres. All these latter processes demand much lower operating pressures than do ordinary drying operations.

Design Methods for Vacuum-Shelf Dryers Heat is transferred to the wet material by conduction through the shelf and bottom of the tray and by radiation from the shelf above. The critical moisture content will not be necessarily the same as for atmospheric tray drying [Ernst, Ridgway, and Tiller, *Ind. Eng. Chem.*, **30**, 1122 (1938)].

TABLE 12-10   Manufacturer's Performance Data for Tray and Tray-Truck Dryers*

Material	Color	Chrome yellow	Toluidine red	Half-finished Titone	Color
Type of dryer	2-truck	16-tray dryer	16-tray	3-truck	2-truck
Capacity, kg product/h	11.2	16.1	1.9	56.7	4.8
Number of trays	80	16	16	180	120
Tray spacing, cm	10	10	10	7.5	9
Tray size, cm	$60 \times 75 \times 4$	$65 \times 100 \times 2.2$	$65 \times 100 \times 2$	$60 \times 70 \times 3.8$	$60 \times 70 \times 2.5$
Depth of loading, cm	2.5 to 5	3	3.5	3	
Initial moisture, % bone-dry basis	207	46	220	223	116
Final moisture, % bone-dry basis	4.5	0.25	0.1	25	0.5
Air temperature, °C	85–74	100	50	95	99
Loading, kg product/m²	10.0	33.7	7.8	14.9	9.28
Drying time, h	33	21	41	20	96
Air velocity, m/s	1.0	2.3	2.3	3.0	2.5
Drying, kg water evaporated/(h·m²)	0.59	65	0.41	1.17	0.11
Steam consumption, kg/kg water evaporated	2.5	3.0	—	2.75	
Total installed power, kW	1.5	0.75	0.75	2.25	1.5

*Courtesy of Proctor & Schwartz, Inc.

During the constant-rate period, moisture is rapidly removed. Often 50 percent of the moisture will evaporate in the first hour of a 6- to 8-h cycle. The drying time has been found to be proportional to between the first and second power of the depth of loading. Shelf vacuum dryers operate in the range of 1 to 25 mmHg pressure. For size-estimating purposes, a heat-transfer coefficient of 20 J/(m²·s·K) may be used. The area employed in this case should be the shelf area in direct contact with the trays. Trays should be maintained as flatly as possible to obtain maximum area of contact with the heated shelves. For the same reason, the shelves should be kept free from scale and rust. Air vents should be installed on steam-heated shelves to vent noncondensable gases. The heating medium should not be applied to the shelves until after the air has been evacuated from the chamber in order to reduce the possibility of the material's overheating or boiling at the start of drying. Casehardening can sometimes be avoided by retarding the rate of drying in the early part of the cycle.

Performance Data for Vacuum-Shelf Dryers The purchase price of a vacuum-shelf dryer depends upon the cabinet size and number of shelves per cabinet. For estimating purposes, typical prices (1985) and auxiliary-equipment requirements are given in Table 12-12. Installed cost of the equipment will be roughly 100 percent of the carbon steel purchase cost.

The thermal efficiency of a vacuum-shelf dryer is usually on the order of 60 to 60 percent. Table 12-13 gives operating data for one

TABLE 12-11   Performance Data for Batch Through-Circulation Dryers*

Kind of material	Granular polymer	Vegetable	Vegetable seeds
Capacity, kg product/h	122	42.5	27.7
Number of trays	16	24	24
Tray spacing, cm	43	43	43
Tray size, cm	91.4×104	91.4×104	85×98
Depth of loading, cm	7.0	6	4
Physical form of product	Crumbs	0.6-cm diced cubes	Washed seeds
Initial moisture content, % dry basis	11.1	669.0	100.0
Final moisture content, % dry basis	0.1	5.0	9.9
Air temperature, °C	88	77 dry-bulb	36
Air velocity, superficial, m/s	1.0	0.6 to 1.0	1.0
Tray loading, kg product/m²	16.1	5.2	6.7
Drying time, h	2.0	8.5	5.5
Overall drying rate, kg water evaporated/(h·m²)	0.89	11.86	1.14
Steam consumption, kg/kg water evaporated	4.0	2.42	6.8
Installed power, kW	7.5	19	19

*Courtesy of Proctor & Schwartz, Inc.

organic color and two inorganic compounds. Labor may constitute 50 percent of the operating cost; maintenance, 20 percent. Annual maintenance costs amount to 5 to 10 percent of the total installed cost. Actual labor costs will depend on drying time, facilities for loading and unloading trays, etc. The power required for these dryers is only that for the vacuum system; for vacuums of 680 to 735 mmHg the power requirements are in the order of 0.06 to 0.12 kW/m² tray surface.

Batch Furnaces, Other Furnace Types, and Kilns Batch furnaces are employed mainly for the heat treating of metals, such as annealing, normalizing, and "drawing" (tempering), and for the drying and calcination of ceramic articles. Many specialized furnaces have been designed for these purposes and may be either batch or continuous in operation. Batch furnaces are used in chemical processing for the same purposes as batch tray and truck dryers, when the drying or process temperature exceeds that which can be tolerated by unlined metal walls; ordinary tray and truck dryers are rarely employed when the circulating-gas temperature will exceed 600 to 700 K. They are employed for small-batch calcination, thermal decompositions, and other chemical reactions; these are the same as those reactions performed on a larger scale in rotary kilns, hearth furnaces, and shaft furnaces.

Design procedures and information on heat release in furnaces are given in Sec. 11. Tables indicating normal operating temperatures in various heating furnaces and the more common process furnaces are also included. Specialized designs of batch furnaces are shown in Figs. 12-50 to 12-53 and are described briefly in the following paragraphs. All may be heated by gas, oil, or electricity. **Standard oven furnaces** are similar in design to the small muffle furnace depicted in Fig. 12-53, but with the muffle housing eliminated.

Forced-convection pit furnaces are employed for heat-treating small metal parts in bulk. Small pieces are suspended in a mesh-bottom basket, while larger pieces are placed on racks. Air heating is by means of Nichrome electric coils set in refractory walls around the periphery of the pit. A high-velocity fan beneath the basket circulates heated air up past the coils and then down through the basket. Some heat is radiated to the outer basket shell, but most is transferred by direct convection from the circulating gas to the solids.

Car-bottom furnaces differ from standard types in that the charge is placed upon movable cars for running into the furnace enclosure. The top of the car is refractory-lined and forms the furnace hearth. The top only is exposed to heat, the lower metal structure being protected by the hearth brick, sand, and water seals at the sides and ends and by the circulation of cooling air around the car structure below the hearth. For use where floor space is limited **elevator furnaces** serve similar purposes.

The **rotary-hearth furnace** consists of a heating chamber lined with refractory brick within which is an annular-shaped refractory-lined rotating hearth. Around the periphery of the rotating hearth, sand or circulating liquid seals are employed to prevent air infiltration. It can be made semicontinuous in operation. The hearth speed can be

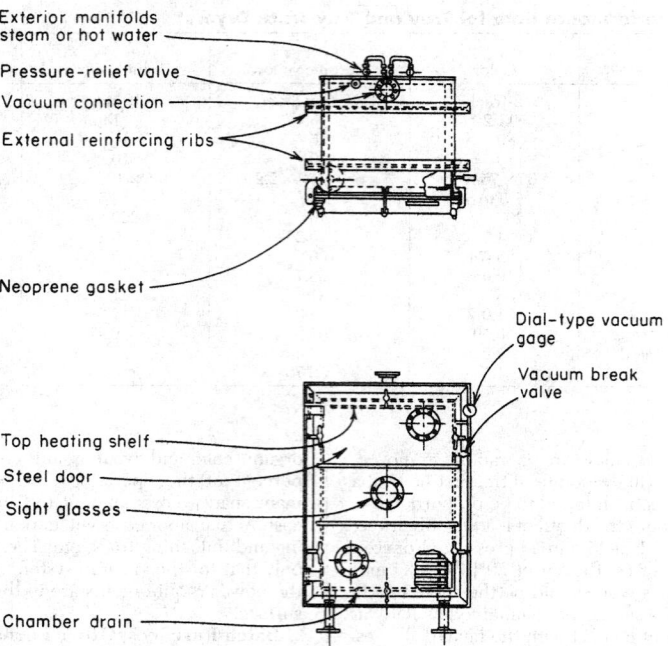

FIG. 12-49 Vacuum-shelf dryer. (*Stokes Equipment Division, Pennwalt Corp.*)

TABLE 12-12   Standard Vacuum-Shelf Dryers*

Shelf area, m²	Floor space, m²	Weight average, kg	Pump capacity, m³/s	Pump motor, kW	Condenser area, m²	Price/m² (1995)	
						Carbon steel	304 stainless steel
0.4–1.1	4.5	540	0.024	1.12	1	$110	$170
1.1–2.2	4.5	680	0.024	1.12	1	75	110
2.2–5.0	4.6	1130	0.038	1.49	4	45	65
5.0–6.7	5.0	1630	0.038	1.49	4	36	65
6.7–14.9	6.4	3900	0.071	2.24	9	27	45
16.7–21.1	6.9	5220	0.071	2.24	9	22	36

*Stokes Vacuum, Inc.

varied to meet changing requirements in size, weight, and load of the charge. For gas and oil heating, the burners fire from the sides of the chamber, tangentially to the hearth.

Standard furnaces are usually direct-heated, in that the burner combustion gases circulate directly over the charge; occasionally the flame may be permitted to impinge on the charge. For bright annealing, tool hardening, powdered-metal sintering, and other work requir-

TABLE 12-13   Performance Data of Vacuum-Shelf Dryers

Material	Sulfur black	Calcium carbonate	Calcium phosphate
Loading, kg dry material/m²	25	17	33
Steam pressure, kPa gauge	410	410	205
Vacuum, mmHg	685–710	685–710	685–710
Initial moisture content, % (wet basis)	50	50.3	30.6
Final moisture content, % (wet basis)	1	1.15	4.3
Drying time, h	8	7	6
Evaporation rate, kg/ (s·m²)	8.9×10^{-4}	7.9×10^{-4}	6.6×10^{-4}

ing protection of the charge by special atmospheres, **muffle-type furnaces** are frequently employed. In these, the charge is separated from the burners and combustion gases by a refractory arch. Heat is transferred by hot-gas radiation and convection to the arch and by radiation from the arch to the charge.

When used for **ceramic heating,** furnaces are called kilns. Operations include drying, oxidation, calcination, and vitrification. These kilns employ horizontal space burners with gaseous, liquid, or solid fuels. If product quality is not injured, ceramic ware may be exposed to flame and combustion gases; otherwise, muffle kilns are employed. Dutch ovens are used frequently for heat generation.

Downdraft kilns are the most common type, being used for brick, pipe, tile, and stoneware. The name is derived from the direction of combustion-gas flow when contacting the charge. The gases then flow up inside the walls to the top of the kiln and chimney. **Updraft kilns** are similar except in direction of gas flow, which is upward past the charge. They are employed commonly for pottery burning. **Stove kilns** are variations of updraft kilns used for burning common brick. The kiln is built of green brick and covered with a layer of burned brick. It is completely dismantled after each burning. **Clamp kilns** are another variation of updraft kilns used for common brick and temporary in nature. They have no tops or flue systems but consist only of sidewalls with arched spaces for combustion.

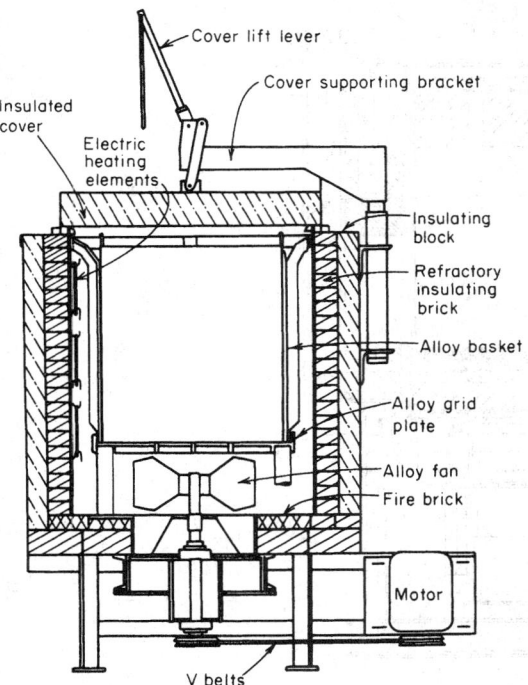

FIG. 12-50 Pit furnace. (*W. S. Rockwell Co.*)

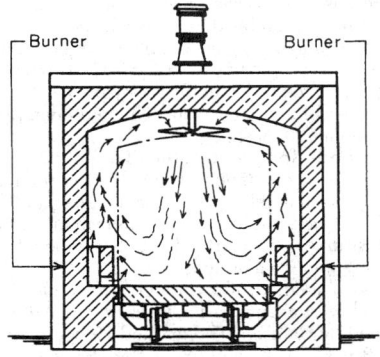

FIG. 12-51 Car-bottom furnace. (*W. S. Rockwell Co.*)

CONTINUOUS TUNNEL DRYERS

Continuous tunnels are in many cases **batch truck** or **tray compartments, operated in series.** The solids to be processed are placed in trays or on trucks which move progressively through the tunnel in contact with hot gases. Operation is **semicontinuous;** when the tunnel is filled, one truck is removed from the discharge end as each new truck is fed into the inlet end. In some cases, the trucks move on tracks or monorails, and they are usually conveyed mechanically, employing chain drives connecting to the bottom of each truck. Schematic diagrams of three typical tunnel arrangements are shown in Fig. 12-54. **Belt-conveyor and screen-conveyor tunnels are truly continuous** in operation, carrying a layer of solids on an endless conveyor.

Air flow can be totally **cocurrent, countercurrent,** or a combination of both as shown in Fig. 12-54. In addition, **cross-flow** designs are employed frequently, with the heating air flowing back and forth across the trucks in series. Reheat coils may be installed after each cross-flow pass to maintain constant-temperature operation; large propeller-type circulating fans are installed at each stage, and air may be introduced or exhausted at any desirable points. Tunnel equipment possesses maximum flexibility for any combination of air flow and temperature staging. When handling granular, particulate solids which do not offer high resistance to air flow, perforated or screen-type belt conveyors are employed with **through circulation** of gas to improve heat- and mass-transfer rates.

In tunnel equipment, the solids are usually heated by direct contact with hot gases. In high-temperature operations, radiation from walls and refractory lining may be significant also. The air in a direct-heat unit may be heated directly or indirectly by combustion or, at temperature below 475 K, by finned steam coils.

Applications of tunnel equipment are essentially the same as for batch tray and compartment units previously described, namely, practically all forms of particulate solids and large solid objects. In operation, they are more suitable for large-quantity production, usually representing investment and installation savings over (multiple) batch compartments. In the case of truck and tray tunnels, labor savings for loading and unloading are not significant compared with batch equipment. Belt and screen conveyors which are truly continuous represent major labor savings over batch operations but require additional investment for automatic feeding and unloading devices.

Auxiliary equipment and the special design considerations discussed for batch trays and compartments apply also to tunnel equipment. For size-estimating purposes, tray and truck tunnels and furnaces can be treated in the same manner as discussed for batch equipment.

Continuous Through-Circulation Dryers Continuous through-circulation dryers operate on the principle of blowing hot air through a permeable bed of wet material passing continuously through the dryer. Dryer rates are high because of the large area of contact and short distance of travel for the internal moisture.

The most widely used type is the **horizontal conveying-screen dryer** in which wet material is conveyed as a layer, 2 to 15 cm deep, on a horizontal mesh screen or perforated apron, while heated air is blown either upward or downward through the bed of material. Its drying characteristics were studied by Marshall and Hougen [*Trans. Am. Inst. Chem. Eng.*, **38**, 91 (1942)]. This dryer consists usually of a number of individual sections, complete with fan and heating coils, arranged in series to form a housing or tunnel through which the conveying screen travels. As shown in the sectional view in Fig. 12-55, the air circulates through the wet material and is reheated before reentering the bed. It is not uncommon to circulate the hot gas upward in the wet end and downward in the dry end. A portion of the air is exhausted continuously by one or two exhaust fans, not shown in the sketch, which handle air from several sections. Since each section can be operated independently, extremely flexible operation is possible, with high temperatures usually at the wet end, followed by lower temperatures; in some cases a unit with cooled or specially humidified air is employed for final conditioning. The **maximum pressure drop** that can be taken through the bed of solids without developing leaks or air bypassing is roughly 50 mm of water.

Through-circulation drying requires that the wet material be in a state of granular or pelleted subdivision so that hot air may be readily blown through it. Many materials meet this requirement without special preparation. Others require special and often elaborate pretreatment to render them suitable for through-circulation drying. The process of converting a wet solid into a form suitable for through circulation of air is called **preforming,** and often the success or failure of this contacting method depends on the preforming step. Fibrous, flaky, and coarse granular materials are usually amenable to drying without preforming. They can be loaded directly onto the conveying screen by suitable spreading feeders of the oscillating-belt or vibrating type or by spiked drums or belts feeding from bins. When materials must be preformed, several methods are available, depending on the physical state of the wet solid.

1. Relatively dry materials such as centrifuge cakes can sometimes be granulated to give a suitably porous bed on the conveying screen.

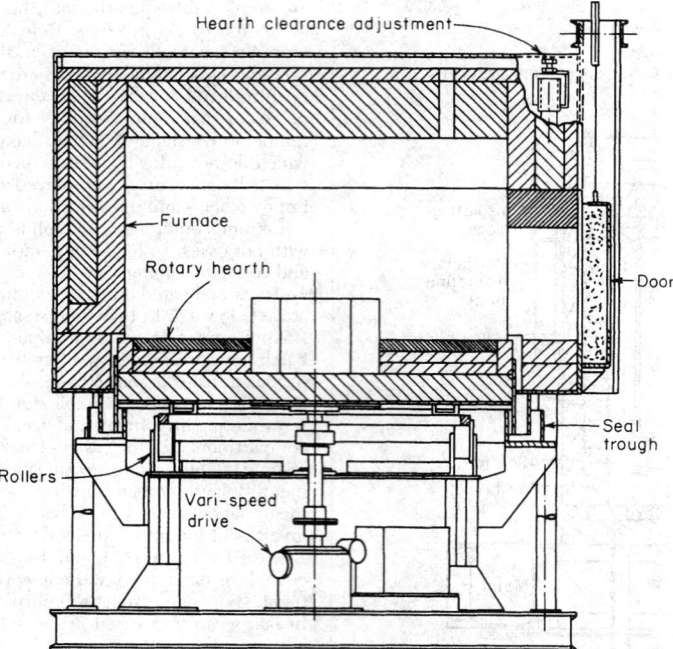

FIG. 12-52   Rotary-hearth furnace. (*W. S. Rockwell Co.*)

2.   Pasty materials can often be preformed by extrusion to form sphaghetti-like pieces, about 6 mm in diameter and several centimeters long.

3.   Wet pastes that cannot be granulated or extruded may be predried and preformed on a steam-heated finned drum. Preforming on a finned drum may be desirable also in that some predrying is accomplished.

4.   Thixotropic filter cakes from rotary vacuum filters that cannot be preformed by any of the above methods can often be scored by knives on the filter, the scored cake discharging in pieces suitable for through-circulation drying.

5.   Material that shrinks markedly during drying is often reloaded during the drying cycle to 2 to 6 times the original loading depth. This is usually done after a degree of shrinkage which, by opening the bed, has destroyed the effectiveness of contact between the air and solids.

6.   In a few cases, powders have been pelleted or formed in briquettes to eliminate dustiness and permit drying by through circula-

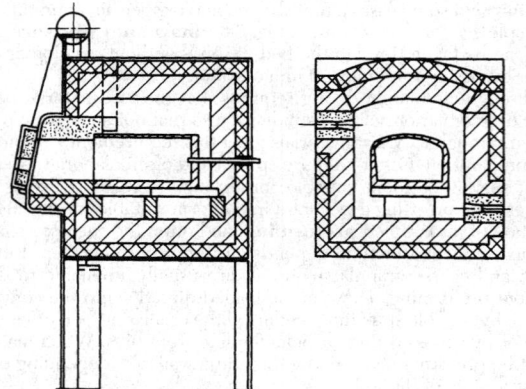

FIG. 12-53   Small muffle furnace. (*W. S. Rockwell Co.*)

tion. Table 12-14 gives a list of materials classified by preforming methods suitable for through-circulation drying.

Steam-heated air is the usual heat-transfer medium employed in these dryers, although combustion gases may be used also. Temperatures above 600 K are not usually feasible because of the problems of lubricating the conveyor, chain, and roller drives. Recirculation of air is in the range of 60 to 90 percent. Conveyors may be made of wire-mesh screen or perforated-steel plate. The minimum practical screen opening size is about 30 mesh.

Design Methods for Continuous Tunnel Dryers   In actual practice, design of a continuous through-circulation dryer is best based upon data taken in pilot-plant tests. Loading and distribution of solids on the screen are rarely as nearly uniform in commercial installations as in test dryers; 50 to 100 percent may be added to the test drying time for commercial design.

A mathematical method of a through-circulation dryer has been developed by Thygeson [*Am. Inst. Chem. Eng. J.*, **16**(5), 749 (1970)]. Results obtained by Gamson, Thodos, and Hougen [*Trans. Am. Inst. Chem. Eng.*, **39**, 1 (1943)] and Wilke and Hougen [ibid., **41**, 444 (1945)] for the rates of adiabatic evaporation of water from packed beds of porous solids are applicable when drying gases flow upward to downward. Use of average additive properties of the drying gas leads to

$$h_c = 0.11 \frac{G^{0.59}}{D_p^{0.41}} \quad \text{for} \quad \frac{D_p G}{\mu} > 350 \quad (12\text{-}49)$$

and

$$h_c = 0.15 \frac{G^{0.49}}{D_p^{0.51}} \quad \text{for} \quad \frac{D_p G}{\mu} < 350 \quad (12\text{-}50)$$

where μ = gas viscosity, lb/(ft·h); D_p = diameter of sphere having the same surface area as particle, ft; and G = mass velocity of drying gas, lb/(h·ft²) [to convert from pounds per foot-hour to newtons per second-square meter, multiply by 4.133×10^{-4}; to convert from feet to meters, multiply by 0.3048; and to convert from pounds per hour-square foot to kilograms per second-square meter, multiply by 1.3562×10^{-3}].

Performance and Cost Data for Continuous Tunnel Dryers
Experimental performance data are given in Table 12-15 for numer-

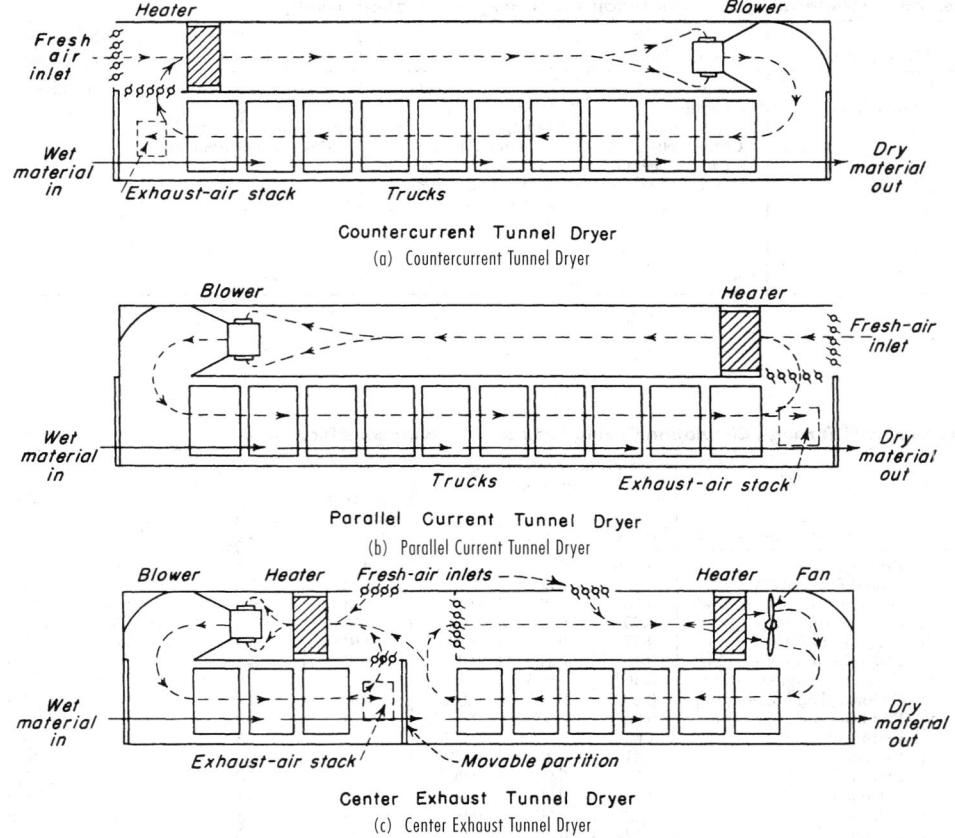

FIG. 12-54 Three types of tunnel dryers. [*Van Arsdale,* Food Ind. *14* (10), 43 (1942).]

ous common materials. Performance data from several commercial through-circulation conveyor dryers are given in Table 12-16. Labor requirements vary depending on the time required for feed adjustments, inspection, etc. These dryers will consume from 0.9 to 1.1 kg of steam/kg of water evaporated. Thermal efficiency is a function of final moisture required and percent air recirculation.

Conveying-screen dryers are fabricated with conveyor widths from 0.3- to 4.4-m sections 1.6 to 2.5 m long. Each section consists of a sheet-metal enclosure, insulated sidewalls and roof, heating coils, a circulating fan, inlet-air distributor baffles, a fines catch pan under the

conveyor, and a conveyor screeen (Fig. 12-56). Table 12-17 gives approximate purchase costs for equipment with type 304 stainless-steel hinged conveyor screens and includes steam-coil heaters, fans, motors, and a variable-speed conveyor drive. Cabinet- and auxiliary-equipment fabrication is of aluminized steel or stainless-steel materials. Prices do not include temperature controllers, motor starters, preform equipment, or auxiliary feed and discharge conveyors. These may add $55,000 to $135,000 to the dryer purchase cost (1995 costs).

A continuous conveyor dryer employing a combination of air impingement and through circulation is shown in Fig. 12-57.

Continuous Furnaces Continuous furnaces are employed for the same general duties cited for batch furnaces. Units are gas, oil, or electrically heated and utilize direct circulation of combustion gases or muffles for heat transfer. Continuous furnaces frequently have an extension added for **cooling** the charge before exposure to atmospheric air.

Conveyors may be of parallel-chain, mat, slat, woven wire-mesh belt, or cast-alloy type. Automatic tensioning devices are used to maintain belt tension during heating and cooling. The product may rest directly on the conveyor or on special supports built into it. Roller-conveyors are used for large pieces. **Flame curtains** are provided for sealing the ends and for protection of special treating atmospheres.

The pusher-type furnace is relatively free from mechanical problems because all mechanical parts are located outside the hot zone. It employs a roller-conveyor usually and will handle charges weighing considerably more per square meter than a belt-conveyor furnace. Pushers are driven by electric motors, compressed air, or hydraulic systems and can be automatically timed and synchronized with door-

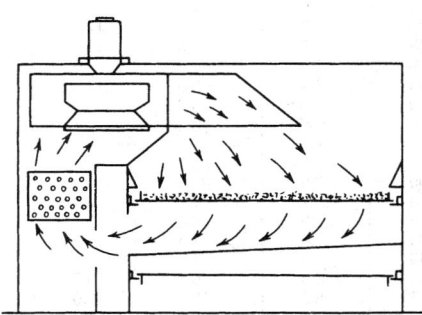

FIG. 12-55 Section view of a continuous through-circulation conveyor dryer. (*Proctor & Schwartz, Inc.*)

TABLE 12-14   Methods of Preforming Some Materials for Through-Circulation Drying

No preforming required	Scored on filter	Granulation	Extrusion	Finned drum	Flaking on chilled drum	Briquetting and squeezing
Cellulose acetate	Starch	Kaolin	Calcium carbonate	Lithopone	Soap flakes	Soda ash
Silica gel	Aluminum hydrate	Cryolite	White lead	Zinc yellow		Cornstarch
Scoured wool		Lead arsenate	Lithopone	Calcium carbonate		Synthetic rubber
Sawdust		Cornstarch	Titanium dioxide	Magnesium carbonate		
Rayon waste		Cellulose acetate	Magnesium carbonate			
Fluorspar		Dye intermediates	Aluminum stearate			
Tapioca			Zinc stearate			
Breakfast food						
Asbestos fiber						
Cotton linters						
Rayon staple						

TABLE 12-15   Experimental Through-Circulation Drying Data for Miscellaneous Materials

Material	Physical form	Moisture contents, kg/kg dry solid			Inlet-air temperature, K	Depth of bed, cm	Loading, kg product/m²	Air velocity, m/s × 10¹	Experimental drying time, s × 10⁻²
		Initial	Critical	Final					
Alumina hydrate	Briquettes	0.105	0.06	0.00	453	6.4	60.0	6.0	18.0
Alumina hydrate	Scored filter cake	9.60	4.50	1.15	333	3.8	1.6	11.0	90.0
Alumina hydrate	Scored filter cake	5.56	2.25	0.42	333	7.0	4.6	11.0	108.0
Aluminum stearate	0.7-cm extrusions	4.20	2.60	0.003	350	7.6	6.5	13.0	36.0
Asbestos fiber	Flakes from squeeze rolls	0.47	0.11	0.008	410	7.6	13.6	9.0	5.6
Asbestos fiber	Flakes from squeeze rolls	0.46	0.10	0.0	410	5.1	6.3	9.0	3.6
Asbestos fiber	Flakes from squeeze rolls	0.46	0.075	0.0	410	3.8	4.5	11.0	2.7
Calcium carbonate	Preformed on finned drum	0.85	0.30	0.003	410	3.8	16.0	11.5	12.0
Calcium carbonate	Preformed on finned drum	0.84	0.35	0.0	410	8.9	25.7	11.7	18.0
Calcium carbonate	Extruded	1.69	0.98	0.255	410	1.3	4.9	14.3	9.0
Calcium carbonate	Extruded	1.41	0.45	0.05	410	1.9	5.8	10.2	12.0
Calcium stearate	Extruded	2.74	0.90	0.0026	350	7.6	8.8	5.6	57.0
Calcium stearate	Extruded	2.76	0.90	0.007	350	5.1	5.9	6.0	42.0
Calcium stearate	Extruded	2.52	1.00	0.0	350	3.8	4.4	10.2	24.0
Cellulose acetate	Granulated	1.14	0.40	0.09	400	1.3	1.4	12.7	1.8
Cellulose acetate	Granulated	1.09	0.35	0.0027	400	1.9	2.7	8.6	7.2
Cellulose acetate	Granulated	1.09	0.30	0.0041	400	2.5	4.1	5.6	10.8
Cellulose acetate	Granulated	1.10	0.45	0.004	400	3.8	6.1	5.1	18.0
Clay	Granulated	0.277	0.175	0.0	375	7.0	46.2	10.2	19.2
Clay	1.5-cm extrusions	0.28	0.18	0.0	375	12.7	100.0	10.7	43.8
Cryolite	Granulated	0.456	0.25	0.0026	380	5.1	34.2	9.1	24.0
Fluorspar	Pellets	0.13	0.066	0.0	425	5.1	51.4	11.6	7.8
Lead arsenate	Granulated	1.23	0.45	0.043	405	5.1	18.1	11.6	18.0
Lead arsenate	Granulated	1.25	0.55	0.054	405	6.4	22.0	10.2	24.0
Lead arsenate	Extruded	1.34	0.64	0.024	405	5.1	18.1	9.4	36.0
Lead arsenate	Extruded	1.31	0.60	0.0006	405	8.4	26.9	9.2	42.0
Kaolin	Formed on finned drum	0.28	0.17	0.0009	375	7.6	44.0	9.2	21.0
Kaolin	Formed on finned drum	0.297	0.20	0.005	375	11.4	56.3	12.2	15.0
Kaolin	Extruded	0.443	0.20	0.008	375	7.0	45.0	10.16	18.0
Kaolin	Extruded	0.36	0.14	0.0033	400	9.6	40.6	15.2	12.0
Kaolin	Extruded	0.36	0.21	0.0037	400	19.0	80.7	10.6	30.0
Lithopone (finished)	Extruded	0.35	0.065	0.0004	408	8.2	63.6	10.2	18.0
Lithopone (crude)	Extruded	0.67	0.26	0.0007	400	7.6	41.1	9.1	51.0
Lithopone	Extruded	0.72	0.28	0.0013	400	5.7	28.9	11.7	18.0
Magnesium carbonate	Extruded	2.57	0.87	0.001	415	7.6	11.0	11.4	17.4
Magnesium carbonate	Formed on finned drum	2.23	1.44	0.0019	418	7.6	13.2	8.6	24.0
Mercuric oxide	Extruded	0.163	0.07	0.004	365	3.8	66.5	11.2	24.0
Silica gel	Granular	4.51	1.85	0.15	400	3.8–0.6	3.2	8.6	15.0
Silica gel	Granular	4.49	1.50	0.215	340	3.8–0.6	3.4	9.1	63.0
Silica gel	Granular	4.50	1.60	0.218	325	3.8–0.6	3.5	9.1	66.0
Soda salt	Extruded	0.36	0.24	0.008	410	3.8	22.8	5.1	51.0
Starch (potato)	Scored filter cake	0.866	0.55	0.069	400	7.0	26.3	10.2	27.0
Starch (potato)	Scored filter cake	0.857	0.42	0.082	400	5.1	17.7	9.4	15.0
Starch (corn)	Scored filter cake	0.776	0.48	0.084	345	7.0	26.4	7.4	54.0
Starch (corn)	Scored filter cake	0.78	0.56	0.098	380	7.0	27.4	7.6	24.0
Starch (corn)	Scored filter cake	0.76	0.30	0.10	345	1.9	7.7	6.7	15.0
Titanium dioxide	Extruded	1.2	0.60	0.10	425	3.0	6.8	13.7	6.3
Titanium dioxide	Extruded	1.07	0.65	0.29	425	8.2	16.0	8.6	6.0
White lead	Formed on finned drum	0.238	0.07	0.001	355	6.4	76.8	11.2	30.0
White lead	Extruded	0.49	0.17	0.0	365	3.8	33.8	10.2	27.0
Zinc stearate	Extruded	4.63	1.50	0.005	360	4.4	4.2	8.6	36.0

TABLE 12-16 *Performance Data for Continuous Through-Circulation Dryers**

	Kind of material						
	Inorganic pigment	Cornstarch	Fiber staple		Charcoal briquettes	Gelatin	Inorganic chemical
Capacity, kg dry product/h	712	4536	1724		5443	295	862
			Stage A	Stage B			
Approximate dryer area, m²	22.11	66.42	57.04	35.12	52.02	104.05	30.19
Depth of loading, cm	3	4			16	5	4
Air temperature, °C	120	115 to 140	130 to 100	100	135 to 120	32 to 52	121 to 82
Loading, kg product/m²	18.8	27.3	3.5	3.3	182.0	9.1	33
Type of conveyor, mm	1.59 by 6.35 slots	1.19 by 4.76 slots	2.57-diameter holes, perforated plate		8.5 × 8.5 mesh screen	4.23 × 4.23 mesh screen	1.59 × 6.35 slot
Preforming method or feed	Rolling extruder	Filtered and scored	Fiber feed		Pressed	Extrusion	Rolling extruder
Type and size of preformed particle, mm	6.35-diameter extrusions	Scored filter cake	Cut fiber		64 × 51 × 25	2-diameter extrusions	6.35-diameter extrusions
Initial moisture content, % bone-dry basis	120	85.2	110		37.3	300	111.2
Final moisture content, % bone-dry basis	0.5	13.6	9		5.3	11.1	1.0
Drying time, min	35	24	11		105	192	70
Drying rate, kg water evaporated/(h·m²)	38.39	42.97	17.09		22.95	9.91	31.25
Air velocity (superficial), m/s	1.27	1.12	0.66		1.12	1.27	1.27
Heat source per kg water evaporated, steam kg/kg gas (m³/kg)	Gas 0.11	Steam 2.0	Steam 1.73		Waste heat	Steam 2.83	Gas 0.13
Installed power, kW	29.8	119.3	194.0		82.06	179.0	41.03

*Courtesy of Proctor & Schwartz, Inc.

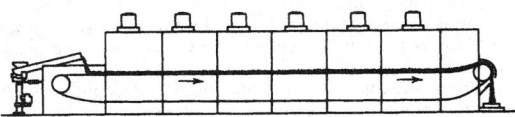

FIG. 12-56 Section assembly of a continuous through-circulation conveyor dryer. (*Proctor & Schwartz, Inc.*)

TABLE 12-17 *Conveyor-Screen-Dryer Costs**

Length	2.4-m-wide conveyor	3.0-m-wide conveyor
7.5 m	$8600/m²	$7110/m²
15 m	$6700/m²	$5600/m²
22.5 m	$6200/m²	$5150/m²
30 m	$5900/m²	$4950/m²

*National Drying Machinery Company, 1996.

opening timers. For small solids, trays of perforated metal alloys are used to carry the product. These carriers ride through the tunnel on rollers, skid rails, and occasionally refractory skids, one tray pushing the next ahead. The charge may travel in a straight line or in counterflow movement in single or multiple chambers.

In counterflow movement, heat from the outgoing solids is transferred directly to cold incoming solids, reducing heat losses and fuel requirements. Continuous conveyor ovens are employed also for drying refractory shapes and for drying and baking enameled pieces. In many of these latter, the parts are suspended from overhead chain conveyors.

Ceramic tunnel kilns handling large irregular-shaped objects must be equipped for precise control of temperature and humidity conditions to prevent cracking and condensation on the product. The internal mechanism causing cracking when drying clay and ceramics has been studied extensively. Information on ceramic tunnel-kiln operation and design is reported fully in publications such as *The American Ceramic Society Bulletin, Ceramic Industry,* and *Transactions of the British Ceramic Society.*

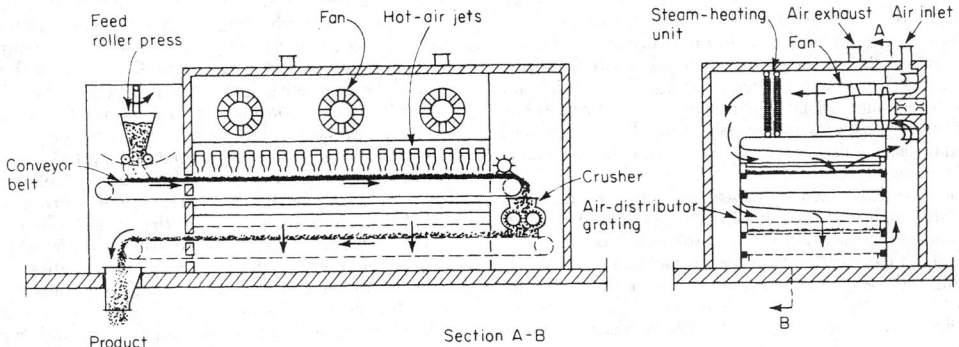

FIG. 12-57 Special conveyor dryer with air jets impinging on surface of bed on first pass. Dried material is crushed and passed again through dryer, with air going through the now-permeable bed. [*Chem. Eng., 192 (June 19, 1967).*]

ROTARY DRYERS

Description   A rotary dryer consists of a cylinder, rotated upon suitable bearings and usually slightly inclined to the horizontal. The length of the cylinder may range from 4 to more than 10 times its diameter, which may vary from less than 0.3 to more than 3 m. Feed solids fed into one end of the cylinder progress through it by virtue of rotation, head effect, and slope of the cylinder and discharge as finished product at the other end. Gases flowing through the cylinder may retard or increase the rate of solids flow, depending upon whether gas flow is countercurrent or cocurrent with solids flow.

Rotary dryers have been classified as **direct, indirect-direct, indirect,** and **special types.** The terms refer to the method of heat transfer, being "direct" when heat is added to or removed from the solids by direct exchange between flowing gas and solids and being "indirect" when the heating medium is separated from physical contact with the solids by a metal wall or tube.

Only totally direct and totally indirect types will be discussed extensively here, as it must be recognized that an infinite number of variations between the two are possible. Their operating characteristics when performing heat- and mass-transfer operations make them suitable for the accomplishment of drying, chemical reactions, solvent recovery, thermal decompositions, mixing, sintering, and agglomeration of solids. The specific types included are the following:

Direct rotary dryer (cooler).   This is usually a bare metal cylinder, with or without flights. It is suitable for low- and medium-temperature operations, the operating temperature being limited primarily by the strength characteristics of the metal employed in fabrication.

Direct rotary kiln.   This is a metal cylinder lined on the interior with insulating block and/or refractory brick. It is suitable for high-temperature operations.

Indirect steam-tube dryer.   This is a bare metal cylinder provided with one or more rows of metal tubes installed longitudinally in the shell. It is suitable for operation up to available steam temperatures or in processes requiring water cooling of the tubes.

Indirect rotary calciner.   This is a bare metal cylinder surrounded on the outside by a fired or electrally heated furnace. It is suitable for operation at medium temperatures up to the maximum which can be tolerated by the metal wall of the cylinder, usually 650 to 700 K for carbon steel and 800 to 1025 K for stainless steel.

Direct Roto-Louvre dryer.   This is one of the more important special types, differing from the direct rotary unit in that true *through circulation* of gas through the solids bed is provided. Like the direct rotary, it is suitable for low- and medium-temperature operation.

Field of Application   Rotating equipment is applicable to batch or continuous processing solids which are relatively free-flowing and granular when discharged as product. Materials which are not completely free-flowing in their feed condition are handled in a special manner, either by recycling a portion of final product and premixing with the feed in an external mixer to form a uniform granular feed to the process or by maintaining a bed of free-flowing product in the cylinder at the feed end and, in essence, performing a premixing operation in the cylinder itself. A properly designed recycle process will permit processing of many forms of slurry and solution feeds in rotating vessels. Direct rotary kilns and indirect calciners without internal flights or other obstructions are often provided with **hanging link chains.** These may serve as surfaces upon which material can accumulate until it is no longer sticky, at which time it will break off as a granular solid and continue its movement through the cylinder. **Scraper chains** may also be provided on indirect calciners to maintain clean internal walls.

As a general rule, the direct-heat units are the simplest and most economical in construction and are employed when direct contact between the solids and flue gases or air can be tolerated. Because the total heat load must be introduced or removed in the gas stream, large gas volumes and high gas velocities are usually required. The latter will be rarely less than 0.5 m/s in an economical design. Therefore, employment of direct rotating equipment with solids containing extremely fine particles is likely to result in excessive entrainment losses in the exit-gas stream.

The indirect forms require only sufficient gas flow through the cylinder to remove vapors or otherwise complete the internal process.

In addition, these can be sealed for processes requiring special gas atmospheres and exclusion of outside air.

Auxiliary Equipment   On direct-heat rotating equipment, a combustion chamber is required for high temperatures and finned steam coils are used for low temperatures. If contamination of the product with combustion gases is undesirable on direct-heat units, indirect gas- or oil-fired air heaters may be employed to achieve temperatures in excess of available steam.

The *method of feeding* rotating equipment depends upon material characteristics and the location and type of upstream processing equipment. When the feed comes from above, a chute extending into the cylinder may be employed. For sealing purposes or if gravity feed is not convenient, a screw feeder is normally used. On cocurrent direct-heat units, cold-water jacketing of the feed chute or conveyor may be desirable if it is contacted by the inlet hot-gas stream. This will prevent overheating of the metal wall with resultant scaling or overheating of heat-sensitive feed materials.

Any type of solids conveyor may be suitable for **recycle** mixing; however, the most universally applicable is the double-shaft pug-mill-type paddle mixer. This conveyor or mixer should be insulated to prevent excessive heat losses from the hot, dry recycle product. To ensure uniformity in the recycle operation, a **surge storage** reserve of recycle solids should be installed for startup purposes and in the event of interruption of product discharge from the cylinder. In recycle operations, 50 to 60 percent product recirculation is found economical in many instances.

One method of feeding direct cocurrent drying equipment utilizes dryer exhaust gases to convey, mix, and predry wet feed. The latter is added to the exhaust gases, at high velocity, from the dryer. The wet feed, mixed with dust entrained from the dryer, separates from the exhaust gases in a cyclone and drops into the feed end of the cylinder. The technique combines pneumatic and rotary drying. High thermal efficiency results from two cocurrent-flow stages operating countercurrently.

Pneumatic conveyors are frequently employed as both dry-product conveyors and coolers. Other types of cooling equipment often used are screw conveyors, vibrating conveyors, and direct or indirect rotating coolers.

Dust entrained in the exit-gas stream is customarily removed in cyclone collectors. This dust may be discharged back into the process or separately collected. For expensive materials or extremely fine particles, bag collectors may follow a cyclone collector, provided fabric temperature stability is not limiting. When toxic gases or solids are present, the exit gas is at a high temperature, the gas is close to saturation as from a steam-tube dryer, or gas recirculation in a sealed system is involved, wet scrubbers may be used independently or following a cyclone. Cyclones and bag collectors in drying applications frequently require insulation and steam tracing. The exhaust fan should be located downstream from the collection system.

Rotating equipment, except brick-lined vessels, operated above ambient temperatures is usually insulated to reduce heat losses. Exceptions are direct-heat units of bare metal construction operating at high temperatures, on which heat losses from the shell are necessary to prevent overheating of the metal. Insulation is particularly necessary on cocurrent direct-heat units. It is not unusual for product cooling or condensation on the shell to occur in the last 10 to 50 percent of the cylinder length if it is not well insulated.

For best operation, the feed rate to rotating equipment should be closely controlled and uniform in quantity and quality. Because solids temperatures are difficult to measure and changes slowly detected, most rotating-equipment operations are controlled by indirect means. Inlet and exit gas temperatures are measured and controlled on direct-heat units such as direct dryers and kilns, steam temperature and pressure and exit-gas temperature and humidity are controlled on steam-tube units, and direct shell temperature measurements are taken on indirect calciners. Product temperature measurements are taken for secondary control purposes only in most instances.

Equipment which is electrically driven and operated with metal temperatures exceeding 425 K should be provided with auxiliary drives or power sources. Loss of rotation of a heated calciner or high-temperature dryer carrying a heavy bed of hot solids will quickly result in sagging of the cylinder due to nonuniform cooling.

Direct-Heat Rotary Dryers   The direct-heat rotary dryer is usually equipped with **flights** on the interior for lifting and showering the solids through the gas stream during passage through the cylinder. These flights are usually offset every 0.6 to 2 m to ensure more continuous and uniform curtains of solids in the gas. The shape of the flights depends upon the handling characteristics of the solids. For free-flowing materials, a radial flight with a 90° lip is employed. For sticky materials, a flat radial flight without any lip is used. When materials change characteristics during drying, the flight design is changed along the dryer length. Many standard dryer designs employ flat flights with no lips in the first one-third of the dryer measured from the feed end, flights with 45° lips in the middle one-third, and flights with 90° lips in the final one-third of the cylinder. **Spiral flights** are usually provided in the first meter or so at the feed end to accelerate forward flow from under the feed chute or conveyor and to prevent leakage over the feed-end retainer ring into the gas seals.

When cocurrent gas-solids flow is used, flights may be left out to the final meter or so at the exit end to reduce entrainment of dry product in the exit gas. Showering of wet feed at the feed end of a countercurrent dryer will, on the other hand, frequently serve as an effective means for scrubbing dry entrained solids from the gas stream before it leaves the cylinder. Some dryers are provided with sawtooth flights to obtain uniform showering, while others use lengths of chain, attached to the underside of the flights, to scrape over and knock the walls of the cylinder, thereby removing sticky solids which might normally adhere to it. In kilns, the chains may contribute significantly to heat transfer; however, their use contributes to maintenance costs when flights are present in direct dryers. Solids sticking on flights and walls are usually removed more efficiently by **external shell knockers.** In dryers of large cross section, internal elements or partitions are sometimes used to increase the effectiveness of material distribution and reduce dusting and impact grinding. Use of internal members increases the difficulty of cleaning and maintenance unless sufficient free area is left between partitions for easy access of a person. Some examples of the more common flight arrangements are shown in Fig. 12-58. Component arrangements of countercurrent direct rotary dryers are shown in Fig. 12-59, and those of a cocurrent unit in Fig. 12-60.

Countercurrent flow of gas and solids gives greater **heat-transfer efficiency** with a given inlet-gas temperature, but cocurrent flow can be used more frequently to dry heat-sensitive materials at higher inlet-gas temperatures because of the rapid cooling of the gas during initial evaporation of surface moisture.

A number of different methods are employed to seal the rotating cylinder and prevent gas leakage through the annular opening between the rotating cylinder and the stationary throat pieces. *None is an effective solids seal,* nor will any function satisfactorily as a gas seal if solids leakage over the retaining ring on the cylinder is permitted.

Three examples of ordinary gas seals are shown in Fig. 12-61. On direct rotary dryers, few gas seals are intended to be completely gastight, but by careful control of the internal pressure, generally between 0.25 and 2.5 mm of water below atmosphere, dusting to the outside is prevented and in-leakage of outside air is minimized.

Figure 12-61 also illustrates three basic types of **trunnion roll-bearing assemblies.** Antifriction pillow blocks are the most common on modern dryers; however, when the dryer load requires larger than a 12.7- to 15.2-cm-diameter bearing on the trunnion shaft, the dead-shaft antifriction bearing is substituted. This represents a considerable cost saving compared with the larger pillow blocks. They are completely sealed and continuously bathed in lubricant. Pillow-block bushings are less often used. The thrust washers are difficult to seal against dust, and they draw more power. Thrust roll mountings are depicted also in Fig. 12-61. These are usually dead-shaft.

Gases are forced through the cylinder by either an exhauster or an exhauster-blower combination. With the latter arrangement it is possible to maintain very precise control of internal pressure even when the total system pressure drop is high. When a low-pressure-drop air heater is employed, however, the exhauster alone is usually sufficient, as the major gas pressure losses are found in the exit-air ductwork and dust collectors. Use of a blower by itself to force gas through the cylinder is an unusual practice, because the internal pressure is above atmospheric and hot air and dust may be blown into the gas seals or out into the surrounding working areas.

Special designs of direct rotary dryers, such as the Renneburg DehydrO-Mat (Edward Renneburg & Sons Co.), are constructed especially to provide lower retention during the falling-rate drying period for the escape of internal moisture from the solids. The DehydrO-Mat is a cocurrent dryer employing a small-diameter shell at the feed end, where rapid evaporation of surface moisture in the stream of initially hot gas is accomplished with low holdup. At the solids- and gas-exit end, the shell diameter is increased to reduce gas velocities and provide increased holdup for the solids while they are exposed to the partially cooled gas stream.

The Louisville type P dryer is a cocurrent dryer developed for the drying of heat-sensitive polymers. It is designed for use on rather finely divided and bulky materials which are easily airborne since its basic design utilizes a discharge cone permitting pneumatic conveying of dried solids from the dryer. Its internal design provides additional retention time by slowing the progress of the material through the dryer cylinder, permitting a comparatively high velocity of the drying medium without excessive blow-through.

The Louisville type H dryer is a modified cocurrent dryer with "flash-drying" characteristics. Its internal arrangements consist of alternating disks and doughnuts which give high differential velocities between the drying medium and the solids being processed to increase the heat transfer and hence the rate of moisture removal.

Design Methods for Direct-Heat Rotary Dryers   Direct drying in a direct-heat rotary dryer is best expressed as a heat-transfer mechanism as follows:

$$Q_t = UaV(\Delta t)_m \tag{12-51}$$

where Q_t = total heat transferred, J/s; Ua = volumetric heat-transfer coefficient, J/(s·m³·K); V = dryer volume, m³; and $(\Delta t)_m$ = true mean temperature difference between the hot gases and material, K. When a considerable quantity of surface moisture is removed from the solids and the solids temperatures are unknown, a good approximation of $(\Delta t)_m$ is the logarithmic mean between the wet-bulb depressions of the drying air at the inlet and exit of the dryer.

Data for evaluating Ua were developed by Miller [*Trans. Am. Inst. Chem. Eng.,* **38,** 841 (1942)], Friedman and Marshall [*Chem. Eng. Prog.,* **45,** 482, 573 (1949)], and Seaman and Mitchell [*Chem. Eng. Prog.,* **50,** 467 (1954)]. These authors all employed relationships that could be reduced to the form $Ua = KG^n/D$, where K = a proportionality constant, G = gas mass velocity, D = diameter, and n = a constant.

P. Y. McCormick [*Chem. Eng. Prog.,* **58**(6), 57 (1962)] compared all available data. The comparisons showed that flight geometry and shell speed should be accounted for in the value of K. He suggested that shell rotational speed and flight number and shape must affect the overall balance; however, data for evaluating these variables separately are not available. Also, it is not believed that the effect of gas velocity

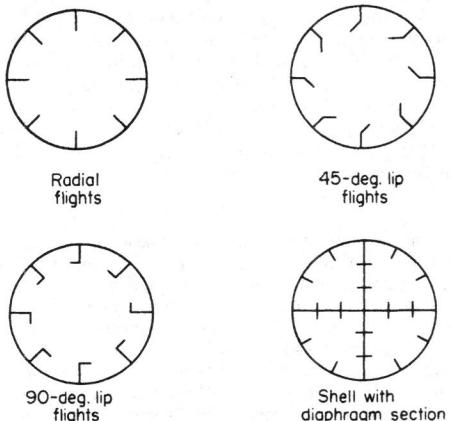

FIG. 12-58   Alternative direct-heat rotary-dryer flight arrangements.

Radial flights

45-deg. lip flights

90-deg. lip flights

Shell with diaphragm section

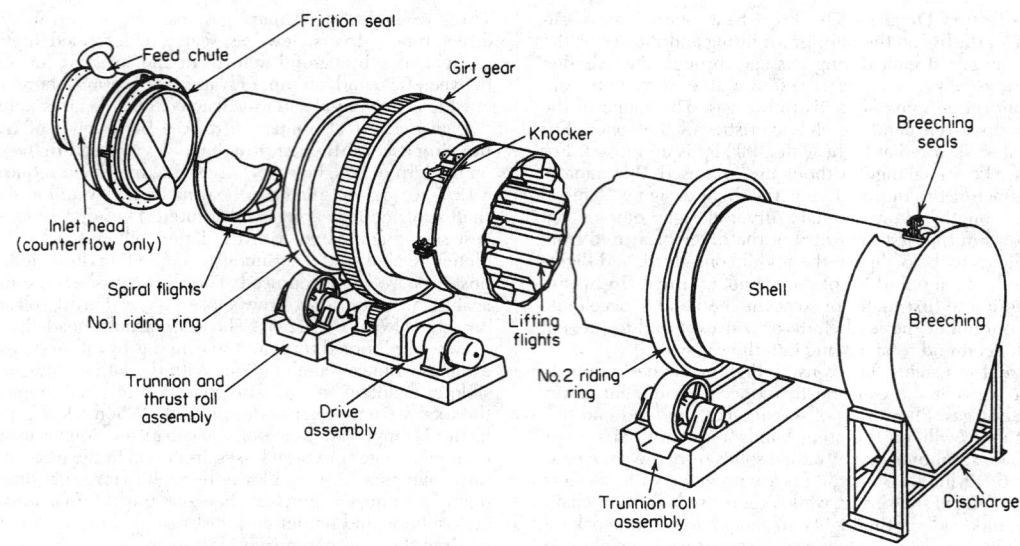

FIG. 12-59 Component arrangements of a countercurrent direct-heat rotary dryer. (*ABB Raymond/Bartlett-Snow TM.*)

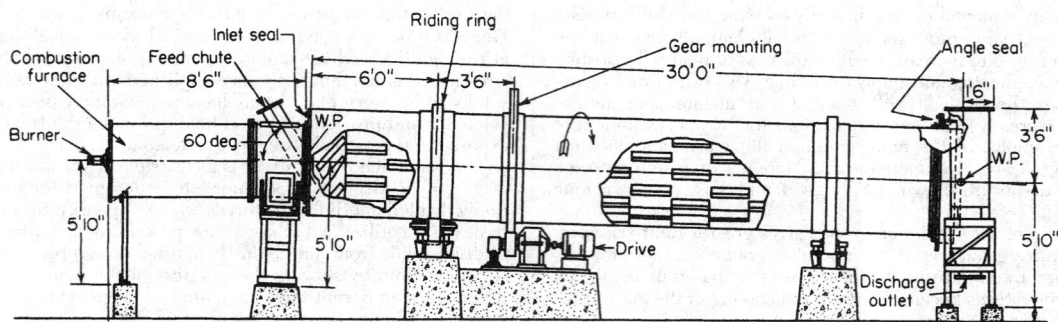

FIG. 12-60 Elevation of a 60-in-diameter by 30-ft-long direct-heat cocurrent rotary dryer. To convert inches to centimeters, multiply by 2.54; to convert feet to meters, multiply by 0.3048. (*ABB Raymond/Bartlett-Snow TM.*)

is toward reducing the surface film thickness around each particle; rather, an increase in gas velocity more likely breaks up more effectively the showering curtains of solids and thereby exposes more solids surface; i.e., the effect is to increase the a in Ua rather than U.

The following relationship is recommended for commercial dryers now manufactured in the United States, which usually have a flight count per circle of 2.4 to 3.0 D and operate at shell peripheral speeds of 60 to 75 ft/min:

$$Q_t = (0.5G^{0.67}/D)V\,\Delta t_m \qquad (12\text{-}52)$$

$$= 0.4LDG^{0.67}\,\Delta t_m \qquad (12\text{-}53)$$

where Q_t = total heat transferred, Btu/h; L = dryer length, ft; D = dryer diameter, ft; G = gas mass velocity, lb/(h·ft² of cross section); and Δt_m = log mean of the drying-gas wet-bulb depression at the inlet end and exit end of the dryer shell.

Typical operating data for cocurrent rotary dryers are given in Table 12-18. (Note that the driving force ΔT_m must be based on wet-bulb depression and not on material temperatures. Use of material temperatures, particularly when the dry solids are superheated after drying, will yield conservative results.)

In most cases, direct-heat rotary dryers are still sized on the basis of pilot-plant tests, because rarely is all the moisture to be removed truly "free" moisture, and residence time for diffusion is frequently needed.

A theoretical model simulating the operation of both cocurrent and countercurrent rotary dryers which relies on pilot and bench scale tests to determine parameters which describe solids transport and drying rate is described by Papadakis, Langrish, Kemp, and Bahu [*Drying Technology—An International Journal,* **12**(1&2), 259 (1994)]. For maximum dryer heat-transfer efficiency, dryer fillage must be sufficient to load the lifting flights fully, as discussed later.

Unless material characteristics limit the gas temperature, the inlet temperature is usually fixed by the heating medium employed; i.e., 400 to 450 K for steam or 800 to 1100 K for gas- and oil-fired burners. The proper exit-gas temperature is largely an economic function. Its value may be determined as follows:

$$N_t = (t_1 - t_2)/(\Delta t)_m \qquad (12\text{-}54)$$

where N_t = number of heat-transfer units based upon the gas; t_1 = initial-gas temperature, K; t_2 = exit-gas temperature allowing for heat losses, K; and $(\Delta t)_m$ is as defined for Eq. (12-51). Equation (12-54) can be used to select an exit-gas temperature since it has been found (empirically) that rotary dryers are most economically operated between $N_t = 1.5$ and $N_t = 2.5$.

The L/D (length-diameter) ratio found most efficient in commercial practice lies between 4 and 10. If the length calculated previously does not fall within these limits, another value of N_t which will place L/D in the proper range may be computed.

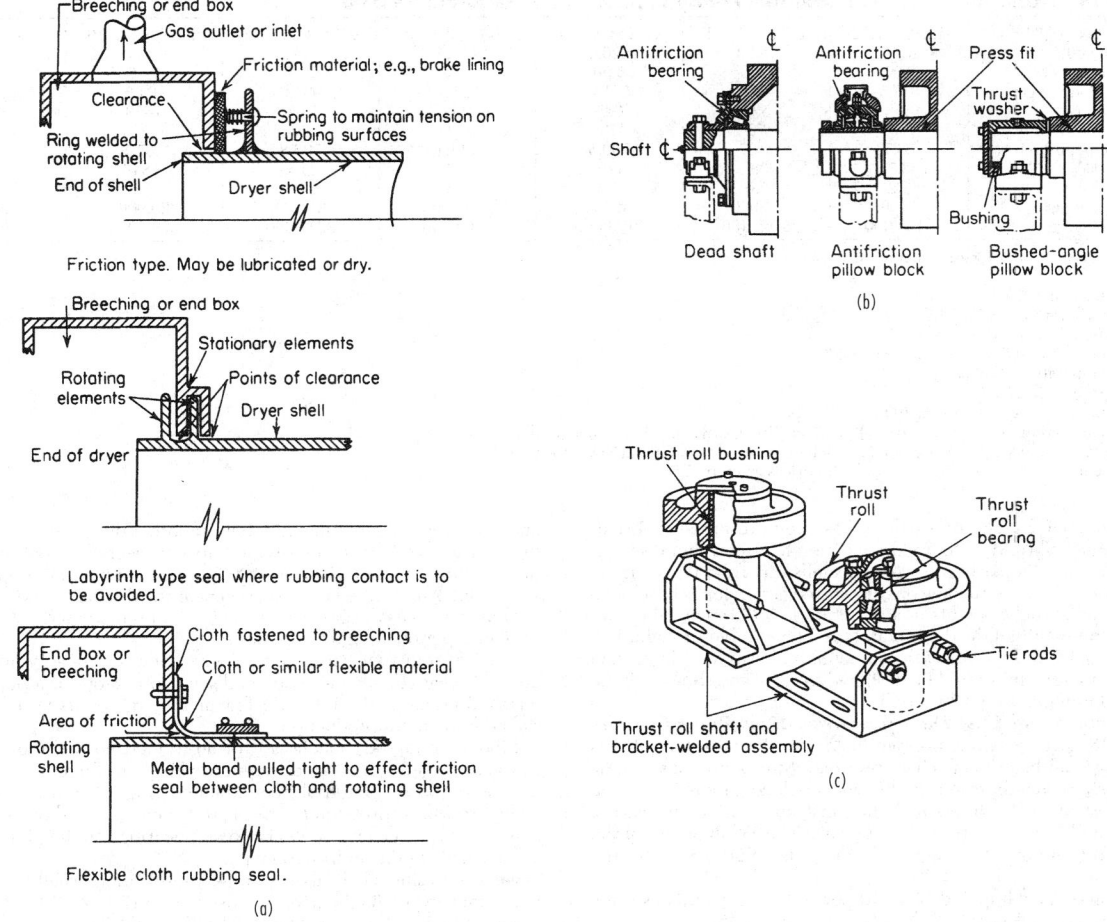

FIG. 12-61 Rotary-dryer components. (*a*) Alternative rotary gas seals. (*b*) Alternative trunnion roll bearings. (*c*) Thrust roll assembly. (*ABB Raymond/Bartlett-Snow TM.*)

Rotary dryers usually operate with 10 to 15 percent of *their volume filled with material*. Lower fillage will be insufficient to utilize the lifters fully, while greater fillage creates the possibility of short-circuiting feed solids across the top of the bed. Under normal fillage conditions, the dryer usually can be made to hold solids long enough to complete the removal of *internal moisture*. If the holdup in the dryer is not great enough, the time of passage may be too short to remove all internal moisture, or because of incomplete flight fillage performance may be erratic. The effect of fillage on retention time and uniformity in rotary dryers has been studied by Miskell and Marshall [*Chem. Eng. Prog.*, **52**, 1 (1956)].

Time of passage is defined as holdup divided by feed rate. It can be measured directly in rotary dryers if holdup and feed rate can be measured directly. Holdup cannot always be measured conveniently on large plant dryers, however, unless a period of shutdown occurs when the dryer can be discharged and its contents weighed. Other methods have been resorted to, one of which consists of adding a pound or two of an inert detectable solid or a radioisotope of a feed constituent to the feed and analyzing for it in the discharged product. The time required for the maximum concentration to occur represents the average time of passage.

The time of passage in rotary dryers can be estimated by the relationships developed by Friedman and Marshall (op. cit.), as given here:

$$\theta = \frac{0.23L}{SN^{0.9}D} \pm 0.6\frac{BLG}{F} \qquad (12\text{-}55)$$

$$B = 5(D_p)^{-0.5} \qquad (12\text{-}56)$$

where B = a constant depending upon the material being handled and approximately defined by Eq. (12-56); D_p = weight average particle size of material being handled, μm; F = feed rate to dryer, lb dry material/(h·ft² of dryer cross section); θ = time of passage, min; S = slope, ft/ft; N = speed, r/min; L = dryer length, ft; G = air-mass velocity, lb/(h·ft²); and D = dryer diameter, ft. The plus sign refers to countercurrent flow and the negative sign to cocurrent flow. To convert from British thermal units per hour to watts, multiply by 0.293; to convert from pounds per hour-square foot of cross section to kilograms per second-square meter, multiply by 0.00135.

Air-mass velocities in rotary dryers usually range from 0.5 to 5.0 kg/(s·m²). It is customary to employ the highest air velocity possible without serious dusting. The amount of dusting occurring during operation is a complex function of the material being dried, its physical state, the air velocity employed, the holdup in the dryer, the number of flights, the rate of rotation, and the construction of the breeching at the end of the dryer. It can be predicted accurately only by experimental tests. An air rate of 1.4 kg/(s·m²) can usually be safely used with 35-mesh solids. Information on the dusting of a number of

TABLE 12-18 Warm-Air Direct-Heat Cocurrent Rotary Dryers: Typical Performance Data*

Dryer size, m × m	$1.219 × 7.62$	$1.372 × 7.621$	$1.524 × 9.144$	$1.839 × 10.668$	$2.134 × 12.192$	$2.438 × 13.716$	$3.048 × 16.767$
Evaporation, kg/h	136.1	181.4	226.8	317.5	408.2	544.3	861.8
Work, 10^5 J/h	3.61	4.60	5.70	8.23	1.12	1.46	2.28
Steam, kg/h at kg/m² gauge	317.5	408.2	521.6	725.7	997.9	131.5	2041
Discharge, kg/h	408	522	685	953	1270	1633	2586
Exhaust velocity, m/min	70	70	70	70	70	70	70
Exhaust volume, m³/min	63.7	80.7	100.5	144.4	196.8	257.7	399.3
Exhaust fan, kW	3.7	3.7	5.6	7.5	11.2	18.6	22.4
Dryer drive, kW	2.2	5.6	5.6	7.5	14.9	18.6	37.3
Shipping weight, kg	7700	10,900	14,500	19,100	35,800	39,900	59,900
Price, FOB Chicago	$158,000	$168,466	$173,066	$204,400	$241,066	$298,933	$393,333

*Courtesy of Swenson Process Equipment Inc.
NOTE:
Material: heat-sensitive solid
Maximum solids temperature: 65°C
Feed conditions: 25 percent moisture, 27°C
Product conditions: 0.5 percent moisture, 65°C
Inlet-air temperature: 165°C
Exit-air temperature: 71°C
Assumed pressure drop in system: 200 mm.
System includes finned air heaters, transition piece, dryer, drive, product collector, duct, and fan.
Prices are for carbon steel construction and include entire dryer system (November, 1994).
For 304 stainless-steel fabrication, multiply the prices given by 1.5.

materials in a 0.3- by 2-m rotary dryer has been presented by Friedman and Marshall (ibid.). Rotary dryers operate at peripheral speeds of 0.25 to 0.5 m/s. Slopes of rotary-dryer shells vary from 0 to 8 cm/m. In some cases of cocurrent-flow operations, negative slopes have been used. The radial flight heights in a direct dryer will range from one-twelfth to one-eighth of the dryer diameter. The number of flights will range from $0.6 D$ to D, where D = diameter, m, for dryers larger than 0.6 m in diameter, and should be designed to carry and shower all the holdup and minimize any kiln action.

Performance and Cost Data for Direct-Heat Rotary Dryers
Table 12-18 gives estimating-price data for direct rotary dryers employing steam-heated air. Higher-temperature operations requiring combustion chambers and fuel burners will cost more. The total installed cost of rotary dryers including instrumentation, auxiliaries, allocated building space, etc., will run from 150 to 300 percent of the purchase cost. Simple erection costs average 10 to 20 percent of the purchase cost.

Operating costs will include 5 to 10 percent of one worker's time, plus power and fuel required. Yearly maintenance costs will range from 50 to 10 percent of total installed costs. Total power for fans, dryer drive, and feed and product conveyors will be in the range of 0.5 D^2 to 1.0 D^2. Thermal efficiency of a high-temperature direct-heat rotary dryer will range from 55 to 75 percent and, with steam-heated air, from 30 to 55 percent.

A representative list of materials dried in direct-heat rotary dryers is given in Table 12-19.

Direct-Heat Rotary Kilns One of the most important of the high-temperature process furnaces is the direct-fired rotary kiln. It replaces the ordinary rotary dryer when the wall temperature exceeds that which can be tolerated by a bare metal shell (650 to 700 K for carbon steel). Rotary-kiln shells are lined in part or for their entire length with a **refractory brick** to prevent overheating of the steel with resulting weakening. Occasionally two linings are used, the one next to the shell being an **insulating brick.** Insulation is infrequently used on the outside of the shell, and caution must be observed not to overheat the shell metal by this confinement. When wet feeds are applied to a kiln lining at the cold end, there may be leakage of liquid through the lining to the shell, which will cause trouble if the liquid is corrosive.

The feed is introduced into the upper end of the kiln by various methods, i.e., inclined chutes, overhung screw conveyors, slurry pipes, etc. Sometimes **ring dams** or chokes of a refractory material are installed within the kiln to build a deeper bed at one or more points, thus changing the flow pattern. The hot product is discharged from the lower end of the kiln into quench tanks, onto conveyors, or into cooling devices which may or may not recover its heat content. These cooling and heat-recovery devices include rotating inclined cylinders, inclined slow-moving grates, shaking grates, etc.

Some kilns have two or three diameters, part of the length being one diameter and the remainder being another diameter. It is claimed that this arrangement increases kiln capacity, decreases fuel consumption, and improves product quality. Two types of kilns are depicted in Fig. 12-62. An enlarged cross section near the discharge end (and hot-gas inlet) reduces the gas velocity and provides increased holdup for a "soaking" period at high temperature.

The first rotary kilns used in the United States were very small, 2 by 20 m. Sizes gradually increased and seemed to stop for a period at a maximum size of 4 by 150 m. A few much larger units have been installed for cement production.

Modern rotary-kiln shells are of all-welded construction. Riding rings are forged or cast steel; support rollers are forged or cast steel and, on rare occasions, tool steel. Main bearings are sleeve-type, normally bronze. Antifriction bearings are frequently used on very small kilns but never on large units. However, bearings on the pinion shafts are normally of the antifriction type.

Gearing is single helical or spur; gear lubrication usually is an automatic spray type. Single drives are used up to 150 kW. Kilns requiring more than 150 kW may be equipped with dual drives, i.e., two driving

TABLE 12-19 Representative Materials Dried in Direct-Heat Rotary Dryers*

Material dried	Moisture content, % (wet basis) Initial	Moisture content, % (wet basis) Final	Heat efficiency, %
High-temperature:			
Sand	10	0.5	61
Stone	6	0.5	65
Fluorspar	6	0.5	59
Sodium chloride (vacuum salt)	3	0.04	70–80
Sodium sulfate	6	0.1	60
Ilmenite ore	6	0.2	60–65
Medium-temperature:			
Copperas	7	1 (moles)	55
Ammonium sulfate	3	0.10	50–60
Cellulose acetate	60	0.5	51
Sodium chloride (grainer salt)	25	0.06	35
Cast-iron borings	6	0.5	50–60
Styrene	5	0.1	45
Low-temperature:			
Oxalic acid	5	0.2	29
Vinyl resins	30	1	50–55
Ammonium nitrate prills	4	0.25	30–35
Urea prills	2	0.2	20–30
Urea crystals	3	0.1	50–55

*Taken from *Chem. Eng.*, June 19, 1967, p. 190, Table III.

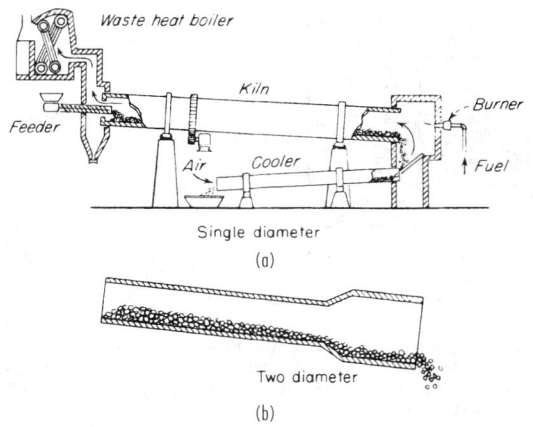

FIG. 12-62   Rotary kilns.

pinions and two motors, both driving one bull gear. In this manner, the power load is split through two separate driving mechanisms, meshing with one and the same gear.

Kiln inclination varies wtih processes from 2 to 6 cm/m. *Speed of rotation* also varies from very slow, i.e., a peripheral speed of 0.15 m/s for a TiO_2 pigment kiln, or 0.22 m/s for a cement kiln, to 0.64 m/s for a unit calcining phosphate materials.

Special features include the discharge end designed for air cooling or kilns that operate at high temperatures, such as cement, dead-burned dolomite, and magnesia. Firing hoods are designed with retractable fronts and large side doors and are mounted on wheels. *Internal heat recuperators* are of numerous designs and are becoming more popular as fuel prices increase. *Thermocouple collector rings* are placed at various points on the shell for indicating and recording internal temperatures.

Scoop systems are provided for introducing collected dust or, in some cases, a feed component through the shell at some intermediate point or points. Ports are installed in the shell for admitting combustion air at points beyond the hot zone; these are used in reducing kilns for burning carbon monoxide and volatiles from materials being processed.

Firing may be accomplished at either end, depending on whether cocurrent or countercurrent flow of the charge and gases is desired. Sometimes a solid fuel is mixed with the charge and burned as it moves down the kiln. Gaseous, liquid, or powdered fuels may be used. The burner may be installed directly at the end of the kiln with combustion occurring inside it. In this case, the discharge-end housing usually consists of a fixed or movable kiln hood through which the fuel pipe enters the kiln. A center position for the fuel pipe is used when the flame is wanted off the charge. Some users prefer an off-center

position toward the trough between the charge chord and the descending kiln lining. The kiln and the hood (combustion chamber) usually have open ends which coincide with each other with the gap being closed by a sliding seal (Fig. 12-63). Sometimes a special offset chamber for the introduction of secondary tempering air is provided on dryers and kilns (Fig. 12-64).

The exhaust gases are generally discharged into dust and fume knockdown equipment to avoid contamination of the atmosphere. Gas-cleaning equipment includes cyclones, settling chambers, scrubbing towers, and electrical precipitators. Heat-recovery devices are utilized both within and outside the kiln. These result in an increase in kiln capacity or a decrease in fuel consumption. Waste-heat boilers, grates, coil systems, and chains are used for this purpose.

The feed end of a rotary kiln is partially closed by a ring-shaped **feed head** which retains the end brick and dams back flow of solids. On the discharge end, a brick retaining-ring casting is made up to suit the application. For low temperatures, segmental alloy-iron rings may be employed. For high-temperature processes, either segmental alloy-steel rings or kiln ends of the air-cooled type are employed; the latter provides longer life for both kiln end and the brick ring.

Efficient **air seals** are essential for the controlled and economical operation of kilns. They reduce outside air entrance; certain types effectively prevent entrance of all outside air. The simplest type of air seal is a floating T-section ring mounted on a wearing pad around the feed end of the kiln and free to slide with expansion of the kiln shell. The web of the T ring is confined within circular retainer plates. Figure 12-65 shows two arrangements. The floating-type discharge-end air seal consists of a circular bar which floats on a wearing pad and which can be moved to provide the desired operating clearance between air seal and firing hood. The floating ring and the fixed portion of these seals can be furnished with renewable wearing surfaces. Air infiltration through this type of seal is usually less than 10 percent (Allis-Chalmers Corporation). For further reduction of air infiltration, lantern-ring-type floating seals, pressurized with inert gas or stack gases, are employed. Accelerated drying of slurries in the feed end of rotary kilns in wet-process operations is achieved by installation of *hanging chains.* Conveying spirals support suspended lengths of chain which are arranged in such a way that they form an effective pattern for drying. With the chain system, slurry is heated in three ways: by direct transfer from chains after suspension in hot gases, by lifting material into the path of hot gases, and by directing the flow of hot gas over the slurry bed in the space formed under the suspended chains. Frequently, the product forms into uniformly sized pellets which progress through the rest of the kiln in that form, resulting in improved heat transfer and reduced dust losses (Fig. 12-66).

Design Methods for Rotary Kilns   In rotary kilns, the material is not showered through the air stream but is retained in the lower part of the cylinder. Gas-solids contacting is much less efficient than in flighted units. Heat transfer is by radiation and convection from the flowing gas to the kiln brick and exposed bed surface and by radiation from the brick to the bed. For units employing separate combustion chambers, it can be assumed that at high temperatures the wall-film

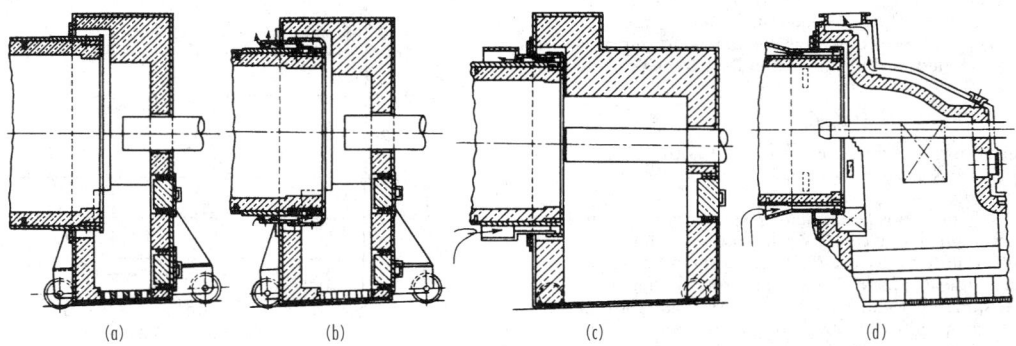

FIG. 12-63   Alternative kiln firing hoods. (*a*) Plain firing hood. (*b*) Hood for high temperatures. (*c*) Hood with enlarged combustion-air passage. (*d*) Hood with cooler connection. (*Allis-Chalmers Corporation.*)

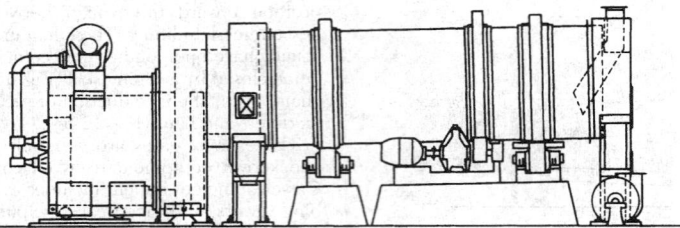

FIG. 12-64 Dryer firing hood with air-tempering chamber. (*ABB Raymond/Bartlett-Snow TM.*)

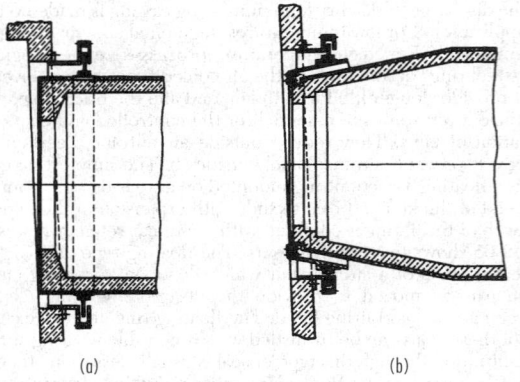

FIG. 12-65 Kiln-seal arrangements. (*a*) Single-floating-type feed end air seal. (*b*) Single-floating-type air seal on an air-cooled tapered feed end. (*Allis-Chalmers Corp.*)

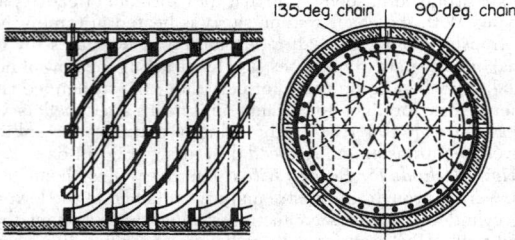

FIG. 12-66 Kiln chain installation (patented). (*Allis-Chalmers Corp.*)

resistance to convection heat transfer from the gas to the brick is limiting and that at any point the bed temperature approaches the wall temperature. Hence, the effective heat-transfer area is the inner kiln surface. For kilns under these conditions, the following empirical relationship is recommended for the convection heat-transfer coefficient from gas to brick:

$$Us = 23.7G^{0.67} \qquad (12\text{-}57)$$

where Us = heat-transfer coefficient, J/(m²·s·K) [(Btu/(h·ft² kiln surface·°F)]; and G = gas mass flow rate, kg/(s·m² kiln cross section) [lb/(h·ft²)].

Equation (12-57) does not account for gas radiation at high temperature when the kiln charge can "see" the burner flame; hence, the method will yield a conservative design. When a kiln is fired internally, the major source of heat transfer is radiation from the flame and hot gases. This occurs directly to both the solids surface and the wall, and from the latter to the product by reradiation (with some conduction).

Generally, a dry-feed kiln will have three zones of heating, and a wet-feed kiln will have four:

1. Drying zone at feed end, when moisture is removed
2. Heating zone, where the charge is heated to the reaction temperature, i.e., the decomposition temperature for limestone or "burning" temperature for cement
3. Reaction zone, in which the charge is burned, decomposed, reduced, oxidized, etc.
4. Soaking zone, where the reacted charge is superheated or "soaked" at temperature or, if desired, cooled before discharge

The rates of heat transfer in each zone will be different.

Rotary kilns operate at various temperatures throughout their length. A graph of approximate gas and charge temperatures for wet-process cement is shown in Fig. 12-67. The maximum charge temperature is 1700 to 1800 K; for the gases, 1800 to 1925 K. Overall heat-transfer rates have been estimated to be in the range of 25 to 60 KJ/(s·m³) on the basis of total kiln volume.

Some commercial performance data for cement and lime kilns are shown in Table 12-20.

Some of the other major uses of direct rotary kilns are in the following processes:

Roasting. Rotary kilns are used for oxidizing and driving off sulfur and arsenic from various ores, including gold, silver, iron, etc. Temperatures employed will vary from 800 to 1600 K.

Chloridizing. Silver ores are chloridized successfully in rotary kilns. Temperatures must be closely controlled between 1030 and 1090 K.

Black ash. Barium sulfide (BaS) is produced by calcining a mixture of barite (BaSO₄) and carbon at a temperature of 1350 K in continuous rotary kilns.

Spodumene. A mixture of quartz, feldspar, and spodumene is being calcined in rotary kilns at 1475 K to produce lithium aluminum silicate.

Vermiculite. A micaceous mineral is roasted to cause exfoliation for use as an insulating material.

Revivification. Temperatures of 800 to 1030 K are used to revivify fuller's and diatomaceous earth, although for some earths lower temperatures are employed.

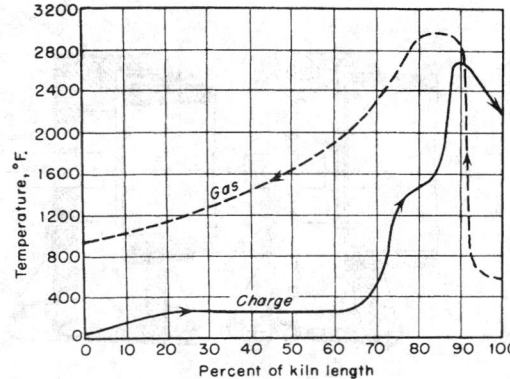

FIG. 12-67 Temperatures in rotary kiln on wet-process cement.

TABLE 12-20 **Typical Rotary-Kiln Installations***

Size, diam. × length	Usual No. of supports	Range of motor hp. to operate†	Nominal 24-hr. capacities‡			
			Portland cement, 376-lb. bbl.		Lime, net tons	
			Dry process	Wet process	Lime sludge	Limestone
5 × 80 ft.	2	5–7.5	140	100	10	16
6 × 70 ft.	2	7.5–15	190	135	15	24
7 × 70 ft.	2	15–20	275	200	20	35
5 ft. 6 in. × 180 ft.	4	15–20	285	250	30	45
7 × 120 ft.	2	15–25	475	340	35	55
7 ft. 6 in. × 125 ft.	2	20–30	575	415	40	70
6 × 220 ft.	4	20–30	420	375	45	65
8 × 140 ft.	2	25–30	750	540	55	90
9 × 160 ft.	2	30–50	1100	800	80	130
8 ft. 6 in. × 185 ft.	4	30–50	1125	810	80	135
10 × 150 ft.	2	40–75	1300	950	—	145
10 × 175 ft.	2	50–75	1500	1100	—	155
8 × 300 ft.	5	50–75	1150	1000	110	160
7 ft. 6 in. × 8 ft. 6 in. × 320 ft.	5	50–75	1175	1020	115	165
7 ft. 6 in. × 10 ft. × 8 ft. 6 in. × 300 ft.	5	50–75	1175	1020		
10 × 11 × 175 ft.	2	50–75	1650	1200	120	180
10 ft. 6 in. × 185 ft.	2	50–75	1800	1300	130	190
11 × 175 ft.	2	60–100	1850	1375	—	205
8 ft. 6 in. × 10 ft. × 8 ft. 6 in. × 300 ft.	5	50–75	1400	1200		
8 × 10 × 300 ft.	5	50–75	1425	1225	140	200
9 ft. 6 in. × 265 ft.	4	60–100	1500	1300	150	215
9 × 10 ft. 6 in. × 9 × 325 ft.	5	60–100	1700	1500		
10 ft. 6 in. × 250 ft.	4	60–100	1750	1525	175	240
9 ft. 6 in. × 11 ft. × 9 ft. 6 in. × 300 ft.	5	60–100	1800	1550		
10 × 300 ft.	5	75–125	1900	1650	190	250
9 ft. 6 in. × 11 ft. × 9 ft. 6 in. × 375 ft.	6	75–125	2025	1800		
11 × 300 ft.	5	75–125	2400	2100	225	300
11 ft. 6 in. × 300 ft.	4	100–150	2600	2250	240	320
10 ft. 6 in. × 375 ft.	5	100–150	2700	2400	250	325
11 ft. 3 in. × 360 ft.	5	125–175	2900	2500	275	350
11 ft. 6 in. × 475 ft.	7	150–250	4000	3500	375	450
12 × 500 ft.	8	200–300	4600	4000	425	500

*Allis-Chalmers Manufacturing Co.

†Power requirements vary according to size of kiln, character of material handled, and method of operation.

‡Capacities indicated are conservative, and apply to normal operation at sea level. Corrections would apply at increased altitudes, and for differing methods of operation.

Zinc. Oxidized ores are calcined to drive off water of hydration and carbon dioxide. The sulfide ore is always roasted before smelting.

Titanium oxide (TiO₂). This is produced from ilmenite ore by mixing ore with carbon and heating in a rotary kiln. Also, the rotary kiln is used in the process of recovery of titanium oxide from hydrated titanium precipitate at about 1250 K.

Roofing granules. Crushed quartz or sand of definite size is treated with various minerals, borax, soda ash, etc., and calcined at temperatures ranging from 1250 to 1600 K. Glass of different colors forms on the surface of the granules at various temperatures. An oxidizing or reducing flame is used to influence the final coloring.

Alumina (Al₂O₃). Alumina is produced by calcining either bauxite or aluminum hydroxide in rotary kilns at temperatures from 1250 to 1600 K. In obtaining the highest-purity alumina, the bauxite is digested with alkali to remove impurities; the resultant aluminum hydroxide [Al₂(OH)₃], of approximately 200-mesh size, is then calcined in rotary kilns at 1350 K.

Potassium salts. In this operation, potassium chloride (KCl) is introduced to the rotary kiln at a fineness of minus 100 mesh and containing 9 percent water. The salt is brought to the fusion temperature of 1048 K.

Magnesium oxide. The natural minerals, i.e., magnesite (MgCO₃), brucite [Mg(OH)₂], etc., after being crushed to predetermined size, are calcined at temperatures varying from 1055 to 2000 K, depending upon whether a caustic or a dead-burned product (periclase) is being produced. Magnesium hydroxide, recovered from seawater or salt brine, is also being treated in a similar manner except that it is added in the form of a sludge.

Sodium aluminum sulfate. This product is now being successfully calcined in rotary kilns. In this process, the salt cake is broken up just before it enters the kiln. Calcination is for the purpose of driving off the combined water (45 percent) and sulfuric acid (3 percent). Temperatures employed are approximately 800 K.

Phosphate rock. In this application, the rotary kiln is used to nodulize the fines in the ore and prepare them for electric-furnace operation. Ore under 5 cm in size and containing 50 percent or more minus 100 mesh is calcined. Ore nodulizes at approximately 1475 to 1500 K.

Mercury. In recovering mercury from cinnabar ores, the ore is crushed to minus 1.5 cm and fed to rotary kilns, where it is calcined to over 800 K. Since the mercury exists as mercuric sulfide (HgS), the sulfur is oxidized to SO₂ and the mercury vaporized. The gases are passed through cooling chambers, where the mercury condenses and is collected. Mercury vaporizes at 625 K.

Gypsum. The rotary kiln is rapidly replacing the kettle in producing plaster of paris. Great care is required, as the temperatures for reaction are low and within narrow limits, 382 to 403 K. Gypsum (CaSO₄·2H₂O) is heated to drive off three-fourths of the water of crystallization to produce plaster of paris [(CaSO₂)·H₂O]. Any overheating drives off all the water, producing gypsite (CaSO₄), which is unsatisfactory.

Clay. To produce lightweight aggregate for concrete, clay is calcined in rotary kilns. Temperatures employed vary from 1350 to 1600 K. The apparent density of the clay is reduced by 50 to 75 percent.

Iron ores. Crushed iron ores are partially reduced in rotary kilns to obtain nodules which are used in blast-furnace charges.

Manganese. Manganese ore, rhodochrosite, or manganese carbonate ($MnCO_2$) is calcined at about 1525 K to produce the oxide (Mn_3O_4). When the oxide ore is available but is in a finely divided state, the rotary kiln is used only for nodulizing.

Petroleum coke. In order to eliminate excess volatile matter, petroleum coke is calcined at temperatures of 1475 to 1525 K. This is a sensitive material, and temperature control is difficult to maintain.

When it is desired to increase the capacity of an existing kiln installation, consideration should be given to the following changes:

1. Increase charge volume held in kiln.
2. Increase temperature and quantity of combustion gases.
3. Decrease quantity of air in excess of combustion needs.
4. Increase speed of rotation of kiln.
5. Install ring dams at intermediate and discharge points.
6. Increase capacity of feeding and discharge mechanisms.
7. Decrease moisture content of feed material.
8. Increase temperature of feed material.
9. Install chains or flights, etc., in feed end.
10. Preheat all combustion air.
11. Reduce leakage of cold air into kiln at hot end.
12. Increase stack draft by increasing height or by use of jets.
13. Install instrumentation to control the kiln at maximum-capacity conditions.

The time of passage in rotary kilns (from which holdup can be calculated) can be estimated by the following formula (U.S. Bur. Mines Tech. Pap. 384, 1927):

$$\theta = 0.19L/NDS \qquad (12\text{-}58)$$

where θ = time of passage in the kiln, min; L = kiln length, ft; N = rotational speed, r/min; S = slope of kiln, ft/ft; and D = diameter inside brick, ft. Other equations for estimating the time of passage employing internal dams and a discharge dam are given by Bayard [*Chem. Metall. Eng.*, **52**(3), 100–102 (1945)].

The total power required to drive a rotary kiln or a dryer with lifters can be calculated by the following formulas (courtesy of ABB Raymond). For a rotary kiln or calciner without lifters,

$$\text{bhp} = \frac{N\,[18.85y\,(\sin B)w + 0.1925DW + 0.33W]}{100{,}000} \qquad (12\text{-}59)$$

For a rotary dryer or section of a kiln with lifters,

$$\text{bhp} = \frac{N\,(4.75dw + 0.1925DW + 0.33W)}{100{,}000} \qquad (12\text{-}60)$$

where bhp = brake horsepower required (1 bhp = 0.75 kW); N = rotational speed, r/min; y = distance between centerline of kiln and the center of gravity of material bed, ft; B = angle of repose of material; W = total rotating load (equipment plus material), lb; w = live load (material), lb; D = riding-ring diameter, ft; and d = shell diameter, ft. For estimating purposes, let $D = (d + 2)$.

Drive motors should be of the high-starting-torque type and selected for 1.33 times maximum rotational speed. For two- or three-diameter kilns, the brake horsepower for the several diameters should be calculated separately and summed. Auxiliary drives should be provided to maintain shell rotation in the event of power failure. These are usually gasoline or diesel engines.

Thermal Efficiency of Rotary Kilns   Kiln length is a major factor in determining thermal efficiency, and kilns with a high ratio of length to diameter have a greater thermal efficiency than those with a low ratio. The use of chains inside the kiln and of heat-recovery equipment on the gases and product leaving the kiln can increase substantially the thermal efficiency of a kiln installation. Efficiencies ranging from 45 to more than 80 percent have been reported. A reasonably satisfactory range based on present fuel prices and construction costs would be 65 to 75 percent utilization and recovery of the heat content of the fuel plus any heat of reaction of the charge. No distinction is made from an efficiency-calculation standpoint between the heat uti-

lized in the kiln and that recovered (or utilized) outside the kiln. With countercurrent flow of the combustion gases and the charge material, an exceptionally long kiln will give high efficiencies within itself. However, good economics may dictate that a shorter kiln be installed with a waste-heat boiler on the hot gases to obtain an equivalent thermal efficiency at a lower investment. The heat in the hot product usually is recovered as preheat in the combustion air.

Size Segregation in Kilns   When an assemblage of solid particles, not very closely screened, is rotated within a cylinder, the solids assume a lunar shape, as shown in Fig. 12-68. This causes serious size segregation. The finest sizes remain at the bottom, in contact with the hot brick. The coarser particles form the upper layer of the agitated mass. As the kiln completes a revolution, the exposed brick, in an upper position, absorbs sensible heat from the gas mass. As the heated brick completes its circuit, it passes under and is in conductive contact with the fine particles. These fines are thus effectively heated by direct solid-to-solid transfer. The larger particles are heated by direct radiation from gas and brick, and become adequately calcined. The particles of size intermediate between the fine and coarse remain, throughout a complete revolution, "sandwiched" between the coarse and fine layers and are protected from heat by the excellent insulation properties of these layers, thus perhaps escaping complete calcination. This factor of segregation is offset by some kiln operators who classify or screen the kiln feed so that only a narrow range of particle size is fed at one time. Also, faster kiln speeds which give a better agitation of the charge are used.

Rotary kilns are usually operated with between 3 and 12 percent of their volume filled with material; 7 percent is considered normal.

Cost Data for Kilns   Purchase prices, weights, and horsepower requirements of typical units are given in Table 12-21. Installed costs will run to from 300 to 500 percent of purchase cost. Maintenance will average 5 to 10 percent of the total installed cost per year but is dependent largely on the life of the refractory lining.

For estimating purposes refractory lining for a 2.7- to 3.4-m-diameter kiln costs $6000 to $15,000 per meter of kiln length installed (50 percent material, 50 percent labor).

A discussion of retention time in rotary kilns is given in *Brit. Chem. Eng.*, 27–29 (January 1966). Rotary-kiln heat control is discussed in detail by Bauer [*Chem. Eng.*, 193–200 (May 1954)] and Zubrzycki [*Chem. Can.*, 33–37 (February 1957)]. Reduction of iron ore in rotary kilns is described by Stewart [*Min. Congr. J.*, 34–38 (December 1958)]. The use of balls to improve solids flow is discussed in [*Chem. Eng.*, 120–222 (March 1956)]. Brisbane examined problems of shell deformation [(*Min. Eng.*, 210–212 (February 1956)]. Instrumentation is discussed by Dixon [*Ind. Eng. Chem. Process Des. Dev.*, 1436–1441 (July 1954)], and a mathematical simulation of a rotary kiln was developed by Sass [*Ind. Eng. Chem. Process Des. Dev.*, 532–535 (October 1967)]. This last paper employed the empirical convection heat-transfer coefficient given previously, and its use is discussed in later correspondence [ibid., 318–319 (April 1968)].

Indirect-Heat Rotary Steam-Tube Dryers   Probably the most common type of indirect-heat rotary dryer is the steam-tube dryer (Fig. 12-69). Steam-heated tubes running the full length of the cylinder are fastened symmetrically in one, two, or three concentric rows inside the cylinder and rotate with it. Tubes may be simple pipe with condensate draining by gravity into the discharge manifold or bayonet-type. Bayonet-type tubes are also employed when units are used

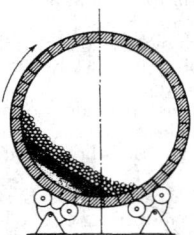

FIG. 12-68   Size segregation of solids in a rotary kiln.

TABLE 12-21   Approximate Purchase Costs and Weight of Rotary Kilns*

Kiln size, diameter × length	Total purchase price includes drive, burner, and controls (not including brick)†
8'0" × 80'0"	$ 448,000
8'0" × 140'0"	600,000
8'0" × 200'0"	960,000
8'0" × 300'0"	1,240,000
9'0" × 250'0"	1,373,000
9'0" × 300'0"	1,545,000
10'0" × 100'0"	682,000
10'0" × 150'0"	965,000
10'0" × 250'0"	1,502,000
10'0" × 300'0"	1,682,000
10'0" × 350'0"	1,779,000
10'6" × 175'0"	1,182,000
10'6" × 250'0"	1,677,000
10'6" × 350'0"	1,942,000
11'0" × 160'0"	1,213,000
11'0" × 250'0"	1,670,000
11'0" × 300'0"	1,858,000
11'0" × 350'0"	2,344,000
11'0" × 400'0"	2,544,000
11'6" × 160'0"	1,251,000
11'6" × 250'0"	1,768,000
11'6" × 350'0"	2,393,000
11'6" × 425'0"	2,676,000
12'0" × 250'0"	1,837,000
12'0" × 325'0"	2,598,000
12'0" × 400'0"	2,645,000
12'0" × 450'0"	3,570,000
13'0" × 500'0"	4,388,000
14'0" × 400'0"	4,155,000
16'6" × 600'0"	8,190,000

*Courtesy of Fuller Co.
†Prices for January 1982

as water-tube coolers. When handling sticky materials, one row of tubes is preferred. These are occasionally shielded at the feed end of the dryer to prevent buildup of solids behind them. Lifting flights are usually inserted behind the tubes to promote solids agitation.

Wet feed enters the dryer through a chute or screw feeder. The product discharges through peripheral openings in the shell in ordinary dryers. These openings also serve to admit purge air to sweep moisture or other evolved gases from the shell. In practically all cases, gas flow is countercurrent to solids flow. To retain a deep bed of material within the dryer, normally 10 to 20 percent fillage, the discharge openings are supplied with removable chutes extending radially into the dryer. These, on removal, permit complete emptying of the dryer.

Steam is admitted to the tubes through a revolving steam joint into the steam side of the manifold (Fig. 12-70). Condensate is removed continuously, by gravity through the steam joint to a condensate receiver and my means of lifters in the condensate side of the manifold. By employing simple tubes, noncondensables are continuously vented at the other ends of the tubes through Sarco-type vent valves mounted on an auxiliary manifold ring, also revolving with the cylinder.

Vapors (from drying) are removed at the feed end of the dryer to the atmosphere through a natural-draft stack and settling chamber or wet scrubber. When employed in simple drying operations with 3.5×10^5 to 10×10^5 Pa steam, draft is controlled by a damper to admit only sufficient outside air to sweep moisture from the cylinder, discharging the air at 340 to 365 K and 80 to 90 percent saturation. In this way, shell gas velocities and dusting are minimized. When used for solvent recovery or other processes requiring a sealed system, sweep gas is recirculated through a scrubber-gas cooler and blower.

Steam manifolds for pressures up to 10×10^5 Pa are of cast iron. For higher pressures, the manifold is fabricated from plate steel, staybolted, and welded. The tubes are fastened rigidly to the manifold face plate and are supported in a close-fitting annular plate at the other end to permit expansion. Packing on the steam neck is normally graphite-asbestos. Ordinary rotating seals are similar in design to those depicted in Fig. 12-61, with allowance for the admission of small quantities of outside air when the dryer is operated under a slight negative internal pressure.

Steam-tube dryers are used for the continuous drying, heating, or cooling of granular or powdery solids which cannot be exposed to ordinary atmospheric or combustion gases. They are especially suitable for fine dusty particles because of the low gas velocities required for purging of the cylinder. Tube sticking is avoided or reduced by employing recycle, shell knockers, etc., as previously described; tube scaling by sticky solids is one of the major hazards to efficient operation. The dryers are suitable for drying, solvent recovery, and chemical reactions. Steam-tube units have found effective employment in

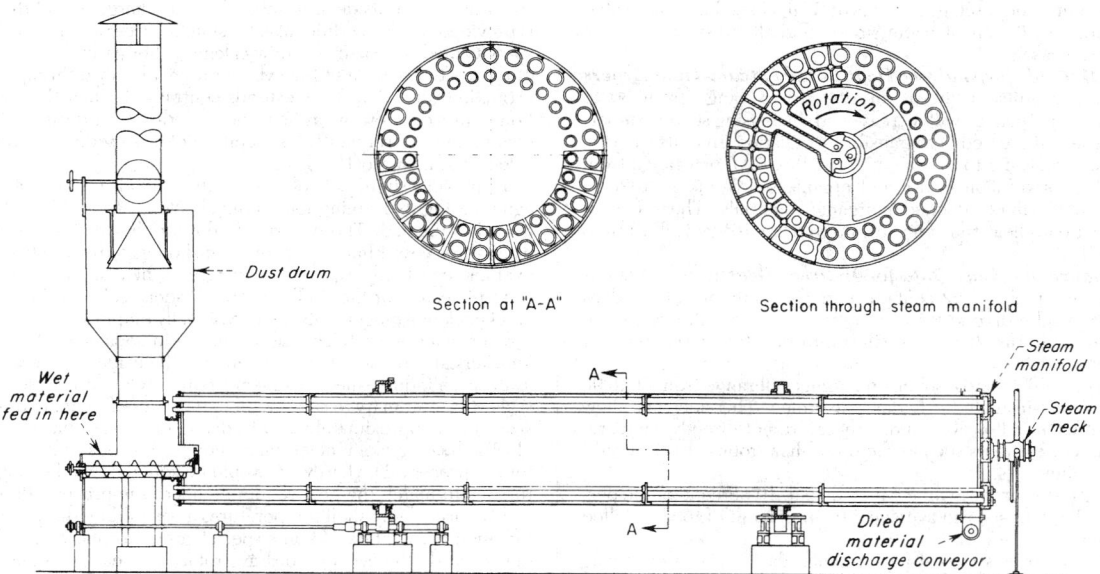

Section at "A-A"

Section through steam manifold

FIG. 12-69   Steam-tube rotary dryer.

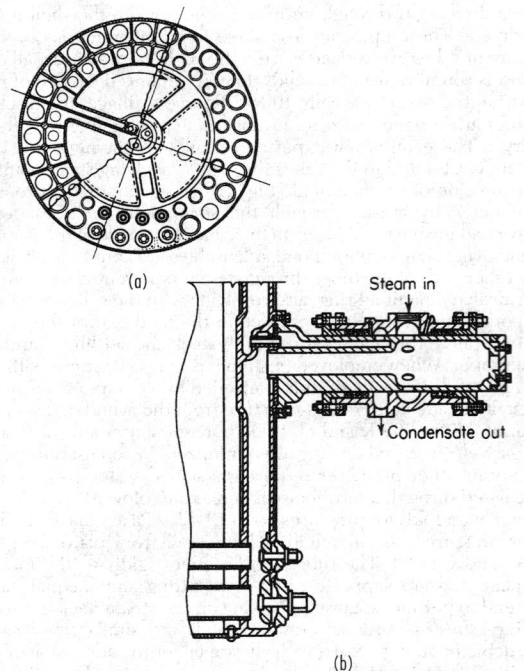

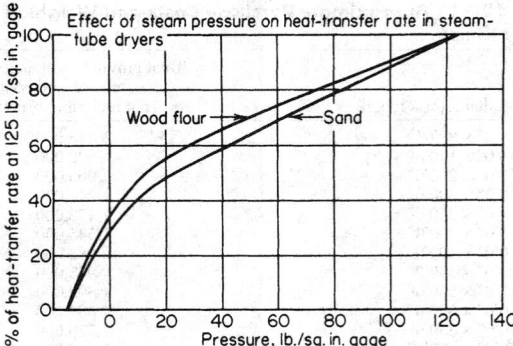

FIG. 12-71   Effect of steam pressure on the heat-transfer rate in steam-tube dryers.

FIG. 12-70   Rotary steam joint for a standard steam-tube dryer. (*a*) Section of cast-steam manifold. (*b*) Section of manifold and steam joint.

soda-ash production, replacing more expensive indirect-heat rotary calciners.

Special types of steam-tube dryers employ packed and purged seals on all rotating joints, with a central solids-discharge manifold through the steam neck to reduce the seal diameter. This manifold contains the product discharge conveyor and a passage for the admission of sweep gas. Solids are removed from the shell by special volute lifters and dropped into the discharge conveyor. Units have been fabricated for operation at 76 mm of water, internal shell pressure, with no detectable air leakage.

Design Methods for Indirect-Heat Rotary Steam-Tube Dryers
Heat-transfer coefficients in steam-tube dryers range from 30 to 85 J/(m²·s·K). Coefficients will increase with increasing steam temperature because of increased heat transfer by radiation. In units carrying saturated steam at 420 to 450 K, the heat flux $U\Delta T$ will range from 6300 J/(m²·s) for difficult-to-dry and organic solids and to 1890 to 3790 J/(m²·s) for finely divided inorganic materials. The effect of steam pressure on heat-transfer rates up to 8.6×10^5 Pa is illustrated in Fig. 12-71.

Performance and Cost Data for Indirect-Heat Rotary Steam-Tube Dryers   Table 12-22 contains data for a number of standard sizes of steam-tube dryers. Prices tabulated are for ordinary carbon steel construction. Installed costs will run from 150 to 300 percent of purchase cost.

The thermal efficiency of steam-tube units will range from 70 to 90 percent, if a well-insulated cylinder is assumed. This does not allow for boiler efficiency, however, and is therefore not directly comparable with direct-heat units such as the direct-heat rotary dryer or indirect-heat calciner.

Operating costs for these dryers include 5 to 10 percent of one person's time. Maintenance will average 5 to 10 percent of total installed cost per year.

Table 12-23 outlines typical performance data from three drying applications in steam-tube dryers.

Indirect-Heat Calciners   Indirect-heat rotary calciners, either batch or continuous, are employed for heat treating and drying at higher temperatures than can be obtained in steam-heated rotating equipment. They generally require a minimum flow of gas to purge the cylinder, to reduce dusting and are suitable for gas-sealed operation with oxidizing, inert, or reducing atmospheres. Indirect calciners are widely utilized and some examples of specific applications are as follows:
1.   Activating charcoal
2.   Reducing mineral high oxides to low oxides
3.   Drying and devolatilizing contaminated soils and sludges
4.   Calcination of alumina oxide-based catalysts
5.   Drying and removal of sulfur from cobalt, copper, and nickel
6.   Reduction of metal oxides in a hydrogen atmosphere
7.   Oxidizing and "burning off" of organic impurities
8.   Calcination of ferrites

This unit consists essentially of a cylindrical retort, rotating within a stationary insulation lined furnace. The latter is arranged so that fuel combustion occurs within the annular ring between the retort and the furnace. The retort cylinder extends beyond both ends of the furnace. These end extensions carry the riding rings and drive gear. Material may be fed continuously at one end and discharged continuously at the other. Feeding and solids discharging are usually accomplished with screw feeders or other positive feeders to prevent leakage of gases into or out of the calciner.

In some cases in which it is desirable to cool the product before removal to the outside atmosphere, the discharge end of the cylinder is provided with an additional extension, the exterior of which is water-spray-cooled. In cocurrent-flow calciners, hot gases from the interior of the heated portion of the cylinder are withdrawn through a special extraction tube. This tube extends centrally through the cooled section to prevent flow of gas near the cooled-shell surfaces and possible condensation. Frequently a separate cooler is used, isolated from the calciner by an air lock.

To prevent sliding of solids over the smooth interior of the shell, agitating flights running longitudinally along the inside wall are frequently provided. These normally do not shower the solids as in a direct-heat vessel but merely prevent sliding so that the bed will turn over and constantly expose new surface for heat and mass transfer. To prevent scaling of the shell interior by sticky solids, cylinder scraper and knocker arrangements are occasionally employed. For example, a scraper chain is fairly common practice in soda-ash calciners, while knockers are frequently utilized on metallic-oxide calciners.

Because indirect-heat calciners frequently require close-fitting gas seals, it is customary to support all parts on a self-contained frame, for sizes up to approximately 2 m in diameter. The furnace can employ electric heating elements or oil and/or gas burners as the heat source for the process. The hardware would be zoned down the length of the furnace to match the heat requirements of the process. Process control is normally by shell temperature, measured by thermocouples or radiation pyrometers. When a special gas atmosphere must be maintained inside the cylinder, positive rotary gas seals, with one or more pressurized and purged annular chambers, are employed. The diaphragm-type seal ABB Raymond (Bartlett-Snow TM) is suitable for pressures up to 5 cm of water, with no detectable leakage.

TABLE 12-22 Standard Steam-Tube Dryers*

Size, diameter × length, m	Tubes			m² of free area	Dryer speed, r/min	Motor size, hp	Shipping weight, kg	Estimated price
	No. OD (mm)		No. OD (mm)					
0.965 × 4.572	14 (114)			21.4	6	2.2	5,500	$152,400
0.965 × 6.096	14 (114)			29.3	6	2.2	5,900	165,100
0.965 × 7.620	14 (114)			36.7	6	3.7	6,500	175,260
0.965 × 9.144	14 (114)			44.6	6	3.7	6,900	184,150
0.965 × 10.668	14 (114)			52.0	6	3.7	7,500	196,850
1.372 × 6.096	18 (114)		18 (63.5)	58.1	4.4	3.7	10,200	203,200
1.372 × 7.620	18 (114)		18 (63.5)	73.4	4.4	3.7	11,100	215,900
1.372 × 9.144	18 (114)		18 (63.5)	88.7	5	5.6	12,100	228,600
1.372 × 10.668	18 (114)		18 (63.5)	104	5	5.6	13,100	243,840
1.372 × 12.192	18 (114)		18 (63.5)	119	5	5.6	14,200	260,350
1.372 × 13.716	18 (114)		18 (63.5)	135	5.5	7.5	15,000	273,050
1.829 × 7.62	27 (114)		27 (76.2)	118	4	5.6	19,300	241,300
1.829 × 9.144	27 (114)		27 (76.2)	143	4	5.6	20,600	254,000
1.829 × 10.668	27 (114)		27 (76.2)	167	4	7.5	22,100	266,700
1.829 × 12.192	27 (114)		27 (76.2)	192	4	7.5	23,800	278,400
1.829 × 13.716	27 (114)		27 (76.2)	217	4	11.2	25,700	292,100
1.829 × 15.240	27 (114)		27 (76.2)	242	4	11.2	27,500	304,800
1.829 × 16.764	27 (114)		27 (76.2)	266	4	14.9	29,300	317,500
1.829 × 18.288	27 (114)		27 (76.2)	291	4	14.9	30,700	330,200
2.438 × 12.192	90 (114)			394	3	11.2	49,900	546,100
2.438 × 15.240	90 (114)			492	3	14.9	56,300	647,700
2.438 × 18.288	90 (114)			590	3	14.9	63,500	736,600
2.438 × 21.336	90 (114)			689	3	22.4	69,900	838,200
2.438 × 24.387	90 (114)			786	3	29.8	75,300	927,100

*Courtesy of Swenson Process Equipment Inc. (prices from November, 1994). Carbon steel fabrication; multiply by 1.75 for 304 stainless steel.

TABLE 12-23 Steam-Tube Dryer Performance Data

	Class 1	Class 2	Class 3
Class of materials handled	High-moisture organic, distillers' grains, brewers' grains, citrus pulp	Pigment filter cakes, blanc fixe, barium carbonate, precipitated chalk	Finely divided inorganic solids, water-ground mica, water-ground silica, flotation concentrates
Description of class	Wet feed is granular and damp but not sticky or muddy and dries to granular meal	Wet feed is pasty, muddy, or sloppy; product is mostly hard pellets	Wet feed is crumbly and friable; product is powder with very few lumps
Normal moisture content of wet feed, % dry basis	233	100	54
Normal moisture content of product, % dry basis	11	0.15	0.5
Normal temperature of wet feed, K	310–320	280–290	280–290
Normal temperature of product, K	350–355	380–410	365–375
Evaporation per product, kg	2	1	0.53
Heat load per lb product, kJ	2250	1190	625
Steam pressure normally used, kPa gauge	860	860	860
Heating surface required per kg product, m²	0.34	0.4	0.072
Steam consumption per kg product, kg	3.33	1.72	0.85

In general, the temperature range of operation for indirect-heated calciners can vary over a wide range, from 475 K at the low end to approximately 1475 K at the high end. All types of carbon steel, stainless, and alloy construction are used, depending upon temperature, process, and corrosion requirements. Fabricated-alloy cylinders can be used over the greater part of the temperature range; however, the greater creep-stress abilities of cast alloys makes their use desirable for the highest calciner-cylinder temperature applications.

Design Methods for Calciners In indirect-heated calciners, heat transfer is primarily by radiation from the cylinder wall to the solids bed. The thermal efficiency ranges from 30 to 65 percent. By utilization of the furnace exhaust gases for preheated combustion air, steam production, or heat for other process steps, the thermal efficiency can be increased considerably. The limiting factors in heat transmission lie in the conductivity and radiation constants of the shell metal and solids bed. If the characteristics of these are known, equipment may be accurately sized by employing the Stefan-Boltzmann radiation equation. Apparent heat-transfer coefficients will range from 17 J/(m²·s·K) in low-temperature operations to 85 J/(m²·s·K) in high-temperature processes.

Cost Data for Calciners Power, operating, and maintenance costs are similar to those previously outlined for direct- and indirect-heat rotary dryers. Estimating purchase costs for preassembled and frame-mounted rotary calciners with carbon steel and type 316 stainless-steel cylinders are given in Table 12-24 together with size, weight, and motor requirements. Sale price includes the cylinder, ordinary angle seals, furnace, drive, feed conveyor, burners, and controls. Installed cost may be estimated, not including building or foundation costs at up to 50 percent of the purchase cost. A layout of a typical continuous calciner with an extended cooler section is illustrated in Fig. 12-72.

Small batch retorts, heated electrically or by combustion, are widely used as carburizing furnaces and are applicable also to chemical processes involving the heat treating of particulate solids. These are mounted on a structural-steel base, complete with cylinder, furnace, drive motor, burner, etc. Units are commercially available in diameters from 0.24 to 1.25 m and lengths of 1 to 2 m. Continuous retorts with helical internal spirals are employed for metal-heat-treating purposes. Precise retention control is maintained in these operations. Standard diameters are 0.33, 0.5, and 0.67 m with effective lengths up

TABLE 12-24 Indirect-Heat Rotary Calciners: Sizes and Purchase Costs*

Diameter, ft	Overall cylinder length	Heated cylinder length	Cylinder drive motor hp	Approximate Shipping weight, lb	Approximate sale price in carbon steel construction†	Approximate sale price in No. 316 stainless construction
4	40 ft	30 ft	7.5	50,000	$275,000	$325,000
5	45 ft	35 ft	10	60,000	375,000	425,000
6	50 ft	40 ft	20	75,000	475,000	550,000
7	60 ft	50 ft	30	90,000	550,000	675,000

* ABB Raymond (Bartlett-Snow™).
† Prices for November, 1994.

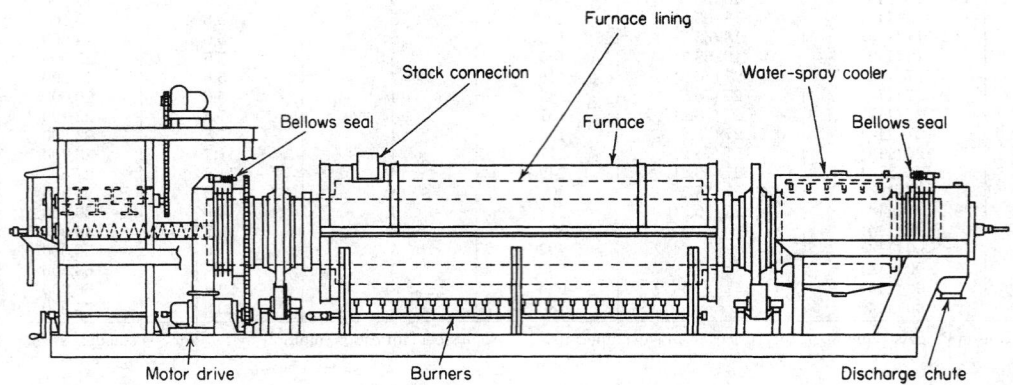

FIG. 12-72 Gas-fired indirect-heat rotary calciner with a water-spray extended cooler and feeder assembly. (*ABB Raymond/Bartlett-Snow™.*)

to 3 m. These vessels are employed in many small-scale chemical-process operations which require accurate control of retention. Their operating characteristics and applications are identical to those of the larger indirect-heat calciners.

Direct-Heat Roto-Louvre Dryer One of the more important special types of rotating equipment is the Roto-Louvre dryer. As illustrated in Fig. 12-73, hot air (or cooling air) is blown through louvers in a double-wall rotating cylinder and up through the bed of solids. The latter moves continuously through the cylinder as it rotates. Constant turnover of the bed ensures uniform gas contacting for heat and mass transfer. The annular gas passage behind the louvers is partitioned so that contacting air enters the cylinder only beneath the solids bed. The number of louvers covered at any one time is roughly 30 percent. Because air circulates through the bed, fillages of 13 to 15 percent or greater are employed.

Roto-Louvre dryers range in size from 0.8 to 3.6 m in diameter and from 2.5 to 11 m long. The largest unit is reported capable of evaporating 5500 kg/h of water. Hot gases from 400 to 865 K may be employed. Because gas flow is through the bed of solids, high pressure drop, from 7 to 50 cm of water, may be encountered within the shell. For this reason, both a pressure inlet fan and an exhaust fan are provided in most applications to maintain the static pressure within the equipment as closely as possible to atmospheric. This prevents excessive in-leakage or blowing of hot gas and dust to the outside. For pressure control, one fan is usually operated under fixed conditions, with an automatic damper control on the other, regulated by a pressure detector-controller.

In heating or drying applications, when cooling of the product is desired before discharge to the atmosphere, cool air is blown through a second annular space, outside the inlet hot-air annulus, and released through the louvers at the solids-discharge end of the shell.

Roto-Louvre dryers are suitable for processing coarse granular solids which do not offer high resistance to air flow, require intimate gas contacting, and do not contain significant quantities of dust.

Heat and mass transfer from the gas to the surface of the solids is extremely efficient; hence the equipment size required for a given duty is frequently less than required when an ordinary direct-heat rotary vessel with lifting flights is used. Purchase-price savings are partially balanced, however, by the more complex construction of the Roto-Louvre unit. A Roto-Louvre dryer will have a capacity roughly 1.5 times that of a single-shell rotary dryer of the same size under equivalent operating conditions. Because of the cross-flow method of heat exchange, the average Δt is not a simple function of inlet and outlet Δt's. There are currently no published data which permit the sizing of equipment without pilot tests as recommended by the manufacturer. Three applications of Roto-Louvre dryers are outlined in Table 12-25. Installation, operating, power, and maintenance costs will be similar to those experienced with ordinary direct-heat rotary dryers. *Thermal efficiency* will range from 30 to 70 percent.

AGITATED DRYERS

Description An agitated dryer is defined as one on which the housing enclosing the process is stationary while solids movement is accomplished by an internal mechanical agitator. Many forms are in use, including batch and continuous versions.

Field of Application Agitated dryers are applicable to process-

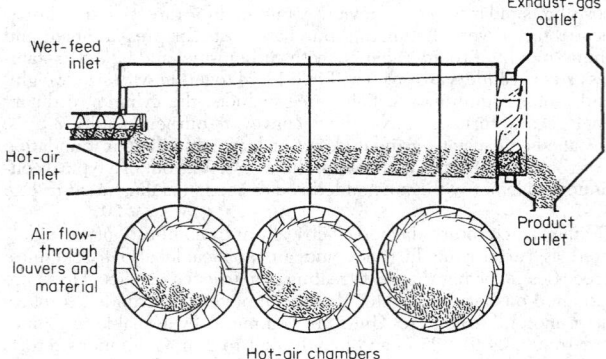

FIG. 12-73 Link-Belt Roto-Louvre dryer. (*Material Handling Systems Division, FMC Corp.*)

TABLE 12-25 Manufacturer's Performance Data for FMC Link-Belt Roto-Louvre Dryers*

Material dried	Ammonium sulfate	Foundry sand	Metallurgical coke
Dryer diameter	2 ft 7 in	6 ft 4 in	10 ft 3 in
Dryer length	10 ft	24 ft	30 ft
Moisture in feed, % wet basis	2.0	6.0	18.0
Moisture in product, % wet basis	0.1	0.5	0.5
Production rate, lb/h	2500	32,000	38,000
Evaporation rate, lb/h	50	2130	8110
Type of fuel	Steam	Gas	Oil
Fuel consumption	255 lb/h	4630 ft³/h	115 gal/h
Calorific value of fuel	837 Btu/lb	1000 Btu/ft³	150,000 Btu/gal
Efficiency, Btu, supplied per lb evaporation	4370	2170	2135
Total power required, hp	4	41	78

*Material Handling Systems Division, FMC Corp. To convert British thermal units to kilojoules, multiply by 1.06; to convert horsepower to kilowatts, multiply by 0.746.

ing solids which are relatively free-flowing and granular when discharged as product. Materials which are not free-flowing in their feed condition can be treated by recycle methods as described in the subsection "Rotary Dryers." In general, agitated dryers have applications similar to those of rotating vessels. Their chief advantages compared with the latter lie in the fact that (1) large-diameter rotary seals are not required at the solids and gas feed and exit points because the housing is stationary, and for this reason gas-leakage problems are minimized. Rotary seals are required only at the points of entrance of the mechanical agitator shaft. (2) Use of a mechanical agitator for solids mixing introduces shear forces which are helpful for breaking up lumps and agglomerates. Balling and pelleting of sticky solids, an occasional occurrence in rotating vessels, can be prevented by special agitator design. The problems concerning dusting of fine particles in direct-heat units are identical to those discussed under "Rotary Dryers."

Batch Vacuum Rotary Dryers   The more common type of vacuum rotary dryer consists of a stationary cylindrical shell, mounted horizontally, in which a set of agitator blades mounted on a revolving central shaft stirs the solids being treated. Heat is supplied by circulation of hot water, steam, or Dowtherm through a jacket surrounding the shell and, in larger units, through the hollow central shaft. The agitator is either a single discontinuous spiral or a double continuous spiral. The outer blades are set as closely as possible to the wall without touching, usually leaving a gap of 0.3 to 0.6 cm. Modern units occasionally employ spring-loaded shell scrapers mounted on the blades. The dryer is charged through a port at the top and emptied through one or more discharge nozzles at the bottom. Vacuum is applied and maintained by any of the conventional methods, i.e., steam jets, vacuum pumps, etc.

Another type of vacuum rotary dryer consists of a rotating horizontal cylindrical shell, suitably jacketed. Vacuum is applied to this unit through hollow trunnions with suitable packing glands. Rotary glands must be used also for admitting and removing the heating medium from the jacket. The inside of the shell may have lifting bars, welded longitudinally, to assist agitation of the solids.

The double-cone rotating vacuum dryer is a more common design. Although it is identical in operating design, the sloping walls of the cones permit more rapid emptying of solids when the dryer is in a stationary position. The older cylinder shape required continuous rotation during emptying to convey product to the discharge nozzles. As a result, a circular dust hood was frequently necessary to enclose the discharge-nozzle turning circle and prevent serious dust losses to the atmosphere during unloading. Several new designs of the double-cone type employ internal tubes or plate coils to provide additional heating surface.

On all rotating dryers, the vapor-outlet tube is stationary; it enters the shell through a rotating gland and is fitted with an elbow and an upward extension so that the vapor inlet, usually protected by a felt dust filter, will be at all times near the top of the shell.

A typical vacuum rotary dryer is illustrated in Fig. 12-74 and a double-cone vacuum dryer in Fig. 12-75.

Vacuum is used in conjunction with drying or other chemical oper-

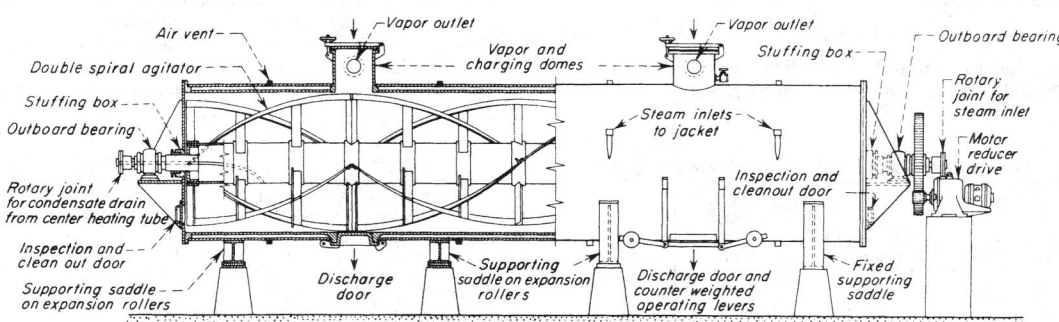

Elevation and partial cross section

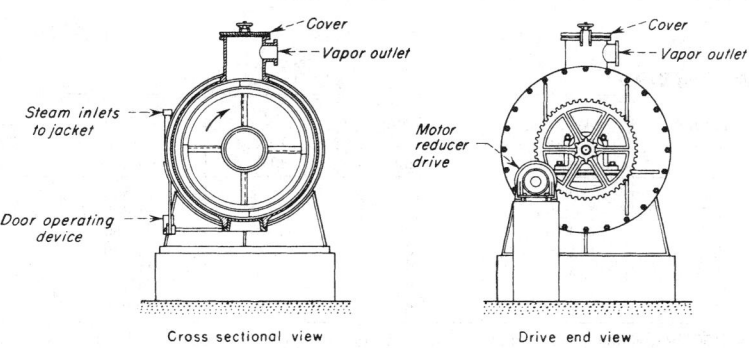

Cross sectional view   Drive end view

FIG. 12-74   A typical vacuum dryer. (*Blaw-Knox Food & Chemical Equipment, Inc.*)

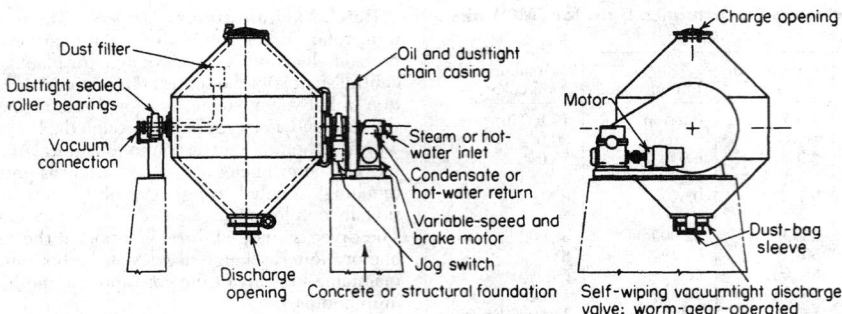

FIG. 12-75 Rotating (double-cone) vacuum dryer. (*Stokes Vacuum, Inc.*)

ations when low solids temperatures must be maintained because heat will cause damage to the product or change its nature, when air combines with the product as it is heated, causing oxidation or an explosive condition, when solvent recovery is required, and when materials must be dried to extremely low moisture levels.

In vacuum processing and drying the objective is to create a large temperature-driving force between the jacket and the product. To accomplish this purpose at fairly low jacket temperatures, it is necessary to reduce the internal process pressure so that the liquid being removed will boil at a lower vapor pressure. It is not always economical, however, to reduce the internal pressure to extremely low levels because of the large vapor volumes thereby created. It is necessary to compromise on operating pressure, considering leakage, condensation problems, and the size of the vapor lines and pumping system. Very few vacuum dryers operate below 5 mmHg pressure on a commercial scale. Air in-leakage through gasket surfaces will be in the range of 0.2 kg/(h·linear m of gasketed surface) under these conditions.

Design Methods for Batch Vacuum Rotary Dryers The rate of heat transfer from the heating medium through the dryer wall to the solids can be expressed by

$$Q = UA\Delta t_m \qquad (12\text{-}61)$$

where Q = heat flux, J/s (Btu/h); U = overall heat-transfer coefficient, J/(m²·s·K) [Btu/(h·ft² jacket area·°F)]; A = total jacket area, m²; and

Δt_m = log-mean-temperature driving force from heating medium to the solids, K.

The overall heat-transfer rate is almost entirely dependent upon the film coefficient between the inner jacket wall and the solids, which depends to a large extent on the solids characteristics. Overall coefficients may range from 30 to 200 J/(m²·s·K), based upon total area if the dryer walls are kept reasonably clean. Coefficients as low as 5 or 10 may be encountered if caking on the walls occurs.

For estimating purposes without tests, a reasonable coefficient for ordinary drying, and without taking the product to absolute dryness, may be assumed at $U = 50$ J/(m²·s·K) for rotary agitator dryers and 35 J/(m²·s·K) for rotating units.

Vacuum dryers are usually filled to 50 to 65 percent of their total shell volume. Agitator speeds range from 3 to 8 r/min. Faster speeds yield a slight improvement in heat transfer but consume more power.

Performance and Cost Data for Batch Vacuum Rotary Dryers Typical performance data for vacuum rotary dryers are given in Table 12-26. Size and cost data for rotary agitator units are given in Table 12-27. Data for double-cone units are in Table 12-28.

Turbo-Tray Dryers The turbo-tray dryer is a continuous dryer consisting of a stack of rotating annular shelves in the center of which turbo-type fans revolve to circulate the air over the shelves. Wet material enters through the roof, falling onto the top shelf as it rotates beneath the feed opening. After completing one revolution, the mate-

TABLE 12-26 Performance Data of Vacuum Rotary Dryers*

Material	Diameter × length, m	Initial moisture, % dry basis	Steam pressure, Pa × 10³	Agitator speed, r/min	Batch dry weight, kg	Final moisture, % dry basis	Pa × 10³	Time, h	Evaporation, kg/(h·m²)
Cellulose acetate	1.5 × 9.1	87.5	97	5.25	610	6	90–91	7	1.5
Starch	1.5 × 9.1	45–48	103	4	3630	12	88–91	4.75	7.3
Sulfur black	1.5 × 9.1	50	207	4	3180	1	91	6	4.4
Fuller's earth/mineral spirit	0.9 × 3.0	50	345	6	450	2	95	8	5.4

*Stokes Vacuum, Inc.

TABLE 12-27 Standard Rotary Vacuum Dryers*

Diameter, m	Length, m	Heating surface, m²	Working capacity, m³†	Agitator speed, r/min	Drive, kW	Weight, kg	Purchase price (1995) Carbon steel	Stainless steel (304)
0.46	0.49	0.836	0.028	7½	1.12	540	$ 43,000	$ 53,000
0.61	1.8	3.72	0.283	7½	1.12	1,680	105,000	130,000
0.91	3.0	10.2	0.991	6	3.73	3,860	145,000	180,000
0.91	4.6	15.3	1.42	6	3.73	5,530	180,000	205,000
1.2	6.1	29.2	3.57	6	7.46	11,340	270,000	380,000
1.5	7.6	48.1	6.94	6	18.7	15,880	305,000	440,000
1.5	9.1	57.7	8.33	6	22.4	19,050	330,000	465,000

*Stokes Vacuum, Inc. Prices include shell, 50-lb/in²-gauge jacket, agitator, drive, and motor; auxiliary dust collectors, condensers.
†Loading with product level on or around the agitator shaft.

TABLE 12-28   Standard (Double-Cone) Rotating Vacuum Dryers*

Working capacity, m³	Total volume, m³	Heating surface, m²	Drive, kW	Floor space, m²	Weight, kg	Purchase cost (1995)	
						Carbon steel	Stainless steel
0.085	0.130	1.11	.373	2.60	730	$ 32,400	$ 38,000
0.283	0.436	2.79	.560	2.97	910	37,800	43,000
0.708	1.09	5.30	1.49	5.57	1810	50,400	57,000
1.42	2.18	8.45	3.73	7.15	2040	97,200	106,000
2.83	4.36	13.9	7.46	13.9	3860	198,000	216,000
4.25	6.51	17.5	11.2	14.9	5440	225,000	243,000
7.08	10.5	°38.7	11.2	15.8	9070	324,000	351,000
9.20	13.9	°46.7	11.2	20.4	9980	358,000	387,000
11.3	16.0	°56.0	11.2	26.0	10,890	378,000	441,000

*Stokes Vacuum, Inc. Price includes dryer, 15-lb/in² jacket, drive with motor, internal filter, and trunnion supports for concrete or steel foundations. Horsepower is established on 65 percent volume loading of material with a bulk density of 50 lb/ft³. Models of 250 ft³, 325 ft³, and 400 ft³ have extended surface area.

rial is wiped by a stationary wiper through radial slots onto the shelf below, where it is spread into a uniform pile by a stationary leveler. The action is repeated on each shelf, with transfers occurring once in each revolution. From the last shelf, material is discharged through the bottom of the dryer (Fig. 12-76). The steel-frame housing consists of removable insulated panels for access to the interior. All bearings and lubricated parts are exterior to the unit with the drives located under the housing. Parts in contact with the product may be of steel or special alloy. The trays can be of any sheet material, such as enameled steel, asbestos-cement composition board, or plastic-glass laminates.

The rate at which each fan circulates air can be varied by changing the pitch of the fan blades. In final drying stages, in which diffusion controls or the product is light and powdery, the circulation rate is considerably lower than in the initial stage, in which high evaporation rates prevail. In the majority of applications, air flows through the

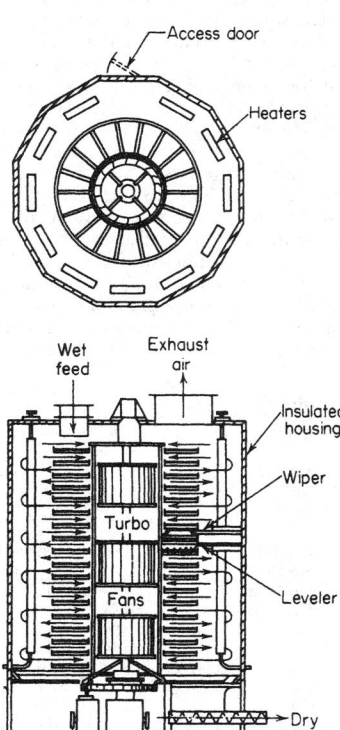

FIG. 12-76   Turbo-tray dryer. (*Wyssmont Company, Inc.*)

dryer upward in counterflow to the material. In special cases, required drying conditions dictate that air flow be cocurrent or both counter-current and cocurrent with the exhaust leaving at some level between solids inlet and discharge. A separate cold-air-supply fan is provided if the product is to be cooled before being discharged.

By virtue of its vertical construction, the turbo-type tray dryer has a stack effect, the resulting draft being frequently sufficient to operate the dryer with natural draft. Pressure at all points within the dryer is maintained close to atmospheric, as low as 0.1, usually less than 0.5 mm of water. Most of the roof area is used as a breeching, lowering the exhaust velocity to settle dust back into the dryer.

Heaters can be located in the space between the trays and the dryer housing, where they are not in direct contact with the product, and thermal efficiencies up to 3500 kJ/kg (1500 Btu/lb) of water evaporated can be obtained by reheating the air within the dryer. Steam is the usual heating medium. The high cost of heating electrically generally restricts its use to relatively small equipment. For materials which have a tendency to foul internal heating surfaces, an external heating system is employed.

The turbo-tray dryer can handle materials from thick slurries [1 million (N·s)/m² (100,000 cP) and over] to fine powders. It is not suitable for fibrous materials which mat or for doughy or tacky materials. Thin slurries can often be handled by recycle of dry product. Filter-press cakes are granulated before feeding. Thixotropic materials are fed directly from a rotary filter by scoring the cake as it leaves the drum. Pastes can be extruded onto the top shelf and subjected to a hot blast of air to make them firm and free-flowing after one revolution.

The turbo-tray dryer is manufactured in sizes from package units 2 m in height and 1.5 m in diameter to large outdoor installations 20 m in height and 11 m in diameter. Tray areas range from 1 for laboratory dryers to 1675 m² for large-scale production in a single unit. The number of shelves in a tray rotor varies according to space available and minimum rate of transfer required, from as few as 12 shelves to as many as 58 in the largest units. Standard construction permits operating temperatures up to 615 K, and high-temperature heaters permit operation at temperatures up to 925 K.

A recent innovation has enabled TURBO-Dryers to operate with very low inert gas make-up. Wyssmont has designed a tank housing that is welded up around the internal structure rather than the column-and-gasket panel design that has been the Wyssmont standard for many years. In field-erected units, the customer does the welding in the field; in packaged units, the tank-type welding is done in the shop. The tank-type housing finds particular application for operation under positive pressure. On the standard design, doors with explosion latches and gang latch operators are used. In the tank-type design, tight-sealing manway-type openings permit access to the interior. Tank-type housing designs have been requested when drying solvent wet materials and for applications where the material being dried is highly toxic and certainty is required that no toxic dust get out.

Design Methods for Turbo-Tray Dryers   The heat- and mass-transfer mechanisms are similar to those in batch tray dryers, except that constant turning over and mixing of the solids significantly improves drying rates. Design must usually be based on previous installations or pilot tests by the manufacturer; apparent heat-transfer

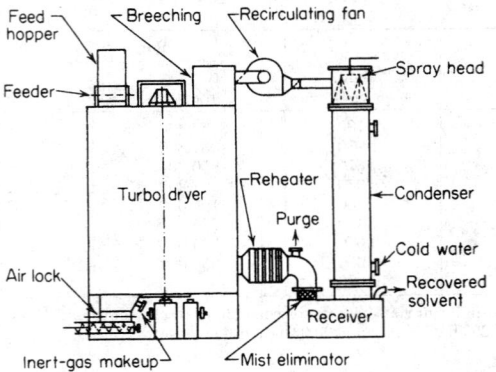

FIG. 12-77 Turbo-tray dryer in closed circuit for continuous drying with solvent recovery. (*Wyssmont Company, Inc.*)

coefficients will range from 28 to 55 J/(m²·s·K) for dry solids to 68 to 115 J/(m²·s·K) for wet solids. Turbo-tray dryers have been employed successfully for the drying and cooling of calcium hypochlorite, urea crystals, calcium chloride flakes, and sodium chloride crystals. The Wyssmont "closed-circuit" system, as shown in Fig. 12-77, consists of the turbo-tray dryer with or without internal heaters, recirculation fan, condenser with receiver and mist eliminators, and reheater. Feed and discharge are through a sealed wet feeder and lock respectively. This method is used for continuous drying without leakage of fumes, vapors, or dust to the atmosphere. A unified approach for scaling up dryers such as turbo-tray, plate, conveyor, or any other dryer type that forms a defined layer of solids next to a heating source is described by C. G. Moyers [*Drying Technology—An International Journal*, **12**(1 and 2), 393 (1994)].

Performance and Cost Data for Turbo-Tray Dryers Performance data for three applications of closed-circuit drying are included in Table 12-29. Operating, labor, and maintenance costs compare favorably with those of direct-heat rotating equipment.

Plate Dryers The plate dryer is an indirect-heated, fully continuous dryer available for three modes of operation: atmospheric, gastight, or full vacuum. The dryer is of vertical design, with horizontal, heated plates mounted inside the housing. The plates are heated by either hot water, steam, or thermal oil, with operating temperatures up to 320°C possible. The product enters at the top and is conveyed through the dryer by a product-transport system consisting of a central-rotating shaft with arms and plows. (See dryer schematic, Fig. 12-78.) The thin product layer (approx. ½ in depth) on the surface of the plates, coupled with frequent product turnover by the conveying system, results in

short-retention times (approx. 5–40 min), true plug flow of the material, and uniform drying. The vapors are removed from the dryer by a small amount of heated purge gas or by vacuum. The material of construction of the plates and housing is normally stainless steel, with special metallurgies also available. The drive unit is located at the bottom of the dryer and supports the central-rotating shaft. Typical speed of the dryer is 1–7 rpm. Full-opening doors are located on two adjacent sides of the dryer for easy access to dryer internals.

The plate dryer may vary in size from 5–35 vertically stacked plates with a heat-exchange area between 3.8–175 m². The largest unit available has overall dimensions of 3 m (w) by 4 m (l) by 10 m (h). Depending upon the loose-bulk density of the material and the overall retention time, the plate dryer can process up to 5,000 kg/hr of wet product.

The plate dryer is limited in its scope of applications only in the consistency of the feed material (the products must be friable, free flowing, and not undergo phase changes) and drying temperatures up to 320°C. Applications include specialty chemicals, pharmaceuticals, foods, polymers, pigments, etc. Initial moisture or volatile level can be as high as 65 percent and the unit is often used as a final dryer to take materials to a bone-dry state, if necessary. The plate dryer can also be used for heat treatment, removal of waters of hydration (bound moisture), solvent removal, and as a product cooler.

The atmospheric plate dryer is a dust-tight system. The dryer housing is an octagonal, panel construction, with operating pressure in the range of ±0.5 kPa gauge. An exhaust-air fan draws the purge air through the housing for removal of the vapors from the drying process. The purge-air velocity through the dryer is in the range of 0.1–0.15 m/sec, resulting in minimal dusting and small dust filters for the exhaust air. The air temperature is normally equal to the plate temperature. The vapor laden exhaust air is passed through a dust filter or a scrubber (if necessary) and is discharged to the atmosphere. Normally, water is the volatile to be removed in this type of system.

The gastight plate dryer, together with the components of the gas-recirculation system, forms a closed system. The dryer housing is semicylindrical and is rated for a nominal pressure of +5 kPa gauge. The flow rate of the recirculating purge gas must be sufficient to absorb the vapors generated from the drying process. The gas temperature must be adjusted according to the specific product characteristics and the type of volatile. After condensation of the volatiles, the purge gas (typically nitrogen) is recirculated back to the dryer via a blower and heat exchanger. Solvents such as methanol, toluene, and acetone are normally evaporated and recovered in the gastight system.

The vacuum plate dryer is provided as part of a closed system. The vacuum dryer has a cylindrical housing and is rated for full-vacuum operation (typical pressure range 3–27 kPa absolute). The exhaust vapor is evacuated by a vacuum pump and is passed through a condenser for solvent recovery. There is **no** purge-gas system required for operation under vacuum. Of special note in the vacuum-drying system

TABLE 12-29 Turbo-Tray Dryer Performance Data in Wyssmont Closed-Circuit Operations*

Material dried	Antioxidant	Water-soluble polymer	Antibiotic filter cake	Petroleum coke
Dried product, kg/h	500	85	2400	227
Volatiles composition	Methanol and water	Xylene and water	Alcohol and water	Methanol
Feed volatiles, % wet basis	10	20	30	30
Product volatiles, % wet basis	0.5	4.8	3.5	0.2
Evaporation rate, kg/h	53	16	910	302
Type of heating system	External	External	External	External
Heating medium	Steam	Steam	Steam	Steam
Drying medium	Inert gas	Inert gas	Inert gas	Inert gas
Heat consumption, J/kg	0.56×10^6	2.2×10^6	1.42×10^6	1.74×10^6
Power, dryer, kW	1.8	0.75	12.4	6.4
Power, recirculation fan, kW	5.6	5.6	37.5	15
Materials of construction	Stainless-steel interior	Stainless-steel interior	Stainless-steel interior	Carbon steel
Dryer height, m	4.4	3.2	7.6	6.5
Dryer diameter, m	2.9	1.8	6.0	4.5
Recovery system	Shell-and-tube condenser	Shell-and-tube condenser	Direct-contact condenser	Shell-and-tube condenser
Condenser cooling medium	Brine	Chilled water	Tower water	Chilled water
Location	Outdoor	Indoor	Indoor	Indoor
Approximate cost of dryer (1995)	$225,000	$115,000	$425,000	$225,000
Dryer assembly	Packaged unit	Packaged unit	Field-erected unit	Field-erected unit

*Courtesy of Wyssmont Company, Inc.

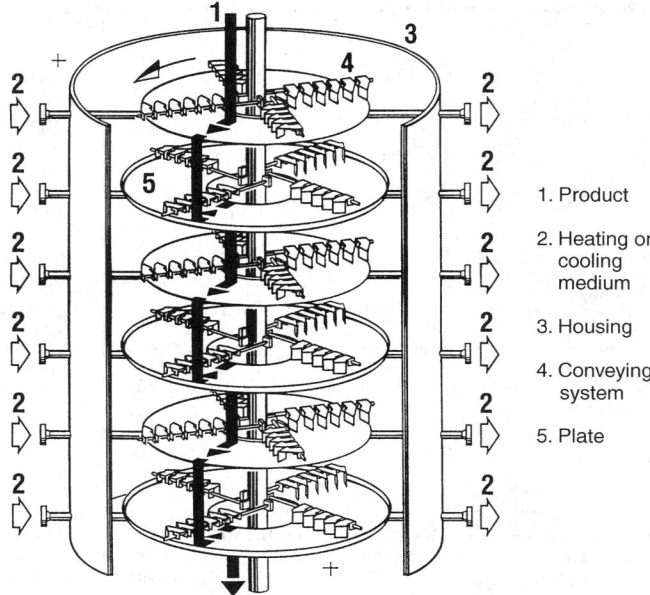

FIG. 12-78 Indirect-heated continuous plate dryer for atmospheric, gastight, or full-vacuum operation. (*Krauss Maffei.*)

1. Product
2. Heating or cooling medium
3. Housing
4. Conveying system
5. Plate

are the vacuum feed and discharge locks, which allow for continuous operation of the plate dryer under full vacuum.

Comparison Data—Plate Dryers Comparative studies have been done on products under both atmospheric and vacuum drying conditions. See Fig. 12-79. These curves demonstrate (1) the improvement in drying achieved with elevated temperature and (2) the impact to the drying process obtained with vacuum operation. Note that curve 4 at 90°C, pressure at 6.7 kPa absolute, is comparable to the atmospheric curve at 150°C. Also, the comparative atmospheric curve at 90°C requires 90 percent more drying time than the vacuum condition. The dramatic improvement with the use of vacuum is important to note for heat-sensitive materials.

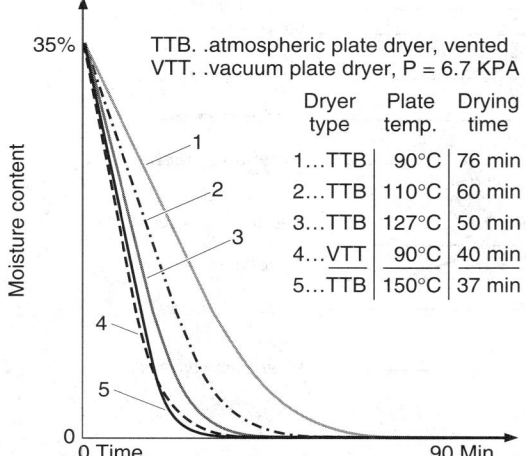

Drying Curve Product "N"

TTB. .atmospheric plate dryer, vented
VTT. .vacuum plate dryer, P = 6.7 KPA

Dryer type	Plate temp.	Drying time
1…TTB	90°C	76 min
2…TTB	110°C	60 min
3…TTB	127°C	50 min
4…VTT	90°C	40 min
5…TTB	150°C	37 min

FIG. 12-79 Plate dryer drying curves demonstrating impact of elevated temperature and/or operation under vacuum. (*Krauss Maffei.*)

The above drying curves have been generated via testing on a plate-dryer simulator. The test unit duplicates the physical setup of the production dryer, therefore linear scale-up from the test data can be made to the full-scale dryer. Because of the thin product layer on each plate, drying in the unit closely follows the normal type of drying curve in which the constant-rate period (steady evolution of moisture or volatiles) is followed by the falling-rate period of the drying process. This results in higher heat-transfer coefficients and specific drying capacities on the upper plates of the dryer as compared to the lower plates. The average specific drying capacity for the plate dryer is in the range of 2–20 kg/m²·hr (based on final dry product). Performance data for typical applications are shown on Table 12-30.

Conical Mixer Dryer The conical mixer dryer is a batch-wise operating unit commonly used in the pharmaceutical and specialty chemical industries for the drying of solvent or water wet, free-flowing powders. The process area is a vertically oriented conical vessel with an internally mounted screw. Figure 12-80 shows a schematic of the bottom drive conical mixer dryer. The dryer utilizes the heatable, internal rotating screw to provide agitation of the batch of material and thus improve the heat and mass transfer of the process. Because it rotates around the full circumference of the vessel, the screw provides a self-cleaning effect for the heated vessel walls.

The distinguishing feature of this dryer is the bottom-screw drive, as opposed to a top-drive unit, thus eliminating any mechanical drive components inside the vessel. The bottom-driven screw rotates about its own axis (speeds up to 100 rpm) and around the interior of the vessel (speeds up to 0.4 rpm). The screw is cantilevered in the vessel and requires no additional support (even in vessel sizes up to 20 m³ operating volume). The dryer is available in a variety of materials of construction, including SS 304 and 316, as well as Hastelloy.

The heatable areas of the dryer are the vessel wall and the screw. The dryer makes maximum use of the product-heated areas—the filling volume of the vessel (up to the knuckle of the dished head) is the usable product loading. The top cover of the vessel is easily heated by either a half-pipe coil or heat tracing, which ensures that no vapor condensation will occur in the process area. In addition to the conical vessel heated area, heating the screw effectively increases the heat exchange area by 15–30 percent. This is accomplished via rotary joints at the base of the screw. The screw can be heated with the same

TABLE 12-30 Plate Dryer Performance Data for Three Applications*

Product	Plastic additive	Pigment	Foodstuff
Volatiles	Methanol	Water	Water
Production rate, dry	362 kg/hr	133 kg/hr	2030 kg/hr
Inlet volatiles content	30%	25%	4%
Final volatiles content	0.1%	0.5%	0.7%
Evaporative rate	155 kg/hr	44 kg/hr	70 kg/hr
Heating medium	Hot water	Steam	Hot water
Drying temperature	70°C	150°C	90°C
Dryer pressure	11 kPa abs	atmospheric	atmospheric
Air velocity	NA	0.1 m/sec	0.2 m/sec
Drying time, min	24	23	48
Heat consumption, kcal/kg dry product	350	480	100
Power, dryer drive	3 kW	1.5 kW	7.5 kW
Material of construction	SS 316L/316Ti	SS 316L/316Ti	SS 316L/316Ti
Dryer height	5 m	2.6 m	8.2 m
Dryer footprint	2.6 m diameter	2.2 m by 3.0 m	3.5 m by 4.5 m
Location	Outdoors	Indoors	Indoors
Dryer assembly	Fully assembled	Fully assembled	Fully assembled
Power, exhaust fan	NA	2.5 kW	15 kW
Power, vacuum pump	20 kW	NA	NA

*Krauss Maffei

medium as the conical jacket (either hot water, steam, or thermal oil). Obviously, the impact of the heated screw will mean shorter batch-drying times, which yields higher productivity and better product quality due to shorter exposure to the drying temperature.

The vessel cover is free of drive components, allowing space for additional process nozzles, manholes, explosion venting, etc., as well as a temperature lance for direct, continuous product-temperature measurement in the vessel. A dust filter is normally mounted directly on the top head of the dryer, thus allowing any entrapped dust to be pulsed back into the process area. Standard cloth-type dust filters are available, along with sintered metal filters. Because there are no drive components in the process area, the risk of batch failures due to contamination from gear lubricants is eliminated. Cleaning of the conical mixer dryer is facilitated with CIP systems and/or the vessel can be

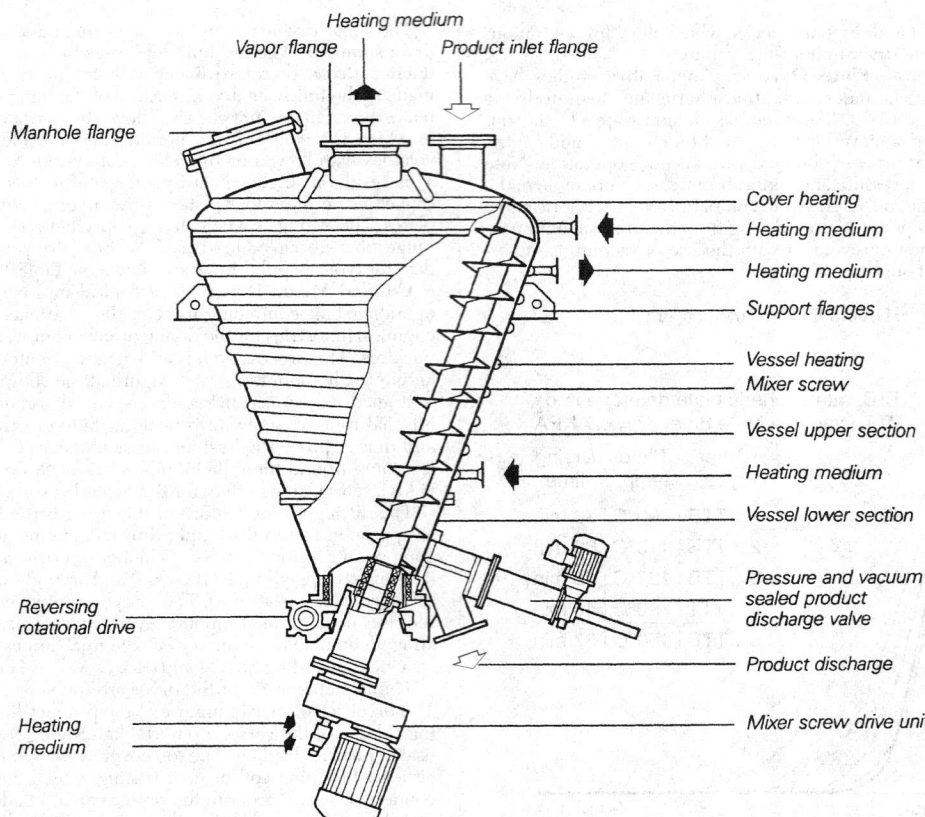

FIG. 12-80 Bottom drive conical mixer dryer. (*Krauss Maffei.*)

completely flooded with water or solvents. The disassembly of the unit is simplified, as all work on removing the screw can be done *without* vessel entry. For disassembly, the screw is simply secured from the top and the drive components removed from the bottom of the dryer.

Since the dryer is a batch-operating unit, it is commonly used in the pharmaceutical industry to maintain batch identification. In addition to pharmaceutical materials, the conical mixer dryer is used to dry polymers, additives, inorganic salts, and many other specialty chemicals.

The conical mixer dryer is an indirect (conduction) dryer designed to operate under full vacuum. The heating medium is either hot water, steam, or thermal oil, with most applications in the temperature range of 50–150°C and pressures in the range of 3–30 kPa absolute. The vapors generated during the drying process are evacuated by a vacuum pump and passed through a condenser for recovery of the solvent.

Design Methods for Conical Mixer Dryers Drying trials are conducted in small pilot dryers (50–100 liter batch units) to determine material handling and drying retention times. Variables such as drying temperature, vacuum level, and screw speed are analyzed during the test trials. Scale-up to larger units is done based upon the area to volume ratio of the pilot unit versus the production dryer. In most applications, the overall drying time in the production models is in the range of 2–24 hours. For design purposes, the heat-transfer coefficient for the conical mixer dryer is 60 $W/m^2 \cdot K$.

Reactotherm The reactotherm is a special thermal processor which has been designed to overcome the problems encountered with standard horizontal rotary-type dryers, such as crust formation on the heat-exchange surfaces and stalling of the rotor during drying. Incorporated in the design of the reactotherm is a high torque drive combined with rugged shaft construction to prevent rotor stall during processing; in addition, stationary mixing elements are installed in the process housing which continually clean the heat-exchange surfaces of the rotor to minimize any crust buildup and ensures an optimum heat-transfer coefficient at all times. Because the unit can handle a wide range of product consistencies (dilute slurries, pastes, friable powders) the reactotherm can be used for processes such as reactions, mixing, drying, cooling, melting, sublimation, distilling, and vaporizing.

The reactotherm is an indirect-heated thermal processor available in both batch and continuous design. Figure 12-81 shows a sketch of the continuous model. The unit is of horizontal design, with both the shell and internal rotor heated to provide high heat-transfer area/product volume ratio. The units are conduction dryers, with operation in atmospheric, pressure and vacuum mode possible. Sizes range from 170 up to 10,000 liter operating volume with heat-exchange area from 3.1 m^2 up to 103.8 m^2. The typical filling volume of each unit is 75 percent of the available operating volume. The material of construction is either carbon steel or 300 series stainless steel, with optional special metallurgy available.

In addition to many varied applications in the chemical industry, the reactotherm is widely used to process waste sludges which must be dried in order to landfill the solid components. In many of these cases, the solvents are valuable components of the waste stream and can be recovered and recycled.

The housing of the continuous model consists of two fully enclosed, cylindrical, flanged-vessel sections. The standard design of the housing is suitable for full vacuum (3–30 kPa abs) or pressures up to 7 bar abs (higher design pressures available). The housing sections, including both end covers, are jacketed for heating or cooling. Hot water, steam, or thermal oil may be used as the heating medium, with maximum temperature up to 350°C possible. Flanged ports and nozzles are provided for product feed and discharge, vapor removal, inspection, and for the stationary mixing elements. The design of the continuous models is such that the ratio of vessel length to housing diameter is approximately 6:1, resulting in a long unit with narrow diameter. Bearing supports are provided at both ends of the unit for shaft support.

The mixing bars in the continuous unit also convey the product through the unit, whereas the product filling level is determined by a weir flanged between the housing sections. The residence time and speed of the mixing shaft are specially matched to the requirements of the process.

The batch-operating units consist of only one jacketed, cylindrical vessel with the necessary nozzles and ports for product inlet and out-

let, vapor, inspection, and mixing elements. The standard design of the housing is suitable for full vacuum or pressures up to 7 bar abs (higher design pressures available), with temperatures up to 350°C possible. The mixing bars are arranged such that a countercurrent flow in a radial direction of the product is achieved in the process vessel. For discharge of the material, a central port is provided at the bottom of the unit. The batch models up to 1080 liters volume are provided with cantilever-mounted mixing shafts. Above 1080 liters, the shaft is supported at both ends with bearings. The design of the batch models is such that the ratio of vessel length to housing diameter is approximately 2.5:1, resulting in a short unit with large diameter.

Design Methods for Reactotherm Product trials are conducted in small pilot dryers (8–60 liter batch or continuous units) to determine material handling and process-retention times. Variables such as drying temperature, pressure level, and shaft speed are analyzed during the test trials. For design purposes, the heat-transfer coefficient for the reactotherm is in the range of 10 $W/m^2 \cdot K$ (light, free-flowing powders) up to 150 $W/m^2 \cdot K$ (dilute slurries).

Hearth Furnaces A special design of a circular hearth furnace is the **Mannheim furnace,** in which sulfuric acid is reacted with sodium chloride to produce salt cake and hydrochloric acid. It consists of a refractory hearth, up to 6 m in diameter, with a silicon carbide arch. Hot flue gases are circulated around the muffle. The major portion of heat is transmitted through the arch and radiated to the product on the hearth. Feed materials are mixed and charged continuously to the center of the hearth, where they are stirred by underdriven rabble arms. The charge is gradually worked toward the periphery as the reaction generates hydrogen chloride gas. The gas is withdrawn through a separate duct to an absorption system. The salt cake is discharged at the periphery. Figure 12-82 shows a diagrammatic cross section of a Mannheim furnace. Combustion-chamber temperatures of about 1475 K are used for heating. The salt cake is discharged from the hearth at about 800 K.

Multiple-Hearth Furnaces Multiple-hearth furnaces are known under various names: the Herreshoff, McDougall, Wedge, Pacific, etc. Figure 12-83 shows a general design. It consists of a number of annular-shaped hearths mounted one above the other. There are rabble arms on each hearth driven from a common center shaft. The feed is charged at the center of the upper hearth. The arms move the charge outward to the periphery, where it falls to the next hearth. Here it is moved again to the center, from which it falls to the next hearth. This continues down the furnace. The hollow center shaft is cooled internally by forced-air circulation.

Burners may be mounted at any of the hearths, and the circulated air is used for combustion. These furnaces handle granular materials and provide a long countercurrent path between the flue gases and the charge material. Industrial sizes are built from 2 to 7 m in diameter and include 4 to 16 hearths. Total hearth areas range from 6.5 to 335 m^2. The furnaces are used for roasting ores, drying and calcining lime, magnesite, and carbonate sludges, reactivation of decolorizing earths, and burning of sulfides to produce sulfur dioxide. The following is a partial list of applications:

1. Lime (*a*) from crushed limestone, (*b*) from oyster or sea shell, and (*c*) from dolomitic limestone
2. Lead and zinc; roasting of sulfides
3. Mercury from cinnabar ores by volatilization
4. Gold and silver: (*a*) chloridizing roast of gold-silver ore, and (*b*) removal of arsenic
5. Sulfuric acid from iron pyrites
6. Paint pigments; roasting of metallic oxides
7. Refractory clays; calcination of refractory clay to reduce shrinkage
8. Foundry sand; removal of carbon from used foundry sand
9. Fuller's earth; calcination of fuller's-earth material
10. Sewage disposal; calcination of sewage slurry

Table 12-31 lists three specific applications with a brief description of the furnaces as to design and operating conditions.

GRAVITY DRYERS

Description A body of solids in which the particles, consisting of granules, pellets, beads, or briquettes, flow downward by gravity at

1. Product inlet
2. Product outlet
3. Vapor outlet
4. Heating/cooling medium inlet
5. Heating/cooling medium outlet
6. Weir plate
7. Mixing shaft
8. Mixing elements
9. Frame
10. Drive
11. Housing

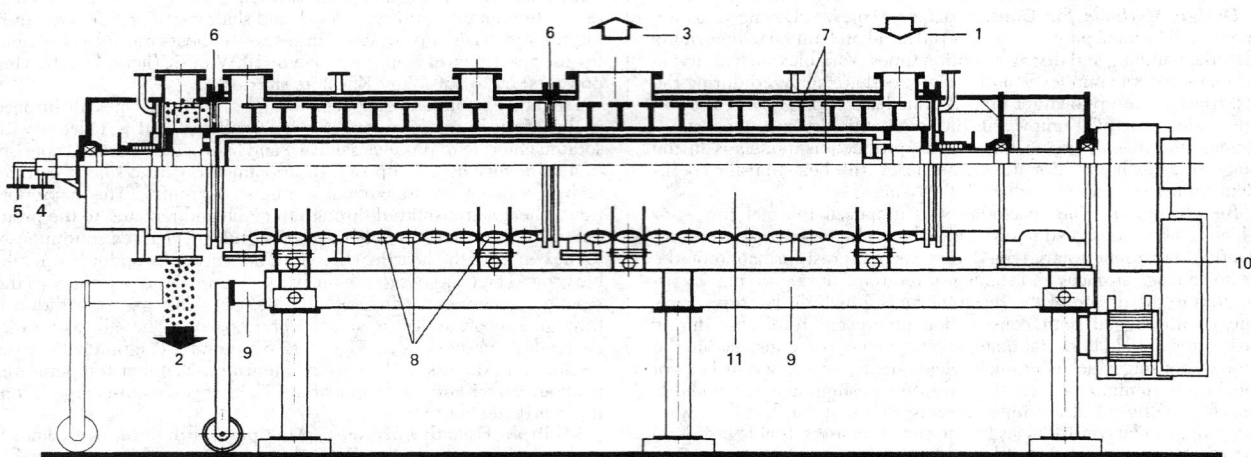

FIG. 12-81   Continuous indirect-heated reactotherm. (*Krauss Maffei.*)

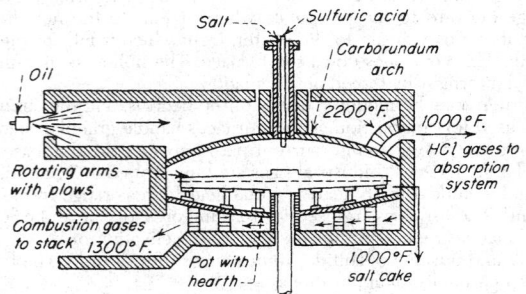

FIG. 12-82   Mannheim-type mechanical hydrochloric acid furnace.

substantially their normal settled bulk density through a vessel in contact with gases is defined frequently as a **moving bed.** Moving-bed equipment finds application in blast furnaces, shaft furnaces, and petroleum refining.

A gravity dryer consists of a stationary vertical, usually cylindrical housing with openings for the introduction of solids (at the top) and removal of solids (at the bottom). Gas flow is through the solids bed and may be cocurrent or countercurrent and, in some instances, crossflow. By definition, the rate of gas flow upward must be less than that required for fluidization.

Fields of Application   One of the major advantages of the gravity-bed technique is that it lends itself well to true intimate countercurrent contacting of solids and gases. This provides for efficient heat transfer and mass transfer. Gravity-bed contacting also permits the use of the solid as a heat-transfer medium, as in pebble heaters.

Gravity vessels are applicable to coarse granular free-flowing solids which are comparatively dust-free. The solids must possess physical properties in size and surface characteristics so that they will not stick together, bridge, or segregate during passage through the vessel. The presence of significant quantities of fines or dust will close the passages among the larger particles through which the gas must penetrate, increasing pressure drop. Fines may also segregate near the sides of the bed or in other areas where gas velocities are low, ultimately completely sealing off these portions of the vessel. The high efficiency of gas-solids contacting in gravity beds is due to the uniform distribution of gas throughout the solids bed; hence choice of feed and its preparation are important factors to successful operation. Preforming techniques such as pelleting and briquetting are employed frequently for the preparation of suitable feed materials.

Gravity vessels are suitable for low-, medium-, and high-temperature operation; in the last case, the housing will be lined completely with refractory brick. Dust-recovery equipment is minimized in this type of operation since the bed actually performs as a dust collector itself, and dust in the bed will not, in a successful application, exist in large quantities.

Other advantages of gravity beds include flexibility in gas and solids flow rates and capacities, variable retention times from minutes to several hours, space economy, ease of startup and shutdown, the potentially large number of contacting stages, and ease of control by using the inlet- and exit-gas temperatures.

Maintenance of a uniform rate of solids movement downward over the entire cross section of the bed is one of the most critical operating problems encountered. For this reason gravity beds are designed to be as high and narrow as practical. In a vessel of large cross section, discharge through a conical bottom and center outlet will usually result in some degree of "ratholing" through the center of the bed. Flow through the center will be rapid while essentially stagnant pock-

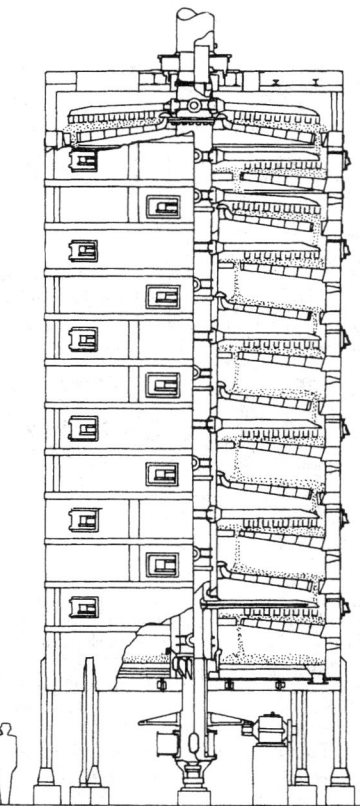

FIG. 12-83 Pacific multiple-hearth furnace.

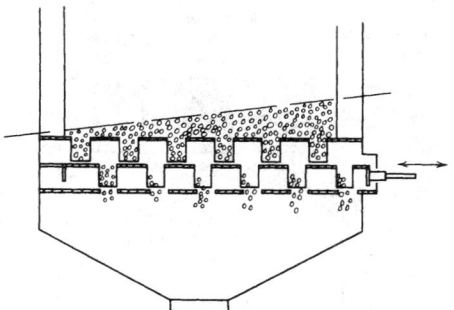

FIG. 12-84 Gravity-bed reactor; solids-discharge mechanism.

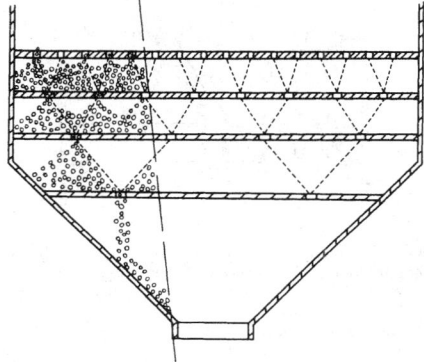

FIG. 12-85 Perforated-tray type of reactor-discharge control.

ets are left around the sides. To overcome this problem, multiple outlets are provided in the center and around the periphery; table unloaders, rotating plows, wide moving grates, and multiple-screw unloaders are employed; insertion of inverted cone baffles in the lower section of the bed, spaced so that flushing at the center is retarded, is also a successful method for improving uniformity of solids movement. Figure 12-84 illustrates a moving tray with multiple downspouts used to remove a precise amount of solids from each increment of area across the base of a gravity-bed reactor. The various pockets are filled at one extremity of its motion and emptied at the other. It is suitable primarily for fine nonabrasive solids. Figure 12-85 depicts a perforated-plate design, taking advantage of the flow characteristics and angle of repose of the solids to control the unloading rate. Still another design of this general type involves the use of a nest of inclined pipes, discharging into a common header, and placed to draw solids at geometrically spaced points across the base of the reactor.

Gas disengaging from the solids may represent another serious operating problem in a gravity bed. One method employs downspouts at the top for solids feeding while leaving an open space in the vessel above the downspout outlet for gas disengaging, as illustrated in Fig. 12-86a. Another uses a series of inverted V-shaped channels inserted into the top of the solids bed. Gas and vapor are collected and removed from under the V's, while the solids flow over the top and around the channels (Fig. 12-86b). These methods for both gas and solids removal were developed originally for petroleum-refining catalytic reactors.

Shaft Furnaces The oldest and most important application of the shaft furnace is the **blast furnace** used for the production of pig iron. Another use is in the manufacture of phosphorus from phosphate rock. Formerly lime was calcined exclusively in this type of furnace. Shaft furnaces are widely used also as gas producers. Chemicals are manufactured in shaft furnaces from briquetted mixtures of the reacting components.

A shaft furnace is a vertical refractory-lined cylinder in which a stationary or descending column of solids is maintained and through

TABLE 12-31 Applications of Multiple-Hearth Furnaces*

Product	Production rate	Furnace size	Special features
Mercury from cinnabar ore	225 tons ore/day (95% recovery)	(2) 18.0 ft. diam., 8 hearth furnaces	Furnaces fired on hearths 3 to 7, inclusive; retention time of 1.0 hr.; furnaces are oil-fired with low-pressure atomizing air burners; all air, both primary and secondary, introduced through the burners; draft control by Monel cold-gas fans downstream from mercury condensers.
Lime from oyster shell	240 tons/day, shell (120 tons/day, lime)	(1) 22 ft., 3 in. diam., 12 hearth furnaces	
Magnesium oxide from magnesium hydroxide	100 tons/day, 50% magnesium hydroxide slurry; yields 50 tons/day magnesium oxide	(2) 22 ft., 3 in., 10 hearth furnaces	Furnace walls of 4.5-in. firebrick, 9-in. insulation for 1550°F. operating temp. Furnace fired on hearths 4 to 10, inclusive

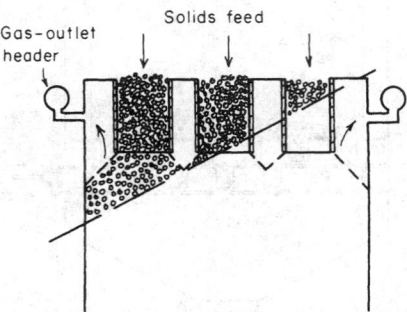

FIG. 12-86a Countercurrent gas-solids flow at the top disengaging section of a moving-bed catalytic reactor.

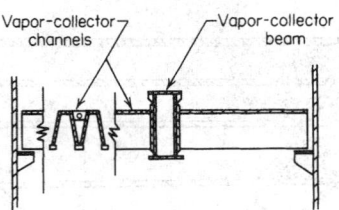

FIG. 12-86b Vapor disengaging tray at the top of a gravity-bed catalytic reactor. This design may also be employed for the addition of gas to a bed of solids.

which an ascending stream of hot gas is forced. Three methods of fuel application may be employed: (1) one in which a solid fuel is added alone or mixed with the reacting solids, (2) one in which the fuel is burned in a separate combustion chamber with the hot gases being blown into the furnace at some level of the column, (3) one in which the fuel is introduced and burned in the bottom of the shaft.

For maximum heat economy, recovered exhaust heat is employed for preheating of the incoming solids and combustion air. The fuels used may be gas, oil, or pulverized coal.

Bucket elevators, skip hoists, and cranes are used for top feeding of the furnace. Retention and downward flow are controlled by timing of the bottom discharge. Gases are propelled by a blower or by induced draft from a stack or discharge fan. In normal operation, the downward flow of solids and upward flow of gas are constant with time, maintaining ideal steady-state conditions.

Figure 12-87 illustrates a shaft lime kiln.

Design Methods for Shaft Furnaces   The size and shape of the charge particles control the amount of surface over which heat may be transmitted to the particle and also the depth of penetration through which the heat must pass to reach the center of each particle. Also, this size and shape control the nature of the random packing in the shaft and the extent of voids for gas passage. As particle size is decreased, the surface area of the particles increases. At the same time, the depth of heat penetration decreases. Both these factors tend to improve performance. With small particle size, however, the charge column presents high resistance to the passage of gas.

With closely screened material, the percentage of voids (usually 37 percent) is independent of particle size. With unscreened particles showing a wide variation in size, the void volume is decreased; irregularity in gas flow results.

There is a large difference between the total surface of the particles (as determined by their size and shape) and the "effective surface" actually exposed to the passing gas stream. In practice, it has been estimated that as little as 10 to 25 percent of the total surface is effective in heat transfer when unscreened particles are treated.

Irregular-shaped particles exhibit greater surface area than regular-shaped cubes and spheres, the amount of this increase being possibly 25 percent. The effect of particle size and size distribution on effective surface, in a shaft employed for calcination of limestone, is shown in

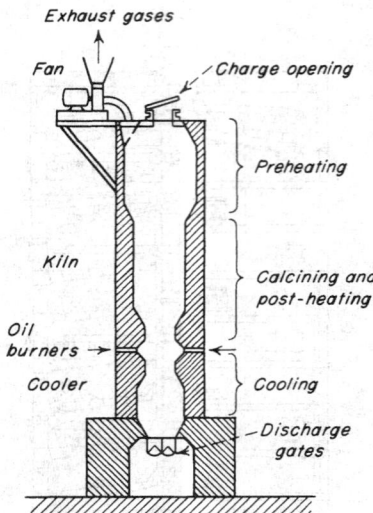

FIG. 12-87 Shaft furnace for lime production.

Fig. 12-88. Curve *A* shows the calculated surface based on an assumed 50 percent void volume and cubical-shaped particles. The *B* set of curves applies to such unscreened irregularly shaped particles as are usually encountered in practice.

The laws governing the flow of fluids through packed beds given in Sec. 6 are applicable to shaft furnaces. Since the pressure drop in a bed is affected by the size and shape of the interstitial voids, the horizontal and vertical nonuniformity of the bed, the changes in gas composition during passage, and other operating factors, test data for a given material are necessary for proper design. In the case of limestone, Fig. 12-89 shows the effect of particle size on the gas-flow friction through the bed, assuming that the friction varies as the square of the gas mass velocity and inversely with the particle size, and utilizing

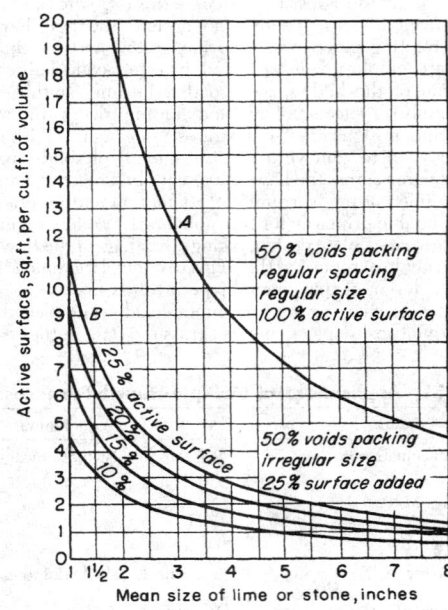

FIG. 12-88 Curve *A* shows surface variations with stone size, 100 percent active surfaces. Curves in group *B* show the effect of irregular stone size.

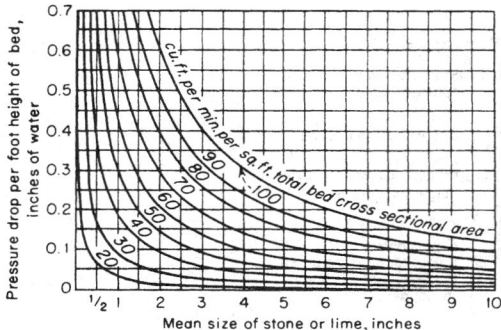

FIG. 12-89 Variation in gas friction with size of stone.

base points established during actual kiln operations. Information on the mathematical treatment of heat transfer in packed beds is included in Sec. 12.

Pellet Coolers and Dryers Gravity beds are employed for the cooling and drying of extruded pellets and briquettes from size-enlargement processes. The rotary cooler illustrated in Fig. 12-90 consists of a stationary steel tank having a wear cylinder at the top for entry of gas and solids (usually from a pneumatic conveyor), with air holes staggered around the outer wall to admit additional air for optimum circulation. The tank encloses a rotating cage for retention of the solids bed. This cage consists of an inner cylinder of wire mesh and an outer perforated shell. Air entering the tank is circulated in cross-flow through the pellet bed and discharges through the center column. Usually a rotary unloading gate and air lock are located underneath the cage.

Another cross-flow design employs a rectangular housing, partitioned into three vertical sections (Fig. 12-91). Solids move downward in the two outer sections, while cooling or drying air is drawn through the louvered outer walls and through the solids bed and is discharged through the center section. Solids are discharged over a baffled shaking shoe. Units of this general type are used for drying wheat and other grain products and numerous forms of pelleted feeds. Gravity-bed dryers are most suitable for drying of granular heat-sensitive products employing moderate air temperatures. These require extended holdup during the falling-rate drying period.

Spouted Beds The spouted-bed technique was developed primarily for solids which are too coarse to be handled in fluidized beds. Although their applications overlap, the methods of gas-solids mixing are completely different. A schematic view of a spouted bed is given in Fig. 12-92. Mixing and gas-solids contacting are achieved first in a fluid "spout," flowing upward through the center of a loosely packed bed of solids. Particles are entrained by the fluid and conveyed to the top of the bed. They then flow downward in the surrounding annulus as in an ordinary gravity bed, countercurrently to gas flow. The mechanisms of gas flow and solids flow in spouted beds were first described by Mathur and Gishler [*Am. Inst. Chem. Eng. J.,* **1**(2), 157–164 (1955)]. Drying studies have been carried out by Cowan [*Eng. J.,* **41,** 5, 60–64 (1958)], and a theoretical equation for predicting the minimum fluid velocity necessary to initiate spouting was developed by Madonna and Lama [*Am. Inst. Chem. Eng. J.,* **4**(4), 497 (1958)]. Investigations to determine maximum spoutable depths and to develop theoretical relationships based on vessel geometry and operating variables have been carried out by Lefroy [*Trans. Inst. Chem. Eng.,* **47**(5), T120–128 (1969)] and Reddy [*Can. J. Chem. Eng.,* **46**(5), 329–334 (1968)].

Gas flow in a spouted bed is partially through the spout and partially through the annulus. About 30 percent of the gas entering the system immediately diffuses into the downward-flowing annulus. Near the top of the bed, the quantity in the annulus approaches 66 percent of the total gas flow; the gas flow through the annulus at any point in the bed equals that which would flow through a loosely packed solids bed under the same conditions of pressure drop. Solids flow in the annulus is both downward and slightly inward. As the fluid spout rises in the bed, it entrains more and more particles, losing velocity and gas into the annulus. The volume of solids displaced by the spout is roughly 6 percent of the total bed.

On the basis of experimental studies, Mathur and Gishler derived an empirical correlation to describe the minimum fluid flow necessary for spouting, in 3- to 12-in-diameter columns:

$$u = \frac{D_p}{D_c}\left(\frac{D_o}{D_c}\right)^{0.33}\left[\frac{2gL(p_s - p_f)}{p_f}\right]^{0.5} \qquad (12\text{-}62)$$

where u = superficial fluid velocity through the bed, ft/s; D_p = particle diameter, ft; D_c = column (or bed) diameter, ft; D_o = fluid-inlet orifice diameter, ft; L = bed height, ft; p_s = absolute solids density, lb/ft^3; p_f = fluid density, lb/ft^3; and g = 32.2 ft/s^2, gravity acceleration. To convert feet per second to meters per second, multiply by 0.305; to convert pounds per cubic foot to kilograms per cubic meter, multiply by 16. g = 9.8 m/s^2 in SI units. The inlet orifice diameter, air rate, bed diameter, and bed depth were all found to be critical and interdependent:

1. In a given-diameter bed, deeper beds can be spouted as the gas-inlet orifice size is decreased. Using air, a 12-in-diameter bed containing 0.125- by 0.250-in wheat can be spouted at a depth of over 100 in with a 0.8-in orifice, but at only 20 in with a 2.4-in orifice.

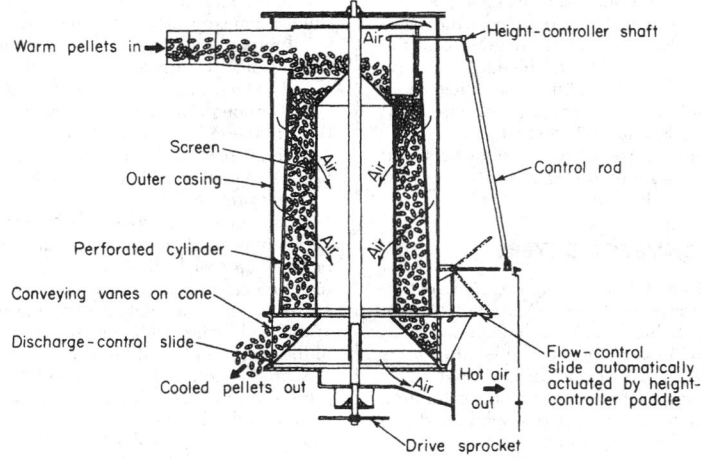

FIG. 12-90 Sprout, Waldron rotary cooler. (*Sprout, Waldron Cos.*)

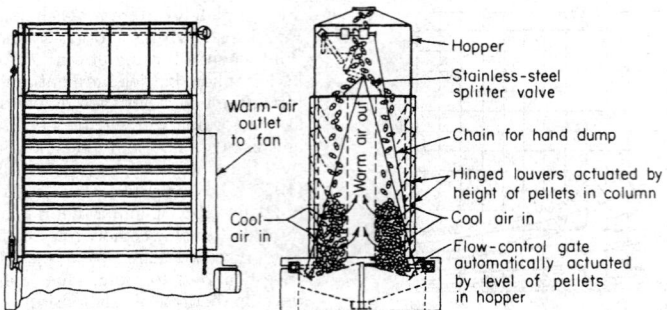

FIG. 12-91   Vertical gravity-bed cooler with louvres. (*Sprout, Waldron Co.*)

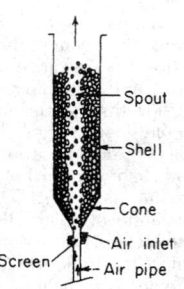

FIG. 12-92   Schematic diagram of a spouted bed. [*Mathur and Gishler, Am. Inst. Chem. Eng. J., 1, 2, 157 (1955).*]

2.   Increasing bed diameter increases spoutable depth. By employing a bed-orifice diameter ratio of 12 for air spouting, a 9-in-diameter bed was spouted at a depth of 65 in while a 12-in-diameter bed was spouted at 95 in.

3.   As indicated by Eq. (12-62) the superficial fluid velocity required for spouting increases with bed depth and orifice diameter and decreases as the bed diameter is increased.

Employing wood chips, Cowan's drying studies indicated that the volumetric heat-transfer coefficient obtainable in a spouted bed is at least twice that in a direct-heat rotary dryer. By using 20- to 30-mesh Ottawa sand, fluidized and spouted beds were compared. The volumetric coefficients in the fluid bed were 4 times those obtained in a spouted bed. Mathur dried wheat continuously in a 12-in-diameter spouted bed, followed by a 9-in-diameter spouted-bed cooler. A drying rate of roughly 100 lb/h of water was obtained by using 450 K inlet air. Six hundred pounds per hour of wheat was reduced from 16 to 26 percent to 4 percent moisture. Evaporation occurred also in the cooler by using sensible heat present in the wheat. The maximum drying-bed temperature was 118°F, and the overall thermal efficiency of the system was roughly 65 percent. Some aspects of the spouted-bed technique are covered by patent (U.S. Patent 2,786,280).

Cowan reported that significant size reduction of solids occurred when cellulose acetate was dried in a spouted bed, indicating its possible limitations for handling other friable particles.

DIRECT-HEAT VIBRATING-CONVEYOR DRYERS

Information on vibrating conveyors and their mechanical construction is given in Sec. 21. The vibrating-conveyor dryer is a modified form of fluidized-bed equipment, in which fluidization is maintained by a combination of pneumatic and mechanical forces. The heating gas is introduced into a plenum beneath the conveying deck through ducts and flexible hose connections, passes up through a screen, perforated, or slotted conveying deck, through the fluidized bed of solids, and into an exhaust hood (Fig. 12-93). If ambient air is employed for cooling, the sides of the plenum may be open and a simple exhaust system

used; however, because the gas-distribution plate may be designed for several inches of water-pressure drop to ensure a uniform velocity distribution through the bed of solids, a combination pressure-blower-exhaust-fan system is desirable to balance the pressure above the deck with the outside atmosphere and prevent gas in-leakage or blowing at the solids feed and exit points.

Units are fabricated in widths from 0.3 to 1.5 m. Lengths are variable from 3 to 50 m; however, most commercial units will not exceed a length of 10 to 16 m per section. Power required for the vibrating drive will be approximately 0.4 kW/m² of deck.

In general, this equipment offers an economical heat-transfer area for first cost as well as operating cost. Capacity is limited primarily by the air velocity which can be used without excessive dust entrainment. Table 12-32 shows limiting air velocities suitable for various solids particles. Usually, the equipment is satisfactory for particles larger than 100 mesh in size. [The use of indirect-heated conveyors eliminates the problem of dust entrainment, but capacity is limited by the heat-transfer coefficients obtainable on the deck (see Sec. 11)].

When a stationary vessel is employed for fluidization, all solids being treated must be fluidized; nonfluidizable fractions fall to the bottom of the bed and may eventually block the gas distributor. The addition of mechanical vibration to a fluidized system offers the following advantages:

1.   Equipment can handle nonfluidizable solids fractions. Although these fractions may drop through the bed to the screen, directional-throw vibration will cause them to be conveyed to the discharge end of the conveyor. Prescreening or sizing of the feed is less critical than in a stationary fluidized bed.

2.   Because of mechanical vibration, incipient channeling is reduced.

3.   Fluidization may be accomplished with lower pressures and gas velocities. This has been evidenced on vibratory units by the fact that fluidization stops when the vibrating drive is stopped.

Vibrating-conveyor dryers are suitable for free-flowing solids containing mainly surface moisture. Retention is limited by conveying speeds which range from 0.02 to 0.12 m/s. Bed depth rarely exceeds 7 cm, although units are fabricated to carry 30- to 46-cm-deep beds; these also employ plate and pipe coils suspended in the bed to provide additional heat-transfer area. Vibrating dryers are not suitable for fibrous materials which mat or for sticky solids which may ball or adhere to the deck.

For estimating purposes for direct-heat drying applications, it can be assumed that the average exit-gas temperature leaving the solids bed will approach the final solids discharge temperature on an ordinary unit carrying a 5- to 15-cm-deep bed. Calculation of the heat load and selection of an inlet-air temperature and superficial velocity (Table 12-32) will then permit approximate sizing, provided an approximation of the minimum required retention time can be made.

Vibrating conveyors employing direct contacting of solids with hot, humid air have also been employed for the agglomeration of fine powders, chiefly for the preparation of agglomerated water-dispersible food products. Control of inlet-air temperature and dew point permits the uniform addition of small quantities of liquids to solids by con-

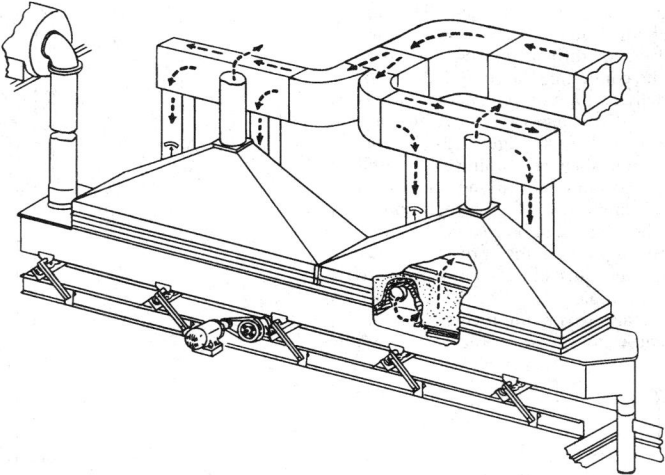

FIG. 12-93 Vibrating conveyor dryer. (*Carrier Division, Rexnord Inc.*)

densation on the cool incoming-particle surfaces. The wetting section of the conveyor is followed immediately by a warm-air-drying section and particle screening.

DISPERSION DRYERS

A gas-solids contacting operation in which the solids phase exists in a dilute condition is termed a **dispersion system.** It is often called a pneumatic system because, in most cases, the quantity and velocity of the gas are sufficient to lift and convey the solids against the force of gravity. Pneumatic systems may be distinguished by two characteristics:

1. Retention of a given solids particle in the system is on the average very short, usually no more than a few seconds. This means that any process conducted in a pneumatic system cannot be diffusion-controlled. The reaction must be mainly a surface phenomenon, or the solids particles must be very small so that heat transfer and mass transfer from the interiors are essentially instantaneous.

2. On an energy-content basis, the system is balanced at all times; i.e., there is sufficient energy in the gas (or solids) present in the system at any time to complete the work on all the solids (or gas) present at the same time. This is significant in that there is no lag in response to control changes or in starting up and shutting down the system; no partially processed residual solids or gas need be retained between runs.

It is for these reasons that pneumatic equipment is especially suitable for processing *heat-sensitive, easily oxidized, explosive,* or *flammable* materials which cannot be exposed to process conditions for extended periods.

Gas flow and solids flow are usually cocurrent, one exception being a countercurrent-flow spray dryer. The method of gas-solids contacting is best described as through circulation; however, in the dilute condition solids particles are so widely dispersed in the gas that they exhibit apparently no effect upon one another, and they offer essentially no resistance to the passage of gas among them.

Pneumatic-Conveyor Dryers A pneumatic-conveyor dryer consists of a long tube or duct carrying a gas at high velocity, a fan to propel the gas, a suitable feeder for addition and dispersion of particulate solids in the gas stream, and a cyclone collector or other separation equipment for final recovery of solids from the gas.

The solids feeder may be of any type; screw feeders, venturi sections, high-speed grinders, and dispersion mills are employed. For pneumatic conveyors, selection of the correct feeder to obtain thorough initial dispersion of solids in the gas is of major importance. For example, by employing an air-swept hammer mill in a drying operation, 65 to 95 percent of the total heat may be transferred within the mill itself if all the drying gas is passed through it. Fans may be of the induced-draft or the forced-draft type. The former is usually preferred because the system can then be operated under a slight negative pressure. Dust and hot gas will not be blown out through leaks in the equipment. Cyclone separators are preferred for low investment. If maximum recovery of dust or noxious fumes is required, the cyclone may be followed by a wet scrubber or bag collector.

In ordinary heating and cooling operations, during which there is no moisture pickup, continuous recirculation of the conveying gas is frequently employed. Also, solvent-recovery operations employing continuously recirculated inert gas with intercondensers and gas reheaters are carried out in pneumatic conveyors.

Pneumatic conveyors are suitable for materials which are granular and free-flowing when dispersed in the gas stream, so they do not stick on the conveyor walls or agglomerate. Sticky materials such as filter cakes may be dispersed and partially dried by an air-swept disintegrator in many cases. Otherwise, dry product may be recycled and mixed with fresh feed, and then the two dispersed together in a disintegrator. Coarse material containing internal moisture may be subjected to fine grinding in a hammer mill. The main requirement in all applications is that the operation be instantaneously completed; internal diffusion of moisture must not be limiting in drying operations, and particle sizes must be small enough so that the thermal conductivity of the solids does not control during heating and cooling operations. Pneumatic conveyors are rarely suitable for abrasive solids. Pneumatic conveying can result in significant particle-size reduction, particularly when crystalline or other friable materials are being handled. This may or may not be desirable but must be recognized if the system is selected. The action is similar to that of a fluid-energy grinder.

TABLE 12-32 Table for Estimating Maximum Superficial Air Velocities through Vibrating-Conveyor Screens*

Mesh size	Velocity, m/s	
	2.0 specific gravity	1.0 specific gravity
200	0.22	0.13
100	0.69	0.38
50	1.4	0.89
30	2.6	1.8
20	3.2	2.5
10	6.9	4.6
5	11.4	7.9

*Carrier Division, Rexnord Inc.

Pneumatic conveyors may be single-stage or multistage. The former is employed for evaporation of small quantities of surface moisture. Multistage installations are used for difficult drying processes, e.g., drying heat-sensitive products containing large quantities of moisture and drying materials initially containing internal as well as surface moisture. Typical single- and two-stage drying systems are illustrated in Figs. 12-94, 12-95a and b, and 12-96. Figure 12-94 incorporates a single-stage dryer with a second stage containing a cage-mill disintegrator. The second stage ensures complete drying after thorough dispersion of lumps and agglomerates. If disintegration is required to disperse the wet feed, the stages can be reversed, or disintegration can be employed in both stages. Systems of the type illustrated are employed for drying synthetic resins, of which low-pressure polyethylene and polypropylene are examples.

Figure 12-96 illustrates a single-stage dryer employing a paddle mixer, recycle, and an ABB Raymond cage mill for fine grinding and

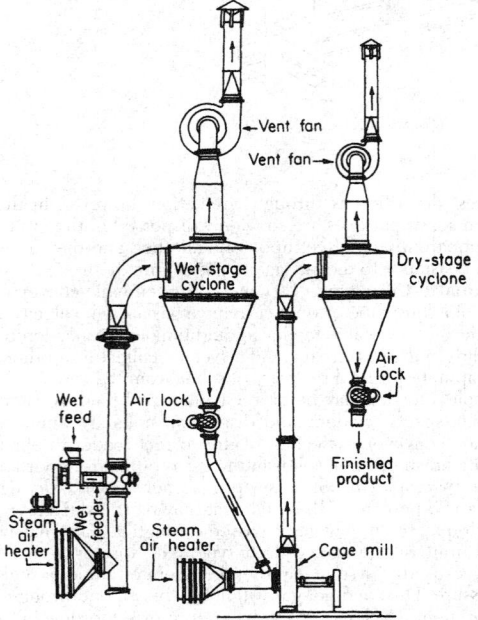

FIG. 12-94 Two-stage air-stream and cage mill, pneumatic-conveyor dryer. (*ABB Raymond.*)

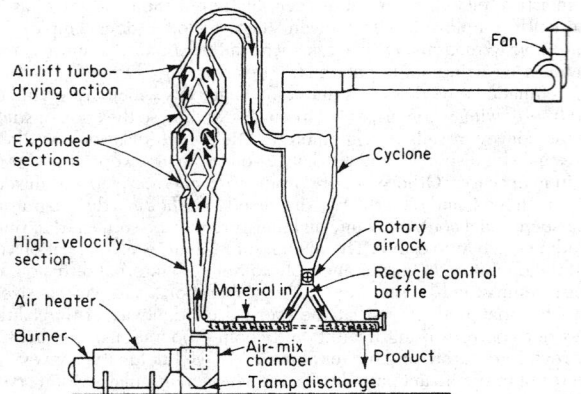

FIG. 12-95a Air-lift pneumatic-conveyor dryer; includes partial recycle of dry product and expanding tube and cone sections to provide longer holdup for coarse particles. (*Bepex Corp.*)

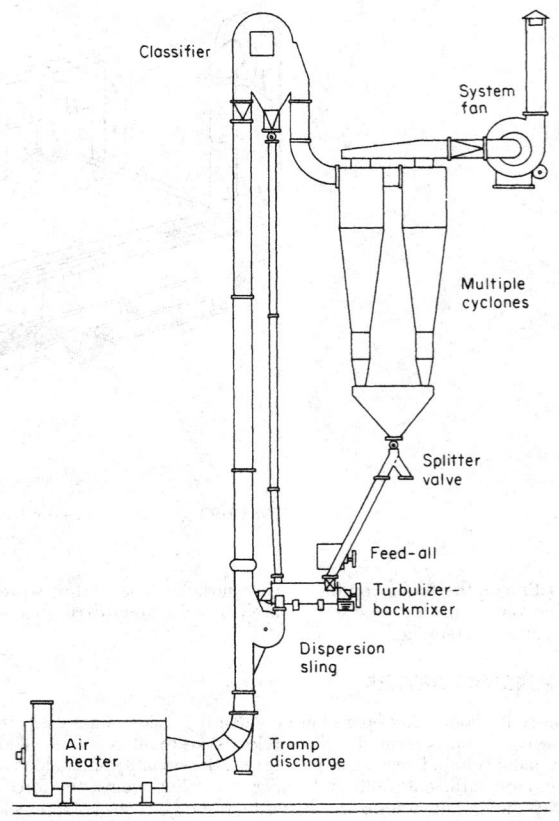

FIG. 12-95b Strong-Scott flash dryer with integral coarse-fraction classifier to separate undried particles for recycle. (*Bepex Corp.*)

dispersion of the mixed feed in the air stream. These units are designed to handle filter and centrifuge cakes and other sticky or pasty feeds. Capacities are given in Table 12-33.

Several typical products dried in pneumatic conveyors are described in Table 12-34. The air-stream type referred to is an ordinary single-stage dryer like the first stage of Fig. 12-94.

Design Methods for Pneumatic-Conveyor Dryers   Depending upon the temperature sensitivity of the product, inlet-air temperatures between 400 and 1000 K are employed. With a heat-sensitive solid, a high initial moisture content should permit use of a high inlet-air temperature. Evaporation of surface moisture takes place at essentially the wet-bulb air temperature. Until this has been completed, by which time the air will have cooled significantly, the surface-moisture film prevents the solids temperature from exceeding the wet-bulb temperature of the air. Pneumatic conveyors are used for solids having initial moisture contents ranging from 3 to 90 percent, wet basis. The air quantity required and solids-to-gas loading are fixed by the moisture load, the inlet-air temperature, and, frequently, the exit-air humidity. If the last is too great to permit complete drying, i.e., if the exit-air humidity is above that in equilibrium with the product at required dryness, the solids-to-gas loading must be reduced together with the inlet-air temperature.

The gas velocity in the conveying duct must be sufficient to convey the largest particle. This may be calculated accurately by methods given in Sec. 12. For estimating purposes, a velocity of 25 m/s, calculated at the exit-air temperature, is frequently employed. If mainly surface moisture is present, the temperature driving force for drying will approach the log mean of the inlet- and exit-gas wet-bulb depressions. (The exit solids temperature will approach the exit-gas dry-bulb temperature.)

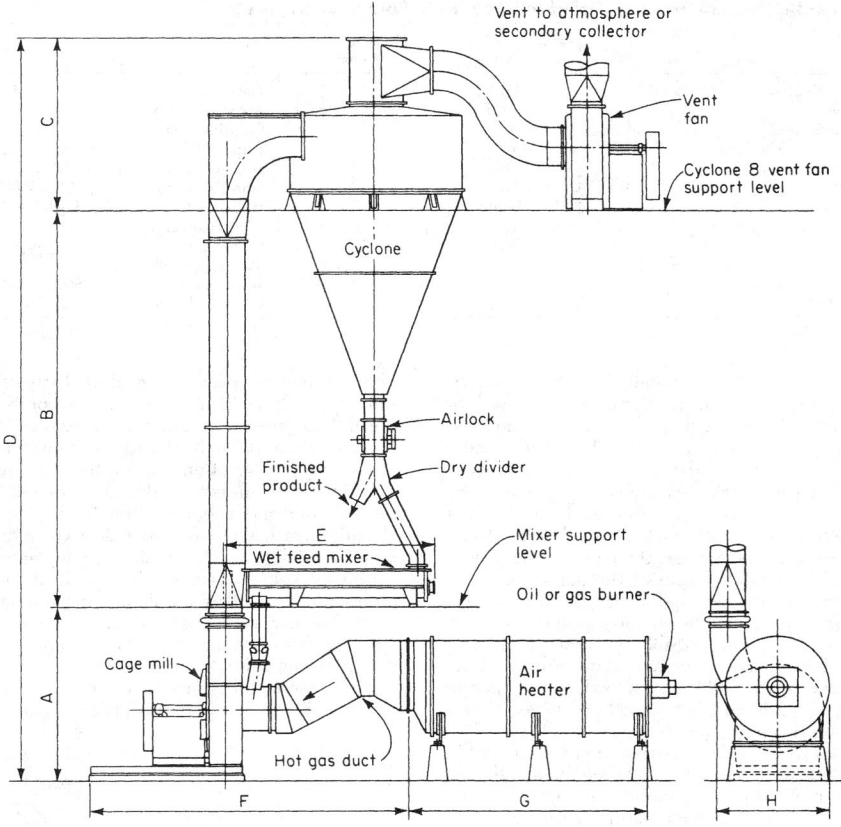

FIG. 12-96   Single-stage pneumatic-conveyor dryer. (*ABB Raymond.*)

Observation of operating conveyors indicates that the solids are rarely uniformly dispersed in the gas phase. With infrequent exceptions, the particles move in a laminar pattern, following a streamline along the duct wall where the flow velocity is at a minimum. Complete or even partial diffusion in the gas phase is rarely experienced even with low-specific-gravity particles. Air velocities may approach 20 to 30 m/s. It is doubtful, however, that even finer and lighter materials reach more than 80 percent of this speed, while heavier and larger fractions may travel at much slower rates [Fischer, *Mech. Eng.*, **81**(11), 67–69 (1959)]. Very little information and operating data on pneumatic-conveyor dryers which would permit a true theoretical basis for design have been published. Therefore, firm design always requires pilot tests. It is believed, however, that the significant velocity effect in a pneumatic conveyor is the difference in velocities between gas and solids, which is the reason why a major part of the total drying actually occurs in the feed section.

TABLE 12-33   Sizes and Capacities of ABB-Raymond Cage-Mill Flash-Drying Systems*

Based upon 1200°F inlet temperatures†

Cyclone	A	B	C	D	E	F	G	Approximate weight, lb H	Evaporation capacity, lb water/h†	Price,‡ FOB shops
18 ft 0 in	18 ft 0 in	38 ft 0 in	20 ft 0 in	76 ft 0 in	19 ft 6 in	49 ft 6 in	10 ft 0 in	115,000	20,000	$550,000
16 ft 0 in	15 ft 0 in	34 ft 0 in	17 ft 4 in	66 ft 4 in	17 ft 0 in	45 ft 6 in	9 ft 9 in	85,000	15,700	475,000
14 ft 0 in	14 ft 0 in	31 ft 0 in	14 ft 4 in	59 ft 4 in	16 ft 6 in	43 ft 4 in	9 ft 6 in	75,000	12,000	400,000
12 ft 0 in	13 ft 0 in	29 ft 0 in	12 ft 6 in	54 ft 6 in	15 ft 3 in	40 ft 4 in	8 ft 2 in	65,000	8,850	375,000
10 ft 0 in	11 ft 6 in	26 ft 0 in	10 ft 9 in	48 ft 3 in	13 ft 3 in	37 ft 8 in	7 ft 1 in	50,000	6,150	325,000
9 ft 0 in	11 ft 0 in	24 ft 0 in	10 ft 3 in	45 ft 3 in	13 ft 3 in	36 ft 8 in	6 ft 11 in	45,000	5,000	300,000
8 ft 0 in	10 ft 6 in	24 ft 0 in	9 ft 6 in	44 ft 0 in	12 ft 6 in	33 ft 11 in	6 ft 6 in	37,000	3,940	285,000
7 ft 0 in	10 ft 0 in	23 ft 0 in	8 ft 9 in	41 ft 9 in	12 ft 6 in	32 ft 3 in	6 ft 3 in	32,000	3,000	270,000
6 ft 0 in	9 ft 6 in	22 ft 0 in	8 ft 0 in	39 ft 6 in	11 ft 0 in	30 ft 9 in	5 ft 7 in	28,000	2,220	250,000
5 ft 0 in	9 ft 0 in	20 ft 0 in	7 ft 3 in	36 ft 3 in	10 ft 9 in	28 ft 4 in	5 ft 5 in	21,000	1,540	230,000
4 ft 0 in	8 ft 6 in	19 ft 0 in	6 ft 3 in	33 ft 9 in	10 ft 9 in	27 ft 1 in	4 ft 0 in	17,000	985	210,000

*ABB Raymond
†With inlet temperature of 600 to 700°F, consider the water evaporation to be one-half of that listed in the table. Considerably lower inlet temperatures are also frequently used for many materials.
‡Price based upon carbon steel construction (motors and secondary dust collectors by others). (1994)

TABLE 12-34 Typical Products Dried in Pneumatic-Conveyor Dryers*

Material	Initial moisture, wet basis, %	Final moisture, wet basis, %	Remarks
Clay, acid-treated	60	18	Cage-mill type
Coal, 1 cm × 0	11.5	1.5	Air-stream type
Corn-gluten feed	65	20	Cage-mill type
Kaolin (H$_2$O washed and partially dried)	10	0.5	Imp-mill type; grind to 99.9%—325 mesh
Gluten (vital wheat)	70	10	Imp-mill type; grind to—80 mesh
Clay, ball	25	0.5	Imp-mill type; grind to 95%—100 mesh
Gypsum, raw	25 total	5	Imp-mill type; grind and calcine to stucco
Pharmaceuticals	15	4	Air-stream type
Silica-gel catalyst	53	10	Cage-mill type
Synthetic resin	50	0.5	Two-stage system
Carboxymethyl cellulose	40	3	Imp-mill type; grind to 80%—200 mesh
Sewage sludge	82	0	Cage-mill-type

*ABB Raymond

One manner in which size may be computed, for estimating purposes, is by employing a volumetric heat-transfer concept as used for rotary dryers. If it is assumed that contacting efficiency is in the same order as that provided by efficient lifters in a rotary dryer and that the velocity difference between gas and solids controls, Eq. (12-52) may be employed to estimate a volumetric heat-transfer coefficient. By assuming a duct diameter of 0.3 m (D) and a gas velocity of 23 m/s, if the solids velocity is taken as 80 percent of this speed, the velocity difference between the two would be 4.6 m/s. If the exit gas has a density of 1 kg/m^3, the relative mass flow rate of the gas G becomes 4.8 kg/(s·m^2); the volumetric heat-transfer coefficient is 2235 J/(m^3·s·K). This is not far different from many coefficients found in commercial installations; however, it is usually not possible to predict accurately the actual difference in velocity between gas and solids. Furthermore, the coefficient is influenced by the solids-to-gas loading and particle size, which control the total solids surface exposed to the gas. Therefore, the figure given is only an approximation.

For estimating purposes, the conveyor cross section is fixed by the assumed air velocity and quantity. The volume, hence the length, can then be calculated by the method just presented, employing the log-mean air wet-bulb depression for the temperature driving force. A conveyor length >50 diameters is rarely required. Scale-up of pneumatic conveying dryers is outlined in a theoretical model which predicts dryer performance but which depends on uncertain parameters such as wall friction and agglomeration. [I. C. Kemp [*Drying Technology—An International Journal*, **12**(1&2), 279 (1994)].

Pressure drop in the system may be computed by methods described in Sec. 6. To prevent excessive leakage into or out of the system, which may have a total pressure drop of 20 to 38 cm of water, rotary air locks or screw feeders are employed at the solids inlet and discharge.

The conveyor and collector parts are thoroughly insulated to reduce heat losses in drying and other heating operations. Operating control is maintained usually by control of the exit-gas temperature, with the inlet-gas temperature varied to compensate for changing feed conditions. A constant solids feed rate must be maintained.

Cost Data for Pneumatic-Conveyor Dryers Purchase costs vary widely; many pneumatic-conveyor installations are assembled units, each component being purchased from a different supplier. Representative prices are given in Table 12-33 for conventional pneumatic-conveyor dryers. These include a cage mill for disintegration and a primary cyclone collector. In general, pneumatic conveyors for similar duties will compete in cost with cocurrent rotary dryers. Space economics may reduce the total installed investment slightly below that of the rotary unit. Operating costs, thermal efficiency, etc., are similar to those of cocurrent rotary units sized for the same duty. When other operations, such as conveying, grinding, or classifying, are simultaneously performed, operating and investment costs may be reduced for the pneumatic-drying process itself by being partially written off on the secondary function. In this situation, a pneumatic conveyor becomes particularly attractive.

Ring Dryers The ring dryer is a development of flash, or pneumatic-conveyor, drying technology, designed to increase the versatility of application of this technology and overcome many of its limitations.

One of the great advantages of flash drying is the very short retention time, typically no more than a few seconds. However, in a conventional flash dryer, residence time is fixed and this limits its application to materials in which the drying mechanism is not diffusion controlled and where a range of moisture within the final product is acceptable. The ring dryer offers two advantages over the flash dryer. Principally there is control of residence time by the use of an adjustable internal classifier which allows fine particles, which dry quickly, to leave, while larger particles, which dry slowly, have an extended residence time within the system. Second, the combination of the classifier with an internal mill can allow simultaneous grinding and drying with control of product particle size and moisture. Available with a range of different feed systems to handle a variety of applications, the ring dryer provide wide versatility.

The essential difference between a conventional flash dryer and the ring dryer is the "manifold centrifugal classifier." The manifold pro-

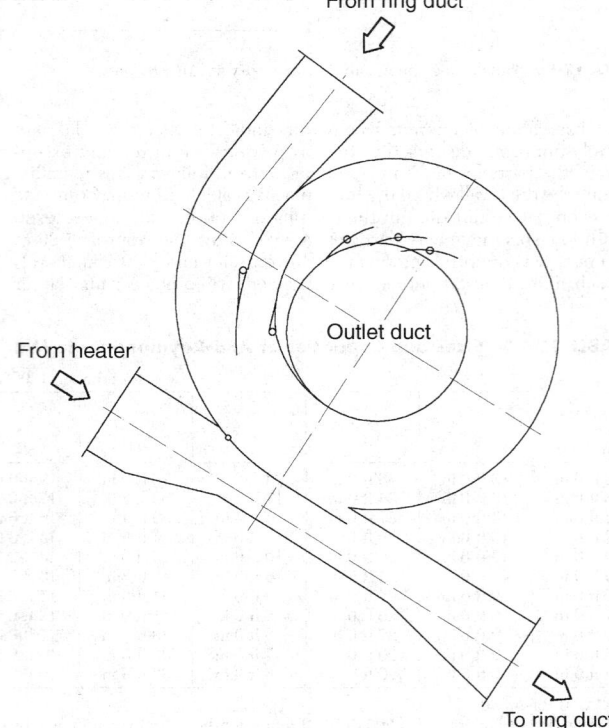

FIG. 12-97 Full manifold classifier for ring dryer. (*Barr—Rosin*)

vides classification of the product about to leave the dryer using differential centrifugal force. The manifold, as shown in Fig. 12-97, uses the centrifugal effect of an air stream passing around the curve to concentrate the product into a moving layer, with the dense material on the bottom and the light material on the top. This enables the adjustable splitter blades within the manifold classifier to segregate the heavier, wetter material and return it for a further circuit of drying. Fine, dried material is allowed to leave the dryer with the exhaust air and pass to the product collection system. This selective extension of residence time ensures a more evenly dried material than possible from a conventional flash dryer. Many materials which have traditionally been regarded as difficult to dry can be processed to the required moisture content in a ring dryer.

The recycle requirements of products in different applications can vary substantially depending upon the scale of operation, the ease of drying, and the finished-product specification. The location of reintroduction of undried material back into the drying medium has a significant impact upon the dryer performance and final-product characteristics.

Three configurations of the ring dryer have been developed to offer flexibility in design and optimal performance:

1. *Single-stage manifold—vertical configuration.* The feed ring dryer (see Fig. 12-98) is similar to a flash dryer but incorporates a single-stage classifier, which diverts 40–60 percent of the product back to the feed point. The feed ring dryer is ideally suited for materials which are neither heat sensitive nor require a high degree of classification. An advantage of this configuration is that it can be manufactured to very large sizes to achieve high-evaporative capacities.

2. *Full manifold—horizontal configuration.* The full ring dryer (see Fig. 12-99) incorporates a multistage classifier which allows much higher recycle rates than the single-stage manifold. This configuration usually incorporates a disintegrator which provides adjustable amounts of product grinding depending upon speed and manifold setting. For sensitive or fine materials, the disintegrator can be omitted. Alternative feed locations are available to suit the material sensitivity and the final-product requirements. The full ring configuration gives a very high degree of control of both residence time and particle size, and is used for a wide variety of applications from small production rates of pharmaceutical and fine chemicals to large production rates of food products, bulk chemicals, and minerals. This is the most versatile configuration of the ring dryer.

3. *P-type manifold—vertical configuration.* The P ring dryer (see Fig. 12-100) incorporates a single-stage classifier and was developed specifically for use with heat-sensitive materials. The undried material is reintroduced into a cool part of the dryer in which it recirculates until it is dry enough to leave the circuit.

An important element in optimizing the performance of a flash or ring dryer is the degree of dispersion at the feed point. Maximizing the product surface area in this region of highest evaporative driving force is a key objective in the design of this type of dryer. Ring dryers are fed using similar equipment to conventional flash dryers. Ring dryers with vertical configuration are normally fed by a flooded screw and a disperser which propels the wet feed into a high-velocity venturi, in which the bulk of the evaporation takes place. The full ring dryer normally employs an air-swept disperser or mill within the drying circuit to provide screenless grinding when required. Together with the manifold classifier this ensures a product with a uniform particle size. For liquid, slurry, or pasty feed materials, backmixing of the feed with a portion of the dry product will be carried out in order to produce a conditioned friable material. This further increases the versatility of the ring dryer, allowing it to handle sludge and slurry feeds with ease.

Dried product is collected in either cyclones or baghouses depending upon the product-particle size. When primary collection is carried out in cyclones, secondary collection in a baghouse or scrubber is usually necessary in order to comply with environmental regulations. A rotary valve is used to provide an airlock at the discharge point. Screws are utilized to combine product from multiple cyclones or large baghouses. If required, a portion of the dried product is separated from the main stream and returned to the feed system for use as backmix.

Design Methods for Ring Dryers Depending on the tempera-

ture sensitivity of the material to be processed, air inlet temperatures as high as 1400°F can be utilized. Even with heat-sensitive solids a high feed-moisture content may permit the use of a high air-inlet temperature since evaporation of surface moisture takes place at the wet-bulb air temperature. Until the surface moisture has been removed it will prevent the solids temperature from exceeding the air wet-bulb temperature, by which time the air will generally have cooled significantly. Ring dryers have been used to process materials with feed-moisture contents between 2–95 percent, weight fraction. The product-moisture content has been controlled to values from 20 percent down to less than 1 percent. The air velocity required and air-to-solids ratio are determined by the evaporative load, the air-inlet temperature, and the exhaust-air humidity. Too high an exhaust-air humidity would prevent complete drying so a lower air-inlet temperature and air-to-solids ratio would be required. The air velocity within the dryer must be sufficient to convey the largest particle, or agglomerate. The air-to-solids ratio must be high enough to convey both the product and backmix, together with internal recycle from the manifold. For estimating purposes a velocity of 5000 ft/min, calculated at dryer exhaust conditions, is appropriate both for pneumatic-conveyor and ring dryers. Sizes, capacity, and costs for full manifold ring dryers are described on Table 12-35.

Spray Dryers A spray dryer consists of a large cylindrical and usually vertical chamber into which material to be dried is sprayed in the form of small droplets and into which is fed a large volume of hot gas sufficient to supply the heat necessary to complete evaporation of the liquid. Heat transfer and mass transfer are accomplished by direct contact of the hot gas with the dispersed droplets. After completion of drying, the cooled gas and solids are separated. This may be accomplished partially at the bottom of the drying chamber by classification and separation of the coarse dried particles. Fine particles are separated from the gas in external cyclones or bag collectors. When only the coarse-particle fraction is desired for finished product, fines may be recovered in wet scrubbers; the scrubber liquid is concentrated and returned as feed to the dryer. Horizontal spray chambers are manufactured with a longitudinal screw conveyor in the bottom of the drying chamber for continuous removal of settled coarse particles.

The principal use of spray dryers is for ordinary drying of water solutions and slurries. They are used also in combined drying and heat-treating operations, and for melt fusion and cooling of molten materials, e.g., ammonium nitrate *prilling*. The latter may be considered a solids size-enlargement process. Spray dryers are employed for wet-agglomeration processes to produce rapidly dispersible forms of concentrated food products, another form of size enlargement. In contacting performance, the spray dryer is similar to a pneumatic conveyor. It differs in application in that the feed material is usually a liquid solution, slurry, or paste capable of being dispersed in a fluidlike spray (rather than being composed of free-flowing particulate solids).

Spray drying involves three fundamental unit processes: (1) liquid atomization, (2) gas-droplet mixing, and (3) drying from liquid droplets. Atomization is accomplished usually by one of three atomizing devices: (1) high-pressure nozzles, (2) two-fluid nozzles, and (3) high-speed centrifugal disks. With these atomizers, thin solutions may be dispersed into droplets as small as 2 μm. The largest drop sizes rarely exceed 500 μm (35 mesh). Because of the large total drying surface and small droplet sizes created, the actual drying time in a spray dryer is measured in seconds. Total residence of a particle in the system is on the average not more than 30 s. A review by Marshall ["Atomization and Spray Drying," *Chem. Eng. Prog. Monogr. Ser.*, **50**, 2 (1954)] considers spray-drying theory in detail as well as the design and operating characteristics of modern spray dryers. A later survey of spray drying, which constitutes a good supplement to Marshall, was published by Masters [*Ind. Eng. Chem.*, **60**(10), 53–63 (1968)]. A more recent article summarizing scale-up of spray dryers has also been published by K. Masters [*Drying Technology—An International Journal,* **12**(1&2), 235 (1994)]. Recently there has been considerable interest in the scale-up and analysis of spray dryers using computational fluid dynamics. D. E. Oakley [*Drying Technology—An International Journal,* **12**(1&2), 217 (1994)]. Liquid atomization and dispersion are discussed in detail in Sec. 14. Atomizers commonly employed on spray dryers are described briefly in the following paragraphs.

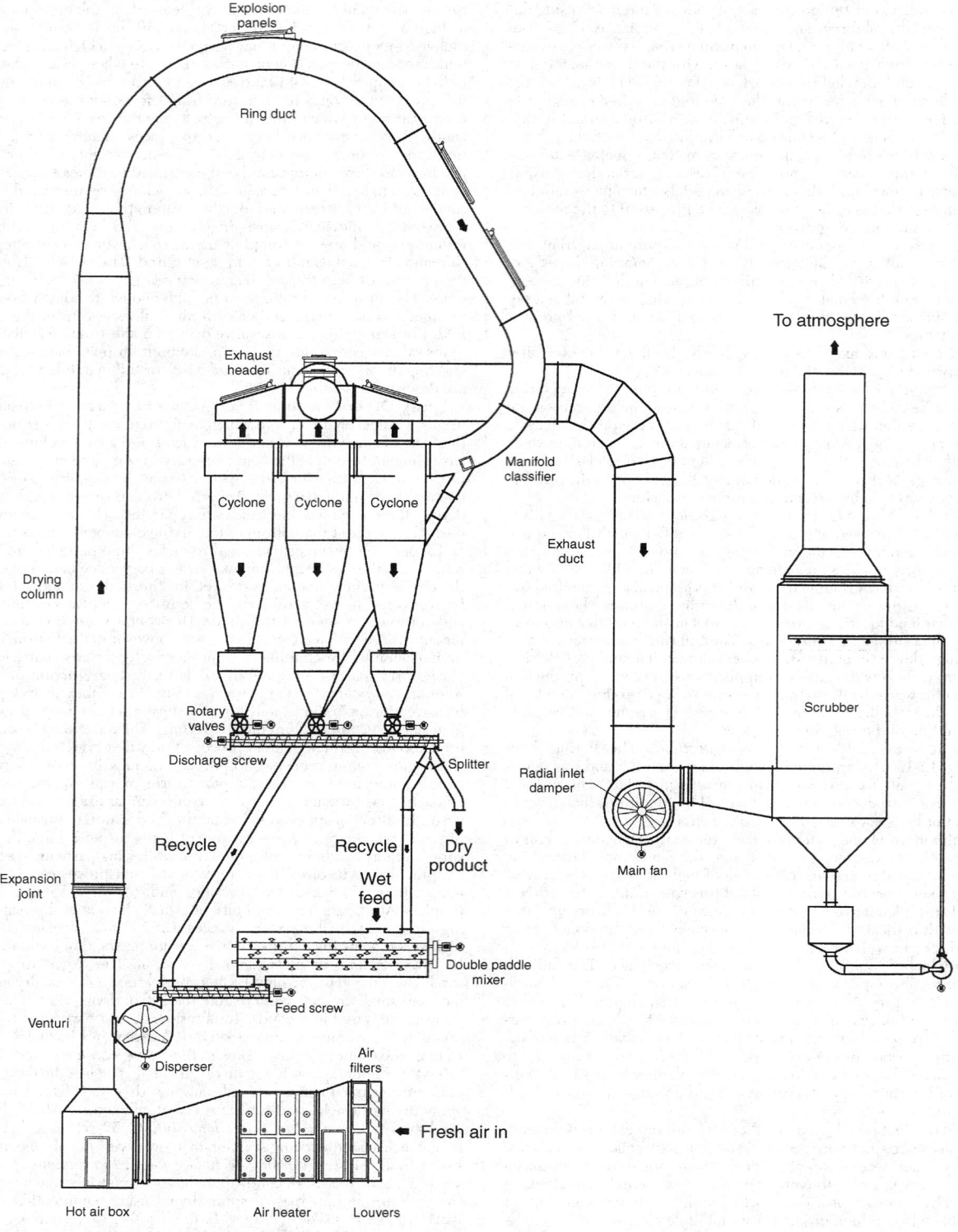

FIG. 12-98 Feed-type ring dryer. (*Barr—Rosin*)

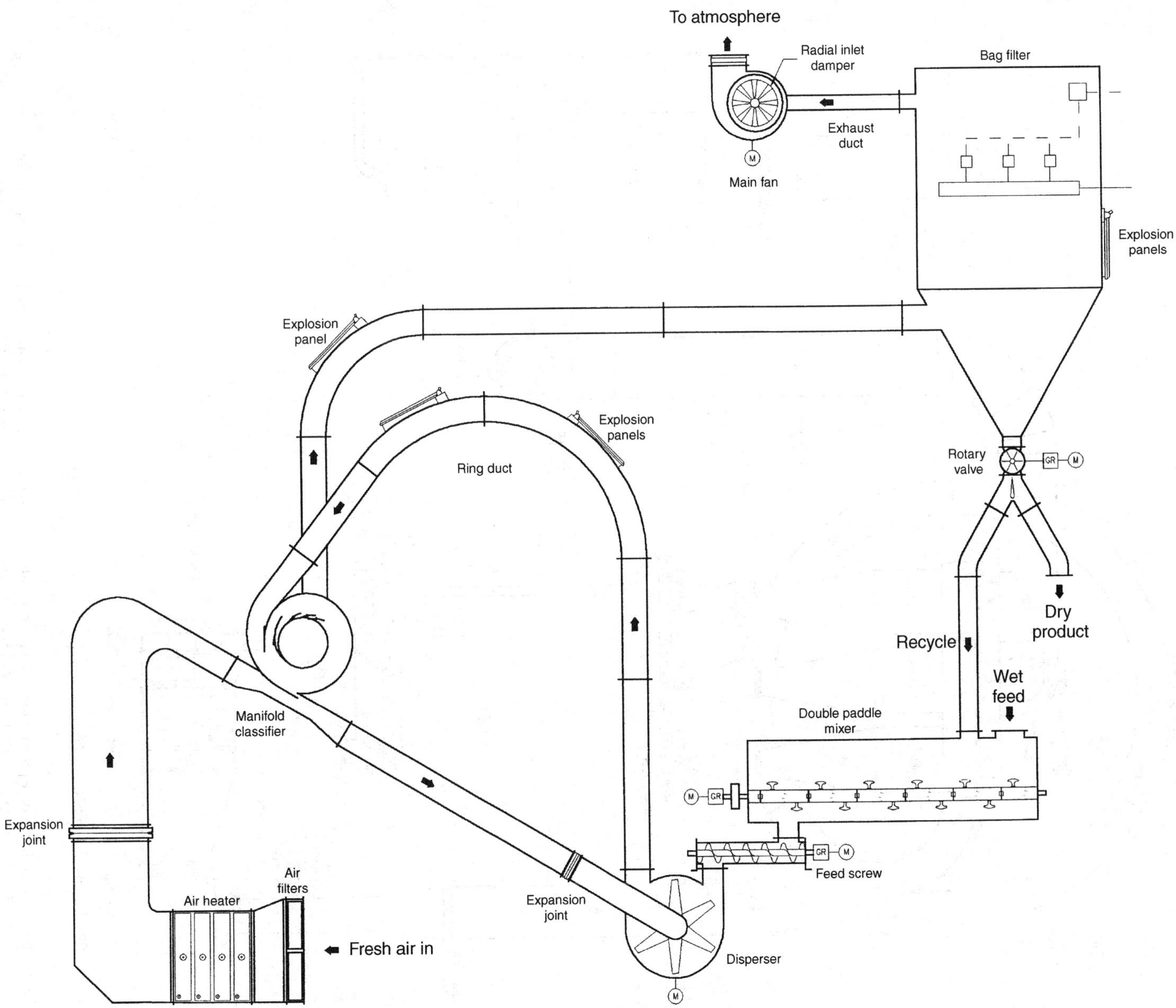

FIG. 12-99 Full manifold-type ring dryer. (*Barr—Rosin*)

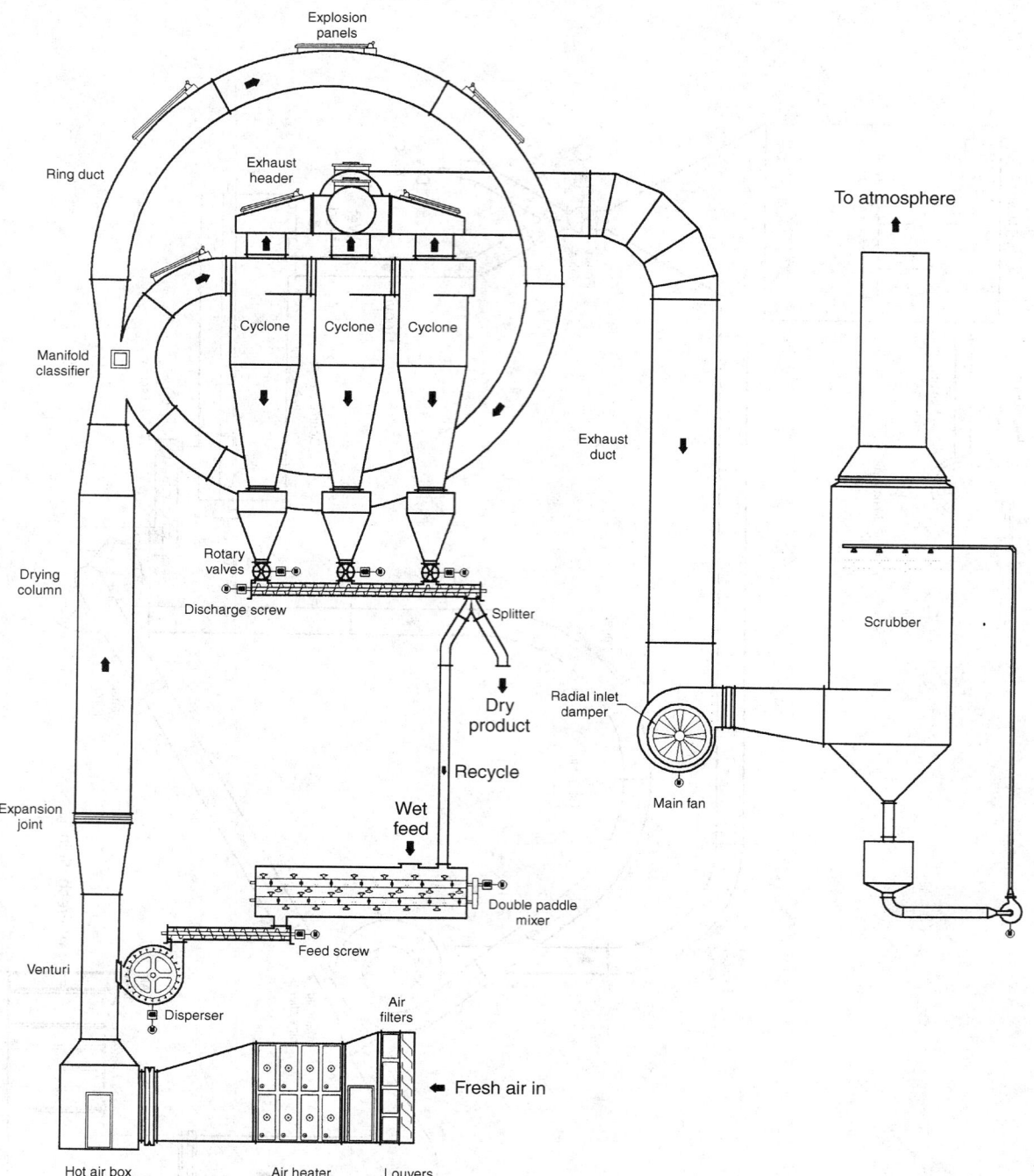

FIG. 12-100   P-type ring dryer. (*Barr—Rosin*)

Special designs of spray dryers may provide for cooling air to enter around the chamber, closed systems for the recovery of solvents, and air sweepers or mechanical rakes to remove dry product from the walls and bottom of the chamber. Some are followed by pneumatic conveyors as depicted in Fig. 12-101, in which drying air is diluted with cool air for product cooling before separation. Spray dryers may operate with cocurrent, mixed, or countercurrent flow of gas and solids. Inlet-gas temperatures may range from 425 to 1100 K.

Pressure nozzles effect atomization by forcing the liquid under high pressure and with a high degree of spin through a small orifice. Pressures may range from 2700 to 69,000 kPa/m², depending on the degree of atomization, capacity, and physical properties. Nozzle ori-

TABLE 12-35   Sizes and Capacities of Full Ring Dryers*

Evaporative capacity (lb/hr)	Nominal area in plan (ft × ft)	Nominal height (ft)	Price FOB shops (US$)
1000	7 × 20	20	125,000
2000	10 × 30	24	200,000
5000	12 × 35	28	300,000
10000	18 × 45	32	500,000
15000	20 × 50	35	600,000
20000	22 × 55	38	700,000
40000	25 × 60	40	1,000,000

*Courtesy of Barr—Rosin.
Based upon 1200°F air-inlet temperature. With an inlet temperature of 700°F the evaporative capacity would be half that given in the table. Prices are based upon carbon steel construction with cyclone collector (January, 1995). Motors and secondary dust collectors excluded.

fices may range in size from 0.25- to 0.4-mm diameter, depending on the pressure desired for a given capacity and the degree of atomization required. For high pressures and when solids are in suspension in the liquid, the nozzle orifice will be subject to wear by erosion, and the orifice should be made of a hard alloy such as tungsten carbide or stellite. Maintenance on pressure nozzles is always a problem since erosion occurs with even the hardest inserts, and once the orifice has become scratched and nonuniform, good atomization is no longer possible. Likewise, incrustation and plugging by particles of foreign matter cause trouble. Piston pumps furnish the liquids at high pressure; erosion of the valves in these pumps is another maintenance problem.

Spray characteristics of pressure nozzles depend on the pressure and nozzle-orifice size. Pressure affects not only the spray characteristics but also the capacity. If it is desired to reduce the amount of liquid sprayed by lowering the pressure, then the spray may become coarser. To correct this, a smaller orifice would be inserted, which might then require a higher pressure to produce the desired capacity, and a spray that would be finer than desired might result. Multiple nozzles tend to overcome this inflexible characteristic of pressure atomization, although several nozzles on a dryer complicate the chamber design and air-flow pattern and risk collision of particles, resulting in nonuniformity of spray and particle size.

Two-fluid nozzles do not operate efficiently at high capacities and consequently are not used widely on plant-size spray dryers. Their chief advantage is that they operate at relatively low pressure, the liquid being 0 to 400 kPa/m² pressure, while the atomizing fluid is usually no more than 700 kPa/m² pressure. The atomizing fluid may be steam or air. Two-fluid nozzles have been employed for the dispersion of thick pastes and filter cakes not previously capable of being handled in ordinary atomizers [Baran, *Ind. Eng. Chem.*, **56**(10), 34–36 (1964); and Turba, *Brit. Chem. Eng.*, **9**(7), 457–460 (1964)].

Centrifugal disks atomize liquids by extending them in thin sheets which are discharged at high speeds from the periphery of the rapidly rotating, specially designed disk. The principal objectives in disk design are to ensure bringing the liquid to disk speed and to obtain a uniform drop-size distribution in the atomized liquid. Disk diameters range from 5 cm in small laboratory models to 35 cm in plant-size dryers. Disk speeds range from 3000 to 50,000 r/min. The high speed is generally used in small-diameter dryers. Usual speeds on plant-size dryers range from 4000 to 20,000 r/min, depending on disk diameter and the degree of atomization desired. The degree of atomization as a function of disk speed is affected by the product of disk diameter and speed, i.e., by peripheral speed as opposed to angular speed. Thus, a 13-cm disk operating at 30,000 r/min would be expected to atomize more finely than a 5-cm disk of the same design running at 50,000 r/min.

Centrifugal-disk atomization is particularly advantageous for atomizing suspensions and pastes that erode and plug nozzles. Thick pastes can be handled if positive-pressure pumps are used to feed them to the disk. Disks are capable of operating over a wide range of feed rates and disk speeds without producing too variable a product. Centrifugal disks may be belt-driven, direct-driven by a high-speed electric motor powered by a frequency changer, or driven by a steam turbine. Direct drive by an electric motor has advantages when very high speeds are required and when closely controlled speed variations are necessary. The life of high-speed bearings in centrifugal-disk atomizers depends on the conditions of operation. Average life may be 2000 h. A spare spray machine should be standard equipment.

The particle-size distribution obtained by any one of the three methods of atomization depends on a number of factors. In general, the size distribution will depend on atomizer design, liquid properties, and degree of atomization. If the finest atomization possible is attempted, a limiting condition is approached, and the particle-size range, regardless of the method of atomization, will be narrow. This is particularly true of pressure nozzles, in which uniformity of size increases with pressure. On the other hand, for the production of a coarse product with a high percentage of large particles, the method of atomization will have a large effect on the particle-size distribution. Production of uniform coarse particles from centrifugal disks frequently can be obtained by careful design.

One of the principal advantages of spray drying is the production of a **spherical particle,** which is usually not obtainable by any other drying method. This spherical particle may be solid or hollow, depending on the material, the feed condition, and the drying conditions. In general, aqueous solutions of materials such as soap, gelatin, and water-soluble polymers which form tough tenuous outer skins on drying will form hollow spherical particles when spray-dried. This is attributed to the formation of a casehardened outer surface on the particle which prevents liquid from reaching the surface from the particle interior. Because of high heat-transfer rates to the drops, the liquid at the center of the particle vaporizes, causing the outer shell to expand and form a hollow sphere. Sometimes the rate of vapor generation within the particle is sufficient to blow a hole through the wall of the spherical shell. Spherical particles may be obtained from true solutions or from slurries and may be produced by any of the previously described atomizers.

The physical properties of spray-dried materials are subject to considerable variation, depending on the direction of flow of the inlet gas and its temperature, the degree and uniformity of atomization, the solids content of the feed, the temperature of the feed, and the degree of aeration of the feed. The properties of the product usually of greatest interest are (1) **particle size,** (2) **bulk density,** and (3) **dustiness.** The particle size is a function of atomizer-operating conditions and also of the solids content, liquid viscosity, liquid density, and feed rate. In general, particle size increases with solids content, viscosity, density, and feed rate.

The bulk density of spray-dried solids is frequently the critical property subject to close control. The bulk density of material from a spray dryer may usually be increased by the following operating changes: (1) reducing droplet size, (2) reducing inlet-air temperature, (3) increasing air throughput, (4) increasing air turbulence, (5) employing countercurrent rather than cocurrent gas flow, and (6) effecting a wide range of size distribution from the atomizer. Chaloud et al. evaluated qualitatively the effects of operating variables on the bulk density of particles from detergent spray dryers [*Chem. Eng. Prog.*, **53,** 12, 593–596 (1957)].

A dusty product is caused by fine atomization or particle degradation after drying. Thin-wall hollow particles are susceptible to breakage during collection. Fine atomization and a high gas temperature contribute to high production rates in small drying chambers; they also generate fine particles and thin-wall spheres. Spray-drying installations yielding exceedingly fine and dusty products are often the result of an honest effort to design equipment for maximum capacity at a minimum investment. Large solids particles or heavy-wall spheres require longer drying cycles, hence larger drying chambers. Careful study in the pilot plant is necessary. In commercial installations, classification of particles and separation of a fine fraction from coarse product may be accomplished by countercurrent flow of gas and solids.

The majority of spray dryers in commercial use employ cocurrent flow of gas and solids. Countercurrent-flow dryers are used primarily for drying soaps and detergents. Their classifying ability is useful in these applications. Air flow is upward, carrying entrained fines from the top of the chamber. The coarse product settles and is removed

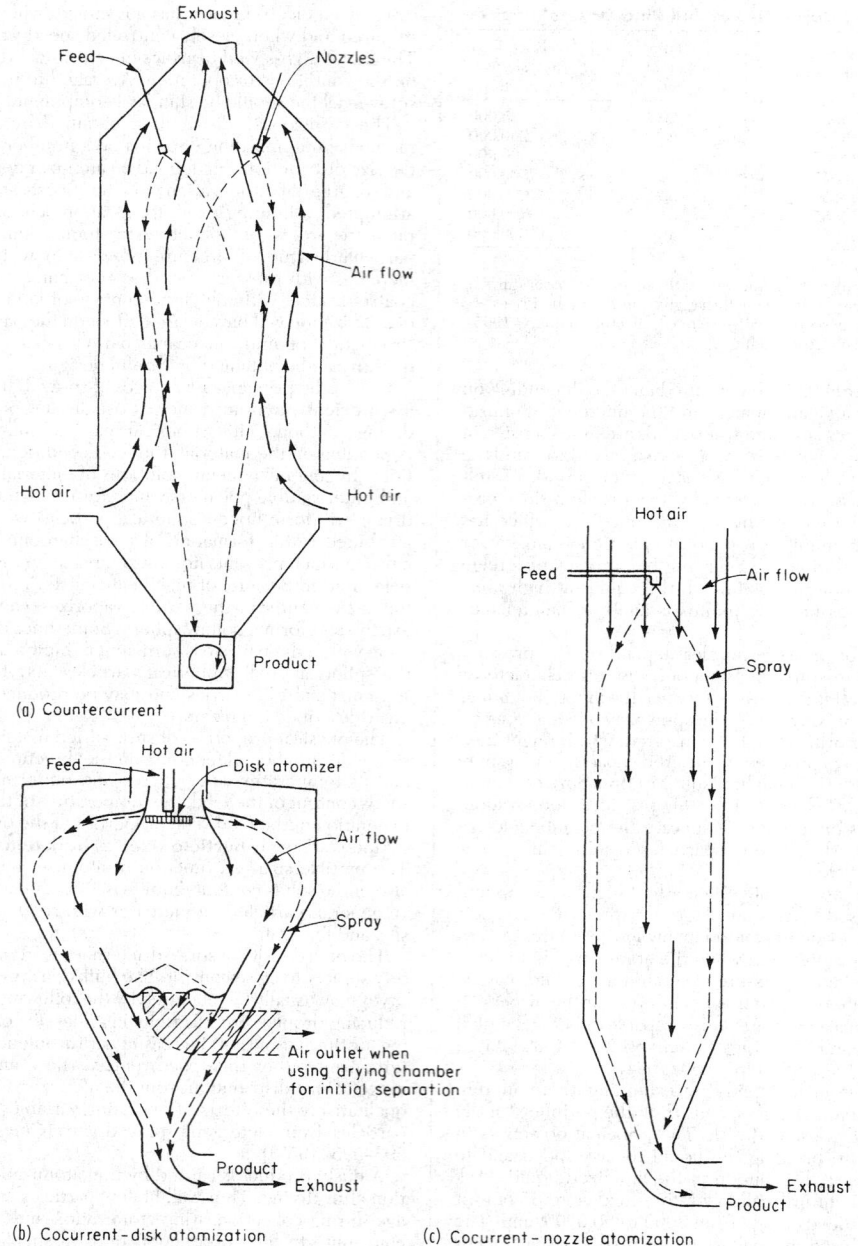

FIG. 12-101 Alternative chambers and gas solids contacting methods in spray dryers.

separately from the bottom of the chamber. Horizontal spray dryers always employ cocurrent flow of gas and solids. A swirling motion is imparted to the air to improve mixing. Many variations of air-flow patterns inside the drying chamber are employed commercially; most are intended primarily to produce turbulence and thorough mixing of gas and droplets and to achieve the most effective use of the chamber volume.

The treatment of drying gas external to the drying chamber may take a variety of forms depending on whether indirect heating (Figs. 12-102 and 103) or direct heating (Figs. 12-104 and 105) is selected; and whether the system is open (Fig. 12-104), semiclosed (Figs. 12-102 and 105), or *totally* closed (Fig. 12-103); and if solvent is flammable, toxic, or total solvent recovery is required.

Applications of Spray Dryers   The major and most successful drying applications of spray dryers are for solutions, slurries, and pastes which (1) cannot be dewatered mechanically, (2) are heat-sensitive and cannot be exposed to high-temperature atmospheres for long periods, or (3) contain ultrafine particles which will agglomerate and fuse if dried in other than a dilute condition. In other applications, spray drying is rarely competitive on a cost basis with two-step dewatering and solids-drying processes. The cost of bag collectors for solids recovery from large volumes of exit gas may double the cost of a spray-

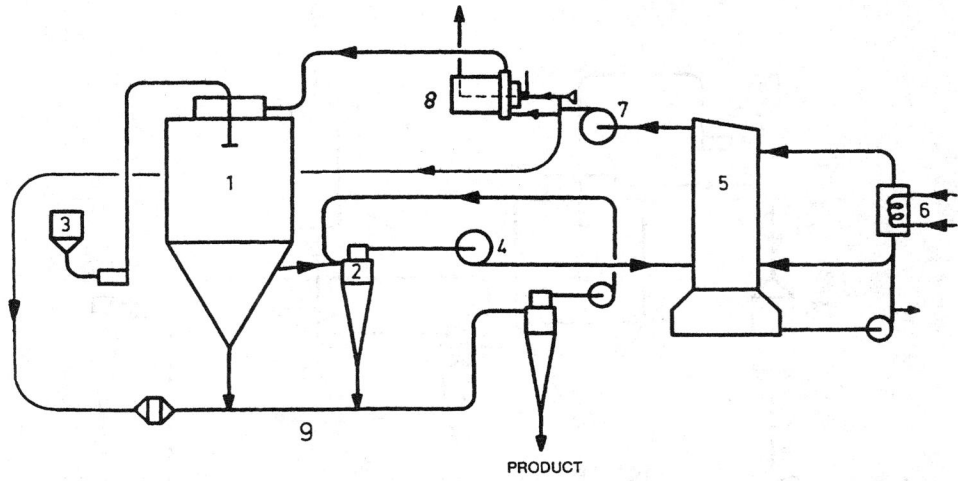

FIG. 12-102   Semiclosed spray-drying cycle with indirect heating. (*NIRO, Inc.*)

1. Drying chamber with atomizer
2. Dry collector
3. Feed system
4. Exhaust fan
5. Scrubber condenser
6. Cooler
7. Supply fan
8. Indirect heater
9. Pneumatic conveying system

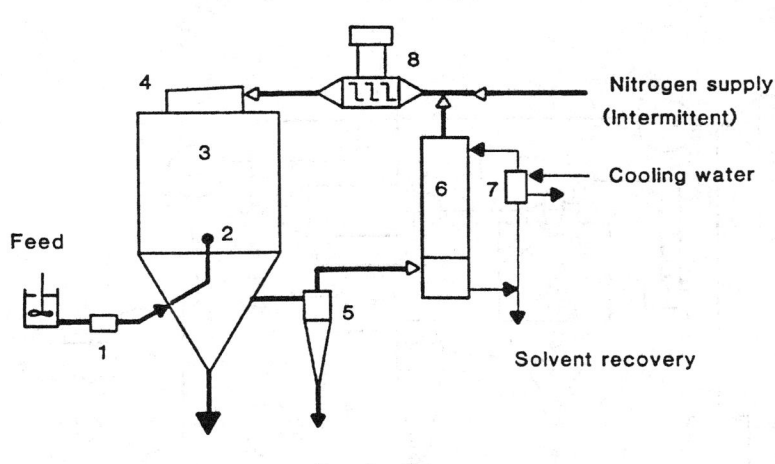

FIG. 12-103   Closed spray-drying cycle with indirect heating. (*NIRO, Inc.*)

1. Feed pump with agitated tanl
2. Nozzle atomizer
3. Drying chamber
4. Gas disperser
5. Cyclone
6. Scrubber condenser
7. Solvent cooler
8. Indirect drying inert gas heater

dryer installation. Additional costs must usually be justified on the basis of some improvement in product quality, such as particle form, size, flavor, color, or heat stability. Spray drying is applicable to heat-sensitive products such as milk powders and other foods and pharmaceuticals because of the short contact time in the dryer hot zone. Further, the water film on the liquid drop protects the solids from high gas temperatures. Drying is carried out at essentially the drying-air wet-bulb temperature. Color pigments are examples of the class of products for which it is desired to maintain as closely as possible the original solids particle size. Table 12-36 lists typical materials which have been successfully spray-dried. One other class of products particularly applicable to spray dryers is solids slurries, containing extremely fine particles, which is nonnewtonian in flow characteristics and remains fluid at very low moisture content. Certain classes of clays are found in this category. Also, spray dryers have been developed for encapsulation processes to convert liquid volatile flavors and perfumes to particulate solids forms [Maleeny, *Soap Chem. Spec.*, **34**, 1, 135–141 (1958)].

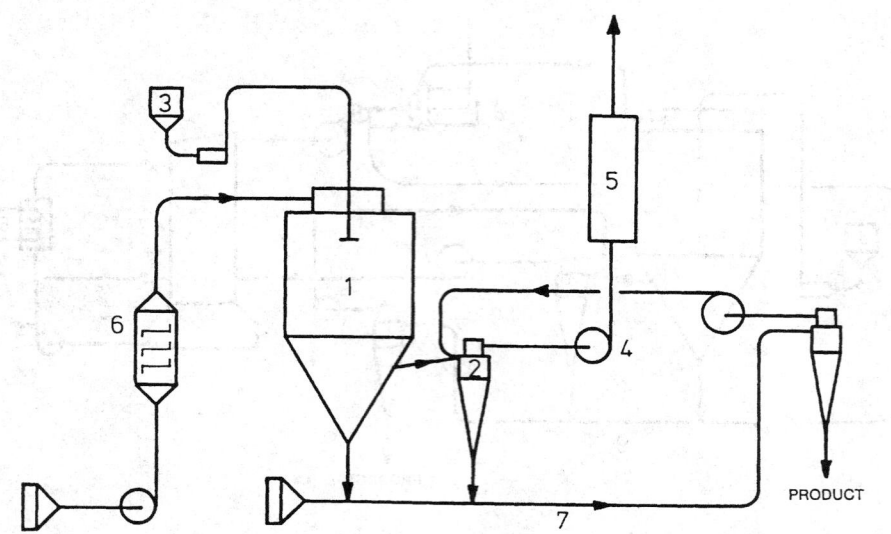

1. Drying chamber with atomizer
2. Dry collector
3. Feed system
4. Exhaust fan
5. Scrubber
6. Heater
7. Pneumatic conveying system

FIG. 12-104   Open spray-drying system with direct-fired heater. (*NIRO, Inc.*)

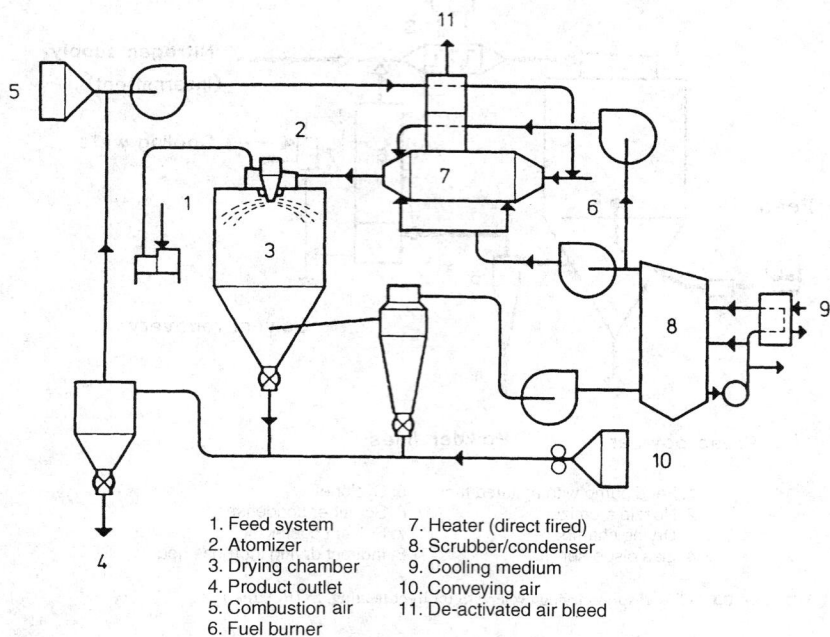

1. Feed system
2. Atomizer
3. Drying chamber
4. Product outlet
5. Combustion air
6. Fuel burner
7. Heater (direct fired)
8. Scrubber/condenser
9. Cooling medium
10. Conveying air
11. De-activated air bleed

FIG. 12-105   Semiclosed spray-drying cycle with direct heating. (*NIRO, Inc.*)

If the product in no way adheres to the dryer parts and simple cyclone collectors are sufficient for gas-solids separation, batch operation of a spray dryer may be considered. Otherwise, the time and costs for cleaning the large equipment parts make them rarely economical for other than continuous processing of a single material.

Design Methods for Spray Dryers   Design variables must be established by experimental tests before final design of a chamber can be carried out. In general, chamber size, atomizer selection, and separation auxiliaries will be determined by the desired physical characteristics of the product. Drying by itself is rarely a problem. An installed spray dryer is relatively inflexible in meeting changing operating requirements while maintaining a constant production rate.

TABLE 12-36 Some Materials That Have Been Successfully Spray-Dried in a 6-m-Diameter by 6-m-High Chamber with a Centrifugal-Disk Atomizer*

Material	Air temperature, K		% water in feed	Evaporation rate, kg/s
	In	Out		
Blood, animal	440	345	65	5.9
Yeast	500	335	86	8.2
Zinc sulfate	600	380	55	10.0
Lignin	475	365	63	6.9
Aluminum hydroxide	590	325	93	19.4
Silica gel	590	350	95	16.9
Magnesium carbonate	590	320	92	18.2
Tanning extract	440	340	46	5.2
Coffee extract	420	355	70	3.8
Detergent A	505	395	50	5.0
Detergent B	510	390	63	6.2
Detergent C	505	395	40	2.6
Manganese sulfate	590	415	50	5.5
Aluminum sulfate	415	350	70	1.7
Urea resin A	535	355	60	3.8
Urea resin B	505	360	70	1.9
Sodium sulfide	500	340	50	2.0
Pigment	515	335	73	13.2

*Courtesy of NIRO, Inc.
NOTE: The fan on this dryer handles about 5.2 m³/s at outlet conditions. The outlet-air temperature includes cold air in-leakage, and the true temperature drop caused by evaporation must therefore be estimated from a heat balance.

Important variables which must be fixed before design of a commercial dryer are the following:
1. The form and particle size of product required
2. The physical properties of the feed: moisture, viscosity, density, etc.
3. The maximum inlet-gas and product temperatures
Theoretical correlations of spray-dryer performance published by Gluckert [*Am. Inst. Chem. Eng. J.*, **8**(4), 460–466 (1962)] may be employed for the scale-up of laboratory dryers and, in some instances, for estimating dryer requirements in the absence of any tests.
Several assumptions are necessary.
1. The largest droplets, which dry most slowly, are the limiting portion of the spray. They determine ultimate chamber dimensions and are employed for the evaluation.
2. The largest droplet in a spray population is 3 times the diameter of the average drop size [see Eq. (12-66)].
3. A droplet Nusselt number = 2, corresponding to pure conduction (Reynolds number = 0) to infinity, is employed for evaluating the coefficient of heat transfer.
4. Drying conditions, because of turbulence and gas mixing, are uniform throughout the chamber; i.e., the entire chamber is at the gas exit temperature—this fact has been well established in many chambers except in the immediate zone of gas inlet and spray atomization.
5. The temperature driving force for drying is the difference between the drying-gas outlet temperature and, in the case of pure water, the gas wet-bulb temperature. In the case of a solution, the adiabatic saturation temperature of the pure saturated solution is employed rather than the wet-bulb temperature.
Methods for calculating average and maximum drop sizes from various atomizers are given by Marshall (op. cit.). For pneumatic nozzles, an expression developed by Nukiyama and Tanasawa is recommended:

$$\overline{X}_{vs} = \frac{1920\sqrt{\alpha}}{V_a\sqrt{\rho_l}} + 597\left(\frac{\mu}{\sqrt{\alpha\rho_l}}\right)^{0.45}\left(\frac{1000 Q_L}{Q_a}\right)^{1.5} \quad (12\text{-}63)$$

where $\overline{X}_{vs}$ = average drop diameter, μm (a drop with the same volume-surface ratio as the total sum of all drops formed)
α = surface tension, dyn/cm
μ = liquid viscosity, P
V_a = relative velocity between air and liquid, ft/s
ρ_l = liquid density, g/cm³

Q_L = liquid volumetric flow rate
Q_a = air volumetric flow rate

For single-fluid pressure nozzles, a rule of thumb is employed:

$$\overline{X}_{vs} = 500/\sqrt[3]{\Delta P} \quad (12\text{-}64)$$

where ΔP = pressure drop across nozzle, lb/in².
For centrifugal disks, the relation of Friedman, Gluckert, and Marshall is employed [*Chem. Eng. Prog.*, **48**, 181 (1952)]:

$$\frac{D_{vs}}{r} = 0.4\left(\frac{\Gamma}{\rho_l N r^2}\right)^{0.6}\left(\frac{\mu}{\Gamma}\right)^{0.2}\left(\frac{\alpha\rho_l L_w}{\Gamma^2}\right)^{0.1} \quad (12\text{-}65)$$

where D_{vs} = average drop diameter, ft
r = disk radius, ft
Γ = spray mass velocity, lb/(min·ft of wetted disk periphery)
ρ_l = liquid density, lb/ft³
N = disk speed, r/min
μ = liquid viscosity, lb/(ft·min)
α = surface tension, lb/min²
L_w = wetted disk periphery, ft

NOTE: All groups are dimensionless. To convert dynes per square centimeter to joules per square meter, multiply by 10^{-3}; to convert poises to newton-seconds per square meter, multiply by 10^{-1}; to convert feet per second to meters per second, multiply by 0.3048; to convert feet to meters, multiply by 0.3048; to convert pounds per minute-foot to kilograms per second-meter, multiply by 0.025; to convert pounds per cubic foot to kilograms per cubic meter, multiply by 16.019; to convert pounds per minute squared to kilograms per second squared, multiply by 1.26×10^{-4}; to convert British thermal units per hour to kilojoules per second, multiply by 2.63×10^{-4}; and to convert British thermal units per hour-square foot-degree Fahrenheit per foot to joules per square meter-second-kelvin per meter, multiply by 1.7307.
Inspection of these relationships will show that the variables are difficult to specify in the absence of tests except when handling pure liquids—which in spray drying is rare indeed. The most useful method for employing these equations is to conduct small-scale drying tests in a chamber under conditions in which wall impingement and sticking are incipient. The maximum particle size can then be back-calculated by using the relationships given in the following paragraphs, and the effects of changing atomizing variables evaluated by using the preceding equations:

$$\overline{X}_m = 3\overline{X}_{vs} \quad (12\text{-}66)$$

where $\overline{X}_m$ = maximum drop diameter, μm.
Gluckert gives the following relationships for calculating heat transfer under various conditions of atomization:
Two-fluid pneumatic nozzles:

$$Q = \frac{6.38 K_f v^{2/3}\,\Delta t}{D_m^2}\frac{w_s}{\rho_s}\sqrt{\frac{\rho_a}{w_a V_a}\frac{w_a + w_s}{w_a}} \quad (12\text{-}67)$$

Single-fluid pressure nozzles:

$$Q = \frac{10.98 K_f v^{2/3}\,\Delta t}{D_m^2}D_s\sqrt{\frac{\rho_l}{\rho_s}} \quad (12\text{-}68)$$

Centrifugal-disk atomizers:

$$Q = \frac{4.19 K_f (R_c - r/2)^2\,\Delta t}{D_m^2 \rho_s}\sqrt{\frac{w_s \rho_l}{rN}} \quad (12\text{-}69)$$

where Q = rate of heat transfer to spray, Btu/h
K_f = thermal conductivity of gas film surrounding the droplet, Btu/(h·ft²)(°F·ft), evaluated at the average between dryer gas and drop temperature
v = volume of dryer chamber, ft³
Δt = temperature driving force (under terminal conditions described above), °F
D_m = maximum drop diameter, ft
w_s = weight rate of liquid flow, lb/h
ρ_s = density of liquid, lb/ft³

w_a = weight rate of atomizing air flow, lb/h
ρ_a = density of atomizing air, lb/ft^3
V_a = velocity of atomizing air at atomizer, ft/h
D_s = diameter of pressure-nozzle discharge orifice, ft
ρ_t = density of dryer gas at exit conditions, lb/ft^3
R_c = radius of drying chamber with centrifugal disk, ft
r = radius of disk, ft
N = rate of disk rotation, r/h

For proper use of the equations, the chamber shape must conform to the spray pattern. With cocurrent gas-spray flow, the angle of spread of single-fluid pressure nozzles and two-fluid pneumatic nozzles is such that wall impingement will occur at a distance approximately four chamber diameters below the nozzle; therefore, chambers employing these atomizers should have vertical height-to-diameter ratios of at least 4 and, more usually, 5. The discharge cone below the vertical portion should have a slope of at least 60°, to minimize settling accumulations, and is used entirely to accelerate gas and solids for entry into the exit duct.

The critical dimension of a centrifugal-disk chamber is the diameter. Vertical height is usually 0.5 to 1.0 times the diameter; the large cone is needed mainly to accelerate to the discharge duct and prevent settling; it contributes little to drying capacity.

Cost Data for Spray Dryers Drying chambers, ductwork, and cyclone separators are usually constructed of stainless steel. Savings of roughly 20 percent may be achieved on the total purchase cost by using carbon steel; the increasing tendency toward the use of heat-resistant and corrosion-resistant plastic coatings (epoxy resins) makes the future appear promising for greater use of carbon steel construction. Wide differences in cost may be experienced in the selection of basic equipment. Air heaters vary in price range according to the selection of steam, electricity, direct-fired, and indirect-fired oil or gas heaters. Dust-collection equipment may consist of cyclone collectors or bag-type filters and may include a wet scrubber. Costs of nozzle and centrifugal atomizers are usually comparable. While the centrifugal atomizer requires mechanical gearing and motor drive, a high-pressure nozzle requires a high-pressure pump, which will usually more than offset the cost of gearing and motor for the centrifugal atomizer. Auxiliary equipment which may be included comprises air filters, drying-chamber insulation, and mechanical or pneumatic cooling conveyors. A minimum of instrumentation consists of indicating and recording thermometers for inlet-air and outlet temperatures, an ammeter for atomizer motor drive (or a pressure gauge for nozzle atomization), a flowmeter, manometers, a high-temperature alarm, and a panelboard with push-button stations for all equipment. The drying process may be completely controlled automatically with some additional instrumentation.

Spray dryers may operate under positive, negative, or neutral pressures. In general, pressure drop in a complete system will range from 15 to 50 cm of water, depending on duct size and separation equipment employed.

Agitated Flash Dryers Agitated flash dryers produce fine powders from feeds with high solids contents, in the form of filter cakes, pastes, or thick, viscous liquids. Many continuous dryers are unable to dry highly viscous feeds. Spray dryers require a pumpable feed. Conventional flash dryers often require backmixing dry product to the feed in order to fluidize. Other drying methods for viscous pastes and filter cakes are well known, such as contact, drum, band and tray dryers. They all require long processing time, large floor space, high maintenance, and after treatment such as milling.

The agitated flash dryer offers a number of process advantages: such as ability to dry pastes, sludges, and filter cakes to a homogeneous, fine powder in a single-unit operation; continuous operation; compact layout; effective heat and mass transfer—short drying times; negligible heat loss—high thermal efficiency; and easy access and cleanability.

The agitated flash dryer (Fig. 12-106) consists of four major components: feed system, drying chamber, heater, and exhaust air system.

Wet feed enters the feed tank which has a slow-rotating impeller to break up large particles. The level in the feed tank is maintained by a

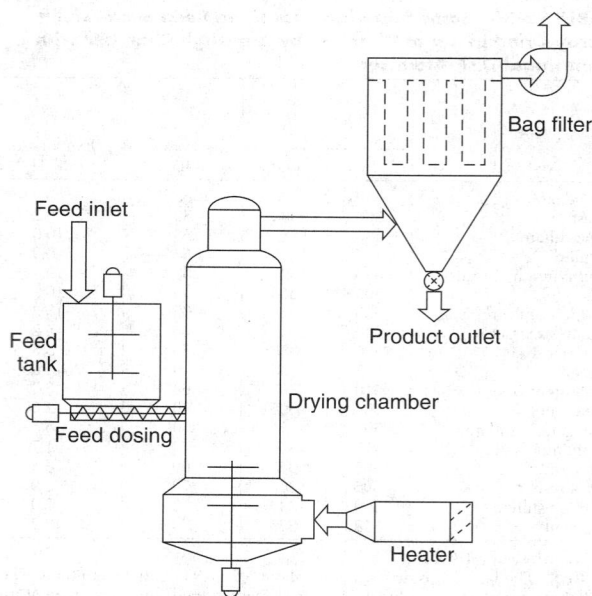

FIG. 12-106 Agitated flash dryer with open cycle. (*NIRO, Inc.*)

level controller. The feed is metered at a constant rate into the drying chamber via a screw conveyor mounted under the feed tank. If the feed is shear thinning and can be pumped, the screw feeder can be replaced by a positive-displacement pump.

The drying chamber is the heart of the system consisting of three important components: air disperser, rotating disintegrator, and drying section. Hot, drying air enters the air disperser tangentially and is introduced into the drying chamber as a swirling airflow. The swirling airflow is established by a guide-vane arrangement. The rotating disintegrator is mounted at the base of the drying chamber. The feed, exposed to the hot, swirling airflow and the agitation of the rotating disintegrator, is broken up and dried. The fine dry particles exit with the exhaust air and are collected in the bag filter. The speed of the rotating disintegrator controls the particles' size. The outlet air temperature controls the product-moisture content.

The drying air is heated either directly or indirectly, depending upon the feed material, powder properties, and available fuel source. The heat sensitivity of the product determines the drying air temperature. The highest possible value is used to optimize thermal efficiency. A bag filter is usually recommended for collecting the fine particles produced. The exhaust fan maintains a slight vacuum in the dryer, to prevent powder leakage into the surroundings. The appropriate process system is selected according to the feed and powder characteristics, available heating source, energy utilization, and operational health and safety requirements.

Open systems use atmospheric air for drying. In cases where products pose a potential for dust explosion, plants are provided with pressure relief or suppression systems. For recycle systems, the drying system medium is recycled and the evaporated solvent recovered as condensate. There are two alternative designs: Self-inertizing mode, where oxygen content is held below 5 percent by combustion control at the heater. This is recommended for products with serious dust-explosion hazards; Inert mode, where nitrogen is the drying gas. This is used when an organic solvent is evaporated or product oxidation during drying must be prevented.

Design Methods The size of the agitated flash dryer is based on the evaporation rate required. The operating temperatures are product-specific. Once established, they determine the airflow requirements. The drying chamber is designed based on air velocity (approximately 3–4 m/s) and residence time (product-specific).

Distillation

J. D. Seader, Ph.D., *Professor of Chemical Engineering, University of Utah, Salt Lake City, Utah; Fellow, American Institute of Chemical Engineers; Member, American Chemical Society; Member, American Society for Engineering Education. (Section Editor°)*

Jeffrey J. Siirola, Ph.D., *Research Fellow, Eastman Chemical Company; Member, National Academy of Engineering; Fellow, American Institute of Chemical Engineers, American Chemical Society, American Association for Artificial Intelligence, American Society for Engineering Education. (Enhanced Distillation)*

Scott D. Barnicki, Ph.D., *Senior Research Chemical Engineer, Eastman Chemical Company. (Enhanced Distillation)*

° Certain portions of this section draw heavily on the work of Buford D. Smith, editor of this section in the fifth edition.

Nomenclature and Units

Symbol	Definition	SI units	U.S. customary units	Symbol	Definition	SI units	U.S. customary units
A	Absorption factor			X	Vector of stage variables		
A	Area	m^2	ft^2	a	Activity		
C	Number of chemical species			b	Component flow rate in bottoms	(kg·mol)/s	(lb·mol)/h
D	Distillate flow rate	(kg·mol)/s	(lb·mol)/h	d	Component flow rate in distillate	(kg·mol)/s	(lb·mol)/h
E	Deviation from set point			e	Rate of heat transfer	kW	Btu/h
E	Residual of heat-transfer expression	kW	Btu/h	f	Fraction of feed leaving in bottoms		
E	Residual of phase equilibrium expression	(kg·mol)/s	(lb·mol)/h	f	Fugacity	Pa	psia
F	Feed flow rate	(kg·mol)/s	(lb·mol)/h	f	Function in homotopy expression		
F	Vector of stage functions			g	Function in homotopy expression		
G	Interlink flow rate	(kg·mol)/s	(lb·mol)/h	g	Residual of energy balance	kW	Btu/h
G	Volume holdup of liquid	m^3	ft^3	h	Height	m	ft
H	Residual of energy balance	kW	Btu/h	h	Homotopy function		
H	Height of a transfer unit	m	ft	ℓ	Component flow rate in liquid	(kg·mol)/s	(lb·mol)/h
H	Enthalpy	J/(kg·mol)	Btu/(lb·mol)	p	Pressure	kPa	psia
K	Vapor-liquid equilibrium ratio (K value)			q	Measure of thermal condition of feed		
K_C	Controller gain			q_c	Condenser duty	kW	Btu/h
K_D	Chemical equilibrium constant for dimerization			q_r	Reboiler duty	kW	Btu/h
K_d	Liquid-liquid distribution ratio			r	Sidestream ratio		
L	Liquid flow rate	(kg·mol)/s	(lb·mol)/h	s	Liquid-sidestream ratio		
M	Residual of component material balance	(kg·mol)/s	(lb·mol)/h	t	Homotopy parameter		
M	Liquid holdup	kg·mol	lb·mol	t	Time	s	h
N	Number of transfer units			v	Component flow rate in vapor	(kg·mol)/s	(lb·mol)/h
N	Number of equilibrium stages			w	Weight fraction		
N_e	Number of relationships			x	Mole fraction in liquid		
N_i	Number of design variables			y	Mole fraction in vapor		
N_m	Minimum number of equilibrium stages			z	Mole fraction in feed		
N_p	Number of phases				Greek symbols		
N_r	Number of repetition variables			α	Relative volatility		
N_o	Number of variables			γ	Activity coefficient		
N	Rate of mass transfer	(kg·mol)/s	(lb·mol)/h	ε	Convergence criterion		
P	Pressure	Pa	psia	ξ	Scale factor		
P	Residual of pressure-drop expression	Pa	psia	η	Murphree-stage efficiency		
P^{sat}	Vapor pressure	Pa	psia	θ	Time for distillation	s	h
Q	Heat-transfer rate	kW	Btu/h	Θ	Parameter in Underwood equations		
Q_c	Condenser duty	kW	Btu/h	Θ	Holland theta factor		
Q_r	Reboiler duty	kW	Btu/h	λ	Eigenvalue		
Q	Residual of phase-equilibrium expression			τ	Sum of squares of residuals		
R	External-reflux ratio			τ	Feedback-reset time	s	h
R_m	Minimum-reflux ratio			Φ	Fugacity coefficient of pure component		
S	Residual of mole-fraction sum			ϕ	Entrainment or occlusion ratio		
S	Sidestream flow rate	(kg·mol)/s	(lb·mol)/h	$\hat{\Phi}$	Fugacity coefficient in mixture		
S	Stripping factor			Φ_A	Fraction of a component in feed vapor that is not absorbed		
S	Vapor-sidestream ratio			Φ_S	Fraction of a component in entering liquid that is not stripped		
T	Temperature	K	°R	Ψ	Factor in Gilliland correlation		
U	Liquid-sidestream rate	(kg·mol)/s	(lb·mol)/h				
V	Vapor flow rate	(kg·mol)/s	(lb·mol)/h				
W	Vapor-sidestream rate	(kg·mol)/s	(lb·mol)/h				

GENERAL REFERENCES: Billet, *Distillation Engineering*, Chemical Publishing, New York, 1979. Fair and Bolles, "Modern Design of Distillation Columns," *Chem. Eng.*, **75**(9), 156 (Apr. 22, 1968). Fredenslund, Gmehling, and Rasmussen, *Vapor-Liquid Equilibria Using UNIFAC, a Group Contribution Method*, Elsevier, Amsterdam, 1977. Friday and Smith, "An Analysis of the Equilibrium Stage Separation Problem—Formulation and Convergence," *Am. Inst. Chem. Eng. J.*, **10**, 698 (1964). Hengstebeck, *Distillation—Principles and Design Procedures*, Reinhold, New York, 1961. Henley and Seader, *Equilibrium-Stage Separation Operations in Chemical Engineering*, Wiley, New York, 1981. Hoffman, *Azeotropic and Extractive Distillation*, Wiley, New York, 1964. Holland, *Fundamentals and Modeling of Separation Processes*, Prentice-Hall, Englewood Cliffs, N.J., 1975. Holland, *Fundamentals of Multicomponent Distillation*, McGraw-Hill, New York, 1981. King, *Separation Processes*, 2d ed., McGraw-Hill, New York, 1980. Kister, *Distillation Design*, McGraw-Hill, New York, 1992. Kister, *Distillation Operation*, McGraw-Hill, New York, 1990. Robinson and Gilliland, *Elements of Fractional Distillation*, 4th ed., McGraw-Hill, New York, 1950. Rousseau, ed., *Handbook of Separation Process Technology*, Wiley-Interscience, New York, 1987. Seader, *The B.C. (Before Computers) and A.D. of Equilibrium-Stage Operations*, Chem. Eng. Educ., Vol. **14**(2), (Spring 1985). Seader, *Chem. Eng. Progress*, **85**(10), 41 (1989). Smith, *Design of Equilibrium Stage Processes*, McGraw-Hill, New York, 1963. Treybal, *Mass Transfer Operations*, 3d ed., McGraw-Hill, New York, 1980. *Ullmann's Encyclopedia of Industrial Chemistry*, Vol. **B3**, VCH, Weinheim, 1988. Van Winkle, *Distillation*, McGraw-Hill, New York, 1967.

CONTINUOUS-DISTILLATION OPERATIONS

GENERAL PRINCIPLES

Separation operations achieve their objective by the creation of two or more coexisting zones which differ in temperature, pressure, composition, and/or phase state. Each molecular species in the mixture to be separated reacts in a unique way to differing environments offered by these zones. Consequently, as the system moves toward equilibrium, each species establishes a different concentration in each zone, and this results in a separation between the species.

The separation operation called *distillation* utilizes vapor and liquid phases at essentially the same temperature and pressure for the coexisting zones. Various kinds of devices such as *random* or *structured packings* and *plates* or *trays* are used to bring the two phases into intimate contact. Trays are stacked one above the other and enclosed in a cylindrical shell to form a *column*. Packings are also generally contained in a cylindrical shell between hold-down and support plates. A typical tray-type distillation column plus major external accessories is shown schematically in Fig. 13-1.

The *feed* material, which is to be separated into fractions, is introduced at one or more points along the column shell. Because of the difference in gravity between vapor and liquid phases, liquid runs down the column, cascading from tray to tray, while vapor flows up the column, contacting liquid at each tray.

Liquid reaching the bottom of the column is partially vaporized in a heated *reboiler* to provide *boil-up*, which is sent back up the column. The remainder of the bottom liquid is withdrawn as *bottoms*, or bottom product. Vapor reaching the top of the column is cooled and condensed to liquid in the *overhead condenser*. Part of this liquid is returned to the column as *reflux* to provide liquid overflow. The remainder of the overhead stream is withdrawn as *distillate*, or overhead product. In some cases only part of the vapor is condensed so that a vapor distillate can be withdrawn.

This overall flow pattern in a distillation column provides countercurrent contacting of vapor and liquid streams on all the trays through the column. Vapor and liquid phases on a given tray approach thermal, pressure, and composition equilibriums to an extent dependent upon the efficiency of the contacting tray.

The *lighter* (lower-boiling) components tend to concentrate in the vapor phase, while the *heavier* (higher-boiling) components tend toward the liquid phase. The result is a vapor phase that becomes richer in light components as it passes up the column and a liquid phase that becomes richer in heavy components as it cascades downward. The overall separation achieved between the distillate and the bottoms depends primarily on the *relative volatilities* of the components, the number of contacting trays, and the ratio of the liquid-phase flow rate to the vapor-phase flow rate.

If the feed is introduced at one point along the column shell, the column is divided into an upper section, which is often called the *rectifying* section, and a lower section, which is often referred to as the *stripping* section. These terms become rather indefinite in *multiple-feed* columns and in columns from which a liquid or vapor *sidestream* is withdrawn somewhere along the column length in addition to the two end-product streams.

EQUILIBRIUM-STAGE CONCEPT

Until recently, energy and mass-transfer processes in an actual distillation column were considered too complicated to be readily modeled in any direct way. This difficulty was circumvented by the *equilibrium-stage model*, developed by Sorel in 1893, in which vapor and liquid streams leaving an equilibrium stage are in complete equilibrium with each other and thermodynamic relations can be used to determine the temperature of and relate the concentrations in the equilibrium streams at a given pressure. A hypothetical column composed of equilibrium stages (instead of actual contact trays) is designed to accomplish the separation specified for the actual column. The number of hypothetical equilibrium stages required is then converted to a number of actual trays by means of *tray efficiencies*, which describe the extent to which the performance of an actual contact tray duplicates the performance of an equilibrium stage. Alternatively and preferably, tray inefficiencies can be accounted for by using rate-based models that are described below.

Use of the equilibrium-stage concept separates the design of a distillation column into three major steps: (1) Thermodynamic data and methods needed to predict equilibrium-phase compositions are assembled. (2) The number of equilibrium stages required to accomplish a specified separation, or the separation that will be accomplished in a given number of equilibrium stages, is calculated. (3) The number of equilibrium stages is converted to an equivalent number of actual contact trays or height of packing, and the column diameter is determined. Much of the third step is eliminated if a rate-based model is used. This section deals primarily with the second step. Section 14 covers the third step. Sections 3 and 4 cover the first step, but a summary of methods and some useful data are included in this section.

COMPLEX DISTILLATION OPERATIONS

All separation operations require energy input in the form of heat or work. In the conventional distillation operation, as typified in Fig. 13-1, energy required to separate the species is added in the form of heat to the reboiler at the bottom of the column, where the temperature is highest. Also, heat is removed from a condenser at the top of the column, where the temperature is lowest. This frequently results

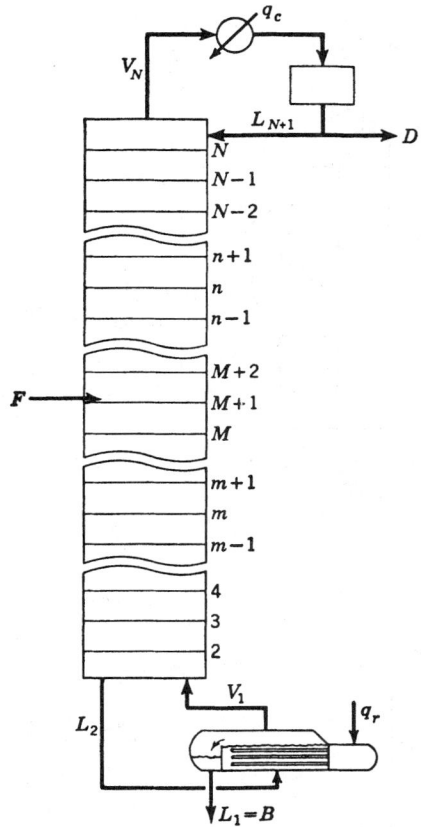

FIG. 13-1 Schematic diagram and nomenclature for a simple distillation column with one feed, a total overhead condenser, and a partial reboiler.

in a large energy-input requirement and low overall thermodynamic efficiency, when the heat removed in the condenser is wasted. Complex distillation operations that offer higher thermodynamic efficiency and lower energy-input requirements have been under intense investigation. In some cases, all or a portion of the energy input is as work.

Complex distillation operations may utilize single columns, as shown in Fig. 13-2 and discussed by Petterson and Wells [*Chem. Eng.*, **84**(20), 78 (Sept. 26, 1977)], Null [*Chem. Eng. Prog.*, **72**(7), 58 (1976)], and Brannon and Marple [*Am. Inst. Chem. Eng. Symp. Ser.* **76, 192**, 10 (1980)], or two or more columns that are thermally linked as shown in Figs. 13-3 and 13-6 and discussed by Petterson and Wells (op. cit.) and Mah, Nicholas, and Wodnik [*Am. Inst. Chem. Eng. J.*, **23**, 651 (1977)].

In Fig. 13-2a, which is particularly useful when a large temperature difference exists between the ends of the column, interreboilers add heat at lower temperatures and/or intercondensers remove heat at higher temperatures. As shown in Fig. 13-2b, these intermediate heat exchangers may be coupled with a heat pump that takes energy from the intercondenser and uses shaft work to elevate this energy to a temperature high enough to transfer it to the interreboiler.

Particularly when the temperature difference between the ends of the column is not large, any of the three heat-pump systems in Fig. 13-2c, d, and e that involve thermal coupling of the overhead condenser and bottoms reboiler might be considered to eliminate external heat transfer almost entirely, substituting shaft work as the prime energy input for achieving the separation. More complex arrangements are considered by Björn, Grén, and Ström [*Chem. Eng. Process.*, **29**, 185 (1991)]. Alternatively, the well-known multiple-

column or split-tower arrangement of Fig. 13-3a, which corresponds somewhat to the energy-saving concept employed in multieffect evaporation, might be used. The feed is split more or less equally among columns that operate in parallel, but at different pressures, in a cascade that decreases from left to right. With proper selection of column-operating pressure, this permits the overhead vapor from the higher-pressure column to be condensed in the reboiler of the lower-pressure column. External heat-transfer media are needed only for the reboiler of the first effect and the condenser of the last effect. Thus, for N effects, utility requirements are of the order 1/N of those for a conventional single-effect column. Wankat [*Ind. Eng. Chem. Res.*, **32**, 894 (1993)] develops a large number of more complex multieffect schemes, some of which show significant reductions in energy requirements.

In another alternative, shown in Fig. 13-3b, the rectifying section may be operated at a pressure sufficiently higher than that of the stripping section such that heat can be transferred between any desired pairs of stages of the two sections. This technique, described by Mah et al. (op. cit.) and referred to as SRV (secondary reflux and vaporization) distillation, can result in a significant reduction in utility requirements for the overhead condenser and bottoms reboiler.

When multicomponent mixtures are to be separated into three or more products, sequences of simple distillation columns of the type shown in Fig. 13-1 are commonly used. For example, if a ternary mixture is to be separated into three relatively pure products, either of the two sequences in Fig. 13-4 can be used. In the direct sequence, shown in Fig. 13-4a, all products but the heaviest are removed one by one as distillates. The reverse is true for the indirect sequence, shown in Fig. 13-4b. The number of possible sequences of simple distillation columns increases rapidly with the number of products. Thus, although only the 2 sequences shown in Fig. 13-4 are possible for a mixture separated into 3 products, 14 different sequences, one of which is shown in Fig. 13-5, can be synthesized when 5 products are to be obtained.

As shown in a study by Tedder and Rudd [*Am. Inst. Chem. Eng. J.*, **24**, 303 (1978)], conventional sequences like those of Fig. 13-4 may not always be the optimal choice, particularly when species of intermediate volatility are present in large amounts in the feed or need not be recovered at high purity. Of particular interest are thermally coupled systems. For example, in Fig. 13-6a, an impure-vapor sidestream is withdrawn from the first column and purified in a side-cut rectifier, the bottoms of which is returned to the first column. The thermally coupled system in Fig. 13-6b, discussed by Stupin and Lockhart [*Chem. Eng. Prog.*, **68**(10), 71 (1972)] and referred to as Petlyuk towers, is particularly useful for reducing energy requirements when the initial feed contains close-boiling species. Shown for a ternary feed, the first column in Fig. 13-6b is a prefractionator, which sends essentially all of the light component and heavy component to the distillate and bottoms respectively, but permits the component of intermediate volatility to be split between the distillate and bottoms. Products from the prefractionator are sent to appropriate feed trays in the second column, where all three products are produced, the middle product being taken off as a sidestream. Only the second column is provided with condenser and reboiler; reflux and boil-up for the prefractionator are obtained from the second column. This concept is readily extended to separations that produce more than three products. Procedures for the optimal design of thermally coupled systems are presented by Triantafyllou and Smith [*Trans. I. Chem. E.*, **70**, Part A, 118 (1992)]. A scheme for combining thermal coupling with heat pumps is developed by Agrawal and Yee [*Ind. Eng. Chem. Res.*, **33**, 2717 (1994)].

RELATED SEPARATION OPERATIONS

The simple and complex distillation operations just described all have two things in common: (1) both rectifying and stripping sections are provided so that a separation can be achieved between two components that are adjacent in volatility; and (2) the separation is effected only by the addition and removal of energy and not by the addition of any mass separating agent (MSA) such as in liquid-liquid extraction.

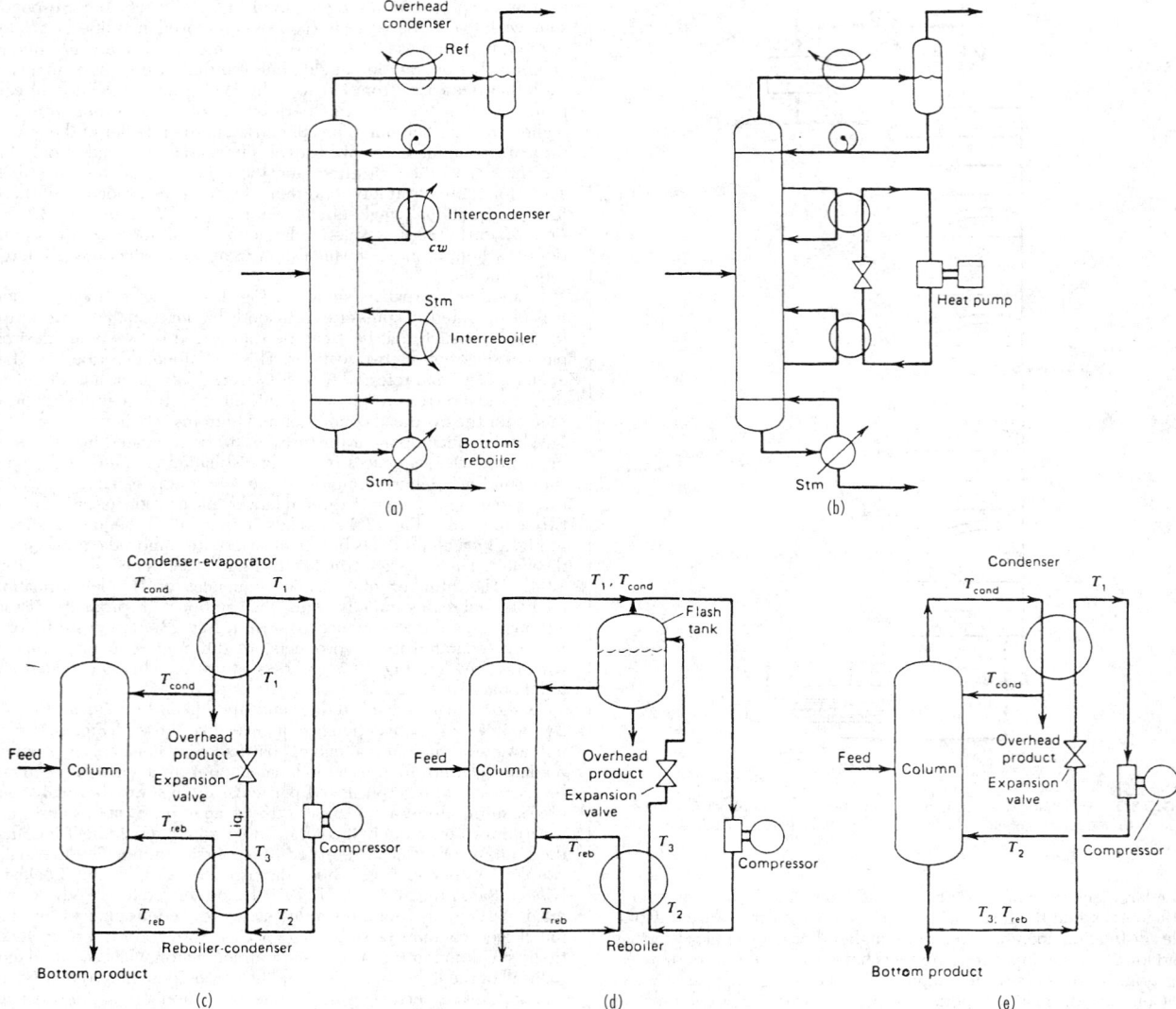

FIG. 13-2 Complex distillation operations with single columns. (*a*) Use of intermediate heat exchangers. (*b*) Coupling of intermediate heat exchangers with heat pump. (*c*) Heat pump with external refrigerant. (*d*) Heat pump with vapor compression. (*e*) Heat pump with bottoms flashing.

Sometimes, alternative single- or multiple-stage vapor-liquid separation operations, of the types shown in Fig. 13-7, may be more suitable than distillation for the specified task.

A single-stage flash, as shown in Fig. 13-7*a*, may be appropriate if (1) the relative volatility between the two components to be separated is very large; (2) the recovery of only one component, without regard to the separation of the other components, in one of the two product streams is to be achieved; or (3) only a partial separation is to be made. A common example is the separation of light gases such as hydrogen and methane from aromatics. The desired temperature and pressure of a flash may be established by the use of heat exchangers, a valve, a compressor, and/or a pump upstream of the vessel used to separate the product vapor and liquid phases. Depending on the original condition of the feed, it may be partially condensed or partially vaporized in a so-called flash operation.

If the recovery of only one component is required rather than a sharp separation between two components of adjacent volatility,

absorption or stripping in a single section of stages may be sufficient. If the feed is vapor at separation conditions, absorption is used either with a liquid MSA absorbent of relatively low volatility as in Fig. 13-7*b* or with reflux produced by an overhead partial condenser as in Fig. 13-7*c*. The choice usually depends on the ease of partially condensing the overhead vapor or of recovering and recycling the absorbent. If the feed is liquid at separation conditions, stripping is used, either with an externally supplied vapor stripping agent of relatively high volatility as shown in Fig. 13-7*d* or with boil-up produced by a partial reboiler as in Fig. 13-7*c*. The choice depends on the ease of partially reboiling the bottoms or of recovering and recycling the stripping agent.

If a relatively sharp separation is required between two components of adjacent volatility, but either an undesirably low temperature is required to produce reflux at the column-operating pressure or an undesirably high temperature is required to produce boil-up, then refluxed stripping as shown in Fig. 13-7*g* or reboiled absorption

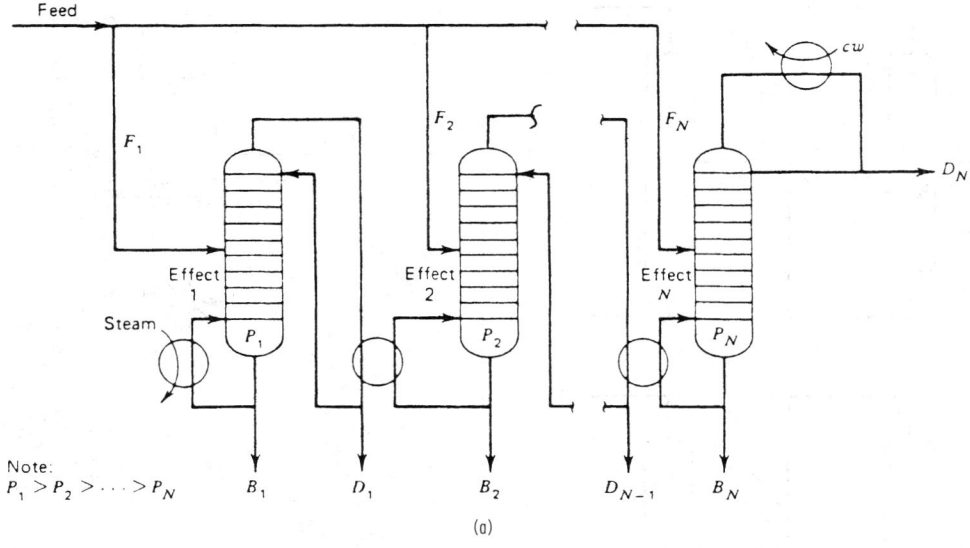

(a)

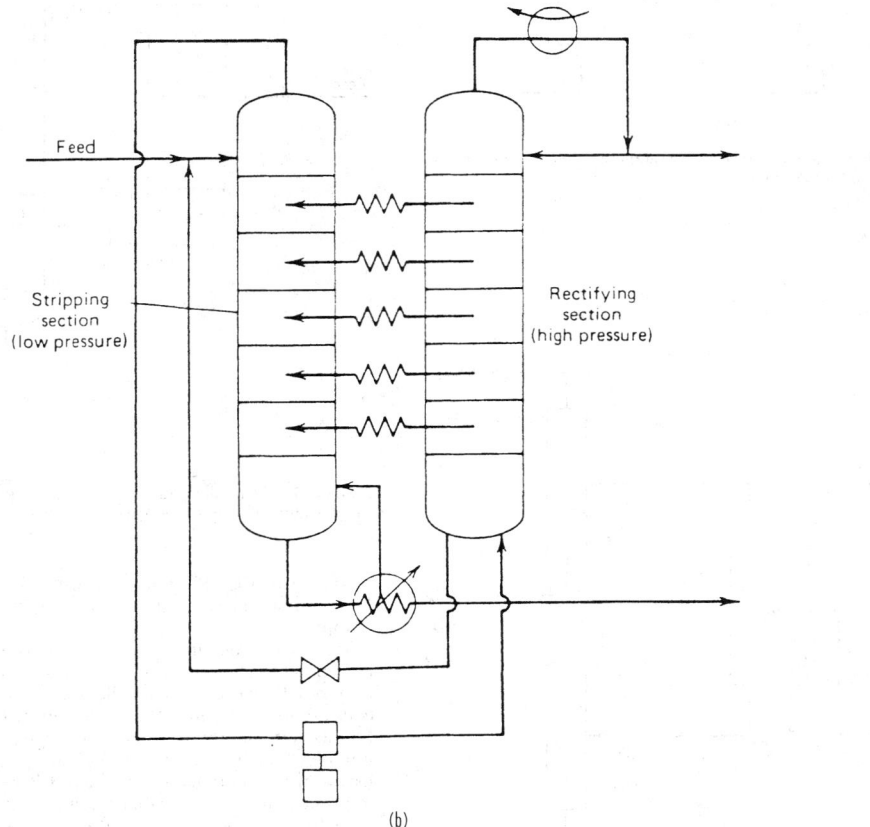

(b)

FIG. 13-3 Complex distillation operations with two or more columns. (*a*) Multieffect distillation. (*b*) SRV distillation.

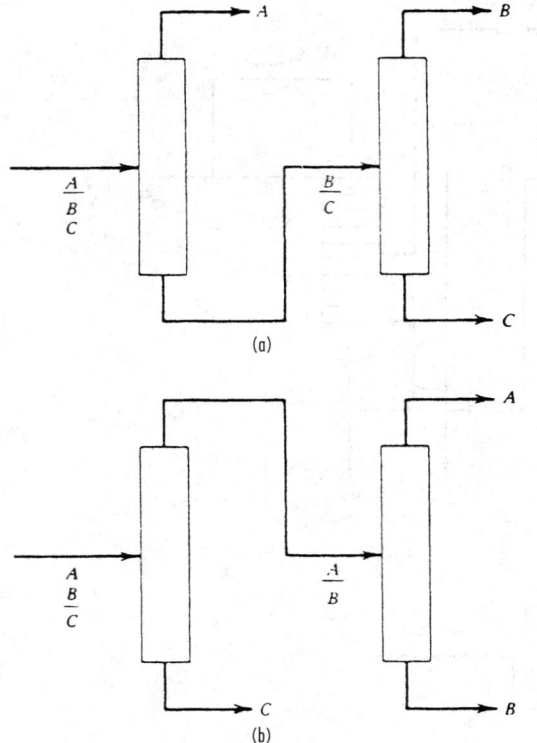

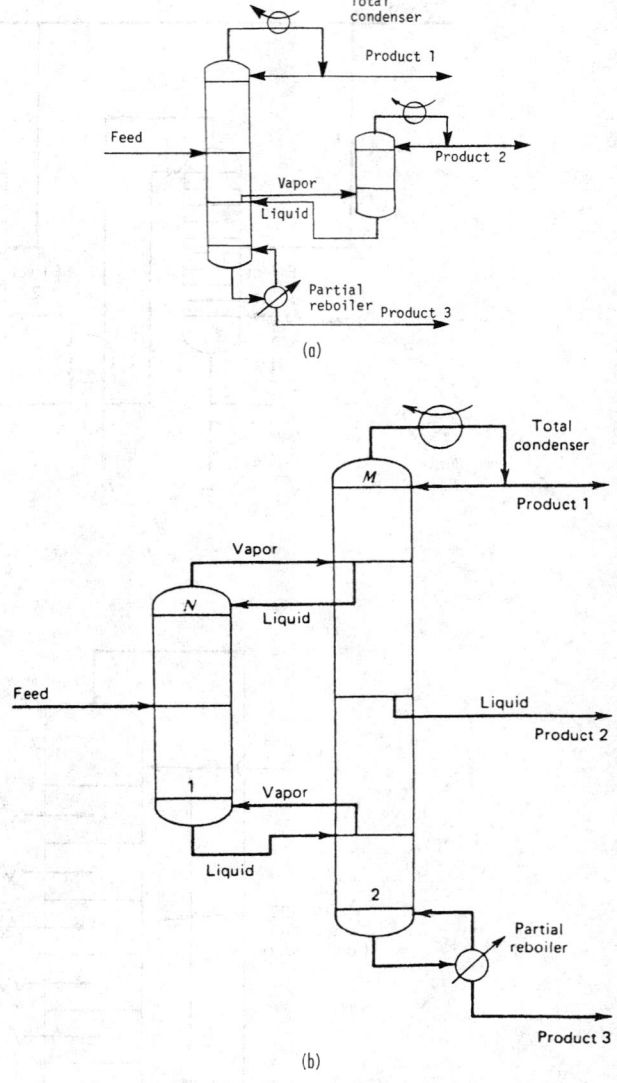

FIG. 13-4   Distillation sequences for the separation of three components. (*a*) Direct sequence. (*b*) Indirect sequence.

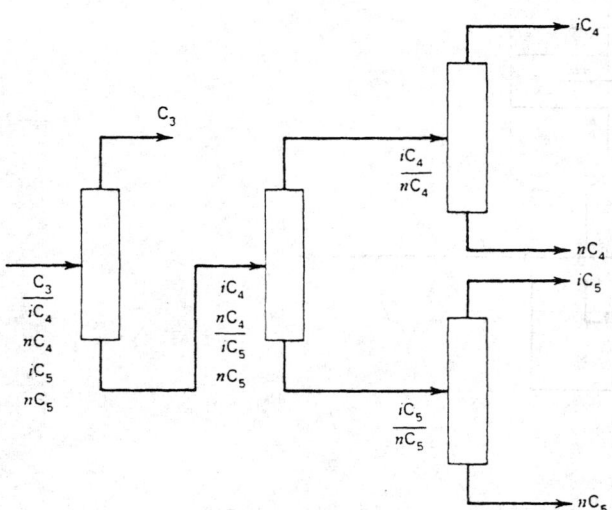

FIG. 13-5   One of 14 different sequences for the separation of a 5-component mixture by simple distillation.

FIG. 13-6   Thermally coupled systems for separation into three products. (*a*) Fractionator with vapor sidestream and side-cut rectifier. (*b*) Petlyuk towers.

as shown in Fig. 13-7*f* may be used. In either case, the choice of MSA follows the same consideration given for simple absorption and stripping.

When the volatility difference between the two components to be separated is so small that a very large number of stages would be required, then extractive distillation, as shown in Fig. 13-7*h*, should be considered. Here, an MSA is selected that increases the volatility difference sufficiently to reduce the stage requirement to a reasonable number. Usually, the MSA is a polar compound of low volatility that leaves in the bottoms, from which it is recovered and recycled. It is introduced in an appreciable amount near the top stage of the column so as to affect the volatility difference over most of the stages. Some reflux to the top stage is utilized to minimize the MSA content in the distillate. An alternative to extractive distillation is azeotropic distillation, which is shown in Fig. 13-7*i* in just one of its many modes. In a common mode, an MSA that forms a heterogeneous minimum-

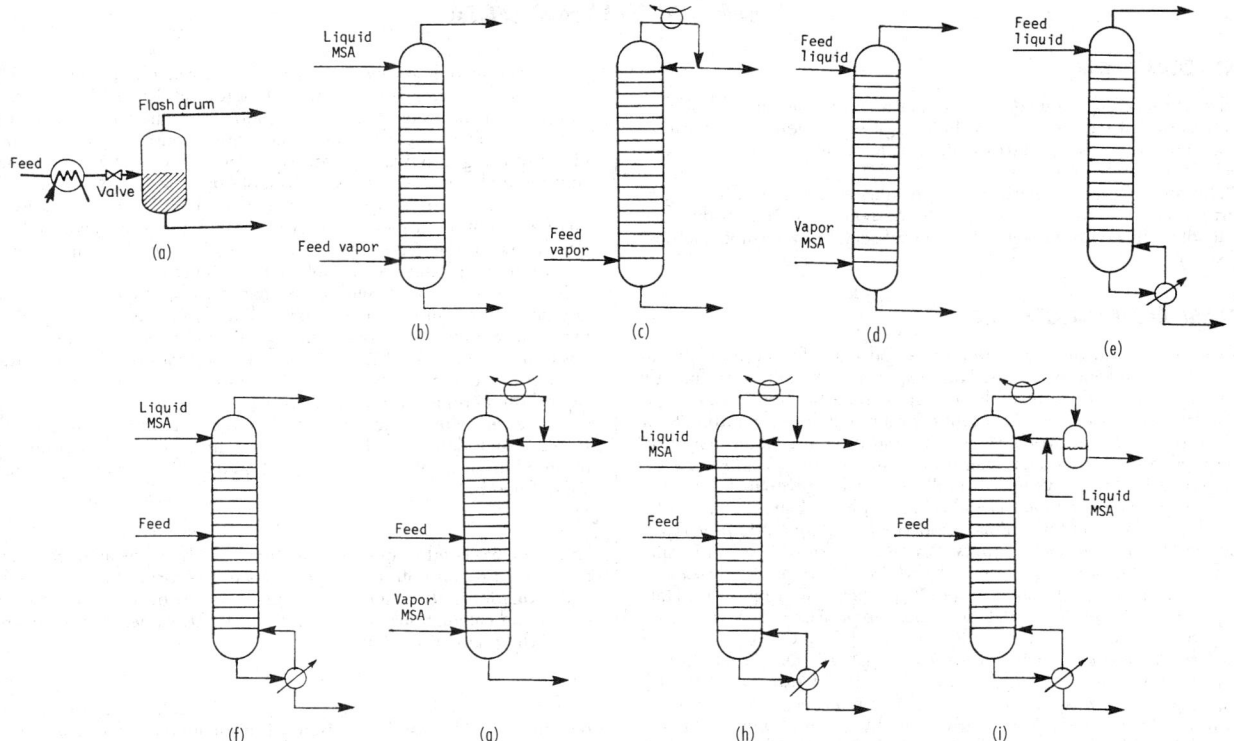

FIG. 13-7   Separation operations related to distillation. (*a*) Flush vaporization or partial condensation. (*b*) Absorption. (*c*) Rectifier. (*d*) Stripping. (*e*) Reboiled stripping. (*f*) Reboiled absorption. (*g*) Refluxed stripping. (*h*) Extractive distillation. (*i*) Azeotropic distillation.

boiling azeotrope with one or more components of the feed is utilized. The azeotrope is taken overhead, and the MSA-rich phase is decanted and returned to the top of the column as reflux.

Numerous other multistaged configurations are possible. One important variation of a stripper, shown in Fig. 13-7*d*, is a refluxed stripper, in which an overhead condenser is added. Such a configuration is sometimes used to steam-strip sour water containing NH₃, H₂O, phenol, and HCN.

All the separation operations shown in Fig. 13-7, as well as the simple and complex distillation operations described earlier, are referred to here as distillation-type separations because they have much in common with respect to calculations of (1) thermodynamic properties, (2) vapor-liquid equilibrium stages, and (3) column sizing. In fact, as will be evident from the remaining treatment of this section, the trend is toward single generalized digital-computer-program packages that compute many or all distillation-type separation operations.

This section also includes a treatment of distillation-type separations from a rate-based point of view that utilizes principles of mass transfer and heat transfer. Section 14 also presents details of that subject as applied to absorption and stripping.

SYNTHESIS OF MULTICOMPONENT SEPARATION SYSTEMS

The sequencing of distillation columns and other types of equipment for the separation of multicomponent mixtures has received much attention in recent years. Although one separator of complex design can sometimes be devised to produce more than two products, more often a sequence of two-product separators is preferable. Often, the sequence includes simple distillation columns. A summary of sequencing methods, prior to 1977, that can lead to optimal or near-optimal designs, is given by Henley and Seader [op. cit.]. More recent methods for distillation-column sequencing are reviewed by Modi and Westerberg [*Ind. Eng. Chem. Res.,* **31,** 839 (1992)], who also present a more generally applicable method based on a marginal price that is the change in price of a separation operation when the separation is carried out in the absence of nonkey components. The synthesis of sequences that consider a wide range of separation operations in a knowledge-based approach is given by Barnicki and Fair for liquid mixtures [*Ind. Eng. Chem. Res.,* **29,** 421 (1990)] and for gas/vapor mixtures [*Ind. Eng. Chem. Res.,* **31,** 1679 (1992)]. A knowledge-based method is also given by Sheppard, Beltramini, and Motard [*Chem. Eng. Comm.,* **106** (1991)] for the synthesis of distillation sequences that involve nonsharp separations where nonkey components distribute. The problem-decomposition approach of Wahnschafft, Le Rudulier, and Westerberg [*Ind. Eng. Chem. Res.,* **32,** 1121 (1993)] is directed to the synthesis of complex separation sequences that involve nonsharp splits and recycle, including azeotropic distillation. The method is applied using a computer-aided separation process designer called *SPLIT.* An expert system, called *EXSEP,* for the synthesis of solvent-based separation trains is presented by Brunet and Liu [*Ind. Eng. Chem. Res.,* **32,** 315 (1993)]. The use of ternary-composition diagrams and residue-curve maps, of the type made popular by Doherty and coworkers, is reviewed and evaluated for application to the synthesis of complex separation sequences by Fien and Liu [*Ind. Eng. Chem. Res.,* **33,** 2506 (1994)].

THERMODYNAMIC DATA

INTRODUCTION

Reliable thermodynamic data are essential for the accurate design or analysis of distillation columns. Failure of equipment to perform at specified levels is often attributable, at least in part, to the lack of such data.

This subsection summarizes and presents examples of phase equilibrium data currently available to the designer. The thermodynamic concepts utilized are presented in the subsection "Thermodynamics" of Sec. 4.

PHASE EQUILIBRIUM DATA

For a binary mixture, pressure and temperature fix the equilibrium vapor and liquid compositions. Thus, experimental data are frequently presented in the form of tables of vapor mole fraction y and liquid mole fraction x for one constituent over a range of temperature T for a fixed pressure P or over a range of pressure for a fixed temperature. A compilation of such data, mainly at a pressure of 101.3 kPa (1 atm, 1.013 bar), for binary systems (mainly nonideal) is given in Table 13-1. More extensive presentations and bibliographies of such data may be found in Hala, Wichterle, Polak, and Boublik [*Vapour-Liquid Equilibrium Data at Normal Pressures,* Pergamon, Oxford (1968)]; Hirata, Ohe, and Nagahama [*Computer Aided Data Book of Vapor-Liquid Equilibria,* Elsevier, Amsterdam (1975)]; Wichterle, Linek, and Hala [*Vapor-Liquid Equilibrium Data Bibliography,* Elsevier, Amsterdam, 1973, Supplement I, 1976, Supplement II, (1979)]; Oe [*Vapor-Liquid Equilibrium Data,* Elsevier, Amsterdam (1989)]; Oe [*Vapor-Liquid Equilibrium Data at High Pressure,* Elsevier, Amsterdam (1990)]; Walas [*Phase Equilibria in Chemical Engineering,* Butterworth, Boston (1985)]; and, particularly, Gmehling and Onken [*Vapor-Liquid Equilibrium Data Collection,* DECHEMA Chemistry Data ser., vol. **1** (parts 1–10), (Frankfort, 1977)].

For application to distillation (a nearly isobaric process), as shown in Figs. 13-8 to 13-13, binary-mixture data are frequently plotted, for a fixed pressure, as y versus x, with a line of 45° slope included for reference, and as T versus y and x. In most binary systems, one of the components is more volatile than the other over the entire composition range. This is the case in Figs. 13-8 and 13-9 for the benzene-toluene system at pressures of both 101.3 and 202.6 kPa (1 and 2 atm), where benzene is more volatile than toluene.

For some binary systems, one of the components is more volatile over only a part of the composition range. Two systems of this type, ethyl acetate–ethanol and chloroform-acetone, are shown in Figs. 13-10 to 13-12. Figure 13-10 shows that for two binary systems chloroform is less volatile than acetone below a concentration of 66 mole percent chloroform and that ethyl acetate is less volatile than ethanol below a concentration of 53 mole percent ethyl acetate. Above these concentrations, volatility is reversed. Such mixtures are known as azeotropic mixtures, and the composition in which the reversal occurs, which is the composition in which vapor and liquid compositions are equal, is the azeotropic composition, or azeotrope. The azeotropic liquid may be homogeneous or heterogeneous (two immiscible liquid phases). Many of the binary mixtures of Table 13-1 form homogeneous azeotropes. Non-azeotrope-forming mixtures such as benzene and toluene in Figs. 13-8 and 13-9 can be separated by simple distillation into two essentially pure products. By contrast, simple distillation of azeotropic mixtures will at best yield the azeotrope and one essentially pure species. The distillate and bottoms products obtained depend upon the feed composition and whether a minimum-boiling azeotrope is formed as with the ethyl acetate–ethanol mixture in Fig. 13-11 or a maximum-boiling azeotrope is formed as with the chloroform-acetone mixture in Fig. 13-12. For example, if a mixture of 30 mole percent chloroform and 70 mole percent acetone is fed to a simple distillation column, such as that shown in Fig. 13-1, operating at 101.3 kPa (1 atm), the distillate could approach pure acetone and the bottoms could approach the azeotrope.

An example of heterogeneous-azeotrope formation is shown in Fig. 13-13 for the water–normal butanol system at 101.3 kPa. At liquid compositions between 0 and 3 mole percent butanol and between 40 and 100 mole percent butanol, the liquid phase is homogeneous. Phase splitting into two separate liquid phases (one with 3 mole percent butanol and the other with 40 mole percent butanol) occurs for any overall liquid composition between 3 and 40 mole percent butanol. A minimum-boiling heterogeneous azeotrope occurs at 92°C (198°F) when the vapor composition and the overall composition of the two liquid phases are 75 mole percent butanol.

For mixtures containing more than two species, an additional degree of freedom is available for each additional component. Thus, for a four-component system, the equilibrium vapor and liquid compositions are only fixed if the pressure, temperature, and mole fractions of two components are set. Representation of multicomponent vapor-liquid equilibrium data in tabular or graphical form of the type shown earlier for binary systems is either difficult or impossible. Instead, such data, as well as binary-system data, are commonly represented in terms of K values (vapor-liquid equilibrium ratios), which are defined by

$$K_i = y_i/x_i \tag{13-1}$$

and are correlated empirically or theoretically in terms of temperature, pressure, and phase compositions in the form of tables, graphs, and equations. K values are widely used in multicomponent-distillation calculations, and the ratio of the K values of two species, called the relative volatility,

$$\alpha_{ij} = K_i/K_j \tag{13-2}$$

is a convenient index of the relative ease or difficulty of separating components i and j by distillation. Rarely is distillation used on a large scale if the relative volatility is less than 1.05, with i more volatile than j.

GRAPHICAL K-VALUE CORRELATIONS

As discussed in Sec. 4, the K value of a species is a complex function of temperature, pressure, and equilibrium vapor- and liquid-phase compositions. However, for mixtures of compounds of similar molecular structure and size, the K value depends mainly on temperature and pressure. For example, several major graphical K-value correlations are available for light-hydrocarbon systems. The easiest to use are the DePriester charts [*Chem. Eng. Prog. Symp. Ser. 7,* **49,** 1 (1953)], which cover 12 hydrocarbons (methane, ethylene, ethane, propylene, propane, isobutane, isobutylene, *n*-butane, isopentane, *n*-pentane, *n*-hexane, and *n*-heptane). These charts are a simplification of the Kellogg charts [*Liquid-Vapor Equilibria in Mixtures of Light Hydrocarbons, MWK Equilibrium Constants, Polyco Data,* (1950)] and include additional experimental data. The Kellogg charts, and hence the DePriester charts, are based primarily on the Benedict-Webb-Rubin equation of state [*Chem. Eng. Prog.,* **47,** 419 (1951); **47,** 449 (1951)], which can represent both the liquid and the vapor phases and can predict K values quite accurately when the equation constants are available for the components in question.

A trial-and-error procedure is required with any K-value correlation that takes into account the effect of composition. One cannot calculate K values until phase compositions are known, and those cannot be known until the K values are available to calculate them. For K as a function of T and P only, the DePriester charts provide good starting values for the iteration. These nomographs are shown in Fig. 13-14*a* and *b*. SI versions of these charts have been developed by Dadyburjor [*Chem. Eng. Prog.,* **74**(4), 85 (1978)].

The Kellogg and DePriester charts and their subsequent extensions and generalizations use the molar average boiling points of the liquid and vapor phases to represent the composition effect. An alternative measure of composition is the convergence pressure of the system, which is defined as that pressure at which the K values for all the components in an isothermal mixture converge to unity. It is analogous to the critical point for a pure component in the sense that the two

TABLE 13-1   Constant-Pressure Liquid-Vapor Equilibrium Data for Selected Binary Systems

Component		Temperature, °C	Mole fraction A in		Total pressure kPa	Reference
A	B		Liquid	Vapor		
Acetone	Chloroform	62.50	0.0817	0.0500	101.3	1
		62.82	0.1390	0.1000		
		63.83	0.2338	0.2000		
		64.30	0.3162	0.3000		
		64.37	0.3535	0.3500		
		64.35	0.3888	0.4000		
		64.02	0.4582	0.5000		
		63.33	0.5299	0.6000		
		62.23	0.6106	0.7000		
		60.72	0.7078	0.8000		
		58.71	0.8302	0.9000		
		57.48	0.9075	0.9500		
Acetone	Methanol	64.65	0.0	0.0	101.3	2
		61.78	0.091	0.177		
		59.60	0.190	0.312		
		58.14	0.288	0.412		
		56.96	0.401	0.505		
		56.22	0.501	0.578		
		55.78	0.579	0.631		
		55.41	0.687	0.707		
		55.29	0.756	0.760		
		55.37	0.840	0.829		
		55.54	0.895	0.880		
		55.92	0.954	0.946		
		56.21	1.000	1.000		
Acetone	Water	74.80	0.0500	0.6381	101.3	3
		68.53	0.1000	0.7301		
		65.26	0.1500	0.7716		
		63.59	0.2000	0.7916		
		61.87	0.3000	0.8124		
		60.75	0.4000	0.8269		
		59.95	0.5000	0.8387		
		59.12	0.6000	0.8532		
		58.29	0.7000	0.8712		
		57.49	0.8000	0.8950		
		56.68	0.9000	0.9335		
		56.30	0.9500	0.9627		
Carbon tetrachloride	Benzene	80.0	0.0	0.0	101.3	4
		79.3	0.1364	0.1582		
		78.8	0.2157	0.2415		
		78.6	0.2573	0.2880		
		78.5	0.2944	0.3215		
		78.2	0.3634	0.3915		
		78.0	0.4057	0.4350		
		77.6	0.5269	0.5480		
		77.4	0.6202	0.6380		
		77.1	0.7223	0.7330		
Chloroform	Methanol	63.0	0.040	0.102	101.3	5
		60.9	0.095	0.215		
		59.3	0.146	0.304		
		57.8	0.196	0.378		
		55.9	0.287	0.472		
		54.7	0.383	0.540		
		54.0	0.459	0.580		
		53.7	0.557	0.619		
		53.5	0.636	0.646		
		53.5	0.667	0.655		
		53.7	0.753	0.684		
		54.4	0.855	0.730		
		55.2	0.904	0.768		
		56.3	0.937	0.812		
		57.9	0.970	0.875		
Ethanol	Benzene	76.1	0.027	0.137	101.3	6
		72.7	0.063	0.248		
		70.8	0.100	0.307		
		69.2	0.167	0.360		
		68.4	0.245	0.390		
		68.0	0.341	0.422		
		67.9	0.450	0.447		
		68.0	0.578	0.478		
		68.7	0.680	0.528		
		69.5	0.766	0.566		

TABLE 13-1   Constant-Pressure Liquid-Vapor Equilibrium Data for Selected Binary Systems (*Continued*)

Component		Temperature, °C	Mole fraction A in		Total pressure, kPa	Reference
A	B		Liquid	Vapor		
		70.4	0.820	0.615		
		72.7	0.905	0.725		
		76.9	0.984	0.937		
Ethanol	Water	95.5	0.0190	0.1700	101.3	7
		89.0	0.0721	0.3891		
		86.7	0.0966	0.4375		
		85.3	0.1238	0.4704		
		84.1	0.1661	0.5089		
		82.7	0.2337	0.5445		
		82.3	0.2608	0.5580		
		81.5	0.3273	0.5826		
		80.7	0.3965	0.6122		
		79.8	0.5079	0.6564		
		79.7	0.5198	0.6599		
		79.3	0.5732	0.6841		
		78.74	0.6763	0.7385		
		78.41	0.7472	0.7815		
		78.15	0.8943	0.8943		
Ethyl acetate	Ethanol	78.3	0.0	0.0	101.3	8
		76.6	0.050	0.102		
		75.5	0.100	0.187		
		73.9	0.200	0.305		
		72.8	0.300	0.389		
		72.1	0.400	0.457		
		71.8	0.500	0.516		
		71.8	0.540	0.540		
		71.9	0.600	0.576		
		72.2	0.700	0.644		
		73.0	0.800	0.726		
		74.7	0.900	0.837		
		76.0	0.950	0.914		
		77.1	1.000	1.000		
Ethylene glycol	Water	69.5	0.0	0.0	30.4	9
		76.1	0.23	0.002		
		78.9	0.31	0.003		
		83.1	0.40	0.010		
		89.6	0.54	0.020		
		103.1	0.73	0.06		
		118.4	0.85	0.13		
		128.0	0.90	0.22		
		134.7	0.93	0.30		
		145.0	0.97	0.47		
		160.7	1.00	1.00		
n-Hexane	Ethanol	78.30	0.0	0.0	101.3	10
		76.00	0.0100	0.0950		
		73.20	0.0200	0.1930		
		67.40	0.0600	0.3650		
		65.90	0.0800	0.4200		
		61.80	0.1520	0.5320		
		59.40	0.2450	0.6050		
		58.70	0.3330	0.6300		
		58.35	0.4520	0.6400		
		58.10	0.5880	0.6500		
		58.00	0.6700	0.6600		
		58.25	0.7250	0.6700		
		58.45	0.7650	0.6750		
		59.15	0.8980	0.7100		
		60.20	0.9550	0.7450		
		63.50	0.9900	0.8400		
		66.70	0.9940	0.9350		
		68.70	1.0000	1.0000		
Methanol	Benzene	70.67	0.026	0.267	101.3	11
		66.44	0.050	0.371		
		62.87	0.088	0.457		
		60.20	0.164	0.526		
		58.64	0.333	0.559		
		58.02	0.549	0.595		
		58.10	0.699	0.633		
		58.47	0.782	0.665		
		59.90	0.898	0.760		
		62.71	0.973	0.907		

TABLE 13-1 Constant-Pressure Liquid-Vapor Equilibrium Data for Selected Binary Systems (*Continued*)

| Component | | Temperature, °C | Mole fraction A in | | Total pressure, kPa | Reference |
A	B		Liquid	Vapor		
Methanol	Ethyl acetate	76.10	0.0125	0.0475	101.3	12
		74.15	0.0320	0.1330		
		71.24	0.0800	0.2475		
		67.75	0.1550	0.3650		
		65.60	0.2510	0.4550		
		64.10	0.3465	0.5205		
		64.00	0.4020	0.5560		
		63.25	0.4975	0.5970		
		62.97	0.5610	0.6380		
		62.50	0.5890	0.6560		
		62.65	0.6220	0.6670		
		62.50	0.6960	0.7000		
		62.35	0.7650	0.7420		
		62.60	0.8250	0.7890		
		62.80	0.8550	0.8070		
		63.21	0.9160	0.8600		
		63.90	0.9550	0.9290		
Methanol	Water	100.0	0.0	0.0	101.3	13
		96.4	0.020	0.134		
		93.5	0.040	0.230		
		91.2	0.060	0.304		
		89.3	0.080	0.365		
		87.7	0.100	0.418		
		84.4	0.150	0.517		
		81.7	0.200	0.579		
		78.0	0.300	0.665		
		75.3	0.400	0.729		
		73.1	0.500	0.779		
		71.2	0.600	0.825		
		69.3	0.700	0.870		
		67.5	0.800	0.915		
		66.0	0.900	0.958		
		65.0	0.950	0.979		
		64.5	1.000	1.000		
Methyl acetate	Methanol	57.80	0.173	0.342	101.3	14
		55.50	0.321	0.477		
		55.04	0.380	0.516		
		53.88	0.595	0.629		
		53.82	0.643	0.657		
		53.90	0.710	0.691		
		54.50	0.849	0.783		
		56.86	1.000	1.000		
1-Propanol	Water	100.00	0.0	0.0	101.3	15
		98.59	0.0030	0.0544		
		95.09	0.0123	0.1790		
		91.05	0.0322	0.3040		
		88.96	0.0697	0.3650		
		88.26	0.1390	0.3840		
		87.96	0.2310	0.3970		
		87.79	0.3110	0.4060		
		87.66	0.4120	0.4280		
		87.83	0.5450	0.4650		
		89.34	0.7300	0.5670		
		92.30	0.8780	0.7210		
		97.18	1.0000	1.0000		
2-Propanol	Water	100.00	0.0	0.0	101.3	16
		97.57	0.0045	0.0815		
		96.20	0.0069	0.1405		
		93.66	0.0127	0.2185		
		87.84	0.0357	0.3692		
		84.28	0.0678	0.4647		
		82.84	0.1330	0.5036		
		82.52	0.1651	0.5153		
		81.52	0.3204	0.5456		
		81.45	0.3336	0.5489		
		81.19	0.3752	0.5615		
		80.77	0.4720	0.5860		
		80.73	0.4756	0.5886		
		80.58	0.5197	0.6033		
		80.52	0.5945	0.6330		
		80.46	0.7880	0.7546		
		80.55	0.8020	0.7680		

TABLE 13-1 **Constant-Pressure Liquid-Vapor Equilibrium Data for Selected Binary Systems** (*Continued*)

Component		Temperature, °C	Mole fraction A in		Total pressure, kPa	Reference
A	B		Liquid	Vapor		
		81.32	0.9303	0.9010		
		81.85	0.9660	0.9525		
		82.39	1.0000	1.0000		
Tetrahydrofuran	Water	73.00	0.0200	0.6523	101.3	17
		66.50	0.0400	0.7381		
		65.58	0.0600	0.7516		
		64.94	0.1000	0.7587		
		64.32	0.2000	0.7625		
		64.27	0.3000	0.7635		
		64.23	0.4000	0.7643		
		64.16	0.5000	0.7658		
		63.94	0.6000	0.7720		
		63.70	0.7000	0.7831		
		63.54	0.8000	0.8085		
		63.53	0.8200	0.8180		
		63.57	0.8400	0.8260		
		63.64	0.8600	0.8368		
		63.87	0.9000	0.8660		
		64.29	0.9400	0.9070		
		65.07	0.9800	0.9625		
		65.39	0.9900	0.9805		
Water	Acetic acid	118.3	0.0	0.0	101.3	18
		110.6	0.1881	0.3063		
		107.8	0.3084	0.4467		
		105.2	0.4498	0.5973		
		104.3	0.5195	0.6580		
		103.5	0.5824	0.7112		
		102.8	0.6750	0.7797		
		102.1	0.7261	0.8239		
		101.5	0.7951	0.8671		
		100.8	0.8556	0.9042		
		100.8	0.8787	0.9186		
		100.5	0.9134	0.9409		
		100.2	0.9578	0.9708		
		100.0	1.0000	1.0000		
Water	1-Butanol	117.6	0.0	0.0	101.3	19
		111.4	0.049	0.245		
		106.7	0.100	0.397		
		102.0	0.161	0.520		
		101.0	0.173	0.534		
		98.5	0.232	0.605		
		96.7	0.288	0.654		
		95.2	0.358	0.693		
		93.6	0.487	0.739		
		93.1	0.551	0.751		
		93.0	0.580	0.752		
		92.9	0.628	0.758		
		92.9	0.927	0.758		
		93.2	0.986	0.760		
		95.2	0.993	0.832		
		96.8	0.996	0.883		
		100.0	1.000	1.000		
Water	Formic acid	102.30	0.0405	0.0245	101.3	20
		104.60	0.1550	0.1020		
		105.90	0.2180	0.1620		
		107.10	0.3210	0.2790		
		107.60	0.4090	0.4020		
		107.60	0.4110	0.4050		
		107.60	0.4640	0.4820		
		107.10	0.5220	0.5670		
		106.00	0.6320	0.7180		
		104.20	0.7400	0.8360		
		102.90	0.8290	0.9070		
		101.80	0.9000	0.9510		
		100.00	1.0000	1.0000		
Water	Glycerol	278.8	0.0275	0.9315	101.3	21
		247.0	0.0467	0.9473		
		224.0	0.0690	0.9563		
		219.2	0.0767	0.9743		
		210.0	0.0901	0.9783		
		202.5	0.1031	0.9724		
		196.5	0.1159	0.9839		

TABLE 13-1   Constant-Pressure Liquid-Vapor Equilibrium Data for Selected Binary Systems (*Concluded*)

| Component | | Temperature, °C | Mole fraction A in | | Total pressure, kPa | Reference |
A	B		Liquid	Vapor		
		175.2	0.1756	0.9899		
		149.3	0.3004	0.9964		
		137.2	0.3847	0.9976		
		136.8	0.3895	0.9878		
		131.8	0.4358	0.9976		
		121.5	0.5633	0.9984		
		112.8	0.7068	0.9993		
		111.3	0.7386	0.9994		
		106.3	0.8442	0.9996		
		100.0	1.0000	1.0000		

NOTE: To convert degrees Celsius to degrees Fahrenheit, °C = (°F − 32)/1.8. To convert kilopascals to pounds-force per square inch, multiply by 0.145.

[1] Kojima, Kato, Sunaga, and Hashimoto, *Kagaku Kogaku*, **32**, 337 (1968).
[2] Marinichev and Susarev, *Zh. Prtkl. Khtm.*, **38**, 378 (1965).
[3] Kojima, Tochigi, Seki, and Watase, *Kagaku Kogaku*, **32**, 149 (1968).
[4] *International Critical Tables*, McGraw-Hill, New York, 1928.
[5] Nagata, *J. Chem. Eng. Data*, **7**, 367 (1962).
[6] Ellis and Clark, *Chem. Age India*, **12**, 377 (1961).
[7] Carey and Lewis, *Ind. Eng. Chem.*, **24**, 882 (1932).
[8] Chu, Getty, Brennecke, and Paul, *Distillation Equilibrium Data*, New York, 1950.
[9] Trimble and Potts, *Ind. Eng. Chem.*, **27**, 66 (1935).
[10] Sinor and Weber, *J. Chem. Eng. Data*, **5**, 243 (1960).
[11] Hudson and Van Winkle, *J. Chem. Eng. Data*, **14**, 310 (1969).
[12] Murti and Van Winkle, *Chem. Eng. Data Ser.*, **3**, 72 (1958).
[13] Dunlop, M.S. thesis, Brooklyn Polytechnic Institute, 1948.
[14] Dobroserdov and Bagrov, *Zh. Prtkl. Kthm. (Leningrad)*, **40**, 875 (1967).
[15] Smirnova, *Vestn. Leningr. Univ. Fiz. Khim.*, 81 (1959).
[16] Kojima, Ochi, and Nakazawa, *Int. Chem. Eng.*, **9**, 342 (1964).
[17] Shnitko and Kogan, *J. Appl. Chem.*, **41**, 1236 (1968).
[18] Brusset, Kaiser, and Hoequel, *Chim. Ind., Gente Chim.* **99**, 207 (1968).
[19] Boublik, *Collect. Czech. Chem. Commun.*, **25**, 285 (1960).
[20] Ito and Yoshida, *J. Chem. Eng. Data*, **8**, 315 (1963).
[21] Chen and Thompson, *J. Chem. Eng. Data*, **15**, 471 (1970).

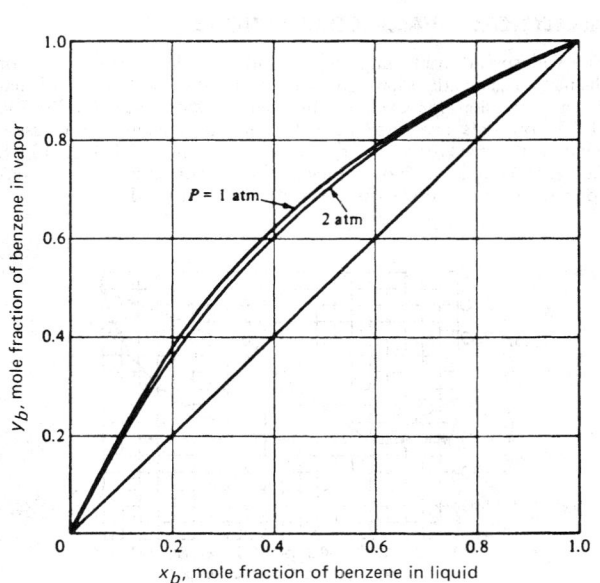

FIG. 13-8 Isobaric *y-x* curves for benzene-toluene. (*Brian, Staged Cascades in Chemical Processing, Prentice-Hall, Englewood Cliffs, NJ, 1972.*)

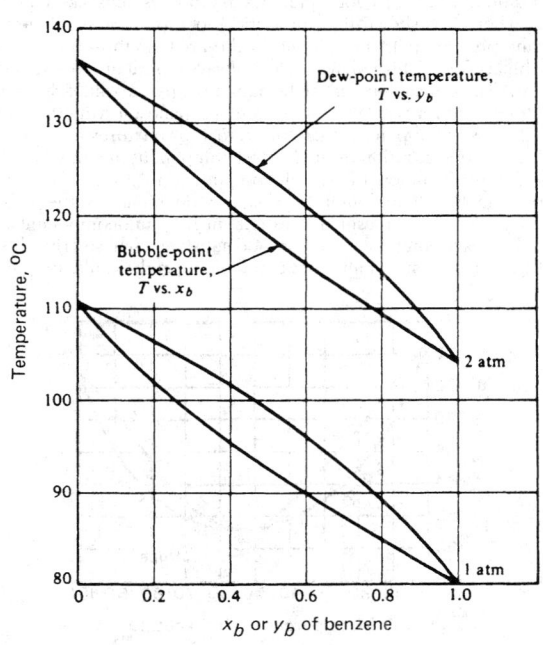

FIG. 13-9 Isobaric vapor-liquid equilibrium data for benzene-toluene. (*Brian, Staged Cascades in Chemical Processing, Prentice-Hall, Englewood Cliffs, NJ, 1972.*)

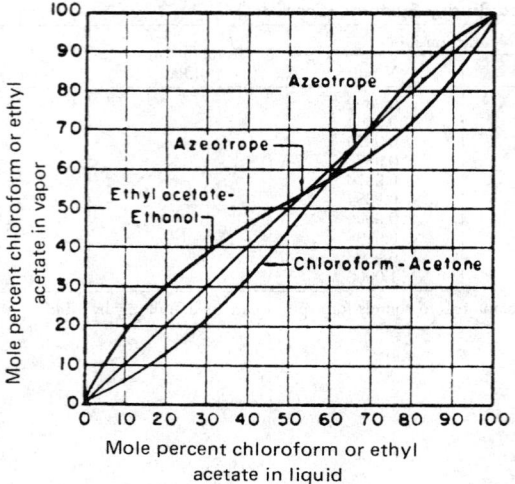

FIG. 13-10 Vapor-liquid equilibriums for the ethyl acetate–ethanol and chloroform-acetone systems at 101.3 kPa (1 atm).

phases become indistinguishable. The behavior of a complex mixture of hydrocarbons for a convergence pressure of 34.5 MPa (5000 psia) is illustrated in Fig. 13-15.

Two major graphical correlations based on convergence pressure as the third parameter (besides temperature and pressure) are the charts published by the Gas Processors Association (GPA, *Engineering Data Book*, 9th ed., Tulsa, 1981) and the charts of the American Petroleum Institute (API, *Technical Data Book—Petroleum Refining*, New York, 1966) based on the procedures from Hadden and Grayson [*Hydrocarbon Process., Pet. Refiner*, **40**(9), 207 (1961)]. The former uses the method proposed by Hadden [*Chem. Eng. Prog. Symp. Ser. 7*, **49**, 53 (1953)] for the prediction of convergence pressure as a function of composition. The basis for Hadden's method is illustrated in Fig. 13-16, where it is shown that the critical loci for various mixtures of methane-propane-pentane fall within the area circumscribed by the three binary loci. (This behavior is not always typical of more nonideal systems.) The critical loci for the ternary mixtures vary linearly, at constant temperature, with weight percent propane on a methane-free basis. The essential point is that critical loci for mixtures are independent of the concentration of the lightest component in a mixture. This permits representation of a multicomponent mixture as a pseudo binary. The light component in this pseudo binary is the lightest species present (to a reasonable extent) in the multicomponent mixture. The heavy component is a pseudo substance whose critical temperature is an average of all other components in the multicomponent

mixture. This pseudocritical point can then be located on a *P-T* diagram containing the critical points for all compounds covered by the charts, and a critical locus can be drawn for the pseudo binary by interpolation between various real binary curves. Convergence pressure for the mixture at the desired temperature is read from the assumed loci at the desired system temperature. This method is illustrated in the left half of Fig. 13-17 for the methane-propane-pentane ternary. Associated *K* values for pentane at 104°C (220°F) are shown to the right as a function of mixture composition (or convergence pressure).

The GPA convergence-pressure charts are primarily for alkane and alkene systems but do include charts for nitrogen, carbon dioxide, and hydrogen sulfide. The charts may not be valid when appreciable amounts of naphthenes or aromatics are present; the API charts use special procedures for such cases. Useful extensions of the convergence-pressure concept to more varied mixtures include the nomographs of Winn [*Chem. Eng. Prog. Symp. Ser. 2*, **48**, 121 (1952)], Hadden and Grayson (op. cit.), and Cajander, Hipkin, and Lenoir [*J. Chem. Eng. Data*, **5**, 251 (1960)].

ANALYTICAL *K*-VALUE CORRELATIONS

The widespread availability and utilization of digital computers for distillation calculations have given impetus to the development of analytical expressions for *K* values. McWilliams [*Chem. Eng.*, **80**(25), 138 (1973)] presents a regression equation and accompanying regression coefficients that represent the DePriester charts of Fig. 13-14. Regression equations and coefficients for various versions of the GPA convergence-pressure charts are available from the GPA.

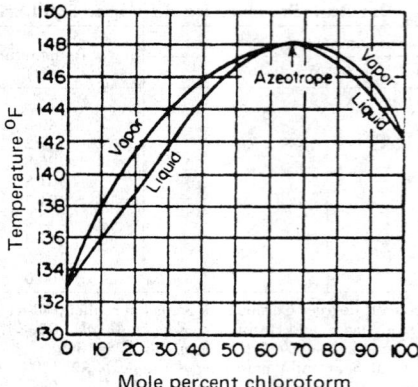

FIG. 13-12 Liquid boiling points and vapor condensation temperatures for maximum-boiling azeotrope mixtures of chloroform and acetone at 101.3 kPa (1 atm) total pressure.

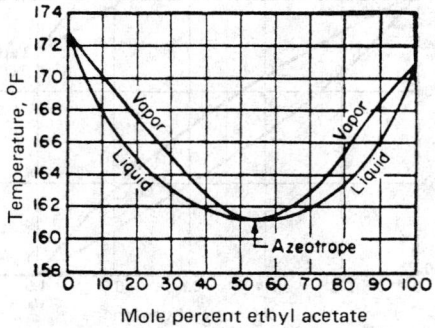

FIG. 13-11 Liquid boiling points and vapor condensation temperatures for minimum-boiling azeotrope mixtures of ethyl acetate and ethanol at 101.3 kPa (1 atm) total pressure.

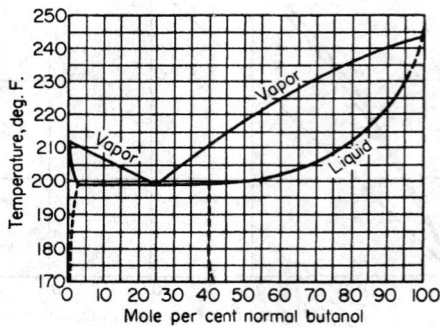

FIG. 13-13 Vapor-liquid equilibrium data for an *n*-butanol–water system at 101.3 kPa (1 atm); phase splitting and heterogeneous-azeotrope formation.

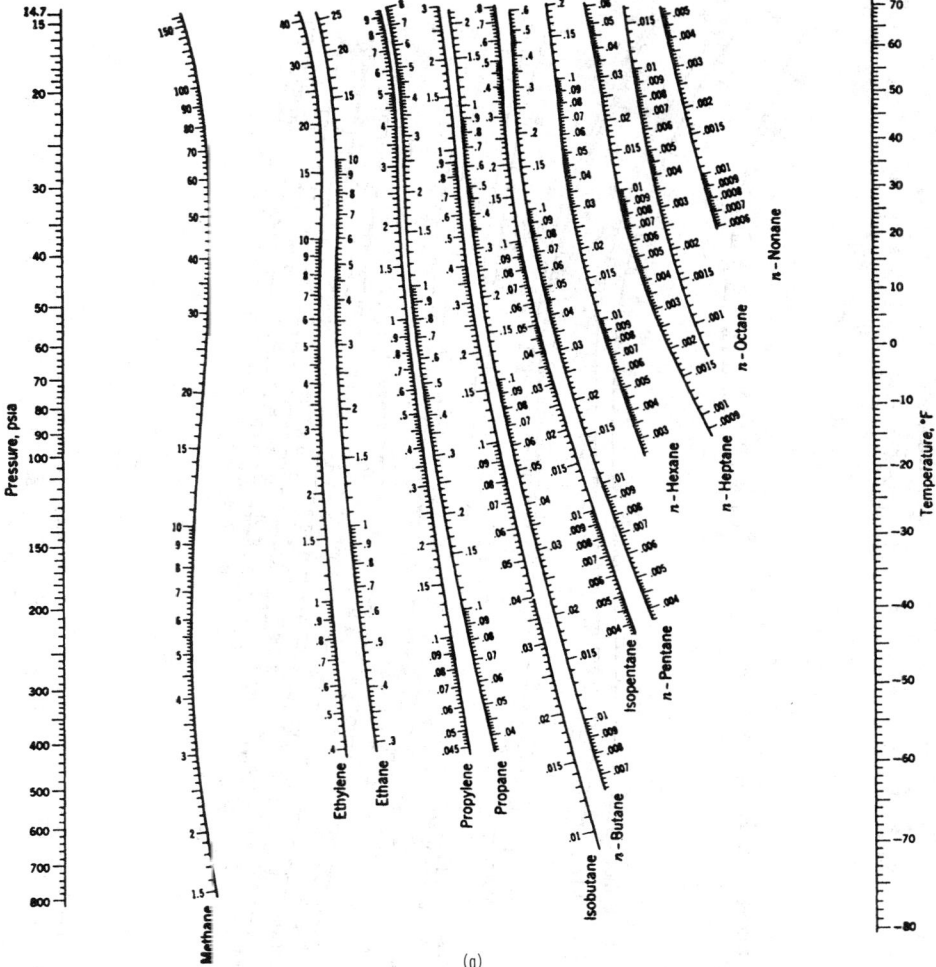

FIG. 13-14 K values ($K = y/x$) in light-hydrocarbon systems. (a) Low-temperature range. [*DePriester*, Chem. Eng. Prog. Symp. Sec. 7, **49**, 1 (1953)]

Preferred analytical correlations are less empirical in nature and most often are theoretically based on one of two exact thermodynamic formulations, as derived in Sec. 4. When a single pressure-volume-temperature (*PVT*) equation of state is applicable to both vapor and liquid phases, the formulation used is

$$K_i = \hat{\Phi}_i^L / \hat{\Phi}_i^V \qquad (13\text{-}3)$$

where the mixture fugacity coefficients $\hat{\Phi}_i^L$ for the liquid and $\hat{\Phi}_i^V$ for the vapor are derived by classical thermodynamics from the *PVT* expression. Consistent equations for enthalpy can similarly be derived.

Until recently, equations of state that have been successfully applied to Eq. (13-3) have been restricted to mixtures of nonpolar compounds, namely, hydrocarbons and light gases. These equations include those of Benedict-Webb-Rubin (BWR), Soave (SRK) [*Chem. Eng. Sci.*, **27**, 1197 (1972)], who extended the remarkable Redlich-Kwong equation, and Peng-Robinson (PR) [*Ind. Eng. Chem. Fundam.*, **15**, 59 (1976)]. The SRK and PR equations belong to a family of so-called cubic equations of state. The Starling extension of the BWR equation (*Fluid Thermodynamic Properties for Light Petroleum Systems*, Gulf, Houston, 1973) predicts *K* values and enthalpies of the normal paraffins up through *n*-octane, as well as isobutane, isopentane, ethylene, propylene, nitrogen, carbon dioxide, and hydrogen sul-

fide, including the cryogenic region. Computer programs for *K* values derived from the SRK, PR and other equations of state are widely available in all computer-aided process design and simulation programs. The ability of the SRK correlation to predict *K* values even when the pressure approaches the convergence pressure is shown for a multicomponent system in Fig. 13-18. Similar results are achieved with the PR correlation. The Wong-Sandler mixing rules for cubic equations of state now permit such equations to be extended to mixtures of organic chemicals, as shown in a reformulated version by Orbey and Sandler [*AIChE J.*, **41**, 683 (1995)].

An alternative *K*-value formulation that has received wide application to mixtures containing polar and/or nonpolar compounds is

$$K_i = \gamma_i^L \Phi_i^L / \hat{\Phi}_i^V \qquad (13\text{-}4)$$

where different equations of state may be used to predict the pure-component liquid fugacity coefficient Φ_i^L and the vapor-mixture fugacity coefficient, and any one of a number of mixture free-energy models may be used to obtain the liquid activity coefficient γ_i^L. At low to moderate pressures, accurate prediction of the latter is crucial to the application of Eq. (13-4).

When either Eq. (13-3) or Eq. (13-4) can be applied, the former is generally preferred because it involves only a single equation of state

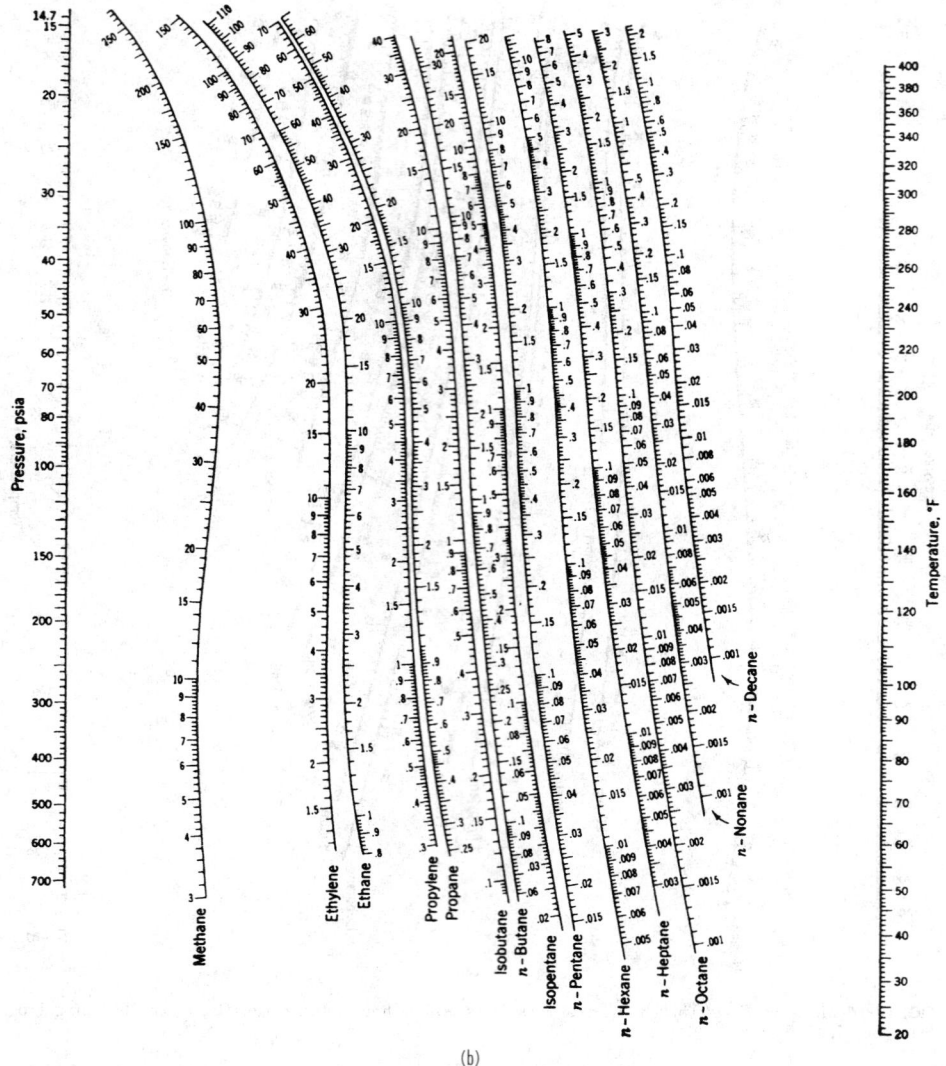

FIG. 13-14   (*Continued*) K values ($K = y/x$) in light-hydrocarbon systems. (*b*) High-temperature range. [*DePriester,* Chem. Eng. Prog. Symp. Sec. 7, **49**, 1 (1953).]

applicable to both phases and thus would seem to offer greater consistency. In addition, the quantity Φ_i^L in Eq. (13-4) is hypothetical for any components that are supercritical. In that case, a modification of Eq. (13-4) that uses Henry's law is sometimes applied.

For mixtures of hydrocarbons and light gases, Chao and Seader (CS) [*AIChE,* **7**, 598 (1961)] applied Eq. (13-4) by using an empirical expression for Φ_i^L based on the generalized corresponding-states *PVT* correlation of Pitzer et al., the Redlich-Kwong equation of state for $\hat{\Phi}_i^V$, and the regular solution theory of Scatchard and Hildebrand for γ_i^L. The predictive ability of the last-named theory is exhibited in Fig. 13-19 for the heptane-toluene system at 101.3 kPa (1 atm). Five pure-component constants for each species (T_c, P_c, ω, δ, and v^L) are required to use the CS method, which when applied within the restrictions discussed by Lenoir and Koppany [*Hydrocarbon Process.*, **46**(11), 249 (1967)] gives good results. Revised coefficients of Grayson and Streed (GS) (Pap. 20-P07, Sixth World Pet. Conf., Frankfurt, June, 1963) for the Φ_i^L expression permit application of

the CS correlation to higher temperatures and pressures and give improved predictions for hydrogen. Jin, Greenkorn, and Chao [*AIChE J,* **41**, 1602 (1995)] present a revised correlation for the standard-state liquid fugacity of hydrogen, applicable from 200 to 730 *K.*

For mixtures containing polar substances, more complex predictive equations for γ^L that involve binary-interaction parameters for each pair of components in the mixture are required for use in Eq. (13-4), as discussed in Sec. 4. Six popular expressions are the Margules, van Laar, Wilson, NRTL, UNIFAC, and UNIQUAC equations. Extensive listings of binary-interaction parameters for use in all but the UNIFAC equation are given by Gmehling and Onken (op. cit.). They obtained the parameters for binary systems at 101.3 kPa (1 atm) from best fits of the experimental *T-y-x* equilibrium data by setting Φ_i^V and Φ_i^L to their ideal-gas, ideal-solution limits of 1.0 and P^{sat}/P respectively, with the vapor pressure P^{sat} given by a three-constant Antoine equation, whose values they tabulate. Table 13-2 lists their parameters for some of the binary systems included in

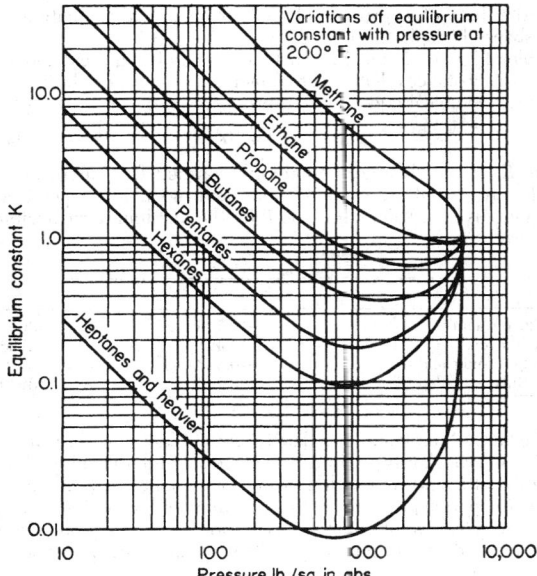

FIG. 13-15 Typical variation of K values with total pressure at constant temperature for a complex mixture. Light hydrocarbons in admixture with crude oil. [*Katz and Hachmuth,* Ind. Eng. Chem., **29,** *1072 (1937).*]

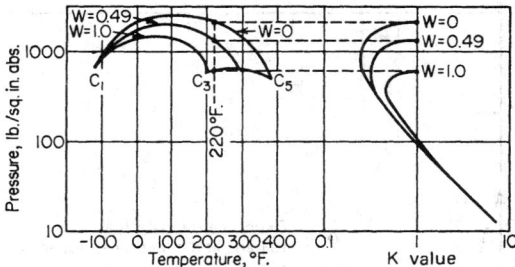

FIG. 13-17 Effect of mixture composition upon K value for *n*-pentane at 104°C (220°F). K values are shown for various values of W, weight fraction propane on a methane-free basis for the methane-propane-pentane system. [*Hadden,* Chem. Eng. Prog. Symp. Sec. 7, **49,** *58 (1953).*]

Table 13-1, based on the binary-system activity-coefficient-equation forms given in Table 13-3. Consistent Antoine vapor-pressure constants and liquid molar volumes are listed in Table 13-4. The Wilson equation is particularly useful for systems that are highly nonideal but do not undergo phase splitting, as exemplified by the ethanol-hexane system, whose activity coefficients are shown in Fig. 13-20. For systems such as this, in which activity coefficients in dilute regions may

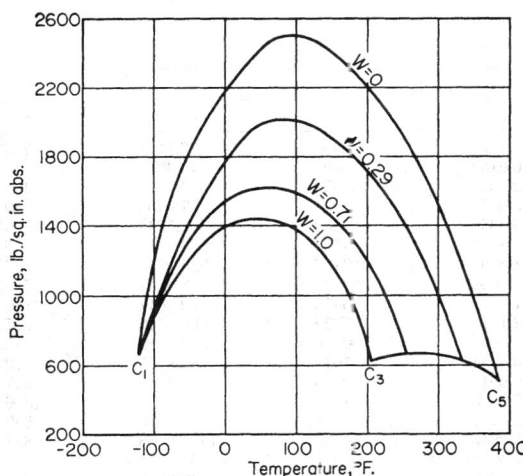

FIG. 13-16 Critical loci for a methane-propane-pentane system according to Hadden. [Chem. Eng. Prog. Symp. Sec. 7, **49,** *53 (1953).*] Parameter W is weight fraction propane on a methane-free basis.

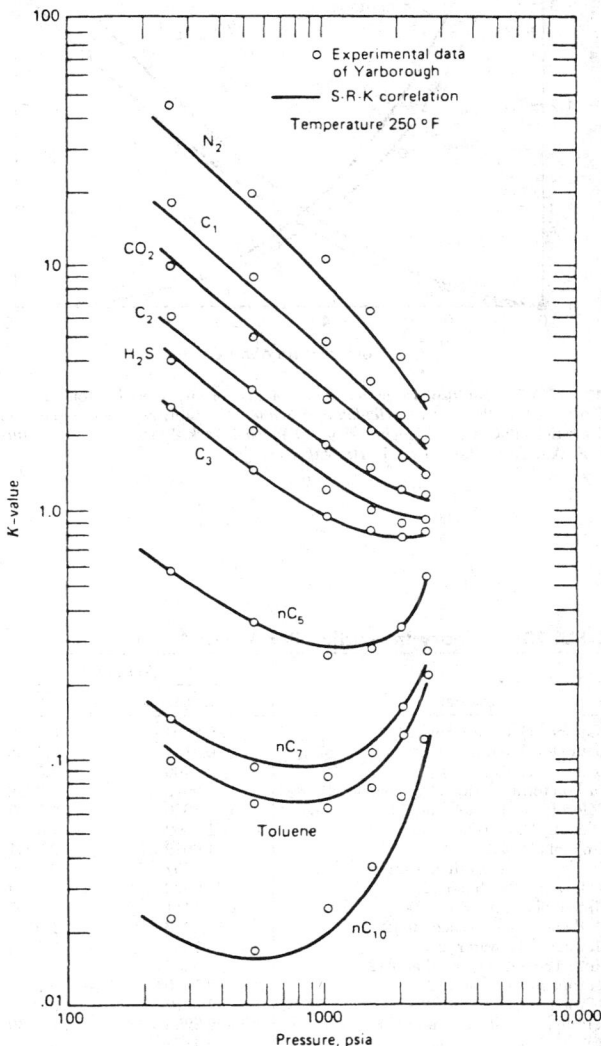

FIG. 13-18 Comparison of experimental K-value data and SRK correlation. [*Henley and Seader,* Equilibrium-Stage Separation Operations in Chemical Engineering, Wiley, New York, 1981; *data of Yarborough,* J. Chem. Eng. Data, **17,** *129 (1972).*]

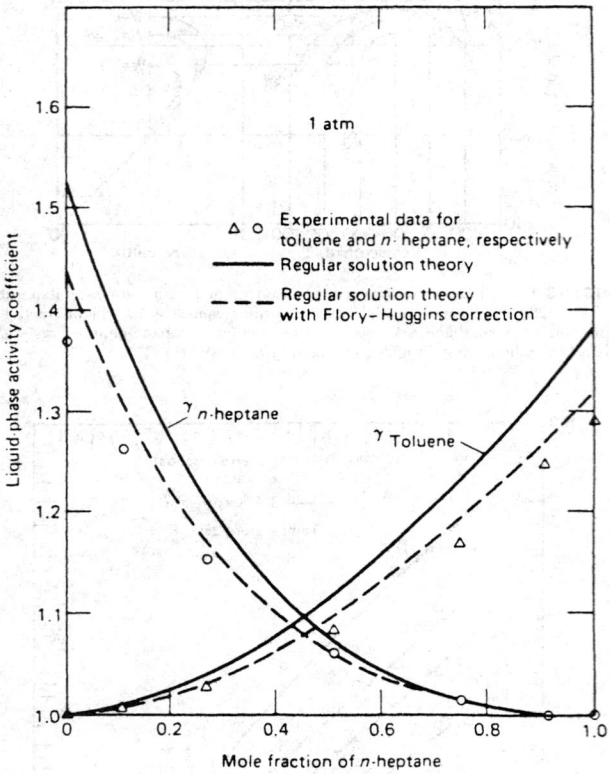

FIG. 13-19 Liquid-phase activity coefficients for an n-heptane-toluene system at 101.3 kPa (1 atm). [*Henley and Seader,* Equilibrium-Stage Separation Operations in Chemical Engineering, *Wiley, New York, 1981; data of Yerazunis et al.,* Am. Inst. Chem. Eng. J., **10**, *660 (1964).*]

exceed values of approximately 7.5, the van Laar equation erroneously predicts phase splitting.

Tables 13-1, 13-2, and 13-4 include data on formic acid and acetic acid, two substances that tend to dimerize in the vapor phase according to the chemical-equilibrium expression

$$K_D = P_D/P_M^2 = 10^{A + B/T} \tag{13-5}$$

where K_D is the chemical-equilibrium constant for dimerization, P_D and P_M are partial pressures of dimer and monomer respectively in torr, and T is in K. Values of A and B for the first four normal aliphatic acids are:

	A	B
Formic acid	−10.743	3083
Acetic acid	−10.421	3166
n-Propionic acid	−10.843	3316
n-Butyric acid	−10.100	3040

As shown by Marek and Standart [*Collect. Czech. Chem. Commun.,* **19**, 1074 (1954)], it is preferable to correlate and utilize liquid-phase activity coefficients for the dimerizing component by considering separately the partial pressures of the monomer and dimer. For example, for a binary system of components 1 and 2, when only compound 1 dimerizes in the vapor phase, the following equations apply if an ideal gas is assumed:

$$P_1 = P_D + P_M \tag{13-6}$$

$$y_1 = (P_M + 2P_D)/P \tag{13-7}$$

These equations when combined with Eq. (13-5) lead to the following equations for liquid-phase activity coefficients in terms of measurable quantities:

$$\gamma_1 = \frac{Py_1}{P_1^{sat}x_1}\left\{\frac{1 + (1 + 4K_D P_1^{sat})^{0.5}}{1 + [1 + 4K_D Py_1(2 - y_1)]^{0.5}}\right\} \tag{13-8}$$

$$\gamma_2 = \frac{Py_1}{P_2^{sat}x_2}\left(\frac{2\{1 - y_1 + [1 + 4K_D Py_1(2 - y_1)]^{0.5}\}}{(2 - y_1)\{1 + [1 + 4K_D Py_1(2 - y_1)]^{0.5}\}}\right) \tag{13-9}$$

Detailed procedures, including computer programs for evaluating binary-interaction parameters from experimental data and then utiliz-

TABLE 13-2 Binary-Interaction Parameters*

System	Margules		van Laar		Wilson (cal/mol)	
	$\overline{A}_{12}$	$\overline{A}_{21}$	A_{12}	A_{21}	$(\lambda_{12} - \lambda_{11})$	$(\lambda_{21} - \lambda_{22})$
Acetone (1), chloroform (2)	−0.8404	−0.5610	−0.8643	−0.5899	116.1171	−506.8519
Acetone (1), methanol (2)	0.6184	0.5788	0.6184	0.5797	−114.4047	545.2942
Acetone (1), water (2)	2.0400	1.5461	2.1041	1.5555	344.3346	1482.2133
Carbon tetrachloride (1), benzene (2)	0.0948	0.0922	0.0951	0.0911	7.0459	59.6233
Chloroform (1), methanol (2)	0.8320	1.7365	0.9356	1.8860	−361.7944	1694.0241
Ethanol (1), benzene (2)	1.8362	1.4717	1.8570	1.4785	1264.4318	266.6118
Ethanol (1), water (2)	1.6022	0.7947	1.6798	0.9227	325.0757	953.2792
Ethyl acetate (1), ethanol (2)	0.8557	0.7476	0.8552	0.7526	58.8869	570.0439
n-Hexane (1), ethanol (2)	1.9398	2.7054	1.9195	2.8463	320.3611	2189.2896
Methanol (1), benzene (2)	2.1411	1.7905	2.1623	1.7925	1666.4410	227.2126
Methanol (1), ethyl acetate (2)	1.0016	1.0517	1.0017	1.0524	982.2689	−172.9317
Methanol (1), water (2)	0.7923	0.5434	0.8041	0.5619	82.9876	520.6458
Methyl acetate (1), methanol (2)	0.9605	1.0120	0.9614	1.0126	−93.8900	847.4348
1-Propanol (1), water (2)	2.7070	0.7172	2.9095	1.1572	906.5256	1396.6398
2-Propanol (1), water (2)	2.3319	0.8976	2.4702	1.0938	659.5473	1230.2080
Tetrahydrofuran (1), water (2)	2.8258	1.9450	3.0216	1.9436	1475.2583	1844.7926
Water (1), acetic acid (2)	0.4178	0.9533	0.4973	1.0623	705.5876	111.6579
Water (1), 1-butanol (2)	0.8608	3.2051	1.0996	4.1760	1549.6600	2050.2569
Water (1), formic acid (2)	−0.2966	−0.2715	−0.2935	−0.2757	−310.1060	1180.8040

*Abstracted from Gmehling and Onken, *Vapor-Liquid Equilibrium Data Collection,* DECHEMA Chemistry Data ser., vol. 1 (parts 1–10), Frankfurt, 1977.

TABLE 13-3 Activity-Coefficient Equations in Binary Form for Use with Parameters and Constants in Tables 13-2 and 13-4

Type of equation	Adjustable parameters	Equations in binary form
Margules	$\bar{A}_{12}$ $\bar{A}_{21}$	$\ln \gamma_1 = [\bar{A}_{12} + 2(\bar{A}_{21} - \bar{A}_{12})x_1]x_2^2$ $\ln \gamma_2 = [\bar{A}_{21} + 2(\bar{A}_{12} - \bar{A}_{21})x_2]x_1^2$
van Laar	A_{12} A_{21}	$\ln \gamma_1 = A_{12}\left(\dfrac{A_{21}x_2}{A_{12}x_1 + A_{21}x_2}\right)^2$ $\ln \gamma_2 = A_{21}\left(\dfrac{A_{12}x_1}{A_{12}x_1 + A_{21}x_2}\right)^2$
Wilson	$\lambda_{12} - \lambda_{11}$ $\lambda_{21} - \lambda_{22}$	$\ln \gamma_1 = -\ln(x_1 + \Lambda_{12}x_2) + x_2\left(\dfrac{\Lambda_{12}}{x_1 + \Lambda_{12}x_2} - \dfrac{\Lambda_{21}}{\Lambda_{21}x_1 + x_2}\right)$ $\ln \gamma_2 = -\ln(x_2 + \Lambda_{21}x_1) - x_1\left(\dfrac{\Lambda_{12}}{x_1 + \Lambda_{12}x_2} - \dfrac{\Lambda_{21}}{\Lambda_{21}x_1 + x_2}\right)$

where $\Lambda_{12} = \dfrac{v_2^L}{v_1^L} \exp\left(-\dfrac{\lambda_{12} - \lambda_{11}}{RT}\right) \quad \Lambda_{21} = \dfrac{v_1^L}{v_2^L} \exp\left(-\dfrac{\lambda_{21} - \lambda_{22}}{RT}\right)$

v_i^L = molar volume of pure-liquid component i
λ_{ij} = interaction energy between components i and j, $\lambda_{ij} = \lambda_{ji}$

ing these parameters to predict K values and phase equilibria, are given in terms of the UNIQUAC equation by Prausnitz et al. (*Computer Calculations for Multicomponent Vapor-Liquid and Liquid-Liquid Equilibria*, Prentice-Hall, Englewood Cliffs, N.J., 1980) and in terms of the UNIFAC group contribution method by Fredenslund, Gmehling, and Rasmussen (*Vapor-Liquid Equilibria Using UNIFAC*, Elsevier, Amsterdam, 1980). Both use the method of Hayden and O'Connell [*Ind. Eng. Chem. Process Des. Dev.*, **14**, 209 (1975)] to compute $\hat{\Phi}_i^V$ in Eq. (13-4). When the system temperature is greater than the critical temperature of one or more components in the mixture, Prausnitz et al. utilize a Henry's-law constant $H_{i,M}$ in place of the product $\gamma_i^L \Phi_i^L$ in Eq. (13-4). Otherwise Φ_i^L is evaluated from vapor-pressure data with a Poynting saturated-vapor fugacity correction. When the total pressure is less than about 202.6 kPa (2 atm) and all components in the mixture have a critical temperature that is greater

than the system temperature, then $\Phi_i^L = P_i^{sat}/P$ and $\hat{\Phi}_i^V = 1.0$. Equation (13-4) then reduces to

$$K_i = \gamma_i^L P_i^{sat}/P \qquad (13\text{-}10)$$

which is referred to as a modified Raoult's-law K value. If, furthermore, the liquid phase is ideal, then $\gamma_i^L = 1.0$ and

$$K_i = P_i^{sat}/P \qquad (13\text{-}11)$$

which is referred to as a Raoult's-law K value that is dependent solely on the vapor pressure P_i^{sat} of the component in the mixture. The UNIFAC method is being periodically updated with new group contributions, with a recent article being that of Hansen et al. [*Ind. Eng. Chem. Res.*, **30**, 2352 (1991)].

TABLE 13-4 Antoine Vapor-Pressure Constants and Liquid Molar Volume*

Species	Antoine constants†			Applicable temperature region, °C	v^L, liquid molar volume, cm³/ g·mol
	A	B	C		
Acetic acid	8.02100	1936.010	258.451	18–118	57.54
Acetone	7.11714	1210.595	229.664	(−13)–55	74.05
Benzene	6.87987	1196.760	219.161	8–80	89.41
1-Butanol	7.36366	1305.198	173.427	89–126	91.97
Carbon tetrachloride	6.84083	1177.910	220.576	(−20)–77	97.09
Chloroform	6.95465	1170.966	226.232	(−10)–60	80.67
Ethanol	7.58670	1281.590	193.768	78–203	58.68
Ethanol	8.11220	1592.864	226.184	20–93	58.68
Ethyl acetate	7.10179	1244.951	217.881	16–76	98.49
Formic acid	6.94459	1295.260	218.000	36–108	37.91
n-Hexane	6.91058	1189.640	226.280	(−30)–170	131.61
Methanol	8.08097	1582.271	239.726	15–84	40.73
Methyl acetate	7.06524	1157.630	219.726	2–56	79.84
1-Propanol	8.37895	1788.020	227.438	(−15)–98	75.14
2-Propanol	8.87829	2010.320	252.636	(−26)–83	76.92
Tetrahydrofuran	6.99515	1202.290	226.254	23–100	81.55
Water	8.07131	1730.630	233.426	1–100	18.07

*Abstracted from Gmehling and Onken, *Vapor-Liquid Equilibrium Data Collection*, DECHEMA Chemistry Data ser., vol. 1 (parts 1–10), Frankfurt, 1977.
†Antoine equation is $\log P^{sat} = A - B/(T + C)$ with P^{sat} in torr and T in °C.
NOTE: To convert degrees Celsius to degrees Fahrenheit, °F = 1.8°C + 32. To convert cubic centimeters per gram-mole to cubic feet per pound-mole, multiply by 0.016.

DEGREES OF FREEDOM AND DESIGN VARIABLES

DEFINITIONS

For separation processes, a design solution is possible if the number of independent equations equals the number of unknowns.

$$N_i = N_v - N_c$$

where N_v is the total number of variables (unknowns) involved in the process under consideration, N_c is the number of restricting relationships among the unknowns (independent equations), and N_i is termed the number of design variables. In the analogous phase-rule analysis, N_i is usually referred to as the degrees of freedom or variance. It is the number of variables that the designer must specify to define one unique operation (solution) of the process.

The variables N_i with which the designer of a separation process must be concerned are:

1. Stream concentrations (e.g., mole fractions)
2. Temperatures
3. Pressures
4. Stream flow rates
5. Repetition variables N_r

The first three are intensive variables. The fourth is an extensive variable that is not considered in the usual phase-rule analysis. The fifth is neither an intensive nor an extensive variable but is a single degree of freedom that the designer utilizes in specifying how often a particular element is repeated in a unit. For example, a distillation-column section is composed of a series of equilibrium stages, and when the designer specifies the number of stages that the section contains, he

or she utilizes the single degree of freedom represented by the repetition variable ($N_r = 1.0$). If the distillation column contains more than one section (such as above and below a feed stage), the number of stages in each section must be specified and as many repetition variables exist as there are sections, that is, $N_r = 2$.

The various restricting relationships N_c can be classified as:

1. Inherent
2. Mass-balance
3. Energy-balance
4. Phase-distribution
5. Chemical-equilibrium

The inherent restrictions are usually the result of definitions and take the form of identities. For example, the concept of the equilibrium stage involves the inherent restrictions that $T^V = T^L$ and $P^V = P^L$ where the superscripts V and L refer to the equilibrium exit streams.

The mass-balance restrictions are the C balances written for the C components present in the system. (Since we will only deal with nonreactive mixtures, each chemical compound present is a phase-rule component.) An alternative is to write $(C - 1)$ component balances and one overall mass balance.

The phase-distribution restrictions reflect the requirement that $f_i^V = f_i^L$ at equilibrium where f is the fugacity. This may be expressed by Eq. (13-1). In vapor-liquid systems, it should always be recognized that all components appear in both phases to some extent and there will be such a restriction for each component in the system. In vapor-liquid-liquid systems, each component will have three such restrictions, but only two are independent. In general, when all components exist in all phases, the number of restricting relationships due to the distribution phenomenon will be $C(N_p - 1)$, where N_p is the number of phases present.

For the analysis here, the forms in which the restricting relationships are expressed are unimportant. Only the number of such restrictions is important.

ANALYSIS OF ELEMENTS

An *element* is defined as part of a more complex *unit*. The unit may be all or only part of an operation or the entire *process*. Our strategy will be to analyze all elements that appear in a separation process and determine the number of design variables associated with each. The appropriate elements can then be combined to form the desired units and the various units combined to form the entire process. Allowance must of course be made for the connecting streams (*interstreams*) whose variables are counted twice when elements or units are joined.

The simplest element is a *single homogeneous stream*. The variables necessary to define it are:

	N_c^e
Concentrations	$C - 1$
Temperature	1
Pressure	1
Flow rate	1
	$C + 2$

There are no restricting relationships when the stream is considered only at a point. Henley and Seader (*Equilibrium-Stage Separation Operations in Chemical Engineering*, Wiley, New York, 1981) count all C concentrations as variables, but then have to include

$$\sum_i x_i = 1.0 \quad \text{or} \quad \sum_i y_i = 1.0$$

as a restriction.

A stream divider simply splits a stream into two or more streams of the same composition. Consider Fig. 13-21, which pictures the division of the condensed overhead liquid L_c into distillate D and

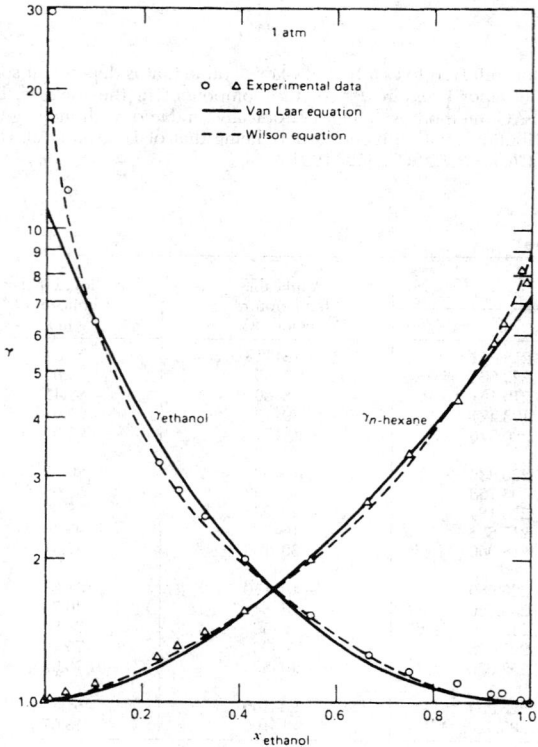

FIG. 13-20 Liquid-phase activity coefficients for an ethanol–*n*-hexane system. [*Henley and Seader*, Equilibrium-Stage Separation Operations in Chemical Engineering, *Wiley, New York, 1931; data of Sinor and Weber*, J. Chem. Eng. Data, **5**, 243–247 (1960).]

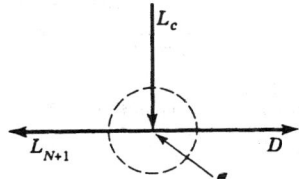

FIG. 13-21 Stream divider.

reflux L_{N+1}. The divider is permitted to operate nonadiabatically if desired. Three mass streams and one possible "energy stream" are involved; so

$$N_v^e = 3(C + 2) + 1 = 3C + 7$$

Each mass stream contributes $C + 2$ variables, but an energy stream has only its rate q as a variable. The independent restrictions are as follows:

	N_c^e
Inherent	
T and P identities between L_{N+1} and D	2
Concentration identities between L_{N+1} and D	$C - 1$
Mass balances	C
Energy balance	1
	$2C + 2$

The number of design variables for the element is given by

$$N_i^e = N_v^e - N_c^e = (3C + 7) - (2C + 2) = C + 5$$

Specification of the feed stream L_c ($C + 2$ variables), the ratio L_{N+1}/D, the "heat leak" q, and the pressure of either stream leaving the divider utilizes these design variables and defines one unique operation of the divider.

A simple equilibrium stage (no feed or sidestreams) is depicted in Fig. 13-22. Four mass streams and a heat-leak (or heat-addition) stream provide the following number of variables:

$$N_v^e = 4(C + 2) + 1 = 4C + 9$$

Vapor and liquid streams V_n and L_n respectively are in equilibrium with each other by definition and therefore are at the same T and P. These two inherent identities when added to C-component balances, one energy balance, and the C phase-distribution relationships give

$$N_c^e = 2C + 3$$

Then
$$N_i^e = N_v^e - N_c^e$$
$$= (4C + 9) - (2C + 3) = 2C + 6$$

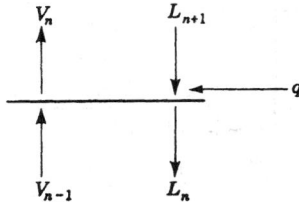

FIG. 13-22 Simple equilibrium stage.

These design variables can be utilized as follows:

Specifications	N_i^e
Specification of L_{n+1} stream	$C + 2$
Specification of V_{n-1} stream	$C + 2$
Pressure of either leaving stream	1
Heat leak q	1
	$2C + 6$

The results of the analyses for all the various elements commonly encountered in distillation processes are summarized in Table 13-5. Details of the analyses are given by Smith (*Design of Equilibrium Stage Processes*, McGraw-Hill, New York, 1967) and in a somewhat different form by Henley and Seader (op. cit.).

ANALYSIS OF UNITS

A "unit" is defined as a combination of elements and may or may not constitute the entire process. By definition

$$N_v^u = N_r + \sum_i N_i^e$$

and
$$N_i^u = N_v^u - N_c^u$$

where N_c^u refers to *new* restricting relationships (identities) that may arise when elements are combined. N_c^u does not include any of the restrictions considered in calculating the N_i^e's for the various elements. It includes only the stream identities that exist in each interstream between two elements. The interstream variables $(C + 2)$ were counted in each of the two elements when their respective N_i^e's were calculated. Therefore, $(C + 2)$ new restricting relationships must be counted for each interstream in the combination of elements to prevent redundancy.

The simple absorber column shown in Fig. 13-23 will be analyzed here to illustrate the procedure. This unit consists of a series of simple equilibrium stages of the type in Fig. 13-22. Specification of the number of stages N utilizes the single repetition variable and

$$N_v^u = N_r + \sum_i N_i^e = 1 + N(2C + 6)$$

since in Table 13-5 $N_i^e = 2C + 6$ for a simple equilibrium stage. There are $2(N - 1)$ interstreams, and therefore $2(N - 1)(C + 2)$ new identities (not previously counted) come into existence when elements are combined. Subtraction of these restrictions from N_v^u gives N_i^u, the design variables that must be specified.

$$N_i^u = N_v^u - N_c^u = N_r + \sum_i N_i^e - N_c^u$$
$$= [1 + N(2C + 6)] - [2(N - 1)(C + 2)]$$
$$= 2C + 2N + 5$$

TABLE 13-5 Design Variables N_i^e for Various Elements

Element	N_v^e	N_c^e	N_i^e
Homogeneous stream	$C + 2$	0	$C + 2$
Stream divider	$3C + 7$	$2C + 2$	$C + 5$
Stream mixer	$3C + 7$	$C + 1$	$2C + 6$
Pump	$2C + 5$	$C + 1$	$C + 4$
Heater	$2C + 5$	$C + 1$	$C + 4$
Cooler	$2C + 5$	$C + 1$	$C + 4$
Total condenser	$2C + 5$	$C + 1$	$C + 4$
Total reboiler	$2C + 5$	$C + 1$	$C + 4$
Partial condenser	$3C + 7$	$2C + 3$	$C + 4$
Partial reboiler	$3C + 7$	$2C + 3$	$C + 4$
Simple equilibrium state	$4C + 9$	$2C + 3$	$2C + 6$
Feed stage	$5C + 11$	$2C + 3$	$3C + 8$
Sidestream stage	$5C + 11$	$3C + 4$	$2C + 7$
Adiabatic equilibrium flash	$3C + 6$	$2C + 3$	$C + 3$
Nonadiabatic equilibrium flash	$3C + 7$	$2C + 3$	$C + 4$

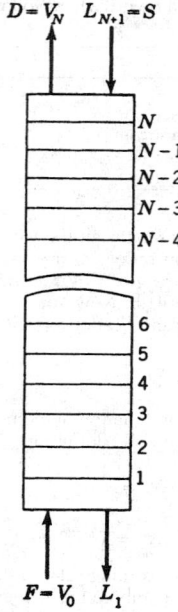

$D = V_N$ $L_{N+1} = S$

N
$N-1$
$N-2$
$N-3$
$N-4$

6
5
4
3
2
1

$F = V_0$ L_1

FIG. 13-23   Simple absorption.

These might be used as follows:

Specifications	N_i^u
Two feed streams	$2C$ $+4$
Number of stages N	1
Pressure of either stream leaving each stage	N
Heat leak for each stage	N
	$2C + 2N + 5$

A more complex unit is shown in Fig. 13-24, which is a schematic diagram of a *distillation column* with one feed, a total condenser, and a partial reboiler. Dotted lines encircle the six connected elements (or units) that constitute the distillation operation. The variables N_v^u that must be considered in the analysis of the entire process are just the sum of the N_i^e's for these six elements since here $N_r = 0$. Using Table 13-5,

Element (or unit)	$N_v^u = \sum_i N_i^e$
Total condenser	C $+4$
Reflux divider	C $+5$
$N - (M + 1)$ equilibrium stages	$2C + 2(N - M - 1)$ $+5$
Feed stage	$3C$ $+8$
$(M - 1)$ equilibrium stages	$2C + 2(M - 1)$ $+5$
Partial reboiler	C $+4$
	$10C + 2N$ $+27$

Here, the two units of $N - (M + 1)$ and $(M - 1)$ stages are treated just like elements. Nine interstreams are created by the combination of elements; so

$$N_c^u = 9(C + 2) = 9C + 18$$

The number of design variables is

$$N_i^u = C + 2N + 9 \; N_v^u - N_c^u = (10C + 2N + 27) - (9C + 18)$$

One set of specifications that is particularly convenient for computer solutions is:

Specifications	N_i^u
Pressure of either stream leaving each stage (including reboiler)	N
Pressure of stream leaving condenser	1
Pressure of either stream leaving reflux divider	1
Heat leak for each stage (excluding reboiler)	$N-1$
Heat leak for reflux divider	1
Feed stream	C $+2$
Reflux temperature	1
Total number of stages N	1
Number of stages below feed M	1
Distillate rate D/F	1
Reflux rate (L_{N+1}/D)	1
	$C + 2N + 9$

Other specifications often used in place of one or more of the last four listed are the fractional recovery of one component in either D or B and/or the concentration of one component in either D or B.

OTHER UNITS AND COMPLEX PROCESSES

In Table 13-6, the number of design variables is summarized for several distillation-type separation operations, most of which are shown in Fig. 13-7. For columns not shown in Figs. 13-1 or 13-7 that

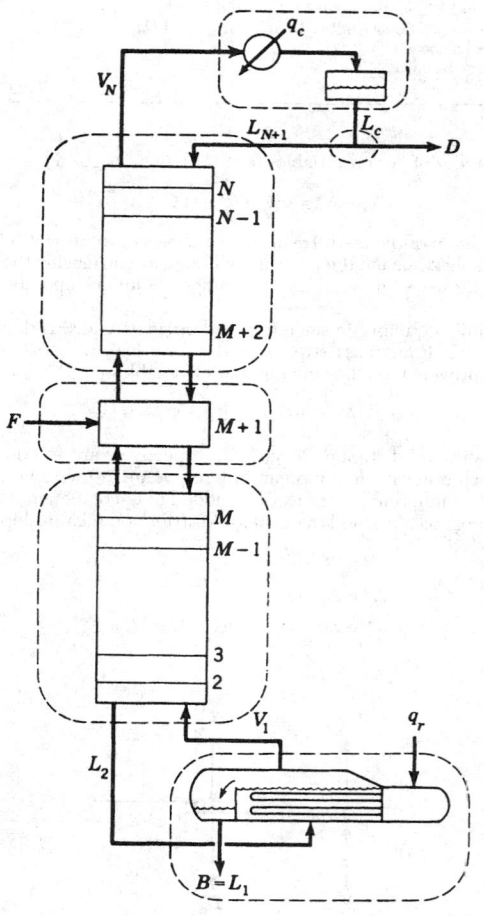

q_c
V_N
L_{N+1} L_c D
N
$N-1$
$M+2$
F $M+1$
M
$M-1$
3
2
V_1
L_2
q_r
$B = L_1$

FIG. 13-24   Distillation column with one feed, a total condenser, and a partial reboiler.

involve additional feeds and/or sidestreams, add $(C + 3)$ degrees of freedom for each additional feed ($C + 2$ to define the feed and 1 to designate the feed stage) and 2 degrees of freedom for each sidestream (1 for the sidestream flow rate and 1 to designate the sidestream-stage location). Any number of elements or units can be combined to form complex processes. No new rules beyond those developed earlier are necessary for their analysis. When applied to the thermally coupled distillation process of Fig. 13-6b, the result is $N_i^u = 2(N + M) + C + 18$. Further examples are given in Henley and Seader (op. cit.). An alternative method for determining the degrees of freedom for equipment and processes is given by Pham [*Chem. Eng. Sci.*, **49**, 2507 (1994)].

TABLE 13-6   Design Variables N_i^u for Separation Units

Unit	N_i^u*
Distillation (partial reboiler–total condenser)	$C + 2N + 9$
Distillation (partial reboiler–partial condenser)	$C + 2N + 6$
Absorption	$2C + 2N + 5$
Rectification (partial condenser)	$C + 2N + 3$
Stripping	$2C + 2N + 5$
Reboiled stripping (partial reboiler)	$C + 2N + 3$
Reboiled absorption (partial reboiler)	$2C + 2N + 6$
Refluxed stripping (total condenser)	$2C + 2N + 9$
Extractive distillation (partial reboiler–total condenser)	$2C + 2N + 12$

*N includes reboiler, but not condenser.

SINGLE-STAGE EQUILIBRIUM-FLASH CALCULATIONS

INTRODUCTION

The simplest continuous-distillation process is the adiabatic single-stage equilibrium-flash process pictured in Fig. 13-25. Feed temperature and the pressure drop across the valve are adjusted to vaporize the feed to the desired extent, while the drum provides disengaging space to allow the vapor to separate from the liquid. The expansion across the valve is at constant enthalpy, and this fact can be used to calculate T_2 (or T_1 to give a desired T_2).

From Table 13-5 it can be seen that the variables subject to the designer's control are $C + 3$ in number. The most common way to utilize these is to specify the feed rate, composition, and pressure ($C + 1$ variables) plus the drum temperature T_2 and pressure P_2. This operation will give one point on the *equilibrium-flash curve* shown in Fig. 13-26. This curve shows the relation at constant pressure between the fraction V/F of the feed flashed and the drum temperature. The temperature at $V/F = 0.0$ when the first bubble of vapor is about to form (saturated liquid) is the *bubble-point* temperature of the feed mixture, and the value at $V/F = 1.0$ when the first droplet of liquid is about to form (saturated liquid) is the *dew-point* temperature.

BUBBLE POINT AND DEW POINT

For a given drum pressure and feed composition, the bubble- and dew-point temperatures bracket the temperature range of the equilibrium flash. At the bubble-point temperature, the total vapor pressure exerted by the mixture becomes equal to the confining drum pressure, and it follows that $\sum y_i = 1.0$ in the bubble formed. Since $y_i = K_i x_i$ and since the x_i's still equal the feed concentrations (denoted by z_i's), calculation of the bubble-point temperature involves a trial-and-error search for the temperature which, at the specified pressure, makes $\sum K_i z_i = 1.0$. If instead the temperature is specified, one can find the bubble-point pressure that satisfies this relationship.

At the dew-point temperature y_i still equals z_i, and the relationship $\sum x_i = \sum z_i / K_i = 1.0$ must be satisfied. As in the case of the bubble point, a trial-and-error search for the dew-point temperature at a specified pressure is involved. Or, if the temperature is specified, the dew-point pressure can be calculated.

ISOTHERMAL FLASH

The calculation for a point on the flash curve that is intermediate between the bubble point and the dew point is referred to as an isothermal-flash calculation because T_2 is specified. Except for an ideal binary mixture, procedures for calculating an isothermal flash are iterative. A popular method is the following due to Rachford and Rice [*J. Pet. Technol.*, **4**(10), sec. 1, p. 19, and sec. 2, p. 3 (October 1952)]. The component mole balance ($Fz_i = Vy_i + Lx_i$), phase-distribution relation ($K_i = y_i/x_i$), and total mole balance ($F = V + L$) can be combined to give

$$x_i = \frac{z_i}{1 + \dfrac{V}{F}(K_i - 1)} \tag{13-12}$$

$$y_i = \frac{K_i z_i}{1 + \dfrac{V}{F}(K_i - 1)} \tag{13-13}$$

Since $\sum x_i - \sum y_i = 0$,

$$f\{V\} = \sum_i \frac{z_i(1 - K_i)}{1 + \dfrac{V}{F}(K_i - 1)} = 0 \tag{13-14}$$

Equation (13-14) is solved iteratively for V/F, followed by the calculation of values of x_i and y_i from Eqs. (13-12) and (13-13) and L from the total mole balance. Any one of a number of numerical root-finding

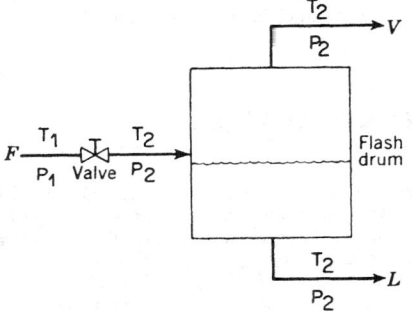

FIG. 13-25   Equilibrium-flash separator.

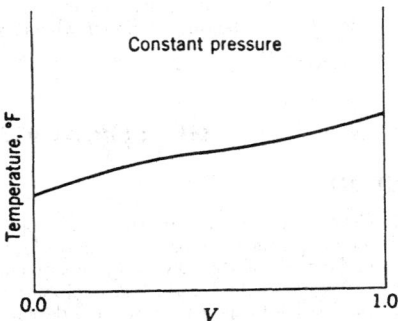

FIG. 13-26   Equilibrium-flash curve.

procedures such as the Newton-Raphson, secant, false-position, or bisection method can be used to solve Eq. (13-14). Values of K_i are constants if they are independent of liquid and vapor compositions. Then the resulting calculations are straightforward. Otherwise, the K_i values must be periodically updated for composition effects, perhaps after each iteration, using prorated values of x_i and y_i from Eqs. (13-12) and (13-13). Generally, the iterations are continued until the calculated value of V/F equals to within ±0.0005 the value of V/F that was used to initiate that iteration. When converged, $\sum x_i$ and $\sum y_i$ will each be very close to a value of 1, and, if desired, T_1 can be computed from an energy balance around the valve if no heat exchanger is used. Alternatively, if T_1 is fixed as mentioned earlier, a heat exchanger must be added before, after, or in place of the valve with the required heat duty being calculated from an energy balance. The limits of applicability of Eqs. (13-12) to (13-14) are the bubble point, at which $V = 0$ and $x_i = z_i$, and the dew point, at which $L = 0$ and $y_i = z_i$, at which Eq. (13-2) reduces to the bubble-point equation

$$\sum_i K_i x_i = 1 \qquad (13\text{-}15)$$

and the dew-point equation

$$\sum_i \frac{y_i}{K_i} = 1 \qquad (13\text{-}16)$$

For a *binary feed*, specification of the flash-drum temperature and pressure fixes the equilibrium-phase concentrations, which are related to the K values by

$$x_1 = (1 - K_2)/(K_1 - K_2) \quad \text{and} \quad y_1 = (K_1 K_2 - K_1)/(K_2 - K_1)$$

The mole balance can be rearranged to

$$\frac{V}{F} = \frac{z_1(K_1 - K_2)/(1.0 - K_2) - 1.0}{K_1 - 1.0}$$

If K_1 and K_2 are functions of temperature and pressure only (ideal solutions), the flash curve can be calculated directly without iteration.

ADIABATIC FLASH

In Fig. 13-25, if P_2 and the feed-stream conditions (i.e., F, z_i, T_1, P_1) are known, then the calculation of T_2, V, L, y_i, and x_i is referred to as an adiabatic flash. In addition to Eqs. (13-12) to (13-14) and the total mole balance, the following energy balance around both the valve and the flash drum combined must be included:

$$H^F F = H^V V + H^L L \qquad (13\text{-}17)$$

Taking a basis of $F = 1.0$ mol and eliminating L with the total mole balance, Eq. (13-17) becomes

$$f_2\{V, T_2\} = H^F - V(H^V - H^L) - H^L = 0 \qquad (13\text{-}18)$$

With T_2 now unknown, Eq. (13-17) becomes

$$f_1\{V, T_2\} = \sum \frac{z_i(1 - K_i)}{1 + V(K_i - 1)} = 0 \qquad (13\text{-}19)$$

A number of iterative procedures have been developed for solving Eqs. (13-18) and (13-19) simultaneously for V and T_2. Frequently, and especially if the feed contains components of a narrow range of volatility, convergence is rapid for a tearing method in which a value of T_2 is

assumed, Eq. (13-19) is solved iteratively by the isothermal-flash procedure, and, using that value of V, Eq. (13-18) is solved iteratively for a new approximation of T_2, which is then used to initiate the next cycle until T_2 and V converge. However, if the feed contains components of a wide range of volatility, it may be best to invert the sequence and assume a value for V, solve Eq. (13-19) for T_2, solve Eq. (13-18) for V, and then repeat the cycle. If K values and/or enthalpies are sensitive to the unknown phase compositions, it may be necessary to solve simultaneously Eqs. (13-18) and (13-19) by a Newton or other suitable iterative technique. Alternatively, the two-tier method of Boston and Britt [*Comput. Chem. Eng.*, **2**, 109 (1978)], which is also suitable for difficult isothermal-flash calculations, may be applied.

OTHER FLASH SPECIFICATIONS

Flash-drum specifications in addition to (P_2, T_2) and $(P_2, \text{adiabatic})$ are also possible but must be applied with care, as discussed by Michelsen [*Comp. Chem. Engng.*, **17**, 431 (1993)]. Most computer-aided process design and simulation programs permit a wide variety of flash specifications.

THREE-PHASE FLASH

Single-stage equilibrium-flash calculations become considerably more complex when an additional liquid phase can form, as from mixtures of water with hydrocarbons. Procedures for computing such situations are referred to as three-phase flash methods, which are given for the general case by Henley and Rosen (*Material and Energy Balance Computations*, Wiley, New York, 1968, chap. 8). When the two liquid phases are almost mutually insoluble, they can be considered separately and relatively simple procedures apply as discussed by Smith (*Design of Equilibrium Stage Processes*, McGraw-Hill, New York, 1963). Condensation of such mixtures may result in one liquid phase being formed before the other. Computer-aided process design and simulation programs all contain a Gibbs free-energy routine that can compute a three-phase flash by minimization of Gibbs free energy. Many difficult aspects of flash calculations are discussed by Michelsen [*Fluid Phase Equil.*, **9**, 1, 21 (1982)].

COMPLEX MIXTURES

Feed analyses in terms of component concentrations are usually not available for complex hydrocarbon mixtures with a final normal boiling point above about 38°C (100°F) (*n*-pentane). One method of handling such a feed is to break it down into pseudo components (narrow-boiling fractions) and then estimate the mole fraction and K value for each such component. Edmister [*Ind. Eng. Chem.*, **47**, 1685 (1955)] and Maxwell (*Data Book on Hydrocarbons*, Van Nostrand, Princeton, N.J., 1958) give charts that are useful for this estimation. Once K values are available, the calculation proceeds as described above for multicomponent mixtures. Another approach to complex mixtures is to obtain an American Society for Testing and Materials (ASTM) or true-boiling point (TBP) curve for the mixture and then use empirical correlations to construct the atmospheric-pressure equilibrium-flash curve (EFV), which can then be corrected to the desired operating pressure. A discussion of this method and the necessary charts are presented in a later subsection entitled "Petroleum and Complex-Mixture Distillation."

GRAPHICAL METHODS FOR BINARY DISTILLATION

INTRODUCTION

Multistage distillation under continuous, steady-state operating conditions is widely used in practice to separate a variety of mixtures. Table 13-7, taken from the study of Mix, Dweck, Weinberg, and Armstrong [*Am. Inst. Chem. Eng. J. Symp. Ser.* **76**, 192, 10 (1980)] lists key components for 27 industrial distillation processes. The design of multiequilibrium-stage columns can be accomplished by graphical techniques when the feed mixture contains only two components. The *x-y* diagram

[McCabe and Thiele, *Ind. Eng. Chem.*, **17**, 605 (1925)] utilizes only equilibrium and mole-balance relationships but approaches rigorousness only for those systems in which energy effects on vapor and liquid rates leaving the stages are negligible. The enthalpy-concentration diagram [Ponchon, *Tech. Mod.*, **13**, 20, 55 (1921); and Savarit, *Arts Metiers*, **65**, 142, 178, 241, 266, 307 (1922)] utilizes the energy balance also and is rigorous when enough calorimetric data are available to construct the diagram without assumptions.

TABLE 13-7   Key Components for Distillation Processes of Industrial Importance

Key components	Typical number of trays
Hydrocarbon systems	
Ethylene-ethane	73
Propylene-propane	138
Propyne–1–3-butadiene	40
1–3 Butadiene–vinyl acetylene	130
Benzene-toluene	34, 53
Benzene–ethyl benzene	20
Benzene–diethyl benzene	50
Toluene–ethyl benzene	28
Toluene-xylenes	45
Ethyl benzene–styrene	34
o-Xylene–m-xylene	130
Organic systems	
Methanol-formaldehyde	23
Dichloroethane-trichloroethane	30
Acetic acid–acetic anhydride	50
Acetic anhydride–ethylene diacetate	32
Vinyl acetate–ethyl acetate	90
Ethylene glycol–diethylene glycol	16
Cumene-phenol	38
Phenol-acetophenone	39, 54
Aqueous systems	
HCN-water	15
Acetic acid–water	40
Methanol-water	60
Ethanol-water	60
Isopropanol-water	12
Vinyl acetate–water	35
Ethylene oxide–water	50
Ethylene glycol–water	16

The availability of computers has decreased our reliance on graphical methods. Nevertheless, diagrams are useful for quick approximations and for demonstrating the effect of various design variables. The x-y diagram is the most convenient for these purposes, and its use is developed in detail here. The use of the enthalpy-concentration diagram is given by Smith (*Design of Equilibrium Stage Processes*, McGraw-Hill, New York, 1963) and Henley and Seader (*Equilibrium-Stage Separation Operations in Chemical Engineering*, Wiley, New York, 1981).

PHASE EQUILIBRIUM DATA

Three types of binary equilibrium curves are shown in Fig. 13-27. The y-x diagram is almost always plotted for the component that is the more volatile (denoted by the subscript 1) in the region where distillation is to take place. Curve A shows the most usual case, in which component 1 remains more volatile over the entire composition range. Curve B is typical of many systems (ethanol-water, for example) in which the component that is more volatile at low values of x_1 becomes less volatile than the other component at high values of x_1. The vapor and liquid compositions are identical for the homogeneous azeotrope where curve B crosses the 45° diagonal. A heterogeneous azeotrope is formed with two liquid phases by curve C.

An azeotrope limits the separation that can be obtained between components by simple distillation. For the system described by curve B, the maximum overhead-product concentration that could be obtained from a feed with $x_1 = 0.25$ is the azeotropic composition. Similarly, a feed with $x_1 = 0.9$ could produce a bottom-product composition no lower than the azeotrope.

The phase rule permits only two variables to be specified arbitrarily in a binary two-phase system at equilibrium. Consequently, the curves in Fig. 13-27 can be plotted at either constant temperature or constant pressure but not both. The latter is more common, and data in Table 13-1 are for that case. The y-x diagram can be plotted in either mole, weight, or volume fractions. The units used later for the phase flow rates must, of course, agree with those used for the equilibrium data. Mole fractions, which are almost always used, are applied here.

It is sometimes permissible to assume constant *relative volatility* in order to approximate the equilibrium curve quickly. Then by applying Eq. (13-2) to components 1 and 2,

$$\alpha = K_1/K_2 = y_1 x_2 / x_1 y_2$$

which, since $x_2 = 1 - x_1$ and $y_2 = 1 - y_1$, can be rewritten as

$$y_1 = \frac{x_1 \alpha}{1 + (\alpha - 1)x_1} \qquad (13\text{-}20)$$

for use in calculating points for the equilibrium curve.

McCABE-THIELE METHOD

Operating Lines   The McCabe-Thiele method is based upon representation of the material-balance equations as operating lines on the y-x diagram. The lines are made straight (and the need for the energy balance obviated) by the assumption of *constant molar overflow*. The liquid-phase flow rate is assumed to be constant from tray to tray in each section of the column between addition (feed) and withdrawal (product) points. If the liquid rate is constant, the vapor rate must also be constant.

The constant-molar-overflow assumption represents several prior assumptions. The most important one is equal molar heats of vaporization for the two components. The other assumptions are adiabatic operation (no heat leaks) and no heat of mixing or sensible heat effects. These assumptions are most closely approximated for close-boiling isomers. The result of these assumptions on the calculation method can be illustrated with Fig. 13-28, which shows two material-balance envelopes cutting through the top section (above the top feed stream or sidestream) of the column. If L_{n+1} is assumed to be identical to L_{n-1} in rate, then $V_n = V_{n-2}$ and the component material balance for both envelopes 1 and 2 can be represented by

$$y_n = (L/V)x_{n+1} + (Dx_D/V) \qquad (13\text{-}21)$$

where y and x have a stage subscript n or $n+1$, but L and V need be identified only with the section of the column to which they apply. Equation (13-21) has the analytical form of a straight line where L/V is the slope and Dx_D/V is the y intercept at $x_1 = 0$.

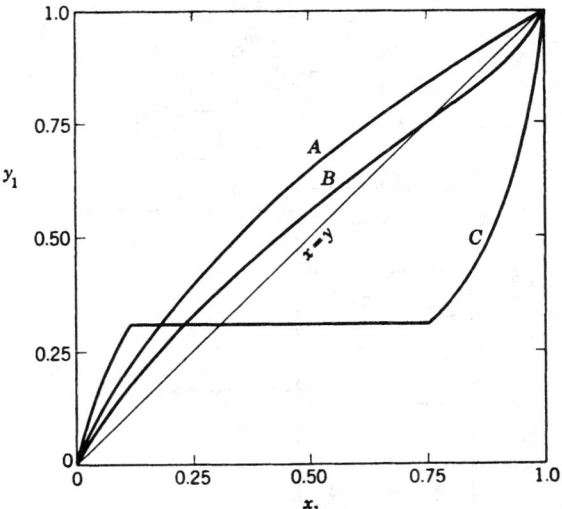

FIG. 13-27   Typical binary equilibrium curves. Curve A, system with normal volatility. Curve B, system with homogeneous azeotrope (one liquid phase). Curve C, system with heterogeneous azeotrope (two liquid phases in equilibrium with one vapor phase).

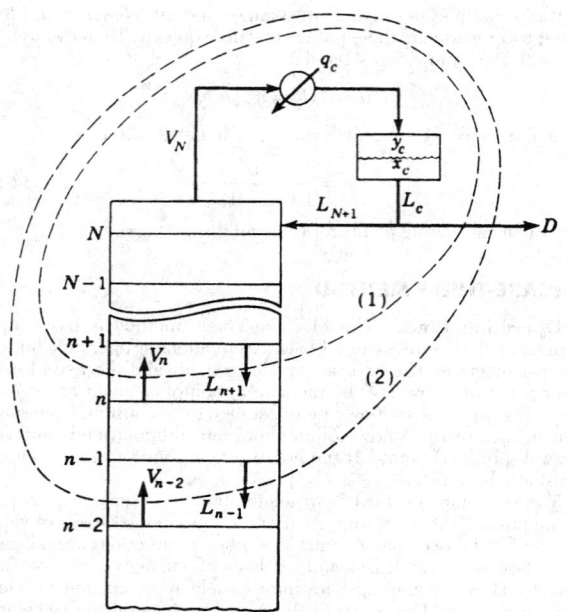

FIG. 13-28   Two material-balance envelopes in the top section of a distillation column.

The effect of a sidestream withdrawal point is illustrated by Fig. 13-29. The material-balance equation for the column section below the sidestream is

$$y_n = \frac{L'}{V'} x_{n+1} + \frac{Dx_D + Sx_S}{V'} \qquad (13\text{-}22)$$

where the primes designate the L and V below the sidestream. Since the sidestream must be a saturated phase, $V = V'$ if a liquid side stream is withdrawn and $L = L'$ if it is a vapor.

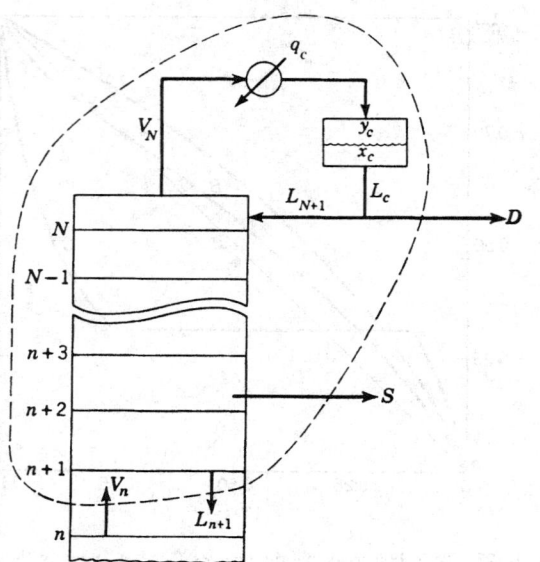

FIG. 13-29   Material-balance envelope which contains two external streams D and S, where S represents a sidestream product withdrawn above the feed plate.

If the sidestream in Fig. 13-29 had been a feed, the balance for the section below the feed would be

$$y_n = \frac{L'}{V'} x_{n+1} + \frac{Dx_D - Fx_F}{V'} \qquad (13\text{-}23)$$

Similar equations can be written for the bottom section of the column. For the envelope shown in Fig. 13-30,

$$y_m = (L''/V'') x_{m+1} - (Bx_B/V) \qquad (13\text{-}24)$$

where the subscript m is used to identify the stage number in the bottom section.

Equations such as (13-21) through (13-24) when plotted on the y-x diagram furnish a set of *operating lines*. A point on an operating line represents two *passing streams*, and the operating line itself is the locus of all possible pairs of passing streams within the column section to which the line applies.

An operating line can be located on the y-x diagram if (1) two points on the line are known or (2) one point and the slope are known. The known points on an operating line are usually its intersection with the y-x diagonal and/or its intersection with another operating line.

The slope L/V of the operating line is termed the *internal-reflux ratio*. This ratio in the operating-line equation for the top section of the column [see Eq. (13-21)] is related to the *external-reflux ratio* $R = L_{N+1}/D$ by

$$\frac{L}{V} = \frac{L_{N+1}}{V_N} = \frac{RD}{(1+R)D} = \frac{R}{1+R} \qquad (13\text{-}25)$$

when the reflux stream L_{N+1} is a saturated liquid.

Thermal Condition of the Feed   The slope of the operating line changes whenever a feed stream or a sidestream is passed. To calculate this change, it is convenient to introduce a quantity q which is defined by the following equations for a feed stream F:

$$L' = L + qF \qquad (13\text{-}26)$$
$$V = V' + (1 - q)F \qquad (13\text{-}27)$$

The primes denote the streams below the stage to which the feed is introduced. The q is a measure of the thermal condition of the feed and represents the moles of saturated liquid formed in the feed stage

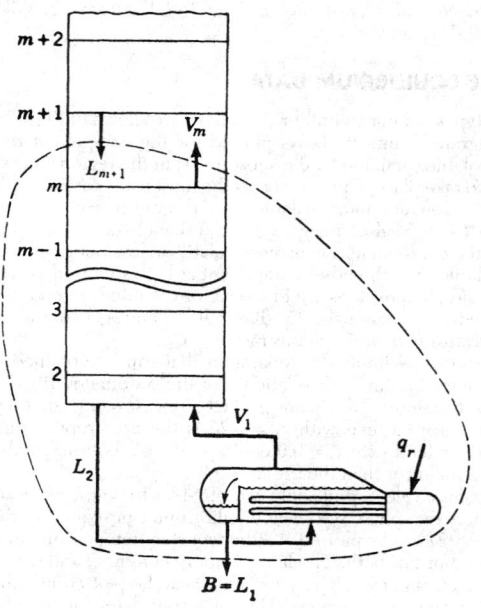

FIG. 13-30   Material-balance envelope around the bottom end of the column. The partial reboiler is equilibrium stage 1.

per mole of feed. It takes on the following values for various possible feed thermal conditions.

Subcooled-liquid feed: $q > 1$
Saturated-liquid feed: $q = 1$
Partially flashed feed: $1 > q > 0$
Saturated-vapor feed: $q = 0$
Superheated-vapor feed: $q < 0$

The q value for a particular feed can be estimated from

$$q = \frac{\text{energy to convert 1 mol of feed to saturated vapor}}{\text{molar heat of vaporization}}$$

Equations analogous to (13-26) and (13-27) can be written for a sidestream, but the q will be either 1 or 0 depending upon whether the sidestream is taken from the liquid or the vapor stream.

The q can be used to derive the "q-line equation" for a feed stream or a sidestream. The q line is the locus of all points of intersection of the two operating lines, which meet at the feed-stream or sidestream stage. This intersection must occur along that section of the q line between the equilibrium curve and the $y = x$ diagonal. At the point of intersection, the same y, x point must satisfy both the operating-line equation above the feed-stream (or sidestream) stage and the one below the feed-stream (or sidestream) stage. Subtracting one equation from the other gives for a feed stage

$$(V - V')y = (L - L')x + Fx_F$$

which when combined with Eqs. (13-26) and (13-27) gives the q-line equation

$$y = \frac{q}{q-1}x - \frac{x_F}{q-1} \qquad (13\text{-}28)$$

A q-line construction for a partially flashed feed is given in Fig. 13-31. It is easily shown that the q line must intersect the diagonal at x_F. The slope of the q line is $q/(q - 1)$. All five q-line cases are shown in Fig. 13-32.

The derivation of Eq. (13-28) assumes a single-feed column and no sidestream. However, the same result is obtained for other column configurations. Typical q-line constructions for sidestream stages are shown in Fig. 13-33. Note that the q line for a sidestream must always intersect the diagonal at the composition (x_1 or x_1) of the sidestream.

Figure 13-33 also shows the intersections of the operating lines with the diagonal construction line. The top operating line must always intersect the diagonal at the overhead-product composition x_D. This

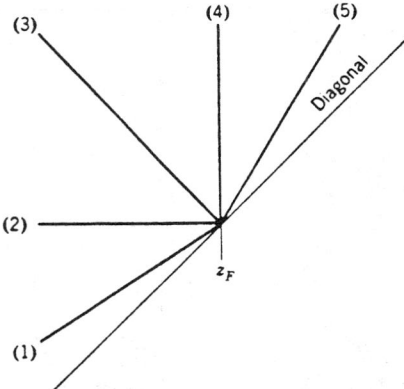

FIG. 13-32 All five cases of q lines; (1) superheated-vapor feed, (2) saturated-vapor feed, (3) partially vaporized feed, (4) saturated-liquid feed, and (5) subcooled-liquid feed. Slope of q line = $q/(q - 1)$.

can be shown by substituting $y = x$ in Eq. (13-21) and using $V - L = D$ to reduce the resulting equation to $x = x_D$. Similarly (except for columns in which open steam is introduced at the bottom), the bottom operating line must always intersect the diagonal at the bottom-product composition x_B.

Equilibrium-Stage Construction The alternate use of the equilibrium curve and the operating line to "step off" equilibrium stages is illustrated in Fig. 13-34. The plotted portions of the equilibrium curve (curved) and the operating line (straight) cover the composition range existing in the column section shown in the lower right-hand corner. If y_n and x_n represent the compositions (in terms of the more volatile component) of the equilibrium vapor and liquid leaving stage n, then point (y_n, x_n) on the equilibrium curve must represent the equilibrium stage n. The operating line is the locus for compositions of all possible pairs of passing streams within the section and therefore a horizontal line (dotted) at y_n must pass through the point (y_n, x_{n+1}) on the operating line since y_n and x_{n+1} represent passing streams. Likewise, a vertical line (dashed) at x_n must intersect the operating line at point (y_{n-1}, x_n). The equilibrium stages above and below stage n can be located by a vertical line through (y_n, x_{n+1}) to find (y_{n+1}, x_{n+1}) and a horizontal line through (y_{n-1}, x_n) to find (y_{n-1}, x_{n-1}). It can be seen that one can work upward or downward through the column by alternating the use of equilibrium and operating lines.

Total-Column Construction The graphical construction for an entire column is shown in Fig. 13-35. The process, pictured in the lower right-hand corner of the diagram, is an existing column with a number of actual trays equivalent to eight equilibrium stages. A partial reboiler (equivalent to an equilibrium stage) and a total condenser are used. This column configuration was analyzed earlier (see Fig. 13-24) and shown to have $C + 2N + 9$ design variables (degrees of freedom) which must be specified to define one unique operation. These may be used as follows as the basis for a graphical solution:

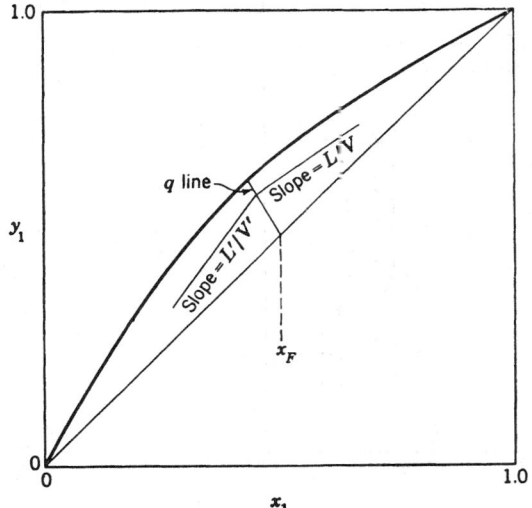

FIG. 13-31 Typical intersection of the two operating lines at the q line for a feed stage. The q line shown is for a partially flashed feed.

Specifications		N_i^u
Stage pressures (including reboiler)		N
Condenser pressure		1
Stage heat leaks (except reboiler)		$N - 1$
Pressure and heat leak in reflux divider		2
Feed stream	C	$+ 2$
Feed-stage location		1
Total number of stages N		1
One overhead purity		1
Reflux temperature		1
External-reflux ratio		1
		$C + 2N + 9$

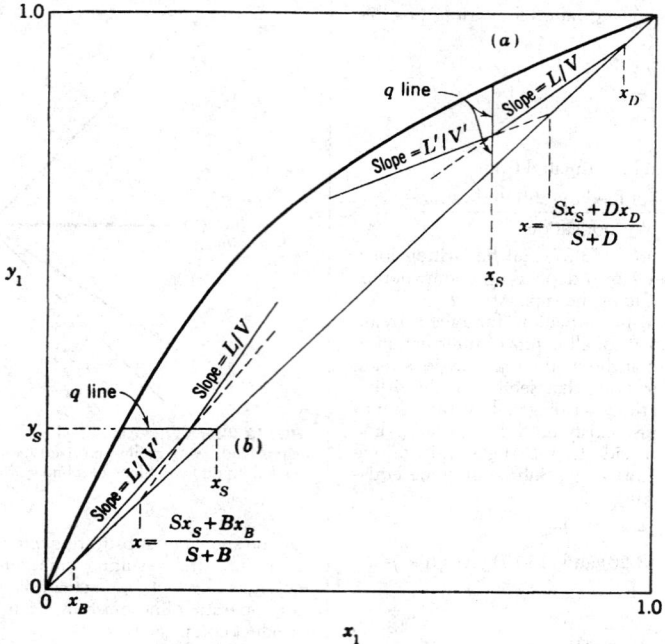

FIG. 13-33   Typical construction for a sidestream showing the intersection of the two operating lines with the q line and with the $x - y$ diagonal. (a) Liquid sidestream near the top of the column. (b) Vapor sidestream near the bottom of the column.

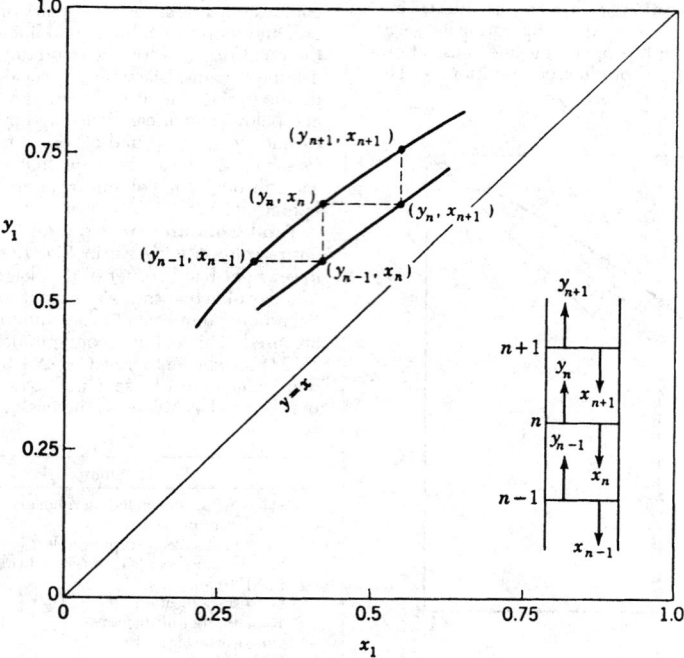

FIG. 13-34   Illustration of how equilibrium stages can be located on the x-y diagram through the alternating use of the equilibrium curve and the operating line.

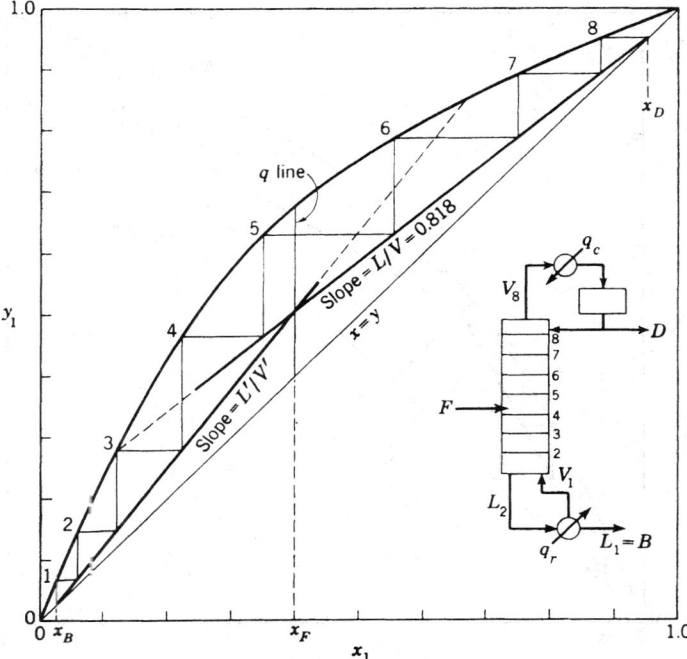

FIG. 13-35 Construction for a column with a bubble-point feed, a total condenser, and a partial reboiler.

Pressures can be specified at any level below the safe working pressure of the column. The condenser pressure will be set at 275.8 kPa (40 psia), and all pressure drops within the column will be neglected. The equilibrium curve in Fig. 13-35 represents data at that pressure. All heat leaks will be assumed to be zero. The feed composition is 40 mole percent of the more volatile component 1, and the feed rate is 0.126 (kg·mol)/s [1000 (lb·mol)/h] of saturated liquid ($q = 1$). The feed-stage location is fixed at stage 4 and the total number of stages at eight.

The overhead purity is specified as $x_D = 0.95$. The reflux temperature is the bubble-point temperature (saturated reflux), and the external-reflux ratio is set at $R = 4.5$.

Answers are desired to the following two questions. First, what bottom-product composition x_B will the column produce under these specifications? Second, what will be the top vapor rate V_N in this operation, and will it exceed the maximum vapor-rate capacity for this column, which is assumed to be 0.252 (kg·mol)/s [2000 (lb·mol)/h] at the top-tray conditions?

The solution is started by using Eq. (13-25) to convert the external-reflux ratio of 4.5 to an internal-reflux ratio of $L/V = 0.818$. The $x_D = 0.95$ value is then located on the diagonal, and the upper operating line is drawn as shown in Fig. 13-35.

If the x_B value were known, the bottom operating line could be immediately drawn from the x_B value on the diagonal up to its required intersection point with the upper operating line on the feed q line. In this problem, since the number of stages is fixed, the x_B which gives a lower operating line that will require exactly eight stages must be found by trial and error. An x_B value is assumed, and the resulting lower operating line is drawn. The stages can be stepped off by starting from either x_B or x_D; x_B was used in this case.

Note that the lower operating line is used until the fourth stage is passed, at which time the construction switches to the upper operating line. This is necessary because the vapor and liquid streams passing each other between the fourth and fifth stages must fall on the upper line.

The x_B that requires exactly eight equilibrium stages is $x_1 = 0.026$. An overall component balance gives $D = 0.051$ (kg·mol)/s [405 (lb·mol)/h]. Then,

$$V_N = V_B = L_{N+1} + D = D(R + 1) = 0.051(4.5 + 1.0)$$
$$= 0.280 \text{ (kg·mol)/s [2230 (lb·mol)/h]}$$

which exceeds the column capacity of 0.252 (kg·mol)/s [2007 (lb·mol)/h]. This means that the column cannot provide an overhead-product yield of 40.5 percent at 95 percent purity. Either the purity specification must be reduced, or we must be satisfied with a lower yield. If the $x_D = 0.95$ specification is retained, the reflux rate must be reduced. This will cause the upper operating line to pivot upward around its fixed point of $x = 0.95$ on the diagonal. The new intersection of the upper line with the q line will lie closer to the equilibrium curve. The x_B value must then move upward along the diagonal because the eight stages will not "reach" as far as formerly. The higher x_B concentration will reduce the recovery of component 1 in the 95 percent overhead product.

Another entire column with a partially vaporized feed, a liquid-sidestream rate equal to D and withdrawn from the second stage from the top, and a total condenser is shown in Fig. 13-36. The specified concentrations are $x_F = 0.40$, $x_B = 0.05$, and $x_D = 0.95$. The specified L/V ratio in the top section is 0.818. These specifications permit the top operating line to be located and the two top stages stepped off to determine the liquid-sidestream composition $x_S = 0.746$. The operating line below the sidestream must intersect the diagonal at the "blend" of the sidestream and the overhead stream. Since S was specified to be equal to D in rate, the intersection point is

$$x = \frac{(1.0)(0.746) + (1.0)(0.95)}{1.0 + 1.0} = 0.848$$

This point plus the point of intersection of the two operating lines on the sidestream q line (vertical at $x_S = 0.746$) permits the location of the middle operating line. (The slope of the middle operating line could

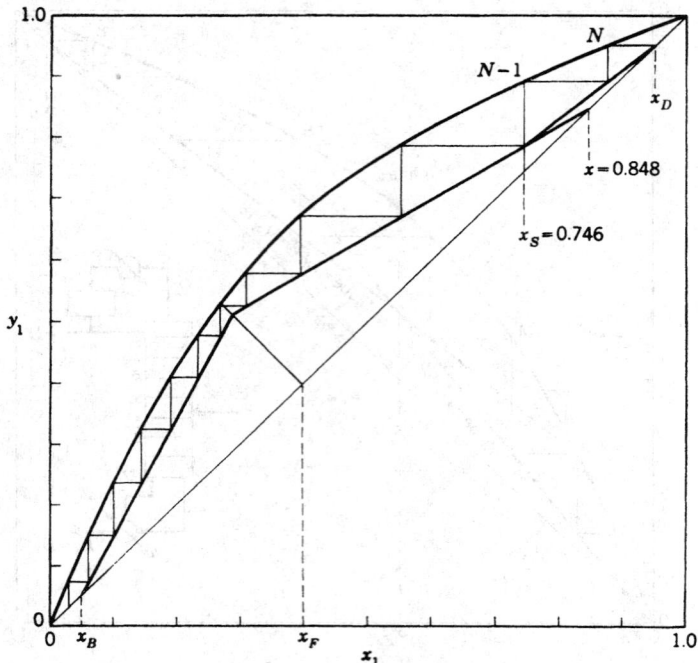

FIG. 13-36 Graphical solution for a column with a partially flashed feed, a liquid side-stream and a total condenser.

also have been used.) The lower operating line must run from the specified x_B value on the diagonal to the required point of intersection on the feed q line. The stages are stepped off from the top down in this case. The sixth stage from the top is the feed stage, and a total of about 11.4 stages is required to reach the specified $x_B = 0.05$.

Fractional equilibrium stages have meaning. The 11.4 will be divided by a tray efficiency, and the rounding to an integral number of actual trays should be done after that division. For example, if the average tray efficiency for the process being modeled in Fig. 13-36 were 80 percent, then the number of actual trays required would be $11.4/0.8 = 14.3$, which would be rounded to 15.

Feed-Stage Location   The *optimum* feed-stage location is that location which, with a given set of other operating specifications, will result in the widest separation between x_D and x_B for a given number of stages. Or, if the number of stages is not specified, the optimum feed location is the one that requires the lowest number of stages to accomplish a specified separation between x_D and x_B. Either of these criteria will always be satisfied if the operating line farthest from the equilibrium curve is used in each step as in Fig. 13-35.

It can be seen from Fig. 13-35 that the optimum feed location would have been the fifth tray for that operation. If a new column were being designed, that would have been the designer's choice. However, when an existing column is being modeled, the feed stage on the diagram should correspond as closely as possible to the actual feed tray in the column. It can be seen that a badly mislocated feed (a feed that requires one to remain with an operating line until it closely approaches the equilibrium curve) can be very wasteful insofar as the effectiveness of the stages is concerned.

Minimum Stages   A column operating at total reflux is diagramed in Fig. 13-37a. Enough material has been charged to the column to fill the reboiler, the trays, and the overhead condensate drum to their working levels. The column is then operated with no feed and with all the condensed overhead stream returned as reflux ($L_{N+1} = V_N$ and $D = 0$). Also all the liquid reaching the reboiler is vaporized and returned to the column as vapor. Since F, D, and B are all zero, $L_{n+1} = V_n$ at all points in the column. With a slope of unity ($L/V = 1.0$), the operating line must coincide with the diagonal throughout the col-

umn. Total-reflux operation gives the minimum number of stages required to effect a specified separation between x_B and x_D.

Minimum Reflux   The minimum-reflux ratio is defined as that ratio which if decreased by an infinitesimal amount would require an infinite number of stages to accomplish a specified separation between two components. The concept has meaning only if a separation between two components is specified and the number of stages is not specified. Figure 13-37b illustrates the minimum-reflux condition. As the reflux ratio is reduced, the two operating lines swing upward, pivoting around the specified x_B and x_D values, until one or both touch the equilibrium curve. For equilibrium curves shaped like the one shown, the contact occurs at the feed q line. Often an equilibrium curve will dip down closer to the diagonal at higher concentrations. In such cases, the upper operating line may make contact before its intersection point on the q line reaches the equilibrium curve. Wherever the contact appears, the intersection of the operating line with the equilibrium curve produces a pinch point which contains a very large number of stages, and a zone of constant composition is formed.

Intermediate Reboilers and Condensers   A distillation column of the type shown in Fig. 13-2a, operating with an interreboiler and an intercondenser in addition to a reboiler and a condenser, is diagramed with the solid lines in Fig. 13-38. The dashed lines correspond to simple distillation with only a bottoms reboiler and an overhead condenser. Total boiling and condensing heat loads are the same for both columns. As shown by Kayihan [*Am. Inst. Chem. Eng. J. Symp. Ser.* **76**, 192, 1 (1980)], the addition of interreboilers and intercondensers increases thermodynamic efficiency but requires additional stages, as is clear from the positions of the operating lines in Fig. 13-38.

Optimum Reflux Ratio   The general effect of the operating reflux ratio on fixed costs, operating costs, and the sum of these is shown in Fig. 13-39. In ordinary situations, the minimum on the total-cost curve will generally occur at an operating reflux ratio of from 1.1 to 1.5 times the minimum $R = L_{N+1}/D$ value, with the lower value corresponding to a value of the relative volatility close to 1.

Difficult Separations   Some binary separations may pose special problems because of extreme purity requirements for one or both products or because of a relative volatility close to 1. The y-x diagram

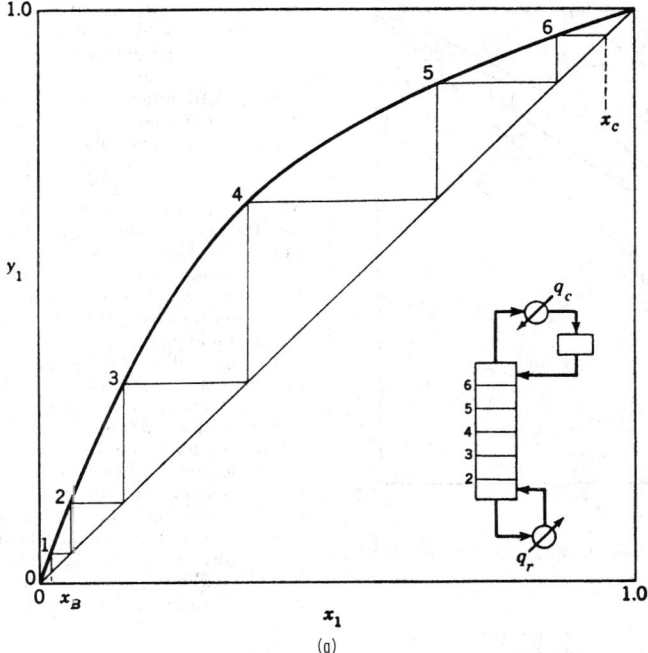

(a)

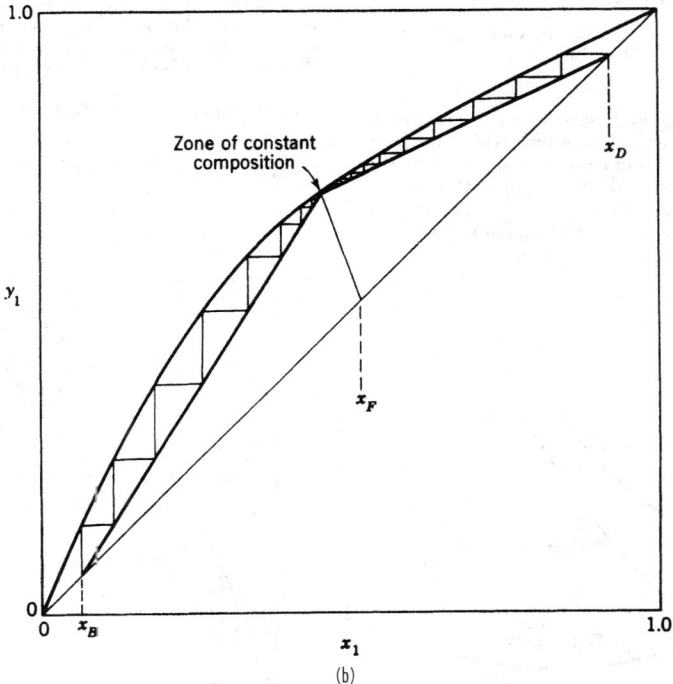

(b)

FIG. 13-37   McCabe-Thiele diagrams for limiting cases. (a) Minimum stages for a column operating at total reflux with no feeds or products. (b) Minimum reflux for a binary system of normal volatility.

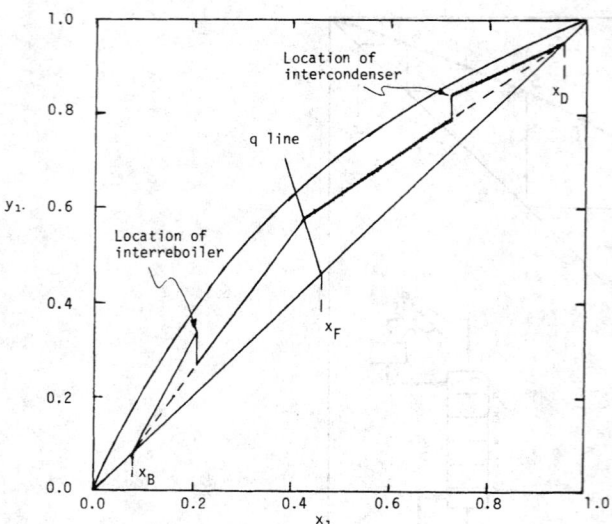

FIG. 13-38 McCabe-Thiele diagram for columns with and without an interreboiler and an intercondenser.

is convenient for stepping off stages at extreme purities if it is plotted on log-log paper. The equilibrium curve at very low x_1 values on ordinary graph paper can usually be assumed to be a straight line with an intercept term of zero which can be expressed as

$$y = (y/x)x + 0.0$$

where the slope y/x is a constant. The necessity for knowing the slope is eliminated by taking the logarithm of both sides

$$\log y = \log x + \log (y/x)$$

and plotting y versus x on a log-log plot to give a straight line with a slope of unity. The slope y/x is now an intercept term which need not be known. One point from the equilibrium curve is sufficient, therefore, to locate the equilibrium curve on the log-log plot. The operating line will be curved on the log-log plot and is located by plotting the appropriate material-balance equation. Both the equilibrium and the operating lines can be extended to any purity desired.

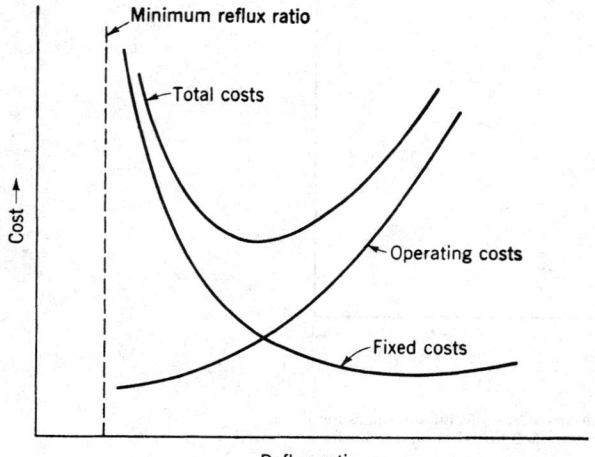

FIG. 13-39 Location of the optimum reflux for a given feed and specified separation.

A system with constant relative volatility can be handled conveniently by the equation of Smoker [*Trans. Am. Inst. Chem. Eng.*, **34**, 165 (1938)]. The derivation of the equation is shown, and its use is illustrated by Smith (op. cit.).

Stage Efficiency   The use of the *Murphree plate efficiency* is particularly convenient on y-x diagrams. The Murphree efficiency is defined for the vapor phase as

$$\eta = (y_n - y_{n-1})/(y_n^\circ - y_{n-1}) \tag{13-29}$$

where y_n° is the composition of the vapor that would be in equilibrium with the liquid leaving stage n and is the value read from the equilibrium curve. The y_{n-1} and y_n are the actual (nonequilibrium) values for vapor streams leaving the $n-1$ and n stages respectively. Note that the y_{n-1} and y_n values assume that vapor streams are completely mixed and uniform in composition. An analogous efficiency can be defined for the liquid phase.

The application of a 50 percent Murphree vapor-phase efficiency on a y-x diagram is illustrated in Fig. 13-40. A "pseudo-equilibrium" curve is drawn halfway (on a vertical line) between the operating lines and the true-equilibrium curve. The true-equilibrium curve is used for the first stage (the partial reboiler is assumed to be an equilibrium stage), but for all other stages the vapor leaving each stage is assumed to approach the equilibrium value y_n° only 50 percent of the way. Consequently, the steps in Fig. 13-40 represent actual trays.

Application of a constant efficiency to each stage as in Fig. 13-40 will not give, in general, the same answer as obtained when the number of equilibrium stages (obtained by using the true-equilibrium curve) is divided by the same efficiency factor.

The prediction and use of stage efficiencies are described in detail in Sec. 14.

Miscellaneous Operations   The y-x diagrams for several other column configurations have not been presented here. The omitted items are *partial condensers, rectifying columns* (feed introduced to the bottom stage), *stripping columns* (feed introduced to the top stage), total reflux in the top section but not in the bottom section, multiple feeds, and introduction of *open steam* to the bottom stage to eliminate the reboiler. These configurations are discussed in Smith (op. cit.) and Henley and Seader (op. cit.), who also describe the more rigorous Ponchon-Savarit method, which is not included here.

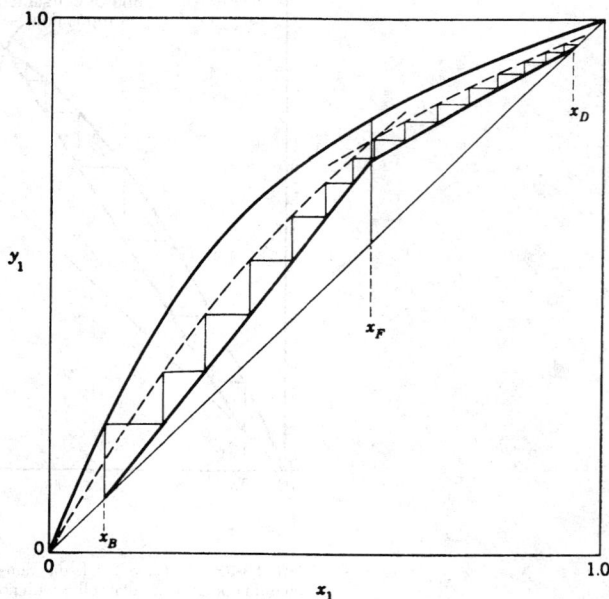

FIG. 13-40 Application of a 50 percent Murphree vapor-phase efficiency to each stage (excluding the reboiler) in the column. Each step in the diagram corresponds to an actual stage.

APPROXIMATE MULTICOMPONENT DISTILLATION METHODS

INTRODUCTION

Some approximate calculation methods for the solution of multicomponent, multistage separation problems continue to serve useful purposes even though computers are available to provide more rigorous solutions. The available phase equilibrium and enthalpy data may not be accurate enough to justify the longer rigorous methods. Or in extensive design and optimization studies, a large number of cases can be worked quickly and cheaply by an approximate method to define roughly the optimum specifications, which can then be investigated more exactly by a rigorous method.

Two approximate multicomponent shortcut methods for simple distillation are the Smith-Brinkley (SB) method, which is based on an analytical solution of the finite-difference equations that can be written for staged separation processes when stages and interstage flow rates are known or assumed and the Fenske-Underwood-Gilliland (FUG) method, which combines Fenske's total-reflux equation and Underwood's minimum-reflux equation with a graphical correlation by Gilliland that relates actual column performance to total- and minimum-reflux conditions for a specified separation between two key components. Thus, the SB and FUG methods are rating and design methods respectively. Both methods work best when mixtures are nearly ideal.

The SB method is not presented here, but is presented in detail in the sixth edition of *Perry's Chemical Engineers' Handbook*. Extensions of the SB method to nonideal mixtures and complex configurations are developed by Eckert and Hlavacek [*Chem. Eng. Sci., 33,* 77 (1978)] and Eckert [*Chem. Eng. Sci., 37,* 425 (1982)] respectively but are not discussed here. However, the approximate and very useful method of Kremser [*Nat. Pet. News,* **22**(21), 43 (May 21, 1930)] for application to absorbers and strippers is discussed at the end of this subsection.

FENSKE-UNDERWOOD-GILLILAND (FUG) SHORTCUT METHOD

In this approach, Fenske's equation [*Ind. Eng. Chem.,* **24,** 482 (1932)] is used to calculate N_m, which is the number of plates required to make a specified separation at total reflux, i.e., the minimum value of N. Underwood's equations [*J. Inst. Pet., 31,* 111 (1945); *32,* 598 (1946); *32,* 614 (1946); and *Chem. Eng. Prog., 44,* 603 (1948)] are used to estimate the minimum-reflux ratio R_m. The empirical correlation of Gilliland [*Ind. Eng. Chem., 32,* 1220 (1940)] shown in Fig. 13-41 then uses these values to give N for any specified R or R for any specified N. Limitations of the Gilliland correlation are discussed by Henley and Seader (*Equilibrium-Stage Separation Operations in Chemical Engineering,* Wiley, New York, 1981). The following equation, developed by Molokanov et al. [*Int. Chem. Eng.,* 12(2), 209 (1972)] satisfies the end points and fits the Gilliland curve reasonably well:

$$\frac{N - N_m}{N + 1} = 1 - \exp\left[\left(\frac{1 + 54.4\Psi}{11 + 117.2\Psi}\right)\left(\frac{\Psi - 1}{\Psi^{0.5}}\right)\right] \quad (13\text{-}30)$$

where $\Psi = (R - R_m)/(R + 1)$.

The *Fenske total-reflux equation* can be written as

$$\left(\frac{x_i}{x_r}\right) = (\alpha_i)^{N_m}\left(\frac{x_i}{x_r}\right)_B \quad (13\text{-}31)$$

or as

$$N_m = \frac{\log\left[\left(\frac{Dx_D}{Bx_B}\right)_i\left(\frac{Bx_B}{Dx_D}\right)_r\right]}{\log \alpha_i} \quad (13\text{-}32)$$

where i is any component and r is an arbitrarily selected reference component in the definition of relative volatilities.

$$\alpha_i = K_i/K_r = y_i x_r/y_r x_i \quad (13\text{-}33)$$

The particular value of α_i is the effective value used in Eqs. (13-36) and (13-34) defined in terms of values for each stage in the column by

$$\alpha^N = \alpha_N \alpha_{N-1} \cdots \alpha_2 \alpha_1 \quad (13\text{-}34)$$

Equations (13-31) and (13-32) are rigorous relationships between the splits obtained for components i and r in a column at total reflux. However, the correct value of α_i must always be estimated, and this is where the approximation enters. It is usually estimated from

$$\alpha = (\alpha_{\text{top}}\alpha_{\text{bottom}})^{1/2} \quad (13\text{-}35)$$

or

$$\alpha = (\alpha_{\text{top}}\alpha_{\text{middle}}\alpha_{\text{bottom}})^{1/3} \quad (13\text{-}36)$$

A reasonably good estimate of the separation that will be accomplished in a plant column often can be obtained by specifying the split of one component (designated as the reference component r), setting N_m equal to from 40 to 60 percent of the number of equilibrium stages (not actual trays), and then using Eq. (13-32) to estimate the splits of all the other components. This is an iterative calculation because the component splits must first be arbitrarily assumed to give end compositions that can be used to give initial end-temperature estimates. The α_{top} and α_{bottom} values corresponding to these end temperatures are used in Eq. (13-35) to give α_i values for each component. The iteration is continued until the α_i values do not change from trial to trial.

The *Underwood minimum-reflux equations* of main interest are those that apply when some of the components do not appear in either the distillate or the bottoms products at minimum reflux. These equations are

$$\sum_i \frac{\alpha_i(x_{i,D})_m}{\alpha_i - \Theta} = R_m + 1 \quad (13\text{-}37)$$

and

$$\sum_i \frac{\alpha_i x_{i,F}}{\alpha_i - \Theta} = 1 - q \quad (13\text{-}38)$$

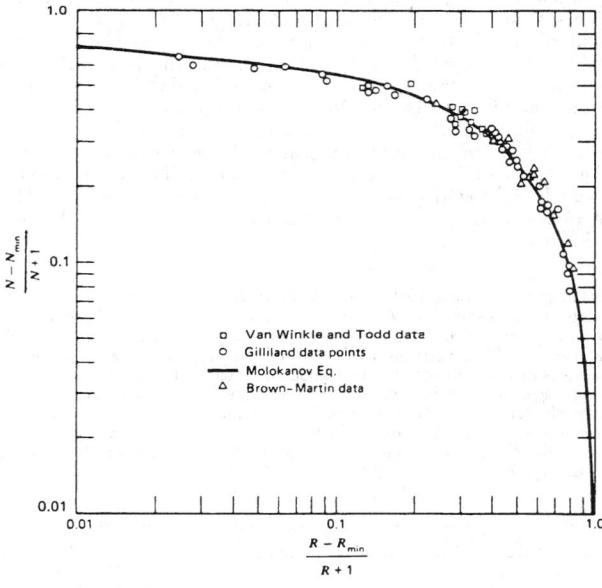

FIG. 13-41 Comparison of rigorous calculations with Gilliland correlation. [*Henley and Seader,* Equilibrium-Stage Separation Operations in Chemical Engineering, *Wiley, New York, 1981; data of Van Winkle and Todd,* Chem. Eng., **78**(21), 136 (Sept. 20, 1971); *data of Gilliland,* Elements of Fractional Distillation, *4th ed., McGraw-Hill, New York, 1950; data of Brown and Martin,* Trans. Am. Inst. Chem. Eng., **35,** 679 (1939).]

The relative volatilities α_i are defined by Eq. (13-33), R_m is the minimum-reflux ratio $(L_{N+1}/D)_{min}$, and q describes the thermal condition of the feed (e.g., 1.0 for a bubble-point feed and 0.0 for a saturated-vapor feed). The $x_{i,F}$ values are available from the given feed composition. The Θ is the common root for the top-section equations and the bottom-section equations developed by Underwood for a column at minimum reflux with separate zones of constant composition in each section. The common root value must fall between α_{hk} and α_{lk}, where hk and lk stand for *heavy key* and *light key* respectively. The *key components* are the ones that the designer wants to separate. In the butane-pentane splitter problem used in Example 1, the light key is n-C_4 and the heavy key is i-C_5.

The α_i values in Eqs. (13-37) and (13-38) are effective values obtained from Eq. (13-35) or Eq. (13-36). Once these values are available, Θ can be calculated in a straightforward iteration from Eq. (13-38). Since the $(\alpha - \Theta)$ difference can be small, Θ should be determined to four decimal places to avoid numerical difficulties.

The $(x_{i,D})_m$ values in Eq. (13-37) are minimum-reflux values, i.e., the overhead concentration that would be produced by the column operating at the minimum reflux with an infinite number of stages. When the light key and the heavy key are adjacent in relative volatility and the specified split between them is sharp or the relative volatilities of the other components are not close to those of the two keys, only the two keys will distribute at minimum reflux and the $(x_{i,D})_m$ values are easily determined. This is often the case and is the only one considered here. Other cases in which some or all of the nonkey components distribute between distillate and bottom products are discussed in detail by Henley and Seader (op. cit.).

The FUG method is convenient for new-column design with the following specifications:
1. R/R_m, the ratio of reflux to minimum reflux
2. Split on the reference component (usually chosen as the heavy key)
3. Split on one other component (usually the light key)

However, the total number of equilibrium stages N, N/N_m, or the external-reflux ratio can be substituted for one of these three specifications. It should be noted that the feed location is automatically specified as the optimum one; this is assumed in the Underwood equations. The assumption of saturated reflux is also inherent in the Fenske and Underwood equations. An important limitation on the Underwood equations is the assumption of constant molar overflow. As discussed by Henley and Seader (op. cit.), this assumption can lead to a prediction of the minimum reflux that is considerably lower than the actual value. No such assumption is inherent in the Fenske equation. An exact calculational technique for minimum reflux is given by Tavana and Hansen [*Ind. Eng. Chem. Process Des. Dev.*, **18**, 154 (1979)]. A computer program for the FUG method is given by Chang [*Hydrocarbon Process.*, **60**(8), 79 (1980)]. The method is best applied to mixtures that form ideal or nearly ideal solutions.

Example 1: Calculation of FUG Method A large butane-pentane splitter is to be shut down for repairs. Some of its feed will be diverted temporarily to an available smaller column, which has only 11 trays plus a partial reboiler. The feed enters on the middle tray. Past experience on similar feeds indicates that the 11 trays plus the reboiler are roughly equivalent to 10 equilibrium stages and that the column has a maximum top-vapor capacity of 1.75 times the feed rate on a mole basis. The column will operate at a condenser pressure of 827.4 kPa (120 psia). The feed will be at its bubble point ($q = 1.0$) at the feed-tray conditions and has the following composition on the basis of 0.0126 (kg·mol)/s [100 (lb·mol)/h]:

Component	Fx_F
C_3	5
i-C_4	15
n-C_4	25
i-C_5	20
n-C_5	35
	100

The original column normally has less than 7 mol percent i-C_5 in the overhead and less than 3 mol percent n-C_4 in the bottoms product when operating at a distillate rate of $D/F = 0.489$. Can these product purities be produced on the smaller column at $D/F = 0.489$?

Pressure drops in the column will be neglected, and the K values will be read at 827 kPa (120 psia) in both column sections from the DePriester nomograph in Fig. 13-14b. When constant molar overflow is assumed in each section, the rates in pound-moles per hour in the upper and lower sections are as follows:

Top section	Bottom section
$D = (0.489)(100) = 48.9$	$B = 100 - 48.9 = 51.1$
$V = (1.75)(100) = 175$	$V' = V = 175$
$L = 175 - 48.9 = 126.1$	$L' = L + F = 226.1$
$V/L = 1.388$	$V'/L' = 0.7739$

$$L/L' = 126.1/226.1 = 0.5577$$
$$R = 126.1/48.9 = 2.579$$

NOTE: To convert pound-moles per hour to kilogram-moles per second, multiply by 1.26×10^{-4}.

Since the feed enters at the middle of the column, $M = 5$ and $M + 1 = 6$. Application of the FUG method is demonstrated on the splitter. Specifications necessary to model the existing column are:
1. $N = 10$, total number of equilibrium stages.
2. Optimum feed location (which may or may not reflect the actual location).
3. Maximum V/F at the top tray of 1.75.
4. Split on one component given in the following paragraphs.

The solution starts with an assumed arbitrary split of all the components to give estimates of top and bottom compositions that can be used to get initial end temperatures. The α_i's evaluated at these temperatures are averaged with an assumed feed-stage temperature (assumed to be the bubble point of the feed) by using Eq. (13-36). The initial assumption for the split on i-C_5 will be $Dx_D/Bx_B = 3.15/16.85$. As mentioned earlier, N_m usually ranges from $0.4N$ to $0.6N$, and the initial N_m value assumed here will be $(0.6)(10) = 6.0$. Equation (13-32) can be rewritten as

$$\left(\frac{Dx_D}{Fx_F - Dx_D}\right)_i = \alpha_i^{6.0}\left(\frac{3.15}{16.85}\right) = \alpha_i^{6.0}(0.1869)$$

or

$$Dx_{i,D} = \frac{0.1869\alpha_i^{6.0}}{1 + 0.1869\alpha_i^{6.0}} Fx_{i,F}$$

The evaluation of this equation for each component is as follows:

Component	α_i	$\alpha_i^{6.0}$	$0.1869\alpha_i^{6.0}$	Fx_F	Dx_D	Bx_B
C_3	5.00			5	5.0	0.0
i-C_4	2.63	330	61.7	15	14.8	0.2
n-C_4	2.01	66	12.3	25	25.1	1.9
i-C_5	1.00	1.00	0.187	20	3.15	16.85
n-C_5	0.843	0.36	0.0672	35	2.20	32.80
				100	48.25	51.75

The end temperatures corresponding to these product compositions are 344 K (159°F) and 386 K (236°F). These temperatures plus the feed bubble-point temperature of 358 K (185°F) provide a new set of α_i's which vary only slightly from those used earlier. Consequently, the $D = 48.25$ value is not expected to vary greatly and will be used to estimate a new i-C_5 split. The desired overhead concentration for i-C_5 is 7 percent; so it will be assumed that $Dx_D = (0.07)(48.25) = 3.4$ for i-C_5 and that the split on that component will be 3.4/16.6. The results obtained with the new α_i's and the new i-C_5 split are as follows:

Component	$\alpha_i^{6.0}$	$0.2048\alpha_i^{6.0}$	Fx_F	Dx_D	Bx_B	x_D	x_B
C_3			5	5.0	0.0	0.102	0.000
i-C_4	322	65.9	15	14.8	0.2	0.301	0.004
n-C_4	68	13.9	25	23.3	1.7	0.473	0.033
i-C_5	1.00	0.205	20	3.4	16.6	0.069	0.327
n-C_5	0.415	0.085	35	2.7	32.3	0.055	0.636
			100	49.2	50.8	1.000	1.000

The calculated i-C_5 concentration in the overhead stream is 6.9 percent, which is close enough to the 7.0 figure for now.

Table 13-8 shows subsequent calculations using the Underwood minimum-reflux equations. The α and x_D values in Table 13-8 are those from the Fenske

TABLE 13-8 Application of Underwood Equations

Component	x_F	α	αx_F	$\theta = 1.36$		$\theta = 1.365$		x_D	αx_D	$\alpha - \theta$	$\dfrac{\alpha x_D}{\alpha - \theta}$
				$\alpha - \theta$	$\dfrac{\alpha x_F}{\alpha - \theta}$	$\alpha - \theta$	$\dfrac{\alpha x_F}{\alpha - \theta}$				
C_3	0.05	4.99	0.2495	3.63	0.0687	3.625	0.0688	0.102	0.5090	3.6253	0.1404
i-C_4	0.15	2.62	0.3930	1.26	0.3119	1.255	0.3131	0.301	0.7886	1.2553	0.6282
n-C_4	0.25	2.02	0.5050	0.66	0.7651	0.655	0.7710	0.473	0.9555	0.6553	1.4581
i-C_5	0.20	1.00	0.2000	−0.36	−0.5556	−0.365	−0.5479	0.069	0.0690	−0.3647	−0.1892
n-C_5	0.35	0.864	0.3024	−0.496	−0.6097	−0.501	−0.6036	0.055	0.0475	−0.5007	−0.0949
	1.00				−0.0196		+0.0014	1.000			$1.9426 = R_m + 1$

Interpolation gives $\theta = 1.3647$.

total-reflux calculation. As noted earlier, the x_D values should be those at minimum reflux. This inconsistency may reduce the accuracy of the Underwood method, but to be useful a shortcut method must be fast, and it has not been shown that a more rigorous estimation of x_D values results in an overall improvement in accuracy. The calculated R_m is 0.9426. The actual reflux assumed is obtained from the specified maximum top vapor rate of 0.022 (kg·mol)/s [175 (lb·mol)/h] and the calculated D of 49.2 (from the Fenske equation).

$$L_{N+1} = V_N - D$$

$$R = V_N/D - 1 = 175/49.2 - 1 = 2.557$$

The $R_m = 0.9426$, $R = 2.557$, and $N = 10$ values are now used with the Gilliland correlation in Fig. 13-41 or Eq. (13-30) to check the initially assumed value of 6.0 for N_m. Equation (13-30) gives $N_m = 6.95$, which differs considerably from the assumed value.

Repetition of the calculations with $N_m = 7.0$ gives $R = 2.519$, $R_m = 0.9782$, and a calculated check value of $N_m = 6.85$, which is close enough. The final-product compositions and the α values used are as follows:

Component	α_i	Dx_D	Bx_B	x_D	x_B
C_3	4.98	5.00	0	0.1004	0.0
i-C_4	2.61	14.91	0.09	0.2996	0.0017
n-C_4	2.02	24.16	0.84	0.4852	0.0168
i-C_5	1.00	3.48	16.52	0.0700	0.3283
n-C_5	0.851	2.23	32.87	0.0448	0.6532
		49.78	50.32	1.0000	1.0000

These results indicate that the 7 percent i-C_5 in D and the 3 percent n-C_4 in B concentrations obtained in the original column can easily be obtained on the smaller column. Unfortunately, this disagrees somewhat with the answers obtained from a rigorous computer solution as shown in the following comparison:

Component	x_D		x_B	
	Rigorous	FUG	Rigorous	FUG
C_3	0.102	0.100	0.0	0.0
i-C_4	0.299	0.300	0.006	0.002
n-C_4	0.473	0.485	0.037	0.017
i-C_5	0.073	0.070	0.322	0.328
n-C_5	0.053	0.045	0.635	0.653
	1.000	1.000	1.000	1.000

KREMSER GROUP METHOD

Starting with the classical method of Kremser (op. cit.), approximate group methods of increasing complexity have been developed to calculate groups of equilibrium stages for a countercurrent cascade, such as is used in simple absorbers and strippers of the type depicted in Fig. 13-7b and d. However, none of these group methods can adequately account for stage temperatures that are considerably higher or lower than the two entering-stream temperatures for absorption and stripping respectively when appreciable composition changes occur. Therefore, only the simplest form of the Kremser method is presented here. Fortunately, rigorous computer methods described later can be applied when accurate results are required. The Kremser method is most useful for making preliminary estimates of absorbent and stripping-agent flow rates or equilibrium-stage requirements. The method can also be used to extrapolate quickly results of a rigorous solution to a different number of equilibrium stages.

Consider the general adiabatic countercurrent cascade of Fig. 13-42 where v and ℓ are molar component flow rates. Regardless of whether the cascade is an absorber or a stripper, components in the entering vapor will tend to be absorbed and components in the entering liquid will tend to be stripped. If more moles are stripped than absorbed, the cascade is a stripper; otherwise, the cascade is an absorber. The Kremser method is general and applies to either case. Application of component material-balance and phase equilibrium equations successively to stages 1 through $N - 1$, 1 through $N - 2$, etc., as shown by Henley and Seader (op. cit.), leads to the following equations originally derived by Kremser. For each component i,

$$(v_i)_N = (v_i)_0 (\Phi_i)_A + (\ell_i)_{N+1}[1 - (\Phi_i)_S] \tag{13-39}$$

where

$$(\Phi_i)_A = \frac{(A_i)_e - 1}{(A_i)_e^{N+1} - 1} \tag{13-40}$$

is the fraction of component i in the entering vapor that is not absorbed.

$$(\Phi_i)_S = \frac{(S_i)_e - 1}{(S_i)_e^{N+1} - 1} \tag{13-41}$$

is the fraction of component i in the entering liquid that is not stripped.

$$(A_i)_e = (L/K_iV)_e \tag{13-42}$$

is the effective or average absorption factor for component i, and

$$(S_i)_e = 1/(A_i)_e \tag{13-43}$$

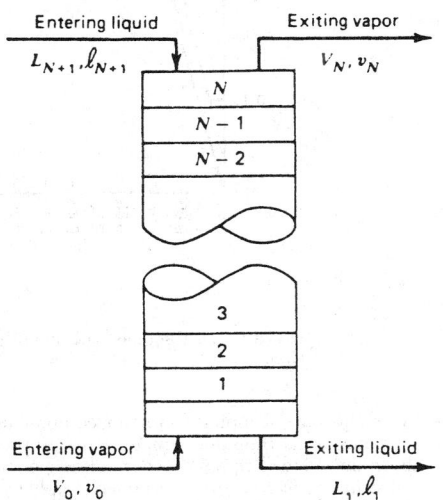

FIG. 13-42 General adiabatic countercurrent cascade for simple absorption or stripping.

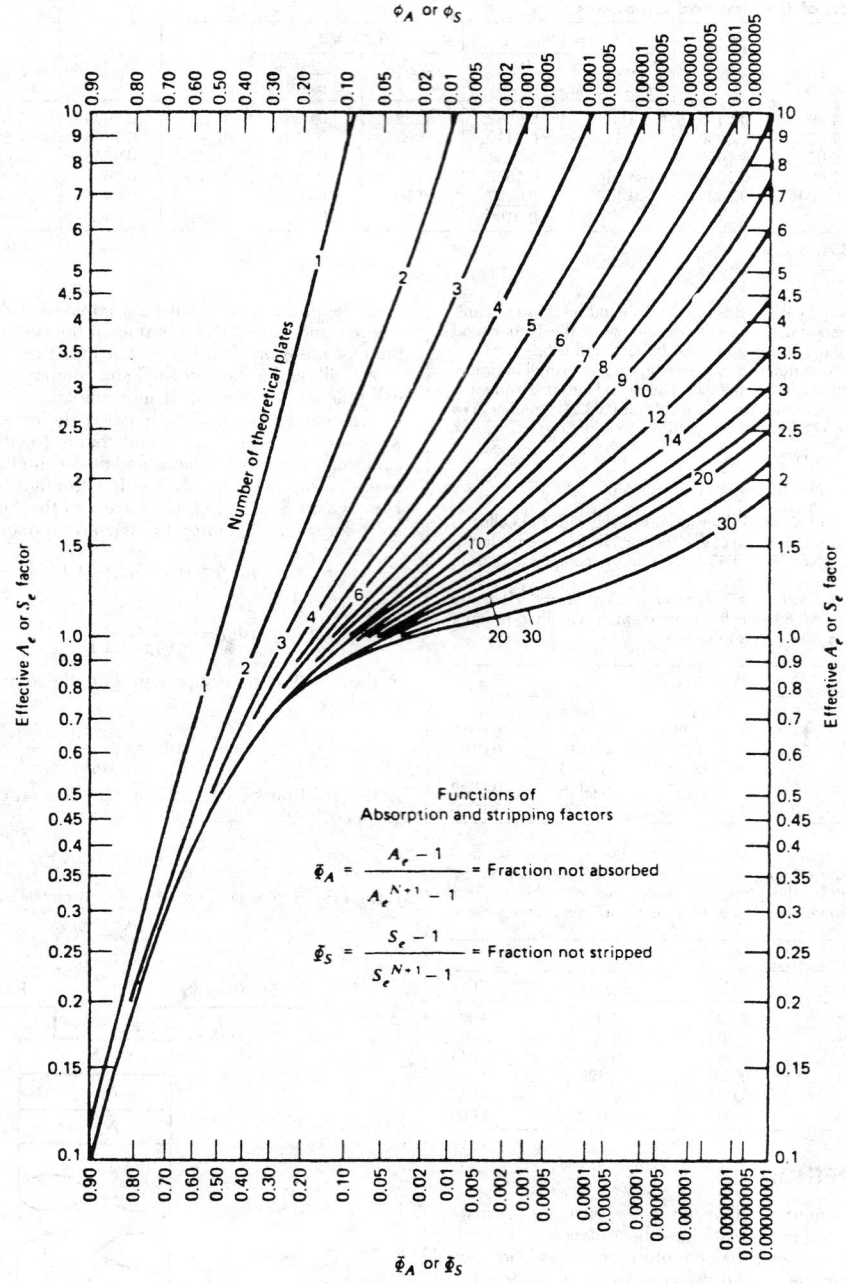

FIG. 13-43 Absorption and stripping factors. [*W. C. Edmister*, Am. Inst. Chem. Eng. J., **3**, 165–171 (1957).]

is the effective or average stripping factor for component i. When the entering streams are at the same temperature and pressure and negligible absorption and stripping occur, effective component absorption and stripping factors are determined simply by entering-stream conditions. Thus, if K values are composition-independent, then

$$(A_i)_e = 1/(S_i)_e = (L_{N+1}/K_i\{T_{N+1}, P_{N+1}\}V_0) \qquad (13\text{-}44)$$

When entering-stream temperatures differ and/or moderate to appreciable absorption and/or stripping occurs, values of A_i and S_i should be based on effective average values of L, V, and K_i in the cascade. However, even then Eq. (13-44) with T_{N+1} replaced by $(T_{N+1} + T_0)/2$ may be able to give a first-order approximation of $(A_i)_e$. In the case of an absorber, $L_{N+1} < L_e$ and $V_0 > V_e$ will be compensated to some extent by $K_i\{(T_{N+1} + T_0)/2, P)\} < K_i\{T_e, P\}$. A similar compensation, but in opposite directions, will occur in the case of a stripper.

Equations (13-40) and (13-41) are plotted in Fig. 13-43. Components having large values of A_e or S_e absorb or strip respectively to a large extent. Coorresponding values of Φ_A and Φ_S approach a value of 1 and are almost independent of the number of equilibrium stages.

An estimate of the minimum absorbent flow rate for a specified amount of absorption from the entering gas of some key component K for a cascade with an infinite number of equilibrium stages is obtained from Eq. (13-40) as

$$(L_{N+1})_{\min} = K_K V_0 [1 - (\Phi_K)_A] \tag{13-45}$$

The corresponding estimate of minimum stripping-agent flow rate for a stripper is obtained as

$$(V_0)_{\min} = L_{N+1} [1 - (\Phi_K)_S]/K_K \tag{13-46}$$

Example 2: Calculation of Kremser Method   For the simple absorber specified in Fig. 13-44, a rigorous calculation procedure as described below gives results in Table 13-9. Values of Φ were computed from component-product flow rates, and corresponding effective absorption and stripping factors were obtained by iterative calculations in using Eqs. (13-40) and (13-41) with $N = 6$. Use the Kremser method to estimate component-product rates if N is doubled to a value of 12.

Assume that values of A_e and S_e will not change with a change in N. Application of Eqs. (13-40), (13-41), and (13-39) gives the results in the last four columns of Table 13-10. Because of its small value of A_e, the extent of absorption of C_1 is unchanged. For the other components, somewhat increased amounts of absorption occur. The degree of stripping of the absorber oil is essentially unchanged. Overall, only an additional 0.5 percent of absorption occurs. The greatest increase in absorption occurs for $n\text{-}C_4$, to the extent of about 4 percent.

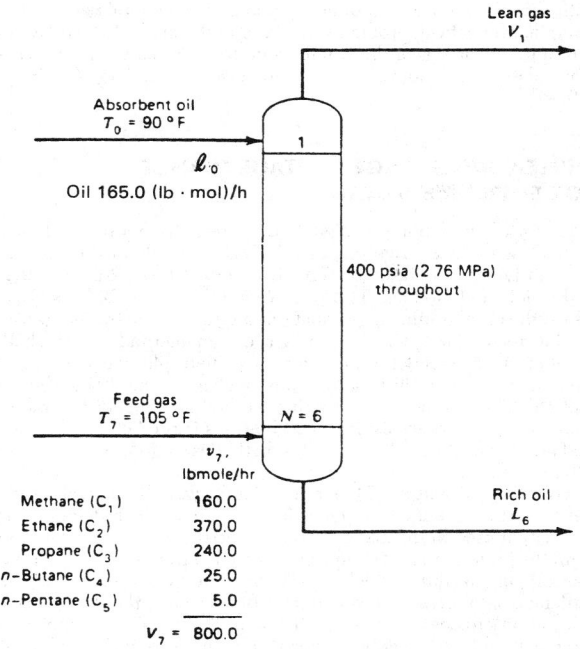

FIG. 13-44   Specifications for the absorber example.

TABLE 13-9   Results of Calculations for Simple Absorber of Fig. 13-44

| Component | $N = 6$ (rigorous method) | | | | | | $N = 12$ (Kremser method) | | | |
| | (lb·mol)/h | | | | | | (lb·mol)/h | | | |
	$(v_i)_6$	$(\ell_i)_1$	$(\Phi_i)_A$	$(\Phi_i)_S$	$(A_i)_e$	$(S_i)_e$	$(v_i)_{12}$	$(\ell_i)_1$	$(\Phi_i)_A$	$(\Phi_i)_S$
C_1	147.64	12.36	0.9228		0.0772		147.64	12.36	0.9228	
C_2	276.03	94.97	0.7460		0.2541		275.98	94.02	0.7459	
C_3	105.42	134.58	0.4393		0.5692		103.46	136.54	0.4311	
nC_4	1.15	23.85	0.0460		1.3693		0.16	24.84	0.0063	
nC_5	0.0015	4.9985	0.0003		3.6		0	5.0	0.0	
Absorber oil	0.05	164.95	———	0.9997	———	0.0003	0.05	164.95	———	0.9997
Totals	530.29	435.71					527.29	437.71		

NOTE: To convert pound-moles per hour to kilogram-moles per hour, multiply by 0.454.

TABLE 13-10   Top-Down Calculations for Example 3

Component	$R+1$	A_{10}	$\dfrac{\ell_{10}}{d}$	$\dfrac{\ell_{10}}{d}+1$	A_9	$\dfrac{\ell_9}{d}$	$\dfrac{\ell_9}{d}+1$	A_8	$\dfrac{\ell_8}{d}$	$\dfrac{\ell_8}{d}+1$	A_7	$\dfrac{\ell_7}{d}$
C_3	3.58	0.260	0.931	1.931	0.232	0.448	1.448	0.212	0.307	1.307	0.198	0.259
$i\text{-}C_4$	3.58	0.522	1.87	2.87	0.450	1.29	2.29	0.405	0.927	1.927	0.371	0.715
$n\text{-}C_4$	3.58	0.693	2.48	3.48	0.590	2.05	3.05	0.526	1.60	2.60	0.484	1.26
$i\text{-}C_5$	3.58	1.44	5.16	6.16	1.22	7.52	8.52	1.05	8.95	9.95	0.936	9.31
$n\text{-}C_5$	3.58	1.72	6.16	7.16	1.46	10.5	11.5	1.23	14.1	15.1	1.09	16.5

RIGOROUS METHODS FOR MULTICOMPONENT DISTILLATION-TYPE SEPARATIONS

INTRODUCTION

Availability of large digital computers has made possible rigorous solutions of equilibrium-stage models for multicomponent, multistage distillation-type columns to an exactness limited only by the accuracy of the phase equilibrium and enthalpy data utilized. Time and cost requirements for obtaining such solutions are very low compared with the cost of manual solutions. Methods are available that can accurately solve almost any type of distillation-type problem quickly and efficiently. The material presented here covers, in some detail, some of the more widely used computer algorithms as well as the classical Thiele-Geddes manual method. All are rating methods, in that the number of equilibrium stages and feed and withdrawal stages are specified. However, a successive-approximation design method that utilizes a rating method is given by Ricker and Grens [*Am. Inst. Chem. Eng. J.*, **20**, 238 (1974).] Those desiring further details are referred to the textbooks by Henley and Seader, King, and Holland cited under "General References" at the beginning of this section. These books, in turn, cite a myriad of references in chemical engineering journals. The mathematics involved is that of dealing with sets of nonlinear algebraic equations. The general nature of the main mathematical problems is presented lucidly by Friday and Smith ["An Analysis of the Equilibrium Stage Separation Problem—Formulation and Convergence," *Am. Inst. Chem. Eng. J.*, **10**, 698 (1964)].

THIELE-GEDDES STAGE-BY-STAGE METHOD FOR SIMPLE DISTILLATION

Prior to the availability of digital computers, the most widely used manual methods for rigorous calculations of simple distillation were those of Lewis and Matheson (LM) [*Ind. Eng. Chem.*, **24**, 496 (1932)] and Thiele and Geddes (TG) [*Ind. Eng. Chem.*, **25**, 290 (1933)], in which the equilibrium-stage equations are solved one by one by using tearing techniques. The former is a design method, in which the number of stages is determined for a specified split between two key components. Thus, it is a rigorous analog of the FUG shortcut method. The TG method is a rating method in which distribution of components between distillate and bottoms is predicted for a specified number of stages. Thus, the TG method is a rigorous analog of the SB method.

Both the LM and the TG methods suffer from numerical difficulties that can prevent convergence in certain cases. The stage-to-stage calculation used in the LM method proceeds from the top down and from the bottom up and is subject to large truncation-error buildup if the components differ widely in volatility. The TG method avoids that difficulty, but numerical instabilities arise as soon as the stage-to-stage calculation crosses a feed stage. Then, a difference term appears in the equations, and sometimes this results in a serious loss of significant digits, making the TG method basically unsuited for multiple-feed columns.

All stage-to-stage methods that work from both ends of the column toward the middle suffer from two other disadvantages. First, the top-down and the bottom-up calculations must "mesh" somewhere in the column. Usually the mesh is made at a feed stage, and if more than one feed stage exists, a choice of mesh point must be made for each component. When the components vary widely in volatility, the same mesh point cannot be used for all components if serious numerical difficulties are to be avoided. Second, arbitrary procedures must be set up to handle *nondistributed* components. (A nondistributed component is one whose concentration in one of the end-product streams is smaller than the smallest number carried by the computer.) In the LM and TG equations, the concentrations for these components do not naturally take on nonzero values at the proper point as the calculations proceed through the column.

Because of all these numerical difficulties, neither the LM nor the TG stage-by-stage method is commonly implemented in modern computer algorithms. Nevertheless, the TG method is very instructive and is developed in the following example. For a single narrow-boiling feed, the TG manual method is quite efficient.

Example 3: Calculation of TG Method The TG method will be demonstrated by using the same example problem that was used above for the approximate methods. The example column was analyzed previously and found to have $C + 2N + 9$ design variables. The specifications to be used in this example were also listed at that time and included the total number of stages ($N = 10$), the feed-plate location ($M = 5$), the reflux temperature (corresponding to saturated liquid), the distillate rate ($D = 48.9$), and the top vapor rate ($V = 175$). As before, the pressure is uniform at 827 kPa (120 psia), but a pressure gradient could be easily handled if desired.

A temperature profile plus a vapor-rate profile through the column must be assumed to start the procedure. These variables are referred to as tear variables and must be iterated on until convergence is achieved in which their values no longer change from iteration to iteration and all equations are satisfied to an acceptable degree of tolerance. Each iteration down and then up through the column is referred to as a column iteration. A set of assumed values of the tear variables consistent with the specifications, plus the component K values at the assumed temperatures, is as follows, using assumed end and middle temperatures and K values from Fig. 13-14b:

Stage	V	L	T	C_3	$i\text{-}C_4$	$n\text{-}C_4$	$i\text{-}C_5$	$n\text{-}C_5$
						K		
10	175	126.1	163.5	2.77	1.38	1.04	0.500	0.420
9		178.5	178.5	3.10	1.60	1.22	0.590	0.495
8			191.3	3.40	1.78	1.37	0.685	0.585
7			202.0	3.63	1.94	1.49	0.770	0.660
6		226.1	210.0	3.84	2.06	1.60	0.825	0.702
5			216.4	4.00	2.21	1.73	0.895	0.765
4			221.7	4.15	2.28	1.80	0.925	0.800
3			226.3	4.28	2.36	1.88	0.965	0.835
2			230.3	4.36	2.43	1.94	1.000	0.870
1		51.1	234.0	4.42	2.50	1.99	1.030	0.890

Stage compositions in the TG method are obtained by stage-to-stage calculations from both ends toward the feed stage. With reference to Fig. 13-1, the calculations work with the ratios v_n/d, ℓ_n/d, v_m/b, and ℓ_m/b instead of v or ℓ directly. The working equations are derived as follows:

In the rectifying section, the equilibrium relationship for component i at any stage n can be expressed in terms of component flow rate in the distillate $d = Dx_D$ and component absorption factor $A_n = L_n/K_n V_n$.

$$x_n = y_n/K_n$$

$$L_n x_n = (L_n/K_n V_n)V_n y_n$$

$$\ell_n = A_n v_n$$

$$\ell_n/d = (v_n/d)A_n \tag{13-47}$$

The general component-i balance around a section of stages from stage n to the top of the column is

$$v_n = \ell_{n+1} + d$$

or

$$v_n/d = (\ell_{n+1}/d) + 1 \tag{13-48}$$

Increasing the subscripts in Eq. (13-47) by 1 and substituting for ℓ_{n+1}/d in Eq. (13-48) gives the following combined equilibrium and material-balance relationship for component i:

$$v_n/d = (v_{n+1}/d)A_{n+1} + 1 \tag{13-49}$$

Or, if v_n/d is eliminated in Eq. (13-48)

$$\frac{\ell_n}{d} = A_n \left(\frac{\ell_{n+1}}{d} + 1 \right) \tag{13-50}$$

Equation (13-50) is used to calculate, from the previous stage, the (ℓ/d) ratio on each stage in the rectifying section. The assumed temperature and phase-rate-profile assumptions conveniently fix all the A_n values for ideal solutions. The calculations are started by writing the equation for stage N:

$$\frac{\ell_N}{d} = A_N \left(\frac{\ell_{N+1}}{d} + 1 \right) \tag{13-51}$$

For a total condenser, $x_D = x_{N+1}$ and

$$\ell_{N+1}/d = L_{N+1}/D = R \tag{13-52}$$

A knowledge of the reflux ratio (obtained from the specified distillate and top vapor rates) permits the calculation of $(\ell_N/d)_i$ from which $(\ell_{m-1}/d)_i$ is obtained, etc. Equation (13-50) is applied to each stage in succession until the ratio ℓ_{M+2}/d in the overflow from the stage above the feed stage is obtained. The calculations are then switched to the stripping section.

The equilibrium relationship for component i in the stripping section can be expressed in terms of component flow rate in the bottoms, $b = Bx_B$, and $S_m = K_m V_m/L_m$ as

$$y_m = K_m x_m$$

$$V_m y_m = (K_m V_m/L_m)L_m x_m \tag{13-53}$$

$$v_m = S_m \ell_m$$

$$v_m/b = (\ell_m/b)S_m$$

Combination with the material balance

$$(\ell_{m+1}/b) = v_m/b + 1 \tag{13-54}$$

gives

$$(\ell_{m+1}/b) = (\ell_m/b)S_m + 1 \tag{13-55}$$

The bottom-up calculations are started by writing Eq. (13-55) for stage 1 as

$$\ell_2/b = V_1 K_1/B + 1 = S_1 + 1 \tag{13-56}$$

The S_m values all are fixed by assumed temperature and phase-rate profiles. Equation (13-55) is applied to each of the stripping stages in sequence until the ratio ℓ_{M+2}/b in the liquid entering the feed stage is obtained.

The manner in which rectifying and stripping-section calculations are meshed at the feed stage depends upon the thermal condition of the feed. Figure 13-45 shows three possible ways in which fresh feed can affect the L and V rates between the feed stage and stage $M + 2$. The superscript bar denotes the stream rate when the stream enters a stage, while the lack of a bar denotes the rate when the stream leaves a stage.

Top-down calculations for the example problem are shown in Table 13-10 and bottom-up calculations in Table 13-11. Top-down and bottom-up calculations have provided values of ℓ_{M+2}/d and ℓ_{M+2}/b respectively. For a bubble-point feed,

$$v_{M+1} = \bar{v}_{M+1}$$

and a combination of Eqs. (13-48) and (13-54) provides for each component i

$$\frac{b}{d} = \frac{v_{M+1}/d}{\bar{v}_{M+1}/b} = \frac{\ell_{M+2}/d + 1}{\ell_{M+2}/b - 1} \tag{13-57}$$

The b/d ratios obtained from this equation can then be used to calculate the individual b and d values as follows. Since

$$d + b = Fx_F \tag{13-58}$$

$$d = \frac{Fx_F}{1 + (b/d)}$$

and

$$b = (b/d)d \tag{13-59}$$

Calculated values of d from the first column iteration in the example problem are as follows:

Component	$\dfrac{\ell_7}{d} + 1$	$\dfrac{\ell_7}{b} - 1$	$\dfrac{b}{d}$	Fx_F	d
C_3	1.26	5450	0.000231	5	5.00
$i\text{-}C_4$	1.71	175	0.00977	15	14.85
$n\text{-}C_4$	2.26	47.7	0.0474	25	23.83
$i\text{-}C_5$	10.3	2.46	0.19	20	3.85
$n\text{-}C_5$	17.5	1.54	11.4	35	2.82
					50.4

The calculated D is 50.4 instead of 48.9. Before these incorrect d (and b) values are used to calculate the stage concentrations, followed by a new set of values of T, V, and L, convergence of the iteration is aided as follows by using the

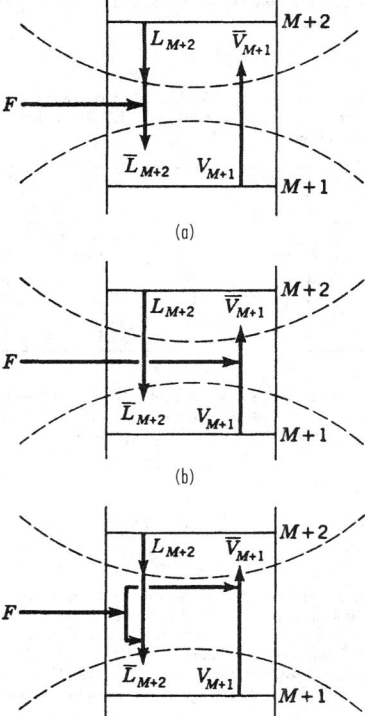

FIG. 13-45 Effect of feed on stream rates just above feed stage $M + 1$. (a) Subcooled or bubble-point feed. (b) Superheated or dew-point feed. (c) Partially flashed feed.

Θ method developed by Holland (*Fundamentals of Multicomponent Distillation*, McGraw-Hill, New York, 1981) and coworkers. A quantity θ is defined by

$$d' = \frac{Fx_F}{1 + (b/d)\Theta} \tag{13-60}$$

where values of d' are the ones that satisfy

$$\sum_i d' = D_{\text{specified}}$$

Comparison of Eqs. (13-57) and (13-60) shows that

$$b' = \Theta (b/d) d' \tag{13-61}$$

TABLE 13-11   Bottom-Up Calculations for Example 3

Component	S_1	$\dfrac{\ell_2}{b}$	S_2	$\dfrac{\ell_2}{b}S_2$	$\dfrac{\ell_3}{b}$	S_3	$\dfrac{\ell_3}{b}S_3$	$\dfrac{\ell_4}{b}$	S_4
C_3	15.1	16.1	3.37	54.3	55.3	3.31	183.0	184.0	3.21
$i\text{-}C_4$	8.56	9.56	1.88	18.0	19.0	1.83	34.8	35.8	1.76
$n\text{-}C_4$	6.81	7.81	1.50	11.7	12.7	1.45	18.4	19.4	1.39
$i\text{-}C_5$	3.53	4.53	0.774	3.51	4.51	0.747	3.37	4.37	0.716
$n\text{-}C_5$	3.05	4.05	0.673	2.73	3.73	0.646	2.41	3.41	0.619

Component	$\dfrac{\ell_4}{b}S_4$	$\dfrac{\ell_5}{b}$	S_5	$\dfrac{\ell_5}{b}S_5$	$\dfrac{\ell_6}{b}$	S_6	$\dfrac{\ell_6}{b}S_6$	$\dfrac{\ell_7}{b}$
C_3	590.6	591.6	3.10	1834	1835	2.97	5450	5451
$i\text{-}C_4$	63.0	64.0	1.71	109.4	110.4	1.59	175	176
$n\text{-}C_4$	27.0	28.0	1.34	37.5	38.5	1.24	47.7	48.7
$i\text{-}C_5$	3.13	4.13	0.693	2.86	3.86	0.638	2.46	3.46
$n\text{-}C_5$	2.11	3.11	0.592	1.84	2.84	0.543	1.54	2.54

The value of θ is found by solving the following nonlinear equation, where D is the specified distillate rate:

$$D - \sum_i \frac{Fx_F}{1 + (b/d)\Theta} = 0 \qquad (13\text{-}62)$$

For the first column iteration, $\Theta = 1.25$ satisfies this equation. The b/d values and the Θ value are used in Eqs. (13-60) and (13-61) to give the following corrected end concentrations:

Component	$\dfrac{b}{d}$	d'	b'	x_D	x_1
C_3	0.000231	5.00	0.00144	0.102	0
$i\text{-}C_4$	0.00977	14.82	0.181	0.303	0.004
$n\text{-}C_4$	0.0474	23.60	1.40	0.482	0.027
$i\text{-}C_5$	4.19	3.21	16.79	0.066	0.329
$n\text{-}C_5$	11.4	2.30	32.7	0.047	0.640
		48.9	51.1	1.000	1.000

Stage-to-stage calculations shown in Tables 13-10 and 13-11 provide ℓ/d and ℓ/b values for each stage. These are used in the following equations to calculate normalized liquid concentrations for each component at each stage:

$$x_n = \frac{(\ell_n/d)d'}{\sum_i (\ell_n/d)d'} \qquad (13\text{-}63)$$

$$x_m = \frac{(\ell_m/b)b'}{\sum_i (\ell_m/b)b'} \qquad (13\text{-}64)$$

Application of these equations gives the results in Table 13-12. A set of T_n is calculated from the normalized x_n by bubble-point calculations. Corresponding values of y_n are obtained from $y_n = K_n x_n$. Once new x_n and T_n are available, new values of V_n are calculated from energy balances by using data from Maxwell (*Data Book on Hydrocarbons,* Van Nostrand, Princeton, N.J., 1950). First, an estimate of condenser duty is computed from an energy balance around the condenser,

$$Q_c = V_N(H_N^V - H_{N+1}^L)$$
$$= 175(18,900 - 10,750) = 1,426,000 \text{ Btu/h } (417.9 \text{ kW}) \qquad (13\text{-}65)$$

The reboiler duty Q_r is obtained from an overall energy balance,

$$Q_r = DH_{N+1}^L + BH_B^L + Q_c - FH_F$$
$$= (48.9)(10,750) + (51.1)(17,080) + 1,426,000 - 100(13,540)$$
$$= 1,465,000 \text{ Btu/h } (429.3 \text{ kW}) \qquad (13\text{-}66)$$

A new set of values of V_m is obtained from energy balances around the bottom section of the column,

$$V_m = \frac{Q_r + B(H_{M+1}^L - H_B^V)}{H_m^V - H_{m+1}^L} \qquad (13\text{-}67)$$

Similar balances around the top section yield a new set of values of V_n. Corresponding values of L_n and L_m are obtained by material balances around the top and bottom sections respectively. The new V, L, and T profiles are listed in Table 13-13. In this example, they do not differ much from the initial guesses in Table 13-10.

It should be noted in Table 13-13 that it is not necessary to list two values of V, L, and T for the feed stage (stage 6) because the TC procedure gives a perfect match at the feed stage in each trial. This completes the first column iteration.

The new temperature and flow-rate profiles (which would be used as the assumptions to begin the second column iteration) are compared in Fig. 13-46 with the final solution. Both profiles are moving toward the final result.

Figure 13-47 shows the concentration profiles from the final solution. Note the discontinuities at the feed stage and the fact that feed-stage composition differs considerably from feed-stream composition. It can be seen in Fig. 13-47 from the $n\text{-}C_4$ and $i\text{-}C_5$ profiles that the separation between the keys improves rapidly with stage number; additional stages would be worthwhile.

Convergence to the final solution is rapid with the TG method for narrow-boiling feeds but may be slow for wide-boiling feeds. Generally, at least four column iterations are required. Convergence is obtained when successive sets of tear variables are identical to approximately four significant digits. This is accompanied by $\Theta = 1.0$, $x = $ normalized x, and nearly identical successive values of Q_c as well as Q_r.

TABLE 13-12   Stage Compositions from the First Trial of Example 3

Component	x_1	$\dfrac{\ell_2}{b}b'$	x_2	$\dfrac{\ell_3}{b}b'$	x_3	$\dfrac{\ell_4}{b}b'$	x_4	$\dfrac{\ell_5}{b}b'$	x_5	$\dfrac{\ell_6}{b}b'$
C_3	0.000	0.0232	0.000	0.0796	0.000	0.279	0.001	0.852	0.004	2.65
$i\text{-}C_4$	0.004	1.73	0.008	3.44	0.016	6.48	0.030	11.6	0.052	20.0
$n\text{-}C_4$	0.027	10.9	0.049	17.8	0.081	27.2	0.124	39.2	0.176	53.9
$i\text{-}C_5$	0.329	76.1	0.344	75.7	0.346	73.4	0.335	69.3	0.311	64.8
$n\text{-}C_5$	0.640	132.4	0.599	122.0	0.557	111.5	0.510	101.7	0.447	92.9
	1.000	221.1	1.000	219.0	1.000	218.9	1.000	226.6	1.000	234.2

Component	x_6	$\dfrac{\ell_7}{d}d'$	x_7	$\dfrac{\ell_8}{d}d'$	x_8	$\dfrac{\ell_9}{d}d'$	x_9	$\dfrac{\ell_{10}}{d}d'$	x_{10}
C_3	0.011	1.295	0.012	1.535	0.013	2.240	0.019	4.66	0.038
$i\text{-}C_4$	0.085	10.6	0.097	13.7	0.120	19.1	0.162	27.7	0.228
$n\text{-}C_4$	0.230	29.7	0.271	37.8	0.331	48.4	0.410	58.5	0.481
$i\text{-}C_5$	0.277	29.9	0.273	28.7	0.252	24.1	0.204	16.6	0.136
$n\text{-}C_5$	0.397	37.9	0.347	32.4	0.284	24.1	0.204	14.2	0.117
	1.000	109.4	1.000	114.1	1.000	118.0	1.000	121.7	1.000

TABLE 13-13   New Temperature and Rate Profiles from the First Trial of Example 3

n	New T	$H_D^L - H_{n+1}^L$	$D(H_D^L - H_{n+1}^L) + Q_c$	$H_N^V - H_{n+1}^L$	V	L
10	160.0	0	1,426,000	8150	175.0	124.0
9	175.0	-1220	1,367,000	7900	172.9	119.6
8	186.0	-2190	1,319,000	7830	168.5	117.6
7	194.0	-3010	1,279,000	7680	166.5	114.2
6	200.0	-3490	1,256,000	7700	163.1	214.3

m	New T	$H_{m+1}^L - H_B^L$	$B(H_{m+1}^L - H_B^L) + Q_r$	$H_m^V - H_{m+1}^L$	V	L
5	211.0	-2480	1,338,000	8200	163.2	214.7
4	220.0	-1780	1,374,000	8400	163.6	215.5
3	228.0	-1130	1,407,000	8560	164.4	221.1
2	233.5	-560	1,438,000	8460	170.0	224.0
1	237.5	-210	1,454,000	8410	172.9	51.1

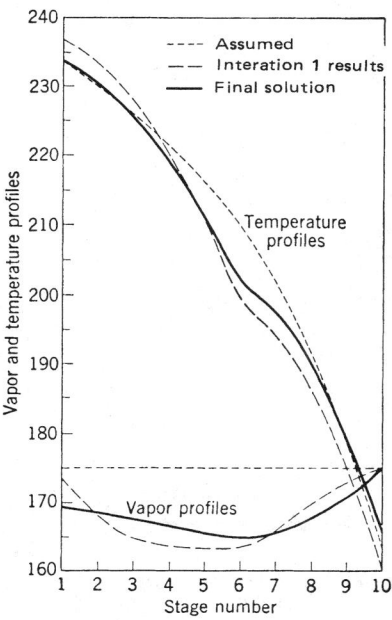

FIG. 13-46 Comparison of the assumed and calculated profiles from the first column iteration in Example 4 with the final computer solution.

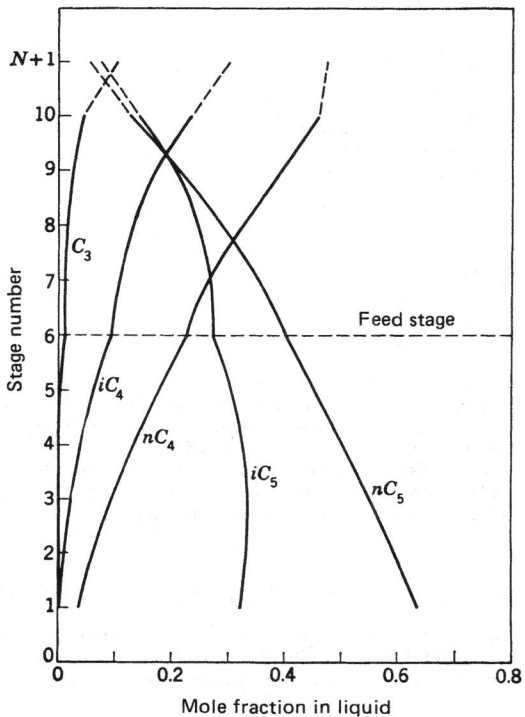

FIG. 13-47 Concentration profiles from the final solution of Example 4. The points at $N + 1$ refer to the reflux composition, which is the same as the overhead vapor.

EQUATION-TEARING PROCEDURES USING THE TRIDIAGONAL-MATRIX ALGORITHM

As seen earlier, the manual Thiele-Geddes method involves solving the equilibrium-stage equations one at a time. More powerful, flexible, and reliable computer programs are based on the application of sparse matrix methods for solving simultaneously all or at least some of the equations. For cases in which combined column feeds represent mixtures that boil within either a narrow range (typical of many distillation operations) or a wide range (typical of absorbers and strippers) and in which great flexibility of problem specifications is not required, equation-tearing procedures that involve solving simultaneously certain subsets of the equations can be applied. Two such equation-tearing procedures are the bubble-point (BP) method for narrow-boiling mixtures suggested by Friday and Smith (op. cit.) and developed in detail by Wang and Henke [*Hydrocarbon Process.*, **45** (8), 155 (1966)], and the sum-rates (SR) method for wide-boiling mixtures proposed by Sujuta [*Hydrocarbon Process.*, **40**(12), 137 (1961)] and further developed by Burningham and Otto [*Hydrocarbon Process.*, **46**(10), 163 (1967)]. Both methods are described here. However, the BP method has been largely superseded by the more reliable and efficient simultaneous-correction and inside-out methods, which do, however, incorporate certain features of the BP method. Both the BP and SR methods start with the same primitive equations for the theoretical model of an equilibrium stage as presented next.

Consider a general, continuous-flow, steady-state, multicomponent, multistage separation operation. Assume that phase equilibrium between an exiting vapor phase and a single exiting liquid phase is achieved at each stage, that no chemical reactions occur, and that neither of the exiting phases entrains the other phase. A general schematic representation of such a stage j is shown in Fig. 13-48. Entering stage j is a single- or two-phase feed at molal flow rate F_j, temperature T_{Fj}, and pressure P_{Fj} and with overall composition in mole fractions $z_{i,j}$. Also entering stage j is interstage liquid from adjacent stage $j - 1$ above at molal flow rate L_{j-1}, temperature T_{j-1}, pressure P_{j-1}, and mole fractions $x_{i,j-1}$. Similarly, interstage vapor from adjacent stage $j + 1$ below enters at molal flow rate V_{j+1}, T_{j+1}, P_{j+1} and mole fractions $y_{i,j+1}$. Heat is transferred from (+) or to (−) stage j at rate Q_j to simulate a condenser, reboiler, intercooler, interheater, etc. Equilibrium vapor and liquid phases leave stage j at T_j and P_j and with mole fractions $y_{i,j}$ and $x_{i,j}$ respectively. The vapor may be partially withdrawn from the column as a sidestream at a molal flow rate W_j, with the remainder V_j sent to adjacent stage $j - 1$ above. Similarly, exiting liquid may be split into a sidestream at a molal flow rate of U_j, with the remainder L_j sent to adjacent stage j below.

For each stage j, the following $2C + 3$ component material-balance (M), phase-equilibrium (E), mole-fraction-summation (S), and energy-balance (H) equations apply, where C is the number of chemical species:

$$L_{j-1}x_{i,j-1} + V_{j+1}y_{i,j+1} + F_jz_{i,j} - (L_j + U_j)x_{i,j} - (V_j + W_j)y_{i,j} = 0 \quad (13\text{-}68)$$

$$y_{i,j} - K_{i,j}x_{i,j} = 0 \quad (13\text{-}69)$$

$$\sum_i y_{i,j} - 1.0 = 0 \quad (13\text{-}70)$$

$$\sum_i x_{i,j} - 1.0 = 0 \quad (13\text{-}71)$$

$$L_{j-1}H_{Lj-1} + V_{j+1}H_{Vj+1} + F_jH_{Fj}$$
$$- (L_j + U_j)H_{Lj} - (V_j + W_j)H_{Vj} - Q_j = 0 \quad (13\text{-}72)$$

In general, K values and molal enthalpies in these MESH equations are complex implicit functions of stage temperature, stage pressure, and equilibrium mole fractions:

$$K_{i,j} = K_{i,j}\{T_j, P_j, x_j, y_j\}$$
$$H_{Vj} = H_{Vj}\{T_j, P_j, y_j\}$$
$$H_{Lj} = H_{Lj}\{T_j, P_j, x_j\}$$

where vectors x_j and y_j refer to all i values of $x_{i,j}$ and $y_{i,j}$ for the particular stage j. As shown in Fig. 13-49, a general countercurrent-flow col-

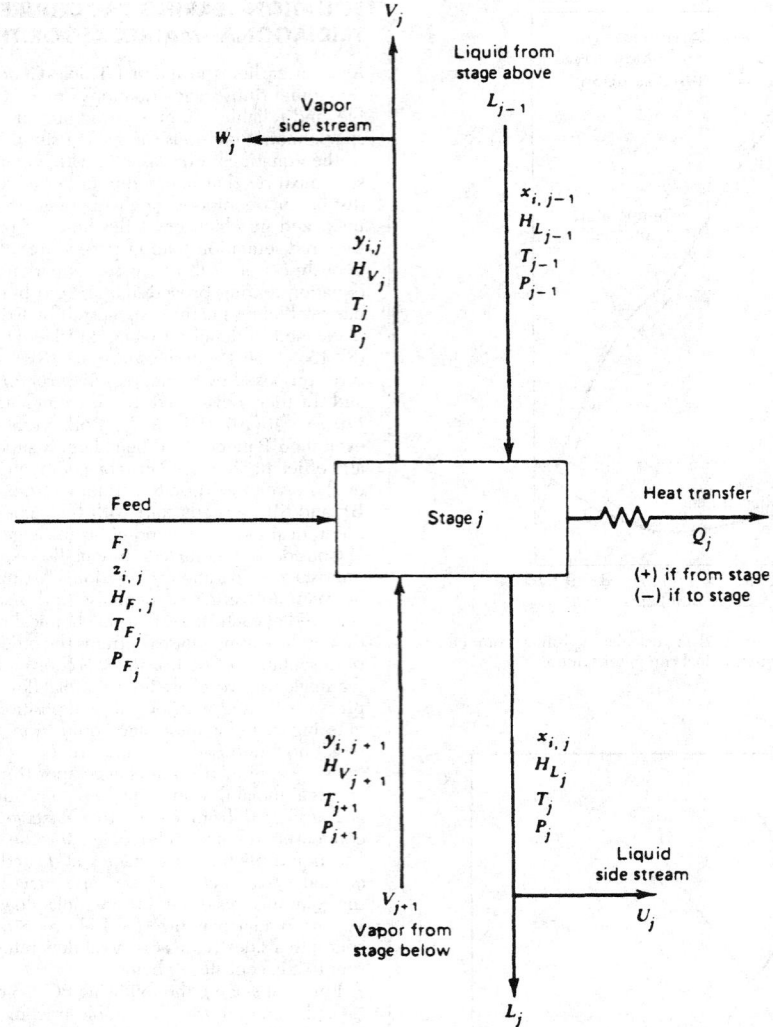

FIG. 13-48   General equilibrium stage.

umn of N stages can be formed from a collection of equilibrium stages of the type in Fig. 13-48. Note that streams L_0, V_{N+1}, W_1 and U_N are zero and do not appear in Fig. 13-49. Such a column is represented by $N(2C + 3)$ MESH equations in $[N(3C + 10) + 1]$ variables, and the difference or $[N(C + 7) + 1]$ variables must be specified. If the specified variables are the value of N and all values of $z_{i,j}$, F_j, T_{Fj}, P_{Fj}, P_j, U_j, W_j, and Q_j, then the remaining $N(2C + 3)$ unknowns are all values of $y_{i,j}$, $x_{i,j}$, L_j, V_j, and T_j. In this case, Eqs. (13-73), (13-74), and (13-77) are nonlinear in the unknowns and the MESH equations can not be solved directly. Even if a different set of variable specifications is made, the MESH equations still remain predominantly nonlinear in the unknowns. For the BP method as applied to distillation, specified variables are those listed except that bottoms rate L_N is specified rather than partial reboiler duty Q_N. This is equivalent by overall material balance to specifying vapor-distillate rate V_1 in the case of a partial condenser or liquid-distillate rate U_1 in the case of a total condenser. Also, reflux rate L_1 is specified rather than condenser duty Q_1. For the SR method as applied to absorption and stripping, the specified variables are those listed without exception.

Tridiagonal-Matrix Algorithm   Both the BP and the SR equation-tearing methods compute liquid-phase mole fractions in the same way by first developing linear matrix equations in a manner shown by Amundson and Pontinen [*Ind. Eng. Chem.*, **50,** 730 (1958)]. Equations (13-69) and (13-68) are combined to eliminate $y_{i,j}$ and $y_{i,j+1}$ (however, the vector y_j still remains implicitly in $K_{i,j}$):

$$L_{j-1}x_{i,j-1} + V_{j+1}K_{i,j+1}x_{i,j+1} + F_j z_{i,j}$$
$$- (L_j + U_j)x_{i,j} - (V_j + W_j)K_{i,j}x_{i,j} = 0 \quad (13\text{-}73)$$

Next, Eq. (13-68) is summed over the C components and over stages 1 through j and combined with Eqs. (13-70), (13-71), and $\sum_i z_{i,j} - 1.0 = 0$ to give a total material balance over stages 1 through j:

$$L_j = V_{j+1} + \sum_{m=1}^{j} (F_m - U_m - W_m) - V_1 \quad (13\text{-}74)$$

By combining Eq. (13-73) with Eq. (13-74), L_j is eliminated to give the following working equations for component material balances:

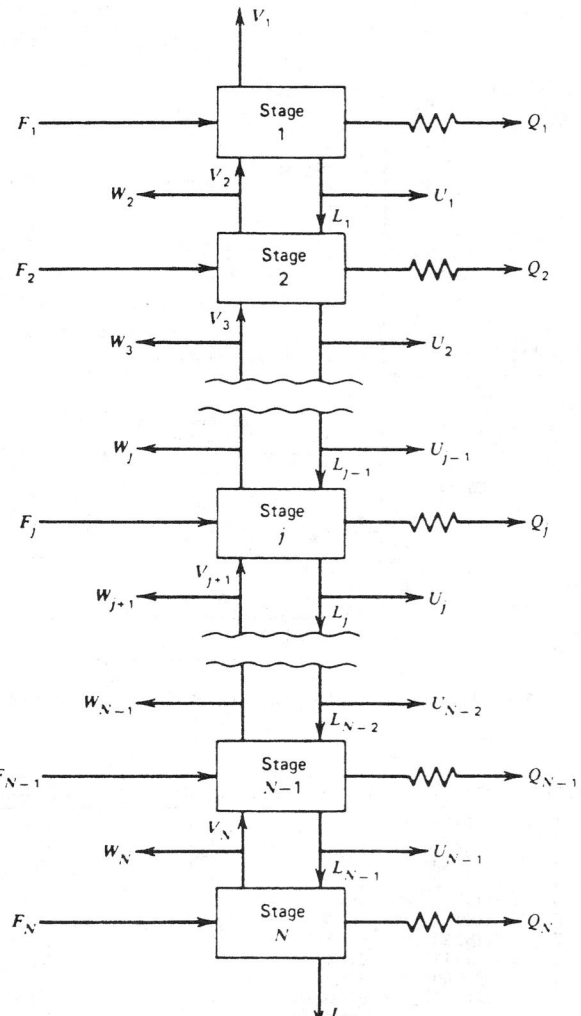

FIG. 13-49   General countercurrent cascade of N stages.

$$A_j x_{i,j-1} + B_{i,j} x_{i,j} + C_{i,j} x_{i,j+1} = D_{i,j} \qquad (13\text{-}75)$$

where

$$A_j = V_j + \sum_{m=1}^{j-1} (F_m - W_m - U_m) - V_i \qquad 2 \le j \le N \quad (13\text{-}76)$$

$$B_{i,j} = -\left[V_{j+1} + \sum_{m=1}^{j} (F_m - W_m - U_m) - V_1 + U_j + (V_j + W_j)K_{i,j} \right]$$

$$1 \ll j \ll N \quad (13\text{-}77)$$

$$C_{i,j} = V_{j+1} K_{i,j+1} \qquad 1 \ll j \ll N - 1 \qquad (13\text{-}78)$$

$$D_{i,j} = -F_j z_{i,j} \qquad 1 \ll j \ll N \qquad (13\text{-}79)$$

The NC equations (13-75) are linearized in terms of the NC unknowns $x_{i,j}$ by selecting unknowns V_j and T_j as tear variables and using values of vectors x_j and y_j from the previous iteration to compute values of $K_{i,j}$ for the current iteration. In this manner all values of A_j, $B_{i,j}$, and $C_{i,j}$ can be estimated. Values of $D_{i,j}$ are fixed by feed specifications. Furthermore, the NC equations (13-75) can be partitioned

into C sets, one for each component, and solved separately for values of x_i, which pertains to all j values of $x_{i,j}$ for the particular species i. Each set of N equations is a special type of sparse matrix equation called a tridiagonal-matrix equation, which has the form shown in Fig. 13-50a for a five-stage example in which, for convenience, the subscript i has been dropped from the coefficients B, C, and D. For this type of sparse matrix equation, we can apply a highly efficient version of the gaussian elimination procedure called the Thomas algorithm, which avoids matrix inversion, eliminates the need to store the zero coefficients in the matrix, almost always avoids buildup of truncation errors, and rarely produces negative values of $x_{i,j}$.

The Thomas algorithm begins by a forward elimination, row by row starting down from the top row ($j = 1$, the condenser stage), to give the following replacements shown in Fig. 13-50b. For row 1:

$$p_1 = C_1/B_1, q_1 = D_1/B_1, B_1 \to 1, C_1 \to p_1, D_1 \to q_1$$

where $\to$ means "is replaced by."
For all subsequent rows:

$$p_j = C_j/(B_j - A_j p_{j-1}), q_j = (D_j - A_j q_{j-1})/(B_j - A_j p_{j-1}),$$
$$A_j \to 0, B_j \to 1, C_j \to p_j, D_j \to q_j$$

At the bottom row for component i, $x_N = q_N$. The remaining values of x_j for species i are computed recursively by backward substitution:

$$x_{j-1} = q_{j-1} - p_{j-1} x_j$$

BP Method for Distillation   The bubble-point method for distillation, particularly when the components involved cover a relatively narrow range of volatility, proceeds iteratively by the following steps, where k is the iteration index for the entire distillation column.

1.   Specify N and all values of $z_{i,j}$, F_j, T_{Fj}, P_{Fj}, P_j, U_j, W_j, and Q_j, except Q_1 and Q_N.
2.   Specify type of condenser. If total ($U_1 \ne 0$), compute L_N from overall material balance; if partial ($U_1 = 0$), specify V_1 and compute L_N from overall material balance.
3.   Specify reflux rate L_1, assuming no subcooling.
4.   Compute $V_2 = V_1 + L_1 + U_1 - F_1$.
5.   Provide initial guesses ($k = 0$) or values of all tear variables T_j and V_j ($j > 2$). Temperature guesses are readily obtained by linear interpolation between estimates of top- and bottom-stage temperatures. The bottom-stage temperature is estimated by making a bubble-point-temperature calculation by using an estimate of bottoms composition at the specified bottom-stage pressure. A similar calculation is made at the top stage by using an estimate of distillate composition; otherwise, for a partial condenser, a dew-point temperature calculation is made. An estimate of the vapor-rate profile is readily obtained by assuming constant molal overflow down the column.
6.   Set index $k = 1$ to initiate the first column iteration.
7.   Using specified stage pressures, current estimates of stage temperatures, and current estimates of stage vapor- and liquid-phase

$$\begin{bmatrix} B_1 & C_1 & 0 & 0 & 0 \\ A_2 & B_2 & C_2 & 0 & 0 \\ 0 & A_3 & B_3 & C_3 & 0 \\ 0 & 0 & A_4 & B_4 & C_4 \\ 0 & 0 & 0 & A_5 & B_5 \end{bmatrix} \begin{bmatrix} x_1 \\ x_2 \\ x_3 \\ x_4 \\ x_5 \end{bmatrix} = \begin{bmatrix} D_1 \\ D_2 \\ D_3 \\ D_4 \\ D_5 \end{bmatrix}$$

$$(a)$$

$$\begin{bmatrix} 1 & p_1 & 0 & 0 & 0 \\ 0 & 1 & p_2 & 0 & 0 \\ 0 & 0 & 1 & p_3 & 0 \\ 0 & 0 & 0 & 1 & p_4 \\ 0 & 0 & 0 & 0 & 1 \end{bmatrix} \begin{bmatrix} x_1 \\ x_2 \\ x_3 \\ x_4 \\ x_5 \end{bmatrix} = \begin{bmatrix} q_1 \\ q_2 \\ q_3 \\ q_4 \\ q_5 \end{bmatrix}$$

$$(b)$$

FIG. 13-50   Tridiagonal-matrix equation for a column with five theoretical stages. (a) Original equation. (b) After forward elimination.

compositions, estimate all $K_{i,j}$ values (for $k = 1$, initial estimates of stage phase compositions may be necessary if $K_{i,j}$ values are sensitive to phase compositions).

8. Compute values of $x_{i,j}$ by solving Eqs. (13-75) through (13-79) by the tridiagonal-matrix algorithm once for each component. Unless all mesh equations are converged, $\sum_i x_{i,j} \neq 1$ for each stage j.

9. To force $\sum_i x_{i,j} = 1$ at each stage j, normalize values by the replacement $x_{i,j} = x_{i,j}/\sum_i x_{i,j}$.

10. Compute a new set of values of $T_j^{(k)}$ tear variables by computing, one at a time, the bubble-point temperature at each stage based on the specified stage pressure and corresponding normalized $x_{i,j}$ values. The equation used is obtained by combining Eqs. (13-69) and (13-70) to eliminate $y_{i,j}$ to give

$$\sum_i K_{i,j}\{T_j, P_j, x_j, y_j\} = 1.0 \qquad (13\text{-}80)$$

which is a nonlinear equation in $T_j^{(k)}$ and must be solved iteratively by some appropriate root-finding method, such as the Newton-Raphson or the Muller method.

11. Compute values of $y_{i,j}$ one at a time from Eq. (13-69).

12. Compute a new set of values of the V_j tear variables one at a time, starting with V_3, from an energy-balance equation that is obtained by combining Eqs. (13-72) and (13-74), eliminating L_{j-1} and L_j to give

$$V_j = (\tilde{C}_{j-1} - \tilde{A}_{j-1} V_{j-1})/\tilde{B}_{j-1} \qquad (13\text{-}81)$$

where $\tilde{A}_{j-1} = H_{L_{j-2}} - H_{V_{j-1}}$

$\tilde{B}_{j-1} = H_{V_j} - H_{L_{j-1}}$

$$\tilde{C}_{j-1} = \left[\sum_{m=1}^{j-2}(F_m - W_m - U_m) - V_1\right](H_{L_{j-1}} - H_{L_{j-2}})$$
$$+ F_{j-1}(H_{L_{j-1}} - H_{F_{j-1}}) + W_{j-1}(H_{V_{j-1}} - H_{L_{j-1}}) + Q_{j-1}$$

13. Check to determine if the new sets of tear variables $T_j^{(k)}$ and $V_j^{(k)}$ are within some prescribed tolerance of sets $T_j^{(k-1)}$ and $V_j^{(k-1)}$ used to initiate the current column iteration. A possible convergence criterion is

$$\sum_{j=1}^N \left[\frac{T_j^{(k)} - T_j^{(k-1)}}{T_j^{(k)}}\right]^2 + \sum_{j=3}^N \left[\frac{V_j^{(k)} - V_j^{(k-1)}}{V_j^{(k)}}\right]^2 \leq 10^{-7}N \qquad (13\text{-}82)$$

but Wang and Henke (op. cit.) use

$$\sum_{j=1}^N [T_j^{(k)} - T_j^{(k-1)}]^2 \leq 0.01N \qquad (13\text{-}83)$$

14. If the convergence criterion is met, compute values of L_j from Eq. (13-74) and values of Q_1 and Q_N from Eq. (13-72). Otherwise, set $k = k + 1$ and repeat steps 7 to 14.

Step 14 implies that if the calculations are not converged, values of $T_j^{(k)}$ computed in step 10 and values of $V_j^{(k)}$ computed in step 12 are used as values of the tear variables to initiate iteration $k + 1$. This is the method of successive substitution, which may require a large number of iterations and/or may result in oscillation. Alternatively, values of $T_j^{(k)}$ and $V_j^{(k)}$ can be adjusted prior to initiating iteration $k + 1$. Experience indicates that values of T_j should be reset if they tend to move outside of specified upper and lower bounds and that negative V_j values be reset to small positive values. Also, damping can be employed to prevent values of absolute T_j and V_j from changing by more than, say, 10 percent on successive iterations. Orbach and Crowe [*Can. J. Chem. Eng.*, **49**, 509 (1971)] show that the dominant eigenvalue method of adjusting values of T_j and V_j can generally accelerate convergence and is a worthwhile improvement to the BP method.

Example 4: Calculation of the BP Method Use the BP method with the SRK equation-of-state for K values and enthalpy departures to compute stage temperatures, interstage vapor and liquid flow rates and compositions, and reboiler and condenser duties for the light-hydrocarbon distillation-column specifications shown in Fig. 13-51 with feed at 260 psia. The specifications are selected to obtain three products: a vapor distillate rich in C_2 and C_3, a vapor sidestream rich in n-C_4, and a bottoms rich in n-C_5 and n-C_6.

Initial estimates provided for the tear variables were as follows compared with final converged values (after 23 iterations), where numbers in parentheses are consistent with specifications:

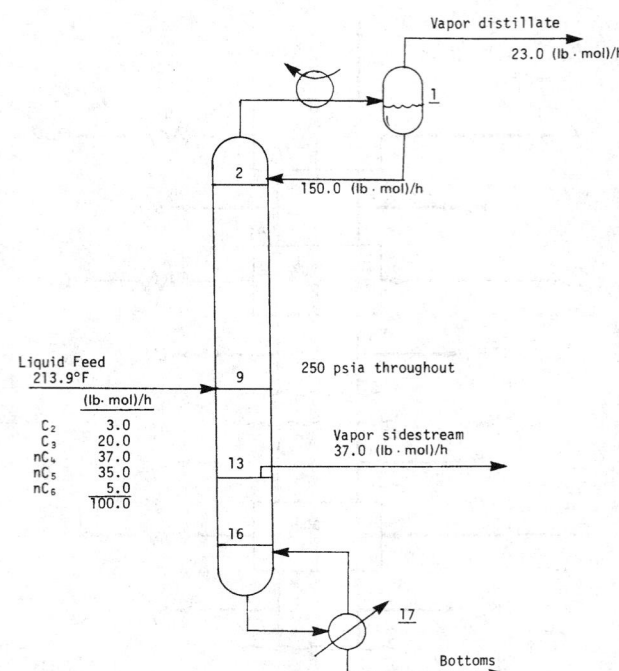

FIG. 13-51 Specifications for the calculation of distillation by the BP method.

Stage	$T^{(0)}$, °F	$T^{(23)}$, °F	$V^{(0)}$, (lb·mol)/h	$V^{(23)}$, (lb·mol)/h
1	110.00	124.4	(23)	23.0
2	121.87	141.0	(173)	173.0
3	133.75	157.4	173	167.3
4	145.62	172.6	173	162.8
5	157.50	184.9	173	160.4
6	169.37	194.5	173	159.5
7	181.25	202.6	173	158.4
8	193.12	211.1	173	155.8
9	205.00	221.9	173	151.4
10	216.87	229.9	173	155.8
11	228.75	236.5	173	160.2
12	240.62	242.9	173	162.9
13	252.50	250.2	173	164.0
14	264.37	258.1	210	200.8
15	276.25	267.4	210	199.4
16	288.12	278.1	210	197.9
17	300.00	290.5	210	196.4

NOTE: To convert degrees Fahrenheit to degrees Celsius, °C = (°F − 32)/1.8. To convert pound-moles per hour to kilogram-moles per second, multiply by 1.26×10^{-4}.

By employing successive substitution of the tear variables and the criterion of Eq. (13-83), convergence was achieved slowly, but without oscillation, in 23 iterations. Computed products are:

Component	Flow rate, (lb·mol)/h		
	Distillate	Sidestream	Bottoms
C_2	3.0	0.0	0.0
C_3	18.4	1.6	0.0
$n\text{C}_4$	1.6	25.7	9.7
$n\text{C}_5$	0.0	9.2	25.8
$n\text{C}_6$	0.0	0.5	4.5
	23.0	37.0	40.0

NOTE: To convert pound-moles per hour to kilogram-moles per second, multiply by 1.26×10^{-4}.

Examination of interstage-composition results showed that maximum nC_4 composition was achieved in the vapor leaving stage 12 rather than stage 13. Therefore, if the sidestream location were moved up one stage, a somewhat higher purity of nC_4 could probably be achieved in that stream. Further improvement in purity of the sidestream as well as the other two products could be achieved by increasing the reflux rate and/or number of stages. Computed condenser and reboiler duties were 268,000 and 418,000 W (914,000 and 1,425,000 Btu/h) respectively.

SR Method for Absorption and Stripping As shown by Friday and Smith (op. cit.), when an attempt is made to apply the BP method to absorption, stripping, or distillation, in which the volatility range of the chemical components in the column is very wide, calculations of stage temperatures from Eq. (13-80) become very sensitive to liquid compositions. This generally causes very oscillatory excursions in temperature from iteration to iteration, making it impossible to obtain convergence. A very successful modification of the BP method for such cases is the sum-rates method, in which new stage temperatures are computed from the energy-balance equation. Interstage vapor rates are computed by material balance from new interstage liquid rates that are obtained by multiplying the previous interstage liquid rates by corresponding unnormalized liquid mole-fraction summations computed from the tridiagonal-matrix algorithm. The SR method proceeds by the following steps:

1. Specify N and all values of $z_{i,j}$, F_j, T_{Fj}, P_{Fj}, P_j, U_j, W_j, and Q_j. For an adiabatic operation, all Q_j are zero.

2. Provide initial guesses ($k = 0$) for values of all tear variables T_j and V_j. Temperature guesses are readily obtained by linear interpolation between estimates of the top- and bottom-stage temperatures, taking the top as that of the liquid feed to the top stage and the bottom as that of the vapor feed to the bottom stage. An estimate of the vapor-rate profile is readily obtained by assuming constant molal overflow working up from the bottom in using the specified vapor feed or feeds. Compute corresponding initial values of L_j from Eq. (13-74).

3. Same as step 6 of the BP method.

4. Same as step 7 of the BP method.

5. Same as step 8 of the BP method.

6. Compute a new set of values of L_j from the sum-rates equation:

$$L_j^{(k+1)} = L_j^{(k)} \sum_i x_{i,j} \qquad (13\text{-}84)$$

7. Compute a corresponding new set of V_j tear variables from the following total material balance, which is obtained by combining Eq. (13-74) with an overall material balance around the column:

$$V_j = L_{j-1} - L_N + \sum_{m=j}^{N} (F_m - W_m - U_m) \qquad (13\text{-}85)$$

8. Same as step 9 of the BP method.

9. Same as step 11 of the BP method.

10. Normalize values of $y_{i,j}$.

11. Compute a new set of values of the T_j tear variables by solving simultaneously the set of N energy-balance equations (13-72), which are nonlinear in the temperatures that determine the enthalpy values. When linearized by a Newton iterative procedure, a tridiagonal-matrix equation that is solved by the Thomas algorithm is obtained. If we set g_j equal to Eq. (13-72), i.e., its residual, the linearized equations to be solved simultaneously are

$$\left(\frac{\partial g_1}{\partial T_1}\right)^{(r)} \Delta T_1 + \left(\frac{\partial g_1}{\partial T_2}\right)^{(r)} \Delta T_2 = -g_1^{(r)} \qquad (13\text{-}86)$$

$$\left(\frac{\partial g_j}{\partial T_{j-1}}\right)^{(r)} \Delta T_{j-1} + \left(\frac{\partial g_j}{\partial T_j}\right)^{(r)} \Delta T_j + \left(\frac{\partial g_j}{\partial T_{j+1}}\right)^{(r)} \Delta T_{j+1} = -g_j^{(r)}$$

$$2 \leq j \leq N-1 \qquad (13\text{-}87)$$

$$\left(\frac{\partial g_N}{\partial T_{N-1}}\right)^{(r)} \Delta T_{N-1} + \left(\frac{\partial g_j}{\partial T_N}\right)^{(r)} \Delta T_N = -g_N^{(r)} \qquad (13\text{-}88)$$

where $\Delta T_j = T_j^{(r+1)} - T_j^{(r)}$, and thus $T_j^{(r+1)} = T_j^{(r)} + \Delta T_j$, and r is the iteration index. The partial derivatives depend upon the enthalpy correlations utilized and may be obtained analytically or numerically. Simultaneous Eqs. (13-86) to (13-88) are solved iteratively until corrections ΔT_j and, therefore, residual values of g_j approach zero.

12. Same as step 13 of the BP method.

13. If the convergence criterion is not met, set $k = k + 1$ and repeat steps 4 to 13.

With the SR method, convergence is often rapid even when successive substitution of T_j and V_j is used from one iteration to the next.

Example 5: Calculation of the SR Method Use the SR method with the PR equation of state for K values and enthalpy departures. The oil was taken as n-dodecane. To compute stage temperatures and interstage vapor and liquid flow rates and compositions for absorber-column specifications shown in Fig. 13-52. Note that a secondary absorber oil is used in addition to the main absorber oil and that heat is withdrawn from the seventh theoretical stage.

Initial estimates provided for the tear variables were as follows compared with final converged values obtained after five iterations:

Stage	$T^{(0)}$, °F	$T^{(5)}$, °F	$V^{(0)}$, (lb·mol)/h	$V^{(5)}$, (lb·mol)/h
1	80.00	84.6	450	323.9
2	81.43	85.5	450	367.7
3	82.86	86.3	450	371.3
4	84.29	85.3	450	374.7
5	85.71	86.0	450	388.6
6	87.14	86.6	450	393.0
7	88.57	85.2	450	398.2
8	90.00	92.7	450	410.4

NOTE: To convert degrees Fahrenheit to degrees Celsius, °C = (°F − 32)/1.8. To convert pound-moles per hour to kilogram-moles per second, multiply by 1.26×10^{-4}.

Convergence was achieved rapidly in five iterations by using Eq. (13-88) as the criterion. Computed compositions for lean gas and rich oil are:

Component	Flow rate, (lb·mol)/h	
	Lean gas	Rich oil
C_1	312.7	60.3
C_2	11.0	32.0
C_3	0.2	28.8
nC_4	0.0	19.0
nC_5	0.0	15.0
Oil	0.0	385.0
	323.9	540.1

NOTE: To convert pound-moles per hour to kilogram-moles per second, multiply by 1.26×10^{-4}.

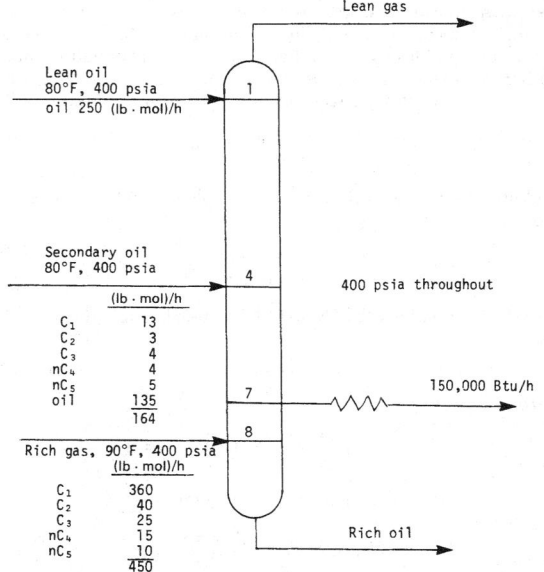

FIG. 13-52 Specifications for the calculation of an absorber by the SR method.

Approximately 0.016 (kg·mol)/s [126 (lb·mol)/h] of vapor is absorbed with an energy liberation of about 198,000 W (670,000 Btu/h), 20 percent of which is removed by the intercooler on stage 7. The temperature profile departs from a smooth curve at stages 4 and 7, where secondary oil enters and heat is removed respectively.

SIMULTANEOUS-CORRECTION PROCEDURES

The BP and SR tearing methods are generally successful only when applied respectively to the distillation of mixtures having a narrow boiling range and to absorbers and strippers. Furthermore, as shown earlier, specifications for these two tearing methods are very restricted. If one wishes to treat distillation of wide-boiling mixtures and other operations shown in Fig. 13-7 such as rectification, reboiled stripping, reboiled absorption, and refluxed stripping, it is usually necessary to utilize other procedures. One class of such procedures involves the solution of most or all of the MESH equations or their equivalent simultaneously by some iterative technique such as a Newton or a quasi-Newton method. Such simultaneous-correction (SC) methods are also useful for separations involving very nonideal liquid mixtures including extractive and azeotropic distillation or for cases in which considerable flexibility in specifications is desired.

The development of an SC procedure involves a number of important decisions: (1) What variables should be used? (2) What equations should be used? (3) How should variables be ordered? (4) How should equations be ordered? (5) How should flexibility in specifications be provided? (6) Which derivatives of physical properties should be retained? (7) How should equations be linearized? (8) If Newton or quasi-Newton linearization techniques are employed, how should the Jacobian be updated? (9) Should corrections to unknowns that are computed at each iteration be modified to dampen or accelerate the solution or be kept within certain bounds? (10) What convergence criterion should be applied?

Perhaps because of these many decisions, a large number of SC procedures have been published. Two quite different procedures that have achieved a significant degree of utilization in solving practical problems include the methods of Naphtali and Sandholm [*Am. Inst. Chem. Eng. J.*, **17**, 148 (1971)] and Goldstein and Stanfield [*Ind. Eng. Chem. Process Des. Dev.*, **9**, 78 (1970)]. The former procedure is of particular interest because, in principle, it can be applied to all cases. However, for situations involving large numbers of components, relatively small numbers of stages, and liquid solutions that are not too highly nonideal, the latter procedure is more efficient computationally.

Naphtali-Sandholm SC Method This method employs the equilibrium-stage model of Figs. 13-48 and 13-49 but reduces the number of variables by $2N$ so that only $N(2C + 1)$ equations in a like number of unknowns must be solved. In place of V_j, L_j, $x_{i,j}$, and $y_{i,j}$, component flow rates are used according to their definitions:

$$v_{i,j} = y_{i,j}V_j \qquad (13\text{-}89)$$

$$\ell_{i,j} = x_{i,j}L_j \qquad (13\text{-}90)$$

In addition, sidestream flow rates are replaced with sidestream flow ratios by

$$s_j = U_j/L_j \qquad (13\text{-}91)$$

$$S_j = W_j/V_j \qquad (13\text{-}92)$$

The MESH equations (13-68) to (13-72) then become the *MEH* functions:

$$M_{i,j} = \ell_{i,j}(1 + s_j) + v_{i,j}(1 + S_j) - \ell_{i,j-1} - v_{i,j+1} - f_{i,j} = 0 \qquad (13\text{-}93)$$

where
$$f_{i,j} = F_j z_{i,j}$$

$$E_{i,j} = K_{i,j}\ell_{i,j}\left(\sum_a v_{a,j} \Big/ \sum_a \ell_{a,j}\right) - v_{i,j} = 0 \qquad (13\text{-}94)$$

$$H_j = H_{Lj}(1 + s_j)\sum_i \ell_{i,j} + H_{Vj}(1 + S_j)\sum_i v_{i,j}$$

$$- H_{Lj-1}\sum_i \ell_{i,j} - H_{Vj+1}\sum_i v_{i,j+1}$$

$$- H_{Fj}\sum_i f_{i,j} - Q_j$$

$$= 0 \qquad (13\text{-}95)$$

where physical properties are not simplified:

$$K_{i,j} = K_{i,j}\{T_j, P_j, \ell_j, v_j\}$$

$$H_{Vj} = H_{Vj}\{T_j, P_j, v_j\}$$

$$H_{Lj} = H_{Lj}\{T_j, P_j, \ell_j\}$$

Let the order of corrections to the unknowns be according to stage number, which in terms of the corresponding unknowns is

$$\overline{X} = [\overline{X}_1, \overline{X}_2, \ldots \overline{X}_j, \ldots \overline{X}_N]^T \qquad (13\text{-}96)$$

where
$$\overline{X}_j = [v_{1,j}, v_{2,j}, \ldots v_{C,j}, T_j, \ell_{1,j}, \ell_{2,j}, \ldots \ell_{C,j}]^T \qquad (13\text{-}97)$$

Let the order of the linearized *MEH* functions also be according to stage number, which in terms of the corresponding nonlinear functions is

$$\overline{F} = [\overline{F}_1, \overline{F}_2, \ldots \overline{F}_j, \ldots \overline{F}_N]^T \qquad (13\text{-}98)$$

where
$$\overline{F}_j = [H_j, M_{1,j}, M_{2,j}, \ldots M_{C,j}, E_{1,j}, \ldots E_{C,j}]^T \qquad (13\text{-}99)$$

Corrections to unknowns for the kth iteration are obtained from

$$\Delta \overline{X}^{(k)} = -\left[\left(\frac{\partial \overline{F}}{\partial X}\right)^{-1}\right]^{(k)} \overline{F}^{(k)} \qquad (13\text{-}100)$$

The next approximations to the unknowns are obtained from

$$\overline{X}^{(k+1)} = \overline{X}^{(k)} + t\,\Delta\overline{X}^{(k)} \qquad (13\text{-}101)$$

where t is a damping ($0 < t < 1$) or acceleration ($t > 1$) factor. By ordering the corrections to the unknowns and the linearized functions in this manner, the resulting Jacobian of partial derivatives of all functions with respect to all unknowns is of a very convenient sparse matrix form of block tridiagonal structure.

$$\left(\frac{d\overline{F}}{dX}\right) = \begin{bmatrix} \overline{\overline{B}}_1 & \overline{\overline{C}}_1 & 0 & 0 & \cdot\cdot & & 0 \\ \overline{\overline{A}}_2 & \overline{\overline{B}}_2 & \overline{\overline{C}}_2 & 0 & \cdot\cdot & & 0 \\ 0° & \overline{\overline{A}}_3 & \overline{\overline{B}}_3\dagger & \overline{\overline{C}}_3\ddagger & & & 0 \\ \cdot\cdot & & & & & & \cdot\cdot \\ \cdot\cdot & & & & & & \cdot\cdot \\ 0 & \cdot\cdot & & & & & 0 \\ 0 & \cdot\cdot & & & 0 & \overline{\overline{A}}_{N-1}° & \overline{\overline{B}}_{N-1}\dagger & \overline{\overline{C}}_{N-1}\ddagger \\ 0 & \cdot\cdot & & & & 0 & 0 & \overline{\overline{A}}_N & \overline{\overline{B}}_N \end{bmatrix}$$

$$(13\text{-}102)$$

Blocks $\overline{\overline{A}}_j$, $\overline{\overline{B}}_j$, and $\overline{\overline{C}}_j$ are $(2C + 1)$ by $(2C + 1)$ submatrices of partial derivatives of the functions on stage j with respect to unknowns on stage $j - 1$, j, and $j + 1$ respectively. The solution to Eq. (13-100) is readily obtained by a matrix-algebra equivalent of the Thomas algorithm for a tridiagonal-matrix equation. Computer storage requirements are minimized by making the following replacements. Starting at top stage 1, using forward-block elimination,

$$\overline{\overline{C}}_1 \rightarrow (\overline{\overline{B}}_1)^{-1}\overline{\overline{C}}_1, \quad \overline{F}_1 \rightarrow (\overline{\overline{B}}_1)^{-1}\overline{F}_1$$

$$\text{and} \qquad \overline{\overline{B}}_1 \rightarrow I \text{ (the identity submatrix)}$$

For stages j from 2 to $(N - 1)$,

$$\overline{\overline{C}}_j \rightarrow (\overline{\overline{B}}_j - \overline{\overline{A}}_j\overline{\overline{C}}_{j-1})^{-1}\overline{\overline{C}}_j,$$

$$\overline{F}_j \rightarrow (\overline{\overline{B}}_j - \overline{\overline{A}}_j\overline{\overline{C}}_{j-1})^{-1}(\overline{F}_j - \overline{\overline{A}}_j\overline{F}_{j-1}), \qquad \overline{\overline{A}}_j \rightarrow 0, \qquad \overline{\overline{B}}_j \rightarrow I$$

For final stage N,

$$\overline{F}_N \rightarrow (\overline{\overline{B}}_N - \overline{\overline{A}}_N\overline{\overline{C}}_{N-1})^{-1}(\overline{F}_N - \overline{\overline{A}}_N\overline{F}_{N-1}), \qquad \overline{\overline{A}}_N \rightarrow 0, \qquad \overline{\overline{B}}_N \rightarrow I$$

This completes the forward steps to give $\Delta\overline{X}_N = -\overline{F}_N$. Remaining values of corrections $\Delta\overline{X}_j$ are obtained by successive backward substitution from $\Delta\overline{X}_j = -\overline{F}_j \rightarrow -(\overline{F}_j - \overline{\overline{C}}_j\overline{F}_{j+1})$. Matrix inversions are best done by *LU* decomposition. Efficiency is best for a small number of components C.

The Newton iteration is initiated by providing reasonable guesses for all unknowns. However, these can be generated from guesses of just T, T_N, and one interstage value of F_j or L_j. Remaining values of T_j are obtained by linear interpolation. By assuming constant molal overflow, calculations are readily made of remaining values of V_j and L_j, from which initial values of $v_{i,j}$ and $\ell_{i,j}$ are obtained from Eqs. (13-89) and (13-90) after obtaining approximations of $x_{i,j}$ and $y_{i,j}$ from steps 4, 5, 8, 9, and 10 of the SR method. Alternatively, a much cruder but often sufficient estimate of $x_{i,j}$ and $y_{i,j}$ is obtained by flashing the combined column feeds at average column pressure and a vapor-to-liquid ratio that approximates the ratio of overhead plus vapor-sidestream flows to bottoms plus liquid-sidestream flows. Resulting compositions are used as the initial estimate for every stage.

At the conclusion of each iteration, convergence is checked by employing an approximate criterion such as

$$\tau = \sum_j \left\{ \left(\frac{H_j}{\xi} \right)^2 + \sum_i [(M_{i,j})^2 + (E_{i,j})^2] \right\} \leq \epsilon \qquad (13\text{-}103)$$

where ξ is a scale factor that is of the order of the average molal heat of vaporization. If we take

$$\epsilon = N(2C+1) \left(\sum_j F_j^2 \right) 10^{-10} \qquad (13\text{-}104)$$

converged values of the unknowns will generally be accurate, on the average, to from four or more significant digits.

During early iterations, particularly when initial estimates of the unknowns are poor, τ and corrections to the unknowns will be very large. It is then preferred to utilize a small value of t in Eq. (13-101) so as to dampen changes to unknowns and prevent wild oscillations. However, the use of values of t much less than 0.25 may slow or prevent convergence.

It is also best to reset to zero or small values any negative values of component flow rates before initiating the next iteration. When the neighborhood of the solution is reached, τ will often decrease by one or more orders of magnitude at each iteration, and it is best to set $t = 1$. Because the Newton method is quadratically convergent in the neighborhood of the solution, usually only three or four additional iterations will be required to reach the convergence criterion. Prior to that, it is not uncommon for τ to increase somewhat from one iteration to the next. If the Jacobian tends toward a singular condition, it may be necessary to restart the procedure with different initial guesses or adjust the Jacobian in some manner.

Standard specifications for the Naphtali-Sandholm method are Q_j (including zero values) at each stage at which heat transfer occurs and sidestream flow ratio s_j or S_j (including zero values) at each stage at which a sidestream is withdrawn. However, the desirable block tridiagonal structure of the jacobian matrix can still be preserved when substitute specifications are made if they are associated with the same stage or an adjacent stage. For example, suppose that for a reboiled absorber, as in Fig. 13-7f, it is desired to specify a boil-up ratio rather than reboiler duty. Equation (13-95) for function H_N is removed from the $N(2C+1)$ set of equations and is replaced by the equation

$$\tilde{H}_N = \sum_i v_{i,N} - (V_N/L_N) \sum_i \ell_{i,N} = 0 \qquad (13\text{-}105)$$

where the value of (V_N/L_N) is specified. Following convergence of the calculations, Q_N is computed from the removed equation.

All of the major computer-aided design and simulation programs have a simultaneous-correction algorithm. A Naphtali-Sandholm type of program, particularly suited for applications to distillation, extractive distillation, and azeotropic distillation, has been published by Fredenslund, Gmehling, and Rasmussen (*Vapor-Liquid Equilibria Using UNIFAC, a Group Contribution Method*, Elsevier, Amsterdam, 1977). Christiansen, Michelsen, and Fredenslund [*Comput. Chem. Eng.*, **3**, 535 (1979)] apply a modified Naphtali-Sandholm type of method to the distillation of natural-gas liquids, even near the critical region, using thermodynamic properties computed from the Soave-Redlich-Kwong equation of state. Block and Hegner [*Am. Inst. Chem. Eng. J.*, **22**, 582 (1976)] extended the Naphtali-Sandholm method to staged separators involving two liquid phases (liquid-liquid extraction) and three coexisting phases (three-phase distillation).

Example 6: Calculation of Naphtali-Sandholm SC Method
Use the Naphtali-Sandholm SC method to compute stage temperatures and interstage vapor and liquid flow rates and compositions for the reboiled-stripper specifications shown in Fig. 13-53. The specified bottoms rate is equivalent to removing most of the nC_5 and nC_6 and some of the nC_7 in the bottoms.

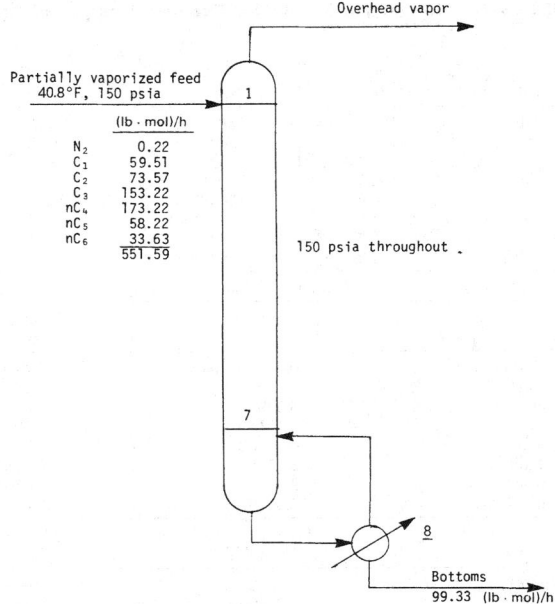

FIG. 13-53 Specifications for the calculation of a reboiled stripper by the Naphtali-Sandholm method.

Calculations were made with the Grayson-Streed modification of the Chao-Seader method for K values and the Lee-Kesler method for enthalpy departures. Initial estimates for stage temperatures and flow rates were as follows, where numbers in parentheses are consistent with specifications:

Stage	T, °F	(lb·mol)/h V	L
1	130	(452.26)	550
8	250		

NOTE: To convert degrees Fahrenheit to degrees Celsius, °C = (°F − 32)/1.8. To convert pound-moles per hour to kilogram-moles per second, multiply by 1.26×10^{-4}.

For specified feed temperature and pressure, an isothermal flash of the feed gave 13.35 percent vaporization.

Convergence was achieved in 3 iterations. Converged values of temperatures, total flows, and component flow rates are tabulated in Table 13-14. Computed reboiler duty is 1,295,000 W (4,421,000 Btu/h). Computed temperature, total vapor flow, and component flow profiles, shown in Fig. 13-54, are not of the shapes that might be expected. Vapor and liquid flow rates for nC_4 change dramatically from stage to stage.

INSIDE-OUT METHODS

The BP, SR, and SC methods described above expend a large percentage of their computational effort during each iteration in the calculation of K values, enthalpies, and derivatives thereof. An algorithm designed to significantly reduce that effort was developed by Boston and Sullivan [*Can. J. Chem. Engr.*, **52**, 52 (1974)]. The MESH equations are solved in an inner loop using simple, approximate equations for K values and enthalpies. The empirical constants in these equations are determined and infrequently updated from the more rigorous, but complex, K value and enthalpy correlations in an outer loop, using calculated compositions and temperatures from the inner loop. Thus, the method is referred to as the inside-out method. The iteration variables for the outer loop are the constants in the approximate thermodynamic-property equations in the inner loop. The iteration variables for the inner loop are related to stage j

TABLE 13-14 Converged Results for Reboiled Stripper of Fig. 13-53

Stg	Temp °F	Pres psia	Net flows Liquid lb·mol/h	Net flows Vapor lb·mol/h	Feeds lb·mol/h	Product lb·mol/h	Duties MMBtu/h
1	130.5	150.00	566.13		551.59	452.26	
2	170.0	150.00	622.63	466.80			
3	184.4	150.00	638.20	523.30			
4	192.7	150.00	637.73	538.87			
5	200.6	150.00	626.86	538.40			
6	211.8	150.00	608.02	527.53			
7	228.5	150.00	586.11	508.69			
8	251.3	150.00		486.78		99.33	4.421

Tray compositions

Stage #1	130.49°F Vap lb·mol/h	150.00 psia Liq lb·mol/h
Nitrogen	0.22000	0.03033
Methane	59.51000	4.72506
Ethane	73.56996	20.85994
Propane	153.18130	115.69593
n-Butane	150.43202	318.78543
n-Pentane	12.75881	70.16827
n-Hexane	2.58982	35.86041
Total lb·mol/h	452.2619	566.1254

Stage #2	170.01°F Vap lb·mol/h	150.00 psia Liq lb·mol/h
Nitrogen	0.03033	0.00517
Methane	4.72506	0.36256
Ethane	20.85991	5.14611
Propane	115.65723	66.94922
n-Butane	295.99744	428.55569
n-Pentane	24.70708	83.84168
n-Hexane	4.82023	37.77268
Total lb·mol/h	466.7973	622.6332

Stage #3	184.42°F Vap lb·mol/h	150.00 psia Liq lb·mol/h
Nitrogen	0.00517	0.00085
Methane	0.36256	0.02460
Ethane	5.14608	1.08705
Propane	66.91052	31.95980
n-Butane	405.76770	466.08041
n-Pentane	38.38049	99.77217
n-Hexane	6.73250	39.27149
Total lb·mol/h	523.3050	638.1964

Stage #4	192.68°F Vap lb·mol/h	150.00 psia Liq lb·mol/h
Nitrogen	0.00085	0.00014
Methane	0.02460	0.00159
Ethane	1.08701	0.21493
Propane	31.92111	14.00705
n-Butane	443.29239	457.42691
n-Pentane	54.31099	124.43126
n-Hexane	8.23131	41.64842
Total lb·mol/h	538.8683	637.7303

Tray compositions

Stage #5	200.64°F Vap lb·mol/h	150.00 psia Liq lb·mol/h
Nitrogen	0.00014	0.00002
Methane	0.00159	0.00010
Ethane	0.21490	0.04043
Propane	13.96836	5.73731
n-Butane	434.63895	410.72699
n-Pentane	78.97008	162.79074
n-Hexane	10.60824	47.56607
Total lb·mol/h	538.4022	626.8616

Stage #6	211.75°F Vap lb·mol/h	150.00 psia Liq lb·mol/h
Nitrogen	0.00002	0.00000
Methane	0.00010	0.00001
Ethane	0.04039	0.00719
Propane	5.69862	2.17265
n-Butane	387.93896	329.74094
n-Pentane	117.32958	212.37617
n-Hexane	16.52589	63.72279
Total lb·mol/h	527.5336	608.0197

Stage #7	228.46°F Vap lb·mol/h	150.00 psia Liq lb·mol/h
Nitrogen	0.00000	0.00000
Methane	0.00001	0.00000
Ethane	0.00716	0.00119
Propane	2.13396	0.74298
n-Butane	306.95294	227.83710
n-Pentane	166.91501	254.54031
n-Hexane	32.68262	102.98384
Total lb·mol/h	508.6917	586.1054

Stage #8	251.31°F Vap lb·mol/h	150.00 psia Liq lb·mol/h
Nitrogen	0.00000	0.00000
Methane	0.00000	0.00000
Ethane	0.00116	0.00003
Propane	0.70429	0.03869
n-Butane	205.04907	22.78802
n-Pentane	209.07915	45.46117
n-Hexane	71.94367	31.04017
Total lb·mol/h	486.7773	99.3281

stripping factors, $K_{ij}V_j/L_j$, for components i, which make use of volatility and energy parameters. Otherwise, the inner-loop calculations utilize computational features of the BP, SR, and SC methods, to compute stage temperatures, compositions, and flow rates. The inside-out method takes advantage of the following observations: (1) relative volatilities vary from iteration to iteration much less than the K values, (2) enthalpy of vaporization varies from iteration to iteration much less than phase enthalpies, and (3) component stripping factors combine effects of temperature and liquid and vapor flows at each stage.

As an example of how the approximate thermodynamic-property equations are handled in the inner loop, consider the calculation of K values. The approximate models for nearly ideal liquid solutions are the following empirical Clausius-Clapeyron form of the K value in terms of a base or reference component, b, and the definition of the relative volatility, α.

$$K_{b,j} = \exp(A_j - B_j/T_j) \qquad (13\text{-}106)$$

$$K_{i,j} = \alpha_{i,j}K_{b,j} \qquad (13\text{-}107)$$

Values of A and B for the base component are back-calculated for each stage in the outer loop from a suitable K-value correlation (e.g. the SRK equation, which is also used to compute the K values of the other components on each of the other stages so that values of α_{ij} can be computed). The values of A, B, and α are passed from the outer loop to the inner loop, where they are used to formulate the phase equilibria equation:

$$v_{i,j} = \alpha_{i,j}S_{b,j}l_{i,j} \qquad (13\text{-}108)$$

where

$$S_{b,j} = K_{ij}V_j/L_j \qquad (13\text{-}109)$$

The initial version of the inside-out method was developed for rapid calculations of simple and complex distillation, absorption, and

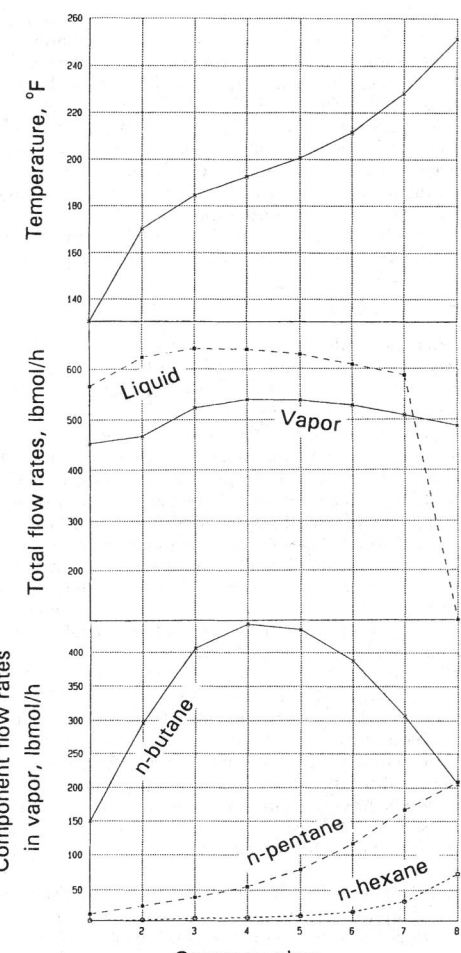

FIG. 13-54 Converged profiles for the reboiled stripper of Fig. 13-53.

stripping operations for hydrocarbon mixtures. However extensions and improvements in the method by Boston and coworkers [*Comput. Chem. Engng.*, **2**, 109 (1978), *ACS Symp. Ser.* No. 124, 135 (1980), *Comput. Chem. Engng.*, **8**, 105 (1984), and *Chem. Eng. Prog.*, **86** (8), 45–54 (1990)] and by Russell [*Chem. Eng.*, **90**, (20), 53 (1983)] and Jelinek [*Comput. Chem. Engng.*, **12**, 195 (1988)] now make it possible to apply the method to reboiled absorption, reboiled stripping, extractive and azeotropic distillation, three-phase systems, reactive distillation, highly nonideal systems, and interlinked distillation systems with pumparounds, bypasses, and external heat exchangers. Inside-out methods are incorporated into most of the computer-aided process design and simulation programs and are now the methods of choice for design and simulation, as stated by Haas, who in Chap. 4 of Kister (op. cit.) presents details of two of the several inside-out algorithms.

Example 7: Calculation of Inside-Out Method For the conditions of the simple distillation column shown in Fig. 13-55, obtain a converged solution by the inside-out method, using the SRK equation-of-state for thermodynamic properties (in the outer loop).

A computer solution was obtained as follows. The only initial assumptions are a condenser outlet temperature of 65°F and a bottoms-product temperature of 165°F. The bubble-point temperature of the feed is computed as 123.5°F. In the initialization procedure, the constants A and B in (13-106) for inner-loop calculations, with T in °R, are determined from the SRK equation, with the following results:

Stage	T, °F	A	B	K_b
1	65	6.870	3708	0.8219
2	95	6.962	4031	0.7374
3	118	7.080	4356	0.6341
4	142	7.039	4466	0.6785
5	165	6.998	4576	0.7205

Values of enthalpy constants for approximate equations are not tabulated here but are also computed for each stage based on the initial temperature distribution.

In the inner-loop calculation sequence, component flow rates are computed from the MESH equations by the tridiagonal matrix method. The resulting bottoms-product flow rate deviates somewhat from the specified value of 50 lb·mol/h. However, by modifying the component stripping factors with a base stripping factor, S_b, in (13-109) of 1.1863, the error in the bottoms flow rate is reduced to 0.73 percent.

The initial inside-loop error from the solution of the normalized energy-balance equations, is found to be only 0.04624. This is reduced to 0.000401 after two iterations through the inner loop.

At this point in the inside-out method, the revised column profiles of temperature and phase compositions are used in the outer loop with the complex SRK thermodynamic models to compute updates of the approximate K and H constants. Then only one inner-loop iteration is required to obtain satisfactory convergence of the energy equations. The K and H constants are again updated in the outer loop. After one inner-loop iteration, the approximate K and H constants are found to be sufficiently close to the SRK values that overall convergence is achieved. Thus, a total of only 3 outer-loop iterations and 4 inner-loop iterations are required.

To illustrate the efficiency of the inside-out method to converge this example, the results from each of the three outer-loop iterations are summarized in the following tables:

Outer-loop iteration	Stage temperatures, °F				
	T_1	T_2	T_3	T_4	T_5
Initial guess	65	—	—	—	165
1	82.36	118.14	146.79	172.66	193.20
2	83.58	119.50	147.98	172.57	192.53
3	83.67	119.54	147.95	172.43	192.43

Outer-loop iteration	Total liquid flows, lb·moles/hr				
	L_1	L_2	L_3	L_4	L_5
Specification	100	—	—	—	—
1	100.00	89.68	187.22	189.39	50.00
2	100.03	89.83	188.84	190.59	49.99
3	100.0	89.87	188.96	190.56	50.00

Outer-loop iteration	Component flows in bottoms product, lb·moles/hr			
	C_3	nC_4	nC_5	L_5
1	0.687	12.045	37.268	50.000
2	0.947	12.341	36.697	49.985
3	0.955	12.363	36.683	50.001

From these tables, it is seen that the stage temperatures and total liquid flows are already close to the converged solution after only one outer-loop iteration. However, the composition of the bottoms product, specifically with respect to the lightest component, C_3, is not close to the converged solution until after two iterations. The inside-out method does not always converge so dramatically, but is usually quite efficient.

HOMOTOPY-CONTINUATION METHODS

Although the SC and inside-out methods are reasonably robust, they are not guaranteed to converge and sometimes fail, particularly for very nonideal liquid solutions and when initial guesses are poor. A much more robust, but more time-consuming, method is differential arclength homotopy continuation, the basic principles and applications of which are discussed by Wayburn and Seader [*Comp. Chem. Engng.*, **11**, 7–25 (1987); *Proceedings Second Intern. Conf. Foundations of Computer-Aided Process Design*, CACHE, Austin, TX, 765–

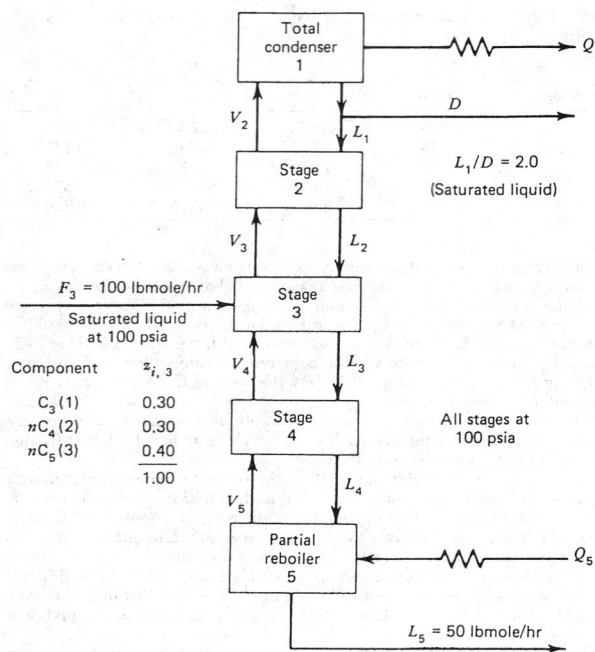

FIG. 13-55 Specifications for distillation column of Example 7.

862 (1984); *AIChE monograph Series*, AIChE, New York, **81,** No. 15 (1985)]. Homotopy methods begin from a known solution of a companion set of equations and follow a path to the desired solution of the set of equations to be solved. In most cases, the path exists and can be followed. In one implementation, the set of equations to be solved, call it $f(x)$, and the companion set of equations, call it $g(x)$, are connected together by a set of mathematical homotopy equations:

$$h(x,t) = t\,f(x) + (1-t)g(x) = 0 \qquad (13\text{-}110)$$

where t is a homotopy parameter. An appropriate function is selected for $g(x)$ such as $f(x) - f(x_0)$, where x_0 are the initial guesses, which can be selected arbitrarily. At the beginning of the path, $t = 0$ and Eq. (13-110) becomes $h(x,t) = f(x) - f(x_0) = 0$ or $f(x) = f(x_0)$. The homotopy parameter is then gradually moved from 0 to 1. At a value of $t = 1$, Eq. (13-110) becomes $h(x,t) = f(x) = 0$, which corresponds to the desired solution.

The movement along the path is accomplished by a predictor-corrector continuation procedure, where the corrector is often a numerical Euler integration step of the differential-arclength form of Eq. (13-110) along the arclength of the path (rather than a step in t), as proposed by Klopfenstein [*J. Assoc. Comput. Mach.*, **8,** 366 (1961)]. The arclength of the path is preferred, over the homotopy parameter, as a continuation parameter because the path may make one or more turns in the homotopy parameter, making it difficult to take an integration step. The predictor step is accompanied by a truncation error that is reduced in the corrector step, which employs Newton's method with Eq. (13-110) to return to the path. If the predictor steps along the path are not too large, the corrector steps always converge.

Another implementation of homotopy-continuation methods is the use of problem-dependent homotopies that exploit some physical aspect of the problem. Vickery and Taylor [*AIChE J.*, **32,** 547 (1986)] utilized thermodynamic homotopies for K values and enthalpies to gradually move these properties from ideal to actual values so as to solve the MESH equations when very nonideal liquid solutions were involved. Taylor, Wayburn, and Vickery [*I. Chem. E. Symp. Ser.* No. 104, B305 (1987)] used a pseudo-Murphree efficiency homotopy to move the solution of the MESH equations from a low efficiency, where little separation occurs, to a higher and more reasonable efficiency.

Continuation methods, also called *imbedding* and *path-following* methods, were first applied to the solution of separation models involving large numbers of nonlinear equations by Salgovic, Hlavacek, and Ilavsky [*Chem. Eng. Sci.*, **36,** 1599 (1981)] and by Byrne and Baird [*Comp. Chem. Engng.*, **9,** 593 (1985)]. Since then, they have been applied successfully to problems involving interlinked distillation (Wayburn and Seader, op. cit.), azeotropic and three-phase distillation [Kovach, III and Seider, *Comp. Chem. Engng.*, **11,** 593 (1987)], and reactive distillation [Chang and Seader, *Comp. Chem. Engng.*, **12,** 1243 (1988)], when SC and inside-out methods have failed. Today, many computer-aided distillation-design and simulation packages include continuation techniques to make the codes more robust.

STAGE EFFICIENCY

The mathematical models presented earlier for rigorous calculations of multistage, multicomponent distillation-type separations assume that equilibrium with respect to both heat and mass transfer is attained at each stage. Unless temperature changes significantly from stage to stage, the assumption that vapor and liquid phases exiting from a stage are at the same temperature is generally valid. However, in most cases, equilibrium with respect to mass transfer is not a valid assumption. If all feed components have the same mass-transfer efficiency, the number of actual stages or trays is simply related to the number of equilibrium stages used in the modeling calculations by an overall stage efficiency. For distillation, as discussed in Sec. 14, this efficiency for well-designed trays typically varies from 40 to 120 percent; the higher value being achieved in some large-diameter towers because of a cross-flow effect. Efficiencies for absorption and extractive distillation can be lower than 40 percent.

When it is desired to compute, with rigorous methods, actual rather than equilibrium stages, Eqs. (13-69) and (13-94) can be modified to include the Murphree vapor-phase efficiency $\eta_{i,j}$, defined by Eq. (13-29). This is particularly desirable for multistage operations involving feeds containing components of a wide range of volatility and/or concentration, in which only a rectification (absorption) or stripping action is provided and all components are not sharply separated. In those cases, the use of a different Murphree efficiency for each component and each tray may be necessary to compute recovery accurately.

Departures from the equilibrium-stage model may also occur when entrainment of liquid droplets in the rising vapor or occlusion of vapor in the liquid flow in the downcomer is significant. The former condition may occur at high vapor loading when flooding is approached. The latter condition is possible at high operating pressures when vapor and liquid densities are not drastically different. Entrainment and occlusion effects are not strictly due to mass-transfer inefficiency and are best taken into account by including entrainment terms in the modeling equations, as shown by Loud and Waggoner [*Ind. Eng. Chem. Process Des. Dev.*, **17,** 149 (1978)].

RATE-BASED MODELS

Although the widely used equilibrium-stage models for distillation, described above, have proved to be quite adequate for binary and close-boiling, ideal and near-ideal multicomponent vapor-liquid mixtures, their deficiencies for general multicomponent mixtures have long been recognized. Even Murphree [*Ind. Eng. Chem.*, **17,** 747–750 and 960–964 (1925)], who formulated the widely used plate efficiencies that carry his name, pointed out clearly their deficiencies for multicomponent mixtures and when efficiencies are small. Later, Walter and Sherwood [*Ind. Eng. Chem.*, **33,** 493 (1941)] showed that experimentally measured efficiencies could cover an enormous range, with some values less than 10 percent, and Krishna et al. [*Trans. Inst. Chem. Engr.*, **55,** 178 (1977)] showed theoretically that the component mass-transfer coupling effects discovered by Toor [*AIChE J.*, **3,** 198 (1957)] could cause the rate of mass transfer for components having small concentration driving forces to be controlled by the other species, with the result that Murphree vapor efficiencies could cover the entire range of values from minus infinity to plus infinity.

The first major step toward the development of a more realistic rate-based (nonequilibrium) model for distillation was taken by Krishnamurthy and Taylor [*AIChE J.*, **31**, 449–465 (1985)]. More recently, Taylor, Kooijman, and Hung [*Comp. Chem. Engng.*, **18**, 205–217 (1994)] extended the initial development so as to add the effects of tray-pressure drop, entrainment, occlusion, and interlinks with other columns. In the augmented MESH equations, which they refer to as the MERSHQ equations, they replace the conventional mass and energy balances around each stage by two balances each, one for the vapor phase and one for the liquid phase. Each of the component-material balances contains a term for the rate of mass transfer between the two phases; the energy balances contain a term for the rate of heat transfer between phases. Thus, each pair of phase balances is coupled by mass or heat-transfer rates, which are estimated from constitutive equations that account, in as rigorous a manner as possible, for bulk transport, species interactions, and coupling effects. The heat and mass-transfer coefficients in these equations are obtained from empirical correlations of experimental data and the Chilton-Colburn analogy. Equilibrium between the two phases is assumed at the phase interface. Thus, the rate-based model deals with both transport and thermodynamics. Although tray efficiencies are not part of the modeling equations, efficiencies can be back-calculated from the results of the simulation. Various options for vapor and liquid flow configurations are employed in the model, including plug flow and perfectly mixed flow on each tray.

A schematic diagram of the nonequilibrium stage for the Taylor et al. model is shown in Fig. 13-56. Entering the stage are the following material streams: F^V = vapor feed; F^L = liquid feed; V_{j+1} = vapor from stage below together with fractional-liquid entrainment, ϕ_{j+1}^L; L_{j-1} = liquid from stage above with fractional-vapor occlusion, ϕ_{j-1}^V; G^V = vapor interlink; and G^L = liquid interlink. Leaving the stage are the following material streams: V_j = vapor with fractional withdrawal as sidestream, r_j^V, and fractional-liquid entrainment, ϕ_j^L; and L_j = liquid with fractional withdrawal as sidestream, r_j^L, and fractional-vapor occlusion, ϕ_j^V. Also leaving the stage are heat-transfer streams, Q_j^V and Q_j^L. The rate of heat transfer from the vapor phase to the liquid phase is E_j and the rate of component mass transfer from the vapor phase to the liquid phase is N_{ij}.

The nonequilibrium-model equations for the stage in Fig. 13-56 are as follows in residual form, where i = component ($i = 1$ to C), j = stage number ($j = 1$ to N), and v = a stage in another column that supplies an interlink.

Material Balances ($2C + 2$ Equations) Component for the vapor phase:

$$M_{ij}^V \equiv (1 + r_j^V + \phi_j^V)V_jy_{ij} - V_{j+1}y_{i,j+1} - \phi_{j-1}^V V_{j-1}y_{i,j-1} - f_{ij}^V - \sum_{v=1}^{n} G_{ijv}^V + N_{ij}$$

$$= 0 \qquad i = 1, 2, \ldots, c \qquad (13\text{-}111)$$

Component for the liquid phase:

$$M_{ij}^L \equiv (1 + r_j^L + \phi_j^L)L_jx_{ij} - L_{j-1}x_{i,j-1} - \phi_{j+1}^L L_{j+1}x_{i,j+1} - f_{ij}^L - \sum_{v=1}^{n} G_{ijv}^L - N_{ij}$$

$$= 0 \qquad i = 1, 2, \ldots, c \qquad (13\text{-}112)$$

Total for the vapor phase:

$$M_{t_j}^V \equiv (1 + r_j^V + \phi_j^V)V_j - V_{j+1} - \phi_{j-1}^L V_{j-1} - F_j^V - \sum_{i=1}^{c}\sum_{v=1}^{n} G_{ijv}^V + N_{tj}$$

$$= 0 \qquad (13\text{-}113)$$

Total for the liquid phase:

$$M_{t_j}^L \equiv (1 + r_j^L + \phi_j^L)L_j - L_{j-1} - \phi_{j+1}^L L_{j+1} - F_j^L - \sum_{i=1}^{c}\sum_{v=1}^{n} G_{ijv}^L - N_{tj}$$

$$= 0 \qquad (13\text{-}114)$$

Energy Balances (3 Equations) For the vapor phase:

$$E_j^V \equiv (1 + r_j^V + \phi_j^V)V_jH_j^V - V_{j+1}H_{j+1}^V - \phi_{j-1}^V V_{j-1}H_{j-1}^V - F_j^VH_j^{VF}$$

$$- \sum_{v=1}^{n} G_{jv}^VH_{jv}^V + Q_j^V + e_j^V = 0 \qquad (13\text{-}115)$$

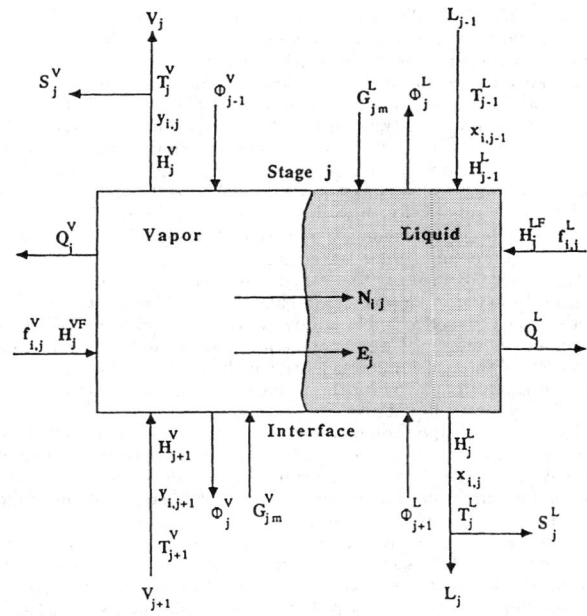

FIG. 13-56 Schematic diagram of a nonequilibrium stage.

For the liquid phase:

$$E_j^L \equiv (1 + r_j^L + \phi_j^L)L_jH_j^L - L_{j-1}H_{j-1}^L - \phi_{j+1}^L L_{j+1}H_{j+1}^L - F_j^LH_j^{LF}$$

$$- \sum_{v=1}^{n} G_{jv}^LH_{jv}^L + Q_j^L - e_j^L = 0 \qquad (13\text{-}116)$$

Continuity across the phase interface:

$$E_j^I \equiv e_j^V - e_j^L = 0 \qquad (13\text{-}117)$$

Mass-Transfer Rates ($2C - 2$ Equations) Component in the vapor phase:

$$R_{ij}^V \equiv N_{ij} - N_{ij}^V = 0 \qquad i = 1, 2, \ldots, c - 1 \qquad (13\text{-}118)$$

Component in the liquid phase:

$$R_{ij}^L \equiv N_{ij} - N_{ij}^L = 0 \qquad i = 1, 2, \ldots, c - 1 \qquad (13\text{-}119)$$

Summation of Mole Fractions (2 Equations) Vapor-phase interface:

$$S_j^{VI} \equiv \sum_{i=1}^{c} y_{ij}^I - 1 = 0 \qquad (13\text{-}120)$$

Liquid-phase interface:

$$S_j^{LI} \equiv \sum_{i=1}^{c} x_{ij}^I - 1 = 0 \qquad (13\text{-}121)$$

Hydraulic Equation for Stage Pressure Drop (1 Equation) Vapor-phase pressure drop:

$$P_j \equiv p_j - p_{j-1} - (\Delta p_{j-1}) = 0 \qquad (13\text{-}122)$$

Interface Equilibrium (C Equations) Component vapor-liquid equilibrium:

$$Q_{ij}^I \equiv K_{ij}x_{ij}^I - y_{ij}^I = 0 \qquad i = 1, 2, \ldots, c \qquad (13\text{-}123)$$

Equations (13-111) to (13-114), (13-118) and (13-119), contain terms, N_{ij}, for rates of mass transfer of components from the vapor phase to the liquid phase (rates are negative if transfer is from the liquid phase to the vapor phase). These rates are estimated from diffusive and bulk-flow contributions, where the former are based on interfacial area, average mole-fraction driving forces, and mass-transfer coefficients that account for coupling effects through binary-pair coefficients. Although the stage shown in Fig. 13-56 appears to apply to a trayed column, the model also applies for a section of a packed column. Accordingly, empirical correlations for the interfacial area and binary-pair mass-transfer coefficients cover bubble-cap trays, sieve trays, valve trays, dumped packings, and structured packings. The average mole-fraction driving forces for diffusion depend upon the assumed vapor and liquid-flow patterns. In the mixed-flow model, both phases are completely mixed. This is the simplest model and is usually suitable for small-diameter trayed columns. In the plug-flow model, both phases move in plug flow. This model is applicable to packed columns and certain trayed columns.

Equations (13-115) to (13-117) contain terms, e_j, for rates of heat transfer from the vapor phase to the liquid phase. These rates are estimated from convective and bulk-flow contributions, where the former are based on interfacial area, average-temperature driving forces, and convective heat-transfer coefficients, which are determined from the Chilton-Colburn analogy for the vapor phase and from the penetration theory for the liquid phase.

The K values (vapor-liquid equilibrium ratios) in Equation (13-123) are estimated from the same equation-of-state or activity-coefficient models that are used with equilibrium-stage models. Tray or packed-section pressure drops are estimated from suitable correlations of the type discussed by Kister (op. cit.).

From the above list of rate-based model equations, it is seen that they total $5C + 6$ for each tray, compared to $2C + 1$ or $2C + 3$ (depending on whether mole fractions or component flow rates are used for composition variables) for each stage in the equilibrium-stage model. Therefore, more computer time is required to solve the rate-based model, which is generally converged by an SC approach of the Newton type.

A potential limitation of the application to design of a rate-based model compared to the equilibrium-stage model is that the latter can be computed independently of the geometry of the column because no transport equations are included in the model. Thus, the sizing of the column is decoupled from the determination of column operating conditions. However, this limitation of the early rate-based models has now been eliminated by incorporating a design mode that simultaneously designs trays and packed sections.

A study of industrial applications by Taylor, Kooijman, and Woodman [IChemE. Symp. Ser. Distillation and Absorption 1992, A415–A427 (1992)] concluded that rate-based models are particularly desirable when simulating or designing: (1) packed columns, (2) systems with strongly nonideal liquid solutions, (3) systems with trace components that need to be tracked closely, (4) columns with rapidly changing profiles, (5) systems where tray-efficiency data are lacking. Besides the extended model just described, a number of other investigators, as summarized by Taylor, Kooijman, and Hung (op. cit.), have developed rate-based models for specific applications and other purposes, including cryogenic distillation, crude distillation, vacuum distillation, catalytic distillation, three-phase distillation, dynamic distillation, and liquid-liquid extraction. Commercial computerized rate-based models are available in two simulation programs: RATEFRAC in ASPEN PLUS from Aspen Technology, Inc., Cambridge, Massachusetts and NEQ2 in ChemSep from R. Taylor and H. A. Kooijman of Clarkson University. Rate-based models could usher in a new era in trayed and packed-column design and simulation.

Example 8: Calculation of Rate-Based Distillation The separation of 655 lb·mol/h of a bubble-point mixture of 16 mol % toluene, 9.5 mol % methanol, 53.3 mol % styrene, and 21.2 mol % ethylbenzene is to be carried out in a 9.84-ft diameter sieve-tray column having 40 sieve trays with 2-inch high weirs and on 24-inch tray spacing. The column is equipped with a total condenser and a partial reboiler. The feed will enter the column on the 21st tray from the top, where the column pressure will be 93 kPa, The bottom-tray pressure is 101 kPa and the top-tray pressure is 86 kPa. The distillate rate will be set at 167 lb·mol/h in an attempt to obtain a sharp separation between toluene-methanol, which will tend to accumulate in the distillate, and styrene and ethylbenzene. A reflux ratio of 4.8 will be used. Plug flow of vapor and complete mixing of liquid will be assumed on each tray. K values will be computed from the UNIFAC activity-coefficient method and the Chan-Fair correlation will be used to estimate mass-transfer coefficients. Predict, with a rate-based model, the separation that will be achieved and back-calculate from the computed tray compositions, the component vapor-phase Murphree-tray efficiencies.

The calculations were made with the RATEFRAC program and comparisons were made with the companion RADFRAC program, which utilizes the inside-out method for an equilibrium-based model.

The rate-based model gave a distillate with 0.023 mol % ethylbenzene and 0.0003 mol % styrene, and a bottoms product with essentially no methanol and 0.008 mol % toluene. Murphree tray efficiencies for toluene, styrene, and ethylbenzene varied somewhat from tray to tray, but were confined mainly between 86 and 93 percent. Methanol tray efficiencies varied widely, mainly from 19 to 105 percent, with high values in the rectifying section and low values in the stripping section. Temperature differences between vapor and liquid phases leaving a tray were not larger than 5°F.

Based on an average tray efficiency of 90 percent for the hydrocarbons, the equilibrium-based model calculations were made with 36 equilibrium stages. The results for the distillate and bottoms compositions, which were very close to those computed by the rate-based method, were a distillate with 0.018 mol % ethylbenzene and less than 0.0006 mol % styrene, and a bottoms product with only a trace of methanol and 0.006 mol % toluene.

ENHANCED DISTILLATION

INTRODUCTION

In distillation operations, separation results from differences in vapor- and liquid-phase compositions arising from the partial vaporization of a liquid mixture or the partial condensation of a vapor mixture. The vapor phase becomes enriched in the more volatile components while the liquid phase is depleted of those same components. In many situations, however, the change in composition between the vapor and liquid phases in equilibrium becomes small (so-called "pinched condition"), and a large number of successive partial vaporizations and partial condensations is required to achieve the desired separation. Alternatively, the vapor and liquid phases may have identical compositions, because of the formation of an azeotrope, and no separation by simple distillation is possible.

Several enhanced distillation-based separation techniques have been developed for close-boiling or low-relative-volatility systems, and for systems exhibiting azeotropic behavior. All of these special techniques are ultimately based on the same differences in the vapor and liquid compositions as ordinary distillation, but, in addition, they rely on some additional mechanism to further modify the vapor-liquid behavior of the key components. These enhanced techniques can be classified according to their effect on the relationship between the vapor and liquid compositions:

1. *Azeotropic distillation and pressure-swing distillation.* Methods that cause or exploit azeotrope formation or behavior to alter the boiling characteristics and separability of the mixture.

2. *Extractive distillation and salt distillation.* Methods that primarily modify liquid-phase behavior to alter the relative volatility of the components of the mixture.

3. *Reactive distillation.* Methods that use chemical reaction to modify the composition of the mixture or, alternatively, use existing vapor-liquid differences between reaction products and reactants to enhance the performance of a reaction.

AZEOTROPISM

At low-to-moderate pressure ranges typical of most industrial applications, the fundamental composition relationship between the vapor and liquid phases in equilibrium can be expressed as a function of the total system pressure, the vapor pressure of each pure component, and the liquid-phase activity coefficient of each component i in the mixture:

$$y_i P = x_i \gamma_i P_i^{sat} \qquad (13\text{-}124)$$

In systems that exhibit ideal liquid-phase behavior, the activity coefficients, γ_i, are equal to unity and Eq. (13-124) simplifies to Raoult's law. For nonideal liquid-phase behavior, a system is said to show negative deviations from Raoult's law if $\gamma_i < 1$, and conversely, positive deviations from Raoult's law if $\gamma_i > 1$. In sufficiently nonideal systems, the deviations may be so large the temperature-composition phase diagrams exhibit extrema, as shown in each of the three parts of Fig. 13-57. At such maxima or minima, the equilibrium vapor and liquid compositions are identical. Thus,

$$y_i = x_i \qquad \text{for all } i = 1, \ldots n \qquad (13\text{-}125)$$

and the system is said to form an azeotrope (from the Greek, meaning *to boil unchanged*). Azeotropic systems show a minimum in the *T-x,y* diagram when the deviations from Raoult's law are positive (Fig. 13-57a) and a maximum in the *T-x,y* diagram when the deviations from Raoult's law are negative (Fig. 13-57b). If at these two conditions, a single liquid phase is in equilibrium with the vapor phase, the azeotrope is homogeneous. If multiple liquid-phase behavior is exhibited at the azeotropic condition, the azeotrope is heterogeneous. For heterogeneous azeotropes, the vapor-phase composition is equal to the overall composition of the two (or more) liquid phases (Fig. 13-57c). Mixtures with only small deviations from Raoult's law may form an azeotrope only if the components are close-boiling. As the boiling-point difference between the components increases, the composition of the azeotrope shifts closer to one of the pure components (toward the lower-boiling pure component for minimum-boiling azeotropes, and toward the higher-boiling pure component for maximum-boiling azeotropes). Mixtures of components whose boiling points differ by more than about 30°C generally do not exhibit azeotropes distinguishable from the pure components even if large deviations from Raoult's law are present. As a qualitative guide to liquid-phase activity-coefficient behavior, Robbins [*Chem. Eng. Prog.*, **76** (10) 58 (1980)] developed a matrix of chemical families, shown in Table 13-15, which indicates expected deviations from Raoult's law.

The formation of two liquid phases within some temperature range for close-boiling mixtures is generally an indication that the system

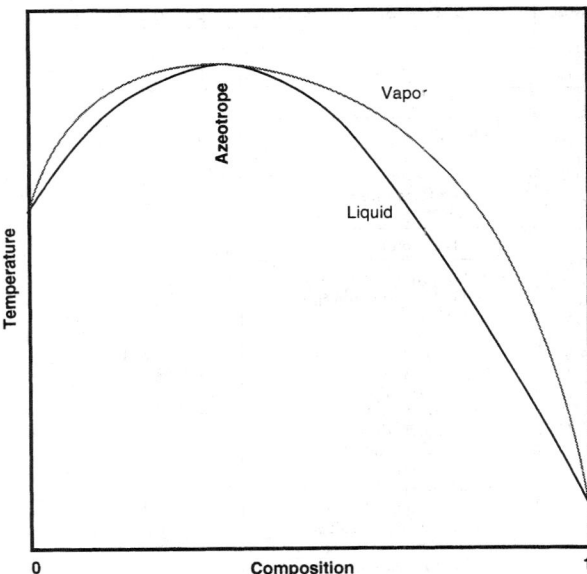

FIG. 13-57 (*Continued*) Schematic isobaric-phase diagrams for binary azeotropic mixtures. (*b*) Homogeneous maximum-boiling azeotrope.

will also exhibit a minimum-boiling azeotrope, since two liquid phases may form when deviations from Raoult's law are extremely positive. The fact that immiscibility does occur, however, does not guarantee that the azeotrope will be heterogeneous. The azeotropic temperature is sometimes outside the range of temperatures at which a system exhibits two liquid phases. Moreover, the azeotropic composition may not necessarily fall within the composition range of the two-liquid-phase region even when within the appropriate temperature range

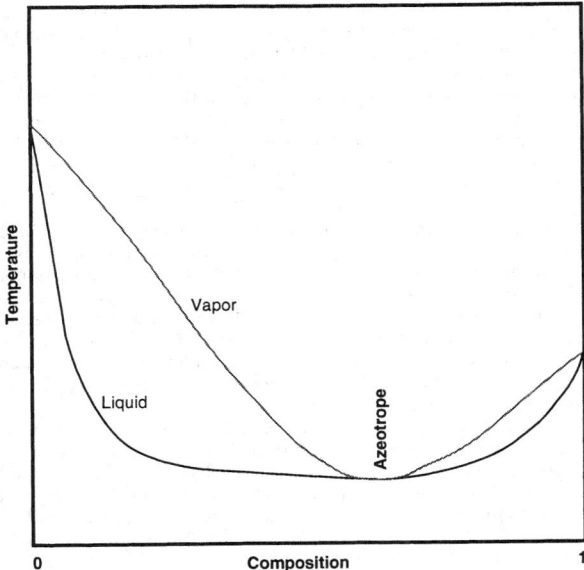

FIG. 13-57 Schematic isobaric-phase diagrams for binary azeotropic mixtures. (*a*) Homogeneous minimum-boiling azeotropes.

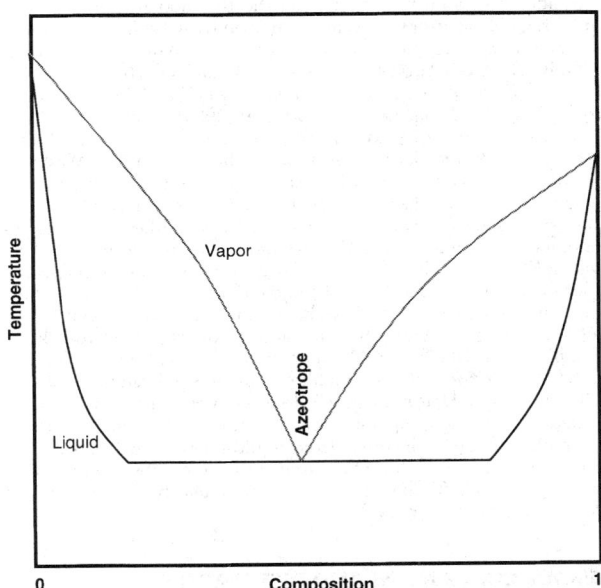

FIG. 13-57 (*Continued*) Schematic isobaric-phase diagrams for binary azeotropic mixtures. (*c*) Heterogeneous azeotrope.

TABLE 13-15 Solute-Solvent Group Interactions

Solute class	Group	Solvent class											
		1	2	3	4	5	6	7	8	9	10	11	12
	H-donor												
1	Phenol	0	0	−	0	−	−	−	−	−	−	−	−
2	Acid, thiol	0	0	−	0	−	−	0	0	0	0	−	−
3	Alcohol, water	−	−	0	+	+	−	0	−	−	−	−	−
4	Active-H on multihalo paraffin	0	0	+	0	−	−	−	−	−	−	0	−
	H-acceptor												
5	Ketone, amide with no H on N, sulfone, phosphine oxide	−	−	+	−	0	+	−	−	−	−	+	−
6	Tertamine	−	−	0	−	+	0	−	−	0	+	0	0
7	Secamine	−	0	−	−	+	+	0	0	0	0	0	−
8	Pri amine, ammonia, amide with 2H on N	−	0	−	−	+	+	0	0	−	+	−	−
9	Ether, oxide, sulfoxide	−	0	+	−	+	0	0	−	0	+	0	−
10	Ester, aldehyde, carbonate, phosphate, nitrate, nitrite, nitrile, intramolecular bonding, e.g., o-nitro phenol	−	0	+	−	+	+	0	0	−	0	−	−
11	Aromatic, olefin, halogen aromatic, multihalo paraffin without active H, monohalo paraffin	+	+	+	0	+	0	0	−	−	0	0	0
	Non-H-bonding												
12	Paraffin, carbon disulfide	+	+	+	+	+	0	+	+	+	+	0	0

SOURCE: Robbins, L. A., *Chem. Eng. Prog.,* **76**(10), 58–61 (1980), by permission.

for liquid-liquid behavior, as is for example the case for the methyl acetate-water and tetrahydrofuran-water systems. Homogeneous azeotropes that are completely miscible at all temperatures usually occur between species with very close boiling points and rather small liquid-phase nonidealities. Moreover, since strong positive deviations from Raoult's law are required for liquid-liquid phase splitting, maximum-boiling azeotropes ($\gamma_i < 1$) are never heterogeneous.

Additional general information on the thermodynamics of phase equilibria and azeotropy is available in Swietoslawski (*Azeotropy and Polyazeotropy*, Pergamon, London, 1963), Van Winkle (*Distillation*, McGraw-Hill, New York, 1967), Smith and Van Ness (*Introduction to Chemical Engineering Thermodynamics*, McGraw-Hill, New York, 1975), Wizniak [*Chem. Eng. Sci.,* **38**, 969 (1983)], and Walas (*Phase Equilibria in Chemical Engineering*, Butterworths, Boston, 1985). Horsley (*Azeotropic Data-III*, American Chemical Society, Washington, 1983) compiled an extensive list of binary and some ternary and higher experimental azeotropic boiling-point and composition data. Another source for azeotrope data and activity coefficient model parameters is the multivolume *Vapor-Liquid Equilibrium Data Collection* (DECHEMA, Frankfort 1977), a compendium of published experimental VLE data. Most of the data have been tested for thermodynamic consistency and have been fit to the Wilson, UNIQUAC, Van Laar, Margules, and NRTL equations. An extensive two-volume compilation of data for 18,800 systems involving 1,700 compounds, entitled *Azeotropic Data* by Gmehling et al., was published in 1994 by VCH Publishers, Deerfield Beach, Florida. A computational method for determining the temperatures and compositions of all azeotropes of a multicomponent mixture, from liquid-phase activity-coefficient correlations, by a differential arclength homotopy continuation method is given by Fidkowski, Malone, and Doherty [*Computers and Chem. Eng.,* **17**, 1141 (1993)].

RESIDUE CURVE MAPS AND DISTILLATION REGION DIAGRAMS

The simplest form of distillation involves boiling a multicomponent liquid mixture batchwise in a single-stage still pot. At any instant in time the vapor being generated and removed from the pot is assumed to be in equilibrium with the remaining liquid (assumed to be perfectly mixed) in the still. Because the vapor is richer in the more volatile components than the liquid, the composition and temperature of the liquid remaining in the still changes continuously over time and moves progressively toward less volatile compositions and higher temperatures until the last drop is vaporized. For some mixtures, this last composition is the highest-boiling pure component in the system. For other mixtures, this final composition may be a maximum-boiling azeotrope. For yet other systems, the final composition varies depending on the initial composition of the mixture charged to the still.

A residue curve is a tracing of this change in perfectly mixed liquid composition for simple single-stage batch distillation with respect to time. Arrows are sometimes added, pointing in the direction of increasing time, increasing temperature, and decreasing volatility. Because simple, batch distillation can be described mathematically by

$$dx_i/d\xi = x_i - y_i \qquad \text{for all } i = 1, \ldots n \qquad (13\text{-}126)$$

where ξ is a nonlinear time scale, residue curves may also be extrapolated backward in time to give more volatile compositions which would produce a residue equal to the specified initial composition. A *residue curve map* (RCM) is generated by varying the initial composition and extrapolating Eq. (13-126) both forward and backward in time [Doherty and Perkins, *Chem. Eng. Sci.,* **33**, 281 (1978)]. Unlike a binary y-x plot, relative-volatility information is not presented. Therefore, it is difficult to determine the ease of separation from a residue curve map alone.

Residue curve maps can be constructed for mixtures of any number of components, but can be pictured graphically only for up to four components. For a binary mixture, a T-x,y diagram suffices; the system is simple enough that vapor-phase information can be included without confusion. With a ternary mixture, liquid-phase compositions are plotted on a triangular diagram, similar to that used in liquid-liquid extraction. Four-component systems can be plotted in a 3-dimensional tetrahedron. The vertices of the triangular diagram or tetrahedron represent the pure components. Any binary, ternary, and quaternary azeotropes are placed at the appropriate compositions on the edges and/or interior of the triangle and tetrahedron.

The simplest form of ternary RCM, as exemplified for the ideal normal-paraffin system of pentane-hexane-heptane, is illustrated in Fig. 13-58*a*, using a right-triangle diagram. Maps for all other non-azeotropic ternary mixtures are qualitatively similar. Each of the infinite number of possible residue curves originates at the pentane vertex, travels toward and then away from the hexane vertex, and terminates at the heptane vertex.

The family of all residue curves that originate at one composition and terminate at another composition defines a region. Systems that do not involve azeotropes have only one region—the entire composition space. However, for many systems, not all residue curves originate or terminate at the same two compositions. Such systems will have more that one region. The demarcation between regions in which adjacent residue curves originate from different compositions or terminate at different compositions is called a *separatrix*. Separatrices are related to the existence of azeotropes. In the composition space for a binary system, the separatrix is a point (the azeotropic composition). With three components, the separatrix becomes a (generally curved) line, with four components the separatrix becomes a surface, and so on.

All pure components and azeotropes in a system lie on region boundaries. Within each region, the most volatile composition on the boundary (either a pure component or a minimum-boiling azeotrope and the origin of all residue curves) is called the *low-boiling node*. The least-volatile composition on the boundary (again either a pure component or a maximum-boiling azeotrope and the terminus of all residue curves) is called the *high-boiling node*. All other pure components and azeotropes are called *intermediate-boiling saddles* (because no residue curves originate or terminate at these compositions). Adjacent regions may share nodes and saddles. Pure components and azeotropes are labeled as nodes and saddles as a result of the boiling points of all of the components and azeotropes in a system. If one species is removed, the labeling of all remaining pure components and azeotropes, particularly those that were saddles, may change. Region-defining separatrices always originate or terminate at saddle azeotropes, but never at saddle-pure components. Saddle-ternary azeotropes are particularly interesting because they are less obvious to determine experimentally (being neither minimum-boiling nor maximum-boiling), and have only recently begun to be recorded in the literature. (Gmehling et al., Azeotropic Data, VCH Publishers, Deer-

field Beach, Florida, 1994). However, their presence in a mixture implies separatrices, which may have an important impact on the design of a separation system.

Both *methylethylketone* (MEK) and *methylisopropylketone* (MIPK) form minimum-boiling azeotropes with water (Fig. 13-58b). In this ternary system, a separatrix connects the binary azeotropes and divides the RCM into two regions. The high-boiling node of Region I is pure water, while the low-boiling node is the MEK-water azeotrope.

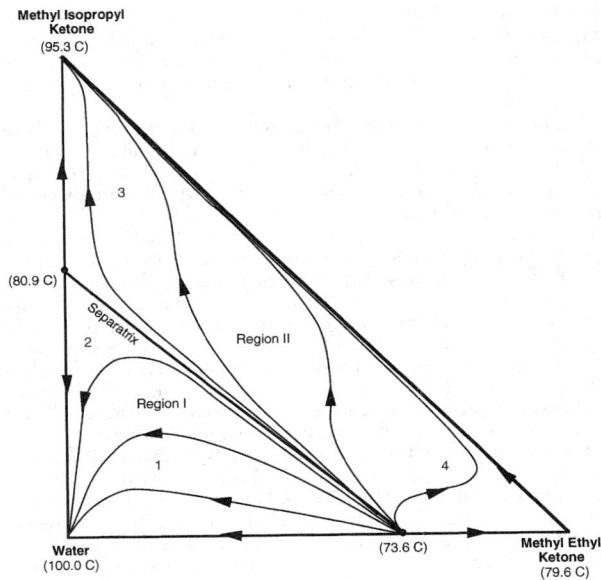

FIG. 13-58 (*Continued*) Residue curve maps. (*b*) MEK-MIPK-water system containing two minumum-boiling binary azeotropes.

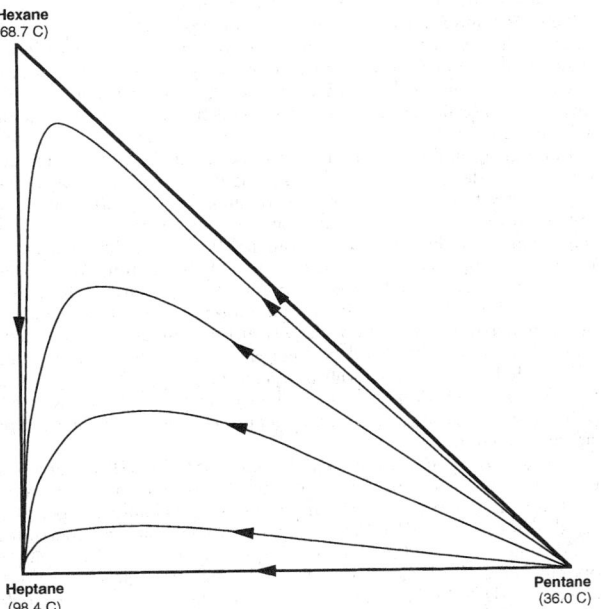

FIG. 13-58 Residue curve maps. (*a*) Nonazeotropic pentane-hexane-heptane system.

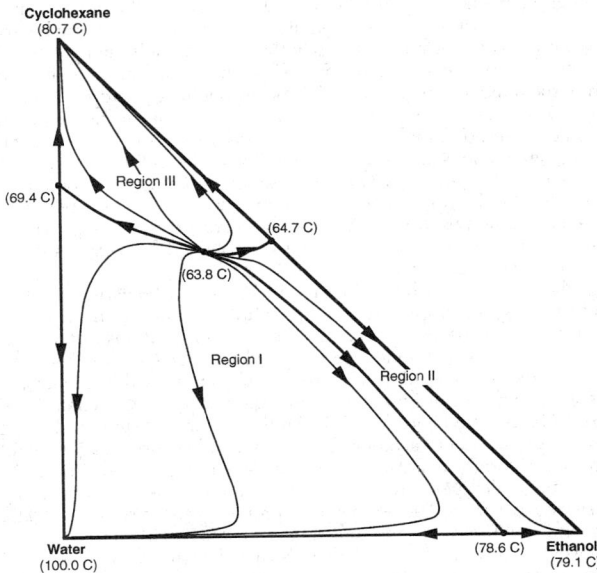

FIG. 13-58 (*Continued*) Residue curve maps. (*c*) Ethanol-cyclohexane-water system containing four minimum-boiling azeotropes and three distillation regions.

In Region II, the high- and low-boiling nodes are MIPK and the MEK-water azeotrope, respectively. The more complicated cyclohexane-ethanol-water system (Fig. 13-58c) has three separatrices and three regions, all of which share the ternary azeotrope as the low-boiling node.

The liquid-composition profiles in continuous staged or packed distillation columns operating at infinite reflux and boilup are closely approximated by simple distillation-residue curves [Van Dongen and Doherty, *Ind. Eng. Chem. Fundam.*, **24**, 454 (1985)]. Residue curves are also indicative of many aspects of the general behavior of continuous columns operating at more practical reflux ratios. For example, to a first approximation, the composition of the distillate and bottoms of a single-feed, continuous distillation column lie on the same residue curve. Therefore, for systems with separatrices and multiple regions, distillation-composition profiles are constrained to lie in specific regions. The precise boundaries of these distillation regions are a function of reflux ratio, but they are closely approximated by the RCM separatrices. If a RCM separatrix exists in a system, a corresponding distillation boundary will also exist. Separatrices and distillation boundaries correspond exactly at all pure components and azeotropes.

Residue curves can be constructed from experimental data or can be calculated analytically if equation-of-state or activity-coefficient expressions are available (e.g., Wilson binary-interaction parameters, UNIFAC groups). However, considerable information on system behavior can still be deduced from a simple semi-qualitative sketch of the RCM separatrices or distillation boundaries based only on pure component and azeotrope boiling-point data and approximate azeotrope compositions. Rules for constructing such qualitative *distillation region diagrams* (DRD) are given by Foucher et al. [*Ind. Eng. Chem. Res.*, **30**, 760–772, 2364 (1991)]. For ternary systems containing no more than one ternary azeotrope, and no more than one binary azeotrope between each pair of components, 125 such DRD are mathematically possible, although only a dozen or so represent most systems commonly encountered in practice.

Figure 13.59 illustrates all of the 125 possible DRD for ternary systems. Azeotropes are schematically depicted generally to have equimolar composition, distillation boundaries are shown as straight lines, and the arrows on the distillation boundaries indicate increasing temperature. These DRD are indexed in Table 13-16 according to a temperature-profile sequence of position numbers, defined in a keyed-triangular diagram at the bottom of the table, arranged by increasing the boiling point. Positions 1, 3, and 5 are the pure components in order of decreasing volatility. Positions 2, 4, and 6 are binary azeotropes at the positions shown in the keyed triangle, and position 7 is the ternary azeotrope. Azeotrope position numbers are deleted from the temperature profile if the corresponding azeotrope is known not to exist. It should be noted that not every conceivable temperature profile corresponds to a thermodynamically consistent system, and such combinations have been excluded from the index. As is evident from the index, some DRD are consistent with more than one temperature profile. Also, some temperature profiles are consistent with more than one DRD. In such cases, the correct diagram for a system must be determined from residue curves obtained from experimental or calculated data.

Schematic DRD shown in Fig. 13-59 are particularly useful in determining the implications of possibly unknown ternary saddle azeotropes by postulating position 7 at interior positions in the temperature profile. It should also be noted that some combinations of binary azeotropes require the existence of a ternary saddle azeotrope. As an example, consider the system acetone (56.4°C), chloroform (61.2°C), and methanol (64.7°C). Methanol forms minimum-boiling azeotropes with both acetone (54.6°C) and chloroform (53.5°C), and acetone-chloroform forms a maximum-boiling azeotrope (64.5°C). Experimentally there are no data for maximum or minimum-boiling ternary azeotropes. The temperature profile for this system is 461325, which from Table 13-16 is consistent with DRD 040 and DRD 042. However, Table 13-16 also indicates that the pure component and binary azeotrope data are consistent with three temperature profiles involving a ternary saddle azeotrope, namely 4671325, 4617325, and 4613725. All three of these temperature profiles correspond to DRD 107. Experimental residue curve trajectories for the acetone-

chloroform-methanol system, as shown in Fig. 13-60, suggest the existence of a ternary saddle azeotrope and DRD 107 as the correct approximation of the distillation regions. Ewell and Welch [*Ind. Eng. Chem.*, **37**, 1224 (1945)] confirm such a ternary saddle at 57.5°C.

APPLICATIONS OF RCM AND DRD

Residue curve maps and distillation region diagrams are very powerful tools for understanding all types of batch and continuous distillation operations, particularly when combined with other information such as liquid-liquid binodal curves. Applications include:

1. *System visualization.* Location of distillation boundaries, azeotropes, distillation regions, feasible products, and liquid-liquid regions.

2. *Evaluation of laboratory data.* Location and confirmation of saddle ternary azeotropes and a check of thermodynamic consistency of data.

3. *Process synthesis.* Concept development, construction of flowsheets for new processes, and redesign or modification of existing process flowsheets.

4. *Process modeling.* Identification of infeasible or problematic column specifications that could cause simulation convergence difficulties or failure, and determination of initial estimates of column parameters including feed-stage location, number of stages in the stripping and enriching sections, reflux ratio, and product compositions.

5. *Control analysis/design.* Analysis of column balances and profiles to aid in control system design and operation.

6. *Process trouble shooting.* Analysis of separation system operation and malfunction, examination of composition profiles, and tracking of trace impurities with implications for corrosion and process specifications.

Material balances for mixing or continuous separation operations are represented graphically on triangular composition diagrams such as residue curve maps or distillation region diagrams by straight lines connecting pertinent compositions. Overall flow rates are found by the inverse-lever-arm rule. Distillation material balance lines are governed by two constraints:

1. The bottoms, distillate, and overall feed compositions must lie on the same straight line.

2. The bottoms and distillate compositions must lie (to a very close approximation) on the same residue curve.

Since residue curves do not by definition cross separatrices, the distillate and bottoms compositions must be in the same distillation region with the mass balance line intersecting a residue curve in two places. Mass balance lines for mixing and for other separations not involving vapor-liquid equilibria, such as extraction and decantation, are of course not limited by distillation boundaries.

For a given multicomponent mixture, a single-feed distillation column can be designed with sufficient stages, reflux, and material balance control to produce separations ranging from the *direct* mode of operation (low-boiling node taken as distillate) to the *indirect* mode (high-boiling node taken as bottoms). The bow-tie shaped set of reachable compositions for single-feed distillation is roughly bounded by the material balance lines corresponding to the sharpest direct separation and the sharpest indirect separation possible. The exact shape of the reachable composition space is further limited by the requirement that the distillate and bottoms lie on the same residue curve [Wahnschafft, et al., *Ind. Eng. Chem. Res.*, **31**, 2345 (1992)]. Since residue curves are deflected by saddles, it is generally not possible to obtain a saddle product (pure component or azeotrope) from a simple single-feed column.

Consider the recovery of MIPK from an MEK-MIPK-water mixture. The bow-tie approximation of reachable compositions for several feeds are shown in Fig 13-61a and the exact reachable compositions are shown in Fig. 13-61b. From Feed F3, which is situated in a different distillation region than the desired product, pure MIPK cannot be obtained at all. With the upper edge of the bow-tie region for Feed F1 along the MEK-MIPK (water-free) face of the composition triangle, and part of the lower edge along the MEK-water (MIPK-free) face, there are conditions under which both the water in the bottoms MIPK product can be driven to low levels (high-product purity) and MIPK in

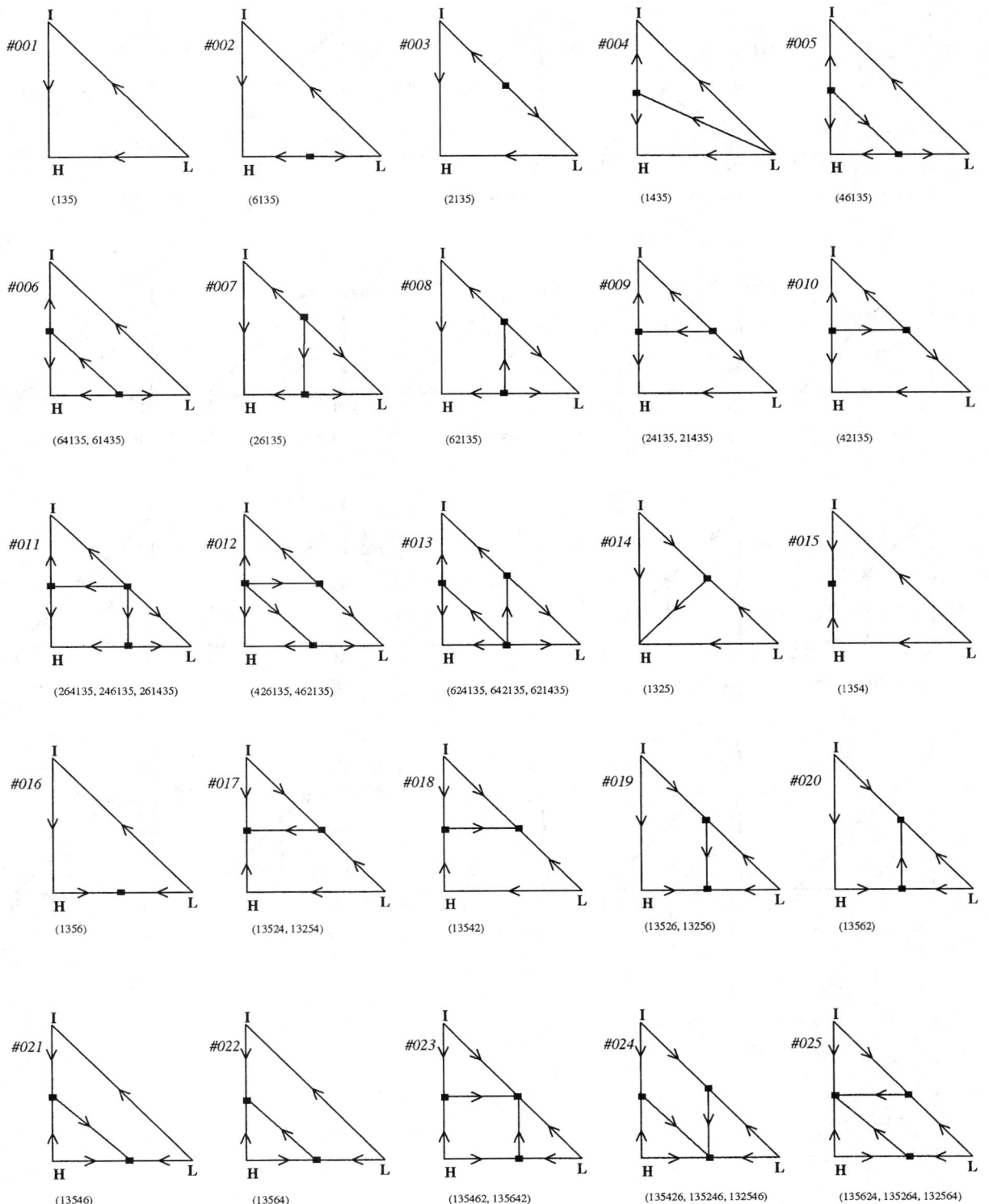

FIG. 13-59 Distillation region diagrams for ternary mixtures.

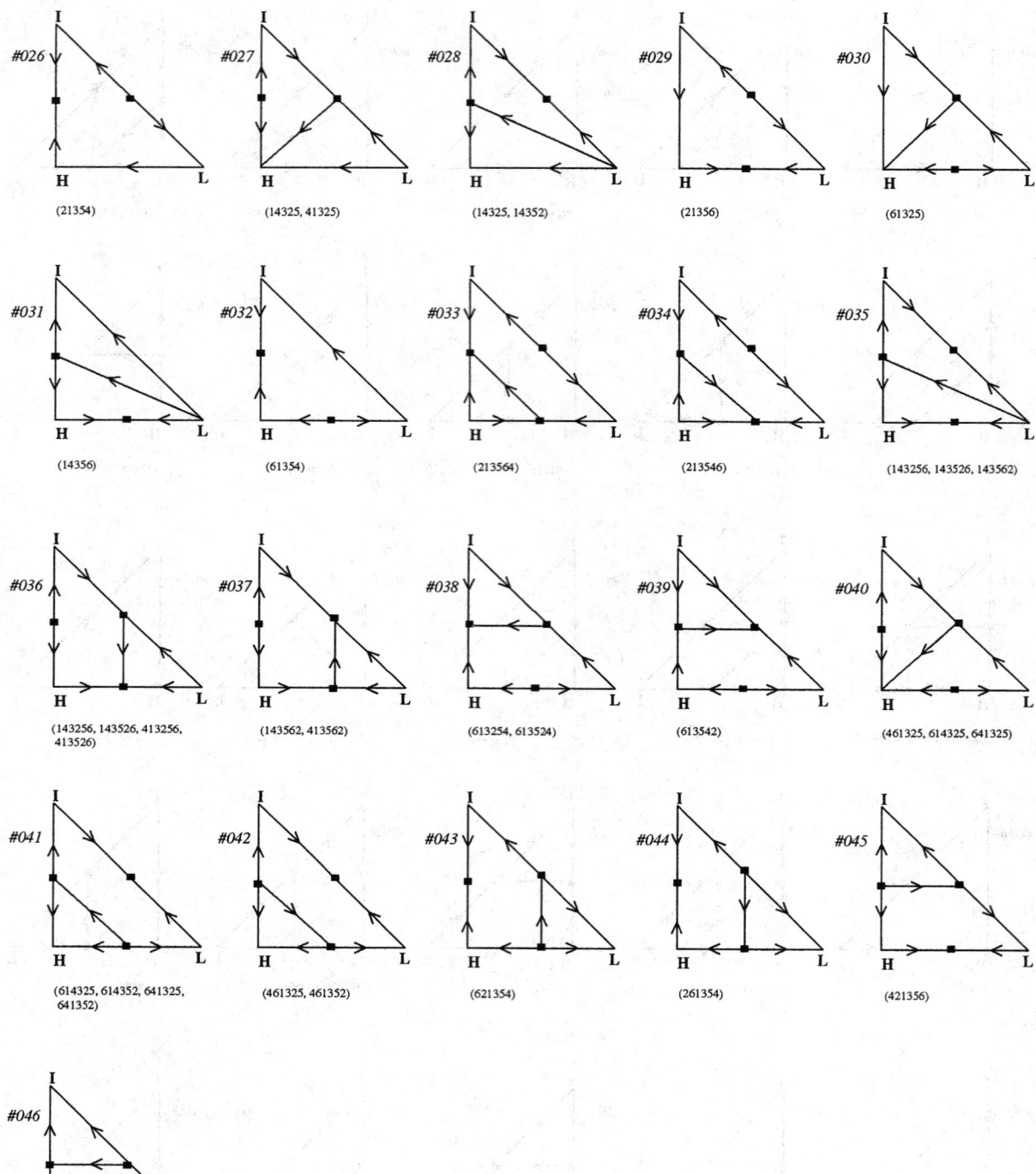

FIG. 13-59 (*Continued*) Distillation region diagrams for ternary mixtures.

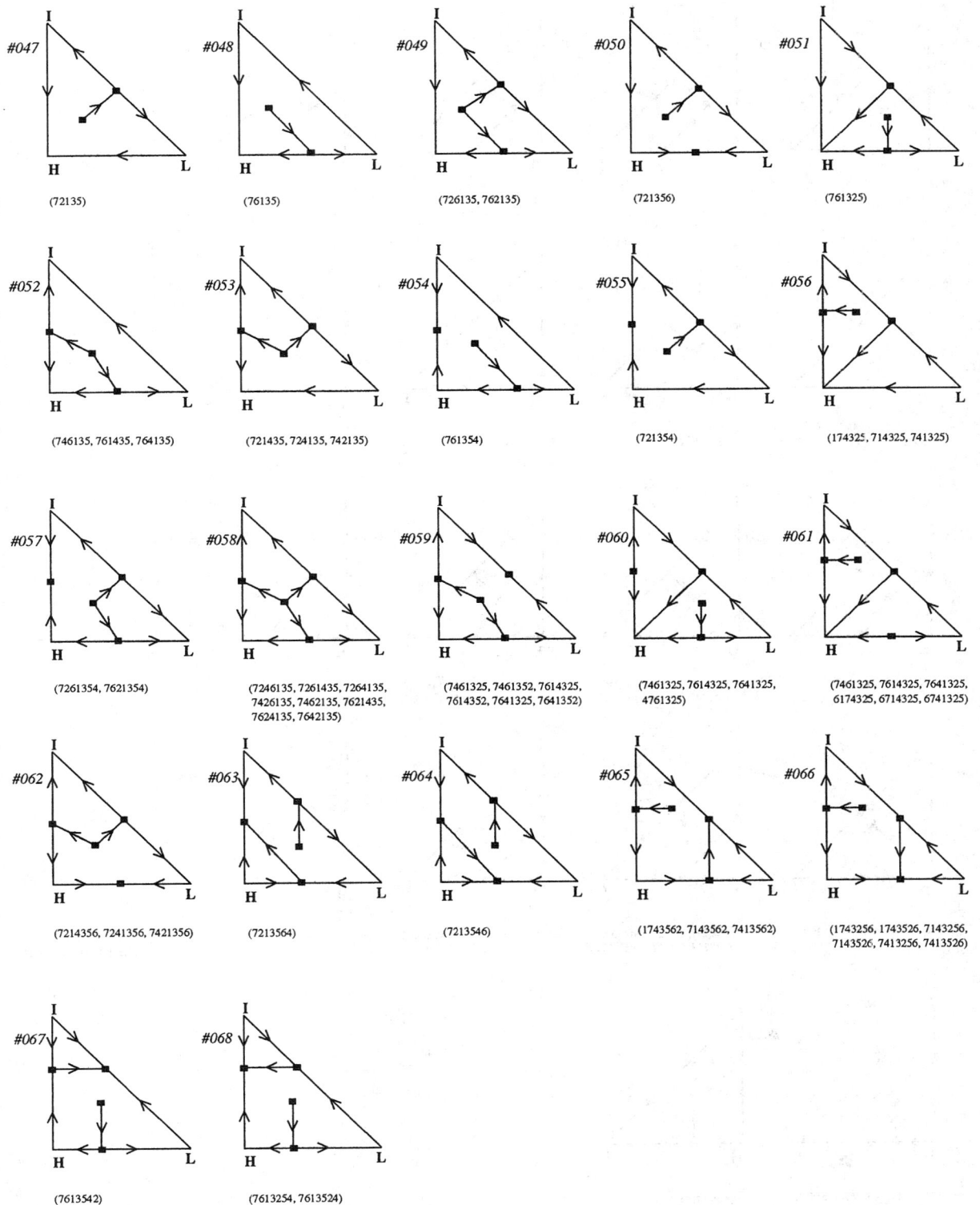

(c)

FIG. 13-59 (*Continued*) Distillation region diagrams for ternary mixtures.

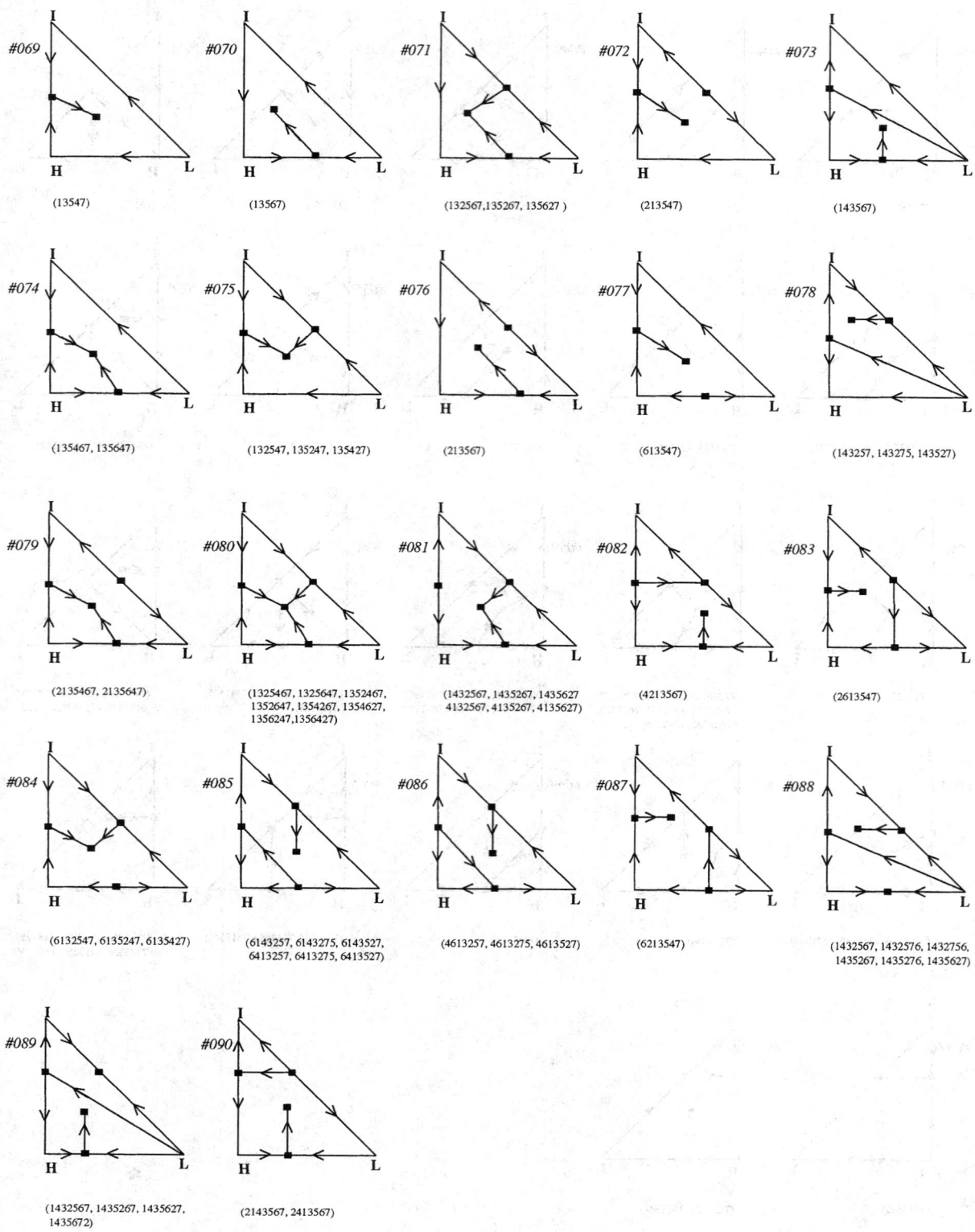

FIG. 13-59   (*Continued*) Distillation region diagrams for ternary mixtures.

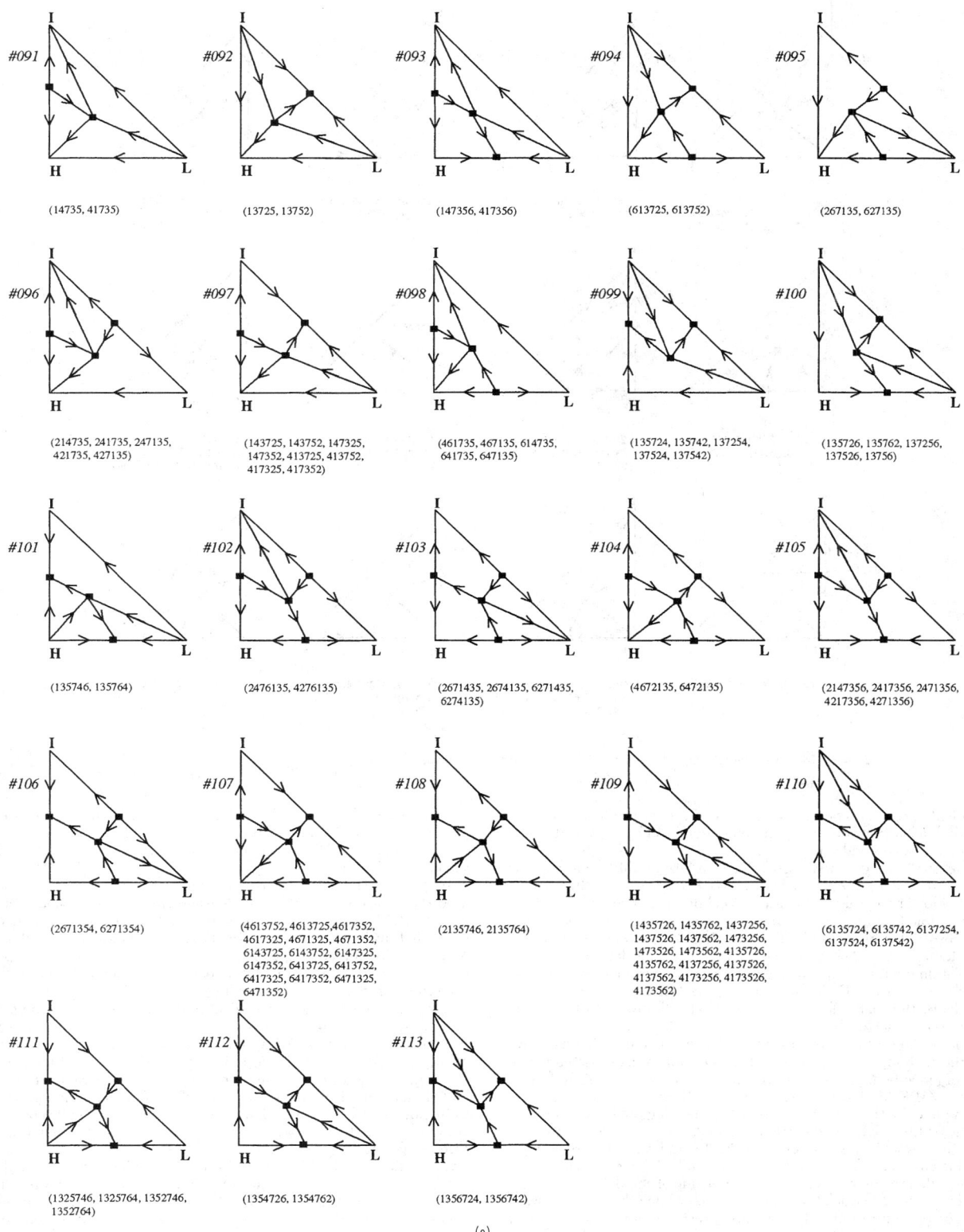

(e)

FIG. 13-59 (*Continued*) Distillation region diagrams for ternary mixtures.

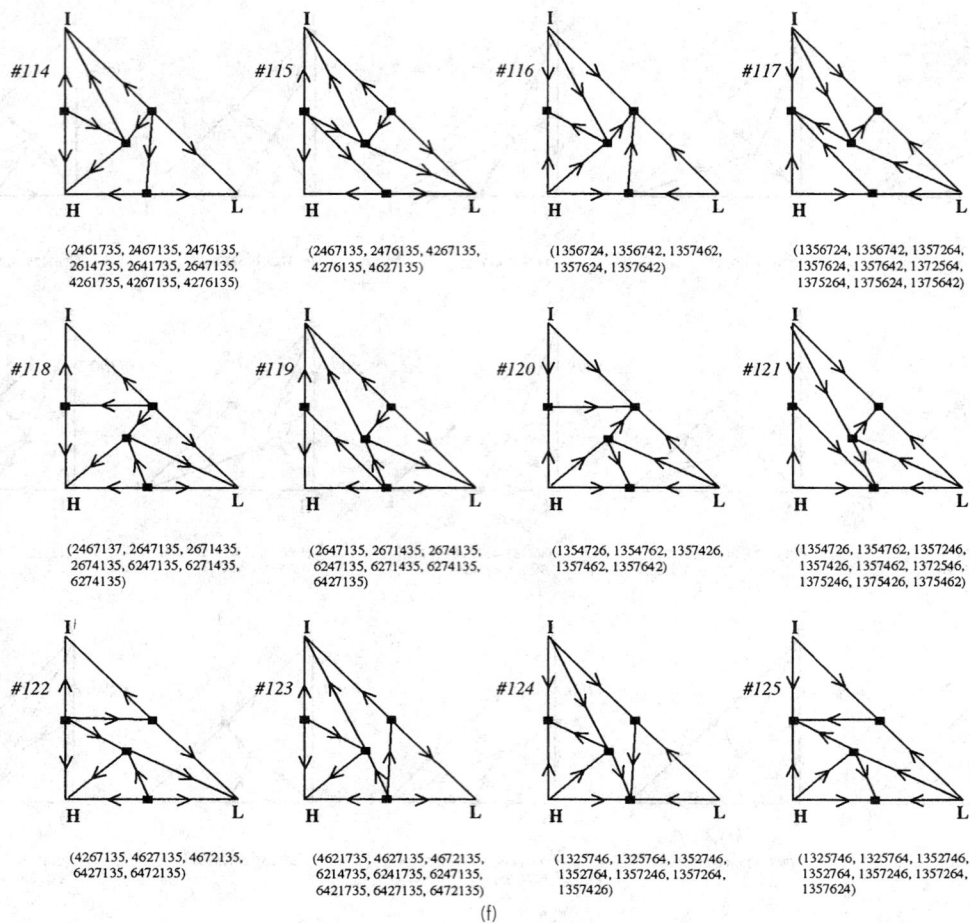

#114 (2461735, 2467135, 2476135, 2614735, 2641735, 2647135, 4261735, 4267135, 4276135)

#115 (2467135, 2476135, 4267135, 4276135, 4627135)

#116 (1356724, 1356742, 1357462, 1357624, 1357642)

#117 (1356724, 1356742, 1357264, 1357624, 1357642, 1372564, 1375264, 1375624, 1375642)

#118 (2467137, 2647135, 2671435, 2674135, 6247135, 6271435, 6274135)

#119 (2647135, 2671435, 2674135, 6247135, 6271435, 6274135, 6427135)

#120 (1354726, 1354762, 1357426, 1357462, 1357642)

#121 (1354726, 1354762, 1357246, 1357426, 1357462, 1372546, 1375246, 1375426, 1375462)

#122 (4267135, 4627135, 4672135, 6427135, 6472135)

#123 (4621735, 4627135, 4672135, 6214735, 6241735, 6247135, 6421735, 6427135, 6472135)

#124 (1325746, 1325764, 1352746, 1352764, 1357246, 1357264, 1357426)

#125 (1325746, 1325764, 1352746, 1352764, 1357246, 1357264, 1357624)

(f)

FIG. 13-59 (*Continued*) Distillation region diagrams for ternary mixtures.

the distillate can also be driven to low levels (high-product recovery), although achieving such an operation depends on having an adequate number of stages and reflux ratio.

The bow-tie region for Feed F2 is significantly different, with the upper edge along the water-MIPK (MEK-free) face of the triangle and the lower edge along the distillation boundary. From this feed it is not possible to achieve a high-purity MIPK specification while simultaneously obtaining high MIPK recovery. If the column is operated to get a high purity of MIPK, then the material balance line runs into the distillation boundary. Alternatively, if the column is operated to obtain a high recovery of MIPK (by removing the MEK-water azeotrope as distillate), the material balance requires the bottoms to lie on the water-MIPK face of the triangle.

The number of saddles in a particular distillation region can have significant impact on column-profile behavior, process stability, and convergence behavior in process simulation of the system. Referring to the MIPK-MEK-water system in Fig. 13-58b, Region I contains one saddle (MIPK-water azeotrope), while Region II contains two saddles (pure MEK and the MIPK-water azeotrope). These are three- and "four-sided" regions respectively. In a three-sided region, all residue curves track toward the solitary saddle. However, in a four (or more) sided region with saddles on either side of a node, some residue curves will tend to track toward one saddle, while others track toward another opposite saddle. For example, residue curve 1 in Region I originates from the MEK-water azeotrope low-boiling node and trav-

els first toward the single saddle of the region (MIPK-water azeotrope) before ending at the water high-boiling node. Likewise, residue curve 2 and all other residue curves in Region I follow the same general path.

In Region II, residue curve 3 originates from the MEK-water azeotrope, travels toward the MIPK-water saddle azeotrope, and ends at pure MIPK. However, residue curve 4 follows a completely different path, traveling toward the pure MEK saddle before ending at pure MIPK. Some multicomponent columns have been designed for operation in four-sided regions with the feed composition adjusted so that both the high-boiling and low-boiling nodes can be obtained simultaneously as products. However, small perturbations in feed composition or reflux can result in feasible operation on many different residue curves that originate and terminate at these compositions. Multiple steady states and composition profiles that shift dramatically from tracking toward one saddle to the other are possible [Kovach, and Seider, *AIChE J.*, **33**, 1300 (1987). Consider a column operating in Region II of the MIPK-MEK-water diagram. Fig. 13-62 shows the composition and temperature profiles for the column operating at three different sets of operating conditions and two feed locations as given in Table 13-17. The desired product specification is 97 mol % MIPK, no more than 3 mol % MEK, and less than 10 ppm residual water. For Case A (Fig. 13-62a), the column profile tracks up the water-free side of the diagram. A pinched zone (area of little change in tray temperature and composition) occurs between the feed tray (tray

TABLE 13-16   Temperature Profile—DRD # Table*

Temp. Profile	DRD #	Temp. Profile	DRD #	Temp. Profile	DRD #	Temp. Profile	DRD #	Temp. Profile	DRD #	Temp. Profile	DRD #
135	001	137524	099	624135	013	1357462	120	2671435	103	6247135	118
1325	014	137526	100	627135	095		121		119		119
1354	015	137542	099	641325	041		116		118		123
1356	016	137562	100		040	1357624	117	2674135	118	6271354	106
1435	004	143256	035	641352	041		116		119	6271435	118
2135	003		036	641735	098		125		103		103
6135	002	143257	078	642135	013	1357642	120	4132567	081		119
13254	017	143275	078	647135	098		117	4135267	081	6274135	103
13256	019	143526	036	714325	056		116	4135627	081		118
13524	017		035	721354	055	1372546	121	4135726	109		119
13526	019	143527	078	721364	050	1372564	117	4135762	109	6413257	085
13542	018	143562	035	721435	053	1375246	121	4137256	109	6413275	085
13546	021		037	724135	053	1375264	117	4137526	109	6413527	085
13547	069	143567	073	726135	049	1375426	121	4137562	109	6413725	107
13562	020	143725	097	741325	056	1375624	117	4173526	109	6413752	107
13564	022	143752	097	742135	053	1375642	117	4173562	109	6417325	107
13567	070	147325	097	746135	052	1432567	089	4213567	082	6417352	107
13725	092	147352	097	761325	051		088	4217356	105	6421735	123
13752	092	147356	093	761354	054		081	4261735	114	6427135	122
14325	028	174325	056	761435	052	1432576	088	4267135	122		123
027		213546	034	762135	049	1432756	088		115		119
14352	028	213547	072	764135	052	1435267	081		114	5471325	107
14356	031	213564	033	1325467	080		089	4271356	105	5471352	107
14735	091	213567	076	1325647	080		088	4276135	114	5472135	104
21354	026	214356	046	1325746	111	1435276	088		102		122
21356	029	214735	096		124	1435627	089		115		123
21435	009	241356	046		125		081	4613257	086	6714325	061
24135	009	241735	096	1325764	111	1435672	089	4613275	086	6741325	061
26135	007	246135	011		124		088	4613527	086	7143256	066
41325	027	247135	096		125	1435726	109	4613752	107	7143526	066
41735	091	261354	044	1352467	080	1435762	109	4617325	107	7143562	065
42135	010	261435	011	1352647	080	1437256	109	4617352	107	7213546	064
46135	005	264135	011	1352746	125	1437526	109	4621735	123	7213564	063
61325	030	267135	095		111	1437562	109	4627135	123	7214356	062
61354	032	413256	036		124	1473256	109		115	7241356	062
61435	006	413526	036	1352764	124	1473526	109		122	7246135	058
62135	008	413562	037		111	1473562	109	4671325	107	7261354	057
64135	006	413725	097		125	1743256	066	4671352	107	7261435	058
72135	047	413752	097	1354267	080	1743526	066	4672135	104	7264135	058
76135	048	417325	097	1354627	080	1743562	065		122	7413256	066
132546	024	417352	097	1354726	120	2135467	079		123	7413526	066
132547	075	417356	093		121	2135647	079	4761325	060	7413562	065
132564	025	421356	045		112	2135746	108	6132547	084	7421356	062
132567	071	421735	096	1354762	121	2135764	108	6135247	084	7426135	058
135246	024	426135	012		120	2143567	090	6135724	110	7461325	060
135247	075	427135	096		112	2147356	105	6135742	110		059
135264	025	461325	040	1356247	080	2413567	090	6137254	110		061
135267	071		042	1356427	080	2417356	105	6137524	110	7461352	059
135426	024	461352	042	1356724	113	2461735	114	6137542	110	7462135	058
135427	075	461735	098		117	2467135	114	6143257	085	7613254	068
135462	023	462135	012		116		115	6143275	085	7613524	068
135467	074	467135	098	1356742	116		117	6143527	085	7613542	067
135624	025	613254	038		117		113	6143725	107	7614325	061
135627	071	613524	038		113	2471356	105	6143752	107		060
135642	023	613542	039	1357246	121	2476135	115	6147325	107		059
135647	074	613547	077		125		114	6147352	107	7614352	059
135724	099	613725	094		124		102	6174325	061	7621354	057
135726	100	613752	094	1357264	117	2613547	083	6213547	087	7621435	058
135742	099	614325	041		125	2614735	114	6214735	123	7624135	058
135746	101	614352	041		124	2641735	114	6214735	123	7641325	061
135762	100	614735	098	1357426	121	2647135	118	6241357	123		060
135764	101	621354	043		120		119				059
137254	099				124		114			7641352	059
137256	100	621435	013			2671354	106			7642135	058

Ternary DRD table lookup procedure:
1. Classify a system by writing down each position number in ascending order of boiling points.
   - A position number is not written down if there is no azeotrope at that position.
   - The resulting sequence of numbers is known as the *temperature profile.*
   - Each temperature profile will have a minimum of three numbers and a maximum of seven numbers.
   - List multiple temperature profiles when you have incomplete azeotropic data.
   - All seven position numbers are shown on the diagram.
2. Using the table, look up the temperature profile(s) to find the corresponding DRD #.

* Table 13-16 and Fig. 13-59 developed by Eric J. Peterson, Eastman Chemical Co.

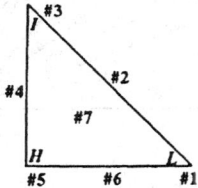

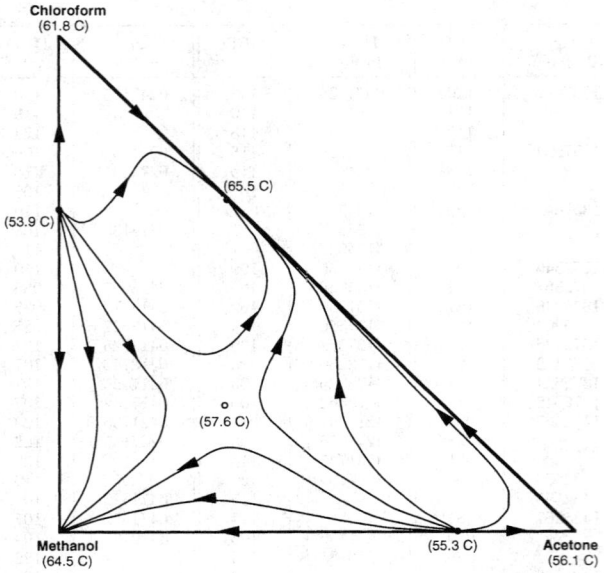

FIG. 13-60 Residue curves for acetone-chloroform-methanol system suggesting a ternary saddle azeotrope.

4) and tray 18. The temperature remains constant at about 93°C throughout the pinch. Product specifications are met.

When the feed composition becomes enriched in water, as with Case B, the column profile changes drastically (Fig. 13-62*b*). At the same reflux and boil-up, the column no longer meets specifications. The MIPK product is lean in MIPK and too rich in water. The profile now tracks generally up the left side of Region II. Note also the dramatic change in the temperature profile. A pinched zone still exists

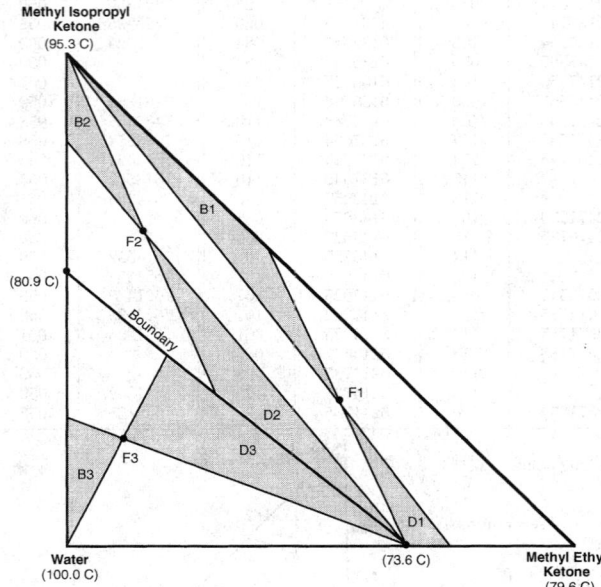

FIG. 13-61 MEK-MIPK-water system. (*a*) Approximate bow-tie reachable compositions by simple distillation.

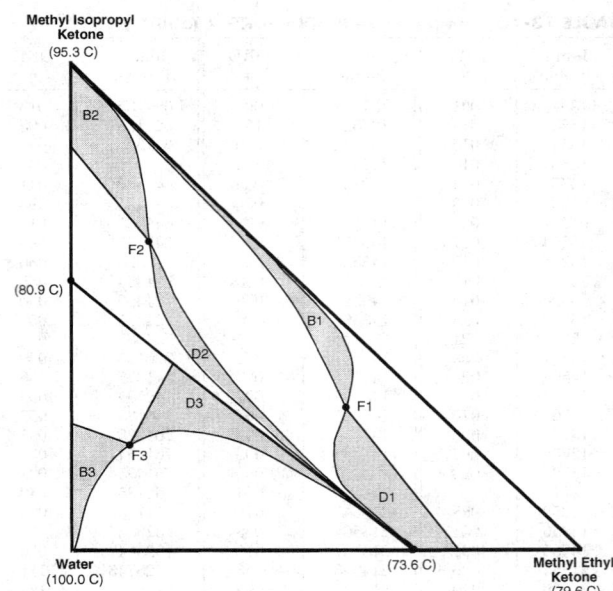

FIG. 13-61 (*Continued*) MEK-MIPK-water system. (*b*) Exact-reachable compositions.

between trays 4 and 18, but the tray temperature in the zone has dropped to 80°C (from 93°C). Most of the trays are required to move through the vicinity of the saddle. Typically, pinches (if they exist) occur close to saddles and nodes.

In Case C (Fig. 13-62*c*), increasing the boil-up ratio to 6 brings the MIPK product back within specifications, but the production rate and recovery have dropped off. In addition, the profile has switched back to the right side of the region; the temperatures on trays in the pinched zone (trays 4–18) are back to 93°C. Such a drastic fluctuation in tray temperature with a relatively minor adjustment of the manipulated variable (boil-up in this case), can make control difficult. This is especially true if the control strategy involves maintaining a constant temperature on one of the trays between tray 4 and 18. If a tray selected that exhibits wide temperature swings, the control system may have a difficult time compensating for disturbances. Such columns are also often difficult to model with a process simulator. Design algorithms often rely on perturbation of a variable (such as reflux or reboil) while checking for convergence of column heat and material balances. In situations where the column profile is altered drastically by minor changes in the perturbed variable, the simulator may be close to a feasible solution, but successive iterations may appear to be very far apart. The convergence routine may continue to oscillate between column profiles and never reach a solution. Likewise, when an attempt is made to design a column to obtain product compositions in different distillation regions, the simulation will never converge.

EXTENSION TO BATCH DISTILLATION

Although batch distillation is covered in a subsequent separate section, it is appropriate to consider the application of RCM and DRD to batch distillation at this time. With a conventional batch-rectification column, a charge of starting material is heated and fractionated, with a vapor product removed continuously. The composition of the vapor product changes continuously and at times drastically as the lighter component(s) are exhausted from the still. Between points of drastic change in the vapor composition, a "cut" is often made. Successive cuts can be removed until the still is nearly dry. The sequence, number, and limiting composition of each cut is dependent on the form of

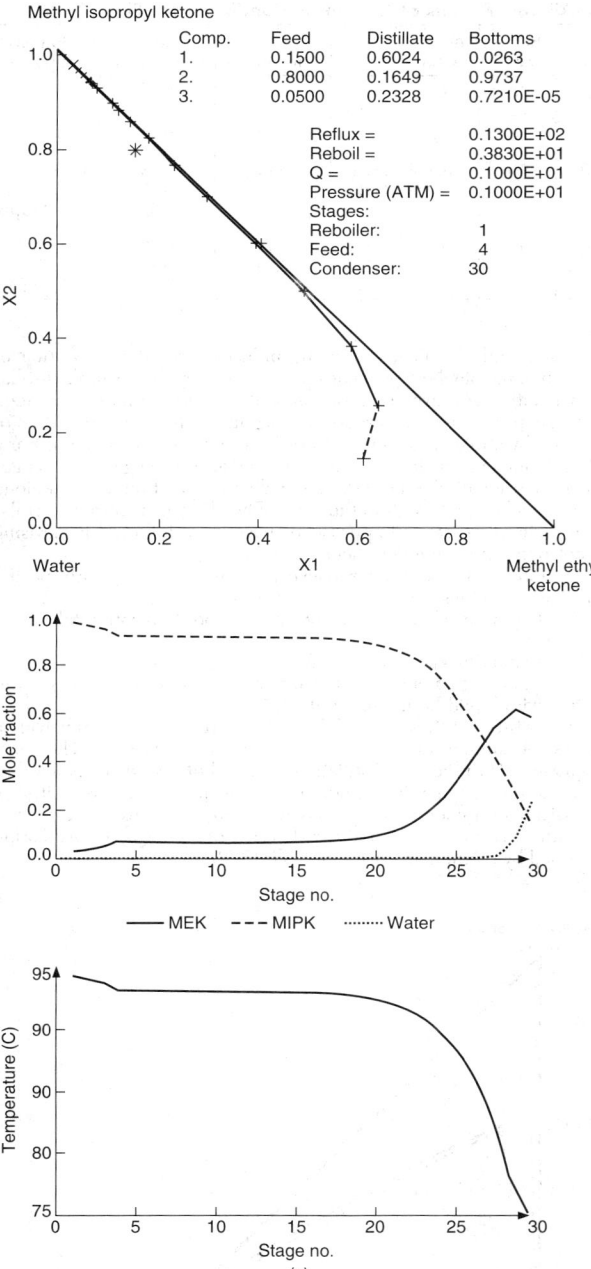

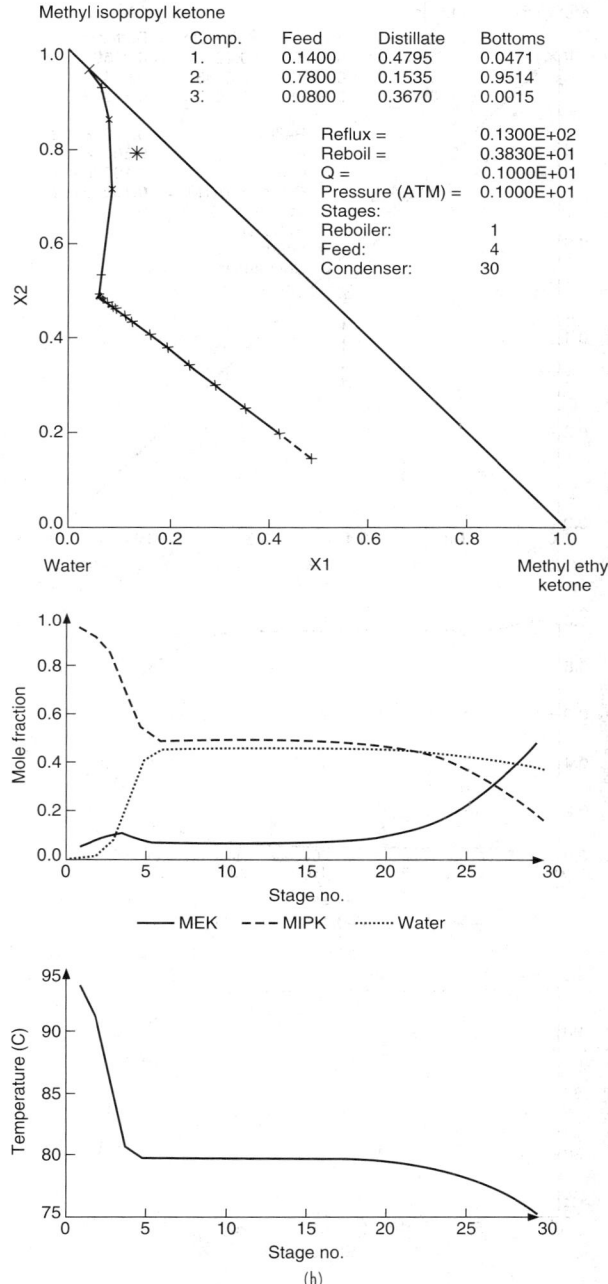

FIG. 13-62 Sensitivity of composition and temperature profiles for MEK-MIPK-water system.

FIG. 13-62 (*Continued*) Sensitivity of composition and temperature profiles for MEK-MIPK-water system.

the residue curve map and the composition of the initial charge to the still. As with continuous distillation operation, the set of reachable products (cuts) for a given charge to a batch distillation is constrained by the residue-curve-map separatrices, which cannot normally be crossed.

Given a sufficient number of stages and reflux, the vapor composition can be made to closely approach direct-mode, continuous operation in which the lowest-boiling species is taken overhead. As the

low-boiling component is removed, the still composition moves along a straight material-balance line through the initial feed composition and the low-boiling node away from the initial composition until it reaches the edge of the composition triangle or a separatrix. The path then follows the edge or separatrix to the high-boiling node of the region. At each turn a new cut is taken. Examples for the acetone-chloroform-methanol and MEK-water-MIPK systems are given in Fig. 13-63 [Bernot et al., *Chem. Eng. Sci.*, **45**, 1207 (1990)].

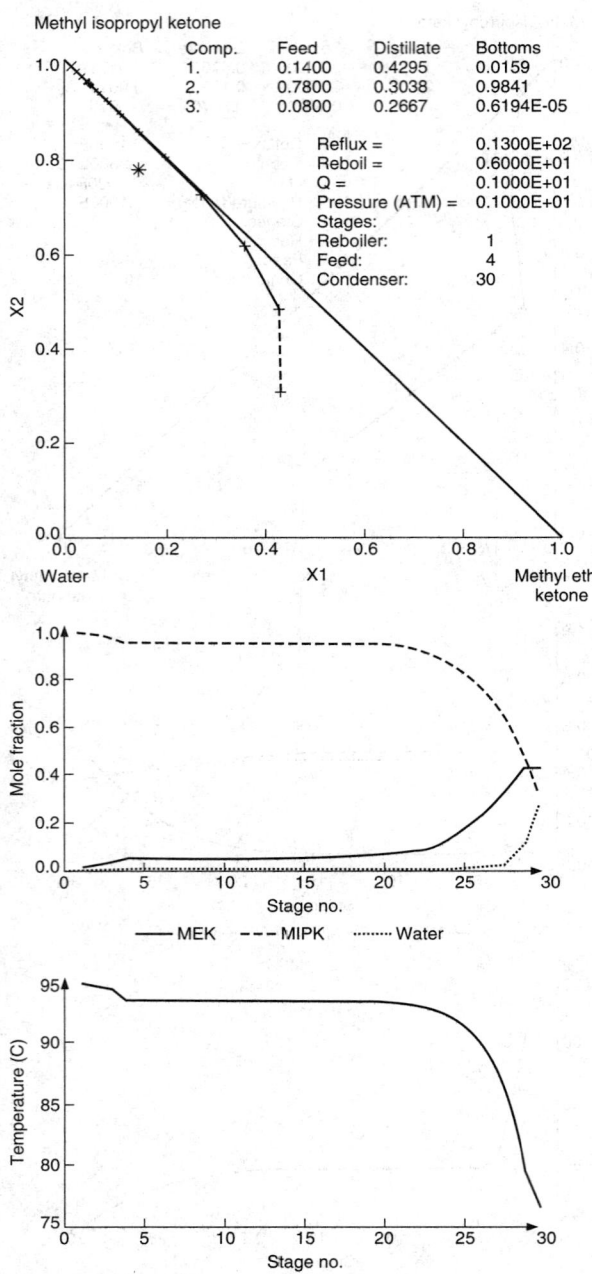

Methyl isopropyl ketone

Comp.	Feed	Distillate	Bottoms
1.	0.1400	0.4295	0.0159
2.	0.7800	0.3038	0.9841
3.	0.0800	0.2667	0.6194E-05

Reflux = 0.1300E+02
Reboil = 0.6000E+01
Q = 0.1000E+01
Pressure (ATM) = 0.1000E+01
Stages:
Reboiler: 1
Feed: 4
Condenser: 30

MEK ——   MIPK – – –   Water ········

(c)

FIG. 13-62 (*Continued*) Sensitivity of composition and temperature profiles for MEK-MIPK-water system.

AZEOTROPIC DISTILLATION

Introduction The term azeotropic distillation has been applied to a broad class of fractional distillation-based separation techniques in that specific azeotropic behavior is exploited to effect a separation. The agent that causes the specific azeotropic behavior, often called the *entrainer,* may already be present in the feed mixture (a self-entraining mixture) or may be an added mass-separation agent. Azeotropic distillation techniques are used throughout the petro-

TABLE 13-17   Sets of Operating Conditions for Fig. 13-62

Case	Reflux ratio	Reboil ratio		Feed composition	Distillate composition	Bottoms composition
A	13	3.8	MEK	0.15	0.60	0.03
			MIPK	0.80	0.16	0.97
			water	0.05	0.24	7 ppm
B	13	3.8	MEK	0.14	0.48	0.05
			MIPK	0.78	0.15	0.95
			water	0.08	0.37	20,000 ppm
C	13	6	MEK	0.14	0.43	0.02
			MIPK	0.78	0.30	0.98
			water	0.08	0.27	6.5 ppm

chemical and chemical processing industries for the separation of close-boiling, pinched, or azeotropic systems for which simple distillation is either too expensive or impossible. With an azeotropic feed mixture, presence of the azeotroping agent results in the formation of a more favorable azeotropic pattern for the desired separation. For a close-boiling or pinched feed mixture, the azeotroping agent changes the dimensionality of the system and allows separation to occur along a less-pinched path. Within the general heading of azeotropic distillation techniques, several approaches have been followed in devising azeotropic distillation flowsheets including:

1.   Choosing an entrainer to give a residue curve map with specific distillation regions and node temperatures.
2.   Exploiting changes in azeotropic composition with total system pressure.
3.   Exploiting curvature of distillation region boundaries.
4.   Choosing an entrainer to cause azeotrope formation in combination with liquid-liquid immiscibility.

The first three of these are solely VLE-based approaches, involving a series of simple distillation operations and recycles. The final approach also relies on distillation (VLE), but also exploits another physical phenomena, liquid-liquid phase formation (phase splitting), to assist in entrainer recovery. This approach is the most powerful and versatile. Examples of industrial uses of azeotropic distillation grouped by method are given in Table 13-18.

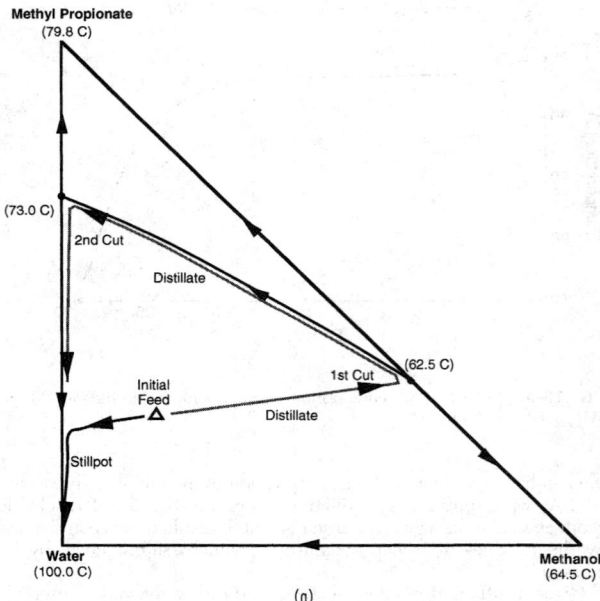

FIG. 13-63 Batch distillation paths. (*a*) Methanol-methyl propionate-water system.

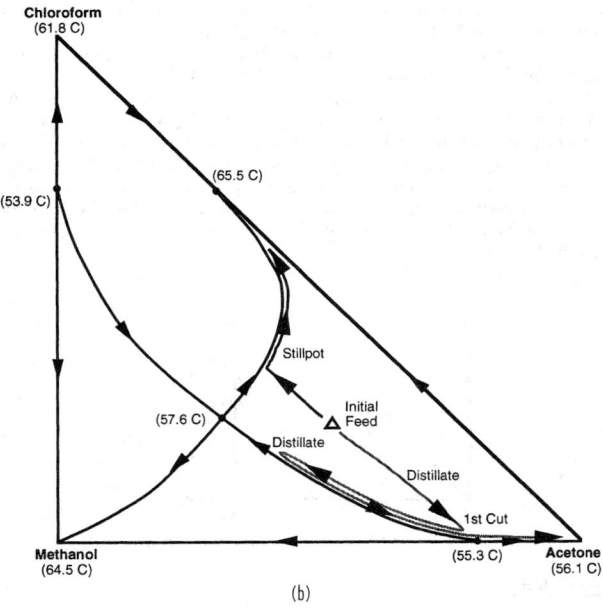

FIG. 13-63 (*Continued*) Batch distillation paths. (*b*) Methanol-acetone-chloroform system.

The choice of the appropriate azeotropic distillation method and the resulting flowsheet for the separation of a particular mixture are strong functions of the separation objective. For example, it may be desirable to recover all constituents of the original feed mixture as pure components, or only some as pure components and some as azeotropic mixtures suitable for recycle. Not every objective may be obtainable by azeotropic distillation for a given mixture and portfolio of candidate entrainers.

Exploitation of Homogeneous Azeotropes Homogeneous azeotropic distillation refers to a flowsheet structure in which azeotrope formation is exploited or avoided in order to accomplish the desired separation in one or more distillation columns. The azeotropes in the system either do not exhibit two-liquid-phase behavior or the liquid-phase behavior is not or cannot be exploited in the separation sequence. The structure of a particular sequence will depend on the geometry of the residue curve map or distillation region diagram for the feed mixture-entrainer system. Two approaches are possible:

1. Selection of an entrainer such that the desired products all lie within a single distillation region (the products may be pure components or azeotropic mixtures).

2. Selection of an entrainer such that although the desired products lie in different regions, some type of boundary-crossing mechanism is employed.

As mentioned previously, ternary mixtures can be represented by 125 different residue curve maps or distillation region diagrams. However, feasible distillation sequences using the first approach can be developed for breaking homogeneous binary azeotropes by the addition of a third component only for those more restricted systems that do not have a distillation boundary connected to the azeotrope and for which one of the original components is a node. For example, from

TABLE 13-18   Examples of Azeotropic Distillation

System	Type	Entrainer(s)	Remark
Exploitation of homogeneous azeotropes			
No known industrial examples			
Exploitation of pressure sensitivity			
THF-water	Minimum boiling azeotrope	None	Alternative to extractive distillation
Methyl acetate-methanol	Minimum boiling azeotrope	None	Element of recovery system for alternative to production of methyl acetate by reactive distillation; alternative to azeotropic, extractive distillation
Alcohol-ketone systems	Minimum boiling azeotropes	None	
Ethanol-water	Minimum boiling azeotrope	None	Alternative to extractive distillation, salt extractive distillation, heterogeneous azeotropic distillation; must reduce pressure to less than 11.5kPa for azeotrope to disappear
Exploitation of boundary curvature			
Hydrochloric acid-water	Maximum boiling azeotrope	Sulfuric acid	Alternative to salt extractive distillation
Nitric acid-water	Maximum-boiling azeotrope	Sulfuric acid	Alternative to salt extractive distillation
Exploitation of azeotropy and liquid phase immiscibility			
Ethanol-water	Minimum boiling azeotrope	Cyclohexane, benzene, heptane, hexane, toluene, gasolene, diethyl ether	Alternative to extractive distillation, pressure-swing distillation
Acetic acid-water	Pinched system	Ethyl acetate, propyl acetate, diethyl ether, dichloroethane, butyl acetate	
Butanol-water	Minimum boiling azeotrope	Self-entraining	
Acetic acid-water-vinyl acetate	Pinched, azeotropic system	Self-entraining	
Methyl acetate-methanol	Minimum boiling azeotrope	Toluene, methyl isobutyl ketone	Element of recovery system for alternative to production of methyl acetate by reactive distillation; alternative to extractive pressure-swing distillation
Diethoxymethanol-water-ethanol	Minimum-boiling azeotropes	Self-entraining	
Pyridine-water	Minimum-boiling azeotrope	Benzene	Alternative to extractive distillation
Hydrocarbon-water	Minimum-boiling azeotrope	Self-entraining	

Fig. 13-59, the following eight residue curve maps are suitable for breaking homogeneous minimum-boiling azeotropes: DRD 002, 027, 030, 040, 051, 056, 060, and 061 as collected in Fig. 13-64a. To produce the necessary distillation region diagrams, an entrainer must be found that is either: (1) an intermediate boiler that forms no azeotropes (DRD 002), or (2) lowest boiling or intermediate boiling and forms a maximum-boiling azeotrope with the lower-boiling original component. In these cases, the entrainer may also optionally form a minimum-boiling azeotrope with the higher boiling of the original components or a minimum-boiling ternary azeotrope. In all cases, after the addition of the entrainer, the higher-boiling original component is a node and is removed as bottoms product from a first column

operated in the indirect mode with the lower-boiling original component recovered as distillate in a second column.

The seven residue curve maps suitable for breaking homogeneous maximum-boiling azeotropes (DRD 028, 031, 035, 073, 078, 088, 089) are shown in Fig. 13-64b. In this case, the entrainer must form a minimum-boiling azeotrope with the higher-boiling original component and either a maximum-boiling azeotrope or no azeotrope with the lower-boiling original component. In all cases, after the addition of the entrainer, the lower-boiling original component is a node and is removed as distillate from a first column operated in the direct mode with the higher-boiling original component recovered as bottoms product in a second column.

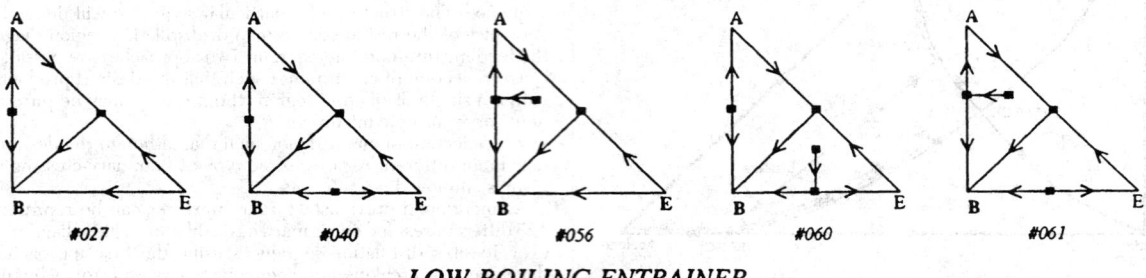

LOW-BOILING ENTRAINER.

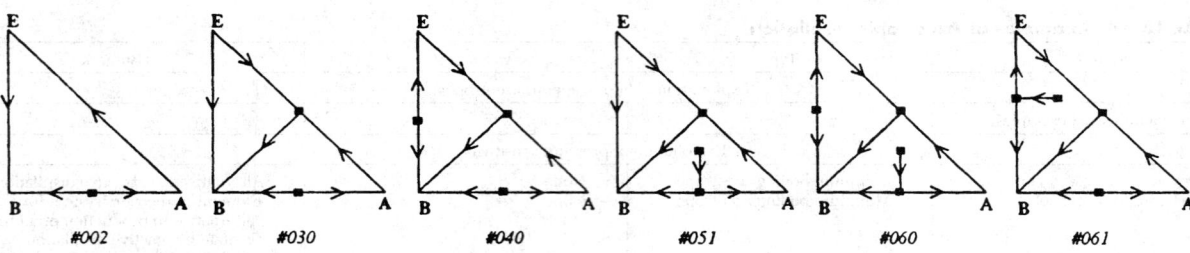

INTERMEDIATE-BOILING ENTRAINER.

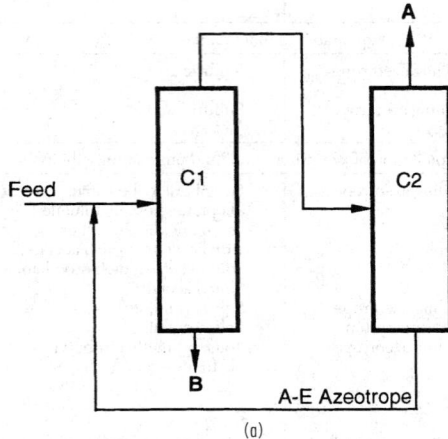

(a)

FIG. 13-64   Feasible distillation region diagrams for breaking homogeneous binary azeotrope A-B. (a) Low-boiling entrances.

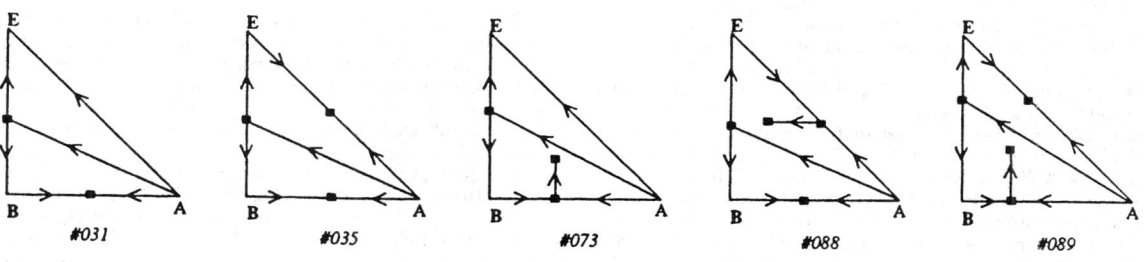

INTERMEDIATE-BOILING ENTRAINER.

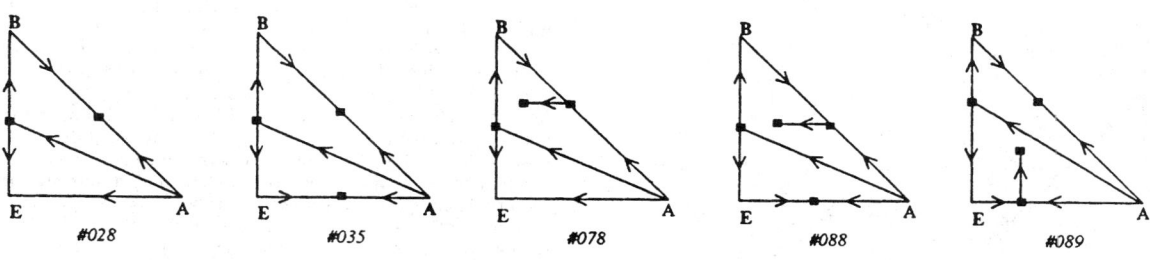

HIGH-BOILING ENTRAINER.

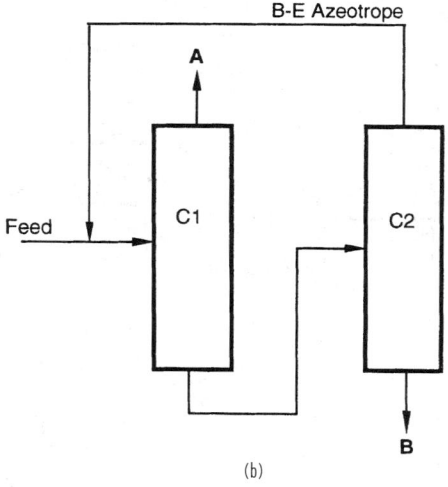

(b)

FIG. 13-64 (*Continued*) Feasible distillation region diagrams for breaking homogeneous binary azeotrope A-B. (*b*) Intermediate-boiling entrances.

In one sense, the restrictions on the boiling point and azeotrope formation of the entrainer act as efficient screening criteria for entrainer selection. Entrainers that do not show appropriate boiling-point characteristics can be discarded without detailed analysis. However, in another sense, although theoretically feasible, the above sequences suffer from serious drawbacks that limit their practical application. DRD 002 requires that the entrainer be an intermediate-boiling component that forms no azeotropes. Unfortunately these are often difficult criteria to meet, as any intermediate boiler will be closer-boiling to both of the original components and, therefore, will be more likely to be at least pinched or even form azeotropes. The remaining feasi-

ble distillation region diagrams require that the entrainer form a maximum-boiling azeotrope with the lower-boiling original component. Because maximum-boiling azeotropes are relatively rare, finding a suitable entrainer may be difficult.

For example, the dehydration of organics that form homogeneous azeotropes with water is a common industrial problem. It is extremely difficult to find an intermediate-boiling entrainer that also does not form an azeotrope with water. Furthermore, the resulting separation is likely to be close-boiling or pinched throughout most of the column, requiring a large number of stages. However, consider the separation of valeric acid (187.0°C) and water. This system exhibits an azeotrope

(99.8°C). Ignoring for the moment potentially severe corrosion problems, formic acid (100.7°C), which is an intermediate boiler and which forms a maximum-boiling azeotrope with water (107.1°C), is a candidate entrainer (DRD 030, Fig. 13-65a). In the conceptual sequence shown in Fig. 13-65b, a recycle of the formic acid-water maximum-boiling azeotrope is added to the original valeric acid-water feed, which may be of any composition. Using the indirect mode of operation, high-boiling node valeric acid is removed in high purity and high recovery as bottoms in a first column, which by mass balance produces a formic acid-water distillate. This binary mixture is fed to a second column that produces pure water as distillate and the formic acid-water azeotrope as bottoms for recycle to the first column. The inventory of formic acid is an important optimization variable in this theoretically feasible but difficult separation scheme.

Exploitation of Pressure Sensitivity   The breaking of homogeneous azeotropes that are part of a distillation boundary (that is, into products in different distillation regions) requires that the boundary be "crossed." This may be done by mixing some external stream with the original feed stream in one region such that the resulting composition is in another region for further processing. However, the external stream must be completely regenerated, and mass-balance observed. For example, it is not possible to break a homogeneous binary azeotrope simply by adding one of the products to cross the azeotropic composition.

The composition of many azeotropes varies with the system pressure (Horsley, *Azeotropic Data-III*, American Chemical Society, Washington, 1983 and Gmehling et al., *Azeotropic Data*, VCH Publishers, Deerfield Beach, Florida, 1994). This effect can be exploited to separate azeotropic mixtures by so-called pressure-swing distillation if at some pressure the azeotrope simply disappears, as for example does the ethanol-water azeotrope at pressures below 11.5 kPa. However, pressure sensitivity can still be exploited if the azeotropic composition and related distillation boundary change sufficiently over a moderate change in total system pressure. A composition in one region under one set of conditions, could be in a different region under a different set of conditions. A two-column sequence for separating a binary maximum-boiling azeotrope is shown in Fig. 13-66 for a system in which the azeotropic composition at pressure P1 is richer in component B than the azeotropic composition at pressure P2. The first column, operating at pressure P1, is fed a mixture of fresh feed and recycle stream from the second column such that the overall composition lies on the A side of the azeotropic composition at P1. Pure component A is recovered as distillate and a mixture near the azeotropic composition is produced as bottoms. The pressure of this bottoms stream is changed to P2 and fed to the second column. This feed is on the B side of the azeotropic composition at P2. Pure component B is now recovered as the distillate and the azeotropic bottoms composition is recycled to the first column. An analogous flowsheet can be used for separating binary-homogeneous minimum-boiling azeotropes. In this case the pure components are recovered as bot-

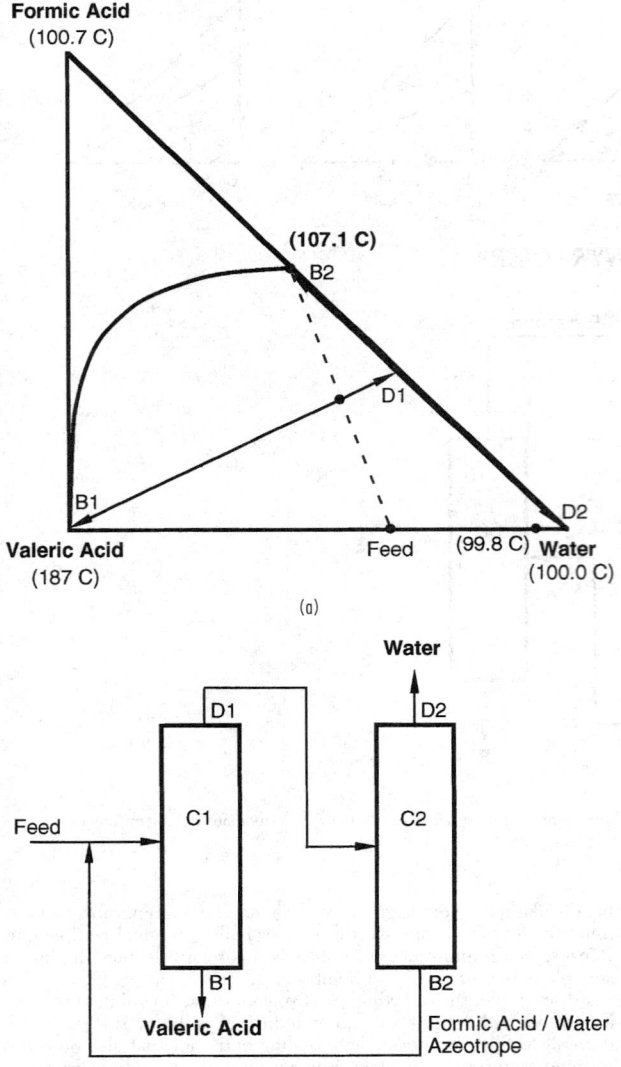

FIG. 13-65   Valeric acid-water separation with formic acid. (*a*) Mass balances on distillation region diagram. (*b*) Conceptual sequence.

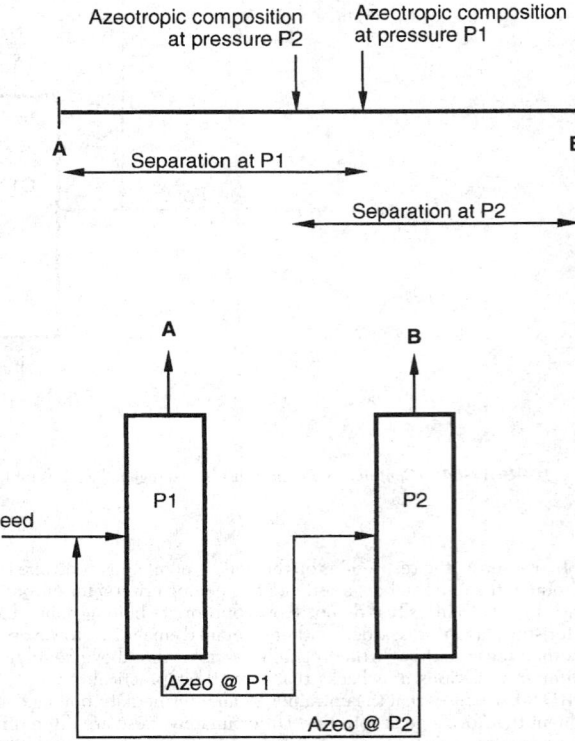

FIG. 13-66   Conceptual sequence for separating maximum-boiling binary azeotrope with pressure swing distillation.

toms in both columns and the distillate from each column is recycled to the other column.

For pressure-swing distillation to be practical, the azeotropic composition must vary at least 5 percent, (preferably 10 percent or more) over a moderate pressure range (not more than ten atmospheres between the two pressures). With a very large pressure range, refrigeration may be required for condensation of the low-pressure distillate or an impractically high reboiler temperature may result in the high-pressure column. The smaller the variation of azeotrope composition over the pressure range, the larger will be the recycle streams between the two columns. In particular, for minimum-boiling azeotropes, the pressure-swing distillation approach requires high energy usage and high capital costs (large-diameter columns) because both recycled azeotropic compositions must be taken overhead. Often one lobe of an azeotropic VLE diagram is pinched regardless of pressure; and, therefore, one of the columns will require a large number of stages to produce the corresponding pure-component product.

General information on pressure-swing distillation can be found in Van Winkle (*Distillation*, McGraw-Hill, New York, 1967), Wankat (*Equilibrium-Staged Separations*, Elsevier, New York, 1988), and Knapp and Doherty [*Ind. Eng. Chem. Res.*, **31**, 346 (1992)]. Only a relatively small fraction of azeotropes are sufficiently pressure sensitive for a pressure-swing process to be economical. Some applications include the minimum-boiling azeotrope tetrahydrofuran-water [Tanabe et al., U.S. Patent 4,093,633 (1978)], and maximum-boiling azeotropes of hydrogen chloride-water and formic acid-water (Horsley, *Azeotropic Data-III*, American Chemical Society, Washington, 1983). Since separatrices also move with pressure-sensitive azeotropes, the pressure-swing principle can also be used for overcoming distillation boundaries in multicomponent azeotropic mixtures.

Exploitation of Boundary Curvature A second approach to boundary crossing exploits boundary curvature in order to produce compositions in different distillation regions. When distillation boundaries exhibit extreme curvature, it may be possible to design a column such that the distillate and bottoms are on the same residue curve in one distillation region, while the feed (which is not required to lie on the column-composition profile) is in another distillation region. In order for such a column to meet material-balance constraints (i.e., bottom, distillate, feed on a straight line), the feed must be located in a region where the boundary is concave.

As an example, Van Dongen [Ph.D. Thesis, University of Massachusetts, (1983)] considered the separation of a methanol-methyl acetate mixture, which forms a homogeneous azeotrope, using *n*-hexane as an entrainer. The separatrices for this system (Fig. 13-67*a*) are somewhat curved. A separation sequence that exploits this boundary curvature is shown in Fig. 13-67*b*. Recycled methanol-methyl acetate binary azeotrope and methanol-methyl acetate-hexane ternary azeotrope are added to the original feed F0 to produce a net-feed composition F1 for column C1 designed to lie on a line between pure methanol and the curved part of the boundary between Regions I and II. C1 is operated in the indirect mode producing the high-boiling node methanol as a bottoms product, and by mass balance, a distillate near the curved boundary. The distillate, although in Region I, becomes feed F2 to column C2 which is operated in the direct mode entirely in Region II, producing the low-boiling node ternary azeotrope as distillate and by mass balance, a methanol-methyl acetate mixture as bottoms. This bottoms mixture is on the opposite side of the methanol-methyl acetate azeotrope at the original feed F0. The bottoms from C2 is finally fed to binary distillation column C3, which produces pure methyl acetate as bottoms product and the methanol-methyl acetate azeotrope as distillate. The distillates from C2 and C3 are recycled to C1. The distillate and bottoms compositions for C2 lie on the same residue curve, and the composition profile lies entirely within Region II, even though its feed composition is in Region I.

Exploitation of boundary curvature for breaking azeotropes is very similar to exploiting pressure sensitivity from a mass-balance point of view, and suffers from the same disadvantages. Separation schemes have large recycle flows, and in the case of minimum-boiling azeotropes, the recycle streams are distillates. However, in the case of maximum-boiling azeotropes, these recycles are underflows and

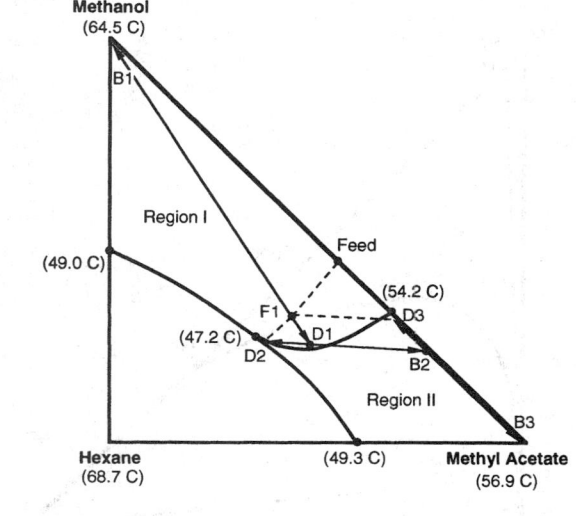

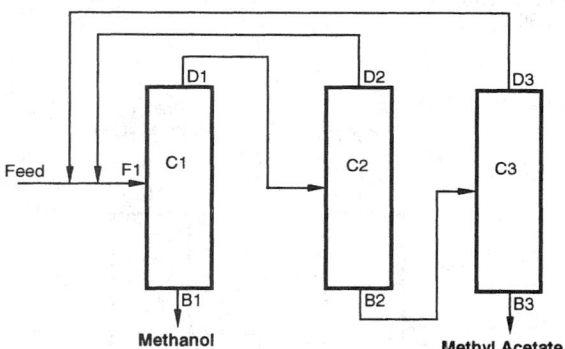

FIG. 13-67 Separation of methanol-methyl acetate by exploitation of distillation boundary curvature.

the economics are improved. One such application, illustrated in Fig. 13-68, is the separation of the nitric acid-water azeotrope by adding sulfuric acid. Recycled sulfuric acid is added to a nitric acid-water mixture near the azeotropic composition to produce a net feed in Region I. The first column, operated in the direct mode, produces a nitric-acid distillate and a bottoms product, by mass balance, near the distillation boundary. In this case, sulfuric acid associates with water so strongly and the separatrix is so curved and nearly tangent to the water-sulfuric acid edge of the composition diagram that the second column operating in the indirect mode in Region II, producing sulfuric acid as bottoms product also produces a distillate close enough to the water specification that a third column is not required (Thiemann et al., in *Ullmann's Encyclopedia of Industrial Chemistry*, Fifth Edition, Volume **A17**, VCH Verlagsgesellschaft mbH, Weinheim, 1991).

Exploitation of Azeotropy and Liquid-Phase Immiscibility One powerful and versatile separation approach exploits several physical phenomena simultaneously including enhanced vapor-liquid behavior, where possible, and liquid-liquid behavior to bypass difficult distillation separations. For example, the overall separation of close-boiling mixtures can be made easier by the addition of an entrainer that forms a heterogeneous minimum-boiling azeotrope with one (generally the lower-boiling) of the key components. Two-liquid-phase formation provides a means of breaking this azeotrope, thus simplifying the entrainer recovery and recycle process. Moreover,

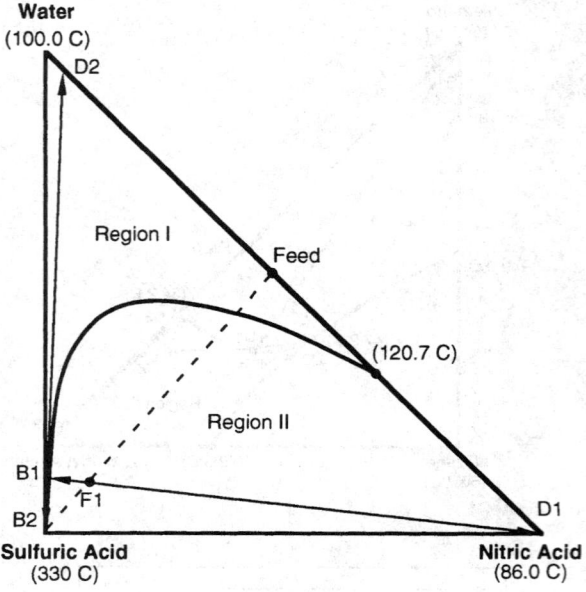

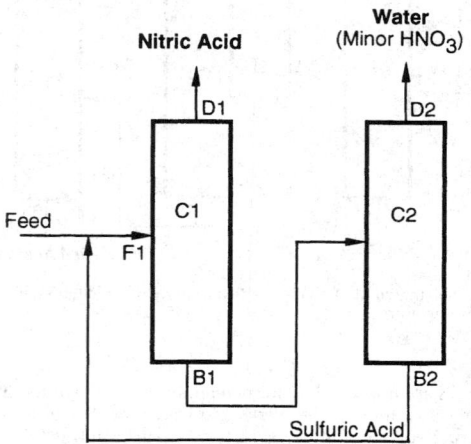

FIG. 13-68 Separation of nitric acid-water system with sulfuric acid in a two-column sequence exploiting extreme boundary curvature.

since liquid-liquid tie lines are unaffected by distillation boundaries (and the separate liquid phases are often located in different distillation regions), liquid-liquid phase splitting is a powerful mechanism for crossing distillation boundaries. The phase separator is usually a simple decanter, but sometimes a multistage extractor is substituted. The decanter or extractor can also be replaced by some other non-VLE-based separation technique such as membrane permeation, chromatography, adsorption, or crystallization. In addition, sequences may include additional separation operations (distillations or other methods) for preconcentration of the feed mixture, entrainer recovery, and final-product purification.

The simplest case of combining VLE and LLE is the separation of a binary heterogeneous azeotropic mixture. One example is the dehydration of 1-butanol, a self-entraining system, in which butanol (117.7°C) and water form a minimum-boiling heterogeneous azeotrope (93.0°C). As shown in Fig. 13-69, the fresh feed may be added

to either column C1 or C2, depending on whether the feed is on the organic-rich side or the water-rich side of the azeotrope. The feed may also be added into the decanter directly if it doesn't move the overall composition of the decanter outside of the two-liquid-phase region. Column C1 produces anhydrous butanol as a bottoms product and a composition close to the butanol-water azeotrope as the distillate. After condensation, the azeotrope rapidly phase separates in the decanter. The upper layer, consisting of 78 wt % butanol, is refluxed totally to C1 for further butanol recovery. The water layer, consisting of 92 wt % water, is fed to C2. This column produces pure water as a bottoms product and, again, a composition close to the azeotrope as distillate for recycle to the decanter. Sparged steam may be used in C2, saving the cost of a reboiler. A similar flowsheet can be used for dehydration of hydrocarbons and other species that are largely immiscible with water.

A second example of the use of liquid-liquid immiscibilities in an azeotropic-distillation sequence is the separation of the ethanol-water minimum-boiling azeotrope. For this separation, a number of entrainers have been proposed, which are usually chosen to be immiscible with water, form a ternary minimum-boiling (preferably heterogeneous) azeotrope with ethanol and water (and, therefore, usually also binary minimum-boiling azeotropes with both ethanol and water). All such systems correspond to DRD 058, although the labeling of the vertices depends on whether the entrainer is lower boiling than ethanol, intermediate boiling, or higher boiling than water. The residue curve map for the case for cyclohexane as entrainer was illustrated in Fig. 13-58c. One three-column distillation sequence is shown in Fig. 13-70. Other two-, three-, or four-column sequences have been described by Knapp and Doherty (*Kirk-Othmer Encyclopedia of Chemical Technology*, Fourth Edition, Vol. **8**, Wiley, New York, 1993).

Fresh aqueous ethanol feed is first preconcentrated to nearly the azeotropic composition in column C3, while producing a water bottoms product. The distillate from C3 is sent to column C1, which is refluxed with the entire organic (entrainer-rich) layer, recycled from a decanter. Mixing of these two streams is the key to this sequence as it allows the overall feed composition to cross the distillation boundary into Region II. column C1 is operated to recover pure high-boiling node ethanol as a bottoms product and to produce a distillate close to the ternary azeotrope. If the ternary azeotrope is heterogeneous (as it

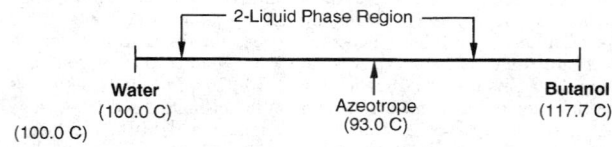

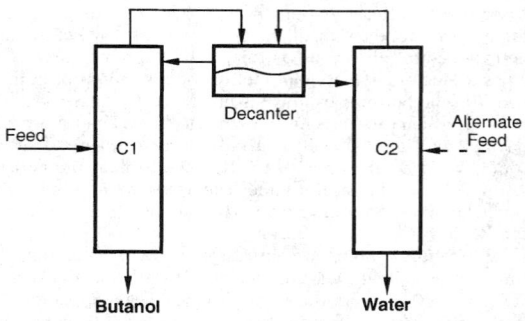

FIG. 13-69 Separation of butanol-water with heterogeneous azeotropic distillation.

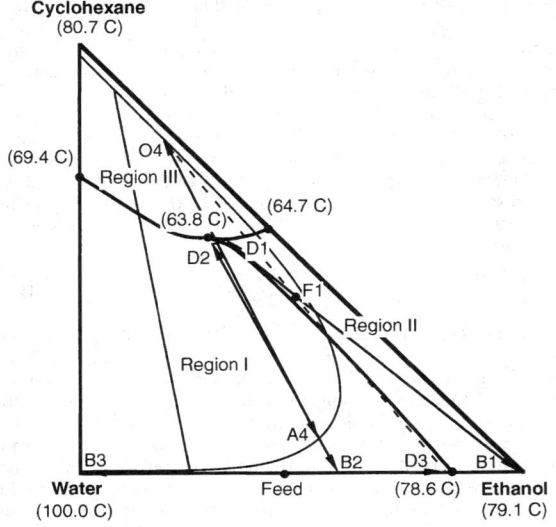

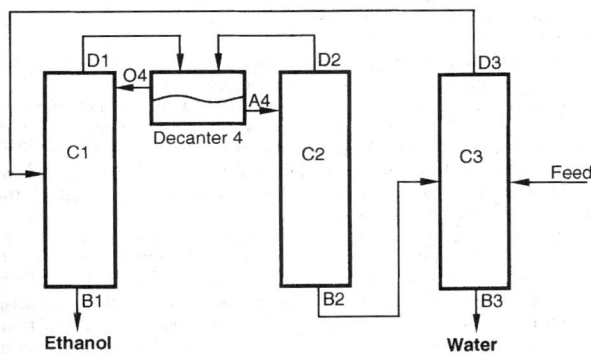

FIG. 13-70 Three-column sequence for ethanol dehydration with cyclohexane (operating column C2 in the direct mode).

is in this case), it is sent to the decanter for phase separation. If the ternary azeotrope is homogeneous (as it is in the alternative case of ethyl acetate as the entrainer) the distillate is first mixed with water before being sent to the decanter. The inventory of entrainer is adjusted to allow C1 to operate essentially between two nodes, although such practice, as discussed previously, is relatively susceptible to instabilities from minor feed or reflux perturbations. Refluxing a fraction of the water-rich decanter layer results in an additional degree of freedom to mitigate against variability in the feed composition. The remaining portion of the water layer from the decanter is stripped of residual cyclohexane in column C2, which may be operated either in the direct mode (producing low-boiling node ternary azeotrope as distillate and, by mass balance, an ethanol-water bottoms for recycle to C3 or, in the indirect mode (producing high-boiling node water as bottoms and, by mass balance, a ternary distillate near the distillation boundary. The distillate may be recycled to the decanter, the top of C2, or C2 feed.

Design and Operation of Azeotropic Distillation Columns Simulation and design of azeotropic distillation columns is a difficult computational problem, but one that is readily handled, in most cases, by widely available commercial computer process simulation packages [Glasscock and Hale, *Chem. Eng.*, **101**(11), 82 (1994)]. Most simula-

tors are capable of modeling the steady state and dynamic behavior of both homogeneous azeotropic distillation systems and those systems involving two-liquid phase behavior within the column, if accurate thermodynamic data and activity-coefficient or equation-of-state models are available. However, VLE and VLLE estimated or extrapolated from binary data or predicted from such methods as UNIFAC may not be able to accurately locate boundaries and predict the extent of liquid immiscibilities. Moreover, different activity-coefficient models fit to the same experimental data often give very different results for the shape of boundaries and liquid-liquid regions. Therefore the design of separation schemes relying on boundary curvature should not be attempted unless accurate, reliable experimental equilibrium data are available.

Two liquid phases can occur within a column in the distillation of heterogeneous systems. Older references, for example Robinson and Gilliland (*Elements of Fractional Distillation*, McGraw-Hill, New York, 1950) state that the presence of two liquid phases in a column should be avoided as much as possible because performance may be reduced. However, more recent studies indicate that problems with two-phase flow have been overstated [Herron et al., *AIChE J.* **34,** 1267 (1988) and Harrison, *Chem. Eng. Prog.*, **86**(11), 80 (1990)]. Based on case-history data and experimental evidence, there is no reason to expect unusual capacity or pressure-drop limitations, and standard correlations for these parameters should give acceptable results. Because of the violent nature of the gas/liquid/liquid mixing on trays, trayed column efficiencies are relatively unaffected by liquid-liquid phase behavior. The falling-film nature of gas/liquid/liquid contact in packing, however, makes that situation more uncertain. Reduced efficiencies may be expected in systems where one of the keys distributes between the phases.

EXTRACTIVE DISTILLATION

Introduction Extractive distillation is a partial vaporization process in the presence of a miscible, high-boiling, non-volatile mass-separation agent, normally called the *solvent*, which is added to an azeotropic or nonazeotropic feed mixture to alter the volatilities of the key components without the formation of any additional azeotropes. Extractive distillation is used throughout the petrochemical- and chemical-processing industries for the separation of close-boiling, pinched, or azeotropic systems for which simple single-feed distillation is either too expensive or impossible. It can also be used to obtain products which are residue-curve saddles, a task not generally possible with single-feed distillation.

Fig. 13-71 illustrates the classical implementation of an extractive distillation process for the separation of a binary system. The configuration consists of a double-feed extractive column and a solvent-recovery column. The components A and B may have a low relative volatility or form a minimum-boiling azeotrope. The solvent is introduced into the extractive column at a high concentration a few stages below the condenser, but above the primary-feed stage. Since the solvent is chosen to be nonvolatile it remains at a relatively high concentration in the liquid phase throughout the sections of the column below the solvent-feed stage.

One of the components, A (not necessarily the most volatile species of the original mixture), is withdrawn as an essentially pure distillate stream. Because the solvent is nonvolatile, at most a few stages above the solvent-feed stage are sufficient to rectify the solvent from the distillate. The bottoms product, consisting of B and the solvent, is sent to the recovery column. The distillate from the recovery column is pure B, and the solvent-bottoms product is recycled back to the extractive column.

Extractive distillation works by the exploitation of the selective solvent-induced enhancements or moderations of the liquid-phase nonidealities of the components to be separated. The solvent selectively alters the activity coefficients of the components being separated. To do this, a high concentration of solvent is necessary. Several features are essential:

1. The solvent must be chosen to affect the liquid-phase behavior of the key components differently, otherwise no enhancement in separability will occur.

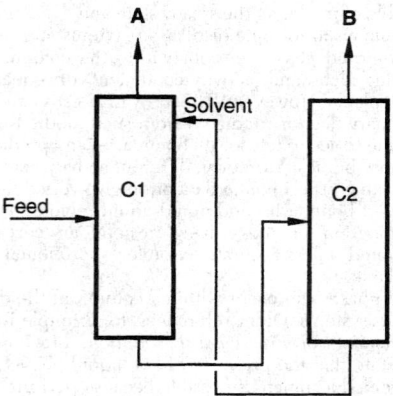

FIG. 13-71 Typical extracting distillation sequence. Component *A* is less associated with the solvent.

2. The solvent must be higher boiling than the key components of the separation and must be relatively nonvolatile in the extractive column, in order to remain largely in the liquid phase.

3. The solvent should not form additional azeotropes with the components in the mixture to be separated.

4. The extractive column must be a double-feed column, with the solvent feed above the primary feed; the column must have an extractive section.

As a consequence of these restrictions, separations of binary systems by extractive distillation correspond to only two possible three-component distillation region diagrams, depending on whether the binary system is pinched or close boiling (DRD 001), or forms a minimum-boiling azeotrope (DRD 003). The addition of high-boiling solvents can also facilitate the breaking of maximum-boiling azeotropes (DRD 014), for example splitting the nitric acid-water azeotrope with sulfuric acid. However, as explained in the section on azeotropic distillation, this type of separation might better be characterized as exploitation of extreme boundary curvature rather than extractive distillation, as the important liquid-phase activity-coefficient modification occurs in the bottom of the column. Although many references show sulfuric acid being introduced high in the column, two separate feeds are in fact not required.

Examples of industrial uses of extractive distillation grouped by distillation region diagram type are given in Table 13-19. Achievable compositions in dual-feed extractive distillation columns are very different from the bow-tie regions for single-feed columns. Either pure component (the higher-boiling of which is a saddle) for close-boiling systems, and either pure component (both of which are saddles) for minimum-boiling azeotropic systems can be obtained as distillate.

Extractive distillation is generally only applicable to systems in which the components to be separated contain one or more different functional groups. Extractive distillation is usually uneconomical for separating stereoisomers, homologs, or homology or structural isomers containing the same functional groups, unless the differences in structure also contribute to significantly different polarity, dipole moment, or hydrophobic character. One such counter-example is the separation of ethanol from isopropanol, where the addition of methyl benzoate raises the relative volatility from 1.09 to 1.27 [Berg et al., *Chem. Eng. Comm.*, **66**, 1 (1988)].

Solvent Effects in Extractive Distillation In the distillation of ideal or nonazeotropic mixtures, the component with the lowest pure-component boiling point is always recovered primarily in the distillate, while the highest boiler is recovered primarily in the bottoms. The situation is not as straightforward for an extractive-distillation operation. With some solvents, the component with the lower pure-component boiling point will be recovered in the distillate as in ordinary distillation. For another solvent, the expected order is reversed, and the component with the higher pure-component boiling point will be

recovered in the distillate. The possibility that the expected relative volatility may be reversed by the addition of solvent is entirely a function of the way the solvent interacts with and modifies the activity coefficients and, thus, the volatility of the components in the mixture.

In normal applications of extractive distillation (i.e., pinched, close-boiling, or azeotropic systems), the relative volatilities between the light and heavy key components will be unity or close to unity. Assuming an ideal vapor phase and subcritical components, the relative volatility between the light and heavy keys of the desired separation can be written as the product of the ratios of the pure-component vapor pressures and activity-coefficient ratios whether the solvent is present or not:

$$\alpha_{L,H} = \left(\frac{P_L^{sat}}{P_H^{sat}} \right) \left(\frac{\gamma_L}{\gamma_H} \right) \tag{13-127}$$

where *L* and *H* denote the lower-boiling and higher-boiling pure component, respectively.

The addition of the solvent has an indirect effect on the vapor-pressure ratio. Because the solvent is high boiling and is generally added at a relatively high mole ratio to the primary-feed mixture, the temperature of an extractive-distillation process tends to increase over that of a simple distillation of the original binary mixture (unless the system pressure is lowered). The result is a corresponding increase in the vapor pressure of both key components. However, the rise in operating temperature generally *does not* result in a significant modification of the relative volatility, because the ratio of vapor pressures often remains approximately constant, unless the slopes of the vapor-pressure curves differ significantly. The ratio of the vapor pressures typically remains greater than unity, following the "natural" volatility of the system.

Since activity coefficients have a strong dependence on composition, the effect of the solvent on the activity coefficients is generally more pronounced. However, the magnitude and direction of change is highly dependent on the solvent concentration, as well as the liquid-phase interactions between the key components and the solvent. The solvent acts to lessen the nonidealities of the key component whose liquid-phase behavior is similar to the solvent, while enhancing the nonideal behavior of the dissimilar key.

The solvent and the key component that show most similar liquid-phase behavior tend to exhibit little molecular interactions. These components form an ideal or nearly ideal liquid solution. The activity coefficient of this key approaches unity, or may even show negative deviations from Raoult's law if solvating or complexing interactions occur. On the other hand, the dissimilar key and the solvent demonstrate unfavorable molecular interactions, and the activity coefficient of this key increases. The positive deviations from Raoult's law are further enhanced by the diluting effect of the high-solvent concentration, and the value of the activity coefficient of this key may approach the infinite dilution value, often a very large number.

The natural relative volatility of the system is enhanced when the activity coefficient of the lower-boiling pure component is increased by the solvent addition (γ_L/γ_H increases and $P_L^{sat}/P_H^{sat} > 1$). In this case, the lower-boiling pure component will be recovered in the distillate as expected. In order for the higher-boiling pure component to be recovered in the distillate, the addition of the solvent must decrease the ratio γ_L/γ_H such that the product of the γ_L/γ_H and P_L^{sat}/P_H^{sat} (i.e., $\alpha_{L,H}$) in the presence of the solvent is less than unity. Generally, the latter is more difficult and requires higher solvent-to-feed ratios. It is normally better to select a solvent that forces the lower-boiling component overhead.

The effect of solvent concentration on the activity coefficients of the key components is shown in Fig. 13-72 for the system methanol-acetone with either water or methylisopropylketone (MIPK) as solvent. For an initial-feed mixture of 50 mol % methanol and 50 mol % acetone (no solvent present), the ratio of activity coefficients of methanol and acetone is close to unity. With water as the solvent, the activity coefficient of the similar key (methanol) rises slightly as the solvent concentration increases, while the coefficient of acetone approaches the relatively large infinite-dilution value. With methylisopropylketone as the solvent, acetone is the similar key and its activity coefficient drops toward unity as the solvent concentration increases, while the activity coefficient of the methanol increases.

TABLE 13-19 Examples of Extractive Distillation, Salt Extractive Distillation

System	Type	Solvent(s)	Remark
Ethanol-water	Minimum-boiling azeotrope	Ethylene glycol, acetate salts for salt process	Alternative to azeotropic distillation, pressure swing distillation
Benzene-cyclohexane	Minimum-boiling azeotrope	Aniline	
Ethyl acetate-ethanol	Minimum-boiling azeotrope	Higher esters or alcohols, aromatics	Process similar for other alcohol-ester systems
THF-water	Minimum-boiling azeotrope	Propylene glycol	Alternative to pressure swing distillation
Acetone-methanol	Minimum-boiling azeotrope	Water, aniline, ethylene glycol	
Isoprene-pentane	Minimum-boiling azeotrope	Furfural, DMF, acetonitrile	
Pyridine-water	Minimum-boiling azeotrope	Bisphenol	
Methyl acetate-methanol	Minimum-boiling azeotrope	Ethylene glycol monomethyl ether	Element of recovery system for alternative to production of methyl acetate by reactive distillation; alternative to azeotropic, pressure, swing distillation
C4 alkenes/C4 alkanes/ C4 dienes	Close-boiling and minimum-boiling azeotropes	Furfural, DMF, acetonitrile, n-methylpyrolidone	
C5 alkenes/C5 alkanes/ C5 dienes	Close-boiling and minimum-boiling azeotropes	Furfural, DMF, acetonitrile, n-methylpyrolidone	
Heptane isomers-cyclohexane	Close-boiling	Aniline, phenol	
Heptane isomers-toluene	Close-boiling and minimum-boiling azeotropes	Aniline, phenol	
Vinyl acetate-ethyl acetate	Close-boiling	Phenol, aromatics	Alternative to simple distillation
Propane-propylene	Close-boiling	Acrylonitrile	Alternative to simple distillation, adsorption
Ethanol-isopropanol	Close-boiling	Methyl benzoate	Alternative to simple distillation
Hydrochloric acid-water	Maximum-boiling azeotrope	Sulfuric acid, calcium chloride for salt process	Sulfuric acid process relies heavily on boundary curvature
Nitric acid-water	Maximum-boiling azeotrope	Sulfuric acid, magnesium nitrate for salt process	Sulfuric acid process relies heavily on boundary curvature

Several methods are available for determining whether the lower- or higher-boiling pure component will be recovered in the distillate. For a series of solvent concentrations, the *y-x* phase diagram of the low-boiling and high-boiling keys can be plotted on a solvent-free basis. At a particular solvent concentration (dependent on the selected solvent and keys), the azeotropic point in the pseudobinary plot disappears at one of the pure component corners. The component corresponding to the corner where the azeotrope disappears is recovered in the distillate [Knapp and Doherty, in *Kirk-Othmer Encyclopedia of Chemical Technology,* Fourth Edition, Vol. **8,** Wiley, New York (1993)]. LaRoche et al. [*Can. J. Chem. Eng.,* **69,** 1302 (1991)] present a related method in which the $\alpha_{L,H} = 1$ line is plotted on the ternary composition diagram. If the $\alpha_{L,H} = 1$ line intersects the lower-boiling pure component-solvent face, then the lower-boiling component will be recovered in the distillate and vice versa if the $\alpha_{L,H} = 1$ line

intersects the higher-boiling pure component-solvent face. A very simple method, if a rigorous residue curve map is available, is to examine the shape and inflection of the residue curves as they approach the pure solvent vertex. Whichever solvent-key component face the residue curves predominantly tend toward as they approach the solvent vertex is the key component that will be recovered in the bottoms with the solvent. In Fig. 13-73*a*, all residue curves approaching the water (solvent) vertex are inflected toward the methanol-water face, with the result that methanol will be recovered in the bottoms and acetone in the distillate. Alternatively, with MIPK as the solvent the residue curves (Fig. 13-73*b*), all residue curves show inflection toward the acetone-MIPK face, indicating that acetone will be recovered in the bottoms and methanol in the distillate.

Extractive Distillation Design and Optimization Extractive distillation column composition profiles have a very characteristic

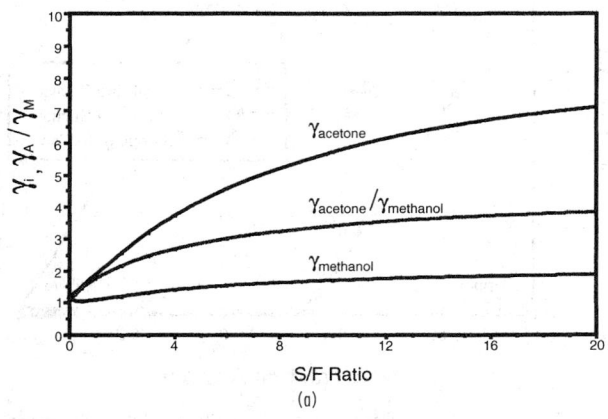

(a)

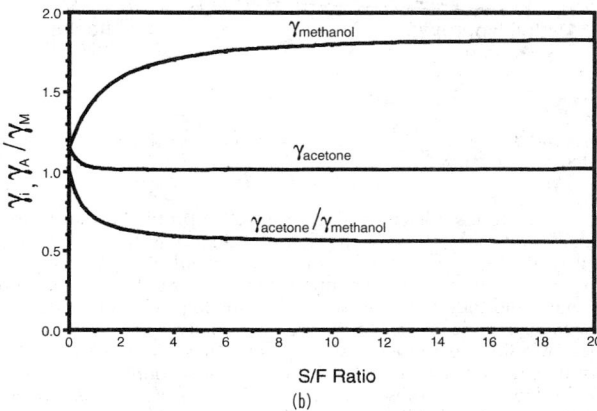

(b)

FIG. 13-72 Effect of solvent concentration on activity coefficients for acetone-methanol system. (*a*) water solvent. (*b*) MIPK solvent.

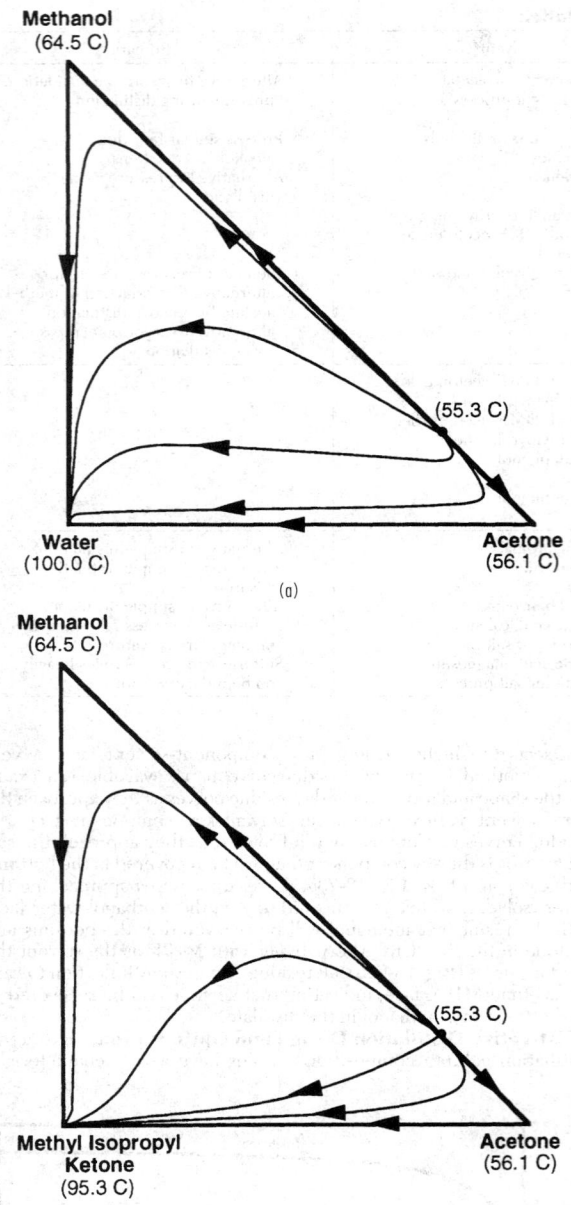

FIG. 13-73 Residue curve maps for acetone-methanol systems. (*a*) With water. (*b*) With MIPK.

shape on a ternary diagram. The composition profile for the separation of methanol-acetone with water is given in Fig. 13-74. Stripping and rectifying profiles start at the bottoms and distillate compositions respectively, track generally along the faces of the composition triangle, and then turn toward the high-boiling (solvent) node and low-boiling node, respectively. For a feasible single-feed design these profiles must cross at some point. However, in an extractive distillation they cannot cross. The extractive-section profile acts at the bridge between these two sections. Most of the key-component separation occurs in this section in the presence of high-solvent concentration.

The variable that has the most significant impact on the economics of an extractive distillation is the solvent-to-feed (S/F) ratio. For close-boiling or pinched nonazeotropic mixtures, no minimum-solvent flow rate is required to effect the separation, as the separation is always theoretically possible (if not economical) in the absence of the solvent. However, the extent of enhancement of the relative volatility is largely determined by the solvent concentration and hence the S/F ratio. The relative volatility tends to increase as the S/F ratio increases. Thus, a given separation can be accomplished in fewer equilibrium stages. As an illustration, the total number of theoretical stages required as a function of S/F ratio is plotted in Fig. 13-75*a* for the separation of the nonazeotropic mixture of vinyl acetate and ethyl acetate using phenol as the solvent.

For the separation of a minimum-boiling binary azeotrope by extractive distillation, there is clearly a minimum-solvent flow rate below which the separation is impossible (due to the azeotrope). For azeotropic separations, the number of equilibrium stages is infinite at or below $(S/F)_{min}$ and decreases rapidly with increasing solvent, and then may asymptote, or rise slowly. The relationship between the total number of stages and the S/F ratio for a given purity and recovery for the azeotropic acetone-methanol system with water as solvent is shown in Fig 13-75*b*. A rough idea of $(S/F)_{min}$ can be determined from a pseudobinary diagram or by plotting the $\alpha_{L,H} = 1$ line on a ternary diagram. The solvent composition at which the azeotrope disappears in a corner of the pseudobinary diagram is an indication of $(S/F)_{min}$ [LaRoche et al., *Can. J. Chem. Eng.*, **69**, 1302 (1991)]. Typically, operating S/F ratios for economically acceptable solvents is between two and five. Higher S/F ratios tend to increase the diameter of both the extractive column and the solvent-recovery columns, but reduce the required number of equilibrium stages and minimum-reflux ratio. Moreover, higher S/F ratios lead to higher reboiler temperatures, resulting in the use of higher-cost utilities, higher utility usages, and greater risk of degradation.

Knight and Doherty [*Ind. Eng. Chem. Fundam.*, **28**, 564 (1989)] have published rigorous methods for computing minimum reflux for extractive distillation, with an operating reflux of 1.2 to 1.5 times the minimum value usually acceptable. Interestingly, unlike other forms of distillation, in extractive distillation the distillate purity or recovery does not increase monotonically with increasing reflux ratio for a given number of stages. Above a maximum-reflux ratio the separation can

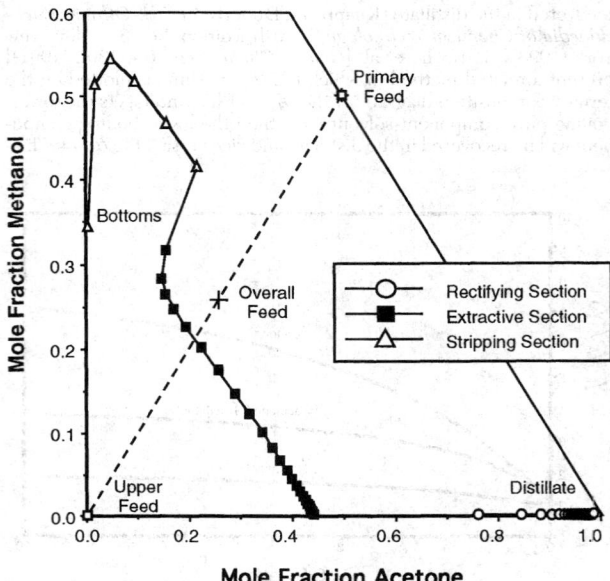

FIG. 13-74 Extractive distillation column composition profile for the separation of acetone-methanol with water.

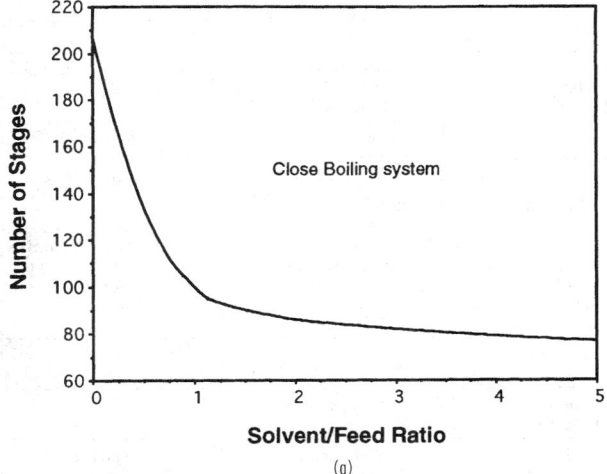

(a)

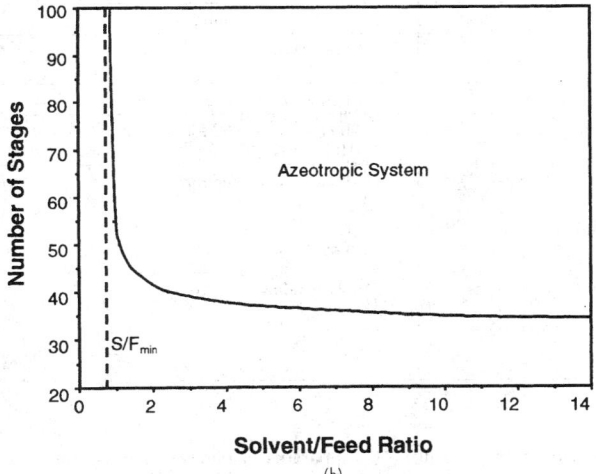

(b)

FIG. 13-75 Number of theoretical stages versus solvent-to-feed ratio for extractive distillation. (a) Close-boiling vinyl acetate-ethyl acetate system with phenol solvent. (b) Azeotropic acetone-methanol system with water solvent.

no longer be achieved and the distillate purity actually decreases for a given number of stages [LaRoche et al., *AIChE J.*, **38**, 1309 (1992)]. The difference between R_{min} and R_{max} increases as the S/F ratio increases. Large amounts of reflux lowers the solvent concentration in the upper section of the column, degrading rather than enhancing column performance. Because the reflux ratio goes through a maximum, the conventional-control strategy of increasing reflux to maintain purity can be detrimental rather than beneficial. However R_{max} generally occurs at impractically high reflux ratios and is typically not of major concern.

The thermal quality of the solvent feed has no effect on the value of $(S/F)_{min}$, but does affect the minimum reflux to some extent, especially as the (S/F) ratio increases. R_{max} occurs at higher values of the reflux ratio as the upper-feed quality decreases; a subcooled upper feed provides additional refluxing capacity and less external reflux is required for the same separation. It is also sometimes advantageous to introduce the primary feed to the extractive distillation column as a vapor to help maintain a higher solvent concentration on the feed tray and the trays immediately below.

Robinson and Gilliland (*Elements of Fractional Distillation*, McGraw-Hill, New York, 1950), Smith (*Design of Equilibrium Stage Processes*, McGraw-Hill, New York, 1963), Van Winkle (*Distillation*, McGraw-Hill, New York, 1967), and Walas (*Chemical Process Equipment*, Butterworths, Boston, 1988) discuss rigorous stage-by-stage design techniques as well as shortcut and graphical methods for determining minimum stages, $(S/F)_{min}$, minimum reflux, and the optimum locations of the solvent and primary feed points. More recently Knapp and Doherty [*AIChE J.*, **40**, 243 (1994)] have published column-design methods based on geometric arguments and fixed-point analysis. Most commercial simulators are capable of solving multiple-feed extractive distillation heat and material balances, but do not include straightforward techniques for calculating $(S/F)_{min}$, minimum or maximum reflux.

Solvent Screening and Selection   Choosing an effective solvent can have the most profound effect on the economics of an extractive distillation process. The approach most often adopted is to first generate a short list of potential solvents using simple qualitative screening and selection methods. Experimental verification is best undertaken only after a list of promising candidate solvents has been generated and some chance at economic viability has been demonstrated via preliminary process modeling.

Solvent selection and screening approaches can be divided into two levels of analysis. The first level focuses on identification of functional groups or chemical families that are likely to give favorable solvent-key component molecular interactions. The second level of analysis identifies and compares individual-candidate solvents. The various methods of analysis are described briefly and illustrated with an example of choosing a solvent for the methanol-acetone separation.

First Level: Broad Screening by Functional Group or Chemical Family

Homologous series.   Select candidate solvents from the high-boiling homologous series of both light and heavy key components. Favor homologs of the heavy key, as this tends to enhance the natural relative volatility of the system. Homologous components tend to form ideal solutions and are unlikely to form azeotropes [Scheibel, *Chem. Eng. Prog.* **44**(12), 927 (1948)].

Robbins chart.   Select candidate solvents from groups in the Robbins Chart (Table 13-15) that tend to give positive (or no) deviations from Raoult's law for the key component desired in the distillate and negative (or no) deviations for the other key.

Hydrogen-bonding characteristics.   Select candidate solvents from groups that are likely to cause the formation of hydrogen bonds with the key component to be removed in the bottoms, or disruption of hydrogen bonds with the key to be removed in the distillate. Formation and disruption of hydrogen bonds are often associated with strong negative and positive deviations, respectively from Raoult's law. Several authors have developed charts indicating expected hydrogen bonding interactions between families of compounds [Ewell et al., *Ind. Eng. Chem.*, **36**, 871 (1944), Gilmont et al., *Ind. Eng. Chem.*, **53**, 223 (1961), and Berg, *Chem. Eng. Prog.*, **65**(9), 52 (1969)]. Table 13-20 presents a hydrogen-bonding classification of chemical families and a summary of deviations from Raoult's law.

Polarity characteristics.   Select candidate solvents from chemical groups that tend to show higher polarity than one key component or lower polarity than the other key. Polarity effects are often cited as a factor in causing deviations from Raoult's law [Hopkins and Fritsch, *Chem. Eng. Prog.*, **51**(8), (1954), Carlson et al., *Ind. Eng. Chem.*, **46**, 350 (1954), and Prausnitz and Anderson, *AIChE J.*, **7**, 96 (1961)]. The general trend in polarity based on the functional group of a molecule is given in Table 13-21. The chart is best for molecules of similar size. A more quantitative measure of the polarity of a molecule is the polarity contribution to the three-term Hansen solubility parameter. A tabulation of calculated three-term solubility parameters is provided by Barton (*CRC Handbook of Solubility Parameters and other Cohesion Parameters*, CRC Press, Boca Raton, 1991), along with a group-contribution method for calculating the three-term solubility parameters of compounds not listed in the reference.

Second Level: Identification of Individual Candidate Solvents

Boiling point characteristics.   Select only candidate solvents that boil at least 30–40°C above the key components to ensure that the sol-

TABLE 13-20 Hydrogen Bonding Classification of Chemical Families

Class	Chemical family			
H-Bonding, Strongly Associative (HBSA)	Water Primary amides Secondary amides	Polyacids Dicarboxylic acids Monohydroxy acids	Polyphenols Oximes Hydroxylamines	Amino alcohols Polyols
H-Bond Acceptor-Donor (HBAD)	Phenols Aromatic acids Aromatic amines Alpha H nitriles	Imines Monocarboxylic acids Other monoacids Peracids	Alpha H nitros Azines Primary amines Secondary amines	n-alcohols Other alcohols Ether alcohols
H-Bond Acceptor (HBA)	Acyl chlorides Acyl fluorides Hetero nitrogen aromatics Hetero oxygen aromatics	Tertiary amides Tertiary amines Other nitriles Other nitros Isocyanates Peroxides	Aldehydes Anhydrides Cyclo ketones Aliphatic ketones Esters Ethers	Aromatic esters Aromatic nitriles Aromatic ethers Sulfones Sulfolanes
π-Bonding Acceptor (π-HBA)	Alkynes Alkenes	Aromatics Unsaturated esters		
H-Bond Donor (HBD)	Inorganic acids Active H chlorides	Active H fluorides Active H iodides	Active H bromides	
Non-Bonding (NB)	Paraffins Nonactive H chlorides	Nonactive H fluorides Sulfides	Nonactive H iodides Disulfides	Nonactive H bromides Thiols

Deviations from Raoult's Law		
H-Bonding classes	Type of deviations	Comments
HBSA + NB HBAD + NB	Alway positive dev., HBSA + NB often limited miscibility	H-bonds broken by interactions
HBA + HBD	Always negative dev.	H-bonds formed by interactions
HBSA + HBD HBAD + HBD	Always positive deviations, HBSA + HBD often limited miscibility	H-bonds broken and formed; dissociation of HBSA or HBAD liquid most important effect
HBSA + HBSA HBSA + HBAD HBSA + HBA HBAD + HBAD HBAD + HBA	Usually positive deviations; some give maximum-boiling azeotropes	H-bonds broken and formed
HBA + HBA HBA + NB HBD + HBD HBD + NB NB + NB	Ideal, quasi-ideal systems; always positive or no deviations; azeotropes, if any, minimum-boiling	No H-bonding involved

NOTE: π-HBA is *enhanced* version of HBA.

vent is relatively nonvolatile and remains largely in the liquid phase. With this boiling point difference, the solvent should also not form azeotropes with the other components.

Selectivity at infinite dilution. Rank candidate solvents according to their selectivity at infinite dilution. The selectivity at infinite dilution is defined as the ratio of the activity coefficients at infinite dilution of the two key components in the solvent. Since solvent effects tend to increase as solvent concentration increases, the infinite-dilution selectivity gives an upper bound on the efficacy of a solvent. Infinite-dilution activity coefficients can be predicted using such methods as UNIFAC, ASOG, MOSCED (Reid et al., *Properties of Gases and Liquids,* Fourth Edition, McGraw-Hill, New York, 1987). They can be found experimentally using a rapid gas-liquid chromatography method based on relative retention times in candidate solvents (Tassios in *Extractive and Azeotropic Distillation,* Advances in Chemistry Series 115, American Chemical Society, Washington, 1972) and they can be correlated to bubble-point data [Kojima and Ochi, *J. Chem. Eng. Japan,* **7**(2), 71 (1974)]. DECHEMA [*Vapor-Liquid Equilibrium Data Collection,* Frankfort (1977)], has also published a compilation of experimental infinite-dilution activity coefficients.

Experimental measurement of relative volatility. Rank candidate solvents by the increase in relative volatility caused by the addition of the solvent. One technique is to experimentally measure the relative volatility of a fixed-composition key component-solvent mixture (often a 1/1 ratio of each key, with a 1/1 to 3/1 solvent/key ratio) for various solvents. [Carlson et al., *Ind. Eng. Chem.,* **46**, 350 (1954)]. The Oth-

mer equilibrium still is the apparatus of choice for these measurements [Zudkevitch, *Chem. Eng. Comm.,* **116**, 41 (1992)].

Methanol and acetone boil at 64.5°C and 56.1°C, respectively and form a minimum-boiling azeotrope at 55.3°C. The natural volatility of the system is acetone > methanol, so the favored solvents most likely will be those that cause the acetone to be recovered in the distillate. However, for the purposes of the example, a solvent that reverses the natural volatility will also be identified. First, examining the polarity of

TABLE 13-21 Relative Polarities of Functional Groups

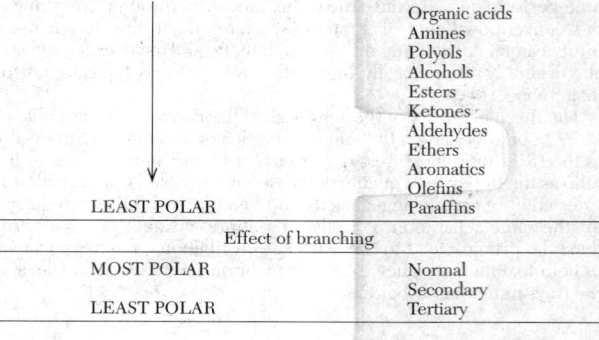

MOST POLAR		Water Organic acids Amines Polyols Alcohols Esters Ketones Aldehydes Ethers Aromatics Olefins
LEAST POLAR		Paraffins
	Effect of branching	
MOST POLAR		Normal
		Secondary
LEAST POLAR		Tertiary

ketones and alcohols (Table 13-21), solvents favored for the recovery of methanol in the bottoms would come from groups more polar than methanol, such as acids, water, and polyols. Turning to the Robbins Chart (Table 13-15), favorable groups are amines, alcohols, polyols, and water since these show expected positive deviations for acetone and zero or negative deviations for methanol. For reversing the natural volatility, solvents should be chosen that are less polar than acetone, such as ethers, hydrocarbons, and aromatics. Unfortunately, both ethers and hydrocarbons are expected to give positive deviations for both acetone and methanol, so should be discarded. Halohydrocarbons and ketones are expected to give positive deviations for methanol and either negative or no deviations for acetone. The other qualitative indicators show that both homologous series (ketones and alcohols) look promising. Thus, after discounting halohydrocarbons for environmental reasons, the best solvents will probably come from alcohols, polyols, and water for recovering methanol in the bottoms and ketones for recovering acetone in the bottoms. Table 13-22 shows the boiling points and experimental or estimated infinite-dilution activity coefficients for several candidate solvents from the aforementioned groups. Methylethylketone boils too low, as does ethanol, and also forms an azeotrope with methanol. These two candidates can be discarded. Other members of the homologous series, along with water and ethylene glycol, have acceptable boiling points (at least 30°C higher than keys). Of these, water (the solvent used industrially) clearly has the largest effect on the activity coefficients, followed by ethylene glycol. Although inferior to water or ethylene glycol, both MIPK and MIBK would probably be acceptable for reversing the natural volatility of the system.

Extractive Distillation by Salt Effects   A second method of modifying the liquid-phase behavior (and thus the relative volatility) of a mixture in order to effect a separation is by the addition of a nonvolatile, soluble, ionic salt. The process is analogous to extractive distillation with a high-boiling liquid. In the simplest case, for the separation of a binary mixture, the salt is fed at the top of the column by dissolving it in the hot reflux stream before introduction into the column. In order to function effectively the salt must be adequately soluble in both components throughout the range of compositions encountered in the column. Since the salt is essentially completely nonvolatile, it remains in the liquid phase on each tray and alters the relative volatility throughout the length of the column. No rectification section is needed above the salt feed. The bottoms product is recovered from the salt solution by evaporation or drying, and the salt is recycled. The ions of a salt are typically capable of causing much larger and more selective effects on liquid-phase behavior than the molecules of a liquid solvent. As a result, salt-to-feed ratios less than 0.1 are typical.

The use of a salting agent presents a number of problems not associated with a liquid solvent, such as the difficulty of transporting and metering a solid or saturated salt solution, slow mixing or dissolution rate of the salt, limits to solubility in the feed components, and potential for corrosion. However, in the limited number of systems for which an effective salt can be found, the energy usage, equipment size, capital investment, and ultimate separation cost can be significantly reduced compared to extractive distillation using a liquid sol-

TABLE 13-22   Comparison of Candidate Solvents for Methanol/Acetone Extractive Distillation

Solvent	Boiling pt. (°C)	Azeotrope formation	$\gamma^\infty_{\text{Acetone}}$	$\gamma^\infty_{\text{MeOH}}$	$\gamma^\infty_{\text{Acetone}}/\gamma^\infty_{\text{MeOH}}$
MEK	79.6	With MeOH	1.01	1.88	0.537
MIPK	102.0	No	1.01	1.89	0.534
MIBK	115.9	No	1.06	2.05	0.517
Ethanol	78.3	No	1.85	1.04	1.78
1-Propanol	97.2	No	1.90	1.20	1.58
1-Butanol	117.8	No	1.93	1.33	1.45
Water	100.0	No	11.77	2.34	5.03
EG	197.2	No	3.71	1.25	2.97

$\gamma^\infty_{\text{Acetone}} = 1.79$ (in MeOH)
$\gamma^\infty_{\text{MeOH}} = 1.81$ (in acetone)

vent [Furter, *Chem. Eng. Commun.*, **116**, 35 (1992)]. Applications of salt extractive distillation include acetate salts to produce absolute ethanol, magnesium nitrate for the production of concentrated nitric acid as an alternative to the sulfuric-acid solvent process, and calcium chloride to produce anhydrous hydrogen chloride. Other examples are noted by Furter [*Can. J. Chem. Eng.*, **55**, 229 (1977)].

One problem limiting the consideration of salt extractive distillation is the fact that the performance and solubility of a salt in a particular system is difficult to predict without experimental data. Some recent advances have been made in modeling the VLE behavior of organic-aqueous-salt solutions using modified UNIFAC, NRTL, UNIQUAC, and other approaches [Kumar, *Sep. Sci. Tech.*, **28**(1), 799 (1993)].

REACTIVE DISTILLATION

Introduction   Reactive distillation is a unit operation in which chemical reaction and distillative separation are carried out simultaneously within a fractional distillation apparatus. Reactive distillation may be advantageous for liquid-phase reaction systems when the reaction must be carried out with a large excess of one or more of the reactants, when a reaction can be driven to completion by removal of one or more of the products as they are formed, or when the product recovery or by-product recycle scheme is complicated or made infeasible by azeotrope formation.

For consecutive reactions in which the desired product is formed in an intermediate step, excess reactant can be used to suppress additional series reactions by keeping the intermediate-species concentration low. A reactive distillation can achieve the same end by removing the desired intermediate from the reaction zone as it is formed. Similarly, if the equilibrium constant of a reversible reaction is small, high conversions can be achieved by use of a large excess of reactant. Alternatively, by Le Chatelier's principle, the reaction can be driven to completion by removal of one or more of the products as they are formed. Typically, reactants can be kept much closer to stoichiometric proportions in a reactive distillation. When a reaction mixture exhibits azeotropism, the recovery of products and recycle of excess reagents can be quite complicated and expensive. Reactive distillation can provide a means of breaking azeotropes by altering or eliminating the condition for azeotrope formation in the reaction zone through the combined effects of vaporization-condensation and consumption-production of the species in the mixture. Alternatively, a reaction may be used to convert the species into components that are more easily distilled. In each of these situations, the conversion and selectivity often can be improved markedly, with much lower-reactant inventories and recycle rates, and much simpler recovery schemes. The capital savings can be quite dramatic. A list of applications of reactive distillation appearing in the literature is given in Table 13-23.

Although reactive distillation has many potential applications, it is not appropriate for all situations. Since it is in essence a distillation process, it has the same range of applicability as other distillation operations. Distillation-based equipment is not designed to effectively handle solids, supercritical components (where no separate vapor and liquid phases exist), gas-phase reactions, or high-temperature or high-pressure reactions such as hydrogenation, steam reforming, gasification, and hydrodealkylation.

Simulation, Modeling, and Design Feasibility   Because reaction and separation phenomena are closely coupled in a reactive distillation process, simulation and design is significantly more complex than that of sequential reaction and separation processes. In spite of the complexity, however, most commercial computer process modeling packages offer reliable and flexible routines for simulating steady-state reactive distillation columns, with either equilibrium or kinetically controlled reaction models. [Venkataraman et al., *Chem. Eng. Prog.*, **86**(6), 45 (1990)]. As with other enhanced distillation processes, the results are very sensitive to the thermodynamics model chosen and the accuracy of the VLE data used to generate model parameters. Of equal, if not more significance is the accuracy of data on reaction rate as a function of temperature. Very different conclusions can be drawn about the feasibility of a reactive distillation if the reaction is assumed to reach chemical equilibrium on each stage of the column or if the reaction is assumed to be kinetically controlled [Barbosa and Doherty, *Chem. Eng. Sci.*, **43**, 541 (1988)]. Tray holdup

TABLE 13-23 Applications of Reactive Distillation

Process	Reaction type	Reference
Methyl acetate from methanol and acetic acid	Esterification	Agreda et al., *Chem. Eng. Prog.*, **86**(2), 40 (1990)
General process for ester formation	Esterification	Simons, "Esterification" in Encyclopedia of Chemical Processing and Design, Vol 19, Dekker, New York, 1983
Dibutyl phthalate from butanol and phthalic acid	Esterification	Berman et al., *Ind. Eng. Chem.*, **40**, 2139 (1948)
Ethyl acetate from ethanol and butyl acetate	Transesterification	Davies and Jeffreys, *Trans. Inst. Chem. Eng.*, **51**, 275 (1973)
Recovery of acetic acid and methanol from methyl acetate by-product of vinyl acetate production	Hydrolysis	Fuchigami, *J. Chem. Eng. Jap.*, **23**, 354 (1990)
Nylon 6,6 prepolymer from adipic acid and hexamethylenediamine	Amidation	Jaswal and Pugi, U.S. Patent 3,900,450 (1975)
MTBE from isobutene and methanol	Etherification	DeGarmo et al., *Chem. Eng. Prog.*, **88**(3), 43 (1992)
TAME frompentenes and methanol	Etherification	Brockwell et al., *Hyd. Proc.*, **70**(9), 133 (1991)
Separation of close boiling 3- and 4-picoline by complexation with organic acids	Acid-base	Duprat and Gau, *Can. J. Chem. Eng.*, **69**, 1320 (1991)
Separation of close-boiling meta and para xylenes by formation of tert-butyl meta-xyxlene	Transalkylation	Saito et al., *J. Chem. Eng. Jap.*, **4**, 37 (1971)
Cumene from propylene and benzene	Alkylation	Shoemaker and Jones, *Hyd. Proc.*, **67**(6), 57 (1987)
General process for the alkylation of aromatics with olefins	Alkylation	Crossland, U.S. Patent 5,043,506 (1991)
Production of specific higher and lower alkenes from butenes	Diproportionation	Jung et al., U.S. Patent 4,709,115 (1987)
4-Nitrochlorobenzene from chlorobenzene and nitric acid	Nitration	Belson, *Ind. Eng. Chem. Res.*, **29**, 1562 (1990)
Production of methylal and high purity formaldehyde		Masamoto and Matsuzaki, *J. Chem. Eng. Jap.*, **27**, 1 (1994)

and stage requirements are two important variables directly affected by the form of the reaction model chosen.

When an equilibrium reaction occurs in a vapor-liquid system, the phase compositions depend not only on the relative volatility of the components in the mixture, but also on the consumption (and production) of species. Thus, the condition for azeotropy in a nonreactive system ($y_i = x_i$, for all i) no longer holds true in a reactive system and must be modified to include reaction stoichiometry:

$$\frac{y_1 - x_1}{v_1 - v_T x_1} = \frac{y_i - x_i}{v_i - v_T x_i} \qquad \text{for all } i = 1, n \qquad (13\text{-}128)$$

where

$$v_T = \sum_{i=1}^{n} v_i$$

x_i = mole fraction of component i in the liquid phase
y_i = mole fraction of component i in the vapor phase
v_i = stoichiometric coefficient of component i (negative for reactants, positive for products)

Phase compositions that satisfy Eq. (13-128) are stationary points on a phase diagram and have been labeled *reactive azeotropes* by Barbosa and Doherty [*Chem. Eng. Sci.*, **43**, 529 (1988)]. At a reactive azeotrope the mass exchange between the vapor and liquid phase and the generation (or consumption) of each species is balanced such that the composition of neither phase changes. Reactive azeotropes show the same distillation properties as ordinary azeotropes and therefore affect what products are achievable. Reactive azeotropes are not easily visualized in conventional y-x coordinates but become apparent upon a transformation of coordinates which depends on the number of reactions, the order of each reaction (e.g., $A + B \leftrightarrow C$ or $A + B \leftrightarrow C + D$), presence of nonreacting components, and the extent of reaction. The general vector-matrix form of the transform for C reacting components, with R reactions, and I nonreacting components has been derived by Ung and Doherty [*Chem. Eng. Sci.*, **50**, 23 (1995)]. For the transformed mole fraction of component i in the liquid phase, X_i, they give

$$X_i = \left[\frac{x_i - v_i^T (v_{\mathrm{Ref}})^{-1} x_{\mathrm{Ref}}}{1 - v_{TOT}^T (v_{\mathrm{Ref}})^{-1} x_{\mathrm{Ref}}} \right], \qquad i = 1, \ldots, C\text{-}R \quad (13\text{-}129)$$

where v_i^T = row vector of stoichiometric coefficients of component i for each reaction
v_{Ref} = square matrix of stoichiometric coefficients for R reference components in R reactions
x_{Ref} = column vector of mole fractions for the R reference components in the liquid phase
v_{TOT}^T = row vector composed of the sum of the stoichiometric coefficients for each reaction

An equation identical to (13-129) defines the transformed mole fraction of component i in the vapor phase, Y_i, where the terms in x are replaced by terms in y.

The transformed variables describe the system composition with or without reaction and sum to unity as do x_i and y_i. The condition for azeotropy becomes $X_i = Y_i$. Barbosa and Doherty have shown that phase and distillation diagrams constructed using the transformed composition coordinates have the same properties as phase and distillation region diagrams for nonreactive systems and similarly can be used to assist in design feasibility and operability studies [*Chem. Eng. Sci.*, **43**, 529, 1523, and 2377 (1988a,b,c)]. A residue curve map in transformed coordinates for the reactive system methanol-acetic acid-methyl acetate-water is shown in Fig. 13-76. Note that the nonreactive azeotrope between water and methyl acetate has disappeared, while the methyl acetate-methanol azeotrope remains intact. Only

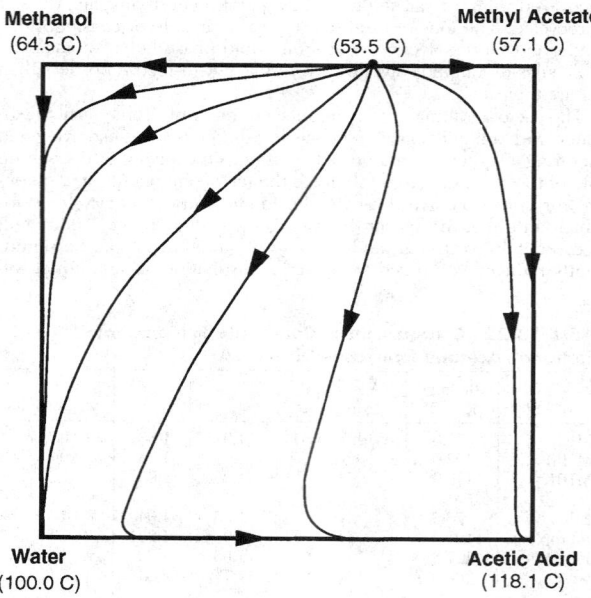

Methanol (64.5 C) (53.5 C) **Methyl Acetate** (57.1 C)

Water (100.0 C) **Acetic Acid** (118.1 C)

FIG. 13-76 Residue curve map for the reactive system methanol-acetic acid-methyl acetate-water in chemical equilibrium.

those azeotropes containing either all the required reactants or products will be altered by the reaction (water and methyl acetate can back-react to form acetic acid and methanol, whereas methanol and methyl acetate cannot further react in the absence of either water or acetic acid). This reactive system consists of only one distillation region in which the methanol-methyl acetate azeotrope is the low-boiling and acetic acid is the high-boiling node.

The situation becomes more complicated when the reaction is kinetically controlled and does not come to complete-chemical equilibrium under the conditions of temperature, liquid holdup, and rate of vaporization in the column reactor. Venimadhavan et al. [*AIChE J.*, **40**, 1814 (1994)] and Rev [*Ind. Eng. Chem. Res.*, **33**, 2174 (1994)] show that the existence and location of reactive azeotropes is a function of approach to equilibrium as well as the evaporation rate.

Mechanical Design and Implementation Issues   The choice of catalyst has a significant impact on the mechanical design and operation of the reactive column. The catalyst must allow the reaction to occur at reasonable rates at the relatively low temperatures and pressures common in distillation operations (typically less than 10 atmospheres and between 50°C and 250°C). Selection of a homogeneous catalyst, such as a high-boiling mineral acid, allows the use of more traditional tray designs and internals (albeit designed with allowance for high-liquid holdups). With a homogeneous catalyst, lifetime is not a problem, as it is added (and withdrawn) continuously. Alternatively, heterogeneous solid catalysts require either complicated mechanical means for continuous replenishment or relatively long lifetimes in order to avoid constant maintenance. As with other multiphase reactors, use of a solid catalyst adds an additional resistance to mass transfer from the bulk liquid (or vapor) to the catalyst surface, which may be the limiting resistance. The catalyst containment system must be designed to ensure adequate liquid-solid contacting and minimize bypassing. A number of specialized column internal designs, catalyst containment methods, and catalyst replenishment systems have been proposed for both homogeneous and heterogeneous catalysts. A partial list of these methods is given in Table 13-24.

Heat management is another important consideration in the implementation of a reactive distillation process. Conventional reactors for highly exothermic or endothermic reactions are often designed as modified shell-and-tube heat exchangers for efficient heat transfer. However, a trayed or packed distillation column is a rather poor mechanical design for the management of the heat of reaction. Although heat can be removed or added in the condenser or reboiler easily, the only mechanism for heat transfer in the column proper is through vaporization (or condensation). For highly exothermic reactions, a large excess of reactants may be required as a heat sink, necessitating high-reflux rates and larger-diameter columns to return the vaporized reactants back to the reaction zone. Often a prereactor of conventional design is used to accomplish most of the reaction and heat removal before feeding to the reactive column for final conversion, as exemplified in most processes for the production of tertiary amyl methyl ether (TAME) [Brockwell et al., *Hyd. Proc.*, **70**(9), 133 (1991)]. Highly endothermic reactions may require intermediate reboilers. None of these heat-management issues preclude the use of reactive distillation, but must be taken into account during the design phase. Comparison of heat of reaction and average heat of vaporization data for a system, as in Fig. 13-77, gives some indication of potential heat imbalances [Sundmacher et al., *Chem. Eng. Comm.*, **127**, 151 (1994)]. The heat neutral systems ($-\Delta H_{react} \approx \Delta H_{vap(avg)}$) such as methyl acetate and other esters can be accomplished in one reactive column, whereas the MTBE and TAME processes, with higher heats of reaction than vaporization, often include an additional prereactor. One exception is the catalytic distillation process for cumene production, which is accomplished without a prereactor. Three moles of benzene reactant are vaporized (and refluxed) for every mole of cumene produced. The relatively high heat of reaction is advantageous in this case as it reduces the overall heat duty of the process by about 30 percent [Shoemaker and Jones, *Hyd. Proc.*, **57**(6), 57 (1987)].

Process Applications   The production of esters from alcohols and carboxylic acids illustrates many of the principles of reactive distillation as applied to equilibrium-limited systems. The equilibrium constants for esterification reactions are usually relatively close to unity. Large excesses of alcohols must be used to obtain acceptable yields with large recycles. In a reactive-distillation scheme, the reac-

TABLE 13-24   Catalyst Systems for Reactive Distillation

Description	Application	Reference
Homogeneous catalysis		
Liquid-phase mineral-acid catalyst added to column or reboiler	Esterifications Dibutyl phlalate Methyl acetate	Keyes, *Ind. Eng. Chem.*, **24**, 1096 (1932) Berman et al., *Ind. Eng. Chem.*, **40**, 2139 (1948) Agreda et al., U.S. Patent 4,435,595 (1984)
Heterogeneous catalysis		
Catalyst-resin beads placed in cloth bags attached to fiberglass strip. Strip wound around helical stainless steel mesh spacer	Etherifications Cumene	Smith et al., U.S. Patent 4,443,559 (1981) Shoemaker and Jones, *Hyd.* **57**(6), 57 (1987)
Ion exchange resin beads used as column packing	Hydrolysis of methyl acetate	Fuchigami, *J. Chem. Eng. Jap.*, **23**, 354 (1990)
Molecular sieves placed in bags or porous containers	Alkylation of aromatics	Crossland, U.S. Patent 5,043,506 (1991)
Ion exchange resins formed into Raschig rings	MTBE	Flato and Hoffman, *Chem. Eng. Tech.*, **15**, 193 (1992)
Granular catalyst resin loaded in corrugated sheet casings	Dimethyl acetals of formaldehyde	Zhang et al., Chinese Patent 1,065,412 (1992)
Trays modified to hold catalyst bed	MTBE	Sanfilippo et al., Eur. Pat. Appl. EP 470,625 (1992)
Distillation trays constructed of porous catalytically active material and reinforcing resins	None specified	Wang et al., Chinese Patent 1,060,228 (1992)
Method described for removing or replacing catalyst on trays as a liquid slurry	None specified	Jones, U.S. Patent, 5,133,942 (1992)
Catalyst bed placed in downcomer, designed to prevent vapor flow through bed	Etherifications, alkylations	Asselineau, Eur. Pat. Appl. EP 547,939 (1993)
Slotted plate for catalyst support designed with openings for vapor flow	None specified	Evans and Stark, Eur. Pat. Appl. EP 571,163 (1993)
Ion exchanger fibers (reinforced ion exchange polymer) used as solid-acid catalyst	Hydrolysis of methyl acetate	Hirata et al., Jap. Patent 05,212,290 (1993)
High-liquid holdup trays designed with catalyst bed extending below tray level, perforated for vapor-liquid contact	None specified	Yeoman et al., Int. Pat. Appl., WO 9408679 (1994)
Catalyst bed placed in downcomer, in-line withdrawal/addition system	None specified	Carland, U.S. Patent, 5,308,451 (1994)

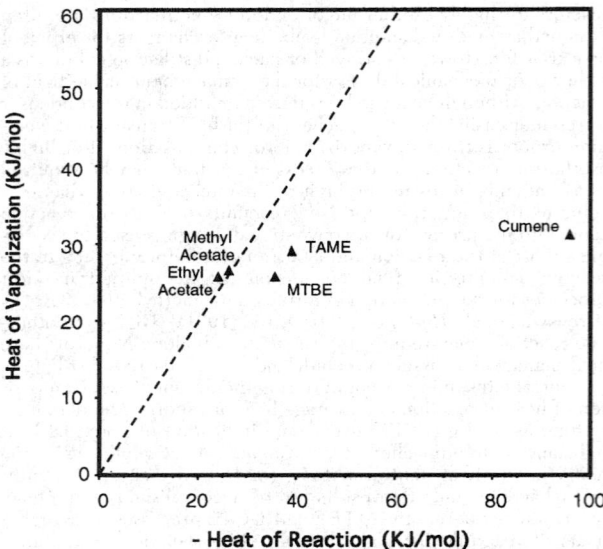

FIG. 13-77 Similarity of heats of reaction and vaporization for compounds made by reactive distillation.

tion is driven to completion by removal of the water of esterification. The method used for removal of the water depends on the boiling points, compositions, and liquid-phase behavior of any azeotropes formed between the products and reactants and largely dictates the structure of the reactive-distillation flowsheet.

When the ester forms a binary low-boiling azeotrope with water or a ternary alcohol-ester-water azeotrope and that azeotrope is heterogeneous (or can be moved into the two-liquid-phase region), the continuous flowsheet illustrated in Fig. 13-78 can be used. Such a flowsheet works for the production of ethyl acetate and higher homologs. In this process scheme, acetic acid and the alcohol are continuously fed to the reboiler of the esterification column, along with a homogeneous strong-acid catalyst. Since the catalyst is largely nonvolatile, the reboiler acts as the primary reaction site. The alcohol is usually fed in slight excess to ensure complete reaction of the acid and to compensate for alcohol losses through distillation of the water-

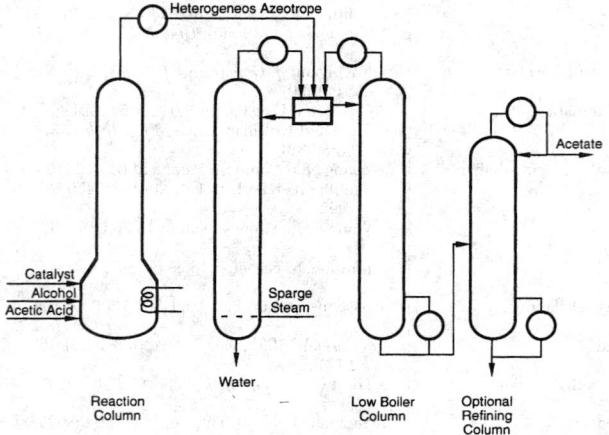

FIG. 13-78 Flowsheet for exters which form a heterogeneous minimum-boiling azeotrope with water.

ester-(alcohol) azeotrope. The esterification column is operated such that the low-boiling, water-laden azeotrope is taken as the distillation product. Upon cooling, the distillate separates into two liquid phases. The aqueous layer is steam-stripped, with the organics recycled to the decanter or reactor. The ester layer from the decanter contains some water and possibly alcohol. Part of this layer may be refluxed to the esterification column. The remainder is fed to a low-boiler column where the water-ester and alcohol-ester azeotropes are removed overhead and recycled to the decanter or reactor. The dry, alcohol-free ester is then optionally taken overhead in a final refining column.

Methyl acetate cannot be produced in high purity using the simple esterification scheme described above. The methyl acetate-methanol-water system does not exhibit a ternary minimum-boiling azeotrope, the methyl acetate-methanol azeotrope is lower boiling than the water-methyl acetate azeotrope, a distillation boundary extends between these two binary azeotropes, and the heterogeneous region does not include either azeotrope, nor does it cross the distillation boundary. Consequently, the water of esterification cannot be removed effectively and methyl acetate cannot be separated from the methanol and water azeotropes by a simple decantation in the same manner as outlined above. Conventional sequential reaction-separation processes rely on large excesses of acetic acid to drive the reaction to higher conversion to methyl acetate, necessitating a capital- and energy-intensive acetic acid-water separation and large recycle streams. The crude methyl acetate product, contaminated with water and methanol, can be purified by a

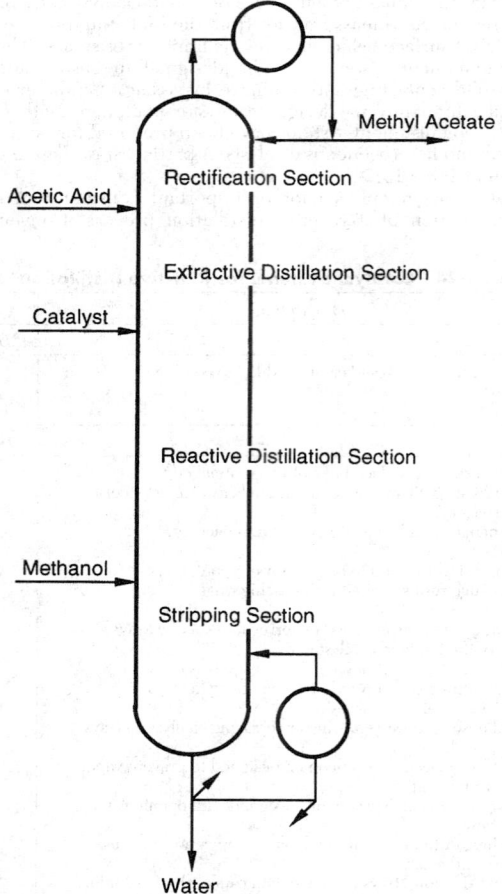

FIG. 13-79 Integrated reactive-extractive distillation column for the production of methyl acetate.

number of the enhanced distillation techniques such as pressure-swing distillation [Harrison, US Patent 2,704,271 (1955)], extractive distillation with ethylene glycol monomethylether as the solvent [Kumerle, German Patent 1,070,165 (1959)], or azeotropic distillation with an aromatic or ketone entrainer [Yeomans, Eur. Patent Appl. 060717 and 060719 (1982)]. The end result is a capital- and energy-intensive process typically requiring multiple reactors and distillation columns.

The reactive-distillation process (Fig. 13-79) provides a mechanism for overcoming both the limitations on conversion due to chemical equilibrium as well as the difficulties in purification imposed by the water-methyl acetate and methanol-methyl acetate azeotropes [Agreda et al., *Chem. Eng. Prog.*, **86**(2), 40 (1990)]. Conceptually, this flowsheet can be thought of as four heat-integrated distillation columns (one of which is also a reactor) stacked on top of each other. The primary reaction zone consists of a series of countercurrent flashing stages in the middle of the column. Adequate residence time for the reaction is provided by high-liquid-holdup bubble-cap trays with specially designed downcomer sumps to further increase tray holdup. A nonvolatile homogeneous catalyst is fed at the top of the reactive section and exits with the underflow water by-product. The extractive-

distillation section, immediately above the reactive section, is critical in achieving high-methyl-acetate purity. As shown in Fig. 13-76, simultaneous reaction and distillation eliminates the water-methyl acetate azeotrope (and the distillation boundary of the nonreactive system). However, pure methyl acetate remains a saddle in the reactive system, and cannot be obtained as a pure component by simple reactive distillation. The acetic acid feed acts as a solvent in an extractive-distillation section placed above the reaction section, breaking the methanol-methyl acetate azeotrope, and yielding a pure methyl acetate distillate product. The uppermost rectification stages serve to remove any acetic acid from the methyl acetate product and the bottommost stripping section removes any methanol and methyl acetate from the water by-product. The countercurrent flow of the reactants results in high local excesses at each end of the reactive section, even though the overall feed to the reactive column is stoichiometric. Therefore, the large excess of acetic acid at the top of the reactive section prevents methanol from reaching the distillate, while, similarly, methanol at the bottom of the reactive section keeps acetic acid from the water bottoms. Temperature and composition profiles for this reactive-extractive-distillation column are shown in Fig. 13-80a and b, respectively.

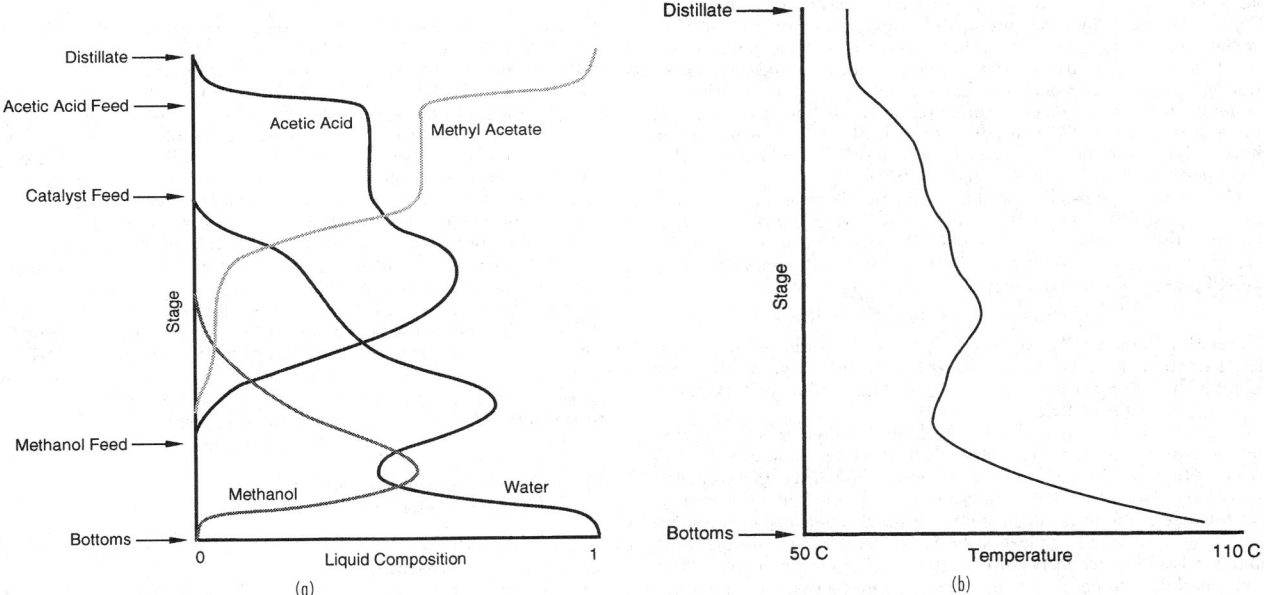

FIG. 13-80   Reactive extracting distillation for methyl acetate production. (*a*) Composition profile. (*b*) Temperature profile.

PETROLEUM AND COMPLEX-MIXTURE DISTILLATION

INTRODUCTION

Although the principles of multicomponent distillation apply to petroleum, synthetic crude oil, and other complex mixtures, this subject warrants special consideration for the following reasons:

1.   Such feedstocks are of exceedingly complex composition, consisting, in the case of petroleum, of many different types of hydrocarbons and perhaps of inorganic and other organic compounds. The number of carbon atoms in the components may range from 1 to more than 50, so that the compounds may exhibit atmospheric-pressure boiling points from −162°C (−259°F) to more than 538°C (1000°F). In a given boiling range, the number of different compounds that exhibit only small differences in volatility multiplies rapidly with

increasing boiling point. For example, 16 of the 18 octane isomers boil within a range of only 12°C (22°F).

2.   Products from the distillation of complex mixtures are in themselves complex mixtures. The character and yields of these products vary widely, depending upon the source of the feedstock. Even crude oils from the same locality may exhibit marked variations.

3.   The scale of petroleum-distillation operations is generally large, and, as discussed in detail by Nelson (*Petroleum Refinery Engineering*, 4th ed., McGraw-Hill, New York, 1958) and Watkins (*Petroleum Refinery Distillation*, 2d ed., Gulf, Houston, 1979), such operations are common in several petroleum-refinery processes including atmospheric distillation of crude oil, vacuum distillation of bottoms residuum obtained from atmospheric distillation, main fractionation of gaseous

effluent from catalytic cracking of various petroleum fractions, and main fractionation of effluent from thermal coking of various petroleum fractions. These distillation operations are conducted in large pieces of equipment that can consume large quantities of energy. Therefore, optimization of design and operation is very important and frequently leads to a relatively complex equipment configuration.

CHARACTERIZATION OF PETROLEUM AND PETROLEUM FRACTIONS

Although much progress has been made in identifying the chemical species present in petroleum, it is generally sufficient for purposes of design and analysis of plant operation of distillation to characterize petroleum and petroleum fractions by gravity, laboratory-distillation curves, component analysis of light ends, and hydrocarbon-type analysis of middle and heavy ends. From such data, as discussed in the *Technical Data Book—Petroleum Refining* [American Petroleum Institute (API), Washington], five different average boiling points and an index of paraffinicity can be determined; these are then used to predict the physical properties of complex mixtures by a number of well-accepted correlations, whose use will be explained in detail and illustrated with examples. Many other characterizing properties or attributes such as sulfur content, pour point, water and sediment content, salt content, metals content, Reid vapor pressure, Saybolt Universal viscosity, aniline point, octane number, freezing point, cloud point, smoke point, diesel index, refractive index, cetane index, neutralization number, wax content, carbon content, and penetration are generally measured for a crude oil or certain of its fractions according to well-specified ASTM tests. But these attributes are of much less interest here even though feedstocks and products may be required to meet certain specified values of the attributes.

Gravity of a crude-oil or petroleum fraction is generally measured by the ASTM D 287 test or the equivalent ASTM D 1298 test and may be reported as specific gravity (SG) 60/60°F [measured at 60°F (15.6°C) and referred to water at 60°F (15.6°C)] or, more commonly, as API gravity, which is defined as

$$\text{API gravity} = 141.5/(\text{SG } 60/60°F) - 131.5 \qquad (13\text{-}130)$$

Water, thus, has an API gravity of 10.0, and most crude oils and petroleum fractions have values of API gravity in the range of 10 to 80. Light hydrocarbons (*n*-pentane and lighter) have values of API gravity ranging upward from 92.8.

The volatility of crude-oil and petroleum fractions is characterized in terms of one or more laboratory distillation tests that are summarized in Table 13-25. The ASTM D 86 and D 1160 tests are reasonably rapid batch laboratory distillations involving the equivalent of approximately one equilibrium stage and no reflux except for that caused by heat losses. Apparatus typical of the D 86 test is shown in Fig. 13-81 and consists of a heated 100-mL or 125-mL Engler flask containing a calibrated thermometer of suitable range to measure the temperature of the vapor at the inlet to the condensing tube, an inclined brass condenser in a cooling bath using a suitable coolant, and a graduated cylinder for collecting the distillate. A stem correction is not applied to the temperature reading. Related tests using similar apparatuses are the D 216 test for natural gasoline and the Engler distillation.

In the widely used ASTM D 86 test, 100 mL of sample is charged to the flask and heated at a sufficient rate to produce the first drop of distillate from the lower end of the condenser tube in from 5 to 15 min, depending on the nature of the sample. The temperature of the vapor at that instant is recorded as the initial boiling point (IBP). Heating is continued at a rate such that the time from the IBP to 5 volume percent recovered of the sample in the cylinder is 60 to 75 s. Again, vapor temperature is recorded. Then, successive vapor temperatures are recorded for from 10 to 90 percent recovered in intervals of 10, and at 95 percent recovered, with the heating rate adjusted so that 4 to 5 mL are collected per minute. At 95 percent recovered, the burner flame is increased if necessary to achieve a maximum vapor temperature referred to as the end point (EP) in from 3 to 5 additional min. The percent recovery is reported as the maximum percent recovered in the cylinder. Any residue remaining in the flask is reported as percent residue, and percent loss is reported as the difference between 100 mL and the sum of the percent recovery and percent residue. If the atmosphere test pressure P is other than 101.3 kPa (760 torr), temperature readings may be adjusted to that pressure by the Sidney Young equation, which for degrees Fahrenheit is

$$T_{760} = T_P + 0.00012(760 - P)(460 + T_P) \qquad (13\text{-}131)$$

Another pressure correction for percent loss can also be applied, as described in the ASTM test method.

Results of a typical ASTM distillation test for an automotive gasoline are given in Table 13-26, in which temperatures have already been corrected to a pressure of 101.3 kPa (760 torr). It is generally assumed that percent loss corresponds to volatile noncondensables that are distilled off at the beginning of the test. In that case, the percent recovered values in Table 13-26 do not correspond to percent evaporated values, which are of greater scientific value. Therefore, it is common to adjust the reported temperatures according to a linear interpolation procedure given in the ASTM test method to obtain corrected temperatures in terms of percent evaporated at the standard intervals as included in Table 13-26. In the example, the corrections are not large because the loss is only 1.5 volume percent.

Although most crude petroleum can be heated to 600°F (316°C) without noticeable cracking, when ASTM temperatures exceed 475°F (246°C), fumes may be evolved, indicating decomposition, which may cause thermometer readings to be low. In that case, the following correction attributed to S. T. Hadden may be applied:

$$\Delta T_{\text{corr}} = 10^{-1.587 + 0.004735T}$$

where T = measured temperature, °F
        ΔT_{corr} = correction to be added to T, °F

At 500 and 600°F (260 and 316°C), the corrections are 6 and 18°F (3.3 and 10°C) respectively.

As discussed by Nelson (op. cit.), virtually no fractionation occurs in an ASTM distillation. Thus, components in the mixture do distill one by one in the order of their boiling points but as mixtures of successively higher boiling points. The IBP, EP, and intermediate points have little theoretical significance, and, in fact, components boiling below the IBP and above the EP are present in the sample. Never-

TABLE 13-25 Laboratory Distillation Tests

Test name	Reference	Main applicability
ASTM (atmospheric)	ASTM D 86	Petroleum fractions or products, including gasolines, turbine fuels, naphthas, kerosines, gas oils, distillate fuel oils, and solvents that do not tend to decompose when vaporized at 760 mmHg
ASTM [vacuum, often 10 torr (1.3 kPa)]	ASTM D 1160	Heavy petroleum fractions or products that tend to decompose in the ASTM D 86 test but can be partially or completely vaporized at a maximum liquid temperature of 750°F (400°C) at pressures down to 1 torr (0.13 kPa)
TBP [atmospheric or 10 torr (1.3 kPa)]	Nelson,° ASTM D 2892	Crude oil and petroleum fractions
Simulated TBP (gas chromatography)	ASTM D 2887	Crude oil and petroleum fractions
EFV (atmospheric, superatmospheric, or subatmospheric)	Nelson†	Crude oil and petroleum fractions

°Nelson, *Petroleum Refinery Engineering*, 4th ed., McGraw-Hill, New York, 1958, pp. 95–99.
†Ibid., pp. 104–105.

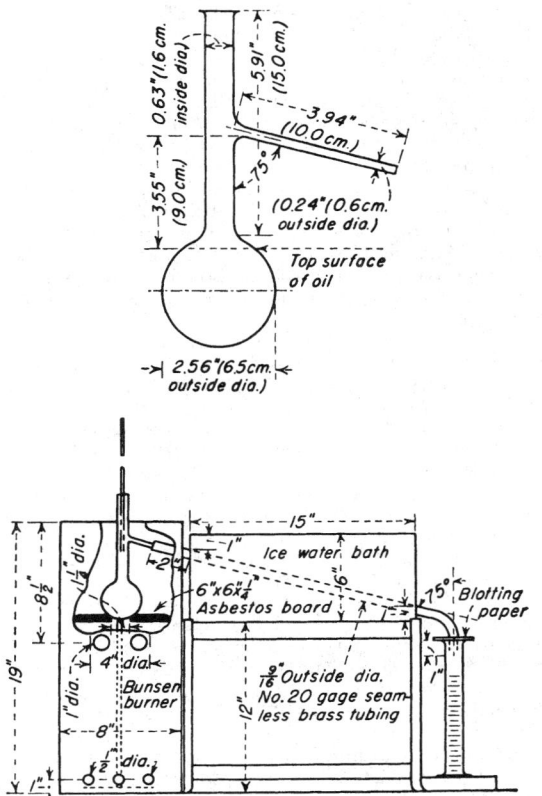

FIG. 13-81 ASTM distillation apparatus; detail of distilling flask is shown in the upper figure.

theless, because ASTM distillations are quickly conducted, have been successfully automated, require only a small sample, and are quite reproducible, they are widely used for comparison and as a basis for specifications on a large number of petroleum intermediates and products, including many solvents and fuels. Typical ASTM curves for several such products are shown in Fig. 13-82.

TABLE 13-26 Typical ASTM D 86 Test Results for Automobile Gasoline Pressure, 760 torr (101.3 kPa)

Percent recovered basis (as measured)			Percent evaporated basis (as corrected)		
Percent recovered	T, °F	Percent evaporated	Percent evaporated	T, °F	Percent recovered
0(IBP)	98	1.5	1.5	98	(IBP)
5	114	6.5	5	109	3.5
10	120	11.5	10	118	8.5
20	150	21.5	20	146	18.5
30	171	31.5	30	168	28.5
40	193	41.5	40	190	38.5
50	215	51.5	50	212	48.5
60	243	61.5	60	239	58.5
70	268	71.5	70	264	68.5
80	300	81.5	80	295	78.5
90	340	91.5	90	334	88.5
95	368	96.5	95	360	93.5
EP	408			408	(EP)

NOTE: Percent recovery = 97.5; percent residue = 1.0; percent loss = 1.5. To convert degrees Fahrenheit to degrees Celsius, °C = (°F − 32)/1.8.

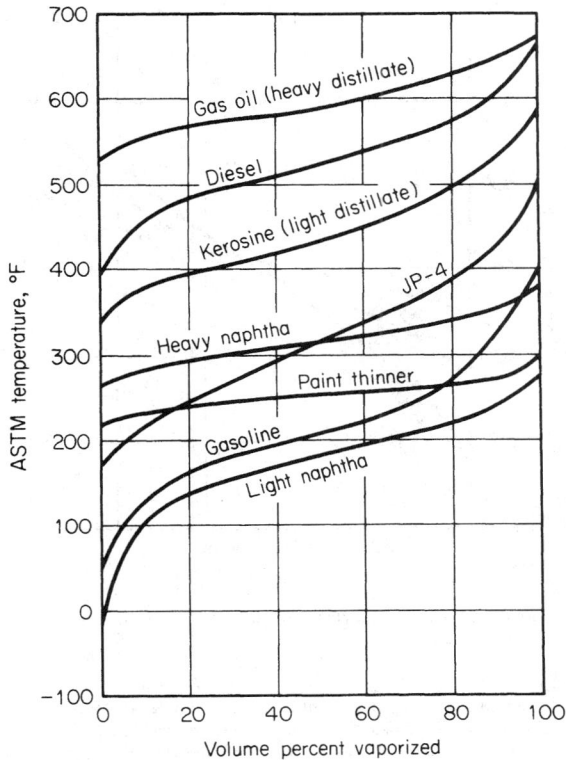

FIG. 13-82 Representative ASTM D 86 distillation curves.

Data from a true-boiling-point (TBP) distillation test provides a much better theoretical basis for characterization. If the sample contains compounds that have moderate differences in boiling points such as in a light gasoline containing light hydrocarbons (e.g., isobutane, n-butane, isopentane, etc.), a plot of overhead-vapor-distillate temperature versus percent distilled in a TBP test would appear in the form of steps as in Fig. 13-83. However, if the sample has a higher average boiling range when the number of close-boiling isomers increases, the steps become indistinct and a TBP curve such as that in Fig. 13-84 results. Because the degree of separation for a TBP distillation test is much higher than for an ASTM distillation test, the IBP is lower and the EP is higher for the TBP method as compared with the ASTM method, as shown in Fig. 13-84.

A standard TBP laboratory-distillation-test method has not been well accepted. Instead, as discussed by Nelson (op. cit., pp. 95–99),

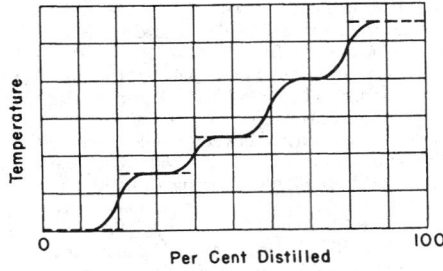

FIG. 13-83 Variation of boiling temperature with percent distilled in true-boiling-point distillation of light hydrocarbons.

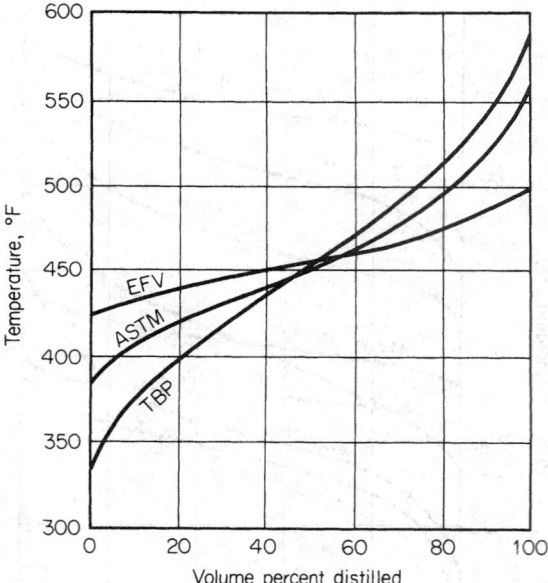

FIG. 13-84 Comparison of ASTM, TBP, and EFV distillation curves for kerosine.

batch distillation equipment that can achieve a good degree of fractionation is usually considered suitable. In general, TBP distillations are conducted in columns with 15 to 100 theoretical stages at reflux ratios of 5 or greater. Thus, the new ASTM D 2892 test method, which involves a column with from 14 to 17 theoretical stages and a reflux ratio of 5, essentially meets the minimum requirements. Distillate may be collected at a constant or a variable rate. Operation may be at 101.3-kPa (760 torr) pressure or at a vacuum at the top of the column as low as 0.067 kPa (0.5 torr) for high-boiling fractions, with 1.3 kPa (10 torr) being common. Results from vacuum operation are extrapolated to 101.3 kPa (760 torr) by the vapor-pressure correlation of Maxwell and Bonner [*Ind. Eng. Chem.*, **49**, 1187 (1957)], which is given in great detail in the API *Technical Data Book—Petroleum Refining* (op. cit.) and in the ASTM D 2892 test method. It includes a correction for the nature of the sample (paraffin, olefin, napthene, and aromatic content) in terms of the UOP characterization factor, UOP-K, as given by

$$UOP\text{-}K = (T_B)^{1/3}/SG \qquad (13\text{-}132)$$

where T_B = mean average boiling point, °R, which is the arithmetic average of the molal average boiling point and the cubic volumetric average boiling point. Values of UOP-K for *n*-hexane, 1-hexene, cyclohexene, and benzene are 12.82, 12.49, 10.99, and 9.73 respectively. Thus, paraffins with their lower values of specific gravity tend to have high values, and aromatics tend to have low values of UOP-K. A movement toward an international TBP standard is discussed by Vercier and Mouton [*Oil Gas J.*, **77**(38), 121 (1979)].

A crude-oil assay always includes a whole crude API gravity and a TBP curve. As discussed by Nelson (op. cit., pp. 89–90) and as shown in Fig. 13-85, a reasonably consistent correlation (based on more than 350 distillation curves) exists between whole crude API gravity and the TBP distillation curve at 101.3 kPa (760 torr). Exceptions not correlated by Fig. 13-85 are highly paraffinic or naphthenic crude oils.

An alternative to TBP distillation is simulated distillation by gas chromatography. As described by Green, Schmauch, and Worman [*Anal. Chem.*, **36**, 1512 (1965)] and Worman and Green [*Anal. Chem.*, **37**, 1620 (1965)], the method is equivalent to a 100-theoretical-plate TBP distillation, is very rapid, reproducible, and easily automated, requires only a small microliter sample, and can better

define initial and final boiling points. The ASTM D 2887 standard test method is based on such a simulated distillation and is applicable to samples having a boiling range greater than 55°C (100°F) for temperature determinations as high as 538°C (1000°F). Typically, the test is conducted with a gas chromatograph having a thermal-conductivity detector, a programmed temperature capability, helium or hydrogen carrier gas, and column packing of silicone gum rubber on a crushed-fire-brick or diatomaceous-earth support.

It is important to note that simulated distillation does not always separate hydrocarbons in the order of their boiling point. For example, high-boiling multiple-ring-type compounds may be eluted earlier than normal paraffins (used as the calibration standard) of the same boiling point. Gas chromatography is also used in the ASTM D 2427 test method to determine quantitatively ethane through pentane hydrocarbons.

A third fundamental type of laboratory distillation, which is the most tedious to perform of the three types of laboratory distillations, is equilibrium-flash distillation (EFV), for which no standard test exists. The sample is heated in such a manner that the total vapor produced remains in contact with the total remaining liquid until the desired temperature is reached at a set pressure. The volume percent vaporized at these conditions is recorded. To determine the complete flash curve, a series of runs at a fixed pressure is conducted over a range of temperature sufficient to cover the range of vaporization from 0 to 100 percent. As seen in Fig. 13-84, the component separation achieved by an EFV distillation is much less than by the ASTM or TBP distillation tests. The initial and final EFV points are the bubble point and the dew point respectively of the sample. If desired, EFV curves can be established at a series of pressures.

Because of the time and expense involved in conducting laboratory distillation tests of all three basic types, it has become increasingly common to use empirical correlations to estimate the other two distillation curves when either the ASTM, TBP, or EFV curve is available. Preferred correlations given in the API *Technical Data Book—Petroleum Refining* (op. cit.) are based on the work of Edmister and Pollock [*Chem. Eng. Prog.*, **44**, 905 (1948)], Edmister and Okamoto [*Pet. Refiner*, **38**(8), 117 (1959); **38**(9), 271 (1959)], Maxwell (*Data Book on*

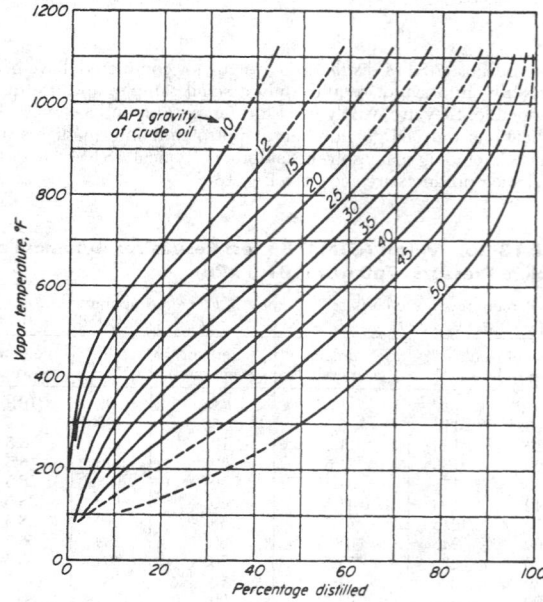

FIG. 13-85 Average true-boiling-point distillation curves of crude oils. (*From W. E. Edmister*, Applied Hydrocarbon Thermodynamics, *vol. 1, 1st ed., 1961 Gulf Publishing Company, Houston, Texas, Used with permission. All rights reserved.*)

Hydrocarbons, Van Nostrand, Princeton, N.J., 1950), and Chu and Staffel [*J. Inst. Pet.,* **41**, 92 (1955)]. Because of the lack of sufficiently precise and consistent data on which to develop the correlations, they are, at best, first approximations and should be used with caution. Also, they do not apply to mixtures containing only a few components of widely different boiling points. Perhaps the most useful correlation of the group is Fig. 13-86 for converting between ASTM D 86 and TBP distillations of petroleum fractions at 101.3 kPa (760 torr). The ASTM D 2889 test method, which presents a standard method for calculating EFV curves from the results of an ASTM D 86 test for a petroleum fraction having a 10 to 90 volume percent boiling range of less than 55°C (100°F), is also quite useful.

APPLICATIONS OF PETROLEUM DISTILLATION

Typical equipment configurations for the distillation of crude oil and other complex hydrocarbon mixtures in a crude unit, a catalytic-cracking unit, and a delayed-coking unit of a petroleum refinery are shown in Figs. 13-87, 13-88, and 13-89. The initial separation of crude oil into fractions is conducted in two main columns, shown in Fig. 13-87. In the first column, called the atmospheric tower or topping still, partially vaporized crude oil, from which water, sediment, and salt have been removed, is mainly rectified, at a feed-tray pressure of no more than about 276 kPa (40 psia), to yield a noncondensable light-hydrocarbon gas, a light naphtha, a heavy naphtha, a light distillate (kerosene), a heavy distillate (diesel oil), and a bottoms residual of components whose TBP exceeds approximately 427°C (800°F). Alternatively, other fractions, shown in Fig. 13-82, may be withdrawn. To control the IBP of the ASTM D 86 curves, each of the sidestreams of the atmospheric tower and the vacuum and main fractionators of Figs. 13-87, 13-88, and 13-89 may be sent to side-cut strippers, which use a partial reboiler or steam stripping. Additional stripping by steam is commonly used in the bottom of the atmospheric tower as well as in the vacuum tower and other main fractionators.

Additional distillate in the TBP range of approximately 427 to 593°C (800 to 1100°F) is recovered from bottoms residuum of the

atmospheric tower by rectification in a vacuum tower, also shown in Fig. 13-87, at the minimum practical overhead condenser pressure, which is typically 1.3 kPa (10 torr). Use of special low-pressure-drop trays or column packing permits feed-tray pressure to be approximately 5.3 to 6.7 kPa (40 to 50 torr) to obtain the maximum degree of vaporization. Vacuum towers may be designed or operated to produce several different products including heavy distillates, gas-oil feedstocks for catalytic cracking, lubricating oils, bunker fuel, and bottoms residua of asphalt (5 to 8° API gravity) or pitch (0 to 5° API gravity). The catalytic-cracking process of Fig. 13-88 produces a superheated vapor at approximately 538°C (1000°F) and 172 to 207 kPa (25 to 30 psia) of a TBP range that covers hydrogen to compounds with normal boiling points above 482°C (900°F). This gas is sent directly to a main fractionator for rectification to obtain products that are typically gas and naphtha [204°C (400°F) ASTM EP approximately], which are often fractionated further to produce relatively pure light hydrocarbons and gasoline; a light cycle oil [typically 204 to 371°C (400 to 700°F) ASTM D 86 range], which may be used for heating oil, hydrocracked, or recycled to the catalytic cracker; an intermediate cycle oil [typically 371 to 482°C (700 to 900°F) ASTM D 86 range], which is generally recycled to the catalytic cracker to extinction; and a heavy gas oil or bottom slurry oil.

Vacuum-column bottoms, bottoms residuum from the main fractionation of a catalytic cracker, and other residua can be further processed at approximately 510°C (950°F) and 448 kPa (65 psia) in a delayed coker unit, as shown in Fig. 13-89 to produce petroleum coke and gas of TBP range that covers methane (with perhaps a small amount of hydrogen) to compounds with normal boiling points that may exceed 649°C (1200°F). The gas is sent directly to a main fractionator that is similar to the type used in conjunction with a catalytic cracker, except that in the delayed-coking operation the liquid to be coked first enters into and passes down through the bottom trays of the main fractionator to be preheated by and to scrub coker vapor of entrained coke particles and condensables for recycling to the delayed coker. Products produced from the main fractionator are similar, except for more unsaturated cyclic compounds, to those produced in a catalytic-cracking unit and include gas and coker naphtha, which are further processed to separate out light hydrocarbons and a coker naphtha that generally needs hydrotreating; and light and heavy coker gas oils, both of which may require hydrocracking to become suitable blending stocks.

DESIGN PROCEDURES

Two general procedures are available for designing fractionators that process petroleum, synthetic crude oils, and complex mixtures. The first, which was originally developed for crude units by Packie [*Trans. Am. Inst. Chem. Eng. J.,* **37,** 51 (1941)], extended to main fractionators by Houghland, Lemieux, and Schreiner [*Proc. API,* sec. III, *Refining,* 385 (1954)], and further elaborated and described in great detail by Watkins (op. cit.), utilizes material and energy balances, with empirical correlations to establish tray requirements, and is essentially a hand-calculation procedure that is a valuable learning experience and is suitable for preliminary designs. Also, when backed by sufficient experience from previous designs, this procedure is adequate for final design.

In the second procedure, which is best applied with a digital computer, the complex mixture being distilled is represented by actual components at the light end and by perhaps 30 pseudo components (e.g., petroleum fractions) over the remaining portion of the TBP distillation curve for the column feed. Each of the pseudo components is characterized by a TBP range, an average normal boiling point, an average API gravity, and an average molecular weight. Rigorous material-balance, energy-balance, and phase equilibrium calculations are then made by an appropriate equation-tearing method as shown by Cecchetti et al. [*Hydrocarbon Process.,* **42**(9), 159 (1963)] or a simultaneous-correction procedure as shown, e.g., by Goldstein and Stanfield [*Ind. Eng. Chem. Process Des. Dev.,* **9,** 78 (1970) and Hess et al. [*Hydrocarbon Process.,* **56**(5), 241 (1977)]. Highly developed procedures of the latter type, suitable for preliminary or final design are included in most com-

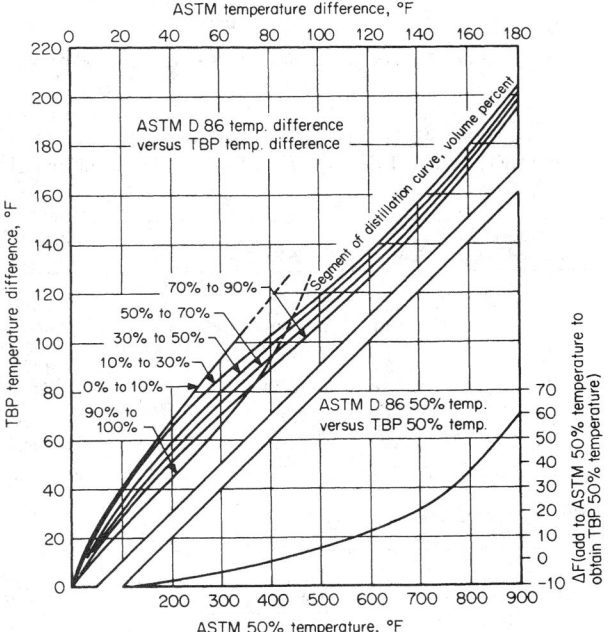

FIG. 13-86 Relationship between ASTM and TBP distillation curves. (*From W. C. Edmister,* Applied Hydrocarbon Thermodynamics, *vol. 1, 1st ed., 1961 Gulf Publishing Company, Houston, Texas. Used with permission. All rights reserved.*)

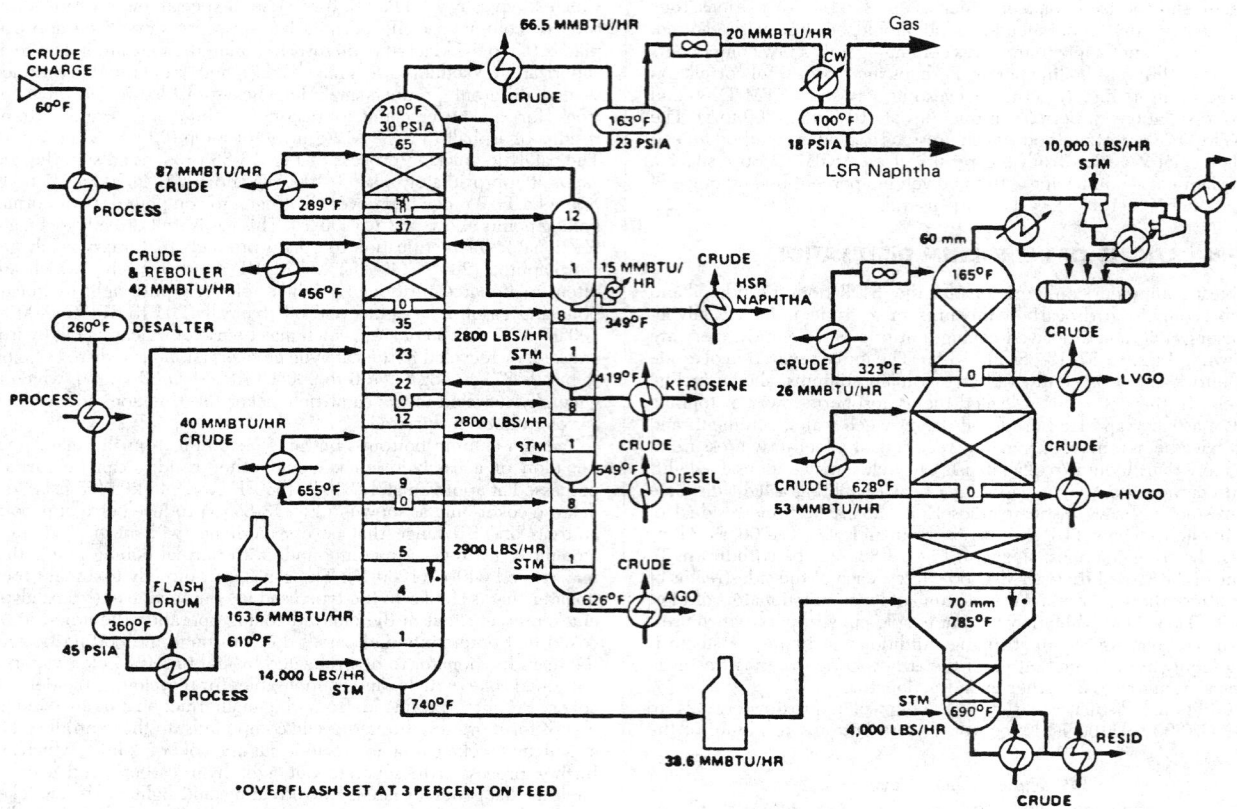

FIG. 13-87 Crude unit with atmospheric and vacuum towers. [*Kleinschrodt and Hammer, "Exchange Networks for Crude Units".* Chem. Eng. Prog., **79**(7), 33 (1983).]

puter-aided steady-state process design and simulation programs as a special case of interlinked distillation, wherein the crude tower or fractionator is converged simultaneously with the sidecut-stripper columns.

Regardless of the procedure used, certain initial steps must be taken for the determination or specification of certain product properties and yields based on the TBP distillation curve of the column feed, method of providing column reflux, column-operating pressure, type of condenser, and type of side-cut strippers and stripping requirements. These steps are developed and illustrated with several detailed examples by Watkins (op. cit.). Only one example, modified from one given by Watkins, is considered briefly here to indicate the approach taken during the initial steps.

For the atmospheric tower shown in Fig. 13-90, suppose distillation specifications are as follows:

Feed: 50,000 bbl (at 42 U.S. gal each) per stream day (BPSD) of 31.6° API crude oil.

Measured light-ends analysis of feed:

Component	Volume percent of crude oil
Ethane	0.04
Propane	0.37
Isobutane	0.27
n-Butane	0.89
Isopentane	0.77
n-Pentane	1.13
	3.47

Measured TBP and API gravity of feed, computed atmospheric pressure EFV (from API *Technical Data Book*), and molecular weight of feed:

Volume percent vaporized	TBP, °F	EFV, °F	°API	Molecular weight
0	−130	179		
5	148	275	75.0	91
10	213	317	61.3	106
20	327	394	50.0	137
30	430	468	41.8	177
40	534	544	36.9	223
50	639	619	30.7	273
60	747	696	26.3	327
70	867	777	22.7	392
80	1013	866	19.1	480

Product specifications:

	ASTM D 86, °F		
Desired cut	5%	50%	95%
Overhead (*OV*)			253
Heavy naphtha (*HN*)	278	314	363
Light distillate (*LD*)	398	453	536
Heavy distillate (*HD*)	546	589	
Bottoms (*B*)			

NOTE: To convert degrees Fahrenheit to degrees Celsius, °C = (°F − 32)/1.8.

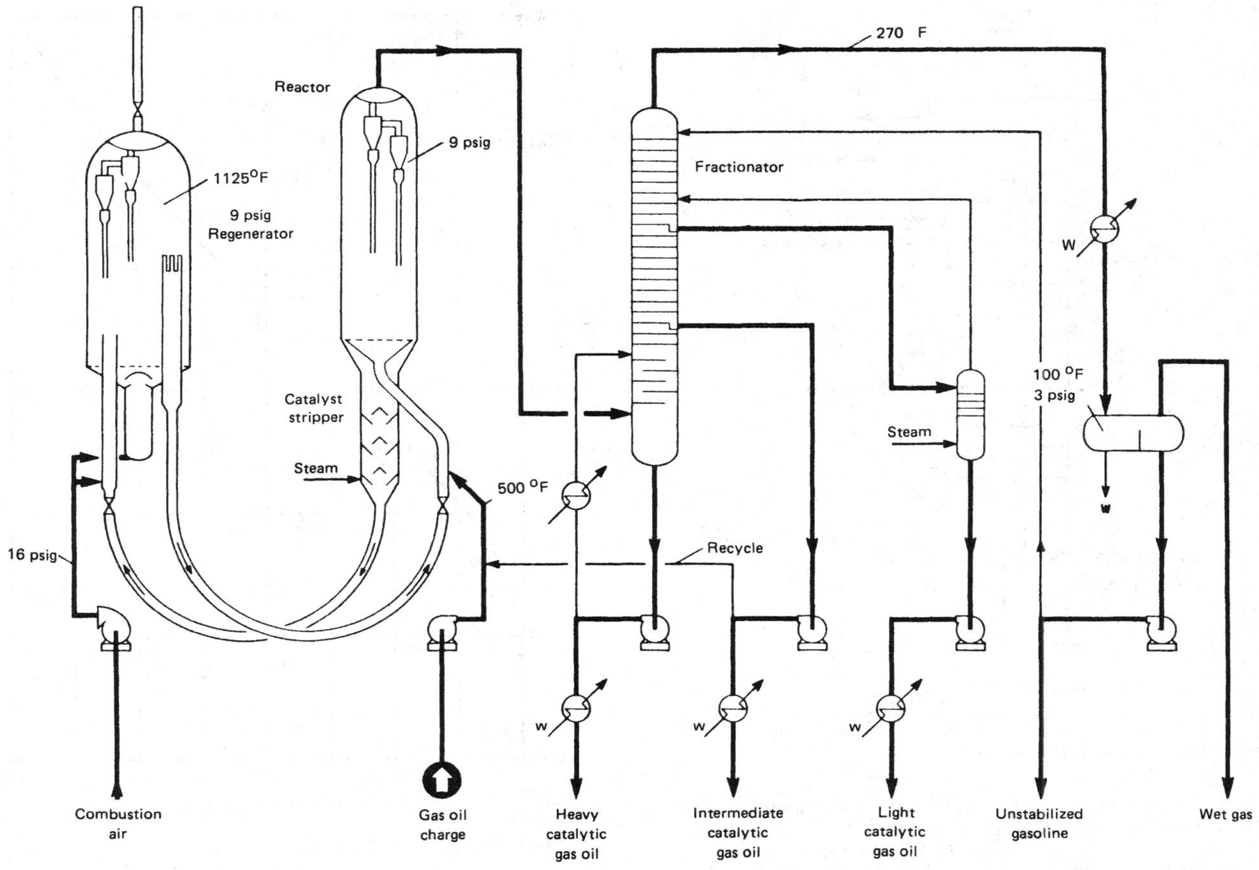

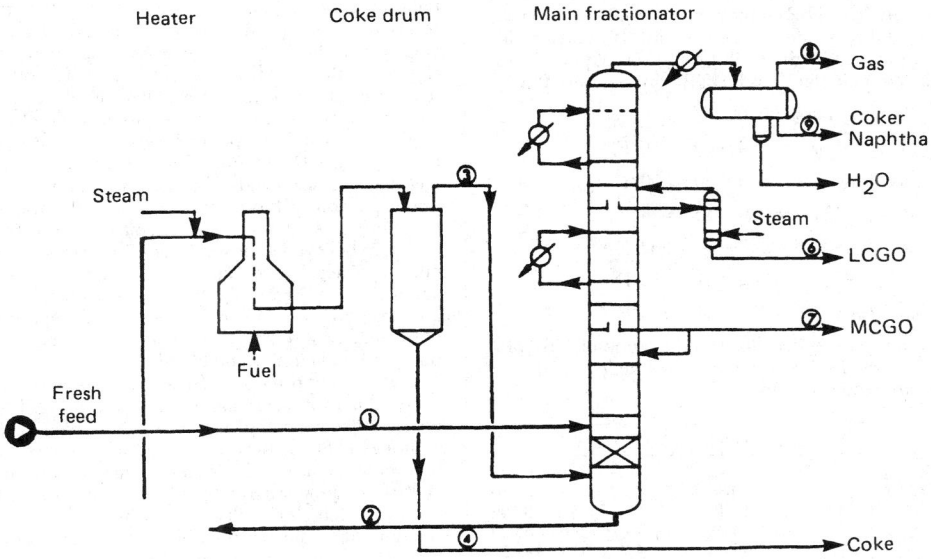

FIG. 13-88   Catalytic cracking unit. (New Horizons, *Lummus Co., New York, 1954.*)

FIG. 13-89   Delayed-coking unit. (*Watkins*, Petroleum Refinery Distillation, *2nd ed., Gulf, Houston, 1979.*)

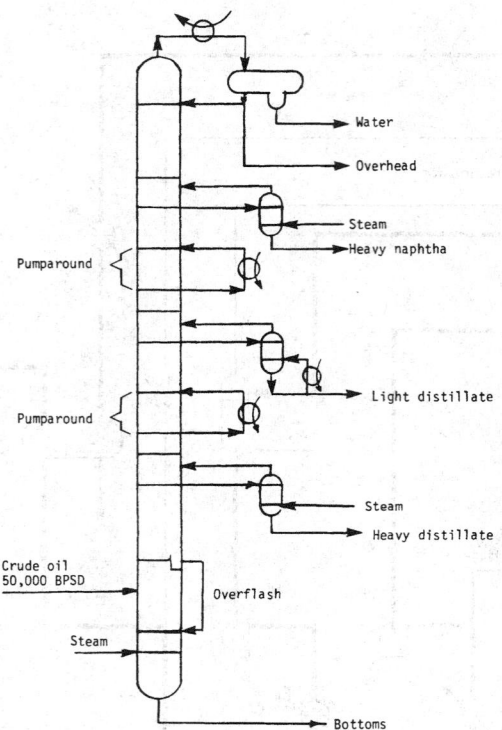

FIG. 13-90   Crude atmospheric tower.

TBP cut point between the heavy distillate and the bottoms = 650°F.
Percent overflash = 2 volume percent of feed.
Furnace outlet temperature = 343°C (650°F) maximum.
Overhead temperature in reflux drum = 49°C (120°F) minimum.

From the product specifications, distillate yields are computed as follows: From Fig. 13-86 and the ASTM D 86 50 percent temperatures, TBP 50 percent temperatures of the three intermediate cuts are obtained as 155, 236, and 316°C (311, 456, and 600°F) for the *HN, LD,* and *HD* respectively. The TBP cut points, corresponding volume fractions of crude oil, and flow rates of the four distillates are readily obtained by starting from the specified 343°C (650°F) cut point as follows, where *CP* is the cut point and *T* is the TBP temperature (°F):

$$CP_{HD,B} = 650°F$$

$$(CP_{HD,B} - T_{HD_{50}}) = 650 - 600 = 50°F$$

$$CP_{LD,HD} = T_{HD_{50}} - 50 = 600 - 50 = 550°F$$

$$(CP_{LD,HD} - T_{LD_{50}}) = 550 - 456 = 94°F$$

$$CP_{HN,LD} = T_{LD_{50}} - 94 = 456 - 94 = 362°F$$

$$(CP_{HN,LD} - T_{HN_{50}}) = 362 - 311 = 51°F$$

$$CP_{OV,HN} = T_{HN_{50}} - 51 = 311 - 51 = 260°F$$

These cut points are shown as vertical lines on the crude-oil TBP plot of Fig. 13-91, from which the following volume fractions and flow rates of product cuts are readily obtained:

Desired cut	Volume percent of crude oil	BPSD
Overhead (*OV*)	13.4	6,700
Heavy naphtha (*HN*)	10.3	5,150
Light distillate (*LD*)	17.4	8,700
Heavy distillate (*HD*)	10.0	5,000
Bottoms (*B*)	48.9	24,450
	100.0	50,000

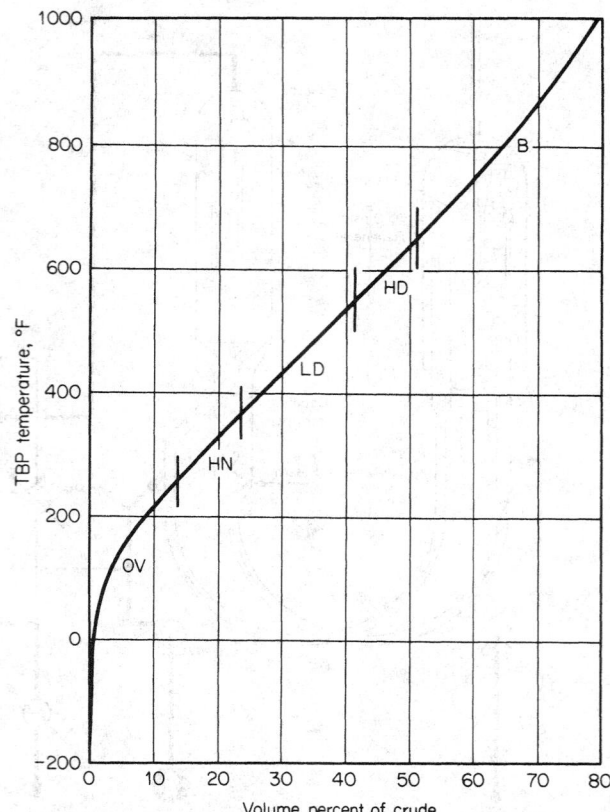

FIG. 13-91   Example of crude-oil TBP cut points.

As shown in Fig. 13-92, methods of providing column reflux include (*a*) conventional top-tray reflux, (*b*) pump-back reflux from side-cut strippers, and (*c*) pump-around reflux. The latter two methods essentially function as intercondenser schemes that reduce the top-tray-reflux requirement. As shown in Fig. 13-93 for the example being considered, the internal-reflux flow rate decreases rapidly from the top tray to the feed-flash zone for case *a.* The other two cases, particularly case *c,* result in better balancing of the column-reflux traffic. Because of this and the opportunity provided to recover energy at a moderate- to high-temperature level, pump-around reflux is the most commonly used technique. However, not indicated in Fig. 13-93 is the fact that in cases *b* and *c* the smaller quantity of reflux present in the upper portion of the column increases the tray requirements. Furthermore, the pump-around circuits, which extend over three trays each, are believed to be equivalent for mass-transfer purposes to only one tray each. Representative tray requirements for the three cases are included in Fig. 13-92. In case *c* heat-transfer rates associated with the two pump-around circuits account for approximately 40 percent of the total heat removed in the overhead condenser and from the two pump-around circuits combined.

Bottoms and three side-cut strippers remove light ends from products and may utilize steam or reboilers. In Fig. 13-92 a reboiled stripper is utilized on the light distillate, which is the largest side cut withdrawn. Steam-stripping rates in side-cut strippers and at the bottom of the atmospheric column may vary from 0.45 to 4.5 kg (1 to 10 lb) of steam per barrel of stripped liquid, depending on the fraction of stripper feed liquid that is vaporized.

Column pressure at the reflux drum is established so as to condense totally the overhead vapor or some fraction thereof. Flash-zone pressure is approximately 69 kPa (10 psia) higher. Crude-oil feed temper-

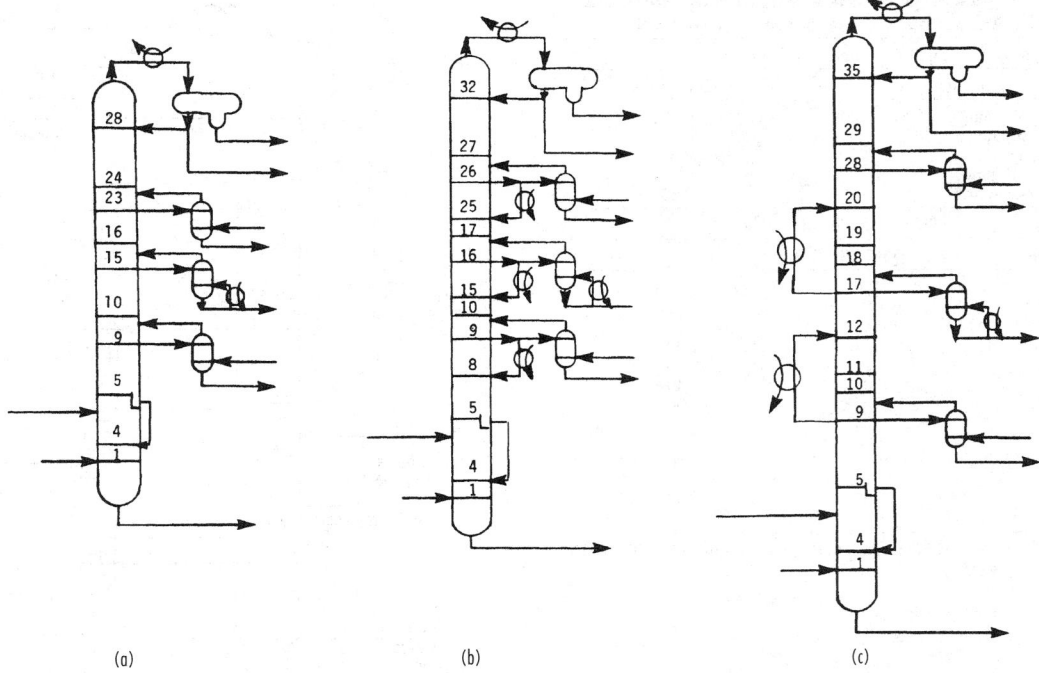

FIG. 13-92   Methods of providing reflux to crude units. (*a*) Top reflux. (*b*) Pump-back reflux. (*c*) Pump-around reflux.

ature at flash-zone pressure must be sufficient to vaporize the total distillates plus the overflash, which is necessary to provide reflux between the lowest sidestream-product draw-off tray and the flash zone. Calculations are made by using the crude-oil EFV curve corrected for pressure. For the example being considered, percent vaporized at the flash zone must be 53.1 percent of the feed.

Tray requirements depend on internal-reflux ratios and ASTM 5-95 gaps or overlaps, and may be estimated by the correlation of Packie (op. cit.) for crude units and the correlation of Houghland, Lemieux, and Schreiner (op. cit.) for main fractionators.

Example 9: Simulation Calculation of an Atmospheric Tower
The ability of a rigorous calculation procedure to simulate operation of an atmospheric tower with its accompanying side-cut strippers may be illustrated by comparing commercial-test data from an actual operation with results computed with the REFINE program of ChemShare Corporation, Houston, Texas. The tower configuration and plant-operating conditions are shown in Fig. 13-94. Light-component analysis and the TBP and API gravity for the feed are given in Table 13-27. Representation of this feed by pseudocomponents is given in Table 13-28 based on 16.7°C (30°F) cuts from 82 to 366°C (180°F to 690°F), followed by 41.7°C (75°F) and then 55.6°C (100°F) cuts. Actual tray numbers are shown in Fig. 13-94. Corresponding theoretical-stage numbers, which were determined by trial and error to obtain a reasonable match of computed- and measured-product TBP distillation curves, are shown in parentheses. Overall tray efficiency appears to be approximately 70 percent for the tower and 25 to 50 percent for the side-cut strippers.

Results of rigorous calculations and comparison to plant data, when possible, are shown in Figs. 13-95, 13-96, and 13-97. Plant temperatures are in good

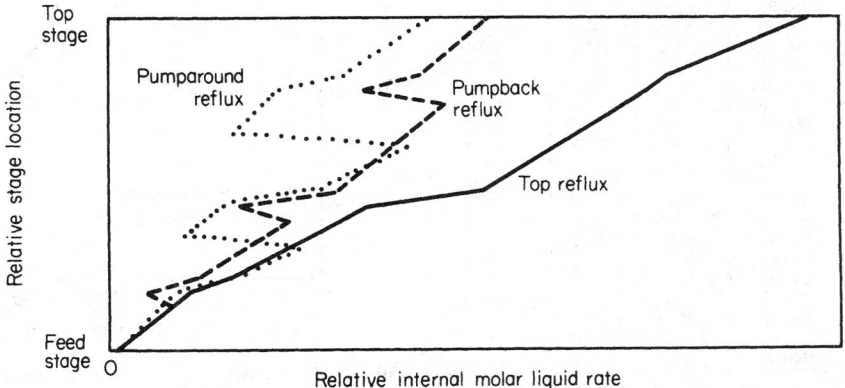

FIG. 13-93   Comparison of internal-reflux rates for three methods of providing reflux.

TABLE 13-27   Light-Component Analysis and TBP Distillation of Feed for the Atmospheric Crude Tower of Fig. 13-94

Light-component analysis	
Component	Volume percent
Methane	0.073
Ethane	0.388
Propane	0.618
n-Butane	0.817
n-Pentane	2.05

TBP distillation of feed		
API gravity	TBP, °F	Volume percent
80	−160.	0.1
70	155.	5.
57.5	242.	10.
45.	377.	20.
36.	499.	30.
29.	609.	40.
26.5	707.	50.
23.	805.	60.
20.5	907.	70.
17.	1054.	80.
10.	1210.	90.
−4.	1303.	95.
−22.	1467.	100.

NOTE: To convert degrees Fahrenheit to degrees Celsius, °C = (°F − 32)/1.8.

TABLE 13-28   Pseudo-Component Representation of Feed for the Atmospheric Crude Tower of Fig. 13-94

No.	Component name	Molecular weight	Specific gravity	API gravity	(lb·mol)/h
1	Water	18.02	1.0000	10.0	.00
2	Methane	16.04	.3005	339.5	7.30
3	Ethane	30.07	.3561	265.8	24.54
4	Propane	44.09	.5072	147.5	37.97
5	n-Butane	58.12	.5840	110.8	43.84
6	n-Pentane	72.15	.6308	92.8	95.72
7	131 ABP	83.70	.6906	73.4	74.31
8	180 ABP	95.03	.7152	66.3	66.99
9	210 ABP	102.23	.7309	62.1	65.83
10	240 ABP	109.78	.7479	57.7	70.59
11	270 ABP	118.52	.7591	54.9	76.02
12	300 ABP	127.69	.7706	52.1	71.62
13	330 ABP	137.30	.7824	49.4	67.63
14	360 ABP	147.33	.7946	46.6	64.01
15	390 ABP	157.97	.8061	44.0	66.58
16	420 ABP	169.37	.8164	41.8	63.30
17	450 ABP	181.24	.8269	39.6	59.92
18	480 ABP	193.59	.8378	37.4	56.84
19	510 ABP	206.52	.8483	35.3	59.05
20	540 ABP	220.18	.8581	33.4	56.77
21	570 ABP	234.31	.8682	31.5	53.97
22	600 ABP	248.30	.8804	29.2	52.91
23	630 ABP	265.43	.8846	28.5	54.49
24	660 ABP	283.37	.8888	27.7	51.28
25	690 ABP	302.14	.8931	26.9	48.33
26	742 ABP	335.94	.9028	25.2	109.84
27	817 ABP	387.54	.9177	22.7	94.26
28	892 ABP	446.02	.9288	20.8	74.10
29	967 ABP	509.43	.9398	19.1	50.27
30	1055 ABP	588.46	.9531	17.0	57.12
31	1155 ABP	665.13	.9829	12.5	50.59
32	1255 ABP	668.15	1.0658	1.3	45.85
33	1355 ABP	643.79	1.1618	−9.7	29.39
34	1436 ABP	597.05	1.2533	−18.6	21.19
		246.90	.8887	27.7	1922.43

NOTE: To convert (lb·mol)/h to (kg·mol)/h, multiply by 0.454.

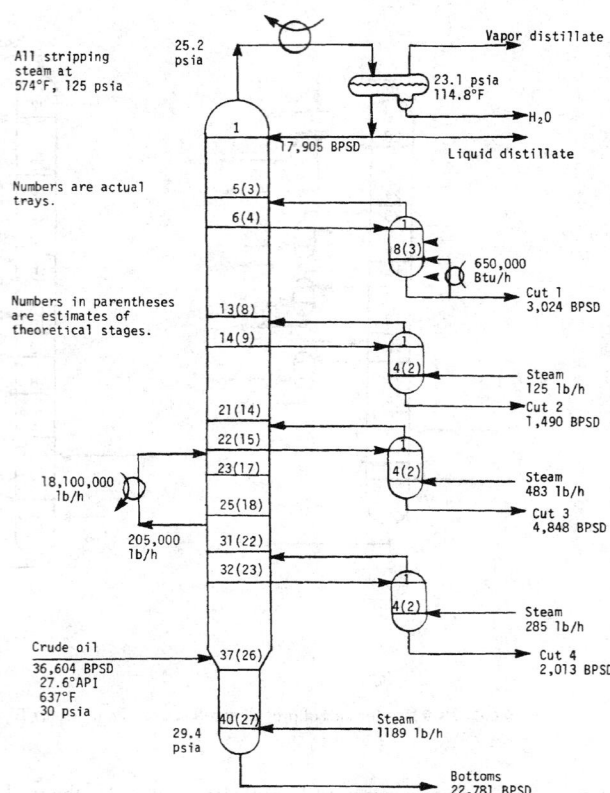

FIG. 13-94   Configuration and conditions for the simulation of the atmospheric tower of crude unit.

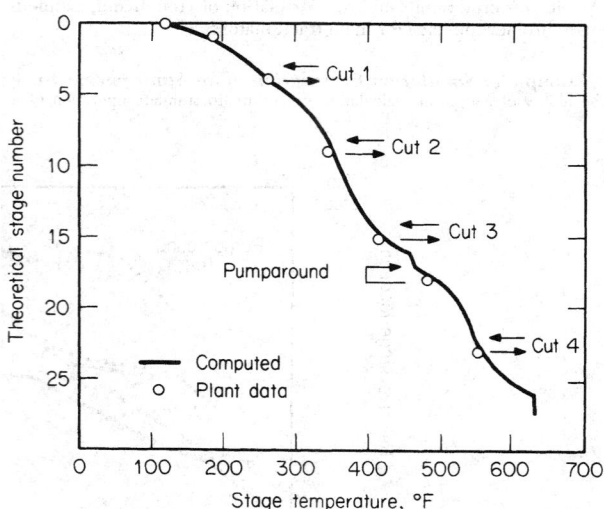

FIG. 13-95   Comparison of computed stage temperatures with plant data for the example of Fig. 13-94.

agreement with computed values in Fig. 13-95. Computed sidestream-product TBP distillation curves are in reasonably good agreement with values converted from plant ASTM distillations as shown in Fig. 13-96. Exceptions are the initial points of all four cuts and the higher-boiling end of the heavy-distillate curve. This would seem to indicate that more theoretical stripping stages should be added and that either the percent vaporization of the tower feed in the simulation is too high or the internal-reflux rate at the lower draw-off tray is too low. The liquid-rate profile in the tower is shown in Fig. 13-97. The use of two or three pump-around circuits instead of one would result in a better traffic pattern than that shown.

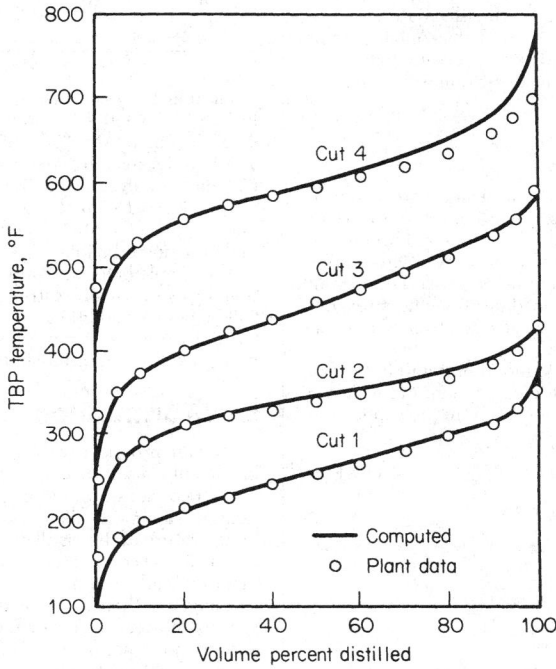

FIG. 13-96 Comparison of computed TBP curves with plant data for the example of Fig. 13-94.

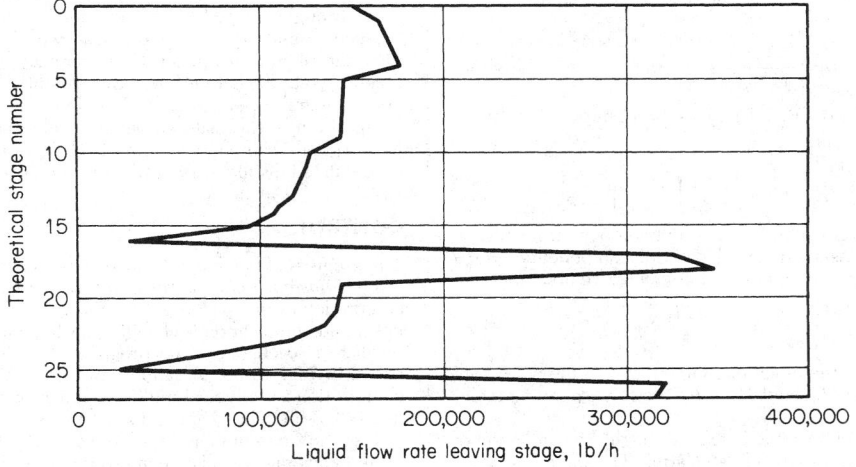

FIG. 13-97 Liquid-rate profile for the example of Fig. 13-94.

BATCH DISTILLATION

Batch distillation, which is the process of separating a specific quantity (the charge) of a liquid mixture into products, is used extensively in the laboratory and in small production units that may have to serve for many mixtures. When there are N components in the feed, one batch column will suffice where $N-1$ simple continuous-distillation columns would be required.

Many larger installations also feature a batch still. Material to be separated may be high in solids content, or it might contain tars or resins that would plug or foul a continuous unit. Use of a batch unit can keep solids separated and permit convenient removal at the termination of the process.

SIMPLE BATCH DISTILLATION

The simplest form of batch still consists of a heated vessel (pot or boiler), a condenser, and one or more receiving tanks. No trays or packing are provided. Feed is charged into the vessel and brought to boiling. Vapors are condensed and collected in a receiver. No reflux is returned. The rate of vaporization is sometimes controlled to prevent "bumping" the charge and to avoid overloading the condenser, but other controls are minimal. This process is often referred to as Rayleigh distillation.

If we represent the moles of vapor by V, moles of liquid in the pot by M, the mole fraction of the more volatile component in this liquid by x, and the mole fraction of the same component in the vapor by y, a material balance yields

$$-y\,dV = d(Mx) \tag{13-133}$$

Since $dV = -dM$, substitution and expansion give

$$y\,dM = M\,dx + x\,dM \tag{13-134}$$

Rearranging and integrating give

$$\ln\frac{M_i}{M_f} = \int_{x_f}^{x_i}\frac{dx}{y - x} \tag{13-135}$$

where subscript i represents the initial condition and f the final condition of the liquid in the still pot. Integration limits have been reversed to obtain a positive integral. If equilibrium is assumed between liquid and vapor, the right-hand side of Eq. (13-135) may be evaluated by plotting $1/(y - x)$ versus x and measuring the area under the curve between limits x_i and x_f. If the mixture is a binary system for which relative volatility α is constant or if an average value that will serve for the range considered can be found, then the relationship that defines relative volatility,

$$\alpha = \frac{y/x}{(1 - y)/(1 - x)} \tag{13-136}$$

can be substituted into Eq. (13-135) and a direct integration can be made:

$$\ln\left(\frac{M_f}{M_i}\right) = \frac{1}{\alpha - 1}\ln\left[\frac{x_f(1 - x_i)}{x_i(1 - x_f)}\right] + \ln\left[\frac{1 - x_i}{1 - x_f}\right] \tag{13-137}$$

For any two components A and B of a multicomponent mixture, if constant α values are assumed for all pairs of components, $-dM_A/-dM_B = y_A/y_B = \alpha_{A,B}(x_A/x_B)$. When this is integrated, we obtain

$$\ln\left(\frac{M_{A(f)}}{M_{A(i)}}\right) = \alpha_{A,B}\ln\left(\frac{M_{B(f)}}{M_{B(i)}}\right) \tag{13-138}$$

where $M_{A(i)}$ and $M_{A(f)}$ are the moles of component A in the pot before and after distillation and $M_{B(i)}$ and $M_{B(f)}$ are the corresponding moles of component B.

A typical application of a simple batch still might be distillation of an ethanol-water mixture at 101.3 kPa (1 atm). The initial charge is 100 mol of ethanol at 18 mole percent, and the mixture must be reduced to a maximum ethanol concentration in the still of 6 mole percent. By using equilibrium data interpolated from Table 13-1,

x	y	$y - x$	$1/(y - x)$
0.18	0.517	0.337	2.97
.16	.502	.342	2.91
.14	.485	.345	2.90
.12	.464	.344	2.90
.10	.438	.338	2.97
.08	.405	.325	3.08
.06	.353	.293	3.41

Plotting $1/(y - x)$ versus x and integrating graphically between the limits of 0.06 and 0.18 for x, the area under the curve is found to be 0.358. Then, $\ln(M_i/M_j) = 0.358$, from which $M_j = 100/1.43 = 70.0$ mol. The liquid remaining consists of $(70.0)(0.06) = 4.2$ mol of ethanol and 65.8 mol of water. By material balance, the total distillate must contain $(18.0 - 4.2) = 13.8$ mol of alcohol and $(82 - 65.8) = 16.2$ mol of water. Total distillate is 30 mol, and distillate composition is $13.8/30 = 0.46$ mole fraction ethanol.

The simple batch still provides only one theoretical plate of separation. Its use is usually restricted to preliminary work in which products will be held for additional separation at a later time, when most of the volatile component must be removed from the batch before it is processed further, or for similar noncritical separations.

BATCH DISTILLATION WITH RECTIFICATION

To obtain products with a narrow composition range, a rectifying batch still is used that consists of a pot (or reboiler), a rectifying column, a condenser, some means of splitting off a portion of the condensed vapor (distillate) as reflux, and one or more receivers. Temperature of the distillate is controlled in order to return the reflux at or near the column temperature to permit a true indication of reflux quantity and to improve column operation. A subcooling heat exchanger is then used for the remainder of the distillate, which is sent to an accumulator or receiver. The column may also operate at elevated pressure or vacuum, in which case appropriate devices must be included to obtain the desired pressure. Equipment-design methods for batch-still components, except for the pot, follow the same principles as those presented for continuous units, but the design should be checked for each mixture if several mixtures are to be processed. It should also be checked at more than one point of a mixture, since composition in the column changes as distillation proceeds. Pot design is based on batch size and required vaporization rate.

In operation, a batch of liquid is charged to the pot and the system is first brought to steady state under total reflux. A portion of the overhead condensate is then continuously withdrawn in accordance with the established reflux policy. Cuts are made by switching to alternate receivers, at which time operating conditions may be altered. The entire column operates as an enriching section. As time proceeds, composition of the material being distilled becomes less rich in the more volatile components, and distillation of a cut is stopped when accumulated distillate attains the desired average composition.

CONTROL

The progress of batch distillation can be controlled in several ways:

1. *Constant reflux, varying overhead composition.* Reflux is set at a predetermined value at which it is maintained for the run. Since pot liquid composition is changing, instantaneous composition of the distillate also changes. The progress of a binary separation is illustrated in Fig. 13-98. Variation with time of instantaneous distillate composition for a typical multicomponent batch distillation is shown in Fig. 13-99. The shapes of the curves are functions of volatility, reflux ratio, and number of theoretical plates. Distillation is continued until the average distillate composition is at the desired value. In the case of a binary, the overhead is then diverted to another receiver, and an intermediate cut is withdrawn until the remaining pot liquor meets the required specification. The intermediate cut is usually added to

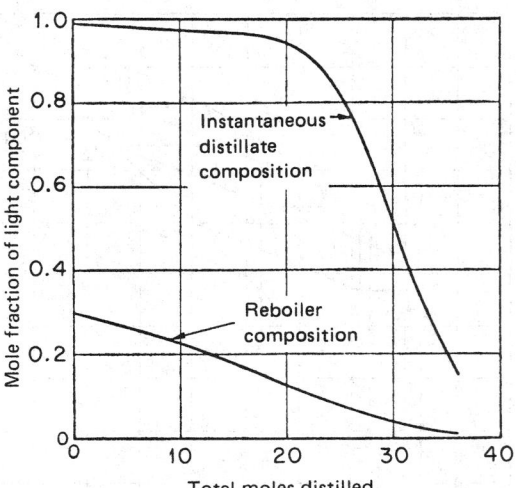

FIG. 13-98 Typical variation in distillate and reboiler compositions with amount distilled in binary batch distillation at a constant-reflux ratio.

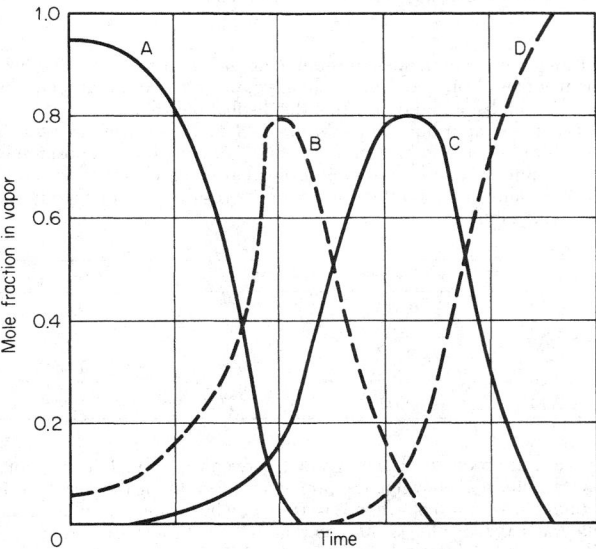

FIG. 13-99 Distillate composition profile for a batch distillation of a four-component mixture.

the next batch. For a multicomponent mixture, two or more intermediate cuts may be taken between product cuts.

2. *Constant overhead composition, varying reflux.* If it is desired to maintain a constant overhead composition in the case of a binary, the amount of reflux returned to the column must be constantly increased throughout the run. As time proceeds, the pot is gradually depleted of the lighter component. Finally, a point is reached at which the reflux ratio has attained a very high value. The receivers are then changed, the reflux is reduced, and an intermediate cut is taken as before. This technique can also be extended to a multicomponent mixture.

3. *Other control methods.* A cycling procedure can be used to set the pattern for column operation. The unit operates at total reflux until equilibrium is established. Distillate is then taken as total draw-

off for a short period of time, after which the column is again returned to total-reflux operation. This cycle is repeated through the course of distillation. Another possibility is to optimize the reflux ratio in order to achieve the desired separation in a minimum of time. Complex operations may involve withdrawal of sidestreams, provision for inter-condensers, addition of feeds to trays, and periodic charge addition to the pot.

APPROXIMATE CALCULATION PROCEDURES FOR BINARY MIXTURES

A useful method for a binary mixture employs an analysis based on the McCabe-Thiele graphical method. In addition to the usual assumptions of adiabatic column and equimolal overflow on the trays, the following procedure assumes negligible holdup of liquid on the trays, in the column, and in the condenser.

As a first step in the calculation, the minimum-reflux ratio should be determined. In Fig. 13-100, point D, representing the distillate, is on the diagonal since a total condenser is assumed and $x_D = y_D$. Point F represents the initial condition in the still pot with coordinates x_{pi}, y_{pi}. Minimum internal reflux is represented by the slope of the line DF,

$$(L/V)_{min} = (y_D - y_{pi})/(x_D - x_{pi}) \qquad (13-139)$$

where L is the liquid flow rate and V is the vapor rate, both in moles per hour. Since $V = L + D$ (where D is distillate rate) and the external-reflux rate R is defined as $R = L/D$, then

$$L/V = R/(R + 1) \qquad (13-140)$$

or

$$R_{min} = \frac{(L/V)_{min}}{1 - (L/V)_{min}} \qquad (13-141)$$

The condition of minimum reflux for an equilibrium curve with an inflection point P is shown in Fig. 13-102. In this case the minimum internal reflux is

$$(L/V)_{min} = (y_D - y_P)/(x_D - x_P) \qquad (13-142)$$

The operating reflux ratio is usually 1.5 to 10 times the minimum. By using the ethanol-water equilibrium curve for 101.3 kPa (1 atm) pressure shown in Fig. 13-101 but extending the line to a convenient point for readability, $(L/V)_{min} = (0.800 - 0.695)/(0.800 - 0.600) = 0.52$ and $R_{min} = 1.083$.

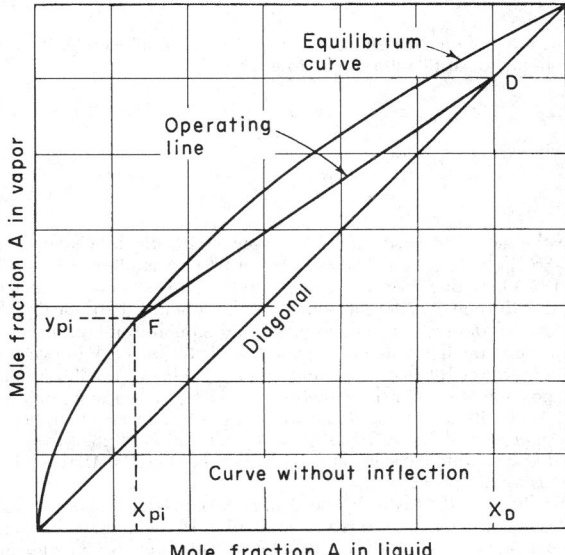

FIG. 13-100 Determination of minimum reflux for normal equilibrium curve.

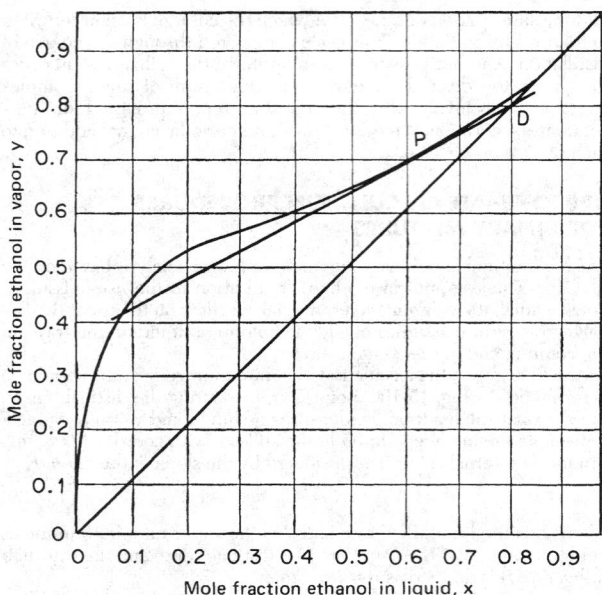

FIG. 13-101 Determination of minimum reflux for equilibrium curve with inflection.

OPERATING METHODS

Batch Rectification at Constant Reflux Using an analysis similar to the simple batch still, Smoker and Rose [*Trans. Am. Inst. Chem. Eng.*, **36**, 285 (1940)] developed the following equation:

$$\ln \frac{M_i}{M_f} = \int_{x_{pf}}^{x_{pi}} \frac{dx_p}{x_D - x_P} \qquad (13\text{-}143)$$

An overall component balance gives the average or accumulated distillate composition $x_{D,\mathrm{avg}}$

$$x_{D,\mathrm{avg}} = \frac{M_i x_{pi} - M_f x_{pf}}{M_i - M_f} \qquad (13\text{-}144)$$

If the integral on the right side of Eq. (13-143) is labeled Q, the time θ in hours for distillation can be found by

$$\theta = (R + 1) \frac{M_i(e^Q - 1)}{Ve^Q} \qquad (13\text{-}145)$$

An alternative equation is

$$\theta = \frac{R + 1}{V} (M_i - M_f) \qquad (13\text{-}146)$$

Development of these equations is given by Block [*Chem. Eng.*, **68**, 88 (Feb. 6, 1961)]. The calculation process is illustrated in Fig. 13-102. Operating lines are drawn with the same slope but intersecting the 45° line at different points. The number of theoretical plates under consideration is stepped off to find equilibrium bottoms composition. In the figure, operating line $L - 1$ with slope L/V drawn from point D_1 where the distillate composition is x_{D1} has an equilibrium pot composition of $x_{p1\text{-}3}$ for three theoretical plates, x_{D2} has an equilibrium pot composition of $x_{p2\text{-}3}$, etc.; performing a graphical integration of the right side of Eq. (13-143) permits calculation of $x_{D,\mathrm{avg}}$ from Eq. (13-144). An iterative calculation is required to find the M_f that corresponds to the specified $x_{D,\mathrm{avg}}$.

To illustrate the use of these equations, consider a charge of 520 mol of an ethanol-water mixture containing 18 mole percent ethanol to be distilled at 101.3 kPa (1 atm). Vaporization rate is 75 mol/h, and the product specification is 80 mole percent ethanol. Let $L/V = 0.75$, corresponding to a reflux ratio $R = 3.0$. If the system has seven theo-

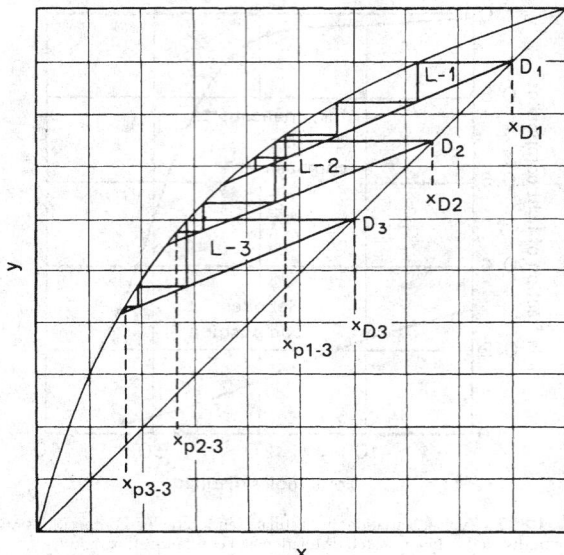

FIG. 13-102 Graphical method for constant-reflux operation.

retical plates, with the pot considered as one of these plates, find how many moles of product will be obtained, what the composition of the residue will be, and the time that the distillation will take.

Using the vapor-liquid equilibrium data, plot a y-x diagram. Draw a number of operating lines at a slope of 0.75. Note the composition at the 45° intersection, and step off seven plates on each to find the equilibrium value of the bottoms. Some of the results are tabulated in the following table:

x_D	x_p	$x_D - x_p$	$1/(x_D - x_p)$
0.800	0.323	0.477	2.097
.795	.245	.550	1.820
.790	.210	.580	1.725
.785	.180	.605	1.654
.780	.107	.673	1.487
.775	.041	.734	1.362

Use a trial procedure by integrating between x_{pi} of 0.18 and various lower limits, and converge the procedure by graphing the results. It is found that $x_{D,\mathrm{avg}} = 0.80$ when $x_{pf} = 0.04$, at which time the value of the integral $= 0.205 = \ln (M_i/M_f)$, so that $M_f = 424$ mol. Product $= M_i - M_f = 520 - 424 = 96$ mol. From Eq. (13-145),

$$\theta = \frac{(4)(520)(e^{0.205} - 1)}{(75)(e^{0.205})} = 5.2 \text{ h}$$

Batch Rectification at Constant Overhead Composition Bogart [*Trans. Am. Inst. Chem. Eng.*, **33**, 139 (1937)] developed the following equation for this situation with column holdup assumed to be negligible:

$$\theta = \frac{M_i(x_D - x_{pi})}{V} \int_{x_{pf}}^{x_{pi}} \frac{dx_p}{(1 - L/V)(x_D - x_p)^2} \qquad (13\text{-}147)$$

where the terms are defined as before. The quantity distilled can then be found by

$$M_i - M_f = \frac{M_i(x_{pi} - x_{pf})}{x_D - x_{pf}} \qquad (13\text{-}148)$$

The progress of a varying-reflux distillation is shown in Fig. 13-103. Distillate composition is held constant by increasing the reflux as pot composition becomes more dilute. Operating lines with varying slopes

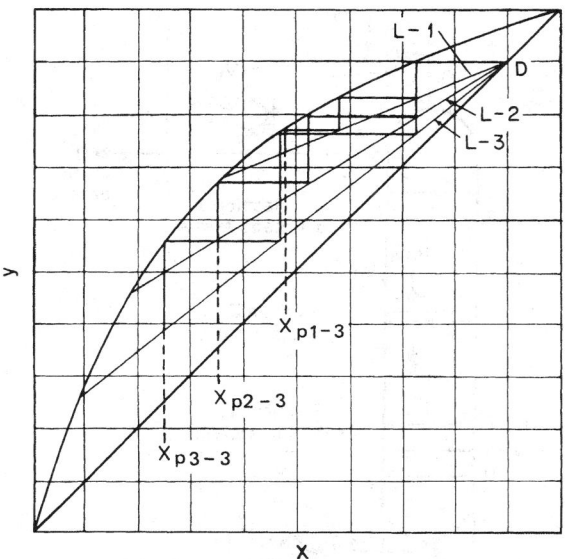

FIG. 13-103   Graphical method for constant-composition operation.

$(= L/V)$ are drawn from the distillate composition, and the appropriate number of plates is stepped off to find the corresponding bottoms composition.

As an example, consider distilling at constant composition the same mixture that was used to illustrate constant reflux. The following table is compiled:

L/V	x_p	$x_D - x_p$	$1/(1 - L/V)(x_D - x_p)^2$
0.600	0.654	0.147	115.7
.700	.453	.348	27.5
.750	.318	.483	17.2
.800	.143	.658	11.5
.850	.054	.747	11.9
.900	.021	.780	16.4

If the right-hand side of Eq. (13-147) is integrated graphically by using a limit for x_{pf} of 0.04, the value of the integral is 1.615 and the time is

$$\theta = \frac{(520)(0.800 - 0.180)(1.615)}{75} = 7.0 \text{ h}$$

The quantity distilled can be found by Eq. (13-148):

$$M_i - M_f = \frac{(520)(0.180 - 0.040)}{0.800 - 0.040} = 96 \text{ mol}$$

Other Operating Methods   A useful control method for difficult industrial or laboratory distillations is *cycling operation*. The most common form of cycling control is operating the column at total reflux until equilibrium is established, taking off the complete distillate for a short period of time, and then returning to total reflux. An alternative scheme is to interrupt vapor flow to the column periodically by the use of a solenoid-operated butterfly valve in the vapor line from the pot. In both cases, equations necessary to describe the system are very complex, as shown by Schrodt et al. [*Chem. Eng. Sci.,* **22,** 759 (1967)]. The most reliable method for establishing the cycle relationships is by experimental trial on an operating column. Several investigators have also proposed that batch distillation can be programmed to attain *time optimization* by proper variation of the reflux ratio. A comprehensive discussion is presented by Coward [*Chem. Eng. Sci.,* **22,** 503 (1967)].

The *choice of operating mode* depends upon characteristics of the

specific system, the product specifications, and the engineer's preference in setting up a control sequence. Probably the most direct and most common method is *constant reflux*. Operation can be regulated by a timed reflux splitter, a ratio controller, or simply a pair of rotameters. Since composition is changing with time, some way must be found to estimate the average accumulated-distillate composition in order to define the end point. This is no problem when the specification is not critical or the change in distillate composition is sharply defined. When the composition of the distillate changes slowly with time, the cut point is more difficult to determine. Operating with *constant composition* (varying reflux), the specification is automatically achieved if control can be linked to concentration or some concentration-sensitive physical variable. The relative advantage, ratewise, of the two systems depends upon the materials being separated and upon the number of theoretical plates in the column. Results of a comparison of distillation rates by using the same initial and final pot composition for the system benzene-toluene are given in Fig. 13-104. Typical control instrumentation is presented in an article by Block [*Chem. Eng.,* **74,** 147 (Jan. 16, 1967)]. Control procedures for *reflux and vapor-cycling* operation and for the *time-optimal* process are largely a matter of empirical trial.

Effect of Column Holdup   When the holdup of liquid on the trays and in the condenser is not negligible compared with the holdup in the pot, the distillate composition at constant-reflux ratio changes with time at a different rate than when the column holdup is negligible because of two separate effects. First, with an appreciable column holdup, composition of the charge to the pot will be higher in the light component than the pot composition at the start of the distillation; the reason for this is that before product takeoff begins, column holdup must be supplied, and its average composition is higher than that of the charge liquid from which it is supplied. Thus, when overhead takeoff begins, the pot composition is lower than it would be if there were no column holdup and separation is more difficult. The second effect of column holdup is to slow the rate of exchange of the components; the holdup exerts an inertia effect, which prevents compositions from changing as rapidly as they would otherwise, and the degree of separation is usually improved. As both these effects occur at the same time and change in importance during the course of distillation, it is difficult, without rigorous calculations, to predict whether the overall effect of holdup will be favorable or detrimental; it is equally difficult to estimate the magnitude of the holdup effect.

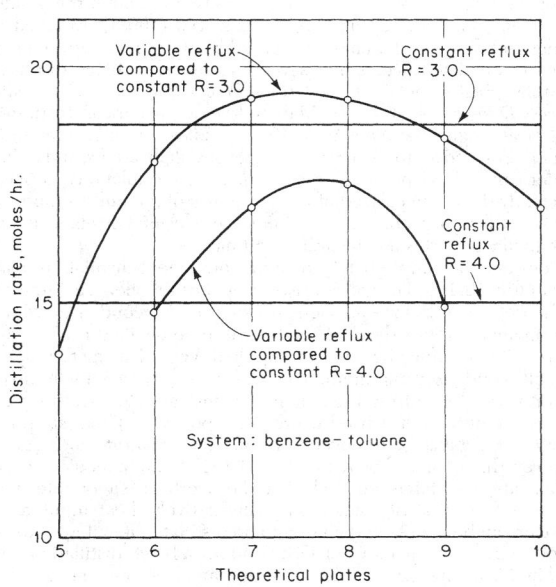

FIG. 13-104   Comparison of operating modes for a batch column.

Although a number of studies were made and approximate methods developed for predicting the effect of liquid holdup in the period of the 1950s and 1960s, as summarized in the 6th edition of *Perry's Chemical Engineers' Handbook,* the complexity of the effect of liquid holdup is such that it is now best to use computer-based batch-distillation algorithms to determine the effect of holdup on a case-by-case basis.

SHORTCUT METHODS FOR MULTICOMPONENT BATCH RECTIFICATION

For preliminary studies of batch rectification of multicomponent mixtures, shortcut methods that assume constant molal overflow and negligible vapor and liquid holdup are useful. The method of Diwekar and Madhaven [*Ind. Eng. Chem. Res.*, **30**, 713 (1991)] can be used for constant reflux or constant overhead rate. The method of Sundaram and Evans [*Ind. Eng. Chem. Res.*, **32**, 511 (1993)] applies only to the case of constant reflux, but is easy to apply. Both methods employ the Fenske-Underwood-Gilliland (FUG) shortcut procedure at successive time steps. Thus, batch rectification is treated as a sequence of continuous, steady-state rectifications.

CALCULATIONAL METHODS

Rigorous Computer-Based Calculation Procedures It is obvious that a set of curves such as shown in Fig. 13-104 for a binary mixture is quite tedious to obtain by hand methods. The curves shown in Fig. 13-99 for a multicomponent batch distillation are extremely difficult to develop by hand methods. Therefore, since the early 1960s, when large digital computers became available, interest has been generated in developing rigorous calculation procedures for binary and multicomponent batch distillation. For binary mixtures of constant relative volatility, Huckaba and Danly [*Am. Inst. Chem. Eng. J.*, **6**, 335 (1960)] developed a computer program that assumed constant-mass tray holdups, adiabatic tray operation, and linear enthalpy relationships but did include energy balances around each tray and permitted use of nonequilibrium trays by means of specified tray efficiencies. Experimental data were provided to validate the simulation. Meadows [*Chem. Eng. Prog. Symp. Ser.* **46**, 59, 48 (1963)] presented a multicomponent-batch-distillation model that included equations for energy, material, and volume balances around theoretical trays. The only assumptions made were perfect mixing on each tray, negligible vapor holdup, adiabatic operation, and constant-volume tray holdup. Distefano [*Am. Inst. Chem. Eng. J.*, **14**, 190 (1968)] extended the model and developed a computer-based-solution procedure that was used to simulate successfully several commercial batch-distillation columns. Boston et al. [*Foundations of Computer-Aided Chemical Process Design*, vol. II, ed. by Mah and Seider, American Institute of Chemical Engineers, New York, 1981, p. 203) further extended the model, provided a variety of practical sets of specifications, and utilized modern numerical procedures and equation formulations to handle efficiently the nonlinear and often stiff nature of the multicomponent-batch-distillation problem. The simpler model of Distefano is used here to illustrate this nonlinear and stiff nature.

Consider the simple batch- or multicomponent-distillation operation in Fig. 13-105. The still consists of a pot or reboiler, a column with N theoretical trays or equivalent packing, and a condenser with an accompanying reflux drum. The mixture to be distilled is charged to the reboiler, to which heat is then supplied. Vapor leaving the top tray is totally condensed and drained into the reflux drum. Initially, no distillate is withdrawn from the system, but instead a total-reflux condition is established at a fixed overhead vapor rate. Then, starting at time $t = 0$, distillate is removed at a constant molal rate and sent to a receiver that is not shown in Fig. 13-105. Simultaneously, a fixed reflux ratio is established such that the overhead vapor rate is not changed from that at total reflux. Alternatively, heat input to the reboiler can be maintained constant and distillate rate allowed to vary accordingly. The equations of Distefano for a batch distillation operated in this manner are as follows (after minor rearrangement), where i,j refers to the ith of C components in the mixture and the jth of N theoretical plates.

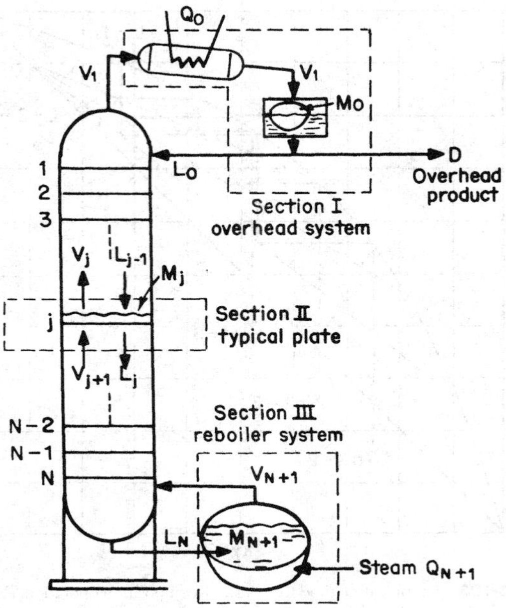

FIG. 13-105 Schematic of a batch-distillation column. [*Distefano*, Am. Inst. Chem. Eng. J., *14, 190 (1963)*.]

Component mole balances for total-condenser-reflux drum, trays, and reboiler, respectively:

$$\frac{dx_{i,0}}{dt} = -\left[\frac{L_0 + D + \dfrac{dM_0}{dt}}{M_0}\right] x_{i,0}$$

$$+ \left[\frac{V_1 K_{i,1}}{M_0}\right] x_{i,1} \qquad i = 1 \text{ to } C \quad (13\text{-}149)$$

$$\frac{dx_{i,j}}{dt} = \left(\frac{L_{j-1}}{M_j}\right) x_{i,j-1} - \left[\frac{L_j + K_{i,j}V_j + \dfrac{dM_j}{dt}}{M_j}\right] x_{i,j}$$

$$+ \left[\frac{K_{i,j+1}V_{j+1}}{M_j}\right] x_{i,j+1} \quad i = 1 \text{ to } C \quad j = 1 \text{ to } N \quad (13\text{-}150)$$

$$\frac{dx_{i,N+1}}{dt} = \left(\frac{L_N}{M_{N+1}}\right) x_{i,N} - \left[\frac{V_{N+1}K_{i,N+1} + \dfrac{dM_{N+1}}{dt}}{M_{N+1}}\right] x_{i,N+1}$$

$$i = 1 \text{ to } C \quad (13\text{-}151)$$

where $L_0 = RD$.

Total mole balance for total-condenser-reflux drum and trays respectively:

$$V_1 = D(R + 1) + \frac{dM_0}{dt} \qquad (13\text{-}152)$$

$$L_j = V_{j+1} + L_{j-1} - V_j - \frac{dM_j}{dt} \qquad j = 1 \text{ to } N \quad (13\text{-}153)$$

Energy balance around jth tray:

$$V_{j+1} = \frac{1}{(H_{V_{j+1}} - H_{L_j})} \left[V_j(H_{V_j} - H_{L_j}) - L_{j-1}(H_{L_{j-1}} - H_{L_j}) + M_j \frac{dH_{L_j}}{dt} \right]$$

$$j = 2 \text{ to } N+1 \quad (13\text{-}154)$$

where H_V and H_L are molar vapor and liquid enthalpies respectively.

Phase equilibriums:

$$y_{i,j} = K_{i,j} x_{i,j} \qquad i = 1 \text{ to } C \qquad j = 1 \text{ to } N+1 \qquad (13\text{-}155)$$

Mole-fraction sum:

$$\sum_i y_{i,j} = \sum_i K_{i,j} x_{i,j} = 1.0 \qquad j = 0 \text{ to } N+1 \qquad (13\text{-}156)$$

Molar holdups in condenser-reflux drum, on trays, and in reboiler:

$$M_0 = G_0 \rho_0 \qquad (13\text{-}157)$$

$$M_j = G_j \rho_j \qquad j = 1 \text{ to } N$$

$$M_{N+1} = M_{N+1}^0 - \sum_{j=0}^{N} M_j - \int_0^t D \, dt \qquad (13\text{-}158)$$

where G is the constant-volume holdup, M_{N+1}^0 is the initial molar charge to reboiler, and ρ is the liquid molar density.

Energy balances around condenser and reboiler respectively:

$$Q_0 = V_1(H_{V_1} - H_{L_0}) - M_0 \frac{dH_{L_0}}{dt} \qquad (13\text{-}159)$$

$$Q_{N+1} = V_{N+1}(H_{V_{N+1}} - H_{L_{N+1}})$$
$$- L_N(H_{L_N} - H_{L_{N+1}}) + M_{N+1}\left(\frac{dH_{L_{N+1}}}{dt}\right) \qquad (13\text{-}160)$$

Equation (13-160) is replaced by the following overall energy-balance equation if Q_{N+1} is to be specified rather than D:

$$D = \frac{Q_{N+1} - H_{V_1}\left(\dfrac{dM_0}{dt}\right) - \displaystyle\sum_{j=1}^{N+1}\left(\dfrac{d(M_j H_{L_j})}{dt}\right)}{(R+1)H_{V_1} - RH_{L_0}} \qquad (13\text{-}161)$$

With D and R specified, Eqs. (13-149) to (13-161) represent a coupled set of $(2CN + 3C + 4N + 7)$ equations constituting an initial-value problem in an equal number of time-dependent unknown variables, namely, $(CN + 2C)x_{i,j}$, $(CN + C)y_{i,j}$, $(N)L_j$, $(N+1)V_j$, $(N+2)T_j$, $(N+2)M_j$, Q_0, and Q_{N+1}, where initial conditions at $t = 0$ for all unknown variables are obtained by determining the total-reflux steady-state condition for specifications on the number of theoretical stages, amount and composition of initial charge, volume holdups, and molar vapor rate leaving the top stage and entering the condenser.

Various procedures for solving Eqs. (13-149) to (13-161), ranging from a complete tearing method to solve the equations one at a time, as shown by Distefano, to a complete simultaneous method, have been studied. Regardless of the method used, the following considerations generally apply:

1. Derivatives or rates of change of tray and condenser-reflux drum liquid holdup with respect to time are sufficiently small compared with total flow rates that these derivatives can be approximated by incremental changes over the previous time step. Derivatives of liquid enthalpy with respect to time everywhere can be approximated in the same way. The derivative of the liquid holdup in the reboiler can likewise be approximated in the same way except when reflux ratios are low.

2. Ordinary differential Eqs. (13-149) to (13-151) for rates of change of liquid-phase mole fractions are nonlinear because the coefficients of $x_{i,j}$ change with time. Therefore, numerical methods of integration with respect to time must be employed. Furthermore, the equations may be difficult to integrate rapidly and accurately because they may constitute a so-called stiff system as considered by Gear (*Numerical Initial Value Problems in Ordinary Differential Equations*, Prentice Hall, Englewood Cliffs, N.J., 1971). The choice of time

step for simple explicit numerical procedures (such as the Euler and Runge-Kutta methods) of integrating sets of ordinary differential equations in initial-value problems may be governed by either stability or truncation-error considerations. Truncation errors in the dependent variables may be scarcely noticeable and generally accumulate gradually with time. Instability generally causes sudden and severe errors that are very noticeable. When the equations are stiff, stability controls and extremely small time steps may be necessary to prevent instability. A common measure of the severity of stiffness is the stiffness ratio $|\lambda|_{\max}/|\lambda|_{\min}$, where λ is an eigenvalue for the jacobian matrix of the set of ordinary differential equations. For Eqs. (13-149) to (13-151), the jacobian matrix is tridiagonal if the equations and variables are arranged by stage (top down) for each component in order.

For a general jacobian matrix pertaining to C components and N theoretical trays, as shown by Distefano [*Am. Inst. Chem. Eng. J.*, **14**, 946 (1968)]. Gerschgorin's circle theorem (Varga, *Matrix Iterative Analysis*, Prentice Hall, Englewood Cliffs, N.J., 1962) may be employed to obtain bounds on the maximum and minimum absolute eigenvalues. Accordingly,

$$|\lambda|_{\max} \le \max_{j=1,N} \left[\left(\frac{L_{j-1}}{M_j}\right) + \left(\frac{L_j + K_{i,j}V_j + \dfrac{dM_j}{dt}}{M_j}\right) + \left(\frac{K_{i,j+1}V_{j+1}}{M_j}\right) \right]$$

The maximum absolute eigenvalue corresponds to the component with the largest K value ($K_{L,j}$) and the tray with the smallest holdup. Therefore, if the derivative term and any variation in L_j, V_j, and $K_{i,j}$ are neglected,

$$|\lambda|_{\max} \simeq 2\left[\frac{L_j + K_{i,j}V_j}{M_j}\right] \qquad (13\text{-}162)$$

In a similar development, the minimum upper limit on the eigenvalue corresponds to the component with the largest K value and to the largest holdup, which occurs in the reboiler. Thus

$$|\lambda|_{\min} = \left[\frac{L_N + K_{L,N}V_N}{M_{N+1}}\right] \qquad (13\text{-}163)$$

Therefore, the lower bound on the stiffness ratio at the beginning of batch distillation is given approximately by

$$\frac{|\lambda|_{\max}}{|\lambda|_{\min}} = 2\left(\frac{M_{N+1}^0}{M_N}\right)$$

where M_{N+1} and M_N are the molar holdups in the reboiler initially and on the bottom tray respectively. In the sample problem presented by Distefano (ibid.) for the smallest charge, the approximate initial-stiffness ratio is of the order of 250, which is not considered to be a particularly large value. Using an explicit integration method, almost 600 time increments, which were controlled by stability criteria, were required to distill 98 percent of the charge.

At the other extreme of Distefano's sample problems, for the largest initial charge, the maximum-stiffness ratio is of the order of 1500, which is considered to be a relatively large value. In this case, more than 10,000 time steps are required to distill 90 percent of the initial change, and the problem is better handled by a stiff integrator.

In Distefano's method, Eqs. (13-149) to (13-161) are solved with an initial condition of total reflux at L_0 equal to $D(R+1)$ from the specifications. At $t = 0$, L_0 is reduced so as to begin distillate withdrawal. The computational procedure is then as follows:

1. Replace L_j^0 by $L_j^0 - D$, but retain V_j^0 and all other initial values from the total-reflux calculation.

2. Replace the holdup derivatives in Eqs. (13-149) to (13-151) by total-stage material-balance equations (e.g., $dM_j/dt = V_{j+1} + L_{j-1} - V_j - L_j$) and solve the resulting equations one at a time by the predictor step of an explicit integration method for a time increment that is determined by stability and truncation considerations. If the mole fractions for a particular stage do not sum to 1, normalize them.

3. Compute a new set of stage temperatures from Eq. (13-156). Calculate a corresponding set of vapor-phase mole fractions from Eq. (13-155).

4. Calculate liquid densities, molar tray and condenser-reflux drum holdups, and liquor and vapor enthalpies. Determine holdup and enthalpy derivatives with respect to time by forward difference approximations.

5. From Eqs. (13-152) to (13-154) compute a new set of values of liquid and vapor molar flow rates.

6. Compute the reboiler molar holdup from Eq. (13-158).

7. Repeat steps 2 through 6 with a corrector step for the same time increment. Repeat again for any further predictor and/or predictor-corrector steps that may be advisable. Distefano (ibid.) discusses and compares a number of suitable explicit methods.

8. Compute condenser and reboiler heat-transfer rates from Eqs. (13-159) and (13-160).

9. Repeat steps 2 through 8 for subsequent time increments until the desired amount of distillate has been withdrawn.

More flexible and efficient methods that can cope with stiffness in batch-distillation calculations utilize stable implicit integration procedures such as the method of Gear (op. cit.). Boston et al. (op. cit.) discuss such a method that also utilizes a two-tier equation-solving technique, referred to as the "inside-cut" algorithm, that can handle both wide-boiling and narrow-boiling charges even when very non-ideal mixtures are formed. In addition to the features of the Distefano model, the Boston et al. model permits multiple feeds, side-stream withdrawals, tray heat transfer, and vapor distillate and divides the batch-distillation process into a sequence of operation steps. At the beginning of each step, the reboiler may receive an additional charge, distillate or sidestream receivers may be dumped, and a feed reservoir may be refilled. Specifications for an operation step include feed, sidestream withdrawal, and tray heat-transfer rates. In addition, any two of the following five variables must be specified: reflux ratio, distillate rate, boil-up rate, condenser duty, and reboiler duty. An operation step is terminated when a specified criterion, selected from the following list, is reached: a time duration; a component purity in the reboiler, distillate, or distillate accumulator; an amount of material in the reboiler or distillate accumulator; or a reboiler or condenser temperature. The purity is specified to be met as the purity is increasing or decreasing. Finally, column configuration and operating conditions (number of stages, holdups, tray pressures, and feed, sidestream, and tray heat-transfer rates) can be changed at the beginning of each operation step. In addition, physical properties may be computed from a wide variety of correlations, including equation-of-state and activity-coefficient models.

Example 10: Calculation of Multicomponent Batch Distillation A charge of 45.4 kg · mol (100 lb·mol) of 25 mole percent benzene, 50 mole percent monochlorobenzene (MCB), and 25 mole percent orthodichlorobenzene (DCB) is to be distilled in a batch still consisting of a reboiler, a column containing 10 theoretical stages, a total condenser, a reflux drum, and a distillate accumulator. Condenser-reflux drum and tray holdups are 0.0056 and

0.00056 m³ (0.2 and 0.02 ft³) respectively. Pressures are 101.3, 107.6, 117.2, and 120.7 kPa (14.696, 15.6, 17, and 17.5 psia) at the condenser outlet, top stage, bottom stage, and reboiler respectively. Initially, the still is to be brought to total-reflux conditions at a boil-up rate of 45.4 (kg·mol)/h [100 (lb·mol)/h] leaving the reboiler. Then, at $t = 0$, the boil-up rate is to be increased to 90.8 (kg·mol)/h [200 (lb·mol)/h], and the reflux ratio is to be set at 3. The batch is then distilled in three steps. The first step, which is designed to obtain a benzene-rich product, is terminated when the mole fraction of benzene in the distillate being sent to the accumulator has dropped to 0.100 or when 2 h have elapsed. The purpose of the second operation step is to recover an MCB-rich product until the mole fraction of MCB in the distillate drops to 0.400 or 2 h have elapsed since the start of this step. The third step is to be terminated when the mole fraction of DCB in the reboiler reaches 0.98 or 2 h have elapsed since the start of this step. Ideal solutions and an ideal gas are assumed such that Raoult's law can be used to obtain K values. Calculations are made by the method of Boston et al. (op. cit.).

First, the total-reflux condition is computed by making several sets of stage-to-stage calculations from the reboiler to the condenser. For the first set, the reboiler composition is assumed to be that of the initial charge. This composition is adjusted by material balance to initiate each subsequent set of calculations until convergence is achieved. Results are shown in Table 13-29. From these data, the initial-stiffness ratio is approximately $[2(99.74)/0.01218] = 16,400$. Thus the equations are quite stiff, and an implicit integration method is preferred. Detailed calculated conditions at the end of the first operation step are given in Table 13-30. A short summary of computed conditions for each of the three operation steps is given in Table 13-31.

From Table 13-31, a total of 394 time increments were necessary to distill all but 22.08 lb·mol of the initial charge of 99.74 lb·mol following the establishment of total-reflux conditions. If this problem had to be solved by an explicit integrator, approximately 25,000 time increments would have been necessary.

Instantaneous distillate (or reflux) composition as a function of total accumulated distillate for all three operation steps is plotted in Fig. 13-106. From these results, an alternative schedule of operation steps can be derived to obtain three relatively rich cuts and two intermediate cuts for recycle to the next batch. One example is as follows:

Cut	Amount, lb·mol	Composition, mole fractions		
		Benzene	MCB	DCB
Benzene-rich	17.750	0.99553	0.00447	0.00000
Recycle 1	15.630	0.44120	0.55880	0.00000
MCB-rich	37.195	0.01164	0.98821	0.00015
Recycle 2	7.155	0.00000	0.55430	0.44570
DCB-rich	22.078	0.00000	0.02000	0.98000
Residual holdup	0.192	0.00000	0.11980	0.88020
Total	100.00	0.25000	0.50000	0.25000

NOTE: To convert pound·moles to kilogram·moles, multiply by 0.454.

From these results, 22.98 lb·mol, or almost 23 percent of the charge, would be recycled for redistillation. All three products are at least 98 mole percent pure.

TABLE 13-29 Total-Reflux Conditions

Stage	T, °F	L, (lb·mol)/h	M, lb·mol	x		
				Benzene	MCB	DCB
Condenser	175.94	116.4	0.1307	1.000	0.525×10^{-7}	0.635×10^{-15}
1	179.46	117.4	0.01304	1.000	0.266×10^{-6}	0.164×10^{-13}
2	180.05	117.5	0.01303	1.000	0.135×10^{-5}	0.424×10^{-12}
3	180.64	117.5	0.01303	1.000	0.683×10^{-5}	0.109×10^{-10}
4	181.22	117.6	0.01302	1.000	0.344×10^{-4}	0.279×10^{-9}
5	181.80	117.7	0.01302	1.000	0.173×10^{-3}	0.711×10^{-8}
6	182.41	117.7	0.01301	0.999	0.871×10^{-3}	0.180×10^{-6}
7	183.14	117.5	0.01300	0.996	0.435×10^{-2}	0.454×10^{-5}
8	184.54	116.4	0.01295	0.979	0.209×10^{-1}	0.112×10^{-3}
9	189.08	111.8	0.01278	0.901	0.965×10^{-1}	0.250×10^{-2}
10	205.91	100.0	0.01218	0.642	0.319	0.389×10^{-1}
Reboiler	250.96	0.0	99.74	0.248	0.501	0.251

NOTE: Reboiler duty = 1,549,000 Btu/h. To convert degrees Fahrenheit to degrees Celsius, °C = (°F − 32)/1.8; to convert pound-moles per hour to kilogram-moles per hour, multiply by 0.454; and to convert pound-moles to kilogram-moles, multiply by 0.454.

TABLE 13-30   Conditions at the End of the First Operation Step

| | | (lb·mol)/h | | | y | | | x | | |
| | | V | L | M, lb·mol | Benzene | MCB | DCB | Benzene | MCB | DCB |
Stage	T, °F									
Condenser	251.58	0	154.6	0.1113	0.0994	0.901	0.293×10^{-5}	0.100	0.900	0.292×10^{-5}
1	267.69	206.1	157.5	0.01992	0.0449	0.955	0.101×10^{-4}	0.276×10^{-1}	0.972	0.124×10^{-4}
2	271.21	209.0	158.0	0.01088	0.0331	0.967	0.312×10^{-4}	0.121×10^{-1}	0.988	0.405×10^{-4}
3	272.48	209.5	158.1	0.01087	0.0306	0.969	0.943×10^{-4}	0.884×10^{-2}	0.991	0.124×10^{-3}
4	273.28	209.6	158.2	0.01086	0.0301	0.970	0.282×10^{-3}	0.818×10^{-2}	0.991	0.373×10^{-3}
5	273.99	209.7	158.2	0.01085	0.0300	0.969	0.839×10^{-3}	0.806×10^{-2}	0.991	0.111×10^{-2}
6	274.75	209.7	158.1	0.01084	0.0300	0.967	0.249×10^{-2}	0.803×10^{-2}	0.989	0.329×10^{-2}
7	275.72	209.6	157.7	0.01083	0.0300	0.963	0.731×10^{-2}	0.801×10^{-2}	0.982	0.969×10^{-2}
8	277.29	209.2	156.6	0.01080	0.0300	0.949	0.211×10^{-1}	0.794×10^{-2}	0.964	0.280×10^{-1}
9	280.50	208.1	154.0	0.01073	0.0301	0.912	0.577×10^{-1}	0.772×10^{-2}	0.915	0.770×10^{-1}
10	287.47	205.4	148.5	0.01057	0.0302	0.829	0.141	0.720×10^{-2}	0.803	0.189
Reboiler	301.67	200.0	0	66.40	0.0304			0.634×10^{-2}	0.617	0.376

NOTE: To convert pound-moles per hour to kilogram-moles per hour, multiply by 0.454; to convert pound-moles to kilogram-moles, multiply by 0.454.

TABLE 13-31   Calculated Conditions for each Operation Step

| | Operation step | | |
	1	2	3
Time of operation step, h	0.5963	0.7944	0.04828
Number of time increments	201	154	39
Accumulated distillate			
Total lb·mol	33.38	41.99	2.361
Mole fractions			
Benzene	0.7360	0.0103	0.74×10^{-10}
MCB	0.2640	0.9537	0.2872
DCB	0.45×10^{-6}	0.036	0.7128
Reboiler holdup			
Total lb·mol	66.40	24.43	22.08
Mole fractions			
Benzene	0.0063	0.63×10^{-11}	0.20×10^{-12}
MCB	0.6172	0.0448	0.0200
DCB	0.3765	0.9552	0.9800
Temperatures, °F			
Condenser outlet	251.58	308.60	330.17
Reboiler outlet	301.67	362.63	366.39
Heat duties, million Btu/h			
Condenser	3.313	3.472	3.456
Reboiler	3.295	3.469	3.433

NOTE: To convert degrees Fahrenheit to degrees Celsius, °C = (°F − 32)/1.8; to convert pound-moles to kilogram-moles, multiply by 0.454; and to convert British thermal units per hour to kilo-joules per hour, multiply by 1.055.

RAPID SOLUTION METHOD

The quasi-steady-state method of Galindez and Fredenslund [*Comput. Chem. Engng.*, **12**, 281 (1988)] is a rapid alternative to the rigorous integration of the stiff differential equations. In their method, which is implemented in the CHEMCAD III process simulator of Chemstations, Inc., Houston, Texas, the transient conditions are simulated, without having to treat stiffness, by a succession of a finite number of continuous steady states of generally a few minutes of batch time each in duration. The calculations are started from an initialization at total-reflux conditions, taking into account holdup. Generally, computed results compare well with rigorous integration methods.

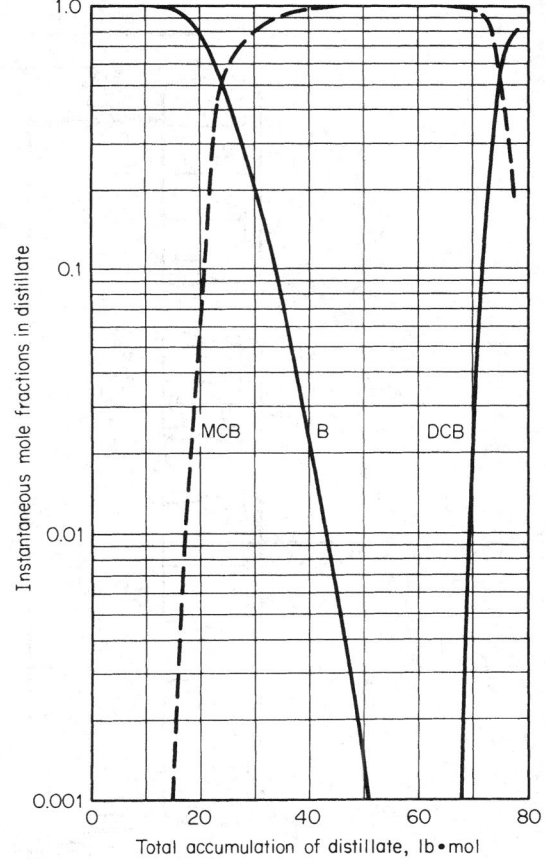

FIG. 13-106 Distillate-composition profile for the multicomponent-batch-distillation example.

DYNAMIC DISTILLATION

INTRODUCTION

As discussed in detail by Archer and Rothfus [*Chem. Eng. Prog. Symp. Series, No. 36*, **57**, 2 (1961)], dynamic or transient behavior of a continuous-distillation operation is important in determining (1) startup and shutdown procedures, (2) transition path between steady states, (3) effect of upsets and fluctuations on controllability, (4) residence times and mass-transfer rates, and (5) operating strategies that may involve deliberate imposition of controlled cyclic fluctuations or oscillations, as summarized by Schrodt [*Ind. Eng. Chem.*, **59**(6), 58 (1967)]. Dynamic behavior may be studied with no controllers in the system to obtain a so-called open-loop response. Alternatively, controllers may be added for certain variables that are to be controlled by manipulating other variables to obtain a so-called closed-loop response. For this latter case, controllers of various levels of complexity [e.g., on-off, proportional (P), proportional with integral action (PI), and proportional with integral and derivative action (PID)] can be considered for various values of tuning parameters, and specific valves of known characteristics may be incorporated if desired.

IDEAL BINARY DISTILLATION

Consider the closed-loop response during the dynamic distillation of an ideal binary mixture in the column shown in Fig. 13-107, under two assumptions of constant relative volatility at a value of 2.0 and constant molar vapor flow for a saturated liquid feed to tray N_s. Following the development by Luyben (op. cit.), it is not necessary to include energy-balance equations for each tray or to treat temperature and pressure as variables. Overhead vapor leaving top tray N_T is totally condensed for negligible liquid holdup with condensate flowing to a reflux drum having constant and perfectly mixed molar liquid holdup M_D. The reflux rate L_{N_T+1} is varied by a proportional-integral (PI) feedback controller to control distillate composition for a set point of 0.98 for the mole fraction x_D of the light component. Holdup of reflux in the line leading back to the top tray is neglected. Under dynamic conditions, y_{N_T} may not equal x_D.

At the bottom of the column, a liquid sump of constant and perfectly mixed molar liquid holdup M_B is provided. A portion of the liquid flowing from this sump passes to a thermosiphon reboiler, with the

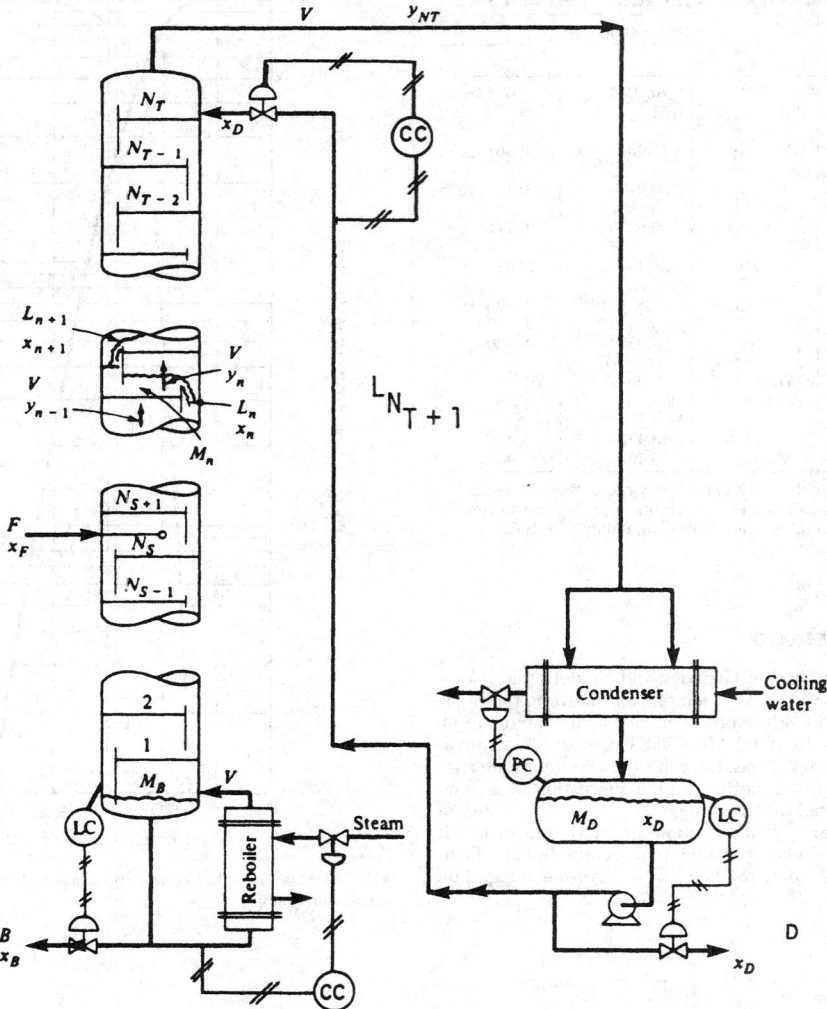

FIG. 13-107   Binary distillation column; dynamic distillation of ideal binary mixture.

remainder taken as bottoms product at a molar flow rate B. Vapor boil-up generated in the reboiler is varied by a PI feedback controller to control bottoms' composition with a set point of 0.02 for the mole fraction x_B of the light component. Liquid holdups in the reboiler and lines leading from the sump are assumed to be negligible. The composition of the boil-up y_B is assumed to be in equilibrium with x_B.

The liquid holdup M_n on each of the N_T equilibrium trays is assumed to be perfectly mixed but will vary as liquid rates leaving the trays vary. Vapor holdup is assumed to be negligible everywhere. Tray molar vapor rates V vary with time but at any instant in time are everywhere equal.

The dynamic material-balance and phase equilibrium equations corresponding to this description are as follows:

All trays, n:

$$dM_n/dt = F_n + L_{n+1} - L_n \qquad (13\text{-}164)$$

$$\frac{d}{dt}(M_n x_n) = F_n x_{F_n} + L_{n+1}x_{n+1} + Vy_{n-1} - L_n x_n - Vy_n \qquad (13\text{-}165)$$

$$L_n = \overline{L}_n + (M_n - \overline{M}_n)/\beta \qquad (13\text{-}166)$$

$$y_n = \frac{\alpha x_n}{1 + (\alpha - 1)x_n} \qquad (13\text{-}167)$$

Where F_n is nonzero only for tray N_s, y and x refer to the light component only such that the corresponding mole fractions for the heavy component are $(1 - y)$ and $(1 - x)$, $\overline{L}_n$ and $\overline{M}_n$ are the initial steady-state values, and β is a constant that depends on tray hydraulics.

For the condenser-reflux-drum combination:

$$D = V - L_{N_T+1} \qquad (13\text{-}168)$$

$$M_D(dx_D/dt) = Vy_{NT} - Vx_D \qquad (13\text{-}169)$$

For the reboiler:

$$B = L_1 - V \qquad (13\text{-}170)$$

$$M_B(dx_B/dt) = L_1 x_1 - Vy_B - Bx_B \qquad (13\text{-}171)$$

$$y_B = \frac{\alpha x_B}{1 + (\alpha - 1)x_B} \qquad (13\text{-}172)$$

The two PI-controller equations are

$$V = \overline{V} - K_{C_B}\left(E_B + \frac{1}{\tau_B}\int E_B \, dt\right) \qquad (13\text{-}173)$$

$$L_{N_T+1} = \overline{L}_{N_T+1} + K_{C_D}\left(E_D + \frac{1}{\tau_D}\int E_D \, dt\right) \qquad (13\text{-}174)$$

where $\overline{V}$ and $\overline{L}_{N_T+1}$ are initial values, K_C and τ are respectively feedback-controller gain and feedback-reset time for integral action, and E is the error or deviation from the set point as given by

$$E_B = x_B^{set} - x_B \qquad (13\text{-}175)$$

$$E_D = x_D^{set} - x_D \qquad (13\text{-}176)$$

These differential equations are readily solved, as shown by Luyben (op. cit.), by simple Euler numerical integration, starting from an initial steady state, as determined, e.g., by the McCabe-Thiele method, followed by some prescribed disturbance such as a step change in feed composition. Typical results for the initial steady-state conditions, fixed conditions, controller and hydraulic parameters, and disturbance given in Table 13-32 are listed in Table 13-33.

MULTICOMPONENT DISTILLATION

Open-loop behavior of multicomponent distillation may be studied by solving modifications of the multicomponent equations of Distefano [*Am. Inst. Chem. Eng. J.*, **14**, 190 (1968)] as presented in the subsection "Batch Distillation." One frequent modification is to include an equation, such as the Francis weir formula, to relate liquid holdup on a tray to liquid flow rate leaving the tray. Applications to azeotropic-distillation towers are particularly interesting because, as discussed by and illustrated in the following example from Prokopakis and Seider

TABLE 13-32 Initial and Fixed Conditions, Controller and Hydraulic Parameters, and Disturbance for Ideal Binary Dynamic-Distillation Example

Other initial conditions		Initial liquid-phase compositions	
$\overline{F} = 100$ (lb·mol)/min		Tray	x_n
$\overline{x}_F = 0.50$		Bottoms	0.02
$\overline{L}_{N_T+1} = 128.01$ (lb·mol)/min		1	0.035
$\overline{V} = 178.01$ (lb·mol)/min		2	0.05719
$\overline{D} = 50$ (lb·mol)/min		3	0.08885
$x_D^{set} = 0.98$		4	0.1318
$\overline{B} = 50$ (lb·mol)/min		5	0.18622
$x_B^{set} = 0.02$		6	0.24951
$M_{n,n=1\,\text{to}\,N_T} = 10$ lb·mol		7	0.31618
		8	0.37948
Fixed conditions		9	0.43391
$N_T = 20$		10	0.47688
$N_S = 10$		11	0.51526
$M_D = 100$ lb·mol		12	0.56295
$M_B = 100$ lb·mol		13	0.61896
Controller and hydraulic parameters		14	0.68052
		15	0.74345
$K_{C_B} = K_{C_D} = 1000$ (lb·mol)/min		16	0.80319
$\tau_B = 1.25$ min		17	0.85603
$\tau_D = 5.0$ min		18	0.89995
$\beta = 0.1$ min		19	0.93458
Disturbance at $t = 0^+$		20	0.96079
$x_F = 0.55$		Distillate	0.98

NOTE: To convert pound-moles per minute to kilogram-moles per minute, multiply by 0.454; to convert pound-moles to kilogram-moles, multiply by 0.454.

[*Am. Inst. Chem. Eng. J.*, **29**, 1017 (1983)], the steep concentration and temperature fronts can be extremely sensitive to small changes in reflux ratio, boil-up rate, product recovery and purity, and feed rate and composition.

Consider azeotropic distillation to dehydrate ethanol with benzene. Initial steady-state conditions are as shown in Fig. 13-108. The overhead vapor is condensed and cooled to 298 K to form two liquid phases that are separated in the decanter. The organic-rich phase is returned to the top tray as reflux together with a portion of the water-rich phase and makeup benzene. The other portion of the water-rich phase is sent to a stripper to recover organic compounds. Ordinarily, vapor from the stripper is condensed and recycled to the decanter, but that coupling is ignored here.

Equations for the decanter are as follows if it is assumed that (1) there are constant holdups in the decanter of both phases in the same ratio as the ratio of the flow rates leaving the decanter, (2) there is a constant decanter temperature, and (3) the two liquid phases in the decanter are in physical equilibrium and each is perfectly mixed.

$$\frac{d}{dt}[M_d(x_i)_d] = V_N y_{i,N} - L_0(x_i)_0 - L_w(x_i)_w \qquad (13\text{-}177)$$

$$dM_d/dt = V_N - L_0 - L_w \qquad (13\text{-}178)$$

$$K_{d_i} = (x_i)_0/(x_i)_w = (\gamma_i)_w/(\gamma_i)_0 \qquad (13\text{-}179)$$

where M_d is the total molar holdup of both phases in the decanter and the total composition in the decanter is

$$(x_i)_d = \frac{(x_i)_0 L_0 + (x_i)_w L_w}{L_0 + L_w} \qquad (13\text{-}180)$$

Combination of Eqs. (13-177), (13-178), and (13-180) gives

$$\frac{d(x_i)_d}{dt} = \frac{V_N}{M_d}[y_{i,N} - (x_i)_d] \qquad (13\text{-}181)$$

These equations, together with those for the tower, constitute a so-called stiff system. They were solved by Prokopakis and Seider (op. cit.), following a prescribed disturbance, using an adaptive semi-implicit Runge-Kutta integration technique by which V_N and the $y_{i,N}$ were obtained by integration of the equations for the tower. Then Eq. (13-181) was integrated to give $(x_i)_d$, which was used with Eqs.

TABLE 13-33 Results for Ideal Binary Dynamic-Distillation Example of Table 13-32*

Time, min	Mole fraction of light component in liquid					Flow rate, (lb·mol)/min	
	Bottoms	Stage 5	Stage 10	Stage 15	Distillate	Reflux	Boli-up
0.00	0.02000	0.18622	0.47688	0.74345	0.98000	128.01	178.01
.50	0.02014	0.19670	0.51310	0.74940	0.98000	128.01	178.16
1.00	0.02107	0.21174	0.52426	0.76049	0.98010	127.91	179.31
1.50	0.02217	0.22038	0.53026	0.76847	0.98034	127.64	181.06
2.00	0.02275	0.22209	0.53229	0.77217	0.98061	127.33	182.65
2.50	0.02268	0.21881	0.53141	0.77222	0.98076	127.11	183.69
3.00	0.02212	0.21287	0.52879	0.76993	0.98077	127.02	184.10
3.50	0.02132	0.20639	0.52560	0.76672	0.98065	127.07	183.99
4.00	0.02051	0.20104	0.52282	0.76381	0.98047	127.19	183.55
4.50	0.01987	0.19777	0.52109	0.76196	0.98030	127.32	182.98
5.00	0.01950	0.19679	0.52057	0.76142	0.98018	127.42	182.47
5.50	0.01939	0.19766	0.52106	0.76198	0.98014	127.45	182.14
6.00	0.01950	0.19956	0.52209	0.76315	0.98016	127.41	182.02
6.50	0.01972	0.20162	0.52320	0.76438	0.98022	127.33	182.08
7.00	0.01995	0.20314	0.52400	0.76525	0.98029	127.24	182.25
7.50	0.02012	0.20380	0.52434	0.76557	0.98034	127.15	182.43
8.00	0.02019	0.20362	0.52422	0.76537	0.98036	127.09	182.56
8.50	0.02016	0.20289	0.52381	0.76484	0.98035	127.07	182.61

*From Luyben, *Process Modeling, Simulation, and Control for Chemical Engineers*, McGraw-Hill, New York, 1973.
NOTE: To convert pound-moles per minute to kilogram-moles per minute, multiply by 0.454.

(13-179) and (13-180) to obtain the equilibrium compositions $(x_i)_0$ and $(x_i)_w$ leaving the decanter. The UNIQUAC equation was used with data from Gmehling and Onken (*Vapor-Liquid Equilibrium Data Collection*, DECHEMA Chemistry Data Ser., Vol. 1 (parts 1–10). Frankfurt, 1977) to obtain the activity coefficients needed in Eq. (13-179). Reboiler and decanter volumetric holdups were assumed constant at 1.0 m³ (35.3 ft³), and volumetric tray holdups were computed from

$$M_n = (\rho_L)_n A_n (h_{w_j} + h_{c_j}) \qquad (13\text{-}182)$$

where $(\rho_L)_n$ is the liquid density; A_n, the cross-sectional area of the active portion of the tray = 0.23 m² (2.48 ft²); h_{w_j}, the weir height = 0.0254 m (0.0833 ft); and h_{c_j}, the weir crest, assumed to be constant at 0.00508 m (0.0167 ft). Accordingly, volumetric tray holdup was constant at 0.007 m³ (0.247 ft³).

Assume that at $t = 0^+$ the feed rate to tray 23 is disturbed by increasing it by 30 percent to 130 mol/min without a change in composition. The resulting ethanol liquid mole fraction on several trays is tracked in Fig. 13-109. Above tray 16, ethanol concentration remains very small. Below tray 9, ethanol concentration initially decreases fairly rapidly but

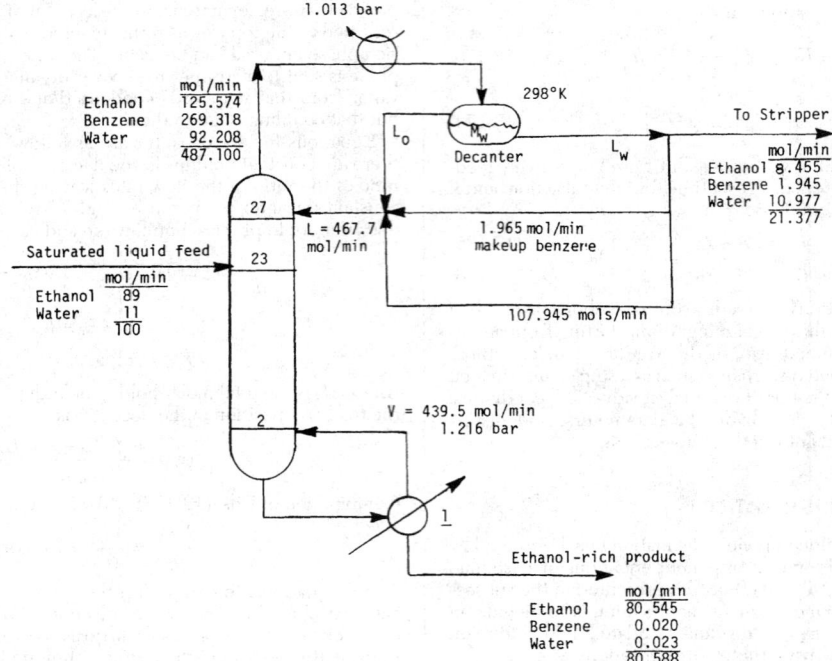

FIG. 13-108 Initial steady state for dynamic azeotropic distillation of ethanol-water with benzene.

then increases slowly and steadies out at significantly higher values than at the initial steady state. Tray 10 is one of the last trays to reach the new steady-state condition, which takes somewhat more than 200 min. This may be compared with initial residence times in the decanter and reboiler of approximately 50 and 250 min respectively. The movement through the column of concentration fronts for all three components is shown in Fig. 13-109. For the first 5 to 10 min, below tray 16, benzene

and ethanol fronts shift downward. Then a reversal occurs, and the fronts shift upward until the new steady state is attained. The upward shift is expected because the increased feed rate increases the water-benzene entrainer ratio. The duration of the initial, temporary downward shift is highly dependent on tray holdup and is due to "wash-out" with the feed liquid. This phenomenon is also observed in the dynamic studies of Peiser and Grover [*Chem. Eng. Prog.*, **58**(9), 65 (1962)].

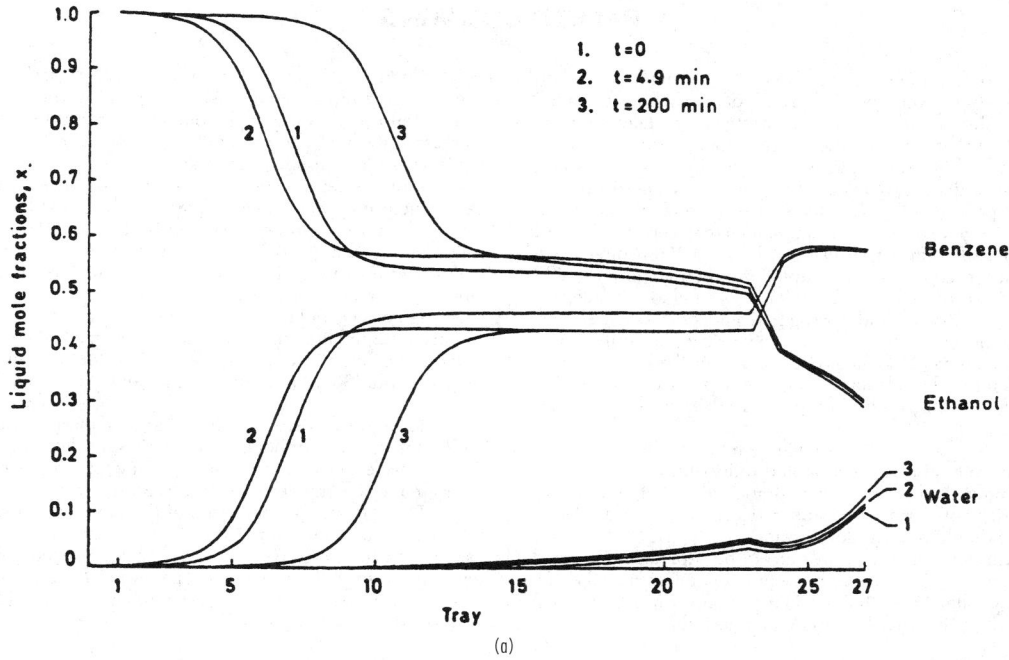

FIG. 13-109*a* Responses after a 30 percent increase in the feed flow rate for the multicomponent-dynamic-distillation example of Fig. 13-100. Profiles of liquid mole fractions at several times.

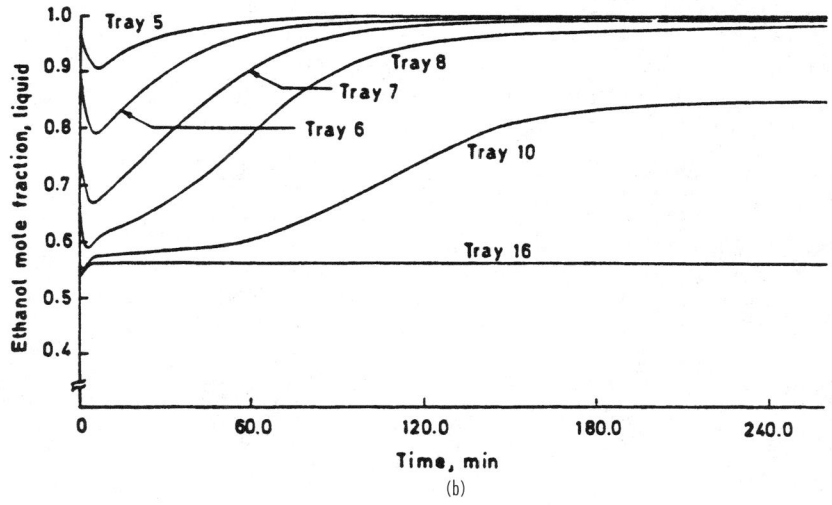

FIG. 13-109*b* Responses after a 30 percent increase in the feed flow rate for the multicomponent-dynamic-distillation example of Fig. 13-100. Alcohol mole fractions on several trays. (*Prokopakis and Seider*, Am. Inst, Chem. Eng. J., **29**, *1017 (1983)*.]

If the feed rate is decreased, the trends of curves in Fig. 13-109 are reversed. The disturbance of other variables such as feed composition, boil-up ratio, and recycle of water-rich effluent from the decanter produces similar shifts in the steep concentration fronts, indicating that azeotropic towers are among the most sensitive separation operations, for which dynamic studies are essential if reliable process control is to be developed. Such studies indicate the importance of adjusting aqueous-phase recycle and reboiler duty to diminish the movement of steep concentration fronts and the possibility of multiple regimes of operation including unstable regimes, as shown by Magnussen et al. [*Inst. Chem. Eng. Symp. Ser.* **56** (1979)].

PACKED COLUMNS

Distillation-type separation operations may be conducted in packed rather than tray-type columns. In prior years, except for small columns, plate columns were heavily favored over packed columns. However, development of more efficient random or structured packing materials and the need to increase capacity, increase efficiency, or reduce pressure drop in many applications has resulted in more extensive use of packed columns in larger sizes in recent years. Both types of contacting devices are discussed extensively in Sec. 14 and by Billet (*Distillation Engineering*, Chemical Publishing, New York, 1979). Packed columns may employ dumped (random) packing, e.g., pall rings, or structured (ordered, arranged, or stacked) packing, e.g., knitted wire and corrugated and perforated sheets. Tray-type columns generally employ valve, sieve, or bubble-cap trays with downcomers. The choice between a packed column and a tray-type column is based mainly on economics when factors of contacting efficiency, loadability, and pressure drop must be considered.

Packed columns must be provided with good initial distribution of liquid across the column cross section and redistribution of liquid at various height intervals that decrease with increasing column diameter. A wide variety of distributors and redistributors are available. Packed columns should be considered when:

1. Temperature-sensitive mixtures are to be separated. To avoid decomposition and/or polymerization, vacuum operation may then be necessary. The smaller liquid holdup and pressure drop theoretical stage of a packed column may be particularly desirable.

2. Ceramic or plastic (e.g., propylene) is a desirable material of construction from a noncorrosion and liquid-wettability standpoint.

3. Refitting of a tray-type column is desired to increase loading, increase efficiency, and/or decrease pressure drop. Structured packing is particularly applicable in this case.

4. Liquid rates are very low and/or vapor rates are high, in which case structured packing may be particularly desirable.

5. The mixture to be separated is clear, nonfouling, and free of solids, and cleaning of column internals will not be necessary.

6. The mixture to be separated tends to form foam, which collapses more readily in a packed column.

7. High recovery of a volatile component by a batch operation is required. Liquid holdup is much lower in a packed column.

Packed columns are almost always used for column diameters less than 762 mm (30 in) but otherwise generally need not be considered when:

1. Multiple feeds, sidestreams, and/or intermediate condensers and/or intermediate reboilers are required or desirable.

2. A wide range of loadability (turndown ratio) is required. Valve trays are particularly desirable in this case.

3. Design data for separation of the particular or similar mixture in a packed column are not available. Design procedures are better established for tray-type columns than for packed columns. This is particularly so with respect to separation efficiency since tray efficiency can be estimated more accurately than packed height equivalent to a theoretical stage (HETP).

Gas Absorption and Gas-Liquid System Design*

James R. Fair, Ph.D., P.E., *Professor of Chemical Engineering, University of Texas; Fellow, American Institute of Chemical Engineers; Member, American Chemical Society, American Society for Engineering Education, National Society of Professional Engineers. (Section Editor, Absorption, Gas-Liquid Contacting)*

D. E. Steinmeyer, M.A., M.S., P.E., *Distinguished Fellow, Monsanto Company; Fellow, American Institute of Chemical Engineers; Member, American Chemical Society. (Liquid-in-Gas Dispersions)*

W. R. Penney, Ph.D., P.E., *Professor of Chemical Engineering, University of Arkansas; Member, American Institute of Chemical Engineers. (Gas-in-Liquid Dispersions)*

B. B. Crocker, S.M., P.E., *Consulting Chemical Engineer; Fellow, American Institute of Chemical Engineers; Member, Air Pollution Control Association. (Phase Separation)*

* Much of the material on absorption is taken from Sec. 14 of the sixth edition, and credit is due to Dr. William M. Edwards, editor of that section.

Nomenclature

Symbol	Description		
a_e	Effective interfacial area	m²/m³	ft²/ft³
A	Cross sectional area	m²	ft²
A_f	Fractional open area	-/-	-/-
A	Absorption factor	-/-	-/-
A_e, A'	Effective absorption factor (Edmister)	-/-	-/-
c	Concentration	kg-moles/m³	lb-mol/ft³
c'	Stokes-Cunningham correction factor for terminal settling velocity	-/-	-/-
C_{sb}, C_s	Flooding coefficient	m/s	ft/s
C_v	Discharge coefficient	-/-	-/-
d	Diameter	m	ft
d_b	Bubble diameter	m	ft
d_h	Hole diameter	m	ft
d_o	Orifice diameter	m	ft
d_{pc}	Cut size of a particle collected in a device, 50% mass efficiency	μm	ft
d_{psd}	Mass median size particle in the pollutant gas	μm	ft
d_{pa50}	Aerodynamic diameter of a real median size particle	μm	ft
D	Diffusion coefficient	m²/s	ft²/s
D_{32}	Sauter mean diameter	m	ft
D_{vm}	Volume mean diameter	m	ft
e	Entrainment, mass liquid/mass gas	kg/kg	lb/lb
E	Plate or stage efficiency, fractional	-/-	-/-
E	Power dissipation per mass	W	btu/lb
E_a	Murphree plate efficiency, with entrainment, gas concentrations, fractional	-/-	-/-
E_g	Point efficiency, gas phase only, fractional	-/-	-/-
E_{oc}	Overall column efficiency, fractional	-/-	-/-
E_{og}	Overall point efficiency, gas concentrations, fractional	-/-	-/-
E_{mv}	Murphree plate efficiency, gas concentrations, fractional	-/-	-/-
f	Fractional approach to flood	-/-	-/-
F	F-factor for gas loading	m/s(kg/m³)^0.5	ft/s(lb/ft³)^0.5
F_{LG}	Flow parameter	-/-	-/-
g	Gravitational constant	m/s²	ft/s²
g_c	Conversion factor	1.0 (kg·m/N·s²)	32.2(lb·ft)/(lbf·s²)
G	Gas phase mass velocity	kg/s-m²	lb/hr-ft²
G_M	Gas phase molar velocity	kg-moles/s-m²	lb-mol/hr-ft²
h	Pressure head	mm	ft
h_f	Height of froth	m	ft
h_T	Height of contacting	m	ft
H	Henry's law constant		
H'	Henry's law constant		
H	Height of a transfer unit	m	ft
H_g	Height of a gas phase transfer unit	m	ft
H_{og}	Height of an overall transfer unit, gas phase concentrations	m	ft
H_{OL}	Height of an overall transfer unit, liquid phase concentrations	m	ft
H_L	Height of a liquid phase transfer unit	m	ft
H'	Henry's law coefficient	kPa/mole fraction	atm/mole fraction
HETP	Height equivalent to a theoretical plate or stage	m	ft
k_1	First order reaction velocity constant	1/s	1/sec
k_2	Second order reaction velocity constant	m³/(s·kmol)	ft³/(h·lb-mol)
k	Individual phase mass transfer coefficient	m/s	ft/sec
k_G	gas phase mass transfer coefficient	m/s	ft/sec
k_L	liquid phase mass transfer coefficient	m/s	ft/sec
K	Vapor-liquid equilibrium ratio	-/-	-/-
K_{OG}, K_G	Overall mass transfer coefficient, gas concentrations	m/s	ft/sec
K_{OL}	Overall mass transfer coefficient, liquid concentrations	m/s	ft/sec
L	Liquid mass velocity	kg/m²-s	lb/ft²-s
L_M	Liquid molar mass velocity	kmoles/m²-s	lb-mol/ft²-s
L_w	Weir length	m	ft
m	Slope of equilibrium curve = dy°/dx	-/-	-/-
M	Molecular weight	kg/kmol	lb/lb-mol
nA	Rate of solute transfer	kmol/s	lb-mol/s
p	Partial pressure	kPa	atm
P, p_T	Total pressure	kPa	atm
q	Volumetric flow rate of liquid	m³/s	ft³/s
Q	Volumetric flow rate of gas or vapor	m³/s	ft³/s
R	Gas constant		
R_h	Hydraulic radius	m	ft
s	Length of corrugation side, structured packing	m	ft
S	Stripping factor	-/-	-/-
S_e, S'	Effective stripping factor (Edmister)	-/-	-/-
T	Absolute temperature	K	°F
TS	Tray or plate spacing	m	ft
U	Linear velocity of gas	m/s	ft/s
U_a	Velocity of gas through active area	m/s	ft/s
U_n	Velocity of gas through net area	m/s	ft/s
U_t	Superficial velocity of gas	m/s	ft/s
x	Mole fraction, liquid phase	-/-	-/-
$x°$	Liquid mole fraction, equilibrium condition	-/-	-/-
y	Mole fraction, gas or vapour phase	-/-	-/-
$y°$	Gas mole fraction, equilibrium condition	-/-	-/-
Z	Height, plate spacing	m	ft

Greek symbols

Symbol	Description		
α	Relative volatility	-/-	-/-
β	Aeration factor	-/-	-/-
ε	Void fraction	-/-	-/-
φ	Relative froth density	-/-	-/-
γ	Activity coefficient	-/-	-/-
Γ	Flow rate per length	kg/(s·m)	lb/(s-ft)
δ	Effective film thickness	m	ft
η	Collection eficiency, fractional	-/-	-/-
λ	Stripping factor = m/(L_M/G_M)	-/-	-/-
μ	Absolute viscosity	Pa-s	lb/(ft-s)
μm	Microns	m	ft
ν	Kinematic viscosity	m²/s	ft²/s
π	3.1416....	-/-	-/-
θ	Residence time	s	s
ρ	Density	kg/m³	lb/ft³
σ	Surface tension	mN/m	dyn/cm
ψ	Fractional entrainment	-/-	-/-

Subscripts

A	Species A
AB	Species A diffusing through species B
B	Species B
e	Effective value
i	Interface value
G	Gas or vapor
L	Liquid
p	Particle
w	water
1	Tower bottom
2	Tower top

Dimensionless Groups

N_{Fr}	Froude number = $(U_t^2)/(Sg)$
N_{Re}	Reynolds number = $(SU_{ga}\rho_G)/(\mu_G)$
N_{Sc}	Schmidt number = $\mu/(\rho D)$
N_{We}	Weber number = $(U_L^2\rho_L S)/(\sigma g_c)$

GENERAL REFERENCES: Astarita, G., *Mass Transfer with Chemical Reaction*, Elsevier, New York, 1967. Astarita, G., D. W. Savage and A. Bisio, *Gas Treating with Chemical Solvents*, Wiley, New York, 1983. Billet, R., *Distillation Engineering*, Chemical Publishing Co., New York, 1979. Danckwerts, P. V., *Gas-Liquid Reactions*, McGraw-Hill, New York, 1970. *Distillation and Absorption 1987*, Rugby, U.K., Institution of Chemical Engineers, 1988. *Distillation and Absorption 1992*, Rugby, U.K., Institution of Chemical Engineers, 1992. Hines, A. L. and R. N. Maddox, *Mass Transfer—Fundamentals and Applications*, Prentice Hall, Englewood Cliffs, New Jersey, 1985. Kister, H. Z., *Distillation Design*, McGraw-Hill, New York, 1992. Lockett, M.J., *Distillation Tray Fundamentals*, Cambridge, U.K., Cambridge Unioversity Press, 1986. Kohl, A. L. and F. C. Riesenfeld, *Gas Purification*, 4th ed., Gulf, Houston, 1985. Sherwood, T. K., R. L. Pigford, C. R. Wilke, *Mass Transfer*, McGraw-Hill, New York, 1975. Treybal, R. E., *Mass Transfer Operations,* McGraw-Hill, New York, 1980.

INTRODUCTION

Definitions   Gas absorption is a unit operation in which soluble components of a gas mixture are dissolved in a liquid. The inverse operation, called stripping or desorption, is employed when it is desired to transfer volatile components from a liquid mixture into a gas. Both absorption and stripping, in common with distillation (Sec. 13), make use of special equipment for bringing gas and liquid phases into intimate contact. This section is concerned with the design of gas-liquid contacting equipment, as well as with the design of absorption and stripping processes.

Equipment   Absorption, stripping, and distillation operations are usually carried out in vertical, cylindrical columns or towers in which devices such as plates or packing elements are placed. The gas and liquid normally flow countercurrently, and the devices serve to provide the contacting and development of interfacial surface through which mass transfer takes place. Background material on this mass transfer process is given in Sec. 5.

Design Procedures   The procedures to be followed in specifying the principal dimensions of gas absorption and distillation equipment are described in this section and are supported by several worked-out examples. The experimental data required for executing the designs are keyed to appropriate references or to other sections of the handbook.

For absorption, stripping, and distillation, there are three main steps involved in design:

1. *Data on the gas-liquid or vapor-liquid equilibrium for the system at hand.* If absorption, stripping, and distillation operations are considered equilibrium-limited processes, which is the usual approach, these data are critical for determining the maximum possible separation. In some cases, the operations are considered rate-based (see Sec. 13) but require knowledge of equilibrium at the phase interface. Other data required include physical properties such as viscosity and density and thermodynamic properties such as enthalpy. Section 2 deals with sources of such data.

2. *Information on the liquid- and gas-handling capacity of the contacting device chosen for the particular separation problem.* Such information includes pressure drop characteristics of the device, in order that an optimum balance between capital cost (column cross section) and energy requirements might be achieved. Capacity and pressure drop characteristics of the available devices are covered later in this Sec. 14.

3. *Determination of the required height of contacting zone for the separation to be made as a function of properties of the fluid mixtures and mass-transfer efficiency of the contacting device.* This determination involves the calculation of mass-transfer parameters such as heights of transfer units and plate efficiencies as well as equilibrium or rate parameters such as theoretical stages or numbers of transfer units. An additional consideration for systems in which chemical reaction occurs is the provision of adequate residence time for desired reactions to occur, or minimal residence time to prevent undesired reactions from occurring. For equilibrium-based operations, the parameters for required height are covered in the present section.

Data Sources in the Handbook   Sources of data for the analysis or design of absorbers, strippers, and distillation columns are manifold, and a detailed listing of them is outside the scope of the presentation in this section. Some key sources within the handbook are shown in Table 14-1.

Equilibrium Data   Finding reliable gas-liquid and vapor-liquid equilibrium data may be the most time-consuming task associated with the design of absorbers and other gas-liquid contactors, and yet it may be the most important task at hand. For gas solubility, an important data source is the set of volumes edited by Kertes et al., *Solubility Data Series*, published by Pergamon Press (1979 ff.). In the introduction to each volume, there is an excellent discussion and definition of the various methods by which gas solubility data have been reported, such as the Bunsen coefficient, the Kuenen coefficient, the Ostwalt coefficient, the absorption coefficient, and the Henry's law coefficient. The fourth edition of *The Properties of Gases and Liquids* by Reid, Prausnitz and Poling (McGraw-Hill, New York, 1987) provides data and recommended estimation methods for gas solubility as well as the broader area of vapor-liquid equilibrium. Finally, the Chemistry Data Series by Gmehling et al., especially the title *Vapor-Liquid Equilibrium Collection* (DECHEMA, Frankfurt, Germany, 1979 ff.), is a rich source of data evaluated against the various models used for interpolation and extrapolation. Section 13 of this handbook presents a good discussion of equilibrium K values.

TABLE 14-1   Directory to Key Data for Absorption and Gas-Liquid Contactor Design

Type of data	Section
Phase equilibrium data	
Gas solubilities	2
Pure component vapor pressures	2
Equilibrium K values	13
Thermal data	
Heats of solution	2
Specific heats	2
Latent heats of vaporization	2
Transport property data	
Diffusion coefficients	
Liquids	2
Gases	2
Viscosities	
Liquids	2
Gases	2
Densities	
Liquids	2
Gases	2
Surface tensions	2
Packed tower data	
Pressure drop and flooding	14
Mass transfer coefficients	5
HTU, physical absorption	5
HTU with chemical reaction	14
Height equivalent to a theoretical plate (HETP)	
Plate tower data	
Pressure drop and flooding	14
Plate efficiencies	14
Costs of gas-liquid contacting equipment	14

DESIGN OF GAS-ABSORPTION SYSTEMS

General Design Procedure   The designer ordinarily is required to determine (1) the best solvent; (2) the best gas velocity through the absorber, namely the vessel diameter; (3) the height of the vessel and its internal members, which is the height and type of packing or the number of contacting trays; (4) the optimum solvent circulation through the absorber and stripper; (5) the temperatures of streams entering and leaving the absorber and the quantity of heat to be removed to account for heat of solution and other thermal effects; (6) the pressures at which the absorber and stripper will operate; and (7) the mechanical design of the absorption and stripping vessels (normally columns or towers), including flow distributors, packing supports, and so on. This section is concerned with all these choices.

The problem presented to the designer of a gas-absorption unit usually specifies the following quantities: (1) gas flow rate; (2) gas composition, at least with respect to the component or components to be absorbed; (3) operating pressure and allowable pressure drop across the absorber; (4) minimum degree of recovery of one or more solutes; and, possibly, (5) the solvent to be employed. Items 3, 4, and 5 may be subject to economic considerations and therefore are sometimes left up to the designer. For determining the number of variables that must be specified in order to fix a unique solution for the design of an absorber one can use the same phase-rule approach described in Sec. 13 for distillation systems.

Recovery of the solvent, sometimes by chemical means but more often by distillation, is almost always required, and the recovery system ordinarily is considered an integral part of the absorption-system process design. A more efficient solvent-stripping operation normally will result in a less costly absorber because of a smaller concentration of residual dissolved solute in the regenerated solvent; however, this may increase the overall cost of solvent recovery. A more detailed discussion of these and other economic considerations is presented later in this section.

Selection of Solvent   When choice is possible, preference is given to liquids with high solubilities for the solute; a high solubility reduces the amount of solvent to be circulated. The solvent should be relatively nonvolatile, inexpensive, noncorrosive, stable, nonviscous, nonfoaming, and preferably nonflammable. Since the exit gas normally leaves saturated with solvent, solvent loss can be costly and may present environmental contamination problems. Thus, low-cost solvents may be chosen over more expensive ones of higher solubility or lower volatility.

Water generally is used for gases fairly soluble in water, oils for light hydrocarbons, and special chemical solvents for acid gases such as CO_2, SO_2, and H_2S. Sometimes a reversible chemical reaction will result in a very high solubility and a minimum solvent rate. Data on actual systems are desirable when chemical reactions are involved, and those available are referenced later under "Absorption with Chemical Reaction."

Selection of Solubility Data   Solubility values determine the liquid rate necessary for complete or economic solute recovery and so are essential to design. Equilibrium data generally will be found in one of three forms: (1) solubility data expressed either as solubility in weight or mole percent or as Henry's-law coefficients, (2) pure-component vapor pressures, or (3) equilibrium distribution coefficients (K values). Data for specific systems may be found in Sec. 2; additional references to sources of data are presented in this section.

In order to define completely the solubility of a gas in a liquid, it generally is necessary to state the temperature, the equilibrium partial pressure of the solute gas in the gas phase, and the concentration of the solute gas in the liquid phase. Strictly speaking, the total pressure on the system also should be stated, but for low total pressures, less than about 507 kPa (5 atm), the solubility for a particular partial pressure of solute gas normally will be relatively independent of the total pressure of the system.

For dilute concentrations of many gases and over a fairly wide range for some gases, the equilibrium relationship is given by Henry's law, which relates the partial pressure developed by a dissolved solute A in a liquid solvent B by one of the following equations:

$$p_A = Hx_A \qquad (14\text{-}1)$$

or

$$P_A = H'c_A \qquad (14\text{-}2)$$

where H is the Henry's law coefficient expressed in kilopascals per mole-fraction solute in liquid and H' is the Henry's law coefficient expressed in kilopascals per kilomole per cubic meter.

Although quite useful when it can be applied, this law should be checked experimentally to determine the accuracy with which it can be used. If Henry's law holds, the solubility is defined by stating the value of the constant H (or H') along with the temperature and the solute partial pressure for which it is to be employed.

For quite a number of gases, Henry's law holds very well when the partial pressure of the solute is less than about 100 kPa (1 atm). For partial pressures of the solute gas greater than 100 kPa, H seldom is independent of the partial pressure of the solute gas, and a given value of H can be used over only a narrow range of partial pressures. There is a strongly nonlinear variation of Henry's-law constants with temperature as discussed by Schulze and Prausnitz [*Ind. Eng. Chem. Fundam.*, **20**, 175 (1981)]. Consultation of this reference is recommended before considering temperature extrapolations of Henry's-law data.

Additional data and information on the applicability of Henry's-law constants can be found in the references cited earlier in the subsection "Directory to Key Gas-Absorption Data." The use of Henry's-law constants is illustrated by the following examples.

Example 1: Gas Solubility   It is desired to find out how much hydrogen can be dissolved in 100 weights of water from a gas mixture when the total pressure is 101.3 kPa (760 torr; 1 atm), the partial pressure of the H_2 is 26.7 kPa (200 torr), and the temperature is 20°C. For partial pressures up to about 100 kPa the value of H is given in Sec. 3 as 6.92×10^6 kPa (6.83×10^4 atm) at 20°C. According to Henry's law,

$$x_{H_2} = p_{H_2}/H_{H_2} = 26.7/6.92 \times 10^6 = 3.86 \times 10^{-6}$$

The mole fraction x is the ratio of the number of moles of H_2 in solution to the total moles of all constituents contained. To calculate the weights of H_2 per 100 weights of H_2O, one can use the following formula, where the subscripts A and w correspond to the solute (hydrogen) and solvent (water):

$$\left(\frac{x_A}{1 - x_A}\right)\frac{M_A}{M_w}100 = \left(\frac{3.86 \times 10^{-6}}{1 - 3.86 \times 10^{-6}}\right)\frac{2.02}{18.02}100$$

$$= 4.33 \times 10^{-5} \text{ weights } H_2/100 \text{ weights } H_2O$$

$$= 0.43 \text{ parts per million weight}$$

Pure-component vapor pressures can be used for predicting solubilities for systems in which **Raoult's law** is valid. For such systems $p_A = p_A^\circ x_A$, where p_A° is the pure-component vapor pressure of the solute and p_A is its partial pressure. Extreme care should be exercised when attempting to use pure-component vapor pressures to predict gas-absorption behavior. Both liquid-phase and vapor-phase nonidealities can cause significant deviations from the behavior predicted from pure-component vapor pressures in combination with Raoult's law. Vapor-pressure data are available in Sec. 3 for a variety of materials.

Whenever data are available for a given system under similar conditions of temperature, pressure, and composition, **equilibrium distribution coefficients** ($K = y/x$) provide a much more reliable tool for predicting vapor-liquid distributions. A detailed discussion of equilibrium K values is presented in Sec. 13.

Calculation of Liquid-to-Gas Ratio   The minimum possible liquid rate is readily calculated from the composition of the entering gas and the solubility of the solute in the exit liquor, saturation being assumed. It may be necessary to estimate the temperature of the exit liquid based on the heat of solution of the solute gas. Values of latent and specific heats and values of heats of solution (at infinite dilution) are given in Sec. 2.

The actual liquid-to-gas ratio (solvent-circulation rate) normally will be greater than the minimum by as much as 25 to 100 percent and may be arrived at by economic considerations as well as by judgment and experience. For example, in some packed-tower applications involving very soluble gases or vacuum operation, the minimum quantity of solvent needed to dissolve the solute may be insufficient to keep the packing surface thoroughly wet, leading to poor distribution of the liquid stream.

When the solute concentration in the inlet gas is low and when nearly all the solute is being absorbed (this is the usual case), the approximation

$$y_1 G_M \doteq x_1 L_M \doteq (y_1^\circ/m) L_M \qquad (14\text{-}3)$$

leads to the conclusion that the ratio $m G_M / L_M$ represents the fractional approach of the exit liquid to saturation with the inlet gas, i.e.,

$$m G_M / L_M \doteq y_1^\circ / y_1 \qquad (14\text{-}4)$$

Optimization of the liquid-to-gas ratio in terms of total annual costs often suggests that the molar liquid-to-gas ratio L_M / G_M should be about 1.2 to 1.5 times the theoretical minimum corresponding to equilibrium at the rich end of the tower (infinite height), provided flooding is not a problem. This would be an alternative to assuming that $L_M / G_M \doteq m/0.7$, for example.

When the exit-liquor temperature rises owing to the heat of absorption of the solute, the value of m changes through the tower, and the liquid-to-gas ratio must be chosen to give reasonable values of $m_1 G_M / L_M$ and $m_2 G_M / L_M$, where the subscripts 1 and 2 refer to the bottom and top of the absorption tower respectively. For this case the value of $m_2 G_M / L_M$ will be taken to be somewhat less than 0.7, so that the value of $m_1 G_M / L_M$ will not approach unity too closely. This rule-of-thumb approach is useful only when low solute concentrations and mild heat effects are involved.

When the solute has a large heat of solution or when the feed gas contains high percentages of the solute, one should consider the use of internal cooling coils or intermediate external heat exchangers in a plate-type tower to remove the heat of absorption. In a packed tower, one could consider the use of multiple packed sections with intermediate liquid-withdrawal points so that the liquid could be cooled by external heat exchange.

Selection of Equipment Packed columns usually are chosen for very corrosive materials, for liquids that foam badly, for either small- or large-diameter towers involving very low allowable pressure drops, and for small-scale operations requiring diameters of less than 0.6 m (2 ft). The type of packing is selected on the basis of resistance to corrosion, mechanical strength, capacity for handling the required flows, mass-transfer efficiency, and cost. Economic factors are discussed later in this section.

Plate columns may be economically preferable for large-scale operations and are needed when liquid rates are so low that packing would be inadequately wetted, when the gas velocity is so low (owing to a very high L/G) that axial dispersion or "pumping" of the gas back down the (packed) column can occur, or when intermediate cooling is desired. Also, plate towers may have a better turndown ratio and are less subject to fouling by solids than are packed towers. Details on the operating characteristics of plate towers are given later in this section.

Column Diameter and Pressure Drop Flooding determines the minimum possible diameter of the absorber column, and the usual design is for 60 to 80 percent of the flooding velocity. Maximum allowable pressure drop may be determined by the cost of energy for compression of the feed gas. For systems having a significant tendency to foam, the maximum allowable velocity will be lower than estimated flooding velocity, especially for plate towers. The safe range of operating velocities should include the velocity one would derive from economic considerations, as discussed later. Methods for predicting flooding velocities and pressure drops are given later in this section.

Computation of Tower Height The required height of a gas-absorption or stripping tower depends on (1) the phase equilibria involved, (2) the specified degree of removal of the solute from the gas, and (3) the mass-transfer efficiency of the apparatus. These same considerations apply both to plate towers and to packed towers. Items 1 and 2 dictate the required number of theoretical stages (plate tower) or transfer units (packed tower). Item 3 is derived from the tray efficiency and spacing (plate tower) or from the height of one transfer unit (packed tower). Solute-removal specifications normally are derived from economic considerations.

For plate towers, the approximate design methods described below may be used in estimating the number of theoretical stages, and the tray efficiencies and spacings for the tower can be specified on the basis of the information given later. Considerations involved in the rigorous design of theoretical stages for plate towers are treated in Sec. 13.

For packed towers, the continuous differential nature of the contact between gas and liquid leads to a design procedure involving the solution of differential equations, as described in the next subsection.

It should be noted that the design procedures discussed in this section are not applicable to reboiled absorbers, which should be designed according to the methods described in Sec. 13.

Caution is advised in distinguishing between systems involving pure physical absorption and those in which a chemical reaction can significantly affect design procedures.

Selection of Stripper-Operating Conditions Stripping involves the removal of one or more volatile components from a liquid by contacting it with a gas such steam, nitrogen, or air. The operating conditions chosen for stripping normally result in a low solubility of the solute (i.e., a high value of m), so that the ratio $m G_M / L_M$ will be larger than unity. A value of 1.4 may be used for rule-of-thumb calculations involving pure physical desorption. For plate-tower calculations the stripping factor $S = K G_M / L_M$, where $K = y^\circ/x$, usually is specified for each tray.

When the solvent from an absorption operation must be regenerated for recycling back to the absorber, one may employ a "pressure-swing concept," a "temperature-swing concept," or a combination of both in specifying stripping conditions. In pressure-swing operation the temperature of the stripper is about the same as that of the absorber, but the stripping pressure is much lower. In temperature-swing operation the pressures are about equal, but the stripping temperature is much higher than the absorption temperature.

In pressure-swing operation a portion of the dissolved gas may be "sprung" from the liquid by the use of a flash drum upstream of the stripping-tower feed point. This type of operation is discussed by Burrows and Preece [*Trans. Inst. Chem. Eng.*, **32**, 99 (1954)] and by Langley and Haselden [*Inst. Chem. Eng. Symp. Ser.* (*London*), no. 28 (1968)]. If the flashing of the feed liquid takes place inside the stripping tower, this effect must be accounted for in the design of the upper section in order to avoid overloading and flooding near the top of the tower.

More often than not the rate at which residual absorbed gas can be driven from the liquid in a stripping tower is limited by the rate of a chemical reaction, in which case the liquid-phase residence time (and hence, the tower liquid holdup) becomes the most important design factor. Thus, many stripper-regenerators are designed on the basis of liquid holdup rather than on the basis of mass transfer rate.

Approximate design equations applicable only to the case of pure physical desorption are developed later in this section for both packed and plate stripping towers. A more rigorous approach using distillation concepts may be found in Sec. 13. A brief discussion of desorption with chemical reaction is given in the subsection "Absorption with Chemical Reaction."

Design of Absorber-Stripper Systems The solute-rich liquor leaving a gas absorber normally is distilled or stripped to regenerate the solvent for recirculation back to the absorber, as depicted in Fig. 14-1. It is apparent that the conditions selected for the absorption step (e.g., temperature, pressure, L_M / G_M) will affect the design of the stripping tower, and, conversely, a selection of stripping conditions will affect the absorber design. The choice of optimum operating conditions for an absorber-stripper system therefore involves a combination of economic factors and practical judgments as to the operability of the system within the context of the overall process flow sheet. Note that in Fig. 14-1 the stripping vapor is provided by a reboiler; alternately, an extraneous stripping gas may be used.

An appropriate procedure for executing the design of an absorber-stripper system is to set up a carefully selected series of design cases and then evaluate the investment costs, the operating costs, and the operability of each case. Some of the economic factors that need to be considered in selecting the optimum absorber-stripper design are discussed later in the subsection "Economic Design of Absorption Systems."

Importance of Design Diagrams One of the first things a designer should try to do is lay out a carefully constructed equilibrium curve, $y^\circ = F(x)$, on an xy diagram, as shown in Fig. 14-2. A horizontal line corresponding to the inlet-gas composition y_1 is then the locus of feasible outlet-liquor compositions, and a vertical line corresponding to the inlet-solvent-liquor composition x_2 is the locus of feasible out-

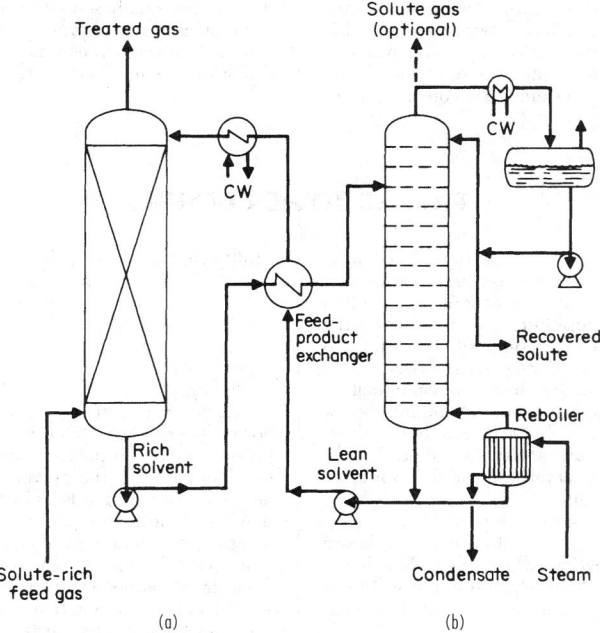

FIG. 14-1 Gas absorber using a solvent regenerated by stripping. (a) Absorber. (b) Stripper.

let-gas compositions. These lines are indicated as $y = y_1$ and $x = x_2$ respectively on Fig. 14-2.

For gas absorption, the region of feasible operating lines lies above the equilibrium curve; for stripping, the feasible region for operating lines lies below the equilibrium curve. These feasible regions are bounded by the equilibrium curve and by the lines $x = x_2$ and $y = y_1$. By inspection, one should be able to visualize those operating lines that are feasible and those that would lead to "pinch points" within the tower. Also, it is possible to determine if a particular proposed design for solute recovery falls within the feasible envelope.

Once the design recovery for an absorber has been established, the operating curve can be constructed by first locating the point x_2, y_2 on the diagram. The intersection of the horizontal line corresponding to the inlet gas composition y_1 with the equilibrium curve $y° = F(x)$

defines the theoretical minimum liquid-to-gas ratio for systems in which there are no intermediate pinch points. The operating line which connects this point with the point x_2, y_2 corresponds to the minimum value of L_M/G_M. The actual design value of L_M/G_M normally should be around 1.2 to 1.5 times this minimum. Thus, the actual design operating line for a gas absorber will pass through the point x_2, y_2 and will intersect the line $y = y_1$ to the left of the equilibrium curve.

For stripping one begins by using the design specification to locate the point x_1, y_1. Then the intersection of the vertical line $x = x_2$ with the equilibrium curve $y° = F(x)$ defines the theoretical minimum gas-to-liquid ratio. The actual value of G_M/L_M is chosen to be about 20 to 50 percent higher than this minimum, so the actual design operating line will intersect the line $x = x_2$ at a point somewhat below the equilibrium curve.

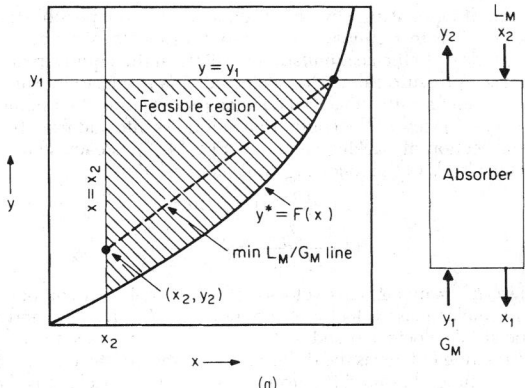

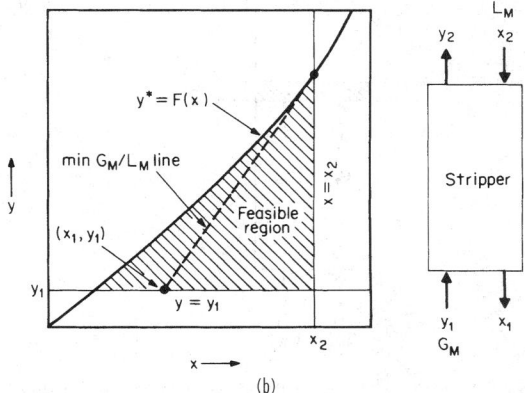

FIG. 14-2 Design diagrams for (a) absorption and (b) stripping.

Design diagrams minimize the possibility of making careless mistakes and allow one to assess easily the effects of operating variable changes on the operability of the system relative to pinch points, etc. Whenever analytical calculations or computer programs are being used for the design of gas-absorption systems, the construction of design diagrams based either on calculation results or on computer printouts may reveal problem areas or even errors in the design concept. It is strongly recommended that design diagrams be employed whenever possible.

PACKED-TOWER DESIGN

Methods for estimating the height of the active section of **counterflow differential contactors** such as packed towers, spray towers, and falling-film absorbers are based on rate expressions representing mass transfer at a point on the gas-liquid interface and on material balances representing the changes in bulk composition in the two phases that flow past each other. The rate expressions are based on the interphase mass-transfer principles described in Sec. 5. Combination of such expressions leads to an integral expression for the number of transfer units or to equations related closely to the number of theoretical plates. The paragraphs which follow set forth convenient methods for using such equations, first in a general case and then for cases in which simplifying assumptions are valid.

Use of Mass-Transfer-Rate Expression   Figure 14-3 shows a section of a packed absorption tower together with the nomenclature that will be used in developing the equations which follow. In a differential section dh, we can equate the rate at which solute is lost from the gas phase to the rate at which it is transferred through the gas phase to the interface as follows:

$$-d(G_M y) = -G_M \, dy - y \, dG_M = N_A a \, dh \qquad (14\text{-}5)$$

When only one component is transferred,

$$dG_M = -N_A a \, dh \qquad (14\text{-}6)$$

Substitution of this relation into Eq. (14-5) and rearranging yields

$$dh = -\frac{G_M \, dy}{N_A a (1-y)} \qquad (14\text{-}7)$$

For this derivation we use the gas-phase rate expression $N_A = k_G(y - y_i)$ and integrate over the tower to obtain

$$h_T = \int_{y_2}^{y_1} \frac{G_M \, dy}{k_G a (1-y)(y-y_i)} \qquad (14\text{-}8)$$

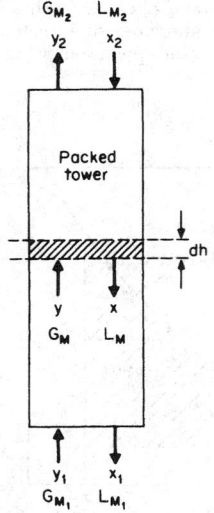

FIG. 14-3   Nomenclature for material balances in a packed-tower absorber or stripper.

Multiplying and dividing by y_{BM} place Eq. (14-8) into the $H_G N_G$ format

$$h_T = \int_{y_2}^{y_1} \left[\frac{G_M}{k_G a y_{BM}} \right] \frac{y_{BM} \, dy}{(1-y)(y-y_i)}$$

$$= H_{G,\text{av}} \int_{y_2}^{y_1} \frac{y_{BM} \, dy}{(1-y)(y-y_i)} = H_{G,\text{av}} N_G \qquad (14\text{-}9)$$

The general expression given by Eq. (14-8) is more complex than normally is required, but it must be used when the mass-transfer coefficient varies from point to point, as may be the case when the gas is not dilute or when the gas velocity varies as the gas dissolves. The values of y_i to be used in Eq. (14-8) depend on the local liquid composition x_i and on the temperature. This dependency is best represented by using the operating and equilibrium lines as discussed later.

Example 3 illustrates the use of Eq. (14-8) for scrubbing chlorine from air with aqueous caustic solution. For this case one can make the simplifying assumption that y_i, the interfacial partial pressure of chlorine over the aqueous caustic solution, is zero owing to the rapid and complete reaction of the chlorine after it dissolves. We note that the feed gas is not dilute.

Example 2: Packed Height Requirement   Let us compute the height of packing needed to reduce the chlorine concentration of 0.537 kg/(s·m²), or 396 lb/(h·ft²), of a chlorine-air mixture containing 0.503 mole-fraction chlorine to 0.0403 mole fraction. On the basis of test data described by Sherwood and Pigford (*Absorption and Extraction,* McGraw-Hill, 1952, p. 121) the value of $k_G a y_{BM}$ at a gas velocity equal to that at the bottom of the packing is equal to 0.1175 kmol/(s·m³), or 26.4 lb·mol/(h·ft³). The equilibrium back pressure y_i can be assumed to be negligible.

Solution.   By assuming that the mass-transfer coefficient varies as the 0.8 power of the local gas mass velocity, we can derive the following relation:

$$\hat{K}_G a = k_G a y_{BM} = 0.1175 \left[\frac{71y + 29(1-y)}{71y_1 + 29(1-y_1)} \left(\frac{1-y_1}{1-y} \right) \right]^{0.8}$$

where 71 and 29 are the molecular weights of chlorine and air respectively. Noting that the inert-gas (air) flow rate is given by $G'_M = G_M(1-y) = 5.34 \times 10^{-3}$ kmol/(s·m²), or 3.94 lb·mol/(h·ft²), and introducing these expressions into the integral gives

$$h_T = 1.82 \int_{0.0403}^{0.503} \left[\frac{1-y}{29+42y} \right]^{0.8} \frac{dy}{(1-y)^2 \ln[1/(1-y)]}$$

This definite integral can be evaluated numerically by the use of Simpson's rule to obtain $h_T = 0.305$ m (1 ft).

Use of Operating Curve   Frequently, it is not possible to assume that $y_i = 0$ as in Example 2, owing to diffusional resistance in the liquid phase or to the accumulation of solute in the liquid stream. When the back pressure cannot be neglected, it is necessary to supplement the equations with a material balance representing the operating line or curve. In view of the countercurrent flows into and from the differential section of packing shown in Fig. 14-3, a steady-state material balance leads to the following equivalent relations:

$$d(G_M y) = d(L_M x) \qquad (14\text{-}10)$$

$$G'_M \frac{dy}{(1-y)^2} = L'_M \frac{dx}{(1-x)^2} \qquad (14\text{-}11)$$

where L'_M = molar mass velocity of the inert-liquid component and G'_M = molar mass velocity of the inert gas. L_M, L'_M, G_M, and G'_M are superficial velocities based on the total tower cross section.

Equation (14-11) is the differential equation of the operating curve, and its integral around the upper portion of the packing is the equation for the operating curve

$$G'_M \left[\frac{y}{1-y} - \frac{y_2}{1-y_2} \right] = L'_M \left[\frac{x}{1-x} - \frac{x_2}{1-x_2} \right] \quad (14\text{-}12)$$

For dilute solutions in which the mole fractions of x and y are small, the total molar flows G_M and L_M will be very nearly constant, and the operating-curve equation is

$$G_M(y - y_2) = L_M(x - x_2) \quad (14\text{-}13)$$

This equation gives the relation between the bulk compositions of the gas and liquid streams at each level in the tower for conditions in which the operating curve can be approximated by a straight line.

Figure 14-4 shows the relationship between the operating curve and the equilibrium curve $y_i = F(x_i)$ for a typical example involving solvent recovery, where y_i and x_i are the interfacial compositions (assumed to be in equilibrium). Once y is known as a function of x along the operating curve, y_i can be found at corresponding points on the equilibrium curve by

$$(y - y_i)/(x_i - x) = k_L/k_G = k'_L \bar{\rho}_L/k'_G p_T = L_M H_G/G_M H_L \quad (14\text{-}14)$$

where L_M = molar liquid mass velocity, G_M = molar gas mass velocity, H_L = height of one transfer unit based on liquid-phase resistance, and H_G = height of one transfer unit based on gas-phase resistance. Thence, the integral in Eq. (14-8) can be evaluated.

Calculation of Transfer Units   In the general case the equations described above must be employed in calculating the height of packing required for a given separation. However, if the local mass-transfer coefficient $k_G a y_{BM}$ is approximately proportional to the first power of the local gas velocity G_M, then the height of one gas-phase transfer unit, defined as $H_G = G_M/k_G a y_{BM}$, will be constant in Eq. (14-9). Similar considerations lead to an assumption that the height of one overall gas-phase transfer unit H_{OG} may be taken as constant. The height of packing required is then calculated according to the relation

$$h_T = H_G N_G = H_{OG} N_{OG} \quad (14\text{-}15)$$

where N_G = number of gas-phase transfer units and N_{OG} = number of overall gas-phase transfer units. When H_G and H_{OG} are not constant, it may be valid to employ averaged values between the top and bottom of the tower and the relation

$$h_T = H_{G,av} N_G = H_{OG,av} N_{OG} \quad (14\text{-}16)$$

In these equations, the terms N_G and N_{OG} are defined by

$$N_G = \int_{y_2}^{y_1} \frac{y_{BM}\, dy}{(1-y)(y-y_i)} \quad (14\text{-}17)$$

and by

$$N_{OG} = \int_{y_2}^{y_1} \frac{y^\circ_{BM}\, dy}{(1-y)(y-y^\circ)} \quad (14\text{-}18)$$

respectively.

Equation (14-18) is the more useful one in practice: it requires either actual experimental H_{OG} data or values estimated by combining

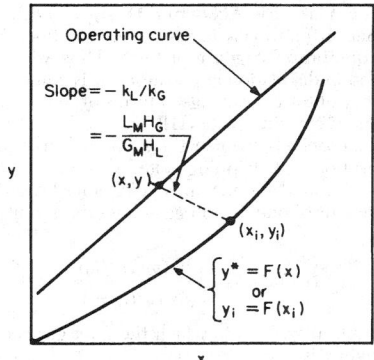

FIG. 14-4   Relationship between equilibrium curve and operating curve in a packed absorber; computation of interfacial compositions.

individual measurements of H_G and H_L by Eq. (14-19). Correlations for predicting H_G, H_L, and H_{OG} in nonreacting systems are presented in Sec. 5.

$$H_{OG} = \frac{y_{BM}}{y^\circ_{BM}} H_G + \frac{mG_M}{L_M} \frac{x_{BM}}{y^\circ_{BM}} H_L \quad (14\text{-}19a)$$

$$H_{OL} = \frac{x_{BM}}{x^\circ_{BM}} H_L + \frac{L_M}{mG_M} \frac{y_{BM}}{x^\circ_{BM}} H_G \quad (14\text{-}19b)$$

On occasion the changes in gas flow and in the mole fraction of inert gas are so small that the inclusion of terms such as $(1 - y)$ and y°_{BM} can be neglected or at least can be included in an approximate way. This leads to some of the simplified procedures described later.

One such simplification was suggested by Wiegand [*Trans. Am. Inst. Chem. Eng.*, **36**, 679 (1940)], who pointed out that the logarithmic-mean mole fraction of inert gas y°_{BM} (or y_{BM}) is often very nearly equal to the arithmetic mean. Thus, substitution of the relation

$$\frac{y^\circ_{BM}}{(1-y)} \doteq \frac{(1-y^\circ) + (1-y)}{2(1-y)} = \frac{y - y^\circ}{2(1-y)} + 1 \quad (14\text{-}20)$$

into the equations presented earlier leads to the simplified forms

$$N_G = \frac{1}{2} \ln \frac{1-y_2}{1-y_1} + \int_{y_2}^{y_1} \frac{dy}{y - y_i} \quad (14\text{-}21)$$

$$N_{OG} = \frac{1}{2} \ln \frac{1-y_2}{1-y_1} + \int_{y_2}^{y_1} \frac{dy}{y - y^\circ} \quad (14\text{-}22)$$

The second (integral) terms represent the numbers of transfer units for an infinitely dilute gas. The first terms, frequently amounting to only small corrections, give the effect of a finite level of gas concentration.

The procedure for applying Eqs. (14-21) and (14-22) involves two steps: (1) evaluation of the integrals and (2) addition of the correction corresponding to the first (logarithmic) term. The discussion which follows deals only with the evaluation of the integral terms (first step).

The simplest possible case occurs when (1) both the operating and the equilibrium lines are straight (i.e., there are dilute solutions), (2) Henry's law is valid ($y^\circ/x = y_i/x_i = m$), and (3) absorption heat effects are negligible. Under these conditions, the integral term in Eq. (14-20) may be computed by Colburn's equation [*Trans. Am. Inst. Chem. Eng.*, **35**, 211 (1939)]:

$$N_{OG} = \frac{1}{1 - (mG_M/L_M)} \ln \left[\left(1 - \frac{mG_M}{L_M} \right) \left(\frac{y_1 - mx_2}{y_2 - mx_2} \right) + \frac{mG_M}{L_M} \right] \quad (14\text{-}23)$$

Figure (14-5) is a plot of Eq. (14-23) from which the value of N_{OG} can be read directly as a function of mG_M/L_M and the ratio of concentrations. This plot and Eq. (14-23) are equivalent to the use of a logarithmic mean of terminal driving forces, but they are more convenient because one does not need to compute the exit-liquor concentration x_1.

In many practical situations involving nearly complete cleanup of the gas, an approximate result can be obtained from the equations just presented even when solutions are concentrated or when absorption heat effects are present. In such cases the driving forces in the upper part of the tower are very much smaller than those at the bottom, and the value of mG_M/L_M used in the equations should be the ratio of the slopes of the equilibrium line m and the operating line L_M/G_M in the low-concentration range near the top of the tower.

Another approach is to divide the tower arbitrarily into a lean section (near the top), where approximate methods are valid, and to deal with the rich section separately. If the heat effects in the rich section are appreciable, consideration could be given to installing cooling units near the bottom of the tower. In any event a design diagram showing the operating and equilibrium curves should be prepared to check on the applicability of any simplified procedure. Figure 14-8, presented in Example 6 is one such diagram for an adiabatic absorption tower.

Stripping Equations   Stripping, or desorption, involves the removal of a volatile component from the liquid stream by contact with an inert gas such as nitrogen or steam. In this case the change in concentration of the liquid stream is of prime importance, and it is more convenient to formulate the rate equation analogous to Eq. (14-6) in terms of the liquid composition x. This leads to the following equations defining numbers of transfer units and heights of transfer units based on liquid-phase resistance:

$$h_T = H_L \int_{x_2}^{x_1} \frac{x_{BM}\, dx}{(1-x)(x_i - x)} = H_L N_L \qquad (14\text{-}24)$$

$$h_T = H_{OL} \int_{x_2}^{x_1} \frac{x_{BM}^\circ\, dx}{(1-x)(x^\circ - x)} = H_{OL} N_{OL} \qquad (14\text{-}25)$$

where, as before, subscripts 1 and 2 refer to the bottom and top of the tower respectively (see Fig. 14-3).

In situations in which one cannot assume that H_L and H_{OL} are constant, these terms must be incorporated inside the integrals in Eqs. (14-24) and (14-25). and the integrals must be evaluated graphically or numerically (by using Simpson's rule, for example). In the normal case involving stripping without chemical reactions, the liquid-phase resistance will dominate, making it preferable to use Eq. (14-25) in conjunction with the relation $H_L \doteq H_{OL}$.

The Wiegand approximations of the above integrals in which arithmetic means are substituted for the logarithmic means x_{BM} and x_{BM}° are

$$N_L = \frac{1}{2} \ln \frac{1-x_1}{1-x_2} + \int_{x_1}^{x_2} \frac{dx}{x - x_i} \qquad (14\text{-}26)$$

$$N_{OL} = \frac{1}{2} \ln \frac{1-x_1}{1-x_2} + \int_{x_1}^{x_2} \frac{dx}{x - x^\circ} \qquad (14\text{-}27)$$

In these equations, the first term is a correction for finite liquid-phase concentrations, and the integral term represents the numbers of transfer units required for dilute solutions. It would be very unusual in practice to find an example in which the first (logarithmic) term is of any significance in a stripper design.

For dilute solutions in which both the operating and the equilibrium lines are straight and in which heat effects can be neglected, the integral term in Eq. (14-27) is

$$N_{OL} = \frac{1}{(1 - L_M/mG_M)} \ln \left[\left(1 - \frac{L_M}{mG_M} \right) \left(\frac{x_2 - y_1/m}{x_1 - y_1/m} \right) + \frac{L_M}{mG_M} \right] \qquad (14\text{-}28)$$

This equation is identical in form to Eq. (14-23). Thus, Fig. 14-5 is applicable if the concentration ratio $(x_2 - y_1/m)/(x_1 - y_1/m)$ is substituted for the abscissa and if the parameter on the curves is identified as L_M/mG_M.

Example 3: Air Stripping of VOCs from Water A 0.45-m diameter packed column was used by Dvorack et al. [*Environ. Sci. Tech.* **20**, 945 (1996)] for removing trichloroethylene (TCE) from wastewater by stripping with atmospheric air. The column was packed with 2.5-cm Pall rings, fabricated from polypropylene, to a height of 3.0 m. The TCE concentration in the entering water was 38 parts per million by weight (ppmw). A molar ratio of entering water to entering air was kept at 23.7. What degree of removal was to be expected? The temperatures of water and air were 20°C. Pressure was atmospheric.

Solution. For TCE in water, the Henry's law coefficient may be taken as 417 atm/mf at 20°C. In this low-concentration region, the coefficient is constant and equal to the slope of the equilibrium line m. The solubility of TCE in water, based on $H = 417$, is 2390 ppm. Because of this low solubility, the entire resistance to mass transfer resides in the liquid phase. Thus, Eq. (14-25) may be used to obtain N_{OL}, the number of overall liquid phase transfer units.

In the equation, the ratio $x_{BM}\cdot/(1-x)$ is unity because of the very dilute solution. It is necessary to have a value of H_L for the packing used, at given flow rates of liquid and gas. Methods for estimating H_L may be found in Sec. 5. Dvorack et al. found $H_{OL} = 0.8$ m. Then, for $h_T = 3.0$ m, $N_L = N_{OL} = 3.0/0.8 = 3.75$ transfer units.

Transfer units may be calculated from Eq. (14-25), replacing mole fractions with ppm concentrations, and since the operating and equilibrium lines are straight,

$$N_{OL} = \frac{38 - (\text{ppm})_{\text{exit}}}{\ln 38/(\text{ppm})_{\text{exit}}} = 3.75$$

Solving, $(\text{ppm})_{\text{exit}} = 0.00151$. Thus, the stripped water would contain 1.51 parts per billion of TCE.

Use of HTU and $K_G a$ Data In estimating the size of a commercial gas absorber or liquid stripper it is desirable to have data on the

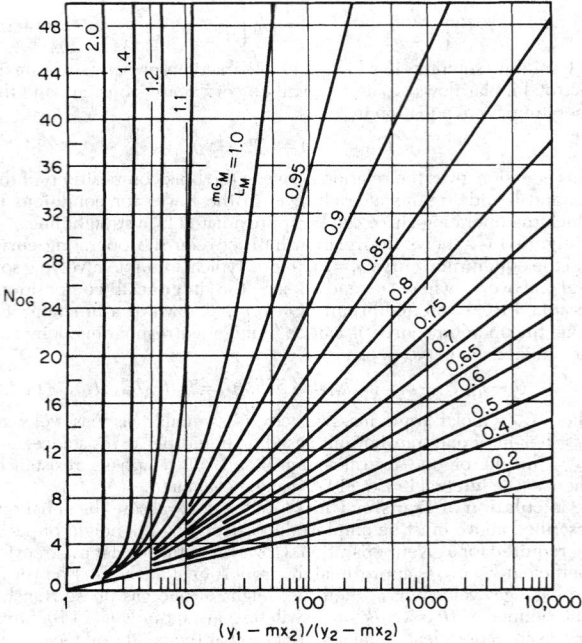

FIG. 14-5 Number of overall gas-phase mass-transfer units in a packed absorption tower for constant mG_M/L_M; solution of Eq. (14-23). (*From Sherwood and Pigford*, Absorption and Extraction, *McGraw-Hill, New York, 1952.*)

overall mass transfer coefficients (or heights of transfer units) for the system of interest, and at the desired conditions of temperature, pressure, solute concentration, and fluid velocities. Such data should be obtained in an apparatus of pilot-plant or semiworks size to avoid abnormalities of scaleup. It must be remembered that values of the mass-transfer parameters are dependent not only on the phase properties and mass throughput in the contactor but also on the type of device used. Within the packing category, there are both random and ordered (structured) type packing elements. Physical characteristics of these devices will be described later.

When no $K_G a$ or HTU data are available, their values may be estimated by means of a generalized model. A summary of useful models is given in Section 5, Table 5-28. The values obtained may then be combined by the use of Eq. 14-19 to obtain values of H_{OG} and H_{OL}. This procedure is not valid, however, when the rate of absorption is limited by a chemical reaction.

Use of HETP Data for Absorber Design Distillation design methods (see Sec. 13) normally involve determination of the number of theoretical equilibrium stages or plates N. Thus, when packed towers are employed in distillation applications, it is common practice to rate the efficiency of tower packings in terms of the height of packing equivalent to one theoretical plate (HETP).

The HETP of a packed-tower section, valid for either distillation or dilute-gas absorption and stripping systems in which constant molal overflow can be assumed and in which no chemical reactions occur, is related to the height of one overall gas-phase mass-transfer unit H_{OG} by the equation

$$\text{HETP} = H_{OG} \frac{\ln (mG_M/L_M)}{(mG_M/L_M - 1)} \qquad (14\text{-}29)$$

For gas-absorption systems in which the inlet gas is concentrated, the correct equation is

$$\text{HETP} = \left(\frac{y_{BM}^\circ}{1 - y} \right)_{\text{av}} H_{OG} \frac{\ln (mG_M/L_M)}{mG_M/L_M - 1} \qquad (14\text{-}30)$$

where the correction term $y_{BM}^\circ/(1 - y)$ is averaged over each individual theoretical plate. The equilibrium compositions corresponding to each theoretical plate may be estimated by the methods described in the subsection "Plate-Tower Design." These compositions are used in conjunction with the local values of the gas and liquid flow rates and the equilibrium slope m to obtain values for H_G, H_L, and H_{OG} corresponding to the conditions on each theoretical stage, and the local values of the HETP are then computed by Eq. (14-30). The total height of packing required for the separation is the summation of the individual HETPs computed for each theoretical stage.

PLATE-TOWER DESIGN

The design of a plate tower for gas-absorption or gas-stripping operations involves many of the same principles employed in distillation calculations, such as the determination of the number of theoretical plates needed to achieve a specified composition change (see Sec. 13). Distillation differs from gas absorption in that it involves the separation of components based on the distribution of the various substances between a gas phase and a liquid phase when all the components are present in both phases. In distillation, the new phase is generated from the original feed mixture by vaporization or condensation of the volatile components, and the separation is achieved by introducing reflux to the top of the tower.

In gas absorption, the new phase consists of an inert nonvolatile solvent (absorption) or an inert nonsoluble gas (stripping), and normally no reflux is involved. The following paragraphs discuss some of the considerations peculiar to gas-absorption calculations for plate towers and some of the approximate design methods that can be employed when simplifying assumptions are valid.

Graphical Design Procedure Construction of design diagrams (xy diagrams showing the equilibrium and operating curves) should be an integral part of any design involving the distribution of a single solute between an inert solvent and an inert gas. The number of theoretical plates can be stepped off rigorously provided the curvatures of the operating and equilibrium lines are correctly accounted for in the diagram. This procedure is valid even though an insoluble inert gas is present in the gas phase and an inert nonvolatile solvent is present in the liquid phase.

Figure 14-6 illustrates the graphical method for a three-theoretical-plate system. Note that in gas absorption the operating line is above the equilibrium curve, whereas in distillation this does not happen. In gas stripping, the operating line will be below the equilibrium curve.

On Fig. 14-6, note that the stepping-off procedure begins on the operating line. The starting point x_f, y_3 represents the compositions of the entering lean wash liquor and of the gas exiting from the top of the tower, as determined by the design specifications. After three steps one reaches the point x_1, y_f, representing the compositions of the solute-rich feed gas y_f and of the solute-rich liquor leaving the bottom of the tower x_1.

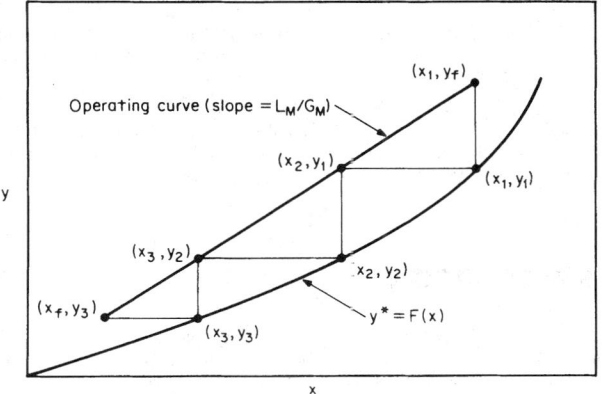

FIG. 14-6 Graphical method for a three-theoretical-plate gas-absorption tower with inlet-liquor composition x_j and inlet-gas composition y_j.

Algebraic Method for Dilute Gases By assuming that the operating and equilibrium curves are straight lines and that heat effects are negligible, Souders and Brown [*Ind. Eng. Chem.*, **24**, 519 (1932)] developed the following equation:

$$(y_1 - y_2)/(y_1 - y_2^\circ) = (A^{N+1} - A)/(A^{N+1} - 1) \tag{14-31}$$

where N = number of theoretical plates, y_1 = mole-fraction solute in the entering gas, y_2 = mole-fraction solute in the leaving gas, $y_2^\circ = mx_2$ = mole-fraction solute in equilibrium with the incoming solvent liquor (zero for a pure solvent), and A = absorption factor = L_M/mG_M. Note that the absorption factor is the reciprocal of the expression given in Eq. (14-4) for packed columns.

When $A = 1$, Eq. (14-31) is indeterminate, and for this case the solution is given by

$$(y_1 - y_2)/(y_1 - y_2^\circ) = N/(N + 1) \tag{14-32}$$

Although Eq. (14-31) is convenient for computing the composition of the exit gas as a function of the number of theoretical stages, an alternative equation derived by Colburn [*Trans. Am. Inst. Chem. Eng.*, **35**, 211 (1939)] is more useful when the number of theoretical plates is the unknown:

$$N = \frac{\ln\left[(1 - A^{-1})(y_1 - y_2^\circ)/(y_2 - y_2^\circ) + A^{-1}\right]}{\ln(A)} \tag{14-33}$$

The numerical results obtained by using either Eq. (14-31) or Eq. (14-33) are identical. Thus, the two equations may be used interchangeably as the need arises.

Comparison of Eqs. (14-33) and (14-23) shows that

$$N_{OG}/N = \ln(A)/(1 - A^{-1}) \tag{14-34}$$

thus revealing the close relationship between theoretical stages in a plate tower and mass-transfer units in a packed tower. Equations (14-23) and (14-33) are related to each other by virtue of the relation

$$h_T = H_{OG}N_{OG} = (\text{HETP})N \tag{14-35}$$

Algebraic Method for Concentrated Gases When the feed gas is concentrated, the absorption factor, which is defined in general as $A = L_M/KG_M$ where $K = y^\circ/x$, can vary throughout the tower owing to changes in the equilibrium K values due to temperature increases. An approximate solution to this problem can be obtained by substitution of the "effective" absorption factors A_e and A' derived by Edmister [*Ind. Eng. Chem.*, **35**, 837 (1943)] into the equation

$$\frac{y_1 - y_2}{y_1} = \left[1 - \frac{1}{A'}\frac{(L_M x)_2}{(G_M y)_1}\right]\frac{A_e^{N+1} - A_e}{A_e^{N+1} - 1} \tag{14-36}$$

where subscripts 1 and 2 refer to the bottom and top of the tower respectively and the absorption factors are defined by the equations

$$A_e = \sqrt{A_1(A_2 + 1) + 0.25} - 0.5 \tag{14-37}$$

$$A' = A_1(A_2 + 1)/(A_1 + 1) \tag{14-38}$$

This procedure has been applied to the absorption of C_5 and lighter hydrocarbon vapors into a lean oil, for example.

Stripping Equations When the liquid feed is dilute and the operating and equilibrium curves are straight lines, the stripping equations analogous to Eqs. (14-31) and (14-33) are

$$(x_2 - x_1)/(x_2 - x_1^\circ) = (S^{N+1} - S)/(S^{N+1} - 1) \tag{14-39}$$

where $x_1^\circ = y_1/m$; $S = mG_M/L_M = A^{-1}$; and

$$N = \frac{\ln\left[(1-A)(x_2 - x_1^\circ)/(x_1 - x_1^\circ) + A\right]}{\ln(S)} \qquad (14\text{-}40)$$

For systems in which the concentrations are large and the stripping factor S may vary along the tower, the following Edmister equations [*Ind. Eng. Chem.*, **35**, 837 (1943)] are applicable:

$$\frac{x_2 - x_1}{x_2} = \left[1 - \frac{1}{S'}\frac{(G_M y)_1}{(L_M x)_2}\right]\frac{S_e^{N+1} - S_e}{S_e^{N+1} - 1} \qquad (14\text{-}41)$$

where

$$S_e = \sqrt{S_2(S_1 + 1) + 0.25} - 0.5 \qquad (14\text{-}42)$$

$$S' = S_2(S_1 + 1)/(S_2 + 1) \qquad (14\text{-}43)$$

and the subscripts 1 and 2 refer to the bottom and top of the tower respectively.

Equations (14-37) and (14-42) represent two different ways of obtaining an effective factor, and a value of A_e obtained by taking the reciprocal of S_e from Eq. (14-42) will not check exactly with a value of A_e derived by substituting $A_1 = 1/S_1$ and $A_2 = 1/S_2$ into Eq. (14-37). Regardless of this fact, the equations generally give reasonable results for approximate design calculations.

It should be noted that throughout this section the subscripts 1 and 2 refer to the bottom and to the top of the apparatus respectively regardless of whether it is an absorber or a stripper. This has been done to maintain internal consistency among all the equations and to prevent the confusion created in some derivations in which the numbering system for an absorber is different from the numbering system for a stripper.

Tray Efficiencies in Plate Absorbers and Strippers Computations of the number of theoretical plates N assume that the liquid on each plate is completely mixed and that the vapor leaving the plate is in equilibrium with the liquid. In actual practice a condition of complete equilibrium cannot exist since interphase mass transfer requires a finite driving-force difference. This leads to the definition of an overall plate efficiency

$$E = N_{\text{theoretical}}/N_{\text{actual}} \qquad (14\text{-}44)$$

which can be correlated to system design variables.

Mass-transfer theory indicates that for trays of a given design the factors most likely to influence E in absorption and stripping towers are the physical properties of the fluids and the dimensionless ratio mG_M/L_M. Systems in which the mass transfer is gas-film-controlled may be expected to have plate efficiencies as high as 50 to 100 percent, whereas plate efficiencies as low as 1 percent have been reported for the absorption of gases of low solubility (large m) into solvents of relatively high viscosity.

The fluid properties are represented by the Schmidt numbers of the gas and liquid phases. For gases, the Schmidt numbers normally are close to unity and are independent of temperature and pressure. Thus, the gas-phase mass-transfer coefficients are relatively independent of the system.

By contrast, the liquid-phase Schmidt numbers range from about 10^2 to 10^4 and depend strongly on the temperature. The effect of temperature on the liquid-phase mass-transfer coefficient is related primarily to changes in the liquid viscosity with temperature, and this derives primarily from the strong dependency of the liquid-phase Schmidt number upon viscosity.

Consideration of the preceding discussion in connection with the relationship between mass-transfer coefficients (see Sec. 5):

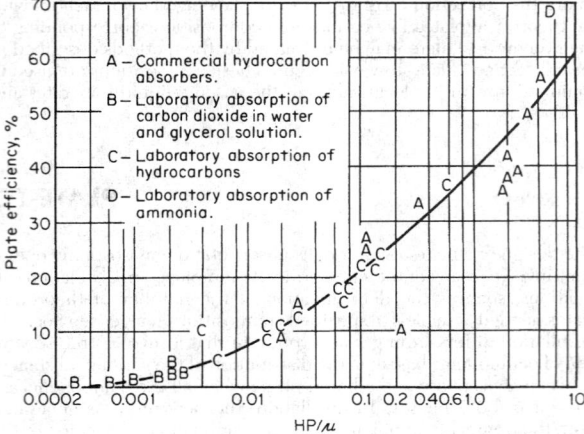

FIG. 14-7 O'Connell correlation for overall column efficiency E_{oc} for absorption. To convert HP/μ in pound-moles per cubic foot-centipoise to kilogram-moles per cubic meter-pascal-second, multiply by 1.60×10^4. [*O'Connell, Trans. Am. Inst. Chem. Eng.*, **42**, 741 (1946).]

$$1/K_G = (1/k_G + m/k_L) \qquad (14\text{-}45)$$

indicates that variations in the overall resistance to mass transfer in absorbers and strippers are related primarily to variations in the liquid-phase viscosity μ and to variations in the slope m. A correlation of the efficiency of plate absorbers in terms of the viscosity of the liquid solvent and the solubility of the solute gas was developed by O'Connell [*Trans. Am. Inst. Chem. Eng.*, **42**, 741 (1946)]. The O'Connell correlation for plate absorbers is presented here as Fig. 14-7.

The best procedure for making plate-efficiency corrections (which obviously can be quite large) is to use experimental-test data from a prototype system that is large enough to be representative of an actual commercial tower.

Example 4: Actual Plates for Steam Stripping The number of actual plates required for steam-stripping an acetone-rich liquor containing 0.573 mole percent acetone in water is to be estimated. The design overhead recovery of acetone is 99.9 percent, leaving 18.5 ppm weight of acetone in the stripper bottoms. The design operating temperature and pressure are 101.3 kPa and 94°C respectively, the average liquid-phase viscosity is 0.30 cP, and the average value of $K = y^\circ/x$ for these conditions is 33.

By choosing a value of $mG_M/L_M = S = A^{-1} = 1.4$ and noting that the stripping medium is pure steam (i.e., $x_1^\circ = 0$), the number of theoretical trays according to Eq. (14-40) is

$$N = \frac{\ln\left[(1 - 0.714)(1000) + 0.714\right]}{\ln(1.4)} = 16.8$$

The O'Connell parameter for gas absorbers is $\rho_L/KM\mu_L$, where ρ_L is the liquid density, lb/ft^3; μ_L is the liquid viscosity, cP; M is the molecular weight of the liquid; and $K = y^\circ/x$. For the present design

$$\rho_L/KM\mu_L = 60.1/(33 \times 18 \times 0.30) = 0.337$$

and according to the O'Connell graph for absorbers (Fig. 14-7) the overall tray efficiency for this case is estimated to be 30 percent. Thus, the required number of actual trays is $16.8/0.3 = 56$ trays.

HEAT EFFECTS IN GAS ABSORPTION

Overview One of the most important considerations involved in designing gas-absorption towers is to determine whether or not temperatures will vary along the length of the tower because of heat effects, since the solubility of the solute gas normally depends strongly upon the temperature. When heat effects can be neglected, computation of the tower dimensions and required flows is relatively straight-

forward, as indicated by the simplified design methods discussed earlier for both packed and plate absorbers and strippers. When heat effects cannot be neglected, the computational problem becomes much more difficult.

Heat effects that may cause temperatures to vary from point to point in a gas absorber are (1) the heat of solution of the solute (including heat of condensation, heat of mixing, and heat of reaction), which can lead to a rise in the liquid temperature; (2) the heat of vaporization or condensation of the solvent; (3) the exchange of sensible heat between the gas and liquid phases; and (4) the loss of sensible heat from the fluids to internal or external cooling coils or to the atmosphere via the tower walls.

Y. T. Shah (*Gas-Liquid-Solid Reactor Design,* McGraw-Hill, New York, 1979, p. 51) has reviewed the literature concerning heat effects in systems involving gas-liquid reactions and concludes that in the majority of the systems involving chemical reactions temperature effects are not very important. For some systems in which large amounts of heat may be liberated, there are compensating effects which decrease the effect on the rate of absorption. For example, increasing temperatures tend simultaneously to increase the rate of chemical reaction and to decrease the solubility of the reactant at the solvent interface. Systems in which compensating effects can occur include absorption of CO_2 in amine solutions and absorption of CO_2 in NaOH solutions.

There are, however, a number of well-known systems in which heat effects definitely cannot be ignored. Examples include absorption of ammonia in water, dehumidification of air with concentrated H_2SO_4, absorption of HCl in water, and absorption of SO_3 in H_2SO_4. Another interesting example is the absorption of acetone in water, in which the heat effects are mild but not negligible.

Some very thorough and knowledgeable discussions of the problems involved in gas absorption with large heat effects have been presented by Coggan and Bourne [*Trans. Inst. Chem. Eng.,* **47,** T96, T160 (1969)], by Bourne, von Stockar, and Coggan [*Ind. Eng. Chem. Process Des. Dev.,* **13,** 115, 124 (1974)], and also by von Stockar and Wilke [*Ind. Eng. Chem. Fundam.,* **16,** 89 (1977)]. The first two of these references discuss plate-tower absorbers and include interesting experimental studies of the absorption of ammonia in water. The third reference discusses the design of packed-tower gas absorbers and includes a shortcut design method based on a semitheoretical correlation of rigorous calculation results. All these authors clearly demonstrate both theoretically and experimentally that when the solvent is volatile, the temperature inside an absorber can go through a maximum. They note that the least expensive of the solvents, water, is capable of exhibiting this unusual "hot-spot" behavior.

From a designer's point of view there are a number of different approaches to be considered in dealing with heat effects, depending on the requirements of the job at hand. For example, one can (1) add internal or external heat-transfer surface to remove heat from the absorber; (2) treat the process as if it were isothermal by arbitrarily assuming that the temperature of the liquid phase is everywhere the same and then add a design safety factor; (3) employ the classical adiabatic model, which assumes that the heat of solution manifests itself only as sensible heat in the liquid phase and that solvent vaporization is negligible; (4) use semitheoretical shortcut methods derived from rigorous calculations; and (5) employ rigorous design procedures requiring the use of a digital computer.

For preliminary-screening work the simpler methods may be adequate, but for final designs one should seriously consider using a more rigorous approach.

Effects of Operating Variables   Conditions that can give rise to significant heat effects are (1) an appreciable heat of solution and (2) absorption of large amounts of solute in the liquid phase. The second of these conditions can arise when the solute concentration in the inlet gas is large, when the liquid flow rate is relatively low (small L_M/G_M), when the solubility of the solute in the liquid phase is high, and/or when the operating pressure is high.

When the solute is absorbed very rapidly, the rate of heat liberation often is largest near the bottom of the tower. This has the effect of causing the equilibrium line to curve upward near the solute-rich end, although it may remain relatively straight near the lean end, corresponding to the temperature of the lean solvent.

If the solute-rich gas entering the bottom of an absorption tower is cold, the liquid phase may be cooled somewhat by transfer of sensible heat to the gas. A much stronger cooling effect can occur when the solvent is volatile and the entering rich gas is not saturated with respect to the solvent. It is possible to experience a condition in which

solvent is being evaporated near the bottom of the tower and condensed near the top. Under these conditions there may develop a pinch point in which the operating and equilibrium curves approach each other at a point inside the tower.

In the references cited previously, the authors discuss the influence of operating variables upon the performance of plate towers when large heat effects are involved. Some general observations are as follows:

Operating Pressure   Raising the pressure may increase the separation efficiency considerably. Calculations involving the absorption of methanol from water-saturated air showed that doubling the pressure doubled the concentration of methanol which could be tolerated in the feed gas while still achieving a preset concentration specification in the off gas.

Temperature of Fresh Solvent   The temperature of the entering solvent has surprisingly little influence upon the degree of absorption or upon the internal-temperature profiles in an absorber when the heat effects are due primarily to heat of solution or to solvent vaporization. In these cases the temperature profile in the liquid phase apparently is dictated solely by the internal-heat effects.

Temperature and Humidity of Rich Gas   Cooling and consequent dehumidification of the feed gas to an absorption tower can be very beneficial. A high humidity (or relative saturation with solvent) limits the capacity of the gas phase to take up latent heat and therefore is unfavorable to absorption. Thus, dehumidification of the inlet gas prior to introducing it into the tower is worth considering in the design of gas absorbers with large heat effects.

Liquid-to-Gas Ratio   The L/G ratio can have a significant influence on the development of temperature profiles in gas absorbers. High L/G ratios tend to result in less strongly developed temperature profiles owing to the high heat capacity of the liquid phase. As the L/G is increased, the operating line moves away from the equilibrium line and there is a tendency for more solute to be absorbed per stage. However, there is a compensating effect in that as more heat is liberated at each stage, the plate temperatures will tend to rise, causing an upward shifting of the equilibrium line.

As the L/G is decreased, the concentration of solute tends to build up in the upper parts of the absorber, and the point of highest temperature tends to move upward in the tower until finally the maximum temperature develops only on the topmost plate. Of course, the capacity of the liquid phase to absorb solute falls progressively as the L/G is reduced.

Number of Stages   When the heat effects combine to produce an extended zone within the tower where little absorption is taking place (i.e., a pinch zone), the addition of plates to the tower will have no useful effect on separation efficiency. Solutions to these difficulties must be sought by increasing the solvent flow, introducing strategically placed coolers, cooling and dehumidifying the inlet gas, and/or raising the tower operating pressure.

Equipment Considerations   When the solute has a large heat of solution and the feed gas contains a large percentage of solute, as in the absorption of HCl in water, the effects of heat release during absorption may be so pronounced that the installation of heat-transfer surface to remove the heat of absorption may be as important as providing sufficient interfacial area for the mass-transfer process itself. The added heat-transfer surface may consist of internal cooling coils on the plates, or else the solvent may be withdrawn from a point intermediate in the tower and passed through an external heat exchanger (intercooler) for cooling.

In many cases the rate of heat liberation is largest near the bottom of the tower, where solute absorption is more rapid, so that cooling surfaces or intercoolers are required only on the first few trays at the bottom of the column. Coggan and Bourne [*Trans. Inst. Chem. Eng.,* **47,** T96, T160 (1969)] found, however, that the optimal position for a single interstage cooler does not necessarily coincide with the position of the maximum temperature or with the center of a pinch. They found that in a 12-plate tower, two strategically placed interstage coolers tripled the allowable ammonia feed concentration for a given off-gas specification. For a case involving methanol absorption, it was found that more separation was possible in a 12-stage column with two intercoolers than in a simple column with 100 stages and no intercoolers.

In the case of HCl absorption, a shell-and-tube heat exchanger often is employed as a cooled wetted-wall vertical-column absorber so that the exothermic heat of reaction can be removed continuously as it is released into the liquid film.

Installation of heat-exchange equipment to precool and dehumidify the feed gas to an absorber also deserves consideration in order to take advantage of the cooling effects created by vaporization of solvent in the lower sections of the tower.

Classical Isothermal Design Method   When the feed gas is sufficiently dilute, the exact design solution may be approximated by the isothermal one over a broad range of L/G ratios, since heat effects generally are less important when washing dilute-gas mixtures. The problem, however, is one of defining the term "sufficiently dilute" for each specific case. For a new absorption duty, the assumption of isothermal operation must be subjected to verification by the use of a rigorous design procedure.

When heat-exchange surface is being provided in the design of an absorber, the isothermal design procedure can be rendered valid by virtue of the exchanger design specifications. With ample surface area and a close approach, isothermal operation can be guaranteed.

For preliminary screening and feasibility studies or for rough cost estimates, one may wish to employ a version of the isothermal method which assumes that the liquid temperatures in the tower are everywhere equal to the inlet-liquid temperature. In their analysis of packed-tower designs, von Stockar and Wilke [*Ind. Eng. Chem. Fundam.* **16,** 89 (1977)] showed that the isothermal method tended to underestimate the required depth of packing by a factor of as much as 1.5 to 2. Thus, for rough estimates one may wish to employ the assumption that the temperature is equal to the inlet-liquid temperature and then apply a design factor to the result.

Another instance in which the constant-temperature method is used involves the direct application of experimental $K_G a$ values obtained at the desired conditions of inlet temperatures, operating pressure, flow rates, and feed-stream compositions. The assumption here is that, regardless of any temperature profiles that may exist within the actual tower, the procedure of "working the problem in reverse" will yield a correct result. One should be cautious about extrapolating such data very far from the original basis and be careful to use compatible equilibrium data.

Classical Adiabatic Design Method   The classical adiabatic method assumes that the heat of solution serves only to heat up the liquid stream and that there is no vaporization of solvent. This assumption makes it feasible to relate increases in the liquid-phase temperature to the solute concentration x by a simple enthalpy balance. The equilibrium curve can then be adjusted to account for the corresponding temperature rise on an xy diagram. The adjusted equilibrium curve will become more concave upward as the concentration increases, tending to decrease the driving forces near the bottom of the tower, as illustrated in Fig. 14-8 in Example 6.

Colburn [*Trans. Am. Inst. Chem. Eng.*, **35,** 211 (1939)] has shown that when the equilibrium line is straight near the origin but curved slightly at its upper end, N_{OG} can be computed approximately by assuming that the equilibrium curve is a parabolic arc of slope m_2 near the origin and passing through the point $x_1, K_1 x_1$ at the upper end. The Colburn equation for this case is

$$N_{OG} = \frac{1}{1 - m_2 G_M / L_M}$$

$$\times \ln\left[\frac{(1 - m_2 G_M / L_M)^2}{1 - K_1 G_M / L_M} \left(\frac{y_1 - m_2 x_2}{y_2 - m_2 x_2} \right) + \frac{m_2 G_M}{L_M} \right] \quad (14\text{-}46)$$

Comparisons by von Stockar and Wilke [*Ind. Eng. Chem. Fundam.*, **16,** 89 (1977)] between the rigorous and the classical adiabatic design methods for packed towers indicate that the simple adiabatic method underestimates the packing depths by as much as a factor of 1.25 to 1.5. Thus, when using the classical adiabatic method, one should consider the possible need to apply a design safety factor.

A slight variation of the above method accounts for increases in the solvent content of the gas stream between the inlet and the outlet of the tower and assumes that the evaporation of solvent tends to cool the liquid. This procedure offsets a part of the temperature rise that

would have been predicted with no solvent evaporation and leads to the prediction of a shorter tower.

Rigorous Design Methods   A detailed discussion of rigorous methods for the design of packed and plate absorbers when large heat effects are involved is beyond the scope of this section. In principle, material and energy balances may be executed under the same constraints as for rigorous distillation calculations (see Sec. 13). The MESH equations are solved, but for absorption or stripping, convergence may be quite sensitive to the relatively large heat effects compared with distillation. Absorption-stripping programs are included in the software packages for process simulation. The paper of von Stockar and Wilke [*Ind. Eng. Chem. Fundam.* **16,** 89 (1977)] presents an interesting shortcut method for the design of packed absorbers which closely approximates rigorous results.

Direct Comparison of Design Methods   The following problem, originally solved by Sherwood, Pigford, and Wilke (*Mass Transfer*, McGraw-Hill, New York, 1975, p. 616), was employed by von Stockar and Wilke as the basis for a direct comparison between the isothermal, adiabatic, semitheoretical shortcut, and rigorous design methods for estimating the height of packed towers.

Example 5: Packed Absorber, Acetone into Water   Inlet gas to an absorber consists of a mixture of 6 mole percent acetone in air saturated with water vapor at 15°C and 101.3 kPa (1 atm). The scrubbing liquor is pure water at 15°C, and the inlet gas and liquid rates are given as 0.080 and 0.190 kmol/s respectively. The liquid rate corresponds to 20 percent over the theoretical minimum as calculated by assuming a value of x_1 corresponding to complete equilibrium between the exit liquor and the incoming gas. H_G and H_L are given as 0.42 and 0.30 m respectively, and the acetone equilibrium data at 15°C are $p_A^0 = 19.7$ kPa (147.4 torr), $\gamma_A = 6.46$, and $m_A = 6.46 \times 19.7/101.3 = 1.26$. The heat of solution of acetone is 7656 cal/gmol (32.05 kJ/gmol), and the heat of vaporization of solvent (water) is 10,755 cal/gmol (45.03 kJ/gmol). The problem calls for determining the height of packing required to achieve a 90 percent recovery of the acetone.

The following table compares the results obtained by von Stockar and Wilke (op. cit.) for the various design methods:

Design method used	N_{OG}	Packed height, m	Design safety factor
Rigorous	5.56	3.63	1.00
Shortcut rigorous	5.56	3.73	0.97
Classical adiabatic	4.01	2.38	1.53
Classical isothermal	3.30	1.96	1.85

It should be clear from this example that there is considerable room for error when approximate design methods are employed in situations involving large heat effects, even for a case in which the solute concentration in the inlet gas was only 6 mole percent.

Example 6: Solvent Rate for Absorption   Let us consider the absorption of acetone from air at atmospheric pressure into a stream of pure water fed to the top of a packed absorber at 25°C. The inlet gas at 35°C contains 2 percent by volume of acetone and is 70 percent saturated with water vapor (4 percent H_2O by volume). The mole-fraction acetone in the exit gas is to be reduced to 1/400 of the inlet value, or 50 ppmv. For 100 kmol of feed-gas mixture, how many kilomoles of fresh water should be fed to provide a positive-driving force throughout the packing? How many transfer units will be needed according to the classical adiabatic method? What is the estimated height of packing required if $H_{OG} = 0.70$ m?

The latent heats at 25°C are 7656 kcal/kmol for acetone and 10,490 kcal/kmol for water, and the differential heat of solution of acetone vapor in pure water is given as 2500 kcal/kmol. The specific heat of air is 7.0 kcal/(kmol·K).

Acetone solubilities are defined by the equation

$$K = y^\circ / x = \gamma_a p_a / p_T \quad (14\text{-}47)$$

where the vapor pressure of pure acetone in mmHg (torr) is given by (Sherwood et al., *Mass Transfer*, McGraw-Hill, New York, 1975, p. 537):

$$p_A^0 = \exp(18.1594 - 3794.06/T) \quad (14\text{-}48)$$

and the liquid-phase-activity coefficient may be approximated for low concentrations ($x \leq 0.01$) by the equation

$$\gamma_a = 6.5 \exp(2.0803 - 601.2/T) \quad (14\text{-}49)$$

Typical values of acetone solubility as a function of temperature at a total pressure of 760 mmHg are shown in the following table:

t, °C	25	30	35	40
γ_a	6.92	7.16	7.40	7.63
p_a, mmHg	229	283	346	422
$K = \gamma_a p_a^0/760$	2.09	2.66	3.37	4.23

For dry gas and liquid water at 25°C, the following enthalpies are computed for the inlet- and exit-gas streams (basis, 100 kmol of gas entering):

Entering gas:

Acetone	$2(2500 + 7656) =$	20,312 kcal
Water vapor	$4(10,490) =$	41,960
Sensible heat	$(100)(7.0)(35 - 25) =$	7,000
		69,272 kcal

Exit gas (assumed saturated with water at 25°C):

Acetone	$(2/400)(94/100)(2500) =$	12 kcal
Water vapor	$94\left(\dfrac{23.7}{760 - 23.7}\right)(10,490) =$	31,600
		31,612 kcal

Enthalpy change of liquid = 69,272 − 31,612 = 37,660 kcal/100 kmol gas. Thus, $\Delta t = t_1 - t_2 = 37,660/18 L_M$, and the relation between L_M/G_M and the liquid-phase temperature rise is

$$L_M/G_M = (37,660)/(18)(100) \, \Delta t = 20.92/\Delta t$$

The following table summarizes the critical values for various assumed temperature rises:

Δt, °C	L_M/G_M	K_1	$K_1 G_M/L_M$	$m_2 G_M/L_M$
0		2.09	0.	0.
2	10.46	2.31	0.221	0.200
3	6.97	2.42	0.347	0.300
4	5.23	2.54	0.486	0.400
5	4.18	2.66	0.636	0.500
6	3.49	2.79	0.799	0.599
7	2.99	2.93	0.980	0.699

Evidently a temperature rise of 7°C would not be a safe design because the equilibrium line nearly touches the operating line near the bottom of the tower, creating a pinch. A temperature rise of 6°C appears to give an operable design, and for this case $L_M = 349$ kmol per 100 kmol of feed gas.

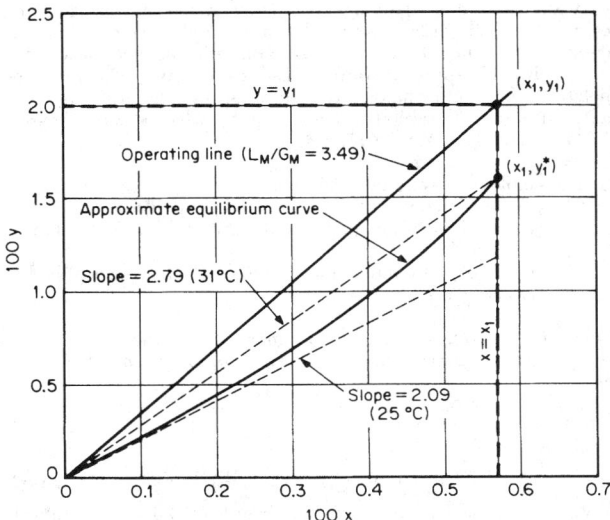

FIG. 14-8 Design diagram for adiabatic absorption of acetone in water, Example 6.

The design diagram for this case is shown in Fig. 14-8, in which the equilibrium curve is drawn with a french curve so that the slope at the origin m_2 is equal to 2.09 and passes through the point $x_1 = 0.02/3.49 = 0.00573$ at $y_1^\circ = 0.00573 \times 2.79 = 0.0160$.

The number of transfer units can be calculated from the adiabatic design equation, Eq. (14-46):

$$N_{OG} = \frac{1}{1 - 0.599} \ln\left[\frac{(1 - 0.599)^2}{(1 - 0.799)}(400) + 0.599\right] = 14.4$$

The estimated height of tower packing by assuming $H_{OG} = 0.70$ m and a design safety factor of 1.5 is

$$h_T = (14.4)(0.7)(1.5) = 15.1 \text{ m (49.6 ft)}$$

For this tower, one should consider the use of two or more shorter packed sections instead of one long section.

MULTICOMPONENT SYSTEMS

When no chemical reactions are involved in the absorption of more than one soluble component from an insoluble gas, the design conditions (pressure, temperature, and liquid-to-gas ratio) normally are determined by the volatility or the physical solubility of the least soluble component for which complete recovery is economical. Components of lower volatility (higher solubility) also will be recovered completely.

The more volatile (i.e., less soluble) components will be only partially absorbed even though the effluent liquid becomes completely saturated with respect to these lighter substances. When a condition of saturation exists, the value of y_1/y_2 will remain finite even for an infinite number of plates or transfer units. This can be seen in Fig. 14-9, in which the asymptotes become vertical for values of mG_M/L_M greater than unity. If the amount of volatile component in the incoming fresh solvent is negligible, then the limiting value of y_1/y_2 for each of the highly volatile components is

$$y_1/y_2 = S/(S - 1) \qquad (14\text{-}50)$$

where $S = mG_M/L_M$ and the subscripts 1 and 2 refer to the bottom and top of the tower respectively.

When the gas stream is dilute, absorption of each constituent can be considered separately as if the other components were absent. The following example illustrates the use of this principle.

Example 7: Multicomponent Absorption, Dilute Case Air entering a tower contains 1 percent acetaldehyde and 2 percent acetone. The liquid-to-gas ratio for optimum acetone recovery is $L_M/G_M = 3.1$ mol/mol when the fresh-solvent temperature is 31.5°C. The value of y°/x for acetaldehyde has been measured as 50 at the boiling point of a dilute solution, 93.5°C. What will the percentage recovery of acetaldehyde be under conditions of optimal acetone recovery?

Solution. If the heat of solution is neglected, y°/x at 31.5°C is equal to 50(1200/7300) = 8.2, where the factor in parentheses is the ratio of pure-acetaldehyde vapor pressures at 31.5 and 93.5°C respectively. Since L_M/G_M is equal to 3.1, the value of S for the aldehyde is $S = mG_M/L_M = 8.2/3.1 = 2.64$, and $y_1/y_2 = S/(S - 1) = 2.64/1.64 = 1.61$. The acetaldehyde recovery is therefore equal to $100 \times 0.61/1.61 = 38$ percent recovery.

In concentrated systems the change in gas and liquid flow rates within the tower and the heat effects accompanying the absorption of all the components must be considered. A trial-and-error calculation from one theoretical stage to the next usually is required if accurate results are to be obtained, and in such cases calculation procedures similar to those described in Sec. 13 normally are employed. A computer procedure for multicomponent adiabatic absorber design has been described by Feintuch and Treybal [*Ind. Eng. Chem. Process Des. Dev.*, **17**, 505 (1978)]. Also see Holland, *Fundamentals and Modeling of Separation Processes*, Prentice Hall, Englewood Cliffs, N.J., 1975.

When two or more gases are absorbed in systems involving chemical reactions, the situation is much more complex. This topic is discussed later in the subsection "Absorption with Chemical Reaction."

Graphical Design Method for Dilute Systems   The following notation for multicomponent absorption calculations has been adapted from Sherwood, Pigford, and Wilke (*Mass Transfer*, McGraw-Hill, New York 1975, p. 415):

L_M^s = moles of solvent per unit time
G_M^0 = moles of rich feed gas to be treated per unit time
X = moles of one solute per mole of solute-free solvent fed to the top of the tower
Y = moles of one solute in the gas phase per mole of rich feed gas to be treated

Subscripts 1 and 2 refer to the bottom and top of the tower respectively, and the material balance for any one component may be written as

$$L_M^s(X - X_2) = G_M^0(Y - Y_2) \qquad (14\text{-}51)$$

or else as

$$L_M^s(X_1 - X) = G_M^0(Y_1 - Y) \qquad (14\text{-}52)$$

For the special case of absorption from lean gases with relatively large amounts of solvent, the equilibrium lines are defined for each component by the relation

$$Y^0 = K'X \qquad (14\text{-}53)$$

Thus, the equilibrium line for each component passes through the origin with slope K', where

$$K' = K(G_M/G_M^0)/(L_M/L_M^s) \qquad (14\text{-}54)$$

and $K = y^\circ/x$. When the system is sufficiently dilute, $K' \doteq K$.

The liquid-to-gas ratio L_M^s/G_M^0 is chosen on the basis of the solubility of the least soluble substance in the feed gas that must be absorbed completely. Each individual component will then have its own operating line with slope equal to L_M^s/G_M^0 (i.e., the operating lines for all the various components will be parallel to each other).

A typical diagram for the complete absorption of pentane and heavier components from a lean gas mixture is shown in Fig. 14-9. The oil used as solvent for this case was assumed to be solute-free (i.e., $X_2 = 0$), and the "key component," butane, was identified as that component absorbed in appreciable amounts whose equilibrium line is most

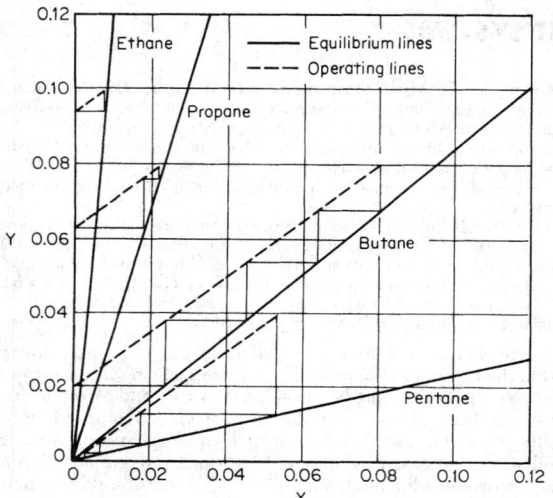

FIG. 14-9   Graphical design method for multicomponent systems; absorption of butane and heavier components in a solute-free lean oil.

nearly parallel to the operating lines (i.e., the K value for butane is approximately equal to L_M^s/G_M^0).

In Fig. 14-9, the composition of the gas with respect to components more volatile than butane will approach equilibrium with the liquid phase at the bottom of the tower. The gas composition with respect to components less volatile (heavier) than butane will approach equilibrium with the oil entering the tower, and since $X_2 = 0$, the components heavier than butane will be completely absorbed.

Four theoretical plates have been stepped off for the key component (butane) on Fig. 14-9 and are sufficient to give a 75 percent recovery of butane. The operating lines for the other components were drawn in with the same slope and were placed so as to give the same number of theoretical plates insofar as possible.

The diagram of Fig. 14-9 shows that for the light components equilibrium is achieved easily in fewer than four theoretical plates and that for the heavier components nearly complete recovery is obtained in four theoretical plates. The diagram also shows that absorption of the light components takes place in the upper part of the tower and absorption of the heavier components takes place in the lower section of the tower.

Algebraic Design Method for Dilute Systems   The design method described above can be performed algebraically by employing the following modified version of the Kremser formula:

$$\frac{Y_1 - Y_2}{Y_1 - mX_2} = \frac{(A^0)^{N+1} - A^0}{(A^0)^{N+1} - 1} \qquad (14\text{-}55)$$

where for dilute gas absorption $A^0 = L_M^s/mG_M^0$ and $m \doteq K = y^0/x$.

The left-hand side of Eq. (14-55) represents the efficiency of absorption of any one component of the feed-gas mixture. If the solvent oil is denuded of solute so that $X_2 = 0$, the left-hand side is equal to the fractional absorption of the component from the rich feed gas. When the number of theoretical plates N and the liquid and gas rates L_M^s and G_M^0 have been fixed, the fractional absorption of each component may be computed directly and the operating lines need not be placed by trial and error as in the graphical approach described earlier.

According to Eq. (14-55), when A^0 is less than unity and N is large,

$$(Y_1 - Y_2)/(Y_1 - mX_2) \doteq A^0 \qquad (14\text{-}56)$$

This equation can be employed for estimating the fractional absorption of the more volatile components whenever the value of A^0 for the component is smaller than the value of A^0 for the key component by a factor of 3 or more.

When A^0 is very much larger than unity and when N is large, the right-hand side of Eq. (14-55) becomes equal to unity. This signifies that the gas will leave the top of the tower in equilibrium with the incoming oil, and when $X_2 = 0$, it corresponds to complete absorption of the component in question. Thus, the least volatile components may be assumed to be at equilibrium with the lean oil at the top of the tower.

When $A^0 = 1$, the right-hand side of Eq. (14-55) becomes indeterminate. The solution for this case is

$$(Y_1 - Y_2)/(Y_1 - mX_2) = N/(N + 1) \qquad (14\text{-}57)$$

For systems in which the absorption factor A^0 for each component is not constant throughout the tower, an effective absorption factor for use in the equations just presented can be estimated by the Edmister formula

$$A_e^0 = \sqrt{A_1^0 (A_2^0 + 1) + 0.25} - 0.5 \qquad (14\text{-}58)$$

This procedure is a reasonable approximation only when no pinch points exist within the tower and when the absorption factors vary in a regular manner between the bottom and the top of the tower.

Example 8: Multicomponent Absorption, Concentrated Case
A hydrocarbon feed gas is to be treated in an existing four-theoretical-tray absorber to remove butane and heavier components. The recovery specification for the key component, butane, is 75 percent. The composition of the exit gas from the absorber and the required liquid-to-gas ratio are to be estimated. The feed-gas composition and the equilibrium K values for each component at the temperature of the (solute-free) lean oil are presented in the following table:

Component	Mole %	K value
Methane	68.0	74.137
Ethane	10.0	12.000
Propane	8.0	3.429
Butane	8.0	0.833
Pentane	4.0	0.233
C_6 plus	2.0	0.065

For $N = 4$ and $Y_2/Y_1 = 0.25$, the value of A^0 for butane is found to be equal to 0.89 from Eq. (14-55) by using a trial-and-error method. The values of A^0 for the other components are then proportional to the ratios of their K values to that of butane. For example, $A^0 = 0.89(0.833/12.0) = 0.062$ for ethane. The values of A^0 for each of the other components and the exit-gas composition as computed from Eq. (14-55) are shown in the following table:

Component	A^0	Y_2, mol/ mol feed	Exit gas, mole %
Methane	0.010	67.3	79.1
Ethane	0.062	9.4	11.1
Propane	0.216	6.3	7.4
Butane	0.890	2.0	2.4
Pentane	3.182	0.027	0.03
C_6 plus	11.406	0.0012	0.0014

The molar liquid-to-gas ratio required for this separation is computed as $L_M^s/G_M^0 = A^0 \times K = 0.89 \times 0.833 = 0.74$.

We note that this example is the analytical solution to the graphical design problem shown in Fig. 14-9, which therefore is the design diagram for this system.

ABSORPTION WITH CHEMICAL REACTION

Introduction Many present-day commercial gas absorption processes involve systems in which chemical reactions take place in the liquid phase. These reactions generally enhance the rate of absorption and increase the capacity of the liquid solution to dissolve the solute, when compared with physical absorption systems.

A necessary prerequisite to understanding the subject of absorption with chemical reaction is the development of a thorough understanding of the principles involved in physical absorption, as discussed earlier in this section and in Section 5. There are a number of excellent references on the subject, such as the book by Danckwerts (*Gas-Liquid Reactions*, McGraw-Hill, New York, 1970) and Astarita et al. (*Gas Treating with Chemical Solvents*, Wiley, New York, 1983).

Recommended Overall Design Strategy When considering the design of a gas-absorption system involving chemical reactions, the following procedure is recommended:

1. Consider the possibility that the physical design methods described earlier in this section may be applicable.

2. Determine whether commercial design overall $\hat{K}_G a$ values are available for use in conjunction with the traditional design method, being careful to note whether or not the conditions under which the $\hat{K}_G a$ data were obtained are essentially the same as for the new design. Contact the various tower-packing vendors for information as to whether $\hat{K}_G a$ data are available for your system and conditions.

3. Consider the possibility of scaling up the design of a new system from experimental data obtained in a laboratory-bench scale or a small pilot-plant unit.

4. Consider the possibility of developing for the new system a rigorous, theoretically based design procedure which will be valid over a wide range of design conditions.

These topics are discussed in the subsections that follow.

Applicability of Physical Design Methods Physical design methods such as the classical isothermal design method or the classical adiabatic design method may be applicable for systems in which chemical reactions are either extremely fast or extremely slow or when chemical equilibrium is achieved between the gas and liquid phases.

If the liquid-phase reaction is extremely fast and irreversible, the rate of absorption may in some cases be completely governed by the gas-phase resistance. For practical design purposes one may assume (for example) that this gas-phase mass-transfer limited condition will exist when the ratio y_i/y is less than 0.05 everywhere in the apparatus.

From the basic mass-transfer flux relationship for species A (Sec. 5),

$$N_A = k_G(y - y_i) = k_L(x_i - x) \tag{14-59}$$

one can readily show that this condition on y_i/y requires that the ratio x/x_i be negligibly small (i.e., a fast reaction) and that the ratio $mk_G/k_L = mk_G/k_L^0\phi$ be less than 0.05 everywhere in the apparatus. The ratio $mk_G/k_L^0\phi$ will be small if the equilibrium back pressure of the solute over the liquid solution is small (i.e., small m; high reactant sol-

ubility), or the reaction-enhancement factor $\phi = k_L/k_L^0$ is very large, or both.

As discussed later, the reaction-enhancement factor ϕ will be large for all extremely fast pseudo-first-order reactions and will be large for extremely fast second-order irreversible reaction systems in which there is a sufficiently large excess of liquid-phase reagent. When the rate of an extremely fast second-order irreversible reaction system $A + vB \rightarrow$ products is limited by the availability of the liquid-phase reagent B, then the reaction-enhancement factor may be estimated by the formula $\phi = 1 + B^0/vc_i$. In systems for which this formula is applicable, it can be shown that the interface concentration y_i will be equal to zero whenever the ratio $k_G y v/k_L^0 B^0$ is less than or equal to unity.

Figure 14-10 illustrates the gas-film and liquid-film concentration profiles one might find in an extremely fast (gas-phase mass-transfer limited) second-order irreversible reaction system. The solid curve for reagent B represents the case in which there is a large excess of bulk-liquid reagent B^0. The dashed curve in Fig. 14-10 represents the case in which the bulk concentration B^0 is not sufficiently large to prevent the depletion of B near the liquid interface and for which the equation $\phi = 1 + B^0/vc_i$ is applicable.

Whenever these conditions on the ratio y_i/y apply, the design can be based upon the physical rate coefficient $\hat{k}_G$ or upon the height of one gas-phase mass-transfer unit H_G. The gas-phase mass-transfer limited condition is approximately valid, for instance, in the following systems: absorption of NH_3 into water or acidic solutions, vaporization of water into air, absorption of H_2O into concentrated sulfuric acid solutions, absorption of SO_2 into alkali solutions, absorption of H_2S from a dilute-

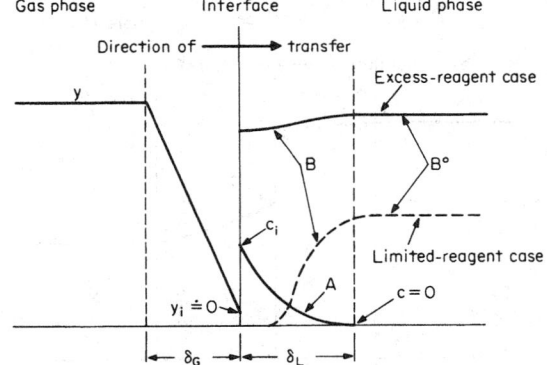

FIG. 14-10 Gas-phase and liquid-phase solute-concentration profiles for an extremely fast (gas-phase mass-transfer limited) irreversible reaction system $A + vB \rightarrow$ products.

gas stream into a strong alkali solution, absorption of HCl into water or alkaline solutions, or absorption of Cl_2 into strong alkali.

When liquid-phase chemical reactions are **extremely slow,** the gas-phase resistance can be neglected and one can assume that the rate of reaction has a predominant effect upon the rate of absorption. In this case the differential rate of transfer is given by the equation

$$dn_A = R_A f_H S \, dh = (k_L^0 a/\rho_L)(c_i - c)S \, dh \tag{14-60}$$

where n_A = rate of solute transfer, R_A = volumetric reaction rate, a function of c and T, f_H = fractional liquid volume holdup in tower or apparatus, S = tower cross-sectional area, h = vertical distance, k_L^0 = liquid-phase mass-transfer coefficient for pure physical absorption, a = effective interfacial mass-transfer area per unit volume of tower or apparatus, ρ_L = average molar density of liquid phase, c_i = solute concentration in liquid at gas-liquid interface, and c = solute concentration in bulk liquid.

Although the right-hand side of Eq. (14-60) remains valid even when chemical reactions are extremely slow, the mass-transfer driving force may become increasingly small, until finally $c \doteq c_i$. For extremely slow first-order irreversible reactions, the following rate expression can be derived from Eq. (14-60):

$$R_A = k_1 c = k_1 c_i / (1 + k_1 \rho_L f_H / k_L^0 a) \tag{14-61}$$

where k_1 = first-order reaction rate coefficient.

For **dilute systems** in countercurrent absorption towers in which the equilibrium curve is a straight line (i.e., $y_i = m x_i$) the differential relation of Eq. (14-60) is formulated as

$$dn_A = -G_M S \, dy = k_1 c f_H S \, dh \tag{14-62}$$

where G_M = molar gas-phase mass velocity and y = gas-phase solute mole fraction.

Substitution of Eq. (14-61) into Eq. (14-62) and integration lead to the following relation for an **extremely slow first-order reaction** in an absorption tower:

$$y_2 = y_1 \exp(-\gamma) \tag{14-63}$$

where
$$\gamma = \frac{k_1 \rho_L f_H h_T / m G_M}{(1 + k_1 \rho_L f_H / k_L^0 a)} \tag{14-64}$$

In Eq. (14-63) the subscripts 1 and 2 refer to the bottom and the top of the tower respectively.

The Hatta number N_{Ha} usually is employed as the criterion for determining whether or not a reaction can be considered extremely slow. For extremely slow reactions a reasonable criterion is

$$N_{Ha} = \sqrt{k_1 D_A} / k_L^0 \le 0.3 \tag{14-65}$$

where D_A = liquid-phase diffusion coefficient of the solute in the sol-

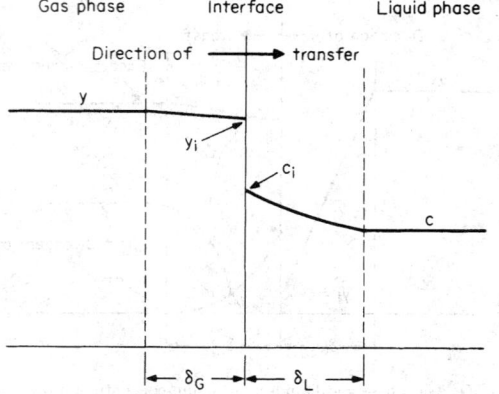

Gas phase   Interface   Liquid phase

Direction of ——→ transfer

FIG. 14-11   Gas-phase and liquid-phase solute-concentration profiles for an extremely slow (kinetically limited) reaction system for which N_{Ha} is less than 0.3.

vent. Figure 14-11 illustrates the concentration profiles in the gas and liquid films for the case of an extremely slow chemical reaction.

We note that when the second term in the denominator of Eq. (14-64) is small, the liquid holdup in the tower can have a significant influence upon the rate of absorption if an extremely slow chemical reaction is involved.

When **chemical equilibrium** is achieved quickly throughout the liquid phase (or can be assumed to exist), the problem becomes one of properly defining the physical and chemical equilibria for the system. It sometimes is possible to design a plate-type absorber by assuming chemical-equilibrium relationships in conjunction with a stage efficiency factor as is done in distillation calculations. Rivas and Prausnitz [*Am. Inst. Chem. Eng. J.*, **25**, 975 (1979)] have presented an excellent discussion and example of the correct procedures to be followed for systems involving chemical equilibria.

Traditional Design Method   The traditionally employed conventional procedure for designing packed-tower gas-absorption systems involving chemical reactions makes use of overall volumetric mass-transfer coefficients as defined by the equation

$$K_G' a = n_A / (h_T S p_T \Delta y_{1m}^\circ) \tag{14-66}$$

where $K_G' a$ = overall volumetric mass-transfer coefficient, n_A = rate of solute transfer from the gas to the liquid phase, h_T = total height of tower packing, S = tower cross-sectional area, p_T = total system pressure, and Δy_{1m}° is defined by the equation

$$\Delta y_{1m}^\circ = \frac{(y - y^\circ)_1 - (y - y^\circ)_2}{\ln[(y - y^\circ)_1/(y - y^\circ)_2]} \tag{14-67}$$

in which subscripts 1 and 2 refer to the bottom and top of the absorption tower respectively, y = mole-fraction solute in the gas phase, and y° = gas-phase solute mole fraction in equilibrium with bulk-liquid-phase solute concentration x. When the equilibrium line is straight, $y^\circ = m x$.

The traditional design method normally makes use of overall $K_G' a$ values even when resistance to transfer lies predominantly in the liquid phase. For example, the CO_2-NaOH system most commonly used for comparing the $K_G' a$ values of various tower packings is a liquid-phase-controlled system. When the liquid phase is controlling, extrapolation to different concentration ranges or operating conditions is not recommended since changes in the reaction mechanism can cause k_L to vary unexpectedly and the overall $K_G' a$ values do not explicitly show such effects.

Overall $K_G' a$ data may be obtained from tower-packing vendors for many of the established commercial gas-absorption processes. Such data often are based either upon tests in large-diameter test units or upon actual commercial operating data. Since extrapolation to untried operating conditions is not recommended, the preferred procedure for applying the traditional design method is equivalent to duplicating a previously successful commercial installation. When this is not possible, then a commercial demonstration at the new operating conditions may be required, or else one could consider using some of the more rigorous methods described later.

Aside from the lack of an explicitly defined liquid-phase-resistance term, the limitations on the use of Eq. (14-66) are related to the fact that its derivation implicitly assumes that the system is dilute ($y_{BM} \doteq 1$) and that the operating and equilibrium curves are straight lines over the range of tower operation. Also, Eq. (14-66) is strictly valid only for the temperature and pressure at which the original test was run even though the total pressure p_T appears in the denominator. The ambiguity of the total pressure effect can be seen by a comparison of the gas-phase- and liquid-phase-controlled cases: when the gas phase controls, the liquid-phase resistance is negligible and $K_G a = K_G' a p_T$ is independent of the total pressure. For this case the coefficient $K_G' a$ is inversely proportional to the total system pressure as shown in Eq. (14-66). On the other hand, when the liquid phase controls, the correct equation is

$$K_G' a = K_G a / p_T = k_L a / H \tag{14-68}$$

where H is the Henry's-law constant defined as $H = p_i / x_i$. This equation indicates that $K_G' a$ will be independent of the total system pressure as long as the Henry's-law constant H does not depend on the

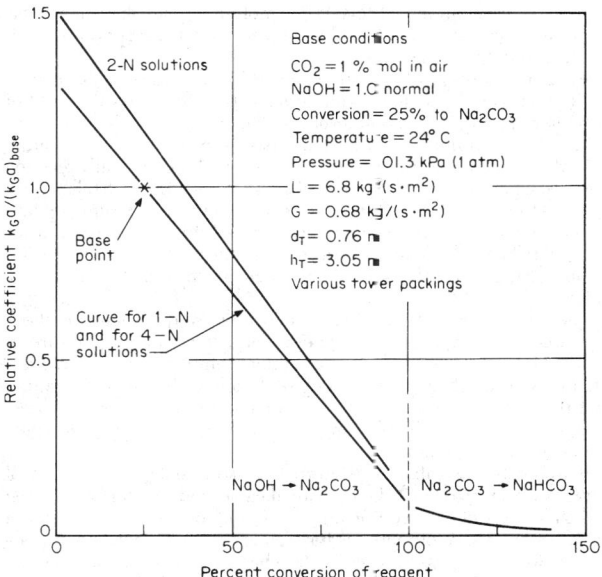

FIG. 14-12 Effects of reagent-concentration and reagent-conversion level upon the relative values of $K_G a$ in the CO_2-NaOH-H_2O system. [*Adapted from Eckert et al.*, Ind. Eng. Chem., **59**(2), 41 (1967)]

total pressure (this will be true only for relatively low pressures). On the basis of this comparison it should be clear that the effects of total system pressure upon $K_G' a$ are not properly defined by Eq. (14-66), especially in cases in which the liquid-phase resistance cannot be neglected.

In using Eq. (14-66), therefore, it should be understood that the numerical values of $K_G' a$ will be a complex function of the pressure, the temperature, the type and size of tower packing employed, the liquid and gas mass flow rates, and the system composition (for example, the degree of conversion of the liquid-phase reactant).

Figure 14-12 illustrates the influence of system composition and **degree of reactant conversion** upon the numerical values of $K_G' a$ for the absorption of CO_2 into sodium hydroxide solutions at constant conditions of temperature, pressure, and type of packing. An excellent experimental study of the influence of operating variables upon overall $K_G' a$ values is that of Field et al. (*Pilot-Plant Studies of the Hot Carbonate Process for Removing Carbon Dioxide and Hydrogen Sulfide*, U.S. Bureau of Mines Bulletin 597, 1962).

Table 14-2 illustrates the observed variations in $\hat{K}_G a$ values for different packing types and sizes for the CO_2-NaOH system at a 25 percent reactant-conversion level for two different liquid flow rates. The lower rate of 2.7 kg/(s·m²) or 2000 lb/(h·ft²) is equivalent to 4 (U.S. gal/min)/ft² and is typical of the liquid rates employed in fume scrubbers. The higher rate of 13.6 kg/(s·m²) or 10,000 lb/(h·ft²) is equivalent to 20 (U.S. gal/min)/ft² and is more typical of absorption towers such as are used in CO_2 removal systems, for example. We note also that two different gas velocities are represented in the table, corresponding to superficial velocities of 0.59 and 1.05 m/s (1.94 and 3.44 ft/s).

Table 14-3 presents a typical range of $\hat{K}_G a$ values for chemically reacting systems. The first two entries in the table represent systems that can be designed by the use of purely physical design methods, for they are completely gas-phase mass-transfer limited. To ensure a negligible liquid-phase resistance in these two tests, the HCl was absorbed into a solution maintained at less than 8 percent weight HCl and the NH_3 was absorbed into a water solution maintained below pH 7 by the addition of acid. The last two entries in Table 14-3 represent liquid-phase mass-transfer limited systems.

The effects of system pressure on these $\hat{K}_G a$ values can be estimated as in Eq. (14-68) by noting that $\hat{K}_G a = K_G a y_{BM}^\circ = K_G' a y_{BM}^\circ p_T$ and recalling that (1) in gas-phase mass-transfer limited systems $\hat{K}_G a = \hat{k}_G a$ and is independent of system pressure, and (2) for liquid-phase mass-transfer limited systems in which H is constant the $\hat{K}_G a$ values can be corrected to other pressures by the relation $\hat{K}_G a$ at $p_2 = (\hat{K}_G a$ at $p_1) \times p_2/p_1$. When both resistances are significant, it is advisable to employ experimentally derived corrections. In any case it is inadvisable to make large pressure corrections by these procedures without experimental verification.

Scaling Up from Laboratory or Pilot-Plant Data For many years it has been thought by practitioners of the art of gas absorption that it would be impossible to carry out an absorption process in a laboratory apparatus or small-scale pilot plant in such a way that the data could be of use in the design of a commercial absorption unit. Indeed, even today most commercial gas-absorption units are designed primarily on the basis of prior commercial experience by using the traditional design methods described previously. Although duplication of a previous commercial design is by far the preferred method, this approach is of little value in developing a completely new process or in attempting to extrapolate an existing design to widely different operating conditions.

Since the early 1960s there have been developed some excellent laboratory experimental techniques, which unfortunately have largely been ignored by the industry. A noteworthy exception was described by Ouwerkerk (*Hydrocarbon Process.*, April 1978, pp. 89–94), in which it was revealed that both laboratory and small-scale pilot-plant data were employed as the basis for the design of an 8.5-m- (28-ft-) diameter commercial Shell Claus off-gas treating (SCOT) plate-type absorber. It is claimed that the cost of developing comprehensive design procedures can be kept to a minimum, especially in the development of a new process, by the use of these modern techniques.

TABLE 14-2 Typical Effects of Packing Type, Size, and Liquid Rate on $\hat{K}_G a$ in a Chemically Reacting System, $\hat{K}_G a$, kmol/(h m³)

Packing size, mm	L = 2.7 kg/(s·m²)				L = 13.6 kg/(s·m²)			
	25	38	50	75–90	25	38	50	75–90
Berl-saddle ceramic	30	24	21		45	38	32	
Raschig-ring ceramic	27	24	21		42	34	30	
Raschig-ring metal	29	24	19		45	35	27	
Pall-ring plastic	29	27	26*	16	45	42	38*	24
Pall-ring metal	37	32	27	21*	56	51	43	27*
Intalox-saddle ceramic	34	27	22	16*	56	43	34	26*
Super-Intalox ceramic	37*		26*		59*		40*	
Intalox-saddle plastic	40*		24*	16*	56*		37*	26*
Intalox-saddle metal	43*	35*	30*	24*	66*	58*	48*	37*
Hy-Pak metal	35	32*	27*	18*	54	50*	42*	27*

Data courtesy of the Norton Company.

Operating conditions: CO_2, 1 percent mole in air; NaOH, 4 percent weight (1 normal); 25 percent conversion to sodium carbonate; temperature, 24°C (75°F); pressure, 98.6 kPa (0.97 atm); gas rate = 0.68 kg/(s·m²) = 0.59 m/s = 500 lb/(h·ft²) = 1.92 ft/s except for values with asterisks, which were run at 1.22 kg/(s·m²) = 1.05 m/s = 900 lb/(h·ft²) = 3.46 ft/s superficial velocity; packed height, 3.05 m (10 ft); tower diameter, 0.76 m (2.5 ft). To convert table values to units of (lb·mol)/(h·ft³), multiply by 0.0624.

TABLE 14-3 Typical $\hat{K}_G a$ Values for Various Chemically Reacting Systems, kmol/(h·m³)

Gas-phase reactant	Liquid-phase reactant	$\hat{K}_G a$	Special conditions
HCl	H_2O	353	Gas-phase limited
NH_3	H_2O	337	Gas-phase limited
Cl_2	NaOH	272	8% weight solution
SO_2	Na_2CO_3	224	11% weight solution
HF	H_2O	152	
Br_2	NaOH	131	5% weight solution
HCN	H_2O	114	
HCHO	H_2O	114	Physical absorption
HBr	H_2O	98	
H_2S	NaOH	96	4% weight solution
SO_2	H_2O	59	
CO_2	NaOH	38	4% weight solution
Cl_2	H_2O	8	Liquid-phase limited

Data courtesy of the Norton Company.
  Operating conditions (see text): 38-mm ceramic Intalox saddles; solute gases, 0.5–1.0 percent mole; reagent conversions = 33 percent; pressure, 101 kPa (1 atm); temperature, 16–24°C; gas rate = 1.3 kg/(s·m²) = 1.1 m/s; liquid rates = 3.4 to 6.8 kg/(s·m²); packed height, 3.05 m; tower diameter, 0.76 m. Multiply table values by 0.0624 to convert to (lb·mol)/(h·ft³).

In 1966, in a paper that now is considered a classic, Danckwerts and Gillham [*Trans. Inst. Chem. Eng.,* **44,** T42 (1966)] showed that data taken in a small stirred-cell laboratory apparatus could be used in the design of a packed-tower absorber when chemical reactions are involved. They showed that if the packed-tower mass-transfer coefficient in the absence of reaction (k_L^0) can be reproduced in the laboratory unit, then the rate of absorption in the laboratory apparatus will respond to chemical reactions in the same way as in the packed column even though the means of agitating the liquid in the two systems might be quite different.

According to this method, it is not necessary to investigate the kinetics of the chemical reactions in detail, nor is it necessary to determine the solubilities or the diffusivities of the various reactants in their unreacted forms. To use the method for scaling up, it is necessary independently to obtain data on the values of the interfacial area per unit volume a and the physical mass-transfer coefficient k_L^0 for the commercial packed tower. Once these data have been measured and tabulated, they can be used directly for scaling up the experimental laboratory data for any new chemically reacting system.

Danckwerts and Gillham did not investigate the influence of the gas-phase resistance in their study (for some processes gas-phase resistance may be neglected). However, in 1975 Danckwerts and Alper [*Trans. Inst. Chem. Eng.,* **53,** 34 (1975)] showed that by placing a stirrer in the gas space of the stirred-cell laboratory absorber, the gas-phase mass-transfer coefficient $\hat{k}_G$ in the laboratory unit could be made identical to that in a packed-tower absorber. When this was done, laboratory data obtained for chemically reacting systems having a significant gas-side resistance could successfully be scaled up to predict the performance of a commercial packed-tower absorber.

If it is assumed that the values of $\hat{k}_G$, k_L^0, and a have been measured for the commercial tower packing to be employed, the procedure for using the laboratory stirred-cell reactor is as follows:
  1. The gas-phase and liquid-phase stirring rates are adjusted so as to produce the same values of $\hat{k}_G$ and k_L^0 as will exist in the commercial tower.
  2. For the reaction system under consideration, experiments are made at a series of bulk-liquid and bulk-gas compositions representing the compositions to be expected at different levels in the commercial absorber (on the basis of a material balance).
  3. The rates of absorption $r_A(c_i, B^0)$ are measured at each pair of gas and liquid compositions.
  For dilute-gas systems one form of the equation to be solved in conjunction with these experimental data is

$$h_T = \frac{G_M}{a} \int_{y2}^{y1} \frac{dy}{r_A} \qquad (14\text{-}69)$$

where h_T = height of commercial tower packing, G_M = molar gas-phase mass velocity, a = effective interfacial area for mass transfer per unit

volume in the commercial tower, y = mole-fraction solute in the gas phase, and r_A = experimentally determined rate of absorption per unit of exposed interfacial area.

By using the series of experimentally measured rates of absorption, Eq. (14-69) can be integrated numerically to determine the height of packing required in the commercial tower.

There are a number of different types of experimental laboratory units that could be used to develop design data for chemically reacting systems. Charpentier [*ACS Symp. Ser.,* **72,** 223–261 (1978)] has summarized the state of the art with respect to methods of scaling up laboratory data and tabulated typical values of the mass-transfer coefficients, interfacial areas, and contact times to be found in various commercial gas absorbers as well as in currently available laboratory units.

The laboratory units that have been employed to date for these experiments were designed to operate at a total system pressure of about 100 kPa (1 atm) and at near-ambient temperatures. In practical situations, it may become necessary to design a laboratory absorption unit that can be operated either under vacuum or at elevated pressures and over a reasonable range of temperatures in order to apply the Danckwerts method.

It would be desirable to reinterpret existing data for commercial tower packings to extract the individual values of the interfacial area a and the mass-transfer coefficients $\hat{k}_G$ and k_L^0 in order to facilitate a more general usage of methods for scaling up from laboratory experiments. Some progress in this direction has already been made, as discussed later in this section. In the absence of such data, it is necessary to operate a pilot plant or a commercial absorber to obtain $\hat{k}_G$, k_L^0, and a as described by Ouwerkerk (op. cit.).

Principles of Rigorous Absorber Design Danckwerts and Alper [*Trans. Inst. Chem. Eng.,* **53,** 34 (1975)] have shown that when adequate data are available for the kinetic-reaction-rate coefficients, the mass-transfer coefficients $\hat{k}_G$ and k_L^0, the effective interfacial area per unit volume a, the physical solubility or Henry's-law constants, and the effective diffusivities of the various reactants, then the design of a packed tower can be calculated from first principles with considerable precision.

For example, the packed-tower design equation for a dilute system in which gas-phase reactant A is being absorbed and reacted with liquid-phase reagent B is

$$r_A a\,dh = \frac{L_M}{\nu\rho_L} dB_h^0 = -G_M\,dy_h \qquad (14\text{-}70)$$

where r_A = specific rate of absorption per unit interfacial area, a = interfacial area per unit volume of packing, h = height of packing, L_M = molar liquid mass velocity, ν = number of moles of B reacting with 1 mol of A, ρ_L = average molar density of liquid phase, B_h^0 = bulk-liquid-phase reagent concentration (a function of h), G_M = molar gas-phase mass velocity, and y_h = mole fraction A in gas phase (a function of h).

For dilute systems it can be assumed that G_M, L_M, and ρ_L are constant, and it normally is assumed that the interfacial area a of the packing is constant and is equal to the value that would exist without reaction. This last assumption needs careful consideration, since different methods for measuring a may give different results. Sharma and Danckwerts [*Br. Chem. Eng.,* **15**(4), 522 (1970)] have reviewed various techniques for measuring interfacial areas.

Under the above assumptions for dilute systems Eq. (14-70) can be integrated as follows:

$$h_T = \frac{L_M}{\nu\rho_L a} \int_{B_1^0}^{B_2^0} \frac{dB_h^0}{r_A} = \frac{G_M}{a} \int_{y2}^{y1} \frac{dy_h}{r_A} \qquad (14\text{-}71)$$

where h_T = total height of packing and the subscripts 1 and 2 refer to the bottom and the top of the tower packing respectively.

The specific absorption rate $r_A = r_A(c_i, B^0)$ is a function of h and may be computed by combining the rate equation

$$r_A = k_L(x_i - x) = (k_L/\rho_L)(c_i - c) \qquad (14\text{-}72)$$

with the material-balance, or operating-curve, equation

$$G_M(y - y_2) = (L_M/\nu\rho_L)(B_2^0 - B^0) \qquad (14\text{-}73)$$

and with the appropriate relation for computing the interfacial concentration x_i of reactant A. In Eq. (14-72) the mass-transfer coefficient k_L is the coefficient with chemical reaction; i.e., $k_L = \phi k_L^0$.

The interfacial concentration x_i is computed by combining the equilibrium relation $y_i = mx_i$ with the equation $k_G(y - y_i) = k_L(x_i - x)$ to obtain

$$x_i = \frac{y/m + (k_L/mk_G)x}{(1 + k_L/mk_G)} \tag{14-74}$$

According to Eq. (14-74), when k_L is very large and the ratio k_L/mk_G is much larger than unity, $x_i - x = yk_G/k_L$ and the specific absorption rate is defined by the equation

$$r_A = k_L(x_i - x) = k_G y \tag{14-75}$$

This is the **gas-phase mass-transfer limited condition,** which can be substituted into Eq. (14-71) to obtain the following equation for calculating the height of packing for a dilute system:

$$h_T = (G_M/k_G a) \ln (y_1/y_2) = H_G \ln (y_1/y_2) \tag{14-76}$$

At the other extreme, when the ratio k_L/mk_G is much smaller than unity, the interfacial concentration of reactant A may be approximated by the equilibrium relation $x_i = y/m$, and the specific absorption rate expression is

$$r_A = k_L(x_i - x) = k_L(y/m - x) \tag{14-77}$$

For **fast chemical reactions** the reactant A is *by definition* completely consumed in the thin film near the liquid interface. Thus, $x = 0$, and

$$r_A = k_L y/m = (k_L/\rho_L)c_i \tag{14-78}$$

This is known as the **liquid-phase mass-transfer limited condition,** as illustrated in Fig. 14-13.

Inspection of Eqs. (14-71) and (14-78) reveals that for fast chemical reactions which are liquid-phase mass-transfer limited the only unknown quantity is the mass-transfer coefficient k_L. The problem of rigorous absorber design therefore is reduced to one of defining the influence of chemical reactions upon k_L. Since the physical mass-transfer coefficient k_L^0 is already known for many tower packings, it

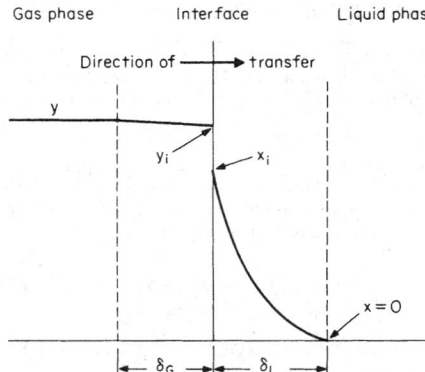

FIG. 14-13 Gas-phase and liquid-phase solute-concentration profiles for a liquid-phase mass-transfer limited reaction system in which N_{Ha} is larger than 3.

often is convenient to work in terms of the ratio k_L/k_L^0 as discussed in the following paragraphs.

Estimation of k_L for Irreversible Reactions Figure 14-14 illustrates the influence of either first- or second-order irreversible chemical reactions on the mass-transfer coefficient k_L as developed by Van Krevelen and Hoftyzer [*Rec. Trav. Chim.*, **67**, 563 (1948)] and as later refined by Perry and Pigford and by Brian et al. [*Am. Inst. Chem. Eng. J.*, **7**, 226 (1961)].

First-order and pseudo-first-order reactions are represented by the upper curve in Fig. 14-14. We note that for first-order reactions when the Hatta number N_{Ha} is larger than about 3, the rate coefficient k_L can be computed by the formula

$$k_L = \sqrt{k_1 D_A} = \sqrt{(k_2 B^0) D_A} \tag{14-79}$$

where k_L = liquid-phase mass-transfer coefficient, k_1 = first-order-reaction-rate coefficient, $k_2 B^0$ = pseudo-first-order-reaction-rate coef-

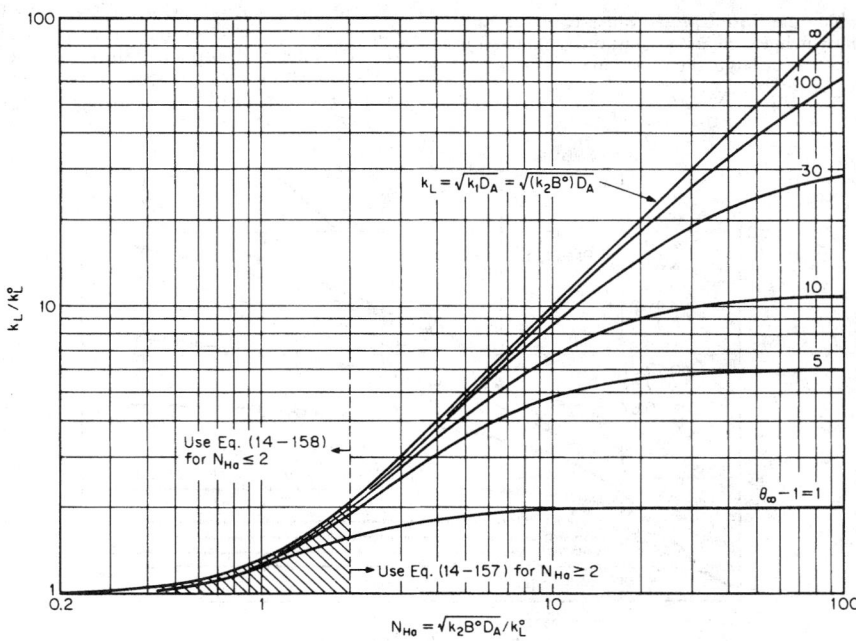

FIG. 14-14 Influence of irreversible chemical reactions on the liquid-phase mass-transfer coefficient k_L. [*Adapted from Van Krevelen and Hoftyzer,* Rec. Trav. Chim., **67**, 563 (1948).]

ficient, and D_A = diffusion coefficient of gaseous reactant A in the liquid phase.

The parameter values for the curves of Fig. 14-14 originally were defined from film theory as $(D_B/D_A)(B^0/vc_i)$ but later were refined by the results of penetration theory to the definition $(\phi_\infty - 1)$, where

$$\phi_\infty = \sqrt{D_A/D_B} + \sqrt{D_B/D_A}(B^0/vc_i) \qquad (14\text{-}80)$$

in which D_B = diffusion coefficient of the liquid-phase reactant B and ϕ_∞ = value of k_L/k_L^0 for large values of N_{Ha} approaching infinity.

For design purposes the entire graph of Fig. 14-14 can be represented by the following pair of equations:

For $N_{Ha} \geq 2$:

$$k_L/k_L^0 = 1 + (\phi_\infty - 1)\{1 - \exp[-(N_{Ha} - 1)/(\phi_\infty - 1)]\} \qquad (14\text{-}81)$$

For $N_{Ha} \leq 2$:

$$k_L/k_L^0 = 1 + (\phi_\infty - 1)\{1 - \exp[-(\phi_\infty - 1)^{-1}]\}\exp[1 - 2/N_{Ha}] \qquad (14\text{-}82)$$

where the Hatta number N_{Ha} is defined as

$$N_{Ha} = \sqrt{k_2 B^0 D_A}/k_L^0 \qquad (14\text{-}83)$$

Equation (14-81) originally was reported by Porter [*Trans. Inst. Chem. Eng.*, **44**(1), T25 (1966)]. Equation (14-82) was derived by the author.

The Van Krevelen-Hoftyzer relationship was tested experimentally for the second-order system in which CO_2 reacts with either NaOH or KOH solutions by Nijsing et al. [*Chem. Eng. Sci.*, **10**, 88 (1959)]. Nijsing's results for the NaOH system are shown in Fig. 14-15 and are in excellent agreement with the second-order-reaction theory. Indeed, these experimental results can be described very well by Eqs. (14-80) and (14-81) when values of $v = 2$ and $D_A/D_B = 0.64$ are employed in the equations.

For fast irreversible chemical reactions, therefore, the principles of rigorous absorber design can be applied by first establishing the effects of the chemical reaction on k_L and then employing the appropriate material-balance and rate equations in Eq. (14-71) to perform the integration to compute the required height of packing.

For an isothermal absorber involving a dilute system in which a **liquid-phase mass-transfer limited** first-order irreversible chemical reaction is occurring, the packed-tower design equation is derived as

$$h_T = (mG_M/\sqrt{k_1 D_A}a)\ln(y_1/y_2) \qquad (14\text{-}84)$$

For a dilute system in which the **liquid-phase mass-transfer limited** condition is valid, in which a very fast second-order reaction is involved, and for which N_{Ha} is very large, the equation

$$k_L/k_L^0 = \phi_\infty = \sqrt{D_A/D_B} + \sqrt{D_B/D_A}(B^0/vc_i) \qquad (14\text{-}85)$$

is valid and results in the following equation for computing the height of packing in a packed tower:

$$h_T = \frac{mG_M}{k_L^0 a}\sqrt{\frac{D_B}{D_A}}\int_{y_2}^{y_1}\frac{dy}{\dfrac{mB^0 D_B}{v\rho_L D_A} + y} \qquad (14\text{-}86)$$

Evaluation of the integral in Eq. (14-86) requires a knowledge of the liquid-phase bulk concentration of B as a function of y. This relationship is obtained by means of a material balance around the tower, as shown in Eq. (14-73). Numerical integration by a quadrature method such as Simpson's rule normally will be required for this calculation.

Estimation of k_L for Reversible Reactions   When the reaction is of the form $A \wedge B$, where B is a nonvolatile product and the equilibrium constant is defined by $c_B = K_{eq}c_A$, the expressions for computing k_L become extremely complex. A good discussion of this situation is given in *Mass Transfer* by Sherwood, Pigford, and Wilke (McGraw-Hill, New York, 1975, p. 317). Three limiting cases are listed below:

1. For very slow reactions,
$$\lim k_L = k_L^0$$
$$k_1 \to 0 \qquad (14\text{-}87)$$

2. For extremely fast reactions where K_{eq} is very large,
$$\lim k_L = \sqrt{k_1 D_A}$$
$$k_1 \to \infty$$
$$K_{eq} = \infty \qquad (14\text{-}88)$$

3. For extremely fast reactions where K_{eq} is finite,
$$\lim k_L = (1 + K_{eq})k_L^0$$
$$k_1 \to \infty$$
$$K_{eq} = \text{finite} \qquad (14\text{-}89)$$

When one of these three conditions is applicable, the appropriate design equation can be obtained by substitution into Eq. (14-71), followed by integration of the resulting relationship.

Some more complex situations involving reversible reactions are discussed in *Mass Transfer* (ibid., pp. 336–343).

Simultaneous Absorption of Two Reacting Gases   In multicomponent physical absorption the presence of one gas often does not affect the rates of absorption of the other gases. When chemical reactions in which two or more gases are competing for the same liquid-phase reagent are involved, selectivity of absorption can be affected by

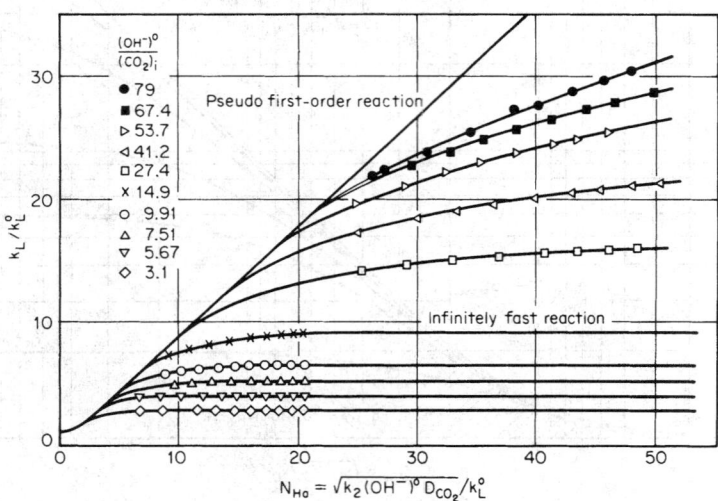

FIG. 14-15 Experimental values of k_L/k_L^0 for absorption of CO_2 into NaOH solutions at 20°C. [*Data of Nijsing et al.,* Chem. Eng. Sci., **10**, 88 (1959).]

the choice of design conditions, and the situation may become extremely complex from a designer's point of view.

The classic work on this subject is that of Ramachandran and Sharma [*Trans. Inst. Chem. Eng., **49**,* 253 (1971)] and is recommended to those needing further details. The following references also are offered as a sampling of the literature on the subject:

• *CO₂ and H₂S.* Danckwerts and Sharma, *Chem. Eng. (London),* CE244–280 (October 1966); Onda, et al., *J. Chem. Eng. Japan,* **5,** 27 (1972); Rivas and Prausnitz, *Am. Inst. Chem. Eng. J.,* **25,** 975 (1979).

• *CO₂ and SO₂.* Goettler and Pigford, *Inst. Chem. Eng. Symp. Ser.,* **28,** 1 (1968); Teramoto et al., *Int. Chem. Eng.,* **18,** 250 (1978).

• *SO₂ and NO₂.* Takeuchi and Yamanaka, *Ind. Eng. Chem. Process Des. Dev.,* **17,** 389 (1978).

Desorption with Chemical Reaction When chemical reactions are involved in a stripping operation, the design problem can become extremely complex. In fact, much less is known about this very important process than is known about absorption. A classic work on this subject is that of Shah and Sharma [*Trans. Inst. Chem. Eng.,* **54,** 1 (1976)], which is recommended to those in need of more details.

In the subsection "Design of Gas-Absorption Systems" it was stated that more often than not the liquid-phase residence time and, hence, the liquid holdup are considered to be the most important design parameters for stripping towers. If Eq. (14-60) is redefined to represent the stripping process for an extremely slow liquid-phase reaction for which $R_A = k_1 c$, then one finds that the liquid holdup will be a factor only when the ratio $k_1 \rho_L f_H / k_L^0 a$ is less than unity. Thus, one can ensure that the liquid-phase reaction rate is not limiting by increasing the temperature and the liquid holdup until this ratio is equal to or greater than unity. The preferred method at present is to base the design on prior commercial experience.

Use of Literature for Specific Systems A large body of experimental data obtained in bench-scale laboratory units and in small-diameter packed towers has been published since the early 1940s. One might wish to consider using such data for a particular chemically reacting system as the basis for scaling up to a commercial design. Extreme caution is recommended in interpreting such data for the purpose of developing commercial designs, as extrapolations of this kind of information can lead to serious errors. Extrapolation to temperatures, pressures, or liquid-phase reagent conversions different from those that were employed by the original investigator definitely should be regarded with caution.

Bibliographies presented in the general references listed at the beginning of this section are an excellent source of information on specific chemically reacting systems. *Gas-Liquid Reactions* by Danckwerts (McGraw-Hill, New York, 1970) contains a tabulation of references to specific chemically reacting systems. *Gas Treating with Chemical Solvents* by Astarita et al. (Wiley, New York, 1983) deals with the absorption of acid gases, and includes an extensive listing of patents. *Gas Purification* by Kohl and Riesenfeld (Gulf Publishing, Houston, 1985) presents data and references for many chemically reacting systems of current commercial interest.

In searching for data on a particular system, a computerized search of *Chemical Abstracts, Engineering Index,* and National Technical Information Service (NTIS) data bases should seriously be considered. Although the NTIS computer contains only information published after 1970, one normally can assume that most pre-1970 publications of merit likely will be referenced in the bibliographies of current articles on the subject.

The experimental data for the system CO_2-$NaOH$-Na_2CO_3 are unusually well known as the result of the work of many experimenters. A serious study of the data and theory for this system therefore is recommended as the basis for developing a good understanding of the kind and quality of experimental information that is needed for design purposes.

In addition to data on CO_2, information can readily be found in the literature for the following systems: O_2, Cl_2, NH_3, NO_2, NO, SO_2, SO_3, H_2S, COS, CS_2, HCl, HBr, HCN, H_2, $COCl_2$, PCl_3, olefins, dienes, and water vapor.

GAS-LIQUID CONTACTING SYSTEMS

Gas-liquid contacting systems are utilized for transferring mass, heat, and momentum between the phases, subject to constraints of physical and chemical equilibrium. Process equipment for such systems is designed to achieve the appropriate transfer operations with a minimum expenditure of energy and capital investment.

In this section emphasis is placed on the transfer of mass. Typical gas-liquid mass-transfer systems are:

Distillation	Evaporation
Flashing	Humidification
Rectification	Dehumidification
Absorption	Dephlegmation
Stripping	Spray drying

Distillation is the separation of the constituents of a liquid mixture via partial vaporization of the mixture and separate recovery of vapor and residue. The process of vaporization is generally of a differential nature.

Flashing is a distillation process in which the total vapor removed approaches phase equilibrium with the residue liquid.

Rectification is the separation of the constituents of a liquid mixture by successive distillations (partial vaporizations and condensations) and is obtained via the use of an integral or differential process. Separations into effectively pure components may be obtained through this procedure.

Stripping or desorption is the transfer of gas, dissolved in a liquid, into a gas stream. The term is also applied to that section of a fractionating column below the feed plate.

Absorption is the transfer of a soluble component in a gas-phase mixture into a liquid absorbent whose volatility is low under process conditions.

Evaporation generally refers to the removal of water, by vaporization, from aqueous solutions of nonvolatile substances.

Humidification and dehumidification refer to the transfer of water between a gas stream and a water stream.

Dephlegmation, or partial condensation, refers to the process in which a vapor stream is cooled to a desired temperature such that a portion of the less volatile components of the stream is removed from the vapor by condensation.

Spray drying is an extension of the evaporative process in which almost all the liquid is removed from a solution of a nonvolatile solid in the liquid.

All these processes are, in common, liquid-gas mass-transfer operations and thus require similar treatment from the aspects of phase equilibrium and kinetics of mass transfer. The fluid-dynamic analysis of the equipment utilized for the transfer also is similar for many types of liquid-gas process systems.

Process equipment utilized for liquid-gas contacting is based on a combination of operating principles of the three categories:

Mode of flow of streams
   Countercurrent
   Cocurrent
   Cross-flow
Gross mechanism of transfer
   Differential
   Integral
Continuous phase
   Gas°
   Liquid

° In this section the terms "gas" and "vapor" are used interchangeably. The latter is often used in distillation, in which the gas phase is represented by an equilibrium vapor.

TABLE 14-4   Characteristics of Liquid-Gas Systems

Equipment designation	Mode of flow	Gross mechanism	Continuous phase	Primary process applications
Plate column	Cross-flow, countercurrent	Integral	Liquid and/or gas	Absorption, rectification, stripping
Packed column	Countercurrent, cocurrent	Differential	Liquid and/or gas	Absorption, rectification, stripping, humidification, dehumidification
Wetted-wall (falling-film) column	Countercurrent, cocurrent	Differential	Liquid and/or gas	Absorption, rectification, stripping, evaporation
Spray chamber	Cocurrent, cross-flow, countercurrent	Differential	Gas	Absorption, stripping, humidification, dehumidification
Heat exchanger	Cocurrent, countercurrent	Differential	Gas	Evaporation, dephlegmation
Agitated vessel	Complete mixing	Integral	Liquid	Absorption
Line mixer	Cocurrent	Differential	Liquid or gas	Absorption, stripping

The combination of these characteristics utilized in the various types of process equipment is indicated in Table 14-4.

PLATE COLUMNS

Plate Types   Plate columns utilized for liquid-gas contacting may be classified according to mode of flow in their internal contacting devices:
1. Cross-flow plates
2. Counterflow plates

The cross-flow plate (Fig. 14-16a) utilizes a liquid downcomer and is more generally used than the counterflow plate (Fig. 14-16b) because of transfer-efficiency advantages and greater operating range. The liquid-flow pattern on a cross-flow plate can be controlled by placement

of downcomers in order to achieve desired stability and transfer efficiency. Commonly used flow arrangements are shown in Fig. 14-17. A guide for the tentative selection of flow pattern is given in Table 14-5.

It should be noted that the fraction of column cross-sectional area available for gas dispersers (perforations, bubble caps) decreases when more than one downcomer is used. Thus, optimum design of the plate involves a balance between liquid-flow accommodation and effective use of cross section for gas flow.

Most new designs of cross-flow plates employ perforations for dispersing gas into liquid on the plate. These perforations may be simple

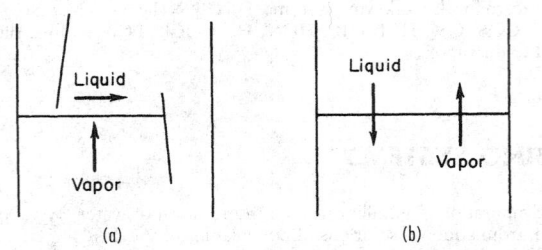

FIG. 14-16   (a) Cross-flow plate (side view). (b) Countercurrent plate (side view).

TABLE 14-5   Selection of Cross-Flow-Plate Flow Pattern*

Estimated tower diam., ft	Range of liquid capacity, gal/min			
	Reverse flow	Cross-flow	Double pass	Cascade double pass
3	0–30	30–200		
4	0–40	40–300		
6	0–50	50–400	400–700	
8	0–50	50–500	500–800	
10	0–50	50–500	500–900	900–1400
12	0–50	50–500	500–1000	1000–1600
15	0–50	50–500	500–1100	1100–1800
20	0–50	50–500	500–1100	1100–2000

*Bolles, chap. 14 in Smith, *Design of Equilibrium Stage Processes*, McGraw-Hill, New York, 1963. To convert feet to meters, multiply by 0.3048; to convert gallons per minute to decimeters per second (liters per second), multiply by 0.06309; and to convert gallons per minute to cubic meters per second, multiply by 6.309×10^{-5}.

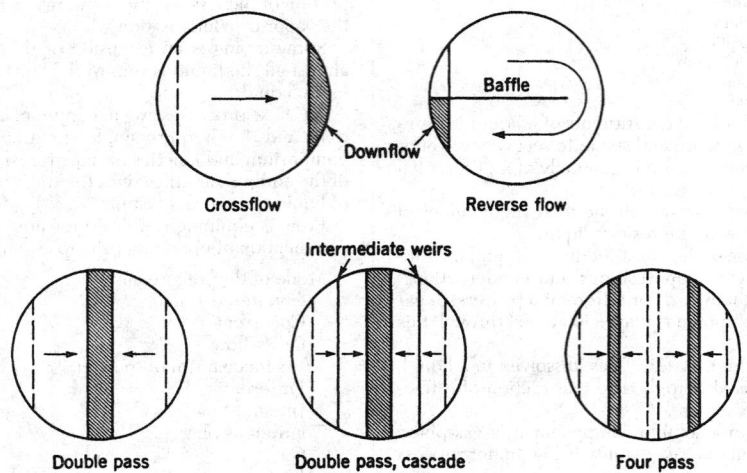

FIG. 14-17   Common liquid-flow patterns, cross-flow plates. (*Smith, Design of Equilibrium Stage Processes, McGraw-Hill, New York, 1963.*)

round orifices, or they may contain movable "valves" that provide variable orifices of noncircular shape. These perforated plates are called sieve plates (Fig. 14-18) or valve plates (Fig. 14-19). For sieve plates, liquid is prevented from flowing through the perforations by the flowing action of the gas; thus, when the gas flow is low, it is possible for some or all of the liquid to drain through the perforations and in effect bypass portions of the contacting zone. The valve plate is designed to minimize this drainage, or "weeping," since the valve tends to close as the gas flow becomes lower, the total orifice area varying to maintain a dynamic-pressure balance across the plate

Historically the most common gas disperser for cross-flow plates has been the bubble cap. This device has a built-in seal which prevents liquid drainage at low gas-flow rates. Typical bubble caps are shown in Fig. 14-20. Gas flows up through a center riser, reverses flow under the cap, passes downward through the annulus between riser and cap, and finally passes into the liquid through a series of openings, or "slots," in the lower side of the cap.

Bubble caps were used almost exclusively as cross-flow-plate dispersers until about 1950, when they were largely displaced by simple or valve-type perforations. Many varieties of bubble-cap design were used (and therefore are extant in many operating columns), but in most cases bell caps of 75- to 150-mm (3- to 6-in) diameter were utilized.

In counterflow plates, liquid and gas utilize the same openings for flow. Thus, there are no downcomers. Openings are usually simple round perforations in the 3- to 13-mm (⅛- to ½-in) range (dual-flow plate) or long slots with widths of 6 to 13 mm (¼ to ½ in) (Turbogrid tray). The plate material can be corrugated (Ripple tray) to segregate partially gas and liquid flow. In general, gas and liquid flow in a pulsating fashion with a particular opening passing both gas and liquid in an intermittent fashion.

A counterflow plate often used for contacting gases with liquids containing solids is the baffle plate, or "shower deck" (Fig. 14-21).

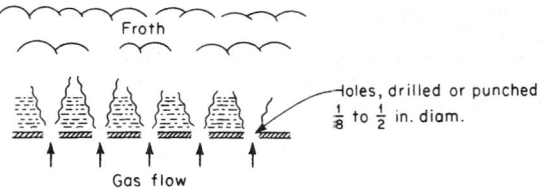

FIG. 14-18   Sieve-plate dispersers. To convert inches to millimeters, multiply by 25.4.

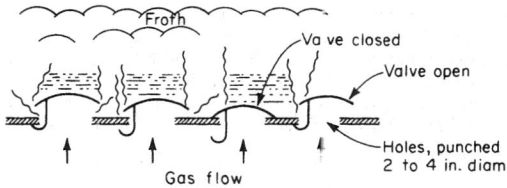

FIG. 14-19   Valve-plate dispersers. To convert inches to millimeters, multiply by 25.4.

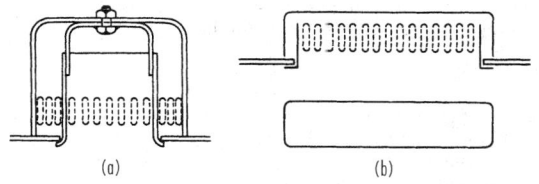

FIG. 14-20   (*a*) Circular or bell cap. (*b*) Tunnel cap.

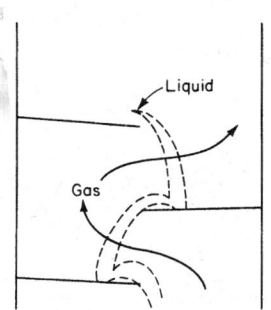

FIG. 14-21   Baffle plate (shower deck).

Typically the plate is half-moon in shape and is sloped slightly in the direction of liquid flow. Gas contacts the liquid as it showers from the plate, and a serrated lip or weir at the edge of the plate can be used to improve the distribution of liquid in the shower.

The baffle plate operates with liquid dispersed and gas as the continuous phase and is used primarily in heat-transfer applications.

In summary, the perforated plate with liquid cross-flow (the "sieve plate") is the most common type specified for new designs. Schematic diagrams of such a plate are shown in Fig. 14-22. Nomenclature items are shown, with heights h_{li}, h_f, h_{lo}, and h_l referring to liquid entering, froth, liquid + froth leaving, and equivalent clear liquid averaged across the plate. For the plan view, area terms are as follows: A_t = tower total cross section; A_a = active area; A_d = area of one downcomer; A_n = net area for vapor flow (usually total cross section minus blocking downcomers); and A_h = area of holes or perforations. For the single cross-flow plate shown,

$$A_t = A_a + 2A_d$$

$$A_n = A_a + A_d = A_t - A_d$$

When downcomers are sloped or when perforations do not occupy essentially all the area between the downcomers, these simple relations do not apply. However, their adaptation should be obvious from the geometry involved.

The term "froth" in Fig. 14-22 suggests aeration in which the liquid phase is continuous. Under certain conditions there can be an inversion to a gas-continuous regime, or "spray." The spray has its phase boundaries equivalent to the boundaries for froth shown in Fig. 14-22.

Plate-Column Capacity   The maximum allowable capacity of a plate for handling gas and liquid flow is of primary importance because it fixes the minimum possible diameter of the column. For a constant liquid rate, increasing the gas rate results eventually in excessive entrainment and flooding. At the flood point it is difficult to obtain net downward flow of liquid, and any liquid fed to the column is carried out with the overhead gas. Furthermore, the column inventory of liquid increases, pressure drop across the column becomes quite large, and control becomes difficult. Rational design calls for operation at a safe margin below this maximum allowable condition.

Flooding may also be brought on by increasing the liquid rate while holding the gas rate constant. Excessive liquid flow can overtax the capacity of downcomers or other passages, with the ultimate result of increased liquid inventory, increased pressure drop, and the other characteristics of a flooded column.

These two types of flooding are usually considered separately when a plate column is being rated for capacity. For identification purposes they are called entrainment flooding (or "priming") and downflow flooding. When counterflow action is destroyed by either type, transfer efficiency is lost and reasonable design limits have been exceeded.

Minimum allowable capacity of a column is determined by the need for effective dispersion and contacting of the phases. The types of plates differ in their ability to permit low flows of gas and liquid. A cross-flow sieve plate can operate at reduced gas flow down to a point where liquid drains through the perforations and gas dispersion is inadequate for good efficiency. Valve plates can be operated at very

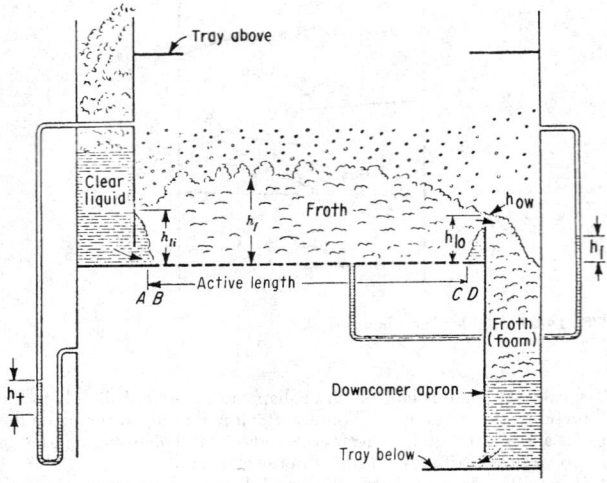

Elevation view

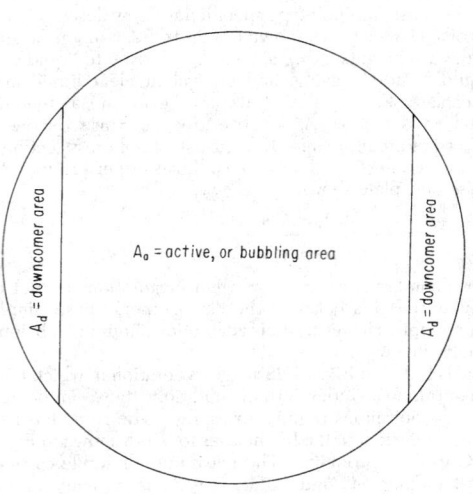

Plan view

FIG. 14-22 Sieve-plate diagram. (*Smith*, *Design of Equilibrium Stage Processes. McGraw-Hill, New York, 1963.*)

low gas rates because of valve closing. Bubble-cap plates can be operated at very low gas rates because of their seal arrangement. All devices have a definite minimum gas rate below which there is inadequate dispersion for intimate contacting. Similarly, there are minimum liquid flows below which good distribution is not possible, although the reverse-flow plate (Fig. 14-17) can accommodate extremely low liquid flows.

For all plate devices a qualitative capacity diagram is shown in Fig. 14-23. The shape and extent of the satisfactory operating zone in Fig. 14-23 vary according to type of plate device. As a specific example, Fig. 14-24 shows actual test data for two cross-flow plates in distillation service at total reflux. The abscissa parameter is called the *F*-factor and is a vapor kinetic energy term. The decline in efficiency of the sieve plate at low *F*-factors is evident and is the result of liquid falling through some of the holes ("weeping"). The decline in efficiency of both devices at high *F*-factors results from liquid entrainment. For new designs, it is the objective of the designer to predict the likely location of performance curves such as those shown.

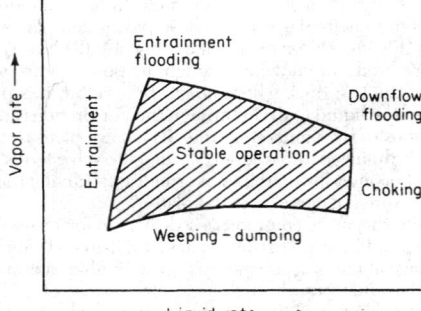

FIG. 14-23 Stable operating region, plates. (*Smith*, *Design of Equilibrium Stage Processes, McGraw-Hill, New York, 1963.*)

Entrainment Flooding The early work of Souders and Brown [*Ind. Eng. Chem.*, **26**, 98 (1934)] based on a force balance on an average suspended droplet of liquid led to the definition of a capacity parameter C_{sb}:

$$C_{sab} = U_n \sqrt{\rho_L/(\rho_t - \rho_g)} \qquad (14\text{-}90)$$

where U_n = linear gas velocity based on net area A_n, m/s
ρ_G = gas density, kg/m^3
ρ_L = liquid density, kg/m^3

For cross-flow plates, net area is the column cross section less that area blocked by the downcomer or downcomers (Fig. 14-22). The vapor velocity in the net area represents an approach velocity and thus controls the level of liquid entrainment. For counterflow plates, net area is the same as the column cross section, since no downcomers are involved.

Maximum allowable values of the capacity parameter are for a flooding condition and are designated C_{sbf}. Experimental values have been correlated against a dimensionless flow parameter F_{LG} as shown in Fig. 14-25. The flow parameter represents a ratio of liquid to vapor kinetic energies:

$$F_{LG} = \frac{L}{G} \left(\frac{\rho_G}{\rho_L} \right)^{0.5} \qquad (14\text{-}91)$$

Low values of F_{LG} indicate vacuum operation, high values indicate operation at higher ressures or at high liquid/vapor loadings as in gas absorption. The liquid/gas ratio L/G is based on mass flow rates. The parameter serves as a criterion for two-phase flow characteristics on the plate, as discussed by Hofhuis and Zuiderweg [*Inst. Chem. Engrs. Symp. Ser. No. 56*, 2.2-1 (1979)]. Notations on Fig. 14-25 are from this source. The correlation in Fig. 14-25 is intended to cover the full range of flow parameters, with the low values of C_{sbf} to the right likely to result from downcomer flow restrictions rather than excessive entrainment. The curves may be expressed in equation form as [Lygeros and Magoulas, *Hydrocarbon Proc.* **65**(12), 43 (1986)]:

$$C_{sbf} = 0.0105 + 8.127 \, (10^{-4})(TS^{0.755}) \exp[-1.463 \, F_{LG}^{0.842}] \qquad (14\text{-}92)$$

where TS = plate spacing, mm.

Figure 14-25 or Eq. (14-92) may be used for sieve plates, valve plates, or bubble-cap plates. The value of the flooding vapor velocity must be considered as approximate, and prudent designs call for approaches to flooding of 75 to 85 percent. The value of the capacity parameter (ordinate term in Fig. 14-25) may be used to calculate the maximum allowable vapor velocity through the net area of the plate:

$$U_{nf} = C_{sb,\text{flood}} \left(\frac{\sigma}{20} \right)^{0.2} \left(\frac{\rho_L - \rho_g}{\rho_g} \right)^{0.5} \qquad (14\text{-}93)$$

where U_{nf} = gas velocity through net area at flood, m/s
C_{sbf} = capacity parameter, m/s
σ = liquid surface tension, mN/m (dyn/cm)
ρ_L = liquid density, kg/m^3
ρ_G = gas density, kg/m^3

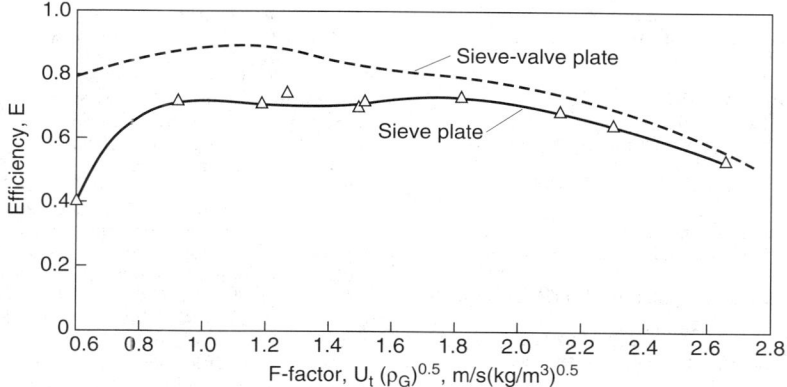

FIG. 14-24   Performance of two crossflow plates operating at 0.13 bar pressure and total reflux. Test mixture: ethylbenzene/styrene. Spacing between plates is 0.50 m, and outlet weir height is 38 mm. U_t = superficial vapor velocity, ρ_G = vapor density. [*Billet, Conrad, and Grubb, I. Chem. E. Symp. Ser. No. 32, 5, 111 (1969).*]

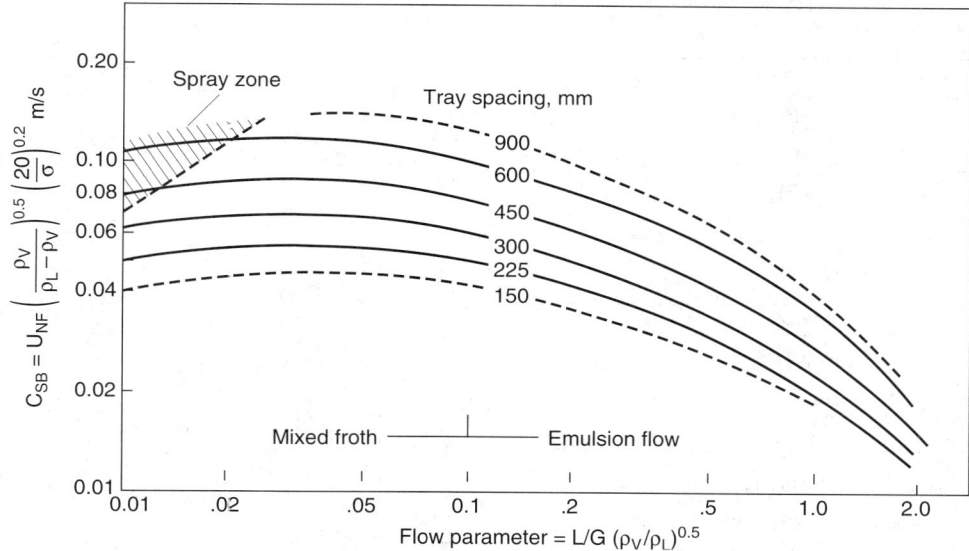

FIG. 14-25   Flooding correlation for columns with crossflow plates (sieve, valve, bubble-cap). [*Fair, Pet/Chem Eng 33(10), 45 (September 1961).*]

Figure 14-25 gives flooding-gas velocities to ±10 percent subject to the following restrictions:

1. System is low or non-foaming
2. Weir height is less than 15 percent of plate spacing
3. Sieve-plate perforations are 13 mm (½ in) or less in diameter
4. Ratio of slot (bubble cap), perforation (sieve), or full valve opening (valve plate) area A_h to active area A_a is 0.1 or greater. Otherwise the value of U_{nf} obtained from Fig. 14-25 should be corrected:

A_h/A_a	$U_{nf}/U_{nf,\text{Fig. 14-25}}$
0.10	1.00
008	090
0 06	0.80

where A_h = total slot, perforated, or open-valve area on plate.

For counterflow plates, the curves of Fig. 14-25 may be used for open areas of 20 percent or greater. Plates with 15 percent open areas have about 85 percent of the curve values, and open areas of less than 15 percent are not recommended. For counterflow-plate columns of the segmental-baffle type, 50 percent cut, allowable C_{sbf} values are about 15 percent greater than those shown in Fig. 14-25, when vertical spacings of the baffles are equal to the tray spacings shown.

An alternate method for predicting the flood point of sieve and valve plates has been reported by Kister and Haas [*Chem. Eng. Progr.,* **86**(9), 63 (1990)] and is said to reproduce a large data base of measured flood points to within ±30 percent. It applies to entrainment flooding only (values of F_{LG} less than about 0.5). The general predictive equation is

$$C_{sbf} = 0.0277(d_h^2\sigma/\rho_L)^{0.125}(\rho_G/\rho_L)^{0.1}(TS/h_{cl})^{0.5} \qquad (14\text{-}94)$$

where d_h = hole diameter, mm
σ = surface tension, mN/m (dyn/cm)
ρ_G, ρ_L = vapor and liquid densities, kg/m³

TS = plate spacing, mm
h_{cl} = clear liquid height at the froth-to-spray transition, mm; obtained from:

$$h_{cl} = h_{cl,H2O} \, (996/\rho_L)^{0.5(1-n)} \qquad (14\text{-}95)$$

$$h_{cl,H2O} = \frac{0.497 \, A_f^{-0.791} \, d_h^{0.833}}{1 + 0.013 \, L^{-0.59} \, A_f^{-1.79}} \qquad (14\text{-}96)$$

$$n = 0.00091 \, d_h/A_f \qquad (14\text{-}97)$$

In Eq. 14-96, L = m³ liquid downflow/(hr-m weir length) and A_f = fractional hole area based on active ("bubbling") area; for instance, $A_f = A_h/A_a$.

For valve trays, adaptations of Eqs. (14-94) to (14-97) are required:

$$d_h = \frac{4 \times (\text{area of opening of one fully open valve})}{\text{wetted perimeter of opening of one fully open valve}} \qquad (14\text{-}98)$$

$$A_f = \frac{\text{no. valves} \times (\text{area of opening of one fully open valve})}{\text{active (bubbling) area}} \qquad (14\text{-}99)$$

Example 9: Loading/Flooding of a Distillation Plate An available sieve plate column of 2.5-m diameter is being considered for an ethylbenzene/styrene separation. An evaluation of loading at the top plate will be made. Key dimensions of the single-crossflow plate are:

Column cross section, m²	4.91
Downcomer area, m²	0.25
Net area, m²	4.66
Active area, m²	4.41
Hole area, m²	0.617
Hole diameter, mm	4.76
Weir length, m	1.50
Weir height, mm	38
Plate spacing, mm	500

Conditions and properties at the top plate are:

Temperature, °C	78
Pressure, torr	100
Vapor flow, kg/h	25,500
Vapor density, kg/m³	0.481
Liquid flow, kg/h	22,000
Liquid density, kg/m³	841
Surface tension, mN/m	25

The dimensions and flow rates are scaled to represent the conditions shown in Fig. 14-24.

Solution. The flow parameter $F_{LG} = 0.021$ (Eq. 14-73). From Fig. 14-25, $C_{sbf} = 0.095$ m/s. Then,

$$U_{nf} = \frac{0.095}{(0.481/841)^{0.5} (20/25)^{0.2}} = 4.15 \text{ m/s}$$

This gives a superficial F-factor at flood = $U_t \rho_G^{0.5}$ = 4.15 (4.66/4.91)(.481)0.5 = 2.73 m/s(kg/m³)0.5, or about 90 percent of a value extrapolated from Fig. 14-24.

The alternate method of Kister and Haas may be applied to the same problem:

$$L = \frac{22,000}{841 \times 1.50} = 17.44 \text{ m}^3/\text{h-m weir}$$

$$A_f = \frac{0.617}{4.41} = 0.14$$

By Eq. 14-96, $h_{cl,H2O} = 7.98$ mm

Eq. 14-97: $n = 0.0309$

Eq. 14-95: $h_{cl} = 8.66$ mm

Finally, by Eq. 14-94,

$$C_{sbf} = 0.0277[(4.762)(25/841)]^{0.125} \times (0.481/841)^{0.1}(500/8.66)^{0.5}$$

$$= 0.0947 \text{ m/s}$$

about the same as the answer obtained from Fig. 14-25.

For the design condition, F-factor is 2.08 m/s(kg/m³)0.5, or about 76 percent of flood. The proposed column is entirely adequate for the service required.

Entrainment Entrainment in a plate column is that liquid which is carried with the vapor from a plate to the plate above. It is detrimental in that the effective plate efficiency is lowered because liquid from a plate of lower volatility is carried to a plate of higher volatility, thereby diluting distillation or absorption effects. Entrainment is also detrimental when nonvolatile impurities are carried upward to contaminate the overhead product from the column.

Many experimental studies of entrainment have been made, but few of them have been made under actual distillation conditions. The studies are often questionable because they are limited to the air-water system, and they do not use a realistic method for collecting and measuring the amount of entrainment. It is clear that the dominant variable affecting entrainment is gas velocity through the two-phase zone on the plate. Mechanisms of entrainment generation are discussed in the subsection "Liquid-in-Gas Dispersions."

For distillations, it is often of more interest to ascertain the effect of entrainment on efficiency than to predict the quantitative amount of liquid entrained. For this purpose, the correlation shown in Fig. 14-26 is useful. The parametric curves in the figure represent approach to the entrainment flood point as measured or as predicted by Fig. 14-25 or some other flood correlation. The abscissa values are those of the flow parameter discussed earlier. The ordinate values ψ are fractions of gross liquid downflow, defined as follows:

$$\psi = \frac{e}{L + e} \qquad (14\text{-}100)$$

where e = absolute entrainment of liquid, moles/time
L = liquid downflow rate without entrainment, moles/time

Figure 14-26 also accepts the validity of the Colburn equation [*Ind. Eng. Chem.*, **28**, 526 (1936)] for the effect of entrainment on efficiency:

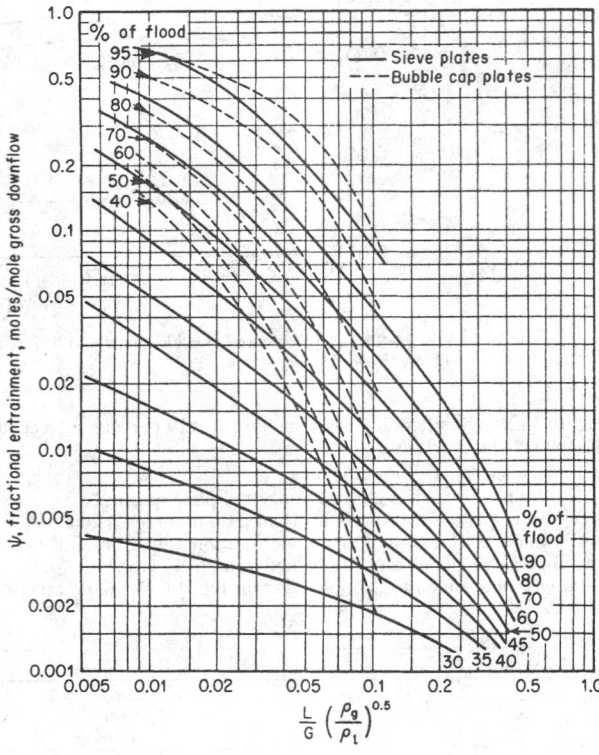

FIG. 14-26 Entrainment correlation. L/G = liquid-gass mass ratio; and ρ_l and ρ_g = liquid and gas densities. [*Fair, Pet./Chem. Eng.*, **33**(10), 45 (*September 1961*).]

$$\frac{E_a}{E_{mv}} = \frac{1}{1 + E_{mv}\left[\psi/(1 - \psi)\right]} \tag{14-101}$$

where E_{mv} = Murphree vapor efficiency [see Eq. (14-28)]
E_a = Murphree vapor efficiency, corrected for recycle of liquid entrainment

The Colburn equation is based on complete mixing on the plate. For incomplete mixing, e.g., liquid approaching plug flow on the plate, Rahman and Lockett [*I. Chem. E. Symp. Ser. No. 61*, 111 (1981)] and Lockett et al. [*Chem. Eng. Sci.*, **38**, 661 (1983)] have provided corrections for Eq. 14-44. Figure 14-26 and Eq. 14-94 may be used to evaluate the effects of entrainment on efficiency.

Example 10: Entrainment Effect on Plate Efficiency For the data shown in Fig. 14-24, estimate the efficiency of the sieve plate at a superficial F-factor of 2.6 m/s(kg/m³).

The data show a midrange dry efficiency of 0.7 (70 percent). They indicate a flood F-factor value of about 3.0. Thus, the approach to flood is 2.6/3.0 = 0.87 (87 percent). The data were taken at total reflux, and thus $F_{LG} = (0.481/841)0.5 = 0.024$ (densities taken from Example 9). From Fig. 14-26, $\psi = 0.19$, and from Eq. (14-76):

$$\frac{E_a}{E_{mv}} = \frac{1}{1 + 0.70[0.19/(1 - 0.19)]} = 0.87$$

Thus, the wet efficiency $E_a = 0.87(0.70) = 0.61$ (61 percent). Figure 14-24 shows that for $F = 2.6$, the measured efficiency is 0.55 (55 percent).

Weeping Liquid flow through sieve-plate perforations occurs when the gas pressure drop through the perforations is not sufficient to create bubble surface and support the static head of froth above the perforations. Weeping can be deleterious in that liquid tends to short-circuit the primary contacting zones. On the other hand, some mass transfer to and from the weeping liquid occurs. Usual practice is to design so that deleterious weeping does not occur, based on a correlation such as that shown in Fig. 14-27.

In Fig. 14-27, h_d = head loss to gas flow through perforations, mm liquid [see Eq. (14-107)], and h_e = head loss due to bubble formation, mm liquid. The latter loss is based on the energy required for bubble formation:

$$h_b = \frac{4\sigma}{d_h} \tag{14-102}$$

with a convenient dimensional form for use in Fig. 14-27 being:

$$h_\sigma = 409\left(\frac{\sigma}{\rho_L d_h}\right) \tag{14-103}$$

where σ = surface tension, mN/m
ρ_L = liquid density, kg/m³
d_h = diameter of a perforation, mm
h_e = head loss due to bubble formation, mm liquid

If design shows a condition *above* the appropriate curve of Fig. 14-27, weeping will not be deleterious to plate performance as measured by a drop in plate efficiency (as in Fig. 14-24 for the sieve plate).

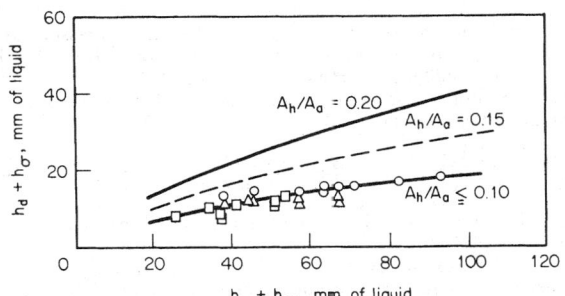

FIG. 14-27 Weeping, sieve plates. To convert millimeters to inches, multiply by 0.0394. (*Smith, Design of Equilibrium Stage Processes, McGraw-Hill, New York, 1963.*)

Downflow Flooding Columns can flood because of their inability to handle large quantities of liquid. For crossbow plates this limit on liquid rate Is evidenced by downcomer backup to the plate above. To avoid downflow flooding one must size the column downcomers such that excessive backup does not occur.

Downcomer backup is calculated from the pressure-balance equation

$$h_{dc} = h_t + h_w + h_{ow} + h_{da} + h_{hg} \tag{14-104}$$

where h_{dc} = height in downcomer, mm liquid
h_t = total pressure drop across the plate, mm liquid
h_w = height of weir at plate outlet, mm liquid
h_{ow} = height of crest over weir, mm liquid
h_{da} = head loss due to liquid flow under downcomer apron, mm liquid
h_{hg} = liquid gradient across plate, mm liquid

The heights of head losses in Eq. (14-104) should be in consistent units, e.g., millimeters or inches of liquid under operating conditions on the plate.

As noted, h_{dc} is calculated in terms of equivalent clear liquid. Actually, the liquid in the downcomer may be aerated and actual backup is

$$h'_{dc} = \frac{h_{dc}}{\phi_{dc}} \tag{14-105}$$

where ϕ_{dc} is an average *relative* froth density (ratio of froth density to liquid density) in the downcomer. Design must not permit h'_{dc} to exceed the value of plate spacing; otherwise, flooding can be precipitated. In fact, plate spacing may be determined by some safe approach to the calculated value of h'_{dc}.

The value of ϕ_{dc} depends upon the tendency for gas and liquid to disengage (froth to collapse) in the downcomer. For cases favoring rapid bubble rise (low gas density, low liquid viscosity, low system foamability) collapse is rapid, and clear liquid fills the bottom of the downcomer (Fig. 14-22). For such cases, it is usual practice to employ a value of $\phi_{dc} = 0.5$. For cases favoring slow bubble rise (high gas density, high liquid viscosity, high system foamability), values of $\phi_{dc} = 0.2$ to 0.3 should be used. As the critical point is approached in high-pressure distillations and absorptions, special precautions with downcomer sizing are mandatory, and sloping of the downcomer apron may be used to provide additional disengaging surface (but at the expense of cross-sectional area for perforations). Even so, some gas can be expected to recycle under the downcomer apron.

Plate Layouts Cross-flow plates, whether bubble-cap, sieve, or valve, are similar in layout (Fig. 14-28). Possible zones on each plate are:

Active vapor-dispersion
Peripheral stiffening and support
Disengaging
Distributing
Downcomer

The downcomer zones generally occupy 10 to 30 percent of the total cross section. For segmental downcomers, weir length ranges from 60 to 80 percent of the column diameter, so that the downcomer zone on each end of the plate occupies from 5 to 15 percent of the total cross section.

The fraction of plate area occupied by disengaging and distributing zones ranges from 5 to 20 percent of the cross section. For most sieve-plate designs, these zones are eliminated completely.

The peripheral stiffening zone (tray ring) is generally 25 to 50 mm (1 to 2 in) wide and occupies 2 to 5 percent of the cross section, the fraction decreasing with increase in plate diameter. Periphery waste (Fig. 14-28) occurs primarily with bubble-cap trays and results from the inability to fit the cap layout to the circular form of the plate. Valves and perforations can be located close to the wall and little dead area results. Typical values of the fraction of the total cross-sectional area available for vapor dispersion and contact with the liquid for cross-flow plates with a chord weir equal to 75 percent of the column diameter are given in Table 14-6.

The plate thickness of bubble-cap and sieve plates is generally established by mechanical design factors and has little effect on pressure drop. For a sieve plate, however, the plate is an integral component of the vapor-dispersion system, and its thickness is important.

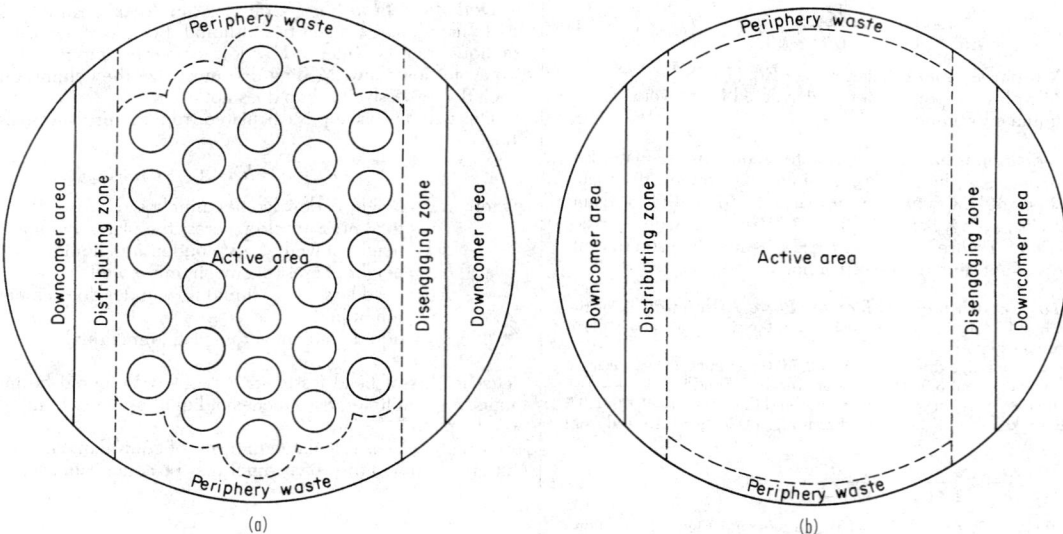

FIG. 14-28   Zone distribution. (*a*) Bubble-cap plate. (*b*) Sieve or valve plate.

For sieve plates, thickness is usually in the 10-to-14 U.S. standard gauge range of 3.58 to 1.98 mm, or 0.141 to 0.078 in. Hardness of metal, size of die, and limits on hole size (for process reasons) lead to the following thickness criterion:

$$0.4 < \frac{\text{plate thickness}}{\text{hole diameter}} < 0.7$$

Bubble caps and valves are generally arranged on an equilateral-triangle layout. Center-to-center spacing should be great enough to minimize impact of vapor streams from adjacent units. Weights of valves on the same plate can be varied to control available flow area at different vapor loadings. Hole sizes for sieve plates range from 1 mm to 25 mm (0.04 to 1 in) diameter, with sizes in the 4- to 12-mm (0.16- to 0.50-in) range being popular. Small holes have less entrainment but may present fouling problems. The smaller hole sizes may lead to punching problems, however, although 1–2 mm holes can be punched from aluminum plate materials. The spacing of the holes, usually on an equilateral-triangle basis, ranges from 2½ to four diameters. Closer

spacings lead to excessive weeping, and greater spacings lead to excessive pressure drop and to entrainment because of high hole velocities.

Countercurrent plates are of perforated or slotted construction and require no downcomers. The vapor and the liquid use the same openings, alternating on an intermittent basis. Layout of such plates is extremely simple. Types of such plates used commercially are
Perforated (dual-flow)
Slotted (Turbogrid)
Perforated-corrugated (Ripple)
The open area for these plates ranges from 15 to 30 percent of the total cross section compared with 5 to 15 percent for sieve plates and 8 to 15 percent for bubble-cap plates. Hole sizes range from 6 to 25 mm (¼ to 1 in), and slot widths from 6 to 12 mm (¼ to ½ in). The Turbogrid and Ripple plates are proprietary devices.

Pressure Drop   Methods for estimating fluid-dynamic behavior of crossflow plates are analogous, whether the plates be bubble-cap, sieve, or valve. The total pressure drop across a plate is defined by the general equation (see Fig. 14-29)

$$h_t = h_d + h_L' \tag{14-106}$$

where h_t = total pressure drop, mm liquid
h_d = pressure drop across the dispersion unit (dry hole for sieve plates; dry valve for valve plates; dry cap, riser, and slot drop for bubble caps, mm liquid
h_L' = pressure drop through aerated mass over and around the disperser, mm liquid

It is convenient and consistent to relate all of these pressure-drop terms to height of equivalent clear liquid (deaerated basis) on the plate, in either millimeters or inches of liquid.

Pressure drop across the disperser is calculated by variations of the standard orifice equation:

$$h_d = K_1 + K_2 \left(\frac{\rho_G}{\rho_L} \right) U_h^2 \tag{14-107}$$

where U_h = linear gas velocity through risers (bubble caps) or perforations (sieve plate), m/s.

For bubble caps, K_1 is the drop through the slots and K_2 is the drop through the riser, reversal, and annular areas. Equations for evaluating these terms for various bubble-cap designs are given by Bolles (in chap. 14 of Smith, *Equilibrium Stage Processes*, McGraw-Hill, New York, 1963), or may be found in previous editions of this handbook.

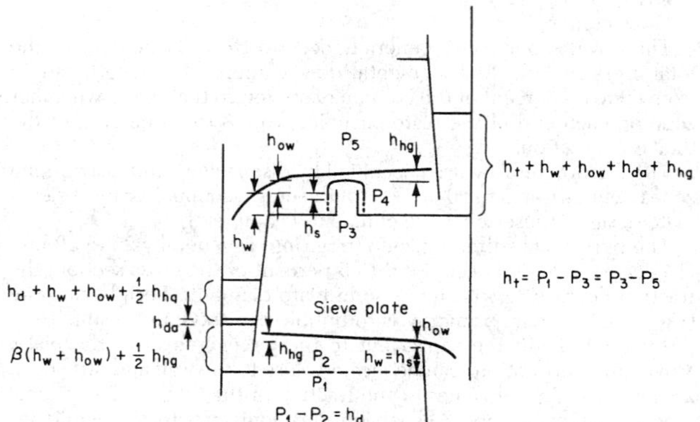

FIG. 14-29   Pressure-drop contributions for cross-flow plates. h_d = pressure drop through cap or sieve, equivalent height of plate liquid; h_w = height of weir; h_{ow} = weir crest; h_s = static liquid seal; h_{hg} = hydraulic gradient; and h_{dc} = loss under downcomer.

TABLE 14-6   Representative Plate Efficiencies

Disperser	System	Column diameter, ft	Tray spacing, in	Pressure, psia	Static submergence, in	Efficiency, % E_{mv}°	Efficiency, % E_{oc}†	Remarks	Ref.
Bubble-cap	Ethanol-water	1.31	10.6	14.7	1.18	83–87			1
		1.31	16.3	14.7	1.18	84–97			
		2.5	14	14.7	1.2	80–85			2
	Methanol-water	3.2	15.7	14.7	1.0	90–95			3
	Ethyl benzene-styrene	2.6	19.7	1.9	0.2	55–68			4
	Cyclohexane-n-heptane	4.0	24	14.7	0.25	65–90			5
				24	4.25	65–90			
				50		65–90			
	Cyclohexane-n-heptane	4.0	24	5	0.6	65–85		Tunnel caps	6
				24		75–100			
	Benzene-toluene	1.5	15.7	14.7	1.5	70–80			7
	Toluene-isooctane	5.0	24	14.7	0.4		60–80		8
Sieve	Methanol-water	3.2	15.7	14.7		70–90		10.8% open	3
	Ethanol-water	2.5	14	14.7	1.0	75–85		10.4%	2
	Methanol-water	3.2	15.7	14.7	1.57	90–100		4.8% open	3
	Ethyl benzene-styrene	2.6	19.7	1.9	0.75	70		12.3% open	9
	Benzene-toluene	1.5	15.7	14.7	3.0	60–75		8% open	14
	Methyl alcohol-n-propyl alcohol-sec-butyl alcohol	6.0	18	18	1.38		64		10
	Mixed xylenes + C$_8$-C$_{10}$ paraffins and naphthenes	13.0	21	25	1.25		86		5
	Cyclohexane-n-heptane	4.0	24	5	2.0	60–70		14% open	13
				24		80		14% open	13
		4.0	24	5	2.0	70–80		8% open	12
	Isobutane-n-butane	4.0	24	165	2.0	110		14% open	13
		4.0	24	165	2.0	120		8% open	12
		4.0	24	300	2.0	110		8% open	12
		4.0	24	400	2.0	100		8% open	12
	n-heptane-toluene	1.5	15.7	14.7	3.0	60–75		8% open	14
	methanol-water	2.0	13.6	14.7	2.0	68–72%		10% open	15
	isopropanol-water	2.0	13.6	14.7	2.0	59–63%			15
	toluene-methylcyclohexane	2.0	13.6	14.7	2.0	70–82%			15
Valve	Methanol-water	3.2	15.7	14.7		70–80		14.7% open	3
	Ethanol-water	2.5	14	14.7	1.0	75–85			2
	Ethyl benzene-styrene	2.6	19.7	1.9	0.75	75–85			4
	Cyclohexane-n-heptane	4.0	24	20	3.0		50–96	Rect. valves	11
	n-Butane-isobutene	4.0	24	165	3.0		104–121	Rect. valves	11

References
1. Kirschbaum, Z. Ver. Dtsch. Ing. Beih. Verfahrenstech., (5), 131 (1938); (3), 69 (1940).
2. Kirschbaum, Distillier-Rektifiziertechnik, 4th ed., Springer-Verlag, Berlin and Heidelberg, 1969.
3. Kastanek and Standart, Sep. Sci. 2, 439 (1967).
4. Billet and Raichle, Chem. Ing. Tech., 38, 825 (1966); 40, 377 (1968).
5. AIChE Research Committee, Tray Efficiency in Distillation Columns, final report, University of Delaware, Newark, 1958.
6. Raichle and Billet, Chem. Ing. Tech., 35, 831 (1963).
7. Zuiderweg, Verburg, and Gilissen, Proc. Intn. Symp., Brighton, England, 1960.
8. Manning, Marple, and Hinds, Ind. Eng. Chem., 49, 2051 (1957).
9. Billet, Proc. Intn. Symp., Brighton, England, 1970.
10. Mayfield, Church, Green, Lee, and Rasmussen, Ind. Eng. Chem., 44, 2238 (1952).
11. Fractionation Research, Inc. "Report of Tests of Nutter Type B Float Valve Tray," July 2, 1964 from Nutter Engineering Co.
12. Sakata and Yanagi, Inst. Chem. Eng. Syrup. Ser., no. 56, 3.2/21 (1979).
13. Yanagi and Sakata, Ind. Eng. Chem. Process Des. Dev., 21, 712 (1982).
14. Zuiderweg and Van der Meer, Chem. Tech. (Leipzig), 24, 10 (1972).
15. Korchinsky, Trans. I. Chem. E., 72, Part A, 472 (1994).
°See Eq. (14-130).
†See Eq. (14-127).
NOTE: To convert feet to meters, multiply by 0.3048; to convert inches to centimeters, multiply by 2.54; and to convert psia to kilopascals, multiply by 6.895.

For sieve plates, $K_1 = 0$ and $K_2 = 50.8/C_v^2$. Values of C_v are taken from Fig. 14-30. Values from Fig. 14-30 may be calculated from

$$C_v = 0.74 (A_h/A_a) + \exp[0.29(t_t/d_h) - 0.56] \quad (14\text{-}108)$$

For valve plates, values of K_1 and K_2 depend on whether the valves are fully open. They also depend on the shape and weight of the valves. Vendors of valve plates make K_1 and K_2 data (or their equivalent) readily available. An analysis of valve plate pressure drop has been reported by Bolles [Chem. Eng. Progr. 72(9), 43 (1976)], and typical dry head loss data, shown in Fig. 14-31, are taken from that work.

Pressure drop through the aerated liquid [h'_L, in Eq. (14-106)] is calculated by

$$h'_L = \beta h_{ds} \quad (14\text{-}109)$$

where β = aeration factor, dimensionless
h_{ds} = calculated height of clear liquid over the dispersers, mm (dynamic seal)

The aeration factor β has been determined for bubble-cap and sieve plates, and a representative correlation of its values is shown in Fig. 14-32. Values of β in the figure may be calculated from

$$\beta = 0.0825 \ln\left(\frac{q}{L_w}\right) - 0.269 \ln F_{vh} + 1.679 \quad (14\text{-}110)$$

where L_w = weir length, m
F_{vh} = F-factor for flow through holes, $F_{vh} = U_h \rho_G^{0.5}$, m/s (kg/m^3)$^{0.5}$

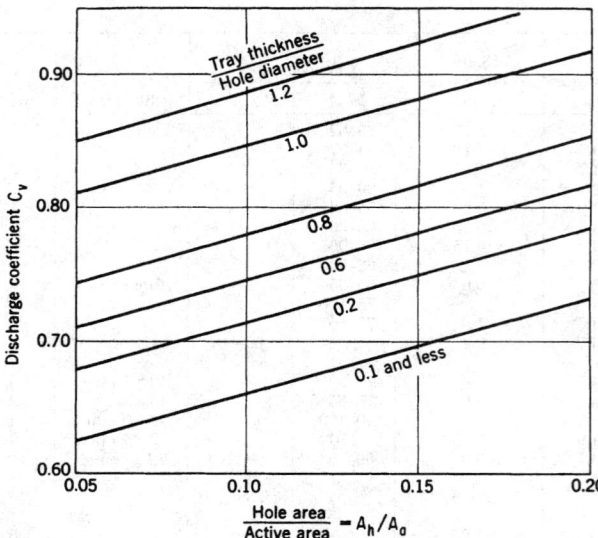

FIG. 14-30 Discharge coefficients for gas flow, sieve plates. [*Liebson, Kelley, and Bullington,* Pet. Refiner, **36**(3), 288 (1957).]

For sieve and valve plates,

$$h_{ds} = h_w + h_{ow} + 0.5h_{hg} \qquad (14\text{-}111)$$

where h_w = weir height, mm
  h_{ow} = height of crest over weir equivalent clear liquid, mm
  h_{hg} = hydraulic gradient across plate height of equivalent clear liquid, mm

The value of weir height crest h_{ow} may be calculated from the Francis weir equation and its modifications for various weir types. For a segmental weir and for height in millimeters of clear liquid,

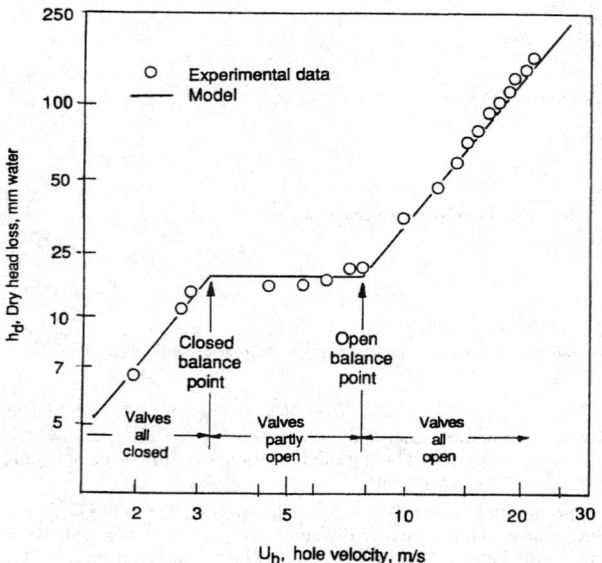

FIG. 14-31 Pressure drop for a valve plate, measured versus model prediction of Bolles [Chem. Eng. Progr. **72**(9), 43 (1976)]. Reproduced with permission of the American Institute of Chemical Engineers. Copyright © 1976 AIChE. All rights reserved.

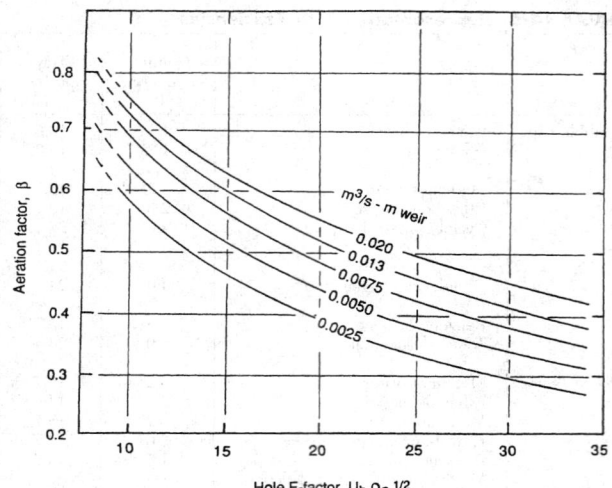

FIG. 14-32 Aeration factor for pressure drop calculation, sieve plates. [*Bolles and Fair,* Encyclopedia of Chemical Processing and Design, *vols. 16, 86. J. M. McKetta (ed.), Marcel Dekker, New York, 1982.*]

$$h_{ow} = 664\left(\frac{q}{L_w}\right)^{2/3} \qquad (14\text{-}112)$$

where q = liquid flow, m³/s
  L_w = weir length, m

For serrated weirs,

$$h_{ow} = 851\left(\frac{q'}{\tan\theta/2}\right)^{0.4} \qquad (14\text{-}113)$$

where q' = liquid flow, m³/s per serration
  θ = angle of serration, degrees

For circular weirs,

$$h_{ow} = 44,300\left(\frac{q}{d_w}\right)^{0.704} \qquad (14\text{-}114)$$

where q = liquid flow, m³/s
  d_w = weir diameter, mm

As noted, the weir crest h_{ow} is calculated on an equivalent clear-liquid basis. A more realistic approach is to recognize that in general a froth or spray flows over the outlet weir (settling can occur upstream of the weir if a large "calming zone" with no dispersers is used). Bennett et al. [*AIChE J.*, **29**, 434 (1983)] allowed for froth overflow in a comprehensive study of pressure drop across sieve plates; their correlation for residual pressure drop h_L' in Eq. (14-87) is represented by Eqs. (14-115) through (14-120):

$$h_L' = h_L + h_\sigma' \qquad (14\text{-}115)$$

where h_L' = pressure drop through the aerated liquid ($= h_L - h_\sigma'$), mm
  h_L = effective clear-liquid height (liquid holdup), mm
  h_σ' = pressure drop for surface generation, mm

$$= \left(\frac{472\sigma}{g\rho_L}\right)\left(\frac{g(\rho_L - \rho_G)}{d_h\sigma}\right)^{1/3} \qquad (14\text{-}116)$$

with σ = surface tension, mN/m.
  First, an effective froth density ϕ_e (dimensionless) is calculated:

$$\phi_e = \exp(-12.55K_s^{0.91}) \qquad (14\text{-}117)$$

where $\phi_e = h_L/h_f$
  h_f = froth height, mm
  $K_s = U_a \cdot [\rho_g/(\rho_L - \rho_G)]^{0.5} = F_{va}/(\Delta\rho)^{0.5} \qquad (14\text{-}118)$

U_a = vapor velocity through the active area, m/s
F_{va} = F-factor based on active area, m/s(kg/m³)$^{0.5}$

Then the liquid holdup is calculated:

$$h_L = \phi_e \left[h_w + 15{,}330 C(q/\phi_e)^{2/3}\right] \qquad (14\text{-}119)$$

where

$$C = 0.0327 + 0.0286 \exp[-0.1378 h_w]. \qquad (14\text{-}120)$$

In these equations, the h terms and d_h (perforation diameter) are in mm, densities are in kg/m³, surface tension is in mN/m, and flow rate q is in m³/s. The gravitational constant g is 9.81 m/s². For total pressure drop across the plate, Eq. (14-96) is used in conjunction with Eq. (14-107) and Fig. 14-30.

For a base of 302 data points covering a wide range of systems and conditions, Eqs. (14-115) through (14-120) gave an average error of ±0.35 percent. For a similar data base, Eqs. (14-106) and (14-109) together with Fig. 14-32 gave an average error of less than 5 percent. Although more difficult to use, the method of Bennett et al. is recommended when determination of pressure drop is of critical importance.

Example 11: Pressure Drop, Sieve Plate For the conditions of Example 10, estimate the pressure drop for flow across one plate. The thickness of the plate metal is 2 mm and the hole diameter is 4.8 mm. The superficial F-factor is 2.08 m/s(kg/m³)$^{1/2}$.

Solution. Method A: Eqs. (14-106, 14-109, 14-110), where $h_t = h_d + \beta(h_w + h_{ow})$. For $F_{vs} = 2.08$, $F_{va} = 2.32$, and $F_{vh} = 16.55$. From Example 9, $L_w = 1.50$ m and $h_w = 38$ mm. For a liquid rate of 22,000 kg/hr, $q = 7.27(10^{-3})$ m³/s, and $q/L_w = 4.8(10^{-3})$. By Eq. (14-110) or Fig. 14-32, $\beta = 0.48$. From Eq. (14-108) or Fig. 14-30, $C_v = 0.78$. Then, by Eq. (14-107), $h_d = 32.6$ mm liquid. Using Eq. (14-112), $h_{ow} = 18.9$ mm. Finally, $h_t = h_d + \beta(h_w + h_{ow}) = 32.6 + 0.48(38 + 18.9) = 60.0$ mm liquid.

Method B: Bennett et al. $h_t = h_d + h_L + h'_\sigma$; h_d = same as for Method A. Eq. (14-116): $h'_\sigma = [(472 \times 25)/(9.81 \times 841)] \, [(9.81 \times 841)/(5 \times 25)]^{1/3} = 5.47$ mm; Eq. (14-118): $K_s = Fva/\Delta\rho^{0.5} = 2.32/(841 - 0.481)^{0.5} = 2.32/(840.5)^{0.5} = 0.080$ m/s; Eq. (14-117): $\phi_e = \exp[-12.55(0.080)^{0.91}] = 0.284$ (effective froth density); Eq. (14-120): $C = 0.0327 + 0.0286 \exp[-0.1378(38)] = 0.0329$; and Eq. (14-119): $h_L = 0.284[38 + 15{,}330(0.0329)(0.00727/0.284]^{2/3} = 23.23$ mm. Finally, $h_t = h_d + h_L + h'_\sigma = 32.6 + 23.23 + 5.47 = 61.3$ mm liquid.

When straight or serrated segmental weirs are used in a column of circular cross section, a correction may be needed for the distorted pattern of flow at the ends of the weirs, depending on liquid flow rate. The correction factor F_w from Fig. 14-33 is used directly in Eq. (14-112) or Eq. (14-114). Even when circular downcomers are utilized, they are often fed by the overflow from a segmental weir. When the weir crest over a straight segmental weir is less than 6 mm (¼ in), it is desirable to use a serrated (notched) weir to provide good liquid distribution. Inasmuch as fabrication standards permit the tray to be 3 mm (⅛ in) out of level, weir crests less than 6 mm (¼ in) can result in maldistribution of liquid flow.

Loss under Downcomer The head loss under the downcomer apron, as millimeters of liquid, may be estimated from

$$h_{da} = 165.2 \left(\frac{q}{A_{da}}\right)^2 \qquad (14\text{-}121)$$

where q = volumetric flow of liquid, m³/s and A_{da} = minimum area of flow under the downcomer apron, m². Although the loss under the downcomer is small, the clearance is significant from the aspect of tray stability and liquid distribution. The seal between the top of the liquid on the plate and the bottom of the downcomer should range between 13 and 38 mm (½ and 1½ in).

Hydraulic Gradient Hydraulic gradient, the head of liquid necessary to overcome the frictional resistance to liquid (froth) passage across the plate, is important for plate stability inasmuch as it is the only liquid head that varies across the length of passage. If the gradient is excessive, the upstream portion of the plate may be rendered inoperative because of increased resistance to gas flow caused by increased liquid head (Fig. 14-34). In general the empirical criterion for stable operation is $h_d > 2.5 h_{hg}$.

Sieve plates usually have negligible hydraulic gradient. Bubble-cap plates can have significant gradient because of the blockage by the caps. Valve plates presumably are intermediate, with hydraulic-gradient characteristics approaching those of sieve plates.

For bubble-cap plates, hydraulic-gradient must be given serious consideration. It is a function of cap size, shape, and density on the plate. Methods for analyzing bubble-cap gradient may be found in the chapter by Bolles (Smith, *Design of Equilibrium Stage Processes*, Chap. 14, McGraw-Hill, New York, 1963) or in previous edition of this handbook.

The hydraulic gradient on sieve plates should be checked in cases of long flow path of liquid. Hughmark and O'Connell (*Chem. Eng. Prog.*, **53**(3), 127 (1957)] presented a correlation for determining sieve-plate hydraulic gradient. Although the correlation does not explicitly indicate an effect of gas velocity, the effect is implicit in the choice of friction factor. The gradient is predicted by the relationship

$$h_{hg} = \frac{1000 f U_f^2 L_f}{g R_h} \qquad (14\text{-}122)$$

where f = friction factor correlated against a Reynolds modulus as in pipe flow

$$N_{Reh} = \frac{R_h U_f \rho_L}{\mu_L} \qquad (14\text{-}123)$$

as shown in Fig. 14-34. In Eqs. (14-122) and (14-123), R_h is the hydraulic radius of the aerated mass, defined as follows:

$$R_h = \frac{\text{cross section}}{\text{wetted perimeter}} = \frac{h_f D_f}{2h_f + 1000 D_f} \qquad (14\text{-}124)$$

where D_f is the arithmetic average between tower diameter and weir

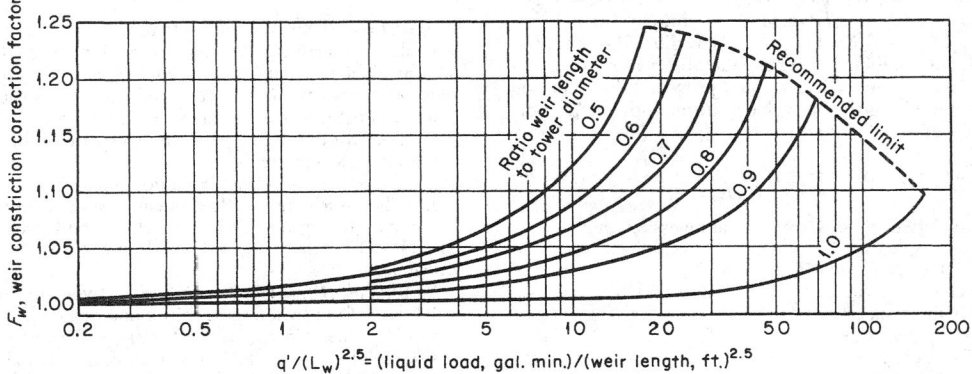

FIG. 14-33 Correction for effective weir length. To convert gallons per minute to cubic meters per second, multiply by 6.309×10^{-5}; to convert feet to meters, multiply by 0.3048. [*Bolles*, Pet. Refiner, **25**, 613 (1946).]

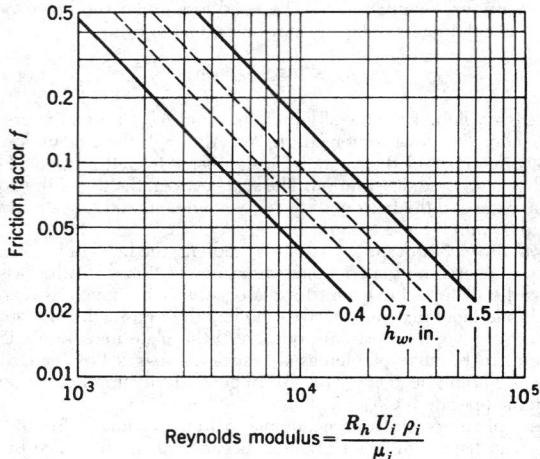

FIG. 14-34 Friction factor for froth crossflow, sieve plates. To convert inches to millimeters, multiply by 25.4. (*Smith, Design of Equilibrium Stage Processes, McGraw-Hill, New York, 1963.*)

length (average width of flow path), and h_f is froth height. The value of h_f is estimated from Eq. (14-117a). U_f is the velocity of the aerated mass, m/s, and is the same as for the clear liquid:

$$U_f = \frac{1000q}{h_f \phi e D_f} = \frac{1000q}{h_L D_f} \tag{14-125}$$

Other terms in Eqs. (14-122) through (14-125) are:

g = acceleration of gravity, m/s²
L_f = length of flow path across plate, m
q = liquid-flow rate, m³/s
ϕe = froth density on plate, dimensionless
μ_L = liquid viscosity, Pa·s or kg/(m·s)
ρ_L = liquid density, kg/m³

Phase Inversion   Normally the two-phase mixture on the plate is in the form of a bubbly, or aerated liquid. This liquid-continuous mixture is called a *froth*. Under high gas rates and low liquid rates, however, the regime can invert to a gas-continuous *spray* comprising a multitude of liquid droplets of varying diameter. Many studies of this froth-to-spray transition have been made, most of them with air and water. The results of one such study, useful for design purposes, are shown in Fig. 14-35. The spray is predicted to prevail above the appropriate curve. Reviews of phase inversion have been provided by Lockett [*Distillation Tray Fundamentals*, Cambridge Univ. Press, Cambridge, U.K., 1986] and Prado et al. [*Chem. Eng. Progr.*, **83**(3), 32 (1987)]. The latter combined experimental observations of inversion with evaluations of plate efficiency, and proposed the following relationship for determining the gas velocity through the active portion of the tray, at the inversion point:

$$U_a^\circ = C_1 \rho_G^{-0.50} \rho_L^{0.692} \sigma^{0.06} A_f^{0.25} \left(\frac{q}{L_w}\right)^{0.05} d_h^{-0.1} \tag{14-126}$$

where U_a° = gas velocity through active area at inversion, m/s
ρ_G = gas density, kg/m³
ρ_L = liquid density, kg/m³
σ = surface tension, mN/m
A_f = hole/active area ratio
q/L_w = liquid flow, m³/s·m weir
d_h = hole diameter, mm
C_1 = 0.0583 for 25.4-mm overflow weirs
      = 0.0568 for 50.4-mm overflow weirs
      = 0.0635 for 101.6-mm overflow weirs

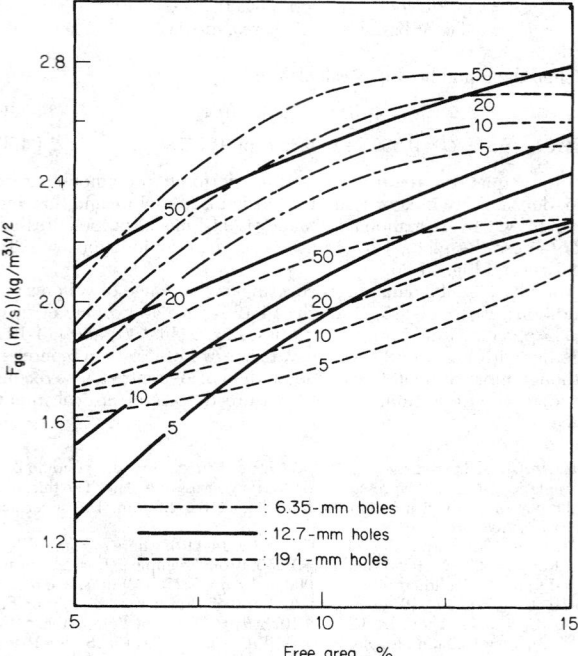

FIG. 14-35   Transition from froth to spray regime for holes of various diameters. Values on curves are liquid loadings, m³/(h·m weir length). To convert cubic meters per hour-meter to cubic feet per hour-foot, multiply by 10.764; to convert (meters per second) (kilograms per cubic meter)^1/2 to (feet per second) (pounds per cubic foot)^1/2, multiply by 0.8197; and to convert millimeters to inches, multiply by 0.0394. [*Loon, Pinczewski, and Fell*, Trans. Inst. Chem. Eng., **51**, 374 (1973).]

Figure 14-25 also provides a means for estimating whether spray or froth might prevail on the tray. As can be seen, low values of the flow parameter F_{LG}, as for vacuum fractionators, can lead to the spray regime.

Plate Efficiency   The efficiency of a plate for mass transfer depends upon three sets of design parameters:
1.   The system—composition and properties
2.   Flow conditions—rates of throughput
3.   Geometry—plate type and dimensions
The designer has little control over the first set but can deal effectively with the other two. Ultimate concern is with *overall column efficiency*:

$$E_{oc} = N_t/N_a \tag{14-127}$$

or the ratio of *theoretical plates* to *actual plates* required to make the separation. In arriving at a value of E_{oc} for design, the designer may rely on plant test data or on judicious use of pilot-plant-efficiency measurements. If such direct information is not available, the designer must resort to predictive methods.

Methods for predicting plate efficiency are of three general types:
1.   Empirical methods
2.   Direct scale-up from laboratory measurements
3.   Theoretical or semitheoretical mass-transfer methods
The first of these gives E_{oc} directly. The second gives a point efficiency [Eq. (18-30)]. The third involves the prediction of individual phase efficiencies.

Empirical Predictive Methods   Two empirical correlations which have found wide use are the one of Drickamer and Bradford [*Trans. Am. Inst. Chem. Eng.*, **39**, 319 (1943)] and a modification of it by O'Connell [*Trans. Am. Inst. Chem. Eng.*, **42**, 741 (1946)]. The latter is shown in Fig. 14-36, the Drickamer-Bradford data are included in the distillation plot.

A semitheoretical method which gives overall efficiency is that of

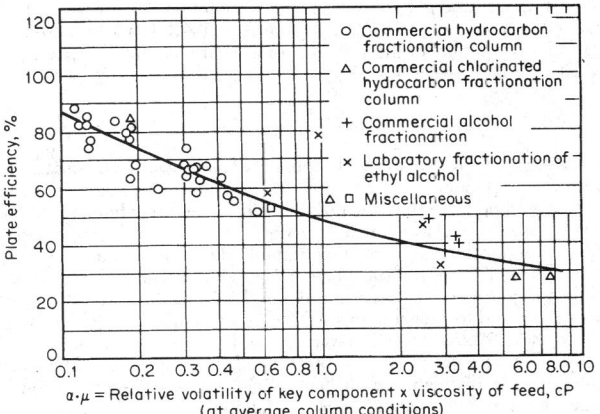

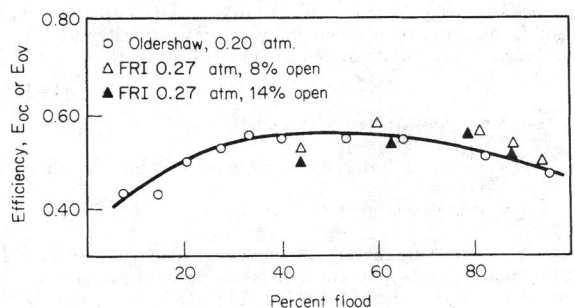

FIG. 14-37 Overall column efficiency of 25-mm Oldershaw column compared with point efficiency of 1.22-m-diameter-sieve sieve-plate column of Fractionation Research, Inc. System = cyclohexane-n-heptane. [(*Fair, Null, and Bolles, Ind. Eng. Chem. Process Des. Dev.,* **22**, 53 (*1982*).]

FIG. 14-36 O'Connell correlation for overall column efficiency E_{oc} for distillation. To convert centipoises to pascal-seconds, multiply by 10^{-3}. [*O'Connell, Trans. Am. Inst. Chem. Eng.,* **42**, 741 (1946).]

Bakowski [*Br. Chem. Eng.,* **8**, 384, 472 (1963); **14**, 945 (1969)]. It is based on the assumption that the mass-transfer rate for a component moving to the vapor phase is proportional to the concentration of the component in the liquid and to its vapor pressure. Also, the interfacial area is assumed proportional to liquid depth, and surface renewal rate is assumed proportional to gas velocity. The resulting general equation for binary distillation is

$$E_{oc} = \frac{1}{1 + 3.7(10^4)\dfrac{KM}{h'\rho_l T}} \tag{14-128}$$

where E_{oc} = overall column efficiency, fractional
   K = vapor-liquid equilibrium ratio, $y°/x$
   $y°$ = gas-phase concentration at equilibrium, mole fraction
   x = liquid-phase concentration, mole fraction
   M = molecular weight
   h' = effective liquid depth, mm
   ρ_l = liquid density, kg/m³
   T = temperature, K

For sieve or valve plates, $h' = h_w$, outlet weir height. For bubble-cap plates, h' = height of static seal. The original references present validations against laboratory and small-commercial-column data. Modifications of the efficiency equation for absorption-stripping are also included.

Direct Scale-Up of Laboratory Distillation Efficiency Measurements   It has been found by Fair, Null, and Bolles [*Ind. Eng. Chem. Process Des. Dev.,* **22**, 53 (1983)] that efficiency measurements in 25- and 50-mm- (1- and 2-in-) diameter laboratory Oldershaw columns closely approach the point efficiencies [Eq. (14-129)] measured in large sieve-plate columns. A representative comparison of scales of operation is shown in Fig. 14-37. Note that in order to achieve agreement between efficiencies it is necessary to ensure that (1) the systems being distilled are the same, (2) comparison is made at the same relative approach to the flood point, (3) operation is at total reflux, and (4) a standard Oldershaw device (a small perforated-plate column with downcomers) is used in the laboratory experimentation. Fair et al. made careful comparisons for several systems, utilizing as large-scale information the published efficiency studies of Fractionation Research, Inc.

Theoretical Predictive Methods   The approach to equilibrium on a plate may be defined as the ratio of the actual change in gas composition as it passes through the plate to the change that would have occurred if the gas had reached a state of equilibrium with the liquid. If a *point* on plate n is considered, this definition leads to the **point efficiency:**

$$E_{og} = \left(\frac{y_n - y_{n-1}}{y_n° - y_{n-1}}\right)_{\text{point}} \tag{14-129}$$

where $y_n°$ is the gas concentration in equilibrium with liquid concentration at the point. This efficiency cannot exceed 1.0 (100 percent). If there are liquid-concentration gradients on the plate (i.e., plate liquid is not completely mixed), then $y°$ will vary and E_{og} may vary from point to point on the plate. It should be noted that an analogous efficiency definition could be expressed on the basis of liquid concentrations. It should be noted also that vaporization efficiency (Holland, *Fundamentals of Multicomponent Distillation,* McGraw-Hill, New York, 1981) could be used:

$$E_v = y_n/y_n° \tag{14-130}$$

For the entire plate and for gas concentrations, the Murphree vapor efficiency is used:

$$E_{mv} = \left(\frac{y_n - y_{n-1}}{y_n° - y_{n-1}}\right)_{\text{plate}} \tag{14-131}$$

where $y_n°$ is gas concentration in equilibrium with the concentration of the liquid leaving the plate (flowing into the downcomer, for a cross-flow plate). Because of concentration gradients in the liquid, E_{mv} can exceed 100 percent.

The best-established theoretical method for predicting E_{oc} is that of the AIChE (*Bubble-Tray Design Manual,* American Institute of Chemical Engineers, New York, 1958). It is based on the sequential prediction of point efficiency, Murphree efficiency, and overall column efficiency:

$$E_{og} \rightarrow E_{mv} \rightarrow E_{oc}$$

with suitable correction of E_{mv} for entrainment. The AIChE model is the basis for the development which follows.

On the basis of the two-film model for mass transfer, and relating all efficiencies to gas-phase concentrations (for convenience only; a similar development can be made on the basis of liquid concentrations), point efficiency can be expressed in terms of transfer units:

$$E_{og} = 1 - e^{-N_{og}} \tag{14-132}$$

where N_{og} = overall transfer units calculated from Eq. (14-133).

$$N_{og} = \frac{1}{1/N_g + \lambda/N_\ell} \tag{14-133}$$

where N_g = gas-phase transfer units
   N_ℓ = liquid-phase transfer units
   $\lambda = mG_m/L_m$ (stripping factor)
   m = slope of equilibrium curve
   G_m = gas rate, mol/s
   L_m = liquid rate, mol/s

Transfer units are dimensionless and are defined further in Sec. 5. According to Eq. (14-133), the evaluation of point efficiencies reduces to the prediction of point values of N_g and N_ℓ plus the evaluation of m, G_m, and L_m for the particular conditions under investigation.

Gas-phase transfer units are obtained from Eq. (14-134)

$$N_g = k_g a \theta_g \qquad (14\text{-}134)$$

where k_g = gas-phase mass-transfer coefficient, (kg·mol)/(s·m²)
         (kg·mol/m³) or m/s
     a = effective interfacial area for mass transfer, m²/m³ froth
        on plate
     θ_g = residence time of gas in froth zone, s

The effect of increasing gas rate is to increase k_g and decrease θ_g, with the result that N_g tends to be constant over a range of gas rates.

For sieve trays, Chan and Fair [*Ind. Eng. Chem. Proc. Des. Dev.*, **23**, 814 (1983)] used a data bank of larger-scale distillation column efficiencies to deduce the following expression for the product $k_G a$:

$$k_g a = \frac{316 D_G^{1/2}\,(1030f + 867f^{2})}{h_L^{1/2}} \qquad (14\text{-}135)$$

where k_g = gas-phase mass-transfer coefficient, m/s
     a = effective interfacial area, m²/m³ froth
     D_G = gas-phase diffusion coefficient, m²/s
     f = approach to flood, fractional
     h_L = liquid holdup on plate, mm

Note that the product of the mass-transfer coefficient and the interfacial area is a volumetric coefficient and obviates the need for a value of the interfacial area. While areas for mass transfer on plates have been measured, the experimental contacting equipment differed significantly from that used for commercial distillation or gas absorption, and the reported areas are considered unreliable for design purposes.

For evaluating the residence time θ_G of gas in the froth, the volume of the froth is taken as $A_a h_f$, where the height of the froth h_f is obtained by first determining effective froth density [Eq. (14-117)]. The dimensionless froth density is defined by

$$\phi = \frac{h_L}{h_f + \varepsilon(\rho_G/\rho_L)} \qquad (14\text{-}136)$$

When $\rho_L \gg \rho_G$, $\phi \sim h_L/h_f$ and $\varepsilon \sim 1 - \phi$ (14-137)

then, residence time(s) may be estimated as

$$\theta_G = \frac{\varepsilon h_f A_a}{10^3\,Q} = \frac{1 - \phi h_L A a}{10^3\,\phi Q} \qquad (14\text{-}138)$$

where Q = volumetric flow of vapor through the plate, m³/s
     h_f = froth height, mm = h_L/ϕ

Liquid phase transfer units are obtained from

$$N_L = k_L a \theta_L \qquad (14\text{-}139)$$

where k_L = liquid phase transfer coefficient, (kg·mol)/(s·m²)
         (kg·mol/m³) = m/s
     a = effective interfacial area for mass transfer, m²/m³ froth
        or spray on the plate
     θ_L = residence time of liquid in the froth or spray zone, s

The mass-transfer coefficient of Eq. (14-139) is carried as a product with interfacial area (giving a volumetric mass transfer coefficient):

Sieve plates:

$$k_L a = (3.875 \times 10^8\,D_L)^{0.5}\,(0.40 U_a \rho_G^{0.5} + 0.17) \qquad (14\text{-}140)$$

Bubble-cap plates:

$$k_L a = (4.127 \times 10^8\,D_L)^{0.5}\,(0.21 U_a \rho_G^{0.5} + 0.15) \qquad (14\text{-}141)$$

where D_L = liquid-phase diffusion coefficient, m²/s (see Sec. 3).

The residence time of liquid in the froth is

$$\theta_L = \frac{(1 - \varepsilon)\,h_f A_a}{10^3\,q} \qquad (14\text{-}142)$$

where q = volumetric flow of liquid across the plate, m³/s.

In summary, the point efficiency E_{og} is computed from Eq. (14-132) using N_{oG} from Eq. (14-113), N_G from Eq. (14-134) and m based on the relative voatility of the system. For a binary mixture and a zone of constant relative volatility,

$$m = \frac{\alpha_{ij}}{[1 + (\alpha_{ij} - 1)\,x_i]^2} \qquad (14\text{-}143)$$

where α_{ij} = relative volatility, component i (more volatile material) relative to component j, and x_i = mole fraction of i in the liquid.

The method for estimating point efficiency, outlined here, is not the only approach available for sieve plates, and more mechanistic methods are under development. For example, Prado and Fair [*Ind. Eng. Chem. Res.*, **29**, 1031 (1990)] have proposed a method whereby bubbling and jetting are taken into account; however the method has not been validated for nonaqueous systems. Chen and Chuang [*Ind. Eng. Chem. Res.*, **32**, 701 (1993)] have proposed a more mechanistic model for predicting point efficiency, but it needs evaluation against a commercial scale distillation data bank. One can expect more development in this area of plate efficiency prediction.

Example 12: Estimation of Plate Efficiency   For the conditions of Examples 9 and 11, estimate the point efficiency of the tray. Additional property data:

Relative volatility, ethylbenzene/styrene	1.40
Stripping factor, λ	1.17
Gas diffusion coefficient, m²/s	2.09 (10⁻⁵)
Liquid diffusion coefficient, m²/s	3.74 (10⁻⁹)

Solution

Gas flow: 25,500/(3600 × 0.481) = 14.73 m³/s
Liquid flow: 22,000/(3600 × 841) = 0.00727 m³/s
Froth density (Example 11), $\phi = 0.284$
Liquid holdup (Example 11), $h_L = 23.23$ mm liquid
Approach to flood (Example 9) = 0.74 = 74%
Gas residence time in froth [Eq. (14-138)] = [(1 − 0.284)(23.23)(4.41)/ [10³ (0.284)(14.73)] = 0.0175 s
Gas volumetric coefficient [Eq. (14-135)] = $k_G a$ = {316 (2.09 × 10⁻⁵)⁰·⁵ [1030(0.74) − 867(0.74²)]}/(23.23)⁰·⁵ = 86.1 s⁻¹
Number of gas-phase transfer units [Eq. (14-134)]: $N_G = k_G a \theta_G$ = 86.1(0.0175) = 1.51
Liquid residence time in froth [Eq. (14-142)] = [(0.284)(81.8)(4.41)]/10³ (0.00727)] = 14.1 s
Liquid volumetric coefficient [Eq. (14-140)] = [3.875(10⁸)(3.74(10⁻⁹))⁰·⁵ [0.40(3.34)(0.481)⁰·⁵ + 0.17] = 1.32 s⁻¹
Number of liquid-phase transfer units [Eq. (14-139)]: $N_L = k_L a \theta_L$ = 1.32 (14.1) = 18.6
Number of overall transfer units [Eq. (14-133)]: N_{OG} = 1/[1/1.51 + 1.17/18.6)] = 1.37
Point efficiency [Eq. (14-132)]: E_{OG} = 1 − exp(−N_{OG}) = 1 − exp(−1.37) = 0.75 (75%)

This value of efficiency is slightly higher than the measured value (Fig. 14-24).

Effects of Gas and Liquid Mixing   As noted previously, it is necessary in most instances to convert point efficiency E_{og} to Murphree plate efficiency E_{mv}. This is true because of incomplete mixing; only in small laboratory or pilot-plant columns, under special conditions, is the assumption $E_{og} = E_{mv}$ likely to be valid. For a crossflow plate with *no* liquid mixing there is plug flow of liquid. For this condition of liquid flow, Lewis [*Ind. Eng. Chem.*, **28**, 399 (1936)] analyzed effects of gas mixing on efficiency. He considered three cases:

1.   Gas enters plate at uniform composition (gas completely mixed between plates).
2.   Gas unmixed; liquid flows in the same direction on successive plates.
3.   Gas unmixed; liquid flows in alternate direction on successive plates.

Case 1 has found the widest application in practice and is represented by the relationship

$$E_{mv} = 1/\lambda\,[\exp\,(\lambda E_{vg}) - 1] \qquad (14\text{-}144)$$

λ is defined as for Eq. (14-133). Equation (14-144) assumes the following in addition to the base conditions:

1. L/V is constant.
2.   Slope of equilibrium curve m is constant.
3.   Point efficiency is constant across the tray.

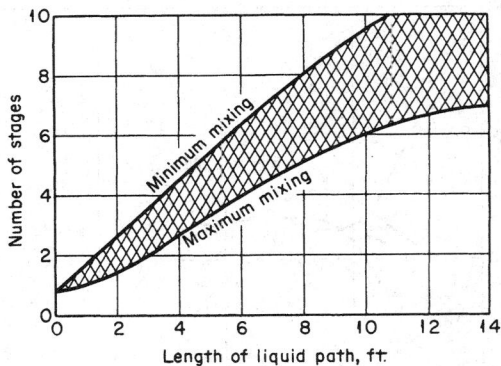

FIG. 14-38 Effect of length of liquid path on number of stages. To convert feet to meters, multiply by 0.3048. [*O'Connell and Gautreaux,* Chem. Eng. Prog., **51**, 236 (1955).]

Most plate columns operate under conditions such that gas is completely mixed as it flows between the plates, but few operate with pure plug flow of liquid. Departure from plug flow of liquid has been studied by Gautreaux and O'Connell [*Chem. Eng. Prog.,* **51**, 232 (1955)] by assuming that liquid mixing can be represented as occurring in a series of stages of completely mixed liquid. For this model,

$$E_{mv} = \frac{1}{\lambda}\left[\left(1 + \lambda\frac{E_{og}}{n}\right)^n - 1\right] \quad (14\text{-}145)$$

where n = number of stages occurring on the tray.

An approximation of the number of stages can be obtained from Fig. 14-38 using the following criteria:

1. Increased liquid rate favors plug flow.

2. Sieve plates have less back mixing than bubble-cap plates because of less obstruction to flow.

3. Increased gas rate increases turbulence and the degree of back mixing of liquid.

An alternative approach is presented in the AIChE *Bubble-Tray Design Manual* and is based on an eddy-diffusion model. According to this model,

$$\frac{E_{mv}}{E_{og}} = \frac{1 - e^{-(n' + N_{Pe})}}{(n' + N_{Pe})\{1 + [(n' + N_{Pe})/n']\}}$$
$$+ \frac{e^n - 1}{n'\{1 + [n'/(n' + N_{Pe})]\}} \quad (14\text{-}146)$$

where $\quad n' = \dfrac{N_{Pe}}{2}\left(\sqrt{1 - 4\lambda\dfrac{E_{og}}{N_{Pe}}} - 1\right)$

N_{Pe} = Péclet number (dimensionless) = $\dfrac{Z_\ell^2}{D_E\theta_l}$

Z_ℓ = length of liquid travel, m; and
λ = stripping factor [see Eq. (14-133)].

The value of θ_l is calculated from Eq. (14-142). The term D_E is an eddy-diffusion coefficient that is obtained from experimental measurements. For sieve plates, Barker and Self [*Chem. Eng. Sci.,* **17,** 541 (1962)] obtained the following correlation:

$$D_E = 6.675(10^{-3})U_a^{1.44} + 0.922(10^{-4})h_\ell - 0.00562 \quad (14\text{-}147)$$

where D_E = eddy-diffusion coefficient, m²/s
    U_a = gas velocity through active area of plate, m/s
    h_ℓ = liquid holdup on plate (Eq. 18-16), mm

For bubble-cap plates, the eddy-diffusion correlation in the AIChE *Bubble-Tray Design Manual* should be used.

The graphical representation of Eq. (14-146) is indicated in Fig. 14-39, where as usual $\lambda = mG_m/L_m$.

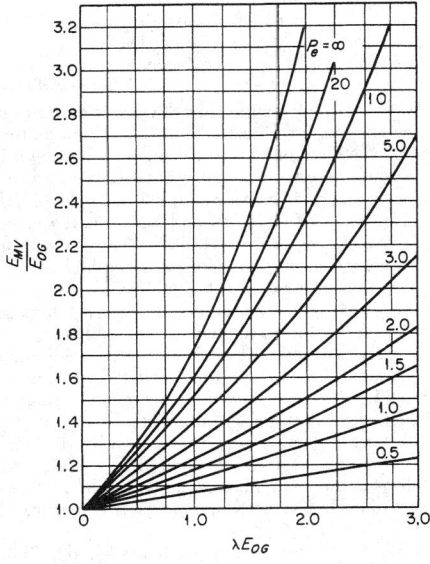

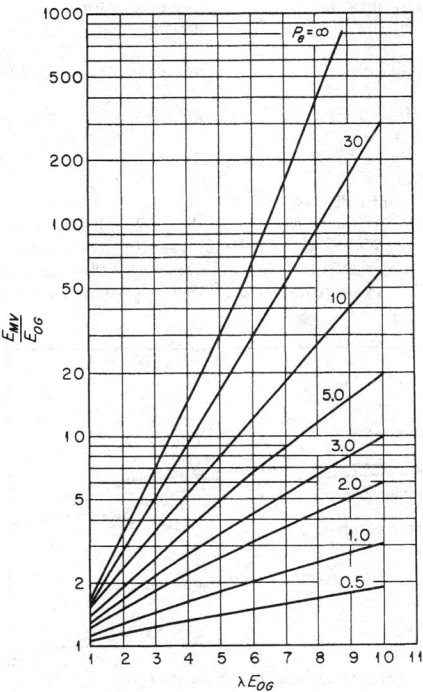

FIG. 14-39 Mixing curves. Pe = Péclet number (N_{Pe}). (Bubble-Tray Design Manual, *American Institute of Chemical Engineers, New York, 1958.*)

The dimensionless Péclet number in Eq. (14-146) is a key parameter in evaluating departure from plug flow of liquid across the plate. The higher its value, the greater the enhancement of point efficiency, as shown in Fig. 14-39. A long liquid flow path Z_L, a high liquid flow rate (low q_L, and a low amount of diffusive backmixing (low value of D_E) contribute to the plug flow effect. For use in the Gautreaux/O'Connell model [Eq. (14-145), Fig. 14-38)], it can be shown that the number of mixing stages n is approximately

$$n = \frac{N_{Pe} + 2}{2} \qquad (14\text{-}148)$$

Overall Column Efficiency   Calculated values of E_{mv} must be corrected for entrainment, if any, by the Colburn equation [Eq. (14-101)]. The resulting corrected efficiency E_a is then converted to column efficiency by the relationship of Lewis [*Ind. Eng. Chem.*, **28**, 399 (1936)]:

$$E_{oc} = \frac{N}{N_t} = \frac{\log\left[1 + E_a(\lambda - 1)\right]}{\log \lambda} \qquad (14\text{-}149)$$

Comparison of Efficiency of Various Plates   Several studies of various plates have been carried out under conditions such that direct and meaningful comparisons are possible. Required conditions include identical system, same pressure, same column diameter, and equivalent submergence. Standart and coworkers [*Br. Chem. Eng.*, **11** (11), 1370 (1966); *Sep. Sci.*, **2**, 439 (1967)] used the methanol-water system at atmospheric pressure in a 1.0-m (3.3-ft) column. For a plate spacing of 0.4 m (15.7 in) they studied the following:
1. Bubble-cap plate—70-mm (2.75-in) round caps, 25.4-mm (1.0-in) submergence
2. Sieve plate—4.0-mm (⁵⁄₃₂ in) holes, hole-active area = 0.048, 40-mm (1.57-in) outlet weir
3. Turbogrid plate—4.6-mm (0.18-in) slot width, 14.7 percent open area
4. Ripple plate—2.85-mm (⁷⁄₆₄ in) holes, 10.8 percent open area
Efficiency data from the work are summarized in Fig. 14-40.

Kirschbaum (*Distillier-Rektifiziertechnik*, 4th ed., Springer-Verlag, Berlin and Heidelberg, 1969) reported on studies of the ethanol-water system at atmospheric pressure, using several columns. For a 0.75-m (2.46-ft) column and 0.35-m (14-in) plate spacing, the following were covered:
1. Bubble-cap plate—90-mm (3.5-in) round caps, 30-mm (1.2-in) static submergence
2. Sieve plate—10-mm (²⁵⁄₆₄ -in) holes, hole/active area = 0.104, 25.4-mm (1.0-in) outlet weir
3. Valve plate—40-mm (1.57-in) holes, 45 valves per plate, 25.4-mm (1.0-in) outlet weir.
Efficiency data are given in Fig. 14-41.

Billet and coworkers [*Chem. Ing. Tech.*, **38**, 825 (1966); *Instn. Chem. Engrs. Symp. Ser. No. 32*, **5**, 111 (1969] used the ethylbenzene/styrene system at 100 torr and a 0.8-m column with 500-mm plate spacing. Two weir heights were used, 19 and 38 mm. Operation was at

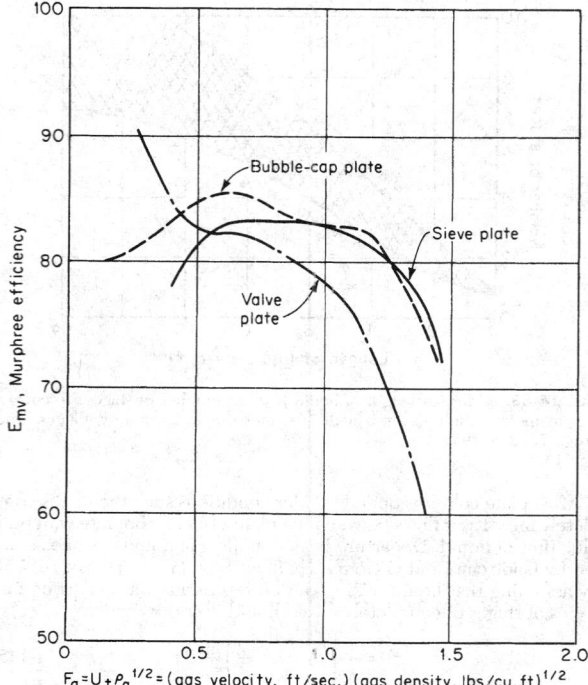

FIG. 14-41   Plate efficiencies, ethanol-water. To convert (feet per second) (pounds per cubic foot)$^{1/2}$ to (meters per second) (kilograms per cubic meter)$^{1/2}$, multiply by 1.2199. (*Kirschbaum, Destillier-Rektifiziertechnik, 4th ed., Springer-Verlag, Berlin and Heidelberg, 1969.*)

total reflux. A variety of plate devices were tested, and Fig. 14-42 shows typical results for the following:

Curve 1	Bubble-cap plate, 35-mm weir
Curve 2	Valve plate with 64 V-1 valves
Curve 3	Sieve plate, 38-mm weir
Curve 4	Sieve-valve plate, 38-mm weir, 49 valves, 140 holes

Testing of plates and other devices is carried out by Fractionation Research, Inc. for industrial sponsors. Some of the test data for sieve plates have been published for the cyclohexane/n-heptane and isobutane/n-butane systems. Representative data are shown in Fig. 14-43. These are taken from Sakata and Yanagi [*Instn. Chem. Engrs. Symp. Ser. No. 56*, 3.2/21 (1979)] and Yanagi and Sakata [*Ind. Eng. Chem. Proc. Des. Devel.*, **21**, 712 (1982)]. The column diameter was 1.2 m, tray spacing was 600 mm, and weir height was 50 mm.

Work at the University of Manchester Institue of Science and Technology (UMIST) has resulted in several papers reporting efficiency data taken in a 0.6-m-diameter column. The systems methanol/water, isopropanoil/water, and toluene methylcyclohexane have been used. The results may be found in Lockett and Ahmed [*Chem. Eng. Res. Des.*, **61**, 110 (1983)], Korchinsky et al. [*Trans. Chem. E.*, **72**, 406 (1994)], and Korchinsky [ibid., 472]

All the foregoing test programs involve distillation of well-defined mixtures under total reflux conditions. The primary value of the results is in the comparative data, but it should be emphasized that the design of each device was not necessarily optimized for the test conditions.

Additional plate-efficiency data are listed in Table 14-6.

PACKED COLUMNS

Introduction   Packed columns for gas-liquid contacting are used extensively for absorption, stripping, and distillation operations. Usually the columns are filled with a randomly oriented packing material, but for an increasing number of applications the packing is very care-

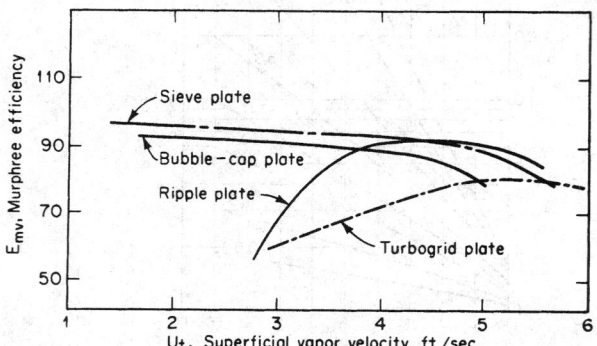

FIG. 14-40   Plate efficiencies, methanol-water. To convert feet per second to meters per second, multiply by 0.3048. [*Standart et al.*, Br. Chem. Eng., **11**, 1370 (1966); Sep. Sci., **2**, 439 (1967).]

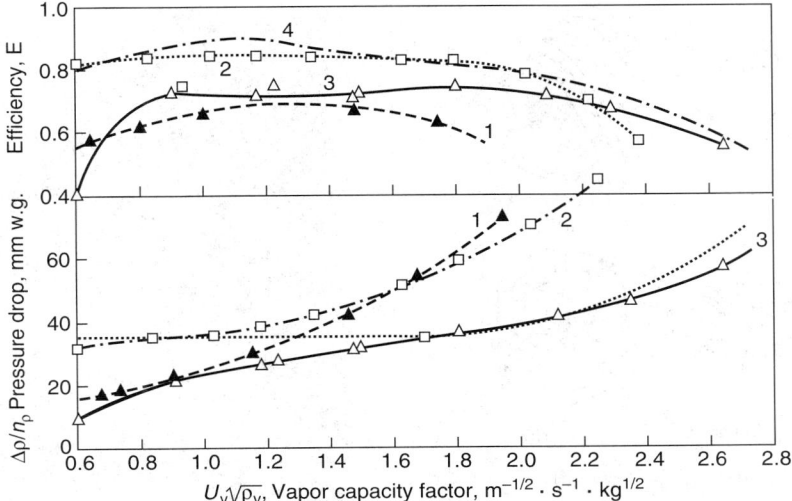

FIG. 14-42 Overall (Murphree) efficiency and pressure drop data for several devices using the same test mixture (ethylbenzene/styrene). See text for details. [*Billet, Conrad, and Grubb,* Instn. Chem. Engrs. Symp. Ser. No. 32, **5,** *111 (1979)*.]

fully positioned in the column. The packed column is characteristically operated with counterflow of the phases.

The packed column is a simple device compared with the plate column (Fig. 14-44). A typical column consists of a cylindrical shell containing a support plate for the packing material and a liquid distributing device designed to provide effective irrigation of the packing. Devices may be added to the packed bed to provide redistribution of liquid that might channel down the wall or otherwise become maldistributed. Several beds, each with liquid distributor and support device, may be used within the same column shell. For example, a distillation column with rectifying and stripping zones, requires a minimum of two beds.

The key issue in packed column design is the selection of the packing material. Such material should provide effective contacting of the phases without excessive pressure drop. Many packings are commercially available, each possessing specific advantages with respect to cost, surface availability, interface regeneration, pressure drop, weight, and corrosion resistance. Packed beds may be divided into two categories: Those containing packing elements that are placed in the column in a random arrangement, usually by dumping; and those containing carefully installed elements designed specifically to fit the column dimensions. The former elements are called random, or dumped, packings. The latter are called ordered, or structured, packings. Typical random and structured packings are shown in Fig. 14-45.

Packed Columns versus Plate Columns Packed columns are usually specified when plate devices would not be feasible because of undesirable fluid characteristics or some special design requirement. Conditions favoring packed columns are:

1. For columns less than 0.6-m (2.0-ft) diameter, packings are usually cheaper than plates unless alloy-metal packings are required.

2. Acids and many other corrosive materials can be handled in

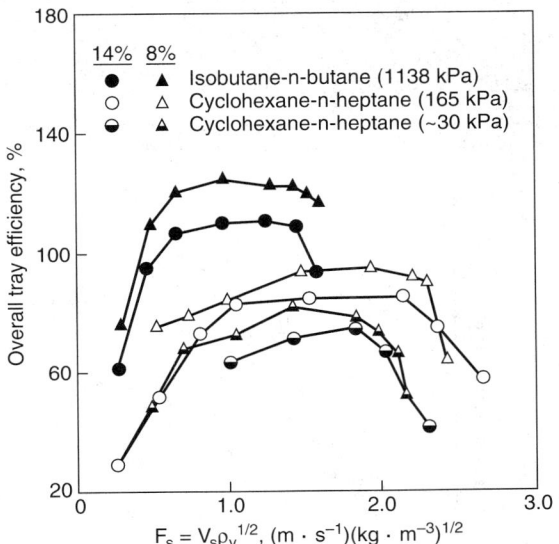

FIG. 14-43 Overall (Murphree) efficiencies of sieve plates with hole/active area ratios of 0.08 and 0.14. Efficiency values greater than 1.0 (100%) result from crossflow effects (Figs. 14-38, 14-39). [*Yanagi and Sakata,* Ind. Eng. Chem., Proc. Des. Devel., **21,** *712 (1982)*.] Reproduced with permission, copyright © 1982 American Chemical Society.

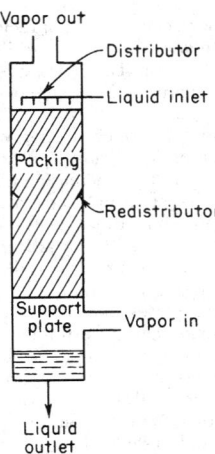

FIG. 14-44 Packed column (schematic).

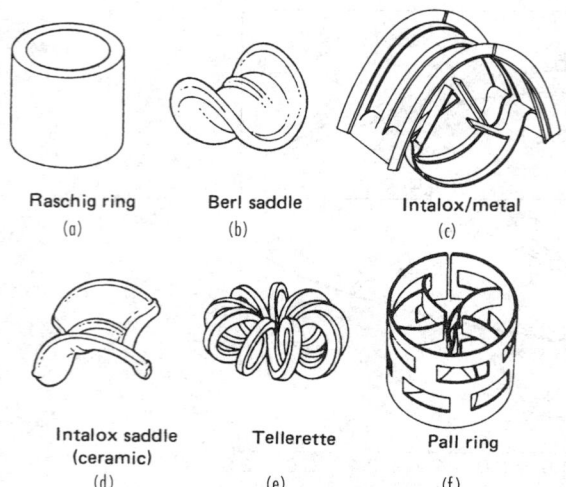

Raschig ring
(a)

Berl saddle
(b)

Intalox/metal
(c)

Intalox saddle
(ceramic)
(d)

Tellerette
(e)

Pall ring
(f)

FIG. 14-45a Representative random packings. Types (c), (e) and (f) are the through-flow variety.

packed columns because construction can be of ceramic, carbon, or other resistant materials.

3. Packings often exhibit desirable efficiency-pressure-drop characteristics for critical vacuum distillations.

4. Liquids tending to foam may be handled more readily in packed columns because of the relatively low degree of liquid agitation by the gas.

5. Holdup of liquid can be quite low in packed columns, an advantage when the liquid is thermally sensitive.

Conditions unfavorable to packed columns are:

1. If solids are present in the liquid or gas, plate columns can be designed to permit easier cleaning.

2. Some packing materials are subject to easy breakage during insertion into the column or resulting from thermal expansion and contraction.

3. High liquid rates can often be handled more economically in plate columns than in packed columns.

4. Cooling coils can be incorporated more readily into plate devices.

5. Low liquid rates lead to incomplete wetting of column packings, thus decreasing contacting efficiency.

Packed-Column Hydraulics Pressure drop of a gas flowing upward through a packing countercurrently to liquid flow, is characterized graphically in Fig. 14-46. At very low liquid rates, the effective open cross section of the packing is not appreciably different from that of dry packing, and pressure drop is due to flow through a series of variable openings in the bed. Thus, pressure drop is proportional approximately to the square of the gas velocity, as indicated in the region AB.

At higher liquid rates the effective open cross section is smaller because of the presence of liquid, and a portion of the energy of the gas stream is used to support an increasing quantity of liquid in the column (region A'B'). For all liquid rates, a zone is reached where pressure drop is proportional to a gas-flow-rate power distinctly higher than 2; this zone is called the **loading zone,** as indicated in Fig. 14-46. The increase in pressure drop is due to the rapid accumulation of liquid in the packing-void volume.

As the liquid holdup increases, one of two changes may occur. If the packing is composed essentially of extended surfaces, the effective orifice diameter becomes so small that the liquid surface becomes continuous across the cross section of the column, generally at the top of the packing. Column instability occurs concomitantly with a rising continuous-phase liquid body in the column. The change in pressure drop is quite great with only a slight change in gas rate (condition C or C'). The phenomenon is called *flooding* and is analogous to entrainment flooding in a plate column.

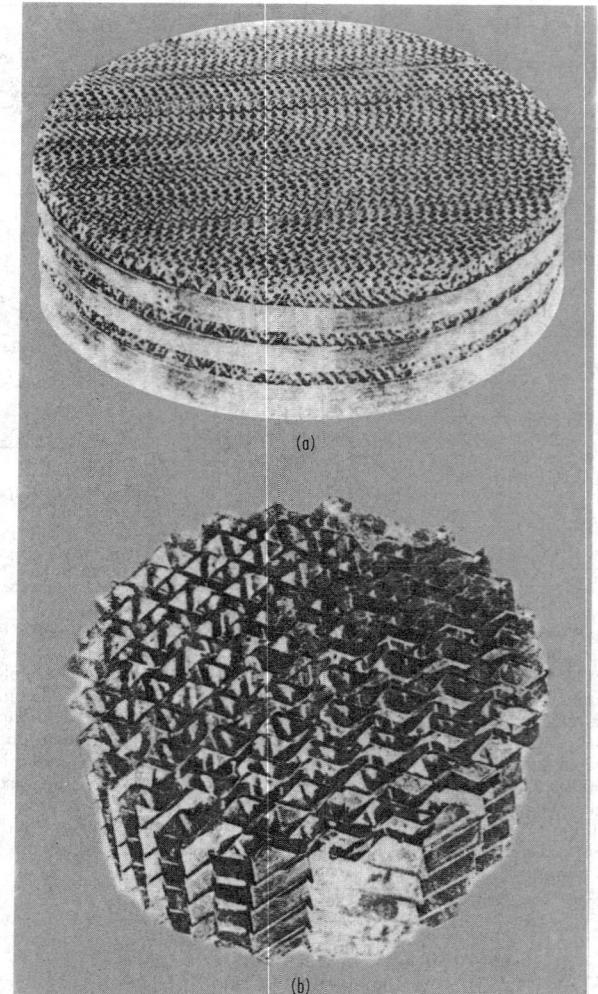

(a)

(b)

FIG. 14-45b Representative arranged-type packings: (a) Koch Sulzer, (b) Flexipac. (*Courtesy Koch Engineering Co., Inc.*)

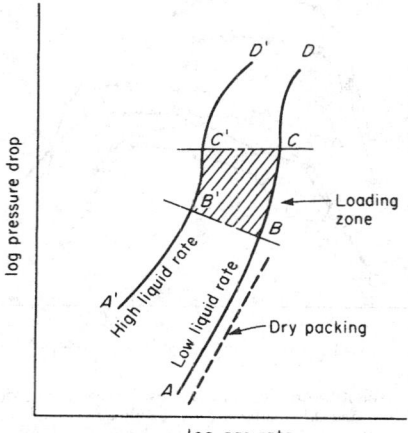

FIG. 14-46 Pressure-drop characteristics of packed columns.

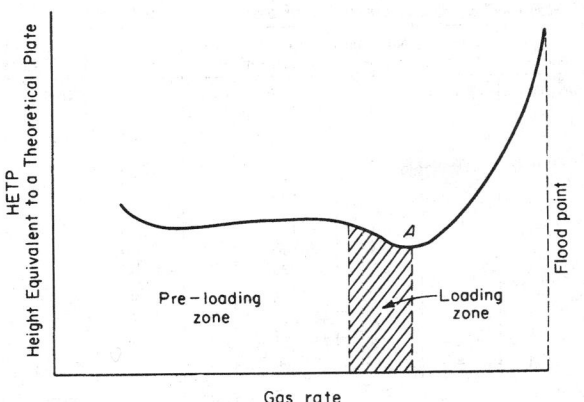

FIG. 14-47 Efficiency characteristics of packed columns (total-reflux distillation.)

If the packing surface is discontinuous in nature, a phase inversion occurs, and gas bubbles through the liquid. The column is not unstable and can be brought back to gas-phase continuous operation by merely reducing the gas rate. Analogously to the flooding condition, the pressure drop rises rapidly as phase inversion occurs.

A stable operating condition beyond "flooding" (region *CD* or *C′D′*) for nonextended surface packing with the liquid as the continuous phase and the gas as the dispersed phase has been reported by Lerner and Grove [*Ind. Eng. Chem.*, **43**, 216 (1951)] and Teller [*Chem. Eng.*, **61**(9), 168 (1954)].

For total-reflux distillations carried out in packed columns, regions of loading and flooding are identified by their effects on mass-transfer efficiency, as shown in Fig. 14-47. Gas and liquid rate increase

together, and a point is reached at which liquid accumulates rapidly (point *A*) and effective surface for mass transfer decreases rapidly.

Flooding and Loading Since flooding or phase inversion normally represents the maximum capacity condition for a packed column, it is desirable to predict its value for new designs. The first generalized correlation of packed-column flood points was developed by Sherwood, Shipley, and Holloway [*Ind. Eng. Chem.*, **30**, 768 (1938)] on the basis of laboratory measurements primarily on the air-water system.

Later work with air and liquids other than water led to modifications of the Sherwood correlation, first by Leva [*Chem. Eng. Prog. Symp. Ser.*, **50**(10), 51 (1954)] and later in a series of papers by Eckert. The generalized flooding-pressure drop chart by Eckert [*Chem. Eng. Progr.*, **66**(3), 39 (1970)], included in the previous edition of this handbook, was modified and simplified by Strigle [*Packed Tower Design and Applications*, Gulf Publ. Co., Houston, 1994] as shown in Fig. 14-48. It includes pressure drop curves, as introduced by Leva [*Chem. Eng. Progr. Symp. Ser. No. 10*, **50**, 51 (1954)], and is often called the generalized pressure drop correlation (GPDC). The ordinate scale term is a capacity parameter related to the Souders-Brown coefficient used for plate columns:

$$C_s = U_t \left[\frac{\rho_G}{(\rho_L - \rho_G)} \right]^{0.50} F_p^{0.5} \, v^{0.05} \qquad (14\text{-}150)$$

where U_t = superficial gas velocity, ft/s
ρ_G, ρ_L = gas and liquid densities, lb/ft^3 or kg/m^3
F_p = packing factor, ft^{-1}
v = kinematic viscosity of liquid, cS

The abscissa scale term is the same flow parameter used for plates (dimensionless):

$$F_{LG} = \frac{L}{G(\rho_G/\rho_L)^{0.5}} \qquad (14\text{-}151)$$

There is not a specific flood curve; a pressure drop of 1.50 in H$_2$O/ft is considered to represent an incipient flooding condition, although

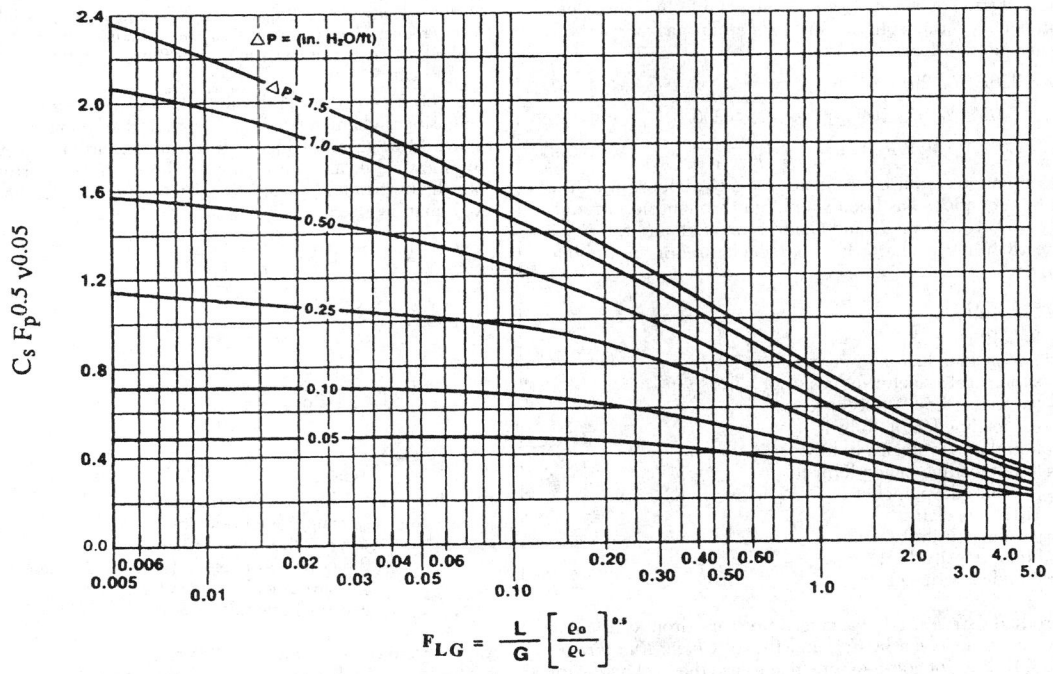

FIG. 14-48 Generalized pressure drop correlation of Eckert/Leva, as modified by Strigle. To convert inches H$_2$O to mm H$_2$O/m, multiply by 83.31. From *Packed Tower Design and Applications* by Ralph E. Strigle, Jr., copyright © 1994 by Gulf Publishing Co., Houston, Texas. Used with permission. All rights reserved.

pressure drops at flooding have been measured in the range of 2.0–2.5 inches H_2O/ft.

The packing factor F_p is empirically determined for each packing type and size. Values of F_p, together with general dimensional data for individual packings, are given in Table 14-7. For pressure drop and flooding estimates, values of F_p should always be used with caution. A detailed analysis of the GPDC approach (Fig. 14-48) was made by Kister and Gill [*Chem. Eng. Progr.*, **87**(2), 32 (1991)] who amassed a large data bank, mostly for random packings. They found it necessary to use separate curves for each packing type and size and also to differentiate between aqueous and nonaqueous systems. An example of their work is shown in Fig. 14-49, for 2-inch (50 mm) metal Pall rings. Note that a packing factor of 27 ft⁻¹ was used for this packing. A complete set of charts for both random and ordered packings is given in Chap. 10 of the book by Kister (*Distillation Design*, McGraw-Hill, New York, 1992).

Pressure Drop   Reference to pressure drop has already been made in connection with the GPDC (Fig. 14-48). For gas flow through dry packings, pressure drop may also be estimated by use of an orifice equation. For irrigated packings, pressure drop increases because of the presence of liquid, which effectively decreases the available cross section for gas flow (Fig. 14-46). In principle, there should be a method for correcting the dry pressure drop for the presence of water. This approach was used by Leva [*Chem. Eng. Progr. Symp. Ser. No. 10*, **50**, 51 (1954)]. A more recent method by by Robbins [*Chem. Eng. Progr.*, **87**(1), 19 (1990)] utilizes the same approach and is described here. The total pressure drop is

$$\Delta P_t = \Delta P_d + \Delta P_L \qquad (14\text{-}152)$$

where ΔP_t = total pressure drop, inches H_2O per foot of packing
ΔP_d = dry pressure drop = $C_3 G_f^2 \, 10^{(C_4 L_f)}$ (14-153)
ΔP_L = pressure drop due to liquid presence
= $0.4[L_f/20,000]^{0.1} [C_3 G_f^2 \, 10^{(C_4 L_f)}]^4$ (14-154)
G_f = gas loading factor = $986 F_s (F_{pd}/20)^{0.5}$ (14-155)
L_f = liquid loading factor = $L(62.4/\rho_L)(F_{pd}/20)^{0.5}\mu_L^{0.1}$ (14-156)

The term F_{pd} is a dimensionless dry packing factor, specific for a given packing type and size. Values of F_{pd} are given in Table 14-7. For operating pressures above atmospheric, and for certain packing sizes, L_f and G_f are calculated differently:

$$G_f = 986 F_s (F_{pd}/20)^{0.5} \, 10^{0.3 \rho_G} \qquad (14\text{-}157)$$

$$L_f = L(62.4/\rho_L)(F_{pd}/20)^{0.5}\mu_L^{0.2} \qquad F_{pd} > 200 \qquad (14\text{-}158)$$

$$= L(62.4/\rho_L)(20/F_{pd})^{0.5}\mu_L^{0.1} \qquad F_{pd} < 200 \qquad (14\text{-}159)$$

The Robbins equations require careful attention to dimensions. However, use of the equations has been simplified through the introduction of Fig. 14-50. The terms L_f and G_f are evaluated, and the ΔP_L is obtained directly from the chart. Basic nomenclature for the Robbins method follows:

$C_3 = 7.4 \, (10)^{-8}$
$C_4 = 2.7 \, (10)^{-5}$
F_{pd} = dry packing factor, dimensionless
F_s = superficial F-factor for gas, $U_s \rho_g^{0.5}$, ft/s(lb/ft³)^{0.5}
G = gas mass velocity = lb/hr·ft²
G_f = gas loading factor, lb/hr·ft²
L = liquid mass velocity, lb/hr·ft²
L_f = liquid loading factor, lb/hr·ft²
ΔP = pressure drop, inches H_2O/ft packing (× 83.3 = mm H_2O/m packing)
ρ_G = gas density, lb/ft³
ρ_L = liquid density, lb/ft³
μ_L = liquid viscosity, cP

For ordered, or structured, packings, pressure-drop estimation methods have been reviewed by Fair and Bravo [*Chem. Eng. Progr.*, **86**(1), 19 (1990)]. It is not common practice to use the packing factor approach for predicting pressure drop or flooding. For operation below the loading point, the model of Bravo et al. [*Hydrocarbon*

TABLE 14-7a   Characteristics of Structured Packings

Name	Material	Nominal size	Surface area, m²/m³	% voids	Packing factor, 1/m	Vendor
Flexipac	S	1	558	91	108	Koch
		2	223	93	72	
		3	134	96	52	
Flexiramic	C	28	282	70	131	Koch
		48	157	74	79	
		88	102	85	49	
Gempak	S	4A	446	92	105	Glitsch
		3A	335	93	69	
		2A	223	95	53	
Intalox	S	1T	315	95	66	Norton
		2T	213	97	56	
		3T	177	97	43	
Max-Pak	S	—	229	95	39	Jaeger
Mellapak	S	125Y	125	97	33	Sulzer
		125X°	125	97		
		250Y	250	95	66	
		250X°	250	95	8	
		350Y	350	93	75	
		500X°	500	91	25	
Montz-Pak	S	B1-125	125	97		Julius Montz
		B1-250	250	95	72	
		BI-350	350	93		
	G	A3-500	500	91		
	E	BSH-250	250	95		
		BSH-500	500	91		
Ralupak	S	250YC°	250	95		Raschig
Sulzer	G	AX°	250	95		Sulzer
		BX°	492	90	69	
		CY	700	85		

NOTES: °60° corrugation angle (with the horizontal); all others 45°.
Packing factors from Kister and Gill [*Chem. Eng. Progr.*, **87**(2), 32 (1991), and Houston AIChE meeting, March 19–23, 1995].
Materials of construction: C = ceramic; E = expanded metal; G = metal gauze; S = sheet metal.
Vendors: Glitsch, Inc., Dallas, Texas; Koch Engineering Co., Wichita, Kansas; Jaeger Products, Inc., Houston, Texas; Julius Montz, Hilden, Germany; Raschig AG, Ludwigshafen, Germany; and Sulzer Bros., Winterthur, Switzerland.

Proc., **65**(3), 45 (1986)] is preferred. To use this and alternate models, dimensional characteristics of structured packing must be defined. Figure 14-51 shows nomenclature and definitions of key dimensions. Not shown, but also important, is the angle the corrugations make with the horizontal (usually 45 or 60°). Then the Rocha et al. predictive equation is:

$$\Delta P_t = \left[0.171 + \left(\frac{92.7}{Re_g}\right)\right]\left(\rho_g \frac{U_{ge}^2}{S}\right)\left[\frac{1}{1 - C_o \, Fr_L^{0.05}}\right]^5 \qquad (14\text{-}160)$$

where $Re_g = \dfrac{SU_{ge}\rho_g}{\mu_g}$

$U_{ge} = \dfrac{U_g}{(\varepsilon \sin \theta)}$

$Fr_L = \dfrac{U_L^2}{(Sg)}$

and where S = length of corrugation side
U_g = superficial velocity of gas
ε = void fraction of packing
θ = corrugation angle (from horizontal), deg.
U_L = superficial liquid velocity
g = gravitational constant

Any consistent set of units may be used.

This model applies in the region below the loading point, and it cannot predict the flood point because it does not include the effects of gas velocity on liquid holdup. The model of Stichlmair et al. [*Gas*

TABLE 14-7b **Characteristics of Random Packings**

Name	Material	Nominal size mm	Nominal size Number	Wall thickness, mm	Bed weight, kg/m³	Area, m²/m³	% voids	Packing factor Fp, m⁻¹	Dry packing factor Fpd, m⁻¹	Vendor
Raschig rings	C	6	—	1.6	960	710	62	—	5250	Various
		13	—	2.4	880	370	64	1900	1705	
		25	—	3.2	670	190	74	587	492	
		50	—	6.4	660	92	74	213	230	
		75	—	9.5	590	62	75	121	—	
Raschig rings	M	19	—	1.6	1500	245	80	984	—	Various
		25	—	1.6	1140	185	86	472	492	
		50	—	1.6	590	95	92	187	223	
		75	—	1.6	400	66	95	105	—	
Pall rings	M	16	—	0.40			92	256	262	Norton, Koch, Glitsch
		25	—	0.51	480	205	94	183	174	
		38	—	0.64	415	130	95	131	91	
		50	—	0.81	385	115	96	89	79	
		90	—	—	270	92	97	59	46	
Cascade mini rings (CMR)	M	—	1	—	389	250	96	131	102	Glitsch
		—	1.5	—	234	144	97	95	—	
		—	2.5	—	195	123	98	72	79	
		—	3	—	58	103	98	46	43	
	P	—	1A	—	71	185	94	98	92	
		—	3A	—	40	74	96	39	33	
Berl saddles	C	6	—	—	900	900	60		2950	Koch
		13	—	—	865	465	62	790	900	
		25	—	—	720	250	68	360	308	
		38	—	—	640	150	71	215	154	
		50	—	—	625	105	72	150	102	
Intalox saddles	C	6	—	—	864	984	65	302	2720	Norton
		13	—	—	736	623	71	—	613	
		25	—	—	672	256	73	302	208	
		50	—	—	608	118	76	131	121	
		75	—	—	576	92	79	72	66	
Fleximax	M	—	300	—	—	141	98	85	—	Koch
		—	400	—	—	85	98	56	—	
Metal Intalox (IMTP)	M	25	—	—	352	230	97	134	141	Norton
		40	—	—	237	154	97	79	85	
		50	—	—	150	98	98	59	56	
		70	—	—	130	56	98	39	—	
Nutter rings	M	—	1	0.30	178	168	98	98	89	Nutter
		—	2	0.45	173	96	98	59	56	
		—	2.5	0.45	145	83	66	52	49	
		—	3.0	0.50	133	66	98	43	36	
Pall rings	P	25	—	—	80	206	90	180	180	Norton
		50	—	—	61	102	92	85	82	
		90	—	—	53	85	92	56	39	
	C	25	—	—				351	—	
		38	—	—				180	—	
		50	—	—				141	—	
Intalox saddles	P	—	1	—	96	207	90	131	131	Norton
		—	2	—	56	108	93	92	85	
Snowflake	P	—	—	—	45	92	95	43	—	Norton
Nor-Pac	P	25	1	—	72	180	92	82	—	NSW
		38	1.5	—	61	144	93	56	—	
		50	2.0	—	53	102	94	39	—	
Tri-Pack	P	25	1	—	72	180	92	82	—	Jaeger
		50	2	—	53	102	94	39	—	
VSP	M	25	1	—	352	206	98	105	—	Jaeger
		50	2	—	296	112	96	69	—	
Tellerettes	P	25	1	—	112	180	87	—	—	Ceilcote
		50	2	—	59	125	93	—	—	

NOTES: M = metal, carbon steel. Other metals available.
P = plastic, polypropylene. Other plastics available.
Packing factor F_p from Kister and Gill [*Chem. Eng. Progr.*, **87**(2), 32 (1991) and Houston AIChE meeting, March 19–23, 1995]; Strigle, *Packed Tower Design and Applications* [Gulf Publishing Co., Houston, Texas, 1994]; dry packing factor F_{pd} from Robbins [*Chem. Eng. Progr.*, **87**(1), 19 (1990)].
Vendors: Ceilcote Co., Berea, Ohio; Glitsch, Inc., Dallas, Texas; Koch Engineering Co., Wichita, Kansas; Jaeger Products, Inc., Houston, Texas; NSW Corp., Roanoke, Virginia; Norton Co., Akron, Ohio; and Nutter Engineering Co., Tulsa, Oklahoma.

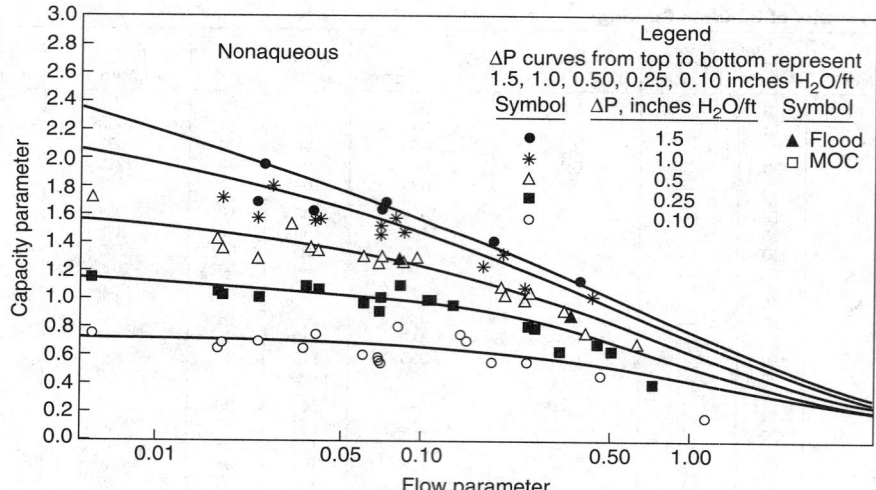

Basic: Fp = 27
Pressure drop measured in inches H₂O/ft

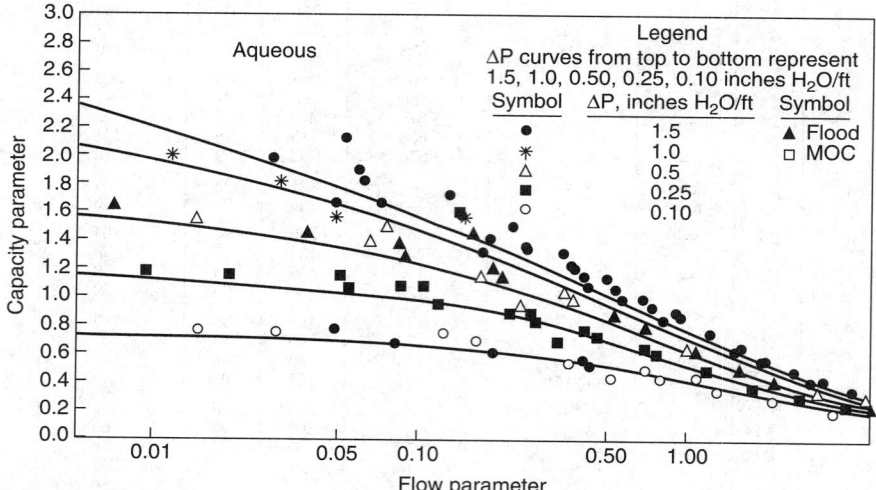

Basic: Fp = 27
Pressure drop measured in inches H₂O/ft

FIG. 14-49   Pressure drop/flooding correlation of Kister and Gill for 2-inch metal Pall rings. The upper chart is for nonaqueous systems, the lower chart for aqueous systems. To convert inches H₂O/ft to mm H₂O/m, multiply by 83.31.

Sepn. Purif., **3**, 19 (1989)] takes holdup into account and applies to random as well as structured packings. It is somewhat cumbersome to use and requires three constants for each packing type and size. Such constants have been evaluated, however, for a number of commonly used packings. A more recent pressure drop and holdup model, suitable for extension to the flood point, has been published by Rocha et al. [*Ind. Eng. Chem. Research*, **35**, 1660 (1996)]. This model takes into account variations in surface texturing of the different brands of packing.

Representative pressure drop data for random and structured packings are given in Figs. 14-52–14-54.

Example 13: Packed Column Pressure Drop   Air and water are flowing countercurrently through a bed of 2-inch metal Pall rings. The air mass velocity is 2.03 kg/s·m² (1500 lbs/hr·ft²), and the liquid mass velocity is 12.20 kg/s·m² (9000 lbs/hr·ft²). Calculate the pressure drop by the generalized pres-

sure drop (GPDC, Fig. 14-48) and the Robbins methods. Properties: $\rho_G = 0.074$ lbs/ft³; $\rho_L = 62.4$ lbs/ft³, $\mu_L = 1.0$ cP, $\nu = 1.0$ cS. The packing factor $F_p = 27$ ft⁻¹. For Robbins, $F_{pd} = 24$ ft⁻¹. The flow parameter $F_{LG} = L/G \ (\rho_G/\rho_L)^{0.5} = (9000/1500) \ (0.072/62.4)^{0.5} = 0.207$. The F-factor $= F_s = U_t\rho_G^{0.5} = G/(\rho_G^{0.5} \ 3600) = 1500/[(0.074)^{0.5} \ (3600)] = 1.53$ ft/s(lb/ft³)^{0.5}

Using the GPDC method, the capacity parameter [by Eq. (14-150)] = $U_t[\rho_G/(\rho_L - \rho_G)]^{0.5} F_p^{0.5} \nu^{0.05}$, which is roughly equivalent to

$$\frac{F_s}{\rho_L^{0.5}} F_p^{0.5} \nu^{0.05} = \left(\frac{1.53}{62.4}\right)^{0.5} 27^{0.5}(1.0)$$

$$= 1.01$$

Referring to Fig. 14-48, the intersection of the capacity parameter and the flow parameter lines gives a pressure drop of 0.38 inches H₂O/ft packing. (The same result is obtained from the Kister/Gill chart, Fig. 14-49.)

Using the Robbins method, $G_f = 986F_s(F_{pd}/20)^{0.5} = 986(1.53)(24/20)^{0.5} = 1653$. $L_f = L \ (62.4/\rho_L)(F_{pd}/20)^{0.5} \mu^{0.1} = 9000 \ (1.0)(1.095)(1.0) = 9859$. $L_f/G_f = 5.96$.

From Fig. 14-50, pressure drop = 0.35 in. H₂O/ft packing.

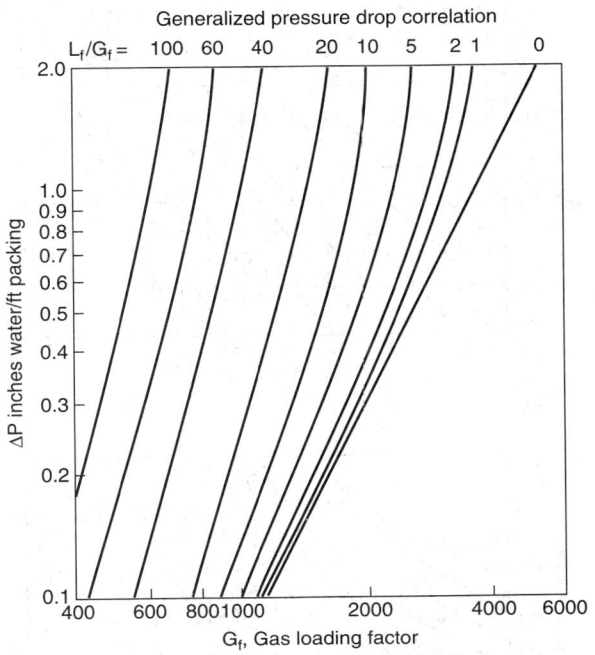

Generalized pressure drop correlation

$L_f/G_f = $ 100 60 40 20 10 5 2 1 0

G_f, Gas loading factor

FIG. 14-50 Pressure drop correlation for random packings, as presented by Robbins. [Chem. Eng. Progr., **87**(1), 19 (1990). *Reproduced with permission of the American Institute of Chemical Engineers. Copyright © 1990 AIChE. All rights reserved.*] To convert inches H_2O/ft to mm H_2O/m, multiply by 83.31.

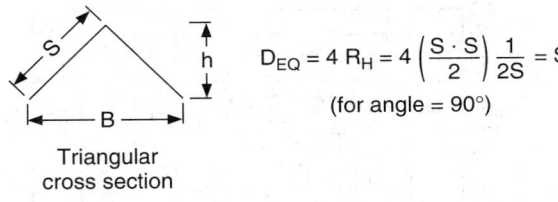

$$D_{EQ} = 4 R_H = 4 \left(\frac{S \cdot S}{2} \right) \frac{1}{2S} = S$$

(for angle = 90°)

Triangular cross section

a. Flow channel cross section

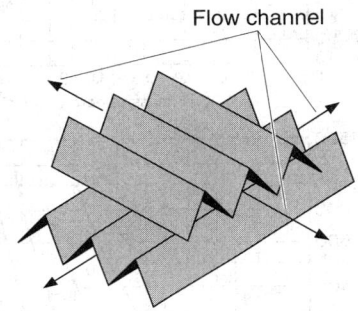

Flow channel

b. Flow channel arrangement

FIG. 14-51 Geometric properties of typical structured packings.

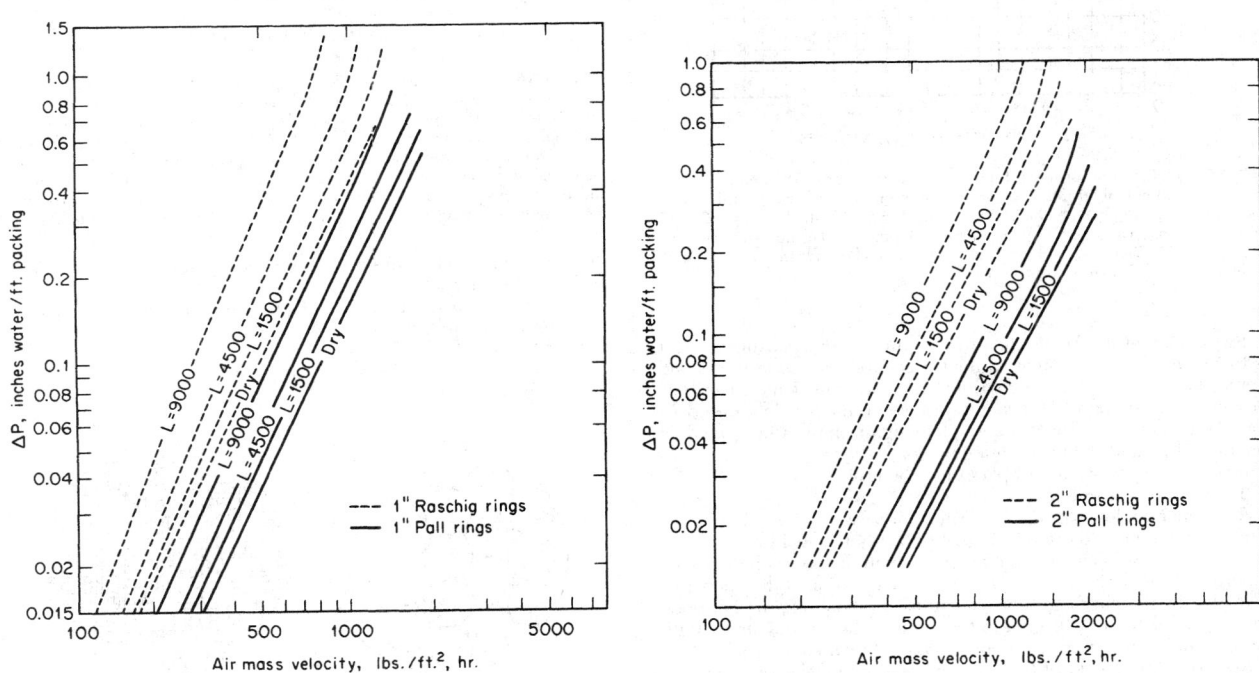

FIG. 14-52 Pressure drop for metal pall rings, 0.024-in wall thickness (1-in size) and 0.036-in wall (2-in size). Metal Raschig rings have 1/16-in wall. L = lb liquid/(h·ft²). To convert inches of water per foot to millimeters of water per meter, multiply by 83.31; to convert inches to millimeters, multiply by 25.4; and to convert pounds per hour-square foot to kilograms per second-square meter, multiply by 0.001356. [*Eckert, Foote, and Huntington*, Chem. Eng. Progr., **54**(1), 70 (1958).]

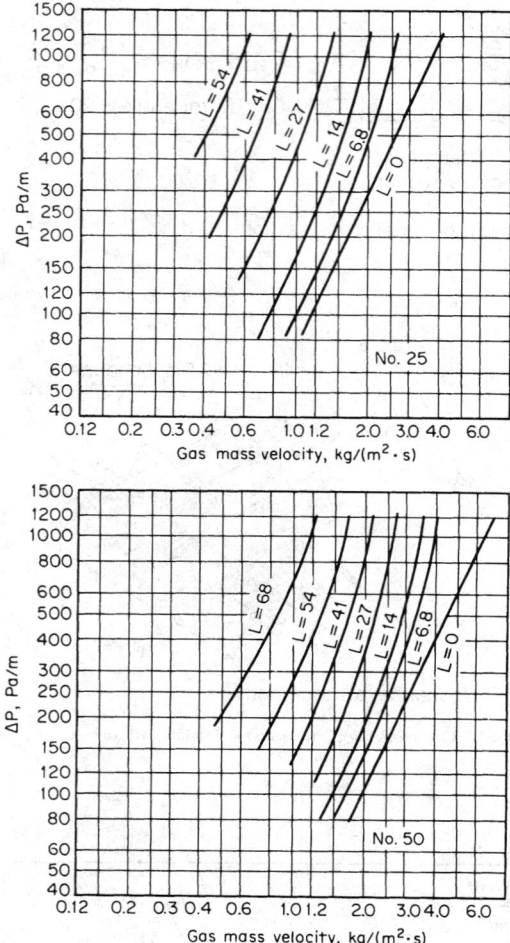

FIG. 14-53 Pressure for metal Intalox saddles, sizes No. 25 (nominal 25 mm) and No. 50 (nominal 50 mm). Air-water system at atmospheric pressure, 760-mm (30-in) column, bed height, 3.05 m (10 ft). L = liquid rate, kg/(s·m²). To convert kilograms per second-square meter to pounds per hour-square foot, multiply by 151.7; to convert pascals per meter to inches of water per foot, multiply by 0.1225. (*Courtesy Norton Company, Akron, Ohio.*)

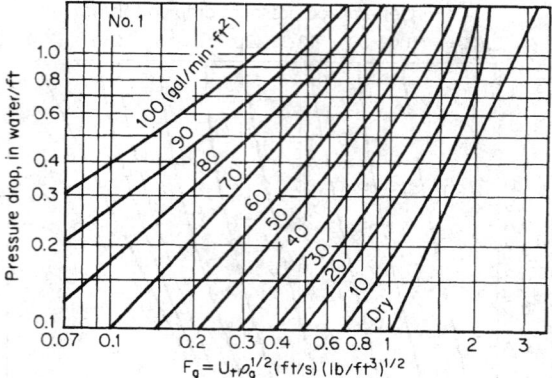

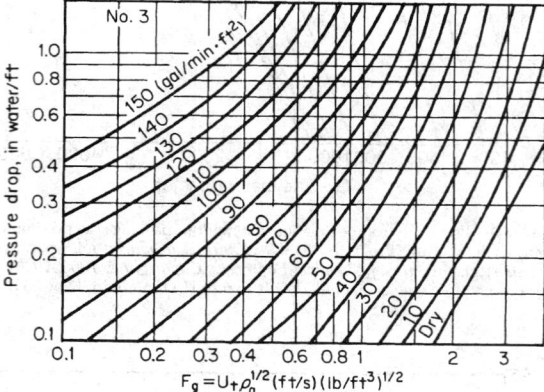

FIG. 14-54 Pressure drop for Flexipac packing, sizes No. 1 and No. 3. Air-water system at atmospheric pressure. Liquid rate in gallons per minute-square foot. To convert (feet per second) (pounds per cubic foot)$^{1/2}$ to (meters per second) (kilograms per cubic meter)$^{1/2}$, multiply by 1.2199; to convert gallons per minute-square foot to pounds per hour-square foot, multiply by 500; to convert inches of water per foot to millimeters of water per meter, multiply by 83.31; and to convert pounds per hour-square foot to kilograms per second-square meter, multiply by 0.001356. (*Courtesy Koch Engineering Co., Wichita, Kansas.*)

Support Plates

Support Plates While the primary purpose of a packing support is to retain a bed of packing without excessive restriction to gas and liquid flow, it also serves to distribute both streams. Unless carefully designed, the support plate can also cause premature column flooding. Thus, design of the support plate significantly affects column pressure drop and stable operating range.

Two basic types of support plates may be utilized:
1. Countercurrent
2. Separate flow passages for liquid and gas

The two types are indicated in Figs. 14-55, 14-56, and 14-57.

The degree of open area on a support plate is the fraction of void inherent in the design of the plate minus that portion of the open area occluded by the packing. To avoid premature flooding, the net open area of the plate must be greater than that of the packing itself. With the countercurrent type of support plate the free area for gas flow can range up to 90 percent of the column cross-sectional area. However, such a plate is easily occluded by the packing pieces resting directly on it.

The separate flow passage devices can be designed for free areas up

to 90 percent, and because of their geometry they will have very little occlusion by the packing.

Liquid Holdup Three modes of liquid holdup in packed columns are recognized:
1. Static, h_s
2. Total, h_t
3. Operating, h_o

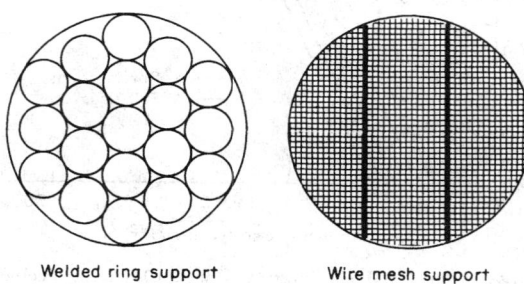

Welded ring support   Wire mesh support

FIG. 14-55 Packing supports (countercurrent).

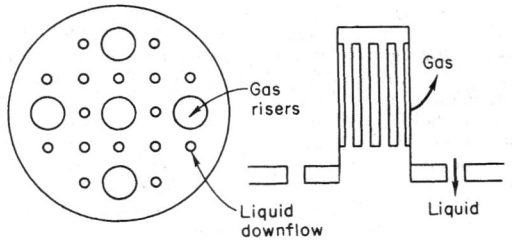

FIG. 14-56   Support plate, cap type (separate flow or "gas-injection" type).

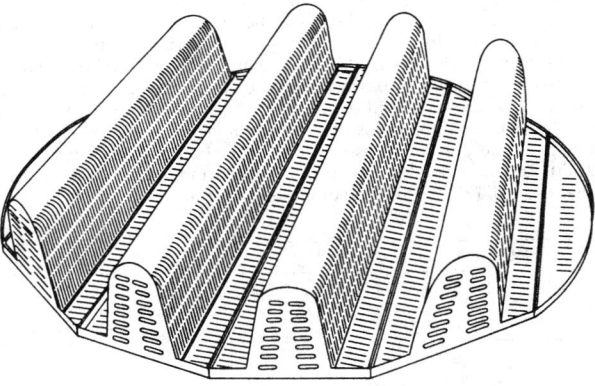

FIG. 14-57   Beam-type "gas-injection" support plate for large columns.

Static holdup is the amount of liquid remaining on packing that has been fully wetted and then drained. Total holdup is the amount of liquid on the packing under dynamic conditions. Operating holdup is the amount of liquid attributed to operation and is measured experimentally as the difference between total and static holdup. Thus,

$$h_t = h_o + h_s \qquad (14\text{-}161)$$

where h values are in volumes of liquid per total volume of bed. The effective void fraction under operating conditions is

$$\varepsilon' = \varepsilon - h_t \qquad (14\text{-}162)$$

Static holdup depends upon the balance between surface-tension forces tending to hold liquid in the bed and gravity or other forces that tend to displace the liquid out of the bed. Estimates of static holdup (for gravity drainage) may be made from the following relationship of Shulman et al. [*Am. Inst. Chem. Eng. J.*, **1**, 259 (1955)]:

$$h_s = 2.79 \frac{C_1 \mu^{C2} \sigma^{C3}}{\rho_\ell^{0.37}} \qquad (14\text{-}163)$$

where μ_ℓ = liquid viscosity, mPa·s
        σ = surface tension, mN/m
        ρ_ℓ = liquid density, kg/m³

and constants are

Packing	C_1	C_2	C_3
1.0-in carbon Raschig rings	0.086	0.02	0.23
1.0-in ceramic Raschig rings	0.00092	0.02	0.99
1.0-in ceramic Berl saddles	0.0055	0.04	0.55

For other packings and for the case in which static holdup is changed by gas flowing through the bed, the method of Dombrowski and Brownell [*Ind. Eng. Chem.*, **46**, 1207 (1954)], which correlates static holdup with a dimensionless capillary number, should be used.

Typical total holdup data for packings are shown in Figs. 14-58 and 14-59. It should be noted that over much of the preloading range gas rate has little effect on holdup.

Operating holdup may be estimated by the dimensionless equation of Buchanan [*Ind. Eng. Chem. Fundam.*, **6**, 400 (1967)]:

$$h_o = 2.2 \left(\frac{\mu_\ell' u_\ell}{g \rho_\ell d_p^2} \right)^{1/3} + 1.8 \left(\frac{u_\ell^2}{g d_p} \right)^{1/2} \qquad (14\text{-}164)$$

where μ_ℓ' = liquid viscosity, Pa·s
        u_ℓ = liquid superficial velocity, m/s
        g = gravitational constant, m/s²
        ρ_ℓ = liquid density, kg/m³
        d_p = nominal packing size, m

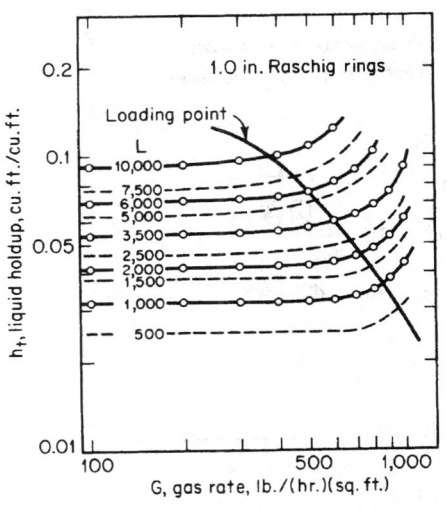

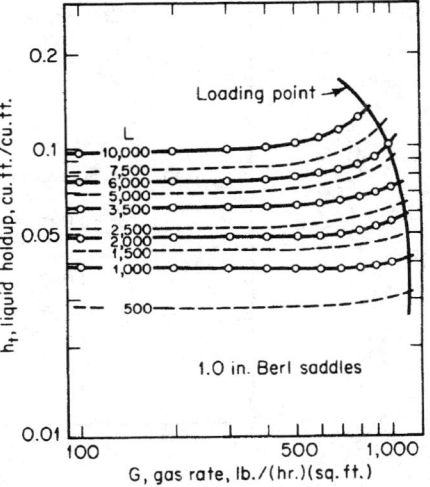

FIG. 14-58   Typical holdup data for random packings and the air-water system. The raschig rings are of ceramic material. To convert pounds per hour per ft² to kilograms per second per m², multiply by 0.001356; to convert inches to millimeters, muultiply by 25.4. [*Shulman et al.,* AIChE J. **1**, 247 (1955).]

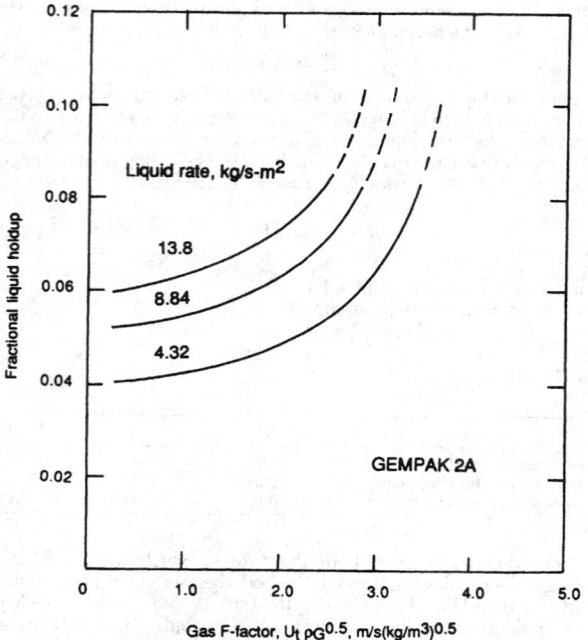

FIG. 14-59 Typical vendor data for liquid holdup of a structured packing, Gempak 2A. [*Courtesy Glitsch, Inc., Dallas, Texas.*]

The first term is a "film number"; the second is the Froude number. The equation applies to ring packings only operating below the load point and correlates all literature data to about ±20 percent.

Liquid holdup has been studied by Stichlmair et al. [*Gas Sepn. Purif.*, **3**, 19 (1989)] for both random and structured packings, and by Rocha et al. [*Ind. Eng. Chem. Res.*, **32**, 641 (1993)] for a variety of

structured packings. The holdup model developed by the latter investigators is shown in Fig. 14-60 in a representative case. It is clear that the modern, through-flow random packings and the structured packings have much less operating liquid holdup of liquid than do the older, traditional bluff-body packings.

Operating holdup contributes effectively to mass-transfer rate, since it provides residence time for phase contact and surface regeneration via agglomeration and dispersion. Static holdup is limited in its contribution to mass-transfer rates, as indicated by Thoenes and Kramers [*Chem. Eng. Sci.*, **8**, 271 (1958)]. In laminar regions holdup in general has a negative effect on the efficiency of separation.

Liquid Distribution   Uniform initial distribution of liquid at the top of the packed bed is essential for efficient column operation. This is accomplished by a device that spreads the liquid uniformly across the top of the packing. Baker, Chilton, and Vernon [*Trans. Am. Inst. Chem. Eng.*, **31**, 296 (1935)] studied the influence of the bed itself as a distributor and found that a single-point distribution in a 305-mm (12-in) column with 19-mm (¾-in) packing required 3.05 m (10 ft) of bed before achieving uniform distribution across the bed. They found also a tendency for liquid to migrate toward the column wall (Fig. 14-61), especially for ratios of column diameter to packing size less than 8. For a multipoint distribution, their recommendation was one liquid stream for each 194-cm² (30-in²) column area. Eckert [*Chem. Eng. Prog.*, **57**, 54 (1961)] recommends the following:

Column diameter, m	Streams/m²
1.2 or greater	40
0.75	170
0.40	340

Silvey and Keller [*I. Chem. E. Symp. Ser. No. 32*, **4**, 18 (1969)] found that a trough-type distributor with 34 streams/m² gave good liquid distribution in a 1.2-m column, up to a bed height of 10.7 m. Ceramic raschig rings of 38 and 76 mm nominal size were used, and efficiency profiles were measured by means of intermediate bed samples. A plot of their data is shown in Fig. 14-62; for the type of plot, a straight line indicates uniform distribution. The plotting technique is based on the Fenske relationship for theoretical stages (Sec. 13). This

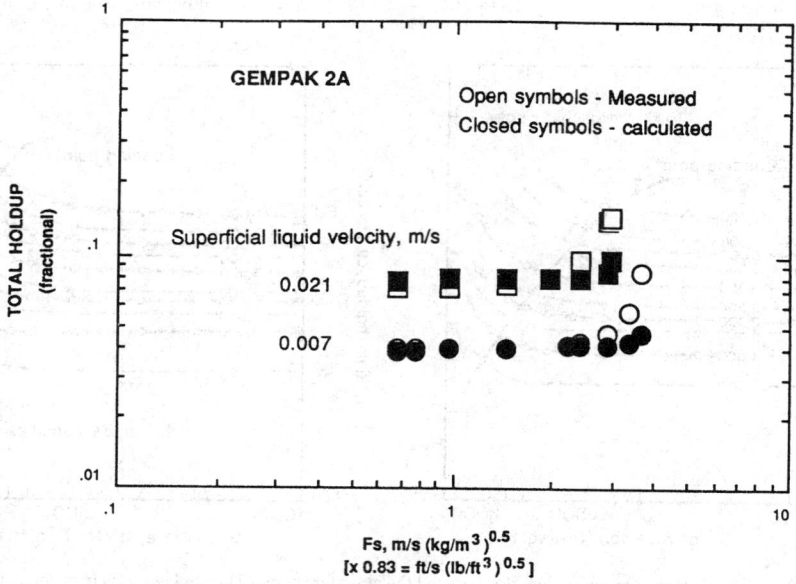

FIG. 14-60 Comparison of measured and calculated values of liquid holdup for Gempak 2A structured packing, air-water system. [*Rocha et al.,* Ind. Eng. Chem., **32**, 641 (1993).] Reproduced with permission. Copyright © 1993 American Chemical Society.

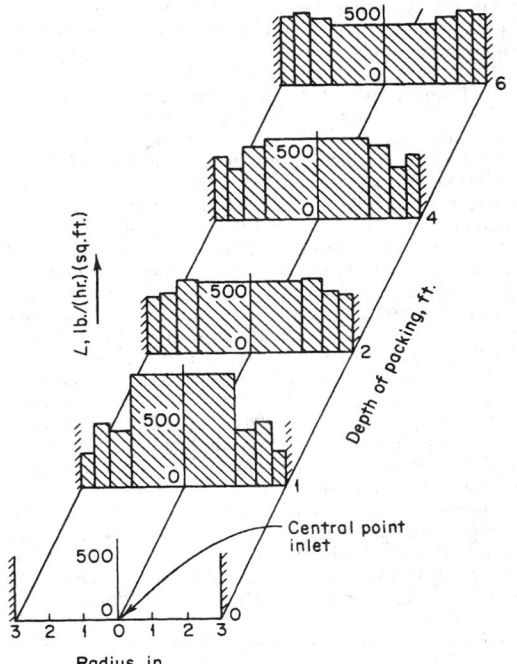

FIG. 14-61 Liquid distribution in a 6-in column packed with ½-in broken-stone packing. Increments of radius represent equal-annual-area segments of tower cross section. Central-point inlet. Water rate = 500 lb/(h·ft²). Air rate = 810 lb/(h·ft²). To convert pounds per hour-square foot to kilograms per second-square meter, multiply by 0.001356; to convert inches to centimeters, multiply by 2.54. (*Data from Baker, Chilton, and Vernon, in Sherwood and Pigford, Absorption and Extraction, 2d ed., McGraw-Hill, New York, 1952.*)

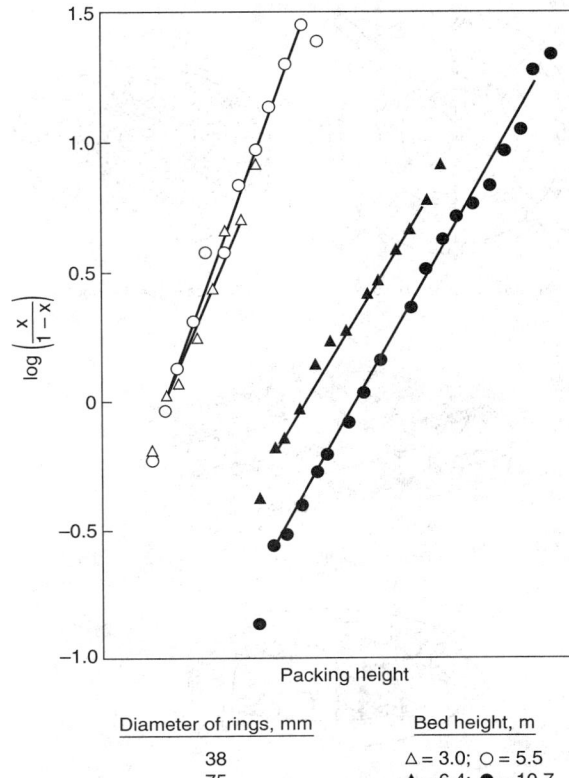

Diameter of rings, mm	Bed height, m
38	△ = 3.0; ○ = 5.5
75	▲ = 6.4; ● = 10.7

FIG. 14-62 Comparison of composition profiles at different bed heights and two sizes of ceramic raschig rings. Column diameter = 1.2 m, cyclohexane/*n*-heptane system at 1.65 bar and total reflux. [*Silvey and Keller*, I. Chem. E. Symp. Ser. No. 32, *1969.*]

same distributor was found to be unsatisfactory by Shariat and Kunesh [*Ind. Eng. Chem. Res.*, **34**, 1273 (1995)] when through-flow pall rings were used. A comparison between the trough distributor and an orifice-riser distributor with 100 streams/m² is shown in Fig. 14-63. In the normal range of operation, 30–70 percent more stages were obtained with the increased number of streams. The column, test mixture (cyclohexane/*n*-heptane) and operating pressure (1.65 bar) were the same as for the earlier Silvey-Keller tests.

The through-flow random packings and the structured packings require careful attention to liquid distribution. Such packings cannot correct an initial poor distribution. Several of the packing vendors maintain elaborate test stands for conducting special distributor evaluations. In general, practical aspects of design limit the number of streams to 100–120 per m², and these values have proven adequate for the newer packings. Kunesh et al. [*Ind. Eng. Chem. Res.*, **26**, 1845 (1987)] have an in-depth treatment of commercial-scale experiments that support designs of packed column liquid distributors.

Several distributor types are available: trough, orifice-rise, and perforated pipe. Examples of these types are shown in Fig. 14-64. The trough distributor provides good distribution under widely varying flow rates of gas and liquid. The liquid may flow through simple V-notch weirs, or it may flow through tubes that extend from the troughs to near the upper level of the packing. Some deposition of solids can be accommodated.

The orifice-riser distributor is designed to lay the liquid carefully onto the bed, with a minimum of contact with gas during the process. It can be designed to provide a large number of liquid streams, with the limit of sufficient liquid head to provide uniform liquid flow through the orifices. The gas risers must be designed to accommodate the expected variations in flow rate, often with a minimum of pressure drop. For very distribution-sensitive packings, it is necessary to include pour points in the vicinity of the column wall (to within 25 mm).

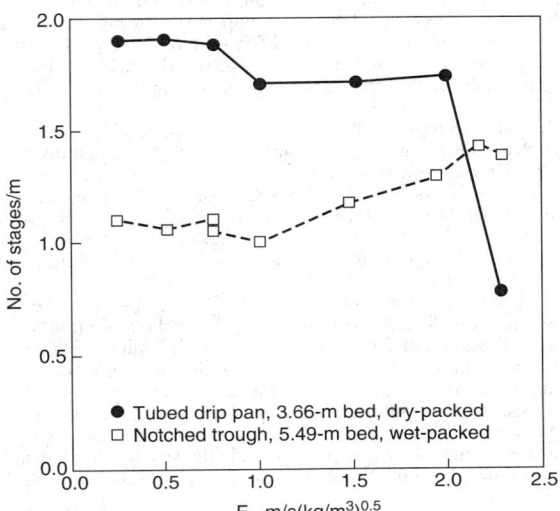

FIG. 14-63 Efficiency of beds of 51 mm Pall rings with two different distributors. Column diameter = 1.2 m, cyclohexane/*n*-heptane system at 1.65 bar and total reflux. [*Shariat and Kunesh*, Ind. Eng. Chem. Res., **34**, 1273 (1995).] Reproduced with permission. Copyright © 1995, American Chemical Society.

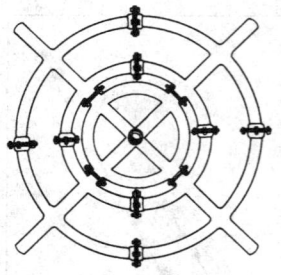
Perforated pipe distributor

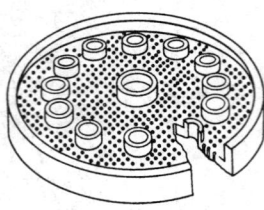
Orifice–type distributor

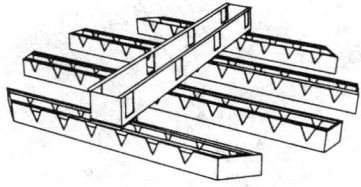
Trough–type distributor

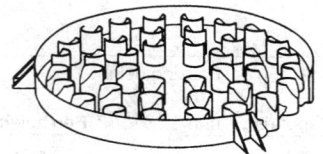
Weir–riser distributor

FIG. 14-64 Typical liquid distributors. (*Courtesy Norton Company, Akron, Ohio.*)

For larger diameter columns, and for low liquid rates, the distributor must be almost exactly level (e.g., within 6 mm for a 3-m diameter) or all pour points will not function. On the other hand, the rises must be high enough to accommodate the backup caused by high liquid rates. The needed head can be estimated from the orifice equation, with a discharge coefficient of 0.5. In some cases the orifices discharge directly into tubes that extend to the packed bed (the "tubed drip-pan distributor").

The perforated pipe distributor comprises a central feed sump and pipes that branch out from the sump to provide the liquid discharge. The level in the sump varies with liquid total flow rate, and the size of the lateral pipes and their perforations must be determined carefully to ensure that the ends of the pipes are not starved for liquid. The orifices are typically 4 to 6 mm diameter, and can be subject to plugging if foreign matter is present. The pipes must be leveled carefully, especially for large diameter columns.

Another type of distributor, not shown in Fig. 14-64, is the spray nozzle. It is usually not recommended for liquid distribution for two reasons. First, except for small columns, it is difficult to obtain a uniform spray pattern for the packing. The full-cone nozzle type is usually used, with the need for a bank of nozzles in larger columns. When there is more than one nozzle, the problem of overlap or underlap arises. A second reason for not using spray nozzles is their tendency toward entrainment by the gas, especially the smaller droplets in the spray size distribution. However, some mass transfer in the spray can be expected.

Maldistribution   Departure from uniform distribution of the phases in a packed column can be caused by:
1. The liquid distributor not dividing the liquid evenly over the column cross section.

2. The liquid moving more easily to the wall than vice versa. The resultant channeling along the wall may be accentuated by vapor condensing because of column heat losses.
3. The packing geometry inhibiting lateral distribution.
4. Void variations due to packing being improperly installed.
5. The column being out of vertical alignment.

Cause 1 has been covered in the preceding subsection. Causes 4 and 5 can be handled through careful design and installation. Causes 2 and 3 bear additional discussion at this point.

The effect of channeling on mass-transfer efficiency has been studied theoretically by Manning and Cannon [*Ind. Eng. Chem.*, **49**, 347 (1957)], Huber and Hiltbrunner [*Chem. Eng. Sci.*, **21**, 819 (1966)], and Meier and Huber (*Proc. Intn. Symp. Distill.*, Brighton, England, 1970). Typical results are shown in Fig. 14-65. The effect of maldistribution can be severe, and efforts are being made to relate a maldistribution index to mass-transfer efficiency. Albright [*Hydrocarbon Proc.*, **63**, 173 (Sept. 1984)] developed a computer model to simulate effects of distributor design on bed distribution. Zuiderweg and coworkers [Hoek et al., *Chem. Eng. Res. Des.*, **64**, 431 (1986); Zuiderweg et al., *I. Chem. E. Symp. Ser. No. 104*, B247 (1987); Zuiderweg et al., *Trans. I. Chem. E.*, **71**, Part A, (1993)] have studied the effect of structured and random packed bed designs on maldistribution and, in turn, its effect of packed column efficiency.

Liquid migration to the wall appears to be favored by small-column diameter-packing diameter ratios (for random packings) and can be corrected by the use of side wipers or redistributors. Inhibition of lateral dispersion can be caused by the geometry of certain types of structured packing, according to Huber and Hiltbrunner (op. cit.). With careful attention given to the five causes of maldistribution, it is possible to design commercial packed columns for heights of 8 to 9 m between redistributors.

End Effects   Analysis of the mass-transfer efficiency of a packed column should take into account that transfer which takes place outside the bed, i.e., at the ends of the packed sections. Inlet gas may very well contact exit liquid below the bottom support plate, and exit gas can contact liquid from some types of distributors (e.g., spray nozzles). The bottom of the column is the more likely place for transfer, and Silvey and Keller [*Chem. Eng. Prog.*, **62**(1), 68 (1966)] found that the

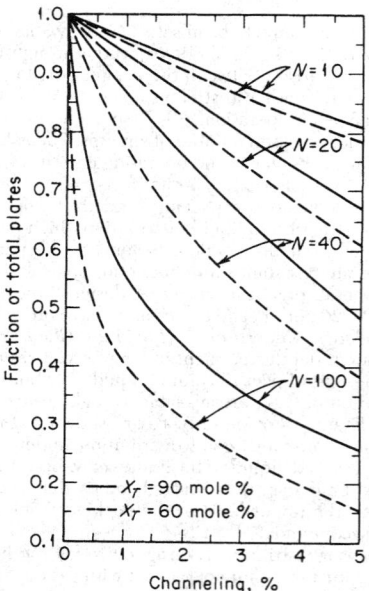

FIG. 14-65 Effect of liquid channeling on column efficiency for a system with a relative volatility of 1.07. Total number of theoretical plates N of 10, 20, 40, and 100 at top liquid composition X of 90 and 60 mole percent. [*Manning and Cannon*, Ind. Eng. Chem., **49**, 347 (1957).]

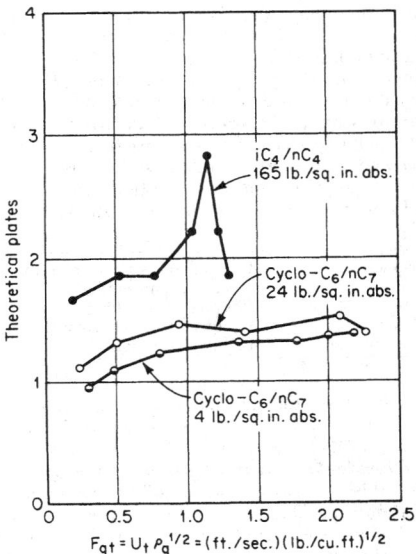

FIG. 14-66 Efficiency of FRI reboiler and space below bottom support plate. To convert pounds per square inch absolute to kilopascals, multiply by 6.8947; to convert (feet per second) (pounds per cubic foot)$^{1/2}$, to (meters per second) (kilograms per cubic meter)$^{1/2}$, multiply by 1.2199. [*Silvey and Keller, Chem. Eng. Prog.*, **62**(*1*), 68 (*1966*).]

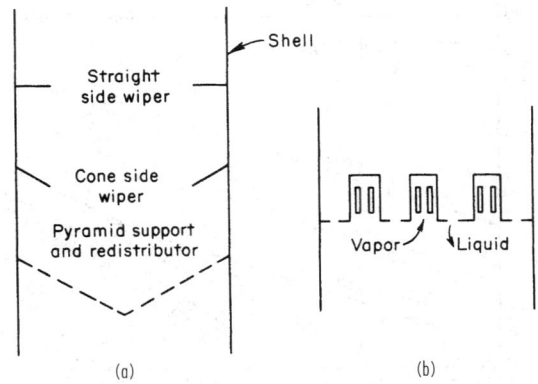

FIG. 14-67 Effective interfacial area based on data of Fellinger. (*a*) 1-in Raschig rings. (*b*) 1-in Berl saddles. To convert square feet per cubic foot to square meters per cubic meter, multiply by 3.28; to convert pounds per hour-square foot to kilograms per second-square meter, multiply by 0.001356. [*Shulman, Am. Inst. Chem. Eng. J.*, **1**, 257 (*1955*).]

reboiler plus the end effect could give up to two or more theoretical plates (Fig. 14-66).

Interfacial Area The effective area of contact between gas and liquid is that area which participates in the gas-liquid mass-exchange process. This area may be less than the actual interfacial area because of stagnant pools where liquid reaches saturation and no longer participates in the transfer process.

Effective area should not be confused with "wetted area." While film flow of liquid across the packing surface is a contributor, effective area includes also contributions from rivulets, drippings, and gas bubbles. Because of this complex physical picture, effective interfacial area is difficult to measure directly.

Weisman and Bonilla [*Ind. Eng. Chem.*, **42**, 1099 (1950)] determined effective area a_i of 25-mm (1-in) Raschig rings indirectly through the relationship $a_i = (k_g a_i)/k_g$. The k_g data were obtained via evaporation of water from presaturated rings by Taecker and Hougen [*Chem. Eng. Prog.*, **45**, 188 (1949)] and the vaporization $k_g a_i$ data from McAdams et al. [*Chem. Eng. Prog.*, **45**, 241 (1949)] for the air-water irrigated system. The authors proposed that

$$a_i/a_t = 0.54 \, G^{0.31} L^{0.07} \qquad (14\text{-}165)$$

for the range of liquid rates from 4 to 17 kg/(s·m^2) [2950 to 12,537 lb/(h·ft^2)], and where a_t = external surface area of the packing (Table 14-7). In Eq. (14-165), both gas and liquid rates, G and L, are in kg/(s·m^2). Areas are in consistent units.

A greater dependency on liquid rate was reported by Shulman et al. [*Am. Inst. Chem. Eng. J.*, **1**, 253 (1955)], who obtained the effective area via vaporization of packing constructed of naphthalene and from calculated ammonia absorption data of Fellinger (Sc.D. thesis, Massachusetts Institute of Technology, 1941), taking account of liquid-phase resistance. On the basis of gross system conditions, the values obtained are indicated in Fig. 14-67 for 25-mm (1-in) Raschig rings and Berl saddles. These packing types in the 12-, 38-, and 50-mm (0.5-, 1.5-, and 2.0-in) sizes were also studied.

Yoshida and Koyanagi [*Am. Inst. Chem. Eng. J.*, **8**, 309 (1962)] used the Weisman-Bonilla approach, accepting the Taecker-Hougen k_g data and making their own $k_g a_i$ measurements under vaporization, absorption, and distillation conditions. They found that the effective

area differs between vaporization and absorption, as shown in Fig. 14-68. For distillation, they found effective areas to be different for systems with different surface tensions (Fig. 14-69) but, upon making the surface-tension correction, concluded that areas for distillation are approximately the same as those for absorption.

For structured packings, Rocha et al. [*Ind. Eng. Chem. Res.*, **35**, 1660 (1996)] adapted the model of Shi and Mersmann [*Ger. Chem. Eng.*, **8**, 87 (1985)] to predict effective wetted area:

$$\frac{a_i}{a_t} = F_{SE} \frac{29.12(We_L Fr_L)^{0.15} S^{0.359}}{Re_L^{0.2} \varepsilon^{0.6} (1 - 0.93 \cos \gamma)(\sin \theta)^{0.3}} \qquad (14\text{-}166)$$

where $We_L = (U_L^2 \rho_L S)/\sigma g_c$ = Weber number for liquid
  $Fr_L = U_L^2/Sg$ = Froude number for liquid
  S = side dimension of corrugation (Fig. 14-51), m
  $Re_L = (U_L S \rho_L)/\mu_L$
  ε = void fraction of packing
  γ = contact angle (degrees)
  θ = angle of corrugation channel with horizontal (degrees)
  F_{SE} = factor for surface enhancement

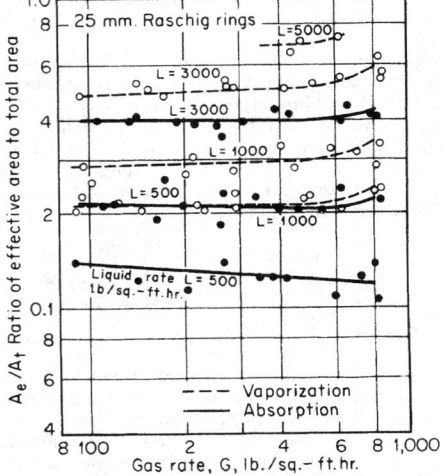

FIG. 14-68 Effective areas for 25-mm Raschig rings. To convert pounds per hour-square foot to kilograms per second-square meter, multiply by 0.001356. [*Yoshida and Koyanagi, Am. Inst. Chem. Eng. J.*, **8**, 309 (*1962*).]

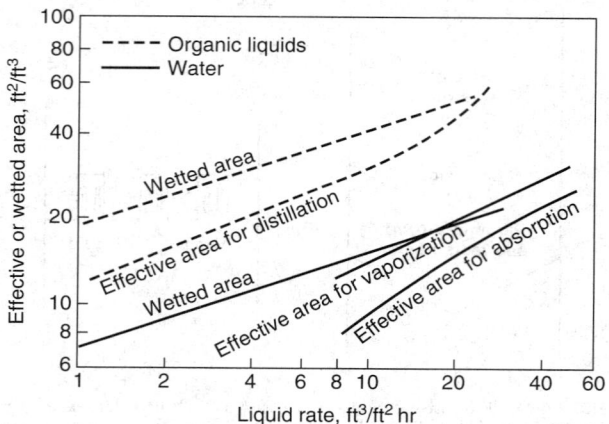

FIG. 14-69 Comparison of effective and wetted areas, 25-mm Raschig rings. To convert cubic feet per square foot-hour to cubic meters per square meter-second, multiply by 8.47×10^{-5}; to convert square feet per cubic foot to square meters per cubic meter, multiply by 3.28. [*Yoshida and Koyanagi*, Am. Inst. Chem. Eng. J., **8**, 309 (1962).]

Values of F_{SE} range from 0.009 for a smooth metal surface with small perforations, to 0.029 for a pierced metal surface. The contact angle is a key parameter for wettability and was related to the surface tension by Shi and Mersmann. This accounts for the differences in wettability found by Yoshida and Koyanagi (Fig. 14-69). In general, aqueous systems will wet ceramic surfaces better than metal surfaces, unless the latter are oxidized.

Mass Transfer Relationships for calculating rates of mass transfer between gas and liquid in packed absorbers, strippers, and distillation columns may be found in Sec. 5 and are summarized in Table 5-28. The two-resistance approach is used, with rates expressed as transfer units:

$$H_{OG} = H_G + \lambda H_L \tag{14-167}$$

where H_{OG} = height of an overall transfer unit, gas concentration basis, m
H_G = height of a gas-phase transfer unit, m
H_L = height of a liquid-phase transfer unit, m
$\lambda = m/(Lm/Gm)$ = slope of equilibrium line/slope of operating line

The various models for predicting values of H_G and H_L are given in Sec. 5. The important parameters in the models include gas rate, liquid rate, gas and liquid properties (density, viscosity, surface tension, diffusivity), packing type and size, and overall bed dimensions.

In design practice, a less rigorous parameter, HETP, is used as an index of packing efficiency. The HETP is the height of packed bed required to achieve a theoretical stage. The terms H_{OG} and HETP may be related under certain conditions:

$$HETP = H_{OG} \left[\frac{\ln \lambda}{(\lambda - 1)} \right] \tag{14-168}$$

and since

$$Z_p = (H_{OG})(N_{OG}) = (HETP)(N_t) \tag{14-169}$$

$$N_{OG} = N_t \left[\ln \lambda/(\lambda - 1) \right] \tag{14-170}$$

Equations (14-168) and (14-170) have been developed for binary mixture separations and hold for cases where the operating line and equilibrium line are straight. Thus, when there is curvature, the equations should be used for sections of the column where linearity can be assumed. When the equilibrium line and operating line have the same slope, HETP = H_{OG} and $N_{OG} = N_t$ (theoretical stages).

In practice, the following procedure is normally used:

1. Determine theoretical stages for the separation: binary or multicomponent.

2. Convert theoretical stages to transfer units using Eq. (14-170).
3. Determine the height of an overall transfer unit H_{OG} using methods given in Sec. 5, Table 5-28. For a multicomponent mixture, the key components are often used in determining H_{OG}.
4. Calculate the required packed bed height by Eq. (14-169).

Behavior of Various Systems and Packings For orientation purposes, it is helpful to have representative data available for packing performance under a variety of conditions. In the preceding edition of this handbook, extensive data on absorption/stripping systems were given. Emphasis was given to the following systems:

Ammonia-air-water	Liquid and gas phases contributing; chemical reaction contributing
Air-water	Gas phase controlling
Sulfur dioxide-air-water	Liquid and gas-phase controlling
Carbon dioxide-air-water	Liquid phase controlling

The reader may refer to the data in the preceding edition. For the current work, emphasis will be given to one absorption system, carbon dioxide-air-caustic, and to several distillation systems.

Carbon Dioxide-Air-Caustic System The vendors of packings have adopted this system as a "standard" for comparing the performance of different packing types and sizes. For tests, air containing 1.0 mol % CO_2 is passed countercurrently to a circulating stream of sodium hydroxide solution. The initial concentration of NaOH in water is 1.0 N (4.0 wt %), and as the circulating NaOH is converted to sodium carbonate it is necessary to make a mass-transfer correction because of reduced mass-transfer rate in the liquid phase. The procedure has been described by Eckert et al. [*Ind. Eng. Chem.*, **59**(2), 41 (1967); *Chem. Eng. Progr.*, **54**(1), 790 (1958)]. An overall coefficient is measured using gas-phase (CO_2) concentrations:

$$K_{OG}a_e = \frac{\text{moles } CO_2 \text{ absorbed}}{\text{time-bed volume-partial pressure } CO_2 \text{ driving force}} \tag{14-171}$$

The coefficients are usually corrected to a hydroxide conversion of 25 percent at 24°C. For other conversions, Fig. 14-15 may be used. Reported values of $K_{OG}a$ for representative random packings are given in Table 14-8. The effect of liquid rate on the coefficient is shown in Figs. 14-70 and 14-71.

While the carbon dioxide/caustic test method has become accepted, one should use the results with caution. The chemical reaction masks the effect of physical absorption, and the relative values in the table may not hold for other cases, especially distillation applications where much of the resistance to mass transfer is in the gas phase. Background on this combination of physical and chemical absorption may be found earlier in the present section, under "Absorption with Chemical Reaction."

Distillation Applications Packings are now routinely considered for distillation columns with diameters up to 10 m or more. The pressure drop advantages of the modern, through-flow random pack-

TABLE 14-8 Overall Coefficients for Representative Packings

	Nominal size, mm	Overall coefficient $K_{OG}a$, kg·moles/(hr·m³·atm)
CO_2-air-caustic system		
Ceramic raschig rings	25	37.0
	50	26.1
Ceramic Intalox saddles	25	45.1
	50	30.1
Metal pall rings	25	49.6
	50	34.9
Metal Intalox saddles (IMTP®)	25	54.8
	50	39.1

NOTE: Basis for reported values: CO_2 concentration in inlet gas, 1.0 vol %; 1N NaOH solution in water, 25 percent NaOH conversion; temperature = 24°C; atmospheric pressure: gas rate = 1.264 kg/(s·m²); liquid rate = 6.78 kg/(s·m²).

SOURCE: Strigle, R. L., *Random Packings and Packed Towers*, Gulf Publ. Co., Houston, 1987.

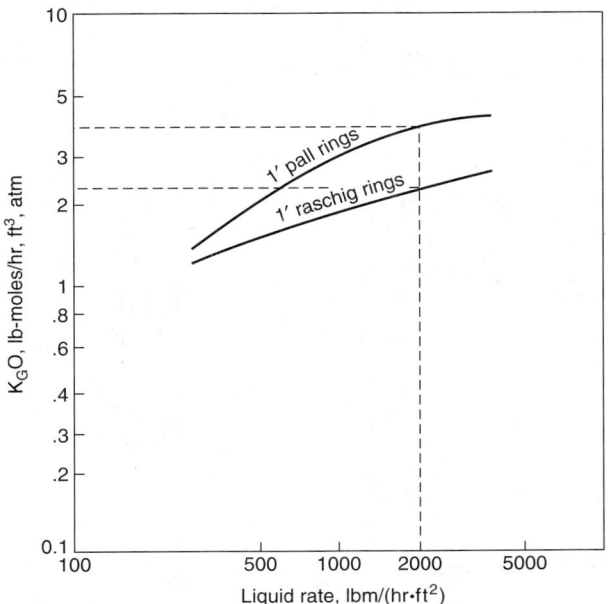

FIG. 14-70 Overall mass transfer coefficients for carbon dioxide absorbed from air by 1N caustic solution. Packing = 1-inch metal raschig and pall rings. Air rate = 0.61 kg/s·m² (450 lb/hr·ft²). To convert from lb/hr·ft² to kg/s·m², multiply by 0.00136; to convert lb·moles/h·ft³ atm to kg·moles/s·m³ atm, multiply by 0.00445. [*Eckert et al.*, Chem. Eng. Progr., **54**(1), 70 (1958).]

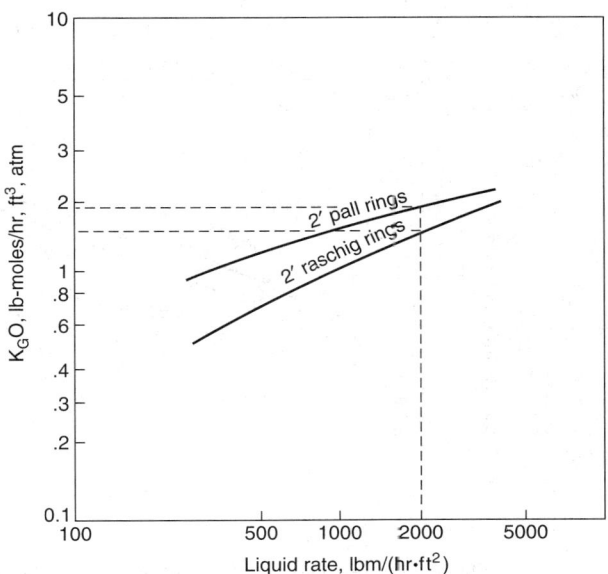

FIG. 14-71 Overall mass transfer coefficients for carbon dioxide absorbed from air by 1N caustic solution. Packing = 2-inch metal raschig and pall rings. Air rate = 0.61 kg/s·m² (450 lb/hr·ft²). To convert from lb/hr·ft² to kg/s·m², multiply by 0.00136. To convert from lb-moles/hr·ft³ atm to kg-moles/s·m³ atm, multiply by 0.0045. [*Eckert et al.*, Chem. Eng. Progr., **54**(1), 70 (1958).]

ings and the ordered, structured packings have made vacuum column applications of particular interest. Test data for total reflux distillations have become available from several sources, such as Fractionation Research, Inc. and The University of Texas at Austin Separations Research Program. Examples of the data from these and other sources are shown in Figs. 14-72–14-75. The comparative efficiency of different sizes of random packing is shown in Fig. 14-72. As would be expected, the smaller packings have higher mass-transfer efficiency (lower values of HETP). However, their capacity is limited. The equivalent pressure drop data are shown in Fig. 14-73, where the smaller packings are shown to have higher pressure drop.

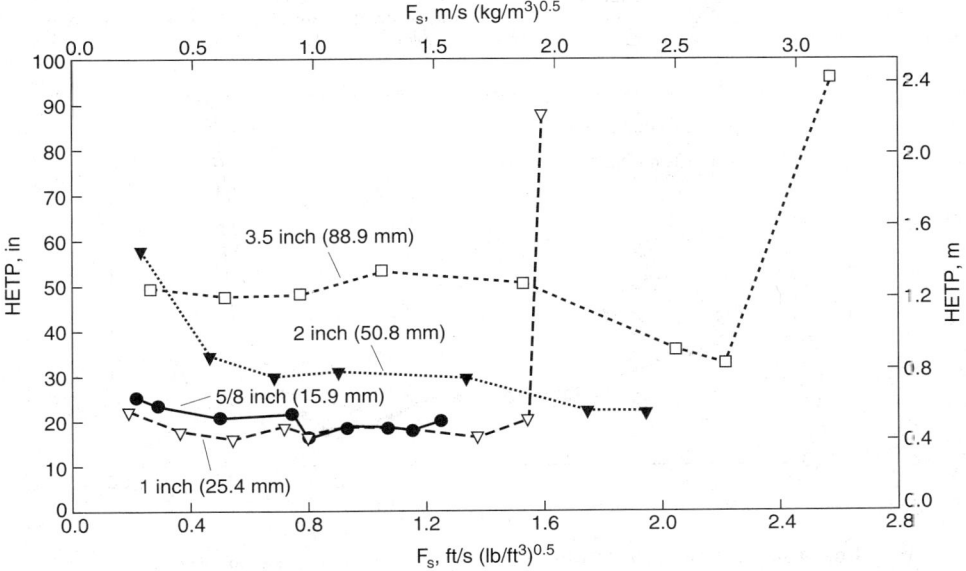

FIG. 14-72 HETP values for four sizes of metal pall rings, vacuum operation. Cyclohexane/*n*-heptane system, total reflux, 35 kPa (5.0 psia). Column diameter = 1.2 m (4.0 ft). Bed height = 3.7 m (12 ft). Distributor = tubed drip pan, 100 streams/m². [*Adapted from Shariat and Kunesh*, Ind. Eng. Chem. Res., **34**, 1273 (1995).] Reproduced with permission. Copyright © 1995 American Chemical Society.

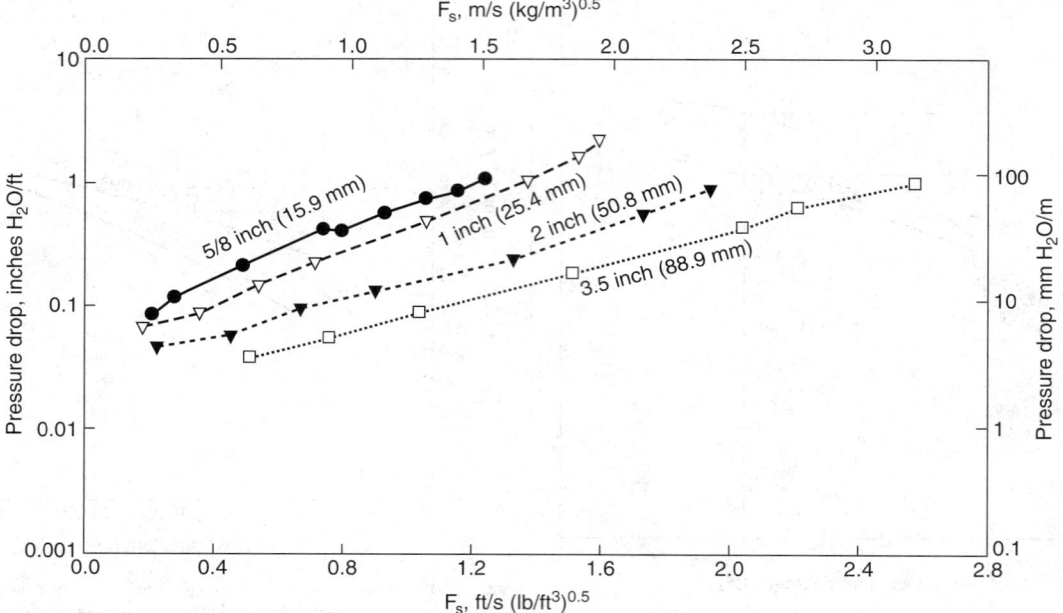

FIG. 14-73 Pressure-drop data for several sizes of pall rings; test conditions same as described for Fig. 14-72. [*Shariat and Kunesh*, Ind. Eng. Chem. Res., **34**, 1273 (1995).] Reproduced with permission. Copyright © 1995 American Chemical Society.

Efficiency data for a representative structured packing at two column diameters are shown in Fig. 14-74. The Max-Pak® packing has a surface area of 246 m²/m³ (75 ft²/ft³). The same test mixture (cyclohexane/*n*-heptane) and operating pressure was used for both tests. It would appear that column diameter does not have an influence in this range of values (0.43 to 1.2 m).

Efficiency and pressure drop data for Sulzer BX metal gauze structured packing and for three test mixtures are shown in Fig. 14-75. For the ethyl benzene/styrene test mixture, the effect of operating pressure is shown. The high viscosity mixture, propylene glycol/ethylene

glycol, has a significantly lower efficiency than the other mixtures but does not appear to have a lower capacity.

LIQUID-DISPERSED CONTACTORS

Introduction   There are two types of gas-liquid contactors where the liquid is deliberately dispersed. In the most common, a spray nozzle is used to generate droplets. A second type is the pipeline contactor, where the entrainment generated by flowing gas generates the droplets.

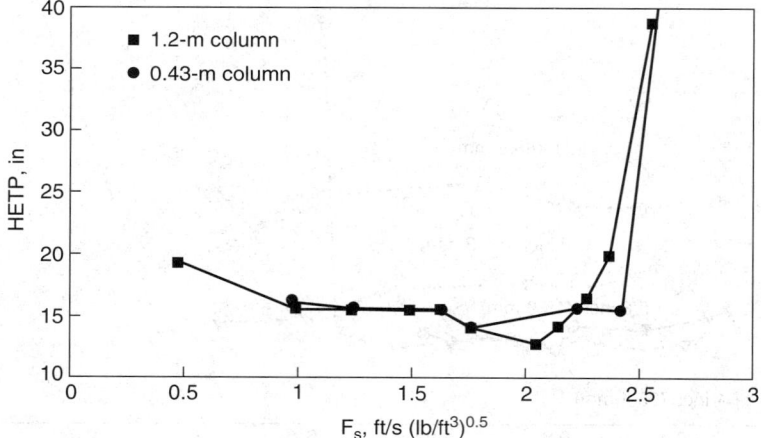

FIG. 14-74 HETP values for Max-Pak structured packing, 35 kPa (5 psia), two column diameters. Cyclohexane/*n*-heptane system, total reflux. For 0.43 m (1.4 ft) column: perforated pipe distributor, 400 streams/m2, 3.05 m (10 ft) bed height. For 1.2 m (4.0 ft) column: tubed drip pan distributor, 100 streams/m², 3.7 m (12 ft) bed height. Smaller column data, University of Texas/Austin; Larger column data, Fractionation Research, Inc. To convert (ft/s)(lb/ft³)⁰·⁵ to (m/s)(kg/m³)⁰·⁵, multiply by 1.2199. (*Courtesy Jaeger Products, Inc., Houston, Texas.*)

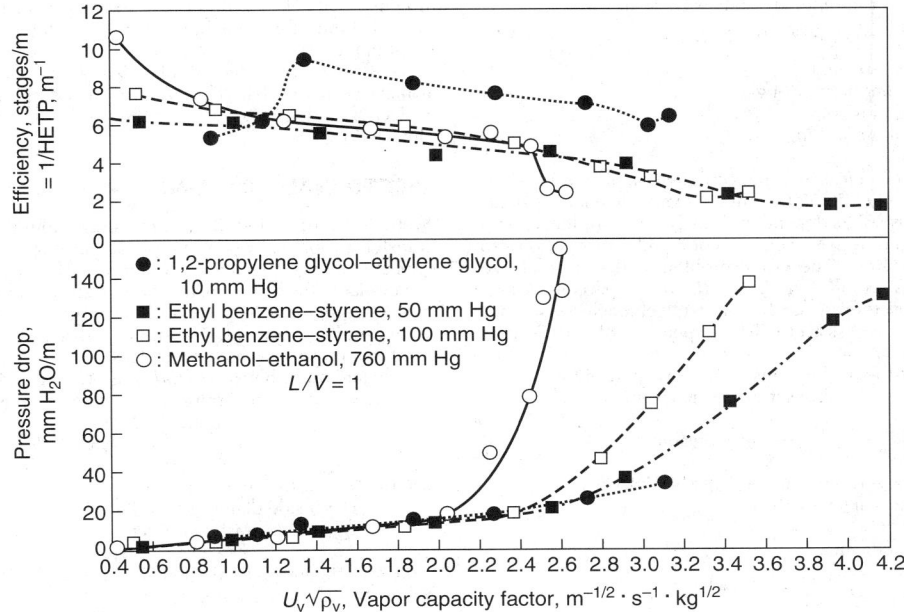

FIG. 14-75 Efficiency and pressure drop data for Sulzer BX structured packing. Test mixtures and operating pressures shown on graph. Total reflux, 0.8 m (2.6 ft) column diameter, ca. 2 m (6.5 ft) bed height, perforated pipe distributor, 400 streams/m². To convert (m/s)(kg/m³)⁰·⁵ to (ft/s)(lb/ft³)⁰·⁵, multiply by 0.8197. [*Billet, Conrad, and Grubb,* I. Chem. E. Symp. Serv. **5**(32), 111 (1969).]

Because of the minimal internals for solids to grow on, both types of contactors are common in fouling services. The spray nozzle based devices are also inherently low pressure drop.

The disadvantage is that volumetric efficiency is usually much less than conventional trays or packed contactors. Applications are usually limited to cases when only a few transfer units or a single equilibrium stage is required. Since many of these applications tend to be in heat-transfer service, the following discussion will be in terms of thermal properties and thermal measures of performance.

Heat-Transfer Applications   Heat-transfer analogs of common mass-transfer terms are:

$$\text{No. of gas-phase transfer units} = \frac{T_{g,\text{out}} - T_{g,\text{in}}}{(T_g - T_i)_{\text{mean}}} = N_g \quad (14\text{-}172)$$

$$\text{No. of liquid-phase transfer units} = \frac{T_{\ell,\text{out}} - T_{\ell,\text{in}}}{(T_\ell - T_i)_{\text{mean}}} = N_\ell \quad (14\text{-}173)$$

$$\text{Gas-phase volumetric transfer rate} = \frac{Gc_p}{H_g} = h_g a \quad (14\text{-}174)$$

$$\text{Diffusivity} = \frac{k}{\rho c_p} = \alpha_T \quad (14\text{-}175)$$

where T_g = gas temperature
T_ℓ = liquid temperature
T_i = interface temperature
H_g = height of a gas-phase transfer unit
G = weight gas flow/area
c_p = specific heat
ρ = density
k = thermal conductivity
α_T = thermal diffusivity

Note that the relative performance of a device can be converted from a mass-transfer basis to a heat-transfer basis by introducing these analogies into the rate equations.

The conversion is simplest when $K_g a$ and $K_\ell a$ are defined in terms of mole-fraction driving force:

$$\frac{h_g a}{K_g a} \text{ or } \frac{h_\ell a}{K_\ell a} = \frac{c_p}{M}\left(\frac{\alpha_T}{D}\right)^{0.5} \quad (14\text{-}176)$$

where $h_\ell a, h_g a$ = volumetric heat-transfer rate
$K_\ell a, K_g a$ = volumetric mass-transfer rate
c_p = heat capacity
M = molecular weight
α_T/D = ratio of thermal diffusivity to molecular diffusivity, dimensionless

For gases, α_T/D is usually close to 1, since the same basic transfer mechanism exists. For liquids, α_T/D is invariably much greater than 1. A simplified model yields the relation

$$\frac{\alpha_T}{D} = 1.9 \times 10^7 (\mu/T) \quad (14\text{-}177)$$

where μ is in Pa·s and T is in K. For a 1-cP (10^{-3}·Pa·s) liquid at 310 K, this gives a ratio of 61. This means that a closer approach to equilibrium can be obtained for heat transfer applications that are liquid-limited (such as in condensation of a pure vapor), than can be obtained for mass transfer liquid-limited applications.

Theoretical Transfer Model   Transfer from single droplets is theoretically well defined for the gas side. For a droplet moving counter to a gas, interfacial area is (in consistent units)

$$a = \left(\frac{G_\ell}{\rho_\ell (U_\ell - U_g)}\right)\left(\frac{6}{D_d}\right) \quad (14\text{-}178)$$

where a = interfacial area/volume
G_ℓ = liquid-flow rate (weight/cross-sectional area)
ρ_ℓ = liquid density
U_ℓ = liquid velocity relative to the gas, often approximately the terminal velocity of droplets (see Sec. 6 for estimation)
U_g = superficial gas velocity
D_d = droplet diameter

The transfer coefficient is defined by (in consistent units)

$$h_g = (k/D_d)(2 + 0.6 N_{Re}^{0.5} N_{Pr}^{0.33}) \qquad (14\text{-}179)$$

where h_g = transfer coefficient
k = thermal conductivity of gas
N_{Re} = Reynolds number = $D_d \bar{U}_t \rho_g / \mu_g$
N_{Pr} = Prandtl number = $(c_p \mu / k)_g$

The volumetric coefficient $h_g a$ from the combination of Eqs. (14-178) and (14-179) is useful in defining the effect of variable changes but is limited in value because of its dependence on D_d. The product of area and coefficient obtained from a given mass of liquid is proportional to $(1/D_d)^2$ for small diameters. The prime problem is that droplet-size estimating procedures are often no better than ±50 percent. A secondary problem is that there is no D_d that truly characterizes either the motion or transfer process for the whole spectrum of particle sizes present. See Eqs. (14-193) and (14-194).

The corresponding theory for transfer in the liquid phase is even less certain. If one had static drops, the transfer would be

$$N_\ell = 0.5 + \frac{4(k/\rho c_p)\pi^2 t}{D_d^2} \quad (\text{dimensionless}) \qquad (14\text{-}180)$$

where N_ℓ = liquid-phase transfer units [Eq. (14-173)]
k = liquid thermal conductivity
ρ = liquid density
c_p = liquid specific heat
D_d = droplet diameter
t = time of contact

However, the static-drop assumption is usually extremely conservative. For example, the high interfacial velocity in the spray from nozzles yields a high degree of internal mixing and much higher transfer.

Countercurrent, Cocurrent, or Backmixed Spray-chamber contactors rarely approach countercurrent performance. Backmixing is hard to prevent and often limiting. Backmixing is so severe that many designers simply limit spray-chamber performance to a single equilibrium stage regardless of height. For a direct-contact heat-transfer device, this means that the temperatures of the exiting gas and liquid would be equal. The main cause of the high degree of backmixing is that there is no stabilizing pressure drop caused by packing of plates. Consequently the chief resistance to gas flow is the rain of drops. Anything less than perfect liquid distribution will induce a dodging action in the opposed vapor flow. The result is the development of large eddies and bypass streams. Other sources of deviation from countercurrent flow are large drops falling faster than small ones and liquid striking the walls About all that can be done is to take special care to obtain a uniform spray pattern with minimum collection at the walls.

Empirical Approach

Sprays Large units generally yield approximately one equilibrium stage even when in nominal counter current flow. See Masters, [*Spray Drying Handbook*, 5th ed., Wiley, New York, (1991)]. For smaller towers, less than one equilibrium stage is typical. Transfer units can be estimated from the data of Pigford and Pyle [*Ind. Eng. Chem.* **43**, 1949 (1951)]. These data show the height of a gas-limited transfer unit to be 1 to 3 m (3 to 12 ft). Pigford and Pyle show much shorter heights for a liquid-limited transfer unit, in the range of 0.5 to 1 m (1.5 to 3 ft). The high liquid transfer rates result from the high energy dissipation in the liquid and the enhanced transfer at the time of droplet formation. The same behavior is indicated by Simpson and Lynn [*Am. Inst. Chem. Eng. J.*, **23**(5), 666 (1977)], who show a 75 to 95 percent approach to equilibrium stripping in a 1.4-m-tall spray contact in a liquid-limited system.

Pipeline Flow (Quenching) For the case of pipeline quenching, the flows are cocurrent. How closely the gas temperature approaches the liquid depends on e^{-Ng}, where N_g varies with $1/(\text{droplet diameter})^2$.

Since the predicted droplet diameter at high velocity pipeline flow varies with $(1/\text{velocity})^{1.2}$, as shown by Eq. (14-201), the volumetric performance is strongly dependent on velocity:

$$N_g = \text{constant} \cdot (\text{velocity})^{2.4} \qquad (14\text{-}181)$$

An empirical point in support of the strong velocity dependence is the rule of thumb for quenching, requiring a high relative velocity [>60 m/s (200 ft/s)].

Venturi scrubbers are similar in that they need high velocity to achieve small droplets. They are primarily employed for mist and dust collection and are discussed further in the mist-collection portion of this section.

WETTED-WALL COLUMNS

Wetted-wall or falling-film columns have found application in mass-transfer problems when high-heat-transfer-rate requirements are concomitant with the absorption process. Large areas of open surface are available for heat transfer for a given rate of mass transfer in this type of equipment because of the low mass-transfer rate inherent in wetted-wall equipment. In addition, this type of equipment lends itself to annular-type cooling devices.

Gilliland and Sherwood [*Ind. Eng. Chem.*, **26**, 516 (1934)] found that, for vaporization of pure liquids in air streams for streamline flow,

$$\frac{k_g D}{D_g} \frac{P_{BM}}{\pi} = 0.023 N_{Re}^{0.83} N_{Sc}^{0.44} \qquad (14\text{-}182)$$

where D_g = diffusion coefficient
D = inside diameter of tube
k_g = mass-transfer coefficient, gas phase

Note that the group on the left side of Eq. (14-182) is dimensionless. When turbulence promoters are used at the inlet-gas section, an improvement in gas mass-transfer coefficient for absorption of water vapor by sulfuric acid was observed by Greenewalt [*Ind. Eng. Chem.*, **18**, 1291 (1926)]. A falling off of the rate of mass transfer below that indicated in Eq. (14-182) was observed by Cogan and Cogan (thesis, Massachusetts Institute of Technology, 1932) when a calming zone preceded the gas inlet in ammonia absorption (Fig. 14-76).

In work with the hydrogen chloride-air-water system, Dobratz, Moore, Barnard, and Meyer [*Chem. Eng. Prog.*, **49**, 611 (1953)] using a cocurrent-flow system found that $k_g \propto G^{1.8}$ (Fig. 14-77) instead of the 0.8 power as indicated by the Gilliland equation. Heat-transfer coefficients were also determined in this study. The radical increase in heat-transfer rate in the range of $G = 30$ kg/(s·m²) [20,000 lb/(h·ft²)] was similar to that observed by Tepe and Mueller [*Chem. Eng. Prog.*, **43**, 267 (1947)] in condensation inside tubes.

Gaylord and Miranda [*Chem. Eng. Prog.*, **53**, 139M (1957)] using a multitube cocurrent-flow falling-film hydrochloric acid absorber for hydrogen chloride absorption found

$$K_g = \frac{1.66(10^{-5})}{M_m^{1.75}} \left(\frac{DG}{\mu} \right) \qquad (14\text{-}183)$$

where K_g = overall mass-transfer coefficient, (kg·mol)/(s·m²·atm)
M_m = mean molecular weight of gas stream at inlet to tube
D = diameter of tube, m
G = mass velocity of gas at inlet to tube, kg/(s·m²)
μ = viscosity of gas, Pa·s

Note that the group DG/μ is dimensionless. This relationship also satisfied the data obtained for this system, with a single-tube falling-film unit, by Coull, Bishop, and Gaylor [*Chem. Eng. Prog.*, **45**, 506 (1949)].

The rate of mass transfer in the liquid phase in wetted-wall columns is highly dependent on surface conditions. When laminar-flow conditions prevail without the presence of wave formation, the laminar-penetration theory prevails. When, however, ripples form at the surface, and they may occur at a Reynolds number exceeding 4, a significant rate of surface regeneration develops, resulting in an increase in mass-transfer rate.

If no wave formations are present, analysis of behavior of the liquid-film mass transfer as developed by Hatta and Katori [*J. Soc. Chem. Ind.*, **37**, 280B (1934)] indicates that

$$k_\ell = 0.422 \sqrt{\frac{D_\ell \Gamma}{\rho B_F^2}} \qquad (14\text{-}184)$$

where $B_F = (3u\Gamma/\rho^2 g)^{1/3}$
D_ℓ = liquid-phase diffusion coefficient, m²/s

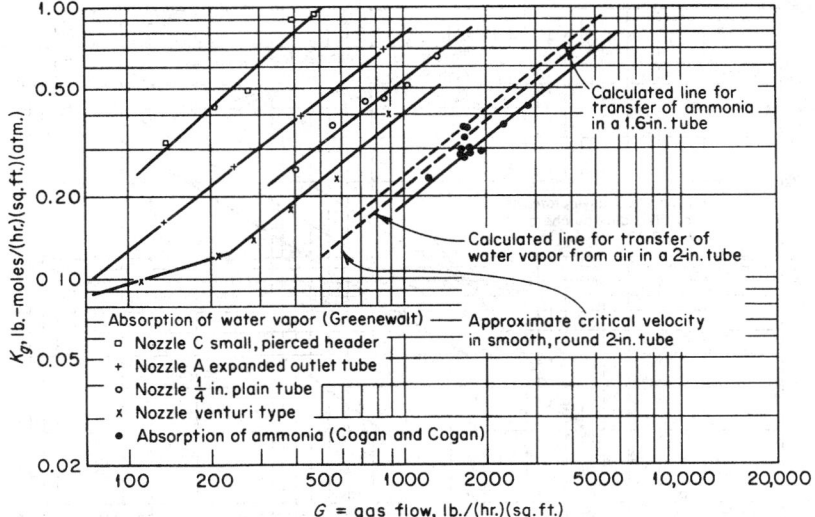

FIG. 14-76 Mass-transfer rates in wetted-wall columns having turbulence promoters. To convert pound-moles per hour-square foot-atmosphere to kilogram-moles per second-square meter-atmosphere, multiply by 0.00136; to convert pounds per hour-square foot to kilograms per second-square meter, multiply by 0.00136; and to convert inches to millimeters, multiply by 25.4. (Data of Greenewalt and Cogan and Cogan, *Sherwood, and Pigford,* Absorption and Extraction, *2d ed., McGraw-Hill, New York, 1952.*)

ρ = liquid density, kg/m³
Z = length of surface, m
k_ℓ = liquid-film-transfer coefficient, (kg·mol)/[(s·m²)(kg·mol)/m³]
Γ = liquid-flow rate, kg/(s·m) based on wetted perimeter
μ = viscosity of liquid, Pa·s
g = gravity acceleration, 9.81 m/s²

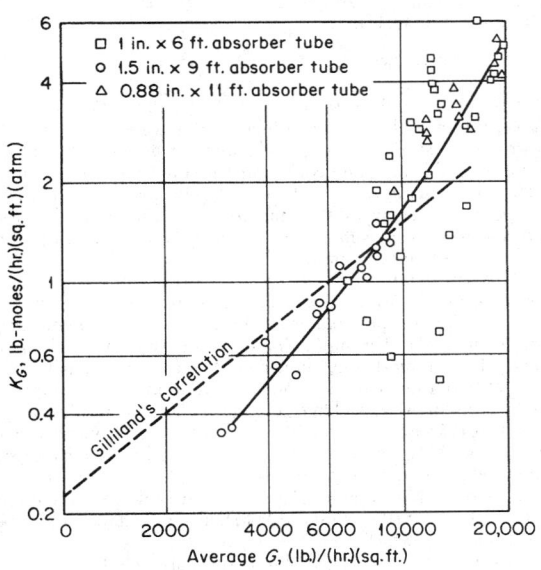

FIG. 14-77 Mass-transfer coefficients versus average gas velocity—HCl absorption, wetted-wall column. To convert pound-moles per hour-square foot-atmosphere to kilogram-moles per second-square meter-atmosphere, multiply by 0.00136; to convert pounds per hour-square foot to kilograms per second-square meter, multiply by 0.00136; to convert feet to meters, multiply by 0.305; and to convert inches to millimeters, multiply by 25.4. [*Dobratz et al., Chem. Eng. Prog.,* **49,** *611 (1953).*]

When Z is large or $\Gamma/\rho B_F$ is so small that liquid penetration is complete,

$$k_\ell = 11.800 \, D_\ell/B_F \quad (14\text{-}185)$$

and

$$H_\ell = 0.95 \, \Gamma B_F/D_\ell \quad (14\text{-}186)$$

A comparison of experimental data for carbon dioxide absorption obtained by Hatta and Katori (op. cit.), Grimley [*Trans. Inst. Chem. Eng.,* **23,** 228 (1945)], and Vyazov [*Zh. Tekh. Fiz.* (U.S.S.R.), **10,** 1519 (1940)] and for absorption of oxygen and hydrogen by Hodgson (S. M. thesis, Massachusetts Institute of Technology, 1949), Henley (B.S. thesis, University of Delaware, 1949), Miller (B.S. thesis, University of Delaware, 1949), and Richards (B.S. thesis, University of Delaware, 1950) was made by Sherwood and Pigford (*Absorption and Extraction,* McGraw-Hill, New York, 1952) and is indicated in Fig. 14-78.

In general, the observed mass-transfer rates are greater than those predicted by theory and may be related to the development of surface rippling, a phenomenon which increases in intensity with increasing liquid path.

Vivian and Peaceman [*Am. Inst. Chem. Eng. J.,* **2,** 437 (1956)] investigated the characteristics of the CO_2-H_2O and Cl_2-HCl, H_2O system in a wetted-wall column and found that gas rate had no effect on the liquid-phase coefficient at Reynolds numbers below 2200. Beyond this rate, the effect of the resulting rippling was to increase significantly the liquid-phase transfer rate. The authors proposed a behavior relationship based on a dimensional analysis but suggested caution in its application concomitant with the use of this type of relationship. Cognizance was taken by the authors of the effects of column length, one to induce rippling and increase of rate of transfer, one to increase time of exposure which via the penetration theory decreases the average rate of mass transfer in the liquid phase. The equation is

$$\frac{k_\ell h}{D_\ell} = 0.433 \left(\frac{\mu_\ell}{\rho_\ell D_\ell}\right)^{1/2} \left(\frac{\rho_\ell^2 g h^3}{\mu_\ell^2}\right)^{1/6} \left(\frac{4\Gamma}{\mu_\ell}\right)^{0.4} \quad (14\text{-}187)$$

where D_ℓ = diffusion coefficient of solute in liquid
g = gravity-acceleration constant
h = length of wetted wall
k_ℓ = mass-transfer coefficient, liquid phase
Γ = mass rate of flow of liquid
μ_ℓ = viscosity of liquid
ρ_ℓ = density of liquid

The equation is dimensionless.

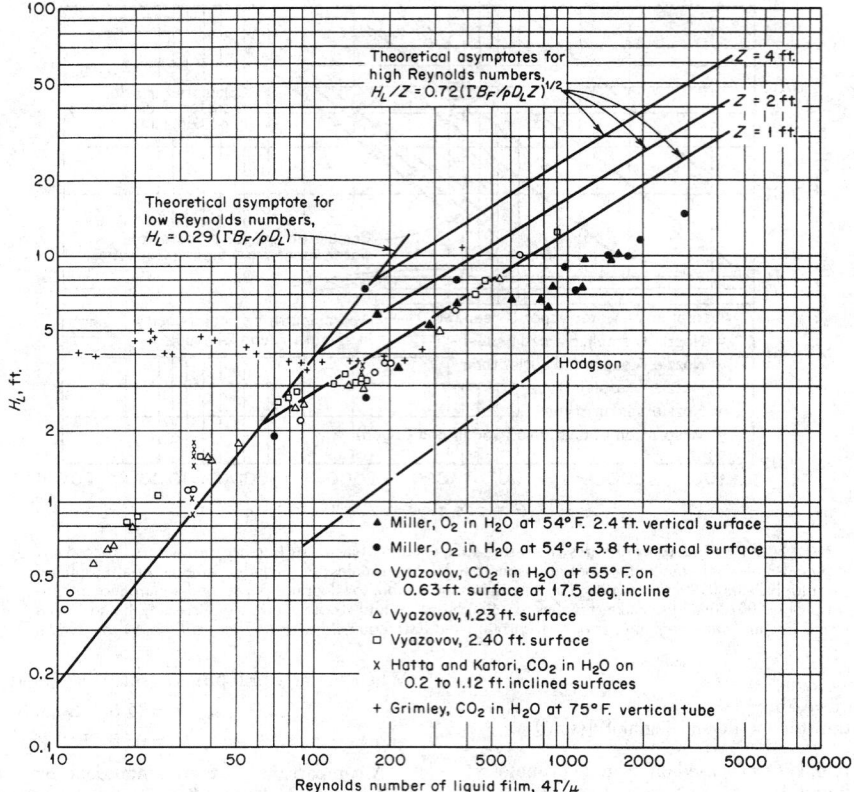

FIG. 14-78 Liquid-film resistance in absorption of gases in wetted-wall columns. Theoretical lines are calculated for oxygen absorption in water at 55°F. To convert feet to meters, multiply by 0.3048; °C = 5/9 (°F − 32). (*Sherwood and Pigford,* Absorption and Extraction, *2d ed., McGraw-Hill, New York, 1952.*)

The effect of chemical reaction in reducing the effect of variation of the liquid rate on the rate of absorption in the laminar-flow regime was illustrated by the evaluation of the rate of absorption of chlorine in ferrous chloride solutions in a wetted-wall column by Gilliland, Baddour, and White [*Am. Inst. Chem. Eng. J.,* **4,** 323 (1958)].

Flooding in Wetted-Wall Columns When gas and liquid are in counterflow in wetted-wall columns, flooding can occur at high gas rates. Diehl and Koppany [*Chem. Eng. Prog.,* **65,** *Symp. Ser.* 42, 77 (1969)] correlated flooding data from a number of sources, including their own work, and developed the following expression:

$$U_f = F_1 F_2 \left(\frac{\sigma}{\rho_g} \right)^{1/2} \qquad (14\text{-}188)$$

where U_f = flooding gas velocity, m/s
F_1 = 1.22 when 3.2 $d_i/\sigma > 1.0$
F_1 = 1.22 (3.2 $d_i/\sigma)^{0.4}$ when 3.2 $d_i/\sigma < 1.0$
$F_2 = (G/L)^{0.25}$
G/L = gas-liquid mass ratio
d_i = inside diameter of column, mm
σ = surface tension, mN/m (dyn/cm)
ρ_g = gas density, kg/m³

The data covered column sizes up to 50-mm (2-in) diameter; the correlation should be used with caution for larger columns.

GAS-LIQUID-COLUMN ECONOMICS

Estimation of column costs for preliminary process evaluations requires consideration not only of the basic type of internals but also of their effect on overall system cost. For a distillation system, for example, the overall system can include the vessel (column), attendant structures, supports, and foundations; auxiliaries such as reboiler, condenser, feed heater, and control instruments; and connecting piping. The choice of internals influences all these costs, but other factors influence them as well. A complete optimization of the system requires a full-process simulation model that can cover all pertinent variables influencing economics.

Cost of Internals Installed costs of plates (trays) may be estimated from Fig. 14-79, with corrections for plate material taken from Table 14-9. For two-pass plates the cost is 15 to 20 percent higher. Approximate costs of dumped (random) packing materials may be obtained from Table 14-10, but it should be recognized that, because of competition, there can be significant variations in these costs from vendor to vendor. Also, packings sold in very large quantities carry discounts. In 1995, costs of structured packings, made from sheet metal, averaged $90–$110 per cubic foot, but the need for special distributors and redistributors can double the cost of structured-packings on a volumetric basis. Note that for Fig. 14-79 and Table 14-9, the effective cost date is January 1990, with the Marshall and Swift cost index being taken as 904.

As indicated above, packed column internals include liquid distributors, packing support plates, redistributors (as needed), and hold-down plates (to prevent movement of packing under flow conditions). Costs of these internals for columns with random packing are given in Fig. 14-80, based on early 1976 prices, and a Marshall and Swift cost index of 460.

Cost of Column The cost of the vessel, including heads, skirt, nozzles, and ladderways, is usually estimated on the basis of weight.

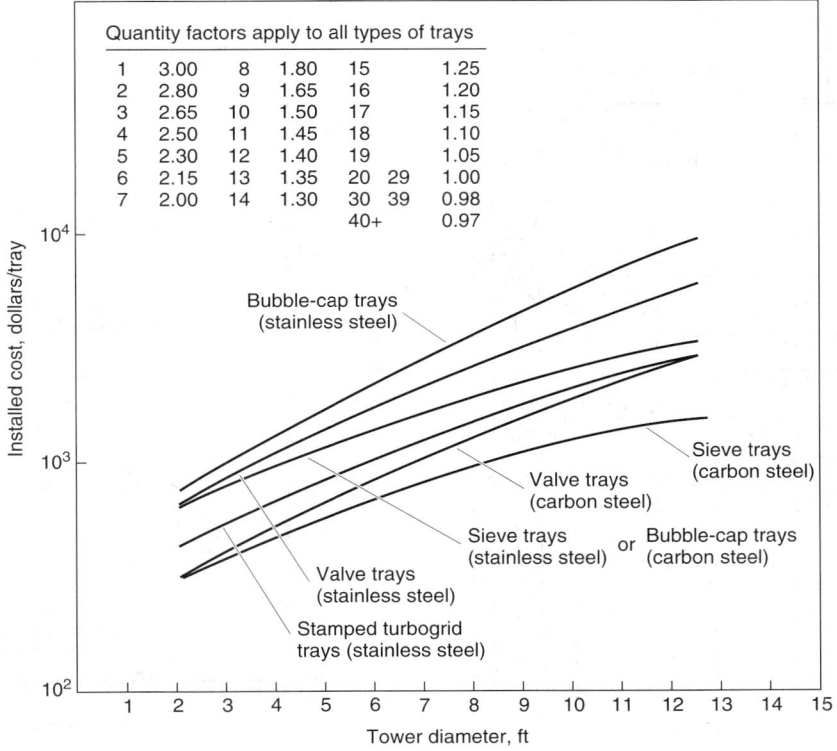

FIG. 14-79 Cost of trays in plate towers. Price includes tray deck, bubble caps, risers, downcomers, and structural-steel parts. The stainless steel designated is type 410 (*Peters and Timmerhaus*, Plant Design and Economics for Chemical Engineers, *4th ed., McGraw-Hill, New York, 1991*).

Figure 14-81 provides early 1990 cost data for the shell and heads, and Fig. 14-82 provides 1990 cost data for connections. For very approximate estimates of complete columns, including internals, Fig. 14-83 may be used. As for Figs. 14-81 and 14-82, the cost index is 904.

Plates versus Packings Bases for using packings instead of plates have been given earlier in this section. In many cases, either type of device may be used. For vacuum fractionations, the low pressure drop characteristics of throughflow random packings and structured packings tend to give them a distinct advantage over plates. On the other hand, structured packings and their many requirements, such as distributors, are more expensive than plates. For atmospheric

and pressure columns, the pressure-drop characteristics of packings are less important, and a decision may be made on installed cost and reliability of design methods. Kister et al. [*Chem. Eng. Progr.*, **90**(2), 23 (1994)] reported a study of the relative capacity and efficiency of plates, modern random packings, and structured packings. They found that, for each device optimally designed for the design require-

TABLE 14-9 Relative Fabricated Cost for Metals Used in Tray-tower Construction*

Materials of construction	Relative cost per ft² of tray area (based on carbon steel = 1)
Sheet-metal trays	
Steel	1
4–6% chrome—½ moly alloy steel	2.1
11–13% chrome type 410 alloy steel	2.6
Red brass	3
Stainless steel type 304	4.2
Stainless steel type 347	5.1
Monel	7.0
Stainless steel type 316	5.5
Inconel	8.2
Cast-iron trays	2.8

*Peters and Timmerhaus, *Plant Design and Economics for Chemical Engineers*, 4th ed., McGraw-Hill, New York, 1991. To convert cost per square foot to cost per square meter, multiply by 10.76.

TABLE 14-10 Costs of Tower Packings, Uninstalled, January, 1990

Prices in dollars per ft³, 100 ft³ orders, f.o.b. manufacturing plant

	Size, in, $/ft³			
	1	1½	2	3
Raschig rings				
Chemical porcelain	12.8	10.3	9.4	7.8
Carbon steel	36.5	23.9	20.5	16.8
Stainless steel	155	117	87.8	—
Carbon	52	46.2	33.9	31.0
Intalox saddles				
Chemical stoneware	17.6	13.0	11.8	10.7
Chemical porcelain	18.8	14.1	12.9	11.8
Polypropylene	21.2	—	13.1	7.0
Berl saddles				
Chemical stoneware	27.0	21.0	—	—
Chemical porcelain	33.5	21.5	15.6	—
Pall rings				
Carbon steel	29.3	19.9	18.2	—
Stainless steel	131	99.0	86.2	—
Polypropylene	21.2	14.4	13.1	—

Peters and Timmerhaus, *Plant Design and Economics for Chemical Engineers*, 4th ed., McGraw-Hill, New York, 1991. To convert cubic feet to cubic meters, multiply by 0.0283; to convert inches to millimeters, multiply by 25.4; and to convert dollars per cubic foot to dollars per cubic meter, multiply by 35.3.

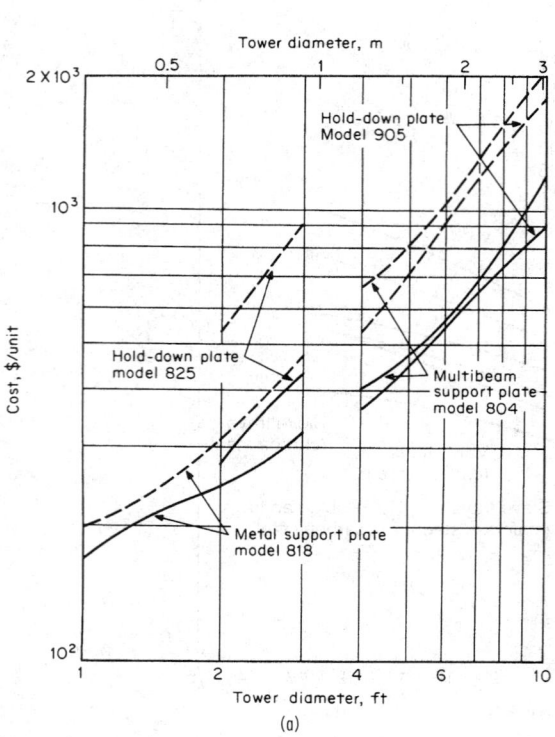

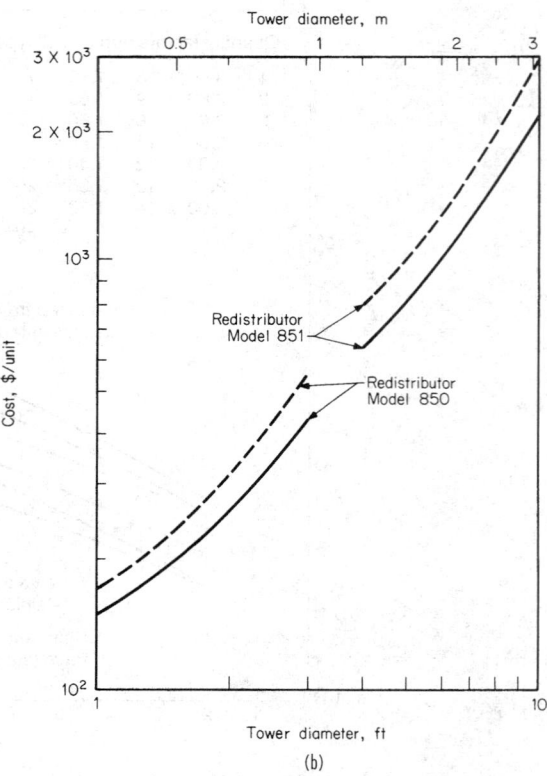

FIG. 14-80 Cost of internal devices for columns containing dumped packings. (*a*) Holddown plates and support plates. (*b*) Redistributors. (*c*) Liquid distributors. [*Pikulik and Diaz*, Chem. Eng., *84*(21), *106* (Oct. 10, 1977).]

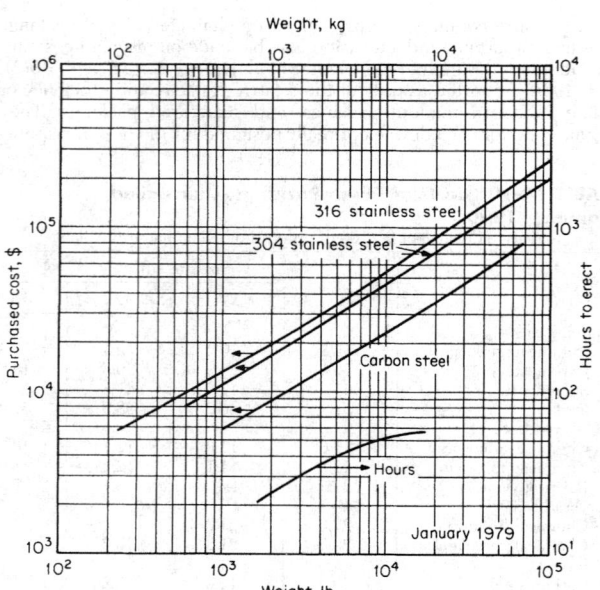

FIG. 14-81 Fabricated costs and installation time of towers. Costs are for shell with two heads and skirt, but without trays, packing, or connections. (*Peters and Timmerhaus*, Plant Design and Economics for Chemical Engineers, *4th ed.*, *McGraw-Hill, New York, 1991.*)

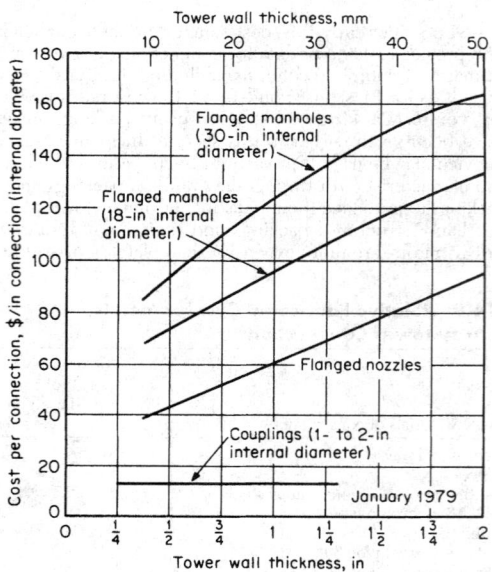

FIG. 14-82 Approximate installed cost of steel-tower connections. Values apply to 2070-kPa connections. Multiply costs by 0.9 for 1035-kPa (150-lb) connections and by 1.2 for 4140-kPa (600-lb) connections. To convert inches to millimeters, multiply by 25.4; to convert dollars per inch to dollars per centimeter, multiply by 0.394. (*Peters and Timmerhaus*, Plant Design and Economics for Chemical Engineers, *New York, McGraw-Hill, 1991.*)

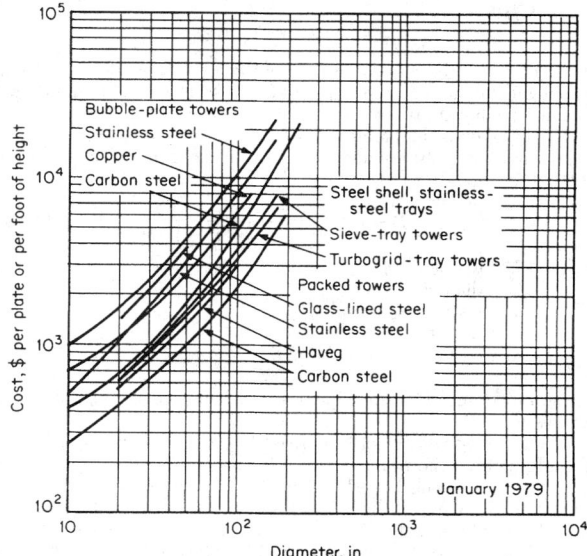

FIG. 14-83 Cost of towers including installation and auxiliaries. To convert inches to millimeters, multiply by 25.4; to convert feet to meters, multiply by 0.305; and to convert dollars per foot to dollars per meter, multiply by 3.28. (*Peters and Timmerhaus,* Plant Design and Economics for Chemical Engineers, 4th ed., *McGraw-Hill, New York, 1991.*)

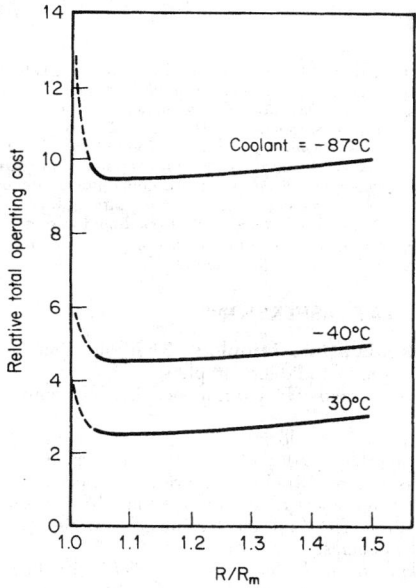

FIG. 14-84 Optimum-reflux-ratio determination. °F = ⁹/₅ °C + 32. [*Fair and Bolles,* Chem. Eng., **75**(4), 156 (1968).]

ments, a rough guide could be developed on the basis of flow parameter $L/G\,(\rho_G/\rho_L)^{0.5}$ (ordinate scale in Figs. 14-25 and 14-26) and the following tentative conclusions could be drawn:

Flow Parameter 0.02–0.1

1. Plates and random packings have much the same efficiency and capacity.
2. Structured packing efficiency is about 1.5 times that of plates or random packing.
3. At a parameter of 0.02, the structured packing has a 1.3–1.4 capacity advantage over random packing and plates. This advantage disappears as the parameter approaches 0.1.

Flow Parameter 0.1–0.3

1. Plates and random packings have about the same efficiency and capacity.
2. Structured packing has about the same capacity as plates and random packings.
3. The efficiency advantage of structured packing over random packings and plates decreases from 1.5 to 1.2 as the parameter increases from 0.1 to 0.3.

Flow Parameter 0.3–0.5

1. The loss of capacity of structured packing is greatest in this range.
2. The random packing appears to have the highest capacity and efficiency, and structured packing the least capacity and efficiency.

Experience indicates that use of structured packings may not have advantages in the higher-pressure (higher-flow-parameter) region. Special considerations for the use of structured packing at higher pressures are detailed by Kurtz et al. [*Chem. Eng. Progr.,* **87**(2), 43 (1991)].

Optimization As stated previously, optimization studies should include the entire system. Such a study was made by Fair and Bolles [*Chem. Eng.,* **75**(9), 156 (1968)], using a light-hydrocarbon system and with the objective of defining optimum reflux ratio. Coolants used were at −87, −40, and +30°C (−125, −40, and +85°F), corresponding to different pressures of operation and associated different condens-

ing temperatures. The results are shown in Fig. 14-84; the optimum reflux ratio is quite close to the calculated minimum reflux ratio.

Colburn (chemical engineering lecture notes, University of Delaware, 1943) proposed that the optimum reflux ratio is

$$R_{\mathrm{opt}} = \frac{N + [(C_2/hG_b + C_3)/C_1] + (dN/dR)}{dN/dR} \qquad (14\text{-}189)$$

where
- R = external reflux ratio = L/D
- N = number of theoretical plates
- G_b = allowable vapor velocity in heat exchangers, $(\text{lb·mol})/(\text{h·ft}^2)$
- h = hours of operation
- C_1 = amortization rate for tower, dollars/(ft²·plate·year)
- C_2 = amortization rate for heat exchangers, dollars/(ft²·year)
- C_3 = cost of utilities per mole of distillate, dollars/mol

Happel [*Chem. Eng.,* **65**(14), 144 (1958)] using a modification of the Colburn relationship found that the optimum number of trays varies from 2 to 3 times the number at total reflux. Gilliland [*Ind. Eng. Chem.,* **32**, 1220 (1940)] from the establishment of an empirical relationship between reflux ratio and theoretical trays based on a study of existing columns indicated that

$$0.1 < \frac{R_{\mathrm{opt}} - R_{\mathrm{min}}}{R_{\mathrm{min}} + 1} < 0.3$$

and correspondingly

$$0.35 < \frac{N_{\mathrm{opt}} - N_{\mathrm{min}}}{N_{\mathrm{min}} + 1} < 0.52$$

The effect of utilities costs on optimum operation was noted by Kiguchi and Ridgway [*Pet. Refiner,* **35**(12), 179 (1956)], who indicated that in petroleum-distillation columns the optimum reflux ratio varies between 1.1 and 1.5 times the minimum reflux ratio. When refrigeration is involved, $1.1R_{\mathrm{min}} < R_{\mathrm{opt}} < 1.2R_{\mathrm{min}}$, and when cooling-tower water is used in the condensers, $1.2R_{\mathrm{min}} < R_{\mathrm{opt}} < 1.4R_{\mathrm{min}}$.

PHASE DISPERSION

GENERAL REFERENCES: For an overall discussion of gas-liquid breakup processes, see Brodkey, *The Phenomena of Fluid Motions*, Addison-Wesley, Reading, Massachusetts, 1967. For a discussion of atomization devices and how they work, see Masters, *Spray Drying Handbook*, 5th ed., Wiley, New York, 1991; and Lefebvre, *Atomization and Sprays*, Hemisphere, New York, 1989. A beautifully illustrated older source is Dombrowski and Munday, *Biochemical and Biological Engineering Science*, vol. 2, Academic Press, London, 1968, pp. 209–320. Bayvel and Orzechowski, *Liquid Atomization*, Taylor and Francis, Washington DC, 1993, provides additional background into atomizer design. For a survey on fog formation, see Amelin, *Theory of Fog Formation*, Israel Program for Scientific Translations, Jerusalem, 1967.

LIQUID-IN-GAS DISPERSIONS

Liquid Breakup into Droplets   There are four basic mechanisms for breakup of liquid into droplets:
- Droplets in a field of high turbulence (i.e., high power dissipation per unit mass)
- Simple jets at low velocity
- Expanding sheets of liquid at relatively low velocity
- Droplets in a steady field of high relative velocity

These mechanisms coexist, and the one that gives the smallest drop size will control. The four mechanisms follow distinctly different velocity dependencies:

1. *Breakup in a highly turbulent field* $(1/\text{velocity})^{1.2}$. This appears to be the dominant breakup process in distillation trays in the spray regime, pneumatic atomizers, and high-velocity pipeline contactors.

2. *Breakup of a low-velocity liquid jet* $(1/\text{velocity})^{0}$. This governs in special applications like prilling towers and is often an intermediate step in liquid breakup processes.

3. *Breakup of a sheet of liquid* $(1/\text{velocity})^{0.67}$. This governs drop size in most hydraulic spray nozzles.

4. *Single-droplet breakup at very high velocity* $(1/\text{velocity})^{2}$. This governs drop size in free fall as well as breakup when droplets impinge on solid surfaces.

Droplet Breakup—High Turbulence   This is the dominant breakup mechanism for many process applications. Breakup results from local variations in turbulent pressure that distort the droplet shape. Hinze [*Am. Inst. Chem. Eng. J.*, **1**, 289–295 (1953)] applied turbulence theory to obtain the form of Eq. (14-171) and took liquid-liquid data to define the coefficient:

$$D_{max} = 0.725(\sigma/\rho_G)^{0.6}/E^{0.4} \qquad (14\text{-}190)$$

where E = (power dissipated)/mass   length²/time³
$\quad\quad\sigma$ = surface tension   mass/time²
$\quad\quad\rho_G$ = gas density   mass/length³

Note that D_{max} comes out with units of length. Since E typically varies with (gas velocity)³, this results in drop size dependence with $(1/\text{velocity})^{1.2}$.

The theoretical requirement for use of Eq. (14-190) is that the microscale of turbulence $\ll D_{max}$. This is satisfied in most gas systems. For example, in three cases,

	(microscale of turbulence)/D_{max}
distillation tray in spray regime	0.007
pipeline @ 40 m/s and atmospheric pressure	0.012
two-fluid atomizer using 100 m/s air	0.03

For these three applications, Eq. (14-190) gives good prediction of drop size when the design variables are used to calculate E, as illustrated by Eqs. (14-198) and (14-201).

Liquid-Column Breakup   Because of increased pressure at points of reduced diameter, the liquid column is inherently unstable. As a result, it breaks into small drops with no external energy input. Ideally, it forms a series of uniform drops with the size of the drops set by the fastest-growing wave. This yields a dominant droplet diameter

about 1.9 times the initial diameter of the jet as shown by Fig. 14-85. As shown, the actual breakup is quite close to prediction, although smaller satellite drops are also formed. The prime advantage of this type of breakup is the greater uniformity of drop size.

For high-viscosity liquids, the drops are larger, as shown by Eq. (14-191):

$$D = 1.9D_j \left[1 + \frac{3\mu_\ell}{(\sigma\rho_\ell D_j)^{1/2}} \right] \qquad (14\text{-}191)$$

where D = diameter of droplet
$\quad\quad D_j$ = diameter of jet
$\quad\quad \mu_\ell$ = viscosity of liquid
$\quad\quad \rho_\ell$ = density of liquid
$\quad\quad \sigma$ = surface tension of liquid

These units are dimensionally consistent; any set of consistent units can be used.

As the velocity of the jet is increased, the breakup process changes and ultimately becomes a mix of various competing effects, such as the capture of small drops by bigger ones in the slowing jet and the "turbulent breakup" of the bigger drops. The high-velocity jet is occasionally used in process applications because of the very narrow spray angle (5–20°) and the high penetration into a gas it can give. The focused stream also aids erosion of a surface.

Liquid-Sheet Breakup   The basic principle of most hydraulic atomizers is to form a thin sheet that breaks via a variety of mechanisms to form ligaments of liquid which in turn yield chains of droplets. See Fig. 14-86.

For a typical nozzle, the drop size varies with $1/(\text{pressure drop})^{1/3}$. When (velocity)² is substituted for pressure drop, droplet size is seen to vary with (velocity)$^{-2/3}$.

Isolated Droplet Breakup—in a Velocity Field   Much effort has focused on defining the conditions under which an isolated drop will break in a velocity field. The criterion for the largest stable drop

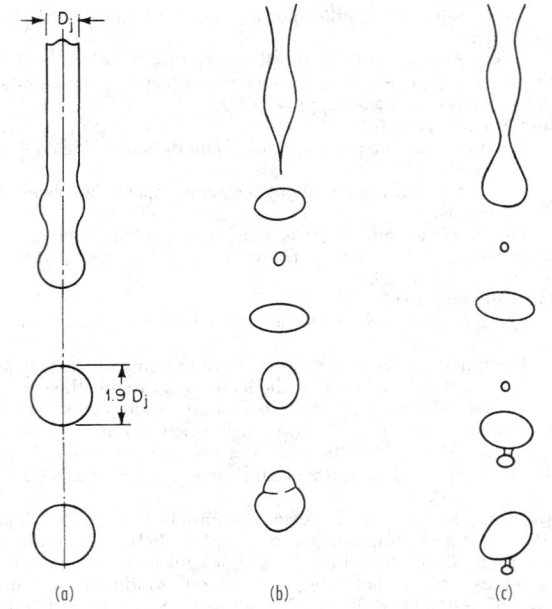

FIG. 14-85   (*a*) Idealized jet breakup suggesting uniform drop diameter and no satellites. (*b*) and (*c*) Actual breakup of a water jet as shown by high-speed photographs. [*From W. R. Marshall, "Atomization and Spray Drying," Chem. Eng. Prog. Monogr. Ser., no. 2 (1954).*]

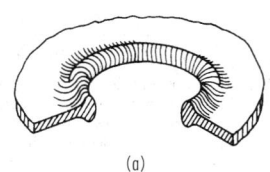

(a)

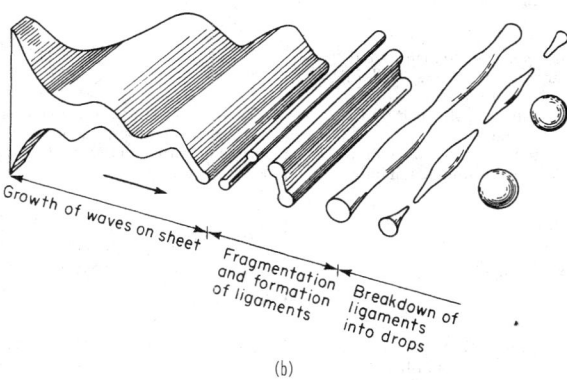

Growth of waves on sheet

Fragmentation and formation of ligaments

Breakdown of ligaments into drops

(b)

FIG. 14-86   Sheet breakup. (a) By perforation. [*After Fraser et al.,* Am. Inst. Chem. Eng. J., *8(5), 672 (1962).*] (b) By sinusoidal wave growth. [*After Dombrowski and Johns,* Chem. Eng. Sci., *18, 203 (1963).*]

size is the ratio of aerodynamic forces to surface-tension forces defined by the Weber number, N_{We} (dimensionless):

$$N_{We\ crit} = \text{constant} = [\rho_G \, (\text{velocity})^2 (D_{max})/(\sigma)] \qquad (14\text{-}192)$$

$N_{We\ crit}$ for low-viscosity fluids commonly ranges from 10 to 20, with the larger value for a free-fall condition and the smaller for a sudden acceleration. High liquid viscosity also increases $N_{We\ crit}$.

Droplet breakup via impingement appears to follow a similar relationship, but much less data is available. This type of breakup can result from impingement on equipment walls or compressor blades. In general, there is less tendency to shatter on wetted surfaces.

Droplet Size Distribution   Instead of the single droplet size implied by the discussion above, a spectrum of droplet sizes is produced. The most common ways to characterize this spectrum are:

• *Volume median (mass median)* D_{vm}. This has no fundamental meaning but is easy to determine since it is at the midpoint of a cumulative-volume plot.

• *Sauter mean* D_{32}.   This has the same ratio of surface to volume as the total drop population. It is typically 70 to 90 percent of D_{vm}. It is frequently used in transport processes and is used here to characterize drop size.

• *Maximum* D_{max}.   This is the largest-sized particle in the population. It is typically 3 to 4 times D_{32} in turbulent breakup processes, per Walzel [*International Chemical Engineering*, **33**, 46, (1993)]. It is the size directly calculated from the power/mass relationship. D_{32} is estimated from D_{max} by

$$D_{32} = 0.3 \cdot D_{max} \qquad (14\text{-}193)$$

and D_{vm} is estimated from it by

$$D_{vm} = 0.4 \cdot D_{max} \qquad (14\text{-}194)$$

However, any average drop size is fictitious, and none is completely satisfactory. For example, there is no way in which the high surface and transfer coefficients in small drops can be made available to the larger drops. Hence, a process calculation based on a given droplet size describes only what happens to that size and gives at best an approximation to the total mass.

There are a variety of ways to describe the droplet population. Figures 14-88 and 14-90 illustrate one of the most common methods, the plot of cumulative volume against droplet size on log-normal graph paper. This satisfies the restraint of not extrapolating to a negative drop size. Its other advantages are that it is easy to plot, the results are easy to visualize, and it yields a nearly straight line at lower drop sizes.

Cumulative volume over the range of 1 to 50 percent can also be shown to vary approximately as D^2. This is equivalent to finding that the number of droplets of a given size is inversely proportional to the droplet area or the surface energy of the droplet.

Atomizers   The common need to disperse a liquid into a gas has spawned a large variety of mechanical devices. The different designs emphasize different advantages such as freedom from plugging, pattern of spray, small droplet size, uniformity of spray, high turndown ratio, and low power consumption.

As shown in Table 14-11, most atomizers fall into three categories:
1. Pressure nozzles (hydraulic)
2. Two-fluid nozzles (pneumatic)
3. Rotary devices (spinning cups, disks, or vaned wheels)
These share certain features such as relatively low efficiency and low cost relative to most process equipment. The energy required to produce the increase in area is typically less than 0.1 percent of the total energy consumption. This is because atomization is a secondary process resulting from high interfacial shear or turbulence. As droplet sizes decrease, this efficiency drops lower.

Other types are available that use sonic energy (from gas streams), ultrasonic energy (electronic), and electrostatic energy, but they are less commonly used in process industries. See Table 14-11 for a summary of the advantages/disadvantages of the different type units. An expanded discussion is given by Masters [*Spray Drying Handbook,* Wiley, New York, (1991)].

Special requirements such as size uniformity in prilling towers can dictate still other approaches to dispersion. Here plates are drilled with many holes to develop nearly uniform columns.

Commonly, the most important feature of a nozzle is the size of droplet it produces. Since the heat or mass transfer that a given dispersion can produce is often proportional to $(1/D_d)^2$, fine drops are usually favored. On the other extreme, drops that are too fine will not settle, and a concern is the amount of liquid that will be entrained from a given spray operation. For example, if sprays are used to contact atmospheric air flowing at 1.5 m/s, drops smaller than 350 µm [terminal velocity = 1.5 m/s (4.92 ft/s)] will be entrained. Even for the relative coarse spray of the hollow-cone nozzle shown in Fig. 14-88, 7.5 percent of the total liquid mass will be entrained.

Hydraulic (Pressure) Nozzles   Manufacturers' data such as shown by Fig. 14-88 are available for most nozzles for the air-water system. In Fig. 14-88, note the much coarser solid-cone spray. The coarseness results from the less uniform discharge.

Effect of Physical Properties on Drop Size   Because of the extreme variety of available geometries, no attempt to encompass this variable is made here. The suggested predictive route starts with air-water droplet size data from the manufacturer at the chosen flow rate. This drop size is then corrected by Eq. (14-195) for different viscosity and surface tension:

$$\frac{D_{vm,\ system}}{D_{vm,\ water}} = \left(\frac{\sigma_{system}}{73}\right)^{0.25} \left(\frac{\mu_\ell}{1.0}\right)^{0.25} \qquad (14\text{-}195)$$

where D_{vm} = volume median droplet diameter
σ = surface tension, mN/m (dyn/cm)
μ_ℓ = liquid viscosity, mPa·s (cP)

The exponential dependencies in Eq. (14-195) represent averages of values reported by a number of studies with particular weight given to Lefebvre ([*Atomization and Sprays,* Hemisphere, New York, (1989)]. Since viscosity can vary over a much broader range than surface tension, it has much more leverage on drop size. For example, it is common to find an oil with 1000 times the viscosity of water, while most liquids fall within a factor of 3 of its surface tension. Liquid density is generally even closer to that of water, and since the data are not clear that a liquid density correction is needed, none is shown in Eq.

TABLE 14-11 Atomizer Summary

Types of atomizer	Design features	Advantages	Disadvantages
Pressure.	Flow $\alpha(\Delta P/\rho_\ell)^{1/2}$. Only source of energy is from fluid being atomized.	Simplicity and low cost.	Limited tolerance for solids; uncertain spray with high-viscosity liquids; susceptible to erosion. Need for special designs (e.g., bypass) to achieve turndown.
1. Hollow cone.	Liquid leaves as conical sheet as a result of centrifugal motion of liquid. Air core extends into nozzle.	High atomization efficiency.	Concentrated spray pattern at cone boundaries.
a. Whirl chamber (see Fig. 14-87*a*).	Centrifugal motion developed by tangential inlet in chamber upstream of orifice.	Minimum opportunity for plugging.	
b. Grooved core.	Centrifugal motion developed by inserts in chamber.	Smaller spray angle than 1*a* and ability to handle flows smaller than 1*a*.	
2. Solid cone (see Fig. 14-87*b*).	Similar to hollow cone but with insert to provide even distribution.	More uniform spatial pattern than hollow cone.	Coarser drops for comparable flows and pressure drops. Failure to yield same pattern with different fluids.
3. Fan (flat) spray.	Liquid leaves as a flat sheet or flattened ellipse.	Flat pattern is useful for coating surfaces and for injection into streams.	Small clearances.
a. Oval or rectangular orifice (see Fig. 14-87*c*). Numerous variants on cavity and groove exist.	Combination of cavity and orifice produces two streams that impinge within the nozzle.		
b. Deflector (see Fig. 14-87*d*).	Liquid from plain circular orifice impinges on curved deflector.	Minimal plugging.	Coarser drops.
c. Impinging jets (see Fig. 14-87*e*).	Two jets collide outside nozzle and produce a sheet perpendicular to their plane.	Different liquids are isolated until they mix outside of orifice. Can produce a flat circular sheet when jets impinge at 180°.	Extreme care needed to align jets.
4. Nozzles with wider range of turndown.			
a. Spill (bypass) (see Fig. 14-87*f*).	A portion of the liquid is recirculated after going through the swirl chamber.	Achieves uniform hollow cone atomization pattern with very high turndown (50:1).	Waste of energy in bypass stream. Added piping for spill flow.
b. Poppet (see Fig. 14-87*g*).	Conical sheet is developed by flow between orifice and poppet. Increased pressure causes poppet to move out and increase flow area.	Simplest control over broad range.	Difficult to maintain proper clearances.
Two-fluid (see Fig. 14-87*h*).	Gas impinges coaxially and supplies energy for breakup.	High velocities can be achieved at lower pressures because the gas is the high-velocity stream. Liquid-flow passages can be large, and hence plugging can be minimized.	Because gas is also accelerated, efficiency is inherently lower than pressure nozzles.
Sonic.	Gas generates an intense sound field into which liquid is directed.	Similar to two-fluid but with greater tolerance for solids.	Similar to two-fluid.
Rotary wheels (see Fig. 14-87*i*) disks, and cups.	Liquid is fed to a rotating surface and spreads in a uniform film. Flat disks, disks with vanes, and bowl-shaped cups are used. Liquid is thrown out at 90° to the axis.	The velocity that determines drop size is independent of flow. Hence these can handle a wide range of rates. They can also tolerate very viscous materials as well as slurries. Can achieve very high capacity in a single unit; does not require a high-pressure pump.	Mechanical complexity of rotating equipment. Radial discharge.
Ultrasound.	Liquid is fed over a surface vibrating at a frequency > 20 kHz.	Fine atomization, small size, and low injection velocity.	Low flow rate and need for ultrasound generator.

(14-195). Vapor density also has an impact on dropsize but the impact is complex, involving conflicts of a number of effects, and vapor density is commonly omitted in atomizer dropsize correlations.

Effect of Pressure Drop and Nozzle Size For a nozzle with a developed pattern, the average drop size can be estimated to fall with rising ΔP (pressure drop) by Eq. (14-196):

$$\frac{D_1}{D_2} = \left(\frac{\Delta P_2}{\Delta P_1}\right)^{1/3} \qquad (14\text{-}196)$$

For similar nozzles and constant ΔP, the drop size will increase with nozzle size as indicated by Eq. (14-197):

$$\frac{D_1}{D_2} = \left(\frac{\text{orifice diameter}_1}{\text{orifice diameter}_2}\right)^{1/2} \qquad (14\text{-}197)$$

Once again, these relationships are averages of a number of reported values and are intended as rough guides.

The normal operating regime is well below turbulent breakup velocity. However the data of Kennedy [*J. of Engineering for Gas Turbines and Power*, **108**, 191, (1986)] at very high pressure drop in large nozzles shows a shift to a higher dependence on pressure drop. This data suggests that turbulent droplet breakup can also be governing with hydraulic spray nozzles, although this is unusual.

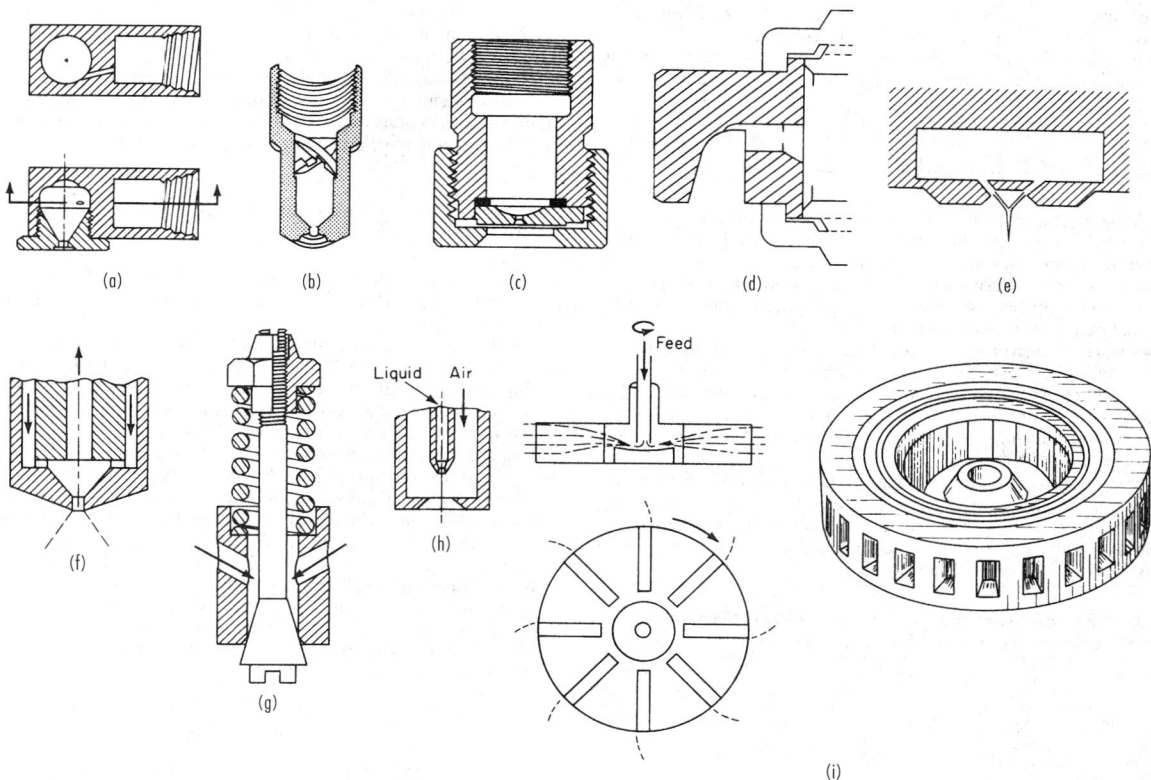

FIG. 14-87 Charactersitic spray nozzles. (*a*) Whirl-chamber hollow cone. (*b*) Solid cone. (*c*) Oval-orifice fan. (*d*) Deflector jet. (*e*) Impinging jet. (*f*) Bypass. (*g*) Poppet. (*h*) Two-fluid. (*i*) Vaned rotating wheel.

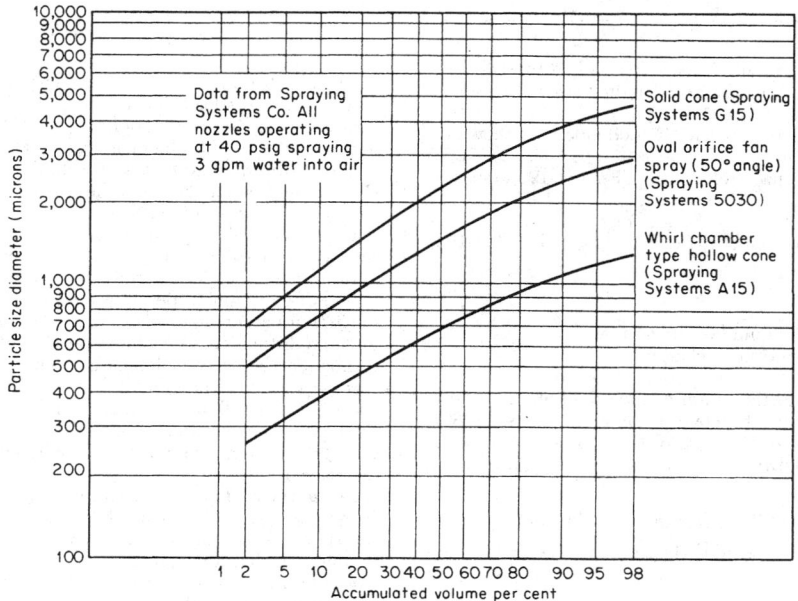

FIG. 14-88 Droplet-size distribution for three different types of nozzles. To convert pounds per square inch gauge to kilopascals, multiply by 6.89; to convert gallons per minute to cubic meters per hour, multiply by 0.227. (*Spraying Systems Inc.*)

Spray Angle A shift to a smaller-angle nozzle gives slightly larger drops for a given type of nozzle because of the reduced tendency of the sheet to thin. Dietrich [*Proc. 1st Int. Conf. Liq. Atomization Spray Systems*, Tokyo, (1978)] shows the following:

Angle	25°	50°	65°	80°	95°
D_{vm}, µm	1459	1226	988	808	771

In calculating the impact point of spray, one should recognize that the spray angle closes in as the spray moves away from the nozzle. This is caused by loss of momentum of the spray to the gas.

At some low flow, pressure nozzles do not develop their normal pattern but tend to approach solid streams. The required flow to achieve the normal pattern increases with viscosity.

Two-Fluid (Pneumatic) Atomizers This general category includes such diverse applications as venturi atomizers and reactor-effluent quench systems in addition to two-fluid spray nozzles. Depending on the manner in which the two fluids meet, several of the breakup mechanisms may be applicable, but the final one is high-level turbulent rupture.

As shown by Table 14-12, empirical correlations for two-fluid atomization show dependence on high gas velocity to supply atomizing energy, usually to a power dependence close to that for turbulent breakup. In addition, the correlations show a dependence on the ratio of gas to liquid and system dimension.

TABLE 14-12 Exponential Dependence of Drop Size on Different Parameters in Two-Fluid Atomization

	Relative velocity	Surface tension	Gas density	$1 + L/G$	Atomizer dimension
Jasuja (empirical for small nozzle)	−0.9	0.45	−0.45	0.5	0.55
El-Shanawany and Lefebvre (empirical for small nozzle)	−1.2	0.6	−0.7	1	0.40
Tatterson, Dallman, and Hanratty (pipe flow)	−1	0.5	−0.5		0.5
Hinze (turbulence theory)	−1.2	0.6	−0.6		
Steinmeyer (extension of turbulence theory)	−1.2	0.6	−0.6	0.4	0.4

Further differences from hydraulic nozzles (controlled by sheet and ligament breakup) are the stronger increase in drop size with increasing surface tension and decreasing gas density.

The similarity of these correlations to the dependencies shown by Eq. (14-190) was noted by Steinmeyer [*Chem. Eng. Progr.*, **91**(7), 72 (1995)] who reformulated Hinze's relationship, Eq. (14-190), into Eq. (14-198) by including atomizer variables.

$$D_{32} = 0.29 \left(\frac{\sigma}{\rho_G} \right)^{0.6} (1/velocity)^{1.2} \left(\frac{1+L}{G} \right)^{0.4} \left(D_{nozzle} \right)^{0.4} \quad (14\text{-}198)$$

where

σ = surface tension
ρ_G = gas density
L/G = mass ratio of liquid flow to gas flow
D_{nozzle} = diameter of the air discharge

This is remarkably similar to the empirical two-fluid atomizer relationships of El-Shanawany and Lefebvre [*J. Energy*, **4**, 184 (1980)] and Jasuja [*Trans. Am. Soc. Mech. Engr.*, **103**, 514 (1981)]. For example, El-Shanawany and Lefebvre give a relationship for a prefilming atomizer:

$$D_{32} = 0.0711(\sigma/\rho_G)^{0.6}(1/velocity)^{1.2}(1 + L/G)(D_{nozzle})^{0.4}(\rho_L/\rho_G)^{0.1}$$
$$+ 0.015[(\mu_L)^2/(\sigma \times \rho_L)]^{0.5}(D_{nozzle})^{0.5}(1 + L/G) \quad (14\text{-}199)$$

where μ_L is liquid viscosity.

According to Jasuja,

$$D_{32} = 0.17(\sigma/\rho_G)^{0.45}(1/velocity)^{0.9}(1 + L/G)^{0.5}(D_{nozzle})^{0.55}$$
$$+ \text{viscosity term} \quad (14\text{-}200)$$

(Eqs. 14-198, 14-199, and 14-200 are dimensionally consistent; any set of consistent units on the right-hand side yields the droplet size in units of length on the left-hand side.)

The second, additive term carrying the viscosity impact in Eq. (14-199) is small at viscosities around 1 centipoise but can become controlling as viscosity increases. For example, for air at atmospheric pressure atomizing water, with nozzle conditions.

$$D_{nozzle} = 0.076 \text{ m (3 inches)}$$
$$velocity = 100 \text{ m/s (328 ft/s)}$$
$$L/G = 1$$

For this case, Steinmeyer's correlation becomes El-Shanawany, Eq. (14-199) predicts 76 microns with the viscosity term contributing less than 1 percent. With the same system and same L/G, but with an oil with 30 times water viscosity, Eq. (14-199) predicts 91 microns, with the viscosity term contributing 54 percent of the total. The measure values for water and oil cases were 70 and 95 microns, respectively. For comparison, Eq. (14-198) predicts 102 microns for the water case.

Rotary Atomizers For rotating wheels, vaneless disks, and cups, there are three regimes of operation. At low rates, the liquid is shed directly as drops from the rim. At intermediate rates, the liquid leaves the rim as threads; and at the highest rate, the liquid extends from the edge as a thin sheet that breaks down similarly to a fan or hollow-cone spray nozzle. As noted in Table 14-12, rotary devices have many unique advantages such as the ability to handle high viscosity and slurries and produce small droplets without high pressures. The prime applications are in spray drying. See Masters [*Spray Drying Handbook*, Wiley, New York (1991)] for more details.

Pipeline Contactors The power dissipation per unit mass for pipeline flow is similar to that for two-fluid nozzles.

$$D_{32} = 0.79 \left(\frac{\sigma}{\rho_G} \right)^{0.6} \left(\frac{1}{velocity} \right)^{1.2} (D_{pipe})^{0.4} \quad (14\text{-}201)$$

(The relation is dimensionally consistent; any set of consistent units on the right-hand side yields the droplet size in units of length on the left-hand side.)

The relationship is similar to the empirical correlation of Tatterson, Dallman, and Hanratty [*Am. Inst. Chem. Eng. J.*, **23**(1), 68 (1977)]

$$D_{32} \sim \left(\frac{\sigma}{\rho_G} \right)^{0.5} (1/velocity)^1 (D_{pipe})^{0.5}$$

Predictions from Eq. (14-201) align well with the Tatterson data. For example, for a velocity of 43 m/s (140 ft/s) in a 0.05-m (1.8-inch) equivalent diameter channel, Eq. (14-201) predicts D_{32} of 490 microns, compared to the measured 460 to 480 microns.

Entrainment Due to Gas Bubbling/Jetting through a Liquid Entrainment generally limits the capacity of distillation trays and is commonly a concern in vaporizers and evaporators. Fortunately, it is readily controllable by simple inertial entrainment capture devices such as wire mesh pads in gravity separators.

In distillation towers, entrainment lowers the tray efficiency, and 1 pound of entrainment per 10 pounds of liquid is sometimes taken as the limit for acceptable performance. However, the impact of entrainment on distillation efficiency depends on the relative volatility of the component being considered. Entrainment has a minor impact on close separations when the difference between vapor and liquid concentration is small, but this factor can be dominant for systems where the liquid concentration is much higher than the vapor in equilibrium with it (i.e., when a component of the liquid has a very low volatility, as in an absorber).

As shown by Fig. 14-90, entrainment droplet sizes span a broad range. The reason for the much larger drop sizes of the upper curve is the short disengaging space. For this curve, over 99 percent of the entrainment has a terminal velocity greater than the vapor velocity. For contrast, in the lower curve the terminal velocity of the largest particle reported is the same as the vapor velocity. For the settling velocity to limit the maximum drop size entrained, at least 0.8 m (30 in) disengaging space is usually required. Note that even for the lower curve, less than 10 percent of the entrainment is in drops of less than

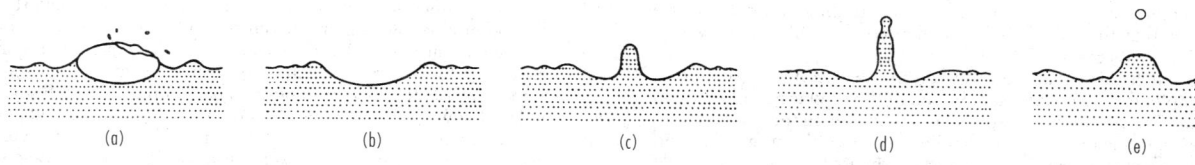

FIG. 14-89 Mechanism of the burst of an air bubble on the surface of water. [*Newitt, Dombrowski, and Knellman,* Trans. Inst. Chem. Eng., **32**, 244 (1954).]

50 μm. The coarseness results from the relatively low power dissipation per mass on distillation trays. This means that it is relatively easy to remove by a device such as a wire mesh pad. Over 50 percent is typically captured by the underside of the next higher tray or by a turn in the piping leaving an evaporator. Conversely, though small on a mass basis, the smaller drops are extremely numerous. On a number basis, more than one-half of the drops in the lower curve are under 5 μm. These can serve as nuclei for fog condensation in downstream equipment.

Entrainment can stem from a variety of sources.

1. Excessive foaming. This is a case of a gas-in-liquid dispersion (covered in the next subsection).

2. Droplets formed from the collapse of the bubble dome (see Fig. 14-89). These are virtually unavoidable. They are generally under 25 μm, which means that their terminal velocities are low and they are invariably entrained. Fortunately, because of their small size, they contribute little on a weight basis (<0.001 kg liquid/kg vapor), although they dominate on a number basis.

3. Droplets from the jet caused by liquid rushing to fill the cavity left by the bubble (see Fig. 14-89). These droplets range up to 1000 μm, their size depending on bubble size. This is important only at modest loadings. Once foam forms over the surface, drop ejection by this mode decreases sharply.

4. At higher vapor loads, the kinetic energy of the vapor rather than the bubble burst supplies the thrust for jets and sheets of liquid that are thrown up as well as the energy from breakup into "spray." This yields much higher levels of entrainment. In distillation trays it is the most common limit to capacity.

The major variable in setting entrainment (E, weight of liquid entrained per weight of vapor) is vapor velocity. As velocity is increased, the dependence of E on velocity steepens. In the lowest velocity regime, E is proportional to velocity. At values of E of about 0.001 (around 10 percent of flood), there is a shift to a region where the dependence is with (velocity)$^{3-5}$. Near flood, the dependence rises to approximately (velocity)^8. In this regime, the kinetic energy of the vapor dominates, and the bulk of the dispersion on the plate is often in the form of a coarse spray.

Pinczewski and Fell [*Trans. Inst. Chem Eng.*, **55**, 46 (1977)] show that the velocity at which vapor jets onto the tray sets the droplet size, rather than the superficial tray velocity. A maximum superficial velocity formulation that incorporates φ, the fractional open area, is logical since the fractional open area sets the jet velocity. Stichlmair and Mersmann [*Int. Chem. Eng.*, **18**(2), 223 (1978)] give such a correlation:

$$U_{max} = \frac{2.5[\phi g(\rho_l - \rho_g)\sigma]^{1/4}}{\rho_g^{1/2}} \qquad (14\text{-}202)$$

(The relation is dimensionally consistent; any set of consistent units on the right-hand side yields velocity units for the left-hand side.)

Stichlmair uses the ratio of actual velocity to this maximum velocity together with a term that increases entrainment as the distance gets small between the liquid-vapor layer and the tray deck above. His correlation spans a 10^5 fold range in entrainment. He shows a sharp increase in entrainment at 60 percent of the maximum velocity and attributes the increase to a shift to the spray regime.

Puppich and Goedecke [*Chem. Eng. Tech.*, **10**, 224 (1987)] test this correlation against a wide range of tray types and find generally good agreement. Bubble caps give about twice as much entrainment as the correlation predicts. Sieve trays give about half as much as the correlation predicts.

Steinmeyer [*Chem. Eng. Progr.*, **91**(7), 72 (1995)] derived a correlation for entrainment utilizing predicted drop size from a turbulent power dissipation per unit mass) breakup in the "spray" regime. His predicted drop size matched the data of Pinczewski and Fell. When this prediction is combined with the estimated fraction of the droplet population that is entrained, the entrainment prediction, Eq. (14-203), results. Note that this matches the empirical experience in predicting entrainment varying with (velocity)^8

$$E = \text{constant } [V]^8 \phi^3 \frac{\rho_G^4}{[g(\rho_L - \rho_G)]^{2.5}\sigma^{1.5}} \qquad (14\text{-}203)$$

If *flood* is defined as the velocity at which E equals 1, this yields (dimensionless):

$$V_{flood} = 1.25 \times \frac{\{\phi^{0.375}[g(\rho_L - \rho_G)]^{0.3125}\sigma^{0.1875}(Z/Z_{base})^{0.125}\}}{\rho_G^{0.5}} \qquad (14\text{-}204)$$

where Z is tray spacing in meters and Z/Z_{base} is the ratio of tray spacing to 0.3 m (1 ft).

Fair's empirical correlation for sieve and bubble-cap trays shown in Fig. 14-26 is similar. Note that Fig. 14-26 incorporates a velocity dependence (velocity)$^{6-8}$ above 90 percent of flood for high-density systems. The correlation implicitly considers the tray design factors such as the open area, tray spacing, and hole diameter through the impact of these factors on percent of flood.

The dependencies of all three correlations are remarkably close, as shown by Table 14-13 and the numeric prediction of flooding velocity is also close.

Correlations can be extended to evaporators at lower velocities by assuming that E declines with (velocity)^2 between 60 percent and 10 percent of the maximum velocity. At velocities below 10 percent of the maximum velocity, E can be assumed to change directly with velocity.

Fog Condensation This is an entirely different way of forming dispersions. Here, the dispersion results from condensation of a vapor rather than mechanical breakup. The particle sizes are usually much finer (0.1 to 30 μm) and are designated as mist or fog.

Fog particles grow because of excess saturation in the gas. Usually this means that the gas is supersaturated (i.e., it is below its dew point). Sometimes, fog can also grow on soluble foreign nuclei at partial pressures below saturation. Increased saturation can occur through a variety of routes:

1. Mixing of two saturated streams at different temperatures. This is commonly seen in the plume from a stack. Since vapor pressure is an exponential function of temperature, the resultant mixture of two saturated streams will be supersaturated at the mixed temperature. Uneven flow patterns and cooling in heat exchangers make this route to supersaturation difficult to prevent.

2. Increased partial pressure due to reaction. An example is the reaction of SO_3 and H_2O to yield H_2SO_4, which has much lower vapor pressure than its components.

3. Isoentropic expansion (cooling) of a gas, as in a steam nozzle.

TABLE 14-13 Dependency of Distillation Flood Velocity on Physical Properties and Tray Open Area

	Power/mass	Fair	Stichlmair/Mersmann
φ	0.375	0.44	0.5
$\rho_L - \rho_G$	0.3125	0.5	0.25
σ	0.1875	0.2	0.25
ρ_G	0.5	0.5	0.5

4. Cooling of a gas containing a condensable vapor. Here the problem is that the gas cools faster than condensable vapor can be removed by mass transfer.

These mechanisms can be observed in many common situations. For example, fog via mixing can be seen in the discharge of breath on a cold day. Fog via adiabatic expansion can be seen in the low-pressure area over the wing of an airplane landing on a humid summer day; and fog via condensation can be seen in the exhaust from an automobile air conditioner (if you follow closely enough behind another car to pick up the ions or NO molecules needed for nucleation). All of these occur at a very low supersaturation and appear to be keyed to an abundance of foreign nuclei. All of these fogs also quickly dissipate as heat or unsaturated gas is added.

The supersaturation in condensers arises for two reasons. First, the condensable vapor is generally of higher molecular weight than the noncondensable gas. This means that the molecular diffusivity of the vapor will be much less than the thermal diffusivity of the gas. Restated, the ratio of N_{Sc}/N_{Pr} is greater than 1. The result is that a condenser yields more heat-transfer units $dT_g/(T_g - T_i)$ than mass-transfer units $d\bar{Y}_g/(Y_g - Y_i)$. Second, both transfer processes derive their driving force from the temperature difference between the gas T_g and the interface T_i. Each incremental decrease in interface temperature yields the same relative increase in temperature driving force. However, the interface vapor pressure can only approach the limit of zero. Because of this, for equal molecular and thermal diffusivities a saturated mixture will supersaturate when cooled. The tendency to supersaturate generally increases with increased molecular weight of the condensable, increased temperature differences, and reduced initial superheating. To evaluate whether a given condensing step yields fog requires rigorous treatment of the coupled heat-transfer and mass-transfer processes through the entire condensation. Steinmeyer [*Chem. Eng. Prog.*, **68**(7), 64 (1972)] illustrates this, showing the impact of foreign-nuclei concentration on calculated fog formation. See Table 14-14. Note the relatively large particles generated for cases 1 and 2 for 10,000 foreign nuclei per cm³. These are large enough to be fairly easily collected. There have been very few documented problems with industrial condensers despite the fact that most calculate to generate supersaturation along the condensing path. The explanation appears to be a limited supply of foreign nuclei.

Ryan et al. [*Chem. Eng. Progr.*, **90**(8), 83 (1994)] show that separate mass and heat transfer-rate modeling of an HCl absorber predicts 2 percent fog in the vapor. The impact is equivalent to lowering the stage efficiency to 20 percent.

Spontaneous (Homogeneous) Nucleation This process is quite difficult because of the energy barrier associated with creation of the interfacial area. It can be treated as a kinetic process with the rate a very steep function of the supersaturation ratio (S = partial pressure of condensable per vapor pressure at gas temperature). For water, an increase in S from 3.4 to 3.9 causes a 10,000-fold increase in the nucleation rate. As a result, below a critical supersaturation (S_{crit}), homogeneous nucleation is slow enough to be ignored. Generally, S_{crit} is defined as that which limits nucleation to one particle produced per cubic centimeter per second. It can be estimated roughly by traditional theory (*Theory of Fog Condensation*, Israel Program for Scientific Translations, Jerusalem, 1967) using the following equation:

$$S_{crit} = \exp\left[0.56\, \frac{M}{\rho_l}\left(\frac{\sigma}{T}\right)^{3/2}\right] \qquad (14\text{-}205)$$

where σ = surface tension, mN/m (dyn/cm)
  ρ_l = liquid density, g/cm³
  T = temperature, K
  M = molecular weight of condensable

Table 14-15 shows typical experimental values of S_{crit} taken from the work of Russel [*J. Chem. Phys.*, **50**, 1809 (1969)]. Since the critical supersaturation ratio for homogeneous nucleation is typically greater than 3, it is not often reached in process equipment.

Growth on Foreign Nuclei As noted above, foreign nuclei are often present in abundance and permit fog formation at much lower supersaturation. For example,

1. *Solids.* Surveys have shown that air contains thousands of particles per cubic centimeter in the 0.1-μm to 1-μm range suitable for nuclei. The sources range from ocean-generated salt spray to combustion processes. The concentration is highest in large cities and industrial regions. When the foreign nuclei are soluble in the fog, nucleation occurs at S values very close to 1.0. This is the mechanism controlling atmospheric water condensation. Even when not soluble, a foreign particle is an effective nucleus if wet by the liquid. Thus, a 1-μm insoluble particle with zero contact angle requires an S of only 1.001 in order to serve as a condensation site for water.

2. *Ions.* Amelin [*Theory of Fog Condensation*, Israel Program for Scientific Translations, Jerusalem, (1967)] reports that ordinary air contains even higher concentrations of ions. These ions also reduce the required critical supersaturation, but by only about 10 to 20 percent, unless multiple charges are present.

3. *Entrained liquids.* Production of small droplets is inherent in the bubbling process, as shown by Fig. 14-90. Values range from near zero to 10,000/cm³ of vapor, depending on how the vapor breaks through the liquid and on the opportunity for evaporation of the small drops after entrainment.

As a result of these mechanisms, most process streams contain enough foreign nuclei to cause some fogging. While fogging has been reported in only a relatively low percent of process partial condensers, it is rarely looked for and volunteers its presence only when yield losses or pollution is intolerable.

Dropsize Distribution Monodisperse (nearly uniform droplet size) fogs can be grown by providing a long retention time for growth. However, industrial fogs usually show a broad distribution, as in Fig. 14-91. Note also that for this set of data, the sizes are several orders of magnitude smaller than those shown earlier for entrainment and atomizers.

The result, as discussed in a later subsection, is a demand for different removal devices for the small particles.

While generally fog formation is a nuisance, it can occasionally be useful because of the high surface area generated by the fine drops. An example is insecticide application.

TABLE 14-14 Simulation of Three Heat Exchangers with Varying Foreign Nuclei

	1	2	3
Weight fraction, noncondensable			
Inlet	0.51	0.42	0.02
Outlet	0.80	0.80	0.32
Molecular weight			
Inert	28	29	29
Condensable	86	99	210
Temperature difference between gas and liquid interface, K			
Inlet	14	24	67
Outlet	4	10	4
Percent of liquid that leaves unit as fog Nuclei concentration in inlet particles/cm³			
100	0.05	1.1	2.2
1,000	0.44	5.6	3.9
10,000	3.2	9.8	4.9
100,000	9.6	11.4	5.1
1,000,000	13.3	11.6	
10,000,000	14.7		
∞	14.7	11.8	5.1
Fog particle size based on 10,000 nuclei/cm³ at inlet, μm	28	25	4

TABLE 14-15 Experimental Critical Supersaturation Ratios

	Temperature, K°	S_{crit}
H_2O	264	4.91
C_2H_5OH	275	2.13
CH_4OH	264	3.55
C_6H_6	253	5.32
CCl_4	247	6.5
$CHCl_3$	258	3.73
C_6H_5Cl	250	9.5

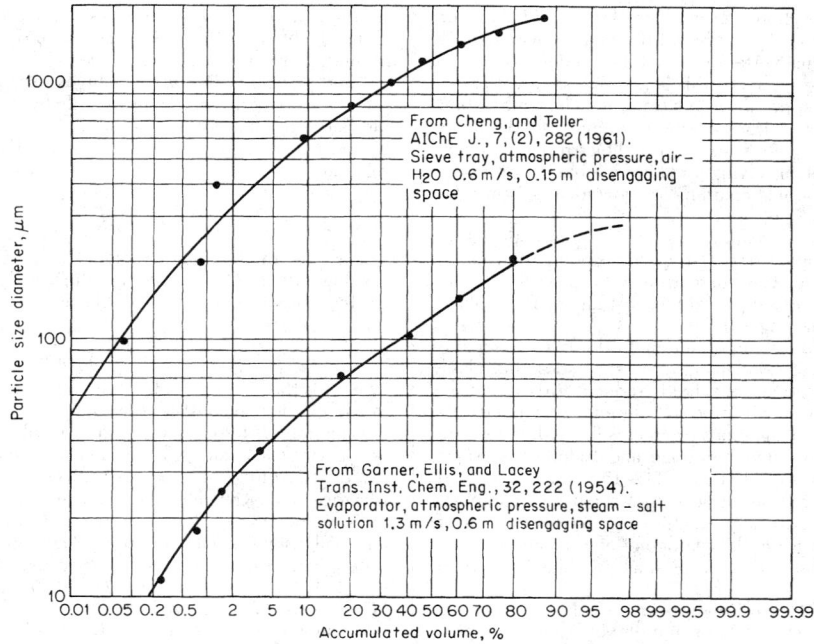

FIG. 14-90   Entrainment droplet-size distribution. To convert meters per second to feet per second, multiply by 3.28, to convert meters to feet multiply by 3.28.

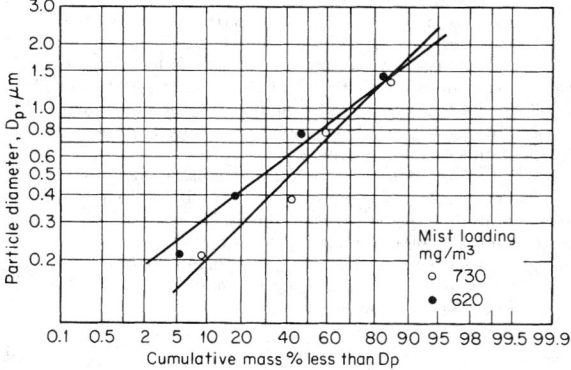

FIG. 14-91   Particle-size distribution and mist loading from absorption tower in a contact H_2SO_4 plant. [*Gillespie and Johnstone,* Chem. Eng. Prog., *51*(2), 74 (1955).]

GAS-IN-LIQUID DISPERSIONS

GENERAL REFERENCES: Comprehensive treatments of bubbles or foams are given by Akers, *Foams: Symposium 1975,* Academic Press, New York, 1973; Bendure, *Tappi,* **58,** 83 (1975); Benfratello, *Energ Elettr.,* **30,** 80, 486 (1953); Berkman and Egloff, *Emulsions and Foams,* Reinhold, New York, 1941, pp. 1 12–152; Bikerman, *Foams,* Springer-Verlag, New York, 1975; *Kirk-Othmer Encyclopedia of Chemical Technology,* 4th ed., Wiley, New York, 1993, pp. 82–145; Haberman and Morton, *Report 802,* David W. Taylor Model Basin, Washington, 1953; Levich, *Physicochemical Hydrodynamics,* Prentice-Hall, Englewood Cliffs, NJ, 1962; and Soo, *Fluid Dynamics of Multiphase Systems,* Blaisdell, Waltham, Massachusetts, 1967. The formation of bubbles is comprehensively treated by Clift, Grace, and Weber, *Bubbles, Drops and Particles,* Academic, New York, 1978; Kumar and Kuloor, *Adv. Chem. Eng,* **8,** 255–368 (1970) and Wilkinson and Van Dierendonck, *Chem. Eng Sci.,* **49,** 1429–1438 (1994). Design methods for units operation in bubble columns and stirred vessels are

covered by Atika and Yoshida, *Ind. Eng Chem. Process Des. Dev.,* **13,** 84 (1974); Calderbank, *The Chem. Eng.* (London), CE209 (October, 1967); and Mixing, vol. II, Academic, New York, 1967, pp. 1–111; Fair, *Chem. Eng,* **74,** 67 (July 3, 1967); Jordan, *Chemical Process Dev.,* Interscience, New York, 1968, part 1, pp. 111–175; Mersmann, *Ger. Chem. Eng,* **1,** 1 (1978); Resnick and Gal-Or, *Adv. Chem. Eng.,* **7,** 295–395 (1968); Valentin, *Absorption in Gas-Liquid Dispersions,* E. & F. N. Spon, London, 1967; Tatterson, *Fluid Mixing and Gas Dispersion in Agitated Tanks,* McGraw-Hill, 1991; and Deckwer and Schumpe, *Chem. Eng. Sci.,* **48,** 889–991 (1993).

The influence of surface-active agents on bubbles and foams is summarized in selected passages from Schwartz and Perry, *Surface Active Agents,* vol. 1, Interscience, New York, 1949; and from Schwartz, Perry, and Berch, *Surface Active Agents and Detergents,* vol. 2, Interscience, New York, 1958. See also Elenkov, *Theor. Found Chem. Eng.,* **1,** 1, 117 (1967); and Rubel, *Antifoaming and Defoaming Agents,* Noyes Data Corp., Park Ridge, NJ, 1972.

A review of foam stability also is given by de Vries, Meded, *Rubber Sticht. Delft.* No. 328, 1957. Foam-separation methodology is discussed by Aguoyo and Lemlich, *Ind. Eng. Chem. Process Des. Dev.,* **13,** 153 (1974); and Lemlich, *Ind. Eng. Chem.,* **60,** 16 (1968). The following reviews of specific applications of gas-to-liquid dispersions are recommended: Industrial fermentations Aiba, Humphrey, and Millis, *Biochemical Engineering,* Academic, New York, 1965. Finn, *Bacteriol. Rev.,* **18,** 254 (1954). Oldshue, "Fermentation Mixing Scale-Up Techniques," in *Biotechnology and Bioengineering,* vol. 8, 1966, pp. 3–24. Aerobic oxidation of wastes: Eckenfelder and McCabe, *Advances in Biological Waste Treatment,* Macmillan, New York, 1963. Eckenfelder and O'Connor, *Biological Waste Treatment,* Pergamon, New York, 1961. McCabe and Eckenfelder, *Biological Treatment of Sewage and Industrial Wastes,* vol. 1, Reinhold, New York, 1955. Proceedings of Industrial Waste Treatment Conference, Purdue University, annually. Zlokarnik, *Adv. Biochem. Eng.,* **11,** 158–180 (1979). *Cellular elastomers:* Fling, *Natural Rubber Latex and Its Applications: The Preparation of Latex Foam Products,* British Rubber Development Board, London, 1954. Gould, in *Symposium on Application of Synthetic Rubbers,* American Society for Testing and Materials, Philadelphia, 1944, pp. 90–103. *Firefighting foams:* Perri, in Bikerman, op. cit., Chap. 12. Ratzer, *Ind. Eng. Chem.,* **48,** 2013 (1956). *Froth-flotation methods and equipment:* Booth, in Bikerman, op. cit., Chap. 13. Gaudin, *Flotation,* McGraw-Hill, New York, 1957. Taggart, *Handbook of Mineral Dressing,* Wiley, New York, 1945, Sec. 12, pp. 52–81. Tatterson, *Fluid Mixing and Gas Dispersion in Agitated Tanks,* McGraw-Hill, New York, 1991.

Objectives of Gas Dispersion   The dispersion of gas as bubbles in a liquid or in a plastic mass is effected for one of the following purposes: (1) gas-liquid contacting (to promote absorption or stripping, with or without chemical reaction), (2) agitation of the liquid phase, or (3) foam or froth production. Gas-in-liquid dispersions also may be produced or encountered inadvertently, sometimes undesirably.

Gas-Liquid Contacting Usually this is accomplished with conventional columns or with spray absorbers (see preceding subsection "Liquid-in-Gas Dispersions"). For systems containing solids or tar likely to plug columns, absorptions accomplished by strongly exothermic reactions, or treatments involving a readily soluble gas or a condensable vapor, however, bubble columns or agitated vessels may be used to your advantage.

Agitation Agitation by a stream of gas bubbles (often air) rising through a liquid is often employed when the extra expense of mechanical agitation is not justified. Gas spargers may be used for simple blending operations involving a liquid of low volatility or for applications where agitator shaft sealing is difficult.

Foam Production This is important in froth-flotation separations; in the manufacture of cellular elastomers, plastics, and glass; and in certain special applications (e.g., food products, fire extinguishers). Unwanted foam can occur in process columns, in agitated vessels, and in reactors in which a gaseous product is formed; it must be avoided, destroyed, or controlled. Berkman and Egloff (*Emulsions and Foams*, Reinhold, New York, 1941, pp. 112–152) have mentioned that foam is produced only in systems possessing the proper combination of interfacial tension, viscosity, volatility, and concentration of solute or suspended solids. From the standpoint of gas comminution, foam production requires the creation of small bubbles in a liquid capable of sustaining foam.

Theory of Bubble and Foam Formation A bubble is a globule of gas or vapor surrounded by a mass or thin film of liquid. By extension, globular voids in a solid are sometimes called bubbles. Foam is a group of bubbles separated from one another by thin films, the aggregation having a finite static life. Although nontechnical dictionaries do not distinguish between foam and froth, a technical distinction is often made. A highly concentrated dispersion of bubbles in a liquid is considered a froth even if its static life is substantially nil (i.e., it must be dynamically maintained). Thus, all foams are also froths, whereas the reverse is not true. The term *lather* implies a froth that is worked up on a solid surface by mechanical agitation; it is seldom used in technical discussions. The thin walls of bubbles comprising a foam are called *laminae* or *lamellae*.

Bubbles in a liquid originate from one of three general sources: (1) They may be formed by desupersaturation of a solution of the gas or by the decomposition of a component in the liquid; (2) They may be introduced directly into the liquid by a bubbler or sparger or by mechanical entrainment; and (3) They may result from the disintegration of larger bubbles already in the liquid.

Generation Spontaneous generation of gas bubbles within a homogeneous liquid is theoretically impossible (Bikerman, *Foams: Theory and Industrial Applications*, Reinhold, New York, 1953, p. 10). The appearance of a bubble requires a gas nucleus as a void in the liquid. The nucleus may be in the form of a small bubble or of a solid carrying adsorbed gas, examples of the latter being dust particles, boiling chips, and a solid wall. A void can result from cavitation, mechanically or acoustically induced. Blander and Katz [*AIChE J.*, **21**, 833 (1975)] have thoroughly reviewed bubble nucleation in liquids.

Theory permits the approximation of the maximum size of a bubble that can adhere to a submerged horizontal surface if the contact angle between bubble and solid (angle formed by solid-liquid and liquid-gas interfaces) is known [Wark, *J. Phys. Chem.*, **37**, 623 (1933); Jakob, *Mech. Eng.*, **58**, 643 (1936)]. Because the bubbles that actually rise from a surface are always considerably smaller than those so calculated and inasmuch as the contact angle is seldom known, the theory is not directly useful.

Formation at a Single Orifice The formation of bubbles at an orifice or capillary immersed in a liquid has been the subject of much study, both experimental and theoretical. Bikerman (op. cit., Secs. 3 to 7), Valentin (op. cit., Chap. 2), Jackson (op. cit.), Soo (op. cit., Chap. 3), Fair (op. cit.), Kumer et al. (op. cit.), Clift et al. (op. cit.) and Wilkinson and Van Dierendonck [*Chem. Eng. Sci.*, **49**, 1429 (1994)] have presented reviews and analyses of this subject.

There are three regimes of bubble production (Silberman in *Pro-*ceedings of the Fifth Midwestern Conference on Fluid Mechanics, Univ. of Michigan Press, Ann Arbor, 1957, pp. 263–284): (1) single-bubble, (2) intermediate, and (3) jet.

Single-Bubble Regime Bubbles are produced one at a time, their size being determined primarily by the orifice diameter d_o, the interfacial tension of the gas-liquid film σ, the densities of the liquid ρ_L and gas ρ_G, and the gravitational acceleration g according to the relation

$$\frac{d_b}{d_o} = \left(\frac{6\sigma}{d_o^2 (\rho_L - \rho_G)} \right)^{1/3} \quad (14\text{-}206)$$

where D_b is the bubble diameter. The bubble size is independent of gas flow rate; the frequency, therefore, is directly proportional to the gas flow rate. Equation (14-206) leads to

$$f = Qg \times \frac{\rho_L - \rho_G}{\pi d_o \sigma} \quad (14\text{-}207)$$

where f is the frequency of bubble formation and Q is the volumetric rate of gas flow in consistent units.

Equations (14-206) and (14-207) result from a balance of bubble buoyancy against interfacial tension. They include no inertia or viscosity effects. At low bubbling rates (<1/s), these equations are quite satisfactory. Van Krevelen and Hoftijzer [*Chem. Eng. Prog.*, **46**, 29 (1950)], Guyer and Peterhaus [*Helv. Chim. Acta*, **26**, 1099 (1943)] and Wilkinson (op. cit.) report good agreement with Eq. (14-185) for water, transformer oil, ether, and carbon tetrachloride for vertically oriented orifices with $0.004 < D < 0.95$ cm. If the orifice diameter becomes too large, the bubble diameter will be smaller than the orifice diameter, as predicted by Eq. (14-206), and instability results; consequently, stable, stationary bubbles cannot be produced.

For bubbles formed in water, the orifice diameter that permits bubbles of about its own size is calculated as 0.66 cm. Davidson and Amick [*AIChE J.*, **2**, 337 (1956)] confirmed this estimate in their observation that stable bubbles in water were formed at a 0.64-cm orifice but could not be formed at a 0.79-cm orifice.

For very thin liquids, Eqs. (14-206) and (14-207) are expected to be valid up to a gas-flow Reynolds number of 200 (Valentin, op. cit., p. 8). For liquid viscosities up to 100 cP, Datta, Napier, and Newitt [*Trans. Inst. Chem. Eng.*, **28**, 14 (1950)] and Siems and Kauffman [*Chem. Eng. Sci.*, **5**, 127 (1956)] have shown that liquid viscosity has very little effect on the bubble volume, but Davidson and Schuler [*Trans. Instn. Chem. Eng.*, **38**, 144 (1960)] and Krishnamurthi et al. [*Ind. Eng. Chem. Fundam.*, **7**, 549 (1968)] have shown that bubble size increases considerably over that predicted by Eq. (14-206) for liquid viscosities above 1000 cP. In fact, Davidson et al. (op. cit.) found that their data agreed very well with a theoretical equation obtained by equating the buoyant force to drag based on Stokes' law and the velocity of the bubble equator at break-off:

$$d_b = \left(\frac{6}{\pi} \right) \left(\frac{4\pi}{3} \right)^{1/4} \left(15 \times \frac{\nu Q}{2g} \right)^{3/4} \quad (14\text{-}208)$$

where ν is the liquid kinematic viscosity and Q is the gas volumetric flow rate. This equation is dimensionally consistent. The relative effect of liquid viscosity can be obtained by comparing the bubble diameters calculated from Eqs. (14-206) and (14-208). If liquid viscosity appears significant, one might want to use the long and tedious method developed by Krishnamurthi et al. (op. cit.) that considers both surface-tension forces and viscous-drag forces.

Intermediate Regime This regime extends approximately from a Reynolds number of 200 to one of 2100. As the gas flow through a submerged orifice increases beyond the limit of the single-bubble regime, the frequency of bubble formation increases more slowly, and the bubbles begin to grow in size. Between the two regimes there may indeed be a range of gas rates over which the bubble size decreases with increasing rate, owing to the establishment of liquid currents that nip the bubbles off prematurely. The net result can be the occurrence of a minimum bubble diameter at some particular gas rate [Mater, *U.S. Bur. Mines Bull.* 260 (1927) and Bikerman, op. cit., p. 4]. At the upper portion of this region, the frequency becomes very nearly constant with respect to gas rate, and the bubble size correspondingly increases with gas rate. The bubble size is affected primarily by (1) orifice diameter, (2) liquid-inertia effects, (3) liquid viscosity, (4) liquid density, and (5) the relationship between the constancy of gas flow and the constancy of pressure at the orifice.

Kumar et al. have done extensive experimental and theoretical work reported in *Ind. Eng. Chem. Fundam.*, **7**, 549 (1968); *Chem. Eng. Sci,* **24**, part 1, 731; part 2, 749; part 3, 1711 (1969) and summarized in *Adv. Chem. Eng.*, **8**, 255 (1970). They, along with other investigators—Swope [*Can. J Chem. Eng.*, **44**, 169 (1972)], Tsuge and Hibino [*J. Chem. Eng. Japan*, **11**, 307 (1972)], Pinczewski [*Chem. Eng. Sci.*, **36**, 405 (1981)], Tsuge and Hibino [*Int. Chem. Eng.*, **21**, 66 (1981)], and Takahashi and Miyahara [ibid., p. 224]—have solved the equations resulting from a force balance on the forming bubble, taking into account buoyancy, surface tension, inertia, and viscous-drag forces for both conditions of constant flow through the orifice and constant pressure in the gas chamber. The design method is complex and tedious and involves the solution of algebraic and differential equations. Although Mersmann [*Ger. Chem. Eng.*, **1**, 1 (1978)] claims that the results of Kumar et al. (loc. cit.) well fit experimental data, Lanauze and Harn [*Chem. Eng. Sci.*, **29**, 1663 (1974)] claim differently:

Further, it has been shown that the mathematical formulation of Kumar's model, including the condition of detachment, cord not adequately describe the experimental situation—Kumar's model has several fundamental weaknesses, the computational simplicity being achieved at the expense of physical reality.

In lieu of careful independent checks of predictive accuracy, the results of the comprehensive theoretical work will not be presented here. Simpler, more easily understood predictive methods, for certain important limiting cases, will be presented. As a check on the accuracy of these simpler methods, it will perhaps be prudent to calculate the bubble diameter from the graphical representation by Mersmann (loc. cit.) of the results of Kumar et al. (loc. cit.).

For conditions approaching constant flow through the orifice, a relationship derived by equating the buoyant force to the inertia force of the liquid [Davidson et al., *Trans. Instn. Chem. Engrs.*, **38**, 335 (1960)] (dimensionally consistent),

$$d_b = 1.378 \times \frac{6Q^{6/5}}{\pi g^{3/5}} \qquad (14\text{-}209)$$

fits experimental data reasonably well. Surface tension and liquid viscosity tend to increase the bubble size—at a low Reynolds number. The effect of surface tension is greater for large orifice diameters. The magnitude of the diameter increase due to high liquid viscosity can be obtained from Eq. (14-208).

For conditions approaching constant pressure at the orifice entrance, which probably simulates most industrial applications, there is no independently verified predictive method. For air at near atmospheric pressure sparged into relatively inviscid liquids (11 ~ 100 cP), the correlation of Kumar et al. [*Can. J. Chem. Eng.*, **54**, 503 (1976)] fits experimental data well. Their correlation is presented here as Fig. 14-92.

Wilkinson et al. (op. cit.) make the following observation about the effect of gas density on bubble size: "The fact that the bubble size decreases slightly for higher gas densities can be explained on the basis of a force balance."

Jet Regime   With further rate increases, turbulence occurs at the orifice, and the gas stream approaches the appearance of a continuous jet that breaks up 7.6 to 10.2 cm above the orifice. Actually, the stream consists of large, closely spaced, irregular bubbles with a rapid swirling motion. These bubbles disintegrate into a cloud of smaller ones of random size distribution between 0.025 cm or smaller and about 1.25 cm, with a mean size for air and water of about 0.4 cm (Leibson et al., loc. cit.). According to Wilkinson et al. (op. cit.), jetting begins when

$$N_{We.g} = \frac{\rho_g d_o U_o^2}{\sigma} \leq 2 \qquad (14\text{-}210)$$

There are many contradictory reports about the jet regime, and theory, although helpful (see, for example, Siberman, loc. cit.), is as yet unable to describe the phenomena observed. The correlation of Kumar et al. (Fig. 14-92) is recommended for air-liquid systems.

Formation at Multiple Orifices   At high velocities, coalescence of bubbles formed at individual orifices occurs; Helsby and Tuson [*Research* (London), **8**, 270 (1955)], for example, observed the frequent coalescence of bubbles formed in pairs or in quartets at an orifice. Multiple orifices spaced by the order of magnitude of the orifice diameter increase the probability of coalescence, and when the magnitude is small (as in a sintered plate), there is invariably some. The broken lines of Fig. 14-92 presumably represent zones of increased coalescence and relatively less effective dispersion as the gas rate through porous-carbon tubes is increased. Savitskaya [*Kolloidn. Zh.*, **13**, 309 (1951)] found that the average bubble size formed at the surface of a porous plate was such as to maintain constancy of the product of bubble specific surface and interfacial tension as the latter was varied by addition of a surfactant. Konig et al. [*Ger. Chem. Eng.*, **1**, 199 (1978)] produced bubble sizes varying from 0.5 to 4 mm by the use of two porous-plate spargers and one perforated-plate sparger with superficial gas velocities from 1 to 8 cm/s. The small bubble sizes were stabilized by adding up to 0.5 percent of various alcohols to water.

At high-flow rates through perforated plates such as those that occur in distillation columns, Calderbank and Rennie [*Trans. Instn. Chem. Engrs.*, **40**, T3 (1962)]; Porter et al. [ibid., **45**, T265 (1967)]; Rennie and Evans [*Br. Chem. Eng*, **7**, 498 (1962)]; and Valentin (op. cit., Chap. 3) have investigated and discussed the effect of the flow conditions through the multiple orifices on the froths and foams that occur above perforated plates.

Entrainment and Mechanical Disintegration   Gas can be entrained into a liquid by a solid or a stream of liquid falling from the gas phase into the liquid, by surface ripples or waves, and by the vertical swirl of a mass of agitated liquid about the axis of a rotating agita-

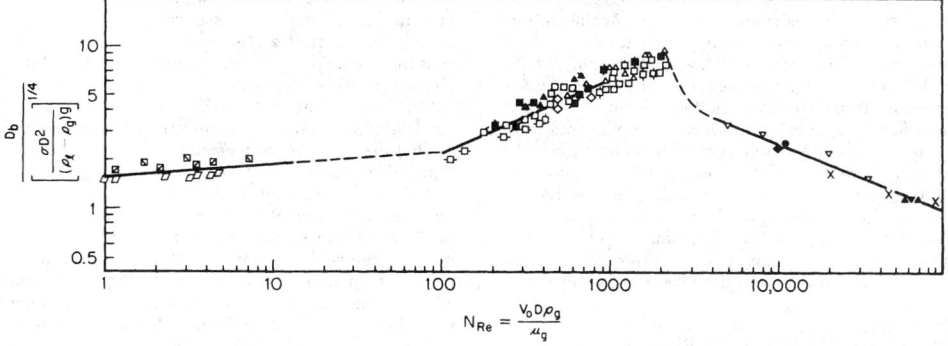

FIG. 14-92   Bubble-diameter correlation for air sparged into relatively inviscid liquids. D_b = bubble diameter, D = orifice diameter, V_o = gas velocity through sparging orifice, P = fluid density, and μ = fluid viscosity. [*From* Can. J. Chem. Eng., **54**, 503 (1976).]

tor. Small bubbles probably form near the surface of the liquid and are caught into the path of turbulent eddies, whose velocity exceeds the terminal velocity of the bubbles. The disintegration of a submerged mass of gas takes place by the turbulent tearing of smaller bubbles away from the exterior of the larger mass or by the influence of surface tension on the mass when it is attenuated by inertial or shear forces into a cylindrical or disk form. A fluid cylinder that is greater in length than in circumference is unstable and tends to break spontaneously into two or more spheres. These effects account for the action of fluid attrition and of an agitator in the disintegration of suspended gas. Quantitative correlations for gas entrainment by liquid jets and in agitated vessels will be given later.

Foams   Two excellent reviews (Shedlovsky, op. cit.; Lemlich, op. cit.) covering the literature pertinent to foams have been published. A foam is formed when bubbles rise to the surface of a liquid and persist for a while without coalescence with one another or without rupture into the vapor space. The formation of foam, then, consists simply of the formation, rise, and aggregation of bubbles in a liquid in which foam can exist. The life of foams varies over many magnitudes—from seconds to years—but in general is finite. Maintenance of a foam, therefore, is a dynamic phenomenon.

Gravitational force favors the separation of gas from liquid in a disperse system, causing the bubbles to rise to the liquid surface and the liquid contained in the bubble walls to drain downward to the main body of the liquid. Interfacial tension favors the coalescence and ultimate disappearance of bubbles; indeed, it is the cause of bubble destruction upon the rupture of the laminae.

The viscosity of the liquid in a film opposes the drainage of the film and its displacement by the approach of coalescing bubbles. The higher the viscosity, the slower will be the film-thinning process; furthermore, if viscosity increases as the film grows thinner, the process becomes self-retarding. The viscosity of films appears to be greater than that of the main body of the parent liquid in many cases. Sometimes this is a simple temperature effect, the film being cooler because of evaporation; sometimes it is a concentration effect, with dissolved or fine suspended solids migrating to the interface and producing classical or anomalous increases in viscosity; at yet other times, the effect seems to occur without explanation.

If the liquid laminae of a foam system can be converted to impermeable solid membranes, the film viscosity can be regarded as having become infinite, and the resulting solid foam will be permanent. Likewise, if the laminae are composed of a gingham plastic or a thixotrope, the foam will be permanently stable for bubbles whose buoyancy does not permit exceeding the yield stress. For other nonnewtonian fluids, however, and for all newtonian ones, no matter how viscous, the viscosity can only delay but never prevent foam disappearance. The popular theory, held since the days of Plateau, that foam life is proportional to surface viscosity and inversely proportional to interfacial tension, is not correct, according to Bikerman (op. cit., p. 161), who points out that it is contradicted by experiment.

The idea that foam films drain to a critical thickness at which they spontaneously burst is also rejected by Bikerman. Foam stability, rather, is keyed to the existence of a surface skin of low interfacial tension immediately overlying a solution bulk of higher tension, latent until it is exposed by rupture of the superficial layer [Maragoni, *Nuovo Cimento*, **2** (5–6), 239 (1871)]. Such a phenomenon of surface elasticity, resulting from concentration differences between bulk and surface of the liquid, accounts for the ability of bubbles to be penetrated by missiles without damage. It is conceivable that films below a certain thickness no longer carry any bulk of solution and hence have no capacity to close surface ruptures, thus becoming vulnerable to mechanical damage that will destroy them. The Maragoni phenomenon is consistent also with the observation that neither pure liquids nor saturated solutions will sustain a foam, since neither extreme will allow the necessary differences in concentration between surface and bulk of solution.

The specific ability of certain finely divided, insoluble solids to stabilize foam has long been known [Berkman and Egloff, op. cit., p. 133; and Bikerman, op. cit., Chap. 11]. Bartsch [*Kolloidchem. Beih.*, **20**, 1 (1925)] found that the presence of fine galena greatly extended the life of air foam in aqueous isoamyl alcohol, and the finer the solids, the

greater the stability. Particles on the order of 50 μm length extended the life from 17 seconds to several hours. This behavior is consistent with theory, which indicates that a solid particle of medium contact angle with the liquid will prevent the coalescence of two bubbles with which it is in simultaneous contact. Quantitative observations of this phenomenon are scanty.

Berkman and Egloff explain that some additives increase the flexibility or toughness of bubble walls, rather than their viscosity, to render them more durable. They cite as illustrations the addition of small quantities of soap to saponin solutions or of glycerin to soap solution to yield much more stable foam. The increased stability with ionic additives is probably due to electrostatic repulsion between charged, nearly parallel surfaces of the liquid film, which acts to retard draining and hence rupture.

Characteristics of Dispersion

Properties of Component Phases   As discussed in the preceding subsection, dispersions of gases in liquids are affected by the viscosity of the liquid, the density of the liquid and of the gas, and the interfacial tension between the two phases. They also may be affected directly by the composition of the liquid phase. Both the formation of bubbles and their behavior during their lifetime are influenced by these quantities as well as by the mechanical aspects of their environment.

Viscosity and density of the component phases can be measured with confidence by conventional methods, as can the interfacial tension between a pure liquid and a gas. The interfacial tension of a system involving a solution or micellar dispersion becomes less satisfactory, because the interfacial free energy depends on the concentration of solute at the interface. Dynamic methods and even some of the so-called static methods involve the creation of new surfaces. Since the establishment of equilibrium between this surface and the solute in the body of the solution requires a finite amount of time, the value measured will be in error if the measurement is made more rapidly than the solute can diffuse to the fresh surface. Eckenfelder and Barnhart (Am. Inst. Chem. Engrs., 42d national meeting, Repr. 30, Atlanta, 1960) found that measurements of the surface tension of sodium lauryl sulfate solutions by maximum bubble pressure were higher than those by DuNouy tensiometer by 40 to 90 percent, the larger factor corresponding to a concentration of about 100 ppm, and the smaller to a concentration of 2500 ppm of sulfate.

Even if the interfacial tension is measured accurately, there may be doubt about its applicability to the surface of bubbles being rapidly formed in a solution of a surface-active agent, for the bubble surface may not have time to become equilibrated with the solution. Coppock and Meiklejohn [*Trans. Instn. Chem. Engrs.*, **29**, 75 (1951)] reported that bubbles formed in the single-bubble regime at an orifice in a solution of a commercial detergent had a diameter larger than that calculated in terms of the measured surface tension of the solution [Eq. (14-206)]. The disparity is probably a reflection of unequilibrated bubble laminae.

One concerned with the measurement of gas-liquid interfacial tension should consult the useful reviews of methods prepared by Harkins [in Chap. 9 of Weissberger, *Techniques of Organic Chemstry*, 2d ed., vol. 1, part 2, Interscience, New York, 1949), Schwartz and coauthors [*Surface Acttve Agents*, vol. 1, Interscience, New York, 1949, pp. 263–271; *Surface Active Agents and Detergents*, vol. 2, Interscience, New York, 1958, pp. 389–391, 417–418], and by Adamson [*Physical Chemistry of Surfaces*, Interscience, New York, 1960].

Dispersion Characteristics   The chief characteristics of gas-in-liquid dispersions, like those of liquid-in-gas suspensions, are heterogeneity and instability. The composition and structure of an unstable dispersion must be observed in the dynamic situation by looking at the mixture, with or without the aid of optical devices, or by photographing it, preferably in nominal steady state; photographs usually are required for quantitative treatment. Stable foams may be examined after the fact of their creation if they are sufficiently robust or if an immobilizing technique such as freezing is employed [Chang et al., *Ind. Eng Chem.*, **48**, 2035 (1956)].

The rate of rise of bubbles has been discussed in many papers, including two that present good reviews of the subject [Benfratello, *Energ Elettr.*, **30**, 80 (1953); Haberman and Morton, *Report 802:*

David W. Taylor Model Basin, Washington, September 1953; Jackson, loc. cit.; Valentin, op. cit., Chap. 2; Soo, op. cit., Chap. 3; Calderbank, loc. cit., p. CE220; and Levich, op. cit., Chap. 8). A comprehensive and apparently accurate predictive method has been published [Jamialahamadi et al., *Trans ICE,* **72,** part A, 119–122 (1994)]. Small bubbles (below 0.2 mm in diameter) are essentially rigid spheres and rise at terminal velocities that place them clearly in the laminar-flow region; hence their rising velocity may be calculated from Stokes' law. As bubble size increases to about 2 mm, the spherical shape is retained, and the Reynolds number is still sufficiently small (<10) that Stokes' law should be nearly obeyed.

As bubble size increases, two effects set in, however, that alter the velocity. At about $N_{Re} = 100$, a wobble begins that can develop into a helical path if the bubbles are not liberated too closely to one another [Houghton, McLean, and Ritchie, *Chem. Eng. Sci.,* **7,** 40 (1957); and Houghton et al., ibid., p. 111]. Furthermore, for bubbles in the range of 1 mm and larger (until distortion becomes serious) internal circulation can set in [Garner and Hammerton, *Chem. Eng. Sci.,* **3,** (1954); and Haberman and Morton, loc. cit.], and according to theoretical analyses by Hadamard and Rybczynski and given by Levich (op. cit.), the drag coefficient for a low-viscosity dispersed phase and a high-viscosity continuous phase will approach two-thirds of the drag coefficient for rigid spheres, namely $C_D = 16/N_{Re}$. The rise velocity of a circulating bubble or drop will thus be 1.5 times that of a rigid sphere. Redfield and Houghton [*Chem. Eng. Sci.,* **20,** 131 (1965)] have found that CO_2 bubbles rising in pure water agree with the theoretical solution for a circulating drop below $N_{Re} = 1$. Many investigators (see Valentin, op. cit.) have found that extremely small quantities of impurities can retard or stop this internal circulation. In this behavior may lie the explanation of the fact that the addition of long-chain fatty acids to water to produce a concentration of 1.5×10^{-4} molar markedly reduces the rate of rise of bubbles [Stuke, *Naturwissenschaften,* **39,** 325 (1952)].

Above diameters of about 2 mm, bubbles begin to change to ellipsoids, and above 1 cm they become lens-shaped, according to Davies and Taylor [*Proc. Roy. Soc.* (London), **A200,** 379 (1950)]. The rising velocity in thin liquids for the size range 1 mm < D_B < 20 mm has been reported as 20 to 30 cm/s by Haberman and Morton (op. cit.) and Davenport, Richardson, and Bradshaw [*Chem. Eng. Sci.,* **22,** 1221 (1967)]. Schwerdtieger [ibid., **23,** 937 (1968)] even found the same for argon bubbles rising in mercury. Surface-active agents have no effect on the rise velocity of bubbles larger than 4 mm in thin liquids (Davenport et al., loc. cit.).

Above a Reynolds number of the order of magnitude of 1000, bubbles assume a helmet shape, with a flat bottom (Eckenfelder and Barnhart, loc. cit.; and Leibson et al., loc. cit.). After bubbles become large enough to depart from Stokes' law at their terminal velocity, behavior is generally complicated and erratic, and the reported data scatter considerably. The rise can be slowed, furthermore, by a wall effect if the diameter of the container is not greater than 10 times the diameter of the bubbles, as shown by Uno and Kintner [*AIChE J.,* **2,** 420 (1956); and Collins, *J. Fluid Mech.,* **28**(1), 97 (1967)]. Work has been done to predict the rise velocity of large bubbles [Rippin and Davidson, *Chem. Eng. Sci.,* **22,** 217 (1967); Grace and Harrison, ibid., 1337; Mendelson, *AIChE J.,* **13,** 250 (1967); Cole, ibid., Lehrer, *J. Chem. Eng. Japan,* **9,** 237; (1976) and Lehrer, *AIChE J.,* **26,** 170 (1980)]. The works of Lehrer present correlations that accurately predict rise velocities for a wide range of system properties. An excellent review of the technical literature concerning the rise of single bubbles and drops has been published by Clift, Grace, and Weber (*Bubbles, Drops and Particles,* Academic, New York, 1978). Mendelson has used a wave theory to predict the terminal velocity, and Cole has checked the theory with additional data. The other authors listed solved some simplified form of the Navier-Stokes equations. Jamialahmadi et al., loc. cit., have developed a single equation predictive method for bubble rise velocity, which covers the entire range of bubble diameters.

When bubbles are produced in clouds, as by a porous disperser, their behavior during rising is further complicated by interaction among themselves. In addition to the tendency for small bubbles to coalesce and large ones to disintegrate, there are two additional opposing influences on the rate of rise of bubbles of any particular size: (1) A "chimney effect" can develop in which a massive current upward appears at the axis of the bubble stream, leading to increased net bubble velocity; and (2) the proximity of the bubbles to one another can result in a hindered-settling condition, leading to reduced average bubble velocity. Figure 14-93 shows the data of Houghton et al. (op. cit.) for clouds of bubbles compared with their single-bubble data for pure water and seawater and of Peebles and Garber [*Chem. Eng. Progr.,* **49,** 88 (1953)] for acetic acid and ethyl acetate. The bubble clouds were produced with a sintered-glass plate of mean pore size (inferred from air wet-permeability data) of 81 μm.

The difference between the curves for pure water and seawater again illustrates the significance of small concentrations of solute with respect to bubble behavior. In commercial bubble columns and agitated vessels coalescence and breakup are so rapid and violent that the rise velocity of a single bubble is meaningless. The average rise velocity can, however, be readily calculated from holdup correlations that will be given later.

The quantitative examination of bubble systems is aided by the use of proper illumination and photography. The formation of bubbles at single sources often is sufficiently periodic to be stopped by stroboscopic light. Clouds of rising bubbles are more difficult to assess and require careful technique. Satisfactory photographic methods have been developed by Vermenlen, Williams, and Langlois [*Chem. Eng. Progr.,* **51,** 85 (1955)] and by Calderbank [*Trans. Instn. Chem. Engrs.,* **36,** 443 (1958)] and are described by these authors. Calderbank's technique resulted in particularly precise measurements that permitted a good estimation of the surface area of the dispersed bubbles.

Methods of Gas Dispersion The problem of dispersing a gas in a liquid may be attacked in several ways: (1) The gas bubbles of the desired size or which grow to the desired size may be introduced directly into the liquid; (2) a volatile liquid may be vaporized by either decreasing the system pressure or increasing its temperature; (3) a chemical reaction may produce a gas; or (4) a massive bubble or stream of gas is disintegrated by fluid shear and/or turbulence in the liquid.

Spargers: Simple Bubblers The simplest method of dispersing gas in a liquid contained in a tank is to introduce the gas through an open-end standpipe, a horizontal perforated pipe, or a perforated plate at the bottom of the tank. At ordinary gassing rates (corresponding to the jet regime), relatively large bubbles will be produced regardless of the size of the orifices.

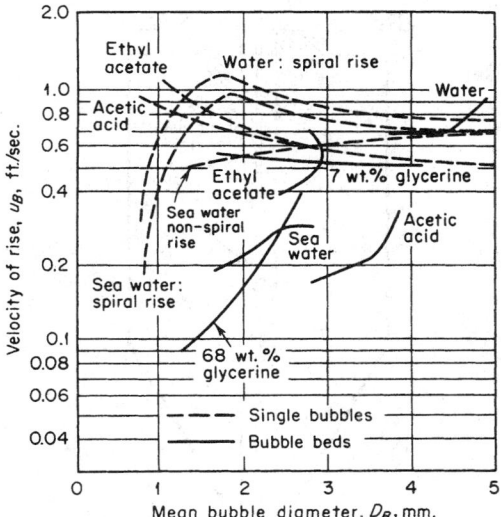

FIG. 14-93 Velocity of rising bubbles, singly and in clouds. To convert feet per second to meters per second, multiply by 0.305. [*From* Chem. Eng. Sci., **7,** 48 (1957).]

Perforated-pipe or -plate spargers usually have orifices 3 to 12 mm in diameter. Effective design methods to minimize maldistribution are presented in the fifth edition of this handbook, p. 5–47, 1973, and by Knaebel [*Chem. Eng.*, 116 (Mar. 9, 1981)]. For turbulent flow conditions into the sparger, the following relationship will allow design of a perforated-pipe sparger for a given degree of maldistribution provided $N_h > 5$ and length/diameter < 300.

$$d_p = 0.95(N_hC_v)^{1/2} \times \left(\frac{d_h}{\Delta U_h/U_h}\right)^{1/4} \qquad (14\text{-}211)$$

where d_p = pipe diameter, d_h = sparging hole diameter, N_h = number of holes in sparger, C_v = orifice coefficient for sparger hole (see *Chemical Engineers' Handbook*, 5th ed., pp. 5–13, 5–34), U_h = average velocity through sparger holes, ΔU_h = difference between maximum and minimum velocities through sparger holes, and $\Delta U_h/U_h$ = fractional maldistribution of flow through sparger holes.

Simple spargers are used as agitators for large tanks, principally in the cement and oil industries. Kauffman [*Chem. Metall. Eng.*, **37**, 178–180 (1930)] reported the following air rates for various degrees of agitation in a tank containing 2.7 m (9 ft) of liquid:

Degree of agitation	Air rate, m³/(m² tank cross section, min)
Moderate	0.0033
Complete	0.0066
Violent	0.016

For a liquid depth of 0.9 m (3 ft), Kauffman recommended that the listed rates be doubled.

An air lift consisting of a sparger jetting into a draft tube with ports discharging at several heights has been recommended by Heiser [*Chem. Eng.*, **55**(1), 135 (1948)] for maintaining agitation in a heavy, coarse slurry, the level of which varies widely. The design is illustrated in Fig. 14-94.

The ability of a sparger to blend miscible liquids might be described in terms of a fictitious diffusivity. Siemes did so, reporting that the agitation produced by a stream of bubbles rising in a tube with a superficial velocity of about 8.2 cm/s corresponded to an apparent diffusion coefficient as large as 75 cm²/s [*Chem. Ing. Tech.*, **29**, 727 (1957)]. The blending rate thus is several orders of magnitude higher than it would be by natural diffusive action. These results are typical of subsequent investigations on back mixing, which will be discussed in more detail later.

Lehrer [*Ind. Eng. Chem. Process Des. Dev.*, **7**, 226 (1968)] conducted liquid-blending tests with air sparging in a 0.61-m-diameter by 0.61-m-tall vessel and found that an air volume equal to about one-half of the vessel volume gave thorough blending of inviscid liquids of equal viscosities. Using an analogy to mechanically agitated vessels in which equal tank turnovers give equal blend times, one would expect this criterion to be applicable to other vessel sizes. Liquids of unequal density would probably require somewhat more air.

Open-end pipes, perforated plates, and ring- or cross-style perforated-pipe spargers are used without mechanical agitation to promote mass transfer, as in chlorinators and biological sewage treatment. In the "quiescent regime" (superficial gas velocity less than 4.57 to 6.1 cm/s [0.15 to 0.2 ft/s]) the previously mentioned spargers are usually operated at orifice Reynolds numbers in excess of 6000 in order to get small bubbles so as to increase the interfacial area and thus increase mass transfer. In the "turbulent regime" (superficial gas velocity greater than 4.57 to 6.1 cm/s), sparger design is not critical because a balance between coalescence and breakup is established very quickly according to Towell et al. [*AIChE Symp. Ser. No. 10*, 97 (1965)]. However, a reasonably uniform orifice distribution over the column cross section is desirable, and according to Fair [*Chem. Eng.*, **74**, 67 (July 3, 1967); 207 (July 17, 1967)] the orifice velocity should be less than 75 to 90 m/s.

Porous Septa In the quiescent regime porous plates, tubes, disks, or other shapes that are made by bonding or sintering together carefully sized particles of carbon, ceramic, polymer, or metal are frequently used for gas dispersion, particularly in foam fractionators. The resulting septa may be used as spargers to produce much smaller bubbles than will result from a simple bubbler. Figure 14-95 shows a com-

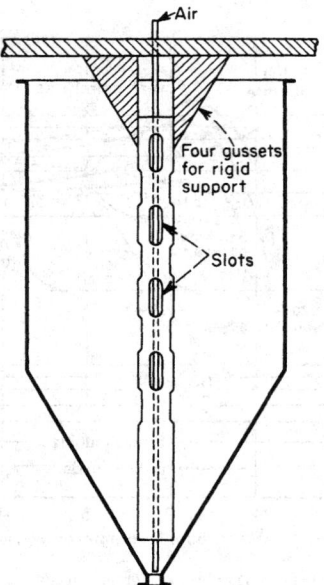

FIG. 14-94 Slotted air lift for agitation of a variable-level charge. [*From* Chem. Eng., **55**(1), 135 (1948).]

FIG. 14-95a Comparison of bubbles from a porous septum and from a perforated-pipe sparger. Air in water at 70°F. Grade 25 porous-carbon diffuser operating under a pressure differential of 13.7 in of water.

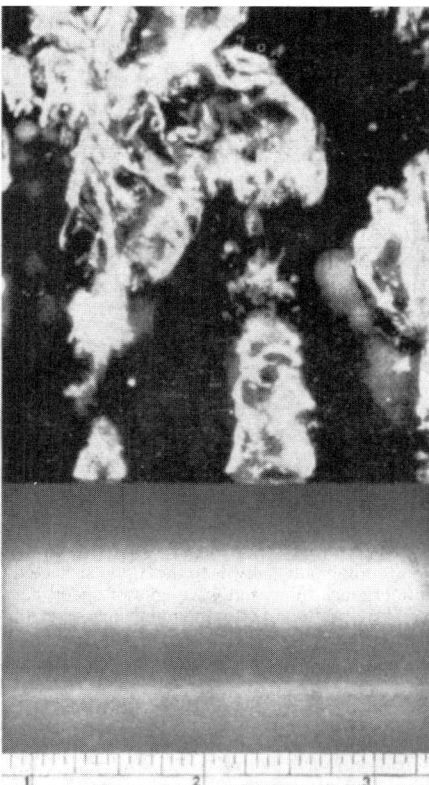

FIG. 14-95b Comparison of bubbles from a porous septum and from a perforated-pipe sparger. Air in water at 70°F. Karbate pipe perforated with 1/16-in holes on 1-in centers. (*National Carbon Co.*) To convert inches to centimeters, multiply by 2.54; °C = 5⁄9 (°F − 32).

TABLE 14-16 Characteristics of Porous Septa

Grade	Avg. % porosity	Avg. pore diam.	Air-permeability data		
			Diaphragm thickness, in	Pressure differential, in water	Air flow, cu ft/ (sq ft)(min)
Alundum porous alumina°					
P2220		25	1	2	0.35
P2120	36	60	1	2	2
P260	35	164	1	2	15
P236	34	240	1	2	40
P216		720	1	2	110
National porous carbon†					
60	48	33	1	2	
45	48	58	1	2	2
25	48	120	1	2	13
Filtros porous silica‡					
Extra fine	26.0	55	1.5	2	1–3
Fine	28.8	110	1.5	2	4–8
Medium fine	31.1	130	1.5	2	9–12
Medium	33.7	150	1.5	2	13–20
Medium coarse	33.8	200	1.5	2	21–30
Coarse	34.5	250	1.5	2	31–59
Extra coarse	36.5	300	1.5	2	60–100
Porous plastic§					
Teflon		9	0.125	1.38	5
Kel-F		15	0.125	1.38	13
Micro Metallic porous stainless steel§,¶					
H	45	5	0.125	1.38	1.8
G	50	10	0.125	1.38	3
F	50	20	0.125	1.38	5
E	50	35	0.125	1.38	18
D	50	65	0.125	1.38	60
C	55	165	0.125	27.7	990

°Data by courtesy of Norton Co., Worcester, Mass. A number of other grades between the extremes listed are available.
†Data by courtesy of National Carbon Co., Cleveland, Ohio.
‡Data by courtesy of Filtros Inc., East Rochester, N.Y.
§Data by courtesy of Pall Corp., Glen Cove, N.Y.
¶Similar septa made from other metals are available.

parison of the bubbles emitted by a perforated-pipe sparger [0.16-cm orifices] and a porous carbon septum (120 μm pores). The gas flux through a porous septum is limited on the lower side by the requirement that, for good performance, the whole sparger surface should bubble more or less uniformly, and on the higher side by the onset of serious coalescence at the surface of the septum, resulting in poor dispersion. In the practical range of fluxes, the size of the bubbles produced depends on both the size of pores in the septum and the pressure drop imposed across it, being a direct function of both.

Table 14-16 lists typical grades of porous carbon, silica, alumina, stainless steel (type 316), and polymers commercially available.

Porous media are also manufactured from porcelain, glass, silicon carbide, and a number of metals: Monel, Inconel, nickel, bronze, Hastelloy C, Stellite L-605, gold, platinum, and many types of stainless steel. The air permeabilities of Table 14-16 indicate the relative flow resistances of the various grades to homogeneous fluid but may not be used in designing a disperser for submerged operation, for the resistance of a septum to the flow of gas increases when it is wet. The air permeabilities for water-submerged porous carbon of some of the grades listed in the table are shown in Fig. 14-96. The data were determined with septa 0.625 inches thick in water at 70°F. Comparable wet-permeability data for 1-in Alundum plates of two grades of fineness are given in Table 14-17.

The gas rate at which coalescence begins to reduce the effectiveness of dispersion appears to depend not only on the pore size and pore structure of the dispersing medium but also on the liquid properties, liquid depth, agitation, and other features of the sparging environment; coalescence is strongly dependent on the concentration of

surfactants capable of forming an electrical double layer and thus produce ionic bubbles, long-chain alcohols in water being excellent examples. For porous-carbon media, the manufacturer suggests that the best dispersion performance will result if the broken-line regions of Fig. 14-96 are avoided. For porous stainless-steel spargers, which

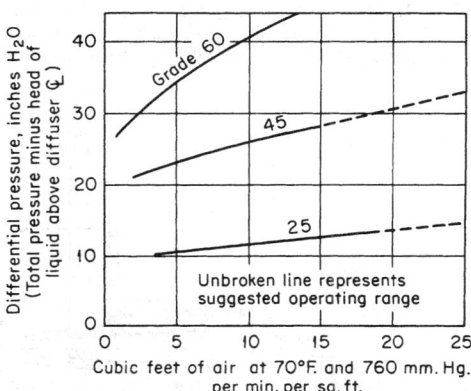

FIG. 14-96 Pressure drop across porous-carbon diffusers submerged in water at 70°F. To convert feet per minute to meters per second, multiply by 0.0051; to convert inches to millimeters, multiply by 25.4; °C = 5⁄9 (°F − 32). (*National Carbon Co.*)

TABLE 14-17 Wet Permeability of Alundum Porous Plates 1 in Thick*

Dry permeability at 2 in of water differential, cu ft/(min)(sq ft)	Pressure differential across wet plate, in of water	Air flow through wet plate, cu ft/(min)(sq ft)
4.3	20.67	2.0
	21.77	3.0
	22.86	4.0
	23.90	5.0
55.0	4.02	1.0
	4.14	2.0
	4.22	3.0
	4.27	4.0
	4.30	5.0

*Data by courtesy of Norton Company, Worcester, Mass. To convert inches to centimeters, multiply by 2.54; to convert feet per minute to meters per second, multiply by 0.0051.

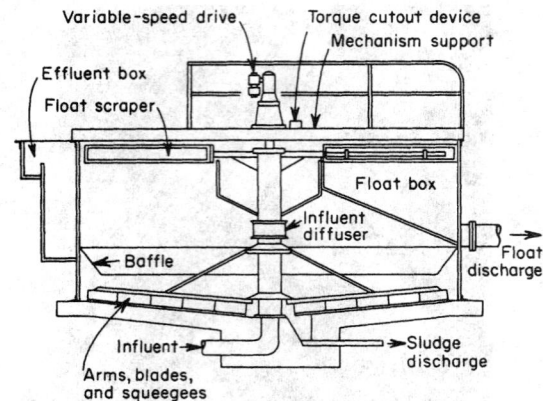

FIG. 14-97 The Flotator dissolved-air flotation thickener. (*Process Engineers, Inc., a division of Etmco Corp., now Envirotech Corporation.*)

extend to a lower pore size than carbon, Micro Metallic Division, Pall Corp., recommends (Release 120A, 1959) a working limit of 8 ft/min (0.044 m/s) to avoid serious coalescence. This agrees with the data reported by Konig et al. (loc. cit.), in which 0.08 m/s was used and bubbles as small as 1 mm were produced from a 5-μm porous sparger.

Slabs of porous material are installed by grouting or welding together to form a diaphragm, usually horizontal. Tubes are prone to produce coalesced gas at rates high enough to cause bubbling from their lower faces, but they have the advantage of being demountable for cleaning or replacement (U.S. Patent 2,328,655). Roe [*Sewage Works J.*, **18**, 878 (1945)] claimed that silicon carbide tubes are superior to horizontal plates, principally because of the wiping action of the liquid circulating past the tube. He reported respective maximum capacities of 2.5 and 1.5 cm²/s of gas/cm² of sparger for a horizontal tube and a horizontal plate of the same material (unspecified grade). Mounting a flat-plate porous sparger vertically instead of horizontally seriously reduces the effectiveness of the sparger for three reasons: (1) The gas is distributed over a reduced cross section; (2) at normal rates, the lower portion of the sparger may not operate because of difference in hydrostatic head; and (3) there is a marked tendency for bubbles to coalesce along the sparger surface. Bone (M.S. thesis in chemical engineering, University of Kansas, 1948) found that the oxygen sulfite solution coefficient for a 3.2- by 10-cm rectangular porous carbon sparger was 26 to 41 percent lower for vertical than for horizontal operation of the sparger, the greatest reduction occurring when the long dimension was vertical.

Precipitation and Generation Methods For a thorough understanding of the phenomena involved, bubble nucleation should be considered. A discussion of nucleation phenomena is beyond the scope of this Handbook; however, excellent coverages are presented by Blander and Katz.

Boerma and Lankester [*Chem. Eng. Sci.*, **23**, 799 (1968)] have measured the surface aeration of a six-bladed disk-type turbine (NOTE: A well-designed pitched-blade turbine will give equal or better perfor-

mance). In a fully baffled vessel, the optimum depth to obtain maximum gas dispersion was 15 percent of the liquid depth. In a vessel with baffles extending only halfway to the liquid surface the optimum impeller submergence increased with agitator speed because of the vortex formed. At optimum depth, the following correlation is recommended for larger vessels:

$$Q \ (\text{m}^3/\text{s}) = 0.3 \left(\frac{\text{impeller speed, rpm}}{500} \right)^{2.5} \left(\frac{\text{impeller diameter, cm}}{25.4} \right)^{4.5}$$

$$(14\text{-}214)$$

Gas dispersion through the free surface by mechanical aerators is commonplace in aerobic waste-treatment lagoons. Surface aerators are generally of three types: (1) large-diameter flow-speed turbines operating just below the free surface of the liquid, often pontoon-mounted; (2) small-diameter high-speed (normally motor-speed) propellers operating in draft tubes, the units of which are always pontoon-mounted; and (3) hollow-tube turbines (Fig. 14-101). An example of the turbine type is illustrated in Fig. 14-102 and the propeller type is illustrated in Fig. 14-103. There are several other styles of the turbine type; for instance, Mixing Equipment Co., Inc., uses an unshrouded 45° axial-flow turbine [see Dykman and Michel, *Chem. Eng.*, **117** (Mar. 10, 1969)], and Infilco makes a unit that has a large-diameter vaned disk operating just below the free surface with a smaller-diameter submerged-disk turbine for additional solids suspension.

Equipment Selection Ideally, selection of equipment to produce a gas-in-liquid dispersion should be made on the basis of a complete economic analysis. The design engineer and especially the pilot-plant engineer seldom have sufficient information or time to do

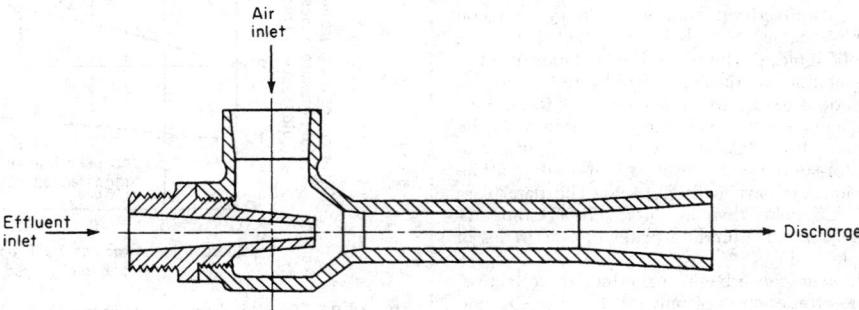

FIG. 14-98 Aeration ejector. (*Penberthy, a division of Houdaille Industries, Inc.*)

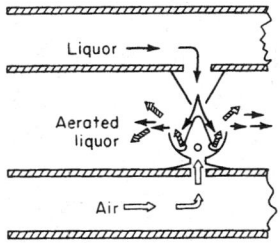

FIG. 14-99 Impingement aerator.

a complete economic analysis. In the following discussion, some guidelines are given as to what equipment might be feasible and what equipment might prove most economical.

For producing foam for foam-separation processes, perforated-plate or porous-plate spargers are normally used. Mechanical agitators are often not effective in the light foams needed in foam fractionation. Dissolved-air flotation, based on the release of a pressurized flow in which oxygen was dissolved, has been shown to be effective some times for particulate removal when sparged air failed because the bubbles formed upon precipitation are smaller—down to 80 μm—than bubbles possible with sparging, typically 1000 μm [Grieves and Ettelt, *AIChE J.*, **13**, 1167 (1967)]. Mechanically agitated surface aerators such as the Wemco-Fagergren flotation unit are used extensively for ore flotation.

To produce foam in batch processes, mechanical agitators are used almost exclusively. The gas can either be introduced through the free surface by the entraining action of the impeller or alternatively sparged beneath the impeller. In such batch operation, the liquid

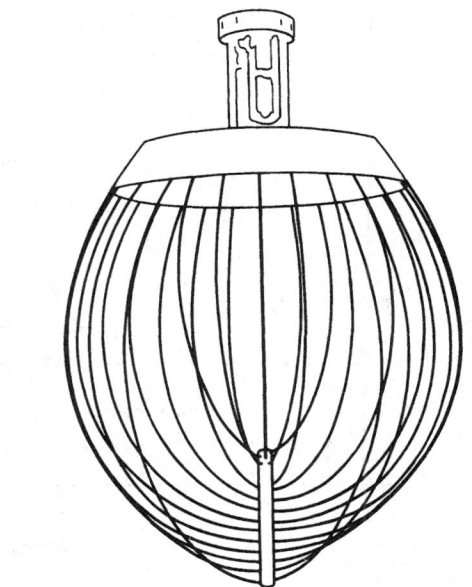

FIG. 14-100 Wire whip.

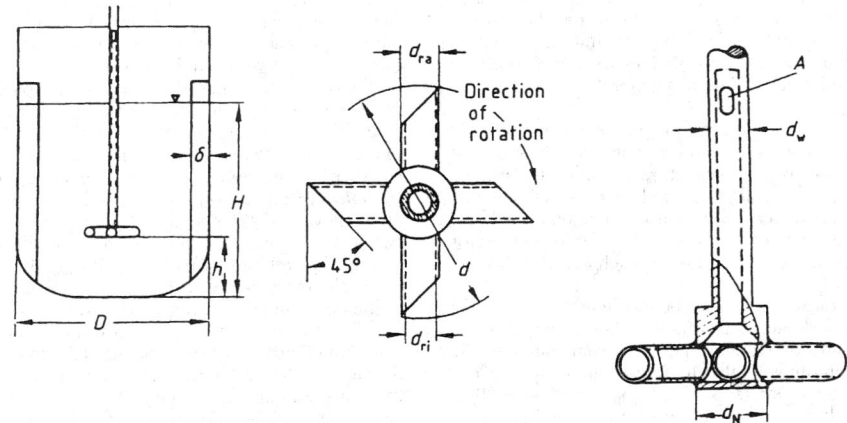

FIG. 14-101 Installation and dimensions of a tube stirrer: $h/d = 1$; $H/D \approx 1$; $D/\delta = 10$; $A = 1.5\, d_w^2$; $D/d_N = 10$; $d/d_N = 3$; $d/d_{ri} = 7.5$; $d/d_{ra} = 6$. [*Zlokarnik, Ullman's Encyclopedia of Industrial Chemistry, Sec. 25, VCH, Weinheim, Germany, 1988.*]

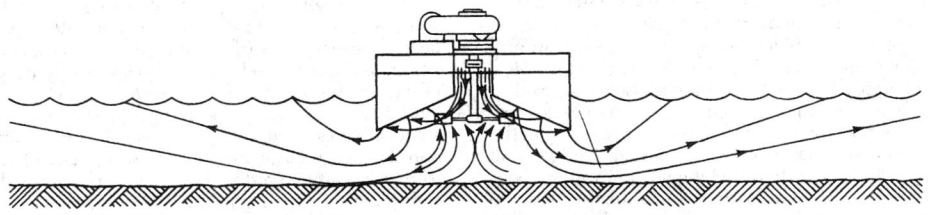

FIG. 14-102 The Cyclox surface aerator. (*Cleveland Mixer Co.*)

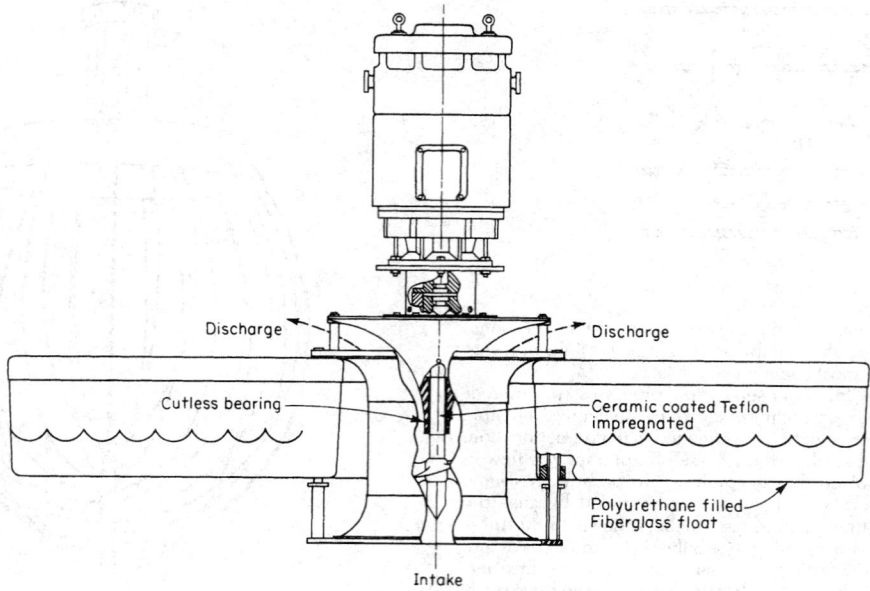

FIG. 14-103   Propeller-type surface aerator. (*Ashbrook-Simon-Hartley Corp.*)

level gradually rises as the foam is generated; thus, squatly impellers such as turbines are rapidly covered with foam and must almost always be sparged from below. Tall impellers such as the previously mentioned wire whips are especially well suited to entrain gas from the vapor space. For a new application, generally some experimentation with different impellers is necessary in order to get the desired fine final bubble size without getting frothing over initially. For producing foams continually, an aspirating venturi nozzle and restrictions in pipes such as baffles and metal gauzes are generally most economical.

For gas absorption, the equipment possibilities are generally packed columns; plate distillation towers, possibly with mechanical agitation on every plate; deep-bed contactors (bubble columns or sparged lagoons); and mechanically agitated vessels or lagoons. Packed towers and plate distillation columns are discussed elsewhere. Generally these devices are used when a relatively large number of stages (more than two or three) is required to achieve the desired result practically.

The volumetric mass-transfer coefficients and heights of transfer units for bubble columns and packed towers have been compared for absorption of CO_2 into water by Houghton et al. [*Chem. Eng. Sci.*, **7**, 26 (1957)]. The bubble column will tolerate much higher vapor velocities, and in the overlapping region (superficial gas velocities of 0.9 to 1.8 cm/s), the bubble column has about three times higher mass-transfer coefficient and about 3 times greater height of transfer unit. The liquid in a bubble column is, for practical purposes, quite well mixed; thus, chemical reactions and component separations requiring significant plug flow of the liquid cannot be carried out with bubble columns. Bubble columns and agitated vessels are the ideal equipment for processes in which the fraction of gas absorbed need not be great, possibly the gas can be recycled, and the liquid can or should be well mixed. The gas phase in bubble columns is not nearly so well back-mixed as the liquid, and often plug flow of the gas is a logical assumption, but in agitated vessels the gas phase is also well mixed.

The choice of a bubble column or an agitated vessel depends primarily on the solubility of the gas in the liquid, the corrosiveness of the liquid (often a gas compressor can be made of inexpensive material, whereas a mechanical agitator may have to be made of exotic, expensive materials), and the rate of chemical reaction as compared with the mass-transfer rate. Bubble columns and agitated vessels are seldom used for gas absorption except in chemical reactors. As a general rule,

if the overall reaction rate is five times greater than the mass-transfer rate in a simple bubble column, a mechanical agitator will be most economical unless the mechanical agitator would have to be made from considerably more expensive material than the gas compressor.

In bubble columns and simply sparged lagoons, selecting the sparger is a very important consideration. In the turbulent regime (superficial gas velocity greater than 4.6 to 6 cm/s), inexpensive perforated-pipe spargers should be used. Often the holes must be placed on the pipe bottom in order to make the sparger free-draining during operation. In the quiescent regime, porous septa will often give considerably higher overall mass-transfer coefficients than perforated plates or pipes because of the formation of tiny bubbles that do not coalesce. Chain and coworkers (*First International Symposium on Chemical Microbiology,* World Health Organization, Monograph Ser. 10, Geneva, 1952) claimed that porous disks are about twice as effective as open-pipe and ring spargers for the air oxidation of sodium sulfite. Eckenfelder [*Chem. Eng. Progr.*, **52**(7), 290 (1956)] has compared the oxygen-transfer capabilities of various devices on the basis of the operating power required to absorb a given quantity of O_2. The installed cost of the various pieces of equipment probably would not vary sufficiently to warrant being including in an economic analysis. Surface mechanical aerators are not included in this comparison. Of the units compared, it appears that porous tubes give the most efficient power usage. Kalinske (*Adv. Biol. Waste Treatment,* 1963, p. 157) has compared submerged sparged aerators with mechanical surface aerators. He has summarized this comparison in *Water Sewage Works,* **33** (January 1968). He indicates that surface aerators are significantly more efficient than subsurface aeration, both for oxygen absorption and for gas-stripping operations.

Zlokarnik and Mann (paper at Mixing Conf., Rindge, New Hampshire, August 1975) have found the opposite of Kalinske, i.e., subsurface diffusers, subsurface sparged turbines, and surface aerators compare approximately as 4:2:1 respectively in terms of O_2 transfer efficiency; however, Zlokarnik [*Adv. Biochem. Eng.,* **11**, 157 (1979)] later indicates that the scale-up correlation used earlier might be somewhat inaccurate. When all available information is considered, it appears that with near-optimum design any of the aeration systems (diffusers, submerged turbines, or surface impellers) should give a transfer efficiency of at least 2.25 kg O_2/kWh. Thus, the final selection should probably be made primarily on the basis of operational reliability, maintenance, and capital costs.

Mass Transfer   Mass transfer in plate and packed gas-liquid contactors has been covered earlier in this subsection. Attention here will be limited to deep-bed contactors (bubble columns and agitated vessels). Theory underlying mass transfer between phases is discussed in Sec. 5 of this handbook.

To design deep-bed contactors for mass-transfer operations, one must have, in general, predictive methods for the following design parameters:
- Flooding (for both columns and agitator impellers)
- Holdup of gas phase
- Agitator power requirements
- Gas-phase and liquid-phase mass-transfer coefficients interfacial area
  - Interfacial resistance
  - Mean driving force for transfer

In most cases, available methods are incomplete or unreliable, and some supporting experimental work is necessary. The methods given here should allow theoretical feasibility studies, help minimize experimentation, and permit a measure of optimization in final design.

Flooding of Agitator Impellers   A review of impeller flooding has been done by Sensel et al. [*AIChE Symp. Series No. 283,* **89** (1993)] and they have offered the following flooding correlation for a six-bladed disk-type turbine.

$$\frac{Q}{ND^3} = 0.0675(ND)T^{-0.4} \qquad \text{for} \quad (ND)T^{-0.4} \le 1.6 \qquad (14\text{-}215a)$$

$$\frac{Q}{ND^3} = 0.0231[(ND)T^{-0.4}]^{3.75} \qquad \text{for} \quad (ND)T^{-0.4} \ge 1.6 \qquad (14\text{-}215b)$$

Where Q = volumetric flow of gas, ft³/s; N = impeller speed, rev/s; D = impeller diameter, ft; and T = tank diameter, ft.

Gassed Impeller Power   Sensel et al. (op. cit.) have developed the following correlation for six-bladed disk impellers:

$$\frac{P_g}{P_o} = 1 - (-0.000715\,\mu_L + 0.723)\tanh\left(\frac{24.54Q}{ND^3}\right)\left(\frac{N^2D}{g}\right)^{0.25} \qquad (14\text{-}216)$$

where P_g = gassed power, P_o = ungassed power, g = gravitational acceleration, and μ_L = liquid viscosity.

Gas Holdup in Agitated Vessels   Sensel et al. (op. cit.) have also developed the following correlation for six-bladed disk-type impellers:

$$\varepsilon = 0.105\left(\frac{Q}{ND^3}\right)\left(\frac{N^2D}{g}\right)^{0.5}\left(\frac{ND^2\rho_L}{\mu_L}\right)^{0.1} \qquad (14\text{-}217)$$

Gas-Phase Mass-Transfer Coefficient   This term is quite high in deep-bed contactors, normally leading to negligible gas-phase resistance.

Interfacial Area   This consideration in agitated vessels has been reviewed and summarized by Tatterson (*op. cit.*). Predictive methods for interfacial area are not presented here because correlations are given for the overall volumetric mass transfer coefficient liquid phase controlling mass transfer.

Overall Mass-Transfer Coefficient   Tatterson (op. cit.) and Zlokarnik (op. cit.) have summarized the literature covering overall mass-transfer coefficients. There is much scatter in the experimental data because the presence of surface-active agents and electrolytes have a significant effect on the mass transfer. The correlation of Van't Riet [*Ind. Eng Chem. Process Des. Dev.,* **18**(3), 357 (1979)] is recommended:

$$k_L a = 0.026\left(\frac{P}{V}\right)^{0.4} U_s^{0.5} \qquad \text{for water} \qquad (14\text{-}218)$$

and

$$k_L a = 0.002\left(\frac{P}{V}\right)^{0.7} U_s^{0.2} \qquad \text{for ionic mixtures} \qquad (14\text{-}219)$$

where P/V = power/volume, W/m³; U_s = superficial gas velocity, m/s; $k_L a$ = volumetric mass-transfer coefficient, s⁻¹

Interfacial Phenomena   These can significantly affect overall mass transfer. In fermentation reactors, small quantities of surface-active agents (especially antifoaming agents) can drastically reduce overall oxygen transfer (Aiba et al., op. cit., pp. 153, 154), and in aerobic mechanically aerated waste-treatment lagoons, overall oxygen transfer has been found to be from 0.5 to 3 times that for pure water from tests with typical sewage streams (Eckenfelder et al., op. cit., p. 105).

One cannot quantitatively predict the effect of the various interfacial phenomena; thus, these phenomena will not be covered in detail here. The following literature gives a good general review of the effects of interfacial phenomena on mass transfer: Goodridge and Robb, *Ind. Eng. Chem. Fund.,* **4**, 49 (1965); Calderbank, *Chem. Eng. (London),* CE 205 (1967); Gal-Or et al., *Ind. Eng. Chem.,* **61**(2), 22 (1969); Kintner, *Adv. Chem. Eng.,* **4** (1963); Resnick and Gal-Or, op. cit., p. 295; Valentin, loc. cit.; and Elenkov, loc. cit., and *Ind. Eng. Chem. Ann. Rev. Mass Transfer,* **60**(1), 67 (1968); **60**(12), 53 (1968); **62**(2), 41 (1970). In the following outline, the effects of the various interfacial phenomena on the factors that influence overall mass transfer are given. Possible effects of interfacial phenomena are tabulated below:

1. Effect on continuous-phase mass-transfer coefficient
   a. Impurities concentrate at interface. Bubble motion produces circumferential surface-tension gradients that act to retard circulation and vibration, thereby decreasing the mass-transfer coefficient.
   b. Large concentration gradients and large heat effects (very soluble gases) can cause interfacial turbulence (the Marangoni effect), which increases the mass-transfer coefficient.
2. Effect on interfacial area
   a. Impurities will lower static surface tension and give smaller bubbles.
   b. Surfactants can electrically charge the bubble surface (produce ionic bubbles) and retard coalescence (soap stabilization of an oil-water emulsion is an excellent example of this phenomenon), thereby increasing the interfacial area.
   c. Large concentration gradients and large heat effects can cause bubble breakup.
3. Effect on mean mass-transfer driving force
   a. Relatively insoluble impurities concentrate at the interface, giving an interfacial resistance. This phenomenon has been used in retarding evaporation from water reservoirs.
   b. The axial concentration variation can be changed by changes in coalescence. The mean driving force for mass transfer is therefore changed.

Gas Holdup (ε) in Bubble Columns   With coalescing systems, holdup may be estimated from a correlation by Hughmark [*Ind. Eng Chem. Process Des. Dev.,* **6**, 218–220 (1967)] reproduced here as Fig. (14-104). For noncoalescing systems, with considerably smaller bub-

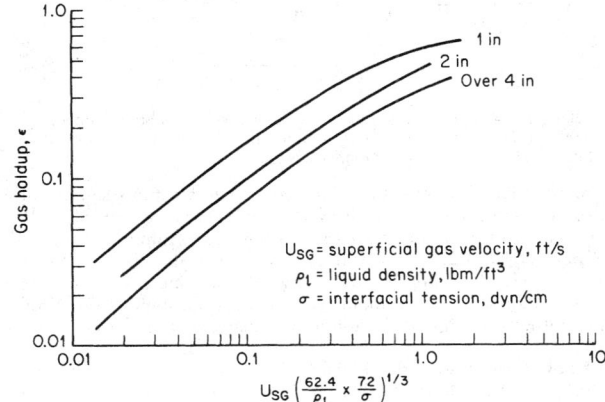

FIG. 14-104   Gas holdup correlation. [Ind. Eng. Chem. Process Des. Dev., **6**, 218 (1967).]

bles, ε can be as great as 0.6 at $U_{sg} = 0.05$ m/s, according to Mersmann [*Ger. Chem. Eng.,* **1**, 1 (1978)].

It is often helpful to use the relationship between ε and superficial gas velocity (U_{sg}) and the rise velocity of a gas bubble relative to the liquid velocity ($U_r + U_L$, with U_L defined as positive upward):

$$\varepsilon = \frac{U_{sg}}{U_r + U_L} \qquad (14\text{-}220)$$

Rise velocities of bubbles through liquids have been discussed previously.

For a better understanding of the interactions between parameters, it is often helpful to calculate the effective bubble rise velocity U_r from measured valves of ε; for example, the data of Mersmann (loc. cit.) indicated $\varepsilon = 0.6$ for $U_{sg} = 0.05$ m/s, giving $U_r = 0.083$ m/s, which agrees with the data reported in Fig. 14-43 for the rise velocity of bubble clouds. The rise velocity of single bubbles, for $d_b \sim 2$ mm, is about 0.3 m/s, for liquids with viscosities not too different from water. Using this value in Eq. (14-220) and comparing with Fig. 14-104, one finds that at low values of U_{sg}, the rise velocity of the bubbles is less than the rise velocity of a single bubble, especially for small-diameter tubes, but that the opposite occurs for large values of U_{sg}.

More recent literature regarding generalized correlational efforts for gas holdup is adequately reviewed by Tsuchiya and Nakanishi [*Chem. Eng Sci.,* **47**(13/14), 3347 (1992)] and Sotelo *et al.* [*Int. Chem. Eng.,* **34**(1), 82–90 (1994)]. Sotelo et al. (op. cit.) have developed a dimensionless correlation for gas holdup that includes the effect of gas and liquid viscosity and density, interfacial tension, and diffuser pore diameter. For systems that deviate significantly from the waterlike liquids for which Fig. 14-104 is applicable, their correlation (the fourth numbered equation in the paper) should be used to obtain a more accurate estimate of gas holdup. Mersmann (op. cit.) and Deckwer et al. (op. cit.) should also be consulted.

Liquid-phase mass-transfer coefficients in bubble columns have been reviewed by Calderbank ("Mixing," loc. cit.), Fair (*Chem. Eng.,* loc. cit.), Mersmann [*Ger. Chem. Eng.* **1**, 1 (1978), *Int. Chem. Eng.,* **32**(3) 397–405 (1991)], Deckwer et al. [*Can. J. Chem. Eng,* **58**, 190 (1980)], Hikita et al. [*Chem. Eng. J.,* **22**, 61 (1981)] and Deckwer and Schumpe [*Chem. Eng. Sci.,* **48**(5), 889–911 (1993)]. The correlation of Ozturk, Schumpe, and Deckwer [*AIChE J.,* **33**, 1473–1480 (1987)] is recommended. Deckwer et al. (op. cit.) have documented the case for using the correlation:

Ozturk et al. (1987) developed a new correlation on the basis of a modification of the Akita-Yoshida correlation suggested by Nakanoh and Yoshida (1980). In addition, the bubble diameter d_b rather than the column diameter was used as the characteristic length as the column diameter has little influence on $k_L a$. The value of d_b was assumed to be approximately constant ($d_b = 0.003$ m). The correlation was obtained by nonlinear regression is as follows:

$$\left(\frac{k_L a\, d_B}{D_L}\right) = 0.62 \left[\frac{\mu_L}{(\rho_L D_L)}\right]^{0.5} \left(\frac{g\rho_L d_B^2}{\sigma}\right)^{0.33} \left(\frac{g\rho_L^2 d_B^3}{\mu_L^2}\right)^{0.29}$$
$$\times \left[\frac{U_G}{(gd_B)^{0.5}}\right]^{0.68} \left(\frac{\rho_G}{\rho_L}\right)^{0.04} \qquad (14\text{-}221)$$

where $k_L a$ = overall mass-transfer coefficient, d_B = bubble diameter = 0.003 m, D_L = diffusivity of gas in liquid, ρ = density, μ = viscosity, σ = interfacial tension, g = gravitational acceleration.

As mentioned earlier, surfactants and ionic solutions significantly affect mass transfer. Normally, surface affects act to retard coalescence and thus increase the mass transfer. For example, Hikita *et al.* [*Chem. Eng. J.,* **22**, 61–69 (1981)] have studied the effect of KCl on mass transfer in water. As KCI concentration increased, the mass transfer increased up to about 35 percent at an ionic strength of 6 gm/l. Other investigators have found similar increases for liquid mixtures.

Axial Dispersion   Backmixing in bubble columns has been extensively studied. An excellent review article by Shah et al. [*AIChE*

J., **24**, 369 (1978)] has summarized the literature prior to 1978. Works by Konig et al. [*Ger. Chem. Eng.,* **1**, 199 (1978)], Lucke et al. [*Trans. Inst. Chem. Eng.,* **58**, 228 (1980)], Riquarts [*Ger. Chem. Eng.,* **4**, 18 (1981)], Mersmann (op. cit.), Deckwer (op. cit.), Yang et al. [*Chem. Eng. Sci.,* **47**(9–11), 2859 (1992)], and Garcia-Calvo and Leton [*Chem. Eng. Sci.,* **49**(21), 3643 (1994)] are particularly useful references.

Axial dispersion occurs in both the liquid and the gas phases. The degree of axial dispersion is affected by vessel diameter, vessel internals, gas superficial velocity, and surface-active agents that retard coalescence. For systems with coalescence-retarding surfactants the initial bubble size produced by the gas sparger is also significant. The gas and liquid physical properties have only a slight effect on the degree of axial dispersion, except that liquid viscosity becomes important as the flow regime becomes laminar. With pure liquids, in the absence of coalescence-inhibiting, surface-active agents, the nature of the sparger has little effect on the axial dispersion, and experimental results are reasonably well correlated by the dispersion model. For the liquid phase the correlation recommended by Deckwer et al. (op. cit.), after the original work by Baird and Rice [*Chem. Eng. J.,* **9**, 171(1975)] is as follows:

$$\frac{E_L}{(DU_G)} = 0.35 \left(\frac{gD}{U_G^2}\right)^{1/3} \qquad (14\text{-}222)$$

where E_L = liquid-phase axial dispersion coefficient, U_G = superficial velocity of the gas phase, D = vessel diameter, and g = gravitational acceleration.

The recommended correlation for the gas-phase axial-dispersion coefficient is given by Field and Davidson (loc. cit.):

$$E_G = 56.4\, D^{1.33} \left(\frac{U_G}{\varepsilon}\right)^{3.56} \qquad (14\text{-}223)$$

where E_G = gas-phase axial-dispersion coefficient, m²/s; D = vessel diameter, m; U_G = superficial gas velocity, m/s; and ε = fractional gas holdup, volume fraction.

The correlations given in the preceding paragraphs are applicable to vertical cylindrical vessels with pure liquids without coalescence inhibitors. For other vessel geometries such as columns of rectangular cross section, packed columns, and coiled tubes, the work of Shah et al. (loc. cit.) should be consulted. For systems containing coalescence-inhibiting surfactants, axial dispersion can be vastly different from that in systems in which coalescence is negligible. Konig et al. (loc. cit.) have well demonstrated the effects of surfactants and sparger type by conducting tests with weak alcohol solutions using three different porous spargers. With pure water, the sparger—and, consequently, initial bubble size—had little effect on back mixing because coalescence produced a dynamic-equilibrium bubble size not far above the sparger. With surfactants, the average bubble size was smaller than the dynamic-equilibrium bubble size. Small bubbles produced minimal back mixing up to $\varepsilon \approx 0.40$; however, above $\varepsilon \approx 0.40$ backmixing increased very rapidly as U_G increased The rapid increase in back mixing as ε exceeds 0.40 was postulated to occur indirectly because a bubble carries upward with it a volume of liquid equal to about 70 percent of the bubble volume, and, for $\varepsilon \approx 0.40$, the bubbles carry so much liquid upward that steady, uniform bubble rise can no longer be maintained and an oscillating, slugging flow develops, which produces fluctuating pressure at the gas distributor and the formation of large eddies. The large eddies greatly increase backmixing. For the air alcohol-water system, the minimum bubble size to prevent unsteady conditions was about 1, 1.5, and 2 mm for U_G = 1, 3, and 5 cm/s, respectively. Any smaller bubble size produced increased backmixing. The results of Konig et al. (loc. cit.) clearly indicate that the interaction of surfactants and sparger can be very complex; thus, one should proceed very cautiously in designing systems for which surfactants significantly retard coalescence. Caution is particularly important because surfactants can produce either much more or much less backmixing than surfactant-free systems, depending on the bubble size, which, in turn, depends on the sparger utilized.

PHASE SEPARATION

Gases and liquids may be intentionally contacted as in absorption and distillation, or a mixture of phases may occur unintentionally as in vapor condensation from inadvertent cooling or liquid entrainment from a film. Regardless of the origin, it is usually desirable or necessary ultimately to separate gas-liquid dispersions. While separation will usually occur naturally, the rate is often economically intolerable and separation processes are employed to accelerate the step.

GAS-PHASE CONTINUOUS SYSTEMS

Practical separation techniques for liquid particles in gases are discussed. Since gas-borne particulates include both liquid and solid particles, many devices used for dry-dust collection (discussed in Sec. 17 under "Gas-Solids Separation") can be adapted to liquid-particle separation. Also, the basic subject of particle mechanics is covered in Sec. 6. Separation of liquid particulates is frequently desirable in chemical processes such as in countercurrent-stage contacting because liquid entrainment with the gas partially reduces true countercurrency. Separation before entering another process step may be needed to prevent corrosion, to prevent yield loss, or to prevent equipment damage or malfunction. Separation before the atmospheric release of gases may be necessary to prevent environmental problems and for regulatory compliance.

GENERAL REFERENCES
G-1. Buonicore and Davis, eds., *Air Pollution Engineering Manual*, Van Nostrand Reinhold, New York, 1992.
G-2. Calvert and Englund, eds., *Handbook of Air Pollution Technology*, Wiley, New York, 1984.
G-3. Cheremisinoff, ed., *Encyclopedia of Environmental Control Technology*, vol. 2, Gulf Pub., Houston, 1989.
G-4. McKetta, *Unit Operations Handbook*, vol. 1–2, Dekker, New York, 1992.
G-5. Wark and Warner, *Air Pollution: Its Origin and Control*, 2d ed., Harper & Row, New York, 1981.
G-6. Hesketh, *Air Pollution Control*, 1979; *Fine Particles in Gaseous Media*, Ann Arbor Science Pubs., Ann Arbor, MI, 1977.
G-7. Stern, *Air Pollution*, 3d ed., vols. 3–5, Academic, Orlando, FL, 1976–77.
G-8. Strauss, *Industrial Gas Cleaning*, 2d ed., Pergamon, New York, 1975.
G-9. Theodore and Buonicore, *Air Pollution Control Equipment; Selection, Design, Operation and Maintenance*, Prentice Hall, Englewood Cliffs, NJ, 1982.
G-10. Wang and Pereira, eds., *Handbook of Environmental Engineering*, vol. 1, Humana, Clifton, NJ 1979.
G-11. Cheremisinoff and Young, *Air Pollution Control and Design Handbook*, parts 1–2, Dekker, New York, 1977.
G-12. Nonhebel, *Gas Purification Processes for Air Pollution Control*, Newnes-Butterworth, London, 1972.

Sampling
R-1. *Code of Federal Regulations*, 40 (CFR 40), subchapter C—Air Programs, parts 50–99, Office of the Federal Register, Washington.
R-2. Ref. G-11, part 1, pp. 65–121.
R-3. Cooper and Rossano, *Source Testing for Air Pollution Control*, Environmental Science Services, Wilton, Connecticut, 1970.
R-4. Ref. G-7, vol. 3, pp. 525–587.
R-5. *Methods of Air Sampling and Analysis*, 2d Ed., American Public Health Assoc., Washington, 1977.
R-6. Stockham and Fochtman, *Particle Size Analysis*, Ann Arbor Science Pubs., Ann Arbor, Michigan, 1977.
R-7. Ref. G-2, Ch. 31, pp. 785–832.
R-8. Ref. G-8, Ch. 2, pp. 39–79.

Specific
R-9. Calvert, Goldchmid, Leith, and Mehta, NTIS Publ. PB-213016, 213017, 1972.
R-10. Calvert, *J. Air Pollut. Control Assoc.* **24,** 929 (1974).
R-11. Calvert, *Chem. Eng.*, **84**(18), 54 (1977).
R-12. Calvert, Yung, and Leung, NTIS Publ. PB-248050, 1975.
R-13. Calvert and Lundgren, *J. Air Pollut. Control Assoc.*, **18,** 677 (1968).
R-14. Calvert, Lundgren, and Mehta, *J. Air Pollut. Control Assoc.*, **22,** 529 (1972).
R-15. Yung, Barbarika, and Calvert, *J. Air Pollut. Control Assoc.*, **27,** 348, (1977).
R-16. Katz, M.S. thesis, Pennsylvania State University, 1958.
R-17. York and Poppele, *Chem. Eng. Prog.*, **59**(6), 45 (1963).
R-18. York, *Chem. Eng. Prog.*, **50,** 421 (1954).
R-19. Ref. G-2, Ch. 10, pp. 215–248.
References with the notation (R-) are cited in the text.

Definitions: Mist and Spray   Little standardization has been adopted in defining gas-borne liquid particles, and this frequently leads to confusion in the selection, design, and operation of collection equipment. Aerosol applies to suspended particulate, either solid or liquid, which is slow to settle by gravity and to particles from the submicrometer range up to 10 to 20 µm. Mists are fine suspended liquid dispersions usually resulting from condensation and ranging upward in particle size from around 0.1 µm. Spray refers to entrained liquid droplets. The droplets may be entrained from atomizing processes previously discussed under "Liquid-in-Gas Dispersions" in this section. In such instances, size will range from the finest particles produced up to a particle whose terminal settling velocity is equal to the entraining gas velocity if some settling volume is provided. Process spray is often created unintentionally, such as by the condensation of vapors on cold duct walls and its subsequent reentrainment, or from two-phase flow in pipes, gas bubbling through liquids, and entrainment from boiling liquids. Entrainment size distribution from sieve trays has been given by Cheng and Teller [*Am. Inst. Chem. Eng. J.*, **7**(2), 282 (1961)] and evaporator spray by Garner et al. [*Trans. Inst. Chem. Eng.*, **32,** 222 (1954)]. In general, spray can range downward in particle size from 5000 µm. There can be overlapping in size between the coarsest mist particles and the finest spray particles, but some authorities have found it convenient arbitrarily to set a boundary of 10 µm between the two. Actually, considerable overlap exists in the region of 5 to 40 µm. Table 14-18 lists typical ranges of particle size created by different mechanisms. The sizes actually entrained can be influenced by the local gas velocity. Figure 14-105 compares the approximate size range of liquid particles with other particulate material and the approximate applicable size range of collection devices. Figure 17-34 gives an expanded chart by Lapple for solid particles. Mist and fog formation has been discussed previously.

Gas Sampling   The sampling of gases containing mists and sprays may be necessary to obtain data for collection-device design, in which case particle-size distribution, total mass loading, and gas volume, temperature, pressure, and composition may all be needed. Other reasons for sampling may be to determine equipment performance, measure yield loss, or determine compliance with regulations.

Location of a sample probe in the process stream is critical especially when larger particles must be sampled. Mass loading in one portion of a duct may be severalfold greater than in another portion as affected by flow patterns. Therefore, the stream should be sampled at a number of points. The U.S. Environmental Protection Agency (R-1) has specified 8 points for ducts between 0.3 and 0.6 m (12 and 24 in) and 12 points for larger ducts, provided there are no flow disturbances for eight pipe diameters upstream and two downstream from the sampling point. When only particles smaller than 3 µm are to be sampled, location and number of sample points are less critical since such particles remain reasonably well dispersed by brownian motion. However, some gravity settling of such particles and even gases of high density have been observed in long horizontal breeching. Isokinetic sampling (velocity at the probe inlet is equal to local duct velocity) is required to get a representative sample of particles larger than 3 µm (error is small for 4- to 5-µm particles). Sampling methods and procedures for mass loading have been developed (R-1 through R-8).

TABLE 14-18   Particle Sizes Produced by Various Mechanisms

Mechanism or process	Particle-size range, µm
Liquid pressure spray nozzle	100–5000
Gas-atomizing spray nozzle	1–100
Gas bubbling through liquid or boiling liquid	20–1000
Condensation processes with fogging	0.1–30
Annular two-phase flow in pipe or duct	10–2000

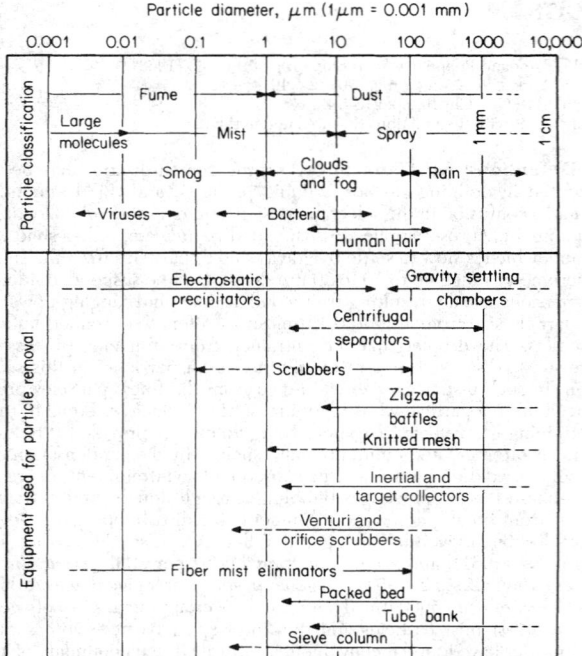

FIG. 14-105 Particle classification and useful collection equipment versus particle size.

Particle-Size Analysis Many particle-size-analysis methods suitable for dry-dust measurement are unsuitable for liquids because of coalescence and drainage after collection. Measurement of particle sizes in the flowing aerosol stream by using a cascade impactor is one of the better means. The impacting principle has been described by Ranz and Wong [*Ind. Eng. Chem.*, **44**, 1371 (1952)] and Gillespie and Johnstone [*Chem. Eng. Prog.*, **51**, 75F (1955)]. The Andersen, Sierra, and University of Washington impactors may be used if the sampling period is kept short so as not to saturate the collection substrate. An impactor designed specifically for collecting liquids has been described by Brink, Kennedy, and Yu [*Am. Inst. Chem. Eng. Symp. Ser.*, **70**(137), 333 (1974)].

Collection Mechanisms Mechanisms which may be used for separating liquid particles from gases are (1) gravity settling, (2) inertial (including centrifugal) impaction, (3) flow-line interception, (4) diffusional (brownian) deposition, (5) electrostatic attraction, (6) thermal precipitation, (7) flux forces (thermophoresis, diffusiophoresis, Stefan flow), and (8) particle agglomeration (nucleation) techniques. Equations and parameters for these mechanisms are given in Table 17-2. Most collection devices rarely operate solely with a single mechanism, although one mechanism may so predominate that it may be referred to, for instance, as an inertial-impaction device.

After collection, liquid particles coalesce and must be drained from the unit, preferably without reentrainment. Calvert (R-12) has studied the mechanism of reentrainment in a number of liquid-particle collectors. Four types of reentrainment were typically observed: (1) transition from separated flow of gas and liquid to a two-phase region of separated-entrained flow, (2) rupture of bubbles, (3) liquid creep on the separator surface, and (4) shattering of liquid droplets and splashing. Generally, reentrainment increased with increasing gas velocity. Unfortunately, in devices collecting primarily by centrifugal and inertial impaction, primary collection efficiency increases with gas velocity; thus overall efficiency may go through a maximum as reentrainment overtakes the incremental increase in efficiency. Prediction of collection efficiency must consider both primary collection and reentrainment.

Procedures for Design and Selection of Collection Devices
Calvert and coworkers (R-9 to R-12 and R-19) have suggested useful design and selection procedures for particulate-collection devices in which direct impingement and inertial impaction are the most significant mechanisms. The concept is based on the premise that the mass median aerodynamic particle diameter d_{p50} is a significant measure of the difficulty of collection of the liquid particles and that the collection device cut size d_{pc} (defined as the aerodynamic particle diameter collected with 50 percent efficiency) is a significant measure of the capability of the collection device. The aerodynamic diameter for a particle is the diameter of a spherical particle (with an arbitrarily assigned density of 1 g/cm³) which behaves in an air stream in the same fashion as the actual particle. For real spherical particles of diameter d_p, the equivalent aerodynamic diameter d_{pa} can be obtained from the equation $d_{pa} = dp(\rho_p C')^{1/2}$, where ρ_p is the apparent particle density (mass/volume) and C' is the Stokes-Cunningham correction factor for the particle size, all in consistent units. If particle diameters are expressed in micrometers, ρ_p can be in grams per cubic centimeter and C' can be approximated by $C' = 1 + A_c(2\lambda/D_p)$, where A_c is a constant dependent upon gas composition, temperature, and pressure ($A_c = 0.88$ for atmospheric air at 20°C) and λ is the mean free path of the gas molecules ($\lambda = 0.10$ μm for 20°C atmospheric air). For other temperatures or pressures, or gases other than air, calculations using these more precise equations may be made: $A_c = 1.257 + 0.4 \exp[-1.1 (d_p/2\lambda)]$ and $\lambda = \mu_g/0.499\rho_g \times \mu_m$ (where μ_g is the gas viscosity, kg/m·h; ρ_g is gas density, g/cm³; and μ_m is the mean molecular speed, m/s. $u_m = [8R_uT/\pi M]^{0.5}$, where R_u is the universal gas constant, 8.315 kJ/kg·mol·K; T is the gas absolute temperature, K; and M is the molar mass or equivalent molecular weight of the gas. (π is the usual geometric constant.) For test purposes (air at 25°C and 1 atm), $\rho_g = 1.183$ kg/m, $\mu_g = 0.0666$ kg/m·h, $\lambda = 0.067$ μm, and $u_m = 467$ m/s. For airborne liquid particles, the assumption of spherical shape is reasonably accurate, and ρ_p is approximately unity for dilute aqueous particles at ambient temperatures. C' is approximately unity at ambient conditions for such particles larger than 1 to 5 μm, so that often the actual liquid particle diameter and the equivalent aerodynamic diameter are identical.

When a distribution of particle sizes which must be collected is present, the actual size distribution must be converted to a mass distribution by aerodynamic size. Frequently the distribution can be represented or approximated by a log-normal distribution (a straight line on a log-log plot of cumulative mass percent of particles versus diameter) which can be characterized by the mass median particle diameter d_{p50} and the standard statistical deviation of particles from the median σ_g. σ_g can be obtained from the log-log plot by $\sigma_g = D_{pa50}/D_{pe}$ at 15.87 percent = D_{pe} at 84.13 percent/D_{pa50}.

The grade efficiency η of most collectors can be expressed as a function of the aerodynamic particle size in the form of an exponential equation. It is simpler to write the equation in terms of the particle penetration P_t (those particles not collected), where the fractional penetration $P_t = 1 - \eta$, when η is the fractional efficiency. The typical collection equation is

$$P_t = e^{(-A_a D_{pa}B)} \qquad (14\text{-}224)$$

where A_a and B are functions of the collection device. Calvert (R-12) has determined that for many devices in which the primary collection mechanism is direct interception and inertial impaction, such as packed beds, knitted-mesh collectors, zigzag baffles, target collectors such as tube banks, sieve-plate columns, and venturi scrubbers, the value of B is approximately 2.0. For cyclonic collectors, the value of B is approximately 0.67. The overall integrated penetration $\overline{P}_t$ for a device handling a distribution of particle sizes can be obtained by

$$\overline{P}_t = \int_0^W \left(\frac{dW}{W}\right) P_t \qquad (14\text{-}225)$$

where (dW/W) is the mass of particles in a given narrow size distribution and P_t is the average penetration for that size range. When the particles to be collected are log-normally distributed and the collection device efficiency can be expressed by Eq. (14-224), the required overall integrated collection efficiency $\overline{P}_t$ can be related to the ratio of the device aerodynamic cut size D_{pc} to the mass median aerodynamic particle size D_{pa50}. This required ratio for a given distribution and collection is designated R_{rL} and these relationships are illustrated graphically in Fig. 14-106. For the many devices for which B is approximately 2.0, a simplified plot (Fig. 14-107) is obtained. From these figures, by knowing the desired overall collection efficiency and particle distribution, the value of R_{rL} can be read. Substituting the mass median particle diameter gives the aerodynamic cut size required from the collection device being considered. Therefore, an experimental plot of aerodynamic cut size for each collection device versus operating parameters can be used to determine the device suitability.

Collection Equipment

Gravity Settlers   Gravity can act to remove larger droplets. Settling or disengaging space above aerated or boiling liquids in a tank or spray zone in a tower can be very useful. If gas velocity is kept low, all particles with terminal settling velocities (see Sec. 6) above the gas

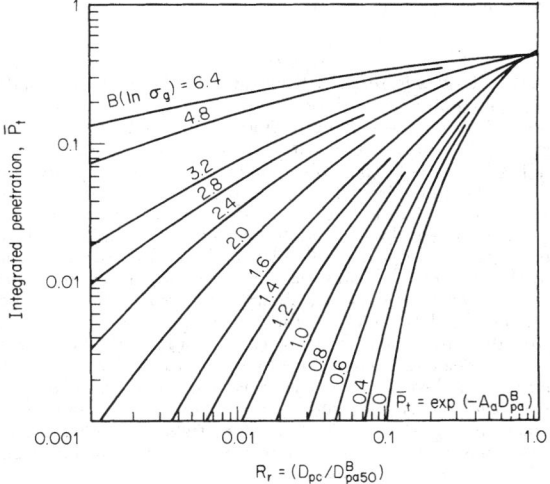

FIG. 14-106   Overall integrated penetration as a function of particle-size distribution and collector parameters. (*Calvert, Yung, and Leung, NTIS Publ. PB-248050, 1975.*)

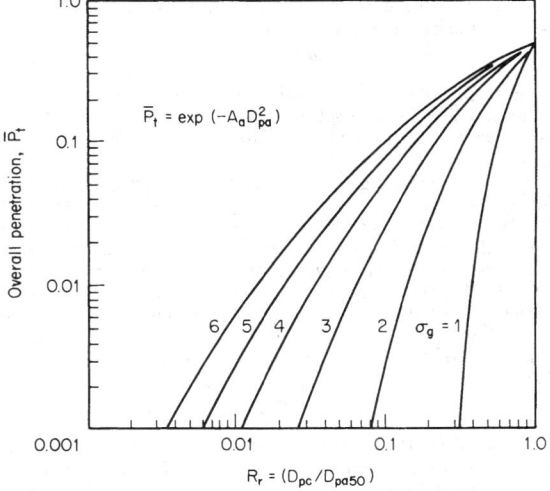

FIG. 14-107   Overall integrated penetration as a function of particle-size distribution and collector cut diameter when $B = 2$ in Eq. (14-224). (*Calvert, Goldshmid, Leith, and Mehta, NTIS Publ. PB-213016, 213017, 1972.*)

velocity will eventually settle. Increasing vessel cross section in the settling zone is helpful. Terminal velocities for particles smaller than 50 μm are very low and generally not attractive for particle removal. Laminar flow of gas in long horizontal paths between trays or shelves on which the droplets settle is another effective means of employing gravity. Design equations are given in Sec. 17 under "Gas-Solids Separations." Settler pressure drop is very low, usually being limited to entrance and exit losses.

Centrifugal Separation   Centrifugal force can be utilized to enhance particle collection to several hundredfold that of gravity. The design of cyclone separators for dust removal is treated in detail in Sec. 17 under "Gas-Solids Separations," and typical cyclone designs are shown in Fig. 17-43. Dimension ratios for one family of cyclones are given in Fig. 17-36. Cyclones, if carefully designed, can be more efficient on liquids than on solids since liquids coalesce on capture and are easy to drain from the unit. However, some precautions not needed for solid cyclones are necessary to prevent reentrainment.

Tests by Calvert (R-12) show high primary collection efficiency on droplets down to 10 μm and in accordance with the efficiency equations of Leith and Licht [*Am. Inst. Chem. Eng. Symp. Ser.*, 68(126), 196–206 (1972)] for the specific cyclone geometry tested if entrainment is avoided. Typical entrainment points are (1) creep along the gas outlet pipe, (2) entrainment by shearing of the liquid film from the walls, and (3) vortex pickup from accumulated liquid in the bottom (Fig. 14-108*a*). Reentrainment from creep of liquid along the top of the cyclone and down the outlet pipe can be prevented by providing the outlet pipe with a flared conical skirt (Fig. 14-108*b*), which provides a point from which the liquid can drip without being caught in the outlet gas. The skirt should be slightly shorter than the gas outlet pipe but extend below the bottom of the gas inlet. The cyclone inlet gas should not impinge on this skirt. Often the bottom edge of the skirt is V-notched or serrated.

Reentrainment is generally reduced by lower inlet gas velocities. Calvert (R-12) reviewed the literature on predicting the onset of entrainment and found that of Chien and Ibele (ASME Pap. 62-WA170) to be the most reliable. Calvert applies their correlation to a liquid Reynolds number on the wall of the cyclone, $N_{Re,L} = 4Q_L/h_i v_L$, where Q_L is the volumetric liquid flow rate, cm³/s; h_i is the cyclone inlet height, cm; and v_L is the kinematic liquid viscosity, cm²/s. He finds that the onset of entrainment occurs at a cyclone inlet gas velocity V_{ci}, m/s, in accordance with the relationship in $V_{ci} = 6.516 - 0.2865$ ln $N_{Re,L}$.

Reentrainment from the bottom of the cyclone can be prevented in several ways. If a typical long-cone dry cyclone is used and liquid is kept continually drained, vortex entrainment is unlikely. However, a vortex breaker baffle in the outlet is desirable, and perhaps a flat disk on top extending to within 2 to 5 cm (0.8 to 2 in) of the walls may be

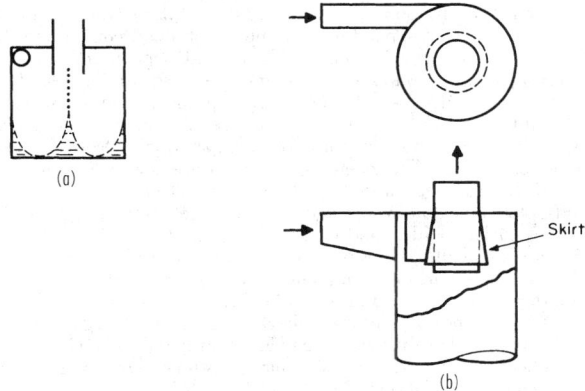

FIG. 14-108   (*a*) Liquid entrainment from the bottom of a vessel by centrifugal flow. (*Rietema and Verver, Cyclones in Industry, Elsevier, Amsterdam, 1961.*) (*b*) Gas-outlet skirt for liquid cyclones. (*Stern et al., Cyclone Dust Collectors, American Petroleum Institute, New York, 1955.*)

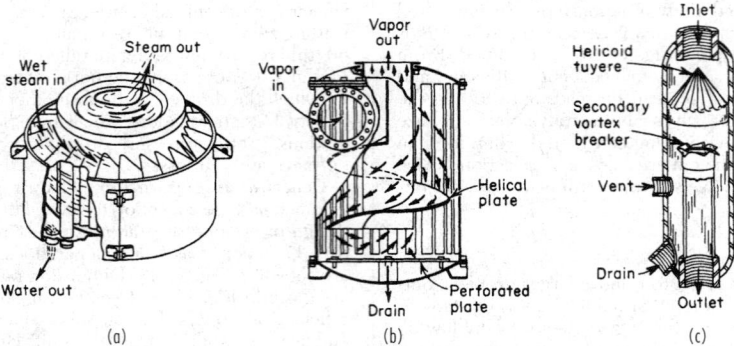

FIG. 14-109 Typical separators using impingement in addition to centrifugal force. (*a*) Hi-eF purifier. (*V. D. Anderson Co.*) (*b*) Flick separator. (*Wurster & Sanger, Inc.*) (*c*) Type RA line separator. (*Centrifix Corp., Bull. 220.*)

beneficial. Often liquid cyclones are built without cones and have dished bottoms. The modifications described earlier are definitely needed in such situations. Stern, Caplan, and Bush (*Cyclone Dust Collectors,* American Petroleum Institute, New York, 1955) and Rietema and Verver (in Tengbergen, *Cyclones in Industry,* Elsevier, Amsterdam, 1961, chap. 7) have discussed liquid-collecting cyclones.

As with dust cyclones, no reliable pressure-drop equations exist (see Sec. 17), although many have been published. A part of the problem is that there is no standard cyclone geometry. Calvert (R-12) experimentally obtained $\Delta P = 0.000513 \, \rho_g (Q_g/h_i W_i)^2 (2.8 h_i w_i/d_o^2)$, where ΔP is in cm of water; ρ_g is the gas density, g/cm³; Q_g is the gas volumetric flow rate, cm³/s; h_i and w_i are cyclone inlet height and width respectively, cm; and d_o is the gas outlet diameter, cm. This equation is in the same form as that proposed by Shepherd and Lapple [*Ind. Eng. Chem.,* **31,** 1246 (1940)] but gives only 37 percent as much pressure drop.

Liquid cyclone efficiency can be improved somewhat by introducing a coarse spray of liquid in the cyclone inlet. Large droplets which are easily collected collide with finer particles as they sweep the gas stream in their travel to the wall. (See subsection "Wet Scrubbers" regarding optimum spray size.) Cyclones may also be operated wet to improve their operation on dry dust. Efficiency can be improved through reduction in entrainment losses since the dust particles become trapped in the water film. Collision between droplets and dust particles aids collection, and adequate irrigation can eliminate problems of wall buildup and fouling. The most effective operation is obtained by spraying countercurrently to the gas flow in the cyclone inlet duct at liquid rates of 0.7 to 2.0 L/m³ of gas. There are also many proprietary designs of liquid separators using centrifugal force, some of which are illustrated in Fig. 14-109. Many of these were originally developed as steam separators to remove entrained condensate. In some designs, impingement on swirl baffles aids separation.

Impingement Separation Impingement separation employs direct impact and inertial forces between particles, the gas streamlines, and target bodies to provide capture. The mechanism is discussed in Sec. 17 under "Gas-Solids Separations." With liquids, droplet coalescence occurs on the target surface, and provision must be made for drainage without reentrainment. Calvert (R-12) has studied droplet collection by impingement on targets consisting of banks of tubes, zigzag baffles, and packed and mesh beds. Figure 14-110 illustrates some other types of impingement-separator designs.

In its simplest form, an impingement separator may be nothing more than a target placed in front of a flow channel such as a disk at the end of a tube. To improve collection efficiency, the gas velocity may be increased by forming the end into a nozzle (Fig. 14-110*a*). Particle collection as a function of size may be estimated by using the target-efficiency correlation in Fig. 17-39. Since target efficiency will be low for systems with separation numbers below 5 to 10 (small particles, low gas velocities), the mist will frequently be subjected to a number of targets in series as in Fig. 14-110*c, d,* and *g.*

The overall droplet penetration is the product of penetration for each set of targets in series. Obviously, for a distribution of particle sizes, an integration procedure is required to give overall collection efficiency. This target-efficiency method is suitable for predicting efficiency when the design effectively prevents the bypassing or short-circuiting of targets by the gas stream and provides adequate time to accelerate the liquid droplets to gas velocity. Katz (R-16) investigated a jet and target-plate entrainment separator design and found the pressure drop less than would be expected to supply the kinetic energy both for droplet acceleration and gas friction. An estimate based on his results indicates that the liquid particles on the average were being accelerated to only about 60 percent of the gas velocity. The largest droplets, which are the easiest to collect, will be accelerated less than the smaller particles. This factor has a leveling effect on collection efficiency as a function of particle size so that experimental results on such devices may not show as sharp a decrease in efficiency with particle size as predicted by calculation. Such results indicate that in many cases our lack of predicting ability results, not from imperfections in the theoretical treatment, but from our lack of knowledge of velocity distributions within the system.

Katz (R-16) also studied *wave-plate impingement separators* (Fig. 14-110*b*) made up of 90° formed arcs with an 11.1-mm (0.44-in) radius and a 3.8-mm (0.15-in) clearance between sheets. The pressure drop is a function of system geometry. The pressure drop for Katz's system and collection efficiency for seven waves are shown in Fig. 14-111. Katz used the Souders-Brown expression to define a design velocity for the gas between the waves:

$$U = K \sqrt{(\rho_l - \rho_g)/\rho_g} \qquad (14\text{-}226)$$

K is 0.12 to give U in ms⁻¹ (0.4 for ft/s), and ρ_l and ρ_g are liquid and gas densities in any consistent set of units. Katz found no change in efficiency at gas velocities from one-half to 3 times that given by the equation.

Calvert (R-12) investigated *zigzag baffles* of a design more like Fig. 14-110*e.* The baffles may have spaces between the changes in direction or be connected as shown. He found close to 100 per collection for water droplets of 10 μm and larger. Some designs had high efficiencies down to 5 or 8 μm. Desirable gas velocities were 2 to 3.5 m/s (6.6 to 11.5 ft/s), with a pressure drop for a six-pass baffle of 2 to 2.5 cm (0.8 to 1.0 in) of water. On the basis of turbulent mixing, an equation was developed for predicting primary collection efficiency as a function of particle size and collector geometry:

$$\eta = 1 - \exp\left[\frac{-u_{te} n W \theta}{57.3 U_g b \tan \theta}\right] \qquad (14\text{-}227)$$

where η is the fractional primary collection efficiency; u_{te} is the drop terminal centrifugal velocity in the normal direction, cm/s; U_g is the superficial gas velocity, cm/s; n is the number of rows of baffles or bends; θ is the angle of inclination of the baffle to the flow path, °; W is the width of the baffle, cm; and b is the spacing between baffles in the same row, cm. For conditions of low Reynolds number ($N_{Re,D} <$

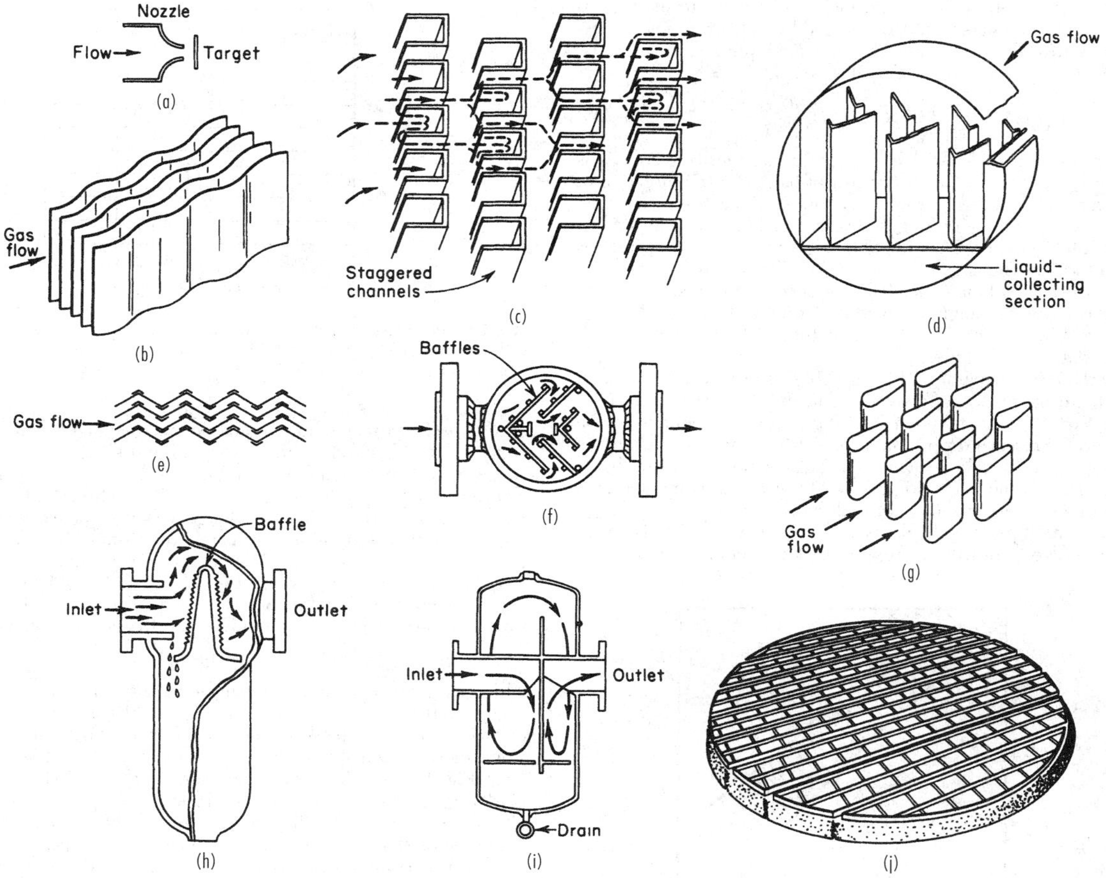

FIG. 14-110   Typical impingement separators. (*a*) Jet impactor. (*b*) Wave plate. (*c*) Staggered channels. (*Blaw-Knox Food & Chemical Equipment, Inc.*) (*d*) Vane-type mist extractor. (*Maloney-Crawford Tank and Mfg. Co.*) (*e*) Peerless line separator. (*Peerless Mfg. Co.*) (*f*) Strong separator. (*Strong Carlisle and Hammond.*) (*g*) Karbate line separator. (*Union Carbide Corporation*) (*h*) Type E horizontal separator. (*Wright-Austin Co.*) (*i*) PL separator. (*Ingersoll Rand.*) (*j*) Wire-mesh demister. (*Otto H. York Co.*)

0.1) where Stokes' law applies, Calvert obtains the value for drop terminal centrifugal velocity of $u_{te} = d_p^2 \rho_p a / 18 \mu_g$, where d_p and ρ_p are the drop particle diameter, cm, and particle density, g/cm³, respectively; μ_g is the gas viscosity, P; and a is the acceleration due to centrifugal force. It is defined by the equation $a = 2U_g^2 \sin \theta / W \cos^3 \theta$. For situations in which Stokes' law does not apply, Calvert recommends substitution in the derivation of Eq. (14-227) for u of drag coefficients from

drag-coefficient data of Foust et al. (*Principles of Unit Operations,* Toppan Co., Tokyo, 1959).

Calvert found that reentrainment from the baffles was affected by the gas velocity, the liquid-to-gas ratio, and the orientation of the baffles. Horizontal gas flow past vertical baffles provided the best drainage and lowest reentrainment. Safe operating regions with vertical baffles are shown in Fig. 14-112. Horizontal baffles gave the poorest drainage

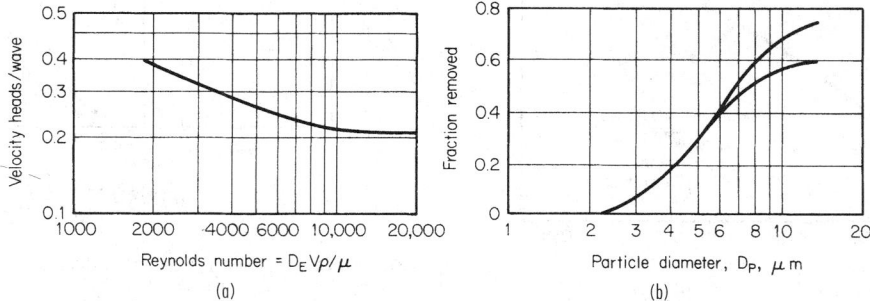

FIG. 14-111   Pressure drop and collection efficiency of a wave-plate separator. (*a*) Pressure drop. (*b*) Efficiency D_E = clearance between sheets. (*Katz, M.S. thesis, Pennsylvania State University, 1958.*)

and the highest reentrainment, with inclined baffles intermediate in performance. Equation (14-228), developed by Calvert, predicts pressure drop across zigzag baffles. The indicated summation must be made over the number of rows of baffles present.

$$\Delta P = \sum_{i=1}^{i=n} 1.02 \times 10^{-3} f_D \rho_g \frac{U_g' A_p}{2 A_t} \qquad (14\text{-}228)$$

ΔP is the pressure drop, cm of water; ρ_g is the gas density, g/cm³; A_p is the total projected area of an entire row of baffles in the direction of inlet gas flow, cm²; and A_t is the duct cross-sectional area, cm². The value f_D is a drag coefficient for gas flow past inclined flat plates taken from Fig. 14-113, while U_g' is the actual gas velocity, cm/s, which is related to the superficial gas velocity U_g by $U_g' = U_g/\cos \theta$. It must be noted that the angle of incidence θ for the second and successive rows of baffles is twice the angle of incidence for the first row. Most of Calvert's work was with 30° baffles, but the method correlates well with other data on 45° baffles.

The Karbate line separator (Fig. 14-110g) is composed of several layers of teardrop-shaped target rods of Karbate. A design flow constant K in Eq. (14-226) of 0.035 m/s (1.0 ft/s) is recommended by the manufacturer. Pressure drop is said to be 5½ velocity heads on the basis of the superficial gas velocity. This value would probably increase at high liquid loads. Figure 14-114 gives the manufacturer's reported grade efficiency curve at the design air velocity.

The use of multiple tube banks as a droplet collector has also been studied by Calvert (R-12). He reports that collection efficiency for

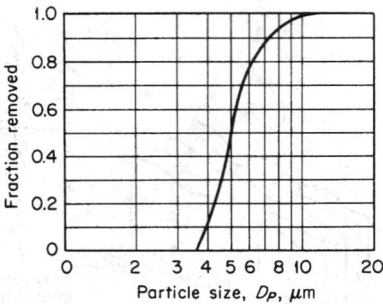

FIG. 14-114   Collection efficiency of Karbate line separator, based on particles with a specific gravity of 1.0 suspended in atmospheric air with a pressure drop of 2.5 cm water gauge. (*Union Carbide Corporation Cat. Sec. S-6900, 1960.*)

closely packed tubes follows equations for rectangular jet impaction which can be obtained graphically from Fig. 14-115 by using a dimensional parameter β which is based on the tube geometry; $\beta = 2l_i/b$, where b is the open distance between adjacent tubes in the row (orifice width) and l_i is the impaction length (distance between orifice and impingement plane), or approximately the distance between centerlines of successive tube rows. Note that the impaction parameter K_p is plotted to the one-half power in Fig. 14-115 and that the radius of the droplet is used rather than the diameter. Collection efficiency overall for a given size of particle is predicted for the entire tube bank by

$$\eta = 1 - (1 - \eta_b)^N \qquad (14\text{-}229)$$

where η_b is the collection efficiency for a given size of particle in one stage of a rectangular jet impactor (Fig. 14-115) and N is the number of stages in the tube bank (equal to one less than the number of rows). For widely spaced tubes, the target efficiency η_g can be calculated from Fig. 17-39 or from the impaction data of Golovin and Putnam [*Ind. Eng. Chem. Fundam.*, **1**, 264 (1962)]. The efficiency of the overall tube banks for a specific particle size can then be calculated from the equation $\eta = 1 - (1 - \eta_t a'/A)^n$, where a' is the cross-sectional area of all tubes in one row, A is the total flow area, and n is the number of rows of tubes.

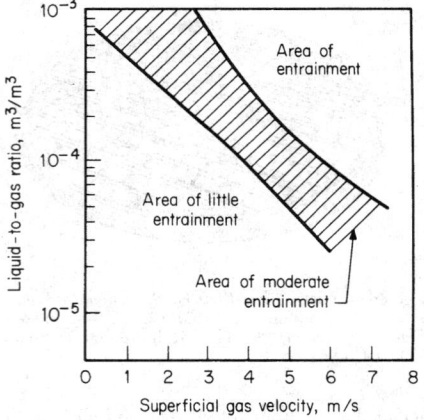

FIG. 14-112   Safe operating region to prevent reentrainment from vertical zigzag baffles with horizontal gas flow. (*Calvert, Yung, and Leung, NTIS Publ. PB-248050, 1975.*)

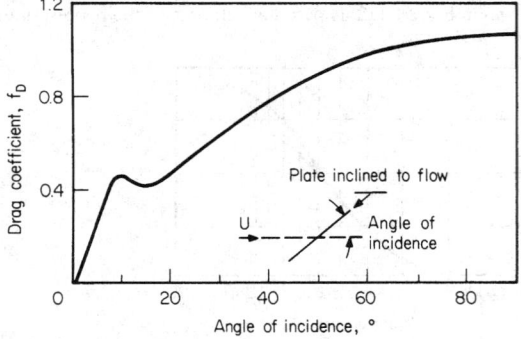

FIG. 14-113   Drag coefficient for flow past inclined flat plates for use in Eq. (14-228). (*Calvert, Yung, and Leung, NTIS Publ. PB-248050; based on Fage and Johansen*, Proc. R. Soc. (London), *116A, 170 (1927).*]

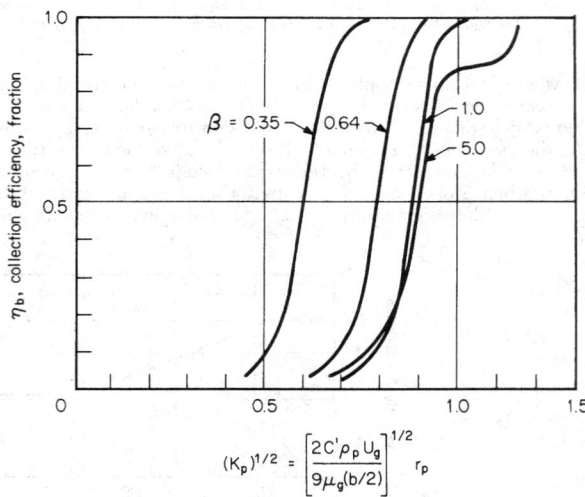

FIG. 14-115   Experimental collection efficiencies of rectangular impactors. C' is the Stokes-Cunningham correction factor; ρ_p, particle density, g/cm³; U_g, superficial gas velocity, approaching the impactor openings, cm/s; and μ_g, gas viscosity, P. [*Calvert, Yung, and Leung, NTIS Publ. PB-248050; based on Mercer and Chow*, J. Coll. Interface Sci., **27**, 75 (1968).]

Calvert reports pressure drop through tube banks to be largely unaffected by liquid loading and indicates that Grimison's correlations in Sec. 6 ("Tube Banks") for gas flow normal to tube banks or data for gas flow through heat-exchanger bundles can be used. However, the following equation is suggested:

$$\Delta P = 8.48 \times 10^{-3} n \rho_g U_g'^2 \qquad (14\text{-}230)$$

where ΔP is cm of water; n is the number of rows of tubes; ρ_g is the gas density, g/cm³; and U_g' is the actual gas velocity between tubes in a row, cm/s. Calvert did find an increase in pressure drop of about 80 to 85 percent above that predicted by Eq. (14-230) in vertical upflow of gas through tube banks due to liquid holdup at gas velocities above 4 m/s. The onset of liquid reentrainment from tube banks can be predicted from Fig. 14-116. Reentrainment occurred at much lower velocities in vertical upflow than in horizontal gas flow through vertical tube banks. While the top of the cross-hatched line of Fig. 14-116a predicts reentrainment above gas velocities of 3 m/s (9.8 ft/s) at high liquid loading, most of the entrainment settled to the bottom of the duct in 1 to 2 m (3.3 to 6.6 ft), and entrainment did not carry significant distances until the gas velocity exceeded 7 m/s (23 ft/s).

Packed-Bed Collectors   Many different materials, including coal, coke, broken solids of various types such as brick, tile, rock, and stone, as well as normal types of tower-packing rings, saddles, and special plastic shapes, have been used over the years in packed beds to remove entrained liquids through impaction and filtration. Separators using natural materials are not available as standard commercial units but are designed for specific applications. Coke boxes were used extensively in the years 1920 to 1940 as sulfuric acid entrainment separators (see *Chemical Engineers' Handbook*, 5th ed., p. 18–87) but have now been largely superseded by more sophisticated and efficient devices.

Jackson and Calvert [*Am. Inst. Chem. Eng. J.*, **12**, 1075 (1966)] studied the collection of fine fuel-oil-mist particles in beds of ½-in glass spheres, Raschig rings, and Berl and Intalox saddles. The mist had a mass median particle diameter of 6 μm and a standard deviation of 2.0. The collection efficiency as a function of particle size and gas

velocity in a 355-mm- (14-in-) diameter by 152-mm- (6-in-) thick bed of Intalox saddles is given in Fig. 14-117. This and additional work have been generalized by Calvert (R-12) to predict collection efficiencies of liquid particles in any packed bed. Assumptions in the theoretical development are that the drag force on the drop is given by Stokes' law and that the number of semicircular bends to which the gas is subjected, η_1, is related to the length of the bed, Z (cm), in the direction of gas flow, the packing diameter, d_c (cm), and the gas-flow channel width, b (cm), such that $\eta_1 = Z/(d_c + b)$. The gas velocity through the channels, U_{gb} (cm/s), is inversely proportional to the bed free volume for gas flow such that $U_{gb} = U_g[1/(\varepsilon - h_b)]$, where U_g is the gas superficial velocity, cm/s, approaching the bed, ε is the bed void fraction, and h_b is the fraction of the total bed volume taken up with liquid which can be obtained from data on liquid holdup in packed beds. The width of the semicircular channels b can be expressed as a fraction j of the diameter of the packing elements, such that $b = jd_c$. These assumptions (as modified by G. E. Goltz, personal communication) lead to an equation for predicting the penetration of a given size of liquid particle through a packed bed:

$$P_t = \exp\left[\frac{-\pi}{2(j + j^2)(\varepsilon - h_b)}\left(\frac{Z}{d_c}\right)K_p\right] \qquad (14\text{-}231)$$

where

$$K_p = \frac{\rho_p d_p^2 U_g}{9\mu_g d_c} \qquad (14\text{-}232)$$

Values of ρ_p and d_p are droplet density, g/cm³, and droplet diameter, cm; μ_g is the gas viscosity, P. All other terms were defined previously. Table 14-19 gives values of j calculated from experimental data of Jackson and Calvert. Values of j for most manufactured packing appear to fall in the range from 0.16 to 0.19. The low value of 0.03 for coke may be due to the porosity of the coke itself.

Calvert (R-12) has tested the correlation in cross-flow packed beds, which tend to give better drainage than countercurrent beds, and has found the effect of gas-flow orientation insignificant. However, the onset of reentrainment was somewhat lower in a bed of 2.5-cm

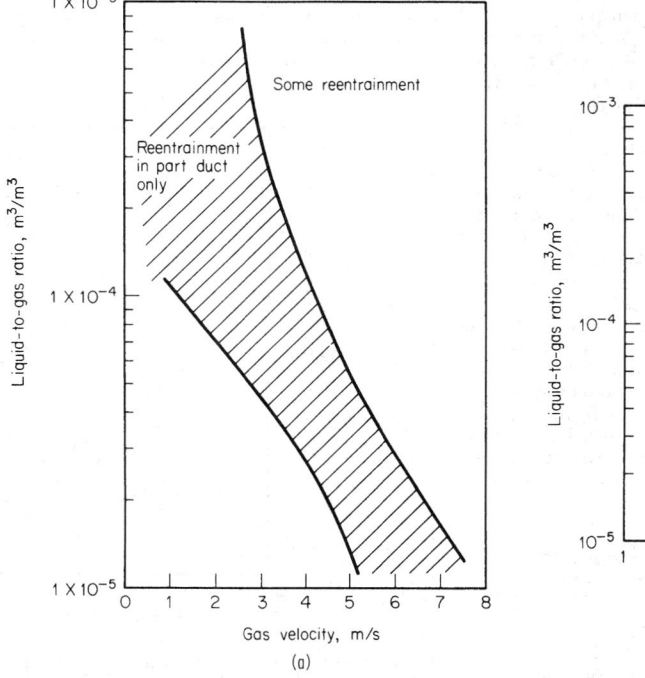

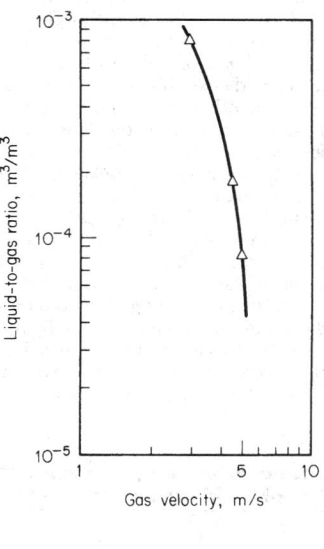

FIG. 14-116   Experimental results showing effect of gas velocity and liquid load on entrainment from (a) vertical tube banks with horizontal gas flow and (b) horizontal tube banks with upflow. To convert meters per second to feet per second, multiply by 3.281. (*Calvert, Yung, and Leung, NTIS Publ. PB-248050.*)

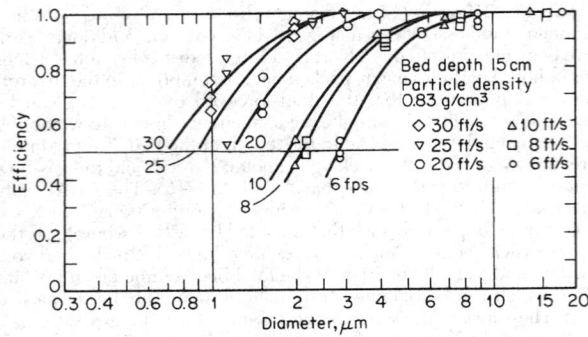

FIG. 14-117 Experimental collection efficiency. ½-in Intalox saddles. To convert feet per second to meters per second, multiply by 0.3048; to convert centimeters to inches, multiply by 0.394; and to convert grams per cubic centimeter to pounds per cubic foot, multiply by 62.43. [*Jackson and Calvert, Am. Inst. Chem. Eng. J., 12, 1975 (1968).*]

(1.0-in) pall rings with gas upflow [6 m/s (20 ft/s)] than with horizontal cross-flow of gas. The onset of reentrainment was independent of liquid loading (all beds were nonirrigated), and entrainment occurred at values somewhat above the flood point for packed beds as predicted by conventional correlations. In beds with more than 3 cm (1.2 in) of water pressure drop, the experimental drop with both vertical and horizontal gas flow was somewhat less than predicted by generalized packed-bed pressure-drop correlations. However, Calvert recommends these correlations for design as conservative.

Calvert's data indicate that packed beds irrigated only with the collected liquid can have collection efficiencies of 80 to 90 percent on mist particles down to 3 μm but have low efficiency on finer mist particles. Frequently, irrigated packed towers and towers with internals will be used with liquid having a wetting capability for the fine mist which must be collected. Tennessee Valley Authority (TVA) experiments with the collection of 1.0-μm mass median phosphoric acid mist in packed towers have shown that the strength of the circulating phosphoric acid is highly important [see Baskerville, *Am. Inst. Chem. Eng. J., 37, 79 (1941); and p. 18–87, 5th ed. of the Handbook*]. Hesketh [*J. Air Pollut. Control Assoc., 24, 942 (1974)*] has reported up to 50 percent improvement in collection efficiency in venturi scrubbers on fine particles with the addition of only 0.10 percent of a low-foaming nonionic surfactant to the scrubbing liquid, and others have experienced similar results in other gas-liquid-contacting devices. Calvert (R-9 and R-10) has reported on the efficiency of various gas-liquid-contacting devices for fine particles. Figure 14-118 gives the particle aerodynamic cut size for a single-sieve-plate gas scrubber as a function of sieve hole size d_h, cm; hole gas velocity u_h, m/s; and froth or foam density on the plate F, g/cm³. This curve is based on standard air and water properties and wettable (hydrophilic) particles. The cut diameter decreases with an increase in froth density, which must be predicted from correlations for sieve-plate behavior (see Fig. 14-32). Equation (14-231) can be used to calculate generalized design curves for collection in packed columns in the same fashion by finding parameters of packing size, bed length, and gas velocity which give collection efficiencies of 50 percent for various size particles. Figure (14-119) illustrates such a plot for three gas velocities and two sizes of packing.

TABLE 14-19   Experimental Values for *j*, Channel Width in Packing as a Fraction of Packing Diameter

Packing size		Type of packing	*j*
cm	in		
1.27	0.5	Berl and Intalox saddles, marbles, Raschig rings	0.192
2.54	1.0	Berl and Intalox saddles, pall rings	0.190
3.8	1.5	Berl and Intalox saddles, pall rings	0.165
7.6–12.7	3–5	Coke	0.03

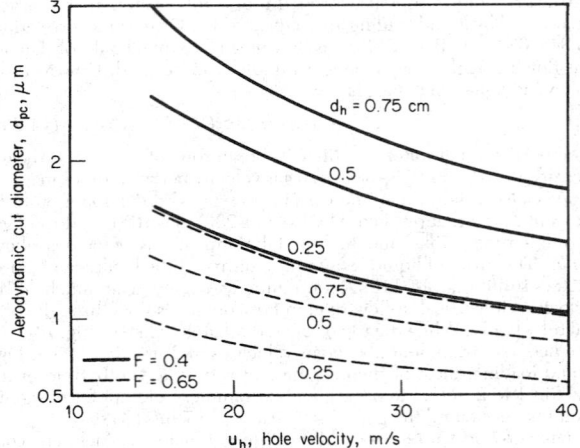

FIG. 14-118 Aerodynamic cut diameter for a single-sieve-plate scrubber as a function of hole size, hole-gas velocity, and froth density, F, g/cm³. To convert meters per second to feet per second, multiply by 3.281; to convert grams per cubic centimeter to pounds per cubic foot, multiply by 62.43. [*Calvert, J. Air Pollut. Control Assoc., 24, 929 (1974).*]

Wire-Mesh Mist Collectors   Knitted mesh of varying density and voidage is widely used for entrainment separators. Its advantage is close to 100 percent removal of drops larger than 5 μm at superficial gas velocities from about 0.2 m/s (0.6 ft/s) to 5 m/s (16.4 ft/s), depending somewhat on the design of the mesh. Pressure drop is usually no more than 2.5 cm (1 in) of water. A major disadvantage is the ease with which tars and insoluble solids plug the mesh. The separator can be made to fit vessels of any shape and can be made of any material which can be drawn into a wire. Stainless-steel and plastic fibers are most common, but other metals are sometimes used. Generally three basic types of mesh are used: (1) layers with a crimp in the same direction (each layer is actually a nested double layer); (2) layers with a crimp in

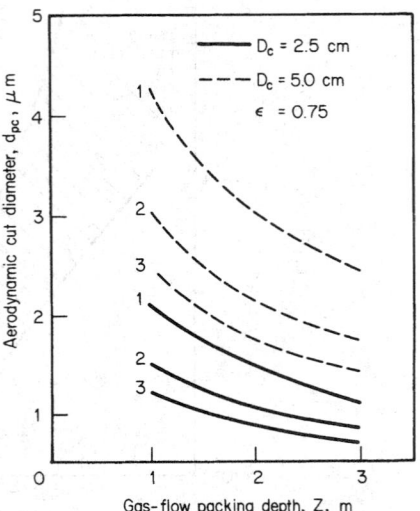

FIG. 14-119 Aerodynamic cut diameter for a typical packed-bed entrainment separator as a function of packing size, bed depth, and three gas velocities: curve 1–1.5 m/s, curve 2–3.0 m/s, and curve 3–4.5 m/s. To convert meters to feet, multiply by 3.281; to convert centimeters to inches, multiply by 0.394. [*Calvert, J. Air Pollut. Control Assoc., 24, 929 (1974).*]

alternate directions, which increases voidage, reduces sheltering and increases target efficiency per layer, and gives a lower pressure drop per unit length; and (3) spiral-wound layers which reduce pressure drop by one-third, but fluid creep may lead to higher entrainment. Some small manufacturers of plastic meshes may offer other weaves claimed to be superior. The filament size can vary from about 0.15 mm (0.006 in) for fine-wire pads to 3.8 mm (0.15 in) for some plastic fibers. Typical pad thickness varies from 100 to 150 mm (4 to 6 in), but occasionally pads up to 300 mm (12 in) thick are used. A typical wire diameter for standard stainless mesh is 0.28 mm (0.011 in), with a finished mesh density of 0.15 g/cm³ (9.4 lb/ft³). A lower mesh density may be produced with standard wire to give 10 to 20 percent higher flow rates.

Figure 14-120 presents an early calculated estimate of mesh efficiency as a fraction of mist-particle size. Experiments by Calvert (R-12) confirm the accuracy of the equation of Bradie and Dickson (*Joint Symp. Proc. Inst. Mech. Eng./Yorkshire Br. Inst. Chem. Eng.,* 1969, pp. 24–25) for primary efficiency in mesh separators:

$$\eta = 1 - \exp(-2/3)\pi a l \eta_i) \qquad (14\text{-}232)$$

where η is the overall collection efficiency for a given-size particle; l is the thickness of the mesh, cm, in the direction of gas flow; a is the surface area of the wires per unit volume of mesh pad, cm²/cm³; and η_i, the target collection efficiency for cylindrical wire, can be calculated from Fig. 17-39 or the impaction data of Golovin and Putnam [*Ind. Eng. Chem.,* **1**, 264 (1962)]. The factor 2/3, introduced by Carpenter and Othmer [*Am. Inst. Chem. Eng. J.,* **1**, 549 (1955)], corrects for the fact that not all the wires are perpendicular to the gas flow and gives the projected perpendicular area. If the specific mesh surface area a is not available, it can be calculated from the mesh void area ε and the mesh wire diameter d_w in cm, $a = 4(1 - \varepsilon)/d_w$.

York and Poppele (R-17) have stated that factors governing maximum allowable gas velocity through the mesh are (1) gas and liquid density, (2) liquid surface tension, (3) liquid viscosity, (4) specific wire surface area, (5) entering-liquid loading, and (6) suspended-solids content. York (R-18) has proposed application of the Souders-Brown equation [Eq. (14-226)] for correlation of maximum allowable gas velocity with values of K for most cases of 0.1067 m/s to give U in m/s (0.35 for ft/s). When liquid viscosity or inlet loading is high or the liquid is dirty, the value of K must be reduced. Schroeder (M.S. thesis, Newark College of Engineering, 1962) found lower values for K necessary when liquid surface tension is reduced such as by the presence of surfactants in water. Ludwig (*Applied Process Design for Chemical and Petrochemical Plants,* 2d ed., vol. I, Gulf, Houston, 1977, p. 157) recommends reduced K values of (0.061 m/s) under vacuum at an absolute pressure of 6.77 kPa (0.98 lbf/in²) and $K = 0.082$ m/s at 54 kPa (7.83 lbf/in²) absolute. Most manufacturers suggest setting the design velocity at three-fourths of the maximum velocity to allow for surges in gas flow.

York and Poppele (R-17) have suggested that total pressure drop through the mesh is equal to the sum of the mesh dry pressure drop

plus an increment due to the presence of liquid. They considered the mesh to be equivalent to numerous small circular channels and used the D'Arcy formula with a modified Reynolds number to correlate friction factor (see Fig. 14-121) for Eq. (14-233) giving dry pressure drop.

$$\Delta P_{\text{dry}} = f l a \rho_g U_g^2 / 981 \, \varepsilon^3 \qquad (14\text{-}233)$$

where ΔP is in cm of water; f is from Fig. (14-121); ρ_g is the gas density, g/cm³; U_g is the superficial gas velocity, cm/s; and ε is the mesh porosity or void fraction; l and a are as defined in Eq. (14-232). Figure 14-121 gives data of York and Poppele for mesh crimped in the same and alternating directions and also includes the data of Satsangee, of Schuring, and of Bradie and Dickson.

The incremental pressure drop for wet mesh is not available for all operating conditions or for mesh of different styles. The data of York and Poppele for wet-mesh incremental pressure drop, ΔP_L in cm of water, are shown in Fig. 14-122 or parameters of liquid velocity L/A, defined as liquid volumetric flow rate, cm³/min per unit of mesh cross-sectional area in cm²; liquid density ρ_L is in g/cm³.

York generally recommends the installation of the mesh horizontally with upflow of gas as in Fig. 14-110f; Calvert (R-12) tested the mesh horizontally with upflow and vertically with horizontal gas flow. He reports better drainage with the mesh vertical and somewhat higher permissible gas velocities without reentrainment, which is contrary to past practice. With horizontal flow through vertical mesh, he found collection efficiency to follow the predictions of Eq. 14-232 up to 4 m/s (13 ft/s) with air and water. Some reentrainment was encountered at higher velocities, but it did not appear serious until velocities exceeded 6.0 m/s (20 ft/s). With vertical upflow of gas, entrainment was encountered at velocities above and below 4.0 m/s (13 ft/s), depending on inlet liquid quantity (see Fig. 14-123). Figure 14-124 illustrates the onset of entrainment from mesh as a function of liquid loading and gas velocity and the safe operating area recommended by Calvert. Measurements of dry pressure drop by Calvert gave values only about one-third of those predicted from Eq. (14-233). He found the pressure drop to be highly affected by liquid load. The pressure drop of wet mesh could be correlated as a function of $U_g^{1.65}$ and parameters of liquid loading L/A, as shown in Fig. 14-125.

As indicated previously, mesh efficiency drops rapidly as particles decrease in size below 5 μm. An alternative is to use two mesh pads in series. The first mesh is made of fine wires and is operated beyond the

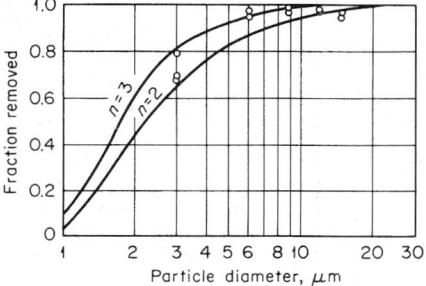

FIG. 14-120 Collection efficiency of wire-mesh separator; 6-in thickness, 98.6 percent free space, 0.006-in-diameter wire used for experiment points. Curves calculated for target area equal to 2 and 3 times the solids volume of packing. To convert inches to millimeters, multiply by 25.4.

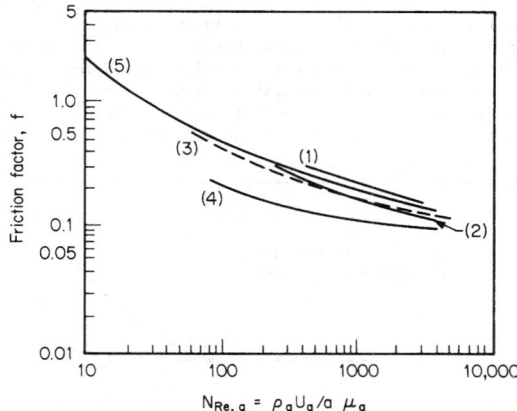

FIG. 14-121 Value of friction factor f for dry knitted mesh for Eq. (14-233). Values of York and Poppele [*Chem. Eng. Prog.,* **50**, 421 (1954)] are given in curve 1 for mesh crimped in the alternating direction and curve 2 for mesh crimped in the same direction. Data of Bradie and Dickson (*Joint Symp. Proc. Inst. Mech. Eng./Yorkshire Br. Inst. Chem. Eng.,* 1969, pp. 24–25) are given in curve 3 for layered mesh and curve 4 for spiral-wound mesh. Curve 5 is data of Satsangee (M.S. thesis, Brooklyn Polytechnic Institute, 1948) and Schurig (D.Ch.E. dissertation, Brooklyn Polytechnic Institute, 1946). (*From Calvert, Yung, and Leung, NTIS Publ. PB-248050, 1975.*)

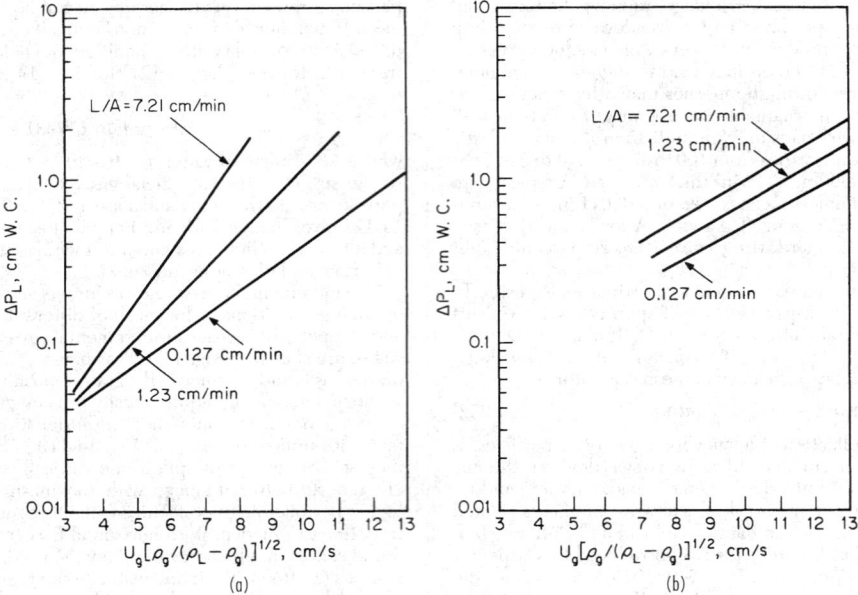

(a)   (b)

FIG. 14-122 Incremental pressure drop in knitted mesh due to the presence of liquid (*a*) with the mesh crimps in the same direction and (*b*) with crimps in the alternating direction, based on the data of York and Poppele [*Chem. Eng. Prog.*, **50**, 421 (1954)]. To convert centimeters per minute to feet per minute, multiply by 0.0328; to convert centimeters per second to feet per second, multiply by 0.0328. (*From Calvert, Yung, and Leung, NTIS Publ. PB-248050, 1975.*)

flood point. It results in droplet coalescence, and the second mesh, using standard wire and operated below flooding, catches entrainment from the first mesh. Coalescence and flooding in the first mesh may be assisted with water sprays or irrigation. Massey [*Chem. Eng. Prog.*, **53**(5), 114 (1959)] and Coykendall et al. [*J. Air Pollut. Control Assoc.*, **18**, 315 (1968)] have discussed such applications. Calvert (R-12) presents data on the particle size of entrained drops from mesh as a function of gas velocity which can be used for sizing the secondary collector. A major disadvantage of this approach is high pressure drop, which can be in the range from 25 cm (10 in) of water to as high as 85 cm (33 in) of water if the mist is mainly submicrometer.

Wet Scrubbers Scrubbers have not been widely used for the collection of purely liquid particulate, probably because they are generally more complex and expensive than impaction devices of the types previously discussed. Further, scrubbers are no more efficient than the former devices for the same energy consumption. However,

scrubbers of the types discussed in Sec. 17 and illustrated in Figs. 17-48 to 17-55 can be used to capture liquid particles efficiently. Their use is primarily indicated when it is desired to accomplish simultaneously another task such as gas absorption or the collection of solid and liquid particulate mixtures.

Table 20-41 [*Chemical Engineers' Handbook*, 5th ed.)], showing

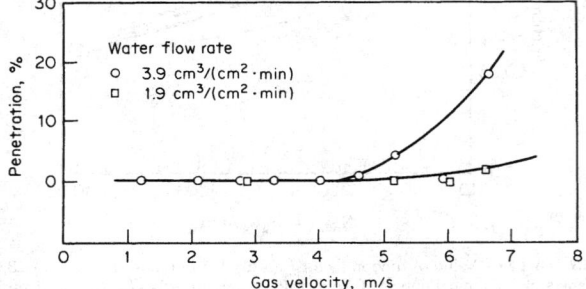

FIG. 14-123 Experimental data of Calvert with air and water in mesh with vertical upflow, showing the effect of liquid loading on efficiency and reentrainment. To convert meters per second to feet per second, multiply by 3.281; to convert cubic centimeters per square centimeter-minute to cubic feet per square foot-minute, multiply by 0.0328. (*Calvert, Yung, and Leung, NTIS Publ. PB-248050, 1975.*)

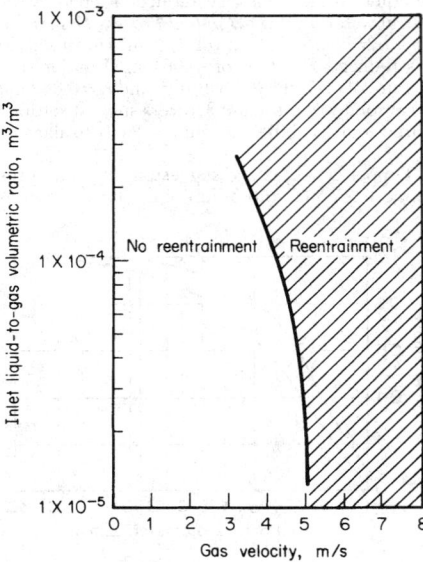

FIG. 14-124 Effect of gas and liquid rates on onset of mesh reentrainment and safe operating regions. To convert meters per second to feet per second, multiply by 3.281. (*Calvert, Yung, and Leung, NTIS Publ. PB-248050, 1975.*)

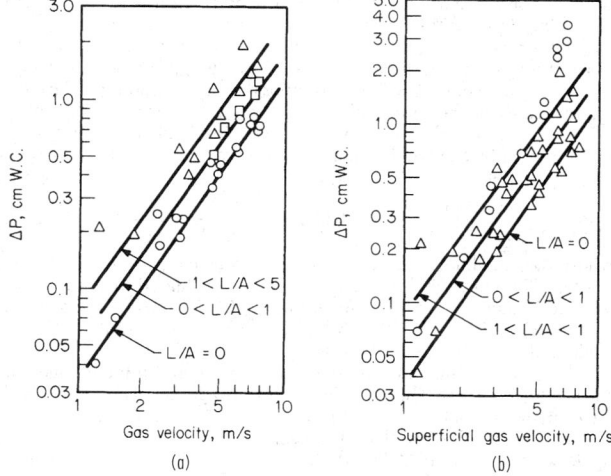

FIG. 14-125 Experimental pressure measured by Calvert as a function of gas velocity and liquid loading for (a) horizontal gas flow through vertical mesh and (b) gas upflow through horizontal mesh. Mesh thickness was 10 cm with 2.8-mm wire and void fraction of 98.2 percent, crimped in alternating directions. To convert meters per second to feet per second, multiply by 3.281; to convert centimeters to inches, multiply by 0.394. (*Calvert, Yung, and Leung, NTIS Publ. PB-248050, 1975.*)

the minimum size of particles collectible in different types of scrubbers at reasonably high efficiencies, is a good selection guide. Cyclonic spray towers (Fig. 17-52) can effectively remove liquid particles down to around 2 to 3 μm. Figures 20-112 and 20-113 (*Chemical Engineers' Handbook,* 5th ed.), giving target efficiency between spray drop size and particle size as calculated by Stairmand or Johnstone and Roberts, should be considered in selecting spray atomization for the most efficient tower operation. Figure 14-126 gives calculated particle cut size as a function of tower height (or length) for vertical countercurrent spray towers and for horizontal-gas-flow, vertical-liquid-flow cross-current spray towers with parameters for liquid drop size. These curves are based on physical properties of standard air and water and should be used under conditions in which these are reasonable

approximations. Lack of uniform liquid distribution or liquid flowing down the walls can affect the performance, requiring empirical correction factors. Calvert (R-10) suggests that a correction factor of 0.2 be used in small-diameter scrubbers to account for the liquid on the walls, i.e., let $Q_L/Q_g = 0.2 (Q_L/Q_g)_{actual}$. Many more complicated wet scrubbers employ a combination of sprays or liquid atomization, cyclonic action, baffles, and targets. These combinations are not likely to be more efficient than similar devices previously discussed that operate at equivalent pressure drop. The vast majority of wet scrubbers operate at moderate pressure drop [8 to 15 cm (3 to 6 in) of water or 18 to 30 cm (7 to 12 in) of water] and cannot be expected to have high efficiency on particles smaller than 10 μm or 3 to 5 μm respectively. Fine and submicrometer particles can be captured efficiently only in wet scrubbers having high energy input such as venturi scrubbers, two-phase eductor scrubbers, and flux-force-condensation scrubbers.

Venturi Scrubbers   One type of venturi scrubber is illustrated in Fig. 17-48. Venturi scrubbers have been used extensively for collecting fine and submicrometer solid particulate, condensing tars and mists, and mixtures of liquids and solids. To a lesser extent, they have also been used for simultaneous gas absorption, although Lundy [*Ind. Eng. Chem.,* **50,** 293 (1958)] indicates that they are generally limited to three transfer units. They have been used to collect submicrometer chemical incinerator fume and mist as well as sulfuric and phosphoric acid mists. The collection efficiency of a venturi scrubber is highly dependent on the throat velocity or pressure drop, the liquid-to-gas ratio, and the chemical nature of wettability of the particulate. Throat velocities may range from 60 to 150 m/s (200 to 500 ft/s). Liquid injection rates are typically 0.67 to 1.4 m³/1000 m³ of gas. A liquid rate of 1.0 m³ per 1000 m³ of gas is usually close to optimum, but liquid rates as high as 2.7 m³ (95 ft³) have been used. Efficiency improves with increased liquid rate but only at the expense of higher pressure drop and energy consumption. Pressure-drop predictions for a given efficiency are hazardous without determining the nature of the particulate and the liquid-to-gas ratio. In general, particles coarser than 1 μm can be collected efficiently with pressure drops of 25 to 50 cm of water. For appreciable collection of submicrometer particles, pressure drops of 75 to 100 cm (30 to 40 in) of water are usually required. When particles are appreciably finer than 0.5 μm, pressure drops of 175 to 250 cm (70 to 100 in) of water have been used.

One of the problems in predicting efficiency and required pressure drop of a venturi is the chemical nature or wettability of the particulate, which on 0.5-μm-size particles can make up to a threefold difference in required pressure drop for its efficient collection. Calvert

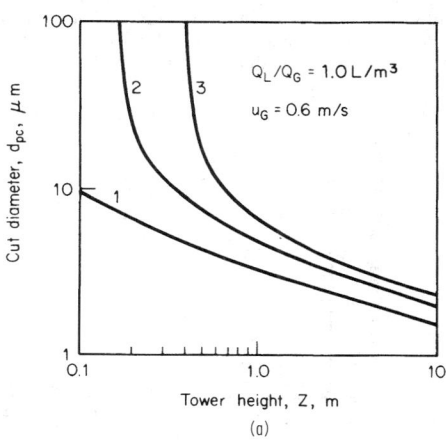

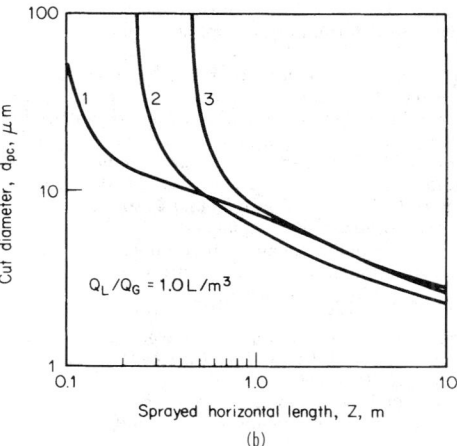

FIG. 14-126 Predicted spray-tower cut diameter as a function of sprayed length and spray droplet size for (a) vertical-countercurrent towers and (b) horizontal-cross-flow towers per Calvert [J. Air Pollut. Control Assoc., **24,** 929 (1974)]. Curve 1 is for 200-μm spray droplets, curve 2 for 500-μm spray, and curve 3 for 1000-μm spray. Q_L/Q_C is the volumetric liquid-to-gas ratio, L liquid/m³ gas, and u_G is the superficial gas velocity in the tower. To convert liters per cubic meter to cubic feet per cubic foot, multiply by 10^{-3}.

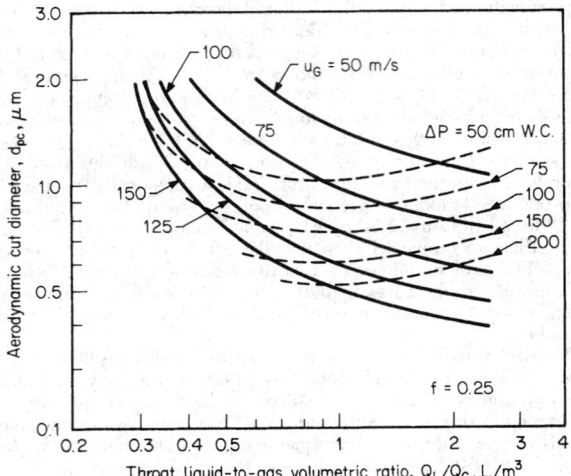

FIG. 14-127 Prediction of venturi-scrubber cut diameter for hydrophobic particles as functions of operating parameters as measured by Calvert [*Calvert, Goldshmid, Leith, and Mehta, NTIS Publ. PB-213016, 213017, 1972; and Calvert, J. Air Pollut. Control Assoc., 24, 929 (1974).*] u_G is the superficial throat velocity, and ΔP is the pressure drop from converging to diverging section. To convert meters per second to feet per second, multiply by 3.281; to convert liters per cubic meter to cubic feet per cubic foot, multiply by 10^{-3}; and to convert centimeters to inches, multiply by 0.394.

(R-9, R-10) has represented this effect by an empirical factor f, which is based on the hydrophobic ($f = 0.25$) or hydrophilic ($f = 0.50$) nature of the particles. Figure 14-127 gives the cut diameter of a venturi scrubber as a function of its operating parameters (throat velocity, pressure drop, and liquid-to-gas ratio) for hydrophobic particles. Figure 14-129 compares cut diameter as a function of pressure drop for an otherwise identically operating venturi on hydrophobic and hydrophilic particles. Calvert (R-9) gives equations which can be used for constructing cut-size curves similar to those of Fig. 14-127 for other values of the empirical factor f. Most real particles are neither completely hydrophobic nor completely hydrophilic but have f values lying between the two extremes. Phosphoric acid mist, on the basis of data of Brink and Contant [*Ind. Eng. Chem.*, 50, 1157 (1958)] appears to have a value of $f = 0.46$. Unfortunately, no chemical-test methods have yet been devised for determining appropriate f values for a particulate in the laboratory.

Pressure drop in a venturi scrubber is controlled by throat velocity. While some venturis have fixed throats, many are designed with variable louvers to change throat dimensions and control performance for changes in gas flow. Pressure-drop equations have been developed by Calvert (R-13, R-14, R-15), Boll [*Ind. Eng. Chem. Fundam.*, 12, 40 (1973)], and Hesketh [*J. Air Pollut. Control Assoc.*, 24, 939 (1974)]. Hollands and Goel [*Ind. Eng. Chem. Fundam.*, 14, 16 (1975)] have developed a generalized pressure-drop equation.

The Hesketh equation is empirical and is based upon a regression analysis of data from a number of industrial venturi scrubbers:

$$\Delta P = U_{gt}^2 \rho_g A_t^{0.155} L^{0.78}/1270 \qquad (14\text{-}234)$$

where ΔP is the pressure drop, in of water; U_{gt} is the gas velocity in the throat, ft/s; ρ_g is the gas density, lb/ft^3; A_t is the throat area, ft^2; and L is the liquid-to-gas ratio, gal/1000 acf.

Calvert (R-15) critiqued the many pressure-drop equations and suggested the following simplified equation as accurate to ±10 percent:

$$\Delta P = \frac{2\rho_\ell U_g^2}{981 g_c}\left(\frac{Q_\ell}{Q_g}\right)[1 - x^2 + \sqrt{(x^4 - x^2)^{0.5}}] \qquad (14\text{-}235)$$

where $\qquad\qquad x = (3l_t C_{Di}\rho_g/16d_l\rho_l) + 1 \qquad (14\text{-}236)$

ΔP is the pressure drop, cm of water; ρ_ℓ and ρ_g are the density of the scrubbing liquid and gas respectively, g/cm^3; U_g is the velocity of the gas at the throat inlet, cm/s; Q_ℓ/Q_g is the volumetric ratio of liquid to gas at the throat inlet, dimensionless; l_t is the length of the throat, cm; C_{Di} is the drag coefficient, dimensionless, for the mean liquid diameter, evaluated at the throat inlet; and d_l is the Sauter mean diameter, cm, for the atomized liquid. The atomized-liquid mean diameter must be evaluated by the Nukiyama and Tanasawa [*Trans. Soc. Mech Eng.* (*Japan*), 4, 5, 6 (1937–1940)] equation:

$$d_\ell = \frac{0.0585}{U_g}\left(\frac{\sigma_\ell}{\rho_\ell}\right)^{0.5} + 0.0597\left[\frac{\mu_\ell}{(\sigma_\ell\rho_\ell)^{0.5}}\right]^{0.45}\left(\frac{Q_\ell}{Q_g}\right)^{1.5} \qquad (14\text{-}237)$$

where σ_ℓ is the liquid surface tension, dyn/cm; and μ_ℓ is the liquid viscosity; P. The drag coefficient C_{Di} should be evaluated by the Dickinson and Marshall [*Am. Inst. Chem. Eng. J.*, 14, 541 (1968)] correlation $C_{Di} = 0.22 + (24/N_{Rei})(1 + 0.15 N_{Rei}^{0.6})$. The Reynolds number, N_{Rei}, is evaluated at the throat inlet considerations as $d_\ell G_g/\mu_g$.

All venturi scrubbers must be followed by an entrainment collector for the liquid spray. These collectors are usually centrifugal and will have an additional pressure drop of several centimeters of water, which must be added to that of the venturi itself.

Other Scrubbers   A liquid-ejector venturi (Fig. 17-53), in which high-pressure water from a jet induces the flow of gas, has been used to collect mist particles in the 1- to 2-µm range, but submicrometer particles will generally pass through an eductor. Power costs for liquid pumping are high if appreciable motive force must be imparted to the gas because jet-pump efficiency is usually less than 10 percent. Harris [*Chem. Eng. Prog.*, 42(4), 55 (1966)] has described their application. Two-phase eductors have been considerably more successful on capture of submicrometer mist particles and could be attractive in situations in which large quantities of waste thermal energy are available. However, the equivalent energy consumption is equal to that required for high-energy venturi scrubbers, and such devices are likely to be no more attractive than venturi scrubbers when the thermal energy is priced at its proper value. Sparks [*J. Air Pollut. Control Assoc.*, 24, 958 (1974)] has discussed steam ejectors giving 99 percent collection of particles 0.3 to 10 µm. Energy requirements were 311,000 J/m^3(8.25 Btu/scf). Gardenier [*J. Air Pollut. Control Assoc.*, 24, 954 (1974)] operated a liquid eductor with high-pressure (6900- to 27,600-kPa) (1000- to 4000-lbf/in^2) hot water heated to 200°C (392°F) which flashed into two phases as it issued from the jet. He obtained 95 to 99 percent collection of submicrometer particulate. Figure 14-128 shows the water-to-gas ratio required as a function of particle size to achieve 99 percent collection.

Effect of Gas Saturation in Scrubbing   If hot unsaturated gas is introduced into a wet scrubber, spray particles will evaporate to cool and saturate the gas. The evaporating liquid molecules moving away from the target droplets will repel particles which might collide with them. This results in the forces of diffusiophoresis opposing particle

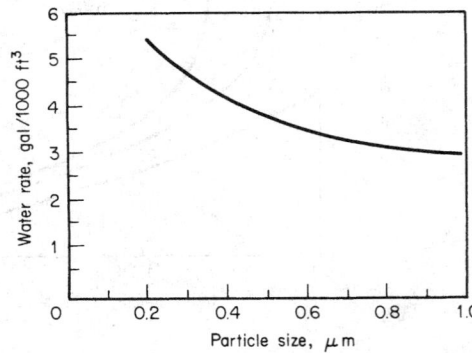

FIG. 14-128 Superheated high-pressure hot-water requirements for 99 percent collection as a function of particle size in a two-phase eductor jet scrubber. To convert gallons per 1000 cubic feet to cubic meters per 1000 cubic meters, multiply by 0.134. [*Gardenier, J. Air Pollut. Control Assoc., 24, 954 (1974).*]

collection. Semrau and Witham (Air Pollut. Control Assoc. Prepr. 75-30.1) investigated temperature parameters in wet scrubbing and found a definite decrease in the efficiency of evaporative scrubbers and an enhancement of efficiency when a hot saturated gas is scrubbed with cold water rather than recirculated hot water. Little improvement was experienced in cooling a hot saturated gas below a 50°C dew point.

Energy Requirements for Inertial-Impaction Efficiency Semrau [*J. Air Pollut. Control Assoc.*, **13**, 587 (1963)] proposed a "contacting-power" principle which states that the collecting efficiency of a given size of particle is proportional to the power expended and that the smaller the particle, the greater the power required. Mathematically expressed, $N_T = \propto P_T^\gamma$, where N_T is the number of particulate transfer units achieved and P_T is the total energy expended within the collection device, including gas and liquid pressure drop and thermal and mechanical energy added in atomizers. N_T is further defined as $N_T = \ln [1/(1 - \eta)]$, where η is the overall fractional collection efficiency. This was intended as a universal principle, but the constants $\propto$ and γ have been found to be functions of the chemical nature of the system and the design of the control device. Others have pointed out that the principle is applicable only when the primary collection mechanism is impaction and direct interception. Calvert (R-10, R-12) has found that plotting particle cut size versus pressure drop (or power expended) as in Fig. 18-129 is a more suitable way to develop a generalized energy-requirement curve for impaction devices. The various curves fall close together and outline an imaginary curve that indicates the magnitude of pressure drop required as particle size decreases bound by the two limits of hydrophilic and hydrophobic particles. By calculating the required cut size for a given collection efficiency, Fig. 14-129. can also be used as a guide to deciding between different collection devices.

Subsequently, Calvert (R-19, p. 228) has combined mathematical modeling with performance tests on a variety of industrial scrubbers and has obtained a refinement of the power-input/cut-size relationship as shown in Fig. 14-130. He considers these relationships sufficiently reliable to use this data as a tool for selection of scrubber type and performance prediction. The power input for this figure is based solely on gas pressure drop across the device.

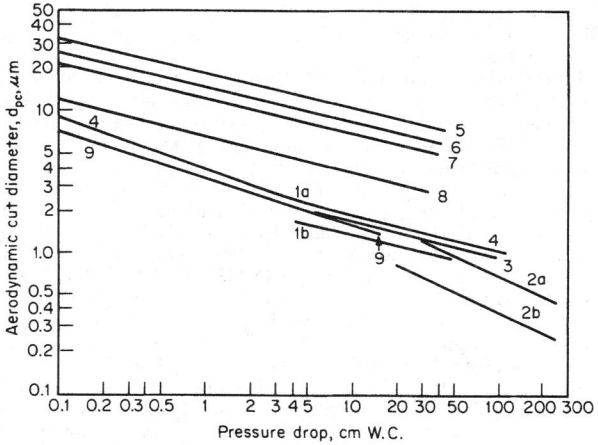

FIG. 14-129 Typical cut diameter as a function of pressure drop for various liquid-particle collectors. Curves 1*a* and *b* are single-sieve plates with froth density of 0.4 g/cm³; 1*a* has sieve holes of 0.5 cm and 1*b* holes of 0.3 cm. Curves 2*a* and *b* are for a venturi scrubber with hydrophobic particles (2*a*) and hydrophilic particles (2*b*). Curve 3 is an impingement plate, and curve 4 is a packed column with 2.5-cm-diameter packing. Curve 5 is a zigzag baffle collector with six baffles at θ = 30°. Curve 7 is for six rows of staggered tubes with 1-cm spacing between adjacent tube walls in a row. Curve 8 is similar, except that tube-wall spacing in the row is 0.3 cm. Curve 9 is for wire-mesh pads. To convert grams per cubic centimeter to pounds per cubic foot, multiply by 62.43; to convert centimeters to inches, multiply by 0.394. [*Calvert, J. Air Pollut. Control Assoc.*, **24**, 929 (1974); and Calvert, Yung, and Leung, NTIS Publ. PB-248050, 1975.]

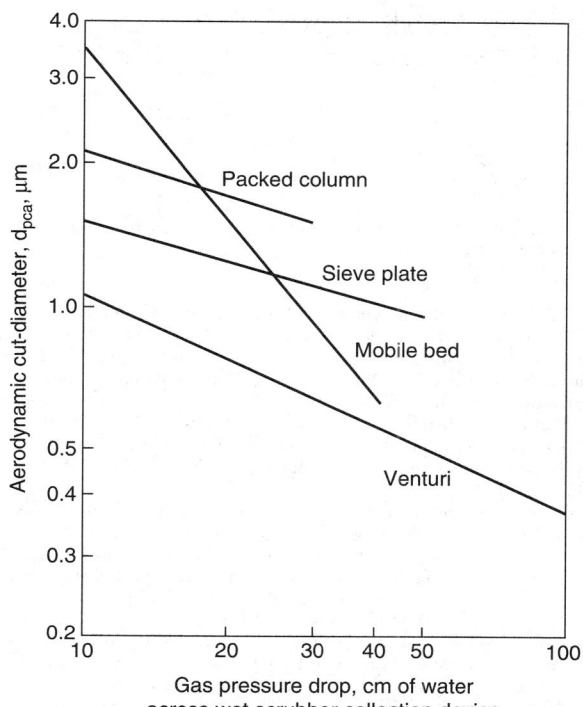

FIG. 14-130 Calvert's refined particle cut-size/power relationship for particle inertial impaction wet collectors. Ref. (R-19) by permission.

Collection of Fine Mists Inertial-impaction devices previously discussed give high efficiency on particles above 5 μm in size and often reasonable efficiency on particles down to 3 μm in size at moderate pressure drops. However, this mechanism becomes ineffective for particles smaller than 3 μm because of the particle gaslike mobility. Only impaction devices having extremely high energy input such as venturi scrubbers and a flooded mesh pad (the pad interstices really become miniature venturi scrubbers in parallel and in series) can give high collection efficiency on fine particles, defined as 2.5 or 3 μm and smaller, including the submicrometer range. Fine particles are subjected to brownian motion in gases, and diffusional deposition can be employed for their collection. Diffusional deposition becomes highly efficient as particles become smaller, especially below 0.2 to 0.3 μm. Table 14-20 shows typical displacement velocity of particles. Randomly oriented fiber beds having tortuous and narrow gas passages are suitable devices for utilizing this collection mechanism. (The diffusional collection is discussed in Sec. 17 under "Gas-Solids Separations.") Other collection mechanisms which are efficient for fine particles are electrostatic forces and flux forces such as thermophoresis and diffusiophoresis. Particle growth and nucleation methods are also applicable. Efficient collection of fine particles is important because particles in the range of 2.0 to around 0.2 μm are the ones which penetrate and are deposited in the lung most efficiently. Hence, particles in this range constitute the largest health hazard.

Fiber Mist Eliminators These devices are produced in various configurations. Generally, randomly oriented glass or polypropylene fibers are densely packed between reinforcing screens, producing fiber beds varying in thickness usually from 25 to 75 mm (1 to 3 in), although thicker beds can be produced. Units with efficiencies as high as 99.9 percent on fine particles have been developed (see *Chemical Engineers' Handbook*, 5th ed., p. 18–88). A combination of mechanisms interacts to provide high overall collection efficiency. Particles larger than 2 to 3 μm are collected on the fibers by inertial impaction

TABLE 14-20 Brownian Movement of Particles*

Particle diameter, μm	Brownian displacement of particle, μm/s
0.1	29.4
0.25	14.2
0.5	8.92
1.0	5.91
2.5	3.58
5.0	2.49
10.0	1.75

*Brink, *Can. J. Chem. Eng.*, **41**, 134 (1963). Based on spherical water particles in air at 21°C and 1 atm.

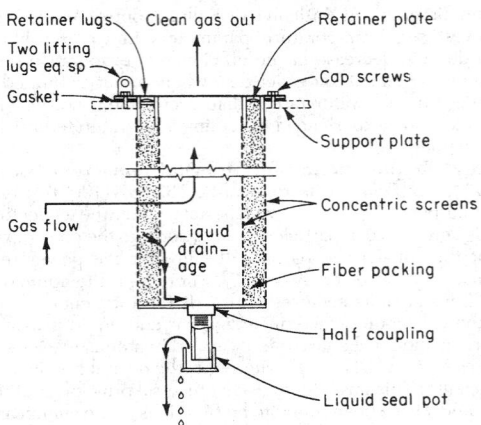

FIG. 14-131 Monsanto high-efficiency fiber-mist-eliminator element. (*Monsanto Company.*)

and direct interception, while small particles are collected by brownian diffusion. When the device is designed to use this latter mechanism as the primary means, efficiency turndown problems are eliminated as collection efficiency by diffusion increases with residence time. Pressure drop through the beds increases with velocity to the first power since the gas flow is laminar. This leads to design capability trade-offs. As pressure drop is reduced and energy is conserved, capital increases because more filtering area is required for the same efficiency.

Three series of fiber mist eliminators are typically available. A spray-catcher series is designed primarily for essentially 100 percent capture of droplets larger than 3 μm. The high-velocity type is designed to give moderately high efficiency on particles down to 1.0 μm as well. Both of these types are usually produced in the form of flat panels of 25- to 50-mm (1- to 2-in) thickness. The high-efficiency type is illustrated in Fig. 14-131. As mist particles are collected, they coalesce into a liquid film which wets the fibers. Liquid is moved horizontally through the bed by the gas drag force and downward by gravity. It drains down the downstream retaining screen to the bottom of the element and is returned to the process through a liquid seal. Table 14-21 gives typical operating characteristics of the three types of collectors. The application of these devices to sulfuric acid plants and other process gases has been discussed by Brink (see *Chemical Engineers' Handbook*, 5th ed., pp. 18–89, 18–90).

Solid particulates are captured as readily as liquids in fiber beds but can rapidly plug the bed if they are insoluble. Fiber beds have frequently been used for mixtures of liquids and soluble solids and with soluble solids in condensing situations. Sufficient solvent (usually water) is atomized into the gas stream entering the collector to irrigate the fiber elements and dissolve the collected particulate. Such fiber beds have been used to collect fine fumes such as ammonium nitrate and ammonium chloride smokes, and oil mists from compressed air.

Electrostatic Precipitators The principles and operation of electrical precipitators are discussed in Sec. 17 under "Gas-Solids Separations." Precipitators are admirably suited to the collection of fine mists and mixtures of mists and solid particulates. Tube-type precipitators have been used for many years for the collection of acid mists and the removal of tar from coke-oven gas. The first practical installation of a precipitator by Cottrell was made on sulfuric acid mist in 1907. Most older installations of precipitators were tube-type rather than plate-type. However, recently two plate-type wet precipitators employing water sprays or overflowing weirs have been introduced by Mikropul Corporation [Bakke, *J. Air Pollut. Control Assoc.*, **25**, 163 (1975)] and by Fluid Ionics. Such precipitators operate on the principle of making all particles conductive when possible, which increases the particle migration velocity and collection efficiency. Under these conditions, particle dielectric strength becomes a much more impor-

tant variable, and particles with a low dielectric constant such as condensed hydrocarbon mists become much more difficult to collect than water-wettable particles. Bakke (U.S.–U.S.S.R. Joint Work. Group Symp.: Fine Particle Control, San Francisco, 1974) has developed equations for particle charge and relative collection efficiency in wet precipitators that show the effect of dielectric constant. Wet precipitators can also be used to absorb soluble gases simultaneously by adjusting the pH or the chemical composition of the liquid spray. The presence of the electric field appears to enhance absorption. Wet precipitators have found their greatest usefulness to date in handling mixtures of gaseous pollutants and submicrometer particulate (either liquid or solid, or both) such as fumes from aluminum-pot lines, carbon anode baking, fiberglass-fume control, coke-oven and metallurgical operations, chemical incineration, and phosphate-fertilizer operations. Two-stage precipitators are used increasingly for moderate-volume gas streams containing nonconductive liquid mists which will drain from the collecting plates. Their application on hydrocarbon mists has been quite successful, but careful attention must be given to fire and explosion hazards.

Electrically Augmented Collectors A new area for enhancing collection efficiency and lowering cost is the combining of electrostatic forces with devices using other collecting mechanisms such as impaction and diffusion. Cooper (Air Pollut. Control Assoc. Prepr. 75-02.1) evaluated the magnitude of forces operating between charged and uncharged particles and concluded that electrostatic attraction is the strongest collecting force operating on particles finer than 2 μm. Nielsen and Hill [*Ind. Eng. Chem. Fundam.*, **15**, 149 (1976)] have quantified these relationships, and a number of practical devices have been demonstrated. Pilat and Meyer (NTIS Publ. PB-252653, 1976) have demonstrated up to 99 percent collection of fine particles in a two-stage spray tower in which the inlet particles and water spray are charged with opposite polarity. The principle has been applied to retrofitting existing spray towers to enhance collection.

Klugman and Sheppard (Air Pollut. Control Assoc. Prepr. 75-30.3) have developed an ionizing wet scrubber in which the charged mist particles are collected in a grounded, irrigated cross-flow bed of Tellerette packing. Particles smaller than 1 μm have been collected

TABLE 14-21 Operating Characteristics of Various Types of Fiber Mist Eliminators as Used on Sulfuric Acid Plants*

	High efficiency	High velocity	Spray catcher
Controlling mechanism for mist collection	Brownian movement	Impaction	Impaction
Superficial velocity, m/s	0.075–0.20	2.0–2.5	2.0–2.5
Efficiency on particles greater than 3 μm, %	Essentially 100	Essentially 100	Essentially 100
Efficiency on particles 3 μm and smaller, %	95–99+	90–98	15–30
Pressure drop, cm H₂O	12–38	15–20	1.0–2.5

*Brink, Burggrabe, and Greenwell, *Chem. Eng. Prog.*, **64**(11), 82 (1968). To convert centimeters to inches, multiply by 0.394.

with 98 percent efficiency by using two units in series. Dembinsky and Vicard (Air Pollut. Control Assoc. Prepr. 78-17.6) have used an electrically augmented low-pressure [5 to 10 cm (2 to 4 in) of water] venturi scrubber to give 95 to 98 percent collection efficiency on submicrometer particles.

Particle Growth and Nucleation   Fine particles may be subjected to conditions favoring the growth of particles either through condensation or through coalescence. Saturation of a hot gas stream with water, followed by condensation on the particles acting as nuclei when the gas is cooled, can increase particle size and ease of collection. Addition of steam can produce the same results. Scrubbing of the humid gas with a cold liquid can bring diffusiophoresis into play. The introduction of cold liquid drops causes a reduction in water-vapor pressure at the surface of the cold drop. The resulting vapor-pressure gradient causes a hydrodynamic flow toward the drop known as Stefan flow which enhances the movement of mist particles toward the spray drop. If the molecular mass of the diffusing vapor is different from the carrier gas, this density difference also produces a driving force, and the sum of these forces is known as diffusiophoresis. A mathematical description of these forces has been presented by Calvert (R-9) and by Sparks and Pilat [*Atmos. Environ.*, **4**, 651 (1970)]. Thermal differences between the carrier gas and the cold scrubbing droplets can further enhance collection through thermophoresis. Calvert and Jhaseri [*J. Air Pollut. Control Assoc.*, **24**, 946 (1974)]; and NTIS Publ. PB-227307, 1973)] have investigated condensation scrubbing in multiple-sieve plate towers.

Submicrometer droplets can be coagulated through brownian diffusion if given ample time. The introduction of particles 50 to 100 times larger in diameter can enhance coagulation, but the addition of a broad range of particle sizes is discouraged. Increasing turbulence will aid coagulation, so fans to stir the gas or narrow, tortuous passages such as those of a packed bed can be beneficial. Sonic energy can also produce coagulation, especially the production of standing waves in the confines of long, narrow tubes. Addition of water and oil mists can sometimes aid sonic coagulation. Sulfuric acid mist [Danser, *Chem. Eng.*, **57**(5), 158 (1950)] and carbon black [Stokes, *Chem. Eng. Prog.*, **46**, 423 (1950)] have been successfully agglomerated with sonic energy. Frequently sonic agglomeration has been unsuccessful because of the high energy requirement. Most sonic generators have very poor energy-transformation efficiency. Wegrzyn et al. (U.S. EPA Publ. EPA-600/7-79-004C, 1979, p. 233) have reviewed acoustic agglomerators. Mednikov (U.S.S.R. Akad. Soc. Moscow, 1963) suggested that the incorporation of sonic agglomeration with electrostatic precipitation could greatly reduce precipitator size.

Other Collectors   Tarry particulates and other difficult-to-handle liquids have been collected on a dry, expendable phenol formaldehyde-bonded glass-fiber mat (Goldfield, *J. Air Pollut. Control Assoc.*, **20**, 466 (1970)] in roll form which is advanced intermittently into a filter frame. Superficial gas velocities are 2.5 to 3.5 m/s (8.2 to 11.5 ft/s), and pressure drop is typically 41 to 46 cm (16 to 18 in) of water. Collection efficiencies of 99 percent have been obtained on submicrometer particles. Brady [*Chem. Eng. Prog.*, **73**(8), 45 (1977)] has discussed a cleanable modification of this approach in which the gas is passed through a reticulated foam filter that is slowly rotated and solvent-cleaned.

In collecting very fine (mainly submicron) mists of a hazardous nature where one of the collectors previously discussed has been used as the primary one (fiber-mist eliminators of the Brownian diffusion type and electrically augmented collectors are primarily recommended), there is the chance that the effluent concentration may still be too high for atmospheric release when residual concentration must be in the range of 1–2 μm. In such situations, secondary treatment may be needed. Probably removal of the residual mist by adsorption will be in order. See "Adsorption," Sec. 16. Another possibility might be treatment of the remaining gas by membrane separation. A separator having a gas-permeable membrane that is essentially nonliquid-permeable could be useful. However, if the gas-flow volumes are appreciable, the device could be expensive. Most membranes have low capacity (requiring high membrane surface area) to handle high gas-permeation capacity. See "Membrane Processes," Sec. 22.

Continuous Phase Uncertain   Some situations exist such as in

two-phase gas-liquid flow where the volume of the liquid phase may approach being equal to the volume of the vapor phase, and where it may be difficult to be sure which phase is the continuous phase. Svrcek and Monnery [*Chem. Eng. Prog.*, **89**(10), 53–60 (Oct. 1993)] have discussed the design of two-phase separation in a tank with gas-liquid separation in the middle, mist elimination in the top, and entrained gas-bubble removal from the liquid in the bottom. Monnery and Svrcek [*Chem. Eng. Prog.*, **90**(9), 29–40 (Sept. 1994)] have expanded the separation to include multiphase flow, where the components are a vapor and two immiscible liquids and these are also separated in a tank. A design approach for sizing the gas-liquid disengaging space in the vessel is given using a tangential tank inlet nozzle, followed by a wire mesh mist eliminator in the top of the vessel for final separation of entrained mist from the vapor. Design approaches and equations are also given for sizing the lower portion of the vessel for separation of the two immiscible liquid phases by settling and separation of discontinuous liquid droplets from the continuous liquid phase.

LIQUID-PHASE CONTINUOUS SYSTEMS

Practical separation techniques for gases dispersed in liquids are discussed. Processes and methods for dispersing gas in liquid have been discussed earlier in this section, together with information for predicting the bubble size produced. Gas-in-liquid dispersions are also produced in chemical reactions and electrochemical cells in which a gas is liberated. Such dispersions are likely to be much finer than those produced by the dispersion of a gas. Dispersions may also be unintentionally created in the vaporization of a liquid.

GENERAL REFERENCES: Adamson, *Physical Chemistry of Surfaces*, 4th ed., Wiley, New York, 1982. Akers, *Foams*, Academic, New York, 1976. Bikerman, *Foams*, Springer-Verlag, New York, 1973. Bikerman, et al., *Foams: Theory and Industrial Applications*, Reinhold, New York, 1953. Cheremisinoff, ed., *Encyclopedia of Fluid Mechanics*, vol. 3, Gulf Publishing, Houston, 1986. Kerner, *Foam Control Agents*, Noyes Data Corp, Park Ridge, NJ, 1976. Rubel, *Antifoaming and Defoaming Agents*, Noyes Data Corp., Park Ridge, NJ, 1972. Rosen, *Surfactants and Interfacial Phenomena*, 2d ed., Wiley, New York, 1989. Sonntag and Strenge, *Coagulation and Stability of Disperse Systems*, Halsted-Wiley, New York, 1972. Wilson, ed., *Foams: Physics, Chemistry and Structure*, Springer-Verlag, London, 1989. "Defoamers" and "Foams," *Encyclopedia of Chemical Technology*, 4th ed., vols. 7, 11, Wiley, New York, 1993–1994.

Types of Gas-in-Liquid Dispersions   Two types of dispersions exist. In one, gas bubbles produce an unstable dispersion which separates readily under the influence of gravity once the mixture has been removed from the influence of the dispersing force. Gas-liquid contacting means such as bubble towers and gas-dispersing agitators are typical examples of equipment producing such dispersions. More difficulties may result in separation when the gas is dispersed in the form of bubbles only a few micrometers in size. An example is the evolution of gas from a liquid in which it has been dissolved or released through chemical reaction such as electrolysis. Coalescence of the dispersed phase can be helpful in such circumstances.

The second type is a stable dispersion, or foam. Separation can be extremely difficult in some cases. A pure two-component system of gas and liquid cannot produce dispersions of the second type. Stable foams can be produced only when an additional substance is adsorbed at the liquid-surface interface. The substance adsorbed may be in true solution but with a chemical tendency to concentrate in the interface such as that of a surface-active agent, or it may be a finely divided solid which concentrates in the interface because it is only poorly wetted by the liquid. Surfactants and proteins are examples of soluble materials, while dust particles and extraneous dirt including traces of nonmiscible liquids can be examples of poorly wetted materials.

Separation of gases and liquids always involves coalescence, but enhancement of the rate of coalescence may be required only in difficult separations.

Separation of Unstable Systems   The buoyancy of bubbles suspended in liquid can frequently be depended upon to cause the bubbles to rise to the surface and separate. This is a special case of gravity settling. The mixture is allowed to stand at rest or is moved along a

Bubble diameter, μm	10	30	50	100	200	300
Terminal velocity, mm/s	0.061	0.488	1.433	5.486	21.95	49.38

*Calculated from Stokes' law. To convert millimeters per second to feet per second, multiply by 0.003281.

flow path in laminar flow until the bubbles have surfaced. Table 14-22 shows the calculated rate of rise of air bubbles at atmospheric pressure in water at 20°C (68°F) as a function of diameter. It will be observed that the velocity of rise for 10-μm bubbles is very low, so that long separating times would be required for gas which is more finely dispersed.

For liquids other than water, the rise velocity can be approximated from Table 14-22 by multiplying by the liquid's specific gravity and the reciprocal of its viscosity (in centipoises). For bubbles larger than 100 μm, this procedure is erroneous, but the error is less than 15 percent for bubbles up to 1000 μm. More serious is the underlying assumption of Table 14-22 that the bubbles are rigid spheres. Circulation within the bubble causes notable increases in velocity in the range of 100 μm to 1 mm, and the flattening of bubbles 1 cm and larger appreciably decreases their velocity. However, in this latter size range the velocity is so high as to make separation a trivial problem.

In design of separating chambers, static vessels or continuous-flow tanks may be used. Care must be taken to protect the flow from turbulence, which could cause back mixing of partially separated fluids or which could carry unseparated liquids rapidly to the separated-liquid outlet. Vertical baffles to protect rising bubbles from flow currents are sometimes employed. Unseparated fluids should be distributed to the separating region as uniformly and with as little velocity as possible. When the bubble rise velocity is quite low, shallow tanks or flow channels should be used to minimize the residence time required.

Quite low velocity rise of bubbles due either to small bubble size or to high liquid viscosity can cause difficult situations. With low-viscosity liquids, separation-enhancing possibilities in addition to those previously enumerated are to sparge the liquid with large-diameter gas bubbles or to atomize the mixture as a spray into a tower. Large gas bubbles rising rapidly through the liquid collide with small bubbles and aid their coalescence through capture. Atomizing of the continuous phase reduces the distance that small gas bubbles must travel to reach a gas interface. Evacuation of the spray space can also be beneficial in promoting small-bubble growth and especially in promoting gas evolution when the gas has appreciable liquid solubility. Liquid heating will also reduce solubility.

Surfaces in the settling zone for bubble coalescence such as closely spaced vertical or inclined plates or tubes are beneficial. When clean low-viscosity fluids are involved, passage of the undegassed liquid through a tightly packed pad of mesh or fine fibers at low velocity will result in efficient bubble coalescence. Problems have been experienced in degassing a water-based organic solution that has been passed through an electrolytic cell for chemical reaction in which extremely fine bubbles of hydrogen gas are produced in the liquid within the cell. Near-total removal of hydrogen gas from the liquid is needed for process safety. This is extremely difficult to achieve by gravity settling alone because of the fine bubble size and the need for a coalescing surface. Utilization of a fine fiber media is strongly recommended in such situations. A low-forward liquid flow through the media is desirable to provide time for the bubbles to attach themselves to the fiber media through Brownian diffusion. Spielman and Goren [*Ind. Eng. Chem.*, **62**(10), (1970)] reviewed the literature on coalescence with porous media and reported their own experimental results [*Ind. Eng. Chem. Fundam.*, **11**(1), 73 (1972)] on the coalescence of oil-water liquid emulsions. The principles are applicable to a gas-in-liquid system. Glass-fiber mats composed of 3.5-, 6-, or 12-μm diameter fibers, varying in thickness from 1.3 to 3.3 mm, successfully coalesced and separated 1- to 7-μm oil droplets at superficial bed velocities of 0.02 to 1.5 cm/s (0.00067 to 0.049 ft/s).

In the deaeration of high-viscosity fluids such as polymers, the material is flowed in thin sheets along solid surfaces. Vacuum is applied to increase bubble size and hasten separation. The Versator (Cornell Machine Co.) degasses viscous liquids by spreading them into a thin film by centrifugal action as the liquids flow through an evacuated rotating bowl.

Separation of Foam   Foam is a colloidal system containing relatively large volumes of dispersed gas in a relatively small volume of liquid. Foams are thermodynamically unstable with respect to separation into their components of gas and vapor, and appreciable surface energy is released in the bursting of foam bubbles. Foams are dynamic systems in which a third component produces a surface layer that is different in composition from the bulk of the liquid phase. The stabilizing effect of such components (often present only in trace amounts) can produce foams of troubling persistence in many operations. (Foams which have lasted for years when left undisturbed have been produced.) Bendure [TAPPI, 58(2), 83 (1975)], Keszthelyi [*J. Paint Technol.*, **46**(11), 31 (1974)], Ahmad [*Sep. Sci.* **10**, 649 (1975)], and Shedlovsky ("Foams," *Encyclopedia of Chemical Technology*, 2d ed., Wiley, New York, 1966) have presented concise articles on the characteristics and properties of foams in addition to the general references cited at the beginning of this subsection.

Foams can be a severe problem in chemical-processing steps involving gas-liquid interaction such as distillation, absorption, evaporation, chemical reaction, and particle separation and settling. It can also be a major problem in pulp and paper manufacture, oil-well drilling fluids, production of water-based paints, utilization of lubricants and hydraulic fluids, dyeing and sizing of textiles, operation of steam boilers, fermentation operations, polymerization, wet-process phosphoric acid concentration, adhesive production, and foam control in products such as detergents, waxes, printing inks, instant coffee, and glycol antifreeze.

Foams, as freshly generated, are gas emulsions with spherical bubbles separated by liquid films up to a few millimeters in thickness. They age rapidly by liquid drainage and form polyhedrals in which three bubbles intersect at corners with angles of approximately 120°. During drainage, the lamellae become increasingly thinner, especially in the center (only a few micrometers thickness), and more brittle. This feature indicates that with some foams if a foam layer can be tolerated, it may be self-limiting, as fresh foam is added to the bottom of the layer with drained foam collapsing on the top. (A quick-breaking foam may reach its maximum life cycle in 6 s. A moderately stable foam can persist for 140 s.) During drainage, gas from small foam bubbles, which is at a high pressure, will diffuse into large bubbles so that foam micelles increase with time. As drainage proceeds, weak areas in the lamella may develop. However, the presence of a higher concentration of surfactants in the surface produces a lower surface tension. As the lamella starts to fail, exposing bulk liquid with higher surface tension, the surface is renewed and healed. This is known as the *Marangoni effect*. If drainage can occur faster than Marangoni healing, a hole may develop in the lamella. The forces involved are such that collapse will occur in milliseconds without concern for rupture propagation. However, in very stable foams, electrostatic surface forces (zeta potential) prevent complete drainage and collapse. In some cases, stable lamella thicknesses of only two molecules have been measured.

Drainage rate is influenced by surface viscosity, which is very temperature-sensitive. At a critical temperature, which is a function of the system, a temperature change of only a few degrees can change a slow-draining foam to a fast-draining foam. This change in drainage rate can be a factor of 100 or more; thus increasing the temperature of foam can cause its destruction. An increase in temperature may also cause liquid evaporation and lamella thinning. As the lamellae become thinner, they become more brittle and fragile. Thus, mechanical deformation or pressure changes, which cause a change in gas-bubble volume, can also cause rupture.

Bendure indicates 10 ways to increase foam stability: (1) increase bulk liquid viscosity, (2) increase surface viscosity, (3) maintain thick walls (higher liquid-to-gas ratio), (4) reduce liquid surface tension, (5) increase surface elasticity, (6) increase surface concentration, (7) reduce surfactant-adsorption rate, (8) prevent liquid evaporation, (9) avoid mechanical stresses, and (10) eliminate foam inhibitors. Obviously, the reverse of each of these actions, when possible, is a way to control and break foam.

Physical Defoaming Techniques Typical physical defoaming techniques include mechanical methods for producing foam stress, thermal methods involving heating or cooling, and electrical methods. Combinations of these methods may also be employed, or they may be used in conjunction with chemical defoamers. Some methods are only moderately successful when conditions are present to reform the foam such as breaking foam on the surface of boiling liquids. In some cases it may be desirable to draw the foam off and treat it separately. Foam can always be stopped by removing the energy source creating it, but this is often impractical.

Thermal Methods Heating is often a suitable means of destroying foam. As indicated previously, raising the foam above a critical temperature (which must be determined experimentally) can greatly decrease the surface viscosity of the film and change the foam from a slow-draining to a fast-draining foam. Coupling such heating with a mechanical force such as a revolving paddle to cause foam deformation is frequently successful. Other effects of heating are expansion of the gas in the foam bubbles, which increases strain on the lamella walls as well as requiring their movement and flexing. Evaporation of solvent may occur causing thinning of the walls. At sufficiently high temperatures, desorption or decomposition of stabilizing substances may occur. Placing a high-temperature bank of steam coils at the maximum foam level is one control method. As the foam approaches or touches the coil, it collapses. The designer should consider the fact that the coil will frequently become coated with solute.

Application of radiant heat to a foam surface is also practiced. Depending on the situation, the radiant source may be electric lamps, Glowbar units, or gas-fired radiant burners. Hot gases from burners will enhance film drying of the foam. Heat may also be applied by jetting or spraying hot water on the foam. This is a combination of methods since the jetting produces mechanical shear, and the water itself provides dilution and change in foam-film composition. Newer approaches might include foam heating with the application of focused microwaves. This could be coupled with continuous or intermittent pressure fluctuations to stress lamella walls as the foam ages.

Cooling can also destroy foam if it is carried to the point of freezing since the formation of solvent crystals destroys the foam structure. Less drastic cooling such as spraying a hot foam with cold water may be effective. Cooling will reduce the gas pressure in the foam bubbles and may cause them to shrink. This is coupled with the effects of shear and dilution mentioned earlier. In general, moderate cooling will be less effective than heating since the surface viscosity is being modified in the direction of a more stable foam.

Mechanical Methods Static or rotating breaker bars or slowly revolving paddles are sometimes successful. Their application in conjunction with other methods is frequently better. As indicated in the theory of foams, they will work better if installed at a level at which the foam has had some time to age and drain. A rotating breaker works by deforming the foam, which causes rupture of the lamella walls. Rapidly moving slingers will throw the foam against the vessel wall and may cause impact on other foam outside the envelope of the slinger. In some instances, stationary bars or closely spaced plates will limit the rise of foam. The action here is primarily one of providing surface for coalescence of the foam. Wettability of the surface, whether moving or stationary, is frequently important. Usually a surface not wetted by the liquid is superior, just as is frequently the case of porous media for foam coalescence. However, in both cases there are exceptions for which wettable surfaces are preferred. Shkodin [*Kolloidn. Zh.,* **14,** 213 (1952)] found molasses foam to be destroyed by contact with a wax-coated rod and unaffected by a clean glass rod.

Goldberg and Rubin [*Ind. Eng. Chem. Process Des. Dev.,* **6** 195 (1967)] showed in tests with a disk spinning vertically to the foam layer that most mechanical procedures, whether centrifugation, mixing, or blowing through nozzles, consist basically of the application of shear stress. Subjecting foam to an air-jet impact can also provide a source

of drying and evaporation from the film, especially if the air is heated. Other effective means of destroying bubbles are to lower a frame of metal points periodically into the foam or to shower the foam with falling solid particles.

Pressure and Acoustic Vibrations These methods for rupturing foam are really special forms of mechanical treatment. Change in pressure in the vessel containing the foam stresses the lamella walls by expanding or contracting the gas inside the foam bubbles. Oscillation of the vessel pressure subjects the foam to repeated film flexing. Parlow [*Zucker,* 3, 468 (1950)] controlled foam in sugar-sirup evaporators with high-frequency air pulses. It is by no means certain that high-frequency pulsing is necessary in all cases. Lower frequency and higher amplitude could be equally beneficial. Acoustic vibration is a similar phenomenon causing localized pressure oscillation by using sound waves. Impulses at 6 kHz have been found to break froth from coal flotation [Sun, *Min. Eng.,* 3, 865 (1958)]. Sonntag and Strenge (*Coagulation and Stability of Disperse Systems,* Halsted-Wiley, New York, 1972, p. 121) report foam suppression with high-intensity sound waves (11 kHz, 150 dB) but indicate that the procedure is too expensive for large-scale application. The Sontrifuge (Teknika Inc., a subsidiary of Chemineer, Inc.) is a commercially available low-speed centrifuge employing sonic energy to break the foam. Walsh [*Chem. Process.,* 29, 91 (1966)], Carlson [*Pap. Trade J.,* 151, 38 (1967)], and Thorhildsen and Rich [*TAPPI,* 49, 95A (1966)] have described the unit.

Electrical Methods As colloids, most foams typically have electrical double layers of charged ions which contribute to foam stability. Accordingly, foams can be broken by the influence of an external electric field. While few commercial applications have been developed, Sonntag and Strenge (op. cit., p. 114) indicate that foams can be broken by passage through devices much like electrostatic precipitators for dusts. Devices similar to two-stage precipitators having closely spaced plates of opposite polarity should be especially useful. Sonntag and Strenge, in experiments with liquid-liquid emulsions, indicate that the colloid structure can be broken at a field strength of the order of 8 to 9×10^5 V/cm.

Chemical Defoaming Techniques Sonntag and Strenge (op. cit., p. 111) indicate two chemical methods for foam breaking. One method is causing the stabilizing substances to be desorbed from the interface, such as by displacement with other more surface-active but nonstabilizing compounds. Heat may also cause desorption. The second method is to carry on chemical changes in the adsorption layer, leading to a new structure. Some defoamers may act purely by mechanical means but will be discussed in this subsection since their action is generally considered to be chemical in nature. Often chemical defoamers act in more than one way.

Chemical Defoamers The addition of chemical foam breakers is the most elegant way to break a foam. Effective defoamers cause very rapid disintegration of the foam and frequently need be present only in parts per million. The great diversity of compounds used for defoamers and the many different systems in which they are applied make a brief and orderly discussion of their selection difficult. Compounds needed to break aqueous foams may be different from those needed for aqueous-free systems. The majority of defoamers are insoluble or nonmiscible in the foam continuous phase, but some work best because of their ready solubility. Lichtman (*Defoamers,* 3d ed., Wiley, New York, 1979) has presented a concise summary of the application and use of defoamers. Rubel (*Antifoaming and Defoaming Agents,* Noyes Data Corp., Park Ridge, N.J., 1972) has reviewed the extensive patent literature on defoamers. Defoamers are also discussed extensively in the general references at the beginning of this subsection.

One useful method of aqueous defoaming is to add a nonfoam stabilizing surfactant which is more surface-active than the stabilizing substance in the foam. Thus a foam stabilized with an ionic surfactant can be broken by the addition of a very surface-active but nonstabilizing silicone oil. The silicone displaces the foam stabilizer from the interface by virtue of its insolubility. However, it does not stabilize the foam because its foam films have poor elasticity and rupture easily.

A major requirement for a defoamer is cost-effectiveness. Accordingly, some useful characteristics are low volatility (to prevent stripping from the system before it is dispersed and does its work), ease of dispersion and strong spreading power, and surface attraction-orientation. Chemical defoamers must also be selected in regard to their possible effect on product quality and their environmental and health suitability. For instance, silicone antifoam agents are effective in textile jet dyeing but reduce the fire retardancy of the fabric. Mineral-oil defoamers in sugar evaporation have been replaced by specifically approved materials. The tendency is no longer to use a single defoamer compound but to use a formulation specially tailored for the application comprising carriers, secondary antifoam agents, emulsifiers, and stabilizing agents in addition to the primary defoamer. Carriers, usually hydrocarbon oils or water, serve as the vehicle to support the release and spread of the primary defoamer. Secondary defoamers may provide a synergistic effect for the primary defoamer or modify its properties such as spreadability or solubility. Emulsifiers may enhance the speed of dispersion, while stabilizing agents may enhance defoamer stability or shelf life.

Hydrophobic silica defoamers work on a basis which may not be chemical at all. They are basically finely divided solid silica particles dispersed in a hydrocarbon or silicone oil which serves as a spreading vehicle. Kulkarni [*Ind. Eng. Chem. Fundam.,* **16**, 472 (1977)] theorizes that this mixture defoams by the penetration of the silica particle into the bubble and the rupture of the wall. Table 14-23 lists major types of defoamers and typical applications.

Other Chemical Methods These methods rely chiefly on destroying the foam stabilizer or neutralizing its effect through methods other than displacement and are applicable when the process will permit changing the chemical environment. Forms stabilized with alkali esters can be broken by acidification since the equivalent free acids do not stabilize foam. Foams containing sulfated and sulfonated ionic detergents can be broken with the addition of fatty-acid soaps and calcium salts. Several theories have been proposed. One suggests that the surfactant is tied up in the foam as double calcium salts of both the sulfonate and the soap. Another suggests that calcium soaps oriented in the film render it inelastic.

Ionic surfactants adsorb at the foam interface and orient with the charged group immersed in the lamellae and their uncharged tails pointed into the gas stream. As the film drains, the charged groups, which repel each other, tend to be moved more closely together. The repulsive force between like charges hinders drainage and stabilizes the film. Addition of a salt or an electrolyte to the foam screens the repulsive effect, permits additional drainage, and can reduce foam stability.

Foam Prevention Chemical prevention of foam differs from defoaming only in that compounds or mixtures are added to a stream prior to processing to prevent the formation of foam either during processing or during customer use. Such additives, sometimes distinguished as antifoam agents, are usually in the same chemical class of materials as defoamers. However, they are usually specifically formulated for the application. Typical examples of products formulated with antifoam agents are laundry detergents (to control excess foaming), automotive antifreeze, instant coffee, and jet-aircraft fuel. Foaming in some chemical processes such as distillation or evaporation may be due to trace impurities such as surface-active agents. An alternative to antifoam agents is their removal before processing such as by treatment with activated carbon [Pool, *Chem. Process.,* **21**(9), 56 (1958)].

Automatic Foam Control In processing materials when foam can accumulate, it is often desirable to measure the height of the foam layer continuously and to dispense defoamer automatically as required to control the foam. Other corrective action can also be taken automatically. Methods of sensing the foam level have included electrodes in which the electrical circuit is completed when the foam touches the electrode [Nelson, *Ind. Eng. Chem.,* **48**, 2183 (1956); and Browne, U.S. Patent 2,981,693, 1961], floats designed to rise in a foam layer (Carter, U.S. Patent 3,154,577, 1964), and change in power input required to turn a foam-breaking impeller as the foam level rises (Yamashita, U.S. Patent 3,317,435, 1967). Timers to control the duration of defoamer addition have also been used. Browne has suggested automatic addition of defoamer through a porous wick when the foam level reaches the level of the wick. Foam control has also been discussed by Kroll [*Ind. Eng. Chem.,* **48**, 2190 (1956)].

TABLE 14-23 Major Types and Applications of Defoamers

Classification	Examples	Applications
Silicones	Dimethyl silicone, trialkyl and tetraalkyl silanes	Lubricating oils; distillation; fermentation; jam and wine making; food processing
Aliphatic acids or esters	Mostly high-molecular-weight compounds; diethyl phthalate; lauric acid	Papermaking; wood-pulp suspensions; water-based paints; food processing
Alcohols	Moderate- to high-molecular-weight monohydric and polyhydric alcohols; octyl alcohol; C-12 to C-20 alcohols; lauryl alcohol	Distillation; fermentation; papermaking; glues and adhesives
Sulfates or sulfonates	Alkali metal salts of sulfated alcohols, sulfonic acid salts; alkyl-aryl sulfonates; sodium lauryl sulfate	Nonaqueous systems; mixed aqueous and nonaqueous systems; oil-well drilling muds; spent H_3SO_4 recovery; deep-fat frying
Amines or amides	Alkyl amines (undecyloctyl and diamyl methyl amine); polyamides (acyl derivatives of piperazine)	Boiler foam; sewage foam; fermentation; dye baths
Halogenated compounds	Fluochloro hydrocarbons with 5 to 50 C atoms; chlorinated hydrocarbons	Lubrication-oil and grease distillation; vegetable-protein glues
Natural products	Vegetable oils; waxes, mineral oils plus their sulfated derivatives (including those of animal oils and fats)	Sugar extraction; glue manufacture; cutting oils
Fatty-acid soaps	Alkali, alkaline earth, and other metal soaps; sodium stearate; aluminum stearate	Gear oils; paper stock; paper sizing; glue solutions
Inorganic compounds	Monosodium phosphate mixed with boric acid and ethyl carbonate, disodium phosphate; sodium aluminate, bentonite and other solids	Distillation; instant coffee; boiler feedwater; sugar extraction
Phosphates	Alkyl-alkalene diphosphates; tributyl phosphate in isopropanol	Petroleum-oil systems; foam control in soap solutions
Hydrophobic silica	Finely divided silica in polydimethyl siloxane	Aqueous foaming systems
Sulfides or thio derivatives	Metallic derivatives of thio ethers and disulfides, usually mixed with organic phosphite esters; long-chain alkyl thienyl ketones	Lubricating oils; boiler water

Liquid-Liquid Extraction Operations and Equipment

Lanny A. Robbins, Ph.D., *Research Fellow, Dow Chemical Company; Member, American Institute of Chemical Engineers*

Roger W. Cusack, Vice President, *Glitsch Process Systems, Inc.; Member, American Institute of Chemical Engineers*

Nomenclature

Symbol	Definition	SI units	U.S. customary units
$A°$	Activity of solute	—	—
A_p	Area of a drop	m^2	ft^2
A_t	Cross-sectional area of tower	m^2	ft^2
a	Specific interfacial surface between liquids	m^2/m^3	ft^2/ft^3
a_p	Specific packing surface	m^2/m^3	ft^2/ft^3
B	Ratio of total length to characteristic length		
b	Constant		
C	Constant		
C_O	Orifice coefficient	Dimensionless	Dimensionless
c	Concentration	$kmol/m^3$	$(lb \cdot mol)/ft^3$
c_p	Heat capacity		
D	Solute diffusivity	m^2/s	ft^2/h
D'	Enhanced diffusivity	m^2/s	ft^2/h
d	Differential operator		
d_F	Packing size	m	ft
d_{FC}	Critical packing size	m	ft
d_i	Impeller diameter	m	ft
d_O	Nozzle, perforation, orifice diameter	m	ft
d_p	Drop diameter, diameter of sphere of same volume per surface	m	ft
d_{pJ}	Drop diameter at jetting if no jet forms	m	ft
d_r	Diameter of rotor	m	ft
d_s	Diameter of stator-ring opening	m	ft
d_t	Tube or tank diameter	m	ft
E	Weight (or mass flow rate) of extract	kg (or kg/s)	lb (or lb/h)
E'	Weight (or mass flow rate) of extraction solvent alone in extract	kg (or kg/s)	lb (or lb/h)
E_f	Fractional efficiency of a single stage (mixer-settlers)		
E_d	Longitudinal dispersion coefficient (differential extractors)	m^2/s	ft^2/h
E_{MD}	Murphree dispersed-phase stage efficiency, fractional		
E_O	Overall stage efficiency of a cascade, fractional		
e	2.7183 (napierian logarithm base)		
e	Extraction factor (slope of equilibrium line/slope of operating line)		
F	Weight (or mass flow rate) of feed	kg (or kg/s)	lb (or lb/h)
F'	Weight (or mass flow rate) of feed solvent alone in feed	kg (or kg/s)	lb (or lb/h)
f_e	Weighting factor		
f_r	Weighting factor		
g	Local acceleration due to gravity	9.83 m/s^2	4.18 E08 ft/h^2
g_c	Gravitational conversion factor	$1(kg \cdot m)/(N \cdot s)$	$4.18 \text{ E08 (lbm·} ft)/(lbf \cdot h^2)$
H_e	Height of a transfer unit attributed to driving force in extract phase	m	ft
H_{or}	Height of a transfer unit based on overall driving force in raffinate concentrations	m	ft
H_r	Height of a transfer unit attributed to driving force in raffinate phase	m	ft

Symbol	Definition	SI units	U.S. customary units
H_{to}	Overall height of a transfer unit	m	ft
HETS	Height equivalent to a theoretical stage	m	ft
h	Head loss due to friction	m	ft
$A°$	Activity of solute	—	—
h_C	Contribution to h due to continuous phase	m	ft
h_D	Contribution to h due to dispersed phase	m	ft
h_o	Contribution to h_D due to orifice	m	ft
h_σ	Contribution to h_D due to interfacial tension	m	ft
K	Overall mass-transfer coefficient	$kmol/(s \cdot m^2)$ $(kmol/m^3)$	$(lb \cdot mol)/(h \cdot ft^2)$ $[(lb \cdot mol)/ft^3]$
K	Partition coefficient in weight fractions	Dimensionless	Dimensionless
$K°$	Partition coefficient in mole fractions	Dimensionless	Dimensionless
K'	Partition coefficient in Bancroft (weight-ratio) coordinates	Dimensionless	Dimensionless
K_C	Mass transfer coefficient for overall driving force in continuous phase concentration units	$kmol/(s \cdot m^2)$ $(kmol/m^3)$	$(lb \; mol)/(h \cdot ft^2)$ $[(lb \cdot mol)/ft^3]$
K_D	Mass transfer coefficient for overall driving force in dispersed phase concentration units	$kmol/(s \cdot m^2)$ $(kmol/m^3)$	$(lb \; mol)/(h \cdot ft^2)$ $[(lb \cdot mol)/ft^3]$
k	Individual-phase mass-transfer coefficient	$kmol/(s \cdot m^2)$ $(kmol/m^3)$	$(lb \cdot mol)/(h \cdot ft^2)$ $[(lb \cdot mol)/ft^3]$
k_t	Thermal conductivity	$W/(m \cdot K)$	$Btu/[h \cdot ft^2 \cdot °F)/ ft]$
L	Superficial mass velocity	$kg/(s \cdot m^2)$	$lbm/(h \cdot ft^2)$
L'	Superficial molar mass velocity	$kmol/(s \cdot m^2)$	$(lb \cdot mol)/(h \cdot ft^2)$
m	Slope of equilibrium distribution curve, dy/dx		
m	Slope of equilibrium line in Bancroft coordinates	Dimensionless	Dimensionless
$m°$	Slope of equilibrium line in mole fractions	Dimensionless	Dimensionless
m_{CD}	Slope of equilibrium line continuous/dispersed phase	Dimensionless	Dimensionless
m'	Slope of equilibrium distribution curve, dc_E/dc_R	$(kmol/m^3)/ (kmol/m^3)$	$[(lb \cdot mol)/ft^3]/ [(lb \cdot mol)/ft^3]$
m'	Slope of equilibrium line in concentration, c, units	Dimensionless	Dimensionless
m'_{CD}	Slope of equilibrium curve, dc_C/dc_D	$(kmol/m^3)/ (kmol/m^3)$	$[(lb \cdot mol)/ft^3]/ [(lb \cdot mol)/ft^3]$
N_s	Impeller speed	r/s	r/h
N_f	Flux of mass transfer	$kmol/(s \cdot m^2)$	$(lb \cdot mol)/(h \cdot ft^2)$
N_{oh}	Number of heat transfer units based on hot phase	Dimensionless	Dimensionless
N_{or}	Number of mass transfer units based on overall driving force in raffinate concentration	Dimensionless	Dimensionless
N_{Pe}	Péclet number for axial dispersion, Vd_F/E_d for packing	Dimensionless	Dimensionless
N_{Po}	Power number, $Pg_c/\rho N^3 d_i^5$	Dimensionless	Dimensionless
N_{Re}	Reynolds number; for pipe flow, $d_t V \rho_{av}/u_{av}$; for an impeller, $d_i^2 N \rho_{av}/u_{av}$; for drops, $d_p V_t u_c/\rho c$	Dimensionless	Dimensionless

Nomenclature (*Concluded*)

Symbol	Definition	SI units	U.S. customary units
N_{Sc}	Schmidt number, $u/\rho D$	Dimensionless	Dimensionless
N_{to}	Number of overall transfer units	Dimensionless	Dimensionless
$N_{We,i}$	Impeller Weber number, $\rho d_i^3 N^2/\sigma g_c$	Dimensionless	Dimensionless
$N_{We,t}$	Pipe Weber number, $\rho d_t V^2/\sigma g_c$	Dimensionless	Dimensionless
NTS	Number of theoretical ¡(equilibrium) stages	Dimensionless	Dimensionless
n	Number of orifices or perforations per plate		
n_d	Number of drops		
P	Power for one real stage	W	(ft·lbf)/h
Q	Total flow rate	m³/s	ft³/h
R	Weight (or mass flow rate) of raffinate	kg (or kg/s)	lb (or lb/h)
R'	Weight (or mass flow rate) of feed solvent alone in raffinate	kg (or kg/s)	lb (or lb/h)
S	Weight (or mass flow rate) of extraction-solvent stream	kg (or kg/s)	lb (or lb/h)
S'	Weight (or mass flow rate) of extraction-solvent alone	kg (or kg/s)	lb (or lb/h)
SPM	Reciprocating speed, strokes/minute		
T	Diameter of mixing vessel or extraction tower	m	ft
T	Temperature in hot (raffinate) phase		
t	Temperature in cold (extract) phase		
U_o	Overall heat-transfer coefficient	W/(m²·K)	Btu/(h·ft²·°F)
V	Superficial velocity	m/s = m³/(s·m²)	ft/h = ft³/(h·ft²)
V_d	Velocity in a down spout	m/s	ft/h
V_K	Characteristic velocity	m/s	ft/h
V_O	Velocity through an orifice or nozzle	m/s	ft/h
V'_O	Velocity through an orifice or nozzle	m/s	ft/h
V'_{OJ}	Jetting velocity	m/s	ft/h
V_S	Slip velocity	m/s	ft/h
V_t	Terminal settling velocity	m/s	ft/h
v	Liquid volume	m³	ft³
v_p	Drop volume	m³	ft³
W	Weight (or mass flow rate) of wash phase or stream	kg (or kg/s)	lb (or lb/h)
W'	Weight (or mass flow rate) of wash solvent alone	kg (or kg/s)	lb (or lb/h)
X	Weight solute/weight feed solvent in feed (raffinate) phase	Dimensionless	Dimensionless
x	Weight-fraction solute in feed (raffinate) phase	Dimensionless	Dimensionless
x°	Mole-fraction solute in feed (raffinate) phase	Dimensionless	Dimensionless
Y	Weight solute/weight extraction solvent in extract	Dimensionless	Dimensionless
y	Weight-fraction solute in extract phase	Dimensionless	Dimensionless
y°	Mole-fraction solute in extract phase	Dimensionless	Dimensionless
Z	Height of liquid in vessel or mixer; for towers, height of packed section	m	ft

Symbol	Definition	SI units	U.S. customary units
Z_H	Height of the heavy phase in decanter	m	ft
Z_i	Height of the interface in decanter	ft m	
Z_L	Height of the light phase in decanter	m	ft
Z_t	Distance between trays	m	ft
Z'_t	Distance between trays	m	in
z	Distance	m	ft
z	Weight-fraction solute in mixture	Dimensionless	Dimensionless

Greek symbols			
α	Relative separation factor (selectivity)		
γ	Activity coefficient of solute		
Δ	Delta (or difference) mixture		
Δp	Pressure drop	Pa	lbf/ft²
$\Delta\rho$	Difference in density	kg/m³	lbm/ft³
δ	Dimensionless amplitude for oscillating drops		
ϵ	Fraction void volume in packed section		
θ	Time of contact	s	h
θ_C	Time between coalescences	s	h
θ_F	Time of drop formation	s	h
λ	Eigenvalue		
μ	Viscosity	Pa·s	lbm/(ft·h)
μ'	Viscosity	Pa·s	cP
ν	Coalescence frequency, fraction of drops coalescing per time	L/s	L/h
π	3.1416		
ρ	Density	kg/m³	lbm/ft³
Σ	Summation		
σ	Interfacial tension	N/m	lbf/ft
σ'	Interfacial tension	N/m	dyn/cm
ϕ	Volume fraction of a liquid in a vessel or extractor's void volume		
ω	Vibration frequency for oscillating drops	L/s	L/h

Additional subscripts	
av	Average
C	Continuous phase
D	Dispersed phase
E	Extract
e	Extract phase or stream
F	Flooding
f	Feed phase or stream
H	Heavy liquid
L	Light liquid
M	Mixture
max	Maximum
o	Organic
plug	Plug flow
R	Raffinate
r	Raffinate phase or stream
s	Extraction solvent phase or stream
w	Water or aqueous liquid
1,2,etc.	Stream leaving stage 1,2,etc.
1	Concentrated end
2	Dilute end

GENERAL REFERENCES: Foust, Wenzel, Clump, Maus, and Anderson, *Principles of Unit Operations,* 2d ed., Wiley, New York, 1980. Schweitzer (ed.), *Handbook of Separation Techniques for Chemical Engineers,* 2d ed., McGraw-Hill, New York, 1988. Sorenson and Arlt, "Liquid-Liquid Equilibrium Data Collection," DECHEMA, Frankfurt, Germany, *Binary Systems* vol V, part 1, 1979, *Ternary Systems* vol V, part 2, 1980, *Ternary & Quaternary Systems,* vol 5, part 3, 1980, Macedo and Rasmussen, *Supplement 1,* vol V, part 4, 1987. Treybal, *Liquid Extraction,* 2d ed., McGraw-Hill, New York, 1963. Treybal, *Mass-Transfer Operations,* 3d ed., McGraw-Hill, New York, 1980. Wisniak and Tamir, *Liquid-Liquid Equilibrium and Extraction: A Literature Source Book,* part A, Elsevier, Amsterdam, 1981. Lo, Baird, and Hanson, *Handbook of Solvent Extraction,* Wiley, New York, 1983. Astarita, *Mass Transfer with Chemical Reactions,* Elsevier, New York, 1967. Calderbank, "Mass Transfer," in Uhl and Gray (eds.), *Mixing,* vol. 2, Academic, New York, 1967. Cremer and Davies, *Chemical Engineering Practice,* vol. 5, Academic, New York, 1958. Davies, "Mass Transfer and Interfacial Phenomena," in Drew, Hoopes, and Vermeulen (eds.), *Advances in Chemical Engineering,* vol. 4, Academic, New York, 1963. Hanson (ed.), *Recent Advances in Liquid-Liquid Extraction,* Pergamon, New York, 1971. Hyman, *Mixing and Agitation,* ibid, vol. 3, 1962. Kalichevsky and Kobe, *Petroleum Refining with Chemicals,* Van Nostrand, Princeton, N.J., 1965. Kintner,

"Drop Phenomena Affecting Liquid Extraction," in Drew et al. (eds.), *Advances in Chemical Engineering,* vol. 4, Academic, New York, 1963. Molyneux, *Chemical Plant Design,* vol. I, Butterworth, Washington, 1965. Olney and Miller, "Liquid Extraction," in Acrivos (ed.) *Modern Chemical Engineering,* vol 1, Reinhold, New York, 1963. Olson and Stout, "Mixing and Chemical Reactions," in Uhl and Gray (eds.), *Mixing,* vol. 2, Academic, New York, 1967. Reitma, "Segregation in Liquid-Liquid Dispersions," in Drew et al. (eds.), *Advances in Chemical Engineering,* vol. 5, Academic, New York, 1964. Rod, Misek, and Sterbacek, *Liquid Extraction,* Statne Nakladatelstor Techniki Literatury, Prague, 1964. Sideman, "Direct-Contact Heat Transfer between Immiscible Liquids," in Drew et al. (eds.), *Advances in Chemical Engineering,* vol. 6, Academic, New York, 1966. Treybal, *Mechanically Aided Liquid Extraction,* ibid., vol. 1, 1956. Ziolkowski, *Liquid Extraction in the Chemical Industry,* Gos. Nauchn. Tekln. Izd. Khim. Lit, Leningrad, 1963. Cusack, Fremeaux, and Glatz, "A Fresh Look at Liquid-Liquid Extraction," part 1, *Extraction Systems, Chemical Engineering,* vol. 98, no. 2, p. 66–67, Feb. 1991. Cusack, and Fremeaux, part 2, "Inside the Extractor," *Chemical Engineering,* vol. 98, no. 3, p. 132–138, Mar. 1991. Cusack and Karr, part 3, "Extractor Design and Specification," *Chemical Engineering,* vol. 98, no. 4, p. 112–120, Apr. 1991.

LIQUID-LIQUID EXTRACTION OPERATIONS

Liquid-liquid extraction is a process for separating components in solution by their distribution between two immiscible liquid phases. Such a process can also be simply referred to as **liquid extraction** or **solvent extraction;** however, the latter term may be confusing because it also applies to the leaching of a soluble substance from a solid.

Since liquid-liquid extraction involves the transfer of mass from one liquid phase into a second immiscible liquid phase, the process can be carried out in many different ways. The simplest example involves the transfer of one component from a binary mixture into a second immiscible liquid phase. One example is liquid-liquid extraction of an impurity from wastewater into an organic solvent. This is analogous to stripping or absorption in which mass is transferred from one phase to another. Transfer of the dissolved component (solute) may be enhanced by the addition of "salting out" agents to the feed mixture or by adding "complexing" agents to the extraction solvent. Or in some cases a chemical reaction can be used to enhance the transfer, an example being the use of an aqueous caustic solution to remove phenolics from a hydrocarbon stream. A more sophisticated concept of liquid-liquid fractionation can be used in a process to separate two solutes completely. A primary extraction solvent is used to extract one of the solutes from a mixture (similar to stripping in distillation), and a wash solvent is used to scrub the extract free from the second solute (similar to rectification in distillation).

USES FOR LIQUID-LIQUID EXTRACTION

Liquid-liquid extraction is used primarily when distillation is impractical or too costly to use. It may be more practical than distillation when the relative volatility for two components falls between 1.0 and 1.2. Likewise, liquid-liquid extraction may be more economical than distillation or steam-stripping a dissolved impurity from wastewater when the relative volatility of the solute to water is less than 4. In one case discussed by Robbins [*Chem. Eng. Prog.,* **76** (10), 58 (1980)], liquid-liquid extraction was economically more attractive than carbon-bed or resin-bed adsorption as a pretreatment process for wastewater detoxification before biotreatment.

In other cases the components to be separated may be heat-

sensitive, like antibiotics, or relatively nonvolatile, like mineral salts, and liquid-liquid extraction may provide the most cost-effective separation process. However, the potential use of distillation should generally be evaluated carefully before considering liquid-liquid extraction. An extraction process usually requires (1) liquid-liquid extraction, (2) solvent recovery, and (3) raffinate desolventizing.

Several examples of cost-effective liquid-liquid extraction processes include the recovery of acetic acid from water (Fig. 15-1), using ethyl ether or ethyl acetate as described by Brown [*Chem. Eng. Prog.,* **59**(10), 65 (1963)], or the recovery of phenolics from water as described by Lauer, Littlewood, and Butler [*Iron Steel Eng.,* **46**(5), 99 (1969)] with butyl acetate, or with isopropyl ether as described by Wurm [*Glückauf,* **12**, 517 (1968)], or with methyl isobutyl ketone as described by Scheibel ["Liquid-Liquid Extraction," in Perry & Weiss-

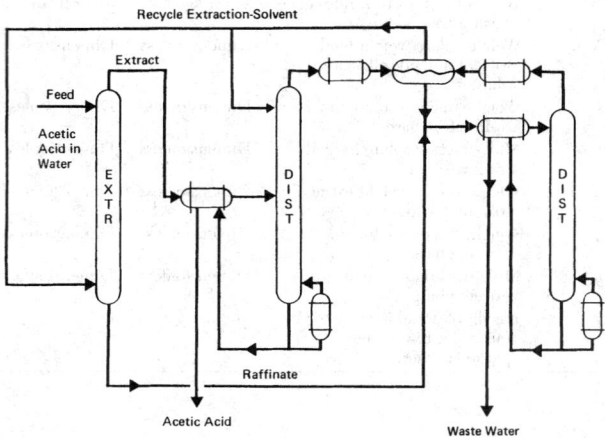

FIG. 15-1 Solvent extraction of acetic acid from water.

burg (eds.), *Separation and Purification,* 3d ed., Wiley, New York, 1978, chap. 3]. The solvent is recovered by distillation, and the raffinate is desolventized by steam stripping. In some cases the extraction solvent may have a higher boiling point than the solute to achieve reduced energy consumption, but a buildup of heavies in the recycle solvent can create another problem.

The Udex process (Fig. 15-2) is a cost-effective liquid-liquid fractionation process for the separation of aromatics from aliphatics as described by Grote [*Chem. Eng. Prog.,* **54**(8), 43 (1958)]. In this process the extraction solvent, diethylene or triethylene glycol, is recovered by steam distillation, and the raffinate and extract streams are desolventized by water extraction. Subsequent process modifications described by Symoniak, Ganju, and Vidueira [*Hydrocarbon Process.,* 139 (September 1981)] use tetraethylene glycol as the extraction solvent and a mixture of light aliphatics and benzene as the wash solvent to the main extractor. Water condensate from the steam distillation is used to extract residual extraction solvent from the raffinate and extract streams, so distillation for drying the extraction solvent has been eliminated. Solids are removed from recycle extraction solvent by filtration, while acids and heavies are removed by a solid adsorbent bed. Other processes similar to this use sulfolane (tetrahydrothiophene-1,1-dioxide) or NMP (N-methyl-pyrrolidone) as the extraction solvent.

Another example of a cost-effective liquid-liquid extraction process is the one used for recovery of uranium from ore leach liquors (Fig. 15-3). In this case the solvents, alkyl phosphates in kerosine, are recovered by liquid-liquid extraction using a strip solution, and the raffinate requires practically no desolventizing because the solubility of the solvents in water is extremely low. Most of the solvent loss occurs because of the entrainment of small droplets in the water. The economic utility of a liquid-liquid extraction process depends strongly on the solvent selected and on the procedures used for solvent recovery and raffinate desolventizing. After these matters have been considered, the selection and design of an extraction device or assembly can be considered in proper perspective.

DEFINITIONS

The feed to a liquid-liquid extraction process is the solution that contains the components to be separated. The major liquid component in the feed can be referred to as the **feed solvent.** Minor components in solution are often referred to as **solutes.** The **extraction solvent,** or just plain **solvent,** is the immiscible liquid added to a process for the purpose of extracting a solute or solutes from the feed. The extraction-solvent phase leaving a liquid-liquid contactor is called the **extract.** The **raffinate** is the liquid phase left from the feed after being contacted by the second phase. A **wash solvent** is a liquid added to a liquid-liquid fractionation process to wash or enrich the purity of a solute in the extract phase.

A **theoretical** or **equilibrium stage** is a device or combination of devices that accomplishes the effect of intimately mixing two immiscible liquids until equilibrium concentrations are reached, then physically separating the two phases into clear layers. **Crosscurrent extraction** (Fig. 15-4) is a cascade, or series of stages, in which the raffinate R from one extraction stage is contacted with additional fresh solvent S in a subsequent stage.

Countercurrent extraction (Fig. 15-5) is an extraction scheme in which the extraction solvent enters the stage or end of the extraction farthest from where the feed F enters and the two phases pass countercurrently to each other. The objective is to transfer one or more components from the feed solution F into the extract E. When a **staged contactor** is used, the two phases are mixed with droplets of one phase suspended in the other, but the phases are separated before leaving each stage. When a **differential contactor** is used, one of the phases can remain dispersed as droplets throughout the contactor as the phases pass countercurrently to each other. The dispersed phase is then allowed to coalesce at the end of the device before being discharged.

Liquid-liquid fractionation, or **fractional extraction** (Fig. 15-6), is a sophisticated scheme for nearly complete separation of one solute from a second solute by liquid-liquid extraction. Two immiscible liquids travel countercurrently through a contactor, with the solutes being fed near the center of the contactor. The ratio of immiscible-liquid flow rates is operated so that one of the phases preferentially moves the first solute to one end of the contactor and the other phase moves the second solute to the opposite end of the contactor. Another way to describe the operation is that a primary solvent S preferentially extracts, or strips, the first solute from the feed F and a wash solvent

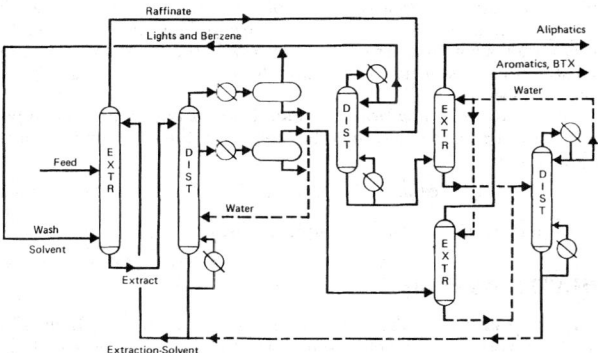

FIG. 15-2 Udex process.

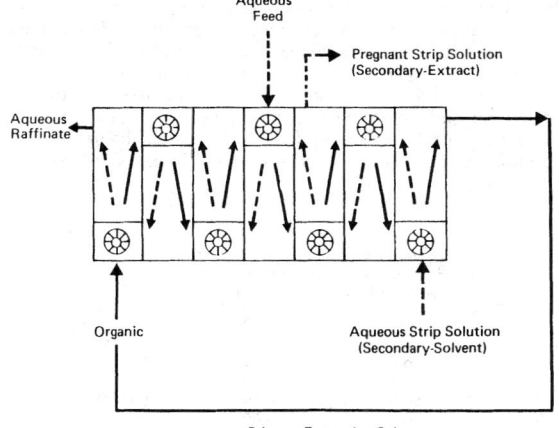

FIG. 15-3 Liquid-liquid extraction of uranium.

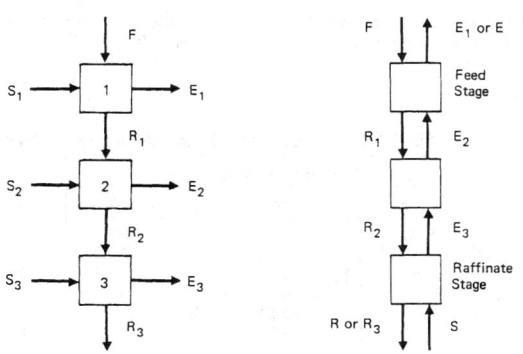

FIG. 15-4 Crosscurrent extraction.   **FIG. 15-5** Countercurrent extraction.

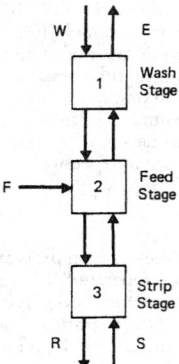

FIG. 15-6   Liquid-liquid fractionation.

W scrubs the extract free from the unwanted second solute. The second solute leaves the contactor in the raffinate stream.

Dissociation extraction is the process of using chemical reaction to force a solute to transfer from one liquid phase to another. One example is the use of a sodium hydroxide solution to extract phenolics, acids, or mercaptans from a hydrocarbon stream. The opposite transfer can be forced by adding an acid to a sodium phenate stream to **spring** the phenolic back to a **free phenol** that can be extracted into an organic solvent. Similarly, primary, secondary, and tertiary amines can be protonated with a strong acid to transfer the amine into a water solution, for example, as an amine hydrochloride salt. Conversely, a strong base can be added to convert the amine salt back to **free base,** which can be extracted into a solvent. This procedure is quite common in pharmaceutical production.

Fractionation dissociation extraction involves both the chemical reaction and the fractionation scheme for the separation of components by their difference in dissociation constants as described by Colby [in Hanson (ed.), *Recent Advances in Liquid-Liquid Extraction*, Pergamon, New York, 1971, chap. 4].

PHASE EQUILIBRIA

The separation of components by liquid-liquid extraction depends primarily on the thermodynamic equilibrium partition of those components between the two liquid phases. Knowledge of these partition relationships is essential for selecting the ratio of extraction solvent to feed that enters an extraction process and for evaluating the mass-transfer rates or theoretical stage efficiencies achieved in process equipment. Since two liquid phases that are immiscible are used, the thermodynamic equilibrium involves considerable evaluation of nonideal solutions. In the simplest case a feed solvent F contains a solute that is to be transferred into an extraction solvent S.

EQUILIBRIUM PARTITION RATIOS

The weight fraction of solute in the extract phase y divided by the weight fraction of solute in the raffinate phase x at equilibrium is called the partition ratio, K [Eq. (15-1)].

$$K = y/x \qquad (15\text{-}1)$$

Thermodynamically the partition ratio $K°$ is derived in mole fractions $y°$ and $x°$ [Eq. (15-2)].

$$K° = y°/x° \qquad (15\text{-}2)$$

For shortcut calculations the partition ratio K' in Bancroft [*Phys. Rev.,* **3**, 120 (1895)] coordinates using the weight ratio of solute to extraction solvent in the extract phase Y and the weight ratio of solute to feed solvent in the raffinate phase X is preferred [Eq. (15-3)].

$$K' = Y/X \qquad (15\text{-}3)$$

In shortcut calculations the slope of the equilibrium line in Bancroft (weight-ratio) coordinates m is also used [Eq. (15-4)].

$$m = dY/dX \qquad (15\text{-}4)$$

For low concentrations in which the equilibrium line is linear the value of K' is equal to m.

The value of K' is one of the main parameters used to establish the minimum ratio of extraction solvent to feed solvent that can be employed in an extraction process. For example, if the partition ratio K' is 4, then a countercurrent extractor would require 0.25 kg or more of extraction-solvent flow to remove all the solute from 1 kg of feed-solvent flow.

The **relative separation,** or **selectivity,** α between two components, b and c, can be described by the ratio of the two partition ratios [Eq. (15-5)].

$$\alpha(b/c) = K_b°/K_c° = K_b/K_c = K_b'/K_c' \qquad (15\text{-}5)$$

This is analogous to relative volatility in distillation.

PHASE DIAGRAMS

Ternary-phase equilibrium data can be tabulated as in Table 15-1 and then worked into an electronic spreadsheet as in Table 15-2 to be presented as a **right-triangular diagram** as shown in Fig. 15-7. The weight-fraction solute is on the horizontal axis and the weight-fraction extraction-solvent is on the vertical axis. The tie-lines connect the points that are in equilibrium. For low-solute concentrations the horizontal scale can be expanded. The water-acetic acid-methylisobutylketone ternary is a **Type I** system where only one of the binary pairs, water-MIBK, is immiscible. In a **Type II** system two of the binary pairs are immiscible, i.e. the solute is not totally miscible in one of the liquids.

Many immiscible-liquid systems exhibit a **critical solution temperature** beyond which the system no longer separates into two liquid phases. This is shown in Fig. 15-8, in which an increase in temperature can change a Type II system to a Type I system above the

TABLE 15-1   Water–Acetic Acid–Methyl Isobutyl Ketone, 25°C*

| Weight % in raffinate | | | | Weight % in extract | | | |
Water	Acetic acid	MIBK	X	Water	Acetic acid	MIBK	Y
98.45	0	1.55	0	2.12	0	97.88	0
95.46	2.85	1.7	0.0299	2.80	1.87	95.33	0.0196
85.8	11.7	2.5	0.1364	5.4	8.9	85.7	0.1039
75.7	20.5	3.8	0.2708	9.2	17.3	73.5	0.2354
67.8	26.2	6.0	0.3864	14.5	24.6	60.9	0.4039
55.0	32.8	12.2	0.5964	22.0	30.8	47.2	0.6525
42.9	34.6	22.5	0.8065	31.0	33.6	35.4	0.9492

*From Sherwood, Evans, and Longcor [*Ind. Eng. Chem.*, **31**, 599 (1939)].

TABLE 15-2 Spreadsheet for Right Triangular Ternary Diagram of Water/Acetic Acid/MIBK Liquid-Liquid-Equilibrium Data at 25°C in Fig. 15-7

Wt. fraction Variable	Acetic acid X	MIBK Y1	MIBK Y2	1 − wf AA Y3
Water	0.0000	0.0155		1.0000
Phase	0.0285	0.0170		0.9715
Line	0.1170	0.0250		0.8830
and	0.2050	0.0380		0.7950
Top of	0.2620	0.0600		0.7380
Triangle	0.3280	0.1220		0.6720
	0.3460	0.2250		0.6540
	1.0000			0.0000
MIBK	0.0000		0.9788	
Phase	0.0187		0.9533	
Line	0.0890		0.8570	
	0.1730		0.7350	
	0.2460		0.6090	
	0.3080		0.4720	
	0.3360		0.3540	
Tie-line 1	0.0285	0.0170		
	0.0187	0.9533		
Tie-line 2	0.1170		0.0250	
	0.0890		0.8570	
Tie-line 3	0.2050	0.0380		
	0.1730	0.7350		
Tie-line 4	0.2620		0.0600	
	0.2460		0.6090	
Tie-line 5	0.3280	0.1220		
	0.3080	0.4720		
Tie-line 6	0.3460		0.2250	
	0.3360		0.3540	

Data from Sherwood, Evans, and Longcor [*Ind. Eng. Chem.,* **31,** 599 (1939)].

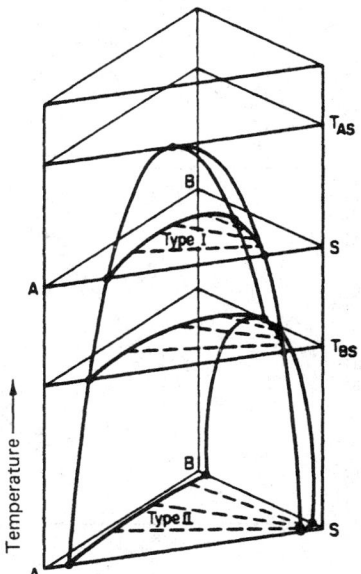

FIG. 15-8 Effect of temperature on ternary liquid-liquid equilibrium. A = feed solvent, B = solute, and S = extraction solvent.

critical temperature of the solute and extraction-solvent binary system T_{BS}. The system becomes totally miscible above the critical temperature of the feed solvent and extraction-solvent binary T_{AS}. Occasionally a system can also have a lower critical solution temperature below which the system will be totally miscible. The methyl ethyl ketone-water binary system provides one example. Changes in pressure ordinarily have a negligible effect on liquid-liquid equilibrium.

For graphical calculation of the number of theoretical stages in a ternary system the **right-triangular diagram** is more convenient to use than an equilateral triangle. The ternary equilibrium data are simply plotted on ordinary rectangular-coordinate graph paper with the weight fraction of the solute on the horizontal axis and the weight fraction of the extraction solvent on the vertical axis. For low-solute concentrations the horizontal scale can be expanded.

For the McCabe-Thiele type of graphical calculations and shortcut methods, the Bancroft (weight-ratio) concentrations can be used on ordinary rectangular-coordinate graph paper. The entire ternary system can be plotted in Bancroft (weight-ratio) concentrations on log-log graph paper as shown by Hand [*J. Phys. Chem.,* **34,** 1961 (1930)], and the equilibrium line can often be correlated by three straight-line segments (Fig. 15-9 and Table 15-3). The plait-point composition for a Type I system can easily be found by using this Hand plot as shown by Treybal, Weber, and Daley [*Ind. Eng. Chem.,* **38,** 817 (1946)]. This type of plot is also helpful for extrapolation and interpolation when data are scarce.

Multicomponent systems containing four or more components become difficult to display graphically. However, process-design calculations can often be made for the extraction of the component with the lowest partition ratio K' and treated as a ternary system. The components with higher K' values may be extracted more thoroughly from the raffinate than the solute chosen for design. Or computer calculations can be used to reduce the tedium of multicomponent, multistage calculations.

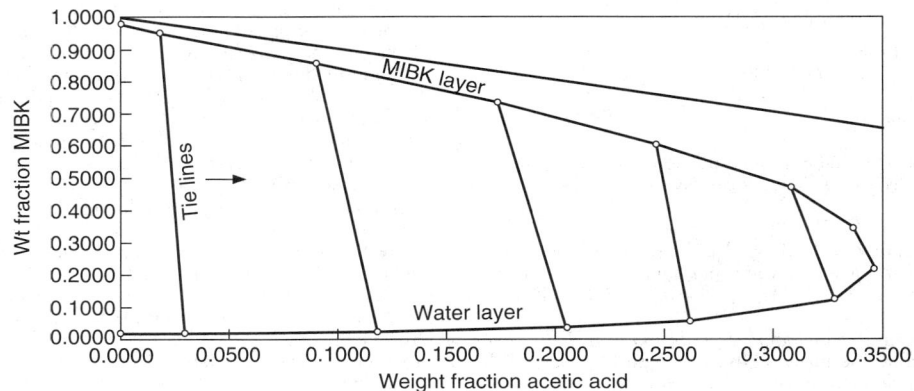

FIG. 15-7 Type I ternary diagram (water-acetic acid-MIBK).

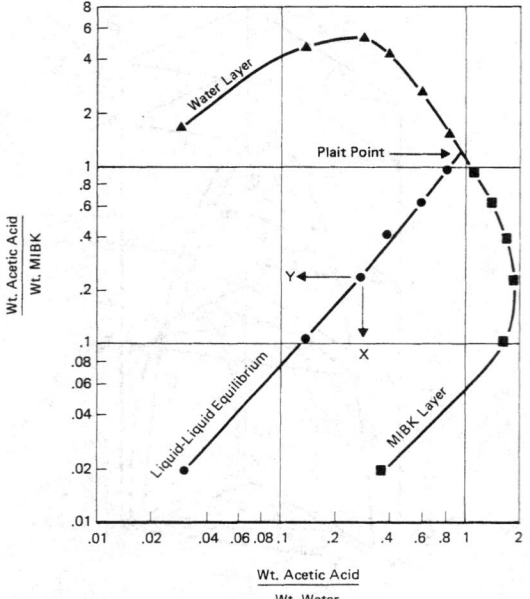

FIG. 15-9 Hand-type ternary diagram for water-acetic acid-methyl isobutyl ketone.

TABLE 15-3 Correlation of Liquid-Liquid Equilibrium Data for Water-Acetic Acid-MIBK

X	Y	$0.93 \cdot X^{1.10}$	$1.27 \cdot X^{1.29}$
0.0299	0.0196	0.0195	
0.1364	0.1039	0.1039	
0.2708	0.2354	0.2210	0.2355
0.3864	0.4039		0.3725
0.5964	0.6525		0.6519
0.8065	0.9492		0.9624

THERMODYNAMIC BASIS OF LIQUID-LIQUID EQUILIBRIA

In a ternary liquid-liquid system, such as the acetic acid–water–MIBK system, all three components are present in both liquid phases. At equilibrium the activity $A°$ of any component is the same in both phases by definition [Eq. (15-6)].

$$A_r° = \gamma_r x° = A_e° = \gamma_e y° \qquad (15-6)$$

where $A°$ = activity of solute
  γ = activity coefficient of solute
  r = raffinate phase
  e = extract phase

Consequently, the partition ratio in mole-fraction units $K°$ is a result of the ratio of activity coefficients in the two layers [Eq. (15-7)].

$$K° = y°/x° = \gamma_r/\gamma_e \qquad (15-7)$$

The **activity coefficient** γ can be defined as the escaping tendency of a component relative to Raoult's law in vapor-liquid equilibrium (see Sec. 4 in this handbook or Null, *Phase Equilibrium in Process Design*, Wiley-Interscience, 1970).

Gmehling and Onken (*Vapor-Liquid Equilibrium Data Collection*, DECHEMA, Frankfurt, Germany, 1979) have reported a large collection of vapor-liquid equilibrium data along with correlations of the resulting activity coefficients. This can be used to predict liquid-liquid equilibrium partition ratios as shown in Example 1.

Example 1: Partition Ratios Let us estimate the partition ratio in weight fractions K for extracting low concentrations of acetone from water into chloroform. The solute is acetone, the feed solvent is water, and the extraction solvent is chloroform in this case.

Gmehling and Onken (op. cit.) give the activity coefficient of acetone in water at infinite dilution γ^∞ as 6.74 at 25°C, depending on which set of vapor-liquid equilibrium data is correlated. From Eqs. (15-1) and (15-7) the partition ratio at infinite dilution of solute can be calculated as follows:

$$K = \frac{\gamma_r}{\gamma_e} \frac{\text{molecular weight of feed solvent}}{\text{molecular weight of extraction solvent}} = \frac{6.74}{0.30} \frac{(18)}{(119.4)} = 3.4$$

Sorenson and Arlt (*Liquid-Liquid Equilibrium Data Collection*, DECHEMA, Frankfurt, Germany, 1979) report several sets of liquid-liquid equilibrium data for the system acetone-water-chloroform, but the lowest solute concentrations reported at 25°C were 3 weight percent acetone in the water layer in equilibrium with 9 weight percent acetone in the chloroform layer. This gives a partition ratio K of 3.0.

This example clearly shows good distribution because of a negative deviation from Raoult's law in the extract layer. The activity coefficient of acetone is less than 1.0 in the chloroform layer. However, there is another problem because acetone and chloroform reach a maximum-boiling-point azeotrope composition and cannot be separated completely by distillation at atmospheric pressure.

A higher-boiling solvent, e.g., 1,1,2-trichloroethane, can be used which still gives acetone a negative deviation from Raoult's law ($\gamma_e = 0.732$ at 2 mole percent acetone) but does not form a maximum-boiling-point azeotrope according to Treybal, Weber, and Daley [*Ind. Eng. Chem.*, **38,** 817 (1946)].

An activity coefficient greater than 1.0 for a solute in solution is generally considered to be a **positive deviation** from Raoult's law; i.e., the escaping tendency is higher than predicted by Raoult's law. Likewise, an activity coefficient less than 1.0 is considered to be a **negative deviation** from Raoult's law; i.e., the escaping tendency is lower than predicted by ideal-solution behavior. "Positive" and "negative" thus refer to the sign of the logarithm of the activity coefficient.

HYDROGEN-BONDING INTERACTIONS

Deviations from Raoult's law in solution behavior have been attributed to many characteristics such as molecular size and shape, but the strongest deviations appear to be due to hydrogen bonding and electron donor-acceptor interactions. Robbins [*Chem. Eng. Prog.*, **76** (10), 58 (1980)] presented a table of these interactions, Table 15-4, that provides a qualitative guide to solvent selection for liquid-liquid extraction, extractive distillation, azeotropic distillation, or even solvent crystallization. The activity coefficient in the liquid phase is common to all these separation processes.

In Example 1 the solute, acetone, contains a ketone carbonyl group which is a hydrogen acceptor, i.e., solute class 5 according to Table 15-4. This solute is to be extracted from water with chloroform solvent which contains a hydrogen donor group, i.e., solvent class 4. The solute class 5 and solvent class 4 interaction in Table 15-4 is shown to give a negative deviation from Raoult's law.

A negative deviation reduces the activity of the solute in the solvent, which enhances the liquid-liquid partition ratio but also leads to maximum-boiling-point azeotropes. Among other classes of solvents shown in Table 15-4 that suppress the escaping tendency of a ketone are classes 1 and 2, i.e., phenolics and acids.

Other ketones, i.e., solvent class 5, are shown to be compatible with acetone, i.e., solute class 5, and tend to give activity coefficients near 1.0, i.e., nearly zero deviation from Raoult's law, and tend to be non-azeotropic. The solvent classes 6 through 12 tend to provide a hostile environment for acetone which increases the escaping tendency, i.e., give activity coefficients greater than 1.0 and tend to form minimum-boiling-point azeotropes. Whenever positive deviations give activity coefficients greater than 7.4, then phase separation, i.e., two liquid phases, can result, as shown by Martin [*Hydrocarbon Process.*, 241 (November 1975)].

Most of the classes in Table 15-4 are self-explanatory, but some can use additional definition. Class 4 includes halogenated solvents that have highly active hydrogens as described by Ewell, Harrison, and Berg [*Ind. Eng. Chem.*, **36,** 871 (1944)]. These are molecules that

TABLE 15-4 Organic-Group Interactions Based on 900 Binary Systems*

Solute class		Solvent class											
		1	2	3	4	5	6	7	8	9	10	11	12
H donor groups													
1	Phenol	0	0	−	0	−	−	−	−	−	−	+	+
2	Acid, thiol	0	0	−	0	−	−	0	0	0	0	+	+
3	Alcohol, water	−	−	0	+	+	0	−	−	+	+	+	+
4	Active H on multihalogen paraffin	0	0	+	0	−	−	−	−	−	−	0	+
H acceptor groups													
5	Ketone, amide with no H on N, sulfone, phosphine oxide	−	−	+	−	0	+	+	+	+	+	+	+
6	Tertiary amine	−	−	0	−	+	0	+	+	0	+	0	0
7	Secondary amine	−	0	−	−	+	+	0	0	0	0	0	+
8	Primary amine, ammonia, amide with 2H on N	−	0	−	−	+	+	0	0	+	+	+	+
9	Ether, oxide, sulfoxide	−	0	+	−	+	0	0	+	0	+	0	+
10	Ester, aldehyde, carbonate, phosphate, nitrate, nitrite, nitrile, intramolecular bonding, e.g., o-nitrophenol	−	0	+	−	+	+	0	+	+	0	+	+
11	Aromatic, olefin, halogen aromatic, multihalogen paraffin without active H, monohalogen paraffin	+	+	+	0	+	0	0	+	0	+	0	0
Non-H-bonding groups													
12	Paraffin, carbon disulfide	+	+	+	+	+	0	+	+	+	+	0	0

*From Robbins, *Chem. Eng. Prog.*, **76**(10), 58–61 (1980), by permission.

have two or three halogen atoms on the same carbon as a hydrogen atom, such as methylene chloride, chloroform, 1,1-dichloroethane and 1,1,2,2-tetrachloroethane. Class 4 also includes molecules that have one halogen on the same carbon atom as a hydrogen atom and one or more halogen atoms on an adjacent carbon atom, such as 1,2-dichloroethane and 1,1,2-trichloroethane. Apparently the halogens interact intramolecularly to leave the hydrogen atom highly active.

Monohalogen paraffins like methyl chloride and ethyl chloride are in class 11 along with multihalogen paraffins and olefins without active hydrogen such as carbon tetrachloride and perchloroethylene. Chlorinated benzenes are also in class 11 because they do not have halogens on the same carbon as a hydrogen atom.

Intramolecular bonding on aromatics is another fascinating interaction which gives a net result that behaves much like an ester group, class 10. Examples of this include *ortho*-nitrophenol and *ortho*-hydroxybenzaldehyde (salicylaldehyde). The intramolecular hydrogen bonding is so strong between the hydrogen donor group (phenol) and the hydrogen acceptor group (nitrate or aldehyde) that the molecule ends up by acting as an ester. One result is its low solubility in hot water. By contrast, the *para* derivative is highly soluble in hot water.

Table 15-4 gives a qualitative indication of interactions between classes of molecules but does not give quantitative differences within each class. Taft et al. [*J. Am. Chem. Soc.*, **91**, 4801 (1969)] have quantified the strength of hydrogen acceptors. The quantitative prediction of activity coefficients for solutions is reviewed by Reid, Prausnitz, and Sherwood (*The Properties of Gases and Liquids*, 3d ed., McGraw-Hill, New York, 1977) for the UNIFAC method, the Perotti, Deal, and Derr [*Ind. Eng. Chem.*, **51**, 95 (1959)] method, and the analytical-solution-of-groups (ASOG) method. Leo, Hansch, and Elkins [*Chem. Rev.*, **71**(6), 525 (1971)] also provide methods for predicting partition ratios for solutes between water and many solvents. Magnussen, Rasmussen, and Fredenslund [*Ind. Eng. Chem. Process Des. Dev.*, **20**(2), 331 (1981)] have presented a UNIFAC parameter table specifically for predicting liquid-liquid equilibrium.

EXPERIMENTAL EQUILIBRIUM DATA

Several large collections of experimental equilibrium data are now available for liquid-liquid systems. Sorenson and Arlt (*Liquid-Liquid Equilibrium Data Collection*, DECHEMA, Frankfurt, Germany, 1979) have reported several volumes of data that have been correlated with activity-coefficient equations.

Wisniak and Tamir (*Liquid-Liquid Equilibrium and Extraction: A Literature Source Book*, Elsevier, Amsterdam, 1980) have listed many references. Leo, Hansch, and Elkins [*Chem. Rev.*, **71**(6), 525 (1971)] have tabulated partition ratios for a large number of solutes between water and solvents. Table 15-5 gives a selected list of partition ratios.

DESIRABLE SOLVENT PROPERTIES

The following properties of a potential solvent should be considered before use in a liquid-liquid extraction process.

1. *Selectivity.* The relative separation, or selectivity, α of a solvent is the ratio of two components in the extraction-solvent phase divided by the ratio of the same components in the feed-solvent phase. The separation power of a liquid-liquid system is governed by the deviation of α from unity, analogous to relative volatility in distillation. A relative separation α of 1.0 gives no separation of the components between the two liquid phases. Dilute solute concentrations generally give the highest relative separation factors.

2. *Recoverability.* The extraction solvent must usually be recovered from the extract stream and also from the raffinate stream in an extraction process. Since distillation is often used, the relative volatility of the extraction-solvent to nonsolvent components should be significantly greater or less than unity. A low latent heat of vaporization is desirable for a volatile solvent.

3. *Partition ratio.* The partition ratio for a solute should preferably be large so that a low ratio of extraction solvent to feed can be used.

4. *Capacity.* This property refers to the loading of solute per weight of extraction solvent that can be achieved in an extract layer at the plait point in a Type I system or at the solubility limit in a Type II system.

5. *Solvent solubility.* A low solubility of extraction solvent in the raffinate generally leads to a high relative volatility in a raffinate stripper or a low solvent loss if the raffinate is not desolventized. A low solubility of feed solvent in the extract leads to a high relative separation and, generally, to low solute-recovery costs.

TABLE 15-5   Selected List of Ternary Systems

Component A = feed solvent, component B = solute, and component S = extraction solvent. K_1 is the partition ratio in weight-fraction solute y/x for the tie line of lowest solute concentration reported. Ordinarily, K will approach unity as the solute concentration is increased.

Component B	Component S	Temp., °C.	K_1	Ref.
A = cetane				
Benzene	Aniline	25	1.290	47
n-Heptane	Aniline	25	0.0784	47
A = cottonseed oil				
Oleic acid	Propane	85	0.150	46
		98.5	0.1272	46
A = cyclohexane				
Benzene	Furfural	25	0.680	44
Benzene	Nitromethane	25	0.397	127
A = docosane				
1,6-Diphenylhexane	Furfural	45	0.980	11
		80	1.100	11
		115	1.062	11
A = dodecane				
Methylnaphthalene	β,β′-Iminodipropionitrile	ca. 25	0.625	92
Methylnaphthalene	β,β′-Oxydipropionitrile	ca. 25	0.377	92
A = ethylbenzene				
Styrene	Ethylene glycol	25	0.190	10
A = ethylene glycol				
Acetone	Amyl acetate	31	1.838	86
Acetone	n-Butyl acetate	31	1.940	86
Acetone	Cyclohexane	27	0.508	86
Acetone	Ethyl acetate	31	1.850	86
Acetone	Ethyl butyrate	31	1.903	86
Acetone	Ethyl propionate	31	2.32	86
A = furfural				
Trilinolein	n-Heptane	30	47.5	15
		50	21.4	15
		70	19.5	15
Triolein	n-Heptane	30	95	15
		50	108	15
		70	41.5	15
A = glycerol				
Ethanol	Benzene	25	0.159	62
Ethanol	Carbon tetrachloride	25	0.0667	63
A = n-heptane				
Benzene	Ethylene glycol	25	0.300	50
		125	0.316	50
Benzene	β,β′-thiodipropionitrile	25	0.350	92
Benzene	Triethylene glycol	25	0.351	89
Cyclohexane	Aniline	25	0.0815	47
Cyclohexane	Benzyl alcohol	0	0.107	29
		15	0.267	29
Cyclohexane	Dimethylformamide	20	0.1320	28
Cyclohexane	Furfural	30	0.0635	78
Ethylbenzene	Dipropylene glycol	25	0.329	90
Ethylbenzene	β,β′-Oxydipropionitrile	25	0.180	101
Ethylbenzene	β,β′-Thiodipropionitrile	25	0.100	101
Ethylbenzene	Triethylene glycol	25	0.140	89
Methylcyclohexane	Aniline	25	0.087	116
Toluene	Aniline	0	0.577	27
		13	0.477	27
		20	0.457	27
		40	0.425	27
Toluene	Benzyl alcohol	0	0.694	29
Toluene	Dimethylformamide	0	0.667	28
		20	0.514	28
Toluene	Dipropylene glycol	25	0.331	90
Toluene	Ethylene glycol	25	0.150	101
Toluene	Propylene carbonate	20	0.732	39
Toluene	β,β′-Thiodipropionitrile	25	0.150	101
Toluene	Triethylene glycol	25	0.289	89
m-Xylene	β,β′-Thiodipropionitrile	25	0.050	101
o-Xylene	β,β′-Thiodipropionitrile	25	0.150	101
p-Xylene	β,β′-Thiodipropionitrile	25	0.080	101
A = n-hexane				
Benzene	Ethylenediamine	20	4.14	23
A = neo-hexane				
Cyclopentane	Aniline	15	0.1259	96
		25	0.311	96
A = methylcyclohexane				
Toluene	Methylperfluorooctanoate	10	0.1297	58
		25	0.200	58

TABLE 15-5 Selected List of Ternary Systems (*Continued*)

Component *B*	Component *S*	Temp., °C.	K_1	Ref.
A = *iso*-octane				
Benzene	Furfural	25	0.833	44
Cyclohexane	Furfural	25	0.1076	44
n-Hexane	Furfural	30	0.083	78
A = perfluoroheptane				
Perfluorocyclic oxide	Carbon tetrachloride	30	0.1370	58
Perfluorocyclic oxide	*n*-Heptane	30	0.329	58
A = perfluoro-*n*-hexane				
n-Hexane	Benzene	30	6.22	80
n-Hexane	Carbon disulfide	25	6.50	80
A = perfluorotri-*n*-butylamine				
Iso-octane	Nitroethane	25	3.59	119
		31.5	2.36	119
		33.7	4.56	119
A = toluene				
Acetone	Ethylene glycol	0	0.286	100
		24	0.326	100
A = triethylene glycol				
α-Picoline	Methylcyclohexane	20	3.87	14
α-Picoline	Diisobutylene	20	0.445	14
α-Picoline	Mixed heptanes	20	0.317	14
A = triolein				
Oleic acid	Propane	85	0.138	46
A = water				
Acetaldehyde	*n*-Amyl alcohol	18	1.43	74
Acetaldehyde	Benzene	18	1.119	74
Acetaldehyde	Furfural	16	0.967	74
Acetaldehyde	Toluene	17	0.478	74
Acetaldehyde	Vinyl acetate	20	0.560	81
Acetic acid	Benzene	25	0.0328	43
		30	0.0984	38
		40	0.1022	38
		50	0.0558	38
		60	0.0637	38
Acetic acid	1-Butanol	26.7	1.613	102
Acetic acid	Butyl acetate	30	0.705	45
			0.391	67
Acetic acid	Caproic acid	25	0.349	73
Acetic acid	Carbon tetrachloride	27	0.1920	91
		27.5	0.0549	54
Acetic acid	Chloroform	*ca.* 25	0.178	70
		25	0.0865	72
		56.8	0.1573	17
Acetic acid	Creosote oil	34	0.706	91
Acetic acid	Cyclohexanol	26.7	1.325	102
Acetic acid	Diisobutyl ketone	25–26	0.284	75
Acetic acid	Di-*n*-butyl ketone	25–26	0.379	75
Acetic acid	Diisopropyl carbinol	25–26	0.800	75
Acetic acid	Ethyl acetate	30	0.907	30
Acetic acid	2-Ethylbutyric acid	25	0.323	73
Acetic acid	2-Ethylhexoic acid	25	0.286	73
Acetic acid	Ethylidene diacetate	25	0.85	104
Acetic acid	Ethyl propionate	28	0.510	87
Acetic acid	Fenchone	25–26	0.310	75
Acetic acid	Furfural	26.7	0.787	102
Acetic acid	Heptadecanol	25	0.312	114
		50	0.1623	114
Acetic acid	3-Heptanol	25	0.828	76
Acetic acid	Hexalin acetate	25–26	0.520	75
Acetic acid	Hexane	31	0.0167	85
Acetic acid	Isoamyl acetate	25–26	0.343	75
Acetic acid	Isophorone	25–26	0.858	75
Acetic acid	Isopropyl ether	20	0.248	31
		25–26	0.429	75
Acetic acid	Methyl acetate		1.273	67
Acetic acid	Methyl butyrate	30	0.690	66
Acetic acid	Methyl cyclohexanone	25–26	0.930	75
Acetic acid	Methylisobutyl carbinol	30	1.058	83
Acetic acid	Methylisobutyl ketone	25	0.657	97
		25–26	0.755	75
Acetic acid	Monochlorobenzene	25	0.0435	77
Acetic acid	Octyl acetate	25–26	0.1805	75
Acetic acid	*n*-Propyl acetate		0.638	67
Acetic acid	Toluene	25	0.0644	131
Acetic acid	Trichloroethylene	27	0.140	91
		30	0.0549	54

TABLE 15-5   Selected List of Ternary Systems (Continued)

Component B	Component S	Temp., °C.	K_i	Ref.
Acetic acid	Vinyl acetate	28	0.294	103
Acetone	Amyl acetate	30	1.228	117
Acetone	Benzene	15	0.940	11
		30	0.862	11
		45	0.725	11
Acetone	n-Butyl acetate		1.127	67
Acetone	Carbon tetrachloride	30	0.238	12
Acetone	Chloroform	25	1.830	43
		25	1.720	3
Acetone	Dibutyl ether	25–26	1.941	75
Acetone	Diethyl ether	30	1.00	54
Acetone	Ethyl acetate	30	1.500	117
Acetone	Ethyl butyrate	30	1.278	117
Acetone	Ethyl propionate	30	1.385	117
Acetone	n-Heptane	25	0.274	112
Acetone	n-Hexane	25	0.343	114
Acetone	Methyl acetate	30	1.153	117
Acetone	Methylisobutyl ketone	25–26	1.910	75
Acetone	Monochlorobenzene	25–26	1.000	75
Acetone	Propyl acetate	30	0.243	117
Acetone	Tetrachloroethane	25–26	2.37	57
Acetone	Tetrachloroethylene	30	0.237	88
Acetone	1,1,2-Trichloroethane	25	1.467	113
Acetone	Toluene	25–26	0.835	75
Acetone	Vinyl acetate	20	1.237	81
		25	3.63	104
Acetone	Xylene	25–26	0.659	75
Allyl alcohol	Diallyl ether	22	0.572	32
Aniline	Benzene	25	14.40	40
		50	15.50	40
Aniline	n-Heptane	25	1.425	40
		50	2.20	40
Aniline	Methylcyclohexane	25	2.05	40
		50	3.41	40
Aniline	Nitrobenzene	25	18.89	108
Aniline	Toluene	25	12.91	107
Aniline hydrochloride	Aniline	25	0.0540	98
Benzoic acid	Methylisobutyl ketone	26.7	76.9°	49
iso-Butanol	Benzene	25	0.989	1
iso-Butanol	1,1,2,2-Tetrachloroethane	25	1.80	36
iso-Butanol	Tetrachloroethylene	25	0.0460	7
n-Butanol	Benzene	25	1.263	126
		35	2.12	126
n-Butanol	Toluene	30	1.176	37
tert-Butanol	Benzene	25	0.401	99
tert-Butanol	tert-Butyl hypochlorite	0	0.1393	130
		20	0.1487	130
		40	0.200	129
		60	0.539	129
tert-Butanol	Ethyl acetate	20	1.74	5
2-Butoxyethanol	Methylethyl ketone	25	3.05	68
2,3-Butylene glycol	n-Butanol	26	0.597	71
		50	0.893	71
2,3-Butylene glycol	Butyl acetate	26	0.0222	71
		50	0.0326	71
2,3-Butylene glycol	Butylene glycol diacetate	26	0.1328	71
		75	0.565	71
2,3-Butylene glycol	Methylvinyl carbinol acetate	26	0.237	71
		50	0.351	71
		75	0.247	71
n-Butylamine	Monochlorobenzene	25	1.391	77
1-Butyraldehyde	Ethyl acetate	37.8	41.3	52
Butyric acid	Methyl butyrate	30	6.75	66
Butyric acid	Methylisobutyl carbinol	30	12.12	83
Cobaltous chloride	Dioxane	25	0.0052	93
Cupric sulfate	n-Butanol	30	0.000501	9
Cupric sulfate	sec-Butanol	30	0.00702	9
Cupric sulfate	Mixed pentanols	30	0.000225	9
p-Cresol	Methylnaphthalene	35	9.89	82
Diacetone alcohol	Ethylbenzene	25	0.335	22
Diacetone alcohol	Styrene	25	0.445	22
Dichloroacetic acid	Monochlorobenzene	25	0.0690	77
1,4-Dioxane	Benzene	25	1.020	8
Ethanol	n-Amyl alcohol	25–26	0.598	75
Ethanol	Benzene	25	0.1191	13
		25	0.0536	115
Ethanol	n-Butanol	20	3.00	26

TABLE 15-5   Selected List of Ternary Systems (*Continued*)

Component B	Component S	Temp., °C.	K_1	Ref.
Ethanol	Cyclohexane	25	0.0157	118
Ethanol	Cyclohexene	25	0.0244	124
Ethanol	Dibutyl ether	25–26	0.1458	75
Ethanol	Di-*n*-propyl ketone	25–26	0.592	75
Ethanol	Ethyl acetate	0	0.0263	5
		20	0.500	5
		70	0.455	41
Ethanol	Ethyl isovalerate	25	0.392	13
Ethanol	Heptadecanol	25	0.270	114
Ethanol	*n*-Heptane	30	0.274	94
Ethanol	3-Heptanol	25	0.783	76
Ethanol	*n*-Hexane	25	0.00212	111
Ethanol	*n*-Hexanol	28	1.00	56
Ethanol	*sec*-Octanol	28	0.825	56
Ethanol	Toluene	25	0.01816	122
Ethanol	Trichloroethylene	25	0.0682	16
Ethylene glycol	*n*-Amyl alcohol	20	0.1159	59
Ethylene glycol	*n*-Butanol	27	0.412	85
Ethylene glycol	Furfural	25	0.315	18
Ethylene glycol	*n*-Hexanol	20	0.275	59
Ethylene glycol	Methylethyl ketone	30	0.0527	85
Formic acid	Chloroform	25	0.00445	72
		56.9	0.0192	17
Formic acid	Methylisobutyl carbinol	30	1.218	83
Furfural	*n*-Butane	51.5	0.712	42
		79.5	0.930	42
Furfural	Methylisobutyl ketone	25	7.10	19
Furfural	Toluene	25	5.64	53
Hydrogen chloride	*iso*-Amyl alcohol	25	0.170	21
Hydrogen chloride	2,6-Dimethyl-4-heptanol	25	0.266	21
Hydrogen chloride	2-Ethyl-1-butanol	25	0.534	21
Hydrogen chloride	Ethylbutyl ketone	25	0.01515	79
Hydrogen chloride	3-Heptanol	25	0.0250	21
Hydrogen chloride	1-Hexanol	25	0.345	21
Hydrogen chloride	2-Methyl-1-butanol	25	0.470	21
Hydrogen chloride	Methylisobutyl ketone	25	0.0273	79
Hydrogen chloride	2-Methyl-1-pentanol	25	0.502	21
Hydrogen chloride	2-Methyl-2-pentanol	25	0.411	21
Hydrogen chloride	Methylisopropyl ketone	25	0.0814	79
Hydrogen chloride	1-Octanol	25	0.424	21
Hydrogen chloride	2-Octanol	25	0.380	21
Hydrogen chloride	1-Pentanol	25	0.257	21
Hydrogen chloride	Pentanols (mixed)	25	0.271	21
Hydrogen fluoride	Methylisobutyl ketone	25	0.370	79
Lactic acid	*iso*-Amyl alcohol	25	0.352	128
Methanol	Benzene	25	0.01022	4
Methanol	*n*-Butanol	0	0.600	65
		15	0.479	65
		30	0.510	65
		45	1.260	65
		60	0.682	65
Methanol	*p*-Cresol	35	0.313	82
Methanol	Cyclohexane	25	0.0156	125
Methanol	Cyclohexene	25	0.01043	124
Methanol	Ethyl acetate	0	0.0589	5
		20	0.238	5
Methanol	*n*-Hexanol	28	0.565	55
Methanol	Methylnaphthalene	25	0.025	82
		35	0.0223	82
Methanol	*sec*-Octanol	28	0.584	55
Methanol	Phenol	25	1.333	82
Methanol	Toluene	25	0.0099	60
Methanol	Trichloroethylene	27.5	0.0167	54
Methyl-*n*-butyl ketone	*n*-Butanol	37.8	53.4	52
Methylethyl ketone	Cyclohexane	25	1.775	48
		30	3.60	85
Methylethyl ketone	Gasoline	25	1.686	64
Methylethyl ketone	*n*-Heptane	25	1.548	112
Methylethyl ketone	*n*-Hexane	25	1.775	112
		37.8	2.22	52
Methylethyl ketone	2-Methyl furan	25	84.0	109
Methylethyl ketone	Monochlorobenzene	25	2.36	68
Methylethyl ketone	Naphtha	26.7	0.885†	6
Methylethyl ketone	1,1,2-Trichloroethane	25	3.44	68
Methylethyl ketone	Trichloroethylene	25	3.27	68
Methylethyl ketone	2,2,4-Trimethylpentane	25	1.572	64
Nickelous chloride	Dioxane	25	0.0017	93

TABLE 15-5   Selected List of Ternary Systems (*Continued*)

Component B	Component S	Temp., °C.	K_1	Ref.
Nicotine	Carbon tetrachloride	25	9.50	34
Phenol	Methylnaphthalene	25	7.06	82
α-Picoline	Benzene	20	8.75	14
α-Picoline	Diisobutylene	20	1.360	14
α-Picoline	Heptanes (mixed)	20	1.378	14
α-Picoline	Methylcyclohexane	20	1.00	14
iso-Propanol	Benzene	25	0.276	69
iso-Propanol	Carbon tetrachloride	20	1.405	25
iso-Propanol	Cyclohexane	25	0.0282	123
iso-Propanol	Cyclohexene	15	0.0583	124
		25	0.0682	124
		35	0.1875	124
iso-Propanol	Diisopropyl ether	25	0.406	35
iso-Propanol	Ethyl acetate	0	0.200	5
		20	1.205	5
iso-Propanol	Tetrachloroethylene	25	0.388	7
iso-Propanol	Toluene	25	0.1296	121
n-Propanol	iso-Amyl alcohol	25	3.34	20
n-Propanol	Benzene	37.8	0.650	61
n-Propanol	n-Butanol	37.8	3.61	61
n-Propanol	Cyclohexane	25	0.1553	123
		35	0.1775	123
n-Propanol	Ethyl acetate	0	1.419	5
		20	1.542	5
n-Propanol	n-Heptane	37.8	0.540	61
n-Propanol	n-Hexane	37.8	0.326	61
n-Propanol	n-Propyl acetate	20	1.55	106
		35	2.14	106
n-Propanol	Toluene	25	0.299	2
Propionic acid	Benzene	30	0.598	57
Propionic acid	Cyclohexane	31	0.1955	84
Propionic acid	Cyclohexene	31	0.303	84
Propionic acid	Ethyl acetate	30	2.77	87
Propionic acid	Ethyl butyrate	26	1.470	87
Propionic acid	Ethyl propionate	28	0.510	87
Propionic acid	Hexanes (mixed)	31	0.186	84
Propionic acid	Methyl butyrate	30	2.15	66
Propionic acid	Methylisobutyl carbinol	30	3.52	83
Propionic acid	Methylisobutyl ketone	26.7	1.949°	49
Propionic acid	Monochlorobenzene	30	0.513	57
Propionic acid	Tetrachloroethylene	31	0.167	84
Propionic acid	Toluene	31	0.515	84
Propionic acid	Trichloroethylene	30	0.496	57
Pyridine	Benzene	15	2.19	110
		25	3.00	105
		25	2.73	120
		45	2.49	110
		60	2.10	110
Pyridine	Monochlorobenzene	25	2.10	77
Pyridine	Toluene	25	1.900	120
Pyridine	Xylene	25	1.260	120
Sodium chloride	iso-Butanol	25	0.0182	36
Sodium chloride	n-Ethyl-sec-butyl amine	32	0.0563	24
Sodium chloride	n-Ethyl-tert-butyl amine	40	0.1792	24
Sodium chloride	2-Ethylhexyl amine	30	0.187	24
Sodium chloride	1-Methyldiethyl amine	39.1	0.0597	24
Sodium chloride	1-Methyldodecyl amine	30	0.693	24
Sodium chloride	n-Methyl-1,3-dimethylbutyl amine	30	0.0537	24
Sodium chloride	1-Methyloctyl amine	30	0.589	24
Sodium chloride	tert-Nonyl amine	30	0.0318	24
Sodium chloride	1,1,3,3-Tetramethyl butyl amine	30	0.072	24
Sodium hydroxide	iso-Butanol	25	0.00857	36
Sodium nitrate	Dioxane	25	0.0246	95
Succinic acid	Ethyl ether	15	0.220	33
		20	0.198	33
		25	0.1805	33
Trimethyl amine	Benzene	25	0.857	51
		70	2.36	51

°Concentrations in lb.-moles/cu. ft.
†Concentrations in volume fraction.

References:

1. Alberty and Washburn, *J. Phys. Chem.*, **49**, 4 (1945).
2. Baker, *J. Phys. Chem.*, **59**, 1182 (1955).
3. Bancroft and Hubard, *J. Am. Chem. Soc.*, **64**, 347 (1942).
4. Barbaudy, *Compt. rend.*, **182**, 1279 (1926).
5. Beech and Glasstone, *J. Chem. Soc.*, **1938**, 67.

6. Berg, Manders, and Switzer, *Chem. Eng. Progr.*, **47**, 11 (1951).
7. Bergelin, Lockhart, and Brown, *Trans. Am. Inst. Chem. Engrs.*, **39**, 173 (1943).
8. Berndt and Lynch, *J. Am. Chem. Soc.*, **66**, 282 (1944).
9. Blumberg, Cejtlin, and Fuchs, *J. Appl. Chem.*, **10**, 407 (1960).

TABLE 15-5   Selected List of Ternary Systems (*Concluded*)

10. Boobar *et al.*, *Ind. Eng. Chem.*, **43**, 2922 (1951).
11. Briggs and Comings, *Ind. Eng. Chem.*, **35**, 411 (1943).
12. Buchanan, *Ind. Eng. Chem.*, **44**, 2449 (1952).
13. Chang and Moulton, *Ind. Eng. Chem.*, **45**, 2350 (1953).
14. Charles and Morton, *J. Appl. Chem.*, **7**, 39 (1957).
15. Church and Briggs, *J. Chem. Eng. Data*, **9**, 207 (1964).
16. Colburn and Phillips, *Trans. Am. Inst. Chem. Engrs.*, **40**, 333 (1944).
17. Conti, Othmer, and Gilmont, *J. Chem. Eng. Data*, **5**, 301 (1960).
18. Conway and Norton, *Ind. Eng. Chem.*, **43**, 1433 (1951).
19. Conway and Phillips, *Ind. Eng. Chem.*, **46**, 1474 (1954).
20. Coull and Hope, *J. Phys. Chem.*, **39**, 967 (1935).
21. Crittenden and Hixson, *Ind. Eng. Chem.*, **46**, 265 (1954).
22. Crook and Van Winkle, *Ind. Eng. Chem.*, **46**, 1474 (1954).
23. Cumming and Morton, *J. Appl. Chem.*, **3**, 358 (1953).
24. Davison, Smith, and Hood, *J. Chem. Eng. Data*, **11**, 304 (1966).
25. Denzler, *J. Phys. Chem.*, **49**, 358 (1945).
26. Drouillon, *J. chim. phys.*, **22**, 149 (1925).
27. Durandet and Gladel, *Rev. Inst. Franc. Pétrole*, **9**, 296 (1954).
28. Durandet and Gladel, *Rev. Inst. Franc. Pétrole*, **11**, 811 (1956).
29. Durandet, Gladel, and Graziani, *Rev. Inst. Franc. Pétrole*, **10**, 585 (1955).
30. Eaglesfield, Kelly, and Short, *Ind. Chemist*, **29**, 147, 243 (1953).
31. Elgin and Browning, *Trans. Am. Inst. Chem. Engrs.*, **31**, 639 (1935).
32. Fairburn, Cheney, and Chernovsky, *Chem. Eng. Progr.*, **43**, 280 (1947).
33. Forbes and Coolidge, *J. Am. Chem. Soc.*, **41**, 150 (1919).
34. Fowler and Noble, *J. Appl. Chem.*, **4**, 546 (1954).
35. Frere, *Ind. Eng. Chem.*, **41**, 2365 (1949).
36. Fritzsche and Stockton, *Ind. Eng. Chem.*, **38**, 737 (1946).
37. Fuoss, *J. Am. Chem. Soc.*, **62**, 3183 (1940).
38. Garner, Ellis, and Roy, *Chem. Eng. Sci.*, **2**, 14 (1953).
39. Gladel and Lablaude, *Rev. Inst. Franc. Pétrole*, **12**, 1236 (1957).
40. Griswold, Chew, and Klecka, *Ind. Eng. Chem.*, **42**, 1246 (1950).
41. Griswold, Chu, and Winsauer, *Ind. Eng. Chem.*, **41**, 2352 (1949).
42. Griswold, Klecka, and West, *Chem. Eng. Progr.*, **44**, 839 (1948).
43. Hand, *J. Phys. Chem.*, **34**, 1961 (1930).
44. Henty, McManamey, and Price, *J. Appl. Chem.*, **14**, 148 (1964).
45. Hirata and Hirose, *Kagaku Kogaku*, **27**, 407 (1963).
46. Hixon and Bockelmann, *Trans. Am. Inst. Chem. Engrs.*, **38**, 891 (1942).
47. Hunter and Brown, *Ind. Eng. Chem.*, **39**, 1343 (1947).
48. Jeffreys, *J. Chem. Eng. Data*, **8**, 320 (1963).
49. Johnson and Bliss, *Trans. Am. Inst. Chem. Engrs.*, **42**, 331 (1946).
50. Johnson and Francis, *Ind. Eng. Chem.*, **46**, 1662 (1954).
51. Jones and Grigsby, *Ind. Eng. Chem.*, **44**, 378 (1952).
52. Jones and McCants, *Ind. Eng. Chem.*, **46**, 1956 (1954).
53. Knight, *Trans. Am. Inst. Chem. Engrs.*, **39**, 439 (1943).
54. Krishnamurty, Murti, and Rao, *J. Sci. Ind. Res.*, **12B**, 583 (1953).
55. Krishnamurty and Rao, *J. Sci. Ind. Res.*, **14B**, 614 (1955).
56. Krishnamurty and Rao, *Trans. Indian Inst. Chem. Engrs.*, **6**, 153 (1954).
57. Krishnamurty, Rao, and Rao, *Trans. Indian Inst. Chem. Engrs.*, **6**, 161 (1954).
58. Kyle and Reed, *J. Chem. Eng. Data*, **5**, 266 (1960).
59. Laddha and Smith, *Ind. Eng. Chem.*, **40**, 494 (1948).
60. Mason and Washburn, *J. Am. Chem. Soc.*, **59**, 2076 (1937).
61. McCants, Jones, and Hopson, *Ind. Eng. Chem.*, **45**, 454 (1953).
62. McDonald, *J. Am. Chem. Soc.*, **62**, 3183 (1940).
63. McDonald, Kluender, and Lane, *J. Phys. Chem.*, **46**, 946 (1942).
64. Moulton and Walkey, *Trans. Am. Inst. Chem. Engrs.*, **40**, 695 (1944).
65. Mueller, Pugsley, and Ferguson, *J. Phys. Chem.*, **35**, 1314 (1931).
66. Murty, Murty, and Subrahmanyam, *J. Chem. Eng. Data*, **11**, 335 (1966).
67. Murti, Venkataratnam, and Rao, *J. Sci. Ind. Res.*, **13B**, 392 (1954).
68. Newman, Hayworth, and Treybal, *Ind. Eng. Chem.*, **41**, 2039 (1949).
69. Olsen and Washburn, *J. Am. Chem. Soc.*, **57**, 303 (1935).
70. Othmer, *Chem. Met. Eng.*, **43**, 325 (1936).
71. Othmer, Bergen, Schlechter, and Bruins, *Ind. Eng. Chem.*, **37**, 890 (1945).
72. Othmer and Ku, *J. Chem. Eng. Data*, **4**, 42 (1959).
73. Othmer and Serrano, *Ind. Eng. Chem.*, **41**, 1030 (1949).
74. Othmer and Tobias, *Ind. Eng. Chem.*, **34**, 690 (1942).
75. Othmer, White, and Treuger, *Ind. Eng. Chem.*, **33**, 1240 (1941).
76. Oualline and Van Winkle, *Ind. Eng. Chem.*, **44**, 1668 (1952).
77. Peake and Thompson, *Ind. Eng. Chem.*, **44**, 2439 (1952).
78. Pennington and Marwill, *Ind. Eng. Chem.*, **45**, 1371 (1953).
79. Pilloton, *A.S.T.M. Spec. Tech. Publ.*, **238**, 5 (1958).
80. Pliskin and Treybal, *J. Chem. Eng. Data*, **11**, 49 (1966).
81. Pratt and Glover, *Trans. Inst. Chem. Engrs.* (*London*), **24**, 52 (1946).
82. Prutton, Walsh, and Desar, *Ind. Eng. Chem.*, **42**, 1210 (1950).
83. Rao, Ramamurty, and Rao, *Chem. Eng. Sci.*, **8**, 265 (1958).
84. Rao and Rao, *J. Appl. Chem.*, **6**, 270 (1956).
85. Rao and Rao, *J. Appl. Chem.*, **7**, 659 (1957).
86. Rao and Rao, *J. Sci. Ind. Res.*, **14B**, 204 (1955).
87. Rao and Rao, *J. Sci. Ind. Res.*, **14B**, 444 (1955).
88. Rao and Rao, *Trans. Indian Inst. Chem. Engrs.*, **7**, 78 (1954–1955).
89. Rifai, *Riv. Combust.*, **11**, 811 (1957).
90. Rifai, *Riv. Combust.*, **11**, 829 (1957).
91. Saletore, Mene, and Warhadpande, *Trans. Indian. Inst. Chem. Engrs.*, **2**, 16 (1950).
92. Saunders, *Ind. Eng. Chem.*, **43**, 121 (1951).
93. Schott and Lynch, *J. Chem. Eng. Data*, **11**, 215 (1966).
94. Schweppe and Lorah, *Ind. Eng. Chem.*, **46**, 2391 (1954).
95. Selikson and Ricci, *J. Am. Chem. Soc.*, **64**, 2474 (1942).
96. Serjian, Spurr, and Gibbons, *J. Am. Chem. Soc.*, **68**, 1763 (1946).
97. Sherwood, Evans, and Longcor, *Ind. Eng. Chem.*, **31**, 1144 (1939).
98. Sidgwick, Pickford, and Wilsdon, *J. Chem. Soc.*, **99**, 1122 (1911).
99. Simonsen and Washburn, *J. Am. Chem. Soc.*, **68**, 235 (1946).
100. Sims and Bolme, *J. Chem. Eng. Data*, **10**, 111 (1965).
101. Skinner, *Ind. Eng. Chem.*, **47**, 222 (1955).
102. Skrzec and Murphy, *Ind. Eng. Chem.*, **46**, 2245 (1954).
103. Smith, *J. Phys. Chem.*, **45**, 1301 (1941).
104. Smith, *J. Phys. Chem.*, **46**, 229 (1942).
105. Smith, *J. Phys. Chem.*, **46**, 376 (1942).
106. Smith and Bonner, *Ind. Eng. Chem.*, **42**, 896 (1950).
107. Smith and Drexel, *Ind. Eng. Chem.*, **37**, 601 (1945).
108. Smith, Foecking, and Barber, *Ind. Eng. Chem.*, **41**, 2289 (1949).
109. Smith and La Bonte, *Ind. Eng. Chem.*, **44**, 2740 (1952).
110. Smith, Stibolt, and Day, *Ind. Eng. Chem.*, **43**, 190 (1951).
111. Taresenkov and Paul'sen, *J. Gen. Chem.* (*U.S.S.R.*), **7**, 2143 (1937).
112. Treybal and Vondrak, *Ind. Eng. Chem.*, **41**, 1761 (1949).
113. Treybal, Weber, and Daley, *Ind. Eng. Chem.*, **38**, 817 (1946).
114. Upchurch and Van Winkle, *Ind. Eng. Chem.*, **44**, 618 (1952).
115. Varteressian and Fenske, *Ind. Eng. Chem.*, **28**, 928 (1936).
116. Varteressian and Fenske, *Ind. Eng. Chem.*, **29**, 270 (1937).
117. Venkataratnam, Rao, and Rao, *Chem. Eng. Sci.*, **7**, 102 (1957).
118. Vold and Washburn, *J. Am. Chem. Soc.*, **54**, 4217 (1932).
119. Vreeland and Dunlap, *J. Phys. Chem.*, **61**, 329 (1957).
120. Vriens and Medcalf, *Ind. Eng. Chem.*, **45**, 1098 (1953).
121. Washburn and Beguin, *J. Am. Chem. Soc.*, **62**, 579 (1940).
122. Washburn, Beguin, and Beckord, *J. Am. Chem. Soc.*, **61**, 1694 (1939).
123. Washburn, Brockway, Graham, and Deming, *J. Am. Chem. Soc.*, **64**, 1886 (1942).
124. Washburn, Graham, Arnold, and Transue, *J. Am. Chem. Soc.*, **62**, 1454 (1940).
125. Washburn and Spencer, *J. Am. Chem. Soc.*, **56**, 361 (1934).
126. Washburn and Strandskov, *J. Phys. Chem.*, **48**, 241 (1944).
127. Weck and Hunt, *Ind. Eng. Chem.*, **46**, 2521 (1954).
128. Weiser and Geankoplis, *Ind. Eng. Chem.*, **47**, 858 (1955).
129. Westwater, *Ind. Eng. Chem.*, **47**, 451 (1955).
130. Westwater and Audrieth, *Ind. Eng. Chem.*, **46**, 1281 (1954).
131. Woodman, *J. Phys. Chem.*, **30**, 1283 (1926).

6. *Density.* The difference in density between the two liquid phases in equilibrium affects the countercurrent flow rates that can be achieved in extraction equipment as well as the coalescence rates. The density difference decreases to zero at a plait point, but in some systems it can become zero at an intermediate solute concentration (isopycnic, or twin-density, tie line) and can invert the phases at higher concentrations. Differential types of extractors cannot cross such a solute concentration, but mixer-settlers can.

7. *Interfacial tension.* A high interfacial tension promotes rapid coalescence and generally requires high mechanical agitation to produce small droplets. A low interfacial tension allows drop breakup with low agitation intensity but also leads to slow coalescence rates. Interfacial tension usually decreases as solubility and solute concentration increase and falls to zero at the plait point (Fig. 15-10).

8. *Toxicity.* Low toxicity from solvent-vapor inhalation or skin contact is preferred because of potential exposure during repair of equipment or while connections are being broken after a solvent transfer. Also, low toxicity to fish and bioorganisms is preferred when extraction is used as a pretreatment for wastewater before it enters a biotreatment plant and with final effluent discharge to a stream or lake. Often solvent toxicity is low if water solubility is high.

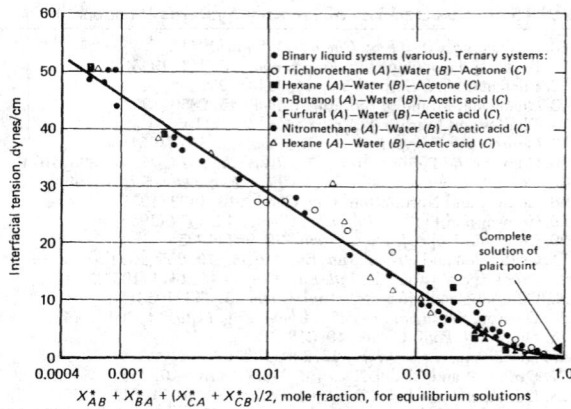

FIG. 15-10 Correlation of interfacial tension with mutual solubility for binary and ternary liquid mixtures. (*From Treybal*, Liquid Extraction, 2d ed., McGraw-Hill, New York, 1963.)

CALCULATION METHODS

SINGLE STAGE

An equilibrium, or theoretical, stage in liquid-liquid extraction as defined earlier is routinely utilized in laboratory procedures. A feed solution is contacted with an immiscible solvent to remove one or more of the solutes from the feed. This can be carried out in a separating funnel, or, preferably, in an agitated vessel that can produce droplets of about 1 mm in diameter. After agitation has stopped and the phases separate, the two clear liquid layers are isolated by decantation.

The equilibrium distribution coefficient can be calculated by material balance, using the weight of the feed F, raffinate R, and extract E, plus the weight-fraction solute in the feed x_f and raffinate x_r, when the weight-fraction solute in the extraction solvent y_s is zero [Eq. (15-8)].

$$K = \frac{y_e}{x_r} = \frac{R}{E}\left[\frac{F}{R}\frac{x_f}{x_r} - 1\right] \tag{15-8}$$

However, an actual analysis of the weight-fraction solute in the extract y_e and raffinate x_r is preferred.

CROSSCURRENT THEORETICAL STAGES

After a single-stage liquid-liquid contact the phase remaining from the feed solution (raffinate) can be contacted with another quantity of fresh extraction solvent. This **crosscurrent extraction scheme** (Fig. 15-4) is an excellent laboratory procedure because the extract and raffinate phases can be analyzed after each stage to generate equilibrium data. Also, the feasibility of solute removal to low levels can be demonstrated.

The number of crosscurrent stages N that are required to reach a specified raffinate composition, in Bancroft coordinates X_n, can be calculated directly if K' is constant, the ratio of extraction solvent to feed solvent S'/F' is kept constant, and fresh extraction solvent $Y_s = 0$ (presaturated with feed solvent) is used in each stage [Eq. (15-9)].

$$N = \frac{\log(X_f/X_n)}{\log(K'S'/F' + 1)} \tag{15-9}$$

The crosscurrent scheme is not generally economically attractive for large commercial processes because solvent usage is high and solute concentration in the combined extract is low.

COUNTERCURRENT THEORETICAL STAGES

The main objective for calculating the number of theoretical stages (or mass-transfer units) in the design of a liquid-liquid extraction process is to evaluate the compromise between the size of the equipment, or number of contactors required, and the ratio of extraction solvent to feed flow rates required to achieve the desired transfer of mass from one phase to the other. In any mass-transfer process there can be an infinite number of combinations of flow rates, number of stages, and degrees of solute transfer. The optimum is governed by economic considerations.

The number of stages that are required can be kept to a minimum by selecting a solvent with a high partition ratio or by operating with a high ratio of extraction solvent to feed. However, a high solvent flow rate usually requires a high operating cost because of the cost of recovering the solvent. A high solvent flow rate should be carefully compared with an increase in capital cost for taller or more equipment to achieve more theoretical stages (or mass-transfer units) and reduce the required flow of solvent. The operating cost of an extractor is generally quite low in comparison with the operating cost of the solvent-recovery distillation column.

The other common objective for calculating the number of countercurrent theoretical stages (or mass-transfer units) is to evaluate the performance of liquid-liquid extraction test equipment in a pilot plant or to evaluate production equipment in an industrial plant. Most liquid-liquid extraction equipment in common use can be designed to achieve the equivalent of 1 to 8 theoretical countercurrent stages, with some designed to achieve 10 to 12 stages.

Right-Triangular Method This method is a rigorous Ponchon-Savarit type of graphical technique for determining the number of countercurrent theoretical stages of a ternary system (Fig. 15-11). The horizontal axis is the concentration of solute in weight fractions x or y. The vertical axis is the weight fraction of extraction solvent. The weight fraction of feed solvent is simply the amount remaining so that all three weight fractions add up to 1.0.

For the system water–acetic acid–MIBK in Fig. 15-11 the raffinate (water) layer is the solubility curve with low concentrations of MIBK, and the extract (MIBK) layer is the solubility curve with high concentrations of MIBK. The dashed lines are **tie lines** which connect the two layers in equilibrium as given in Table 15-1. Example 2 describes the right-triangular method of calculating the number of theoretical stages required.

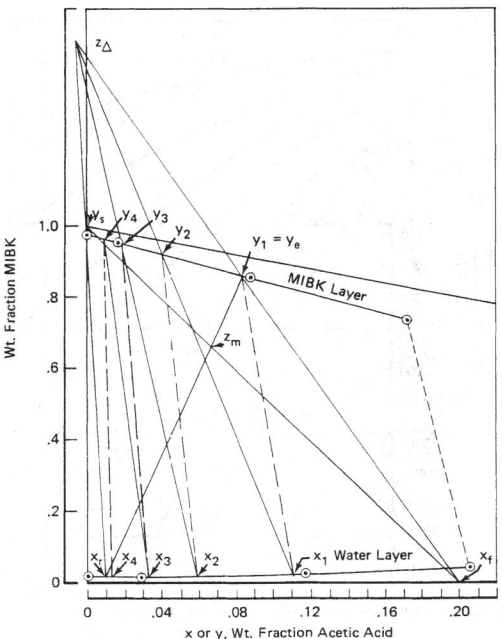

FIG. 15-11   Right-triangular graphical stages.

Example 2: Stage and Composition Calculation   A 100-kg/h feed stream containing 20 weight percent acetic acid in water is to be extracted with 200 kg/h of recycle MIBK that contains 0.1 percent acetic acid and 0.01 percent water. The aqueous raffinate is to be extracted down to 1 percent acetic acid. How many theoretical stages will be required and what will the extract composition be?

The solute concentration in the feed, $x_f = 0.20$, in the raffinate, $x_r = 0.01$, and in the extraction solvent, $y_s = 0.001$, can be located on the diagram. Then the mix point z_m can be calculated from the feed, $F = 100$ kg/h, and the solvent, $S = 200$ kg/h, entering the extractor [Eq. (15-10)].

$$z_m = (Fx_f + Sy_s)/(F + S) \qquad (15\text{-}10)$$

The mix point, $z_m = 0.0673$, falls on a straight line connecting x_f and y_s. The extract composition is then determined by drawing a straight line from x_r through z_m until the line intersects the extract line at the final extract composition, $y_e = 0.084$. The delta point z_Δ is then found at the intersection of two lines. One line connects the feed and extract compositions x_f and y_e. The other line connects the raffinate and solvent compositions x_r and y_s.

The graphical stepping off of theoretical stages starts at the extract composition y_e, and a tie line is drawn (parallel to the nearest one) to the raffinate composition leaving stage 1, $x_1 = y_e/K = 0.084 \, (0.117/0.089) = 0.1104$. The size of the extract stream can be calculated by the material balance $E = (F + S)(z_m - x_r)/(y_e - x_r)$. A straight line is drawn between x_1 and z_Δ to find the extract composition leaving stage 2, $y_2 = 0.0415$. Another tie line is drawn to find the raffinate composition leaving stage 2, x_2, and the stepwise procedure continues until the final raffinate composition, $x_r = 0.01$, is achieved. This requires four theoretical stages plus a fraction. Additional details on the derivation of this procedure are provided by Foust, Wenzel, Clump, Maus, and Anderson (*Principles of Unit Operations*, 2d ed, Wiley, New York, 1980) and Treybal (*Mass-Transfer Operations*, 3d ed., McGraw-Hill, New York, 1980).

Shortcut Methods   These methods are often preferred for repetitive calculations of pilot-plant data and numerous design conditions. In distillation calculations the assumption of constant molar vapor and liquid flow rates gave rise to the McCabe-Thiele stepwise calculation method with straight operating lines and a curved equilibrium line. A similar concept can be achieved in liquid-liquid extraction by assuming a constant flow rate of feed solvent F' and a constant flow rate of extraction solvent S' through the extractor. The solute concentrations are then given as the weight ratio of solute to feed solvent X and the weight ratio of solute to extraction solvent Y, i.e., Bancroft coordi-

nates. These concentrations and coordinates will essentially give a straight operating line on an XY diagram for stages 2 through $r - 1$ in Fig. 15-12. Equilibrium data using these weight ratios have already been shown to follow straight-line segments on a log-log plot (see Fig. 15-9). The main problem, then, is to evaluate the primary ratio of extraction solvent to feed solvent passing through the extractor in stages 2 through $r - 1$.

Robbins ("Liquid-Liquid Extraction," in Schweitzer, *Handbook of Separation Techniques for Chemical Engineers*, McGraw-Hill, New York, 1979, sec. 1.9) reported that most liquid-liquid extraction systems can be treated as having either (*A*) immiscible solvents, (*B*) partially miscible solvents with a low solute concentration in the extract, or (*C*) partially miscible solvents with a high solute concentration in the extract.

In case *A* the solvents are immiscible, so the rate of feed solvent alone in the feed stream F' is the same as the rate of feed solvent alone in the raffinate stream R'. In like manner, the rate of extraction solvent alone is the same in the stream entering S' as in the extract stream leaving E' (Fig. 15-12). The ratio of extraction-solvent to feed-solvent flow rates is therefore $S'/F' = E'/R'$. A material balance can be written around the feed end of the extractor down to any stage n (see Fig. 15-12) and then rearranged to a McCabe-Thiele type of operating line with a slope of F'/S' [Eq. (15-11)].

$$Y_{n+1} = \frac{F'}{S'} X_n + \frac{E'Y_e - F'X_f}{S'} \qquad (15\text{-}11)$$

Similarly, the same operating line can be derived from a material balance around the raffinate end of the extractor up to stage n [Eq. (15-12)].

$$Y_n = \frac{F'}{S'} X_{n-1} + \frac{S'Y_s - R'X_r}{S'} \qquad (15\text{-}12)$$

The overall extractor material balance is given by Eq. (15-13).

$$Y_e = \frac{F'X_f + S'Y_s - R'X_r}{E'} \qquad (15\text{-}13)$$

The end points of the operating line on an XY plot (Fig. 15-13) are X_r, Y_s and X_f, Y_e, and the number of theoretical stages can be stepped off graphically. The equilibrium curve is taken from the Hand type of correlation shown earlier (Fig. 15-9). When the equilibrium line is straight, its intercept is zero, and the operating line is straight, the number of theoretical stages can be calculated with one of the Kremser equations [Eqs. (15-14) and (15-15)]. When the intercept of the equilibrium line is not zero, the value of Y_s/K_s' should be used

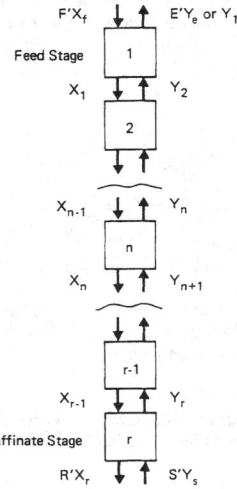

FIG. 15-12   Countercurrent extraction cascade.

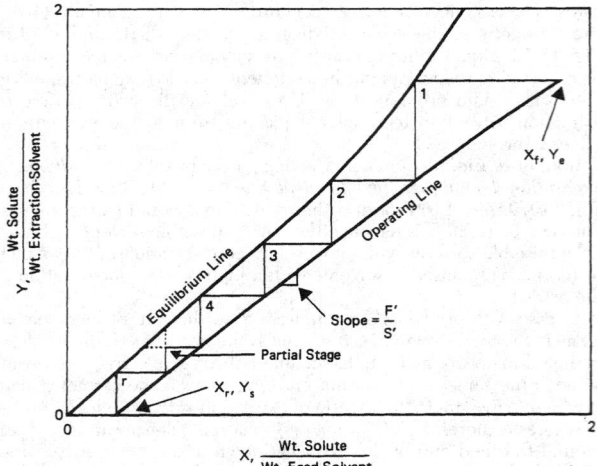

FIG. 15-13 Graphical calculation of countercurrent stages (Bancroft coordinates).

instead of Y_s/m, where K'_s is the partition ratio in Bancroft coordinates at Y_s.

When $\mathscr{E} \neq 1.0$,

$$N = \frac{\ln\left[\left(\dfrac{X_f - Y_s/m}{X_r - Y_s/m}\right)\left(1 - \dfrac{1}{\mathscr{E}}\right) + \dfrac{1}{\mathscr{E}}\right]}{\ln \mathscr{E}} \qquad (15\text{-}14)$$

When $\mathscr{E} = 1.0$,

$$N = \frac{X_f - Y_s/m}{X_r - Y_s/m} - 1 \qquad (15\text{-}15)$$

The value of m is the slope of the equilibrium line dY/dX [Eq. (15-4)]. This is equal to K' [Eq. (15-3)] at low concentrations where the equilibrium line is straight. The value of $\mathscr{E}$, the **extraction factor,** is calculated by dividing the slope of the equilibrium line m by the slope of the operating line F'/S' [Eq. (15-16)].

$$\mathscr{E} = mS'/F' \qquad (15\text{-}16)$$

The solution to the Kremser equation is shown graphically in Fig. 15-14. When a system responds with a constant number of theoretical stages N, the solute concentration in the raffinate X_r can readily be evaluated as the result of changing the ratio of solvent to feed [Eqs. (15-17) and (15-18)].

When $\mathscr{E} \neq 1.0$,

$$\frac{X_r - Y_s/m}{X_f - Y_s/m} = \frac{\mathscr{E} - 1}{\mathscr{E}^{N+1} - 1} \qquad (15\text{-}17)$$

When $\mathscr{E} = 1.0$,

$$\frac{X_r - Y_s/m}{X_f - Y_s/m} = \frac{1}{N+1} \qquad (15\text{-}18)$$

When the equilibrium line is not straight, Treybal (*Liquid Extraction,* 2d ed., McGraw-Hill, New York, 1963) recommends that the geometric mean value of m be used. The geometric mean of the slope of the equilibrium line at the concentration leaving the feed state m_1 and at the raffinate concentration leaving the raffinate stage m_r is $\sqrt{m_1 m_r}$.

Example 3: Shortcut Calculation, Case A   Let us solve the problem in Example 2 by using the shortcut calculation method assuming immiscible solvents, case A.

From the problem,

$$F' = 100(1 - 0.2) = 80 \text{ kg water/h}$$

$$X_f = 0.2/0.8 = 0.25 \text{ kg acetic acid/kg water}$$

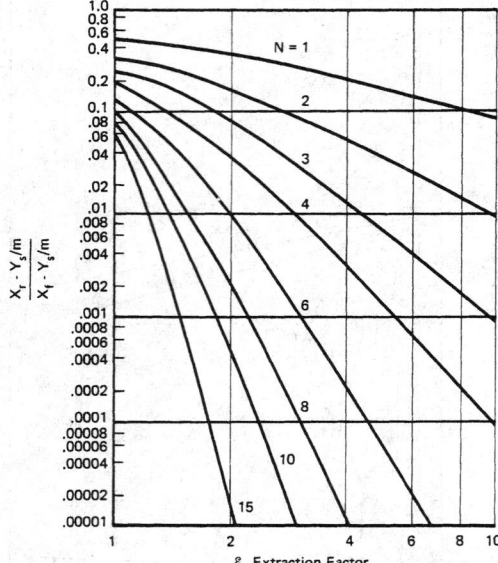

FIG. 15-14 Graphical solution to the Kremser equation.

$$X_r = 0.01/0.99 = 0.01 \text{ kg acetic acid/kg water}$$

$$S' = 200(1 - 0.001) = 199.8 \text{ kg MIBK/h}$$

$$Y_s = 0.2/199.8 = 0.001 \text{ kg acetic acid/kg MIBK}$$

If we assume $R' = F'$ and $E' = S'$, calculate Y_e from Eq. (15-13):

$$Y_e = \frac{80(0.25) + (199.8)(0.001) - 80(0.01)}{199.8} = 0.097 \frac{\text{kg acetic acid}}{\text{kg MIBK}}$$

From the correlation of equilibrium data (Table 15-3), $Y = 0.930(X)^{1.10}$, for X between 0.03 and 0.25.

Calculate $X_1 = (0.097/0.930)^{1/1.10} = 0.128$:

$$m = dY/dX = (0.930)(1.10)(X)^{0.1}, \text{ for } X \text{ between 0.03 and 0.25}$$

$$m_1 = 0.833 \text{ at } X = 0.128$$

$$m_r = dY/dX = K' = 0.656, \text{ for } X \text{ below 0.03}$$

$$K'_s = 0.656 \text{ at } Y_s = 0.001$$

$$\mathscr{E} = \sqrt{m_1 m_r} \, S'/F' = (0.739)(199.8)/80 = 1.85$$

N is determined from Fig. 15-14, Eq. (15-14), or the McCabe-Thiele type of plot (Fig. 15-13):

$$N = \frac{\ln\left[\left(\dfrac{0.25 - 0.001/0.656}{0.01 - 0.001/0.656}\right)\left(1 - \dfrac{1}{1.85}\right) + \dfrac{1}{1.85}\right]}{\ln 1.85} = 4.3$$

From solubility data at $Y = 0.1039$ (Table 15-1) the extract layer contains $5.4/85.7 = 0.0630$ kg water/kg MIBK and $y_e = (0.097)/(1 + 0.097 + 0.063) = 0.084$ weight-fraction acetic acid in the extract.

For cases B and C, Robbins ("Liquid-Liquid Extraction," in Schweitzer, *Handbook of Separation Techniques for Chemical Engineers,* McGraw-Hill, New York, 1979, sec. 1.9) developed the concept of pseudo solute concentrations for the feed and solvent streams entering the extractor that will allow the Kremser equations to be used.

In case B the solvents are partially miscible, and the miscibility is nearly constant through the extractor. This frequently occurs when all solute concentrations are relatively low. The feed stream is assumed to dissolve extraction solvent only in the feed stage and to retain the same amount throughout the extractor. Likewise, the extraction solvent is assumed to dissolve feed solvent only in the raffinate stage. With these assumptions the primary extraction-solvent rate moving through the extractor is assumed to be S', and the primary feed-solvent rate is assumed to be F'. The extract rate E' is less than S', and the raffinate rate R' is less than F' because of solvent solubilities.

The slope of the operating line is F'/S', just as in Eqs. (15-11) and (15-12), but only stages 2 through $r-1$ will fall directly on the operating line. And one knows that X_1 will be on the equilibrium line in equilibrium with Y_e by definition (see Fig. 15-12). One can calculate a pseudo feed concentration X_f that will fall on the operating line [Eq. (15-11)] at $Y_{n+1} = Y_e$ [Eq. (15-19)].

$$X_f^B = X_f + \frac{S' - E'}{F'} Y_e \qquad (15\text{-}19)$$

Likewise, one knows that Y_r will be on the equilibrium line with X_r (see Fig. 15-12). One can therefore calculate a pseudo concentration of solute in the inlet extraction solvent Y_s^B that will fall on the operating line [Eq. (15-12)] where $X_{n-1} = X_r$ [Eq. (15-20)].

$$Y_s^B = Y_s + \frac{F' - R'}{S'} X_r \qquad (15\text{-}20)$$

For case B, the two pseudo inlet concentrations X_f^B and Y_s^B can be used in the Kremser equation with the actual value of X_r and $\mathscr{E} = mS'/F'$ to calculate rapidly the number of theoretical stages required. The graphical stepwise solution shown in Fig. 15-13 can also be used. The operating line will go through points X_r, Y_s^B and X_f^B, Y_e with a slope of F'/S'. In one example studied by Robbins [*Chem. Eng. Prog.*, **76**(10), 58 (1980)], the actual feed and extract compositions gave a point to the left of the equilibrium line on an XY graph like Fig. 15-13 because the solubility of the solvent was so high. But the use of the pseudo feed composition still gave an accurate calculation of the number of theoretical stages as confirmed by a right-triangular graphical calculation.

Example 4: Shortcut Calculation, Case B   Let us solve the problem in Example 2 by assuming case B. The solute (acetic acid) concentration is low enough in the extract so that we may assume that the mutual solubilities of the solvents remain nearly constant. The material balance can be calculated by an iterative method.

From equilibrium data (Table 15-1) the extraction-solvent (MIBK) loss in the raffinate will be about $0.016/0.984 = 0.0163$ kg MIBK/kg water, and the feed-solvent (water) loss in the extract will be about $5.4/85.7 = 0.0630$ kg water/kg MIBK.

First iteration: assume $R' = F' = 80$ kg water/h. Then, extraction solvent in raffinate $= (0.0163)(80) = 1.30$ kg MIBK/h. Estimate $E' = 199.8 - 1.3 = 198.5$ MIBK/h. Then feed solvent in extract $= (0.063)(198.5) = 12.5$ kg water/h.

Second iteration: calculate $R' = 80 - 12.5 = 67.5$ kg water/h. $E' = 199.8 - (0.0163)(67.5) = 198.7$ kg MIBK/h.

Third iteration: converge $R' = 80 - (0.063)(198.7) = 67.5$ kg water/h. Y_e is calculated from the overall extractor material balance [Eq. (15-13)]:

$$Y_e = \frac{(80)(0.25) + (199.8)(0.001) - (67.5)(0.01)}{198.7} = 0.0983 \ \frac{\text{kg acetic acid}}{\text{kg MIBK}}$$

$$Y_e = \frac{0.0983}{1 + 0.0983 + 0.0630} = 0.0846$$

weight fraction acetic acid in extract.

From the correlation of equilibrium data (Table 15-3),

$$Y_e = 0.930(X)^{1.10}, \text{ for } X \text{ between } 0.03 \text{ and } 0.25$$

The raffinate composition leaving the feed (first stage) is calculated:

$$X_1 = (0.0983/0.930)^{1/1.10} = 0.130$$
$$m_1 = dY/dX = (0.930)(1.10)(X)^{0.1}$$
$$m_r = dY/dX = K' = 0.656$$
$$m_1 = 0.834 \text{ at } X_1 = 0.13$$
$$m_r = 0.656 \text{ at } X_r = 0.01$$
$$K_s' = 0.656 \text{ at } Y_s = 0.001$$
$$\mathscr{E} = \sqrt{m_1 m_r}\, S'/F' = (0.740)(199.8)/80 = 1.85$$

X_f^B is calculated from Eq. (15-19):

$$X_f^B = 0.25 + \frac{(199.8 - 198.7)(0.0983)(0.0983)}{80} = 0.251$$

Y_s^B is calculated from Eq. (15-20):

$$Y_s^B = 0.001 + \frac{(80 - 67.5)(0.01)}{199.8} = 0.0016$$

N is determined from Fig. 15-13, Eq. (15-14), or a McCabe-Thiele type of plot (Fig. 15-13) for case B.

$$N = \frac{\ln\left[\left(\dfrac{0.251 - 0.0016/0.656}{0.01 - 0.0016/0.656}\right)\left(1 - \dfrac{1}{1.85}\right) + \dfrac{1}{1.85}\right]}{\ln 1.85}$$

$$= 4.5 \text{ theoretical stages}$$

A less frequent situation, case C, can occur when the solute concentration in the extract is so high that a large amount of feed solvent is dissolved in the extract stream in the "feed stage" but a relatively small amount of feed solvent (say one-tenth as much) is dissolved by the extract stream in the "raffinate stage." The feed stream is assumed to dissolve the extraction solvent only in the feed stage just as in case B. But the extract stream is assumed to dissolve a large amount of feed solvent leaving the feed stage and a negligible amount leaving the raffinate stage. With these assumptions the primary feed-solvent rate is assumed to be R', so the slope of the operating line for case C is R'/S'. Again the extract rate E' is less than S', and the raffinate rate R' is less than F'.

The pseudo feed concentration for case C, X_f^C, can be calculated from Eq. (15-21).

$$X_f^C = \frac{F'}{R'} X_f + \frac{S' - E'}{R'} Y_e \qquad (15\text{-}21)$$

And the value of Y_s will fall on the operating line for case C. The extraction factor for case C is calculated from Eq. (15-22).

$$\mathscr{E}^C = mS'/R' \qquad (15\text{-}22)$$

On an XY diagram for case C the operating line will go through points X_r, Y_s and X_f^C, Y_e with a slope of R'/S' similar to Fig. 15-13. When using the Kremser equation for case C, one uses the pseudo feed concentration X_f^C from Eq. (15-21) and the stripping factor $\mathscr{E}^C$ from Eq. (15-22). One uses the raffinate concentration X_r and inlet solvent concentration Y_s without modification.

For the first time through a liquid-liquid extraction problem, the right-triangular graphical method may be preferred because it is completely rigorous for a ternary system and reasonably easy to understand. However, the shortcut methods with the Bancroft coordinates and the Kremser equations become valuable time-savers for repetitive calculations and for data reduction from experimental runs. The calculation of pseudo inlet compositions and the use of the McCabe-Thiele type of stage calculations lend themselves readily to programmable calculator or computer routines with a simple correlation of equilibrium data.

COUNTERCURRENT MASS-TRANSFER-UNIT CALCULATIONS

The concept of a mass-transfer unit was developed many years ago to represent more rigorously what happens in a differential contactor rather than a stagewise contactor. For a straight operating line and a straight equilibrium line with an intercept of zero, the equation for calculating the number of mass-transfer units based on the overall raffinate phase N_{or} is identical to the Kremser equation except for the denominator when the extraction factor is not equal to 1.0 [Eq. (15-23)].

When $\mathscr{E} \neq 1.0$,

$$N_{or} = \frac{\ln\left[\left(\dfrac{X_f - Y_s/m}{X_r - Y_s/m}\right)\left(1 - \dfrac{1}{\mathscr{E}}\right) + \dfrac{1}{\mathscr{E}}\right]}{1 - 1/\mathscr{E}} \qquad (15\text{-}23)$$

The number of mass-transfer units N_{or} is identical to the number of theoretical stages when the extraction factor $\mathscr{E}$ is 1.0 [Eq. (15-24)].

When $\mathscr{E} = 1.0$,

$$N_{or} = [(X_f - Y_s/m)/(X_r - Y_s/m)] - 1 \qquad (15\text{-}24)$$

The differences become pronounced when values of the extraction factor are high [Eq. (15-25)].

$$N_{or} = N \ln \mathscr{E}/(1 - 1/\mathscr{E}) \qquad (15\text{-}25)$$

Even staged equipment may be modeled best by the number of mass-transfer units when the extraction factor is much higher than 1.5, especially if the stage efficiencies are low.

The response of solute concentration in the raffinate X_r to the solvent-to-feed ratio S'/F' can be calculated by Eqs. (15-26) and (15-27) for a constant number of transfer units based on the overall raffinate phase N_{or}.

When $\mathscr{E} \neq 1.0$,

$$\frac{X_r - Y_s/m}{X_f - Y_s/m} = \frac{1 - 1/\mathscr{E}}{e^{N_{or}(1 - 1/\mathscr{E})} - 1/\mathscr{E}} \qquad (15\text{-}26)$$

When $\mathscr{E} = 1.0$,

$$\frac{X_r - Y_s/m}{X_f - Y_s/m} = \frac{1}{N_{or} + 1} \qquad (15\text{-}27)$$

The solution to these equations is shown graphically in Fig. 15-15. Note that the raffinate composition is not reduced appreciably when the extraction factor $\mathscr{E}$ is increased from 5 to infinity. This is true because mass transfer from the raffinate phase limits the performance. This is typical of the performance of many devices including actual staged equipment. However, if there is sufficient residence time in each stage of a staged device so that high stage efficiencies can be achieved, then the raffinate can be reduced substantially by increasing the extraction factor above 5 (see Fig. 15-14). However, the solute concentration in the extract stream would be quite dilute.

Example 5: Number of Transfer Units Let us calculate the number of transfer units required to achieve the separation in Example 3. The solution to the problem is the same as in Example 3 except that the denominator is changed in the final equation [Eq. (15-25)]:

$$N_{or} = 4.5 \; \frac{\ln 1.85}{1 - 1/1.85} = 6.0 \text{ transfer units}$$

STAGE EFFICIENCY AND HEIGHT OF A THEORETICAL STAGE OR TRANSFER UNIT

The overall stage efficiency of a staged extraction system is simply the number of theoretical stages divided by the number of actual stages times 100 [Eq. (15-28)].

$$\text{Percent stage efficiency} = 100N/\text{number of actual stages} \qquad (15\text{-}28)$$

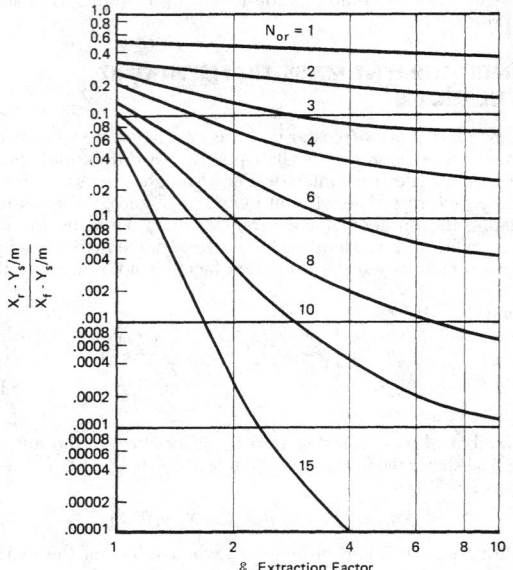

FIG. 15-15 Graphical solution to the mass-transfer-unit equations.

A similar term of number of transfer units per actual stage could also be envisioned.

The height equivalent to a theoretical stage (HETS) in an extraction tower is simply the height of the tower Z_t divided by the number of theoretical stages achieved [Eq. (15-29)].

$$\text{HETS} = Z_t/N \qquad (15\text{-}29)$$

Likewise, the height of a transfer unit based on raffinate-phase compositions H_{or} is the height of tower divided by the number of transfer units [Eq. (15-30)].

$$H_{or} = Z_t/N_{or} \qquad (15\text{-}30)$$

The contribution to the height of a transfer unit overall based on the raffinate-phase compositions is the sum of the contribution from the resistance to mass transfer in the raffinate phase H_r plus the contribution from the resistance to mass transfer in the extract phase H_e, divided by the extraction factor $\mathscr{E}$ [Eq. (15-31)].

$$H_{or} = H_r + H_e/\mathscr{E} \qquad (15\text{-}31)$$

At high extraction factors the height of a transfer unit is mostly dependent on the resistance to the transfer of solute from the raffinate phase.

Prediction methods attempt to quantify the resistances to mass transfer in terms of the raffinate rate R and the extract rate E, per tower cross-sectional area A_t, and the mass-transfer coefficient in the raffinate phase k_r and the extract phase k_e, times the interfacial (droplet) mass-transfer area per volume of tower a [Eqs. (15-32) and (15-33)].

$$H_r = R/A_t k_r a \qquad (15\text{-}32)$$

$$H_e = E/A_t k_e a \qquad (15\text{-}33)$$

The mass-transfer coefficients depend on complex functions of diffusivity, viscosity, density, interfacial tension, and turbulence. Similarly, the mass-transfer area of the droplets depends on complex functions of viscosity, interfacial tension, density difference, extractor geometry, agitation intensity, agitator design, flow rates, and interfacial rag deposits. Only limited success has been achieved in correlating extractor performance with these basic principles. The lumped parameter H_{or} deals directly with the ultimate design criterion, which is the height of an extraction tower.

FRACTIONATION STAGES

One of the most sophisticated separations achievable by liquid-liquid extraction is fractionation. Two solutes can be separated almost completely by isolating one solute b into the extraction solvent S' and another c into a wash solvent W' (Fig. 15-16). The bottom section of a fractionation extraction is about the same as the countercurrent extractions described earlier, with the extraction solvent S' entering the bottom and extracting, i.e., stripping, one of the solutes b almost completely from the raffinate R'. As the extract stream moves above the feed stage, it is contacted countercurrently with a wash solvent W' that scrubs the unwanted solute c out of the extract stream. This in effect purifies the solute b that is being extracted. The stripping section and the washing (enriching) section will each have its own operating line on a McCabe-Thiele type of XY diagram (Fig. 15-17). The overall material balance must be met at the feed stage.

For the case in which the extraction solvent can be assumed to be totally immiscible with the wash solvent and there is no solvent in the feed, the extraction factor $\mathscr{E}$ must be greater than 1.0 for component b and less than 1.0 for component c [Eq. (15-34)].

$$\mathscr{E} = mS'/W' \qquad (15\text{-}34)$$

For a symmetrical separation of component b from c, Brian (*Staged Cascades in Chemical Processing*, Prentice-Hall, Englewood Cliffs, N.J., 1972) reported that the ratio of wash solvent to extraction solvent W'/S' should be set equal to the geometric mean of the two slopes of the equilibrium lines [Eq. (15-35)].

$$W'/S' = \sqrt{m_b m_c} \qquad (15\text{-}35)$$

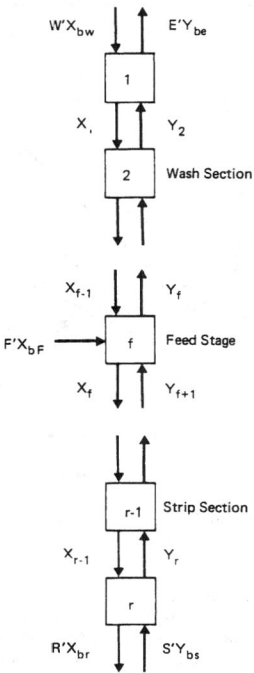

FIG. 15-16   Liquid-liquid fractionation cascade.

The ratio of wash solvent to extraction solvent is the same in the enriching section as in the stripping section if no solvent is added in the feed. The degree of separation to be achieved can be chosen for the process design, such as 99 percent of component b into the extract stream and 99 percent of component c into the raffinate stream. Then the feed rate can be chosen so that the solute loadings in the extract stream and

the raffinate stream are reasonable. This becomes especially critical near the feed stage, where the solute loadings are highest.

An overall material balance can be calculated around the extractor, and then an XY plot can be constructed for each solute (Figs. 15-17 and 15-18). The solute concentrations at the raffinate end of the extractor, X_{br} and Y_{bs}, can be plotted for component b, and the operating line can be drawn with a slope of W'/S' with no solvent in the feed. The solute concentrations at the extract end of the extractor, X_{bw} and Y_{be}, can also be plotted for component b with the enriching-section operating line also having the slope W'/S' if no solvents were added with the feed. The theoretical stages can be stepped off for each section of the extractor by starting at the extract end, stage 1, and stepping toward the feed stage f, then restarting at the raffinate end, stage r, and stepping toward the feed stage f (Fig. 15-17). A similar procedure is repeated for component c (Fig. 15-18).

The feed-stage number is found by matching the concentrations and stage number. This occurs at the point where the feed should be introduced (see Treybal, *Mass-Transfer Operations,* 3d ed., McGraw-Hill, New York, 1980). The procedure for matching concentrations is carried out by plotting the stage number on the vertical axis and the raffinate concentration X for each component (Fig. 15-19). The concentrations are matched when the rectangle *HJLK* can be drawn as shown. The number of stages in the wash section including the feed stage is determined from the position of line *HJ*. The total number of stages r is determined from the position of *LK*, which is also at the feed stage.

The solute concentrations can be seen to be highest at the feed stage (Figs. 15-17 and 15-18). Also the solute concentrations increase as the number of theoretical stages is increased. For a given flow rate of feed, the flow rates of the solvents entering the extraction must be sufficiently high so that neither solubility limits nor a plait point is exceeded, nor a pinch point is reached between the operating lines and the equilibrium lines. The presence of solvents in the feed stream will change the slope of one or both of the operating lines, and several ratios of extraction solvent to wash solvent may have to be evaluated to find the optimum. The final optimization is usually carried out in pilot-plant equipment. Theoretically the use of solute reflux to the ends of the extraction cascade can reduce the number of theoretical stages required by a factor of 2 according to Brian (*Staged Cascades in Chemical Processing,* Englewood Cliffs, N.J., 1972), but again the amount of solvent flow rates may have to be increased to avoid a pinch point or plait point near the feed stage.

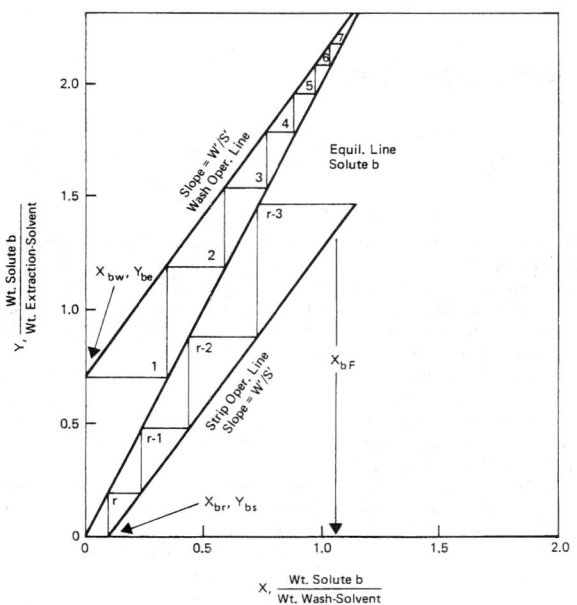

FIG. 15-17   Graphical calculation of fractionation stages for solute b.

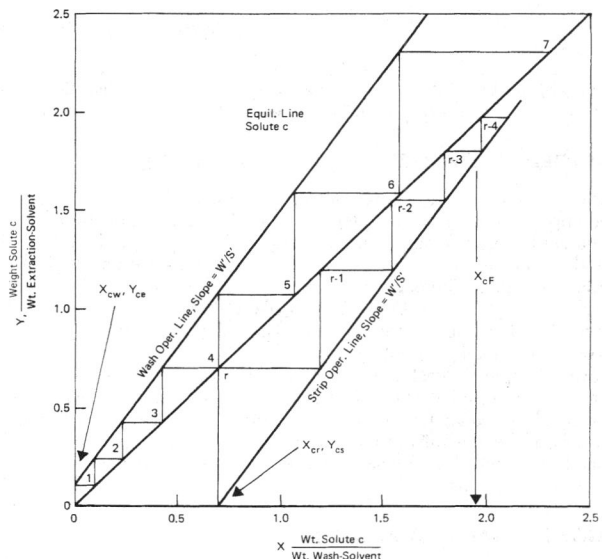

FIG. 15-18   Graphical calculation of fractionation stages for solute c.

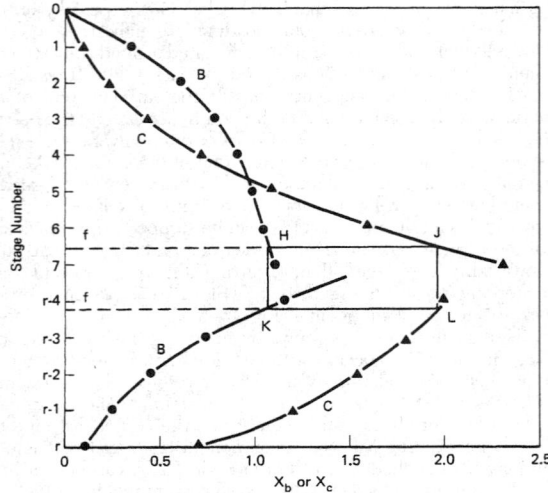

FIG. 15-19   Matching concentrations at the feed stage.

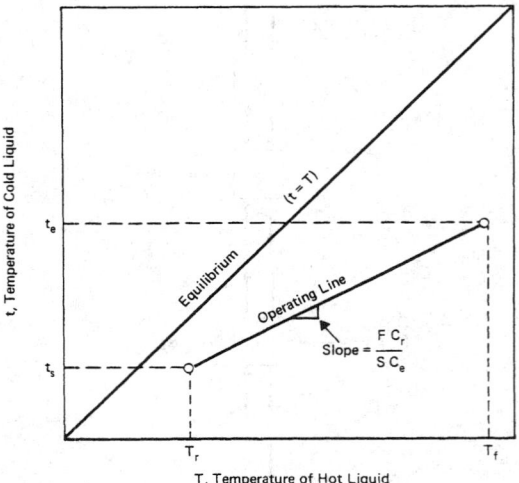

FIG. 15-20   Countercurrent heat transfer.

COUNTERCURRENT LIQUID-LIQUID HEAT TRANSFER

Heat may be transferred between two insoluble liquids in countercurrent flow through an extractor, and the performance can be evaluated in the same general manner as in mass transfer (Fig. 15-20). For a differential contactor the number of overall heat-transfer units based on the hot phase N_{oh} can be derived from the same equations used for the number of mass-transfer units based on the feed (raffinate) phase [Eq. (15-36)].

$$N_{oh} = \int_{T_r}^{T_f} \frac{dT}{T - t} = \frac{Z_t}{N_{oh}} = \frac{Z_t U_o a_r A_t}{FC_r} \qquad (15\text{-}36)$$

where T = temperature of the hot (raffinate) phase, t = temperature of

the cold (extract) phase, Z_t = height of tower, H_{oh} = height of an overall heat-transfer unit based on the hot (raffinate) phase, U_o = overall heat-transfer coefficient, a_r = heat-transfer surface of the droplets per volume, A_t = cross-sectional area of tower, F = hot feed rate, and C_r = heat capacity of raffinate and feed. The solution to the integral in Eq. (15-36) is identical with Eqs. (15-23) and (15-24), where $X_f = T_f$, $X_r = T_r$, $Y_s = T_s$, $\mathcal{E} = SC_e/FC_r$, S = cold solvent rate, and C_e = heat capacity of solvent and extract. The slope of the equilibrium line $m = dT/dt = 1.0$ since $t = T$ at equilibrium. The height of a heat transfer unit H_{oh} is reported by Von Berg [in C. Hanson (ed.), *Recent Advances in Liquid-Liquid Extraction*, Pergamon, New York, 1971, chap. 11] to be shorter than the height of a mass-transfer unit H_{or} by a factor of 3 to 20. As an alternative, the Kremser equations [Eqs. (15-14) and (15-15)] could be used to calculate the number of theoretical heat-transfer stages.

LIQUID-LIQUID EXTRACTION EQUIPMENT

Liquid-liquid contacting equipment may be generally classified into two categories: **stagewise** and **continuous (differential) contact.**

STAGEWISE EQUIPMENT (MIXER-SETTLERS)

The function of a stage is to contact the liquids, allow equilibrium to be approached, and to make a mechanical separation of the liquids. The contacting and separating correspond to mixing the liquids, and settling the resulting dispersion; so these devices are usually called **mixer-settlers.** The operation may be carried out in batch fashion or with continuous flow. If batch, it is likely that the same vessel will serve for both mixing and settling, whereas if continuous, separate vessels are usually but not always used.

In principle, at least, any mixer may be coupled with any settler to provide the complete stage. There are several combinations which are especially popular. Continuously operated devices usually, but not always, place the mixing and settling functions in separate vessels. Batch-operated devices may use the same vessel alternately for the separate functions.

RATES OF MASS TRANSFER

Measurements simply of the extent of extraction in an agitated vessel lead to the overall "volumetric" mass-transfer coefficients, $K_C a_{av}$ or

$K_D a_{av}$, or the equivalent stage efficiency. The coefficients K_C and K_D are made up of the coefficients for the individual liquids, k_C and k_D:

$$\frac{1}{K_D} = \frac{1}{k_D} + \frac{1}{m'_{CD}k_C}; \quad \frac{1}{K_C} = \frac{1}{k_C} + \frac{m'_{CD}}{k_D} \qquad (15\text{-}37)$$

The evidence is that the coefficients k_C and k_D and the interfacial area a_{av} depend differently upon operating variables. For purposes of design, therefore, it is ultimately necessary to have separate information on the quantities k_C, k_D, and a_{av}. The role of an additional surface resistance is emphasized by the studies of Kishinevski and Moehalova [*Zh. Prikl. Khim.*, **33**, 2049 (1960)].

Information on the coefficients is relatively undeveloped. They are evidently strongly influenced by rate of drop coalescence and breakup, presence of surface-active agents, "interfacial turbulence" (Marangoni effect), drop-size distribution, and the like, none of which can be effectively evaluated at this time.

Continuous-Phase Coefficients   There have been a large number of measurements of k_C for solid particles and gas bubbles suspended in agitated liquids [for review, see Miller, *Ind. Eng. Chem.*, **56**(10), 18 (1964)]. A typical correlation of these data is that of Calderbank and Moo-Young [*Chem. Eng. Sci.*, **16**, 39 (1961)]:

$$k_C N_{Sc}^{2/3} = 0.13(P\mu_c g_c/v\rho^2)^{1/4} \qquad (15\text{-}38)$$

Schindler and Treybal [*Am. Inst. Chem. Eng. J.*, **14**, 790 (1968)], however, found that for liquid dispersions of ethyl acetate saturated with water, agitated in water by flat-blade turbine impellers, k_C was appreciably larger than that given by Eq. (15-38) for baffled vessels and even higher for unbaffled vessels (no air-liquid interface). The increase was attributed to the rate of coalescence of the droplets as the dispersion emerged from the impeller and recirculated through the tank and to their redispersion at the impeller. It was described by an expression of the form

$$k_C = k_S + C(D_C/\theta_C)^{0.5} \qquad (15\text{-}39)$$

where k_S was calculated from Harriott's data for small-diameter solids [Harriott, *Am. Inst. Chem. Eng. J.*, **8**, 93 (1962)]. The continuous phase was found to be completely back-mixed and of uniform composition throughout for both baffled and unbaffled vessels.

Dispersed-Phase Coefficients   There have been no direct measurements of k_D for liquid dispersions in agitated vessels. If the drops are small (as they usually are), internal circulation causes them to behave like rigid spheres with an enhanced diffusivity D'_D. In stirred vessels, the ratio D'_D/D_D has been estimated to lie in the range of about 1:2 [Olney, *Am. Inst. Chem. Eng. J.*, **7**, 348 (1961); Treybal *Liquid Extraction*, 2d ed., McGraw-Hill, New York, 1963]. For a pump-mix impeller (Fig. 15-28), Coughlin and von Berg [*Chem. Eng. Sci.*, **21**, 3 (1966)], on the other hand, estimate k_D to be higher than that for circulating drops but not so large as that for oscillating drops (see below). These estimates do not take into account drop coalescence, interfacial turbulence, etc.; they are based on an assumed value for k_C and measured overall coefficients.

Overall Coefficients and Stage Efficiency   If it is assumed that values of a_{av}, k_C, k_D (and therefore K_D) can somehow be estimated, the stage efficiency can be calculated through

$$E_{MD} = 1 - \exp\left(-\frac{K_D a_{av} Z}{V_D}\right) = 1 - \exp\left[-\frac{K_D a_{av}\theta(V_C + V_D)}{V_D}\right] \quad (15\text{-}40)$$

See also Treybal [*Am. Inst. Chem. Eng. J.*, **4**, 202 (1958); **6**, 5M (1960)] and Olander [*Chem. Eng. Sci.*, **18**, 47 (1963); **19**, 275 (1964)]. The remaining discussion is confined to measured stage efficiency or volumetric overall coefficients. These are largely of value only for the particular systems studied. For this reason, one fairly complete study will be described, and the others will only be mentioned.

Figure 15-21 summarizes the results for the extraction of *n*-butylamine from kerosine into water in a continuously operated mixer [$T = 0.37$ m (1.23 ft); $Z = 0.48$ m (1.562 ft)] fed cocurrently upward, with and without four wall baffles and with a variety of impellers [Overcashier, Kingsley, and Olney, *Am. Inst. Chem. Eng. J.*, **2**, 529 (1956)]. When unbaffled, the vessel was full and without an air-liquid interface. E_O represents the overall countercurrent efficiency of a single stage. E_O at zero agitator speed was 0.18 at a liquid residence time of 1.08 min. The improved performance in the absence of baffles may be attributed to the reduction in back mixing and to the reduced power requirement for a given impeller speed. In the absence of baffles, vertical location of the impeller is immaterial. With baffles, the best performance is given with the impeller at 0.667 Z from the bottom, the worst at 0.25 Z from the bottom. For the spiral turbine, wall baffles and stator-ring baffles produced the same power-efficiency relationship. Off-center unbaffled operation at a propeller was intermediate between centered baffled and centered unbaffled operation. The data for propellers, spiral turbines, and flat-blade turbines, $d_i = 0.10$ to 0.25 m (0.333 to 0.833 ft), in both unbaffled and baffled tanks, with a flow rate to produce a residence time $\theta = 0.18$ h, kerosine-water ratio = 1.57 by volume, are empirically correlated by

$$E_O = 1 - \frac{0.318(10^{15})(d_i/d_t)^b}{N_{Re}^{3.2} N_{Po}^{1.37}} \qquad (15\text{-}41)$$

where $b = 0$ for baffled operation and 1.6 for unbaffled operation.

Other detailed studies are the following:

1. Hixson and Smith [*Ind. Eng. Chem.*, **41**, 973 (1949)]. Batch extraction of iodine from water into carbon tetrachloride; unbaffled vessels, propeller agitated. Log $(1 - E)$ is linear with time.
2. Karr and Scheibel [*Chem. Eng. Prog.*, **50**, Symp. Ser. 10, 73 (1954)]. Continuous extraction of acetic acid between methyl isobutyl ketone and water, and

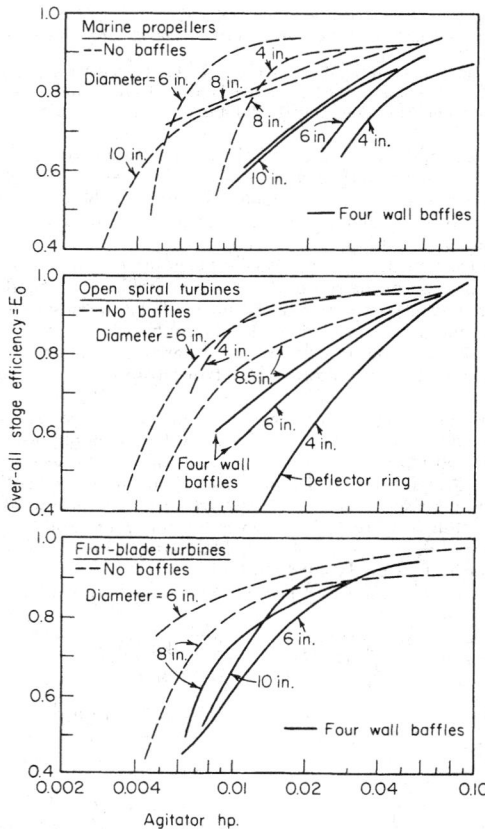

FIG. 15-21   Continuous extraction of *n*-butylamine from kerosine into water. $T = 1.23$ ft, $Z = 1.56$ ft, no air-liquid interface, impellers centered, $V_R/V_E \times 1.57$, residence time $\times 1.08$ min. To convert feet to meters, multiply by 0.3048; to convert inches to centimeters, multiply by 2.54; and to convert horsepower to kilowatts, multiply by 0.746. [*Overcashier, Kingsley, and Olney*, Am. Inst. Chem. Eng. J., **2**, 529 (1956), *with permission.*]

xylene and water, and of acetone between xylene and water; unbaffled vessels consisting of the unpacked section of the extractor of Fig. 15-44. Rate of extraction is larger when organic liquid is dispersed in the extractant than with other arrangements.

3. Flynn and Treybal [*Am. Inst. Chem. Eng. J.*, **1**, 324 (1955)]. Continuous extraction of benzoic acid from toluene and kerosine into water; baffled vessels, turbine agitators. Stage efficiency is correlated with agitator energy per unit of liquid treated.
4. Mottel and Colvin (U.S. AEC DP-254, 1957). Continuous heat transfer between kerosine and water; vessel of Pump-Mix design (Fig. 15-28).
5. Ryon, Daley, and Lowrie [*Chem. Eng. Prog.*, **55**(10), 70, (1959), U.S. AFC ORNL-2951, 1960]. Continuous extraction of uranium from sulfate-ore-leach liquors and kerosine + tributyl phosphate and di(2-ethylhexyl)-phosphoric acid; baffled vessels, turbine agitated. There is strong evidence of the influence of a slow chemical reaction.
6. Ryon and Lowrie (U.S. AEC ORNL-3381, 1960). Batch and continuous extraction of uranium from aqueous sulfate solutions into kerosine + amines, stripping of extract with aqueous sodium carbonate; baffled vessels, turbine agitated. A detailed process study.
7. David and Colvin [*Am. Inst. Chem. Eng. J.*, **7**, 72 (1961)]. Continuous heat transfer between kerosine and water; unbaffled vessel. Open impellers (paddles and propellers) are better than closed (centrifugal and disk impellers) at the same tip speed.
8. Simard et al. [*Can. J. Chem. Eng.*, **39**, 229 (1961)]. Continuous extraction of uranium from aqueous nitrate solutions into kerosine + tributyl phosphate and from sulfate solutions containing tricaprylamine; unbaffled vessel, propeller agitated. Process details for high recovery and low reagent costs.
9. Rushton, Nagata, and Rooney [*Am. Inst. Chem. Eng. J.*, **10**, 298 (1964)].

Batch extraction of octanoic acid from water and corn syrup into xylene, paraffin oil, and their mixtures; baffled vessel, turbine impeller. $K_C a_{av}$ proportional to $N^{2.1} \mu_C^{-0.6} \mu_D^{-0.55}$.

10. Coughlin and von Berg [*Chem. Eng. Sci.*, **21**, 3 (1966)]. Continuous heat transfer and extraction of ethylbutyric acid between kerosine and water; unbaffled vessel, Pump-Mix design (Fig. 15-28). Interfacial area measured.

Scale-Up of Mixers   For the details associated with the design and scale-up of agitated vessels, the reader is referred to Section 18 which covers this topic in great detail. The intention here is to provide only some of the general principles involved which have particular application to liquid-liquid extraction.

For extraction, the mixing usually takes place either in a vessel which also serves as the settler (these can be baffled or unbaffled), or a separate mixing compartment (usually baffled if there is a gas-liquid interface, and usually unbaffled if it is liquid filled).

The most common impellers are the marine impeller or disc flat-blade turbine; the flow patterns which typically result are illustrated in Fig. 15-22.

The power for agitation of two-phase mixtures in vessels such as these is given by the curves in Fig. 15-23. At low levels of power input, the dispersed phase holdup in the vessel (ϕ_D) can be less than the value in the feed (ϕ_{DF}); it will approach the value in the feed as the agitation is increased. Treybal (*Mass Transfer Operations*, 3d ed., McGraw-Hill, New York, 1980) gives the following correlations for estimation of the dispersed phase holdup based on power and physical properties for disc flat-blade turbines:

Baffled vessels, impeller power/vessel volume > 105 W/m³ = 2.2 ft lb$_f$/ft³·s:

$$\frac{\phi_D}{\phi_{DF}} = 0.764 \left(\frac{P q_D \mu_C^2}{v_L \sigma^3 g_c} \right)^{0.300} \left(\frac{\mu_C^3}{q_D \rho_C^2 \sigma g_c} \right)^{0.178} \left(\frac{\rho_C}{\Delta \rho} \right)^{0.0741}$$
$$\times \left(\frac{\sigma^3 \rho_C g_c^3}{\mu_C^4 g} \right)^{0.276} \left(\frac{\mu_D}{\mu_c} \right)^{0.136} \quad (15\text{-}42)$$

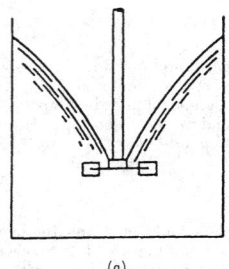 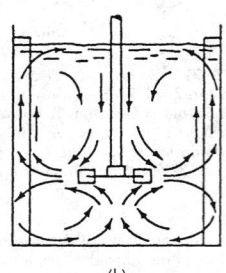

(a)   (b)

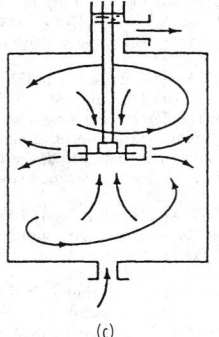

(c)

FIG. 15-22   Liquid agitation by a disc flat blade turbine in the presence of a gas-liquid interface (*a*) without wall baffles, (*b*) with wall baffles, and (*c*) in full vessels without a gas-liquid interface (continuous flow) and without baffles. [*Courtesy Treybal*, Mass Transfer Operations, *3rd ed., p. 148, McGraw-Hill, NY, (1980).*]

Unbaffled vessels, full, no gas-liquid surface, no vortex:

$$\frac{\phi_D}{\phi_{DF}} = 3.39 \left(\frac{P q_D \mu_C^2}{v_L \sigma^3 g_c} \right)^{0.247} \left(\frac{\mu_C^3}{q_D \rho_C^2 \sigma g_c} \right)^{0.427} \left(\frac{\rho_C}{\Delta \rho} \right)^{0.430}$$
$$\times \left(\frac{\sigma^3 \rho_C g_c^3}{\mu_C^4 g} \right)^{0.401} \left(\frac{\mu_D}{\mu_c} \right)^{0.0987} \quad (15\text{-}43)$$

He then recommends that the following relationships be used to estimate the mixture average density and viscosity for the power calculations:

Baffled vessels:

$$\rho_M = \rho_C \phi_C + \rho_D \phi_D \quad (15\text{-}44)$$

$$\mu_M = \frac{\mu_C}{\phi_C} \left(1 + \frac{1.5 \mu_D \phi_D}{\mu_C + \mu_D} \right) \quad (15\text{-}45)$$

Unbaffled vessels, no gas-liquid interface, no vortex:

$$\mu_M = \begin{cases} \dfrac{\mu_w}{\phi_w} \left(1 + \dfrac{6\mu_o \phi_o}{\mu_o + \mu_w} \right); & \phi_w > 0.4 \quad (15\text{-}46a) \\[2ex] \dfrac{\mu_o}{\phi_o} \left(1 - \dfrac{1.5 \mu_w \phi_w}{\mu_o + \mu_w} \right); & \phi_w < 0.4 \quad (15\text{-}46b) \end{cases}$$

In scale-up, there are three types of similarity to be considered:

1. *Geometric similarity*. Two vessels are geometrically similar if the ratio of *all* corresponding dimensions is the same

2. *Kinematic similarity*. Two vessels are kinematically similar if they are first geometrically similar *and* have the same ratio of velocities in corresponding positions of the vessel

3. *Dynamic similarity*. Two vessels are dynamically similar if they are first kinematically similar *and* all force ratios are equal in corresponding positions of the vessel

For most liquid-liquid extraction applications, the mixing section is usually scaled up on the principle of geometric similarity, and the power is based on maintaining the same power per unit volume. Treybal [*Chem. Eng. Prog.*, **62**(9), 67 (1966)] demonstrates that, for geometrically similar vessels with equal holding time and power per unit volume, the stage efficiency for liquid extraction is likely to increase on scale-up, so this is generally a conservative approach.

Because of the difficulty in obtaining good data on mass-transfer coefficients and interfacial area as outlined earlier, it is necessary that bench or pilot scale experiments be performed to obtain the data needed for scale-up. The usual procedure is to determine a suitable range of residence times at various power inputs for a given mixer geometry. Most extractions are mass-transfer limited, so relatively short residence times are adequate (in the range of 1–3 minutes). However in some cases (such as metal extractions), there is actually a reactive-extraction taking place, and residence time becomes more critical; times in the range of 10–15 minutes are not unusual.

Besides looking at just the mixing, it is important at this time to also consider the settling time of the phases after mixing since this will impact on the settler design. Higher intensity of mixing may decrease the residence time for mass transfer, but at the same time create fine dispersions which are difficult to settle.

With the batch data, Slater and Godfrey in Lo, Baird, and Hanson, *Handbook of Solvent Extraction*, Wiley, New York, 1983, recommend that an approach to equilibrium be used to provide the fundamental basis for scale-up; they define the approach to equilibrium (E_f) as:

$$E = \frac{C_i - C_t}{C_i - C_e} \quad (15\text{-}47)$$

It has been found that this data can be correlated for batch extraction using the following correlation:

$$1 - E_b = e^{(-k t_b)} \quad (15\text{-}48)$$

Once the value of k is obtained from the batch data, it can be related to a continuous extraction via the correlation:

$$E_f = \frac{k t_c}{1 + k t_c} \quad (15\text{-}49)$$

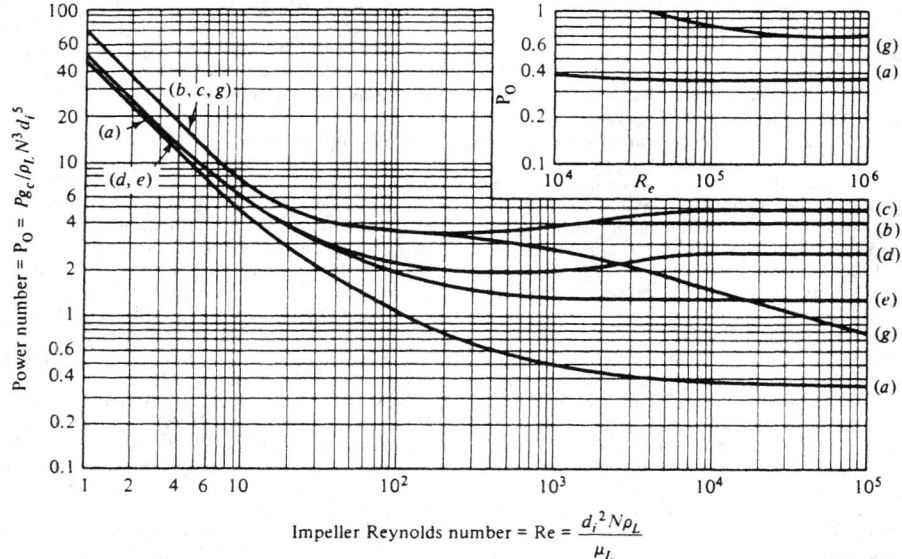

FIG. 15-23   Power for agitation impellers immersed in single-phase liquids, baffled vessels with a gas-liquid surface [except curves (c) and (g)]. Curves correspond to: (a) marine impellers, (b) flat-blade turbines, $w = d_i/5$, (c) disk flat-blade turbines with and without a gas-liquid surface, (d) curved-blade turbines, (e) pitched-blade turbines, (g) flat-blade turbines, no baffles, no gas-liquid interface, no vortex.

Notes on Fig. 15-23.

1.   The power P is only that imparted to the liquid by the impeller. It is not that delivered to the motor drive, which additionally includes losses in the motor and speed-reducing gear. These may total 30 to 40 percent of P. A stuffing box where the shaft enters a covered vessel causes additional losses.

2.   All the curves are for axial impeller shafts, with liquid depth Z equal to the tank diameter d_t.

3.   Curves a to e are for open vessels, with a gas-liquid surface, fitted with four baffles, $b = d_t/10$ to $d_t/12$.

4.   Curve a is for marine propellers, $d_i/d_t \approx \frac{1}{3}$, set a distance $C = d_i$ or greater from the bottom of the vessel. The effect of changing d_i/d_t is apparently felt only at very high Reynolds numbers and is not well established.

5.   Curves b to e are for turbines located at a distance $C = d_i$ or greater from the bottom of the vessel. For disk flat-blade turbines, curve c, there is essentially no effect of d_i/d_t in the range 0.15 to 0.50. For open types, curve b, the effect of d_i/d_t may be strong, depending upon the group nb/d_t.

6.   Curve g is for disk flat-blade turbines operated in unbaffled vessels filled with liquid, covered, so that no vortex forms. If baffles are present, the power characteristics at high Reynolds numbers are essentially the same as curve b for baffled open vessels, with only a slight increase in power.

7.   For very deep tanks, two impellers are sometimes mounted on the same shaft, one above the other. For the flat-blade disk turbines, at a spacing equal to $1.5d_i$ or greater, the combined power for both will approximate twice that for a single turbine.

8.   SOURCE: Treybal, *Mass Transfer Operations,* 3d ed., p. 152, McGraw-Hill, NY, 1963.

where E = approach to equilibrium of a single-stage contact
C_i = initial concentration
C_t = concentration of time t
C_e = concentration of equilibrium
E_b = approach to equilibrium for a batch process
E_f = approach to equilibrium for a continuous process
t_b = mixing time for a batch process
t_c = residence time for a continuous process

Having established the residence time and power input, the scale-up can be now done using the principle of geometric similarity together with equal power per unit volume discussed earlier.

The above covers most conventional mixers; there is another class of mixers, called pump-mix impellers, where the impeller serves not only to mix the fluids, but also to move the fluids through the extraction stages. These are specialized designs, often used in the metals extraction industries. For these types of impellers, a knowledge of the power characteristics for pumping is required in addition to that for mixing. For a more detailed treatment of these special cases, the reader is referred to Lo et al.

Recycling some of one of the settled liquids back to the agitated vessel sometimes improves settling of the dispersion. In addition, the stage efficiency of a stirred vessel can be considerably enhanced by recycling the liquid favored by solute distribution, whereas recycling the other liquid reduces the stage efficiency. When solute distribution favors the dispersed phase and mass-transfer rates are poor, recycling the settled dispersed phase can result in minimizing the volume of a cascade of extraction vessels [Treybal, *Ind. Eng. Chem. Fundam.,* **3,** 185 (1964)]. See also Gel'perin et al. (*Khim. Neft. Mashinostr.,* 1966, 23).

Extractive reaction, in which a solvent extracts one of the products to enhance the yield, is considered by Piret, Trambouze, et al. [*Am. Inst. Chem. Eng. J.,* **6,** 394, 574 (1960); **7,** 138 (1961)]. See also Schmitz and Amundsen [*Chem. Eng. Sci.,* **18,** 265, 415, 447 (1963)].

SETTLERS

Emulsions and Dispersions   The mixture of liquids leaving a mixer is a cloudy dispersion which must be settled, coalesced, and separated into its liquid phases in order to be withdrawn as separate liquids from a stage. For a dispersion to "break" into separate phases, both sedimentation and coalescence of the drops of the dispersed phase must occur. Unstable dispersions usually have droplet diameters of about 1 mm or larger and settle rapidly. Stable dispersions, or emulsions, are generally characterized by droplet diameters of about 1 μm or less. The unstable dispersions are preferred in liquid-liquid-extraction operations and chemical-reaction systems involving two liquid phases that ultimately need to be separated. Dispersions and emulsions are usually characterized by the terms **water-in-oil** (meaning aqueous liquid droplets dispersed in organic liquid continuous phase) and **oil-in-water** (organic droplets in aqueous liquid). **Dual emulsions** and **liquid-membrane systems** are those in which the continuous phase is also present as very small droplets within larger drops of the other liquid. See Becher, *Emulsions: Theory and Practice*, ACS Monogr. 175, Reinhold, New York, 1957; and Li and Shrier, in Li (ed.), *Recent Developments in Separation Science*, vol. I, CRC Press, Cleveland, 1972, p. 163.

The "breaking" of a dispersion in a batch settler may be divided into two periods: (1) primary break, or rapid settling and coalescence of most of the dispersed phase, which often leaves a fog of very small droplets suspended as parts per million in the majority phase; and (2) secondary break, which represents the slow settling of the fog. Most industrial settlers are designed for the primary break since the slow secondary break would require much longer residence times. The small amount of entrainment to a subsequent stage seldom influences stage efficiency in a multistage cascade. However, for conserving solvent and desolventizing the effluent streams from the final stages of a cascade, it may be necessary to clarify as completely as possible, including the use of coalescers to eliminate secondary fog.

Sedimentation Isolated droplets, settling or rising in a stagnant liquid under the force of gravity, generally move more rapidly than solid spheres. The rate of settling or rising is more rapid for large droplet size, large density difference between phases, and low viscosity of the continuous phase. Felix and Holder [*Am. Inst. Chem. Eng. J.*, **1**, 296 (1955)] show considerably shorter settling time of petroleum-oil dispersions in water and phenol by reducing the continuous-phase viscosity simply by raising the temperature.

Coalescence The coalescence of droplets can occur whenever two or more droplets collide and remain in contact long enough for the continuous-phase film to become so thin that a hole develops and allows the liquid to become one body. A clean system with a high interfacial tension will generally coalesce quite rapidly. Particulates and polymeric films tend to accumulate at droplet surfaces and reduce the rate of coalescence. This can lead to the buildup of a "rag" layer at the liquid-liquid interface in an extractor. Rapid drop breakup and rapid coalescence can significantly enhance the rate of mass transfer between phases.

Gravity Settlers; Decanters These are tanks in which a liquid-liquid dispersion is continuously settled and coalesced and from which the settled liquids are continuously withdrawn. They can be either horizontal or vertical. Figure 15-24 shows some typical horizontal decanters. For an uninstrumented decanter the height of the heavy-phase-liquid leg above the interface is balanced against the height of the light-liquid phase above the interface, Eq. 15-50.

$$(Z_h - Z_i)\rho_h = (Z_L - Z_i)\rho_L \qquad (15\text{-}50)$$

The velocity of the liquid entering the decanter should be kept low to minimize disturbance of the interface. Sometimes an impingement baffle, or "picket fence," has been used. In other cases, opposing inlets as in Fig. 15-24c and d have been used. For an external jackleg shown in Fig. 15-24a the heavy-liquid takeoff requires a siphon break to prevent emptying the vessel by siphoning. Some problems can occur because of pressure drop through the outlet piping and variable levels under flow conditions. The horizontal weirs in Fig. 15-24b and the circular weirs in Fig. 15-24c can be designed for a very low crest height at maximum design flow rates. When rag builds up at the interface, sometimes it can be purged by withdrawing a small stream, filtering out the solids, and returning the liquids to the decanter. The decanter can also be instrumented with an interface detector and automatic control valve on the heavy-phase flow. The light phase can still overflow from the vessel.

For general reviews, see Ingersoll [*Pet. Refiner*, **30**(6), (1951)] and Hart [*Pet. Process.*, **2**, 282, 471, 513, 632 (1947)]. In the petroleum industry, settler volumes have frequently been sufficiently large so as to provide a holding time from 0.5 to 1.0 h, which in most cases is probably excessive and costly. For most thin liquids, in which unusual emulsification problems do not occur, 5 to 10 min is ample. The size of the settler seems to be set by the rate of flow per unit of horizontal cross-sectional area as well as holding time [Williams et al., *Trans. Inst. Chem. Eng. (London)*, **36**, 464 (1958)]; Ryon, Daley, and Lowrie [*Chem. Eng. Prog.*, **55**(10), 70 (1959)], for the settling of aqueous uranium solutions and kerosine-alkyl phosphate solvents, used decanters of the type shown in Fig. 15-24b. The depth of the decanter having been chosen, these authors recommend that the horizontal cross section for the prevailing flow rate be set at twice the value which would give a dispersion-band thickness equal to the depth of the tank. In this manner dispersions of 9.08 m³/h aqueous + 15.90 m³/h solvent (40 gal/min aqueous + 70 gal/min solvent) were successfully settled in a decanter of 1.4-min holding time.

Gravity settlers, basically of the type shown in Fig. 15-24a, were used by Wilke et al. (UCRL-10625, 1963; UCRL-11182, 1964) to settle water and Aroclor (specific gravity, 1.36). The dispersion may

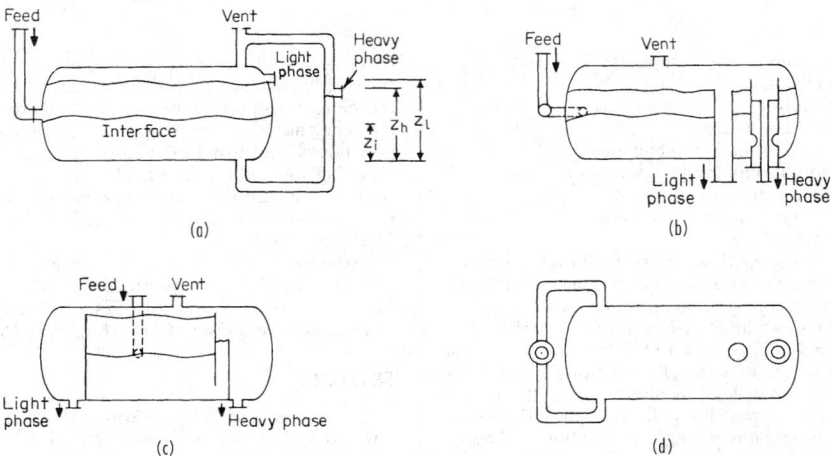

FIG. 15-24 Gravity decanters. (*a*) External jackleg, side view. (*b*) Straight weirs, side view. (*c*) Circular weirs, side view. (*d*) Circular weirs, top view.

occupy a wedge-shaped volume in the region of the interface at low flow rates instead of covering the entire interface as for higher flow rates. Important variables influencing the performance are (1) the value of φ_D in the entering liquid mixture, (2) whether the dispersion is introduced above or below the interface, and (3) the distance of an impact baffle from the inlet pipe. The length of the dispersion wedge for kerosine-water (dispersed) is proportional to V_D/d_p^3 (Jeffreys and Pitt, paper at AIChE meeting, Salt Lake City, May 1967). Higher temperatures (to decrease viscosity) and longer residence times within each phase improve the settling of water (dispersed)-coconut fatty acids [Manchanda and Woods, *Ind. Eng. Chem. Process Des. Dev.*, **7**, 182 (1968)].

In the extraction of uranium from ore leach liquors with kerosine-reagent solvents, there is a saving in the cost of thickeners and filters if the aqueous liquors are not clarified before extraction. If such slurries are extracted, however, it is necessary to increase the solvent-aqueous ratio in the extractor in order to make the organic phase continuous; otherwise, unsettleable emulsions are produced. Table 15-6 gives the data of Shaw and Long [*Chem. Eng.*, **64**(11), 251 (1957)] for settling areas required for such extractions. The high organic-aqueous ratios are obtained by recycling settled organic phase from the settler to the mixer. Entrainment of organic solvent with the settled solids represents a serious problem in such operations.

TABLE 15-6 Settling of Aqueous Uranium Leach Liquors with Kerosine–Alkyl Phosphate Solvent*

Nature of aqueous feed	Organic-aqueous ratio required	Permissible settler flow rate, U.S. gal/(min·ft²) horizontal area
Clear liquor	4	1.4–1.6
Slimes (5% solids)	8	0.6
Dense pulps (50–60% solids)	10	0.3

*Shaw and Long, *Chem. Eng.*, **64**(11), 251 (1957).
To convert gallons per minute–square foot to cubic meters per hour–square meter, multiply by 2.44.

CYCLONES

Cyclones have been suggested as simple means of enhancing by centrifugal force the rate of settling of liquid dispersions. Tepe and Woods (U.S. AEC AECD-2864, 1943) report a few data for the separation of isobutanol-water dispersions in such devices, but the results were poor. The most thorough studies are those of Simkin and Olney [*Am. Inst. Chem. Eng. J.*, **2**, 545 (1956)] and of Hitchon [At. Energy Res. Estab. (Gt. Brit.) CE/R-2777, 1959], who conclude that high extraction efficiencies (requiring high degrees of dispersion) and good clarification of both effluents cannot be obtained in one stage. Tepe and Woods (loc. cit.) also tried *helical coils of pipe* for separating isobutanol-water mixtures, with poor results.

CENTRIFUGES

Mechanical centrifuges, high-speed machines, have been used for many years for separating liquid-liquid dispersions, for example, in the separation of caustic solutions and oils in the soap-making process, more recently in uranium extractions, and in many others. By enhancing the settling rate (without, however, influencing coalescence), they reduce the settling time considerably. See, for example, Landis [*Chem. Eng. Prog.*, **61**(10), 58 (1965)]. For details see Section 19.

SETTLER AUXILIARIES

These include the use of coalescers, separating membranes, and electrical devices and the addition of emulsion-breaking reagents. These last are used for treating permanent emulsions and will not be discussed here.

Coalescers The small drops of a fine dispersion may be caused to coalesce and thus become larger by passing the dispersion through a coalescer. The enlarged drops then settle more rapidly. Coalescers are mats, beds, or layers of porous or fibrous solids whose properties are

especially suited for the purpose at hand. In an extensive study, Sareen et al. [*Am. Inst. Chem. Eng. J.*, **12**, 1045 (1966)] found, in part, that (1) coalescence is promoted by decreased fiber diameter, (2) a minimum bed density is required to achieve complete coalescence, dependent upon the system characteristics, (3) wetting of the fibers by droplets of dispersed phase is not necessary for good coalescence, (4) a fibrous bed of medical cotton can be made to coalesce almost any kind of liquid dispersed in another except if $\sigma' < 3$ mN/m (dyn/cm), (5) cotton fibers are best supported from collapse by mixing with fibers of glass or Dynel [see also Langdon et al., *Petro/Chem Eng.*, **1963**(11), 35], (6) the optimum bed thickness of a mixed bed depends on the ratio of cotton to support (0.75 in for 50 percent cotton), (7) the maximum velocity through the bed with effective coalescing increases with bed depth, but increased pressure drop causes redispersion, presumably at values depending upon the liquid system, and (8) some surfactants interfere with coalescence, but others do not. For tests on petroleum-brine emulsions and Fiberglas, see Burtis and Kirkbride [*Trans. Am. Inst. Chem. Eng.*, **42**, 413 (1946)] and Hayes et al. [*Chem. Eng. Prog.*, **45**, 235 (1949)]. Beds of granular solids such as sand, etc., and bats of excelsior, steel wool, and the like have also been used.

Separating Membranes If the capillary size of a porous substance is very small, the liquid which preferentially wets the solid may flow through the capillaries readily but strong interfacial films block the capillaries for flow of nonwetting liquid. Sufficient pressure will cause disruption of the films and permit passage of the nonwetting liquid, but regulation of the pressure commensurate with the pore size permits perfect phase separation. Separating membranes of this type are made of a variety of materials such as porcelain, paper which has been coated with special resins, and the like and may be either hydrophilic or hydrophobic in character. They are made thin so as to permit maximum passage of the wetting liquid [see Jordan, *Trans. Am. Soc. Mech. Eng.*, **77**, 393 (1955); and Belk, *Chem. Eng. Prog.*, **61**(10), 72 (1965)]. In practice, the dispersion is usually first passed through a coalescer so as to permit settling of the bulk of the dispersed phase before the mixture is presented to the separating membrane, thus relieving the load on the membrane.

Figure 15-25 shows a combination device containing coalescers and both hydrophobic and hydrophilic separating membranes. Coalescers

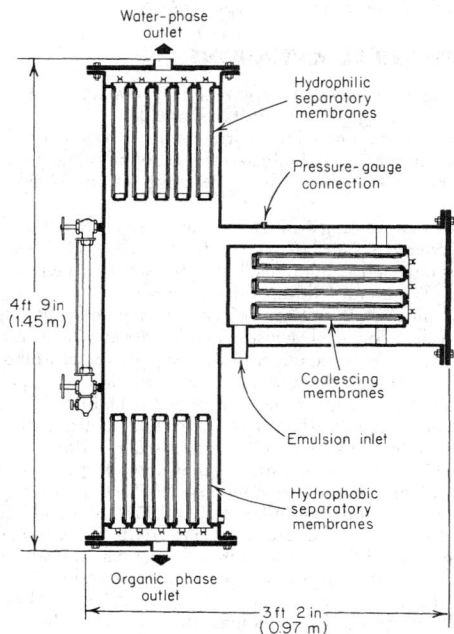

FIG. 15-25 Combination coalescer, settler, and membrane separator. (*Courtesy of Selas Corporation of America.*)

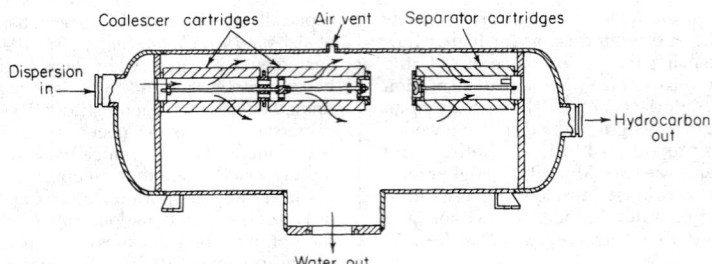

FIG. 15-26   Fuel-water separator. (*Courtesy of Warner-Lewis Co. Division, Fram Corp.*)

and separating membranes are fashioned in the form of hollow cylinders, and flow is radially through the wall. After passing through the coalescers, the bulk of the liquids settles in the vertical member of the device, and then the settled phases are passed through their respective separating membranes. Devices of this type are designed to handle 0.57 to 6.81 m³/h (150 to 1800 gal/h), delivering completely separated phases; and further settling is unnecessary. Figure 15-26 shows another design for removing dispersed water from jet fuel or gasoline, available in sizes to handle from 68 to 250 m³/h (300 to 1100 gal/min) and delivering clear effluents. In this case only a hydrophobic membrane is required [Redmon, *Chem. Eng. Prog.*, **59**(9), 87 (1963)].

Electrical Devices   Subjecting electrically conducting emulsions or dispersions to high-voltage electric fields may cause rupture of the protective film about a droplet and thus induce coalescence. Dispersions of low conductivity are subject, in an electric field, to forces between particles resulting from acquired induced dipoles, which induce coalescence. These phenomena have been used particularly for the desalting of petroleum emulsified with brine, and for similar applications. Devices have been built to handle 828 m³/h (125,000 bbl/day) of crude oil, at costs of approximately 0.1 to 0.5 cent/bbl [Waterman, *Chem. Eng. Prog.*, **61**(10), 51 (1965)]. For a detailed study see Sjoblom and Goren [*Ind. Eng. Chem. Fundam.*, **5**, 519 (1966)] and Brown and Hanson [*Trans. Faraday Soc.*, **61**, 1754 (1965)]. Figure 15-27 shows schematically the flow through a typical device.

MIXER-SETTLER COMBINATIONS

Any mixer and settler can be combined to produce a stage, and the stages in turn arranged in a multistage cascade.

A great many commonly used arrangements have been developed in an effort to reduce or eliminate interstage pumping and to reduce costs generally. Only a few of the more commonly used types are mentioned here.

A compact alternating arrangement of mixers and settlers has been adopted in many of the "box-type" extractors developed originally for processing radioactive solutions, but now used in principle for many processes, with literally dozens of modifications. An example is the Pump-Mix mixer-settler (Fig. 15-28), in which adjacent stages have common walls [Coplan, Davidson, and Zebroski, *Chem. Eng. Prog.*, **50**, 403 (1954)]. The impellers in this case pump as well as mix by drawing the heavy liquid upward through the hollow impeller shaft and discharging it at a higher level through the hollow impeller. These extractors or variants of them have been built not only in relatively large sizes but also in miniature for bench-scale work.

Figure 15-29 represents still further modification for low cost [Hazen and Henrickson, *Min. Eng.*, 994 (1957); Quinn, Trefoil (Denver Equipment Company) Bull. M4-B90, 1957]. At *a* and *b* in the figure, the settler is a circular tank $d_t = 4.9$ m (16 ft), $Z = 2.1$ m (7 ft), with the mixing vessel, 1.2 by 1.2 m (4 by 4 ft), contained inside. Agitators are turbines, $d_i = 0.46$ m (1.5 ft), operated at 150 r/min (1.12 kW) and 200 r/min (2.02 kW). The aqueous feed is 22.7-m³/h (100-gal/min) uranium-bearing ore leach liquor, the organic solvent 4.5-m³/h (20-gal/min) alkyl phosphate solutions in kerosine. Adjacent stages are at 0.3-m (1-ft) elevation difference, allowing gravity flow of the aqueous liquor, while the organic phase is pumped in countercurrent by air

lifts. Provision is made for recycle of settled organic phase by overflow to the mixer, the amount of which can be adjusted by changing the height of the organic-overflow pipe. The vanadium extractor at *c* in the figure is a box type, built into a circular tank, $d_t = 9.8$ m (32 ft), $Z = 2.1$ m (7 ft). The 0.46-m- (18-in-) diameter turbines draw 5.6 kW (7.5 hp). Other modifications of the box-type mixer-settler (Denver Equipment Company, Denver, Bull. A1-B6), with capacities of from 0.23- to 5700-m³/h (1- to 25,000-gal/min) liquid flow, have been extensively used in a great variety of metal separations in process metallurgy. These provide for intrastage recycle of liquids, particularly advantageous when very low solvent-feed ratios typical of good solvents must be used and when it is desired to make the minority liquid continuous in order to improve settling characteristics. See also Williams et al. [*Trans. Inst. Chem. Eng. (London)*, **36**, 464 (1958)] and Hanson and Kaye [*Chem. Process Eng.*, **44**, 27, 654 (1963); **45**, 413 (1964)].

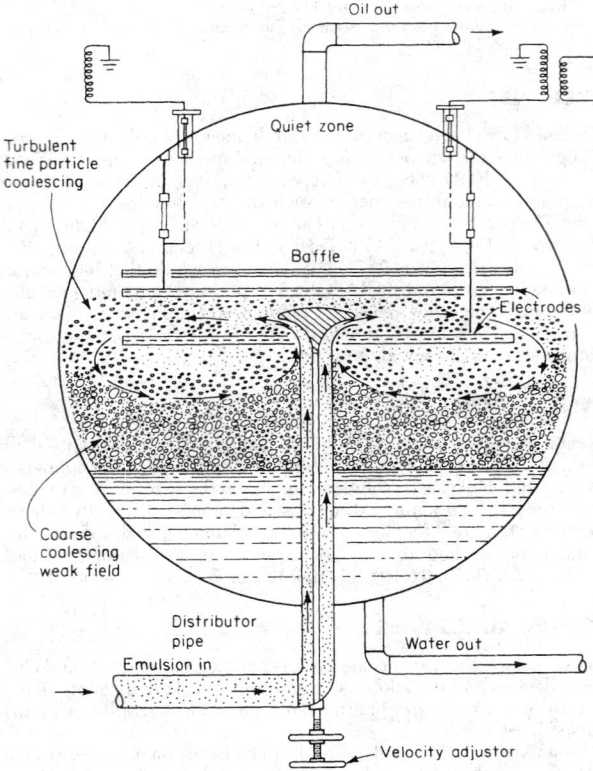

FIG. 15-27   Internal circulation and electric field, Petreco Cylectric coalescer (schematic). [*Waterman*, Chem. Eng., **61**(10), 51 (1965), *with permission.*]

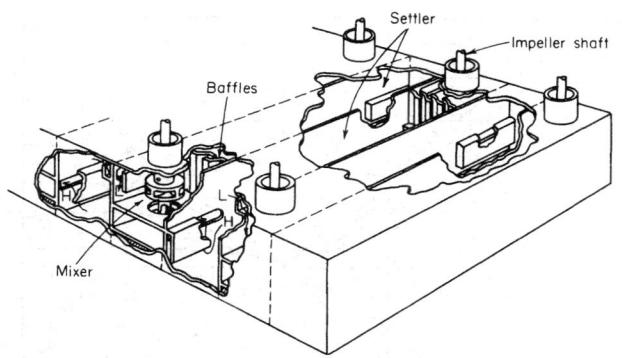

FIG. 15-28   Pump-Mix mixer-settler. [*Coplan, Davidson, and Zebroski,* Chem. Eng. Prog., **50**, *403 (1954), with permission.*]

Vertical arrangement of the stages is desirable, since then a single drive may be used for agitators and the floor-space requirement of a cascade is reduced to that of a single stage. See, for example, Hanson and Kaye, loc. cit. In the Lurgi extractor, the mixer and settlers are in separate vertical shells interconnected with piping [Guccione, *Chem. Eng.,* 78 (July 4, 1966)].

A great many other devices are known. The Fenske extractor [Fenske and Long, *Chem. Eng. Prog.,* **51**, 194 (1955); *Ind. Eng. Chem.,* **53**, 791 (1961); *Ind. Eng. Chem. Fundam.,* **1**, 152 (1962)] is a vertical stack of mixer-settler stages, with mixing done by a vertically moving reciprocating plate in each mixer. One very successful device, particularly in the extraction of radioactive solutions, uses a centrifuge instead of a settler to separate the mixed liquids, and the pump-mixer

and centrifuge of each stage operate on a common shaft [Clark, U.S. AEC DP-752 (1962); Kisbaugh, ibid., DP-841 (1963)]. See also Goncharenko et al. (*Tr. Vses. Khemosorbtsii Nauchn-Tekhn. Sovesch. Protessy Zhidkostnoi Ekstraktsii Khemosorbtsii,* 2d, Leningrad, **1964,** 75) and Berestovoi et al. (ibid., 171). Still another uses a cyclone (hydroclone) for separating [Whatley and Woods, U.S. AEC ORNL-3533 (1964); Finsterwalder, ibid., ORNL-4088 (1967)]. The Graesser extractor (Coleby, U.S. Patent 3,017,253, 1962) is a horizontal shell filled with stratified settled liquids, with a series of buckets revolving around the inner periphery which rain droplets of one liquid through the other. It has been used primarily in Europe for easily emulsified liquids.

Overall Stage Efficiencies   The mixer-settler extractors described have generally produced overall stage efficiencies in excess of 80 percent, usually nearly 90 to 95 percent.

CONTINUOUS (DIFFERENTIAL) CONTACT EQUIPMENT

Equipment in this category is usually arranged for multistage countercurrent contact of the insoluble liquids, without repeated complete separation of the liquids from each other between stages or their equivalent. Instead, the liquids remain in continuous contact throughout their passage through the equipment.

General Characteristics   Countercurrent flow is maintained by virtue of the difference in densities of the liquids and either the force of gravity (vertical towers) or centrifugal force (centrifugal extractors). Only one of the liquids may be pumped through the equipment at any desired velocity. The maximum velocity for the second is then fixed; if it is attempted to exceed this limit, the second liquid will be rejected and the extractor will be **flooded.**

It cannot be overemphasized that knowledge of the characteristics of such equipment is surprisingly underdeveloped. The number of quantities that influence the rate of extraction is very large, and many of them are not well understood. Most of the available data were taken

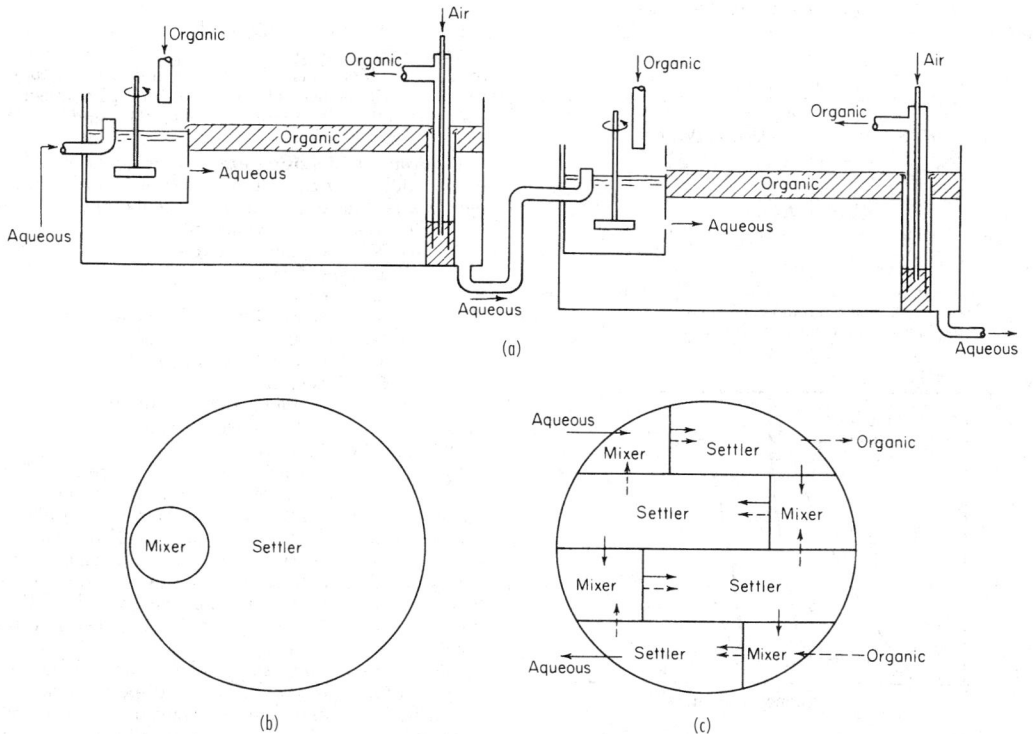

FIG. 15-29   Kerr-McGee multistage mixer-settler. (*a*) and (*b*) For uranium. (*c*) For vanadium extraction.

from small laboratory devices, frequently only a few inches in diameter and a few feet high. For these reasons the generalizations given here should be used only for very rough estimates, with allowance for generous factors of safety.

Axial Dispersion   The devices in this category are subject to axial (longitudinal) dispersion within both liquids or departure from strictly "plug," countercurrent flow. The result of this axial mixing is to decrease the effective concentration driving force in the contactor as illustrated in Fig. 15-30. As a result, the towers must be taller than simple application of the plug-flow numbers of transfer units would indicate. The problem has been extensively studied by Sleicher [*Am. Inst. Chem. Eng. J.*, **5**, 145 (1959)] and Vermeulen et al. [U.S. AEC UCRL-3911, 1958; suppl., 1958; 10928, 1963; *Ind. Eng. Chem. Fundam.*, **2**, 113, 304 (1963); *Chem. Eng. Prog.*, **62**(9), 95 (1966)]. The two studies lead to essentially the same results although they are expressed somewhat differently. For a review, see Li and Ziegler [*Ind. Eng. Chem.*, **59**(3), 30 (1967)]. The subject of axial mixing has also been treated extensively by Pratt and Baird (*Handbook of Solvent Extraction*, Wiley, NY, 1983, pp. 197–247). It will not be possible to outline in detail here all the considerations taken into account; for these the original papers should be consulted. For present purposes the procedure to be used in design, as developed by Vermeulen et al., will be outlined. It is limited to cases in which flow rates, distribution coefficients, and mass-transfer coefficients are constant throughout the extractor. For the final design it is important to conduct pilot tests in equipment which closely resembles that being considered for the full-size plant, and at conditions which simulate the conditions expected in the full-size plant.

1. Obtain $N_{tOR,plug}$ from Colburn's equation [*Trans. Am. Inst. Chem. Eng.*, **35**, 211 (1938)]:

$$N_{tOR,plug} = \frac{1}{1 - V_R/m'V_E} \ln\left[\left(\frac{c_{R1} - c_{E2}/m'}{c_{R2} - c_{E2}/m'}\right)\left(1 - \frac{V_R}{m'V_E}\right) + \frac{V_R}{m'V_E}\right] \quad (15\text{-}51)$$

2. Obtain H_{tOR} (from data correlations, etc.) and Z_{plug}:

$$Z_{plug} = N_{tOR,plug}H_{tOR} \quad (15\text{-}52)$$

3. Solve Eqs. (15-53) to (15-55) together with Fig. 15-31 simultaneously by trial and error to obtain N_{tOR}:

$$1/N_{tOR} = 1/N_{tOR,plug} - 1/N'_{tOR} \quad (15\text{-}53)$$

$$N'_{tOR} = (N_{Pe}B)_E + \frac{\ln(V_R/m'V_E)}{V_R/m'V_E - 1} \quad (15\text{-}54)$$

Equation (15-54) is applicable only for cases in which $N_{tOR}V_R/m'V_E$ and $(N_{Pe}B)_E \geq 1.0$.

$$(N_{Pe}B)_E = \left(\frac{V_R/m'V_E}{f_R N_{Pe,R}B} + \frac{1}{f_E N_{Pe,E}B}\right)^{-1} \quad (15\text{-}55)$$

4. The final height of the effective portion of the tower, Z, is then

$$Z = Z_{plug}(N_{tOR}/N_{tOR,plug}) \quad (15\text{-}56)$$

In these expressions, $B = Z/d$, $N_{Pe,E} = dV_E/E_E$, $N_{Pe,R} = dV_R/E_R$, where $d =$ some characteristic length such as d_F for packed towers or T for spray towers. E_E and E_R are the longitudinal dispersion coefficients, which must ultimately be deter-

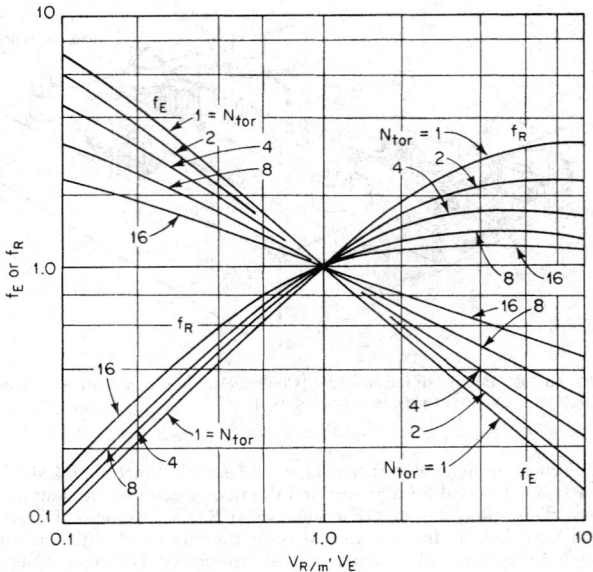

FIG. 15-31   Factors for Eq. 15-67. [*Vermeulen et al.*, Chem. Eng. Prog., **62**(9), 95 (1966), with permission.]

mined experimentally. They are usually reported as E_C, E_D, $N_{Pe,C}$, and $N_{Pe,D}$, since they are more characteristic of the continuous or dispersed nature of the liquid than whether the liquid is extract or raffinate. For plug flow, $E = 0$; for complete mixing, $E = \infty$. In using these expressions, H_{tOR} should represent data that have been corrected for axial dispersion; unfortunately, very few data have been so corrected. Rod [*Br. Chem. Eng.*, **9**, 300 (1964)] presents a graphical calculation suitable even for curvilinear equilibrium curves.

Devices that are stagelike in character (sieve trays, compartmented extractors, etc.) are perhaps better treated by a somewhat different procedure which space does not permit outlining here. See Sleicher [*Am. Inst. Chem. Eng. J.*, **6**, 529 (1960)], Miyauchi and Vermeulen [*Ind. Eng. Chem. Fundam.*, **2**, 304 (1963)], and Van der Laan [*Chem. Eng. Sci.*, **7**, 187 (1958)].

Equipment Classification   Equipment can be broadly classified into the following categories, generally in order of increasing complexity of internal construction. Those most generally used are:

I. Gravity-operated extractors
    A. No mechanical agitation
        1. Spray towers
        2. Packed towers
        3. Perforated-plate (sieve-plate) towers
    B. Mechanically agitated extractors
        1. Towers with rotating stirrers
        2. Pulsed towers
            a. Liquid contents pulsed
            b. Reciprocating plates
II. Centrifugal extractors

Spray Towers   These are simple gravity extractors, consisting of empty towers with provisions for introducing and removing liquids at the ends (see Fig. 15-32). The interface can be run above the top distributor, below the bottom distributor, or in the middle, depending on where the best performance is achieved. Because of severe axial back mixing, it is difficult to achieve the equivalent of more than one or two theoretical stages or transfer units on one side of the interface. For this reason they have only rarely been applied in extraction applications.

Distributors   Most spray columns operate with the drops being formed at the ends of jets from the dispersed phase inlet distributor. The orifices or nozzles for introducing the dispersed phase are usually not smaller than 0.13 cm (0.05 in) in diameter in order to avoid clogging, nor larger than 0.64 cm (0.25 in) in order to avoid formation of

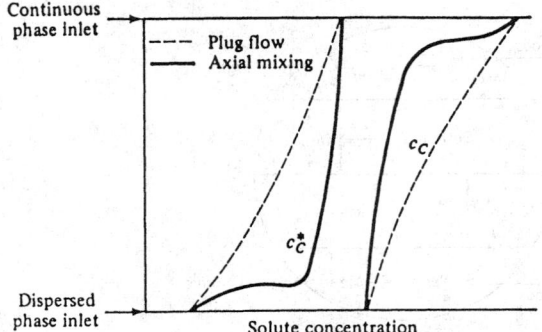

FIG. 15-30   Effect of axial mixing on concentration profiles in towers subject to axial mixing.

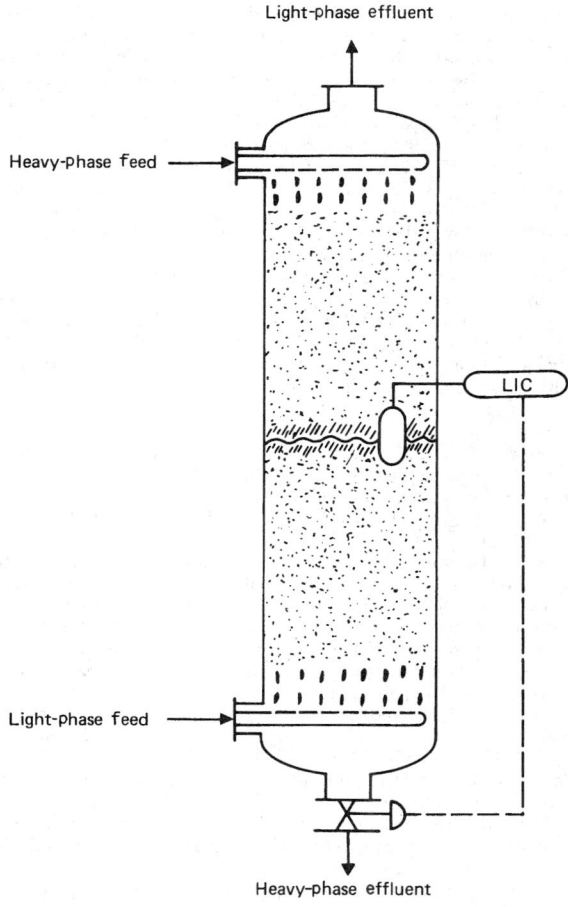

FIG. 15-32 Spray tower with both phases dispersed.

excessively large drops. They should be designed to eliminate wetting by the dispersed liquid. The following equation is recommended for calculating the velocity at which a jet appears:

$$V'_{oj} = \left[\frac{3\sigma(1 - d_o/d_{vj})}{\rho_d d_o}\right]^{0.3} \tag{15-57}$$

The value of d_{vj} can be found from the relationship:

$$d_{vj} = \frac{d_o}{\rho + q(\phi)^m} \qquad \text{where} \qquad \phi = \frac{d_o}{(\sigma/\Delta\rho\ g)^{0.5}} \tag{15-58}$$

where
V_{oj} = hole velocity where a jet appears
σ = interfacial tension
d_o = hole diameter
d_{vj} = drop diameter in jetting
ρ_d = dispersed phase density
$\Delta\rho$ = density difference
g = gravitational constant
ϕ, q, m = constants

For
$\phi < 0.785$: $\rho = 1.0$, $q = 0.485$, $m = 2$

$\phi > 0.785$: $\rho = 0.12$, $q = 1.51$, $m = 1$

The distributor should be sized so that the hole velocity is greater than the jet formation velocity. As the velocity is increased, the jet reaches a maximum length at which it breaks into drops of approximately uniform size. It has been found that at this velocity the drop surface area

produced is also approximately at its maximum. Typical nozzle velocities are in the range of 0.1–0.25 m/s.

Holdup and Flooding   At this point it is useful to introduce the concepts of holdup and flooding in column contactors. It is normal practice to select the phase which preferentially wets the internals of the column as the **continuous phase.** This then allows the **dispersed phase** to exist as discrete droplets within the column. If the dispersed phase were to preferentially wet the internals, this could cause the dispersion to prematurely coalesce and pass through the column as rivulets or streams which would decrease interfacial area and therefore column efficiency.

The volume of droplets within the contactor at any time is referred to as the operational holdup of the dispersed phase, generally expressed as a fraction of the contactor volume.

In a countercurrent-type column contactor, stable operation is possible as long as the rate of arrival of droplets in any section does not exceed the coalescence rate at the main interface; once this value is exceeded, droplet backup will occur at the interface and slowly build back into the column active area, a condition known as flooding. This is an inoperable condition.

Besides starting at the interface and building back into the column, flooding can also start in other sections of the column depending on local hydrodynamic conditions. If for any reason the velocity of the continuous phase is increased, this will increase the drag force on the droplets and cause the smaller droplets to rise (or fall) more slowly. As the continuous phase velocity is increased further, there is reached a point where a significant number of droplets stop rising (or falling), forming a dense region which eventually coalesces and forms a second interface in the column; this also is an inoperable condition. This same phenomenon can be caused at constant continuous phase velocity by inducing the formation of smaller droplets, such as by increased agitation. These smaller droplets can no longer overcome the continuous phase velocity drag force and holdup, thus inducing flooding.

The concepts of slip velocity and characteristic velocity are useful in defining the flooding point and operational regions of different types of column contactors. The slip (or relative) velocity is given by the equation:

$$V_s = \frac{V_c}{1 - \phi_d} + \frac{V_d}{\phi_d} \tag{15-59}$$

From this has been derived the concept of characteristic velocity which is defined by the general equation:

$$V_k = \frac{V_s}{(1 - \phi_d)} \tag{15-60}$$

The value of V_k may be identified with the average terminal velocity of the droplets within the contactor. Each different type of contactor will have a different and unique characteristic velocity.

As flooding is approached, the slip velocity continues to decrease until at the flood point is zero and the following relationship applies;

$$\left(\frac{\delta V_c}{\delta\phi_d}\right)_f = \left(\frac{\delta V_d}{\delta\phi_d}\right)_f = 0 \tag{15-61}$$

Applying the above to the relationship for slip velocity yields the following relationships at flooding:

$$V_{df} = 2V_k\phi_{df}(1 - \phi_{df}) \tag{15-62}$$

$$V_{cf} = V_k(1 - \phi_{df})^2(1 - 2\phi_{df}) \tag{15-63}$$

$$\phi_{df} = \frac{[1 + 8(V_c/V_d)_f]^{0.5} - 3}{4(V_c/V_d)_f - 4} \tag{15-64}$$

where
V_s = slip velocity
V_c = continuous phase velocity
V_d = dispersed phase velocity
ϕ_d = dispersed phase holdup
V_k = characteristic velocity
V_{cf} = continuous phase velocity of flooding
V_{df} = dispersed phase velocity of flooding

These relationships are not restricted to any type of contactor; they can be used to predict either the flooding velocity at a given holdup, or the holdup at flooding. However, it requires the knowledge of the characteristic velocity under actual mass-transfer conditions, and this has not been easy to obtain. As a result, column contactors still require pilot tests for reliable scale-up to full size.

There are also other factors to be aware of which can have a significant impact on the holdup and flooding characteristics. One of the most important of these is the direction of mass transfer. When transfer is from the dispersed to continuous phases, there occurs what is known as the Marangoni effect which causes rapid interdroplet coalescence with resulting decrease in holdup, sometimes by as much as 50 percent. Also, changes in phase densities, particularly in the continuous phase, can have significant impact on axial mixing with resulting effects on extraction efficiency. Finally, the presence of contaminents can affect interfacial properties (in particular interfacial tension) as well as inhibit mass transfer.

Flooding can be estimated theoretically by setting $\partial V_c/\partial \phi_d = \partial V_d/\partial \phi_d = 0$ [Thornton, *Chem. Eng. Sci.,* **5**, 201 (1956)], using Eq. (15-61). On the basis of purely statistical comparison of observed and calculated data, the empirical correlation of Minard and Johnson [*Chem. Eng. Prog.,* **48**, 62 (1952)], slightly modified, is recommended:

$$V_{cf} = \frac{10,000\Delta\rho^{0.28}}{[0.453\mu_c^{0.075}\rho_c^{0.5} + d_p^{0.56}\rho_d^{0.5}(Q_d/Q_c)^{0.5}]^2} \quad (15\text{-}65)$$

Use U.S. customary units only in this equation. In sizing the column diameter, it is ususlly assumed that the continuous phase velocity will set at 40 percent of this value, and therefore the column diameter is calculated by:

$$d_t = \sqrt{\frac{4Q_c}{0.4\pi V_{cf}}} \quad (15\text{-}66)$$

where d_t = column diameter
Q_c = volumetric flow rate of the continuous phase
V_{cf} = velocity of the continuous phase of flooding

Mass Transfer   As mentioned earlier, spray columns rarely develop more than 1 theoretical stage due to the axial mixing in the column. Nevertheless, it is necesary to determine what column height will give this theoretical stage. It is recommended by Cavers in Lo et al. *Handbook of Solvent Extraction* p. 323 and p. 327, John Wiley & Sons, New York, 1983 that the following equation be used to estimate the overall efficiency coefficient:

$$K_c a = \frac{\phi_d(1-\phi_d)\left(\dfrac{g^3\,\Delta\rho^3}{\sigma\rho_c^2}\right)^{0.25}}{(N_{Sc,c})^{0.5} + \dfrac{1}{m}(N_{Sc,d})^{0.5}} \quad (15\text{-}67)$$

where $K_c a$ = overall mass transfer capacity coefficient based on the continuous phase
ϕ_d = dispersed phase holdup
$\Delta\rho$ = density difference of phases
σ = interfacial tension
ρ_c = density of the continuous phase
$N_{Sc,c}$ = Schmidt number—continuous phase
$N_{Sc,d}$ = Schmidt number—dispersed phase
m = distribution coefficient

With this value, the height of a transfer unit, $(HTU)_{oc}$ can be estimated from:

$$(HTU)_{oc} = \frac{V_c}{K_c a} \quad (15\text{-}68)$$

where $(HTU)_{oc}$ = overall height of a transfer unit based on the continuous phase
V_c = continuous phase superficial velocity

On top of this should be put a safety factor of 30 percent due to the unreliability of the correlations.

There are not many data on the scale-up of spray columns from pilot to industrial scale, so these types of calculations must be used for the column design. As mentioned earlier, because of its limitations, the spray column is only rarely used in industrial applications.

Heat Transfer   Heat-transfer rates are generally large despite severe axial dispersion, with Ua_{av} frequently observed in the range 18.6 to 74.5 and even to 130 kW/(m³·K) [1000 to 4000 and even to 7000 Btu/(h·ft³·°F)][see Bauerle and Ahlert, *Ind. Eng. Chem. Process Des. Dev.,* **4**, 225 (1965); and Greskovich et al., *Am. Inst. Chem. Eng. J.,* **13**, 1160 (1967); Sideman, in Drew et al. (eds.), *Advances in Chemical Engineering,* vol. 6, Academic, New York, 1966, p. 207, reviewed earlier work]. In the absence of specific heat-transfer correlations, it is suggested that rates be estimated from mass-transfer correlations via the heat–mass-transfer analogy.

Axial Dispersion   For low values of ϕ_d and in the absence of interdrop coalescence, axial dispersion for the dispersed phase is evidently very small ($E_d \sim 0$). For the continuous phase, low μ_c and ϕ_d, Vermeulen et al. [*Chem. Eng. Prog.,* **62**(9), 95 (1966)] reviewed the available data and concluded that, for d_t = 3.6 to 15.2 cm (0.117 to 0.5 ft), E_c is given empirically by

$$E_c = c(V_d\,d_t)^{1/2} \quad (15\text{-}69)$$

where c = 23.6 for U.S. customary units and 7.2 for SI units. For treatment of heat transfer, particularly for high values of ϕ_d when axial dispersion evidently is the controlling factor, see Letan and Kehat [*Am. Inst. Chem. Eng. J.,* **11**, 804 (1965); **14**, 398 (1968)] and Mixon et al. [ibid., **13**, 21 (1967)].

Packed Towers   For a packed-tower liquid-liquid extractor the empty shell of a spray tower is filled with packing to reduce the vertical circulation of the continuous phase. The standard commercial packings used in vapor-liquid systems are also used in liquid-liquid systems. This includes Raschig and pall rings, Berl and Intalox saddles, and other random-dumped packings as well as the newer structured packings. The packing reduces the available free space for flow but also significantly reduces the height required for mass transfer. However, Nemunaitis, Eckert, Foote, and Rollinson [*Chem. Eng. Prog.,* **67**(11), 60 (1971)] reported little benefit from a packed height greater than 3.05 m (10 ft) and recommended redistributing the dispersed phase about every 1.52 to 3.05 m (5 to 10 ft) to generate new droplets and mass-transfer surfaces. From this perspective the packing allows a wider spacing between sieve plates than described for a conventional sieve-plate tower.

The pieces of random-dumped packing should be no larger than one-eighth of the tower diameter to minimize the wall effect which gives larger voids at the wall. The packing support can be an open grid or multiarch support if the dispersed phase is distributed to the top of the bed. But the packing support may also be a sieve plate with multiple light-liquid risers if the heavy phase is to be redispersed onto a lower bed. Or the packing support may be a sieve plate with multiple heavy-phase downcomers if the light phase is to be dispersed up into the bed. The streams of dispersed phase should be far enough apart to avoid coalescence at the dispersion plate, and the dispersed phase should not preferentially wet the packing. If the droplets wet the packing, they will coalesce and stream along the packing as rivulets. Eckert [*Hydrocarbon Process.,* 117 (March 1976)] recommends the use of packed towers when the interfacial tension is below 10 mN/m (dyn/cm).

Holdup   It is recognized that the dispersed-phase holdup may be placed in two categories: a smaller portion which is permanent and a larger portion, free, which moves through the packing and enters into mass-transfer operations when a solute is transferred between phases. Vignes [*Chem. Ind. Genie Chim.,* **95**, 307 (1966)] further classifies the moving holdup into "free" and "semifree." The total is ϕ_d, which here refers to the volume of dispersed phase expressed as a fraction of the void space in the packed section. See Beckmann et al. [*Am. Inst. Chem. Eng. J.,* **1**, 426 (1955); **3**, 223 (1957)].

What follows is a very brief summary of the extensive work of Pratt and his coworkers, Dell, Gayler, Lewis, Jones, Roberts, and White [*Trans. Inst. Chem. Eng. (London),* **29**, 89, 110, 126 (1951); **31**, 57, 69 (1953); *Chem. Ind. (London),* 1952, p. 358]. For the standard commercial packings of 1.27-cm (½-in) size and larger, at low values of V_d, ϕ_d varies linearly with V_d up to values of ϕ_d = 0.10. With further increase of V_d, ϕ_d increases sharply up to a "lower transition point," resembling "loading" in gas-liquid contact. At still higher values of V_d an upper transition point occurs, the drops of dispersed phase tend to coalesce, and V_d can increase without a corresponding increase in ϕ_d. This regime ends in flooding. Drops of the dispersed phase reach a characteristic size after leaving the distributor nozzles regardless of their initial size. For each system there is a critical packing size above which the mean drop size is a minimum. For smaller packing, the drop size is larger (and the interfacial area smaller). The critical size of packing, usually 1.27 cm ($\frac{1}{2}$ in) or more, is given by

$$d_{FC} = 2.42(\sigma g_c/\Delta\rho g)^{0.5} \qquad (15\text{-}70)$$

For packing larger than d_{FC}, the characteristic drop diameter, for liquids that are in concentration equilibrium, is given by

$$d_p = 0.92(\sigma g_c/\Delta\rho g)^{0.5}(V_k\varepsilon\phi_d/V_d) \qquad (15\text{-}71)$$

For liquids that are not in concentration equilibrium and when an unequilibrated solute is present, the characteristic drop size will generally be larger. If the drops formed at the distributor nozzle are smaller than this, there may be a tendency to flood until they grow to size. Thornton [*Ind. Chem.*, **39**, 632 (1963)] finds that large drops decay in exponential fashion to their final size. It is therefore best to design the nozzles to give drop sizes which are larger than that given by Eq. (21-67). V_k is a characteristic drop velocity (at $V_c = 0$, V_d approaching 0), and is given by Fig. 15-33. Below the upper transition point, the holdup is given by

$$\frac{V_d}{\phi_d} + \frac{V_c}{1 - \phi_d} = \varepsilon V_k(1 - \phi_d) \qquad (15\text{-}72)$$

Additional holdup correlations are offered by Sitarmayya and Laddha [*Chem. Eng. Sci.*, **13**, 263 (1961)] and Ghosal et al. [*Trans. Indian Inst. Chem. Eng.*, **11**, 23 (1958–1959)]. The interfacial area is given by

$$a = 6\varepsilon\phi_d/d_p$$

It is generally desirable to design for ϕ_d in the range 0.15 to 0.25 (the lower value for $V_d/V_c < 0.5$).

Flooding   Many correlations are available. By a comparison of the observed and calculated velocities at flooding for all available data, those of Crawford and Wilke [*Chem. Eng. Prog.*, **47**, 423 (1951)] and Hoffing and Lockhart [*Chem. Eng. Prog.*, **50**, 94 (1954)] are best and about equally effective. The Crawford-Wilke correlation is the simpler and is given in Fig. 15-34. Nemunaitis, Eckert, Foote, and Rollinson [*Chem. Eng. Prog.*, **67**(11), 60 (1971)] updated the correlation using packing factors. See also Dell and Pratt [*Trans. Inst. Chem. Eng.* (*London*), **29**, 89, 270 (1951)], Fujita et al. [*Chem. Eng.* (*Japan*), **17**, 230 (1957)], Sakiadis and Johnson [*Ind. Eng. Chem.*, **46**, 1229 (1954)], and Kafarov and Dytnerskii [*Zh. Prikl. Khim.*, **30**, 1698 (1957)]. For very small packings, see Rao and Rao [*Chem. Eng. Sci.*, **9**, 170 (1958)] and Venkatoramen and Laddha [*Am. Inst. Chem. Eng. J.*, **6**, 355 (1960)]. It is recommended that flow rates be set at no more than 50 percent of the flooding values, less if the interfacial tension of the liquids is high.

Mass Transfer   Extraction rates for packed towers are usually excellently correlated for a given situation on the coordinate system of Fig. 15-35. Treybal [*Chem. Eng. Prog.*, **62**(9), 67 (1966)] has suggested means whereby overall H_{tO}'s may be resolved into constituent

H_t's. In connection with the data on this figure, it should be noted that economical values of $m'V_E/V_R$ will usually lie in the range between 1 and 2, so that overall heights of transfer units are not too unreasonable even for this high-interfacial-tension system. For lower interfacial tensions, H_{tOC} will ordinarily be appreciably less.

The number of variables that are known to influence the rate of extraction is exceedingly large, and includes at least the following:

Size, shape, and material of packing
Tower diameter
Packing depth
Dispersed-phase distributor design
Which liquid is dispersed
Direction of extraction, whether from dispersed to continuous, organic liquid to water, or the reverse
Dispersed-phase holdup
Flow rates and flow ratio of the liquids

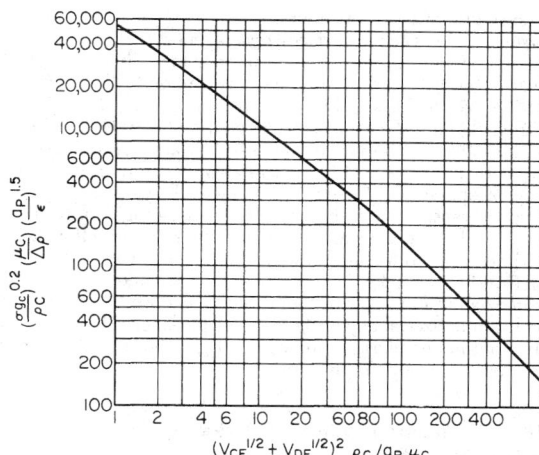

FIG. 15-34   Flooding in packed towers. Use only customary units in the variables. [*Crawford and Wilke*, Chem. Eng. Prog., **47**, 423 (1951), with permission.]

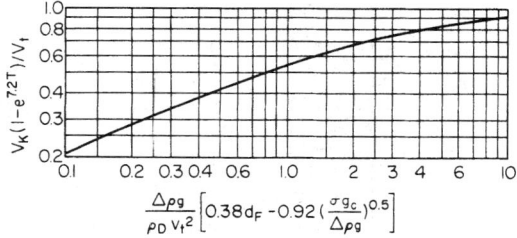

FIG. 15-33   Characteristic drop velocity for packed towers, for equilibrium liquids $d_F > d_{FC}$ and $T > 0.25$ ft. [*Pratt*, Ind. Chem., **31**, 552 (1955), with permission.]

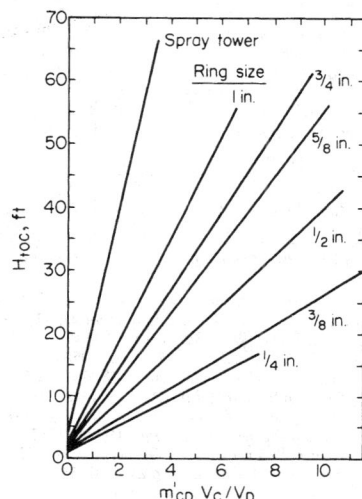

FIG. 15-35   Extraction of diethylamine from water into toluene (dispersed) in towers packed with unglazed porcelain Raschig rings. To convert feet to meters, multiply by 0.3048; to convert inches to centimeters, multiply by 2.54. [*Leibson and Beckman*, Chem. Eng. Prog., **49**, 405 (1953), with permission.]

Physical properties of the liquids
Presence or absence of surface-active agents

Although many attempts have been made to establish a method for estimating the extraction rates [see, for example, Ellis, *Ind. Chem.*, **28**, 483 (1952); Jeffreys and Ellis, *Congr. Chem. Eng. Des.*, **1962**, 65; and Treybal, *Liquid Extraction*, 2d ed., McGraw-Hill, New York, 1963], it is still most important to pilot-plant any new process. About the most that can be said is that, for a given system, packing, and method of operation, H_{td} should be practically constant for all flow rates up to transition and that H_{tc} should vary roughly as $C(V_c/V_d)^n$, where C and n are constants, and to both H_t's correction must be applied on scale-up for axial dispersion [Treybal, *Chem. Eng. Prog.*, **62**(9), 67 (1966)]. Table 15-7 lists additional selected data sources.

Axial Dispersion Vermeulen et al. [*Chem. Eng. Prog.*, **62**(9), 95 (1966)] summarized many of the data for packings. Their correlation for the continuous phase is shown in Fig. 15-36. For the dispersed phase, their correlation is given by

TABLE 15-7 Selected Sources of Packed-Tower Mass-Transfer Data

System	Tower diameter, in.	Packing	Ref.
Water–acetic acid–ethyl acetate, cyclohexane, methylcyclohexane, ethyl acetate + benzene	1	0.25-in. saddles	b
Water–acetic acid–methyl isobutyl ketone	1.95	0.23-in. rings	g
	3	0.375-in. plastic spheres	j
		0.375-in. plastic, ceramic rings	k
		0.5-in. plastic, ceramic saddles	k
Water-acetone-hydrocarbon	1.88	0.25-, 0.375-in. rings, 6-mm. beads	o
	2–4	0.5-, 0.75-in. rings	a
Water–adipic acid–ethyl ether	6	0.5-, 0.75-in. rings, 0.375-in. spheres	e
Water–benzoic acid–carbon tetrachloride	1.95	0.25-in. rings	f
Water-diethylamine-toluene	3, 4, 6	0.25–1-in. rings	i
	3	0.375-in. rings	m
Water–ethyl acetate	4	0.5-in. rings	c
Water-methylisobutyl-carbinol	4	0.5-in. rings	n
Water–methyl ethyl ketone	4	0.5-in. rings	n
Water–propionic acid–methyl isobutyl ketone	1.88	0.25-, 0.375-in. rings, 6-mm. beads	o
Acetone (aq.)–soybean oil, linseed oil	2	0.25-in. saddles, 0.5-in. rings	p
Petroleum-furfural	2	0.25-in. rings	d
	1.2	0.16-in. rings	l
Toluene–heptane–diethylene glycol	1.4, 2.25	Glass and brass rings	h

a Degaleesan and Laddha, *Chem. Eng. Sci.*, **21**, 199 (1966); *Indian Chem. Eng.*, **8**(1), 6 (1966).
b Eaglesfield, Kelly, and Short, *Ind. Chem.*, **29**, 147, 243 (1953).
c Gaylor and Pratt, *Trans. Inst. Chem. Eng. (London)*, **31**, 78 (1953).
d Garwin and Barber, *Pet. Refiner*, **32**(1), 144 (1953).
e Gier and Hougen, *Ind. Eng. Chem.*, **45**, 1362 (1953).
f Guyer, Guyer, and Mauli, *Helv. Chim. Acta*, **38**, 790 (1955).
g Guyer, Guyer, and Mauli, *Helv. Chim. Acta*, **38**, 955 (1955).
h Kishinevskii and Mochalova, *Zh. Prikl. Khim.*, **33**, 2344 (1960).
i Liebson and Beckmann, *Chem. Eng. Prog.*, **49**, 405 (1953).
j Moorhead and Himmelblau, *Ind. Eng. Chem. Fundam.*, **1**, 68 (1962).
k Osmon and Himmelblau, *J. Chem. Eng. Data*, **6**, 551 (1961).
l Sef and Moretu, *Nafta (Zagreb)*, **5**, 125 (1954).
m Shih and Kraybill, *Ind. Eng. Chem. Process. Des. Dev.*, **5**, 260 (1966).
n Smith and Beckmann, *Am. Inst. Chem. Eng. J.*, **4**, 180 (1958).
o Rao and Rao, *J. Chem. Eng. Data*, **6**, 200 (1961).
p Young and Sullans, *J. Am. Oil Chem. Soc.*, **32**, 397 (155).
NOTE: To convert inches to centimeters, multiply by 2.54.

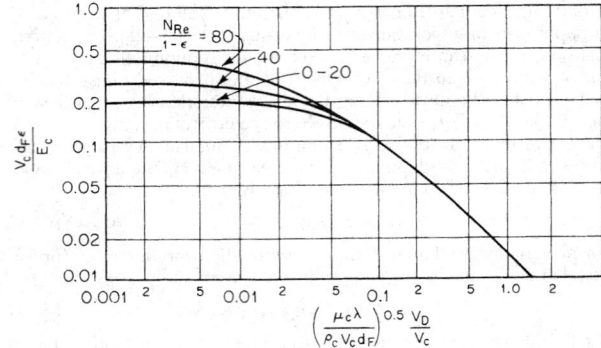

FIG. 15-36 Axial dispersion for the continuous phase in packed towers. Spheres (0.75-in, $\varepsilon = 0.32$ to 0.41; 0.50-in, $\varepsilon = 0.62$), Raschig rings (0.50-in, $\varepsilon = 0.62$; 0.75-in, $\varepsilon = 0.65$), Berl saddles (1.0-in, $\varepsilon = 0.67$). Use customary units in the variables. [*Vermeulen et al.*, Chem. Eng. Prog., **62**(9), 95 (1966), with permission.]

1. Nonwetted carbon rings and wetted Berl saddles:

$$\log \frac{E_d}{V_d d_f} = 0.046 \frac{V_c}{V_d} + 0.301 \qquad (15\text{-}73)$$

2. Wetted ceramic rings:

$$\log \frac{E_d}{V_d d_f} = 0.161 \frac{V_c}{V_d} + 0.347 \qquad (15\text{-}74)$$

The measurements were made with kerosine or diisobutyl ketone dispersed in water. Additional work is reported by Komasawa et al. [*Kagaku Kogaku*, **30**, 237, 450, 928, 1103 (1966); English version, **4**, 288, 363 (1966); **5**, 125, 182 (1967)], and Olbrich et al. [*Trans. Inst. Chem. Eng. (London)*, **44**, T207 (1966)].

GENERAL REFERENCES: Bussolari, Schiff, and Treybal, *Ind. Eng. Chem.*, **45**, 2413 (1953). Fujita and Tanizawa, *Chem. Eng. (Japan)*, **17**, 111 (1953). Garner, Ellis, and Hill, *Am. Inst. Chem. Eng. J.*, **1**, 185 (1955); *Trans. Inst. Chem. Eng. (London)*, **34**, 223 (1956). Major and Hertzog, *Chem. Eng. Prog.*, **51**, 17 (1955). Mayfield and Church, *Ind. Eng. Chem.*, **44**, 2253 (1952). Planovski and Bulatov, *Khim. Mashinostr.*, **1960**(2), 10; (3), 9. Pyle, Duffey, and Colburn, *Ind. Eng. Chem.*, **42**, 1042 (1950).

Perforated-Plate (Sieve-Plate) Towers A schematic diagram for the most common design of perforated-plate, or sieve-plate, tower, arranged for light liquid dispersed, is shown in Fig. 15-37. The light liquid flows through the perforations of each plate and is thereby dispersed into drops which rise through the continuous phase. The continuous liquid flows horizontally across each plate and passes to the plate beneath through the down spout. For heavy liquid dispersed,

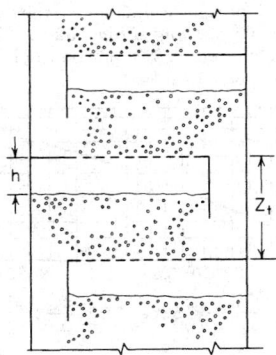

FIG. 15-37 Portion of a perforated-tray tower, arranged for light liquid dispersed.

the same design may be used, but turned upside down. The plates serve to eliminate essentially completely the vertical recirculation of continuous phase characteristic of the spray tower. Furthermore, extraction rates are enhanced by the repeated coalescence and redispersion into droplets of the dispersed phase. Towers of the simple design suggested by Fig. 15-38 have been used successfully in a great variety of services and for petroleum-refining processes have commonly been built to diameters of 3.66 m (12 ft). With careful design, these towers may have excellent flow capacities, and with systems of low interfacial tension equally excellent mass-transfer characteristics.

Many variations in design have been suggested and tried, for example, the use of tower packing in the down spouts to prevent entrainment of dispersed phase, arrangements in which both liquids must pass through perforations at each plate, arrangements with vertical perforated plates, etc. As examples of these, see Bradley (U.S. Patent 2,642,341, 1953), Williams (U.S. Patent 2,652,316, 1953), Maycock and Hartwig (U.S. Patent 2,729,550, 1956), and Pohlenz (U.S. Patent

2,872,295, 1959). Data are available only for arrangements of the sort shown in Fig. 15-39. In general, caplike sieve plates, bubble caps, and vertical perforated plates have not been as satisfactory as horizontal plates.

Sieve-Plate Design   For best tray efficiency, it is well established that the dispersed phase must issue cleanly from the perforations. This requires that the material of the plates be preferentially wet by the continuous phase (requiring the use of plastics or plastic-coated plates in some instances) or that the dispersed phase issue from nozzles projecting beyond the plate surface. These may be formed by punching the holes and leaving the burr in place or otherwise forming the jets (see Mayfield and Church, loc. cit.). The liquid flowing at the larger volume rate should be dispersed.

Perforations are usually 0.32 to 0.64 cm (1/8 to 1/4 in) in diameter, set 1.27 to 1.81 cm (1/2 to 3/4 in) apart, on square or triangular pitch. There appears to be relatively little effect of hole size on extraction rate, except that with systems of high interfacial tension smaller holes will

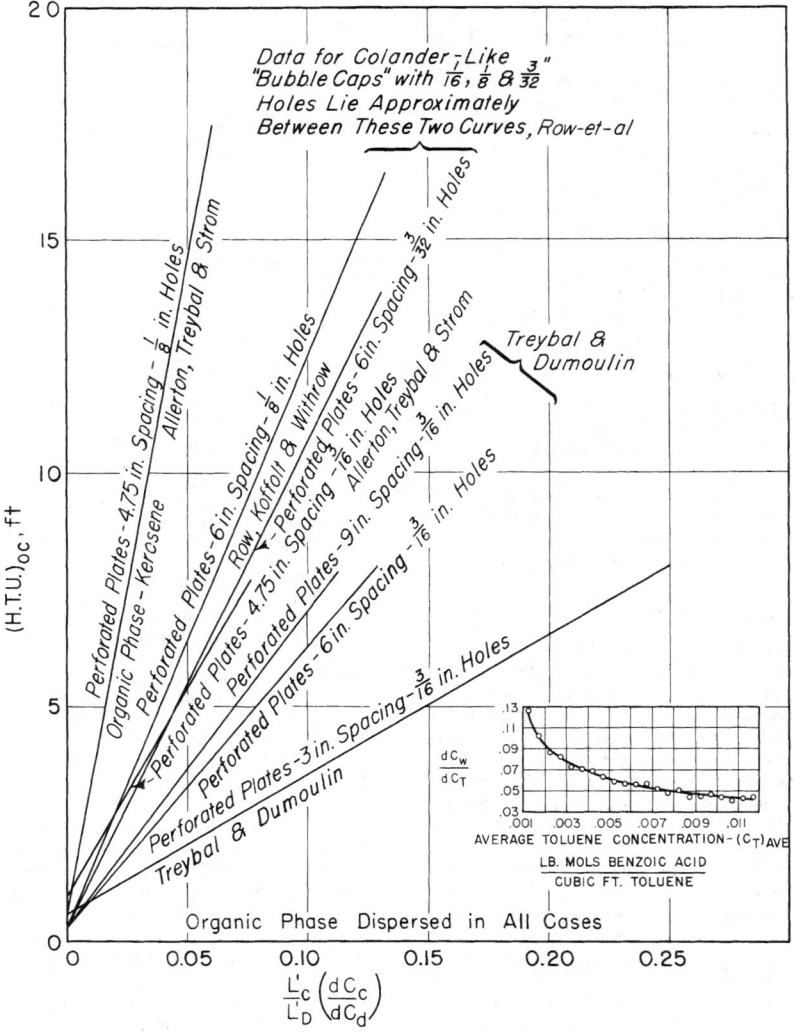

FIG. 15-38   Extraction rates for sieve-plate and modified bubble-plate columns. System: benzoic acid–water–toluene, except where noted. To convert feet to meters, multiply by 0.3048; to convert inches to centimeters, multiply by 2.54. [*Allerton, Strom, and Treybal, Trans. Am. Inst. Chem. Eng.*, **39**, *361* (1943); *Row, Koffolt, and Withrow, ibid.*, **37**, *559* (1941); *Treybal and Dumoulin, Ind. Eng. Chem.*, **34**, *709* (1942).*]

produce somewhat better rates. The entire hole area is suitably set at 15 to 25 percent of the column cross section, subject, however, to check through calculations as outlined below. The velocity through the holes should be such that drops do not form slowly at the holes, but rather that the dispersed phase streams through the openings to be broken up into droplets at a slight distance from the plate. This generally requires average linear velocities through the holes of from 15.2 to 30.5 cm/s (0.5 to 1.0 ft/s). The plate area directly opposite down spouts is kept free of perforations. A scum or "interface-rag" bypass can be incorporated in the trays (see Mayfield and Church, op. cit.) at the expense of tray efficiency, or provision may be made for periodic withdrawal of accumulations through the side of the tower between plates.

Down spouts (or up spouts) are best set flush with the plate from which they lead, with no weir as in gas-liquid contact. The velocity of the continuous phase in the down spout V_d, which sets the down-spout cross section, should be set at a value lower than the terminal velocity of some arbitrarily small droplet of dispersed phase, say, 0.08 or 0.16 cm (1/32 or 1/16 in) in diameter; otherwise, recirculation of entrained dispersed phase around a plate will result in flooding. The down spouts should extend beyond the accumulated layer of dispersed phase on the plate.

The depth of dispersed liquid h accumulating on each plate is determined by the pressure drop required for flow of the liquids,

$$h = h_C + h_D \qquad (15\text{-}75)$$

For the dispersed phase,

$$h_D = h_\sigma + h_O \qquad (15\text{-}76)$$

The available data indicate that, for the orifice effect,

$$h_O = \frac{(V_O^2 - V_D^2)\rho_D}{2g(0.67)^2 \, \Delta\rho} = \frac{V_O'^2 \rho_D}{28.9 \, \Delta\rho} \qquad (15\text{-}77)$$

and that h_σ to overcome interfacial-tension effects may be estimated for drop formation at a low velocity through the holes,

$$h_\sigma = 6\sigma g_c / d_{p0.1} \, \Delta\rho g \qquad (15\text{-}78)$$

where $d_{p0.1}$ = drop diameter produced by flow of dispersed phase at $V_O = 109$ m/h ($V_O' = 0.03$ m/s) [360 ft/h ($V_O' = 0.1$ ft/s)] through the perforations. At hole velocities of 0.3 m/s (1100 m/h) [1 ft/s (3600 ft/h)] or more, h_σ should be omitted, and $h_D = h_O$.

The head required for flow of continuous phase h_C includes losses due to (1) friction in the down spout, which should be negligible, (2) contraction and expansion upon entering and leaving the down spout, and (3) two abrupt changes in direction. These total 4.5 velocity heads:

$$h_C = 4.5 V_d^2 \rho_C / 2g \, \Delta\rho \qquad (15\text{-}79)$$

The distance between trays Z_t should be larger than h, sufficient so that (1) the "streamers" of dispersed liquid from the holes break up into drops before coalescing into the layer of liquid on the next plate, (2) the linear velocity of continuous liquid is not greater than that in the down spout to avoid excessive entrainment, and (3) the tower may be entered through handholes or manholes in the sides for cleaning.

Mass Transfer Mass-transfer rates may be expressed in terms of overall heights of transfer units and successfully correlated for any tower and system in Fig. 15-38. No significance in terms of individual heights of transfer units for the separate phases should be given to the slope and intercept of such lines. The advantage gained by dispersing the liquid flowing at the larger rate, which results in low values for the abscissa of Fig. 15-38 and consequently low transfer-unit heights, is clear. Alternatively, since the plates resemble and basically behave in the manner of stages, the performance is frequently expressed in terms of stage efficiency, either overall E_O for the entire tower or, more satisfactorily, as Murphree efficiencies for each tray.

The system of Fig. 15-38 is one of high interfacial tension, so that the heights of transfer units are relatively high and stage efficiency low. For systems of low interfacial tension, on the other hand, stage efficiencies may be very much improved. Table 15-8 lists sources of mass-transfer data.

Treybal (*Liquid Extraction*, 2d ed., McGraw-Hill, New York, 1963)

TABLE 15-8 Mass-Transfer Data for Perforated-Tray Towers

System	Tower diameter, in.	Tray spacing, in.	Ref.
Benzene–acetic acid–water	1.97	3.9–6.3	t
	1.97	3.2–6.3	s
	2.2	2.8–6.3	r
	1.6 × 3.2	5.9	p
Benzene-acetone-water	3	4, 8	m
Benzene–benzoic acid–water	3	4	m
Benzene–monochloroacetic acid–water	1.97	3.9–6.3	t
Benzene–propionic acid–water	1.97	3.2–6.3	s
Carbon tetrachloride–propionic acid–water	1.97	3.9–6.3	t
Ethyl acetate–acetic acid–water	2	8–24	j
Ethyl ether–acetic acid–water	8.63	4–7.2	n
Gasoline–methyl ethyl ketone–water	3.75	4.5, 6	k
Kerosene-acetone-water	3	4, 8	m
Kerosene–benzoic acid–water	3.63	4.75	a
Isopar-H–benzyl alcohol, methyl benzyl alcohol, acetophenone–water	2 × 12	24	b
Methylisobutylcarbinol–acetic acid–water	3	6	l
Methyl isobutyl ketone–adipic acid–water	4.18	6	e
Methyl isobutyl ketone–butyric acid–water	4.8	6–23	g
Pegasol–propionic acid–water	4.8	6–11	g
Toluene–benzoic acid–water	8.75	6	o
	3.63	4.75	a
	3.56	3–9	q
	3	6	l
	2.72	9	f
	2	24	j
Toluene-diethylamine-water	4.18	6	c, d
2,2,4-Trimethylpentane–methyl ethyl ketone–water	3.75	4.5, 6	k

a Allerton, Strom, and Treybal, *Trans. Am. Inst. Chem. Eng.*, **39**, 361 (1943).
b Angelo and Lightfoot, *Am. Inst. Chem. Eng. J.*, **14**, 53 (1968).
c Garner, Ellis, and Fosbury, *Trans. Inst. Chem. Eng.* (*London*), **31**, 348 (1953).
d Garner, Ellis, and Hill, *Am. Inst. Chem. Eng. J.*, **1**, 185 (1955).
e Garner, Ellis, and Hill, *Trans. Inst. Chem. Eng.* (*London*), **34**, 223 (1956).
f Goldberger and Benenati, *Ind. Eng. Chem.*, **51**, 641 (1959).
g Krishnamurty and Rao, *Indian J. Technol.*, **5**, 205 (1967).
h Krishnamurty and Rao, *Ind. Eng. Chem. Process Des. Dev.*, **7**, 166 (1968).
i Lodh and Rao, *Indian J. Technol.*, **4**, 163 (1966).
j Mayfield and Church, *Ind. Eng. Chem.*, **44**, 2253 (1952).
k Moulton and Walkey, *Trans. Am. Inst. Chem. Eng.*, **40**, 695 (1944).
l Murali and Rao, *J. Chem. Eng. Data*, **7**, 468 (1962).
m Nandi and Ghosh, *J. Indian Chem. Soc., Ind. News Ed.*, **13**, 93, 103, 108 (1950).
n Pyle, Duffey, and Colburn, *Ind. Eng. Chem.*, **42**, 1042 (1950).
o Row, Koffolt, and Withrow, *Trans. Am. Inst. Chem. Eng.*, **37**, 559 (1941).
p Shirotsuka and Murakami, *Kagaku Kogaku*, **30**, 727 (1966).
q Treybal and Dumoulin, *Ind. Eng. Chem.*, **34**, 709 (1942).
r Ueyama and Koboyashi, *Bull. Univ. Osaka Prefect.*, **A7**, 113 (1959).
s Zheliznyak, *Zh. Prikl. Khim.*, **40**, 689 (1967).
t Zheliznyak and Brounshtein, *Zh. Prikl. Khim.*, **40**, 584 (1967).
NOTE: To convert inches to centimeters, multiply by 2.54.

has shown that good estimates of the rate of extraction, or stage efficiency, may be made by computing the rates of extraction for drop formation, drop rise (by computing dispersed-phase holdup and drop velocity and by considering the continuous phase to be of uniform concentration vertically), and drop coalescence (see the subsections "Single Drops Immersed in Immiscible Liquids" and "Spray Towers." See also Skelland and Cornish [*Can. J. Chem. Eng.*, **43**, 302 (1965)]. Specifically, Angelo and Lightfoot [*Am. Inst. Chem. Eng. J.*, **14**, 531 (1968)] have had good success in applying the surface-stretch theory to drop formation and drop rise for oscillating drops on a perforated-tray extractor. Zheleznyak and Brounshtein [*Zh. Prikl. Khim.*, **40**, 584, 689 (1967)] have shown that if the mass-transfer resistance lies within the drop phase, the approach to equilibrium of that phase produced by an extractor is simply related to the approach reached in one section.

The following empirical expression (Treybal, *Liquid Extraction*, 2d ed., McGraw-Hill, New York, 1963) has been found to represent all the available data reasonably well, considering the great variety of circumstances and the considerable scatter in many of the original data:

$$E_O = \frac{89{,}500 Z_t^{0.5}}{\sigma g_c} \left(\frac{V_D}{V_C}\right)^{0.42} = \frac{0.9 Z_t'^{0.5}}{\sigma'} \left(\frac{V_D}{V_C}\right)^{0.42} \quad (15\text{-}80)$$

Use only U.S. customary units in this equation. Krishnamurty and Rao [*Ind. Eng. Chem. Process Des. Dev.*, **7**, 166 (1968)] suggest that Eq. (21–77) is improved if the right-hand side is multiplied by $0.1123/d_O^{0.35}$.

Mechanically Agitated Gravity Devices Owing to the usual small density differences between the contacted liquids, the energy available from simple counterflow under the force of gravity is insufficient to disperse one liquid in the other and to establish turbulence levels to the extent necessary for rapid mass transfer, particularly for systems of high interfacial tension. Application of energy, mechanically applied through stirring devices, pulsations, etc., assists. The devices of major importance are considered below in order of increasing complexity of design.

Rotary-Disk Contactors (RDC)

GENERAL REFERENCES: Logsdail, Thornton, and Pratt, *Trans. Inst. Chem. Eng.* (*London*), **35**, 301 (1957). Misek. *Collect. Czech. Commun.*, **28**, 426, 570, 1631 (1963); **32**, 4018 (1967) (in English); *Ratacni Diskove Extraktory a Jejich Vypocty*, SNTL, Prague, 1964. Olney et al., *Am. Inst. Chem. Eng. J.*, **8**, 252 (1962); **10**, 827 (1964). Reman et al., U.S. Patent 2,601,674 (1952); *Chem. Eng. Prog.*, **51**, 141 (1955); **62**(9), 56 (1966); *Joint Symposium: Scaling-Up Chemical Plant and Processes*, London, 1957, p. 26.

Refer to Fig. 15-39. The tower is formed into compartments by horizontal doughnut-shaped or annular baffles, and within each compartment agitation is provided by a rotating, centrally located, horizontal disk. Somewhat similar devices have been known for some time. The features here are that the rotating disk is smooth and flat and of a diameter less than that of the opening in the stationary baffles, which facilitates fabrication and apparently improves extraction rates. The typical proportions of the internals of the RDC are as follows:

$$d_s/d_t = 0.7$$

$$d_r/d_t = 0.6$$

Z_t/d_t—the following table applies

For $\quad\quad\quad 0 < d_t < 0.1 \text{ m} \quad\quad Z_t = (d_t)^{0.5}$

$\quad\quad\quad\quad\quad\quad 0.1 < d_t < 1.0 \text{ m} \quad Z_t/d_t = 0.15$

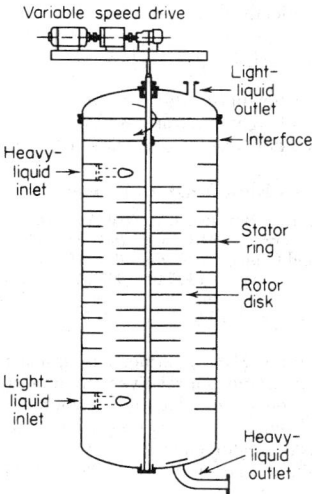

FIG. 15-39 Rotating-disk (RDC) extractor. (*Courtesy of Glitsch Process Systems Inc.*)

$\quad\quad 1.0 < d_t < 1.5 \text{ m} \quad\quad Z_t/d_t = 0.12$

$\quad\quad 1.5 < d_t < 2.5 \text{ m} \quad\quad Z_t/d_t = 0.10$

$\quad\quad 2.5 < d_t > 2.5 \text{ m} \quad\quad Z_t/d_t = 0.08$

where $\quad d_s$ = stator diameter
$\quad\quad\quad d_t$ = tower diameter
$\quad\quad\quad d_r$ = rotor diameter
$\quad\quad\quad Z_t$ = stage height

The general proportions may be varied from one end of the tower to the other to accommodate changing liquid volumes and physical properties. These towers have been used in diameters ranging from a few inches for laboratory work up to 2.4 m (8 ft) in diameter by 12.2 m (40 ft) tall for purposes of deasphalting petroleum. Other commercial services include furfural extraction of lubricating oils, desulfurization of gasoline, phenol recovery from wastewaters, and many others. Columns up to 4.5 m in diameter and up to 50 m in height have been constructed.

A reliable design procedure for new systems, without the necessity for laboratory work, is not yet established. The data available show that the flow capacity increases with (1) decreased rotor speed, (2) decreased diameter of rotating disks, (3) increased diameter of opening in the stationary baffles, and (4) increased compartment height. Logsdail et al. (loc. cit.) have proposed that the slip velocity of Eq. 15-83, in the absence of mass transfer, can be set equal to $V_K(1 - \varphi_D)$, where V_K is a "characteristic" velocity which can be related to the liquid properties, speed of agitation, and tower geometry. Kung and Beckmann [*Am. Inst. Chem. Eng. J.*, **7**, 319 (1961)] and Olney et al. (loc. cit.) have also used this. Misek (loc. cit.), however, has had considerable success by setting the slip velocity equal to $V_K(1 - \varphi_D) \exp [\varphi_D(z - 4.1)]$, where z is a "coalescence coefficient" which depends on the liquid properties. Evidently mass transfer has a profound effect, as a result of drop coalescence; variation in the flooding rate from −15 to +200 percent has been noted in the extraction of acetone to and from water, respectively, with organic solvents. See also Kagan et al., *Izv. Vyssh. Uchebn. Zaved. Khim. Khim. Tekhnol.*, **9**, 836 (1966). Drop-size distribution which has an important influence on axial dispersion in the dispersed phase has been studied extensively by Misek and Olney (loc. cit.).

The value of HETS becomes smaller with (1) increased rotor speed but passes through a minimum, (2) increased diameter of rotating disks, (3) decreased diameter of stationary baffle opening, and (4) decreased compartment height. Reman and Olney [*Chem. Eng. Prog.*, **51**, 141 (1955)] show a correlation of stage height for two sizes of RDCs with the system water-kerosine-butylamine, as in Fig. 15-40. That such correlations cannot be general is indicated by these authors' data on caustic extraction of gasoline, which show quite different

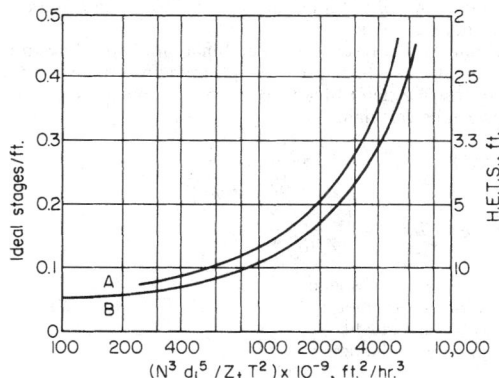

FIG. 15-40 Extraction in RDC columns, water-butylamine-kerosine (continuous). $d_t = 0.33$ and 1.33 ft. Curve A: $V_D = 50.7$, $V_C = 78.9$ ft/h. Curve B: $V_D = 25.4$, $V_C = 78.9$ ft/h. Use customary units in the variables. [*Data of Reman and Olney*, Chem. Eng. Prog., **51**, 141 (1955).]

curves. Logsdail, Thornton, and Pratt (loc. cit.) tentatively suggest that data can be correlated through

$$\frac{H_{tOC}}{V_C}\left(\frac{g^2\rho_C}{\mu_C}\right)^{1/3}\phi_D = C\left[\frac{\mu_C g}{V_K^3(1-\phi_D)^3\rho_C}\right]^{2\beta/3}\left(\frac{\Delta\rho}{\rho_C}\right)^{2(\beta-1)/3} \quad (15\text{-}81)$$

the constants C and β to be determined for each system. For toluene-water-acetone, $\beta = 0.13$; for butyl acetate-water-acetone, $\beta = 0.4$; in both cases, transfer was from water to organic solvent. For transfer in the reverse direction, V_K could not be computed (see above).

A large number of studies of axial mixing have been made [Gel'perin et al., *Teor. Osn. Khim. Tekhnol.*, **1**, 666 (1967); Kagan et al., *Zh. Prikl. Khim.*, **39**, 88 (1966); Miyauchi et al., *Am. Inst. Chem. Eng. J.*, **12**, 508 (1966); Stainthorp and Sudall, *Trans. Inst. Chem. Eng. (London)*, **42**, 198 (1964); Stemerding and Zuiderweg, *Chem. Ing. Tech.*, **35**, 844 (1963); and Strand et al., *Am. Inst. Chem. Eng. J.*, **8**, 252 (1962)]. Reman [*Chem. Eng. Prog.*, **62**(9), 56 (1966)] recommends, for the continuous phase in columns 0.08 to 2.13 m (3 in to 7 ft) in diameter,

$$E_C = 0.5Z_t V_C + 0.012 d_i NZ_t(d_s/d_t)^2 \quad (15\text{-}82)$$

For the dispersed phase firm relationships have not been established, but at high rotor speeds, E_D may be 1 to 3 times E_C. In any event, axial mixing for the liquid flowing at the lower rate becomes very severe for extreme flow ratios (>10).

Costs are given by Clerk (*Chem. Eng.*, 232 (Oct. 12, 1964).

Several modifications of the design have appeared. Modifications of the rotors include perforation of the disk [Krishnara et al., *Br. Chem. Eng.*, **12**, 719 (1967)] and radially supported arc plates [Nakamura and Hiratsuka, *Kagaku Kogaku*, **30**, 1003 (1966)]. An "asymmetric" modification, with off-center rotors and arrangement of settling spaces for the liquids between dispersions (Misek, loc. cit.) is available in Europe.

As stated above, the design of an RDC contactor usually involves the performance of pilot tests due to the large number of factors which can influence performance. These pilot plant data must then be scaled-up to full commercial size. The following procedure is recommended.

1. Pilot plant tests are conducted using the actual plant materials since small amounts of contaminents can have significant effects on throughput and efficiency. These tests are usually conducted in columns ranging from 0.075–0.15 m diameter; the column height (and therefore number of compartments) should be sufficient to accomplish the separation desired; this may require several iterations on column height.

2. The column is run over a range of total throughputs ($V_o + V_c$) and agitation speeds; at each condition the concentrations of the streams are measured after equilibrium is reached; the holdup is also measured by stopping the agitation, isolating the column, and measuring the change in the interface level. The flooding point is determined at each specific throughput by increasing the agitation speed until the column floods.

3. From the above data, the combination of specific throughput and agitation speed which gives the optimum performance in terms of separation can be determined. At this condition the following relationships can be calculated:

Slip Velocity:
$$V_s = \frac{V_d}{\phi_d} + \frac{V_c}{1 - \phi_d} \quad (15\text{-}83)$$

Specific Power Input $= \dfrac{(N_s)^3(d_r)^5}{(Z_t)(d_t)^2} \quad (15\text{-}84)$

where N_s = rotational speed
d_r = rotor diameter
Z_t = stage height
d_t = tower diameter

Max Continuous Phase Velocity at Flooding

$$V_{cf} = \frac{V_s e^{(-\phi_{df})}}{[V_d/V_c]/\phi_{df}} \quad (15\text{-}85)$$

where V_{cf} = velocity of the continuous phase at flooding
ϕ_{df} = holdup of the dispersed phase at flooding
V_d = dispersed phase velocity
V_s = slip velocity

4. For design, the slip velocity is derated to 70–80 percent of the calculated value to give some margin of safety; this sets the design value of the continuous phase velocity (V_c). The column cross sectional area (and therefore diameter) is set by Q_c/V_c. With the diameter set, the other dimensions can be set using the ratios given above.

5. The rotor speed of the scaled up tower is based on maintaining the same specific power input number as used on the pilot column; it can be determined by substituting the specific values into the relationship:

$$\frac{N_s^3 R_r^5}{Z_t d_t^2}$$

6. For the column height, the pilot plant data must be corrected for the effect of axial mixing. The height of a transfer unit (HTU) can be determined from the pilot plant data; to this must be added the height of a diffusion unit (HDU). This is done by determining the axial mixing coefficients of the continuous and dispersed phases according the the following relationships:

$$E_c = 0.5V_c Z_t + 0.012 d_r N_s Z_t(d_s/d_t)^2 \quad (15\text{-}86)$$

$$E_d = E_c\left[\frac{4.2 \times 10^5}{d_t^2}\left(\frac{V_d}{\phi_d}\right)^{3.3}\right] \quad (15\text{-}87)$$

where E_c = diffusion coefficient, continuous phase; E_d = diffusion coefficient, dispersed phase; V_c, Z_t, d_r, N_s, Z_t, d_t, ϕ_d = same as defined previously; see Nomenclature list.

From these the continuous and dispersed phase Péclet numbers can be determined from the relationships:

$$\frac{1}{(Pe)_c} = \frac{E_c(1 - \phi_d)}{Z_t V_c} \quad (15\text{-}88)$$

$$\frac{1}{(Pe)_d} = \left[\frac{1}{(Pe)_c}\right]\left[\frac{\phi_d V_c}{(1 - \phi_d)V_d}\right]\left[\frac{4.2 \times 10^5}{d_t^2}\left(\frac{V_d}{\phi_d}\right)^{3.3}\right] \quad (15\text{-}89)$$

where $(Pe)_c$ = Péclet number, continuous phase
$(Pe)_d$ = Péclet number, dispersed phase

The HDU is then calculated from the relationship:

$$\text{HDU} = \frac{Z_t}{(Pe)_c} + \frac{Z_t}{(Pe)_d} \quad (15\text{-}90)$$

And finally, the effective height of a transfer unit is calculated from:

$$(\text{HTU})_{\text{eff}} = \text{HTU} + \text{HDU} \quad (15\text{-}91)$$

where HDU = height of a diffusion unit
HTU = height of a transfer unit
$(\text{HTU})_{\text{eff}}$ = effective height of a transfer unit

Lightnin Mixer (Oldshue-Rushton) Tower

GENERAL REFERENCES: Bibaud and Treybal, *Am. Inst. Chem. Eng. J.*, **12**, 472 (1966). Dykstra, Thompson, and Clouse, *Ind. Eng. Chem.*, **50**, 161 (1958). Gustison, Treybal, and Capps, *Chem. Eng. Prog.*, **58**, Symp. Ser. **39**, 8 (1962). Gutoff, *Am. Inst. Chem. Eng. J.*, **11**, 712 (1965). Oldshue and Rushton, *Chem. Eng. Prog.*, **48**, 297 (1952). Miyauchi et al., *Am. Inst. Chem. Eng. J.*, **12**, 508 (1966).

The Oldshue-Rushton (Mixco) extractor is similar in construction to the RDC in the fact that it is a relatively open design, consisting of a series of compartments separated by horizontal stator baffles. The major difference from the RDC is that the height/diameter ratio of the compartments is greater, each compartment is fitted with vertical baffles, and the mixing is accomplished by means of a turbine impeller rather than a disc.

Refer to Fig. 15-41. The extractor is an extension of the simple baffled mixing vessel into a multistage column. Although commercial

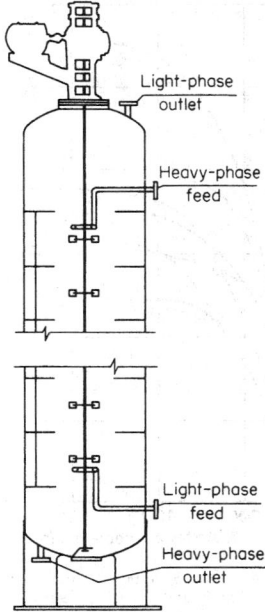

FIG. 15-41 Mixco (Oldshue-Rushton) extractor.

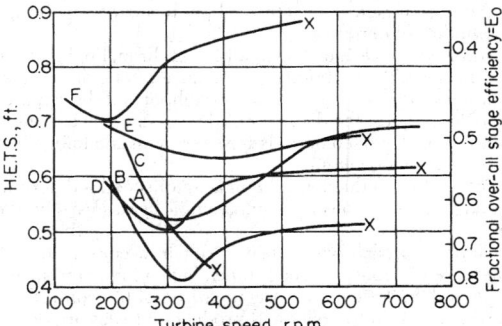

FIG. 15-42 Extraction in Mixco columns, methyl isobutyl ketone-acetic acid-water (continuous). $d_t = 0.5$ ft, $Z_t = 0.333$ ft, X = flooded condition. To convert feet to meters, multiply by 0.348; to convert feet per hour to meters per hour, multiply by 0.3048.

application has been made, data are scarce and are limited to towers of small diameter. The preferred proportions are $Z_t = 0.5d_t$, $d_s > d_t$.

For water (continuous) and toluene or kerosine (dispersed), in a tower with $d_t = 0.152$ m (0.5 ft), $Z_t = 0.082$ m (0.27 ft), $d_i = 0.051$ m (0.1667 ft), dispersed-phase holdup is given by Eq. (15-70) with $V_s = V_k(1 - \phi_d)$ and the following relationship by Wong (M.Ch.E. thesis, New York University, 1963):

$$V_k \mu_c / \sigma g_c = 1.77(10^{-4})(g/d_i N^2)(\Delta\rho/\rho_c)^{0.9} \qquad (15\text{-}92)$$

For the same liquids axial mixing is described by (Bibaud and Treybal, loc. cit.)

$$E_c\phi_c/V_c Z_t = -0.1400 + 0.0268(d_i N\phi_c/V_c) \qquad (15\text{-}93)$$

$$\frac{d_i^2 N}{E_d} = 0.393(10^{-8})\left(\frac{d_i^3 N^2\rho_c}{\sigma g_c}\right)^{1.54}\left(\frac{\rho_c}{\Delta\rho}\right)^{4.18}\left(\frac{d_i^2 N\rho_c}{\mu_c}\right)^{0.61} \qquad (15\text{-}94)$$

See also Miyauchi et al. (loc. cit.), who express the axial mixing in terms of interstage flow. For the continuous phase with no dispersed-phase flow, see Bibaud and Treybal, and Gutoff (loc. cit.).

Figure 15-42 presents some of the data of Oldshue and Rushton (loc. cit.) which show an optimum agitator speed for each configuration studied. The optimum would be expected to vary with physical properties of the liquids contacted. HETS is improved, although capacity is decreased, by smaller openings in the stationary baffles. The effect of stage openings of efficiency and throughput for the system MIBK-acetic acid-water in a 6-inch (150 mm) diameter column is shown in Table 15-9. In the more difficult (because of high interfacial tension) extraction of uranium between kerosine-diluted solvents and aqueous solutions, Dykstra et al. (loc. cit.) have also shown the development of an optimum impeller speed. Gustison et al. (loc. cit) have found it possible to correlate the stage efficiency with the ratio of flow rates (V_d/V_c) and the distribution coefficient, which varies considerably with concentration in the extraction of uranium. They also found it possible to scale up performance from 0.152- to 0.305-m (6- to 12-in) diameter geometrically, on the assumption that the continuous phase was thoroughly mixed in each compartment, by applying equal power per unit volume of liquids treated on the large and the small scale and using the same mass velocities of flow. Bibaud (loc. cit.) found that, for butylamine extracted from kerosine (dispersed) into

water, the extraction rates corrected for axial mixing in either phase were described by assuming the drops to be rigid spheres, with Thornton's correlation [*Ind. Chem.*, **39**, 298 (1963)] for drop size.

A somewhat related design has been studied by Nagata, Eguchi, and coworkers [*Chem. Eng.* (*Japan*), **17**, 20 (1953); **20**, 2 (1956); *Mem. Fac. Eng., Kyoto Univ.*, **19**, 102 (1957); *Kagaku Kogaku*, **22**, 483 (1958)]. This column is characterized by the relatively small, separate openings between compartments for passage of liquids and the eccentric location of the impeller shaft. In a pilot-plant column, $d_t = 0.3$ m (0.983 ft), phenol was extracted from water [$V_c = 11.6$ m/h (38.1 ft/h)] into benzene [$V_d = 6.4$ m/h (21 ft/h)] at a stage efficiency of 0.618.

Because of the above limitations in prediction of column performance based on correlations, the design of an Oldshue-Rushton column must also be based on pilot-plant tests. The minimum column diameter which can be used to give reliable scale-up data is 6 inches (150 mm); it is usuasly fitted with stages 3 inches (75 mm) high, and the stage opening is 2.4 inches (60 mm). The column should be high enough to accomplish the complete extraction; if it is not it will be necessary to "rerun" the extract and raffinate phases to examine effects in the dilute regions of the column.

The following procedure is followed:

1. The column is run over a range of total throughputs ($V_o + V_c$) and agitation speeds; at each condition the concentrations of the streams are measured after equilibrium has been reached. The flood-

TABLE 15-9 Effect of Size of Opening between Compartments*

Compartment opening, mm	Maximum stage efficiency	Minimum HETS, mm	Flow rate, kg s⁻¹ m⁻²
Constant flow rate			
0	100	2560	0†
54	83	3098	2.9†
82	52	4953	2.9
152	38	6731	2.9
At maximum efficiency			
0	100	2560	0†
54	83	3098	2.9†
82	67	3860	5.4†
152	38	6731	6.0ᵇ

*Typical data for operation with methyl isobutyl ketone, water, acetic acid; four stages; 101.6-mm stage height, 152-mm-diameter column; extraction, water → ketone.

†Optimum flow rate.

Oldshue in Lo, Baird, Hanson, *Handbook of Solvent Extraction*, p. 436, John Wiley & Sons, NY, 1983. Used with permission.

ing point is determined at each throughput by increasing the agitation speed until the column floods.

2. From the above data, the condition of throughput and agitation speed which gives the optimum performance can be determined.

3. Based on this design-specific throughput and the required production column rates, the diameter of the commercial column can be calculated. The stage geometry is next set by maintaining geometric similarity to the pilot column.

4. Finally, the production column agitator speed is determined by maintaining the same power per unit volume as was used on the pilot column.

The above approach will usually result in a conservative design, since the stage efficiency is usually much higher in the production column than in the pilot column. A comparison of the controlling parameters which exist in the pilot and production scales are depicted in Fig. 15-43.

Scheibel Extraction Towers   The original Scheibel tower design [*Chem. Eng. Prog.*, **44**, 681, 771 (1948); U.S. Patent 2,493,265, 1950] used knitted-mesh packed sections in a tower for coalescence with a centrally located impeller between the packed sections for drop breakup. Scheibel and Karr [*Ind. Eng. Chem.*, **42**, 1048 (1950)] presented data on a 0.305-m- (12-in-) diameter column of this design (Fig. 15-44) for systems which are difficult to extract because of high interfacial tension or easy because of low interfacial tension. Excellent values of HETS were obtained with a wide variety of conditions. Low throughput and ratios of flow rates greatly different from unity required high agitator speeds for best results. Both direction of extraction and which phase was dispersed influenced the rates. The liquids of Fig. 15-44 were also used in tests involving the mixing sections alone (see operating characteristics of mechanically agitated vessels). Honekamp and Burkhart [*Ind. Eng. Chem. Process Des. Dev.*, **1**, 176 (1962)] found very little change in drop size to occur within the knitted-wire mesh and measured extraction rates in the mesh zone for one system.

A second Scheibel tower design [*Am. Inst. Chem. Eng. J.*, **2**, 74 (1956); U.S. Patent 2,850,362, 1958] reduced HETS and permitted more direct scale-up. The impellers are surrounded by stationary shroud baffles to direct the flow of droplets as they are discharged from the tips of the impellers. Data taken from a 0.305-m- (12-in-)

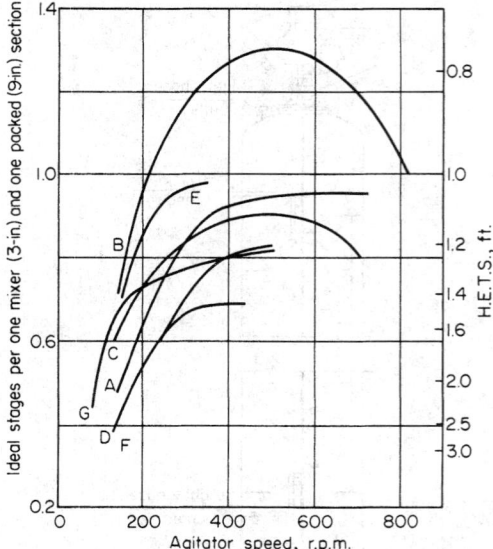

FIG. 15-44 Extraction in first Scheibel column. $T = 0.94$ ft, $d_i = 0.333$ ft, height of mixer section = 3 in, height of packed section = 9 in. To convert inches to centimeters, multiply by 2.54; to convert feet to meters, multiply by 0.3048; and to convert feet per hour to meters per hour, multiply by 0.3048. [*Data of Scheibel and Karr*, Ind. Eng. Chem., **42**, 1048 (1950).]

Curve	System	V_D, ft/h	V_C, ft/h
A	MIBK(C)–water(D,E)–acetic acid	41.7	41.7
	MIBK(D)–water(C,E)–acetic acid		
B	MIBK(C,E)–water(D)–acetic acid	41.7	41.7
C	MIBK(C)–water(D,E)–acetic acid	23.2	23.2
	MIBK(C,E)–water(D)–acetic acid		
	MIBK(D)–water(C,E)–acetic acid		
D	o–Xylene(D)–water(C,E)–acetone	25.9	17.3
E	o–Xylene(D,E)–water(C)–acetone	22.1	21.2
F	o–Xylene(C)–water(D,E)–acetone	25.9	17.3
G	o–Xylene(C,E)–water(D)–acetone	21.1	22.1

MIBK = methyl isobutyl ketone; C = continuous; D = dispersed; E = extractant.

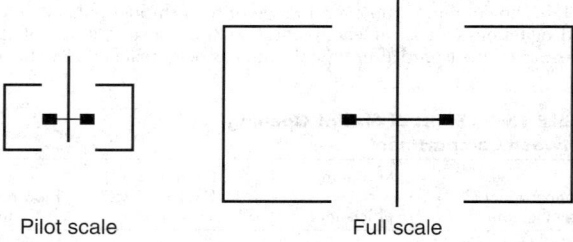

Full scale compared to pilot scale

Residence time	Higher
Blend time, undispersed	Longer
Interstage mixing, undisperesed	Different
Interstage mixing, disp.	Different
Concentration gradient, disp.	Higher
Max. impeller zone shear rate	Higher
Ave. impeller zone shear rate	Lower
Ave. tank zone shear rate	Lower
Turbulent shear rates	Different

FIG. 15-43 Mixing factors compared for pilot and full scale. [*Oldshue in Lo, Baird, and Hanson*, Handbook of Solvent Extraction, *John Wiley & Sons, NY, 1983. Used with permission.*]

diameter tower are shown in Fig. 15-46 and correlated in terms of the power applied per unit volume of liquids handled per compartment. For the impeller used, the power number at turbulent Reynolds numbers is $N_{Po} = 1.85$. The data show that while packing in alternate sections may increase mass-transfer rates, it decreases flow capacity. For many industrial systems, the knitted mesh was not used because of fouling (Fig. 15-45). Towers up to 3 m (9.8 ft) in diameter are in service. A third design by Scheibel (U.S. Patent 3,389,970, 1968) uses closed impellers plus horizontal baffles in the tower.

Scheibel (Ref. 2) has shown that the efficiency of a mixing stage can be correlated to the power per unit of throughput, and is related to the ratio of dispersed/continuous phase flow rates; this is shown in Fig. 15-47.

This figure shows an optimum power input; below this value efficiency drops off due to reduced interfacial area; above this value efficiency decreases due to increased axial mixing of the continuous and dispersed phases.

Scheibel has found that the power input can be correlated by the following equation:

$$P = 1.85 \left[\frac{(d_i)^5 \, \rho (N_s)^3}{g_c} \right] \tag{15-95}$$

where P = Power input per mixing stage
 d_i = impeller diameter
 ρ = average stage density

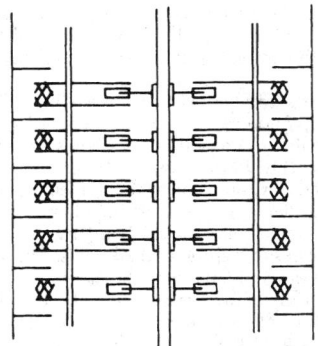

FIG. 15-45 Second Scheibel extractor with horizontal baffles and no wire-mesh packing between stages. [*Reprinted with permission of* Am. Inst. Chem. Eng. J., **2**, *74 (1956)*].

N_s = impeller rotational speed
g_c = gravitational constant

As with the design of the other columns described above, the design of a Scheibel column must be based on pilot plant tests and scale-up. The following procedure is recommended:

1. Pilot tests are usually conducted in 0.075-m diameter columns; the column should contain a sufficient number of stages to complete the extraction; this may require several iterations on column height.

2. The column is run over a range of throughputs ($V_d + V_c$) and agitation speeds; at each condition the concentrations of the streams are measured after equilibrium is reached (usually 3–5 turnovers of column volume). At each throughput the flood point is determined by

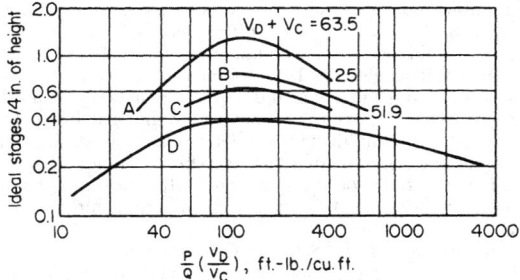

FIG. 15-46 Extraction in second Scheibel column. $T = 0.94$ ft, $d_i = 0.333$ ft, height of packed section = height of mixer section = 2 in. Use customary units in the variables. [*Data of Scheibel,* Am. Inst. Chem. Eng. J., **2**, *74 (1956).*]

Curve	System
A, B°	Methyl isobutyl ketone–water–acetic acid
C°	o–Xylene–water–acetic acid
D†	o–Xylene–water–phenol
	Methyl isobutyl ketone–water–acetic acid
	o–Xylene–water–acetic acid

°Alternate mixing and packed sections.
†Packing omitted. Agitators in alternate and also every section.

increasing the agitation until flooding is induced. A minimum of three throughput ranges are examined in this manner.

3. From the above data, the combination of specific throughput and agitation speed which gives the optimum performance in terms of separation can be determined. This determines the design specific throughput value (m3/m2-h) and agitation speed (RPM).

4. Unlike the RDC and Oldshue-Rushton columns where the

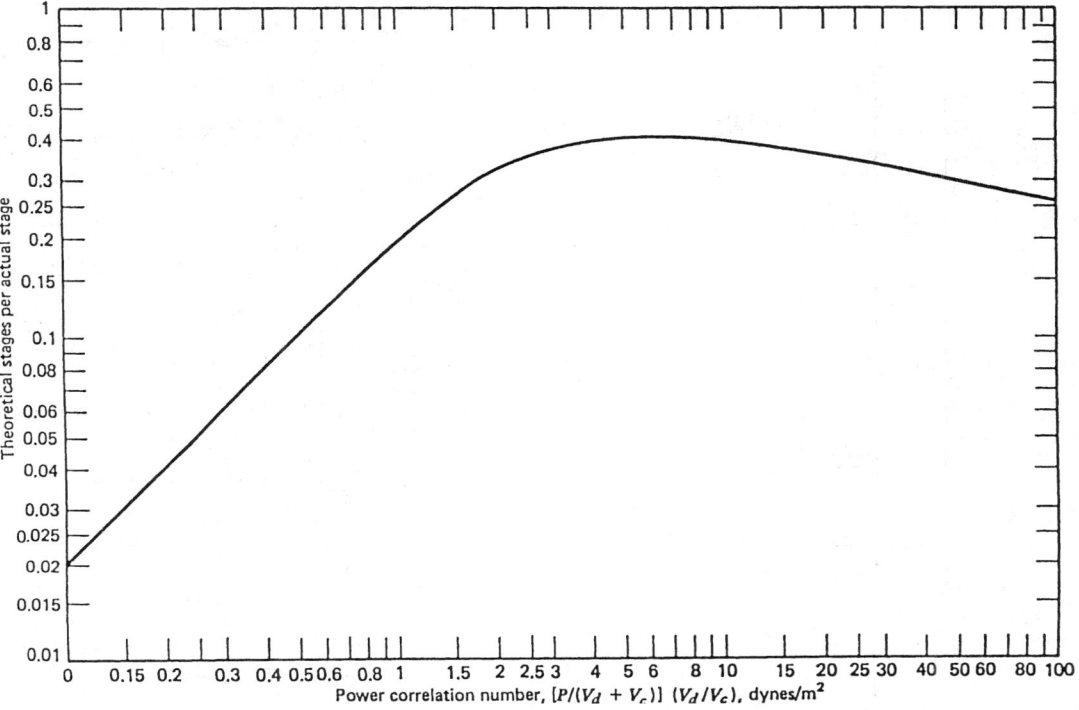

FIG. 15-47 Correlation of mixing-stage efficiency with power input and liquid flow rates. [*Scheibel in Lo, Baird, Hanson,* Handbook of Solvent Extraction, *p. 428 John Wiley & Sons, NY, 1983. Used with permission.*]

specific throughput of the scaled-up version is the same as the pilot column, it is the characteristic of the Scheibel column that each throughput of the scaled-up column is on the order of 3–5 times greater than that acheived on the pilot column. The reason for this is that the restricted geometry of the 0.075 m diameter column limits throughput; these restrictions are removed in the scaled-up columns.

5.   Once the column diameter is determined, the stage geometry can be fixed. The geometry of a stage is a complex function of the column diameter; in the pilot (0.075 m) column the stage height to diameter ratio is on the order of 1:3; on a 3-m diameter column it is on the order of 1:8.

6.   The principle of the Scheibel Column scale-up is to maintain the efficiency of the stage. Therefore, the scaled-up column will have the same number of actual stages as the pilot column. The only difference is that the stages will be taller to take into account the effect of axial mixing. With the agitator dimensions determined, the speed is then calculated to give the same power input per unit of throughput.

The scale-up of the Scheibel column is still considered proprietary, and therefore the vendor (Glitsch Process Systems Inc.) should be consulted for the final design. From pilot tests in 0.075-m diameter column, industrial columns up to 3 m in diameter and containing 90 actual stages have been provided.

Because of its internal baffling which controls the mixing patterns on the stages, the Scheibel column has proven to be one of the more efficient extractors in terms of height of a theoretical stage; this makes it ideally suited for applications which require a large number of stages, or are located indoors with headroom restrictions. Holmes, Karr, and Cusack (*Solvent Extraction and Ion Exchange*, vol. 8, no. 3, pp. 515–528, 1990) have published results comparing the efficiency of the Scheibel column to other extractors on the system toluene-acetone-water.

Kühni Tower   The extraction towers designed at Kühni [see Mögli and Bühlmann, in Lo, Baird, and Hanson (eds.), *Handbook of Solvent Extraction*, Wiley-Interscience, New York, 1983, sec. 13.5] use shrouded (closed) impellers on a central shaft in the tower (Fig. 15-48). The droplet size can be controlled by the speed and diameter

of the impeller, while the circulation rate can be controlled by the design of the width of the impeller. A perforated plate between each stage can control the droplet holdup by the percentage of open area in the plate.

Treybal Tower   Treybal [U.S. Patent 3,325,255, 1967); *Chem. Eng. Prog.*, **60**(5), 77 (1964)] adapted a mixer-settler cascade in tower form in which the liquids are settled between stages.

Karr Reciprocating Plate Tower   Up to this point, the agitated columns presented have all imparted their energy to the fluids by means of rotating elements (discs or impellers). However, there is another class of agitated columns which impart their energy by means of reciprocating plates or pulsing of the liquids. This results in a more uniform drop-size distribution due to the fact that the shear forces are more uniform over the entire cross section of the column.

The reciprocating plate extractor developed by Karr [*Am. Inst. Chem. Eng. J.*, **5**, 446 (1959)] is a mechanically agitated tower using dual-flow plates with 50 to 60 percent open area, mounted on a central shaft and reciprocated vertically (Fig. 15-49). Typical perforated plates and baffle plates for a 35 ⅝-in (0.9-m) diameter column are shown in Fig. 15-50. A typical stroke length is 2.54 cm (1 in) with a speed of 10 to 400 strokes per minute and a plate spacing of 5 to 15 cm (2 to 6 in). Scale-up relationships by Karr (*Sep. Sci. Technol.*, **15**(4), 877 (1980)] show that HETS increases with tower diameter to the 0.38 power in the most difficult case. Laboratory columns of 2.54- and 5.08-cm- (1- and 2-in-) diameter are used to scale up to towers as large as 1.5 to 2.0-m (5 to 6.5-ft) in diameter. A high volumetric efficiency is achieved as measured by total volumetric throughput per cross-sectional area divided by HETS.

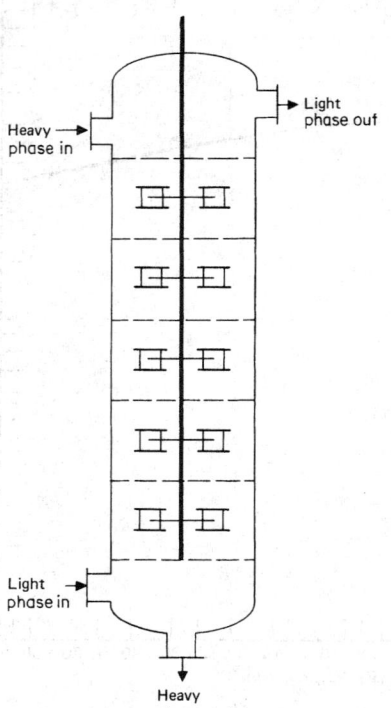

FIG. 15-48   Kühni tower.

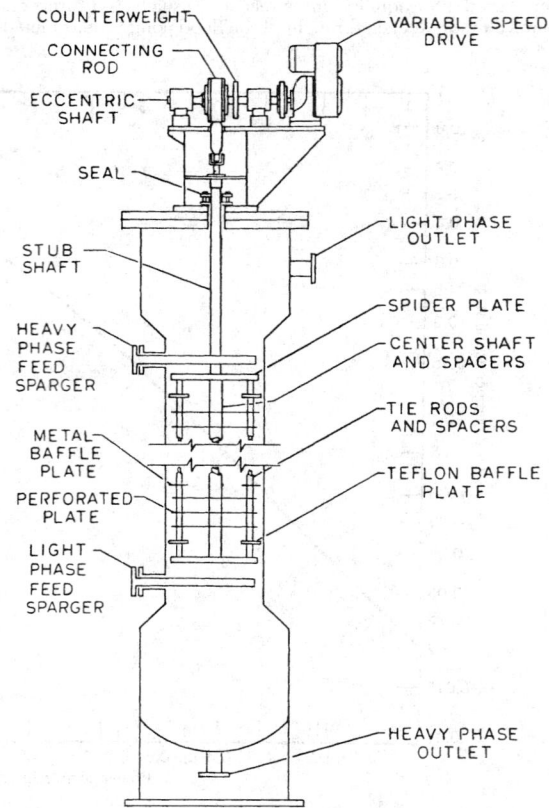

FIG. 15-49   Schematic arrangement of the 900-mm (36-in) reciprocating-plate column. (*Courtesy of Glitsch Process Systems Inc.*)

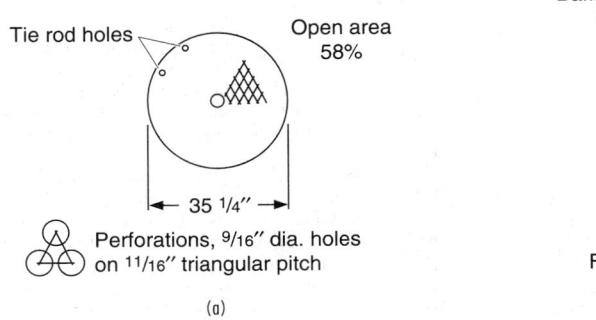

FIG. 15-50 (a) Typical perforated plate. (b) Typical baffle plate.

One of the chief characteristics of the Karr column is its high-volumetric efficiency when compared to other extractors. Volumetric efficiency is defined as:

$$\text{Volumetric efficiency} = \frac{V_d + V_c}{\text{HETS}} \qquad (15\text{-}96)$$

Karr, Holmes, and Cusack have given comparisons of the Karr column volumetric efficiency with other types of extractors. In Table 15-10 are data showing the values of HETS and volumetric efficiency over a range of column diameters from 1–36 in (0.025–0.9 m); Fig. 15-51 shows how the HETS varies with agitation, again over a range of diameters but at relatively constant total throughput. It was from

these data that Karr and Lo developed the scale-up procedure for this type of column.

As with the other extractors, the final design of a Karr column depends on the scale-up from a pilot test. The following procedure is recommended.

1. For scale-up up to 2 m in diameter, testing in a pilot column of 0.025 m is sufficient; if the anticipated scaled-up diameter is greater than 2 m, then the pilot tests should be conducted in a 0.050-m diameter. The column should be tall enough to accomplish the complete extraction; this may require several iterations on column height.

2. The column is first optimized with regard to plate spacing; what is desired is for the tendency to flood to be equal over the entire

TABLE 15-10 Summary of Minimum HETS Values and Volumetric Efficiencies for a Reciprocating-Plate Column*

Column diameter in.	Amplitude, in.	Plate spacing, in.	Agitator speed, strokes/min	Extractant	Dispersed phase	Minimum HETS	Throughput, gal hr⁻¹ ft⁻²	Volumetric efficiencies V_t/HETS, h⁻¹
MIBK-acetic acid-water system								
1	½	1	360	MIBK	Water	3.1	572	296
			401			2.8	913	523
1	½	1	278	Water	MIBK	4.2	459	175
			152			8.1	1030	204
3	½	1	330	MIBK	Water	4.9	600	196
	½	1	245			6.3	1193	304
	½	2	355			7.5	1837	393
	½	1	320	Water	Water	4.3	548	205
	½	1	230			6.7	1168	280
	½	2	367	Water	Water	5.0	1172	376
			240			7.75	1707	353
12 (with baffle)	½	1	430	Water	MIBK	5.8	547	151
			285			5.7	1167	328
	½	1	244	MIBK	MIBK	4.4	599	218
			170			5.6	1193	342
	½	1	250	MIBK	Water	7.2	602	134
			225			7.2	1200	268
			150			14.0	1821	208
	½	1	225	Water	Water	7.0	555	127
			200			9.5	1170	197
			150			11.05	1694	246
	½	1	275	Water	MIBK	9.5	1179	199
	½	1	200	MIBK	MIBK	7.8	595	123
			150			6.2	1202	311
Xylene-acetic acid-water system								
3	1	1	267	Water	Water	9.1	424	75
3	½	1	537	Water	Water	8.2	424	83
3	¼	1	995	Water	Water	7.7	424	88
3	1	2	340	Water	Water	9.1	804	142
36	1	1	168	Water	Water	23.3	425	29†
36	1	1	168	Xylene	Water	20.0	442	36†

*Lo, Baird, Hanson, *Handbook of Solvent Extraction*, John Wiley & Sons, NY, p. 37, 1983.
†Because of instrumentation limits, the maximum volumetric efficiencies have not been explored. Used with permission.

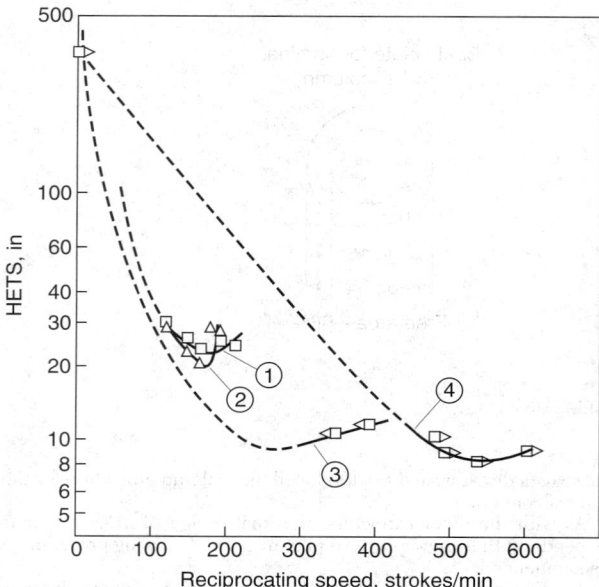

FIG. 15-51 Effect of reciprocating speed on HETS, o-xylene-acetic acid-water system. (*Lo and Prochazka in Lo et al., p. 377.*)

Curve No.	Column diam, in	Phase dispersed	Phase extractant	Double amplitude, in	Plate spacing, in	Total throughput gal/(h)(ft²)
1	36	Water	Water	1	1	425
2	36	Water	Xylene	1	1	442
3	3	Water	Water	1	1	424
4	3	Water	Water	½	1	424

Predicted minimum based on exponents of 0.36 in Eq. (15-98) and 0.14 in Eq. (15-99).

column length. If one particular section apopears to be limiting the throughput, then the plate spacing should be increased in this area; this will decrease the power input into that section. Likewise, in sections which appear to be undermixed, plate spacing should be decreased. It has been found that the following correlation can be used to estimate the relative plate spacing in the column:

$$l \propto \frac{1}{(\Delta\rho)^{5/3}(\sigma)^{3/2}} \qquad (15\text{-}97)$$

where l = relative plate spacing
 $\Delta\rho$ = density difference of the two phases
 σ = interfacial tension

3.   Once the plate spacing is optimized, the column is run over a range of total throughputs ($V_d + V_c$) and agitation speeds. There should be a minimum of three throughput levels, and at each throughput three agitation speeds. After equilibrium is attained at each condition (usually 3–5 turnovers of column volume), samples are taken and separation measured. At each condition the flood point is also determined. It is a characteristic of the Karr column that on the small diameters, the optimum efficiency usually occurs just before the flood point.
4.   From these data, plots are made of volumetric efficiency and agitation speed at each throughput level; from these plots the condition which gives the maximum volumetric efficiency is selected for scale-up.
5.   The following parameters are kept constant on the scale-up: total throughput per unit area, plate spacing, and stroke length. The height and agitation speed of the scaled-up column is then calculated from the following relationships:

$$Z_2/Z_1 = (d_{t2}/d_{t1})^{0.38} \qquad (15\text{-}98)$$

$$(SPM_2/SPM_1) = (d_{t1}/d_{t2})^{0.14} \qquad (15\text{-}99)$$

where Z_1 = plate stack height in pilot column
 Z_2 = plate stack height in scaled-up column
 d_{t1} = diameter of pilot column
 d_{t2} = diameter of scaled up column
 SPM_1 = reciprocating speed of pilot column
 SPM_2 = reciprocating speed of scaled-up column

6.   For the scaled-up column, suitable baffle plates are required to control axial mixing. For the final column layout the equipment vendor (Glitsch Process Systems Inc.) should be consulted.

The Karr column is particularly well suited for systems which tend to emulsify since its uniform shear characteristics tend to minimize emulsion formation. It is also particularly well suited for corrosive systems (since the plates can be constructed of non-metals) or for systems containing significant solids (due to its large open area and hole size on the plates). Slurries containing up to 30 percent solids have been successfully processed in Karr columns.

Pulsed Columns   These are extractors in which a rapid reciprocating motion of relatively short amplitude is applied to the liquid contents. The agitation so produced has been found to give improved rates of extraction. The principle originated with van Dijck (U.S. Patent 2,011,186, 1935). Because agitation was necessary to reduce tower heights and consequently the expense of massive shielding, and because pulsing provided a means of agitation not requiring moving parts, bearings, and the like in contact with highly corrosive, dangerously radioactive liquids, pulsed columns have been freely applied in the extraction and separation of metals from solutions of atomic energy operations. With very few exceptions, applications appear thus far to be limited to this area. There are two major types of columns: (1) ordinary (spray, packed, etc.) extractors on which pulsations are imposed and (2) a special sieve-plate design. Their characteristics are quite different.

Pulsing Devices   Refer to Fig. 15-52. At *a*, a reciprocating plunger or piston pump from which the check valves have been removed is connected to the space containing continuous phase, as shown. This arrangement suffers the disadvantages that (1) the corrosive liquid may be in direct contact with the piston and (2) too rapid pulsing, especially with volatile organic liquids, may cause cavitation. The pipe connecting column and pulser may be of any length to pass through shielding, barriers, and the like, but high pressure drop in the transfer pipe contributes to cavitation difficulties. An alternative arrangement using an air pulse is shown at *b* in the figure [Thornton, *Chem. Eng. Prog.,* **50,** *Symp. Ser.* 13, 39 (1954); U.S. Patent 2,818,324, 1957]. This keeps corrosive liquids out of contact with the pulsing device and obviates the cavitation problem but because of the compressibility of the gas requires greater application of pulsing power for the same results. For design, see Week and Knight [*Ind. Eng. Chem. Process Des. Dev.,* **6,** 480 (1967); **7,** 156 (1968)]. For pulsing at the natural frequency of the column, Baird [*Proc. Am. Inst. Chem. Eng.-Inst. Chem. Eng. Joint Meeting,* London, **1956**(6), 53] connected the liquid space to a volume of gas which acts as a spring. Flexible bellows or diaphragms of reinforced rubber, plastic, or metal in contact with the liquids may be flexed mechanically or by an electromagnetic transducer (Thornton, loc. cit.). If hydraulically activated, these may have a life of up to 30,000,000 cycles or more [Jealous and Johnson, *Ind. Eng. Chem.,* **47,** 1159 (1955)]. With suitable cam mechanisms, pulsations whose amplitude-time characteristics appear as sine, square, or sawtooth wave shapes are possible.

Pressure at the pulsing device and the conditions for cavitation and "water hammer" may be estimated by the methods of Williams and Little [*Trans. Inst. Chem. Eng. (London),* **32,** 174 (1954)] provided the pressure-drop characteristics of the tower internals are known. Jealous and Johnson (loc. cit) have had good success in computing the power required for pulsing. Since power requirement alternates, the use of a flywheel on the pulse mechanism to act as an energy reservoir is suggested as a means of reducing power requirements. Alternatively, two columns could be pulsed 180° out of phase with one pulse generator (Griffith, Jasney, and Tupper, U.S. AEC AECD-3440,

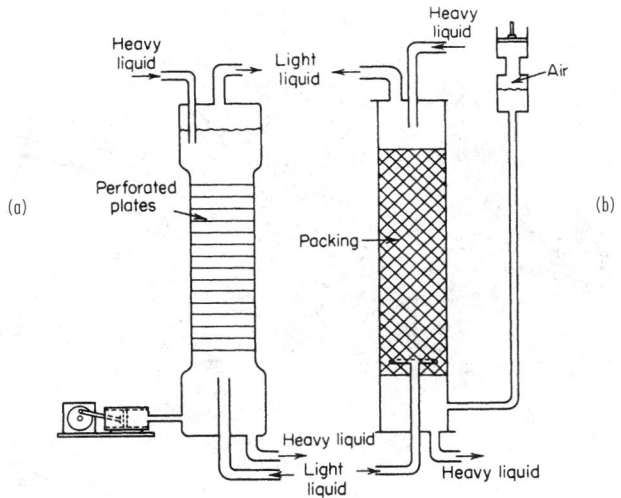

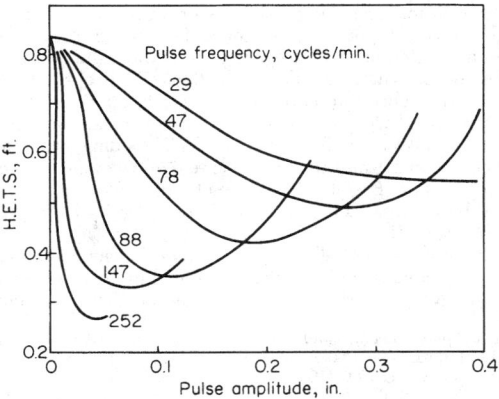

FIG. 15-52 Pulsed columns: (a) Perforated-plate column with pump pulse generator. (b) Packed column with air pulser.

FIG. 15-53 Effect of pulsing on extraction in a packed column: methyl isobutyl ketone-acetic acid-water (continuous). Tower diameter = 1.58 in, 27-in depth of ¼-in Raschig rings. $V_D = V_C = 7.5$ to 10. To convert inches to centimeters, multiply by 2.54. [*Data of Chantry, von Berg, and Wiegandt*, Ind. Eng. Chem., **47**, 1153 (1955), with permission.]

1952). Irvine (U.S. AEC ORNL-2377, 1957) devised a pulse pump to utilize part of the pulse energy. Concatenated columns (long extractors built as several short columns, with liquids led from one to the other in strictly countercurrent fashion) may be pulsed by a single pulse generator to advantage, since less power is required owing to reduced static head [Jealous and Lieberman, *Chem. Eng. Prog.*, **52**, 366 (1956)].

The following terms are generally used to describe the pulse action: **Frequency** is the rate of application of the pulse action, cycles/time. **Amplitude** is the linear distance between extreme positions of the liquid in the column (not of the pulser) produced by pulsing. **Pulsed volume** = amplitude × frequency × column crosssectional area = volumetric rate of movement of liquid, expressed as volume/time or volume/(time-area).

Pulsed Spray Columns Billerbeck et al. [*Ind. Eng. Chem.*, **48**, 183 (1956)] applied pulsing to a laboratory [3.8-cm- (1.5-in-) diameter] column. At pulse amplitude 1.11 cm (7⁄16 in), rates of mass transfer improved slightly with increased frequency up to 400 cycles/min, but the effect was relatively small. Shirotsuka [*Kagaku Kogaku*, **22**, 687 (1958)] provides additional data. There is not believed to be commercial application.

Pulsed Packed Columns Any of the ordinary packings may be used, although random packings tend to orient on pulsing, which may lead to channeling. For this reason, Thornton [*Chem. Eng. Prog.*, **50**, Symp. Ser. 13, 39 (1954); *Br. Chem. Eng.*, **3**, 247 (1958)] recommends fixed packing made from plates of corrugated expanded metal. Polyethylene packing, not wet by aqueous solutions, provides higher flow capacities and mass-transfer rates than ceramic (wetted) packing [Jackson, Holman, and Grove, *Am. Inst. Chem. Eng. J.*, **8**, 659 (1952)]. Pulsing reduces the size of dispersed-phase droplets, increases holdup, and increases interfacial area for mass transfer. There is a greater tendency toward emulsification, and maximum throughput is decreased, but HETS is reduced considerably, by the pulsing. Pulsing can be applied on existing nonpulsed packed towers to good mass-transfer advantage, provided limiting flow rates are not exceeded.

Figure 15-53 is perhaps typical of the results obtainable, although no generalizations have been devised for estimating the mass-transfer rates in the absence of experiment. For additional data, see Crico [*Genie Chim.*, **73**, 57 (1955)], Feich and Anderson [*Ind. Eng. Chem.*, **44**, 404 (1952)], Karpacheva et al. [*Khim. Masinostr.*, **1959**(3), 6; **1960**(2), 13; *Khim. Prom.*, **1960**, 469], Honda et al. [*Kagaku Kikai*, **21**, 645 (1957); *Kagaku Kogaku*, **22**, 97 (1958)], Oyama and Yamaguchi [*Kagaku Kogaku*, **22**, 668 (1958)], Potnis et al. [*Ind. Eng. Chem.*, **51**, 645 (1959)], Widmer [*Chem. Ing. Tech.*, **39**, 900 (1967)];

Worall and Thwaites [*Br. Chem. Eng.*, **10**, 158 (1965)], Ziolkowski and Naumowicz [*Chem. Stosow.*, **2**, 457 (1958); **3**, 475 (1959); **5**, 363 (1961)].

A small perforated-plate column of conventional design was pulsed by Goldberger and Benenati [*Ind. Eng. Chem.*, **51**, 641 (1959)] with marked improvement in mass-transfer rates.

Pulsed Sieve-Plate Columns The standard arrangement (see Fig. 15-52a) consists of a tower fitted with horizontal sieve plates which occupy the entire cross section of the columns. There are *no down spouts* as in ordinary sieve-plate columns. Typical arrangements use 0.32-cm- (⅛-in-) diameter perforations sufficient to provide 20 to 25 percent free space, with 5.08-cm (2-in) plate spacing, pulse amplitudes in the range 0.64 to 2.5 cm (0.25 to 1 in), and frequencies of 100 to 250 cycles/min, although the pulse characteristics will depend upon the system and flow rates under consideration. Plates are usually of metal, but Sobotik and Himmelblau [*Am. Inst. Chem. Eng. J.*, **6**, 619 (1960)] indicate that for certain services plates which are not wet by water (polyethylene) may be advantageous.

Sege and Woodfield [*Chem. Eng. Prog.*, **50**, Symp. Ser. 13, 179 (1954)] provide a good description of the operational characteristics. Refer to Fig. 15-54. Since in many cases the perforations are too small to permit flow owing to interfacial tension of the liquids, the total pulsed volume must ordinarily approximate the volumetric rate of flow of the liquids [Edwards and Beyer, *Am. Inst. Chem. Eng. J.*, **2**, 148 (1956), show that slightly higher rates than $V_d + V_c$ = pulsed volume may be obtained]. In region 1 of the figure, the column is flooded because of insufficient pulsed volume. In region 2, discrete layers of liquid appear between plates during the quiet portion of the pulse cycle. During upward pulsing, the light liquid is forced through the perforations and forms drops which rise to the plate above. During downward pulsing, the heavy liquid behaves similarly. Flow is stable,

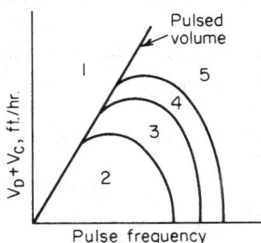

FIG. 15-54 Pulsed column characteristics. [*Sege and Woodfield*, Chem. Eng. Prog., **50**, Symp. Ser. 13, 179 (1954).]

but mass-transfer rates are generally poor. In region 3 there is little change in phase dispersion throughout the pulse cycle, and a fairly uniform dispersion of small droplets persists throughout. This region provides the best mass-transfer rates. Region 4 is characterized by irregular coalescence into fairly large drops, and periodic reversal of the continuous phase (local flooding). Extraction rates are generally poor. Further increase in frequency results in flooding owing to emulsification, region 5. Transition between regions is gradual and continuous, not abrupt. Excellent photographs of these phenomena are provided by Defives, Durandet, and Gladel [*Rev. Inst. Fr. Pet. Ann. Combust. Liq.*, **11**, 231 (1956)].

The literature is unusually large. In view of the fact that application of these extractors is almost entirely confined to processes related to atomic energy, only a brief listing of sources of data is presented here.

Dispersed-phase holdup and flooding. Groenier, McAllister, and Ryon [U.S. AEC ORNL-3890, 1966; *Chem. Eng. Sci.*, **22**, 931 (1967)]; Babb et al. [*Ind. Eng. Chem.*, **51**, 1005 (1959); *Ind. Eng. Chem. Process Des. Dev.*, **2**, 38 (1963)]; Gel'perin et al. [*Khim. Prom.*, **42**, 607 (1966)]; Thornton and Logsdail [*Trans. Inst. Chem. Eng. (London)*, **35**, 316, 331 (1957)].

Longitudinal mixing. Babb et al. [*Ind. Eng. Chem.*, **51**, 1011 (1959); *Ind. Eng. Chem. Process Des. Dev.*, **3**, 210 (1964)]; Burger and Swift (U.S. AEC HW-29010, 1953); Miyauchi et al. [*Am. Inst. Chem. Eng. J.*, **11**, 395 (1965); *Kagaku Kogaku*, **30**, 895 (1966)]; Otake and Komasawa [ibid., **32**(6), 19 (1968)].

Mass-transfer rates. Correlations are offered by Smoot, Mar, and Babb [*Ind. Eng. Chem.*, **51**, 1005 (1959); *Ind. Eng. Chem. Fundam.*, **1**, 93 (1962)] and Zwolkowski and Kubica [*Chem. Stosow.*, Ser. **B2**, 392 (1965)].

Controlled Cycling The compartmental character of sieve-plate columns described above lends itself particularly well to this technique, which is, however, not confined to these devices [Cannon, *Oil Gas J.*, **51**, 268 (1952); **55**, 68 (1956); Szabo et al., *Chem. Eng. Prog.*, **60**(1), 66 (1964); Belter and Speaker, *Ind. Eng. Chem. Process Des.*, **6**, 36 (1967); Horn, ibid., **6**, 30 (1967); Robinson and Engel, *Ind. Eng. Chem.*, **59**(3), 22 (1967); and Lövland, *Ind. Eng. Chem. Process Des. Dev.*, **7**, 65 (1968)]. A cycle is completed by the following sequence of events: (1) a light-phase flow period, during which the heavy phase does not flow; (2) a coalescing period, during which neither phase flows; (3) a heavy-phase flow period, during which the light phase does not flow; and (4) a repeat of the coalescing period. The net result can be an increased flow capacity (in the case of sieve-plate pulsed columns) and stage efficiency, such that the effect of 2N stages may be obtained with a column of N stages, provided the total holdup of each phase is displaced during each cycle.

Centrifugal Extractors The force of gravity for counterflow of liquids of different density may be replaced and in effect increased (many thousandfold if desired) by centrifugal machines. These then become especially useful for handling liquids of low density difference and those with tendencies to form emulsions.

Podbielniak Extractor (Podbielniak, U.S. Patent 2,044,996, 1935, and other patents) This is the most important of the group. Refer to Fig. 15-55. Rotation is about a horizontal shaft. The body of the extractor is a cylindrical drum containing concentric perforate cylinders. The liquids are introduced through the rotating shaft with the help of special mechanical seals; the light liquid is led internally to the drum periphery and the heavy liquid to the axis of the drum. Rapid rotation (up to several thousand revolutions per minute, depending on size) causes radial counterflow of the liquids, which are then led out through the shaft. Materials of construction include steel,

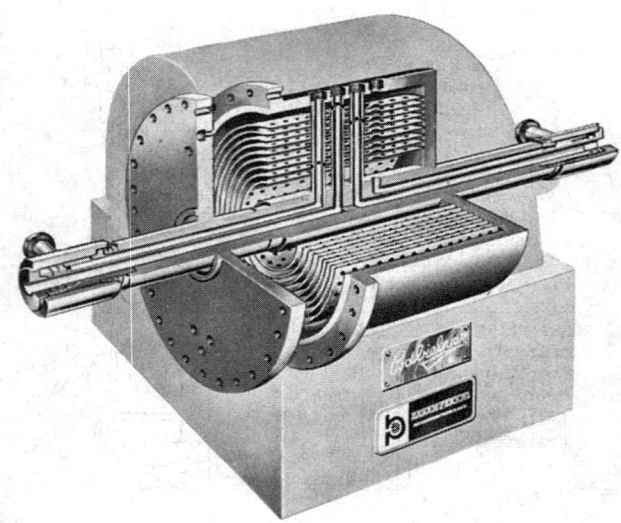

FIG. 15-55 Podbielniak centrifugal extractor. (*Courtesy of Baker Perkins Inc.*)

stainless steel, Hastelloy, and other corrosion-resistant alloys. The machines are particularly characterized by extremely low holdup of liquid per stage, and this led to their extensive use in the extraction of antibiotics, such as penicillin and the like, for which multistage extraction and phase separation must be done rapidly to avoid chemical destruction of the product under conditions of extraction. They have been used extensively in all phases of pharmaceutical manufacture and are increasingly being used in other fields: petroleum processing, both solvent refining and acid treating, dephenolization of wastewaters, extraction of uranium from ore leach liquors, as well as for clarification and phase-separation work. See Kaiser, *Sewage Ind. Wastes*, **27**, 311 (1955); Podbielniak, Gavin, and Kaiser, *J. Am. Oil Chem. Soc.*, **36**, 238 (1959); Doyle and Rauch, *Pet. Eng.*, **27**(5), C-49 (1955); Anderson and Lau, *Chem. Eng. Prog.*, **51**, 507 (1955); Todd and Podbielniak, ibid., **61**(5), 69 (1965); and Todd, ibid., **62**(8), 119 (1966). The last contains data on interstage back mixing. Table 15-11 lists some of the characteristics of the machines.

With a laboratory model [0.55 m (18 in) in diameter, 5.08 cm (2 in) wide, 18 concentric cylinders slotted at 180° intervals], Barson and Beyer [*Chem. Eng. Prog.*, **49**, 243 (1953)] obtained from two to eight ideal stages with isoamyl alcohol–boric acid–water at 5000 r/min. The number of stages increased with ratio of light-to-heavy-liquid flow but with varying position of the interface and consequently varying fraction of the machine devoted to light-liquid-dispersed. At constant flow rate, the number of stages was essentially independent of rotational speed. Jacobson and Beyer [*Am. Inst. Chem. Eng. J.*, **2**, 283 (1956)] obtained about the same results. Alexandre and Gentilini [*Rev. Inst. Fr. Pet. Ann. Combust. Liq.*, **11**, 389 (1956)] similarly obtained five ideal stages with benzene–acetic acid–water, and 3.4 to 12.5 ideal stages with methyl isobutyl ketone–acetic acid–water. Anderson and

TABLE 15-11 Podbielniak Centrifugal Extractors*

Model number	Over-all dimensions, in.				Horsepower		Flow capacity, gal./min.	
	Width	Height	Length (incl. drive)	Total wt., lb.	Connected	Continuous	Multistage extraction	Neutralization, acid treating, extraction of fermentation broths
A-1	16	12	30	150	3.0	2.5	1.0	0.5
B-10	55.5	33	67.5	2,700	7.5	6.7	30	30
D-18	76	45	85	8,600	15	10	150	75
D-36	94	45	85	10,250	25	15	300	150
E-48	113	59	107	21,500	40	22	500	300

*Courtesy Baker Perkins Inc. To convert inches to centimeters, multiply by 2.54; to convert pounds to kilograms, multiply by 0.454; to convert horsepower to kilowatts, multiply by 0.746; and to convert gallons per minute to cubic meters per hour, multiply by 0.227.

Lau [*Chem. Eng. Prog.*, **51**, 507 (1955) describe a model handling 10 to 15 percent suspended solids in the liquids, and report a fraction to two ideal stages when extracting penicillin and chloromycetin, 7.04 to 8.71 m³/h (1860 to 2300 gal/h) total flow rate.

Quadronics (Liquid Dynamics) Extractor   (Doyle et al., U.S. Patent 3,114,707, 1963, and others; paper at AIChE meeting, St. Louis, February 1968) This is a horizontally rotated device, a variant of the Podbielniak extractor, in which either fixed or adjustable orifices may be inserted as a package radially. These permit control of the mixing intensity as the liquids pass radially through the extractor. Flow capacities, depending on machine size, range from 0.34 to 340 m³/h (1.5 to 1500 gal/min).

Luwesta (Centriwesta) Extractor   This is a development from Coutor (U.S. Patent 2,036,924, 1936). See also Eisenlohr [*Ind. Chem.*, **27**, 271 (1951); *Chem. Ing. Tech.*, **23**, 12 (1951); *Pharm. Ind.*, **17**, 207 (1955); *Trans. Indian Inst. Chem. Eng.*, **3**, 7 (1949–1950)] and Husain et al. [*Chim. Ind. (Milan)*, **82**, 435 (1959)]. This centrifuge revolves about a vertical axis and contains three actual stages. It operates at 3800 r/min and handles approximately 4.92 m³/h (1300 gal/h) total liquid flow at 12-kW power requirement. Provision is made in the machine for the accumulation of solids separated from the liquids, for periodic removal. It is used, more extensively in Europe than in the United States, for the extraction of acetic acid, pharmaceuticals, and similar products.

De Laval Extractor   (Palmqvist and Beskow, U.S. Patent 3,108,953, 1959) This machine contains a number of perforated cylinders revolving about a vertical shaft. The liquids follow a spiral path about 25 m (82 ft) long, in countercurrent fashion radially, and mix when passing through the perforations. There are no published performance data.

Adsorption and Ion Exchange

M. Douglas LeVan, Ph.D., *Centennial Professor, Department of Chemical Engineering, Vanderbilt University; Member, American Institute of Chemical Engineers, American Chemical Society, International Adsorption Society. (Section Editor)*

Giorgio Carta, Ph.D., *Professor, Department of Chemical Engineering, University of Virginia; Member, American Institute of Chemical Engineers, American Chemical Society, International Adsorption Society.*

Carmen M. Yon, M.S., *Development Associate, UOP, Des Plaines, IL; Member, American Institute of Chemical Engineers. (Process Cycles, Equipment)*

Nomenclature and Units

a	specific external surface area per unit bed volume, m^2/m^3	t	time, s
a_v	surface area per unit particle volume, m^2/m^3 particle	t_c	cycle time, s
A	surface area of solid, m^2/kg	t_f	feed time, s
A_s	chromatography peak asymmetry factor	t_r	chromatographic retention time, s
b	correction factor for resistances in series	T	absolute temperature, K
c	fluid-phase concentration, mol/m^3 fluid	u_f	fluid-phase internal energy, J/mol
c_p	pore fluid-phase concentration, mol/m^3	u_s, u_{sol}	stationary-phase and sorbent solid internal energy, J/kg
c^s	fluid-phase concentration at particle surface, mol/m^3	v	interstitial velocity, m/s
C_{pf}°	ideal gas heat capacity, J/(mol K)	V_f	extraparticle fluid volume, m^3
C_s	heat capacity of sorbent solid, J/(kg K)	W	volume adsorbed as liquid, m^3;
d_p	particle diameter, m		baseline width of chromatographic peak, s
D	fluid-phase diffusion coefficient, m^2/s	x	adsorbed-phase mole fraction;
D_e	equivalent diffusion coefficient, m^2/s		particle coordinate, m
D_L	axial dispersion coefficient, m^2/s	y	fluid-phase mole fraction
D_p	pore diffusion coefficient, m^2/s	z	bed axial coordinate, m; ionic valence
D_s	adsorbed-phase (solid, surface, particle, or micropore) diffusion coefficient, m^2/s		

Greek letters	
D_0	diffusion coefficient corrected for thermodynamic driving force, m^2/s

$\overline{D}$	ionic self-diffusion coefficient, m^2/s	α	separation factor
F	fractional approach to equilibrium	β	scaling factor in Polanyi-based models;
F_v	volumetric flow rate, m^3/s		slope in gradient elution chromatography
h	enthalpy, J/mol;	Δ	peak width at half height, s
	reduced height equivalent to theoretical plate	ε	void fraction of packing (extraparticle);
htu	reduced height equivalent to a transfer unit		adsorption potential in Polanyi model, J/mol
HETP	height equivalent to theoretical plate, m	ε_p	particle porosity (intraparticle void fraction)
HTU	height equivalent to a transfer unit, m	ε_b	total bed voidage (inside and outside particles)
J	mass-transfer flux relative to molar average velocity, $mol/(m^2 \cdot s)$;	γ	activity coefficient
	J function	Γ	surface excess, mol/m^2
k	rate coefficient, s^{-1}	κ	Boltzmann constant
k_a	forward rate constant for reaction kinetics, $m^3/(mol\ s)$	Λ	partition ratio
k_c	rate coefficient based on fluid-phase concentration driving force, $m^3/(kg \cdot s)$	Λ^{∞}	ultimate fraction of solute adsorbed in batch
		μ	fluid viscosity, kg/(m s)
k_f	external mass-transfer coefficient, m/s	μ_0	zero moment, mol s/m^3
k_n	rate coefficient based on adsorbed-phase concentration driving force, s^{-1}	μ_1	first moment, s
		ν	kinematic viscosity, m^2/s
k'	retention factor	Ω	cycle-time dependent LDF coefficient
K	isotherm parameter	φ	volume fraction or mobile-phase modulator concentration, mol/m^3
K^c	molar selectivity coefficient	π	spreading pressure, N/m
K'	rational selectivity coefficient	ψ	LDF correction factor
L	bed length, m	Ψ	mechanism parameter for combined resistances
m	isotherm exponent	ρ	subparticle radial coordinate, m
M_r	molecular weight, kg/kmol	ρ_b	bulk density of packing, kg/m^3
M_s	mass of adsorbent, kg	ρ_p	particle density, kg/m^3
n	adsorbed-phase concentration, mol/kg adsorbent	ρ_s	skeletal particle density, kg/m^3
n^s	ion-exchange capacity, g-equiv/kg	σ^2	second central moment, s^2
N	number of transfer or reaction units; $k_f a L/(\varepsilon v^{ref})$ for external mass transfer; $15(1 - \varepsilon)\varepsilon_p D_p L/(\varepsilon v^{ref} R_p^2)$ for pore diffusion; $15 \Lambda D_s L/(\varepsilon v^{ref} R_p^2)$ for solid diffusion; $k_n \Lambda L/(\varepsilon v^{ref})$ for linear driving-force approximation; $k_a c_{ref} \Lambda L/[(1 - R)\varepsilon v^{ref}]$ for reaction kinetics	τ	dimensionless time
		τ_1	throughput parameter
		τ_p	tortuosity factor
		ξ	particle dimensionless radial coordinate
		ζ	dimensionless bed axial coordinate
N_p	number of theoretical plates		

Subscripts	
p	partial pressure, Pa

N_{Pe}	$v^{ref}L/D_L$, bed Peclet number (number of dispersion units)	a	adsorbed phase
P	pressure, Pa	f	fluid phase
Pe	particle-based Peclet number, $d_p v/D_L$	i, j	component index
q^{st}	isosteric heat of adsorption, J/mol	tot	total

Superscripts	
Q_i	amount of component i injected with feed, mol

r, R	separation factor; particle radial coordinate, m	$-$	an averaged concentration
r_c	column internal radius, m	$\wedge$	a combination of averaged concentrations
r_m	hydrodynamic radius of molecule, m	$\circ$	dimensionless concentration variable
r_p	particle radius, m	e	equilibrium
r_{pore}	pore radius, m	ref	reference
r_s	radius of subparticles, m	s	saturation
$\Re$	gas constant, $Pa \cdot m^3/(mol\ K)$	SM	service mark
Re	Reynolds number based on particle diameter, $d_p \varepsilon v/\nu$	TM	trademark
s	UNILAN isotherm parameter	0	initial fluid concentration in batch
Sc	Schmidt number, ν/D	0'	initial adsorbed-phase concentration in batch
Sh	Sherwood number, $k_f d_p/D$	∞	final state approached in batch

GENERAL REFERENCES

1. Adamson, *Physical Chemistry of Surfaces,* Wiley, New York, 1990.
2. Barrer, *Zeolites and Clay Minerals as Adsorbents and Molecular Sieves,* Academic Press, New York, 1978.
3. Breck, D. W., *Zeolite Molecular Sieves,* Wiley, New York, 1974.
4. Cheremisinoff and Ellerbusch, *Carbon Adsorption Handbook,* Ann Arbor Science, Ann Arbor, 1978.
5. Dorfner (ed.), *Ion Exchangers,* W. deGruyter, Berlin, 1991.
6. Dyer, *An Introduction to Zeolite Molecular Sieves,* Wiley, New York, 1988.
7. EPA, *Process Design Manual for Carbon Adsorption,* U.S. Envir. Protect. Agency., Cincinnati, 1973.
8. Gembicki, Oroskar, and Johnson, "Adsorption, Liquid Separation" in *Kirk-Othmer Encyclopedia of Chemical Technology,* 4th ed., Wiley, 1991.
9. Guiochon, Golsham-Shirazi, and Katti, *Fundamentals of Preparative and Nonlinear Chromatography,* Academic Press, Boston, Massachusetts, 1994.
10. Gregg and Sing, *Adsorption, Surface Area and Porosity,* Academic Press, New York, 1982.
11. Helfferich, *Ion Exchange,* McGraw-Hill, New York, 1962; reprinted by University Microfilms International, Ann Arbor, Michigan.
12. Helfferich and Klein, *Multicomponent Chromatography,* Marcel Dekker, New York, 1970.
13. Jaroniec and Madey, *Physical Adsorption on Heterogeneous Solids,* Elsevier, New York, 1988.
14. Karge and Ruthven, *Diffusion in Zeolites and Other Microporous Solids,* Wiley, New York, 1992.
15. Keller, Anderson, and Yon, "Adsorption" in Rousseau (ed.), *Handbook of Separation Process Technology,* Wiley-Interscience, New York, 1987.
16. Rhee, Aris, and Amundson, *First-Order Partial Differential Equations:*
Volume 1. Theory and Application of Single Equations; Volume 2. Theory and Application of Hyperbolic Systems of Quasi-Linear Equations, Prentice Hall, Englewood Cliffs, New Jersey, 1986, 1989.
17. Rodrigues, LeVan, and Tondeur (eds.), *Adsorption: Science and Technology,* Kluwer Academic Publishers, Dordrecht, The Netherlands, 1989.
18. Rudzinski and Everett, *Adsorption of Gases on Heterogeneous Surfaces,* Academic Press, San Diego, 1992.
19. Ruthven, *Principles of Adsorption and Adsorption Processes,* Wiley, New York, 1984.
20. Ruthven, Farooq, and Knaebel, *Pressure Swing Adsorption,* VCH Publishers, New York, 1994.
21. Sherman and Yon, "Adsorption, Gas Separation" in *Kirk-Othmer Encyclopedia of Chemical Technology,* 4th ed., Wiley, 1991.
22. Streat and Cloete, "Ion Exchange," in Rousseau (ed.), *Handbook of Separation Process Technology,* Wiley, New York, 1987.
23. Suzuki, *Adsorption Engineering,* Elsevier, Amsterdam, 1990.
24. Tien, *Adsorption Calculations and Modeling,* Butterworth-Heinemann, Newton, Massachusetts, 1994.
25. Valenzuela and Myers, *Adsorption Equilibrium Data Handbook,* Prentice Hall, Englewood Cliffs, New Jersey, 1989.
26. Vermeulen, LeVan, Hiester, and Klein, "Adsorption and Ion Exchange" in Perry, R. H. and Green, D. W. (eds.), *Perry's Chemical Engineers' Handbook* (6th ed.), McGraw-Hill, New York, 1984.
27. Wankat, *Large-Scale Adsorption and Chromatography,* CRC Press, Boca Raton, Florida, 1986.
28. Yang, *Gas Separation by Adsorption Processes,* Butterworth, Stoneham, Massachusetts, 1987.
29. Young and Crowell, *Physical Adsorption of Gases,* Butterworths, London, 1962.

DESIGN CONCEPTS

INTRODUCTION

Adsorption and ion exchange share so many common features in regard to application in batch and fixed-bed processes that they can be grouped together as sorption for a unified treatment. These processes involve the transfer and resulting equilibrium distribution of one or more solutes between a fluid phase and particles. The partitioning of a single solute between fluid and sorbed phases or the selectivity of a sorbent towards multiple solutes makes it possible to separate solutes from a bulk fluid phase or from one another.

This section treats batch and fixed-bed operations and reviews process cycles and equipment. As the processes indicate, fixed-bed operation with the sorbent in granule, bead, or pellet form is the predominant way of conducting sorption separations and purifications. Although the fixed-bed mode is highly useful, its analysis is complex. Therefore, fixed beds including chromatographic separations are given primary attention here with respect to both interpretation and prediction.

Adsorption involves, in general, the accumulation (or depletion) of solute molecules at an interface (including gas-liquid interfaces, as in foam fractionation, and liquid-liquid interfaces, as in detergency). Here we consider only gas-solid and liquid-solid interfaces, with solute distributed selectively between the fluid and solid phases. The accumulation per unit surface area is small; thus, highly porous solids with very large internal area per unit volume are preferred. Adsorbent surfaces are often physically and/or chemically heterogeneous, and bonding energies may vary widely from one site to another. We seek to promote physical adsorption or *physisorption,* which involves van der Waals forces (as in vapor condensation), and retard chemical adsorption or *chemisorption,* which involves chemical bonding (and often dissociation, as in catalysis). The former is well suited for a regenera-ble process, while the latter generally destroys the capacity of the adsorbent.

Adsorbents are natural or synthetic materials of amorphous or microcrystalline structure. Those used on a large scale, in order of sales volume, are activated carbon, molecular sieves, silica gel, and activated alumina [Keller et al., gen. refs.].

Ion exchange usually occurs throughout a polymeric solid, the solid being of gel-type, which dissolves some fluid-phase solvent, or truly porous. In ion exchange, ions of positive charge in some cases (cations) and negative charge in others (anions) from the fluid (usually an aqueous solution) replace dissimilar ions of the same charge initially in the solid. The ion exchanger contains permanently bound functional groups of opposite charge-type (or, in special cases, notably weak-base exchangers act as if they do). Cation-exchange resins generally contain bound sulfonic acid groups; less commonly, these groups are carboxylic, phosphonic, phosphinic, and so on. Anionic resins involve quaternary ammonium groups (strongly basic) or other amino groups (weakly basic).

Most ion exchangers in large-scale use are based on synthetic resins—either preformed and then chemically reacted, as for polystyrene, or formed from active monomers (olefinic acids, amines, or phenols). Natural zeolites were the first ion exchangers, and both natural and synthetic zeolites are in use today.

Ion exchange may be thought of as a reversible reaction involving chemically equivalent quantities. A common example for cation exchange is the familiar water-softening reaction

$$Ca^{++} + 2NaR \rightleftharpoons CaR_2 + 2Na^+$$

where R represents a stationary univalent anionic site in the polyelectrolyte network of the exchanger phase.

TABLE 16-1   Classification of Sorptive Separations

Type of interaction	Basis for separation	Examples
Adsorption	Equilibrium	Numerous purification and recovery processes for gases and liquids Activated carbon-based applications Desiccation using silica gels, aluminas, and zeolites Oxygen from air by PSA using 5A zeolite
	Rate Molecular sieving	Nitrogen from air by PSA using carbon molecular sieve Separation on n- and iso-parafins using 5A zeolite Separation of xylenes using zeolite
Ion exchange (electrostatic)	Equilibrium	Deionization Water softening Rare earth separations Recovery and separation of pharmaceuticals (e.g., amino acids, proteins)
Ligand exchange	Equilibrium	Chromatographic separation of glucose-fructose mixtures with Ca-form resins Removal of heavy metals with chelating resins Affinity chromatography
Solubility	Equilibrium	Partition chromatography
None (purely steric)	Equilibrium partitioning in pores	Size exclusion or gel permeation chromatography

Table 16-1 classifies sorption operations by the type of interaction and the basis for the separation. In addition to the normal sorption operations of adsorption and ion exchange, some other similar separations are included. Applications are discussed in this section in "Process Cycles."

Example 1: Surface Area and Pore Volume of Adsorbent A simple example will show the extent of internal area in a typical granular adsorbent. A fixed bed is packed with particles of a porous adsorbent material. The bulk density of the packing is 500 kg/m³, and the interparticle void fraction is 0.40. The intraparticle porosity is 0.50, with two-thirds of this in cylindrical pores of diameter 1.4 nm and the rest in much larger pores. Find the surface area of the adsorbent and, if solute has formed a complete monomolecular layer 0.3 nm thick inside the pores, determine the percent of the particle volume and the percent of the total bed volume filled with adsorbate.

From surface area to volume ratio considerations, the internal area is practically all in the small pores. One gram of the adsorbent occupies 2 cm³ as packed and has 0.4 cm³ in small pores, which gives a surface area of 1150 m²/g (or about 1 mi² per 5 lb or 6.3 mi²/ft³ of packing). Based on the area of the annular region filled with adsorbate, the solute occupies 22.5 percent of the internal pore volume and 13.5 percent of the total packed-bed volume.

DESIGN STRATEGY

The design of sorption systems is based on a few underlying principles. First, knowledge of *sorption equilibrium* is required. This equilibrium, between solutes in the fluid phase and the solute-enriched phase of the solid, supplants what in most chemical engineering separations is a fluid-fluid equilibrium. The selection of the sorbent material with an understanding of its equilibrium properties (i.e., capacity and selectivity as a function of temperature and component concentrations) is of primary importance. Second, because sorption operations take place in batch, in fixed beds, or in simulated moving beds, the processes have *dynamical character.* Such operations generally do not run at steady state, although such operation may be approached in a simulated moving bed. Fixed-bed processes often approach a periodic condition called a periodic state or cyclic steady state, with several different feed steps constituting a cycle. Thus, some knowledge of how transitions travel through a bed is required. This introduces both time and space into the analysis, in contrast to many chemical engineering operations that can be analyzed at steady state with only a spatial dependence. For good design, it is crucial to understand fixed-bed performance in relation to adsorption equilibrium and rate behavior. Finally, many *practical aspects* must be included in design so that a process starts up and continues to perform well, and that it is not so overdesigned that it is wasteful. While these aspects are process-specific, they include an understanding of dispersive phenomena at the bed scale and, for regenerative processes, knowledge of aging characteristics of the sorbent material, with consequent changes in sorption equilibrium.

Characterization of Equilibria Phase equilibrium between fluid and sorbed phases for one or many components in adsorption or two or more species in ion exchange is usually the single most important factor affecting process performance. In most processes, it is much more important than mass and heat transfer rates; a doubling of the stoichiometric capacity of a sorbent or a significant change in the shape of an isotherm would almost always have a greater impact on process performance than a doubling of transfer rates.

A difference between adsorption and ion exchange with completely ionized resins is indicated in the *variance* of the systems. In adsorption, part of the solid surface or pore volume is vacant. This diminishes as the fluid-phase concentration of solute increases. In contrast, for ion exchange the sorbent has a fixed total capacity and merely exchanges solutes while conserving charge. Variance is defined as the number of independent concentration variables in a sorption system at equilibrium—that is, variables that one can change separately and thereby control the values of all others. Thus, it also equals the difference between the total number of concentration variables and the number of independent relations connecting them. Numerous cases arise in which ion exchange is accompanied by chemical reaction (neutralization or precipitation, in particular), or adsorption is accompanied by evolution of sensible heat. The concept of variance helps greatly to assure correct interpretations and predictions.

The working capacity of a sorbent depends on fluid concentrations and temperatures. Graphical depiction of sorption equilibrium for single component adsorption or binary ion exchange (monovariance) is usually in the form of isotherms $[n_i = n_i(c_i)$ or $n_i(p_i)$ at constant $T]$ or isosteres $[p_i = p_i(T)$ at constant $n_i]$. Representative forms are shown in Fig. 16-1. An important dimensionless group dependent on adsorption equilibrium is the **partition ratio** (see Eq. 16-125), which is a measure of the relative affinities of the sorbed and fluid phases for solute.

Historically, isotherms have been classified as *favorable* (concave downward) or *unfavorable* (concave upward). These terms refer to the spreading tendencies of transitions in fixed beds. A favorable isotherm gives a compact transition, whereas an unfavorable isotherm leads to a broad one.

Example 2: Calculation of Variance In mixed-bed deionization of a solution of a single salt, there are 8 concentration variables: 2 each for cation, anion, hydrogen, and hydroxide. There are 6 connecting relations: 2 for ion exchange and 1 for neutralization equilibrium, and 2 ion-exchanger and 1 solution electroneutrality relations. The variance is therefore $8 - 6 = 2$.

Adsorbent/Ion Exchanger Selection Guidelines for sorbent selection are different for regenerative and nonregenerative systems. For a nonregenerative system, one generally wants a high capacity and a strongly favorable isotherm for a purification and additionally high selectivity for a separation. For a regenerative system, high overall capacity and selectivity are again desired, but needs for cost-effective

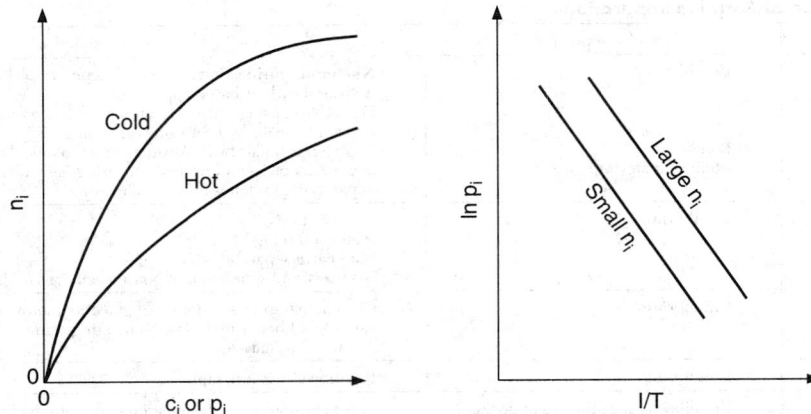

FIG. 16-1 Isotherms (left) and isosteres (right). Isosteres plotted using these coordinates are nearly straight parallel lines, with deviations caused by the dependence of the isosteric heat of adsorption on temperature and loading.

regeneration leading to a reasonable working capacity influence what is sought after in terms of isotherm shape. For separations by pressure swing adsorption (or vacuum pressure swing adsorption), generally one wants a linear to slightly favorable isotherm (although purifications can operate economically with more strongly favorable isotherms). Temperature-swing adsorption usually operates with moderately to strongly favorable isotherms, in part because one is typically dealing with heavier solutes and these are adsorbed fairly strongly (e.g., organic solvents on activated carbon and water vapor on zeolites). Exceptions exist, however; for example, water is adsorbed on silica gel and activated alumina only moderately favorably, with some isotherms showing unfavorable sections. Equilibria for ion exchange separations generally vary from moderately favorable to moderately unfavorable; depending on feed concentrations, the alternates often exist for the different steps of a regenerative cycle. Other factors in sorbent selection are mechanical and chemical stability, mass transfer characteristics, and cost.

Fixed-Bed Behavior The number of transitions occurring in a fixed bed of initially uniform composition before it becomes saturated by a constant composition feed stream is generally equal to the variance of the system. This introductory discussion will be limited to single transition systems.

Methods for analysis of fixed-bed transitions are shown in Table 16-2. Local equilibrium theory is based solely of stoichiometric concerns and system nonlinearities. A transition becomes a "simple wave" (a gradual transition), a "shock" (an abrupt transition), or a combination of the two. In other methods, mass-transfer resistances are incorporated.

The asymptotic behavior of transitions under the influence of mass-transfer resistances in long, "deep" beds is important. The three basic asymptotic forms are shown in Fig. 16-2. With an unfavorable isotherm, the breadth of the transition becomes proportional to the depth of bed it has passed through. For the linear isotherm, the breadth becomes proportional to the square root of the depth. For the favorable isotherm, the transition approaches a constant breadth called a *constant pattern*.

Design of nonregenerative sorption systems and many regenerative ones often relies on the concept of the mass-transfer zone or MTZ, which closely resembles the constant pattern [Collins, *Chem. Eng. Prog. Symp. Ser. No. 74,* **63,** 31 (1974); Keller et al., gen. refs.]. The length of this zone (depicted in Fig. 16-3) together with stoichiometry can be used to predict accurately how long a bed can be utilized prior to breakthrough. Upstream of the mass-transfer zone, the adsorbent is in equilibrium with the feed. Downstream, the adsorbent is in its initial state. Within the mass-transfer zone, the fluid-phase concentration drops from the feed value to the initial, presaturation state. Equilibrium with the feed is not attained in this region. As a result, because an adsorption bed must typically be removed from service shortly after breakthrough begins, the full capacity of the bed is not utilized. Obviously, the broader that the mass-transfer zone is, the greater will be the extent of unused capacity. Also shown in the figure is the length of the equivalent equilibrium section (LES) and the length of equivalent unused bed (LUB). The length of the MTZ is divided between these two.

Adsorption with strongly favorable isotherms and ion exchange between strong electrolytes can usually be carried out until most of the stoichiometric capacity of the sorbent has been utilized, corresponding to a thin MTZ. Consequently, the total capacity of the bed is

TABLE 16-2 Methods of Analysis of Fixed-Bed Transitions

Method	Purpose	Approximations
Local equilibrium theory	Shows wave character—simple waves and shocks Usually indicates best possible performance Better understanding	Mass and heat transfer very rapid Dispersion usually neglected If nonisothermal, then adiabatic
Mass-transfer zone	Design based on stoichiometry and experience	Isothermal MTZ length largely empirical Regeneration often empirical
Constant pattern and related analyses	Gives asymptotic transition shapes and upper bound on MTZ	Deep bed with fully developed transition
Full rate modeling	Accurate description of transitions Appropriate for shallow beds, with incomplete wave development General numerical solutions by finite difference or collocation methods	Various to few

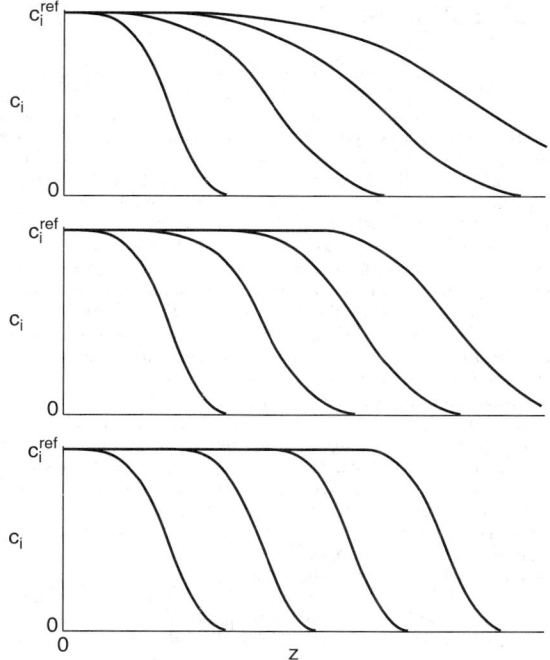

FIG. 16-2   Limiting fixed-bed behavior: simple wave for unfavorable isotherm (top), square-root spreading for linear isotherm (middle), and constant pattern for favorable isotherm (bottom). [*From LeVan in Rodrigues et al. (eds.),* Adsorption: Science and Technology, *Kluwer Academic Publishers, Dordrecht, The Netherlands, 1989; reprinted with permission.*]

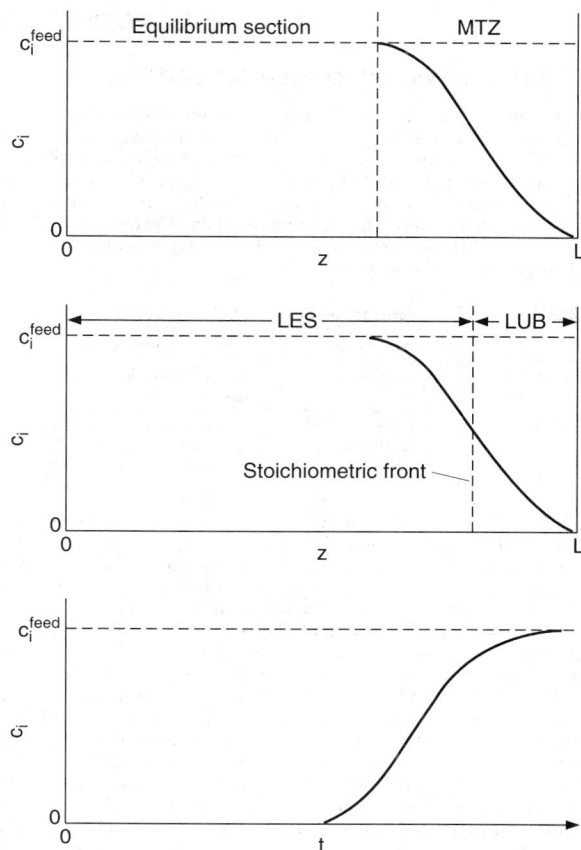

FIG. 16-3   Bed profiles (top and middle) and breakthrough curve (bottom). The bed profiles show the mass-transfer zone (MTZ) and equilibrium section at breakthrough. The stoichiometric front divides the MTZ into two parts with contributions to the length of equivalent equilibrium section (LES) and the length of equivalent unused bed (LUB).

practically constant regardless of the composition of the solution being treated.

The effluent concentration history is the breakthrough curve, also shown in Fig. 16-3. The effluent concentration stays at or near zero or a low residual concentration until the transition reaches the column outlet. The effluent concentration then rises until it becomes unacceptable, this time being called the breakthrough time. The feed step must stop and, for a regenerative system, the regeneration step begins.

Two dimensionless variables play key roles in the analysis of single transition systems (and some multiple transition systems). These are the **throughput parameter** [see Eq. (16-129)] and the **number of transfer units** (see Table 16-13). The former is time made dimensionless so that it is equal to unity at the stoichiometric center of a breakthrough curve. The latter is, as in packed tower calculations, a measure of mass-transfer resistance.

Cycles   Design methods for cycles rely on mathematical modeling (or empiricism) and often extensive pilot plant experiments. Many cycles can be easily analyzed using the methods described above applied to the collection of steps. In some cycles, however, especially those operated with short cycle times or in shallow beds, transitions may not be very fully developed, even at a periodic state, and the complexity may be compounded by multiple sorbates.

A wide variety of complex process cycles have been developed. Systems with many beds incorporating multiple sorbents, possibly in layered beds, are in use. Mathematical models constructed to analyze such cycles can be complex. With a large number of variables and nonlinear equilibria involved, it is usually not beneficial to make all variables in such models dimensionless; doing so does not help appreciably in making comparisons with other largely dissimilar systems. If dimensionless variables are used, these usually begin with a dimensionless bed length and a dimensionless time, which is often different from the throughput parameter.

Practical Aspects   There are a number of process-specific concerns that are accounted for in good design. In regenerable systems, sorbents age, losing capacity because of fouling by heavy contaminants, loss of surface area or crystallinity, oxidation, and the like. Mass-transfer resistances may increase over time. Because of particle shape, size distribution, or column packing method, dispersion may be more pronounced than would normally be expected. The humidity of an entering stream will usually impact a solvent recovery application. Safety, including the possibility of a fire, may be a concern. For gas-phase adsorption, scale-up from an isothermal laboratory column to a nonisothermal pilot plant column to a largely adiabatic process column requires careful judgment. If the MTZ concept is utilized, the length of the MTZ cannot be reliably determined solely from knowledge on other systems. Experience plays the key role in accounting for these and other such factors.

ADSORBENTS AND ION EXCHANGERS

CLASSIFICATIONS AND CHARACTERIZATIONS

Adsorbents Table 16-3 classifies common adsorbents by structure type and water adsorption characteristics. Structured adsorbents take advantage of their crystalline structure (zeolites and silicalite) and/or their molecular sieving properties. The hydrophobic (nonpolar surface) or hydrophilic (polar surface) character may vary depending on the competing adsorbate. A large number of zeolites have been identified, and these include both synthetic and naturally occurring (e.g., mordenite and chabazite) varieties.

TABLE 16-3   Classification of Common Adsorbents

	Amorphous	Structured
Hydrophobic	Activated carbon Polymers	Carbon molecular sieves Silicalite

Hydrophilic	Silica gel
	Activated alumina
		Common zeolites: 3A (KA),
		  4A (NaA), 5A (CaA), 13X (NaX),
		  Mordenite, Chabazite, etc.

The classifications in Table 16-3 are intended only as a rough guide. For example, a carbon molecular sieve is truly amorphous but has been manufactured to have certain structural, rate-selective properties. Similarly, the extent of hydrophobicity of an activated carbon will depend on its ash content and its level of surface oxidation.

Zeolites are crystalline aluminosilicates. Zeolitic adsorbents have had their water of hydration removed by calcination to create a structure with well-defined openings into crystalline cages. The molecular sieving properties of zeolites are based on the size of these openings. Two crystal types are common: type A (with openings formed by 4 sodalite cages) and type X or Y (with openings formed by 6 sodalite cages). Cations balancing charge and their locations determine the size of opening into a crystal unit cell. Nominal openings sizes for the most common synthetic zeolites are 0.3 nm for KA, 0.4 nm for NaA, 0.5 nm for CaA, and 1.0 nm for NaX. Further details, including effective molecular diameters, are widely available [Barrer; Breck; Ruthven; Yang, gen. refs.].

Many adsorbents, particularly the amorphous adsorbents, are characterized by their pore size distribution. The distribution of small pores is usually determined by analysis, using one of several available methods, of a cryogenic nitrogen adsorption isotherm, although other probe molecules are also used. Russell and LeVan [*Carbon*, **32**, 845 (1994)] compare popular methods using a single nitrogen isotherm measured on activated carbon and provide numerous references. The distribution of large pores is usually determined by mercury porisimetry [Gregg and Sing, gen. refs.].

Table 16-4 shows the IUPAC classification of pores by size. Micropores are small enough that a molecule is attracted to both of the opposing walls forming the pore. The potential energy functions for these walls superimpose to create a deep well, and strong adsorption results. Hysteresis is generally not observed. (However, water vapor adsorbed in the micropores of activated carbon shows a large hysteresis loop, and the desorption branch is sometimes used with the Kelvin equation to determine the pore size distribution.) Capillary condensation occurs in mesopores and a hysteresis loop is typically found. Macropores form important paths for molecules to diffuse into a particle; for gas-phase adsorption, they do not fill with adsorbate until the gas phase becomes saturated.

TABLE 16-4   Classification of Pore Sizes

Type	Slit Width° (w)	Characteristic
Micropore†	$w < 2$ nm	Superimposed wall potentials
Mesopore	2 nm $< w < 50$ nm	Capillary condensation
Macropore	$w > 50$ nm	Effectively flat walled until $p \to P^\circ$

°Or pore diameter.
†Further subdivided into ultramicropores and supermicropores (Gregg and Sing, gen. refs.).

Ion Exchangers Ion exchangers are classified according to (1) their functionality and (2) the physical properties of the support matrix. Cation and anion exchangers are classified in terms of their ability to exchange positively or negatively charged species. Strongly acidic and strongly basic ion exchangers are ionized and thus are effective at nearly all pH values (pH 0–14). Weakly acidic exchangers are typically effective in the range of pH 5–14. Weakly basic resins are effective in the range of pH 0–9. Weakly acidic and weakly basic exchangers are often easier to regenerate, but leakage due to incomplete exchange may occur. Chelating resins containing iminodiacetic acid form specific metal complexes with metal ions with complex stability constants that follow the same order as those for EDTA. However, depending on pH, they also function as weak cation exchangers. The achievable ion-exchange capacity depends on the concentration of ionogenic groups and their availability as an exchange site. The latter is a function of the support matrix.

Polymer-based, synthetic ion-exchangers known as resins are available commercially in gel type or truly porous forms. Gel-type resins are not porous in the usual sense of the word, since their structure depends upon swelling in the solvent in which they are immersed. Removal of the solvent usually results in a collapse of the three-dimensional structure, and no significant surface area or pore diameter can be defined by the ordinary techniques available for truly porous materials. In their swollen state, gel-type resins approximate a true molecular-scale solution. Thus, we can identify an internal porosity ε_p only in terms of the equilibrium uptake of water or other liquid. When crosslinked polymers are used as the support matrix, the internal porosity so defined varies in inverse proportion to the degree of crosslinking, with swelling and therefore porosity typically being more pronounced in solvents with a high dielectric constant. The ion held by the exchanger also influences the resin swelling. Thus, the size of the resin particles changes during the ion-exchange process as the resin is changed from one form to another, and this effect is more dramatic for resins with a lower degree of crosslinking. The choice of degree of crosslinking is dependent on several factors including: the extent of swelling, the exchange capacity, the intraparticle diffusivity, the ease of regeneration, and the physical and chemical stability of the exchanger under chosen operating conditions. The concentration of ionogenic groups determines the capacity of the resin. Although the capacity per unit mass of dry resin is insensitive to the degree of crosslinking, except for very highly crosslinked resins, the exchange capacity per unit volume of swollen resin increases significantly with degree of crosslinking, so long as the mesh size of the polymer network allows the ions free access to functional groups within the interior of the resin. The degree of crosslinking also affects the rate of ion exchange. The intraparticle diffusivity decreases nearly exponentially with the mesh size of the matrix. As a result, resins with a lower degree of crosslinking are normally required for the exchange of bulky species, such as organic ions with molecular weight in excess of 100. The regeneration efficiency is typically greater for resins with a lower degree of crosslinking. Finally, the degree of crosslinking also affects the long-term stability of the resin. Strongly acidic and strongly basic resins are subject to irreversible oxidative degradation of the polymer and thermal and oxidative decomposition of functional groups. Generally, more highly crosslinked resins are less prone to irreversible chemical degradation but may be subject to osmotic breakage caused by volume changes that occur during cyclic operations. In general, experience shows that an intermediate degree of crosslinking is often preferred. However, readers are referred to manufacturers' specifications for resin stability data at different operating conditions.

Truly porous, synthetic ion exchangers are also available. These materials retain their porosity even after removal of the solvent and have measurable surface areas and pore size. The term macroreticular is commonly used for resins prepared from a phase separation technique, where the polymer matrix is prepared with the addition of a liquid that is a good solvent for the monomers, but in which the polymer is insoluble. Matrices prepared in this way usually have the appearance of a conglomerate of gel-type microspheres held together to form an interconnected porous network. Macroporous resins possessing a more continuous gellular structure interlaced with a pore network have also been obtained with different techniques and are commercially available. Since higher degrees of crosslinking are typically used to produce truly porous ion-exchange resins, these materials tend to be more stable under highly oxidative conditions, more attrition-resistant, and more resistant to breakage due to osmotic shock than their gel-type counterparts. Moreover, since their porosity does not depend entirely on swelling, they can be used in solvents with low dielectric constant where gel-type resins can be ineffective. Porous ion-exchange resins are also useful for the recovery and separation of high-molecular-weight substances such as proteins or colloidal particles. Specialty resins with very large pore sizes (from 30 nm to larger than 1000 nm) are available for these applications. In general, compared to gel-type resins, truly porous resins typically have somewhat lower capacities and can be more expensive. Thus, for ordinary ion-exchange applications involving small ions under nonharsh conditions, gel-type resins are usually preferred.

PHYSICAL PROPERTIES

Selected data on commercially available adsorbents and ion exchangers are given in Tables 16-5 and 16-6. The purpose of the tables is twofold: to assist the engineer or scientist in identifying materials suitable for a needed application, and to supply typical physical property values.

Excellent sources of information on the characteristics of adsorbents or ion exchange products for specific applications are the manufacturers themselves. The names, addresses, and phone numbers of suppliers may be readily found in most libraries in sources such as the *Thomas Register* and *Dun & Bradstreet*. Additional information on adsorbents and ion exchangers is available in many of the general references and in several articles in *Kirk-Othmer Encyclopedia of Chemical Technology*. A recent comprehensive summary of commercial ion-exchangers, including manufacturing methods, properties, and applications, is given by Dorfner (*Ion Exchangers*, de Gruyter, New York, 1991).

TABLE 16-5   Physical Properties of Adsorbents

Material and uses	Shape* of particles	Size range, U.S. standard mesh†	Internal porosity, %	Bulk dry density, kg/L	Average pore diameter, nm	Surface area, km²/kg	Sorptive capacity, kg/kg (dry)
Aluminas							
Low-porosity (fluoride sorbent)	G, S	8–14, etc.	40	0.70	~7	0.32	0.20
High-porosity (drying, separations)	G	Various	57	0.85	4–14	0.25–0.36	
							0.25–0.33
Desiccant, CaCl₂-coated	G	3–8, etc.	30	0.91	4.5	0.2	0.22
Activated bauxite	G	8–20, etc.	35	0.85	5		0.1–0.2
Chromatographic alumina	G, P, S	80–200, etc.	30	0.93			~0.14
Silicates and aluminosilicates							
Molecular sieves	S, C, P	Various					
Type 3A (dehydration)			~30	0.62–0.68	0.3	~0.7	0.21–0.23
Type 4A (dehydration)			~32	0.61–0.67	0.4	~0.7	0.22–0.26
Type 5A (separations)			~34	0.60–0.66	0.5	~0.7	0.23–0.28
Type 13X (purification)			~38	0.58–0.64	1.0	~0.6	0.25–0.36
Silicalite (hydrocarbons)	S, C, P	Various		0.64–0.70	0.6	~0.4	0.12–0.16
Dealumininated Y (hydrocarbons)	S, C, P	Various		0.48–0.53	0.8	0.5–0.8	0.28–0.42
Mordenite (acid drying)				0.88	0.3–0.8		0.12
Chabazite (acid drying)				0.72	0.4–0.5		0.20
Silica gel (drying, separations)	G, P	Various	38–48	0.70–0.82	2–5	0.6–0.8	0.35–0.50
Magnesium silicate (decolorizing)	G, P	Various	~33	~0.50		0.18–0.30	
Calcium silicate (fatty-acid removal)	P		75–80	~0.20		~0.1	
Clay, acid-treated (refining of petroleum, food products)	G	4–8		0.85			
Fuller's earth (same)	G, P	<200		0.80			
Diatomaceous earth	G	Various		0.44–0.50		~0.002	
Carbons							
Shell-based	G	Various	60	0.45–0.55	2	0.8–1.6	0.40
Wood-based	G	Various	~80	0.25–0.30		0.8–1.8	~0.70
Petroleum-based	G, C	Various	~80	0.45–0.55	2	0.9–1.3	0.3–0.4
Peat-based	G, C, P	Various	~55	0.30–0.50	1–4	0.8–1.6	0.5
Lignite-based	G, P	Various	70–85	0.40–0.70	3	0.4–0.7	0.3
Bituminous-coal-based	G, P	8–30, 12–40	60–80	0.40–0.60	2–4	0.9–1.2	0.4
Synthetic polymer based (pyrolyzed)	S	20–100	40–70	0.49–0.60		0.1–1.1	
Carbon molecular sieve (air separation)		Various	35–50	0.5–0.7	0.3–0.6		0.5–0.20
Organic polymers							
Polystyrene (removal of organics, e.g., phenol; antibiotics recovery)	S	20–60	40–60	0.64	4–20	0.3–0.7	
Polyacrylic ester (purification of pulping wastewaters; antibiotics recovery)	G, S	20–60	50–55	0.65–0.70	10–25	0.15–0.4	
Phenolic (also phenolic amine) resin (decolorizing and deodorizing of solutions)	G	16–50	45	0.42		0.08–0.12	0.45–0.55

*Shapes: C, cylindrical pellets; F, fibrous flakes; G, granules; P, powder; S, spheres.
†U.S. Standard sieve sizes (given in parentheses) correspond to the following diameters in millimeters: (3) 6.73, (4) 4.76, (8) 2.98, (12) 1.68, (14) 1.41, (16) 1.19, (20) 0.841, (30) 0.595, (40) 0.420, (50) 0.297, (60) 0.250, (80) 0.177, (200) 0.074.

TABLE 16-6 Physical Properties of Ion-Exchange Materials

Material	Shape° of particles	Bulk wet density (drained), kg/L	Moisture content (drained), % by weight	Swelling due to exchange, %	Maximum operating temperature,† °C	Operating pH range	Exchange capacity Dry, equivalent/kg	Exchange capacity Wet, equivalent/L
Cation exchangers: strongly acidic								
Polystyrene sulfonate								
Homogeneous (gel) resin	S				120–150	0–14		
4% cross-linked		0.75–0.85	64–70	10–12			5.0–5.5	1.2–1.6
6% cross-linked		0.76–0.86	58–65	8–10			4.8–5.4	1.3–1.8
8–10% cross-linked		0.77–0.87	48–60	6–8			4.6–5.2	1.4–1.9
12% cross-linked		0.78–0.88	44–48	5			4.4–4.9	1.5–2.0
16% cross-linked		0.79–0.89	42–46	4			4.2–4.6	1.7–2.1
20% cross-linked		0.80–0.90	40–45	3			3.9–4.2	1.8–2.0
Porous structure								
12–20% cross-linked	S	0.81	50–55	4–6	120–150	0–14	4.5–5.0	1.5–1.9
Sulfonated phenolic resin	G	0.74–0.85	50–60	7	50–90	0–14	2.0–2.5	0.7–0.9
Sulfonated coal	G							
Cation exchangers: weakly acidic								
Acrylic (pK 5) or methacrylic (pK 6)								
Homogeneous (gel) resin	S	0.70–0.75	45–50	20–80	120	4–14	8.3–10	3.3–4.0
Macroporous	S	0.67–0.74	50–55	10–100	120		~8.0	2.5–3.5
Phenolic resin	G	0.70–0.80	~50	10–25	45–65	0–14	2.5	1.0–1.4
Polystyrene phosphonate	G, S	0.74	50–70	<40	120	3–14	6.6	3.0
Polystyrene iminodiacetate	S	0.75	68–75	<100	75	3–14	2.9	0.7
Polystyrene amidoxime	S	~0.75	58	10	50	1–11	2.8	0.8–0.9
Polystyrene thiol	S	~0.75	45–50		60	1–13	~5	2.0
Cellulose								
Phosphonate	F						~7.0	
Methylene carboxylate	F, P, G						~0.7	
Greensand (Fe silicate)	G	1.3	1–5	0	60	6–8	0.14	0.18
Zeolite (Al silicate)	G	0.85–0.95	40–45	0	60	6–8	1.4	0.75
Zirconium tungstate	G	1.15–1.25	~5	0	>150	2–10	1.2	1.0
Anion exchangers: strongly basic								
Polystyrene-based								
Trimethyl benzyl ammonium (type I)								
Homogeneous, 8% CL	S	0.70	46–50	~20	60–80	0–14	3.4–3.8	1.3–1.5
Porous, 11% CL	S	0.67	57–60	15–20	60–80	0–14	3.4	1.0
Dimethyl hydroxyethyl ammonium (type II)								
Homogeneous, 8% CL	S	0.71	~42	15–20	40–80	0–14	3.8–4.0	1.2
Porous, 10% CL	S	0.67	~55	12–15	40–80	0–14	3.8	1.1
Acrylic-based								
Homogeneous (gel)	S	0.72	~70	~15	40–80	0–14	~5.0	1.0–1.2
Porous	S	0.67	~60	~12	40–80	0–14	3.0–3.3	0.8–0.9
Cellulose-based								
Ethyl trimethyl ammonium	F				100	4–10	0.62	
Triethyl hydroxypropyl ammonium					100	4–10	0.57	
Anion exchangers: intermediately basic (pK 11)								
Polystyrene-based	S	0.75	~50	15–25	65	0–10	4.8	1.8
Epoxy-polyamine	S	0.72	~64	8–10	75	0–7	6.5	1.7
Anion exchangers: weakly basic (pK 9)								
Aminopolystyrene								
Homogeneous (gel)	S	0.67	~45	8–12	100	0–7	5.5	1.8
Porous	S	0.61	55–60	~25	100	0–9	4.9	1.2
Acrylic-based amine								
Homogeneous (gel)	S	0.72	~63	8–10	80	0–7	6.5	1.7
Porous	S	0.72	~68	12–15	60	0–9	5.0	1.1
Cellulose-based								
Aminoethyl	P						1.0	
Diethyl aminoethyl	P						~0.9	

°Shapes: C, cylindrical pellets; G, granules; P, powder; S, spheres.
†When two temperatures are shown, the first applies to H form for cation, or OH form for anion, exchanger; the second, to salt ion.

Several densities and void fractions are commonly used. For adsorbents, usually the **bulk density** ρ_b, the weight of clean material per unit bulk volume as packed in a column, is reported. The dry **particle density** ρ_p is related to the (external) **void fraction of packing** ε by

$$\rho_p = \frac{\rho_b}{1-\varepsilon} \qquad (16\text{-}1)$$

The **skeletal density** ρ_s of a particle (or crystalline density for a pure chemical compound) is given in terms of **internal porosity** ε_p by

$$\rho_s = \frac{\rho_p}{1-\varepsilon_p} \qquad (16\text{-}2)$$

For an adsorbent or ion exchanger, the **wet density** ρ_w of a particle is related to these factors and to the liquid density ρ_f by

$$\rho_w = \rho_p + \varepsilon_p \rho_f \qquad (16\text{-}3)$$

The **total voidage** ε_b in a packed bed (outside and inside particles) is

$$\varepsilon_b = \varepsilon + (1-\varepsilon)\varepsilon_p \qquad (16\text{-}4)$$

SORPTION EQUILIBRIUM

The quantity of a solute adsorbed can be given conveniently in terms of moles or volume (for adsorption) or ion-equivalents (for ion exchange) per unit mass or volume (dry or wet) of sorbent. Common units for adsorption are mol/(m³ of fluid) for the fluid-phase concentration c_i and mol/(kg of clean adsorbent) for adsorbed-phase concentration n_i. For gases, partial pressure may replace concentration.

Many models have been proposed for adsorption and ion exchange equilibria. The most important factor in selecting a model from an engineering standpoint is to have an accurate mathematical description over the entire range of process conditions. It is usually fairly easy to obtain correct capacities at selected points, but isotherm shape over the entire range is often a critical concern for a regenerable process.

GENERAL CONSIDERATIONS

Forces Molecules are attracted to surfaces as the result of two types of forces: dispersion-repulsion forces (also called London or van der Waals forces) such as described by the Lennard-Jones potential for molecule-molecule interactions; and electrostatic forces, which exist as the result of a molecule or surface group having a permanent electric dipole or quadrupole moment or net electric charge.

Dispersion forces are always present and in the absence of any stronger force will determine equilibrium behavior, as with adsorption of molecules with no dipole or quadrupole moment on nonoxidized carbons and silicalite.

If a surface is polar, its resulting electric field will induce a dipole moment in a molecule with no permanent dipole and, through this polarization, increase the extent of adsorption. Similarly, a molecule with a permanent dipole moment will polarize an otherwise nonpolar surface, thereby increasing the attraction.

For a polar surface and molecules with permanent dipole moments, attraction is strong, as for water adsorption on a hydrophilic adsorbent. Similarly, for a polar surface, a molecule with a permanent quadrupole moment will be attracted more strongly than a similar molecule with a weaker moment; for example, nitrogen is adsorbed more strongly than oxygen on zeolites (Sherman and Yon, gen. refs.).

Surface Excess With a Gibbs dividing surface placed at the surface of the solid, the surface excess of component i, Γ_i (mol/m²), is the amount per unit area of solid contained in the region near the surface, above that contained at the fluid-phase concentration far from the surface. This is depicted in two ways in Fig. 16-4. The quantity adsorbed per unit mass of adsorbent is

$$n_i = \Gamma_i A \qquad (16\text{-}6)$$

where A (m²/kg) is the surface area of the solid.

For a porous adsorbent, the amount adsorbed in the pore structure per unit mass of adsorbent, based on surface excess, is obtained by the difference

$$n_i = n_i^{\text{tot}} - V_p c_i^{\infty} \qquad (16\text{-}6)$$

where n_i^{tot} (mol/kg) is the total amount of component i contained within the particle's pore volume V_p (m³/kg), and c_i^{∞} is the concentration outside of the particle. If thermodynamics is a concern and n_i differs significantly from n_i^{tot} (as it will for weakly adsorbed species), then

it is important to consider adsorbed-phase quantities in terms of surface excesses.

A second convention is the placement of an imaginary envelope around the outermost boundary of a porous particle, so that all solute and nonadsorbing fluid contained within the pores of the particle is considered adsorbed.

Classification of Isotherms by Shape Representative isotherms are shown in Fig. 16-5, as classified by Brunauer and coworkers.

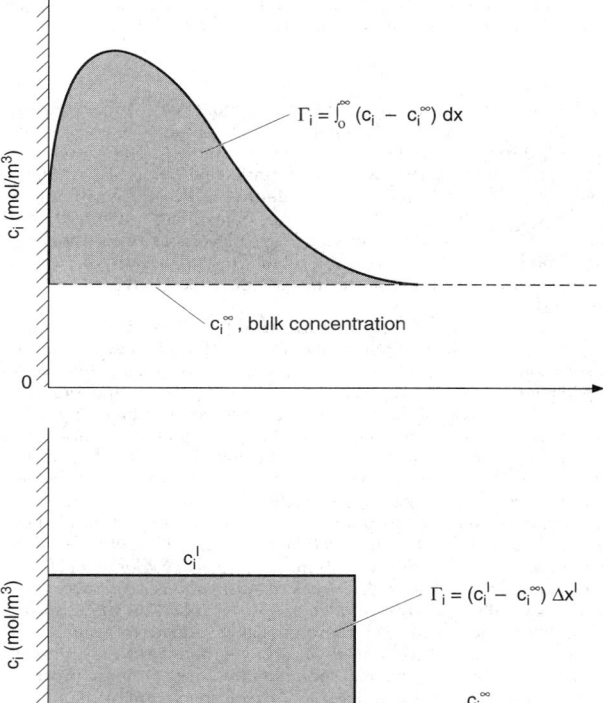

FIG. 16-4 Depictions of surface excess Γ_i. *Top:* The force field of the solid concentrates component i near the surface; the concentration c_i is low at the surface because of short-range repulsive forces between adsorbate and surface. *Bottom:* Surface excess for an imagined homogeneous surface layer of thickness Δx^l.

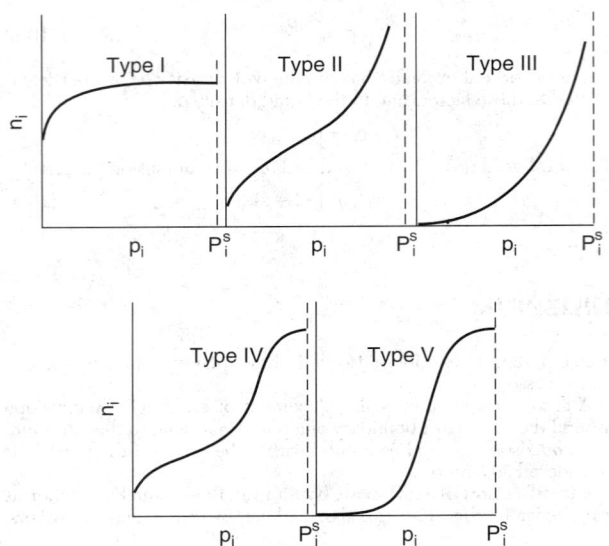

FIG. 16-5 Representative isotherm types. p_i and P_i^s are pressure and vapor pressure of the solute. [*Brunauer*, J. Am. Chem. Soc., **62**, *1723 (1940); reprinted with permission.*]

Curves that are concave downward throughout (type I) have historically been designated as "favorable," while those that are concave upward throughout (type III) are "unfavorable." Other isotherms (types II, IV, and V) have one or more inflection points. The designations "favorable" and "unfavorable" refer to fixed-bed behavior for the uptake step, with a favorable isotherm maintaining a compact wave shape. A favorable isotherm for uptake is unfavorable for the discharge step. This becomes particularly important for a regenerative process, in which a favorable isotherm may be too favorable for regeneration to occur effectively.

Categorization of Equilibrium Models Historically, sorption equilibrium has been approached from different viewpoints. For adsorption, many models for flat surfaces have been used to develop explicit equations and equations of state for pure components and mixtures, and many of the resulting equations are routinely applied to porous materials. Explicit equations for pore filling have also been proposed, generally based on the Polanyi potential theory. Ion exchange adds to these approaches concepts of absorption or dissolution (absorption) and exchange reactions. Statistical mechanics and molecular dynamics contribute to our understanding of all of these approaches (Steele, *The Interaction of Gases with Solid Surfaces*, Pergamon, Oxford, 1974; Nicholson and Parsonage, *Computer Simulation and the Statistical Mechanics of Adsorption*, Academic Press, New York, 1982). Mixture models are often based on the adsorbed solution theory, which uses thermodynamic equations from vapor-liquid equilibria with volume replaced by surface area and pressure replaced by a two-dimensional spreading pressure. Other approaches include lattice theories and mass-action exchange equilibrium.

Heterogeneity Adsorbents and ion exchangers can be physically and chemically heterogeneous. Although exceptions exist, solutes generally compete for the same sites. Models for adsorbent heterogeneity have been developed for both discrete and continuous distributions of energies [Ross and Olivier, *On Physical Adsorption*, Interscience, New York, 1964; Jaroniec and Madey, Rudzinski and Everett, gen. refs.].

Isosteric Heat of Adsorption The most useful heat of adsorption for fixed-bed calculations is the *isosteric heat of adsorption*, which is given by the Clausius-Clapeyron type relation

$$q_i^{st} = \Re T^2 \left. \frac{\partial \ln p_i}{\partial T} \right|_{n_i, n_j} \tag{16-7}$$

where the n_j can be dropped for single-component adsorption. q_i^{st} is positive by convention. If isosteres are straight lines when plotted as $\ln p_i$ versus T^{-1} (see Fig. 16-1), then Eq. (16-7) can be integrated to give

$$\ln p_i = f(n_i) - \frac{q_i^{st}}{\Re T} \tag{16-8}$$

where $f(n_i)$ is an arbitrary function dependent only on n_i. Many other heats of adsorption have been defined and their utility depends on the application (Ross and Olivier; Young and Crowell, gen. refs.).

From Eq. (16-8), if a single isotherm is known in the pressure explicit form $p_i = p_i(n_i, T)$ and if q_i^{st} is known at least approximately, then equilibria can be estimated over a narrow temperature range using

$$\ln \frac{p_i}{p_i^{ref}} = \frac{q_i^{st}}{\Re} \left(\frac{1}{T^{ref}} - \frac{1}{T} \right) \qquad (\text{const } n_i) \tag{16-9}$$

Similarly, Eq. (16-9) is used to calculate the isosteric heat of adsorption from two isotherms.

Experiments Sorption equilibria are measured using apparatuses and methods classified as volumetric, gravimetric, flow-through (frontal analysis), and chromatographic. Apparatuses are discussed by Yang (gen. refs.). Heats of adsorption can be determined from isotherms measured at different temperatures or measured independently by calorimetric methods.

Dimensionless Concentration Variables Where appropriate, isotherms will be written here using the dimensionless system variables

$$c_i^\circ = \frac{c_i}{c_i^{ref}} \qquad n_i^\circ = \frac{n_i}{n_i^{ref}} \tag{16-10}$$

where the best choice of reference values depends on the operation.

In some cases, to allow for some preloading of the adsorbent, it will be more convenient to use the dimensionless transition variables

$$c_i^\circ = \frac{(c_i - c_i')}{(c_i'' - c_i')} \qquad n_i^\circ = \frac{(n_i - n_i')}{(n_i'' - n_i')} \tag{16-11}$$

where single and double primes indicate initial and final concentrations, respectively. Figure 16-6 shows n° plotted versus c° for a sample system. Superimposed are an upward transition (loading) and a downward transition (unloading), shown by the respective positions of (c_i', n_i') and (c_i'', n_i'').

SINGLE COMPONENT OR EXCHANGE

The simplest relationship between solid-phase and fluid-phase concentrations is the **linear isotherm**

$$n_i = K_i c_i \qquad \text{or} \qquad n_i = K_i' p_i \tag{16-12}$$

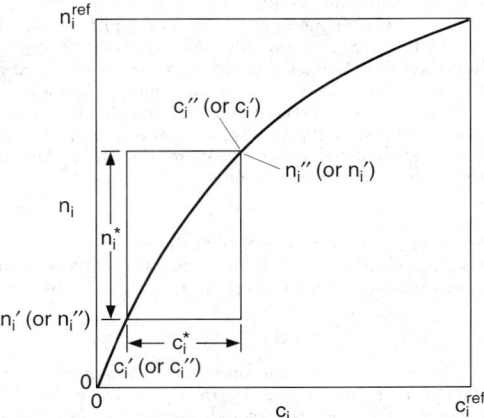

FIG. 16-6 Isotherm showing concentration variables for a transition from (c_i', n_i') to (c_i'', n_i'').

Thermodynamics requires that a linear limit be approached in the Henry's law region for all isotherm equations.

Flat Surface Isotherm Equations The classification of isotherm equations into two broad categories for flat surfaces and pore filling reflects their origin. It does not restrict equations developed for flat surfaces from being applied successfully to describe data for porous adsorbents.

The classical isotherm for a homogeneous flat surface, and most popular of all nonlinear isotherms, is the **Langmuir isotherm**

$$n_i = \frac{n_i^s K_i c_i}{1 + K_i c_i} \tag{16-13}$$

where n_i^s is the monolayer capacity approached at large concentrations and K_i is an equilibrium constant. These parameters are often determined by plotting $1/n_i$ versus $1/c_i$. The derivation of the isotherm assumes negligible interaction between adsorbed molecules.

The classical isotherm for multilayer adsorption on a homogeneous, flat surface is the **BET isotherm** [Brunauer, Emmett, and Teller, *J. Am. Chem. Soc.*, **60**, 309 (1938)]

$$n_i = \frac{n_i^s K_i p_i}{[1 + K_i p_i - (p_i/P_i^s)][1 - (p_i/P_i^s)]} \tag{16-14}$$

where p_i is the pressure of the adsorbable component and P_i^s is its vapor pressure. It is useful for gas-solid systems in which condensation is approached, fitting type-II behavior.

For a heterogeneous flat surface, a classical isotherm is the **Freundlich isotherm**

$$n_i = K_i c_i^{m_i} \tag{16-15}$$

where m_i is positive and generally not an integer. The isotherm corresponds approximately to an exponential distribution of heats of adsorption. Although it lacks the required linear behavior in the Henry's law region, it can often be used to correlate data on heterogeneous adsorbents over wide ranges of concentration.

Several isotherms combine aspects of both the Langmuir and Freudlich equations. One that has been shown to be effective in describing data mathematically for heterogeneous adsorbents is the **Tóth isotherm** [*Acta Chim. Acad. Sci. Hung.*, **69**, 311 (1971)]

$$n_i = \frac{n_i^s p_i}{[(1/K_i) + p_i^{m_i}]^{1/m_i}} \tag{16-16}$$

This three-parameter equation behaves linearly in the Henry's law region and reduces to the Langmuir isotherm for $m = 1$. Other well-known isotherms include the **Radke-Prausnitz isotherm** [Radke and Prausnitz, *Ind. Eng. Chem. Fundam.*, **11**, 445 (1972); *AIChE J.*, **18**, 761 (1972)]

$$n_i = \frac{n_i^s K_i p_i}{(1 + K_i p_i)^m} \tag{16-17}$$

and the **Sips isotherm** [Sips, *J. Chem. Phys.*, **16**, 490 (1948); Koble and Corrigan, *Ind. Eng. Chem.*, **44**, 383 (1952)] or **loading ratio correlation** with prescribed temperature dependence [Yon and Turnock, *AIChE Symp. Ser.*, **67**(117), 75 (1971)]

$$n_i = \frac{n_i^s (K_i p_i)^m}{1 + (K_i p_i)^m} \tag{16-18}$$

Another three-parameter equation that often fits data well and is linear in the Henry's law region is the **UNILAN equation** [Honig and Reyerson, *J. Phys. Chem.*, **56**, 140 (1952)]

$$n_i = \frac{n_i^s}{2s_i} \ln \frac{1 + K_i e^{s_i} p_i}{1 + K_i e^{-s_i} p_i} \tag{16-19}$$

which reduces to the Langmuir equation as $s_i \to 0$.

Equations of state are also used for pure components. Given such an equation written in terms of the two-dimensional spreading pressure π, the corresponding isotherm is easily determined, as described later for mixtures [see Eq. (16-42)]. The two-dimensional equivalent of an ideal gas is an ideal surface gas, which is described by

$$\pi A = n_i \Re T \tag{16-20}$$

which readily gives the linear isotherm, Eq. (16-12). Many more com-

plicated equations of state are available, including two-dimensional analogs of the virial equation and equations of van der Waals, Redlich-Kwong, Peng-Robinson, and so forth [Adamson, gen. refs.; Patrykiejewet et al., *Chem. Eng. J.*, **15**, 147 (1978); Haydel and Kobayashi, *Ind. Eng. Chem. Fundam.*, **6**, 546 (1967)].

Pore-Filling Isotherm Equations Most pore-filling models are grounded in the **Polanyi potential theory.** In Polanyi's model, an attracting potential energy field is assumed to exist adjacent to the surface of the adsorbent and concentrates vapors there. Adsorption takes place whenever the strength of the field, independent of temperature, is great enough to compress the solute to a partial pressure greater than its vapor pressure. The last molecules to adsorb form an equipotential surface containing the adsorbed volume. The strength of this field, called the adsorption potential ε (J/mol), was defined by Polanyi to be equal to the work required to compress the solute from its partial pressure to its vapor pressure

$$\varepsilon = \Re T \ln\left(\frac{P_i^s}{p_i}\right) \tag{16-21}$$

The same result is obtained by considering the change in chemical potential. In the basic theory, W (m³/kg), the volume adsorbed as saturated liquid at the adsorption temperature, is plotted versus ε to give a characteristic curve. Data measured at different temperatures for the same solute-adsorbent pair should fall on this curve. Using the method, it is possible to use data measured at a single temperature to predict isotherms at other temperatures. Data for additional, homologous solutes can be collapsed into a single "correlation curve" by defining a scaling factor β, the most useful of which has been V/V^{ref}, the adsorbate molar volume as saturated liquid at the adsorption temperature divided by that for a reference compound. Thus, by plotting W versus ε/β or ε/V for data measured at various temperatures for various similar solutes and a single adsorbent, a single curve should be obtained. Variations of the theory are often used to evaluate properties for components near or above their critical points [Grant and Manes, *Ind. Eng. Chem. Fund.*, **5**, 490 (1966)].

The most popular equations used to describe the shape of a characteristic curve or a correlation curve are the two-parameter **Dubinin-Radushkevich** (DR) equation

$$\frac{W}{W_0} = \exp\left[-k\left(\frac{\varepsilon}{\beta}\right)^2\right] \tag{16-22}$$

and the three-parameter **Dubinin-Astakhov** (DA) equation

$$\frac{W}{W_0} = \exp\left[-k\left(\frac{\varepsilon}{\beta}\right)^m\right] \tag{16-23}$$

where W_0 is micropore volume and m is related to the pore size distribution [Gregg and Sing, gen. refs.]. Neither of these equations has correct limiting behavior in the Henry's law regime.

Ion Exchange A useful tool is provided by the **mass action law** for describing the general exchange equilibrium in fully ionized exchanger systems as

$$z_B A + z_A \overline{B} \rightleftharpoons z_B \overline{A} + z_A B$$

where overbars indicate the ionic species in the ion exchanger phase and z_A and z_B are the valences of ions A and B. The associated equilibrium relation is of the form

$$K_{A,B}^c = K_{A,B} \left(\frac{\gamma_A}{\overline{\gamma}_A}\right)^{z_B} \left(\frac{\overline{\gamma}_B}{\gamma_B}\right)^{z_A} = \left(\frac{n_A}{c_A}\right)^{z_B} \left(\frac{c_B}{n_B}\right)^{z_A} \tag{16-24}$$

where $K_{A,B}^c$ is the apparent equilibrium constant or molar selectivity coefficient, $K_{A,B}$ is the thermodynamic equilibrium constant based on activities, and the γs are activity coefficients. Often it is desirable to represent concentrations in terms of equivalent ionic fractions based on solution normality c_{tot} and fixed exchanger capacity n^s as $c_i^\circ = z_i c_i / c_{\text{tot}}$ and $n_i^\circ = z_i n_i / n_{\text{tot}}$, where $c_{\text{tot}} = \sum z_j c_j$ and $n_{\text{tot}} = \sum z_j n_j = n^s$ with the summations extended to all counter-ion species. A rational selectivity coefficient is then defined as

$$K_{A,B}' = K_{A,B}^c \left(\frac{n_{\text{tot}}}{c_{\text{tot}}}\right)^{z_A - z_B} = \left(\frac{n_A^\circ}{c_A^\circ}\right)^{z_B} \left(\frac{c_B^\circ}{n_B^\circ}\right)^{z_A} \tag{16-25}$$

For the exchange of ions of equal valence ($z_A = z_B$), $K_{A,B}^c$ and $K_{A,B}'$ are coincident and, to a first approximation, independent of concentration. When $z_A > z_B$, $K'_{A,B}$ decreases with solution normality. This reflects the fact that ion exchangers exhibit an increasing affinity for ions of lower valence as the solution normality increases.

An alternate form of Eq. (16-25) is

$$n_A^\circ = K_{A,B}^c \left(\frac{n_{tot}}{c_{tot}}\right)^{(z_A - z_B)/z_B} \left(\frac{1 - n_A^\circ}{1 - c_A^\circ}\right)^{z_A/z_B} c_A^\circ \qquad (16\text{-}26)$$

When B is in excess ($c_A^\circ, n_A^\circ \to 0$), this reduces to

$$n_A^\circ = K_{A,B}^c \left(\frac{n_{tot}}{c_{tot}}\right)^{(z_A - z_B)/z_B} c_A^\circ = K_A c_A^\circ \qquad (16\text{-}27)$$

where the linear equilibrium constant $K_A = K_{A,B}^c (n_{tot}/c_{tot})^{(z_A/z_B)-1}$ decreases with solution normality c_{tot} when $z_A > z_B$ or increases when $z_A < z_B$.

Table 16-7 gives equilibrium constants for crosslinked cation and anion exchangers for a variety of counterions. The values given for cation exchangers are based on ion A replacing Li$^+$, and those given for anion exchangers are based on ion A replacing Cl$^-$. The selectivity for a particular ion generally increases with decreasing hydrated ion size and increasing degree of crosslinking. The selectivity coefficient for any two ions A and D can be obtained from Table 16-7 from values of $K_{A,B}$ and $K_{D,B}$ as

$$K_{A,D} = K_{A,B}/K_{D,B} \qquad (16\text{-}28)$$

The values given in this table are only approximate, but they are adequate for process screening purposes with Eqs. (16-24) and (16-25). Rigorous calculations generally require that activity coefficients be accounted for. However, for the exchange between ions of the same valence at solution concentrations of 0.1 N or less, or between any ions at 0.01 N or less, the solution-phase activity coefficients prorated to unit valence will be similar enough that they can be omitted.

Models for ion exchange equilibria based on the mass-action law taking into account solution and exchanger-phase nonidealities with equations similar to those for liquid mixtures have been developed by several authors [see Smith and Woodburn, *AIChE J.*, **24**, 577 (1978); Mehablia et al., *Chem. Eng. Sci.*, **49**, 2277 (1994)]. Thermodynamics-based approaches are also available [Soldatov in Dorfner, gen. refs.; Novosad and Myers, *Can J. Chem. Eng.*, **60**, 500 (1982); Myers and Byington in Rodrigues, ed., *Ion Exchange Science and Technology*, NATO ASI Series, No. 107, Nijhoff, Dordrecht, 1986, pp. 119–145] as

well as approaches for the exchange of macromolecules, taking into account steric-hindrance effects [Brooks and Cramer, *AIChE J.*, **38**, 12 (1992)].

Example 3: Calculation of Useful Ion-Exchange Capacity An 8 percent crosslinked sulfonated resin is used to remove calcium from a solution containing 0.0007 mol/l Ca^{+2} and 0.01 mol/l Na$^+$. The resin capacity is 2.0 equiv/l. Estimate the resin capacity for calcium removal.

From Table 16-7, we obtain $K_{Ca,Na} = 5.16/1.98 = 2.6$. Since $n_{tot} = 2$ equiv/l and $c_{tot} = 2 \times 0.0007 + 0.01 = 0.024$ equiv/l, $K'_{Ca,Na} = 2.6 \times 2/0.011 = 470$. Thus, with $c_{Ca}^\circ = 2 \times 0.0007/0.011 = 0.13$, Eq. (16-25) gives $470 = (n_{Ca}^\circ/0.13)[(1 - 0.13)/(1 - n_{Na}^\circ)]^2$ or $n_{Ca}^\circ = 0.9$. The available capacity for calcium is $n_{Ca} = 0.9 \times 2.0/2 = 0.9$ mol/l.

Donnan Uptake The uptake of an electrolyte as a neutral ion pair of a salt is called Donnan uptake. It is generally negligible at low ionic concentrations. Above 0.5 g·equiv/l with strongly ionized exchangers (or at lower concentrations with those more weakly ionized), the resin's fixed ion-exchange capacity is measurably exceeded as a result of electrolyte invasion. With only one coion species Y (matching the charge sign of the fixed groups in the resin), its uptake n_Y equals the total excess uptake of the counterion. Equilibrium is described by the mass-action law. For the case of a resin in A-form in equilibrium with a salt AY, the excess counterion uptake is given by [Helfferich, gen. refs., pp. 133–147]

$$z_Y n_Y = \left[\frac{n^{s2}}{4} + (z_Y c_Y)^2 \left(\frac{\gamma_{AY}}{\bar\gamma_{AY}}\right)^2 \left(\frac{\bar a_w}{a_w}\right)^{\bar v_{AY}/\bar v_w}\right]^{1/2} - \frac{n^s}{2} \qquad (16\text{-}29)$$

where γs are activity coefficients, as water activities, and vs partial molar volumes. For dilute conditions, Eq. (16-29) predicts a squared dependence of n_Y on c_Y. Thus, the electrolyte sorption isotherm has a strong positive curvature. Donnan uptake is more pronounced for resins of lower degree of crosslinking and for counterions of low valence.

Separation Factor By analogy with the mass-action case and appropriate for both adsorption and ion exchange, a **separation factor** r can be defined based on dimensionless system variables [Eq. (16-10)] by

$$r = \frac{c_i^\circ(1 - n_i^\circ)}{n_i^\circ(1 - c_i^\circ)} \quad \text{or} \quad r = \frac{n_B^\circ/c_B^\circ}{n_A^\circ/c_A^\circ} \qquad (16\text{-}30)$$

This term is analogous to relative volatility or its reciprocal (or to an equilibrium selectivity). Similarly, the assumption of a constant sepa-

TABLE 16-7 Equilibrium Constants for Polystyrene DVB Cation and Anion Exchangers

	Strong acid sulfonated cation exchangers (Li$^+$ reference ion)						
	Degree of crosslinking				Degree of crosslinking		
Counterion	4% DVB	8% DVB	16% DVB	Counterion	4% DVB	8% DVB	16% DVB
Li$^+$	1.00	1.00	1.00	Mg^{++}	2.95	3.29	3.51
H$^+$	1.32	1.27	1.47	Zn^{++}	3.13	3.47	3.78
Na$^+$	1.58	1.98	2.37	Co^{++}	3.23	3.74	3.81
NH$_4^+$	1.90	2.55	3.34	Cu^{++}	3.29	3.85	4.46
K$^+$	2.27	2.90	4.50	Cd^{++}	3.37	3.88	4.95
Rb$^+$	2.46	3.16	4.62	Ni^{++}	3.45	3.93	4.06
Cs$^+$	2.67	3.25	4.66	Ca^{++}	4.15	5.16	7.27
Ag$^+$	4.73	8.51	22.9	Pb^{++}	6.56	9.91	18.0
Tl$^+$	6.71	12.4	28.5	Ba^{++}	7.47	11.5	20.8

	Strong base anion exchangers, 8% DVB (Cl$^-$ reference ion)				
Counterion	Type I resin[*]	Type II resin[†]	Counterion	Type I resin[(a)]	Type II resin[(b)]
Salicylate	32	28	Cyanide	1.6	1.3
Iodide	8.7	7.3	Chloride	1.0	1.0
Phenoxide	5.2	8.7	Hydroxide	0.05–0.07	0.65
Nitrate	3.8	3.3	Bicarbonate	0.3	0.5
Bromide	2.8	2.3	Formate	0.2	0.2
Nitrite	1.2	1.3	Acetate	0.2	0.2
Bisulfite	1.3	1.3	Fluoride	0.09	0.1
Cyanide	1.6	1.3	Sulfate	0.15	0.15

[*]Trimethylamine
[†]Dimethyl-hydroxyethylamine
Data from Bonner and Smith, *J. Phys. Chem.*, **61**, 326 (1957) and Wheaton and Bauman, *Ind. Eng. Chem.*, **45**, 1088 (1951).

ration factor is a useful assumption in many sorptive operations. [It is constant for the Langmuir isotherm, as described below, and for mass-action equilibrium with $m = 1$ in Eq. (16-24).] This gives the constant separation factor isotherm

$$n_i^\circ = \frac{c_i^\circ}{r + (1 - r)c_i^\circ} \qquad (16\text{-}31)$$

The separation factor r identifies the equilibrium increase in n_i° from 0 to 1, which accompanies an increase in c_i° from 0 to 1. For a concentration change over only part of the isotherm, a separation factor R can be defined for the dimensionless transition variables [Eq. (16-11)]. This separation factor is

$$R = \frac{n_i''/c_i''}{n_i'/c_i'} = \frac{c_i^\circ(1 - n_i^\circ)}{n_i^\circ(1 - c_i^\circ)} \qquad (16\text{-}32)$$

and gives an equation identical to Eq. (16-31) with R replacing r.

Figure 16-7 shows constant separation factor isotherms for a range of r (or R) values. The isotherm is linear for $r = 1$, favorable for $r < 1$, rectangular (or irreversible) for $r = 0$, and unfavorable for $r > 1$. As a result of symmetry properties, if r is defined for adsorption of component i or exchange of ion A for B, then the reverse process is described by $1/r$.

The Langmuir isotherm, Eq. (16-13), corresponds to the constant separation factor isotherm with

$$r = 1/(1 + K_i c_i^{\text{ref}}) \qquad (16\text{-}33)$$

for system variables Eq. (16-10) or

$$R = \frac{r + (1 - r)(c_i'/c_i^{\text{ref}})}{r + (1 - r)(c_i''/c_i^{\text{ref}})} \qquad (16\text{-}34)$$

for transition variables [Eq. (16-11)]. Vermeulen et al. [gen. refs.] give additional properties of constant separation factor isotherms.

Example 4: Application of Isotherms Thomas [*Ann. N.Y. Acad. Sci.*, **49**, 161 (1948)] provides the following Langmuir isotherm for the adsorption of anthracene from cyclohexane onto alumina:

$$n_i = \frac{22c_i}{1 + 375c_i}$$

with n_i in mol anthracene/kg alumina and c_i in mol anthracene/l liquid.
a. What are the values of K_i and n_i^s according to Eq. (16-13)?

$$K_i = 375 \text{ l/mol}$$

$$n_i^s = \frac{22}{K_i} = 0.0587 \text{ mol/kg}$$

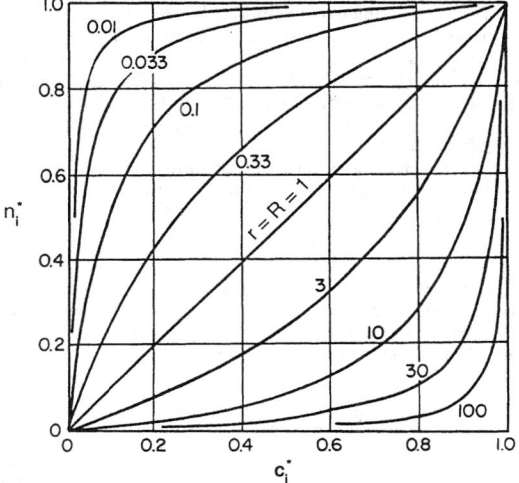

FIG. 16-7 Constant separation factor isotherm as a function of the separation factor r (or interchangeably R). Each isotherm is symmetric about the perpendicular line connecting $(0,1)$ and $(1,0)$. Isotherms for r and $1/r$ are symmetric about the 45° line.

b. For a feed concentration of 8.11×10^{-4} mol/l, what is the value of r?

$$r = \frac{1}{1 + K_i(8.11 \times 10^{-4})} = 0.766 \qquad \text{[from Eq. (16-33)]}$$

c. If the alumina is presaturated with liquid containing 2.35×10^{-4} mol/l and the feed concentration is 8.11×10^{-4} mol/l, what is the value of R?

$$R = 0.834 \qquad \text{[from Eq. (16-32) or (16-34)]}$$

MULTIPLE COMPONENTS OR EXCHANGES

When more than one adsorbed species or more than two ion-exchanged species interact in some manner, equilibrium becomes more complicated. Usually, thermodynamics provides a sound basis for prediction.

Adsorbed-Solution Theory The common thermodynamic approach to multicomponent adsorption treats adsorption equilibrium in a way analogous to fluid-fluid equilibrium. The theory has as its basis the Gibbs adsorption isotherm [Young and Crowell, gen. refs.], which is

$$A d\pi = \sum_i n_i \, d\mu_i \qquad (\text{const } T) \qquad (16\text{-}35)$$

where μ is chemical potential. For an ideal gas ($d\mu = \Re T d \ln p_i$), if it is assumed that an adsorbed solution is defined with a pure-component standard state (as is common for a liquid solution), then Eq. (16-35) can be integrated to give [Rudisill and LeVan, *Chem. Eng. Sci.*, **47**, 1239 (1992)]

$$p_i = \gamma_i x_i P_i^{\text{ref}}(T, \pi) \qquad (16\text{-}36)$$

where γ_i and x_i are the adsorbed-phase activity coefficient and mole fraction of component i and P_i^{ref} is the standard state, specified to be at the temperature and spreading pressure of the mixture.

Equation (16-36) with $\gamma_i = 1$ provides the basis for the ideal adsorbed-solution theory [Myers and Prausnitz, *AIChE J.*, **11**, 121 (1965)]. The spreading pressure for a pure component is determined by integrating Eq. (16-35) for a pure component to obtain

$$\frac{\pi A}{\Re T} = \int_0^{P_i^{\text{ref}}} \frac{n_i}{p_i} \, dp_i \qquad (16\text{-}37)$$

where n_i is given by the pure-component isotherm. Also, since $\sum x_i = 1$, Eq. (16-36) with $\gamma_i = 1$ gives $\sum (p_i/P_i^{\text{ref}}) = 1$. With no area change on mixing for the ideal solution, the total number of moles adsorbed per unit weight of adsorbent is determined using a two-dimensional form of Amagat's law:

$$\frac{1}{n_{\text{tot}}} = \sum \frac{x_i}{n_i^{\text{ref}}} \qquad (16\text{-}38)$$

where $n_{\text{tot}} = \sum n_i$ and n_i^{ref} is given by the pure-component isotherm at P_i^{ref}. Adsorbed-phase concentrations are calculated using $n_i = x_i n_{\text{tot}}$. Generally, different values of π [or $\pi A/(\Re T)$] must be tried until the one is found that satisfies Eqs. (16-37) and $\sum x_i = 1$.

Example 5: Application of Ideal Adsorbed-Solution Theory Consider a binary adsorbed mixture for which each pure component obeys the Langmuir equation, Eq. (16-13). Let $n_1^s = 4$ mol/kg, $n_2^s = 3$ mol/kg, $K_1 p_1 = K_2 p_2 = 1$. Use the ideal adsorbed-solution theory to determine n_1 and n_2.

Substituting the pure component Langmuir isotherm

$$n_i = \frac{n_i^s K_i p_i}{1 + K_i p_i}$$

into Eq. (16-37) and integrating gives

$$\frac{\pi A}{\Re T} = n_i^s \ln(1 + K_i P_i^{\text{ref}})$$

which can be solved explicitly for $K_i P_i^{\text{ref}}$. Values are guessed for $\pi A/(\Re T)$, values of $K_i P_i^{\text{ref}}$ are calculated from the equation above, and $\sum x_i = \sum K_i p_i/(K_i P_i^{\text{ref}}) = 1$ is checked to see if it is satisfied. Trial and error gives $\pi A/(\Re T) = 3.8530$ mol/kg, $K_1 P_1^{\text{ref}} = 1.6202$, $K_2 P_2^{\text{ref}} = 2.6123$, and $x_1 = 0.61720$. Evaluating the pure-component isotherms at the reference pressures and using Eq. (16-38) gives $n_{\text{tot}} = 2.3475$ mol/kg, and finally $n_i = x_i n_{\text{tot}}$ gives $n_1 = 1.4489$ mol/kg and $n_2 = 0.8986$ mol/kg.

Other approaches to account for various effects have been developed. Negative deviations from Raoult's law (i.e., $\gamma_i < 1$) are frequently found due to adsorbent heterogeneity [e.g., Myers, *AIChE J.*, **29**, 691 (1983)]. Thus, contributions include accounting for adsorbent heterogeneity [Valenzuela et al., *AIChE J.*, **34**, 397 (1988)] and excluded pore-volume effects [Myers, in Rodrigues et al., gen. refs.]. Several activity coefficient models have been developed to account for nonideal adsorbate-adsorbate interactions including a spreading pressure-dependent activity coefficient model [e.g., Talu and Zwiebel, *AIChE J.*, **32**, 1263 (1986)] and a vacancy solution theory [Suwanayuen and Danner, *AIChE J.*, **26**, 68, 76 (1980)].

Langmuir-Type Relations   For systems composed of solutes that individually follow Langmuir isotherms, the traditional multicomponent Langmuir equation, obtained via a kinetic derivation, is

$$n_i = \frac{n_i^s K_i p_i}{1 + \sum K_i p_i} \tag{16-39}$$

This equation has been criticized on thermodynamic grounds because it does not satisfy the Gibbs adsorption isotherm unless all monolayer capacities n_i^s are equal.

To satisfy the Gibbs adsorption isotherm for unequal monolayer capacities, explicit isotherms can be obtained in the form of a series expansion [LeVan and Vermeulen, *J. Phys. Chem.*, **85**, 3247 (1981)]. A two-term form is

$$n_1 = \frac{(n_1^s + n_2^s)K_1 p_1}{2(1 + K_1 p_1 + K_2 p_2)} + \frac{(n_1^s - n_2^s)K_1 p_1 K_2 p_2}{(K_1 p_1 + K_2 p_2)^2} \ln(1 + K_1 p_1 + K_2 p_2) \tag{16-40}$$

where the subscripts may be interchanged. Multicomponent forms are also available [Frey and Rodrigues, *AIChE J.*, **40**, 182 (1994)].

Example 6: Comparison of Binary Langmuir Isotherms   Use the numerical values in Example 5 to evaluate the binary Langmuir isotherms given by Eqs. (16-39) and (16-40) and compare results with the exact answers given in Example 5.

Equation (16-39) gives $n_1 = 1.3333$ mol/kg and $n_2 = 1.0000$ mol/kg for an average deviation from the exact values of approximately 10 percent. Equation (16-40) gives $n_1 = 1.4413$ mol/kg and $n_2 = 0.8990$ mol/kg for an average deviation of about 0.6 percent.

Freundlich-Type Relations   A binary Freundlich isotherm, obtained from the ideal adsorbed solution theory in loading-explicit closed form [Crittenden et al., *Environ. Sci. Technol.*, **19**, 1037 (1985)], is

$$c_i = \frac{n_i}{\sum_{j=1}^N n_j} \left(\frac{\sum_{j=1}^N (n_j/m_j)}{K_i/m_i} \right)^{1/m_i} \tag{16-41}$$

Equations of State   If an equation of state is specified for a multicomponent adsorbed phase of the form $\pi A/(\Re T) = f(n_1, n_2, \dots)$, then the isotherms are determined using [Van Ness, *Ind. Eng. Chem. Fundam.*, **8**, 464–473 (1969)]

$$\ln \left(\frac{K_i p_i}{n_i/M_s} \right) = \int_A^\infty \left[\frac{\partial[\pi A/(\Re T)]}{\partial n_i} \bigg|_{T,A,n_j} - 1 \right] \frac{dA}{A} \tag{16-42}$$

where, because integration is over A, M_s is mass of adsorbent, units for n and A are mol and m² (rather than mol/kg and m²/kg), and $K_i p_i/(n_i/M_s) = 1$ is the linear lower limit approached for the ideal surface gas [see Eqs. (16-12) and (16-20)].

Ion Exchange—Stoichiometry   In most applications, except for some weak-electrolyte and some concentrated-solution cases, the following summations apply:

$$\sum_i z_i c_i = c_{\text{tot}} = \text{const} \qquad \sum_i z_i n_i = n_{\text{tot}} = \text{const} \tag{16-43}$$

In equivalent-fraction terms, the sums become

$$\sum_i c_i^\circ = 1 \qquad \sum_i n_i^\circ = 1 \tag{16-44}$$

Mass Action   Here the equilibrium relations, consistent with Eq. (16-25), are

$$K_{ij}' = \left(\frac{n_i^\circ}{c_i^\circ} \right)^{z_j} \left(\frac{c_j^\circ}{n_j^\circ} \right)^{z_i} \tag{16-45}$$

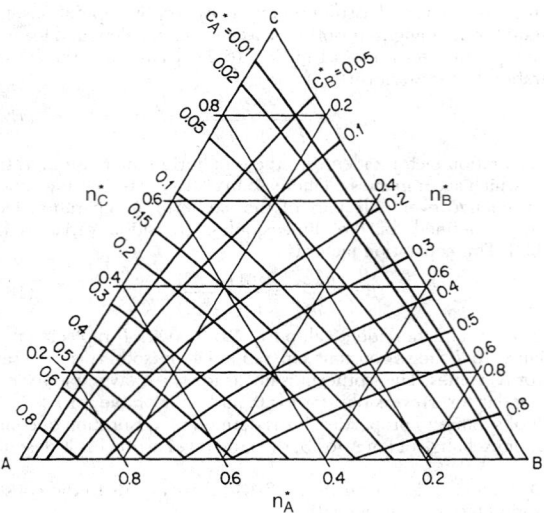

FIG. 16-8   Ideal mass-action equilibrium for three-component ion exchange with unequal valences. $K_{A,C}' = 8.06$; $K_{B,C}' = 3.87$. Duolite C-20 polystyrenesulfonate resin, with Ca as *A*, Mg as *B*, and Na as *C*. [*Klein et al., Ind. Eng. Chem. Fund.*, **6**, 339 (1967); reprinted with permission.]

For an N-species system, with N c°s (or n°s) known, the N n°s (or c°s) can be found by simultaneous solution of the $N - 1$ independent i,j combinations for Eq. (16-45) using Eq. (16-43); one n_j°/c_j° is assumed, the other values can be calculated using Eq. (16-45), and the sum of the trial n°s (or c°s) is compared with Eq. (16-44).

Because an N-component system has $N - 1$ independent concentrations, a three-component equilibrium can be plotted in a plane and a four-component equilibrium in a three-dimensional space. Figure 16-8 shows a triangular plot of c° contours in equilibrium with the corresponding n° coordinates.

Improved models for ion-exchange equilibria based on the mass-action law that take into account solution and exchanger-phase non-idealities with equations similar to those for liquid mixtures have been developed by several authors including Smith and Woodburn [*AIChE J.*, **24**, 577 (1978)] and Mehablia et al. [*Chem. Eng. Sci.*, **49**, 2277 (1994)]. Thermodynamics-based approaches are also available [Soldatov in Dorfner, gen. refs.; Novosad and Myers, *Can J. Chem. Eng.*, **60**, 500 (1982); Myers and Byington in Rodrigues, ed., *Ion Exchange: Science and Technology*, NATO ASI Series, No. 107, Nijhoff, Dordrecht, 1986, pp. 119–145] as well as approaches for the exchange of macromolecules taking into account steric-hindrance effects [Brooks and Cramer, *AIChE J.*, **38**, 12 (1992)].

Constant Separation-Factor Treatment   If the valences of all species are equal, the separation factor α_{ij} applies, where

$$\alpha_{ij} = K_{ij} = \frac{n_i^\circ c_j^\circ}{c_i^\circ n_j^\circ} \tag{16-46}$$

For a binary system, $r = \alpha_{BA} = 1/\alpha_{AB}$. The symbol r applies primarily to the process, while α is oriented toward interactions between pairs of solute species. For each binary pair, $r_{ij} = \alpha_{ji} = 1/\alpha_{ij}$.

Equilibrium then is given explicitly by

$$n_i^\circ = \frac{c_i^\circ}{\sum_j \alpha_{ji} c_j^\circ} = \frac{\alpha_{iN} c_i^\circ}{\sum_j \alpha_{jN} c_j^\circ} \tag{16-47}$$

and

$$c_i^\circ = \frac{n_i^\circ}{\sum_j \alpha_{ij} n_j^\circ} = \frac{\alpha_{Ni} n_i^\circ}{\sum_j \alpha_{Nj} n_j^\circ} \tag{16-48}$$

For the constant separation factor case, the c° contours in a plot like Fig. 16-8 are linear.

CONSERVATION EQUATIONS

Material balances, often an energy balance, and occasionally a momentum balance are needed to describe an adsorption process. These are written in various forms depending on the specific application and desire for simplicity or rigor. Reasonably general material balances for various processes are given below. An energy balance is developed for a fixed bed for gas-phase application and simplified for liquid-phase application. Momentum balances for pressure drop in packed beds are given in Sec. 6.

MATERIAL BALANCES

At a microscale, a sorbable component exists at three locations—in a sorbed phase, in pore fluid, and in fluid outside particles. As a consequence, in material balances time derivatives must be included of terms involving n_i, c_{pi} (the pore concentration), and c_i (the extraparticle concentration). Let $\bar{n}_i$ represent n_i averaged over particle volume, and let $\bar{c}_{pi}$ represent c_{pi} averaged over pore fluid volume.

For batch or stirred tank processes, in terms of the mass of adsorbent M_s (kg), extraparticle volume of fluid V_f (m^3), and volumetric flow rates F_v (m^3/s) in and out of a tank, the material balance on component i is

$$M_s \frac{d\hat{n}_i}{dt} + \frac{d(V_f c_i)}{dt} = F_{v,\text{in}} c_{\text{in},i} - F_{v,\text{out}} c_i \qquad (16\text{-}49)$$

with

$$\hat{n}_i = \bar{n}_i + (\varepsilon_p/\rho_p)\bar{c}_{pi} = \bar{n}_i + [(1-\varepsilon)\varepsilon_p/\rho_b]\bar{c}_{pi} \qquad (16\text{-}50)$$

where ρ_p and ρ_b are particle and bulk densities, and ε and ε_p are void fraction (extra particle volume fraction) and particle porosity, respectively.

For a fixed-bed process, the material balance for component i is

$$\rho_b \frac{\partial \hat{n}_i}{\partial t} + \varepsilon \frac{\partial c_i}{\partial t} + \varepsilon \frac{\partial (vc_i)}{\partial z} = \varepsilon D_L \frac{\partial}{\partial z}\left(c \frac{\partial y_i}{\partial z}\right) \qquad (16\text{-}51)$$

where v is interstitial fluid velocity, D_L is a Fickian axial dispersion coefficient, and $y_i = c_i/c$ is the fluid-phase mole fraction of component i. An alternative form, grouping together fluid-phase concentrations rather than intraparticle concentrations, is

$$\rho_b \frac{\partial \bar{n}_i}{\partial t} + \varepsilon_b \frac{\partial \hat{c}_i}{\partial t} + \varepsilon \frac{\partial (vc_i)}{\partial z} = \varepsilon D_L \frac{\partial}{\partial z}\left(c \frac{\partial y_i}{\partial z}\right) \qquad (16\text{-}52)$$

where, noting Eq. (16-4), $\hat{c}_i$ is defined by

$$\varepsilon_b \hat{c}_i = \varepsilon c_i + (1-\varepsilon)\varepsilon_p \bar{c}_{pi} \qquad (16\text{-}53)$$

For moving-bed processes, we add a term to Eq. (16-51) to obtain

$$\rho_b\left[\frac{\partial \hat{n}_i}{\partial t} + v_s \frac{\partial \hat{n}_i}{\partial z}\right] + \varepsilon\left[\frac{\partial c_i}{\partial t} + \frac{\partial (vc_i)}{\partial z} - D_L \frac{\partial}{\partial z}\left(c \frac{\partial y_i}{\partial z}\right)\right] = 0 \quad (16\text{-}54)$$

where v_s is the solid-phase velocity (opposite in sign to v for a countercurrent process).

ENERGY BALANCE

Many different forms of the energy balance have been used in fixed-bed adsorption studies. The form chosen for a particular study depends on the process considered (e.g., temperature swing adsorption or pressure swing adsorption) and on the degree of approximation that is appropriate.

The energy balance for a general fixed-bed process, ignoring dispersion, is

$$\rho_b \frac{\partial u_s}{\partial t} + \frac{\partial(\varepsilon_b c u_f)}{\partial t} + \frac{\partial(\varepsilon v c h_f)}{\partial z} = -\frac{2h_w(T-T_w)}{r_c} \qquad (16\text{-}55)$$

where h_w is a heat transfer coefficient for energy transfer with the column wall and r_c is the radius of the column. The second term of Eq. (16-55) combines contributions from both pore and extraparticle fluid.

Thermodynamic paths are necessary to evaluate the enthalpy (or internal energy) of the fluid phase and the internal energy of the stationary phase. For gas-phase processes at low and modest pressures, the enthalpy departure function for pressure changes can be ignored and a reference state for each pure component chosen to be ideal gas at temperature T^{ref}, and a reference state for the stationary phase (adsorbent plus adsorbate) chosen to be adsorbate-free solid at T^{ref}. Thus, for the gas phase we have

$$h_f = \sum_i y_i h_{fi} = \sum_i y_i\left[h_{fi}^{\text{ref}} + \int_{T^{\text{ref}}}^T C_{pfi}^\circ \, dT\right] \qquad (16\text{-}56)$$

$$u_f = h_f - P/c \qquad (16\text{-}57)$$

and for the stationary phase

$$u_s = u_{\text{sol}} + nu_a \approx u_{\text{sol}} + nh_a \qquad (16\text{-}58)$$

$$u_{\text{sol}} = u_{\text{sol}}^{\text{ref}} + \int_{T^{\text{ref}}}^T C_s \, dT \qquad (16\text{-}59)$$

The enthalpy of the adsorbed phase h_a is evaluated along a path for which the gas-phase components undergo temperature change from T_{ref} to T and then are adsorbed isothermally, giving

$$h_a = \sum_i x_i h_{fi} - \left(\frac{1}{n}\right)\sum_i \int_0^{n_i} q_i^{st}(n_i, n_j, T) \, dn_i \qquad (16\text{-}60)$$

The isoteric heat of adsorption q_i^{st} is composition-dependent, and the sum of integrals Eq. (16-60) is difficult to evaluate for multicomponent adsorption if the isoteric heats indeed depend on loading. Because each isosteric heat depends on the loadings of all components, the sum must be evaluated for a path beginning with clean adsorbent and ending with the proper loadings of all components. If the isosteric heat of adsorption is constant, as is commonly assumed, then the energy balance (Eq. 16-55) becomes

$$\left[\rho_b\left(C_s + \sum_i n_i C_{pfi}^\circ\right) + \varepsilon_b c C_{pf}^\circ\right]\frac{\partial T}{\partial t} - \rho_b \sum_i q_i^{st}\frac{\partial n_i}{\partial t} - \frac{\partial(\varepsilon_b P)}{\partial t}$$
$$+ \varepsilon v c C_{pf}^\circ \frac{\partial T}{\partial z} = -\frac{2h_w(T-T_w)}{r_c} \qquad (16\text{-}61)$$

where Eq. (16-51) with $D_L = 0$ has been used. Equation (16-61) is a popular form of the energy balance for fixed-bed adsorption calculations. Often the first summation on the left-hand side, which involves gas-phase heat capacities, is neglected, or the gas-phase heat capacities are replaced by adsorbed-phase heat capacities.

Nonisothermal liquid-phase processes may be driven by changes in feed temperature or heat addition or withdrawal through a column wall. For these, heats of adsorption and pressure effects are generally of less concern. For this case a suitable energy balance is

$$(\rho_b C_s + \varepsilon_b c C_{pf}^\circ)\frac{\partial T}{\partial t} + \varepsilon v c C_{pf}^\circ \frac{\partial T}{\partial z} = -\frac{2h_w(T-T_w)}{r_c} \qquad (16\text{-}62)$$

RATE AND DISPERSION FACTORS

The performance of adsorption processes results in general from the combined effects of thermodynamic and rate factors. It is convenient to consider first thermodynamic factors. These determine the process performance in a limit where the system behaves ideally; i.e. without mass transfer and kinetic limitations and with the fluid phase in perfect piston flow. **Rate factors** determine the efficiency of the real process in relation to the ideal process performance. Rate factors include heat- and mass-transfer limitations, reaction kinetic limitations, and hydrodynamic dispersion resulting from the velocity distribution across the bed and from mixing and diffusion in the interparticle void space.

TRANSPORT AND DISPERSION MECHANISMS

Figure 16-9 depicts porous adsorbent particles in an adsorption bed with sufficient generality to illustrate the nature and location of individual transport and dispersion mechanisms. Each mechanism involves a different driving force and, in general, gives rise to a different form of mathematical result.

Intraparticle Transport Mechanisms Intraparticle transport may be limited by *pore diffusion, solid diffusion, reaction kinetics* at phase boundaries, or two or more of these mechanisms together.

1. *Pore diffusion in fluid-filled pores.* These pores are sufficiently large that the adsorbing molecule escapes the force field of the adsorbent surface. Thus, this process is often referred to as *macropore diffusion.* The driving force for such a diffusion process can be approximated by the gradient in mole fraction or, if the molar concentration is constant, by the gradient in concentration of the diffusing species within the pores.

2. *Solid diffusion in the adsorbed phase.* Diffusion in pores sufficiently small that the diffusing molecule never escapes the force field of the adsorbent surface. In this case, transport may occur by an activated process involving jumps between adsorption sites. Such a process is often called *surface diffusion* or, in the case of zeolites, *micropore* or *intracrystalline* diffusion. The driving force for the process can thus be approximated by the gradient in concentration of the species in its adsorbed state. Phenomenologically, the process is not distinguishable from that of homogeneous diffusion that occurs inside a sorbent gel or in a pore-filling fluid that is immiscible with the external fluid. The generic term *solid diffusion* is used here to encompass the general traits of these physically different systems.

3. *Reaction kinetics at phase boundaries.* Rates of adsorption and desorption in porous adsorbents are generally controlled by mass transfer within the pore network rather than by the kinetics of sorption at the surface. Exceptions are the cases of chemisorption and affinity-adsorption systems used for biological separations, where the kinetics of bond formation can be exceedingly slow.

Intraparticle convection can also occur in packed beds when the adsorbent particles have very large and well-connected pores. Although, in general, bulk flow through the pores of the adsorbent particles is only a small fraction of the total flow, intraparticle convection can affect the transport of very slowly diffusing species such as macromolecules. The driving force for convection, in this case, is the pressure drop across each particle that is generated by the frictional resistance to flow experienced by the fluid as this flows through the packed bed [Rodrigues et al., *Chem. Eng. Sci.,* **46,** 2765 (1991); Carta et al., *Sep. Technol.,* **2,** 62 (1992); Frey et al., *Biotechnol. Progr.,* **9,** 273 (1993); Liapis and McCoy, *J. Chromatogr.,* **599,** 87 (1992)]. Intraparticle convection can also be significant when there is a total pressure difference between the center of the particle and the outside, such as is experienced in pressurization and depressurization steps of pressure swing adsorption or when more gas is drawn into an adsorbent to equalize pressure as adsorption occurs from the gas phase within a porous particle [Lu et al., *AIChE J.,* **38,** 857 (1992); Lu et al., *Gas Sep. Purif.,* **6,** 89 (1992)].

Extraparticle Transport and Dispersion Mechanisms Extraparticle mechanisms are affected by the design of the contacting device and depend on the hydrodynamic conditions outside the particles.

4. *External mass transfer* between the external surfaces of the adsorbent particles and the surrounding fluid phase. The driving force is the concentration difference across the boundary layer that surrounds each particle, and the latter is affected by the hydrodynamic conditions outside the particles.

5. *Mixing,* or lack of mixing, between different parts of the contacting equipment. This may occur through the existence of a velocity distribution or dead zones in a packed bed or through inefficient mixing in an agitated contactor. In packed-bed adsorbers, mixing is often described in terms of an axial dispersion coefficient whereby all mechanisms contributing to axial mixing are lumped together in a single effective coefficient.

Heat Transfer Since adsorption is generally accompanied by the evolution of heat, the rate of heat transfer between the adsorbent particles and the fluid phase may be important. In addition, heat transfer can occur across the column wall in small diameter beds and is important in energy applications of adsorption. In gas adsorption systems, even with highly porous particles, the controlling heat transfer resistance is generally associated with extraparticle transport [Lee and Ruthven, *Can. J. Chem. Eng.,* **57,** 65 (1979)], so that the temperature within the particles is essentially uniform. In liquid-phase adsorption, intraparticle and extraparticle heat transfer resistances are generally comparable. However, in this case the heat capacity of the fluid phase is sufficiently high that temperature effects may be negligible except in extreme cases. General discussions of heat-transfer effects in adsorbents and adsorption beds are found in Suzuki (gen. refs., pp. 187–208 and pp. 275–290) and in Ruthven (gen. refs., pp. 189–198 and pp. 215–219).

INTRAPARTICLE MASS TRANSFER

The phenomenological aspects of diffusional mass transfer in adsorption systems can be described in terms of Fick's law:

$$J_i = -D_i(c_i)\,\frac{\partial c_i}{\partial x} \tag{16-63}$$

This expression can be used to describe both pore and solid diffusion so long as the driving force is expressed in terms of the appropriate concentrations. Although the driving force should be more correctly expressed in terms of chemical potentials, Eq. (16-63) provides a qualitatively and quantitatively correct representation of adsorption systems so long as the diffusivity is allowed to be a function of the adsorbate concentration. The diffusivity will be constant only for a thermodynamically ideal system, which is only an adequate approximation for a limited number of adsorption systems.

Results for several individual mechanisms will now be considered. The equations that follow refer to local conditions within the contacting equipment that may apply to the average concentrations in the neighborhood of a single adsorbent particle. Generally, they apply to particles that can be approximated as spherical and of a uniform size and properties. An appropriately chosen mean particle size must be used in these equations when dealing with adsorbents having a broad particle size distribution. The appropriate average depends on the controlling mass-transfer mechanism. For intraparticle mass-transfer mechanisms, the volume or mass-average particle size usually provides the best prediction.

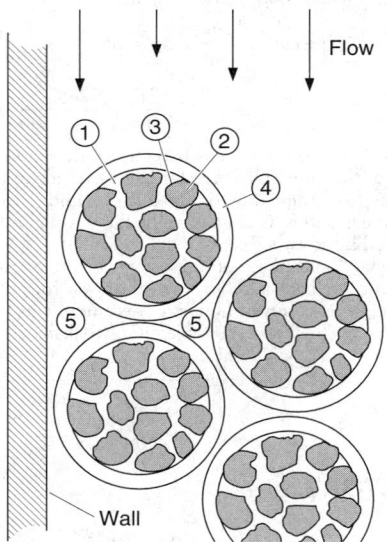

Flow

Wall

FIG. 16-9 General scheme of adsorbent particles in a packed bed showing the locations of mass transfer and dispersive mechanisms. Numerals correspond to numbered paragraphs in the text: 1, pore diffusion; 2, solid diffusion; 3, reaction kinetics at phase boundary; 4, external mass transfer; 5, fluid mixing.

Pore Diffusion When fluid transport through a network of fluid-filled pores inside the particles provides access for solute adsorption sites, the diffusion flux can be expressed in terms of a pore diffusion coefficient D_{pi} as:

$$J_i = -\varepsilon_p D_{pi} \frac{\partial c_{pi}}{\partial r} \tag{16-64}$$

D_{pi} is smaller than the diffusivity in a straight cylindrical pore as a result of the random orientation of the pores, which gives a longer diffusion path, and the variation in the pore diameter. Both effects are commonly accounted for by a tortuosity factor τ_p such that $D_{pi} = D_i/\tau_p$. In principle, predictions of the tortuosity factor can be made if the pore structure, pore size, and shape distributions are known (see Dullien, *Porous Media: Fluid Transport and Pore Structure*, Academic Press, NY, 1979). In some cases, approximate prediction can be obtained from the following equations.

Mackie and Meares, *Proc. Roy. Soc.*, **A232**, 498 (1955):

$$\tau_p = \frac{(2 - \varepsilon_p)^2}{\varepsilon_p} \tag{16-65a}$$

Wakao and Smith, *Chem. Eng. Sci.*, **17**, 825 (1962):

$$\tau_p = \frac{1}{\varepsilon_p} \tag{16-65b}$$

Suzuki and Smith, *Chem. Eng. J.*, **3**, 256 (1972):

$$\tau_p = \varepsilon_p + 1.5(1 - \varepsilon_p) \tag{16-65c}$$

In practice, however, the predictive value of these equations is rather uncertain, and vastly different results are obtained from each. All of them, on the other hand, predict that τ_p increases as the porosity decreases.

For catalyst particles, Satterfield (*Heterogeneous Catalysis in Practice*, McGraw-Hill, 1980) recommends the use of a value of $\tau_p = 4$ when no other information is available, and this can be used for many adsorbents. In general, however, it is more reliable to treat the tortuosity as an empirical constant that is determined experimentally for any particular adsorbent.

For adsorbent materials, experimental tortuosity factors generally fall in the range 2–6 and generally decrease as the particle porosity is increased. Higher apparent values may be obtained when the experimental measurements are affected by other resistances, while values much lower than 2 generally indicate that surface or solid diffusion occurs in parallel to pore diffusion.

Ruthven (gen. refs.) summarizes methods for the measurement of effective pore diffusivities that can be used to obtain tortuosity factors by comparison with the estimated pore diffusion coefficient of the adsorbate. Molecular diffusivities can be estimated with the methods in Sec. 6.

For gas-phase diffusion in small pores at low pressure, the molecular mean free path may be larger than the pore diameter, giving rise to Knudsen diffusion. Satterfield (*Mass Transfer in Heterogeneous Catalysis*, MIT, Cambridge, MA, 1970, p. 43), gives the following expression for the pore diffusivity:

$$D_{pi} = \frac{1}{\tau_p} \left[\frac{3}{4r_{pore}} \left(\frac{\pi M_{ri}}{2RT} \right)^{1/2} + \frac{1}{D_i} \right]^{-1} \tag{16-66}$$

where r_{pore} is the average pore radius, T the absolute temperature, and M_{ri} the molecular weight.

For liquid-phase diffusion of large adsorbate molecules, when the ratio $\lambda_m = r_m/r_{pore}$ of the molecule radius r_m to the pore radius is significantly greater than zero, the pore diffusivity is reduced by steric interactions with the pore wall and hydrodynamic resistance. When $\lambda_m < 0.2$, the following expressions derived by Brenner and Gaydos [*J. Coll. Int. Sci.*, **58**, 312 (1977)] for a hard sphere molecule (a particle) diffusing in a long cylindrical pore, can be used

$$D_{pi} = \frac{D_i}{\tau_p} (1 - \lambda_m)^{-2} \left[1 + \frac{9}{8} \lambda_m \ln \lambda_m - 1.539 \lambda_m \right] \tag{16-67}$$

r_m is the Stokes-Einstein radius of the solute that can be determined from the free diffusivity as

$$r_m = \frac{\kappa T}{6\pi\mu D_i} \tag{16-68}$$

where κ is the Boltzmann constant. When $\lambda_m > 0.2$, the centerline approximation [Anderson and Quinn, *Biophys. J.*, **14**, 130, (1974)] can be used instead of Eq. (16-67)

$$D_{pi} = \frac{D_i}{\tau_p} (1 - 2.1044\lambda_m + 2.089\lambda_m^3 - 0.984\lambda_m^5) \times 0.865 \tag{16-69}$$

The 0.865 factor is used to match this equation to the Brenner and Gaydos expression for $\lambda_m = 0.2$. In these cases, the pore concentration C_{pi} is related to the external concentration C_i by the partition ratio $(1 - \lambda_m)^2$.

Solid Diffusion In the case of pore diffusion discussed above, transport occurs within the fluid phase contained inside the particle; here the solute concentration is generally similar in magnitude to the external fluid concentration. A solute molecule transported by pore diffusion may attach to the sorbent and detach many times along its path. In other cases, attachment can be essentially permanent, but in both cases, only detached molecules undergo transport. In contrast, the following four instances illustrate cases where diffusion of adsorbate molecules occurs in their adsorbed state within phases that are distinct from the pore fluid:

1. Movement of mobile adsorbed solute molecules along pore surfaces, without detaching
2. Transport in a homogeneously dissolved state, as for a neutral molecule inside a sorbent gel or in a pore filled with a liquid which is immiscible with the external fluid
3. Ion transport in charged ion-exchange resins
4. Advance of an adsorbate molecule from one cage to another within a zeolite crystal

In these cases, the diffusion flux may be written in terms of the adsorbed solute concentration as

$$J_i = -\rho_p D_{si} \frac{\partial n_i}{\partial r} \tag{16-70}$$

The diffusion coefficient in these phases D_{si} is usually considerably smaller than that in fluid-filled pores; however, the adsorbate concentration is often much larger. Thus, the diffusion rate can be smaller or larger than can be expected for pore diffusion, depending on the magnitude of the fluid/solid partition coefficient.

Numerical values for solid diffusivities D_{si} in adsorbents are sparse and disperse. Moreover, they may be strongly dependent on the adsorbed phase concentration of solute. Hence, locally conducted experiments and interpretation must be used to a great extent. Summaries of available data for surface diffusivities in activated carbon and other adsorbent materials and for micropore diffusivities in zeolites are given in Ruthven, Yang, Suzuki, and Karger and Ruthven (gen. refs.).

Surface diffusivities are generally strongly dependent on the fractional surface coverage and increase rapidly at surface coverage greater than 80 percent [see for example Yang et al., *AIChE J.*, **19**, 1052 (1973)]. For estimation purposes, the correlation of Sladek et al. [*Ind. Eng. Chem. Fundam.*, **13**, 100 (1974)] can be used to predict surface diffusivities for gas-phase adsorption on a variety of adsorbents.

Zeolite crystallite diffusivities for sorbed gases range from 10^{-7} to 10^{-14} cm²/s. These diffusivities generally show a strong increase with the adsorbate concentration that is accounted for by the Darken thermodynamic correction factor

$$D_{si} = D_{0i} \frac{d \ln a_i}{d \ln n_i} \tag{16-71}$$

where D_{0i} is the corrected diffusivity, a_i the thermodynamic activity of the species in the adsorbed phase, and n_i the adsorbed phase solute concentration. Corrected diffusivities D_{0i} calculated according to this equation are often found to be essentially independent of concentration. If the adsorption equilibrium isotherm obeys the Langmuir equation [Eq. (16-13)], Eq. (16-71) yields:

$$D_{si} = D_{0i} \left(1 - \frac{n_i}{n_i^s} \right)^{-1} \tag{16-72}$$

The effect of temperature on diffusivities in zeolite crystals can be expressed in terms of the Eyring equation (see Ruthven, gen. refs.).

In ion-exchange resins, diffusion is further complicated by electrical coupling effects. In a system with M counterions, diffusion rates are described by the Nernst-Planck equations (Helfferich, gen. refs.). Assuming complete Donnan exclusion, these equations can be written as:

$$J_i = -\rho_p \frac{1}{z_i} \sum_{j=1}^{M-1} \overline{D}_{i,j} \frac{\partial z_j n_j}{\partial r} \qquad (16\text{-}73)$$

with

$$\overline{D}_{i,j} = -\frac{\overline{D}_i(\overline{D}_j - \overline{D}_M)z_i^2 n_i}{\sum_{k=1}^{M} \overline{D}_k z_k^2 n_k} \qquad (16\text{-}74a)$$

$$\overline{D}_{i,i} = \overline{D}_i - \frac{\overline{D}_i(\overline{D}_i - \overline{D}_M)z_i^2 n_i}{\sum_{k=1}^{M} \overline{D}_k z_k^2 n_k} \qquad (16\text{-}74b)$$

which are dependent on the **ionic self diffusivities** $\overline{D}_i$ of the individual species. As a qualitative rule, ionic diffusivities of inorganic species in crosslinked polystyrene-DVB ion-exchange resins compared with those in water are 1:10 for monovalent ions, 1:100 for divalent ions, and 1:1000 for trivalent ions. Table 16-8 shows typical ionic diffusivities of inorganic ions in cation and anion exchange resins; larger organic ions, however, can have ionic diffusivities much smaller than inorganic ions of the same valence [see, for example, Jones and Carta, *Ind. Eng. Chem. Research*, **32**, 117 (1993)].

For mixtures of unlike ions (the usual case), the apparent diffusivity will be intermediate between these values because of the electrical coupling effect. For a system with two counterions A and B, with charge z_A and z_B, Eqs. (16-73) and (16-74) reduce to:

$$J_A = -\rho_p \overline{D}_{A,B} \frac{\partial n_A}{\partial r} = -\rho_p \frac{\overline{D}_A \overline{D}_B [z_A^2 n_A + z_B^2 n_B]}{z_A^2 \overline{D}_A n_A + z_B^2 \overline{D}_B n_B} \frac{\partial n_A}{\partial r} \qquad (16\text{-}75)$$

which shows that the apparent diffusivity $\overline{D}_{A,B}$ varies between $\overline{D}_A$ when the ionic fraction of species A in the resin is very small and $\overline{D}_B$ when the ionic fraction of A in the resin approaches unity, indicating that the ion present in smaller concentration has the stronger effect on the local interdiffusion rate.

Combined Pore and Solid Diffusion In porous adsorbents and ion-exchange resins, intraparticle transport can occur with pore and solid diffusion in parallel. The dominant transport process is the faster one, and this depends on the relative diffusivities and concentrations in the pore fluid and in the adsorbed phase. Often, equilibrium between the pore fluid and the solid phase can be assumed to exist locally at each point within a particle. In this case, the mass-transfer flux is expressed by:

$$J_i = -\left[\varepsilon_p D_{pi} + \rho_p D_{si} \frac{dn_i^e}{dc_i} \right] \frac{\partial c_{pi}}{\partial r} = -D_{ei}(c_{pi}) \frac{\partial c_{pi}}{\partial r} \qquad (16\text{-}76)$$

where dn_i^e/dc_i is the derivative of the adsorption isotherm and it has been assumed that at equilibrium $c_{pi} = c_i$. This equation suggests that

in such an adsorbent, pore and solid diffusivities can be obtained by determining the apparent diffusivity D_{ei} for conditions of no adsorption ($dn_i^e/dc_i = 0$) and for conditions of strong adsorption, where dn_i^e/dc_i is large. If the adsorption isotherm is linear over the range of experimental measurement:

$$D_{ei} = \varepsilon_p D_{pi} + \rho_p K_i D_{si} \qquad (16\text{-}77)$$

Thus, a plot of the apparent diffusivity versus the linear adsorption equilibrium constant should be linear so long as D_{pi} and D_{si} remain constant.

In a particle having a **bidispersed pore structure** comprising spherical adsorptive subparticles of radius r_s forming a macroporous aggregate, separate flux equations can be written for the macroporous network in terms of Eq. (16-64) and for the subparticles themselves in terms of Eq. (16-70) if solid diffusion occurs.

EXTERNAL MASS TRANSFER

Because of the complexities encountered with a rigorous treatment of the hydrodynamics around particles in industrial contactors, mass transfer to and from the adsorbent is described in terms of a mass-transfer coefficient k_f. The flux at the particle surface is:

$$N_i = k_f(c_i - c_i^s) \qquad (16\text{-}78)$$

where c_i and c_i^s are the solute concentrations in the bulk fluid and at the particle surface, respectively. k_f can be estimated from available correlations in terms of the Sherwood number $Sh = k_f d_p/D_i$ and the Schmidt number $Sc = v/D_i$. For packed-bed operations, the correlations in Table 16-9 are recommended. A plot of these equations is given in Fig. 16-10 for representative ranges of Re and Sc with $\varepsilon = 0.4$.

External mass-transfer coefficients for particles suspended in agitated contactors can be estimated from equations in Levins and Glastonbury [*Trans. Instn. Chem. Eng.*, **50**, 132 (1972)] and Armenante and Kirwan [*Chem. Eng. Sci.*, **44**, 2871 (1989)].

AXIAL DISPERSION IN PACKED BEDS

The axial dispersion coefficient [cf. Eq. (16-51)] lumps together all mechanisms leading to axial mixing in packed beds. Thus, the axial dispersion coefficient must account not only for molecular diffusion and convective mixing but also for nonuniformities in the fluid velocity across the packed bed. As such, the axial dispersion coefficient is best determined experimentally for each specific contactor.

The effects of **flow nonuniformities**, in particular, can be severe in gas systems when the ratio of bed-to-particle diameters is small; in liquid systems when viscous fingering occurs as a result of large viscosity gradients in the adsorption bed; when very small particles (<50 μm) are used, such as in high performance liquid chromatography systems; and in large-diameter beds. A lower bound of the axial

TABLE 16-8 Self Diffusion Coefficients in Polystyrene-divinylbenzene Ion Exchangers (units of 10^{-7} cm²/s)*

Temperature	0.3 °C					25 °C				
Crosslinking, %	4	6	8	12	16	4	6	8	12	16
Cation exchangers (sulfonated): Dowex 50										
Na^+	6.7		3.4	1.15	0.66	14.1		9.44		2.40
Cs^+			6.6		1.11			13.7		3.10
Ag^+			2.62		1.00			6.42		2.75
Zn^{2+}			0.21		0.03			0.63		0.14
La^{3+}		0.30	0.03		0.002			0.092		0.005
Anion exchangers (dimethyl hydroxyethylamine): Dowex 2										
Cl^-		1.25						3.54		
Br^-	(1.8)	1.50	0.63		0.06	(4.3)		3.87	2.04	0.26
I^-		0.35						1.33		
BrO_3^-		1.76						4.55		
WO_4^{2-}		0.60						1.80		
PO_4^{3-}		0.16						0.57		

*Data from Boyd and Soldano, *J. Am. Chem. Soc.*, **75**, 6091 (1953).

TABLE 16-9 Recommended Correlations for External Mass Transfer Coefficients in Adsorption Beds ($Re = \varepsilon v d_p/\nu$, $Sc = \nu/D$)

Equation	Re	Phase	Ref.
$Sh = 1.15\left(\dfrac{Re}{\varepsilon}\right)^{0.5} Sc^{0.33}$	$Re > 1$	Gas/liquid	Carberry, *AIChE J.*, **6**, 460 (1960)
$Sh = 2.0 + 1.1\,Re^{0.6} Sc^{0.33}$	$3 < Re < 10^4$	Gas/liquid	Wakao and Funazkri, *Chem. Eng. Sci.*, **33**, 1375 (1978)
$Sh = 1.85\left(\dfrac{1-\varepsilon}{\varepsilon}\right)^{0.33} Re^{0.33} Sc^{0.33}$	$Re < 40$	Liquid	Kataoka et al., *J. Chem. Eng. Japan*, **5**, 132 (1972)
$Sh = \dfrac{1.09}{\varepsilon}\,Re^{0.33} Sc^{0.33}$	$0.0015 < Re < 55$	Liquid	Wilson and Geankoplis, *Ind. Eng. Chem. Fundam.*, **5**, 9 (1966)
$Sh = \dfrac{0.25}{\varepsilon}\,Re^{0.69} Sc^{0.33}$	$55 < Re < 1050$	Liquid	Wilson and Geankoplis, *Ind. Eng. Chem. Fundam.*, **5**, 9 (1966)

dispersion coefficient can be estimated for well-packed beds from correlations that follow.

Neglecting flow nonuniformities, the contributions of molecular diffusion and turbulent mixing arising from stream splitting and recombination around the sorbent particles can be considered additive [Langer et al., *Int. J. Heat and Mass Transfer*, **21**, 751 (1978)]; thus, the axial dispersion coefficient D_L is given by:

$$\frac{D_L}{D_i} = \gamma_1 + \gamma_2 \frac{d_p v}{D_i} = \gamma_1 + \gamma_2 \frac{(Re)(Sc)}{\varepsilon} \qquad (16\text{-}79)$$

or, in terms of a particle-based Peclet number ($Pe = d_p v/D_L$), by:

$$\frac{1}{Pe} = \frac{\gamma_1 \varepsilon}{(Re)(Sc)} + \gamma_2 \qquad (16\text{-}80)$$

The first term in Eqs. (16-79) and (16-80) accounts for molecular diffusion, and the second term accounts for mixing. For the first term, Wicke [*Ber. Bunsenges*, **77**, 160 (1973)] has suggested:

$$\gamma_1 = 0.45 + 0.55\varepsilon \qquad (16\text{-}81)$$

which, for typical void fractions, $\varepsilon = 0.35 - 0.5$ gives $\gamma_1 = 0.64 - 0.73$ (Ruthven, gen. refs.). Expressions for the axial mixing term, γ_2 in Eq. (16-79) are given in Table 16-10. The expression of Wakao and Funazkri includes an axial diffusion term, γ_1, that varies from 0.7 for nonporous particles to $20/\varepsilon$, depending on the intraparticle mass-transfer mechanism. For strongly adsorbed species, Wakao and Funazkri suggest that the effective axial dispersion coefficient is much larger than that predicted on the basis of nonporous, nonadsorbing particles. The Gunn expression includes a term σ_v^2 accounting for deviations from plug flow. σ_v^2 is defined as the dimensionless variance of the distribution of the ratio of velocity to average velocity over the cross section of the bed. The parameter values included in this equation are valid for spherical particles. Values for nonspherical particles can be found in the original reference.

Figure 16-11 compares predicted values of D_L/D_i for $\sigma_v = 0$ and $\varepsilon = 0.4$ with $Sc = 1$ (gases at low pressure), and $Sc = 1000$ (liquids), based on the equations in Table 16-10.

Correlations for axial dispersion in beds packed with very small particles ($<50\ \mu$m) that take into account the holdup of liquid in a bed are discussed by Horvath and Lin [*J. Chromatogr.*, **126**, 401 (1976)].

RATE EQUATIONS

Rate equations are used to describe interphase mass transfer in batch systems, packed beds, and other contacting devices for sorptive processes and are formulated in terms of fundamental transport properties of adsorbent and adsorbate.

General Component Balance For a spherical adsorbent particle:

$$\varepsilon_p \frac{\partial c_{pi}}{\partial t} + \rho_p \frac{\partial n_i}{\partial t} = \frac{1}{r^2} \frac{\partial}{\partial r}(-r^2 N_i) \qquad (16\text{-}82)$$

For particles that have no macropores, such as gel-type ion-exchange resins, or when the solute holdup in the pore fluid is small, ε_p may be

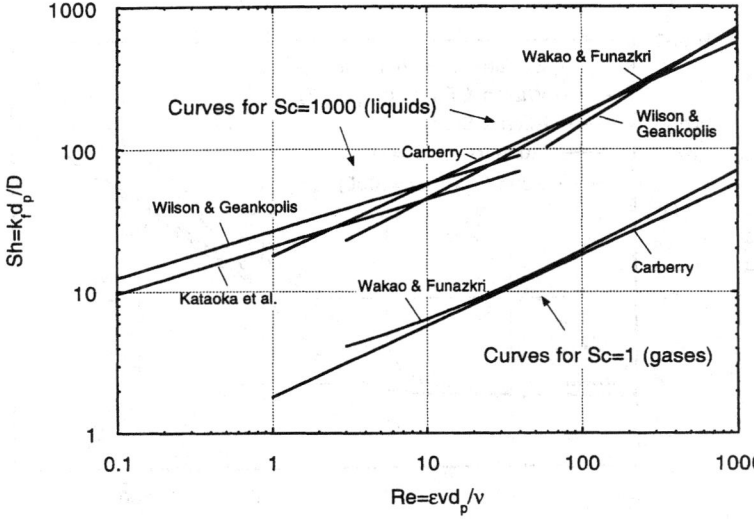

FIG. 16-10 Sherwood number correlations for external mass-transfer coefficients in packed beds for $\varepsilon = 0.4$ (adapted from Suzuki, gen. refs.).

TABLE 16-10 Coefficients for Axial Dispersion Correlations in Packed Beds Based on Eq. (16-79)

γ_1	γ_2	Ref.
0.73	$0.5\left(1 + \dfrac{13\gamma_1\varepsilon}{ReSc}\right)^{-1}$	Edwards and Richardson, *Chem. Eng. Sci.*, **23**, 109 (1968)
Nonporous particles: 0.7 Porous particles: $\leq 20/\varepsilon$	0.5	Wakao and Funazkri, *Chem. Eng. Sci.*, **33**, 1375 (1978)
1	$\dfrac{3}{4}\varepsilon + \dfrac{\pi^2\varepsilon(1-\varepsilon)}{6}\ln(ReSc)$	Koch and Brady, *J. Fluid Mech.*, **154**, 399 (1985)
0.714	$\dfrac{\sigma_v^2}{2} + (1+\sigma_v^2)\left\{\gamma(1-p)^2 + \gamma^2\,p(1-p)^3\left[e^{-\frac{1}{\gamma p(1-p)}} - 1\right]\right\}$ with $\gamma = 0.043\,ReSc/(1-\varepsilon)$ $p = 0.33\exp(-24/Re) + 0.17$	Gunn, *Chem. Eng. Sci.*, **2**, 363 (1987)

taken as zero. Ignoring bulk flow terms, the fluxes N_i and J_i are equal. In this case, coupling the component balance with the flux expressions previously introduced gives the rate equations in Table 16-11. Typical boundary conditions are also included in this table.

Linear Driving Force Approximation   Simplified expressions can also be used for an approximate description of adsorption in terms of rate coefficients for both extraparticle and intraparticle mass transfer controlling. As an approximation, the rate of adsorption on a particle can be written as:

$$\frac{\partial \hat{n}_i}{\partial t} = kf(n_i, c_i) \qquad (16\text{-}83)$$

where k is a rate coefficient, and the function $f(n_i, c_i)$ is a driving force relationship. The variables k_c and k_n are used to denote rate coefficients based on fluid-phase and adsorbed-phase concentration driving forces, respectively.

Commonly used forms of this rate equation are given in Table 16-12. For adsorption bed calculations with constant separation factor systems, somewhat improved predictions are obtained using correction factors ψ_s and ψ_p defined in Table 16-12 is the partition ratio defined by Eq. (16-25).

The linear driving force (LDF) approximation is obtained when the driving force is expressed as a concentration difference. It was originally developed to describe packed-bed dynamics under linear equilibrium conditions [Glueckauf, *Trans. Far. Soc.*, **51**, 1540 (1955)]. This form is exact for a nonlinear isotherm only when external mass transfer is controlling. However, it can also be used for nonlinear systems with pore or solid diffusion mechanisms as an approximation, since it provides qualitatively correct results.

Alternate driving force approximations, item 2B in Table 16-12, for solid diffusion, and item 3B in Table 16-12, for pore diffusion, provide somewhat more accurate results in constant pattern packed-bed calculations with pore or solid diffusion controlling for constant separation factor systems.

The reaction kinetics approximation is mechanistically correct for systems where the reaction step at pore surfaces or other fluid-solid interfaces is controlling. This may occur in the case of chemisorption on porous catalysts and in affinity adsorbents that involve very slow binding steps. In these cases, the mass-transfer parameter k is replaced by a second-order reaction rate constant k_a. The driving force is written for a constant separation factor isotherm (column 4 in Table 16-12). When diffusion steps control the process, it is still possible to describe the system by its apparent second-order kinetic behavior, since it usually provides a good approximation to a more complex exact form for single transition systems (see "Fixed Bed Transitions").

Combined Intraparticle Resistances   When solid diffusion and pore diffusion operate in parallel, the effective rate is the sum of these two rates. When solid diffusion predominates, mass transfer can be represented approximately in terms of the LDF approximation, replacing k_n in column 2 of Table 16-12 with

$$k_n^c = \frac{15\psi_s D_{si}}{r_p^2} + \frac{15(1-\varepsilon)\psi_p\varepsilon_p D_{pi}}{\Lambda r_p^2} \qquad (16\text{-}84)$$

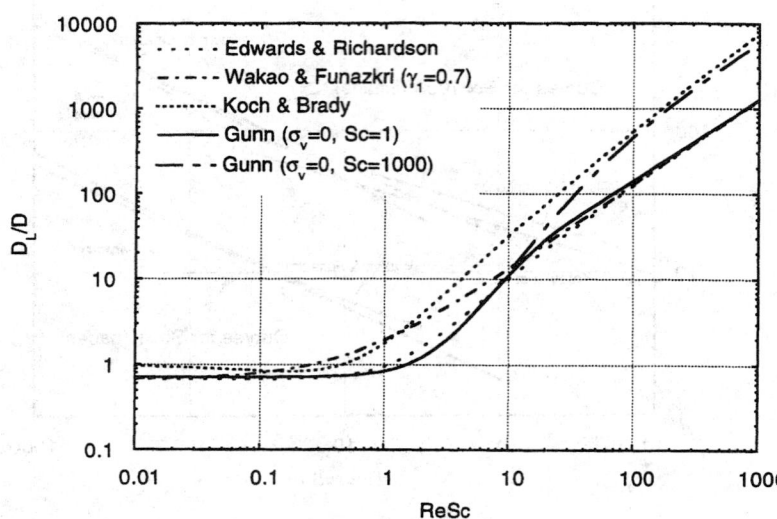

FIG. 16-11   Axial dispersion coefficient correlations for well-packed beds for $\varepsilon = 0.4$.

TABLE 16-11   Rate Equations for Description of Mass Transfer in Spherical Adsorbent Particles

Mechanism	Flux equation	Rate equation
A. Pore diffusion	16-64	$\varepsilon_p \dfrac{\partial c_{pi}}{\partial t} + \rho_p \dfrac{\partial n_i}{\partial t} = \dfrac{1}{r^2}\dfrac{\partial}{\partial r}\left(\varepsilon_p D_{pi} r^2 \dfrac{\partial c_{pi}}{\partial r}\right)$ $(\partial c_{pi}/\partial r)_{r=0} = 0,\ (\varepsilon_p D_{pi}\partial c_{pi}/\partial r)_{r=r_p} = k_f(c_i - c_{pi|_{r=r_p}})$ or $(c_{pi})_{r=r_p} = c_i$ for no external resistance
B. Solid diffusion	16-70	$\dfrac{\partial n_i}{\partial t} = \dfrac{1}{r^2}\dfrac{\partial}{\partial r}\left(D_{si} r^2 \dfrac{\partial n_i}{\partial r}\right)$ $(\partial n_i/\partial r)_{r=0} = 0,\ (\rho_p D_{si}\partial n_i/\partial r)_{r=r_p} = k_f(c_i - c_i^s)$ or $(n_i)_{r=r_p} = n_i^e(c_i)$ for no external resistance
C. Parallel pore and solid diffusion (local equilibrium between pore and adsorbed phase)	16-76	$\left(\varepsilon_p + \rho_p \dfrac{dn_i^e}{dc_i}\right)\dfrac{\partial c_{pi}}{\partial t} = \dfrac{1}{r^2}\dfrac{\partial}{\partial r}\left[r^2\left(\varepsilon_p D_{pi} + \rho_p D_{si}\dfrac{dn_i^e}{dc_i}\right)\dfrac{\partial c_{pi}}{\partial r}\right]$ $(\partial c_{pi}/\partial r)_{r=0} = 0,\ [(\varepsilon_p D_{pi} + \rho_p D_{si} dn_i^e/dc_i)\partial c_{pi}/\partial r]_{r=r_p} = k_f(c_i - c_{pi|_{r=r_p}})$ or $(c_{pi})_{r=r_p} = c_i$ for no external resistance
D. Diffusion in bidispersed particles (no external resistance)	16-64 and 16-70	$\dfrac{\partial n_i}{\partial t} = \dfrac{1}{\rho^2}\dfrac{\partial}{\partial \rho}\left(D_{si}\rho^2 \dfrac{\partial n_i}{\partial \rho}\right),\ (\partial n_i/\partial \rho)_{\rho=0} = 0,\ (n_i)_{\rho=r_s} = n_i^e(c_{pi})$ $\bar{n}_i(r,t) = \dfrac{3}{r_s^3}\displaystyle\int_0^{r_s} \rho^2 n_i\, d\rho$ $\varepsilon_p \dfrac{\partial c_{pi}}{\partial t} + \rho_p \dfrac{\partial \bar{n}_i}{\partial t} = \dfrac{1}{r^2}\dfrac{\partial}{\partial r}\left(\varepsilon_p D_{pi} r^2 \dfrac{\partial c_{pi}}{\partial r}\right),\ (\partial c_{pi}/\partial r)_{r=0} = 0,\ (c_{pi})_{r=r_p} = c_i$

TABLE 16-12   Expressions for Rate Coefficient k and Driving Force Relationships for Eq. 16-83

Mechanism	1. External film	2. Solid diffusion	3. Pore diffusion	4. Reaction kinetics
Expression for rate coefficient, k	$k_c = \dfrac{k_f a}{\rho_b} = \dfrac{3(1-\varepsilon)k_f}{\rho_b r_p}$	$k_n = \dfrac{15\psi_s D_{si}}{r_p^2}$	$k_n = \dfrac{15\psi_p(1-\varepsilon)\varepsilon_p D_{pi}}{\Lambda r_p^2}$	k_a
A. Linear driving force (LDF)	$c - c_i^e$	$n_i^e - \bar{n}_i$	$n_i^e - \bar{n}_i$	—
LDF for constant R	$c_i - \dfrac{Rc^{\mathrm{ref}}\bar{n}_i/n^{\mathrm{ref}}}{1-(R-1)\bar{n}_i/n^{\mathrm{ref}}}$	$\dfrac{n^{\mathrm{ref}}c_i/c^{\mathrm{ref}}}{R+(R-1)c_i/c^{\mathrm{ref}}} - \bar{n}_i$	$\dfrac{n^{\mathrm{ref}}c_i/c^{\mathrm{ref}}}{R+(R-1)c_i/c^{\mathrm{ref}}} - \bar{n}_i$	—
Correction factors ψ for constant R	—	$\dfrac{0.894}{1-0.106R^{0.5}}$	$\dfrac{0.775}{1-0.225R^{0.5}}$	—
B. Alternate driving force for constant R	—	$\dfrac{n_i^{e2}-\bar{n}_i^2}{2\bar{n}_i}$	$\dfrac{n_i^e - \bar{n}_i}{[1-(R-1)\bar{n}_i/n^{\mathrm{ref}}]^{0.5}}$	—
Correction factors ψ for alternate driving force	—	$\dfrac{0.590}{1-0.410R^{0.5}}$	$\dfrac{0.548}{1-0.452R^{0.5}}$	—
C. Reaction kinetics for constant R	—	—	—	$\dfrac{c_i(n^{\mathrm{ref}}-\bar{n}_i)-R\bar{n}_i(c^{\mathrm{ref}}-c_i)}{1-R}$

References: 1A. Beaton and Furnas, *Ind. Eng. Chem.*, **33**, 1500 (1941); Michaels, *Ind. Eng. Chem.*, **44**, 1922 (1952)
2A,3A. Glueckauf and Coates, *J. Chem. Soc.*, 1315 (1947); *Trans. Faraday Soc.*, **51**, 1540 (1955); Hall et al., *Ind. Eng. Chem. Fundam.*, **5**, 212 (1966)
2B. Vermeulen, *Ind. Eng. Chem.*, **45**, 1664 (1953)
3B. Vermeulen and Quilici, *Ind. Eng. Chem. Fundam.*, **9**, 179 (1970)
4C. Hiester and Vermeulen, *Chem. Eng. Progr.*, **48**, 505 (1952)

When pore diffusion predominates, use of column 3 in Table 16-12 is preferable, with k_n^c replacing k_n.

For particles with a **bidispersed pore structure,** the mass-transfer parameter k_n in the LDF approximation (column 2 in Table 16-12) can be approximated by the series-combination of resistances as:

$$\frac{1}{k_n^c} = \frac{1}{b_s}\left[\frac{\Lambda r_p^2}{15(1-\varepsilon)\psi_p \varepsilon_p D_{pi}} + \frac{r_s^2}{15\psi_s D_{si}}\right] \qquad (16\text{-}85)$$

where b_s is a correction to the driving force that is described below. In the limiting cases where the controlling resistance is diffusion through the particle pores or diffusion within the subparticles, the rate coefficients $k_n = 15(1-\varepsilon)\psi_p \varepsilon_p D_{pi}/\Lambda r_p^2$ and $k_n = 15\psi_s D_{si}/r_s^2$ are obtained.

Overall Resistance   With a linear isotherm ($R = 1$), the overall mass transfer resistance is the sum of intraparticle and extraparticle resistances. Thus, the overall LDF coefficient for use with a particle-side driving force (column 2 in Table 16-12) is:

$$\frac{1}{k_n^o} = \frac{\Lambda r_p}{3(1-\varepsilon)k_f} + \frac{1}{k_n^c} \qquad (16\text{-}86)$$

or

$$\frac{1}{k_c^o} = \frac{\rho_b r_p}{3(1-\varepsilon)k_f} + \frac{\rho_b}{\Lambda k_n^c} \qquad (16\text{-}87)$$

for use with a fluid-phase driving force (column 1 in Table 16-12).

In either equation, k_n^c is given by Eq. (16-84) for parallel pore and surface diffusion or by Eq. (16-85) for a bidispersed particle. For nearly linear isotherms $(0.7 < R < 1.5)$, the same linear addition of resistance can be used as a good approximation to predict the adsorption behavior of packed beds, since solutions for all mechanisms are nearly identical. With a highly favorable isotherm $(R \to 0)$, however, the rate at each point is controlled by the resistance that is locally greater, and the principle of additivity of resistances breaks down. For approximate calculations with intermediate values of R, an overall transport parameter for use with the LDF approximation can be calculated from the following relationship for solid diffusion and film resistance in series

$$\frac{\Lambda r_p}{3(1-\varepsilon)k_f} + \frac{r_p^2}{15\psi_s D_{si}} = \frac{b_s}{k_n^o}$$

$$= \frac{b_f \Lambda}{\rho_b k_c^o} \qquad (16\text{-}88)$$

b_s and b_f are correction factors that are given by Fig. 16-12 as a function of the separation factor R and the mechanism parameter

$$\Psi = \frac{10\psi_s D_{si}\Lambda}{D_i} \frac{1}{Sh} \qquad (16\text{-}89)$$

Axial Dispersion Effects In adsorption bed calculations, axial dispersion effects are typically accounted for by the axial diffusionlike term in the bed conservation equations [Eqs. (16-51) and (16-52)]. For nearly linear isotherms $(0.5 < R < 1.5)$, the combined effects of axial dispersion and mass-transfer resistances on the adsorption behavior of packed beds can be expressed approximately in terms of an apparent rate coefficient k_c for use with a fluid-phase driving force (column 1, Table 16-12):

$$\frac{1}{k_c} = \frac{\rho_b}{\varepsilon} \frac{D_L}{v^2} + \frac{1}{k_c^o} \qquad (16\text{-}90)$$

which extends the linear addition principle to combined axial dispersion and mass-transfer resistances. Even for a highly nonlinear isotherm $(R = 0.33)$, the linear addition principle expressed by this equation provides a useful approximation except in the extreme case of low mass-transfer resistance and large axial dispersion, when $D_L \rho_b k_c^o / v^2 \varepsilon \gg 5$ [Garg and Ruthven, *Chem. Eng. Sci.*, **30**, 1192 (1975)]. However, when the isotherm is irreversible $(R \to 0)$, the linear addition principle breaks down and axial dispersion has to be taken into account by explicit models (see "Fixed Bed Transitions").

Rapid Adsorption-Desorption Cycles For rapid cycles with particle diffusion controlling, when the cycle time t_c is much smaller than the time constant for intraparticle transport, the LDF approximation becomes inaccurate. The generalized expression

$$\frac{\partial \hat{n}_i}{\partial t} = \Omega k_n (n_i^e - \overline{n}_i) \qquad (16\text{-}91)$$

can be used for packed-bed calculations when the parameter Ω is defined to be a function of the cycle time such that the amount of solute adsorbed and desorbed during a cycle is equal to that obtained by solution of the complete particle diffusion equations. Graphical and analytical expressions for Ω in the case of a single particle, usable for very short beds, are given by Nakao and Suzuki [*J. Chem. Eng. Japan*, **16**, 114 (1983)] and Carta [*Chem. Eng. Sci.*, **48**, 622 (1993)]. With equal adsorption and desorption times, $t_a = t_d = t_c/2$, Ω approaches the value $\pi^2/15$ for long cycle times and the asymptote $\Omega = 1.877/\sqrt{t_c k_n}$ for short cycle times [Alpay and Scott, *Chem. Eng. Sci.*, **47**, 499 (1992)]. However, other results by Raghavan et al. [*Chem. Eng. Sci.*, **41**, 2787 (1986)] indicate that a limiting constant value of Ω (larger than 1) is obtained for very short cycles, when calculations are carried out for beds of finite length.

Determination of Controlling Rate Factor The most important physical variables determining the controlling dispersion factor are particle size and structure, flow rate, fluid- and solid-phase diffusivities, partition ratio, and fluid viscosity. When multiple resistances and axial dispersion can potentially affect the rate, the spreading of a concentration wave in a fixed bed can be represented approximately

in terms of the single rate parameter k. In customary separation-process calculations, the height of an adsorption bed can be calculated approximately as the product of the number of transfer units times the height of one fluid-phase transfer unit (HTU). The HTU is related to the LDF rate parameters k_c and k_n by:

$$\text{HTU} = \frac{\varepsilon v}{\rho_b k_c} = \frac{\varepsilon v}{\Lambda k_n} \qquad (16\text{-}92)$$

Figure 16-13 is a plot of the dimensionless HTU (htu = HTU/d_p) multiplied times the correction factor b_f (between 1 and 2) as a function of the dimensionless velocity $(Re)(Sc) = \varepsilon v d_p/D$ and a ratio of the controlling diffusivity to the fluid-phase diffusivity, generated on the basis of results of Vermeulen et al. (gen. ref.) using typical values of the individual physical factors likely to be found in adsorption beds. This figure can be used to determine the controlling rate factor from a knowledge of individual physical parameters. If fluid-side effects control, the dimensionless HTU is given by the bottom curve (dotted for gas and solid for liquid-phase systems). If particle-side diffusivities control, the dimensionless HTU is given by a point above the lower envelope on the appropriate diffusional contour (through the ψs, the contour value depends slightly on the separation factor R). If pore and solid diffusion occur in parallel, the reciprocal of the HTU is the sum of the reciprocals of the HTU values for the two mechanisms. Near the intersections of the diffusional contours with the envelope, the dimensionless HTU is the sum of the HTU values for fluid-side and particle-side resistances.

Example 7: Estimation of Rate Coefficient An adsorption bed is used to remove methane from a methane-hydrogen mixture at 10 atm (abs.) (10.1 bar) and 25°C (298 K), containing 10 mol % methane. Activated carbon particles having a mean diameter $d_p = 0.17$ cm, a surface area $A = 1.1 \times 10^7$ cm²/s, a bulk density $\rho_b = 0.509$ g/cm³, a particle density $\rho_p = 0.777$ g/cm³, and a skeletal density $\rho_s = 2.178$ g/cm³ is used as the adsorbent. Based on data of Grant et al. [*AIChE J.*, **8**, 403 (1962)], adsorption equilibrium is represented by $n = 2.0 \times 10^{-3} K_A p_A / (1 + K_A p_A)$ mol/g adsorbent, with $K_A = 0.346$ atm⁻¹. Estimate the rate coefficient and determine the controlling rate factor for a superficial velocity of 30 cm/s.

1. The intraparticle void fraction is $\varepsilon_p = (0.777^{-1} - 2.178^{-1})/(0.777^{-1}) = 0.643$ and the extraparticle void fraction is $\varepsilon = (0.509^{-1} - 0.777^{-1})/(0.509^{-1}) = 0.345$. The pore radius is estimated from $r_p = 2\varepsilon_p/(A\rho_p) = 1.5 \times 10^{-7}$ cm.

2. The fluid phase diffusivity is $D = 0.0742$ cm²/s. The pore diffusivity is estimated from Eq. (16-66) with a tortuosity factor $\tau_p = 4$; $D_p = 1.45 \times 10^{-3}$ cm²/s.

3. The fluid-side mass transfer coefficient is estimated from Fig. 16-10. For these conditions, $v = 0.108$ cm²/s, $Re = 30 \times 0.17/0.108 = 47$, and $Sc = 0.108/0.0742 = 1.5$. From Fig. 16-10 or equations in Table 16-9, $Sh \sim 13$.

4. The isotherm parameters based on the feed concentration are $R = 1/(1 + K_A p_A) = 0.4$ and $\Lambda = 0.509 \times 7.68 \times 10^{-4}/4.09 \times 10^{-5} = 9.56$. For pore diffusion, $\psi_p = 0.961$ from item 3A in Table 16-12. Thus, $(1 - \varepsilon)\psi_p \varepsilon_p D_p/D = (1 - 0.345) \times 0.961 \times 0.643 \times 1.45 \times 10^{-3}/0.0742 = 7.9 \times 10^{-3}$. From Fig. 16-13 at $ReSc = 69$, b htu ~ 150. b is found from Fig. 16-12. However, since the mechanism parameter Ψ is very small, $b \sim 1$. Thus, $k_n = \varepsilon v/(\text{htu } d_p\Lambda) = 0.12$ s⁻¹. This value applies to the driving force $n_i^e - \overline{n}_i$. Since pore diffusion is dominant, this value is very close to the value $k_n = 15(1 - \varepsilon)\psi_p \varepsilon_p D_p/(r_p^2\Lambda) = 0.13$ s⁻¹ obtained directly from Table 16-12. It should be noted that surface diffusion is neglected in this estimation. Its occurrence could significantly affect the overall mass transfer rate (see Suzuki, gen. refs., pp. 70–85).

Example 8: Estimation of Rate Coefficient Estimate the rate coefficient for flow of a 0.01-M water solution of NaCl through a bed of cation exchange particles in hydrogen form with $\varepsilon = 0.4$. The superficial velocity is 0.2 cm/s and the temperature is 25°C. The particles are 600 μm in diameter, and the diffusion coefficient of sodium ion is 1.2×10^{-5} cm²/s in solution and 9.4×10^{-7} cm²/s inside the particles (cf. Table 16-8). The bulk density is 0.7 g dry resin/cm³ of bed, and the capacity of the resin is 4.9 mequiv/g dry resin. The mass action equilibrium constant is 1.5.

1. Estimate the fluid-side mass transfer coefficient; $Re = \varepsilon v d_p/v = 0.2 \times 0.06/0.00913 = 1.3$, $Sc = v/D = 0.00913/1.2 \times 10^{-5} = 761$. From Fig. 16-10 or Table 16-9, $Sh \sim 23$, $k_f = D$ $Sh/d_p = 4.5 \times 10^{-3}$ cm/s.

2. From the equilibrium constant, $R = 1/K_{Na,H} = 0.67$. Thus, from Table 16-12, item 2A, $\psi_s = 0.979$. Using $n^{ref} = 4.9$ mequiv/g and $c^{ref} = 0.01$ mmole/cm³, $\Lambda = \rho_b n^{ref}/c^{ref} = 343$. Thus, $\psi_s D_s\Lambda/D = 26$ and the external mass-transfer resistance is controlling (cf. Fig. 16-13).

3. The rate coefficient for use with a fluid-phase driving force is $k_c = 3(1 - \varepsilon)k_f/(\rho_b r_p) = 0.39$ cm³/(g·s).

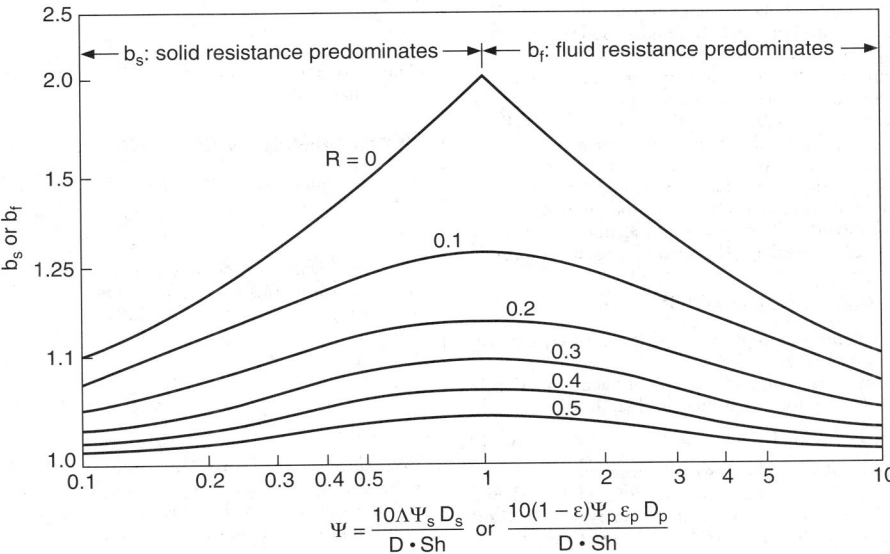

FIG. 16-12   Correction factors for addition of mass-transfer resistances, relative to effective overall solid phase or fluid phase rates, as a function of the mechanism parameter. Each curve corresponds to both b_s and b_f over its entire range.

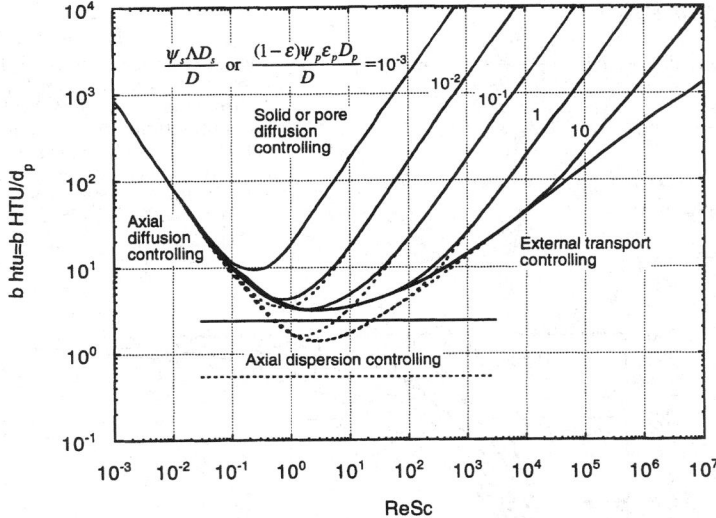

FIG. 16-13   Effect of *ReSc* group, distribution ratio, and diffusivity ratio on height of a transfer unit. Dotted lines for gas and solid lines for liquid-phase systems.

BATCH ADSORPTION

In this section, we consider the transient adsorption of a solute from a dilute solution in a constant-volume, well-mixed batch system or, equivalently, adsorption of a pure gas. The solutions provided can approximate the response of a stirred vessel containing suspended adsorbent particles, or that of a very short adsorption bed. Uniform, spherical particles of radius r_p are assumed. These particles, initially of uniform adsorbate concentration, are assumed to be exposed to a step change in concentration of the external fluid.

In general, solutions are obtained by coupling the basic conservation equation for the batch system, Eq. (16-49) with the appropriate rate equation. Rate equations are summarized in Table 16-11 and 16-12 for different controlling mechanisms.

Solutions are provided for external mass-transfer control, intraparticle diffusion control, and mixed resistances for the case of constant V_f and $F_{v,\,in} = F_{v,\,out} = 0$. The results are in terms of the fractional approach to equilibrium $F = (\hat{n}_i - \hat{n}_i^{0'})/(\hat{n}_i^{\infty} - \hat{n}_i^{0'})$, where $\hat{n}_i^{0'}$ and $\hat{n}_i^{\infty}$ are the initial and ultimate solute concentrations in the adsorbent. The solution concentration is related to the amount adsorbed by the material balance $c_i = c_i^0 - (\hat{n}_i - \hat{n}_i^0)M_s/V_f$.

Two general cases are considered: (1) adsorption under conditions of constant or nearly constant external solution concentration (equivalent to infinite fluid volume); and (2) adsorption in a batch with finite volume. In the latter case, the fluid concentration varies from c_i^0 to c_i^∞ when equilibrium is eventually attained. $\Lambda^\infty = (c_i^0 - c_i^\infty)/c_i^0 = M_s(\hat{n}_i^\infty - \hat{n}_i^0)/(V_f c_i^0)$ is a partition ratio that represents the fraction of adsorbate that is ultimately adsorbed. It determines which general case should be considered in the analysis of experimental systems. Generally, when $\Lambda^\infty \geq 0.1$, solutions for the second case are required.

EXTERNAL MASS-TRANSFER CONTROL

The intraparticle concentration is uniform, and the rate equation is given by column 1 in Table 16-12.

For a **Langmuir isotherm** with negligible solute accumulation in the particle pores, the solution for an infinite fluid volume:

$$(1-R)(1-n_i^{0'}/n_i^0)F - R\ln(1-F) = (3k_f t/r_p)(c_i^0/\rho_p n_i^0) \quad (16\text{-}93)$$

where $n_i^0 = n_i^\infty = n_i^s K_i c_i^0/(1 + K_i c_i^0)$ is the adsorbate concentration in the particle at equilibrium with the fluid concentration. The predicted behavior is shown in Fig. 16-14 for $n_i^{0'} = 0$. In the **irreversible limit** ($R = 0$), F increases linearly with time; and in the **linear limit** ($R = 1$), $1 - F$ decreases exponentially with time.

For a finite fluid volume ($\Lambda^\infty > 0$), the fractional approach to equilibrium is given by:

$$\left[1 - \frac{b'(1-R^\infty)}{2c'}\right]\frac{1}{q'}\ln\frac{(2c'F - b' - q')(-b' + q')}{(2c'F - b' + q')(-b' - q')}$$

$$- \frac{1-R^\infty}{2c'}\ln(1 - b'F + c'F^2) = \frac{3k_f t}{r_p}\frac{c_i^0}{\rho_p n_i^\infty} \quad (16\text{-}94)$$

where

$$b' = \frac{1-R^\infty}{1-R^0} + \Lambda^\infty \quad (16\text{-}95a)$$

$$c' = \Lambda^\infty(1 - R^\infty) \quad (16\text{-}95b)$$

$$q' = (b'^2 - 4c')^{0.5} \quad (16\text{-}95c)$$

$$R^0 = \frac{1}{1 + K_i c_i^0} \quad (16\text{-}95d)$$

$$R^\infty = \frac{1}{1 + Kc_i^\infty} \quad (16\text{-}95e)$$

The predicted behavior is shown in Fig. 16-15 for $R^0 = 0.5$ with different values of Λ^∞.

SOLID DIFFUSION CONTROL

For a constant diffusivity and an infinite fluid volume the solution is:

$$F = 1 - \frac{6}{\pi^2}\sum_{n=1}^{\infty}\frac{1}{n^2}\exp\left(-\frac{n^2\pi^2 D_{si}t}{r_p^2}\right) \quad (16\text{-}96)$$

For short times, this equation does not converge rapidly. The following approximations can be used instead (Helfferich and Hwang, in Dorfner, gen. refs., pp. 1277–1309):

$$F = \frac{6}{r_p}\left(\frac{D_{si}t}{\pi}\right)^{0.5}, \qquad F < 0.2 \quad (16\text{-}97)$$

$$F = \frac{6}{r_p}\left(\frac{D_s t}{\pi}\right)^{0.5} - \frac{3D_{si}t}{r_p^2}, \qquad F < 0.8 \quad (16\text{-}98)$$

For values of $F > 0.8$, the first term ($n = 1$) in Eq. (16-96) is generally sufficient. If the controlling resistance is diffusion in the subparticles of a bidispersed adsorbent, Eq. (16-96) applies with r_s, replacing r_p.

For a finite fluid volume the solution is:

$$F = 1 - 6\sum_{n=1}^{\infty}\frac{\exp(-p_n^2 D_{si}t/r_p^2)}{9\Lambda^\infty/(1-\Lambda^\infty) + (1-\Lambda^\infty)p_n^2} \quad (16\text{-}99)$$

where the p_ns are the positive roots of

$$\frac{\tan p_n}{p_n} = \frac{3}{3 + (1/\Lambda^\infty - 1)p_n^2} \quad (16\text{-}100)$$

The predicted behavior is shown in Fig. 16-16. F is calculated from Eq. (16-96) for $\Lambda^\infty = 0$ and from Eq. (16-99) for $\Lambda^\infty > 0$. Significant deviations from the $\Lambda^\infty = 0$ curve exist for $\Lambda^\infty > 0.1$.

For nonconstant diffusivity, a numerical solution of the conservation equations is generally required. In molecular sieve zeolites, when equilibrium is described by the Langmuir isotherm, the concentration dependence of the intracrystalline diffusivity can often be approximated by Eq. (16-72). The relevant rate equation is:

$$\frac{\partial n_i}{\partial t} = \frac{D_{0i}}{r^2}\frac{\partial}{\partial r}\left(\frac{r^2}{1 - n_i/n_i^s}\frac{\partial n_i}{\partial r}\right) \quad (16\text{-}101)$$

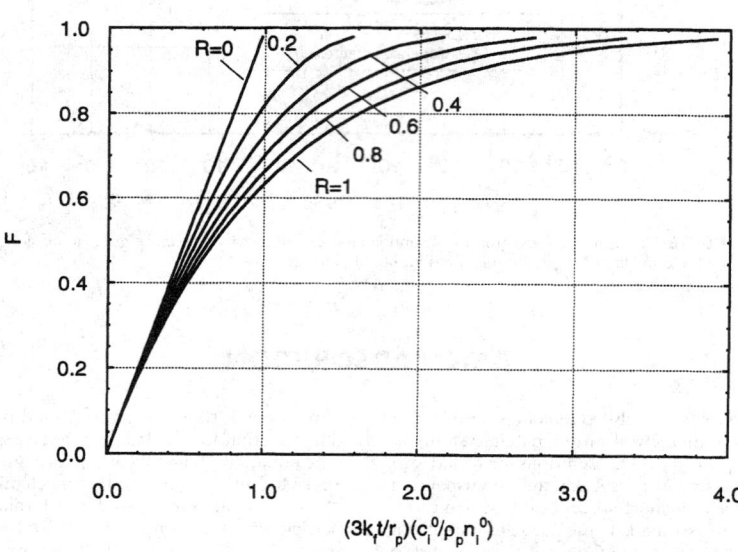

FIG. 16-14 Constant separation factor batch adsorption curves for external mass-transfer control with an infinite fluid volume and $n_i^0 = 0$.

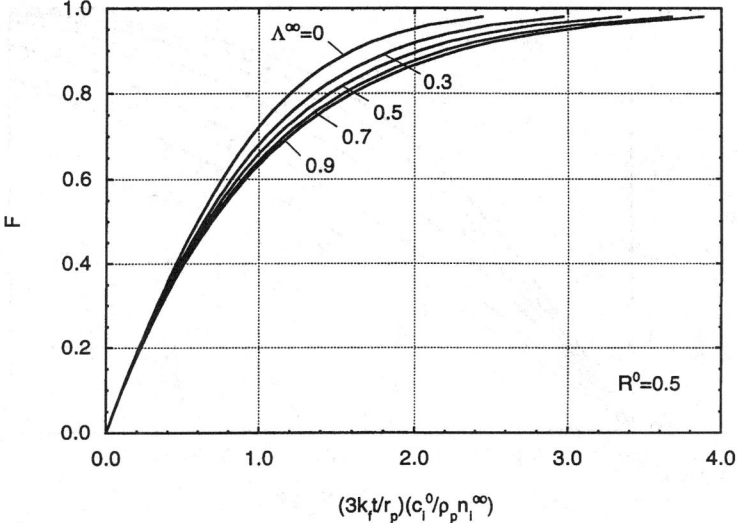

FIG. 16-15   Constant separation factor batch adsorption curves for external mass-transfer control with a finite fluid volume, $n_i^{0'} = 0$ and $R^0 = 0.5$.

A numerical solution of this equation for a constant surface concentration (infinite fluid volume) is given by Garg and Ruthven [*Chem. Eng. Sci.*, **27**, 417 (1972)]. The solution depends on the value of $\lambda = (n_i^0 - n_i^{0'})/(n_i^s - n_i^{0'})$. Because of the effect of adsorbate concentration on the effective diffusivity, for large concentration steps adsorption is faster than desorption, while for small concentration steps, when D_s can be taken to be essentially constant, adsorption and desorption curves are mirror images of each other as predicted by Eq. (16-96); see Ruthven, gen. refs., p. 175.

In binary ion-exchange, intraparticle mass transfer is described by Eq. (16-75) and is dependent on the ionic self diffusivities of the exchanging counterions. A numerical solution of the corresponding conservation equation for spherical particles with an infinite fluid volume is given by Helfferich and Plesset [*J. Chem. Phys.*, **66**, 28, 418

(1958)]. The numerical results for the case of two counterions of equal valence where a resin bead, initially partially saturated with A, is completely converted to the B form, is expressed by:

$$F = \{1 - \exp [\pi^2(f_1(\alpha')\tau_D + f_2(\alpha')\tau_D^2 + f_3(\alpha')\tau_D^3)]\}^{1/2} \quad (16\text{-}102)$$

with $f_1(\alpha') = -(0.570 + 0.430\alpha'^{0.775})^{-1}$ $\qquad$ (16-103a)

$f_2(\alpha') = (0.260 + 0.782\alpha')^{-1}$ $\qquad\qquad$ (16-103b)

$f_3(\alpha') = -(0.165 + 0.177\alpha')^{-1}$ $\qquad\quad$ (16-103c)

where $\tau_D = \overline{D}_A t/r_p^2$ and $\alpha' = 1 + (\overline{D}_A/\overline{D}_B - 1)n_A^{0'}/n^s$ for $0.1 \leq \alpha' \leq 10$. The predicted behavior is shown in Fig. 16-17. When $\alpha' = 1$ (equal ion diffusivities or $n_A^{0'} \sim 0$), the solution coincides with Eq. (16-96). For $\alpha' \neq 1$, the exchange rate is faster or slower depending on which coun-

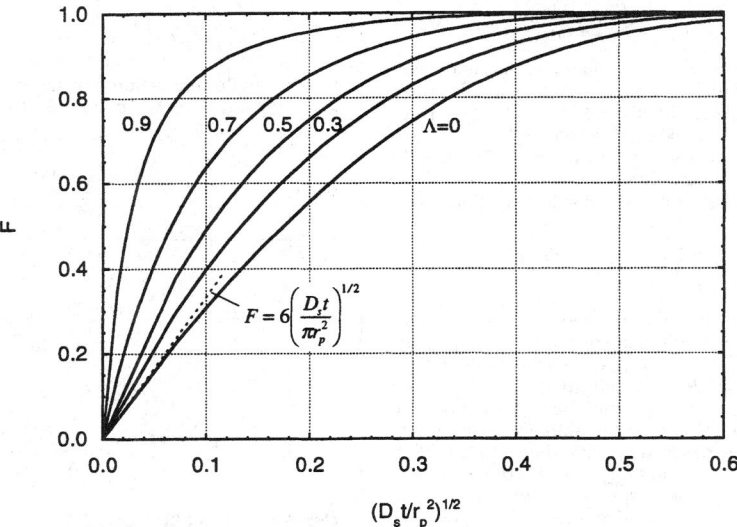

FIG. 16-16   Batch adsorption curves for solid diffusion control. The curve for $\Lambda^\infty = 0$ corresponds to an infinite fluid volume (adapted from Ruthven, gen. refs., with permission).

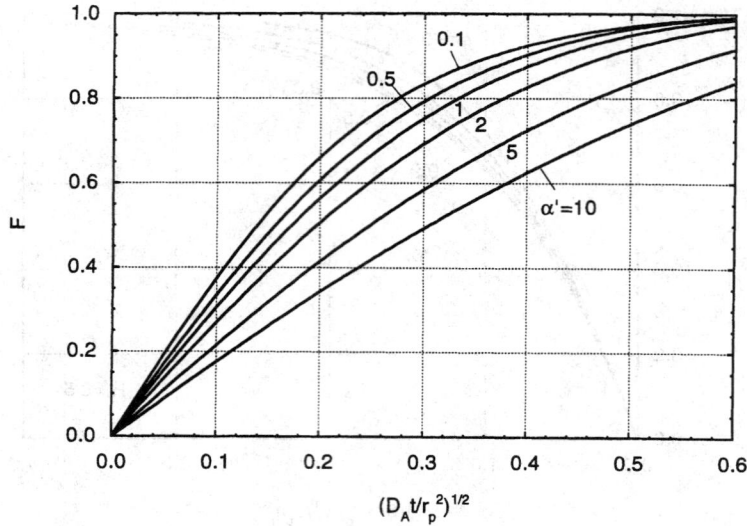

FIG. 16-17 Batch ion exchange for two equal-valence counterions. The exchanger is initially uniformly loaded with ion A in concentration n_A^{0r} and is completely converted to the B form. $\alpha' = 1 + (\overline{D_A}/\overline{D_B} - 1)n_A^{0r}/n_A^s$.

terion is initially present in the ion exchanger and on the initial level of saturation.

For an initially fully saturated particle, the exchange rate is faster when the faster counterion is initially in the resin, with the difference in rate becoming more important as conversion from one form to the other progresses. Helfferich (gen. refs., pp. 270–271) gives explicit expressions for the exchange of ions of unequal valence.

PORE DIFFUSION CONTROL

The rate equation is given by item A in Table 16-11. With pore fluid and adsorbent at equilibrium at each point within the particle and for a constant diffusivity, the rate equation can be written as:

$$\frac{\partial c_{pi}}{\partial t} = \frac{\varepsilon_p D_{pi}}{\varepsilon_p + \rho_p \, dn_i^e/dc_i} \frac{1}{r^2} \frac{\partial}{\partial r}\left(r^2 \frac{\partial c_{pi}}{\partial r}\right) \quad (16\text{-}104)$$

For a **linear isotherm** ($n_i = K_i c_i$), this equation is identical to the conservation equation for solid diffusion, except that the solid diffusivity D_{si} is replaced by the equivalent diffusivity $D_{ei} = \varepsilon_p D_{pi}/(\varepsilon_p + \rho_p K_i)$. Thus, Eqs. (16-96) and (16-99) can be used for pore diffusion control with infinite and finite fluid volumes simply by replacing D_{si} with D_{ei}.

When the adsorption isotherm is nonlinear, a numerical solution is generally required. For a **Langmuir system** with negligible solute holdup in the pore fluid, item A in Table 16-11 gives:

$$\frac{\partial n_i}{\partial t} = \frac{\varepsilon_p D_{pi}}{\rho_p n_i^s K_i} \frac{1}{r^2} \frac{\partial}{\partial r}\left[\frac{r^2}{(1 - n_i/n_i^s)^2} \frac{\partial n_i}{\partial r}\right] \quad (16\text{-}105)$$

This equation has the same form of that obtained for solid diffusion control with D_{si} replaced by the equivalent concentration-dependent diffusivity $D_{ei} = \varepsilon_p D_{pi}/[\rho_p n_i^s K_i(1 - n_i/n_i^s)^2]$. Numerical results for the case of adsorption on an initially clean particle are given in Fig. 16-18 for different values of $\lambda = n_i^0/n_i^s = 1 - R$. The uptake curves become increasingly steeper, as the nonlinearity of the isotherm, measured by the parameter λ, increases. The desorption curve shown for a particle with $n_i^{0r}/n_i^s = 0.9$ shows that for the same step in concentration, adsorption occurs much more quickly than desorption. This difference, however, becomes smaller as the value of λ is reduced and in the linear region of the adsorption isotherm ($\lambda \to 0$), adsorption and desorption curves are mirror images. The solution in Fig. 16-18 is applicable to a nonzero initial adsorbent loading by redefining λ as $(n_i^0 - n_i^{0r})/(n_i^s - n_{0r})$

and the dimensionless time variable as $[\varepsilon_p D_{pi}t/\rho_p(1 - n_i^{0r}/n_i^s)^2 n_i^s K_i r_p^2]^{1/2}$ (Ruthven, gen. refs.).

In the **irreversible limit** ($R < 0.1$), the adsorption front within the particle approaches a shock transition separating an inner core into which the adsorbate has not yet penetrated from an outer layer in which the adsorbed phase concentration is uniform at the saturation value. The dynamics of this process is described approximately by the shrinking-core model [Yagi and Kunii, *Chem. Eng. (Japan)*, **19**, 500 (1955)]. For an infinite fluid volume, the solution is:

$$\frac{\varepsilon_p D_{pi}t}{r_p^2} \frac{c_i^0}{\rho_p n_i^s} = \frac{1}{2} - \frac{1}{3}F - \frac{1}{2}(1 - F)^{2/3} \quad (16\text{-}106)$$

or, in explicit form [Brauch and Schlunder, *Chem. Eng. Sci.*, **30**, 540 (1975)]:

$$F = 1 - \left\{\frac{1}{2} + \cos\left[\frac{\pi}{3} + \frac{1}{3}\cos^{-1}\left(1 - \frac{12\varepsilon_p D_{pi}t}{r_p^2} \frac{c_i^0}{\rho_p n_i^s}\right)\right]\right\}^3 \quad (16\text{-}107)$$

For a finite fluid volume with $0 < \Lambda^\infty \leq 1$, the solution is [Teo and Ruthven, *Ind. Eng. Chem. Process Des. Dev.*, **25**, 17 (1986)]:

$$\frac{\varepsilon_p D_{pi}t}{r_p^2} \frac{c_i^0}{\rho_p n_i^s} = I_2 - I_1 \quad (16\text{-}108)$$

where

$$I_1 = \frac{1}{\lambda'\Lambda^\infty\sqrt{3}}\left[\tan^{-1}\frac{2\eta - \lambda'}{\lambda'\sqrt{3}} - \tan^{-1}\frac{2 - \lambda'}{\lambda'\sqrt{3}}\right] + \frac{1}{6\lambda'\Lambda^\infty}\ln\left[\frac{\lambda'^3 + \eta^3}{\lambda'^3 + 1}\left(\frac{\lambda' + 1}{\lambda' + \eta}\right)^3\right]$$

$$(16\text{-}109a)$$

$$I_2 = \frac{1}{3\Lambda^\infty}\ln\frac{\lambda'^3 + \eta^3}{\lambda'^3 + 1} \quad (16\text{-}109b)$$

$$\eta = (1 - F)^{1/3} \quad (16\text{-}109c)$$

$$\lambda' = \left(\frac{1}{\Lambda^\infty} - 1\right)^{1/3} \quad (16\text{-}109d)$$

COMBINED RESISTANCES

In general, exact analytic solutions are available only for the linear ($R = 1$) and irreversible limits ($R \to 0$). Intermediate cases require

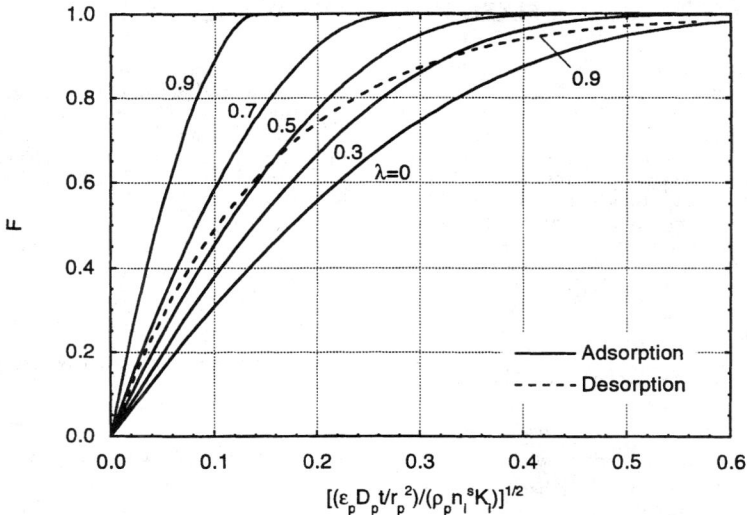

FIG. 16-18   Constant separation factor batch adsorption curves for pore diffusion control with an infinite fluid volume. λ is defined in the text.

numerical solution or use of approximate driving force expressions (see "Rate and Dispersion Factors").

Parallel Pore and Solid Diffusion Control   With a **linear isotherm,** assuming equilibrium between the pore fluid and the solid adsorbent, batch adsorption can be represented in terms of an equivalent solid diffusivity $D_{ei} = (\varepsilon_p D_{pi} + \rho_p D_{si})/(\varepsilon_p + \rho_p K_i)$. Thus, Eqs. (16-96) and (16-99) can be used for this case with D_{si} replaced by D_{ei}.

External Mass Transfer and Intraparticle Diffusion Control   With a **linear isotherm,** the solution for combined external mass transfer and pore diffusion control with an infinite fluid volume is (Crank, *Mathematics of Diffusion*, 2d ed., Clarendon Press, 1975):

$$F = 1 - \sum_{n=1}^{\infty} \frac{6Bi^2 \exp\left[-(p_n^2 \varepsilon_p D_{pi} t/r_p^2)/(\varepsilon_p + \rho_p K_i)\right]}{p_n^2[p_n^2 + Bi(Bi - 1)]} \quad (16\text{-}110)$$

where $Bi = k_f r_p/\varepsilon_p D_{pi}$ is the **Biot number** and the p_ns are the positive roots of

$$p_n \cot p_n = 1 - Bi \quad (16\text{-}111)$$

For a finite fluid volume the solution is:

$$F = 1 - 6 \sum_{n=1}^{\infty} \frac{\exp\left[-(p_n^2 \varepsilon_p D_{pi} t/r_p^2)/(\varepsilon_p + \rho_p K_i)\right]}{\dfrac{9\Lambda^\infty}{1 - \Lambda^\infty} + (1 - \Lambda^\infty)p_n^2 - (5\Lambda^\infty + 1)\dfrac{p_n^2}{Bi} + (1 - \Lambda^\infty)\dfrac{p_n^4}{Bi^2}} \quad (16\text{-}112)$$

where the p_ns are the positive roots of

$$\frac{\tan p_n}{p_n} = \frac{3 - \dfrac{1 - \Lambda^\infty}{\Lambda^\infty}\dfrac{p_n^2}{Bi}}{3 + \dfrac{1 - \Lambda^\infty}{\Lambda^\infty}\dfrac{(Bi - 1)p_n^2}{Bi}} \quad (16\text{-}113)$$

These expressions can also be used for the case of external mass transfer and solid diffusion control by substituting D_{si} for $\varepsilon_p D_{pi}/(\varepsilon_p + \rho_p K_i)$ and $k_f r_p/(\rho_p K_i D_{si})$ for the Biot number.

In the **irreversible limit,** the solution for combined external resistance and pore diffusion with infinite fluid volume is (Yagi and Kunii):

$$\frac{\varepsilon_p D_{pi} t}{r_p^2} \frac{c_i^0}{\rho_p n_i^s} = \frac{1}{2} - \frac{1}{3}\left(1 - \frac{1}{Bi}\right)F - \frac{1}{2}(1 - F)^{2/3} \quad (16\text{-}114)$$

For a finite fluid volume the solution is (Teo and Ruthen):

$$\frac{\varepsilon_p D_{pi} t}{r_p^2} \frac{c_i^0}{\rho_p n_i^s} = \left(1 - \frac{1}{Bi}\right)I_2 - I_1 \quad (16\text{-}115)$$

where I_1 and I_2 are given be Eqs. (16-109a) and (16-109b).

Bidispersed Particles   For particles of radius r_p comprising adsorptive subparticles of radius r_s that define a macropore network, conservation equations are needed to describe transport both within the macropores and within the subparticles and are given in Table 16-11, item D. Detailed equations and solutions for a linear isotherm are given in Ruthven (gen. refs., p. 183) and Ruckenstein et al. [*Chem. Eng. Sci.*, **26**, 1306 (1971)]. The solution for a **linear isotherm** with no external resistance and an infinite fluid volume is:

$$F = 1 - \frac{18}{\beta + 3\alpha} \sum_{m=1}^{\infty} \sum_{n=1}^{\infty} \left(\frac{n^2\pi^2}{p_{n,m}^4}\right)$$

$$\times \frac{\exp(-p_{n,m}^2 D_{si} t/r_s^2)}{\alpha + \dfrac{\beta}{2}\left[1 + \dfrac{\cot p_{n,m}}{p_{n,m}}(p_{n,m} \cot p_{n,m} - 1)\right]} \quad (16\text{-}116)$$

where the $p_{n,m}$ values are the roots of the equation

$$\alpha p_{n,m}^2 - n^2\pi^2 = \beta(p_{n,m} \cot p_{n,m} - 1) \quad (16\text{-}117)$$

and

$$\alpha = \frac{D_{si}/r_s^2}{D_{pi}/r_p^2} \quad (16\text{-}118a)$$

$$\beta = \frac{3\alpha\rho_p K_i}{\varepsilon_p} \quad (16\text{-}118b)$$

In these equations, D_{si} is the diffusivity in the subparticles, and D_{pi} is the diffusivity in the pore network formed by the subparticles.

For large K_i values, the uptake curve depends only upon the value of the parameter β representing the ratio of characteristic time constants for diffusion in the pores and in the subparticles. For small β values, diffusion in the subparticles is controlling and the solution coincides with Eq. (16-96) with r_s replacing r_p. For large β values, pore diffusion is controlling, and the solution coincides with Eq. (16-96) with $\varepsilon_p D_{pi}/(\varepsilon_p + \rho_p K_i)$ replacing D_{si}.

Lee [*AIChE J.*, **24**, 531 (1978)] gives the solution for batch adsorption with bidispersed particles for the case of a finite fluid volume.

FIXED-BED TRANSITIONS

As discussed in "Design Concepts," a large fraction of adsorption and ion-exchange processes takes place in fixed beds. Two classical methods for analyzing fixed-bed transitions are described here. First, local equilibrium theory is presented. In this, all mass-transfer resistances are ignored to focus on the often dominating role of isotherm shape. Second, results of constant pattern analysis are presented. This gives the maximum breadth to which a mass-transfer zone will spread for various rate mechanisms. It is therefore conservative for design purposes. Both of these methods pertain to behavior in deep beds. For shallow beds, the equations given below must be solved for the particular case of interest.

DIMENSIONLESS SYSTEM

For the methods, we consider Eq. (16-52), the material balance for a fixed bed, written in the form

$$\rho_b \frac{\partial \overline{n}_i}{\partial t} + \varepsilon_b \frac{\partial c_i}{\partial t} + \varepsilon \frac{\partial (v c_i)}{\partial z} = \varepsilon D_L \frac{\partial}{\partial z}\left(c \frac{\partial y_i}{\partial z}\right) \quad (16\text{-}119)$$

where it has been assumed that D_L is constant and that $\hat{c}_i \approx c_i$ (or that the second term in the balance is small compared to the first—usually a good assumption).

Dimensionless variables can be defined for time, the axial coordinate, and velocity:

$$\tau = \frac{\varepsilon v^{\text{ref}} t}{L} \quad (16\text{-}120)$$

$$\zeta = \frac{z}{L} \quad (16\text{-}121)$$

$$v^\circ = \frac{v}{v^{\text{ref}}} \quad (16\text{-}122)$$

where L is bed length, v^{ref} is the interstitial velocity at the bed inlet, and τ is equal to the number of empty bed volumes of feed passed into the bed. The material balance becomes

$$\rho_b \frac{\partial \overline{n}_i}{\partial \tau} + \varepsilon_b \frac{\partial c_i}{\partial \tau} + \frac{\partial (v^\circ c_i)}{\partial \zeta} = \frac{1}{N_{Pe}} \frac{\partial}{\partial \zeta}\left(c \frac{\partial y_i}{\partial \zeta}\right) \quad (16\text{-}123)$$

where $N_{Pe} = \varepsilon v^{\text{ref}} L / D_L$ is a Peclet number for the bed or a **number of dispersion units.** Equation (16-123) or a similar equation is often the material balance used in nonisothermal problems, in problems involving adsorption of nontrace components, and in calculations of cycles.

For a trace, isothermal system, we have $v^\circ = 1$, and using the dimensionless system variables for concentrations [Eq. (16-10)], Eq. (16-123) becomes

$$\Lambda \frac{\partial \overline{n}_i^\circ}{\partial \tau} + \varepsilon_b \frac{\partial c_i^\circ}{\partial \tau} + \frac{\partial c_i^\circ}{\partial \zeta} = \frac{1}{N_{Pe}} \frac{\partial^2 c_i^\circ}{\partial \zeta^2} \quad (16\text{-}124)$$

where Λ is the **partition ratio** defined by

$$\Lambda = \frac{\rho_b n_i^{\text{ref}}}{c_i^{\text{ref}}} \quad (16\text{-}125)$$

This important dimensionless group is the volumetric capacity of the bed for the sorbable component divided by the concentration of the sorbable component in the feed. The stoichiometric capacity of the bed for solute is exactly equal to Λ empty bed volumes of feed (to saturate the sorbent at the feed concentration) plus a fraction of a bed volume of feed to fill the voids outside and inside the particles. Alternatively, we also obtain Eq. (16-124) using the dimensionless transition variables for concentrations [Eq. (16-11)], but now the partition ratio in the first term of Eq. (16-124) pertains to the transition and is given by

$$\Lambda = \rho_b \frac{n_i'' - n_i'}{c_i'' - c_i'} \quad (16\text{-}126)$$

Equation (16-124) is a commonly used form of material balance for a fixed-bed adsorber.

If the system under consideration involves use of the sorbent for only a single feed step or reuse after uniform regeneration, as in many applications with activated carbons and ion exchangers, then one of two paths is often followed at this point to simplify Eq. (16-124) further. The second term on the left-hand side of the equation is often assumed to be negligibly small (usually a good assumption), and time is redefined as

$$\tau_1 = \frac{\tau}{\Lambda} \quad (16\text{-}127)$$

to give

$$\frac{\partial \overline{n}_i^\circ}{\partial \tau_1} + \frac{\partial c_i^\circ}{\partial \zeta} = \frac{1}{N_{Pe}} \frac{\partial^2 c_i^\circ}{\partial \zeta^2} \quad (16\text{-}128)$$

Alternatively, in the absence of axial dispersion, a variable of the form

$$\tau_1 = \frac{\tau - \varepsilon_b \zeta}{\Lambda} \quad (16\text{-}129)$$

can be defined to reduce Eq. (16-124) directly to

$$\frac{\partial \overline{n}_i^\circ}{\partial \tau_1} + \frac{\partial c_i^\circ}{\partial \zeta} = 0 \quad (16\text{-}130)$$

The variable τ_1 defined by Eq. (16-127) or (16-129) is a **throughput parameter,** equal to unity (hence, the "1" subscript) at the time when the stoichiometric center of the concentration wave leaves the bed. This important group, in essence a dimensionless time variable, essentially determines the location of the stoichiometric center of the transition in the bed at any time.

LOCAL EQUILIBRIUM THEORY

In local equilibrium theory, fluid and sorbed phases are assumed to be in local equilibrium with one another at every axial position in the bed. Thus, because of uniform concentrations, the overbar on n_i° is not necessary and we have $\hat{c}_i \approx c_i$ [note Eqs. (16-52) and (16-119)].

Single Transition System For a system described by a single material balance, Eq. (16-130) gives

$$\frac{d\tau_1}{d\zeta} = -\frac{\partial c_i^\circ / \partial \zeta}{\partial c_i^\circ / \partial \tau_1} = \frac{dn_i^\circ}{dc_i^\circ} \quad (16\text{-}131)$$

where $d\tau_1 / d\zeta$ is the reciprocal of a concentration velocity. Equation (16-131) is the equation for a **simple wave** (or gradual transition or proportionate pattern). If a bed is initially uniformly saturated, then $d\tau_1 / d\zeta = \tau_1 / \zeta$. Thus, for the dimensionless system, the reciprocal of the velocity of a concentration is equal to the slope of the isotherm at that concentration. Furthermore, from Eq. (16-131), the depth of penetration of a given concentration into the bed is directly proportional to time, so the breadth of a simple wave increases in direct proportion to the depth of its penetration into the bed (or to time). Thus, for the simple wave, the length of the MTZ is proportional to the depth of the bed through which the wave has passed. Consideration of isotherm shape indicates that a simple wave occurs for an unfavorable dimensionless isotherm ($d^2 n_i^\circ / dc_i^{\circ 2} > 0$), for which low concentrations will go faster than high concentrations. Equation (16-131) also pertains to a linear isotherm, in which case the wave is called a **contact discontinuity,** because it has neither a tendency to spread nor sharpen. If mass-transfer resistance is added to the consideration of wave character for unfavorable isotherms, the wave will still asymptotically approach the simple wave result given by Eq. (16-131).

For a favorable isotherm ($d^2 n_i^\circ / dc_i^{\circ 2} < 0$), Eq. (16-131) gives the impossible result that three concentrations can coexist at one point in the bed (see example below). The correct solution is a **shock** (or abrupt transition) and not a simple wave. Mathematical theory has been developed for this case to give "weak solutions" to conservation laws. The form of the solution is

$$\text{shock speed} = \frac{\text{change in flux}}{\text{change in accumulated quantity}}$$

where the changes are jump discontinuities across the shock. The reciprocal of this equation, using Eq. (16-130), is

$$\frac{d\tau_1}{d\zeta} = \frac{\Delta n_i^\circ}{\Delta c_i^\circ} \qquad (16\text{-}132)$$

where the differences are taken across the shock.

It is also possible to have a **combined wave,** which has both gradual and abrupt parts. The general rule for an isothermal, trace system is that in passing from the initial condition to the feed point in the isotherm plane, the slope of the path must not decrease. If it does, then a shock chord is taken for that part of the path. Referring to Fig. 16-19, for a transition from (0,0) to (1,1), the dashes indicate shock parts, which are connected by a simple wave part between points P_1 and P_2.

Example 9: Transition Types   For the constant separation-factor isotherm given by Eq. (16-31), determine breakthrough curves for $r = 2$ and $r = 0.5$ for transitions from $c_i^\circ = 0$ to $c_i^\circ = 1$.

Using Eq. (16-131), we obtain

$$\frac{\tau_1}{\zeta} = \frac{r}{[r + (1-r)c_i^\circ]^2}$$

This equation, evaluated at $\zeta = 1$, is plotted for $r = 2$ and $r = 0.5$ in Fig. 16-20. Clearly, the solution for $r = 0.5$ is not physically correct. Equation (16-132), with $d\tau_1/d\zeta = \tau_1/\zeta$, is applied to this case to give the shock indicated by the dashed line. Alternatively, we could have obtained bed profiles by evaluating equations at $\tau_1 = $ const.

Multiple Transition System   Local equilibrium theory for multiple transitions begins with some combination of material and energy balances, written

$$\rho_b \frac{\partial n_i}{\partial \tau} + \varepsilon_b \frac{\partial c_i}{\partial \tau} + \frac{\partial(v^\circ c_i)}{\partial \zeta} = 0 \qquad (i = 1, 2, \ldots) \quad (16\text{-}133)$$

$$\rho_b \frac{\partial u_s}{\partial \tau} + \varepsilon_b \frac{\partial(cu_f)}{\partial \tau} + \frac{\partial(v^\circ ch_f)}{\partial \zeta} = 0 \qquad (16\text{-}134)$$

which are Eq. (16-123) written with no dispersion and Eq. (16-55) written for an adiabatic bed.

For a simple wave, application of the method of characteristics (hodograph transformation) gives

$$\frac{d\tau}{d\zeta} = \frac{d(\rho_b n_1 + \varepsilon_b c_1)}{d(v^\circ c_1)} = \frac{d(\rho_b n_2 + \varepsilon_b c_2)}{d(v^\circ c_2)} = \cdots = \frac{d(\rho_b u_s + \varepsilon_b cu_f)}{d(v^\circ ch_f)} \quad (16\text{-}135)$$

where the derivatives are taken along the path of a transition (i.e., directional derivatives).

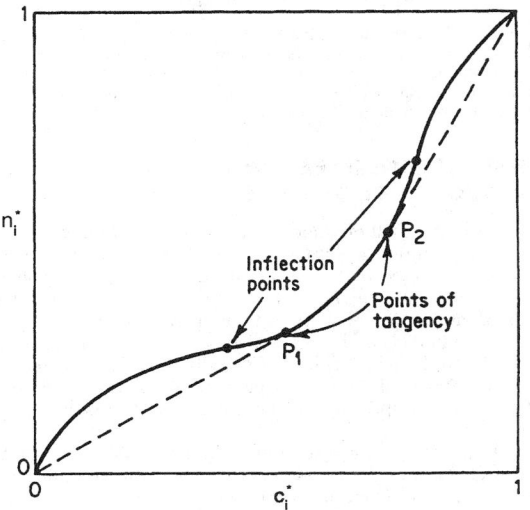

FIG. 16-19   Path in isotherm plane for a combined wave (After Tudge).

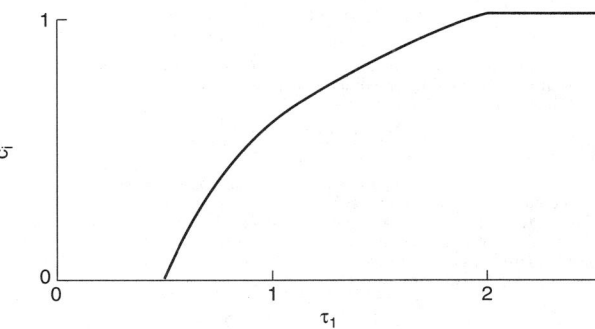

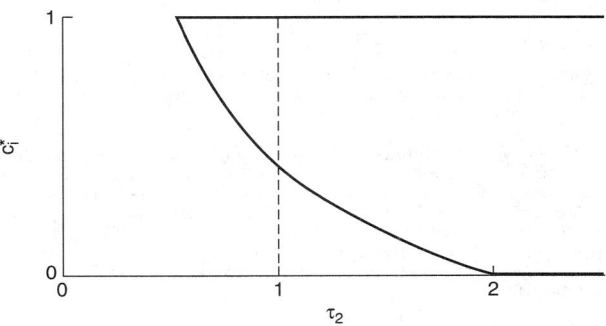

FIG. 16-20   Breakthrough curves for $r = 2$ (top) and $r = 0.5$ (bottom) for Example 9.

If a simple wave is not possible on physical grounds, then it (or part of it) is replaced by a shock, given by

$$\frac{d\tau}{d\zeta} = \frac{\Delta(\rho_b n_1 + \varepsilon_b c_1)}{\Delta(v^\circ c_1)} = \frac{\Delta(\rho_b n_2 + \varepsilon_b c_2)}{\Delta(v^\circ c_2)} = \cdots = \frac{\Delta(\rho_b u_s + \varepsilon_b cu_f)}{\Delta(v^\circ ch_f)} \quad (16\text{-}136)$$

Extensions   When more than two conservation equations are to be solved simultaneously, matrix methods for eigenvalues and left eigenvectors are efficient [Jeffrey and Taniuti, *Nonlinear Wave Propagation,* Academic Press, New York, 1964; Jacob and Tondeur, *Chem. Eng. J.,* **22,** 187 (1981), **26,** 41 (1983); Davis and LeVan, *AIChE J.,* **33,** 470 (1987); Rhee et al., gen. refs.].

Nontrace isothermal systems give the "adsorption effect" (i.e., significant change in fluid velocity because of loss or gain of solute). Criteria for the existence of simple waves, contact discontinuities, and shocks are changed somewhat [Peterson and Helfferich, *J. Phys. Chem.,* **69,** 1283 (1965); LeVan et al., *AIChE J.,* **34,** 996 (1988); Frey, *AIChE J.,* **38,** 1649 (1992)].

Local equilibrium theory also pertains to adsorption with axial dispersion, since this mechanism does not disallow existence of equilibrium between stationary and fluid phases across the cross section of the bed [Rhee et al., *Chem. Eng. Sci.,* **26,** 1571 (1971)]. It is discussed below in further detail from the standpoint of the constant pattern.

Example 10: Two-Component Isothermal Adsorption   Two components present at low mole fractions are adsorbed isothermally from an inert fluid in an initially clean bed. The system is described by $\rho_b = 500$ kg/m³, $\varepsilon_b = 0.7$, and the binary Langmuir isotherm

$$n_i = \frac{n_i^s K_i c_i}{1 + K_1 c_1 + K_2 c_2} \qquad (i = 1, 2)$$

with $n_1^s = n_2^s = 6$ mol/kg, $K_1 = 40$ m³/mol, and $K_2 = 20$ m³/mol. The feed is $c_1 = c_2 = 0.5$ mol/m³. Find the bed profile.

Using the isotherm to calculate loadings in equilibrium with the feed gives $n_1 = 3.87$ mol/kg and $n_2 = 1.94$ mol/kg. An attempt to find a simple wave solution for this problem fails because of the favorable isotherms (see the next example for the general solution method). To obtain the two shocks, Eq. (16-136) is written

$$\frac{d\tau}{d\zeta} = \frac{\Delta(\rho_b n_1 + \varepsilon_b c_1)}{\Delta c_1} = \frac{\Delta(\rho_b n_2 + \varepsilon_b c_2)}{\Delta c_2}$$

The concentration of one of the components will drop to zero in the shock nearest the bed inlet. If it is component 1, then using feed values and the equation above, that shock would be at

$$\frac{\tau}{\zeta} = \frac{\rho_b n_1 + \varepsilon_b c_1}{c_1} \approx \frac{\rho_b n_1}{c_1} = 3870$$

Similarly, if the second component were to disappear in the first shock, we would have $\tau/\zeta = 1940$. Material balance considerations require that we accept the shorter distance, so component 1 disappears in the first shock.

The concentrations of component 2 on the plateau downstream of the first shock are then calculated from

$$\frac{\tau}{\zeta} = \frac{\Delta(\rho_b n_2 + \varepsilon_b c_2)}{\Delta c_2} \approx \frac{\rho_b \Delta n_2}{\Delta c_2} = 3870$$

and its pure component isotherm, giving $c_2 = 0.987$ mol/m^3 and $n_2 = 5.71$ mol/kg. The location of this shock is determined using these concentrations and

$$\frac{\tau}{\zeta} = \frac{\rho_b n_2 + \varepsilon_b c_2}{c_2} \approx \frac{\rho_b n_2}{c_2}$$

which gives $\tau/\zeta = 2890$. The bed profile is plotted in Fig. 16-21 using ζ/τ as the abscissa. (This example can also be worked with the h-transformation described in this section under chromatography.)

Example 11: Adiabatic Adsorption and Thermal Regeneration

An initially clean activated carbon bed at 320 K is fed a vapor of benzene in nitrogen at a total pressure of 1 MPa. The concentration of benzene in the feed is 6 mol/m^3. After the bed is uniformly saturated with feed, it is regenerated using benzene-free nitrogen at 400 K and 1 MPa. Solve for both steps. For simplicity, neglect fluid-phase accumulation terms and assume constant mean heat capacities for stationary and fluid phases and a constant velocity. The system is described by

$$\rho_b n_1 = \frac{\rho_b n_1^s K_1 c_1}{1 + K_1 c_1}$$

$$\rho_b n_1^s = 2750 \text{ mol/m}^3$$

$$K_1 = 3.88 \times 10^{-8} \sqrt{T} \exp\left[q_1^{st}/(\Re T)\right] \text{ m}^3/(\text{mol K}^{1/2})$$

$$\rho_b C_{sm} = 850 \text{ kJ/(m}^3 \text{ K)}$$

$$c C_{pfm} = 11.3 \text{ kJ/(m}^3 \text{ K)}$$

$$q_1^{st} = 43.5 \text{ kJ/mol}$$

$$T^{ref} = 320 \text{ K}$$

Extensive analysis has been made of this system [Rhee et al., *Chem. Eng. J.*, **1**, 241 (1970) and gen. refs.; LeVan, in Rodrigues et al., gen. refs.].

To obtain the concentration and temperature profiles, the two transitions are first assumed to be gradual. Equation (16-135) is written in the form

$$\frac{\tau}{\zeta} = \rho_b \frac{dn_1}{dc_1} = \rho_b \frac{du_s}{d(ch_f)}$$

where, from Eqs. (16-56) to (16-60),

$$\rho_b u_s = \rho_b C_{sm}(T - T^{ref}) - \rho_b n_1 q_1^{st}$$

$$ch_f = c C_{pfm}(T - T^{ref})$$

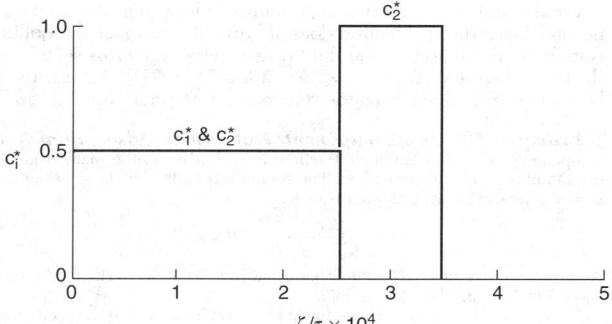

FIG. 16-21 Bed profiles for two-component isothermal adsorption, Example 10.

This equation is expanded in terms of c_1 and T to obtain

$$\frac{\tau}{\zeta} = \rho_b \left(\frac{\partial n_1}{\partial c_1} + \frac{\partial n_1}{\partial T} \frac{dT}{dc_1}\right) = \frac{\rho_b C_{sm}}{c C_{pfm}} - \frac{\rho_b q_1^{st}}{c C_{pfm}}\left(\frac{\partial n_1}{\partial T} + \frac{\partial n_1}{\partial c_1} \frac{dc_1}{dT}\right)$$

Solving the rightmost equality for the directional derivative using the quadratic formula gives

$$\frac{dc_1}{dT} = \frac{-b \pm \sqrt{b^2 - 4ad}}{2a}$$

with

$$a = \left[\frac{\rho_b q_1^{st}}{(c C_{pfm})}\right] \frac{\partial n_1}{\partial c_1}$$

$$b = \left[\frac{\rho_b q_1^{st}}{(c C_{pfm})}\right] \frac{\partial n_1}{\partial T} - \frac{\rho_b C_{sm}}{(c C_{pfm})} + \frac{\rho_b \partial n_1}{\partial c_1}$$

$$d = \frac{\rho_b \partial n_1}{\partial T}$$

The plus sign corresponds to the downstream transition and the minus sign to the upstream one. This equation is solved along each path beginning at the respective end points—the initial condition of the bed for the downstream transition and the feed condition for the upstream transition. If either path fails to evolve continuously in the expected direction, the difference form, from Eq. (16-136),

$$\frac{d\tau}{d\zeta} = \rho_b \frac{\Delta n_1}{\Delta c_1} = \rho_b \frac{\Delta u_s}{\Delta(ch_f)}$$

is used for that path (or part thereof, if appropriate); two solutions of this equation pass through each composition point, and care must be taken to ensure that the correct path is taken. The two correct paths found intersect to give the composition and temperature of an intermediate plateau region.

Letting $c_1^{ref} = 6$ mol/m^3, the isotherm gives $\rho_b n_1^{ref} = 2700$ mol/m^3, and the partition ratio is $\Lambda = 450$.

In the figures, $\Gamma(k)$ and $\Sigma(k)$ symbolize simple waves and shocks, respectively, with $k = 1$ downstream and $k = 2$ upstream.

Adiabatic Adsorption The construction is shown in Figs. 16-22 and 16-23. The first path begins at the initial condition, point A ($T = 320$ K, $c_1 = 0$ mol/m^3). Since $\partial n_1/\partial T = 0$ there, we obtain $dc_1/dT = 0$ and $\tau/\zeta = \rho_b C_{sm}/(c C_{pfm}) = 75.2$ (or $\tau/\Lambda = 0.167$ at $\zeta = 1$), corresponding to a pure thermal wave along the $c_1 = 0$ axis of Fig. 16-22. The second path begins at the feed condition, point B ($T = 320$ K, $c_1 = 6$ mol/m^3). A $\Gamma(2)$ fails, so a $\Sigma(2)$ is calculated and plotted in the two figures. The $\Gamma(1)$ and $\Sigma(2)$ intersect at point C ($T = 348$ K, $c_1 = 0$ mol/m^3) Breakthrough curves of temperature and concentration are shown in Fig. 16-24.

Thermal Regeneration The bed, initially at point B, is fed with pure nitrogen. The feed, at point D ($T = 400$ K, $c_1 = 0$ mol/m^3), provides a successful $\Gamma(2)$. The $\Gamma(1)$ from point B fails, so a $\Sigma(1)$, which differs imperceptibly from $\Gamma(1)$, is determined. The $\Sigma(1)$ and $\Gamma(2)$ meet at point E ($T = 320.8$ K, $c_1 = 8.93$ mol/m^3), where $\rho_b n_1 = 2716$ mol/m^3, which is greater than the initial loading, indicating "roll-up," a term attributed to Basmadjian et al. [*Ind. Eng. Chem. Process Des. Dev.*, **14**, 328 (1975)], of the adsorbed-phase concentration. Breakthrough curves are shown in Fig. 16-25. The $\Sigma(1)$ leaves the bed at $\tau/\Lambda = 0.011$. The $\Gamma(2)$ begins to emerge at $\tau/\Lambda = 0.169$, but regeneration is not complete until $\tau/\Lambda = 2.38$.

It is possible (with lower initial bed temperature, higher initial loading, or higher regeneration temperature or pressure) for the transition paths to contact the saturated vapor curve in Fig. 16-22 rather than intersect beneath it. For this case, liquid benzene condenses in the bed, and the effluent vapor is saturated during part of regeneration [Friday and LeVan, *AIChE J.*, **30**, 679 (1984)].

CONSTANT PATTERN BEHAVIOR FOR FAVORABLE ISOTHERMS

With a favorable isotherm and a mass-transfer resistance or axial dispersion, a transition approaches a **constant pattern,** which is an asymptotic shape beyond which the wave will not spread. The wave is said to be "self-sharpening." (If a wave is initially broader than the constant pattern, it will sharpen to approach the constant pattern.) Thus, for an initially uniformly loaded bed, the constant pattern gives the maximum breadth of the MTZ. As bed length is increased, the constant pattern will occupy an increasingly smaller fraction of the bed. (Square-root spreading for a linear isotherm gives this same qualitative result.)

The treatment here is restricted to the Langmuir or constant separation factor isotherm, single-component adsorption, dilute systems, isothermal behavior, and mass-transfer resistances acting alone. References to extensions are given below. Different isotherms have been considered, and the theory is well understood for general isotherms.

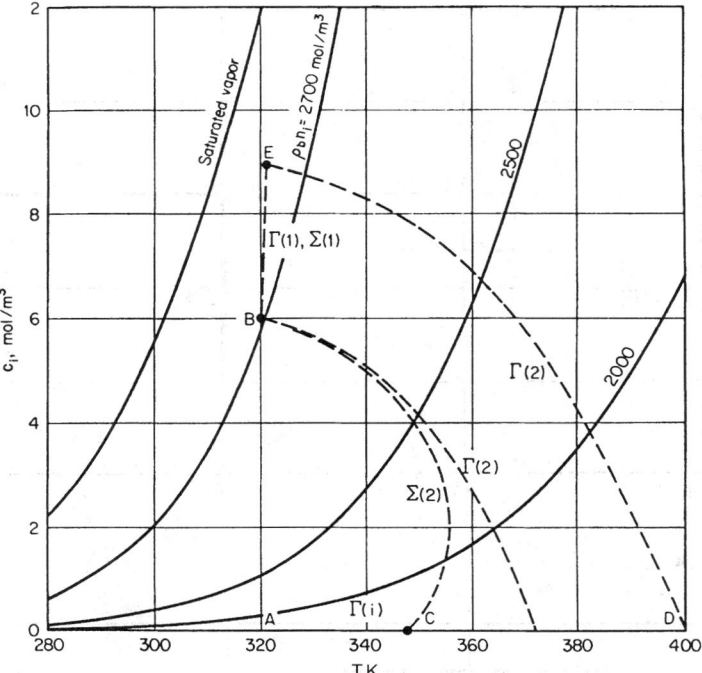

FIG. 16-22 Transition paths in c_1, T plane for adiabatic adsorption and thermal regeneration.

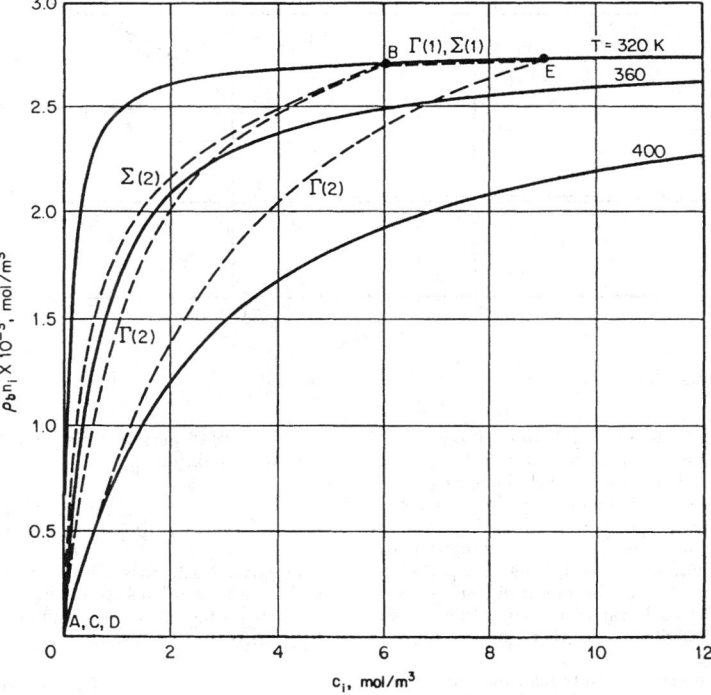

FIG. 16-23 Transition paths in isotherm plane for adiabatic adsorption and thermal regeneration.

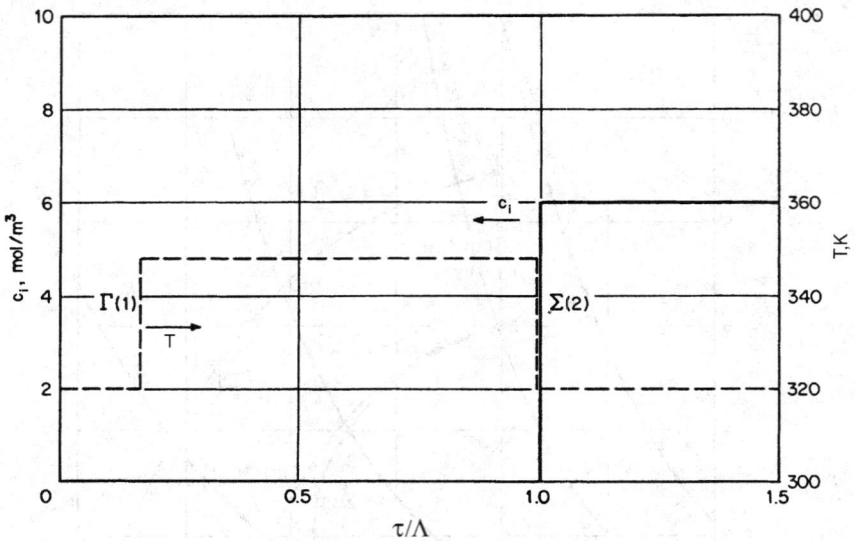

FIG. 16-24   Breakthrough curves for adiabatic adsorption.

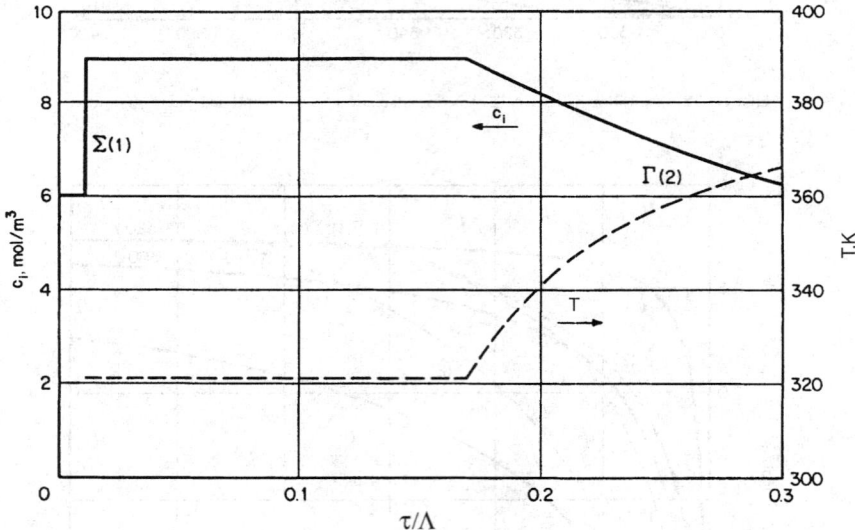

FIG. 16-25   Breakthrough curves for thermal regeneration.

Asymptotic Solution   Rate equations for the various mass-transfer mechanisms are written in dimensionless form in Table 16-13 in terms of a number of transfer units, $N = L/\text{HTU}$, for particle-scale mass-transfer resistances, a number of reaction units for the reaction kinetics mechanism, and a number of dispersion units, N_{Pe}, for axial dispersion. For pore and solid diffusion, $\xi = r/r_p$ is a dimensionless radial coordinate, where r_p is the radius of the particle. If a particle is bidisperse, then r_p can be replaced by r_s, the radius of a subparticle. For preliminary calculations, Fig. 16-13 can be used to estimate N for use with the LDF approximation when more than one resistance is important.

In constant pattern analysis, equations are transformed into a new coordinate system that moves with the wave. Variables are changed from (ζ, τ_1) to $(\zeta - \tau_1, \tau_1)$. The new variable $\zeta - \tau_1$ is equal to zero at the stoichiometric center of the wave. Equation (16-130) for a bed

with no axial dispersion, when transformed to the $(\zeta - \tau_1, \tau_1)$ coordinate system, becomes

$$-\frac{\partial \overline{n}_i^\circ}{\partial(\zeta - \tau_1)} + \frac{\partial \overline{n}_i^\circ}{\partial \tau_1} + \frac{\partial c_i^\circ}{\partial(\zeta - \tau_1)} = 0 \qquad (16\text{-}137)$$

The constant pattern is approached as the τ_1 dependence in this equation disappears. Thus, discarding the derivative with respect to τ_1 and integrating, using the condition that $\overline{n}_i^\circ$ and c_i° approach zero as $N(\zeta - \tau_1) \to \infty$ [or approach unity as $N(\zeta - \tau_1) \to -\infty$], gives simply

$$\overline{n}_i^\circ = c_i^\circ \qquad (16\text{-}138)$$

For adsorption with axial dispersion, the material balance transforms to

TABLE 16-13   Constant Pattern Solutions for Constant Separation Factor Isotherm ($R < 1$)

Mechanism	N	Dimensionless rate equation[1]	Constant pattern	Refs.
Pore diffusion	$\dfrac{15(1-\varepsilon)\varepsilon_p D_p L}{\varepsilon v^{\text{ref}} R_p^2}$	$\dfrac{\partial n_i^\circ}{\partial \tau_1} = \dfrac{N}{15}\dfrac{1}{\xi^2}\dfrac{\partial}{\partial \xi}\left(\xi^2 \dfrac{\partial c_i^\circ}{\partial \xi}\right)$	Numerical	A
Solid diffusion	$\dfrac{15 \Lambda D_s L}{\varepsilon v^{\text{ref}} R_p^2}$	$\dfrac{\partial n_i^\circ}{\partial \tau_1} = \dfrac{N}{15}\dfrac{1}{\xi^2}\dfrac{\partial}{\partial \xi}\left(\xi^2 \dfrac{\partial n_i^\circ}{\partial \xi}\right)$	Numerical	B
External mass transfer	$\dfrac{k_f a L}{\varepsilon v^{\text{ref}}}$	$\dfrac{\partial \overline{n}_i^\circ}{\partial \tau_1} = N\,(c_i^\circ - c_i^{e\circ})$	$\dfrac{1}{1-R}\ln\left[\dfrac{(1-c_i^\circ)^R}{c_i^\circ}\right] - 1 = N\,(\zeta - \tau_1)$	C
Linear driving force	$\dfrac{k_n \Lambda L}{\varepsilon v^{\text{ref}}}$	$\dfrac{\partial \overline{n}_i^\circ}{\partial \tau_1} = N\,(n_i^{e\circ} - \overline{n}_i^\circ)$	$\dfrac{1}{1-R}\ln\left[\dfrac{1-c_i^\circ}{c_i^{\circ R}}\right] + 1 = N\,(\zeta - \tau_1)$	D
Reaction kinetics	$\dfrac{k_a c^{\text{ref}} \Lambda L}{(1-R)\varepsilon v^{\text{ref}}}$	$\dfrac{\partial \overline{n}_i^\circ}{\partial \tau_1} = N\,[(1-\overline{n}_i^\circ)c_i^\circ - R\overline{n}_i^\circ(1-c_i^\circ)]$	$\dfrac{1}{1-R}\ln\left[\dfrac{1-c_i^\circ}{c_i^\circ}\right] = N\,(\zeta - \tau_1)$	E
Axial dispersion	$\dfrac{v^{\text{ref}} L}{D_L}$	Eq. (16-128)	$\dfrac{1}{1-R}\ln\left[\dfrac{1-c_i^\circ}{c_i^{\circ R}}\right] = N\,(\zeta - \tau_1)$	F

1: Dimensional rate equations are given in Table 16-11 and 16-12.
A: Hall et al., *Ind. Eng. Chem. Fundam.*, **5**, 212 (1966).
B: Hall et al., *Ind. Eng. Chem. Fundam.*, **5**, 212 (1966); Garg and Ruthven, *Chem. Eng. Sci.*, **28**, 791, 799 (1973).
C: Michaels, *Ind. Eng. Chem.*, **44**, 1922 (1952).
D: Glueckauf Coates, *J. Chem. Soc.*, **1947**, 1315 (1947); Vermeulen, *Advances in Chemical Engineering*, **2**, 147 (1958); Hall et al., *Ind. Eng. Chem. Fundam.*, **5**, 212 (1966); Miura and Hashimoto, *J. Chem. Eng. Japan*, **10**, 490 (1977).
E: Walter, *J. Chem. Phys.*, **13**, 229 (1945); Hiester and Vermeulen, *Chem. Eng. Progress*, **48**, 505 (1952).
F: Acrivos, *Chem. Eng. Sci.*, **13**, 1 (1960); Coppola and LeVan, *Chem. Eng. Sci.*, **36**, 967 (1981).

$$-\frac{\partial n_i^\circ}{\partial(\zeta - \tau_1)} + \frac{\partial n_i^\circ}{\partial \tau_1} + \frac{\partial c_i^\circ}{\partial(\zeta - \tau_1)} = \frac{1}{N_{Pe}}\frac{\partial^2 c_i^\circ}{\partial(\zeta - \tau_1)^2} \quad (16\text{-}139)$$

The partial derivative with respect to τ_1 is discarded and the resulting equation integrated once to give

$$-n_i^\circ + c_i^\circ = \frac{1}{N_{Pe}}\frac{dc_i^\circ}{d(\zeta - \tau_1)} \quad (16\text{-}140)$$

After eliminating n_i° or c_i° using the adsorption isotherm, Eq. (16-140) can be integrated directly to obtain the constant pattern.

For other mechanisms, the particle-scale equation must be integrated. Equation (16-140) is used to advantage. For example, for external mass transfer acting alone, the dimensionless rate equation in Table 16-13 would be transformed into the $(\zeta - \tau_1, \tau_1)$ coordinate system and derivatives with respect to τ_1 discarded. Equation (16-138) is then used to replace c_i° with $\overline{n}_i^\circ$ in the transformed equation. Furthermore, for this case there are assumed to be no gradients within the particles, so we have $\overline{n}_i^\circ = n_i^\circ$. After making this substitution, the transformed equation can be rearranged to

$$-\frac{dn_i^\circ}{n_i^\circ - c_i^\circ} = d(\zeta - \tau_1) \quad (16\text{-}141)$$

Since n_i° and c_i° are related by the adsorption isotherm, Eq. (16-141) can be integrated.

The integration of Eq. (16-140) or (16-141) as an indefinite integral will give an integration constant that must be evaluated to center the transition properly. The material balance depicted in Fig. 16-26 is used. The two shaded regions must be of equal area if the stoichiometric center of the transition is located where the throughput parameter is unity. Thus, we have

$$\int_{-\infty}^{0}(1 - n_i^\circ)\,d(\zeta - \tau_1) = \int_{0}^{\infty} n_i^\circ\,d(\zeta - \tau_1) \quad (16\text{-}142)$$

Integrating Eq. (16-142) by parts gives

$$\int_{0}^{1}(N - NT)\,dn_i^\circ = 0 \quad (16\text{-}143)$$

For all mechanisms except axial dispersion, the transition can be centered just as well using c_i°, because of Eq. (16-138). For axial dispersion, the transition should be centered using n_i°, provided the fluid-phase accumulation term in the material balance, Eq. (16-124),

can be neglected. If fluid-phase accumulation is important, then the transition for axial dispersion can be centered by taking into account the relative quantities of solute held in the fluid and adsorbed phases.

Constant pattern solutions for the individual mechanisms and constant separation factor isotherm are given in Table 16-13. The solutions all have the expected dependence on R—the more favorable the isotherm, the sharper the profile.

Figure 16-27 compares the various constant pattern solutions for $R = 0.5$. The curves are of a similar shape. The solution for reaction kinetics is perfectly symmetrical. The curves for the axial dispersion fluid-phase concentration profile and the linear driving force approximation are identical except that the latter occurs one transfer unit further down the bed. The curve for external mass transfer is exactly that for the linear driving force approximation turned upside down [i.e., rotated 180° about $c_i^\circ = n_i^\circ = 0.5$, $N(\zeta - \tau_1) = 0$]. The linear driving force approximation provides a good approximation for both pore diffusion and surface diffusion.

Because of the close similarity in shape of the profiles shown in Fig. 16-27 (as well as likely variations in parameters; e.g., concentration-dependent surface diffusion coefficient), a controlling mechanism cannot be reliably determined from transition shape. If reliable correlations are not available and rate parameters cannot be measured in independent experiments, then particle diameters, velocities, and other factors should be varied and the observed impact considered in relation to the definitions of the numbers of transfer units.

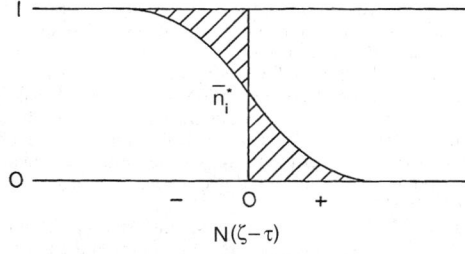

FIG. 16-26 Material balance for centering profile [*from LeVan in Rodrigues et al. (eds), Adsorption: Science and Technology, Kluwer Academic Publishers, Dordrecht, The Netherlands, 1989; reprinted with permission.*]

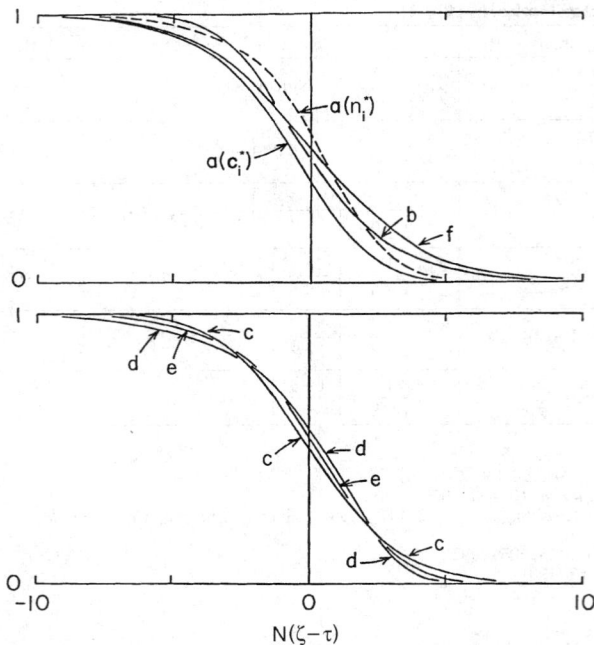

FIG. 16-27 Constant pattern solutions for $R = 0.5$. Ordinant is c_i° or n_i° except for axial dispersion for which individual curves are labeled: a, axial dispersion; b, external mass transfer; c, pore diffusion (spherical particles); d, surface diffusion (spherical particles); e, linear driving force approximation; f, reaction kinetics. [*from LeVan in Rodrigues et al. (eds.), Adsorption: Science and Technology, Kluwer Academic Publishers, Dordrecht, The Netherlands, 1989; reprinted with permission.*]

Breakthrough Behavior for Axial Dispersion Breakthrough behavior for adsorption with axial dispersion in a deep bed is not adequately described by the constant pattern profile for this mechanism. Equation (16-128), the partial differential equation of the second order Fickian model, requires two boundary conditions for its solution. The constant pattern pertains to a bed of infinite depth—in obtaining the solution we apply the downstream boundary condition $c_i^\circ \to 0$ as $N_{Pe}\zeta \to \infty$. Breakthrough behavior presumes the existence of a bed outlet, and a boundary condition must be applied there.

The full mathematical model for this problem is Eq. (16-128) with boundary conditions

$$c_i^\circ - \frac{1}{N_{Pe}}\frac{\partial c_i^\circ}{\partial \zeta} = 1 \quad \text{at} \quad \zeta = 0 \qquad (16\text{-}144)$$

$$\frac{\partial c_i^\circ}{\partial \zeta} = 0 \quad \text{at} \quad \zeta = 1 \qquad (16\text{-}145)$$

and an initial condition. Equation (16-144) specifies a constant flux at the bed inlet, and Eq. (16-145), the Danckwerts-type boundary condition at the bed outlet, is appropriate for fixed-bed adsorption, provided that the partition ratio is large.

The solution to this model for a deep bed indicates an increase in velocity of the fluid-phase concentration wave during breakthrough. This is most dramatic for the rectangular isotherm—the instant the bed becomes saturated, the fluid-phase profile jumps in velocity from that of the adsorption transition to that of the fluid, and a near shock-like breakthrough curve is observed [Coppola and LeVan, *Chem. Eng. Sci.*, **36**, 967 (1981)].

Extensions Existence, uniqueness, and stability criteria have been developed for the constant pattern [Cooney and Lightfoot, *Ind. Eng. Chem. Fundam.*, **4**, 233 (1965); Rhee et al., *Chem. Eng. Sci.*, **26**, 1571 (1971); Rhee and Amundson, *Chem. Eng. Sci.*, **26**, 1571 (1971), **27**, 199 (1972), **29**, 2049 (1974)].

The rectangular isotherm has received special attention. For this, many of the constant patterns are developed fully at the bed inlet, as shown for external mass transfer [Klotz, *Chem. Revs.*, **39**, 241 (1946)], pore diffusion [Vermeulen, *Adv. Chem. Eng.*, **2**, 147 (1958); Hall et al., *Ind. Eng. Chem. Fundam.*, **5**, 212 (1966)], the linear driving force approximation [Cooper, *Ind. Eng. Chem. Fundam.*, **4**, 308 (1965)], reaction kinetics [Hiester and Vermeulen, *Chem. Eng. Progress*, **48**, 505 (1952); Bohart and Adams, *J. Amer. Chem. Soc.*, **42**, 523 (1920)], and axial dispersion [Coppola and LeVan, *Chem. Eng. Sci.*, **38**, 991 (1983)].

Multiple mass-transfer resistances have been considered in many studies [Vermeulen, *Adv. in Chem. Eng.*, **2**, 147 (1958); Vermeulen et al., Ruthven, gen. refs.; Fleck et al., *Ind. Eng. Chem. Fundam.*, **12**, 95 (1973); Yoshida et al., *Chem. Eng. Sci.*, **39**, 1489 (1984)].

Treatments of constant pattern behavior have been carried out for multicomponent adsorption [Vermeulen, *Adv. in Chem. Eng.*, **2**, 147 (1958); Vermeulen et al., Ruthven, gen. refs.; Rhee and Amundson, *Chem. Eng. Sci.*, **29**, 2049 (1974); Cooney and Lightfoot, *Ind. Eng. Chem. Fundam.*, **5**, 25 (1966); Cooney and Strusi, *Ind. Eng. Chem. Fundam.*, **11**, 123 (1972); Bradley and Sweed, *AIChE Symp. Ser. No. 152*, **71**, 59 (1975)]. The behavior is such that coexisting compositions advance through the bed together at a uniform rate; this is the *coherence* concept of Helfferich and coworkers [gen. refs.].

Nontrace systems have been considered [Sircar and Kumar, *Ind. Eng. Chem. Proc. Des. Dev.*, **22**, 271 (1983)].

Constant patterns have been developed for adiabatic adsorption [Pan and Basmadjian, *Chem. Eng. Sci.*, **22**, 285 (1967); Ruthven et al., *Chem. Eng. Sci.*, **30**, 803 (1975); Kaguei et al., *Chem. Eng. Sci.*, **42**, 2964 (1987)].

The constant pattern concept has also been extended to circumstances with nonplug flows, with various degrees of rigor, including flow profiles in tubes [Sartory, *Ind. Eng. Chem. Fundam.*, **17**, 97 (1978); Tereck et al., *Ind. Eng. Chem. Res.*, **26**, 1222 (1987)], wall effects [Vortmeyer and Michael, *Chem. Eng. Sci.*, **40**, 2135 (1985)], channeling [LeVan and Vermeulen in Myers and Belfort (eds.), *Fundamentals of Adsorption*, Engineering Foundation, New York (1984), pp. 305–314, *AIChE Symp. Ser. No. 233*, **80**, 34 (1984)], networks [Avilés and LeVan, *Chem. Eng. Sci.*, **46**, 1935 (1991)], and general structures of constant cross section [Rudisill and LeVan, *Ind. Eng. Chem. Res.*, **29**, 1054 (1991)].

SQUARE ROOT SPREADING FOR LINEAR ISOTHERMS

The simplest isotherm is $n_i^\circ = c_i^\circ$, corresponding to $R = 1$. For this isotherm, the rate equation for external mass transfer, the linear driving force approximation, or reaction kinetics, can be combined with Eq. (16-130) to obtain

$$\frac{\partial \overline{n}_i^\circ}{\partial \tau_1} = -\frac{\partial c_i^\circ}{\partial \zeta} = N(c_i^\circ - \overline{n}_i^\circ) \qquad (16\text{-}146)$$

The solution to this equation, with initial condition $n_i^\circ = 0$ at $\tau_1 = 0$ and boundary condition $c_i^\circ = 1$ at $\zeta = 0$, originally obtained for an analogous heat transfer case [Anzelius, *Z. Angew Math. Mech.*, **6**, 291 (1926); Schumann, *J. Franklin Inst.*, **208**, 405 (1929)], is

$$c_i^\circ = J(N\zeta, N\tau_1) \qquad n_i^\circ = 1 - J(N\tau_1, \zeta) \qquad (16\text{-}147)$$

where the J function is [Hiester and Vermeulen, *Chem. Eng. Prog.*, **48**, 505 (1952)]

$$J(s, t) = 1 - \int_0^s e^{-t-\xi} I_0(2\sqrt{t\xi})\, d\xi \qquad (16\text{-}148)$$

where I_0 is the modified Bessel function of the first kind of order zero. This linear isotherm result can be generalized to remove the assumption that $\hat{c}_i^\circ \approx c_i^\circ$ if the throughput parameter is redefined as

$$\tau_1 = \frac{\tau - \varepsilon\zeta}{(1-\varepsilon)(\rho_p K_i + \varepsilon_p)}$$

The J function is plotted in Fig. 16-28 and tables are available (e.g., Sherwood et al., *Mass Transfer*, McGraw-Hill, New York, 1975). Vermeulen et al. (gen. refs.) discuss several approximations of the J function. For large arguments it approaches

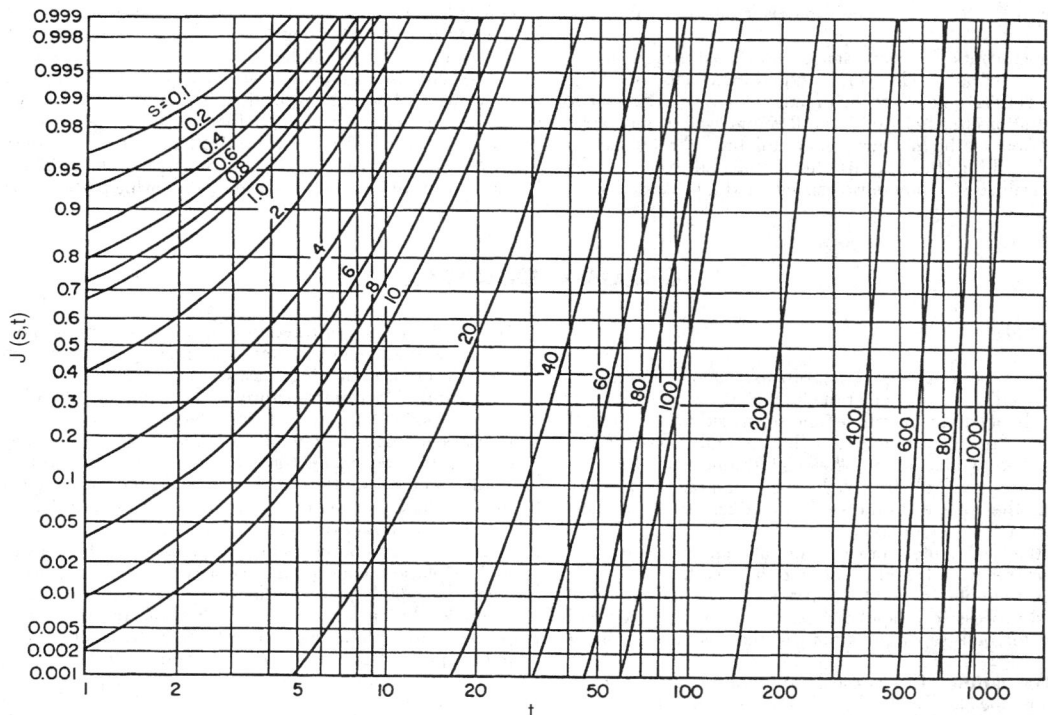

FIG. 16-28 Plot of the function $J(s, t)$ defined by Eq. (16-148).

$$J(s, t) = \frac{1}{2} \operatorname{erfc} \left[\sqrt{s} - \sqrt{t} \right] \qquad (16\text{-}149)$$

A derivation for particle-phase diffusion accompanied by fluid-side mass transfer has been carried out by Rosen [*J. Chem. Phys.*, **18**, 1587 (1950); ibid., **20**, 387 (1952); *Ind. Eng. Chem.*, **46**, 1590 (1954)] with a limiting form at $N > 50$ of

$$c_i^\circ = \frac{1}{2} \operatorname{erfc} \left[\frac{\sqrt{N}}{2} (\zeta - \tau_1) \right] \qquad (16\text{-}150)$$

For axial dispersion in a semi-infinite bed with a linear isotherm, the complete solution has been obtained for a constant flux inlet boundary condition [Lapidus and Amundson, *J. Phys. Chem.*, **56**, 984 (1952); Brenner, *Chem. Eng. Sci.*, **17**, 229 (1962); Coates and Smith, *Soc. Petrol. Engrs. J.*, **4**, 73 (1964)]. For large N, the leading term is

$$c_i^\circ = \frac{1}{2} \operatorname{erfc} \left[\frac{\sqrt{N}}{2\sqrt{\tau_1}} (\zeta - \tau_1) \right] \qquad (16\text{-}151)$$

All of these solutions are very similar and show, for large N, a wave with breadth proportional to the square root of the bed depth through which it has passed.

COMPLETE SOLUTION FOR REACTION KINETICS

In general, full time-dependent analytical solutions to differential equation-based models of the above mechanisms have not been found for nonlinear isotherms. Only for reaction kinetics with the constant separation factor isotherm has a full solution been found [Thomas, *J. Amer. Chem. Soc.*, **66**, 1664 (1944)]. Referred to as the *Thomas solution*, it has been extensively studied [Amundson, *J. Phys. Colloid Chem.*, **54**, 812 (1950); Hiester and Vermeulen, *Chem. Eng. Progress*, **48**, 505 (1952); Gilliland and Baddour, *Ind. Eng. Chem.*, **45**, 330 (1953); Vermeulen, *Adv. in Chem. Eng.*, **2**, 147 (1958)]. The solution to Eqs. (16-130) and (16-130) for the same boundary conditions as Eq. (16-146) is

$$c_i^\circ = \frac{J(RN\zeta, N\tau_1)}{J(RN\zeta, N\tau_1) + e^{-(R-1)N(\zeta-\tau_1)}[1 - J(N\tau_1, RN\zeta)]}$$

$$n_i^\circ = \frac{1 - J(N\tau_1, RN\zeta)}{J(RN\zeta, N\tau_1) + e^{-(R-1)N(\zeta-\tau_1)}[1 - J(N\tau_1, RN\zeta)]} \qquad (16\text{-}152)$$

The solution gives all of the expected asymptotic behaviors for large N—the proportionate pattern spreading of the simple wave if $R > 1$, the constant pattern if $R < 1$, and square root spreading for $R = 1$.

NUMERICAL METHODS AND CHARACTERIZATION OF WAVE SHAPE

For the solution of sophisticated mathematical models of adsorption cycles including complex multicomponent equilibrium and rate expressions, two numerical methods are popular. These are finite difference methods and orthogonal collocation. The former vary in the manner in which distance variables are discretized, ranging from simple backward difference stage models (akin to the plate theory of chromatography) to more involved schemes exhibiting little numerical dispersion. Collocation methods are often thought to be faster computationally, but oscillations in the polynomial trial function can be a problem. The choice of best method is often the preference of the user.

For both the finite difference and collocation methods a set of coupled ordinary differential equations results which are integrated forward in time using the method of lines. Various software packages implementing Gear's method are popular.

The development of mathematical models is described in several of the general references [Guiochon et al., Rhee et al., Ruthven, Ruthven et al., Suzuki, Tien, Wankat, and Yang]. See also Finlayson [*Numerical Methods for Problems with Moving Fronts*, Ravenna Park, Washington, 1992; Holland and Liapis, *Computer Methods for Solving Dynamic Separation Problems*, McGraw-Hill, New York, 1982; Villadsen and Michelsen, *Solution of Differential Equation Models by*

Polynomial Approximation, Prentice Hall, Englewood Cliffs, New Jersey, 1978].

For the characterization of wave shape and breakthrough curves, three methods are popular. The MTZ method [Michaels, *Ind. Eng. Chem.,* **44,** 1922 (1952)] measures the breadth of a wave between two chosen concentrations (e.g., $c_i^o = 0.05$ and 0.95 or $c_i^o = 0.01$ and 0.99). Outside of a laboratory, the measurement of full breakthrough curves is rare, so the breadth of the MTZ is often estimated from an independently determined stoichiometric capacity and a measured small concentration in the "toe" of the breakthrough curve. A second method for characterizing wave shape is by the slope of the breakthrough curve at its midheight (i.e., $c_i^o = 0.5$ [Vermeulen et al., gen. refs.]). The use of moments of the slope of breakthrough curves is a third means for characterization. They can often be used to extract numerical values for rate coefficients for linear systems [Ruthven, Suzuki, gen. refs.; Nauman and Buffham, *Mixing in Continuous Flow Systems,* Wiley-Interscience, New York, 1983]. The method of moments is discussed further in the following part of this section.

CHROMATOGRAPHY

CLASSIFICATION

Chromatography is a sorptive separation process where a portion of a solute mixture (feed) is introduced at the inlet of a column containing a selective adsorbent (stationary phase) and separated over the length of the column by the action of a carrier fluid (mobile phase) that is continually supplied to the column following introduction of the feed. The mobile phase is generally free of the feed components, but may contain various other species introduced to modulate the chromatographic separation. The separation occurs as a result of the different partitioning of the feed solutes between the stationary phase; the separated solutes are recovered at different times in the effluent from the column. Chromatography is used both in analysis of mixtures and in preparative and process-scale applications. It can be used for both trace-level and for bulk separations both in the gas and the liquid phase.

Modes of Operation The classical modes of operation of chromatography as enunciated by Tiselius [*Kolloid Z.,* **105,** 101 (1943)] are: elution chromatography, frontal analysis, and displacement development. Basic features of these techniques are illustrated in Fig. 16-29. Often, each of the different modes can be implemented with the same equipment and stationary phase. The results are, however, quite different in the three cases.

Elution Chromatography The components of the mobile phase supplied to the column after feed introduction have less affinity for the stationary phase than any of the feed solutes. Under trace conditions, the feed solutes travel through the column as bands or zones at different velocities that depend only on the composition of the mobile phase and the operating temperature and that exit from the column at different times.

Two variations of the technique exists: **isocratic elution,** when the mobile phase composition is kept constant, and **gradient elution,** when the mobile phase composition is varied during the separation. Isocratic elution is often the method of choice for analysis and in process applications when the retention characteristics of the solutes to be separated are similar and not dramatically sensitive to very small changes in operating conditions. Isocratic elution is also generally practical for systems where the equilibrium isotherm is linear or nearly linear. In all cases, isocratic elution results in a dilution of the separated products.

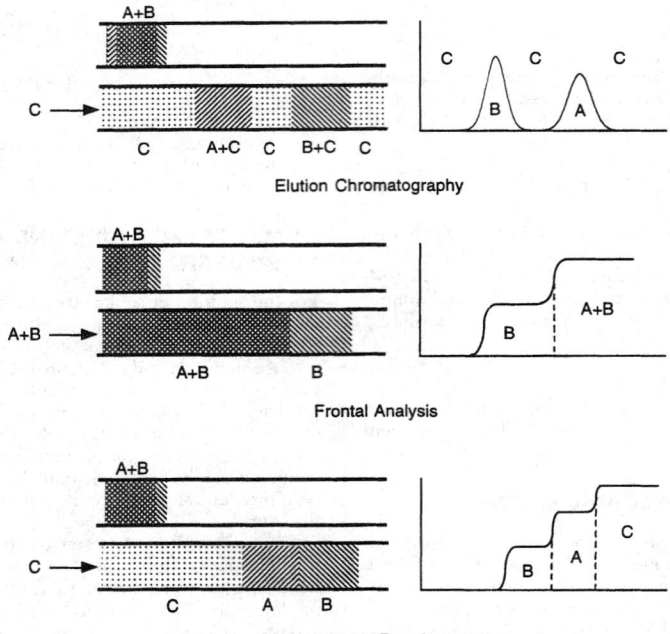

FIG. 16-29 Modes of operation of chromatography for the separation of a mixture of two components A and B. Figures on the left represent a schematic of the column with sample passing through it. Top diagrams show column at end of feed step. Figures on the right show the corresponding effluent concentrations. C is either the eluent or the displacer (adapted from Ettre, 1980).

In **gradient elution,** the eluting strength of the mobile phase is gradually increased after supplying the feed to the column. In liquid chromatography, this is accomplished by changing the mobile phase composition. The gradient in eluting strength of the mobile phase that is generated in the column is used to modulate the separation allowing control of the retention time of weakly and strongly retained components. A similar effect can be obtained in gas chromatography by modulating the column temperature. In either case, the column has to be brought back to the initial conditions before the next cycle is commenced.

Generally, gradient elution is best suited for the separation of complex mixtures that contain both species that interact weakly with the stationary phase and species that interact strongly. Since the eluting strength of the mobile phase is adjusted continuously, weakly retained components of a mixture are separated in the initial phase when the relative elution strength of the mobile phase is low, while strongly retained components are separated later in the gradient when the elution strength is high. In addition, the technique is used to obtain reproducible chromatographic separations when the solute retention characteristics are extremely sensitive to the operating conditions, as in the case of the chromatography of biopolymers, such as proteins. These molecules are often found to be very strongly retained in an extremely small range of mobile phase compositions and completely unretained elsewhere, making it practically impossible to obtain reproducible isocratic separations.

Frontal Analysis The feed mixture to be separated is continuously supplied to the column where the mixture components are competitively retained by the stationary phase. These solutes are partially separated in a series of fronts, preceded downstream by the least strongly retained species forming a pure component band, and upstream by the feed mixture. The technique is best suited for the removal of strongly adsorbed impurities present in trace amounts from an unretained or weakly retained product of interest. In this case, a large amount of feed can be processed before the impurities begin to break through. When this point is reached, the bed is washed to remove any desired product from the interstitial voids, and the adsorbent is regenerated. The method can only provide a single component in pure form, but avoids product dilution completely. Multicomponent separations require a series of processing steps; either a

series of frontal analysis separations, or a combination of elution and displacement separations. Example 10 illustrates bed concentration profiles for the frontal analysis separation of two components.

Displacement Development The column is partially loaded with the feed mixture as in frontal analysis, usually for conditions where all solutes of interest are strongly and competitively retained by the stationary phase. The feed supply is then stopped and a mobile phase containing the **displacer,** a component that has an affinity for the stationary phase stronger than any of the feed components, is fed to the column. The advancement of the displacer front through the column causes desorption of the feed components and their competitive readsorption downstream of the displacer front. As in frontal chromatography, the less strongly retained species tends to migrate faster down the column concentrating in a band farthest from the displacer front, while the most strongly adsorbed solute tends to move more slowly concentrating in a band adjacent to the displacer front. If the column is sufficiently long, all feed components eventually become distributed into a pattern of adjacent pure component bands where each upstream component acts as a displacer for each downstream species located in the band immediately downstream. When this occurs, all bands in the displacement train move at the same velocity which is equal to the velocity of the displacer front and the bed concentration profile is called an **isotachic pattern.**

The various operational steps of a displacement development separation are shown in Fig. 16-30. Ideally, the separated species exit the column as adjacent rectangular bands in order of increasing affinity for the stationary phase as shown in this figure. In practice, dispersion effects result in a partial mixing of adjacent bands requiring recycling of portions of the effluent that do not meet purity requirements. Following separation, the displacer has to be removed from the column with a suitable regenerant and the initial conditions of the column restored before the next cycle. Column regeneration may consume a significant portion of the cycle, when removal of the displacer is difficult.

Displacement chromatography is suitable for the separation of multicomponent bulk mixtures. For dilute multicomponent mixtures it allows a simultaneous separation and concentration. Thus, it permits the separation of compounds with extremely low separation factors without the excessive dilution that would be obtained in elution techniques.

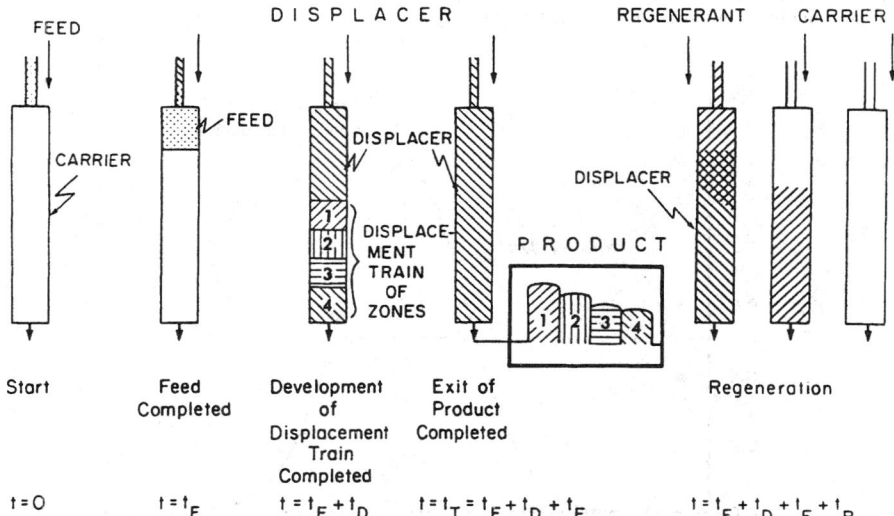

FIG. 16-30 Operational steps in displacement chromatography. The column, initially equilibrated with a carrier solvent at time 0, is loaded with feed until time t_F and supplied with displacer for a time $t_D + t_E$. Development of the displacement train occurs during the time t_D and elution of the separated products ends at time t_E. t_R is the time required to remove the displacer from the column and restore the initial conditions. Components are numbered in order of decreasing affinity for the stationary phase. [*Reference: Horvath et al., J. Chromatogr.,* **218,** 365 (1981). *Reprinted with permission of* J. Chromatogr.]

Other modes of operation, including recycle and flow reversal schemes and continuous chromatography, are discussed in Ganetsos and Barker (*Preparative and Production Scale Chromatography*, Marcel Dekker, New York, 1993).

CHARACTERIZATION OF EXPERIMENTAL CHROMATOGRAMS

METHOD OF MOMENTS

Method of Moments   The first step in the analysis of chromatographic systems is often a characterization of the column response to small pulse injections of a solute under trace conditions in the Henry's law limit. For such conditions, the statistical moments of the response peak are used to characterize the chromatographic behavior. Such an approach is generally preferable to other descriptions of peak properties which are specific to Gaussian behavior, since the statistical moments are directly correlated to equilibrium and dispersion parameters. Useful references are Schneider and Smith [*AIChE J.*, **14**, 762 (1968)], Suzuki and Smith [*Chem. Eng. Sci.*, **26**, 221 (1971)], and Carbonell et al. [*Chem. Eng. Sci.*, **9**, 115 (1975); **16**, 221 (1978)].

The relevant moments are:

$$\mu_0 = \int_0^\infty c_i \, dt \tag{16-153}$$

$$\mu_1 = \frac{1}{\mu_0} \int_0^\infty c_i t \, dt \tag{16-154}$$

$$\sigma^2 = \frac{1}{\mu_0} \int_0^\infty c_i (t - \mu_1)^2 \, dt \tag{16-155}$$

where c_i is the peak profile. μ_0 represents the area, μ_1 the mean residence time, and σ^2 the variance of the response peak. Moments can be calculated by numerical integration of experimental profiles.

The **retention factor** is defined as a dimensionless peak locator as:

$$k_i' = \frac{\mu_1 - \mu_1^0}{\mu_1^0} \tag{16-156}$$

where μ_1^0 is the first moment obtained for an inert tracer which is excluded from the stationary phase. k_i' is a partition ratio representing the equilibrium ratio of the amount of solute in the stationary phase (including any pores) and the amount in the external mobile phase. Another commonly used definition of the retention factor uses as a reference the first moment of an inert that has access to all the pores.

The number of plates, N_p, and the height equivalent to a theoretical plate, HETP, are defined as measures of dispersion effects as:

$$N_p = \frac{L}{\text{HETP}} = \frac{\mu_1^2}{\sigma^2} \tag{16-157}$$

$$\text{HETP} = \frac{\sigma^2 L}{\mu_1^2} \tag{16-158}$$

A high number of plates and a low HETP indicate a high column efficiency.

Higher moments can also be computed and used to define the skewness of the response peak. However, difficulties often arise in such computations as a result of drifting of the detection system.

In practice, experimental peaks can be affected by extracolumn retention and dispersion factors associated with the injector, connections, and any detector. For linear chromatography conditions, the apparent response parameters are related to their corresponding true column value by

$$\mu_1^{app} = \mu_1 + \mu_1^{inj} + \mu_1^{conn} + \mu_1^{det} \tag{16-159}$$

$$\sigma^{2app} = \sigma^2 + \sigma^{2inj} + \sigma^{2conn} + \sigma^{2det} \tag{16-160}$$

Approximate Methods   For certain conditions, symmetrical, Gaussian-like peaks are obtained experimentally. Such peaks may be empirically described by:

$$c_i = \frac{Q_i/F_v}{\sigma\sqrt{2\pi}} \exp\left[-\left(\frac{t - t_{Ri}}{\sigma\sqrt{2}}\right)^2\right] = \frac{Q_i/F_v}{t_{Ri}} \sqrt{\frac{N_p}{2\pi}} \exp\left[-\frac{N_p}{2}\left(\frac{t}{t_{Ri}} - 1\right)^2\right] \tag{16-161}$$

where Q_i is the amount of solute injected, F_v is the volumetric flow rate, and t_{Ri} is the peak apex time. The relationships between the moments and other properties of such peaks are shown in Fig. 16-31. For such peaks, approximate calculations of the number of plates can be done with the following equations:

$$N_p = 5.54\left(\frac{t_{Ri}}{\Delta}\right)^2 \tag{16-162}$$

$$N_p = 16\left(\frac{t_{Ri}}{W}\right)^2 \tag{16-163}$$

$$N_p = 2\pi\left(\frac{c_i^{max} t_{Ri}}{\mu_0}\right)^2 \tag{16-164}$$

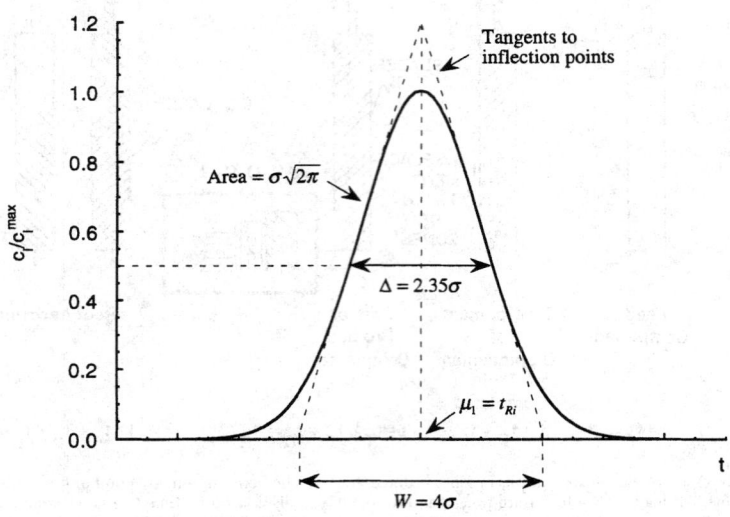

FIG. 16-31   Properties of a Gaussian peak. c_i^{max} is the peak height; t_{Ri}, the peak apex time; σ, the standard deviation; Δ, the peak width at midheight; and W, the distance between the baseline intercepts of the tangents to the peak.

where Δ is the peak width at half peak height and W is the distance between the baseline intercepts of the tangents to the inflection points of the peak.

In general, Gaussian behavior can be tested by plotting the cumulative fractional recovery $\int_0^t c_i \, dt / \mu_0$ versus time on probability-linear coordinates; if the plot is linear, Gaussian behavior is confirmed. For nearly Gaussian peaks, calculations of N_p based on Eqs. (16-162) to (16-164) provide results close to those obtained with a rigorous calculation of moments. When deviations from Gaussian behavior are significant, however, large errors can be obtained using these expressions.

Tailing Peaks Tailing peaks can be obtained experimentally when the column efficiency is very low, when there are large extracolumn dispersion effects, when the stationary phase is heterogeneous (in the sense that it contains different adsorption sites), or when the adsorption equilibrium deviates from the Henry's law limit. Asymmetrical tailing peaks can some times be described empirically by an exponentially modified Gaussian (EMG) defined as the convolute integral of a Gaussian constituent with mean time t_G and standard deviation σ_G and an exponential decay with time constant τ_G [Grushka, *Anal. Chem.*, **44**, 1733 (1972)]:

$$c_i = \frac{Q_i / F_v}{\sigma_G \tau_G \sqrt{2\pi}} \int_0^\infty \exp\left[-\left(\frac{t - t_G - t'}{\sigma_G \sqrt{2}}\right)^2 - \frac{t'}{\tau_G}\right] dt' \quad (16\text{-}165)$$

Although a numerical integration is required to compute the peak profile, the moments are calculated directly as $\mu_1 = t_G + \tau_G$ and $\sigma^2 = \sigma_G^2 + \tau_G^2$ and the **peak skew** as:

$$\text{Peak skew} = \frac{1}{\mu_0 \sigma^3} \int_0^\infty c_i (t - \mu_1)^3 \, dt \sim \frac{2(\tau_G / \sigma_G)^3}{[1 + (\tau_G / \sigma_G)^2]^{3/2}} \quad (16\text{-}166)$$

For EMG peaks, peak skew increases with the ratio τ_G / σ_G. Figure 16-32 illustrates the characteristics of such a peak calculated for $\tau_G / \sigma_G = 1.5$. In general, with $\tau_G / \sigma_G > 1$ (peak skew > 0.7), a direct calculation of the moments is required to obtained a good approximation of the true value of N_p, since other methods give a large error (Yau et al., *Moderns Size-Exclusion Liquid Chromatography*, Wiley, New York, 1979). Alternatively, Eq. (16-165) can be fitted to experimental peaks to determine the optimum values of t_G, σ_G, and τ_G.

In practice, the calculation of peak skew for highly tailing peaks is rendered difficult by baseline errors in the calculation of third moments. The **peak asymmetry factor,** $A_s = b/a$, at 10 percent of peak height (see Fig. 16-32) is thus frequently used. An approximate relationship between peak skew and A_s for tailing peaks, based on data in Yau et al. is: Peak skew = $[0.51 + 0.19/(A_s - 1)]^{-1}$. Values of $A_s < 1.25$ (corresponding to peak skew < 0.7) are generally desirable for an efficient chromatographic separation.

Resolution The chromatographic separation of two components, A and B, under trace conditions with small feed injections can be characterized in terms of the resolution, R_s. For nearly Gaussian peaks:

$$R_s = \frac{2(t_{R,A} - t_{R,B})}{W_A + W_B} \sim \frac{\Delta t_R}{4\sigma_{AB}} \quad (16\text{-}167)$$

where Δt_R is the difference in retention time of the two peaks and $\sigma_{AB} = (\sigma_A + \sigma_B)/2$ is the average of their standard deviations. When Eq. (16-161) is applicable, the resolution for two closely spaced peaks is approximated by:

$$R_s = \frac{1}{2} \frac{\alpha - 1}{\alpha + 1} \frac{\bar{k}_i'}{1 + \bar{k}_i'} \sqrt{N_p}$$

$$\sim \frac{\alpha - 1}{4} \frac{k_A'}{1 + k_A'} \sqrt{N_p}, \quad \text{for } \alpha \sim 1 \quad (16\text{-}168)$$

where $\alpha = k_A' / k_B'$ and $\bar{k}' = (k_A' + k_B')/2$.

Equation (16-168) shows that the resolution is the result of independent effects of the separation selectivity (α), column efficiency (N_p), and capacity ($\bar{k}'$). Generally, peaks are essentially completely resolved when $R_s = 1.5$ (>99.5 percent separation). In practice, values of $R_s \sim 1$, corresponding to 98 percent separation, are often considered adequate.

The preceding equations are accurate to within about 10 percent for feed injections that do not exceed 40 percent of the final peak width. For large, rectangular feed injections, the baseline width of the response peak is approximated by:

$$W \sim 4\sigma + t_F \quad (16\text{-}169)$$

where 4σ is the baseline width obtained with a small pulse injection and t_F is the duration of feed injection. In this case, the resolution is defined as (see Ruthven, gen. refs., pp. 324–331):

$$R_s = \frac{t_{R,A} - t_{R,B} - t_F}{4\sigma_{AB}} \quad (16\text{-}170)$$

For strongly retained components ($\bar{k}' \gg 1$), the number of plates required to obtain a given resolution with a finite feed injection is approximated by:

$$N_p = 4R_s^2 \left(\frac{\alpha + 1}{\alpha - 1}\right)^2 \left(1 + \frac{t_F}{4\sigma_{AB}R_s}\right)^2 \quad (16\text{-}171)$$

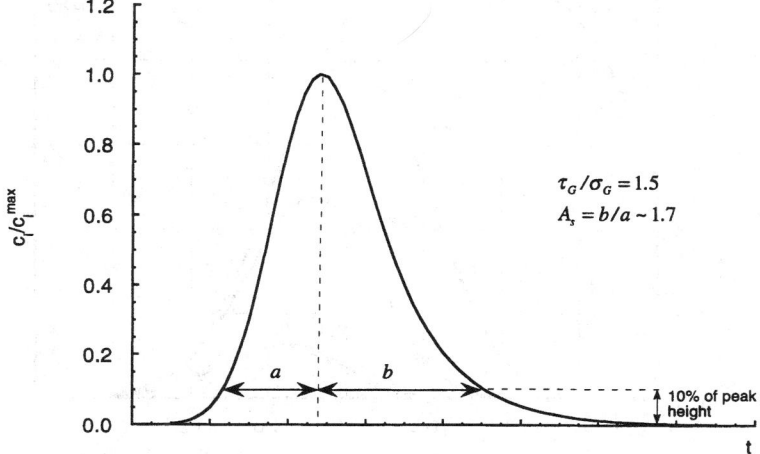

FIG. 16-32 Exponentially modified Gaussian peak with $\tau_G / \sigma_G = 1.5$. The graph also shows the definition of the peak asymmetry factor A_s at 10 percent of peak height.

PREDICTION OF CHROMATOGRAPHIC BEHAVIOR

The conservation equations and the rate models described in the "Rate and Dispersion Factors" can normally be used for a quantitative description of chromatographic separations. Alternatively, **plate models** can be used for an approximate prediction, lumping together all dispersion contributions into a single parameter, the HETP or the number of plates [Sherwood et al., *Mass Transfer,* McGraw-Hill, New York, 1975, p. 576; Dondi and Guiochon, *Theoretical Advancements in Chromatography and Related Techniques,* NATO-ASI, Series C: Mathematical and Physical Sciences, vol. 383, Kluwer, Dordrecht, 1992, pp. 1–61]. Exact analytic solutions are generally available for linear isocratic elution under trace conditions (see Dondi and Guiochon, ibid., Ruthven [gen. refs., pp. 324–335], Suzuki [gen. refs., pp. 224–243]). Other cases generally require numerical solution (see Guiochon et al., gen. refs.) or approximate treatments with simplified rate models.

Isocratic Elution In the simplest case, feed with concentration c_i^F is applied to the column for a time t_F followed by the pure carrier fluid. Under trace conditions, for a linear isotherm with external mass-transfer control, the linear driving force approximation or reaction kinetics (see Table 16-12), solution of Eq. (16-146) gives the following expression for the dimensionless solute concentration at the column outlet:

$$c_i^\circ = J(N, N\tau_1) - J(N, N\tau_1') \qquad (16\text{-}172)$$

where N is the number of transfer units given in Table 16-13 and $\tau_1 = (\varepsilon v t/L - \varepsilon)/[(1 - \varepsilon)(\rho_p K_i + \varepsilon_p)]$ the throughput parameter (see "Square Root Spreading for Linear Isotherms" in "Fixed Bed Transitions"). τ_1' represents the value of τ_1 with time measured from the end of the feed step. Thus, the column effluent profile is obtained as the difference between a breakthrough profile for a feed started at $t = 0$ and another for a feed started at $t = t_F$.

The behavior predicted by this equation is illustrated in Fig. 16-33 with $N = 80$. $\tau_F = (\varepsilon v t_F/L)/[(1 - \varepsilon)(\rho_p K_i + \varepsilon_p)]$ is the dimensionless duration of the feed step and is equal to the amount of solute fed to the column divided by the sorption capacity. Thus, at $\tau_F = 1$, the column has been supplied with an amount of solute equal to the stationary phase capacity. The graph shows the transition from a case where complete saturation of the bed occurs before elution ($\tau_F = 1$) to incomplete saturation as τ_F is progressively reduced. The lower curves with $\tau_F \leq 0.4$ are seen to be nearly Gaussian and centered at a dimensionless time $\tau_m \sim (1 - \tau_F/2)$. Thus, as $\tau_F \to 0$, the response curve approaches a Gaussian centered at $\tau_1 = 1$.

When τ_F is small ($<<0.4$), the solution for a feed pulse represented by a Dirac's delta function at $\zeta = 0$ can be used in lieu of Eq. (16-172). In terms of the dimensionless concentration $c_i^\circ = c_i/c_i^F$ (Sherwood et al., ibid., pp. 571–577):

$$c_i^\circ = \frac{\tau_F N}{\sqrt{\tau_1}} e^{-N} e^{-N\tau_1} I_1(2N\sqrt{\tau_1}) \qquad (16\text{-}173)$$

where I_1 is the Bessel function of imaginary argument. When N is larger than 5, this equation is approximated by:

$$c_i^\circ = \frac{\tau_F}{2\sqrt{\pi}} \sqrt{\frac{N}{\tau_1}} \frac{\exp[-N(\sqrt{\tau_1} - 1)^2]}{(\tau_1)^{1/4}} \qquad (16\text{-}174)$$

The behavior predicted by Eq. (16-174) is shown in Fig. 16-34 as $c_i^\circ/(\tau_F\sqrt{N})$ versus τ_1 for different values of N. For $N > 50$, the response peak is symmetrical, the peak apex occurs at $\tau_1 = 1$, and the dimensionless peak height is $c_i^{\circ\,\text{max}} = \tau_F \sqrt{N/4\pi}$. A comparison of this equation with Eq. (16-172) with $N = 80$ is shown in Fig. 16-33 for $\tau_F = 0.05$.

The moments of the response peak predicted by Eq. (16-173) are

$$\mu_1 = \frac{L}{v}\left[1 + \frac{1-\varepsilon}{\varepsilon}(\varepsilon_p + \rho_p K_i)\right] = \frac{L}{v}(1 + k_i') \qquad (16\text{-}175)$$

$$\sigma^2 = \frac{2\mu_1^2}{N}\left(\frac{k_i'}{1 + k_i'}\right)^2 \qquad (16\text{-}176)$$

where $k_i' = (1 - \varepsilon)(\varepsilon_p + \rho_p K_i)/\varepsilon$. Correspondingly, the number of plates and the HETP are:

$$N_p = \frac{N}{2}\left(\frac{1 + k_i'}{k_i'}\right)^2 \qquad (16\text{-}177)$$

$$\text{HETP} = \frac{2L}{N}\left(\frac{k_i'}{1 + k_i'}\right)^2 \qquad (16\text{-}178)$$

Since the term $(1 + k_i')/k_i'$ approaches unity for large k'-value, the number of plates is equal to one half the number of transfer units for a strongly retained component. For these conditions, when $N_p = N/2$, Eq. (16-174) and Eq. (16-161) produce the same peak retention time, peak spreading, and predict essentially the same peak profile.

In the general case of axially dispersed plug flow with bidispersed particles, the first and second moment of the pulse response are [Haynes and Sarma, *AIChE J.,* **19**, 1043 (1973)]:

$$\mu_1 = \frac{L}{v}\left[1 + \frac{1-\varepsilon}{\varepsilon}(\varepsilon_p + \rho_p K_i)\right] = \frac{L}{v}(1 + k_i') \qquad (16\text{-}179)$$

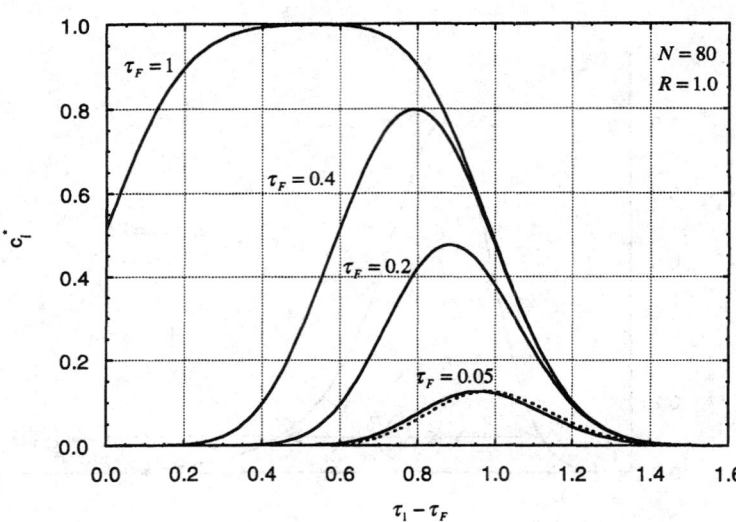

FIG. 16-33 Elution curves under trace linear equilibrium conditions for different feed loading periods and $N = 80$. Solid lines, Eq. (16-172); dashed line, Eq. (16-174) for $\tau_F = 0.05$.

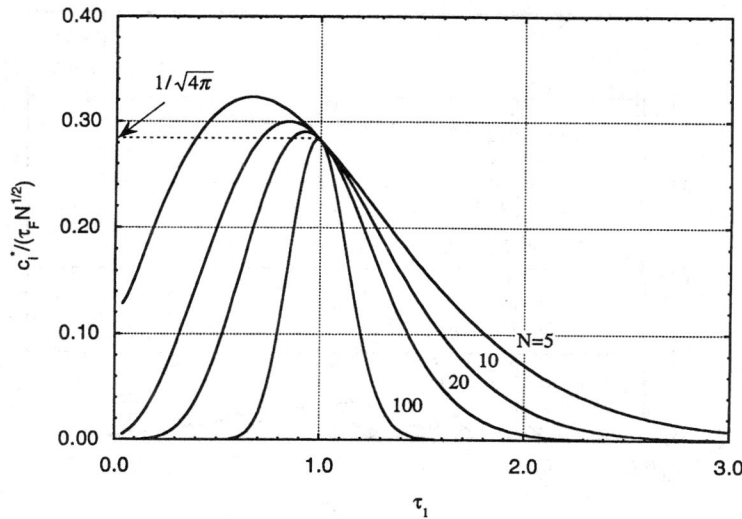

FIG. 16-34   Elution curves under trace linear equilibrium conditions with a pulse feed from Eq. (16-174).

$$\sigma^2 = \frac{2LD_L}{v^3}\,(1+k_i')^2$$

$$+ 2\,\frac{L}{v}\,\frac{\varepsilon k_i'^2}{1-\varepsilon}\left[\frac{r_p}{3k_f}+\frac{r_p^2}{15\varepsilon_p D_{pi}}+\frac{\rho_p K_i}{(\varepsilon_p+\rho_p K_i)^2}\,\frac{r_s^2}{15D_{si}}\right] \quad (16\text{-}180)$$

Correspondingly, the number of plates and the plate height are:

$$N_p = \left\{\frac{2D_L}{vL}+\frac{2\varepsilon}{1-\varepsilon}\,\frac{v}{L}\left(\frac{k_i'}{1+k_i'}\right)^2\right.$$

$$\left.\times\left[\frac{r_p}{3k_f}+\frac{r_p^2}{15\varepsilon_p D_{pi}}+\frac{\rho_p K_i}{(\varepsilon_p+\rho_p K_i)^2}\,\frac{r_s^2}{15D_{si}}\right]\right\}^{-1} \quad (16\text{-}181)$$

$$\text{HETP} = \frac{2D_L}{v}+\frac{2\varepsilon v}{1-\varepsilon}\left(\frac{k_i'}{1+k_i'}\right)^2$$

$$\times\left[\frac{r_p}{3k_f}+\frac{r_p^2}{15\varepsilon_p D_{pi}}+\frac{\rho_p K_i}{(\varepsilon_p+\rho_p K_i)^2}\,\frac{r_s^2}{15D_{si}}\right] \quad (16\text{-}182)$$

In dimensionless form, a **reduced HETP**, $h = \text{HETP}/d_p$, analogous to the reduced HTU (cf. Fig. 16-13), is obtained as a function of the dimensionless velocity $ReSc$:

$$h = \frac{b}{ReSc}+a+cReSc \quad (16\text{-}183)$$

where $b = 2\varepsilon\gamma_1$ $\qquad\qquad\qquad\qquad\qquad (16\text{-}184a)$

$a = 2\gamma_2$ $\qquad\qquad\qquad\qquad\qquad\qquad (16\text{-}184b)$

$$c = \frac{1}{30}\,\frac{1}{1-\varepsilon}\left(\frac{k_i'}{1+k_i'}\right)^2\left[\frac{10}{Sh}+\frac{\tau_p}{\varepsilon_p}+\frac{\rho_p K_i}{(\varepsilon_p+\rho_p K_i)^2}\,\frac{r_s^2}{D_{si}}\,\frac{D_{pi}}{r_p^2}\right]$$

$$(16\text{-}184c)$$

Equation 16-183 is qualitatively the same as the **van Deemter equation** [van Deemter and Zuiderweg, *Chem. Eng. Sci.*, **5**, 271 (1956)] and is equivalent to other empirical reduced HETP expressions such as the **Knox equation** [Knox, *J. Chromatogr. Sci.*, **15**, 352 (1977)].

The Sherwood number, Sh, is estimated from Table 16-9, and the dispersion parameters γ_1 and γ_2 from Table 16-10 for well-packed columns. Typical values are $a = 1$–4 and $b = 0.5$–1. Since HETP $\sim$ 2HTU, Fig. 16-13 can also be used for approximate calculations.

Concentration Profiles   In the general case but with a linear isotherm, the concentration profile can be found by numerical inversion of the Laplace-domain solution of Haynes and Sarma [see

Lenhoff, *J. Chromatogr.*, **384**, 285 (1987)] or by direct numerical solution of the conservation and rate equations. For the special case of no-axial dispersion with external mass transfer and pore diffusion, an explicit time-domain solution, useful for the case of time-periodic injections, is also available [Carta, *Chem. Eng. Sci.*, **43**, 2877 (1988)]. In most cases, however, when $N > 50$, use of Eq. (16-161), or (16-172) and (16-174) with $N = 2N_p$ calculated from Eq. (16-181) provides an approximation sufficiently accurate for most practical purposes.

When the adsorption equilibrium is nonlinear, skewed peaks are obtained, even when N is large. For a constant separation-factor isotherm with $R < 1$ (favorable), the leading edge of the chromatographic peak is steeper than the trailing edge. When $R > 1$ (unfavorable), the opposite is true.

Figure 16-35 portrays numerically calculated chromatographic peaks for a constant separation factor system showing the effect of feed loading on the elution profile with $R = 0.5$ ($\tau_1 - \tau_F = [\varepsilon v(t-t_F)/L - \varepsilon]/\Lambda$). When the dimensionless feed time $\tau_F = 1$, the elution curve comprises a sharp leading profile reaching the feed concentration followed by a gradual decline to zero. As τ_F is reduced, breakthrough of the leading edge occurs at later times, while the trailing edge, past the peak apex, continues to follow the same profile. As the amount of feed injected approaches zero, mass-transfer resistance reduces the solute concentration to values that fall in the Henry's law limit of the isotherm, and the peak retention time gradually approaches the value predicted in the infinite dilution limit for a linear isotherm.

For high feed loads, the shape of the diffuse trailing profile and the location of the leading front can be predicted from local equilibrium theory (see "Fixed Bed Transitions"). This is illustrated in Fig. 16-35 for $\tau_F = 0.4$. For the diffuse profile (a "simple wave"), Eq. (16-131) gives:

$$c_i^\circ = \frac{R}{1-R}\left[\left(\frac{1}{R(\tau_1-\tau_F)}\right)^{1/2}-1\right] \quad (16\text{-}185)$$

Thus, the effluent concentration becomes zero at $\tau_1 - \tau_F = 1/R$. The position of the leading edge (a "shock front") is determined from Eq. (16-132):

$$\tau_{1s} = \tau_F + \frac{1}{R}\left[1-\sqrt{(1-R)\tau_F}\right]^2 \quad (16\text{-}186)$$

and the peak highest concentration by [Golshan-Shirazi and Guiochon, *Anal. Chem.*, **60**, 2364 (1988)]:

$$c_i^{\circ\,\max} = \frac{R}{1-R}\,\frac{\sqrt{(1-R)\tau_F}}{1-\sqrt{(1-R)\tau_F}} \quad (16\text{-}187)$$

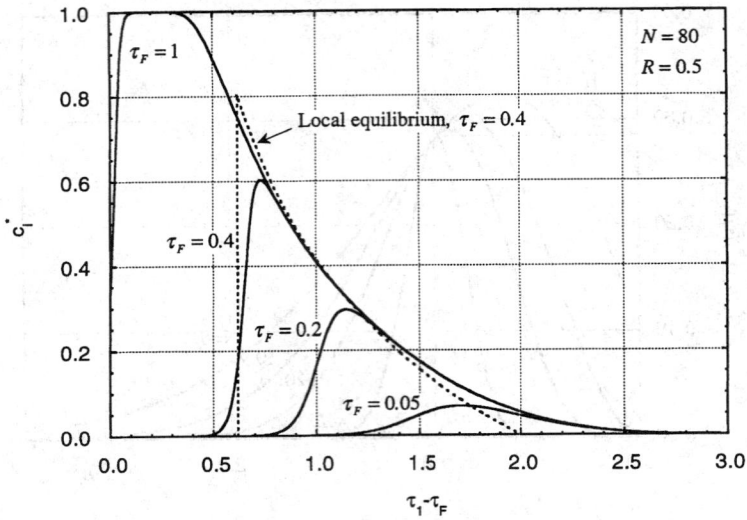

FIG. 16-35 Elution curves under trace conditions with a constant separation factor isotherm for different feed loadings and $N = 80$. Solid lines, rate model; dashed line, local equilibrium theory for $\tau_F = 0.4$.

The local equilibrium curve is in approximate agreement with the numerically calculated profiles except at very low concentrations when the isotherm becomes linear and near the peak apex. This occurs because band-spreading, in this case, is dominated by adsorption equilibrium, even if the number of transfer units is not very high. A similar treatment based on local equilibrium for a two-component mixture is given by Golshan-Shirazi and Guiochon [*J. Phys. Chem.*, **93**, 4143 (1989)].

Prediction of multicomponent nonlinear chromatography accounting for rate factors requires numerical solution (see Guiochon et al., gen. refs., and "Numerical Methods and Characterization of Wave Shape" in "Fixed Bed Transitions").

Linear Gradient Elution Analytical solutions are available for special cases under trace conditions with a linear isotherm. Other situations normally require a numerical solution. General references are Snyder [in Horvath (ed.), *High Performance Liquid Chromatography: Advances and Perspectives*, vol. 1, Academic Press, 1980, p. 208], Antia and Horvath [*J. Chromatogr.*, **484**, 1 (1989)], Yamamoto et al. [*Ion Exchange Chromatography of Proteins*, Marcel Dekker, 1988], Guiochon et al. (gen. refs.).

The most commonly used gradients are linear gradients where the starting solvent is gradually mixed with a second gradient-forming solvent at the column entrance to yield a volume fraction φ of the mobile phase modulator that increases linearly with time:

$$\varphi = \varphi_0 + \beta t \qquad (16\text{-}188)$$

When the mobile phase modulator is a dilute solute in a solvent, as when the modulator is a salt, φ indicates molar concentration.

Under *trace conditions*, the retention of the modulator in the column is independent of the presence of any solutes. The modulator concentration at the column exit is approximated by

$$\varphi = \varphi_0 + \beta\left[t - \frac{L}{v}\left(1 + k'_M\right)\right] \qquad (16\text{-}189)$$

where k'_M is the retention factor of the modulator.

For a small feed injection, the modulator concentration φ_R at which a feed solute is eluted from the column is obtained from the following integral relationship

$$G(\varphi_R) = \frac{\beta L}{v} = \int_{\varphi_0}^{\varphi_R} \frac{d\varphi}{k'_i(\varphi) - k'_M} \qquad (16\text{-}190)$$

where $k'_i(\varphi)$ is the solute retention factor as a function of the mobile

phase modulator concentration. Note that the solute retention time in the column is affected by the steepness of the gradient at the column entrance.

$k'_i(\varphi)$ can be obtained experimentally from isocratic elution experiments at different φ values, or from linear gradient elution experiments where the ratio $G = \beta L/v$ is varied. In the latter case, the retention factor is obtained by differentiation of Eq. (16-190) from $k'_i(\varphi_R) = k'_M + (dG/d\varphi_R)^{-1}$.

Table 16-14 gives explicit expressions for chromatographic peak properties in isocratic elution and linear gradient elution for two cases.

In **reversed-phase chromatography** (RPC), the mobile phase modulator is typically a water-miscible organic solvent, and the stationary phase is a hydrophobic adsorbent. In this case, the logarithm of solute retention factor is commonly found to be linearly related to the volume fraction of the organic solvent.

In **ion-exchange chromatography** (IEC), the mobile phase modulator is typically a salt in aqueous solution, and the stationary phase is an ion-exchanger. For dilute conditions, the solute retention factor is commonly found to be a power-law function of the salt normality [cf. Eq. (16-27) for ion-exchange equilibrium].

Band broadening is also affected by the gradient steepness. This effect is expressed in Table 16-14 by a **band compression factor** C, which is a function of the gradient steepness and of equilibrium parameters. Since $C < 1$, gradient elution yields peaks that are sharper than those that would be obtained in isocratic elution at $\varphi = \varphi_R$.

Other cases, involving an arbitrary relationship between the solute retention factor and the modulator concentration can be handled analytically using the approaches of Frey [*Biotechnol. Bioeng.*, **35**, 1055 (1990)] and Carta and Stringfield [*J. Chromatogr.*, **605**, 151 (1992)].

Displacement Development A complete prediction of displacement chromatography accounting for rate factors requires a numerical solution since the adsorption equilibrium is nonlinear and intrinsically competitive. When the column efficiency is high, however, useful predictions can be obtained with the local equilibrium theory (see "Fixed Bed Transitions").

For constant-separation factor systems, the **h-transformation** of Helfferich and Klein (gen. refs.) or the method of Rhee et al. [*AIChE J.*, **28**, 423 (1982)] can be used [see also Helfferich, *Chem. Eng. Sci.*, **46**, 3320 (1991)]. The equations that follow are adapted from Frenz and Horvath [*AIChE J.*, **31**, 400 (1985)] and are based on the h-transformation. They refer to the separation of a mixture of $M - 1$

TABLE 16-14   Expressions for Predictions of Chromatographic Peak Properties in Linear Gradient Elution Chromatography under Trace Conditions with a Small Feed Injection and Inlet Gradient Described by $\varphi = \varphi_o + \beta t$ (Adapted from Refs. A and B).

Parameter	Isocratic	Gradient elution—RPC (Ref. A)	Gradient elution—IEC (Ref. B)
Dependence of retention factor on modulator concentration	—	$k' - k'_M = \alpha_s e^{-S\varphi}$	$k' - k'_M = \alpha_Z \varphi^{-z}$
Retention time	$t_R = \dfrac{L}{v}(1 + k'_o)$	$t_R = \dfrac{L}{v}(1 + k'_M) + \dfrac{1}{S\beta}\ln\left[\dfrac{S\beta L}{v}(k'_o - k'_M) + 1\right]$	$t_R = \dfrac{L}{v}(1 + k'_M) + \dfrac{1}{\beta}\left[\dfrac{\alpha_Z\beta L}{v}(Z+1) + \varphi_o^{Z+1}\right]^{1/(Z+1)} - \dfrac{\varphi_o}{\beta}$
Mobile phase composition at column exit	$\varphi_R = \varphi_o$	$\varphi_R = \varphi_o + \beta\left[t_R - \dfrac{L}{v}(1 + k'_M)\right]$	$\varphi_R = \varphi_o + \beta\left[t_R - \dfrac{L}{v}(1 + k'_M)\right]$
Retention factor at peak elution	$k'_R = k'_o$	$k'_R = \dfrac{k'_o + (k'_o - k'_M)S\beta L/v}{1 + (k'_o - k'_M)S\beta L/v}$	$k'_R = \alpha_Z\left[\dfrac{\alpha_Z\beta L}{v}(Z+1) + \varphi_o^{Z+1}\right]^{-Z/(Z+1)} + k'_M$
Peak standard deviation	$\sigma = \dfrac{L/v}{\sqrt{N_P}}(1 + k'_o)$	$\sigma = C\dfrac{L/v}{\sqrt{N_{pR}}}(1 + k'_R)$	$\sigma = C\dfrac{L/v}{\sqrt{N_{pR}}}(1 + k'_R)$
Band compression factor	—	$C = \dfrac{(1 + p + p^2/3)^{1/2}}{1 + p}$ $p = \dfrac{(k'_o - k'_M)(1 + k'_M)S\beta L/v}{1 + k'_o}$	$C = \begin{cases} \sqrt{M'} & \text{for } M' < 0.25 \\ \dfrac{3.22M'}{1 + 3.13M'} & \text{for } 0.25 < M' < 0.25 \\ 1 & \text{for } M' > 120.25 \end{cases}$ $M' = \dfrac{1}{2}\dfrac{1 + k'_R}{1 + k'_M}\dfrac{Z + 1}{Z}$

References: A. Snyder in Horvath (ed.), *High Performance Liquid Chromatography: Advances and Perspectives,* vol. 1, Academic Press, 1980, p. 208.
B. Yamamoto, *Biotechnol. Bioeng.,* **48,** 444 (1995).
Solute equilibrium parameters: α_s, S for RPC and α_Z, Z for IEC
Solute retention factor for initial mobile phase: k'_o
Retention factor of mobile phase modulator: k'_M
Plate number obtained for $k' = k'_R$: N_{pR}

components with a displacer (component 1) that is more strongly adsorbed than any of the feed solutes. The **multicomponent Langmuir isotherm** [Eq. (16-39)] is assumed valid with equal monolayer capacities, and components are ranked numerically in order of decreasing affinity for the stationary phase (i.e., $K_1 > K_2 > \cdots K_M$).

The development of component bands is predicted by mapping the trajectories of the **h-function roots** h_i, which are obtained from the solution of

$$\sum_{i=1}^{M} \frac{K_i c_i}{h\alpha_{i,1} - 1} = 1 \qquad (16\text{-}191)$$

where $\alpha_{i,j} = K_i/K_j$.

Trivial roots also exist for each component with zero concentration. In displacement chromatography, for each of the transitions at the column entrance, a new set of M roots is generated. These roots are given in Table 16-15. The change from the initial solvent (carrier) to feed changes $M - 1$ roots, generating $M - 1$ boundaries or transitions that move through the column. These boundaries are all self-sharpening, characterized by upstream h''_i values larger than the corresponding downstream h'_i values.

The switch from feed to displacer changes M roots, generating M boundaries. All of these boundaries are self-sharpening, except the boundary associated with the transition from h_{MF} to $\alpha_{1,M+1}$ that can be self-sharpening or diffuse depending on the relative value of these

roots. Note that only one root changes value across a particular boundary.

The **adjusted propagation velocity** of each self-sharpening boundary is:

$$u_{si} = \frac{d\zeta}{d\tau_1} = h'_i h''_i P_i \qquad (16\text{-}192)$$

$$P_i = \prod_{j=1}^{i-1} h'_j \prod_{j=i+1}^{M} h''_j \prod_{j=1}^{M+1} \alpha_{j,1} \qquad (16\text{-}193)$$

and that of each diffuse boundary:

$$u_{hi} = h_i^2 P_i \qquad (16\text{-}194)$$

where $\zeta = z/L$ and $\tau_1 = (\varepsilon v t/L - \varepsilon_b \zeta)/\Lambda$ with $\Lambda = \rho_b n_1^D/c_1^D$ equal to the partition ratio for the displacer.

The root trajectories are mapped as follows on the ζ-τ_1 plane:

1. The starting points are the beginning and end of the feed step. From these points trajectories are traced according to Eqs. (16-192) and (16-194).

2. At the points of intersection of trajectories, a change in boundary velocity occurs: When two boundaries associated with a change in the same root number intersect, they combine into a single boundary with velocity given by Eq. (16-192) or (16-194), intermediate in value between those of the intersecting trajectories. When two trajectories associated with different root numbers intersect, they both continue beyond the point of intersection with new velocities calculated from Eqs. (16-192) and (16-194), reflecting any change in root values.

3. After constructing the trajectories, the following equation is used to calculate the band profiles:

$$c_j = \frac{\prod_{i=1}^{M}(h_i\alpha_{j,1} - 1)}{K_j \prod_{i=1}^{M}(\alpha_{j,i} - 1)} \qquad (16\text{-}195)$$

Example 12: Calculation of Band Profiles in Displacement Chromatography An equimolar mixture of two components (concentrations $c_2^F = c_3^F = 1$ arbitrary unit) is separated with a displacer with concentration $c_1^F = 2$. The equilibrium isotherm is:

TABLE 16-15   Concentrations and h-Function Roots for Displacement Chromatography of a Mixture of M-1 Components Numbered in Order of Decreasing Affinity for the Stationary Phase (Adapted from Frenz and Horvath, 1985).

Solution	$c_i = 0$		$c_i > 0$	
	i	h_i	i	h_i
Carrier	1, 2, ... M	$\alpha_{1,1}, ... \alpha_{1,M}$	—	
Feed	1	$\alpha_{1,1}$	2, ... M	$h_{2F}, ...h_{MF}$°
Displacer	2, 3, ... M	$\alpha_{1,2}, ... \alpha_{1,M}$	1	$\alpha_{1,M+1} = 1 + K_1 c_1^D$

°Roots calculated from the solution of Eq. (16-191).

$$n_i = \frac{n^s K_i c_i}{1 + K_1 c_1 + K_2 c_2 + K_3 c_3} \qquad (i = 1, 2, 3)$$

with $K_1 = 2$, $K_2 = 1$, and $K_3 = 0.667$. The dimensionless feed time is $\tau_F = 0.2$. The separation factors used for calculating the trivial roots are $\alpha_{1,1} = 1$, $\alpha_{1,2} = 2$, $\alpha_{1,3} = 3$, $\alpha_{1,4} = 1 + K_1 c_1^D = 5$. The remaining h-function roots are calculated by solving Eq. (16-191) for the carrier, feed, and displacer concentrations. The results are summarized in Table 16-16.

TABLE 16-16   Values of h-Roots for Example 16-12

State	h_1	h_2	h_3
Carrier	1	2	3
Feed	1	2.44	6.56
Displacer	2	3	5

The switch from carrier to feed generates self-sharpening boundaries, since the h_i''s (upstream) are larger than the h_i's (downstream). The switch from feed to displacer generates two self-sharpening boundaries (corresponding to h_1 and h_2) and a diffuse boundary (corresponding to h_3) since $h_3'' < h_3'$. The velocity of each boundary is calculated from Eq. (16-192) for self-sharpening boundaries and from Eq. (16-194) for diffuse boundaries. The root trajectories are shown in Fig. 16-36 (top). From one boundary to the next, the root values change one at a time from the values corresponding to the carrier to the values corresponding to the displacer. Concentration profiles calculated from Eq. (16-195) are shown in Fig. 16-36 (bottom) for two different values of ζ. The concentrations in the z and t coordinates are easily reconstructed from the definitions of τ_1 and ζ.

In general, for a constant separation factor system, complete separation is obtained when:

$$c_1^D > \frac{1}{K_1} (\alpha_{1,M} - 1) \qquad (16-196)$$

Then, band profiles eventually develop into the **isotachic pattern** of pure component bands moving at the velocity of the displacer front. In Example 12, this occurs at $\zeta = 0.765$.

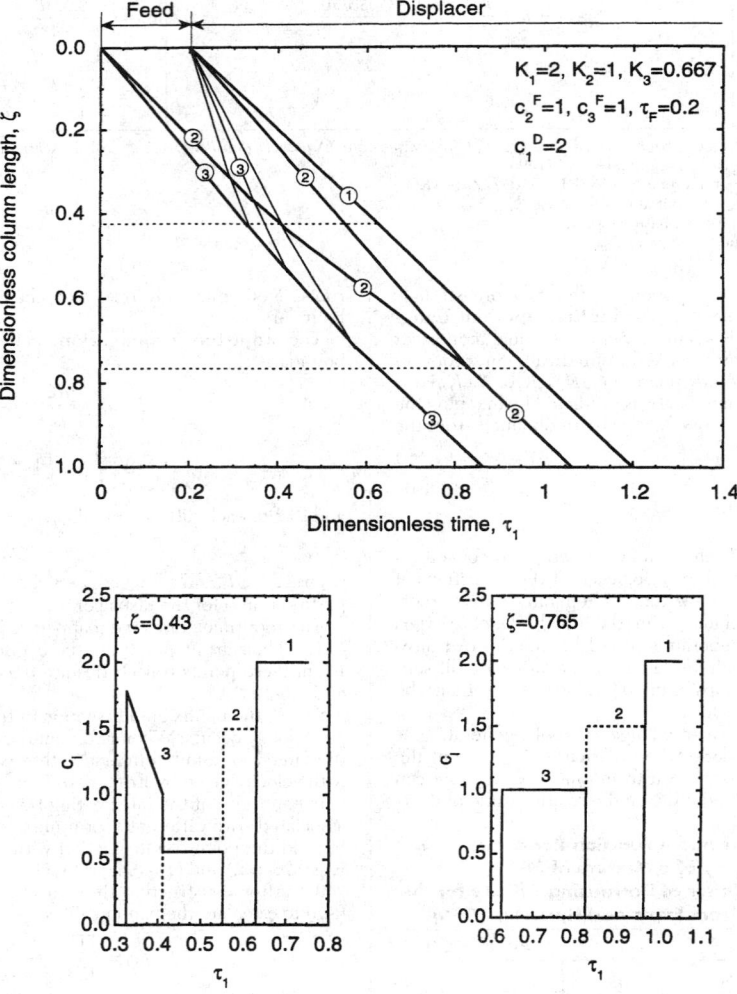

FIG. 16-36 Dimensionless time-distance plot for the displacement chromatography of a binary mixture. The darker lines indicate self-sharpening boundaries and the thinner lines diffuse boundaries. Circled numerals indicate the root number. Concentration profiles are shown at intermediate dimensionless column lengths $\zeta = 0.43$ and $\zeta = 0.765$. The profiles remain unchanged for longer column lengths.

The **isotachic concentrations** c_i^I in the fully developed train are calculated directly from the single component isotherms using:

$$\frac{n_1^e}{c_1^I} = \frac{n_2^e}{c_2^I} = \cdots = \frac{n_M^e}{c_M^I} \qquad (16\text{-}197)$$

or

$$c_j^I = c_1^D - \frac{1}{K_j}(1 - \alpha_{j,1}) \qquad (16\text{-}198)$$

The graphical interpretation of Eq. (16-197) is shown in Fig. 16-37 for the conditions of Example 12. An operating line is drawn from the origin to the point of the pure displacer isotherm at $c_1 = c_1^D$. For displacement to occur, the operating line must cross the pure component isotherms of the feed solutes. The product concentrations in the isotachic train are found where the operating line crosses the isotherms. When this condition is met, the feed concentrations do not affect the final product concentrations.

Analyses of displacement chromatography by the method of characteristics with non-Langmuirian systems is discussed by Antia and Horvath [*J. Chromatogr.*, **556**, 199 (1991)] and Carta and Dinerman [*AIChE J.*, **40**, 1618 (1994)]. Optimization studies and analyses by computer simulations are discussed by Jen and Pinto [*J. Chromatogr.*, **590**, 47 (1992)], Katti and Guiochon [*J. Chromatogr.*, **449**, 24 (1988)], and Phillips et al. [*J. Chromatogr.*, **54**, 1 (1988)].

DESIGN FOR TRACE SOLUTE SEPARATIONS

The design objectives of the analyst and the production line engineer are generally quite different. For analysis the primary concern is typically resolution. Hence operating conditions near the minimum value of the HETP or the HTU are desirable (see Fig. 16-13).

In preparative chromatography, however, it is generally desirable to reduce capital costs by maximizing the **productivity,** or the amount of feed processed per unit column volume, subject to specified purity requirements. This reduction, however, must be balanced against operating costs that are determined mainly by the mobile phase flow rate and the pressure drop. In practice, preparative chromatography is often carried out under **overload conditions,** i.e., in the nonlinear region of the adsorption isotherm. Optimization under these conditions is discussed in Guiochon et al. (gen. refs.). General guidelines for trace-level, isocratic binary separations, in the Henry's law limit of the isotherm are:

1. The stationary phase is selected to provide the maximum selectivity. Where possible, the retention factor is adjusted (by varying the mobile phase composition, temperature, or pressure) to an optimum value that generally falls between 2 and 10. Resolution is adversely affected when $k' \ll 2$, while product dilution and separation time

increase greatly when $k' \gg 10$. When this is not possible for all feed components and large differences exist among the k'-values of the different solutes, gradient elution should be considered.

2. The average feed mixture charging rate, molar or volumetric, is fixed by the raw material supply or the demand for finished product.

3. The value of N_p required to achieve a desired resolution is determined by Eq. (16-168) or (16-171). Since $N = L/\text{HTU} \sim 2N_p = 2L/\text{HETP}$, Fig. 16-13 or Eq. (16-183) can be used to determine the range of the dimensionless velocity $ReSc$ that maximizes N_p for a given particle diameter and column length.

4. The allowable pressure drop influences the choice of the particle size and helps determine the column length. Equations for estimating the pressure drop in packed beds are given in Section 6.

5. When only a few solutes are separated, they may occupy only a small portion of the total column volume at any given instant. In such cases, the productivity is improved by cyclic feed injections, timed so that the most strongly retained component from an injection elutes just before the least strongly retained component from the following injection (see Fig. 16-57). For a mixture of two components with $k' > 1$, when the same resolution is maintained between bands of the same injections and bands of successive injections, the cycle time t_C and the plate number requirement are:

$$t_C = 2(t_{R,A} - t_{R,B}) = \frac{2L}{v}(k_A' - k_B') \qquad (16\text{-}199)$$

$$N_p = 4R_s^2 \left(\frac{\alpha+1}{\alpha-1}\right)^2 (1 - 2\phi)^{-2} \qquad (16\text{-}200)$$

where $\phi = t_F/t_C$ is the fraction of the cycle time during which feed is supplied to the column. The productivity, $P = $ volume of feed/(time × bed volume), is:

$$P = \frac{1}{4R_s^2}\left(\frac{\alpha-1}{\alpha+1}\right)^2 \frac{\varepsilon v}{\text{HETP}}\,\phi(1-2\phi)^2$$

$$= \frac{1}{4R_s^2}\left(\frac{\alpha-1}{\alpha+1}\right)^2 \frac{D_i}{d_p^2}\left(\frac{ReSc}{b/ReSc + a + cReSc}\right)\phi(1-2\phi)^2 \qquad (16\text{-}201)$$

For a given resolution, P is maximized when $\phi = 1/6$ (i.e., feed is supplied for one sixth of the cycle time), and by the use of small particle sizes. The function $ReSc/(b/ReSc + a + cReSc)$ generally increases with $ReSc$, so that productivity generally increases with the mobile phase velocity. For typical columns, however, this function is within about 10 percent of its maximum value ($\sim 1/c$) when $ReSc$ is in the range 30–100. Thus, increasing the velocity above this range must be balanced against the costs associated with the higher pressure drop.

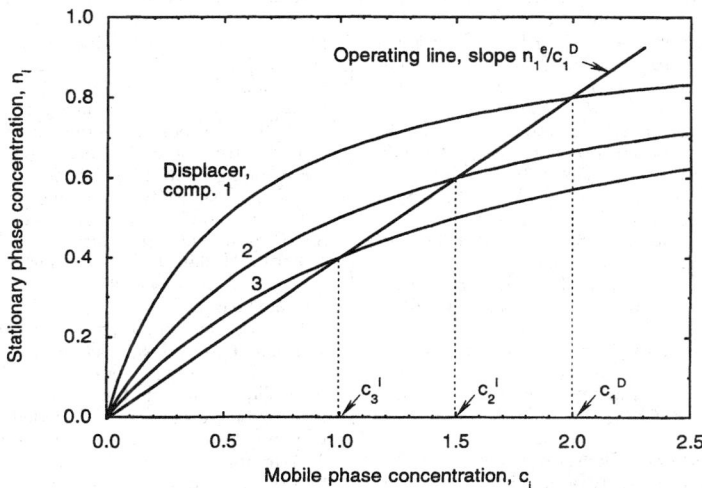

FIG. 16-37 Schematic showing the intersection of the operating line with the pure-component isotherms in displacement chromatography. Conditions are the same as in Fig. 16-36.

PROCESS CYCLES

GENERAL CONCEPTS

Some applications of adsorption and ion exchange can be achieved by sorbent-fluid contact in batch equipment. Batch methods are well adapted to laboratory use and have also been applied on a larger scale in several specific instances. In a batch run for either adsorption or ion exchange, a sorbent is added to a fluid, completely mixed, and subsequently separated. Batch treatment is adopted when the capacity and equilibrium of the sorbent are large enough to give nearly complete sorption in a single step, as in purifying and decoloration of laboratory preparations with carbons and clays. Batch runs are useful in the measurement of equilibrium isotherms and also of adsorptive diffusion rate. Batch tests or measurements are often conducted in portions of adsorbent or ion-exchange material intended for larger-scale use. For example, either the equilibrium sorption of a solid and/or its ultimate sorption capacity can be determined in this way. The solid is first equilibrated with a fluid at the concentration of interest; after separating the phases, gravimetric and chemical analyses can be used to determine the sorbed amount and composition.

Like the laboratory batch use, some commercial applications use the adsorbent on a throwaway basis. Reasons for using sorption nonregeneratively are usually: (1) low cost of the sorbent, (2) high value of the product, (3) very low dosage (sorbent-to-fluid ratio), and (4) difficulty in desorbing the sorbates. Magnesium perchlorate and barium oxide are used for drying, iron sponge (hydrated iron oxide on wood chips) is used to remove hydrogen sulfide, and sodium or potassium hydroxide is applied to removal of sulfur compounds or carbon dioxide. In wastewater treatment, powdered activated carbon (PAC) is added to enhance biological treatment but is not regenerated; instead, it remains with the sludge. Silica gel is used as a desiccant in packaging. Activated carbon is used in packaging and storage to adsorb other chemicals for preventing the tarnishing of silver, retarding the ripening or spoiling of fruits, "gettering" (scavenging) out-gassed solvents from electronic components, and removing odors. Synthetic zeolites, or blends of zeolites with silica gel, are used in dual-pane windows to adsorb water during initial dry-down and any in-leakage and to adsorb organic solvents emitted from the sealants during their cure; this prevents fogging between the sealed panes that could result from the condensation of water or the solvents [Ausikaitis, *Glass Digest*, **61**, 69–78 (1982)]. Activated carbon is used to treat recirculated air in office buildings, apartments, and manufacturing plants using thin filter-like frames to treat the large volumes of air with low pressure drop. On a smaller scale, activated carbon filters are in kitchen hoods, air conditioners, and electronic air purifiers. On the smallest scale, gas masks containing carbon or carbon impregnated with promoters are used to protect individual wearers from industrial odors, toxic chemicals, and gas-warfare chemicals. Activated carbon fibers have been formed into fabrics for clothing to protect against vesicant and percutaneous chemical vapors [Macnair and Arons in Cheremisinoff and Ellerbusch (eds.), gen. refs.].

Ion exchangers are sometimes used on a throwaway basis also. In the laboratory, ion exchangers are used to produce deionized water, purify reagents, and prepare inorganic sols. In medicine, they are used as antacid, for sodium reduction, for the sustained release of drugs, in skin-care preparations, and in toxin removal.

Although there are many practical applications for which the sorbent is discarded after one use, most applications of interest to chemical engineers involve the removal of adsorbates from the sorbent (i.e., regeneration). This allows the adsorbent to be reused and the adsorbates to be recovered.

The maximum efficiency that a cyclic adsorption process can approach for any given set of operating conditions is defined by the adsorptive loading in equilibrium with the feed fluid. There are several factors that reduce the practical (or "operating") adsorption: mass-transfer resistance (see above), deactivation (see above), and incomplete regeneration (or *desorption*). The severity of regeneration influences how closely the dynamic capacity of an adsorbent resembles that of fresh, virgin material. Regeneration, or reversal of the

adsorption process, requires a reduction in the driving force for adsorption. This is accomplished by increasing the equilibrium driving force for the adsorbed species to desorb from the solid to the surrounding fluid.

TEMPERATURE SWING ADSORPTION

A temperature-swing, or thermal-swing, adsorption (TSA) process cycle is one in which desorption takes place at a temperature much higher than adsorption. The elevation of temperature is used to shift the adsorption equilibrium and affect regeneration of the adsorbent. Figure 16-38 depicts a simplified (and ideal) TSA cycle. The feed fluid containing an adsorbate at a partial pressure of p_1 is passed through an adsorbent at temperature T_1. This adsorption step continues until the equilibrium loading n_1 is achieved with p_1. Next the adsorbent temperature is raised to T_2 (heating step) so that the partial pressure in equilibrium with n_1 is increased to p_2, creating a partial pressure driving force for desorption into any fluid containing less than p_2 of the adsorbate. By means of passing a purge fluid across the adsorbent, adsorbate is swept away, and the equilibrium proceeds down the isotherm to some point such as p_1, n_2. (This point need not coincide with the feed partial pressure; it is selected for illustrative purposes. Also, in some applications, roll-up of the adsorbed-phase concentration can occur during heating such that in some regions of the bed p_2 reaches the condensation pressure of the component, causing a condensed liquid phase to form temporarily in particles [Friday and LeVan, *AIChE J.*, **31**, 1322 (1985)].) During a cooling step, the adsorbent temperature is returned to T_1. The new equilibrium p_3, n_2 represents the best-quality product that can be produced from the adsorbent at a regenerated loading of n_2 in the simplest cycle. The adsorption step is now repeated. The differential loading, $n_1 - n_2$, is the maximum loading that can be achieved for a TSA cycle operating between a feed containing p_1 at temperature T_1, regeneration at T_2, and a product containing a partial pressure p_3 of the adsorbate. The regeneration fluid will contain an average partial pressure between p_2 and p_1 and will therefore have accomplished a concentration of the adsorbate in the regenerant fluid. For liquid-phase adsorption, the partial pressure can be replaced by the fugacity of the adsorbate. Then, the entire discussion above is applicable whether the regeneration is by a fluid in the gas or liquid phase.

In a TSA cycle, the heating step must provide the thermal energy necessary to raise the adsorbate, adsorbent, and adsorber temperatures, to desorb the adsorbate, and to make up for heat losses. TSA regeneration is classified as (1) heating-limited (or stoichiometric-limited) when transfer of energy to the system is limiting, or (2) stripping-limited (or equilibrium-limited) when transferring adsorbate away is limiting. Heating is accomplished by either direct contact of the adsorbent by the heating medium (external heat exchange to a purge gas) or indirect means (heating elements, coils, or panels inside the adsorber). Direct heating is the most commonly used, especially for stripping-limited heating. Indirect heating can be considered for stripping-limited heating, but the complexity of indirect heating limits its practicality to heating-limited regeneration where purge gas is in short supply. Microwave fields [Benchanaa, Lallemant, Simonet-Grange, and Bertrand, *Thermochim. Acta*, **152**, 43–51 (1989)] and dielectric fields [Burkholder, Fanslow, and Bluhm, *Ind. Eng. Chem. Fundam.*, **25**, 414–416 (1986)] are also used to supply indirect heating.

Because high temperatures can be used, resulting in thorough desorption, TSA cycles are characterized by low residual loadings and thus high operating loadings. These high capacities at low concentrations allow long cycle times for reasonably sized adsorbers (hours to days). Long cycle time is needed because particles of adsorbent respond slowly to changes in gas temperature. Most applications of TSA are for systems in which adsorbates are present at low concentration (purification), such as drying, and in which species are strongly adsorbed, such as sweetening, CO_2 removal, and pollution control.

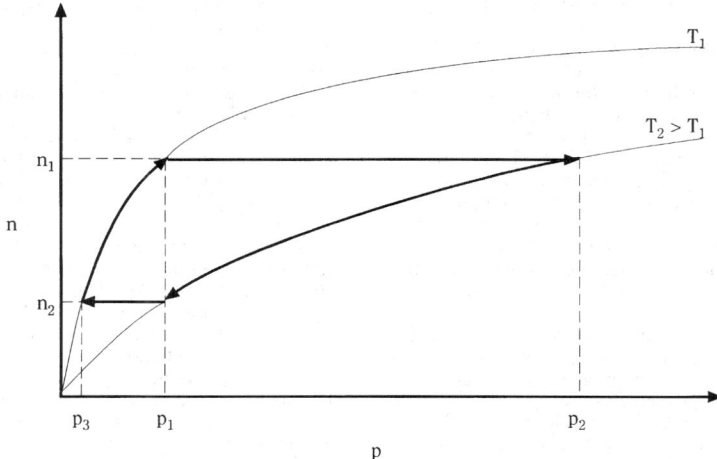

FIG. 16-38 Ideal temperature swing cycle. (*Reprinted with permission of UOP.*)

Other Cycle Steps Besides the necessary adsorption and heating steps, TSA cycles may employ additional steps. A purge or sweep gas removes the thermally desorbed components from the adsorbent, and cooling returns it to adsorption temperature. Although the cooling is normally accomplished as a separate step after the heating, sometimes adsorption is started on a hot bed. If certain criteria are met [Basmadjian, *Can. J. Chem. Eng.*, **53**, 234–238 (1975)], the dynamic adsorption efficiency is not significantly affected by the lack of a cooling step.

For liquid-phase adsorption cycles when the unit is to treat a product of significant value, there must be a step to remove the liquid from the adsorbent and one to displace any liquid regenerant thoroughly before filling with the valuable fluid. Because adsorbents and ion exchangers are porous, some retention of the product is unavoidable but needs to be minimized to maximize recovery. When regeneration is by a gas, removal and recovery is accomplished by a drain (or pressure-assisted drain) step using the gas to help displace liquid from the sorbent before heating. When the regenerant is another liquid, the feed or product can be displaced out of the adsorbent. When heating and cooling are complete, liquid feed must be introduced again to the adsorbent with a corresponding displacement of the gas or liquid regenerant. In ion exchange, these steps for draining and filling are commonly referred to as "sweetening off" and "sweetening on," respectively.

Applications Drying is the most common gas-phase application for TSA. The natural gas, chemical, and cryogenics industries all use adsorbents to dry streams. Zeolites, activated alumina, and silica gel are used for drying pipeline natural gas. Alumina and silica gel are used because they have higher equilibrium capacity and are more easily regenerated with waste-level heat [Crittenden, *Chem. Engr.*, **452**, 21–24 (1988); Goodboy and Fleming, *Chem. Eng. Progr.*, **80**, 63–68 (1984); Ruthven, *Chem. Eng. Progr.*, **84**, 42–50 (1988)]. The low dew-point that can be achieved with zeolites is especially important when drying cryogenic-process feed-streams to prevent freeze-up. Zeolites dry natural gas before liquefaction to liquefied natural gas (LNG) and before ethane recovery utilizing the cryogenic turboexpander process [Anderson in Katzer (ed.), *Molecular Sieves—II*, Am. Chem. Soc. *Symp. Ser.*, **40**, 637–649 (1977); Brooking and Walton, *The Chem. Engr.*, **257**, 13–17 (1972)]. The feed air to be cryogenically separated into N_2, O_2, and argon is purified of both water and CO_2 with 13X zeolites using TSA cycles. Zeolites, silica gel, and activated alumina are used to dry synthesis gas, inert gas, hydrocracker gas, rare gases, and reformer recycle H_2. Because 3A and pore-closed 4A zeolites size-selectively adsorb water but exclude hydrocarbons, they are used extensively to dry reactive streams such as cracked gas in order to prevent coke formation on the adsorbent. This molecular sieving increases the recovery of hydrocarbons by reducing the coadsorption that would otherwise cause them to be desorbed and lost with the water.

Another area of application for TSA processes is in sweetening. H_2S, mercaptans, organic sulfides and disulfides, and carbonyl sulfide all must be removed from natural gas, H_2, biogas, and refinery streams in order to prevent corrosion and catalyst poisoning. Natural gas feed to steam methane reforming is sweetened in order to protect the sulfur-sensitive, low-temperature shift catalyst. Well-head natural gas is treated by TSA to prevent pipeline corrosion using 4A zeolites to remove sulfur compounds without the coadsorption of CO_2 that would cause shrinkage. Sweetening and drying of refinery hydrogen streams are needed to prevent poisoning of reformer catalysts. Adsorption can be used to dry and sweeten these in the same unit operation.

TSA processes are applied to the removal of many inorganic pollutants. CO_2 is removed from base-load and peak-shaving natural-gas liquefaction facilities using 4A zeolite in a TSA cycle. The Sulfacid and Hitachi fixed-bed processes, the Sumitomo and BF moving-bed processes, and the Westvaco fluidized-bed process all use activated carbon adsorbents to remove SO_2 from flue gases and sulfuric acid plant tail gases [Juentgen, *Carbon*, **15**, 273–283 (1977)]. Activated carbon with a catalyst is used by the Unitaka process to remove NO_x by reacting with ammonia, and activated carbon has been used to convert NO to NO_2, which is removed by scrubbing. Mercury vapor from air and other gas streams is removed and recovered by activated carbon impregnated with elemental sulfur; the Hg is then recovered by thermal oxidation in a retort [Lovett and Cunniff, *Chem. Eng. Progr.*, **70**, 43–47 (1974)]. Applications for HCl removal from Cl_2, chlorinated hydrocarbons, and reformer catalyst gas streams use TSA with mordenite and clinoptilolite zeolites [Dyer, gen. refs., pp. 102–105]. Activated aluminas are also used for HCl adsorption as well as fluorine and boron-fluorine compounds from alkylation processes [Crittenden, ibid.].

PRESSURE-SWING ADSORPTION

A pressure-swing adsorption (PSA) process cycle is one in which desorption takes place at a pressure much lower than adsorption. Reduction of pressure is used to shift the adsorption equilibrium and affect regeneration of the adsorbent. Figure 16-39 depicts a simplified pressure-swing cycle. Feed fluid containing an adsorbate at a molar concentration of $y_1 = p_1/P_1$ is passed through an adsorbent at conditions T_1, P_1, and the adsorption step continues until the equilibrium loading n_1 is achieved with y_1. Next, the total pressure is reduced to P_2 during the depressurization (or blowdown) step. Now, although the partial pressure in equilibrium with n_1 is still p_1, there is a concentration driving force of $y_2 = p_1/P_2 > y_1$ for desorption into any fluid containing less than y_2. By passing a fluid across the adsorbent in a purge step, adsorbate is swept away, and the equilibrium proceeds down the isotherm to some point such as y_1, n_2. (The choice of y_1 is arbitrary and need not coincide with feed composition.) At this time, the adsorbent is repressurized to P_1. The new equilibrium y_3, n_2 represents the best quality product that can be produced from the adsorbent at a regenerated loading of n_2. The adsorption step is now repeated. The differential loading, $n_1 - n_2$, is the maximum loading that can be achieved for a pressure-swing cycle operating between a feed containing y_1 and a product containing a molar concentration y_3 of the adsorbate. The regeneration fluid will contain an average concentration between y_2 and y_1 and will therefore have accomplished a concentration of the adsorbate in the regenerant gas. There is no analog for a liquid-phase PSA process cycle.

Thus, in a PSA process cycle, regeneration is achieved by a depressurization that must reduce the partial pressure of the adsorbates to allow desorption. These cycles operate at constant temperature, requiring no heating or cooling steps. Rather, they use the exothermic heat of adsorption remaining in the adsorbent to supply the endothermic heat needed for desorption. Pressure-swing cycles are classified as: (1) PSA, which, although used broadly, usually swings between a high superatmospheric and a low superatmospheric pressure; (2) VSA (vacuum-swing adsorption), which swings from a superatmospheric pressure to a subatmospheric pressure; and (3) PSPP (pressure-swing parametric pumping) and RPSA (rapid pressure-swing adsorption), which operate at very fast cycle times such that significant pressure gradients develop in the adsorbent bed (see the subsection on parametric pumping). Otherwise, the broad principles remain the same.

Low pressure is not as effective in totally reversing adsorption as is temperature elevation unless very high feed to purge pressure ratios are applied (e.g., deep vacuum). Therefore, most PSA cycles are characterized by high residual loadings and thus low operating loadings. These low capacities at high concentrations require that cycle times be short for reasonably sized beds (seconds to minutes). These short cycle times are attainable because particles of adsorbent respond quickly to changes in pressure. Major uses for PSA processes include purification as well as applications where contaminants are present at high concentration (bulk separations).

Other Cycle Steps A PSA cycle may have several other steps in addition to the basic adsorption, depressurization, and repressurization. Cocurrent depressurization, purge, and pressure-equalization steps are normally added to increase efficiency of separation and recovery of product. At the end of the adsorption step, the more weakly adsorbed species have been recovered as product, but there is still a significant amount held up in the bed in the inter- and intra-particle void spaces. A cocurrent depressurization step can be added before the blowdown step, which is countercurrent to adsorption. This increases the amount of product produced each cycle. In some applications, the purity of the more strongly adsorbed components has also been shown to be heavily dependent on the cocurrent depressurization step [Cen and Yang, *Ind. Eng. Chem. Fundam.*, **25**, 758–767 (1986)]. This cocurrent blowdown is optional because there is always a countercurrent one. Skarstrom developed criteria to determine when the use of both is justified [Skarstrom in Li, *Recent Developments in Separation Science*, vol. II, CRC Press, Boca Raton, pp. 95–106 (1975)].

Additional stripping of the adsorbates from the adsorbent and purging of them from the voids can be accomplished by the addition of a purge step. The purge can begin toward the end of the depressurization or immediately afterward. Purging is accomplished with a flow of product countercurrent to adsorption to provide a lower residual at the product effluent end of the bed.

The repressurization step that returns the adsorber to feed pressure and completes the steps of a PSA cycle should be completed with pressure equalization steps to conserve gas and compression energy. Portions of the effluent gas during depressurization, blowdown, and enrichment purge can be used for repressurization to reduce the quantity of feed or product gas needed to pressurize the beds. The most efficient cycle is one that most closely matches available pressures and adsorbate concentration to the appropriate portion of the bed at the proper point in the cycle.

Applications PSA cycles are used primarily for purification of wet gases and of hydrogen. One of the earliest applications was the original Skarstrom two-bed cycle (adsorption, countercurrent blowdown, countercurrent purge, and cocurrent repressurization) to dry air stream to less than 1 ppm H_2O [Skarstrom, ibid.]. Instrument-air dryers still use a PSA cycle similar to Skarstrom's with activated alumina or silica gel [Armond, in Townsend, *The Properties and Applications of Zeolites*, The Chemical Society, London, pp. 92–102 (1980)].

The hydrocarbon exclusion by small-pore zeolites allows PSA to achieve a 10 to 30 K dewpoint depression in air-brake compressors, even at high discharge air temperatures in the presence of compressor oil [Ausikaitis, in Katzer, *Molecular Sieves—II, Am. Chem. Soc. Symp. Ser.*, **40**, pp. 681–695 (1977)]. The high-purity hydrogen employed in processes such as hydrogenation, hydrocracking, and ammonia and methanol production is produced by PSA cycles with adsorbent beds compounded of activated carbon, zeolites, and carbon molecular sieves [Martin, Gotzmann, Notaro, and Stewart, *Adv. Cryog. Eng.*, **31**, 1071–1086 (1986)]. The impurities to be removed include ammonia, carbon oxides, nitrogen, oxygen, methane, and heavier hydrocarbons. In order to be able to produce purities as high as 99.9999

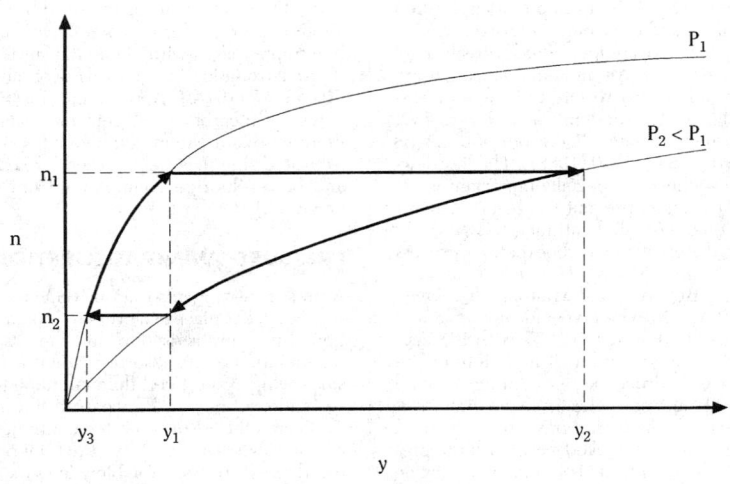

FIG. 16-39 Ideal pressure swing cycle. (*Reprinted with permission of UOP.*)

percent, systems such as a UOP™ Polybed™ separation unit use six to ten adsorbers with complex cycles (see below) [Fuderer and Rudelstorfer, U.S. Patent 3,986,849 (1976)].

Air separation, methane enrichment, and iso/normal separations are the major bulk separations utilizing PSA; recovery of CO and CO_2 are also growing uses. PSA process cycles are used to produce oxygen and/or nitrogen from air. Synthetic zeolites, clinoptilolite, mordenite, and carbon molecular sieves are all used in various PSA, VSA, and RPSA cycles. The 85 to 95 percent purity oxygen produced is employed for electric furnace steel, waste water treating, solid waste combustion, and kilns [Martin et al., ibid.]. Small PSA oxygen units are used for patients requiring inhalation therapy in the hospital and at home [Cassidy and Holmes, *AIChE Symp. Ser.*, **80** (1984), pp. 68–75] and for pilots on board aircraft [Tedor, Horch, and Dangieri, *SAFE J.*, **12**, 4–9 (1982)]. Lower purity oxygen (25 to 55 percent) can be produced to enhance combustion, chemical reactions, and ozone production [Sircar, in Rodrigues et al., gen. refs., pp. 285–321]. High purity nitrogen (up to 99.99 percent) for inert blanketing is produced in PSA and VSA processes using zeolites and carbon molecular sieves [Kawai and Kaneko, *Gas Sep. & Purif.*, **3**, 2–6 (1989); Ruthven, ibid.]. Methane is upgraded to natural gas pipeline quality by another PSA process. The methane is recovered from fermentation gases of landfills and wastewater purification plants and from poor-quality natural gas wells and tertiary oil recovery. Carbon dioxide is the major bulk contaminant but the gases contain water and other "garbage" components such as sulfur and halogen compounds, alkanes, and aromatics [Kumar and VanSloun, *Chem. Eng. Progr.*, **85**, 34–40 (1989)]. These impurities are removed by TSA using activated carbon or carbon molecular sieves and then the CO_2 is adsorbed using a PSA cycle. The cycle can use zeolites or silica gel in an equilibrium-selective separation or carbon molecular sieve in a rate-selective separation [Kapoor and Yang, *Chem. Eng. Sci.*, **44**, 1723–1733 (1989); Richter, Erdoel Kohle, Erdgas, *Petrochem.*, **40**, 432–438 (1987)]. The pore-size selectivity of zeolite 5A is employed to adsorb straight-chain molecules while excluding branched and cyclic species in the UOP IsoSiv^SM process. This PSA process separates C_5 to C_9 range hydrocarbons into a normal-hydrocarbon fraction of better than 95 percent purity, and a higher-octane isomer fraction with less than 2 percent normals [Cassidy and Holmes, ibid.].

PURGE/CONCENTRATION SWING ADSORPTION

A purge-swing adsorption cycle is usually considered to be one in which desorption takes place at the same temperature and total pressure as adsorption. Desorption is accomplished either by partial-pressure reduction using an inert gas purge or by adsorbate displacement with another adsorbable component. Purge cycles operate adiabatically at nearly constant inlet temperature and require no heating or cooling steps. As with PSA, they utilize the heat of adsorption remaining in the adsorbent to supply the heat of desorption. Purge processes are classified as (1) inert or (2) displacement.

Inert Purge   In inert-purge desorption cycles, *inert* refers to the fact that the purge is not adsorbable at the cycle conditions. Inert purging desorbs the adsorbate solely by partial pressure reduction. Regeneration of the adsorbent can be achieved while maintaining the same temperature and pressure by the introduction of an inert purge fluid. Figure 16-40 depicts a simplified inert-purge swing cycle utilizing a nonadsorbing purge fluid. As before, the feed stream containing an adsorbate at a partial pressure of p_1 is passed through an adsorbent at temperature T_1, and the adsorption step continues until equilibrium n_1 is achieved. Next the nonadsorbing fluid is introduced to reduce the partial pressure below p_1 by dilution. Therefore, there is a partial pressure driving force for desorption into the purge fluid, and the equilibrium proceeds down the isotherm to the point p_2, n_2, where p_2 represents the best quality product that can be produced from the adsorbent at a regenerated loading of n_2. The adsorption step is now repeated, and the differential loading is $n_1 - n_2$. The regeneration fluid will contain an average partial pressure between p_2 and p_1, and the cycle will therefore not have accomplished a concentration of the adsorbate in the regenerant fluid. But it will have transferred the adsorbates to a fluid from which it may be more easily separated by means such as distillation.

Like PSA cycles, inert-purge processes are characterized by high residual loadings, low operating loadings, and short cycle times (minutes). Bulk separations of contaminants not easily separable at high concentration and of weakly adsorbed components are especially suited to inert-purge-swing adsorption. Another version of UOP's Iso-Siv process employs H_2 in an inert-purge cycle for separating C_5 to C_9 naphtha by adsorbing straight-chain molecules and excluding branched and cyclic species on size selective 5A zeolite [Cassidy and Holmes, ibid.]. Automobiles made in the United States have canisters of activated carbon to adsorb gasoline vapors lost from the carburetor or the gas tank during running, from the tank during diurnal cycling, and from carburetor hot-soak losses; the vapors are desorbed by an inert purge of air that is drawn into the carburetor as fuel when the engine is running [Clarke, Gerrard, Skarstrom, Vardi and Wade, *S.A.E. Trans.*, **76**, 824–842 (1968)]. UOP's Adsorptive Heat Recovery drying system has been commercialized for drying azeotropic ethanol to be blended with gasoline into gasohol; the process uses a closed loop of N_2 as the inert purge to desorb the water [Garg and Yon, *Chem. Eng. Progr.*, **82**, 54–60 (1986)].

Displacement Purge   Isothermal, isobaric regeneration of the

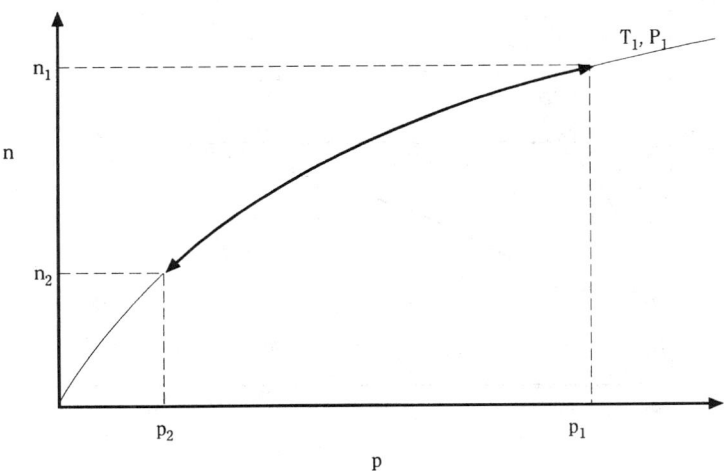

FIG. 16-40   Ideal inert-purge swing cycle. (*Reprinted with permission of UOP.*)

adsorbent can also be accomplished by using a purge fluid that can adsorb. In displacement-purge stripping, *displacement* refers to the displacing action of the purge fluid caused by its ability to adsorb at the cycle conditions. Figure 16-41 depicts a simplified displacement-purge swing cycle utilizing an adsorbable purge. Again, the feed stream containing an adsorbate at a partial pressure of p_1 is passed through an adsorbent at temperature T_1, and the adsorption step continues until equilibrium n_1 is achieved. Next the displacement fluid, B, is introduced. The presence of another adsorbable species reduces the adsorptivity of the key adsorbate, A. Therefore, there exists a partial pressure driving force for desorption into the purge fluid, and the equilibrium proceeds down the isotherm to some point such as p_1, n_2 (again arbitrary.) Next the adsorbent is recharged with a fluid that contains no component B, shifting the effective isotherm to that where the equilibrium of component A is p_3. The new equilibrium p_3, n_2 represents the best quality product that can be produced from the adsorbent at a regenerated loading of n_2. The adsorption step is now repeated. The differential loading $(n_1 - n_2)$ is the maximum loading that can be achieved for a pressure-swing cycle operating between a feed containing y_1 and a product containing a partial pressure p_3 of the adsorbate. The regeneration fluid will contain an average partial pressure between p_2 and p_1 and will therefore have accomplished a concentration of the adsorbate in the regenerant gas. Displacement-purge cycles are not as dependent on the heat of adsorption remaining on the adsorbent, because the adsorption of purge releases most or all of the energy needed to desorb the adsorbate. It is best if the adsorbate is more selectively adsorbed than the displacement purge, so that the adsorbates can easily desorb the purge fluid during adsorption. The displacement purge must be carefully selected, because it contaminates both the product stream and the recovered adsorbate, and requires separation for recovery (e.g., distillation).

Displacement-purge processes are more efficient for less selective adsorbate/adsorbent systems, while systems with high equilibrium loading of adsorbate will require more purging [Sircar and Kumar, *Ind. Eng. Chem. Proc. Des. Dev.*, **24**, 358–364 (1985)]. Several displacement-purge-swing processes have been commercialized for the separation of branched and cyclic C_{10}–C_{18} from normals using the molecular-size selectivity of 5A zeolite: Exxon's Ensorb, UOP's IsoSiv, Texaco Selective Finishing (TSF), Leuna Werke's Parex, and the Shell Process [Ruthven, ibid.]. All use a purge of normal paraffin or light naphtha with a carbon number of two to four less than the feed stream except for Ensorb, which uses ammonia [Yang, gen. refs.]. UOP has also developed a similar process, OlefinSiv, which separates isobutylene from normal butenes with displacement purge and a size-selective zeolite [Adler and Johnson, *Chem. Eng. Progr.*, **75**, 77–79

(1979)]. Solvent extraction to regenerate activated carbon is another example of a displacement-purge cycle; the adsorbent is then usually steamed to remove the purge fluid [Martin and Ng, *Water Res.*, **18**, 59–73 (1984)]. The best use of solvent regeneration is for water phase adsorption where the separation of water from carbon would use too much steam and where purge and water are easily separated; and for vapor-phase where the adsorbate is highly nonvolatile but soluble. Air Products has developed a process for separating ethanol/water on activated carbon using acetone as a displacement agent and adding a water rinse to improve the recovery of two products [Sircar, U.S. Patent 5,026,482 (1991)].

Displacement-purge forms the basis for most simulated continuous countercurrent systems (see hereafter) such as the UOP Sorbex^SM processes. UOP has licensed close to one hundred Sorbex units for its family of processes: Parex^SM to separate *p*-xylene from C_8 aromatics, Molex^SM for *n*-paraffin from branched and cyclic hydrocarbons, Olex^SM for olefins from paraffin, Sarex^SM for fructose from dextrose plus polysaccharides, Cymex^SM for *p*- or *m*-cymene from cymene isomers, and Cresex^SM for *p*- or *m*-cresol from cresol isomers. Toray Industries' Aromax^SM process is another for the production of *p*-xylene [Otani, *Chem. Eng.*, **80**(9), 106–107, (1973)]. Illinois Water Treatment [*Making Waves in Liquid Processing*, Illinois Water Treatment Company, IWT Adsep System, Rockford, IL, **6**(1), (1984)] and Mitsubishi [Ishikawa, Tanabe, and Usui, U.S. Patent 4,182,633 (1980)] have also commercialized displacement-purge processes for the separation of fructose from dextrose.

Chromatography   Chromatography is a sorptive separation technique that allows multicomponent separations in both gas and liquid phase. As a preparative tool, it is often used as a displacement-purge process, although many applications employ an inert-displacement mode, especially for use in analysis. General characteristics and operating modes are discussed in a separate part of this section.

ION EXCHANGE

Except in very small-scale applications, ion-exchangers are used in cyclic operations involving sorption and desorption steps. A typical ion-exchange cycle used in water-treatment applications involves (a) *backwash*—used to remove accumulated solids obtained by an upflow of water to expand (50–80 percent expansion is typical) and fluidize the exchanger bed; (b) *regeneration*—a regenerant is passed slowly through the used to restore the original ionic form of the exchanger; (c) *rinse*—water is passed through the bed to remove regenerant from the void volume and, in the case of porous exchangers, from the resin pores; (d) *loading*—the fresh solution to be treated is passed through

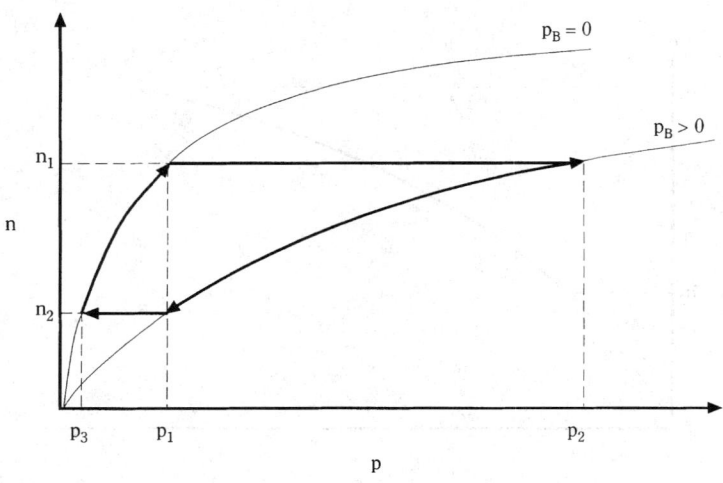

FIG. 16-41   Ideal displacement-purge swing cycle. (*Reprinted with permission of UOP.*)

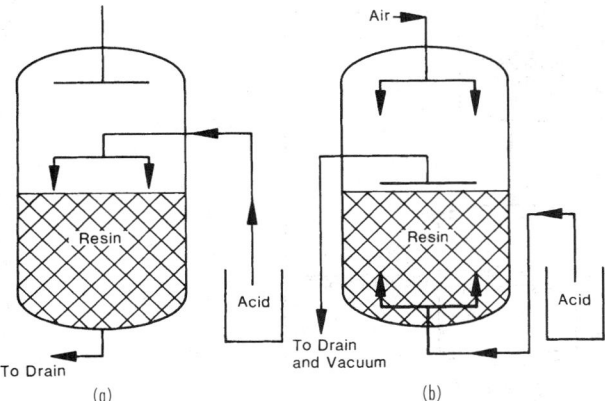

FIG. 16-42 Ion-exchanger regeneration. (*a*) Conventional. Acid is passed downflow through the cation-exchange resin bed. (*b*) Counterflow. Regenerant solution is introduced upflow with the resin bed held in place by a dry layer of resin.

the bed until leakage begins to occur. Water softening is practiced in this way with a cation exchange column in sodium form. At the low ionic strength used in the loading step, calcium and magnesium are strongly preferred over sodium, allowing nearly complete removal. Since the selectivity for divalent cations decreases sharply with ionic concentration, regeneration is carried out effectively with a concentrated sodium chloride solution. Removal of sulfates from boiler feed water is done by similar means with anion exchangers in chloride form.

Many ion-exchange columns operate downflow and are regenerated in the same direction (Fig. 16-42*a*). However, a better regeneration and lower leakage during loading can be achieved by passing the regenerant countercurrently to the loading flow. Specialized equipment is available to perform countercurrent regeneration (see "Equipment" in this section). One approach (Fig. 16-42*b*) is to apply a vacuum to remove the regenerant at the top of the bed.

Complete deionization with ion-exchange columns is the classical method of producing ultrapure water for boiler feed, in electronics manufacture, and for other general uses in the chemical and allied industries. Deionization requires use of two exchangers with opposite functionality to remove both cations and anions. These can be in separate columns, packed in adjacent layers in the same column, or, more frequently, in a mixed bed. In the latter case, the two exchangers are intimately mixed during the loading step. For regeneration, back-

washing separates the usually lighter anion exchanger from the usually denser cation exchanger. The column typically has a screened distributor at the interface between the two exchangers, so that they may be separately regenerated without removing them from the column. The most common cycle (Fig. 16-43) permits sequential regeneration of the two exchangers, first with alkali flowing downward through the anion exchanger to the interface distributor and then acid flowing downward from the interface distributor through the cation exchanger. After regeneration and rinsing, the exchangers are remixed by compressed air. To alleviate the problem of intermixing of the two different exchangers and chemical penetration through the wrong one, an inert material of intermediate density can be used to provide a buffer zone between layers of cation and anion exchangers.

When recovery of the sorbed solute is of interest, the cycle is modified to include a displacement step. In the manufacture of pharmaceuticals, ion-exchangers are used extensively in recovery and separation. Many of these compounds are amphoteric and are positively or negatively charged depending on the solution pH. Thus, using for example a cation exchanger, loading can be carried out at a low pH and displacement at a high pH. Differences in selectivity for different species can be used to carry out separations during the displacement [Carta et al., *AIChE Symp. Ser.*, **84**, 54–61 (1988)]. Multibed cycles are also used to facilitate integration with other chemical process operations. Fig. 16-44 shows a two-bed ion exchange system using both cation and anion exchangers to treat and recover chromate from rinse water in plating operations. The cation exchanger removes trivalent chromium, while the anion exchanger removes hexavalent chromium as an anion. Regeneration of the cation exchanger with sulfuric acid produces a concentrated solution of trivalent chromium as the sulfate salt. The hexavalent chromium is eluted from the anion exchanger with sodium hydroxide in a concentrated solution. This solution is recycled to the plating tank by passing it through a second cation exchange column in hydrogen form to convert the sodium chromate to a dilute chromic acid solution that is concentrated by evaporation.

PARAMETRIC PUMPING

The term *parametric pumping* was coined by Wilhelm et al. [Wilhelm, Rice, and Bendelius, *Ind. Eng. Chem. Fundam.*, **5**, 141–144 (1966)] to describe a liquid-phase adsorption process in which separation is achieved by periodically reversing not only flow but also an intensive thermodynamic property such as temperature, which influences adsorptivity. Moreover, they considered the concurrent cycling of pressure, pH, and electrical and magnetic fields. A lot of research and development has been conducted on thermal, pressure, and pH driven cycles, but to date only gas-phase pressure-swing parametric pumping has found much commercial acceptance.

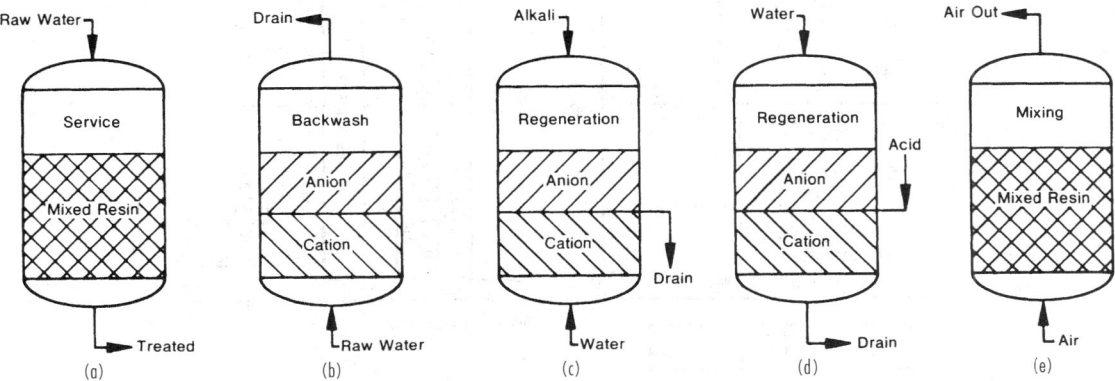

FIG. 16-43 Principles of mixed-bed ion exchange. (*a*) Service period (loading). (*b*) Backwash period. (*c*) Caustic regeneration. (*d*) Acid regeneration. (*e*) Resin mixing.

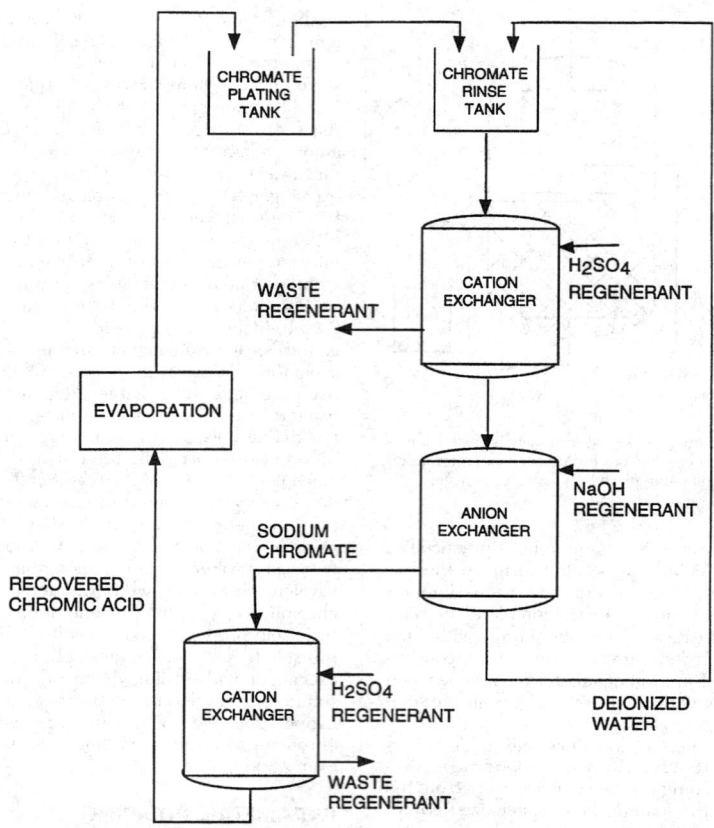

FIG. 16-44 Multicomponent ion-exchange process for chromate recovery from plating rinse water. (*Adapted from Rhom & Haas with permission.*)

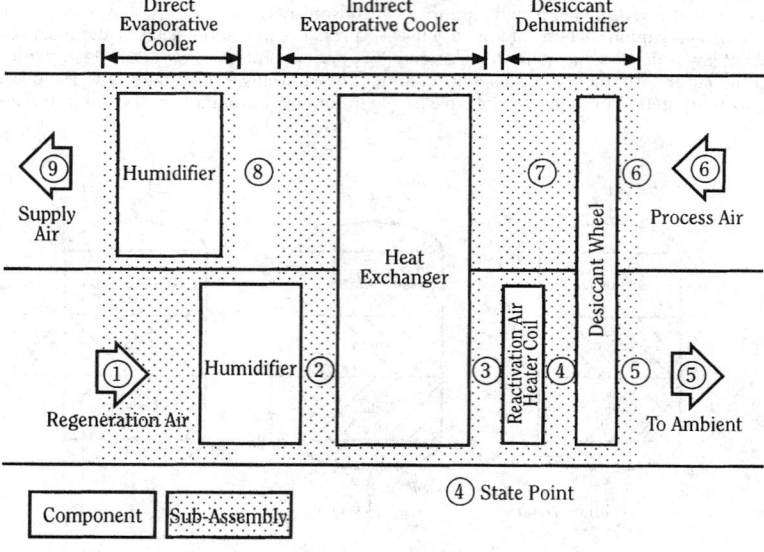

FIG. 16-45 Flow diagram of desiccant cooling cycle. [*Reprinted with permission of American Society of Heating, Refrigeration and Air Conditioning Engineers, Inc. (ASHRAE). Reference: Collier, Cohen, and Slosberg in Harrimam,* Desiccant Cooling and Dehumidification, *ASHRAE, Atlanta, 1992.*]

Temperature Two modes of temperature parametric-pumping cycles have been defined—direct and recuperative. In direct mode, an adsorbent column is heated and cooled while the fluid feed is pumped forward and backward through the bed from reservoirs at each end. When the feed is a binary fluid, one component will concentrate in one reservoir and one in the other. In recuperative mode, the heating and cooling takes place outside the adsorbent column. Parametric pumping, thermal and pH modes, have been widely studied for separation of liquid mixtures. However, the primary success for separating gas mixtures in thermal mode has been the separation of propane/ethane on activated carbon [Jencziewski and Myers, *Ind. Eng. Chem. Fundam.*, **9**, 216–221 (1970)] and of air/SO$_2$ on silica gel [Patrick, Schrodt, and Kermode, *Sep. Sci.*, **7**, 331–343 (1972)]. The difficulty with applying the thermal mode to gas separation is that in a fixed-volume gas-pressure increases during the hot step, which defeats the desorption purpose of this step. No thermal parametric-pumping cycle has yet been practiced commercially.

Pressure Another approach to parametric pumping is accomplished by pressure cycling of an adsorbent. An adsorbent bed is alternately pressurized with forward flow and depressurized with backward flow through the column from reservoirs at each end. Like TSA parametric pumping, one component concentrates in one reservoir and one in the other. The pressure mode of parametric pumping has been called pressure-swing parametric pumping (PSPP) and rapid pressure swing adsorption (RPSA). It was developed to minimize process complexity and investment at the expense of product recovery. RPSA is practiced in single-bed [Keller and Jones in Flank, *Adsorption and Ion Exchange with Synthetic Zeolites,* **135** (1980), pp. 275–286] and multiple-bed [Earls and Long, U.S. Patent number 4,194,892, 1980] implementations. Adsorbers are short (about 0.3 to 1.3 m), and particle sizes are very small (about 150 to 400 mm). The total cycle time including adsorption, dead time, countercurrent purge, and sometimes a second dead time, ranges from a few to about 30 seconds. The feature of RPSA that differentiates it from traditional PSA is the existence of axial pressure profiles throughout the cycle much as temperature gradients are present in TSA parametric-pumping. Whereas PSA processes have essentially constant pressure through the bed at any given time, the flow resistance of the very small adsorbent particles produce substantial pressure drops in the bed. These pressure dynamics are key to the attainment of separation performance. RPSA has been commercialized for the production of oxygen and for the recovery of ethylene and chlorocarbons (the selectively adsorbed species) in an ethylene-chlorination process while purging nitrogen (the less selectively adsorbed specie).

OTHER ADSORPTION CYCLES

Steam When steam is used for regeneration of activated carbon, it is desorbing by a combination thermal swing and displacement purge (described earlier in this section). The exothermic heat released when the steam is adsorbed supplies the thermal energy much more efficiently than is possible with heated gas purging. Slightly superheated steam at about 130°C is introduced into the bed countercurrent to adsorption; for adsorbates with high boiling points, the steam temperature must be higher. Adsorbates are desorbed and purged out of the bed with the steam. Steam and desorbates then go to a condenser for subsequent separation. The water phase can be further cleaned by air stripping, and the sorbate-laden air from that stripper can be recycled with the feed to the adsorption bed.

Steam regeneration is most commonly applied to activated carbon that has been used in the removal and/or recovery of solvents from gases. At volatile organic compound (VOC) concentration levels from 500 to 15,000 ppm, recovery of the VOC from the stream used for regeneration is economically justified. Below about 500 ppm, recovery is not economically justifiable, but environmental concerns often dictate adsorption followed by destruction. While activated carbon is also used to remove similar chemicals from water and wastewater, regeneration by steam is not usual. The reason is that the water-treatment carbon contains 1 to 5 kg of water per kg of adsorbent that must be removed by drying before regeneration or an excessive amount of superheated steam will be needed. In water treatment,

there can also be significant amounts of nonvolatile compounds that do not desorb during steam regeneration and that residual will reduce the adsorption working capacity. There is a growing use of reticulated styrene-type polymeric resins for VOC removal from air [Beckett, Wood, and Dixon, *Environ. Technol.,* **13**, 1129–1140 (1992); Heinegaard, *Chem.-Ing.-Tech.,* **60**, 907–908 (1988)]. LeVan and Schweiger [in Mersmann and Scholl (eds.), *Fundamentals of Adsorption*, United Engineering Trustees, New York (1991), pp. 487–496] tabulate reported steam utilizations (kg steam/kg adsorbate recovered) for a number of processes.

Energy Applications Desiccant cooling is a means for more efficiently providing air conditioning for enclosures such as supermarkets, ice rinks, hotels, and hospitals. Adsorbers are integrated with evaporative and electric vapor compression cooling equipment into an overall air handling system. Air conditioning is comprised of two cooling loads, latent heat for water removal and sensible heat for temperature reduction. The energy savings derive from shifting the latent heat load from expensive compression cooling (chilling) to cooling tower load. Early desiccant cooling used adsorption wheels (see hereafter) impregnated with the hygroscopic salt, LiCl. More recently, these wheels are being fabricated with zeolite and/or silica gel. They are then incorporated into a system such as the example shown in Fig. 16-45 [Collier, Cohen, and Slosberg, in Harrimam, *Desiccant Cooling and Dehumidification*, ASHRAE, Atlanta (1992)]. Process air stream 6, to be conditioned, passes through the adsorbent wheel, where it is dried. This is a nonisothermal process due to the release of heat of adsorption and transfer of heat from a wheel that may be above ambient temperature. The dry but heated air (7) is cooled in a heat exchanger that can be a thermal wheel. This stream (8) is further cooled, and the humidity adjusted back up to a comfort range by direct contact evaporative cooling to provide supply air. Regeneration air stream 1, which can be ambient air or exhausted air, is evaporatively cooled to provide a heat sink for the hot, dry air. This warmed air (3) is heated to the desired temperature for regeneration of the adsorbent wheel and exhausted to the atmosphere. Many other combinations of drying and cooling are used to accomplish desiccant cooling [Belding, in *Proceedings of AFEAS Refrigeration and Air Conditioning Workshop*, Breckenridge, CO (June 23–25, 1993)].

Heat pumps are another developing application of adsorbents. Zeolite/water systems have been evaluated as a means for transferring heat from a low temperature to a higher, more valuable level. Both natural (chabazite and clinoptilolite) and synthetic (NaX and high silica NaY) zeolites have favorable properties for sorption heat pumps. Data have demonstrated that hydrothermally stable Na-mordenite and dealuminated NaY can be used with water in chemical heat pumps to upgrade 100°C heat sources by 50 to 80°C using a 20°C heat sink [Fujiwara, Suzuki, Shin, and Sato, *J. Chem. Eng. Japan,* **23**, 738–743 (1990)]. Other work has shown that integration of two adsorber beds can achieve heating coefficients of performance of 1.56 for the system NaX/water, upgrading 150°C heat to 200°C with a 50°C sink [Douss, Meunier, and Sun, *Ind. Eng. Chem. Res.,* **27**, 310–316 (1988)].

Energy Conservation Techniques The major use of energy in an adsorption cycle is associated with the regeneration step, whether it is thermal energy for TSA or compression energy for PSA. Since the regeneration energy per pound of adsorbent tends to be constant, the first step in minimizing consumption is to maximize the operating loading. When the mass-transfer zone (MTZ) is a large portion of the adsorber relative to the equilibrium section, the fraction of the bed being fully utilized is small. Most fixed-bed adsorption systems have two adsorbers so that one is on stream while the other is being regenerated. One means of improving adsorbent utilization is to use a lead/trim (or cascade, or merry-go-round) cycle. Two (or more) adsorbent beds in series treat the feed. The feed enters the lead bed first and then the trim bed. The trim bed is the one that has most recently been regenerated. The MTZ is allowed to proceed through the lead beds but not to break through the trim bed. In this way the lead bed can be almost totally utilized before being regenerated. When a lead bed is taken out of service, the trim bed is placed in lead position, and a regenerated bed is placed in the trim position.

A thermal pulse cycle is a means of conserving thermal energy in heating-limited desorption. A process cycle that is heat-limited needs only a very small time (dwell) at temperature to achieve satisfactory desorption. If the entire bed is heated before the cooling is begun, every part of the bed will dwell at temperature for the entire time it takes the cooling front to traverse the bed. Thus, much of the heat in the bed at the start of cooling would be swept from the bed. Instead, cooling is begun before any heat front has exited the bed, creating a thermal pulse that moves through the bed. The pulse expends its thermal energy for desorption so that only a small temperature peak remains at the end of the regeneration and no excess heat has been wasted. If the heating step is stripping-limited, a thermal pulse is not applicable.

A series cool/heat cycle is another way in which the heat that is purged from the bed during cooling can be conserved. Sometimes the outlet fluid is passed to a heat sink where energy is stored to be reused to preheat heating fluid, or cross exchanged against the purge fluid to recover energy. However, there is also a process cycle that accomplishes the same effect. Three adsorbers are used, with one on adsorption, one on heating, and one on cooling. The regeneration fluid flows in series, first to cool the bed just heated and then to heat the bed to be desorbed. Thus all of the heat swept from the adsorber during heating can be reused to reduce the heating requirement. Unlike thermal pulse, this cycle is applicable to both heat- and stripping-limited heating.

Process Selection The preceding sections present many process cycles and their variations. It is important to have some guidelines for design engineers to narrow their choice of cycles to the most economical for a particular separation. Keller and coworkers [Keller et al., gen. refs.] have presented a method for choosing appropriate adsorp-

TABLE 16-17 Process Descriptors

Number	Statement
1	Feed is a vaporized liquid or a gas.
2	Feed is a liquid that can be fully vaporized at less than about 200°C.
3	Feed is a liquid that cannot be fully vaporized at 200°C.
4	Adsorbate concentration in the feed is less than about 3 wt %.
5	Adsorbate concentration in the feed is between about 3 and 10 wt %.
6	Adsorbate concentration in the feed is greater than about 10 wt %.
7	Adsorbate must be recovered in high purity (> than 90–99% rejection of nonadsorbed material).
8	Adsorbate can be desorbed by thermal regeneration.
9	Practical purge or displacement agents cannot be easily separated from the adsorbate.

Keller, Anderson, and Yon in Rousseau (ed.), *Handbook of Separation Process Technology*, John Wiley & Sons, Inc., New York, 1987; reprinted with permission.

tion processes. Their procedure considers the economics of capital, energy, labor, and other costs. Although these costs can vary from site to site, the procedure is robust enough to include most scenarios. In Table 16-17, nine statements are made about the character of the separation being considered. The numbers of the statements that are true (i.e., applicable) are used in the matrix in Table 16-18. A "no" for any true statement under a given process should remove that process from further consideration. Any process having all "yes" answers for true statements deserves strong consideration. Entries other than "yes" or "no" provide a means of prioritizing processes when more than one cycle is satisfactory.

TABLE 16-18 Process Selection Matrix

Statement number, Table 16-17	Gas- or vapor-phase processes					Liquid-phase processes		
	Temperature swing	Inert purge	Displacement purge	Pressure swing	Chromatography	Temperature swing°	Simulated moving bed	Chromatography
1	Yes	Yes	Yes	Yes	Yes	No	No	No
2	Not likely	Yes	Yes	Yes	Yes	Yes	Yes	Yes
3	No	No	No	No	No	Yes	Yes	Yes
4	Yes	Yes	Not likely	Not likely	Not likely	Yes	Not likely	Maybe
5	Yes	Yes	Yes	Yes	Yes	No	Yes	Yes
6	No	Yes	Yes	Yes	Yes	No	Yes	Yes
7	Yes	Yes	Yes	Maybe†	Yes	Yes	Yes	Yes
8	Yes	No	No	No	No	Yes‡	No	No
9	Maybe§	Not likely	Not likely	N/A	Not likely	Maybe§	Not likely	Not likely

°Includes powdered, fixed-bed, and moving-bed processes.
†Very high ratio of feed to desorption pressure (>10:1) will be required. Vacuum desorption will probably be necessary.
‡If adsorbate concentration in the feed is very low, it may be practical to discard the loaded adsorbent or reprocess off-site.
§If it is not necessary to recover the adsorbate, these processes are satisfactory.
Keller, Anderson, and Yon in Rousseau (ed.), *Handbook of Separation Process Technology*, John Wiley & Sons, Inc., New York, 1987; reprinted with permission.

EQUIPMENT

ADSORPTION

General Design Adsorbents are used in adsorbers with fixed inventory, with intermittent solids flow, or with continuous-moving solids flow. The most common are fixed beds operating as batch units or as beds of adsorbent through which the feed fluid passes, with periodic interruption for regeneration. Total systems consist of pressure vessels or open tanks along with the associated piping, valves, controls, and auxiliary equipment needed to accomplish regeneration of the adsorbent. Gas treating equipment includes blowers or compressors with a multiplicity of paths to prevent dead-heading. Liquid treating equipment includes pumps with surge vessels as needed to assure continuous flow.

Adsorber Vessel The most frequently used method of fluid-solid contact for adsorption operations is in cylindrical, vertical vessels, with

the adsorbent particles in a fixed and closely but randomly packed arrangement. The adsorbers must be designed with consideration for pressure drop and must contain a means of supporting the adsorbent and a means of assuring that the incoming fluid is evenly distributed to the face of the bed. There are additional design considerations for adsorbers when the streams are liquid and for high-performance separation applications using very small particles (<0.05 mm) such as in HPLC.

For most large-scale processes, adsorbent particle size varies from 0.06 to 6 mm (0.0025 to 0.25 in), but the adsorbent packed in a fixed bed will have a fairly narrow particle size range. Pressure drop in adsorbers can be changed by changing the diameter to bed depth ratio and by changing the particle size (see Sec. 5). Adsorbent size also determines separation performance of adsorbent columns—increasing efficiency with decreasing particle size. In liquid-phase processing, total cost of the adsorption step can sometimes be reduced by designing for overall pressure drops as large as 300 to 600 kPa (45 to 90 psi) because pumping is not the major utility cost. In special, high-resolution applications (HPLC), pressure drops as high as 5000–25,000 kPa (800–4000 psi) are sometimes used requiring special pumping and column hardware [Colin in Ganetsos and Barker, *Preparative and Production Scale Chromatography*, Marcel Dekker, New York, 1993, pp. 11–45]. However, the cost of compressing gases is significant. Since blowers are limited to about 5 kPa (20 in wc) of lift, atmospheric gas applications are typically designed with adsorbent pressure drops of 1 to 4 kPa (4 to 16 in wc). To keep compression ratio low, compressed gas adsorption pressure drops are 5 to 100 kPa (0.7 to 15 psi) depending on the pressure level.

Besides influencing how much pressure drop is allowable, the operating pressure determines other design features. When adsorption and/or regeneration is to be performed at pressures above atmospheric, the adsorber vessels are designed like process pressure vessels (Fig. 16-46 [EPA, gen. refs.]). Their flow distributors can consume more gas momentum at higher pressure. On the other hand, for applications near atmospheric pressure, any pressure drop can be costly, and most design choices are made in the direction of minimizing head loss. Beds have large face areas and shallow depth. Many times, the choice is to fabricate a horizontal (horizontal axis) vessel where flows are radial rather than axial as in conventional vertical beds. Figure 16-47 [Leatherdale in Cheremisinoff and Ellerbusch, gen. refs.] depicts how a rectangular, shallow adsorber bed is oriented in a horizontal vessel. Flow distributors, especially for large units, are often elaborate in order to evenly divide the flow rather than consume precious head.

There are two types of support systems used for fixed beds of adsorbent. The first is a progressive series of grid- and screen-layers. In this system, each higher layer screen has successively smaller openings, with the last small enough to prevent particles from passing through. Each lower layer has greater strength. A series of I-beams can be used to support a layer of subway grating that, in turn, supports several layers of screening. In other cases, special support grills such as Johnson™ screens may rest on the I-beams or on clips at the vessel wall and thus directly support the adsorbent. The topmost screen must retain the original size particles and some reasonable size of broken particles. The second type of support is a graded system of particles such as ceramic balls or gravel that rests directly on the bottom of the adsorber. A typical system might consist of 100 mm (4 in) of 50 mm (2 in) diameter material, covered by succeeding layers of 25, 12, and 6 mm (1, ½, ¼ in) of support material for a 3 mm (⅛ in) adsorbent. In water treatment, the support may actually start with filter blocks and have an upper layer of sand (see Fig. 16-48 [EPA, gen. refs.]).

If flow is not evenly distributed throughout the bed of adsorbent, there will be less than maximum utilization of the adsorbent during adsorption and of the desorption fluid during regeneration. Incoming fluids from the nozzles is at a much higher velocity than the average through the bed and may have asymmetric momentum components due to the piping manifold. The simplest means of allowing flow to redistribute across the face of the bed is to employ ample plenum space above and below the fixed bed. A much more cost-effective

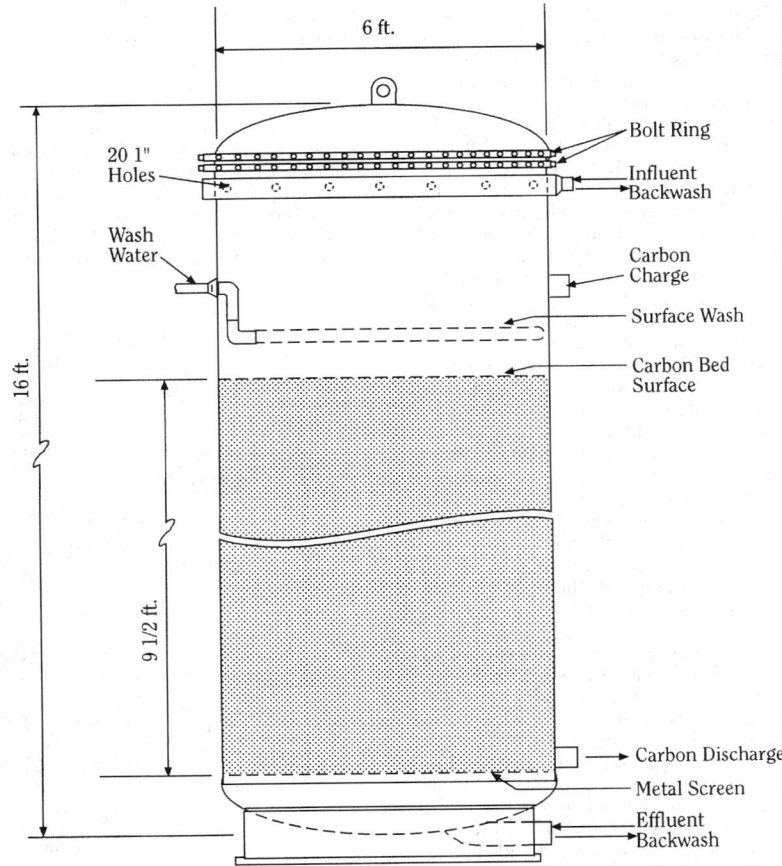

FIG. 16-46 Pressurized adsorber vessel. (*Reprinted with permission of EPA. Reference: EPA, Process Design Manual for Carbon Adsorption, U.S. Envir. Protect. Agency., Cincinnati, 1973.*)

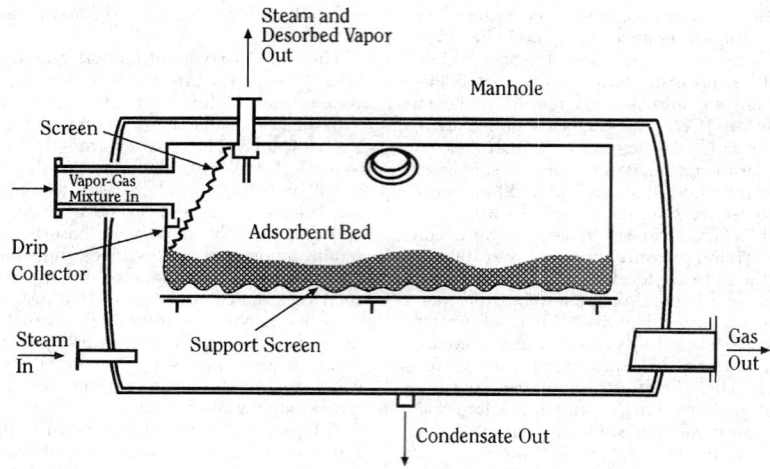

FIG. 16-47   Ambient pressure adsorber vessel. (*Reprinted with permission of Ann Arbor Science. Reference: Leatherdale in Cheremisinoff and Ellerbusch,* Carbon Adsorption Handbook, Ann Arbor Science, Ann Arbor, 1978.)

method is to install simple baffle plates with symmetrically placed inlet and outlet nozzles. The solid, or perforated, baffles are designed to break the momentum of the incoming fluid and redistribute it to prevent direct impingement on the adsorbent. When graded bed support is installed at the bottom, the baffles should be covered by screening to restrain the particles. An alternative to screened baffles is slotted metal or Johnson™ screen distributors. Shallow horizontal beds often have such a large flow area that multiple inlet and outlet nozzles are required. These nozzle headers must be carefully designed to assure balanced flow to each nozzle. In liquid systems, a single inlet may enter the vessel and branch into several pipes that are often perforated along their length (Fig. 16-46). Such "spiders" and "Christmas trees" often have holes that are not uniformly spaced and sized but are distributed to provide equal flow per bed area.

Although allowable pressure loss with liquids is not a restricting factor, there are special considerations for liquid treating systems. Activated carbon adsorbers used in water and wastewater treatment are designed and constructed using the same considerations used for turbidity removal by sand or multilayer filters. A typical carbon bed is shown in Fig. 16-48. Such contactors for liquids at ambient pressure are often nothing more than open tanks or concrete basins with flow distribution simply an overflow weir. In liquid treating, the adsorbers must be designed with a means for liquid draining and filling occasionally or during every cycle when a gas is used for regeneration. Draining is by gravity, sometimes assisted by a 70–140 kPa (10–20 psig) pressure pad. Even with time to drain, there will be significant liquid holdup to recover. As much as 40 cc of liquid per 100 g of adsorbent is retained in the micro- and macropores and bridged between particles. When drain is cocurrent to adsorption, the drained liquid can be recovered as treated product. When drain is countercurrent, drained fluid must be returned to feed surge. Minimizing other holdup in dead volume is especially important for liquid separation processes such as chromatography, because it adversely affects product recovery and regeneration efficiency. In filling an adsorber, there must be sufficient time for any gas trapped in the pores to escape. The fill step is preferably upflow to sweep the vapor out and to prevent gas pockets that could cause product contamination, bed lifting, or flow maldistribution. In liquid upflow, the buoyancy force of the liquid plus the pressure drop must not exceed the gravitational forces if bed lifting is to be prevented. Because there is very little increase in pressure drop beyond the lifting (or fluidization) velocity, some liquid systems are designed with bed expansion to limit pressure drop. Upflow-adsorption expanded-beds are also preferred when the liquid contains any suspended solids, so that the bed does not act as a filter and

become plugged. Since increased expansion causes the adsorbent to become increasingly well mixed, with accompanying drop in removal efficiency, expansion is usually limited to about 10 percent. Higher velocities also tend to cause too much particle turbulence, abrasion, attrition, and erosion.

Regeneration Equipment   Sometimes it is economically justified to remove the adsorbent from the adsorber when it is exhausted and have an outside contractor regenerate it rather than install on-site regeneration equipment. This is feasible only if the adsorbent can treat feed for weeks or months rather than only hours or days. In other cases, the process conditions during regeneration are so much more severe than those for adsorption that a single regenerator with materials of construction capable of handling the conditions is more cost-effective than to make all adsorbers the expensive material. This is

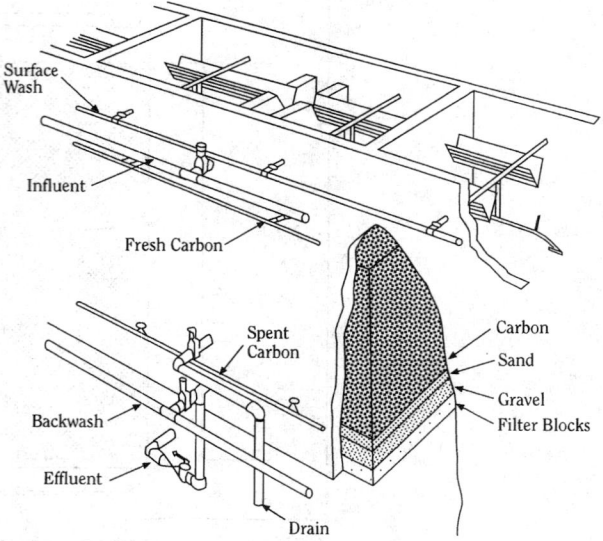

FIG. 16-48   Adsorber vessel with graded support system. (*Reprinted with permission of EPA. Reference: EPA,* Process Design Manual for Carbon Adsorption, U.S. Envir. Protect. Agency., Cincinnati, 1973.)

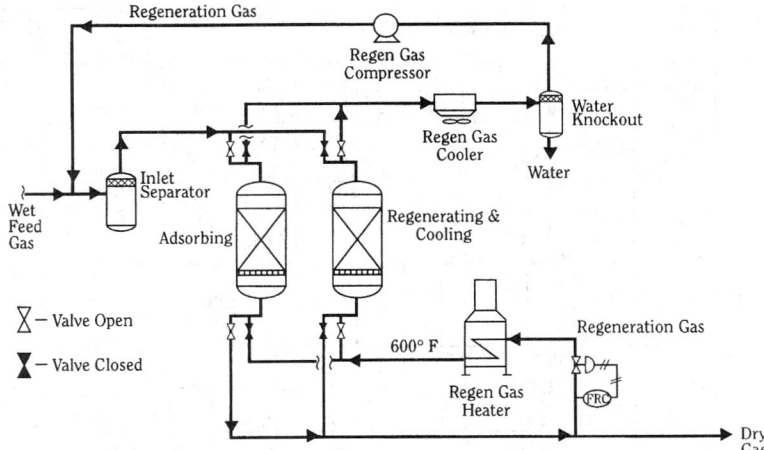

FIG. 16-49 Two-bed TSA system with regeneration equipment. (*Reprinted with permission of GPSA. Reference:* Engineering Data Book, *10th ed., Gas Processors Suppliers Association, Tulsa, 1988, Sec. 20, p. 22.*)

true for most water and wastewater treatment with thermally reactivated carbon. Otherwise, desorption is conducted in situ with any additional equipment connected to the adsorbers.

Figure 16-49 [*Engineering Data Book,* 10th ed., Gas Processors Suppliers Association, Tulsa, 1988, Sec. 20, p. 22] depicts the flow scheme for a typical two-bed TSA dryer system showing the auxiliary equipment associated with regeneration. Some of the dry product gas is externally heated and used countercurrently to heat and desorb water from the adsorber not currently drying feed. The wet, spent regeneration gas is cooled; the water is condensed out; and the gas is recycled to feed for recovery.

The thermal reactivation of spent activated carbon may require the same high temperatures and reaction conditions used for its manufacture. Although the exact conditions to be used for reactivation depend on the type of carbon and the adsorbates to be removed, the objective is to remove the adsorbed material without altering the carbon structure. This occurs in four stages: (a) drying, (b) desorption, (c) pyrolysis and carbonization, and (d) burnoff. Each of these steps is associated with a particular temperature range and is carried out in a multiple hearth furnace such as that depicted in Fig. 16-50. This six-hearth system is a typical configuration for water treatment carbons. The gas temperature ranges from 100°C on the top hearth to 950°C at the bottom. The rotating rabble arms rake the carbon from hearth to hearth. Wet, spent carbon is fed to the top, where it is dried by the top two hearths. Pyrolysis (desorption and decomposition) is accomplished on the next hearth. The bottom four hearths are for reactivation. Varying amounts of air, steam, and/or fuel are added to the different hearths to maintain the conditions established for the particular reactivation. A hearth area requirement of 0.1 m²/kg/hr (0.5 ft²/lb/hr) is adequate for cost estimating, but there are also detailed design procedures available [von Dreusche in Cheremisinoff and Ellerbusch, gen. refs.].

In addition to the multiple-hearth furnace, the reactivation system is comprised of additional equipment to transport, store, dewater, and quench the carbon.

A typical two-bed PSA dryer process flow scheme is very similar to Fig. 16-49 without the regeneration gas heater. A portion of the dry product gas is used countercurrently to purge the water from the adsorber not currently drying feed. Again, the wet, spent purge gas is cooled, and the water is condensed out and recycled to feed for recovery. If the cycle is VSA, then instead of the regeneration gas compressor, there is a vacuum compressor at the purge outlet upstream of the regeneration cooler. Some PSA cycles such as in air-brake dryers operate with one adsorber, a surge volume, and a three-way valve to

switch between pressurizing and countercurrent blowdown. The other extreme of complexity is demonstrated by the UOP nine-bed Polybed PSA H₂ unit shown in Fig. 16-51 [Fuderer and Rudelstorfer, U.S. Patent number 3,986,849, 1976]. This cycle requires nine adsorbers sequenced by fifty-five valves to maximize purity and recovery.

Cycle Control Valves are the heart of cycle control for cyclic adsorption systems. These on/off valves switch flows among beds so

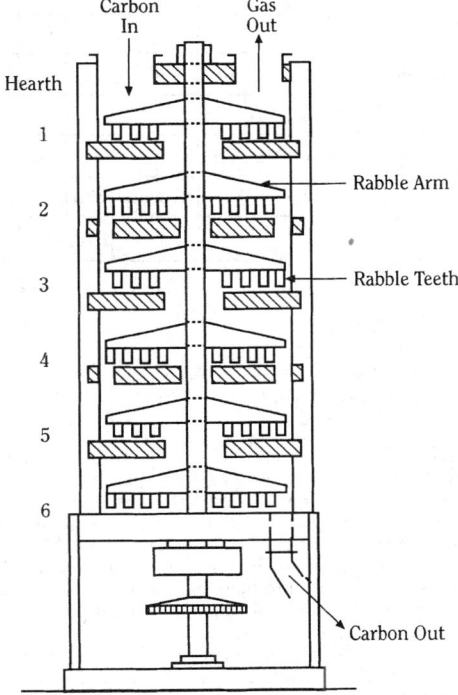

FIG. 16-50 Multiple hearth furnace for carbon reactivation. (*Reprinted with permission of EPA. Reference:* EPA, Process Design Manual for Carbon Adsorption, *U.S. Envir. Protect. Agency., Cincinnati, 1973.*)

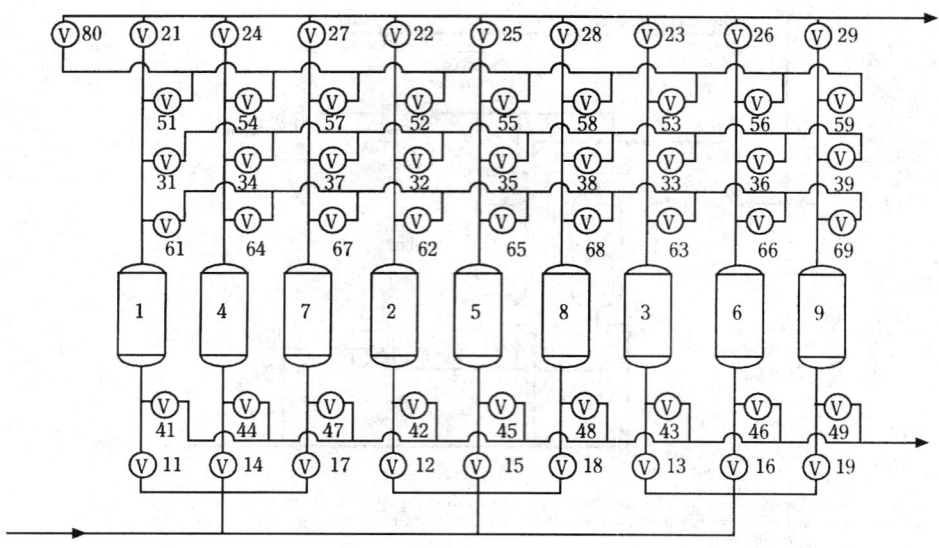

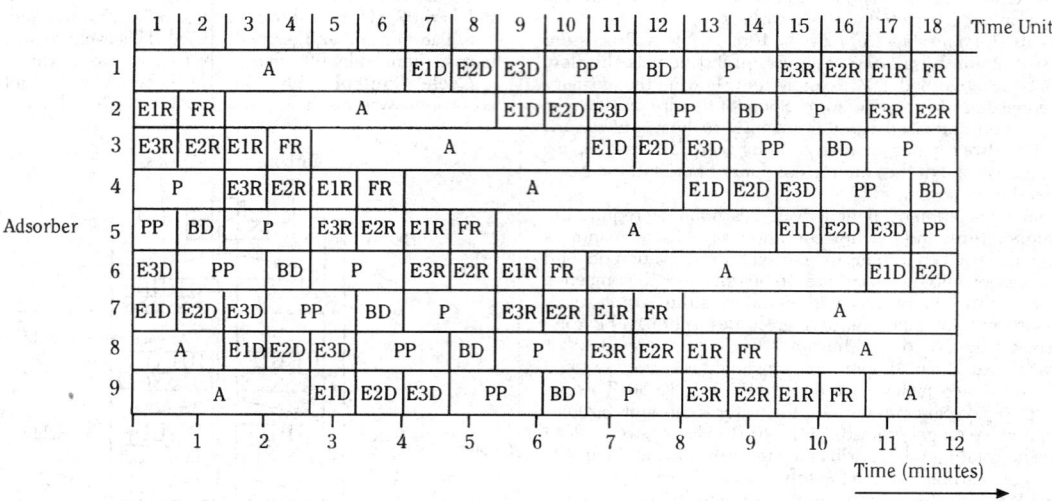

FIG. 16-51 UOP nine-bed polybed PSA H_2 unit: (*a*) flow scheme; (*b*) cycle diagram. (*Reference: Fuderer and Rudelstorfer, U.S. Patent number 3,986,849, 1976.*)

that external to the system it appears as if operation is continuous. In general, one valve is needed for each bed at each end for each step that is performed (e.g., for a two-bed system with an adsorption step plus heating and cooling step [carried out in the same direction and during the same step], only $2 \times 2 \times 2 = 8$ valves would be needed [see Fig. 16-49]). In some cycles such as pressure-swing systems, it may be possible to use valves for more than one function (e.g., repressurization with feed gas using the same manifold as adsorption feed). Without multiple use, the cycle in Fig. 16-51 would need $9 \times 2 \times 5 = 90$ valves instead of 55 to accommodate the five steps of adsorption, purge to product, blowdown, purge, and feed repressurization (even without the equalization). For some applications, 3- and 4-way valves can replace two and four valves, respectively. The ultimate integration of switching valves is the UOP rotary valve discussed below. For long step times (8 hours or more) it is possible for the valves to be manually switched by operators. For most systems, it is advantageous for the opening and closing of the valves be controlled by automatic timers. The same controller can be responsible for maintaining flows and pressure, logic for proceeding only on completion of events, and safety bypass or shutdown. Automatic control can provide for a period of parallel flow paths to assure transitions. In some applications, process analyzers can interface with the controller to initiate bed-switching when adsorbate is detected breaking through into the effluent.

Continuous Countercurrent Systems Most adsorption systems use fixed-bed adsorbers. However, if the fluid to be separated and that used for desorption can be countercurrently contacted by a moving bed of the adsorbent, there are significant efficiencies to be realized. Because the adsorbent leaves the adsorption section essentially in equilibrium with the feed composition, the inefficiency of the mass-transfer zone is eliminated. The adsorption section only needs to contain a MTZ length of adsorbent compared to a MTZ plus an equilibrium section in a fixed bed. Likewise, countercurrent regeneration is more efficient. Since the adsorbent is moved from an adsorption chamber to another chamber for regeneration, only the regeneration section is designed for the often severe conditions. Countercurrent adsorption can take advantage of an exceptionally favorable equilibrium in water softening; and the regeneration step can be made favorable by the use of relatively concentrated eluent. Continuous units generally require more headroom but much less footprint. The foremost problems to be overcome in the design and operation of continuous countercurrent sorption operations are the mechanical complexity of equipment and the attrition of the sorbent.

An early example of a commercial countercurrent adsorption processes is the hypersorption process developed by Union Oil Company for the recovery of propane and heavier components from natural gas [Berg, *Chem. Eng. Progr.*, **47**, 585–590 (1951)]. Hypersorption used an activated carbon adsorbent moving as a dense bed continuously downward through a rising gas stream (Fig. 16-52 [Berg, ibid.]). However, this process proved to be less economical than cryogenic distillation and had excessive carbon losses resulting from attrition. The commercialization by Kureha Chemical Co. of Japan of a new, highly attrition-resistant, activated-carbon adsorbent, allowed development of a process employing fluidized-bed adsorption and moving bed desorption for removal of VOC compounds from air. The process has been marketed as GASTAK^SM in Japan and as PuraSiv^SM HR (Fig. 16-53 [Anon., *Chem. Eng.*, **84**(18), 39–40 (1977)]) in the United States, and now as SOLDACS by Daikin Industries, Ltd. A

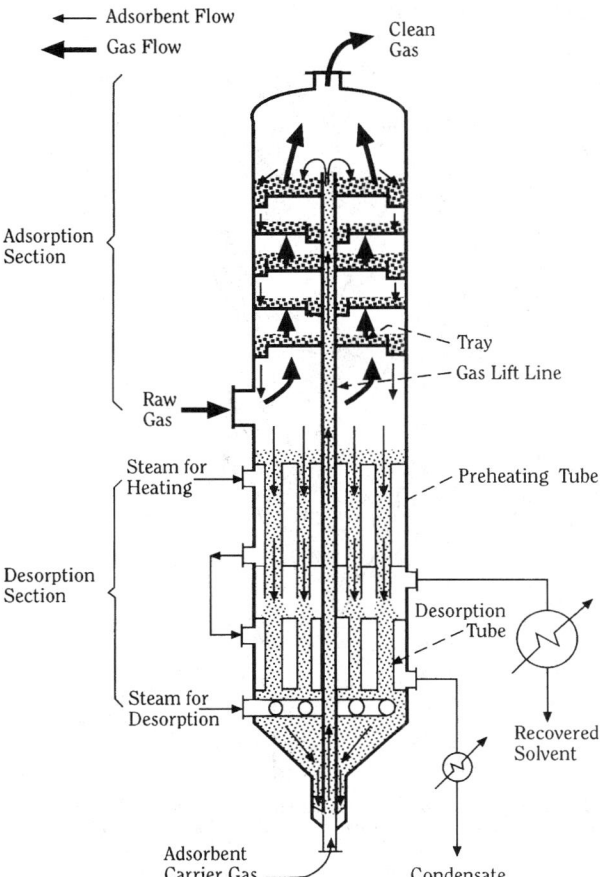

FIG. 16-53 PuraSiv HR adsorber vessel. [*Reprinted with permission of Chemical Engineering. Reference: Anon., Chem. Eng., 84(18), 39–40 (1977).*]

similar process using beaded polymeric resin is offered by Chematur [Heinegaard, *Chem.-Ing.-Tech.*, **60**, 907–908 (1988)]. The recent discovery [Acharya and BeVier, U.S. Patent number 4,526,877, 1985] that the graphite coating of zeolites can dramatically improve attrition resistance without significantly impairing adsorption performance could allow the extension of moving-bed technology to other separations. A good review of continuous ion-exchange applications is presented by Wankat [gen. refs.].

Continuous Cross-Flow Systems There are at least three implementations of moving-bed adsorption that are cross flow rather than fixed beds or countercurrent flow: (1) panel beds, (2) adsorbent wheels, and (3) rotating annular beds. By *cross flow* is meant that the adsorbent is moving in a direction perpendicular to the fluid flow. All of these employ moving adsorbent—the first, a down-flowing solid; and the others, a constrained solid. Panel beds of activated carbon like those depicted in Fig. 16-54 [Lovett and Cunniff, *Chem. Eng. Progr.*, **70**(5), 43–47 (1974)] have been applied to odor control [Lovett and Cunniff, ibid.] and to the desulfurization of waste gas [Richter, Knoblauch, and Juntgen, *Verfahrenstechnik (Mainz)*, **14**, 338–342 (1980)]. The spent solid falls from the bottom panel into a load-out bin, and fresh regenerated carbon is added to the top; gas flows through as shown.

The heart of an adsorbent wheel system is a rotating cylinder containing the adsorbent. Figure 16-55 illustrates two types: horizontal and vertical. In some adsorbent wheels, the adsorbent particles are placed in basket segments (a multitude of fixed beds) to form a hori-

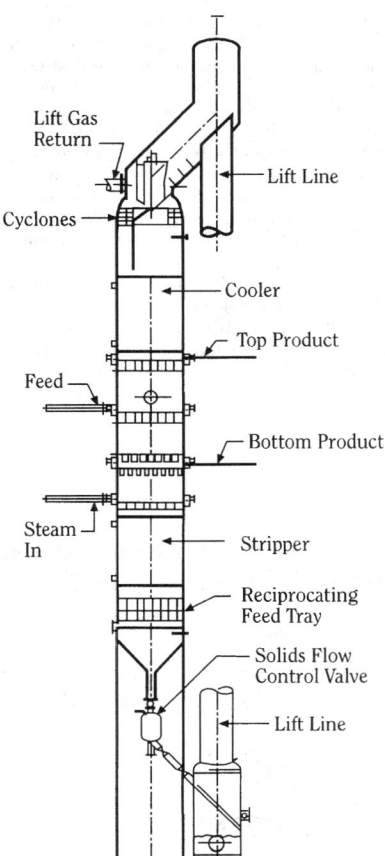

FIG. 16-52 Hypersorption adsorber vessel [*Reprinted with permission of AIChE. Reference: Berg, Chem. Eng. Progr., 47, 585–590 (1951).*]

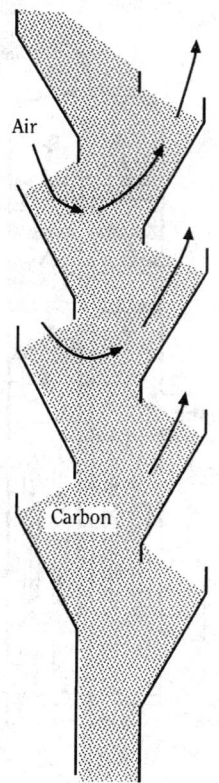

FIG. 16-54 Cross-flow panel-bed adsorber. [*Reprinted with permission of AIChE. Reference: Lovett and Cunniff,* Chem. Eng. Progr., **70***(5), 43–47 (1974).*]

zontal wheel that rotates around a vertical axis. In other instances, the adsorbent is integral to the monolithic wheel or coated onto a metal, paper, or ceramic honeycomb substrate. These monolithic or honeycomb structures rotate around either a vertical or a horizontal axis. The gas to be treated usually flows through the wheel parallel to the axis of rotation, although some implementations use radial flow con-

figurations. Most of the wheel is removing adsorbates. The remaining (smaller) portion of the wheel is undergoing thermal regeneration—usually countercurrently. The wheel constantly rotates to provide a continuous treated stream and a steady concentrated stream. Adsorbent wheels are most often used to treat ambient air because they have very low pressure drop. One application of wheels is the removal of VOC where the regeneration stream is usually sent to an incinerator for destruction of VOC. Another use is in desiccant cooling (see previously). They do suffer from low efficiency due to the short contact time, mechanical leakage at seals, and the tendency to allow the wheel to exceed breakthrough in order to get better adsorbent utilization. Some adsorbent wheels are operated in an intermittent manner such that the wheel periodically indexes to a new position; this is particularly true of radial flow wheels.

The rotating annular bed system for liquid chromatographic separation (two-dimensional chromatography) is analogous to the horizontal adsorbent wheel for gases. Feed to be separated flows to a portion of the top face of an annular bed of sorbent. A displacement purge in the form of a solvent or carrier gas flows to the rest of the annulus. The less strongly adsorbed components travel downward through the sorbent at a higher rate. Thus, they will exit at the bottom of the annulus at a smaller angular distance from the feed sector. The more strongly adsorbed species will exit at a greater angle. The mechanical and packing complexities of such an apparatus have been overcome for a pressurized system by workers at the Oak Ridge National Laboratory shown in Fig. 16-56 [Canon, Begovich, and Sisson, *Sep. Sci. Technol.,* **15**, 655–678 (1980)]. Several potential applications are reviewed by Carta and Byers [*Chromatographic and Membrane Processes in Biotechnology,* NATO ASI Proceeding, Kluwer, 1991].

Simulated Continuous Countercurrent Systems   Because of the problems associated with moving adsorbent in true continuous countercurrent mode, most commercial application of continuous countercurrent processes has been accomplished by using flow schemes that simulates the flow of adsorbent and process fluid. There are at least four types of implementation: (1) units designed to intermittently move the adsorbent rather than continuously, (2) continuously operating chromatographic columns to which the feed is periodically pulsed, (3) enhancement of the chromatographic separation by dividing the column and using recycles, and (4) moving the point of introduction of the feed and regeneration point along a series of beds.

Pulsed beds of activated carbon are used in water and wastewater treatment systems. The adsorber tank is usually a vertical cylindrical pressure vessel, with fluid distributors at top and bottom, similar to the arrangement of an ion exchanger. The column is filled with granular carbon. Fluid flow is upward, and carbon is intermittently dis-

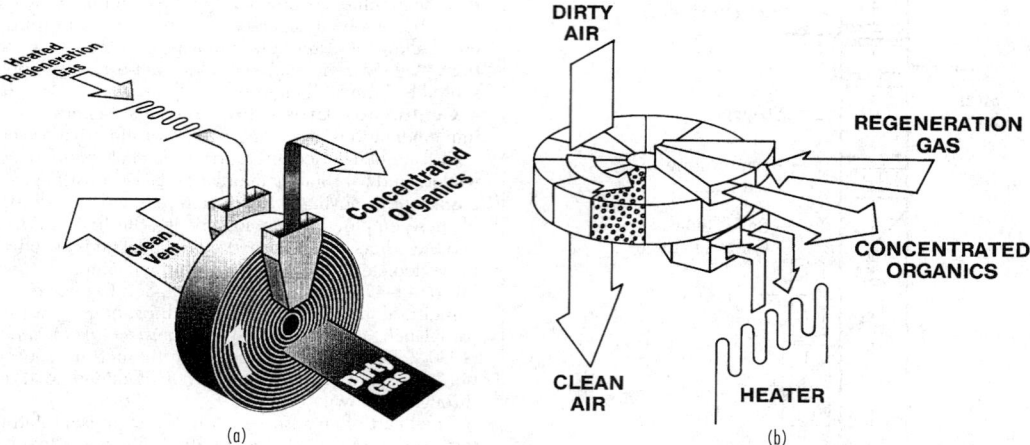

FIG. 16-55   Adsorbent wheels for gas separation: (*a*) horizontal with fixed beds; (*b*) vertical monolith. (*Reprinted with permission of UOP.*)

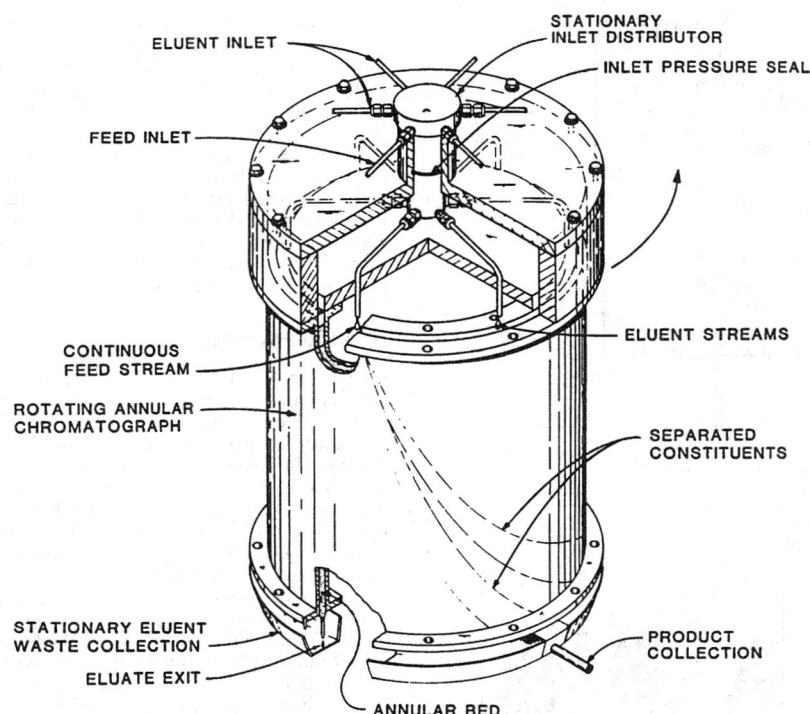

FIG. 16-56   Annular bed for liquid separation. [*Reprinted with permission of* Marcel Dekker. *Reference: Canon, Begovich, and Sisson*, Sep. Sci. Technol., **15**, 655–678 (1980).]

placed downward by opening a valve at the bottom and injecting a measured slug of carbon into the top of the vessel. The exhausted slug (a small fraction of the total charge in order to approximate fully countercurrent operation very closely) is transferred to the sweeten-off tank, where residual product is displaced. It next is dewatered and fed to the regeneration furnace. The carbon is eventually returned to the adsorber where fluid flow is interrupted briefly to permit carbon transfer in, as in continuous ion exchanger systems.

Liquid chromatography has been used commercially to separate glucose from fructose and other sugar isomers, for recovery of nucleic acids, and other uses. A patent to Sanmatsu Kogyo Co., Ltd [Yoritomi, Kezuka, and Moriya, U.S. Patent number 4,267,054, 1981] presents an improved chromatographic process that is simpler to build and operate than simulated moving-bed processes. Figure 16-57 [Keller et al., gen. refs.] diagrams its use for a binary separation. It is a displacement-purge cycle where pure component cuts are recovered, while cuts that contain both components are recycled to the feed end of the column.

The UOP Cyclesorb^SM is another adsorptive separation process with semicontinuous recycle. It utilizes a series of chromatographic columns to separate fructose from glucose. A series of internal recycle streams of impure and dilute portions of the chromatograph are used to improve the efficiency [Gerhold, U.S. Patent numbers 4,402,832, 1983; and 4,478,721, 1984]. A schematic diagram of a six-vessel UOP Cyclesorb process is shown in Fig. 16-58 [Gembicki et al., gen. refs., p. 595]. The process has four external streams and four internal recycles: Dilute raffinate and impure extract are like displacement steps; and impure raffinate and dilute extract are recycled from the bottom of an adsorber to its top. Feed and desorbent are fed to the top of each column in a predetermined sequence. The switching of the feed and desorbent are accomplished by the same rotary valve used for Sorbex switching (see hereafter). A chromatographic profile is established in each column that is moving from top to bottom, and all portions of an adsorber are performing a useful function at any time.

The concept of a simulated moving bed (SMB) was originally used in a process developed and licensed under the name UOP Sorbex process [Broughton, Bieser, and Anderson, *Pet. Int. (Milan)*, **23**(3), p. 91 (1976); Broughton, Bieser, and Persak, *Pet. Int. (Milan)*, **23**(5), p. 36 (1976)]. The following discussion is based on that process, but the concepts can be generally applied. In a moving-bed system for continuous countercurrent effect, solids move continuously in a closed loop past fixed points of introduction and withdrawal of feed and regenerant. The same effect can be achieved by holding the bed stationary and periodically moving the points of introduction and withdrawal of the various streams. Shifting those points in the direction of fluid flow simulates the movement of solid in the opposite direction. Since moving the positions continuously is impractical, a finite number of access lines are provided to a limited series of adsorbent beds. The Sorbex commercial application of this concept is portrayed in Fig. 16-59, which shows the adsorbent as a stationary bed and the auxiliary distillation columns needed to separate the desorbent from raffinate and extract so that it can be recycled to the process. In this example, feed flows through the three beds of Zone 1, where the most strongly adsorbed component A is adsorbed and depleted from the raffinate. Desorbent flows through the three beds of Zone 3 to displace component A into the extract. The less strongly adsorbed component B is adsorbed slightly in the two beds of Zone 4 between raffinate outlet and desorbent, inlet to prevent it from contaminating extract in Zone 3. Component B is desorbed in the four beds of Zone 2 between the extract outlet and the feed inlet and most leaves with the raffinate. A pump draws liquid from the bottom outlet to the top inlet of the adsorbent chamber. All flows in the beds are downward, simulating an upward flow of solid. The four active port positions are likewise moved downward by the selection of the rotary valve. The rotary valve functions in the same manner as a multiport stopcock or chromatography valve, sequencing the four streams on the right to the lines connected to the inlet/outlet nozzles on the adsorber. The next position of the rotary valve will direct desorbent to

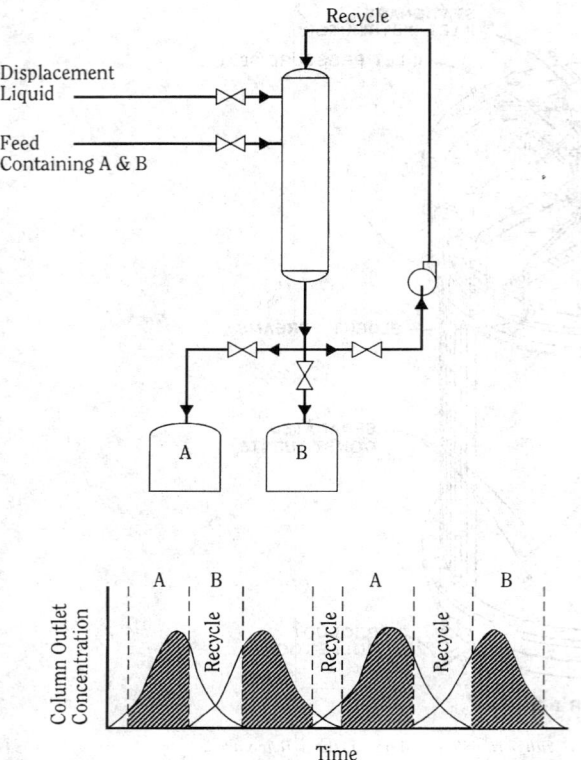

FIG. 16-57 Sanmatsu Kogyo chromatographic process. (*Reprinted with permission of Wiley. Reference: Keller, Anderson, and Yon, Chap. 12 in Rousseau, Handbook of Separation Process Technology, John Wiley & Sons, Inc., New York, 1987.*)

line 3, extract from 6, and feed to 10; and raffinate moves to the top of the column at line 1. The liquid flow rate in each of the four zones is different because of the addition or withdrawal of the various streams and is controlled by the circulating pump. The circulating pump must be programmed to pump at four different rates in order to keep the flow in each zone constant. The variables available for fine tuning performance are the cycle time, measured as the time required for one rotation of the rotary valve, and the liquid flow rate in Zones 2, 3, and 4. Chemical analyses at the liquid circulating pump can trace the performance of the entire bed and are used for changing operating conditions. Other versions of the SMB system have been used commercially by Toray Industries, Illinois Water Treatment and Mitsubishi (see "Displacement Purge").

ION EXCHANGE

A typical fixed-bed ion exchanger consists of a vertical cylindrical vessel of lined steel or stainless steel. Linings are usually of natural or synthetic rubber. Spargers are provided at the top and bottom, and frequently a separate distributor is used for the regenerant solution. The resin bed, consisting of a meter or more of ion-exchange particles, is supported by the screen of the bottom distributor. Externally, the unit is provided with a valve manifold to permit downflow loading, upflow backwash, regeneration, and rinsing. For deionization, a two-exchanger assembly comprising a cation and an anion exchanger, is a common configuration (Fig. 16-60).

Column hardware designed to allow countercurrent, upflow regeneration of ion-exchange resins is available. An example is given in Fig. 16-61. During upflow of the regenerant, bed expansion is prevented by withdrawing the effluent through the application of vacuum. A layer of drained particles is formed at the top of the bed while the rest of the column functions in the usual way.

Typical design data for fixed-bed ion-exchange columns are given in Table 16-19. These should be used for preliminary evaluation purposes only. Characteristic design calculations are presented and illustrated by Applebaum [*Demineralization by Ion Exchange*, Academic, New York, 1968]. Large amounts of data are available in published literature and in bulletins of manufacturers of ion exchangers. In general, however, laboratory testing and pilot-plant work is advisable to determine usable exchanger capacities, regenerant quantities,

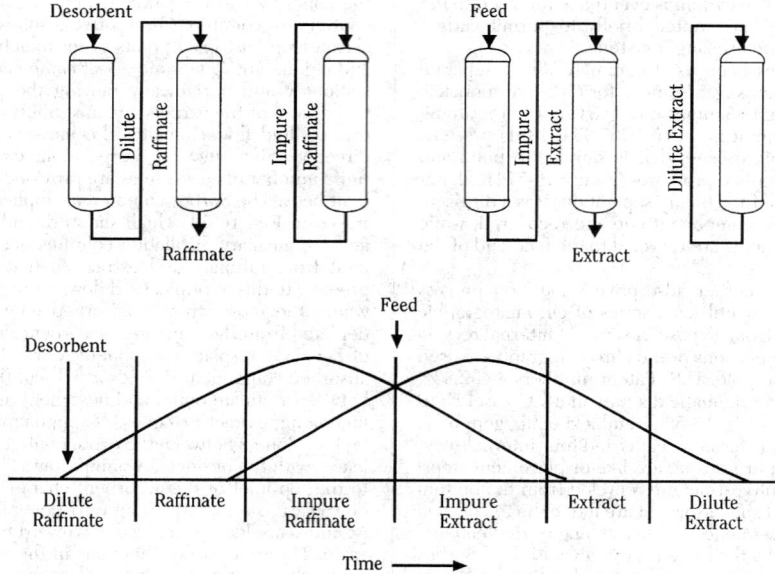

FIG. 16-58 UOP Cyclesorb process. (*Reprinted with permission of John Wiley & Sons, Inc. Reference: Gembicki, Oroskar, and Johnson, "Adsorption, Liquid Separation," in Kirk-Othmer Encyclopedia of Chemical Technology, 4th ed., Wiley, New York, 1991.*)

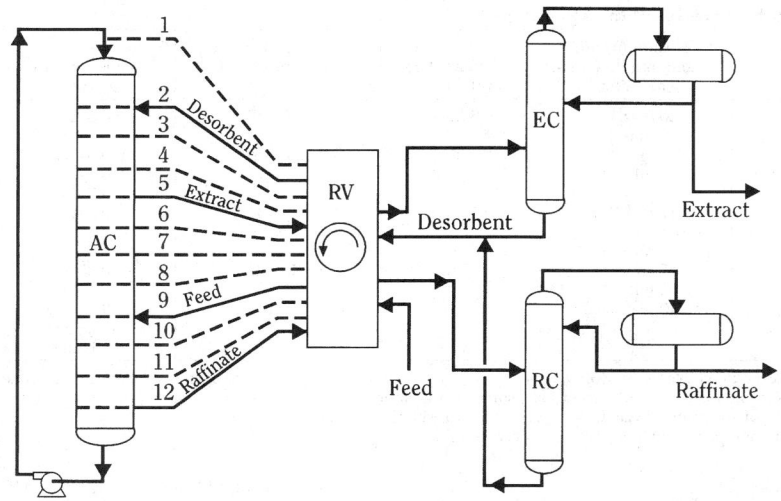

FIG. 16-59   UOP Sorbex process. (*Reprinted with permission of John Wiley & Sons, Inc. Reference: Gembicki, Oroskar, and Johnson, "Adsorption, Liquid Separation," in Kirk-Othmer Encyclopedia of Chemical Technology, 4th ed., John Wiley & Sons, Inc., New York, 1991.*)

exchanger life, and quality of product for any application not already thoroughly proven in other plants for similar conditions. Firms that manufacture ion exchangers and ion-exchange equipment will often cooperate in such tests.

For larger-scale applications, a number of continuous or semicontinuous ion-exchange units are also available. The Higgins contactor (Fig. 16-62) was originally developed to recover uranium from leach slurries at the Oak Ridge National Laboratory. More recently, it has been adapted to a wide variety of applications, including large-volume water softening.

The Asahi process (Fig. 16-63) is used principally for high-volume water treatment. The liquid to be treated is passed upward through a resin bed in the adsorption tank. The upward flow at 30–40 m/h [12–16 gal/(min ft²)] keeps the bed packed against the top. After a preset time, 10 to 60 min, the flow is interrupted for about 30 s, allowing the entire bed to drop. A small portion (10 percent or less) of the ion-exchange resin is removed from the bottom of the adsorption tank and transferred hydraulically to the hopper feeding the regeneration tank.

The process is then resumed. Meanwhile, regeneration is occurring by a similar flow system in the regeneration tank, from which the regenerated ion exchanger is transferred periodically to the hopper above the water-rinse tank. In the latter, the resin particles are fluidized to flush away fines and accumulated foreign matter before the resin is returned to the adsorption tank.

Another continuous ion-exchange system is described by Himsley and Farkas (see Fig. 16-64). Such a system is used to treat 1590 m³/h (7000 gal/min) of uranium-bearing copper leach liquor using fiberglass-construction columns 3.7 m (12 ft) in diameter. The adsorption column is divided vertically into stages. Continuous transfer of batches of resin occurs from stage to stage without any interruption of flow. This is accomplished by pumping solution from one stage (A) to the stage (B) immediately above by means of external pumping in such a

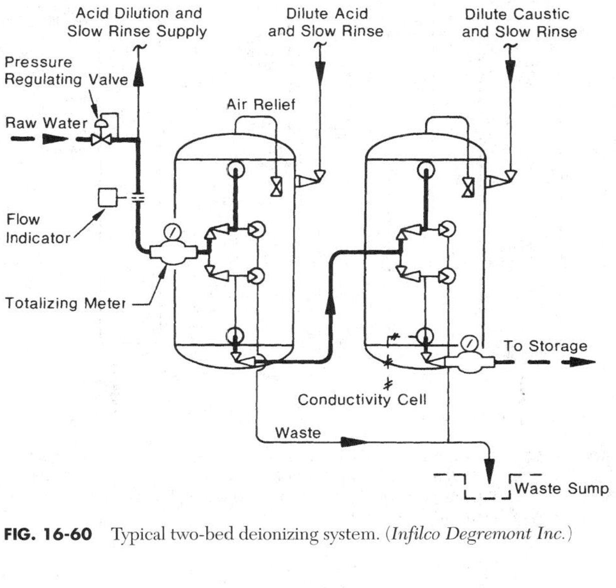

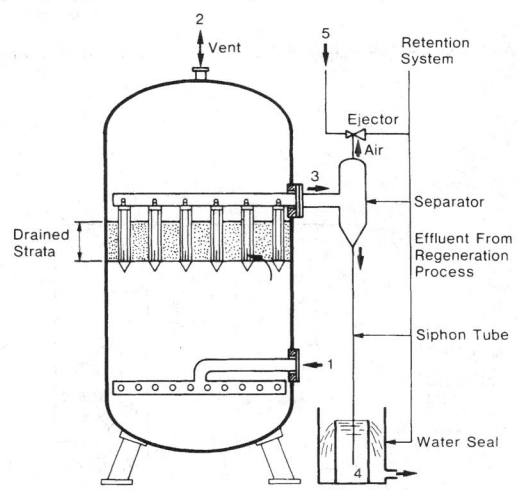

1 - Regenerant Inlet
2 - Vent To Atmosphere
3 - Outlet Air + Effluent From Regeneration Cycle
4 - Outlet Effluent From Regeneration Cycle
5 - Inlet Injected Water

FIG. 16-60   Typical two-bed deionizing system. (*Infilco Degremont Inc.*)

FIG. 16-61   Internals of an upflow regenerated unit. (*Infilco Degremont Inc.*)

TABLE 16-19 Design Data for Fixed-Bed Ion Exchanger*

Type of resin	Maximum and minimum flow, m/h [gal/(min·ft²)]	Minimum bed depth, m (in)	Maximum operating temperatures, °C (°F)	Usable capacity, g-equivalent/L†	Regenerant, g/L resin‡
Weak acid cation	20 max. (8) 3 min. (1)	0.6 (24)	120 (248)	0.5–2.0	110% theoretical (HCl or H₂SO₄)
Strong acid cation	30 max. (12) 3 min. (1)	0.6 (24)	120 (248)	0.8–1.5 0.5–1.0 0.7–1.4	80–250 NaCl 35–200 66° Bé. H₂SO₄ 80–500 20° Bé. HCl
Weak and intermediate base anions	17 max. (7) 3 min. (1)	0.75 (30)	40 (104)	0.8–1.4	35–70 NaOH
Strong-base anions	17 max. (7) 3 min. (1)	0.75 (30)	50 (122)	0.35–0.7	70–140 NaOH
Mixed cation and strong-base anion (chemical-equivalent mixture)	40 max. (16)	1.2 (47)	50 (122)	0.2–0.35 (based on mixture)	Same as cation and anion individually

*These figures represent the usual ranges of design for water-treatment applications. For chemical-process applications, allowable flow rates are generally somewhat lower than the maximums shown, and bed depths are usually somewhat greater.
†To convert to capacity in terms of kilograms of $CaCO_3$ per cubic foot of resin, multiply by 21.8.
‡To convert to pounds of regenerant per cubic foot of resin, multiply by 0.0625.

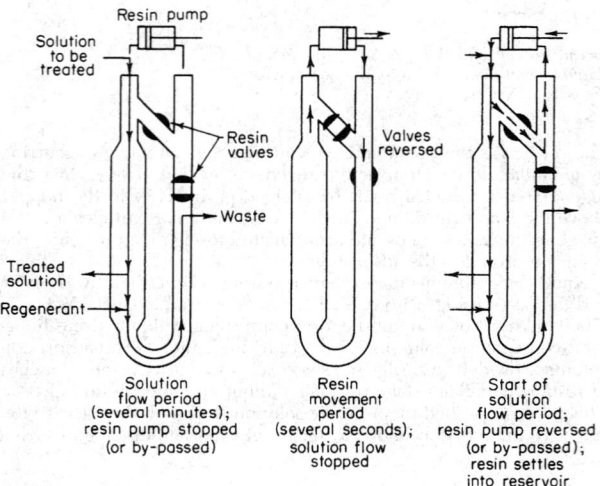

FIG. 16-62 Mode of operation of the Higgins contractor. (*ORNL-LR-Dwg. 27857R.*)

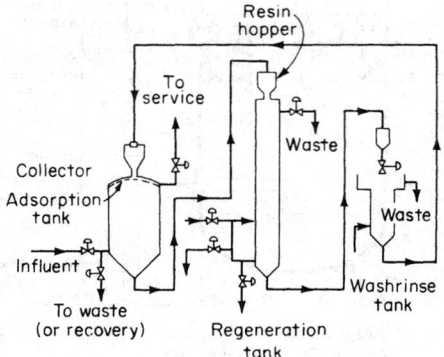

FIG. 16-63 Asahi countercurrent ion-exchange process. [*Gilwood, Chem. Eng.,* **74**(26), 86 (1967); copyright 1967 by McGraw-Hill, Inc., New York. Excerpted with special permission of McGraw-Hill.]

manner that the net flow through stage B is downward, carrying with it all of the resin in that stage B. When transfer of the ion exchanger is complete, the resin in stage C above is transferred downward in a similar manner. The process continues until the last stage (F) is empty. The regenerated resin is then transferred from the elution column to the empty stage (F). Elution of the sorbed product is carried out in the elution column in a moving packed bed mode. The countercurrent contact achieved in this column yields a concentrated eluate with a minimum consumption of regenerant.

Disadvantages of these continuous countercurrent systems are associated primarily with the complexity of the equipment required and with the attrition resulting from the transport of the ion exchanger. An effective alternative for intermediate scale processes is the use of merry-go-round systems and SMB units employing only packed-beds with no movement of the ion-exchanger.

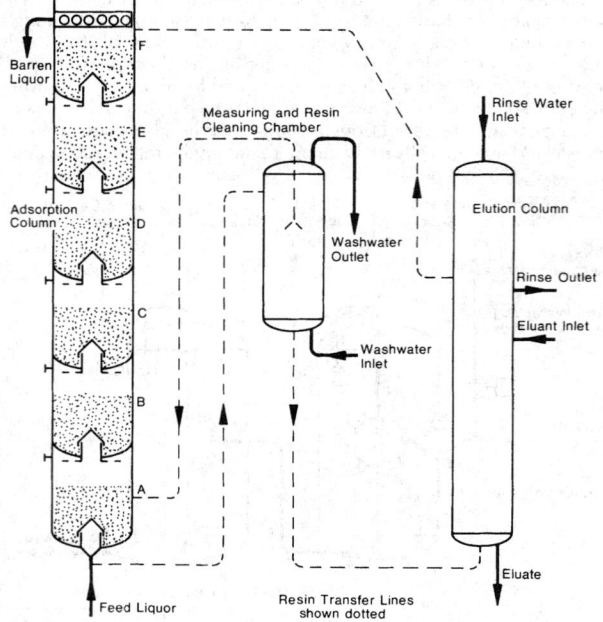

FIG. 16-64 Himsley continuous ion-exchange system. (*Himsley and Farkas, "Operating and Design Details of a Truly Continuous Ion Exchange System," Soc. Chem. Ind. Conf., Cambridge, England, July 1976. Used by permission of the Society of Chemical Industry.*)

Gas-Solid Operations and Equipment

Mel Pell, Ph.D., *Senior Consultant, E. I. duPont de Nemours & Co.; Fellow, American Institute of Chemical Engineers. (Section Editor, Fluidized Bed Systems)*

James B. Dunson, M.S., *Principal Consultant, E. I. duPont de Nemours & Co.; Member American Institute of Chemical Engineers; Registered Professional Engineer. (Delaware) (Gas-Solid Separations)*

FLUIDIZED-BED SYSTEMS

Fluidization, or fluidizing, converts a bed of solid particles into an expanded, suspended mass that has many properties of a liquid. This mass has zero angle of repose, seeks its own level, and assumes the shape of the containing vessel.

Fluidized beds are used successfully in a multitude of processes both catalytic and noncatalytic. Among the catalytic uses are hydrocarbon cracking and re-forming, oxidation of naphthalene to phthalic anhydride, and ammoxidation of propylene to acrylonitrile. A few of the noncatalytic uses are roasting of sulfide ores, coking of petroleum residues, calcination of ores, incineration of sewage sludge, drying, and classification. Considerable effort and interest are now centered in the areas of coal and waste combustion to raise steam.

The size of solid particles which can be fluidized varies greatly from less than 1 µm to 6 cm (2½ in). It is generally concluded that particles distributed in sizes between 150 µm and 10 µm are the best for smooth fluidization (least formation of large bubbles). Large particles cause instability and result in slugging or massive surges. Small particles (less than 20 µm) frequently, even though dry, act as if damp, forming agglomerates or fissures in the bed, or spouting. Adding finer-sized particles to a coarse bed or coarser-sized particles to a bed of fines usually results in better fluidization.

The upward velocity of the gas is usually between 0.15 m/s (0.5 ft/s) and 6 m/s (20 ft/s). This velocity is based upon the flow through the empty vessel and is referred to as the **superficial velocity.**

For details beyond the scope of this subsection, reference should be made to Kunii and Levenspiel, *Fluidization Engineering*, 2d ed., Butterworth Heinemann, Boston, 1991; Pell, *Gas Fluidization*, Elsevier, New York, 1990; D. Geldart (ed.), *Gas Fluidization Technology*, Wiley, New York, 1986; and the vast number of papers published in periodicals, transcripts of symposia, and the American Institute of Chemical Engineers symposium series.

GAS-SOLID SYSTEMS

Several workers in the field have systematized the various types of fluidization. Several of these types are discussed in the following subsections because each adds another dimension to the understanding of the phenomena.

Types of Solids Geldart [*Powder Technol., **7**, 285–292 (1973)] has characterized four groups of solids that exhibit different properties when fluidized with a gas. Figure 17-1 shows the division of the classes as a function of mean particle size, $\bar{d}_{sv}$, µm, and density difference, $(\rho_s - \rho_f)$, g/cm³, where ρ_s = particle density and ρ_f = fluid density

$$\bar{d}_{sv} = \frac{1}{\sum (x_i/d_{pi})}$$

where $\bar{d}_{sv}$ = surface volume diameter of particle and x_i = weight fraction of particles of size d_{pi}. $\bar{d}_{sv}$ is the preferred particle averaging method for fluid bed applications.

When gas is passed upward through a bed of particles of groups A, B, or D, friction causes a pressure drop expressed by the Carman-Kozeny fixed-bed correlation. As the gas velocity is increased, the pressure drop increases until it equals the weight of the bed divided by the cross-sectional area. This velocity is called minimum fluidizing velocity, U_{mf}. When this point is reached, the bed of group A particles will expand uniformly until at some higher velocity gas bubbles will form (minimum bubbling velocity, U_{mb}). For group B and group D particles U_{mf} and U_{mb} are essentially equal. Group C particles exhibit cohesive tendencies, and as the gas flow is further increased, usually "rathole"; the gas opens channels that extend from the gas distributor to the surface. If channels are not formed, the whole bed will lift as a piston. At higher velocities or with mechanical agitation or vibration,

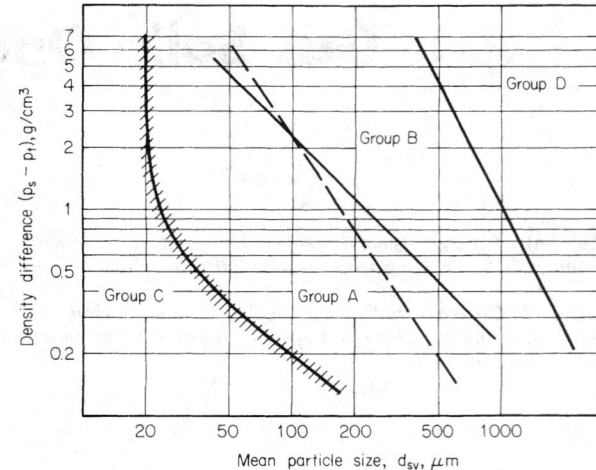

FIG. 17-1 Powder-classification diagram for fluidization by air (ambient conditions). [*From Geldart*, Powder Technol., **7**, 285–292 (1973).]

this type of particle will fluidize but with the appearance of clumps or clusters of particles. For all groups of powder (A, B, C, and D) as the gas velocity is further increased, bed density is decreased and turbulence increased. In smaller-diameter beds, especially with group B and D powders, slugging will occur as the bubbles increase in size to greater than half of the bed diameter. Bubbles grow by vertical and lateral merging. Bubbles also increase in size as the gas velocity is increased [Whitehead, in Davidson and Harrison (eds.), *Fluidization*, Academic, London and New York, 1971]. As the gas velocity is increased further and bubbles tend to disappear and streamers of solids and gas prevail, pressure fluctuations in the bed are greatly reduced. Further increase in velocity results in dilute-phase pneumatic transport.

Phase Diagram (Zenz and Othmer) Zenz and Othmer (op. cit.) have graphically represented (Fig. 17-2) all gas-solid systems in which the gas is flowing counter to gravity as a function of pressure drop per unit of height versus velocity. Note that line OAB in Fig. 17-2 is the pressure-drop versus gas-velocity curve for a packed bed and BD the curve for a fluid bed. Zenz indicates an instability between D and H because with no solids flow all the particles will be entrained from the bed; however, if solids are added to replace those entrained, system IJ prevails. The area $DHIJ$ will be discussed further.

Phase Diagram (Grace) Grace [*Can. J. Chem. Eng., **64**, 353–363 (1986); Fig. 17-3] has correlated the various types of gas-solid systems in which the gas is flowing counter to gravity in a status graph using the parameters of the Archimedes number (Ar) for the particle size and a nondimensional velocity (U°) for the gas effects. By means of this plot, the regime of fluidization can be predicted.

Regime Diagram (Grace) Grace [*Can. J. Chem. Eng., **64**, 353–363 (1986); Fig. 17-4] has also sketched the way a fluid bed appears in the different regimes of fluidization resulting from increasing velocity from packed-bed operation to a transport reactor. The area that Zenz and Othmer consider to be discontinuous is called the fast fluid bed regime. Note that since beds in turbulent and faster regimes operate above the terminal velocity of some or all of the particles, solids return is necessary to maintain a bed.

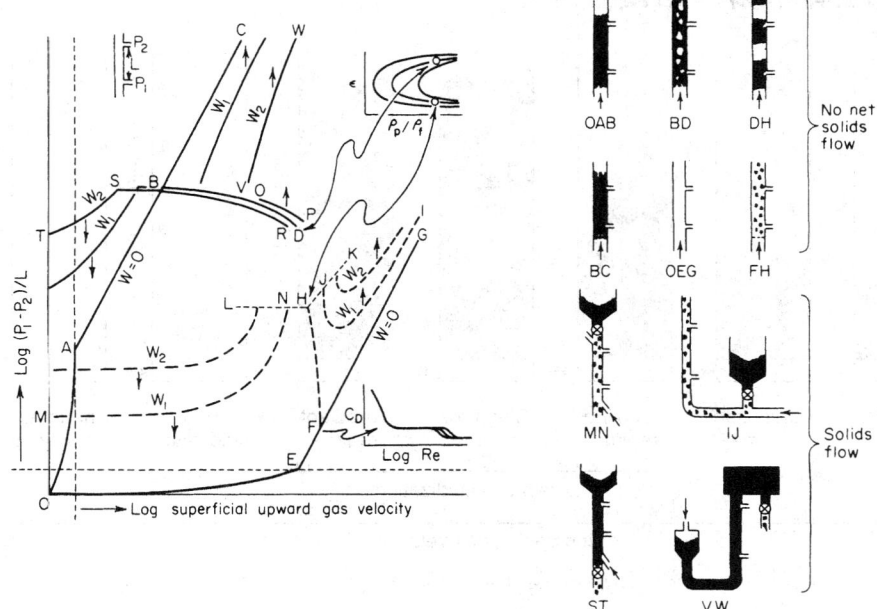

FIG. 17-2 Schematic phase diagram in the region of upward gas flow. W = mass flow solids, lb/(h · ft²); ε = fraction voids; ρ_p = particle density, lb/ft³; ρ_f = fluid density, lb/ft³; Cd = drag coefficient; Re = modified Reynolds number. (*Zenz and Othmer,* Fluidization and Fluid Particle Systems, *Reinhold, New York, 1960.*)

Key:

OAB = packed bed	IJ = cocurrent flow	AC = packed bed	FH = dilute phase
BD = fluidized bed	(dilute phase)	(restrained at top)	MN = countercurrent flow
DH = slugging bed	ST = countercurrent flow	OEG = fluid only	(dilute phase)
	(dense phase)	(no solids)	VW = cocurrent flow
			(dense phase)

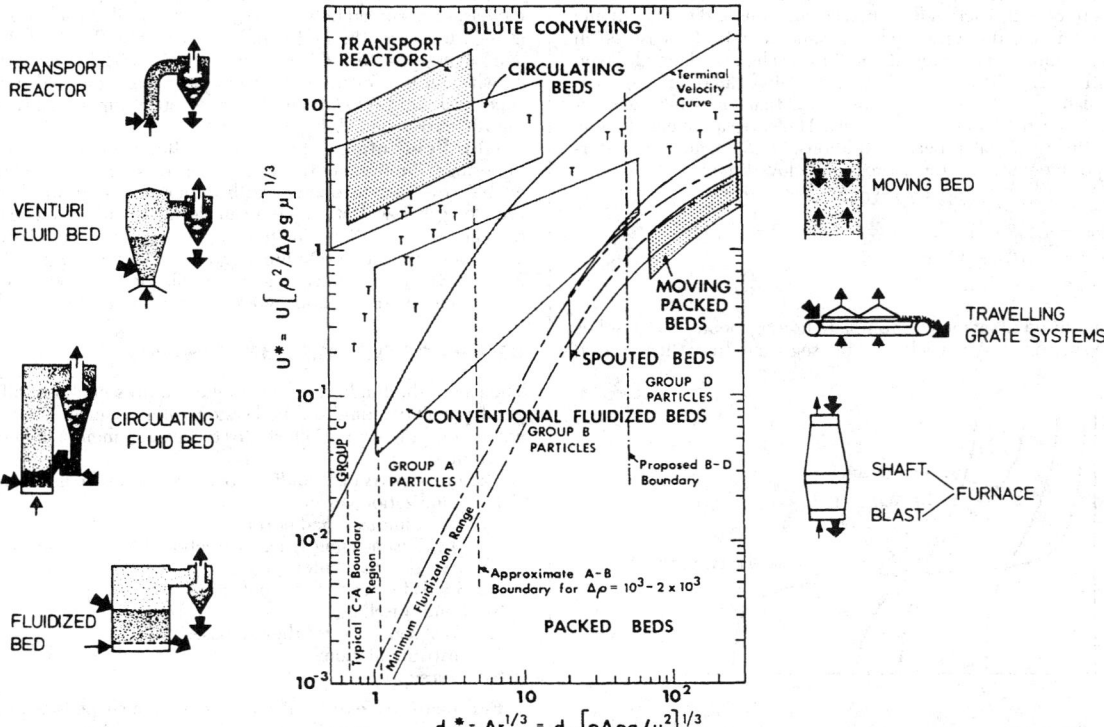

FIG. 17-3 Simplified fluid-bed status graph [*From Grace,* Can. J. Chem. Eng., **64**, 353–363 (1986); *sketches from Reh,* Ger. Chem. Eng., **1**, 319–329 (1978).]

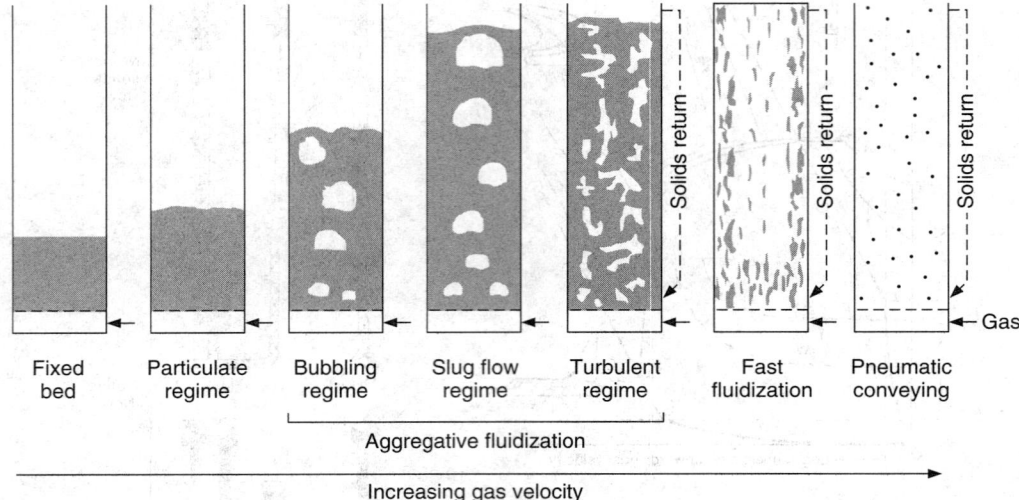

FIG. 17-4   Fluidization regimes. [*Adapted from Grace, Can J. Chem. Eng., **64**, 353–363 (1986).*]

Solids Concentration versus Height   From the foregoing it is apparent that there are several regimes of fluidization. These are, in order of increasing gas velocity, particulate fluidization (Geldart group A), bubbling (aggregative), turbulent, fast, and transport. Each of these regimes has characteristic solids concentration profiles as shown in Fig. 17-5.

Equipment Types   Fluidized-bed systems take many forms. Figure 17-6 shows some of the more prevalent concepts with approximate ranges of gas velocities.

Minimum Fluidizing Velocity U_{mf}, the minimum fluidizing velocity, is frequently used in fluid-bed calculations and in quantifying one of the particle properties. This parameter is best measured in small-scale equipment at ambient conditions. The correlation by Wen and Yu [*A.I.Ch.E.J.*, 610–612 (1966)] given below can then be used to back calculate d_p. This gives a particle size that takes into account effects of size distribution and sphericity. The correlation can then be used to estimate U_{mf} at process conditions. If U_{mf} cannot be determined experimentally, use the expression below directly.

$$Re_{mf} = (1135.7 + 0.0408Ar)^{0.5} - 33.7$$

where $Re_{mf} = \bar{d}_p\rho_f U_{mf}/\mu$
$Ar = \bar{d}_p^3\rho_f(\rho_s - \rho_f)g/\mu^2$
$\bar{d}_p = 1/\sum (x_i/d_{pi})$

The flow required to maintain a complete homogeneous bed of solids in which coarse or heavy particles will not segregate from the fluidized

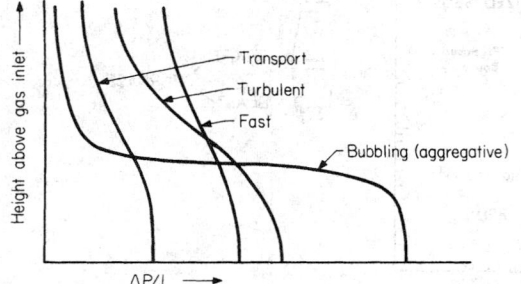

FIG. 17-5   Solids concentration versus height above distributor for regimes of fluidization.

portion is very different from the minimum fluidizing velocity. See Nienow and Chiba, *Fluidization*, 2d ed., Wiley, 1985, pp. 357–382, for a discussion of segregation or mixing mechanism as well as the means of predicting this; also see Baeyens and Geldart, *Gas Fluidization Technology*, Wiley, 1986, 97–122.

Particulate Fluidization   Fluid beds of Geldart class A powders that are operated at gas velocities above the minimum fluidizing velocity (U_{mf}) but below the minimum bubbling velocity (U_{mb}) are said to be particulately fluidized. As the gas velocity is increased above U_{mf}, the bed further expands. Decreasing ($\rho_s - \rho_f$), d_p and/or increasing μ_f increases the spread between U_{mf} and U_{mb} until at some point, usually at high pressure, the bed is fully particulately fluidized. Richardson and Zaki [*Trans. Inst. Chem. Eng.*, **32**, 35 (1954)] showed that $U/U_i = \varepsilon^n$, where n is a function of system properties, ε = void fraction, U = superficial fluid velocity, and U_i = theoretical superficial velocity from the Richardson and Zaki plot when $\varepsilon = 1$.

Vibrofluidization   It is possible to fluidize a bed mechanically by imposing vibration to throw the particles upward cyclically. This enables the bed to operate with either no gas upward velocity or reduced gas flow. Entrainment can also be greatly reduced compared to unaided fluidization. The technique is used commercially in drying and other applications [Mujumdar and Erdesz, *Drying Tech.*, **6**, 255–274 (1988)], and chemical reaction applications are possible. See Sec. 12 for more on drying applications of vibrofluidization.

DESIGN OF FLUIDIZED-BED SYSTEMS

The use of the fluidization technique requires in almost all cases the employment of a fluidized-bed system rather than an isolated piece of equipment. Figure 17-7 illustrates the arrangement of components of a system.

The major parts of a fluidized-bed system can be listed as follows:
1. Fluidization vessel
   a. Fluidized-bed portion
   b. Disengaging space or freeboard
   c. Gas distributor
2. Solids feeder or flow control
3. Solids discharge
4. Dust separator for the exit gases
5. Instrumentation
6. Gas supply

Fluidization Vessel   The most common shape is a vertical cylinder. Just as for a vessel designed for boiling a liquid, space must be provided for vertical expansion of the solids and for disengaging

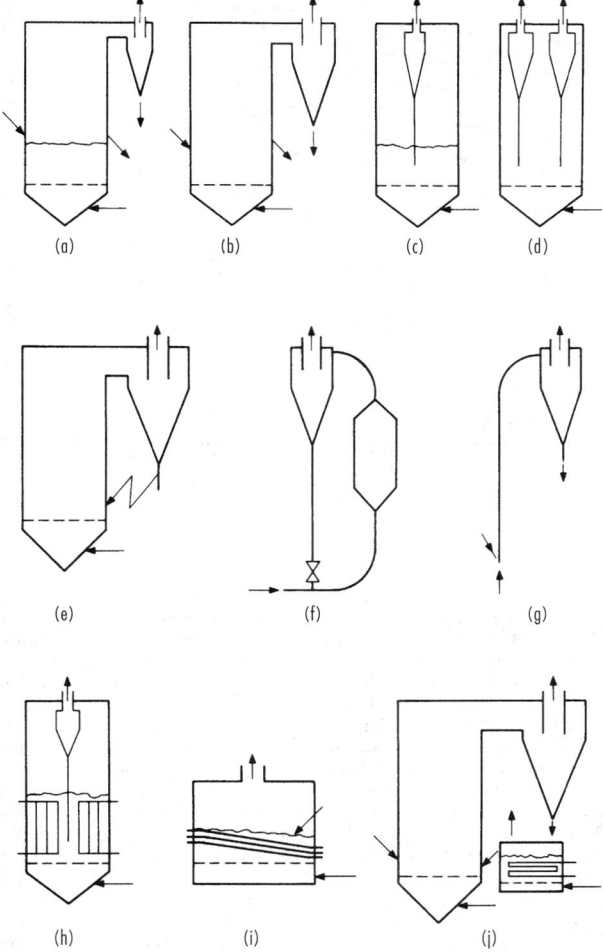

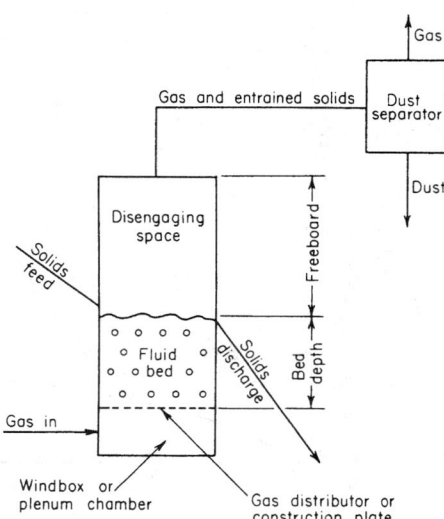

FIG. 17-7 Noncatalytic fluidized-bed system.

FIG. 17-6 Fluidized-bed systems. (*a*) Bubbling bed, external cyclone, $U < 20 \times U_{mf}$. (*b*) Turbulent bed, external cyclone, $20 \times U_{mf} < U < 200 \times U_{mf}$. (*c*) Bubbling bed, internal cyclones, $U < 20 \times U_{mf}$. (*d*) Turbulent bed, internal cyclones, $20 \times U_{mf} < U < 200 \times U_{mf}$. (*e*) Circulating (fast) bed, external cyclones, $U > 200 \times U_{mf}$. (*f*) Circulating bed, $U > 200 \times U_{mf}$. (*g*) Transport, $U > U_T$. (*h*) Bubbling or turbulent bed with internal heat transfer, $2 \times U_{mf} < U < 200 \times U_{mf}$. (*i*) Bubbling or turbulent bed with internal heat transfer, $2 \times U_{mf} < U < 100 \times U_{mf}$. (*j*) Circulating bed with external heat transfer, $U > 200 \times U_{mf}$.

splashed and entrained material. The volume above the bed is called the disengaging space. The cross-sectional area is determined by the volumetric flow of gas and the allowable or required fluidizing velocity of the gas at operating conditions. In some cases the lowest permissible velocity of gas is used, and in others the greatest permissible velocity is used. The maximum flow is generally determined by the carry-over or entrainment of solids, and this is related to the dimensions of the disengaging space (cross-sectional area and height).

Bed Bed height is determined by a number of factors, either individually or collectively, such as:
1. Gas-contact time
2. *L/D* ratio required to provide staging
3. Space required for internal heat exchangers
4. Solids-retention time

Generally, bed heights are not less than 0.3 m (12 in) or more than 15 m (50 ft).

Although the reactor is usually a vertical cylinder, there is no real limitation on shape. The specific design features vary with operating

conditions, available space, and use. The lack of moving parts lends toward simple, clean design.

Many fluidized-bed units operate at elevated temperatures. For this use, refractory-lined steel is the most economical design. The refractory serves two main purposes: (1) it insulates the metal shell from the elevated temperatures, and (2) it protects the metal shell from abrasion by the bed and particularly the splashing solids at the top of the bed resulting from bursting bubbles. Depending on specific conditions, several different refractory linings are used [Van Dyck, *Chem. Eng. Prog.*, 46–51 (December 1979)]. Generally, for the moderate temperatures encountered in catalytic cracking of petroleum, a reinforced-gunnite lining has been found to be satisfactory. This also permits the construction of larger units than would be permissible if self-supporting ceramic domes were to be used for the roof of the reactor.

When heavier refractories are required because of operating conditions, insulating brick is installed next to the shell and firebrick is installed to protect the insulating brick. Industrial experience in many fields of application has demonstrated that such a lining will successfully withstand the abrasive conditions for many years without replacement. Most serious refractory wear occurs with coarse particles at high gas velocities and is usually most pronounced near the operating level of the fluidized bed.

Gas leakage behind the refractory has plagued a number of units. Care should be taken in the design and installation of the refractory to reduce the possibility of the formation of "chimneys" in the refractories. A small flow of solids and gas can quickly erode large passages in soft insulating brick or even in dense refractory. Gas stops are frequently attached to the shell and project into the refractory lining. Care in design and installation of openings in shell and lining is also required.

In many cases, cold spots on the reactor shell will result in condensation and high corrosion rates. Sufficient insulation to maintain the shell and appurtenances above the dew point of the reaction gases is necessary. Hot spots can occur where refractory cracks allow heat to permeate to the shell. These can sometimes be repaired by pumping castable refractory into the hot area from the outside.

The violent motion of a fluidized bed requires ample foundations and sturdy supporting structure for the reactor. Even a relatively small differential movement of the reactor shell with the lining will materially shorten refractory life. The lining and shell must be designed as a unit. Structural steel should not be supported from a vessel that is subject to severe vibration.

Freeboard and Entrainment The freeboard or disengaging height is the distance between the top of the fluid bed and the gas-exit

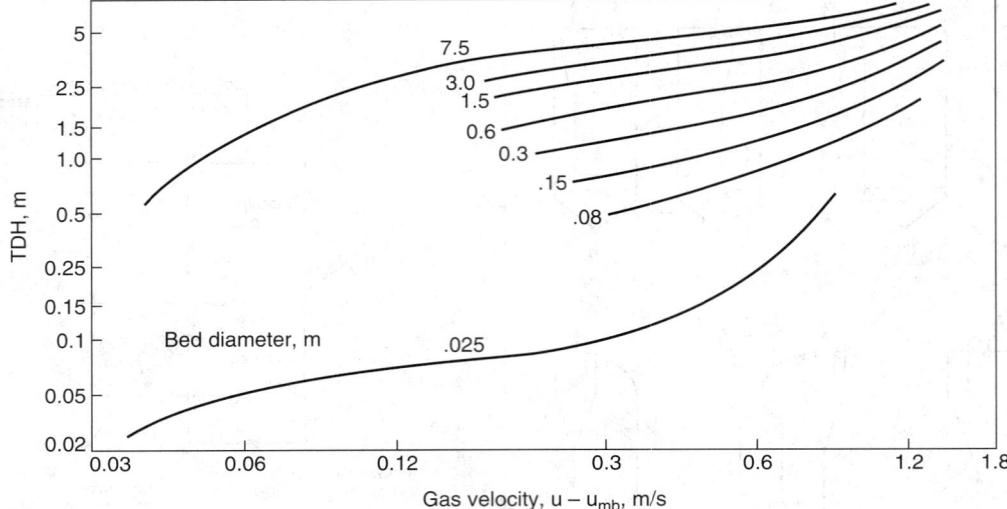

FIG. 17-8 Estimating transport disengaging height (TDH).

nozzle in bubbling- or turbulent-bed units. The distinction between bed and freeboard is difficult to determine in fast and transport units (see Fig. 17-5).

At least two actions can take place in the freeboard: classification of solids and reaction of solids and gases.

As a bubble reaches the upper surface of a fluidized bed, the gas breaks through the thin upper envelope composed of solid particles entraining some of these particles. The crater-shaped void formed is rapidly filled by flowing solids. When these solids meet at the center of the void, solids are geysered upward. The downward pull of gravity and the upward pull of the drag force of the upward-flowing gas act on the particles. The larger and denser particles return to the top of the bed, and the finer and lighter particles are carried upward. The distance above the bed at which the entrainment becomes constant is the transport disengaging height, TDH. Cyclones and vessel gas outlets are usually located above TDH. Figure 17-8 graphically estimates TDH as a function of velocity and bed size.

The higher the concentration of an entrainable component in the bed, the greater its rate of entrainment. Finer particles have a greater rate of entrainment than coarse ones. These principles are embodied in the method of Geldart (*Gas Fluidization Tech.*, Wiley, 1986, pp. 123–153) via the equation, $E(i) = K^\circ(i)x(i)$, where $E(i)$ = entrainment rate for size i, kg/m^2 s; $K^\circ(i)$ = entrainment rate constant for particle size i; and $x(i)$ = weight fraction for particle size i. K° is a function of operating conditions given by $K^\circ(i)/(P_f u) = 23.7 \exp[-5.4 \, U_t(i)/U]$. The composition and the total entrainment are calculating by summing over the entrainable fractions. An alternative is to use the method of Zenz as reproduced by Pell (*Gas Fluidization*, Elsevier, 1990, pp. 69–72).

In batch classification, the removal of fines (particles less than any arbitrary size) can be correlated by treating as a second-order reaction $K = (F/\theta)[1/x(x - F)]$, where K = rate constant, F = fines removed in time θ, and x = original concentration of fines.

Gas Distributor The gas distributor has a considerable effect on proper operation of the fluidized bed. Basically there are two types: (1) for use when the inlet gas contains solids and (2) for use when the inlet gas is clean. In the latter case, the distributor is designed to prevent back flow of solids during normal operation, and in many cases it is designed to prevent back flow during shutdown. In order to provide distribution, it is necessary to restrict the gas or gas and solids flow so that pressure drops across the restriction amount to from 0.5 kPa (2 in of water) to 20 kPa (3 lbf/in^2).

The bubbling action of the bed produces local pressure fluctua-

tions. A temporary condition of low pressure in one section of the bed due to bubbles can combine with higher pressure in another section with fewer bubbles to produce permanent flow maldistribution. The grid should be designed for a pressure drop of at least ⅓ the bed weight for upflow distributors and about ¹⁄₁₀ the bed weight for downflow spargers. In units with shallow beds such as dryers or where distribution is less crucial, lower distributor pressure drops can be used.

When both solids and gases pass through the distributor, such as in catalytic-cracking units, a number of variations are or have been used, such as concentric rings in the same plane, with the annuli open (Fig. 17-9a); concentric rings in the form of a cone (Fig. 17-9b); grids of T bars or other structural shapes (Fig. 17-9c); flat metal perforated plates supported or reinforced with structural members (Fig. 17-9d); dished and perforated plates concave both upward and downward (Fig. 17-9e and f). The last two forms are generally more economical.

In order to generate the required pressure drop, a high velocity through the grid openings may be needed. It is best to limit this velocity to less than 70 m/s to minimize attrition of the bed material. It is common industrial practice to put a shroud of pipe over the opening so that the velocity can be kept high for pressure drop but is reduced for entry to the bed. The technique is applied to both plate and pipe spargers.

Pressure drop through a pipe or a drilled plate is given by:

$$\Delta P = \frac{u^2 \rho_f}{0.64 \times 2g}$$

where u = velocity in the hole at inlet conditions
ρ_f = fluid density in the hole at inlet conditions
ΔP = pressure drop in consistent units (may be lbs/ft^2)

Experience has shown that the concave-upward type is a better arrangement than the concave-downward type, as it tends to increase the flow of gases in the outer portion of the bed. This counteracts the normal tendency of higher gas flows in the center of the bed.

Structurally, distributors must withstand the differential pressure across the restriction during normal and abnormal flow. In addition, during a shutdown all or a portion of the bed will be supported by the distributor until sufficient back flow of the solids has occurred both to reduce the weight of solids above the distributor and to support some of this remaining weight by transmitting the force to the walls and bottom of the reactor. During startup considerable upward thrust can be exerted against the distributor as the settled solids under the distributor are carried up into the normal reactor bed.

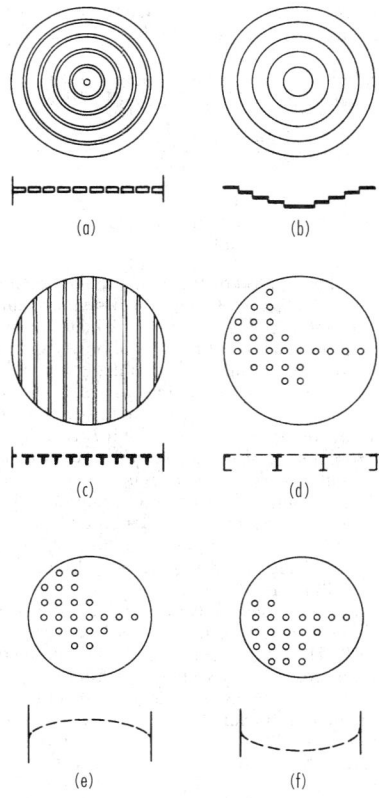

FIG. 17-9   Gas distribution for gases containing solids.

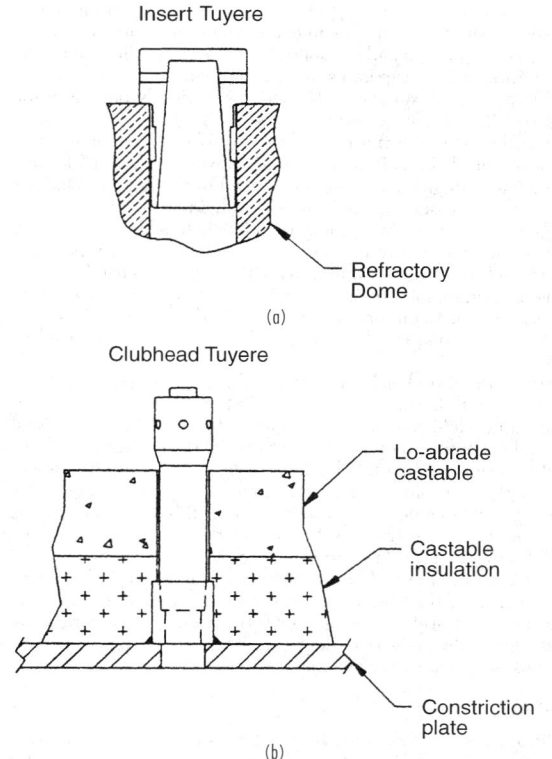

FIG. 17-10   Gas inlets designed to prevent backflow of solids. (*a*) Insert tuyere; (*b*) clubhead tuyere. (*Dorr-Oliver, Inc.*)

When the feed gas is devoid of or contains only small quantities of fine solids, more sophisticated designs of gas distributors can be used to effect economies in initial cost and maintenance. This is most pronounced when the inlet gas is cold and noncorrosive. When this is the case, the plenum chamber gas distributor, and distributor supports can be fabricated of mild steel by using normal temperature design factors. The first commercial fluidized-bed ore roaster [Mathews, *Trans. Can. Inst. Min. Metall.*, **L11,** 97 (1949)], supplied by the Dorr Co. (now Dorr-Oliver Inc.) in 1947 to Cochenour-Willans, Red Lake, Ontario, was designed with a mild-steel constriction plate covered with castable refractory to insulate the plate from the calcine and also to provide cones in which refractory balls were placed to act as ball checks. The balls eroded unevenly, and the castable cracked. However, when the unit was shut down by closing the air-control valve, the runback of solids was negligible because of bridging. If, however, the unit was shut down by deenergizing the centrifugal blower motor, the higher pressure in the reactor would relieve through the blower and fluidizing gas plus solids would run back through the constriction plate. Figure 17-10 illustrates two designs of gas inlets which have been successfully used to prevent flowback of solids. For best results, irrespective of the design, the gas flow should be stopped and pressure-relieved from the bottom upward through the bed.

Some units have been built and successfully operated with simple slot-type distributors made of heat-resistant steel. This requires a heat-resistant plenum chamber but eliminates the frequently encountered problem of corrosion caused by condensation of acids and water vapor on the cold metal of the distributor.

When the inlet gas is hot, such as in dryers or in the upper distributors of multibed units, ceramic arches or heat-resistant metal grates are generally used.

Self-supporting ceramic domes have been in successful use for many years as gas distributors when temperatures range up to 1100°C. Some of these domes are fitted with alloy-steel orifices to regulate air distribution. However, the ceramic arch presents the same problem as the dished head positioned concave downward. Either the holes in the center must be smaller so that the sum of the pressure drops through the distributor plus the bed is constant across the whole cross section, or the top of the arch must be flattened so that the bed depth in the center and outside is equal. This is especially important when shallow beds are used.

It is important to consider thermal effects in design of the grid to

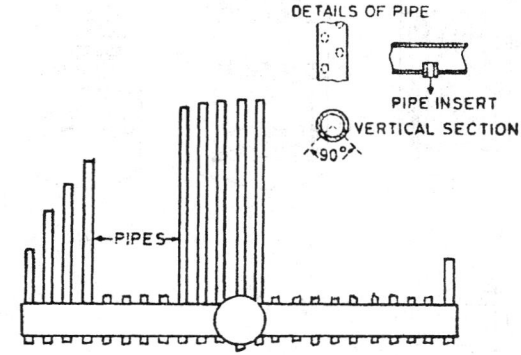

FIG. 17-11   Multiple-pipe gas distributor. [*From Stemerding, de Groot, and Kuypers,* Soc. Chem. Ind. J. Symp. Fluidization Proc., *35–46, London (1963).*]

shell seal. Bypassing of the grid at the seal point is a common problem caused by situations such as uneven expansion of metal and ceramic parts, a cold plenum and hot solids in contact with the grid plate at the same time, and startup and shutdown scenarios.

When the atmosphere in the bed is sufficiently benign, a sparger-type distributor may be used. See Fig. 17-11.

In some cases, it is impractical to use a plenum chamber under the constriction plate. This condition arises when a flammable and explosive mixture of gases is being introduced to the reactor. One solution is to pipe the gases to a multitude of individual gas inlets in the floor of the reactor. In this way it may be possible to maintain the gas velocities in the pipes above the flame velocity or to reduce the volume of gas in each pipe to the point at which an explosion can be safely contained. Another solution is to provide separate inlets for the different gases and depend on mixing in the fluidized bed. The inlets should be fairly close to one another, as lateral gas mixing in fluidized beds is poor.

Much attention has been given to the effect of gas distribution on bubble growth in the bed and the effect of this on catalyst utilization, space-time yield, etc. It would appear that the best gas distributor would be a porous membrane. This type of distributor is seldom practical for commercial units because of both structural limitations and the need for absolutely clean gas. Practically, the limitations on hole spacing are dependent on particle size of solids, materials of construction, and type of distributor. If easily worked metals are used, punching, drilling, and welding are not expensive operations and permit the use of large numbers of holes. The use of tuyeres or bubble caps permits horizontal distribution of the gas so that a smaller number of gas-inlet ports can still achieve good gas distribution. If a ceramic arch is used, generally only one hole per brick is permissible and brick dimensions must be reasonable.

Scale-Up

Bubbling or Turbulent Beds   Scale-up of noncatalytic fluidized beds when the reaction is fast, as in roasting or calcination, is straightforward and is usually carried out on an area basis. Small-scale tests are made to determine physical limitations such as sintering, agglomeration, solids-holdup time required, etc. Slower ($k < 1/s$) catalytic or more complex reactions in which several gas interchanges are required are usually scaled up in several steps, from laboratory to commercial size. The hydrodynamics of gas-solids flow and contacting is quite different in small-diameter high-L/D fluid beds as compared with large-diameter moderate-L/D beds. In small-diameter beds, bubbles tend to be small and cannot grow larger than the vessel diameter. In larger, deeper units, bubbles can grow very large. The large

bubbles have less surface for mass transfer to the solids than the same volume of small bubbles. The large bubbles also rise through the bed more quickly.

The size of a bubble as a function of height was given by Darton et al. [*Trans. Inst. Chem. Eng.*, **55**, 274–280 (1977)] as

$$d_b = \frac{0.54(u - u_{mb})^{0.4}(h + 4\sqrt{A_t/N_o})^{0.8}}{g^{0.2}}$$

where
d_b = bubble diameter, m
h = height above the grid, m
A_t/N_o = grid area per hole

Bubble growth will be limited by the containing vessel and the bubble hydrodynamic stability. Bubbles in group-B systems can grow to several meters in diameter. Bubbles in group-A materials with high fines may reach a maximum stable bubble size of only several cm.

Furthermore, the solids and gas back mixing is much less in high-L/D beds, either slugging or bubbling, as compared with low-L/D beds. Thus, the conversion or yield in large unstaged reactors is sometimes considerably lower than in small high-L/D units. To overcome some of the problems of scale-up, staged units are used (see Fig. 17-12). It is generally concluded than an unstaged 1-m- (40-in-) diameter unit will achieve about the same conversion as a large industrial unit. The validity of this conclusion is dependent on many variables, including bed depth, particle size, and size distribution, temperature, and pressure. A brief history of fluidization, fluidized-bed scale-up, and modeling will illustrate the problems.

Fluidized beds were used in Europe in the 1920s to gasify coal. Scale-up problems either were insignificant or were not publicized. During World War II, catalytic cracking of oil to produce gasoline was successfully commercialized by scaling up from pilot-plant size (a few centimeters in diameter) to commercial size (several meters in diameter). It is fortunate that the kinetics of the cracking reactions are fast, that the ratio of crude oil to catalyst is determined by thermal balance and the required catalyst circulation rates, and that the crude feed point was in the plug-flow riser. The first experience of problems with scale-up was associated with the production of gasoline from natural gas by using the Fischer-Tropsch process. Some 0.10-m- (4-in-), 0.20-m- (8-in-), and 0.30-m- (12-in-) diameter pilot-plant results were scaled to a 7-m-diameter commercial unit, where the yield was only about 50 percent of that achieved in the pilot units. The Fischer-Tropsch synthesis is a relatively slow reaction; therefore, gas-solid contacting is very important. Since this unfortunate experience or perhaps because of it, much effort has been given to the scale-up of fluidized beds. Many models have been developed; these basically are

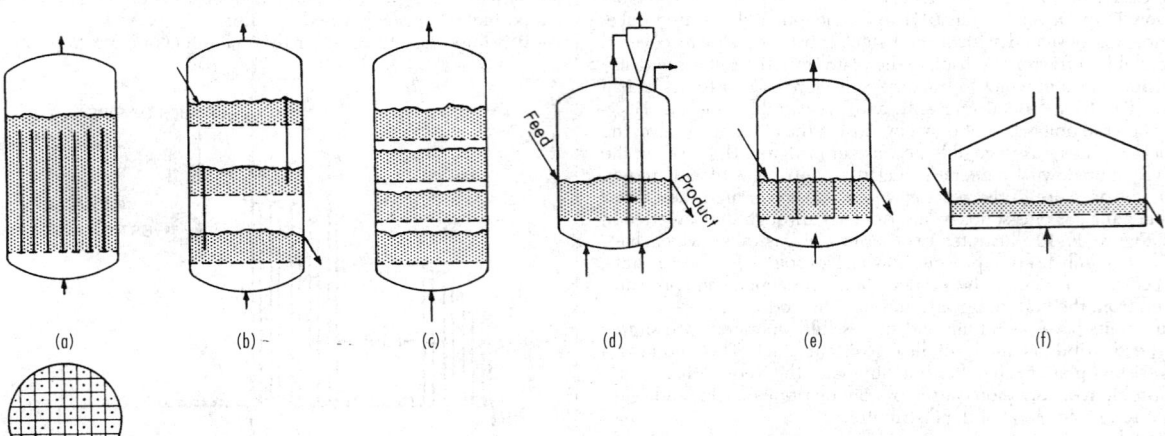

(a)   (b)   (c)   (d)   (e)   (f)

FIG. 17-12   Methods of providing staging in fluidized beds.

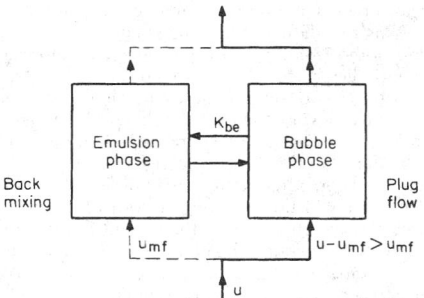

FIG. 17-13 Two-phase model according to May [Chem. Eng. Prog., **55,** 12, 5, 49–55 (1959)] and Van Deemter [Chem. Eng. Sci., **13,** 143–154 (1961)]. U = superficial velocity, U_{mf} = minimum fluidizing velocity, E = axial dispersion coefficient, and K_{be} = mass-transfer coefficient.

of two types, the two-phase model [May, *Chem. Eng. Prog.,* **55,** 12, 5, 49–55 (1959); and Van Deemter, *Chem. Eng. Sci.,* **13,** 143–154 (1961)] and the bubble model (Kunii and Levenspiel, *Fluidization Engineering,* Wiley, New York, 1969). The two-phase model according to May and Van Deemter is shown in Fig. 17-13. In these models all or most of the gas passes through the bed in plug flow in the bubbles which do not contain solids (catalyst). The solids form a dense suspension-emulsion phase in which gas and solids mix according to an axial dispersion coefficient (E). Cross flow between the two phases is predicted by a mass-transfer coefficient.

Conversion of a gaseous reactant can be given by C/C_0 = exp $[-Na \times Nr/(Na + Nr)]$ where C = the exit concentration, C_0 = the inlet concentration, Na = diffusional driving force and Nr = reaction driving force. Conversion is determined by both reaction and diffusional terms. It is possible for reaction to dominate in a lab unit with small bubbles and for diffusion to dominate in a plant size unit. It is this change of limiting regime that makes scaleup so difficult. Refinements of the basic model and predictions of mass-transfer and axial-dispersion coefficients are the subject of many papers [Van Deemter, *Proc. Symp. Fluidization,* Eindhoven (1967); de Groot, ibid.; Van Swaaij; and Zuidiweg, *Proc. 5th Eur. Symp. React. Eng., Amsterdam,* B9–25 (1972); DeVries, Van Swaaij, Mantovani, and Heijkoop, ibid., B9–59 (1972); Werther, *Ger. Chem. Eng.,* **1,** 243–251 (1978); and Pell, *Gas Fluidization,* Elsevier, 75–81 (1990).

The bubble model (Kunii and Levenspiel, *Fluidization Engineering,* Wiley, New York, 1969; Fig. 17-14) assumes constant-sized bubbles (effective bubble size d_b) rising through the suspension phase. Gas is transferred from the bubble void to the mantle and wake at

mass-transfer coefficient K_{bc} and from the mantle and wake to the emulsion phase at mass-transfer coefficient K_{ce}. Experimental results have been fitted to theory by means of adjusting the effective bubble size. As mentioned previously, bubble size changes from the bottom to the top of the bed, and thus this model is not realistic though of considerable use in evaluating reactor performance. Several bubble models using bubbles of increasing size from the distributor to the top of the bed and gas interchange between the bubbles and the emulsion phase according to Kunii and Levenspiel have been proposed [Kato and Wen, *Chem. Eng. Sci.,* **24,** 1351–1369 (1969); and Fryer and Potter, in Keairns (ed.), *Fluidization Technology,* vol. I, Hemisphere, Washington, 1975, pp. 171–178].

There are several methods available to reduce scaleup loss. These are summarized in Fig. 17-15. The efficiency of a fluid bed reactor usually decreases as the size of the reactor increases. This can be minimized by the use of high velocity, fine solids, staging methods, and a high L/D. High velocity maintains the reactor in the turbulent mode, where bubble breakup is frequent and backmixing is infrequent. A fine catalyst leads to smaller maximum bubble sizes by promoting instability of large bubbles. Maintaining high L/D minimizes backmixing, as does the use of baffles in the reactor. By these techniques, Mobil was able to scale up its methanol to gasoline technology with little difficulty. [Krambeck, Avidan, Lee and Lo, *A.I.Ch.E.J.,* 1727–1734 (1987)].

Another way to examine scaleup of hydrodynamics is to build a cold or hot scale model of the commercial design. Validated scaling criteria have been developed and are particularly effective for group B and D materials [Glicksman, Hyre and Woloshun, *Powder Tech.,* 177–199 (1993)].

Circulating or Fast Beds The circulating or fast fluidized bed carries the principles given above to the maximum extent. Velocity is increased until solids are entrained from the bed at a massive rate. Thus, solids extend to the exit and may constitute up to 10 percent of the freeboard volume. There are no bubbles, mass transfer rates are high, and there is little backmixing. The high velocity means high gas throughput, minimizing reactor cost. Scaleup is less of a problem than with bubbling beds.

The system is characterized by an external cyclone return system that is usually as large as the reactor itself. The axial solids density profile is relatively flat, as indicated in Figs. 17-4 and 17-5. There is a radial solids density profile called core annular flow. In the center of the reactor, gas flow may be double the average, and the solids are in dilute flow, traveling at $U_g - U_t$, their expected slip velocity. Near the wall, the solids are close to their minimum bubbling density and are likely to be in downflow. Engineering methods for evaluating the hydrodynamics of the circulating bed are given by Kunii and Levenspiel (*Fluidization Engineering,* 2d ed., Butterworth, 1991, pp. 195–

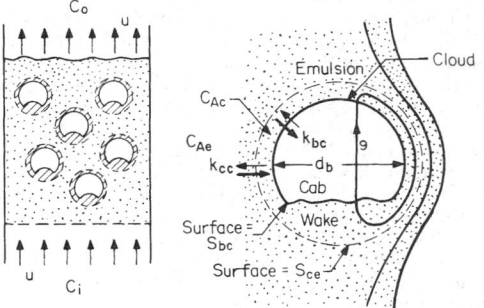

FIG. 17-14 Bubbling-bed model of Kunii and Levenspiel. d_b = effective bubble diameter, C_{Ab} = concentration of A in bubble, C_{Ac} = concentration of A in cloud, C_{Ae} = concentration of A in emulsion, q = volumetric gas flow into or out of bubble, k_{bc} = mass-transfer coefficient between bubble and cloud, and k_{ce} = mass-transfer coefficient between cloud and emulsion. (*From Kunii and Levenspiel,* Fluidization Engineering, *Wiley, New York, 1969, and Krieger, Malabar, Fla., 1977.*)

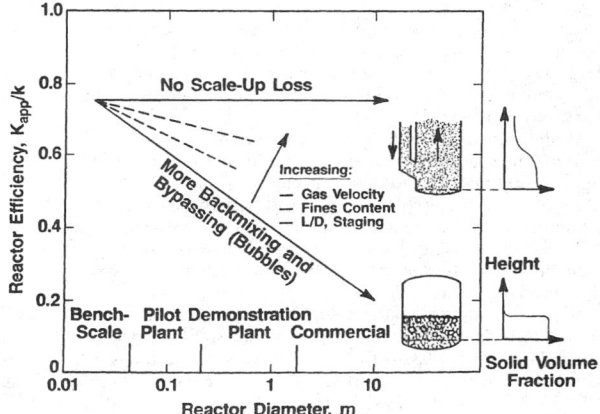

FIG. 17-15 Reducing scaleup loss (*From Krambeck, Avidan, Lee and Lo,* A.I.Ch.E.J., *1727–1734, 1987.*)

209) and Werther (*Circulating Fluid Bed Technology IV*, in press, 1994).

Transport Transport units can be scaled up on the principles of pneumatic conveying. Mass and heat transfer can be predicted on both the slip velocity during acceleration and the slip velocity at full acceleration. The slip velocity is increased as the solids concentration is increased.

Heat Transfer Heat-exchange surfaces have been used to provide means of removing or adding heat to fluidized beds. Usually, these surfaces are provided in the form of vertical tubes manifolded at top and bottom or in trombone shape manifolded exterior to the vessel.

Other shapes such as horizontal bayonets have been used. In any such installations adequate provision must be made for abrasion of the exchanger surface by the bed.

The prediction of the heat-transfer coefficient is covered in Secs. 5 and 11. Normally, the transfer rate is between 5 and 25 times that for the gas alone.

Heat transfer from solids to gas and gas to solids usually results in a coefficient of about 6 to 20 J/(m²·s·K) [3 to 10 Btu/(h·ft²·°F)]. However, the large area of the solids per cubic foot of bed, 5000 m²/m³ (15,000 ft²/ft³) for 60-μm particles of 600 kg/m³ (40 lb/ft³) bulk density, results in the rapid approach of gas and solids temperatures. With a fairly good distributor, essential equalization of temperatures occurs within 2 to 6 cm (1 to 3 in) of the top of the distributor.

Bed thermal conductivities in the vertical direction have been measured in the laboratory in the range of 40 to 60 kJ/(m²·s·K) [20,000 to 30,000 Btu/(h·ft²·°F·ft)]. Horizontal conductivities for 3-mm (⅛-in) particles in the range of 2 kJ/(m²·s·K) [1000 Btu/(h·ft²·°F·ft)] have been measured in large-scale experiments.

Except in extreme L/D ratios, the temperature in the fluidized bed is uniform—generally the temperature at any point being within 5 K (10°F) of any other point.

Temperature Control Because of the rapid equalization of temperatures in fluidized beds, temperature control can be accomplished in a number of ways.

1. *Adiabatic.* Control gas flow and/or solids feed rate so that the heat of reaction is removed as sensible heat in off gases and solids or heat supplied by gases or solids.

2. *Solids circulation.* Remove or add heat by circulating solids.

3. *Gas circulation.* Recycle gas through heat exchangers to cool or heat.

4. *Liquid injection.* Add volatile liquid so that the latent heat of vaporization equals excess energy.

5. *Cooling or heating surfaces in bed.*

Solids Mixing Solids are mixed in fluidized beds by means of solids entrained in the lower portion of bubbles, and the shedding of these solids from the wake of the bubble (Rowe and Patridge, "Particle Movement Caused by Bubbles in a Fluidized Bed," Third Congress of European Federation of Chemical Engineering, London, 1962). Thus, no mixing will occur at incipient fluidization, and mixing increases as the gas rate is increased. Naturally, particles brought to the top of the bed must displace particles toward the bottom of the bed. Generally, solids upflow is greater in the center of the bed and downward at the wall.

At high ratios of fluidizing velocity to minimum fluidizing velocity, tremendous solids circulation from top to bottom of the bed assures rapid mixing of the solids. For all practical purposes, beds with L/D ratios of from 4 to 0.1 can be considered to be completely mixed continuous-reaction vessels insofar as the solids are concerned.

Batch mixing using fluidization has been successfully employed in many industries. In this case there is practically no limitation to vessel dimensions.

All the foregoing pertains to solids of approximately the same physical characteristics. There is evidence that solids of widely different characteristics will classify one from the other at certain gas flow rates [Geldart, Baeyens, Pope, and van de Wijer, *Powder Technol.*, **30**(2), 195 (1981)]. Two fluidized beds, one on top of the other, may be formed, or a lower static bed with a fluidized bed above may result. The latter frequently occurs when agglomeration takes place because of either fusion in the bed or poor dispersion of sticky feed solids.

Increased gas flows sometimes overcome the problem; however, improved feeding techniques or a change in operating conditions may be required. Another solution is to remove agglomerates either continuously or periodically from the bottom of the bed.

Gas Mixing The mixing of gases as they pass vertically up through the bed has never been considered a problem. However, horizontal mixing is very poor and requires effective distributors if two gases are to be mixed in the fluidized bed.

In bubbling beds operated at velocities of less than about 5 to 11 times U_{mf} the gases will flow upward in both the emulsion and the bubble phases. At velocities greater than about 5 to 11 times U_{mf} the downward velocity of the emulsion phase is sufficient to carry the contained gas downward. The back mixing of gases increases as U/U_{mf} is increased until the circulating or fast regime is reached where the back mixing decreases as the velocity is further increased.

Size Enlargement Under proper conditions, particles of solids can be caused to grow. That is sometimes advantageous and at other times disadvantageous. Growth is associated with the liquefaction or softening of some portion of the bed material (i.e., addition of soda ash to calcium carbonate feed in lime reburning, tars in fluidized-bed coking, or lead or zinc roasting cause agglomeration of dry particles in much the same way as binders act in rotary pelletizers). The motion of the particles, one against the other, in the bed results in spherical pellets. If the size of these particles is not controlled, segregation of the large particles from the bed will occur. Control can be achieved by crushing a portion of the bed product and recycling it to form nuclei for new growth.

In drying solutions or slurries of solutions, the location of the feed-injection nozzle (spray nozzle) has a great effect on the size of particle formed in the bed. Also of importance are the operating temperature, relative humidity of the off gas, and gas velocity. Particle growth can occur as agglomeration or as an "onion skinning."

Size Reduction Three major size-reduction mechanisms occur in the fluidized bed. These are attrition, impact, and thermal decrepitation.

Because of the random motion of the solids, some abrasion of the surface occurs. This is generally quite small, usually amounting to about 0.25 to 1 percent of the solids per day.

In areas of high gas velocities, greater rates of attrition will occur as well as fracture of particles by impact. This type of jet grinding is employed in some of the coking units to control particle size [Dunlop, Griffin and Moser, *J. Chem. Eng. Prog.*, **54**, 39–43, (1958)]. It also occurs to a lesser degree at the point of gas introduction when the pressure drop to assure gas distribution is taken across an orifice or pipe that discharges directly into the bed.

Thermal decrepitation occurs frequently when crystals are rearranged because of transition from one form to another or when new compounds are formed (i.e., calcination of limestone). Sometimes the strains in cases such as this are sufficient to reduce the particle to the basic crystal size.

All these mechanisms will cause completion of fractures that were started before the introduction of the solids into the fluidized bed.

Standpipes, Solids Feeders, and Solids Flow Control In the case of catalytic-cracking units in which the addition of catalyst is small and need not be steady, the makeup catalyst may be fed from pressurized hoppers into one of the conveying lines. The main solids-flow-control problem is to maintain balanced inventories of catalyst in and controlled flow from and to the reactor and regenerator. This flow of solids from an oxidizing atmosphere to a reducing one, or vice versa, usually necessitates stripping gases from the interstices of the solids as well as gases adsorbed by the particles. Steam is usually used for this purpose. The point of removal of the solids from the fluidized bed is usually under a lower pressure than the point of feed introduction into the carrier gas.

The pressure is higher at the bottom of the solids draw-off pipe due to the relative flow of gas counter to the solids flow. The gas may either be flowing downward more slowly than the solids or upward. The standpipe may be fluidized, or the solids may be in moving packed bed flow with no expansion. Gas is introduced at the bottom (best for group B) or at about 3-m intervals along the standpipe (best for group A). The increasing pressure causes gas inside and between

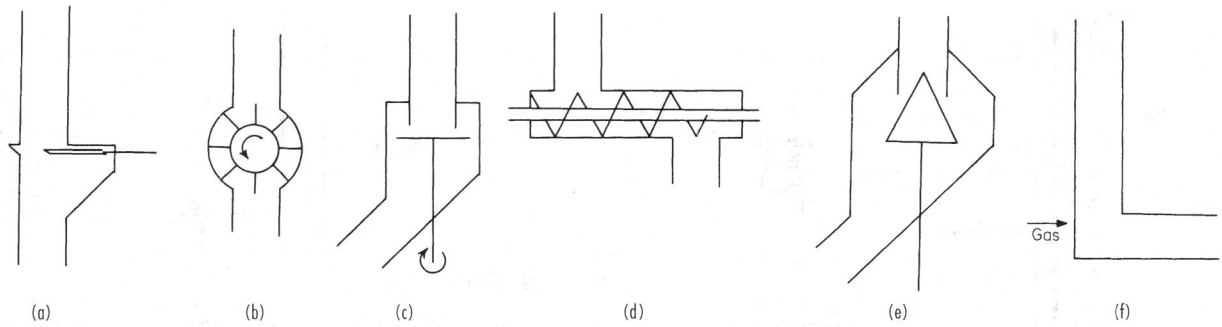

(a)   (b)   (c)   (d)   (e)   (f)

FIG. 17-16   Solids-flow-control devices. (*a*) Slide valve. (*b*) Rotary valve. (*c*) Table feeder. (*d*) Screw feeder. (*e*) Cone valve. (*f*) L Valve.

the particles to be compressed. Unless aeration gas is added, the solids could defluidize and become a moving fixed bed with a lower-pressure head than that of fluidized solids. In any event, the pressure drop across the solids control valve should be designed for 3 psi or more to safely prevent backflow. See Zenz, *Powder Tech.*, 105–113 (1986).

Several designs of valves for solids flow control are used. These should be chosen with care to suit the specific conditions. Usually, block valves are used in conjunction with the control valves. Figure 17-16 shows schematically some of the devices used for solids flow control. Not shown in Fig. 17-16 is the flow-control arrangement used in the Exxon Research & Engineering Co. model IV catalytic-cracking units. This device consists of a U bend. A variable portion of regenerating air is injected into the riser leg. Changes in air-injection rate change the fluid density in the riser and thereby achieve control of the solids flow rate. Catalyst circulation rates of 1200 kg/s (70 tons/min) have been reported.

When the solid is one of the reactants, such as in ore roasting, the flow must be continuous and precise in order to maintain constant conditions in the reactor. Feeding of free-flowing granular solids into a fluidized bed is not difficult. Standard commercially available solids-weighing and -conveying equipment can be used to control the rate and deliver the solids to the feeder. Screw conveyors, dip pipes, seal legs, and injectors are used to introduce the solids into the reactor proper (Fig. 17-16). Difficulties arise and special techniques must be used when the solids are not free-flowing, such as is the case with most filter cakes. One solution to this problem was developed at Cochenour-Willans. After much difficulty in attempting to feed a wet and sometimes frozen filter cake into the reactor by means of a screw feeder, experimental feeding of a water slurry of flotation concentrates was attempted. This trial was successful, and this method has been used in almost all cases in which the heat balance, particle size of solids, and other considerations have permitted. Gilfillan et al. (*J. Chem. Metall. Min. Soc. S. Afr.*, May 1954) and Soloman and Beal (*Uranium in South Africa, 1946–56*) present complete details on the use of this system for feeding.

When slurry feeding is impractical, recycling of solids product to mix with the feed, both to dry and to achieve a better-handling material, has been used successfully. Also, the use of a rotary table feeder mounted on top of the reactor, discharging through a mechanical disintegrator, has been successful. The wet solids generally must be broken up into discrete particles of very fine agglomerates either by mechanical action before entering the bed or by rapidly vaporizing water. If lumps of dry or semidry solids are fed, the agglomerates do not break up but tend to fuse together. As the size of the agglomerate is many times the size of the largest individual particle, these agglomerates will segregate out of the bed, and in time the whole of the fluidized bed may be replaced with a static bed of agglomerates.

Solids Discharge   The type of discharge mechanism utilized is dependent upon the necessity of sealing the atmosphere inside the fluidized-bed reactor and the subsequent treatment of the solids. The simplest solids discharge is an overflow weir. This can be used only when the escape of fluidizing gas does not present any hazards due to

nature or dust content or when the leakage of gas into the fluidized-bed chamber from the atmosphere into which the bed is discharged is permitted. Solids will overflow from a fluidized bed through a port even though the pressure above the bed is maintained at a slightly lower pressure than the exterior pressure. When it is necessary to restrict the flow of gas through the opening, a simple flapper valve is frequently used. Overflow to combination seal and quench tanks (Fig. 17-17) is used when it is permissible to wet the solids and when disposal or subsequent treatment of the solids in slurry form is desirable. The FluoSeal is a simple and effective way of sealing and purging gas from the solids when an overflow-type discharge is used (Fig. 17-18).

Either trickle (flapper) or star (rotary) valves are effective sealing devices for solids discharge. Each functions with a head of solids above it. Bottom of the bed discharge is also acceptable via a slide valve with a head of solids.

Seal legs are frequently used in conjunction with solids-flow-control valves to equalize pressures and to strip trapped or adsorbed gases from the solids. The operation of a seal leg is shown schematically in Fig. 17-19. The solids settle by gravity from the fluidized bed into the seal leg or standpipe. Seal and/or stripping gas is introduced near the bottom of the leg. This gas flows both upward and downward. Pressures indicated in the illustration have no absolute value but are only relative. The legs are designed for either fluidized or settled solids.

The L valve is shown schematically in Fig. 17-20. It can act as a seal and as a solids-flow control valve. However, control of solids rate is only practical for solids that deaerate quickly (Geldart B and D). The

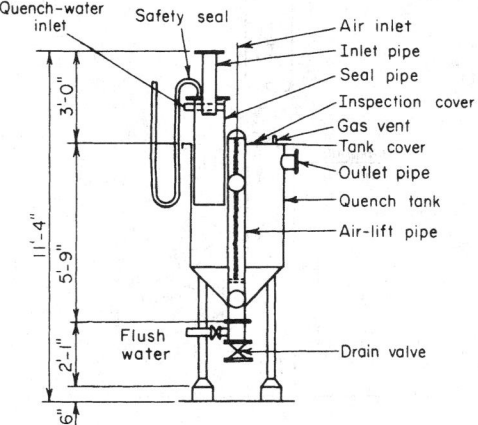

FIG. 17-17   Quench tank for overflow or cyclone solids discharge. [*Gilfillan et al., "The FluoSolids Reactor as a Source of Sulphur Dioxide,"* J. Chem. Metall. Min. Soc. S. Afr. (*May 1954*).]

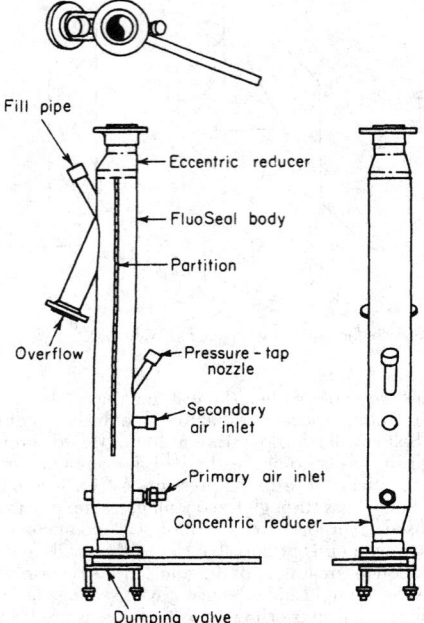

FIG. 17-18 Dorrco FluoSeal, type UA. (*Dorr-Oliver Inc.*)

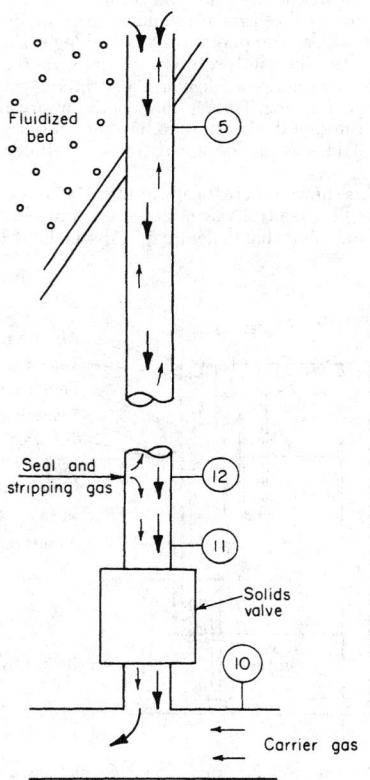

FIG. 17-19 Fluidized-bed seal leg.

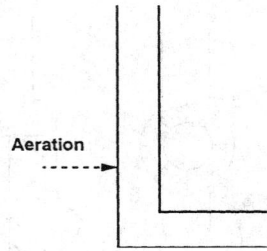

FIG. 17-20 L valve.

height at which aeration is added in Fig. 17-20 is usually one exit pipe diameter above the centerline of the exit pipe. For L-valve design equations, see Yang and Knowlton [*Powder Tech.*, **77**, 49–54 (1993)].

In the sealing mode, the leg is usually fluidized. Gas introduced below the normal solids level and above the discharge port will flow upward and downward. The relative flow in each direction is self-adjusting, depending upon the differential pressure between the point of solids feed and discharge and the level of solids in the leg. The length and diameter of the discharge spout are selected so that the undisturbed angle of repose of the solids will prevent discharge of the solids. As solids are fed into the leg, height H of solids increases. This in turn reduces the flow of gas in an upward direction and increases the flow of gas in a downward direction. When the flow of gas downward and through the solids-discharge port reaches a given rate, the angle of repose of the solids is upset and solids discharge commences. Usually, the level of solids above the point of gas introduction will float. When used as a flow controller, the vertical leg is best run in the packed bed mode. The solids flow rate is controlled by varying the aeration gas flow.

In most catalytic-reactor systems, no solids removal is necessary as the catalyst is retained in the system and solids loss is in the form of fines that are not collected by the dust-recovery system.

Dust Separation   It is usually necessary to recover the solids carried by the gas leaving the disengaging space or freeboard of the fluidized bed. Generally, cyclones are used to remove the major portion of these solids (see "Gas-Solids Separation"). However, in a few cases, usually on small-scale units, filters are employed without the use of cyclones to reduce the loading of solids in the gas. For high-temperature usage, either porous ceramic or sintered metal has been employed. Multiple units must be provided so that one unit can be blown back with clean gas while one or more are filtering.

Cyclones are arranged generally in any one of the arrangements shown in Fig. 17-21. The effect of cyclone arrangement on the height of the vessel and the overall height of the system is apparent. Details regarding cyclone design and collection efficiencies are to be found in another portion of this section.

Discharging of the cyclone into the fluidized bed requires some care. It is necessary to seal the bottom of the cyclone so that the collection efficiency of the cyclone will not be impaired by the passage of appreciable quantities of gas up through the solids-discharge port. This is usually done by sealing the dip leg in the fluid bed. Experience has shown, particularly in the case of deep beds, that the bottom of the dip pipe must be protected from the action of large gas bubbles which, if allowed to pass up the leg, would carry quantities of fine solids up into the cyclone and cause momentarily high losses. This can be done by attaching a plate larger in diameter than the pipe to the bottom (see Fig. 17-22*e*).

Example 1: Length of Seal Leg   The length of the seal leg can be estimated as shown in the following example.

Given: Fluid density of bed at 0.3-m/s (1-ft/s) superficial gas velocity = 1100 kg/m³ (70 lb/ft³).

Fluid density of cyclone product at 0.15 m/s (0.5 ft/s) = 650 kg/m³ (40 lb/ft³).
Settled bed depth = 1.8 m (6 ft)
Fluidized-bed depth = 2.4 m (8 ft)
Pressure drop through cyclone = 1.4 kPa (0.2 lbf/in²)

In order to assure seal at startup, the bottom of the seal leg is 1.5 m (5 ft) above the constriction plate or submerged 0.9 m (3 ft) in the fluidized bed.

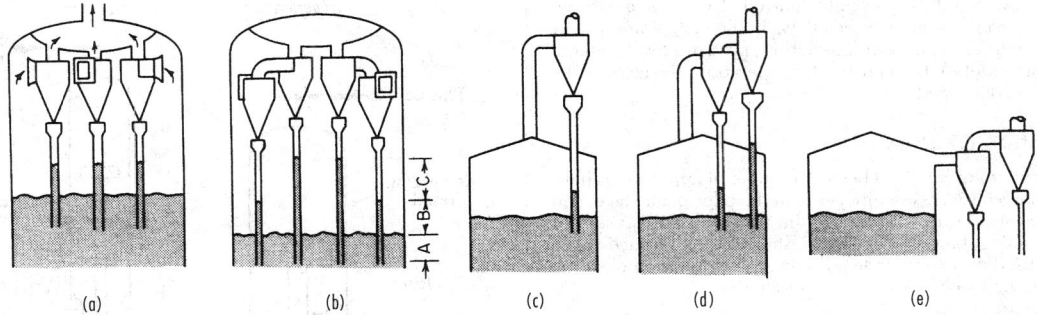

FIG. 17-21 Fluidized-bed cyclone arrangements. (*a*) Single-stage internal cyclone. (*b*) Two-stage internal cyclone. (*c*) Single-stage external cyclone; dust returned to bed. (*d*) Two-stage external cyclone; dust returned to bed. (*e*) Two-stage external cyclone; dust collected externally.

The pressure at the solids outlet of a gas cyclone is usually about 0.7 kPa (0.1 lbf/in²) lower than the pressure at the discharge of the leg. Total pressure to be balanced by the fluid leg in the cyclone dip leg is

$$(0.9 \times 1100 \times 9.81)/1000 + 1.4 + 0.7 = 11.8 \text{ Kpa}$$

$$[(3 \times 70)/144 + 0.2 + 0.1 = 1.7 \text{ lb/in}^2]$$

Height of solids in dip leg = $(11.8 \times 1000)/(650 \times 9.81) = 1.9$ m $[(1.7 \times 144)/40 = 6.1$ ft]; therefore, the bottom of the separator pot on the cyclone must be at least $1.9 + 1.5$ or 3.4 m $(6.1 + 5$ or 11.1 ft) above the gas distributor. To allow for upsets, changes in size distribution, etc., use 4.6 m (15 ft).

In addition to the open dip leg, various other devices have been used to seal cyclone solids returns, especially for second-stage cyclones. A number of these are shown in Fig. 17-22. One of the most frequently used is the trickle valve (17-22*a*). There is no general agreement as to whether this valve should discharge below the bed level or in the freeboard. In any event, the legs must be large enough to carry momentarily high rates of solids and must provide seals to overcome cyclone pressure drops as well as to allow for differences in fluid density of bed and cyclone products. It has been reported that, in the case of catalytic-cracking catalysts, the fluid density of the solids collected by the primary cyclone is essentially the same as that in the fluidized bed because the particles in the bed are so small, nearly all are entrained. However, as a general rule the fluidized density of solids collected by the first cyclone is less than the fluidized density of the bed. Each succeeding cyclone collects finer and less dense solids.

As cyclones are less effective as the particle size decreases, secondary collection units are frequently required, i.e., filters, electrostatic precipitators, and scrubbers. When dry collection is not required, elimination of cyclones is possible if allowance is made for heavy solids loads in the scrubber (see "Gas-Solids Separations"; see also Sec. 14).

Instrumentation

Temperature Measurement This is usually simple, and standard temperature-sensing elements are adequate for continuous use. Because of the high abrasion wear on horizontal protection tubes, vertical installations are frequently used. In highly corrosive atmospheres in which metallic protection tubes cannot be used, short, heavy ceramic tubes have been used successfully.

Pressure Measurement Although successful pressure-measurement probes or taps have been fabricated by using porous materials, the most universally accepted pressure tap consists of a purged tube projecting into the bed as nearly vertically as possible. Minimum internal diameters are 1 to 2 cm (½ to 1 in). A purge rate of at least 0.9 m/s (3 ft/s) is usually required. Pressure measurements taken at various heights in the bed are used to determine bed level.

Bed density is determined directly from $\Delta P/L$, the pressure drop inside the bed itself ($\Delta P/L$ in units of weight/area $\times L$). The overall bed weight is obtained from ΔP taken between the grid and the freeboard. Nominal bed height is the length required to have the overall weight at the measured density. Of course, splashing and entrainment will place solids well above the nominal bed height in most cases.

The pressure-drop signal is noisy due to bubble effects and the generally statistical nature of fluid bed flow parameters. A fast fourier transform of the pressure drop signal transforms the perturbations to a frequency-versus-amplitude plot with a maximum at about 5 Hz and frequencies generally tailing off above 20 Hz. Changes in frequency and amplitude are associated with changes in the quality of the fluidization. Experienced operators can frequently predict performance from changes in the ΔP signal.

Flow Measurement Measurement of flow rates of clean gases presents no problem. Flow measurement of dirty gases is usually

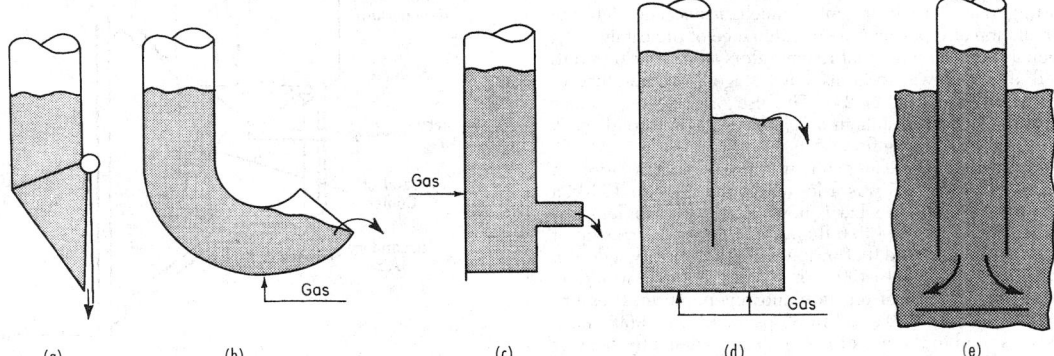

FIG. 17-22 Cyclone solids-return seals. (*a*) Trickle valve. (*Ducon Co., Inc.*) (*b*) J valve. (*c*) L valve. (*d*) Fluid-seal pot. (*e*) "Dollar" plate. *a*, *b*, *c*, and *d* may be used above the bed; *a* and *e* are used below the bed.

avoided. The flow of solids is usually controlled but not measured except externally to the system. Solids flows in the system are usually adjusted on an inferential basis (temperature, pressure level, catalyst activity, gas analysis, etc.). In many roasting operations the color of the calcine indicates solids feed rate.

USES OF FLUIDIZED BEDS

There are many uses of fluidized beds. A number of applications have become commercial successes; others are in the pilot-plant stage, and others in bench-scale stage. Generally, the fluidized bed is used for gas-solids contacting; however, in some instances the presence of the gas or solid is used only to provide a fluidized bed to accomplish the end result. Uses or special characteristics follow:

I. Chemical reactions
   A. Catalytic
   B. Noncatalytic
      1. Homogeneous
      2. Heterogeneous
II. Physical contacting
   A. Heat transfer
      1. To and from fluidized bed
      2. Between gases and solids
      3. Temperature control
      4. Between points in bed
   B. Solids mixing
   C. Gas mixing
   D. Drying
      1. Solids
      2. Gases
   E. Size enlargement
   F. Size reduction
   G. Classification
      1. Removal of fines from solids
      2. Removal of fines from gas
   H. Adsorption-desorption
   I. Heat treatment
   J. Coating

Chemical Reactions

Catalytic Reactions   This use has provided the greatest impetus for use, development, and research in the field of fluidized solids. Some of the details pertaining to this use are to be found in the preceding pages of this section. Reference should also be made to Sec. 23.

Cracking.   The evolution of fluidized catalytic cracking (FCC) since the early 1940s has resulted in several configurations depending upon the particular use and designer.

The high rate of transfer of solids between the regenerator and the reactor permits a balancing of the exothermic burning of carbon and tars in the regenerator and the endothermic cracking of petroleum in the reactor, so that temperature in both units can usually be controlled without resorting to auxiliary heat-control mechanisms. The high rate of catalyst circulation also permits the maintenance of the catalyst at a constantly high activity. The original regenerators were considered to be backmixed units. Newer systems have staged regenerators to improve conversion (see Fig. 17-23). The use of the riser reactor (transport or fast fluid bed) results in much lower gas and solids back mixing and an approach to plug flow.

The first fluid catalytic-cracking unit was placed in operation in Baytown, Texas, in 1942. This was a low-pressure, 115- to 120-kPa (2- to 3-psig) unit operating in what is now called the turbulent fluidization mode, 1.2 to 1.8 m/s (4 to 6 ft/s). Even before the startup of the first model I, it was realized that by lowering the velocity, a dense, aggregative fluidized bed, 300 to 400 kg/m³ (20 to 25 lb/ft³), would be formed, allowing completion of reaction and regeneration. Pressure was increased to 240 to 320 kPa (20 to 30 psig). In the 1960s more active catalysts resulted in the use of riser cracking. Recently, heavier crude feedstocks have produced higher coke yields. This has necessitated addition of catalyst cooling to the regeneration step as shown in Fig. 17-24. Many companies participated in the development of the

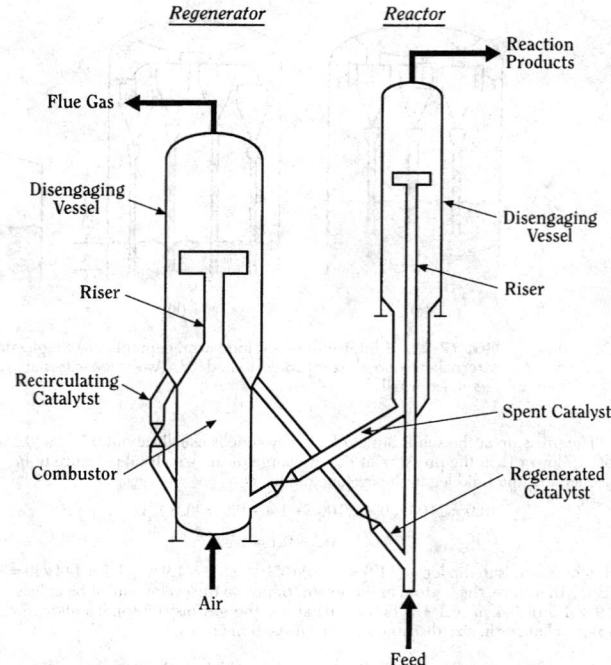

FIG. 17-23   UOP fluid cracking unit. (*Reprinted with permission of UOP.*)

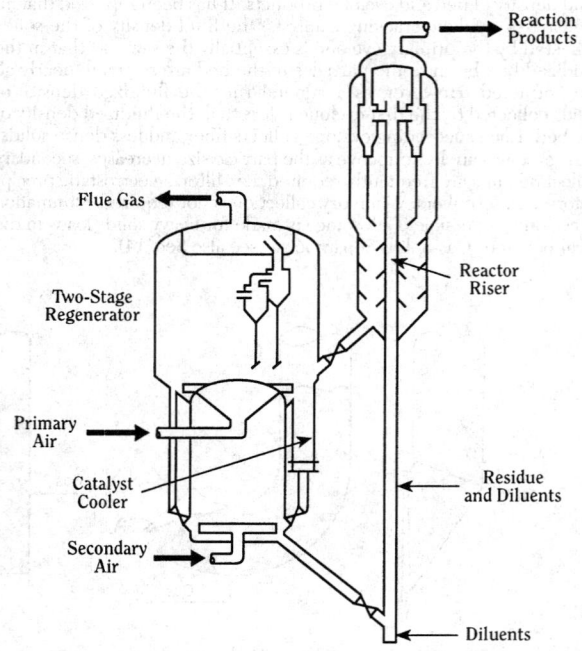

FIG. 17-24   Modern FCC unit configured for high-efficiency regeneration and extra catalyst cooling. (*Reprinted with permission of UOP. RCC is a service mark of Ashland Oil Inc.*)

cat cracker, including Exxon Research & Engineering Co., Universal Oil Products Companies, Kellogg Co., Texaco Development Corp., Gulf Research Development Co., and Shell Oil Company. Many of the companies provide designs and/or licenses to operate to others. For further details, see Luckenbach, Reichle, Gladrow, and Worley, "Cracking, Catalytic," in McKetta (ed.), *Encyclopedia of Chemical Processing and Design*, vol. 13, Marcel Dekker, New York, 1981, pp. 1–132.

Alkyl chlorides. Olefins are chlorinated to alkyl chlorides in a single fluidized bed. HCl reacts with O_2 over a copper chloride catalyst to form chlorine. The chlorine reacts with the olefin to form the alkyl chloride. The process developed by the Shell Development Co. uses a recycle of catalyst fines in aqueous HCl to control the temperature [*Chem. Proc.*, **16**, 42 (1953)].

Phthalic anhydride. Naphthalene is oxidized by air to phthalic anhydride in a bubbling fluidized reactor. Even though the naphthalene feed is in liquid form, the reaction is highly exothermic. Temperature control is achieved by removing heat through vertical tubes in the bed to raise steam [Graham and Way, *Chem. Eng. Prog.*, **58**, 96 (January 1962)].

Acrylonitrile. Acrylonitrile is produced by reacting propylene, ammonia, and oxygen (air) in a single fluidized bed of a complex catalyst. Known as the SOHIO process, this process was first operated commercially in 1960. In addition to acrylonitrile, significant quantities of HCN and acetonitrile are also produced. This process is also exothermic. Temperature control is achieved by raising steam inside vertical tubes immersed in the bed [Veatch, *Hydrocarbon Process. Pet. Refiner*, **41**, 18 (November 1962)].

Fischer-Tropsch synthesis. The scale-up of a bubbling-bed reactor to produce gasoline from CO and H_2 was unsuccessful (see "Design of Fluidized-Bed Systems: Scale-Up"). However, Kellogg Co. developed a successful Fischer-Tropsch synthesis reactor based on a dilute-phase or transport-reactor concept. Kellogg, in its design, prevented gas bypassing by using the transport reactor and maintained temperature control of the exothermic reaction by inserting heat exchangers in the transport line. This process has been very successful and repeatedly expanded at the South African Synthetic Oil Limited (SASOL) plant in the Republic of South Africa, where politics and economics favor the conversion of coal to gasoline and other hydrocarbons. Refer to Jewell and Johnson, U.S. Patent 2,543,974, Mar. 6, 1951. Recently, the process has been modified to a simpler, less expensive turbulent bed catalytic reactor system [Silverman et al., *Fluidization V*, Engineering Foundation 1986, pp. 441–448).

The first commercial fluidized bed polyethylene plant was constructed by Union Carbide in 1968. Modern units operate at 100°C and 32 MPa (300 psig). The bed is fluidized with ethylene at about 0.5 m/s and probably operates near the turbulent fluidization regime. The excellent mixing provided by the fluidized bed is necessary to prevent hot spots, since the unit is operated near the melting point of the product. A model of the reactor (Fig. 17-25) that couples kinetics to the hydrodynamics was given by Choi and Ray, *Chem. Eng. Sci.*, **40**, 2261, 1985.

Additional catalytic processes. Nitrobenzene is hydrogenated to aniline (U.S. Patent 2,891,094). Melamine and isophthalonitrile are produced in catalytic fluidized-bed reactors. Badger has announced a process to produce maleic anhydride by the partial oxidation of butane (Schaffel, Chen, and Graham, "Fluidized Bed Catalytic Oxidation of Butane to Maleic Anhydride," presented at Chemical Engineering World Congress, Montreal, 1981). Dupont has announced a circulating bed process for production of maleic anhydride (Contractor, *Circulating Fluidized Bed Tech. II*, Pergamon, 1988, pp. 467–474). Mobil Oil has developed a commercial process to convert methanol to gasoline (Grimmer et al., *Methane Conversion*, Elsevier, 1988, pp. 273–291).

Noncatalytic Reactions

Homogeneous reactions. Homogeneous noncatalytic reactions are normally carried out in a fluidized bed to achieve mixing of the gases and temperature control. The solids of the bed act as a heat sink or source and facilitate heat transfer from or to the gas or from or to heat-exchange surfaces. Reactions of this type include chlorination of hydrocarbons or oxidation of gaseous fuels.

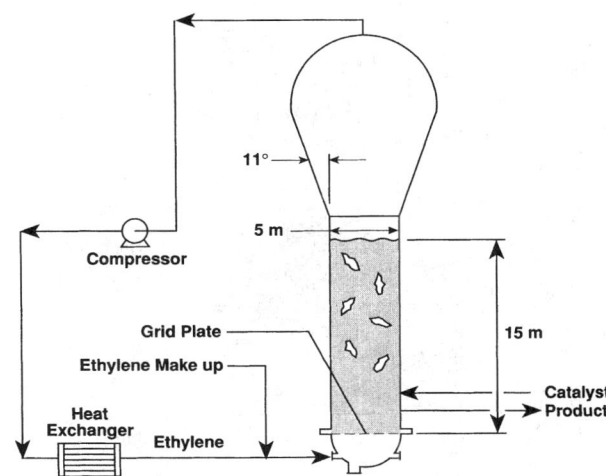

FIG. 17-25 High-pressure polyethylene reactor.

Heterogeneous reactions. This category covers the greatest commercial use of fluidized beds other than petroleum cracking. The roasting of sulfide, arsenical, and/or antimonial ores to facilitate the release of gold or silver values; the roasting of pyrite, pyrrhotite, or naturally occurring sulfur ores to provide SO_2 for sulfuric acid manufacture; and the roasting of copper, cobalt, and zinc sulfide ores to solubilize the metal values are the major metallurgical uses. Figure 17-26 shows basic items in the system.

Thermally efficient **calcination** of lime dolomite and clay can be carried out in a multicompartment fluidized bed (Fig. 17-27). Fuels are burned in a fluidized bed of the product to produce the required heat. Bunker C oil, natural gas, and coal are used in commercial units. Temperature control is accurate enough to permit production of lime of very high availability with close control of slaking characteristics. Also, half calcination of dolomite is an accepted practice. The requirement of large crystal size for the limestone limits application. Small-sized crystals in the limestone result in low yields due to high dust losses.

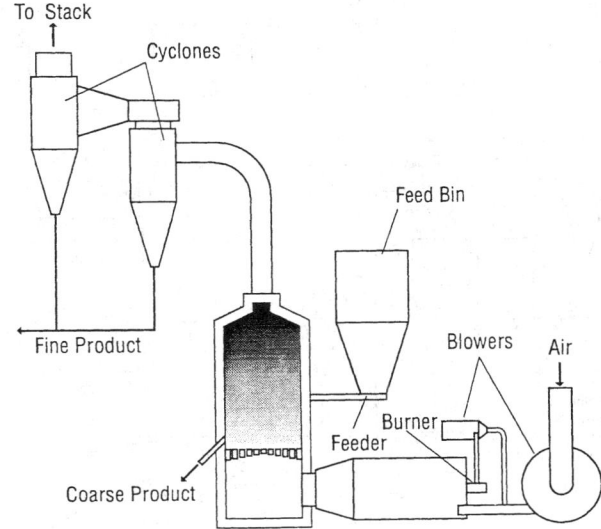

FIG. 17-26 Single-stage FluoSolids roaster or dryer. (*Dorr-Oliver, Inc.*)

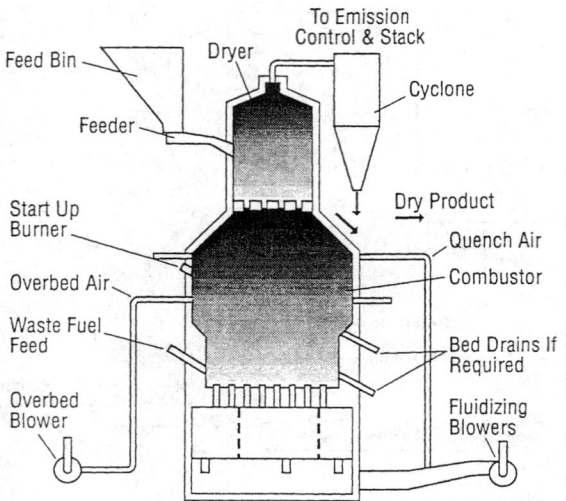

FIG. 17-27   FluoSolids multicompartment fluidized bed. (*Dorr-Oliver, Inc.*)

Phosphate rock is calcined to remove carbonaceous material before being digested with sulfuric acid. Several different fluidization processes have been commercialized for the direct reduction of hematite to high-iron, low-oxide products. Foundry sand is also calcined to remove organic binders and release fines.

The calcination of $Al(OH)_3$ to Al_2O_3 in a circulating fluidized process produces a high-grade product. The process combines the use of circulating, bubbling, and transport beds to achieve high thermal efficiency. See Fig. 17-28.

An interesting feature of these high-temperature-calcination applications is the direct injection of either heavy oil, natural gas, or fine coal into the fluidized bed. Combustion takes place at well below flame temperatures without atomization. Considerable care in the design of the fuel- and air-supply system is necessary to take full advantage of the fluidized bed, which serves to mix the air and fuel.

Coal can be burned in fluidized beds in an environmentally acceptable manner by adding limestone or dolomite to the bed to react with the SO_2 to form $CaSO_4$. Because of moderate combustion temperature, about 800 to 900°C, NO_x, which results from the oxidation of nitrogen compounds contained in the coal, is kept at a low level. NO_x is increased by higher temperatures and higher excess oxygen. Two-stage air addition reduces NO_x.

Several concepts of fluidized-bed combustion have been or are being developed. Atmospheric fluidized-bed combustion (AFBC), in which most of the heat-exchange tubes are located in the bed, is illustrated in Fig. 17-29. This type of unit is most commonly used for industrial applications up to about 50 metric tons/h of steam generation. Larger units are generally of the circulating fluidized bed type as shown in Fig. 17-30. **Circulating fluidized bed combustors** have many advantages. The velocity is significantly higher for greater throughput. Since all the solids are recycled, fine limestone and coal can be fed, which gives better limestone utilization and greater laitude in specifying coal sizing. Because of erosion due to high velocity coarse solids, heat-transfer surface is usually not designed into the bottom of the combustion zone. Figure 17-30 shows a commercial 110 MWe unit. Pressurized fluidized-bed combustion (PFBC) is, as the name implies, operated at above atmospheric pressures. The beds and heat-transfer surface are stacked to conserve space and to reduce the size of the pressure vessel. This type of unit is usually conceived as a cogeneration unit. Steam raised in the boilers would be employed to drive turbines or for other uses, and the hot pressurized gases after cleaning would be let down through an expander coupled to a compressor to supply the compressed combustion air and/or electric generator. A 71 MWe unit is shown in Fig. 17-31. Also see Sec. 27, "Energy Sources: Conversion and Utilization."

Incineration   The majority of over 400 units in operation are used for the incineration of biological sludges. These units can be designed to operate autogenously with wet sludges containing as little as 6 MJ/kg (2600 Btu/lb) heating value (Fig. 17-32). Depending on the calorific value of the feed, heat can be recovered as steam either by means of waste-heat boilers or by a combination of waste-heat boilers and the heat-exchange surface in the fluid bed.

Several units are used for sulfite-paper-mill waste-liquor disposal. At least six units are used for oil-refinery wastes, which sometimes include a mixture of liquid sludges, emulsions, and caustic waste

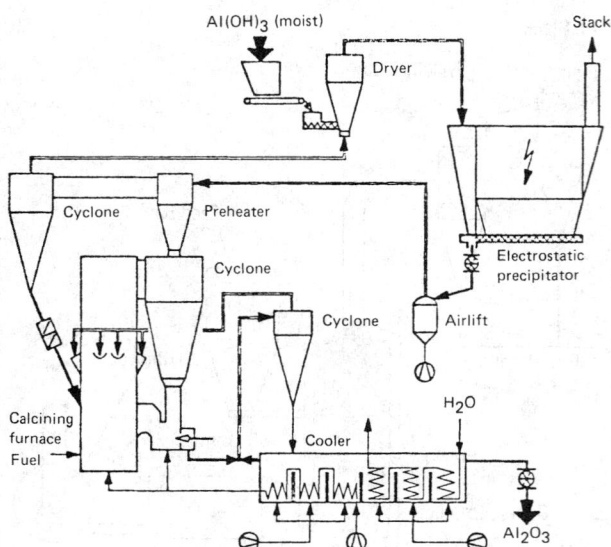

FIG. 17-28   Circulating-fluid-bed calciner. (*Lurgi Corp.*)

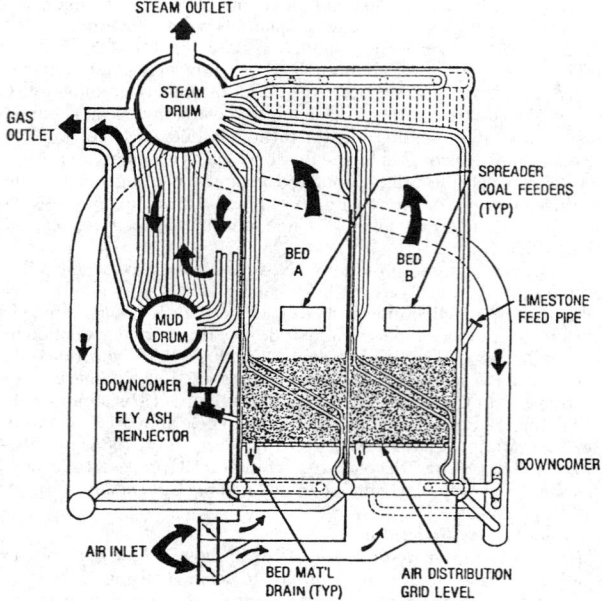

FIG. 17-29   Fluidized-bed steam generator at Georgetown University; 12.6-kg/s (100,000-lb/h) steam at 4.75-MPa (675-psig) pressure. (*From Georgetown Univ. Q. Tech. Prog. Rep. METC/DOE/10381/135, July–September 1980.*)

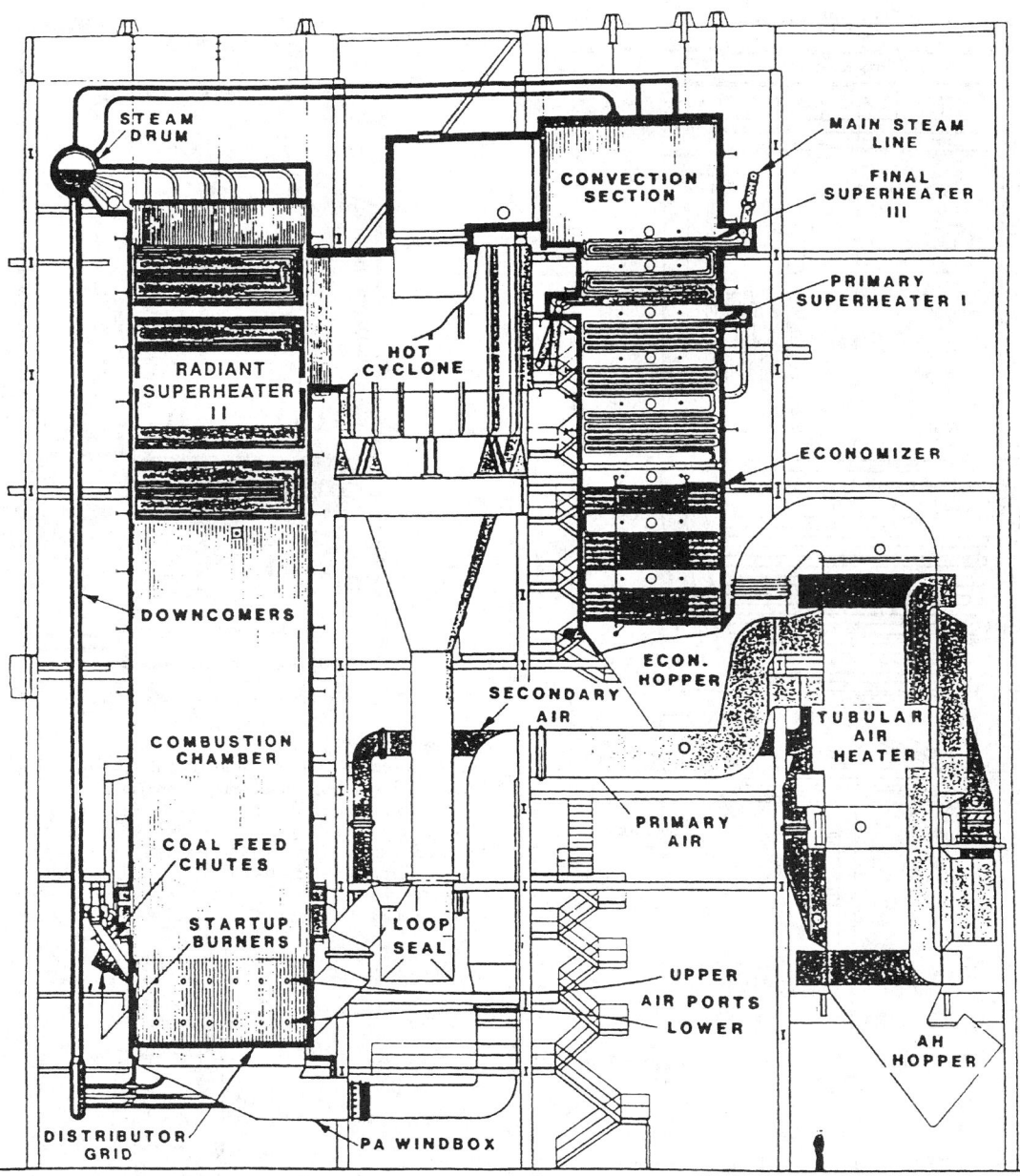

FIG. 17-30 Side view of 110 MWe Nucla CFB boiler. (*Source: Pyropower Corporation.*)

[Flood and Kernel, *Chem. Proc.* (Sept. 8, 1973)]. Miscellaneous uses include the incineration of sawdust, carbon-black waste, pharmaceutical waste, grease from domestic sewage, spent coffee grounds, and domestic garbage.

Toxic or hazardous wastes can be disposed of in fluidized beds by either chemical capture or complete destruction. In the former case, bed material, such as limestone, will react with halides, sulfides, metals, etc., to form stable compounds which can be landfilled. Contact times of up to 5 or 10 s at 1200 K (900°C) to 1300 K (1000°C) assure complete destruction of most compounds.

Physical Contacting

Drying Fluidized-bed units for drying solids, particularly coal, cement, rock, and limestone, are in general acceptance. Economic considerations make these units particularly attractive when large tonnages of solids are to be handled. Fuel requirements are 3.3 to 4.2 MJ/kg (1500 to 1900 Btu/lb of water removed), and total power for blowers, feeders, etc., is about 0.08 kWh/kg of water removed. The maximum-sized feed is 6 cm (1½ in) × 0 coal. One of the major advantages of this type of dryer is the close control of conditions so that a predetermined amount of free moisture may be left with the solids to

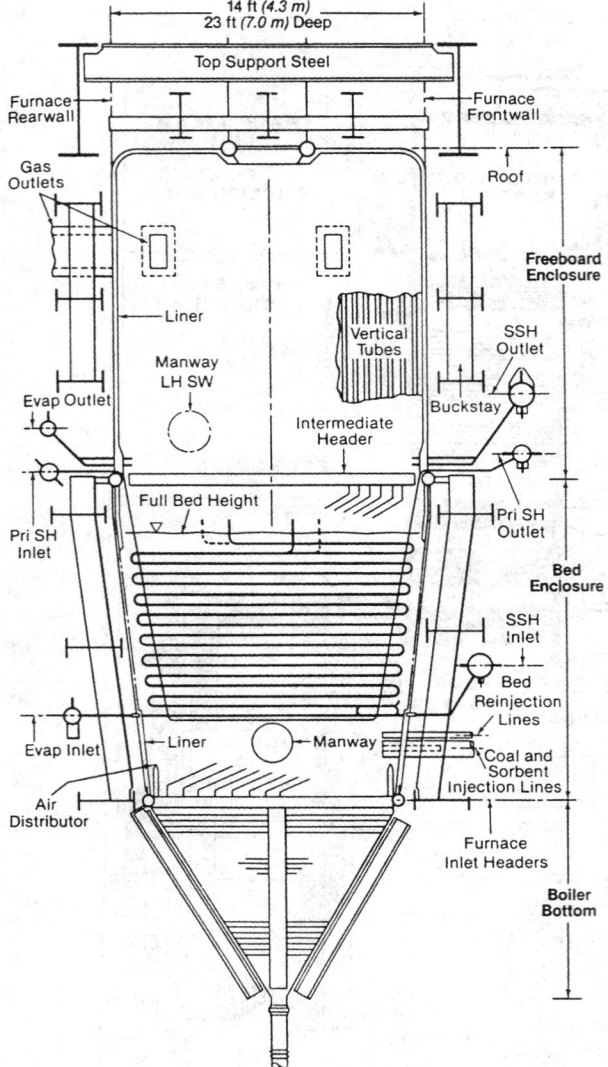

FIG. 17-31 71 MWe PFBC unit. (*From* Steam, *40th ed., 29-9, Babcock & Wilcox, 1992*).

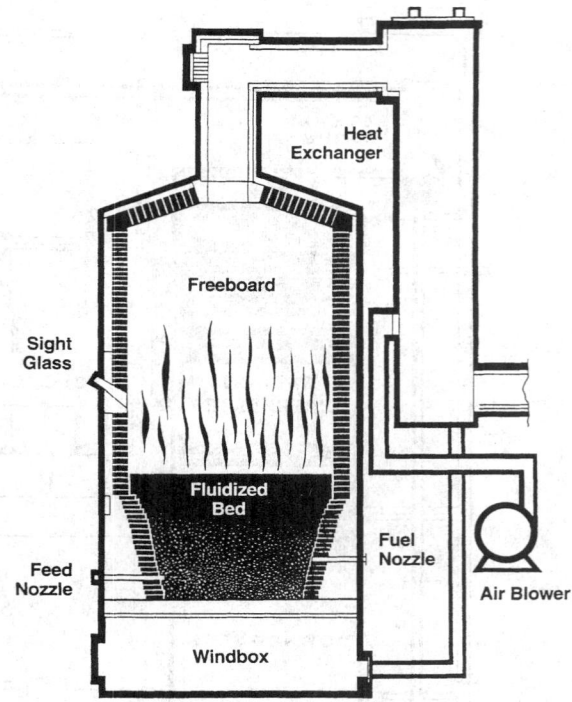

FIG. 17-32 Hot windbox incinerator/reactor with air preheating. (*Dorr-Oliver, Inc.*)

Classification The separation of fine particles from coarse can be effected by use of a fluidized bed (see "Drying"). However, for economic reasons (i.e., initial cost, power requirements for compression of fluidizing gas, etc.), it is doubtful except in special cases if a fluidized-bed classifier would be built for this purpose alone.

It has been proposed that fluidized beds be used to remove fine solids from a gas stream. This is possible under special conditions.

Adsorption-Desorption An arrangement for gas fractionation is shown in Fig. 17-33.

The effects of adsorption and desorption on the performance of fluidized beds are discussed elsewhere. Adsorption of carbon disulfide

prevent dusting of the product during subsequent material-handling operations. The fluidized-bed dryer is also used as a classifier so that both drying and classification operations are accomplished simultaneously.

Wall and Ash [*Ind. Eng. Chem.,* **41,** 1247 (1949)] state that, in drying 4.8-mm (−4 mesh) dolomite with combustion gases at a superficial velocity of 1.2 m/s (4 ft/s), the following removals of fines were achieved:

Particle size	% removed
−65 + 100 mesh	60
−100 + 150 mesh	79
−150 + 200 mesh	85
−200 + 325 mesh	89
−325 mesh	89

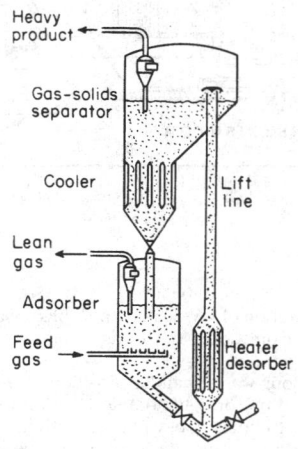

FIG. 17-33 Fluidized bed for gas fractionation. [*Sittig,* Chem. Eng. (*May 1953*).]

vapors from air streams as great as 300 m³/s (540,000 ft³/min) in a 17-m- (53-ft-) diameter unit has been reported by Avery and Tracey ("The Application of Fluidized Beds of Activated Carbon to Recover Solvent from Air or Gas Streams," Tripartate Chemical Engineering Conference, Montreal, Sept. 24, 1968).

Heat Treatment   Heat treatment can be divided into two types, treatment of fluidizable solids and treatment of large, usually metallic objects in a fluid bed. The former is generally accomplished in multi-compartment units to conserve heat (Fig. 17-27). The heat treatment of large metallic objects is accomplished in long, narrow heated beds.

The objects are conveyed through the beds by an overhead conveyor system. Fluid beds are used because of the high heat-transfer rate and uniform temperature. See Reindl, "Fluid Bed Technology," *American Society for Metals,* Cincinnati, Sept. 23, 1981; Fennell, *Ind. Heat.,* **48,** 9, 36 (September 1981).

Coating   Fluidized beds of thermoplastic resins have been used to facilitate the coating of metallic parts. A properly prepared, heated metal part is dipped into the fluidized bed, which permits complete immersion in the dry solids. The heated metal fuses the thermoplastic, forming a continuous uniform coating.

GAS-SOLIDS SEPARATIONS

This subsection is concerned with the application of particle mechanics (see Sec. 5, "Fluid and Particle Mechanics") to the design and application of dust-collection systems. It includes wet collectors, or scrubbers, for particle collection. Scrubbers designed for purposes of mass transfer are discussed in Secs. 14 and 18. Equipment for removing entrained liquid mist from gases is described in Sec. 18.

Nomenclature

Except where otherwise noted here or in the text, either consistent system of units (SI or U.S. customary) may be used. Only SI units may be used for electrical quantities, since no comparable electrical units exist in the U.S. customary system. When special units are used, they are noted at the point of use.

Symbols	Definition	SI units	U.S. customary units	Special units
B_c	Width of rectangular cyclone inlet duct	m	ft	
B_e	Spacing between wire and plate, or between rod and curtain, or between parallel plates in electrical precipitators	m	ft	
B_s	Width of gravity settling chamber	m	ft	
$C°$	Dry scrubber pollutant gas equilibrium concentration over sorbent			
C_1	Dry scrubber pollutant gasinlet concentration			
C_2	Dry scrubber pollutant gasoutlet concentration			
c_d	Dust concentration in gas stream	g/m³		grains/ft³
c_h	Specific heat of gas	J/(k·K)	Btu/(lbm·°F)	
c_{hb}	Specific heat of collecting body	J/(kg·K)	Btu/(lbm·°F)	
c_{hp}	Specific heat of particle	J/(kg·K)	Btu/(lbm·°F)	
D_b	Diameter or other representative dimension of collector body or device	m	ft	
D_{b1},D_{b2}	Other characteristic dimensions of collector body or device	m	ft	
D_c	Cyclone diameter	m	ft	
D_d	Outside diameter of wire or discharge electrode of concentric-cylinder type of electrical precipitator	m	ft	
D_e	Diameter of cyclone gas exit duct	m	ft	
D_o	Volume/surface-mean-drop diameter			μm
D_p	Diameter of particle	m	ft	μm
D_{pth}	Cut diameter, diameter of particles of which 50% of those present are collected	m	ft	μm
d_P	Particle diameter of fraction number c'	m	ft	
D_t	Inside diameter of collecting tube of concentric-cylinder type of electrical precipitator	m	ft	
D_v	Diffusion coefficient for particle	m²/s	ft²/s	
DI	Decontamination index $= \log_{10}[1/(1 - \eta)]$	Dimensionless	Dimensionless	
e	Natural (napierian) logarithm base	2.718 . . .	2.718 . . .	
E	Potential difference	V		
E_c	Potential difference required for corona discharge to commence	V		
E_d	Voltage across dust layer	V		
E_s	Potential difference required for sparking to commence	V		
E_L	Cyclone collection efficiency at actual loading			
E_O	Cyclone collection efficiency at low loading			
F_E	Effective friction loss across wetted equipment in scrubber	kPa		in water
F_k	Packed bed friction loss			
g_c	Conversion factor		32.17 (lbm/lbf)(ft/s²)	
g_L	Local acceleration due to gravity	m/s²	ft/s²	
H_c	Height of rectangular cyclone inlet duct	m	ft	
H_s	Height of gravity settling chamber	m	ft	
I	Electrical current per unit of electrode length	A/m		

Nomenclature (*Continued*)

Symbols	Definition	SI units	U.S. customary units	Special units
j	Corona current density at dust layer	A/m^2		
k_p	Density of gas relative to its density at 0°C, 1 atm	Dimensionless	Dimensionless	Dimensionless
k_t	Thermal conductivity of gas	$W/(m \cdot K)$	$Btu/(s \cdot ft \cdot °F)$	
k_{tb}	Thermal conductivity of collecting body	$W/(m \cdot K)$	$Btu/(s \cdot ft \cdot °F)$	
k_{tp}	Thermal conductivity of particle	$W/(m \cdot K)$	$Btu/(s \cdot ft \cdot °F)$	
K	Empirical proportionality constant for cyclone pressure drop or friction loss	Dimensionless	Dimensionless	
K_1	Resistance coefficient of "conditioned" filter fabric	$kPa/(m/min)$	in water/(ft/min)	
K_2	Resistance coefficient of dust cake on filter fabric	$\dfrac{kPa}{(m/min)(g/m^2)}$	$\dfrac{in\ water}{(ft/min)(lbm/ft^2)}$	
K_a	Proportionality constant, for target efficiency of a single fiber in a bed of fibers	Dimensionless	Dimensionless	
K_c	Resistance coefficient for "conditioned" filter fabric			$\dfrac{in\ water}{(ft/min)(cP)}$
K_d	Resistance coefficient for dust cake on filter fabric			$\dfrac{in\ water}{(ft/min)(gr/ft^2)(cP)}$
K_e	Electrical-precipitator constant	s/m	s/ft	
K_F	Resistance coefficient for clean filter cloth			$\dfrac{in\ water}{(ft/min)(cP)}$
K_o	"Energy-distance" constant for electrical discharge in gases	m		
K_m	Stokes-Cunningham correction factor	Dimensionless	Dimensionless	Dimensionless
L	Thickness of fibrous filter or of dust layer on surface filter	m	ft	
L_e	Length of collecting electrode in direction of gas flow	m	ft	
L_s	Length of gravity settling chamber in direction of gas flow	m	ft	
ln	Natural logarithm (logarithm to the base e)	Dimensionless	Dimensionless	Dimensionless
M	Molecular weight	kg/mol	lbm/mol	
n	Exponent	Dimensionless	Dimensionless	Dimensionless
N_{Kn}	Knudsen number $= \lambda_m/D_b$	Dimensionless	Dimensionless	
N_{Ma}	Mach number	Dimensionless	Dimensionless	
N_o	Number of elementary electrical charges acquired by a particle	Dimensionless	Dimensionless	
N_{Re}	Reynolds number $= (D_p \rho V_o/\mu)$ or $(D_p \rho u_t/\mu)$	Dimensionless	Dimensionless	
N_{sc}	Interaction number $= 18\ \mu/K_m \rho_p D_v$	Dimensionless	Dimensionless	
N_{sd}	Diffusional separation number	Dimensionless	Dimensionless	
N_{sec}	Electrostatic-attraction separation number	Dimensionless	Dimensionless	
N_{sei}	Electrostatic-induction separation number	Dimensionless	Dimensionless	
N_{sf}	Flow-line separation number	Dimensionless	Dimensionless	
N_{sg}	Gravitational separation number	Dimensionless	Dimensionless	
N_{si}	Inertial separation number	Dimensionless	Dimensionless	
N_{st}	Thermal separation number	Dimensionless	Dimensionless	
N_t	Number of transfer units $= \ln\,[1/(1 - \eta)]$	Dimensionless	Dimensionless	
N_s	Number of turns made by gas stream in a cyclone separator	Dimensionless	Dimensionless	
Δp	Gas pressure drop	kPa	lbf/ft^2	in water
Δp_i	Gas pressure drop in cyclone or filter			in water
p_F	Gauge pressure of water fed to scrubber	kPa		lbf/in^2
P_G	Gas-phase contacting power	$MJ/1000\ m^3$		$hp/(1000\ ft^3/min)$
P_L	Liquid-phase contacting power	$MJ/1000\ m^3$		$hp/(1000\ ft^3/min)$
P_M	Mechanical contacting power	$MJ/1000\ m^3$		$hp/(1000\ ft^3/min)$
P_T	Total contacting power	$MJ/1000\ m^3$		$hp/(1000\ ft^3/min)$
q	Gas flow rate	m^3/s	ft^3/s	
Q_G	Gas flow rate		ft^3/s	ft^3/min
Q_L	Liquid flow rate		ft^3/s	gal/min
Q_p	Electrical charge on particle	C		
r	Radius; distance from centerline of cyclone separator; distance from centerline of concentric-cylinder electrical precipitator	m	ft	
t_m	Time			min
T	Absolute gas temperature	K	$°R$	
T_b	Absolute temperature of collecting body	K	$°R$	

Nomenclature (*Concluded*)

Symbols	Definition	SI units	U.S. customary units	Special units
u_s	Velocity of migration of particle toward collecting electrode	m/s	ft/s	
u_t	Terminal settling velocity of particle under action of gravity	m/s	ft/s	ft/s
v_m	Average cyclone inlet velocity, based on area A_c	m/s	ft/s	ft/s
v_p	Actual particle velocity	m/s	ft/s	
V_f	Filtration velocity (superficial gas velocity through filter)	m/min		ft/min
V_o	Gas velocity	m/s	ft/s	
V_s	Average gas velocity in gravity settling	m/s	ft/s	
V_{ct}	Tangential component of gas velocity in cyclone	m/s	ft/s	
w	Loading of collected dust on filter	g/m^2	lbm/ft^2	gr/ft^2

Greek symbols

Symbols	Definition	SI units	U.S. customary units	Special units
α	Empirical constant in equation of scrubber performance curve	$\left[\dfrac{MJ}{1000\ m^3}\right]^{-\gamma}$		$\left[\dfrac{hp}{100\ ft^3/min}\right]^{-\gamma}$
γ	Empirical constant in equation of scrubber performance curve	Dimensionless		Dimensionless
δ	Dielectric constant	Dimensionless		
δ_g	Dielectric constant at 0°C, 1 atm	Dimensionless		
δ_o	Permittivity of free space	F/m		
δ_b	Dielectric constant of collecting body	Dimensionless		
δ_p	Dielectric constant of particle	Dimensionless		
Δ	Fractional free area (for screens, perforated plates, grids)	Dimensionless	Dimensionless	
ε	Elementary electrical charge	1.60210×10^{-19} C		
ε_b	Characteristics potential gradient at collecting surface	V/m		
ε_v	Fraction voids in bed of solids	Dimensionless	Dimensionless	Dimensionless
ζ	$= 1 + 2\dfrac{(\delta - 1)}{(\delta + 2)}$ ranges from a value of 1 for materials with a dielectric constant of 1 to 3 for conductors	Dimensionless		
η	Collection efficiency, weight fraction of entering dispersoid collected	Dimensionless	Dimensionless	Dimensionless
η_o	Target efficiency of an isolated collecting body, fraction of dispersoid in swept volume collected on body	Dimensionless	Dimensionless	Dimensionless
η_t	Target efficiency of a single collecting body in an array of collecting bodies, fraction of dispersoid in swept volume collected on body	Dimensionless	Dimensionless	Dimensionless
λ_i	Ionic mobility of gas	(m/s)/(V/m)		
λ_p	Particle mobility $= u_e/\mathscr{E}$	(m/s)/(V/m)		
μ	Gas viscosity	Pa·s	lbm/(s·ft)	cP
μ_L	Liquid viscosity			cP
ρ	Gas density	g/m^3	lb/ft^3	
ρ_d	Resistivity of dust layer	$\Omega \cdot$m		
ρ_L	Liquid density		lbm/ft^3	lbm/ft^3
ρ_s	True (*not* bulk) density of solids or liquid drops	kg/m^3	lbm/ft^3	lbm/ft^3
ρ'	Density of gas relative to its density at 25°C, 1 atm	Dimensionless	Dimensionless	Dimensionless
σ	Ion density	Number/m^3		
σ_{avg}	Average ion density	Number/m^3		
σ_L	Liquid surface tension			dyn/cm
ϕ_s	Particle shape factor = (surface of sphere)/(surface of particle of same volume)	Dimensionless	Dimensionless	Dimensionless

Script symbols

Symbols	Definition	SI units	U.S. customary units	Special units
$\mathscr{E}$	Potential gradient	V/m		
$\mathscr{E}_c$	Potential gradient required for corona discharge to commence	V/m		
$\mathscr{E}_i$	Average potential gradient in ionization stage	V/m		
$\mathscr{E}_o$	Electrical breakdown constant for gas	V/m		
$\mathscr{E}_p$	Average potential gradient in collection stage	V/m		
$\mathscr{E}_s$	Potential gradient required for sparking to commence	V/m		

GENERAL REFERENCES: Burchsted, Kahn, and Fuller, *Nuclear Air Cleaning Handbook,* ERDA 76-21, Oak Ridge, Tenn., 1976. Cadle, *The Measurement of Airborne Particles,* Wiley, New York, 1975. Davies, *Aerosol Science,* Academic, New York, 1966. Davies, *Air Filtration,* Academic, New York, 1973. Dennis, *Handbook on Aerosols,* ERDA TID-26608, Oak Ridge, Tenn., 1976. Drinker and Hatch, *Industrial Dust,* 2d ed., McGraw-Hill, New York, 1954. Friedlander, *Smoke, Dust, and Haze,* Wiley, New York, 1977. Fuchs, *The Mechanics of Aerosols,* Pergamon, Oxford, 1964. Green and Lane, *Particulate Clouds: Dusts, Smokes, and Mists,* Van Nostrand, New York, 1964. Lapple, *Fluid and Particle Mechanics,* University of Delaware, Newark, 1951. Licht, *Air Pollution Control Engineering—Basic Calculations for Particle Collection,* Marcel Dekker, New York, 1980. Liu, *Fine Particles—Aerosol Generation, Measurement, Sampling, and Analysis,* Academic, New York, 1976. Lunde and Lapple, *Chem. Eng. Prog.,* **53,** 385 (1957). Lundgren et al., *Aerosol Measurement,* University of Florida, Gainesville, 1979. Mercer, *Aerosol Technology in Hazard Evaluation,* Academic, New York, 1973. Nonhebel, *Processes for Air Pollution Control,* CRC Press, Cleveland, 1972. Shaw, *Fundamentals of Aerosol Science,* Wiley, New York, 1978. Stern, *Air Pollution: A Comprehensive Treatise,* vols. 3 and 4, Academic, New York, 1977. Strauss, *Industrial Gas Cleaning,* 2d ed., Pergamon, New York, 1975. Theodore and Buonicore, *Air Pollution Control Equipment: Selection, Design, Operation, and Maintenance,* Prentice-Hall, Englewood Cliffs, N.J., 1982. White, *Industrial Electrostatic Precipitation,* Addison-Wesley, Reading, Mass., 1963. White and Smith, *High-Efficiency Air Filtration,* Butterworth, Washington, 1964. ASME Research Committee on Industrial and Municipal Wastes, *Combustion Fundamentals for Waste Incineration,* American Society of Mechanical Engineers, 1974. Buonicore and Davis (eds.), *Air Pollution & Waste Management Association,* Van Nostrand Reinhold, 1992. Burchsted, Fuller, and Kahn, *Nuclear Air Cleaning Handbook,* ORNL for the U.S. Energy Research and Development Administration, NTIS Report ERDA 76-21, 1976. Dennis (ed.), *Handbook on Aerosols,* GCA for the U.S. Energy Research and Development Administration, NTIS Report TID-26608, 1976. Stern, *Air Pollution,* 3d ed., Academic Press, 1977 (supplement 1986).

PURPOSE OF DUST COLLECTION

Dust collection is concerned with the removal or collection of solid dispersoids in gases for purposes of:

1. Air-pollution control, as in fly-ash removal from power-plant flue gases
2. Equipment-maintenance reduction, as in filtration of engine-intake air or pyrites furnace-gas treatment prior to its entry to a contact sulfuric acid plant
3. Safety- or health-hazard elimination, as in collection of siliceous and metallic dusts around grinding and drilling equipment and in some metallurgical operations and flour dusts from milling or bagging operations
4. Product-quality improvement, as in air cleaning in the production of pharmaceutical products and photographic film
5. Recovery of a valuable product, as in collection of dusts from dryers and smelters
6. Powdered-product collection, as in pneumatic conveying; the spray drying of milk, eggs, and soap; and the manufacture of high-purity zinc oxide and carbon black

PROPERTIES OF PARTICLE DISPERSOIDS

An understanding of the fundamental properties and characteristics of gas dispersoids is essential to the design of industrial dust-control equipment. Figure 17-34 shows characteristics of dispersoids and other particles together with the types of gas-cleaning equipment that are applicable to their control. Two types of solid dispersoids are shown: (1) dust, which is composed of particles larger than 1 μm; and (2) fume, which consists of particles generally smaller than 1 μm. Dusts usually result from mechanical disintegration of matter. They may be redispersed from the settled, or bulk, condition by an air blast. Fumes are submicrometer dispersoids formed by processes such as combustion, sublimation, and condensation. Once collected, they cannot be redispersed from the settled condition to their original state of dispersion by air blasts or mechanical dispersion equipment.

The primary distinguishing characteristic of gas dispersoids is particle size. The generally accepted unit of particle size is the micrometer, μm. (Prior to the adoption of the SI system, the same unit was known as the micron and was designated by μ.) The particle size of a gas dispersoid is usually taken as the diameter of a sphere having the same mass and density as the particle in question. Another common method is to designate the screen mesh that has an aperture corresponding to the particle diameter; the screen scale used must also be specified to avoid confusion.

From the standpoint of collector design and performance, the most important size-related property of a dust particle is its dynamic behavior. Particles larger than 100 μm are readily collectible by simple inertial or gravitational methods. For particles under 100 μm, the range of principal difficulty in dust collection, the resistance to motion in a gas is viscous (see Sec. 6, "Fluid and Particle Mechanics"), and for such particles, the most useful size specification is commonly the Stokes settling diameter, which is the diameter of the spherical particle of the same density that has the same terminal velocity in viscous flow as the particle in question. It is yet more convenient in many circumstances to use the "aerodynamic diameter," which is the diameter of the particle of unit density (1 g/cm^3) that has the same terminal settling velocity. Use of the aerodynamic diameter permits direct comparisons of the dynamic behavior of particles that are actually of different sizes, shapes, and densities [Raabe, *J. Air Pollut. Control Assoc.,* **26,** 856 (1976)].

When the size of a particle approaches the same order of magnitude as the mean free path of the gas molecules, the settling velocity is greater than predicted by Stokes' law because of molecular slip. The slip-flow correction is appreciable for particles smaller than 1 μm and is allowed for by the Cunningham correction for Stokes' law (Lapple, op. cit.; Licht, op. cit.). The Cunningham correction is applied in calculations of the aerodynamic diameters of particles that are in the appropriate size range.

Although solid fume particles may range in size down to perhaps 0.001 μm, fine particles effectively smaller than about 0.1 μm are not of much significance in industrial dust and fume sources because their aggregate mass is only a very small fraction of the total mass emission. At the concentrations present in such sources (e.g., production of carbon black) the coagulation, or flocculation, rate of the ultrafine particles is extremely high, and the particles speedily grow to sizes of 0.1 μm or greater. The most difficult collection problems are thus concerned with particles in the range of about 0.1 to 2 μm, in which forces for deposition by inertia are small. For collection of particles under 0.1 μm, diffusional deposition becomes increasingly important as the particle size decreases.

In a gas stream carrying dust or fume, some degree of particle flocculation will exist, so that both discrete particles and clusters of adhering particles will be present. The discrete particles composing the clusters may be only loosely attached to each other, as by van der Waals forces [Lapple, *Chem. Eng.,* **75**(11), 149 (1968)]. Flocculation tends to increase with increases in particle concentration and may strongly influence collector performance.

PARTICLE MEASUREMENTS

Measurements of the concentrations and characteristics of dust dispersed in air or other gases may be necessary (1) to determine the need for control measures, (2) to establish compliance with legal requirements, (3) to obtain information for collector design, and (4) to determine collector performance.

Atmospheric-Pollution Measurements The dust-fall measurement is one of the common methods for obtaining a relative long-period evaluation of particulate air pollution. Stack-smoke densities are often graded visually by means of the Ringelmann chart. Plume opacity may be continuously monitored and recorded by a photoelectric device which measures the amount of light transmitted through a stack plume. Equipment for local atmospheric-dust-concentration measurements fall into five general types: (1) the impinger, (2) the hot-wire or thermal precipitator, (3) the electrostatic precipitator, (4) the filter, and (5) impactors and cyclones. The filter is the most widely used, in the form of either a continuous tape, or a number of filter disks arranged in an automatic sequencing device, or a single, short-term, high-volume sampler. Samplers such as these are commonly used to obtain mass emission and particle-size distribution. Impactors and small cyclones are commonly used as size-discriminating samplers and are usually followed by filters for the determination of the finest

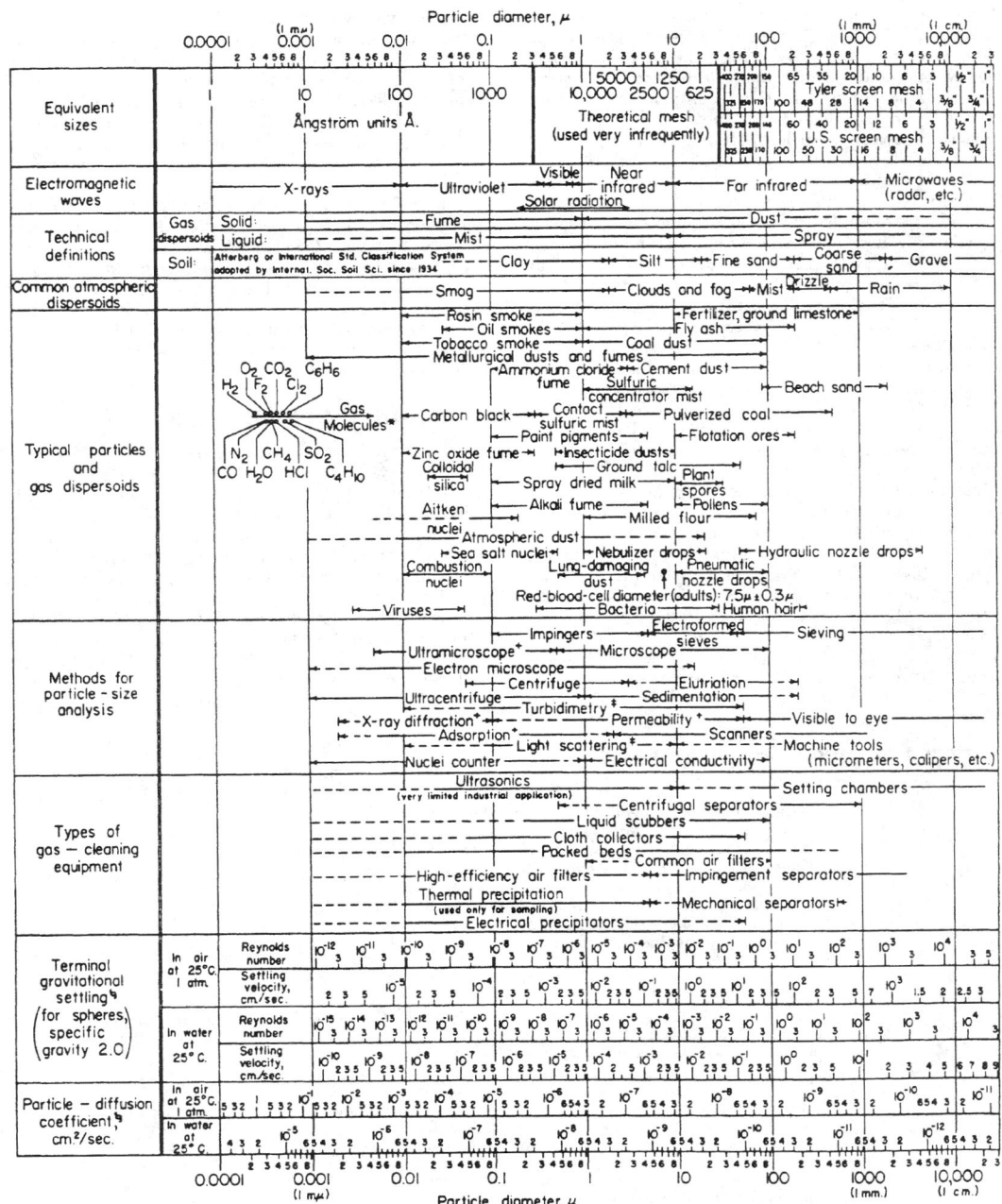

FIG. 17-34 Characteristics of particles and particle dispersoids. (*Courtesy of the Stanford Research Institute; prepared by C. E. Lapple.*)

fraction of the dust (Lundgren et al., *Aerosol Measurement*, University of Florida, Gainesville, 1979; and Dennis, *Handbook on Aerosols*, U.S. ERDA TID-26608, Oak Ridge, Tenn., 1976).

Process-Gas Sampling In sampling process gases either to determine dust concentration or to obtain a representative dust sample, it is necessary to take special precautions to avoid inertial segregation of the particles. To prevent such classification, a traverse of the duct may be required, and at each point the sampling nozzle must face directly into the gas stream with the velocity in the mouth of the nozzle equal to the local gas velocity at that point. This is called "isokinetic sampling." If the sampling velocity is too high, the dust sample will contain a lower concentration of dust than the mainstream, with a greater percentage of fine particles; if the sampling velocity is too low, the dust sample will contain a higher concentration of dust with a greater percentage of coarse particles [Lapple, *Heat. Piping Air Cond.*, **16,** 578 (1944); *Manual of Disposal of Refinery Wastes*, vol. V, American Petroleum Institute, New York, 1954; and Dennis, op. cit.].

Particle-Size Analysis Methods for particle-size analysis are shown in Fig. 17-34, and examples of size-analysis methods are given in Table 17-1. More detailed information may be found in Lapple, *Chem. Eng.*, **75**(11), 140 (1968); Lapple, "Particle-Size Analysis," in *Encyclopedia of Science and Technology*, 5th ed., McGraw-Hill, New York, 1982; Cadle, *The Measurement of Airborne Particles*, Wiley, New York, 1975; Lowell, *Introduction to Powder Surface Area*, 2d ed., Wiley, New York, 1993; and Allen, *Particle Size Measurement*, 4th ed, Chapman and Hall, London, 1990. Particle-size distribution may be presented on either a frequency or a cumulative basis; the various methods are discussed in the references just cited. The most common method presents a plot of particle size versus the cumulative weight percent of material larger or smaller than the indicated size, on logarithmic-probability graph paper.

For determination of the aerodynamic diameters of particles, the most commonly applicable methods for particle-size analysis are those based on inertia: aerosol centrifuges, cyclones, and inertial impactors (Lundgren et al., *Aerosol Measurement*, University of Florida, Gainesville, 1979; and Liu, *Fine Particles—Aerosol Generation, Measurement, Sampling, and Analysis*, Academic, New York, 1976). Impactors are the most commonly used. Nevertheless, impactor measurements are subject to numerous errors [Rao and Whitby, *Am. Ind. Hyg. Assoc. J.*, **38,** 174 (1977); Marple and Willeke, "Inertial Impactors," in Lundgren et al., *Aerosol Measurement*; and Fuchs, "Aerosol Impactors," in Shaw, *Fundamentals of Aerosol Sci-*

ence, Wiley, New York, 1978]. Reentrainment due to particle bouncing and blowoff of deposited particles makes a dust appear finer than it actually is, as does the breakup of flocculated particles. Processing cascade-impactor data also presents possibilities for substantial errors (Fuchs, *The Mechanics of Aerosols*, Pergamon, Oxford, 1964) and is laborious as well. Lawless (Rep. No. EPA-600/7-78-189, U.S. EPA, 1978) discusses problems in analyzing and fitting cascade-impactor data to obtain dust-collector efficiencies for discrete particle sizes.

The measured diameters of particles should as nearly as possible represent the effective particle size of a dust as it exists in the gas stream. When significant flocculation exists, it is sometimes possible to use measurement methods based on gravity settling.

For dust-control work, it is recommended that a preliminary qualitative examination of the dust first be made without a detailed particle count. A visual estimate of particle-size distribution will often provide sufficient guidance for a preliminary assessment of requirements for collection equipment.

MECHANISMS OF DUST COLLECTION

The basic operations in dust collection by any device are (1) separation of the gas-borne particles from the gas stream by deposition on a collecting surface; (2) retention of the deposit on the surface; and (3) removal of the deposit from the surface for recovery or disposal. The separation step requires (1) application of a force that produces a differential motion of a particle relative to the gas and (2) a gas retention time sufficient for the particle to migrate to the collecting surface. The principal mechanisms of aerosol deposition that are applied in dust collectors are (1) gravitational deposition, (2) flow-line interception, (3) inertial deposition, (4) diffusional deposition, and (5) electrostatic deposition. Thermal deposition is only a minor factor in practical dust-collection equipment because the thermophoretic force is small. Table 17-2 lists these six mechanisms and presents the characteristic

TABLE 17-1   Particle Size Analysis Methods and Equipment

Method	Brand names	Size range	Sample size, g
Quantitative image analysis	American Innovation Videometric, Analytical Measuring Systems Quickstep & Optomax, Artec Omnicon Automatix, Boeckeler, Buehler Omnimat, Compix Imaging Systems, Data Teanslation— Global Lab, Image, Hamamatsu C-1000, Hitech Olympus Cue-3, Joyce-Loebl Magiscan, Leco AMF System, Leico Quantimet, LeMont Oasys, Millipore_MC, Nachet 1500, Nicon Microphot, Oncor Instrument System, Optomax V, Outokumpu Imagist, Shapespeare Juliet, Tracor Northern, Carl Zeiss Videoplan	1–1000 μm	0.001
Sieves:			
Punched Plate	Many	4–100 mm	500
Woven wire	Many	20 μm–125 mm	25–200
Micromesh	Buckbee Mears, Veco, Endecottes	5–500 + μm	1–5
Sieving machines (Air jet, sonic wet and dry)	Alpine, ATM, Gilson, Gradex, Hosokawa, Retsch, Seishin		
Sedimentation	*Pipette*—Gilson, *Photosedimentometer—gravitational*, Paar; *centrifugal*, Joyce Loebl, Brookhaven, Horiba, Seishin, Shimadzu; *x-ray absorption*—gravitational, Quantachrome, Micromeretics; *centrifugal*, Brookhaven	Gravitational 1–100 μm Centrifugal 0.05–5 mm	0.1–5+ g
Classification	Air-Bahco, Water-Warmain Cyclosizer	5–500 μm	5–100 g
Field scanning (light)	Cilas, Coulter, Insitec, Fritsch, Horiba, Leeds & Northrup (Microtrac), Malvern, Nitto, Seishin, Shimadzu, Sympatec	0.04–3500 μm	<1 g (wet) >20 g (on-line)
Field scanning (ultrasonics)	Sympatec OPUS, Pen-Kem		On-line
Stream scanning	Brinkmann, Climet, Coulter, Dantec, Erdco, Faley, Flowvision, Hiac/Royco, Kowa, Lasentec, Malvern, Met One, Particle Measuring Systems, Polytec, Procedyne, Rion, Spectrex	0.2–10,000 μm	0.1–10 g (also on-line)
Zeta potential Photon correlation Spectroscopy	Zeta Plus, Micromeretics, Zeta sizer Malvern, Nicomp, Brookhaven, Coulter, Photol	0.001–30 μm	0.1–1 μm

NOTE: This table was compiled with the assistance of T. Allen, DuPont Particle Science and Technology, and is not intended to be comprehensive. Many other fine suppliers of particle analysis equipment are available.

TABLE 17-2 Summary of Mechanisms and Parameters in Aerosol Deposition

Deposition	Origin of force field	Deposition mechanism measureable in terms of		System parameters
		Basic parameter	Specific modifying parameters	
Flow-line interception°	Physical gradient°	$N_{sf} = \left(\dfrac{D_p}{D_b}\right)$		Geometry: $(D_{b1}/D_b), (D_{b2}/D_b)$, etc. ϵ_v Δ
Inertial deposition	Velocity gradient	$N_{st} = \left(\dfrac{K_m \rho s D_p^2 V_o}{18\mu D_b}\right)$	$N_{sc} = \left(\dfrac{N_{sf}^2}{N_{st} N_{sd}}\right)$ $= \left(\dfrac{18\mu}{K_m \rho_p D_v}\right)^\dagger$	
Diffusional deposition	Concentration gradient	$N_{sd} = \left(\dfrac{D_v}{V_o D_b}\right)$		
Gravity settling	Elevation gradient	$N_{sg} = \left(\dfrac{u_t}{V_o}\right)$		Flow pattern: N_{Re}‡ N_{Ma} N_{Kn}
Electrostatic precipitation	Electric-field gradient§ *a.* Attraction *b.* Induction	$N_{sec} = \left(\dfrac{K_m Q_p \epsilon_b}{\mu D_p V_o}\right)$ $N_{sei} = \left(\dfrac{\delta_p - 1}{\delta_p + 2}\right)\left(\dfrac{K_m D_p^2 \delta_o \epsilon_b 2}{\mu D_b V_o}\right)$	δ_p, δ_b¶	Surface accommodation
Thermal precipitation	Temperature gradient	$N_{st} =$ $\left(\dfrac{T - T_b}{T}\right)\left(\dfrac{\mu}{K_m \rho D_b V_o}\right)\left(\dfrac{k_t}{2k_t + k_{tp}}\right)$	$(T_b/T), (T_p/T),\dagger (N_{Pr}),$ $(k_{tp}/k_t), (k_{tb}/k_t),¶ (c_{hp}/c_h), (c_{hb}/c_h)$¶	

SOURCE: Lunde and Lapple, *Chem. Eng. Prog.*, **53**, 385 (1957).

°This has also commonly been termed "direct interception" and in conventional analysis would constitute a physical boundary condition imposed upon the particle path induced by action of other forces. By itself it reflects deposition that might result with a hypothetical particle having finite size but no mass or elasticity.

†This parameter is an alternative to N_{sf}', N_{si}, or N_{ad} and is useful as a measure of the interactive effect of one of these on the other two. It is comparable with the Schmidt number.

‡When applied to the inertial deposition mechanism, a convenient alternative is $(K_m \rho_s/18\rho) = N_{si}/(N_{sf}^2 N_{Re})$.

§In cases in which the body charge distribution is fixed and known, ϵ_b may be replaced with Q_{bs}/δ_o.

¶Not likely to be significant contributions.

parameters of their operation [Lunde and Lapple, *Chem. Eng. Prog.*, **53**, 385 (1957)]. The actions of the inertial-deposition, flow-line-interception, and diffusional-deposition mechanisms are illustrated in Fig. 17-35 for the case of a collecting body immersed in a particle-laden gas stream.

Two other deposition mechanisms, in addition to the six listed, may be in operation under particular circumstances. Some dust particles may be collected on filters by sieving when the pore diameter is less than the particle diameter. Except in small membrane filters, the sieving mechanism is probably limited to surface-type filters, in which a layer of collected dust is itself the principal filter medium.

The other mechanism appears in scrubbers. When water vapor diffuses from a gas stream to a cold surface and condenses, there is a net hydrodynamic flow of the noncondensable gas directed toward the surface. This flow, termed the Stefan flow, carries aerosol particles to the condensing surface (Goldsmith and May, in Davies, *Aerosol Science*, Academic, New York, 1966) and can substantially improve the performance of a scrubber. However, there is a corresponding Stefan flow directed away from a surface at which water is evaporating, and this will tend to repel aerosol particles from the surface.

In addition to the deposition mechanisms themselves, methods for preliminary conditioning of aerosols may be used to increase the effectiveness of the deposition mechanisms subsequently applied. One such conditioning method consists of imposing on the gas high-intensity acoustic vibrations to cause collisions and flocculation of the aerosol particles, producing large particles that can be separated by simple inertial devices such as cyclones. This process, termed "sonic (or acoustic) agglomeration," has attained only limited commercial acceptance.

Another conditioning method, adaptable to scrubber systems, consists of inducing condensation of water vapor on the aerosol particles as nuclei, increasing the size of the particles and making them more susceptible to collection by inertial deposition.

Most forms of dust-collection equipment use more than one of the collection mechanisms, and in some instances the controlling mechanism may change when the collector is operated over a wide range of conditions. Consequently, collectors are most conveniently classified by type rather than according to the underlying mechanisms that may be operating.

PERFORMANCE OF DUST COLLECTORS

The performance of a dust collector is most commonly expressed as the collection efficiency η, the weight ratio of the dust collected to the dust entering the apparatus. However, the collection efficiency is usually related exponentially to the properties of the dust and gas and the operating conditions of most types of collectors and hence is an insensitive function of the collector operating conditions as its value approaches 1.0. Performance in the high-efficiency range is better expressed by the penetration $1 - \eta$, the weight ratio of the dust escaping to the dust entering. Particularly in reference to collection of radioactive aerosols, it is common to express performance in terms of the reciprocal of the penetration $1/(1 - \eta)$, which is termed the "decontamination factor." The number of transfer units N_t, which is equal to $\ln [1/(1 - \eta)]$ in the case of dust collection, was first proposed for use by Lapple (Wright, Stasny, and Lapple, "High Velocity Air Filters," WADC Tech. Rep. 55-457, ASTIA No. AD-142075, October 1957) and is more commonly used than the DI. Because of the exponential form of the relationship between efficiency and process variables for most dust collectors, the use of N_t (or DI) is particularly suitable for correlating collector performance data.

In comparing alternative collectors for a given service, a figure of merit is desirable for ranking the different devices. Since power consumption is one of the most important characteristics of a collector, the ratio of N_t to power consumption is a useful criterion. Another is the ratio of N_t to capital investment.

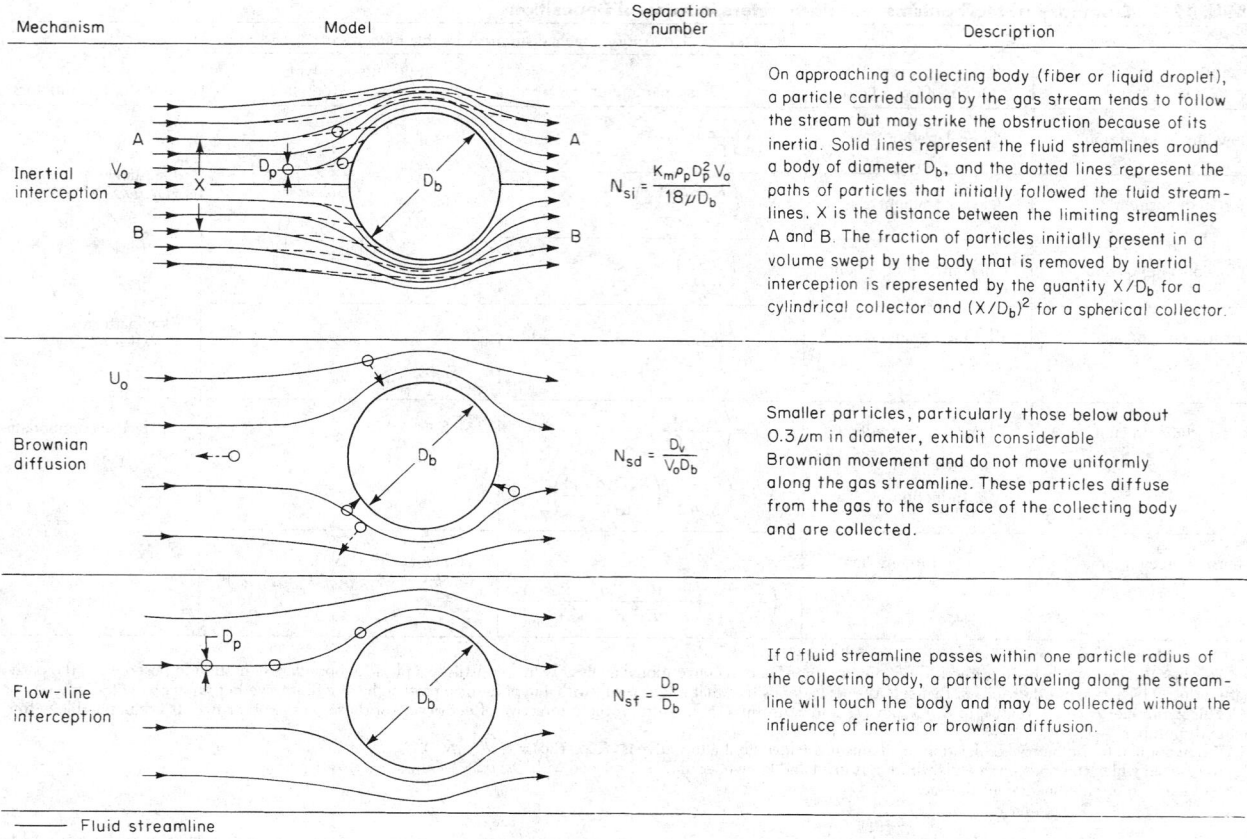

Mechanism	Model	Separation number	Description
Inertial interception		$N_{si} = \dfrac{K_m \rho_p D_p^2 V_0}{18 \mu D_b}$	On approaching a collecting body (fiber or liquid droplet), a particle carried along by the gas stream tends to follow the stream but may strike the obstruction because of its inertia. Solid lines represent the fluid streamlines around a body of diameter D_b, and the dotted lines represent the paths of particles that initially followed the fluid stream- lines. X is the distance between the limiting streamlines A and B. The fraction of particles initially present in a volume swept by the body that is removed by inertial interception is represented by the quantity X/D_b for a cylindrical collector and $(X/D_b)^2$ for a spherical collector.
Brownian diffusion		$N_{sd} = \dfrac{D_v}{V_0 D_b}$	Smaller particles, particularly those below about 0.3 μm in diameter, exhibit considerable Brownian movement and do not move uniformly along the gas streamline. These particles diffuse from the gas to the surface of the collecting body and are collected.
Flow-line interception		$N_{sf} = \dfrac{D_p}{D_b}$	If a fluid streamline passes within one particle radius of the collecting body, a particle traveling along the stream- line will touch the body and may be collected without the influence of inertia or brownian diffusion.

——— Fluid streamline
- - - - Particle path

FIG. 17-35 Particle deposition on collector bodies.

DUST-COLLECTOR DESIGN

In dust-collection equipment, most or all of the collection mechanisms may be operating simultaneously, their relative importance being determined by the particle and gas characteristics, the geometry of the equipment, and the fluid-flow pattern. Although the general case is exceedingly complex, it is usually possible in specific instances to determine which mechanism or mechanisms may be controlling. Nevertheless, the difficulty of theoretical treatment of dust-collection phenomena has made necessary simplifying assumptions, with the introduction of corresponding uncertainties. Theoretical studies have been hampered by a lack of adequate experimental techniques for verification of predictions. Although theoretical treatment of collector performance has been greatly expanded in the period since 1960, few of the resulting performance models have received adequate experimental confirmation because of experimental limitations.

The best-established models of collector performance are those for fibrous filters and fixed-bed granular filters, in which the structures and fluid-flow patterns are reasonably well defined. These devices are also adapted to small-scale testing under controlled laboratory conditions. Realistic modeling of full-scale electrostatic precipitators and scrubbers is incomparably more difficult. Confirmation of the models has been further limited by a lack of monodisperse aerosols that can be generated on a scale suitable for testing equipment of substantial sizes. When a polydisperse test dust is used, the particle-size distributions of the dust both entering and leaving a collector must be determined with extreme precision to avoid serious errors in the determination of the collection efficiency for a given particle size.

The design of industrial-scale collectors still rests essentially on empirical or semiempirical methods, although it is increasingly guided by concepts derived from theory. Existing theoretical models frequently embody constants that must be evaluated by experiment and that may actually compensate for deficiencies in the models.

DUST-COLLECTION EQUIPMENT

Gravity Settling Chambers The gravity settling chamber is probably the simplest and earliest type of dust-collection equipment, consisting of a chamber in which the gas velocity is reduced to enable dust to settle out by the action of gravity. Its simplicity lends it to almost any type of construction. Practically, however, its industrial utility is limited to removing particles larger than 325 mesh (43-μm diameter). For removing smaller particles, the required chamber size is generally excessive.

Gravity collectors are generally built in the form of long, empty, horizontal, rectangular chambers with an inlet at one end and an outlet at the side or top of the other end. By assuming a low degree of turbulence relative to the settling velocity of the dust particle in question, the performance of a gravity settling chamber is given by

$$\eta = \frac{u_t L_s}{H_s V_s} = \frac{u_t B_s L_s}{q} \qquad \text{(for } \eta \leqq 1.0) \qquad (17\text{-}1)$$

where V_s = average gas velocity. Expressing u_t in terms of particle size (equivalent spherical diameter), the smallest particle that can be completely separated out corresponds to $\eta = 1.0$ and, assuming Stokes' law, is given by

$$D_{p,\text{min}} = \sqrt{\frac{18\mu H_s V_s}{g_L L_s (\rho_s - \rho)}}$$

$$= \sqrt{\frac{18\mu q}{g_L B_s L_s (\rho_s - \rho)}} \qquad (17\text{-}2)$$

where ρ = gas density and ρ_s = particle density. For a given volumetric air-flow rate, collection efficiency depends on the total plan cross section of the chamber and is independent of the height. The height need be made only large enough so that the gas velocity V_s in the chamber is not so high as to cause reentrainment of separated dust. Generally V_s should not exceed about 3 m/s (10 ft/s).

Horizontal plates arranged as shelves within the chamber will give a marked improvement in collection. This arrangement is known as the Howard dust chamber (Fume Arrester, U.S. Patent 896,111, 1908). The disadvantage of the unit is the difficulty of cleaning owing to the close shelf spacing and warpage at elevated temperatures.

The pressure drop through a settling chamber is small, consisting primarily of entrance and exit losses. Because low gas velocities are used, the chamber is not subject to abrasion and may therefore be used as a precleaner to remove very coarse particles and thus minimize abrasion on subsequent equipment.

Impingement Separators   Impingement separators are a class of inertial separators in which particles are separated from the gas by inertial impingement on collecting bodies arrayed across the path of the gas stream, as shown on Fig. 17-35. Fibrous-pad inertial impingement separators for the collection of wet particles are the main application in current technology, as is described in Sec. 18 and Fig. 17-48. With the growing need for very high performance dust collectors, there is little application anymore for impingement collectors that catch large amounts of dry dust.

Cyclone Separators   The most widely used type of dust-collection equipment is the cyclone, in which dust-laden gas enters a cylindrical or conical chamber tangentially at one or more points and leaves through a central opening (Fig. 17-36). The dust particles, by virtue of their inertia, will tend to move toward the outside separator wall, from which they are led into a receiver. A cyclone is essentially a settling chamber in which gravitational acceleration is replaced by centrifugal acceleration. At operating conditions commonly employed, the centrifugal separating force or acceleration may range from 5 times gravity in very large diameter, low-resistance cyclones, to 2500 times gravity in very small, high-resistance units. The immediate entrance to a cyclone is usually rectangular.

Fields of Application   Within the range of their performance capabilities, cyclone collectors offer one of the least expensive means of dust collection from the standpoint of both investment and operation. Their major limitation is that unless very small units are used, their efficiency is low for collection of particles smaller than 5 μm. Although cyclones may be used to collect particles larger than 200 μm, gravity settling chambers or simple inertial separators (such as gas-reversal chambers) are usually satisfactory and less subject to abrasion. In special cases in which the dust is highly flocculated or high dust concentrations (over 230 g/m³, or 100 gr/ft³) are encountered, cyclones will remove dusts having small particle sizes. In certain instances efficiencies as high as 98 percent have been attained on dusts having ultimate particle sizes of 0.1 to 2.0 μm because of the predominant effect of flocculation.

Cyclones are used to remove both solids and liquids from gases and have been operated at temperatures as high as 1000°C and pressures as high as 50,700 kPa (500 atm).

Flow Pattern   In a cyclone the gas path involves a double vortex with the gas spiraling downward at the outside and upward at the inside. When the gas enters the cyclone, its velocity undergoes a redistribution so that the tangential component of velocity increases with decreasing radius as expressed by $V_{ct} \sim r^{-n}$. The spiral velocity in a cyclone may reach a value several times the average inlet-gas velocity. Theoretical considerations indicate that n should be equal to 1.0 in the absence of wall friction. Actual measurements [Shepherd and Lapple, *Ind. Eng. Chem.*, **31**, 972 (1939); **32**, 1246 (1940)], however, indicate that n may range from 0.5 to 0.7 over a large portion of the cyclone

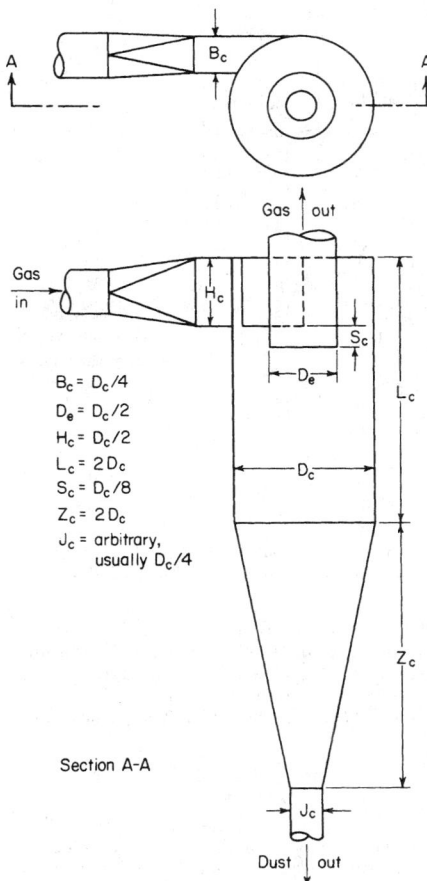

$B_c = D_c/4$
$D_e = D_c/2$
$H_c = D_c/2$
$L_c = 2 D_c$
$S_c = D_c/8$
$Z_c = 2 D_c$
J_c = arbitrary, usually $D_c/4$

Section A-A

FIG. 17-36   Cyclone-separator proportions.

radius. Ter Linden [*Inst. Mech. Eng. J.*, **160**, 235 (1949)] found n to be 0.52 for tangential velocities measured in the cylindrical portion of the cyclone at positions ranging from the radius of the gas-outlet pipe to the radius of the collector. Although the velocity approaches zero at the wall, the boundary layer is sufficiently thin that pitot-tube measurements show relatively high tangential velocities there, as shown in Fig. 17-37. The radial velocity V_r is directed toward the center throughout most of the cyclone, except at the center, where it is directed outward.

Superimposed on the "double spiral," there may be a "double eddy" [Van Tongeran, *Mech. Eng.*, **57**, 753 (1935); and Wellmann, *Feuerungstechnik*, **26**, 137 (1938)] similar to that encountered in pipe coils. Measurements on cyclones of the type shown in Fig. 17-36 indicate, however, that such double-eddy velocities are small compared with the spiral velocity (Shepherd and Lapple, op. cit.). Recent analyses of flow patterns can be found in Hoffman et al., *Powder Tech.*, **70**, 83 (1992); and Trefz and Muschelknautz, *Chem. Eng. Tech.*, **16**, 153 (1993).

Cyclone Efficiency   The methods described below for pressure drop and efficiency calculations were given by Zenz in *Manual on Disposal of Refinery Wastes—Atmospheric Emissions*, chap. 11 (1975), American Petroleum Institute Publ. 931 and improved by the Particulate Solid Research Inc. (PSRI), Chicago.

Cyclones work by using centrifugal force to increase the gravity field experienced by the solids. They then settle to the wall under the influence of their increased weight. Settling is improved as the path the solids traverse under centrifugal flow is increased. This path is

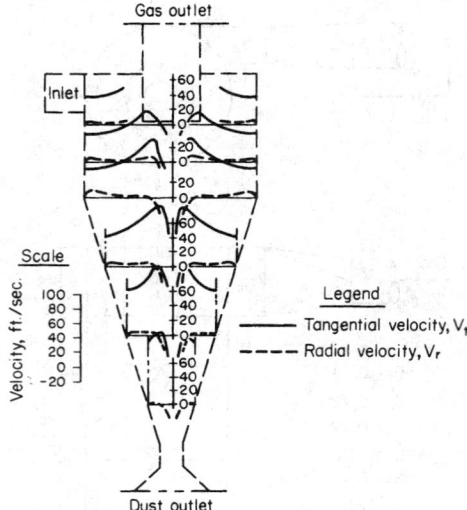

FIG. 17-37 Variation of tangential velocity and radial velocity at different points in a cyclone. [*Ter Linden,* Inst. Mech. Eng. J., **160**, 235 (1949).]

equated with the number of spirals the solids make in the cyclone barrel. Figure 17-38 gives the number of spirals N_s as a function of the maximum velocity in the cyclone. The maximum velocity may be seen either at the inlet or outlet depending on design.

The equation for D_{pth}, the theoretical size particle removed by the cyclone, is

$$D_{pth} = \sqrt{\frac{9\mu_g B_c}{\pi N_s v_{in}(\rho_p - \rho_g)}}$$

When consistent units are used, the particle size will either be in meters or feet. The equation contains effects of cyclone size, velocity, viscosity, and density of solids. In practice, a design curve as given in Fig. 17-39 uses D_{pth} as the size at which 50 percent of solids of a given size are collected by the cyclone. The material entering the cyclone is divided into fractional sizes, and the collection efficiency for each size is determined. The total efficiency of collection is the sum of the collection efficiencies of the cuts.

The above applies for very dilute systems, usually on the order of 1 grain/ft³, or 2.3 g/m³ where a grain equals 1/7000 of a pound. When an appreciable amount of solids are present, the efficiency increases

**Effective Number of Spiral Paths Taken
By the Gas Within the Body of a Cyclone**

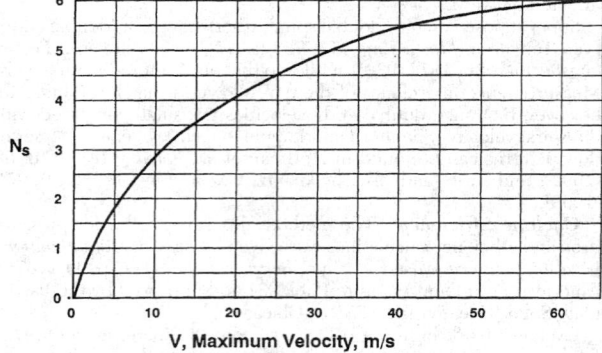

FIG. 17-38 N_s versus velocity—where the larger of either the inlet or outlet velocity is used.

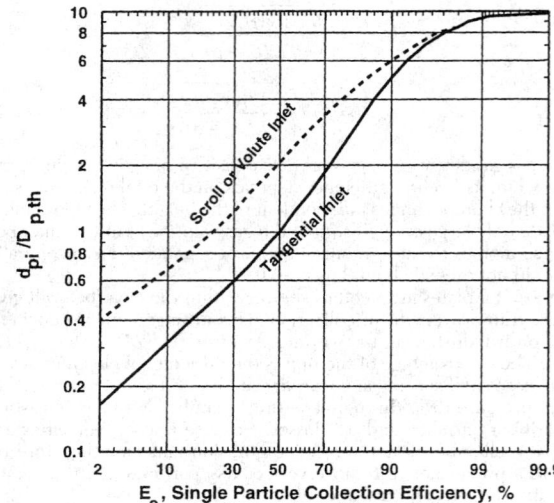

FIG. 17-39 Single particle collection efficiency curve. (*Courtesy of PSRI, Chicago.*)

dramatically. This may be due to the coarse particles colliding with fines as they settle, which takes the fines to the wall more quickly. Other explanations are that the solids have a lower drag coefficient or tend to flocculate in multiparticle environments. At very high loadings, it is believed the gas simply cannot hold that much solid material in suspension at high gravities, and the bulk of the solids simply "condenses" out of the gas stream.

The phenomenon is represented by Figs. 17-40 and 17-41 for Geldart-type A and B solids, respectively (see beginning of Sec. 17). The initial efficiency of a particle size cut is found on the chart, and the parametric line is followed to the proper overall solids loading. The efficiency for that cut size is then read from the graph.

Pressure Drop Pressure drop is first determined by summing five flow losses associated with the cyclone.

1. Inlet contraction,

$$\Delta P = 0.5\rho_g(v_{in}^2 - v_{vessel}^2 + Kv_{in}^2)$$

where K is taken from Table 17-3. Using SI units gives the pressure drop in Pa. In English units, the factor of 32.2 for g must be included. This loss is primarily associated with cyclones inside a vessel. If the cyclone is connected outside a vessel, the dp tap may measure acceleration, and this term should not be used for total dp.

TABLE 17-3   K versus Area Ratio

Area ratio	K
0	.50
0.1	.47
0.2	.43
0.3	.395
0.4	.35

2. Particle acceleration,

$$\Delta P = Lv_{in}(v_{pin} - v_{pvessel})$$

For small particles, the velocity is taken as equal to the gas velocity and L is the loading, kg/m³.

3. Barrel friction. The inlet diameter d_{in} is taken as 4 × (inlet area)/ inlet perimeter. Then,

$$\Delta P = \frac{2f\rho_g v_{in}^2 \pi D_c N_s}{d_{in}}$$

where the Reynolds number for determining the friction factor f is based on the inlet area.

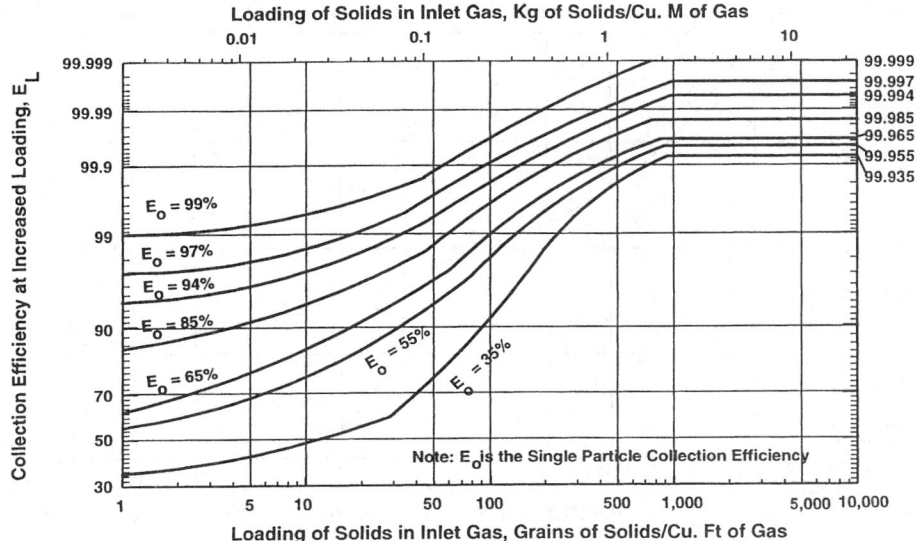

FIG. 17-40   Effect of inlet loading on collection efficiency for Geldart Group A and Group C particles. (*Courtesy of PSRI, Chicago.*)

4.   Gas flow reversal,

$$\Delta P = \frac{\rho_g v_{in}^2}{2}$$

5.   Exit contraction,

$$\Delta P = 0.5\rho_g(v_{exit}^2 - v_c^2 + Kv_{exit}^2)$$

where K is determined from Table 17-3 based on the area ratio of barrel and exit tube of the cyclone.

The total pressure drop is the sum of the 5 individual pressure drops.

However, the actual pressure drop observed turns out to be a function of the solids loading. The pressure drop is high when the gas is free of solids and then *decreases* as the solids loading increases up to about 3 kg/m³ (0.2 lb/ft³). The cyclone *dp* then begins to increase with

loading. The cause of the initial decline is that the presence of solids decreases the tangential velocity of the gas [Yuu, *Chem. Eng. Sci.,* **33,** 1573 (1978)]. Figure 17-42 gives the actual pressure drop based on the loading. When solids are absent, the observed pressure drop can be 2.5 times the calculated pressure drop.

Cyclone Design Factors   Cyclones are generally designed to meet specified pressure-drop limitations. For ordinary installations, operating at approximately atmospheric pressure, fan limitations generally dictate a maximum allowable pressure drop corresponding to a cyclone inlet velocity in the range of 8 to 30 m/s (25 to 100 ft/s). Consequently, cyclones are usually designed for an inlet velocity of 15 m/s (50 ft/s), though this need not be strictly adhered to.

In the removal of dusts, collection efficiency can be changed by only a relatively small amount by a variation in operating conditions.

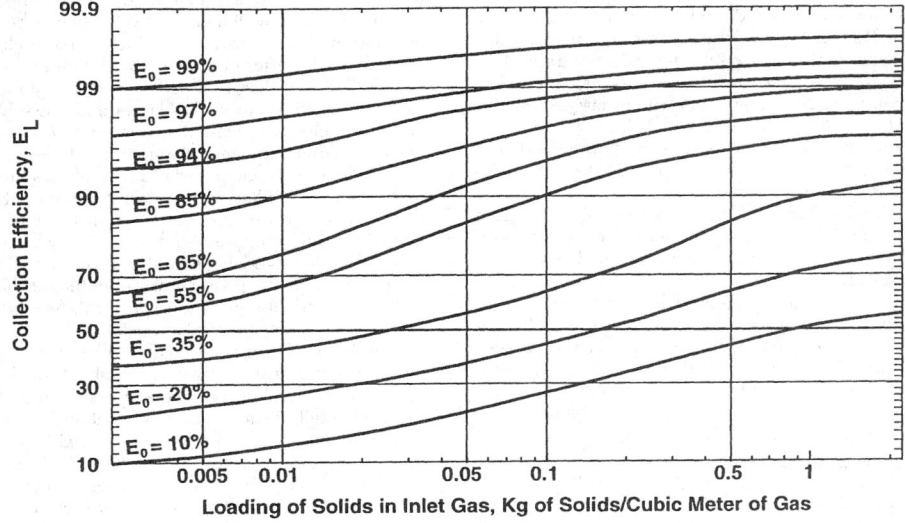

FIG. 17-41   Effect of inlet loading on collection efficiency (Geldart Group B and Group D) particles. (*Courtesy of PSRI, Chicago.*)

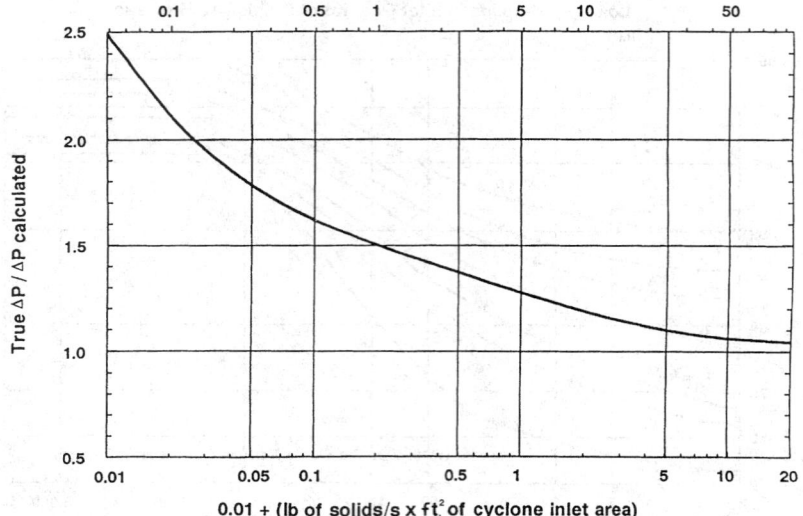

FIG. 17-42   Effect of cyclone inlet loading on pressure drop. (*Courtesy of PSRI, Chicago.*)

The primary design factor that can be utilized to control collection efficiency is the cyclone diameter, a smaller-diameter unit operating at a fixed pressure drop having the higher efficiency [Anderson, *Chem. Metall.*, **40**, 525 (1933); Drijver, *Wärme*, **60**, 333 (1937); and Whiton, *Power*, **75**, 344 (1932); *Chem. Metall.*, **39**, 150 (1932)]. Small-diameter cyclones, however, will require a multiple of units in parallel for a specified capacity. In such cases the individual cyclones can discharge the dust into a common receiving hopper [Whiton, *Trans. Am. Soc. Mech. Eng.*, **63**, 213 (1941)]. The final design involves a compromise between collection efficiency and complexity of equipment. It is customary to design a single cyclone for a given capacity, resorting to multiple parallel units only if the predicted collection efficiency is inadequate for a single unit.

Reducing the gas outlet diameter will increase both collection efficiency and pressure drop. To exit the cyclone, gas must enter the cyclonic flow associated with the outlet tube. If the outlet diameter is reduced, the outlet vortex increases in length to compensate. Therefore, when the outlet area is less than the inlet area, the length of the cyclone must increase. Too short a cyclone is associated with erosion of the cone and reentrainment of solids into the exit flow. Table 17-4 below gives this increase as a function of outlet-to-inlet area. The length is measured centrally along a cylinder 10 cm larger than the inner diameter of the outlet tube to prevent interference with the cone. If the cone interferes, the barrel must be lengthened. The minimum cone angle should be 60° or greater with steeper angles appropriate to materials that are cohesive.

The inlet is usually rectangular and sometimes circular. In either case, projection of the flow path should never interfere with the outlet tube. If a very heavy solids loading is anticipated, the barrel diameter should be increased slightly.

Collection efficiency is normally increased by increasing the gas throughput (Drijver, op. cit.). However, if the entering dust is flocculated, increased gas velocities may cause deflocculation in the cyclone, so that efficiency remains the same or actually decreases. Also, variations in design proportions that result in increased collection effi-

ciency with dispersed dusts may be detrimental with flocculated dusts. Kalen and Zenz [*Am. Inst. Chem. Eng. Symp. Ser.*, **70** (137), 388 (1974)] report that collection efficiency increases with increasing gas inlet velocity up to a minimum tangential velocity at which dust is either reentrained or not deposited because of saltation. Koch and Licht [*Chem. Eng.*, **84**(24), 80 (1977)] estimate that for typical cyclones the saltation velocity is consistent with cyclone inlet velocities in the range of 15 to 27 m/s (50 to 90 ft/s). C. E. Lapple (private communication) reports that in cyclone tests with talc dust collection efficiency increased steadily as the inlet velocity was increased up to the maximum of 52 m/s (170 ft/s). With ilmenite dust, which was much more strongly flocculated, efficiency decreased over the same inlet-velocity range. In later experiments with well-dispersed talc dust, collection efficiency continued to increase at inlet velocities up to the maximum used, 82 m/s (270 ft/s). Another effect of increasing the cyclone inlet gas velocity is that friable materials may disintegrate as they hit the cyclone wall at high velocity. Thus, the increase in efficiency associated with increased velocity may be more than lost due to generation of fine attrited material that the cyclone cannot contain.

Cyclones in series may be justified under some circumstances:

1. The dust has a broad size distribution, including particles under 10 to 15 μm as well as larger and possibly abrasive particles. A large low-velocity cyclone may be used to remove the coarse particles ahead of a unit with small-diameter multiple tubes.

2. The dust is composed of fine particles but is highly flocculated or tends to flocculate in preceding equipment and in the cyclones themselves. Efficiencies predicted on the basis of ultimate particle size will be highly conservative.

3. The dust is relatively uniform, and the efficiency of the second-stage cyclone is not greatly lower than that of the first stage.

4. Dependable operation is critical. Second-stage or even third-stage cyclones may be used as backup.

A cyclone will operate equally well on the suction or pressure side of a fan if the dust receiver is airtight. Probably the greatest single cause for poor cyclone performance, however, is the leakage of air into the dust outlet of the cyclone. A slight air leak at this point can result in a tremendous drop in collection efficiency, particularly with fine dusts. For a cyclone under pressure, air leakage at this point is objectionable primarily because of the local dust nuisance created. For batch operation, an airtight hopper or receiver may be used. For continuous withdrawal of collected dust, a rotary star valve, a double-lock valve, or a screw conveyor may be used, the latter only with fine dusts. A collapsible open-ended rubber tube can be used for cyclones operating under slight negative pressure; mechanical flapgate valves and

TABLE 17-4   Required Cyclone Length as a Function of Area Ratio

A_{out}/A_{in}	Length below outlet tube/D_c
>1.0	2.0
0.8	2.2
0.6	2.6
0.4	3.2

fluidized seal legs can be used (see "Fluidized-Bed Systems: Solids Discharge"). Special pneumatic unloading devices can also be used with dusts. In any case it is essential that sufficient unloading and receiving capacity be provided to prevent collected material from accumulating in the cyclone.

Generally cone-and-disk baffles, helical guide vanes, etc., placed inside a cyclone, will have a detrimental effect on performance. A few of these devices do have some merit, however, under special circumstances. Although an inlet vane will reduce pressure drop, it causes a correspondingly greater reduction in collection efficiency. Its use is recommended only when collection efficiency is normally so high as to be a secondary consideration and when it is desired to decrease the resistance of an existing cyclone system for purposes of increased air-handling capacity or when floor-space or headroom requirements are controlling factors. If an inlet vane is used, it is advantageous to increase the gas-exit-duct length inside the cyclone chamber. A disk or cone baffle located beneath the gas-outlet duct may be beneficial if air in-leakage at the dust outlet cannot be avoided. A heavy chain suspended from the gas-outlet duct has been found beneficial to minimize dust buildup on the cyclone walls. Such a chain should be suspended from a swivel so that it is free to rotate without twisting. At present there are no known devices that will recover the gas spiral-velocity energy in the gas-outlet duct. Substantially all devices that

have been reported to reduce pressure drop do so by reducing spiral velocities in the cyclone chamber and consequently result in reduced collection efficiency.

At low dust loadings the pressure in the dust receiver of a single cyclone will generally be lower than in the gas-outlet duct. Increased dust loadings will increase the pressure in the dust receiver. Such devices as cones, disks, and inlet vanes will generally cause the pressure in the dust receiver to exceed that in the gas-outlet duct. A cyclone will operate as well in a horizontal position as in a vertical position. However, departure from the normal vertical position results in an increasing tendency to plug the dust outlet. If the dust outlet becomes plugged, collection efficiency will, of course, be low. If the cyclone exit duct must be reduced to tie in with proposed duct sizes, the transition should be made at least five diameters downstream from the cyclone and preferably after a bend. In the event that the transition must be made closer to the cyclone, a Greek cross should be installed in the transition piece in order to avoid excessive pressure drop.

Commercial Equipment   Simple cyclones are available in a wide variety of shapes ranging from long, slender units to short, large-diameter units. The body may be conical or cylindrical, and entrances may be involute or tangential and round or rectangular.

In Fig. 17-43 are shown some of the special types of commercial cyclones. In the Multiclone a spiral motion is imparted to the gas by

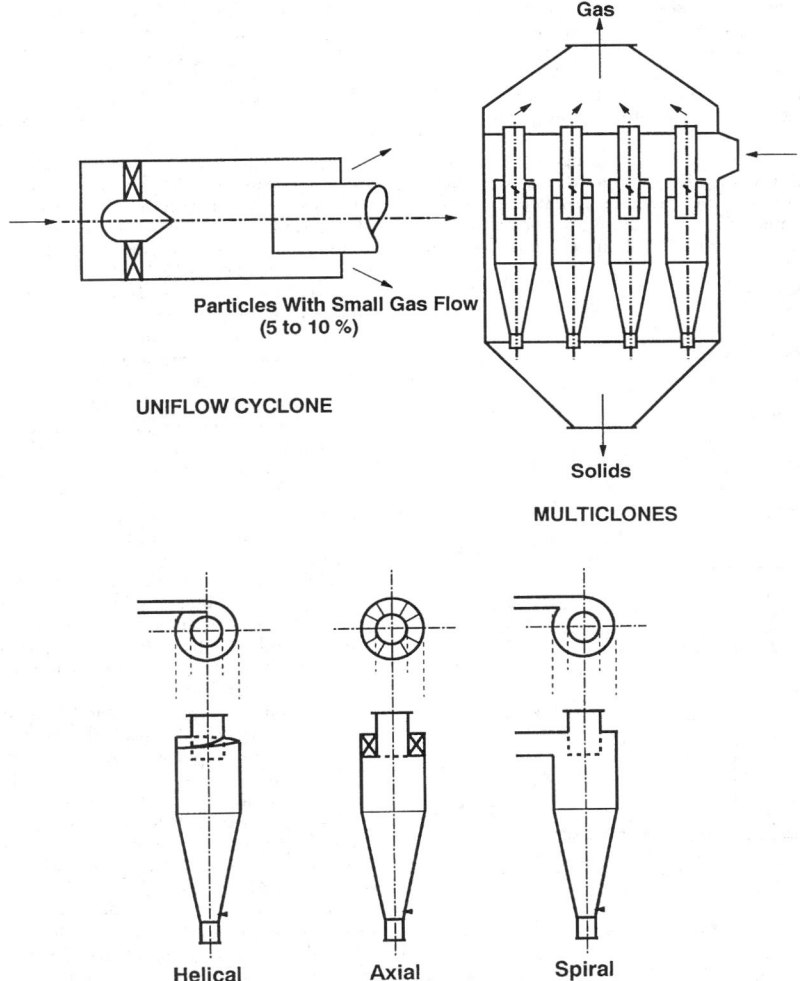

FIG. 17-43   Some commercial cyclone design alternatives. (*Courtesy of PSRI, Chicago.*)

annular vanes, and it is furnished in multiple units of 15.2- and 22.9-cm (6- and 9-in) diameter. Its largest field of application has been in the collection of fly ash from steam boilers. The tubes are commonly constructed of cast iron and other abrasion-resistant alloys.

In addition to the conventional reverse-flow cyclones, some use is made of uniflow, or straight-through, cyclones, in which the gas and solids discharge at the same end (Fig. 17-44). These devices act as concentrators; the concentrated dust, together with 5 to 20 percent of the inlet gas, is discharged at the periphery, while the clean gas passes out axially. The purge gas and concentrated dust enter a conventional cyclone for final separation. The straight-through cyclones are usually multiple-tube units.

Mechanical Centrifugal Separators   A number of collectors in which the centrifugal field is supplied by a rotating member are commercially available. In the typical unit shown in Fig. 17-45, the exhauster or fan and dust collector are combined as a single unit. The blades are especially shaped to direct the separated dust into an annular slot leading to the collection hopper while the cleaned gas continues to the scroll.

Although no comparative data are available, the collection efficiency of units of this type is probably comparable with that of the single-unit high-pressure-drop-cyclone installation. The clearances are smaller and the centrifugal fields higher than in a cyclone, but these advantages are probably compensated for by the shorter gas path and the greater degree of turbulence with its inherent reentrainment tendency. The chief advantage of these units lies in their compactness, which may be a prime consideration for large installations or plants requiring a large number of individual collectors. Caution should be exercised when attempting to apply this type of unit to a dust that shows a marked tendency to build up on solid surfaces, because of the high maintenance costs that may be encountered from plugging and rotor unbalancing.

Particulate Scrubbers   Wet collectors, or scrubbers, form a class of devices in which a liquid (usually water) is used to assist or accomplish the collection of dusts or mists. Such devices have been in use for well over 100 years, and innumerable designs have been or are offered commercially or constructed by users. Wet-film collectors logically

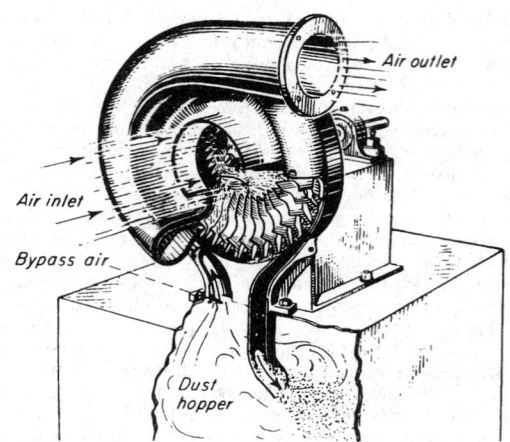

FIG. 17-45   Typical mechanical centrifugal separator. Type D Rotoclone (cutaway view). (*American Air Filter Co., Inc.*)

form a separate subcategory of devices. They comprise inertial collectors in which a film of liquid flows over the interior surfaces, preventing reentrainment of dust particles and flushing away the deposited dust. Wetted-wall cyclones are an example [Stairmand, *Trans. Inst. Chem. Eng.*, **29**, 356 (1951)]. Wet-film collectors have not been studied systematically but can probably be expected to perform much as do equivalent dry inertial collectors, except for the benefit of reduced reentrainment.

In particulate scrubbers, the liquid is dispersed into the gas as a spray, and the liquid droplets are the principal collectors for the dust particles. Depending on their design and operating conditions, particulate scrubbers can be adapted to collecting fine as well as coarse particles. Collection of particles by the drops follows the same principles illustrated in Fig. 17-35. Various investigations of the relative contributions of the various mechanisms have led to the conclusion that the predominant mechanism is inertial deposition. Flow-line interception is only a minor mechanism in the collection of the finer dust particles by liquid droplets of the sizes encountered in scrubbers. Diffusion is indicated to be a relatively minor mechanism for the particles larger than 0.1 μm that are of principal concern. Thermal deposition is negligible. Gravitational settling is ineffective because of the high gas velocities and short residence times used in scrubbers. Electrostatic deposition is unlikely to be important except in cases in which the dust particles or the water, or both, are being deliberately charged from an external power source to enhance collection. Deposition produced by Stefan flow can be significant when water vapor is condensing in a scrubber.

Despite numerous claims or speculations that wetting of dust particles by the scrubbing liquid plays a major role in the collection process, there is no unequivocal evidence that this is the case. The issue is whether wetting is an important factor in the adherence of a particle to a collecting droplet upon impact. From the body of general experience, it can be inferred that wettable particles probably are not collected much, if any, more readily than nonwettable particles of the same size. However, the available experimental techniques have not been adequate to permit any direct test to resolve the question. Changing from a wettable to a nonwettable test aerosol or from one scrubbing liquid to another is virtually certain to introduce other (and possibly unknown) factors into the scrubbing process. The most informative experimental studies appear to be some by Weber [*Staub*, English trans., **28**, 37 (November 1968); **29**, 12 (July 1969)], who bombarded single drops of various liquids with dust particles at different velocities and studied the behavior at impact by means of high-speed photography. Dust particles hitting the drops were invariably retained by the latter, regardless of their wettability by the liquid used.

The use of wetting agents in scrubbing water is equally controversial, and there has been no clear demonstration that it is beneficial.

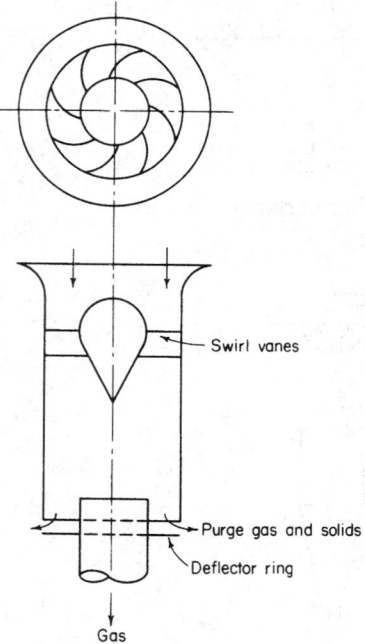

FIG. 17-44   Uniflow cyclone. [*Ter Linden*, Inst. Mech. Eng. J., **160**, 233 (1949).]

A particulate scrubber may be considered as consisting of two parts: (1) a contactor stage, in which a spray is generated and the dust-laden gas stream is brought into contact with it; and (2) an entrainment separation stage, in which the spray and deposited dust particles are separated from the cleaned gas. These two stages may be separate or physically combined. The contactor stage may be of any form intended to bring about effective contacting of the gas and spray. The spray may be generated by the flow of the gas itself in contact with the liquid, by spray nozzles (pressure-atomizing or pneumatic-atomizing), by a motor-driven mechanical spray generator, or by a motor-driven rotor through which both gas and liquid pass.

Entrainment separation is accomplished with inertial separators, which are usually cyclones or impingement separators of various forms. If properly designed, these devices can remove virtually all droplets of the sizes produced in scrubbers. However, reentrainment of liquid can take place in poorly designed or overloaded separators.

Scrubber Types and Performance   The diversity of particulate scrubber designs is so great as to defy any detailed and self-consistent system of classification based on configuration or principle of operation. However, it is convenient to characterize scrubbers loosely according to prominent constructional features, even though the modes of operation of different devices in a group may vary widely.

A relationship of power consumption to collection efficiency is characteristic of all particulate scrubbers. Attaining increased efficiency requires increased power consumption, and the power consumption required to attain a given efficiency increases as the particle size of the dust decreases. Experience generally indicates that the power consumption required to provide a specific efficiency on a given dust does not vary widely even with markedly different devices. The extent to which this generalization holds true has not been fully explored, but the known extent is sufficient to suggest that the underlying collection mechanism may be essentially the same in all types of particulate scrubbers.

Since some relationship of power consumption to performance appears to be a universal characteristic of particulate scrubbers, it is useful to characterize such devices broadly according to the source from which the energy is supplied to the gas-liquid contacting process. The energy may be drawn from (1) the gas stream itself, (2) the liquid stream, or (3) a motor driving a rotor. For convenience, devices in these classes may be termed respectively (1) gas-atomized spray scrubbers, (2) spray, or preformed-spray, scrubbers, and (3) mechanical scrubbers. In the spray scrubbers, all the energy may be supplied from the liquid, using a pressure nozzle, but some or all may be provided by compressed air or steam in a two-fluid nozzle or by a motor driving a spray generator.

Particulate scrubbers may also be classed broadly into low-energy and high-energy scrubbers. The distinction between the two classes is arbitrary, since the devices are not basically different and the same device may fall into either class depending on the amount of power it consumes. However, some differences in configuration are sometimes necessary to adapt a device for high-energy service. No specific level of power consumption is commonly agreed upon as the boundary between the two classes, but high-energy scrubbers may be regarded as those using sufficient power to give substantial efficiencies on submicrometer particles.

Scrubber Performance Models   A number of investigators have made theoretical studies of the performance of venturi scrubbers and have sought to produce performance models, based on first principles, that can be used to design a unit for a given duty without recourse to experimental data other than the particle size and size distribution of the dust. Among these workers are Calvert [*Am. Inst. Chem. Eng. J.*, **16**, 392 (1970); *J. Air Pollut. Control Assoc.*, **24**, 929 (1974)], Boll [*Ind. Eng. Chem. Fundam.*, **12**, 40 (1973)], Goel and Hollands [*Atmos. Environ.*, **11**, 837 (1977)], and Yung et al. [*Environ. Sci. Technol.*, **12**, 456 (1978)]. Comparatively few efforts have been made to model the performance of scrubbers of types other than the venturi, and a number of such models are summarized by Yung and Calvert (U.S. EPA-600/8-78-005b, 1978).

The various venturi-scrubber models embody a variety of assumptions and approximations. The solutions of the equations for particulate collection must in general be determined numerically, although

Calvert et al. [*J. Air Pollut. Control Assoc.*, **22**, 529 (1972)] obtained an explicit equation by making some simplifying assumptions and incorporating an empirical constant that must be evaluated experimentally; the constant may absorb some of the deficiencies in the model. Although other models avoid direct incorporation of empirical constants, use of empirical relationships is necessary to obtain specific estimates of scrubber collection efficiency. One of the areas of greatest uncertainty is the estimation of droplet size.

Most of the investigators have assumed the effective drop size of the spray to be the Sauter (surface-mean) diameter and have used the empirical equation of Nukiyama and Tanasawa [*Trans. Soc. Mech. Eng., Japan,* **5**, 63 (1939)] to estimate the Sauter diameter:

$$D_o = \frac{1920\sqrt{\sigma_L}}{V_o\sqrt{\rho_L/62.3}} + 75.4 \left(\frac{\mu_L}{\sqrt{\sigma_L \rho_L/62.3}} \right)^{0.45} \left(\frac{1000 Q_L}{Q_G} \right)^{1.5} \quad (17\text{-}3)$$

where D_o = drop diameter, μm; V_o = gas velocity, ft/s; σ_L = liquid surface tension, dyn/cm; ρ_L = liquid density, lb/ft³; μ_L = liquid viscosity, cP; Q_L = liquid flow rate, ft³/s; and Q_G = gas flow rate, ft³/s.

The Nukiyama-Tanasawa equation, which is not dimensionally homogeneous, was derived from experiments with small, internal-mix pneumatic atomizing nozzles with concentric feed of air and liquid (Lapple et al., "Atomization: A Survey and Critique of the Literature," Stanford Res. Inst. Tech. Rep. No. 6, AD 821-314, 1967; Lapple, "Atomization," in *McGraw-Hill Encyclopedia of Science and Technology,* 5th ed., vol. 1, McGraw-Hill, New York, 1982, p. 858). The effect of nozzle size on the drop size is undefined. Even within the range of parameters for which the relationship was derived, the drop sizes reported by various investigators have varied by twofold to threefold from those predicted by the equation (Boll, op. cit.). The Nukiyama-Tanasawa equation has, nevertheless, been applied to large venturi and orifice scrubbers with configurations radically different from those of the atomizing nozzles for which the equation was originally developed.

Primarily because of the lack of adequate experimental techniques (particularly, the production of appropriate monodisperse aerosols), there has been no comprehensive experimental test of any of the venturi-scrubber models over wide ranges of design and operating variables. The models for other types of scrubbers appear to be essentially untested.

Contacting Power Correlation   A scrubber design method that has achieved wide acceptance and use is based on correlation of the collection efficiency with the power dissipated in the gas-liquid contacting process, which is termed "contacting power." The method originated from an investigation by Lapple and Kamack [*Chem. Eng. Prog.*, **51**, 110 (1955)] and has been extended and refined in a series of papers by Semrau and coworkers [*Ind. Eng. Chem.*, **50**, 1615 (1958); *J. Air Pollut. Control Assoc.*, **10**, 200 (1960); **13**, 587 (1963); U.S. EPA-650/2-74-108, 1974; U.S. EPA-650/2-77-234, 1977; *Chem. Eng.*, **84**(20), 87 (1977); and "Performance of Particulate Scrubbers as Influenced by Gas-Liquid Contactor Design and by Dust Flocculation," EPA-600/9-82-005c, 1982, p. 43]. Other workers have made extensive independent studies of the correlation method [Walker and Hall, *J. Air Pollut. Control Assoc.*, **18**, 319 (1968)], and numerous studies of narrower scope have been made. The major conclusion from these studies is that the collection efficiency of a scrubber on a given dust is essentially dependent only on the contacting power and is affected to only a minor degree by the size or geometry of the scrubber or by the way in which the contacting power is applied. This contacting-power rule is strictly empirical, and the full extent of its validity has still not been explored. It has been best verified for the class of gas-atomized spray scrubbers, in which the contacting power is derived from the gas stream and takes the form of gas pressure drop. Tests of the equivalence of contacting power supplied from the liquid stream in pressure spray nozzles have been far less extensive and are strongly indicative but not yet conclusive. Evidence for the equivalence of contacting power from mechanically driven devices is also indicative but extremely limited in quantity.

Contacting power is defined as the power per unit of volumetric gas flow rate that is dissipated in gas-liquid contacting and is ultimately converted to heat. In the simplest case, in which all the energy is obtained from the gas stream in the form of pressure drop, the contacting power is equivalent to the friction loss across the wetted equipment, which is termed "effective friction loss," F_E. The pressure drop may reflect kinetic-energy changes rather than energy dissipation, and pressure drops that result solely from kinetic-energy changes in the gas stream do not correlate with performance. Likewise, any friction losses taking place across equipment that is operating dry do not contribute to gas-liquid contacting and do not correlate with performance. The gross power input to a scrubber includes losses in motors, drive shafts, fans, and pumps that obviously should be unrelated to scrubber performance.

The effective friction loss, or "gas-phase contacting power," is easily determined by direct measurements. However, the "liquid-phase contacting power," supplied from the stream of scrubbing liquid, and the "mechanical contacting power," supplied by a mechanically driven rotor, are not directly measurable; the theoretical power inputs can be estimated, but the portions of these quantities effectively converted to contacting power can only be inferred from comparison with gas-phase contacting power. Such data as are available indicate that the contributions of contacting power from different sources are directly additive in their relation to scrubber performance.

Contacting power is variously expressed in units of MJ/1000 m³ (SI), kWh/1000 m³ (meter-kilogram-second system), and hp/(1000 ft³/min) (U.S. customary). Relationships for conversion to SI units are

$$1.0 \text{ kWh/1000 m}^3 = 3.60 \text{ MJ/1000 m}^3$$

$$1.0 \text{ hp/(1000 ft}^3\text{/min)} = 1.58 \text{ MJ/1000 m}^3$$

The gas-phase contacting power P_G may be calculated from the effective friction loss by the following relationships:
SI units:

$$P_G = 1.0F_E \qquad (17\text{-}4)$$

where F_E = kPa.
U.S. customary units:

$$P_G = 0.1575F_E \qquad (17\text{-}5)$$

where F_E = in of water.

The power input from a liquid stream injected with a hydraulic spray nozzle may usually be taken as approximately equal to the product of the nozzle feed pressure p_F and the volumetric liquid rate. The liquid-phase contacting power P_L may then be calculated from the following formulas:
SI units:

$$P_L = 1.0p_F(Q_L/Q_G) \qquad (17\text{-}6)$$

where p_F = kPa gauge, and Q_L and Q_G = m³/s.
U.S. customary units:

$$P_L = 0.583p_F(Q_L/Q_G) \qquad (17\text{-}7)$$

where p_G = lbf/in² gauge, Q_L = gal/min, and Q_G = ft³/min.

The correlation of efficiency data is based on the total contacting power P_T, which is the sum of P_G, P_L, and any power P_M that may be supplied mechanically by a power-driven rotor.

In general, the liquid-to-gas ratio does not have an influence independent of contacting power on the collection efficiency of scrubbers of the venturi type. This is true at least of operation with liquid-to-gas ratios above some critical lower value. However, several investigations [Semrau and Lunn, "Performance of Particulate Scrubbers as Influenced by Gas-Liquid Contactor Design and by Dust Flocculation," EPA-600/9-82-005c, 1982, p. 43; and Muir et al., *Filtr. Sep.*, **15**, 332 (1978)] have shown that at low liquid-to-gas ratios relatively poor efficiencies may be obtained at a given contacting power. Such regions of operation are obviously to be avoided.

It has sometimes been asserted that multiple gas-liquid contactors in series will give higher efficiencies at a given contacting power than will a single contacting stage. However, there is little experimental evidence to support this contention. Lapple and Kamack (op. cit.) obtained slightly higher efficiencies with a venturi and an orifice in series than they did with a venturi alone. Muir and Mihisei [*Atmos. Environ.*, **13**, 1187 (1979)] obtained somewhat higher efficiencies on two redispersed dusts when using two venturis in series rather than

one. The improvement obtained with two-stage scrubbing was greatest with the coarser of the two dusts and was relatively small with the finer dust. Flocculation or deflocculation of the dusts may have been responsible for some of the behavior encountered. Semrau et al. (EPA-600/2-77-234, 1977) compared the performance of a four-stage multiple-orifice contactor with that of a single-orifice contactor, using well-dispersed aerosols generated from ammonium fluorescein. The multiple-orifice contactor gave about the same efficiency as the single-orifice in the upper range of contacting power but lower efficiencies in the lower range. The deviations in performance in this case were probably characteristic of the particular multiple-orifice contactor rather than of multistage contacting as such.

Most scrubbers actually incorporate more than one stage of gas-liquid contacting even though these may not be identical (e.g., the contactor and the entrainment separator). The preponderance of evidence indicates that multiple-stage contacting is not inherently either more or less efficient than single-stage contacting. However, two-stage contacting may have practical benefits in dealing with abrasive or flocculated dusts.

Some investigators have proposed, mostly on the basis of mathematical modeling, to optimize the design of scrubbers to obtain a given efficiency with a minimum power consumption (e.g., Goel and Hollands, op. cit.). In fact, no optimum in performance appears to exist; apart from some avoidable regions of unfavorable operation, increased contacting power yields increased efficiency.

Scrubber Performance Curves   The scrubber performance curve, which shows the relationship of scrubber efficiency to the contacting power, has been found to take the form

$$N_t = \alpha P_T^\gamma \qquad (17\text{-}8)$$

where α and γ are empirical constants that depend primarily on the aerosol (dust or mist) collected. In a log-log plot of N_t versus P_T, γ is the slope of the performance curve and α is the intercept at $P_T = 1$. Figure 17-46 shows such a performance curve for the collection of coal fly ash by a pilot-plant venturi scrubber (Raben "Use of Scrubbers for Control of Emissions from Power Boilers," United States–U.S.S.R. Symposium on Control of Fine-Particulate Emissions from Industrial Sources, San Francisco, 1974). The scatter in the data reflects not merely experimental errors but actual variations in the particle-size characteristics of the dust. Because the characteristics of an industrial dust vary with time, the scrubber performance curve necessarily must represent an average material, and the scatter in the data is frequently greater than is shown in Fig. 17-46. For best definition, the curve should cover as wide a range of contacting power as possible. Obtaining the data thus requires pilot-plant equipment with the flexibility to operate over a wide range of conditions. Because scrubber performance is not greatly affected by the size of the unit, it is feasible to conduct the tests with a unit handling no more than 170 m³/h (100 ft³/min) of gas.

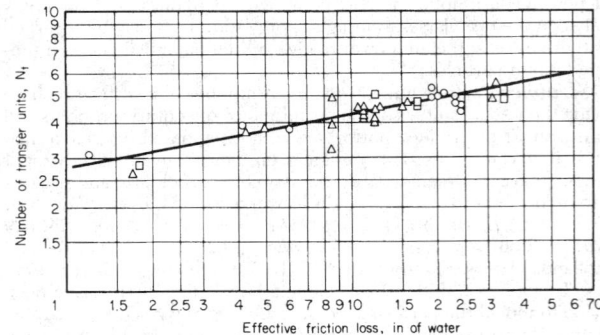

FIG. 17-46   Performance of pilot-plant venturi scrubber on fly ash. Liquid-to-gas ratio, gal/1000 ft³: ○, 10; △, 15; □, 20. (*Raben, United States–U.S.S.R. Symposium on Fine-Particulate Emissions from Industrial Sources, San Francisco, 1974.*)

A clear interpretation of γ, the slope of the curve, is still lacking. Presumably, it should be related to the particle-size distribution of the dust. Because scrubbing preferentially removes the coarser particles, the fraction of the dust removed (or the increment of N_t) per unit of contacting power should decrease as the contacting power and efficiency increase, so that the value of γ should be less than unity. In fact, the value of γ has been less than unity for most dusts. Nevertheless, some data in the literature have displayed values of γ greater than unity when plotted on the transfer-unit basis, indicating that the residual fraction of dust became more readily collectible as contacting power and efficiency increased. More recent studies by Semrau et al. (EPA-650/2-74-108 and EPA-600/2-77-237) have revealed performance curves having two branches (typified by Fig. 17-47), the lower having a slope greater than unity and the upper a slope less than unity. This suggests that had the earlier tests been extended into higher contacting-power ranges, performance curves with flatter slopes might have appeared in those ranges.

Among the aerosols that gave performance curves with γ > 1, the only obvious common characteristic was that a large fraction of each was composed of submicrometer particles.

Cut-Power Correlation Another design method, also based on scrubber power consumption, is the cut-power method of Calvert [*J. Air Pollut. Control Assoc.*, **24**, 929 (1974); *Chem. Eng.*, **84**(18), 54 (1977)]. In this approach, the cut diameter (the particle diameter for which the collection efficiency is 50 percent) is given as a function of the gas pressure drop or of the power input per unit of volumetric gas flow rate. The functional relationship is presented as a log-log plot of the cut diameter versus the pressure drop (or power input). In principle, the function could be constructed by experimentally determining scrubber performance curves for discrete particle sizes and then plotting the particle sizes against the corresponding pressure drops necessary to give efficiencies of 50 percent. In practice, Calvert and coworkers evidently have in most cases constructed the cut-power functions for various scrubbers by modeling (Yung and Calvert, U.S. EPA-600/8-78-005b, 1978). They show a variety of curves, whereas empirical studies have indicated that different types of scrubbers generally have about the same performance at a given level of power consumption.

Condensation Scrubbing The collection efficiency of scrubbing can be increased by the simultaneous condensation of water vapor from the gas stream. Water-vapor condensation assists in particle removal by two entirely different mechanisms. One is the deposition of particles on cold-water droplets or other surfaces as the result of

Stefan flow. The other is the condensation of water vapor on particles as nuclei, which enlarges the particles and makes them more readily collected by inertial deposition on droplets. Both mechanisms can operate simultaneously. However, for the buildup of particles by condensation to be effective, there must be adequate time for the particles to grow substantially before the principal gas-liquid-contacting operation takes place. Hence, if particle buildup is to be sought, the scrubber should be preceded by an appropriate gas-conditioning section. On the other hand, particle collection by Stefan flow can be induced simply by scrubbing the hot, humid gas with sufficient cold water to bring the gas below its initial dew point. Any practical method of inducing condensation on the dust particles will incidentally afford opportunities for the operation of the Stefan-flow mechanism. The hot gas stream must, of course, have a high initial moisture content, since the magnitude of the effects obtained is related to the quantity of water vapor condensed.

Although there is a considerable body of literature on particle collection by condensation mechanisms, most of it is either theoretical or, if experimental, treats basic phenomena in simplified cases. Few studies have been made to determine what performance may be expected from condensation scrubbing under practical conditions in industrial applications. In a series of studies, Calvert and coworkers investigated several types of equipment for condensation scrubbing, generally emphasizing the use of the condensation center effect to build up the particles for collection by inertial deposition (Calvert and Parker, EPA-600/8-78-005c, 1978). From early estimates, they predicted that a condensation scrubber would require only about one-third or less of the power required by a conventional high-energy scrubber. A subsequent demonstration-plant scrubber system consisted of a direct-contact condensing tower fed with cold water followed by a venturi scrubber fed with recirculated water (Chmielewski and Calvert, 600/7-81-148, 1981). The condensation and particle buildup took place in the cooling tower. In operation on humidified iron-foundry-cupola gas, this system still required about 65 percent as much power as for conventional high-energy scrubbing.

Semrau and coworkers [*Ind. Eng. Chem.*, **50**, 1615 (1958); *J. Air Pollut. Control Assoc.*, **13**, 587 (1963); EPA-650/2-74-108, 1974] investigated condensation scrubbing in pilot-plant studies in the field and, later, under laboratory conditions. Hot, humid gases were scrubbed directly with cold water under conditions that were favorable for the Stefan-flow mechanism but offered little or no opportunity for particle buildup. Some of the field studies indicated a contacting-power saving of as much as 50 percent for condensation scrubbing of Kraft-recovery-furnace fume. Laboratory tests on a predominantly submicrometer synthetic aerosol showed contacting-power savings of up to 40 percent with condensation scrubbing.

In the scrubbing of hot gases with high water content, condensation reduces contacting power and affords a direct power saving through the reduction of the gas volume by cooling and water-vapor condensation, but it incurs other costs for power and equipment for heat transfer and water cooling. However, condensation scrubbing may offer a net economic advantage if recovery of low-level heat is practical. It should also be advantageous when a hot gas must be not only cleaned but cooled and dehumidified as well; examples are the cleaning of blast-furnace gas for use as fuel and of SO_2-bearing waste gases for feed to a sulfuric acid plant.

Entrainment Separation The entrainment separator is a critical element of a scrubber, since the collection efficiency of the scrubber depends on essentially total removal of the spray from the gas stream. The sprays generated in scrubbers are generally large enough in droplet size that they can be readily removed by properly designed inertial separators. Primary collection of the spray is seldom the critical limitation on separator performance, but reentrainment is a common problem. In dust scrubbers it is essential that the entrainment separator not be of a form readily subject to blockage by solids deposits and that it be readily cleared of deposits if they should occur. Cyclone separators are advantageous in this respect and are widely used with venturi contactors. However, they cannot readily be made integral with scrubbers of some other configurations, which can be more conveniently fitted with various forms of impingement separators. Although separator design can be important, the most common

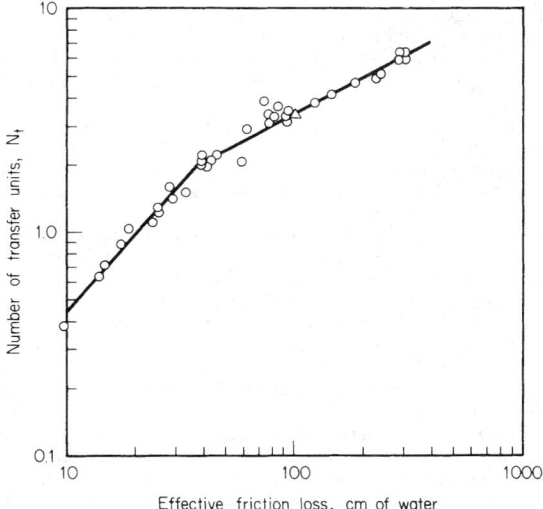

FIG. 17-47 Performance curve for orifice scrubber collecting ammonium fluorescin aerosol. (*Semrau et al., EPA 600/2-77-237, 1977.*)

cause of reentrainment is simply the use of excessive gas velocities, and few data are available on the gas-handling capacities of separators. In the absence of good data, there is a frequent tendency to underdesign separators in an effort to reduce costs.

Venturi Scrubbers The venturi scrubber is one of the most widely used types of particulate scrubbers. The designs have become generally standardized, and units are manufactured by a large number of companies. Venturi scrubbers may be used as either high- or low-energy devices but are most commonly employed as high-energy units. The units originally studied and used were designed to the proportions of the classical venturis used for metering, but since it was discovered that these proportions have no special merits, simpler and more practical designs have been adopted. Most "venturi" contactors in current use are in fact not venturis but variable orifices of one form or another. Any of a wide range of devices can be used, including a simple pipe-line contactor. Although the venturi scrubber is not inherently more efficient at a given contacting power than other types of devices, its simplicity and flexibility favor its use. It is also useful as a gas absorber for relatively soluble gases, but because it is a cocurrent contactor, it is not well suited to absorption of gases having low solubilities.

Current designs for venturi scrubbers generally use the vertical downflow of gas through the venturi contactor and incorporate three features: (1) a "wet-approach" or "flooded-wall" entry section, to avoid dust buildup at a wet-dry junction; (2) an adjustable throat for the venturi (or orifice), to provide for adjustment of the pressure drop; and (3) a "flooded elbow" located below the venturi and ahead of the entrainment separator, to reduce wear by abrasive particles. The venturi throat is sometimes fitted with a refractory lining to resist abrasion by dust particles. The entrainment separator is commonly, but not invariably, of the cyclone type. An example of the "standard form" of venturi scrubber is shown in Fig. 17-48. The wet-approach entry section has made practical the recirculation of slurries. Various forms of adjustable throats, which may be under manual or automatic control,

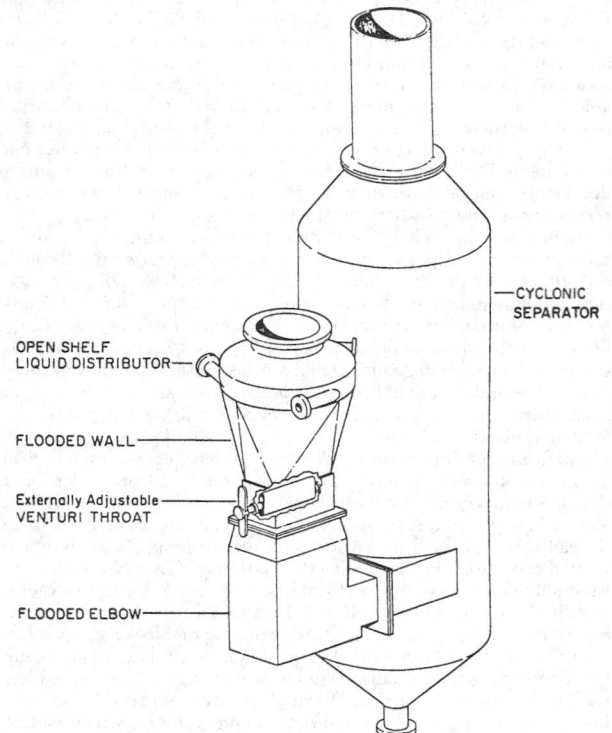

OPEN SHELF
LIQUID DISTRIBUTOR

CYCLONIC
SEPARATOR

FLOODED WALL

Externally Adjustable
VENTURI THROAT

FLOODED ELBOW

FIG. 17-48 Venturi scrubber. (*Neptune AirPol.*)

permit maintaining a constant pressure drop and constant efficiency under conditions of varying gas flow.

Self-Induced Spray Scrubbers Self-induced spray scrubbers form a category of gas-atomized spray scrubbers in which a tube or a duct of some other shape forms the gas-liquid-contacting zone. The gas stream flowing at high velocity through the contactor atomizes the liquid in essentially the same manner as in a venturi scrubber. However, the liquid is fed into the contactor and later recirculated from the entrainment separator section by gravity instead of being circulated by a pump as in venturi scrubbers. The scheme is well illustrated in Fig. 17-49a. A great many such devices using contactor ducts of various shapes, as in Fig. 17-49b are offered commercially. Although self-induced spray scrubbers can be built as high-energy units and sometimes are, most such devices are designed for only low-energy service.

The principal advantage of self-induced spray scrubbers is the elimination of a pump for recirculation of the scrubbing liquid. However, the designs for high-energy service are somewhat more complex and less flexible than those for venturi scrubbers.

Plate Towers Plate (tray) towers are countercurrent gas-atomized spray scrubbers using one or more plates for gas-liquid contacting. They are essentially the same as, if not identical to, the devices used for gas absorption and are frequently employed in applications in which gases are to be absorbed simultaneously with the removal of dust. Except possibly in cases in which condensation effects are involved, countercurrent operation is not significantly beneficial in dust collection.

The plates may be any of several types, including sieve, bubble-cap, and valve trays. The impingement baffle plate (Fig. 17-50) is commonly used for dust collection applications. Impingement on the baffles is not the controlling mechanism of particle collection; the principal collecting bodies are the droplets produced from the liquid by the gas as it flows through the perforations and around the baffles. The slot stage (Fig. 17-50) is in effect a miniature venturi contactor. Valve trays constitute multiple self-adjusting orifices that provide nearly constant gas pressure drop over considerable ranges of variation in gas flow. The gas pressure drop that can be taken across a single plate is necessarily limited, so that units designed for high contacting power must use multiple plates.

Plate towers are more subject to plugging and fouling than venturi-type scrubbers that have large passages for gas and liquid.

Packed-Bed Scrubbers Packed-bed scrubbers of the types used for gas absorption may also be used for dust collection but are subject to plugging by deposits of insoluble solids. Random packings, such as dumped Raschig rings and Berl saddles, are most seriously affected by plugging. Regular packings, such as stacked grids, are better in dust-collection service. When both a gas and particulate matter are to be collected, it is advisable to use a primary-stage scrubber of the venturi or similar type to collect the particulate matter ahead of a packed gas absorber.

Packed-bed scrubbers may be constructed for either vertical or horizontal gas flow. Vertical-flow units (packed towers) commonly use countercurrent flow of gas and liquid, although cocurrent flow is sometimes used. Packed scrubbers using horizontal gas flow usually employ cross-flow of liquid.

Scrubber packings are too large to serve as collecting bodies for any except very large dust particles. In the collection of fine particles, the packings serve primarily to promote fluid turbulence that aids the deposition of the dust particles on droplets. In a packed tower operating below the flooding point, with most of the liquid flowing in films and little spray formation, the relative efficiency in collection of particles may possibly be lower than that of a venturi-type scrubber operating at the same contacting power. However, no data are available to resolve the question.

Mobile-Bed Scrubbers Mobile-bed scrubbers (Fig. 17-51) are constructed with one or more beds of low-density spheres that are free to move between upper and lower retaining grids. The spheres are commonly 1.0 in (2.5 cm) or more in diameter and made from rubber or a plastic such as polypropylene. The plastic spheres may be solid or hollow. Gas and liquid flows are countercurrent, and the spherical packings are fluidized by the upward-flowing gas. The movement of

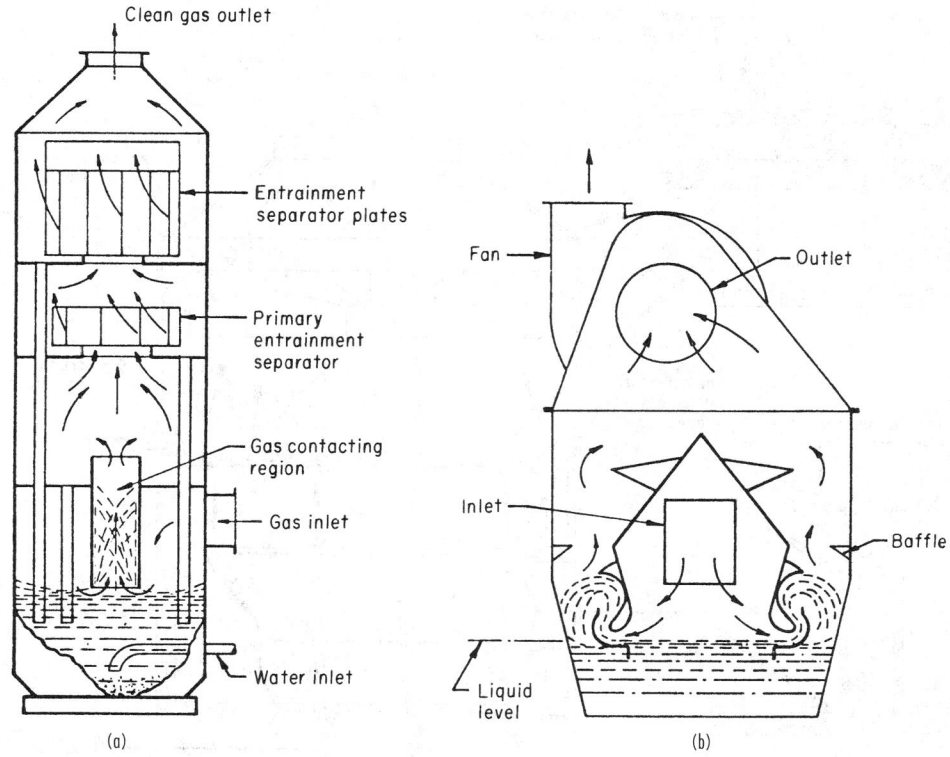

FIG. 17-49   Self-induced spray scrubbers. (*a*) Blaw-Knox Food & Chemical Equipment, Inc. (*b*) American Air Filter Co., Inc.

the packings is intended to minimize fouling and plugging of the bed. Mobile-bed scrubbers were first developed for absorbing gases from gas streams that also carry solid or semisolid particles.

The spherical packings are too large to serve as effective targets for the deposition of fine dust particles. In dust-collection service, the packings actually serve as turbulence promoters, while the dust particles are collected primarily by the liquid droplets.

The gas pressure drop through the scrubber may be increased by increasing the gas velocity, the liquid-to-gas ratio, the depth of the bed, the density of the packings, and the number of beds in series. In an experimental study, Yung et al. (EPA-600/7-79-071, 1979) determined that the collection efficiency of a mobile-bed scrubber was dependent only on the gas pressure drop and was not influenced independently by the gas velocity, the liquid-to-gas ratio, or the number of beds except as these factors affected the pressure drop. Yung et al. also reported that the mobile-bed scrubber was less efficient at a given pressure drop than scrubbers of the venturi type, but without offering comparable experimental supporting evidence.

Spray Scrubbers   Spray scrubbers consist of empty chambers of some simple form in which the gas stream is contacted with liquid droplets generated by spray nozzles. A common form is a spray tower, in which the gas flows upward through a bank or successive banks of spray nozzles. Similar arrangements are sometimes used in spray chambers with horizontal gas flow. Such devices have very low gas pressure drops, and all but a small part of the contacting power is derived from the liquid stream. The required contacting power is obtained from an appropriate combination of liquid pressure and flow rate. Most spray scrubber are low-energy units. Collection of fine particles is possible but may require very high liquid-to-gas ratios, liquid feed pressures, or both. Plugging of the nozzles can be a persistent maintenance problem. Entrainment separators are necessary to prevent carry-over of spray into the exit gas.

Cyclone Scrubbers   The vessels of cyclone scrubbers are all in the form of cyclones, which provide for entrainment separation. However, the gas-liquid-contacting devices may be of either the gas-atomized-spray or the preformed-spray type. The cyclone-spray scrubber shown in Fig. 17-52*a* has an axial spray tree, or manifold, equipped with hydraulic spray nozzles. Similar units are available with the spray nozzles mounted in the wall of the cyclone, discharging inward. This latter arrangement makes the nozzles more accessible for maintenance. In the cyclone scrubber shown in Fig. 17-52*b*, most of the gas-liquid contacting is accomplished in the swirl vanes, with the energy being supplied from the gas stream in the form of pressure drop. The swirl vanes serve the same function as do the trays in a tray tower. Higher contacting power can be provided by using additional sets of swirl vanes in series.

Ejector-Venturi Scrubbers   In the ejector-venturi scrubber (Fig. 17-53) the cocurrent water jet from a spray nozzle serves both to scrub the gas and to provide the draft for moving the gas. No fan is required, but the equivalent power must be supplied to the pump that delivers water to the ejector nozzle. The water must be supplied in sufficient volume and at high enough pressure to provide both adequate draft and enough contacting power for the required scrubbing operation. Considered as a gas pump, the ejector is not a very efficient device, but the dissipated energy that is not effective in pumping does serve in gas-liquid contacting. The energy equivalent to any gas pressure rise across the scrubber is not part of the contacting power (Semrau et al., EPA-600/2-77-234, 1974).

The ejector-venturi scrubber is widely used as a gas absorber, but the combinations of water pressure and flow rate that are sufficient to provide the required draft usually do not also yield enough contacting power to give high collection efficiency on submicrometer particles. Other types of ejectors have been employed to provide higher contacting-power levels. In one, superheated water is discharged through

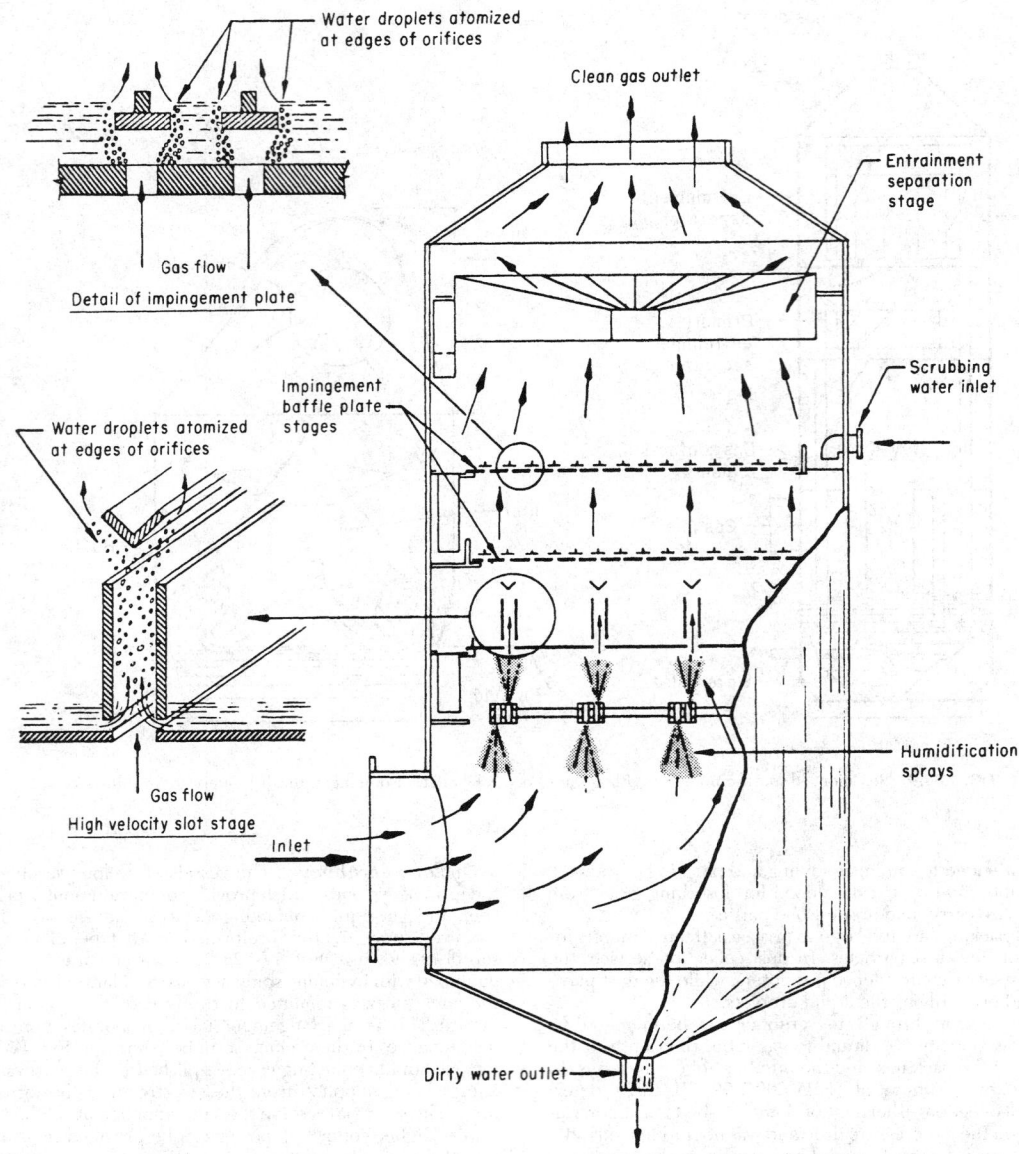

FIG. 17-50   Impingement-plate scrubber. (*Peabody Engineering Corp.*)

the nozzle, and part flashes to steam, increasing the mechanical energy available for scrubbing [Gardenier, *J. Air Pollut. Control Assoc.*, **24**, 954 (1974)]. Some units use two-fluid nozzles, with either compressed air or steam as the compressible fluid [Sparks, *J. Air Pollut. Control Assoc.*, **24**, 958 (1974)]. Most of the energy for gas movement and for atomizing the liquid and scrubbing the gas is derived from the compressed air or steam. In some ejector-venturi-scrubbers installations, part of the draft is supplied by a fan [Williams and Fuller, TAPPI, **60**(1), 108 (1977)].

Mechanical Scrubbers   Mechanical scrubbers comprise those devices in which a power-driven rotor produces the fine spray and the contacting of gas and liquid. As in other types of scrubbers, it is the droplets that are the principal collecting bodies for the dust particles. The rotor acts as a turbulence producer. An entrainment separator must be used to prevent carry-over of spray. Among potential mainte-

nance problems are unbalancing of the rotor by buildup of dust deposits and abrasion by coarse particles.

The simplest commercial devices of this type are essentially fans upon which water is sprayed. The unit shown in Fig. 17-54 is adapted to light duty, and heavy dust loads are avoided to minimize buildup on the rotor.

Fiber-Bed Scrubbers   Fibrous-bed structures are sometimes used as gas-liquid contactors, with cocurrent flow of the gas and liquid streams. In such contactors, both scrubbing (particle deposition on droplets) and filtration (particle deposition on fibers) may take place. If only mists are to be collected, small fibers may be used, but if solid particles are present, the use of fiber beds is limited by the tendency of the beds to plug. For dust-collection service, the fiber bed must be composed of coarse fibers and have a high void fraction, so as to minimize the tendency to plug. The fiber bed may be made from metal or

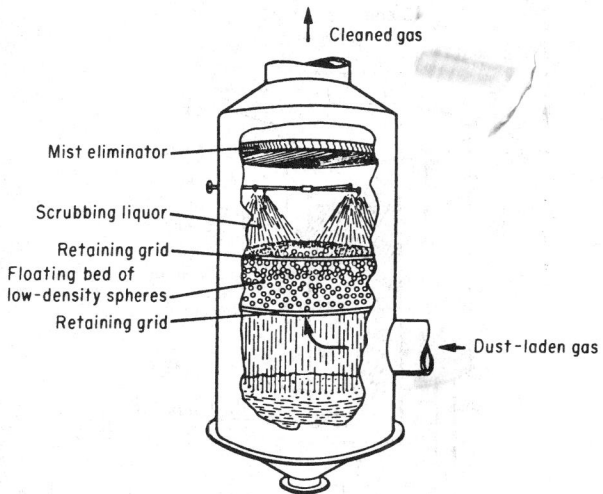

FIG. 17-51 Mobile-bed scrubber. (*Air Correction Division, UOP.*)

plastic fibers in the form of knitted structures, multiple layers of screens, or random-packed fibers. However, the bed must have sufficient dimensional stability so that it will not be compacted during operation.

Lucas and Porter (U.S. Patent 3,370,401, 1967) developed a fiber-bed scrubber in which the gas and scrubbing liquid flow vertically upward through a fiber bed (Fig. 17-55). The beds tested were composed of knitted structures made from fibers with diameters ranging from 89 to 406 μm. Lucas and Porter reported that the fiber-bed scrubber gave substantially higher efficiencies than did venturi-type scrubbers tested with the same dust at the same gas pressure drop. In similar experiments, Semrau (Semrau and Lunn, op. cit.) also found that a fiber-bed contactor made with random-packed steel-wool fibers gave higher efficiencies than an orifice contactor. However, there

were indications that the fiber bed would have little advantage in the collection of submicrometer particles, presumably because of the large fiber size feasible for dust-collection service.

Despite their potential for increased collection efficiency, fiber-bed scrubbers have had only limited commercial acceptance for dust collection because of their tendency to become plugged. Their principal use has been in small units such as engine-intake-air cleaners, for which it is feasible to remove the fiber bed for cleaning at frequent intervals.

Electrically Augmented Scrubbers In some types of wet collectors, attempts are made to apply the electrostatic-deposition mechanism by charging the dust particles, the water droplets, or both. The objective is to combine in a scrubber high efficiency in collecting fine particles and the moderate power consumption characteristic of an electrical precipitator. Successful devices of this type have been essentially wet electrical precipitators and should properly be discussed in that category (see "Electrical Precipitators"). So far, there has been no clear demonstration of a device that combines the small size, compactness, and high efficiency of a high-energy scrubber with the relatively low power consumption of an electrical precipitator.

Dry Scrubbing *Dry scrubbing* is an umbrella term used to associate several different unit operations and types of hardware that can be used in combinations to accomplish the unit process of dry scrubbing. They all utilize scrubbing, in which mass transfer takes place between the gas phase and an active liquidlike surface, and they all discharge the resulting products separately as a gas and a solidlike "dry" product for reuse or disposal.

It is helpful to recognize three phases in the development of systems using dry scrubbing. The first phase used mainly good contactors and relatively expensive sorbents, which were tailored to catch targeted compounds with high efficiency (>99 percent). The second phase used mainly minimal contactors and cheap sorbents in order to get enough sulfur dioxide capture (>60 percent) to justify using cheaper higher sulfur coal. The third phase is ongoing; it uses mainly better contactors, sorbents, and fabric filters in order to get not only high SO_2 capture (>90 percent), but also substantial capture of other regulated emissions such as metals and dioxins.

The principle of dry scrubbing was first recognized by Lamb and Wilson [*I&EC,* **11**(5), 420 (1919)] and Wilson [*I&EC,* **12**(10), 1000 (1920)] at MIT during their World War I program to develop better

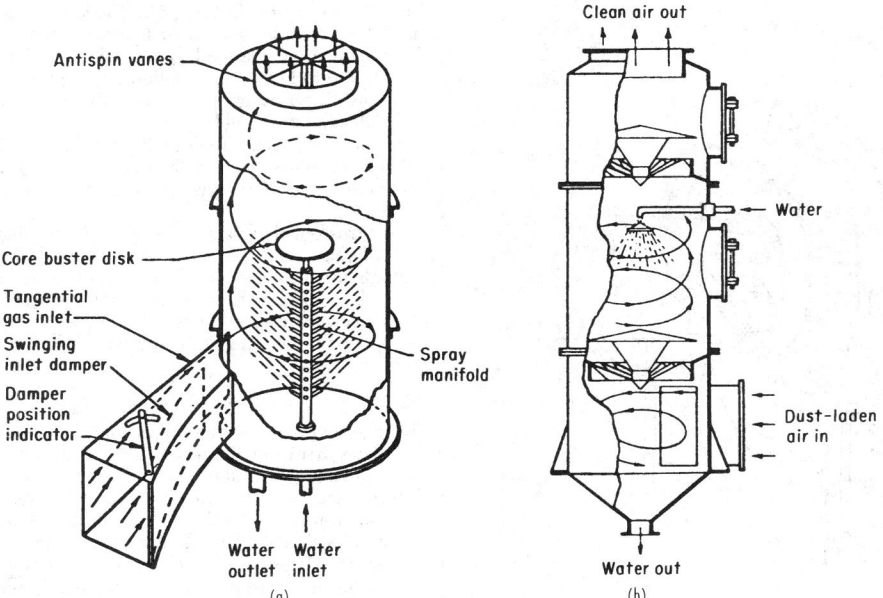

FIG. 17-52 Cyclone scrubbers. (*a*) Chemico Air Pollution Control Corp. (*b*) Ducon Co., Inc.

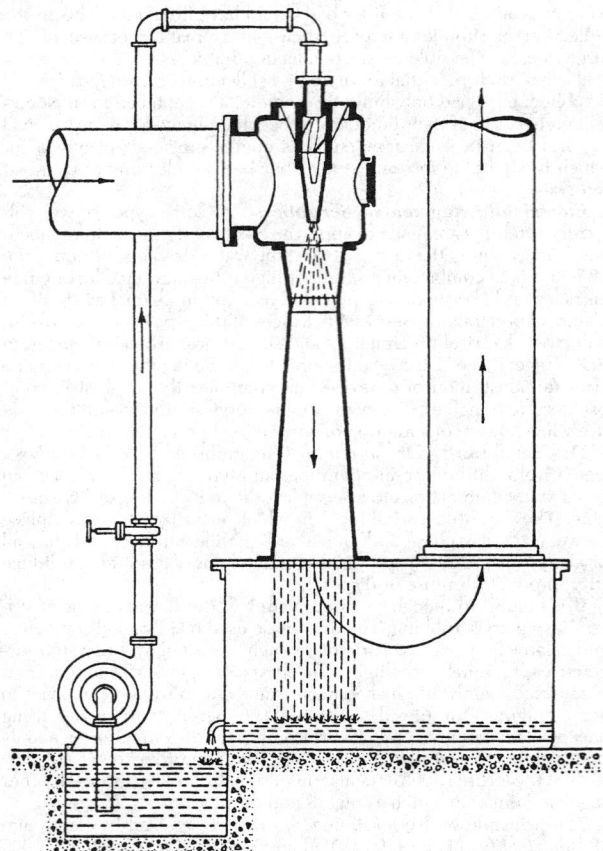

FIG. 17-53   Ejector-venturi scrubber. (*Schutte & Koerting Division, Amtek, Inc.*)

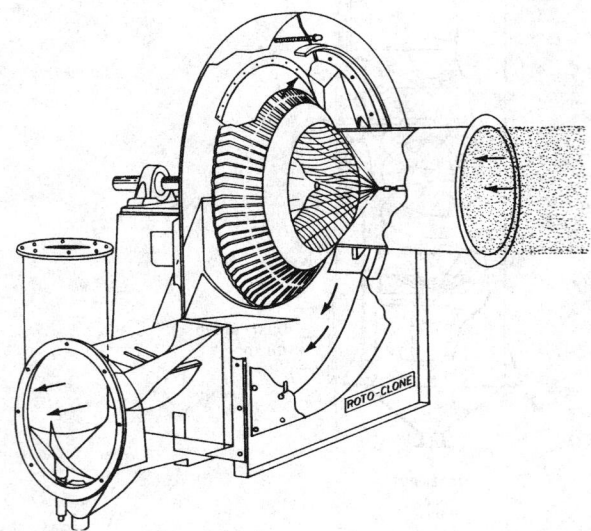

FIG. 17-54   Mechanical scrubber. (*American Air Filter Co., Inc.*)

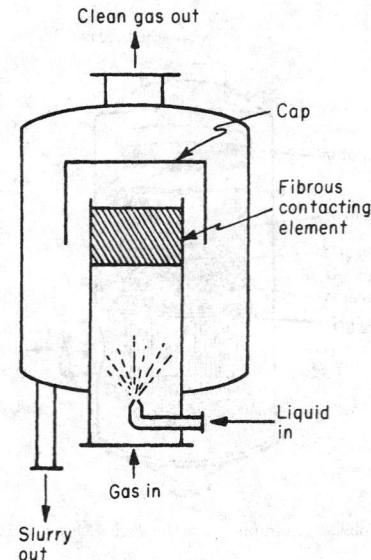

FIG. 17-55   Fibrous-bed scrubber. (*Lucas and Porter, U.S. Patent 3,370,401, 1967.*)

media for military gas masks. The preferred medium turned out to be a mixture of granular activated carbon with a new kind of porous granular hydrated lime, which they dubbed *activated soda-lime*. What activated it was a small but critical amount of sodium hydroxide, which was enough to maintain in use on French battlefields an equilibrium moisture film a few molecules thick on the surface of pores within lime particles. That film enhanced surface hydrolysis, diffusion, and neutralization of acid gases such as Cl_2 and SO_2, to the point that their removal became controlled by fast external gas-film diffusion, up to substantial utilization of the potential stoichiometric capacity. The better varieties of army activated soda-lime used in gas masks captured more than 99 percent of Cl_2 or SO_2 even after using up more than 10 percent of alkalinity. Gas-mask-canister sorbent beds typically used 8×14 mesh granules 10 cm deep, and tests were run with 1 percent acid gas at space velocities of 3000 to 7400 l/hr. Carbon dioxide was tested to higher penetrations; its capture was more than 50 percent even after using up more than 50 percent of available alkalinity.

With good dry scrubbing sorbents, the controlling resistance for gas cleaning is external turbulent diffusion, which also depends on energy dissipated by viscous and by inertial mechanisms. It turns out to be possible to correlate mass-transfer rate as a function of the friction factor.

It turns out that packed beds much less than a hundred particles thick behave as if they were well-stirred due to the entrance effect. Although it may seem odd that a packed bed can behave as if well-stirred, it typically takes at least a 100-particle bed depth in order for a plug-flow concentration wave to develop.

Integration of the Ergun equation for well stirred flow gives

$$\frac{C_2 - C_1}{C^\circ - C_2} = F_k \frac{\rho D_v (1 - \varepsilon_v) L}{\mu \varepsilon_v D_p} \tag{17-9}$$

Recognition of dry scrubbing as a mechanism useful in treating large gas flows first came in the 1960s as regulations were tightening on fly ash emissions from utility boilers. Dry scrubbing turns out to be the appropriate way to describe the interactions of sulfuric acid vapor with fly ash from pulverized coal combustion. On one hand, powdered sodium bicarbonate was added to some flue gases in order to protect fabric filter bags from attack by sulfuric acid films on ash from high sulfur coal combustion. On the other hand, sulfur trioxide was added to other flue gases in order to generate sulfuric acid films that would improve the performance of electrostatic precipitators collecting inherently high resistivity ash from low sulfur coal. From Eq. (17-9), it is readily apparent that the way to get more gas cleaning per unit pressure drop is to optimize the particle size and void fraction, and use relatively fewer particles operating at higher velocity. Such an arrangement runs out of stoichiometric capacity quickly, which leads naturally to transported bed contactors fed with fresh sorbent.

The actual name "dry scrubbing" was first publicized by Teller [U.S. Patent no. 3,721,066 (1973)]. He worked both with classical Army-type soda-lime and with his patented water-activated form of the alkaline feldspar nepheline syenite as a flow agent and feedstock sorbent for HF and SO_2 in hot, sticky fumes from glass melting furnaces. He claimed capture of more than 99 percent of 180 ppm HF and SO_2 for more than 20 hours in a packed bed of 200×325 mesh hydrated nepheline syenite at 42,000/hr.

Teller [U.S. Patent no. 3,997,652 (1976)] also patented a number of "chromatographic sorbents" using semivolatile liquid coatings similar to those used in capillary columns for gas chromatographs. For simple acid gases, Teller generally used a flash-drying-type venturi injector and pneumatic transport system to disperse fresh sorbent into hot gas to be treated and to provide time for turbulent mixing. He then typically used a baghouse to collect the loaded sorbent for inclusion with other feeds into the glass melting system.

Aluminum producers were also early users of another type of dry scrubbing for the difficult cleaning problem of acid gases in hot, sticky fumes from aluminum reduction potlines. Bazhenov et al. [*Tsvetn. Met.—Russia*, (6) 44 (1975)] described the use of a fluidized bed of feedstock aluminum oxide to clean potline offgas in an arrangement where there was an integral freeboard baghouse. Loaded sorbent was purged from the bed for feed into the aluminum reduction system. In that anhydrous system there was no water film as such inside the pores, but there was high specific surface, having liquidlike mobile active sites with a strong affinity for binding fluoride. The potential for achieving significant gas cleaning at very high space velocities led to dry scrubbing being applied both for incremental desulfurization of flue gases from coal-fired utility boilers and for removal of trace acid gases from solid waste incineration. A number of different hardware arrangements were tested all the way to full scale in order to determine accurately the various effects that must be optimized in order to achieve minimum net cost to be carried by the rate-paying public.

It was apparent from the very earliest tests that control of thin moisture films on the surface of reactive particles was the key to success. The main three competing arrangements, as compared by Statnick et al. [4th Annual Pitt. Coal Conf. (1987)] involved slurry spray dryers, where lime and water were injected together, versus systems where the gas was humidified by water injection before or after injection of limey dry powder reagents. It turns out that there are tradeoffs among the costs of hardware, reagent, and water dispersion and reagent purchase and disposal. Systems where water evaporates in the presence of active particles are usually less expensive overall.

The evaporation lifetimes of pure water droplets are described by Marshall and Seltzer [*CEP*, **46**(10), 501 and **46**(11), 575 (1950)] and by Duffie and Marshall [*CEP*, **49**(8), 417 (1953) and **49**(9), 480 (1953)]. Figure 17-56 is easy to use to estimate the necessary spray dispersion. For typical humidifier conditions of 300°F gas inlet temperature and a 20°F approach to adiabatic saturation at the outlet, a 60-micrometer droplet will require about 0.9 seconds to evaporate, while a 110-micrometer droplet will require about 3 seconds. If the time available for drying is no more than 3 seconds, it therefore follows that the largest droplets in a humidifier spray should be no larger than 100 micrometers. If the material being dried contains solids, droplet top size will need to be smaller due to slower evaporation rates. Droplets this small not only require considerable expensive power to generate, but since they have inherent penetration distances of less than a meter, they require expensive dispersion arrangements to get good mixing into large gas flows without allowing damp, sticky particles to reach walls.

The economics seem to be better for systems where dry powdered fresh lime plus ground recycled lime is injected along with a relatively coarse spray which impinges on and dries out from the reagent, as described by Stouffer et al. [*I&EC Res.*, **28**(1) 20 (1989)]. Withum et al. [9th Ann. Pitt. Coal Prep. Util. Env. Control Contractors Conf. (1993)] describes an advanced version of that system that has been further optimized to the point that it is competitive with wet limestone scrubbing for >90 percent flue gas desulfurization.

The most popular dry scrubbing systems for incinerators have involved the spray drying of lime slurries, followed by dry collection in electrostatic precipitators or fabric filters. Moller and Christiansen [Air Poll. Cont. Assoc. 84-9.5 (1984)] published data on early European technology. Moller et al. [U.S. Patent no. 4,889,698 (1989)] describe the newer extension of that technology to include both spray-dryer absorption and dry scrubbing with powdered, activated carbon injection. They claim greatly improved removal of mercury, dioxins, and NOx.

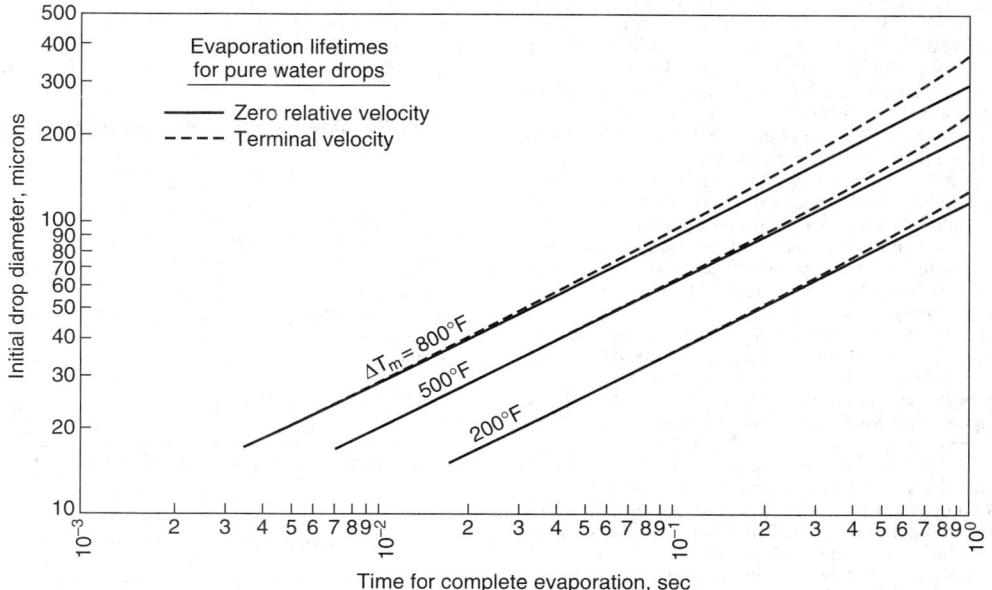

FIG. 17-56 Effect of drop diameter on time for complete evaporation of water drops.

Optimized modern dry scrubbing systems for incinerator gas cleaning are much more effective (and expensive) than their counterparts used so far for utility boiler flue gas cleaning. Brinckman and Maresca [ASME Med. Waste Symp. (1992)] describe the use of dry hydrated lime or sodium bicarbonate injection followed by membrane filtration as preferred treatment technology for control of acid gas and particulate matter emissions from modular medical waste incinerators, which have especially high dioxin emissions.

Kreindl and Brinckman [Air Waste Man. Assoc., paper 93-RP-154.04 (1993)] describe the three conventional flue-gas cleaning-process concepts (standard dry scrubbing, quasi-dry scrubbing, and wet scrubbing) as being inherently inadequate to meet the emerging European incinerator metal and organic emissions limits. Standard dry scrubbing can be upgraded by using powdered carbon along with lime, and switching to membrane filter media. Quasi-dry scrubbing can be upgraded by using powdered carbon along with lime spray drying [as also described by Moller (1989)], and switching to membrane filter media. Wet scrubbing can be upgraded by post-treatment following the wet scrubber with a circulating fluidized bed of granular activated carbon, again followed by a membrane filter.

Control of metal and organic emissions by dry scrubbing involves more than the simple acid gas neutralization discussed so far. Capture of metals other than mercury is mainly a matter of using a fabric filter having the capability of retaining unagglomerated submicron particles that are relatively enriched in toxic metals such as cadmium. Use of membrane-coated filter media is almost essential to do that well. Capture of mercury and dioxins is mainly achieved with activated carbon, which is more effective when used in combination with chlorine inhibitors. Karasek [U.S. Patent no. 4,793,270 (1988)] describes the inhibiting effects of sulfur compounds, which is why dioxins are rarely a problem in flue gases from coal-fired combustion. Naikwadi and Karasek [Chemosphere, **27**(1–3), 335 (1993)] describe the potent inhibiting effect of amines. That is presumably why German experience is finding that dioxins are rarely a problem in flue gases that have undergone some form of ammonia deNOx treatment, as described by Hahn [Chemosphere, **25**(1–2), 57 (1992)]. Hahn, like Kreindl and Brinckman, also argues that classical wet scrubbers will need to be upgraded with modern dry scrubbers and fabric filters.

Fabric Filters Fabric filters, commonly termed "bag filters" or "baghouses," are collectors in which dust is removed from the gas stream by passing the dust-laden gas through a fabric of some type (e.g., woven cloth, felt, or porous membrane). These devices are "surface" filters in that dust collects in a layer on the surface of the filter medium, and the dust layer itself becomes the effective filter medium. The pores in the medium (particularly in woven cloth) are usually many times the size of the dust particles, so that collection efficiency is low until sufficient particles have been collected to build up a "precoat" in the fabric pores (Billings and Wilder, Handbook of Fabric Filter Technology, vol. I, EPA No. APTD-0690, NTIS No. PB-200648, 1979). During this initial period, particle deposition takes place mainly by inertial and flow-line interception, diffusion, and gravity. Once the dust layer has been fully established, sieving is probably the dominant deposition mechanism, penetration is usually extremely low except during the fabric-cleaning cycle, and only limited additional means remain for influencing collection efficiency by filter design. Filter design is related mainly to choices of gas filtration velocities and pressure drops and of fabric-cleaning cycles.

Because of their inherently high efficiency on dusts in all particle-size ranges, fabric filters have been used for collection of fine dusts and fumes for over 100 years. The greatest limitation on filter application has been imposed by the temperature limits of available fabric materials. The upper limit for natural fibers is about 90°C (200°F). The major new developments in filter technology that have been made since 1945 have followed the development of fabrics made from glass and synthetic fibers, which has extended the temperature limits to about 230 to 260°C (450 to 500°F). The capabilities of available fibers to resist high temperatures are still among the most severe limitations on the possible applications of fabric filters.

Gas Pressure Drops The filtration, or superficial face, velocities used in fabric filters are generally in the range of 0.3 to 3 m/min (1 to 10 ft/min), depending on the types of fabric, fabric supports, and cleaning methods used. In this range, gas pressure drops conform to Darcy's law for streamline flow in porous media, in which the pressure drop is directly proportional to the flow rate. The pressure drop across the fabric and the collected dust layer may be expressed (Billings and Wilder, op. cit.) by

$$\Delta p = K_1 V_f + K_2 w V_f \qquad (17\text{-}10)$$

where Δp = kPa, or in of water; V_f = superficial velocity through filter, m/min, or ft/min; w = dust loading on filter, g/m^2, or lbm/ft^2; and K_1 and K_2 are resistance coefficients for the "conditioned" fabric and the dust layer respectively. The conditioned fabric is that fabric in which a relatively consistent dust load remains deposited in depth following cycles of filtration and cleaning. K_1, expressed in units of kPa/(m/min) or in water/(ft/min), may be more than 10 times the value of the resistance coefficient for the original clean fabric. If the depth of the dust layer on the fabric is greater than about 0.2 cm ($\frac{1}{16}$ in), corresponding to a fabric dust loading on the order of 500 g/m^2 (0.1 lbm/ft^2), the pressure drop across the fabric (including the dust in the pores) is usually negligible relative to that across the dust layer.

The specific resistance coefficient for the dust layer K_2 was originally defined by Williams et al. [Heat. Piping Air Cond., **12**, 259 (1940)], who proposed estimating values of the coefficient by use of the Kozeny-Carman equation [Carman, Trans. Inst. Chem. Eng. (London), **15**, 150 (1937)]. In practice, K_1 and K_2 are measured directly in filtration experiments. The K_1 and K_2 values can be corrected for temperature by multiplying by the ratio of the gas viscosity at the desired condition to the gas viscosity at the original experimental conditions. Values of K_2 determined for certain dusts by Williams et al. (op. cit.) are presented in Table 17-5.

Lapple (in Perry, Chemical Engineers' Handbook, 3d ed., McGraw-Hill, New York, 1950) presents an alternative form of Eq. (17-10) in which the gas-viscosity term is explicit instead of being incorporated into the resistance coefficients:

$$\Delta p_i = K_c \mu V_f + K_d \mu w V_f \qquad (17\text{-}11)$$

where Δp_i = in water, μ = cP, V_f = ft/min, w = gr/ft^2, K_c = cloth resistance coefficient = (in water)/(cP)(ft/min), and K_d = dust-layer resistance coefficient = (in water)/(cP)(gr/ft^2)(ft/min). K_d may be expressed in the same units, using the Kozeny-Carman equation:

$$K_d = \frac{160.0(1 - \varepsilon_v)}{\phi_s^2 D_p^2 \rho_s \varepsilon_v^3} \qquad (17\text{-}12)$$

where D_p = μm and ρ_s = lbm/ft^3.

Data sufficient to permit reasonable predictions of K_d from Eq. (17-12) are seldom available. The range in the values of K_d that may be encountered in practice is illustrated in Fig. 17-57 in which avail-

TABLE 17-5 Specific Resistance Coefficients for Certain Dusts*

	K_2† for particle size less than						
Dust	20 mesh	140 mesh	375 mesh	90 μm	45 μm	20 μm	2 μm
Granite	1.58	2.20				19.8	
Foundry	0.62	1.58	3.78				
Gypsum			6.30			18.9	
Feldspar			6.30			27.3	
Stone	0.96			6.30			
Lampblack							47.2
Zinc oxide							15.7‡
Wood				6.30			
Resin (cold)		0.62				25.2	
Oats	1.58			9.60	11.0		
Corn	0.62		1.58	3.78	8.80		

*Data from Williams et al., Heat. Piping Air Cond., **12**, 259–263 (1940).

$$\dagger K_2 = \frac{\Delta p_i}{V_f w}, \quad \frac{\text{in water}}{\text{(ft/min)(lbm/ft}^2)}.$$

NOTE: These data were obtained when filtering air at ambient conditions. For gases other than atmospheric air, the Δp_i values predicted from Table 17-5 should be multiplied by the actual gas viscosity divided by the viscosity of atmospheric air.

‡Flocculated material not dispersed; size actually larger.

able experimental determinations of K_d reported in the literature for a variety of dusts are plotted against particle size. In most cases no accurate particle-size data were reported, and the curves represent the estimated range of particle size involved. The data of Williams et al. (op. cit.) are related to a wide variety of dusts, and only the approximate limits enclosing these data are shown. Also included are curves predicted from Eq. (17-12) for specific values of ϕ_s, ρ_s, and ε_v. It is apparent from these curves that smaller particles tend toward higher values of ε_v, which is consistent with the observation that fine dusts (particularly those smaller than 10 μm) have lower bulk densities than coarser fractions, apparently because of the effects of surface forces. For coarse dusts, K_d varies approximately inversely as the square of the particle diameter, which implies that the void fraction (or bulk density) does not change with particle size. However, for particles finer than 10 μm, the value of K_d appears to become constant, increased voids compensating for the reduction in size. In addition to the increased voidages encountered with small particles, slip flow also contributes to the relative constancy of the experimental values of K_d for particle sizes under 5 μm.

Because of the assumptions underlying its derivation, the Kozeny-Carman equation is not valid at void fractions greater than 0.7 to 0.8 (Billings and Wilder, op. cit.). In addition, in situ measurement of the void fraction of a dust layer on a filter fabric is extremely difficult and has seldom even been attempted. The structure of the layer is dependent on the character of the fabric surface as well as on the characteristics of the dust, whereas the application of Eq. (17-12) implicitly assumes that K_2 is dependent only on the properties of the dust. A smooth fabric surface permits the dust to become closely packed, leading to a relatively high value of K_2. If the surface is napped or has numerous extended fibrils, the dust cake formed will be more porous and have a lower value of K_2 [Billings and Wilder, op. cit.; Snyder and Pring, *Ind. Eng. Chem.*, **47**, 960 (1955); and K. T. Semrau, unpublished data, SRI International, Menlo Park, Calif., 1952–1953].

Equation (17-10) indicates that for filtration at a given velocity the pressure drop is a linear function of the fabric dust loading w. In some cases, particularly with smooth-surfaced fabrics, this is at least approx-

imately the case, but in other instances the function displays an upward curvature with increases in w, indicating compression of the dust layer, the fabric, or both, and a consequent increase in K_2 (Snyder and Pring, op. cit.; Semrau, op. cit.). Several investigations have shown K_2 to be increased by increases in the filtration velocity [Billings and Wilder, op. cit.; Spaite and Walsh, *Am. Ind. Hyg. Assoc. J.*, **24**, 357 (1968)]. However, the various investigators do not agree on the magnitude of the velocity effects. Billings and Wilder suggest assuming as an approximation that K_2 is directly proportional to the filtration velocity, but the actual relationship is probably dependent on the nature of the fabric and fabric surface, the characteristics of the dust, the dust loading on the fabric, and the pressure drop.

Clearly, the factors determining K_2 are far more complex than is indicated by a simple application of the Kozeny-Carman equation, and when possible, filter design should be based on experimental determinations made under conditions approximating those expected in the planned installation.

Types of Filters Current fabric-filter designs fall into three types, depending on the method of cleaning used: (1) shaker-cleaned, (2) reverse-flow-cleaned, and (3) reverse-pulse-cleaned. The shaker-cleaned filter is the earliest form of bag filter (Fig. 17-58). The open lower ends of the bags are fastened over openings in the tube sheet that separates the lower dirty-gas inlet chamber from the upper clean-gas chamber. The bag supports from which the bags are suspended are connected to a shaking mechanism. The dirty gas flows upward into the filter bags, and the dust collects on the inside surfaces of the bags. When the gas pressure drop rises to a chosen upper limit as the result of dust accumulation, the gas flow is stopped and the shaker is operated, giving a whipping motion to the bags. The dislodged dust falls into the dust hopper located below the tube sheet. If the filter is to be operated continuously, it must be constructed with multiple compartments, so that the individual compartments can be sequentially taken off line for cleaning while the other compartments continue in operation (Fig. 17-59).

Shaker-cleaned filters are available as standard commercial units, although large baghouses for heavy-duty service are commonly custom-designed and -fabricated. The oval or round bags used in the standard units are usually 12 to 20 cm (5 to 8 in) in diameter and 2.5 to 5 m (8 to 17 ft) long. The large, heavy-duty baghouses may use bags up to 30 cm (12 in) in diameter and 9 m (30 ft) long. The bags must be made of woven fabrics to withstand the flexing and stretching involved in shaking. The fabrics may be made from natural fibers (cotton or wool) or synthetic fibers. Fabrics of glass or mineral fibers are generally too fragile to be cleaned by shaking and are usually used in reverse-flow-cleaned filters.

Large units (other than custom units) are usually built up of standardized rectangular sections in parallel. Each section contains on the order of 1000 to 2000 ft^2 of cloth, and the sections are assembled in the field to form a single filter housing. In this manner, the filter can be partitioned so that one or more sections at a time can be cut out of service for shaking or general maintenance.

Ordinary shaker-cleaned filters may be shaken every 1/4 to 8 h, depending on the service. A manometer connected across the filter is useful in determining when the filter should be shaken. Fully automatic filters may be shaken as frequently as every 2 min, but bag maintenance will be greatly reduced if the time between shakings can be increased to 15 or 20 min without developing excessive pressure drop. Cleaning may be actuated automatically by a differential-pressure switch. It is essential that the gas flow through the filter be stopped when shaking in order to permit the dust to fall off. With very fine dust, it may even be necessary to equalize the pressure across the cloth [Mumford, Markson, and Ravese, *Trans. Am. Soc. Mech. Eng.*, **62**, 271 (1940)]. In practice this can be accomplished without interrupting the operation by cutting one section out of service at a time, as shown in Fig. 17-59. In automatic filters this operation involves closing the dampers, shaking the filter units either pneumatically or mechanically, sometimes with the addition of a reverse flow of cleaned gas through the filter, and lastly reopening the dampers. For compressed-air-operated automatic filters, this entire operation may take only 2 to 10 s. For ordinary mechanical filters equipped for automatic control, the operation may take as long as 3 min.

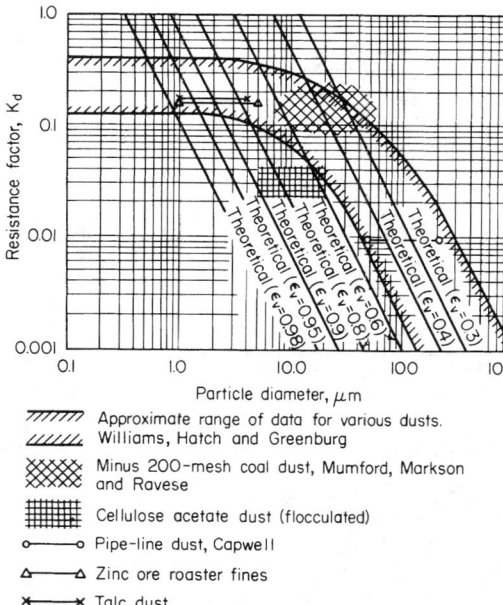

FIG. 17-57 Resistance factors for dust layers. Theoretical curves given are based on Eq. (20-78) for a shape factor of 0.5 and a true particle specific gravity of 2.0. [*Williams, Hatch, and Greenburg, Heat. Piping Air. Cond., 12, 259 (1940); Mumford, Markson, and Ravese, Trans. Am. Soc. Mech. Eng., 62, 271 (1940); Capwell, Gas, 15, 31 (August 1939)*].

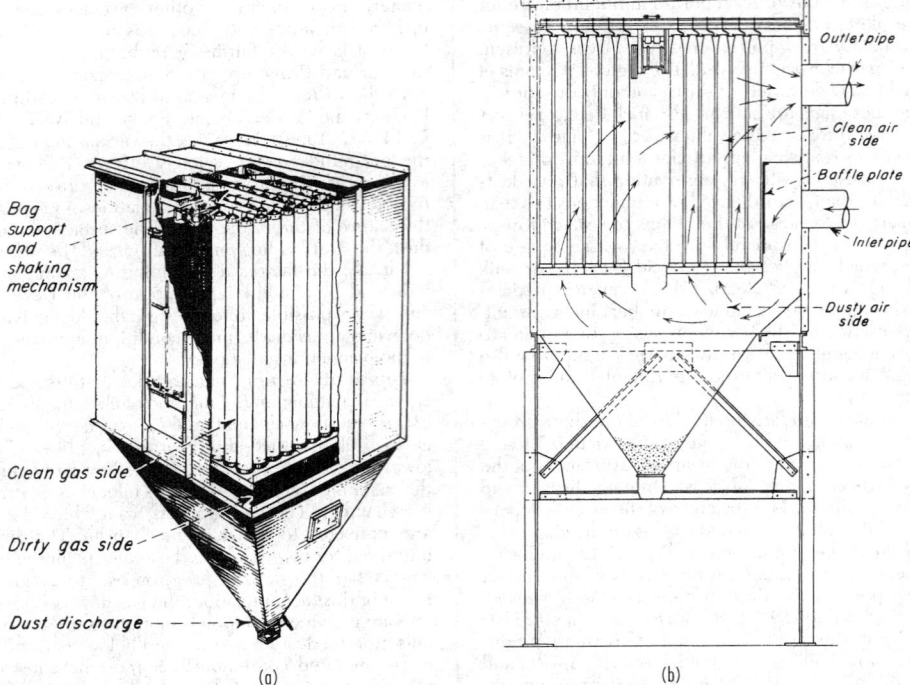

FIG. 17-58 Typical shaker-type fabric filters. (*a*) Buell Norblo (cutaway view). (*b*) Wheelabrator-Frye Inc. (sectional view).

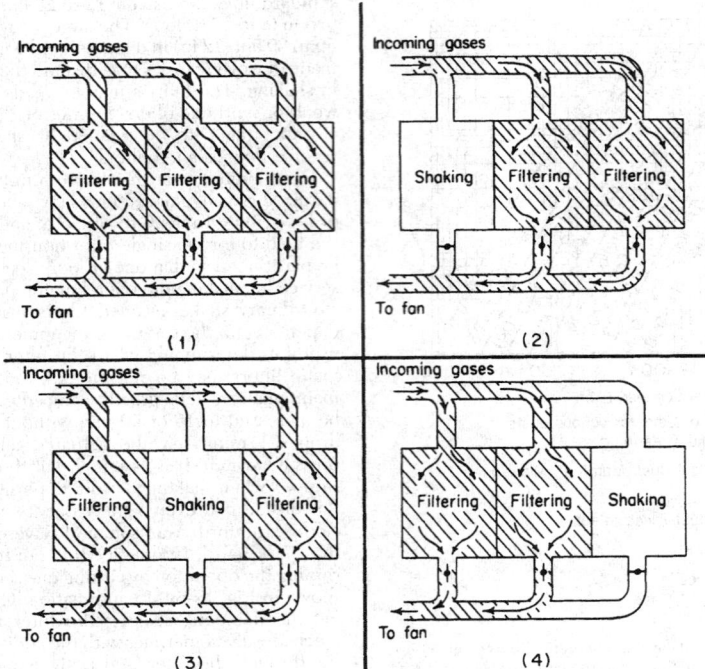

FIG. 17-59 Three-compartment bag filter at various stages in the cleaning cycle. (*Wheelabrator-Frye Inc.*)

Equation (17-11) may be rewritten as

$$\Delta p_i = K_d \mu c_d V_f^2 t_m \qquad (17\text{-}13)$$

where c_d = dust concentration in dirty gas, gr/ft³; and t_m = filtration time, min. This shows that the pressure drop due to dust accumulation varies as the square of the gas velocity through the filter. (The actual effect of velocity on pressure drop may be even greater in some instances.) Greater cloth area and reduced filtration velocity therefore afford substantial reductions in shaking frequency and in bag wear. Consequently, it is generally economical to be conservative in specifying cloth area. Shaker-cleaned filters are generally operated at filtration velocities of 0.3 to 2.5 m/min (1 to 8 ft/min) and at pressure drops of 0.5 to 1.5 kPa (2 to 6 in water). For very fine dusts or high dust concentrations, filtration velocities should not exceed 1 m/min (3 ft/min). For fine fumes and dusts in heavy-duty installations, filtration velocities of 0.3 to 0.6 m/min (1 to 2 ft/min) have long been accepted on the basis of operating experience.

Cyclone precleaners are sometimes used to reduce the dust load on the filter or to remove large hot cinders or other materials that might damage the bags. However, reducing the dust load on the filter by this means may not reduce the pressure drop, since the increase in K_2 produced by the reduction in average particle size may compensate for the decrease in the fabric dust loading.

In filter operation, it is essential that the gas be kept above its dew point to avoid water-vapor condensation on the bags and resulting plugging of the bag pores. However, fabric filters have been used successfully in steam atmospheres, such as those encountered in vacuum dryers. In such cases, the housing is generally steam-traced.

Reverse-flow-cleaned filters are generally similar to the shaker-cleaned filters except for the elimination of the shaker. After the flow of dirty gas has stopped, a fan is used to force clean gas through the bags from the clean-gas side. This flow of gas partly collapses the bags and dislodges the collected dust, which falls to the dust hopper. Rings are usually sewn into the bags at intervals along the length to prevent complete collapse, which would obstruct the fall of the dislodged dust. The principal applications of reverse-flow cleaning are in units using fiberglass fabric bags for dust collection at temperatures above 150°C (300°F). Collapsing and reinflation of the bags can be made sufficiently gentle to avoid putting excessive stresses on the fiberglass fabrics [Perkins and Imbalzano, "Factors Affecting the Bag Life Performance in Coal-Fired Boilers," 3d APCA Specialty Conference on the User and Fabric Filtration Equipment, Niagara Falls, N.Y., 1978; and Miller, *Power*, **125**(8), 78 (1981)]. As with shaker-cleaned filters, compartments of the baghouse are taken off line sequentially for bag cleaning. The gas for reverse-flow cleaning is commonly supplied in

an amount necessary to give a superficial velocity through the bags of 0.5 to 0.6 m/min (1.5 to 2.0 ft/min), which is the same range as the filtration velocities frequently used.

In the reverse-pulse filter (frequently termed a reverse-jet filter), the filter bag forms a sleeve that is drawn over a wire cage, which is usually cylindrical (Fig. 17-60). The cage supports the fabric on the clean-gas side, and the dust is collected on the outside of the bag. A venturi nozzle is located in the clean-gas outlet from the bag. For cleaning, a jet of high-velocity air is directed through the venturi nozzle and into the bag, inducing a flow of cleaned gas to enter the bag and flow through the fabric to the dirty-gas side. The high-velocity jet is released in a sudden, short pulse (typical duration 100 ms or less) from a compressed-air line by a solenoid valve. The pulse of air and clean gas expands the bag and dislodges the collected dust. Rows of bags are cleaned in a timed sequence by programmed operation of the solenoid valves. The pressure of the pulse is sufficient to dislodge the dust without cessation of the gas flow through the filter unit.

It has been a common practice to clean the bags on line (i.e., without stopping the flow of dirty gas into the filter), and reverse-pulse bag filters have been built without division into multiple compartments. However, investigations [Leith et al., *J. Air Pollut. Control Assoc.*, **27**, 636 (1977)] and experience have shown that, with online cleaning of reverse-pulse filters, a large fraction of the dust dislodged from the bag being cleaned may redeposit on neighboring bags rather than fall to the dust hopper. As a result, there is a growing trend to off-line cleaning of reverse-pulse filters. The baghouse is sectionalized so that the outlet-gas plenum serving the bags in a section can be closed off from the clean-gas exhaust, thereby stopping the flow of inlet gas through the bags. On the dirty-gas side of the tube sheet, the bags of the section are separated by partitions from the neighboring sections, where filtration is continuing. Sections of the filter are cleaned in rotation, as in shaker and reverse-flow filters.

Some manufacturers are using relatively low-pressure air (100 kPa, or 15 lbf/in², instead of 690 kPa, or 100 lbf/in²) and are eliminating the venturi tubes for clean-gas induction. Others have eliminated the separate jet nozzles located at the individual bags and use a single jet to inject a pulse into the outlet-gas plenum.

Reverse-pulse filters are typically operated at higher filtration velocities (air-to-cloth ratios) than shaker or reverse-flow filters designed for the same duty. Filtration velocities may range from 1 to 4.5 m/min (3 to 15 ft/min), depending on the dust being collected, but for most dusts the commonly used range is about 1.2 to 2.5 m/min (4 to 8 ft/min). The frequency of cleaning is also dependent on the nature and concentration of the dust, with the intervals between pulses varying from about 2 to 15 min.

The cleaning action of the pulse is so effective that the dust layer may be completely removed from the surface of the fabric. Consequently, the fabric itself must serve as the principal filter medium for at least a substantial part of the filtration cycle. Woven fabrics are unsuitable for such service, and felts of various types must be used. The bulk of the dust is still removed in a surface layer, but the felt ensures that an adequate collection efficiency is maintained until the dust layer has formed.

Filter Fabrics   The cost of the filter bags represents a substantial part of the erected cost of a bag filter—typically 5 to 20 percent, depending on the bag material [Reigel and Bundy, *Power*, **121**(1), 68 (1977)]. The cost of bag repair and replacement is the largest component of the cost of bag-filter maintenance. Consequently, the proper choice of filter fabric is critical to both the technical performance and the economics of operating a filter. With the advent of synthetic fibers, it has become possible to produce fabrics having a wide range of properties. However, demonstrating the acceptability of a fabric still depends on experience with prolonged operation under the actual or simulated conditions of the proposed application. The choice of a fabric material for a given service is necessarily a compromise, since no single material possesses all the properties that may be desired. Following the choice of material, the type of fabric construction is critical.

Two principal types of fabric are adaptable to filter use: woven fabrics, which are used in shaker and reverse-flow filters; and felts, which are used in reverse-pulse filters. The felts made from synthetic fibers are needle felts (i.e., felted on a needle loom) and are normally rein-

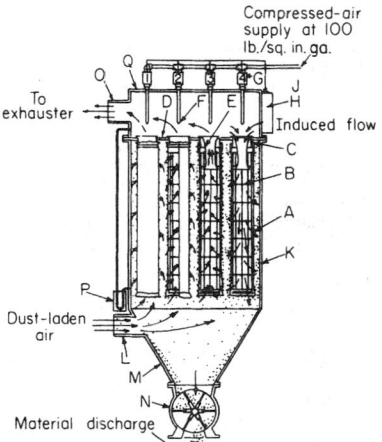

FIG. 17-60 Reverse-pulse fabric filter: (*a*) filter cylinders; (*b*) wire retainers; (*c*) collars; (*d*) tube sheet; (*e*) venturi nozzle; (*f*) nozzle or orifice; (*g*) solenoid valve; (*h*) timer; (*j*) air manifold; (*k*) collector housing; (*l*) inlet; (*m*) hopper; (*n*) air lock; (*o*) upper plenum. (*Mikropul Division, U.S. Filter Corp.*)

forced with a woven insert. The physical properties and air permeabilities of some typical woven and felt filter fabrics are presented in Tables 17-6 and 17-7. The "air permeability" of a filter fabric is defined as the flow rate of air in cubic feet per minute (at 70°F, 1 atm) that will pass through 1 ft² of clean fabric under an applied differential pressure of ½ in water. The resistance coefficient K_F of the clean fabric is defined by the equation in Table 17-6, which may be used to calculate the value of K_F from the air permeability. If Δp_i is taken as 0.5 in water, μ as 0.0181 cP (the viscosity of air at 70°F and 1 atm), and V_f as the air permeability, then $K_F = 27.8/$air permeability.

Collection Efficiency   The inherent collection efficiency of fabric filters is usually so high that, for practical purposes, the precise

level has not commonly been the subject of much concern. Furthermore, for collection of a given dust, the efficiency is usually fixed by the choices of filter fabric, filtration velocity, method of cleaning, and cleaning cycle, leaving few if any controllable variables by which efficiency can be further influenced. Inefficiency usually results from bags that are poorly installed, torn, or stretched from excessive dust loading and pressure drop. Of course, certain types of fabrics may simply be unsuited for filtration of a particular dust, but usually this will soon become obvious.

Few basic studies of the efficiency of bag filters have been made. Increased dust penetration immediately following cleaning has been readily observed while the dust layer is being reestablished. However,

TABLE 17-6   Resistance Factors and Air Permeabilities for Typical Woven Fabrics

Cloth	Pore size,° in	Threads/in	Weight, oz/yd²	Thread° diameter, in	K_F†	Air permeability, (ft³/min)/ft² at Δp_i = ½ in H₂O
Osnaburg cotton	0.01	32 × 28		0.02	0.51	55
Osnaburg cotton (soiled)‡		32 × 28			4.80	5.8
Drill cotton	0.01	68 × 40	5.28	0.01	0.093	300
Cotton§		46 × 56			1.39	20
Cotton§		104 × 68			1.54	18
Cotton sateen (unnapped)	0.007	96 × 56	6.88	0.009	0.27	103
Cotton sateen (unnapped)	0.005	96 × 64	8.23	0.01	0.88	32
Cotton sateen (unnapped)		96 × 60		0.012	1.63	17
Cotton sateen (unnapped)	0.004	96 × 56	10.2	0.011	1.12	25
Wool					0.25	111
Wool		40 × 50	11.5	0.014	0.33	84
Wool, white§		36 × 32			0.15	185
Wool, black§		28 × 30			0.25	110
Wool§		30 × 26			0.51	55
Vinyon§		37 × 37			0.12	23
Nylon tackle twill		72 × 196		0.010	0.66	42
Nylon sailcloth		130 × 130		0.007	1.66	17
Nylon§		37 × 37			1.74	16
Nylon§					3.71	7.5
Asbeston§					0.56	50
Orlon§		72 × 72			0.66	42
Orlon§		74 × 38			0.75	37
Orlon§					1.16	24
Orlon§					1.98	14
Smoothtex nickel screen		(300 mesh)			0.16	174
Glass		32 × 28		0.03	1.60	17
Dacron		60 × 40	5.8		0.84	33
Dacron		76 × 48	13.4		0.29	9.5
Teflon		76 × 70	8.7		1.39	20

°Estimates based on microscopic examination.
†Measured with atmospheric air. This value will be constant only for streamline flow, which is the case for values of $\rho V_f/\mu$ of less than approximately 100.

$$K_F = \Delta p_i/\mu V_f$$

where Δp_i = pressure drop, in water; μ = gas viscosity, cP; V_f = superficial gas velocity through cloth, ft/min; and ρ = gas density, lb/ft³.
‡Cloth, similar to previous one, that had been in service and contained dust in pores although free of surface accumulation.
§Data from Pring, *Air Pollution,* McGraw-Hill, New York, 1952, p. 280.

TABLE 17-7   Physical Properties of Selected Felts for Reverse-Pulse Filters

Fiber	Weight, oz/yd²	Thickness, in	Breaking strength, lbf/in width	Elongation, % to rupture	Air permeability, (ft³/min)/ft² at Δp_i = ½ in water	K_F
Wool	23.1	0.135			27.1	1.03
Wool	21.2	0.129			29.8	0.93
Orlon°	10.9	0.045	65	18	20–25	1.11–1.39
Orlon°	17.9	0.088	85	18	15–20	1.39–1.85
Orlon°	24	0.125	110	60	10–20	1.39–2.78
Acrilan°	17.9	0.075	100	22	15–20	1.39–1.85
Dynel°	24	0.125	60	80	30–40	0.70–0.93
Dacron°	17.9	0.080	125	22	15–20	1.39–1.85
Dacron°	9.9	0.250	20	150	200–225	0.11–0.14
Dacron°	24	0.125	175	80	20–30	0.93–1.39
Nylon°	24	0.125	100	100	30–40	0.70–0.93
Arnel°	24	0.125	60	80	30–40	0.70–0.93
Teflon	15.6	0.053			82.5	0.34
Teflon	43.5	0.119			21.6	1.29

°These data courtesy of American Felt Co.

field and laboratory studies have indicated that during the rest of the filtration cycle the effluent-dust concentration tends to remain constant regardless of the inlet concentration [Dennis, *J. Air Pollut. Control Assoc.*, **24**, 1156 (1974)]. In addition, there has been little indication that the penetration is strongly related to dust-particle size, except possibly in the low-submicrometer range. These observations appear to be generally consistent with sieving being the principal collection mechanism.

Leith and First [*J. Air Pollut. Control Assoc.*, **27**, 534 (1977); **27**, 754 (1977)] studied the collection efficiency of reverse-pulse filters and concluded that once the dust cake has been established, "straight-through" penetration by dust particles that pass through the filter without being stopped is negligible by comparison with penetration by dust that actually deposits initially and then "seeps" through the fabric to be reentrained into the exit air stream. They also noted that "pinholes" may form in the dust cake, particularly over pores between yarns in a woven fabric, and that particles may subsequently penetrate straight through at the pinholes. The formation of pinholes, or "cake puncture," had been observed earlier by Stephan et al. [*Am. Ind. Hyg. Assoc. J.*, **21**, 1 (1960)], but without measurement of the associated loss of collection efficiency. When a supported flat filter medium with extremely fine pores (e.g., glass-fiber paper, membrane filter) was used, no cake puncture took place even with very high pressure differentials across the cake. However, puncture did occur when a cotton-sateen filter fabric was used as the cake support. The formation of pinholes with certain combinations of dusts, fabrics, and filtration conditions was also observed by Koscianowski et al. (EPA-600/7-78-056, 1978). Evidently puncture occurs when the local cake structure is not strong enough to maintain a bridge over the aperture represented by a large pore and the portion of the cake covering the pore is blown through the fabric. This suggests that formation of pinholes will be highly dependent on the strength of the surface forces between particles that produce flocculation of dusts. The seepage of a dust through a filter is probably also closely related to the strength of the surface forces.

Surface pores can be greatly reduced in size by coating what will become the dusty side of the filter fabric with a thin microporous membrane that is supported by the underlying fabric. That has the effect of decreasing the effective penetration, both by eliminating cake pinholes, and by preventing the seepage of dust that is dragged through the fabric by successive cleanings. A variety of different membrane-forming polymers can be used in compatible service. The most versatile and effective surface filtration membranes are microfibrous Teflon as already described by Brinckman and Maresca [*ASME Med. Waste Symp.* (1992)] in the section on dry scrubbing.

Granular-Bed Filters Granular-bed filters may be classified as "depth" filters, since dust particles deposit in depth within the bed of granules. The granules themselves present targets for the deposition of particles by inertia, diffusion, flow-line interception, gravity, and electrostatic attraction, depending on the dust and filter characteristics and the operating conditions. Other deposition mechanisms are minor at most. Although it is physically possible under some circumstances for a dust layer to form on the inlet face of the filter, the practical limits of gas pressure drop will normally have been reached long before a surface dust layer can be established.

Granular-bed filters may be divided into three classes:

1. *Fixed-bed, or packed-bed, filters.* These units are not cleaned when they become plugged with deposited dust particles but are broken up for disposal or simply abandoned. If they are constructed from fine granules (e.g., sand filters), they may be designed to give high collection efficiencies on fine dust particles. However, if such a filter is to have a reasonable operating life, it can be used only on a gas containing a low concentration of dust particles.

2. *Cleanable granular-bed filters.* In these devices provisions are made to separate the collected dust from the granules either continuously or periodically, so that the units can operate continuously on gases containing moderate to high dust concentrations. The necessity for cleaning and recycling the granules generally restricts the practical lower granule size to about 3 to 10 mm. This in turn makes it difficult to attain high collection efficiencies on fine particles with granule beds of reasonable depth and gas pressure drop.

3. *Fluidized-bed filters.* Fluidized beds of granules have received considerable study on theoretical and experimental levels but have not been applied on a practical commercial scale.

Fixed Granular-Bed Filters Fixed-bed filters composed of granules have received considerable theoretical and experimental study [Thomas and Yoder, *AMA Arch. Ind. Health*, **13**, 545 (1956); **13**, 550 (1956); Knettig and Beeckmans, *Can. J. Chem. Eng.*, **52**, 703 (1974); Schmidt et al., *J. Air Pollut. Control Assoc.*, **28**, 143 (1978); Tardos et al., *J. Air Pollut. Control Assoc.*, **28**, 354 (1978); and Gutfinger and Tardos, *Atmos. Environ.*, **13**, 853 (1979)]. The theoretical approach is the same as that used in the treatment of deep-bed fibrous filters.

Fibers for filter applications can be produced with diameters smaller than it is practical to obtain with granules. Consequently, most concern with filtration of fine particles has been focused on fibrous-bed rather than granular-bed filters. However, for certain specialized applications granular beds have shown some superior properties, such as greater dimensional stability. Granular-bed filters of special design (deep-bed sand filters) have been used since 1948 for removing radioactive particles from waste air and gas streams in atomic energy plants (Lapple, "Interim Report—200 Area Stack Contamination," U.S. AEC Rep. HDC-743, Oct. 11, 1948; Juvinall et al., "Sand-Bed Filtration of Aerosols: A Review of Published Information," U.S. AEC Rep. ANL-7683, 1970; and Burchsted et al., *Nuclear Air Cleaning Handbook,* U.S. ERDA 76-21, 1976). The filter characteristics needed included high collection efficiency on fine particles, large dust-holding capacity to give long operating life, and low maintenance requirements. The sand filters are as much as 2.7 m (9 ft) in depth and are constructed in graded layers with about a 2:1 variation in the granule size from one layer to the next. The air-flow direction is upward, and the granules decrease in size in the direction of the air flow. The bottom layer is composed of rocks about 5 to 7.5 cm (2 to 3 in) in diameter, and granule sizes in successive layers decrease to 0.3 to 0.6 mm (50 to 30 mesh) in the finest layer. With superficial face velocities of about 1.5 m/min (5 ft/min), gas pressure drops of clean filters have ranged from 1.7 to 2.8 kPa (7 to 11 in water). Collection efficiencies of up to 99.98 percent with a polydisperse dioctyl phthalate aerosol of 0.7-μm mean diameter have been reported (Juvinall et al., op. cit.). Operating lives of 5 years or more have been attained.

Cleanable Granular-Bed Filters The principal objective in the development of cleanable granular-bed filters is to produce a device that can operate at temperatures above the range that can be tolerated with fabric filters. In some of the devices, the granules are circulated continuously through the unit, then are cleaned of the collected dust and returned to the filter bed. In others, the granular bed remains in place but is periodically taken out of service and cleaned by some means, such as backflushing with air.

A number of moving-bed granular filters have used cross-flow designs. One form of cross-flow moving-granular-bed filter, produced by the Combustion Power Company (Fig. 17-61), is currently in commercial use in some applications. The granular filter medium consists of 1/8- to 1/4-in (3- to 6-mm) pea gravel. Gas face velocities range from 30 to 46 m/min (100 to 150 ft/min), and reported gas pressure drops are in the range of 0.5 to 3 kPa (2 to 12 in water). The original form of the device [Reese, *TAPPI*, **60**(3), 109 (1977)] did not incorporate electrical augmentation. Collection efficiencies for submicrometer particles were low, and the electrical augmentation was added to correct the deficiency (Parquet, "The Electroscrubber Filter: Applications and Particulate Collection Performance," EPA-600/9-82-005c, 1982, p. 363). The electrostatic grid immersed in the bed of granules is charged to a potential of 20,000 to 30,000 V, producing an electric field between the grid and the inlet and outlet louvers that enclose the bed. No ionizing electrode is used to charge particles in the incoming gas; reliance is placed on the existence of natural charges on the dust particles. Individual dust particles commonly carry positive or negative charges even though the net charge on the dust as a whole is normally neutral. Depending on their charges, dust particles are attracted or repelled by the electrical field and are therefore caused to deposit on the rocks in the bed.

Self et al. ("Electrical Augmentation of Granular Bed Filters," EPA-600/9-80-039c, 1980, p. 309) demonstrated in theoretical studies and

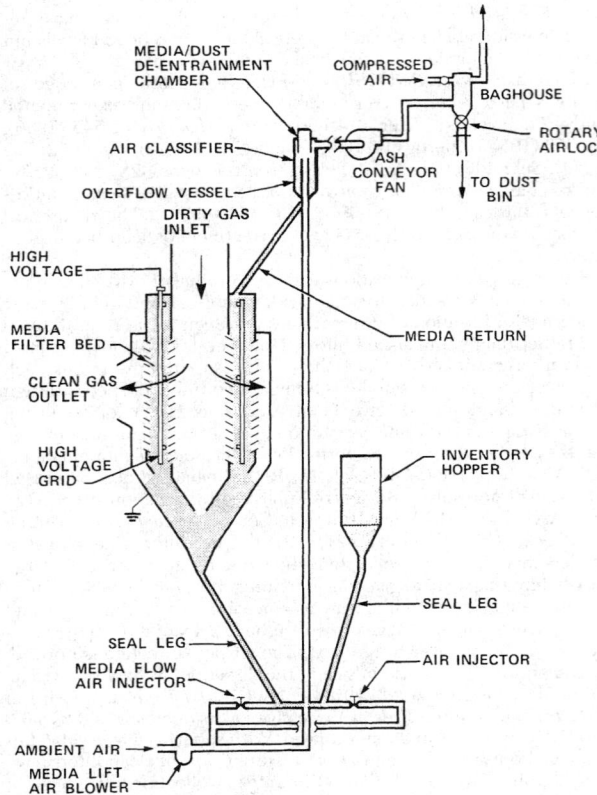

FIG. 17-61 Electrically augmented granular-bed filter. (*Combustion Power Company.*)

TABLE 17-8 Average Atmospheric-Dust Concentrations*

1 gr/1000 ft³ = 2.3 mg/m³

Location	Dust concentration, gr/1000 ft³
Rural and suburban districts	0.02–0.2
Metropolitan districts	0.04–0.4
Industrial districts	0.1–2.0
Ordinary factories or workrooms	0.2–4.0
Excessive dusty factories or mines	4.0–400

*Heating Ventilating Air Conditioning Guide, American Society of Heating, Refrigerating and Air-Conditioning Engineers, New York, 1960, p. 77.

ferent quantities of dust. Process-dust concentrations may run as high as several hundred grams per cubic meter (or grains per cubic foot) but usually do not exceed 45 g/m³ (20 gr/ft³). Atmospheric-dust concentrations that may be expected in various types of locations are shown in Table 17-8 and are generally below 12 mg/m³ (5 gr/1000 ft³).

The most frequent application of air filters is in cleaning atmospheric air for building ventilation, which usually requires only moderately high collection-efficiency levels. However, a variety of industrial operations developed mostly since the 1940s require air of extreme cleanliness, sometimes for pressurizing enclosures such as clean rooms and sometimes for use in a process itself. Examples of applications include the manufacture of antibiotics and other pharmaceuticals, the production of photographic film, and the manufacture and assembly of semiconductors and other electronic devices. Air cleaning at the necessary efficiency levels is accomplished by the use of high-efficiency fibrous filters that have been developed since the 1940s.

Air filters are also used to protect internal-combustion engines and gas turbines by cleaning the intake air. In some locations and applications, the atmospheric-dust concentrations encountered are much higher than those normally encountered in air-conditioning service.

High-efficiency air filters are sometimes used for emission control when particulate contaminants are low in concentration but present special hazards; cleaning of ventilation air and other gas streams exhausted from nuclear plant operations is an example.

Air-Filtration Theory Current high-efficiency air- and gas-filtration methods and equipment have resulted largely from the development of filtration theory since about 1930 and particularly since the 1940s. Much of the theoretical advance was originally encouraged by the requirements of the military and atomic energy programs. The fibrous filter has served both as a practical device and as a model for theoretical and experimental investigation. Extensive reviews and new treatments of air-filtration theory and experience have been presented by Chen [*Chem. Rev.*, **55**, 595 (1955)], Dorman ("Filtration," in Davies, *Aerosol Science*, Academic, New York, 1966), Pich (*Theory of Aerosol Filtration by Fibrous and Membrane Filters*, in ibid.), Davies (*Air Filtration*, Academic, New York, 1973), and Kirsch and Stechkina ("The Theory of Aerosol Filtration with Fibrous Filters," in Shaw, *Fundamentals of Aerosol Science*, Wiley, New York, 1978). The theoretical treatment of filtration starts with the processes of dust-particle deposition on collecting bodies, as outlined in Fig. 17-35 and Table 17-2. All the mechanisms shown in Table 17-2 may come into play, but inertial deposition, flow-line interception, and diffusional deposition are usually dominant. Electrostatic precipitation may become a major mechanism if the collecting body, the dust particle, or both, are charged. Gravitational settling is a minor influence for particles in the size range of usual interest. Thermal precipitation is nil in the absence of significant temperature gradients. Sieving is a possible mechanism only when the pores in the filter medium are smaller than or approximately equal to the particle size and will not be encountered in fibrous filters unless they are loaded sufficiently for a surface dust layer to form.

The theoretical prediction of the efficiency of collection of dust particles by a fibrous filter consists of three steps (Chen, op. cit.):

1. Calculation of the target efficiency η_o of an isolated fiber in an air stream having a superficial velocity the same as that in the filter

2. Determining the difference between the target efficiency of the isolated fiber and that of an individual fiber in the filter array η_f

laboratory experiments that such an augmentation system should yield substantial increases in the collection efficiency for fine particles if the particles carry significant charges. Significant improvements in the performance of the Combustion Power units with electrical augmentation have been reported by the manufacturer (Parquet, op. cit.).

Another type of gravel-bed filter, developed by GFE in Germany, has had limited commercial application in the United States [Schueler, *Rock Prod.*, **76**(7), 66 (1973); **77**(11), 39 (1974)]. After precleaning in a cyclone, the gas flows downward through a stationary horizontal filter bed of gravel. When the bed becomes loaded with dust, the gas flow is cut off, and the bed is backflushed with air while being stirred with a double-armed rake that is rotated by a gear motor. The backflush air also flows backward through the cyclone, which then acts as a dropout chamber. Multiple filter units are constructed in parallel so that individual units can be taken off the line for cleaning. The dust dislodged from the bed and carried by the backflush air is flocculated, and part is collected in the cyclone. The backflush air with the remaining suspended dust is cleaned in the other gravel-bed filter units that are operating on line. Performance tests made on one installation for the U.S. Environmental Protection Agency (EPA-600/7-78-093, 1978) did not give clear results but indicated that collection efficiencies were low on particles under 2 μm and that some of the dust in the backflush air was redispersed sufficiently to penetrate the operating filter units.

Air Filters The types of equipment previously described are intended primarily for the collection of process dusts, whereas air filters comprise a variety of filtration devices designed for the collection of particulate matter at low concentrations, usually atmospheric dust. The difference in the two categories of equipment is not in the principles of operation but in the adaptations required to deal with the dif-

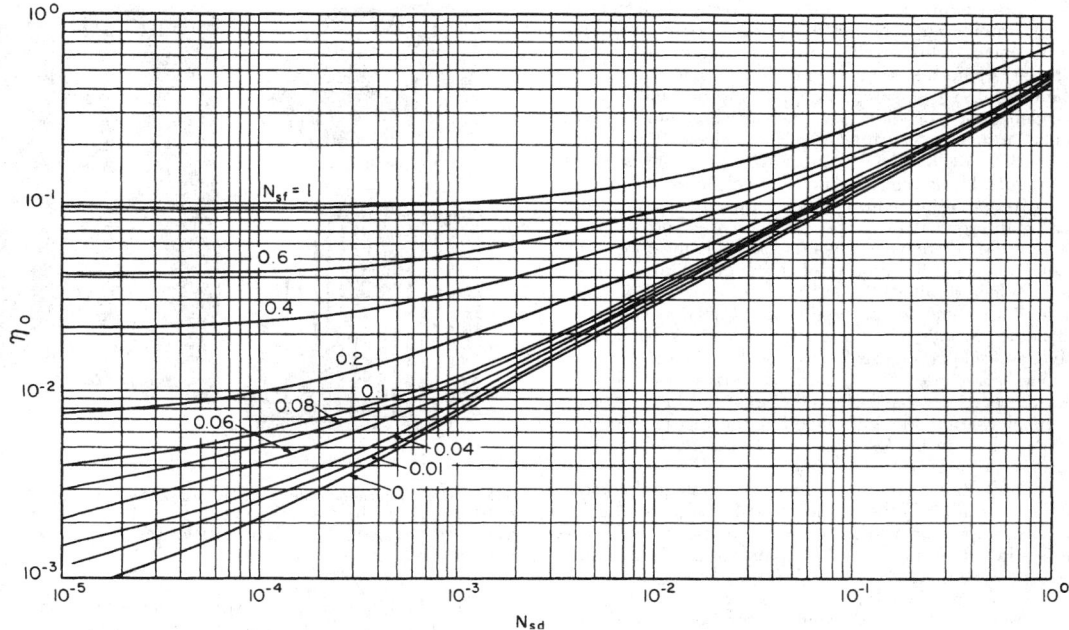

FIG. 17-62 Isolated fiber efficiency for combined diffusion and interception mechanism at $N_{Re} = 10^{-2}$. [*Chen*, Chem. Rev., **55**, 595 (1955).]

3.   Determining the collection efficiency of the filter η from the target efficiency of the individual fibers

The results of computations of η_o for an isolated fiber are illustrated in Figs. 17-62 and 17-63. The target efficiency η_t of an individual fiber in a filter differs from η_o for two main reasons (Pich, op. cit.): (1) the average gas velocity is higher in the filter, and (2) the velocity field around the individual fibers is influenced by the proximity of neighboring fibers. The interference effect is difficult to determine on a purely theoretical basis and is usually evaluated experimentally. Chen (op. cit.) expressed the effect with an empirical equation:

$$\eta_t = \eta_o[1 + K_\alpha(1 - \varepsilon_v)] \tag{17-14}$$

This indicates that the target efficiency of the fiber is increased by the proximity of other fibers. The value of K_α averaged 4.5 for values of the void fraction ε_v, ranging from 0.90 to 0.99. Extending use of the equation to values of ε_v lower than 0.90 may result in large errors.

The collection efficiency of the filter may be calculated from the fiber target efficiency and other physical characteristics of the filter (Chen, op. cit.):

$$N_t = \frac{4\eta_t L(1 - \varepsilon_v)}{\pi D_b \varepsilon_v} \tag{17-15}$$

where D_b = fiber diameter and L = filter thickness. The derivation of Eq. (17-15) assumes that (1) η_t is the same throughout the filter, (2) all fibers are of the same diameter D_b, are cylindrical and are normal to the direction of the gas flow, (3) the fraction of the particles deposited in any one layer of fiber is small, and (4) the gas passing through the filter is essentially completely remixed after it leaves one layer of the filter and before it enters the next. The first assumption requires that Eq. (17-15) apply only for particles of a single size for which there are corresponding values of η_t, η, and N_t.

For filters of high porosity, ε_v approaches unity and Eq. (17-15) reduces to the expression used by Wong et al. [*J. Appl. Phys.*, **27**, 161 (1956)] and Thomas and Lapple [*Am. Inst. Chem. Eng. J.*, **7**, 203 (1961)]:

$$N_t = \frac{4\eta_t L(1 - \varepsilon_v)}{\pi D_b} \tag{17-16}$$

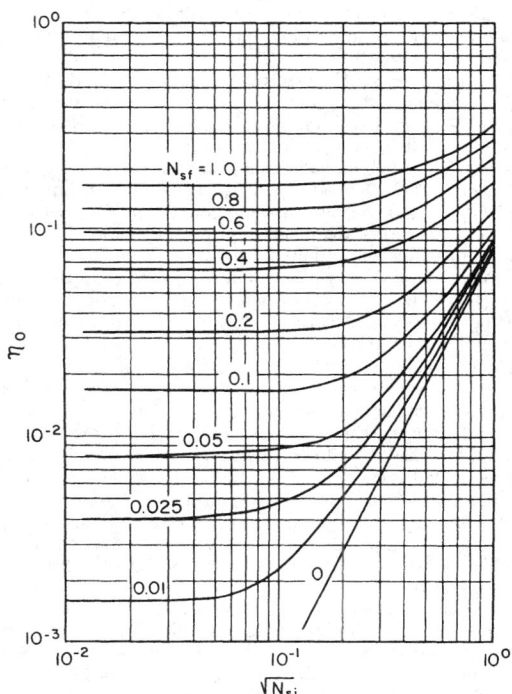

FIG. 17-63 Isolated fiber efficiency for combined inertia and interception mechanisms at $N_{Re} = 0.2$. [*Chen*, Chem. Rev., **55**, 595 (1955).]

The foregoing procedure is commonly employed in reverse to determine or confirm fiber target efficiencies from the experimentally determined efficiencies of fibrous filter pads.

Filtration theory assumes that a dust particle that touches a collector body adheres to it. This assumption appears to be valid in most cases, but evidence of nonadherence, or particle bouncing, has appeared in some instances. Wright et al. ("High Velocity Air Filters," WADC TR 55-457, ASTIA Doc. AD-142075, 1957) investigated the performance of fibrous filters at filtration velocities of 0.091 to 3.05 m/s (0.3 to 10 ft/s), using 0.3-μm and 1.4-μm supercooled liquid aerosols and a 1.2-μm solid aerosol. The collection efficiencies agreed well with theoretical predictions for the liquid aerosols and apparently also for the solid aerosol at filtration velocities under 0.3 m/s (1 ft/s). But at filtration velocities above 0.3 m/s some of the solid particles failed to adhere. With a filter composed of 30-μm glass fibers and a filtration velocity of 9.1 m/s (30 ft/s), there were indications that 90 percent of the solid aerosol particles striking a fiber bounced off.

Bouncing may be regarded as a defect in the particle-deposition process. However, particles that have been deposited in filters may subsequently be blown off and reentrained into the air stream (Corn, "Adhesion of Particles," in Davies, *Aerosol Science,* Academic, New York, 1966; and Davies, op. cit.).

The theories of filtration by a fibrous filter relate only to the initial efficiency of the clean filter in the "static" period of filtration before the deposition of any appreciable quantity of dust particles. The deposition of particles in a filter increases the number of targets available to intercept particles, so that collection efficiency increases as the filter loads. At the same time, the filter undergoes clogging and the pressure drop increases. No theory is available for dealing with the "dynamic" period of filtration in which collection efficiency and pressure drop vary with the loading of collected dust. The theoretical treatment of this filtration period is incomparably more complex than that for the "static" period. Investigators have noted that both the increase in collection efficiency and the increase in pressure drop are exponential functions of the loading of collected dust or are at least roughly so (Davies, op. cit.). Some empirical relationships have been derived for correlating data in particular instances.

The dust particles collected by a fibrous filter do not deposit in uniform layers on fibers but tend to deposit preferentially on previously deposited particles (Billings, "Effect of Particle Accumulation in Aerosol Filtration," Ph.D. dissertation, California Institute of Technology, Pasadena, 1966), forming chainlike agglomerates termed "dendrites." The growth of dendritic deposits on fibers has been studied experimentally [Billings, op. cit.; Bhutra and Payatakes, *J. Aerosol Sci.,* **10,** 445 (1979)], and Payatakes and coworkers [Payatakes and Tien, *J. Aerosol Sci.,* **7,** 85 (1976); Payatakes, *Am. Inst. Chem. Eng. J.,* **23,** 192 (1977); and Payatakes and Gradon, *Chem. Eng. Sci.,* **35,** 1083 (1980)] have attempted to model the growth of dendrites and its influence on filter efficiency and pressure drop.

Air-Filter Types Air filters may be broadly divided into two classes: (1) panel, or unit, filters; and (2) automatic, or continuous, filters. Panel filters are constructed in units of convenient size (commonly 20- by 20-in or 24- by 24-in face area) to facilitate installation, maintenance, and cleaning. Each unit consists of a cleanable or replaceable cell or filter pad in a substantial frame that may be bolted to the frames of similar units to form an airtight partition between the source of the dusty air and the destination of the cleaned air.

Panel filters may use either viscous or dry filter media. Viscous filters are so called because the filter medium is coated with a tacky liquid of high viscosity (e.g., mineral oil and adhesives) to retain the dust. The filter pad consists of an assembly of coarse fibers (now usually metal, glass, or plastic). Because the fibers are coarse and the media are highly porous, resistance to air flow is low and high filtration velocities can be used.

Dry filters are usually deeper than viscous filters. The dry filter media use finer fibers and have much smaller pores than the viscous media and need not rely on an oil coating to retain collected dust. Because of their greater resistance to air flow, dry filters must use lower filtration velocities to avoid excessive pressure drops. Hence, dry media must have larger surface areas and are usually pleated or arranged in the form of pockets (Fig. 17-64), generally sheets of cellulose pulp, cotton, felt, or spun glass.

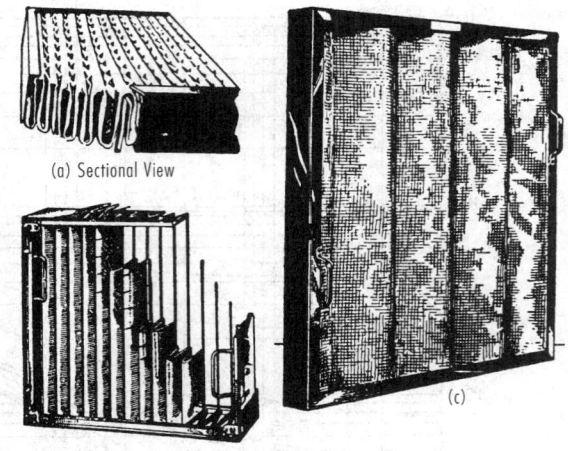

(a) Sectional View

(b) Cutaway View

(c)

FIG. 17-64 Typical dry filters. (*a*) Throwaway type, Airplex. (*Davies Air Filter Corporation.*) (*b*) Replaceable medium type, Airmat PL-24, cutaway view. (*American Air Filter Co., Inc.*) (*c*) Cleanable type, Amirglass sawtooth. (*Amirton Company.*)

Automatic filters are made with either viscous-coated or dry filter media. However, the cleaning or disposal of the loaded medium is essentially continuous and automatic. In most such devices the air passes horizontally through a movable filter curtain. As the filter loads with dust, the curtain is continuously or intermittently advanced to expose clean media to the air flow and to clean or dispose of the loaded medium. Movement of the curtain can be provided by a hand crank or a motor drive. Movement of a motor-driven curtain can be actuated automatically by a differential-pressure switch connected across the filter.

High-Efficiency Air Cleaning Air-filter systems for nuclear facilities and for other applications demanding extremely high standards of air purity require filtration efficiencies well beyond those attainable with the equipment described above. The *Nuclear Air Cleaning Handbook* (Burchsted et al., op. cit.) presents an extensive treatment of the requirements for and the design of such air-cleaning facilities. Much of the material is pertinent to high-efficiency air-filter systems for applications to other than nuclear facilities.

HEPA (high-efficiency particulate air) filters were originally developed for nuclear and military applications but are now widely used and are manufactured by numerous companies. By definition, an HEPA filter is a "throwaway, extended-medium dry-type" filter having (1) a minimum particle-removal efficiency of not less than 99.97 percent for 0.3-μm particles, (2) a maximum resistance, when clean, of 1.0 in water when operated at rated air-flow capacity, and (3) a rigid casing extending the full depth of the medium (Burchsted et al., op. cit.). The filter medium is a paper made of submicrometer glass fibers in a matrix of larger-diameter (1- to 4-μm) glass fibers. An organic binder is added during the papermaking process to hold the fibers and give the paper added tensile strength. Filter units are made in several standard sizes (Table 17-9).

TABLE 17-9 Standard HEPA Filters*

Face dimensions, in	Depth, less gaskets, in	Design air-flow capacity at clean-filter resistance of 1.0 in water (standard ft³/min)
24 × 24	11½	1000
24 × 24	5⅞	500
12 × 12	5⅞	125
8 × 8	5⅞	20
8 × 8	3 1/16	25

*Burchsted et al., *Nuclear Air Cleaning Handbook,* ERDA 76-21, Oak Ridge, Tenn., 1976.

Because HEPA filters are designed primarily for high efficiency, their dust-loading capacities are limited, and it is common practice to use prefilters to extend their operating lives. In general, HEPA filters should be protected from (1) lint, (2) particles larger than 1 to 2 μm in diameter, and (3) dust concentrations greater than 23 mg/m³ (10 gr/1000 ft³). Air filters used in nuclear facilities as prefilters and building-supply air filters are classified as shown in Table 17-10. The standard of the American Society of Heating, Refrigerating and Air-Conditioning Engineers (*Method of Testing Air Cleaning Devices Used in General Ventilation for Removing Particulate Matter*, ASHRAE 52-68, 1968) requires both a dust-spot (dust-stain) efficiency test made with atmospheric dust and a weight-arrestance test made with a synthetic test dust. A more precise comparison of the different groups of filters, based on removal efficiencies for particles of specific sizes, is presented in Table 17-11.

Table 17-12 presents the relative performance of Group I, II, and III filters with respect to air-flow capacity, resistance, and dust-holding capacity. The dust-holding capacities correspond to the manufacturers' recommended maximum allowable increases in air-flow resistance. The values for dust-holding capacity are based on tests with a synthetic dust and hence are relative. The actual dust-holding capacity in a specific application will depend on the characteristics of the dust encountered. In some instances it may be appropriate to use two or more stages of precleaning in air-filter systems to achieve a desired combination of operating life and efficiency. In very dusty locations, inertial devices such as multiple small cyclones may be used as first-stage separators.

Electrical Precipitators When particles suspended in a gas are exposed to gas ions in an electrostatic field, they will become charged and migrate under the action of the field. The functional mechanisms of electrical precipitation may be listed as follows:

1. Gas ionization
2. Particle collection
   a. Production of electrostatic field to cause charging and migration of dust particles
   b. Gas retention to permit particle migration to a collection surface
   c. Prevention of reentrainment of collected particles
   d. Removal of collected particles from the equipment

TABLE 17-10   Classification of Common Air Filters*

Group	Efficiency	Filter type	Stain test efficiency, %	Arrestance, %
I	Low	Viscous impingement, panel type	<20†	40–80†
II	Moderate	Extended medium, dry type	20–60†	80–96†
III	High	Extended medium, dry type	60–98‡	96–99†
HEPA	Extreme	Extended medium, dry type	100§	100†

*Burchsted et al., *Nuclear Air Cleaning Handbook,* ERDA 76-21, Oak Ridge, Tenn., 1976.
†Test using synthetic dust.
‡Stain test using atmospheric dust.
§ASHRAE/52-68, American Society of Heating, Refrigerating and Air-Conditioning Engineers.

TABLE 17-11   Comparison of Air Filters by Percent Removal Efficiency for Various Particle Sizes*

Group	Efficiency	Removal efficiency, %, for particle size of			
		0.3 μm	1.0 μm	5.0 μm	10.0 μm
I	Low	0–2	10–30	40–70	90–98
II	Moderate	10–40	40–70	85–95	98–99
III	High	45–85	75–99	99–99.9	99.9
HEPA	Extreme	99.97 min	99.99	100	100

*Burchsted et al., *Nuclear Air Cleaning Handbook,* ERDA 76-21, Oak Ridge, Tenn., 1976.

TABLE 17-12   Air-Flow Capacity, Resistance, and Dust-Holding Capacity of Air Filters*

Group	Efficiency	Air-flow capacity, ft³/(min·ft² of frontal area)	Resistance, in water		Dust-holding capacity, g/(1000 ft³·min of air-flow capacity)
			Clean filter	Used filter	
I	Low	300–500	0.05–0.1	0.3–0.5	50–1000
II	Moderate	250–750	0.1–0.5	0.5–1.0	100–500
III	High	250–750	0.20–0.5	0.6–1.4	50–200

*Burchsted et al., *Nuclear Air Cleaning Handbook,* ERDA 76-21, Oak Ridge, Tenn., 1976.

There are two general classes of electrical precipitators: (1) single-stage, in which ionization and collection are combined; (2) two-stage, in which ionization is achieved in one portion of the equipment, followed by collection in another. Various types in each class differ essentially in the details by which each function is accomplished.

The underlying theory presented in the following paragraphs assumes that the dust concentration is small, since only very incomplete evaluations for conditions of high dust concentration have been made.

Field Strength Whereas the applied potential or voltage is the quantity commonly known, it is the field strength that determines behavior in an electrostatic field. When the current flow is low (i.e., before the onset of spark or corona discharge), these are related by the following equations for two common forms of electrodes:

Parallel plates:

$$\mathscr{E} = E/B_e \tag{17-17}$$

Concentric cylinders (wire-in-cylinder):

$$\mathscr{E} = \frac{E}{r \ln (D_t/D_d)} \tag{17-18}$$

The field strength is uniform between parallel plates, whereas it varies in the space between concentric cylinders, being highest at the surface of the central cylinder. After corona sets in, the current flow will become appreciable. The field strength near the center electrode will be less than given by Eq. (17-18) and that in the major portion of the clearance space will be greater and more uniform [see Eqs. (17-23) and (17-24)].

Potential and Ionization In order to obtain gas ionization it is necessary to exceed, at least locally, the electrical breakdown strength of the gas. Corona is the name applied to such a local discharge that fails to propagate itself. Sparking is essentially an advanced stage of corona in which complete breakdown of the gas occurs along a given path. Since corona represents a local breakdown, it can occur only in a nonuniform electrical field (Whitehead, *Dielectric Phenomena—Electrical Discharge in Gases,* Van Nostrand, Princeton, N.J., 1927, p. 40). Consequently, for parallel plates, only sparking occurs at a field strength or potential difference given by the empirical expressions

$$\mathscr{E}_s = \mathscr{E}_o k_\rho \left[1 + \left(\frac{K_o}{k_\rho B_e} \right) \right] \tag{17-19}$$

$$E_s = \mathscr{E}_o k_\rho B_e + K_o \mathscr{E}_o \tag{17-20}$$

For air in the range of $k_\rho B_e$ from 0.1 to 2, $\mathscr{E}_o = 111.2$ and $K_o = 0.048$. Thornton [*Phil. Mag.,* **28**(7), 666 (1939)] gives values for other gases. For concentric cylinders (Loeb, *Fundamental Processes of Electrical Discharge in Gases,* Wiley, New York, 1939; Peek, *Dielectric Phenomena in High-Voltage Engineering,* McGraw-Hill, New York, 1929; and Whitehead, op. cit.), corona sets in at the central wire when

$$\mathscr{E}_c = \mathscr{E}_o k_\rho \left(1 + \sqrt{\frac{K_o}{k_\rho D_d}} \right) \tag{17-21}$$

$$E_c = \left(\frac{\mathscr{E}_o k_\rho D_d}{2} \right) \left(1 + \sqrt{\frac{K_o}{k_\rho D_d}} \right) \ln \left(\frac{D_t}{D_d} \right) \tag{17-22}$$

TABLE 17-13 Sparking Potentials* (Small Wire Concentric in Pipe)

Pipe diameter, in	Sparking potential,† volts	
	Peak	Root mean square
4	59,000	45,000
6	76,000	58,000
9	90,000	69,000
12	100,000	77,000

*Data reported by Anderson in Perry, "Chemical Engineers' Handbook," 2d ed., p. 1873, McGraw-Hill, New York, 1941.

†For gases at atmospheric pressure, 100°F, containing water vapor, air, CO_2, and mist, and negative-discharge-electrode polarity.

For air approximate values are $\mathscr{E}_o = 110$, $K_o = 0.18$. Corona, however, will set in only if $(D_t/D_d) > 2.718$. If this ratio is less than 2.718, no corona occurs, and only sparking will result, following the laws given by Eqs. (17-21) and (17-22) (Peek, op. cit.).

In practice, precipitators are usually operated at the highest voltage practicable without sparking, since this increases both the particle charge and the electrical precipitating field. The sparking potential is generally higher with a negative charge on the discharge electrode and is less erratic in behavior than a positive corona discharge. It is the consensus, however, that ozone formation with a positive discharge is considerably less than with a negative discharge. For these reasons negative discharge is generally used in industrial precipitators, and a positive discharge is utilized in air-conditioning applications. In Table 17-13 are given some typical values for the sparking potential for the case of small wires in pipes of various sizes. The sparking potential varies approximately directly as the density of the gas but is very sensitive to the character of any material collected on the electrodes. Even small amounts of poorly conducting material on the electrodes may markedly lower the sparking voltage. For positive polarity of the discharge electrode, the sparking voltage will be very much lower. The sparking voltage is greatly affected by the temperature and humidity of the gas, as shown in Fig. 17-65.

Current Flow Corona discharge is accompanied by a relatively small flow of electric current, typically 0.1 to 0.5 mA/m² of collecting-electrode area (projected, rather than actual area). Sparking usually involves a considerably larger flow of current which cannot be tolerated except for occasional periods of a fraction of a second duration, and then only when suitable electrical controls are provided to limit the current. However, when suitable controls are provided, precipitators have been operated continuously with a small amount of sparking

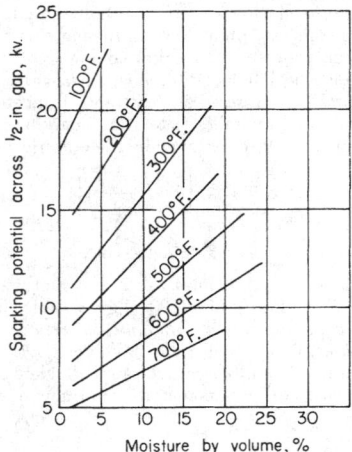

FIG. 17-65 Sparking potential for negative point-to-plane ½-in (1.3-cm) gap as a function of moisture content and temperature of air at 1-atm (101.3-kPa) pressure. [*Sproull and Nakada*, Ind. Eng. Chem., **43**, 1356 (1951).]

to ensure that the voltage is in the correct range to ensure corona. Besides disruptive effects on the electrical equipment and electrodes, sparking will result in low collection efficiency because of reduction in applied voltage, redispersion of collected dust, and current channeling. Although an exact calculation can be made for the current flow for a direct-current potential applied between concentric cylinders, the following simpler expression, based on the assumption of a constant space charge or ion density, gives a good approximation of corona current [Ladenburg, *Ann. Phys.*, **4**(5), 863 (1930)]:

$$I = \frac{8\lambda_i E(E - E_c)}{D_t^2 \ln (D_t/D_d)} \qquad (17\text{-}23)$$

and the average space charge is given by (Whitehead, op. cit.)

$$\sigma_{avg} = \frac{4(E - E_c)}{\pi D_t^2 \varepsilon} \qquad (17\text{-}24)$$

In the space outside the immediate vicinity of corona discharge, the field strength is sensibly constant, and an average value is given by

$$\mathscr{E} = \sqrt{2I/\lambda_i} \qquad (17\text{-}25)$$

which applies if the potential difference is above the critical potential required for corona discharge so that an appreciable current flows.

Ionic mobilities are given by Loeb (*International Critical Tables*, vol. 6, McGraw-Hill, New York, 1929, p. 107). For air at 0°C, 760 mmHg, $\lambda_i = 624$ (cm/s)/(statV/cm) for negative ions. Positive ions usually have a slightly lower mobility. Loeb (*Basic Processes of Gaseous Electronics*, University of California Press, Berkeley and Los Angeles, 1955, p. 53) gives a theoretical expression for ionic mobility of gases which is probably good to within ±50 percent:

$$\lambda_i = \frac{100.0}{k_\rho \sqrt{(\delta_g - 1)M}} \qquad (17\text{-}26)$$

In general, ionic mobilities are inversely proportional to gas density. Ionic velocities in the usual electrostatic precipitator are on the order of 30.5 m/s (100 ft/s).

Electric Wind By virtue of the momentum transfer from gas ions moving in the electrical field to the surrounding gas molecules, a gas circulation, known as the "electric" or "ionic" wind, is set up between the electrodes. For conditions encountered in electrical precipitators, the velocity of this circulation is on the order of 0.6 m/s (2 ft/s). Also, as a result of this momentum transfer, the pressure at the collecting electrode is slightly higher than at the discharge electrode (Whitehead, op. cit., p. 167).

Charging of Particles [Deutsch, *Ann. Phys.*, **68**(4), 335 (1922); **9**(5), 249 (1931); **10**(5), 847 (1931); Ladenburg, op. cit.; and Mierdel, *Z. Tech. Phys.*, **13**, 564 (1932).] Three forces act on a gas ion in the vicinity of a particle: attractive forces due to the field strength and the ionic image; and repulsive forces due to the Coulomb effect. For spherical particles larger than 1-μm diameter, the ionic image effect is negligible, and charging will continue until the other two forces balance according to the equation

$$N_o = \left(\frac{\zeta \mathscr{E} D_p^2}{4\varepsilon} \right) \left(\frac{\pi \sigma \varepsilon t \lambda_i}{1 + \pi \sigma \varepsilon \lambda_i t} \right) \qquad (17\text{-}27)$$

The ultimate charge acquired by the particle is given by

$$N_o = \zeta \mathscr{E} D_p^2 / 4\varepsilon \qquad (17\text{-}28)$$

and is very nearly attained in a fraction of a second. For particles smaller than 1-μm diameter, the initial charging will occur according to Eq. (17-27). However, owing to the ionic-image effect, the ultimate charge will be considerably greater because of penetration resulting from the kinetic energy of the gas ions. For charging times of the order encountered in electrical precipitation, the ultimate charge acquired by spherical particles smaller than about 1-μm diameter may be approximated (±30 percent) by the empirical expression

$$N_o = 3.4 \times 10^3 D_p T \qquad (17\text{-}29)$$

Values of N_o for various sized particles are listed in Table 17-14 for 70°F, $\zeta = 2$, and $\mathscr{E} = 10$ statV/cm.

Particle Mobility By equating the electrical force acting on a

TABLE 17-14 Charge and Motion of Spherical Particles in an Electric Field

For $\zeta = 2$, and $\varepsilon = \varepsilon_i = \varepsilon_p = 10$ statV/cm

Particle diam., μ	Number of elementary electrical charges, N_0	Particle migration velocity,° u_e, ft/sec
0.1	10	0.27
.25	25	.15
.5	50	.12
1.0	105	.11
2.5	655	.26
5.0	2,620	.50
10.0	10,470	.98
25.0	65,500	2.40

NOTE: To convert feet per second to meters per second, multiply by 0.3048.

TABLE 17-15 Performance Data on Typical Single-Stage Electrical Precipitator Installations*

Type of precipitator	Type of dust	Gas volume, cu ft/min	Average gas velocity, ft/sec	Collecting electrode area, sq ft	Over-all collection efficiency, %	Average particle migration velocity, ft/sec
Rod curtain	Smelter fume	180,000	6	44,400	85	0.13
Tulip type	Gypsum from kiln	25,000	3.5	3,800	99.7	.64
Perforated plate	Fly ash	108,000	6	10,900	91	.40
Rod curtain	Cement	204,000	9.5	26,000	91	.31

*Research-Cottrell, Inc. To convert cubic feet per minute to cubic meters per second, multiply by 0.00047; to convert feet per second to meters per second, multiply by 0.3048; and to convert square feet to square meters, multiply by 0.0929.

particle to the resistance due to air friction, as expressed by Stokes' law, the particle velocity or mobility may be expressed by

1. For particles larger than 1-μm diameter:

$$\lambda_p = \left(\frac{u_e}{\mathscr{E}_p}\right) = \frac{\zeta D_p \mathscr{E}_i K_m}{12\pi\mu} \qquad (17\text{-}30)$$

2. For particles smaller than 1-μm diameter:

$$\lambda_p = \left(\frac{u_e}{\mathscr{E}_p}\right) = \frac{360 K_m \varepsilon T}{\mu} \qquad (17\text{-}31)$$

For single-stage precipitators, $\mathscr{E}_i$ and $\mathscr{E}_p$ may be considered as essentially equal. It is apparent from Eq. (17-31) that the mobility in an electric field will be almost the same for all particles smaller than about 1-μm diameter, and hence, in the absence of reentrainment, collection efficiency should be almost independent of particle size in this range. Very small particles will actually have a greater mobility because of the Stokes-Cunningham correction factor. Values of u_e are listed in Table 17-14 for 70°F, $\zeta = 2$, and $\mathscr{E} = \mathscr{E}_i = \mathscr{E}_p = 10$ statV/cm.

Collection Efficiency Although actual particle mobilities may be considerably greater than would be calculated on the basis given in the preceding paragraph because of the action of the electric wind in single-stage precipitators, the latter acts in a compensating fashion, and the overall effect of the electric wind is probably to provide an equalization of particle concentration between the electrodes similar to the action of normal turbulence (Mierdel, op. cit.). On this basis Deutsch (op. cit.) has derived the following equations for collection efficiency, the form of which had previously been suggested by Anderson on the basis of experimental data:

$$\eta = 1 - e^{-(u_e A_e/q)} = 1 - e^{-K_e t_e} \qquad (17\text{-}32)$$

For the concentric-cylinder (or wire-in-cylinder) type of precipitator, $K_e = 4L_e/D_t V_e$; for rod-curtain or wire-plate types, $K_e = L_e/B_e V_e$. Strictly speaking, Eq. (17-32) applies only for a given particle size, and the overall efficiency must be obtained by an integration process for a specific dust distribution, as described in the subsection "Cyclone Separators." However, over limited ranges of performance conditions, Eq. (17-32) has been found to give a good approximation of overall collection efficiency, with the term for particle migration velocity representing an empirical average value. Such values, calculated from overall collection-efficiency measurements, are given in Table 17-15 for specific installations.

For two-stage precipitators with close collecting-plate spacings (Figs. 20-152, 20-153), the gas flow is substantially streamline, and no electric wind exists. Consequently, with reentrainment neglected, collection efficiency may be expressed as [Penny, *Electr. Eng.*, **56**, 159 (1937)]

$$\eta = u_e L_e / V_e B_e \qquad (17\text{-}33)$$

which holds for values of $\eta \leq 1.0$. In practice, however, extraneous factors may cause the actual efficiency to approach a relationship of the type given by Eq. (17-32).

Application The theoretical considerations that have been expounded should be used only for order-of-magnitude estimates, since a number of extraneous factors may enter into actual performance. In actual installations rectified alternating current is em-

ployed. Hence the electric field is not fixed but varies continuously, depending on the waveform of the rectifier, although Schmidt and Anderson [*Electr. Eng.*, **57**, 332 (1938)] report that the waveform is not a critical factor. Allowances for high dust concentrations have not been fully studied, although Deutsch (op. cit.) has presented a theoretical approach. In addition, irregularities on the discharge electrode will result in local discharges. Such irregularities can readily result from dust incrustation on the discharge electrodes due to charging of particles with opposite polarity within the thin but appreciable flow or ionization layer surrounding this electrode. Very high dust loadings increase the potential difference required for corona and reduce the current due to the space charge of the particles. This tends to reduce the average particle charge and reduces collection efficiency. This can be compensated for by increasing the potential difference when high dust loadings are involved.

Several investigators have attempted to modify the basic Deutsch equation so that it would more nearly describe precipitator performance. Cooperman ("A New Theory of Precipitator Efficiency," Pap. 69-4, APCA meeting, New York, 1969) introduced correction factors for diffusional forces arising from variations in particle concentration along the precipitator length and also perpendicular to the collecting surface. Robinson [*Atmos. Environ.* **1**(3), 193 (1967)] derived an equation for collection efficiency in which two erosion or reentrainment terms are introduced.

An analysis of precipitator performance based on theoretical considerations was undertaken by the Southern Research Institute for the National Air Pollution Control Administration (Nichols and Oglesby, "Electrostatic Precipitator Systems Analysis," AIChE annual meeting, 1970). A mathematical model was developed for calculating the particle charge, electric field, and collection efficiency based on the Deutsch-Anderson equation. The system diagram is shown in Fig. 17-66. This system-analysis method, using high-speed computers, makes it possible to analyze what takes place in each increment of precipitator length. Collection efficiency versus particle size is computed for each 1 ft (0.3 m) of gas travel, and the inlet particle-size distribution is modified accordingly. Computed overall efficiencies compare well with measured values on three precipitators. The model assumes that field charging is the only charging mechanism. The authors considered the addition of several refinements to the program: the influence of diffusion charging; reentrainment effects due to rapping and erosion; and loss of efficiency due to maldistribution of gas, dust resistivity, and gas-property effects. The modeling technique appeared promising, but much more work was needed before it could be used for design. The same authors prepared a general treatise (Oglesby and Nichols, *A Manual of Electrostatic Precipitator Technology*, parts I and II, Southern Research Institute, Birmingham, Ala., U.S. Government Publications PB196360, 196381, 1970).

High-Pressure-High-Temperature Electrostatic Precipitation In general, increased pressure increases precipitation efficiency, although a somewhat higher potential is required, because it reduces ion mobility and hence increases the potential required for corona

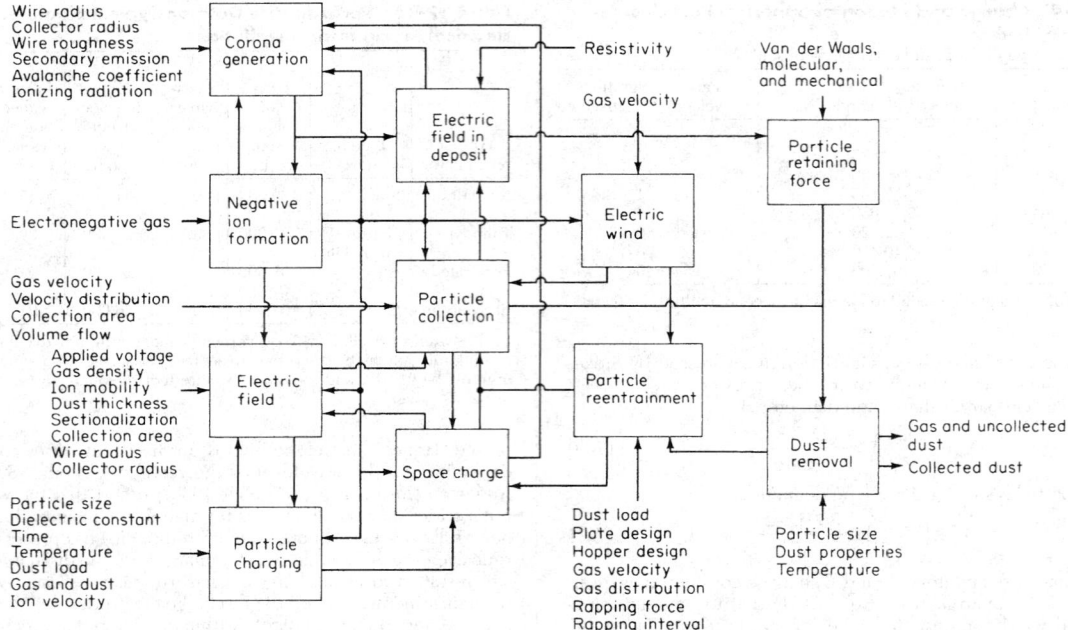

FIG. 17-66   Electrostatic-precipitator-system model. (*Nichols and Oglesby, "Electrostatic Precipitator Systems Analysis," AIChE annual meeting, 1970.*)

and sparking. Increased temperature reduces collection efficiency because ion mobility is increased, lowering critical potentials, and because gas viscosity is increased, reducing migration velocities.

Precipitators have been operated at pressures up to 5.5 MPa (800 psig) and temperatures to 800°C.

The effect of increasing gas density on sparkover voltage has been investigated by Robinson [*J. Appl. Phys.*, **40**, 5107 (1969); *Air Pollution Control*, part 1, Wiley-Interscience, New York, 1971, chap. 5]. Figure 17-67 shows the effect of gas density on corona-starting and sparkover voltages for positive and negative corona in a pipe precipitator. The sparkover voltages are experimental and are given by the solid points. Experimental corona-starting voltages are given by the hollow points. The solid lines are corona-starting voltage curves calculated from Eq. (17-33). This is an empirical relationship developed by Robinson.

$$\frac{E_c}{\rho'} = A \frac{B}{\sqrt{D_d \, \rho'/2}} \qquad (17\text{-}34)$$

E_c is the corona-starting field, kV/cm. ρ' is the relative gas density, equal to the actual gas density divided by the density of air at 25°C, 1 atm. D_d is the diameter of the ionizing wire, cm. A and B are constants which are characteristics of the gas. In dry air, $A = 32.2$ kV/cm and $B = 8.46$ kV/cm$^{1/2}$. Agreement between experimental and calculated starting voltages is good for the case of positive corona, but in the case of negative corona the calculated line serves as an upper limit for the data. This lower-than-expected starting-voltage characteristic of negative corona is confirmed by Hall et al. [*Oil Gas J.*, **66**, 109 (1968)] in a report of an electrostatic precipitator which removes lubricating-oil mist from natural gas at 5.5 MPa (800 psig) and 38°C (100°F). The use of electrostatic precipitators at elevated pressure is expected to in-

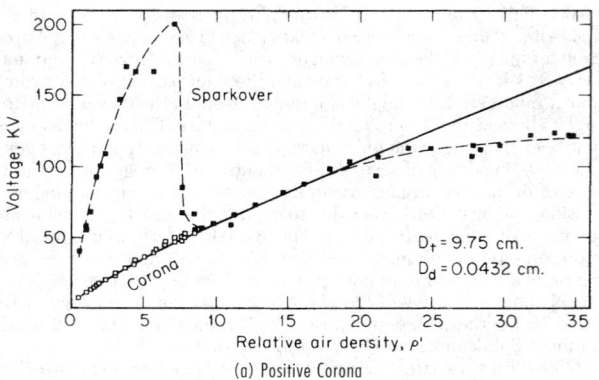

(a) Positive Corona

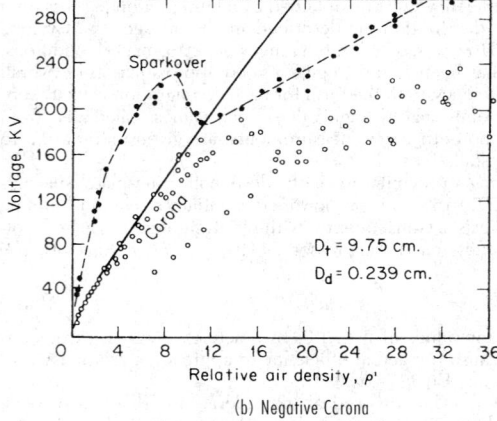

(b) Negative Corona

FIG. 17-67   Corona-starting and sparkover voltages for coaxial wire-pipe electrodes in air (25°C). D_t and D_d are the respective pipe and wire diameters. The voltage is unvarying direct current. (*Robinson, Air Pollution Control, part 1, Wiley-Interscience, New York, 1971, chap. 5.*)

crease, because the method requires very low pressure drop [approximately 69 Pa (0.1 lbf/in²)]. This results from the fact that the electric separation forces are applied directly to the particles themselves rather than to the entire mass of the gas, as in inertial separators. The use of electrostatic precipitators at temperatures up to 400°C is well developed for the powerhouse fly-ash application, but in the range of 600 to 800°C they are still in the experimental phase. The U.S. Bureau of Mines has tested a pilot-scale tubular precipitator for fly ash. See Shale [*Air Pollut. Control Assoc. J.*, **17**, 159 (1967)] and Shale and Fasching (*Operating Characteristics of a High-Temperature Electrostatic Precipitator*, U.S. Bur. Mines Rep. 7276, 1969). It operated over a temperature range of 27 to 816°C (80 to 1500°F) and a pressure range of 552 kPa (35 to 80 psig). Initial collection efficiencies ranged from 90 to 98 percent at 793°C (1460°F), 552 kPa (80 psig), but continuous operation was not achieved because of excessive thermal expansion of internal parts.

Resistivity Problems Optimum performance of electrostatic precipitators is achieved when the electrical resistivity of the collected dust is sufficiently high to result in electrostatic pinning of the particles to the collecting surface, but not so high that dielectric breakdown of the dust layer occurs as the corona current passes through it. The optimum resistivity range is generally considered to be from 10^8 to 10^{10} Ω·cm, measured at operating conditions. As the dust builds up on the collecting electrode, it impedes the flow of current, so that a voltage drop is developed across the dust layer:

$$E_d = j\rho_d L_d \qquad (17\text{-}35)$$

If E_d/L_d exceeds the dielectric strength of the dust layer, sparks occur in the deposit and form back-corona craters. Ions of both polarities are formed. Positive ions formed in the craters are attracted to the negatively charged particles in the gas stream, whose charge level is reduced so that collection efficiency decreases. Some of the positive ions neutralize part of the negative-space-charge cloud normally present near the wire, thereby increasing total current. Collection efficiency under these conditions will not correlate with total power input (Owens, E. I. du Pont de Nemours & Co. internal communication, 1971). Under normal conditions, collection efficiency is an exponential function of corona power (White, *Industrial Electrostatic Precipitation*, Addison-Wesley, Reading, Mass., 1963). With typical ion density in the range of 10^9/cm³, overall voltage gradient would be about 4000 V/cm, and current about 1 μA/cm². Dielectric breakdown of the dust layer (at about 10,000 V/cm) would therefore be expected for dusts with resistivities above 10^{10} Ω·cm.

Problems due to high resistivity are of great concern in **fly-ash** precipitation because air-pollution regulations require that coals have low (<1 percent) sulfur content. Figure 17-68 shows that the resistivity of low-sulfur coal ash exceeds the threshold of 10^{10} Ω·cm at common operating temperatures. This has resulted in the installation of a number of precipitators which have failed to meet guaranteed performance. This has occurred to an alarming extent in the United States but has also been encountered in Australia, where the sulfur content is typically 0.3 to 0.6 percent. Maartmann (Pap. EN-34F, 2d International Clean Air Congress, Washington, 1970) reports the installation of a number of precipitators which performed below guarantees, so that the Electricity Commission of New South Wales decided that each manufacturer wishing to bid on a new station must first make pilot tests to prove performance on the actual coal to be burned in that station. Problems of back corona and excessive sparking with low-sulfur coal usually require that the operating voltage be reduced. This reduces the migration velocity and leads to larger precipitators. Ramsdell (*Design Criteria for Precipitators for Modern Central Station Power Plants*, American Power Conference, Chicago, 1968) developed the curves in Fig. 17-69. They show the results of extensive field tests by the Consolidated Edison Co. In another paper ("Antipollution Program of Consolidated Edison Co. of New York," ASCE, May 13–17, 1968), Ramsdell traces the remarkable growth in the size of precipitators required for high efficiency on low-sulfur coals. The culmination of this work was the precipitator at boiler 30 at Ravenswood Station, New York. Resistivity problems were avoided by operating at high temperature [343°C (650°F)]. The mechanical

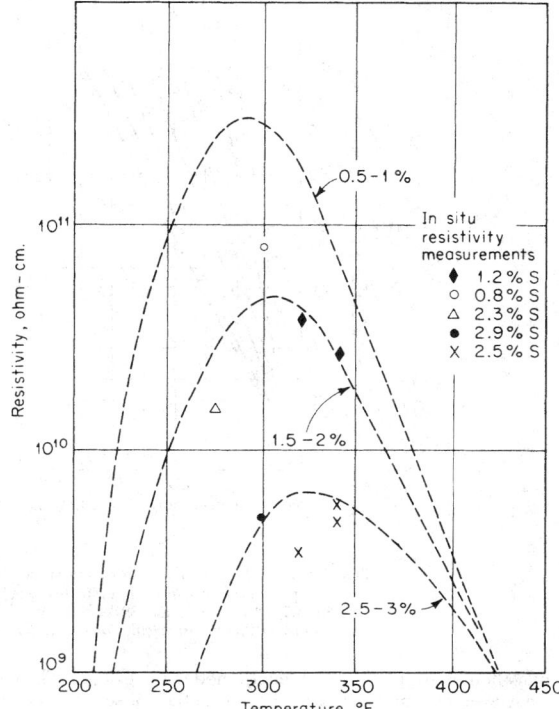

FIG. 17-68   Trends in resistivity of fly ash with variations in flue-gas temperature and coal sulfur content. °C = (°F − 32) × ⁵⁄₉. (*Oglesby and Nichols*, A Manual of Electrostatic Precipitator Technology, *part II, Southern Research Institute, Birmingham, Ala., 1970.*)

(cyclone) collector was installed after the precipitator to clean up puffs due to rapping.

Maartmann (op. cit.) agrees that sulfur content is important but feels that it should not be the sole criterion for the determination of collecting surface. He points to specific collecting-surface requirements as high as 500 ft²/(1000 ft³·min) for 95 percent collection efficiency with high-resistivity Australian ash.

Schmidt and Anderson (op. cit.) and Anderson [*Physics*, **3**, 23 (July 1932)] claim that resistivity of the collected dust may be a controlling factor which is very sensitive to moisture. They state that an increase in relative humidity of 5 percent may double the precipitation rate because of its effect on the conductivity of the collected dust layer.

Conditioning agents have been added to the flue gas to alter dust resistivity. Steam, sodium chloride, sulfur trioxide, and ammonia have all been successfully used. Research by Chittum and others [Schmidt, *Ind. Eng. Chem.*, **41**, 2428 (1949)] led to a theory of conditioning by alteration of the moisture-adsorption properties of dust surfaces. Chittum proposed that an intermediate chemical-adsorption film, which was strongly bound to the particle and which in turn strongly adsorbed water, would be an effective conditioner. This explains how acid conditioners, such as SO_3, help resistivity problems associated with basic dusts, such as many types of fly ash, whereas ammonia is a good additive for acidic dusts such as alumina. Moisture alone can be used as a conditioning agent. This is shown in Fig. 17-65. Moisture is beneficial in two ways: it reduces the electrical resistivity of most dusts (an exception is powdered sulfur, which apparently does not absorb water), and it increases the voltage which may safely be employed without sparking, as in Fig. 17-65.

Low resistivity can sometimes be a problem. If the resistivity is below 10^4 Ω·cm, the collected particles are so conductive that their charges leak to ground faster than they are replenished by the corona. The particles are no longer electrostatically pinned to the plate, and

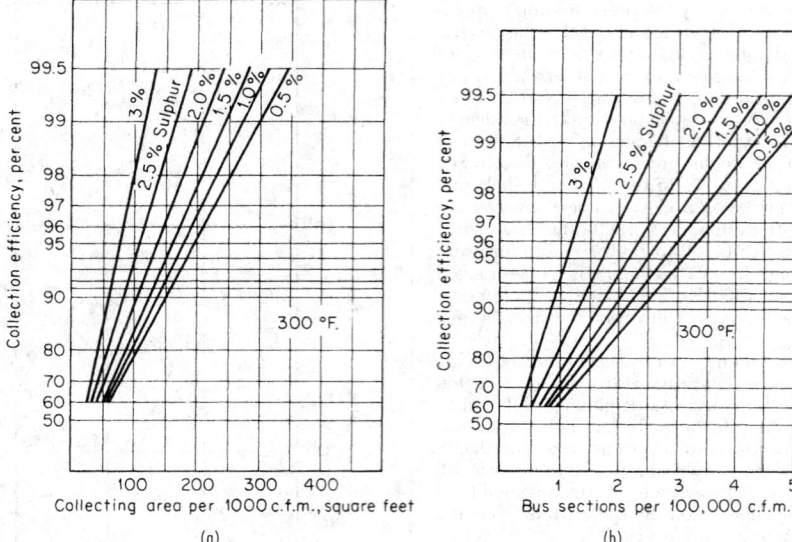

FIG. 17-69 Design curves for electrostatic precipitators for fly ash. Collection efficiency for various levels of percent sulfur in coal versus (*a*) specific collecting surface, and (*b*) bus sections per 100,000 ft³/min (4.7 m³/s). °C = (°F − 32) × ⅝. (*Ramsdell,* Design Criteria for Precipitators for Modern Central Station Power Plants, *American Power Conference, Chicago, Ill., 1968.*)

they may then be swept away and reentrained in the exit gas. The particles may even pick up positive charges from the collecting plate and then be repelled. Low-resistivity problems are common with dusts of high carbon content and may also occur in fly-ash precipitators which handle the ash from high-sulfur coal and operate at low gas temperatures. Low resistivity in this case results from excessive condensation of electrically conductive sulfuric acid.

Single-Stage Precipitators   The single-stage type of unit, commonly known as a Cottrell precipitator, is most generally used for dust or mist collection from industrial-process gases. The corona discharge is maintained throughout the precipitator and, besides providing initial ionization, serves to prevent redispersion of precipitated dust and recharges neutralized or discharged particle ions. Cottrell precipitators may be divided into two main classes, the so-called plate type (Fig. 17-70), in which the collecting electrodes consist of parallel plates, screens, or rows of rods, chains, or wires; and the pipe type (Fig. 17-71), in which the collecting electrodes consist of a nest of parallel pipes which may be square, round, or of any other shape. The discharge or precipitating electrodes in each case are wires or rods, either round or edged, which are placed midway between the collecting electrodes or in the center of the pipes and may be either parallel

or perpendicular to the gas flow in the case of plate precipitators. When the collecting electrodes are screens or rows of rods or wires, the gases are usually passed parallel to the plane of each but may also be passed through it. In pipe precipitators, the gas flow is generally vertical up through the pipe, although downflow is not unusual. The pipe-type precipitator is usually used for the removal of liquid particles and volatilized fumes [Beaver, op. cit.; and Cree, *Am. Gas J.*, **162**, 27 (March 1945)], and the plate type is used mainly on dusts. In the pipe type, the discharge electrodes are usually suspended from an insulated support and kept taut by a weight at the bottom. Cree (op.

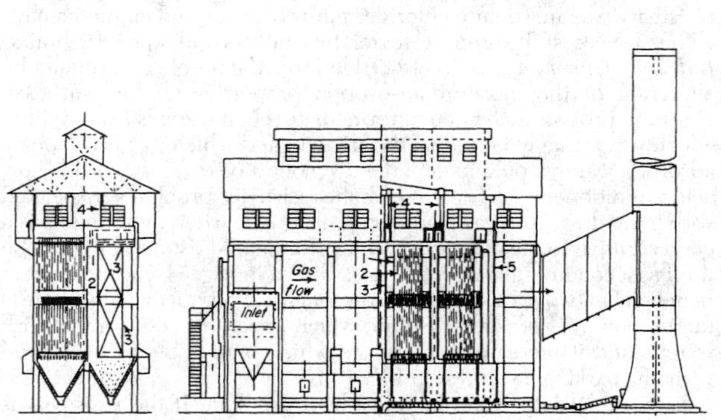

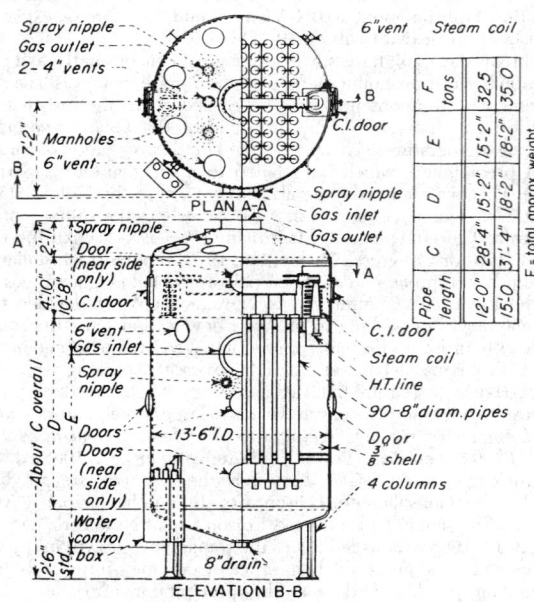

FIG. 17-70 Horizontal-flow plate precipitator used in a cement plant. (*Western Precipitation Division, Joy Manufacturing Company.*)

FIG. 17-71 Blast-furnace pipe precipitator. (*Research-Cottrell, Inc.*)

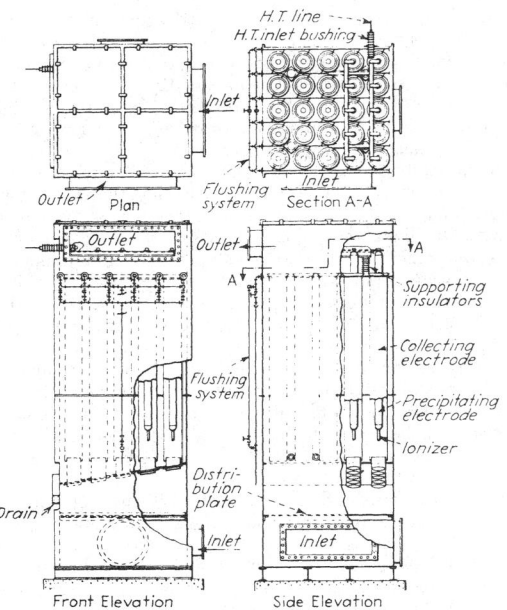

FIG. 17-72 Two-stage water-film pipe precipitator. (*Western Precipitation Division, Joy Manufacturing Company.*)

cit.) discusses the application of electrical precipitators to tar removal in the gas industry.

Rapping Except when liquid dispersoids are being collected or, in the case of film precipitators, when a liquid is circulated over the collecting-electrode surface (Fig. 17-72), thus continuously removing the precipitated material, the collected dust is dislodged from the electrodes either periodically or continuously by mechanical rapping or scraping, which may be performed automatically or manually. Automatic rapping with either impact-type or vibrator-type rappers is common practice. White (op. cit.) recommends fairly continuous rapping with magnetic-impulse rappers. Rapping with excessive force leads to dust reentrainment and possible mechanical failure of the plates, while insufficient rapping leads to excessive dust buildup with poor electrical operation and reduced collection efficiency. Intermittent rapping at intervals of an hour or more causes heavy puffs of reentrained dust. Sproull [*Air Pollut. Control Assoc. J.*, **15**, 50 (1965)] reports the importance of electrode acceleration and shows that it varies with the type of dust, whether the electrode is rapped perpendicularly (normally) to the plate or parallel to it. Figure 17-73 shows the accelerations required for rapping normally to the plate. Difficult dusts may require as much as 100 G acceleration for 90 percent removal, and even higher accelerations are required when the vibrating force is applied in the plane of the plate.

Perforated-plate or rod-curtain precipitators are frequently rapped without shutting off the gas flow and with the electrodes energized. This procedure, however, results in a tendency for reentrainment of collected dust. Sectional or composite-plate collecting electrodes (sometimes known as hollow, pocket, or tulip electrodes) are used to minimize this tendency in the continuous removal of the precipitated material, provided that it is free-flowing. These are generally designed for vertical gas flow and comprise a collecting electrode containing a dead air space and provided with horizontal protruding slots that guide the dust into this space (see Fig. 17-74), although some types use horizontal flow.

The choice of size, shape, and type of electrode is based on economic considerations and is usually determined by the characteristics of the gas and suspended matter and by mechanical considerations such as flue arrangement, the available space, and previous experience with the electrodes on similar problems. The spacing between collecting electrodes in plate-type precipitators and the pipe diameter

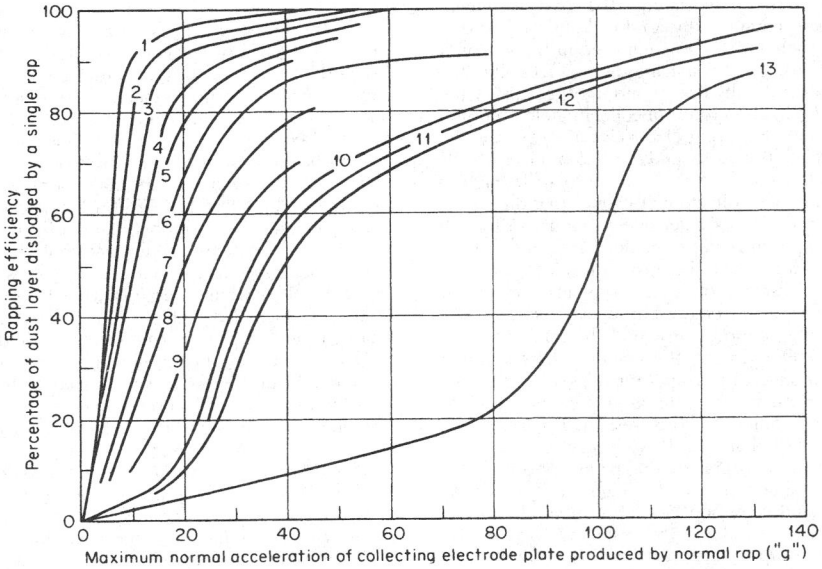

FIG. 17-73 Normal (perpendicular) rapping efficiency for various precipitated dust layers having about 0.03 g dust/cm² (0.2 g dust/in²) as a function of maximum acceleration in multiples of g. Curve 1, fly ash, 200 or 300°F, power off. Curve 2, fly ash, 70°F, power off; also 200 or 300°F, power on. Curve 3, fly ash, 70°F, power on. Curve 4, cement-kiln feed, 300°F, power off. Curve 5, cement dust, 300°F, power off. Curve 6, same as 5, except power on. Curve 7, cement-kiln feed, 300°F, power on. Curve 8, cement dust, 200°F, power off. Curve 9, same as 8, except power on. Curve 10, cement-kiln feed, 200°F, power off. Curve 11, same as 10, except at 70°F. Curve 12, cement-kiln feed, 200°F, power on. Curve 13, cement-kiln feed, 70°F, power on. °C = (°F − 32) × ⁵⁄₉. [*Sproull, Air Pollut. Control Assoc. J.*, **15**, 50 (1965).]

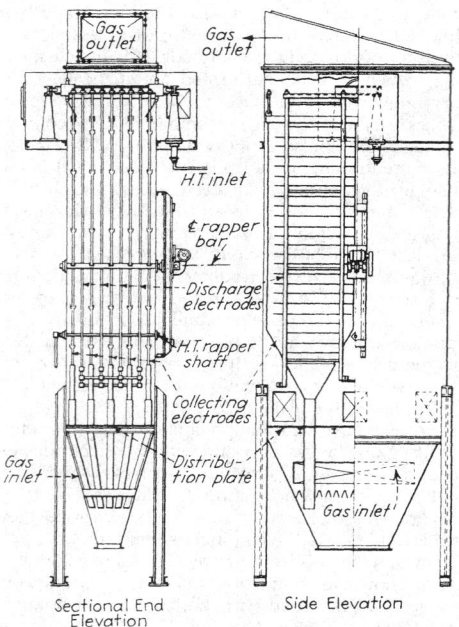

FIG. 17-74 Vertical-flow heavy-duty plate precipitator. (*Western Precipitation Division, Joy Manufacturing Company.*)

in pipe-type precipitators usually ranges from 15 to 38 cm (6 to 15 in). The smaller the spacing, the lower the necessary voltage and overall equipment size, but the greater the difficulties involved in maintaining proper alignment and resulting from disturbances due to collected material. Large spacings are usually associated with high dust concentration in order to minimize sparkover due to dust buildup. For very high dust concentrations, such as those encountered in fluid-catalyst plants, it is advantageous to use greater spacings in the first half of the precipitator than in the second half. Precipitators, especially of the plate type, are frequently built with groups of collecting electrodes in series in a common housing. Collecting electrodes are generally on the order of 0.9 to 1.8 m (3 to 6 ft) wide and 3 to 5.5 m (10 to 18 ft) high in plate-type precipitators and 1.8 to 4.6 m (6 to 15 ft) high in pipe types. It is essential for good collection efficiency that the gas be evenly distributed across the various electrode elements. Although this can be achieved by proper gas-inlet transitions and guide vanes, perforated plates or screens located on the upstream side of the electrodes are generally used for distribution. Perforated plates or screens located on the downstream side may be used in special cases.

Electrical precipitators are generally designed for collection efficiency in the range of 90 to 99.9 percent. It is essential, however, that the units be properly maintained in order to achieve the required collection efficiency. Electric power consumption is generally 0.2 to 0.6 kW/(1000 ft³·min) of gas handled, and the pressure drop across the precipitator unit is usually less than 124 Pa (0.5 in water), ranging from 62 to 248 Pa (¼ to 1 in) and representing primarily distributor and entrance-exit losses. Applied potentials range from 30,000 to 100,000 V. Gas velocities and retention times are generally in the range of 0.9 to 3 m/s (3 to 10 ft/s) and 1 to 15 s respectively. Velocities are kept low in conventional precipitators to avoid reentrainment of dust. There are, however, precipitator installations on carbon black in which the precipitator acts to flocculate the dust so that it may be subsequently collected in multiple small-diameter cyclone collectors. By not attempting to collect the particles in the precipitator, higher velocities may be used with a correspondingly lower investment cost.

Power Supply Electrical precipitators are generally energized by rectified alternating current of commercial frequency. The voltage is stepped up to the required value by means of a transformer and then rectified. The rectifying equipment has undergone an evolution which began with the synchronous mechanical rectifier in 1904 and was followed by mercury-vapor rectifiers in the 1920s; the first solid-state selenium rectifiers were introduced about 1939. Silicon rectifiers are the latest and most widely used type, since they provide high efficiency and reliability. Automatic controls commonly are tied to voltage, current, spark rate, or some combination of these parameters. Modern precipitators use control circuits similar to those on Fig. 17-75. A high-voltage silicon rectifier is used together with a saturable reactor and means for limiting current and controlling voltage and/or spark rate. One popular method adjusts the voltage to give a specified sparking frequency (typically 50 to 150 sparks per minute per bus section). Half-wave rectification is sometimes used because of its lower equipment requirements and power consumption. It also has the advantage of longer decay periods for sparks to extinguish between current pulses.

Electrode **insulators** must also be designed for a particular service. The properties of the dust or mist and gas determine their design as well as the physical details of the installation. Conducting mists require special allowances such as oil seals, energized shielding cups, or air bleeds. With saturated gas, steam coils are frequently used to prevent condensation on the electrodes.

Typical applications in the chemical field (Beaver, op. cit.) include detarring of manufactured gas, removal of acid mist and impurities in contact sulfuric acid plants, recovery of phosphoric acid mists, removal of dusts in gases from roasters, sintering machines, calciners, cement and lime kilns, blast furnaces, carbon-black furnaces, regenerators on fluid-catalyst units, chemical-recovery furnaces in soda and sulfate pulp mills, and gypsum kettles. Figure 17-74 shows a vertical-flow steel-plate-type precipitator similar to a type used for catalyst-dust collection in certain fluid-catalyst plants.

A development of interest to the chemical industry is the tubular precipitator of reinforced-plastic construction (Wanner, *Gas Cleaning Plant after TiO₂ Rotary Kilns,* technical bulletin, Lurgi Corp., Frankfurt, Germany, 1971). Tubes made of polyvinyl chloride plastic are reinforced on the outside with polyester-fiber glass. The use of modern economical materials of construction to replace high-maintenance materials such as lead has been long awaited for corrosive applications.

Electrical precipitators are probably the most versatile of all types of dust collectors. Very high collection efficiencies can be obtained regardless of the fineness of the dust, provided that the precipitators are given proper maintenance. The chief disadvantages are the high initial cost and, in some cases, high maintenance costs. Furthermore, caution must be exercised with dusts that are combustible in the carrier gas.

Two-Stage Precipitators In two-stage precipitators, corona discharge takes place in the first stage between two electrodes having a nonuniform field (see Fig. 17-76). This is generally obtained by a fine-wire discharge electrode and a large-diameter receiving electrode. In this stage the potential difference must be above that required for corona discharge. The second stage involves a relatively uniform electrostatic field in which charged particles are caused to migrate to a collecting surface. This stage usually consists of either alternately charged parallel plates or concentric cylinders with relatively close clearances compared with their diameters. The only voltage requirement in this stage is that no sparking occur, though higher voltages will result in increased collection efficiency. Since collection occurs in the absence of corona discharge, there is no way of recharging reentrained and discharged particles. Consequently, some means must be provided for avoiding reentrainment of particles from the collecting surface. It is also essential that there be sufficient time and mixing between the first and second stages to secure distribution of gas ions across the gas stream and proper charging of the dust particles.

A unit is available in which electrostatic precipitation is combined with a dry-air filter of the type shown in Fig. 17-64b. In another unit an electrostatic field is superimposed on an automatic filter. In this case the ionizer wires are located on the leading face of the unit, and the collecting electrodes consist of alternate stationary and rotating parallel plates. Cleaning in this case is automatic and continuous.

Although intended primarily for air-conditioning applications, these units have been successfully applied to the collection of relatively nonconducting mists such as oil. However, other process applications

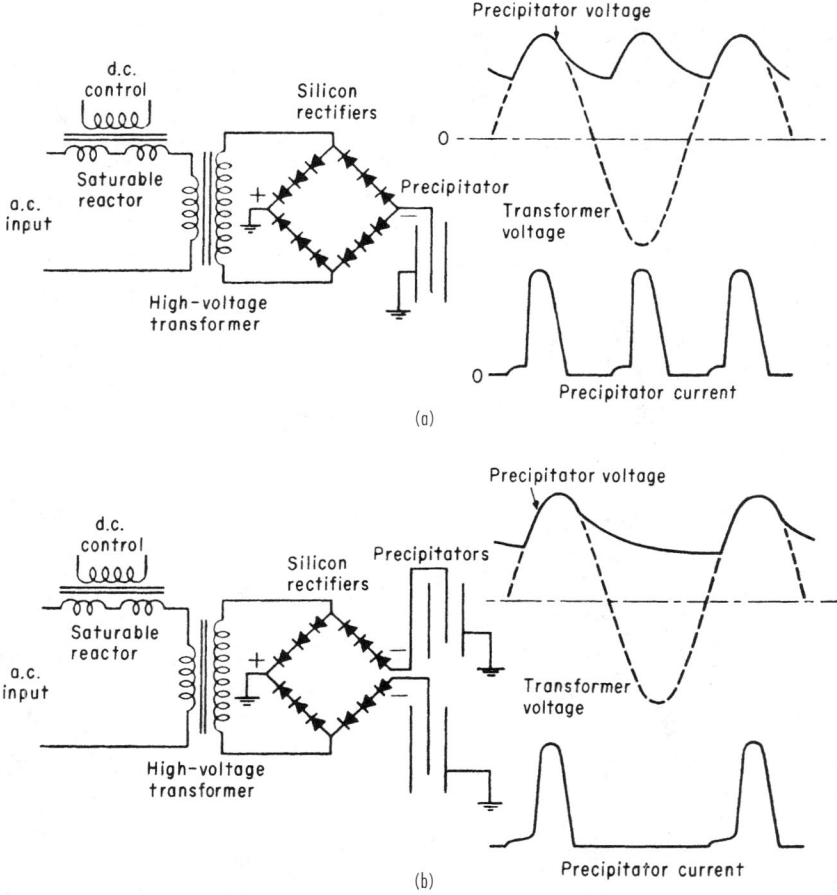

FIG. 17-75 Schematic circuits for silicon rectifier sets with saturable reactor control. (*a*) Full-wave silicon rectifier. (*b*) Half-wave silicon rectifier. (*White*, Industrial Electrostatic Precipitation, *Addison-Wesley, Reading, Mass., 1963.*)

have been limited largely to experimental installations. The large cost advantage of these units over the Cottrell precipitator lies in the smaller equipment size made possible by the close plate spacing, in the lower power consumption due to the two-stage operation, and primarily in the mass production of standardized units. In process applications, the close plate spacing is objectionable because of the relatively high dust concentrations involved. Special material or weight requirements for the structural members may eliminate the mass-production advantage except for individual wide applications.

Alternating-Current Precipitators High-voltage alternating current may be employed for electrical precipitation. Corona discharge will result in a net rectification, provided that no spark gaps are used in series with the precipitator. However, the equipment capacity for a given efficiency is considerably lower than for direct current. In addition, difficulties due to induced high-frequency currents may be encountered. The simplicity of an ac system, on the other hand, has permitted very satisfactory adaptation for laboratory and sampling purposes [Drinker, Thomson, and Fitchet, *J. Ind. Hyg.*, **5**, 162 (September 1923)].

Some promising work with alternating current has been undertaken at the University of Karlsruhe. Lau [*Staub*, English ed., **29**, 10 (1969)] and coworkers found that ac precipitators operated at 50 Hz were more effective than dc precipitators for dusts with resistivities higher than 10^{11} Ω·cm. An insulating screen covering the collecting electrode permitted higher-voltage operation without sparkover.

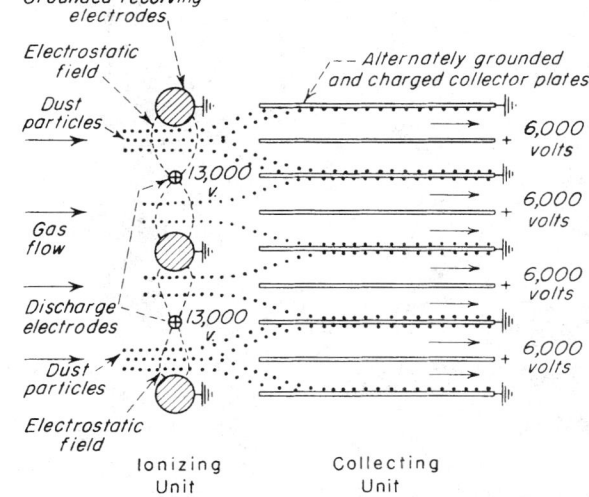

FIG. 17-76 Two-stage electrical-precipitation principle.

Liquid-Solid Operations and Equipment

Donald A. Dahlstrom, Ph.D., *Research Professor, Chemical and Fuels Engineering Department and Metallurgical Engineering Department, University of Utah; Member, National Academy of Engineering, American Institute of Chemical Engineers (AIChE), American Chemical Society (ACS), Society of Mining, Metallurgical, and Exploration Engineers (SME) of the American Institute of Mining, Metallurgical, and Petroleum Engineers (AIME), American Society of Engineering Education (ASEE). (Section Editor)*

Richard C. Bennett, B.S. Ch.E., *Registered Professional Engineer, Illinois; Member, American Institute of Chemical Engineers (AIChE), President of Crystallization Technology, Inc.; Former President of Swenson Process Equipment, Inc. (Crystallization)*

Robert C. Emmett, Jr., B.S., Ch.E., *Senior Process Consultant, EIMCO Process Equipment Co.; Member, American Institute of Chemical Engineers (AIChE), American Institute of Mining, Metallurgical, and Petroleum Engineers (AIME), Society of Mining, Metallurgical, and Exploration Engineers (SME. Gravity Sedimentation Operations; Filtration)*

Peter Harriott, Ph.D., *Professor, School of Chemical Engineering, Cornell University; Member, American Institute of Chemical Engineering (AIChE), American Chemical Society (ACS). (Selection of a Solids-Liquid Separator)*

Tim Laros, M.S., *Mineral Processing, Senior Process Consultant, EIMCO Process Equipment Co.; Member, Society of Mining, Metallurgy, and Exploration (SME of AIME). (Gravity Sedimentation Operations; Filtration)*

Wallace Leung, Sc.D., *Director, Process Technology, Bird Machine Company; Member, American Filtration and Separation Society (Director). (Centrifuges)*

Chad McCleary, *EIMCO Process Equipment Company, Process Consultant. (Vacuum filters)*

Shelby A. Miller, Ph.D., *Retired Sr. Eng., Argonne National Laboratories; Member, American Association for the Advancement of Science (AAAS), American Chemical Society (ACS), American Institute of Chemical Engineering (AIChE), American Institute of Chemists, Filtration Society, New York Academy of Sciences; Registered Professional Engineer, New York. (Leaching)*

Booker Morey, Ph.D., *Senior Consultant, SRI International; Member, Society of Mining, Metallurgy, and Exploration (SME of AIME), Filtration Society, Air and Waste Management, Association, Registered Professional Engineer, California and Massachusetts. (Expression)*

James Y. Oldshue, Ph.D., *President, Oldshue Technologies International, Inc.; Member, National Academy of Engineering; Adjunct Professor of Chemical Engineering at Beijing Institute of Chemical Technology, Beijing, China; Member, American Chemical Society (ACS), American Institute of Chemical Engineering (AIChE), Traveler Century Club; Member of Executive Committee on the Transfer of Appropriate Technology for the World Federation of Engineering Organizations. (Agitation of Low-Viscosity Particle Suspensions)*

George Priday, B.S., Ch.E.,   *EIMCO Process Equipment Co.; Member, American Institute of Chemical Engineering (AIChE), Instrument Society of America (ISA). (Gravity Sedimentation Operations)*

Charles E. Silverblatt, M.S., Ch.E.,   *Peregrine International Associates, Inc.; Consultant to WesTech Engineering, Inc.; American Institute of Mining, Metallurgical, and Petroleum Engineers (AIME). (Gravity Sedimentation Operations; Filtration)*

J. Stephen Slottee, M.S., Ch.E.,   *Manager, Technology and Development, EIMCO Process Equipment Co.; Member, American Institute of Chemical Engineers (AIChE). (Filtration)*

Julian C. Smith, B. Chem., Ch.E.,   *Professor Emeritus, Chemical Engineering, Cornell University; Member, American Chemical Society, (ACS), American Institute of Chemical Engineers (AIChE). (Selection of Solids-Liquid Separator)*

David B. Todd, Ph.D.,   *President, Todd Engineering; Member, American Association for the Advancement of Science (AAAS), American Chemical Society (ACS), American Institute of Chemical Engineering (AIChE), American Oil Chemists Society (ADCS), Society of Plastics Engineers (SPE), and Society of the Plastics Industry (SPI); Registered Professional Engineer, Michigan. (Paste and Viscous Material Mixing)*

EXPRESSION

SELECTION OF A SOLIDS-LIQUID SEPARATOR

Nomenclature

Symbol	Definition	SI units	U.S. customary units
c	Specific heat	J/(kg·k)	Btu/(lb·°F)
C	Constant		
C_o	Orifice coefficient	Dimensionless	Dimensionless
d_o	Orifice diameter	m	in
$d_{p, max}$	Drop diameter	m	ft
d_t	Pipe diameter	m	in
d_t	Tube diameter	m	ft
D	Impeller diameter	m	ft
D_a	Impeller diameter	m	ft
D_j	Diameter of jacketed vessel	m	ft
D_T	Tank diameter	m	ft
g	Acceleration	m/s^2	ft/s^2
g_c	Dimensional constant	$g_c = 1$ when using SI units	32.2 (ft·lb)/(lbf·s^2)
h	Local individual coefficient of heat transfer, equals $dq/(dA)(\Delta T)$	J/(m^2·s·K)	Btu/(h·ft^2·°F)
H	Velocity head	m	ft
k	Thermal conductivity	J/(m·s·K)	(Btu·ft)/(h·ft^2·°F)
L_p	Diameter of agitator blade	m	ft
N	Agitator rotational speed	s^{-1}, (r/s)	s^{-1}, (r/s)
N_{JS}	Agitator speed for just suspension	s^{-1}	s^{-1}
N_{Re}	$D_a^2 N\rho/\mu$ impeller Reynolds number	Dimensionless	Dimensionless
N_p	Power number $= (q_c P)/\rho N^3 D_a^5$	Dimensionless	Dimensionless
N_Q	Impeller pumping coefficient $= Q/ND_a^3$	Dimensionless	Dimensionless
N_r	Impeller speed	s^{-1}	s^{-1}
N_t	Impeller speed	s^{-1}	s^{-1}
P	Power	(N·m/s)	ft·lb$_f$/s
Q	Impeller flow rate	m^3/s	ft^3/s
T	Tank diameter	m	ft
v	Average fluid velocity	m/s	ft/s
v'	Fluid velocity fluctuation	m/s	ft/s
V	Bulk average velocity	m/s	ft/s
Z	Liquid level in tank	m	ft
Greek symbols			
γ	Rate of shear	s^{-1}	s^{-1}
Δp	Pressure drop across orifice		lbf/ft^2
μ	Viscosity of liquid at tank temperature	Pa·s	lb/(ft·s)
μ	Stirred liquid viscosity	Pa·s	lb/(ft·s)
μ_b	Viscosity of fluid at bulk temperature	Pa·s	lb/(ft·s)
μ_c	Viscosity, continuous phase	Pa·s	lb/(ft·s)
μ_D	Viscosity of dispersed phase	Pa·s	lb/(ft·s)
μ_f	Viscosity of liquid at mean film temperature	Pa·s	lb/(ft·s)
μ_{wt}	Viscosity at wall temperature	Pa·s	lb/(ft·s)
ρ	Stirred liquid density	g/m^3	lb/ft^3
ρ	Density of fluid	kg/m^3	lb/ft^3
ρ_{av}	Density of dispersed phase	kg/m^3	lb/ft^3
ρ_c	Density	kg/m^3	lb/ft^3
σ	Interfacial tension	N/m	lbf/ft
Φ_D	Average volume fraction of discontinuous phase	Dimensionless	Dimensionless

PHASE CONTACTING AND LIQUID-SOLID PROCESSING: AGITATION OF LOW-VISCOSITY PARTICLE SUSPENSIONS

GENERAL REFERENCES: Harnby, N., M. F. Edwards, and A. W. Neinow (eds.), *Mixing in the Process Industries*, Butterworth, Stoneham, Mass., 1986. Lo, T. C., M. H. I. Baird, and C. Hanson, *Handbook of Solvent Extraction*, Wiley, New York, 1983. Nagata, S., *Mixing: Principles and Applications*, Kodansha Ltd., Tokyo, Wiley, New York, 1975. Oldshue, J. Y., *Fluid Mixing Technology*, McGraw-Hill, New York, 1983. Tatterson, G. B., *Fluid Mixing and Gas Dispersion in Agitated Tanks*, McGraw-Hill, New York, 1991. Uhl, V. W., and J. B. Gray (eds.), *Mixing*, vols. I and II, Academic Press, New York, 1966; vol. III, Academic Press, Orlando, Fla., 1992. Ulbrecht, J. J., and G. K. Paterson (eds.), *Mixing of Liquids by Mechanical Agitation*, Godon & Breach Science Publishers, New York, 1985.

PROCEEDINGS: *Fluid Mixing*, vol. I, Inst. Chem. Eng. Symp., Ser. No. 64 (Bradford, England), The Institute of Chemical Engineers, Rugby, England, 1984. *Mixing—Theory Related to Practice*, AIChE, Inst. Chem. Eng. Symp. Ser. No. 10 (London), AIChE and The Institute of Chemical Engineers, London, 1965. *Proc. First (1974), Second (1977), Third (1979), Fourth (1982), Fifth (1985), and Sixth (1988) European Conf. on Mixing*, N. G. Coles (ed.), (Cambridge, England) BHRA Fluid Eng., Cranfield, England. *Process Mixing, Chemical and Biochemical Applications*, G. B. Tatterson, and R. V. Calabrese (eds.), AIChE Symp. Ser. No. 286, 1992.

FLUID MIXING TECHNOLOGY

Fluid mixers cut across almost every processing industry including the chemical process industry; minerals, pulp, and paper; waste and water treating and almost every individual process sector. The engineer working with the application and design of mixers for a given process has three basic sources for information. One is published literature, consisting of several thousand published articles and several currently available books, and brochures from equipment vendors. In addition, there may be a variety of in-house experience which may or may not be cataloged, categorized, or usefully available for the process application at hand. Also, short courses are currently available in selected locations and with various course objectives, and a large body of experience and information lies in the hands of equipment vendors.

In the United States, it is customary to design and purchase a mixer from a mixing vendor and purchase the vessel from another supplier. In many other countries, it is more common to purchase the vessel and mixer as a package from one supplier.

In any event, the users of the mixer can issue a mechanical specification and determine the speed, diameter of an impeller, and power with in-house expertise. Or they may issue a processes specification which describes the engineering purpose of the mixing operation and the vendor will supply a description of the mixer process performance as well as prepare a mechanical design.

This section describes fluid mixing technology and is referred to in other sections in this handbook which discuss the use of fluid mixing equipment in their various operating disciplines. This section does not describe paste and dough mixing, which may require planetary and extruder-type mixers, nor the area of dry solid-solid mixing.

It is convenient to divide mixing into five pairs (plus three triplets and one quadruplicate combination) of materials, as shown in Table 18-1. These five pairs are blending (miscible liquids), liquid-solid, liquid-gas, liquid-liquid (immiscible liquids), and fluid motion. There are also four other categories that occur, involving three or four phases. One concept that differentiates mixing requirements is the difference between physical criteria listed on the left side of Table 18-1, in which some degree of sampling can be used to determine the character of the mixture in various parts in the tank, and various definitions of mixing requirements can be based on these physical descriptions. The other category on the right side of Table 18-1 involves chemical and mass-transfer criteria in which rates of mass transfer or chemical reaction are of interest and have many more complexities in expressing the mixing requirements.

The first five classes have their own mixing technologies. Each of these 10 areas has its own mixing technology. There are relationships for the optimum geometry of impeller types, D/T ratios, and tank geometry. They each often have general, overall mixing requirements and different scale-up relationships based on process definitions. In addition, there are many subclassifications, some of which are based on the viscosity of fluids. In the case of blending, we have blending in the viscous region, the transition region, and the turbulent region. Since any given mixer designed for a process may be required to do several different parts of these 10 categories, it must be a compromise of the geometry and other requirements for the total process result and may not optimize any one particular process component. If it turns out that one particular process requirement is so predominant that all the other requirements are satisfied as a consequence, then it is possible to optimize that particular process step. Often, the only process requirement is in one of these 10 areas, and the mixer can be designed and optimized for that one step only.

As an example of the complexity of fluid mixing, many batch processes involve adding many different materials and varying the liquid level over wide ranges in the tank, have different temperatures and shear rate requirements, and obviously need experience and expert attention to all of the requirements. Superimpose the requirements for sound mechanical design, including drives, fluid seals, and rotating shafts, means that the concepts presented here are merely a beginning to the overall, final design.

A few general principles are helpful at this point before proceeding to the examination of equipment and process details. For any given impeller geometry, speed, and diameter, the impeller draws a certain amount of power. This power is 100 percent converted to heat. In low-viscosity mixing (defined later), this power is used to generate a *macro-scale* regime in which one typically has the visual observation of flow pattern, swirls, and other surface phenomena. However, these flow patterns are primarily energy transfer agents that transfer the power down to the micro scale. The macro-scale regime involves the pumping capacity of the impeller as well as the total circulating capacity throughout the tank and it is an important part of the overall mixer design. The micro-scale area in which the power is dissipated does not care much which mixer is used to generate the energy dissipation. In contrast, in high-viscosity processes, there is a continual progress of energy dissipation from the macro scale down to the micro scale.

There is a wide variety of impellers using *fluidfoil principles*, which are used when flow from the impeller is predominant in the process requirement and macro- or micro-scale shear rates are a subordinate issue.

Scale-up involves selecting mixing variables to give the desired performance in both pilot and full scale. This is often difficult (sometimes

TABLE 18-1 Classification System for Mixing Processes

Physical	Components	Chemical, mass transfer
Blending	Blending	Chemical reactions
Suspension	Solid-liquid	Dissolving, precipitation
Dispersion	Gas-liquid	Gas absorption
	Solid-liquid-gas	
Emulsions	Liquid-liquid	Extraction
	Liquid-liquid-solid	
	Gas-liquid-liquid	
	Gas-liquid-liquid-solid	
Pumping	Fluid motion	Heat transfer

impossible) using geometric similarity, so that the use of nongeo-metric impellers in the pilot plant compared to the impellers used in the plant often allows closer modeling of the mixing requirements to be achieved.

Computational fluid mixing allows the modeling of flow patterns in mixing vessels and some of the principles on which this is based in current techniques are included.

INTRODUCTORY FLUID MECHANICS

The fluid mixing process involves three different areas of viscosity which affect flow patterns and scale-up, and two different scales within the fluid itself: macro scale and micro scale. Design questions come up when looking at the design and performance of mixing processes in a given volume. Considerations must be given to proper impeller and tank geometry as well as the proper speed and power for the impeller. Similar considerations come up when it is desired to scale up or scale down, and this involves another set of mixing considerations.

If the fluid discharge from an impeller is measured with a device that has a high-frequency response, one can track the velocity of the fluid as a function of time. The velocity at a given point in time can then be expressed as an average velocity v plus fluctuating component v'. Average velocities can be integrated across the discharge of the impeller, and the pumping capacity normal to an arbitrary discharge plane can be calculated. This arbitrary discharge plane is often defined as the plane bounded by the boundaries of the impeller blade diameter and height. Because there is no casing, however, an additional 10 to 20 percent of flow typically can be considered as the primary flow from an impeller.

The velocity gradients between the average velocities operate only on larger particles. Typically, these larger-size particles are greater than 1000 μm. This is not a proven definition, but it does give a feel for the magnitudes involved. This defines macro-scale mixing. In the turbulent region, these macro-scale fluctuations can also arise from the finite number of impeller blades. These set up velocity fluctuations that can also operate on the macro scale.

Smaller particles see primarily only the fluctuating velocity component. When the particle size is much less than 100 μm, the turbulent properties of the fluid become important. This is the definition of the physical size for micro-scale mixing.

All of the power applied by a mixer to a fluid through the impeller appears as heat. The conversion of power to heat is through viscous shear and is approximately 2542 Btu/h/hp. Viscous shear is present in turbulent flow only at the micro-scale level. As a result, the power per unit volume is a major component of the phenomena of micro-scale mixing. At a 1-μm level, in fact, it doesn't matter what specific impeller design is used to supply the power.

Numerous experiments show that power per unit volume in the zone of the impeller (which is about 5 percent of the total tank volume) is about 100 times higher than the power per unit volume in the rest of the vessel. Making some reasonable assumptions about the fluid mechanics parameters, the root-mean-square (rms) velocity fluctuation in the zone of the impeller appears to be approximately 5 to 10 times higher than in the rest of the vessel. This conclusion has been verified by experimental measurements.

The ratio of the rms velocity fluctuation to the average velocity in the impeller zone is about 50 percent with many open impellers. If the rms velocity fluctuation is divided by the average velocity in the rest of the vessel, however, the ratio is on the order of 5 percent. This is also the level of rms velocity fluctuation to the mean velocity in pipeline flow. There are phenomena in micro-scale mixing that can occur in mixing tanks that do not occur in pipeline reactors. Whether this is good or bad depends upon the process requirements.

Figure 18-1 shows velocity versus time for three different impellers. The differences between the impellers are quite significant and can be important for mixing processes.

All three impellers are calculated for the same impeller flow Q and the same diameter. The A310 (Fig. 18-2) draws the least power and has the least velocity fluctuations. This gives the lowest micro-scale turbulence and shear rate. The A200 (Fig. 18-3) shows increased velocity

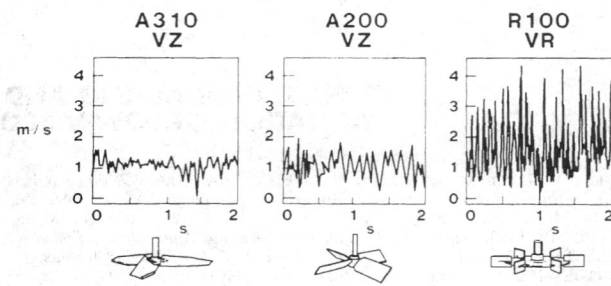

FIG. 18-1 Velocity fluctuations versus time for equal total pumping capacity from three different impellers.

fluctuations and draws more power. The R100 (Fig. 18-4) draws the most power and has the highest micro-scale shear rate. The proper impeller should be used for each individual process requirement.

Scale-up/Scale-down Two aspects of scale-up frequently arise. One is building a model based on pilot-plant studies that develop an understanding of the process variables for an existing full-scale mixing installation. The other is taking a new process and studying it in the pilot plant in such a way that pertinent scale-up variables are worked out for a new mixing installation.

There are a few principles of scale-up that can indicate which approach to take in either case. Using geometric similarity, the macro-scale variables can be summarized as follows:

- Blend and circulation times in the large tank will be much longer than in the small tank.

FIG. 18-2 An A310 impeller.

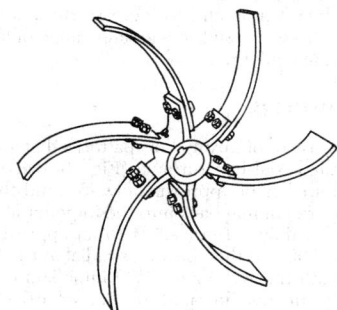

FIG. 18-4 Flat-blade turbine.

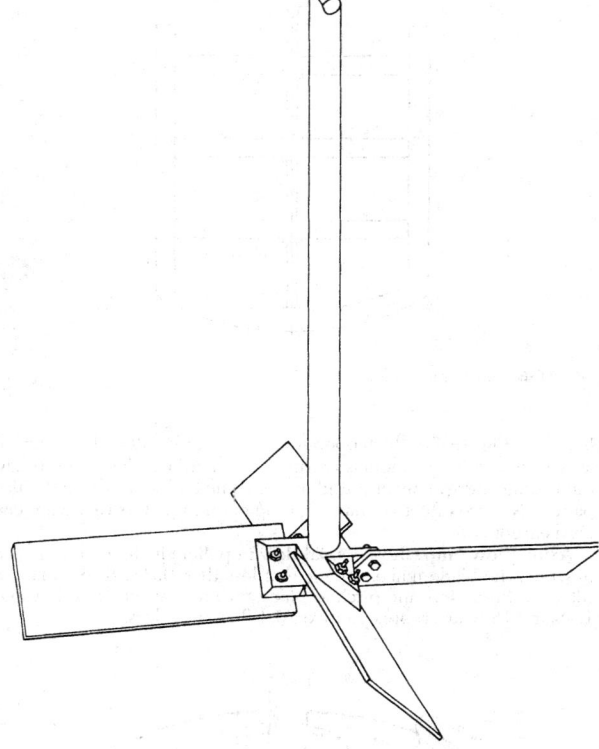

FIG. 18-3 Pitched-blade turbine.

• Maximum impeller zone shear rate will be higher in the larger tank, but the average impeller zone shear rate will be lower; therefore, there will be a much greater variation in shear rates in a full-scale tank than in a pilot unit.

• Reynolds numbers in the large tank will be higher, typically on the order of 5 to 25 times higher than those in a small tank.

• Large tanks tend to develop a recirculation pattern from the impeller through the tank back to the impeller. This results in a behavior similar to that for a number of tanks in a series. The net result is that the mean circulation time is increased over what would be predicted from the impeller pumping capacity. This also increases the standard deviation of the circulation times around the mean.

• Heat transfer is normally much more demanding on a large scale. The introduction of helical coils, vertical tubes, or other heat-transfer devices causes an increased tendency for areas of low recirculation to exist.

• In gas-liquid systems, the tendency for an increase in the gas superficial velocity upon scale-up can further increase the overall circulation time.

What about the micro-scale phenomena? These are dependent primarily on the energy dissipation per unit volume, although one must also be concerned about the energy spectra. In general, the energy dissipation per unit volume around the impeller is approximately 100 times higher than in the rest of the tank. This results in an rms velocity fluctuation ratio to the average velocity on the order of 10:1 between the impeller zone and the rest of the tank.

Because there are thousands of specific processes each year that involve mixing, there will be at least hundreds of different situations requiring a somewhat different pilot-plant approach. Unfortunately, no set of rules states how to carry out studies for any specific program, but here are a few guidelines that can help one carry out a pilot-plant program.

• For any given process, one takes a qualitative look at the possible role of fluid shear stresses. Then one tries to consider pathways related to fluid shear stress that may affect the process. If there are none, then this extremely complex phenomenon can be dismissed and the process design can be based on such things as uniformity, circulation time, blend time, or velocity specifications. This is often the case in the blending of miscible fluids and the suspension of solids.

• If fluid shear stresses are likely to be involved in obtaining a process result, then one must qualitatively look at the scale at which the shear stresses influence the result. If the particles, bubbles, droplets, or fluid clumps are on the order of 1000 μm or larger, the variables are macro scale and average velocities at a point are the predominant variable.

When macro-scale variables are involved, every geometric design variable can affect the role of shear stresses. They can include such items as power, impeller speed, impeller diameter, impeller blade shape, impeller blade width or height, thickness of the material used to make the impeller, number of blades, impeller location, baffle location, and number of impellers.

Micro-scale variables are involved when the particles, droplets, baffles, or fluid clumps are on the order of 100 μm or less. In this case, the critical parameters usually are power per unit volume, distribution of power per unit volume between the impeller and the rest of the tank, rms velocity fluctuation, energy spectra, dissipation length, the smallest micro-scale eddy size for the particular power level, and viscosity of the fluid.

• The overall circulating pattern, including the circulation time and the deviation of the circulation times, can never be neglected. No matter what else a mixer does, it must be able to circulate fluid throughout an entire vessel appropriately. If it cannot, then that mixer is not suited for the task being considered.

Qualitative and, hopefully, quantitative estimates of how the process result will be measured must be made in advance. The evaluations must allow one to establish the importance of the different steps in a process, such as gas-liquid mass transfer, chemical reaction rate, or heat transfer.

• It is seldom possible, either economically or timewise, to study every potential mixing variable or to compare the performance of many impeller types. In many cases, a process needs a specific fluid regime that is relatively independent of the impeller type used to generate it. Because different impellers may require different geometries to achieve an optimum process combination, a random choice of only one diameter of each of two or more impeller types may not tell what is appropriate for the fluid regime ultimately required.

• Often, a pilot plant will operate in the viscous region while the commercial unit will operate in the transition region, or alternatively, the pilot plant may be in the transition region and the commercial unit in the turbulent region. Some experience is required to estimate the difference in performance to be expected upon scale-up.

• In general, it is not necessary to model Z/T ratios between pilot and commercial units.

• In order to make the pilot unit more like a commercial unit in macro-scale characteristics, the pilot unit impeller must be designed to lengthen the blend time and to increase the maximum impeller zone shear rate. This will result in a greater range of shear rates than is normally found in a pilot unit.

MIXING EQUIPMENT

There are three types of mixing flow patterns that are markedly different. The so-called axial-flow turbines (Fig. 18-3) actually give a flow coming off the impeller of approximately 45°, and therefore have a recirculation pattern coming back into the impeller at the hub region of the blades. This flow pattern exists to an approximate Reynolds number of 200 to 600 and then becomes radial as the Reynolds number decreases. Both the R100 and A200 impellers normally require four baffles for an effective flow pattern. These baffles typically are 1/12 of the tank diameter and width.

Radial-flow impellers include the flat-blade disc turbine, Fig. 18-4, which is labeled an R100. This generates a radial flow pattern at all Reynolds numbers. Figure 18-17 is the diagram of Reynolds number/power number curve, which allows one to calculate the power knowing the speed and diameter of the impeller. The impeller shown in Fig. 18-4 typically gives high shear rates and relatively low pumping capacity.

The current design of fluidfoil impellers includes the A310 (Fig. 18-2), as well as several other impellers of that type commonly referred to as *high-efficiency impellers, hydrofoil,* and other descriptive names to illustrate that they are designed to maximize flow and minimize shear rate. These impellers typically require two baffles, but are normally used with three, since three gives a more stable flow pattern. Since most industrial mixing processes involve pumping capacity and, to a lesser degree, fluid shear rate, the fluidfoil impellers are now used on the majority of the mixer installations. There is now an additional family of these fluidfoil impellers, which depend upon different solidity ratios to operate in various kinds of fluid mixing systems. Figure 18-5 illustrates four of these impellers. The solidity ratio is the ratio of total blade area to a circle circumscribing the impeller and, as viscosity increases, higher values of the solidity ratios are more effective in providing an axial flow pattern rather than a radial flow pattern. Also the A315-type provides an effective area of preventing gas bypassing through the hub of the impeller by having exceptionally wide blades. Another impeller of that type is the Prochem Maxflo T.

Small Tanks   For tanks less than 1.8 m in diameter, the clamp or flanged mounted angular, off-center axial-flow impeller without baffles should be used for a wide range of process requirements (refer to Fig. 18-14). The impellers currently used are the fluidfoil type. Since small impellers typically operate at low Reynolds numbers, often in the transition region, the fluidfoil impeller should be designed to give good flow characteristics over a range of Reynolds numbers, probably on the order of 50 to 500. The Z/T ratio should be 0.75 to 1.5. The volume of liquid should not exceed 4 m³.

Close-Clearance Impellers   There are two close-clearance impellers. They are the *anchor impeller* (Fig. 18-6) and the *helical*

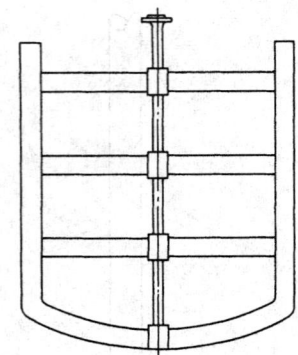

FIG. 18-6   Anchor impeller.

impeller (Fig. 18-7), which operate near the tank wall and are particularly effective in pseudoplastic fluids in which it is desirable to have the mixing energy concentrated out near the tank wall where the flow pattern is more effective than with the open impellers that were covered earlier.

Axial-Flow Impellers   Axial-flow impellers include all impellers in which the blade makes an angle of less than 90° with the plane of rotation. Propellers and pitched-blade turbines, as illustrated in Figs. 18-8 and 18-3, are representative axial-flow impellers.

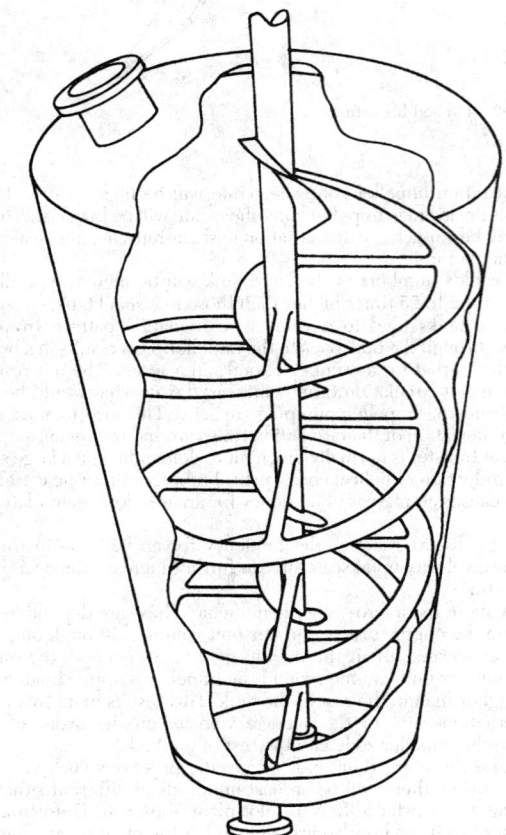

FIG. 18-7   Helical mixer for high-viscosity fluid.

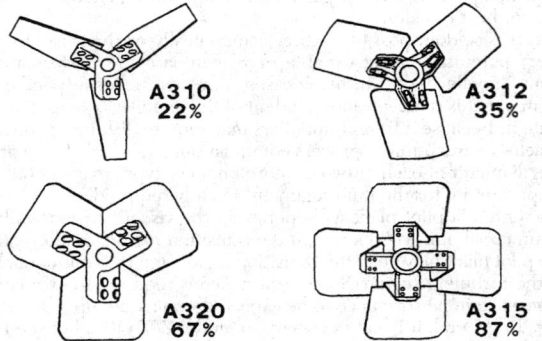

FIG. 18-5   The solidity ratio for four different impellers of the axial-flow fluidfoil type.

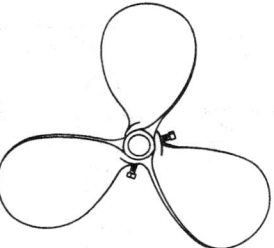

FIG. 18-8 Marine-type mixing impeller.

Portable mixers may be clamped on the side of an open vessel in the angular, off-center position shown in Fig. 18-14 or bolted to a flange or plate on the top of a closed vessel with the shaft in the same angular, off-center position. This mounting results in a strong top-to-bottom circulation.

Two basic speed ranges are available: 1150 or 1750 r/min with direct drive and 350 or 420 r/min with a gear drive. The high-speed units produce higher velocities and shear rates (Fig. 18-9) in the impeller discharge stream and a lower circulation rate throughout the vessel than the low-speed units. For suspension of solids, it is common to use the gear-driven units, while for rapid dispersion or fast reactions the high-speed units are more appropriate.

Axial-flow impellers may also be mounted near the bottom of the cylindrical wall of a vessel as shown in Fig. 18-10. Such side-entering agitators are used to blend low-viscosity fluids [<0.1 Pa·s (100 cP)] or to keep slowly settling sediment suspended in tanks as large as some 4000 m³ (10⁶ gal). Mixing of paper pulp is often carried out by side-entering propellers.

Pitched-blade turbines (Fig. 18-3) are used on top-entering agitator shafts instead of propellers when a high axial circulation rate is desired and the power consumption is more than 2.2 kW (3 hp). A pitched-blade turbine near the upper surface of liquid in a vessel is effective for rapid submergence of floating particulate solids.

Radial-Flow Impellers   Radial-flow impellers have blades which are parallel to the axis of the drive shaft. The smaller multiblade ones are known as *turbines*; larger, slower-speed impellers, with two or four blades, are often called *paddles*. The diameter of a turbine is normally

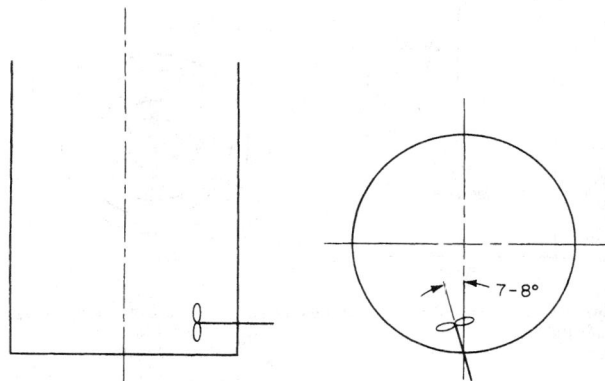

FIG. 18-10 Side-entering propeller mixer.

between 0.3 and 0.6 of the tank diameter. Turbine impellers come in a variety of types, such as curved-blade and flat-blade, as illustrated in Fig. 18-4. Curved blades aid in starting an impeller in settled solids.

For processes in which corrosion of commonly used metals is a problem, glass-coated impellers may be economical. A typical modified curved-blade turbine of this type is shown in Fig. 18-11.

Close-Clearance Stirrers   For some pseudoplastic fluid systems stagnant fluid may be found next to the vessel walls in parts remote from propeller or turbine impellers. In such cases, an "anchor" impeller may be used (Fig. 18-6). The fluid flow is principally circular or helical (see Fig. 18-7) in the direction of rotation of the anchor. Whether substantial axial or radial fluid motion also occurs depends on the fluid viscosity and the design of the upper blade-supporting spokes. Anchor agitators are used particularly to obtain improved heat transfer in high-consistency fluids.

Unbaffled Tanks   If a low-viscosity liquid is stirred in an unbaffled tank by an axially mounted agitator, there is a tendency for a swirling flow pattern to develop regardless of the type of impeller. Figure 18-12 shows a typical flow pattern. A vortex is produced owing to centrifugal force acting on the rotating liquid. In spite of the presence of a vortex, satisfactory process results often can be obtained in an unbaffled vessel. However, there is a limit to the rotational speed that may be used, since once the vortex reaches the impeller, severe air entrainment may occur. In addition, the swirling mass of liquid often generates an oscillating surge in the tank, which coupled with the deep vortex may create a large fluctuating force acting on the mixer shaft.

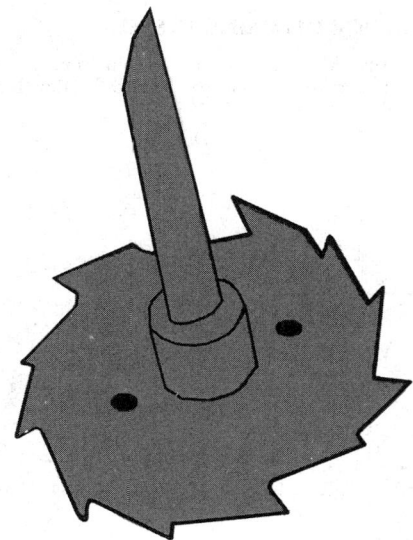

FIG. 18-9 High shear rate impeller.

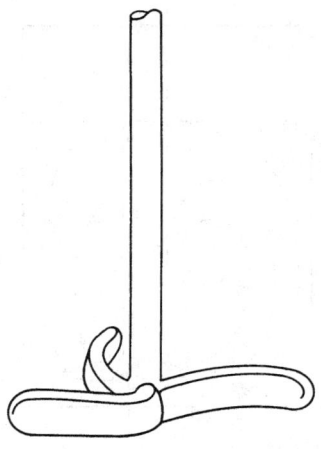

FIG. 18-11 Glass-steel impeller. (*The Pfaudler Company.*)

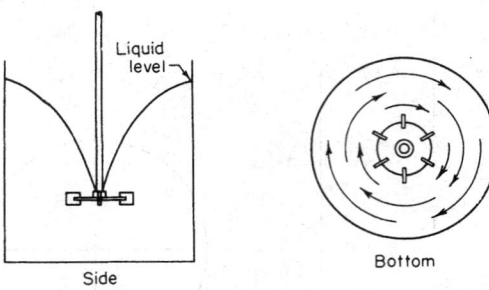

FIG. 18-12 Typical flow pattern for either axial- or radial-flow impellers in an unbaffled tank.

Vertical velocities in a vortexing low-viscosity liquid are low relative to circumferential velocities in the vessel. Increased vertical circulation rates may be obtained by mounting the impeller off center, as illustrated in Fig. 18-13. This position may be used with either turbines or propellers. The position is critical, since too far or too little off center in one direction or the other will cause greater swirling, erratic vortexing, and dangerously high shaft stresses. Changes in viscosity and tank size also affect the flow pattern in such vessels. Off-center mountings have been particularly effective in the suspension of paper pulp.

With axial-flow impellers, an angular off-center position may be used. The impeller is mounted approximately 15° from the vertical, as shown in Fig. 18-14.

The angular off-center position used with fluidfoil units is usually limited to impellers delivering 2.2 kW (3 hp) or less. The unbalanced fluid forces generated by this mounting can become severe with higher power.

Baffled Tanks For vigorous agitation of thin suspensions, the tank is provided with baffles which are flat vertical strips set radially along the tank wall, as illustrated in Figs. 18-15 and 18-16. Four baffles are almost always adequate. A common baffle width is one-tenth to one-twelfth of the tank diameter (radial dimension). For agitating slurries, the baffles often are located one-half of their width from the vessel wall to minimize accumulation of solids on or behind them.

For Reynolds numbers greater than 2000 baffles are commonly used with turbine impellers and with on-centerline axial-flow impellers. The flow patterns illustrated in Figs. 18-15 and 18-16 are quite different, but in both cases the use of baffles results in a large top-to-bottom circulation without vortexing or severely unbalanced fluid forces on the impeller shaft.

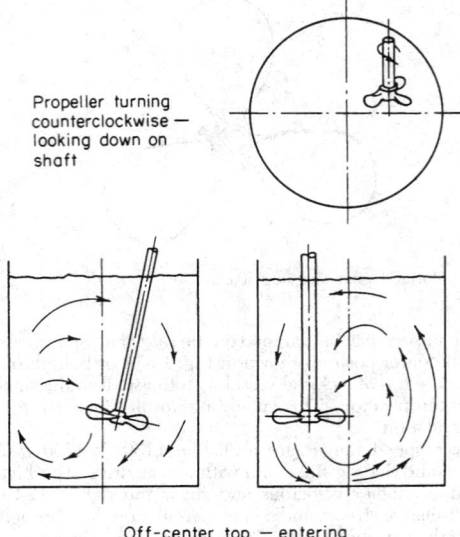

FIG. 18-14 Typical flow pattern with a propeller in angular off-center position without baffles.

In the transition region [Reynolds numbers, Eq. (18-1), from 10 to 10,000], the width of the baffle may be reduced, often to one-half of standard width. If the circulation pattern is satisfactory when the tank is unbaffled but a vortex creates a problem, partial-length baffles may be used. These are standard-width and extend downward from the surface into about one-third of the liquid volume.

In the region of laminar flow ($N_{Re} < 10$), the same power is consumed by the impeller whether baffles are present or not, and they are seldom required. The flow pattern may be affected by the baffles, but not always advantageously. When they are needed, the baffles are usually placed one or two widths radially off the tank wall, to allow fluid to circulate behind them and at the same time produce some axial deflection of flow.

FLUID BEHAVIOR IN MIXING VESSELS

Impeller Reynolds Number The presence or absence of turbulence in an impeller-stirred vessel can be correlated with an impeller Reynolds number defined

$$N_{Re} = \frac{D_a^2 N \rho}{\mu} \qquad (18\text{-}1)$$

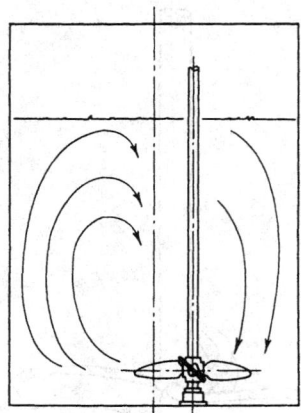

FIG. 18-13 Flow pattern with a paper-stock propeller, unbaffled; vertical off-center position.

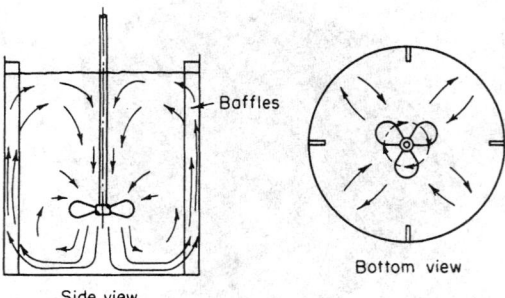

FIG. 18-15 Typical flow pattern in a baffled tank with a propeller or an axial-flow turbine positioned on center.

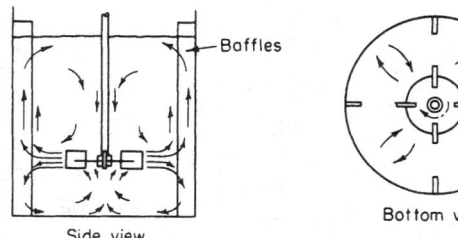

Side view

Bottom view

FIG. 18-16 Typical flow pattern in a baffled tank with a turbine positioned on center.

where N = rotational speed, r/s; D_a = impeller diameter, m (ft); ρ = fluid density, kg/m³ (lb/ft³); and μ = viscosity, Pa·s [lb/(ft·s)]. Flow in the tank is turbulent when $N_{Re} > 10{,}000$. Thus viscosity alone is not a valid indication of the type of flow to be expected. Between Reynolds numbers of 10,000 and approximately 10 is a transition range in which flow is turbulent at the impeller and laminar in remote parts of the vessel; when $N_{Re} < 10$, flow is laminar only.

Not only is the type of flow related to the impeller Reynolds number, but also such process performance characteristics as mixing time, impeller pumping rate, impeller power consumption, and heat- and mass-transfer coefficients can be correlated with this dimensionless group.

Relationship between Fluid Motion and Process Performance Several phenomena which can be used to promote various processing objectives occur during fluid motion in a vessel.

1. Shear stresses are developed in a fluid when a layer of fluid moves faster or slower than a nearby layer of fluid or a solid surface. In laminar flow, the shear stress is equal to the product of fluid viscosity and velocity gradient or rate of shear. Under laminar-flow conditions, shear forces are larger than inertial forces in the fluid.

With turbulent flow, shear stress also results from the behavior of transient random eddies, including large-scale eddies which decay to small eddies or fluctuations. The scale of the large eddies depends on equipment size. On the other hand, the scale of small eddies, which dissipate energy primarily through viscous shear, is almost independent of agitator and tank size.

The shear stress in the fluid is much higher near the impeller than it is near the tank wall. The difference is greater in large tanks than in small ones.

2. Inertial forces are developed when the velocity of a fluid changes direction or magnitude. In turbulent flow, inertia forces are larger than viscous forces. Fluid in motion tends to continue in motion until it meets a solid surface or other fluid moving in a different direction. Forces are developed during the momentum transfer that takes place. The forces acting on the impeller blades fluctuate in a random manner related to the scale and intensity of turbulence at the impeller.

3. The interfacial area between gases and liquids, immiscible liquids, and solids and liquids may be enlarged or reduced by these viscous and inertia forces when interacting with interfacial forces such as surface tension.

4. Concentration and temperature differences are reduced by bulk flow or circulation in a vessel. Fluid regions of different composition or temperature are reduced in thickness by bulk motion in which velocity gradients exist. This process is called bulk diffusion or Taylor diffusion (Brodkey, in Uhl and Gray, op. cit., vol. 1, p. 48). The turbulent and molecular diffusion reduces the difference between these regions. In laminar flow, Taylor diffusion and molecular diffusion are the mechanisms of concentration- and temperature-difference reduction.

5. Equilibrium concentrations which tend to develop at solid-liquid, gas-liquid, or liquid-liquid interfaces are displaced or changed by molecular and turbulent diffusion between bulk fluid and fluid adjacent to the interface. Bulk motion (Taylor diffusion) aids in this mass-transfer mechanism also.

Turbulent Flow in Stirred Vessels Turbulence parameters such as intensity and scale of turbulence, correlation coefficients, and energy spectra have been measured in stirred vessels. However, these characteristics are not used directly in the design of stirred vessels.

Fluid Velocities in Mixing Equipment Fluid velocities have been measured for various turbines in baffled and unbaffled vessels. Typical data are summarized in Uhl and Gray, op. cit., vol. 1, chap. 4. Velocity data have been used for calculating impeller discharge and circulation rates but are not employed directly in the design of mixing equipment.

Impeller Discharge Rate and Fluid Head for Turbulent Flow When fluid viscosity is low and flow is turbulent, an impeller moves fluids by an increase in momentum from the blades which exert a force on the fluid. The blades of rotating propellers and turbines change the direction and increase the velocity of the fluids.

The pumping rate or discharge rate of an impeller is the flow rate perpendicular to the impeller discharge area. The fluid passing through this area has velocities proportional to the impeller peripheral velocity and velocity heads proportional to the square of these velocities at each point in the impeller discharge stream under turbulent-flow conditions. The following equations relate velocity head, pumping rate, and power for geometrically similar impellers under turbulent-flow conditions:

$$Q = N_Q N D_a^3 \tag{18-2}$$

$$H = \frac{N_p N^2 D_a^2}{N_Q g} \tag{18-3}$$

$$P = N_p \rho N^3 \left(\frac{D_a^5}{g_c} \right) \tag{18-4}$$

$$P = \frac{\rho H Q g}{g_c} \tag{18-5}$$

where Q = impeller discharge rate, m³/s (ft³/s); N_Q = discharge coefficient, dimensionless; H = velocity head, m (ft); N_p = power number, dimensionless; P = power, (N·m)/s [(ft·lbf)/s]; g_c = dimensional constant, 32.2 (ft·lb)/(lbf·s²)(g_c = 1 when using SI units); and g = gravitational acceleration, m/s² (ft/s²).

The discharge rate Q has been measured for several types of impellers, and discharge coefficients have been calculated. The data of a number of investigators are reviewed by Uhl and Gray (op. cit., vol. 1, chap. 4): N_Q is 0.4 to 0.5 for a propeller with pitch equal to diameter at $N_{Re} = 10^5$. For turbines, N_Q ranges from 0.7 to 2.9, depending on the number of blades, blade-height-to-impeller-diameter ratio, and impeller-to-vessel-diameter ratio. The effects of these geometric variables are not well defined.

Power consumption has also been measured and correlated with impeller Reynolds number. The velocity head for a mixing impeller can be calculated, then, from flow and power data, by Eq. (18-3) or Eq. (18-5).

The velocity head of the impeller discharge stream is a measure of the maximum force that this fluid can exert when its velocity is changed. Such inertia forces are higher in streams with higher discharge velocities. Shear rates and shear stresses are also higher under these conditions in the smallest eddies. If a higher discharge velocity is desired at the same power consumption, a smaller-diameter impeller must be used at a higher rotational speed. According to Eq. (18-4), at a given power level $N \propto D_a^{-5/3}$ and $ND_a \propto D_a^{-2/3}$. Then, $H \propto D_a^{-4/3}$ and $Q \propto D_a^{4/3}$.

An impeller with a high fluid head is one with high peripheral velocity and discharge velocity. Such impellers are useful for (1) rapid reduction of concentration differences in the impeller discharge stream (rapid mixing), (2) production of large interfacial area and small droplets in gas-liquid and immiscible-liquid systems, (3) solids deagglomeration, and (4) promotion of mass transfer between phases.

The impeller discharge rate can be increased at the same power consumption by increasing impeller diameter and decreasing rotational speed and peripheral velocity so that $N^3 D_a^5$ is a constant (Eq. 18-4)]. Flow goes up, velocity head and peripheral velocity go down, but impeller torque T_Q goes up. At the same torque, $N^2 D_a^5$ is constant, $P \propto D_a^{-5/2}$, and $Q \propto D_a^{1/2}$. Therefore, increasing impeller diameter at constant torque increases discharge rate at lower power consumption. At the same discharge rate, ND_a^3 is constant, $P \propto D_a^{-4}$, and $T_Q \propto D_a^{-1}$.

Therefore, power and torque decrease as impeller diameter is increased at constant Q.

A large-diameter impeller with a high discharge rate is used for (1) short times to complete mixing of miscible liquid throughout a vessel, (2) promotion of heat transfer, (3) reduction of concentration and temperature differences in all parts of vessels used for constant-environment reactors and continuous averaging, and (4) suspension of particles of relatively low settling rate.

Laminar Fluid Motion in Vessels When the impeller Reynolds number is less than 10, the flow induced by the impeller is laminar. Under these conditions, the impeller drags fluid with it in a predominantly circular pattern. If the impeller blades curve back, there is a viscous drag flow toward the tips of these blades. Under moderate-viscosity conditions in laminar flow, centrifugal force acting on the fluid layer dragged in a circular path by the rotating impeller will move fluid in a radial direction. This centrifugal effect causes any gas accumulated behind a rotating blade to move to the axis of impeller rotation. Such radial-velocity components are small relative to tangential velocity.

For turbines at Reynolds numbers less than 100, toroidal stagnant zones exist above and below the turbine periphery. Interchange of liquid between these regions and the rest of the vessel is principally by molecular diffusion.

Suspensions of fine solids may have pseudoplastic or plastic-flow properties. When they are in laminar flow in a stirred vessel, motion in remote parts of the vessel where shear rates are low may become negligible or cease completely. To compensate for this behavior of slurries, large-diameter impellers or paddles are used, with $(D_a/D_T) > 0.6$, where D_T is the tank diameter. In some cases, for example, with some anchors, $D_a > 0.95\, D_T$. Two or more paddles may be used in deep tanks to avoid stagnant regions in slurries.

In laminar flow ($N_{Re} < 10$), $N_p \propto 1/N_{Re}$ and $P \propto \mu N^2 D_a^3$. Since shear stress is proportional to rotational speed, shear stress can be increased at the same power consumption by increasing N proportionally to $D_a^{-3/2}$ as impeller diameter D_a is decreased.

Fluid circulation probably can be increased at the same power consumption and viscosity in laminar flow by increasing impeller diameter and decreasing rotational speed, but the relationship between Q, N, and D_a for laminar flow from turbines has not been determined.

As in the case of turbulent flow, then, small-diameter impellers ($D_a < D_T/3$) are useful for (1) rapid mixing of dry particles into liquids, (2) gas dispersion in slurries, (3) solid-particle deagglomeration, and (4) promoting mass transfer between solid and liquid phases. If stagnant regions are a problem, large impellers must be used and rotational speed and power increased to obtain the required results. Small continuous-processing equipment may be more economical than batch equipment in such cases.

Likewise, large-diameter impellers ($D_a > D_T/2$) are useful for (1) avoiding stagnant regions in slurries, (2) short mixing times to obtain uniformity throughout a vessel, (3) promotion of heat transfer, and (4) laminar continuous averaging of slurries.

Vortex Depth In an unbaffled vessel with an impeller rotating in the center, centrifugal force acting on the fluid raises the fluid level at the wall and lowers the level at the shaft. The depth and shape of such a vortex (Rieger, Ditl, and Novak, *Chem. Eng. Sci.*, **34**, 397 (1978)] depend on impeller and vessel dimensions as well as rotational speed.

Power Consumption of Impellers Power consumption is related to fluid density, fluid viscosity, rotational speed, and impeller diameter by plots of power number ($g_c P/\rho N^3 D_a^5$) versus Reynolds number ($D_a^2 N \rho/\mu$). Typical correlation lines for frequently used impellers operating in newtonian liquids contained in baffled cylindrical vessels are presented in Fig. 18-17. These curves may be used also for operation of the respective impellers in unbaffled tanks when the Reynolds number is 300 or less. When N_{Re} is greater than 300, however, the power consumption is lower in an unbaffled vessel than indicated in Fig. 18-17. For example, for a six-blade disk turbine with $D_T/D_a = 3$ and $D_a/W_i = 5$, $N_p = 1.2$ when $N_{Re} = 10^4$. This is only about one-fifth of the value of N_p when baffles are present.

Additional power data for other impeller types such as anchors, curved-blade turbines, and paddles in baffled and unbaffled vessels are available in the following references: Holland and Chapman, op.

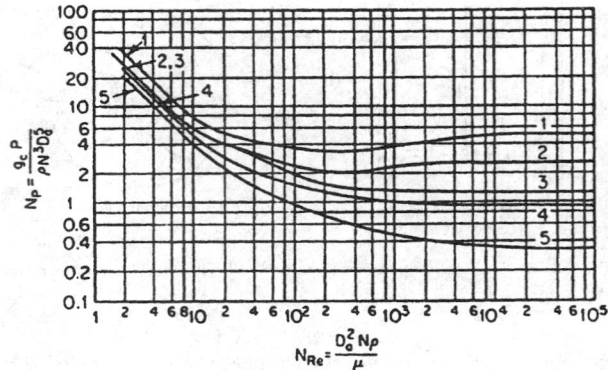

FIG. 18-17 Impeller power correlations: curve 1, six-blade turbine, $D_a/W_i = 5$, like Fig. 18-4 but with six blades, four baffles, each $D_T/12$; curve 2, vertical-blade, open turbine with six straight blades, $D_a/W_i = 8$, four baffles each $D_T/12$; curve 3, 45° pitched-blade turbine like Fig. 18-3 but with six blades, $D_a/W_i = 8$, four baffles, each $D_T/12$; curve 4, propeller, pitch equal to $2D_a$, four baffles, each $0.1D_T$, also same propeller in angular off-center position with no baffles; curve 5, propeller, pitch equal to D_a, four baffles each $0.1D_T$, also same propeller in angular off-center position as in Fig. 18-14 with no baffles. D_a = impeller diameter, D_T = tank diameter, g_c = gravitational conversion factor, N = impeller rotational speed, P = power transmitted by impeller shaft, W_i = impeller blade height, μ = viscosity of stirred liquid, and ρ = density of stirred mixture. Any set of consistent units may be used, but N must be rotations (rather than radians) per unit time. In the SI system, g_c is dimensionless and unity. [Curves 4 and 5 from Rushton, Costich, and Everett, *Chem. Eng. Prog.*, **46**, 395, 467 (1950), by permission; curves 2 and 3 from Bates, Fondy, and Corpstein, *Ind. Eng. Chem. Process Des. Dev.*, **2**, 310 (1963), by permission of the copyright owner, the American Chemical Society.]

cit., chaps. 2, 4, Reinhold, New York, 1966; and Bates, Fondy, and Fenic, in Uhl and Gray, op. cit., vol. 1, chap. 3.

Power consumption for impellers in pseudoplastic, Bingham plastic, and dilatant nonnewtonian fluids may be calculated by using the correlating lines of Fig. 18-17 if viscosity is obtained from viscosity-shear rate curves as described here. For a pseudoplastic fluid, viscosity decreases as shear rate increases. A Bingham plastic is similar to a pseudoplastic fluid but requires that a minimum shear stress be exceeded for any flow to occur. For a dilatant fluid, viscosity increases as shear rate increases.

The appropriate shear rate to use in calculating viscosity is given by one of the following equations when a propeller or a turbine is used (Bates et al., in Uhl and Gray, op. cit., vol. 1, p. 149):

For dilatant liquids,

$$\dot{\gamma} = 13N\left(\frac{D_a}{D_T}\right)^{0.5} \tag{18-6}$$

For pseudoplastic and Bingham plastic fluids,

$$\dot{\gamma} = 10N \tag{18-7}$$

where $\dot{\gamma}$ = average shear rate, s^{-1}.

The shear rate calculated from impeller rotational speed is used to identify a viscosity from a plot of viscosity versus shear rate determined with a capillary or rotational viscometer. Next N_{Re} is calculated, and N_p is read from a plot like Fig. 18-17.

DESIGN OF AGITATION EQUIPMENT

Selection of Equipment The principal factors which influence mixing-equipment choice are (1) the process requirements, (2) the flow properties of the process fluids, (3) equipment costs, and (4) construction materials required.

Ideally, the equipment chosen should be that of the lowest total cost which meets all process requirements. The total cost includes depreciation on investment, operating cost such as power, and maintenance costs. Rarely is any more than a superficial evaluation based on this principle justified, however, because the cost of such an evaluation often exceeds the potential savings that can be realized. Usually optimization is based on experience with similar mixing operations. Often the process requirements can be matched with those of a similar operation, but sometimes tests are necessary to identify a satisfactory design and to find the minimum rotational speed and power.

There are no satisfactory specific guides for selecting mixing equipment because the ranges of application of the various types of equipment overlap and the effects of flow properties on process performance have not been adequately defined. Nevertheless, what is frequently done in selecting equipment is described in the following paragraphs.

Top-Entering Impellers For vessels less than 1.8 m (6 ft) in diameter, a clamp- or flange-mounted, angular, off-center fluidfoil impeller with no baffles should be the initial choice for meeting a wide range of process requirements (Fig. 18-14). The vessel straight-side-height-to-diameter ratio should be 0.75 to 1.5, and the volume of stirred liquid should not exceed 4 m³ (about 1000 gal).

For suspension of free-settling particles, circulation of pseudoplastic slurries, and heat transfer or mixing of miscible liquids to obtain uniformity, a speed of 350 or 420 r/min should be stipulated. For dispersion of dry particles in liquids or for rapid initial mixing of liquid reactants in a vessel, an 1150- or 1750- r/min propeller should be used at a distance $D_T/4$ above the vessel bottom. A second propeller can be added to the shaft at a depth D_a below the liquid surface if the submergence of floating liquids or particulate solids is otherwise inadequate. Such propeller mixers are readily available up to 2.2 kW (3 hp) for off-center sloped-shaft mounting.

Propeller size, pitch, and rotational speed may be selected by model tests, by experience with similar operations, or, in a few cases, by published correlations of performance data such as mixing time or heat transfer. The propeller diameter and motor power should be the minimum which meet process requirements.

If agitation is required for a vessel less than 1.8 m (6 ft) in diameter and the same operations will be scaled up to a larger vessel ultimately, the equipment type should be the same as that expected in the larger vessel.

Axial-Flow Fluidfoil Impellers For vessel volumes of 4 to 200 m³ (1000 to 50,000 gal), a turbine mixer mounted coaxially within the vessel with four or more baffles should be the initial choice. Here also the vessel straight-side-height-to-diameter ratio should be 0.75 to 1.5. Four vertical baffles should be fastened perpendicularly to the vessel wall with a gap between baffle and wall equal to $D_T/24$ and a radial baffle width equal to $D_T/12$.

For suspension of rapidly settling particles, the impeller turbine diameter should be $D_T/3$ to $D_T/2$. A clearance of less than one-seventh of the fluid depth in the vessel should be used between the lower edge of the turbine blade tips and the vessel bottom. As the viscosity of a suspension increases, the impeller diameter should be increased. This diameter may be increased to $0.6\,D_T$ and a second impeller added to avoid stagnant regions in pseudoplastic slurries. Moving the baffles halfway between the impeller periphery and the vessel wall will also help avoid stagnant fluid near the baffles.

As has been shown, power consumption is decreased and turbine discharge rate is increased as impeller diameter is increased at constant torque (in the completely turbulent regime). This means that for a stipulated discharge rate, more efficient operation is obtained (lower power and torque) with a relatively large impeller operating at a relatively low speed ($N \propto D_a^{-3}$). Conversely, if power is held constant, decreasing impeller diameter results in increasing peripheral velocity and decreasing torque. Thus at a stipulated power level the rapid, efficient initial mixing of reactants identified with high peripheral velocity can be achieved by a relatively small impeller operating at a relatively high speed ($N \propto D_a^{-5/3}$).

For circulation and mixing to obtain uniformity, the impeller should be located at one-third of the liquid depth above the vessel bottom unless rapidly settling material or a need to stir a nearly empty vessel requires a lower impeller location.

Side-Entering Impellers For vessels greater than 4 m³ (1000 gal), a side-entering propeller agitator (Fig. 18-9) may be more economical than a top-mounted impeller on a centered vertical shaft.

For vessels greater than 38 m³ (10,000 gal), the economic attractiveness of side-entering impellers increases. For vessels larger than 380 m³ (100,000 gal), units may be as large as 56 kW (75 hp), and two or even three may be installed in one tank. For the suspension of slow-settling particles or the maintenance of uniformity in a viscous slurry of small particles, the diameter and rotational speed of a side-entering agitator must be selected on the basis of model tests or experience with similar operations.

When abrasive solid particles must be suspended, maintenance costs for the submerged shaft seal of a side-entering propeller may become high enough to make this type of mixer an uneconomical choice.

Jet Mixers Continuous recycle of the contents of a tank through an external pump so arranged that the pump discharge stream appropriately reenters the vessel can result in a flow pattern in the tank which will produce a slow mixing action [Fossett, *Trans. Inst. Chem. Eng.*, **29**, 322 (1951)].

Large Tanks Most large vessels (over 4 m³) require a heavy-duty drive. About two-thirds of the mixing requirements industrially involve flow, circulation, and other types of pumping capacity requirements, including such applications as blending and solid suspension. There often is no requirement for any marked level of shear rate, so the use of the fluidfoil impellers is most common. If additional shear rate is required over what can be provided by the fluidfoil impeller, the axial-flow turbine (Fig. 18-3) is often used, and if extremely high shear rates are required, the flat-blade turbine (Rushton turbine) (Fig. 18-4) is required. For still higher shear rates, there is an entire variety of high-shear-rate impellers, typified by that shown in Fig. 18-10 that are used.

The fluidfoil impellers in large tanks require only two baffles, but three are usually used to provide better flow pattern asymmetry. These fluidfoil impellers provide a true axial flow pattern, almost as though there was a draft tube around the impeller. Two or three or more impellers are used if tanks with high D/T ratios are involved. The fluidfoil impellers do not vortex vigorously even at relatively low coverage so that if gases or solids are to be incorporated at the surface, the axial-flow turbine is often required and can be used in combination with the fluidfoil impellers also on the same shaft.

BLENDING

If the blending process is between two or more fluids with relatively low viscosity such that the blending is not affected by fluid shear rates, then the difference in blend time and circulation between small and large tanks is the only factor involved. However, if the blending involves wide disparities in the density of viscosity and surface tension between the various phases, then a certain level of shear rate may be required before blending can proceed to the required degree of uniformity.

The role of viscosity is a major factor in going from the turbulent regime, through the transition region, into the viscous regime and the change in the role of energy dissipation discussed previously. The role of non-newtonian viscosities comes into the picture very strongly since that tends to markedly change the type of influence of impellers and determines the appropriate geometry that is involved.

There is the possibility of misinterpretation of the difference between circulation time and blend time. Circulation time is primarily a function of the pumping capacity of the impeller. For axial-flow impellers, a convenient parameter, but not particularly physically accurate, is to divide the pumping capacity of the impeller by the cross-sectional area of the tank to give a superficial liquid velocity. This is sometimes used by using the total volume of flow from the impeller including entrainment of the tank to obtain a superficial liquid velocity.

As the flow from an impeller is increased from a given power level, there will be a higher fluid velocity and therefore a shorter circulation time. This holds true when dealing with any given impeller. This is shown in Fig. 18-18, which shows that circulation time versus D/T decreases. A major consideration is when increasing D/T becomes too large and actually causes the curve to reverse. This occurs somewhere around 0.45, ±0.05, so that using impellers of D/T ratios of 0.6 to 0.8

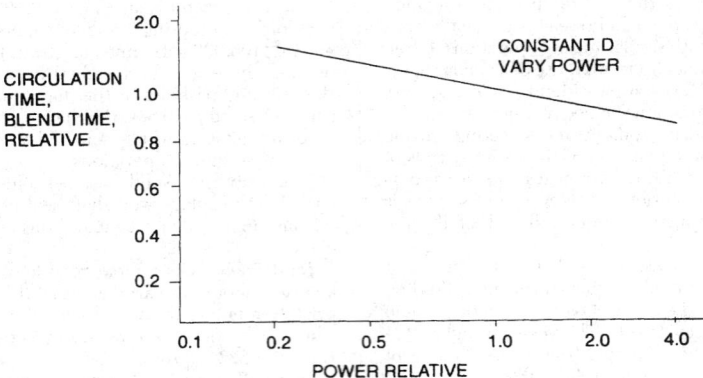

FIG. 18-18   Effect of D/T ratio on two different impellers on the circulation time and the blend time.

is often counterproductive for circulation time. They may be useful for the blending or motion of pseudoplastic fluids.

When comparing different impeller types, an entirely different phenomenon is important. In terms of circulation time, the phenomena shown in Figs. 18-18 and 18-19 still apply with the different impellers shown in Fig. 18-5. When it comes to blending another factor enters the picture. When particles A and B meet each other as a result of shear rates, there has to be sufficient shear stress to cause A and B to blend, react, or otherwise participate in the process.

It turns out that in low-viscosity blending the actual result does depend upon the measuring technique used to measure blend time. Two common techniques, which do not exhaust the possibilities in reported studies, are to use an acid-base indicator and inject an acid or base into the system that will result in a color change. One can also put a dye into the tank and measure the time for color to arrive at uniformity. Another system is to put in a conductivity probe and inject a salt or other electrolyte into the system. With any given impeller type at constant power, the circulation time will increase with the D/T ratio of the impeller. Figure 18-18 shows that both circulation time and blend time decrease as D/T increases. The same is true for impeller speed. As impeller speed is increased with any impeller, blend time and circulation time are decreased (Fig. 18-19).

However, when comparing different impeller types at the same power level, it turns out that impellers that have a higher pumping capacity will give decreased circulation time, but all the impellers, regardless of their pumping efficiency, give the same blend time at the

same power level and same diameter. This means that circulation time must be combined with shear rate to carry out a blending experiment which involves chemical reactions or interparticle mixing (Fig. 18-20).

For other situations in low-viscosity blending, the fluid in tanks may become stratified. There are few studies on that situation, but Oldshue (op. cit.) indicates the relationship between some of the variables. The important difference is that blend time is inversely proportional to power, not impeller flow, so that the exponents are quite different for a stratified tank. This situation occurs more frequently in the petroleum industry, where large petroleum storage tanks become stratified either by filling techniques or by temperature fluctuations.

There is a lot of common usage of the terms *blend time, mixing time,* and *circulation time*. There are differences in concept and interpretation of these different "times." For any given experiment, one must pick a definition of *blend time* to be used. As an example, if one is measuring the fluctuation of concentration after an addition of material to the tank, then one can pick an arbitrary definition of blending such as reducing the fluctuations below a certain level. This often is chosen as a fluctuation equal to 5% of the original fluctuation when the feed material is added. This obviously is a function of the size of the probe used to measure these fluctuations, which often is on the order of 500 to 1000 μm.

At the micro-scale level, there really is no way to measure concentration fluctuations. Resort must be made to other qualitative interpretation of results for either a process or a chemical reaction study.

High-Viscosity Systems   All axial-flow impellers become radial flow as Reynolds numbers approach the viscous region. Blending in

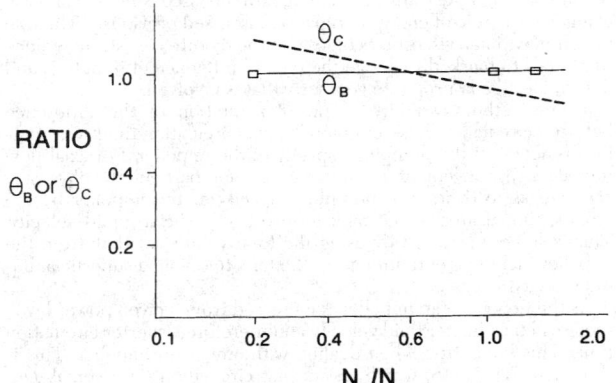

FIG. 18-19   Effect of impeller speed and power for the same diameter on circulation time and blend time for a particular impeller.

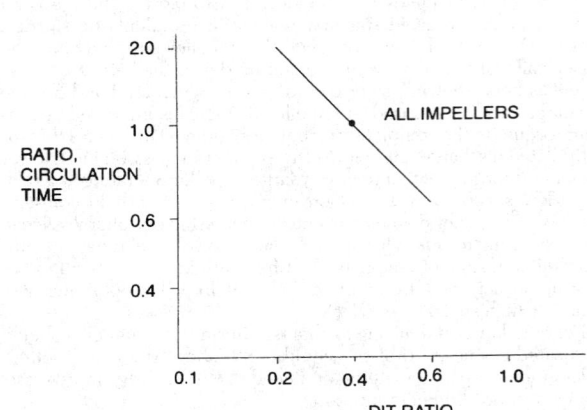

FIG. 18-20   At constant power and constant impeller diameter, three different impellers give the same blend time but different circulation times.

the transition and low-viscosity system is largely a measure of fluid motion throughout the tank. For close-clearance impellers, the anchor and helical impellers provide blending by having an effective action at the tank wall, which is particularly suitable for pseudoplastic fluids.

Figure 18-21 gives some data on the circulation time of the helical impeller. It has been observed that it takes about three circulation times to get one blend time being the visual uniformity of a dye added to the material. This is a macro-scale blending definition.

Axial-flow turbines are often used in blending pseudoplastic materials, and they are often used at relatively large D/T ratios, from 0.5 to 0.7, to adequately provide shear rate in the majority of the batch particularly in pseudoplastic material. These impellers develop a flow pattern which may or may not encompass an entire tank, and these areas of motion are sometimes referred to as *caverns*. Several papers describe the size of these caverns relative to various types of mixing phenomena. An effective procedure for the blending of pseudoplastic fluids is given in Oldshue (op. cit.).

Chemical Reactions   Chemical reactions are influenced by the uniformity of concentration both at the feed point and in the rest of the tank and can be markedly affected by the change in overall blend time and circulation time as well as the micro-scale environment. It is possible to keep the ratio between the power per unit volume at the impeller and in the rest of the tank relatively similar on scale-up, but many details need to be considered when talking about the reaction conditions, particularly where they involve selectivity. This means that reactions can take different paths depending upon chemistry and fluid mechanics, which is a major consideration in what should be examined. The method of introducing the reagent stream can be projected in several different ways depending upon the geometry of the impeller and feed system.

Chemical reactions normally occur in the micro-scale range. In turbulent flow, almost all of the power dissipation occurs eventually in the micro-scale regime because that is the only place where the scale of the fluid fluctuations is small enough that viscous shear stress exists. At approximately 100 μm, the fluid does not know what type of impeller is used to generate the power; continuing down to 10 μm and, even further, to chemical reactions, the actual impeller type is not a major variable as long as the proper macro-scale regime has been provided throughout the entire tank. The intensity of the mixing environment in the micro-scale regime can be related to a series of variables in an increasing order of complexity. Since all of the power is ultimately dissipated in the micro-scale regime, the power per unit volume throughout the tank is one measure of the overall measure of micro-scale mixing and the power dissipation at individual volumes in the tank is another way of expressing the influence. In general, the power per unit

volume dissipated around an impeller zone can be 100 times higher than the power dissipated throughout the remainder of the tank.

The next level of complexity is to look at the rms velocity fluctuation, which is typically 50 percent of the mean velocity around the impeller zone and about 5 percent of the mean velocity in the rest of the vessel. This means that the feed introduction point for either a single reactant or several reactants can be of extreme importance. It seems that the selectivity of competing or consecutive chemical reactions can be a function of the rms velocity fluctuations in the feed point if the chemical reactants remain constant and involve an appropriate relationship to the time between the rms velocity fluctuations. There are three common ways of introducing reagents into a mixing vessel. One is to let them drip on the surface. The second is to use some type of introduction pipe to bring the material into various parts of the vessel. The third is to purposely bring them in and around the impeller zone. Generally, all three methods have to be tried before determining the effect of feed location.

Since chemical reactions are on a scale much below 1 μm, and it appears that the Komolgoroff scale of isotropic turbulence turns out to be somewhere between 10 and 30 μm, other mechanisms must play a role in getting materials in and out of reaction zones and reactants in and out of those zones. One cannot really assign a shear rate magnitude to the area around a micro-scale zone, and it is primarily an environment that particles and reactants witness in this area.

The next level of complexity looks at the kinetic energy of turbulence. There are several models that are used to study the fluid mechanics, such as the Kε model. One can also put the velocity measurements through a spectrum analyzer to look at the energy at various wave numbers.

In the viscous regime, chemical reactants become associated with each other through viscous shear stresses. These shear stresses exist at all scales (macro to micro) and until the power is dissipated continuously through the entire spectrum. This gives a different relationship for power dissipation than in the case of turbulent flow.

Solid-Liquid   The most-used technique to study solid suspension, as documented in hundreds of papers in the literature, is called the *speed for just suspension*, N_{JS}. The original work was done in 1958 by Zwietering and this is still the most extensive range of variables, although other investigators have added to it considerably.

This particular technique is suitable only for laboratory investigation using tanks that are transparent and well illuminated. It does not lend itself to evaluation of the opaque tanks, nor is it used in any study of large-scale tanks in the field. It is a very minimal requirement for uniformity, and definitions suggested earlier are recommended for use in industrial design.

Some Observations on the Use of N_{JS} With D/T ratios of less than 0.4, uniformity throughout the rest of the tank is minimal. In D/T ratios greater than 0.4, the rest of the tank has a very vigorous fluid motion with marked approach to complete uniformity before N_{JS} is reached.

Much of the variation in N_{JS} can be reduced by using P_{JS}, which is the power in the just-suspended state. This also gives a better feel for the comparison of various impellers based on the energy requirement rather than speed, which has no economic relevance.

The overall superficial fluid velocity, mentioned earlier, should be proportional to the settling velocity of the solids if that were the main mechanism for solid suspension. If this were the case, the requirement for power if the settling velocity were doubled should be eight times. Experimentally, it is found that the increase in power is more nearly four times, so that some effect of the shear rate in macro-scale turbulence is effective in providing uplift and motion in the system.

Picking up the solids at the bottom of the tank depends upon the eddies and velocity fluctuations in the lower part of the tank and is a different criterion from the flow pattern required to keep particles suspended and moving in various velocity patterns throughout the remainder of the vessel. This leads to the variables in the design equation and a relationship that is quite different when these same variables are studied in relation to complete uniformity throughout the mixing vessel.

Another concern is the effect of multiple particle sizes. In general, the presence of fine particles will affect the requirements of suspen-

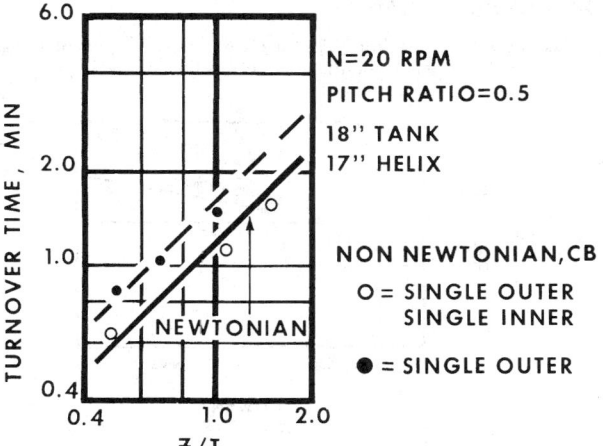

FIG. 18-21   Effect of impeller speed on circulation time for a helical impeller in the Reynolds number arranged less than 10.

sion of larger particles. The fine particles act largely as a potential viscosity-increasing agent and give a similar result to what would happen if the viscosity of the continuous phase were increased.

Another phenomenon is the increase in power required with percent solids, which makes a dramatic change at approximately 40 percent by volume, and then dramatically changes again as we approach the ultimate weight percent of settled solids. This phenomenon is covered by Oldshue (op. cit.), who describes conditions required for mixing slurries in the 80 to 100 percent range of the ultimate weight percent of settled solids.

Solids suspension in general is not usually affected by blend time or shear-rate changes in the relatively low to medium solids concentration in the range from 0 to 40 percent by weight. However, as solids become more concentrated, the effect of solids concentration on power required gives a change in criterion from the settling velocity of the individual particles in the mixture to the apparent viscosity of the more concentrated slurry. This means that we enter into an area where the blending of non-newtonian fluid regions affects the shear rates and plays a marked role.

The suspension of a single solid particle should depend primarily on the upward velocity at a given point and also should be affected by the uniformity of this velocity profile across the entire tank cross section. There are upward velocities in the tank and there also must be corresponding downward velocities.

In addition to the effect of the upward velocity on a settling particle, there is also the random motion of the micro-scale environment, which does not affect large particles very much but is a major factor in the concentration and uniformity of particles in the transition and micro-scale size range.

Using a draft tube in the tank for solids suspension introduces another, different set of variables. There are other relationships that are very much affected by scale-up in this type of process, as shown in Fig. 18-22. Different scale-up problems exist whether the impeller is pumping up or down within the draft tube.

Solid Dispersion   If the process involves the dispersion of solids in a liquid, then we may either be involved with breaking up agglomerates or possibly physically breaking or shattering particles that have a low cohesive force between their components. Normally, we do not think of breaking up ionic bonds with the shear rates available in mixing machinery.

If we know the shear stress required to break up a particle, we can then determine the shear rate required from the machinery by various viscosities with the equation:

$$\text{Shear stress} = \text{viscosity (shear rate)}$$

The shear rate available from various types of mixing and dispersion devices is known approximately and also the range of viscosities in which they can operate. This makes the selection of the mixing equipment subject to calculation of the shear stress required for the viscosity to be used.

In the equation referred to above, it is assumed that there is 100 percent transmission of the shear rate in the shear stress. However, with the slurry viscosity determined essentially by the properties of the slurry, at high concentrations of slurries there is a slippage factor. Internal motion of particles in the fluids over and around each other can reduce the effective transmission of viscosity efficiencies from 100 percent to as low as 30 percent.

Animal cells in biotechnology do not normally have tough skins like those of fungal cells and they are very sensitive to mixing effects. Many approaches have been and are being tried to minimize the effect of increased shear rates on scale-up. These include encapsulating the organism in or on microparticles and/or conditioning cells selectively to shear rates. In addition, traditional fermentation processes have maximum shear-rate requirements in which cells become progressively more and more damaged until they become motile.

Solid-Liquid Mass Transfer   There is potentially a major effect of both shear rate and circulation time in these processes. The solids can either be fragile or rugged. We are looking at the slip velocity of the particle and also whether we can break up agglomerates of particles which may enhance the mass transfer. When the particles become small enough, they tend to follow the flow pattern, so the slip velocity necessary to affect the mass transfer becomes less and less available.

What this shows is that, from the definition of off-bottom motion to complete uniformity, the effect of mixer power is much less than from going to on-bottom motion to off-bottom suspension. The initial increase in power causes more and more solids to be in active communication with the liquid and has a much greater mass-transfer rate than that occurring above the power level for off-bottom suspension, in which slip velocity between the particles of fluid is the major contributor (Fig. 18-23).

Since there may well be chemical or biological reactions happening on or in the solid phase, depending upon the size of the process participants, macro- or micro-scale effects may or may not be appropriate to consider.

In the case of living organisms, their access to dissolved oxygen throughout the tank is of great concern. Large tanks in the fermentation industry often have a Z/T ratio of 2:1 to 4:1; thus, top-to-bottom blending can be a major factor. Some biological particles are facultative and can adapt and reestablish their metabolisms at different dissolved-oxygen levels. Other organisms are irreversibly destroyed by sufficient exposure to low dissolved-oxygen levels.

GAS-LIQUID SYSTEMS

Gas-Liquid Dispersion   This involves physical dispersion of gas bubbles by the impeller, and the effect of gas flow on the impeller.

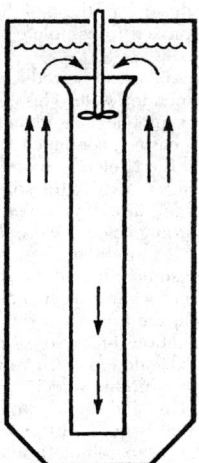

FIG. 18-22   Typical draft tube circulator, shown here for down-pumping mode for the impeller in the draft tube.

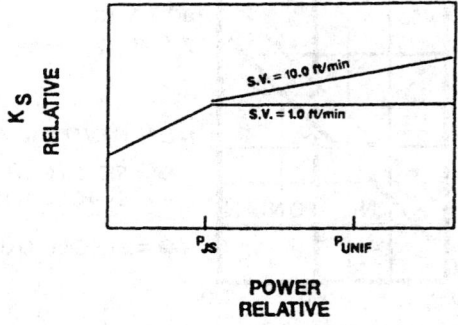

FIG. 18-23   Relative change in solid-liquid mass-transfer ratio with three different suspension levels, i.e., on-bottom motion, off-bottom motion, and complete uniformity.

The observation of the physical appearance of a tank undergoing gas-liquid mass transfer can be helpful but is not a substitute for mass-transfer data on the actual process. The mixing vessel can have four regimes of visual comparisons between gas bubbles and flow patterns. A helpful parameter is the ratio between the power given up by the gas phase and the power introduced by the mixing impeller. In general, if the power in the gas stream (calculated as the expansion energy from the gas expanding from the sparging area to the top of the tank, shown in Fig. 18-24) is greater, there will be considerable blurping and entrainment of liquid drops by a very violent explosion of gas bubbles at the surface. If the power level is more than the expanding gas energy, then the surface action will normally be very coalescent and uniform by comparison, and the gas will be reasonably well distributed throughout the remainder of the tank. With power levels up to 10 to 100 times the gas energy, the impeller will cause a more uniform and vigorous dispersion of the gas bubbles and smaller gas bubbles in the vessel.

In the 1960s and before, most gas-liquid operations were conducted using flat-blade turbines as shown in Fig. 18-4. These impellers required input of approximately three times the energy in the gas stream before they completely control the flow pattern. This was usually the case, and the mass-transfer characteristics were comparable to what would be expected. One disadvantage of the radial-flow impeller is that it is a very poor blending device so blend time is very long compared to that in pilot-scale experiments and compared to the fluidfoil impeller types often used currently. Using curvature of the blades to modify the tendency of gas bubbles to streamline the back of the flat-blade turbine gives a different characteristic to the power drawn by the impeller at a given gas rate compared to no gas rate, but it seems to give quite similar mass transfer at power levels similar to those of the flat-blade design. In order to improve the blending and solid-suspension characteristics, fluidfoil impellers (typified by the A315, Fig. 18-25) have been introduced in recent years and they have many of the advantages and some of the disadvantages of the flat-blade turbine. These impellers typically have a very high solidity ratio, on the order of 0.85 or more, and produce a strong axial downflow at low gas rate. As the gas rate increases, the flow pattern becomes more radial due to the upflow of the gas counteracting the downward flow of the impeller.

Mass-transfer characteristics on large-scale equipment seem to be quite similar, but the fluidfoil impellers tend to release a larger-

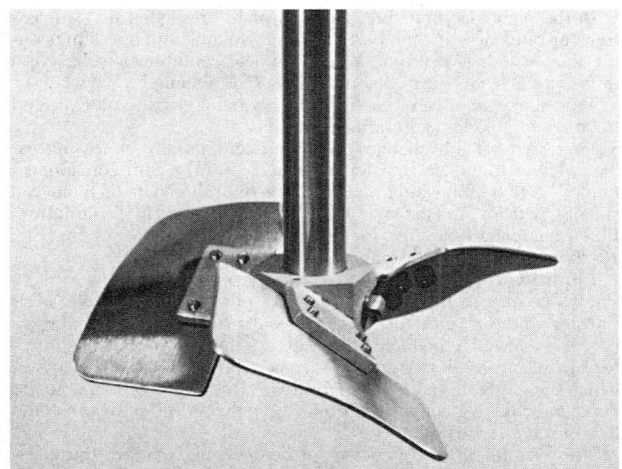

FIG. 18-25 An impeller designed for gas-liquid dispersion and mass transfer of the fluidfoil type, i.e., A315.

diameter bubble than is common with the radial-flow turbines. The blend time is one-half or one-third as long, and solid-suspension characteristics are better so that there have been notable improved process results with these impellers. This is particularly true if the process requires better blending and there is solid suspension. If this is not the case, the results from these impellers can be negative compared to radial-flow turbines.

It is very difficult to test these impellers on a small scale, since they provide better blending on a pilot scale where blending is already very effective compared to the large scale. Caution is recommended if it is desirable to study these impellers in pilot-scale equipment.

Gas-Liquid Mass Transfer   Gas-liquid mass transfer normally is correlated by means of the mass-transfer coefficient $K_g a$ versus power level at various superficial gas velocities. The superficial gas velocity is the volume of gas at the average temperature and pressure at the midpoint in the tank divided by the area of the vessel. In order to obtain the partial-pressure driving force, an assumption must be made of the partial pressure in equilibrium with the concentration of gas in the liquid. Many times this must be assumed, but if Fig. 18-26 is obtained in the pilot plant and the same assumption principle is used in evaluating the mixer in the full-scale tank, the error from the assumption is limited.

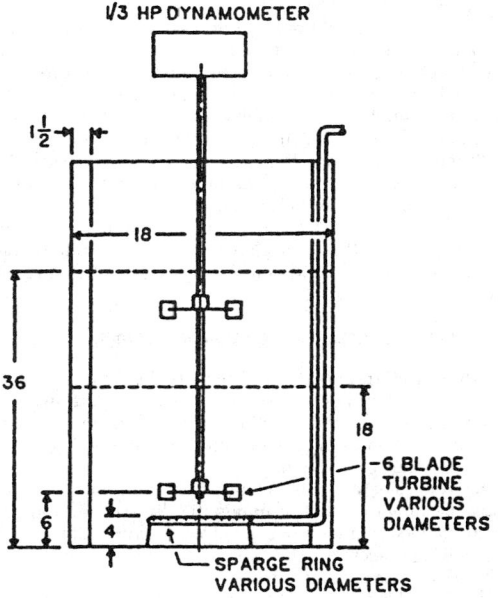

FIG. 18-24 Typical arrangement of Rushton radial-flow R100 flat-blade turbine with typical sparge ring for gas-liquid mass transfer.

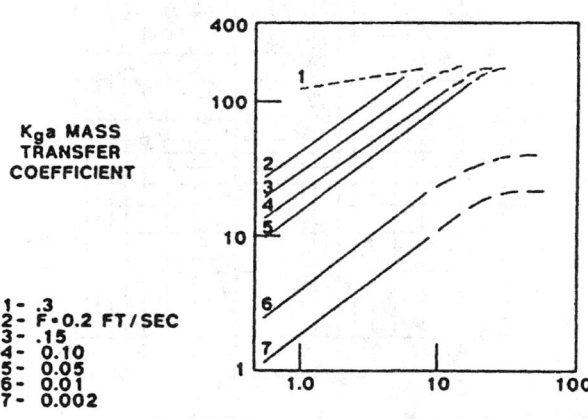

FIG. 18-26 Typical curve for mass transfer coefficient $K_g a$ as a function of mixer power and superficial gas velocity.

In the plant-size unit, Fig. 18-26 must be translated into a mass-transfer-rate curve for the particular tank volume and operating condition selected. Every time a new physical condition is selected, a different curve similar to that of Fig. 18-27 is obtained.

Typical exponents on the effect of power and gas rate on $K_g a$ tend to be around 0.5 for each variable, ±0.1.

Viscosity markedly changes the picture and, usually, increasing viscosity lowers the mass-transfer coefficient. For the common application of waste treating and for some of the published data on biological slurries, data for $k_L a$ (shown in Fig. 18-28) is obtained in the literature. For a completely new gas or liquid of a liquid slurry system, Fig. 18-26 must be obtained by an actual experiment.

Liquid-Gas-Solid Systems Many gas-liquid systems contain solids that may be the ultimate recipient of the liquid-gas-solid mass transfer entering into the process result. Examples are biological processes in which the biological solids are the user of the mass transfer of the mixing-flow patterns, various types of slurries reactors in which the solids either are being reactive or there may be extraction or dissolving taking place, or there may be polymerization or precipitation of solids occurring.

Normally there must be a way of determining whether the mass-transfer rate with the solids is the key controlling parameter or the gas-liquid mass transfer rate.

In general, introduction of a gas stream to a fluid will increase the blend time because the gas-flow patterns are counterproductive to the typical mixer-flow patterns. In a similar vein, the introduction of a gas stream to a liquid-solid suspension will decrease the suspension uniformity because the gas-flow pattern is normally counterproductive to the mixer-flow pattern. Many times the power needed for the gas-liquid mass transfer is higher than the power needed for solid suspension, and the effect of the gas flow on the solid suspensions are of little concern. On the other hand, if power levels are relatively low and solid-suspension characteristics are critical—examples being the case of activated sludge reactors in the waste-treating field or biological solid reactors in the hydrometallurgical field—then the effect of the gas-flow pattern of the mixing system can be quite critical to the overall design.

Another common situation is batch hydrogenation, in which pure hydrogen is introduced to a relatively high pressure reactor and a decision must be made to recycle the unabsorbed gas stream from the top of the reactor or use a vortexing mode for an upper impeller to incorporate the gas from the surface.

Loop Reactors For some gas-liquid-solid processes, a recirculating loop can be an effective reactor. These involve a relatively high horsepower pumping system and various kinds of nozzles, baffles, and turbulence generators in the loop system. These have power levels

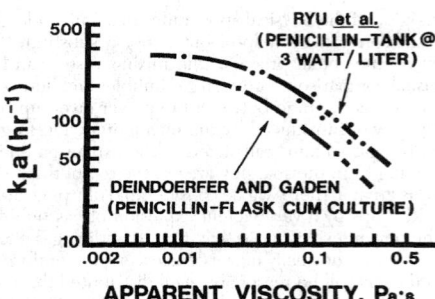

FIG. 18-28 Usually, the gas-liquid mass-transfer coefficient, $K_g a$, is reduced with increased viscosity. This shows the effect of increased concentration of microbial cells in a fermentation process.

anywhere from 1 to 10 times higher than the power level in a typical mixing reactor, and may allow the retention time to be less by a factor of 1 to 10.

LIQUID-LIQUID CONTACTING

Emulsions Almost every shear rate parameter affects liquid-liquid emulsion formation. Some of the effects are dependent upon whether the emulsion is both dispersing and coalescing in the tank, or whether there are sufficient stabilizers present to maintain the smallest droplet size produced for long periods of time. Blend time and the standard deviation of circulation times affect the length of time it takes for a particle to be exposed to the various levels of shear work and thus the time it takes to achieve the ultimate small particle size desired.

The prediction of drop sizes in liquid-liquid systems is difficult. Most of the studies have used very pure fluids as two of the immiscible liquids, and in industrial practice there almost always are other chemicals that are surface-active to some degree and make the prediction of absolute drop sizes very difficult. In addition, techniques to measure drop sizes in experimental studies have all types of experimental and interpretation variations and difficulties so that many of the equations and correlations in the literature give contradictory results under similar conditions. Experimental difficulties include dispersion and coalescence effects, difficulty of measuring actual drop size, the effect of visual or photographic studies on where in the tank you can make these observations, and the difficulty of using probes that measure bubble size or bubble area by light or other sample transmission techniques which are very sensitive to the concentration of the dispersed phase and often are used in very dilute solutions.

It is seldom possible to specify an initial mixer design requirement for an absolute bubble size prediction, particularly if coalescence and dispersion are involved. However, if data are available on the actual system, then many of these correlations could be used to predict relative changes in drop size conditions with changes in fluid properties or impeller variables.

STAGEWISE EQUIPMENT: MIXER-SETTLERS

Introduction Insoluble liquids may be brought into direct contact to cause transfer of dissolved substances, to allow transfer of heat, and to promote chemical reaction. This subsection concerns the design and selection of equipment used for conducting this type of liquid-liquid contact operation.

Objectives There are four principal purposes of operations involving the direct contact of immiscible liquids. The purpose of a particular contact operation may involve any one or any combination of the following objectives:

1. *Separation of components in solution.* This includes the ordinary objectives of liquid extraction, in which the constituents of a solution are separated by causing their unequal distribution between two insoluble liquids, the washing of a liquid with another to remove small

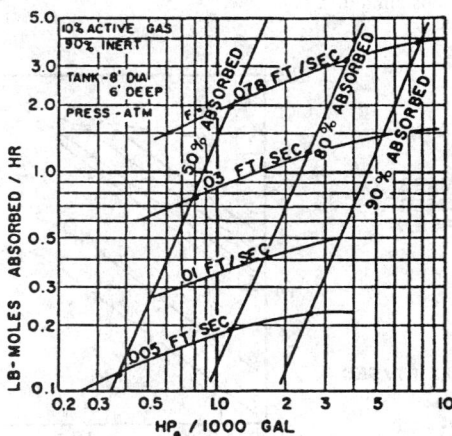

FIG. 18-27 Example of a specific chart to analyze the total mass-transfer rate in a particular tank under a process condition obtained from basic $K_g a$ data shown in Fig. 18-25.

amounts of a dissolved impurity, and the like. The theoretical principles governing the phase relationships, material balances, and number of ideal stages or transfer units required to bring about the desired changes are to be found in Sec. 15. Design of equipment is based on the quantities of liquids and the efficiency and operating characteristics of the type of equipment selected.

2. *Chemical reaction.* The reactants may be the liquids themselves, or they may be dissolved in the insoluble liquids. The kinetics of this type of reaction are treated in Sec. 4.

3. *Cooling or heating a liquid by direct contact with another.* Although liquid-liquid-contact operations have not been used widely for heat transfer alone, this technique is one of increasing interest. Applications also include cases in which chemical reaction or liquid extraction occurs simultaneously.

4. *Creating permanent emulsions.* The objective is to disperse one liquid within another in such finely divided form that separation by settling either does not occur or occurs extremely slowly. The purpose is to prepare the emulsion. Neither extraction nor chemical reaction between the liquids is ordinarily sought.

Liquid-liquid contacting equipment may be generally classified into two categories: **stagewise** and **continuous** (**differential**) contact.

The function of a stage is to contact the liquids, allow equilibrium to be approached, and to make a mechanical separation of the liquids. The contacting and separating correspond to mixing the liquids, and settling the resulting dispersion; so these devices are usually called **mixer-settlers.** The operation may be carried out in batch fashion or with continuous flow. If batch, it is likely that the same vessel will serve for both mixing and settling, whereas if continuous, separate vessels are usually but not always used.

Mixer-Settler Equipment   The equipment for extraction or chemical reaction may be classified as follows:

I. Mixers
    A. Flow or line mixers
        1. Mechanical agitation
        2. No mechanical agitation
    B. Agitated vessels
        1. Mechanical agitation
        2. Gas agitation
II. Settlers
    A. Nonmechanical
        1. Gravity
        2. Centrifugal (cyclones)
    B. Mechanical (centrifuges)
    C. Settler auxiliaries
        1. Coalescers
        2. Separator membranes
        3. Electrostatic equipment

In principle, at least, any mixer may be coupled with any settler to provide the complete stage. There are several combinations which are especially popular. Continuously operated devices usually, but not always, place the mixing and settling functions in separate vessels. Batch-operated devices may use the same vessel alternately for the separate functions.

Flow or Line Mixers

Definition   Flow or line mixers are devices through which the liquids to be contacted are passed, characterized principally by the very small time of contact for the liquids. They are used only for continuous operations or semibatch (in which one liquid flows continuously and the other is continuously recycled). If holding time is required for extraction or reaction, it must be provided by passing the mixed liquids through a vessel of the necessary volume. This may be a long pipe of large diameter, sometimes fitted with segmental baffles, but frequently the settler which follows the mixer serves. The energy for mixing and dispersing usually comes from pressure drop resulting from flow.

There are many types, and only the most important can be mentioned here. [See also Hunter, in Dunstan (ed.), *Science of Petroleum,* vol. 3, Oxford, New York, 1938, pp. 1779–1797.] They are used fairly extensively in treating petroleum distillates, in vegetable-oil, refining, in extraction of phenol-bearing coke-oven liquors, in some metal extractions, and the like. Kalichevsky and Kobe (*Petroleum Refining*

with Chemicals, Elsevier, New York, 1956) discuss detailed application in the refining of petroleum.

Jet Mixers   These depend upon impingement of one liquid on the other to obtain a dispersion, and one of the liquids is pumped through a small nozzle or orifice into a flowing stream of the other. Both liquids are pumped. They can be used successfully only for liquids of low interfacial tension. See Fig. 18-29 and also Hunter and Nash [*Ind. Chem.,* **9,** 245, 263, 317 (1933)]. Treybal (*Liquid Extraction,* 2d ed., McGraw-Hill, New York, 1963) describes a more elaborate device. For a study of the extraction of antibiotics with jet mixers, see Anneskova and Boiko, *Med. Prom. SSSR,* **13**(5), 26 (1959). Insonation with ultrasound of a toluene-water mixture during methanol extraction with a simple jet mixer improves the rate of mass transfer, but the energy requirements for significant improvement are large [Woodle and Vilbrandt, *Am. Inst. Chem. Eng. J.,* **6,** 296 (1960)].

Injectors   The flow of one liquid is induced by the flow of the other, with only the majority liquid being pumped at relatively high velocity. Figure 18-30 shows a typical device used in semibatch fashion for washing oil with a recirculated wash liquid. It is installed directly in the settling drum. See also Hampton (U.S. Patent 2,091,709, 1933), Sheldon (U.S. Patent 2,009,347, 1935), and Ng (U.S. Patent 2,665,975, 1954). Folsom [*Chem. Eng. Prog.,* **44,** 765 (1948)] gives a good review of basic principles. The most thorough study for extraction is provided by Kafarov and Zhukovskaya [*Zh. Prikl. Khim.,* **31,** 376 (1958)], who used very small injectors. With an injector measuring 73 mm from throat to exit, with 2.48-mm throat diameter, they extracted benzoic acid and acetic acid from water with carbon tetrachloride at the rate of 58 to 106 L/h, to obtain a stage efficiency E = 0.8 to 1.0. Data on flow characteristics are also given. Boyadzhiev and Elenkov [*Collect. Czech. Chem. Commun.,* **31,** 4072 (1966)] point out that the presence of surface-active agents exerts a profound influence on drop size in such devices.

Orifices and Mixing Nozzles   Both liquids are pumped through constrictions in a pipe, the pressure drop of which is partly utilized to create the dispersion (see Fig. 18-31). Single nozzles or several in series may be used. For the orifice mixers, as many as 20 orifice plates each with 13.8-kPa (2-lb/in²) pressure drop may be used in series [Morell and Bergman, *Chem. Metall. Eng.,* **35,** 211 (1928)]. In the Dualayer process for removal of mercaptans from gasoline, 258 m³/h (39,000 bbl/day) of oil and treating solution are contacted with 68.9-kPa (10-lb/in²) pressure drop per stage [Greek et al., *Ind. Eng. Chem.,* **49,** 1938 (1957)]. Holland et al. [*Am. Inst. Chem. Eng. J.,* **4,** 346 (1958); **6,** 615 (1960)] report on the interfacial area produced between two immiscible liquids entering a pipe (diameter 0.8 to 2.0 in) from an orifice, γ_D = 0.02 to 0.20, at flow rates of 0.23 to 4.1 m³/h (1 to 18 gal/min). At a distance 17.8 cm (7 in) downstream from the orifice,

$$a_{av} = \frac{0.179}{\sigma g_c} (C_O^2 \, \Delta p)^{0.75} \left(\frac{\sigma \sqrt{g_c} \rho_{av}}{\mu_D} \right)^{0.158} \left[\left(\frac{d_t}{d_O} \right)^4 - 1 \right]^{0.117} \gamma_D^{0.878} \quad (18\text{-}8)$$

where a_{av} = interfacial surface, cm²/cm³; C_O = orifice coefficient, dimensionless; d_t = pipe diameter, in; d_O = orifice diameter, in; g_c =

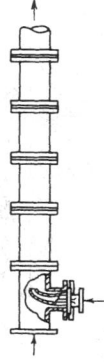

FIG. 18-29   Elbow jet mixer.

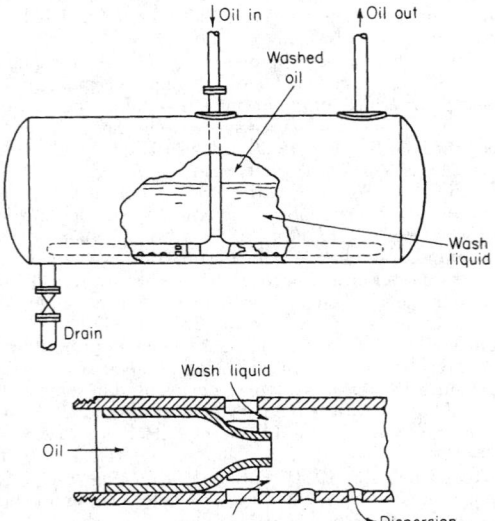

FIG. 18-30 Injector mixer. (*Ayres, U.S. Patent 2,531,547, 1950.*)

gravitational conversion factor, (32.2 lbm·ft)/(lbf·s²); Δp = pressure drop across orifice, lbf/ft²; μ_D = viscosity of dispersed phase, lbm/(ft·s); ρ_{av} = density of dispersed phase, lbm/ft; and σ = interfacial tension, lbf/ft. See also Shirotsuka et al. [*Kagaku Kogaku*, **25**, 109 (1961)].

Valves Valves may be considered to be adjustable orifice mixers. In desalting crude petroleum by mixing with water, Hayes et al. [*Chem. Eng. Prog.*, **45**, 235 (1949)] used a globe-valve mixer operating at 110- to 221-kPa (16- to 32-lb/in²) pressure drop for mixing 66 m³/h (416 bbl/h) oil with 8 m³/h (50 bbl/h) water, with best results at the lowest value. Simkin and Olney [*Am. Inst. Chem. Eng. J.*, **2**, 545 (1956)] mixed kerosine and white oil with water, using 0.35- to 0.62-kPa (0.05- to 0.09-lb/in²) pressure drop across a 1-in gate valve, at 22-m³/h (10-gal/min) flow rate for optimum separating conditions in a cyclone, but higher pressure drops were required to give good extractor efficiencies.

Pumps Centrifugal pumps, in which the two liquids are fed to the suction side of the pump, have been used fairly extensively, and they offer the advantage of providing interstage pumping at the same time. They have been commonly used in the extraction of phenols from coke-oven liquors with light oil [Gollmar, *Ind. Eng. Chem.*, **39**, 596, 1947; Carbone, *Sewage Ind. Wastes*, **22**, 200 (1950)], but the intense shearing action causes emulsions with this low-interfacial-tension system. Modern plants use other types of extractors. Pumps are useful in

the extraction of slurries, as in the extraction of uranyl nitrate from acid-uranium-ore slurries [*Chem. Eng.*, **66**, 30 (Nov. 2, 1959)]. Shaw and Long [*Chem. Eng.*, **64**(11), 251 (1957)] obtain a stage efficiency of 100 percent ($E = 1.0$) in a uranium-ore-slurry extraction with an open impeller pump. In order to avoid emulsification difficulties in these extractions, it is necessary to maintain the organic phase continuous, if necessary by recycling a portion of the settled organic liquid to the mixer.

Agitated Line Mixer See Fig. 18-32. This device, which combines the features of orifice mixers and agitators, is used extensively in treating petroleum and vegetable oils. It is available in sizes to fit ½- to 10-in pipe. The device of Fig. 18-33, with two impellers in separate stages, is available in sizes to fit 4- to 20-in pipe.

Packed Tubes Cocurrent flow of immiscible liquids through a packed tube produces a one-stage contact, characteristic of line mixers. For flow of isobutanol-water° through a 0.5-in diameter tube packed with 6 in of 3-mm glass beads, Leacock and Churchill [*Am. Inst. Chem. Eng. J.*, **7**, 196 (1961)] find

$$k_C a_{av} = c_1 L_C^{0.5} L_D \tag{18-9}$$

$$k_D a_{av} = c_2 L_C^{0.75} L_D^{0.75} \tag{18-10}$$

where $c_1 = 0.00178$ using SI units and 0.00032 using U.S. customary units; and $c_2 = 0.0037$ using SI units and 0.00057 using U.S. customary units. These indicate a stage efficiency approaching 100 percent. Organic-phase holdup and pressure drop for larger pipes similarly packed are also available [Rigg and Churchill, *ibid.*, **10**, 810 (1964)].

Pipe Lines The principal interest here will be for flow in which one liquid is dispersed in another as they flow cocurrently through a pipe (stratified flow produces too little interfacial area for use in liquid extraction or chemical reaction between liquids). Drop size of dispersed phase, if initially very fine at high concentrations, increases as the distance downstream increases, owing to coalescence [see Holland, loc. cit.; Ward and Knudsen, *Am. Inst. Chem. Eng. J.*, **13**, 356 (1967)]; or if initially large, decreases by breakup in regions of high shear [Sleicher, *ibid.*, **8**, 471 (1962); *Chem. Eng. Sci.*, **20**, 57 (1965)]. The maximum drop size is given by (Sleicher, loc. cit.)

° Isobutanol dispersed: $L_D = 3500$ to 27,000; water continuous; $L_C = 6000$ to 32,000 in pounds-mass per hour-square foot (to convert to kilograms per second-square meter, multiply by 1.36×10^{-3}).

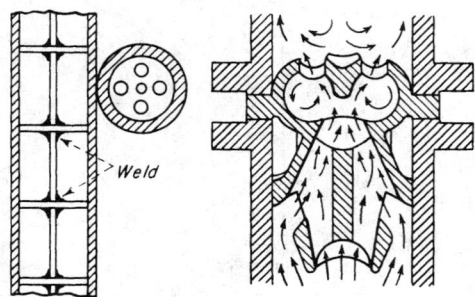

FIG. 18-31 Orifice mixer and nozzle mixer.

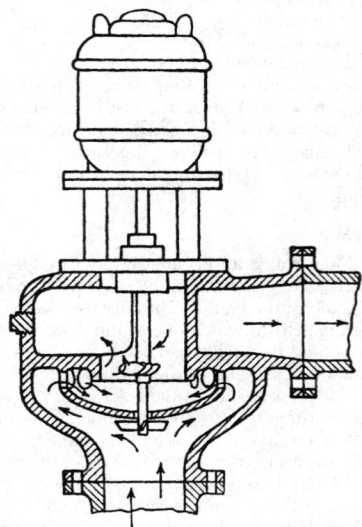

FIG. 18-32 Nettco Corp. Flomix.

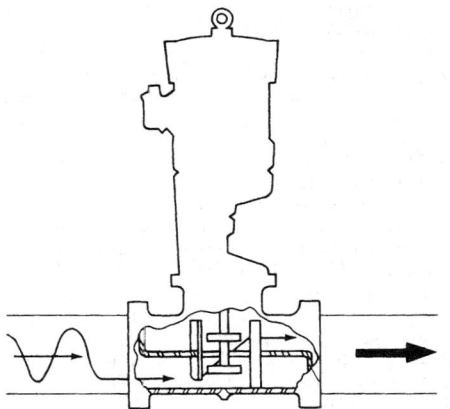

FIG. 18-33 Lightnin line blender. (*Mixing Equipment Co., Inc., with permission.*)

$$\frac{d_{p,\max}\rho_C V^2}{\sigma g_c}\sqrt{\frac{\mu_C V}{\sigma g_c}} = C\left[1 + 0.7\left(\frac{\mu_D V}{\sigma g_c}\right)^{0.7}\right] \qquad (18\text{-}11)$$

where $C = 43$ ($d_t = 0.013$ m or 0.0417 ft) or 38 ($d_t = 0.038$ m or 0.125 ft), with $d_{p,\mathrm{av}} = d_{p,\max}/4$ for high flow rates and $d_{p,\max}/13$ for low velocities.

Extensive measurements of the rate of mass transfer between *n*-butanol and water flowing in a 0.008-m (0.314-in) ID horizontal pipe are reported by Watkinson and Cavers [*Can. J. Chem. Eng.*, **45**, 258 (1967)] in a series of graphs not readily reproduced here. Length of a transfer unit for either phase is strongly dependent upon flow rate and passes through a pronounced maximum at an organic-water phase ratio of 0.5. In energy (pressure-drop) requirements and volume, the pipe line compared favorably with other types of extractors. Boyadzhiev and Elenkov [*Chem. Eng. Sci.*, **21**, 955 (1966)] concluded that, for the extraction of iodine between carbon tetrachloride and water in turbulent flow, drop coalescence and breakup did not influence the extraction rate. Yoshida et al. [*Coal Tar* (Japan), **8**, 107 (1956)] provide details of the treatment of crude benzene with sulfuric acid in a 1-in diameter pipe, $N_{\mathrm{Re}} = 37{,}000$ to 50,000. Fernandes and Sharma [*Chem. Eng. Sci.*, **23**, 9 (1968)] used cocurrent flow downward of two liquids in a pipe, agitated with an upward current of air.

The pipe has also been used for the transfer of heat between two immiscible liquids in cocurrent flow. For hydrocarbon oil-water, the heat-transfer coefficient is given by

$$\frac{Ua_{\mathrm{av}}d_t^2}{vk_{to}} = \frac{\gamma_D N_{\mathrm{We},t}^{6/5}}{\dfrac{k_{to}}{0.415 k_{tC}} + \dfrac{k_{to}}{0.173 k_{tD}}} \qquad (18\text{-}12)$$

for $\gamma_D = 0$ to 0.2. Additional data for $\gamma_D = 0.4$ to 0.8 are also given. Data for stratified flow are given by Wilke et al. [*Chem. Eng. Prog.*, **59**, 69 (1963)] and Grover and Knudsen [*Chem. Eng. Prog.*, **51**, *Symp. Ser.* **17**, 71 (1955)].

Mixing in Agitated Vessels Agitated vessels may frequently be used for either batch or continuous service and for the latter may be sized to provide any holding time desired. They are useful for liquids of any viscosity up to 750 Pa·s (750,000 cP), although in contacting two liquids for reaction or extraction purposes viscosities in excess of 0.1 Pa·s (100 cP) are only rarely encountered.

Mechanical Agitation This type of agitation utilizes a rotating impeller immersed in the liquid to accomplish the mixing and dispersion. There are literally hundreds of devices using this principle, the major variations being found when chemical reactions are being carried out. The basic requirements regarding shape and arrangement of the vessel, type and arrangement of the impeller, and the like are essentially the same as those for dispersing finely divided solids in liquids, which are fully discussed in Sec. 18.

The following summary of operating characteristics of mechanically agitated vessels is confined to the data available on liquid-liquid contacting.

Phase Dispersed There is an ill-defined upper limit to the volume fraction of dispersed liquid which may be maintained in an agitated dispersion. For dispersions of organic liquids in water [Quinn and Sigloh, *Can. J. Chem. Eng.*, **41**, 15 (1963)],

$$\gamma_{Do,\max} = \gamma' + \left(\frac{C}{N^3}\right) \qquad (18\text{-}13)$$

where γ' is a constant, asymptotic value, and C is a constant, both depending in an unestablished manner upon the system physical properties and geometry. Thus, inversion of a dispersion may occur if the agitator speed is increased. With systems of low interfacial tension ($\sigma' = 2$ to 3 mN/m or 2 to 3 dyn/cm), γ_D as high as 0.8 can be maintained. Selker and Sleicher [*Can. J. Chem. Eng.*, **43**, 298 (1965)] and Yeh et al. [*Am. Inst. Chem. Eng. J.*, **10**, 260 (1964)] feel that the viscosity ratio of the liquids alone is important. Within the limits in which either phase can be dispersed, for *batch operation* of baffled vessels, that phase in which the impeller is immersed when at rest will normally be continuous [Rodger, Trice, and Rushton, *Chem. Eng. Prog.*, **52**, 515 (1956); Laity and Treybal, *Am. Inst. Chem. Eng. J.*, **3**, 176 (1957)]. With water dispersed, *dual emulsions* (continuous phase found as small droplets within larger drops of dispersed phase) are possible. In *continuous operation*, the vessel is first filled with the liquid to be continuous, and agitation is then begun, after which the liquid to be dispersed is introduced.

Uniformity of Mixing This refers to the gross uniformity throughout the vessel and not to the size of the droplets produced. For *unbaffled vessels, batch, with an air-liquid interface*, Miller and Mann [*Trans. Am. Inst. Chem. Eng.*, **40**, 709 (1944)] mixed water with several organic liquids, measuring uniformity of mixing by sampling the tank at various places, comparing the percentage of dispersed phase found with that in the tank as a whole. A power application of 200 to 400 W/m³ [(250 to 500 ft·lb)/(min·ft³)] gave maximum and nearly uniform performance for all. See also Nagata et al. [*Chem. Eng. (Japan)*, **15**, 59 (1951)].

For *baffled vessels operated continuously, no air-liquid interface*, flow upward, light liquid dispersed [Treybal, *Am. Inst. Chem. Eng. J.*, **4**, 202 (1958)], the average fraction of dispersed phase in the vessel $\gamma_{D,\mathrm{av}}$ is less than the fraction of the dispersed liquid in the feed mixture, unless the impeller speed is above a certain critical value which depends upon vessel geometry and liquid properties. Thornton and Bouyatiotis [*Ind. Chem.*, **39**, 298 (1963); *Inst. Chem. Eng. Symp. Liquid Extraction*, Newcastle-upon-Tyne, April 1967] have presented correlations of data for a 17.8-cm (7-in) vessel, but these do not agree with observations on 15.2- and 30.5-cm (6- and 12-in) vessels in Treybal's laboratory. See also Kovalev and Kagan [*Zh. Prikl. Khim.*, **39**, 1513 (1966)] and Trambouze [*Chem. Eng. Sci.*, **14**, 161 (1961)]. Stemerding et al. [*Can. J. Chem. Eng.*, **43**, 153 (1965)] present data on a large mixing tank [15 m³ (530 ft³)] fitted with a marine-type propeller and a draft tube.

Drop Size and Interfacial Area The drops produced have a size range [Sullivan and Lindsey, *Ind. Eng. Chem. Fundam.*, **1**, 87 (1962); Sprow, *Chem. Eng. Sci.*, **22**, 435 (1967); and Chen and Middleman, *Am. Inst. Chem. Eng. J.*, **13**, 989 (1967)]. The average drop size may be expressed as

$$d_{p,\mathrm{av}} = \frac{\sum n_i d_{pi}^3}{\sum n_i d_{pi}^2} \qquad (18\text{-}14)$$

and if the drops are spherical,

$$a_{\mathrm{av}} = \frac{6\gamma_{D,\mathrm{av}}}{d_{p,\mathrm{av}}} \qquad (18\text{-}15)$$

The drop size varies locally with location in the vessel, being smallest at the impeller and largest in regions farthest removed from the impeller owing to coalescence in regions of relatively low turbulence intensity [Schindler and Treybal, *Am. Inst. Chem. Eng. J.*, **14**, 790 (1968); Vanderveen, U.S. AEC UCRL-8733, 1960]. Interfacial area and hence average drop size have been measured by light transmittance, light scattering, direct photography, and other means. Typical of the resulting correlations is that of Thornton and Bouyatiotis (*Inst. Chem. Eng. Symp. Liquid Extraction*, Newcastle-upon-Tyne, April 1967) for a 17.8-cm- (7-in-) diameter baffled vessel, six-bladed flat-blade turbine, d_i = 6.85 cm (0.225 ft), operated full, for organic liquids (σ' = 8.5 to 34, ρ_D = 43.1 to 56.4, μ_D = 1.18 to 1.81) dispersed in water, in the absence of mass transfer, and under conditions giving nearly the vessel-average $d_{p,av}$:

$$\frac{d_{p,av}}{d_p^0} = 1 + 1.18\phi_D \left(\frac{\sigma^2 g_c^2}{d_p^0 \mu_c^2 g} \right) \left(\frac{\mu_c^4 g}{\Delta\rho \, \sigma^3 g_c^3} \right)^{0.62} \left(\frac{\Delta\rho}{\rho_c} \right)^{0.05} \quad (18\text{-}16)$$

where d_p^0 is given by

$$\frac{(d_p^0)^3 \rho_C^2 g}{\mu_c^2} = 29.0 \left(\frac{P^3 g_c^3}{v^3 \rho_c^2 \mu_c g^4} \right)^{-0.32} \left(\frac{\rho_C \sigma^3 g_c^3}{\mu_c^4 g} \right)^{0.14} \quad (18\text{-}17)$$

Caution is needed in using such correlations, since those available do not generally agree with each other. For example, Eq. (21-28) gives $d_{p,av}$ = 4.78(10⁻⁴) ft for a liquid pair of properties a' = 30, ρ_C = 62.0, ρ_D = 52.0, μ_C = 2.42, μ_D = 1.94, $\gamma_{D,av}$ = 0.20 in a vessel $T = Z = 0.75$, a turbine impeller d_i = 0.25 turning at 400 r/min. Other correlations provide 3.28(10⁻⁴) [Thornton and Bouyatiotis, *Ind. Chem.*, **39**, 298 (1963)], 8.58(10⁻⁴) [Calderbank, *Trans. Inst. Chem. Eng. (London)*, **36**, 443 (1958)], 6.1(10⁻⁴) [Kafarov and Babinov, *Zh. Prikl. Khim*, **32**, 789 (1959)], and 2.68(10⁻³) (Rushton and Love, paper at AIChE, Mexico City, September 1967). See also Vermeulen et al. [*Chem. Eng. Prog.*, **51**, 85F (1955)], Rodgers et al. [ibid., **52**, 515 (1956); U.S. AEC ANL-5575 (1956)], Rodrigues et al. [*Am. Inst. Chem. Eng. J.*, **7**, 663 (1961)], Sharma et al. [*Chem. Eng. Sci.*, **21**, 707 (1966); **22**, 1267 (1967)], and Kagan and Kovalev [*Khim. Prom.*, **42**, 192 (1966)]. For the effect of absence of baffles, see Fick et al. (U.S. AEC UCRL-2545, 1954) and Schindler and Treybal [*Am. Inst. Chem. Eng. J.*, **14**, 790 (1968)]. The latter have observations during mass transfer.

Coalescence Rates   The droplets coalesce and redisperse at rates that depend upon the vessel geometry, N, $\gamma_{D,av}$, and liquid properties. The few measurements available, made with a variety of techniques, do not as yet permit quantitative estimates of the coalescence frequency v. Madden and Damarell [*Am. Inst. Chem. Eng. J.*, **8**, 233 (1962)] found for baffled vessels that v varied as $N^{2.2}\gamma_{D,av}^{0.5}$, and this has generally been confirmed by Groothius and Zuiderweg [*Chem. Eng. Sci.*, **19**, 63 (1964)], Miller et al. [*Am. Inst. Chem. Eng. J.*, **9**, 196 (1963)], and Howarth [ibid., **13**, 1007 (1967)], although absolute values of v in the various studies are not well related. Hillestad and Rushton (paper at AIChE, Columbus, Ohio, May 1966), on the other hand, find v to vary as $N^{0.73}\gamma_{D,av}$ for impeller Weber numbers $N_{We,i}$ below a certain critical value and as $N^{-3.5}\gamma_{D,av}^{1.58}$ for higher Weber numbers. The influence of liquid properties is strong. There is clear evidence [Groothius and Zuiderweg, loc. cit.; *Chem. Eng. Sci.*, **12**, 288 (1960)] that coalescence rates are enhanced by mass transfer from a drop to the surrounding continuum and retarded by transfer in the reverse direction. See also Howarth [*Chem. Eng. Sci.*, **19**, 33 (1964)]. For a theoretical treatment of drop breakage and coalescence and their effects, see Valentas and Amundsen [*Ind. Eng. Chem. Fundam.*, **5**, 271, 533 (1966); **7**, 66 (1968)], Gal-Or and Walatka [*Am. Inst. Chem. Eng. J.*, **13**, 650 (1967)], and Curl [ibid., **9**, 175 (1963)].

In calculating the power required for mixers, a reasonable estimate of the average density and viscosity for a two-phase system is satisfactory.

Solids are often present in liquid streams either as a part of the processing system or as impurities that come along and have to be handled in the process. One advantage of mixers in differential contact equipment is the fact that they can handle slurries in one or both phases. In many industrial leaching systems, particularly in the minerals processing industry, coming out of the leach circuit is a slurry with a desired material involved in the liquid but a large amount of solids contained in the stream. Typically, the solids must be separated out by filtration or centrifugation, but there has always been a desire to try a direct liquid-liquid extraction with an immiscible liquid contact with this often highly concentrated slurry leach solution. The major prob-

lem with this approach is loss of organic material going out with the highly concentrated liquid slurry.

Data are not currently available on the dispersion with the newer fluidfoil impellers, but they are often used in industrial mixer-settler systems to maintain dispersion when additional resonance time holdup is required, after an initial dispersion is made by a radial- or axial-flow turbine.

Recent data by Calabrese[5] indicates that the sauter mean drop diameter can be correlated by equation and is useful to compare with other predictions indicated previously.

As an aside, when a large liquid droplet is broken up by shear stress, it tends initially to elongate into a dumbbell shape, which determines the particle size of the two large droplets formed. Then, the neck in the center between the ends of the dumbbell may explode or shatter. This would give a debris of particle sizes which can be quite different than the two major particles produced.

Liquid-Liquid Extraction   The actual configuration of mixers in multistage mixer-settlers and/or multistage columns is summarized in Section 15. A general handbook on this subject is *Handbook of Solvent Extraction* by Lowe, Beard, and Hanson. This handbook gives a comprehensive review of this entire operation as well.

In the liquid-liquid extraction area, in the mining industry, coming out of the leach tanks is normally a slurry, in which the desired mineral is dissolved in the liquid phase. To save the expense of separation, usually by filtration or centrifugation, attempts have been made to use a resident pump extraction system in which the organic material is contacted directly with the slurry. The main economic disadvantage to this proposed system is the fact that considerable amounts of organic liquid are entrained in the aqueous slurry system, which, after the extraction is complete, is discarded. In many systems this has caused an economic loss of solvent into this waste stream.

LIQUID-LIQUID-SOLID SYSTEMS

Many times solids are present in one or more phases of a solid-liquid system. They add a certain level of complexity in the process, especially if they tend to be a part of both phases, as they normally will do. Approximate methods need to be worked out to estimate the density of the emulsion and determine the overall velocity of the flow pattern so that proper evaluation of the suspension requirements can be made. In general, the solids will behave as though they were a fluid of a particular average density and viscosity and won't care much that there is a two-phase dispersion going on in the system. However, if solids are being dissolved or precipitated by participating in one phase and not the other, then they will be affected by which phase is dispersed or continuous, and the process will behave somewhat differently than if the solids migrate independently between the two phases within the process.

FLUID MOTION

Pumping   Some mixing applications can be specified by the pumping capacity desired from the impeller with a certain specified geometry in the vessel. As mentioned earlier, this sometimes is used to describe a blending requirement, but circulation and blending are two different things. The major area where this occurs is in draft tube circulators or pump-mix mixer settlers. In draft tube circulators (shown in Fig. 18-22), the circulation occurs through the draft tube and around the annulus and for a given geometry, the velocity head required can be calculated with reference to various formulas for geometric shapes. What is needed is a curve for head versus flow for the impeller, and then the system curve can be matched to the impeller curve. Adding to the complexity of this system is the fact that solids may settle out and change the character of the head curve so that the impeller can get involved in an unstable condition which has various degrees of erratic behavior depending upon the sophistication of the impeller and inlet and outlet vanes involved. These draft tube circulators often involve solids, and applications are often for precipitation or crystallization in these units. Draft tube circulators can either have the impeller pump up in the draft tube and flow down the annulus or just the reverse. If the flow is down the annulus, then the flow has to make a 180° turn where it comes back at the bottom of the tank into the draft tube again. This is a very sensitive area, and special baffles must be used to carefully determine how the fluid will make this turn since many areas of constriction are involved in making this change in direction.

When pumping down the draft tube, flow normally makes a more troublefree velocity change to a flow going up the annulus. Since the area of the draft tube is markedly less than the area of the annulus, pumping up the draft tube requires less flow to suspend solids of a given settling velocity than does pumping down the draft tube.

Another example is to eliminate the interstage pump between mixing and settling stages in the countercurrent mixer-settler system. The radial-flow impeller typically used is placed very close to an orifice at the bottom of the mixing tank and can develop heads from 12 to 18 in. All the head-loss terms in the mixer and settler circuit have to be carefully calculated because they come very close to that 12- to 18-in range when the passages are very carefully designed and streamlined. If the mixing tank gets much above 10 ft in depth, then the heads have to be higher than the 12- to 18-in range and special designs have to be worked on which have the potential liability of increasing the shear rate acting on the dispersed phase to cause more entrainment and longer settling times. In these cases, it is sometimes desirable to put the mixer system outside the actual mixer tank and have it operate in a single phase or to use multiple impellers, each one of which can develop a portion of the total head required.

Heat Transfer   In general, the fluid mechanics of the film on the mixer side of the heat transfer surface is a function of what happens at that surface rather than the fluid mechanics going on around the impeller zone. The impeller largely provides flow across and adjacent to the heat-transfer surface and that is the major consideration of the heat-transfer result obtained. Many of the correlations are in terms of traditional dimensionless groups in heat transfer, while the impeller performance is often expressed as the impeller Reynolds number.

The fluidfoil impellers (shown in Fig. 18-2) usually give more flow for a given power level than the traditional axial- or radial-flow turbines. This is also thought to be an advantage since the heat-transfer surface itself generates the turbulence to provide the film coefficient and more flow should be helpful. This is true to a limited degree in jacketed tanks (Fig. 18-34), but in helical coils (Fig. 18-35), the

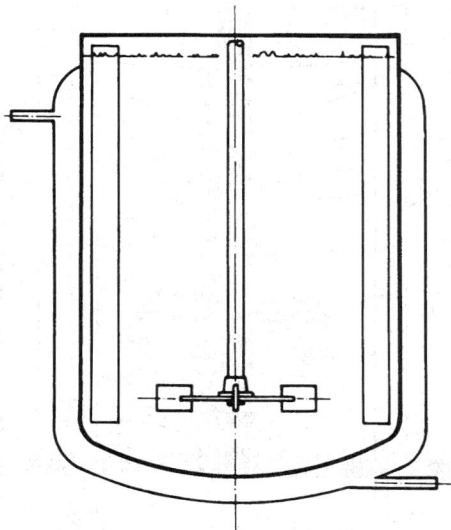

FIG. 18-34   Typical jacket arrangement for heat transfer.

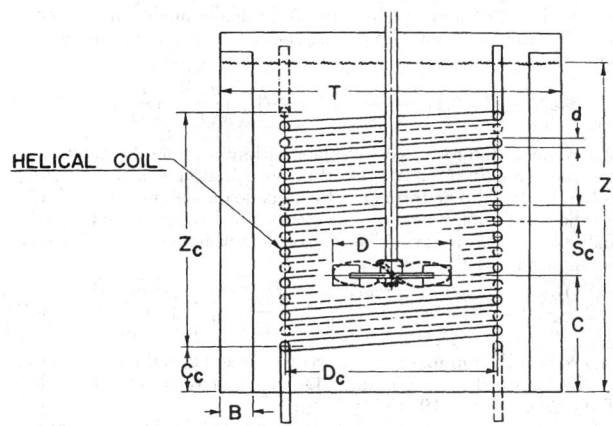

FIG. 18-35   Typical arrangement of helical coil at mixing vessel for heat transfer.

extreme axial flow of these impellers tends to have the first or second turn in the coil at the bottom of the tank blank off the flow from the turns above it in a way that (at the same power level) the increased flow from the fluidfoil impeller is not helpful. It best gives the same coefficient as with the other impellers and on occasion can cause a 5 to 10 percent reduction in the heat-transfer coefficient over the entire coil.

JACKETS AND COILS OF AGITATED VESSELS

Most of the correlations for heat transfer from the agitated liquid contents of vessels to jacketed walls have been of the form:

$$\frac{hD_j}{k} = a \left(\frac{L_p^2 N_r \rho}{\mu} \right)^b \left(\frac{c\mu}{k} \right)^{1/3} \left(\frac{\mu_b}{\mu_w} \right)^m \tag{18-18}$$

The film coefficient h is for the inner wall; D_j is the inside diameter of the mixing vessel. The term $L_p^2 N_r \rho / \mu$ is the Reynolds number for mixing in which L_p is the diameter and N_r the speed of the agitator. Recommended values of the constants a, b, and m are given in Table 18-2.

A wide variety of configurations exists for coils in agitated vessels. Correlations of data for heat transfer to helical coils have been of two forms, of which the following are representative:

$$\frac{hD_j}{k} = 0.87 \left(\frac{L_p^2 N_t \rho}{\mu} \right)^{0.62} \left(\frac{c\mu}{k} \right)^{1/3} \left(\frac{\mu_b}{\mu_w} \right)^{0.14} \tag{18-19}$$

TABLE 18-2   Values of Constants for Use in Eq. (18-18)

Agitator	a	b	m	Range of Reynolds number
Paddle[a]	0.36	⅔	0.21	300–3 × 10⁵
Pitched-blade turbine[b]	0.53	⅔	0.24	80–200
Disk, flat-blade turbine[c]	0.54	⅔	0.14	40–3 × 10⁵
Propeller[d]	0.54	⅔	0.14	2 × 10³ (one point)
Anchor[b]	1.0	½	0.18	10–300
Anchor[b]	0.36	⅔	0.18	300–40,000
Helical ribbon[e]	0.633	½	0.18	8–10⁵

[a]Chilton, Drew, and Jebens, *Ind. Eng. Chem.*, **36**, 510 (1944), with constant *m* modified by Uhl.
[b]Uhl, *Chem. Eng. Progr., Symp. Ser.* 17, **51**, 93 (1955).
[c]Brooks and Su, *Chem. Eng. Progr.*, **55**(10), 54 (1959).
[d]Brown et al., *Trans. Inst. Chem. Engrs.* (*London*), **25**, 181 (1947).
[e]Gluz and Pavlushenko, *J. Appl. Chem. U.S.S.R.*, **39**, 2323 (1966).

where the agitator is a paddle, the Reynolds number range is 300 to 4×10^5 [Chilton, Drew, and Jebens, *Ind. Eng. Chem.*, **36**, 510 (1944)], and

$$\frac{hD_o}{k} = 0.17 \left(\frac{L_p^2 N_r \rho}{\mu}\right)^{0.67} \left(\frac{c\mu}{k}\right)^{0.37} \left(\frac{L_p}{D_j}\right)^{0.1} \left(\frac{D_o}{D_j}\right)^{0.5} \quad (18\text{-}20)$$

where the agitator is a disk flat-blade turbine, and the Reynolds number range is 400 to $(2)(10^5)$ [Oldshue and Gretton, *Chem. Eng. Prog.*, **50**, 615 (1954)]. The term D_o is the outside diameter of the coil tube.

The most comprehensive correlation for heat transfer to vertical baffle-type coils is for a disk flat-blade turbine over the Reynolds number range 10^3 to $(2)(10^6)$:

$$\frac{hD_o}{k} = 0.09 \left(\frac{L_p^2 N_r \rho}{\mu}\right)^{0.65} \left(\frac{c\mu}{k}\right)^{1/3} \left(\frac{L_p}{D_j}\right)^{1/3} \left(\frac{2}{n_b}\right)^{0.2} \left(\frac{\mu}{\mu_f}\right)^{0.4} \quad (18\text{-}21)$$

where n_b is the number of baffle-type coils and μ_f is the fluid viscosity at the mean film temperature [Dunlop and Rushton, *Chem. Eng. Prog. Symp. Ser.* 5, **49**, 137 (1953)].

Chapman and Holland (*Liquid Mixing and Processing in Stirred Tanks*, Reinhold, New York, 1966) review heat transfer to low-viscosity fluids in agitated vessels. Uhl ["Mechanically Aided Heat Transfer," in Uhl and Gray (eds.), *Mixing: Theory and Practice*, vol. I, Academic, New York, 1966, chap. V] surveys heat transfer to low- and high-viscosity agitated fluid systems. This review includes scraped-wall units and heat transfer on the jacket and coil side for agitated vessels.

LIQUID-LIQUID-GAS-SOLID SYSTEMS

This is a relatively unusual combination, and one of the more common times it exists is in the fermentation of hydrocarbons with aerobic microorganisms in an aqueous phase. The solid phase is a microorganism which is normally in the aqueous phase and is using the organic phase for food. Gas is supplied to the system to make the fermentation aerobic. Usually the viscosities are quite low, percent solids is also modest, and there are no special design conditions required when this particular gas-liquid-liquid-solid combination occurs. Normally, average properties for the density of viscosity of the liquid phase are used. In considering that the role the solids play in the system is adequate, there are cases of other processes which consist of four phases, each of which involves looking at the particular properties of the phases to see whether there are any problems of dispersion, suspension, or emulsification.

COMPUTATIONAL FLUID DYNAMICS

There are several software programs that are available to model flow patterns of mixing tanks. They allow the prediction of flow patterns based on certain boundary conditions. The most reliable models use accurate fluid mechanics data generated for the impellers in question and a reasonable number of modeling cells to give the overall tank flow pattern. These flow patterns can give velocities, streamlines, and localized kinetic energy values for the systems. Their main use at the present time is to look at the effect of making changes in mixing variables based on doing certain things to the mixing process. These programs can model velocity, shear rates, and kinetic energy, but probably cannot adapt to the actual chemistry of diffusion or mass-transfer kinetics of actual industrial process at the present time.

Relatively uncomplicated transparent tank studies with tracer fluids or particles can give a similar feel for the overall flow pattern. It is important that a careful balance be made between the time and expense of calculating these flow patterns with computational fluid dynamics compared to their applicability to an actual industrial process. The future of computational fluid dynamics appears very encouraging and a reasonable amount of time and effort put forth in this regard can yield immediate results as well as potential for future process evaluation.

Figures 18-36, 18-37, and 18-38 show some approaches. Figure 18-36 shows velocity vectors for an A310 impeller. Figure 18-37 shows contours of kinetic energy of turbulence. Figure 18-38 uses a particle trajectory approach with neutral buoyancy particles.

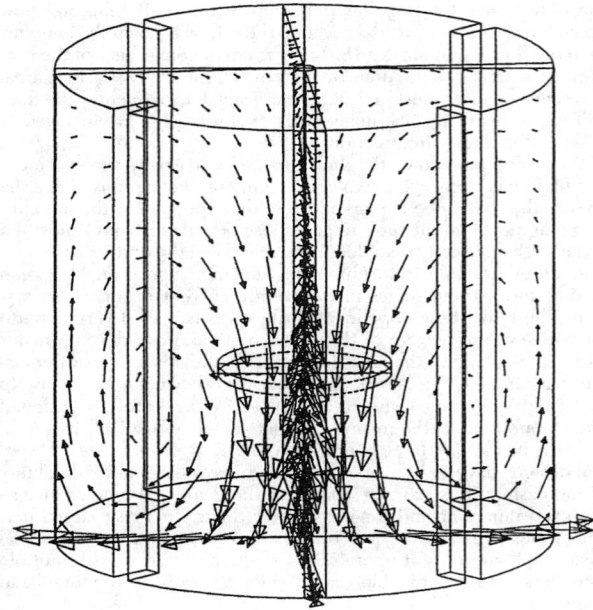

FIG. 18-36   Laser scan.

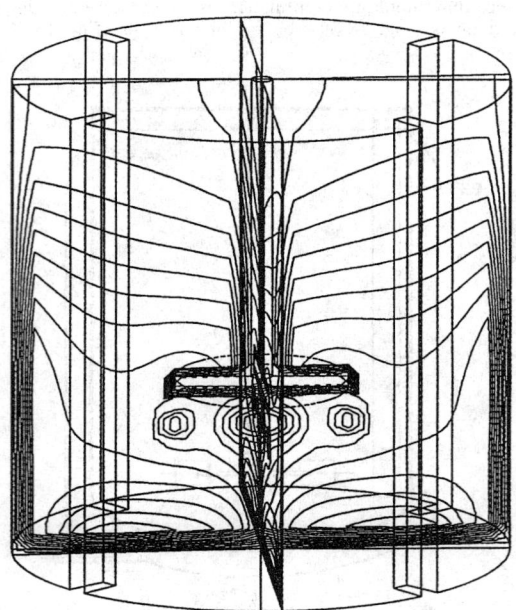

FIG. 18-37   Laser scan.

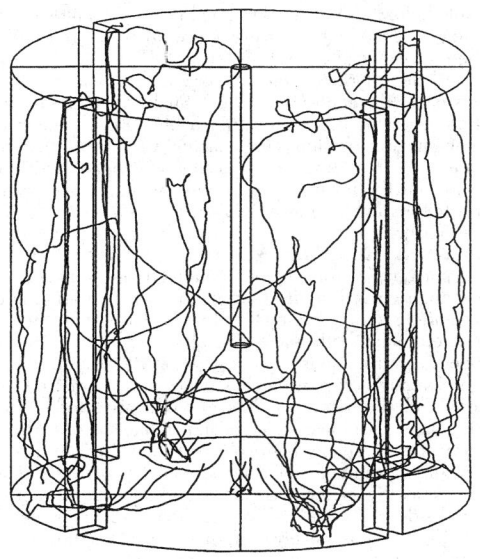

FIG. 18-38   Laser scan.

Numerical fluid mechanics can define many of the fluid mechanics parameters for an overall reactor system. Many of the models break up the mixing tank into small microcells. Suitable material and mass-transfer balances between these cells throughout the reactor are then made. This can involve long and massive computational requirements. Programs are available that can give reasonably acceptable models of experimental data taken in mixing vessels. Modeling the three-dimensional aspect of a flow pattern in a mixing tank can require a large amount of computing power.

Most modeling codes are a time-averaging technique. Depending upon the process, a time-dependent technique may be more suitable. Time-dependent modeling requires much more computing power than does time averaging.

GENERAL REFERENCES

1. J. Y. Oldshue, "Mixing '89," *Chemical Engineering Progress,* **85**(5): 33–42 (1989).
2. J. C. Middleton, *Proc. 3d European Conf. on Mixing,* 4/89, BHRA, pp. 15–36.
3. J. Y. Oldshue, T. A. Post, R. J. Weetman, "Comparison of Mass Transfer Characteristics of Radial and Axial Flow Impellers," *BHRA Proc. 6th European Conf. on Mixing,* 5/88.
4. A. W. Neinow, B. Buckland, R. J. Weetman, *Mixing XII Research Conference,* Potosi, Mo., 8/89.
5. R. Calabrese et al., *AIChE J.* **32:** 657, 677 (1986).
6. T. N. Zwietering, *Chemical Engineering Science,* **8**(3): 244–253 (1958).
7. J. Y. Oldshue, *Chemical Engineering Progress,* "Mixing of Slurries Near the Ultimate Settled Solids Concentration," **77**(5): 95–98 (1981).

PASTE AND VISCOUS-MATERIAL MIXING

GENERAL REFERENCES: *Dry Solids, Paste and Dough Mixing Equipment Testing Procedure,* American Institute of Chemical Engineers, New York, 1979. Fischer, *Chem. Eng.* **69**(3): 52 (1962). Irving and Saxton, "Mixing of High Viscosity Materials," in Uhl and Gray, *Mixing Theory and Practice,* vol. 2, Academic, New York, 1967, chap. 8. Mohr, in Bernhardt, *Processing of Thermoplastic Materials,* Reinhold, New York, 1967, chap. 3. Parker, *Chem. Eng.* **71**(12): 166 (1964). Rauwendaal, *Mixing in Polymer Processing,* Marcel Dekker, 1991. Tadmor and Gogos, *Principles of Polymer Processing,* Wiley, New York, 1979.

INTRODUCTION

Thick mixtures with viscosities greater than 10 Pa·s are not readily mixed in conventional stirred pots with either propeller or turbine agitators. The high viscosity may be due to that of the matrix fluid itself, to a high slurry concentration, or to interactions between components.

Because of the high viscosity, the mixing Reynolds number (Re = $D^2\rho N/\mu$) may be well below 100. Mixing occurs as a consequence of laminar shearing and stretching forces, and turbulence plays no part. Relative motion of an agitator stretches and deforms the material between itself and the vessel wall. As a layer of fluid gets stretched into thinner layers, the striations diminish and the shear forces tear solid agglomerates apart until apparent homogeneity is obtained. When solids are present, it may be necessary to reduce the particle size of the solids, even down to submicron size, such as in pigment dispersion.

Mixers for high-viscosity materials generally have a small high shear zone (to minimize dissipative heat effects) and rely on the impeller to circulate all of the mixer contents past the high shear zone. It may be necessary to take special steps to eliminate any stagnant zones or perhaps to avoid material riding around on mixing blades without being reincorporated into the matrix.

Most paste and high-viscosity mixtures are nonnewtonian, with viscosity dropping with shear rate. Consequently, increasing impeller speed may be counterproductive, as the transmitted shear drops rapidly and an isolated hole may be created in the central mass without circulation of the bulk material in the remainder of the vessel.

The relative viscosities of the materials being mixed can be very important. Karam and Bellinger [*Ind. Eng. Chem. Fund.* **7**: 576 (1981)] and Grace [*Engr. Fndn. Mixing Research Conf.* Andover, N.H., 1973] have shown that there is a maximum viscosity ratio (μ dispersed/μ continuous equal to about 4) above which droplets can no longer be dispersed by shear forces. The only effective dispersing mechanism for high-viscosity-ratio mixtures is elongational flow.

Equipment for viscous mixing usually has a small clearance between impeller and vessel walls, a relatively small volume, and a high power per unit volume. Intermeshing blades or stators may be present to prevent material from cylindering on the rotating impeller. Blade shape can have a significant impact on the mixing process. A scraping profile will be useful if heat transfer is important, whereas a smearing profile will be more effective for dispersion. Ease of cleaning and ease of discharge may also be important.

BATCH MIXERS

Change-Can Mixers   Change-can mixers are vertical batch mixers in which the container is a separate unit easily placed in or removed from the frame of the machine. They are available in capacities of about 4 to 1500 L (1 to 400 gal). The commonest type is the pony mixer. Separate cans allow the batch to be carefully measured or weighed before being brought to the mixer itself. The mixer also may serve to transport the finished batch to the next operation or to storage. The identity of each batch is preserved, and weight checks are easily made.

Change cans are relatively inexpensive. A good supply of cans allows cleaning to be done in a separate department, arranged for efficient cleaning. In paint and ink plants, where mixing precedes milling or grinding and where there may be a long run of the same formulation and color, the cans may be used for an extended period without cleaning, as long as no drying out or surface oxidation occurs.

In most change-can mixers, the mixing elements are raised from the can by either a vertical lift or a tilting head; in others, the can is dropped away from the mixing elements. After separation the mixing elements drain into the can, and the blades can be wiped down. With the can out of the way, complete cleaning in the blades and their supports is simple. If necessary, blades may be cleaned by rotating them in solvent.

Intimate mixing is accomplished in change-can mixers in two ways. One method is to have the mixing-unit assembly revolve in a planetary motion so that the rotating blades sweep the entire circumference of the can (Fig. 18-39). The other is to mount the can on a rotating turntable so that all parts of the can wall pass fixed scraper blades or the agitator blades at a point of minimum clearance.

A type of heavy-duty planetary change-can mixer has been used extensively for processing critical solid-propellant materials. In a design offered by APV Chemical Machinery Inc., the mixer blades pass through all portions of the can volume, and the two blades wipe each other, thus assuring no unmixed portions. A dead spot under the mixer blades is avoided by having both blades off the centerline of the can. The cans fit tightly enough so that mixing can be achieved under vacuum or pressure. There is no contact between the glands and the material being mixed. Charging ports located in the housing directly above the mixing can make it possible to charge materials to the mixer with the can in the operating position. This type of mixer is available in sizes of 0.5 L (about 1 pt) to 1.6 m³ (420 gal), involving power input of 0.2 to 75 kW (0.25 to 100 hp).

Helical-Blade Mixers   Helical mixers are now available in a variety of configurations. The mixing element may be in the form of a conical or a cylindrical helix. It may be a ribbon spaced radially from the shaft by spokes or a screw consisting of a helical surface that is continuous from the shaft to the periphery of the helix. A venerable example of the latter type is the soap crutcher, in which the screw is mounted in a draft tube. Close screw-tube clearance and a high rotational speed result in rapid motion of the material and high shear. The screw lifts the material through the tube, and gravity returns it to the bottom of the tank. If the tank has well-rounded corners, this kind of mixer may be used for fibrous materials. Heavy paper pulp containing 16 to 18 percent solids is uniformly bleached in large mixers of this type.

A double helix shortens mixing time but requires more power. The disadvantages of the higher torque requirement are frequently offset by the better mixing and heat transfer.

A vertical helical ribbon blender can be combined with an axial screw of smaller diameter (Fig. 18-25). Such mixers are used in polymerization reactions in which uniform blending is required but in which high-shear dispersion is not a factor. Addition of the inner flight contributes little more turnover in mixing newtonian fluids but significantly shortens the mixing time in nonnewtonian systems and adds negligibly to the impeller power [Coyle et al., *Am. Inst. Chem. Eng. J.,* **15**, 903 (1970)].

Double-Arm Kneading Mixers   The universal mixing and kneading machine consists of two counterrotating blades in a rectangular trough curved at the bottom to form two longitudinal half cylinders and a saddle section (Fig. 18-40). The blades are driven by gearing at either or both ends. The oldest style empties through a bottom door or valve and is still in use when 100 percent discharge or thorough cleaning between batches is not an essential requirement. More commonly, however, double-arm mixers are tilted for discharge. The tilting mechanism may be manual, mechanical, or hydraulic.

A variety of blade shapes have evolved. The mixing action is a combination of bulk movement, smearing, stretching, folding, dividing, and recombining as the material is pulled and squeezed against blades, saddle, and sidewalls. The blades are pitched to achieve end-to-end circulation. Rotation is usually such that material is drawn down over the saddle. Clearances are as close as 1 mm (0.04 in).

The blades may be tangential or overlapping. Tangential blades are run at different speeds, with the advantages of faster mixing from constant change of relative position, greater wiped heat-transfer area per unit volume, and less riding of material above the blades. Overlapping blades can be designed to avoid buildup of sticky material on the blades.

The agitator design most widely used is the sigma blade (Fig. 18-41*a*). The sigma-blade mixer is capable of starting and operating with either liquids or solids or a combination of both. Modifications in blade-face design have been introduced to increase particular effects, such as shredding or wiping. The sigma blade has good mixing action, readily discharges materials which do not stick to the blades, and is relatively easy to clean when sticky materials are being processed.

The dispersion blade (Fig. 18-41*b*) was developed particularly to provide compressive shear higher than that achieved with standard

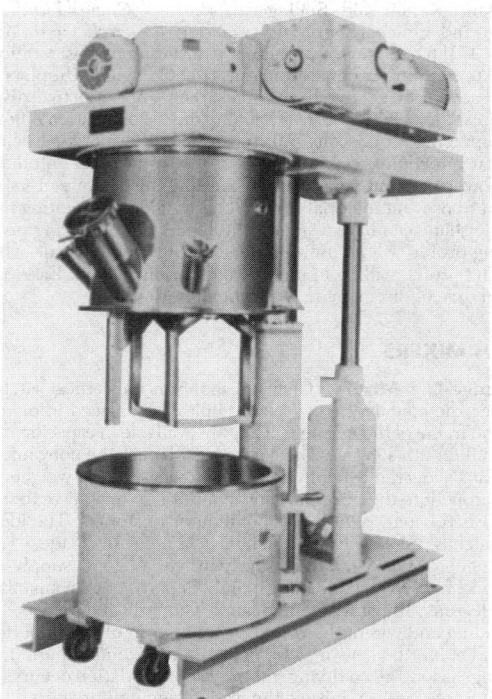

FIG. 18-39   Change-can mixer. (*Charles Ross & Son Co.*)

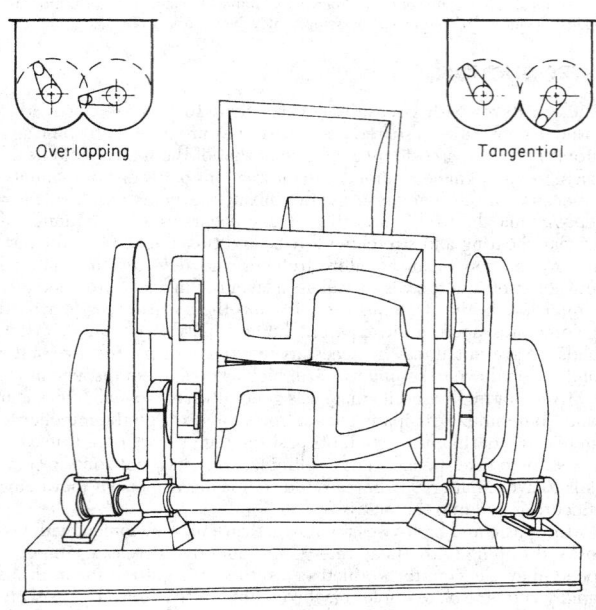

Overlapping   Tangential

FIG. 18-40   Double-arm kneader mixer. (*APV Baker Perkins Inc.*)

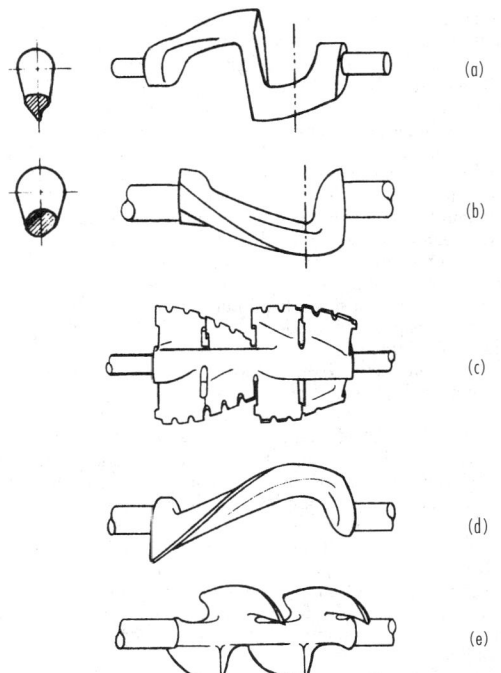

FIG. 18-41 Agitator blades for double-arm kneaders: (*a*) sigma; (*b*) dispersion; (*c*) multiwiping overlap; (*d*) single-curve; (*e*) double-naben. (*APV Baker Perkins, Inc.*)

TABLE 18-3 Characteristics of Double-Arm Kneading Mixers*

Size number	Capacity, U.S. gal		Typical supplied horsepower		Floor space, ft
	Working	Maximum	Sigma blade, MWOL blade	Dispersion blade	
4	0.7	1	1	2	1 × 3
6	2.3	3.5	2	5	2 × 3
8	4.5	7	5	7.5	3 × 4
11	10	15	15	20	5 × 6
12	20	30	25	40	6 × 6
14	50	75	30	60	6 × 8
15	100	150	50	100	8 × 10
16	150	225	60	150	9 × 11
17	200	300	75	200	9 × 13
18	300	450	100	—	10 × 14
20	500	750	150	—	11 × 16
21	600	900	175	—	12 × 16
22	750	1125	225	—	12 × 17
23	1000	1500	300	—	14 × 18

*Data from APV Baker Perkins, Inc. To convert feet to meters, multiply by 0.3048; to convert gallons to cubic meters, multiply by 3.78×10^{-3}; and to convert horsepower to kilowatts, multiply by 0.746.

mixing process. At discharge time, the direction of rotation of the screw is reversed and the mixed material is extruded through suitable die openings in the side of the machine. The discharge screw is driven independently of the mixer blades by a separate drive.

Working capacities range from 4 to 3800 L (1 to 1000 gal), with up to 300 kW (400 hp). This type of kneader is offered by most of the double-arm-kneader manufacturers. It is particularly suitable when a heel from the prior batch can be left without detriment to succeeding batches.

Intensive Mixers

Banbury Mixer   Preeminent in the field of high-intensity mixers, with power input up to 6000 kW/m³ (30 hp/gal), is the Banbury mixer, made by the Farrel Co. (Fig. 18-43). It is used mainly in the plastics and rubber industries. The top of the charge is confined by an air-operated ram cover mounted so that it can be forced down on the charge. The clearance between the rotors and the walls is extremely small, and it is

sigma blades. The blade face wedges material between itself and the trough, rather than scraping the trough, and is particularly suited for dispersing fine particles in a viscous mass. Rubbery materials have a tendency to ride the blades, and a ram is frequently used to keep the material in the mixing zone.

Multiwiping overlapping (MWOL) blades (Fig. 18-41*c*) are commonly used for mixtures which start tough and rubberlike, inasmuch as the blade cuts the material into small pieces before plasticating it.

The single-curve blade (Fig. 18-41*d*) was developed for incorporating fiber reinforcement into plastics. In this application, the individual fibers (e.g., sisal or glass) must be wetted with polymer without incurring undue fiber breakage.

Many other blade designs have been developed for specific applications. The double-naben blade (Fig. 18-41*e*) is a good blade for mixes which "ride," that is, form a lump which bridges across the sigma blade.

Double-arm mixers are available from several suppliers (e.g., Paul O. Abbe, Inc.; Littleford Day; Jaygo, Inc.; Charles Ross & Son Co.; Teledyne Readco). Options include vacuum design, cored blades, jacketed trough, choice of cover design, and a variety of seals and packing glands. Power requirements vary from ⅙ to 2 hp/gal of capacity. Table 18-3 lists specifications and space requirements for typical tilting-type, double-ended-drive, double-arm mixers. The working capacity is generally at or near the top of the blades, and the total capacity is the volume contained when the mixer is filled level with the top of the trough.

Figure 18-42 provides a guide for typical applications. Individual formulation changes may require more power than indicated in the figure. Parker [*Chem. Eng.*, **72**(18), 121 (1965)] has described in greater detail how to select double-arm mixers.

Screw-Discharge Batch Mixers   A variant of the sigma-blade mixer is now available with an extrusion-discharge screw located in the saddle section. During the mixing cycle the screw moves the material within the reach of the mixing blades, thereby accelerating the

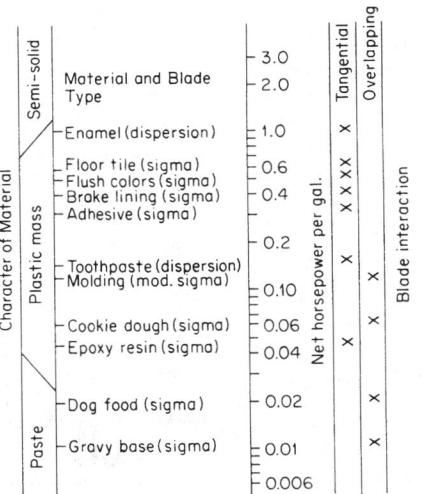

FIG. 18-42 Typical application and power for double-arm kneaders. To convert horsepower per gallon to kilowatts per cubic meter, multiply by 197.3. [*Parker,* Chem. Eng. **72**(18): 125 (1965); *excerpted by special permission of the copyright owner, McGraw-Hill, Inc.*]

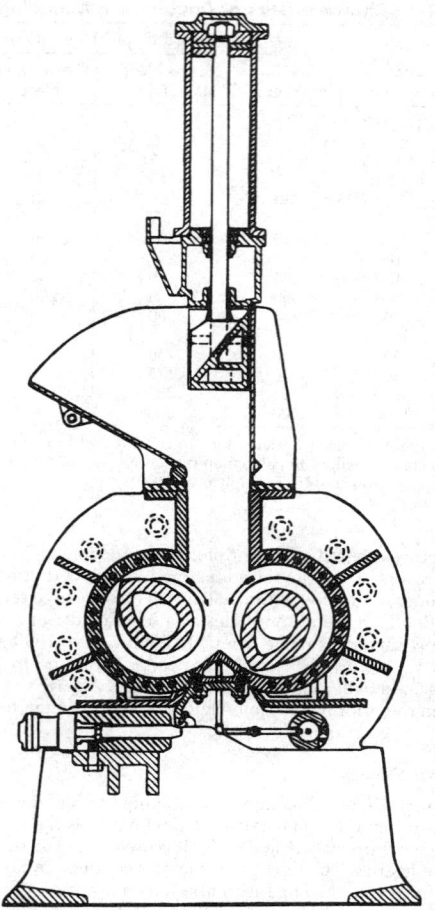

FIG. 18-43   Banbury mixer. (*Farrel Co.*)

here that the mixing action takes place. The operation of the rotors of a Banbury at different speeds enables one rotor to drag the stock against the rear of the other and thus help clean ingredients from this area.

The extremely high power consumption of the machines operating at speeds of 40 r/min or lower calls for rotor shafts of large diameter. The combination of heavy shafts, stubby blades, close clearances, and the confined charge limits the Banbury mixer to small batches. The production rate is increased as much as possible by using powerful drives and rotating the blades at the highest speed that the material will stand. The friction produced in the confined space is great, and with heat-sensitive materials cooling may be the limiting factor. Recent innovations include a drop door, four-wing rotors which can provide 30 percent greater power than the older two-wing rotors, and separable gear housings. Equipment is available from laboratory size to a mixer capable of handling a 450-kg (1000-lb) charge and applying 2240 kW (3000 hp).

High-Intensity Mixer.   Mixers such as that shown in Fig. 18-44 combine a high shear zone with a fluidized vortex mixing action. Blades at the bottom of the vessel scoop the batch upward at peripheral speeds of about 40 m/s (130 ft/s). The high shear stress (to $20,000 \ s^{-1}$) and blade impact easily reduce agglomerates and aid intimate dispersion. Since the energy input is high [200 kW/m³ (about 8 hp/ft³)], even powdery material is heated rapidly.

Mixers of this type are available in sizes from 4 to 2000 L (1 to 500 gal), consuming from 1.5 to 375 kW (2 to 500 hp).

These mixers are particularly suited for rapid mixing of powders and granules with liquids, for dissolving resins or solids in liquids, or for removal of volatiles from pastes under vacuum. Scale-up is usually on the basis of maintaining constant peripheral velocity of the impeller.

Roll Mills   Roll mills can provide exceedingly high localized shear while retaining extended surface for temperature control.

Two-Roll Mills   These mills contain two parallel rolls mounted in a heavy frame with provision for accurately regulating the pressure and distance between the rolls. As one pass between the rolls does little blending and only a small amount of work, the mills are practically always used as batch mixers. Only a small amount of material is in the high-shear zone at any one time.

To increase the wiping action, the rolls are usually operated at different speeds. The material passing between the rolls is returned to the feed point by the rotation of the rolls. If the rolls are at different temperatures, the material usually will stick to the hotter roll and return to the feed point as a thick layer.

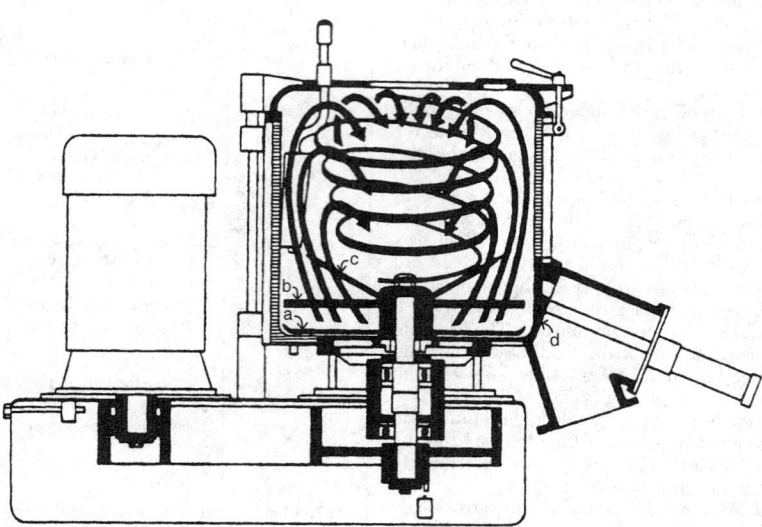

FIG. 18-44   High-intensity mixer: (*a*) bottom scraper; (*b*) fluidizing tool; (*c*) horn tool; (*d*) flush-mounted discharge valve. (*Henschel Mixers America, Inc.*)

At the end of the period of batch mixing, heavy materials may be discharged by dropping between the rolls, while thin mixes may be removed by a scraper bar pressing against the descending surface of one of the rolls.

Two-roll mills are used mainly for preparing color pastes for the ink, paint, and coating industries. There are a few applications in heavy-duty blending of rubber stocks, for which corrugated and masticating rolls are often used.

Miscellaneous Batch Mixers

Bulk Blenders   Many of the mixers used for solids blending (Sec. 19) are also suitable for some liquid-solids blending. **Ribbon blenders** can be used for such tasks as wetting out or coating a powder. When the final paste product is not too fluid, other solids-handling equipment finds frequent use.

Plow Mixers   Plow mixers such as the **Littleford** (Littleford Day) and the **Marion** (Rapids Machinery Co.) machines can be used for either batch or continuous mixing. Plow-shaped heads arranged on the horizontal shaft rotate at high speed, hurling the material throughout the free space of the vessel. Additional intermixing and blending occur as the impellers plow through the solids bed. Special high-speed choppers (3600 r/min) can be installed to break up lumps and aid liquid incorporation. The choppers also disperse fine particles throughout viscous materials to provide a uniform suspension. The mixer is available in sizes of 40 L to 40 m³ (10 to 10,000 gal) of working capacity.

Cone and Screw Mixers   The **Nauta mixer** of Fig. 18-45 (Day Mixing) utilizes an orbiting action of a helical screw rotating on its own axis to carry material upward, while revolving about the centerline of the cone-shaped shell near the wall for top-to-bottom circulation. Reversing the direction of screw rotation aids discharge of pasty materials. Partial batches are mixed in the Nauta type of mixer as efficiently as full loads. These mixers, available in sizes of 40 L to 40 m³ (10 to 10,000 gal), achieve excellent low-energy blending, with some hydraulic shear dispersion. At constant speed, both mixing time and power scale up with the square root of volume.

Pan Muller Mixers   These mixers can be used if the paste is not too fluid or too sticky. The main application of muller mixers is now in the foundry industry, in mixing small amounts of moisture and binder materials with sand particles for both core and molding sand. In paste processing, pan-and-plow mixers are used principally for mixing putty and clay pastes, while muller mixers handle such diversified materials as clay, storage-battery paste, welding-rod coatings, and chocolate coatings.

In muller mixers the rotation of the circular pan or of the plows brings the material progressively into the path of the mullers, where the intensive action takes place. Figure 18-46 shows one type of mixer, in which the mullers and plows revolve around a stationary turret in a stationary pan. The outside plow moves material from the crib wall to the path of the following muller; the inside plow moves it from the central turret to the path of the other muller. The mullers crush the material, breaking down lumps and aggregates.

Standard muller mixers range in capacity from a fraction of a cubic foot to more than 1.8 m³ (60 ft³), with power requirements ranging from 0.2 to 56 kW (⅓ to 75 hp). A continuous muller design employs two intersecting and communicating cribs, each with its own mullers and plows. At the point of intersection of the two crib bodies, the outside plows give an approximately equal exchange of material from one crib to the other, but material builds up in the first crib until the feed rate and the discharge rate of material from the gate in the second crib are equal. The residence time is regulated by adjusting the outlet gate.

CONTINUOUS MIXERS

Some of the batch mixers previously described can be converted for continuous processing. Product uniformity may be poor because of the broad residence time distribution. If the different ingredients can be accurately metered, various continuous mixers are available. Continuous mixers generally consist of a closely fitted agitator element rotating within a stationary housing.

Single-Screw Mixers   Because of the growth of the plastics industry, use of extruders such as that depicted in Fig. 18-47 are now

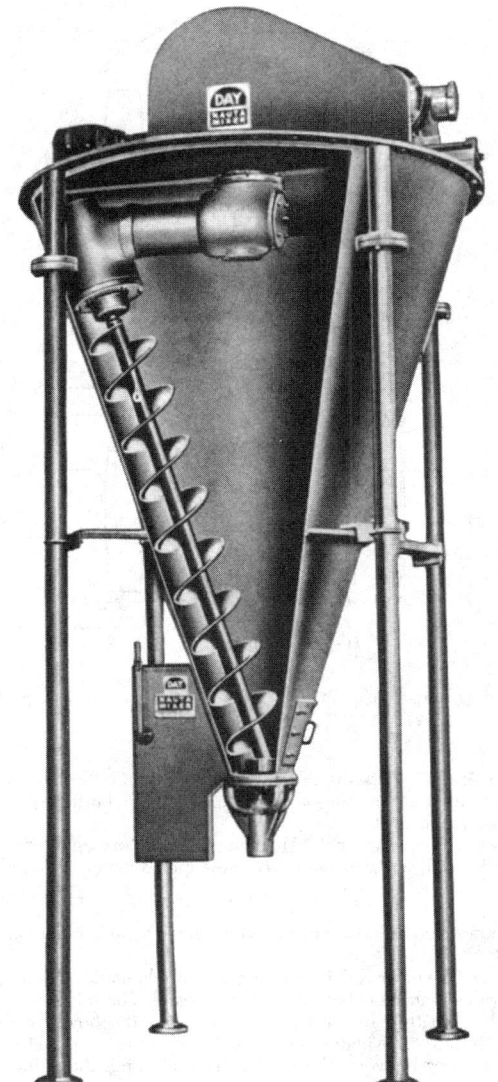

FIG. 18-45   Nauta mixer. (*Littleford Day.*)

very widespread. The quality and usefulness of the product may well depend on how uniformly the various additives, stabilizers, fillers, etc., have been incorporated. The extruder combines the process functions of melting the base resin, mixing in the additives, and developing the pressure required for shaping the product into pellets, sheet, or profiles. Dry ingredients, sometimes premixed in a batch blender, are fed into the feed throat where the channel depth is deepest. As the root diameter of the screw is increased, the plastic is melted by a combination of friction and heat transfer from the barrel. Shear forces can be very high, especially in the melting zone, and the mixing is primarily a laminar shearing action.

Single-screw extruders can be built with a long length-to-diameter ratio to permit a sequence of process operations, such as staged addition of various ingredients. Capacity is determined by diameter, length, and power. Although the majority of extruders are in the 25- to 200-mm-diameter range, much larger units have been made for specific applications such as polyethylene homogenization. Mixing

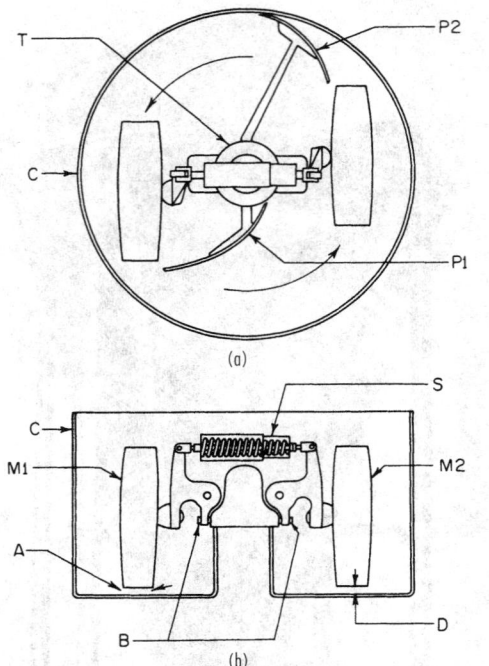

FIG. 18-46   Pan muller: (*a*) plan view; (*b*) sectional elevation. [*Bullock*, Chem. Eng. Prog. **51**: 243 (1955), *by permission.*]

enhancers (Fig. 18-48) are utilized to provide both elongational reorienting and shearing action to provide for both dispersive and distributive mixing.

The maximum power (*P*, kW) being supplied for single-screw extruders varies with screw diameter (*D*, mm) approximately as follows:

$$P = 5.3 \times 10^{-3} \, D^{2.25}$$

The power required for most polymer mixing applications varies from 0.15 to 0.3 kWh/kg.

Rietz Extruder   This extruder, shown schematically in Fig. 18-49, has orifice plates and baffles along the vessel. The rotor carries multiple blades with a forward pitch, generating the head for extrusion through the orifice plates as well as battering the material to break up agglomerates between the baffles. Typical applications include wet

granulation of pharmaceuticals, blending color in bar soap, and mixing and extruding cellulose materials. The Extructor is available in rotor diameters up to 600 mm (24 in) and in a power range of 5 to 112 kW (7 to 150 hp).

Twin-Screw Extruders   With two screws in a figure-eight barrel, advantage can be taken of the interaction between the screws as well as between the screw and the barrel. Twin-screw machines are used in continuous melting, mixing, and homogenizing of different polymers with various additives, or to carry out the intimate mixing required for reactions in which at least one of the components is of high viscosity. The screws can be tangential or intermeshing, and the latter either co- or counterrotating. Tangential designs allow variability in channel depth and permit longer lengths.

Counterrotating intermeshing screws provide a dispersive milling action between the screws and behave like a positive displacement device with the ability to generate pressure more efficiently than any other extruder.

The most common type of twin-screw mixing extruder is the co-rotating intermeshing variety (APV, Berstorff, Davis Standard, Leistritz, Werner & Pfleiderer). The two keyed or splined shafts are fitted with pairs of slip-on kneading or conveying elements, as shown in Fig. 18-50. The arrays of these elements can be varied to provide a wide range of mixing effects. The barrel sections are also segmented to allow for optimum positioning of feed ports, vents, barrel valves, etc. The barrels may be heated electrically or with oil or steam and cooled with air or water.

Each pair of kneading paddles causes an alternating compression and expansion effect which massages the contents and provides a combination of shearing and elongational mixing actions. The corotating twin-screw mixers are available in sizes ranging from 15 to 300 mm, with length-to-diameter ratios up to 50 and throughput capacities up to 25,000 kg/h. Screw speeds can be high (to 500 r/min in the smaller-production-size extruders), and residence times for mixing are usually under 2 min.

Farrel Continuous Mixer   This mixer (Fig. 18-51) consists of rotors similar in cross section to the Banbury batch mixer. The first section of the rotor acts as a screw conveyor, propelling the feed ingredients to the mixing section. The mixing action is a combination of intensive shear, between rotor and chamber wall, kneading between the rotors, and a rolling action of the material itself. The amount and quality of mixing are controlled by adjustment of speed, feed rates, and discharge-orifice opening. Units are available in five sizes with mixing-chamber volumes ranging up to 0.12 m³ (4.2 ft³). At 200 r/min, the power range is 5 to 2200 kW (7 to 3000 hp).

Miscellaneous Continuous Mixers

Trough-and-Screw Mixers   These mixers usually consist of single or twin rotors which continually turn the feed material over as it

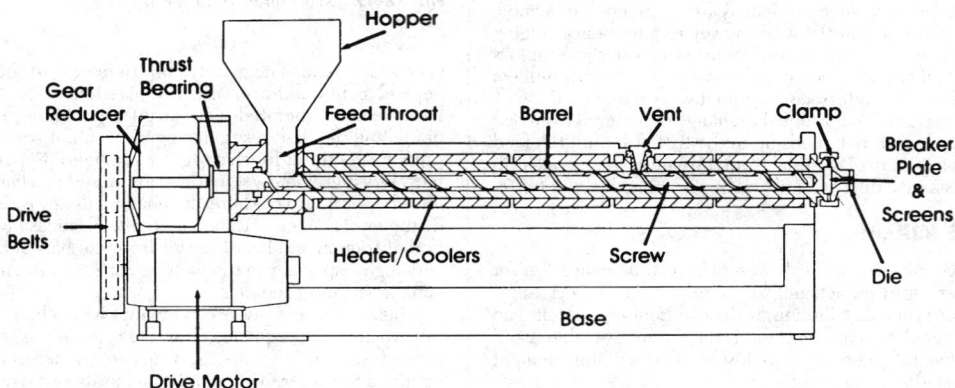

FIG. 18-47   Single-screw extruder. (*Davis Standard.*)

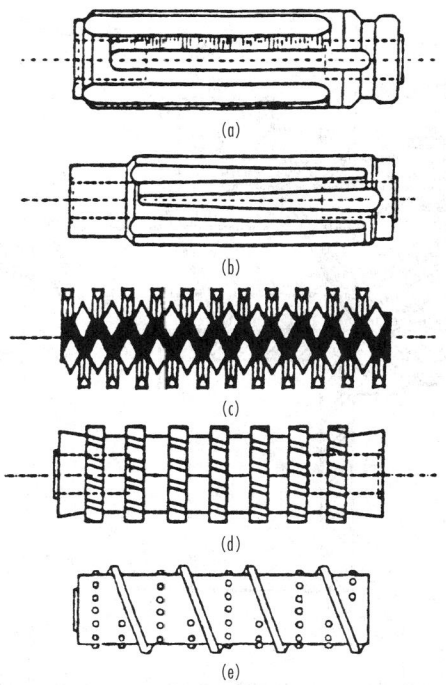

FIG. 18-48 Mixing enhancers for single-screw extruders: (*a*) Maddock (straight); (*b*) Maddock (tapered); (*c*) pineapple; (*d*) gear; (*e*) pin.

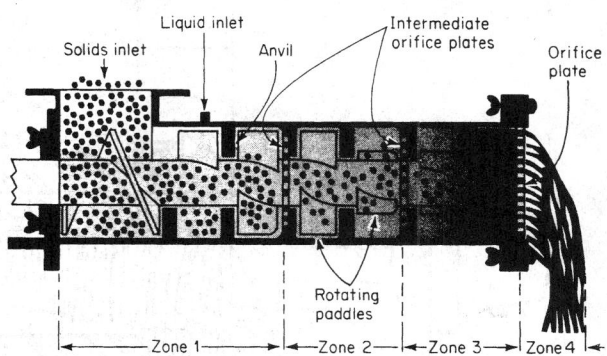

FIG. 18-49   Rietz Extruder. (*Bepex Corporation.*)

progresses toward the discharge end. Some have been designed with extensive heat-transfer area. The continuous-screw **Holo-Flite Processor** (Fig. 18-52; Denver Sala) is used primarily for heat transfer, since the hollow screws present extended surface without contributing much shear. Two or four screws may be used. Bethlehem Corp.'s **Porcupine Processor** also has heat-transfer media going through the flights of the rotor, but the agitator flights are cut to provide a folding action on the process mass. Breaker-bar assemblies, consisting of fingers extending toward the shaft, are frequently used to improve agitation.

Another type of trough-and-screw mixer with a large volume available for mixing is the AP Conti (List AG) machine shown in Fig. 18-53. These self-cleaning-type mixers are particularly appropriate when

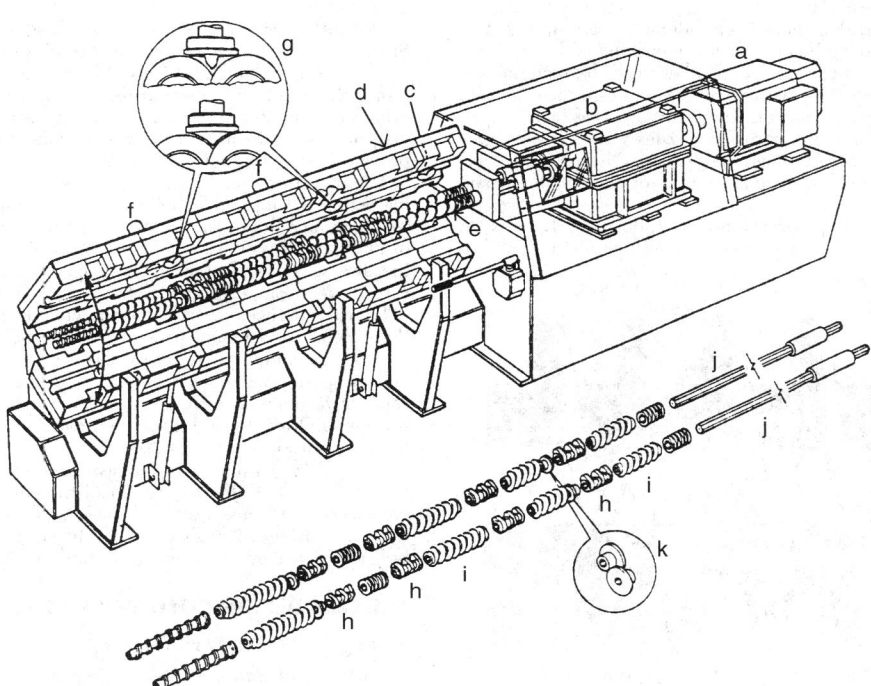

FIG. 18-50   Intermeshing corotating twin screw extruder: (*a*) drive motor; (*b*) gearbox; (*c*) feed port; (*d*) barrel; (*e*) assembled rotors; (*f*) vent; (*g*) barrel valve; (*h*) kneading paddles; (*i*) conveying screws; (*j*) splined shafts; (*k*) blister rings. (*APV Chemical Machinery, Inc.*)

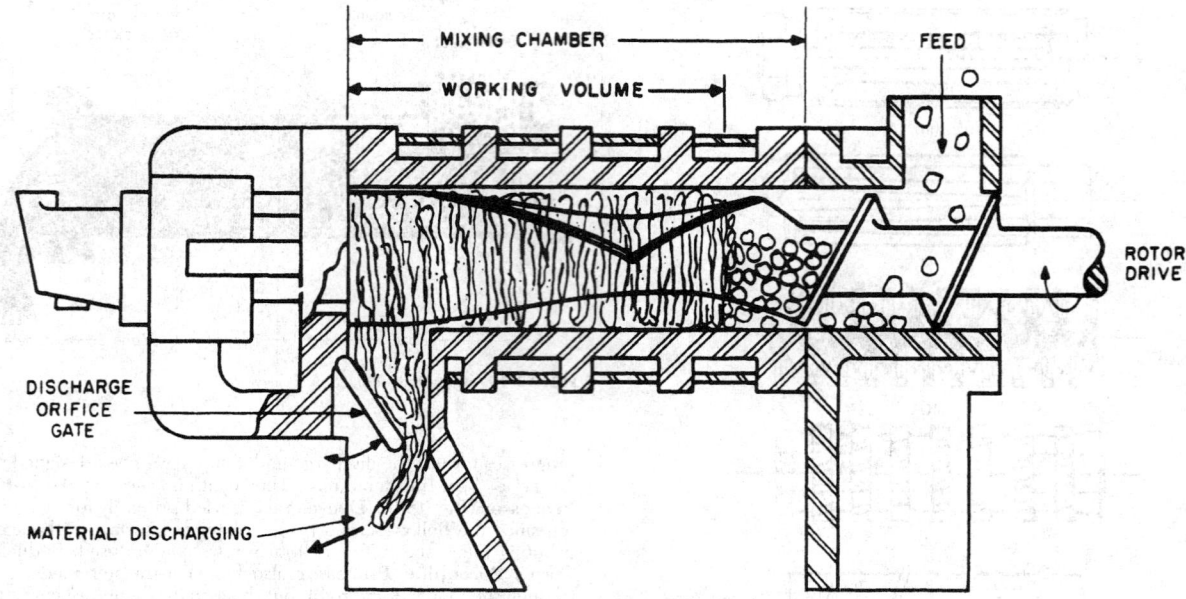

FIG. 18-51   Farrel continuous mixer. (*Farrel Co.*)

the product being handled goes through a sticky stage, which could plug the mixer or foul the heat-transfer surfaces.

Pug Mills   A pug mill contains one or two shafts fitted with short, heavy paddles, mounted in a cylinder or trough which holds the material being processed. In two-shaft mills the shafts are parallel and may be horizontal or vertical. The paddles may or may not intermesh. Clearances are wide so that there is considerable mass mixing. Unmixed or partially mixed ingredients are fed at one end of the machine, which is usually totally enclosed. The paddles push the material forward as they cut through it, and carry the charge toward the discharge end as it is mixed. Product may discharge through one or two open ports or through one or more extrusion nozzles which give roughly shaped, continuous strips. Automatic cutters may be used to make blocks from the strips. Pug mills are most used for mixing mineral and clay products.

Kneadermaster   This mixer (Patterson Industries, Inc.) is an adaptation of a sigma-blade mixer for continuous operation. Each two

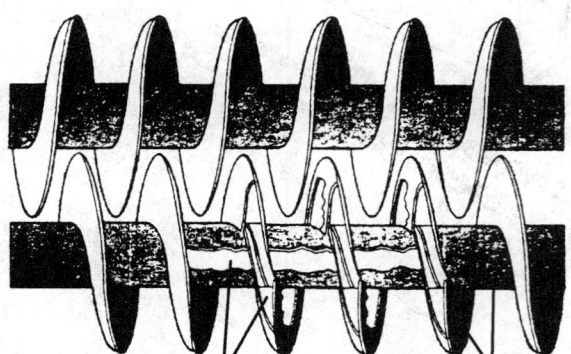

Heat transfer Fluid
(hot oil, water, or steam)

Indirect heat transfer-
fluid never comes in
contact with product

FIG. 18-52   Holo-Flite Processor. (*Denver Sala.*)

pairs of blades establish a mixing zone, the first pair pushing materials toward the discharge end of the trough and the second pair pushing them back. Forwarding to the next zone is by displacement with more feed material. Control of mixing intensity is by variation in rotor speed. Cored blades supplement the heat-transfer area of the jacketed trough.

Motionless Mixers   An alternative to agitated equipment for continuous viscous mixing involves the use of stationary shaped diverters inside conduits which force the fluid media to mix themselves through a progression of divisions and recombinations, forming striations of ever-decreasing thickness until uniformity is achieved. Simple diverters, such as the **Kenics static mixer** (Chemineer, Inc.; Fig. 18-54), provide 2^n layerings per n diverters.

The power consumed by a motionless mixer in producing the mixing action is simply that delivered by a pump to the fluid which it moves against the resistance of the diverter conduit. For a given rate of pumping, it is substantially proportional to that resistance. When the diverter consists of several passageways, as in the **Sulzer static mixer** (Koch Engineering Co., Inc.) shown in Fig. 18-55, the number of layerings (hence, the rate of mixing) per diverter is increased, but at the expense of a higher pressure drop. The pressure drop, usually expressed as a multiple K of that of the empty duct, is strongly dependent upon the hydraulic radius of the divided flow passageway. The value of K, obtainable from the mixer supplier, can range from 6 to several hundred, depending on the Reynolds number and the geometry of the mixer.

Motionless mixers continuously interchange fluid elements between the walls and the center of the conduit, thereby providing enhanced heat transfer and relatively uniform residence times. Distributive mixing is usually excellent; however, dispersive mixing may be poor, especially when viscosity ratios are high.

PROCESS DESIGN CONSIDERATIONS

Scaling Up Mixing Performance

Scale-Up of Batch Mixers   The prime basis of scale-up of batch mixers has been equal power per unit volume, although the most desirable practical criterion is equal blending per unit time. As size is increased, mechanical-design requirements may limit the larger mixer to lower agitator speeds; if so, blend times will be longer in the larger

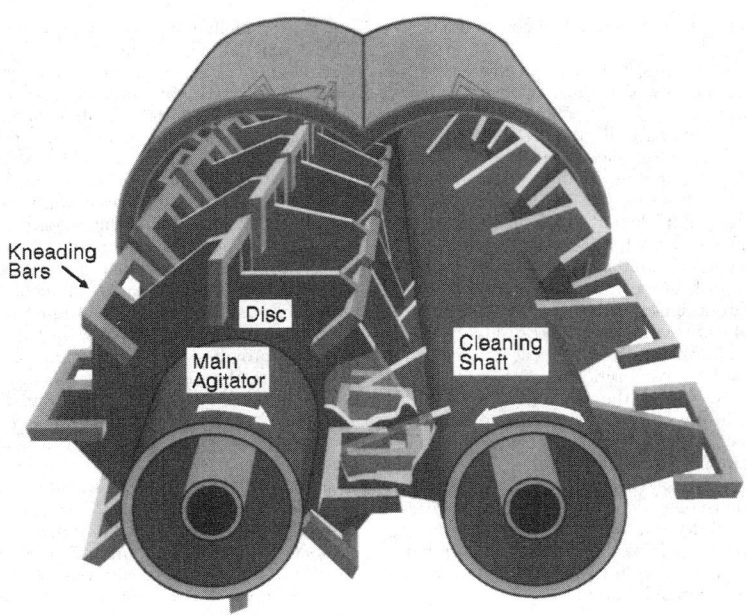

FIG. 18-53   AP Conti paste mixer. (*LIST.*)

mixer than in the smaller prototype. If the power is high, the lower surface-to-volume ratio as size is increased may make temperature buildup a limiting factor. Since the impeller in a paste mixer generally comes close to the vessel wall, it is not possible to add cooling coils. In some instances, the impeller blades can be cored for additional heat-transfer area.

Experience with double-arm mixers indicates that power is proportional to the product of blade radius, blade-wing depth, trough length, and average of the speeds of the two blades (Irving and Saxton, loc. cit.). The mixing time scales up inversely with blade speed. Goodness of mixing is dependent primarily on the number of revolutions that the blades have made. As indicated previously, the minimum possible mixing time may become dependent on heat-transfer rate.

Frequently, the physical properties of a paste vary considerably during the mixing cycle. Even if one knew exactly how power depended upon density and viscosity, it might be better to predict the requirements for a large paste mixer from the power-time curve observed in the prototype mixer rather than to try to calculate or measure all intermediate properties during the processing sequence (i.e., the prototype mixer may be the best instrument to use to measure the effective viscosity).

Scale-Up of Continuous Mixers   Although scaling up on the basis of constant power per unit feed rate [kWh/kg or (hp·h)/lb] is usually a good first estimate, several other factors may have to be considered. As the equipment scale is increased, geometric similarity being

at least approximated, there is a loss in surface-to-volume ratio. As size is increased, changing shear rate or length-to-diameter ratio may be required because of equipment-fabrication limitations. Furthermore, even if a reliable method of scaling up power exists, the determination of net power in small-scale test equipment is frequently difficult and inaccurate because of fairly large no-load power.

As a matter of fact, geometrical similarity usually cannot be maintained exactly as the size of the model is increased. In single-screw extruders, for example, channel depth in the flights cannot be increased in proportion to screw diameter because the distribution of heat generated by friction at the barrel wall requires more time as channel depth becomes greater. With constant retention time, therefore, nonhomogeneous product would be discharged from the scaled-up model. As the result of the departure from geometrical similarity, the throughput rate of single-screw extruders scales up with diameter

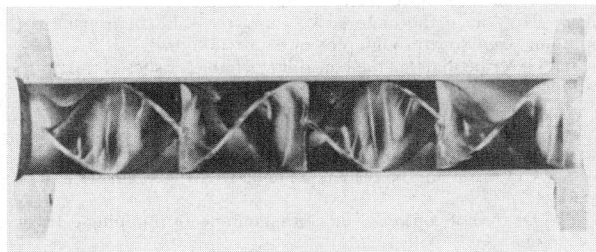

FIG. 18-54   Kenics static mixer. (*Chemineer, Inc.*)

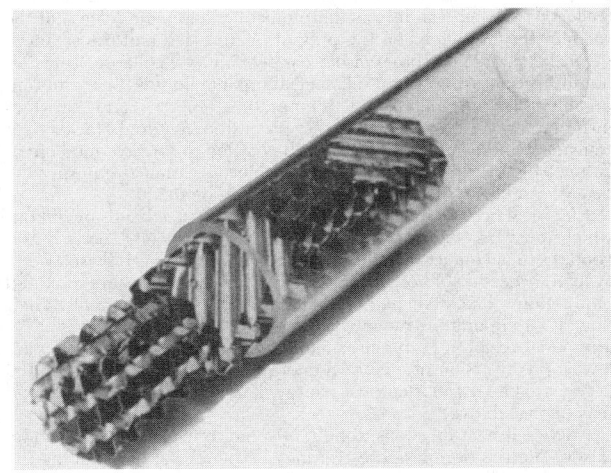

FIG. 18-55   Sulzer static mixer. (*Koch Engineering Co., Inc.*)

to the power 2.0 to 2.5 (instead of diameter cubed) at constant length-to-diameter ratio and screw speed.

The throughput rate of intermeshing twin-screw extruders (Fig. 18-56) and the Farrel continuous mixer (Fig. 18-51) is scaled up with diameter to about the 2.6 power.

If the process can be operated adiabatically, the production capacity is scaled up as the cube of diameter since geometry, shear rate, residence time, and power input per unit volume all can be held constant.

Residence-time distributions.   For flow through a conduit, the extent of axial dispersion can be characterized either by an axial-diffusion coefficient or by analogy to a number of well-mixed stages in series. Retention time can control the performance of a mixing system. As the number of apparent stages increases, there is greater assurance that all the material will have the required residence time. Under conditions requiring uniform retention time, it is imperative that the feed streams be fed in the correct ratio on a time scale much shorter than the average residence time of the mixer; otherwise, a perturbation in the feed will produce a comparable perturbation in the product. The mixing impellers in continuous mixers can be designed to cover the full range from minimum axial mixing (plug flow) to maximum (to damp out feed irregularities). Residence-time distributions and effective Peclet numbers have been determined for a wide variety of twin-screw configurations [Todd and Irving, *Chem. Eng. Prog.*, **65**(9), 84 (1969)]. Conventional single-screw extruder mixers have Peclet numbers about equal to the length-to-diameter ratio, or an equivalent number of stages equal to one-half of that.

HEATING AND COOLING MIXERS

Heat Transfer   Pastes are often heated or cooled by heat transfer through the walls of the container or hollow mixing arms. Good agitation, a large ratio of transfer surface to mixer volume, and frequent removal of material from the surface are essential for high rates of heat transfer. Sometimes evaporation of part of the mix is used for cooling.

In most mixers, the metal wall has a negligible thermal resistance. The paste film, however, usually has high resistance. It is important, therefore, while minimizing the resistance of the heating or cooling medium, to move the paste up to and away from the smooth wall surface as steadily and rapidly as possible. This is best achieved by having the paste flow so as to follow a close-fitting scraper which wipes the film from the wall with each rotation. Typical overall heat-transfer coefficients are between 25 and 200 $J/(m^2 \cdot s \cdot K)$ [4 to 35 Btu/(h·ft²·°F)].

Heating Methods   The most economical heating method varies with plant location and available facilities. Direct firing is rarely used, since it does not permit good surface-temperature control and may cause scorching of the material on the vessel walls. Steam heating is the most widely used method. It is economical, safe, and easily controlled. With thin-wall mixers there must be automatic release of the vacuum that results when the pressure is reduced and the steam in the jacket condenses; otherwise, weak sections will collapse. Transfer-liquid heating using water, oil, special organic liquids, or molten inorganic salts permits good temperature control and provides insurance against overheating the processed material. Jackets for transfer-liquid heating usually must be baffled to provide good circulation. Higher temperatures can be achieved without requiring the heavy vessel construction otherwise required by steam.

Electrical heating is accomplished with resistance bands or ribbons which must be electrically insulated from the machine body but in good thermal contact with it. The heaters must be carefully spaced to avoid a succession of hot and cold areas. Sometimes they are mounted in aluminum blocks shaped to conform to the container walls. Their effective temperature range is 150 to 500°C (about 300 to 930°F). Temperature control is precise, maintenance and supervision costs are low, and conversion of electrical energy to useful heat is almost 100 percent. The cost of electrical energy is usually large, however, and may be prohibitive.

Frictional heat develops rapidly in some units such as a Banbury mixer. The first temperature rise may be beneficial in softening the materials and accelerating chemical reactions. High temperatures detrimental to the product may easily be reached, however, and pro-

vision for cooling or frequent stopping of the machine must be made. Frictional heating may be lessened by reducing the number of working elements, their area, and their speed. Cooling thus is facilitated, but at the expense of increased mixing time.

Cooling Methods   In air cooling, air may be blown over the machine surfaces, the area of which is best extended with fins. Air or cooled inert gas may also be blown over the exposed surface of the mix, provided contamination or oxidation of the charge does not result. Evaporation of excess water or solvent under vacuum or at atmospheric pressure provides good cooling. A small amount of evaporation produces a large amount of cooling. Removing too much solvent, however, may damage the charge. Direct addition of ice to the mixer provides rapid, convenient cooling, but the resulting dilution of the mix must be permissible. Addition of dry ice is more expensive but results in lower temperatures, the mix is not diluted, and the CO_2 gas evolved provides a good inert atmosphere. Many mixers are cooled by circulation of water or refrigerants through jackets or hollow agitators. In general, this is the least expensive method, but it is limited by the magnitude of heat-transfer coefficient obtainable.

Selection of Equipment   If a new product is being considered, the preliminary study must be highly detailed. Laboratory or pilot-plant work must be done to establish the controlling factors. The problem is then to select and install equipment which will operate for quantity production at minimum overall cost. Most equipment vendors have pilot equipment available on a rental basis or can conduct test runs in their own customer-demonstration facilities.

One approach to proper equipment specification is by analogy. What current product is most similar to the new one? How is this material produced? What difficulties are being experienced?

In other situations the following procedure is recommended:

1.   List carefully all materials to be handled at the processing point and describe their characteristics, such as:

a.   How received at the processing unit: in bags, barrels, or drums, in bulk, by pipe line, etc.?

b.   Must storage and/or weighing be done at the site?

c.   Physical form.

d.   Specific gravity and bulk characteristics.

e.   Particle size or size range.

f.   Viscosity.

g.   Melting or boiling point.

h.   Corrosive properties.

i.   Abrasive characteristics.

j.   Is material poisonous?

k.   Is material explosive?

l.   Is material an irritant to skin, eyes, or lungs?

m.   Is material sensitive to exposure of air, moisture, or heat?

2.   List pertinent data covering production:

a.   Quantity to be produced per 8-h shift.

b.   Formulation of finished product.

c.   What accuracy of analysis is required?

d.   Will changes in color, flavor, odor, or grade require frequent cleaning of equipment?

e.   Is this operation independent, or does it serve other process stages with which it must be synchronized?

f.   Is there a change in physical state during processing?

g.   Is there a chemical reaction? Is it endothermic or exothermic?

h.   What are the temperature requirements?

i.   What is the form of the finished product?

j.   How must the material be removed from the apparatus (by pumping, free flow through pipe or chute, dumping, etc.)?

3.   Describe in detail the controlling characteristics of the finished product:

a.   Permanence of the emulsion or dispersion.

b.   Degree of blending of aggregates or of ultimate particles.

c.   Ultimate color development required.

d.   Uniformity of the dispersion of active ingredients, as in a drug product.

e.   Degree of control of moisture content for pumping extrusion, and so on.

Preparation and Addition of Materials   To ensure maximum production of high-grade mixed material, the preliminary preparation of the ingredients must be correct and they must be added in the proper order. There are equipment implications to these considerations.

Some finely powdered materials, such as carbon black, contain much air. If possible they should be compacted or wet out before being added to the mix. If a sufficient quantity of light solvent is a part of the formula, it may be used to wet the powder and drive out the air. If the powder cannot be wet, it may be possible to densify it somewhat by mechanical means. Removal of adsorbed gas under vacuum is sometimes necessary.

Not uncommonly, critical ingredients that are present in small proportion (e.g., vulcanizers, antioxidants, and antiacids) tend to form aggregates when dry. Before entering the mixer, they should be fluffed, either by screening if the aggregates are soft or by passage through a hammer mill, roll mill, or muller if they are hard. The mixing time is cut down and the product is more uniform if all ingredients are freed from aggregates before mixing.

If any solids present in small amounts are soluble in a liquid portion of the mix, it is well to add them as a solution, making provision to distribute the liquid uniformly throughout the mass. When a trace of solid material which is not soluble in any other ingredients is to be added, it may be expedient to add it as a solution in a neutral solvent, with provision to evaporate the solvent at the end of the mixing cycle.

It may be advisable to consider master batching, in which a low-proportion ingredient is separately mixed with part of some other ingredient of the mix, this premix then being added to the rest of the mix for final dispersion. Master batching is especially valuable in adding tinting colors, antioxidants, and the like. The master batch may be made up with laboratory accuracy, while at the mixing station weighing errors are minimized by the dilution of the important ingredient.

Considerations such as these may make it desirable to consider automatic weighing and batch accumulation, metering of liquid ingredients, and automatic control of various time cycles.

CRYSTALLIZATION FROM SOLUTION

GENERAL REFERENCES: *AIChE Testing Procedures: Crystallizers,* American Institute of Chemical Engineers, New York, 1970; *Evaporators,* 1961. Bennett, *Chem. Eng. Prog.,* **58**(9), 76 (1962). Buckley, *Crystal Growth,* Wiley, New York, 1951. Campbell and Smith, *Phase Rule,* Dover, New York, 1951. De Jong and Jancic (eds.), *Industrial Crystallization,* North-Holland Publishing Company, Amsterdam, 1979. "Crystallization from Solution: Factors Influencing Size Distribution," *Chem. Eng. Prog. Symp. Ser.,* **67**(110), (1971). Mullin (ed.), *Industrial Crystallization,* Plenum, New York, 1976. Newman and Bennett, *Chem. Eng. Prog.,* **55**(3), 65 (1959). Palermo and Larson (eds.), "Crystallization from Solutions and Melts," *Chem. Eng. Prog. Symp. Ser.,* **65**(95), (1969). Randolph (ed.), "Design, Control and Analysis of Crystallization Processes," *Am. Inst. Chem. Eng. Symp. Ser.,* **76**(193), (1980). Randolph and Larson, *Theory of Particulate Processes,* Academic, New York, 2d ed., 1988. Rousseau and Larson (eds.), "Analysis and Design of Crystallization Processes," *Am. Inst. Chem. Eng. Symp. Ser.,* **72**(153), (1976). Seidell, *Solubilities of Inorganic and Metal Organic Compounds,* American Chemical Society, Washington, 1965. Myerson (ed.), *Handbook of Industrial Crystallization,* Butterworth, 1993.

Crystallization is important as an industrial process because of the number of materials that are and can be marketed in the form of crystals. Its wide use is probably due to the highly purified and attractive form of a chemical solid which can be obtained from relatively impure solutions in a single processing step. In terms of energy requirements, crystallization requires much less energy for separation than do distillation and other commonly used methods of purification. In addition, it can be performed at relatively low temperatures and on a scale which varies from a few grams up to thousands of tons per day.

Crystallization may be carried out from a vapor, from a melt, or from a solution. Most of the industrial applications of the operation involve crystallization from solutions. Nevertheless, crystal solidification of metals is basically a crystallization process, and much theory has been developed in relation to metal crystallization. This topic is so specialized, however, that it is outside the scope of this subsection, which is limited to crystallization from solution.

PRINCIPLES OF CRYSTALLIZATION

Crystals   A crystal may be defined as a solid composed of atoms arranged in an orderly, repetitive array. The interatomic distances in a crystal of any definite material are constant and are characteristic of that material. Because the pattern or arrangement of the atoms is repeated in all directions, there are definite restrictions on the kinds of symmetry that crystals can possess.

There are five main types of crystals, and these types have been arranged into seven crystallographic systems based on the crystal interfacial angles and the relative length of its axes. The treatment of the description and arrangement of the atomic structure of crystals is the science of **crystallography.** The material in this discussion will be limited to a treatment of the growth and production of crystals as a unit operation.

Solubility and Phase Diagrams   Equilibrium relations for crystallization systems are expressed in the form of solubility data which are plotted as phase diagrams or solubility curves. Solubility data are ordinarily given as parts by weight of anhydrous material per 100 parts by weight of total solvent. In some cases these data are reported as parts by weight of anhydrous material per 100 parts of solution. If water of crystallization is present in the crystals, this is indicated as a separate phase. The concentration is normally plotted as a function of temperature and has no general shape or slope. It can also be reported as a function of pressure, but for most materials the change in solubility with change in pressure is very small. If there are two components in solution, it is common to plot the concentration of these two components on the X and Y axes and represent the solubility by isotherms. When three or more components are present, there are various techniques for depicting the solubility and phase relations in both three-dimension and two-dimension models. For a description of these techniques, refer to Campbell and Smith (loc. cit.). Shown in Fig. 18-56 is a phase diagram for magnesium sulfate in water. The line p–a represents the freezing points of ice (water) from solutions of magne-

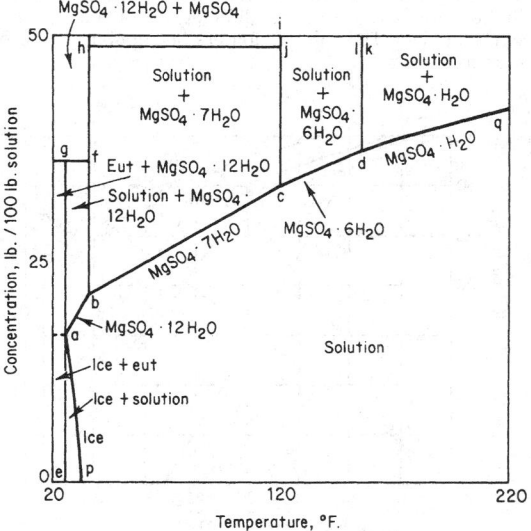

FIG. 18-56   Phase diagram. $MgSO_4 \cdot H_2O$. To convert pounds to kilograms, divide by 2.2; K = (°F + 459.7)/1.8.

sium sulfate. Point *a* is the eutectic, and the line *a–b–c–d–q* is the solubility curve of the various hydrates. Line *a–b* is the solubility curve for $MgSO_4 \cdot 12H_2O$, *b–c* is the solubility curve for $MgSO_4 \cdot 7H_2O$, *c–d* is the solubility curve for $MgSO_4 \cdot 6H_2O$, and *d–q* is the portion of the solubility curve for $MgSO_4 \cdot H_2O$.

As shown in Fig. 18-57, the mutual solubility of two salts can be plotted on the *X* and *Y* axes with temperatures as isotherm lines. In the example shown, all the solution compositions corresponding to 100°C with solid-phase sodium chloride present are shown on the line *DE*. All the solution compositions at equilibrium with solid-phase KCl at 100°C are shown by the line *EF*. If both solid-phase KCl and NaCl are present, the solution composition at equilibrium can only be represented by point *E*, which is the invariant point (at constant pressure). Connecting all the invariant points results in the mixed-salt line. The locus of this line is an important consideration in making phase separations.

There are numerous solubility data in the literature; the standard reference is by Seidell (loc. cit.). Valuable as they are, they nevertheless must be used with caution because the solubility of compounds is often influenced by pH and/or the presence of other soluble impurities which usually tend to depress the solubility of the major constituents. While exact values for any system are frequently best determined by actual composition measurements, the difficulty of reproducing these solubility diagrams should not be underestimated. To obtain data which are readily reproducible, elaborate pains must be taken to be sure the system sampled is at equilibrium, and often this means holding a sample at constant temperature for a period of from 1 to 100 h. While the published curves may not be exact for actual solutions of interest, they generally will be indicative of the shape of the solubility curve and will show the presence of hydrates or double salts.

Heat Effects in a Crystallization Process   The heat effects in a crystallization process can be computed by two methods: (1) a heat balance can be made in which individual heat effects such as sensible heats, latent heats, and the heat of crystallization can be combined into an equation for total heat effects; or (2) an enthalpy balance can be made in which the total enthalpy of all leaving streams minus the total enthalpy of all entering streams is equal to the heat absorbed from external sources by the process. In using the heat-balance method, it is necessary to make a corresponding mass balance, since the heat effects are related to the quantities of solids produced through the heat of crystallization. The advantage of the enthalpy-concentration-diagram method is that both heat and mass effects are taken into account simultaneously. This method has limited use because of the difficulty in obtaining enthalpy-concentration data. This information has been published for only a few systems.

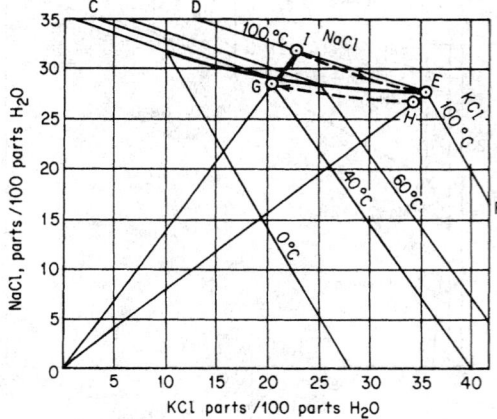

FIG. 18-57   Phase diagram, KCl – NaCl – H_2O. K = °C + 273.2.

With compounds whose solubility increases with increasing temperature there is an absorption of heat when the compound dissolves. In compounds with decreasing solubility as the temperature increases, there is an evolution of heat when solution occurs. When there is no change in solubility with temperature, there is no heat effect. The solubility curve will be continuous as long as the solid substance of a given phase is in contact with the solution, and any sudden change in the slope of the curve will be accompanied by a change in the heat of solution and a change in the solid phase. Heats of solution are generally reported as the change in enthalpy associated with the dissolution of a large quantity of solute in an excess of pure solvent. Tables showing the heats of solution for various compounds are given in Sec. 2.

At equilibrium the heat of crystallization is equal and opposite in sign to the heat of solution. Using the heat of solution at infinite dilution as equal but opposite in sign to the heat of crystallization is equivalent to neglecting the heat of dilution. With many materials the heat of dilution is small in comparison with the heat of solution and the approximation is justified; however, there are exceptions. Relatively large heat effects are usually found in the crystallization of hydrated salts. In such cases the total heat released by this effect may be a substantial portion of the total heat effects in a cooling-type crystallizer. In evaporative-type crystallizers the heat of crystallization is usually negligible when compared with the heat of vaporizing the solvent.

Yield of a Crystallization Process   In most cases the process of crystallization is slow, and the final mother liquor is in contact with a sufficiently large crystal surface so that the concentration of the mother liquor is substantially that of a saturated solution at the final temperature in the process. In such cases it is normal to calculate the yield from the initial solution composition and the solubility of the material at the final temperature. If evaporative crystallization is involved, the solvent removed must be taken into account in determining the final yield. If the crystals removed from solution are hydrated, account must be taken of the water of crystallization in the crystals, since this water is not available for retaining the solute in solution. The yield is also influenced in most plants by the removal of some mother liquor with the crystals being separated from the process. Typically, with a product separated on a centrifuge or filter, the adhering mother liquor would be in the range of 2 to 10 percent of the weight of the crystals.

The actual yield may be obtained from algebraic calculations or trial-and-error calculations when the heat effects in the process and any resultant evaporation are used to correct the initial assumptions on calculated yield. When calculations are made by hand, it is generally preferable to use the trial-and-error system, since it permits easy adjustments for relatively small deviations found in practice, such as the addition of wash water, or instrument and purge water additions. The following calculations are typical of an evaporative crystallizer precipitating a hydrated salt. If SI units are desired, kilograms = pounds × 0.454; K = (°F + 459.7)/1.8.

Example 1: Yield from a Crystallization Process   A 10,000-lb batch of a 32.5 percent $MgSO_4$ solution at 120°F is cooled without appreciable evaporation to 70°F. What weight of $MgSO_4 \cdot 7H_2O$ crystals will be formed (if it is assumed that the mother liquor leaving is saturated)?

From the solubility diagram in Fig. 18-56 at 70°F the concentration of solids is 26.3 lb $MgSO_4$ per 100-lb solution.

The mole weight of $MgSO_4$ is 120.38.

The mole weight of $MgSO_4 \cdot 7H_2O$ is 246.49.

For calculations involving hydrated salts, it is convenient to make the calculations based on the hydrated solute and the "free water."

$$0.325 \text{ weight fraction} \times \frac{246.94}{120.38} = 0.6655 \text{ } MgSO_4 \cdot 7H_2O \text{ in the feed solution}$$

$$0.263 \times \frac{246.94}{120.38} = 0.5385 \text{ } MgSO_4 \cdot 7H_2O \text{ in the mother liquor}$$

Since the free water remains constant (except when there is evaporation), the final amount of soluble $MgSO_4 \cdot 7H_2O$ is calculated by the ratio of

$$\frac{0.538 \text{ lb } MgSO_4 \cdot 7H_2O}{(1 - 0.538) \text{ lb free water}}$$

	Total	$MgSO_4 \cdot 7H_2O$	Free water	$\dfrac{MgSO_4 \cdot 7H_2O}{Free\ water}$
Feed	10,000	6655	3345	1.989
Mother liquor	7249	3904*	3345	1.167
Yield	2751	2751		

*$3345 \times (0.538/0.462) = 3904$

A formula method for calculation is sometimes used where

$$P = R\,\frac{100W_0 - S(H_0 - E)}{100 - S(R-1)}$$

where

P = weight of crystals in final magma, lb
R = mole weight of hydrate/mole weight of anhydrous = 2.04759
S = solubility at mother-liquor temperature (anhydrous basis) in lb per 100 lb solvent. $[0.263/(1 - 0.263)] \times 100 = 35.68521$
W_0 = weight of anhydrous solute in the original batch. $10,000(0.325) = 3250$ lb
H_0 = total weight of solvent at the beginning of the batch. $10,000 - 3250 = 6750$ lb
E = evaporation = 0

$$P = 2.04\,\frac{(100)(3250) - 35.7(6750)}{100 - 35.7(2.04 - 1)} = 2751\ \text{lb}$$

Note that taking the difference between large numbers in this method can increase the chance for error.

Fractional Crystallization When two or more solutes are dissolved in a solvent, it is often possible to (1) separate these into the pure components or (2) separate one and leave the other in the solution. Whether or not this can be done depends on the solubility and phase relations of the system under consideration. Normally alternative 2 is successful only when one of the components has a much more rapid change in solubility with temperature than does the other. A typical example which is practiced on a large scale is the separation of KCl and NaCl from water solution. A phase diagram for this system is shown in Fig. 18-57. In this case the solubility of NaCl is plotted on the Y axis in parts per 100 parts of water, and the solubility of KCl is plotted on the X axis. The isotherms show a marked decrease in solubility for each component as the amount of the other is increased. This is typical for most inorganic salts. As explained earlier, the mixed-salt line is CE, and to make a separation of the solutes into the pure components it is necessary to be on one side of this line or the other. Normally a 95 to 98 percent approach to this line is possible. When evaporation occurs during a cooling or concentration process, this can be represented by movement away from the origin on a straight line through the origin. Dilution by water is represented by movement in the opposite direction.

A typical separation might be represented as follows: Starting at E with a saturated brine at 100°C a small amount of water is added to dissolve any traces of solid phase present and to make sure the solids precipitated initially are KCl. Evaporative cooling along line HG results in the precipitation of KCl. During this evaporative cooling, part of the water evaporated must be added back to the solution to prevent the coprecipitation of NaCl. The final composition at G can be calculated by the NaCl/KCl/H₂O ratios and the known amount of NaCl in the incoming solution at E. The solution at point G may be concentrated by evaporation at 100°C. During this process the solution will increase in concentration with respect to both components until point I is reached. Then NaCl will precipitate, and the solution will become more concentrated in KCl, as indicated by the line IE, until the original point E is reached. If concentration is carried beyond point E, a mixture of KCl and NaCl will precipitate.

Example 2: Yield from Evaporative Cooling Starting with 1000 lb of water in a solution at H on the solubility diagram in Fig. 18-57, calculate the yield on evaporative cooling and concentrate the solution back to point H so the cycle can be repeated, indicating the amount of NaCl precipitated and the evaporation and dilution required at the different steps in the process.

In solving problems of this type, it is convenient to list the material balance and the solubility ratios. The various points on the material balance are calculated by multiplying the quantity of the component which does not precipitate from solution during the transition from one point to another (normally the

NaCl in cooling or the KCl in the evaporative step) by the solubility ratio at the next step, illustrated as follows:

Basis. 1000 lb of water at the initial conditions.

				Solubility ratios		
Solution component	KCl	NaCl	Water	KCl	NaCl	Water
H	343	270	1000	34.3	27.0	100
G(a)	194	270	950	20.4	28.4	100
KCl yield	149					
Net evaporation			50			
I(b)	194	270	860	22.6	31.4	100
E(c)	194	153	554	35.0	27.5	100
NaCl yield		117				
Evaporation			306			
Dilution			11			
H'	194	153	565	34.3	27.0	−100

The calculations for these steps are:

a. 270 lb NaCl (100 lb water/28.4 lb NaCl) = 950 lb water
   950 lb water (20.4 lb KCl/100 lb water) = 194 lb KCl
b. 270 lb NaCl (100 lb water/31.4 lb NaCl) = 860 lb water
   860 lb water (22.6 lb KCl/100 lb water) = 194 lb KCl
c. 194 lb KCl (100 lb water/35.0 lb KCl) = 554 lb water
   554 lb water (27.5 lb NaCl/100 lb water) = 153 lb NaCl

Note that during the cooling step the maximum amount of evaporation which is permitted by the material balance is 50 lb for the step shown. In an evaporative-cooling step, however, the actual evaporation which results from adiabatic cooling is more than this. Therefore, water must be added back to prevent the NaCl concentration from rising too high; otherwise, coprecipitation of NaCl will occur.

Inasmuch as only mass ratios are involved in these calculations, kilograms or any other unit of mass may be substituted for pounds without affecting the validity of the example.

Although the figures given are for a step-by-step process, it is obvious that the same techniques will apply to a continuous system if the fresh feed containing KCl and NaCl is added at an appropriate part of the cycle, such as between steps G and I for the case of dilute feed solutions.

Another method of fractional crystallization, in which advantage is taken of different crystallization rates, is sometimes used. Thus, a solution saturated with borax and potassium chloride will, in the absence of borax seed crystals, precipitate only potassium chloride on rapid cooling. The borax remains behind as a supersaturated solution, and the potassium chloride crystals can be removed before the slower borax crystallization starts.

Crystal Formation There are obviously two steps involved in the preparation of crystal matter from a solution. The crystals must first form and then grow. The formation of a new solid phase either on an inert particle in the solution or in the solution itself is called **nucleation.** The increase in size of this nucleus with a layer-by-layer addition of solute is called **growth.** Both nucleation and crystal growth have supersaturation as a common driving force. Unless a solution is supersaturated, crystals can neither form nor grow. Supersaturation refers to the quantity of solute present in solution compared with the quantity which would be present if the solution were kept for a very long period of time with solid phase in contact with the solution. The latter value is the equilibrium solubility at the temperature and pressure under consideration. The supersaturation coefficient can be expressed

$$S = \frac{\text{parts solute/100 parts solvent}}{\text{parts solute at equilibrium/100 parts solvent}} \geq 1.0 \quad (18\text{-}22)$$

Solutions vary greatly in their ability to sustain measurable amounts of supersaturation. With some materials, such as sucrose, it is possible to develop a supersaturation coefficient of 1.4 to 2.0 with little danger of nucleation. With some common inorganic solutions such as sodium chloride in water, the amount of supersaturation which can be generated stably is so small that it is difficult or impossible to measure.

Certain qualitative facts in connection with supersaturation, growth, and the yield in a crystallization process are readily apparent.

If the concentration of the initial solution and the final mother liquor are fixed, the total weight of the crystalline crop is also fixed if equilibrium is obtained. The particle-size distribution of this weight, however, will depend on the relationship between the two processes of nucleation and growth. Considering a given quantity of solution cooled through a fixed range, if there is considerable nucleation initially during the cooling process, the yield will consist of many small crystals. If only a few nuclei form at the start of the precipitation and the resulting yield occurs uniformly on these nuclei without secondary nucleation, a crop of large uniform crystals will result. Obviously, many intermediate cases of varying nucleation rates and growth rates can also occur, depending on the nature of the materials being handled, the rate of cooling, agitation, and other factors.

When a process is continuous, nucleation frequently occurs in the presence of a seeded solution by the combined effects of mechanical stimulus and nucleation caused by supersaturation (heterogeneous nucleation). If such a system is completely and uniformly mixed (i.e., the product stream represents the typical magma circulated within the system) and if the system is operating at steady state, the particle-size distribution has definite limits which can be predicted mathematically with a high degree of accuracy, as will be shown later in this section.

Geometry of Crystal Growth Geometrically a crystal is a solid bounded by planes. The shape and size of such a solid are functions of the interfacial angles and of the linear dimension of the faces. As the result of the constancy of its interfacial angles, each face of a growing or dissolving crystal, as it moves away from or toward the center of the crystal, is always parallel to its original position. This concept is known as the "principle of the parallel displacement of faces." The rate at which a face moves in a direction perpendicular to its original position is called the translation velocity of that face or the rate of growth of that face.

From the industrial point of view, the term **crystal habit** or **crystal morphology** refers to the relative sizes of the faces of a crystal. The crystal habit is determined by the internal structure and external influences on the crystal such as the growth rate, solvent used, and impurities present during the crystallization growth period. The crystal habit of commercial products is of very great importance. Long, needlelike crystals tend to be easily broken during centrifugation and drying. Flat, platelike crystals are very difficult to wash during filtration or centrifugation and result in relatively low filtration rates. Complex or twinned crystals tend to be more easily broken in transport than chunky, compact crystal habits. Rounded or spherical crystals (caused generally by attrition during growth and handling) tend to give considerably less difficulty with caking than do cubical or other compact shapes.

Internal structure can be different in crystals that are chemically identical, even though they may be formed at different temperatures and have a different appearance. This is called **polymorphism** and can be determined only by X-ray diffraction. For the same internal structure, very small amounts of foreign substances will often completely change the crystal habit. The selective adsorption of dyes by different faces of a crystal or the change from an alkaline to an acidic environment will often produce pronounced changes in the crystal habit. The presence of other soluble anions and cations often has a similar influence. In the crystallization of ammonium sulfate, the reduction in soluble iron to below 50 ppm of ferric ion is sufficient to cause significant change in the habit of an ammonium sulfate crystal from a long, narrow form to a relatively chunky and compact form. Additional information is available in the patent literature and Table 18-4 lists some of the better-known additives and their influences.

Since the relative sizes of the individual faces of a crystal vary between wide limits, it follows that different faces must have different translational velocities. A geometric law of crystal growth known as the **overlapping principle** is based on those velocity differences: in growing a crystal, only those faces having the lowest translational velocities survive; and in dissolving a crystal, only those faces having the highest translational velocities survive.

For example, consider the cross sections of a growing crystal as in Fig. 18-58. The polygons shown in the figure represent varying stages in the growth of the crystal. The faces marked A are slow-growing faces (low translational velocities), and the faces marked B are fast-growing (high translational velocities). It is apparent from Fig. 18-58 that the faster B faces tend to disappear as they are overlapped by the slower A faces.

Hartman and Perdok (1955) predicted that crystal habit or crystal morphology was related to the internal structure based on energy considerations and speculated it should be possible to predict the growth shape of crystals from the slice energy of different flat faces. Later, Hartman was able to predict the calculated attachment energy for various crystal species. Recently computer programs have been developed that predict crystal morphology from attachment energies. These techniques are particularly useful in dealing with organic or molecular crystals and rapid progress in this area is being made by companies such as Molecular Simulations of Cambridge, England.

Purity of the Product If a crystal is produced in a region of the phase diagram where a single-crystal composition precipitates, the crystal itself will normally be pure provided that it is grown at relatively low rates and constant conditions. With many products these purities approach a value of about 99.5 to 99.8 percent. The difference between this and a purity of 100 percent is generally the result of small pockets of mother liquor called occlusions trapped within the crystal. Although frequently large enough to be seen with an ordinary microscope, these occlusions can be submicroscopic and represent dislocations within the structure of the crystal. They can be caused by either attrition or breakage during the growth process or by slip planes within the crystal structure caused by interference between screw-type dislocations and the remainder of the crystal faces. To increase the purity of the crystal beyond the point where such occlusions are normally expected (about 0.1 to 0.5 percent by volume), it is generally necessary to reduce the impurities in the mother liquor itself to an acceptably low level so that the mother liquor contained within these occlusions will not contain sufficient impurities to cause an impure product to be formed. It is normally necessary to recrystallize material from a solution which is relatively pure to surmount this type of purity problem.

In addition to the impurities within the crystal structure itself, there is normally an adhering mother-liquid film left on the surface of the crystal after separation in a centrifuge or on a filter. Typically a centrifuge may leave about 2 to 10 percent of the weight of the crystals as adhering mother liquor on the surface. This varies greatly with the size and shape or habit of the crystals. Large, uniform crystals precipitated from low-viscosity mother liquors will retain a minimum of mother liquor, while nonuniform or small crystals precipitated from viscous solutions will retain a considerably larger proportion. Comparable statements apply to the filtration of crystals, although normally the amounts of mother liquor adhering to the crystals are considerably larger. It is common practice when precipitating materials from solutions which contain appreciable quantities of impurities to wash the crystals on the centrifuge or filter with either fresh solvent or feed solution. In principle, such washing can reduce the impurities quite substantially. It is also possible in many cases to reslurry the crystals in fresh solvent and recentrifuge the product in an effort to obtain a longer residence time during the washing operation and better mixing of the wash liquors with the crystals.

Coefficient of Variation One of the problems confronting any user or designer of crystallization equipment is the expected particle-size distribution of the solids leaving the system and how this distribution may be adequately described. Most crystalline-product distributions plotted on arithmetic-probability paper will exhibit a straight line for a considerable portion of the plotted distribution. In this type of plot the particle diameter should be plotted as the ordinate and the cumulative percent on the log-probability scale as the abscissa.

It is common practice to use a parameter characterizing crystal-size distribution called the coefficient of variation. This is defined as follows:

$$CV = 100 \, \frac{PD_{16\%} - PD_{84\%}}{2PD_{50\%}} \qquad (18\text{-}23)$$

where CV = coefficient of variation, as a percentage
PD = particle diameter from intercept on ordinate axis at percent indicated

TABLE 18-4 Some Impurities Known to Be Habit Modifiers

Material crystallized	Additive(s)	Effect	Concentration	References
$Ba(NO_2)^2$	Mg, Te^{+4}	Helps growth	—	1
$CaSO_4\cdot2H_2O$	Citric, succinic, tartaric acids	Helps growth	Low	
	Sodium citrate	Forms prisms	—	5
$CuSO_4\cdot5H_2O$	H_2SO_4	Chunky crystals	0.3%	5
KCl	$K_4Fe(CN)_6$	Inhibits growth, dendrites	1000 ppm	4
	Pb, Bi, Sn^{+2}, Ti, Zr, Th, Cd, Fe, Hg, Mg	Helps growth	Low	1
$KClO_4$	Congo red (dye)	Modifies the 102 face	50 ppm	6
K_2CrO_4	Acid magenta (dye)	Modifies the 010 face	50 ppm	6
KH_2PO_4	$Na_2B_4O_7$	Aids growth	—	1
KNO_2	Fe	Helps growth	Low	1
KNO_3	Acid magenta (dye)	Tabular crystals		7
	Pb, Th, Bi	Helps growth	Low	1
K_2SO_4	Acid magenta (dye)	Forms plates	2000 ppm	6
	Cl, Mn, Mg, Bi, Cu, Al, Fe	Helps growth	Low	1
	Cl_3	Reduces growth rate	1000 ppm	4
	$(NH_4)_3Ce(NO_3)_6$	Reduces growth rate	1000 ppm	4
$LiCl\cdot H_2O$	Cr·Mn^{+2}, Sn^{+2}, Co, Ni, Fe^{+3}	Helps growth	Low	1
$MgSO_4\cdot7H_2O$	Borax	Aids growth	5%	1
$Na_2B_4O_7\cdot10H_2O$	Sodium oleate	Reduces growth & nuc.	5 ppm	
	Casein, gelatin	Promotes flat crystals	—	2, 5
	NaOH, Na_2CO_3	Promotes chunky crystals	—	
$Na_2CO_3\cdot H_2O$	$SO_4^=$	Reduces L/D ratio	0.1–1.0%	Canadian Patent 812,685
	Ca^{+2} and Mg^{+2}	Increase bulk density	400 ppm	U.S. Patent 3,459,497
$NaCO_3\cdot NaHCO_3\cdot2H_2O$	D-40 detergent	Aids growth	20 ppm	U.S. Patent 3,233,983
NaCl	$Na_4Fe(CN)_6$, CdBr	Forms dendrites	100 ppm	4
	Pb, Mn^{+2}, Bi, Sn^{+2}, Ti, Fe, Hg	Helps growth	Low	1
	Urea, formamide	Forms octahedra	Low	2
	Tetraalkyl ammon. salts	Helps growth & hardness	1–100 ppm	U.S. Patent 3,095,281
	Polyethylene-oxy compounds	Helps growth & hardness	—	U.S. Patent 3,000,708
$NaClO_3$	Na_2SO_4, $NaClO_4$	Tetrahedrons	—	3
$NaNO_3$	Acid green (dye)	Flattened rhombahedra		7
Na_2SO_4	NH_4SO_4 @ pH 6.5	Large single crystals	Low	
	$CdCl_2$	Inhibits growth	1000 ppm	4
	Alkyl aryl sulfonates	Aids growth	—	2
	Calgon	Aids growth	100 ppm	
NH_4Cl	Mn, Fe, Cu, Co, Ni, Cr	Aid growth	Low	1
	Urea	Forms octahedra		5
NH_4ClO_4	Azurine (dye)	Modifies the 102 face	22 ppm	6
NF_4F	Ca	Helps growth	Low	1
$(NH_4)NO_3$	Acid magenta (dye)	Forms 010 face plates	1%	6
$(NH_4)_2HPO_4$	H_2SO_4	Reduces L/D ratio	7%	
$NH_4H_2PO_4$	Fe^{+3}, Cr, Al, Sn	Helps growth	Traces	1
$(NH_4)_2SO_4$	Cr^{+3}, Fe^{+3}, Al^{+3}	Promotes needles	50 ppm	
	H_2SO_4	Promotes needles	2–6%	U.S. Patent 2,092,073
	Oxalic acid, citric acid	Promotes chunky crystals	1000 ppm	U.S. Patent 2,228,742
	H_3PO_4, SO_2	Promotes chunky crystals	1000 ppm	
$ZnSO_4\cdot7H_2O$	Borax	Aids growth	—	1
Adipic acid	Surfactant-SDBS	Aids growth	50–100 ppm	2
Fructose	Glucose, difructose	Affects growth		8
L-asparagine	L-glutamic acid	Affects growth		8
Naphthalene	Cyclohexane (solvent)	Forms needles	—	2
	Methanol (solvent)	Forms plates		
Pentaerythritol	Sucrose	Aids growth	—	1
	Acetone (solvent)	Forms plates		2
Sodium glutamate	Lysine, CaO	Affects growth		8
Sucrose	Raffinose, KCl, NaBr	Modify growth rate		
Urea	Biuret	Reduces L/D & aids growth	2–7%	
	NH_4Cl	Reduces L/D & aids growth	5–10%	

1. Gillman, *The Art and Science of Growing Crystals,* Wiley, New York, 1963.
2. Mullin, *Crystallization,* Butterworth, London, 1961.
3. Buckley, *Crystal Growth,* Wiley, New York, 1961.
4. Phoenix, L., *British Chemical Engineering,* vol. II, no. 1 (Jan. 1966), pp. 34–38.
5. Garrett, D. E., *British Chemical Engineering,* vol. I, no. 12 (Dec. 1959), pp. 673–677.
6. Buckley, *Crystal Growth,* (Faraday Soc.) Butterworths, 1949, p. 249.
7. Butchart and Whetstone, *Crystal Growth,* (Faraday Soc.) Butterworths, 1949, p. 259.
8. Nyvlt, J., *Industrial Crystallization,* Verlag Chemie Publishers, New York, 1978, pp. 26–31.

In order to be consistent with normal usage, the particle-size distribution when this parameter is used should be a straight line between approximately 10 percent cumulative weight and 90 percent cumulative weight. By giving the coefficient of variation and the mean particle diameter, a description of the particle-size distribution is obtained which is normally satisfactory for most industrial purposes. If the product is removed from a mixed-suspension crystallizer, this coefficient of variation should have a value of approximately 50 percent (Randolph and Larson, op. cit., chap. 2).

Crystal Nucleation and Growth

Rate of Growth Crystal growth is a layer-by-layer process, and since growth can occur only at the face of the crystal, material must be transported to that face from the bulk of the solution. Diffusional

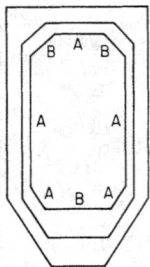

FIG. 18-58   Overlapping principle.

resistance to the movement of molecules (or ions) to the growing crystal face, as well as the resistance to integration of those molecules into the face, must be considered. As discussed earlier, different faces can have different rates of growth, and these can be selectively altered by the addition or elimination of impurities.

If L is a characteristic dimension of a crystal of selected material and shape, the rate of growth of a crystal face that is perpendicular to L is, by definition,

$$G \equiv \lim_{\Delta L \to O} \frac{\Delta L}{\Delta t} = \frac{dL}{dt} \qquad (18\text{-}24)$$

where G is the growth rate over time internal t. It is customary to measure G in the practical units of millimeters per hour. It should be noted that growth rates so measured are actually twice the facial growth rate.

The delta L law.   It has been shown by McCabe [*Ind. Eng. Chem.,* **21**, 30, 112 (1929)] that all geometrically similar crystals of the same material suspended in the same solution grow at the same rate if growth rate is defined as in Eq. (18-24). The rate is independent of crystal size, provided that all crystals in the suspension are treated alike. This generalization is known as the delta L law. Although there are some well-known exceptions, they usually occur when the crystals are very large or when movement of the crystals in the solution is so rapid that substantial changes occur in diffusion-limited growth of the faces.

It is emphasized that the delta L law does not apply when similar crystals are given preferential treatment based on size. It fails also when surface defects or dislocations significantly alter the growth rate of a crystal face. Nevertheless, it is a reasonably accurate generalization for a surprising number of industrial cases. When it is, it is important because it simplifies the mathematical treatment in modeling real crystallizers and is useful in predicting crystal-size distribution in many types of industrial crystallization equipment.

Important exceptions to McCabe's growth-rate model have been noted by Bramson, by Randolph, and by Abegg. These are discussed by Canning and Randolph, *Am. Inst. Chem. Eng. J.,* **13**, 5 (1967).

Nucleation   The mechanism of crystal nucleation from solution has been studied by many scientists, and recent work suggests that—in commercial crystallization equipment, at least—the nucleation rate is the sum of contributions by (1) homogeneous nucleation and (2) nucleation due to contact between crystals and (a) other crystals, (b) the walls of the container, and (c) the pump impeller. If B^0 is the net number of new crystals formed in a unit volume of solution per unit of time,

$$B^0 = B_{ss} + B_e + B_c \qquad (18\text{-}25)$$

where B_e is the rate of nucleation due to crystal-impeller contacts, B_c is that due to crystal-crystal contacts, and B_{ss} is the homogeneous nucleation rate due to the supersaturation driving force. The mechanism of the last-named is not precisely known, although it is obvious that molecules forming a nucleus not only have to coagulate, resisting the tendency to redissolve, but also must become oriented into a fixed lattice. The number of molecules required to form a stable crystal nucleus has been variously estimated at from 80 to 100 (with ice), and the probability that a stable nucleus will result from the simultaneous collision of that large number is extremely low unless the supersatura-

tion level is very high or the solution is supersaturated in the absence of agitation. In commercial crystallization equipment, in which supersaturation is low and agitation is employed to keep the growing crystals suspended, the predominant mechanism is contact nucleation or, in extreme cases, attrition.

In order to treat crystallization systems both dynamically and continuously, a mathematical model has been developed which can correlate the nucleation rate to the level of supersaturation and/or the growth rate. Because the growth rate is more easily determined and because nucleation is sharply nonlinear in the regions normally encountered in industrial crystallization, it has been common to assume

$$B^0 = ks^i \qquad (18\text{-}26)$$

where s, the supersaturation, is defined as $(C - C_s)$, C being the concentration of the solute and C_s its saturation concentration; and the exponent i and dimensional coefficient k are values characteristic of the material.

While Eq. (18-26) has been popular among those attempting correlations between nucleation rate and supersaturation, recently it has become commoner to use a derived relationship between nucleation rate and growth rate by assuming that

$$G = k's \qquad (18\text{-}27)$$

whence, in consideration of Eq. (18-26),

$$B^0 = k''G^i \qquad (18\text{-}28)$$

where the dimensional coefficient k' is characteristic of the material and the conditions of crystallization; and $k'' = k/(k')^i$. Feeling that a model in which nucleation depends only on supersaturation or growth rate is simplistically deficient, some have proposed that contact nucleation rate is also a power function of slurry density and that

$$B^0 = k_n G^i M_T^j \qquad (18\text{-}29)$$

where M_T is the density of the crystal slurry, g/L.

Although Eqs. (18-28) and (18-29) have been adopted by many as a matter of convenience, they are oversimplifications of the very complex relationship that is suggested by Eq. (18-25); Eq. (18-29) implicitly and quite arbitrarily combines the effects of homogeneous nucleation and those due to contact nucleation. They should be used only with caution.

In work pioneered by Clontz and McCabe [*Chem. Eng. Prog. Symp. Ser.,* **67**(110), 6 (1971)] and subsequently extended by others, contact nucleation rate was found to be proportional to the input of energy of contact, as well as being a function of contact area and supersaturation. This observation is important to the scaling up of crystallizers: at laboratory or bench scale, contact energy level is relatively low and homogeneous nucleation can contribute significantly to the total rate of nucleation; in commercial equipment, on the other hand, contact energy input is intense and contact nucleation is the predominant mechanism. Scale-up modeling of a crystallizer, therefore, must include its mechanical characteristics as well as the physiochemical driving force.

Nucleation and Growth   From the preceding, it is clear that no analysis of a crystallizing system can be truly meaningful unless the simultaneous effects of nucleation rate, growth rate, heat balance, and material balance are considered. The most comprehensive treatment of this subject is by Randolph and Larson (op. cit.), who developed a mathematical model for continuous crystallizers of the mixed-suspension or circulating-magma type [*Am. Inst. Chem. Eng. J.,* **8**, 639 (1962)] and subsequently examined variations of this model that include most of the aberrations found in commercial equipment. Randolph and Larson showed that when the total number of crystals in a given volume of suspension from a crystallizer is plotted as a function of the characteristic length as in Fig. 18-59, the slope of the line is usefully identified as the crystal population density, n:

$$n = \lim_{\Delta L \to O} \frac{\Delta N}{\Delta L} = \frac{dN}{dL} \qquad (18\text{-}30)$$

where N = total number of crystals up to size L per unit volume of magma. The population density thus defined is useful because it char-

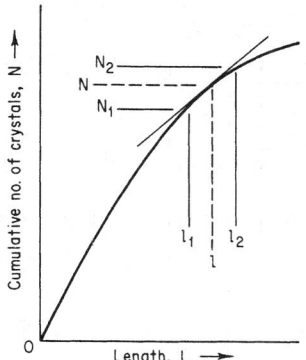

FIG. 18-59 Determination of the population density of crystals.

acterizes the nucleation-growth performance of a particular crystallization process or crystallizer.

The data for a plot like Fig. 18-60 are easily obtained from a screen analysis of the total crystal content of a known volume (e.g., a liter) of magma. The analysis is made with a closely spaced set of testing sieves, as discussed in Sec. 19, Table 19-6, the cumulative number of particles smaller than each sieve in the nest being plotted against the aperture dimension of that sieve. The fraction retained on each sieve is weighed, and the mass is converted to the equivalent number of particles by dividing by the calculated mass of a particle whose dimension is the arithmetic mean of the mesh sizes of the sieve on which it is retained and the sieve immediately above it.

In industrial practice, the size-distribution curve usually is not actually constructed. Instead, a mean value of the population density for any sieve fraction of interest (in essence, the population density of the particle of average dimension in that fraction) is determined directly as $\Delta N / \Delta L$, ΔN being the number of particles retained on the sieve and ΔL being the difference between the mesh sizes of the retaining sieve and its immediate predecessor. It is common to employ the units of $(\text{mm} \cdot L)^{-1}$ for n.

For a steady-state crystallizer receiving solids-free feed and containing a well-mixed suspension of crystals experiencing negligible breakage, a material-balance statement degenerates to a particle balance (the Randolph-Larson general-population balance); in turn, it simplifies to

$$\frac{dn}{dL} + \frac{n}{Gt} = 0 \qquad (18\text{-}31)$$

if the delta L law applies (i.e., G is independent of L) and the drawdown (or retention) time is assumed to be invariant and calculated as $t = V/Q$. Integrated between the limits n^0, the population density of nuclei (for which L is assumed to be zero), and n, that of any chosen crystal size L, Eq. (18-31) becomes

$$\int_{n^0}^{n} \frac{dn}{n} = -\int_{0}^{L} \frac{dL}{Gt} \qquad (18\text{-}32)$$

$$\ln n = \frac{-L}{Gt} + \ln n^0 \qquad (18\text{-}33a)$$

or

$$n = n^0 e^{-L/Gt} \qquad (18\text{-}33b)$$

A plot of $\ln n$ versus L is a straight line whose intercept is $\ln n^0$ and whose slope is $-1/Gt$. (For plots on base-10 log paper, the appropriate slope correction must be made.) Thus, from a given product sample of known slurry density and retention time it is possible to obtain the nucleation rate and growth rate for the conditions tested if the sample satisfies the assumptions of the derivation and yields a straight line. A number of derived relations which describe the nucleation rate, size distribution, and average properties are summarized in Table 18-5.

If a straight line does *not* result (Fig. 18-60), at least part of the explanation may be violation of the delta L law (Canning and Ran-

dolph, loc. cit.). The best current theory about what causes size-dependent growth suggests what has been called growth dispersion or "Bujacian behavior" [Mullen (ed.), op. cit., p. 23]. In the same environment different crystals of the same size can grow at different rates owing to differences in dislocations or other surface effects. The graphs of "slow" growers (Fig. 18-60, curve A) and "fast" growers (curve B) sum to a resultant line (curve C), concave upward, that is described by Eq. (18-34) (Randolph, in deJong and Jancic, op. cit., p. 295):

$$n = \sum \frac{B^0 i}{G_i} e^{(-L/G_i t)} \qquad (18\text{-}34)$$

Equation (18-31) contains no information about the crystallizer's influence on the nucleation rate. If the crystallizer is of a mixed-suspension, mixed-product-removal (MSMPR) type, satisfying the criteria for Eq. (18-31), and if the model of Clontz and McCabe is valid, the contribution to the nucleation rate by the circulating pump can be calculated [Bennett, Fiedelman, and Randolph, *Chem. Eng. Prog.*, **69**(7), 86 (1973)]:

$$B_e = K_e \left(\frac{I^2}{P} \right) \rho G \int_{0}^{\infty} n L^4 \, dL \qquad (18\text{-}35)$$

where I = tip speed of the propeller or impeller, m/s
ρ = crystal density, g/cm³
P = volume of crystallizer/circulation rate (turnover), m³/(m³/s) = s

Since the integral term is the fourth moment of the distribution (m_4), Eq. (18-35) becomes

$$B_e = K_e \rho G \left(\frac{I^2}{P} \right) m_4 \qquad (18\text{-}36)$$

Equation (18-36) is the general expression for impeller-induced nucleation. In a fixed-geometry system in which only the speed of the circulating pump is changed and in which the flow is roughly proportional to the pump speed, Eq. (18-36) may be satisfactorily replaced with

$$B_e = K_e'' \rho G(S_R)^3 m_4 \qquad (18\text{-}37)$$

where S_R = rotation rate of impeller, r/min. If the maximum crystal-impeller impact stress is a nonlinear function of the kinetic energy, shown to be the case in at least some systems, Eq. (18-37) no longer applies.

In the specific case of an MSMPR exponential distribution, the fourth moment of the distribution may be calculated as

$$m_4 = 4! n^0 (Gt)^5 \qquad (18\text{-}38)$$

Substitution of this expression into Eq. (18-36) gives

$$B_e = k n^0 G(S_R)^3 L_D^5 \qquad (18\text{-}39)$$

where $L_D = 3Gt$, the dominant crystal (mode) size.

Equation (18-39) displays the competing factors that stabilize secondary nucleation in an operating crystallizer when nucleation is due mostly to impeller-crystal contact. Any increase in particle size produces a fifth-power increase in nucleation rate, tending to counteract the direction of the change and thereby stabilizing the crystal-size distribution. From dimensional argument alone the size produced in a mixed crystallizer for a (fixed) nucleation rate varies as $(B^0)^{1/3}$. Thus, this fifth-order response of contact nucleation does not wildly upset the crystal size distribution but instead acts as a stabilizing feedback effect.

Nucleation due to crystal-to-crystal contact is greater for equal striking energies than crystal-to-metal contact. However, the viscous drag of the liquid on particle sizes normally encountered limits the velocity of impact to extremely low values. The assumption that only the largest crystal sizes contribute significantly to the nucleation rate by crystal-to-crystal contact permits a simple computation of the rate:

$$B_c = K_c \rho G m_j^2 \qquad (18\text{-}40)$$

where m_j = the fourth, fifth, sixth, or higher moments of the distribution.

A number of different crystallizing systems have been investigated by using the Randolph-Larson technique, and some of the published

TABLE 18-5 Common Equations for Population-Balance Calculations

Name	Symbol	Units	Systems without fines removal	Systems with fines removal		References
				Fines stream	Product stream	
Drawdown time (retention time)	t	h	$t = V/Q$	$t_F = V_{liquid}/Q_F$	$t = V/Q$	
Growth rate	G	mm/h	$G = dL/dt$	$G = dL/dt$	$G = dL/dt$	
Volume coefficient	K_v	1/no. (crystals)	$K_v = \dfrac{\text{volume of one crystal}}{L^3}$	$K_v = \dfrac{\text{volume of one crystal}}{L^3}$	$K_v = \dfrac{\text{volume of one crystal}}{L^3}$	
Population density	n	No. (crystals)/mm	$n = dN/dL$	$n = dN/dL$	$n = dN/dL$	1
Nuclei population density	n^o	No. (crystals)/mm	$n^o = K_M M^i G^{-1}$			2
Population density	n	No. (crystals)/mm	$n = n^o e^{-L/Gt}$	$n_F = n^o e^{-L/Gt_F}$	$n = n^{o-L/Gt_F} e^{-L/Gt}$	1, 3
Nucleation rate	B_0	No. (crystals)/h	$B_0 = Gn^o = K_M M^i G^i$	$B_0 = G_n^o$		4
Dimensionless length	x	None	$x = \dfrac{L}{Gt}$	$x_F = \dfrac{L}{Gt_F},\ L_o \to L_f$	$x = \dfrac{L}{Gt},\ L_f \to L$	1
Mass/unit volume (slurry density)	M_T	g/L	$M_t = K_v\rho \displaystyle\int_0^\infty nL^3\, dL$ $M_t = K_v\rho 6\, n^o (Gt)^4$	$M_{T_F} = K_v\rho \displaystyle\int_0^{L_f} n^o e^{-L/Gt_F} L^3\, dL$	$M_T = K_v\rho \displaystyle\int_{L_f}^\infty n^o e^{-L/Gt_F} e^{-L/Gt} L^3\, dL$	1
Cumulative mass to x — Total mass	W_x	None	$W_x = 1 - e^{-x}\left(\dfrac{x^3}{6} + \dfrac{x^2}{2} + x + 1\right)$	$W_F = \dfrac{e^{-x}(x^3 + 3x^2 + 6x + 6) - 6}{e^{-c}(x_c^3 + 3x_c^2 + 6x_c + 6) - 6}$	$W = \dfrac{6K_v\rho n^o e^{-L/Gt_F}(Gt)^4 \left[1 - e^{-x}\left(\dfrac{x^3}{6} + \dfrac{x^2}{2} + x + 1\right)\right]}{\text{Slurry density } M,\ g/L}$ when $L_c \approx 0$, compared with L_a	5
Dominant particle	L_d	mm	$L_d = 3Gt$			
Average particle, weight	L_a	mm	$L_a = 3.67Gt$			6
Total number of crystals	N_T	No./L	$N_T = \displaystyle\int_0^\infty n\, dl$	$N_F = \displaystyle\int_0^{L_f} n_F\, dL$	$N_T = \displaystyle\int_{L_f}^\infty n\, dL$	1, 3

1. Randolph and Larson, *Am. Inst. Chem. Eng. J.*, **8**, 639 (1962).
2. Timm and Larson, *Am. Inst. Chem. Eng. J.*, **14**, 452 (1968).
3. Larson, private communication.
4. Larson, Timm, and Wolff, *Am. Inst. Chem. Eng. J.*, **14**, 448 (1968).
5. Larson and Randolph, *Chem. Eng. Prog. Symp. Ser.*, **65**(95), 1 (1969).
6. Schoen, *Ind. Eng. Chem.*, **53**, 607 (1961).

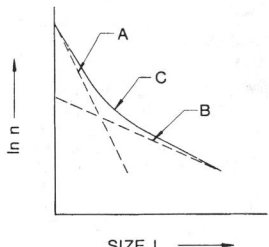

FIG. 18-60 Population density of crystals resulting from Bujacian behavior.

growth rates and nucleation rates are included in Table 18-6. Although the usefulness of these data is limited to the conditions tested, the table gives a range of values which may be expected, and it permits resolution of the information gained from a simple screen analysis into the fundamental factors of growth rate and nucleation rate. Experiments may then be conducted to determine the independent effects of operation and equipment design on these parameters.

Although this procedure requires laborious calculations because of the number of samples normally needed, these computations and the determination of the best straight-line fit to the data are readily programmed for digital computers.

Example 3: Population, Density, Growth and Nucleation Rate
Calculate the population density, growth, and nucleation rates for a crystal sample of urea for which there is the following information. These data are from Bennett and Van Buren [*Chem. Eng. Prog. Symp. Ser.*, **65**(95), 44 (1969)].

Slurry density = 450 g/L
Crystal density = 1.335 g/cm^3
Drawdown time t = 3.38 h
Shape factor k_v = 1.00

Product size:

−14 mesh, +20 mesh	4.4 percent
−20 mesh, +28 mesh	14.4 percent
−28 mesh, +35 mesh	24.2 percent
−35 mesh, +48 mesh	31.6 percent
−48 mesh, +65 mesh	15.5 percent
−65 mesh, +100 mesh	7.4 percent
−100 mesh	2.5 percent

n = number of particles per liter of volume
14 mesh = 1.168 mm, 20 mesh = 0.833 mm, average opening 1.00 mm
Size span = 0.335 mm = ΔL

$$n_{20} = \frac{(450 \text{ g/L})(0.044)}{(1.335/1000) \text{ g/mm}^3(1.00^3 \text{ mm}^3/\text{particle})(0.335 \text{ mm})(1.0)}$$

$$n_{20} = 44{,}270$$

$$\ln n_{20} = 10.698$$

Repeating for each screen increment:

Screen size	Weight, %	k_v	$\ln n$	L, average diameter, mm
100	7.4	1.0	18.099	0.178
65	15.5	1.0	17.452	0.251
48	31.6	1.0	16.778	0.356
35	24.2	1.0	15.131	0.503
28	14.4	1.0	13.224	0.711
20	4.4	1.0	10.698	1.000

Plotting $\ln n$ versus L as shown in Fig. 18-61, a straight line having an intercept at zero length of 19.781 and a slope of −9.127 results. As mentioned in discussing Eq. (18-24), the growth rate can then be found.

Slope = $-1/Gt$ or $-9.127 = -1/[G(3.38)]$

or $\quad G = 0.0324$ mm/h

and $\quad B_0 = Gn^0 = (0.0324)(e^{19.781}) = 12.65 \times 10^6 \dfrac{n^0}{\text{L·h}}$

and $\quad L_a = 3.67(0.0324)(3.38) = 0.40$ mm

An additional check can be made of the accuracy of the data by the relation

$$M_T = 6k_v \rho n^0 (Gt)^4 = 450 \text{ g/L}$$

$$M_T = (6)(1.0)\frac{1.335 \text{ g/cm}^3}{1000 \text{ mm}^3/\text{cm}^3}\, e^{19.78}[(0.0324)(3.38)]^4$$

$$M_T = 455 \text{ g/L} \approx 450 \text{ g/L}$$

Had only the growth rate been known, the size distribution of the solids could have been calculated from the equation

$$W_f = 1 - e^{-x}\left(\frac{x^3}{6} + \frac{x^2}{2} + x + 1\right)$$

where W_f is the weight fraction *up to* size L and $x = L/Gt$.

$$x = \frac{L}{(0.0324)(3.38)} = \frac{L}{0.1095}$$

Screen size	L, mm	x	$W_f°$	Cumulative % retained 100 $(1 - W_f)$	Measured cumulative % retained
20	0.833	7.70	0.944	5.6	4.4
28	0.589	5.38	0.784	21.6	18.8
35	0.417	3.80	0.526	47.4	43.0
48	0.295	2.70	0.286	71.4	74.6
65	0.208	1.90	0.125	87.5	90.1
100	0.147	1.34	0.048	95.2	97.5

°Values of W_f as a function of x may be obtained from a table of Wick's functions.

Note that the calculated distribution shows some deviation from the measured values because of the small departure of the actual sample from the theoretical coefficient of variation (i.e., 47.5 versus 52 percent).

Here it can be seen that the nucleation rate is a decreasing function of growth rate (and supersaturation). The physical explanation is believed to be the mechanical influence of the crystallizer on the growing suspension and/or the effect of Bujacian behavior.

Had sufficient data indicating a change in n^0 for various values of M at constant G been available, a plot of $\ln n^0$ versus $\ln M$ at corresponding G's would permit determination of the power j.

Crystallizers with Fines Removal In Example 3, the product was from a forced-circulation crystallizer of the MSMPR type. In many cases, the product produced by such machines is too small for commercial use; therefore, a separation baffle is added within the crystallizer to permit the removal of unwanted fine crystalline material from the magma, thereby controlling the population density in the machine so as to produce a coarser crystal product. When this is done, the product sample plots on a graph of $\ln n$ versus L as shown in line P, Fig. 18-62. The line of steepest slope, line F, represents the particle-size distribution of the fine material, and samples which show this distribution can be taken from the liquid leaving the fines-separation baffle. The product crystals have a slope of lower value, and typically there should be little or no material present smaller than L_f, the size which the baffle is designed to separate. The effective nucleation rate for the product material is the intersection of the extension of line P to zero size.

As long as the largest particle separated by the fines-destruction baffle is small compared with the mean particle size of the product, the seed for the product may be thought of as the particle-size distribution corresponding to the fine material which ranges in length from zero to L_f, the largest size separated by the baffle.

The product discharged from the crystallizer is characterized by the integral of the distribution from size L_f to infinity:

$$M_T = k_v \rho \int_{L_f}^{\infty} n^0 \exp\left(-L_f/Gt_f\right) \exp\left(L/GT\right) L^3 \, dL \quad (18\text{-}41)$$

The integrated form of this equation is shown in Table 18-5.

For a given set of assumptions it is possible to calculate the characteristic curves for the product from the crystallizer when it is operated at various levels of fines removal as characterized by L_f. This has been done for an ammonium sulfate crystallizer in Fig. 18-63. Also shown in that figure is the actual size distribution obtained. In calculating theoretical size distributions in accordance with the Eq. (18-41), it is

TABLE 18-6 Growth Rates and Kinetic Equations for Some Industrial Crystallized Products

Material crystallized	G, m/s $\times 10^8$	Range t, h	Range M_T, g/L	Temp., °C	Scale°	Kinetic equation for B_0 no./(L·s)	References†
$(NH_4)_2SO_4$	1.67	3.83	150	70	P	$B_0 = 6.62 \times 10^{-25} G^{0.82} p^{-0.92} m_2^{2.05}$	Bennett and Wolf, *AIChE, SFC*, 1979.
$(NH_4)_2SO_4$	0.20	0.25	38	18	B	$B_0 = 2.94(10^{10})G^{1.03}$	Larsen and Mullen, *J. Crystal Growth* **20:** 183 (1973).
$(NH_4)_2SO_4$	—	0.20		34	B	$B_0 = 6.14(10^{-11})S_R^{7.84}M_T^{0.98}G^{1.22}$	Youngquist and Randolph, *AIChE J.* **18:** 421 (1972).
$MgSO_4 \cdot 7H_2O$	3.0–7.0	—	—	25	B	$B_0 = 9.65(10^{12})M_T^{0.67}G^{1.24}$	Sikdar and Randolph, *AIChE J.* **22:** 110 (1976).
$MgSO_4 \cdot 7H_2O$	—	—	Low	29	B	$B_0 = f\,(N, L^4, N^{4.2}, S^{2.5})$	Ness and White, *AIChE Symposium Series 153*, vol. 72, p. 64.
KCl	2–12		200	32	P	$B_0 = 7.12(10^{39})M_T^{0.14}G^{4.99}$	Randolph et al., *AIChE J.* **23:** 500 (1977).
KCl	3.3	1–2	100	37	B	$B_0 = 5.16(10^{22})M_T^{0.91}G^{2.77}$	Randolph et al., *Ind. Eng. Chem. Proc. Design Dev.* **20:** 496 (1981).
KCl	0.3–0.45	—	50–147	25–68	B	$B_0 = 5 \times 10^{-3} G^{2.78}(M_T TIP^2)^{1.2}$	Qian et al., *AIChE J.* **33**(10): 1690 (1987).
KCr_2O_7	1.2–9.1	0.25–1	14–42	—	B	$B_0 = 7.33(10^4)M_T^{0.6}G^{0.5}$	Desari et al., *AIChE J.* **20:** 43 (1974).
KCr_2O_7	2.6–10	0.15–0.5	20–100	26–40	B	$B_0 = 1.59(10^{-3})S_R^3 M_T G^{0.48}$	Janse, Ph.D. thesis, Delft Technical University, 1977.
KNO_3	8.13	0.25–0.050	10–40	20	B	$B_0 = 3.85(10^{16})M_T^{0.5}G^{2.06}$	Jurazsek and Larson, *AIChE J.* **23:** 460 (1977).
K_2SO_4	—	0.03–0.17	1–7	30	B	$B_0 = 2.62(10^3)S_R^{2.5}M_T^{0.5}G^{0.54}$	Randolph and Sikdar, *Ind. Eng. Chem. Fund.* **15:** 64 (1976).
K_2SO_4	2–6	0.25–1	2–20	10–50	B	$B_0 = 4.09(10^6)\exp\left(\dfrac{10900}{RT}\right)M_T G^{0.5}$	Jones, Budz, and Mullin, *AIChE J.* **33:** 12 (1986).
K_2SO_4	0.8–1.6	—	—	—	B	$\dfrac{G}{G_0} = 1 + 2L^{2/3}$ (L in μm)	White, Bendig, and Larson, *AIChE Mtg.*, Washington, D.C., Dec. 1974.
NaCl	4–13	0.2–1	25–200	50	B	$B_0 = 1.92(10^{10})S_R^2 M_T G^2$	Asselbergs, Ph.D. thesis, Delft Technical University, 1978.
NaCl	—	0.6	35–70	55	P	$B_0 = 8 \times 10^{10}N^2 G^2 M_T$	Grootscholten et al., *Chem. Eng. Design* **62:** 179 (1984).
NaCl	0.5	1–2.5	70–190	72	P	$B_0 = 1.47(10^2)\left(\dfrac{I^2}{P}\right)m_4^{0.84}G^{0.98}$	Bennett et al., *Chem. Eng. Prog.* **69**(7): 86 (1973).
Citric acid	1.1–3.7	—	—	16–24	B	$B_0 = 1.09(10^{10})m_4^{0.084}G^{0.84}$	Sikdar and Randolph, *AIChE J.* **22:** 110 (1976).
Fructose	0.1–0.25	—	—	50	B	—	Shiau and Berglund, *AIChE J.* **33:** 6 (1987).
Sucrose	—	—		80	B	$B_0 = 5 \times 10^6 N^{0.7}M_T^{0.3}G^{0.4}$	Berglund and deJong, *Separations Technology* **1:** 38 (1990).
Sugar	2.5–5	0.375	50	45	B	$B_0 = 4.38(10^6)M_T^{1.01}(\Delta C - 0.5)^{1.42}$	Hart et al., *AIChE Symposium Series 193*, vol. 76, 1980.
Urea	0.4–4.2	2.5–6.8	350–510	55	P	$B_0 = 5.48(10^{-1})M_T^{-3.87}G^{1.66}$	Bennett and Van Buren, *Chem. Eng. Prog. Symposium Series* **95**(7): 65 (1973).
Urea	—	—		3–16	B	$B_0 = 1.49(10^{-31})S_R^{2.3}M_T^{1.07}G^{-3.54}$	Lodaya et al., *Ind. Eng. Chem. Proc. Design Dev.* **16:** 294 (1977).

°B = bench scale; P = pilot plant.
†Additional data on many components are in Garside and Shah, *Ind. Eng. Chem. Proc. Design Dev.*, **19**, 509 (1980).

assumed that the growth rate is a constant, whereas in fact larger values of L_f will interact with the system driving force to raise the growth rate and the nucleation rate. Nevertheless, Fig. 18-63 illustrates clearly the empirical result of the operation of such equipment, demonstrating that the most significant variable in changing the particle-size distribution of the product is the size removed by the baffle. Conversely, changes in retention time for a given particle-removal size L_f make a relatively small change in the product-size distribution.

It is implicit that increasing the value of L_f will raise the supersaturation and growth rate to levels at which mass homogeneous nucleation can occur, thereby leading to periodic upsets of the system or cycling [Randolph, Beer, and Keener, *Am. Inst. Chem. Eng. J.*, **19**, 1140 (1973)]. That this could actually happen was demonstrated experimentally by Randolph, Beckman, and Kraljevich [*Am. Inst. Chem. Eng. J.*, **23**, 500 (1977)], and that it could be controlled dynamically by regulating the fines-destruction system was shown by Beckman and Randolph [ibid., (1977)]. Dynamic control of a crystallizer with a fines-destruction baffle and fine-particle-detection equipment

employing a light-scattering (laser) particle-size-measurement instrument is described in U.S. Patent, 4,263,010 and 5,124,265.

CRYSTALLIZATION EQUIPMENT

Whether a vessel is called an evaporator or a crystallizer depends primarily on the criteria used in arriving at its sizing. In an evaporator of the salting-out type, sizing is done on the basis of vapor release. In a crystallizer, sizing is normally done on the basis of the volume required for crystallization or for special features required to obtain the proper product size. In external appearance, the vessels could be identical. Evaporators are discussed in Sec. 11.

In the discussion which follows, crystallization equipment has been classified according to the means of suspending the growing product. This technique reduces the number of major classifications and segregates those to which Eq. (18-31) applies.

Mixed-Suspension, Mixed-Product-Removal Crystallizers
This type of equipment, sometimes called the circulating-magma

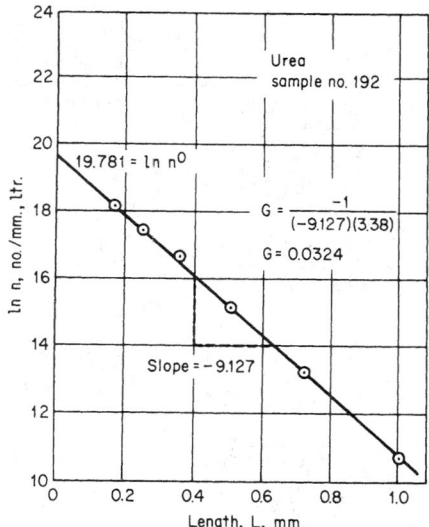

FIG. 18-61   Population density plot for Example 3.

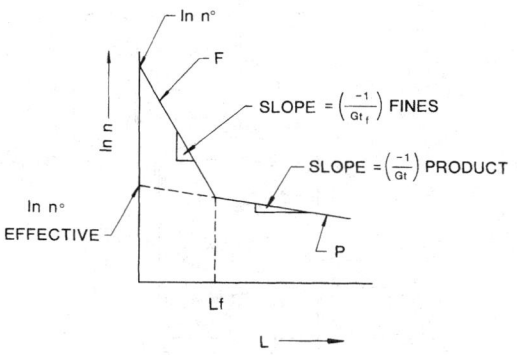

FIG. 18-62   Plot of Log N against L for a crystallizer with fines removal.

crystallizer, is by far the most important in use today. In most commercial equipment of this type, the uniformity of suspension of product solids within the crystallizer body is sufficient for the theory [Eqs. (18-31) to (18-33b)] to apply. Although a number of different varieties and features are included within this classification, the equipment operating with the highest capacity is the kind in which the vaporization of a solvent, usually water, occurs.

Although surface-cooled types of MSMPR crystallizers are available, most users prefer crystallizers employing vaporization of solvents or of refrigerants. The primary reason for this preference is that heat transferred through the critical supersaturating step is through a boiling-liquid-gas surface, avoiding the troublesome solid deposits that can form on a metal heat-transfer surface.

Forced-Circulation Evaporator-Crystallizer   This crystallizer is shown in Fig. 18-64. Slurry leaving the body is pumped through a circulating pipe and through a tube-and-shell heat exchanger, where its temperature increases by about 2 to 6°C (3 to 10°F). Since this heating is done without vaporization, materials of normal solubility should produce no deposition on the tubes. The heated slurry, returned to the body by a recirculation line, mixes with the body slurry and raises its temperature locally near the point of entry, which causes boiling at the liquid surface. During the consequent cooling and

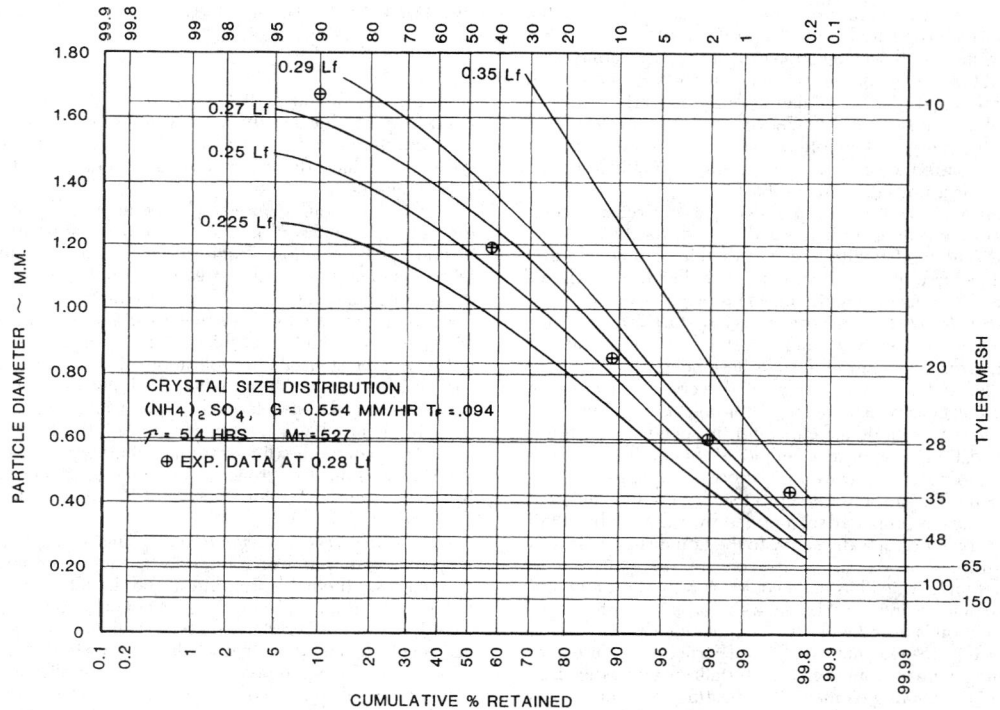

FIG. 18-63   Calculated product-size distribution for a crystallizer operation at different fine-crystal-separation sizes.

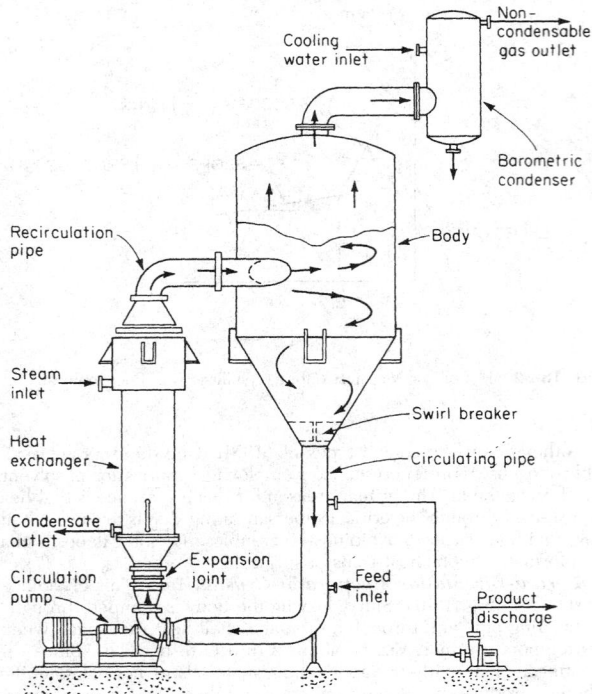

FIG. 18-64   Forced-circulation (evaporative) crystallizer. (*Swenson Process Equipment, Inc.*)

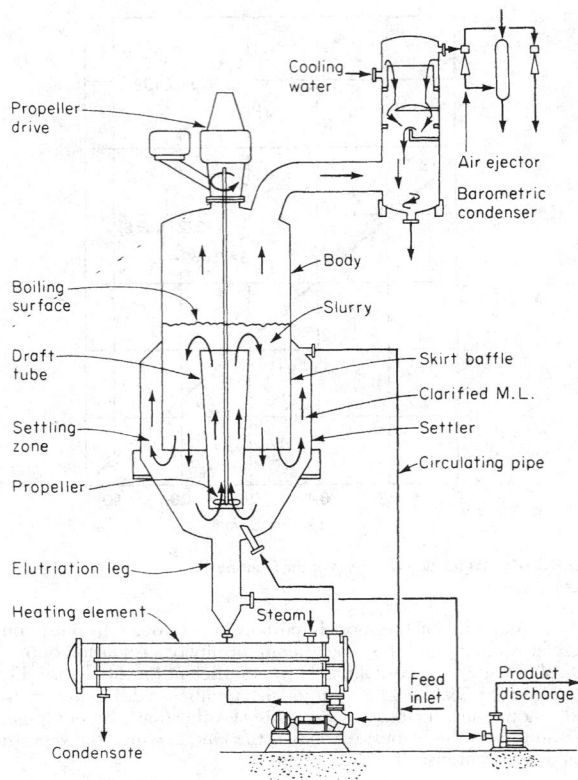

FIG. 18-65   Draft-tube-baffle (DTB) crystallizer. (*Swenson Process Equipment, Inc.*)

vaporization to achieve equilibrium between liquid and vapor, the supersaturation which is created causes deposits on the swirling body of suspended crystals until they again leave via the circulating pipe. The quantity and the velocity of the recirculation, the size of the body, and the type and speed of the circulating pump are critical design items if predictable results are to be achieved. A further discussion of the parameters affecting this type of equipment is given by Bennett, Newman, and Van Buren [*Chem. Eng. Prog.*, **55**(3), 65 (1959); *Chem. Eng. Prog. Symp. Ser.*, **65**(95), 34, 44 (1969)].

If the crystallizer is not of the evaporative type but relies only on *adiabatic evaporative cooling* to achieve the yield, the heating element is omitted. The feed is admitted into the circulating line after withdrawal of the slurry, at a point sufficiently below the free-liquid surface to prevent flashing during the mixing process.

Draft-Tube-Baffle (DTB) Evaporator-Crystallizer   Because mechanical circulation greatly influences the level of nucleation within the crystallizer, a number of designs have been developed that use circulators located within the body of the crystallizer, thereby reducing the head against which the circulator must pump. This technique reduces the power input and circulator tip speed and therefore the rate of nucleation. A typical example is the draft-tube-baffle (DTB) evaporator-crystallizer (Swenson Process Equipment, Inc.) shown in Fig. 18-65. The suspension of product crystals in maintained by a large, slow-moving propeller surrounded by a draft tube within the body. The propeller directs the slurry to the liquid surface so as to prevent solids from short-circuiting the zone of the most intense supersaturation. Slurry which has been cooled is returned to the bottom of the vessel and recirculated through the propeller. At the propeller, heated solution is mixed with the recirculating slurry.

The design of Fig. 18-65 contains a fines-destruction feature comprising the settling zone surrounding the crystallizer body, the circulating pump, and the heating element. The heating element supplies sufficient heat to meet the evaporation requirements and to raise the temperature of the solution removed from the settler so as to destroy

any small crystalline particles withdrawn. Coarse crystals are separated from the fines in the settling zone by gravitational sedimentation, and therefore this fines-destruction feature is applicable only to systems in which there is a substantial density difference between crystals and mother liquor.

This type of equipment can also be used for applications in which the only heat removed is that required for adiabatic cooling of the incoming feed solution. When this is done and the fines-destruction feature is to be employed, a stream of liquid must be withdrawn from the settling zone of the crystallizer and the fine crystals must be separated or destroyed by some means other than heat addition—for example, either dilution or thickening and physical separation.

In some crystallization applications it is desirable to increase the solids content of the slurry within the body above the natural consistency, which is that developed by equilibrium cooling of the incoming feed solution to the final temperature. This can be done by withdrawing a stream of mother liquor from the baffle zone, thereby thickening the slurry within the growing zone of the crystallizer. This mother liquor is also available for removal of fine crystals for size control of the product.

Draft-Tube (DT) Crystallizer   This crystallizer may be employed in systems in which fines destruction is not needed or wanted. In such cases the baffle is omitted, and the internal circulator is sized to have the minimum nucleating influence on the suspension.

In DTB and DT crystallizers the circulation rate achieved is generally much greater than that available in a similar forced-circulation crystallizer. The equipment therefore finds application when it is necessary to circulate large quantities of slurry to minimize supersaturation levels within the equipment. In general, this approach is required to obtain long operating cycles with material capable of growing on the walls of the crystallizer. The draft-tube and draft-tube-baffle designs are commonly used in the production of granular materials such as ammonium sulfate, potassium chloride, photographic hypo, and other inorganic and organic crystals for which product in the range 8 to 30 mesh is required.

Surface-Cooled Crystallizer For some materials, such as sodium chlorate, it is possible to use a forced-circulation tube-and-shell exchanger in direct combination with a draft-tube-crystallizer body, as shown in Fig. 18-66. Careful attention must be paid to the temperature difference between the cooling medium and the slurry circulated through the exchanger tubes. In addition, the path and rate of slurry flow within the crystallizer body must be such that the volume contained in the body is "active." That is to say, crystals must be so suspended within the body by the turbulence that they are effective in relieving supersaturation created by the reduction in temperature of the slurry as it passes through the exchanger. Obviously, the circulating pump is part of the crystallizing system, and careful attention must be paid to its type and its operating parameters to avoid undue nucleating influences.

The use of the internal baffle permits operation of the crystallizer at a slurry consistency other than that naturally obtained by the cooling of the feed from the initial temperature to the final mother-liquor temperature. The baffle also permits fines removal and destruction.

With most inorganic materials this type of equipment produces crystals in the range 30 to 100 mesh. The design is based on the allowable rates of heat exchange and the retention required to grow the product crystals.

Direct-Contact-Refrigeration Crystallizer For some applications, such as the freezing of ice from seawater, it is necessary to go to such low temperatures that cooling by the use of refrigerants is the only economical solution. In such systems it is sometimes impractical to employ surface-cooled equipment because the allowable temperature difference is so small (under 3°C) that the heat-exchanger surface becomes excessive or because the viscosity is so high that the mechanical energy put in by the circulation system requires a heat-removal rate greater than can be obtained at reasonable temperature differences. In such systems, it is convenient to admix the refrigerant with the slurry being cooled in the crystallizer, as shown in Fig. 18-67, so that the heat of vaporization of the refrigerant cools the slurry by direct contact. The successful application of such systems requires that the refrigerant be relatively immiscible with the mother liquor and be capable of separation, compression, condensation, and subsequent recycle into the crystallizing system. The operating pressures and temperatures chosen have a large bearing on power consumption.

This technique has been very successful in reducing the problems associated with buildup of solids on a cooling surface. The use of direct-

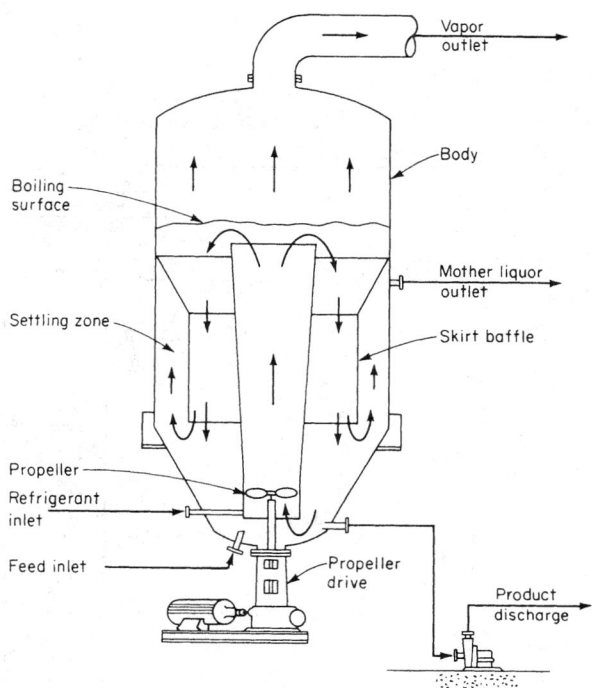

FIG. 18-67 Direct-contact-refrigeration crystallizer (DTB type). (*Swenson Process Equipment, Inc.*)

contact refrigeration also reduces overall process-energy requirements, since in a refrigeration process involving two fluids a greater temperature difference is required on an overall basis when the refrigerant must first cool some intermediate solution, such as calcium chloride brine, and that solution in turn cools the mother liquor in the crystallizer.

Equipment of this type has been successfully operated at temperatures as low as −59°C (−75°F).

Reaction-Type Crystallizers In chemical reactions in which the end product is a solid-phase material such as a crystal or an amorphous solid the type of equipment described in the preceding subsections or shown in Fig. 18-68 may be used. By mixing the reactants in a large circulated stream of mother liquor containing suspended solids of the equilibrium phase, it is possible to minimize the driving force created during their reaction and remove the heat of reaction through the vaporization of a solvent, normally water. Depending on the final particle size required, it is possible to incorporate a fines-destruction baffle as shown in Fig. 18-68 and take advantage of the control over particle size afforded by this technique. In the case of ammonium sulfate crystallization from ammonia gas and concentrated sulfuric acid, it is necessary to vaporize water to remove the heat of reaction, and this water so removed can be reinjected after condensation into the fines-destruction stream to afford a very large amount of dissolving capability.

Other examples of this technique are where a solid material is to be decomposed by mixing it with a mother liquor of a different composition, as shown in Fig. 18-69. Carnallite ore ($KCl \cdot MgCl_2 \cdot 4H_2O$) can be added to a mother liquor into which water is also added so that decomposition of the ore into potassium chloride (KCl) crystals and magnesium chloride–rich mother liquor takes place. Circulated slurry in the draft tube suspends the product crystals as well as the incoming ore particles until the ore can decompose into potassium chloride crystals and mother liquor. By taking advantage of the fact that water must be added to the process, the fines-bearing mother liquor can be removed behind the baffle and then water added so that the finest particles are dissolved before being returned to the crystallizer body.

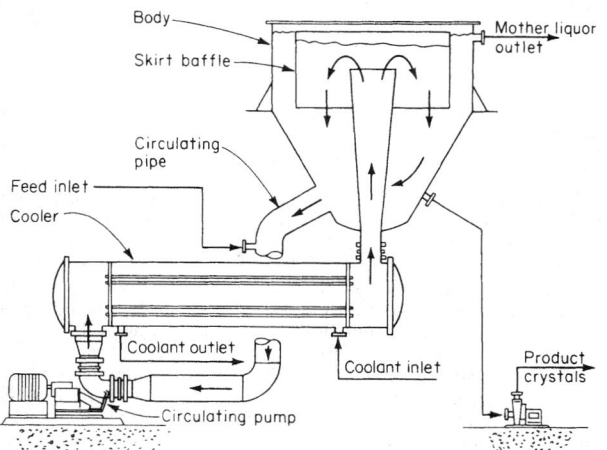

FIG. 18-66 Forced-circulation baffle surface-cooled crystallizer. (*Swenson Process Equipment, Inc.*)

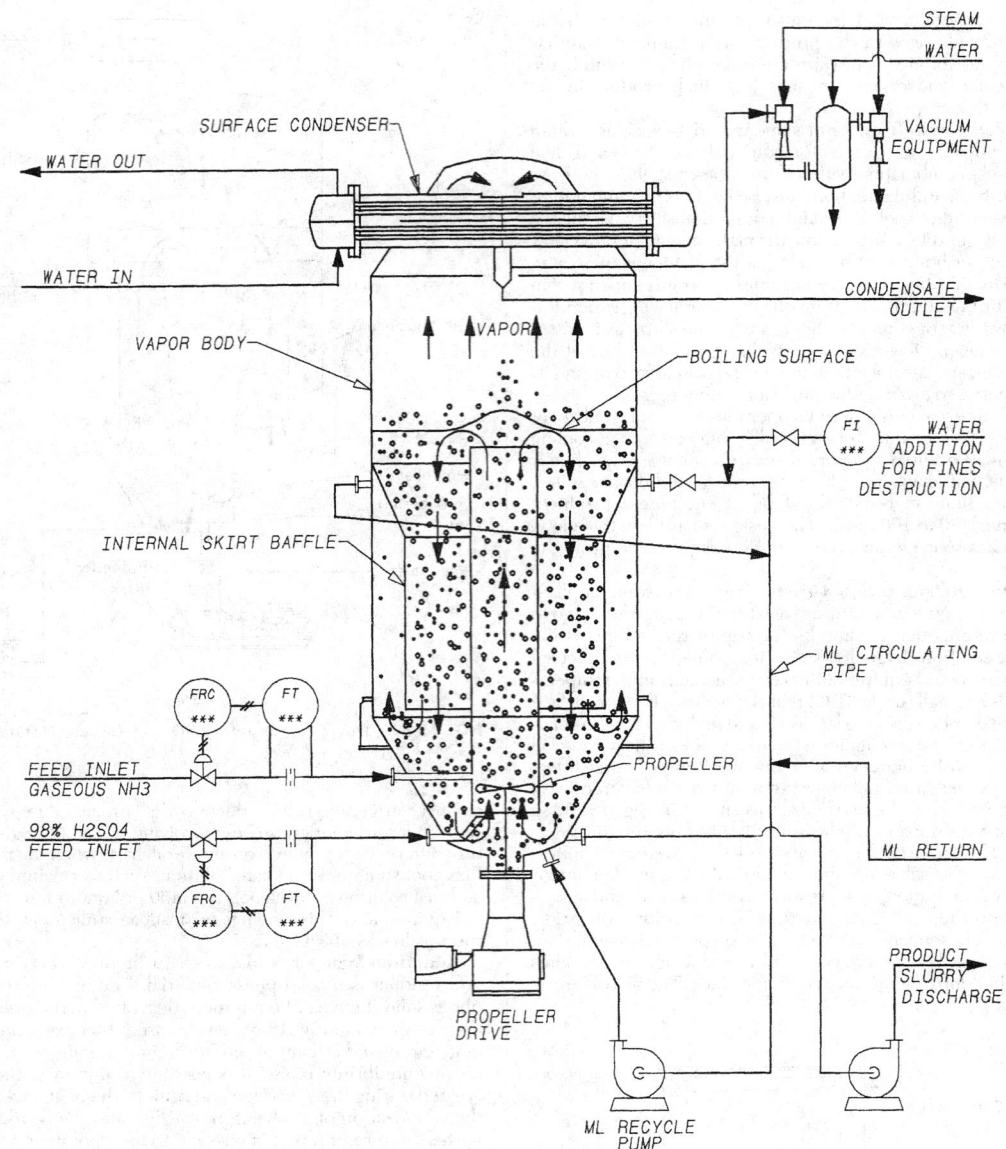

FIG. 18-68 Swenson reaction type DTB crystallizer. (*Swenson Process Equipment, Inc.*)

Other examples of this technique involve neutralization reactions such as the neutralization of sulfuric acid with calcium chloride to result in the precipitation of gypsum.

Mixed-Suspension, Classified-Product-Removal Crystallizers Many of the crystallizers just described can be designed for classified-product discharge. Classification of the product is normally done by means of an elutriation leg suspended beneath the crystallizing body as shown in Fig. 18-66. Introduction of clarified mother liquor to the lower portion of the leg fluidizes the particles prior to discharge and selectively returns the finest crystals to the body for further growth. A relatively wide distribution of material is usually produced unless the elutriation leg is extremely long. Inlet conditions at the leg are critical if good classifying action or washing action is to be achieved.

If an elutriation leg or other product-classifying device is added to a crystallizer of the MSMPR type, the plot of the population density versus L is distorted in the region of largest sizes. Also the incorporation of an elutriation leg destabilizes the crystal-size distribution and under some conditions can lead to cycling. The theoretical treatment of both the crystallizer model and the cycling relations is discussed by Randolph, Beer, and Keener (loc. cit.). Although such a feature can be included on many types of classified-suspension or mixed-suspension crystallizers, it is most common to use this feature with the forced-circulation evaporative-crystallizer and the DTB crystallizer.

Classified-Suspension Crystallizer This equipment is also known as the **growth** or **Oslo crystallizer** and is characterized by the production of supersaturation in a circulating stream of liquor. Supersaturation is developed in one part of the system by evaporative cooling or by cooling in a heat exchanger, and it is relieved by passing the liquor through a fluidized bed of crystals. The fluidized bed may be contained in a simple tank or in a more sophisticated vessel arranged

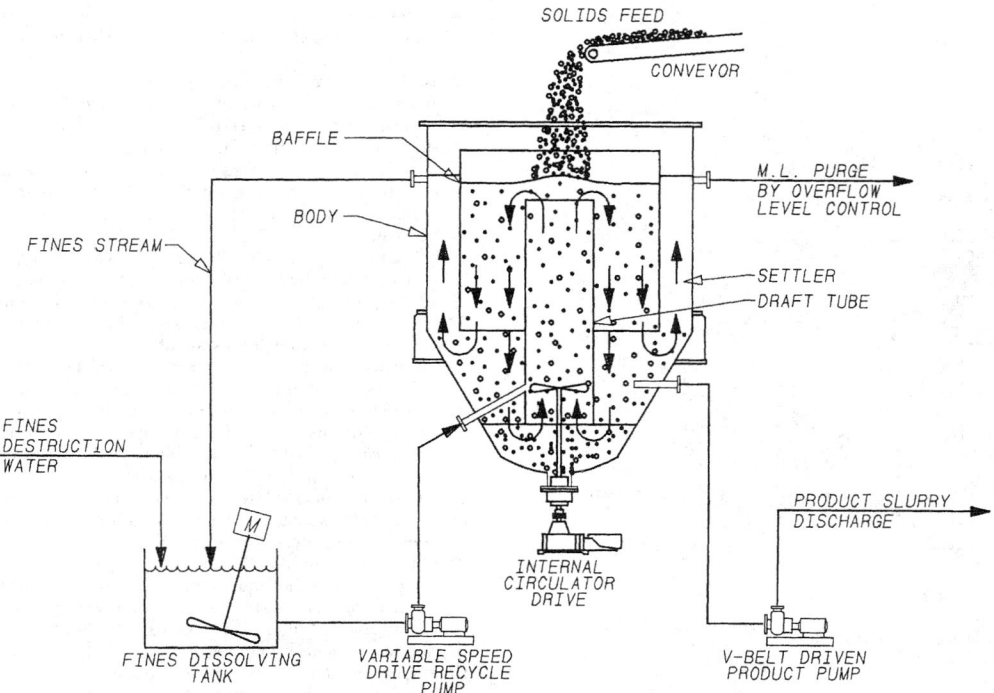

FIG. 18-69 Swenson atmospheric reaction–type DTB crystallizer. (*Swenson Process Equipment, Inc.*)

for a pronounced classification of the crystal sizes. Ideally this equipment operates within the metastable supersaturation field described by Miers and Isaac, *J. Chem. Soc.,* **1906,** 413.

In the **evaporative crystallizer** of Fig. 18-70, solution leaving the vaporization chamber at *B* is supersaturated slightly within the metastable zone so that new nuclei will not form. The liquor contacting the bed at *E* relieves its supersaturation on the growing crystals and leaves through the circulating pipe *F.* In a cooling-type crystallization hot feed is introduced at *G,* and the mixed liquor flashes when it reaches the vaporization chamber at *A.* If further evaporation is required to produce the driving force, a heat exchanger is installed between the circulating pump and the vaporization changer to supply the heat for the required rate of vaporization.

The transfer of supersaturated liquor from the vaporizer (point *B,* Fig. 18-69) often causes salt buildup in the piping and reduction of the operating cycle in equipment of this type. The rate of buildup can be reduced by circulating a thin suspension of solids through the vaporizing chamber; however, the presence of such small seed crystals tends to rob the supersaturation developed in the vaporizer, thereby lowering the efficiency of the recirculation system.

An **Oslo surface-cooled crystallizer** is illustrated in Fig. 18-71. Supersaturation is developed in the circulated liquor by chilling in the cooler *H.* This supersaturated liquor is contacted with the suspension of crystals in the suspension chamber at *E.* At the top of the suspension chamber a stream of mother liquor *D* can be removed to be used for fines removal and destruction. This feature can be added on either type of equipment. Fine crystals withdrawn from the top of the suspension are destroyed, thereby reducing the overall number of crystals in the system and increasing the particle size of the remaining product crystals.

Scraped-Surface Crystallizer For relatively small-scale applications a number of crystallizer designs employing direct heat exchange between the slurry and a jacket or double wall containing a cooling medium have been developed. The heat-transfer surface is scraped or agitated in such a way that the deposits cannot build up.

The scraped-surface crystallizer provides an effective and inexpensive method of producing slurry in equipment which does not require expensive installation or supporting structures.

Double-Pipe Scraped-Surface Crystallizer This type of equipment consists of a double-pipe heat exchanger with an internal agitator fitted with spring-loaded scrapers that wipe the wall of the inner pipe. The cooling liquid passes between the pipes, this annulus being dimensioned to permit reasonable shell-side velocities. The scrapers prevent the buildup of solids and maintain a good film coefficient of heat transfer. The equipment can be operated in a continuous or in a recirculating batch manner.

Such units are generally built in lengths to above 12 m (40 ft). They can be arranged in parallel or in series to give the necessary liquid velocities for various capacities. Heat-transfer coefficients have been reported in the range of 170 to 850 W/(m²·K) [30 to 150 Btu/(h·ft²·°F)] at temperature differentials of 17°C (30°F) and higher [Garrett and Rosenbaum, *Chem. Eng.,* **65**(16), 127 (1958)]. Equipment of this type is marketed as the **Votator** and the **Armstrong crystallizer.**

Batch Crystallization Batch crystallization has been practiced longer than any other form of crystallization in both atmospheric tanks, which are either static or agitated, as well as in vacuum or pressure vessels. It is still widely practiced in the pharmaceutical and fine chemical industry or in those applications where the capacity is very small. The integrity of the batch with respect to composition and history can be maintained easily and the inventory management is more precise than with continuous processes. Batch crystallizers can be left unattended (overnight) if necessary and this is an important advantage for many small producers.

In any batch process the common mode of operation involves charging the crystallizer with concentrated or near-saturated solution, producing supersaturation by means of cooling the batch or evaporating solvent from the batch, seeding the batch by means of injecting seed crystals into the batch or by allowing homogeneous nucleation to occur, reaching the final mother-liquor temperature and concentration by some time-dependent means of control, and stopping the cycle

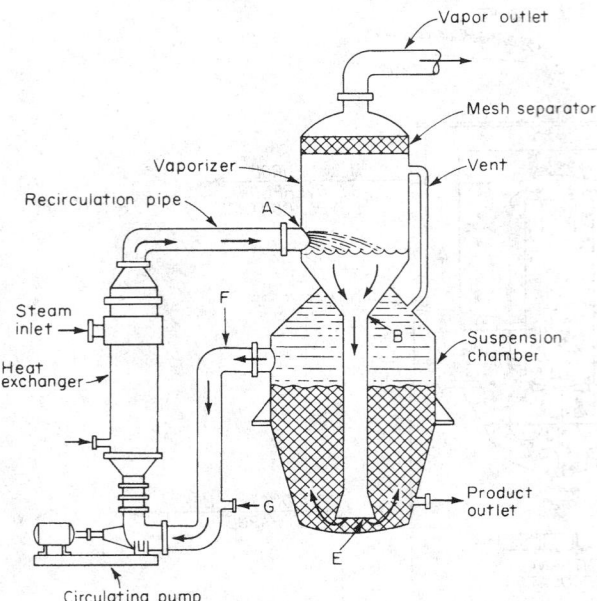

FIG. 18-70   OSLO evaporative crystallizer.

so that the batch may be dumped into a tank for processing by successive steps, which normally include centrifugation, filtration, and/or drying. In some cases, a small "heal" of slurry is left in the batch crystallizer to act as seed for the next batch.

Control of a batch crystallizer is almost always the most difficult part and very often is not practiced except to permit homogeneous nucleation to take place when the system becomes supersaturated. If control is practiced, it is necessary to have some means for determining when the initial solution is supersaturated so that seed of the appropriate size, quantity, and habit may be introduced into the batch. Following seeding, it is necessary to limit the cooling or evaporation in

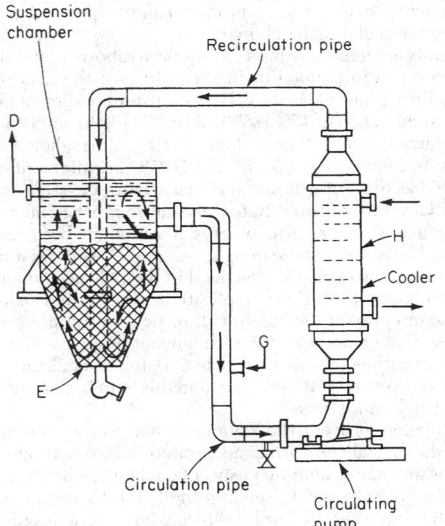

FIG. 18-71   OSLO surface-cooled crystallizer.

the batch to that which permits the generated supersaturation to be relieved on the seed crystals. This means that the first cooling or evaporation following seeding must be at a very slow rate, which is increased nonlinearly in order to achieve the optimum batch cycle. Frequently, such controls are operated by cycle timers or computers so as to achieve the required conditions. Sugar, many pharmaceutical products, and many fine chemicals are produced this way. Shown in Fig. 18-72 is a typical batch crystallizer comprising a jacketed closed tank with top-mounted agitator and feed connections. The tank is equipped with a short distillation column and surface condenser so that volatile materials may be retained in the tank and solvent recycled to maintain the batch integrity. Provisions are included so that the vessel may be heated with steam addition to the shell or cooling solution circulated through the jacket so as to control the temperature. Tanks of this type are intended to be operated with a wide variety of chemicals under both cooling and solvent evaporation conditions.

Recompression Evaporation-Crystallization   In all types of crystallization equipment wherein water or some other solvent is vaporized to produce supersaturation and/or cooling, attention should be given to the use of mechanical vapor recompression, which by its nature permits substitution of electrical energy for evaporation and solvent removal rather than requiring the direct utilization of heat energy in the form of steam or electricity. A typical recompression crystallizer flowsheet is shown in Fig. 18-73, which shows a single-stage evaporative crystallizer operating at approximately atmospheric pressure. The amount of heat energy necessary to remove 1 kg of water to produce the equivalent in crystal product is approximately 550 kilocalories. If the water evaporated is compressed by a mechanical compressor of high efficiency to a pressure where it can be condensed in the heat exchanger of the crystallizer, it can thereby supply the energy needed to sustain the process. Then the equivalent power for this compression is about 44 kilocalories (Bennett, *Chem. Eng. Progress*, 1978, pp. 67–70).

Although this technique is limited economically to those large-scale cases where the materials handled have a relatively low boiling point elevation and in those cases where a significant amount of heat is required to produce the evaporation for the crystallization step, it nevertheless offers an attractive technique for reducing the use of heat energy and substituting mechanical energy or electrical energy in those cases where there is a cost advantage for doing so. This technique finds many applications in the crystallization of sodium sulfate, sodium carbonate monohydrate, and sodium chloride. Shown in Fig. 18-74 is the amount of vapor compressed per kilowatt-hour for water vapor at 100°C and various ΔTs. The amount of water vapor compressed per horsepower decreases rapidly with increasing ΔT and, therefore, normal design considerations dictate that the recompression evaporators have a relatively large amount of heat-transfer surface so as to minimize the power cost. Often this technique is utilized only with the initial stages of evaporation where concentration of the solids is relatively low and, therefore, the boiling-point elevation is negligible. In order to maintain adequate tube velocity for heat transfer and suspension of crystals, the increased surface requires a large internal recirculation within the crystallizer body, which consequently lowers the supersaturation in the fluid pumped through the tubes. One benefit of this design is that with materials of flat or inverted solubility, the use of recompression complements the need to maintain low ΔTs to prevent fouling of the heat-transfer surface.

INFORMATION REQUIRED TO SPECIFY A CRYSTALLIZER

The following information regarding the product, properties of the feed solution, and required materials of construction must be available before a crystallizer application can be properly evaluated and the appropriate equipment options identified. Is the crystalline material being produced a hydrated or an anhydrous material? What is the solubility of the compound in water or in other solvents under consideration, and how does this change with temperature? Are other compounds in solution which coprecipitate with the product being crystallized, or do these remain in solution, increasing in concentration until some change in product phase occurs? What will be the influence of impurities in the solution on the crystal habit, growth, and

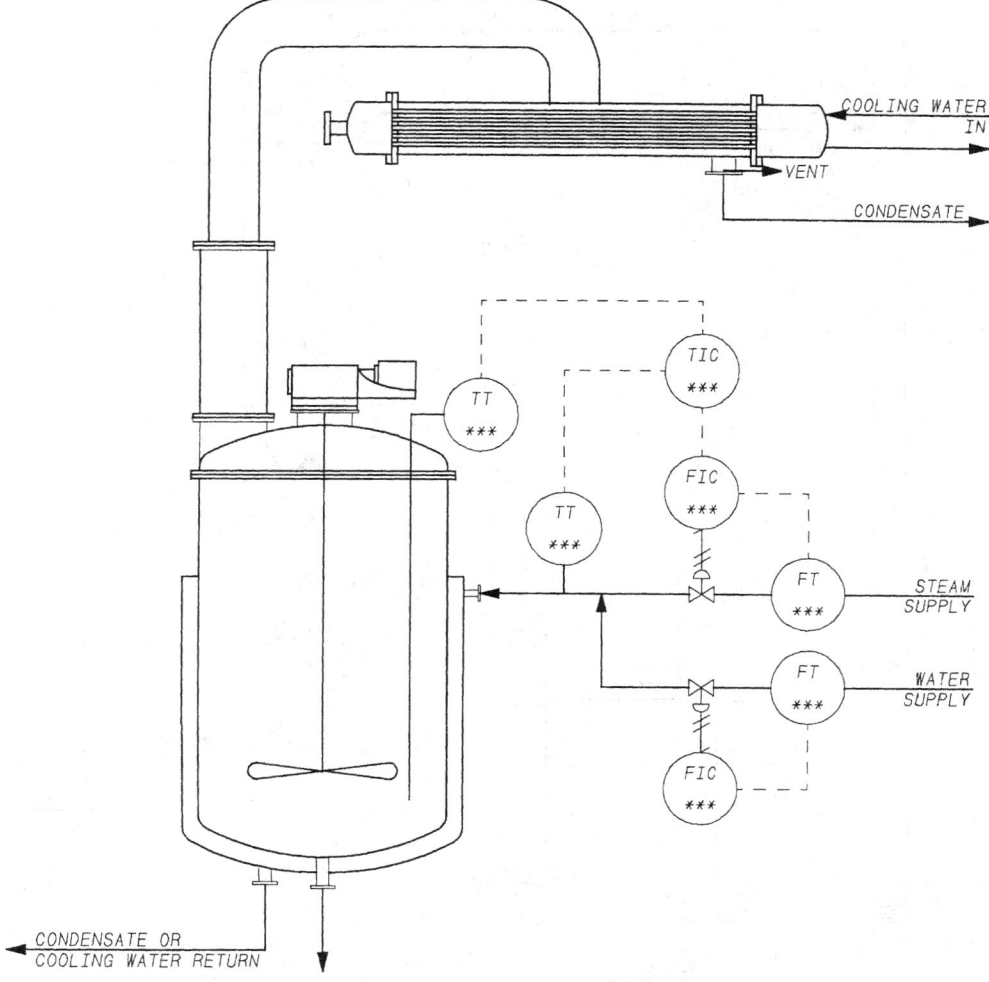

FIG. 18-72 Typical agitated batch crystallizer. (*Swenson Process Equipment, Inc.*)

nucleation rates? What are the physical properties of the solution and its tendency to foam? What is the heat of crystallization of the product crystal? What is the production rate, and what is the basis on which this production rate is computed? What is the tendency of the material to grow on the walls of the crystallizer? What materials of construction can be used in contact with the solution at various temperatures? What utilities will be available at the crystallizer location, and what are the costs associated with the use of these utilities? Is the final product to be blended or mixed with other crystalline materials or solids? What size of product and what shape of product are required to meet these requirements? How can the crystalline material be separated from the mother liquor and dried? Are there temperature requirements or wash requirements which must be met? How can these solids or mixtures of solids be handled and stored without undue breakage and caking?

Another basic consideration is whether crystallization is best carried out on a batch basis or on a continuous basis. The present tendency in most processing plants is to use continuous equipment whenever possible. Continuous equipment permits adjusting of the operating variables to a relatively fine degree in order to achieve the best results in terms of energy usage and product characteristics. It allows the use of a smaller labor force and results in a continuous util-

ity demand, which minimizes the size of boilers, cooling towers, and power-generation facilities. It also minimizes the capital investment required in the crystallizer and in the feed-storage and product-liquor-storage facilities.

Materials that have a tendency to grow readily on the walls of the crystallizer require periodic washout, and therefore an otherwise continuous operation would be interrupted once or even twice a week for the removal of these deposits. The impact that this contingency may have on the processing-equipment train ahead of the crystallizer must be considered.

The batch handling of wet or semidry crystalline materials is substantially more difficult than the storing and handling of dry crystalline materials. A batch operation has economic application only on a relatively small scale or when temperature or product characteristics require unusual precautions.

CRYSTALLIZER OPERATION

Crystal growth is a layer-by-layer process, and the retention time required in most commercial equipment to produce crystals of the size normally desired is on the order of 2 to 6 h. On the other hand, nucleation in a supersaturated solution can be generated in a fraction

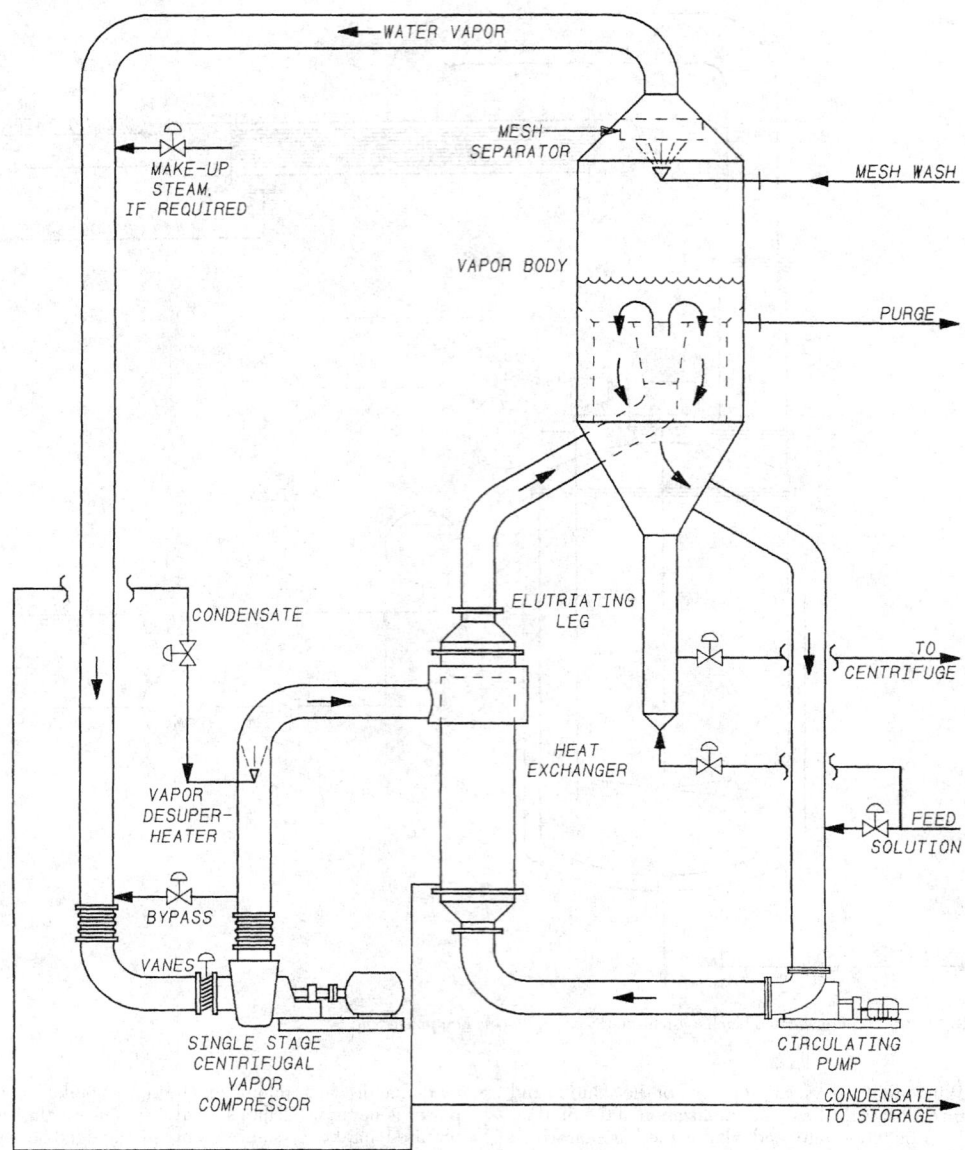

FIG. 18-73 Swenson single-stage recompression evaporator. (*Swenson Process Equipment, Inc.*)

of a second. The influence of any upsets in operating conditions, in terms of the excess nuclei produced, is very short-term in comparison with the total growth period of the product removed from the crystallizer. In a practical sense, this means that steadiness of operation is much more important in crystallization equipment than it is in many other types of process equipment.

It is to be expected that four to six retention periods will pass before the effects of an upset will be damped out. Thus, the recovery period may last from 8 to 36 h.

The **rate of nuclei formation** required to sustain a given product size decreases exponentially with increasing size of the product. Although when crystals in the range of 100 to 50 mesh are produced, the system may react quickly, the system response when generating large crystals in the 14-mesh size range is quite slow. This is because a single pound of 150-mesh seed crystals is sufficient to provide the

total number of particles in a ton of 14-mesh product crystals. In any system producing relatively large crystals, nucleation must be carefully controlled with respect to all internal and external sources. Particular attention must be paid to preventing seed crystals from entering with the incoming feed stream or being returned to the crystallizer with recycle streams of mother liquor coming back from the filter or centrifuge.

Experience has shown that in any given body operating at a given production rate, control of the magma (slurry) density is important to the control of crystal size. Although in some systems a change in slurry density does not result in a change in the rate nucleation, the more general case is that an increase in the magma density increases the product size through reduction in nucleation and increased retention time of the crystals in the growing bed. The reduction in supersaturation at longer retention times together with the smaller distance

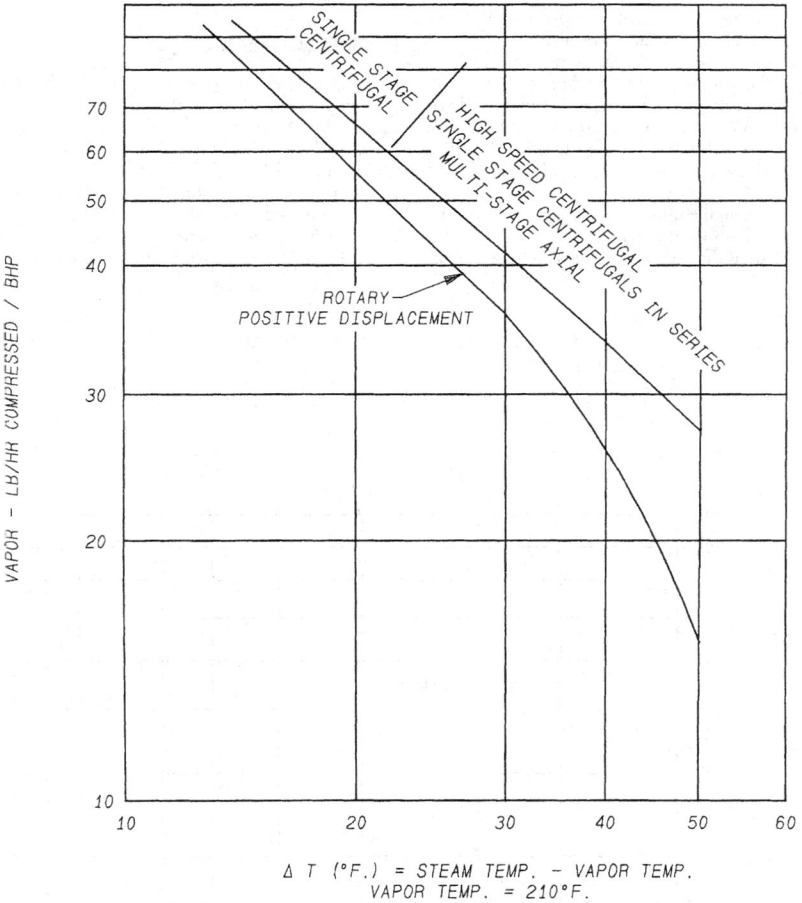

LB/HR OF VAPOR COMPRESSED PER BHP
VS TEMPERATURE DIFFERENCE

Δ T (°F.) = STEAM TEMP. – VAPOR TEMP.
VAPOR TEMP. = 210°F.

FIG. 18-74 Recompression evaporator horsepower as a function of overall ΔT.

between growing crystals, which lowers the driving force that is required to transport material from the liquid phase to the growing solids (**propinquity effect**), appears to be responsible for the larger product.

A reduction in the magma density will generally increase nucleation and decrease the particle size. This technique has the disadvantage that crystal formation on the equipment surfaces increases because lower slurry densities create higher levels of supersaturation within the equipment, particularly at the critical boiling surface in a vaporization-type crystallizer.

High levels of supersaturation at the liquid surface or at the tube walls in a surface-cooled crystallizer are the dominant cause of wall salting. Although some types of crystallizers can operate for several months continuously when crystallizing KCl or $(NH_4)_2SO_4$, most machines have much shorter operating cycles. Second only to control of particle size, the extension of operating cycles is the most difficult operating problem to be solved in most installations.

In the forced-circulation-type crystallizer (Fig. 19-43) primary control over particle size is exercised by the designer in selecting the circulating system and volume of the body. From the operating standpoint there is little that can be done to an existing unit other than supply external seed, classify the discharge crystals, or control the slurry

density. Nevertheless, machines of this type are frequently carefully controlled by these techniques and produce a predictable and desirable product-size distribution.

When crystals cannot be grown sufficiently large in forced-circulation equipment to meet product-size requirements, it is common to employ one of the designs that allow some influence to be exercised over the population density of the finer crystals. In the DTB design (Fig. 18-69) this is done by regulating the flow in the circulating pipe so as to withdraw a portion of the fines in the body in the amount of about 0.05 to 0.5 percent by settled volume. The exact quantity of solids depends on the size of the product crystals and on the capacity of the fines-dissolving system. If the machine is not operating stably, this quantity of solids will appear and then disappear, indicating changes in the nucleation rate within the circuit. At steady-state operation, the quantity of solids overflowing will remain relatively constant, with some solids appearing at all times. Should the slurry density of product crystals circulated within the machine rise to a value higher than about 50 percent settled volume, large quantities of product crystals will appear in the overflow system, disabling the fines-destruction equipment. Too high a circulating rate through the fines trap will produce this same result. Too low a flow through the fines circuit will remove insufficient particles and result in a smaller product-size crystal. To operate effectively, a crystallizer of the type employing fines-destruction techniques requires more sophisticated control than does operation of the simpler forced-circulation equipment.

The classifying crystallizer (Fig. 18-70) requires approximately the same control of the fines-removal stream and, in addition, requires control of the fluidizing flow circulated by the main pump. This flow must be adjusted to achieve the proper degree of fluidization in the suspension chamber, and this quantity of flow varies as the crystal size varies between start-up operation and normal operation. As with the draft-tube-baffle machine, a considerably higher degree of skill is required for operation of this equipment than of the forced-circulation type.

While most of the industrial designs in use today are built to reduce the problems due to excess nucleation, it is true that in some crystallizing systems a deficiency of seed crystals is produced and the prod-uct crystals are larger than are wanted or required. In such systems nucleation can be increased by increasing the mechanical stimulus created by the circulating devices or by seeding through the addition of fine crystals from some external source.

CRYSTALLIZER COSTS

Because crystallizers can come with such a wide variety of attachments, capacities, materials of construction, and designs, it is very difficult to present an accurate picture of the costs for any except certain specific types of equipment, crystallizing specific compounds. This is illustrated in Fig. 18-75, which shows the prices of equipment for crystallizing two different compounds at various production rates, one of the compounds being produced in two alternative crystallizer modes. Installed cost (including cost of equipment and accessories, foundations and supporting steel, utility piping,

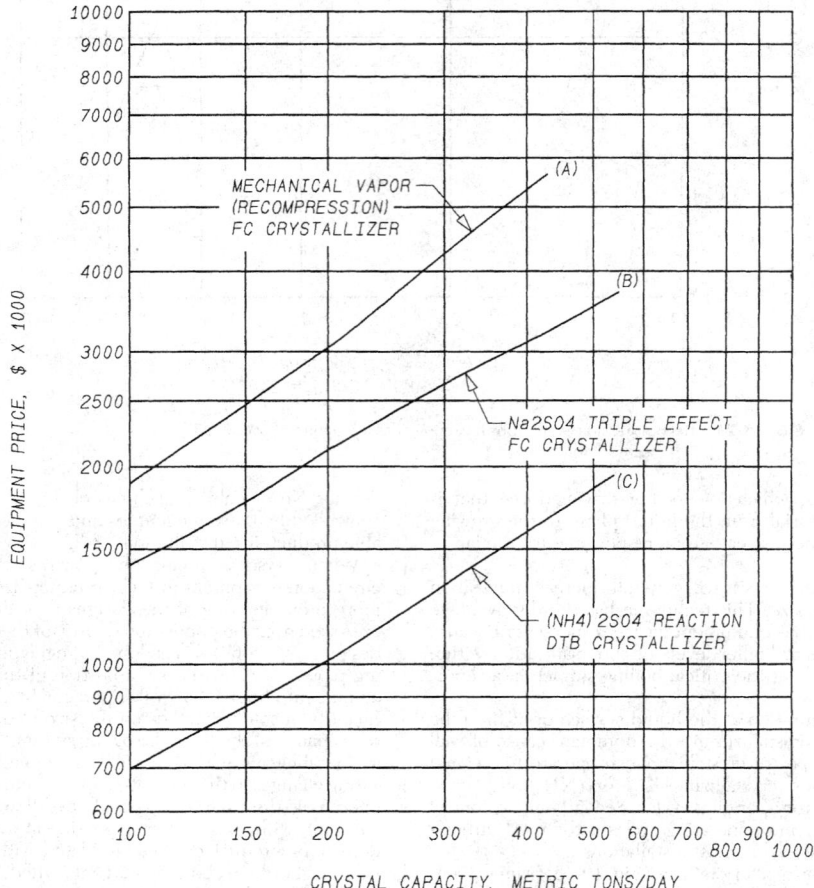

FIG. 18-75 Equipment prices, FOB point of fabrication, for typical crystallizer systems. Prices are for crystallizer plus accessories including vacuum equipment. (*Data supplied by Swenson Process Equipment, Inc., effective January, 1995.*)

process piping and pumps, electrical switchgear, instrumentation, and labor, but excluding cost of a building) will be approximately twice these price figures.

It should be ever present in the reader's mind that for every particular case the appropriate crystallizer manufacturers should be consulted for reliable price estimates. Most crystallization equipment is custom-designed, and costs for a particular application may vary greatly from those illustrated in Fig. 18-75. Realistic estimation of installation costs also requires reference to local labor rates, site-specific factors, and other case specifics.

LEACHING

GENERAL REFERENCES: Cculson and Richardson, *Chemical Engineering*, 4th ed., vol. 2, Pergamon Press, Oxford, 1991. Duby, "Hydrometallurgy," in *Kirk-Othmer Encyclopedia of Chemical Technology*, 4th ed., vol. 16, Wiley, New York, 1995, p. 338. McCabe, Smith, and Harriott, *Unit Operations of Chemical Engineering*, 5th ed., McGraw-Hill, New York, 1993, chaps. 17 and 20. Prub-hudesal, in Schweitzer, *Handbook of Separation Techniques for Chemical Engineers*, McGraw-Hill, New York, 1979, sec. 5.1. Rickles, *Chem. Eng.* **72**(6): 157 (1965). Schwartzberg (chap. 10, "Leaching—Organic Materials") and Wadsworth (chap. 9, "Leaching—Metals Applications") in Rousseau, *Handbook of Separation Process Technology*, Wiley, New York, 1987. Wakeman, "Extraction (Liquid-Solid)" in *Kirk-Othmer Encyclopedia of Chemical Technology*, 4th ed., vol. 10, Wiley, New York, 1993, p. 186.

DEFINITION

Leaching is the removal of a soluble fraction, in the form of a solution, from an insoluble, permeable solid phase with which it is associated. The separation usually involves selective dissolution, with or without diffusion, but in the extreme case of simple washing it consists merely of the displacement (with some mixing) of one interstitial liquid by another with which it is miscible. The soluble constituent may be solid or liquid; and it may be incorporated within, chemically combined with, adsorbed upon, or held mechanically in the pore structure of the insoluble material. The insoluble solid may be massive and porous; more often it is particulate, and the particles may be openly porous, cellular with selectively permeable cell walls, or surface-activated.

It is common practice to exclude from consideration as leaching the elution of surface-adsorbed solute. This process is treated instead as a special case of the reverse operation, adsorption. Also usually excluded is the washing of filter cakes, whether in situ or by reslurrying and refiltration.

Because of its variety of applications and its importance to several ancient industries, leaching is known by a number of other names. Among those encountered in chemical engineering practice are extraction, solid-liquid extraction, lixiviation, percolation, infusion, washing, and decantation-settling. The stream of solids being leached and the accompanying liquid is known as the underflow; in hydrometallurgical practice it is called pulp. The solid content of the stream is sometimes called marc (particularly by oil seed processors). The stream of liquid containing the leached solute is the overflow. As it leaves the leaching process it has several optional names: extract, solution, lixiviate, leachate, or miscella.

Mechanism The mechanism of leaching may involve simple physical solution or dissolution made possible by chemical reaction. The rate of transport of solvent into the mass to be leached, or of soluble fraction into the solvent, or of extract solution out of the insoluble material, or some combination of these rates may be significant. A membranous resistance may be involved. A chemical-reaction rate may also affect the rate of leaching.

Inasmuch as the overflow and underflow streams are not immiscible phases but streams based on the same solvent, the concept of equilibrium for leaching is not the one applied in other mass-transfer separations. If the solute is not adsorbed on the inert solid, true equilibrium is reached only when all the solute is dissolved and distributed uniformly throughout the solvent in both underflow and overflow (or when the solvent is uniformly saturated with the solute, a condition never encountered in a properly designed extractor). The practical interpretation of leaching equilibrium is the state in which the overflow and underflow liquids are of the same composition; on a *y-x* diagram, the equilibrium line will be a straight line through the origin with a slope of unity. It is customary to calculate the number of ideal (equilibrium) stages required for a given leaching task and to adjust the number by applying a stage efficiency factor, although local efficiencies, if known, can be applied stage by stage.

Usually, however, it is not feasible to establish a stage or overall efficiency or a leaching rate index (e.g., overall coefficient) without testing small-scale models of likely apparatus. In fact, the results of such tests may have to be scaled up empirically, without explicit evaluation of rate or quasi-equilibrium indices.

Methods of Operation Leaching systems are distinguished by operating cycle (batch, continuous, or multibatch intermittent); by direction of streams (cocurrent, countercurrent, or hybrid flow); by staging (single-stage, multistage, or differential-stage); and by method of contacting (sprayed percolation, immersed percolation, or solids dispersion). In general, descriptors from all four categories must be assigned to stipulate a leaching system completely (e.g., the Bollman-type extractor is a continuous hybrid-flow multistage sprayed percolator).

Whatever the mechanism and the method of operation, it is clear that the leaching process will be favored by increased surface per unit volume of solids to be leached and by decreased radial distances that must be traversed within the solids, both of which are favored by decreased particle size. Fine solids, on the other hand, cause slow percolation rate, difficult solids separation, and possible poor quality of solid product. The basis for an optimum particle size is established by these characteristics.

LEACHING EQUIPMENT

It is classification by contacting method that provides the two principal categories into which leaching equipment is divided: (1) that in which the leaching is accomplished by percolation and (2) that in which particulate solids are dispersed into a liquid and subsequently separated from it. Each includes batch and continuous units. Materials which disintegrate during leaching are treated in equipment of the second class.

A few designs of continuous machines fall in neither of these major classes.

Percolation In addition to being applied to ores and rock in place and by the simple technique of heap leaching (usually on very large scale; see Wadsworth, loc. cit.); percolation is carried out in batch tanks and in continuous or dump extractors (usually on smaller scale).

Batch Percolators The **batch tank** is not unlike a big nutsche filter; it is a large circular or rectangular tank with a false bottom. The solids to be leached are dumped into the tank to a uniform depth. They are sprayed with solvent until their solute content is reduced to an economic minimum and are then excavated. Countercurrent flow of solvent through a series of tanks is common, with fresh solvent entering the tank containing most nearly exhausted material. In a typical ore-dressing operation the tanks are 53 by 20 by 5.5 m (175 by 67 by 18 ft) and extract about 8200 Mg (9000 U.S. tons) of ore on a 13-day cycle. Some tanks operate under pressure, to contain volatile solvents or increase the percolation rate. A series of pressure tanks operating with countercurrent solvent flow is called a **diffusion battery.**

Continuous Percolators Coarse solids are also leached by percolation in moving-bed equipment, including single-deck and multideck rake classifiers, bucket-elevator contactors, and horizontal-belt conveyors.

The Bollman-type extractor shown in Fig. 18-76 is a bucket-elevator unit designed to handle about 2000 to 20,000 kg/h (50 to 500 U.S. tons/day) of flaky solids (e.g., soybeans). Buckets with perforated

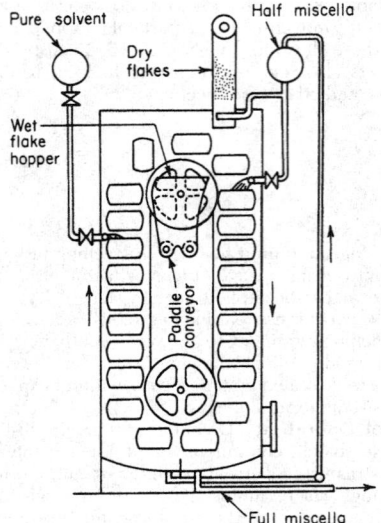

FIG. 18-76 Bollman-type extractor. (*McCabe, Smith, and Harriott,* Unit Operations of Chemical Engineering, *5th ed., p. 616. Copyright 1993 by McGraw-Hill, Inc., New York. Used with permission of McGraw-Hill Book Company.*)

bottoms are held on an endless moving belt. Dry flakes, fed into the descending buckets, are sprayed with partially enriched solvent ("half miscella") pumped from the bottom of the column of ascending buckets. As the buckets rise on the other side of the unit, the solids are sprayed with a countercurrent stream of pure solvent. Exhausted flakes are dumped from the buckets at the top of the unit into a paddle conveyor; enriched solvent, the "full miscella," is pumped from the bottom of the casing. Because the solids are unagitated and because the final miscella moves cocurrently, the Bollman extractor permits the use of thin flakes while producing extract of good clarity. It is only partially a countercurrent device, however, and it sometimes permits channeling and consequent low stage efficiency. Perhaps for

this reason, it is being displaced in the oil extraction industry by horizontal basket, pan, or belt percolators (Schwartzberg, loc. cit.).

In the **horizontal-basket design,** illustrated by the **Rotocel extractor** (Fig. 18-77), walled compartments in the form of annular sectors with liquid-permeable floors revolve about a central axis. The compartments successively pass a feed point, a number of solvent sprays, a drainage section, and a discharge station (where the floor opens to discharge the extracted solids). The discharge station is circumferentially contiguous to the feed point. Countercurrent extraction is achieved by feeding fresh solvent only to the last compartment before dumping occurs and by washing the solids in each preceding compartment with the effluent from the succeeding one. The Rotocel is simple and inexpensive, and it requires little headroom. This type of equipment is made by a number of manufacturers. Horizontal table and tilting-pan vacuum filters, of which it is the gravity counterpart, are used as extractors for leaching processes involving difficult solution-residue separation.

The **endless-belt percolator** (Wakeman, loc. cit.) is similar in principle, but the successive feed, solvent spray, drainage, and dumping stations are linearly rather than circularly disposed. Examples are the **de Smet belt extractor** (uncompartmented) and the **Lurgi frame belt** (compartmented), the latter being a kind of linear equivalent of the Rotocel. Horizontal-belt vacuum filters, which resemble endless-belt extractors, are sometimes used for leaching.

The **Kennedy extractor** (Fig. 18-78), also requiring little headroom, operates substantially as a percolator that moves the bed of solids through the solvent rather than the conventional opposite. It comprises a nearly horizontal line of chambers through each of which in succession the solids being leached are moved by a slow impeller enclosed in that section. There is an opportunity for drainage between stages when the impeller lifts solids above the liquid level before dumping them into the next chamber. Solvent flows countercurrently from chamber to chamber. Because the solids are subjected to mechanical action somewhat more intense than in other types of continuous percolator, the Kennedy extractor is now little used for fragile materials such as flaked oil seeds.

Dispersed-Solids Leaching Equipment for leaching fine solids by dispersion and separation includes batch tanks agitated by rotating impellers or by air and a variety of continuous devices.

Batch Stirred Tanks Tanks agitated by coaxial impellers (turbines, paddles, or propellers) are commonly used for batch dissolution of solids in liquids and may be used for leaching fine solids. Insofar as the controlling rate in the mass transfer is the rate of transfer of mate-

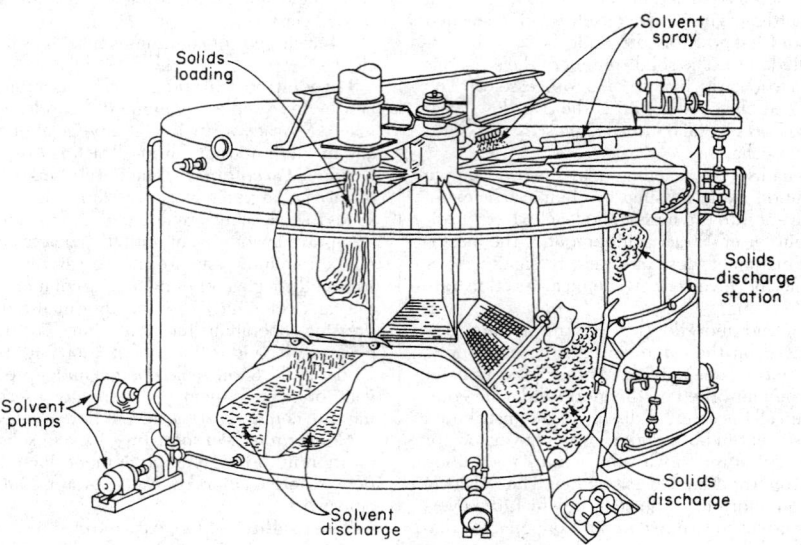

FIG. 18-77 Rotocel extractor. [*Rickles, Chem. Eng.* **72**(6): 164 (1965). *Used with permission of McGraw-Hill, Inc.*]

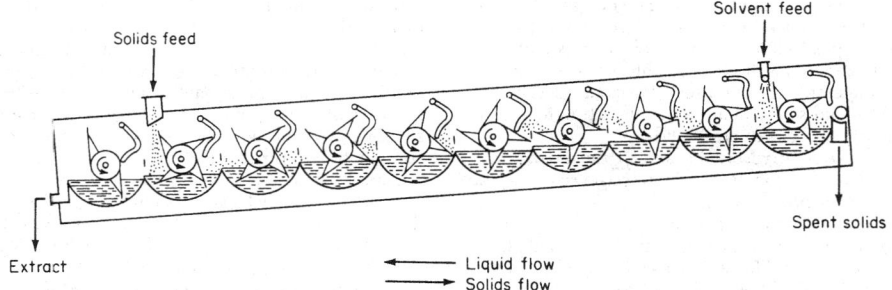

FIG. 18-78   Kennedy extractor. (*Vulcan Cincinnati, Inc.*)

rial into or from the interior of the solid particles rather than the rate of transfer to or from the surface of particles, the main function of the agitator is to supply unexhausted solvent to the particles while they reside in the tank long enough for the diffusive process to be completed. The agitator does this most efficiently if it just gently circulates the solids across the tank bottom or barely suspends them above the bottom.

The leached solids must be separated from the extract by settling and decantation or by external filters, centrifuges, or thickeners, all of which are treated elsewhere in Sec. 18. The difficulty of solids-extract separation and the fact that a batch stirred tank provides only a single equilibrium stage are its major disadvantages.

Pachuca Tanks   Ores of gold, uranium, and other metals are commonly batch-leached in large air-agitated vessels known as Pachuca tanks. A typical tank is a vertical cylinder with a conical bottom section usually with a 60° included angle, 7 m (23 ft) in diameter and 14 m (46 ft) in overall height. In some designs air is admitted from an open pipe in the bottom of the cone and rises freely through the tank; more commonly, however, it enters through a central vertical tube, characteristically about 46 cm (18 in) in diameter, that extends from the bottom of the tank to a level above the conical section—in some cases, almost to the liquid surface. Before it disengages at the liquid surface, the air induces in and above the axial tube substantial flow of pulp, which then finds its way down the outer part of the tank, eventually reentering the riser. The circulation rate in Pachuca tanks is discussed by Lamont [*Can. J. Chem. Eng.,* **36,** 153 (1958)].

Continuous Dispersed-Solids Leaching

Vertical-plate extractor.   Exemplified by the **Bonotto extractor** (Fig. 18-79), this consists of a column divided into cylindrical compartments by equispaced horizontal plates. Each plate has a radial opening staggered 180° from the openings of the plates immediately above and below it, and each is wiped by a rotating radial blade. Alternatively, the plates may be mounted on a coaxial shaft and rotated past stationary blades. The solids, fed to the top plate, thus are caused to fall to each lower plate in succession. The solids fall as a curtain into solvent which flows upward through the tower. They are discharged by a screw conveyor and compactor. Like the Bollman extractor, the Bonotto has been virtually displaced by horizontal belt or tray percolators for the extraction of oil seeds.

Gravity sedimentation tanks.   Operated as thickeners, these tanks can serve as continuous contacting and separating devices in which fine solids may be leached continuously. A series of such units properly connected permit true continuous countercurrent washing of fine solids. If appropriate, a mixing tank may be associated with each thickener to improve the contact between the solids and liquid being fed to that stage. Gravity sedimentation thickeners are described under "Gravity Sedimentation Operations." Of all continuous leaching equipment, gravity thickeners require the most area, and they are limited to relatively fine solids.

The Dorr agitator (Coulson and Richardson, loc. cit.) consolidates in one unit the principles of the thickener and the Pachuca tank. Resembling a rake-equipped thickener, it differs in that the rake is driven by a hollow shaft through which the solids-liquid suspension is lifted and circulated by an air stream. The rake moves the pulp to the center, where it can be entrained by the air stream. The unit may be operated batchwise or continuously.

Impeller-agitated tanks.   These can be operated as continuous leaching tanks, singly or in a series. If the solids feed is a mixture of particles of different settling velocities and if it is desirable that all particles reside in the leaching tank the same lengths of time, design of a continuous stirred leach tank is difficult and uncertain.

Screw-Conveyor Extractors   One type of continuous leaching equipment, employing the screw-conveyor principle, is strictly speaking neither a percolator nor a dispersed-solids extractor. Although it is often classed with percolators, there can be sufficient agitation of the solids during their conveyance by the screw that the action differs from an orthodox percolation.

The **Hildebrandt total-immersion extractor** is shown schematically in Fig. 18-80. The helix surface is perforated so that solvent can pass through countercurrently. The screws are so designed to compact

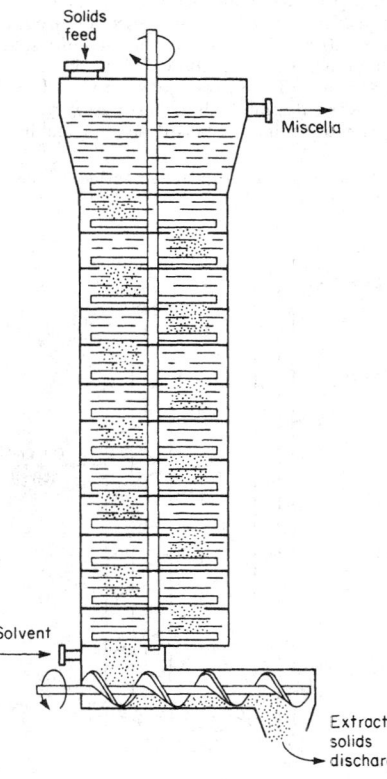

FIG. 18-79   Bonotto extractor. [*Rickles,* Chem. Eng. *72*(6): 163 (1965); copyright 1965 by McGraw-Hill, Inc., New York. Excerpted with special permission of McGraw-Hill.]

the solids during their passage through the unit. The design offers the obvious advantages of countercurrent action and continuous solids compaction, but there are possibilities of some solvent loss and feed overflow, and successful operation is limited to light, permeable solids.

A somewhat similar but simpler design uses a horizontal screw section for leaching and a second screw in an inclined section for washing, draining, and discharging the extracted solids.

In the **De Danske Sukkerfabriker,** the axis of the extractor is tilted to about 10° from the horizontal, eliminating the necessity of two screws at different angles of inclination.

Sugar-beet cossettes are successfully extracted while being transported upward in a vertical tower by an arrangement of inclined plates or wings attached to an axial shaft. The action is assisted by staggered guide plates on the tower wall. The shell is filled with water that passes downward as the beets travel upward. This configuration is employed in the **BMA diffusion tower** (Wakeman, loc. cit.).

Schwartzberg (loc. cit.) reports that screw-conveyor extractors, once widely employed to extract flaked oil seeds, have fallen into disuse for this application because of their destructive action on the fragile seed flakes.

Tray Classifier A hybrid like the screw-conveyor classifier, the tray classifier rakes pulp up the sloping bottom of a tank while solvent flows in the opposite direction. The solvent is forced by a baffle to the bottom of the tank at the lower end before it overflows. The solids must be rugged enough to stand the stress of raking.

SELECTION OR DESIGN OF A LEACHING PROCESS*

At the heart of a leaching plant design at any level—conceptual, preliminary, firm engineering, or whatever—is unit-operations and process design of the extraction unit or line. The major aspects that are particular for the leaching operation are the selection of process and operating conditions and the sizing of the extraction equipment.

Process and Operating Conditions The major parameters that must be fixed or identified are the solvent to be used, the temperature, the terminal stream compositions and quantities, leaching cycle (batch or continuous), contact method, and specific extractor choice.

Choice of Solvent The solvent selected will offer the best balance of a number of desirable characteristics: high saturation limit and selectivity for the solute to be extracted, capability to produce

---

° Portions of this subsection are adaptations from the still-pertinent article by Rickles (loc. cit.).

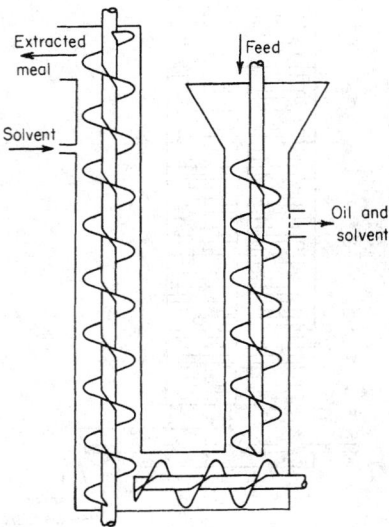

FIG. 18-80 Hildebrandt extractor. (*McCabe, Smith, and Harriott,* Unit Operations of Chemical Engineering, *5th ed., p. 616. Copyright 1993 by McGraw-Hill, Inc., New York. Used with permission by McGraw-Hill Book Company.*)

extracted material of quality unimpaired by the solvent, chemical stability under process conditions, low viscosity, low vapor pressure, low toxicity and flammability, low density, low surface tension, ease and economy of recovery from the extract stream, and price. These factors are listed in an approximate order of decreasing importance, but the specifics of each application determine their interaction and relative significance, and any one can control the decision under the right combination of process conditions.

Temperature The temperature of the extraction should be chosen for the best balance of solubility, solvent-vapor pressure, solute diffusivity, solvent selectivity, and sensitivity of product. In some cases, temperature sensitivity of materials of construction to corrosion or erosion attack may be significant.

Terminal Stream Compositions and Quantities These are basically linked to an arbitrary given: the production capacity of the leaching plant (rate of extract production or rate of raw-material purification by extraction). When options are permitted, the degree of solute removal and the concentration of the extract stream chosen are those that maximize process economy while sustaining conformance to regulatory standards.

Leaching Cycle and Contact Method As is true generally, the choice between continuous and intermittent operation is largely a matter of the size and nature of the process of which the extraction is a part. The choice of a percolation or solids-dispersion technique depends principally on the amenability of the extraction to effective, sufficiently rapid percolation.

Type of Reactor The specific type of reactor that is most compatible (or least incompatible) with the chosen combination of the preceding parameters seldom is clearly and unequivocally perceived without difficulty, if at all. In the end, however, that remains the objective. As is always true, the ultimate criteria are reliability and profitability.

Extractor-Sizing Calculations For any given throughput rate (which fixes the cross-sectional area and/or the number of extractors), the size of the units boils down to the number of stages required, actual or equivalent. In calculation, this resolves into determination of the number of ideal stages required and application of appropriate stage efficiencies. The methods of calculation resemble those for other mass-transfer operations (see Secs. 13, 14, and 15), involving equilibrium data and contact conditions, and based on material balances. They are discussed briefly here with reference to countercurrent contacting.

Composition Diagrams In its elemental form, a leaching system consists of three components: inert, insoluble solids; a single nonadsorbed solute, which may be liquid or solid; and a single solvent.° Thus, it is a ternary system, albeit an unusual one, as already mentioned, by virtue of the total mutual "insolubility" of two of the phases and the simple nature of equilibrium.

The composition of a typical system is satisfactorily presented in the form of a diagram. Those diagrams most frequently employed are a right-triangular plot of mass fraction of solvent against mass fraction of solute (Fig. 18-81a) and a plot suggestive of a Ponchon-Savarit diagram, with inerts taking the place of enthalpy (Fig. 18-81b). A third diagram, less frequently used, is a modified McCabe-Thiele plot in which the overflow solution (inerts-free) and the underflow solution (traveling out of a stage with the inerts) are treated as pseudo phases, the mass fraction of solute in overflow, y, being plotted against the mass fraction of solute in underflow, x. (An additional representation, the equilateral-triangular diagram frequently employed for liquid-liquid ternary systems, is seldom used because the field of leaching data is confined to a small portion of the triangle.)

With reference to Fig. 18-81 (both graphs), EF represents the locus of overflow compositions for the case in which the overflow stream contains no inert solids. $E'F'$ represents the overflow streams containing some inert solids, either by entrainment or by partial solubility in the overflow solution. Lines GF, GL, and GM represent the loci of underflow compositions for the three different conditions indicated on the diagram. In Fig. 18-81a, the constant underflow line GM is parallel to EF, the hypotenuse of the triangle, whereas GF passes through

---

° The solubility of the inert, adsorption of solute on the inert, and complexity of solvent and extracted material can be taken into account if necessary. Their consideration is beyond the scope of this treatment.

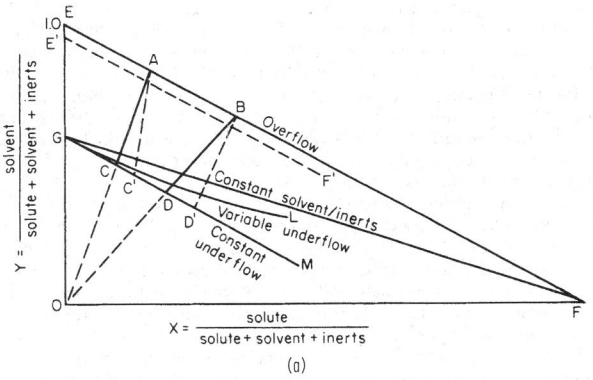

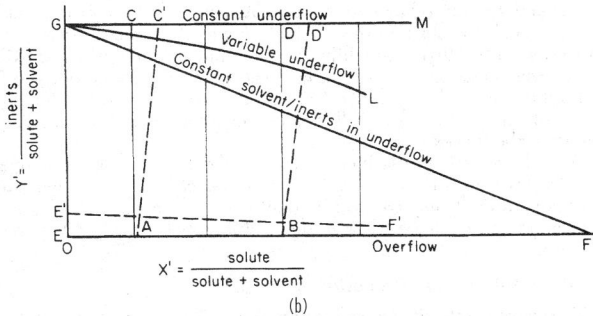

FIG. 18-81 Composition diagrams for leaching calculations: (*a*) right-triangular diagram; (*b*) modified Ponchon-Savarit diagram.

the right-hand vertex representing 100 percent solute. In Fig. 18-81*b*, underflow line *GM* is parallel to the abscissa, and *GF* passes through the point on the abscissa representing the composition of the clear solution adhering to the inert solids.

Compositions of overflow and underflow streams leaving the same stage are represented by the intersection of the composition lines for those streams with a tie line (*AC*, *AC′*, *BD*, *BD′*). Equilibrium tie lines (*AC*, *BD*) pass through the origin (representing 100 percent inerts) in Fig. 18-81*a*, and are vertical (representing the same inert-free solution composition in both streams) in Fig. 18-81*b*. For nonequilibrium conditions with or without adsorption or for equilibrium conditions with selective adsorption, the tie lines are displaced, such as *AC′* and *BD′*. Point *C′* is to the right of *C* if the solute concentration in the overflow solution is less than that in the underflow solution adhering to the solids. Unequal concentrations in the two solutions indicate insufficient contact time and/or preferential adsorption of one of the components on the inert solids. Tie lines such as *AC′* may be considered as "practical tie lines" (i.e., they represent actual rather than ideal stages) if data on underflow and overflow composition have been

obtained experimentally under conditions simulating actual operation, particularly with respect to contact time, agitation, and particle size of solids.

The illustrative construction lines of Fig. 18-81 have been made with the assumption of constant underflow. In the more realistic case of variable underflow, the points *C*, *C′*, *D*, *D′* would lie along line *GL*. Like the practical tie lines, *GL* is a representation of experimental data.

Algebraic Computation   This method starts with calculation of the quantities and compositions of all the terminal streams, using a convenient quantity of one of the streams as the basis of calculation. Material balance and stream compositions are then computed for a terminal ideal stage at either end of an extraction battery (i.e., at Point *A* or Point *B* in Fig. 18-81), using equilibrium and solution-retention data. Calculations are repeated for each successive ideal stage from one end of the system to the other until an ideal stage which corresponds to the desired conditions is obtained. Any solid-liquid extraction problem can be solved by this method.

For certain simplified cases it is possible to calculate directly the number of stages required to attain a desired product composition for a given set of feed conditions. For example, if equilibrium is attained in all stages and if the underflow mass rate is constant, both the equilibrium and operating lines on a modified McCabe-Thiele diagram are straight, and it is possible to calculate directly the number of ideal stages required to accommodate any rational set of terminal flows and compositions (McCabe, Smith, and Harriott, op. cit.):

$$N = \frac{\log\,[(y_b - x_b)/(y_a - x_a)]}{\log\,[(y_b - y_a)/(x_b - x_a)]} \qquad (18\text{-}42)$$

Even when the conditions of equilibrium in each stage and constant underflow obtain, Eq. (18-42) normally is not valid for the first stage because the unextracted solids entering that stage usually are not premixed with solution to produce the underflow mass that will leave. This is easily rectified by calculating the exit streams for the first stage and using those values in Eq. (18-42) to calculate the number of stages required after stage 1.

Graphical Method   This method of calculation is simply a diagrammatic representation of all the possible compositions in a leaching system, including equilibrium values, on which material balances across ideal (or, in some cases, nonideal) stages can be evaluated in the graphical equivalent of the stage-by-stage algebraic computation. It normally is simpler than the hand calculation of the algebraic solution, and it is viewed by many as helpful because it permits visualization of the process variables and their effect on the operation. Any of the four types of composition diagrams described above can be used, but modified Ponchon-Savarit or right-triangular plots (Fig. 18-81) are most convenient for leaching calculations.

The techniques of graphical solution, in fact, are not unlike those for distillation and absorption (binary) problems using McCabe-Thiele, Ponchon-Savarit, and right-triangular diagrams and are similar to those described in Sec. 15 for solvent-extraction (ternary) systems. More detailed explanations of the application of the several graphical conventions to leaching are presented by: Coulson and Richardson, right triangle; Rickles, modified Ponchon-Savarit; McCabe, Smith, and Harriott, modified McCabe-Thiele; and Schwartzberg, equilateral ternary diagram; all in the publications cited as general references. (See also Treybal, *Mass Transfer Operations*, 3d ed., McGraw-Hill, New York, 1980.)

GRAVITY SEDIMENTATION OPERATIONS

GENERAL REFERENCES: Albertson, *Fluid/Particle Separation Journal,* **7**(1), 18 (1994). Coe and Clevenger, *Trans. Am. Inst. Min. Eng.,* **55,** 356 (1916). Comings, Pruiss, and DeBord, *Ind. Eng. Chem.,* **46,** 1164 (1954). Counselmann, *Trans. Am. Inst. Min. Eng.,* **187,** 223 (1950). Fitch, *Ind. Eng. Chem.,* **58**(10), 18 (1966). *Trans. Soc. Min. Eng. AIME,* **223,** 129 (1962). Kynch, *Trans. Faraday Soc.,* **48,** 166 (1952). Pearse, *Gravity Thickening Theories: A Review,* Warren Spring Laboratory, Hertfordshire, 1977. Purchas, *Solid/Liquid Separation Equipment Scale-Up,* Uplands Press, Croydon, England, 1977. Sankey and Payne, *Chemical Reagents in the Mineral Processing Industry,* **245,** SME (1985). Talmage and Fitch, *Ind. Eng. Chem.,* **47,** 38 (1955). Wilhelm and Naide, *Min. Eng. (Littleton, Colo.),* **33,** 1710 (1981).

Sedimentation is the partial separation or concentration of suspended solid particles from a liquid by gravity settling. This field may be divided into the functional operations of thickening and clarification. The primary purpose of thickening is to increase the concentration of suspended solids in a feed stream, while that of clarification is to remove a relatively small quantity of suspended particles and produce a clear effluent. These two functions are similar and occur simultaneously, and the terminology merely makes a distinction between the primary process results desired. Generally, thickener mechanisms are designed for the heavier-duty requirements imposed by a large quantity of relatively concentrated pulp, while clarifiers usually will include features that ensure essentially complete suspended-solids removal, such as greater depth, special provision for coagulation or flocculation of the feed suspension, and greater overflow-weir length.

CLASSIFICATION OF SETTLEABLE SOLIDS; SEDIMENTATION TESTS

The types of sedimentation encountered in process technology will be greatly affected not only by the obvious factors—particle size, liquid viscosity, solid and solution densities—but also by the characteristics of the particles within the slurry. These properties, as well as the process requirements, will help determine both the type of equipment which will achieve the desired ends most effectively and the testing methods to be used to select the equipment.

Figure 18-82 illustrates the relationship between solids concentration, interparticle cohesiveness, and the type of sedimentation that may exist. "Totally discrete" particles include many mineral particles (usually greater in diameter than 20 μm), salt crystals, and similar substances that have little tendency to cohere. "Flocculent" particles generally will include those smaller than 20 μm (unless present in a dispersed state owing to surface charges), metal hydroxides, many chemical precipitates, and most organic substances other than true colloids.

At low concentrations, the type of sedimentation encountered is called particulate settling. Regardless of their nature, particles are sufficiently far apart to settle freely. Faster-settling particles may collide with slower-settling ones and, if they do not cohere, continue downward at their own specific rate. Those that do cohere will form floccules of a larger diameter that will settle at a rate greater than that of the individual particles.

There is a gradual transition from particulate settling into the zone-settling regime, where the particles are constrained to settle as a mass. The principal characteristic of this zone is that the settling rate of the mass, as observed in batch tests, will be a function of its solids concentration (for any particular condition of flocculation, particle density, etc.).

The solids concentration ultimately will reach a level at which particle descent is restrained not only by hydrodynamic forces but also partially by mechanical support from the particles below; therefore,

the weight of particles in mutual contact can influence the rate of sedimentation of those at lower levels. This compression, as it is termed, will result in further solids concentration because of compaction of the individual floccules and partial filling of the interfloc voids by the deformed floccules. Accordingly, the rate of sedimentation in the compression regime is a function of both the solids concentration and the depth of pulp in this particular zone. As indicated in Fig. 18-82, granular, nonflocculent particles may reach their ultimate solids concentration without passing through this regime.

As an illustration, coarse-size (45 μm) the aluminum oxide trihydrate particles produced in the Bayer process would be located near the extreme left of Fig. 18-82. These solids settle in a particulate manner, passing through a zone-settling regime only briefly, and reach a terminal density or ultimate solids concentration without any significant compressive effects. At this point, the solids concentration may be as much as 80 percent by weight. The same compound, but of the gelatinous nature it has when precipitated in water treatment as aluminum hydroxide, would be on the extreme right-hand side of the figure. This flocculent material enters into a zone-settling regime at a low concentration (relative to the ultimate concentration it can reach) and gradually thickens. With sufficient pulp depth present, preferably aided by gentle stirring or vibration, the compression-zone effect will occur; this is essential for the sludge to attain its maximum solids concentration, around 10 percent. Certain fine-size (1- to 2-μm) precipitates of this compound will possess characteristics intermediate between the two extremes.

A feed stream to be clarified or thickened can exist at any state represented within this diagram. As it becomes concentrated owing to sedimentation, it may pass through all the regimes, and the settling rate in any one may be the size-determining factor for the required equipment.

Sedimentation-Test Procedures

Determination of Clarification-Zone Requirements   In the treatment of solids suspensions which are in the particulate-settling regime, the usual objective will be the production of a clear effluent and test methods limited to this type of settling will be the normal sizing procedure, although the area demand for thickening should be verified. With particulate or slightly flocculent matter, any method that measures the rate of particle subsidence will be suitable, and either long-tube or short-tube procedures (described later) may be used. If the solids are strongly flocculent and particles cohere easily during sedimentation, the long tube test will yield erratic data, with better clarities being observed in samples taken from the lower taps (i.e., clarity appears to improve at higher settling rates). In these instances, time alone usually is the principal variable in clarification, and a simple detention test is recommended.

Long-Tube Method   A transparent tube 2 to 4 m long and at least 100 mm in diameter (preferably larger), fitted with sampling taps every 200 to 300 mm, is used in this test. The tube is mounted vertically and filled with a representative sample of feed suspension. At timed intervals approximately 100-mL samples are withdrawn from successive taps, beginning with the uppermost one. The time intervals will be determined largely by the settling rate of the particles and should be chosen so that a series of at least four time intervals will produce samples that bracket the desired solids-removal target. Also, this procedure will indicate whether or not detention time is a factor in the rate of clarification. Typically, intervals may be 30 min long, the last series of samples representing the results obtainable with 2-h detention. The samples are analyzed for suspended-solids concentration by any suitable means, such as filtration through membranes or centrifugation with calibrated tubes.

A plot is made of the suspended-solids concentration in each of the samples as a function of the nominal settling velocity of the sample, which is determined from the corresponding sample-tap depth divided by the elapsed time between the start of the test and the time of sampling. For each sampling series after the first, the depth will have changed because of the removal of the preceding samples, and this must be taken into account. With particulate solids having little or no tendency to cohere, the data points generally will fall on one line irrespective of the detention time. This indicates that the settling area

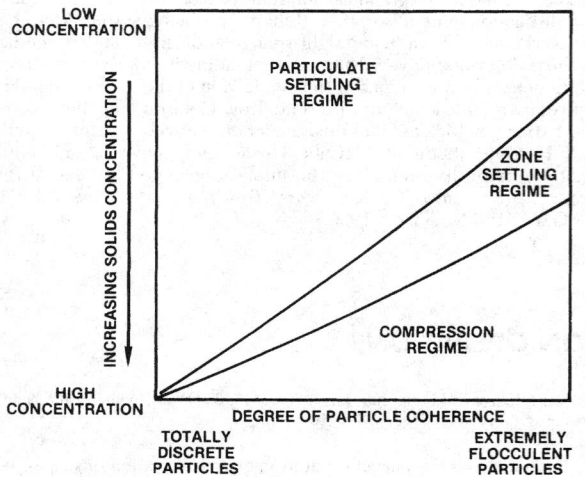

FIG. 18-82 Combined effect of particle coherence and solids concentration on the settling characteristics of a suspension.

available in the basin will determine the degree of solids removal, and the depth will have little bearing on the results, except as it may affect clarification efficiency.

Since the solids concentration in the 100-mL samples withdrawn at different depths does not correspond to the average concentration in the fluid above the sample point (which would be equivalent to the overflow from the clarifier at that particular design rate), an adjustment must be made in calculating the required area. Accordingly, an approximate average of the solids concentration throughout the column of liquid above a given tap can be obtained by summing the values obtained from all the higher taps and the one which is being sampled and dividing by the total number of taps sampled. This value is then used in a plot of overflow suspended-solids concentration versus nominal settling velocity.

Areal efficiencies for properly designed clarifiers in which detention time is not a significant factor range from 65 to 80 percent, and the surface area should be increased accordingly to reduce the overflow rate for scale-up.

Should the particles have a tendency to cohere slightly during sedimentation, each sampling time, representing a different nominal detention time in the clarifier, will produce different suspended-solids concentrations at similar rates. These data can be plotted as sets of curves of concentration versus settling rate for each detention time by the means just described. Scale-up will be similar, except that detention time will be a factor, and both depth and area of the clarifier will influence the results. In most cases, more than one combination of diameter and depth will be capable of producing the same clarification result.

These data may be evaluated by selecting different nominal overflow rates (equivalent to settling rates) for each of the detention-time values, and then plotting the suspended-solids concentrations for each nominal overflow rate (as a parameter) against the detention time. For a specified suspended-solids concentration in the effluent, a curve of overflow rate versus detention time can be prepared from this plot and used for optimizing the design of the equipment.

Short-Tube Method   This test is suitable in cases in which detention time does not change the degree of particle flocculation and hence has no significant influence on particle-settling rates. It is also useful for hydroseparator tests where the sedimentation device is to be used for classification (see Sec. 19). A tube 50 to 75 mm in diameter and 300 to 500 mm long is employed. A sample placed in the tube is mixed to ensure uniformity, and settling is allowed to occur for a measured interval. At the end of this time, the supernatant liquid is siphoned off quickly down to a chosen level, and the collected sample is analyzed for suspended-solids concentration. The level selected usually is based on the relative expected volumes of overflow and underflow. The suspended-solids concentration measured in the siphoned sample will be equivalent to the "averaged" values obtained in the long-tube test, at a corresponding settling rate.

In hydroseparator tests, it is necessary to measure solids concentrations and size distributions of both the supernatant sample withdrawn and the fraction remaining in the cylinder. The volume of the latter sample should be such as to produce a solids concentration that would be typical of a readily pumped underflow slurry.

Detention Test   This test utilizes a 1- to 4-L beaker or similar vessel. The sample is placed in the container, flocculated by suitable means if required, and allowed to settle. Small samples for suspended-solids analysis are withdrawn from a point approximately midway between liquid surface and settled solids interface, taken with sufficient care that settled solids are not resuspended. Sampling times may be at consecutively longer intervals, such as 5, 10, 20, 40, and 80 min.

The suspended-solids concentration can be plotted on log-log paper as a function of the sampling (detention) time. A straight line usually will result, and the required static detention time t to achieve a certain suspended-solids concentration C in the overflow of an ideal basin can be taken directly from the graph. If the plot is a straight line, the data are described by the equation

$$C = Kt^m \qquad (18\text{-}43)$$

where the coefficient K and exponent m are characteristic of the particular suspension.

Should the suspension contain a fraction of solids which can be considered "unsettleable," the data are more easily represented by using the so-called second-order procedure. This depends on the data being reasonably represented by the equation

$$Kt = \frac{1}{C - C_\infty} - \frac{1}{C_0 - C_\infty} \qquad (18\text{-}44)$$

where C_∞ is the unsettleable-solids concentration and C_0 is the concentration of suspended solids in the unsettled (feed) sample. The residual-solids concentration remaining in suspension after a sufficiently long detention time (C_∞) must be determined first, and the data then plotted on linear paper as the reciprocal concentration function $1/(C - C_\infty)$ versus time.

Bulk-settling test.   In cases involving detention time only, the overflow rate must be considered by other means. This is done by carrying out a settling test in which the solids are first concentrated to a level at which zone settling just begins. This is usually marked by a very diffuse interface during initial settling. Its rate of descent is measured with a graduated cylinder of suitable size, preferably at least 1 L, and the initial straight-line portion of the settling curve is used for specifying a bulk-settling rate. The design overflow rate generally should not exceed half of the bulk-settling rate.

Detention efficiency.   Conversion from the ideal basin sized by detention-time procedures to an actual clarifier requires the inclusion of an efficiency factor to account for the effects of turbulence and nonuniform flow. Efficiencies vary greatly, being dependent not only on the relative dimensions of the clarifier and the means of feeding but also on the characteristics of the particles. The curve shown in Fig. 18-83 can be used to scale up laboratory data in sizing circular clarifiers. The static detention time determined from a test to produce a specific effluent solids concentration is divided by the efficiency (expressed as a fraction) to determine the nominal detention time, which represents the volume of the clarifier above the settled pulp interface divided by the overflow rate. Different diameter-depth combinations are considered by using the corresponding efficiency factor. In most cases, area may be determined by factors other than the bulk-settling rate, such as practical tank-depth limitations.

Thickener-Basin Area   The area requirements for thickeners frequently are based on the solids flux rates measured in the zone-settling regime. Theory holds that, for any specific sedimentation condition, a critical concentration which will limit the solids throughput rate will exist in the thickener. This critical concentration will be evidenced as a pulp bed of variable depth in which the solids concentration is fairly uniform from top to bottom. Since the underflow concentration usually will be higher than this, gradually increasing concentrations will be found with progressing depth in the region beneath this constant-concentration zone. As the concentration within this critical zone represents a steady-state condition, its vertical extent may vary continually, responding to minor changes in the feed rate, underflow withdrawal rate, or flocculant dosage. In thickeners operating at relatively high underflow concentrations, with long

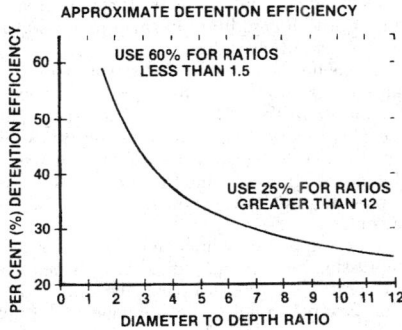

FIG. 18-83   Efficiency curve for scale-up of batch clarification data to determine nominal detention time in a continuous clarifier.

solids-detention times and lower throughput, this zone generally will not be present.

Many batch-test methods which are based on determining the solids flux rate at this critical concentration have been developed. Most methods recognize that as the solids enter compression, thickening behavior is no longer a function only of solids concentration. Hence, these methods attempt to utilize the "critical" point dividing these two zones and size the area on the basis of the settling rate of a layer of pulp at this concentration. The difficulty lies in discerning where this point is located on the settling curve.

Many procedures have been developed, but two in particular have been more widely used: the Coe and Clevenger approach and the Kynch method as defined by Talmage and Fitch (op. cit.).

The former requires measurement of the initial settling rate of a pulp at different solids concentrations varying from feed to final underflow value. The area requirement for each solids concentration tested is calculated by equating the net overflow rate to the corresponding interfacial settling rate, as represented by the following equation for the unit area:

$$\text{Unit area} = \frac{1/C_i - 1/C_u}{v_i} \qquad (18\text{-}45)$$

where C_i is the solids concentration at the interfacial settling velocity v_i and C_u is the underflow concentration, both concentrations being expressed in terms of mass of solids per unit volume of slurry. Using kg/L for the concentrations and m/day for the settling velocity yields a unit area value in m²/(ton/day).

These unit area values, plotted as a function of the feed concentration, will describe a maximum value that can be used to specify the thickener design unit area for the particular underflow concentration C_u employed in Eq. (18-45).

The method is applicable for *unflocculated* pulps or those in which the ionic characteristics of the solution produce a flocculent structure. If polymeric flocculants are used, the floccule size will be highly dependent on the feed concentration, and an approach based on the Kynch theory is preferred. In this method, the test is carried out at the expected feed solids concentration and is continued until underflow concentration is achieved in the cylinder. To determine the unit area, Talmage and Fitch (op. cit.) proposed an equation derived from a relationship equivalent to that shown in Eq. (18-45):

$$\text{Unit area} = \frac{t_u}{C_0 H_0} \qquad (18\text{-}46)$$

where t_u is the time, days; C_0 is the initial solids concentration in the feed, t/m³; and H_0 is the initial height of the slurry in the test cylinder, m. The term t_u is taken from the intersection of a tangent to the curve at the critical point and a horizontal line representing the depth of pulp at underflow concentration. There are various means for selecting this critical point, all of them empirical, and the unit area value determined cannot be considered precise. The review by Pearse (op. cit.) presents many of the different procedures used in applying this approach to laboratory settling test data.

Because this method in itself does not adequately address the effect of different underflow concentrations on the unit area, a thickener sized using this approach could be too small in a case where a relatively concentrated underflow is desired.

Two other approaches avoid using the critical point by computing the area requirements from the settling conditions existing at the underflow concentration. The Wilhelm and Naide procedure (op. cit.) applies zone-settling theory (Kynch) to the entire thickening regime. Tangents drawn to the settling curve are used to calculate the settling velocity at all concentrations obtained in the test. This permits construction of a plot (Figure 18-84) showing unit area as a function of underflow concentration.

A second, "direct" approach which yields a similar result, since it also takes compression into account, utilizes the value of settling time t_x taken from the settling curve at a particular underflow concentration. This value is used to solve the Talmage and Fitch equation (18-46) for unit area.

Compression bed depth will have a significant effect on the overall settling rate (increasing compression zone depth reduces unit area).

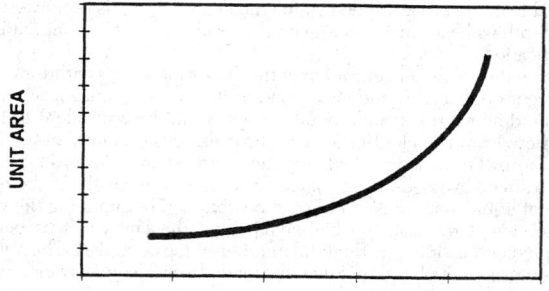

UNDERFLOW SOLIDS CONCENTRATION

FIG. 18-84 Characteristic relationship between thickener unit area and underflow solids concentration (fixed flocculant dosage and pulp depth).

Therefore, in applying either of these two procedures it is necessary to run the test in a vessel having an average bed depth close to that expected in a full-scale thickener. This requires a very large sample, and it is more convenient to carry out the test in a cylinder having a volume of 1 to 4 liters. The calculated unit area value from this test can be extrapolated to full-scale depth by carrying out similar tests at different depths to determine the effect on unit area. Alternatively, an empirical relationship can be used which is effective in applying a depth correction to laboratory cylinder data over normal operating ranges. The unit area calculated by either the Wilhelm and Naide approach or the direct method is multiplied by a factor equal to $(h/H)^n$, where h is the average depth of the pulp in the cylinder, H is the expected full-scale compression zone depth, usually taken as 1 m, and n is the exponent calculated from Fig. 18-85. For conservative design purposes, the minimum value of this factor that should be used is 0.25.

It is essential to use a slow-speed (approximately 0.1 r/min) picket rake in all cylinder tests to prevent particle bridging and allow the sample to attain the underflow density which is obtainable in a full-scale thickener.

Continuously operated, small-scale or pilot-plant thickeners, ranging from 75 mm diameter by 400 mm depth to several meters in diameter, are also effectively used for sizing full-scale equipment. This approach requires a significantly greater volume of sample, such as

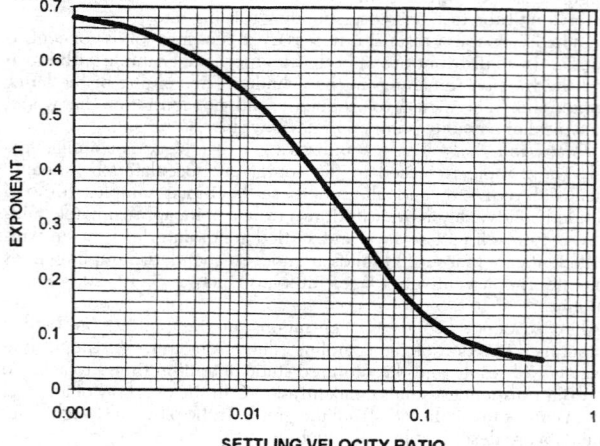

FIG. 18-85 Depth correction factor to be applied to unit areas determined with Wilhelm-Naide and "direct" methods. Velocity ratio calculated using tangents to settling curve at a particular settled solids concentration and at start of test.

would be available in an operating installation or a pilot plant. Continuous units and batch cylinders will produce equivalent results if proper procedures are followed with either system.

Thickener-Basin Depth   The pulp depth required in the thickener will be greatly affected by the role that compression plays in determining the rate of sedimentation. If the zone-settling conditions define the area needed, depth of pulp will be unimportant and can be largely ignored, as the "normal" depth found in the thickener will be sufficient. On the other hand, with the compression zone controlling, depth of pulp will be significant, and it is essential to measure the sedimentation rate under these conditions.

To determine the compression-zone requirement in a thickener, a test should be run in a deep cylinder in which the average settling pulp depth approximates the depth anticipated in the full-scale basin. The average density of the pulp in compression is calculated and used in Eq. (18-47) to determine the required compression-zone volume:

$$V = \frac{\theta_c(\rho_s - \rho_l)}{\rho_s(\rho_{sl} - \rho_l)} \qquad (18\text{-}47)$$

where V is the volume, m^3, required per ton of solids per day; θ_c is the compression time, days, required in the test to reach underflow concentration; and ρ_s, ρ_l, ρ_{sl}, are the densities of the solids, liquid, and slurry (average), respectively, ton/m^3. This value divided by the average depth of the pulp during the period represents the unit area defined from compression requirements. If it exceeds the value determined from the zone-settling tests, it is the quantity to be used.

The side depth of the thickener is determined as the sum of the depths needed for the compression zone and for the clear zone. Normally, 1.5 to 2 m of clear liquid depth above the expected pulp level in a thickener will be sufficient for stable, effective operation. When the location of the pulp level cannot be predicted in advance or it is expected to be relatively low, a thickener sidewall depth of 2 to 3 m is usually safe. Greater depth may be used in order to provide better clarity, although in most thickener applications the improvement obtained by this means will be marginal.

Scale-up Factors   Factors used in thickening will vary, but, typically, a 1.2 to 1.3 multiplier applied to the unit area calculated from laboratory data is sufficient if proper testing procedures have been followed and the samples are representative.

Flocculation   Flocculants are commonly used because of the considerable reduction in equipment size and capital cost that can be effected with very nominal reagent dosages. Selection of the reagents usually involves simple bench-scale comparison tests on small samples of pulp for rough screening, followed by larger-scale tests in cylinders or in continuous pilot-plant thickeners. Determination of the optimum dosage is complicated; it involves a number of economic factors such as reagent and capital costs, cost of a shutdown or loss in availability due to failure in the flocculation system, and consideration of possible future increases in plant capacity.

Polymeric flocculants are available in various chemical compositions and molecular weight ranges, and they may be nonionic in character or may have predominantly cationic or anionic charges. The range of application varies; but, in general, nonionics are well suited to acidic suspensions, anionic flocculants work well in neutral or alkaline environments, and cationics are most effective on organic material and colloidal matter.

Colloidal solids (e.g., as encountered in waste treatment) may require initial treatment with a chemical having strong ionic properties, such as acid, lime, alum, or ferric sulfate. The latter two will precipitate at neutral pH and produce a gelatinous, flocculent structure which further helps collect extremely small particles. Some cationic polymers also may be effective in flocculation of particles of this type. (This action is commonly termed *coagulation*.) Prolonged, gentle agitation improves the degree and rate of flocculation under these conditions. If the solids concentration is relatively low, e.g., <500 mg/L, results can usually be improved by recirculation of settled material to the flocculation zone to produce an optimum concentration for floc growth.

With polymer flocculation of slurries, however, extended agitation after the addition of the polymer may be detrimental. The reagent should be added to the slurry under conditions which promote rapid dispersion and uniform, complete mixing with a minimum of shear. In cylinder tests, this can be accomplished by simultaneously injecting and mixing flocculant with the slurry, using an apparatus consisting of a syringe, a tube, and an inverted rubber stopper. The stopper, with a diameter approximately three-fourths of that of the cylinder, provides sufficient turbulence as it is moved gently up and down through the sample to cause good blending of reagent and pulp.

Flocculant solution preparation and use should take into account specific properties of these reagents. During mixing and distribution, excessive shear should be avoided since it can reduce reagent effectiveness and result in higher consumption. Flocculant solution should be prepared at as high a concentration—generally, 0.25 to 1%—as can be handled by the metering pump and agitator, and added to the pulp at the maximum dilution possible. Dry polymers require sufficient aging time after initial mixing with water in order to develop their full effectiveness. Liquid polymers—emulsions and dispersions—contain additives such as carriers, activators, and other components which can have an effect on the process and other equipment, and this should be investigated before their use. Care should be taken that the water employed for dissolving as well as for dilution will not affect reagent activity, as this could increase consumption. Prior to design of the system, laboratory tests should be conducted with both plant and tap water to determine if there is a detrimental effect using process water.

Torque Requirements   Sufficient torque must be available in the raking mechanism of a full-scale thickener to allow it to move through the slurry and assist solids movement to the underflow outlet. Granular, particulate solids that settle rapidly and reach a terminal solids concentration without going through any apparent compression or zone-settling region require a maximum raking capability, as they must be moved to the outlet solely by the mechanism. At the other end of the spectrum, extremely fine materials, such as clays and precipitates, require a minimum of raking, for most of the solids may reach the underflow outlet hydrodynamically. The rakes prevent a gradual buildup of some solids on the bottom, however, and the gentle stirring action from the rake arm often aids the thickening process. As the underflow concentration approaches its ultimate limit, the consistency will increase greatly, resulting in a higher raking requirement and an increase in torque.

For most materials, the particle size lies somewhere between these two extremes, and the torques required in two properly designed thickeners of the same size but in distinct applications can differ greatly. Unfortunately, test methods to specify torque from small-scale tests are of questionable value, since it is difficult to duplicate actual conditions. Manufacturers of sedimentation equipment select torque ratings from experience with similar substances and will recommend a torque capability on this basis. Definitions of *operating torque* vary with the manufacturer, and the user should ask the supplier to specify the *B-10 life* for bearings and to reference appropriate mechanical standards for continuous operation of the selected gear set at specific torque levels. This will provide guidelines for plant operators and help avoid premature failure of the mechanism. Abnormal conditions above the normal operating torque are inevitable, and a thickener should be provided with sufficient torque capability for short-term operation at higher levels in order to ensure continuous performance.

Underflow Pump Requirements   Many suspensions will thicken to a concentration higher than that which can be handled by conventional slurry pumps. Thickening tests should be performed with this in mind, for, in general, the unit area to produce the maximum concentration that can be pumped is the usual design basis. Determination of this ultimate pumpable concentration is largely a judgmental decision requiring some experience with slurry pumping; however, the behavior of the thickened suspension can be used as an approximate guide to pumpability. The supernatant should be decanted following a test and the settled solids repulped in the cylinder to a uniform consistency. Repulping is done easily with a rubber stopper fastened to the end of a rigid rod. If the bulk of the repulped slurry can be poured from the cylinder when it is tilted 10 to 30° above the horizontal, the corresponding thickener underflow can be handled by most types of slurry pumps. But if the slurry requires cylinder shaking or other mechanical means for its removal, it should be diluted to a more fluid condition, if conventional pump systems are to be employed.

THICKENERS

The primary function of a continuous thickener is to concentrate suspended solids by gravity settling so that a steady-state material balance is achieved, solids being withdrawn continuously in the underflow at the rate they are supplied in the feed. Normally, an inventory of pulp is maintained in order to achieve the desired concentration. This volume will vary somewhat as operating conditions change; on occasion, this inventory can be used for storage of solids when feed and underflow rates are reduced or temporarily suspended.

A thickener has several basic components: a tank to contain the slurry, feed piping and a feedwell to allow the feed stream to enter the tank, a rake mechanism to assist in moving the concentrated solids to the withdrawal points, an underflow solids-withdrawal system, and an overflow launder. The basic design of a bridge-supported thickener mechanism is illustrated in Fig. 18-86.

High-Rate Thickeners   Flocculants are commonly used in thickeners, and this practice has resulted in thickener classification as either *conventional* or *high-rate*. These designations can be confusing in that they imply that there is a sharp distinction between the two,

which *is not* the case. The greater capacity expected from a high-rate thickener is due solely to the effective use of flocculant to maximize throughput, as shown in Fig. 18-87. In most applications, there is a threshold dosage at which a noticeable increase in capacity begins to occur. This effect will continue up to a limit, at which point the capacity will be a maximum unless a lower underflow concentration is accepted, as illustrated in Fig. 18-84. Since flocculant is usually added to a thickener either in the feed line or the feedwell, there are a number of proprietary feedwell designs which are used in high-rate thickeners in order to help optimize flocculation. De-aeration systems may be included in some cases to avoid air entrainment in the flocculated slurry. The other components of these units are not materially different from those of a conventional thickener.

High-Density Thickeners   Thickeners can be designed to produce underflows having a very high apparent viscosity, permitting disposal of waste slurries at a concentration that avoids segregation of fines and coarse particles or formation of a free-liquid pond on the surface of the deposit. This practice is applied in *dry-stacking* systems and underground *paste-fill* operations for disposal of mine tailings and similar materials. The thickener mechanism generally will require a special

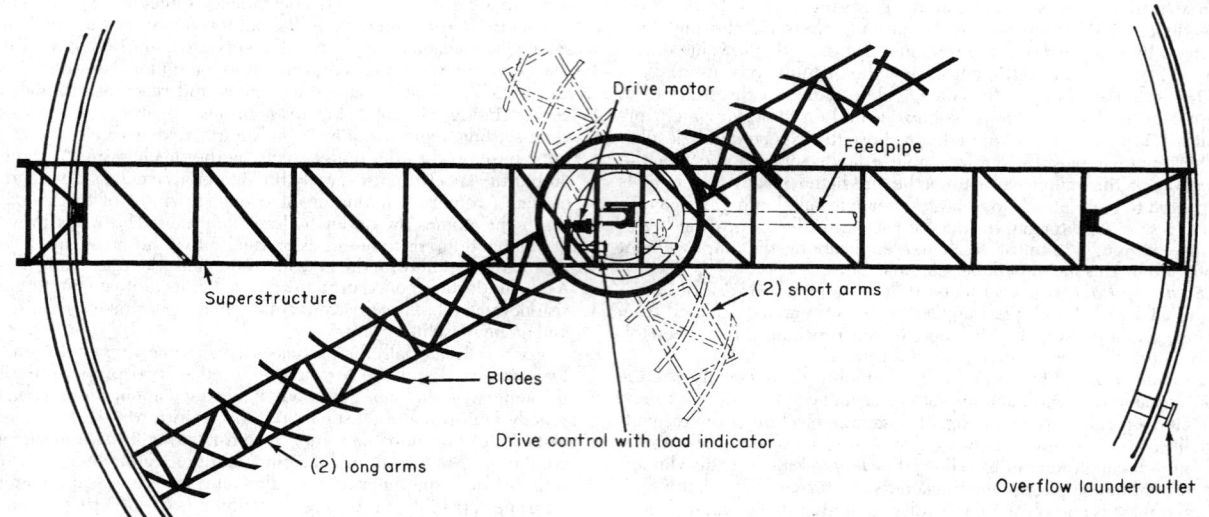

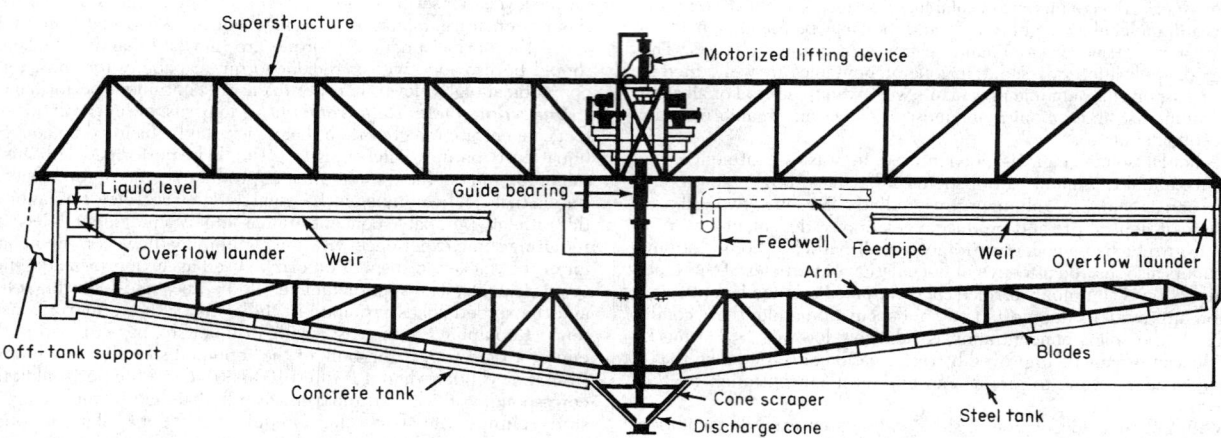

FIG. 18-86   Unit thickener with bridge-supported mechanism. (*EIMCO Process Equipment Co.*)

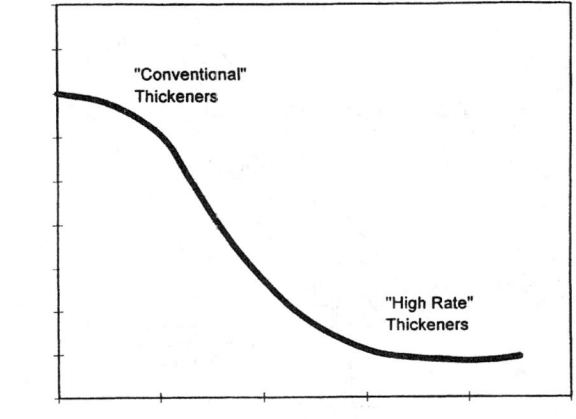

FIG. 18-87   Unit area vs. flocculant dose, illustrating the relationship between conventional and high-rate thickeners.

rake design and have a torque capability 3 to 4 times, or more, the normal for that particular diameter. Underflow slurries usually will be at a solids concentration 5 to 10% lower than that of a vacuum filter cake formed from the same material. Special pumping requirements are necessary if the slurry is to be transferred a significant distance, with line pressure drop typically in the range of 3 to 4 kPa/m of pipeline.

Design Features   There are three classes of thickeners, each differentiated by its drive mechanism: (1) bridge-supported, (2) center-column supported, and (3) traction drives. The diameter of the tank will range from 2 to 150 m (6.5 to 492 ft), and the support structure often is related to the size required. These classes are described in detail in the subsection "Components and Accessories for Sedimentation Units."

Operation   When operated correctly, thickeners require a minimum of attention and, if the feed characteristics do not change radically, can be expected to maintain design performance consistently. In this regard, it is usually desirable to monitor feed and underflow rates and solids concentrations, flocculant dosage rate, and pulp interface level, preferably with dependable instrumentation systems. Process variations are then easily handled by changing the principal operating controls—underflow rate and flocculant dose—to maintain stability.

Starting up a thickener is usually the most difficult part of the operation, and there is more potential for mechanical damage to the mechanism at this stage than at any other time. In general, two conditions require special attention at this point: underflow pumping and mechanism torque. If possible, the underflow pump should be in operation as soon as feed enters the system, recirculating underflow slurry at a reduced rate if the material is relatively fine or advancing it to the next process step (or disposal) if the feed contains a considerable quantity of coarse solids, e.g., more than 20 percent + 75 μm particles. At this stage of the operation, coarse solids separate from the pulp and produce a difficult raking and pumping situation. Torque can rise rapidly if this material accumulates faster than it is removed. If the torque reaches a point where the automatic control system raises the rakes, it is usually preferable to reduce or cut off the feed completely until the torque drops and the rakes are returned to the lowest position. As the fine fraction of the feed slurry begins to thicken and accumulate in the basin, providing both buoyancy and fluidity, torque will drop and normal feeding can be continued. This applies whether the thickener tank is empty at start-up or filled with liquid. The latter approach contributes to coarse-solids raking problems but at the same time provides conditions more suited to good flocculation, with the result that the thickener will reach stable operation much sooner.

As the solids inventory in the thickener reaches a normal level—usually about 0.5 to 1.0 m below the feedwell outlet—with underflow

slurry at the desired concentration, the torque will reach a normal operating range. Special note should be made of the torque reading at this time. Subsequent higher torque levels while operating conditions remain unchanged can almost always be attributed to island formation, and corrective action can be taken early, before serious problems develop. *Island* is the name given to a mass of semisolidified solids that have accumulated on or in front of the rakes, often as a result of excessive flocculant use. This mass usually will continue to grow in size, eventually producing a torque spike that can shut down the thickener and often resulting in lower underflow densities than would otherwise be achievable.

An island is easily detected, usually by the higher-than-normal, gradually increasing torque reading. Probing the rake arms near the thickener center with a rigid rod will confirm this condition—the mass is easily distinguished by its cohesive, claylike consistency. At an early stage, the island is readily removed by raising the rakes until the torque drops to a minimum value. The rakes are then lowered gradually, a few centimeters at a time, so as to shave off the mass of solids and discharge this gelled material through the underflow. This operation can take several hours, and if island formation is a frequent occurrence, the procedure should be carried out on a regular basis, typically once a day, preferably with an automatic system to control the entire operation.

Stable thickener performance can be maintained by carefully monitoring operating conditions, particularly the pulp interface level and the underflow rate and concentration. As process changes occur, the pulp level can vary; regulation of the underflow pumping rate will keep the level within the desired range. If the underflow varies in concentration, this can be corrected by adjusting the flocculant dosage. Response will not be immediate, of course, and care should be taken to make only small step changes at any one time. Procedures for use of automatic control are described in the section on instrumentation.

CLARIFIERS

Continuous clarifiers generally are employed with dilute suspensions, principally industrial process streams and domestic municipal wastes, and their primary purpose is to produce a relatively clear overflow. They are basically identical to thickeners in design and layout except that they employ a mechanism of lighter construction and a drive head with a lower torque capability. These differences are permitted in clarification applications because the thickened pulp produced is smaller in volume and appreciably lower in suspended solids concentration, owing in part to the large percentage of relatively fine (smaller than 10 μm) solids. The installed cost of a clarifier, therefore, is approximately 5 to 10 percent less than that of a thickener of equal tank size, as given in Fig. 18-94.

Rectangular Clarifiers   Rectangular clarifiers are employed primarily in municipal water and waste treatment plants, as well as in certain industrial plants, also for waste streams. The raking mechanism employed in many designs consists of a chaintype drag, although suction systems are used for light-duty applications. The drag moves the deposited pulp to a sludge hopper located on one end by means of scrapers fixed to endless chains. During their return to the sludge raking position, the flights may travel near the water level and thus act as skimming devices for removal of surface scum. Rectangular clarifiers are available in widths of 2 to 10 m (6 to 33 ft). The length is generally 3 to 5 times the width. The larger widths have multiple raking mechanisms, each with a separate drive.

This type of clarifier is used in applications such as preliminary oil-water separations in refineries and clarification of waste streams in steel mills. When multiple units are employed, common walls are possible, reducing construction costs and saving on floor space. Overflow clarities, however, generally are not as good as with circular clarifiers, due primarily to reduced overflow weir length for equivalent areas.

Circular Clarifiers   Circular units are available in the same three basic types as single-compartment thickeners: bridge, center-column, and peripheral-traction. Because of economic considerations, the bridge-supported type is limited generally to tanks less than 20 m in diameter.

A circular clarifier often is equipped with a surface-skimming device, which includes a rotating skimmer, scum baffle, and scum-box assembly. In sewage and organic-waste applications, squeegees normally are provided for the rake-arm blades, as it is desirable that the bottom be scraped clean to preclude accumulation of organic solids, with resultant septicity and flotation of decomposing material.

Center-drive mechanisms are also installed in square tanks. This mechanism differs from the standard circular mechanism in that a hinged corner blade is provided to sweep the corners which lie outside the path of the main mechanism.

Clarifier-Thickener   Clarifiers can serve as thickeners, achieving additional densification in a deep sludge sump adjacent to the center that extends a short distance radially and provides adequate retention time and pulp depth to compact the solids to a high density. Drive mechanisms on this type of clarifier usually must have higher torque capability than would be supplied on a standard clarifier.

Industrial Waste Secondary Clarifiers   Many plants which formerly discharged organic wastes to the sewer have turned to using their own treatment facilities in order to reduce municipal treatment plant charges. For organic wastes, the waste-activated sludge process is a preferred approach, using an aeration basin for the bio-oxidation step and a secondary clarifier to produce a clear effluent and to concentrate the biomass for recycling to the basin. To produce an acceptable effluent and achieve sufficient concentration of the low-density solids that make up the biomass, certain design criteria must be followed. If pilot-plant data are unavailable, the design procedures proposed by Albertson (op. cit.) can be used to specify tank diameter, depth, feedwell dimensions, feed inlet configuration, and rake-blade design for a unit which will meet the specifications for many waste-activated sludge plants. Typical design parameters recommended by Albertson include the following:

Feed pipe velocity: ≤ 1.2 m/s.
Energy-dissipating feed entry velocity (tangential): ≤ 0.5 m/s.
Downward velocity from feedwell: ≤ 0.5–0.75 (peak) m/min.
Feedwell depth: Entry port depth +1 m.
Radial velocity below feedwell: $\leq 90\%$ of downward velocity.
Tank depth: Clear water zone above dense sludge blanket is determined largely by clarification requirements; typically 3–5 m.
Tank diameter: Largely a function of clear water depth; the maximum overflow rate in m/h = $0.278 \times$ clear water depth. Resulting underflow design settling velocity should be less than 1.0 m/h for 1% solids and 120 mL/g SVI.

Tilted-Plate Clarifiers   Lamella or tilted-plate separators have achieved increased use for clarification. They contain a multiplicity of plates inclined at 45 to 60° from the horizontal. Various feed methods are employed so that the influent passes into each inclined channel at about one-third of the vertical height from the bottom. This results in the solids having to settle only a short distance in each channel before sliding down the base to the collection zone beneath the plates. The clarified liquid passes in the opposite direction beneath the ceiling of each channel to the overflow connection.

The area that is theoretically available for separation is equal to the sum of the projected areas or all channels on the horizontal plane. Figure 18-88 shows the horizontally projected area A, of a single channel in a clarifier of unit width. If X is the uniform distance between plates (measured perpendicularly to the plate surface), the clarifier will contain $\sin \alpha/X$ channels per unit length and an effective collection area per unit clarifier length of A, $\sin \alpha/X$, where α is the angle of inclination of the plates to the horizontal. It follows that the total horizontally projected plate area per unit volume of sludge in the clarifier A_s is

$$A_s = \cos \alpha/X \qquad (18\text{-}48)$$

As α and X are decreased, A_s is increased. However, α must be larger than the angle of repose of the sludge so that it will slide down the plate, and the most common range is 55 to 60°. Plate spacing must be large enough to accommodate the opposite flows of liquid and sludge while reducing interference and preventing plugging and to provide enough residence time for the solids to settle to the bottom plate. Usual X values are 50 to 75 mm (2 to 3 in).

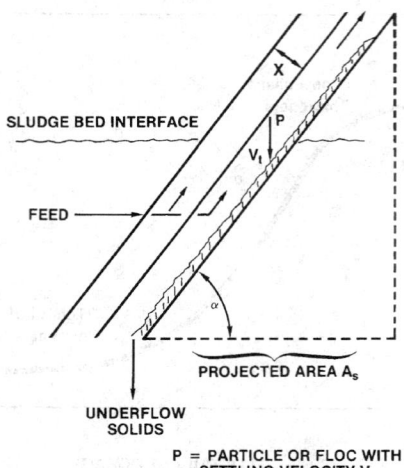

FIG. 18-88   Basic concept of the Lamella-type clarifier.

Many different designs are available, the major difference among them being in feed-distribution methods and plate configurations. Operating capacities range from 1 to 3 m³ of feed/h/m² of projected horizontal area [0.4 to 1.2 gal/(min·ft)].

The principal advantage of the tilted-plate clarifier is the increased capacity per unit of plane area. Major disadvantages are an underflow solids concentration that generally is lower than in other gravity clarifiers and difficulty of cleaning when scaling or deposition occurs. The lower underflow composition is due primarily to the reduced compression-zone volume relative to the large settling area. When flocculants are employed, flocculating equipment and tankage preceding the separator are required, as the design does not permit internal flocculation.

Solids-Contact Clarifiers   When desirable, mixing, flocculation, and sedimentation all may be accomplished in a single tank. Of the various designs available, those employing mechanically assisted mixing in the reaction zone are the most efficient. They generally permit the highest overflow rate at a minimum chemical dosage while producing the best effluent quality. The unit illustrated in Fig. 18-89 consists of a combination dual drive which moves the rake mechanism at a very slow speed as it rotates a high-pumping-rate, low-shear turbine located in the top portion of a center reaction well at a very much higher speed. The influent, dosed with chemicals as it enters, is contacted with previously settled solids in a recirculation draft tube within the reaction well by means of the pumping action of the turbine, resulting in a thorough mixing of these streams. Owing to the higher concentration of solids being recirculated, all chemical reactions are more rapid and more nearly complete, and flocculation is improved. The mixture passes out of the contacting and reaction well into the clarification area, where the flocculated particles settle out. They are raked to the center to be used again in the recirculation process, with a small amount being discharged through the sludge pump. When floccules are too heavy to be circulated up through the draft tube (as in the case of metallurgical pulps), a modified design using external recirculation of a portion of the thickened underflow is chosen. These units employ a special mixing impeller in a large feed well with a small-diameter central outlet.

Solids-contact clarifiers are advantageous for clarifying turbid waters or slurries that require coagulation and flocculation for the removal of bacteria, suspended solids, or color. Applications include softening water by lime addition; clarifying industrial-process streams, sewage, and industrial wastewaters; tertiary treatment for removal of phosphates, BOD$_5$, and turbidity; and silica removal from geothermal brines or from surface water for cooling-tower makeup.

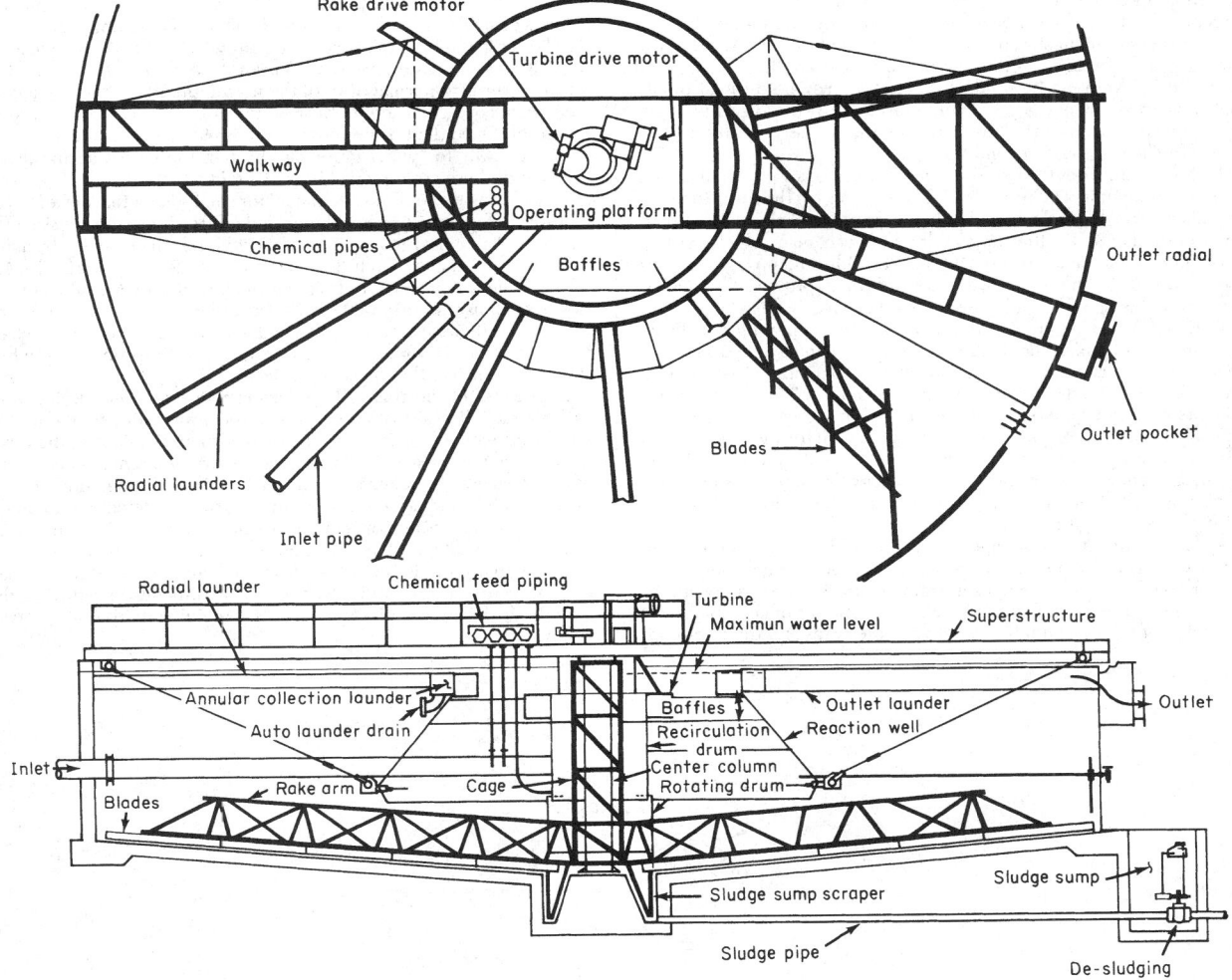

FIG. 18-89   Reactor-clarifier of the high-rate solids-contact type. (*EIMCO Process Equipment Co.*)

COMPONENTS AND ACCESSORIES FOR SEDIMENTATION UNITS

Sedimentation systems consist of a collection of components, each of which can be supplied in a number of variations. The basic components are the same, whether the system is for thickening or clarifying: tank, drive-support structure, drive unit and lifting device, rake structure, feedwell, overflow arrangement, underflow arrangement, instrumentation, and flocculation facilities.

Tanks   Tanks or basins are constructed of such materials as steel, concrete, wood, compacted earth, plastic sheeting, and soil cement. The selection of the materials of construction is based on cost, availability, topography, water table, ground conditions, climate, operating temperature, and chemical-corrosion resistance. Typically, industrial tanks up to 30 m (100 ft) in diameter are made of steel. Concrete generally is used in municipal applications and in larger industrial applications. Extremely large units employing earthen basins with impermeable liners have proved to be economical.

Drive-Support Structures   There are three basic drive mechanisms. These are (1) the bridge-supported mechanism, (2) the center-column-supported mechanism, and (3) the traction-drive thickener containing a center-column-supported mechanism with the driving arm attached to a motorized carriage at the tank periphery.

Bridge-Supported Thickeners   These thickeners (Fig. 18-86) are common in diameters up to 30 m, the maximum being about 45 m (150 ft). They offer the following advantages over a center-column-supported design: (1) ability to transfer loads to the tank periphery; (2) ability to give a denser and more consistent underflow concentration with the single draw-off point; (3) a less complicated lifting device; (4) fewer structural members subject to mud accumulation; (5) access to the drive from both ends of the bridge; and (6) lower cost for units smaller than 30 m in diameter.

Center-Column-Supported Thickeners   These thickeners are usually 20 m (65 ft) or more in diameter. The mechanism is supported by a stationary steel or concrete center column, and the raking arms are attached to a driving cage which rotates around the center column.

Traction Thickeners   These thickeners are most adaptable to tanks larger than 60 m (200 ft) in diameter. Maintenance generally is less difficult than with other types of thickeners, which is an advantage in remote locations. The installed cost of the traction thickener may be more than that of a center-driven unit primarily because of the cost of constructing the heavy concrete wall required to support the drive carriage. Disadvantages of the traction thickener are that (1) no practical lifting device can be used, (2) operation may be difficult in climates where snow and ice are common, and (3) the driving-torque effort must be transmitted from the tank periphery to the center, where the heaviest raking conditions occur.

Drive Assemblies The drive assembly is the key component of a sedimentation unit. The drive assembly provides (1) the force to move the rakes through the thickened pulp and to move settled solids to the point of discharge, (2) the support for the mechanism which permits it to rotate, (3) adequate reserve capacity to withstand upsets and temporary overloads, and (4) a reliable control which protects the mechanism from damage when a major overload occurs.

Drives usually have steel or iron main spur gears mounted on bearings, alloy-steel pinions, and either a bronze or a malleable iron worm gear driven by a hardened-steel worm or a planetary gear system. Direct-drive hydraulic systems are also employed. The gearing components preferably are enclosed for maximum service life. The drive typically includes a torque-measuring system with torque indicated on the mechanism and often transmitted to a remote indicator. If the torque becomes excessive, it can automatically activate such safeguards against structural damage as sounding an alarm, raising the rakes, and stopping the drive.

Rake-Lifting Mechanisms These should be provided when abnormal thickener operation is probable. Abnormal thickener operation or excessive torque may result from insufficient underflow pumping, surges in the solids feed rate, excessive amounts of large particles, sloughing of solids accumulated between the rakes and the bottom of the tank or on structural members of the rake mechanism, or miscellaneous obstructions falling into the thickener. The lifting mechanism may be set to raise the rakes automatically when a specific torque level (e.g., 40 percent of design) is encountered, continuing to lift until the torque returns to normal or until the maximum lift height is reached. Generally, corrective action must be taken to eliminate the cause of the upset. Once the torque returns to normal, the rake mechanism is lowered slowly to "plow" gradually through the excess accumulated solids until these are removed from the tank.

Rake-lifting devices can be manual for small-diameter thickeners or motorized for larger ones. Manual rake-lifting devices consist of a handwheel and a worm to raise or lower the rake mechanism by a distance usually ranging from 30 to 60 cm (1 to 2 ft). Motorized rake-lifting devices typically are designed to allow for a vertical lift of the rake mechanism of up to 90 cm (3 ft). A platform-type lifting device lifts the entire drive and rake mechanism up to 2.5 m (8 ft) and is used for applications in which excessive torque is most probable or when storage of solids in the thickener is desired.

Figure 18-90 illustrates the cable-arm design. This design uses cables attached to a truss above or near the liquid surface to move the rake arms, which are hinged to the drive structure, allowing the rakes to be raised when excessive torque is encountered. A major advantage of this design is the relatively small surface area of the raking mechanism, which reduces the solids accumulation and downtime in applications in which scaling or island formation can occur.

One disadvantage of this or any hinged-arm or other self-lifting design is that there is very little lift at the center, where the overload usually occurs. A further disadvantage is the difficulty of returning the rakes to the lowered position in settlers containing solids that compact firmly.

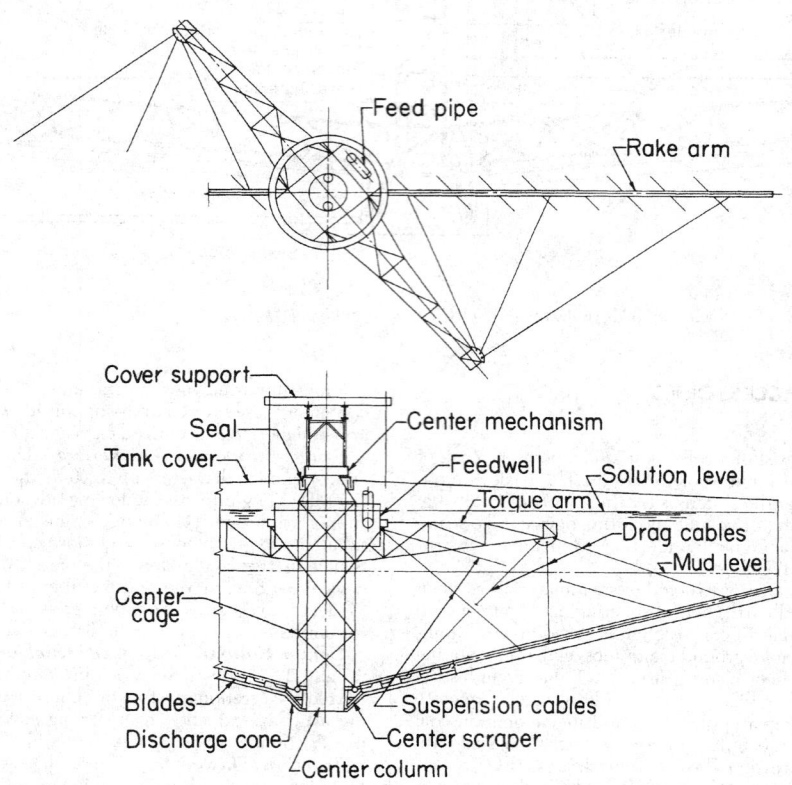

FIG. 18-90 Cable-arm design. (*Dorr-Oliver, Inc.*)

Rake Mechanism   The rake mechanism assists in moving the settled solids to the point of discharge. It also aids in thickening the pulp by disrupting bridged floccules, permitting trapped fluid to escape and allowing the floccules to become more consolidated. Rake mechanisms are designed for specific applications, usually having two long rake arms with an option for two short rake arms for bridge-supported and center-column-supported units. Traction units usually have one long arm and three short arms.

Figure 18-91 illustrates three types of rake-arm designs. The conventional design typically is used in bridge-supported units, while the dual-slope design is used for units of larger diameter. The "thixo post" design employs rake blades on vertical posts extending below the truss to keep the truss out of the thickest pulp and thereby prevent the collection of solids that might cause an "island" to form.

Rake blades can have attached spikes or serrated bottoms to cut into solids that have a tendency to compact. Lifting devices typically are used with these applications.

Rake-speed requirements depend on the type of solids entering the thickener. Peripheral speed ranges used are, for slow-settling solids, 3 to 8 m/min (10 to 25 ft/min); for fast-settling solids, 8 to 12 m/min (25 to 40 ft/min); and for coarse solids or crystalline materials, 12 to 30 m/min (40 to 100 ft/min).

Feedwell   The feedwell is designed to allow the feed to enter the thickener with minimum turbulence and uniform distribution while dissipating most of its kinetic energy. Feed slurry enters the feedwell, which is usually located in the center of the thickener, through a pipe or launder suspended from the bridge. To avoid excess velocity, an open launder normally has a slope no greater than 1 to 2 percent. Pulp should enter the launder at a velocity that prevents sanding at the inlet. With nonsanding pulps, the feed may also enter upward through the center column from a pipeline installed beneath the tank.

The standard feedwell for a thickener is designed for a maximum vertical outlet velocity of about 1.5 m/min (5 ft/min). High turbidity

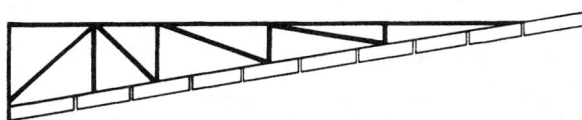

Conventional design (raking inward)

Design used for sloping bottoms

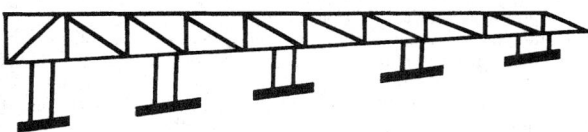

Arms used for thixotropic slimes to prevent "donut" formations and attain maximum underflow densities

FIG. 18-91   Rake-mechanism designs employed for specific duties. (*EIMCO Process Equipment Co.*)

caused by short-circuiting the feed to the overflow can be reduced by increasing the depth of the feedwell. When overflow clarity is important or the solids specific gravity is close to the liquid specific gravity, deep feedwells of large diameter are used, and measures are taken to reduce the velocity of the entering feed slurry.

Shallow feedwells may be used when overflow clarity is not important, the overflow rate is low, and/or solids density is appreciably greater than that of water. Some special feedwell designs used to dissipate entrance velocity and create quiescent settling conditions split the feed stream and allow it to enter the feedwell tangentially on opposite sides. The two streams shear or collide with one another to dissipate kinetic energy.

When flocculants are used, often it will be found that the optimum solids concentration for flocculation is considerably less than the normal concentration, and significant savings in reagent cost will be made possible by dilution of the very dense feed prior to flocculation. This can be achieved by recycling overflow or more efficiently by feedwell modifications, including openings in the feedwell rim. These will allow supernatant to enter the feedwell, and flocculant can be added at this point or injected below the surface of the pulp in the feedwell. Another effective means of achieving this dilution prior to flocculant addition is illustrated in Fig. 18-92. This approach utilizes the energy available in the incoming feed stream to achieve the dilution by momentum transfer and requires no additional energy expenditure to dilute this slurry by as much as three to four times.

Overflow Arrangements   Clarified effluent typically is removed in a peripheral launder located inside or outside the tank. The effluent enters the launder by overflowing a V-notch or level flat weir, or through submerged orifices in the bottom of the launder. Uneven overflow rates caused by wind blowing across the liquid surface in large thickeners can be better controlled when submerged orifices or V-notch weirs are used. Radial launders are used when uniform upward liquid flow is desired in order to improve clarifier detention efficiency. This arrangement provides an additional benefit in reducing the effect of wind, which can seriously impair clarity in applications that employ basins of large diameter.

The hydraulic capacity of a launder must be sufficient to prevent flooding, which can cause short-circuiting of the feed and deterioration of overflow clarity. Standards are occasionally imposed on weir overflow rates for clarifiers used in municipal applications; typical rates are 3.5 to 15 m³/(h·m) [7000 to 30,000 gal/(day·ft)], and they are highly dependent on clarifier side-water depth. Industrial clarifiers may have higher overflow rates, depending on the application and the desired overflow clarity. Launders can be arranged in a variety of configurations to achieve the desired overflow rate. Several alternatives to improve clarity include an annular launder inside the tank (the liquid overflows both sides), radial launders connected to the peripheral launder (providing the very long weir that may be needed when abnormally high overflow rates are encountered and overflow clarity is important), and Stamford baffles, which are located below the launder to direct flow currents back toward the center of the clarifier.

In many thickener applications, on the other hand, complete peripheral launders are not required, and no difference in either overflow clarity or underflow concentration will result through the use of launders extending over only a fraction (e.g., one-fifth) of the perimeter. For design purposes, a weir-loading rate in the range of 7.5 to 30.0 m³/(h·m) [10 to 40 gpm/ft] can be used, the higher values being employed with well-flocculated, rapidly settling slurries. The overflow launder required may occupy only a single section of the perimeter rather than consisting of multiple, shorter segments spaced uniformly around the tank.

Underflow Arrangements   Concentrated solids are removed from the thickener by use of centrifugal or positive displacement pumps or, particularly with large-volume flows, by gravity discharge through a flow control valve or orifice suitable for slurry applications. Due to the risk to the thickener operation of a plugged underflow pipe, it is recommended that duplicate underflow pipes and pumps be installed in all thickening applications. Provision to recycle underflow slurry back to the feedwell is also useful, particularly if solids are to be stored in the thickener.

FIG. 18-92   E-Duc® Feed dilution system installed on 122-m-diameter thickener. (*EIMCO Process Equipment Co.*)

There are several basic underflow arrangements: (1) the underflow pump adjacent to the thickener sidewall with buried piping from the discharge cone, (2) the underflow pump under the thickeners or adjacent to the sidewall with the piping from the discharge cone in a tunnel, (3) the underflow pump adjacent to the thickener sidewall with a peripheral discharge from the tank sidewall, and (4) the underflow pump located in the center of the thickener on the bridge close to the drive mechanism, or at the perimeter, using piping up through the center column.

Pump Adjacent to Thickener with Buried Piping   This arrangement of buried piping from the discharge cone is the least expensive system but the most susceptible to plugging. It is used only when the solids do not compact to an unpumpable slurry and can be easily backflushed if plugging occurs. Typically, two or more underflow pipes are installed from the discharge cone to the underflow pump so that solids removal can continue if one of the lines plugs. Valves should be installed to permit flushing with water and compressed air in both directions to remove blockages.

Tunnel   A tunnel may be constructed under the thickener to provide access to the discharge cone when underflow slurries are difficult to pump and have characteristics that cause plugging. The underflow pump may be installed underneath the thickener or at the perimeter. Occasionally thickeners are installed on legs or piers, making tunneling for access to the center unnecessary. A tunnel or an elevated thickener is more expensive than the other underflow arrangements, but there are certain operational and maintenance advantages. Of course, the hazards of working in a tunnel (flooding and interrupted ventilation, for example) and related safety regulations must be considered.

Peripheral Discharge   Peripheral discharge sometimes is used to permit the reduced installation cost of a flat-bottom tank on compacted soil. Because more torque is required to rake the solids to the perimeter of the tank, this arrangement is not suitable for service involving coarse solids or solids that become nonfluid at high concentrations.

Center-Column Pumping   This arrangement may be used instead of a tunnel. Several designs are available. The commonest is a bridge-mounted pump with a suction line through a wet or dry center column. The pump selection may be limiting, requiring special attention to priming, net positive suction head, and the maximum density that the pump can handle. One design has the underflow pump located in a room under the thickener mechanism and connected to openings in the column. Access is through the drive gear at the top of the column.

INSTRUMENTATION

Thickener control philosophies are usually based on the idea that the underflow density obtained is the most important performance criterion. The overflow clarity is also a consideration, but this is generally not as critical. Additional factors which must be considered are optimization of flocculant usage and protection of the raking mechanism.

Automated control schemes employ one or more sets of controls, which will fit into three categories: (1) control loops which are used to regulate the addition of flocculant, (2) control loops to regulate the withdrawal of underflow, and (3) rake drive controls. Usually, the feed to a thickener is not controlled and most control systems have been designed with some flexibility to deal with changes in feed characteristics, such as an increase in volume or alteration in the nature of the solids themselves. In severe cases, some equalization of the feed is required in order to allow the control system to perform effectively.

Flocculant addition rate can be regulated in proportion to the thickener volumetric feed rate or solids mass flow in a feed-forward mode, or in a feed-back mode on either rake torque, underflow density, settling solids (sludge) bed level, or solids settling rate. Of these, feed-forward on mass flow or feed-back on bed level are probably the most common. In some feed-forward schemes, the ratio multiplier is trimmed by one of the other parameters.

Underflow is usually withdrawn continuously on the bases of bed level, rake torque, or underflow solids concentration in a feed-back mode. Most installations incorporate at least two of these parameters in their underflow withdrawal control philosophy. For example, the continuous withdrawal may be based on underflow solids density with an override to increase the withdrawal rate if either the rake torque or the bed level reaches a preset value. In some cases, underflow withdrawal has been regulated in a feed-forward mode on the basis of thickener feed solids mass flow rate. Any automated underflow pumping scheme should incorporate a lower limit on volumetric flow rate as a safeguard against line pluggage.

It is also important to consider the level of the sludge bed in the thickener. Although this can be allowed to increase or decrease within moderate limits, it must be controlled enough to prevent solids from overflowing the thickener or from falling so low that the underflow density becomes too dilute. The settling slurry within the sludge bed is normally free flowing and will disperse to a consistent level across the thickener diameter.

Rake drive controls protect the drive mechanism from damage and usually incorporate an alarm to indicate high torque with an interlock to shut down the drive at a higher torque level. They can have an automated rake raising and lowering feature with a device to indicate the elevation of the rakes.

A complete automated control scheme incorporates controls from each of the three categories. It is important to consider the interaction of the various controls, especially of the flocculant addition and underflow withdrawal control loops, when designing a system. The lag and dead times of any feedback loops as well as the actual response of the system to changes in manipulated variables must be considered. For example, in some applications it is possible that excessive flocculant addition may produce an increase in the rake torque (due to island formation or viscosity increase) without a corresponding increase in underflow density. Additionally, sludge bed level sensors generally require periodic cleaning to produce a reliable signal. In many cases, it has not been possible to effectively maintain the sludge bed level sensors, requiring a change in the thickener control logic after start-up. Some manufacturers offer complete thickener control packages.

Control philosophies for clarifiers are based on the idea that the overflow is the most important performance criterion. Underflow density or suspended solids content is a consideration, as is optimal use of flocculation and pH control reagents. Automated controls are of three basic types: (1) control loops that optimize coagulant, flocculant, and pH control reagent additions; (2) those that regulate underflow removal; and (3) rake drive controls. Equalization of the feed is provided in some installations, but the clarifier feed is usually not a controlled variable with respect to the clarifier operation.

Automated controls for flocculating reagents can use a feedforward mode based on feed turbidity and feed volumetric rate, or a feed-back mode incorporating a streaming current detector on the flocculated feed. Attempts to control coagulant on the basis of overflow turbidity generally have been less successful. Control for pH has been accomplished by feed-forward modes on the feed pH and by feed-back modes on the basis of clarifier feedwell or external reaction tank pH. Control loops based on measurement of feedwell pH are useful for control in applications in which flocculated solids are internally recirculated within the clarifier feedwell.

Automated sludge withdrawal controls are usually based on the sludge bed level. These can operate in on-off or continuous modes and will use either single-point or continuous sludge level indication sensors. In many applications, automated control of underflow withdrawal does not provide an advantage, since so few settled solids are produced that it is only necessary to remove sludge for a short interval once a day, or even less frequently. In applications in which the underflow is recirculated internally within the feedwell, it is necessary to maintain sufficient sludge inventory for the recirculation turbine to pull from. This could be handled in an automated system with a single-point *low* sludge bed level sensor in conjunction with a low-level alarm or pump shutoff solenoid. Some applications require continuous external recirculation of the underflow direct to the feedwell or external reaction tanks, and an automated control loop could be used to maintain recirculation based on flow measurement, with a manually adjusted setpoint.

Control philosophies applied to continuous countercurrent decantation (CCD) thickeners are similar to those used for thickeners in other applications, but have emphasis on maintaining the CCD circuit in balance. It is important to prevent any one of the thickeners from pumping out too fast, otherwise an upstream unit could be starved of wash liquor while at the same too much underflow could be placed in a downstream unit too quickly, disrupting the operation of both units as well as reducing the circuit washing efficiency. Several control configurations have been attempted, and the more successful schemes

have linked the solids mass flow rate of underflow pumping to that of the upstream unit or to the CCD circuit solids mass feed rate. Wide variations in the solids feed rate to a CCD circuit will require some means of dampening these fluctuations if design wash efficiency is to be maintained.

The following types of devices are commonly applied to measure the various operational parameters of thickeners and clarifiers. They have been used in conjunction with automatic valves and variable-speed pumps to achieve automatic operation as well as to simply provide local or remote indications.

Sludge Bed Level Sensors   Most of the more commonly used devices for detection of the sludge bed level operate on principles of ultrasound, gamma radiation, infrared or visible light, or simply with a float that is carefully weighted to float on the sludge bed level interface. There are basically two types of sensors, those that provide a single-point indication at a fixed height within the thickener or clarifier, and those that follow the sludge bed depth over a distance and provide a continuous indication of the level. Some devices are available that can provide a profile of the slurry concentration within the sludge bed.

Flowmeters   These are used to measure flocculant addition, underflow, and feed flow rates. For automatic control, the more commonly used devices are magnetic flowmeters and Doppler effect flowmeters.

Density Gauges   These are used to measure the density or suspended solids content of the feed and underflow streams. Gamma radiation devices are the most commonly used for automatic control, but ultrasonic devices are effective in the lower range of slurry density. Marcy pulp density scales are an effective manually operated device. A solids "mass flow" indication is usually obtained by combining a density gauge output with the output from a flowmeter.

Turbidity Gauges   These operate with visible light beams and detectors. They are used to monitor feed and effluent turbidity.

Streaming Current Detectors   These units produce a measurement closely related to the zeta potential of a suspension and are used successfully in optimizing the coagulant dose in clarification applications.

CONTINUOUS COUNTERCURRENT DECANTATION

The system of separation of solid-phase material from an associated solution by repeated stages of dilution and gravity sedimentation is adapted for many industrial-processing applications through an operation known as continuous countercurrent decantation (CCD). The flow of solids proceeds in a direction countercurrent to the flow of solution diluent (water, usually), with each stage composed of a mixing step followed by settling of the solids from the suspension. The number of stages ranges from 2 to as many as 10, depending on the degree of separation required, the amount of wash fluid added (which influences the final solute concentration in the first-stage overflow), and the underflow solids concentration attainable. Applications include processes in which the solution is the valuable component (as in alumina extraction), or in which purified solids are sought (magnesium hydroxide from seawater), or both (as frequently encountered in the chemical-processing industry and in base-metal hydrometallurgy).

The factors which may make CCD a preferred choice over other separation systems include the following: rapidly settling solids, assisted by flocculation: relatively high ratio of solids concentration between underflow and feed; moderately high wash ratios allowable (2 to 4 times the volume of liquor in the thickened underflows); large quantity of solids to be processed; and the presence of fine-size solids that are difficult to concentrate by other means. A technical feasibility and economic study is desirable in order to make the optimum choice.

Flow-Sheet Design   Thickener-sizing tests, as described earlier, will determine unit areas, flocculant dosages, and underflow densities for the various stages. For most cases, unit areas will not vary significantly throughout the circuit; similarly, underflow concentrations should be relatively constant. In practice, the same unit area is generally used for all thickeners in the circuit to simplify construction.

Equipment   The equipment selected for CCD circuits may consist of multiple-compartment washing-tray thickeners or a train of individual unit thickeners. The washing-tray thickener consists of a vertical array of coaxial trays connected in series, contained in a single tank. The advantages of this design are smaller floor-area requirements, less pumping equipment and piping, and reduced heat losses in circuits operating at elevated temperatures. However, operation is generally more difficult and user preference has shifted toward unit thickeners despite the larger floor-area requirement and greater initial cost.

Underflow Pumping   Diaphragm pumps with open discharge are employed in some cases, primarily because underflow densities are readily controlled with these units. Disadvantages include the generally higher maintenance and initial costs than for other types and their inability to transfer the slurry any great distance. Large flows often are best handled with variable-speed, rubber-lined centrifugal pumps, utilizing automatic control to maintain the underflow rate and density.

Overflow Pumps   These can be omitted if the thickeners are located at increasing elevations from first to last so that overflows are transferred by gravity or if the mixture of underflow and overflow is to be pumped. Overflow pumps are necessary, however, when maximum flexibility and control are sought.

Interstage Mixing Efficiencies   Mixing or stage efficiencies rarely achieve the ideal 100 percent, in which solute concentrations in overflow and underflow liquor from each thickener are identical. Part of the deficiency is due to insufficient blending of the two streams, and attaining equilibrium will be hampered further by heavily flocculated solids. In systems in which flocculants are used, interstage effi-

TABLE 18-7   Typical Thickener and Clarifier Design Criteria and Operating Conditions

	Percent solids		Unit area, $m^2/(t/d)$[*][†]	Overflow rate, $m^3/(m^2 \cdot h)$[*]
	Feed	Underflow		
Alumina Bayer process				
Red mud, primary	3–4	10–25	2–5	
Red mud, washers	6–8	15–35	1–4	
Hydrate, fine or seed	1–10	20–50	1.2–3	0.07–0.12
Brine purification	0.2–2.0	8–15		0.5–1.2
Coal, refuse	0.5–6	20–40	0.5–1	0.7–1.7
Coal, heavy-media (magnetic)	20–30	60–70	0.05–0.1	
Cyanide, leached-ore	16–33	40–60	0.3–1.3	
Flue dust, blast-furnace	0.2–2.0	40–60		1.5–3.7
Flue dust, BOF	0.2–2.0	30–70		1–3.7
Flue-gas desulfurization sludge	3–12	20–45	0.3–3†	
Magnesium hydroxide from brine	8–10	25–40	5–10	
Magnesium hydroxide from seawater	1–4	15–20	3–10	0.5–0.8
Metallurgical				
Copper concentrates	14–50	40–75	0.2–2	
Copper tailings	10–30	45–65	0.4–1	
Iron ore				
Concentrate (magnetic)	20–35	50–70	0.01–0.08	
Concentrate (nonmagnetic), coarse: 40–65% −325	25–40	60–75	0.02–0.1	
Concentrate (nonmagnetic), fine: 65–100% −325	15–30	60–70	0.15–0.4	
Tailings (magnetic)	2–5	45–60	0.6–1.5	1.2–2.4
Tailings (nonmagnetic)	2–10	45–50	0.8–3	0.7–1.2
Lead concentrates	20–25	60–80	0.5–1	
Molybdenum concentrates	10–35	50–60	0.2–0.4	
Nickel, $(NH_4)_2CO_3$ leach residue	15–25	45–60	0.3–0.5	
Nickel, acid leach residue	20	60	0.8	
Zinc concentrates	10–20	50–60	0.3–0.7	
Zinc leach residue	5–10	25–40	0.8–1.5	
Municipal waste				
Primary clarifier	0.02–0.05	0.5–1.5		1–1.7
Thickening				
Primary sludge	1–3	5–10	8	
Waste-activated sludge	0.2–1.5	2–3	33	
Anaerobically digested sludge	4–8	6–12	10	
Phosphate slimes	1–3	5–15	1.2–18	
Pickle liquor and rinse water	1–8	9–18	3.5–5	
Plating waste	2–5	5–30		1.2
Potash slimes	1–5	6–25	4–12	
Potato-processing waste	0.3–0.5	5–6		1
Pulp and paper				
Green-liquor clarifier	0.2	5		0.8
White-liquor clarifier	8	35–45	0.8–1.6	
Kraft waste	0.01–0.05	2–5		0.8–1.2
Deinking waste	0.01–0.05	4–7		1–1.2
Paper-mill waste	0.01–0.05	2–8		1.2–2.2
Sugarcane defecation			0.5†	
Sugar-beet carbonation	2–5	15–20	0.03–0.07‡	
Uranium				
Acid-leached ore	10–30	25–65	0.02–1	
Alkaline-leached ore	20	60	1	
Uranium precipitate	1–2	10–25	5–12.5	
Water treatment				
Clarification (after 30-min flocculation)				1–1.3
Softening lime-soda (high-rate, solids-contact clarifiers)				3.7
Softening lime-sludge	5–10	20–45	0.6–2.5	

[*]$m^2/(t/d) \times 9.76 = ft^2/(short\ ton/day)$; $m^3/(m^2 \cdot h) \times 0.41 = gal/(ft^2 \cdot min)$; 1 t = 1 metric ton.
†High-rate thickeners using required flocculant dosages operate at 10 to 50 percent of these unit areas.
‡Basis: 1 t of cane or beets.

ciencies often will drop gradually from first to last thickener, and typical values will range from 98 percent to as low as 70 percent. In some cases, operators will add the flocculant to an overflow solution which is to be blended with the corresponding underflow. While this is very effective for good flocculation, it can result in reflocculation of the solids before the entrained liquor has had a chance to blend completely with the overflow liquor. The preferable procedure is to recycle a portion of the overflow back to the feed line of the same thickener, adding the reagent to this liquor.

The usual method of interstage mixing consists of a relatively simple arrangement in which the flows from preceding and succeeding stages are added to a feed box at the thickener periphery. A nominal detention time in this mixing tank of 30 to 60 s and sufficient energy input to avoid solids settling will ensure interstage efficiencies greater than 95 percent.

The performance of a CCD circuit can be estimated through use of the following equations, which assume 100 percent stage efficiency:

$$R = \frac{O/U[(O/U)^N - 1]}{[(O/U)^{N+1} - 1]} \tag{18-49}$$

$$R = 1 - \left(\frac{U}{O'}\right)^N \tag{18-50}$$

for O/U and $U/O' \neq 1$. R is the fraction of dissolved value in the feed which is recovered in the overflow liquor from the first thickener, O and U are the overflow and underflow liquor volumes per unit weight of underflow solids, and N is the number of stages. Equation (18-49) applies to a system in which the circuit receives dry solids with which the second-stage thickener overflow is mixed to extract the soluble component. In this instance, O' refers to the overflow volume from the thickeners following the first stage.

For more precise values, computer programs can be used to calculate soluble recovery as well as solution compositions for conditions that are typical of a CCD circuit, with varying underflow concentrations, stage efficiencies, and solution densities in each of the stages. The calculation sequence is easily performed by utilizing material-balance equations around each thickener.

DESIGN OR STIPULATION OF A SEDIMENTATION UNIT

Selection of Type of Thickener or Clarifier   Selection of the type of unit thickener or clarifier depends primarily on the optimization of performance requirements, installation cost and operating cost. For example, the inclined-plate type of clarifier provides for less solids-holding capacity than a circular or rectangular clarifier, but at a lower installation cost. The high-density thickener maximizes underflow solids concentration, requiring a higher torque rating than conventional thickeners.

Most manufacturers have overlapping sizes in the bridge-supported, center-column-supported, and traction types even though an optimum economical size range exists for each type. Bridge-supported thickeners are generally selected for diameters less than 30 m (100 ft) and center-column-supported thickeners are selected for greater diameters. Traction units often are least expensive in sizes over 75 m (250 ft) if ground conditions permit installing proper supporting walls to carry the loads.

Special conditions affect the choice of thickener or clarifier type. For example, if a unit must be covered to conserve heat, the bridge-supported type may be more economical up to about 45 m (150 ft) in diameter, although 30 m (100 ft) may be the economic limit for an uncovered unit.

Materials of Construction   A wide variety of materials is available for tanks, as indicated earlier. Most mechanisms are made of steel; however, submerged parts may be made of wood, stainless steel, rubber-covered or coated steel, or special alloys.

Design Sizing Criteria   Table 18-7, which lists typical design sizing criteria and operating conditions for a number of thickener and clarifier applications, is presented for purposes of illustration or preliminary estimate. Actual thickening and clarification performance is dependent on particle size distribution, specific gravity, sludge bed compaction characteristics, and other factors. Final designs should be based on bench-scale tests involving the methods previously discussed.

If a solids-contact clarifier is required, the surface-area requirement must exclude the area taken up by the reaction chamber. The reaction chamber itself is normally sized for a detention time of 15 to 45 min, depending on the type of treatment and the design of the unit.

Torque Rating   The choice of torque rating has been discussed earlier. Torque is a function of such factors as quantity and quality of underflow (therefore, of such parameters as particle characteristics and flocculant dosage that affect underflow character), unit area, and rake speed; but, in the final analysis, torque must be specified on the basis of experience modified by these factors. Unless one is experienced in a given application, it is wise to consult a thickener or clarifier manufacturer.

THICKENER COSTS

Equipment   Costs vary widely for a given diameter because of the many types of construction. As a general rule, the total installed cost will be about 3 to 4 times the cost of the raking mechanism (including drivehead and lift), plus walkways and bridge or centerpier cage, railings, and overflow launders. Figure 18-93 shows the approximate installed costs of thickeners up to 107 m (350 ft) in diameter. These costs are to be used only as a guide. They include the erection of mechanism and tank plus normal uncomplicated site preparation, excavation, reinforcing bar placement, backfill, and surveying. The price does not include any electrical work, pumps, piping, instrumentation, walkways, or lifting mechanisms. Special design modifications, which are not in the price, could include elevated tanks (for underflow handling); special feedwell designs to control dilution, entrance velocity, and turbulence; electrical and drive enclosures required because of climatic conditions; and mechanism designs required because of scale buildup tendencies.

Operating Costs   Power cost for a continuous thickener is an almost insignificant item. For example, a unit thickener 60 m (200 ft) in diameter with a torque rating of 1.0 MN·m (8.8 Mlbf·in) will normally require 12 kW (16 hp). The low power consumption is due to the very slow rotative speeds. Normally, a mechanism will be designed for a peripheral speed of about 9 m/min (0.5 ft/s), which corresponds to only 3 r/h for a 60-m (200-ft) unit. This low speed also means very low maintenance costs. Operating labor is low because little attention is normally required after initial operation has balanced the feed and underflow. If chemicals are required for flocculation, the chemical cost frequently dwarfs all other operating costs.

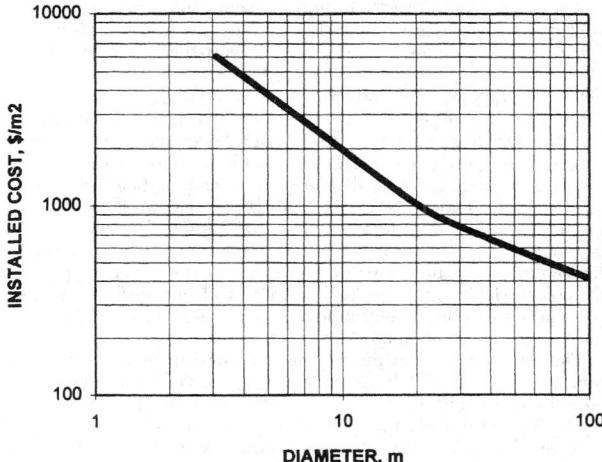

FIG. 18-93   Approximate installed cost of single-compartment thickeners (1995 dollars). To convert to ft and ft² units, multiply diameter in meters by 3.28 and divide cost by 10.76.

FILTRATION

GENERAL REFERENCES: Moir, *Chem. Eng.,* **89**(15), 46 (1982). Brown, ibid., 58; also published as McGraw-Hill Repr. A078. Cheremisinoff and Azbel, *Liquid Filtration,* Ann Arbor Science, Woburn, Mass., 1983. Orr (ed.), *Filtration: Principles and Practice,* part I, Marcel Dekker, New York, 1977; part II, 1979. Purchas (ed.), *Solid/Liquid Separation Equipment Scale-Up,* Uplands Press, Croydon, England, 1977. Schweitzer (ed.), *Handbook of Separation Techniques for Chemical Engineers,* part 4, McGraw-Hill, New York, 1979. Shoemaker (ed.), "What the Filter Man Needs to Know about Filtration," *Am. Inst. Chem. Eng. Symp. Ser.,* **73**(171), (1977). Talcott et al., in *Kirk-Othmer Encyclopedia of Chemical Technology,* 3d ed., vol. 10, Wiley, New York, 1980, p. 284. Tiller et al., *Chem. Eng.,* **81**(9), 116–136 (1974); also published as McGraw-Hill Repr. R203.

DEFINITIONS AND CLASSIFICATION

Filtration is the separation of a fluid-solids mixture involving passage of most of the fluid through a porous barrier which retains most of the solid particulates contained in the mixture. This subsection deals only with the filtration of solids from liquids; gas filtration is treated in Sec. 17. **Filtration** is the term for the unit operation. A filter is a piece of unit-operations equipment by which filtration is performed. The **filter medium** or **septum** is the barrier that lets the liquid pass while retaining most of the solids; it may be a screen, cloth, paper, or bed of solids. The liquid that passes through the filter medium is called the **filtrate.**

Filtration and filters can be classified several ways:

1. *By driving force.* The filtrate is induced to flow through the filter medium by hydrostatic head (gravity), pressure applied upstream of the filter medium, vacuum or reduced pressure applied downstream of the filter medium, or centrifugal force across the medium. Centrifugal filtration is closely related to centrifugal sedimentation, and both are discussed later under "Centrifuges."

2. *By filtration mechanism.* Although the mechanism for separation and accumulation of solids is not clearly understood, two models are generally considered and are the basis for the application of theory to the filtration process. When solids are stopped at the surface of a filter medium and pile upon one another to form a cake of increasing thickness, the separation is called **cake filtration.** When solids are trapped within the pores or body of the medium, it is termed **depth, filter-medium,** or **clarifying filtration.**

3. *By objective.* The process goal of filtration may be dry solids (the cake is the product of value), clarified liquid (the filtrate is the product of value), or both. Good solids recovery is best obtained by cake filtration, while clarification of the liquid is accomplished by either depth or cake filtration.

4. *By operating cycle.* Filtration may be intermittent (batch) or continuous. Batch filters may be operated with constant-pressure driving force, at constant rate, or in cycles that are variable with respect to both pressure and rate. Batch cycle can vary greatly, depending on filter area and solids loading.

5. *By nature of the solids.* Cake filtration may involve an accumulation of solids that is compressible or substantially incompressible, corresponding roughly in filter-medium filtration to particles that are deformable and to those that are rigid. The particle or particle-aggregate size may be of the same order of magnitude as the minimum pore size of most filter media (1 to 10 μm and greater), or may be smaller (1 μm down to the dimension of bacteria and even large molecules). Most filtrations involve solids of the former size range; those of the latter range can be filtered, if at all, only by filter-medium-type filtration or by ultrafiltration unless they are converted to the former range by aggregation prior to filtration.

These methods of classification are not mutually exclusive. Thus filters usually are divided first into the two groups of cake and clarifying equipment, then into groups of machines using the same kind of driving force, then further into batch and continuous classes. This is the scheme of classification underlying the discussion of filters of this subsection. Within it, the other aspects of operating cycle, the nature of the solids, and additional factors (e.g., types and classification of filter media) will be treated explicitly or implicitly.

FILTRATION THEORY

While research has developed a significant and detailed filtration theory, it is still so difficult to define a given liquid-solid system that it is both faster and more accurate to determine filter requirements by performing small-scale tests. Filtration theory does, however, show how the test data can best be correlated, and extrapolated when necessary, for use in scale-up calculations.

In cake or surface filtration, there are two primary areas of consideration: continuous filtration, in which the resistance of the filter cake (deposited process solids) is very large with respect to that of the filter media and filtrate drainage, and batch pressure filtration, in which the resistance of the filter cake is not very large with respect to that of the filter media and filtrate drainage. Batch pressure filters are generally fitted with heavy, tight filter cloths plus a layer of precoat and these represent a significant resistance that must be taken into account. Continuous filters, except for precoats, use relatively open cloths that offer little resistance compared to that of the filter cake.

Simplified theory for both batch and continuous filtration is based on the time-honored Hagen-Poiseuille equation:

$$\frac{1}{A}\frac{dV}{d\Theta} = \frac{P}{\mu(\alpha w V/A + r)} \tag{18-51}$$

where V is the volume of filtrate collected, Θ is the filtration time, A is the filter area, P is the total pressure across the system, w is the weight of cake solids/unit volume of filtrate, μ is the filtrate viscosity, α is the cake-specific resistance, and r is the resistance of the filter cloth plus the drainage system.

CONTINUOUS FILTRATION

Since testing and scale-up are different for batch and continuous filtration, discussion in this section will be limited to continuous filtration.

It is both convenient and reasonable in continuous filtration, except for precoat filters, to assume that the resistance of the filter cloth plus filtrate drainage is negligible compared to the resistance of the filter cake and to assume that both pressure drop and specific cake resistance remain constant throughout the filter cycle. Equation (18-51), integrated under these conditions, may then be manipulated to give the following relationships:

$$W = \sqrt{\frac{2wP\Theta_f}{\mu(\alpha w V/A + r)}} \tag{18-52}$$

$$V_f = \sqrt{\frac{2P\Theta_f}{\mu\alpha w}} \tag{18-53}$$

$$\Theta_w = \frac{WV_w\mu\alpha}{P_w} \tag{18-54}$$

$$\Theta_W \propto NW^2 \tag{18-55}$$

$$\frac{\Theta_W}{\Theta_f} = 2\frac{V_w}{V_f} \tag{18-56}$$

where W is the weight of dry filter cake solids/unit area, V_f is the volume of cake formation filtrate/unit area, V_w is the volume of cake wash filtrate/unit area, Θ_f is the cake formation time, Θ_w is the cake wash time, and N is the wash ratio, the volume of cake wash/volume of liquid in the discharged cake.

As long as the suspended solids concentration in the feed remains constant, these equations lead to the following convenient correlations:

$$\log W \text{ vs. } \log \Theta_f \tag{18-57}$$

$$\log V_f \text{ vs. } \log \Theta_f \tag{18-58}$$

$$\Theta_w \text{ vs. } WV_w \tag{18-59}$$

$$\Theta_w \text{ vs. } NW^2 \tag{18-60}$$

$$\Theta_{w/f} \text{ vs. } V_w/V_f \tag{18-61}$$

There are two other useful empirical correlations as follows:

$$W \text{ vs. cake thickness} \qquad (18\text{-}62)$$

$$\log R \text{ vs. } N \qquad (18\text{-}63)$$

where R is percent remaining—the percent of solute in the unwashed cake that remains after washing.

FACTORS INFLUENCING SMALL-SCALE TESTING

[Purchas (ed.), *Solid/Liquid Separation Equipment Scale-Up,* Uplands Press, Croydon, England, 1977.]

Vacuum or Pressure   The vast majority of all continuous filters use vacuum to provide the driving force for filtration. However, if the feed slurry contains a highly volatile liquid phase, or if it is hot, saturated, and/or near the atmospheric pressure boiling point, the use of pressure for the driving force may be required. Pressure filtration might also be used where the required cake moisture content is lower than that obtainable with vacuum.

The objective of most continuous filters is to produce a dry or handleable cake. Most vacuum filters easily discharge a "dry" consolidated cake as they are usually operated in an open or semiopen environment. However, whenever the filter must operate under pressure or within a vapor-tight enclosure, either because of the need for a greater driving force or because of the vapor pressure of the liquid phase, a dry cake discharge becomes difficult. The problem of removing a dry cake from a pressurized enclosure has precluded the use of continuous-pressure filters in many cases where this is a requirement. Applications which do discharge dry cake from a "sealed" enclosure are restricted to relatively dry, friable cakes that will "flow" through double valves which form a "vapor lock."

Cake Discharge   For any filter application to be practical, it must be possible to produce a cake thick enough to discharge. Table 18-8 tabulates the minimum acceptable cake thickness required for discharge for various types of filters and discharge mechanisms. The experimenter, when running small-scale tests, should decide early in the test program which type of discharge is applicable and then tailor the data collected to fit the physical requirements of that type of unit. Note, however, that the data correlations recommended later are sufficiently general in nature to apply to most equipment types.

Feed Slurry Temperature   Temperature can be both an aid and a limitation. As temperature of the feed slurry is increased, the viscosity of the liquid phase is decreased, causing an increase in filtration rate and a decrease in cake moisture content. The limit to the benefits of increased temperature occurs when the vapor pressure of the liquid phase starts to materially reduce the allowable vacuum. If the liquid phase is permitted to flash within the filter internals, various undesired results may ensue: disruption in cake formation adjacent to the medium, scale deposit on the filter internals, a sharp rise in pressure drop within the filter drainage passages due to increased vapor flow, or decreased vacuum pump capacity. In most cases, the vacuum system should be designed so that the liquid phase does not boil.

In some special cases, steam filtration can be used to gain the advantages of temperature without having to heat the feed slurry.

Where applicable, dry steam is passed through the deliquored cake to raise the temperature of the residual moisture, reduce its viscosity, and lower its content. The final drying or cooling period which follows steam filtration uses the residual heat left in the cake to evaporate some additional moisture.

Cake Thickness Control   Sometimes the rate of cake formation with bottom feed–type filters is rapid enough to create a cake too thick for subsequent operations. Cake thickness may be controlled by adjusting the bridge-blocks in the filter valve to decrease the effective submergence, by reducing the slurry level in the vat, and by reducing the vacuum level in the cake formation portion of the filter valve. If these measures are inadequate, it may be necessary to use a top-loading filter.

Cake thickness must frequently be restricted when cake washing is required or the final cake moisture content is critical. Where the time required for cake washing is the rate-controlling step in the filter cycle, maximum filtration rate will be obtained when using the minimum cake thickness that gives good cake discharge. Where minimum cake moisture content is the controlling factor, there is usually some leeway with respect to cake thickness, although the minimum required for cake discharge is controlling in some cases. Since a relatively constant quantity of moisture is transferred from the medium to the filter cake when the vacuum is released prior to cake discharge, very thin cakes will sometimes be wetter than thicker cakes.

The effect of an increase in cake thickness on the time required for washing is easy to see if one considers what happens when the cake thickness is doubled. Assume that two cakes have the same permeability and that the quantity of washing fluid to cake solids is to remain constant. Doubling of the cake thickness doubles the resistance to flow of the washing fluid through the cake. At the same time, the quantity of washing fluid per unit area is also doubled. Thus, the time required for the washing fluid to pass through the cake is increased by the square of the ratio of the cake thicknesses. In this particular example, the washing time would be increased by a factor of four, while cake production would only be doubled.

Filter Cycle   Each filter cycle is composed of cake formation plus one more of the following operations: deliquoring (dewatering or drying), washing, thermal drying, steam drying, and cake discharge. The number of these operations required by a given filtration operation depends upon the process flowsheet. It is neither possible nor necessary to consider all of these operations at once. The basic testing program is designed to look at each operation individually. The requirements for each of the steps are then fit into a single filter cycle.

All filters utilizing a rotary filter valve have their areas divided into a number of sections, sectors, or segments (see Fig. 18-122). When a drainage port passes from one portion of the filter valve to another, the change at the filter medium does not occur instantaneously nor does it occur at some precise location on the filter surface. The change is relatively gradual and occurs over an area, as the drainage port at the filter valve first closes by passing onto a stationary bridge-block and then opens as it passes off that bridge-block on the other side.

On a horizontal belt filter, the equivalent sections extend across the filter in narrow strips. Therefore, changes in vacuum do occur rapidly and may be considered as happening at a particular point along the length of the filter.

Representative Samples   The results which are obtained in any bench-scale testing program can be only as good as the sample which is tested. It is absolutely essential that the sample used be representative of the slurry in the full-scale plant and that it be tested under the conditions that prevail in the process. If there is to be some significant time between taking or producing the sample and commencing the test program, due consideration must be given to what effect this time lapse may have on the characteristics of the slurry. If the slurry is at a temperature different from ambient, the subsequent heating and/or cooling could change the particle size distribution. Even sample age itself may exert a significant influence on particle size. If there is likely to be an effect, the bench-scale testing program should be carried out at the plant or laboratory site on fresh material.

Whenever a sample is to be held for some time or shipped to a distant laboratory for testing, some type of characterizing filtration test should

TABLE 18-8   Minimum Cake Thickness for Discharge

Filter type	Minimum design thickness	
	mm	in
Drum		
Belt	3–5	⅛–³⁄₁₆
Roll discharge	1	¹⁄₃₂
Std. scraper	6	¼
Coil	3–5	⅛–³⁄₁₆
String discharge	6	¼
Precoat	0–3 max.	0–⅛ max.
Horizontal belt	3–5	⅛–³⁄₁₆
Horizontal table	20	¾
Tilting pan	20–25	¾–1
Disk	10–13	⅜–½

be run on the fresh sample and then duplicated at the time of the test program. A comparison of the results of the two tests will indicate how much of a change there has been in the sample. If the change is too great, there would be no point in proceeding with the tests, and it would be necessary to make arrangements to work on a fresh sample. Any shipped sample, especially during the winter months, must be protected from freezing, as freezing can substantially change the filtration characteristics of a slurry, *particularly hydrated* materials.

The slurry should always be defined as completely as possible by noting suspended solids concentration, particle size distribution, viscosity, density of solids and liquid, temperature, chemical composition, and so on.

Feed Solids Concentration   Feed slurries that are so dilute that they settle rapidly usually yield reduced solids filtration rates and produce stratified cakes with higher moisture contents than would normally be obtained with a homogeneous cake. It is well known that an increase in feed solids concentration is generally an effective means of increasing solids filtration rate, assisting in forming a homogeneous suspension and thereby minimizing cake moisture content, and so on. Equipment required to concentrate a slurry sample and the tests needed to predict how far a slurry will thicken are discussed elsewhere in this section.

Pretreatment Chemicals   Even though the suspended solids concentration of the slurry to be tested may be correct, it is frequently necessary to modify the slurry in order to provide an acceptable filtration rate, washing rate, or final cake moisture content. The most common treatment, and one which may provide improvement in all three of these categories, is the addition of flocculating agents, either inorganic chemicals or natural or synthetic polymers. The main task at this point is to determine which is the most effective chemical and the quantity of chemical which should be used.

It is usually difficult to observe visually a change in floc structure in a concentrated slurry. The two best indications that an effective quantity of chemical has been added is a sudden thickening or increase in viscosity of the slurry and the formation of *riverlets* on the surface of a spatula when treated slurry is shaken from it. It is generally necessary to exceed a threshold quantity of chemical before there is a measurable improvement. The proper dosage becomes an economic balance between the cost of additional chemicals and the savings resulting from a reduction in filter area.

Screening tests are used to determine the best chemical and its approximate dosage. It is usually convenient to use small, graduated beakers and sample quantities in the range of 50 to 150 mL. The chemical may be added with a syringe or medicine dropper and a note made of the quantity used, together with the results. The experimenter should filter and wash the flocculated sample on any convenient, small, top-loading filter with good filtrate drainage. The cake formation and wash times obtained from these micro tests are not intended to provide sizing data, but they do provide an excellent indication of the relative effectiveness of various chemicals and treatment levels.

With any chemical treatment system, the main task is one of getting the chemical thoroughly mixed with the solids without degrading the flocs which are formed. For those slurries that are relatively fluid, the chemical can frequently be added and mixed satisfactorily using a relatively wide spatula. However, for those thick, relatively viscous slurries, a power mixer will be required. In this case, the mixer should be stopped about one second after the last of the flocculant is added. Should this approach be required, it means that a suitably designed addition system must be supplied with the full-scale installation in order to do an effective job of flocculation.

While the volume of chemical used should be minimized, the experimenter must use good judgment based on the viscosities of both the slurry to be treated and the chemical used. If both are relatively viscous, use of a power mixer is indicated.

There are a number of commercially available surfactants that can be employed as an aid in filter cake moisture reduction. These reagents can be added to the filter feed slurry or to the filter cake wash water, if washing is used. Since these reagents have a dispersing effect, flocculation may be required subsequently. Typical moisture reduc-

tions of 2 to 4 percentage points are obtained at reagent dosages of 200 to 500 g/mt of solids.

Cloth Blinding   Continuous filters, except for precoats, generally use some type of medium to effect the separation of the solid and filtrate phases. Since the medium is in contact with the process solids, there is always the danger, and almost invariably the actual occurrence, of medium blinding. The term **blinding** refers to blockage of the fabric itself, either by the wedging of process solids or by solids precipitated in and around the yarn.

The filter medium chosen should be as open as possible yet still able to maintain the required filtrate clarity. Those fabrics which will produce a clear filtrate and yet do not have rapid blinding tendencies are frequently light in weight (woven from thin filaments or yarn) and will not wear as long as some of the heavier, more open fabrics (woven from heavy filaments or yarn). Whenever the filter follows a gravity thickening or clarification step, it is advisable to return the filtrate to the thickener or clarifier so that the filtrate clarity requirements may be relaxed in favor of using a heavier, more open cloth with reduced blinding tendencies. Excessively dirty filtrates should be avoided as the solids may be abrasive and detrimental to the internals of the filter or perhaps may cut the fabric yarn.

It should be noted at this point that an **absolutely** clear filtrate can rarely be obtained on a cloth-covered continuous filter. The passages through the medium are invariably larger than some of the solids in the slurry, and there will be some amount of solids passing through the medium. Once the pores of the fabric have been bridged, the solids themselves form the septum for the remaining particles, and the filtrate becomes clear. It is this bridging action of the solids that permits the use of a relatively open filter medium, while at the same time maintaining a reasonably clear filtrate.

Filters with media in the form of an endless belt have greatly reduced the concern about blinding. Most synthetic fabrics can be successfully cleaned of process solids by washing the medium after cake discharge, and the rate of blinding due to chemical precipitation also can be drastically reduced. Current practice suggests that the belt-type filter with continuous-medium washing be the first choice unless experience has shown that medium blinding is not a factor or if the belt-type system cannot be successfully applied.

Sealing of the belt along the edges of the filter drum is never perfect, and some leakage should be expected. If good clarity is essential, it may be preferable to use a drum filter with the cloth caulked in place and design the system to contend with the effects of blinding.

The one exception to the points noted above is the continuous precoat filter. Here the purpose of the filter medium is to act as a support for the sacrificial bed of precoat material. Thus, the medium should be tight enough to retain the precoat solids and prevent bleeding of the precoat solids through the filter medium during operation, yet open enough to permit easy cleaning at the end of each cycle. Lightweight felt media work well in these respects.

Homogeneous Cake   Accurate test results and optimum filter performance require the formation of a homogeneous cake and thus the maintenance of a similarly homogeneous suspension. Settling in the sample container during a bottom-feed test program can usually be detected by comparing the back-calculated feed solids concentration (based on filtrate, wet cake, and dry cake weights) with the slurry solids concentration as prepared. It is normal to find that the back-calculated concentration is slightly lower than the prepared concentration. This difference is normally within 2 percentage points and should never be greater than 5 percentage points. Since this difference does exist, it means that the slurry sample will concentrate to some extent as the tests continue. Adding fresh slurry to the sample container after each test can counteract this condition, as the system will reach an equilibrium similar to that found in a full-scale machine.

If a more positive check is required on the quality of the filter cake, particle size analyses may be run and compared with the sample as prepared.

In a top-feed filter test, the filter cake will contain all of the solids, provided they are all emptied from the sample container. The danger in this type of test is that the solids will stratify, particularly if the cake formation time is prolonged. Close examination of the filter cake will indicate whether or not this has happened. If there has been significant stratification, the feed slurry should be modified by thickening and/or flocculation in order to increase the dry solids filtration rate and permit formation of a homogeneous cake. Another possibility, but not necessarily the best, is to use a thinner, but still dischargeable, cake to avoid stratification.

Agitation of Sample   All slurries used in bottom-feed tests must be agitated by hand (if slurry characteristics permit) to check whether or not the solids are settling out around the edges of the container and to determine the degree of agitation required to maintain the solids in suspension. Generally speaking, if the solids can be maintained in suspension by hand agitation, the slurry can be processed by a bottom-feed-type filter.

Agitation by a wide spatula may be substituted for hand agitation, but only after it has been determined by feel that the spatula will provide the needed agitation. If this cannot be done, then confirmation of proper agitation must be based on back-calculated feed solids concentrations and/or particle size analyses of the filter cakes.

If it is not possible to maintain a uniform suspension, the sample should be thickened and the flowsheet modified to provide the required thickening.

Mechanical agitation of a sample is very difficult to use effectively. Generally speaking, if enough room is left in the sample container for the leaf and the agitator, the agitation is not sufficient to prevent settling out in the corners of the container. If sufficient agitation is used to maintain suspension in all parts of the container, then it is highly probable that the velocity of the slurry across the face of the test leaf will wash the solids from the leaf and give very erroneous results. Furthermore, the tendency is to leave the agitator running continuously, or at least for so long a period of time that there is attrition of the solids and, therefore, inaccurate results. Mechanical agitation during testing can generally be justified only in a most unusual and exceptional circumstance.

Use of Steam or Hot Air   It was indicated earlier that the cycle might include steam filtration or thermal drying using hot air. While effective use is made of both steam and hot air, the applications are rather limited, and testing procedures are difficult and specialized. As a general rule, steam application will reduce cake moisture 2 to 4 percentage points. Hot-air drying can produce a bone-dry cake, but generally it is practical only if the air rate is high, greater than about 1800 $m^3/m^2 \cdot h$ (98 cfm/ft^2). Both systems require a suitable hood which must contact the dam on the leaf during the drying cycle, allowing the steam or hot air to pass through the cake without dilution by cold air. The end of the operation can be determined by a noticeable increase in the temperature of the gas leaving the leaf.

SMALL-SCALE TEST PROCEDURES

[Purchas (ed.), *Solid/Liquid Separation Equipment Scale-Up*, Uplands Press, Croydon, England, 1977.]

Apparatus   There are several variations of the bench-scale test leaf that may be used, but they all have features similar to the one discussed below.

One typical test leaf is a circular disk with a plane area of 92.9 cm^2 (0.1 ft^2). One face of the leaf is grooved to provide large filtrate drainage passages and a support for the filter medium. A threaded drainage connection is provided on the center of the other face of the leaf. The test leaf is fitted with a filter medium and a dam, and the assembly clamped together as shown in Fig. 18-94. The depth of the dam for bottom-feed tests should be no greater than the depth of the maximum cake thickness, except where cake washing tests are to be performed. In this case, the dam depth should be about 3 mm (⅛ in) greater than the maximum expected cake thickness. Excessive dam depth will interfere with slurry agitation and can result in the formation of a nonhomogeneous cake.

It is absolutely necessary that a dam be used in all cases, except for roll discharge applications which do not involve cake washing or where the maximum cake thickness is on the order of 2 mm or less. If a dam is not used, filter cake will form past the edge of the leaf in the general shape of a mushroom. When this happens, the total filter area is some unknown value, greater than the area of the leaf, that constantly increases with time during cake formation.

The back of the leaf assembly and the joint where the dam overlaps must be sealed with some suitable material so that the filtrate volume collected accurately represents the liquid associated with the deposited cake solids.

Figure 18-95 also contains a schematic layout of the equipment which is required for all bottom-feed leaf tests. Note that there are no valves in the drainage line between the test leaf and the filtrate receiver, nor between the filtrate receiver and the vacuum pump.

At the start of the leaf test run, the hose between the test leaf and filtrate receiver should be crimped by hand to bring the filtrate receiver to the operating vacuum level. The use of a valve at this point is not only less convenient but very frequently results in a hydraulic

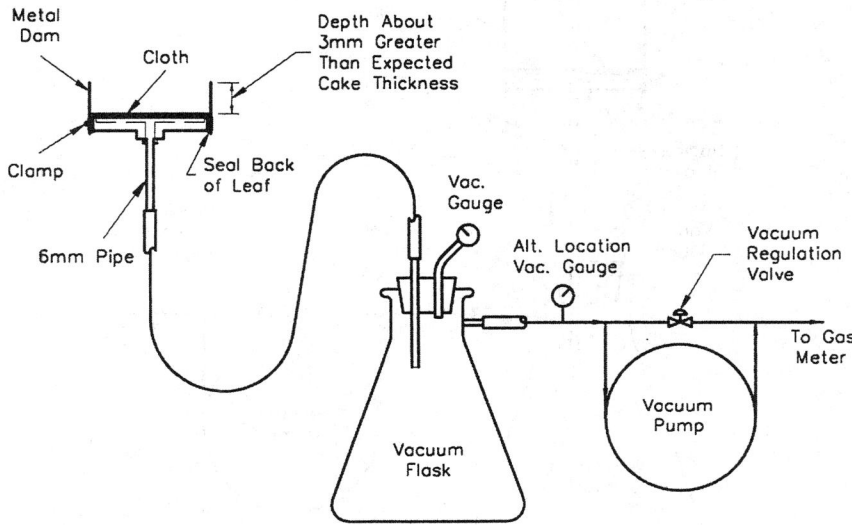

FIG. 18-94   Typical bottom-feed leaf test setup.

restriction. The net result, then, is a measurement of flow through the valve rather than the rate at which the filter cake is capable of forming. Hydraulic restriction is something which should always be kept in mind. If the filtrate runs at a high and full pipe flow rate into the filtrate receiver, it is quite likely that there is some degree of hydraulic restriction, and larger tubing and piping should be considered. When very high air flow rates are obtained, the experimenter must be satisfied that the rates being measured are limited by cake resistance and not by pressure drop through the equipment.

There will be many times when the quantity of sample is limited. While it is best to use the 92.9 cm² (0.1 ft²) area leaf in order to minimize edge effects and improve accuracy, when the sample volume is limited it is much better to have several data points with a smaller leaf than only one or two using the larger leaf. Data from leaves as small as 23.2 cm² (0.025 ft²) are reasonably accurate and can be used to scale up to commercially sized units. However, it is usually prudent to employ a more conservative scale-up factor.

For top-feed applications, the most convenient assembly is that shown in Fig. 18-95. The depth of the dam must, of course, be sufficient to contain the total quantity of feed slurry required for the test. Since the test leaf is mounted on top of the vacuum receiver, it is necessary to provide a valve between the test leaf and the receiver so that the desired operating vacuum may be obtained in the receiver before the start of a test run. It is imperative, however, that there be no restriction in this valve. The preferred choice is a ball valve with the full bore of the drainage piping.

Test Program   Figure 18-96 is a suggested data sheet which contains spaces for most of the information which should be taken during a leaf test program, together with space for certain calculated values. Additional data which may be required include variations in air flow rate through the cake during each dewatering period and chemical and physical data for those tests involving cake washing.

It is difficult to plan a filtration leaf test program until one test has been run. In the case of a bottom-feed test, the first run is normally started with the intention of using a 30-s cake formation time. However, if the filtrate rate is very high, it is usually wise to terminate the run at the end of 15 s. Should the filtrate rate be very low, the initial form period should be extended to at least 1 min. If cake washing is to

be employed, it is useful to apply a quantity of wash water to measure its rate of passage through the cake. The results of this first run will give the experimenter an approximation of cake formation rate, cake washing rate, and the type of cake discharge that must be used. The rest of the leaf test program can then be planned accordingly.

In any leaf test program there is always a question as to what vacuum level should be used. With very porous materials, a vacuum in the range of 0.1 to 0.3 bar (3 to 9 in Hg) should be used, and, except for thermal-drying applications using hot air, the vacuum level should be adjusted to give an air rate in the range of 450 to 900 m³/m²·h (30 to 40 cfm/ft²) measured at the vacuum.

For materials of moderate to low porosity, a good starting vacuum level is 0.6 to 0.7 bar (18 to 21 in Hg), as the capacity of most vacuum pumps starts to fall off rapidly at vacuum levels higher than 0.67 bar (20 in Hg). Unless there is a critical moisture content which requires the use of higher vacuums, or unless the deposited cake is so impervious that the air rate is extremely low, process economics will favor operation at vacuums below this level. When test work is carried out at an elevation above sea level different than that of the plant, the elevation at the plant should be taken into account when determining the vacuum system capacity for high vacuum levels (>0.5 bar).

Generalized correlations are available for each of the operations which make up the full filter cycle. This means that simulated operating conditions can be varied to obtain a maximum of information without requiring an excessive number of test runs. The minimum number of test runs required for a given feed will, of course, vary with the expertise of the experimenter and the number of operations performed during the filter cycle. If, for example, the operation involves only the dewatering of a slurry which forms a cake of relatively low to moderate porosity, frequently sufficient data can be obtained in as little as six runs. For more difficult tests, more runs are usually advisable, and the novice certainly should make a larger number of runs as there is likely to be more data scatter.

Bottom-Feed Procedure   The procedure for collecting data using bottom-feed leaf test techniques is as follows:

1.   Fit the test leaf with a filter cloth expected to give reasonable results and seal the back of the leaf and side of the dam with silicone or other suitable material.

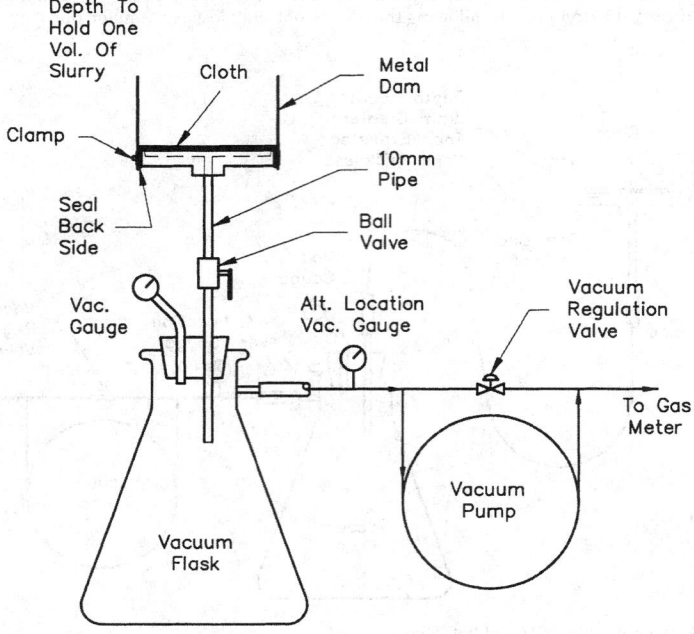

FIG. 18-95   Typical top-feed leaf test setup.

FILTRATION LEAF TEST DATA SHEET – VACUUM AND PRESSURE

Company _____   Mat'l as Received: Date _____   Test No. _____

Address _____   Solids: _____ %   Date Tested _____

_____   Analysis _____   By _____

_____   Liquid: _____ %   Location _____

Filter Type _____ Leaf Size _____ Ft.^{2} Analysis _____

Used Shim: No _____ Yes _____   Precoat Forming Liquid _____   Temp. _____ °F/°C

Run No.	Filter Module and/or Precoat Type	Feed Temp., °F/°C	% Solids in Food		Vacuum = in. Hg. Pressure = PSI.				TIME, MIN.					Air Flow (1)	Filtrate			Wash				Cake Weights			Dish No.
			As Prepared	Back Calculated	Form	Wash	Dry	After Cake Cracks	Form	Dewater	Wash	Dry	To Crack or Gas Breakthrough After Form/Wash		Temp., °F/°C	ML.	Clarity	Precoat Penetration	ML.	Temp., °F/°C	Cake/Precoat Thickness, In.	Dia. of Shared Area, In.	Tare GMS.	Wet & Tare GMS.	Dry & Tare GMS.	

CAKE DISCHARGE		REAGENT TREATMENT	
RUNS	COMMENT	RUNS	COMMENT

REMARKS: (1) Record Basis of Observation in Space Provided.

FIG. 18-96   Sample data sheet.

2. Hand-crimp the hose in back of the test leaf, and then turn on the vacuum pump and regulate the bypass valve on the pump to give the desired vacuum level in the receiver.

3. Agitate the slurry by hand or with a wide spatula to maintain a homogeneous suspension. Immerse the test leaf face downward to approximately one-half the depth of the slurry.

4. Simultaneously start the timer and release the crimped hose to begin cake formation. Maintain agitation during cake formation and move the leaf as may be required to ensure that solids do not settle out in any part of the container. It is not necessary to try to simulate the velocity with which the full-scale unit's filtration surface passes through the slurry in the filter tank.

5. Remove the leaf from the slurry at the end of the cake-formation period and note the time. If the slurry is particularly thick and viscous, the leaf may be gently shaken to remove excess slurry and prevent the dam from scooping up extra material. Maintain the leaf in an upright position (cake surface on top) and elevated so that liquid within the drainage passages may pass to the receiver. Tilt and rotate the leaf to help the filtrate reach the drain outlet. Continue this dewatering period until:

a. the preselected time has elapsed, or
b. the cake cracks.

6. If the cake is to be washed, apply a measured quantity of wash fluid and note the time required for free fluid to disappear from the surface of the cake. Pour the wash fluid onto a deflecting baffle, such as a bent spatula, to prevent the cake from being gouged. Washing must begin before cake cracking occurs. In particular, observe that there is no crack along the edge between the cake and the dam.

7. Continue with the various operations in the predetermined sequence.

8. During each of the operations record all pertinent information such as vacuum level, temperature, time required for the cake to crack, filtrate foaming characteristics, air flow rate during the drying periods, etc.

9. At the end of the run, measure and record the filtrate volume (and weight, if appropriate), cake thickness, final cake temperature (if appropriate), wet cake weight, and note the cake discharge characteristics (roll, sticks to media, etc.).

10. For runs involving cake dewatering only, it is usually convenient to dry the total cake sample, if the associated solution contains little or no dissolved solids.

11. When cake washing is involved, it is usually convenient to weigh the wet cake and then repulp it in a known quantity of distilled water or in water at the same pH as the filtrate, if precipitation of

solute could occur in distilled water. The resultant slurry is then filtered using a clean dry filter and flask and a sample of the clear liquid analyzed for the reference constituent.

Should the mother liquor contain a significant quantity of dissolved solids, the filter cake should be thoroughly washed (after the sample for analysis has been taken) so that the final dry weight of the cake will represent suspended solids only. The quantity of reference constituent in the final washed cake can be readily calculated from the wet and dry cake weights and the known amount of distilled water used for repulping.

In cake-washing tests, it is important that the feed slurry liquid be analyzed for total dissolved solids and density as well as the reference constituent.

Top-Feed Procedure   The sequence of operations with a top-feed leaf test is the same as in a bottom-feed test, except that the leaf is not immersed in the slurry. The best method for transferring the slurry to the top-feed leaf is, of course, a function of the characteristics of the slurry. If the particles in the slurry do not settle rapidly, the feed can usually be transferred to the leaf from a beaker. If, however, the particles settle very rapidly, it is virtually impossible to pour the slurry out of a beaker satisfactorily. In this case, the best method is to make use of an Erlenmeyer flask, preferably one made of plastic. The slurry is swirled in the flask until it is completely suspended and then abruptly inverted over the leaf. This technique will ensure that all of the solids are transferred to the leaf.

When the solids involved are coarse and fast settling, the vacuum should be applied an instant after the slurry reaches the surface of the filter medium.

Precoat Procedure   Precoat filtration tests are run in exactly the same manner as bottom-feed tests except that the leaf must first be precoated with a bed of diatomaceous earth, perlite, or other shaveable inert solids. Some trial and error is involved in selecting a grade of precoat material which will retain the filtered solids to be removed on the surface of the bed without any significant penetration. During this selection process, relatively thin precoat beds of 1 to 2 cm are satisfactory. After a grade has been selected, bench-scale tests should be run using precoat beds of the same thickness as expected on the full-scale unit.

Where the resistance of the precoat bed is significant in comparison to the resistance of the deposited solids, the thickness of the precoat bed effectively controls the filtration rate. In some instances, the resistance of the deposited solids is very large with respect to even a thick precoat bed. In this case, variations in thickness through the life of the precoat bed have relatively little effect on filtration rate. This type of information readily becomes apparent when the filtration rate data are correlated.

The depth of cut involved in precoat filtration is a very important economic factor. There is some disagreement as to the method required to accurately predict the minimum permissible depth of cut. Some investigators maintain that the depth of cut can be evaluated only in a qualitative manner during bench-scale tests by judging whether the process solids remain on the surface of the precoat bed. This being so, they indicate that it is necessary to run a continuous pilot-plant test to determine the minimum permissible depth of cut. The use of a continuous pilot-plant filter is a very desirable approach and will provide accurate information under a variety of operating conditions.

However, it is not always possible to run a pilot-plant test in order to determine the depth of cut. A well-accepted alternative approach makes use of the more sophisticated test leaf illustrated in Fig. 18-97. This test leaf is designed so that the cake and precoat are extruded axially out the open end of the leaf. The top of the retaining wall on this end is a machined surface which serves as a support for a sharp discharge knife. This approach permits variable and known depths of cut to be made so that the minimum depth of cut may be determined. Test units are available from Rutland Tool and Supply Company, City Of Industry, California, (800) 289-4787.

Lacking the above-described actual data, it is possible to estimate precoat consumption by using these values: nonpenetrating solids, 0.06-mm cut/drum revolution (0.0024 in); visible penetration, 0.15- to 0.20-mm cut/drum revolution (0.006 to 0.008 in); precoat bed density, 4.2 kg/m²·cm of bed depth (2.2 lb/ft²·in) for diatomaceous earth or 2.1 to 3.0 kg/m²·cm (1.1 to 1.6 lb/ft²·in) for perlite.

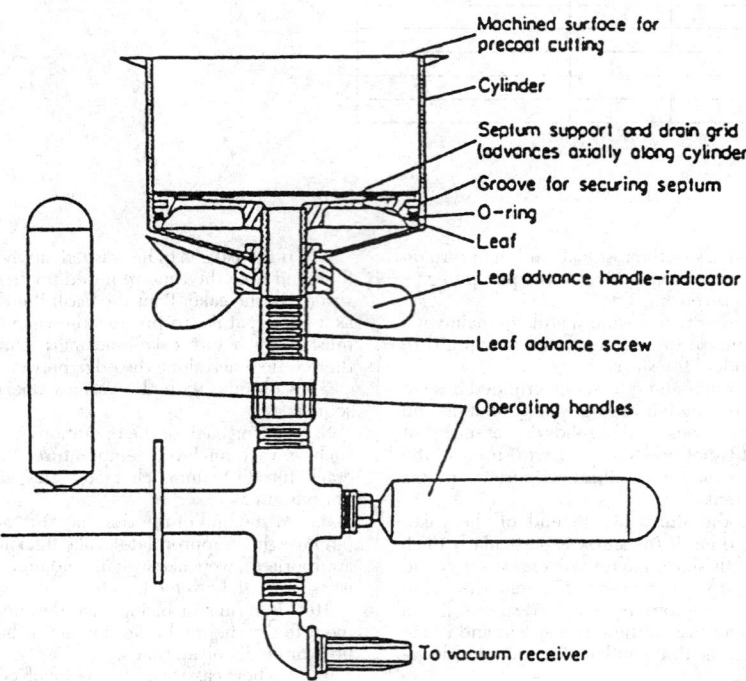

FIG. 18-97   Special test leaf for precoat filtration. (*Rutland Tool and Supply Company.*)

DATA CORRELATION

[Purchas (ed.), *Solid/Liquid Separation Equipment Scale-Up,* Uplands Press, Croydon, England, 1977.]

The correlations used are based partly on theoretical consideration and partly on empirical observations. The basic filtration data are correlated by application of the classic cake-filtration equation, aided by various simplifying assumptions which are sufficiently valid for many (but not all) situations. Washing and drying correlations are of a more empirical nature but with strong experimental justification. If steam or thermal drying is being examined, additional correlations are required beyond those summarized below; for such applications, it is advisable to consult an equipment manufacturer or refer to published technical papers for guidance.

Dry Cake Weight vs. Thickness   It is convenient to convert the test dry cake weight to the weight of dry cake per unit area per cycle (W), and plot these values as a function of cake thickness (Fig. 18-98). Cake weight is measured quite accurately, while cake thickness measurements are subject to some variation. By plotting the data, variations in thickness measurements are averaged. The data usually give a straight line passing through the origin. However, with compressible material, sometimes a slightly curved line best represents the data, since thinner cakes are usually compressed more than thicker cakes.

Dry Solids or Filtrate Rate   Filtration rate, expressed either in terms of dry solids or filtrate volume, may be plotted as a function of time on log-log paper. However, it is more convenient to delay the rate calculation until the complete cycle of operations has been defined.

It is most useful to plot either dry cake weight (weight of dry solids/unit area/cycle) or filtrate volume (volume/unit area/cycle) as a function of time on log-log paper. These data should give straight-line plots for constant operating conditions in accordance with Eqs. (18-52) and (18-53). The expected slope of the resultant rate/time plots is +0.50, as in Fig. 18-99. In practice, the vast majority of slopes range from +0.50 to +0.35. Slopes steeper than +0.5 indicate that there is some significant resistance other than that of the cake solids, such as a hydraulic restriction in the equipment or an exceptionally tight filter cloth.

Data from precoat tests, however, generally produce filtrate curves with much steeper slopes. The precoat bed has a greater resistance than most filter fabrics, and the particles which are separated on a continuous precoat usually form a cake which has a relatively low resistance when compared to that of the precoat bed. Once the thickness of the deposited solids becomes significant, their resistance increases. Thus, at very short form times, the slope of the filtrate curve

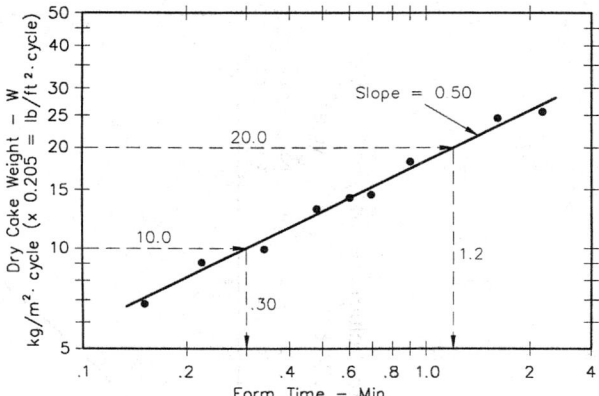

FIG. 18-99   Dry cake weight vs. form time.

may be close to 1.0, but as form time increases, the slope of the curve will decrease and will approach +0.5 (Fig. 18-100).

There are some solids, however, which form a less permeable cake, even in very thin layers. With these solids, the resistance of the deposited cake will be very high when compared to that of the precoat bed, and the slope of the filtrate curve will be +0.5 for all values of form time.

Effect of Time on Flocculated Slurries   Flocculated slurries usually show significant decreases in filterability with time (Fig. 18-101). The rate of degradation may be established by running a series of repetitive leaf tests at frequent intervals on a flocculated slurry, starting as soon as practical after the addition of the flocculant. If there is little change in the filtration rate, this factor need be given no more consideration. However, it is usually found that there is significant degradation.

When a flocculated feed is added to a filter tank, there is a definite time lag before this material reaches the surface of the filter medium. Since this lag time is not known at the time of testing, a lag time of 8 to 10 minutes should be allowed before starting the first leaf test on a flocculated slurry. Two, or perhaps three, tests can be run before the elapsed time exceeds the probable retention time in the full-scale filter tank. With knowledge of the elapsed time after flocculation and data relating to the rate of degradation, the rates obtained on the leaf test runs can be adjusted to some constant lag time consistent with the anticipated full-scale design.

Cake Moisture   Results on a wide variety of materials have shown that the following factor is very useful for correlating cake moisture content data:

$$\text{Correlating factor} = (m^3/m^2\cdot h)(P_c/W)(\Theta_D/\mu), \qquad (18\text{-}64)$$

where $m^3/m^2\cdot h$ = air rate through filter cake measured at downstream pressure or vacuum
P_c = pressure drop across cake
W = dry cake weight/unit area/cycle
Θ_D = dry time per cycle
μ = viscosity of liquid phase

For a more rigorous discussion of cake moisture correlation, the reader is referred to an earlier article by Nelson and Dahlstrom [*Chem. Eng. Progress,* **53**, 7, 1957]. Fig. 18-102 shows the general shape of the curve obtained when using the cake moisture correlating factor. The value of the correlating factor chosen for design should be somewhere past the knee of the curve. Values at and to the left of the knee are in an unstable range where a small change in operating conditions can result in a relatively large change in cake moisture content.

It is not always necessary to use all of the terms in the correlating factor, and those conditions which are held constant throughout the testing may be dropped from the correlating factor. Many times, air rate data are not available and reasonable correlations can be obtained

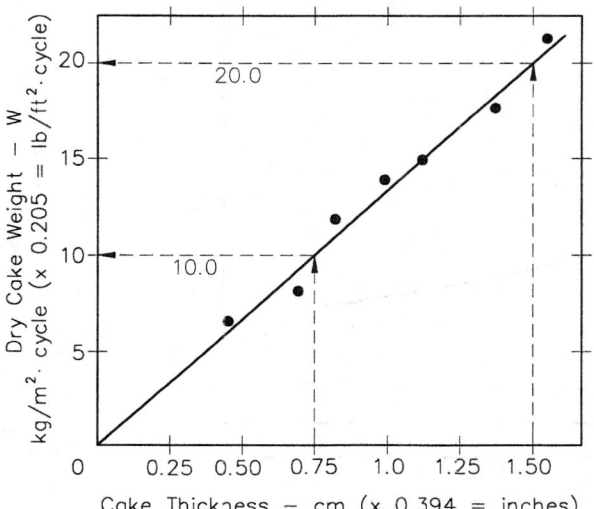

FIG. 18-98   Dry cake weight vs. cake thickness.

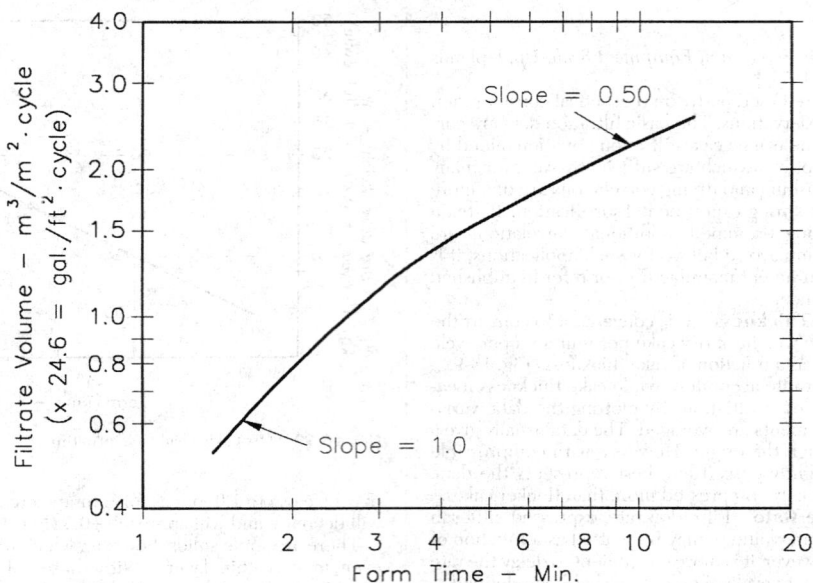

FIG. 18-100 Filtrate volume per cycle vs. form time.

without this information, particularly if the cakes are relatively low in permeability. By dropping these terms, the correlating factor is reduced to the simplified version, Θ_D/W, involving only drying time and cake weight per unit area per revolution. While this is a very convenient factor to use, care must be taken in its application, as it is no longer a generalized factor, and there will be a tendency for the data to produce different curves for changes in operating conditions such as vacuum level and cake thickness.

Cake Washing Wash efficiency data are most conveniently represented by a semilog plot of percent remaining R as a function of wash ratio N as shown in Fig. 18-103. Percent remaining refers to that portion of the solute in the dewatered but unwashed cake which is left in the washed and dewatered cake. Since a cake-washing operation involves the displacement of one volume of liquid by another volume, the removal of solute is related to the ratio of the volume of washing fluid divided by the volume of liquid in the cake. This ratio is defined as wash ratio N.

Practical experience has shown that the most convenient and best means of expressing R is in terms of the solute concentrations in the washed cake liquid, the feed liquid (or unwashed cake liquid), and the cake wash liquid. Furthermore, the wash ratio N may also be expressed either as a volume or weight ratio.

Percent remaining is defined as follows:

$$\frac{R}{100} = \frac{C_2 - C_w}{C_1 - C_w} \tag{18-65}$$

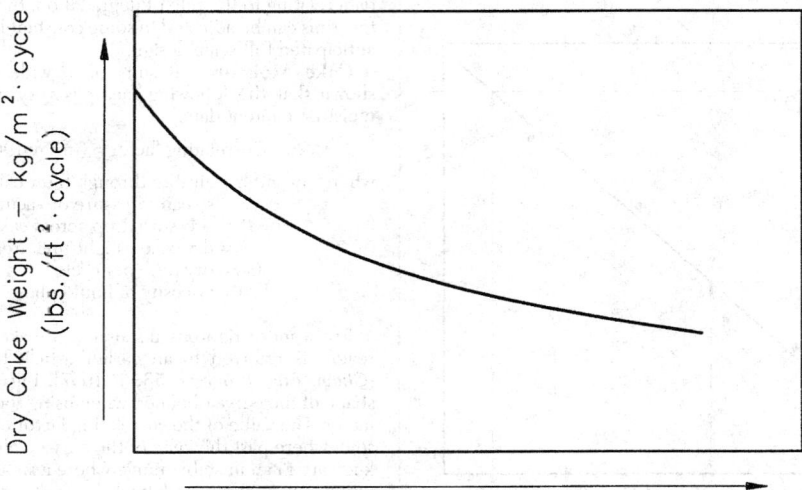

Time After Flocculant Addition

FIG. 18-101 Degradation of flocculation with time.

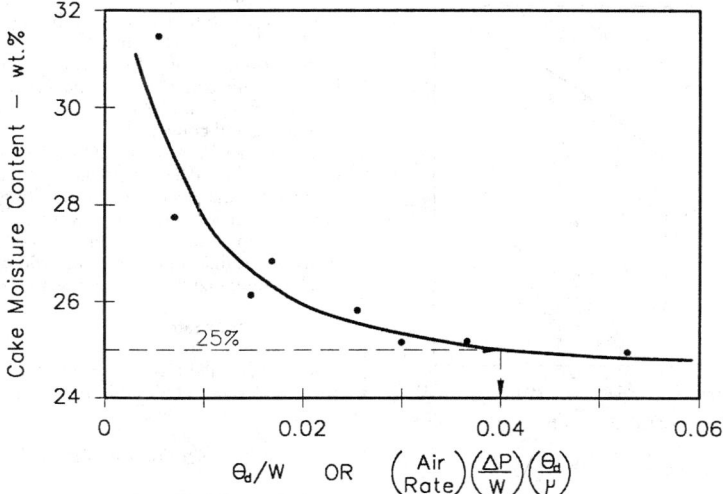

FIG. 18-102   Cake moisture correlation.

where R = % solute remaining after washing
C_2 = solute concentration in washed cake liquid
C_1 = solute concentration in unwashed cake liquid
C_w = solute concentration in wash liquid

If the cake is washed with solute-free liquid, percent remaining is readily calculated by dividing the solute concentration in the liquid remaining in the washed cake by the solute concentration in the liquid in the original feed.

The residence time of the cake-washing fluid within the cake is relatively short and is not normally considered useful for any kind of leaching operation. Therefore, it is assumed that all of the solute is in solution.

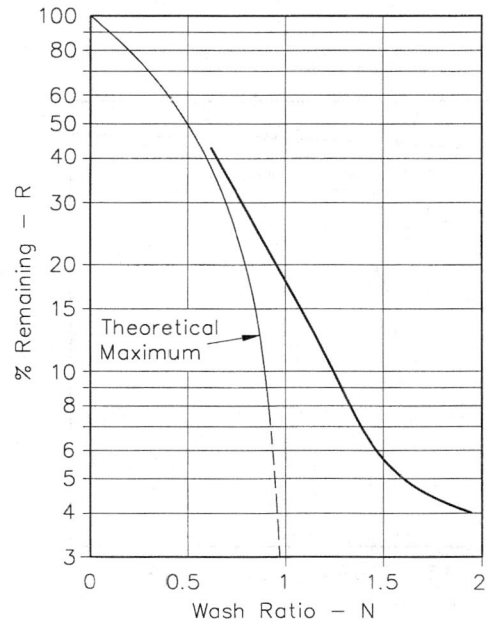

FIG. 18-103   Wash effectiveness.

If it were possible to obtain a perfect slug displacement wash, the fraction remaining would be numerically equal to 1 minus the wash ratio. This ideal condition is represented by the maximum theoretical line as shown in Fig. 18-103. Since it represents the best that can be done, no data point should fall to the left of this curve. Most, but not all, cake-washing curves tend to fall along the heavy solid line shown. In the absence of actual data, one may estimate washing results by using this curve.

The quantity of wash water to be used in a given operation is dictated by flowsheet considerations and the required solute content of the washed cake. Generally speaking, the maximum wash water quantity should be equivalent to a wash ratio of 1.5 to 2.5. Where high solute removals are required, it is frequently necessary to use a two-stage filtration system with intermediate repulping. These two stages may involve countercurrent flow of wash water, or fresh wash water may be used on both filters and for the intermediate repulping step.

Wash Time   Cake-washing time is the most difficult of the filtration variables to correlate. It is obviously desirable to use one which provides a single curve for all of the data. Filtration theory suggests three possible correlations [Eqs. (18-59) to (18-61)]. These are listed below, beginning with the easiest to use:

1.   Wash time vs. WV_w
2.   Wash time vs. NW^2
3.   Wash time/form time vs. wash volume/form volume

where W = weight dry cake/unit area/cycle
V_w = volume of cake wash/unit area/cycle
N = wash ratio
= volume of wash/volume of liquid in discharged cake

Fortunately, the easiest correlation to use usually gives satisfactory results. This curve is usually a straight line passing through the origin, but frequently falls off as the volume of wash water increases (Fig. 18-104). If for some reason this correlation is not satisfactory, one of the other two should be tried.

Air Rate   Air rate through the cake, and thus vacuum pump capacity, can be determined from measurements of the air flow for various lengths of dry time. Figure 18-105 represents instantaneous air rate data. The total volume of gas passing through the cake during a dry period is determined by integrating under the curve.

Note that the air rate at the beginning of each drying or dewatering period within a given cycle starts at zero and then increases with time. The shape of this curve will be a function of the permeability of the deposited cake.

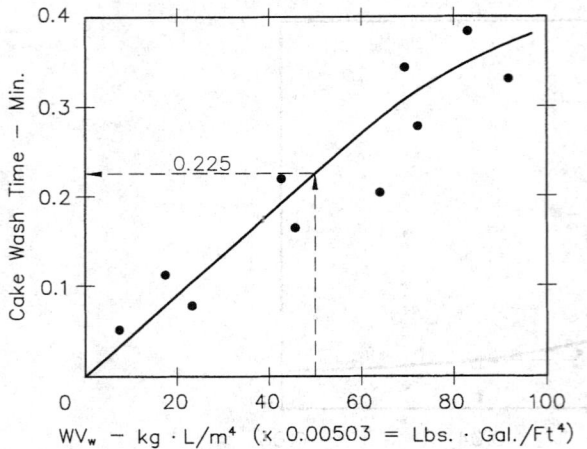

FIG. 18-104 Cake wash time correlation.

Vacuum pump capacity is conventionally based on the total cycle and expressed as m³/h·m² (cfm/ft²) of filter area measured at pump inlet conditions. Thus, the gas volumes per unit area passing during each dry period in the cycle are totaled and divided by the cycle time to arrive at the design air rate. Since air rate measurements in the test program are based on pressure drop across the cake and filter medium only, allowance must be made for additional expansion due to pressure drop within the filter and auxiliary piping system in arriving at vacuum pump inlet conditions.

Air rate measurements made during a leaf test program account only for gas flow through the cake. An operating filter has drainage passages that must also be evacuated. The extra air flow may be conveniently accounted for in most cases by multiplying the leaf test rate by 1.10. With horizontal belt filters, one must also add for the leakage that occurs along the sliding seal and, depending upon the type of fil-

ter cloth used, edge leakage due to lateral permeability of the cloth. Typically, the total leakage can be significant, amounting to about 35 to 50 m³/h·m² (2 to 3 cfm/ft²).

Adjustment may also be required for differences in altitude between the test site and the commercial installation. In general terms, if the plant elevation is higher, the vacuum pump size must be increased, and conversely.

Darcy's law has been used to derive an expression which reflects not only the effect of a change in elevation, but also provides a means for estimating changes in air rate resulting from changes in vacuum level and cake thickness (or cake weight per unit area). In order for this relationship to hold for changes in vacuum and cake thickness, it must be assumed that both cakes have the same specific resistance.

The generalized equation is as follows:

$$(\text{air rate})_{b2} = (\text{air rate})_{a2} \frac{(W_a)}{(W_b)} \frac{(P_{b1}^2 - P_{b2}^2)}{(P_{a1}^2 - P_{a2}^2)} \frac{(P_{a2})}{(P_{b2})} \quad (18\text{-}66)$$

where P_1 = inlet absolute pressure
P_2 = outlet absolute pressure
a = base condition
b = revised condition
W = weight of dry cake solids/unit area/cycle

SCALE-UP FACTORS

[Purchas (ed.), *Solid/Liquid Separation Equipment Scale-Up*, Uplands Press, Croydon, England, 1977.]

The overall scale-up factor used to convert a rate calculated from bench-scale data to a design rate for a commercial installation must incorporate separate factors for each of the following:

Scale-up on rate
Scale-up on cake discharge
Scale-up on actual area

Special note should be made that these scale-up factors are *not* safety factors to allow for additional plant capacity at some future date. Their purpose is to account for differences in scale, including such things as minor deviations in the plant slurry from the sample tested, edge effects due to the size of the test equipment, close control of operating conditions during the leaf test program, and long-term medium blinding.

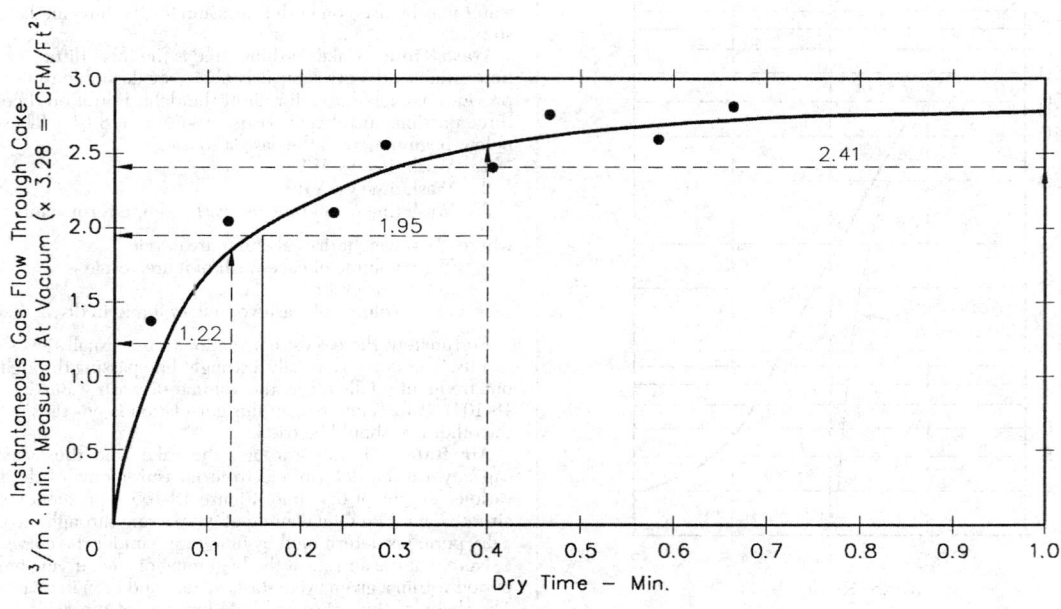

FIG. 18-105 Air flow through cake.

Scale-Up on Rate   Filtration rates calculated from bench-scale data should be multiplied by a factor of 0.8 for all types of commercial units which do not employ continuous washing of the filter medium and on which there is a possibility of filter-medium blinding. For those units which employ continuous filter-medium washing, belt-type drum and horizontal units, the scale-up factor may be increased to 0.9.

The use of this scale-up factor assumes the following:
Complete cake discharge
Nominal filter area approximately equal to actual area
A representative sample
Suitable choice of filter medium
Operating conditions equal to those used in testing
Normal cloth conditioning during testing and operation
The scale-up factor on rate specifically does not allow for:
Changes in slurry filterability
Changes in feed rate
Changes in operating conditions
Cloth blinding
Where there is any doubt about some of the conditions listed above (except for cake discharge and actual area which should be handled separately) a more conservative scale-up factor should be used.

Scale-Up on Cake Discharge   When a filter is selected for a particular application, it is intended that the unit be capable of discharging essentially 100 percent of the cake which is formed. There are, however, many applications which are marginal, regardless of the type of discharge mechanism used. In these cases, the experimenter must judge the percent of cake discharge to be expected and factor the design rate accordingly.

Scale-Up on Actual Area   The nominal area of a filter as used by equipment manufacturers is based upon the overall dimensions of the filtering surface. The fraction of this total area that is active in filtration is a function of the filterability of the material being handled and any special treatment which the surface may receive.

The filtering surface is divided into a number of sections by division strips, radial rods, or some other impervious separator. Material which forms a thin, rather impervious cake will not form across the dividers, and thus the actual area is somewhat less than the nominal. Where relatively thick cakes of at least 1.5 cm are formed, the cake tends to form across the dividers due to cross-drainage in both the filter cake and the filter medium. In this case, the effective area is relatively close to the nominal area.

For most applications, the actual area of a drum filter will generally be no less than 94 to 97 percent of the nominal area, depending upon the size and number of sections. This variation is generally not accounted for separately and is assumed to be taken care of in the scale-up factor on filtration rate.

There are, however, certain special applications where the filter medium around the edge of the section may be deliberately blinded by painting in order to improve cake discharge. This technique is most frequently used on disk filters, with the result that the actual area may be only 75 to 85 percent of the nominal area. This is a significant deviation from the nominal area and must be considered separately.

Overall Scale-Up Factor   The final design filtration rate is determined by multiplying the bench-scale filtration rate by each of the scale-up factors discussed above. While this approach may seem to be ultraconservative, one must realize that the experimenter maintains careful control over the various steps during the filter cycle while running a bench-scale test, whereas a commercial filter operates with a minimum of attendance and at average conditions which are chosen to provide a satisfactory result in a production context.

FULL-SCALE FILTER PERFORMANCE EVALUATION

The correlations which have been presented for the evaluation of bench-scale data are the same correlations which should be used to evaluate the performance of a commercial installation.

A few random samples taken from a commercial installation most probably will not provide enough insight to determine that the filter is performing as expected. However, by making use of reasonable variations in the most important parameters, the desired correlations can be developed. Bench-scale tests should be run on representative feed samples taken at the same time test runs are made on the commercial unit. The bench-scale tests can be varied over a much wider range to provide a sound basis for both the location and shape of the appropriate correlation. A comparison of these results with the data taken from the commercial installation provides a good measure for efficiency of the commercial unit and a basis for identifying problem areas on the full-scale unit.

FILTER SIZING EXAMPLES

[Purchas (ed.), *Solid/Liquid Separation Equipment Scale-Up*, Uplands Press, Croydon, England, 1977.]

The examples which follow show how data from the correlations just presented and a knowledge of the physical characteristics of a particular filter are used to determine a filtration cycle and, subsequently, the size of the filter itself. The three examples which follow involve a disk, a drum belt, and a horizontal belt filter.

Example 1: Sizing a Disk Filter   Equipment physical factors, selected from Table 18-9: Maximum effective submergence = 28%; maximum portion of filter cycle available for dewatering = 45%. (High submergence versions require trunnion seals, and their use is limited to specific applications.)

Scale-up factors: On rate = 0.8; On area = 0.8; On discharge = 0.9. (Scale-up on discharge may be increased to 0.97 if based on previous experience or to 0.95 if the total filter area is based on the measured effective area of the disk.)

Objective: Determine the filter size and vacuum system capacity required to dewater 15 mtph (metric tons per hour) of dry solids and produce a cake containing an average moisture content of 25 wt %.

Calculation procedure:
1. Choose cake thickness = 1.5 cm, slightly thicker than minimum value listed in Table 18-8.
2. From Fig. 18-98, $W = 20.0$ kg dry cake/($m^2 \times$ rev.).
3. From Fig. 18-99, form time, $_f = 1.20$ min.
4. Use simplified moisture content correlating factor in Fig. 18-102.
   Choose $_d/W = 0.04$ at avg. moisture content of 25 wt %.
   Dry time $= _d = 0.04 \times 20.0 = 0.80$ min.
5. Calculate cycle time CT both on the basis of form time and dry time to determine which is controlling:
   $$CT_{form} = 1.20/.28 = 4.29 \text{ mpr (min/rev.)}.$$

TABLE 18-9   Typical Equipment Factors for Cycle Design

	% of cycle					
	Submergence		Total under active vac. or pres.	Max. for washing	Max. for dewatering only	Req'd. for cake discharge
Filter type	Apparent	Max. effective				
Drum						
Standard scraper	35	30	80	29	50–60	20
Roll discharge	35	30	80	29	50–60	20
Belt	35	30	75	29	45–50	25
Coil or string	35	30	75	29	45–50	25
Precoat	35,55	35,55	93	30	60,40	5
Horizontal belt	As req'd.	As req'd.	Lengthen as req'd.	As req'd.	As req'd.	0
Horizontal table	As req'd.	As req'd.	80	As req'd.	As req'd.	20
Tilting pan	As req'd.	As req'd.	75	As req'd.	As req'd.	25
Disk	35	28	75	None	45	25

$CT_{dry} = 0.80/.45 = 1.78$ mpr.

Therefore, cake formation rate is controlling and a cycle time of 4.29 mpr must be used.

6. Overall scale-up factor based on the factors presented previously = $0.8 \times 0.8 \times 0.9 = 0.58$

7. Design filtration rate = $(20.0/4.29)(60 \times 0.58) = 162$ kg/h × m²

8. Area required to filter 15 metric tons of dry solids per hour = $15 \times 1000/162 = 92.6$ m². The practical choice would then be the nearest commercial size of filter corresponding to this calculated area.

9. Dry time = 45% of CT = $0.45 \times 4.29 = 1.93$ min. This is a much longer dry time than required. Therefore, reduce the dry time to 1.00 min by proper bridging in the filter valve.

10. From Fig. 18-105, the average gas flow rate during the 1.00 min drying period was found by graphical integration to be 2.41 m³/m² × min.

11. Total volume of air flowing during dry period = $2.41 \times 1.00 = 2.41$ m³ per m² per cycle. Add 10% to allow for evacuation of drainage passages. Total flow = 2.65 m³/m² × cycle.

12. Required vacuum pump capacity = $2.65/4.29 = 0.62$ m³/min × m² of total filter area. Allow for pressure drop within system when specifying the vacuum pump. See next example.

Example 2: Sizing a Drum Belt Filter with Washing Equipment physical factors, selected from Table 18-9: Maximum effective submergence = 30%; max. apparent subm. = 35%; max. arc for washing = 29%; portion of cycle under vacuum = 75%.

Scale-up factors: On rate = 0.9; On area = 1.0; On discharge = 1.0.

Process data: Sp. gr. of feed liquid = 1.0; TDS (total dissolved solids) in feed liquid = 4.0 wt %; fresh water used for washing; vacuum level = 18 in Hg; final cake liquid content = 25 wt %.

Objective: Determine the filter size and vacuum capacity required to dewater and wash 15 mtph of dry solids while producing a final washed cake with a moisture content of 25 wt % and containing 0.10 wt % TDS based on dry cake solids.

Calculation procedure:

1. Choose cake thickness = 0.75 cm, slightly thicker than the minimum in Table 18-8.

2. From Fig. 18-98, W = 10 kg/m² × cycle.

3. From Fig. 18-99, form time = 0.30 min.

4. From Fig. 18-102, $_d/W = 0.04$ for 25 wt % residual moisture.

5. Dry time = $_d = W \times 0.04 = 10.0 \times 0.04 = 0.40$ min.

6. Determine required wash quantity:

Calculated TDS concentration in washed cake liquor:
Liquid in final cake = $10 \times 0.25/0.75 = 3.33$ kg/m² × cycle.
TDS in dry washed solids = $10 \times 0.001/0.999 = 0.010$ kg/m² × cycle.
TDS in final washed liquor = $(0.010/3.33)100 = 0.300$ wt %.
Percent remaining, $R = ((C_2 - C_w)/(C_1 - C_w))100$.
Since $C_w = 0$,
Required percent remaining, $R = (C_2/C_1)100 = (0.300/4.00)100 = 7.5\%$.
From Fig. 18-103, required wash ratio N = 1.35.
For design, add 10% → N = $1.35 \times 1.1 = 1.49$.
Wash vol. = $V_w = 1.49 \times 3.33/1.00 = 4.96$ L/m² × cycle.

7. Determine wash time:
$WV_w = 10.0 \times 4.96 = 49.6$ kgL/m⁴.
From Fig. 18-104, wash time = $_w = 0.225$ min.

8. Summary of minimum times for each operation:
Form (step 3) = 0.30 min.
Wash (step 6) = 0.225 min.
Final dry (step 5) = 0.40 min.

9. Maximum washing arc = horizontal centerline to 15° past top dead center, or 29% of total cycle. Minimum percent of cycle between end of form and earliest start of wash = area between horizontal centerline and maximum apparent submergence = $(50\% - 35\%)/2 = 7.5\%$.

10. Maximum percentage of cycle for wash + final dry = $75 - 30 - 7.5 = 37.5\%$.

11. Determine cycle time based on the rate-controlling operation:
    a. $CT_{form} = 0.30/.30 = 1.00$ mpr.
    b. $CT_{wash} = 0.225/.29 = 0.77$ mpr.
    c. $CT_{wash + dry} = (0.225 + 0.40)/.375 = 1.67$ mpr.

Therefore, the cake wash + final dry rate is controlling and a cycle time of 1.67 mpr must be used.

12. Since (c) is larger than (a) in the previous step, too thick a cake will be formed and it will not wash or dry adequately unless the effective submergence is artificially restricted to yield the design cake thickness. This may be accomplished by proper bridge-block adjustment or by vacuum regulation within the form zone of the filter valve.

13. The required washing arc of $(0.225/1.67)360 = 48.5°$ is assumed to start at the horizontal center line. Careful control of the wash sprays will be required to minimize runback into the slurry in the vat.

14. Overall scale-up factor = $0.9 \times 1.0 \times 1.0 = 0.9$.

15. Design filtration rate = $(10.0/1.67)(60 \times 0.9)$.
$= 323.3$ kg/h × m².

16. Total filter area required = $15 \times 1000/323.3 = 46.4$ m². Nearest commercial size for a single unit could be a 10 ft dia. × 16 ft long with a total area of 502 ft² = 46.7 m².

17. Determine required vacuum capacity:
Initial dry time = $1.67 \times 0.075 = 0.125$ min.
Calculate gas vol. through cake using data from Fig. 18-105:
Initial dry = $0.125 \times 1.22 = 0.153$ m³/m² × rev.
Final dry = $0.40 \times 1.95 = 0.780$ m³/m² × rev.
Total, including 10% for evacuation of drainage passages = $0.933 \times 1.10 = 1.03$ m³/m² × rev.
Air rate based on total cycle = $1.03/1.67 = 0.62$ m³/min × m² measured at 18 in Hg vacuum.
If pressure drop through system = 1.0 in Hg and barometric pressure = 30 in Hg, design air rate = $0.62 \times 12/11 = 0.68$ m³/min × m² measured at 19 in Hg vacuum.

Horizontal Belt Filter Since the total cycle of a horizontal belt filter occurs on a single, long horizontal surface, there is no restriction with respect to the relative portions of the cycle. Otherwise, scale-up procedures are similar.

BATCH FILTRATION

Since most batch-type filters operate under pressure rather than vacuum, the following discussion will apply primarily to pressure filtration and the various types of pressure filters.

To use Eq. (18-50) one must know the pattern of the filtration process, i.e., the variation of the flow rate and pressure with time. Generally the pumping mechanism determines the filtration flow characteristics and serves as a basis for the following three categories° [Tiller and Crump, *Chem. Eng. Prog.*, **73**(10), 65 (1977)]:

1. *Constant-pressure filtration.* The actuating mechanism is compressed gas maintained at a constant pressure.

2. *Constant-rate filtration.* Positive-displacement pumps of various types are employed.

3. *Variable-pressure, variable-rate filtration.* The use of a centrifugal pump results in this pattern: the discharge rate decreases with increasing back pressure.

Flow rate and pressure behavior for the three types of filtration are shown in Fig. 18-106. Depending on the characteristics of the centrifugal pump, widely differing curves may be encountered, as suggested by the figure.

Constant-Pressure Filtration For constant-pressure filtration Eq. (18-51) can be integrated to give the following relationships between total time and filtrate measurements:

$$\frac{\theta}{V/A} = \frac{\mu\alpha}{2P}\frac{W}{A} + \frac{\mu r}{P} \qquad (18\text{-}67)$$

$$\frac{\theta}{V/A} = \frac{\mu\alpha w}{2P}\frac{V}{A} + \frac{\mu r}{P} \qquad (18\text{-}68)$$

$$\frac{\theta}{V/A} = \frac{\mu\alpha\rho c}{2P(1-mc)}\frac{V}{A} + \frac{\mu r}{P} \qquad (18\text{-}69)$$

For a given constant-pressure filtration, these may be simplified to

$$\frac{\theta}{V/A} = K_p\frac{W}{A} + C = K_p'\frac{V}{A} + C \qquad (18\text{-}70)$$

where K_p, K_p', and C are constants for the conditions employed. It should be noted that K_p, K_p', and C depend on filtering pressure not only in the obvious explicit way but also in the implicit sense that α, m, and r are generally dependent on P.

Constant-Rate Filtration For substantially incompressible cakes, Eq. (18-51) may be integrated for a constant rate of slurry feed to the filter to give the following equations, in which filter-medium resistance is treated as the equivalent constant-pressure component to be deducted from the rising total pressure drop to

---

° A combination of category 2 followed by category 1 as parts of the same filtration cycle is considered by some as a fourth category. For a method of combining the constant-rate and constant-pressure equations for such a cycle, see Brown, loc. cit.

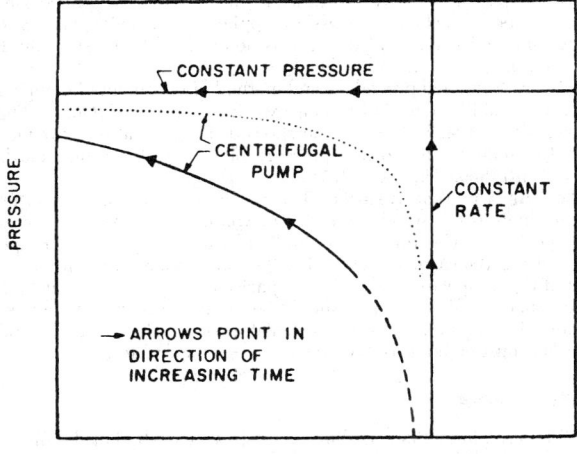

FIG. 18-106 Typical filtration cycles. [*Tiller and Crump*, Chem. Eng. Prog. **73**(10), 72(1977), *by permission.*]

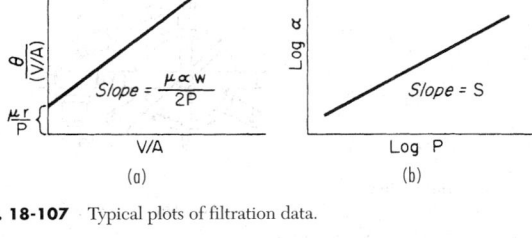

FIG. 18-107 Typical plots of filtration data.

The effect of the change of any variable not affecting α or r can now be estimated. It should be remembered that α and r usually depend on P and may be affected by w.

The symbol α represents the average specific cake resistance, which is a constant for the particular cake in its immediate condition. In the usual range of operating conditions it is related to the pressure by the expression

$$\alpha = \alpha' P^s \qquad (18\text{-}74)$$

where α' is a constant determined largely by the size of the particles forming the cake; s is the cake compressibility, varying from 0 for rigid, incompressible cakes, such as fine sand and diatomite, to 1.0 for very highly compressible cakes. For most industrial slurries, s lies between 0.1 and 0.8. The symbol r represents the resistance of unit area of filter medium but includes other losses (besides those across the cake and the medium) in the system across which P is the pressure drop.

It should be noted also that the intercept is difficult to determine accurately because of large potential experimental error in observing the time of the start of filtration and the time-volume correspondence during the first moments when the filtration rate is high. The value of r calculated from the intercept may vary appreciably from test to test, and will almost always be different from the value measured with clean medium in a permeability test.

To determine the effect of a change in pressure, it is necessary to run tests at three or more pressures, preferably spanning the range of interest. Plotting α or r against P on log-log paper (or log α or log r against the log P on cartesian coordinates) results in an approximate straight line (Fig. 18-107b) from which one may estimate values of α or r at interpolated or reasonably extrapolated magnitudes of P. The slope of the line is the index of a power relationship between α and P or r and P.

Not uncommonly r is found to be only slightly dependent on pressure. When this is true and especially when the filter-medium resistance is, as it should be, relatively small, an average value may be used for all pressures.

It is advisable to start a constant-pressure filtration test, like a comparable plant operation, at a low pressure, and smoothly increase the pressure to the desired operating level. In such cases, time and filtrate-quantity data should not be taken until the constant operating pressure is realized. The value of r calculated from the extrapolated intercept then reflects the resistance of both the filter medium and that part of the cake deposited during the pressure-buildup period. When only the total mass of dry cake is measured for the total cycle time, as is usually true in vacuum leaf tests, at least three runs of different lengths should be made to permit a reliable plot of θ/V against W. If rectification of the resulting three points is dubious, additional runs should be made.

give the variable pressure through the filter cake [*Ruth, Ind. Eng. Chem.*, **27**, 717 (1935)]:

$$\frac{\theta}{V/A} = \frac{1}{\text{rate per unit area}} = \frac{\mu\alpha}{P-P_1}\frac{W}{A} \qquad (18\text{-}71)$$

which may also be written

$$\frac{\theta}{V/A} = \frac{\mu\alpha w}{P-P_1}\frac{V}{A} = \frac{\mu\alpha\rho c}{(P-P_1)(1-mc)}\frac{V}{A} \qquad (18\text{-}72)$$

In these equations P_1 is the pressure drop through the filter medium.

$$P_1 = \mu r(V/A\theta)$$

For a given constant-rate run, the equations may be simplified to

$$V/A = P/K_r + C' \qquad (18\text{-}73)$$

where K_r and C' are constants for the given conditions.

Variable-Pressure, Variable-Rate Filtration The pattern of this category complicates the use of the basic rate equation. The method of Tiller and Crump (loc. cit.) can be used to integrate the equation when the characteristic curve of the feed pump is available.

In the filtration of small amounts of fine particles from liquid by means of bulky filter media (such as absorbent cotton or felt) it has been found that the preceding equations based upon the resistance of a cake of solids do not hold, since no cake is formed. For these cases, in which filtration takes place on the surface or within the interstices of a medium, analogous equations have been developed [Hermans and Bredée, *J. Soc. Chem. Ind.*, **55T**, 1 (1936)]. These are usefully summarized, for both constant-pressure and constant-rate conditions, by Grace [*Am. Inst. Chem. Eng. J.*, **2**, 323 (1956)]. These equations often apply to the clarification of such materials as sugar solutions, viscose and other spinning solutions, and film-casting dopes.

If a **constant-pressure test** is run on a slurry, care being taken that not only the pressure but also the temperature and the solid content remain constant throughout the run and that time readings begin at the exact start of filtration, one can observe values of filtrate volume or weight and corresponding elapsed time. With the use of the known filtering area, values of $\theta/(V/A)$ can be calculated for various values of V/A which, when plotted with $\theta/(V/A)$ as the ordinate and V/A as the abscissa (Fig. 18-107a), result in a straight line having the slope $\mu\alpha w/2P$ and an intercept on the vertical axis of $\mu r/P$. Since μ, w, and P are known, α and r can be calculated from

$$\alpha = 2P/\mu w \times (\text{slope})$$

and

$$r = P/\mu \times (\text{vertical intercept})$$

Pressure Tests

Leaf Tests A bomb filter is used for small-scale leaf tests to simulate the performance of pressure-leaf (leaf-in-shell) filters. The equipment used is a small [50.8- by 50.8-mm (2- by 2-in)] leaf, covered with appropriate filter medium, suspended in a cell large enough to contain sufficient slurry to form the desired cake (Fig. 18-108). The slurry may be agitated gently, for example, by an air sparger.

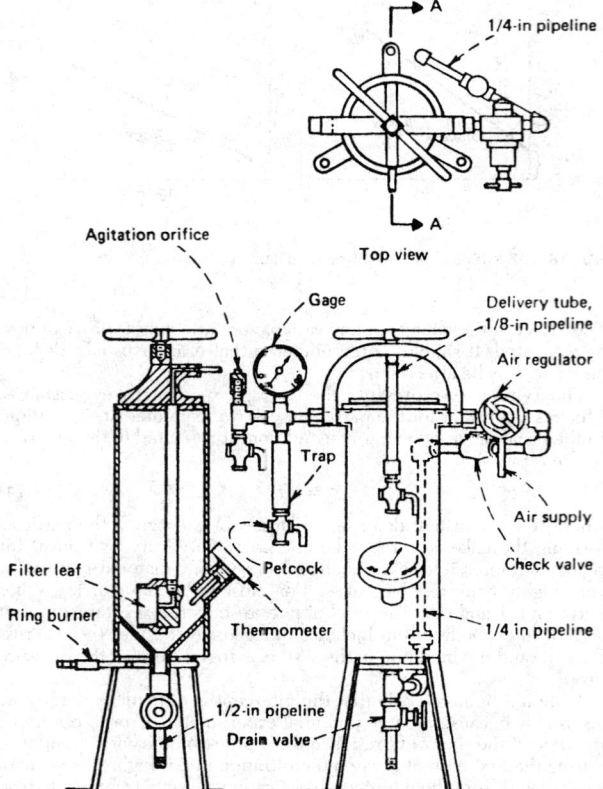

FIG. 18-108   Bomb filter for small-scale pressure filtration tests. [*Silverblatt et al.*, Chem. Eng., **81**(9), 132 (1974), *by permission.*]

Although incremental time and filtrate volume may be taken during a cake-forming cycle at a selected pressure to permit a plot like Fig. 18-107a from a single run, it may be more satisfactory to make several successive quick runs at the same pressure but for different lengths of time, recording only the terminal values of filtrate volume, time, and cake mass. Operation of the commercial unit should be kept in mind when the test cycles are planned. Displacement washing and air blowing of the cake should be tried if appropriate. Wet discharge can be simulated by opening the cell and playing a jet of water on the cake; dry discharge, by applying a gentle air blast to the filtrate-discharge tube. Tests at several pressures must be conducted to determine the compressibility of the cake solids.

Plate-and-Frame Tests   These tests should be conducted if the use of a filter press in the plant is anticipated; at least a few confirming tests are advisable after preliminary leaf tests, unless the slurry is very rapidly filtering. A laboratory-size filter press consisting of two plates and a single frame may be used. It will permit the observation of solids-settling, cake-packing, and washing behavior, which may be quite different for a frame than for a leaf.

Compression-Permeability Tests   Instead of model leaf tests, compression-permeability experiments may be substituted with advantage for appreciably compressible solids. As in the case of constant-rate filtration, a single run provides data equivalent to those obtained from a series of constant-pressure runs, but it avoids the data-treatment complexity of constant-rate tests.

The equipment consists of a cylindrical cell with a permeable bottom and an open top, into which is fitted a close-clearance, hollow, cylindrical piston with a permeable bottom. Slurry is poured into the cell, and a cake is formed by applying gentle vacuum to the filtrate discharge line. The cell is then filled with filtrate, and the counter-

weighted piston is allowed to descend to the cake level. Successive increments of mechanical stress are applied to the solids, at each of which the permeability of the cake is determined by passing filtrate through the piston under low head.

The experimental procedure and method of treatment of compression-permeability data have been explained by Grace [*Chem. Eng. Prog.*, **49**, 303, 427 (1953)], who showed that the values of α measured in such a cell and in a pressure filter were the same, and by Tiller [*Filtr. Sep.*, **12**, 386 (1975)].

Scaling Up Test Results   The results of small-scale tests are determined as dry weight of solids or volume of filtrate per unit of area per cycle. This quantity multiplied by the number of cycles per day permits the calculation of either the filter area required for a stipulated daily capacity or the daily capacity of a specified plant filter. The scaled-up filtration area should be increased by 25 percent as a factor of uncertainty. In the calculation of cycle length, proper account must be made of the downtime of a batch filter.

FILTER MEDIA

All filters require a filter medium to retain solids, whether the filter is for cake filtration or for filter-medium or depth filtration. Specification of a medium is based on retention of some minimum particle size at good removal efficiency and on acceptable life of the medium in the environment of the filter. The selection of the type of filter medium is often the most important decision in success of the operation. For cake filtration, medium selection involves an optimization of the following factors:

1.   Ability to bridge solids across its pores quickly after the feed is started (i.e., minimum propensity to bleed)
2.   Low rate of entrapment of solids within its interstices (i.e., minimum propensity to blind)
3.   Minimum resistance to filtrate flow (i.e., high production rate)
4.   Resistance to chemical attack
5.   Sufficient strength to support the filtering pressure
6.   Acceptable resistance to mechanical wear
7.   Ability to discharge cake easily and cleanly
8.   Ability to conform mechanically to the kind of filter with which it will be used
9.   Minimum cost

For filter-medium filtration, attributes 3, 4, 5, 8, and 9 of the preceding list apply and must have added to them (*a*) ability to retain the solids required, (*b*) freedom from discharge of lint or other adulterant into the filtrate, and (*c*) ability to plug slowly (i.e., long life).

Filter-medium selection embraces many types of construction: fabrics of woven fibers, felts, and nonwoven fibers, porous or sintered solids, polymer membranes, or particulate solids in the form of a permeable bed. Media of all types are available in a wide choice of materials.

Fabrics of Woven Fibers   For cake filtration these fabrics are the most common type of medium. A wide variety of materials are available; some popular examples are listed in Table 18-10, with ratings for chemical and temperature resistance. In addition to the material of the fibers, six construction characteristics describe the filter cloth: (1) weave, (2) style number, (3) weight, (4) count, (5) ply, and (6) yarn number. Of the many types of weaves available, only four are extensively used as filter media: plain (square) weave, twill, chain weave, and satin.

All these weaves may be made from any textile fiber, natural or synthetic. They may be woven from spun staple yarns, multifilament continuous yarns, or monofilament yarns. The performance of the filter cloth depends on the weave and the type of yarn.

A recently developed medium known as a **double weave** incorporates different yarns in warp and fill in order to combine the specific advantages of each type. An example of this is Style 99FS, made by SCAPA Filtration (formerly P&S Filtration), in which multifilament warp yarns provide good cake release properties and spun staple fill yarns contribute to greater retentivity.

Metal Fabrics or Screens   These are available in several types of weave in nickel, copper, brass, bronze, aluminum, steel, stainless steel, Monel, and other alloys. In the plain weave, 400 mesh is the closest

TABLE 18-10   Characteristics of Filter-Fabric Materials*

Generic name and description	Breaking tenacity, g/denier	Abrasion resistance	Resistance to acids	Resistance to alkalies	Resistance to oxidizing agents	Resistance to solvents	Specific gravity	Maximum operating temperature, °F†
Acetate—cellulose acetate. When not less than 92% of the hydroxyl groups are acetylated, "triacetate" may be used as a generic description.	1.2–1.5	G	F	P	G	G	1.33	210
Acrylic—any long-chain synthetic polymer composed of at least 85% by weight of acrylonitrile units.	2.0–4.8	G	G	F	G	E	1.18	300
Glass—fiber-forming substance is glass.	3.0–7.2	P	E	P	E	E	2.54	600
Metallic—composed of metal, metal-coated plastic, plastic-coated metal, or a core completely covered by metal.	—	G						
Modacrylic—fiber-forming substance is any long-chain synthetic polymer composed of less than 85% but at least 35% by weight of acrylonitrile units.	2.5–3.0	G	G	G	G	G	1.30	180
Nylon—any long-chain synthetic polyamide having recurring amide groups as an integral part of the polymer chain.	3.8–9.2	E	F–P	G	F–P	G	1.14	225
Polyester—any long-chain synthetic polymer composed of at least 85% by weight of an ester of a dihydric alcohol and terephthalic acid (p—HOOC—C_6H_4—COOH).	2.2–7.8	E–G	G	G–F	G	G	1.38	300
Polyethylene—long-chain synthetic polymer composed of at least 85% weight of ethylene.	1.0–7.0	G	G	G	F	G	0.92	165‡
Polypropylene—long-chain synthetic polymer composed of at least 85% by weight of propylene.	4.8–8.5	G	E	E	G	G	0.90	250§
Cotton—natural fibers.	3.3–6.4	G	P	F	G	E–G	1.55	210
Fluorocarbon—long-chain synthetic polymer composed of tetrafluoroethylene units.	1.0–2.0	F	E	E	E	G	2.30	550¶

*Adapted from Mais, *Chem. Eng.*, **78**(4), 51 (1971). Symbols have the following meaning: E = excellent, G = good, F = fair, P = poor.
†°C = (°F − 32)/1.8; K = (°F + 459.7)/1.8.
‡Low-density polymer. Up to 230°F for high-density.
§Heat-set fabric; otherwise lower.
¶Requires ventilation because of release of toxic gases above 400°F.

wire spacing available, thus limiting use to coarse crystalline slurries, pulps, and the like. The "Dutch weaves" employing relatively large, widely spaced, straight warp wires and relatively small crimped filling wires can be woven much more closely, providing a good medium for filtering fine crystals and pulps. This type of weave tends to plug readily when soft or amorphous particles are filtered and makes the use of filter aid desirable. Good corrosion and high temperature resistance of properly selected metals makes filtrations with metal media desirable for long-life applications. This is attractive for handling toxic materials in closed filters to which minimum exposure by maintenance personnel is desirable.

Pressed Felts and Cotton Batting   These materials are used to filter gelatinous particles from paints, spinning solutions, and other viscous liquids. Filtration occurs by deposition of the particles in and on the fibers throughout the mat.

Nonwoven media consist of web or sheet structures which are composed primarily of fibers or filaments bonded together by thermal, chemical, or mechanical (such as needlepunching) means. Needled felts are the most commonly used nonwoven fabric for liquid filtration. Additional strength often is provided by including a scrim of woven fabric encapsulated within the nonwoven material. The surface of the medium can be calendered to improve particle retention and assist in filter cake release. Weights range from 270 to 2700 gm/m² (8 to 80 oz/yd²). Because of their good retentivity, high strength, moderate cost, and resistance to blinding, nonwoven media have found

wide acceptance in filter press use, particularly in mineral concentrate filtration applications. They are used frequently on horizontal belt filters where their dimensional stability reduces or eliminates wrinkling and biasing problems often encountered with woven belts.

Filter Papers   These papers come in a wide range of permeability, thickness, and strength. As a class of material, they have low strength, however, and require a perforated backup plate for support.

Rigid Porous Media   These are available in sheets or plates and tubes. Materials include sintered stainless steel and other metals, graphite, aluminum oxide, silica, porcelain, and some plastics—a gamut that allows a wide range of chemical and temperature resistance. Most applications are for clarification.

Polymer Membranes   These are used in filtration applications for fine-particle separations such as microfiltration and ultrafiltration (clarification involving the removal of 1-μm and smaller particles). The membranes are made from a variety of materials, the commonest being cellulose acetates and polyamides. Membrane filtration, discussed in Sec. 22, has been well covered by Porter (in Schweitzer, op. cit., sec. 2.1).

Media made from woven or nonwoven fabrics coated with a polymeric film, such as Primapor, made by SCAPA Filtration, and Gore-Tex, made by W. L. Gore and Associates, combine the high retentivity characteristics of a membrane with the strength and durability of a thick filter cloth. These media are used on both continuous and batch filters where excellent filtrate clarity is required.

Granular Beds of Particulate Solids   Beds of solids like sand or coal are used as filter media to clarify water or chemical solutions containing small quantities of suspended particles. Filter-grade grains of desired particle size can be purchased. Frequently beds will be constructed of layers of different materials and different particle sizes.

Various types of filter media and the materials of which they are constructed are surveyed extensively by Purchas (*Industrial Filtration of Liquids,* CRC Press, Cleveland, 1967, chap. 3), and characterizing measurements (e.g., pore size, permeability) are reviewed in detail by Rushton and Griffiths (in Orr, op. cit., chap. 3). Briefer summaries of classification of media and of practical criteria for the selection of a filter medium are presented by Shoemaker (op. cit., p. 26) and Purchas [*Filtr. Sep.,* **17,** 253, 372 (1980)].

FILTER AIDS

Use of filter aids is a technique frequently applied for filtrations in which problems of slow filtration rate, rapid medium blinding, or unsatisfactory filtrate clarity arise. Filter aids are granular or fibrous solids capable of forming a highly permeable filter cake in which very fine solids or slimy, deformable flocs may be trapped. Application of filter aids may allow the use of a much more permeable filter medium than the clarification would require to produce filtrate of the same quality by depth filtration.

Filter aids should have low bulk density to minimize settling and aid good distribution on a filter-medium surface that may not be horizontal. They should also be porous and capable of forming a porous cake to minimize flow resistance, and they must be chemically inert to the filtrate. These characteristics are all found in the two most popular commercial filter aids: diatomaceous silica (also called diatomite, or diatomaceous earth), which is an almost pure silica prepared from deposits of diatom skeletons; and expanded perlite, particles of "puffed" lava that are principally aluminum alkali silicate. Cellulosic fibers (ground wood pulp) are sometimes used when siliceous materials cannot be used but are much more compressible. The use of other less effective aids (e.g., carbon and gypsum) may be justified in special cases. Sometimes a combination of carbon and diatomaceous silica permits adsorption in addition to filter-aid performance. Various other materials, such as salt, fine sand, starch, and precipitated calcium carbonate, are employed in specific industries where they represent either waste material or inexpensive alternatives to conventional filter aids.

Diatomaceous Silica   Filter aids of diatomaceous silica have a dry bulk density of 128 to 320 kg/m^3 (8 to 20 lb/ft^3), contain particles mostly smaller than 50 μm, and produce a cake with porosity in the range of 0.9 (volume of voids/total filter-cake volume). The high porosity (compared with a porosity of 0.38 for randomly packed uniform spheres and 0.2 to 0.3 for a typical filter cake) is indicative of its filter-aid ability. Different methods of processing the crude diatomite result in a series of filter aids having a wide range of permeability.

Perlite   Perlite filter aids are somewhat lower in bulk density (48 to 96 kg/m^3, or 3 to 6 lb/ft^3) than diatomaceous silica and contain a higher fraction of particles in the 50- to 150-μm range. Perlite is also available in a number of grades of differing permeability and cost, the grades being roughly comparable to those of diatomaceous silica. Diatomaceous silica will withstand slightly more extreme pH levels than perlite, and it is said to be somewhat less compressible.

Filter aids are used in two ways: (1) as a precoat and (2) mixed with the slurry as a "body feed." Precoat filtration, employing a thin layer of about 0.5 to 1.0 kg/m^2 (0.1 to 0.2 lb/ft^2) deposited on the filter medium prior to beginning feed to the filter, is in wide use to protect the filter medium from fouling by trapping solids before they reach the medium. It also provides a finer matrix to trap fine solids and assure filtrate clarity. Body-feed application is the continuous addition of filter aid to the filter feed to increase the porosity of the cake. The amount of addition must be determined by trial, but in general, the quantity added should at least equal the amount of solids to be removed. For solids loadings greater than 1000 ppm this may become a significant cost factor. An acceptable alternative might be to use a rotary vacuum precoat filter [Smith, *Chem. Eng.,* **83**(4), 84 (1976)]. Further details of filter-aid filtration are set forth by Cain (in

Schweitzer, op. cit., sec. 4.2) and Hutto [*Am. Inst. Chem. Eng. Symp. Ser.,* **73**(171), 50 (1977)]. Figure 18-109 shows a flow sheet indicating arrangements for both precoat and body-feed applications. Most filter aid is used on a one-time basis, although some techniques have been demonstrated to reuse precoat filter aid on vertical-tube pressure filters.

FILTRATION EQUIPMENT

Cake Filters   Filters that accumulate appreciable visible quantities of solids on the surface of a filter medium are called cake filters. The slurry feed may have a solids concentration from about 1 percent to greater than 40 percent. The filter medium on which the cake forms is relatively open to minimize flow resistance, since once the cake forms, it becomes the effective filter medium. The initial filtrate therefore may contain unacceptable solids concentration until the cake is formed. This situation may be made tolerable by recycling the filtrate until acceptable clarity is obtained or by using a downstream polishing filter (clarifying type).

Cake filters are used when the desired product of the operation is the solids, the filtrate, or both. When the filtrate is the product, the degree of removal from the cake by washing or blowing with air or gas becomes an economic optimization. When the cake is the desired product, the incentive is to obtain the desired degree of cake purity by washing, blowing, and sometimes mechanical expression of residual liquid.

Implicit in cake filtration is the removal and handling of solids, since the cake is usually relatively dry and compacted. Cakes can be sticky and difficult to handle; therefore, the ability of a filter to discharge the cake cleanly is an important equipment-selection criterion.

In the operational sense, some filters are batch devices, whereas others are continuous. This difference provides the principal basis for classifying cake filters in the discussion that follows. The driving force by which the filter functions—hydrostatic head ("gravity"), pressure imposed by a pump or a gas blanket, or atmospheric pressure ("vacuum")—will be used as a secondary criterion.

Batch Cake Filters

Nutsche Filters   A nutsche is one of the simplest batch filters. It is a tank with a false bottom, perforated or porous, which may either support a filter medium or act as the filter medium. The slurry is fed into the filter vessel, and separation occurs by gravity flow, gas pressure, vacuum, or a combination of these forces. The term "nutsche" comes from the German term for sucking, and vacuum is the common operating mode.

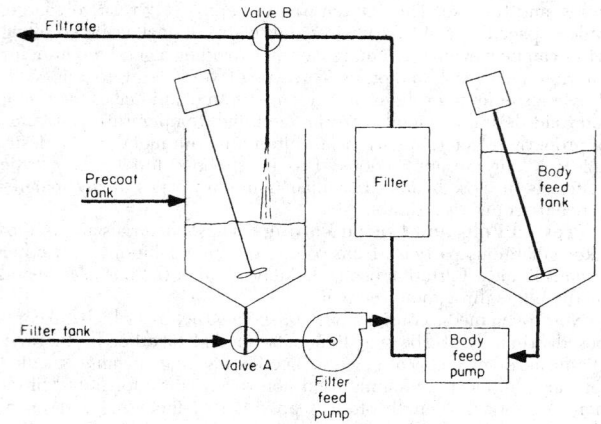

FIG. 18-109   Filter-aid filtration system for precoat or body feed. (*Schweitzer, Handbook of Separation Techniques for Chemical Engineers, p. 4-12. Copyright 1979 by McGraw-Hill, Inc. Used with permission of McGraw-Hill Book Company.*)

The design of most nutsche filters is very simple, and they are often fabricated by the user at low cost. The filter is very frequently used in laboratory, pilot-plant, or small-plant operation. For large-scale processing, however, the excessive floor area encumbered per unit of filtration area and the awkwardness of cake removal are strong deterrents. For small-scale operations, cake is manually removed. For large-scale applications, cake may be further processed by reslurrying or redissolving; or it may be removed manually (by shovel) or by mechanical discharge arrangements such as a movable filter medium belt.

Thorough displacement washing is possible in a nutsche if the wash solvent is added before the cake begins to be exposed to air displacement of filtrate. If washing needs to be more effective, an agitator can be provided in the nutsche vessel to reslurry the cake to allow adequate diffusion of solute from the solids.

Horizontal Plate Filter   The horizontal multiple-plate pressure filter consists of a number of horizontal circular drainage plates and guides placed in a stack in a cylindrical shell (Fig. 18-110). In normal practice the filtering pressure is limited to 345 kPa (50 psig), although special filters have been designed for shell pressures of 2.1 MPa (300 psig) or higher.

Filter Press   The filter press, one of the most frequently used filters in the early years of the chemical industry, is still widely employed. Often referred to generically (in error) as the plate-and-frame filter, it has probably over 100 design variations. Two basic popular designs are the flush-plate, or plate-and-frame, design and the recessed-plate press. Both are available in a wide range of materials: metals, coated metals, plastics, or wood.

Plate-and-frame press.   This press is an alternate assembly of plates covered on both sides with a filter medium, usually a cloth, and hollow frames that provide space for cake accumulation during filtration. The frames have feed and wash manifold ports, while the plates have filtrate drainage ports. The plates and frames usually are rectangular, although circles and other shapes also are used (Fig. 18-111). They are hung on a pair of horizontal support bars and pressed together during filtration to form a watertight closure between two end plates, one of which is stationary. The press may be closed manually, hydraulically, or by a motor drive. Several feed and filtrate discharge arrangements are possible. In the most popular, the feed and discharge of the several elements of the press are manifolded via some of the holes that are in the four corners of each plate and frame (and filter cloth) to form continuous longitudinal channels from the stationary end plate to the other end of the press. Alternatively, the filtrate may be drained from each plate by an individual valve and spigot

(for open discharge) or tubing (for closed). Top feed to and bottom discharge from the chambers provide maximum recovery of filtrate and maximum mean cake dryness. This arrangement is especially suitable for heavy fast-settling solids. For most slurries, bottom feed and top filtrate discharge allow quick air displacement and produce a more uniform cake.

Two wash techniques are used in plate-and-frame filter presses, illustrated in Fig. 18-112. In simple washing, the wash liquor follows the same path as the filtrate. If the cake is not extremely uniform and highly permeable, this type of washing is ineffective in a well-filled press. A better technique is thorough washing, in which the wash is introduced to the faces of alternate plates (with their discharge channels valved off). The wash passes through the entire cake and exits through the faces of the other plates. This improved technique requires a special design and the assembly of the plates in proper order. Thorough washing should be used only when the frames are well filled, since an incomplete fill of cake will allow cake collapse during the wash entry. The remainder of the wash flow will bypass through cracks or channels opened in the cake.

Filter presses are made in plate sizes from 10 by 10 cm (4 by 4 in) to 1.5 by 1.8 m (61 by 71 in). Frame thickness ranges from 0.3 to 20 cm (0.125 to 8 in). Operating pressures up to 689 kPa (100 psig) are common, with some presses designed for 6.9 MPa (1000 psig). Some metal units have cored plates for steam or refrigerant. Maximum pressure for wood or plastic frames is 410 to 480 kPa (60 to 70 psig).

The filter press has the advantage of simplicity, low capital cost, flexibility, and ability to operate at high pressure in either a cake-filter or a clarifying-filter application. Floor-space and headroom needs per unit of filter area are small, and capacity can be adjusted by adding or removing plates and frames. Filter presses are cleaned easily, and the filter medium is easily replaced. With proper operation a denser, drier cake compared with that of most other filters is obtained.

There are several serious disadvantages, including imperfect washing due to variable cake density, relatively short filter-cloth life due to the mechanical wear of emptying and cleaning the press (often involving scraping the cloth), and high labor requirements. Presses frequently drip or leak and thereby create housekeeping problems, but the biggest problem arises from the requirement to open the filter for cake discharge. The operator is thus exposed routinely to the contents of the filter, and this is becoming an increasingly severe disadvantage as more and more materials once believed safe are given restricted exposure limits.

Recessed-plate filter press.   This press is similar to the plate-and-frame press in appearance but consists only of plates (Fig. 18-113).

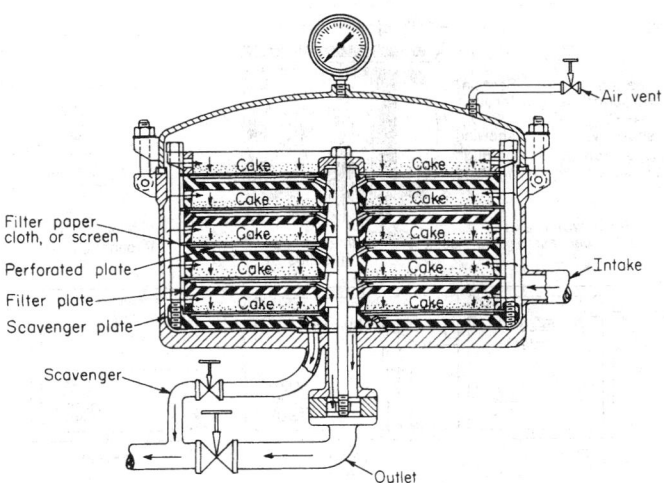

FIG. 18-110   Elevation section of a Sparkler horizontal plate filter. (*Sparkler Filters, Inc.*)

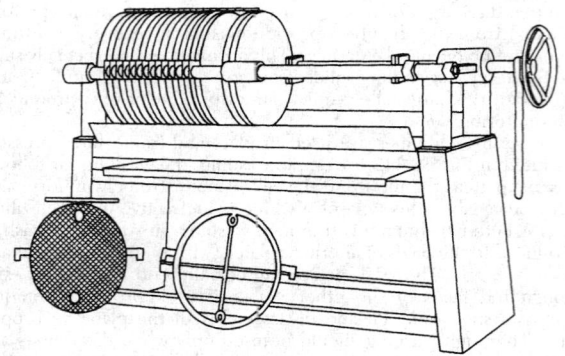

FIG. 18-111 Circular-plate fabricated-metal filter press. (*Star Systems Filtration Division.*)

Both faces of each plate are hollowed to form a chamber for cake accumulation between adjacent plates. This design has the advantage of about half as many joints as a plate-and-frame press, making a tight closure more certain. Figure 18-114 shows some of the features of one type of recessed-plate filter which has a gasket to further minimize leaks. Air can be introduced behind the cloth on both sides of each plate to assist cake removal.

Some interesting variations of standard designs include the ability to roll the filter to change from a bottom to a top inlet or outlet and the ability to add blank dividers to convert a press to a multistage press for further clarification of the filtrate or to do two separate filtrations simultaneously in the same press. Some designs have rubber membranes between plates which can be expanded when filtration is finished to squeeze out additional moisture. Some designs feature automated opening and cake-discharge operations to reduce labor requirements. Examples of this type of pressure filter include Larox, Vertipress, and Oberlin.

Internal Cake Tube Filters or Liquid Bag Filters This type of filter, such as manufactured by Industrial Filter and Pump Mfg. Co., Rosedale, Illinois, and many others, utilizes one or more perforated tubes supported by a tube sheet or by the lip of the pressure vessel. A cylindrical filter bag sealed at one end is inserted into the perforated

tube. The open end of the filter bag generally has a flange or special seal ring to prevent leakage.

Slurry under pressure is admitted to the chamber between the head of the shell and the tube sheet, whence it enters and fills the tubes. Filtration occurs as the filtrate passes radially outward through the filter medium and the wall of each tube into the shell and on out the filtrate discharge line, depositing cake on the medium. The filtration cycle is ended when the tubes have filled with cake or when the media have become plugged. The cake can be washed (if it has not been allowed to fill the tubes completely) and air-blown. The filter has a removable head to provide easy access to the tube sheet and mouth of the tubes; thus "sausages" of cake can be removed by taking out the filter bags or each tube and bag assembly together. The tubes themselves are easily removed for inspection and cleaning.

The advantages of the tubular filter are that it uses an easily replaced filter medium, its filtration cycle can be interrupted and the shell can be emptied of prefill at any time without loss of the cake, the cake is readily recoverable in dry form, and the inside of the filter is conveniently accessible. There is also no unfiltered heel. Disadvantages are the necessity and attendant labor requirements of emptying by hand and replacing the filter media and the tendency for heavy solids to settle out in the header chamber. Applications are as a scavenger filter to remove fines not removed in a prior-filtration stage with a different kind of equipment, to handle the runoff from other filters, and in semiworks and small-plant operations in which the filter's size, versatility, and cleanliness recommend it.

External-Cake Tubular Filters Several filter designs are available with vertical tubes supported by a filtrate-chamber tube sheet in a vertical cylindrical vessel (Fig. 18-115). The tubes may be made of wire cloth; porous ceramic, carbon, plastic, or metal; or closely wound wire. The tubes may have a filter cloth on the outside. Frequently a filter-aid precoat will be applied to the tubes. The prefilt slurry is fed near the bottom of the vertical vessel. The filtrate passes from the outside to the inside of the tubes and into a filtrate chamber at the top or the bottom of the vessel. The solids form a cake on the outside of the tubes with the filter area actually increasing as the cake builds up, partially compensating for the increased flow resistance of the thicker cake. The filtration cycle continues until the differential pressure reaches a specified level, or until about 25 mm (1 in) of cake thickness is obtained.

Cake-discharge methods are the chief distinguishing feature among the various designs. That of the Industrial Filter & Pump Hydro-Shoc, for example, removes cake from the tubes by filtrate backflushing assisted by the "shocking" action of a compressed-gas pocket formed

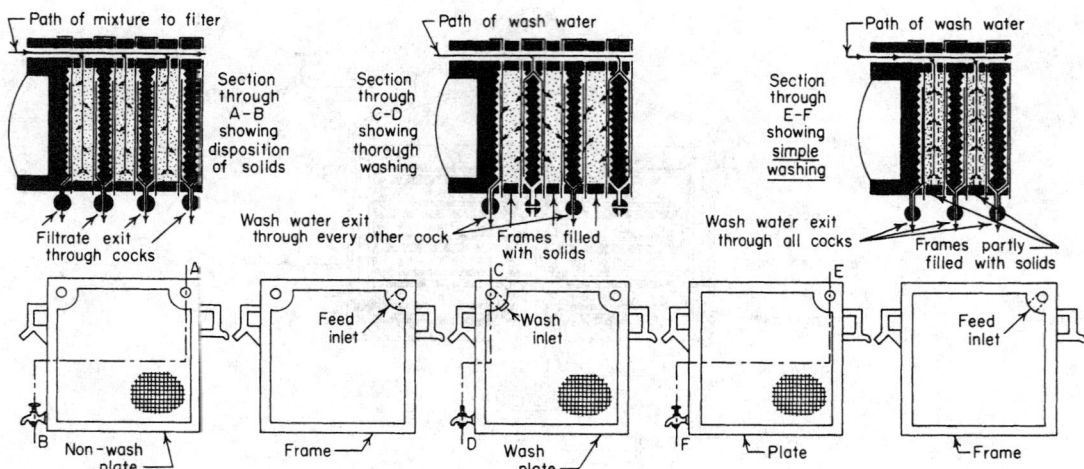

FIG. 18-112 Filling and washing flow patterns in a filter press. (*D. R. Sperry & Co.*)

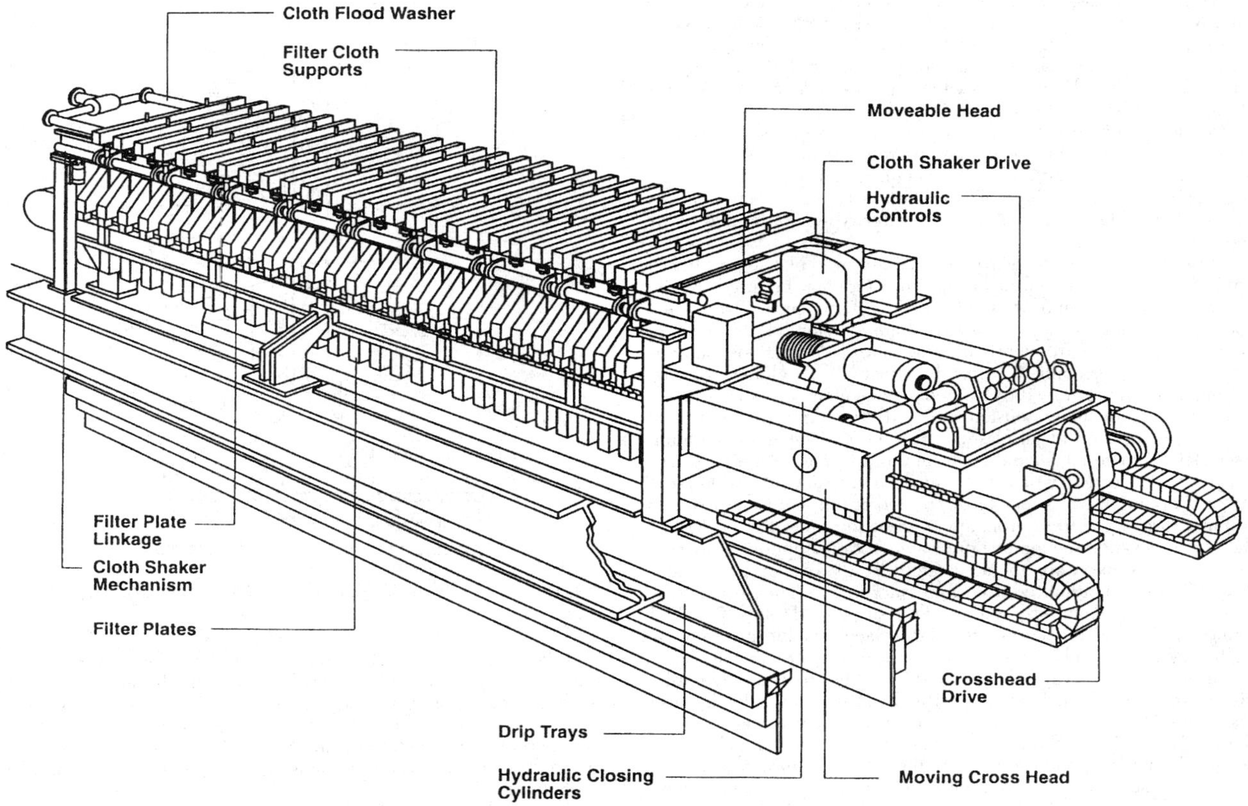

FIG. 18-113 Automated recessed-plate filter press used in mineral applications. (*EIMCO Process Equipment Co.*)

in the filtrate chamber at the top of the vertical vessel. Closing the filtrate outlet valve while continuing to feed the filter causes compression of the gas volume trapped in the dome of the vessel until, at the desired gas pressure, quick-acting valves stop the feed and open a bottom drain. The compressed gas rapidly expands, forcing a rush of filtrate back across the filter medium and dislodging the cake, which drains out the bottom with the flush liquid. Of course, this technique may be used only when wet-cake discharge is permitted.

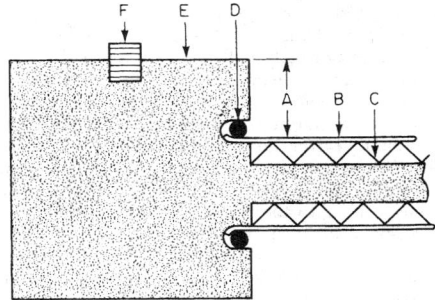

FIG. 18-114 Section detail of a caulked-gasketed-recessed filter plate: (*a*) cake recess; (*b*) filter cloth; (*c*) drainage surface of plate; (*d*) caulking strip; (*e*) plate joint; (*f*) sealing gasket. (*EIMCO Process Equipment Co.*)

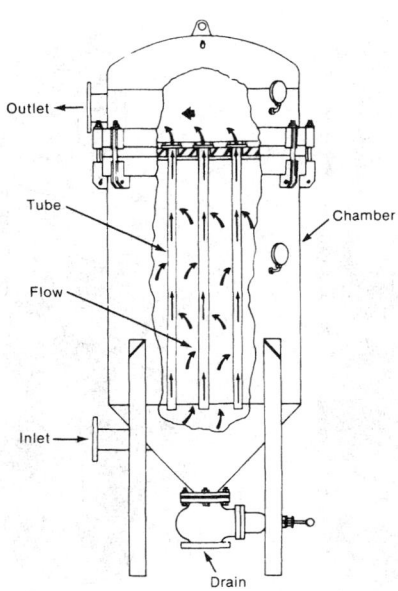

FIG. 18-115 Top-outlet tubular filter. (*Industrial Filter & Pump Mfg. Co.*)

Dry cake discharge can be achieved with a Fundabac candle-type filter manufactured by DrM, Dr. Müller, AG, of Switzerland. This filter uses a candle made up of six small-diameter tubes around a central filtrate delivery tube. This design allows the filter cloth to be flexed outward upon blowback, easily achieving an effective dry cake discharge (Fig. 18-116).

Pressure Leaf Filters   Sometimes called tank filters, they consist of flat filtering elements (leaves) supported in a pressure shell. The leaves are circular, arc-sided, or rectangular, and they have filtering surfaces on both faces. The shell is a cylindrical or conical tank. Its axis may be horizontal or vertical, and the filter type is described by its shell axis orientation.

A filter leaf consists of a heavy screen or grooved plate over which a filter medium of woven fabric or fine wire cloth may be fitted. Textile fabrics are more commonly used for chemical service and are usually applied as bags that may be sewed, zippered, stapled, or snapped. Wire-screen cloth is frequently used for filter-aid filtrations, particularly if a precoat is applied. It may be attached by welding, riveting, bolting, or caulking or by the clamped engagement of two 180° bends in the wire cloth under tension, as in Multi Metal's Rim-Lok leaf. Leaves may also be of all-plastic construction. The filter medium, regardless of material, should be as taut as possible to minimize sagging when it is loaded with a cake; excessive sag can cause cake cracking or dropping. Leaves may be supported at top, bottom, or center and may discharge filtrate from any of these locations. Figure 18-117 shows the elevation section of a precoated bottom-support wire leaf.

Pressure leaf filters are operated batchwise. The shell is locked, and the prefilt slurry is admitted from a pressure source. The slurry enters in such a way as to minimize settling of the suspended solids. The shell is filled, and filtration occurs on the leaf surfaces, the filtrate discharging through an individual delivery line or into an internal manifold, as the filter design dictates. Filtration is allowed to proceed only until a cake of the desired thickness has formed, since to overfill will cause cake consolidation with consequent difficulty in washing and discharge. The decision of when to end the filtering cycle is largely a matter of experience, guided roughly by the rate in a constant-pressure filter or pressure drop in a constant-rate filter. This judgment may be supplanted by the use of a detector which "feels" the thickness of cake on a representative leaf.

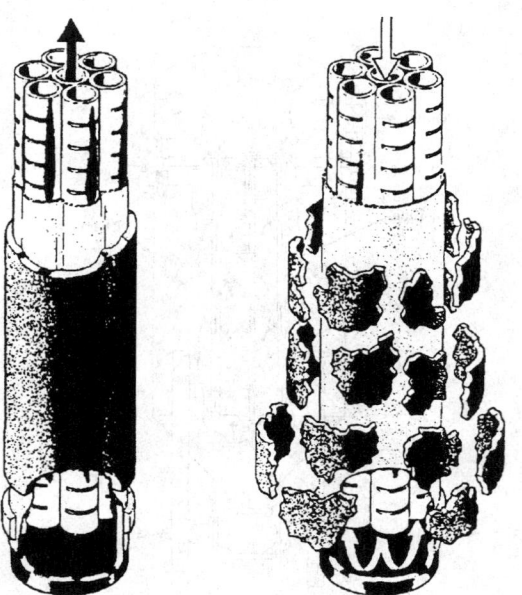

FIG. 18-116   Cake formation and discharge with the Fundabac filter element. (*DrM, Dr. Müller AG, Switzerland.*)

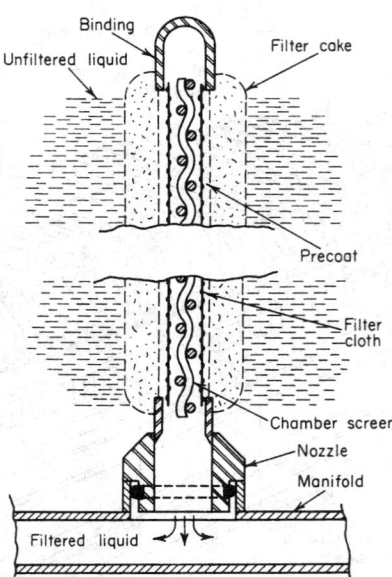

FIG. 18-117   Section of precoated wire filter leaf. (*Multi Metal Wire Cloth.*)

If the cake is to be washed, the slurry heel can be blown from the filter and wash liquor can be introduced to refill the shell. If the cake tends to crack during air blowing, it may be necessary to displace the slurry heel with wash gradually so as never to allow the cake to dry. Upon the completion of filtration and washing, the cake is discharged by one of several methods, depending on the shell and leaf configuration.

Horizontal pressure leaf filters.   In these filters the leaves may be rectangular leaves which run parallel to the axis and are of varying sizes since they form chords of the shell; or they may be circular or square elements parallel to the head of the shell, and all of the same dimension. The leaves may be supported in the shell from an independent rack, individually from the shell, or from a filtrate manifold. Horizontal filters are particularly suited to dry-cake discharge.

Most of the currently available commercial horizontal pressure filters have leaves parallel to the shell head. Cake discharge may be wet or dry; it can be accomplished by sluicing with liquid sprays, vibration of the leaves, or leaf rotation against a knife, wire, or brush. If a wet-cake discharge is allowable, the filters will probably be sluiced with high-pressure liquid. If the filter has a top or bottom filtrate manifold, the leaves are usually in a fixed position, and the spray header is rotated to contact all filter surfaces. If the filtrate header is center-mounted, the leaves are generally rotated at about 3 r/min and the spray header is fixed. Some units may be wet-cake-discharged by mechanical vibration of the leaves with the filter filled with liquid. Dry-cake discharge normally will be accomplished by vibration if leaves are top- or bottom-manifolded and by rotation of the leaves against a cutting knife, wire, or brush if they are center-manifolded.

In many designs the filter is opened for cake discharge, and the leaf assembly is separated from the shell by moving one or the other on rails (Fig. 18-118). For processes involving toxic or flammable materials, a closed filter system can be maintained by sloping the bottom of the horizontal cylinder to the drain nozzle for wet discharge or by using a screw conveyor in the bottom of the shell for dry discharge (Fig. 18-119).

Vertical pressure leaf filters.   These filters have vertical, parallel, rectangular leaves mounted in an upright cylindrical pressure tank. The leaves usually are of such different widths as to allow them to conform to the curvature of the tank and to fill it without waste space. The leaves often rest on a filtrate manifold, the connection being sealed by an O ring, so that they can be lifted individually from the top of the fil-

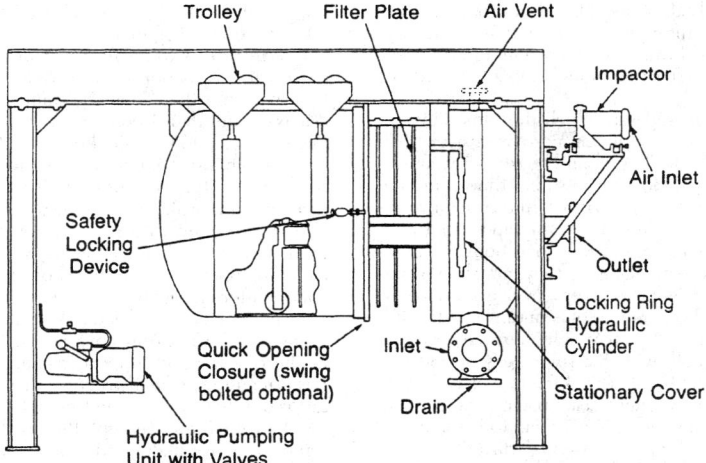

FIG. 18-118 Horizontal-tank pressure leaf filter designed for dry cake discharge. (*Sparkler Filter, Inc.*)

ter for inspection and repair. A scavenger leaf frequently is installed in the bottom of the shell to allow virtually complete filtration of the slurry heel at the end of a cycle.

Vertical filters are not convenient for the removal of dry cake, although they can be used in this service if they have a bottom that can be retracted to permit the cake to fall into a bin or hopper below. They are adapted rather to wet-solids discharge, a process that may be assisted by leaf vibration, air or steam sparging of a filter full of water, sluicing from fixed, oscillating, or traveling nozzles, and blowback.

They are made by many companies, and they enjoy their widest use for filter-aid precoat filtration.

Advantages and uses. The advantages of pressure leaf filters are their considerable flexibility (up to the permissible maximum, cakes of various thickness can be formed successfully), their low labor charges, particularly when the cake may be sluiced off or the dry cake discharged cleanly by blowback, the basic simplicity of many of the designs, and their adaptability to quite effective displacement washing. Their disadvantages are the requirement of exceptionally intelligent

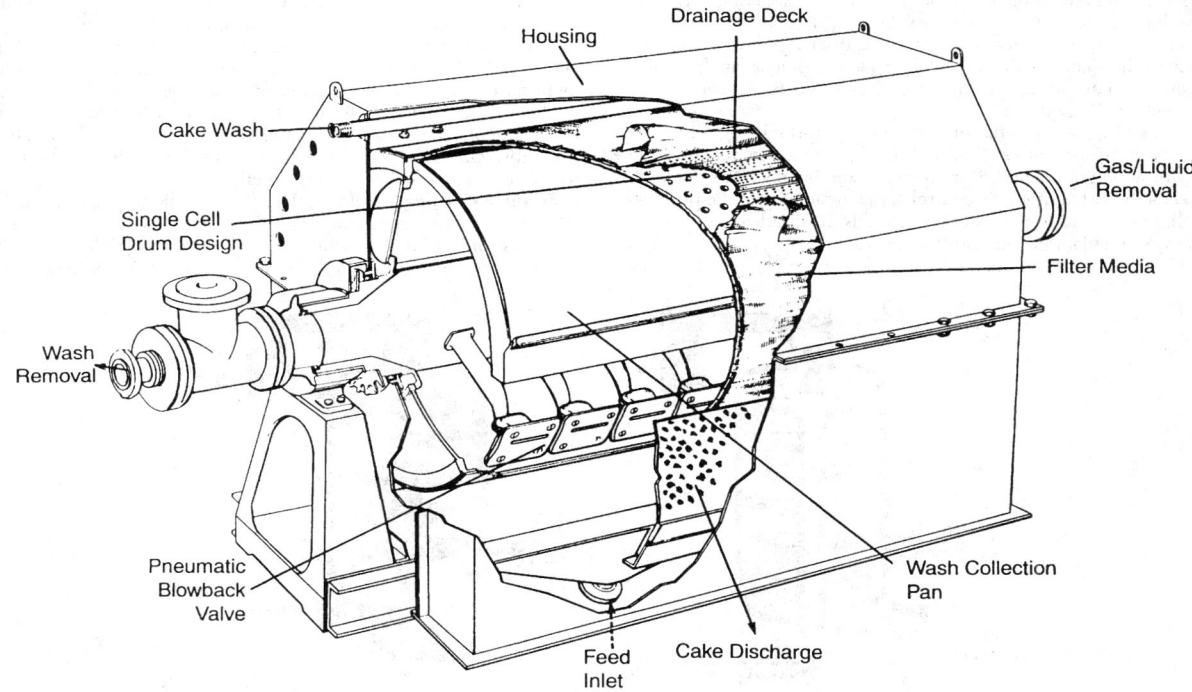

FIG. 18-119 Cutaway of a horizontal tank filter with dry-cake-discharge features. (*United States Filter.*)

and watchful supervision to avoid cake consolidation or dropping, their inability to form as dry a cake as a filter press, their tendency to classify vertically during filtration and to form misshapen nonuniform cakes unless they rotate, and the restriction of most models to 610 kPa (75 psig) or less.

Pressure leaf filters are used to separate much the same kinds of slurries as are filter presses and are used much more extensively than filter presses for filter-aid filtrations. They should be seriously considered whenever uniformity of production permits long-time operation under essentially constant filtration conditions, when thorough washing with a minimum of liquor is desired, or when vapors or fumes make closed construction desirable. Under such conditions, if the filter medium does not require frequent changing, they may show a considerable advantage in cycle and labor economy over a filter press, which has a lower initial cost, and advantages of economy and flexibility over continuous vacuum filters, which have a higher first cost.

Pressure leaf filters are available with filtering areas of 930 cm² (1 ft²) (laboratory size) up to about 440 m² (4734 ft²) for vertical filters and 158 m² (1700 ft²) for horizontal ones. Leaf spacings range from 5 to 15 cm (2 to 6 in) but are seldom less than 7.5 cm (3 in) since 1.3 to 2.5 cm (0.5 to 1 in) should be left open between surfaces.

Centrifugal-Discharge Filter Horizontal top-surface filter plates may be mounted on a hollow motor-connected shaft that serves both as a filtrate-discharge manifold and as a drive shaft to permit centrifugal removal of the cake. An example is the **Funda filter** (marketed in the United States by Steri Technologies), illustrated schematically in Fig. 18-120. The filtering surface may be a textile fabric or a wire screen, and the use of a precoat is optional. The Funda filter is driven from the top, leaving the bottom unobstructed for inlet and drainage lines; a somewhat similar machine that employs a bottom drive, providing a lower center of mass and ground-level access to the drive system, is the German-made **Schenk filter** (marketed in the United States by Schenk Filtersystems).

During filtration, the vessel that coaxially contains the assembly of filter plates is filled with prefilt under pressure, the filtrate passes through the plates and out the hollow shaft, and cake is formed on the top surfaces of the plates. After filtration, the vessel is drained, or the heel may be filtered by recirculation through a cascade ring at the top of the filter. The cake may be washed—or it may be extracted, steamed, air-blown, or dried by hot gas. It is discharged, wet or dry, by rotation of the shaft at sufficiently high speed to sling away the solids. If flushing is permitted, the discharge is assisted by a backwash of appropriate liquid.

The operating advantages of the centrifugal-discharge filter are those of a horizontal-plate filter and, further, its ability to discharge cake without being opened. It is characterized by low labor demand, easy adaptability to automatic control, and amenability to the processing of hazardous, noxious, or sterile materials. Its disadvantages are its complexity and maintenance (stuffing boxes, high-speed drive) and its cost. The Funda filter is made in sizes that cover the filtering area

range of 1 to 50 m³ (11 to 537 ft²). The largest Schenk filter provides 100 m² (1075 ft²) of area.

Continuous Cake Filters Continuous cake filters are applicable when cake formation is fairly rapid, as in situations in which slurry flow is greater than about 5 L/min (1 to 2 gal/min), slurry concentration is greater than 1 percent, and particles are greater than 0.5 μm in diameter. Liquid viscosity below 0.1 Pa·s (100 cP) is usually required for maintaining rapid liquid flow through the cake. Some designs of continuous filters can compromise some of these guidelines by sacrificial use of filter aid when the cake is not the desired product.

Rotary Drum Filters The rotary drum filter is the most widely used of the continuous filters. There are many design variations, including operation as either a pressure filter or a vacuum filter. The major difference between designs is in the technique for cake discharge, to be discussed later. All the alternatives are characterized by a horizontal-axis drum covered on the cylindrical portion by filter medium over a grid support structure to allow drainage to manifolds. Basic materials of construction may be metals or plastics. Sizes (in terms of filter areas) range from 0.37 to 186 m² (4 to 2000 ft²).

All drum filters (except the single-compartment filter) utilize a rotary-valve arrangement in the drum-axis support trunnion to facilitate removal of filtrate and wash liquid and to allow introduction of air or gas for cake blowback if needed. The valve controls the relative duration of each cycle as well as providing "dead" portions of the cycle through the use of bridge blocks. A typical valve design is shown in Fig. 18-121. Internal piping manifolds connect the valve with various sections of the drum.

Most drum filters are fed by operating the drum with about 35 percent of its circumference submerged in a slurry trough, although submergence can be set for any desired amount between zero and almost total. Some units contain an oscillating rake agitator in the trough to aid solids suspension. Others use propellers, paddles, or no agitator.

Slurries of free-filtering solids that are difficult to suspend are sometimes filtered on a top-feed drum filter or filter-dryer. An example application is in the production of table salt. An alternative for slurries of extremely coarse, dense solids is the internal drum filter. In the chemical-process industry both top-feed and internal drums (which are described briefly by Emmett in Schweitzer, op. cit., p. 4-41) have largely been displaced by the horizontal vacuum filter (q.v.).

Most drum filters operate at a rotation speed in the range of 0.1 to 10 r/min. Variable-speed drives are usually provided to allow adjustment for changing cake-formation and drainage rates.

Drum filters commonly are classified according to the feeding arrangement and the cake-discharge technique. They are so treated in this subsection. The characteristics of the slurry and the filter cake usually dictate the cake-discharge method.

Scraper-Discharge Filter The filter medium is usually caulked into grooves in the drum grid, with cake removal facilitated by a scraper blade just prior to the resubmergence of the drum (Fig. 18-122). The scraper serves mainly as a deflector to direct the cake, dislodged by an

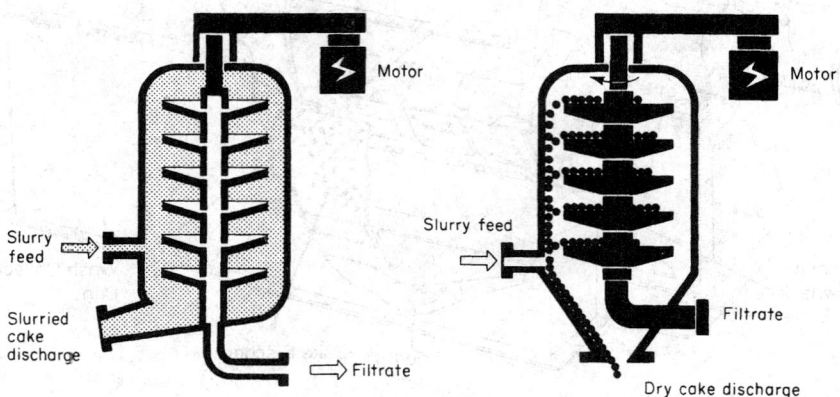

FIG. 18-120 Schematic of a centrifugal-discharge filter. (*Steri Technologies.*)

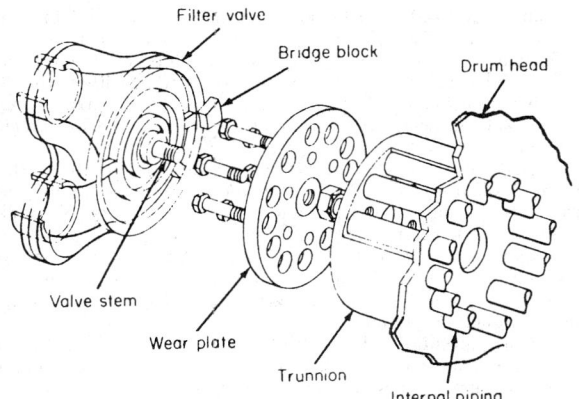

FIG. 18-121 Component arrangement of a continuous-filter valve. (*EIMCO Process Equipment Co.*)

air blowback, into the discharge chute, since actual contact with the medium would cause rapid wear. In some cases the filter medium is held by circumferentially wound wires spaced 50 mm (2 in) apart, and a flexible scraper blade may rest lightly against the wire winding. A taut wire in place of the scraper blade may be used in some applications in which physical dislodging of sticky, cohesive cakes is needed.

For a given slurry, the maximum filtration rate is determined by the minimum cake thickness which can be removed—the thinner the cake, the less the flow resistance and the higher the rate. The minimum thickness is about 6 mm (0.25 in) for relatively rigid or cohesive cakes of materials such as mineral concentrates or coarse precipitates like gypsum or calcium citrate. Solids that form friable cakes composed of less cohesive materials such as salts or coal will usually require a cake thickness of 13 mm (0.5 in) or more. Filter cakes composed of fine precipitates such as pigments and magnesium hydroxide, which often produce cakes that crack or adhere to the medium, usually need a thickness of at least 10 mm (0.38 in).

String-Discharge Filter   A system of endless strings or wires spaced about 13 mm (0.5 in) apart pass around the filter drum but are

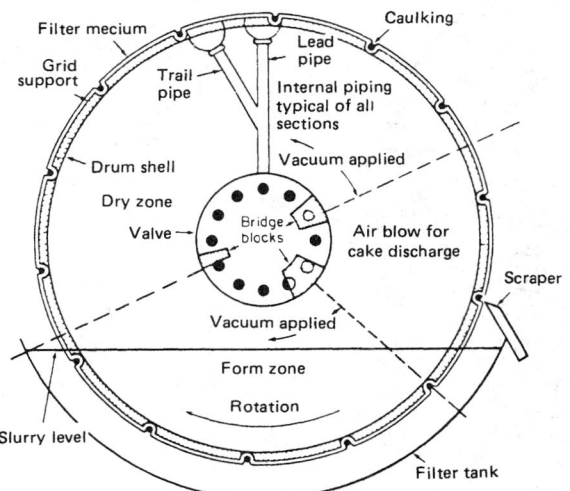

FIG. 18-122 Schematic of a rotary-drum vacuum filter with scraper discharge, showing operating zones. (*Schweitzer, Handbook of Separation Techniques for Chemical Engineers, p. 4-38. Copyright 1979 by McGraw-Hill, Inc. Used with permission of McGraw-Hill Inc.*)

separated tangentially from the drum at the point of cake discharge, lifting the cake off as they leave contact with the drum. The strings return to the drum surface guided by two rollers, the cake separating from the strings as they pass over the rollers. If it has the required body, a thinner cake (5 mm or about ³⁄₁₆ in) than can be handled by drum filters is feasible, allowing more difficult materials to be filtered. This is done at the expense of greater dead area on the drum. Success depends on the ability of the cake to be removed with the strings and must be determined experimentally. Applications are mainly in the starch and pharmaceutical industries, with some in the metallurgical field.

Removable-Medium Filters   Some drum filters provide for the filter medium to be removed and reapplied as the drum rotates. This feature permits the complete discharge of thin or sticky cake and provides the regenerative washing of the medium to reduce blinding. Higher filtration rates are possible because of the thinner cake and clean medium, but this is compromised by a less pure filtrate than normally produced by a nonremovable medium.

Belt-discharge filter.   This is a drum filter carrying a fabric that is removed, passed over rollers, washed, and returned to the drum. Figure 18-123 shows the path of the medium while it is off the drum. A special aligning device keeps the medium wrinkle-free and in proper line during its travel. Thin cakes of difficult solids which may be slightly soluble are good applications. When acceptable, a sluice discharge makes cakes as thin as 1.5 to 2 mm (about ¹⁄₁₆ in) feasible. Several manufacturers offer belt-discharge filters.

Coilfilter.   The **Coilfilter** (Komline-Sanderson Engineering Corp.) is a drum filter with a medium consisting of one or two layers of stainless-steel helically coiled springs, about 10 mm (0.4 in) in diameter, placed in a corduroy pattern around the drum. The springs follow the drum during filtration with cake forming the coils. They are separated from the drum to discharge the cake and undergo washing; if two layers are used, the coils of each layer are further separated from those of the other, passing over different sets of rolls. The use of stainless steel in spring form provides a relatively permanent medium that is readily cleaned by washing and flexing. Filtrate clarity is poorer than with most other media, and a relatively large vacuum pump is needed to handle greater air leakage than is characteristic of fabric media. Material forming a slimy, matlike cake (e.g., raw sewage) is the typical application.

Roll-Discharge Filters   A roll in close proximity to the drum at the point of cake discharge rotates in the opposite direction at a peripheral speed equal to or slightly faster than that of the drum (Fig. 18-124). If the cake on the drum is adequately tacky and cohesive for this discharge technique, it adheres to cake on the smaller roll and separates from the drum. A blade or taut wire removes the material from the discharge roll. This design is especially good for thin, sticky cakes. If necessary, a slight air blow may be provided to help release the cake from the drum. Typical cake thickness is 1 to 10 mm (0.04 to 0.4 in).

Single-Compartment Drum Filter

Bird-Young filter.   This filter (Bird Machine Co.) differs from most drum filters in that the drum is not compartmented and there is no internal piping or rotary valve. The entire inside of the drum is sub-

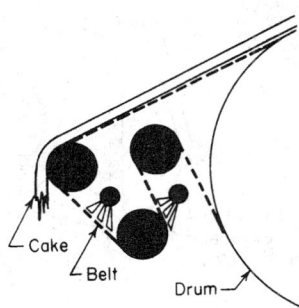

FIG. 18-123 Cake discharge and medium washing on an EIMCO belt filter. (*EIMCO Process Equipment Co.*)

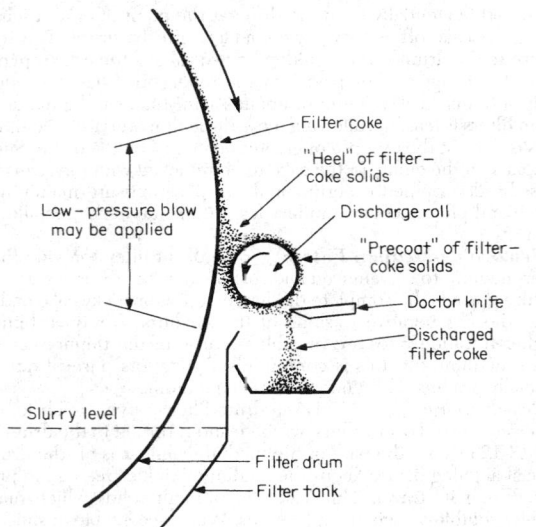

FIG. 18-124 Operating principles of a roll-discharge mechanism. (*Schweitzer,* Handbook of Separation Techniques for Chemical Engineers, *p. 4-40. Copyright 1979 by McGraw-Hill, Inc. Used with permission of McGraw-Hill Book Co.*)

jected to vacuum, with its surface perforated to pass the filtrate. Cake is discharged by an air blowback applied through a "shoe" that covers a narrow discharge zone on the inside surface of the drum to interrupt the vacuum, as illustrated in Fig 18-125. The internal drum surface must be machined to provide close clearance of the shoe to avoid leakage. The filter is designed for high filtration rates with thin cakes.

Rotation speeds to 40 r/min are possible with cakes typically 3 to 6 mm (0.12 to 0.24 in) thick. Filter sizes range from 930 cm² to 19 m² (1 to 207 ft²) with 93 percent of the area active. The slurry is fed into a conical feed tank designed to prevent solids from settling without the use of mechanical agitators. The proper liquid level is maintained by overflow, and submergence ranges from 5 to 70 percent of the drum circumference.

The perforated drum cylinder is divided into sections about 50 to 60 mm (2 to 2.5 in) wide. The filter medium is positioned into tubes between the sections and locked into place by round rods. No caulking, wires, or other fasteners are needed.

Wash sprays may be applied to the cake, with collection troughs or pans inserted inside the drum to keep the wash separate from the filtrate. Filtrate is removed from the lower section of the drum by a pipe passing through the trunnions.

The major advantages of the Bird-Young filter are its ability to handle thin cakes and operate at high speeds, its washing effectiveness, and its low internal resistance to air and filtrate flow. An additional advantage is the possibility of construction as a pressure filter with up to 1.14-MPa (150-psig) operating pressure to handle volatile liquids. The chief disadvantages are its high cost and the limited flexibility imposed by not having an adjustable rotary valve. Best applications are on free-draining nonblinding materials such as paper pulp or crystallized salts.

Continuous Pressure Filters These filters consist of conventional drum or disk filters totally enclosed in pressure vessels. Filtration takes place with the vessel pressurized up to 6 bar and the filtrate discharging either at atmospheric pressure or into a receiver maintained at a suitable backpressure. Cake discharge is facilitated through a dual valve and lock-hopper arrangement in order to maintain vessel pressure. Alternatively, the discharged filter cake can be reslurried within the filter or in an adjoining pressure vessel and removed through a control valve.

One variation in design, the Ceramec, offered by Outokumpu Mintec, employs "gasless" ceramic media instead of traditional filter fabrics, relying partly on capillary action to achieve low moistures.

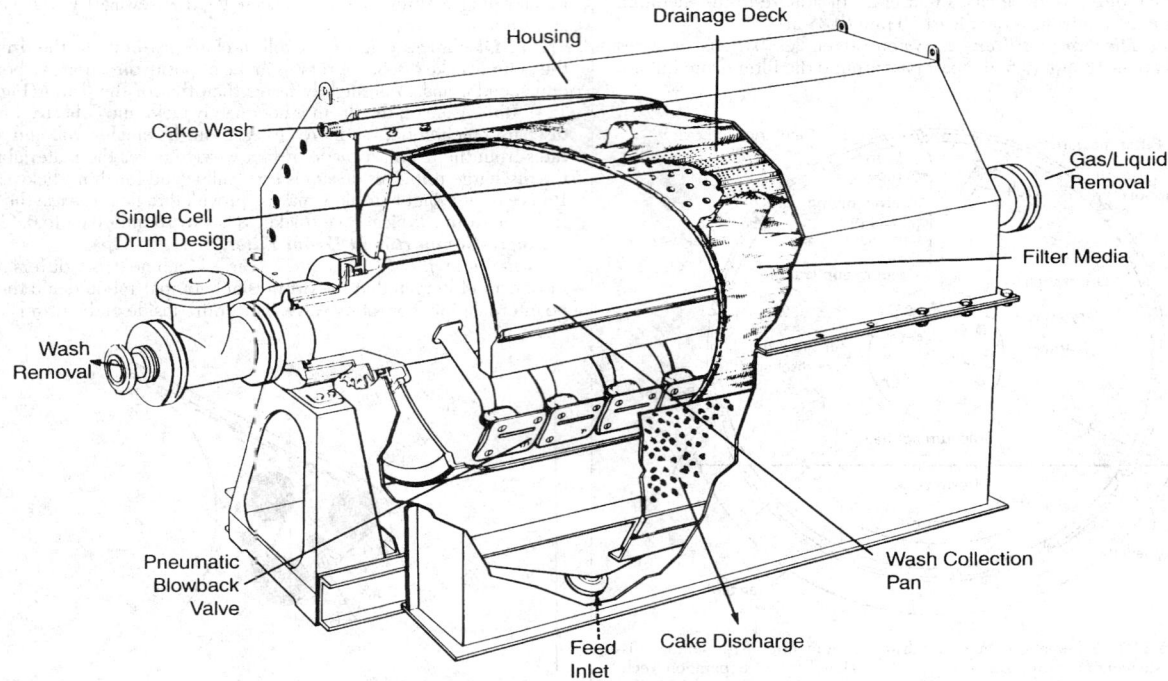

FIG. 18-125 Cutaway of the single-compartment drum filter. (*Bird Machine Co.*)

This results in a significant drop in power consumption by greatly reducing the compressed air requirements.

Continuous Precoat Filters These filters may be operated as either pressure or vacuum filters, although vacuum operation is the prevailing one. The filters are really not continuous but have an extremely long batch cycle (1 to 10 days). Applications are for continuous clarification of liquids from slurries containing 50 to 5000 ppm of solids when only very thin unacceptable cakes would form on other filters and where "perfect" clarity is required.

Construction is similar to that of other drum filters, except that vacuum is applied to the entire rotation. Before feeding slurry a precoat layer of filter aid or other suitable solids, 75 to 125 mm (3 to 5 in) thick, is applied. The feed slurry is introduced and trapped in the outer surface of the precoat, where it is removed by a progressively advancing doctor knife which trims a thin layer of solids plus precoat (Fig. 18-126). The blade advances 0.05 to 0.2 mm (0.002 to 0.008 in) per revolution of the drum. When the precoat has been cut to a predefined minimum thickness, the filter is taken out of service, washed, and freshly precoated. This turnaround time may be 1 to 3 h.

Disk Filters A disk filter is a vacuum filter consisting of a number of vertical disks attached at intervals on a continuously rotating horizontal hollow central shaft (Fig. 18-127). Rotation is by a gear drive. Each disk consists of 10 to 30 sectors of metal, plastic, or wood, ribbed on both sides to support a filter cloth and provide drainage via an outlet nipple into the central shaft. Each sector may be replaced individually. The filter medium is usually a cloth bag slipped over the sectors and sealed to the discharge nipple. For some heavy-duty applications on ores, stainless-steel screens may be used.

The disks are typically 30 to 50 percent submerged in a troughlike vessel containing the slurry. Another horizontal shaft running beneath the disks may contain agitator paddles to maintain suspension of the solids, as in the EIMCO **Agidisc filter.** In some designs, feed is distributed through nozzles below each disk. Vacuum is supplied to the sectors as they rotate into the liquid to allow cake formation. Vacuum is maintained as the sectors emerge from the liquid and are exposed to air. Wash may be applied with sprays, but most applications are for dewatering only. As the sectors rotate to the discharge point, the vacuum is cut off, and a slight air blast is used to loosen the cake. This allows scraper blades to direct the cake into discharge chutes positioned between the disks. Vacuum and air blowback is controlled by an automatic valve as in rotary-drum filters.

Of all continuous filters, the vacuum disk is the lowest in cost per unit area of filter when mild steel, cast iron, or similar materials of construction may be used. It provides a large filtering area with minimum floor space, and it is used mostly in high-tonnage dewatering applications in sizes up to about 300 m² (3300 ft²) of filter area.

The main disadvantages are the inadaptability to have effective wash and the difficulty of totally enclosing the filter for hazardous-material operations.

Horizontal Vacuum Filters These filters are generally classified into two broad classes: rotary circular and belt-type units. Regardless of geometry, they have similar advantages and limitations. They pro-

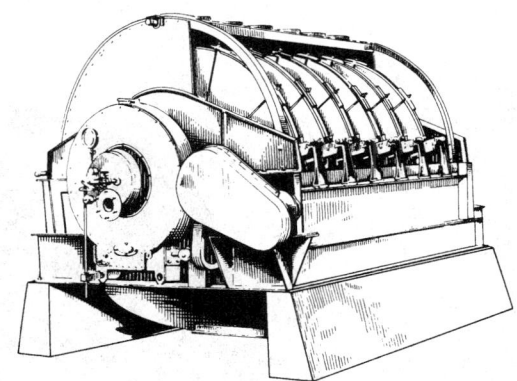

FIG. 18-127 Rotary disk filter. (*Dorr-Oliver, Inc.*)

vide flexibility of choice of cake thickness, washing time, and drying cycle. They effectively handle heavy, dense solids, allow flooding of the cake with wash liquor, and are easily designed for true countercurrent leaching or washing. The disadvantages are they are more expensive to build than drum or disk filters, they use a large amount of floor space per filter area, and they are difficult to enclose for hazardous applications.

Horizontal-Table, Scroll-Discharge, and Pan Filters These are all basically revolving annular tables with the top surface a filter medium (Fig. 18-128). The table is divided into sectors, each of which is a separate compartment. Vacuum is applied through a drainage chamber beneath the table that leads to a large rotary valve. Slurry is fed at one point, and cake is removed after completing more than three-fourths of the circle, by a horizontal scroll conveyor which elevates the cake over the rim of the filter. A clearance of about 10 mm (0.4 in) is maintained between the scroll and the filter medium to prevent damage to the medium. Residual cake on the medium may be loosened by an air blow from below or with high-velocity liquid sprays from above. This residual cake is a disadvantage peculiar to this type of filter. With material that can cause blinding, frequent shutdowns for thorough cleaning may be needed. Unit sizes range from about 0.9 to 7 m (3 to 24 ft) in diameter, with about 80 percent of the surface available for filtration.

Tilting-Pan Filter This is a modification of the table or pan filter in which each of the sectors is an individual pan pivoted on a radial axis to allow its inversion for cake discharge, usually assisted by an air blast. Filter-cake thicknesses of 50 to 100 mm (2 to 4 in) are common. Most applications involve free-draining inorganic-salt dewatering. In addition to the advantages and disadvantages common to all horizontal continuous filters, tilting-pan filters have the relative advantages of complete wash containment per sector, good cake discharge, filter-medium washing, and feasibility of construction in very large sizes, up to about 25 m (80 ft) in diameter, with about 75 percent of the area usable. Relative disadvantages are high capital cost (especially in smaller sizes) and mechanical complexity leading to higher maintenance costs.

Horizontal-Belt Filter This filter consists of a slotted or perforated elastomer drainage belt driven as a conveyor belt carrying a filter fabric belt (Fig. 18-129). Both belts are supported by and pass across a lubricated support deck. A vacuum pan, aligned with the slots in the elastomer belt, forms a continuous vacuum surface which may include multiple zones for cake formation, washing and final dewatering. Several manufacturers provide horizontal-belt filters, the major differences among which lie in the construction of the drainage belt, the method of retaining the slurry/cake on the belt, and the method of maintaining the alignment of the filter medium. The filters are rated according to the available active filtration area. *Indexing horizontal-belt filters* do away with the elastomer drainage belt of the original design in favor of large drainage pans directly beneath the filter medium. Either the pans or the filter medium is indexed to provide a

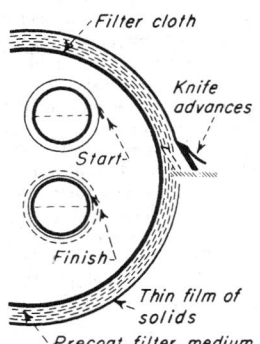

FIG. 18-126 Operating method of a vacuum precoat filter. (*Dorr-Oliver, Inc.*)

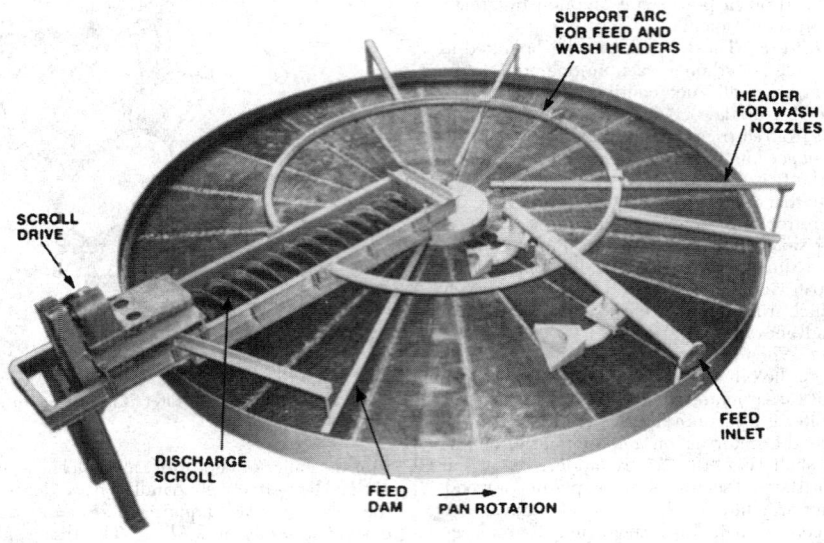

FIG. 18-128   Continuous horizontal vacuum table filter. (*Dorr-Oliver, Inc.*)

pseudo-continuous filtration operation. The applied vacuum is cycled with the indexing operation to minimize wear to the sliding surfaces. As a result the indexing filter must be de-rated for the indexing cycle. The indexing horizontal-belt filter avoids the problem of process compatibility with the elastomer drainage belt. The major differences among the indexing machines of several manufacturers lie in the method of indexing and the method of cycling the applied vacuum.

The method of feeding, washing, dewatering and discharging is essentially the same with all horizontal-belt filters. Slurry is fed at one end by overflow weirs or a fantail chute; wash liquor, if required, is applied by sprays or weirs at one or more locations as the formed cake moves along the filter. Wiping dams and separations in the drainage pan(s) provide controlled wash application. The cake is discharged as the filter-medium belt passes over the end pulley after separation

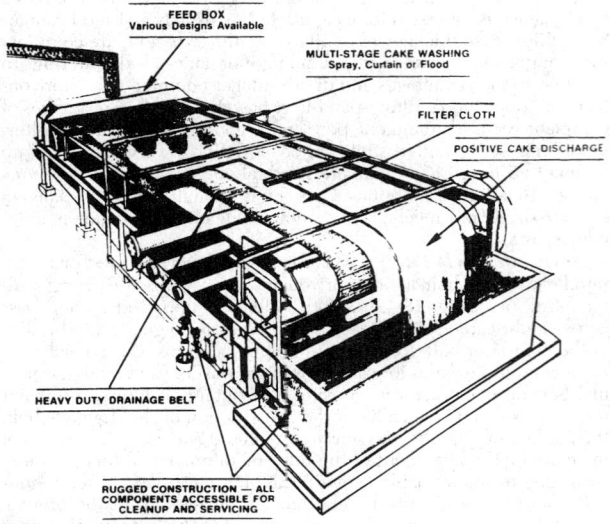

FIG. 18-129   Horizontal-belt filter. (*EIMCO Process Equipment Co.*)

from the drainage surface. Separating the filter-medium from the drainage surface allows thorough spray cleaning of the filter-medium belt. The duration of the filtration cycle is controlled by belt speed which may be as high as 1 m/s (3.3 ft/s) and is typically variable. The minimum possible cake thickness, at a given solids loading, which can be effectively discharged limits the belt speed from a process point of view. The maximum cake thickness is dependent on the method used to retain the slurry during cake formation and can be 100 to 150 mm (4 to 6 in) with fast draining materials.

Some of the advantages of horizontal-belt filters are the precise control of the filtration cycle including the capability for countercurrent washing of the cake, effective cake discharge and thorough cleaning of the filter medium belt. The horizontal-belt filter's primary disadvantage is that at least half of the filtration medium is always idle during the return loop. This contributes to a significantly higher capital cost which can be two to four times that of a drum or disk filter with equal area. Horizontal-belt filters with active filtration area ranging from 0.18 m² to 120 m² (2 ft² to 1300 ft²) on a single machine have been installed.

Additional equipment is sometimes integrated with horizontal-belt filters to further dewater the cake through expression. The addition of such equipment shouldn't be confused with expression equipment that utilizes filter medium belts. Belt type expression equipment is described later in the "Expression" subsection.

Filter Thickeners   Thickeners are devices which remove a portion of the liquid from a slurry to increase the concentration of solids in suspension. Thickening is done to prepare a dilute slurry for more economical filtration or to change the consistency or concentration of the slurry for process reasons. The commonest method of thickening is by gravity sedimentation, discussed earlier in this section. Occasions may arise, however, in which a filter may be called upon for thickening service. Many of the filters previously discussed as cake filters can be operated as thickeners: the filter press with special plates containing flow channels that keep velocity high enough to prevent cake buildup, cycled tube or candle filters with the cake discharge into the filter tank, and continuous leaf filters which use rotating elements adjacent to the filtering surfaces to limit filter cake buildup. Examples of these filters include the Shriver Thickener, the Industrial Hydra-Shoc Filter employing Back Pulse Technology, the DrM, Dr. Müller AG, Contibac Thickener, and the Ingersoll Rand Continuous Pressure Filter.

Clarifying Filters   Clarifying filters are used to separate liquid mixtures which contain only very small quantities of solids. When the solids are finely divided enough to be observed only as a haze, the filter which removes them is sometimes called a polishing filter. The prefilt slurry generally contains no more than 0.10 percent solids, the size of which may vary widely (0.01 to 100 μm). The filter usually produces no visible cake, sometimes because the amount of solids removed is so small, sometimes because the particles are removed by being entrapped within rather than upon the filter medium. Compared with cake filters, clarifying filters are of minor importance to pure chemical-process work, their greatest use being in the fields of beverage and water polishing, pharmaceutical filtration, fuel- and lubricating-oil clarification, electroplating-solution conditioning, and dry-cleaning-solvent recovery. They are essential, however, to the processes of fiber spinning and film extrusion; the spinning solution or dope must be free of particles above a certain size to maintain product quality and to prevent the clogging of spinnerets.

Most cake filters can be so operated as to function as clarifiers, although not necessarily with efficiency. On the other hand, a number of clarifying filters which can be used for no purpose other than clarifying or straining have been developed. In general, clarifying filters are less expensive than cake filters. Clarifying filters may be classified as disk and plate presses, cartridge clarifiers, precoat pressure filters, deep-bed filters, and miscellaneous types. Membrane filters constitute a special class of plate presses and cartridge filters. Simple strainers sometimes are used as clarifiers of liquids containing very large particles. Because they more closely resemble wet screens than filters and because they have little primary process application, they are not discussed here.

Disk Filters and Plate Presses   Filters employing asbestos-pulp disks, cakes of cotton fibers (filtermasse), or sheets of paper or other media are used widely for the polishing of beverages, plating solutions, and other low-viscosity liquids containing small quantities of suspended matter. The term **disk filter** is applied to assemblies of pulp disks made of asbestos and cellulose fibers and sealed into a pressure case. The disks may be preassembled into a self-supporting unit (Fig. 18-130), or each disk may rest on an individual screen or plate against which it is sealed as the filter is closed (Fig. 18-131). The liquid flows through the disks, and into a central or peripheral discharge manifold. Flow rates are on the order of 122 L/(min·m²) [3 gal/(min·ft²)], and the operating pressure does not normally exceed 345 kPa (50 psig) (usually it is less). Disk filters are almost always operated as pressure filters. Individual units deliver up to 378 L/min (6000 gal/h) of low-viscosity liquid.

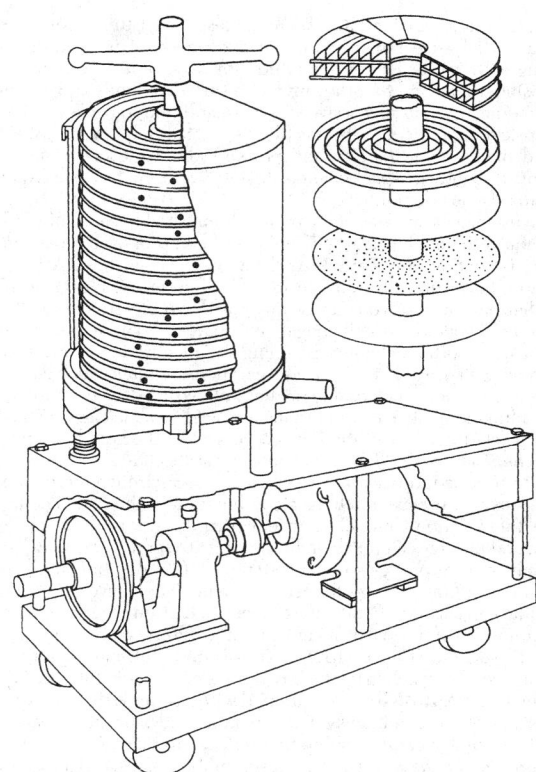

FIG. 18-131   Disk-and-plate clarifying-filter assembly. (*Alsop Engineering Co.*)

Disk-and-plate assemblies somewhat resemble horizontal-plate pressure filters, which, in fact, may be used for polishing. In one design (**Sparkler VR filter**) both sides of each plate are used as filtering surfaces, having paper or other media clamped against them.

Pulp filters.   These filters employ one or more packs of filter-masse (cellulose fibers compressed to a compact cylinder) stacked into a pressure case. The packs are sometimes supported in individual trays which provide drainage channels and sometimes rest on one another with a loose spacer plate between each two packs and with a drainage screen buried in the center of each pack. The liquid being clarified flows under a pressure of 345 kPa (50 psig) or less through the pulp packs and into a drainage manifold. Flow rates are somewhat less than for disk filters, on the order of 20 L/(min·m²) [0.5 gal/(min·ft²)]. Pulp filters are used chiefly to polish beverages. The filter-masse may be washed in special washers and re-formed into new cakes.

Plate presses.   Sometimes called sheet filters, these are assemblies of plates, sheets of filter media, and sometimes screens or frames. They are essentially modified filter presses with practically no cake-holding capacity. A press may consist of many plates or of a single filter sheet between two plates, the plates may be rectangular or circular, and the sheets may lie in a horizontal or vertical plane. The operation is similar to that of a filter press, and the flow rates are about the same as for disk filters. The operating pressure usually does not exceed 138 kPa (20 psig). The presses are used most frequently for low-viscosity liquids, but an ordinary filter press with thin frames is commonly used as a clarifier for 100-Pa·s (1000-P) rayon-spinning solution. Here the filtration pressure may be 6900 kPa (1000 psig).

Disk, pulp, and sheet filters accomplish extreme clarification. Not infrequently their mission is complete removal of particles above a

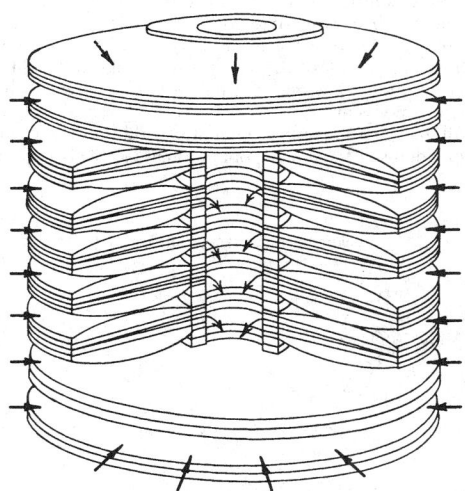

FIG. 18-130   Preassembled pack of clarifying-filter disks. (*Alsop Engineering Co.*)

stipulated cut size, which may be much less than 1 μm. They operate over a particle-size range of four to five orders of magnitude, contrasting with two orders of magnitude for most other filters. It is not surprising, therefore, that they involve a variety of kinds and grades of filter media, often in successive stages. In addition to packs or disks of cellulosic, polymeric, or asbestos fiber, sheets of pulp, paper, asbestos, carded fiber, woven fabrics, and porous cellophane or polymer are employed. Sandwich-pack composites of several materials have been used for viscous-dope filtration.

The use of asbestos has been greatly diminished because of its identification with health hazards. There have been proposed replacement materials such as the **Zeta Plus** filter media from the AMF Cuno Division, consisting of a composite of cellulose and inorganic filter aids that have a positive charge and provide an electrokinetic attraction to hold colloids (usually negatively charged). These media therefore provide both mechanical straining and electrokinetic adsorption.

Cartridge Clarifiers   Cartridge clarifiers are units which consist of or use one or more replaceable or renewable cartridges containing the active filter element. The unit usually is placed in a line carrying the liquid to be clarified; clarification thus occurs while the liquid is in transit.

Mechanical or edge filters.   These consist of stacks of disks separated by precise intervals by spacer plates, or a wire wound on a cage in grooves of a precise pitch, or a combination of the two. The liquid to be filtered flows radially between the disks, wires, or layers of paper, and particles larger than the spacing are screened out. Edge filters can remove particles down to 0.001 in (25 μm) but more often have a minimum spacing of twice this value. They have small solids-retaining capacity and hence must be cleaned often to avoid plugging. Continuous cleaning is provided in some filters. For example, the Cuno Flo-Klean (Fig. 18-132), a wire-wound unit, employs a slowly rotating nozzle which backwashes the element with filtered liquid; and the Cuno **Auto-Klean** is equipped with a scraper that fits into the interdisk slots to comb away accumulated solids. In either case, the dislodged solids fall into a sump that may be drained at intervals.

Micronic clarifiers.   The greatest number of cartridge clarifiers are of the micronic class, with elements of fiber, resin-impregnated filter paper, porous stone, or porous stainless steel of controlled porosity. Other rustless metals are also available. The elements may be chosen to remove particles larger than a fraction of a micrometer, although many are made to pass 10-μm solids and smaller. By proper choice of multiple-cylinder cartridges or multiple cartridges in parallel any desired flow rate can be obtained at a reasonable pressure drop, often less than 138 kPa (20 psig).

When the pressure rises to the permissible maximum, the cartridge must be opened and the element replaced. Micronic elements of the fiber type cannot be cleaned and are so priced that they can be discarded or the filter medium replaced economically. Stone elements usually must be cleaned, a process best accomplished by the manufacturer of the porous ceramic or in accordance with the manufacturer's directions. The user can clean stainless-steel elements by chemical treatment.

Flexibility.   Cartridge filters are flexible: cartridges of different ratings and materials of construction can be interchanged, permitting

ready accommodation to shifting conditions. They have the disadvantage of very limited solid-handling capability so that the maximum solid concentrations in the feed are limited to about 0.01 percent solids. The biggest limitation for modern process-plant operation is the need to open the filter to replace cartridges, which makes their use for the processing of hazardous materials undesirable. Some manufacturers—for example, the Hydraulic Research Division of Textron Inc. and the Fluid Dynamics Division of Brunswick Corp.—have designed cartridges of bonded metal fibers that can be back-flushed or chemically cleaned without opening the unit. These filters, which can operate at temperatures to 482°C (900°F) and at pressures of 33 MPa (325 atm) or greater, are particularly useful for filtering polymers.

Granular Media Filters   Many types of granular media filters are used for clarification, operating either as gravity or pressure filters. Gravity filters rely on a difference in elevation between inlet and outlet to provide the driving force necessary to force the liquid through the granular media. Pressure filters employ enclosed vessels operating at relatively low pressure differentials, in the order of 50 to 70 kPa (7 to 10 psig), which may function in either an upflow or a downflow mode.

The media may be a single material, such as sand, but more often will consist of two or even three layers of different materials, such as anthracite coal in the top layer and sand in the lower one. Solids are captured throughout the bed depth, rather than on the surface, and the gradient in void size provides substantially more solids-holding capacity. The anthracite layer, typically employing 1-mm grain size, serves as a roughing filter and also provides a flocculating action which helps the finer sand, ~0.5-mm particle size, to serve as an effective polishing zone. Media depths vary, but 0.7 to 1.0 m is typical of a dual media installation. Deeper beds of up to 2.5 m (8 ft) are employed in some cases involving special applications where greater solids-holding capacity is desired.

Filtrate is collected in the underdrain system, which may be as simple as a network of perforated pipes covered by graded gravel or a complex structure with slotted nozzles or conduits that will retain the finest sand media while maintaining high flow rates. This latter design allows the use of both air and liquid for the backwashing and cleaning operations.

Backwashing usually is carried out when a limiting pressure drop is reached and before the bed becomes nearly filled with solids, which would lead to a deterioration in filtrate clarity. Cleaning the media is greatly aided by the use of an air scour which helps break loose the trapped solids and provides efficient removal of this material in the subsequent backflushing step. The filtration action tends to agglomerate the filtered solids and, as a result, these generally will settle out readily from the backwash fluid. If the filter is handling a clarifier overflow, usually it is possible to discharge the backwash liquid into the clarifier without risk of these solids returning to the filter. Filter media consumption is low, with normal replacement usually being less than 5 percent per year.

These filters are best applied on relatively dilute suspensions, <150 mg/L suspended solids, allowing operation at relatively high rates, 7.5

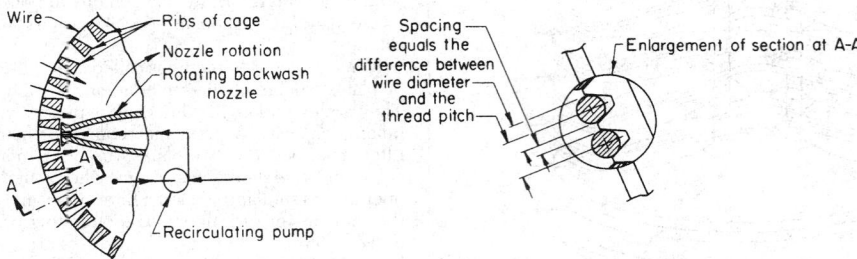

FIG. 18-132   Sectional view of the Cuno Flo-Klean backwashing edge filter. Fluid pumped through the nozzle loosens solids from the filter surface and clears the filtering area. The pump draws filtered fluid from the filter discharge and returns it to the system through the nozzle. Thus, there is no loss of backwash fluid. (*Cuno Division, AMF, Inc.*)

to 15.0 m³/m²·h (3 to 6 gpm/ft²). Solids capture will range from 90 to 98 percent in a well-designed system. Typical operating cycles range from 8 to 24 h of filtration (and up to 48 h in municipal water treatment), followed by a backwash interval of 15 to 30 min. Applications are principally in municipal and wastewater treatment, but granular media filters also have been employed in industrial uses such as pulp and paper plant inlet water treatment; removal of oil, grease, and scale from steelmaking process wastewater; and clarification of electrolyte in copper electrowinning operations.

United States Filter Corp. Maxi-Flo Filter.   The **Maxi-Flo Filter** is an example of the upflow closed-vessel design. Filtration rates to 0.0081 m³/(m²·s) [12 gal/(ft²·min)] and filter cross-section areas up to 10.5 m² (113 ft²) are possible. Deep-bed filtration has been reviewed

by Tien and Payatakes [*Am. Inst. Chem. Eng. J.*, **25,** 737 (1970)] and by Oulman and Baumann [*Am. Inst. Chem. Eng. Symp. Ser.*, **73**(171), 76 (1977)].

Dyna Sand Filter.   A filter that avoids batch backwashing for cleaning, the **Dyna Sand Filter** is available from Parkson Corporation. The bed is continuously cleaned and regenerated by recycling solids internally through an air-lift pipe and a sand washer. Thus a constant pressure drop is maintained across the bed, and the need for parallel filters to allow continued on-stream operation, as with conventional designs, is avoided.

Miscellaneous Clarifiers   Various types of filters such as cartridge, magnetic, and bag filters are widely used in polishing operations, generally to remove trace amounts of suspended solids

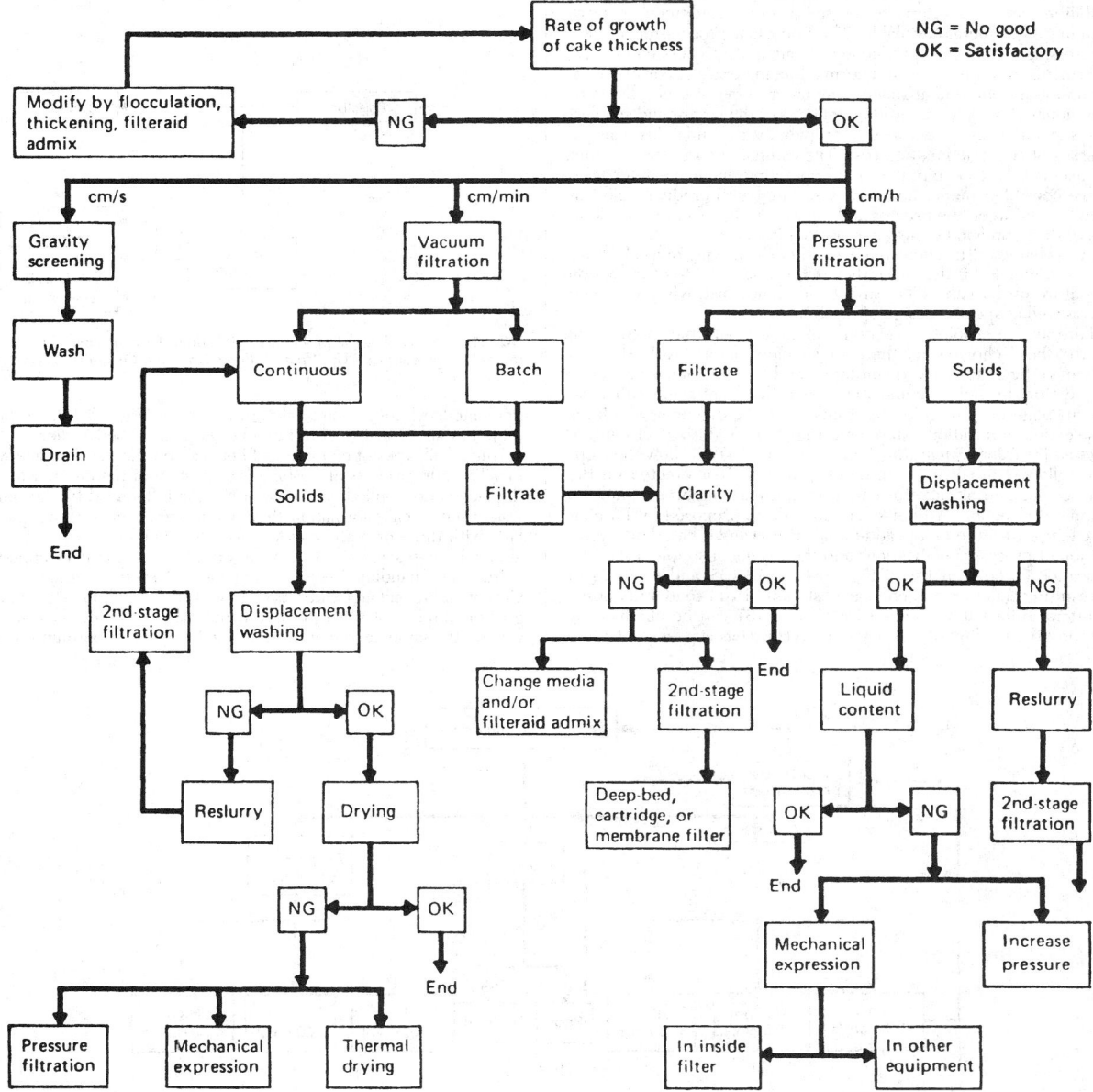

FIG. 18-133   Decision pattern for solving a filtration problem. [*Tiller,* Chem. Eng., **81**(9), 118 (1974), *by permission.*]

remaining from prior unit operations. A thorough discussion of cartridge and felt strainer bag filters is available in Schweitzer, op. cit., Section 4.6 (Nickolaus) and Section 4.7 (Wrotnowski).

SELECTION OF FILTRATION EQUIPMENT

If a process developer who must provide the mechanical separation of solids from a liquid has cleared the first decision hurdle by determining that filtration is the way to get the job done (see the final subsection of Sec. 18, "Selection of a Solids-Liquid Separator")—or that it must remain in the running until some of the details of equipment choice have been settled—choosing the right filter and right filtration conditions may still be difficult. Much as in the broader determination of which unit operation to employ, the selection of filtration equipment involves the balancing of process specifications and objectives against capabilities and characteristics of the various equipment choices (including filter media) available. The important process-related factors are slurry character, production throughput, process conditions, performance requirements, and permissible materials of construction. The important equipment-related factors are type of cycle (batch or continuous), driving force, production rates of the largest and smallest units, separation sharpness, washing capability, dependability, feasible materials of construction, and cost. The estimated cost must account for installed cost, equipment life, operating labor, maintenance, replacement filter media, and costs associated with product-yield loss (if any). In between the process and equipment factors are considerations of slurry preconditioning and use of filter aids.

Slurry characteristics determine whether a clarifying or a cake filter is appropriate; and if the latter, they determine the rate of formation and nature of the cake. They affect the choice of driving force and cycle as well as specific design of machine.

There are no absolute selection techniques available to come up with the "best" choice since there are so many factors involved, many of them difficult to make quantitative and, not uncommonly, some contradictory in their demands. However, there are some published general suggestions to guide the thinking of the engineer who faces the selection of filtration equipment. Figure 18-133 is a decision tree designed by Tiller [*Chem. Eng.*, **81**(9), 118 (1974)] to show the steps to be followed in solving a filtration problem. It is erected on the premise that rate of cake formation is the most important guide to equipment selection. A filter-selection process proposed by Purchas (op. cit., pp. 10–14) employs additional criteria and is based on a combination of process specifications and the results of simple tests. The application is coded by use of Figs. 18-134, 18-135, and 18-136, and the resulting codes are matched against Table 18-11 to identify possible filters. Information needed for Fig. 18-137 can be obtained by observing the settling of a slurry sample (Purchas suggests 1 L) in a

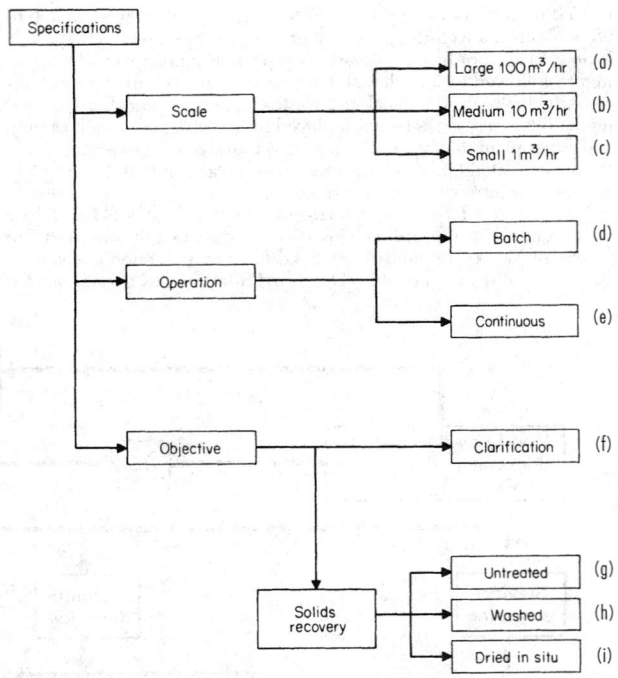

FIG. 18-134   Coding the problem specification. (*Purchas*, Solid/Liquid Separation Equipment Scale-Up, *Uplands Press, Croydon, England, 1977, p. 10, by permission.*)

graduated cylinder. Filter-cake-growth rate (Fig. 18-136) is determined by small-scale leaf or funnel tests as described earlier.

Almost all types of continuous filters can be adapted for cake washing. The effectiveness of washing is a function of the number of wash displacements applied, and this, in turn, is influenced by the ratio of wash time to cake-formation time. Countercurrent washing, particularly with three or more stages, is usually limited to horizontal filters, although a two-stage countercurrent wash sometimes can be applied on a drum filter handling freely filtering material, such as crystallized salts. Cake washing on batch filters is commonly done, although, generally, a greater number of wash displacements may be required in order to achieve the same degree of washing obtainable on a continuous filter.

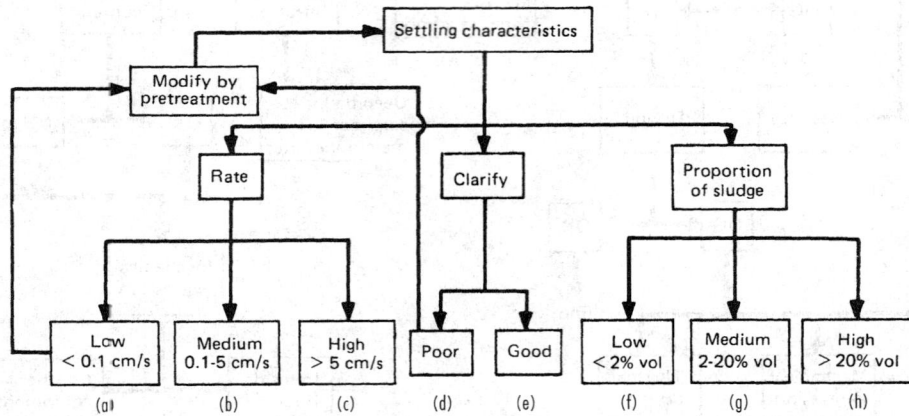

FIG. 18-135   Coding the settling characteristics of a slurry. (*Purchas*, Solid/Liquid Separation Equipment Scale-Up, *Uplands Press, Croydon, England, 1977, p. 11, by permission.*)

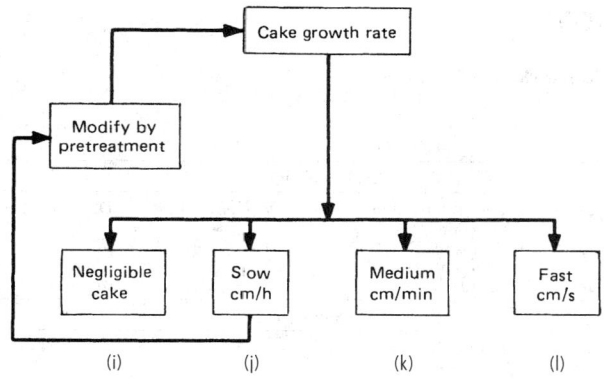

FIG. 18-136 Coding the filtration characteristics of a slurry. (*Purchas, Solid/Liquid Separation Equipment Scale-Up,* Uplands Press, Croydon, *England, 1977, p. 12, by permission.*)

Continuous filters are most attractive when the process application is a steady-state continuous one, but the rate at which cake forms and the magnitude of production rate are sometimes overriding factors. A rotary vacuum filter, for example, is a dubious choice if a 3-mm (0.12-in) cake will not form under normal vacuum in less than 5 min and if less than 1.4 m³/h (50 ft³/h) of wet cake is produced. Upper production-rate limits to the practicality of batch units are harder to

establish, but any operation above 5.7 m³/h (200 ft³/h) of wet cake should be considered for continuous filtration if it is at all feasible. Again, however, other factors such as the desire for flexibility or the need for high pressure may dictate batch equipment.

For estimating filtration rate (therefore, operating pressure and size of the filter), washing characteristics, and other important features, small-scale tests such as the leaf or pressure bomb tests described earlier are usually essential. In the conduct and interpretation of such tests, and for advice on labor requirements, maintenance schedule, and selection of accessory equipment the assistance of a dependable equipment vendor is advisable.

FILTER PRICES

As indicated, one of the factors affecting the selection of a filter is total cost of carrying out the separation with the selected machine. An important component of this cost item is the installed cost of the filter, which starts with the purchase price.

From a survey of early 1982, prices of a number of widely used types of process filter were collated by Hall and coworkers [*Chem. Eng.,* **89**(7), 80 (1982)]. These data are drawn together in Fig. 18-137, updated to 1995 prices. They have a claimed accuracy of ±10 percent, but they should be used confidently only with study-level cost estimations (±25 percent) at best. Cost of delivery to the plant can be approximated as 3 percent of the FOB price [Pikulik and Diaz, *Chem. Eng.,* **84**(21), 106 (1977)].

The cost of the filter station includes not only the installed cost of the filter itself but also that of all the accessories dedicated to the filtration operation. Examples are feed pumps and storage facilities, precoat tanks, vacuum systems (often a major cost factor for a vacuum filter station), and compressed-air systems. The delivered cost of the accessories plus the cost of installation of filter and accessories generally is of the same order of magnitude as the delivered filter cost and commonly is several times as large. Installation costs, of course, must be estimated with reference to local labor costs and site-specific considerations.

The relatively high prices of pulp and paper filters reflect the construction features that accommodate the very high hydraulic capacity that is required. The absence of data for some common types of filters, in particular the filter press, is explained by Hall as due to the complex variety of individual features and materials of construction. For information about missing filters and for firmer estimates for those types presented, vendors should be consulted. In all cases of serious interest, consultation should take place early in the evaluation procedure so that it can yield timely advice on testing, selection, and price.

TABLE 18-11   Classification of Filters According to Duty and Slurry-Separation Characteristics*

Type of equipment	Suitable for duty specification†	Required slurry-separation characteristics‡	
		Slurry-settling characteristics	Slurry-filtering characteristics
Deep-bed filters	a or b	A	T
	e	D	
	f	F	
Cartridges	b or c	A or B	
	d	D or E	
	f	F	
Batch filters			
Pressure vessel with vertical elements	a, b, or c	A or B	I or J
	d	D or E	
	f, g, h, or i	F or G	
Pressure vessel with horizontal elements	b or c	A or B	J or K
	d	D or E	
	g or h	F or G	
Filter presses	a, b, or c	A (or B)	I or J
	d	D or E	
	f, g, h, or i	F, G, or H	
Variable-volume filters	a, b, or c	A (or B)	J or K
	d or e	D or E	
	g (or h)	G or H	
Continuous filters			
Bottom-fed drum or belt drum	a, b, or c	A or B	I, J, K, or L
	e	D or E	
	f, g, h, or i	F, G, or H	
Top-fed drum	a, b, or c	C	L
	e	E	
	g, i (or h)	G or H	
Disk	a, b, or c	A or B	J or K
	e	D or E	
	g	G or H	
Horizontal belt, pan, or table	a, b, or c	A, B, or C	J, K, or L
	d or e	D or E	
	g or h	F, G, or H	

*Adapted from Purchas, *Solid/Liquid Separation Equipment Scale-Up,* Uplands Press, Croydon, England, 1977, p. 13, by permission.
†Symbols are identified in Fig. 18-135.
‡Symbols are identified in Figs. 18-136 and 18-137.

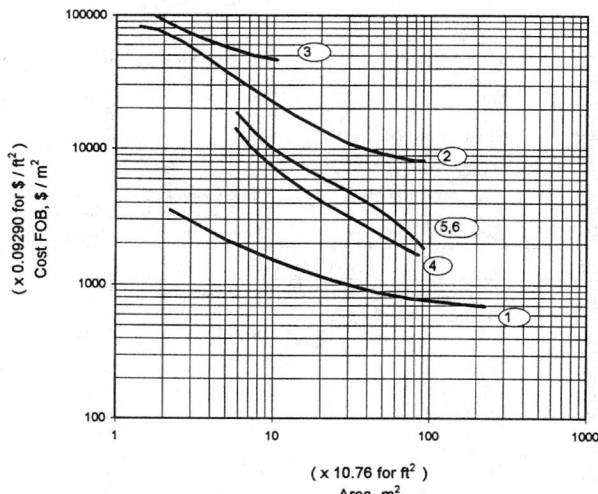

FIG. 18-137 Price of filters installed, FOB point of manufacture. (*EIMCO Process Equipment Co.*)

CENTRIFUGES

GENERAL REFERENCES: Ambler, in McKetta, *Encyclopedia of Chemical Processing and Design*, vol. 7, Marcel Dekker, New York, 1978; also in Schweitzer, *Handbook of Separation Techniques for Chemical Engineers*, McGraw-Hill, New York, 1979, sec. 4. Ambler and Keith, in Perry and Weissberger, *Separation and Purification Techniques of Chemistry*, 3d ed., vol. 12, Wiley, New York, 1978. Flood, Porter, and Rennie, *Chem. Eng.*, **73**(13), 190 (1966). Greenspan, *J. of Fluid Mech.*, **127**(9), 91 (1983). Hultsch and Wilkesmann, in Purchas, *Solid/Liquid Separation Equipment Scale-Up*, Uplands Press, Croydon, England, 2d ed., 1986, chap. 12. Gerl, Stadager, and Stahl, *Chemical Eng. Progress*, **91**, 48–54, (May 1995). Leung, *Chem. Eng.* (1990). Leung, *Fluid-Particle Sep. J.*, **5**(1), 44 (1992). Leung, *10th Pittsburgh Coal Conf. Proceed.* (1993). Leung and Shapiro, *Filtration and Separation Journal*, Sept. and Oct. 1996. Leung and Shapiro, U.S. Patents 5,520,605 (May 28, 1996), 5,380,266 (Jan. 10, 1995), and 5,401,423 (March 28 1995). Mayer and Stahl, *Aufbereitungs-Technik*, **11**, 619 (1988). Moyers, *Chem. Eng.*, **73**(13), 182 (1966). Records, in Purchas, op. cit., chap. 6. Smith, *Ind. Eng. Chem.*, **43**, 439 (1961). Sullivan and Erikson, ibid., p. 434. Svarovsky, in *Solid-Liquid Separation*, 3d ed., Butterworths, 1990, chap. 7. Tiller, *AICHE J.*, **33**(1), (1987). Zeitsch, in Svarovsky, op. cit., chap. 14.

Nomenclature

Symbol	Definition	SI units	U.S. customary units
a	Acceleration	m/s^2	ft/s^2
B_o	Bond number	Dimensionless	Dimensionless
b	Basket axial length	m	ft
C_f	Frictional coefficient	Dimensionless	Dimensionless
D	Bowl/basket diameter	m	in
d	Particle diameter	m	ft
Ek	Ekman number	Dimensionless	Dimensionless
F	Cumulative fraction	Dimensionless	Dimensionless
G	Centrifugal gravity	m/s^2	ft/s^2
g	Earth gravity	m/s^2	ft/s^2
h	Cake height	m	ft
K	Cake permeability	m^2	ft^2
L	Length	m	ft
M	Mass	kg	lb
m	Bulk mass rate	kg/s	lb/h
N_c	Capillary number	Dimensionless	Dimensionless
P	Power	kw	hp
Q	Flow rate	L/s	gpm
R_m	Filter media resistance	m^{-1}	ft^{-1}
Ro	Rossby number	Dimensionless	Dimensionless
r	Radius	m	ft
Rec	solids recovery	Dimensionless	Dimensionless
S	Liquid saturation in cake (=volume of liquid/ volume cake void)	Dimensionless	Dimensionless
sg	Specific gravity	Dimensionless	Dimensionless
t	Time	s	s
t_d	Time	Dimensionless	Dimensionless
u	Velocity	m/s	ft/s
Vθ	Circumferential velocity	m/s	ft/s
V_c	Bulk cake volume	m^3	ft^3
V	Velocity	m/s	ft/s
W	Weight fraction of solids	Dimensionless	Dimensionless
Y	Yield	Dimensionless	Dimensionless
Z	Capture efficiency	Dimensionless	Dimensionless
Greek symbols			
ε	Cake void volume fraction	Dimensionless	Dimensionless
ε_s	Cake solids volume fraction	Dimensionless	Dimensionless
μ	Liquid viscosity	Pa·s	P
ρ_s	Solid density	kg/m^3	lb/ft^3
ρ_L	Liquid density	kg/m^3	lb/ft^3
σ	Surface tension	N/m	lbf/ft
σ_h	Hoop stress	Pa	psi
θ	Angle	Radian	degree
Σ	Scale-up factor (equivalent sedimentation area)	m^2	ft^2
ξ	Time	Dimensionless	Dimensionless

Nomenclature (Concluded)

Symbol	Definition	SI units	U.S. customary units
Greek symbols (Cont.)			
ϕ	Feed solids volume fraction	Dimensionless	Dimensionless
τ_y	Yield stress	Pa	psi
Δ	Differential speed	1/s	r/min
Ω	Angular speed	1/s	r/min
Subscripts			
b	Bowl or basket		
c	Cake		
e	Centrate		
f	Feed		
f	Filtrate		
acc	Acceleration		
p	Pool		
con	Conveyance		
t	Tangential		
L	Liquid		
s	Solid		

INTRODUCTION

Centrifuges for the separation of solids from liquids are of two general types: (1) sedimenting centrifuges, which require a difference in density of the two phases (solid-liquid or liquid-liquid) and (2) filtering centrifuges (for solid-liquid separation), in which the solid phase is retained by the filter medium through which the liquid phase is free to pass. The following discussion is focused on solid-liquid separation for both types of centrifuges; however, a dispersed liquid phase in another continuous liquid phase as used in sedimenting centrifuges exhibits similar behavior as solid in liquid, and therefore the results developed are generally applicable. The use of centrifuges covers a broad range of applications, from separation of fine calcium carbonate particles of less than 10 μm to coarse coal of 0.013 m (½ in).

GENERAL PRINCIPLES

Centripetal and Centrifugal Acceleration   A centripetal body force is required to sustain a body of mass moving along a curve trajectory. The force acts perpendicular to the direction of motion and is directed radially inward. The centripetal acceleration, which follows the same direction as the force, is given by the kinematic relationship:

$$a = \frac{V_\theta^2}{r} \tag{18-75}$$

where V_θ is the tangential velocity at a given point on the trajectory and r is the radius of curvature at that point. This analysis holds for the motion of a body in an inertial reference frame, for example, a stationary laboratory. It is most desirable to consider the process in a centrifuge, and the dynamics associated with such, in a noninertial reference frame such as in a frame rotating at the same angular speed as the centrifuge. Here, additional forces and accelerations arise, some of which are absent in the inertial frame. Analogous to centripetal acceleration, an observer in the rotating frame experiences a centrifugal acceleration directed radially outward from the axis of rotation with magnitude:

$$a = \Omega^2 r \tag{18-76}$$

where Ω is the angular speed of the rotating frame and r is the radius from the axis of rotation.

Solid-Body Rotation   When a body of fluid rotates in a solid-body mode, the tangential or circumferential velocity is linearly proportional to radius:

$$V_\theta = \Omega r \tag{18-77}$$

as with a system of particles in a rigid body. Under this condition, the magnitude of the centripetal acceleration, Eq. (18-75), equals that of the centrifugal acceleration, Eq. (18-76), despite the fact that these accelerations are considered in two different reference frames. Hereafter, the rotating frame attached to the centrifuge is adopted. Therefore, centrifugal acceleration is exclusively used.

G-Level Centrifugal acceleration G is measured in multiples of earth gravity g:

$$\frac{G}{g} = \frac{\Omega^2 r}{g} \qquad (18\text{-}78)$$

With the speed of the centrifuge Ω in r/min and D the diameter of the bowl,

$$\frac{G}{g} = 0.000559\Omega^2 D, \; D(m) \qquad (18\text{-}79)$$

With D in inches, the constant in Eq. (18-79) is 0.0000142. G can be as low as $100g$ for slow-speed, large basket units to as much as $10,000g$ for high-speed, small decanter centrifuges and $15,000g$ for disk centrifuges. Because G is usually very much greater than g, the effect due to earth's gravity is negligible. In analytical ultracentrifuges used to process small samples, G can be as much as $500,000g$ to effectively separate two phases with very small density difference.

Coriolis Acceleration The Coriolis acceleration arises in a rotating frame, which has no parallel in an inertial frame. When a body moves at a linear velocity u in a rotating frame with angular speed Ω, it experiences a Coriolis acceleration with magnitude:

$$a = 2\Omega u \qquad (18\text{-}80)$$

The Coriolis vector lies in the same plane as the velocity vector and is perpendicular to the rotation vector. If the rotation of the reference frame is anticlockwise, then the Coriolis acceleration is directed 90° clockwise from the velocity vector, and vice versa when the frame rotates clockwise. The Coriolis acceleration distorts the trajectory of the body as it moves rectilinearly in the rotating frame.

Effect of Fluid Viscosity and Inertia The dynamic effect of viscosity on a rotating liquid slurry as found in a sedimenting centrifuge is confined in very thin fluid layers, known as **Ekman layers.** These layers are adjacent to rotating surfaces which are perpendicular to the axis of rotation, such as bowl heads, flanges, and conveyor blades, etc. The thickness of the Ekman layer δ is of the order

$$\delta = \sqrt{\frac{(\mu/\rho)}{\Omega}} \qquad (18\text{-}81)$$

where μ/ρ is the kinematic viscosity of the liquid. For example with water at room temperature, μ/ρ is 1×10^{-6} m^2/s and for a surface rotating at $\Omega = 3000$ r/min, δ is 0.05 mm! These layers are very thin; nevertheless, they are responsible for transfer of angular momentum between the rotating surfaces to the fluid during acceleration and deceleration. They worked together with the larger-scale inviscid bulk flow transferring momentum in a rather complicated way. This is demonstrated by the teacup example in which the content of the cup is brought to speed when it is stirred and it is brought to a halt after undergoing solid-body rotation. The viscous effect is characterized by the dimensionless Ekman number:

$$\text{Ek} = \frac{\mu/\rho}{\Omega L^2} \qquad (18\text{-}82)$$

where L is a characteristics length. It measures the scale of the viscous effect to that of the bulk flow.

The effect of fluid inertia manifests during abrupt change in velocity of the fluid mass. It is quantified by the Rossby number:

$$\text{Ro} = \frac{u}{\Omega L} \qquad (18\text{-}83)$$

Typically, Ro is small to the order of 1 with the high end of the range showing possible effect due to inertia, whereas the Ek number is usually very small, 10^{-6} or smaller. Therefore, the viscous effect is confined to thin boundary layers with thickness $\text{Ek}^{1/2}L$.

Sedimenting and Filtering Centrifuges Under centrifugal force, the solid phase assumed to be denser than the liquid phase set-

tles out to the bowl wall—sedimentation. Concurrently, the lighter, more buoyant liquid phase is displaced toward the smaller diameter—flotation. This is illustrated in Fig. 18-138a. Some centrifuges run with an air core, i.e., with free surface, whereas others run with slurry filled to the center hub or even to the axis in which pressure can be sustained.

In a sedimenting centrifuge, the separation can be in the form of **clarification,** wherein solids are separated from the liquid phase in which clarity of the liquid phase is of prime concern. For biological sludge, polymers are used to agglomerate fine solids to facilitate clarification. Separation can also be in the form of **classification** and degritting at which separation is effected by means of particle size and density. Typically, the finer solids (such as kaolin) of smaller size and/or density in the feed slurry are separated in the centrate stream as product (for example 90 percent of particles less than 1 μm, etc.), whereas the larger and/or denser solids are captured in cake as reject. Furthermore, separation can be in the form of **thickening,** where solids settle under centrifugal force to form a stream with concentrated solids. In **dewatering** or **deliquoring,** the objective is to produce dry cake with high solids consistency by centrifugation.

In a filtering centrifuge, separating solids from liquid does not require a density difference between the two phases. Should a density difference exist between the two phases, sedimentation is usually at a much more rapid rate compared to filtration. In both cases, the solid and liquid phases move toward the bowl under centrifugal force. The solids are retained by the filter medium, while the liquid flows through the cake solids and the filter. This is illustrated in Fig. 18-138b.

Performance Criteria Separation of a given solid-liquid slurry is usually measured by the purity of the separated liquid phase in the centrate (or liquid effluent) in sedimenting mode or filtrate in filtering mode, and the separated solids in the cake. In addition, there are other important considerations. Generally, a selected subset of the following criteria are used, depending on the objectives of the process:

Cake dryness or moisture content
Total solids recovery
Polymer dosage
Size recovery and yield
Volumetric and solids throughput
Solid purity and wash ratio
Power consumption

Cake Dryness In dewatering, usually the cake needs to be as dry as possible. Cake dryness is commonly measured by the solids fraction by weight W or by volume ε_s. The moisture content is measured by the complement of W or ε_s. The volume fraction of the pores and void in the wet cake is measured by the cake porosity $\varepsilon (= 1 - \varepsilon_s)$; whereas the volume fraction of the liquid in the pores of the cake is measured by the saturation S. For well-defined solids in the cake with solid density (bone dry) ρ_s and liquid density ρ_L, and given that the cake volume $\dot{V}_c$

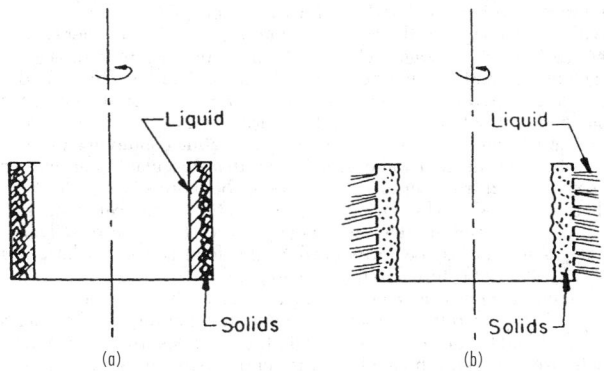

FIG. 18-138 Principles of centrifugal separation and filtration: (*a*) sedimentation in rotating imperforate bowl; (*b*) filtration in rotating perforate basket.

and the mass of solids in the cake w_s are known, the cake porosity is determined by

$$\varepsilon = 1 - \frac{w_s}{\rho_s V_c} \qquad (18\text{-}84)$$

For undersaturated cake with $S < 1$, saturation can be inferred from the weight fraction of solids and the porosity of the cake, together with the solid and liquid densities:

$$S = \left(\frac{1-W}{W}\right)\left(\frac{1-\varepsilon}{\varepsilon}\right)\frac{\rho_s}{\rho_L} \qquad (18\text{-}85)$$

When the cake is saturated $S = 1$, the cake porosity can be determined from Eq. (18-86) as

$$\varepsilon = \left(1 - \frac{\rho_L}{\rho_s}\frac{W}{1-W}\right)^{-1} \qquad (18\text{-}86)$$

Cake dewatering by compression and rearrangement of the solids in the cake matrix reduce ε, yet the cake is still saturated with $S = 1$. (Assuming cake solids are ideal spheres of uniform size, the maximum packing, in rhombohedral arrangement, is such that $\varepsilon_s = 74$ percent or $\varepsilon = 26$ percent.) Drainage of liquid within the cake by centrifugation further reduces S to be less than 1. There is a lower limit on S which is determined by the cake height, dewatering time, centrifugal force as compared to the capillary and surface forces, as well as the surface roughness and porosity of the particles.

Total Solids Recovery   In clarification, the clarity of the effluent is measured indirectly by the total solids recovered in the cake as

$$\text{Rec} = \frac{m_c W_c}{m_f W_f} \qquad (18\text{-}87)$$

where subscripts c and f denote, respectively, the cake and the feed. m is the bulk mass flow rate in kg/s (lb/h).

Under steady state, the mass balance on both solids and liquid yield, respectively:

$$m_f W_f = m_c W_c + m_e W_e \qquad (18\text{-}88)$$

$$m_f = m_c + m_e \qquad (18\text{-}89)$$

From the above, it follows that

$$\text{Rec} = \frac{1 - (W_e/W_f)}{1 - (W_e/W_c)} \qquad (18\text{-}90)$$

where subscript e represents liquid centrate. Stringent requirements on centrate quality or capture of valuable solid product often require the recovery to exceed 90 percent and, in some cases, 99+ percent. In such cases, the centrate solids are typically measured in ppm.

Polymer Dosage   Cationic and anionic polymers have been commonly used to coagulate and flocculate fine particles in the slurry. This is especially pertinent to biological materials such as are found in wastewater treatment. In the latter, cationic polymers are often used to neutralize the negative-charge ions left on the surface of the colloidal particles. Polymer dosage is measured by kg of dry polymer/1000 kg of dry solids cake (lbm of dry polymer/ton of dry solids cake). With liquid polymers, the equivalent (active) dry solid polymer is used to calculate the dosage. There is a minimum polymer dosage to agglomerate and capture the fines in the cake. Overdose can be undesirable to recovery and cake dryness. The range of optimal dosage is dictated by the type of solids in the slurry, slurry physical properties such as pH, ionic strength, etc., and the operating condition and characteristics of the centrifuge. It is known that flocculated particles or flocs obtained from certain polymers may be more sensitive to shear than others, especially during feed acceleration in the centrifuges. A more gentle feed accelerator is beneficial for this type of polymer. Also, polymers can be introduced to the feed at various locations either within or external to the centrifuge.

Size Recovery and Yield   Centrifuges have been applied to classify polydispersed fine particles. The size distribution of the particles is quantified by the cumulative weight fraction F less than a given particle size d for both the feed and the centrate streams. It is measured by a particle size counter which operates based on principles such as sedimentation or optical scattering.

In kaolin classification, the product is typically measured with a certain percentage less than a given size (example 90 percent or 95 percent less than 1 or 2 μm). Each combination of percent and size cut represents a condition by which the centrifuge would have to tune to yield the product specification.

The yield Y is defined as the fraction of feed particles of a given size below which they report to the centrate product. Thus,

$$Y = \frac{m_e W_e F_e}{m_f W_f F_f} \qquad (18\text{-}91)$$

From material balance, the particle size distribution of the feed and centrate, as well as the total solids recovery, determine the yield,

$$Y(d) = \frac{F_e(d)}{F_f(d)}(1 - \text{Rec}) \qquad (18\text{-}92)$$

The complementary is the cumulative capture efficiency Z ($= 1 - Y$), which is defined as the feed particles of a given size and smaller which are captured in the cake, which in most dewatering applications is the product stream.

Volumetric and Solids Throughput   The maximum volumetric and solids throughput to a centrifuge are dictated by one or several governing factors, the most common ones are the centrate solids, cake dryness, and capacity (torque and power) of drive/gear unit. The solids throughput is also governed by other factors such as solids conveyance and discharge mechanisms for continuous and batch centrifuges. The settling rate, as may be significantly reduced by increasing feed solids concentration, also becomes crucial to solids throughput, especially if it has to meet a certain specification on centrate quality.

Solid Purity and Wash Ratio   Cake washing is used to remove impurities in cake solids. It is most effective with filtering centrifuges; however, washing can also be done on some sedimenting centrifuges. The wash ratio is defined as the volumetric amount of wash liquid per unit void volume in the cake or per cake solids amount if the cake porosity is not available. Wash ratio of 0.5:4 is typical, depending on the effectiveness of the wash. Separation and repulping also proves to be effective in removing dissolved contaminants in the cake.

Power Consumption   Power is consumed to overcome windage and bearing (and seal) friction, to accelerate feed stream from zero speed to full tangential speed at the pool so as to establish the required G-force for separation, and to convey and discharge cake. The power to overcome windage and bearing friction is usually established through tests for a given centrifuge geometry at different rotation speeds. It is proportional to the mass of the centrifuge, to the first power of the speed for the bearing friction, and to the second power of speed for windage. It is also related to the bearing diameter. The seal friction is usually small.

The horsepower for feed acceleration is given by

$$P_{\text{acc}} = 5.984(10^{-10})\text{sg}Q(\Omega r_p)^2 \qquad (18\text{-}93)$$

where sg is the specific gravity of the feed slurry, Q the volumetric flow rate of feed in gpm(l/s), Ω the speed in r/min, r_p in meters corresponds to the radius of the pool surface for sedimenting centrifuge, or to the radius of the cake surface for filtering centrifuge. *Note:* To convert horsepower to kilowatts, multiply by 0.746.

The horsepower for cake conveyance for scroll centrifuge is

$$P_{\text{con}} = 1.587(10^{-5})\,T\Delta \qquad (18\text{-}94)$$

where Δ is the differential speed in r/min (s^{-1}) between the scroll conveyor and the bowl, and T is the conveyance torque in in·lb_f (N·m). For centrifuge where cake is discharged differently, the conveyance power is simply

$$P_{\text{con}} = MGC_f V \qquad (18\text{-}95)$$

where M is the mass of the cake, G the centrifugal acceleration, C_f is the coefficient of friction, and V is the cake velocity. Comparing Eqs. (18-94) and (18-95) the conveyance torque is inversely related to the differential speed and directly proportional to the G acceleration, cake velocity, and cake mass.

Stress in the Centrifuge Rotor   The stress in the centrifuge rotor is quite complex. Analytical methods, such as the finite element method, are used to analyze the mechanical integrity of a given rotor design. Without getting into an involved analysis, some useful knowledge can be gained from a simple analysis of the hoop stress of a rotating bowl under load. At equilibrium, the tensile hoop stress σ_h of the cylindrical bowl wall with thickness t is balanced by the centrifugal body force due to the mass of the bowl wall with density ρ_m and its contents (cake or slurry or liquid) with equivalent density ρ_L. Consider a circular wall segment with radius r, unit subtended angle, and unit axial length. A force balance requires

$$\sigma_h = \rho_m V_t^2 \left[1 + 0.5 \frac{\rho_L}{\rho_m} \left(1 - \frac{r_s^2}{r_b^2} \right) \frac{r_b}{t} \right] \quad (18\text{-}96)$$

$V_t = \Omega r_b$ is the tip speed of the bowl. The term in the bracket is typically of order 1. If the maximum allowed σ_h is designed to be no more than 60 percent of the yield stress of the bowl material, which for steel is about $2.07(10^8)$ $\mathrm{Nm^{-2}}$ (30,000 $\mathrm{lb_f/in^2}$), and there is no liquid load, then $(V_t)_{\max} = \{\sigma_h/\rho_m\}^{1/2} = 126$ m/s (412 ft/s). With additional liquid load, $\rho_L = 1000$ kg/m³ (0.0361 $\mathrm{lbm/in^3}$), $r_b/t = 10$, and further assuming the worst case with liquid filling to the axis, the term in the curly bracket is 1.636. Using Eq. (18-96), $(V_t)_{\max} = \{\sigma_h/\rho_m/1.636\}^{1/2} = 98$ m/s (322 ft/s). Indeed, almost all centrifuges are designed with top rim speed about 91 m/s (300 ft/s). With special construction materials for the rotor, such as Ferralium, with higher yield stress, the maximum rim speed under full load can be over 122 m/s (400 ft/s).

G-Force vs. Throughput   The G-acceleration can be expressed as

$$G = \Omega^2 r_b = \frac{(V_t)_{\max}^2}{r_b} \quad (18\text{-}97)$$

Figure 18-139 shows the range of diameter of commercial centrifuges and the range of maximum G developed in each type. It demonstrates an inverse relationship between G and r_b at $V_t = (V_t)_{\max}$, which is constant for a given material. Figure 18-140 shows a log-log plot of G versus Ω for various bowl diameters, [Eq. (18-97)]. Also, the limiting conditions as delineated by $G = \Omega^2 R = \Omega(V_t)_{\max}$ with various $(V_t)_{\max}$, are superimposed on these curves. These two sets of curves dictate the operable speed and G for a given diameter and a given construction material for the bowl. The throughput capacity of a machine, depending on the process need, is roughly proportional to the nth power of the bowl radius,

$$Q = C_1 r_b^n \quad (18\text{-}98)$$

where n is normally between 2 to 3, depending on clarification, classification, thickening, or dewatering. Thus,

$$G = c_2 (V_t)_{\max}^2 Q^{1/n} \quad (18\text{-}99)$$

where c_1 and c_2 are constants. It follows that large centrifuges can deliver high flow rate but separation is at lower G-force; vice versa, smaller centrifuges can deliver lower flow rate but separation is at higher G-force. Also, using higher-strength material for construction

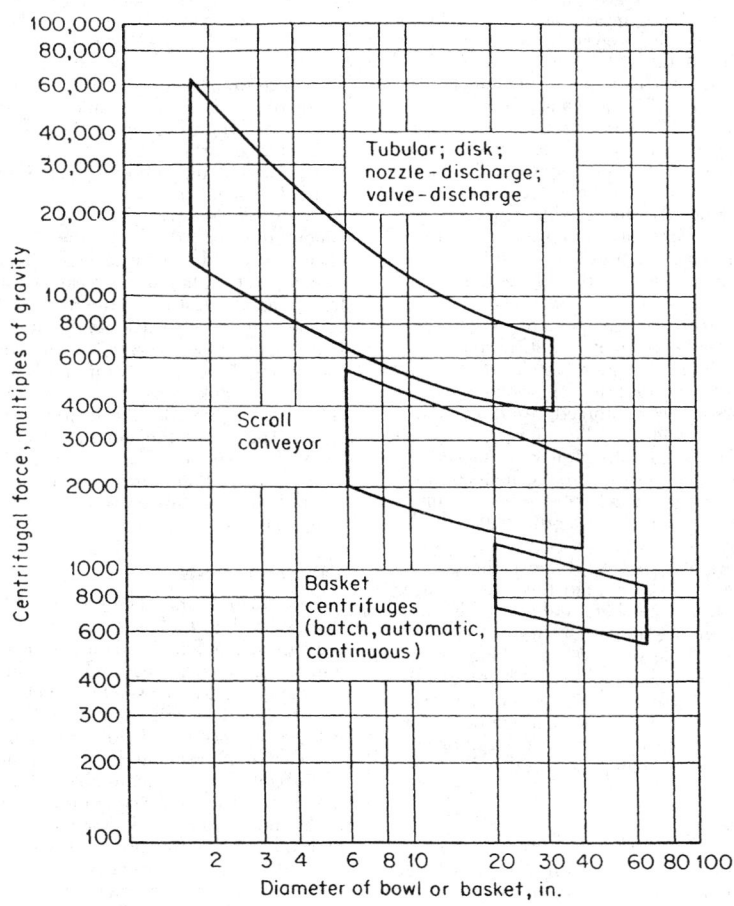

FIG. 18-139   Variation of centrifugal force with diameter in industrial centrifuges.

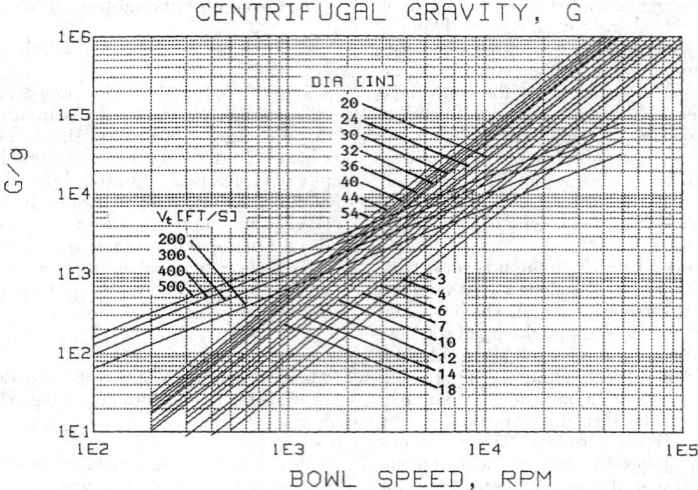

FIG. 18-140 Variation of centrifugal force with r/min.

of the rotating assembly permits higher maximum tip speed, thus allowing higher G-force for separation at a given feed rate.

Centrifuge bowls are made of almost every machinable alloy of reasonably high strength. Preference is given to those alloys having 1 percent elongation to minimize the risk of cracking at stress-concentration points. Typically, the list includes carbon steel, types 304, 316, and 317 stainless steels, alloy 20 (Carpenter) stainless steel, Monel Metal, Inconel, nickel, Hastelloy B, titanium, and alloyed aluminum. Vertical-basket centrifuges are frequently constructed of carbon steel or stainless steel coated with rubber, neoprene, Penton, or Kynar. Casings and feed, rinse, and discharge lines that are stationary and lightly stressed may be constructed of any suitable rigid corrosion-resistant material. Wear-resistant materials—tungsten and cermaic carbide, hard-facing, and others—are often used to protect the bare metal surfaces in high-wear areas such as the blade tips of the decanter centrifuge.

Critical Speeds In the design of any high-speed rotating machinery, attention must be paid to the phenomenon of critical speed. This is the speed at which the frequency of rotation matches the natural frequency of the rotating part. At this speed, any vibration induced by slight unbalance in the rotor is strongly reinforced, resulting in large deflections, high stresses, and even failure of the equipment. Speeds corresponding to harmonics of the natural frequency are also critical speeds but give relatively small deflections and are much less troublesome than the fundamental frequency. The critical speed of simple shapes may be calculated from the moment of inertia; with complex elements such as a loaded centrifuge bowl, it is best found by tests.

Nearly all centrifuges operate at speeds well above the primary critical speed and therefore must pass through this speed during acceleration and deceleration. To permit them to do so safely, some degree of damping in their mounting must be provided. This may result from the design of the spindle or driveshaft alone, spring-loading of the spindle bearing nearest the rotor, elastic loading of the suspension, or a combination of these. Smaller and medium-sized centrifuges of the cream-separator and bottle-centrifuge design are frequently mounted on elastic cushions. Horizontal decanter centrifuges are mounted on isolators with dampers to reduce vibration transmitted to the foundation.

SEDIMENTING CENTRIFUGES

When a spherical particle of diameter d settles in a viscous liquid under earth gravity g, the terminal velocity V_s is determined by the weight of the particle-balancing buoyancy and the viscous drag on the particle in accordance to Stokes' law. In a rotating flow, Stokes' law is modified by the "centrifugal gravity" $G = \Omega^2 r$, thus

$$V_s = \frac{1}{18\mu}\Omega^2 r(\rho_s - \rho_L)d^2 \qquad (18\text{-}100)$$

In order to have good separation or high settling velocity, a combination of the following conditions is generally sufficient:
1. High centrifuge speed
2. Large particle size
3. Large density difference between solid and liquid
4. Large radius
5. Small viscosity

Among the five parameters, the settling velocity is very sensitive to change in speed and particle size. It varies as the square of both parameters. The rotation speed can be increased to the limit as dictated by the maximum stress of the rotor or the periphery equipment, such as gear unit. If the particles in the slurry are too small, coagulation and flocculation by polymers is very effective to build up bigger particles for settling. Large density difference leads to higher settling rate and vice versa. Unlike separation under a constant gravitational field, the settling velocity under a centrifugal field increases linearly with the radius. The greater the radius at which separation takes place, the better is the separation. Sedimentation of particles is favorable in a less viscous liquid. Some processes are run under elevated temperature where liquid viscosity drops to a fraction of its original value at room temperature.

Laboratory Tests

Spin-Tube Tests The objective of the spin-tube test is to check the settleability of solids in a slurry under centrifugation. The clarity of the supernatant liquid and the solid concentration in the sediment can also be evaluated. A small and equal amount of feed is introduced into two diametrically opposite test tubes (typically plastic tubes in a stainless steel holder) with a volume of 15 to 50 mL. The samples are centrifuged at a given G and for a period of time t. The supernatant liquid is decanted off from the spin tubes from which the clarity (in the form of turbidity or any measurable solids, dissolved and suspended) is measured. The integrity—more precisely, the yield stress—of the cake can be determined approximately by the amount of penetration of a rod into the cake under its weight and accounting for the buoyancy effect due to the wet cake. It is further assumed that the rod does not lean on the sides of the tube. The yield stress τ_y of the centrifuged cake can be determined from:

$$\tau_y = \frac{1}{2}\left(\frac{\rho_r L}{h} - \rho_L\right)gr_d \qquad (18\text{-}101)$$

where r_d and L are, respectively, the radius and length of the solid circular rod; ρ_r is the density of the rod; and h is the penetration of the rod into the cake. By using rods of various sizes and densities, yield stress which is indicative of cake handling and integrity can be measured for a wide range of conditions.

The solids recovery in the cake can be inferred from measurements using Eq. (18-90). It is shown as a function of G-seconds for different feed solids concentration in Fig. 18-141; see also theory discussed below.

Imperforate Bowl Tests The amount of supernant liquid from spin tubes is usually too small to warrant accurate gravimetric analysis. A fixed amount of slurry is introduced at a controlled rate into a rotating imperforate bowl to simulate a continuous sedimentation centrifuge. The liquid is collected as it overflows the ring weir. The test is stopped when the solids in the bowl build up to a thickness which affects centrate quality. The solid concentration of the centrate is determined similarly to that of the spin tube.

Transient Centrifugation Theory As in gravitational sedimentation, there are three layers which exist during batch settling of a slurry in a centrifuge: a clarified liquid layer closest to the axis, a middle feed slurry layer with suspended solids, and a cake layer adjacent to the bowl wall with concentrated solids. Unlike with constant gravity g, the centrifugal gravity G increases linearly with radius. It is highest near the bowl and is zero at the axis of rotation. Also, the cylindrical surface area through which the particle has to settle increases linearly with radius. Both of these effects give rise to some rather unexpected results.

Consider the simple initial condition $t = 0$ where the solid concentration ϕ_{so} is constant across the entire slurry domain $r_L \leq r \leq r_b$ where r_L and r_b are, respectively, the radii of the slurry surface and the bowl. At a later time $t > 0$, three layers coexist: the top clarified layer, a middle slurry layer, and a bottom sediment layer. The air-liquid interface remains stationary at radius r_L, while the liquid-slurry interface with radius r_s expands radially outward, with t with r_s given by:

$$\frac{r_s}{r_L} = \sqrt{\frac{\phi_{so}}{\phi_s}} \qquad (18\text{-}102)$$

Eq. (18-102) can be derived from conservation of angular momentum as applied to the liquid-slurry interface.

Interestingly, the solid concentration in the slurry layer ϕ_s does not remain constant with time as in gravitational sedimentation. Instead, ϕ_s decreases with time uniformly in the entire slurry layer in accordance to:

$$\frac{\phi_s}{\phi_{smax}} = 1 - \left[1 - \left(\frac{\phi_{so}}{\phi_{smax}}\right)\right]e^{2\xi} \qquad (18\text{-}103)$$

where ξ is a dimensionless time variable:

$$\xi = \left(\frac{V_{go}t}{r_b}\right)\left(\frac{G}{g}\right) \qquad (18\text{-}104)$$

In Eqs. (18-103) and (18-104), under hindered settling and 1 g, the solids flux $\phi_s V_s$ is assumed to be a linear function of ϕ_s decreasing at a rate of V_{go}. Also, the solids flux is taken to be zero at the "maximum" solids concentration ϕ_{smax}. As $G/g \gg 1$, this solids flux behavior based on 1 g is assumed to be ratioed by G/g.

Concurrent with the liquid-slurry interface moving radially outward, the cake layer builds up with the cake-slurry interface moving radially inward, with radial position given by:

$$\frac{r_c}{r_b} = \sqrt{\frac{(\varepsilon_s - \phi_{so})}{(\varepsilon_s - \phi_s)}} \qquad (18\text{-}105)$$

where ε_s is a constant cake solids concentration. Sedimentation stops when the growing cake-slurry interface meets the decreasing slurry-liquid interface with $r_c = r_s$. This point is reached at $\phi_s = \phi_s^\circ$ and $t = t^\circ$ when

$$\frac{1}{\phi_s^\circ} = \frac{1}{\varepsilon_s} + \left(\frac{r_b}{r_L}\right)^2\left(\frac{1}{\phi_{so}} - \frac{1}{\varepsilon_s}\right) \qquad (18\text{-}106)$$

$$t^\circ = \frac{1}{2}\left(\frac{gr_b}{GV_{go}}\right)\ln\left(\frac{\phi_{smax} - \phi_s^\circ}{\phi_{smax} - \phi_{so}}\right) \qquad (18\text{-}107)$$

Example
Calcium carbonate–water slurry
  $G/g = 2667$
  $V_{go} = 1.31 \times 10^{-6}$ m/s (5.16×10^{-5} in/s)

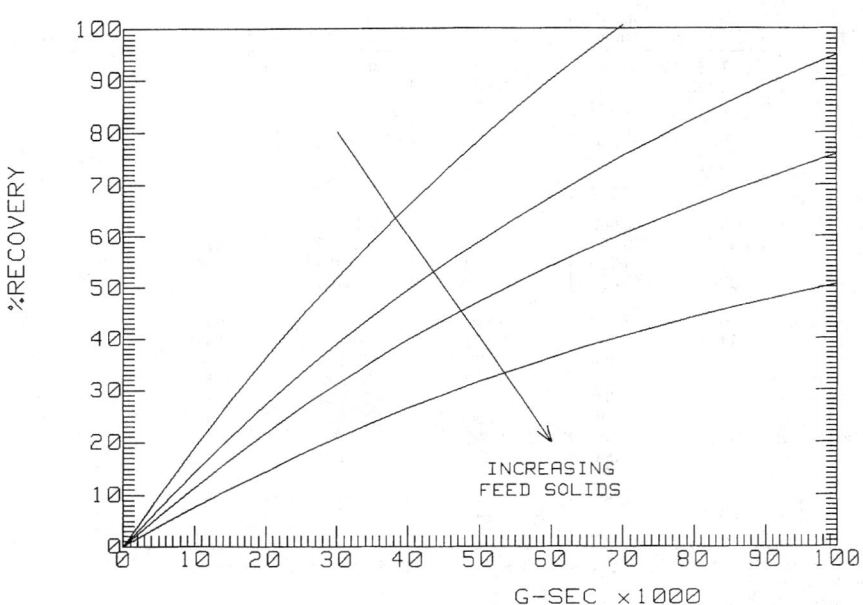

FIG. 18-141 Recovery as a function of G-seconds for centrifugal sedimentation.

$\phi_{smax} = 0.26$ (with $\phi_s V_g = 0$)
$\phi_{so} = 0.13$
$r_L = 0.0508$ m (2 in)
$r_b = 0.1016$ m (4 in)
$\xi = 2667 \, (1.31 \times 10^{-6}) \, (1/0.1016) = 0.0344$
$\varepsilon_s = 0.52$

$t(s)$	ξ	ϕ_s/ϕ_{smax}	r_s (m)	r_c (m)
0.0	0	0.50	0.051	0.102
1.0	0.034	0.46	0.053	0.100
5.0	0.173	0.29	0.067	0.095
7.6	0.261	0.16	0.091	0.091

There are six types of industrial sedimenting centrifuges:
Tubular-bowl centrifuges
Multichamber centrifuges
Skimmer pipe/knife-discharge centrifuges
Disk centrifuges
Decanter centrifuges
Screenbowl centrifuges
The first three types, including the manual-discharge disk, are batch-feed centrifuges, whereas the latter three, including the intermittent and nozzle-discharge disks, are continuous centrifuges.

Tubular-Bowl Centrifuges The tubular-bowl centrifuge is widely employed for purifying used lubricating and other industrial oils and in the food, biochemical, and pharmaceutical industries. Industrial models have bowls 102 to 127 mm (4 to 5 in) in diameter and 762 mm (30 in) long (Table 18-12). It is capable of delivering up to 18,000g. The smallest size, 44 mm × 229 mm (1.75 in × 9 in bowl), is a laboratory model capable of developing up to 65,000g. It is also used for separating difficult-to-separate biological solids with very small density difference, such as cells and virus.

The bowl is suspended from an upper bearing and drive (electric or turbine motor) assembly through a flexible-drive spindle with a loose guide in a controlled damping assembly at the bottom. The unit finds its axis of rotation if it becomes slightly unbalanced due to process load.

The feed slurry is introduced into the lower portion of the bowl through a small orifice. Immediately downstream of the orifice is a distributor and a baffle assembly which distribute and accelerate the feed to circumferential speed. The centrate discharges from the top end of the bowl by overflowing a ring weir. Solids that have sedimented against the bowl wall are removed manually from the centrifuge when the buildup of solids inside the bowl is sufficient to affect the centrate clarity.

The liquid-handling capacity of the tubular-bowl centrifuge varies with use. The low end shown in Table 18-11 corresponds to stripping small bacteria from a culture medium. The high end corresponds to purifying transformer oil and restoring its dielectric value. The solids-handling capacity of this centrifuge is limited to 4.5 kg (10 lb) or less. Typically, the feed stream solids should be less than 1 percent in practice.

Multichamber Centrifuges While the tubular has a high aspect ratio (i.e., length-to-diameter ratio) of 5 to 7, the multichamber centrifuges have aspect ratios of 1 or less. The bowl driven from below consists of a series of short tubular sections of increasing diameter nested to form a continuous tubular passage of stepwise increasing diameter for the liquid flow. The feed is introduced at the center tube and gradually finds its way to tubes with larger diameters. The larger and denser particles settle out in the smaller-diameter tube, while the smaller and lighter particles settle out in the larger-diameter tubes. Classification of particles can be conveniently carried out. Clarification may be significantly improved by spacing especially the outer tubes more closely together to reduce the settling distance, a concept which is fully exploited by the disk-centrifuge design. This also serves to maintain a constant velocity of flow between adjacent tubes. As much as six chambers can be accommodated. The maximum solids-holding capacity is 0.064 m³ (17 gal). The most common use is for clarifying fruit juices, wort, and beer. For these services it is equipped with a centripetal pump at effluent discharge to minimize foaming and contact with air.

Knife-Discharge Centrifugal Clarifiers Knife-discharge centrifuges with solid instead of perforated bowls are used as sedimenting

TABLE 18-12   Specifications and Performance Characteristics of Typical Sedimenting Centrifuges

Type	Bowl diameter	Speed, r/min	Maximum centrifugal force × gravity	Throughput		Typical motor size, hp
				Liquid, gal/min	Solids, tons/h	
Tubular	1.75	50,000°	62,400	0.05–0.25		°
	4.125	15,000	13,200	0.1–10		2
	5	15,000	15,900	0.2–20		3
Disk	7	12,000	14,300	0.1–10		1/3
	13	7,500	10,400	5–50		6
	24	4,000	5,500	20–200		7½
Nozzle discharge	10	10,000	14,200	10–40	0.1–1	20
	16	6,250	8,900	25–150	0.4–4	40
	27	4,200	6,750	40–400	1–11	125
	30	3,300	4,600	40–400	1–11	125
Helical conveyor	6	8,000	5,500	To 20	0.03–0.25	5
	14	4,000	3,180	To 75	0.5–1.5	20
	18	3,500	3,130	To 100	1–3	50
	24	3,000	3,070	To 250	2.5–12	125
	30	2,700	3,105	To 350	3–15	200
	36	2,250	2,590	To 600	10–25	300
	44	1,600	1,600	To 700	10–25	400
	54	1,000	770	To 750	20–60	250
Knife discharge	20	1,800	920	†	1.0‡	20
	36	1,200	740	†	4.1‡	30
	68	900	780	†	20.5‡	40

°Turbine drive, 100 lb/h (45 Kg/h) of steam at 40 lbf/in² gauge (372 KPa) or equivalent compressed air.
†Widely variable.
‡Maximum volume of solids that the bowl can contain, ft³.
NOTE: To convert inches to millimeters, multiply by 25.4; to convert revolutions per minute to radians per second, multiply by 0.105; to convert gallons per minute to liters per second, multiply by 0.063; to convert tons per hour to kilograms per second, multiply by 0.253; and to convert horsepower to kilowatts, multiply by 0.746.

centrifuges. The liquid flow is usually continuous until the settled solids start to interfere with the effluent liquid. The feed enters the hub end and is accelerated to speed before introducing to the separation pool. The solids settle out to the bowl wall and the clarified liquid overflows the ring weir or discharges through a skimmer pipe. In some designs, internal baffles in the bowl are required to stop wave action primarily along the axial direction. When sufficient thick solid layer has built up inside the bowl, the supernatant liquid is skimmed off by moving the opening of the skimmer pipe radially inward. After the liquid is sucked out, the solids are knifed out as with centrifugal filters. However, unlike centrifugal filters, the cake is always fully saturated with liquid, $S = 100\%$. These centrifuges are used for coarse, fast-settled solids. When greater clarification effectiveness is required, the operation may be totally batchwise with prolonged spinning of each batch. If the solids content in the feed is low, several batches may be successively charged and the resulting supernatant liquor skimmed off before unloading of the accumulated solids.

Commercial centrifuges of this type have bowl diameters ranging from 0.3 to 2.4 m (12 to 96 in). The large sizes are used on heavy-duty applications such as coal dewatering and are limited by stress considerations to operate at 300g. The intermediate sizes for chemical process service develop up to 1000g (see Table 18-10).

Disk Centrifuges   One of the commonest clarifier centrifuges is the vertically mounted disk machine, as illustrated in Fig. 18-142a, b. Feed is introduced proximate to the axis of the bowl, accelerated to speed typically by a radial vane assembly, and flows through a stack of closely spaced conical disks in the form of truncated cones. Generally 50 to 150 disks are used. They are spaced 0.4 to 3 mm (0.015 to 0.125 in) apart to reduce the distance for solid/liquid separation. The angle made by conical disks with the horizontal is typically between 40 to 55° to facilitate solids conveyance. Under centrifugal force the solids settle against the underside of the disk surface and move down to the large end of the conical disk and subsequently to the bowl wall. Concurrently, the clarified liquid phase moves up the conical channel. Each disk carries several holes spaced uniformly around the circumference. When the disk stack is assembled, the holes provide a continuous upward passage for the lighter clarified liquid released from each conical channel. The liquid collects at the top of the disk stack and discharges through overflow ports. To recover the kinetic energy and avoid foaming due to discharging of a high-velocity jet against a stationary casing, the rotating liquid is diverted to a stationary impeller from which the kinetic energy of the stream is converted to hydrostatic pressure. Unlike most centrifuges operating with a slurry pool in contact with a free surface, disk centrifuges with a rotary seal arrangement can operate under high pressure. The settled solids at the bowl wall are discharged in different forms, depending on the type of disk centrifuges.

Manual Discharge Disk Centrifuges   In the simplest design shown in Fig. 18-142a, the accumulated solids must be removed manually on a periodic basis, similar to that for the tubular-bowl centrifuge. This requires stopping and disassembling the bowl and removing the disk stack. Although the individual disks rarely require cleaning, manual removal of solids is economical only when the fraction of solids in the feed is very small.

Intermittent-Discharge Disk Centrifuges   In the intermittent-discharge-disk centrifuge, solids discharge through ports at the periphery of the bowl, the ports opening on a timed cycle while the bowl is at speed. In another design, the bottom of the bowl drops, exposing an annular opening where accumulated solids are discharged. The bowl closes and the cycle repeats.

Nozzle-Discharge Disk Centrifuges   In the nozzle-discharge disk centrifuge, solids are discharged continuously, along with a portion of the liquid phase, through nozzles spaced around the periphery of the bowl, which are tapered radially outward, providing a space for solids storage (see Fig. 18-142b). The angle of repose of the sedimented solids determines the slope of the bowl walls for satisfactory operation. Clarification efficiency is seriously impaired if the buildup of solids between nozzles reaches into the disk stack. The nozzle diameter should be at least twice the diameter of the largest particle to be processed, and prescreening is recommended of extraneous solids. Typically, nozzle diameters range from 0.6 to 3 mm (0.25 to

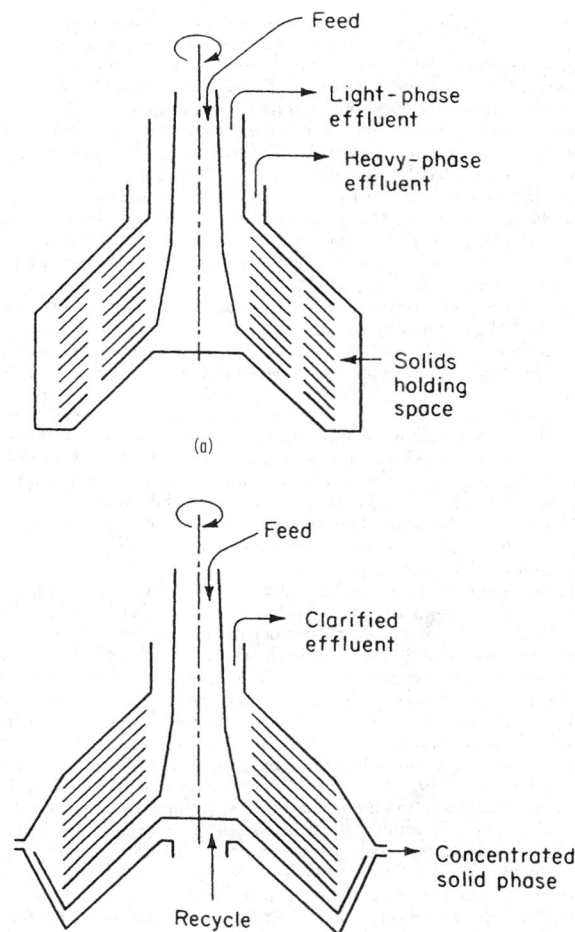

FIG. 18-142   Disk-centrifuge bowls: (a) separator, solid wall; (b) recycle clarifier, nozzle discharge.

0.125 in). Large disk centrifuges may have as much as 24 nozzles spaced out at the bowl.

For clarification of a single liquid phase with controlled concentration of the discharged slurry, a centrifuge which provides recirculation is used (see Fig. 18-142b). A fraction of the sludge discharged out of the machine is returned to the bowl to the area adjacent to the nozzles through lines external to the machine as well as built-in annular passages at the periphery of the bowl. This has the effect of preloading the nozzles with sludge which has already been separated and reduces the net flow of liquid with the newly sedimented solids from the feed. Increased concentration can be obtained alternatively by recycling a portion of the sludge to the feed, but this increases solids loading at the disk stack, with a corresponding sacrifice in the effluent clarity for a given feed rate.

With proper rotary seals, the pressure in the machine can be contained up to 1.1 MPa (150 psig) or higher. Also, operating temperature can be as high as 315°C (about 600°F). The rotating parts are made of stainless steel with the high-wear nozzles made of tungsten carbide. The bowls may be underdriven or suspended and range from several centimeters to over 1 m (3.3 ft) in outer diameter. The largest size capable of clarifying up to 1920 L/m (500 gpm) requires 112 kW (150 hp).

Decanter Centrifuges   The decanter centrifuge (also known as the **solid-bowl** or **scroll** centrifuge) consists of a solid bowl with a screw or scroll conveyor between the solid- and the liquid-bowl heads, or hubs (see Fig. 18-143). Both the bowl and the conveyor rotate at a high speed, yet there is a difference in speed between the two, which is responsible for conveying the sediment along the machine from the cylinder to the conical discharge end. The rotating assembly is commonly mounted horizontally with bearings on each end. Some centrifuges are vertically mounted with the weight of the rotating assembly supported by a single bearing at the bottom or with the entire machine suspended from the top. With the former configuration, the weight of the rotating assembly provides a good sealing surface at the bearing for high-pressure applications. The bowl may be conical in shape or, in most instances, it has combined conical and cylindrical sections (see Fig. 18-143).

Slurry is introduced into the feed accelerator through a stationary pipe located proximate to the axis of the machine. The feed slurry is accelerated through contact with the rotating surfaces to angular speed before discharging to the separation pool through a series of ports in the conveyor hub. In the separation pool, under centrifugal gravity the solids which are heavier compared to the liquid settle toward the bowl wall, while the clarified liquid moves radially toward the pool surface. Subsequently, the liquid flows along the helical channel (or channels, if the screw conveyor has multiple leads) formed by adjacent blades of the conveyor to the liquid bowl head, from which it discharges over the weirs. The annular pool can be changed by adjusting the radial position of the weir openings, which take the form of circular holes or crescent-shaped slots.

The cake solids adjacent to the bowl wall are transported by the differential speed from the cylinder up the cone, also known as the **beach.** The half cone angle ranges between 5° and 20°. The cake is submerged in the pool when it is in the cylinder and at the beginning of the beach. In this region, liquid buoyancy helps to reduce the effective weight of the cake under centrifugal gravity, resulting in lower conveyance torque. Farther up the beach, the cake emerges above the pool and moves along the "dry beach," where buoyancy force is absent, resulting in more difficult conveyance and higher torque. But it is also in this section that the cake is dewatered, with expressed liquid returned back to the pool. The centrifugal force helps to dewater, yet at the same time hinders the transport of the cake in the dry beach. Therefore, a balance in cake conveyance and cake dewatering is the key in setting the pool and the *G*-force for a given application. Also, clarification is important in dictating this decision.

The cylindrical section provides clarification under high centrifugal gravity. In some cases, the pool should be shallow to maximize the *G*-force for separation. In other cases, when the cake layer is too thick inside the cylinder, the settled solids—especially the finer particles at the cake surface—entrain into the fast-moving liquid stream above, which eventually ends up in the centrate. A slightly deeper pool

becomes beneficial in these cases because there is a thicker buffer liquid layer to ensure settling of resuspended solids. This can be at the expense of cake dryness due to reduction of the dry beach. Consequently, there is again a compromise between centrate clarity and cake dryness. Another reason for the tradeoff of centrate clarity with cake dryness is that, in losing fine solids to the centrate (i.e., classification), the cake with larger particles, having less surface-to-volume ratio, can dewater more effectively, resulting in drier cake. It is best to determine the optimal pool for a given application through tests.

The speed with which the cake transports is controlled by the differential speed. High differential speed facilitates high solids throughput where the cake thickness is kept to a minimum so as not to impair centrate quality due to entrainment of fine solids. Also, cake dewatering is improved due to a reduction in the drainage path with smaller cake height; however, this is offset by the fact that higher differential speed also reduces cake residence time, especially in the dry beach. The opposite holds for low differential speed. Therefore, an optimal differential speed is required to balance centrate clarity and cake dryness. The desirable differential speed is usually maintained using a two-stage planetary gearbox, the housing of which rotates with the bowl speed, with a fixed first-stage pinion shaft. In some applications, the pinion is driven by an electrical backdrive (dc or ac), hydraulic backdrive, or braked by an eddy-current device at a fixed rotation speed. The differential speed is then the difference in speed between the bowl and the pinion divided by the gear ratio. This also applies to the case when the pinion arm is held stationary, in which the pinion speed is zero. The torque at the spline of the conveyor, conveyance torque, is equal to the product of the pinion torque and the gear ratio. Higher gear ratio gives lower differential speed, and vice versa; lower gear ratio gives higher differential for higher solids capacity. The torque at the pinion shaft has been used to control the feed rate or to signal an overload condition by shearing of a safety pin. Under this condition, both the bowl and the conveyor are bound to rotate at the same speed (zero differential) with no conveyance torque and no load at the pinion.

Soft solids, most of which are biological waste such as sewage, are difficult to convey up the beach. Annular baffles or dams have been commonly used to provide a pool-level difference wherein the pool is deeper upstream of the baffle toward the clarifier and lower downstream of the baffle toward the beach. The pool-level difference across the baffle, together with the differential speed, provide the driving force to convey the compressible sludge up the beach. This has been used effectively in thickening of waste-activated sludge and in some cases of fine clay with dilatant characteristics.

High solids decanter centrifuges have been used to dewater mixed raw sewage sludge (with volume ratio of primary to waste-activated sludge such as 50 percent to 50 percent or 40 percent to 60 percent, etc.), aerobically digested sludge, and anaerobically digested sludge.

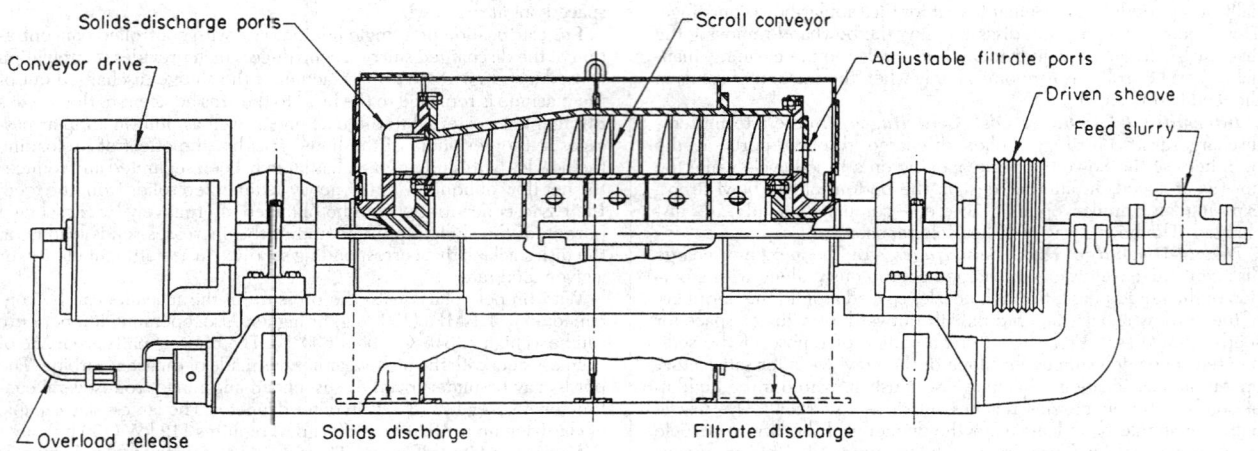

FIG. 18-143   Cylindrical-conical solid-bowl centrifuge. (*Bird Machine Co.*)

Cake solids as dry as 28 percent to 35 percent by weight are obtained for raw mixed sludge and 20 percent to 28 percent for the digested sludges, with the aerobic sludge at the lower end of the range. The typical characteristics of high-solids applications are: low differential speed (0.5 to 3 r/min), high conveyance torque, high polymer dosage (10 to 30 lb dry polymer/ton dry solids, depending on the feed sewage), and slightly lower volumetric throughput rate. An electrical (dc or ac with variable-frequency drive) or hydraulic backdrive on the conveyor with high torque capacity is essential to operate these conditions at steady state.

The horizontal decanter centrifuge is operated below its critical speed. The bowl is mounted between fixed bearings anchored to a rigid frame. The gearbox is cantilevered outboard of one of these bearings, and the feed pipe enters the rotating assembly through the other end. Particularly in the larger sizes, the frame is connected to ground through vibration isolators. In the vertical configuration, the bowl and the gearbox are suspended from the drive head, which is connected to the frame and casing through vibration isolators. A clearance bushing at the bottom limits the excursion of the bowl during start-up and shutdown but does not provide the radial constraint of a bearing under normal operating conditions.

Decanter centrifuges with mechanical shaft-to-casing seals are available for pressure containment up to 1.1 MPa (150 psig), similar to the nozzle-disk centrifuge. They can be built to operate at temperatures from −87 to +260°C (−125 to +500°F).

When abrasive solids are processed, the points of wear are protected with a replaceable hard facing such as Colmonoy, Hastelloy, tungsten carbide, or ceramic tiles. These high-wear areas include the feed zone; the conveyor blade tip, especially the pressure and pushing face; the conical beach; and the solids discharge ports. Transport of solids is encouraged in some applications by longitudinal strips or grooves at the inner diameter of the bowl, especially at the beach, to enhance the frictional characteristics between the sediment and the bowl surface, and by polished conveyor faces to reduce frictional drag. For fluidlike sediment cake, by using the strips in the beach, a much tighter gap between the conveyor blade tip and the bowl surface is possible with a cake heel layer trapped by the strips. This reduces leakage of the fluid sediment flowing through an otherwise larger gap opening to the pool. Gypsum coating on the bowl wall at the beach section has been used to achieve the same objective.

Washing in a continuous decanter is fairly effective on solid particles larger than 80 μm (200 mesh), provided the particles are reasonably uniform in size with porous structure. Otherwise, the wash flows across the cake surface with little penetration because the pores at the cake surface are plugged by fines. Rinsing efficiency, the proportion of soluble impurities displaced from the solids, is in the range of 50 to 80

percent, depending on cake porosity, permeability, and mass-transfer rate. A wash-solids weight ratio of up to 0.25 is required.

Various bowl configurations with a wide range of aspect ratios—i.e., length-to-diameter ratio—from less than 1 to 4 are available for specific applications, depending on whether the major objective is maximum clarification, classification, or solids dryness. Generally, the movement of liquid and solids is in countercurrent directions, but in the cocurrent design the movement of liquid is in the same direction as that of the solids. In this design, the feed is introduced at the large end of the machine and the centrate is taken by a skimmer at the beach-cylinder junction. The settled solids transverse the entire machine and discharge at the beach exit. Compound angle beaches are used in specific applications such as washing and drying of polystyrene beads. The pool level is located at the intersection of the two angles at the beach, the steeper angle being under the pool and the shallower angle above the pool (i.e., dry beach), allowing a longer dewatering time. The wash is applied at the pool side of the beach-angle intersection and functions as a continuously replenished annulus of wash liquid through which the solids are conveyed. The size of decanter centrifuge ranges from 6 in diameter to 54 in. The larger the machine, the slower is the speed, and less is the G-force (Fig. 18-141). However, it provides a much higher throughput capacity, which cannot be accommodated with smaller machines (see Table 18-10).

Screenbowl Centrifuges   The screenbowl centrifuge consists of a solid-bowl decanter to which, at the smaller conical end, a cylindrical screen has been added (see Fig. 18-144). The scroll spans the entire bowl, conforming to the profile of the bowl. It combines a sedimenting centrifuge together with a filtering centrifuge. Therefore, the solids which are processed are typically larger than 23 to 44 μm.

As in a decanter, an accelerated feed is introduced to the separation pool. The denser solids settle toward the bowl wall and the effluent escapes through the ports at the large end of the machine. The sediment is scrolled toward the beach, typically with a steeper angle compared to the decanter centrifuge. As the solids are conveyed to the screen section, the liquid in the sediment cake further drains through the screen, resulting in drier cake. Washing of the sediment in the first half of the screen section is very effective in removing impurities, with the second half of the screen section reserved for dewatering of mother liquor and wash liquid.

The screen is typically constructed of a wedge-bar with an aperture between adjacent bars, which opens up to a larger radius. This prevents solids from blinding the screen as well as reduces conveyance torque. For abrasive materials such as coal, the screens are made of wear-resistant materials such as tungsten carbide.

Continuous Centrifugal Sedimentation Theory   The Stokes settling velocity of a spherical particle under centrifugal field is given

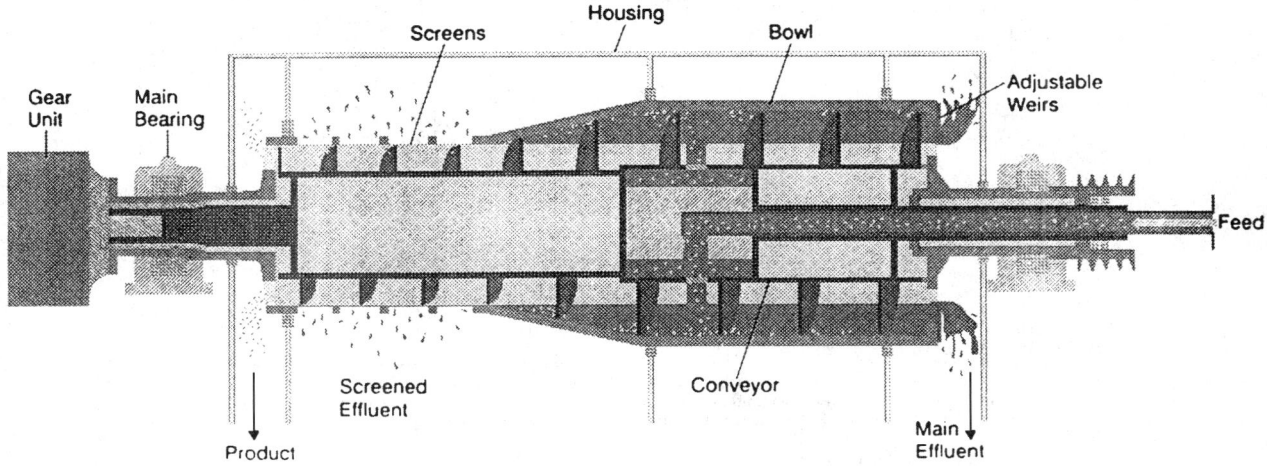

FIG. 18-144 Cylindrical-conical screen-bowl centrifuge. (*Bird Machine Co.*)

by Eq. (18-100). Useful relationships have been established on continuous sedimentation by studying the kinematics of settling of a spherical particle of diameter d in an annular rotating pool. Equating the time rate of change in a radial position to the settling velocity, and the rate of change in an axial position to bulk-flow velocity, thus gives

$$\frac{dr}{dt} = crd^2 \qquad (18\text{-}108)$$

$$\frac{dx}{dt} = \frac{Q}{\pi\,(r_b^2 - r_p^2)} \qquad (18\text{-}109)$$

where $c = (\rho_s - \rho_L)\,\Omega^2/18\mu$, x is distance along the axis of the bowl, Q is the volumetric feed rate, r_b and r_p are, respectively, the bowl and pool surface radii. For a particle located at one end of the bowl at radius r with $r_p < r < r_b$, after transversing the full bowl length, it settles out and is captured by the bowl wall. Solving the above equations with these boundary conditions, the limiting trajectory is:

$$\frac{r}{r_b} = \exp\left[\frac{-\pi\,cL(r_b^2 - r_p^2)d^2}{Q}\right] \qquad (18\text{-}110)$$

If the same size particle d is located at an initial starting radius less than r given by Eq. (18-110) it is assumed to escape from being captured by the bowl, whereas it would have been captured if it had been at an initial radius greater than r. Assuming that the number of particles with size d is uniformly distributed across the annular pool, the recovery Rec_d (known also as **grade efficiency**) is the differential of the cumulative recovery $Z = 1 - Y$, with Y given in Eq. (18-92) for particles with size d, as the ratio of the two annular areas:

$$\mathrm{Rec}_d = \frac{r_b^2 - r^2}{r_b^2 - r_p^2} \qquad (18\text{-}111)$$

Combining Eqs. (18-110) and (18-111), the maximum Q to the centrifuge, so as to meet a given recovery Rec_d of particles with diameter d, is

$$\frac{Q_d}{2V_{gd}} = \left(\frac{\pi\,\Omega^2 L}{g}\right)\left(\frac{r_p^2 - r_b^2}{\ln\{1 - \mathrm{Rec}_d[1 - (r_p/r_b)^2]\}}\right) = \Sigma_{\mathrm{Rec}_d} \qquad (18\text{-}112)$$

Note in Eq. (18-112) that V_{gd} is the settling rate under $1g$, and it is a function of the particle size and density and fluid properties. The ratio $Q_d/2V_{gd}$ is then related only to the operating speed and geometry of the centrifuge, as well as to the size recovery. It measures the *required*

surface area for settling under centrifugal gravity to meet a specified Rec_d. When the size recovery Rec_d is set at 50 percent, the general result, Eq. (18-112), reduces to the special case, which is the well-known Ambler's Sigma factor, which for a straight rotating bowl (applicable to bottle centrifuge, decanter centrifuge, etc.) is:

$$\Sigma = \left[\frac{\pi\,\Omega^2 L}{g}\right]\left[\frac{r_b^2 - r_p^2}{\ln\left[2r_b^2/(r_b^2 + r_p^2)\right]}\right] \qquad (18\text{-}113)$$

It can be simplified to:

$$\Sigma = \pi\,\Omega^2 L\,\frac{(3r_b^2 + r_p^2)}{2g} \qquad (18\text{-}114)$$

For a disk centrifuge, a similar derivation results in

$$\Sigma = \frac{2\pi\,\Omega^2\,(N-1)\,(r_2^3 - r_1^3)}{3g\tan\theta} \qquad (18\text{-}115)$$

where N is the number of disks in the stack, r_1 and r_2 are the outer and inner radii of the disk stack, and θ is the conical half-angle.

Typical Σ factors for the three types of sedimenting centrifuges are given in Table 18-13. In scale-up from laboratory tests, sedimentation performance should be the same if the value of Q/Σ is the same for the two machines. This is a widely used criterion for the comparison of centrifuges of similar geometry and liquid-flow patterns developing approximately the same G; however, it should be used with caution when comparing centrifuges of different configurations. In general, the shortcomings of the theory are due to the oversimplified assumptions being made, such as (1) there is an idealized plug-flow pattern; (2) sedimentation abides by Stokes law as extended to many g's; (3) feed solids are uniformly distributed across the surface of the bowl head at one end of the clarifier and capture implies that the particles' trajectory intersect the bowl wall; (4) the feed reaches full tangential speed as it is introduced to the pool; (5) the recovery of given-size particles is at 50 percent; (6) this does not account for possible entrainment of already settled particles in the liquid stream; (7) there is absence of entrance and exit effects.

Experience in using the Σ concept has demonstrated that the calculated Σ factor should be modified by an efficiency factor to account for some of the aforementioned effects which are absent in the theory and, as such, this factor depends on the type of centrifuge. It is nearly 100 percent for simple spin-tube bottle centrifuge, 80 percent for tubular centrifuge, and less than 55 percent for disk centrifuges. The

TABLE 18-13 Scale-up Factors for Sedimenting Centrifuges

Type of centrifuge	Inside diameter, in	Disk diameter, in/number of disks	Speed, r/min	Σ value, units of 10^4 ft²	Recommended scale-up factors*
Tubular	1.75	—	23,000	0.32	1†
Tubular	4.125	—	15,000	2.7	21
Tubular	4.90	—	15,000	4.2	33
Disk	—	4.1/33	10,000	1.1	1
Disk	—	9.5/107	6,500	21.5	15
Disk	—	12.4/98	6,250	42.5	30
Disk	—	13.7/132	4,650	39.3	25
Disk	—	19.5/144	4,240	105	73
Helical conveyor	6	—	6,000	0.27	1
Helical conveyor	14	—	4,000	1.34	5
Helical conveyor	14‡	—	4,000	3.0	10
Helical conveyor	18	—	3,450	3.7	12.0
Helical conveyor	20	—	3,350	4.0	13.3
Helical conveyor	25	—	3,000	6.1	22
Helical conveyor	25	—	2,700	8.6	31

*These scale-up factors are relative capacities of centrifuges of the same type but different sizes when performing at the same level of separation achievement (e.g., same degree of clarification). These factors must not be used to compare the capacities of different types of centrifuges.

†Approaches 2.5 at rates below mL/min.

‡Long bowl configuration.

NOTE: To convert inches to millimeters, multiply by 25.4; to convert revolutions per minute to radians per second, multiply by 0.105; and to convert 10^4 square feet to square meters, multiply by 929.

efficiency varies widely for decanter centrifuges, depending on cake conveyability and other factors.

FILTERING CENTRIFUGES

As in sedimenting centrifuges, heavier solids settle to the bowl to form a cake layer. In filtering centrifuges, the bowl wall is a screen, a perforated surface, or, in general, a filtering medium. Under a centrifugal field, the liquid above, as well as that trapped in the cake, flows through the cake and the filter. Because the solids are coarser as compared to those processed by sedimenting centrifuges, settling under centrifugal gravity is quick, leaving filtration as the limiting step of the process. Washing and subsequent dewatering of the cake are very common for filtering centrifuges. Filtering centrifuges are also known as centrifugal filters, **wringers, extractors,** or **dryers.** There is a difference in the various types of machines according to whether the feed is batch, intermittent, or continuous, and in the manner in which the cake solids are removed from the basket. The operating range of filtering centrifuges is shown in Table 18-14.

Batch Centrifuges Despite the tendency of the industry to be in favor of continuous filtering centrifuges to reduce downtime and increase productivity, batch-filtering centrifuges still share a significant market due to improvement in modern control and measurement technology, allowing certain operational flexibilities which are unique in the batch machines.

Unlike continuous-filtering centrifuges, the duration and centrifugal gravity for filtration, washing, and final dewatering in batch centrifuges are adjustable. This facilitates qualitative optimization of the processed products and adjustment to varying product requirements. Surge tanks with controlled inlet and a bypass loop permit integration of batch centrifuges into continuous processes.

Variable-Speed Basket Centrifuges A typical cycle of a variable-speed basket centrifuge is shown in Fig. 18-145. The basket accelerates to medium speed, after which feed slurry is introduced. Subsequently, both the slurry and the basket are further accelerated to a higher operating speed. Under the high G-force, the cake is washed and spun dry. Afterwards the basket decelerates to a low speed at which the cake is unloaded. The cycle is repeated. The rate of acceleration depends on the inertia of the basket and load, as well as the available torque from the drive; whereas the rate of deceleration depends on the inertia of the rotating mass and the available braking force. The unloading time depends on the torque available for turning the basket at unloading speed, the rheological properties of the cake to be unloaded, and the design of the unloading knife. The remainder of the cycle depends on the processing characteristics of the slurry to be centrifuged—the filtration rate of the mother liquor, the wash and the drain rate of the wash liquor, and the final drying of the cake to a residual moisture level. The operating cycle may be either manual or fully automatic through a sequence of programmed steps employing reset timers, speed sensors, and limit switches.

Variable-speed basket centrifuges rotate on a vertical axis. The cylindrical perforated basket, connected at one end to a bowl head or hub, is driven from below or suspended from above by a driveshaft. It is capped at the other end with a weir ring. The hub may be a solid piece or with an annular opening. If the hub is a solid piece, the slurry is introduced into the basket at low speed despite the fact that the earth's gravity may influence the annular pool. The settled solids can be removed from the top only when the machine is at rest. Otherwise,

the feed has to be introduced when the machine is at high rotation speed so that an annular pool forms instantly. With an open bottom, settled solids are removed from the bottom either manually while the basket is at rest or by an unloader knife while the basket is rotating at a low speed of less than 100 r/min. The filter medium can be screen or filter cloth with a support.

Solid-Bottom Basket Centrifuges Smaller-scale, solid-bottom batch basket centrifuges are available for small test samples, when the sample cannot tolerate mechanical handling or when the traces of solids remaining in a more automated centrifuge would be subject to decomposition or spoilage.

Base-Bearing Centrifuges The driveshaft of the centrifuge is supported from below on a thrust bearing, which is often held and pivoted in a ball joint. The shaft is centered by radial springs or rubber in compression. This provides damped freedom of motion of the axis of rotation to compensate for the out-of-balance condition of a basket load. This type of centrifuge is used extensively in **chip wringers** that recover excess oil from metal chips and turnings. The basket wall is usually solid and tapered radially outward toward the top, ending in an annular lid wherein the oil is freely discharged but the chip is withheld. A typical unit is 660 mm (26 in) in diameter at the top and 584 mm (23 in) at the bottom, holds up to 0.15 m^3 (5 ft^3) or about 225 kg (500 lb) of crushed steel chips and is driven at 1025 r/min (370g) by a 7.5-kW (10-hp) motor.

Link-Suspended Basket Centrifuges In centrifuges with diameters larger than 762 mm (30 in), the basket, curb, curb cover, and drive form a rigid assembly flexibly suspended from three fixed posts (also known as a **three-column centrifuge**). The three suspension members may be either chain links or stiff rods in ball-and-socket joints and are spring-loaded. The suspended assembly has restrained freedom to oscillate to compensate for a normal out-of-balance condition. The drive is vertical with more efficient power transmission compared to the base-bearing type.

This type of centrifuge is usually loaded at zero speed. Therefore, loading should be uniform inside the basket, after which the machine is brought up to operating speed and maintained for a period until the free liquid has drained off through an opening in the bottom of the curb. The basket is brought to rest and the cover lid is opened for unloading. This may be facilitated if the filter medium is in the form of a bag contoured to fit the inside of the basket.

Link-suspended centrifuges with solid bottoms are available with inside diameters ranging from 305 to 2743 mm (12 to 108 in).

Open-Bottom Basket Centrifuges These centrifuges are made in top-suspended and link-suspended configurations. In both, the bottom bowl head consists of three functional components contained in a single fabrication: (1) the central nave by which the basket is attached to the driveshaft, (2) an outer ring to which the cylindrical shell is attached and whose inside diameter is less than that of the liquid ring weir on the opposite end of the basket, and (3) spokes connecting the nave to the outer ring. A typical cycle follows that shown in Fig. 18-145. The control of the cycle may be manual, semiautomatic, or fully automatic. The drive may be a variable-speed electric motor, either direct or through V-belts; a high-pressure, fixed-volume hydraulic motor receiving its energy from a constant-speed variable-volume pump; or, infrequently in modern practice, a steam or water turbine.

The unloading may be accomplished manually with a hand-driven plow or knife mounted on the curb cover or automatically with a double- or single-acting knife with hydraulic or pneumatic piston actua-

TABLE 18-14 Operating Range of Filtering Centrifuges

Type of centrifuge	G/g	Minimum feed solids concentration by wt.	Minimum mean particle size, μm	Minimum V_{fo}, m/s
Vibratory	30–120	50	300	5×10^{-4}
Tumbler	50–300	50	200	2×10^{-4}
Screen scroll	500–2000	35	75	1×10^{-5}
Pusher	300–1200	40	120	5×10^{-5}
Screen bowl	500–2000	20	45	2×10^{-6}
Peeler	300–1600	10	10	2×10^{-7}
Pendulum	200–1200	10	5	1×10^{-7}

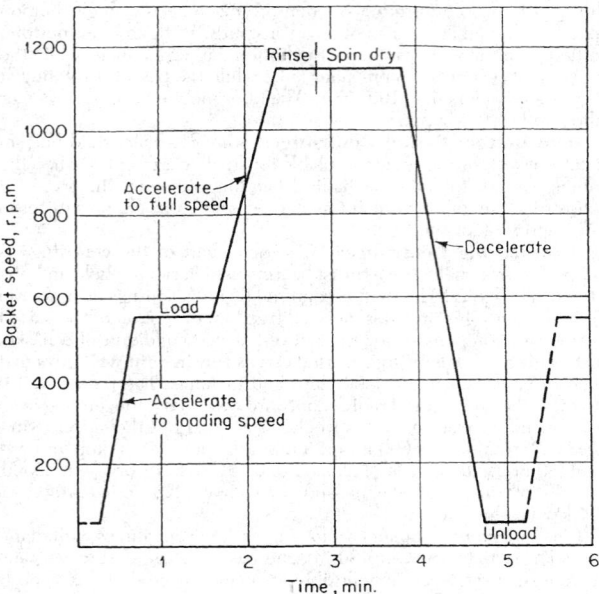

FIG. 18-145 Typical operating cycle, batch-filtering centrifuge.

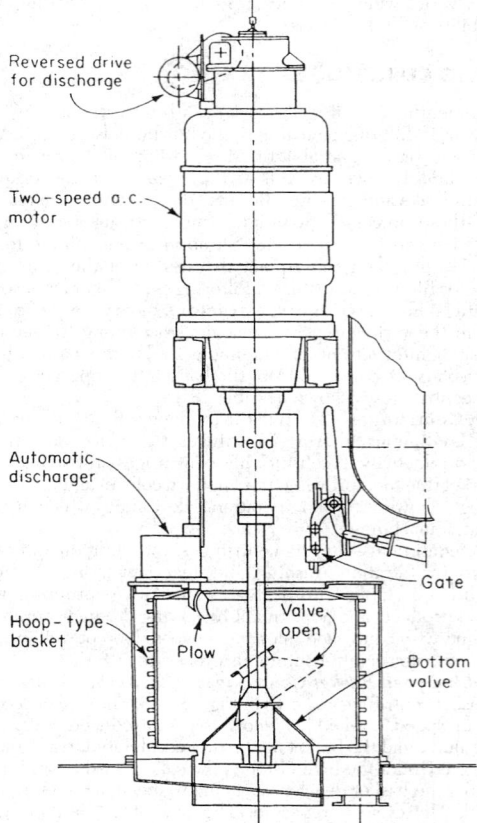

FIG. 18-146 Top-suspended filtering centrifuge. (*Western States Machine Co.*)

tors. In some designs, the opening through the basket is covered with a plate which is lifted during unloading.

Top-Suspended Centrifuges The top-suspended centrifuge (also known as a **pendulum centrifuge,** see Fig. 18-146) is widely used for purging molasses from crystallized sugar, as well as for many other applications. Conventionally, the drive is suspended from a horizontal bar supported at both ends from two A-frames. The drive head, which is connected to the motor or a driven pulley through a flexible coupling, carries the thrust and radial bearings that support the shaft and its load. The cylindrical bowl has a ring weir at the top. The stationary casing is attached to the A-frame. The filtrate is collected and diverted to an outlet in the casing.

The entire weight of the motor is carried on the frame with no component of force other than its rotation reacting on the centrifuge proper. This permits the use of very large special motors for rapid cycling. On white-sugar service, up to 24 cycles/h with 364 kg (800 lb) of sugar per load are provided by wound-rotor variable-speed ac motors with a power rating of 112 kW (150 hp).

Typical pilot-plant top-suspended baskets are 305 mm (12 in) diameter by 127 mm (5 in) deep. Commercial machines are available in sizes from 508-mm (20-in) diameter by 305-mm (12-in) depth to 1524-mm (60-in) diameter by 1016-mm (40-in) depth and develop up to 1800g in the smaller and intermediate sizes. Except in the sugar application, operation with a two-speed motor (half speed for loading and full speed for purging) is typical. Hydraulic drives with variable-speed capability are commonly used in the chemical industry. To maximize the number of cycles per hour, a combination of electrical and mechanical braking is employed to minimize the deceleration period, which is a transition period of no value to the process.

Link-Suspended Bottom-Discharge Centrifuges With the top-suspended design, the only limit to the size and weight of the drive motor is the strength of the supporting structure, which can be made as strong as possible. With the link-suspended design, the weight of the side-mounted motor constitutes an overturning moment, and therefore this weight is limited to a proportion of the remaining suspending components. For a basket with diameter 1219 mm (48 in) and depth 762 mm (30 in), only a relatively lightweight 45-kW (60-hp) motor is employed. Typically it is a two-speed operation, as with other basket centrifuges—half speed for loading and full speed for deli-

quoring. Greater torque transmission and more flexibility in speed control are possible with hydraulic drives.

Batch-Automatic Basket Centrifuges Some batch centrifuges operate automatically on a preprogrammed cycle and are known as **batch-automatic** or **semiautomatic** centrifugal filters. The peeler, the rotary siphon centrifuge, and the pressurized-siphon centrifuges all belong to this category.

Peeler Centrifuges The peeler centrifuges operate at a constant speed during the entire sequence of feeding, deliquoring, and cake discharging, so that no process time is lost in acceleration and deceleration. All peeler centrifuges operate on a horizontal axis of rotation, with the driveshaft supported by fixed bearings. The basket may be cantilevered at one end of the driveshaft, with the driven pulley at the other end. In another design, the driveshaft extends through the basket and is also supported by an outboard bearing. The through-shaft design may carry two baskets with a common hub to allow feeding one basket while the other is spinning to smooth out the power demand.

A cake distributor may be provided to level the load during feeding, to indicate cake depth, and to activate the feed-valve closure when the desired cake depth has been reached. After the cake has been spun dry, the cake is peeled out by rotating an unloader knife into it to discharge the solids down a chute extending through the opening in the ring weir at the front of the basket. In other designs, the unloader knife is smaller and double-acting, and in very large sizes, the discharge of the cake through the front of the basket is facilitated with a horizontal screw conveyor. Because the unloader knife cannot be allowed to contact the filter medium, a heel of product remains in the basket after each unloading. This serves as a precoat to prevent loss of fines through the screen during the next cycle. The disadvantage is that it also adds resistance to filtration similar to the filter medium. The heel may become glazed and impervious from the rubbing action of the knife, and a rinse or backwash of the screen is frequently required to restore the permeability.

The peeler centrifuges are used to dewater insoluble solids such as coal, starch, and citric acid gypsum. Their widest application is for dewatering and washing solids such as specialty chemicals, pharmaceuticals, and thermoplastics having medium to fast drain rates, typically 50 to 200 mesh (297 to 74 μm). The drive is usually an electric motor and must be large enough to bring the empty basket to operating speed and also to accelerate the feed slurry. Usually, for optimum performance, the feed rate should match the cake drain rate so that a minimum of free mother liquor is left on the surface of the cake when the feed valve closes. Some modern versions are equipped with gastight design for fume control.

Diameters of the peelers range from 609 mm (24 in) to 1422 mm (56 in) with widths of 406 mm (16 in) to 1422 mm (56 in). The small units can achieve 3000g, whereas the large units operate at 1000g.

Rotary-Siphon Peeler Centrifuges In this type of centrifuge, a partial vacuum is drawn on the outer diameter of the filter such that the filtrate flows through the cake under both centrifugal force as well as a positive pressure difference of about 1 atm or less. Thus, a higher rate of filtration takes place due to the increased driving force.

In the siphon peeler (Fig. 18-147), the filtrate leaving the basket is collected at the solid bowl. It flows farther axially along the bowl through an annulus to a chamber with a larger diameter as compared to the solid bowl. In this chamber, an annular baffle is used to maintain a difference in liquid head across the baffle. The liquid head is higher upstream toward the basket and is maintained at a lower level downstream by a stationary skimmer tube which opens tangentially along the pool surface. The kinetic energy of the filtrate sustains the liquid to flow through the skimmer tube and connecting tube and into a storage tank. Part of the liquid is returned to the rotating chamber of the centrifuge, whereas the remaining portion overflows the storage tank and exits the system.

The suction pressure generated is equal to the liquid head difference across the annular baffle as magnified by G. The maximum suction that can be realized, assuming negligible vapor pressure, is 1 atm. For example, with a differential liquid level of 2 cm (0.79 in) of water across the annulus under 500g, the suction generated by the siphon is 0.98 atm.

Pressurized Siphon Basket Centrifuges In addition to a siphoning action downstream of the filter, an additional pressure difference of as much as 5 to 6 atm is imposed across the filter to further enhance dewatering. Usually before the high pressure is imposed, the centrifuge reduces speed so that the cake formed is less compact and the positive pressure gradient provides a better filtration on an otherwise low-permeability centrifuge cake. The basket is pressurized; so is the feed pipe. In one design, the filter is inverted during cake discharge to remove the heel cake left on the filter. In another design, dewatering of the centrifuged cake is enhanced by thermal drying.

Continuous-Filtering Centrifuges The trend toward continuous processing and reduction in labor has resulted in more demanding requirements on continuous-processing equipment, including centrifuges. Despite the elimination of downtime, which translates to lower solids throughput, the continuous centrifuge is less flexible than the batch counterpart because there is no separate control on filtering, dry spinning, washing, and final dewatering for both time and G-force, which for different functions may be different for optimal result. There are several types of continuous-filtering centrifuges, the principal distinction among which is the solids conveyance mechanism.

Conical-Screen Centrifuges When a conical screen in the form of a frustum is rotated about its axis, the component of the centrifugal force normal to the screen surface impels the liquid to filter through the cake and the screen, whereas the component of the centrifugal force parallel to the screen in the longitudinal direction conveys the cake to the screen at a larger diameter. The sliding of the solids on the cone is favored by smooth perforated plates or wedge-wire sections with slots parallel to the axis of rotation, rather than woven wire mesh.

Wide-angle conical screen centrifuges. If the half-angle of the cone screen is greater than the angle of repose of the solids, the solids will slide across it with a velocity which depends on frictional properties of the cake but not on feed rate. The frictional property of the cake depends on the solid property, such as shape and size, as well as on moisture content. If the half-angle of the cone greatly exceeds the angle of repose, the cake slides across the screen at a high velocity, thereby reducing the retention time for dewatering. The angle selected is therefore highly critical with respect to performance on a specific application. Wide-angle and compound-angle centrifuges are used to dewater coarse coal and rubber crumb and to dewater and wash crude sugar and vegetable fibers such as from corn and potatoes.

Shallow-angle conical screen centrifuges. By selecting a half-angle for the conical screen that is less than the angle of repose of the cake and providing supplementary means for the controlled conveyance of the cake across the conical screen from the small to large diameter, longer retention time is available for cake dewatering. Three methods are in common use for cake conveyance:

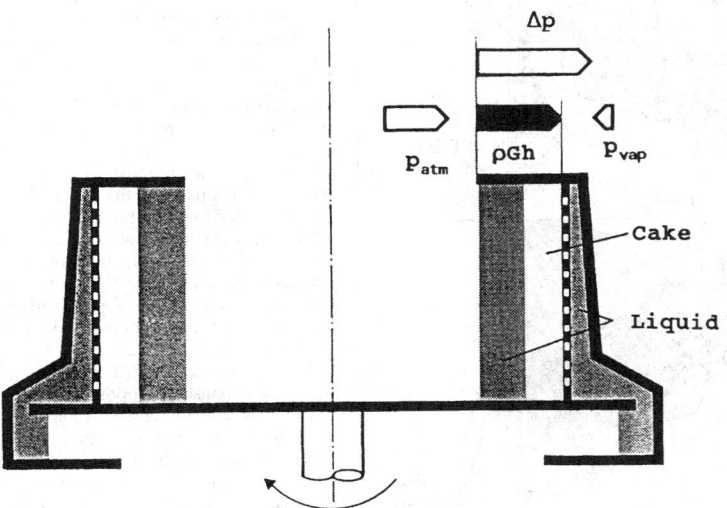

FIG. 18-147 Filtration with a rotary-siphon centrifuge.

1. *Vibrational conveyance.* This is referred as the vibratory centrifuge. A relatively high frequency force is superimposed on the rotating assembly. This can be either in-line with the axis of rotation or torsional, around the driveshaft. In either case, the cake under inertial force from the vibration is partly "fluidized" and propelled down the screen under a somewhat steady pace toward the large end, where it is discharged.

2. *Oscillating or "tumbling" conveyance.* This is commonly known as the tumbling centrifuge. The driveshaft is supported at its lower end on a pivot point. A supplementary power source causes the shaft and the rotating bracket it carries to gyrate about the pivot at a controlled amplitude and at a frequency lower than the rate of rotation of the basket. The inertia force generated also provides partial fluidization of the bed of solids in the basket, causing the cake to convey toward the large end, as in the vibrational conveyance.

3. *Scroll conveyance.* The screen-scroll centrifuge, Fig. 18-148, is the most common shallow-angle conical screen centrifuge. The cake conveyance is controlled by either a continuous helical screw conveyor or with a discrete set of profiled scraper blades, usually in sets of eight. The conveyor guides the cake solids down the conical screen by the differential rotation between the conveyor and the screen. The conveyor is driven by a cyclogear or a gearbox.

The several types of conical-screen centrifuges are constructed with both vertical and horizontal axes of rotation for various applications and installation requirements. Their energy requirement is low, for example, as little as 792 J/kg (0.2 kWh/ton) to dewater −19 +6 mm (−¾ +¼ in) stoker coal at the rate of 38 kg/s (150 ton/h) to 6 percent surface moisture on a large oscillatory centrifuge. The screen-scroll centrifuge can handle finer-size coal with the lower end at 0.5 mm (28 mesh). Their liquid-handling capacity is limited, and 50% feed-slurry concentration is recommended to obtain best performance and capacity. For this reason, the amount of cake wash that can be applied is restricted and, therefore, the wash efficiency is relatively low. As with any filtering centrifuge, performance is also optimized with operation on large and uniformly sized solid particles. Screen thickness, and, hence, screen life, is a function of the openings that will support the solids of the size being centrifuged, the conveyance mechanism, and, finally, whether the feed slurry is accelerated to speed before it lies on the screen at the small diameter. Underaccelerated feed slurry, when introduced to the small conical area, wears out the screen as it slips on the screen surface.

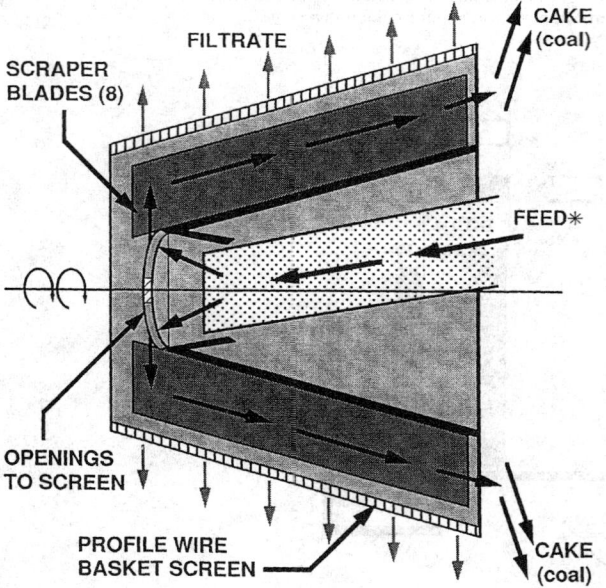

FIG. 18-148   Screen-scroll centrifuge. (*Bird Machine Co.*)

Recently, the performance of the screen-scroll centrifuges benefit from a more effective feed accelerator technology. Not only is the screen wear at the small end of the screen being significantly reduced, the liquid-handling capacity has also increased. For fine coal application, a 900-mm (36-in) diameter centrifuge operating at a speed of 700 r/min used to process 10 to 15 kg/s (40 to 60 ton/h) can handle with a better feed accelerator 20 to 25 kg/s (80 to 100 ton/h), yielding the same cake moisture of 5 percent.

The screen-scroll centrifuge is somewhat more flexible than the others in terms of ability to handle lower feed solids concentration and provide fair to good washing efficiency. This type, together with the wide-angle centrifuge, are used for the dewatering of cellulosic fibers. The screen-scroll centrifuge has also applied to the process crystal form of edible lactose with size 150 to 165 mesh (100 to 240 μm). With a 450-mm (18-in) outer diameter screen, the machine can process 5 ton/h at 50 percent solids. The wash rate is 3.5 to 7×10^{-8} m³/s/kg (0.5 to 1 gpm/ton). After passing through the centrifuge, the lactose solids are dried to 4 to 8 percent moisture.

Pusher Centrifuges

Single-stage pusher. The pusher centrifuge consists of a rotating, perforated, cylindrical basket lined with wedge-wire screen with the slots parallel to the axis of rotation. The cylinder is open with no ring weir at the solids-discharge end and is cantilevered at the other end through its hub to a hollow driveshaft (Fig. 18-149*a*). Inside the basket and fitting close to the cylindrical screen is an annular pusher plate. This is mounted on its own shaft inside the driveshaft. It rotates at the same speed as the basket and concurrently reciprocates axially via a hydraulic mechanism. The slurry is fed near the centerline of the machine and is distributed and accelerated in the rotating feed chamber, which usually takes the form of a cone or a disk. As the pusher plate retracts, a clean surface of screen is exposed for bulk drainage of the incoming feed slurry. As the pusher plate advances, the incremental annulus of the cake thus formed transmits its "pressure" to the annulus of the cake already in the basket, causing an equivalent amount to discharge out of the basket at the open end. Cake wash is applied as a spray through which the cake advances stepwise. Pusher stroke, usually less than 50 mm (2 in), and stroke rate, usually under 100 cpm, depending on the size of the unit and the load requirement, are usually controllable from outside the machine.

The wedge-bar screen construction minimizes friction between the screen and the advancing cake to permit the use of a relatively long screen and correspondingly long retention time for the cake without buckling.

In one modification, the pusher plate consists of a conical screen with an angle slightly greater than the angle of repose of the cake solids. This accelerates the feed slurry and provides extra area for bulk drainage, permitting operation over a wider range of feed concentrations.

Multistage pusher. In a multistage variation, as shown in Fig. 18-149*b*, the basket consists of a series of concentric stages of baskets that increase in diameter in the direction of solids discharge. The first (smallest-diameter) stage and each alternate stage, which are fixed to the inner shaft, rotate and reciprocate. The final stage (largest-diameter) and alternate stages, which are fixed to the outer shaft, rotate but do not reciprocate. Each screen section is relatively short, with its own pusher action from the preceding stage so that there is less tendency for cake buckling with even a relatively long bowl. Bulk filtration, with liquid above and in the cake, occurs in the small-diameter section with minimum power consumption, and film drainage takes place in the large-diameter section under maximum G-force. In one design, the second-stage basket is conical to facilitate cake transport, has a thinner cake layer, and a higher G-force toward the solid discharge end.

Single-stage vs. multistage. Below a limiting feed rate, the cake thickness is constant, depending on the length of the basket. The cake advance velocity increases with increasing feed rate. At higher feed rate, the velocity reaches a maximum (stroke frequency × stroke length) and stays constant. Any further increase in feed rate results in a thicker cake at the same maximum conveyance velocity. As the feed rate further increases, it reaches a point at which the feed rate is greater than the bulk filtration rate, resulting in a slurry pool forming above the cake surface, which in effect drives a higher filtration rate due to the extra liquid

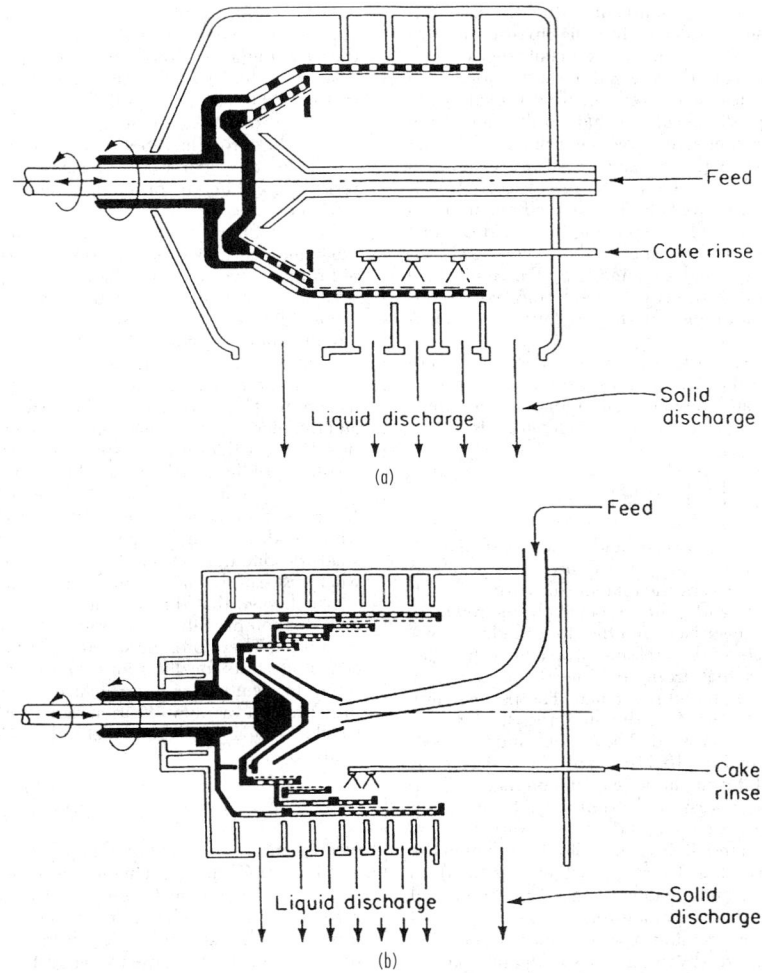

FIG. 18-149 Schematic of pusher centrifuge: (*a*) single-stage pusher; (*b*) multistage pusher.

head. If this increase in filtration rate does not offset the increase in feed rate, a swallowing limit is reached in which the liquid slurry runs over the product cake, which heretofore acts as a dam for the slurry.

Another limit plays a role when the cake at the feed zone has not reached a consistency to be conveyed, in which case it slops over on the preceding cake in the basket. This is especially true for cake which has a low filtration rate. The stroke frequency has to be reduced accordingly.

The velocity of the cake is less than the stroke per unit cycle time, the ratio of which is the **pushing efficiency.** Depending on the moisture of the cake as well as the frictional characteristics of the screen, a 90 percent efficiency can be attained in proper operation.

By having a short basket in each stage for a multistage pusher, cake buckling is reduced, as it may be a problem for a long, single-stage pusher. Because the cake thickness in each stage can be smaller (cake thickness is proportional to the basket length), a thinner cake results in more effective washing and dewatering. In addition, there is no cake heel, as the cake tumbles in transition from one stage to the next. This also helps to eliminate moisture retained by capillary rise in an otherwise undisturbed cake. Effective cake washing takes place at the transition between stages. Despite these advantages for the multistage, the single-stage can handle better bulk filtration than a multistage, in which it has to be accomplished in the first stage where

length is limited. Because a thicker cake is formed on a one-stage basket, solids recovery is higher for the single-stage than for the multistage pusher. With either design, it is important that the cake thickness be circumferentially and axially uniform, with no axial ridges and valleys on the cake surface, which otherwise result in poor washing and dewatering.

Pusher centrifuges are used for dewatering and washing crystals and other particulate solids, including short fibers. The crystals in the feed should be 100 mesh (149 μm) or larger for good operation and to minimize loss of fine solids to filtrate. Feed-slurry concentration should be 35 percent by weight for conventional pushers. Lower concentrations can be tolerated on the cone-screen and multistage types, and especially the ones with better feed accelerators, which can handle higher liquid filtration. Washing efficiency of pusher centrifuges can be excellent, frequently in excess of 95 percent displacement of mother-liquor impurities with a wash-crystal weight ratio of only 1:10. Pusher centrifuges range in size from 300 to 900 mm (12 to 36 in), 4300 to 1300 mm long. They are operated at 400 to 900g. The upper range corresponds to a smaller unit.

Theory of Centrifugal Filtration   Theoretical predictions of the behavior of solid-liquid mixtures in a filtering centrifuge are more difficult compared to pressure and gravity filtration. The area of flow and driving force are both proportional to the radius, and the specific resistance and porosity may also change markedly within the cake. Filtering centrifuges are nearly always selected by scale-up from lab tests on materials to be processed, such as using bucket centrifuges where a wide range of test conditions (cake thickness, time, and G-force) can be controlled. Although tests with the bucket centrifuge provide some quantitative data to scale-up, the results include wall effect from buckets, which are not representative of actual cylindrical basket geometry. A modified version of the buckets or even a cylindrical perforated basket can be used. In the latter, there is less control of cake depth and circumferential uniformity. The desired quantities to measure are filtration rate, washing rate, spinning time, and residual moisture. Also, with filtering centrifuges such as the screen-bowl centrifuge, screen-scroll centrifuge, and to some extent in multistage pushers, the cake is constantly disturbed by the scroll conveyor or conveyance mechanism; liquid saturation due to capillary rise as measured in bucket tests is absent.

Filtration Rate When the centrifuge cake is submerged in a pool of liquid (Fig. 18-139b), as in the case of a fast-sedimenting, solids-forming cake almost instantly, and the rate of filtration becomes limiting, the bulk filtration rate Q for a basket with axial length b is:

$$Q = \frac{\pi\, b\, \rho K \Omega^2 \,(r_b^2 - r_p^2)}{\mu \left(\ln \left[\dfrac{r_b}{r_c} \right] + \dfrac{KR_m}{r_b} \right)} \qquad (18\text{-}116a)$$

where μ and ρ are, respectively, the viscosity and density of the liquid; Ω is the angular speed; K is the average permeability of the cake and is related to the specific resistance α by the relationship $\alpha K \rho_s = 1$, with ρ_s being the solids density; r_p, r_c and r_b are, respectively, the radius of the liquid pool surface, the cake surface, and the filter medium adjacent to the perforated bowl. Here, the pressure drop across the filter medium, which also includes that from the cake heel, is $\Delta p_m = \mu R_m(Q/A)$ with R_m being the combined resistance. The permeability K has a unit m^2, α m/kg, and R_m m^{-1}. The driving force is due to the hydrostatic pressure difference across the bowl wall and the pool surface—i.e., the numerator of Eq. (18-116a), and the resistance is due to the cake layer and the filter medium—i.e., denominator of Eq. (18-116a). Fig. 18-150 shows the pressure distribution in the cake and the liquid layer above. The pressure (gauge) rises from zero to a maximum at the cake surface; thereafter, it drops monotonically within the cake in overcoming resistance to flow. There is a further pressure drop across the filter medium, the magnitude dependent on the combined resistance of the medium and the heel at a given flow rate. This scenario holds, in general, for incompressible as well as for compressible cake. For the latter, the pressure distribution also depends on the compressibility of the cake.

For incompressible cake, the pressure distribution and the rate depend on the resistance of the filter medium and the permeability of the cake. Figure 18-150 shows several possible pressure profiles in the cake with increasing filtration rates through the cake. It is assumed that $r_c/r_b = 0.8$ and $r_p/r_b = 0.6$. The pressure at $r = r_b$ corresponds to pressure drop across the filter medium Δp_m with the ambient pressure taken to be zero. The filtration rate as well as the pressure distribution depend on the medium resistance and that of the cake. High medium resistance or blinding of the medium results in greater penalty on filtration rate.

In most filtering centrifuges, especially the continuous-feed ones, the liquid pool above the cake surface should be minimum to avoid liquid running over the cake. Setting $r_p = r_c$ in Eq. 18-116a, the dimensionless filtration flux is plotted in Fig. 18-151 against r_c/r_b for different ratios of filter-medium resistance to cake resistance, KR_m/r_b. For negligible medium resistance, the flux is a monotonic decreasing function with increasing cake thickness, i.e., smaller r_c. With finite medium resistance, the flux curve for a range of different cake thicknesses has a maximum. This is because for thin cake the driving liquid head is small and the medium resistance plays a dominating role, resulting in lower flux. For very thick cake, despite the increased driving liquid head, the resistance of the cake becomes dominant; therefore, the flux decreases again. The medium resistance to cake resistance should be small, with $KR_m/r_b < 5$ percent (see Fig. 18-152). However, the cake thickness, which is directly proportional to the throughput, should not be too small, despite the fact that the machine may have to operate at somewhat less than the maximum flux condition.

It is known that the specific resistance for centrifuge cake, especially for compressible cake, is greater than that of the pressure or vacuum filter. Therefore, the specific resistance has to be measured from centrifuge tests for different cake thicknesses so as to scale up accurately for centrifuge performance. It cannot be extrapolated from pressure and vacuum filtration data. For cake thickness that is much smaller compared to the basket radius, Eq. (18-116a) can be approximated by

$$V_f = V_{fo} \left(\frac{h}{h_c} \right) \qquad (18\text{-}116b)$$

where $h = r_b - r_p$ is the liquid depth, $h_c = r_b - r_c$ is the cake thickness, and $v_{fo} = (\rho GK/\mu)$ is a characteristic filtration velocity. Table 18-14 shows some common filtering centrifuges and the application with respect to the G-level, minimum feed-solids concentration, minimum mean particle size, and typical filtration velocity. The vibratory and tumbler centrifuges have the largest filtration rate of 5×10^{-4} m/s (0.02 in/s) for processing 200-μm or larger particles, whereas the pen-

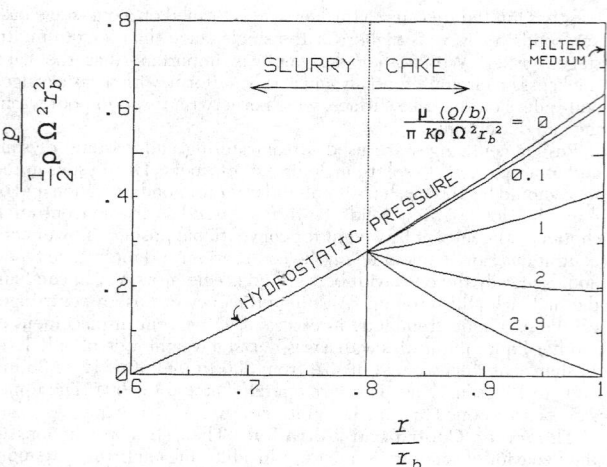

FIG. 18-150 Pressure distribution in a basket centrifuge under bulk filtration.

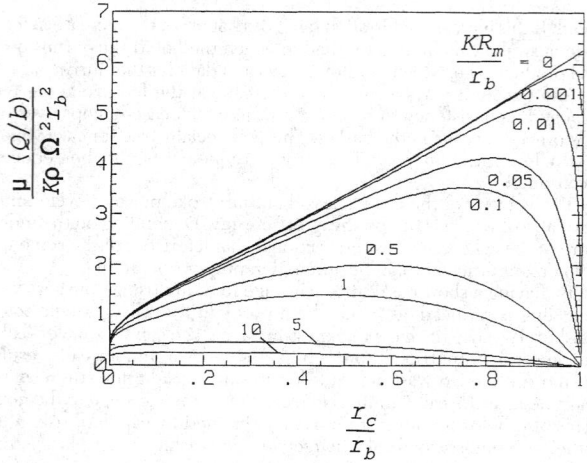

FIG. 18-151 Centrifugal filtration rate as a function of both cake and medium resistance.

dulum has the lowest filtration rate of 1×10^{-7} m/s (4×10^{-6} in/s) for processing 5-μm particles with increased cycle time. The screen-scroll, pusher, screen-bowl, and peeler centrifuges are in between.

Film Drainage and Residual Moisture Content   Desaturation of the liquid cake ($S < 1$) begins as the bulk filtration ends, at which point the liquid level starts to recede below the cake surface. Liquids are trapped in: (1) cake pores between particles that can be drained with time (free liquid), (2) particle contact points (pendular liquid), (3) fine pores forming continuous capillaries (capillary rise), (4) particle pores or they are bound by particles (bound liquid). Numbers (1) to (3) can be removed by centrifugation, and, as such, each of these components depends on G to a different extent. Only desaturation of the free liquid, and to a much lesser entent the liquid at contact points, is a function of time. The wet cake starts from a state of being fully saturated, $S = 1$, to a point where $S < 1$, depending on the dewatering time. At a very large amount of time, it approaches an equilibrium point S_∞, which is a function of G, capillary force, and the amount of bound liquid trapped inside or externally attached to the particles.

The following equations, which have been tested in centrifugal dewatering of granular solids, prove useful:

Total saturation:

$$S_{total} = S_\infty + S_T(t) \qquad (18\text{-}117)$$

Equilibrium component:

$$S_\infty = S_c + (1 - S_c)(S_p + S_z) \qquad (18\text{-}118)$$

Transient component:

$$S_T(t) = (1 - S_c)(1 - S_p - S_z)S_t(t) \qquad (18\text{-}119)$$

The details of the mathematical model of these four components are given below.

Drainage of free liquid in thin film:

$$S_t(t) = \left(\frac{4}{3}\right)\left(\frac{1}{t_d^n}\right), \quad t_d > 0 \qquad (18\text{-}120)$$

where for smooth-surface particles, $n = 0.5$, and for particles with rough surfaces, n can be as low as 0.25.

Bound liquid saturation:

$$S_p = \text{function(particle characteristics)} \qquad (18\text{-}121)$$

Pendular saturation:

$$S_z = 0.075, \quad N_c \leq 5 \qquad (18\text{-}122a)$$

$$S_z = \frac{5}{(40 + 6N_c)}, \quad 5 \leq N_c \leq 10 \qquad (18\text{-}122b)$$

$$S = \frac{0.5}{N_c}, \quad N_c \geq 10 \qquad (18\text{-}122c)$$

Frequently when $N_c < 10$, S_p and S_z are combined for convenience, the sum of which is typically 0.075 for smooth particles and can be as high as 0.35 for rough-surface particles. This has to be determined from tests.

Saturation due to capillary rise:

$$S_c = \frac{4}{B_o} \qquad (18\text{-}123)$$

where the dimensionless time t_d, capillary number N_c, and Bond number B_o are, respectively:

$$t_d = \frac{\rho\, G d_h^2 t}{\mu\, H} \qquad (18\text{-}124)$$

$$N_c = \frac{\rho\, G d_h^2}{\sigma \cos \theta} \qquad (18\text{-}125)$$

$$B_o = \frac{\rho\, G H d_h}{\sigma \cos \theta} \qquad (18\text{-}126)$$

where ρ and μ are, respectively, the density and viscosity of the liquid; θ is the wetting angle of the liquid on the solid particles; σ is the inter-

facial tension; H is the cake height; d is the mean particle size; and t is the dewatering time. The hydraulic diameter of the particles can be approximated by either $d_h = 0.667\, \varepsilon d/(1 - \varepsilon)$ or $d_h = 7.2(1 - \varepsilon)K^{1/2}/\varepsilon^{3/2}$, where ε is the cake porosity.

Example

Given: $\rho = 1000$ kg/m³, $\rho_s = 1200$ kg/m³, $\mu = 0.004$ N·s/m², $\sigma \cos \theta = 0.068$ N/m, $H = 0.0254$ m, $d = 0.0001$ m, $\varepsilon = 0.4$, $G/g = 2000$, $t = 2$ s, $S_p = 0.03$.

Calculate: $d_h = 4.4 \times 10^{-5}$ m, $t_d = 748$, $N_c = 0.56$, $B_o = 322$, $S_t = 0.048$, $S_z = 0.075$, $S_c = 0.012$, $S_\infty = 0.116$, $S_T = 0.043$, $S_{total} = 0.158$, $W = 0.919$.

Note that W is the solids fraction by weight and is determined indirectly from Eq. (18-86). The moisture weight fraction is 0.081.

The transient component depends not only on G, cake height, and cake properties, but also on dewatering time, which ties to solids throughput for a continuous centrifuge and cycle time for batch centrifuge. If the throughput is too high or the dewatering cycle is too short, the liquid saturation can be high and becomes limiting. Given that time is not the limiting factor, dewatering of the liquid lens at particle contact points requires a much higher G-force. The residual saturation depends on the G-force to the capillary force, as measured by N_c, the maximum of which is about 7.5%, which is quite significant. If the cake is not disturbed (scrolled and tumbled) during conveyance and dewatering, liquid can be further trapped in fine capillaries due to liquid rise, the amount of which is a function of B_o, which weighs the G-force to the capillary force. This amount of liquid saturation is usually smaller as compared to capillary force associated with liquid-lens (also known as pendular) saturation. Lastly, liquid can be trapped by chemical force at the particle surface or physical capillary or interfacial force in the pores within the particles. Because the required desaturating force is extremely high, this portion of moisture cannot be removed by mechanical centrifugation. Fortunately, for most applications it is a small percentage, if it exists.

SELECTION OF CENTRIFUGES

Table 18-15 summarizes the several types of commercial centrifuges, their manner of liquid and solids discharge, their unloading speed, and their relative volumetric capacity. When either the liquid or the solids discharge is not continuous, the operation is said to be cyclic. Cyclic or batch centrifuges are often used in continuous processes by providing appropriate upstream and downstream surge capacity.

Sedimenting Centrifuges   These centrifuges frequently are selected on the basis of tests on tubular, disk, or helical-conveyor centrifuges of small size. The centrifuge should be of a configuration similar to that of the commercial centrifuge it is proposed to be used for. The results in terms of capacity for a given performance (effluent clarity and solids concentration) may be scaled up by using the sigma concept of Eqs. (18-114) to (18-116). Spin-tube tests may be used for information on systems containing well-dispersed solids. Such tests are totally unreliable on systems containing a dispersed phase that agglomerates or flocculates during the time of centrifugation.

Filtering Centrifuges   These filters often can be selected on the basis of batch tests on a laboratory unit, preferably one at least 12 in (305 mm) in diameter. A bucket centrifuge test would be helpful to study the effect of G, cake height, and dewatering time. Caution has to be taken in correcting for capillary saturation, which may be absent in large continuous centrifuges with scrolling conveyances.

Unless operating data on similar material are available from other sources, continuous centrifuges should be selected and sized only after tests on a centrifuge of identical configuration.

It seems needless to state but is frequently overlooked that test results are valid only to the extent that the slurry and the test conditions duplicate what will exist in the operating plant. This may involve testing on a small scale (or even on a large one) with a slipstream from an existing unit, but the dependability of the data is often worth the extra effort involved. Most centrifuge manufacturers provide testing services and demonstration facilities in their own plants and maintain a supply of equipment for field-testing in the customer's plant, such as with a pilot centrifuge mounted on a mobile truck or trailer.

TABLE 18-15 *Characteristics of Commercial Centrifuges*

Method of separation	Rotor type	Centrifuge type	Manner of liquid discharge	Manner of solids discharge or removal	Centrifuge speed for solids discharge	Capacity°
Sedimentation	Batch	Ultracentrifuge				1 mL
		Laboratory, clinical	Batch	Batch manual	Zero	To 6 L
	Tubular	Supercentrifuge	Continuous†	Batch manual	Zero	To 1,200 gal/h
		Multipass clarifier	Continuous†	Batch manual	Zero	To 3,000 gal/h
	Disk	Solid wall	Continuous†	Batch manual	Zero	To 30,000 gal/h
		Light-phase skimmer	Continuous	Continuous for light-phase solids	Full	To 1,200 gal/h
		Peripheral nozzles	Continuous	Continuous	Full	To 24,000 gal/h
		Peripheral valves	Continuous	Intermittent	Full	To 3,000 gal/h
		Peripheral annulus	Continuous	Intermittent	Full	To 12,000 gal/h
	Solid bowl	Constant-speed horizontal	Continuous†	Cyclic	Full (usually)	To 60 ft³
		Variable-speed vertical	Continuous†	Cyclic	Zero or reduced	To 16 ft³
		Continuous decanter	Continuous	Continuous screw conveyor		To 54,000 gal/h
					Full	To 100 tons/h solids
Sedimentation and filtration		Screen-bowl decanter	Continuous	Continuous	Full	To 60,000 gal/h
						To 125 tons/h solids
Filtration	Conical screen	Wide-angle screen	Continuous	Continuous	Full	To 40 tons/h solids
		Differential conveyor	Continuous	Continuous	Full	To 80 tons/h solids
		Vibrating, oscillating, and tumbling screens	Continuous	Essentially continuous	Full	To 250 tons/h solids
	Cylindrical screen	Reciprocating pusher	Continuous	Essentially continuous	Full	Limited data
		Reciprocating pusher, single and multistage	Continuous	Essentially continuous	Full	To 50 tons/h solids
		Horizontal	Cyclic	Intermittent, automatic	Full (usually)	To 25 tons/h solids
		Vertical, underdriven	Cyclic	Intermittent, automatic, or manual	Zero or reduced	To 6 tons/h solids
		Vertical, suspended	Cyclic	Intermittent, automatic, or manual	Zero or reduced	To 10 tons/h solids

°To convert gallons per hour to liters per second, multiply by 0.00105; to convert tons per hour to kilograms per second, multiply by 0.253; and to convert cubic feet to cubic meters, multiply by 0.0283.
†Feed and liquid discharge interrupted while solids are unloaded.

COSTS

Neither the investment cost nor the operating cost of a centrifuge can be directly correlated with any single characteristic of a given type of centrifuge. The costs depend on the features of the centrifuge tailored toward the physical and chemical nature of the materials being separated, the degree and difficulty of separation, the flexibility and capability of the centrifuge and its auxiliary equipment, the environment in which the centrifuge is located, and many other nontechnical factors, including market competition. The cost figures presented herewith represent centrifuges only for use in the process industries as of 1996. In any particular installations, the costs may be somewhat less or much greater than those presented here.

The useful parameter for value analysis is the installed cost of the number of centrifuges required to produce the demanded separative effect (end product) at the specified capacity of the plant. The possible benefits of adjustments in the upstream and downstream components of the plant and the process should be carefully examined in order to minimize the total overall plant costs; the systems approach should be used.

Purchase Price Typical purchase prices, including drive motors, of tubular and disk sedimenting centrifuges are given in Table 18-16. The price will vary upward with the use of more exotic materials of construction, the need for explosion-proof electrical gear, the type of enclosure required for vapor containment, and the degree of portability, and this holds for all types of centrifuges.

The average purchase prices of continuous-feed, solid-bowl centrifuges made, respectively, of 316 stainless steel and steel are shown in Fig. 18-152. The average purchase prices of continuous-feed filtering centrifuges are shown in Fig. 18-153. This chart includes a comparison of prices on screen-bowl, pusher, screen-scroll, and oscillating conical baskets. On average, the screen bowl is approximately 10 percent higher in price than the solid bowl of the same diameter and length. This incremental cost results from the added complexity of the screen section, bowl configuration, and casing differences. Prices for both the solid-bowl and the filtering centrifuges do not include the drive motor,

TABLE 18-16 *Typical Purchase Prices, Including Drive Motors, of Tubular and Disk Sedimenting Centrifuges, 1996*

Type	Bowl diameter, in (mm)	Approximate Σ value, units of 10⁴ ft² (10³ m²)	Designation	Purchase price, 1996 $
Tubular	4 (102)	2.7 (2.5)	Oil purifier	60,000–80,000
	4 (102)	2.7 (2.5)	Chemical separation	60,000–80,000
	5 (127)	4.2 (3.9)	Blood fractionation	100,000–140,000
Manual discharge disk	13.5 (343)	21 (20)	Hermetic	100,000–130,000
	24 (610)	95 (88)	Centripetal pump	150,000–300,000
Continuous nozzle-discharge disk	12 (305)	12 (11)	Clarifier	100,000–130,000
	18 (457)	25 (23)	Separator	150,000–200,000
	30 (762)	100 (93)	Recycle clarifier	270,000–300,000
Intermittent discharge disk	14 (356)	13 (12)	Centripetal pump	130,000–150,000
	18 (457)	22 (20)	Centripetal pump	170,000–200,000
	24 (610)	38 (35)	Centripetal pump	250,000–300,000

°NOTE: All prices quoted are for stainless steel construction with the exception of the oil purifier noted.

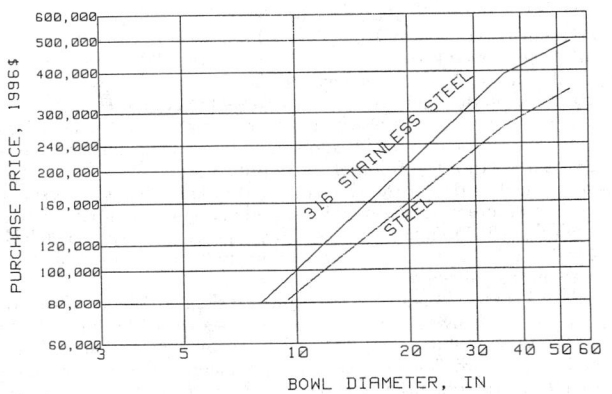

FIG. 18-152 Costs of continuous-feed solid-bowl.

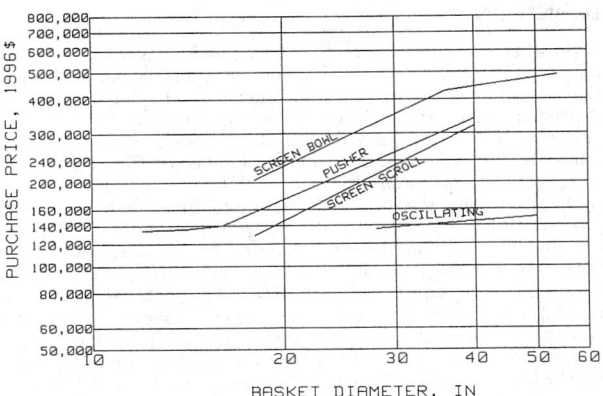

FIG. 18-153 Costs of continuous baskets (316 stainless steel).

which typically adds another 5 to 25 percent to the cost. The higher end of this range represents a variable-speed-type drive. If a variable-speed backdrive is used instead of the gear unit, the incremental cost is about another 10 to 15 percent, depending on the capability.

The average prices of the batch centrifuge are shown in Fig. 18-154. All the models include the drive motor and control. In Fig. 18-154, the inverting filter, horizontal peeler, and the advanced vertical peeler are the premium baskets especially used for specialty chemicals and pharmaceuticals. Control versatility with the use of programmable logic control (PLC), automation, and cake-heel removal are the key features which are responsible for the higher price. The under-driven, top-driven, and pendulum baskets are less expensive with fewer features.

Installation Costs   Installation costs of centrifuges vary over an extremely wide range, depending on the type of centrifuge, on the area and kind of structure in which it is installed, and on the details of installation. Some centrifuges, such as portable tubular and disk oil purifiers, are shipped as package units and require no foundation and a minimum of connecting piping and electrical wiring. Others, such as large batch automatic and continuous scroll-type centrifuges, may require substantial foundations and even building reinforcement, extensive interconnecting piping with required flexibility, auxiliary feed and discharge tanks and pumps and other facilities, and elaborate electrical and process-control equipment. Minimum installation costs, covering a simple foundation and minimum piping and wiring, are about 5 to 10 percent of purchase price for tubular and disk centrifuges; 10 to 25 percent for bottom drive, batch automatic, and continuous-scroll centrifuges; and up to 30 percent for top-suspended basket centrifuges. If the cost of all auxiliaries—special foundations, tanks, pumps, conveyors, electrical and control equipment, etc.—is included, the installation cost may well range from 1 to 3 times the purchase price of the centrifuge itself.

Maintenance Costs   Because of the care with which centrifuges are designed and built, their maintenance costs are in line with those of other slower-speed separation equipment, averaging in the range of 5 to 10 percent per year of the purchase price for centrifuges in light to moderate duty. For centrifuges in severe service and on highly cor-

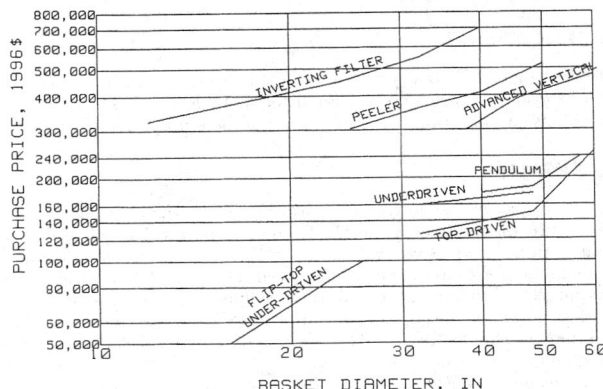

FIG. 18-154 Costs of batch baskets (316 stainless steel).

rosive fluids, the maintenance cost may be several times this value. Maintenance costs are likely to vary from year to year, with lower costs for general maintenance and periodic large expenses for major overhaul. Centrifuges are subject to erosion from abrasive solids such as sand, minerals, and grits. When these solids are present in the feed, the centrifuge components are subject to wear. Feed and solids discharge ports, unloader knives, helical scroll blade tips, etc., should be protected with replaceable wear-resistant materials. Excessive out-of-balance forces strongly contribute to maintenance requirements and should be avoided.

Operating Labor   Centrifuges run the gamut from completely manual control to fully automated operation. For the former, one operator can run several centrifuges, depending on their type and the application. Fully automatic centrifuges usually require little direct operation attention.

EXPRESSION

GENERAL REFERENCES: F. M. Tiller and N. B. Hsyung, "Compaction of Filter Cakes," *Advances in Filtration Separation Technology,* **5,** pp. 327–331 (1992). F. M. Tiller and C. S. Yeh, "Relative Liquid Removal in Filtration and Expression," *Filtr. Sep.,* **27,** p. 129 (1990). F. M. Tiller and C. S. Yeh, "The Role of Porosity in Filtration Part XI: Filtration Followed by Expression," *AIChE Journal,* **33**(8), p. 1241 (August 1987). F. M. Tiller, C. S. Yeh, and W. F. Leu, "Compressibility of Particulate Structures in Relation to Thickening, Filtration and Expression—A Review," *Separation Science and Technology,* **22**(2) and (3), pp. 1037–1063 (1987). Shirato, M. et al., "Deliquoring by Expression," in R. J. Wakeman (ed.), *Progress in Filtration,* **4,** Elsevier, pp. 181–287 (1986). M. Shirato et al., "Fundamental Studies on Continuous Extrusion Using a Screw Press," *International Chemical Engineering,* **18**(4), pp. 680–687 (October 1978).

DEFINITION

Expression, possibly followed by air (or steam) displacement, is the last stage in mechanically dewatering° compressible solids. Expression is used to wring out the last remaining liquid before resorting to thermal drying or solvent (chemical) extraction of the remaining liquids from the solids.† The goal of this stage of dewatering is maximum removal of liquid rather than creation of a solids-free liquid. The operating costs for expression are much lower than those for heat or solvent recovery, and the former is used in preference to the two latter processes. Tiller estimates that for pressures up to 1000 kPa (10 atm) the mechanical energy required for liquid removal is 400 times less than the thermal energy required for evaporation [Tiller, Yeh, and Leu, *Separation Science and Technology*, **22**, pp. 1037–1063 (1987)]. For sludges intended for incineration, expression can often dewater the material sufficiently to eliminate the need for auxiliary fuel. For example, in wastewater treatment plants, particularly those that include some form of thermal treatment, the degree of dewatering has a greater impact on the energy balance than any other single unit process [Campbell and Plaisier, *Advances in Filtration and Separation Technology*, **7** (System Approach to Separation and Filtration Process Equipment), pp. 583–586 (1993)].

APPLICATIONS

Typical materials that are or can be processed in various types of expression dewatering devices are listed in Table 18-17. The table also provides an indication of the efficiency of the equipment.

EQUIPMENT

In the past, expression presses were used in many processes for extracting oil and juice, generally from seeds and fruits such as olives. Batch presses were typically used in these applications, and hand unloading of the pressed cake was often required. Batch presses that require hand unloading or extensive cleaning between pressings are rarely used now; descriptions of various types are presented in earlier editions of this handbook. This section, therefore, describes mainly continuous presses.

Continuous is used here to include not only presses that take feed and produce dewatered product in a constant stream, but also automated batch presses that require little or no operator intervention during operating cycles. Continuous-expression presses, under this broader operational definition, include variable-volume filter presses (q.v.), which inflate a membrane against the cake to press additional liquid from the solids within the press; screw presses, which compress a generally fibrous or polymerized feed in a screw whose diameter increases as the feed progresses through the barrel; belt presses, which constrain the feed between two porous belts that are mechanically squeezed together; and rigid perforated disk or roll presses, which squeeze solids through a nip formed by converging rotating disks or in the nip between two parallel, closely spaced, porous drums. Many other types of presses are also available, but they are used less frequently. Among them is the tube press, which can exert higher overall pressure on filter cake than any other device. Although the tube press is technically a batch device like the variable-volume filter press, it can also be automated for continuous cycling.

Screw Presses Figures 18-155 and 18-156 show screw presses. The rotating screw shown in Fig. 18-156 is of constant pitch and has a constant diameter of about 0.3 m (12 in). Pressure to squeeze juice from fruit placed in the press comes from restricting the opening at the end of the barrel with a hydraulically adjustable cone and by making the spindle of the screw thicker toward the discharge end.

° Technically, the term should be *deliquoring* because this article concerns only separation of fluid from solids. However, in many applications the devices are intended for water removal, and *dewatering* is a common term. Therefore, the term *dewatering* will be used loosely in this article.

† Incompressible solids can be further mechanically dewatered by passing air or steam through a filter or centrifuge cake; for these materials, expression offers virtually no benefit.

The press in Fig. 18-156 is widely used in dewatering synthetic rubber crumb. With 200 to 250 connected horsepower, capacities range from 900 to 5000 kg/h (2000 to 11,000 lb/h) of solids. The hydraulic cylinders at the discharge end control the choke, or the space that the dewatered product must pass through to be discharged. The setting of the choke opening controls the dewatering that will be achieved, as well as the throughput. The approximate weight of the machine is 13.5 tonnes (15 tons). In other screw presses, the pitch of the screw can be tightened near the discharge end to apply higher pressures. The largest screw presses, used for sugar beets have 750-hp drives and weigh about 165 tonnes (185 tons). Processing capacity is 180,000 kg/h (400,000 lb/h). A few screw presses are built with a vertical shaft; these presses take up less floor space.

Disk Presses Figure 18-157 shows a disk press. The two disks, or press wheels, converge to a very narrow space at the bottom. This is the point of maximum compression, which can be more than 14 times the feed pressure. The press wheels have channels to carry the liquid from the dewatered product, and they are covered with a screen plate. Wheels 1.5 m (5 ft) in diameter are used on a large press that requires about 80 connected horsepower and produces just under 1 tonne/h (0.9 ton/h) of solids (dry basis). Typical applications are fibrous materials such as coffee grounds, pineapple and citrus peels and wastes, alfalfa, and brewer's spent grain.

Roll Presses The twin roll press shown in Fig. 18-158 is intended for paper pulp dewatering, specifically for one-step dewatering of an incoming slurry that starts at 2 to 5 percent solids and is dewatered to 50 percent. Other uses include citrus processing. The two rolls are perforated, and the mat to be dewatered forms on them. They rotate toward each other. Throughput and degree of dewatering depend on the size of the gap between the rolls, as well as on feed rate and solids consistency. The bottom vat is sealed so that the slurry can be pumped beneath the rolls under pressures up to 140 kPa (20 psig), and the rolls act as pressure filters until the cake is compressed between them as it is lifted out of the slurry by rotating the roll. The largest machines, with rolls 1.5 m (4.5 ft) in diameter and 7 m (21 ft) long, can produce up to 37 tonnes/h (41 tons/h), dry basis. Such machines weigh about 160 tonnes (180 tons) and require up to 450 connected horsepower.

Belt Presses Belt presses were fully described in the section on filtration. The description here is intended to cover only the parts and designs that apply expression pressure by a mechanism in addition to the normal compression obtained from tensioning the belts and pulling them over rollers of smaller and smaller diameters. The tension on the belt produces a squeezing pressure on the filter cake proportional to the diameter of the rollers. Normally, that static pressure is calculated as $P = 2T/D$, where P is the pressure (psi), T is the tension on the belts (lb/linear in), and D is the roller diameter. This calculation results in values about one-half as great as the measured values because it ignores pressure created by drive torque and some other forces [Laros, *Advances in Filtration and Separation Technology*, **7** (System Approach to Separation and Filtration Process Equipment), pp. 505–510 (1993)].

Expression belt presses (such as the Parkson Magnum Press) add an extra section to the path of the filter belts; an additional tensioned belt compresses the belts containing the filter cake to achieve a pressure of about 18 kg/cm (100 lb/linear in) instead of the pressure of 6 kg/cm (35 lb/linear in) achieved by a conventional dewatering belt press. Other expression dewatering belt presses, such as the Eimco Expressor Press shown in Figure 18-159, are designed with a large central drum and many smaller press rolls mounted radially around it. The smaller press rolls directly compress the two belts, and the filter cake enclosed between them is compressed against the larger perforated drum. This design can apply over 3500 kPa (500 psi), or 10 or more times as much pressure as that achieved by belt tensioning alone. Horsepower requirements are in the range of 10 to 40, and a large unit weighs about 11.5 tonnes (13 tons). A large press will process up to 13 tonnes/h (15 tons/h) of apples for juice extraction.

Variable-Volume Filter Presses These membrane filter presses are covered in the section on filtration. Two designs are used for the presses: (1) the typical variable-volume filter press has a normal vertical-leaf design; (2) other presses, such as those provided by Filtra-Systems and Larox, are designed with horizontally arranged leaves.

TABLE 18-17 Materials Suitable for Dewatering in Expression Devices and Typical Performance

Type of press	Material	Feed	Product
Food and dairy products			
Oilseeds	Single screw		20% oil recovery
	Twin screw		94% oil recovery
Cocoa butter	Screw		
Alfalfa for silage	Screw	10–15	25–30
	Disk	10–15	25–35
Brewers' spent grain	Screw	20–25	30–35
	Disk	20–25	35–40
Corn silage	Screw		
Starch	Screw		
Citrus or pineapple peel, unlimed	Screw	15	20–25
	Disk	15	25–30
Citrus peel, limed for pectin	Screw	15	20–25
	Disk	15	25–30
Coffee grounds	Disk	20–25	45–50
Fish waste			
Sugar beets			
Sugarcane mud			
Spent tea	Expression belt	22	48
Fruit for juice	Expression belt	3	19
Slaughterhouse, tanning, and rendering waste and by-products			
Sewage sludge	Expression belt	4	36
Coagulated suspensions of rubber or polymers	Expression belt	48	90
		40–60	85–92
Pulp and paper mill sludges°	Expression belt	3	43
	Twin roll	3	35–50
	Screw	3	35
Fibrous materials (peat)°	Expression belt	18	35
	Variable-volume filter press	3	37
Fiberglass manufacturing scrubber sludge			
Secondary and synthetic fiber			
Chemicals			
Calcium carbonate	Expression belt	43	71
Calcium hypochlorite			
Calcium stearate			
Magnesium hydroxide	Expression belt	48	68
Biogums: xanthan, carrageenan	Screw	50–55	
Textile dyeing wastes	Screw	3	25–30
Paint and pigments	Expression belt	33	57
Ceramics, talc, and clay			
Paper and plastic fillers			
Metal ore concentrates	Variable-volume filter press	46–55	81–93°
Fine coal refuse			
Metal tailings			
Refinery sludges			
Wastewater treatment plant sludges			
Potable water treatment sludges			
Plating and other metal hydroxide sludges	Screw	5–6	30

°Final moisture content depends on operating conditions; longer cycles produce higher values.
°The results for these different presses are not directly comparable.
SOURCE: Various manufacturers.

Tube Presses The tube press (Denver Equipment Co.; Alfa-Dyne Div., Aquilla Eng.) achieves pressures of up to 13,800 kPa (2,000 psi). A tube with a central porous candle is sealed at the top, where the inlet for slurry is located, and at the space at the bottom between the tube shell and the porous candle. Feed is pumped into the annular space between the candle and inner wall of the tube, and clear liquid flows through the candle and out the bottom. When the pump is no longer capable of adding feed to the tube, the inlet is sealed, and a rubber bladder that lines the inside wall of the tube is expanded by pumping a fluid into the space between the tube and the rubber bladder. This forces the solids and the bladder against the central candle, completing the expression dewatering. After the bladder retracts, the bottom of the tube and the candle are dropped away from the tube to allow discharge of the dewatered, and usually quite hard, filter cake. Then the cycle starts again. Used on clays, calcium carbonate, mineral concentrates, and other inorganic solids, capacity ranges up to 1 tonne/h per tube.

OPTIMIZATION

The range of feed materials that are dewatered by expression devices is expanding, and the degree of dewatering achieved by each such devices is improving. Although new equipment design is responsible for most of the improvements, pretreatment of various feeds plays a critical role in allowing the expression presses to work on many feeds. Until recently, expression has been viewed as a brute force method; users generally ignored the properties of the feed. When feed materials would react to physical compression or shearing in the press by flowing, breaking down, or in some other way allowing the solids to follow the liquid out of the press, expression presses were not used.

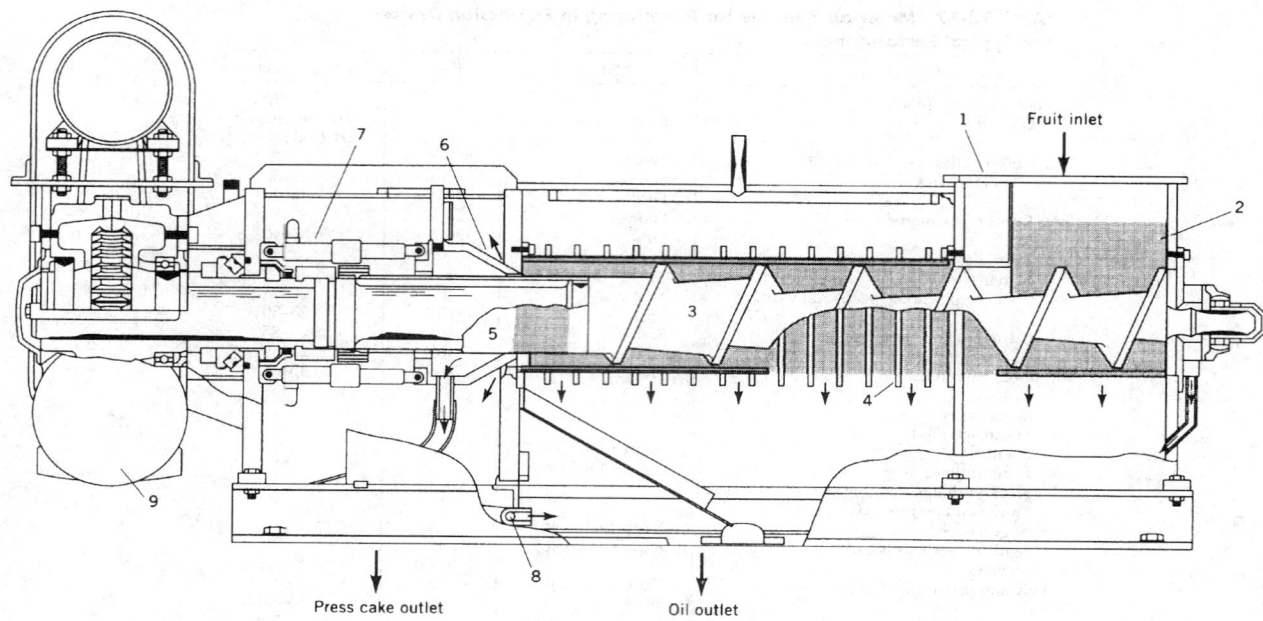

FIG. 18-155   Cross section of screw press used for fruit juice (32): (1) hopper, (2) perforated sheets, (3) main shaft, (4) perforated cage, (5) draining cylinder, (6) cone, (7) hydraulic cylinder, (8) draining cylinder oil, (9) gearbox. (*Courtesy of the French Oil Mill Machinery Co.*)

About the only treatment given to make a feed stay in the press while it was being dewatered was to add a **body feed.** The body feed would provide a network of pores and add some strength to the feed during the pressing cycle. Typical body feeds, used at 10 to 50 percent of the volume of the original dry solids content of the feed, include rice hulls, sawdust, bagasse, diatomaceous earth, kiln dust, flyash, or perlite. Fibrous body feeds are very effective in aiding screw-press performance. A recent trend is the addition of coagulants and synthetic floc-culants. These additives make both screw and belt presses effective for dewatering sewage sludges and other fine-grained hydrophilic feeds.

In the past decade adjustments in many of the more subtle variables that affect the feed to a filter have begun to be used to control dewatering presses and improve their performance. These variables affect the permeability, compressibility, and rheological properties of the feed and the resulting cake. For example, pH, streaming potential,

FIG. 18-156   Screw press. (*Anderson International Corp.*)

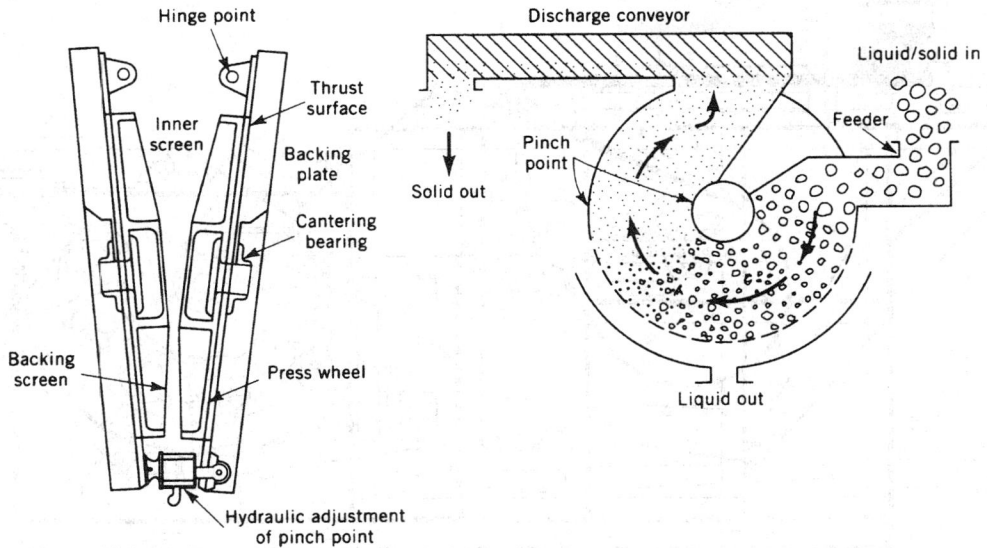

FIG. 18-157 Schematic of a disk press (41). (*Courtesy of Bepex Corp., a subsidiary of Berwind Corp.*)

viscosity, temperature, contact angle, and surface tension all affect filtration, and they can be adjusted to enhance expression dewatering as well. One indirect, continuous, online system controls polymer dosing by observing the reflectivity of the filter cake in the gravity drainage section of a belt filter press. Higher moisture content leads to high reflectivity, a condition that a human operator often uses to monitor performance [Ho, in B. M. Moudgil and B. J. Scheiner (eds.), "Floc-

culation and Dewatering," *Eng. Found. Conf.*, pp. 433–444 (Jan. 1988)]. Online control appears to be difficult to achieve [Campbell and Plaisier, *Advances in Filtration and Separation Technology,* **7** (System Approach to Separation and Filtration Process Equipment), pp. 583–586 (1993)].

Screw presses were traditionally used for seeds and fruits that had to be mechanically ruptured to release the liquid in the seeds or cells

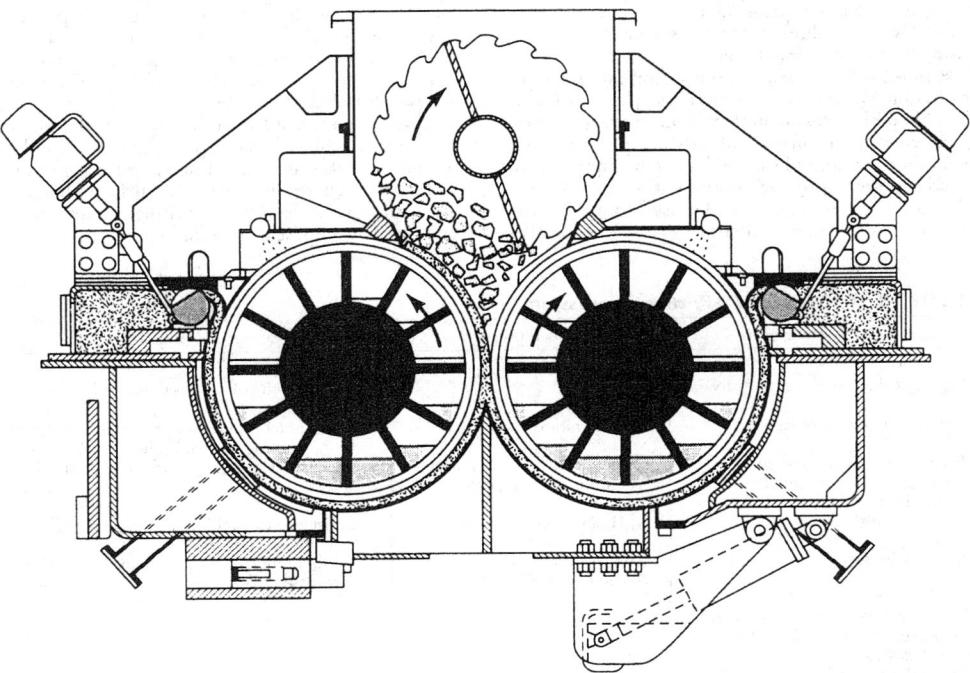

FIG. 18-158 Schematic of the Vari-Nip press (43). (*Courtesy of Ingersoll-Rand Co.*)

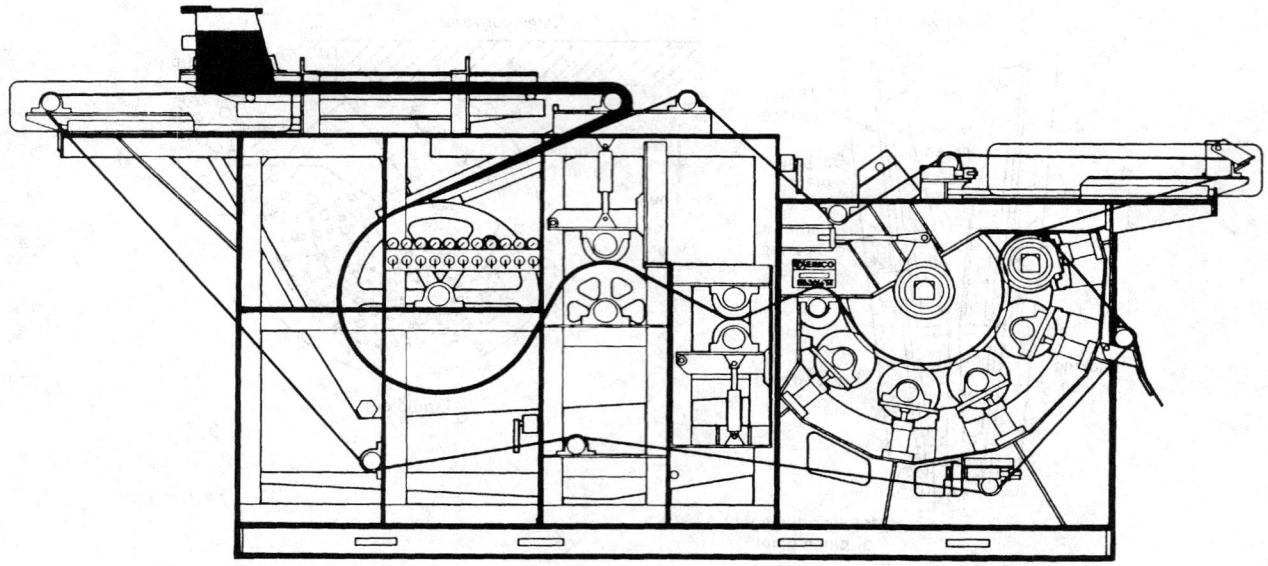

FIG. 18-159   Expressor press. (*Eimco Process Equipment Co.*)

of the fruit. Shearing was often required for effective rupturing of cells in those presses. However, when the feeds to the presses are agglomerated fine solids that are held together only by added floccu-lants and coagulants, shearing is not beneficial. Thus, designs of the presses for these two types of feeds have diverged. Designs for seeds emphasize shearing by placing screws nearly in contact with the per-forated barrel or by using twin screws in a press. Designs for express-ing without shearing emphasize the way the pitch and diameter in the screw press are varied in order to create squeezing with less shearing. Much less shearing occurs in a belt press, and still less occurs in an expression belt press. Variable-volume filter presses or tube presses give the least shearing for the most fragile cakes.

There is no substitute for fully testing various expression devices with specific feeds to determine their effectiveness. The only short-cuts are to seek the help of the press manufacturers and reagent sup-pliers in testing. Because performance of existing presses often depends critically on pretreatment of the feed, reagent suppliers are a good source for advice for improving continuing operations. For belt presses, small-scale test equipment can be used to predict the full-scale results of polymer selection and dosage [Novak, Knocke et

al., *Water Sci. Technol.*, **28**(1), pp. 11–19 (1993); Kaesler, Connelly, and Richardson in B. M. Moudgil and B. J. Scheiner, op. cit., pp. 473–490].

THEORY

The variables affecting expression are known, can be described and modeled in a variety of ways, and are useful in understanding the prin-ciples of the mechanism of expression. They are not particularly use-ful in sizing operations or in choosing one type of press over another. The variables include those listed in Table 18-18.

The response of a cake to these properties is often not linear, and it can vary as conditions in the press change. This variability significantly complicates efforts to understand the process. For example, a highly compressible cake of latex in water easily dewaters with pressure, up to a point at which the latex particles deform and essentially create a barrier to further movement of water through the cake. Particle size and surface charge interact, but surface charge affects only small (<0.1-μm) particles. Not surprisingly, these interrelationships are described by empirical equations covering restrictive ranges.

TABLE 18-18   Variables Affecting Expression

Cake properties	Slurry feed properties	Equipment properties
Thickness	Liquid/solid ratio	Pressure
Specific cake resistance (porosity)	Viscosity	Homogeneity of cake as deposited on the fil-ter medium
Compressibility, permeability	Temperature	Effect of mechanical forces on cake structure (e.g., shearing or axial loading)
Particle size, shape, degree of aggregation, and capillary pore size distribution	Surface tension, interfacial tension (contact angle)	Time the cake is subject to the highest pressures
Surface charge	pH, ionic concentration	Resistance to liquid flow through the filter medium and support
Size distribution	Rate of change in compressibility over time	Velocity required for the liquid to escape the expressed cake
Particle surface characteristics		
Type of solid (in terms of internal liquid content): gel, flocculated, hard particle		
Strength of particle (resistance to deformation under pressure)		

SELECTION OF A SOLIDS-LIQUID SEPARATOR

A good solids-liquid separator performs well in service, both initially and over time. It operates reliably day after day, with enough flexibility to accommodate to normal fluctuations in process conditions, and does not require frequent maintenance and repair. Selection of such a separator begins with a preliminary listing of a number of possible devices, which may solve the problem at hand, and usually ends with the purchase and installation of one or more commercially available machines of a specific type, size, and material of construction. Rarely is it worthwhile to develop a new kind of separator to fill a particular need.

In selecting a solids-liquid separator, it is important to keep in mind the capabilities and limitations of commercially available devices. Among the multiplicity of types on the market, many are designed for fairly specific applications, and unthinking attempts to apply them to other situations are likely to meet with failure. The danger is the more insidious because failure often is not of the clean no-go type; rather it is likely to be in the character of underproduction, subspecification product, or excessively costly operation—the kinds of limping failure that may be slowly detected and difficult to analyze for cause. In addition, it should be recognized that the performance of mechanical separators—more, perhaps, than most chemical-processing equipment—strongly depends on preceding steps in the process. A relatively minor upstream process change, one that might be inadvertent, can alter the optimal separator choice.

PRELIMINARY DEFINITION AND SELECTION

The steps in solving a solids-liquid separation problem, in general, are:
1. Define the overall problem, with expert assistance if necessary.
2. Establish process conditions.
3. Identify appropriate separator types; make preliminary selections.
4. Develop a test program.
5. Take representative samples.
6. Make simple tests.
7. Modify process conditions if necessary.
8. Consult equipment manufacturers.
9. Make final selection; obtain quotations.

Problem Definition   Intelligent selection of a separator requires a careful and complete statement of the nature of the separation problem. Focusing narrowly on the specific problem, however, is not sufficient, especially if the separation is to be one of the steps in a new process. Instead, the problem must be defined as broadly as possible, beginning with the chemical reactor or other source of material to be separated and ending with the separated materials in their desired final form. In this way the influence of preceding and subsequent process steps on the separation step will be illuminated. Sometimes, of course, the new separator is proposed to replace an existing unit; the new separator must then fit into the current process and accept feed materials of more or less fixed characteristics. At other times the separator is only one item in a train of new equipment, all parts of which must work in harmony if the separator is to be effective.

Assistance in problem definition and in developing a test program should be sought from persons experienced in the field. If your organization has a consultant in separations of this kind, by all means make use of the expertise available. If not, it may be wise to employ an outside consultant, whose special knowledge and guidance can save time, money, and headaches. It is important to do this early; after the separation equipment has been installed, there is little a consultant can do to remedy the sometimes disastrous effects of a poor selection. Often it is best to work with established equipment manufacturers throughout the selection process, unless the problem is unusually sensitive or confidential. Their experience with problems similar to yours may be most helpful and avoid many false starts.

Preliminary Selections   Assembling background information permits tentative selection of promising equipment and rules out clearly unsuitable types. If the material to be processed is a slurry or pumpable suspension of solids in a liquid, several methods of mechanical separation may be suitable, and these are classified into settling and filtration

methods as shown in Fig. 18-160. If the material is a wet solid, removal of liquid by various methods of expression should be considered.

Settling does not give a complete separation: one product is a concentrated suspension and the other is a liquid which may contain fine particles of suspended solids. However, settling is often the best way to process very large volumes of a dilute suspension and remove most of the liquid. The concentrated suspension can then be filtered with smaller equipment than would be needed to filter the original dilute suspension, and the cloudy liquid can be clarified if necessary. Settlers can also be used for classifying particles by size or density, which is usually not possible with filtration.

Solid-liquid separation by screening is possible for some suspensions of coarse particles, but it is not widely applicable. For separation of fine solids from liquids, cake filtration or the newer systems for crossflow filtration should be considered. Crossflow filtration includes ultrafiltration, where the solids are very fine particles or macromolecules, and microfiltration, where the particle size generally ranges from 0.05 to 2 μm. In microfiltration, a suspension is passed at high velocity and moderate pressure (10 to 30 lb_f/in^2 gauge) parallel to a semipermeable membrane in sheet or tubular form. The solid particles are too large to enter the pores of the membrane and tend to accumulate at the membrane surface as the fluid passes through. However, high shear at the surface moves particles back into the fluid stream, and the solid is discharged as a concentrated suspension. This product suspension is similar to that produced in a settler, but the microfiltration equipment is much smaller for the same capacity. Also, by proper membrane selection, nearly clean liquid can be obtained as the permeate product.

SAMPLES AND TESTS

Once the initial choice of promising separator types is made, representative liquid-solid samples should be obtained for preliminary tests. At this point, a detailed test program should be developed, preferably with the advice of a specialist.

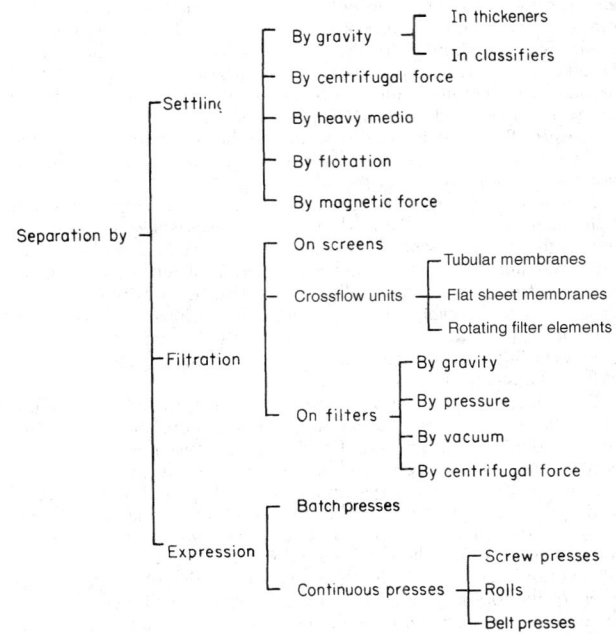

FIG. 18-160   Main paths to solids-liquid separation.

Establishing Process Conditions   Step 2 is taken by defining the problem in detail. Properties of the materials to be separated, the quantities of feed and products required, the range of operating variables, and any restrictions on materials of construction must be accurately fixed, or reasonable assumptions must be made. Accurate data on the concentration of solids, the average particle size or size distribution, the solids and liquid densities, and the suspension viscosity should be obtained *before* selection is made, not after an installed separator fails to perform. The required quantity of the liquid and solid may also influence separator selection. If the solid is the valuable product and crystal size and appearance are important, separators that minimize particle breakage and permit nearly complete removal of fluid may be required. If the liquid is the more valuable product, can minor amounts of solid be tolerated, or must the liquid be sparkling clear? In some cases, partial or incomplete separation is acceptable and can be accomplished simply by settling or by crossflow filtration. Where clarity of the liquid is a key requirement, the liquid may have to be passed through a cartridge-type clarifying filter after most of the solid has been removed by the primary separator.

Table 18-19 lists the pertinent background information that should be assembled. It is typical of data requested by manufacturers when they are asked to recommend and quote on a solid-liquid separator. The more accurately and thoroughly these questions can be answered, the better the final choice is likely to be.

Representative Samples   For meaningful results, tests must be run on representative samples. In liquid-solids systems good samples are hard to get. Frequently a liquid-solid mixture from a chemical process varies significantly from hour to hour, from batch to batch, or from week to week. A well-thought-out sampling program over a prolonged period, with samples spaced randomly and sufficiently far apart, under the most widely varying process conditions possible, should be formulated. Samples should be taken from all shifts in a continuous process and from many successive batches in a batch process. The influence of variations in raw materials on the separating characteristics should be investigated, as should the effect of reactor or crystallizer temperature, intensity of agitation, or other process variables.

Once samples are taken, they must be preserved unchanged until tested. Unfortunately, cooling or heating the samples or the addition of preservatives may markedly change the ease with which solids may be separated from the liquid. Sometimes they make the separation easier, sometimes harder; in either case, tests made on deteriorated samples give a false picture of the capabilities of separation equipment. Even shipping of the samples can have a significant effect. Often it is so difficult to preserve liquid-solids samples without deterioration that accurate results can be obtained only by incorporating a test separation unit directly in the process stream.

Simple Tests   It is usually profitable, however, to make simple preliminary tests, recognizing that the results may require confirmation through subsequent large-scale studies.

Preliminary gravity settling tests are made in a large graduated cylinder in which a well-stirred sample of slurry is allowed to settle, the height of the interface between clear supernatant liquid and concentrated slurry being recorded as a function of settling time. Centrifugal settling tests are normally made in a bottle centrifuge in which the slurry sample is spun at various speeds for various periods of time, and the volume and consistency of the settled solids are noted. In gravity settling tests in particular, it is important to evaluate the effects of flocculating agents on settling rates.

Preliminary filtration tests may be made with a Büchner funnel or a small filter leaf, covered with canvas or other appropriate medium and connected to a vacuum system. Usually the suspension is poured carefully into the vacuum-connected funnel, whereas the leaf is immersed in a sample of the slurry and vacuum is applied to pull filtrate into a collecting flask. The time required to form each of several cakes in the range of 3 to 25 mm (⅛ to 1 in) thick under a given vacuum is noted, as is the volume of the collected filtrate. Properly conducted tests with a Büchner or a filter leaf closely simulate the action of rotary vacuum filters of the top- and bottom-feed variety, respectively, and may give the experienced observer enough information for complete specification of a plant-size filter. Alternatively, they may point to pressure-filter tests or, indeed, to a search for an alternative to filtration. Centrifugal

TABLE 18-19   Data for Selecting a Solids-Liquid Separator*

1. Process
    a. Describe the process briefly. Make up a flowsheet showing places where liquid-solid separators are needed.
    b. What are the objections to the present process?
    c. Briefly, what results are expected of the separator?
    d. Is the process batch or continuous?
    e. Number the following objectives in order of importance in your problem: (*a*) separation of two different solids _____ ; (*b*) removal of solids to recover valuable liquor as overflow _____ ; (*c*) removal of solids to recover the solids as thickened underflow _____ or as "dry" cake _____ ; (*d*) washing of solids _____ ; (*e*) classification of solids _____ ; (*f*) clarification or "polishing" of liquid _____ ; (*g*) concentration of solids _____ .
    f. List the available power and current characteristics.

2. Feed
    a. Quantity of feed:
        Continuous process: _____ gal/min; _____ h/day; _____ lb/h of dry solids.
        Batch process: volume of batch: _____ ; total batch cycle: _____ h.
    b. Feed properties: temp. _____ ; pH _____ ; viscosity _____ .
    c. What maximum feed temperature is allowable?
    d. Chemical analysis and specific gravity of carrying liquid.
    e. Chemical analysis and specific gravity of solids.
    f. Percentage of solids in feed slurry.
    g. Screen analysis of solids: wet _____ dry _____
    h. Chemical analysis and concentration of solubles in feed.
    i. Impurities: form and probable effect on separation.
    j. Is there a volatile component in the feed? _____ Should the separator be vapor-tight? _____ Must it be under pressure? _____ If so, how much? _____

3. Filtration and settling rates
    a. Filtration rate on Büchner funnel: _____ gal/(min)(ft²) of filter area under a vacuum of _____ in Hg. Time required to form a cake _____ in thick: _____ s.
    b. At what rate do the solids settle by gravity?
    c. What percentage of the total feed volume do the settled solids occupy after settling is complete? After how long?

4. Feed preparation
    a. If the feed tends to foam, can antifoaming agents be used? If so, what type?
    b. Can flocculating agents be used? If so, what agents?
    c. Can a filter aid be used?
    d. What are the process steps immediately preceding the separation? Can they be modified to make the separation easier?
    e. Could another carrying liquid be used?

5. Washing
    a. Is washing necessary?
    b. What are the chemical analysis and specific gravity of wash liquid?
    c. Purpose of wash liquid: to displace residual mother liquor or to dissolve soluble material from the solids?
    d. Temperature of wash liquid.
    e. Quantity of wash allowable, in lb/lb of solids.

6. Separated solids
    a. What percentage of solids is desired in the cake or thickened underflow?
    b. Is particle breakage important?
    c. Amount of residual solubles allowable in solids.
    d. What further processing will have to be carried out on the solids?

7. Separated liquids
    a Clarity of liquor: what percentage of solids is permissible?
    b. Must the filtrate and spent wash liquid be kept separate?
    c. What further processing will be carried out on the filtrate and/or spent wash?

8. Materials of construction
    a. What metals look most promising?
    b. What metals must not be used?
    c. What gasket and packing materials are suitable?

*U.S. customary engineering units have been retained in this data form. The following SI or modified-SI units might be used instead: centimeters = inches × 2.54; kilograms per kilogram = pounds per pound × 1.0; kilograms per hour = pounds per hour × 0.454; liters per minute = gallons per minute × 3.785; liters per second-square meter = gallons per minute·square foot × 0.679; and pascals = inches mercury × 3377.

filter tests are made in a perforated basket centrifugal 254 or 305 mm (10 or 12 in) in diameter lined with a suitable filter medium. Slurry is poured into the rotating basket until an appropriately thick cake—say, 25 mm (1 in)—is formed. Filtrate is recycled to the basket at such a rate that a thin layer of liquid is just visible on the surface of the cake. The discharge rate of the liquor under these conditions is the draining rate. The test is repeated with cakes of other thicknesses to establish the productive capacity of the centrifugal filter.

Batch tests of crossflow filtration can be carried out in small pressurized cells with a polymer, ceramic, or porous metal membrane at the bottom and a magnetic stirrer to provide high shear at the membrane surface. In other test cells, a pump circulates the suspension through a rectangular channel whose top or bottom surface is the selective membrane. Membranes with a wide range of permeabilities and pore sizes are available, and, to minimize plugging, the pore size is generally selected to be much smaller than the size of the particles in the suspension. The permeate flux (flow rate per unit area) is measured for different applied pressures and slurry concentrations, and the tests are continued for several hours to check for any flux decline caused by fouling of the membrane.

More detailed descriptions of small-scale sedimentation and filtration tests are presented in other parts of this section. Interpretation of the results and their conversion into preliminary estimates of such quantities as thickener size, centrifuge capacity, filter area, sludge density, cake dryness, and wash requirements also are discussed. Both the tests and the data treatment must be in experienced hands if error is to be avoided.

Modification of Process Conditions   Relatively small changes in process conditions often markedly affect the performance of specific solids-liquid separators, making possible their application when initial test results indicated otherwise or vice versa. Flocculating agents are an example; many gravity settling operations are economically feasible only when flocculants are added to the process stream. Changes in precipitation or crystallization steps may greatly enhance or diminish filtration rates and hence filter capacity. Changes in the temperature of the process stream, the solute content, or the chemical nature of the suspending liquid also influence solids-settling rates. Occasionally it is desirable to add a heavy, finely divided solid to form a pseudo-liquid suspending medium in which the particles of the desired solid will rise to the surface. Attachment of air bubbles to solid particles in a flotation cell, using a suitable flotation agent, is another way of changing the relative densities of liquid and solid.

Consulting the Manufacturer   Early in the selection campaign—certainly no later than the time at which the preliminary tests are completed—manufacturers of the more promising separators should be asked for assistance. Additional tests may be made at a manufacturer's test center; again a major problem is to obtain and preserve representative samples. As much process information as tolerable should be shared with the manufacturers to make full use of their experience with their particular equipment. Full-scale plant tests, although expensive, may well be justified before final selection is made. Such tests demonstrate operation on truly representative feed, show up long-term operating problems, and give valuable operating experience.

In summary, separator selection calls for clear problem definition, in broad terms; thorough cataloging of process information; and preliminary and tentative equipment selection, followed by refinement of the initial selections through tests on an increasingly larger scale. Reliability, flexibility of operation, and ease of maintenance should be weighed heavily in the final economic evaluation; rarely is purchase price, by itself, a governing factor in determining the suitability of a liquid-solids separator.

Solid-Solid Operations and Equipment

Kalanadh V. S. Sastry, Ph.D., *Professor, Department of Materials Science and Mineral Engineering, University of California, Berkeley, CA; Member, American Institute of Chemical Engineers, Society for Mining, Metallurgy and Exploration. (Section Editor)*

Harrison Cooper, Ph.D., *Harrison R. Cooper Systems, Inc., Salt Lake City, Utah. (Sampling of Solids and Slurries)*

Richard Hogg, Ph.D., *Professor, Department of Mineral Engineering, The Pennsylvania State University, University Park, PA. (Mixing)*

T. L. P. Jespen, M.S., Min. Proc., *Metallurgical Engineer, Basic, Inc., Gabbs, NV. (Dense Media Separation)*

Frank Knoll, M.S., Min. Proc., *President, Carpco, Inc., Jacksonville, FL. (Super Conducting Magnetic Separation and Electrostatic Separation)*

Bhupendra Parekh, Ph.D., *Associate Director, Center for Applied Energy Research, University of Kentucky, Lexington, KY. (Other Techniques)*

Raj K. Rajamani, Ph.D., *Professor, Department of Metallurgy and Metallurgical Engineering, University of Utah, Salt Lake City, Utah. (Wet Classification)*

Thomas Sorenson, M.B.A., Min. Eng., *President, Galigher Ash (Canada) Ltd. (Flotation)*

Ionel Wechsler, M.S., Min. and Met., *Vice President, Sala Magnetics, Inc., Cambridge, MA. (Magnetic Separation)*

INTRODUCTION

SAMPLING OF DRY SOLIDS AND SLURRIES OF SOLIDS

INTRODUCTION

GENERAL REFERENCES: *Proceedings of XIX International Mineral Processing Congress,* SME, Littleton, CO, 1995. Gaudin, *Principles of Mineral Dressing,* McGraw-Hill, New York, 1939. Kelly and Spottiswood, *Introduction to Mineral Processing,* Wiley, 1980. Roberts, Stavenger, Bowdersox, Walton, and Mehta, *Chem. Eng.,* **78**(4), 89 (Feb. 15, 1971). Taggart, *Handbook of Mineral Dressing,* 2d ed., Wiley, New York, 1964. Weiss, *SME Mineral Processing Handbook,* SME, Littleton, CO, 1985.

Most of the process industries deal with solid-solid systems which belong to the class of particulate systems. Particulate systems are composed of discrete solids known as particles dispersed in a gaseous or liquid phase. Solids dispersed in liquids are known as slurry systems. Thus, the processing of particulate solids might be carried out in either dry or wet state. Processing of particulate solids involves basically two kinds of operations: *mixing* leading to the generation of a homogeneous product, and *separation* in order to produce valuable solid components and to discard undesired less valuable solids.

The control of processes involving the treatment of solids generally requires means for careful sampling and analysis of solids and slurries at various points in an operation. Unlike liquids, particulate solids are not homogeneous. The composition of individual particles will vary with particle size and particle density. It follows that care must be

exercised to take a sample that represents the entire solids mixture at the point of interest in the process. If the solids are not sampled in a representative manner, process and product control will not be reliable. The first subsection presents various aspects of sampling of solids and slurries including the underlying theory and details of different sampling equipment and their selection.

Mixing of solids is an important unit operation in the production of solids with consistent properties. A number of properties of the solid particles influence the mixing process, the design, and selection of mixing equipment. The second subsection elaborates on the theory of mixing, types of mixing equipment, and their operation.

Various techniques are available to separate the different types of particles that may be present in a solid mixture. The choice depends on the physicochemical nature of the solids and on site-specific considerations (for example, wet versus dry methods). A key consideration is the extent of the "liberation" of the individual particles to be separated. Particles attached to each other obviously cannot be separated by direct mechanical means except after the attachment has been broken. In ore processing, the mineral values are generally liberated by size reduction (see Sec. 20). Rarely is liberation complete at any one size, and a physical-separation flow sheet will incorporate a sequence of operations that often are designed first to reject as much

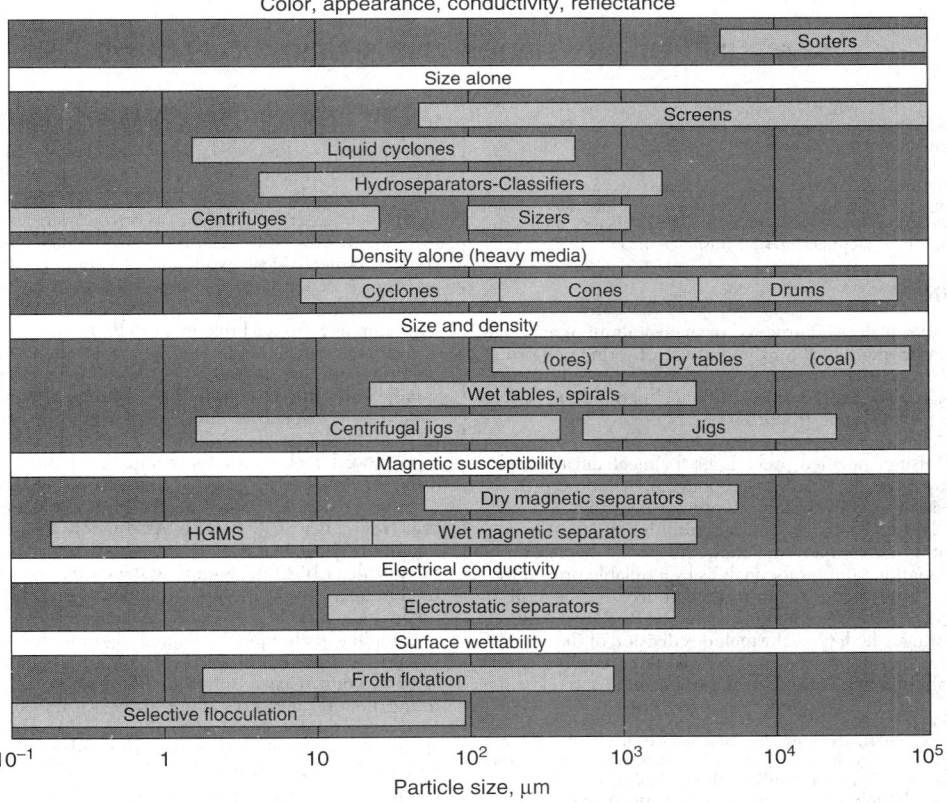

FIG. 19-1 Particle-size range as a guide to the range of applications of various solid-solid operations.

unwanted material as is possible at a coarse size and subsequently to recover the values after further size reduction.

Any difference in physical properties of the individual solids can be used as the basis for separation. Differences in density, size, shape, color, and electrical and magnetic properties are used in successful commercial separation processes. An important factor in determining the techniques that can be practically applied is the particle-size range of the mixture. A convenient guide to the application of different solid-solid separation techniques in relation to the particle-size range is presented in Fig. 19-1, which is a modification of an original illustration by Roberts et al.

The classification of solids by particle size is carried out for a number of reasons. Size classification can facilitate subsequent processing steps. An example is the scalping of tramp oversize material to avoid clogging a piece of processing apparatus. Similarly, better efficiency is achieved by removing fines before size reduction in crushers or ball or rod mills. Finished products generally are required to meet particle-size limits. Size separation is accomplished either in the dry condition or with the solids in suspension as a slurry. Wet classification allows higher process rates, particularly for materials of very fine sizes. Classification often is an integral part of a unit operation, as in closed-circuit grinding. Air classification methods for dry size classification in conjunction with size-reduction operations is covered in Sec. 20.

Gravity concentration is one of the oldest of the solids-separation techniques and the most important mineral-dressing method for obtaining ore concentrates. It is used mainly now for coal cleaning, yet Mills ["Process Design, Scale-Up and Plant Design for Gravity Concentration," in Mular and Bhappu (eds.), *Mineral Processing Plant Design,* 2d ed., Society of Mining Engineers, AIME, New York, 1980] notes that still more tonnage and greater values of material are concentrated by gravity methods than by a method such as froth flotation. The major unit operations which comprise gravity separation are jigging, tabling, spiral concentration, and dense-media separation. For high-capacity treatment of finer-sized low-grade ore materials, particularly the heavy mineral sands, the Reichert cone is becoming an industry standard [Ferree, "An Expanded Role in Minerals Processing Is Seen for the Reichert Cone," *Min. Eng.,* **25**(3), 29 (1973)].

Solids separation based on density loses its effectiveness as the particle size decreases. For particles below 100 microns, separation methods make use of differences in the magnetic susceptibility (magnetic separation), electrical conductivity (electrostatic separation), and in the surface wettability (flotation and selective flocculation). Treatment of ultrafine solids, say smaller than 10 microns can also be achieved by utilizing differences in dielectric and electrophoretic properties of the particles.

Physical separation methods are most widely used for the processing of coal and ore materials, and their basic development was designed for that purpose. Tremendous tonnages of solids are processed routinely at costs often as low at $1 per ton of material separated. The methods are applicable for other than ore processing, and solid-separation technology has become a more integral part of chemical-process operations. Recent requirements to recover values from various solid wastes have emphasized the need to adapt the relatively low-cost physical separation techniques of the ore processor, and as the needs to treat new types of materials and to improve recovery efficiency are constantly increasing, new designs are being developed.

The following subsections discuss the basic considerations involved in various unit operations of solid-solid separation and describe present industrial practice and equipment in general use.

SAMPLING OF DRY SOLIDS AND SLURRIES OF SOLIDS

REFERENCES: Society of Mining Engineers, *Minerals Processing Handbook,* Norman L. Weiss, ed., chap. 30, "Sampling and Testing", part 2. "Theory and Practice of Incremental Sampling", Littleton, Colorado, 1985. Gy, Pierre M., *Sampling of Particulate Materials—Theory and Practice,* Elsevier Scientific Publishing Co., New York, 1979. Pitard, Francis F., *Pierre Gy's Sampling Theory and Sampling Practice,* CRC Press, Inc., Boca Raton, Florida, 1995. Gayle, G. B., *Theoretical Precision of Screen Analysis,* Report of Investigations No. 4993, U.S. Bureau of Mines, Dept. of Interior, Washington, D.C., 1952.

INTRODUCTION*

Sampling is a statistically derived process—a small amount of material S is taken from a large quantity B for the purpose of estimating properties of B. If S is an accurate sample (or stated more correctly, is representative of B according to a defined statistical parameter), it is a suitable estimator for the properties of B.

Sampling is typically required for quality control, wherein statistical data are compiled using specified procedures for mechanical sample collection and sample testing. Another sampling application is providing data for process control. A key factor in process-control sampling is minimizing time delays in making data available for use. Automatic analysis equipment is often employed, and the role of mechanical sampling becomes presenting samples for analysis by a reliable procedure.

The process of sample taking encompasses several steps, beginning with (1) taking a gross sample S from bulk materials B; (2) preparation of sample S for testing, which typically includes division of the sample and possible further substeps according to whether sampling is for analysis, size distribution, moisture, ash, and so on; and (3) the testing

(analysis) step itself to determine properties of interest. Each step of the process contributes statistical error to the final result.

Estimations based on statistics can be made for total accuracy, precision, and reproducibility of results related to the sampling procedure being applied. Statistical error is expressed in terms of variance. Total sampling error is the sum of error variance from each step of the process. However, discussions herein will take into consideration only step (1)—mechanical extraction of samples. Mechanical-extraction accuracy is dependent on design reflecting mechanical and statistical factors in carrying out efficient and practical collection of representative samples S from a bulk quantity B.

Although mechanical sampling methods are to be the focus of attention, manual sampling methods are also employed for practical sample collection in commerce. Techniques of mechanical sampling should be emulated as closely as possible for best results with sampling by manual procedures.

Approved techniques for manual and mechanical sampling are often documented for various commodities handled in commerce by industry groups. Examples are the International Standards Organization (ISO), British Standards Association (BSA), Japan Institute of Standards (JIS), American Society for Testing Materials (ASTM), and the Fertilizer Institute. Sampling standards developed for use in specified industry applications frequently include instructions for laboratory work in sample preparation and analysis—steps (2) and (3) above.

Sampling techniques are more rigorous for materials with large variations in particle size and density compared to sampling of fine-sized powders. Coarse solids are often comprised of substantially differing mineral and crystalline forms within complex solids matrix. Fine-sized solid materials typically are relatively uniform in terms of chemical and physical characteristics with particle-size distributions and mineral densities usually within narrow ranges. Solids of organic chemical derivation and many commercial chemical materials, such as fertilizers, generally follow patterns of property distributions typical of powdered-mineral solids.

---

* Sampling of slurries and solids, differs fundamentally from sampling a completely mixed liquid or gas. A bulk quantity of solids incorporates characteristic heterogeneity—that is, a sample S_1 differs inherently from a sample S_2 when both are taken from a thoroughly mixed load of solids as a result of property variances embodied in solids. In contrast, all individual samples from a completely mixed liquid or gas container are statistically identical.

The following discussion centers on sampling applications for powder solids comprised of small particulate sizes and equivalents in dry form or slurries. Sampling applications involving coarser solids (⅜ inch or 10 mm nominal size) as encountered in mineral products, typical ores, coal, and quarry rock for cement manufacture, are given more complete discussion in the *Mineral Processing Handbook* published by the Society of Mining Engineers and in other references (Pitard, Gy). "Nominal" particle size implies 95 percent through-screen particle size.

THEORY OF SAMPLING

Two principal topics are considered under theory of sampling. First is theory accounting for physical properties of material to be sampled. Second is the process of mechanical sample extraction. The theory predicts accuracy of sample taking—how much sample to take and how to take it to meet an accuracy specification.

Theory related to material characteristics states that a minimum quantity of sample is predicated as that amount required to achieve a specified limit of error in the sample-taking process. Theory of sampling in its application acknowledges sample preparation and testing as additional contributions to total error, but these error sources are placed outside consideration of sampling accuracy in theory of sample extraction.

Variations in measurable properties existing in the bulk material being sampled are the underlying basis for sampling theory. For samples that correctly lead to valid analysis results (of chemical composition, ash, or moisture as examples), a fundamental theory of sampling is applied. The fundamental theory as developed by Gy (see references) employs descriptive terms reflecting material properties to calculate a minimum quantity to achieve specified sampling error. Estimates of minimum quantity assumes completely mixed material. Each quantity of equal mass withdrawn provides equivalent representation of the bulk.

The theory enables a reasonable estimate of sample quantity needed to attain specified accuracy of a composition variable. The result is an ideal quantity—not realized in practice. Actual quantities for practical estimation are larger by an appropriate multiple to account for the reality that material is incompletely mixed when stored in stockpiles or carried on conveyors. Sample quantity to accommodate incompletely mixed solids can be specified through evaluating variance by autocorrelation of data derived with a series of stockpile samples, or from multiple sample extractions taken from a moving stream (Gy, Pitard).

In addition to composition factors, a sampling theory is available in sampling for size distribution. Quantity of sample needed to reach a specified error in determining size fraction retained on a designated screen is estimated by application of the binomial theorem (Gayle).

The second topic in theory of sampling pertains to mechanical sample taking. Design of mechanical sampling must conform to established criteria for sample-taking error to be minimal. This ensures error variance introduced by mechanical sample extraction is statistically insignificant compared to physical factors of sampling arising from heterogeniety, sample preparation, and sample testing sources of error.

Estimating Minimum Sample Quantity for Analysis The fundamental theory of sampling error variance can be applied to estimating a minimum quantity required from a completely mixed lot of solids for attaining an objective level of accuracy (Gy):

$$V = \left[\frac{1}{W_S} - \frac{1}{W_B}\right]\left[\frac{1-F}{F}\{(1-F)A_m + FA_g\}\right]fgbd^3$$

where V is the objective sampling-error variance (weight fraction), W_S is weight of the sample, W_B is weight of the bulk-solids lot, F is weight fraction mineral or other measurable quantity in the solids, A_m is density of mineral, and A_g is density of the nonmineral matrix.

Remaining terms to right of the bracket relate to properties to be measured within the matrix. The factor f is adjusted from 0 to 1 in relationship to the purpose of testing. A low value of f is indicated for scarce elements such as precious metals in electronic-source scrap. Moisture content has a high f value. The factor g is adjusted from 0 to

1 according to the degree of particulate classification. A high degree of size classification, as in a case of fine powders from screening, indicates values of 0.5 or higher. Unclassified fine solids from crushing have a value assigned to g of 0.25 or less. The factor b relates to size of elemental or crystal particles in bulk-solids particulate and degree of liberation ranging from 0 to 1. The term d is nominally the largest particle size. Estimated values employed in calculations rely on sampling experience and from solids-property investigation according to development of the theory, as described in related publications (Gy).

Example 1: Sample Quantity for Composition Quality Control Testing An example is sampling for quality control of a 1,000 metric ton (W_B) trainload of ⅜ in (9.4 mm) nominal top-size bentonite. The specification requires silica to be determined with an accuracy of plus or minus three percent for two standard errors (s.e.). With one s.e. of 1.5 percent, V is 0.000225 (one s.e. weight fraction of 0.015 squared). The problem to be solved is thus calculating weight of sample to determine silica with the specified error variance.

Bentonite has expected silica content of 0.5 weight percent (F is 0.005). Silica density (A_m) is 2.4 gm per cu cm, and bentonite (A_g) is 2.6. The calculation requires knowledge of mineral properties described by the factor ($fgbd^3$). Value of the factor can be established from fundamental data (Gy) or be derived from previous experience. In this example, data from testing a shipment of bentonite of 10 mesh top-size screen analysis determined value of the mineral factor to be 0.28. This value is scaled by the cube of diameter to ⅜-in screen size of the example shipment. The mineral factor is scaled from 0.28 to 52 by multiplying 0.28 with the ratio of cubed 9.4 mm (⅜-in screen top-size of the shipment to be tested) and cubed 1.65 mm (equivalent to 10 mesh).

Minimum weight W_S of sample is 110 kg from

$$0.000225 = \left[\frac{1}{W_S} - \frac{1}{10^6}\right]\left[\frac{1-0.005}{0.005}[(1-0.005)2.6 + (0.005)2.4]\right]52$$

noting dimension of d^3 (particle diameter) is cubic mm requiring division by 1000 to rationalize with cubic cm of density. Sample weight in grams (from density) is divided by 1000 in converting to kg.

Estimating Change of Sampling Error with Change in Sample Size Increased accuracy in estimating a quality parameter by sampling through larger sample quantity can be estimated using the simplified Gy sampling equation

$$W_1V_1 = W_2V_2$$

W and V are values for sample weights and variances of parameter measurements at states 1 and 2 respectively.

Example 2: Calculation of Error with Doubled Sample Weight Repeated measurements from a lot of anhydrous alumina for loss on ignition established test standard error of 0.15 percent for sample weight of 500 grams, noting V is the square of s.e. Calculation of variance V and s.e. for a 1000 gram sample is

$$V = \frac{(0.15)^2(500)}{(1000)} = 0.01125; \text{ standard error} = 0.11 \text{ percent}$$

Estimating Minimum Sample Quantity for Size Distribution Testing A simplistic approach to specifying minimum sample size for estimating particle distributions within allowed variance is based on a screening process in terms of binomial distribution. Each screening event is an outcome of two possibilities—particles either pass the screen or not. A relationship according to this principle presented by Gayle (loc. cit.) is employed in the example. Further development of sampling concepts for particle-size distribution is provided in the references (Pitard).

Example 3: Calculating Sample Weight for Screen-Size Measurement Weight W of bulk sample for screen analysis is calculated by the Gayle model for percent retained on a specified screen with relative standard error s.e. in percent

$$W = \frac{G(100-G)w}{V}; \quad \text{example } W = \frac{5.5(100-5.5)0.0120}{1.56} = 4.0$$

where G is the weight percent of the sample retained on the given screen either as determined by testing or defined per specification, and w is the weight of a particle of the size retained on that screen.

Sample weight estimated in this example is for two standard errors of 2.5 percent (resulting in V of 1.56) for testing iron ore (hematite) retained on a ½-in screen. Estimate of G is 5.5 for 94.5 percent of weight passing. Particle weight

w retained on a ½-in opening screen assuming spherical shape and 5.1 specific gravity is 0.0120 lb. The calculation yields 4.0 lb as a minimum sample size, W.

Estimating Minimum Sample Quantity for Moisture Measurement   Estimates of material quantity for testing moisture content depend on mechanisms of moisture distribution in the material. Moisture is physically retained on particle surfaces, chemically adsorbed on surfaces and within pores of particulate solids, and contained as an internal constituent of solids. Significant internal moisture is most often encountered in organic and agricultural source materials.

Sample quantity to estimate moisture for specific material is influenced to various levels of significance by properties such as particle-size range as well as relative amounts of moisture distributed among denoted forms of retention. Practical sample size estimates require background knowledge of parameters derived from experience for specific materials. More detailed examination of moisture-sampling aspects is provided in reference texts (Pitard).

Example 4: Calculation of Sample Weight for Surface Moisture Content   An example is given with reference to material with minimal internal or pore-retained moisture such as mineral concentrates wherein physically adhering moisture is the sole consideration. With this simplification, a moisture coefficient K is employed as multiplier of nominal top-size particle size d taken to the third power to account for surface area. Adapting fundamental sampling theory to moisture sampling, variance is of a minimum sample quantity is expressed as

$$V = \left[\frac{1}{W_S}\right]\left[\frac{1-F}{F}\right]Kd^3; \text{ example } 0.0000562 = \frac{0.95\ 5\ 0.00633}{W_S\ 0.05\ 1728}$$

where V is variance in weight fraction, W_S is minimum weight of sample, F is nominal weight fraction moisture, and K is a constant with dimension mass per unit volume. In absence of prior knowledge for material surface moisture characteristics, a value of K equal to 5 lb/ft^3 can be used for typical mineral concentrates and other nonabsorbing fine materials. This relationship is applied in an example of a crystalline product—hydrated sodium sulfate (Glaubers salt) with d of minus 4 mesh (0.185 in). Standard material moisture content is 5 percent by weight, with required sampling error of 1.5 percent relative to total weight for two s.e. Variance for this value in weight fraction is 0.0000562 in calculating 6.1 lb as sample weight (1728 converts in^3 to ft^3).

MECHANICAL DELIMITATIONS OF SAMPLING

Sample increment extraction requires a cutter to move through (traverse) a flowing stream being sampled while meeting accepted criteria of design and operation. Two methods of mechanical sampling for materials in flow regime are employed. A preferred first method is sample extraction from material in gravity free fall, such as from trajectory discharge at the pulley of a conveyor or gravity flow down an enclosed chute. Cutter motion can be linear or rotational with constant speed while taking samples by traversing a gravity free-fall flow stream.

Sampling is required to meet the principle of mechanical sample extraction in maintaining statistical validity. This principle states that the cutter must take through-stream extractions during each traverse of the flow stream being sampled such that each particle in the flow stream at any place in the stream has equal probability of being extracted into sample. The diagram of Fig. 19-2 illustrates a typical arrangement meeting criteria (sampling delimitations) for a linear-traversing cutter installation extracting from a free-fall stream of material.

An alternative method is sampling directly from a moving or stationary conveyor with cutter traverse through the complete material bed carried on the conveyor. The alternative method cannot assure executing complete extractions, or through-stream sampling, because in many applications residual fines from the material stream remain on the conveyor surface.

The alternative method of sample extraction is termed the *cross-stream* sampling method, or *cross-belt* when used in conjunction with a belt conveyor. Sample extraction typically take place with a belt conveyor in motion. However, with a rotary table-feeder conveyor, extractions are made with the table stopped. A cutter can perform extractions by this means from a machined flat surface with negligible

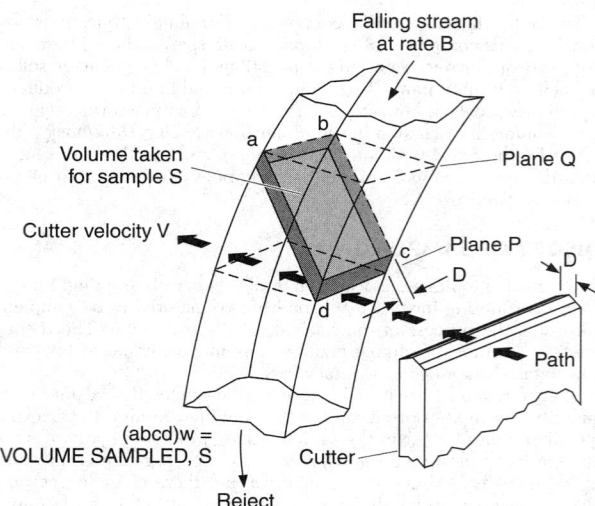

FIG. 19-2   Through-stream linear sampling. (*Courtesy of Harrison R. Cooper Systems, Inc. Salt Lake City, Utah.*)

residual fines left out of the sample. When sampling from a moving belt conveyor, residual fines become more significant resulting in loss of accuracy in extractions. This is due to clearances necessary between cutter edges and the conveyor belt, and also due to belt surface irregularities.

CRITERIA FOR SAMPLER DESIGN

Operation of a traversing sampler for *gravity flow* of material for through-stream sampling is required to meet the following design factors:

1.   The cutter moves at constant speed (or constant rotation rate in the case of a rotary-motion sample cutter) such that the entire flow of material is traversed by the cutter, with the further requirement that the stopped position of the cutter at either limit of traverse (out of stream) is at sufficient distance from the stream so that no material from the stream enters the cutter while it is held stationary between traversing operations.

2.   The sample cutter opening is set to specified width according to a multiple of the maximum (nominal) size of particulate being sampled and selected speed of the cutter. A minimum width of 10 mm or 0.375 in is recommended unless material is moist or has other properties to induce bridging of the cutter, suggesting need for a wider opening for practical operation. Experiments have determined that a cutter opening of a multiple of three times the nominal largest particle size and an 18-inches-per-second cutter speed (0.46 meters per second) is optimum to minimize sample extraction quantity with negligible delimitation error for fine-sized materials.

3.   Cutter blade length extends beyond the material stream width on either side of the stream and volume of the cutter is sufficient to ensure all material taken into sample can be contained in the cutter body. Cutter blades are parallel, and are beveled to a sharp edge in the case of linear-motion traverse. For rotary-motion sample cutters, sharp edges of the cutter blades are radial to the center point of rotation.

Criteria for mechanical delimitations in sampling by the alternative cross-stream method to fulfill through-stream extraction requirements are revised from gravity-sampling criteria in the following respects:

1.   The cutter opening is to exceed maximum (nominal) particle size by sufficient clearance to ensure that a large particle will not wedge into the opening. Sampling error due to free-fall deflection is avoided as a factor in setting cutter opening width. A 2 inch minimum cutter opening, required for practical operation, is recommended.

2.   The cutter length should be approximately equal to the width of the material load carried on the conveyor.

3.   When sampling from moving belt conveyors, the cutter operates in a radial mode with the belt surface contoured at the point of sampling by idlers, fixing radial curvature to the outer radius of the cutter. Clearance is minimized between outer edges of cutter blades and belt surface by cutter-shaft adjustment in the drive-clamping bracket.

4.   Cutter speed at the outer radius is recommended at twice the conveyor belt speed for through-stream extractions from moving belts. The cutter is adjusted in a lateral angle to a 30-degree position, matching the cutter extraction path through the material bed on the belt at specified speed.

Cross-stream sampling from flat surfaces with material handled on a linear conveyor or rotary table is best carried out with the conveyor stopped. Sample extraction is then performed by linear traverse.

MECHANICAL SAMPLING EQUIPMENT

Repeating an axiom stated earlier, mechanical samplers are designed to extract increments of sample from a bulk quantity of material B in a manner that increments S are representative within statistical bounds of the bulk B. Further, the sampler is designed and constructed in conformance to criteria stated previously under "Mechanical Delimitations of Sampling" to assure that negligible errors arise from mechanical influence.

Many designs of equipment purported for sample extraction have been offered to industry or placed into service for sampling that fail to meet accepted mechanical standards. Extracted increments often have bias—inaccuracies found from tests on increments showing deviations usually with more or less fixed offset from true median values, or otherwise producing inconsistent and statistically poor test data compared to true values. Extraction increments using nonconforming designs may best be regarded as specimens of bulk B, but not samples in the statistical sense.

Mechanical sampling procedures further discussed are limited to sampling of flowing materials. Dry solid flows carried on a conveyor or in chute gravity fall are subject to through-stream sampling designed to extract correctly defined increments. Slurry gravity flows in launders and sloped pipes are sampled at the point of discharge, or slurries are sampled at open discharge from a vertical gravity pipe.

Static sampling methods with mechanical systems to operate thief-pipe sampling of solids taking increments from railroad hopper cars, trucks, or bins are seen in use. These are considered manual sampling methods operated mechanically. Applying criteria of through-stream sample extraction is infeasible, and it is inherently understood that bulk materials to be sampled in this manner are not perfectly mixed. An assured mode of sampling is providing through-stream sample extraction of bulk materials as they are loaded into bins, rail cars, trucks, and so on.

Various static thief or pipe samplers, often including pumps for stream transfers, are employed in slurry flows as well. These lack validity in terms of through-stream extraction capability. A pressure-thief sampler mounted on a pump discharge flange can be an approximation to through-stream sampling with assumption of complete mixing in flow from the pump if time lapse for flow to the thief from a pump is minimal, and pipe bends or other elements inducing classification are absent.

SELECTING A SAMPLER

Mechanical samplers meeting delimitation criteria are available in two basic designs for sampling material in gravity free fall. The basic designs are sampling with linear cutter motion and sampling with radial cutter motion (see Fig. 19-3 and Fig. 19-4 respectively). The net result is the same with either when equipment is properly designed and operated.

Selection of linear or radial (rotary cutter) sampling is made according to mechanical installation factors often on a basis of flow quantity. Smaller flows can be sampled in a cost-effective manner by rotary cutter samplers (frequently termed "vezin" design samplers, see Fig. 19-4).

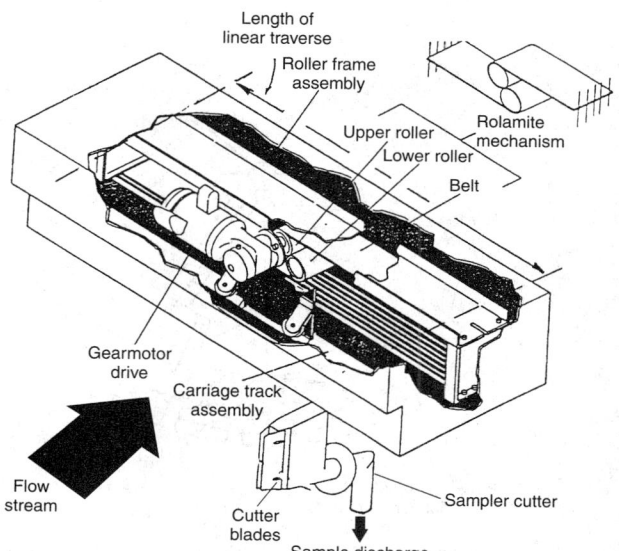

FIG. 19-3   Traversing sampler. (*Courtesy of Harrison R. Cooper Systems, Inc. Salt Lake City, Utah.*)

Sampling directly from material lying on the conveyor using a cross-stream cutter for extracting sample increments is diagrammed in Fig. 19-5 for moving conveyor belts and in Fig. 19-6 for a rotary table application. Cross-stream sampling can frequently be applied with acceptable delimitation error to materials of relatively low particle size and minimal variation, and also to materials with moisture content sufficient to avoid fines classification onto conveyor surfaces. A brush fixed to the cutter trailing edge aids in fines extraction to minimize residual sample remaining on the belt surface following cutter traverse.

In Fig. 19-5, the conveyor belt is radially profiled at the point of sample extraction with contouring idlers set to match the path of the cutter moving from its driveshaft rotation axis. Cutter edges are posi-

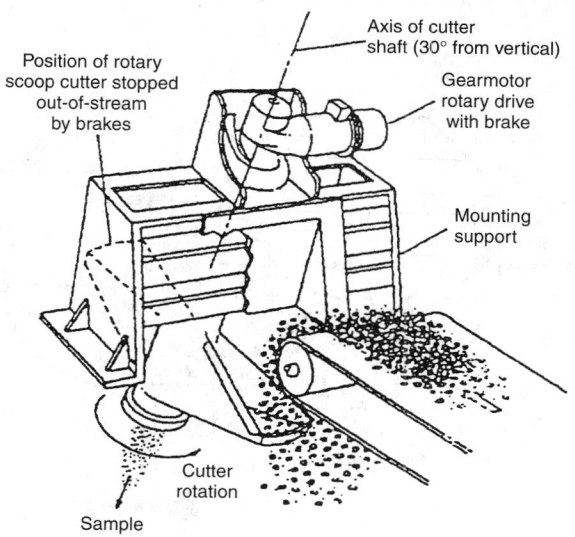

FIG. 19-4   Rotary sampler. (*Courtesy of Harrison R. Cooper Systems, Inc. Salt Lake City, Utah.*)

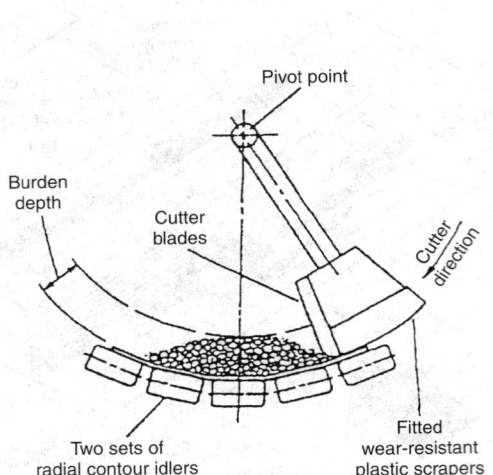

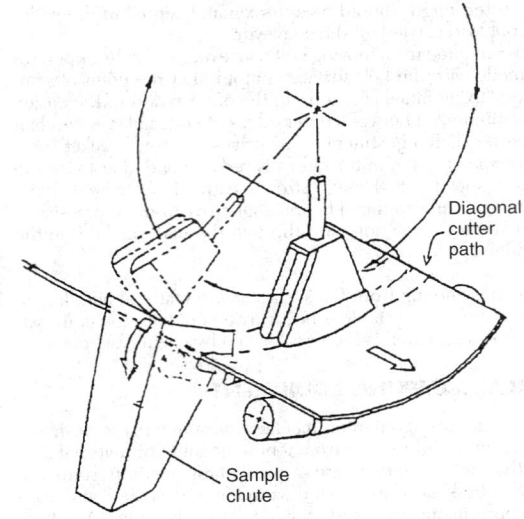

FIG. 19-5 Cross-belt sampler. (*Courtesy of Harrison R. Cooper Systems, Inc. Salt Lake City, Utah.*)

tioned with minimum clearance from the belt surface as is reasonably accomplished without contact of the cutter with the belt surface. Cutter blades are angled 30 degrees from the conveyor belt direction in positioning the cutter to its path through the conveyor belt load for cutter speed twice conveyor speed.

Extractions performed with the conveyor stopped allow more assured accuracy by the certainty of including fines in the sample increment. Sampler design to extract increments from a flat belt or rotary table sampler while the conveyor is stopped minimizes potential for residual fine particles remaining on the conveyor surface in carrying out extractions. See Fig. 19-6 for rotary table sampler extraction diagram.

Composite Samples Obtained by Multiple Sample Extractions Material flow streams are sampled in practice by combining extractions taken at successive time intervals into a composite sample. Multiple increment collection to obtain representative composite samples for specified bulk-material flows is performed according to a

designated process in accommodating the presence of material property variations.

The requirement is to obtain proportional samples from the flowing material. This is accomplished in a technically accurate procedure by extractions taken on fixed time intervals. Variable time intervals with intervals determined from random selection are optionally employed to avoid bias error in sampling when characteristic periodic effects are known to be present in the stream of material. Possibilities for fixed sampling intervals to systematically coincide with periodicities are avoided by random time interval selection. Setting sampling intervals to material flow quantity, as in using belt weigh scale readings, opens potential for nonproportionalities and error in the composite sample.

Sampling of specified material flows to obtain representative composite quantities is a common practice for material accounting and quality control. A typical case is composite sampling of a shipload or trainload cargo transfer for either receiving or delivering materials. Another frequently used specification is eight-hour shift production quantities to be sampled to generate composite samples for testing.

Industry standards are frequently applicable in designing sampling procedures for many commodities in commerce transferred by ship cargoes or trainloads. Standards for iron ore, coal, metallurgical concentrates, and similar materials are often to be observed. Standards are likely to give details on sampling specifications necessary for acceptance-based material characteristics and lot size to mandate minimum number and weights of increments, gross (combined) sample weights, and other factors.

Selection of appropriate time intervals for increment extractions relates to property variation (inhomogeneity) within material flow streams. Ten minute extraction intervals are generally adequate to obtain suitably representative samples from material flows under practical circumstances. Precise determination of extraction intervals consistent with individual applications can be calculated through autocorrelation of historical sampling data, a statistical method described in references (Gy, Pitard).

Sample Quantity Reduction As sample increments are accumulated by multiple extractions from a bulk flow of material, according to the parameters of sampling to accommodate material stratification and nonhomogeneous composition, gross sample quantities (primary sample) often become quite extensive. Large primary sample volumes are subject to mechanical resampling to obtain final samples of practical, reduced quantities for testing. The same principles of sampling applied to primary sampling are used to design resampling to accomplish sample reduction without loss of sample statistical validity.

FIG. 19-6 Rotary-table sampler. (*Courtesy of Harrison R. Cooper Systems, Inc. Salt Lake City, Utah.*)

Sample reduction in successive stages—primary to secondary, secondary to tertiary, etc.—can be fulfilled using automatic sampling equipment while observing design principles of statistical sampling. Alternatively, sample quantity reduction may be carried out in a laboratory.

Sample reduction by mechanical procedures in automatic on-line mode encompasses (1) particle-size reduction preceding a following stage of resampling, and (2) multiple secondary increments taken for each primary increment when resampling without particle-size reduction. Particle-size reduction implies crushing or grinding the sample before resampling. A sampling-unit design incorporating primary and successive stages of sampling, with particle-size reduction and controlled flow of sample through intermediate stages, is developed in accord with application requirements while maintaining specified standards of sample accuracy.

Calculation of Sample Extraction Increments   Sample quantities taken in an extraction increment are calculated in accord with the mechanical sampler employed. The following three examples illustrate calculations for three commonly used sampling methods.

Example 5: Solids Sampling by Linear Traversing Trajectory Cutter   Increment weight S by a linear traversing cutter from bulk flow of fine powder B expressed in unit weight per unit time is calculated by

$$S = \frac{B \times D}{V} \quad \text{example } S = 1.38 = \frac{120 \times 2000 \times 0.375}{3600 \times 18}$$

where V is cutter velocity and D is cutter opening. For S given in 120 short tons per hour converted to lbs, 0.375-in cutter opening, and 18-inches-per-second cutter speed, 1.38 lb. For consistent units, tons per hour is multiplied by 2,000 for lbs per hour, and divided by 3,600 for lbs per second.

Example 6: Slurry Sampling by Rotary Traverse of Gravity Flow   Increment volume, quantity of slurry extracted by one cutter rotation, is S from bulk slurry flow B expressed in volume-per-unit time. R is cutter rotation per minute. D is cutter angle opening, with D/360 extraction ratio for continuous cutter rotation.

$$S = \frac{D \times B}{360 \times R} \quad \text{example } S = 0.055 = \frac{2.5 \times 200}{360 \times 25}$$

with S gallons per extraction for 200 gallons per minute, 2.5 degree cutter opening, and 25 RPM cutter rotation rate.

Example 7: Cross-Belt Sampling of Solids from Conveyors   Increment weight is S from bulk material flow B expressed in unit weight per unit time. J is belt speed in length-per-unit time. J × 2 is cutter speed. Therefore, cutter angle determining path length through material loads on the conveyor belt is 30 deg for traversing perpendicular to direction of conveyor movement. Extraction weight is corrected by csc(30 deg) to account for diagonal path of the cutter. Solids weight-per-unit length of conveyor belt is B/J. With cutter width D of 50 mm, minimum recommended for fine powders, increment weight S is

$$S = \frac{B \times D \times 1.16}{J} \quad \text{example } S = 1.93 = \frac{120 \times 1000 \times 50 \times 1.16}{3600 \times 1 \times 1000}$$

for S given in kg at 120 metric tons per hour and 1 meter per second conveyor belt speed. Consistent units require tons per hour be multiplied by 1,000 for kg per hour, and divided by 3,600 for kg per second. Cutter opening is divided by 1,000 for meters.

Sampling Trajectory Stream from Conveyor-Belt Discharge   Conveyor-belt speeds above approximately 300 ft per minute (1.5 meters per second) impart sufficient momentum to material discharging at its head pulley to cause lifting of material streams in a trajectory from the head pulley. A trajectory is illustrated in Fig. 19-7. Blades of the sample cutter are positioned to intersect the trajectory. See Fig. 19-7 for an example of a linear-traversing bottom-dump cutter installation. Calculation of trajectory profiles are described in the Conveyor Equipment Manufacturers Association publications and similar references.

SAMPLING EQUIPMENT COST DATA

The cost of an electric-drive rotary-cutter sample of the smallest size manufactured—suitable for gravity sampling of fine particulate solids or slurry flow—including timer and control unit was approximately $5,000 in 1996.

An electric-drive linear-traversing sampler of minimum standard manufactured size with cutter and controls will range upwards of $8,000.

Pneumatic as well as electric-drive samplers are available. Generally, pneumatic-drive samplers are lower in cost.

Cross-belt samplers of minimum size for 24-in (600-mm) conveyors cost approximately $15,000 with controls using an electric drive, and about $12,500 with pneumatic drive.

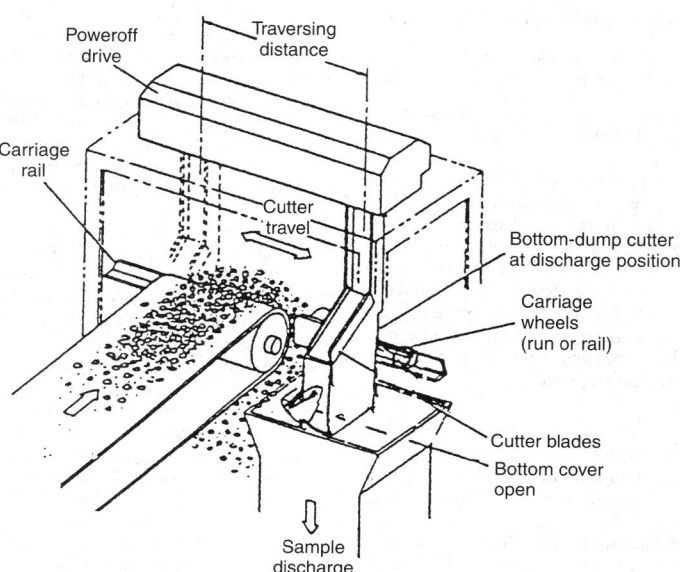

FIG. 19-7   Traversing linear bottom-dump sampler. (*Courtesy of Harrison R. Cooper Systems, Inc. Salt Lake City, Utah.*)

Hydraulic-drive samplers are also available, but cost factors tend to be substantially greater than electromechanical units. Recent use of hydraulic-drive systems has diminished with the availability of increased strength and durability electric-motor linear-drive units capable of reliable operation in high-capacity applications.

Sampling systems for multiple-stage sample reduction incorporating components such as crushing units, interstage feeders, reject handling, and others range up to several hundred thousand dollars in cost. A requirement would be rarely encountered in fine-powder applications.

SOLID-SOLID SYSTEMS

MIXING

GENERAL REFERENCES

1. *AIChE Standard Testing Procedure for Solids Mixing Equipment*, American Institute of Chemical Engineers, New York.
2. Bullock, *Chem. Eng.*, **66**, 177 (Apr. 20, 1959).
3. Danckwerts, *Research*, **6**, 355–361 (1953).
4. Fischer, *Chem. Eng.*, **67**, 107 (Aug. 8, 1960).
5. Kirk and Othmer (eds), *Encyclopedia of Chemical Technology*, Wiley-Interscience, New York: Rushton, Boutros, and Selheimer, "Mixing and Agitating," vol. 9, 1st ed., pp. 133–166; Rushton and Boutros, "Mixing and Blending," vol. 13, 2d ed., pp. 577–613; Oldshue and Todd, "Mixing and Blending," vol. 15, 3d ed., pp. 604–637.
6. Lacey, *J. Appl. Chem.*, **4**, 257–268 (1954).
7. Quillen, *Chem. Eng.*, **61**, 178 (June 1954).
8. Scott, in Cremer and Davies (eds.), *Chemical Engineering Practice*, vol. 3, Butterworth, London, 1957, p. 362.
9. Weidenbaum, in Drew and Hoopes (eds.), *Advances in Chemical Engineering*, vol. II, Academic, New York, 1958, chap. on mixing of solids.
10. Work, *Chem. Eng. Prog.*, **50**(9), (September 1954).
11. Vance, "Statistical Properties of Dry Blends," *Ind. Eng. Chem.*, **58**, 37–44 (1966).
12. Gren, "Solids Mixing: A Review of Theory," *Br. Chem. Eng.*, **12**, 1733–1738 (1967).
13. Ashton and Valentin, "Mixing of Powders and Particles in Industrial Mixers," *Trans. Inst. Chem. Eng. (London)*, **44**, t165–t188 (1966).
14. Valentin, "Mixing of Powders and Pastes: Basic Concepts," *Chem. Eng. (London)*, **45**, CE99–CE106 (1967).
15. Uhl and Gray (eds.), *Mixing*, vols. I and II, Academic, New York, 1966–1967.
16. Bhatia and Cheremisinoff (eds.), *Solids Separation and Mixing*, vol. 1, Technomic Publishing Co., Westport, Conn., 1979.
17. Beddow, *Particulate Science and Technology*, Chemical Publishing, New York, 1980.
18. Fan, Chen and Lai, "Recent Developments in Solids Mixing," *Powder Technol.* **61**, 255–287 (1990).
19. Harnby, N., "The Estimation of the Variance of Samples Withdrawn from a Random Mixture of Multi-Sized Particles," *Chem. Eng.* No. 214: CE270–71 (1967).
20. Stange, K., "Mixing Quality in a Random Mixture of Three or More Components," *Chem. Ing. Tech.*, **35**:580–82 (1963).

A comprehensive bibliography is available in Ref. 9. A more recent update can be found in Ref. 18. Equipment photographs and details are available in Refs. 2, 4, 5, 7, 8, and 10. References 3 and 6 give excellent theoretical work. Reference 5 gives a tabulation and summary of many mixer types and applications. References 8 and 9 are book chapters dealing with mixing of solids and cover both the theoretical and the equipment aspects. Interpretive summaries of the literature in various areas (state of mixedness, theoretical frequency distributions, rate equations, and equipment) are included in Refs. 9 and 18. Reference 1 gives a procedure for testing solids-mixing equipment.

Fundamentals

Objectives Equipment in which solid materials are mixed may be used for a number of operations. Blending of ingredients may be the main objective, as, for example, in the preparation of feeds, insecticides, fertilizer, glass batches, packaged foods, and cosmetics. Other objectives may include cooling or heating such as in the cooling of limestone or sugar or the preheating of plastic prior to calendering. Drying or roasting of the solids is sometimes desired. In some applications, such as polymerization of plastics, catalyst manufacture, or the preparation of cereal products, the solids mixture may be reacted.

Coating is desired in some cases, as in the manufacture of pigments, dyes, minerals, candy, and other food products and in the preparation of feeds. In certain of these cases, small amounts of liquid may be added, but the end product is a solids mixture. Sometimes agglomerates are desired, as in the preparation of food products, pharmaceuticals, detergents, and fertilizer. Often size reduction is desired while solids are being mixed. In all cases, the mixing of solids occurs. However, in some of these operations, the details of the equipment to accomplish operations other than pure blending may become a major problem. This portion of Sec. 19 will deal with equipment whose major function is to give a thorough mixture of solids. Specialized equipment to perform the other functions is discussed in other sections of the *Handbook* and will not be dealt with here. Thus, for example, Sec. 8 is devoted to size reduction and enlargement, although equipment mentioned there may also accomplish mixing.

Properties Affecting Solids Mixing Wide differences among properties such as particle-size distribution, density, shape, and surface characteristics (such as electrostatic charge) may make blending very difficult. In fact, the properties of the ingredients dominate the mixing operation. The most commonly observed characteristics of solids are as follows:

1. *Particle-size distribution.* This tells the percentages of the material in different size ranges.
2. *Bulk density.* This is the weight per unit of volume of a quantity of solid particles, usually expressed in kilograms per cubic meter (pounds per cubic foot). It is not a constant and can be decreased by aeration and increased by vibration or mechanical packing.
3. *True density.* The true density of the solid material is usually expressed in kilograms per cubic meter (pounds per cubic foot). This, divided by the density of water, equals specific gravity.
4. *Particle shape.* Some types are pellets, egg shapes, blocks, spheres, flakes, chips, rods, filaments, crystals, or irregular shapes.
5. *Surface characteristics.* These include surface area and tendency to hold a static charge.
6. *Flow characteristics.* Angle of repose and flowability are measurable characteristics for which standard tests are available (e.g., ASTM Test B213-48, Flow Rate of Metal Powders, etc.). A steeper angle of repose would indicate less flowability. The term "lubricity" has sometimes been used for solid particles to correspond roughly to viscosity of a fluid.
7. *Friability.* (Also see "Grindability," Sec. 8.) This is the tendency of the material to break into smaller sizes in the course of handling. There are quantitative tests specially devised for certain materials such as coal which can be used to estimate this property. Abrasiveness of one ingredient upon another should also be considered.
8. *State of agglomeration.* This refers to whether the particles exist independently or adhere to one another in clusters. The kind and degree of energy employed during mixing and the friability of the agglomerates will affect the extent of agglomerate breakdown and particle dispersion.
9. *Moisture or liquid content of solids.* Often a small amount of liquid is added for dust reduction or special requirements (such as oils for cosmetics). The resultant material may still have the appearance of a dry solid rather than a paste.
10. *Density, viscosity, and surface tension.* These are properties at operating temperature of any liquid added.
11. *Temperature limitations of ingredients.* Any unusual effects due to temperature changes which might occur (such as heat of reaction) should be noted.

A look at these properties for the ingredients to be mixed is a first step toward selecting mixing equipment.

Measuring Uniformity Except for cases in which a coating of one ingredient with another takes place, the theoretical end result of mixing will not be an arrangement in which one type of particle is directly next to a different type. Rather, the theoretical end result when random tumbling takes place will be a random mixture along the lines shown in Fig. 19-8.

The variation among spot samples of known size can be predicted theoretically for a random mixture and used as a guide to determine how closely random blending of the ingredients has been approached. Various types of analyses can be made on spot samples to determine batch uniformity. These could include x-ray fluorescence, flame spectrometry, polarography, emission spectroscopy, and so on, depending on the powder being examined. Radio-tracing techniques may also be appropriate. As many spot samples as possible should be analyzed. These should be taken at random from different locations in the batch. Sample size is an important consideration and is discussed below.

Evaluation Statistical tests can be used to evaluate relative homogeneity based on observed variations in spot sample composition. For a simple binary mixture such as that shown in Fig. 19-8, it can be shown (see Ref. 9) that the expected variance among samples containing n particles each is given by

$$\sigma^2 = \frac{p(1-p)}{n} \qquad (19\text{-}1)$$

where p is the overall fraction of black (or white) particles in the mixture. The observed sample variance can be computed using

$$S^2 = \frac{1}{m-1}\left[\sum_{i=1}^{m} p_i^2 - \frac{1}{m}\left(\sum_{i=1}^{m} p_i\right)^2\right] \qquad (19\text{-}2)$$

where p_i is the fraction of black (or white) in the ith sample and m is the total number of samples taken. The expected and observed variances can be compared using the statistical F-test (see Sec. 2 or any standard reference on statistics) which determines the likelihood that the F-ratio (S^2/σ^2) could be obtained from a random mixture, purely by chance.

The procedure can be readily extended to multicomponent systems by applying the test to each component in turn. In real systems, it is generally convenient to take samples of fixed volume or mass rather than fixed number of particles. In such cases, the expected variance can be computed using (see Refs. 19 and 20)

$$\sigma^2 = \frac{f_{ij}(1-f_{ij})w_{ij} + f_{ij}^2(\overline{w}-w_{ij})}{M} \qquad (19\text{-}3)$$

where f_{ij} is the overall mass fraction of size i composition j material in the mixture, M is the sample mass, w_{ij} is the mass of a single particle of size i composition j and $\overline{w}$ is the mean particle mass:

$$\overline{w} = \sum_i \sum_j f_{ij} w_{ij} \qquad (19\text{-}4)$$

FIG. 19-8 Random arrangement of black and white particles. [*Lacey*, Trans. Inst. Chem. Eng. (London), **21**, 52 (1943).]

The test for homogeneity is based on the probability of including different kinds of particles in a sample. For large samples, containing many particles, the expected variance given by Eq. (19-3) becomes extremely small and will often be exceeded by the variance due to experimental (analytical) error. The approach described above is, therefore, appropriate only for evaluating homogeneity at a scale approaching the size of the individual particles. If information at that scale is needed, it is necessary to use extremely small samples, containing no more than some hundreds of particles each. For very fine powders, this may seriously limit the choice of analytical techniques.

The use of very small samples to evaluate fine-scale homogeneity will often tend to mask long-range but small variations in composition. The use of somewhat larger samples is appropriate for detecting and quantifying such variations. In such cases, the sample variance can be compared, using the F-test, with an experimental variance S_E^2 obtained from replicate testing of the analytical procedure used to determine sample composition.

In general, a two-level procedure is recommended in which very small samples are used to evaluate microhomogeneity at the individual particle scale and larger samples are employed to investigate longer range variability. The actual sample sizes should be chosen such that microhomogeneity is evaluated from samples for which σ^2, as calculated using Eq. (19-3), is substantially less than the experimental (analytical) variance S_E^2 while macrohomogeneity is tested using samples with $\sigma^2 \gg S_E^2$.

Whether the desired end product is satisfactory can also be used as a practical criterion of the adequacy of the solids mixture. A further consideration is the effect of the solids mixture on the overall economics of the manufacturing process. Studies of the type mentioned in the preceding subsection *may* be part of such an evaluation. When the solids mixture is made directly into a product, as in the case of feed pellets or pharmaceutical tablets, uniformity tests on these items will speak for themselves. If the solids mixture must be further processed, as in the manufacture of glass or plastics, the efficiency and costs of the subsequent operations can often be related to the starting solids mixture. In such cases, knowledge of the homogeneity of the solids mixture is needed to determine its effect on the manufacturing process.

Regardless of the method of evaluating the solids mixture, the sampling procedure is vital. Often a sampling thief, or other special device, is used to remove samples from the mixture without excessive disturbance of the batch. If an easier method of sampling is obvious and will bring less contamination to the batch, it should be used.

Method of sampling, location, size and number of samples, method of sample analysis, and fraction of the batch removed for sampling all contribute to how well the sampling study reflects the actual conditions.

A standard testing procedure for solids-mixing equipment is available (Ref. 1). This contains details and references pertaining to sampling from solids mixtures for both batch and continuous mixing.

Segregation Problems Previously it was pointed out that wide differences among properties may make blending very difficult. For example, natural segregating tendencies will be observed with extreme differences in specific gravity, size, or shape. The heavier, smaller, or smoother and rounder particles tend to sink through the lighter, larger, or jagged ones respectively. In some cases, preparation of the materials to avoid extreme differences in such ingredient properties can avoid segregation problems.

There are also other factors which can cause segregation.

Electrostatic charges may cause particles to repel each other. When continued blending may cause such charges to build up, it is important to determine the precise blending time required and not to overblend.

Loss of material as dust must be considered as a possible means of segregation and should not be aggravated by too strong suction in the dust-collection apparatus.

If there are smeary particles which have an almost pastelike behavior and barely flow (high angle of repose), frictional anchorage of these onto the other particles in the mixture may be necessary in order to achieve good mixing.

If a batch ingredient is in agglomerate form, some device to break up the agglomerates should be used to prevent them from segregating from the rest of the mixture and to ensure the intimate dispersion of this ingredient throughout the mixture.

The use of a liquid such as water (possibly with a surface-active agent) can have remarkable effects in overcoming segregation which may appear inevitable otherwise.

Although these statements apply to the actual solids-mixing operation, thought must also be given to the subsequent processing steps. Thus, the solids-mixing operation must be checked from the point of view of delivering a well-mixed batch to a certain point. The system must be scrutinized for possible segregating points such as transfer points, long drops, flow through silos, and vibratory equipment. Where a liquid is used, the amount that can be added without getting into caking problems which may upset the later processing of the solids mixture should be determined.

Equipment

Mixing Mechanisms   There are several basic mechanisms by which solid particles are mixed. These include small-scale random motion (diffusion), large-scale random motion (convection), and shear.

Motions which increase the mobility of the individual particles will promote diffusive mixing. If there are no opposing segregating effects, this diffusive mixing will in time lead to a high degree of homogeneity. Diffusive mixing occurs when particles are distributed over a freshly developed surface and when individual particles are given increased internal mobility. A plain tumbler gives the former, while an impact mill gives the latter.

For most rapid mixing, in addition to diffusive (fine-scale) mixing, there should be a means by which large groups of particles are intermixed. This can be accomplished by either the convective or the shear mechanism. A ribbon mixer illustrates the former, whereas a plain tumbler gives the latter.

The diffusion mechanism occurs readily for free-flowing powders in which individual particles are highly mobile, but is inhibited by cohesion among particles. It follows that cohesive powders, containing fine material or liquid phases, are relatively difficult to mix. At the same time, reduced particle mobility inhibits demixing so that once mixed, cohesive powders tend to remain so. Free-flowing powders, on the other hand are prone to demixing during any transport/handling operation. The beneficial effects, noted above, of liquid addition presumably result from increased cohesion.

Types of Solids-Mixing Machines   There are several types of solids-mixing machines. In some machines the container moves. In others a device rotates within a stationary container. In some cases, a combination of rotating container and rotating internal device is used.

Sometimes baffles or blades are present in the mixer. Most types can be quite effective for free-flowing powders, bearing in mind that segregation may also be favored. Highly cohesive powders generally require high shear (velocity gradient) to achieve a high degree of microhomogeneity. Table 19-1 classifies solids-mixing machines via the characteristics given in the column headings. Illustrations of several of the machines listed there are shown in Fig. 19-9. The various types listed in Table 19-1 will be briefly discussed, with paragraph numbers referring to the columns.

1. *Tumbler.*   Suitable for gentle blending; capable of handling large volumes; easily cleaned; suitable for dense powders and abrasive materials. Not for breaking up agglomerates.

Figure 19-9a and b (without broken-line portions) shows some unbaffled tumblers.

Figure 19-9c and d shows some baffled tumblers.

2. *Tumbler with agglomerate breaker.*   See Sec. 20: "Tumbling Mills," for ball mill, rod mill, and vibratory pebble mill which will accomplish mixing along with size reduction.

Several tumblers are available with separately driven internal rotating devices for breaking up agglomerates. The tumbler itself can be used for gentle blending if agglomerate breakdown is not required.

The broken-line portions of Fig. 19-9a and b show some types of agglomerate-breaking devices for tumblers.

Table 19-2 includes impact velocities for some internal rotating devices in tumblers as well as other mixers. Contamination and wear problems of internal rotating devices are discussed under "Performance Characteristics."

3. *Stationary shell or trough.*   There are a number of different types of mixers in which the container is stationary and material displacement is accomplished by single or multiple rotating inner mixing devices.

a. Ribbon mixer (Fig. 19-9e).   Within this subgroup there are several types. Ribbon cross section and pitch, clearances between outer ribbon and shell, and number of spirals on the ribbon are some features which can be varied to accommodate materials ranging from low-density finely divided materials that aerate rapidly to fibrous or sticky materials that require positive discharge aid. Other construction variations are center or end discharge and the mounting of paddles or cutting blades on the center shaft. A broad ribbon can be used for lifting as well as for conveying, while a narrow one will cut through the material while conveying. The ribbon is adaptable to batch or continuous mixing.

b. Vertical screw mixer.   This subgroup also has several variations. One type is shown in Fig. 19-9f. In this type, the screw rotates

TABLE 19-1   Types of Solids-Mixing Machines*

Tumbler (1)	Tumbler with internal agglomerate breaker (2)	Stationary shell or trough (3)	Both shell and internal device rotate (4)	Impact mixing (5)	Process steps which can affect solids mixing‡ (6)

Without baffles:
  Drum, either horizontal
    or inclined
  Double cone
	Ball mill
	Pebble mill
	Rod mill
		Ribbon
		
		Stationary pan, rotating muller turret†
			Countercurrent, muller turret and pan rotate in opposite directions	Hammer mill
				
				Impact mill
					Filling of hoppers
					
					Fluidization

Twin shell Cube Mushroom type	Vibratory pebble mill Double cone Twin shell Cube	Vertical screw Single rotor Twin rotor Turbine Paddle mixer Sifter (turbosifter)	Planetary types	Cage mill Jet mill Attrition mill	Screw feeders Conveyor-belt loading Elevator loading Pneumatic conveying Vibrating

With baffles:
  Horizontal drum
  Double cone revolving
    around long axis					

* Diagrammatic sketches of many of these machines are shown in Fig. 19-9.
† There is also a muller in which the turret is stationary but the pan rotates.
‡ Although these steps, when carefully selected, can aid mixing, caution must be exercised with pneumatic conveying and vibrating, as they may tend to separate materials.

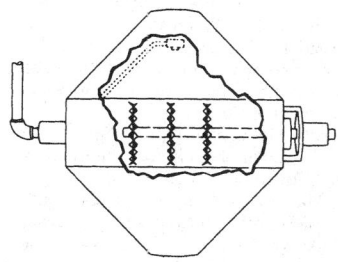

(a) Double cone

Agglomerate breaking device shown in broken line. Spray nozzle shown in dotted line. Tumblers of this type available plain or with either or both of the above features.

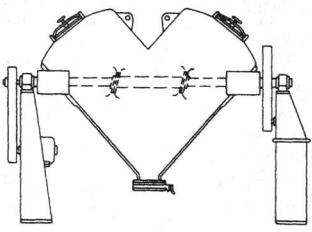

(b) Twin shell (Vee)

Agglomerate breaking and liquid feeding device shown in broken line. Where no liquid feeding is necessary, a pin-type agglomerate breaking device is used. Tumblers of this type are available plain or with any of the above features.

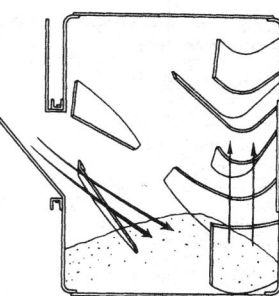

(c) Horizontal drum (with baffles)

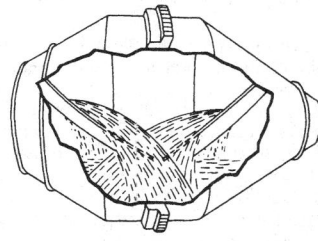

(d) Double-cone revolving around long axis (with baffles)

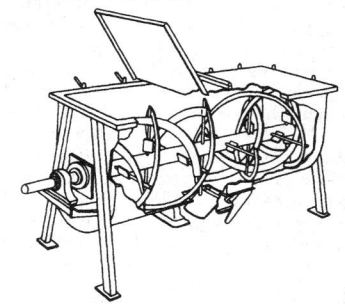

(e) Ribbon

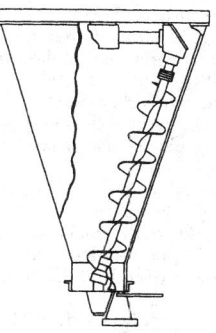

(f) Vertical screw (orbiting type)

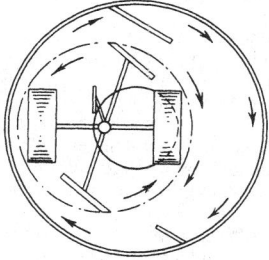

(g) Batch muller

Three types are available:
(1) pan is stationary and muller turret rotates;
(2) muller turret is stationary and pan rotates;
(3) pan rotates clockwise, muller turret rotates counterclockwise.
Type 3 is illustrated above

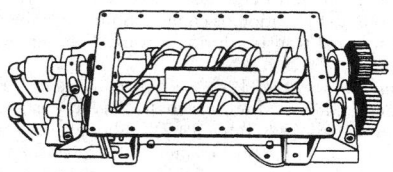

(h) Continuous muller (stationary shell)

(i) Twin rotor (adapted to heat transfer-jacketed body and hollow screws)

(j) Single rotor

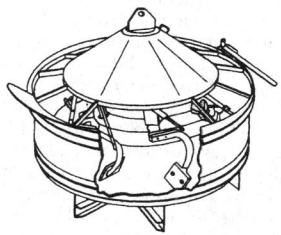

(k) Turbine

FIG. 19-9 Several types of solids-mixing machines. (See Table 19-1.)

TABLE 19-2 Approximate Impact Velocities of Some Rotating Internal Devices in Mixers*

Type of mixer (see Table 19-1)	Tip speed, ft/min
Ribbon	280
Turbine	600
Twin-shell tumbler with	
Pin-type intensifier	1700
Liquid-feed bar	3300
Twin rotor	Up to 1300
Single rotor	6000–9000
Mills of various types	2500–20,000

*To convert feet per minute to meters per second, multiply by 0.00508.

about its own axis while also orbiting around the center axis of the conical tank. In another variation, the screw does not orbit but remains in the center of the conical tank and is tapered so that the swept area steadily increases with increasing height. In another type, the central screw is contained in an inner cylindrical casing. This type of mixer is primarily suitable for free-flowing dry solids.

c. Muller mixer. The stationary-pan muller with rotating turret is one of several types. Other muller types are the countercurrent type, in which the pan and muller turret rotate in opposite directions, and the rotating-pan type, in which the muller turret is stationary.

The heavy, wide roller rides over the material. There is some skidding action where the rollers engage the mass of materials. This gives local shearing plus coarse-scale mixing which is aided by the plows and scrapers.

The muller is useful for mixing problems requiring certain types of aggregate breakdown, frictional anchorage of particles to one another, and densification of the final mix. Materials which are excessively fluid or sticky should be avoided. The muller mixer is generally used for batch operations (Fig. 19-9g), although Fig. 19-9h shows a continuous muller.

d. Twin rotor (Fig. 19-9i). This consists of two shafts with either paddles or screws encased in a cylindrical shell. There are various types available with shaft speeds ranging from moderately low to relatively high (see Table 19-2). The twin rotor is useful for continuously mixing non-free-flowing solids; liquids can be added, there is minor product attrition, and materials can be added beyond the inlet. It is easily adaptable to heating or cooling. Some machines are specifically designed for heat transfer during mixing. The pug mill is one type of twin rotor.

e. Single rotor (Fig. 19-9j). This consists of a single shaft with paddles encased in a cylindrical shell. This type is available with relatively high speeds (see Table 19-2), although in certain cases lower speeds are used. A high-speed single rotor gives the maximum impact short of a grinding mill. It is used for intensive dispersion and disintegration. The type is available with split casing and is suitable for heating or cooling and for small amounts of liquid addition.

f. Turbine mixer (Fig. 19-9k). This is a circular trough with a housing in the center around which revolves a spider or a series of legs with plowshares or moldboards on each leg. The moldboards spin around through the circular trough. This mixer is suitable for free-flowing dry materials or semiwet materials which do not flow well and is also adaptable to liquid-solid mixing and coating problems.

4. *Shell and internal device rotate.* The countercurrent muller (Fig. 19-9g), which is in this category, is mentioned under "Muller mixer." This machine has a clockwise rotating mixing pan with a counterclockwise rotating mixing tool head mounted off center of the pan, thus providing a planetary mixing pattern. For the mixing of free-flowing solids not requiring the shearing and compressive action of mullers, plows are sometimes used alone. When used with mullers, plows deflect material into their path. Special mixing tools are also available.

5. *Impact mixing.* This process, which includes size reduction, is covered in Sec. 20.

The process steps listed in Table 19-1 can sometimes be used to promote mixing. However, they are primarily for functions other than solids mixing. (Note precautions for pneumatic conveying and vibrating in Table 19-1.)

Since paste mixing is not within the scope of this section, such widely used paste mixers as the sigma blade and banbury types will not be covered here but instead are discussed in Sec. 18.

Performance Characteristics Before selecting solids-mixing equipment, a careful study should be made of various performance characteristics. These are given here.

Uniformity of Mixture The proper type of mixer should be chosen to assure the desired degree of batch homogeneity. This cannot be compromised for other conveniences. Information is given under "Types of Solids-Mixing Machines" about the special abilities of various kinds of machines to blend different types of materials.

Care should be taken to avoid mixing too long, as in some cases this will result in a poorer blend. A graph of degree of mixing versus time should be made to select the proper mixing time quantitatively.

Mixing Time The actual time during which the batch is being mixed is usually less than 15 min if the proper type of machine and working capacity have been chosen. In some cases much more lengthy mixing times are tolerated so as to avoid the cost of purchasing more efficient equipment. However, there is usually a machine that can properly homogenize almost any type of mixture in less than 15 min provided one is willing to pay the price. In fact, proper mixer design in most instances will produce the desired blend in a few minutes.

Besides actual mixing time, however, the total cycle time should be optimized.

Charging and Discharging The total handling system must be considered in order to obtain optimum charging and discharging conditions. This includes the efficient use of weigh hoppers and surge bins, minor-ingredient premixing, location of discharge gates, and so on.

Power In general, power requirements are not a major consideration in choosing a solids mixer since other requirements usually predominate. However, sufficient power must be supplied to handle the maximum needs should there be changes during the mixing operation. Also, when a variety of mixes may be required, power must be sufficient for the heaviest bulk-density materials. If the loaded mixer is to be started from rest, there should be sufficient power for this. When speed variation may be desirable, this should be taken into account in planning power requirements.

Horsepower requirements of several types of mixers are listed in Table 19-3.

Cleaning The ease, frequency, and thoroughness of cleaning may be crucial considerations when incompatible batches are to be mixed at different times in the same machine. Plain tumbling vessels are easy to clean provided that adequate openings are available. Areas that may present cleaning problems are (1) seals or stuffing boxes, (2) crevices at baffle supports, (3) any corners, and (4) discharge arrangement. If cleaning between different batches may be time-consuming, several small mixers should be considered. Special sanitary construction can usually be provided at extra expense.

Agglomerate Breakdown and Attrition The two methods of producing agglomerate breakdown and attrition are as follows:

1. *Impact.* The major factor is the peripheral speed of the rotating internal device. Table 19-2 gives impact-velocity data for various mixers.

2. *Shearing and compressive action.* In mullers this depends upon the clearance between muller and pan and the muller weight or spring load respectively.

When an attrition device is necessary to break down aggregates but may also produce too much size reduction on other batch ingredients, tolerable attrition should be determined by tests.

Dust Formation Loss of dust can seriously affect batch composition, particularly when vital minor ingredients are lost. Methods of minimizing dust formation are: (1) Use of less dusty but equally satisfactory batch ingredients. Sometimes a pelletized form of an extremely dusty material is available. (2) Proper venting so as to enable filtering of displaced air rather than unregulated loss of dust-laden air. (3) Dust-tight arrangements for loading and unloading the mixer. (4) Addition of liquids if tolerable. Not only is water effective in minimizing dust upon discharging from the mixer, but if properly added it will render the batch less dusty in subsequent handling steps. The addition of a small quantity of surface-active agent will improve the penetration of the water throughout the batch and enable

TABLE 19-3 Horsepower Requirements and Speeds of Rotation for Some Commercial Solids Mixers

[Approximately 1.5 m³ (50 ft³) Working Capacity]

Type of solids-mixing machine	Approximate working capacity, ft³	Horsepower, hp		Rotational speed, r/min		Comments
		Shell	Internal device	Shell	Internal-device shaft speed	
1. Tumbler						
Without baffles						
Double cone	54	7½		18		Based on 100-lb/ft³ material.
Twin shell	50	5		13.7		Maximum bulk density of material = 55 lb/ft³.
With baffles						
Horizontal drum						
Manufacturer E	50	20		11.1		Heavy-duty (material 100 lb/ft³). For extremely heavy duty (150–200-lb/ft³ material), the maximum working capacity with 20-hp motor is 35 ft³.
Manufacturer F	50	10		14		For material of 40-lb/ft³ maximum bulk density.
Double cone revolving about horizontal axis	56	25		11.5		Mixer can be tilted. Rear end charger. Capacity based on mixed concrete.
2. Tumbler with agglomerate breaker						
Double cone	54	7½	See Comments.	18	See Comments.	Horsepower requirement for internal device depends on character of material, type, and speed of agitator. These are to be determined by adequate testing.
Twin shell	50	5	5 (pin-type intensifier bar) 7½ (liquid-solids intensifier bar)	13.7	945 (1730-ft/min tip speed) 1055 (3320-ft/min tip speed)	Maximum bulk density of material = 55 lb/ft³.
3. Stationary shell or trough						
Ribbon						
Manufacturer C	50		12		28	Horsepower required based on material of 50–60-lb/ft³ bulk density, medium free-flowing, using 10 hp/ton for average mix cycle of 3–10 min (depending on material, range can be 3–18 hp/ton).
Manufacturer A	46		10		37	Based on material of 30-lb/ft³ bulk density.
Manufacturer D	50		15		45	Based on material of 40–50-lb/ft³ bulk density.
Three-shaft ribbon	50		Blender shaft 20 Feeder shaft 7½ (total)		Variable-speed drives on all shafts	This blender is rated at 300 ft³/h on batch-mixing basis; 900 ft³/h on continuous-mixing basis. Materials rated at 70-lb/ft³ bulk density.
Vertical screw	52.9		5		Screw, 64.4 Orbit, 2.2	Horsepower based on 37-lb/ft³ bulk density. This may vary with different materials. Maximum hp = 10, maximum weight = 4410 lb.
Muller:						
Batch; stationary pan, rotating turret	40		60		24 (turret speed)	Based on material of 60–75-lb/ft³ bulk density.
Continuous; stationary pan, rotating turret						Basically, the continuous mullers are merely two-batch mullers joined together at the cribs, making a figure-8 design. Thus, the 40-ft³ batch muller rated at 60 hp becomes an 80-ft³ working-capacity continuous muller requiring 125 hp. This would give 125 tons/h with a 2½-min residence time. Turret speeds are 24 r/min.
Single rotor			See Comments.			In this continuous unit, output can range from 25–600 lb/min with hp from 5 to 100 and r/min of 500 to 4000, depending on the materials mixed.
Double rotor			See Comments.			In this continuous unit the output can range from 200–500 lb/min with hp from 5 to 40 and r/min from 200 to 300, depending on the materials mixed.
Twin-rotor heat-exchanger mixer	49.2		5–15		20–100	Amount of conveying and mixing action affected by amount of pitch and type of ribbons mounted on exterior of hollow screws.
Turbine	50		50		Peripheral speed of 600 ft/min	
4. Both shell and internal device rotate						
Countercurrent muller	45 60–90†	° 20	° 25	6.75–8.75 6.65	28–35 20	

°One 25-hp motor drives both the shell (mixing pan) and the internal device (mixing star).

†Batch-capacity range depends on nature of materials to be mixed.

NOTE: To convert cubic feet to cubic meters, multiply by 0.02832; to convert horsepower to kilowatts, multiply by 0.7457; to convert pounds per cubic foot to kilograms per cubic meter, multiply by 16.02; to convert tons per hour to kilograms per second, multiply by 0.252; to convert revolutions per minute to radians per second, multiply by 0.1047; to convert pounds per minute to kilograms per minute, multiply by 0.4535; and to convert horsepower per ton to kilowatts per metric ton, multiply by 0.8352.

it to wet even such materials as coal dust. The method of adding water is important (see "Method of Adding Liquids").

Care should be taken to avoid powerful suction or air flow on the mixer or the weigh hopper from which the ingredients feed into the mixer. If the dust-collection suction on the mixer is too strong, vital ingredients may be sucked out. If the dust-collection suction on the weighing system is too strong, errors in weighing may result.

Electrostatic Charge   Certain batch materials such as plastics tend to accumulate a charge easily. Work input will affect the charge on the batch. Coating of the inside of the mixer shell or rotating elements may occasionally result because of electrostatic charge. This can present a cleaning problem. Possible aids in overcoming this are (1) addition of special solid materials with very high surface area to weight ratios, (2) addition of liquids (see "Dust Formation" and "Method of Adding Liquids"), (3) proper choice of material of construction of the mixer, (4) controlling humidity, (5) preparation of the batch ingredients so as to minimize accumulated charge.

Equipment Wear   Simple tumbling mixers give the least wear. Attrition devices in tumblers may present serious abrasion problems with certain materials such as sand and abrasive grinding-wheel grains. Abrasion-resistant coating such as rubber coating, special alloys, or platings should be considered for these cases. An internal agitator device may wear even though its speed is low. Particularly when highly abrasive materials are to be mixed, the benefits of an agglomerate-breaking device must be weighed against potential contamination and replacement and maintenance costs.

Contamination of Product   This has been partially covered under "Cleaning" and "Equipment Wear." Other sources of contamination are lubricants and repair materials. Types which are not compatible with the batches to be mixed should be avoided.

Heating or Cooling   Nearly all commercial mixers can be heated or cooled. Some can be provided with heated or cooled agitators. If temperature rise during mixing is detrimental, cooling facilities should be provided. The various manufacturers can provide details on the means of heating their machines. Most common heating means are (1) water or steam in the jacket and in hollow-screw or paddle-type internal agitator, (2) hot oil, (3) Dowtherm liquid or vapor, (4) electric heaters, contact or radiant, (5) hot air in direct contact with product (suitable only for revolving-drum-type mixers), (6) exterior heating of drum by direct or indirect firing. For cooling, the most common means are (1) water or refrigerated fluid in the jacket and in hollow-screw or paddle-type internal agitator, (2) an evaporant such as liquid ammonia, (3) direct air contact (for rotating-shell mixers), and (4) oil or Dowtherm (or its equivalent) for cooling high-temperature materials.

Flexibility   When batches of widely different size must be mixed, flexibility of operating capacity may enable use of fewer mixers. Certain features may necessitate a nonflexible capacity requirement. For example, ordinarily an internal agitating device in a tumbling mixer does not function effectively unless the batch is loaded to a certain level. The need for such features must be weighed against the limitations imposed by a narrow operating-capacity range when choosing equipment for an operation in which batch size will vary considerably.

In general, the effect of percentage of mixer volume occupied by the batch on the adequacy of mixing should be borne in mind, particularly when any change from the recommended volume percent is considered.

Vacuum or Pressure   Most tumbling mixers can have provision for vacuum or pressure. Mixers which cannot be adapted to these conditions are mullers with rotating pans. Continuous mixers introduce problems of sealing the charge and discharge ends.

Method of Adding Liquids   When the addition of liquids may be desirable (see "Dust Formation" and "Electrostatic Charge"), this should be considered when designing the mixing system rather than hastily improvised. The purpose of the liquid should be considered, whether for (1) dust suppression, (2) product, or (3) heating and cooling. If a viscous liquid must be well distributed, this requirement should be considered when choosing the mixer.

Liquid should be directed into the batch materials and not onto bare mixer surface since this could cause buildup. Nozzle spray pressure should be sufficient to penetrate the batch but not so high as to cause heavy splashing. The liquid should be added to the well-mixed batch. In particular, when premature addition of liquid could impair the adequacy of blending, both the time during which it is added in the mixing cycle and the time taken to add the liquid are important.

Automated equipment for the addition of liquids can be worked into the overall mixing plant when necessary. For dust-reduction purposes, a volumetric method of metering is satisfactory. However, should a critical batch ingredient be added in liquid form, a more precise method of metering may be necessary.

Other considerations are (1) proper ventilation and discharge enclosures, (2) provision for relief of internal explosion, (3) vibration isolation (shock mounts), (4) remote operation of charge and discharge, (5) noise during operation.

Equipment Selection   Types of mixers and performance characteristics have been given. Segregating tendencies among solid materials have also been described. A sound approach to solids-mixer selection starts with a careful examination of these areas. However, mixer selection should also involve consideration of the mixer's place in the overall process. Possible consolidation of many solids-processing steps or the opposite (splitting one operation into several) deserves scrutiny at this time. If no one standard machine has all the necessary requirements, thought should be given to which machine can best be modified to achieve the most desirable combination of features. One should look at the overall process objectives as well as at equipment details when selecting a solids mixer.

Pilot Tests   In some cases, it is possible to perform pilot tests on a small-scale version of the equipment to be used in production. Much useful information can be found here but the following must be borne in mind:

1.   In general, the larger the pilot unit, the more reliable the prediction of large-scale performance. The pilot unit should be a prototype with all dimensions properly scaled down.

2.   Published solids-mixing scale-up data are rare. Equipment suppliers can provide scale-up information for their particular types of equipment on the basis of experience. With geometrically similar tumblers, if the speeds are adjusted to give comparable motion and the mixer volume fraction occupied by the charge is the same, scale-up of results will be straightforward. The presence of a rotating internal device presents problems in the scaling up of clearances, blade area to mixture volume, and sizes and speeds of the rotating devices. For agglomerate breakers, the key factor in scaling up is impact velocity. Scale-up in cylinders is discussed on pages 290–292 of Ref. 9. Solids-processing scale-up is discussed in a paper by Sterret (*Chem. Eng.*, Sept. 21, 1959).

3.   The actual process materials should be used if possible. If substitute materials are used, they should have the same mixing characteristics. Tests with differently colored but otherwise identical beads can be misleading, and so can tracers. The reason is that the flow properties of the specific materials to be mixed in the plant may not be the same as these demonstration materials. Regardless of how the mixer contents appear to be moved around, the properties of the actual batch ingredients may cause segregation or other problems.

4.   Differences in materials of construction between the pilot unit and the production unit should be considered. These may have a bearing on caking, abrasion, and electrostatic effects.

Continuous Mixing   Although batch mixing has been the predominant method of mixing solids, consideration is being given to the use of continuous mixing in many industries. There are two types of continuous-mixing operations. The first type has a low holdup volume and will provide fine-scale blending of the particles via impact and shear elements such as are used in grinding machines. Some machines of this type are hammer, impact, cage, and jet mills. It is essential that the feed to these machines be properly proportioned and premixed to achieve a uniform product.

The second type of continuous mixer involves high holdup machines which contain agitating and conveying mechanisms. These rearrange the individual particles and also displace large volumes of material and move the batch through the machine. Mixers of this type can produce both fine-scale and coarse-scale blending. The ribbon-type mixer is frequently used for continuous mixing, although this is also used for batch mixing. A continuous muller mixer has been developed as shown in Fig. 19-9h.

The average composition of the stream leaving a continuous mixer is the same as the average of the added entering streams. Variations in proportions of the entering streams will be damped out by the mixing action of a continuous mixer. These effluent-stream variations will become smaller as average solids residence time is increased and the frequency of the variations increases.

Certain general criteria can be used to determine whether continuous flow will be beneficial. Continuous flow is worth consideration if (1) a single formulation can be run for an extended period, (2) the fluctuations of the outgoing product are within process requirements, (3) sufficiently accurate metering of ingredients can be achieved, (4) the rest of the process warrants continuous mixing. Continuous flow is of doubtful benefit if (1) frequent changes of formulations are anticipated, (2) fluctuations of product composition will be outside the permitted range, (3) the ingredients cannot be metered with the necessary level of accuracy, (4) complex temperature or pressure cycles are involved.

Sometimes a system of mixing and dispersing is composed of one or more batch units providing a feed to a continuous intensive dispersion unit. Another possibility would be a batch mixer and surge bin which provide a continuous feed to a final dispersion unit. Various combinations of this type with adequate sampling at the proper points may be used when continuous flow would be beneficial provided that certain features could be overcome.

OPTICAL SORTING

Difference in optical properties can be used as the basis to separate solids in a mixture. Optical properties include color, light reflectance, opacity, and fluorescence excited by ultraviolet rays or x-rays. Differences in electrical conductance can also be used for separation. With appropriate sensing, the particles in a moving stream can be sorted by using an air jet or other means to deflect certain particles away from the mainstream (Fig. 19-10). The lower limit of particle size is about 0.003 m (⅛ in); below this limit the process rate would be slow and equipment costs become exceedingly high.

In a typical **optical sorting** installation, the mixture of particles is fed from a hopper onto a vibrating feeder. Dust may be removed by dry screening or by water spray. The solids then enter a troughed conveyor belt and align the flow and cause the particles to be projected in a continuous stream along their free-fall trajectory. They are viewed in midair during fall through an optical chamber by a series of cameras arranged to view the entire surface of each particle. The color or reflectivity of the surface of each particle sets up a characteristic voltage pattern in the output circuit of a light-sensing photomultiplier. The patterns are analyzed electronically and compared with a preset reflectance level. When appropriate, a reject signal delayed electronically will activate on air jet to deflect a particle from the main stream.

An example of throughput is given in Fig. 19-11. For the machine referenced in Fig. 19-10, electronic sensing is capable of inspecting 80 particles per second.

Color sorting has been used for the recovery of glass from nontransparent materials in municipal solid waste. The separation of mixed glasses from opaques is illustrated in Figs. 19-12 and 19-13. As the throughput is increased, the glass recovery falls; and for any given installation there will be a breakeven point for the optimum number of machines against the amount of glass recovered. Color sorting can also separate mixed colored glasses to produce amber and green products that meet a specification of 10 percent contamination of one type of glass in the other.

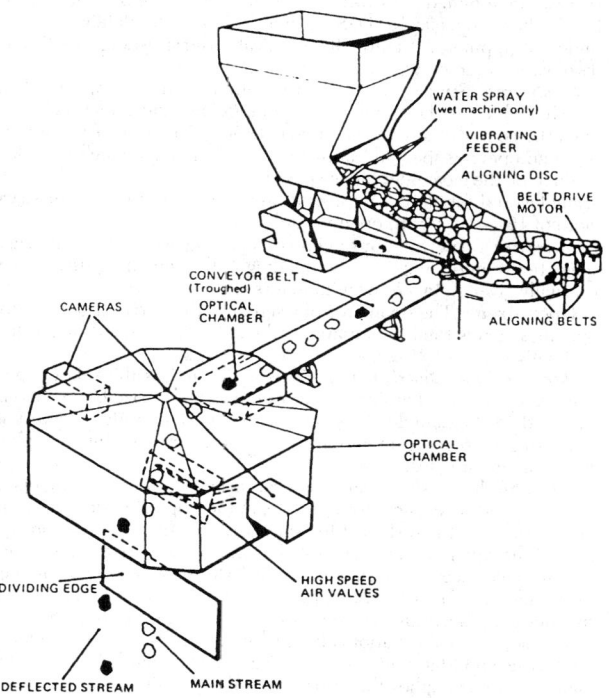

FIG. 19-10 Sortex 711M optical separator. (*Courtesy Gunson's Sortex Ltd.*)

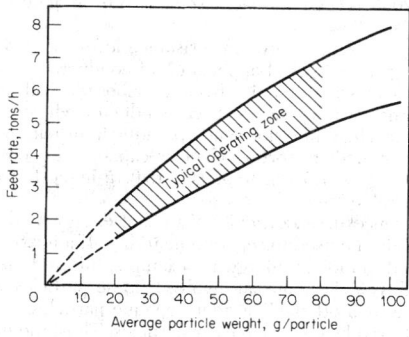

FIG. 19-11 Sortex 711M feed-rate characteristics. (*Courtesy Gunson's Sortex Ltd.*)

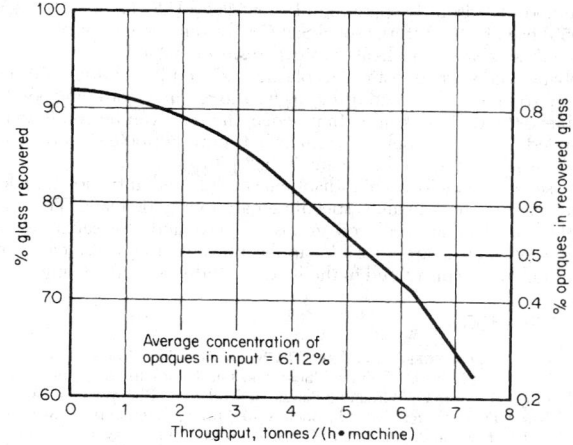

FIG. 19-12 Optical separation of mixed glasses: separation of opaques from glasses. (*Courtesy Gunson's Sortex Ltd.*)

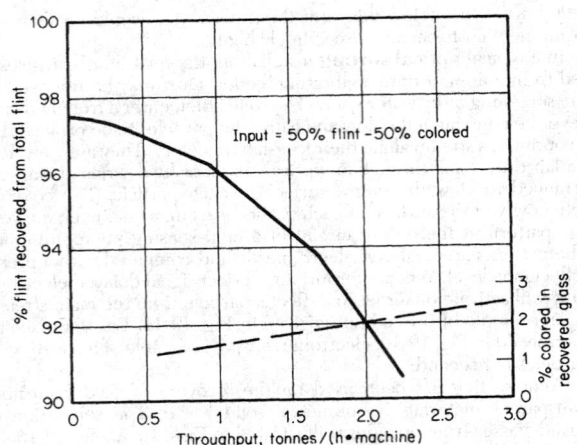

FIG. 19-13   Optical separation of mixed glasses: separation of flint from colored. (*Courtesy Gunson's Sortex Ltd.*)

When **electrical conductivity** is used as the basis of the sorting process, contact of the particles is made by a brush type of electrode to generate the signal for analysis. Materials having a resistance difference of 2000 kΩ can be readily separated from material of 100 kΩ resistance.

However, separation between resistance levels of 1000 to 300 kΩ may be marginal. A typical application of conductivity sorting is the separation of massive ilmenite from anorthosite. Both are compact rocks, but ilmenite is a good electrical conductor, whereas anorthosite is an insulator. The dimensions and operating information for the Sortex CS-03 conductivity sorter, which is capable of processing up to about 25,000 kg/h (27.5 tons/h) of 0.05- to 0.15-in mesh size (2 to 6 in), are given in Table 19-4.

The differences in absorptivity of radiant energy by different substances can also be used to separate materials. If a mixture of materials is exposed to radiant energy, for example, infrared, depending on the properties of the material involved, some particles will become heated more than others. The more opaque particles will be heated more by infrared heating than clear particles. Thus, the more opaque will become hotter. By spreading the irradiated material onto a surface coated with a low-melting thermoplastic or a heat-sensitive polymer, the higher-temperature particles will adhere, while the cooler particles will not. This is the basis for the **thermoadhesive-separation** method of Brison ("Separation of Materials," U.S. Patent 2,907,456, 1959) used by the International Salt Co. for removing impurities from mined rock salt. The heat-sensitive resin employed is a mixture of polymerized styrene resins, Piccolastic A-25 and Piccolastic A-50. The proportion of each was adjusted to give the required softening point to achieve the desired results. In practice, the resin can be continuously applied to a moving belt by brush or hot spray. Periodic scrapping and redressing are required to maintain belt performance.

Use of specific forms of radiant energy, infrared, ultraviolet, dielectric heating, etc., can allow specific separations to be made. The separation of clear and colored grains of glass and the separation of different metals are possible applications of the thermoadhesive method being considered in the field of solid-waste processing.

SCREENING

GENERAL REFERENCES: Beddow, "Dry Separation Techniques," *Chem. Eng.,* **88,** 70 (Aug. 10, 1981). Colman, "Selection Guidelines for Size and Type of Vibrating Screens in Ore Crushing Plants," in Mular and Bhappu (eds.), *Mineral Processing Plant Design,* 2d ed., Society of Mining Engineers, AIME, New York, 1980. Kuenhold, "Factors to Consider in Vibrating Screen Installations," *Min. Eng.,* 650–653 (June 1957). Matthews, *Chem. Eng.,* deskbook issue, **99** (Feb. 15, 1971). Matthews, "Screening," *Chem. Eng.,* **79,** 76 (July 10, 1972). Moir, "Recent Developments in Mineral Processing and Their Implications,"

TABLE 19-4   Specifications for Sortex Conductivity Sorter CS-03*

Dimensions	Height (excluding feeder), 1.7 m (70 in)
	Width, 1.9 m (76 in)
	Length, 2.5 m (100 in)
	Diameter of disk, 1.50 m (49 in)
Electric power	380–440 V, three-phase 50/60 Hz; consumption, 2 kW (excluding vibrating feeder)
Air consumption (depending on rejection rate)	7 m³/min at 6 bar (250 ft³/min at 80 lbf/in²)
Water supply	4 L/min (1 gal/min) for cleansing the light source
Net weight	560 kg (22 cwt) approximately

*Courtesy of Gunson's Sortex Ltd.

Economics of Mineral Engineering Mining Journal, Books Ltd., London, 1976, p. 125. Mular, *Mineral Processing Equipment Costs and Preliminary Capital Cost Estimations,* spec. vol. 18, Canadian Institute of Mining and Metallurgy, Montreal, 1978. Pryor, *Mineral Processing,* 3d ed., Elsevier, New York, 1965. Reed, "The Story behind the New Sieve Specifications," *Test. World,* October 1959. Taggart, *Handbook of Mineral Dressing,* 2d ed., Wiley, New York, 1945.

Definitions

Screening   Screening is the separation of a mixture of various sizes of grains into two or more portions by means of a screening surface, the screening surface acting as a multiple go-no-go gauge and the final portions consisting of grains of more uniform size than those of the original mixture.

Material that remains on a given screening surface is the oversize or plus material, material passing through the screening surface is the undersize or minus material, and material passing one screening surface and retained on a subsequent surface is the intermediate material.

The screening surface may consist of woven-wire, silk, or plastic cloth, perforated or punched plate, grizzly bars, or wedge wire sections.

Classification of screening operations and the range of separations that can be attained with various screens were given in concise form by Matthews (op. cit., 1971). See Table 19-5. Further details are given under "Equipment." Figure 19-14 indicates the size-range applicability of various screen types.

Mesh and Space Cloth   Wire cloth is generally specified by "mesh," which is the number of openings per linear inch counting from the center of any wire to a point exactly 25.4 mm (1 in) distant, or by an opening specified in inches or millimeters, which is understood to be the clear opening or space between the wires. Mesh is generally favored for cloth 2 mesh and finer and clear opening for space cloth of 12.7-mm (½-in) opening and coarser.

Aperture   Aperture, or screen-size opening, is the minimum clear space between the edges of the opening in the screening surface and is usually given in inches or millimeters.

Open Area   The open area of a screen is the percentage of actual openings versus total screen area and can be determined by the formulas given in Fig. 19-22.

Particle-Size Distribution   This is defined as the relative percentage by weight of grains of each of the different size fractions represented in the sample. It is one of the most important factors in evaluating a screening operation and is best determined by a complete size analysis using testing sieves.

Sieve Scale   A sieve scale is a series of testing sieves having openings in a fixed succession; for example, in the original basic Tyler standard sieve scale the widths of the successive openings have a constant ratio of the square root of 2, or 1.414, while the areas of the successive openings have a constant ratio of 2. The Tyler scale has been enlarged to include intermediate openings so that the entire scale has successive openings according to the fourth root of 2, or 1.189. The sieve series adopted by the National Bureau of Standards, American Society for Testing and Materials, American National Standards Institute, and many countries applies the fourth-root-of-2 principle, and the openings are fully compatible with the Tyler standard scale even though the sieve designations may vary (Table 19-6).

TABLE 19-5 Types of Screening Operations

Operation and description	Type of screen commonly employed
Scalping—Strictly, the removing of a small amount of oversize from a feed which is predominantly fines. Typically, the removal of oversize from a feed with approximately a maximum of 5% oversize and a minimum of 50% half-size.	Coarse (grizzly); fine, same as fine separation; ultrafine, same as ultrafine separation.
Separation (coarse)—Making a size separation at 4 mesh and larger.	Vibrating screen, horizontal or inclined.
Separation (fine)—Making a size separation smaller than 4 mesh and larger than 48 mesh.	Vibrating screen, horizontal or inclined; high-speed low-amplitude vibrating screens; sifter screens; static sieves; centrifugal screens.
Separation (ultrafine)—Making a size separation smaller than 48 mesh.	High-speed low-amplitude vibrating screen; sifter screens; static sieves; centrifugal screens.
Dewatering—Removal of free water from a solids-water mixture. Generally limited to 4 mesh and above.	Horizontal vibrating screen; inclined vibrating screens (about 10°); centrifugal screen.
Trash removal—Removal of extraneous foreign matter from a processed material. Essentially a form of scalping operation. Type of screen employed will depend on size range of processed material—coarse, fine, or ultrafine.	Vibrating screen, horizontal or inclined; sifter screens; static sieves; centrifugal screen.
Other applications: Desliming—Removal of extremely fine particles from a wet material by passing it over a screening surface. *Conveying*—In some instances transport of the material may be as important as the operation. *Media recovery*—A combination washing and dewatering operation.	Vibrating screens, inclined and horizontal; oscillating screens; centrifugal screens.

Testing Sieves Many product specifications require size of material in terms of given percentages passing or retained on specified test sieves. Test sieves are also generally used to determine the efficiency of screening devices and the work of crushing and grinding machinery.

It is essential that standard sieves, with standard-size openings be used for sieve analyses. The time of screening and the method of agitating the material on the sieve should also be standard, and in many industries the practice of specifying the test-sieve designation and the time and method of sieving is followed. An excellent overview of the theory and use of standard testing sieves is given in the *Testing Sieve Handbook*, No. 53, published by W. S. Tyler, Inc., Mentor, Ohio.

U.S. Sieve Series The American Society for Testing and Materials in cooperation with the National Bureau of Standards and the American National Standards Institute has further refined the U.S. sieve series, combining the former coarse and fine series into a single series series with a fourth-root-of-2 ratio (Table 19-6). The openings in the individual sieves have remained unchanged except for minor adjustments in sieves coarser than 0.00673 m (6.73 mm). In the revised series, sieves 1 mm and coarser are identified by opening in millimeters, and those finer than 1 mm by their openings in microns.

Tyler Standard Sieve Series Many users base their tests on Tyler standard testing sieves (Table 19-6). The only difference between the U.S. sieves and the Tyler screen scale sieves is the identification method. Tyler screen scale sieves are identified by nominal meshes per linear inch while the U.S. sieves are identified by millimeters or micrometers or by an arbitrary number which does not necessarily mean the mesh count. The Tyler standard sieve scale series has

as its base a 200-mesh screen in which the opening is 7.37×10^{-5} m (0.0029 in) and the wire diameter 5.35×10^{-5} m (0.0021 in).

International Test Sieve Series The International Organization for Standardization (ISO) has been intensifying its efforts to establish an international test sieve series. At a meeting held at The Hague in October 1959, the ISO provisionally recommended for adoption as an international standard 19 sieves as shown by the data in Table 19-6. These sieves correspond to every alternating sieve in the fourth-root-of-2 U.S. sieve series from 0.022-m (8/8-in) opening to 325 mesh. The ISO has prepared a manual of sieving procedures that is available through the American Society for Testing and Materials.

The Ro-Tap testing sieve shaker (Fig. 19-15) manufactured by the W. S. Tyler, Inc., is the standard machine for automatically carrying out sieve-test procedures with accuracy and dependability. This device is built to hold a series of 0.203-m- (8-in-) diameter Tyler standard scale testing sieves and imparts to the sieves both a circular and a tapping motion. In effect, it reproduces the circular and tapping motion given testing sieves in hand sieving but does it with a uniform mechanical action. An important feature of the Ro-Tap is that both speed and stroke are fixed and not adjustable. This ensures the comparability between a number of sieve tests not only in a manufacturer's plant but between tests of a supplier and a customer.

The Ro-Tap is equipped to handle from 1 to 13 sieves at a time and is equipped with a timer that automatically terminates the test after any predetermined time.

Another mechanical shaker is the End-Shak, made by the Newark Wire Cloth Co. Sieves used are Newark test sieves, made to conform with the U.S. standard series.

A number of less expensive sieve shakers are on the market, such as the Dynamic, by Soiltest Inc., Chicago; the Cenco-Meinzer, by Central Scientific Co., Chicago; the Tyler portable, by W. S. Tyler, Inc., Mentor, Ohio; and also a number of electromagnetic vibratory shakers. The latter should be used only when strict comparability with other tests is not required, since it is difficult to be sure that identical intensity of vibration was present in the tests being compared.

Equipment Screening machines may be divided into five main classes: grizzlies, revolving screens, shaking screens, vibrating screens, and oscillating screens. Grizzlies are used primarily for scalping at 0.05 m (2 in) and coarser, while revolving screens and shaking screens are generally used for separations above 0.013 m (½ in). Vibrating screens cover this coarse range and also down into the fine meshes. Oscillating screens are confined in general to the finer meshes below 4 mesh.

Grizzly Screens These consist of a set of parallel bars held apart by spacers at some predetermined opening. Bars are frequently made of manganese steel to reduce wear. A grizzly is widely used before a primary crusher in rock- or ore-crushing plants to remove the fines before the ore or rock enters the crusher. It can be a stationary set of bars or a vibrating screen.

Stationary grizzlies. These are the simplest of all separating devices and the least expensive to install and maintain. They are normally lim-

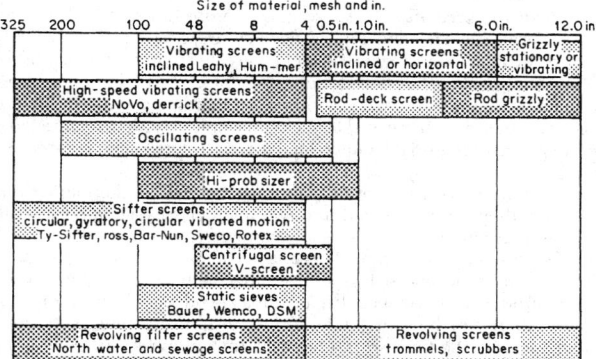

FIG. 19-14 Range of separations that can be obtained with various kinds of screens. To convert inches to meters, multiply by 0.0254. [*Matthews, Chem. Eng. (Feb. 15, 1971).*]

TABLE 19-6   U.S. Sieve Series and Tyler Equivalents
(ASTM—E-11-61)

| Sieve designation | | Sieve opening | | Nominal wire diam. | | Tyler equivalent designation |
Standard	Alternate	mm	in (approx. equivalents)	mm	in (approx. equivalents)	
107.6 mm	4.24 in	107.6	4.24	6.40	0.2520	
101.6 mm	4 in†	101.6	4.00	6.30	.2480	
90.5 mm	3½ in	90.5	3.50	6.08	.2394	
76.1 mm	3 in	76.1	3.00	5.80	.2283	
64.0 mm	2½ in	64.0	2.50	5.50	.2165	
53.8 mm	2.12 in	53.8	2.12	5.15	.2028	
50.8 mm	2 in†	50.8	2.00	5.05	.1988	
45.3 mm	1¾ in	45.3	1.75	4.85	.1909	
38.1 mm	1½ in	38.1	1.50	4.59	.1807	
32.0 mm	1¼ in	32.0	1.25	4.23	.1665	
26.9 mm	1.06 in	26.9	1.06	3.90	.1535	1.050 in
25.4 mm	1 in†	25.4	1.00	3.80	.1496	
22.6 mm°	⅞ in	22.6	0.875	3.50	.1378	0.883 in
19.0 mm°	¾ in	19.0	.750	3.30	.1299	.742 in
16.0 mm°	⅝ in	16.0	.625	3.00	.1181	.624 in
13.5 mm	0.530 in	13.5	.530	2.75	.1083	.525 in
12.7 mm	½ in†	12.7	.500	2.67	.1051	
11.2 mm°	7/16 in	11.2	.438	2.45	.0965	.441 in
9.51 mm°	⅜ in	9.51	.375	2.27	.0894	.371 in
8.00 mm°	5/16 in	8.00	.312	2.07	.0815	2½ mesh
6.73 mm	0.265 in	6.73	.265	1.87	.0736	3 mesh
6.35 mm	¼ in†	6.35	.250	1.82	.0717	
5.66 mm°	No. 3½	5.66	.223	1.68	.0661	3½ mesh
4.76 mm°	No. 4	4.76	.187	1.54	.0606	4 mesh
4.00 mm°	No. 5	4.00	.157	1.37	.0539	5 mesh
3.36 mm	No. 6	3.36	.132	1.23	.0484	6 mesh
2.83 mm°	No. 7	2.83	.111	1.10	.0430	7 mesh
2.38 mm°	No. 8	2.38	.0937	1.00	.0394	8 mesh
2.00 mm°	No. 10	2.00	.0787	0.900	.0354	9 mesh
1.68 mm	No. 12	1.68	.0661	.810	.0319	10 mesh
1.41 mm°	No. 14	1.41	.0555	.725	.0285	12 mesh
1.19 mm°	No. 16	1.19	.0469	.650	.0256	14 mesh
1.00 mm°	No. 18	1.00	.0394	.580	.0228	16 mesh
841 micron°	No. 20	0.841	.0331	.510	.0201	20 mesh
707 micron°	No. 25	.707	.0278	.450	.0177	24 mesh
595 micron°	No. 30	.595	.0234	.390	.0154	28 mesh
500 micron°	No. 35	.500	.0197	.340	.0134	32 mesh
420 micron°	No. 40	.420	.0165	.290	.0114	35 mesh
354 micron°	No. 45	.354	.0139	.247	.0097	42 mesh
297 micron°	No. 50	.297	.0117	.215	.0085	48 mesh
250 micron°	No. 60	.250	.0098	.180	.0071	60 mesh
210 micron°	No. 70	.210	.0083	.152	.0060	65 mesh
177 micron°	No. 80	.177	.0070	.131	.0052	80 mesh
149 micron°	No. 100	.149	.0059	.110	.0043	100 mesh
125 micron°	No. 120	.125	.0049	.091	.0036	115 mesh
105 micron°	No. 140	.105	.0041	.076	.0030	150 mesh
88 micron°	No. 170	.088	.0035	.064	.0025	170 mesh
74 micron°	No. 200	.074	.0029	.053	.0021	200 mesh
63 micron°	No. 230	.063	.0025	.044	.0017	250 mesh
53 micron°	No. 270	.053	.0021	.037	.0015	270 mesh
44 micron°	No. 325	.044	.0017	.030	.0012	325 mesh
37 micron°	No. 400	.037	.0015	.025	.0010	400 mesh

°These sieves correspond to those proposed as an international (I.S.O.) standard. It is recommended that wherever possible these sieves be included in all sieve analysis data or reports intended for international publication.

†These sieves are not in the fourth-root-of-2 series, but they have been included because they are in common usage.

ited to the scalping or rough screening of dry material at 0.05 m (2 in) and coarser and are not satisfactory for moist and sticky material. The slope, or angle with the horizontal, will vary between 20 and 50°. Stationary grizzlies require no power and little maintenance. It is, of course, difficult to change the opening between the bars, and the separation may not be sufficiently complete.

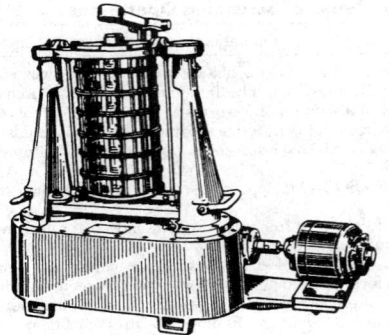

FIG. 19-15   Ro-Tap testing sieve shaker. (*W. S. Tyler, Inc.*)

Flat grizzlies. These, in which the parallel bars are in a horizontal plane, are used on tops of ore and coal bins and under unloading trestles. This type of grizzly is used to retain occasional pieces too large for the following plant equipment. These lumps must then be broken up or removed manually.

Vibrating grizzlies. These are simply bar grizzlies mounted on eccentrics so that the entire assembly is given a back-and-forth movement or a positive circle throw. These are made by companies such as Allis-Chalmers, Hewitt Robins, Nordberg, Link-Belt, Simplicity, and Tyler.

Revolving Screens   Revolving screens, or trommel screens, once widely used, are being largely replaced by vibrating screens. They consist of a cylindrical frame surrounded by wire cloth or perforated plate, open at both ends, and inclined at a slight angle. The material to be screened is delivered at the upper end, and the oversize is discharged at the lower end. The desired product falls through the wire-cloth openings. The screens revolve at relatively low speeds of 15 to 20 r/min. Their capacity is not great, and efficiency is relatively low.

Mechanical Shaking Screens   These screens consist of a rectangular frame which holds wire cloth or perforated plate and is slightly inclined and suspended by loose rods or cables or supported from a base frame by flexible flat springs. The frame is driven with a reciprocating motion. The material to be screened is fed at the upper end and is advanced by the forward stroke of the screen while the finer particles pass through the openings. In many screening operations such devices have given way to vibrating screens.

Shaking screens, such as the mechanical-conveyor type made by Syntron Co., may be used for both screening and conveying.

The advantages of this type are low headroom and low power requirement. The disadvantages are the high cost of maintenance of the screen and the supporting structure owing to vibration and low capacity compared with inclined high-speed vibrating screens.

Vibrating Screens   These screens are used as standard practice when large capacity and high efficiency are desired. The capacity, especially in the finer sizes, is so much greater than that of any of the other screens that they have practically replaced all other types when efficiency of the screen is an important factor. Advantages include accuracy of sizing, increased capacity per unit area, low maintenance cost per ton of material handled, and a saving in installation space and weight.

There are a great number of vibrating screens on the market, but basically they can be divided into two main classes: (1) mechanically vibrated screens and (2) electrically vibrated screens.

Mechanically Vibrated Screens   The most versatile vibration for medium to coarse sizing is generally conceded to be the vertical circle produced by an eccentric or unbalanced shaft, but other types of vibration may be more suitable for certain screening operations, particularly in the finer sizes. One well-known *four-bearing mechanically vibrated screen*, installed in an inclined position, is the Ty-Rock (Fig. 19-16). This is a balanced circle-throw machine mounted on a base frame, having a full-floating body mounted on shear rubber mounting units which absorb the shocks of heavy material and allow the shaft to revolve around its own natural center of rotation.

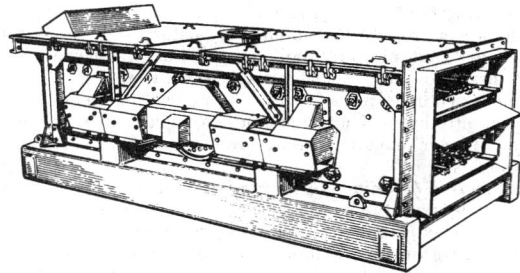

FIG. 19-16 Ty-Rock screen with air-seal enclosure. (*W. S. Tyler, Inc.*)

Two-bearing screens, of which there are many types, have the same screen body as the four-bearing type but without the two outer bearings and the base frame. The gyrating motion is caused by eccentric weights on the shaft, and the screen itself is supported by overhead cables or springs on the floor.

Screening machines actuated by rotating unbalanced weights have a symmetrical shaft through the screen body with an unbalanced flywheel on each end. Counterweights on each flywheel, which may be moved in relation to the shaft, permit adjustment of the amplitude of vibration. On some makes of machines the complete shaft assembly is contained in a unit bolted to the top of the screen body.

The horizontal-type screen is actuated by an enclosed mechanism consisting of off-center weights geared together on short horizontal shafts. The mechanism is usually mounted between the side plates and above the screen body (Fig. 19-17).

Electrically Vibrated Screens These screens are particularly useful in the chemical industry. They handle very successfully many light, fine, dry materials and metal powders from approximately 4 mesh to as fine as 325 mesh. Most of these screens have an intense, high-speed (25 to 120 vibrations/s) low-amplitude vibration supplied by means of an electromagnet.

Typical of these is the Hum-mer screen used throughout the chemical industry. Figure 19-18 shows one used throughout the fertilizer industry for handling mixed chemical fertilizers.

Oscillating Screens These screens are characterized by low-speed oscillations [5 to 7 oscillations per second (300 to 400 r/min)] in a plane essentially parallel to the screen cloth.

Screens in this group are usually used from 0.013 m (½ in) to 60 mesh. Some light free-flowing materials, however, can be separated at 200 to 300 mesh. Silk cloths are often used.

Reciprocating Screens These screens have many applications in chemical work. An eccentric under the screen supplies oscillation, ranging from gyratory [about 0.05-m (2-in) diameter] at the feed end to reciprocating motion at the discharge end. Frequency is 8 to 10

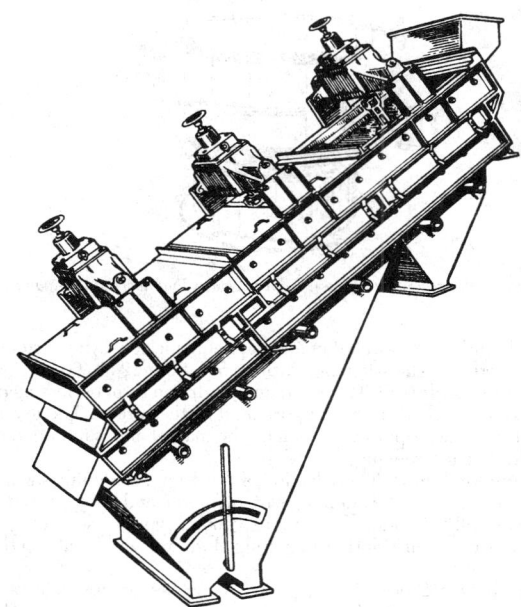

FIG. 19-18 Type 38 Hum-mer screen. (*W. S. Tyler, Inc.*)

oscillations per second (500 to 600 r/min), and since the screen is inclined about 5°, a secondary high-amplitude normal vibration of about 0.0025 m (¹⁄₁₀ in) is also set up. Further vibration is caused by balls bouncing against the lower surface of the screen cloth.

These screens are used extensively in the United States and are standard equipment in many chemical and processing plants for handling fine separations even down to 300 mesh. They are used to handle a variety of chemicals, usually dry, light, or bulky materials, light metal powders, powdered foods, and granular materials. They are not designed for handling heavy tonnages of materials like rock or gravel. Machines of this type are exemplified by Fig. 19-19.

Gyratory Screens These are boxlike machines, either round or square, with a series of screen cloths nested atop one another. Oscillation, supplied by eccentrics or counterweights, is in a circular or near-circular orbit. In some machines a supplementary whipping action is set up. Most gyratory screens have an auxiliary vibration caused by balls bouncing against the lower surface of the screen cloth. A typical machine is shown in Fig. 19-20. Machines of this type are operated con-

FIG. 19-17 Mechanically vibrated horizontal screen. (*Courtesy of Deister Concentrator Company, Inc.*)

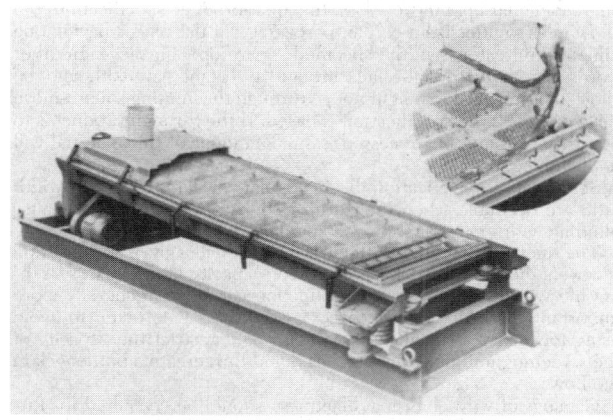

FIG. 19-19 Reciprocating screen. (*Courtesy of Rotex Corp.*)

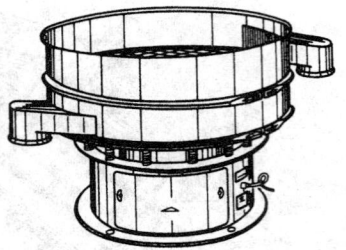

FIG. 19-20   Vibro-energy screen. (*Southwestern Engineering Company.*)

tinuously and can be located in line in pneumatic conveying systems as scalping screens. The size ranges from 0.6 to 1.5 m (24 to 60 in).

Gyratory Riddles   These screens are driven in an oscillating path by a motor attached to the support shaft of the screen. The gyratory riddle is the least expensive screen on the market and is intended normally for batch screening.

Screen Surfaces   The selection of the proper screening surface is very important, and the opening, wire diameter, and open area should all be carefully considered. The four general types of screening surfaces are woven-wire cloth, silk bolting cloth, punched plate, and bar or rod screens.

Woven-Wire Cloth   This type has by far the greatest selection as to screen opening, wire diameter, and percentage of open area. Thousands of specifications are available from over 0.10 m (4 in) clear opening to 500 mesh. Woven-wire screens are obtainable in a variety of metals and alloys. Steel and high-carbon steel are generally favored for the coarser openings because of their abrasion-resistant qualities, and other materials, such as phosphor bronze, Monel, and stainless steel, are used for their corrosion-resisting or noncontamination qualities.

Square-mesh cloth is the conventional type of screen cloth, but there are many types of cloth with an oblong weave. This latter construction provides greater open area and capacity and in addition makes it possible to use stronger wire for the same size of screen opening and for the same percentage of open area.

In choosing a wire-cloth specification there must be a compromise between sharpness of separation, capacity, freedom from blinding, and life of the wire cloth. The square-mesh cloth will give the closest control of the maximum size particle in the undersize material; but the effective size of the openings will be reduced, because of the foreshortening when used at an angle of inclination, with consequent reduction in capacity. It should be realized that it is often necessary to use a cloth specification with an aperture larger than the smallest-size material acceptable in the oversize in order to ensure thorough removal of the undersize. A screen with a rectangular opening will increase the capacity with but little loss of sharpness when handling rounded or cubical grains. Slabby or flat material may also be handled on rectangular-opening cloth if the final-product specification will allow in the undersize a certain percentage of flat pieces having one dimension greater than the specified square-opening sieve. In other words pieces that might fall through a rectangular cloth might be allowed in the product might not go through the limiting square-mesh sieve on which the specification is based. If the through product is to be further ground or processed, a small amount of this material will not be objectionable.

Screen-cloth specifications having a relatively large length-to-width ratio are desirable when moisture or sticky material tends to cause blinding with square or short rectangular openings.

The finer the diameter of the wire from which a given specification is woven, the greater will be its screening capacity, although its screening life will be shorter. Since production capacity is generally more important than screen-surface cost, care should be taken to avoid using too heavy a specification which might restrict the capacity of the screening unit on which it is used and thus create a bottleneck in the flow.

Catalogs of wire-cloth manufacturers should be consulted for further study of the different types of wire-cloth specifications.

Silk Bolting Cloth   This material originated in Switzerland and is generally woven from twisted multistrand-natural silk. The system of numbers and grades for both bolting cloth and gritz gauze has been handed down from the original Swiss weavers. In recent years, nylon and similar synthetic materials woven largely from monofilaments have been introduced. The nylon grades are generally designated by their micrometer opening and are available in light, standard, and heavy weights.

Comparative Openings of Screening Cloths   In screening any material, the size of the particles going through the screen is determined by the actual opening and not by the number of meshes per linear unit. As a rule, the lighter grades of wire-screen cloth, having greater percentages of open area, screen more freely and accurately and should be used whenever they will give satisfactory length of service. Tables of comparative openings are available for selecting a screen specification with a specific opening or for picking a specification having a heavier or lighter wire but the same opening.

Punched Plates   These are available in a variety of perforations including round, square, hexagonal, and elongated openings. Punched metal will generally wear longer than wire cloth and has more rigidity, which is an advantage in certain applications. However, it usually does not give the capacity per unit area that wire cloth does and is generally heavier. Its use is normally limited to the coarser separations.

Bar Screens   These screens are generally used in handling large and heavy pieces of material. They are formed from rails, rods, or bars, suitably shaped; made from rolled steel or castings; fixed in parallel position and held by crossbars and spacers. Bars, which taper in thickness from top to bottom and may also taper in width from one end to the other, are recommended because they tend to avoid blinding.

Rod decks, composed of spring-steel rods approximately 0.6 m (2 ft) long, sprung into position between molded-rubber blocks, and held in position by means of rubber spacers, are also available.

Probability Screening Principle   Probability screening uses the fact that particles moving almost at right angles to a screening surface are not likely to pass through when the particle size is greater than about half of the distance between the screening elements. The screens utilizing the probability principle are manufacturing by Dutch State Mines (DSM), Bartles (CTS), and Morgensen. The last-named incorporates multiple decks. Higher throughput, longer screen life, and lower capital costs are claimed for these screening systems. The performance of several types of probability screens was reviewed by Moir (op cit.).

Factors in Selecting Screening Equipment   In attempting to pick a screening machine for a specific screening problem it should be emphasized that generalized formulas and charts used to predict screen capacity can give only an approximation because of the many variables which may affect performance. Screen consultants will readily admit that they must depend largely on laboratory tests and field experience. However, two governing factors bear mention: generally width of screen relates to capacity and length of screen relates to efficiency. Width is necessary to reduce the bed thickness to a practical maximum and length to allow the undersize to be removed without an inordinate amount of fines in the oversize. In attempting to choose a screening machine for a particular screening application the customer and manufacturer should consider the following: (1) Full description of material involved, including the name and type of material, bulk density, and physical characteristics such as hardness, particle shape, flow characteristics (free-flowing, sluggish, or sticky), percent of moisture and temperature. (2) Normal and maximum total rate of feed to screen. (3) Complete sieve analysis of screen feed, including maximum lump size, and sieve analysis of desired product. (4) Separation or separations required and the purpose of screening. Can slotted or rectangular openings be used in place of square openings? (5) Is screening to be accomplished dry or wet, and what amount of water is available? (6) Other important factors include method of delivering feed to the screen, open or closed circuit, open or enclosed screens, previous screening experience with the material, flow sheet or description of related equipment, operating hours per day, power available, and space limitations.

Variables in Screening Operations   It will readily be seen that many variables in a screening operation can easily be changed in the field, and practical operators will always be trying to improve their operations or adapt them to new products or processes. Capacity and efficiency in screening operations are closely related. Capacity may be large if low efficiency is not objectionable. Usually, as the tonnage to a screen is increased, efficiency is decreased.

Method of Feed   The screening machine must be fed properly in order to obtain maximum capacity and efficiency. The feed should be spread evenly over the full width of the screen cloth and approach the screen surface in a direction parallel to the longitudinal axis of the screen and at as low a practical velocity as is possible.

Screening Surfaces   It is generally agreed that the most efficient screening results when a series of single-deck screens is used. This is true because lower decks of multiple-deck screens are not fed so that their entire area is used and because each separation requires a different combination of angle, speed, and amplitude of vibration for maximum performance.

Angle of Slope   The optimum slope of inclined vibrating screens is that which will handle the greatest volume of oversize and still remove the available undersize required by the standards of the particular operation. To separate a material into coarse and fine fractions, the bed thickness must be limited so that vibration can stratify the load and allow fines to work their way to the screen surface and pass through the opening. Increased slope naturally increases the rate of travel, and at a given rate it reduces the bed thickness.

In the oscillating screen the angle of inclination must be coordinated with the speed and stroke for best results.

Direction of Rotation   In circle-throw screens somewhat greater efficiency can be obtained by counterflow rotation, that is, having the material move down the screen against the rotation. Screens rotating with the flow of material will handle greater tonnage and operate at a lower angle.

Vibration Amplitude and Frequency   Speed and amplitude of vibration should be designed to convey the material properly and to prevent blinding of the cloth. They are somewhat dependent upon the size and weight of the material being handled and are related to the angle of installation and the type of screen surface. The object, of course, is to see that the feed is properly stratified for the most efficient separation.

Noise and Safety   Noise is generated in screening due to the impact of the feed material on the screen surface. The drive mechanism also generates noise. Rubber and rubber bearings reduce substantially feed-impact noise with the added longer life of the decks. Noise from the drive mechanism is reduced by enclosing the mechanism in a box or by adding rubber linings to the side plates to dampen the noise.

Depending on the feed materials, the dust generated during operation may be hazardous because of possible emissions and toxicity. These hazards must be carefully evaluated before proper design of the facility and selecting the apparatus.

Performance Formulas

Screen Efficiency   There is confusion concerning the meaning of screen efficiency, as a uniform method for figuring efficiency has never been established. A sound method of evaluating screen performance is given by W. S. Tyler, Inc., Mentor, Ohio, in its *Sieve Handbook*, no. 53. In this formula, when material put through the screen is the desired product, "efficiency" is the ratio of the amount of undersize obtained to the amount of undersize in the feed.

$$E = (R \times d)/b \qquad (19\text{-}5)$$

where E = efficiency, R = percent of fines through the screen, d = percent finer than the designated size in screen fines, and b = percent finer than the designated size in screen feed.

When the object is to recover an oversize product from the screen, efficiency may be expressed as a ratio of the amount of oversize obtained to the amount of true oversize:

$$E = (O \times c)/a \qquad (19\text{-}6)$$

where O = percent of oversize over the screen, c = percent coarser than the designated size in screen oversize, and a = percent coarser than the designated size in screen feed.

Other formulas for the derivation of screen efficiency are used. Taggart (*Handbook of Mineral Dressing*) gives the formula

$$E = 100 \times \frac{100(e - v)}{e(100 - v)}$$

where E is the efficiency, e is the percentage of undersize in the feed, and v is the percentage of undersize in the screen oversize.

Graphical methods of evaluating efficiency, using sieve analyses, are also employed and are recommended when serious research on screening is done.

Estimating Screen Capacity   Various methods of predicting screening capacity have been proposed, and each has its limitations. The throughflow method of Matthews uses the following equation:

$$A = 0.4C_t/C_u F_{oa} F_s \qquad (19\text{-}7)$$

where A = screen area
C_t = throughflow rate
C_u = unit capacity
F_{oa} = open-area factor
F_s = slotted-area factor

The unit capacity C_u can be determined from Fig. 19-21. Figure 19-22 can be used to determine the open-area factor F_{oa}, and the slotted-opening factor F_s for various screen types is given in Table 19-7.

WET CLASSIFICATION

GENERAL REFERENCES: Dyakowski, T., and Williams, R. A., "Modelling Turbulent Flow within a Small Diameter Hydrocyclone," *Chemical Engineering Science*, vol. 48, p. 1143, 1993. Heiskanen, K., "Particle Classification," Scarlett, B., (ed.), *Powder Technology Series*, Chapman & Hall, 1993. Fitch, B., "Gravity Sedimentation Operations," McKetta J. J., (ed.), *Unit Operations Handbook*, vol. 2, Mechanical Separation and Material Handling, p. 51, Marcel Dekker, Inc., New York, 1993. Neal Abernathy, M. W., "Gravity Settlers, Design" McKetta J. J., (ed.), *Unit Operations Handbook*, vol. 2, Mechanical Separation and Material Handling, p. 127, Marcel Dekker, Inc., New York, 1993. Zanker, A., "Gravity Settlers, Sizing of Decanters," McKetta J. J., (ed.), *Unit Operations Handbook*, vol. 2, Mechanical Separation and Material Handling, p. 136, Marcel Dekker, Inc., New York, 1993. Rajamani, R. K., and Milin L., "Fluid-Flow Model of the Hydrocyclone for Concentrated Slurry Classification" Svarovsky, L., and Thew M. T., (eds.), *Hydrocyclones Analysis and Applications*, p. 95, Kluwer Academic Publishers, Dordrecht, The Netherlands, 1992. Dahlstrom D. A., "Fundamental of Solid-Liquid Separation," Mular A. L., and Anderson, M. A., (eds.), *Design and Installation of Concentration and Dewatering Circuits*, p. 103, SME, New York, 1986. Kelly, E. G., and Spottiswood, D. J., *Introduction to Mineral Processing*. John Wiley & Sons, New York, 1982. Tarr, D. T., Jr., "Hydrocyclones," Weiss, N. L. (ed.), *SME Mineral Processing Handbook*, vol. 1., p. 3D-10, SME, New York, 1985. Devulapalli, B., and Rajamani, R. K., "A comprehensive CFD model for particle size classification in industrial hydrocyclones," Claxton, D., Svarovsky, L., and Thew, M. (eds.), Hydrocyclone '96, p. 83–104, *Mechanical Engineering Publications Limited*, London and Bury St Edmunds, UK, 1996.

Introduction   Wet classification is defined here as that art of separating the solid particles in a mixture of solids and liquid into fractions according to particle size or density by methods other than screening. In general, the products resulting are (1) a partially drained fraction containing the coarse material (called the underflow) and (2) a fine fraction along with the remaining portion of the liquid medium (called the overflow).

The classifying operation is carried out in a pool of fluid pulp confined in a tank arranged to allow the coarse solids to settle out, whereupon they are removed by gravity, mechanical means, or induced pressure. Solids which do not settle report as overflow. Mesh of sepa-

TABLE 19-7   Slotted-Opening Factors

Screen type	Length-to-width ratio
Square and slightly rectangular openings	Less than 2
Rectangular openings	Equal to or greater than 2 but less than 4
Slotted openings	Equal to or greater than 4 but less than 25
Parallel-rod decks	Equal to or greater than 25

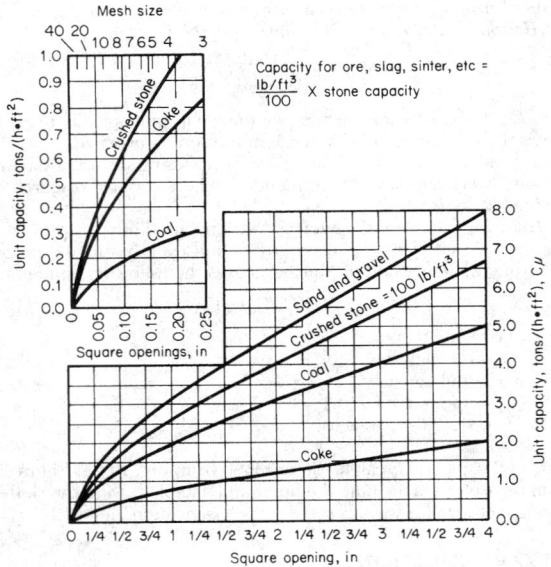

FIG. 19-21 Unit capacity (C_u) for square-opening screens. To convert inches to meters, multiply by 0.0254; to convert tons per hour-square foot to kilograms per second-square meter, multiply by 2.7182.

Aperture	Formula	
Rectangular openings	$F_{oa} = \dfrac{a_1 a_2}{(a_1 + d_1)(a_2 + d_2)} \times 100$	
	F_{oa} is open area, %, d is diameter of wire, or horizontal width of bar (for plate); a is clear opening dimension	(21-4)
Square openings Specified by opening size	$F_{oa} = 100 \left(\dfrac{a}{a+d}\right)^2$ $a_1 = a_2 = a$ $d_1 = d_2 = d$	(21-5)
Square openings Specified in mesh, m	$F_{oa} = 100\, a^2 m^2$ $m = \dfrac{1}{a+d}$	(21-6)
Parallel-rod decks 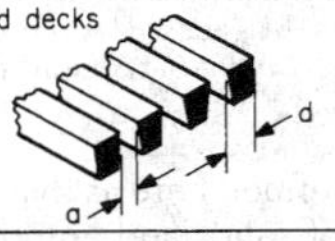	$F_{oa} = \dfrac{100a}{(a+d)}$	(21-7)
Special weaves 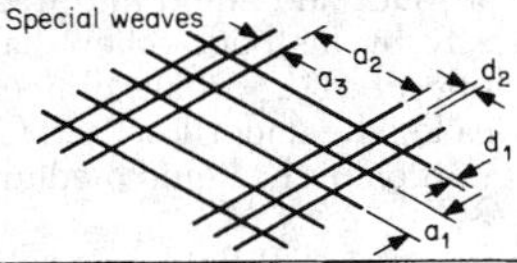	Assuming $a_3 = a_1$; $F_{oa} = 100 \left[\dfrac{a_1(a_2 + 2a_1)}{(a_2 + 2a_1 + 3d_2)(a_1 + d_1)}\right]$	(21-8)

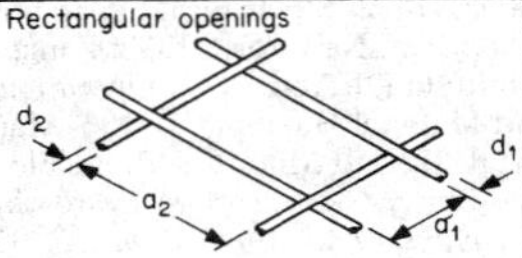

FIG. 19-22 Open-area factor (F_{oa}) for flow-through screen-capacity calculation.

ration as used in this text is the screen size retaining 1½ percent of the overflow solids.

All wet classifiers depend on the difference in settling rate between coarse and fine or heavy- and light-specific gravity particles to be separated. Rates can be controlled to some extent by mild agitation, providing for hindered settling, and centrifugal force versus gravity in centrifuging types of units.

Several fundamental laws on classification are:

1. Coarse particles have a relatively faster settling velocity than fine particles of the same specific gravity.

2. Heavy-gravity particles have a relatively faster settling velocity than light-gravity particles of the same size. High solids concentration increases the viscosity and density of the fluid medium.

3. Settling rates of solid particles become progressively slower as the viscosity or density of the fluid medium increases.

  a. There is a point (called critical dilution) where the lowering of density or viscosity by addition of more liquid creates a velocity effect which overcomes normal classification settling velocity, thereby coarsening the separation.

  b. Conversely, at this point less liquid will cause a viscosity and buoyancy effect which will also coarsen the separation.

Typical problems to be solved by wet-classification means fall into several broad categories such as (1) to effect a simple sand-slime separation resulting in two products; (2) to effect a concentration of smaller heavy-gravity particles in a product containing larger light-gravity particles; (3) to obtain a washing effect by successive dewatering, repulping in weaker solution, and further dewatering; (4) to sort solids having a full range of screen sizes into a number of partials each having a short range of screen sizes; and (5) to achieve closed-circuit control of grinding mills.

Classification is by definition used preponderantly in the treatment of raw materials. However, these raw materials find their way into chemical processing per se and thus become of interest to the chemical engineer, particularly when the products to be treated react better when of a defined cleanliness, size, gravity, or moisture content.

Classifier types fall into two basic categories: (1) gravitational and (2) centrifugal classifiers. Gravitational classifiers can be subdivided into (1) sedimentation and (2) hydraulic classifiers. Furthermore each type falls into mechanical and nonmechanical types.

There are numerous machines and machine types to obtain a number of different particle-size classes from solids having a full range of sizes, and there is much overlapping in the possibilities. Usually, one type will provide optimum economy for the specific problem involved.

The quick reference Table 19-8 will help by way of rapid elimination of poor possibilities. Following that the brief comments and illustrations will help pinpoint most probable selections. Further study of the more elaborate data in the references and contact with the usual suppliers are recommended, as there are many possible modifications of equipment which can improve operating results from any type of machine finally selected.

Nonmechanical Classifiers

Cone Type Cone classifiers are one of the oldest types but are still used for relatively crude work because of low cost of installation. They are limited in diameter because of high headroom requirements caused by the ±60° sloping sides. Units are simple and are often fabricated locally with millwright ingenuity fashioning the apex opening arrangement for adjustment or control of the spigot coarse product. Operating attention is often necessary to a greater degree than for the more positive mechanical types. Cost figures are not available.

Hydrocyclone The wet cyclone classifier has rapidly achieved prominence since the 1950s and continues to gain popularity throughout chemical and ore-dressing industries. Standout virtues are its low capital cost and ability to make extremely fine separations by proper adjustment of design/operating condition. See Fig. 19-23.

In simplest terms the unit has a top cylindrical section and a lower conical section terminating in an apex opening, often adjustable. The unit operates under pressure induced by a static hydraulic head or by means of a pump forcing new feed into the cylindrical portion tangentially, thus producing centrifuging action and vortexing. The cover has a downward-extending pipe to cut the vortex and remove the overflow product called vortex finder. Coarse solids travel down the sides of the steeply sided cone section and are removed in a partially dewatered form at the apex.

Hydrocyclones are available in numerous sizes and types ranging from pencil-sized 10-mm diameters of plastic to the 1.2-m (48-in) diameter of rubber-protected mild or stainless steel. Porcelain units 25 to 100 mm (1 to 4 in) in diameter are becoming popular, and in the 150-mm (6-in) size the starch industry has standardized on special molded nylon types. Small units for fine-size separations are usually manifolded in multiple units in parallel with up to 480 ten-mm

TABLE 19-8 The Major Types of Classifiers

Classifier	(Type°)	Description	Size (m) Width Diameter Max. length	Limiting size (max. feed size)	Feed rate (t/hr)	Vol. % solids Feed overflow underflow	Power (kW)	Suitability and applications
Sloping tank classifier (spiral, rake, drag)	(M-S)	Classification occurs near deep end of sloping, elongated pool. Spiral, rake or drag mechanism lifts sands from pool.	0.3 to 7.0 (spiral) 2.4 (spiral) 14	1 mm to 45 μm (25 mm)	5 to 850	Not critical 2 to 20 45 to 65	0.4 to 110	Used for closed circuit grinding, washing and dewatering, desliming; particularly where clean dry underflow is important. (Drag classifier sands not so clean.) In closed circuit grinding discharge mechanism (spirals especially) may give enough lift to eliminate pump.
Log washer	(M-S)	Essentially a spiral classifier with paddles replacing the spiral.	0.8 to 2.6 0.6 to 1.1 4.6 to 11	(100 mm)	40 to 450		7.5 to 60	Used for rough separations such as removing trash, clay from sand. Also to remove or break down agglomerates.
Bowl classifier	(M-S)	Extension of sloping tank classifiers, with settling occurring in large circular pool, which has rotating mechanism to scrape sands inwards (outwards in Bowl Desiltor) to discharge rake or spiral.	0.5 to 6.0 1.2 to 15 12	150 μm to 45 μm (12 mm)	5 to 225	Not critical 0.4 to 8 50 to 60 (15 to 25 in Bowl Desiltor)	Bowl: 0.75 to 7.5 Rake: 0.75 to 20	Used for closed circuit grinding (particularly regrind circuits) where clean underflow is necessary. Bowl Desiltor has finer separations. Bowl Desiltor has larger pools (and capacities). Relatively expensive.
Hydraulic bowl classifier	(M-F)	Basically a hydraulic bowl classifier. Vibrating plate replaces rotating mechanism in pool. Hydraulic water passes through perforations in plate and fluidizes sands.	–1.2 to 3.7 1.2 to 4.3 12	1 mm to 100 μm (12 mm)	5 to 225	Not critical 2 to 15 50 to 65	Vib: 2.2 to 7.5 Rake: 3.7 to 15	Gives very clean sands and has relatively low hydraulic water requirements (0.5 t/t underflow). One of the most efficient single-stage classifiers available for closed circuit grinding and washing. Relatively expensive.
Cylindrical tank classifier	(M-S)	Effectively an overloaded thickener. Rotating rake feeds sands to central underflow.	— 3 to 45 —	150 μm to 45 μm (6 mm)	5 to 625	Not critical 0.4 to 8 15 to 25	0.75 to 11	Simple, but gives relatively inefficient separation. Used for primary dewatering where the separations involve large feed volumes, and underflow drainage is not critical.

°M: Mechanical transport of sands to discharge
N: Nonmechanical (gravity or pressure) discharge of underflow
S: Sedimentation classifier
F: Fluidized bed classifier
From Kelley, E. G. and D. J. Spottiswood, *Introduction to Mineral Processing*, John Wiley & Sons, New York, 1982, pp. 200–201, with permission.

TABLE 19-8 The Major Types of Classifiers (Concluded)

Classifier	(Type°)	Description	Size (m) Width Diameter Max. length	Limiting size (max. feed size)	Feed rate (t/hr)	Vol. % solids Feed overflow underflow	Power (kW)	Suitability and applications
Hydraulic cylindrical tank classifier	(M-F)	Hydraulic form of overloaded thickener. Siphon-Sizer (N-F) uses siphon to discharge underflow instead of rotating rake.	— 1.0 to 40 —	1.4 mm to 45 μm (25 mm)	1 to 150	Not critical 0.4 to 15 20 to 35	0.75 to 11	Two-product device giving very clean underflow. Requires relatively little hydraulic water (2 t/t solids feed). Used for washing, desliming, and closed circuit grinding.
Cone classifier	(N-S)	Similar to cylindrical tank classifier, except tank is conical to eliminate need for rake.	— 0.6 to 3.7 —	600 μm to 45 μm (6 mm)	2 to 100	Not critical 5 to 30 35 to 60	None	Low cost (simple enough to be made locally), and simplicity can justify relatively inefficient separation. Used for desliming and primary dewatering. Solids buildup can be a problem.
Hydraulic cone classifier	(M-F)	Open cylindrical upper section with conical lower section containing slowly rotating mechanism.	— 0.6 to 1.6 —	400 μm to 100 μm (6 mm)	10 to 120	Not critical 2 to 15 30 to 50	3 to 7.5	Used primarily in closed circuit grinding to reclassify hydrocyclone underflow.
Hydrocyclone	(N-S)	(Pumped) pressure feed generates centrifugal action to give high separating forces, and discharge.	— 0.01 to 1.2 —	300 μm to 5 μm (1400 μm to 45 μm)	to 20 m³/min	4 to 35 2 to 15 30 to 50	35 to 400 kN/m² pressure head	Small cheap device, widely used for closed circuit grinding. Gives relatively efficient separations of fine particles in dilute suspensions.
Air separator	(N-S)	Similar shape to hydrocyclone, but higher included angle. Internal impellor induces recycle within classifier.	— 0.5 to 7.5 —	2 mm to 38 μm	to 2100		4 to 500	Used where solids must be kept dry, such as cement grinding. Air classifiers may be integrated into grinding mill structure.

Equipment	Type	Description						Remarks
Solid bowl centrifuge	(M-S)	Power generates high settling forces. Slurry centrifuged against rotating bowl, and removed by slower rotating helical screw conveyor within bowl.	— 0.3 to 1.4 1.8	74 μm to 1 μm (6 mm)	0.04 to 2.5 m³/min	2 to 25 0.4 to 20 5 to 50	11 to 110	Relatively expensive, but high capacity for a given floor space; used for finer separations.
Scrubber	(M-S)	Essentially a rotating drum mounted on slight incline.	— 1.5 to 3.5 3 to 10	(450 mm)	to 700		1 to 55	Similar applications to log washer, but lighter action. Tumbling (85% critical speed) provides attrition to remove clay from sand. Also removes trash.
Countercurrent classifier	(M-F)	One form based on scrubber, another on spiral classifier. They have wash water added to flow essentially horizontally in opposite direction to underflow which is conveyed and *resuspended* by some form of spiral.	— 0.5 to 3.3 (spiral type) 12 (spiral type)	2 mm to 40 μm	3 to 600	Not critical 2 to 15 50 to 65	0.2 to 19	Very clean coarse product, but relatively low capacity for a given size.
Elutriator	(N-F)	Basically a tube with hydraulic water fed near bottom to produce hindered settling. Underflow withdrawn through valve at base. Column may be filled with network to even out flow.	— 1.2 to 4.3 —	2.4 mm to 100 μm (7.5 mm)	4 to 120	15 to 35 0.4 to 5 20 to 35	0.75 for valves	Simple and relatively efficient separation. Normally a two-product device but may be operated in series to give a range of size fractions.
Pocket classifier	(N-F)	A series of classification pockets, with decreasing quantities of hydraulic water in each, producing a range of product sizes.	— 0.5 to 6.0 12	2.4 mm to 100 μm (10 mm)	4 to 120	15 to 35 0.4 to 5 20 to 35		Efficient separations, but requires 3 t hydraulic water/t solids feed. Used to produce exceptionally clean underflow fractioned into narrow size ranges.

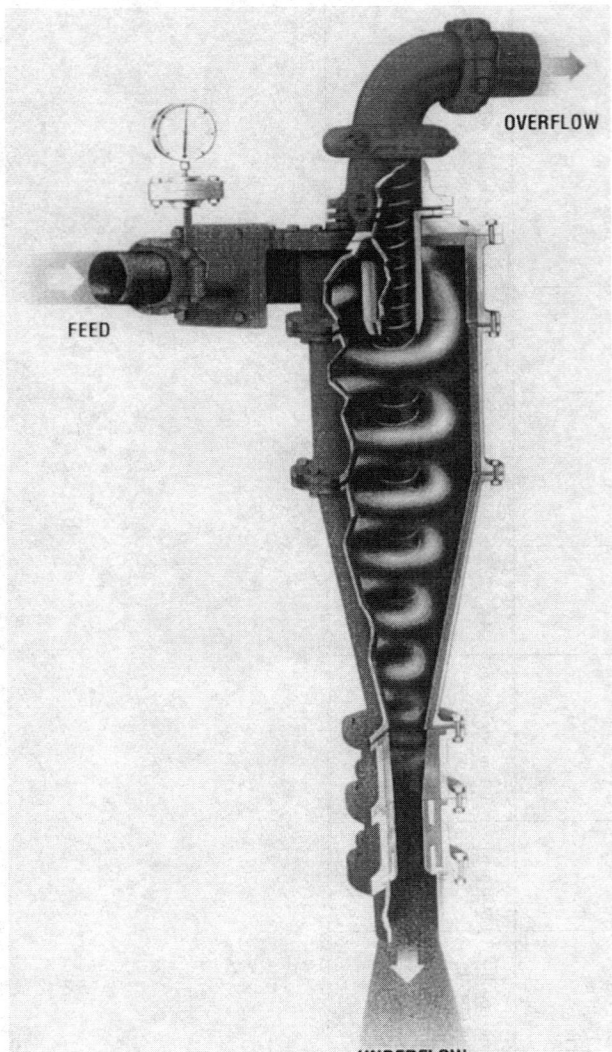

FIG. 19-23 Hydrocyclone. (*Courtesy Krebs Engineers.*)

cyclones in a single case. Larger sizes may be used singly or manifolded by outside piping.

The hydrocyclone has mostly replaced other classifiers in closed-circuit grinding.

Typical uses more in line with chemical applications are degritting milk of lime and of red mud in alumina production, removal of carbonaceous material in upgrading gypsum produced in making phosphoric acid, open-circuit washing of fine uranium pulps, classification of crystal magma such as lactose and sodium bisulfite, and classifying pigment and plastic beads into size ranges.

Mechanical Classifiers

Drag Classifiers Single endless-belt or chain suspensions with cross flights running in an inclined trough have long been used for draining and classifying. Many styles, sizes, and shapes have resulted from locally built units, and operating results on a scientific basis are meager. In general, they have served their purpose consistent with the type of engineering and cost included.

The Hardinge Overdrain° classifier is of the belt type, but it embodies the innovation of allowing entrapped water and slimes to escape through holes in the belt just uphill of the cross flights in an upward direction and thence flow down on top of the belt into the pool without again intermingling with the coarse product being advanced by the cross flights. Coarse product with lower moisture and fines content result from this action. Modern design and materials of construction permit sizes up to 3 m (10 ft) wide and 12.5 m (41 ft) long on steeper than average slopes and for very high tonnages.

Rake and Spiral Classifiers Rake-type classifiers such as the Dorr† classifier and spiral types such as the Akins‡ have been the workhorses for general-classification problems for half a century, and their names describe the mechanisms installed in sloping-bottom tanks. See Fig. 19-24. Mechanically the devices are powerfully built, and functionally they are versatile and flexible. They were the first classifiers used successfully for closed-circuit grinding. Separations as fine as 325 mesh can be accomplished at reduced tonnage rates.

Control of water into the classifiers is important since separation into fine and coarse products is made largely by the buoyancy, viscosity, and degree of agitation in the pool.

Both types of devices will produce rake products of consistent moisture content even with considerable variation in feed tonnage or volume.

Bowl Classifier The bowl classifier was developed to provide more separation area necessary for fine separations consistent with high tonnage. In essence a shallow bowl with revolving plows is superimposed over a rake or screw dewatering section. Feed enters at the center of the bowl, and fine solids overflow at the periphery. Coarse solids collected on the bowl bottom are raked to the center for discharge into the dewatering compartment below where wash water may be added for counterflow.

Hydrocyclones are rapidly taking over the functions formerly handled by bowl classifiers because of lower capital costs and floor-area requirements.

Bowl Desilter The bowl desilter provides for separation areas well beyond areas possible in bowl classifiers, in which larger sizes are limited by mechanical design. Its use is in operations involving large flow volumes and fine separations. Rake tonnages can be great or small with a dewatering compartment to suit the conditions.

In the bowl desilter the rotating blades in the bowl plow outward and discharge settled coarse material at the periphery, where it drops into the drainage compartment. This configuration does away with the long cantilevered rake construction necessary in bowl classifiers.

Widest application has been for the recovery of and drainage of very fine material overflowing coarser washing units in glass sand, concrete sand, coal, and limestone processing plants.

Hydroseparator The hydroseparator is merely a thickener-type machine receiving more flow than can be clarified in the area provided. Thus the overflow contains fine solids, and the greater the feed rate per unit of area the coarser the solids in the overflow.

Classification efficiency of the hydroseparator compares with that of the cone classifier and is appreciably lower than that obtained from mechanical or hydraulic units. The chief virtue of the hydroseparator is its ability to receive and slough off great quantities of water at low per-unit-volume cost.

Typical applications include primary dewatering of phosphate rock matrix and silica sand products following wet screening. In ore dressing it is used mainly to protect large-diameter thickeners by scalping out +65-mesh material.

Solid-Bowl Centrifuge The Bird solid-bowl centrifuge uses power instead of gravity and can develop centrifugal forces up to 1800 times the force of gravity. It is therefore a unique type in classification practice.

The unit consists essentially of two rotating elements, the outer being a solid-shell conical-shaped bowl and the inner comprising a helical-screw conveyor revolving at a speed slightly lower than that of the bowl. Raw feed slurry is delivered through a stationary feed pipe

° Trademark of Koppers Co., Inc.
† Trademark of Dorr-Oliver Inc.
‡ Trademark of Mine & Smelter.

FIG. 19-24 Spiral type classifier. Wemco S-H 78-in classifiers in closed-circuit grinding operation at St. Joseph Lead Co., Indian Creek plant. (*Courtesy Wemco Div., Envirotech Corp.*)

to the conveyor, where, urged by centrifugal force, it is transferred to the revolving bowl. A circumferential classifying area is formed and contained at the larger diameter of the cone shell. The ports for oversize material are located closer to the axis of rotation than the ports for the overflow to effect a beach line and drainage.

Centrifugal force deposits the oversize particles against the bowl wall, from which they are conveyed by the helix. The overflow fractions flow around the helix to the liquid-discharge ports. Size of separation is controlled by feed rate and degree of centrifugal force.

Several prime features of this totally enclosed unit are its high capacity per unit of floor area, small volume of material in process, high degree of separation, and shear action for dispersion of solids. Typical applications are desliming to upgrade cement rock, sizing of abrasives, fractionating for reagent control, and classification of pigments.

Countercurrent Classifier The countercurrent classifier is an inclined, slowly rotating cylindrical drum with continuous spiral flights attached to the interior of the shell forming helical troughs. Direction of rotation is such that material in the troughs is impelled toward the higher end. The lower end of the shell is closed except for a central overflow opening. Attached to the upper end is a coarse solid dewatering elevator which rotates with the shell. Wash water introduced at the upper end drains from the lifting flights above the normal water level and progresses countercurrently to the sand toward the overflow.

Usual application is for sand-slime separations, washing and for closed construction restricting escape of heat and chemical fumes, easy start-up after shutdown, and general simplicity. Weights range from 500 to 55,000 kg (1100 to 120,000 lb).

Hydraulic Classifiers

Jet Sizer° and SuperSorter† The Jet Sizer and SuperSorter are multicompartment and, therefore, multiproduct classifiers operating on the basis of hindered settling. The classification pockets are arranged in series for throughflow with parallel pockets to take care of high tonnage size fractions. Each compartment is served with low-pressure hydraulic water.

Hydraulic classification ensures the highest separating efficiency obtainable by wet-classification means. The amount of hydraulic water is controlled so that in each succeeding compartment the coarsest particles are maintained in hindered-settling condition and the finer fractions pass along for similar treatment. Two compartments will normally capture 90 percent of a two-screen-size fraction. Spigot discharge is controlled by air-actuated valves in the Jet Sizer and motor-driven pincer-type valves in the SuperSorter. Solid fractions can be taken from single or combinations of compartments as desired.

Typical applications include careful sizing of silica-glass sand, washing phosphate rock, sizing of abrasives, smokeless powder, sodium aluminate, etc.

D-O SiphonSizer° The D-O SiphonSizer (Fig. 19-25) is a high-efficiency hydraulic classifier developed originally for the washing and sizing of phosphate rock. In ore-dressing work it is normally a two-product unit; but by use of an upper column sealed at the top and

° Trademark of Dorr-Oliver Inc.
† Trademark of Deister Concentrator Company, Inc.

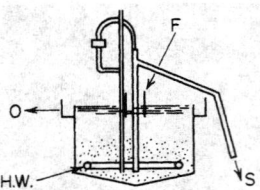

FIG. 19-25 D-O SiphonSizer.

open at the bottom, three products are possible: coarse, intermediate, and fine fractions.

Feed to be sized is put into hindered-settling condition by hydraulic water in quantity only sufficient to teeter the smallest particle wanted in the coarse product. The finer fractions report to the overflow or pass into the upper column for removal in a three-product unit.

Coarse solids are discharged by siphons extending to the bottom of the hindered-settling zone. Siphon control is obtained by a novel hydrostatically actuated valve which makes or breaks the siphon to flow only when the teeter zone is in correct condition. Discharge by an intermediate fraction from the upper column is by means of additional siphons. Hydraulic-water consumption is considerably lower than required for multipocket sizers.

SiphonSizers vary so widely in configuration that general cost data are not meaningful.

JIGGING

GENERAL REFERENCES: Aplan, "Gravity Concentration," in Kirk and Othmer (eds.), *Encyclopedia of Chemical Technology*, vol. 12, 3d ed., Wiley, New York, 1980, pp. 1–29. Bogert, "Fine Coal Cleaning with the Feldspar Jig," *Min. Congr. J.*, **46**, 42 (July 1960). *Mineral Dressing*. McGraw-Hill. Burt, *Gravity Concentration Technology*, Elsevier (1985). Green, "Designers improve jig efficiency," *Coal Age*, **89**, 50 (1984). Hasse and Wasmuth, "Use of air-pulsated Battac jigs for production of high-grade lump ore and sinter feed from intergrown hematite iron ores," *Proc. XVI International Mineral Processing Congress*, A, 1063, Elsevier (1988). Kirchberg and Hentzschel, "A Study of the Behavior of Particles in Jigging," *Trans. Int. Miner. Dressing Congr.*, 1957, Almquiste, Wiksell, Stockholm, 1958, pp. 193–215. Knelson, "The Knelson Concentrator. Metamorphosis from crude beginning to sophisticated worldwide acceptance," *Minerals Engineering*, **5**, 10 (1992). Krantzavelos and Frangiscos, "Contribution to the modeling of the jigging process," *Control '84: Minerals/Metallurgical Processing*, **97**, SME, Littleton, CO (1984). Mayer, "Fundamentals of Potential Theory of Jigging," 7th International Mineral Processing Congress, New York, 1964. Miller, "Design and operating experience with the Goldsworthy Mining Limited Batac jig and spiral concentrator iron ore beneficiation plant," *Minerals Engineering*, **4**, 411 (1991). Zimmerman, "Performance of the Batac jig for cleaning fine and coarse coal sizes," *Trans. SME*, **258**, 199 (1975).

Introduction A jig is a mechanical device used for separating materials of different specific gravities by the pulsation of a stream of liquid flowing through a bed of materials. The liquid pulsates, or "jigs" up and down, causing the heavy material to work down to the bottom of the bed and the lighter material to rise to the top. Each product is then drawn off separately.

Jigging is one of the oldest processes used for concentrating heavy mineral particles from the light. Jigging is best suited for coarse material that is unlocked in the size range 20 mesh and coarser and when there is a considerable difference between the effective specific gravity (sp gr mineral minus sp gr water) of the valuable and the waste material. Jigs are simple in operation. Water consumption is high, and the tailings losses on metallic ores are usually high. Also, because of the scarcity of still-available ore deposits having coarse mineralization, the jigs are used to a limited extent, mostly to treat iron ores, a few lead-zinc ores, and some heavy nonmetallic ores like barite and diamonds. Jigging is widely employed for the concentration of coal. Over 50 million tons of coal is concentrated by jigs annually in the United States. High-speed types of jigs are used for the recovery of fine-grained heavy minerals from placer deposits, gold, tin, and tungsten, and for recovering a portion of coarse metallic values liberated in ball-mill grinding circuits. Jigging has been superseded in many milling

operations by the adoption of the dense-media process or by fine grinding followed by flotation.

Principles of Operation The principle of jig operation can easily be understood by taking a 10-mesh laboratory sieve, placing a 1 cm thick bed of a mixture of heavy and light particles, immersing the sieve in a bucket of water, and oscillating it up and down under water. The pulsations will dilate the bed of material and make the particles settle as the larger and denser particles forming the lower layers with the finer and lighter particles on the top.

The motion of the mixture of particles during jigging is modulated by the amplitude and frequency of jigging strokes and these strokes result in displacement of particle bed in a harmonic wave (Fig. 19-26*a*). During the pulsation stroke the original bed (Fig. 19-26*b*) dilates resulting in the bed as shown in Fig. 19-26*c*. During the suction stroke the bed of particles undergoes differential initial acceleration followed by hindered settling and consolidation trickling (Wills, op. cit.). It is found that the initial acceleration of the particles is independent of size and dependent only on the densities of the solids and fluid, thus causing the heavy particles to settle faster than the lighter as illustrated in Fig. 19-26*d*. The hindered settling on the other hand is controlled by both size and density of particles with smaller particles settling less and heavier settling more (Fig. 19-26*e*). Finally, during the consolidation trickling the bed begins to compact, the larger particles interlock and allow the smaller grains to move downwards (Fig. 19-26*f*).

Types of Jigs A jig is essentially an open tank filled with water and provided with a horizontal screen on the top and a *hutch* compartment fitted with a spigot (Fig. 19-27*a*). A layer of coarse, heavy particles, known as *ragging*, is placed on the top of the screen onto which the feed slurry is introduced. The feed moves over the ragging and the separation takes place as the bed is pulsated by a different mechanical device. The heavy particles are collected into the hutch compartment and removed through the spigot while the lighter particles are made to overflow from the top of the tank.

Several types of jigs are currently available with the main differences being in the pulsating mechanism and the stroke modification. Figures 19-27*b* through Fig. 19-27*e* illustrate four different designs that are commonly used. One of the earliest designs of jigs is the *Harz* and it uses reciprocating plunger with differential piston action (Fig. 19-27*b*). The Harz jig is commonly used in the treatment of gold, tungsten, and chromite ores. Remer jig (Fig. 19-27*c*) is an improvement over Harz by providing a driving mechanism that has two motions, a normal jig pulse of 80 to 120 strokes per minute on which imposed a fast pulse in the range of 200 to 300 per minute. This kind of jig is commonly used in concentrating such materials as iron and barite ores and in removing impurities such as wood, shale, and lignite from sand and gravel. In contrast, *Baum* and *Batac* jigs make use of air pulsations and are widely used in the coal-preparation industry to reduce the ash content of the run-of-mine coals. The standard Baum jig (Fig. 19-27*d*) operates by forcing air under pressure at about 17.2 kPa (2.5 lbf/in^2) into a large air chamber on one side of the jig vessel to pulsate the jig water which in turn pulsate the bed of particles fed onto the screen. Several design variations exist in the removal of lighter coal and heavier ash fractions (Green, op. cit). The Batac jig (Fig. 19-27*e*) is a modification of the Baum jig in that it employs multiple air chambers under the screen with electronic controls for air input and exhaust. This design is found to provide a uniform flow across the whole bed and a wide control of the speed and length of the jigging strokes. Batac jig is reported to treat both coarse- and fine-size coals satisfactorily (Chen, 1980) and has become an industry standard for coal cleaning (Zimmerman).

Jig Feed In coal washing jigging is practiced on unsized material as coarse as 175 mm (7 in). In metal-milling practice jigging is now seldom employed on material coarser than 20 mm (¾ in). Float-and-sink methods have largely superseded jigs as a way of concentrating metallic ores in the minus 75 to plus 10-mm (3 to plus ½-in) range. Shaking tables usually are considered more efficient than jigs for treating ores finer than 2 mm (10 mesh). Jigs are used in some plants to obtain flow-sheet simplicity since they can handle a wide range of sizes. Jigs, except when extremely heavy minerals such as gold, galena, cassiterite, or tungsten minerals are treated, recover only a small percentage of the sizes finer than 65 mesh (¼ mm).

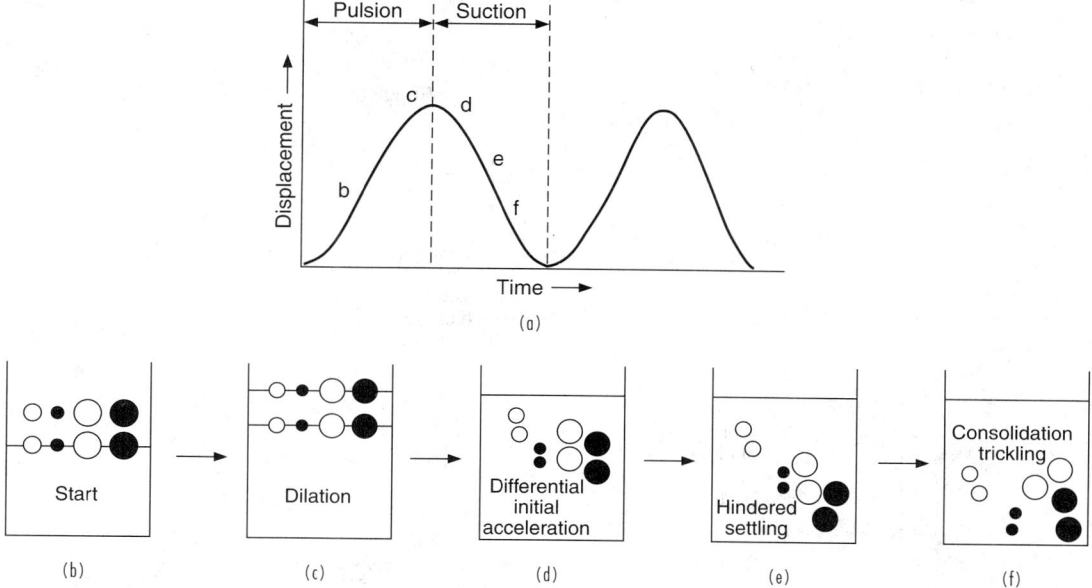

FIG. 19-26 Movement of particles in a jig. (*a*) Displacement of the bed as a function of time. (*b*) Starting position of particles. (*c*) After dilation. (*d*) After differential initial acceleration. (*e*) After hindered settling. (*f*) After consolidation trickling.

Capacity The Jeffrey-Baum will treat minus 100-mm (4-in) coal at the rate of 8 kg/(s·m²) [3 tons/(h·ft²)] of active screen area. For fine sizes capacity decreases. A standard 1.52- by 4.87-m (5- by 16-ft) Wemco-Remer jig will treat minus 9.5-mm (⅜-in) iron ore at the rate of 7.5 to 11.3 kg/s (30 to 45 tons/h). A Cooley jig, a variation of the Harz jig consisting of six compartments 1.07 by 1.22 m (42 by 48 in), will handle 6 to 7.5 kg/s (25 to 30 tons/h) of minus 19-mm (½-in) Mid-Continent zinc ore. The largest commercially available jig is the IHC Cleveland 25, a circular jig of 7.5-m (24.6-ft) diameter with a nominal capacity range of 30 to 60 kg/s (130 to 260 tons/h) of coal.

Power Requirements The power required in jigging depends on the screen area, the size of material treated, the percentage of opening in the jig screen, the depth of the bed, the length of stroke, and the choke frequency. The power required for plunger-type jigs treating 12.7-mm (½-in) material is about 7 W/m² (0.1 hp/ft²) jig screen surface.

Water Consumption Jigs require much water. In most installations, the Harz-type jig uses 0.006 to 0.01 m³ water/kg (1500 to 2500 gal/ton) material treated. Water requirements for treating minus 10-mm (⅜-in) iron ore in a Wemco-Remer rougher-cleaner jig circuit are approximately 0.005 m³ water/kg (1200 gal/ton) of material processed.

TABLING

GENERAL REFERENCES: "Automation Keys Two Stage Precision Washing at Moss No. 3," *Coal Age,* **64**, 80 (July 1959). Coghill, DeVaney, Clemmer, and Cooke, *Concentration of Potash Ores of Carlsbad, N.M., by Ore Dressing Methods,* U.S. Bur. Mines Rep. Invest. 3271, 1935. Dickson, Trepp, and Nichols, "Virginia Plant Concentrates Sulphide Ore with Air Tables," *Eng. Min. J.,* **160**(4), (April 1959). Kirchberg and Berger, *Trans. Int. Miner. Process. Congr.,* London, 1960, p. 537. "Linka Mill Added to Nevada WO₅ Output," *Min. World,* **18**, 52 (June 1956). Manser et al., "The shaking table concentrator—the influence of operating conditions and table parameters on mineral separation—the development of a mathematical model for normal operating conditions," *Minerals Engineering,* **4**, 411 (1991). McLeod, "Tungsten Milling and Current Metallurgy at Canadian Exploration Limited," *Can. Min. Metall. Bull.,* **50**, 137 (March 1957). Mitchell, "The Recovery of Pyrite from Coal Mine Refuse," *Min. Technol.,* **8**(4), 2 (1944). Norman and O'Meara, *Froth Flotation and Agglomerate Tabling of Mica,* U.S. Bur. Mines Rep. Invest. 3558, 1941. O'Meara, Norman, and Hammond, "Froth Flotation and Agglomerate Tabling of Feldspars," *Bull. Am. Ceram. Soc.,* **18**, 286 (1939). Sivamohan and Forssberg, "Principles of Tabling," *Int. J. Min. Proc.,* **15**, 281 (1985). Stockett, "Milling Practice of the St. Joseph Lead Co.," *Min. Technol.,* **7**(3), 1 (1943). "Upgrading Fragile Coal to Premium Metallurgical Product," *Coal Age,* **64**, 94 (July 1959). Wills, "Laboratory simulation of shaking table performance," *Mining Magazine,* 489 (June 1981).

Wet Tabling Tabling is a concentration process whereby a separation between two or more minerals is effected by flowing a pulp across a riffled plane surface inclined slightly from the horizontal, differentially shaken in the direction of the long axis, and washed with an even flow of water at right angles to the direction of motion. A separation between two or more minerals depends mainly on the difference in specific gravity between the minerals and to a lesser degree on the shape and size of the particles. The process is best suited for the concentration of ore and coal where there is a considerable difference between the effective specific gravity (sp gr mineral minus sp gr water) of the valuable and the waste material. Tables treat metallic ores effectively in the size range from 6 to 150 mesh but can be used to treat lighter materials such as coal of a considerably larger size.

Tabling is best suited for the treatment of material containing only one valuable mineral that is free at a granular size and when a considerable difference exists between the effective specific gravities of the mineral constituents. Flotation has been found to be best in treating complex ores containing several valuable minerals, those requiring fine grinding for liberation, and those having small gravity differentials.

The heaviest particles in a table feed are the least affected by the current of water washing down over the tables, and they collect in the riffles along which they move to the end of the table. The lighter materials ride above the heavy minerals and tend to be washed over the riffles to the low side of the table. Suitable launders are placed at the end of the low side of the table to catch the various products as they are discharged. These launders are provided with movable dividing devices to separate the concentrates from the middlings and the middlings from the tailings. It seldom is possible in tabling to make a sharp separation of the feed into a high-grade concentrate and a low-grade tailing with one pass. Some material of intermediate grade is almost invariably present as a band between these products, and it is customary to return such middlings either with or without additional grinding to the head of the circuit for retreatment. The amount of middling recirculated may amount to 25 percent of weight of the feed to the table.

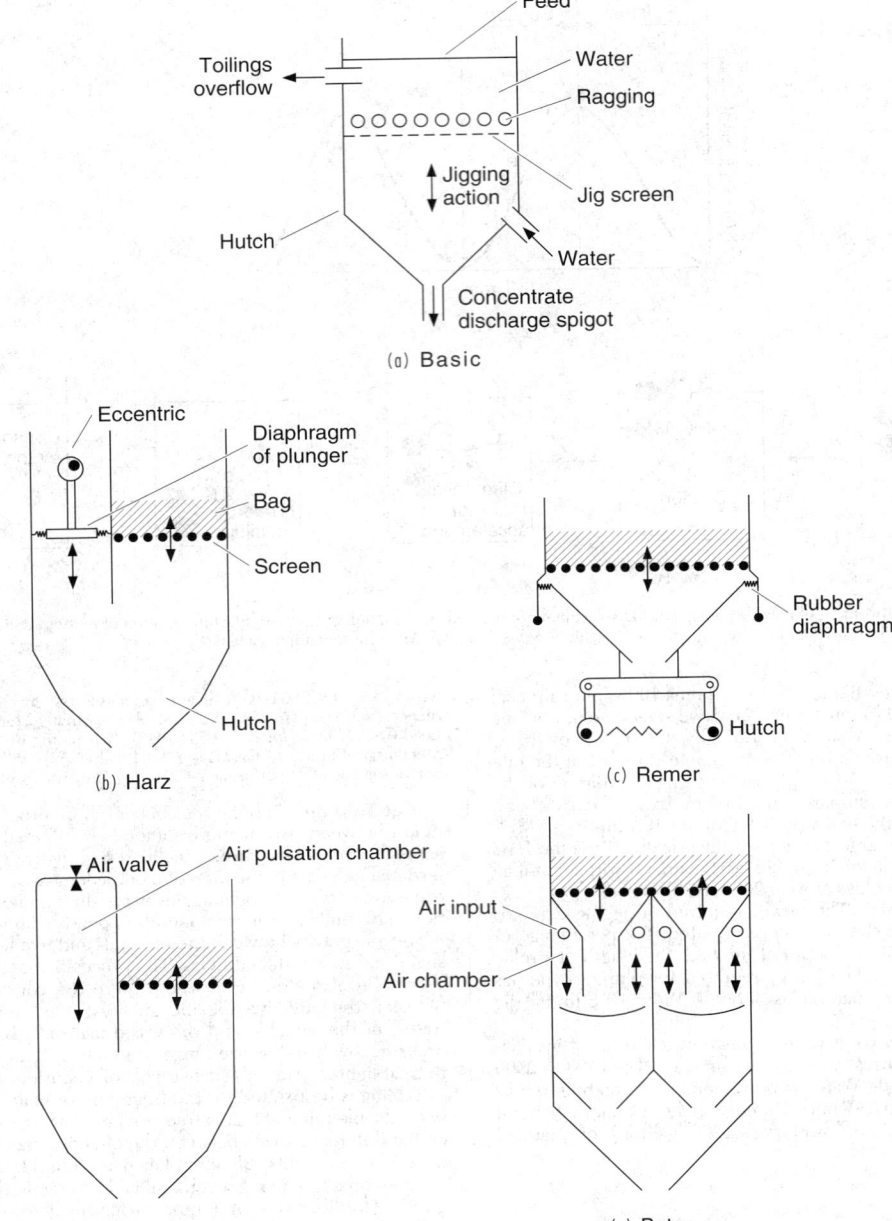

FIG. 19-27   Schematic diagrams of the jigs. (*a*) Basic, (*b*) Denver/Harz, (*c*) Remer, (*d*) Baum, and (*e*) Batac.

Tables usually are surfaced either with heavy battleship linoleum or with rubber. The riffles may be a clear grade of sugar pine or may be rubber strips. Such riffles usually taper from the feed end of the table to the discharge end. Almost all mill operators employ different styles of riffling table, which they believe best for their particular separations. The usual method of riffling is shown in Fig. 19-28.

If the object of tabling is to produce as clean a concentrate as possible, a diagonal area in the upper discharge side corner is left unriffled. This area is known as the cleaning deck. If the table is to be used in making only a rough concentrate and a finished tailing, the riffling is extended by many operators. Tables are provided with adjustable tilt-

ing devices so that the transverse slope may be varied. The head motion is such that the deck reverses its direction with a maximum velocity at one end and a minimum velocity at the other end of the stroke. It is the quickness of the return that causes the material to migrate toward the discharge end. The length of stroke may be adjusted. This will vary from 0.03 m (1¼ in) for coarse material to 0.01 m (½ in) for fines. Modern tables operate from 4 Hz (270 strokes/min) for coarse to 6 Hz (350 strokes/min) for fines.

Present table practice is to use multiple decks. Multiple-deck tables consisting of from two to three decks effect space saving proportionate to the number of decks employed. They also have the advantage in

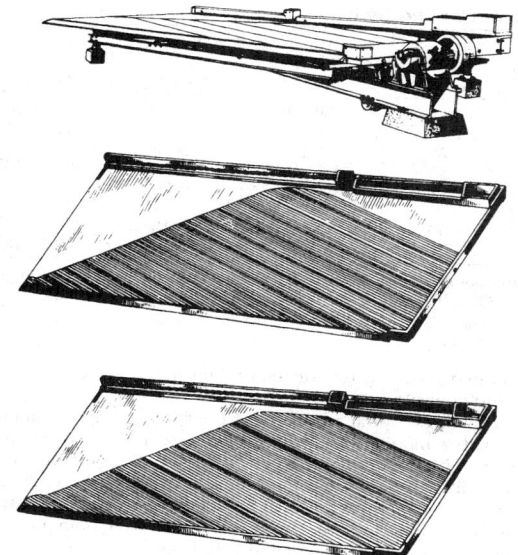

FIG. 19-28 Deister-Overstrom diagonal deck table. Center, diagonal deck with pool riffle system for sand; bottom, diagonal deck with pool riffle system for fine sand and slime.

that no heavy floor supports need be supplied since such tables are supported by suspended mountings. Multiple-deck installations reduce capital expenditures since a single motor and less piping and fewer launders are required than for a comparable number of single-deck installations. A two-deck configuration is shown in Fig. 19-29.

General information for standard-size tables operating on various-sized feeds is shown in Table 19-9. The No. 6 table of the Deister Concentrator Company, Fort Wayne, Indiana, has a diagonal deck approximately 1.83 m (6 ft) wide and 4.27 m (14 ft) long. The No. 7 table used primarily for coal work is approximately 2.44 m (8 ft) wide and 4.88 m (16 ft) long. The figures given apply to single-deck installations. In modern practice, each table, whether it be a single-deck or a multiple-deck installation, is driven by a single motor which is connected to the actuating mechanism by a V-belt drive. The installed

FIG. 19-29 Two-deck concentrating table. (*Courtesy Deister Concentrator Company, Inc.*)

horsepower for the large No. 7 deck is 1120 W (1.5 hp). A comparable figure for the smaller No. 6 deck is 746 W (1 hp) per deck. The actual power consumed in operation is somewhat less.

An essential factor for good table operation is that the rate of feed must be uniform, both as to tonnage and as to physical properties. No one factor will cause more trouble to the table operator than to have a surging feed. The feed to tables may be unsized, or it may be either screened or hydraulically classified. For treating fine coals a common procedure is to use hydrocyclones both to deslime the material and to give a cyclone underflow of about 40 percent solids, which constitutes the table feed.

Tabling is a relatively cheap operation. If the feed is uniform, one operator can take care of many tables. In a modern coal plant with multiple-deck tables, a single operator can handle the tabling of as much as 300 kg/s (1200 tons/h). In an ore-tabling plant such as a lead or zinc operation, a table operator can watch as many as 50 tables with a total capacity in the order of 50 kg/s (200 tons/h). Labor is the principal item of cost. Power requirements and maintenance are both low. The installed cost of a table including supports and launders is from $8000 to $15,000 per deck. In the past, one of the disadvantages of table installation was the relatively large floor space required for the tonnage treated. This disadvantage has now largely been overcome by the use of multiple-deck tables. Their main advantage is that, in the size range for which they are suited, tabling is a cheap and effective method of concentrating simple ores and coal.

Dry Tabling   Tabling may be done dry as well as wet, and for such use tables of special design are used. The Sutton, Steele and Steele table is an example of this type of equipment. It has a shaking motion somewhat similar to that of a wet table, except that the direction of motion is inclined upward from the horizontal, and instead of water acting as the medium of distribution, a blast of air is driven through a perforated deck. The table has application when it is desirable to treat material dry, either because of water shortage or because it is undesirable to wet the materials. An advantage of this table is the ability to handle material coarser than that treated on most wet tables. Ores as coarse as 0.006 m (¼ in) and coal as coarse as 0.076 m (3 in) can be treated.

Close sizing is necessary to give good results, and until recently this has militated against adoption of the table for fine sizes, owing to the difficulties of screening most ores dry below about 40 mesh. The development of improved dry methods for sizing fine material by the use of various cyclonelike devices has tended to increase the use of this apparatus on finer sizes.

Dry tables are used commercially in the separation of many types of minerals. Their greatest use is in the treatment of coal, but ilmenite, various tungsten ores, and even copper ores are so treated. Another important use is the cleaning of industrial materials such as seeds, cork, bagasse, fiber, nuts, wood chips, and coffee. One interesting use is in the sorting of silicon carbide by grain shapes. Flat and splintery grains are removed from others of more nearly equal dimensions.

Agglomeration Tabling   Agglomeration tabling is a process whereby selective flocculation or agglomeration of grains of one mineral in an aggregate is caused by the addition of an agglomerating agent in a conditioning cell or in the ball-mill circuit, the slurry containing the agglomerated grains then being fed across gravity tables. The larger size, the oil-filmed surface, and the feathery texture of the floccules cause them to be washed over the side of the table by the current of cross water, while the unflocculated discrete particles remain on the table and are carried off the end in the position followed normally by the concentrate in the usual table feed. An oiled particle will tend to ride on the surface of the water and thus be more readily carried across the side of the table than an unoiled particle. Agglomeration tabling has had more application in the concentration of phosphate minerals than in any other field, although successful tests have been run on limestone, potash, mica, and other ores.

The process is limited to granular material in the size range from 10 to 100 mesh. In this respect it differs from flotation, which functions best on material 48 mesh and finer. For best results the material should be well deslimed and should be conditioned with the agglomerating reagents at a high percentage of solids, 65 percent or greater. A collector is used that will selectively film the mineral to be agglomerated. In phosphate and limestone practice, this collector is usually a

TABLE 19-9 Generalized Operating Data for Superduty Diagonal-Deck Concentrating Table

Table No.	Feed	Feed size	Feed capacity, tons/hr	Speed, rpm	Stroke, in	Water with feed, gal/min	Dressing water, gal/min	Size of deck
6	Ore	¼ in.–35 mesh	2.0–10.0	275	1.25	30–150	10–100	6′5″ × 14′1″
6	Ore	35–150 mesh	1.0–2.5	285	0.75	16–40	5–20	6′5″ × 14′1″
6	Ore	Minus 150 mesh	0.25–1.0	300	.50	3–12	3–10	6′5″ × 14′1″
7	Coal	1½ in	15.0–25.0	270	1.25	125–210	55–90	8′¼″ × 16′9¼″
7	Coal	¾ in	10.0–15.0	280	1.00	60–85	20–35	8′¼″ × 16′9¼″
7	Coal	½ in	7.5–12.0	285	1.00	42–65	18–31	8′¼″ × 16′9¼″
7	Coal	⅛ in	5.0–7.5	290	0.75	28–42	12–18	8′¼″ × 16′9¼″
7	Coal	1/16 in	3.0–5.0	290	.75	15–28	9–12	8′¼″ × 16′9¼″

NOTE: To convert inches to meters, multiply by 0.0254; to convert tons per hour to kilograms per second, multiply by 0.252; to convert revolutions per minute to hertz, multiply by 0.0167; to convert gallons per minute to cubic meters per second, multiply 6.309×10^{-5}; and to convert feet to meters, multiply by 0.3048.

cheap fatty acid such as tall oil. In potash separation long-chain amines are used to film sylvite (KCl).

A bulk oil is always used in addition to the collector to give body to the film and to assist in forming agglomerates. In Florida practice, it is customary to use 0.14 to 0.23-kg/ton (0.3- to 0.5-lb/ton) tall oil and 1.8 to 23 kg/ton (4 to 5 lb/ton) of a 22°Bé fuel oil. Operating data for the agglomerate tabling of phosphate and potash ore are shown in Table 19-10.

Agglomerate tabling works best on simple ores consisting of two free minerals. It has several advantages over the usual tabling method in that it can be used to separate two minerals the difference in specific gravity of which is so small that an effective separation cannot be made by gravity separation alone. Tables treating an agglomerated feed have a considerably larger capacity than tables using untreated feeds, since the capacity of a table treating an agglomerated feed is limited only by the carrying capacity of the riffles. Disadvantages of the method that must be considered are the cost of the reagents used and the fact that if the mineral fraction filmed is the one to be sold, the oily film may be objectionable and must be removed.

SPIRAL CONCENTRATION

GENERAL REFERENCES: Adair, "New Method for Recovery of Flake Mica," *Min. Eng.,* **3,** 252 (1951). Brown, "Humphreys Spiral Concentration on Mesabi Range Ores," *Trans. Am. Inst. Min. Metall. Pet. Eng., Min. Branch,* **184,** 187 (1949). Gleeson, "Why the Humphreys Spiral Works," *Eng. Min. J.,* **146**(3), 85 (1945). Burt, *Gravity Concentration Technology,* Elsevier (1985). Lenhart, "Spiral Concentrators for Gravity Separation of Minerals," *Rock Prod.,* **54**(12), 92, 131 (1951). Miller, "Design and operating experience with the Goldsworthy Mining Limited Batac jig and spiral concentrator iron ore beneficiation plant," *Minerals Engineering,* **4,** 411 (1991). Roberts, "How New Highland Plant Recovers Titaniferous Minerals," *Min. World,* **17**(11), 52, 72 (1955). Roe, *Iron Ore Beneficiation,* Minerals Publishing Company, Lake Bluff, Ill., 1957. Sivamohan and Forssberg, "Principles of spiral concentration," *Int. J. Min. Proc.,* **15,** 173 (1985). Thompson, "The Humphreys Spiral: Some Present and Potential Applications," *Eng. Min. J.,* **151**(8), 87 (1950). Thompson, "The Humphreys Spiral Concentrator: Its Place in Ore Dressing," *Min. Eng.,* **10**(1), 84 (1958).

Principle of Operation Spiral concentration of ores and industrial materials is based primarily on the specific-gravity differentials of the materials to be separated. The shape factor of the feed material is also important, and utilization of reagentized feed can change the apparent specific gravity of component minerals by forced attachment of air bubbles to mineral flocs. The best known spiral-type concentrator is the Humphreys spiral concentrator, which first proved its commercial feasibility in 1943. In that year an Oregon plant successfully

demonstrated ability to recover chromium minerals from low-grade beach sand deposits.

The Humphreys spiral concentrator is a spirally shaped channel or launder with a modified semicircular cross section, as illustrated in Fig. 19-30. The standard spiral consists of five complete turns, but three-turn units are used in some instances when an unusually rapid and clean separation takes place, as in second-stage or cleaner spirals. There is a drop of 0.34 m (13.5 in)/turn as the flowing pulp progresses from the top to the bottom of the spiral. One spiral concentrator occupies about 0.37 m² (4 ft²) of floor space and about 2.1 m (7 ft) of headroom measured from feed to discharge box. The optimum particle-size range of feed particles for spirals is about 10 to 200 mesh (2 to 0.074 mm).

As the feed slurry flows down the spiral channel, the particles with the highest specific gravity sink to the bottom and move inward toward the inside of the channel. The lighter-weight particles move to the outside and are carried away by the faster, more dilute pulp stream. At 120° intervals circular concentrate "ports," or openings, appear in the bottom of the channel near the inside edge, as illustrated in Fig. 19-31. There are 15 ports in a five-turn spiral, but usually more than half of them are blocked off with smooth stainless-steel disks in order to allow proper configuration of the concentrate stream and good washing of the concentrate. Wash water is available along the entire inside edge of the spiral, where it flows at the rate of 0.2 to 0.6 L/s (3 to 10 gal/min) in a separate wash-water channel. Thus the spiral provides repeated washing stages as the pulp flows down the channel. Generally the richest concentrate is withdrawn from the concentrate ports near the top end of the spiral. Concentrate ports are fitted with very simple stainless-steel "belt-disk splitters," which can split out the desired portion of the concentrate stream. As the gradually impoverished pulp flows down the spiral, wash water is proportioned from the wash-water channel by a series of notches and directed so as to wash repeatedly across the concentrate band and sweep out unwanted gangue particles. The lowest-specific-gravity solids wash outward, and the finest particles actually climb the sloping wall of pulp on the outside of the channel. The concentrate withdrawn from ports near the bottom end of the spiral is usually low-grade and, if liberated, may be recirculated to obtain additional recovery of values and a higher grade of concentrate.

Although the spiral concentrator is mechanically a very simple piece of equipment, the separating action taking place is complex. It involves centrifugal force, friction against the spiral surface, gravity, and the drag of the water.

TABLE 19-10 Operating Data for Agglomerate Tabling of Phosphate and Potash Ore

Type of table	Feed size	Feed capacity, tons/h	Table speed, r/min	Table stroke, in	Water with feed, gal/min	Dressing water, gal/min	Size of deck
No. 6 superduty diagonal deck	10–48 mesh	2.5–3.5	295	1.0	20–40	8–15	6′5″ × 14′1″

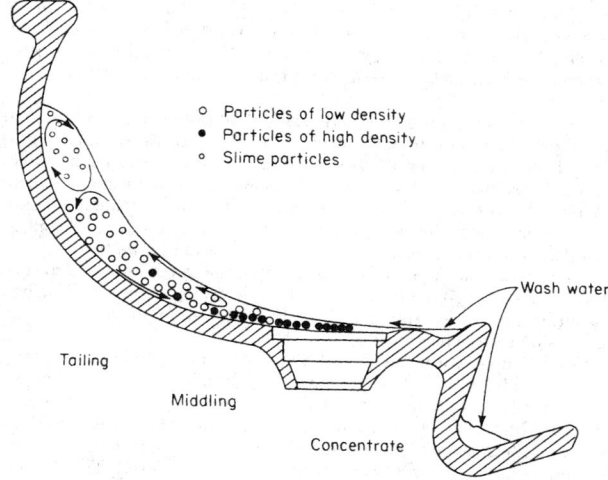

FIG. 19-30 Heavy-mineral separation in the Humphreys spiral concentrator.

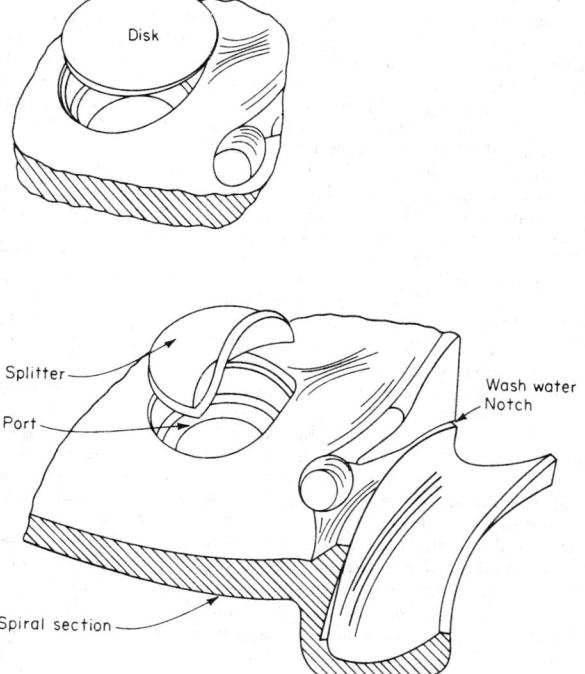

FIG. 19-31 Disks and splitters as used in the Humphreys spiral concentrator.

Basic Requirements for Spiral Concentration Minerals or materials of different specific gravity can usually be concentrated on spirals if the heavy particles do not exceed 10 mesh (2 mm) or are not finer than 200 mesh (0.074 mm). The size of the low-specific-gravity component is not critical when the values to be recovered are in the heavier-particle fraction. In this case the size of the light particles may range from 4 mesh (4.76 mm) to zero. The quantity of locked grain components (referred to as middlings in the mineral industry) that are present in a given pulp can be critical because this material is fre-

quently recirculated and may eventually accumulate to a degree that will inhibit the separation operations. One solution to such a problem is continual removal of all or part of the middling stream to a grinding mill, followed by separate recovery of values in another spiral circuit or other concentrating machines.

Examples of good feed materials for spiral concentration are (1) beach sands that are processed for recovery of chromite, ilmenite, rutile, zircon, tin, and iron-ore minerals; (2) hard-rock iron ores in which good liberation of iron values occurs in the 10- to 200-mesh size range; (3) some mica and phosphate ores; (4) tailings from concentrating plants that contain heavy mineral components not recovered by flotation and other concentrating methods; and (5) some fractions of coal [minus 6-mm (¼-in) sizes] that can be upgraded by spiral concentration. The spiral used for coal cleaning has six complete turns with a more gradual slope [a 0.25-m (10-in) pitch]. The six turns require about the same headroom as the conventional five-turn spiral.

The spiral concentrator has shown unusual capability in the gravity processing of tailing streams from conventional magnetic and froth-flotation types of ore-processing plants. There are a number of minerals plants where some of the iron values are first recovered by spirals and the tailings are then sent to magnetic separators. There are also iron plants in which the reverse order of processing is used. In spirals the nonmagnetic iron minerals can be efficiently recovered as a high-grade product. An outstanding example of tailings processing in spirals is illustrated by a Colorado molybdenum-ore treatment plant. Spirals recover salable tungsten, pyrite, and tin concentrates from thousands of tons of flotation-plant waste every 24 h. There is no other known ore-processing method that can economically recover tin and tungsten values from this source. The crude ore contains only 0.03 percent tungstic oxide and a trace of tin. The tin occurs as the mineral cassiterite and the tungsten as the mineral hubnerite.

Operating Characteristics Spiral capacity can range from 0.12 to over 0.5 kg/s (0.5 to over 2 tons/h) of new feed. The grade of concentrate produced can be adversely affected by either too low or too high a feed rate. A good average feed rate for most spiral installations is 1.5 short tons of new feed per hour. The pulp density of spiral feed may range from 10 to 50 percent solids. If the values are contained in coarse heavy minerals, a high pulp density is preferred, whereas if the values are in fine-sized heavy minerals, it is better to use low pulp densities. Generally 20 to 30 percent solids by weight will constitute a suitable pulp feed.

Water Requirement The water requirement per spiral can range from 1.0 to 2.5 L/s (15 to 40 gal/min); this includes 0.2 to 0.6 L/s (3 to 10 gal/min) of water used in the wash-water channel. An attractive feature of the spiral is that reclaimed water can generally be used in all except the very final upgrading step.

Maintenance The only moving parts in spiral concentrators are those in the pumps that supply the feed and recirculate intermediate products. However, there are sometimes minor maintenance problems associated with the spiral trough itself. Some ores contain sharp particles of very abrasive minerals. The presence of these minerals in some ore causes rapid formation of deep grooves in the surface of cast-iron spirals. Wear grooves can be patched with a variety of plastic and metallic cements. Most spirals presently in service are made of cast iron with molded and vulcanized liners. These liners have successfully solved most wear problems.

Other than the wear problems, actual in-plant maintenance usually involves removal of wood, pieces of blasting wire, and other trash from the ports. When a reagentized feed is used, layers of oily reagents can build up on the spiral surface and sometimes require scrubbing for removal. With feeds containing oily reagents that attack rubber, abrasion-resistant alloy spiral sections are used.

Spirals have been manufactured from concrete, plastics, solid rubber, iron, and special iron alloys.

Operating Costs The operating cost of a spiral concentrator plant will be among the lowest costs of any ore-processing plant handling similar feed material. The only moving equipment parts involved are the pumps included in the flow sheet for the purpose of elevating feed and water to the spirals. One large pump can feed 100 or more spirals in a large plant. At a Canadian iron-ore plant twelve 0.14 m³/s (2200-gal/min) pumps provide feed for 1152 rougher (or first-stage) spirals. In many plants the second- and third-stage spirals are gravity-fed. Thus maintenance is largely limited to pump repair. When unusually abrasive ores are processed, maintenance of worn sections of the spirals can be extensive unless rubber coating or other abrasion-resistant materials are used on the wearing surfaces.

Labor requirements for a spiral plant are low and are governed primarily by the type of material being fed to the spirals. For example, a phosphate ore containing roots, leaves, and other trash will contain sufficient fibrous material to block concentrate ports, and generally prevent good spiral operation. When such an ore is processed, considerably more labor is required for cleaning and adjustments. Generally metallic ores are quite free of fibrous material that will hang up in the spirals, and one person can operate 100 or more spirals.

Power requirements for spiral plants are low, consisting primarily of pumping energy and possibly a thickener or other pulp-handling equipment associated with the flow sheet.

A typical summary of an approximate range of spiral concentration plant direct costs is given in Table 19-11.

DENSE-MEDIA SEPARATION

GENERAL REFERENCES: Aplan and Spedden, "Viscosity Control in Heavy Media Suspension," *Proc. 7th Int. Miner. Process. Congr.*, New York, Sept. 20, 1964, Gordon and Breach, New York, 1965, p. 103. Browning, *Heavy Liquids and Procedures for Laboratory Separation of Minerals*, U.S. Bur. Mines Inf. Circ. 8007. Burton, "The economic impact of modern dense medium systems," *Minerals Engineering*, **4**, 225 (1991). Deurbrouck and Hudy, *Performance Characteristics of Coal-Washing Equipment: Dense-Medium Cyclones*, U.S. Bur. Mines Rep. Invest. 7673. Doyle, "The Sink-Float Process in Lead-Zinc Concentration," *AIME Symp. Lead Zinc*, St. Louis, 1970. "Mineral Engineering Techniques," *Chem. Eng. Prog. Symp. Ser.*, **50**(13), (1954). Mular and Bhappu (eds.), *Mineral Processing Plant Design*, 2d ed., Society of Mining Engineers, AIME, New York, 1980. Rodis and Cremer, "Why an Atomized Ferrosilicon?" *Min. World*, **22**(3), 36 (March 1960). Ruff, "New developments in dynamic dense-medium systems," *Mine & Quarry*, **13**, 24 (1984). Tippin and Browning, *Heavy Liquid Cyclone Concentration of Minerals*, U.S. Bur. Mines Rep. Invest. 6969 and 7134. Volin and Valentyik, "Control of Heavy Media Plants," *Pit Quarry*, **62**, 111 (December 1969). Walker and Allen, "Beneficiation of Industrial Minerals by Heavy Media Separation," *Trans. Am. Inst. Min. Metall. Pet. Eng., Min. Branch*, **184**, 17 (1949). Williams and Kelsall, "Degradation of ferrosilicon media in dense medium separation circuits," *Minerals Engineering*, **5**, 1 (1992).

Dense-media separation, also known as heavy-media or sink-float processing, is an adaptation of the common laboratory procedure for separating solids of differing specific gravities by immersing them in a heavy liquid of specific gravity intermediate between those of the solids, thereby causing the lighter particles to float while the heavier sink. However, in dense-media separation, the parting liquid is produced by dispersing relatively fine-grained solids of a high specific gravity in water and maintaining this pulp in suspension by light agitation. The method is very effective and can be used to separate solids with differences in specific gravity of as little as 0.005. It is often the only process needed for the removal of deleterious wastes from coal. The method is used extensively for beneficiating ore minerals, and it is finding increasing use for the processing of shredded automobile scrap and for the recovery of values from solid municipal waste.

Dense-media separation may be used to produce either a finished concentrate or an upgraded feed for subsequent processing. In the latter case, it provides a low-cost means to reject a significant amount of essentially barren waste at a coarse size.

Sink-float plants are usually custom-designed for each individual application. However, for coal beneficiation modular units are available. For most large mineral-processing applications the plants will be permanently located for easy access of feed and disposal of waste, but for smaller coal and aggregate operations the plants are often constructed with the anticipation of relocation when the deposits have been depleted.

The response of any given feed to sink-float processing can be accurately established in the laboratory by testing with various heavy liquids. The liquids generally used for this purpose are listed in Table 19-12. These halogenated hydrocarbons are mutually miscible, which enables the preparation of almost any pulp density attainable in a commercial plant. Heavy-liquid test work provides the basis for specifying the optimum screen size for the preparation of the feed.

Continuous pilot-plant test runs are generally recommended to verify the laboratory results and to establish criteria for plant design. Facilities for these runs are available at a number of minerals-processing research centers.

Feed Preparation and Feed Size   The ability to achieve a separation of different solid particles on the basis of density, as in all physical separation, depends on the degree to which the particles are liberated (detached) from each other. Liberation can be achieved by breaking the material in a manner that causes it to fracture and free the individual grains of the constituents to be recovered. The degree of separation that can be realized by the dense-media process will depend on the degree of liberation of the individual grains.

There will be an optimum size reduction of the feed material for the dense-media process. This size range can depend on the overall objectives of its use. For example, if the process is to be used in conjunction with a subsequent separation method such as flotation, the intent may be more the rejection of barren waste material at a relatively coarse size and at high recovery of the values, although the resulting grade from the dense-media operation may still be low. On the other hand, if the concentrate grade must be high, a finer degree of liberation will be needed at some loss in recovery. The initial test work can establish the so-called grade versus recovery limitations of the dense-media operation for the specific material of interest. This testing should recognize that the dense-media process is not effective for treating material which contains any substantial amount of particles smaller in size than about 0.5 mm (20 to 28 Tyler mesh).

The largest size that can be treated depends mostly on the dimensions of the separating vessel; coal up to 0.3 m (12 in) has been successfully processed in a drum separator of the type illustrated in Fig. 19-32. Complete removal of fines is usually necessary to ensure proper viscosity of the media. Fines increase viscosity and slow the separation process.

The feed-preparation screen between crusher and separatory vessel may be of either the revolving or the vibrating type. Wash water is applied only to the feed end of the screen so that the process feed will enter the separator moist but without any free water, which would lower pulp density. In a few instances it has been found advantageous to provide for surge storage between screen and separator to drain off further excess water.

A typical flow sheet is shown on Fig. 19-33.

Preparation of the Media   Various solid materials have been used to prepare the media. In the initial development of the process, a suspension of sand and also mixtures of barite and clay were used for separating coal from slate. Galena (lead sulfide mineral) was also used

TABLE 19-11   Approximate Range of Direct Costs for Spiral Concentration

Cost element	Cents per short ton of spiral feed
Labor	3.0–5.0
Power	1.6–3.0
Maintenance	2.0–3.0
Depreciation	2.8–4.0
Total	9.4–15.0

TABLE 19-12   Liquids Used to Test Feeds

Name	Specific gravity, 25°C
Methylene iodide	3.33
Tetrabromoethane	2.96
Bromoform	2.89
Tribromoethane	2.61
Methylene bromide	2.48
Ethylene dibromide	2.17
Methylene chlorobromide	1.92
Pentachloroethane	1.67
Carbon tetrachloride	1.59
Trichloroethylene	1.46
Ethylene dichloride	1.26

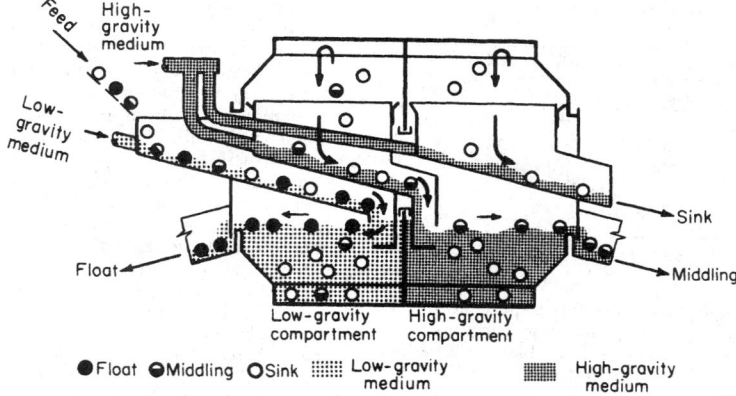

FIG. 19-32 Revolving-drum-type dense-media separatory vessel. (*Courtesy of Western Machinery Co.*)

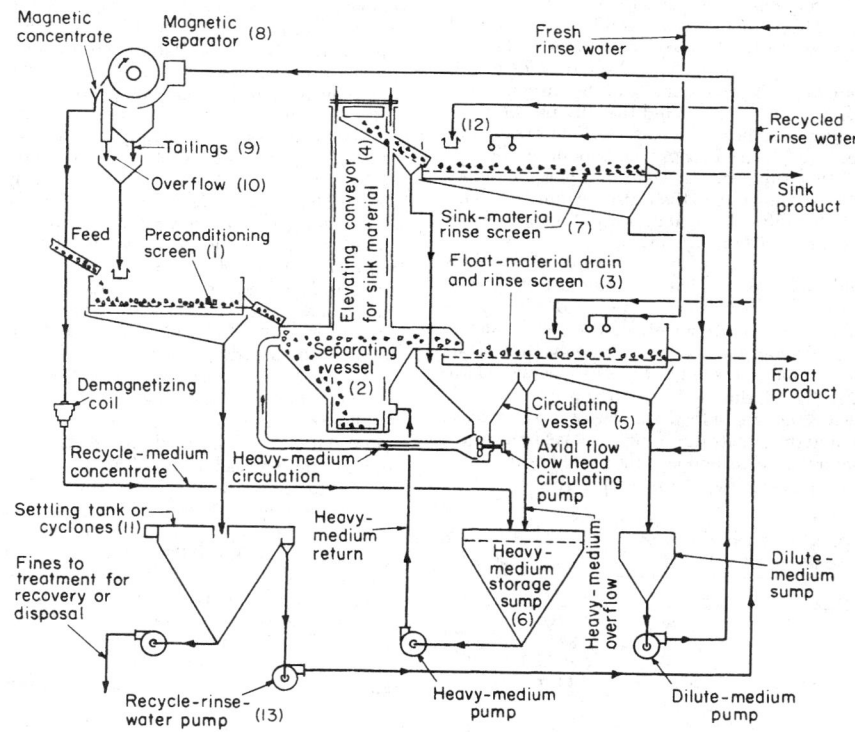

FIG. 19-33 Typical dense-media flow sheet for a coal-cleaning plant. (*Courtesy of Process Machinery Division, Arthur G. McKee Co.*)

to achieve a higher pulp density. In present processing, iron-based particles such as magnetite and ferrosilicon are preferred because they offer suitable density, high resistance to attrition, and ease of recovery by magnetic methods. With magnetite, a pulp density 2.5 times that of water can be obtained. Ferrosilicon can provide a density factor of 3.3, which is effective for separating most gangue constituents from metallic ores. Both materials might be used to obtain intermediate media densities.

Media-Particle Size   The size of the media particle is important. A relatively coarse medium (minus 100 mesh) is commonly used in larger-volume static-type separators such as cones. However, in dynamic sep-

arators, a much finer size is desirable. Ground magnetite or atomized ferrosilicon is advantageous in this application. The latter is produced by pouring molten ferrosilicon into an atomizing chamber, where a jet of steam forms the alloy into spherical particles. These particles are more resistant to wear than are the particles in a ground product and cause less abrasion to the equipment. Atomized ferrosilicon permits pulp densities above about 3.4-density factor.

Custom-ground natural magnetite is available in many size ranges. Table 19-13 gives a typical specification sheet. Cost is based on truckload quantities in 45.4-kg (100-lb) paper bags, FOB Frazer, Pennsylvania.

TABLE 19-13 Typical Size Distribution of Ground Natural Magnetite*

Product grade	Percent retained by weight for mesh size				1978 cost
	100	200	325	Less than 325	
A	0.6	12.0	17.8	69.6	$72
B	0.1	1.0	7.6	91.4	$74
C	5.0	22.0	23.0	50.0	$72
D	6.0	29.5	22.9	36.3	$71
E	0.1	0.4	1.9	97.7	$85
G	0.2	6.2	15.5	78.1	$74

*Foote Mineral Company.

Pulverized ferrosilicon containing approximately 15 percent silicon is available from the Foote Mineral Company and from Carborundum Co. in the sizes and at the prices shown in Table 19-14. Cost is based on truckload quantities in 227-kg (500-lb) steel drums, FOB Keokuk, Iowa, and Niagara Falls, Ontario.

Atomized ferrosilicon is at present available only from West Germany through American Hoechst Corp. in the sizes shown on Table 19-15. Costs vary with the exchange of U.S. dollars to deutsche marks but will be around $770 per metric ton, FOB Germany (1978 estimate).

Chemical Additives The use of chemical additives in sink-float processing is not common except for the use of lime to prevent oxidation and decomposition of the medium. A small amount of clay is sometimes added to improve the kinetic stability of the suspension.

Considerable laboratory work has indicated that the use of a dispersant such as sodium hexametaphosphate may assist in the stabilization of the medium; more recent data report the beneficial effect of the addition of polymers that reduce media viscosity while simultaneously producing a very low settling rate of the ferrous compound. This should be of great value for difficult separations, but at present no data are available from commercial operations.

Separating Vessels Many different types of separating vessels have been proposed and used for sink-float separation. For applications at coarse sizes involving a high ratio of float to sink or a high gravity differential, as in the case of coal, trough-shaped vessels (as shown in Fig. 19-34) or rotating-wheel separators are commonly used. However, for the beneficiation of most ores, other types of separators have found general acceptance. The optimum design for any given ore will depend on such variables as the rate of feed, the size of feed particles, the ratio of float to sink, and the gravity differential between the solids to be separated. Separators are classified as either static or dynamic, depending on whether or not centrifugal force is applied.

TABLE 19-14 Typical Size Distribution of Pulverized Ferrosilicon

Particle size; mesh size less than	1978 cost range
48	$245–279
65	$248–282
100	$252–291
200	$289–400

Drum Separators Very coarse solids, up to 0.3 m (12 in), are often processed in a drum separator of the type shown in Fig. 19-32. This is similar to a ball-mill shell with lifters permanently attached to the wall. Medium and feed enter at one end, and the float product flows out through the discharge trunnion, while the sink is lifted by the rotation of the drum to a stationary launder, through which it is flushed out. Modifications of this type include division of the shell into two compartments, which permits simultaneous operation at two different pulp densities resulting in various grades of products. The two-compartment revolving drum is illustrated in Fig. 19-32.

Drum separators have capacities up to 250 kg/s (900 t/h), cone separators to about 125 kg/s (500 t/h), and dynamic separators a maximum of 28 kg/s (100 t/h); however, these can readily be manifolded for any required tonnage. Economics will dictate the minimum tonnage for which a plant would be justified; several plants of 2.8-kg/s (10-t/h) capacity have been built.

Cone Separators Feed materials in the intermediate sizes, 0.1 to 0.01 m (4 to 0.5 in), may be processed in cone separators as shown in Fig. 19-35. These have a large surface area and increased volume pulp, which permit longer retention time than that of most other types of separators; this is a great advantage when separating solids of small gravity differential. The feed is introduced into the cone at a point below the pool surface and as far from the overflow baffle as possible. Slow-moving scrapers prevent a buildup of medium on the cone wall, provide the necessary agitation to prevent settling of the medium, and push the float particles toward the overflow weir. The sink product is removed from the cone bottom by rock pump, internal air lift, or external air lift. Mechanical elevators of the screw or bucket type have also been used but require more maintenance.

Cyclone Separators Finer feed solids, from 0.04 to 0.0005 m (1.5 in to 28 mesh), may be treated in dynamic separators of the Dutch State Mines cyclone type (Fig. 19-36). In cyclone separators, the medium and the feed enter the separator together tangentially at the feed inlet (1); the short cylindrical section (2) carries the central vortex finder (3), which prevents short circuiting within the cyclone. Separation is made in the cone-shaped part of the cyclone (4) by the action of centrifugal and centripetal forces. The heavier portion of the feed leaves the cyclone at the apex opening (5), and the lighter portion leaves at the overflow top orifice (6).

The sharpness of separation of the mineral from the gangue is dependent on (1) the stability of the suspension, which is influenced by the size of the medium; (2) the specific gravity of the medium; (3) the cleanliness of the medium; (4) the cone angle; (5) the size and ratios of the internal openings in the cyclone (inlet, apex, and vortex); and (6) the pressure at which the pulp is introduced into the cyclone. A 20° cone angle is the most common. Cyclone diameter will be determined by the separation to be made as well as by the capacity required. The 0.5- and 0.6-m (20- and 24-in) cyclones are most common in coal plants, whereas multiple cones of 0.25- or 0.3-m (10- or 12-in) diameter are used in higher-gravity separations.

Dense-media cyclones are generally operated in the $(0.7–1.0) \times 10^6 \cdot Pa$ (10–15-lbf/in²) range. It is not advisable to go below $(0.4–0.56) \times 10^6$ Pa because the recovery of low-specific-gravity material and the rejection of impurity are improved at higher pressures,

TABLE 19-15 Size Distribution of Atomized Ferrosilicons

Particle size, greater than Tyler mesh	Manufacturers' grade distribution					
	Extra coarse	Coarse	Fine	Cyclone 60	Cyclone 40	Cyclone 20
48	15–0	5–0				
65	25–7	8–2	4–0			
100	42–20	22–7	8–2	3–0		
150	55–35	35–17	22–5	8–0	3–0	
200	70–50	50–35	30–15	20–5	8–3	2–0
Particle size, less than Tyler mesh						
200	30–50	50–65	70–85	80–95	92–97	98–100
325				40–60	70–85	90–100
625				10–20	40–55	70–80

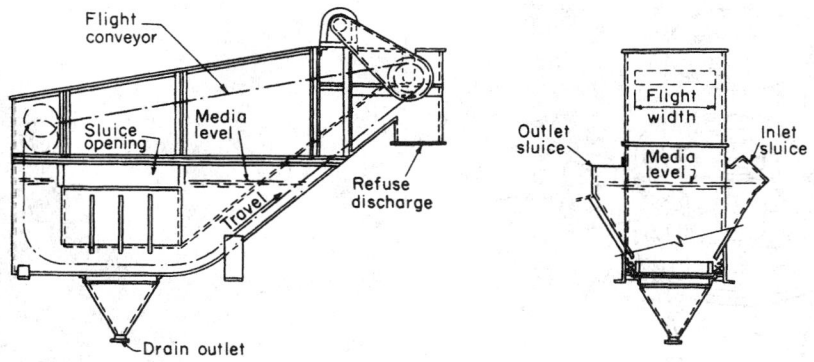

FIG. 19-34   Drag-tank-type dense-media separatory vessel. (*Courtesy of Link-Belt Co.*)

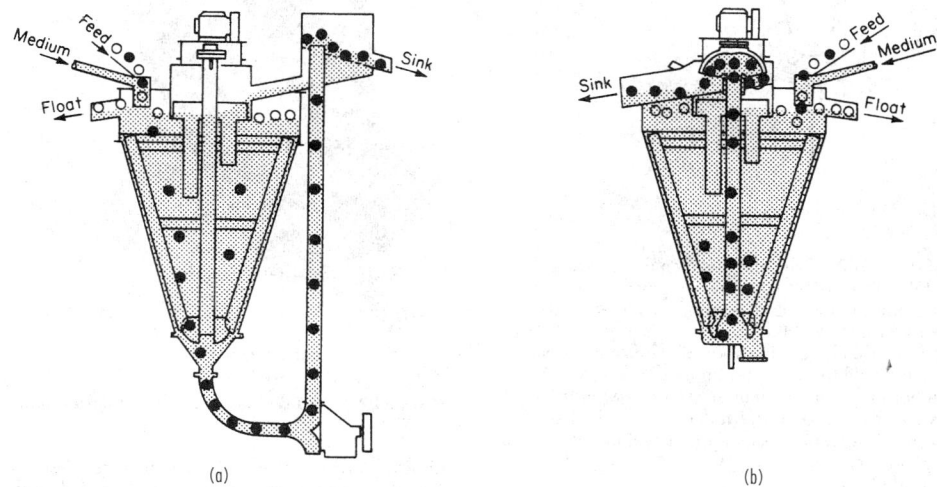

(a)

(b)

FIG. 19-35   Dense-media cone-vessel arrangements. (*a*) Single-gravity two-product system with pump sink removal. (*b*) Single-gravity two-product system with compressed-air sink removal. (*Courtesy of Process Machinery Division, Arthur G. McKee Co.*)

especially for the finer sizes. Pressures as high as 2.5×10^6 Pa (36 lbf/in²) have been used, and they increase capacity but accelerate wear. Residence time of the ore particles is very short in the cyclone, and a large volume of medium is circulated for each ton of feed treated in the cone. Loss of media is higher in cyclone plants because of the finer media required and the additional volume encountered in these plants. Media loss may be 2 to 5 kg/ton (5 to 10 lb/ton) of ore treated in cyclone plants, as compared with 0.2 to 0.8 kg/ton (0.5 to 1.5 lb/ton) in coarse, static heavy-medium circuits. Cyclone-plant labor requirements are low and efficiency is high. A 0.6-m (24-in) heavy-medium cyclone can handle 75 tons of coal per hour.

Dyna Whirlpool   A unique vessel design for capacities up to 100 t/h has been developed by the American Zinc Co. The separation occurs in a cylindrical-shaped separatory vessel maintained in an inclined position from horizontal. This system, known as the Dyna Whirlpool (DWP) process, provides for separate entry of the medium and the feed solids, as illustrated in Fig. 19-37. A distinct feature of this separator is that the feed enters the separator via gravity flow. Feed size may range from 0.05 to 0.0002 m (2 in to 65 mesh). Magnetite or ferrosilicon is generally used.

Process Control   As is the case in all concentration processes, optimum results will be obtained under steady operating conditions.

Because of the simplicity of the dense-media process, these can readily be maintained.

Uniformity of the rate of feed will be ensured by a constant-weight feeder; density control may be automatically obtained through a measuring probe on the media-return line that adjusts delivery of the necessary volume of media from the densifier or media thickener; the viscosity can be controlled automatically by continuously testing a predetermined volume of return media and adjusting the divider under the drainage screen for media cleaning as needed; pH control can be automated by conventional methods.

Notwithstanding the possibility of such automation, many successful operations depend almost entirely on manual sampling. Density determinations of the pulp on the media-return line and on each of the drainage screens are made at scheduled intervals, and the operator adjusts the media flow as needed.

Costs   Because sink-float processing is applied to relatively coarse particles and is a single-pass operation, capital and operating costs are usually considerably lower than would be required for a flotation or a gravity mill of the same capacity. A large flow of water is required for feed preparation and for media recovery, but almost total recovery for recirculation is possible. A minimum of two job-trained operators per shift is generally required by law, but these would be able to attend several separators at almost any feed rate.

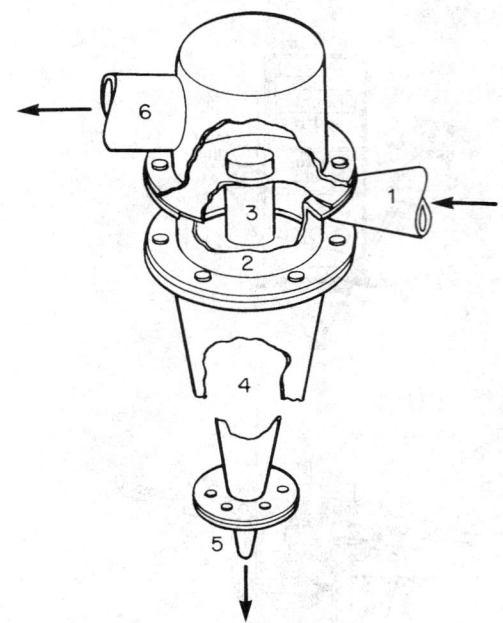

FIG. 19-36 Dutch State Mines cyclone separator.

Estimates for a 30-kg/s (100-ton/h) plant using a dynamic separator is approximately $350,000 (1978), exclusive of power, water, compressed air, crushing, foundations, and housing; installed power for such a plant will be about 298 kW (400 hp).

Direct operating costs usually vary between 30 cents and $1 per ton of feed, depending on hourly tonnage and on media recovery. Media losses are usually higher in a dynamic than in a static separator, mainly because a finer particle size is required. Media loss may be from 2 to 20 kg/t (5 to 10 lb/ton) of mixture processed in a cyclone plant compared with a static plant, which generally operates with media losses of an order of magnitude lower.

MAGNETIC SEPARATION

GENERAL REFERENCES: Kolm, Oberteuffer, and Kelland, "High-Gradient Magnetic Separation," *Sci. Am.,* **233**(5), 46–54 (November 1975). Lawyer and Hopstock, "Wet Magnetic Separation of Weakly Magnetic Materials," *Min. Sci. Eng.,* **6**(3), 154–172 (July 1974). Marston, "The Use of Electromagnetic Fields for the Separation of Minerals," World Electrical Congress, Moscow, 1977. Taggart, *Handbook of Mineral Dressing,* 2d ed., Wiley, New York, 1945.

The principles of magnetic separation have been applied commercially for nearly 100 years. Applications range from the removal of coarse tramp iron to more sophisticated separations, such as the elimination of weakly magnetic iron-stained particulates from paper-coating clays. The application of magnetic-separation methods to weakly magnetic particles has been made possible by recent advances in separator design. Magnetic separators now have a great many industrial applications and range in size from small laboratory-scale devices to those capable of processing hundreds of tons hourly.

Selecting the best separator for a specific application requires an understanding of basic principles of magnetism plus an evaluation of separator capability on the basis of design and application variables such as type of material to be processed, wet or dry processing, particle-size range, magnetic characteristics of the feed, desired throughput rate, etc.

Principles of Magnetic Separation Any particle introduced into a magnetic field will become magnetized to some extent and act as a magnetic dipole. Depending on the magnetic characteristics of the material, it can be classified as ferromagnetic, paramagnetic (mag-

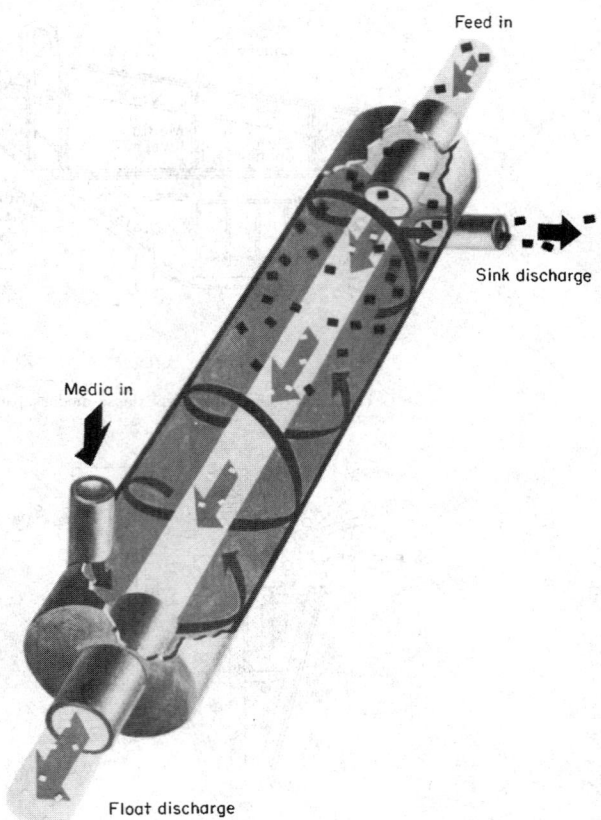

FIG. 19-37 Dyna Whirlpool separator. (*Courtesy of American Zinc Co.*)

netically attracted), or diamagnetic (repelled by a magnetic field). Ferromagnetic substances (e.g., iron, nickel, and cobalt) may be permanently magnetized and have strong magnetic moments per unit volume. Paramagnetic substances are further classified as strongly or weakly magnetic according to the strength of the magnetic moment produced per unit volume in the external magnetic field.

A magnetic field and magnetic-field gradients are produced in a variety of ways and vary in both field geometry and strength. The magnetic field of a magnet is the space through which its influence extends. It is mapped by the lines of magnetic force. A magnetic field is considered uniform or homogeneous when these lines are parallel and equally spaced. It can be noted in Fig. 19-38*a* and *b* that neither the bar (permanent) magnet nor the coils plus iron magnet typical of the C-frame magnet type can produce a uniform magnetic field.

The intensity of the magnetic field H is measured in amperes per meter. For a single-layer solenoid, at any point along its axis the magnetic field intensity is

$$H = \tfrac{1}{2}NI(\cos \theta_2 - \cos \theta_1) \qquad (19\text{-}8)$$

where H is measured in amperes per meter, A/m; N is the number of turns per unit length or the number of turns per meter; I is the current per turn in amperes, A; θ_1 and θ_2 are the angles included between the axis and the lines drawn from the measured point to its near and far edges.

The magnetic flux density is

$$B = \mu_0(H + M) \qquad (19\text{-}9)$$

Flux density is calculated as the permeability of free space times the sum of the magnetic-field intensity and the induced magnetization

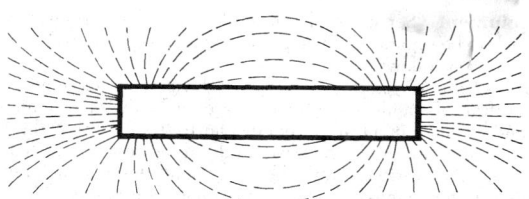

FIG. 19-38a Lines of force surrounding a bar-type magnet.

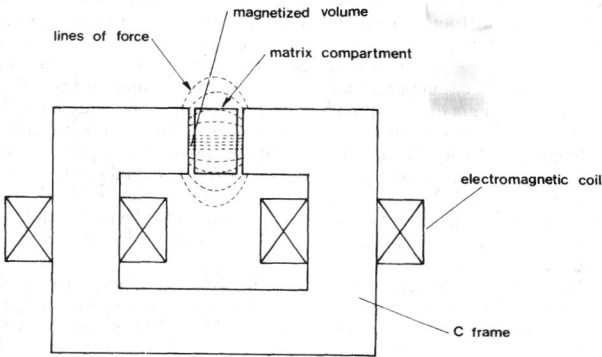

FIG. 19-38b Lines of force produced by a C-frame magnet (coils and iron-magnet surface).

TABLE 19-16 Magnetic Susceptibility of Elements and Minerals

Substance	Susceptibility, 10^{-6} cgs	Substance	Susceptibility, 10^{-6} cgs
Aluminum	+10.5	Ferberite	+39.3
Al_2O_3	−37.0	Galena	−0.4
Apatite	+1.0 to +18.0	Garnierite	+30.7
Aragonite	−0.4	Gold	−28.0
Asbolan	+150.0	Ilmenite	+15.45 to 70.0
Azurite	+12.2 to +19.0	Lead	−23
Anatase	+0.96 to +5.60	Malachite	+10.5 to +14.5
Beryl	+0.4	Millerite	+0.21 to +3.85
Braunite	+35.0 to +150.0	Molybdenite	+4.93 to +7.07
Biotite	+40.0	Molybdenum	+89.0
Barite	+10.0	Platinum	+201.9
Barite (pure)	−71.3	Rutile	+0.85 to +4.78
Brannerite	+3.5	Scheelite	+0.13 to +0.27
Chromium	+180.0	Siderite	+65.19 to +103.81
Chromite	+125.6 to +450.0	Titanium	+150.0
Cobalt	Ferromagnetic	Tungsten	+59.0
Cobaltine	+2.0	Uranium	+395.0
Cobaltite	+0.34 to +0.64	Vanadium	+255.0
Columbite	+32.55 to +37.20	Vanadinite	−0.2 to +0.27
Copper	−0.1	Wolframite	+42.2
Chalcopyrite	+1.0 to +5.0		

NOTE: Extensive listing of magnetic susceptibilities of elements and organic and inorganic compounds can be found in G. Foex, *Tables de constantes et donnes numériques*, Massou et Cie., Paris, 1957.

observed whenever a magnetic material is placed in a magnetic field; it is measured in teslas, T.

The strength of the induced magnetization is equivalent to M dipoles per cubic meter where μ_0 is the permeability of free space, equal to $4\pi \times 10^{-7}$, N/A^2; and M is the magnetization, A/m.

Another method for calculating B is

$$B = \mu H \qquad (19\text{-}10)$$

where μ is the permeability of the material.

The magnetic susceptibility of a material (χ, volume susceptibility) is dimensionless and is defined as the ratio of induced magnetization to magnetic field intensity. It is expressed as

$$\chi = M/H \qquad (19\text{-}11)$$

Thus,

$$B = \mu_0 H(1 + \chi) \qquad (19\text{-}12)$$

Specific magnetic susceptibility (ψ) is

$$\psi = \chi/\rho \qquad (19\text{-}13)$$

where ρ is material density.

Paramagnetic substances have positive susceptibilities, and induced magnetization augments the magnetic-flux density within the substance. Diamagnetic materials have negative susceptibilities, and an induced field in this case cancels part of the magnetic-field intensity.

Permeability (μ), which is often used, albeit imprecisely, in referring to ferromagnetic substances, is the ratio of the magnetic-flux density to the magnetic-field density.

$$\mu = B/H \qquad (19\text{-}14)$$

Relative permeability is μ/μ_0.

In cgs units,

$$B = H + 4\pi I \qquad (19\text{-}15)$$

$$K = I/H \qquad (19\text{-}16)$$

$$K = \chi/4\pi \qquad (19\text{-}17)$$

Table 19-16 shows the magnetic susceptibility of minerals and elements. Magnetization of various materials is directly dependent on

two factors: (1) the degree of magnetic susceptibility and (2) the applied magnetic-field intensity. It can be seen in Fig. 19-39 that ferromagnetic materials quickly become magnetically saturated and that an increase in magnetic-field intensity will have no effect after a certain point. For paramagnetic materials (e.g., hematite), which are more difficult to magnetize, the magnetic-flux density is directly proportional to the magnetic-field intensity, and some of these substances, practically speaking, cannot be saturated.

In addition, the magnetic characteristics of a material can change as a function of stress (e.g., unannealed series 316 stainless steel can be magnetic after machining), temperature, pressure, and physical and chemical treatment. Therefore, when two paramagnetic materials with similar magnetic susceptibilities are to be separated, the possibility that pretreatment will facilitate subsequent separation should be studied.

A magnetic field exerts a force on each of the two poles of a dipole (particle), forcing it to align itself with the lines of magnetic force.

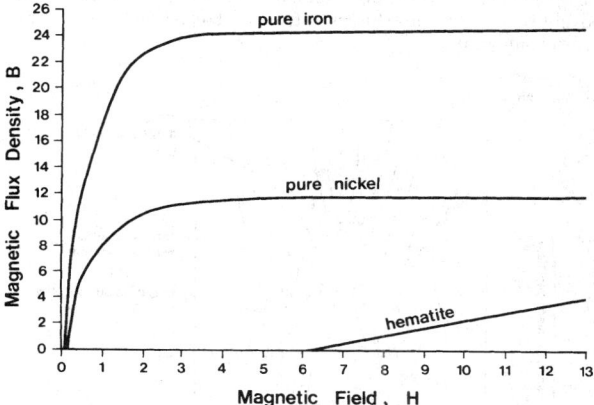

FIG. 19-39 Magnetization curves for ferromagnetic and paramagnetic materials.

These are exerted in opposite directions, and if the magnetic field is uniform, they will be equal. Therefore, the net force on the dipole will be zero. However, if the field varies in space (has a gradient), the force on the dipole will be greater in the direction of the higher field and will be proportional to the magnetic-dipole moment and the magnitude of the magnetic-field gradient.

A model of the forces operating in such a case is shown in Fig. 19-40, where

$$F_m = m \, dB/dz \qquad (19\text{-}18)$$

$$F_d = 3\pi\eta bv \qquad (19\text{-}19)$$

where
F_m = magnetic tractive force
F_d = hydrodynamic drag force
F_g = gravitational force
m = magnetization characterization of the particle
  $(m = \chi HV)$
H = magnetic-field intensity
dB/dz = magnetic-field gradient
η = fluid viscosity
b = particle diameter
v = fluid velocity
V = particle volume
χ = magnetic permeability of particle

In order to retain the magnetic fraction of the material in the collection volume of the separator it is necessary that

$$F_m \geq F_d + F_g \qquad (19\text{-}20)$$

The preceding analysis shows that density is important and cannot be influenced and that particle size is of extreme importance, as the magnetic force F_m is directly dependent on the b^3 of the particle while the drag force F_d is dependent on b. This means that separations between particles with close magnetic susceptibilities can be successfully performed only if the particle-size distribution is within a relatively narrow range. The influence of gravitational force is dependent on the relative direction of the slurry flow. [Buoyant forces are neglected in the relationship expressed in Eq. (19-20).]

It should be noted that effective magnetic separation requires that particles of different species be liberated from each other. It is also important that the finest possible matrices (filamentary type) be used because these produce the highest magnetic-field gradients. The use of high-gradient-producing matrices can substantially reduce the magnetic-field intensity (magnet strength) required to gain the same separation results, thus lowering both capital investment and process costs. The most economical results are achieved when the diameters of the matrix filaments are matched to the size of the particulates being processed.

Factors which adversely influence the separation of very fine particle systems are brownian motion and London forces. However, it is possible to counter these forces by the use of dispersants, temperature control, and so on.

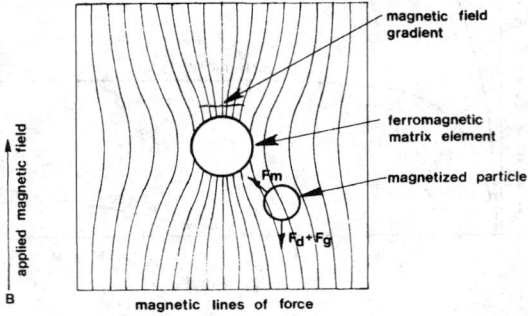

FIG. 19-40 Model of particle-capture forces.

Equipment Separator designs differ for the various types of materials to be separated. In general, magnetic separation devices can be grouped as follows:

Grate-Type Magnets This type of device consists of a series of tubes (often of stainless steel) which are packed with ceramic magnets and installed in a trap perpendicular to the fluid-flow direction. Grate magnets are used for the wet or dry removal of tramp coarse or fine iron. The various available equipment designs include self-cleaning grates, wing-and-drawer-type magnetic grates, permanent magnetic grates, vibratory grates, and rota-grates. Each of the designs is manufactured in a range of sizes, with single or multiple rows (banks) of magnetic tubes. Applications include ferrous traps for slurries such as detergents (e.g., in chemical plants), sugar and candy (e.g., in food plants), ink recycling (e.g., in printing operations), or pulp in paper mills.

Grates may be installed in all circuits of dry, pulverized material where contamination or accidents may occur from tramp or fine iron.

Plate Magnets and Magnetic Humps These devices are used to remove tramp iron from materials being conveyed pneumatically or falling in gravity flow. Tramp iron is removed by being trapped against a magnetized plate. This type of magnet must be cleaned periodically. A chute angle of 45° is recommended. The plate magnet should be close to the feed point to eliminate the influence of velocity. Complete lines of plate magnets and magnetic humps are offered by many manufacturers.

Plate magnets, which are used in chutes, can be either permanent magnets or electromagnets. For the permanent type, magnet width extends to 1.23 m; there is a maximum width of 2.85 m for electromagnets. Capacities are approximately 250 m³/h for each meter of width, at a 45° angle, with the magnet located in the bottom of the chute approximately 0.6 m from the top and material introduced at a slow velocity. Capacity varies with the size of the tramp-iron particles to be removed and with the angle of the chute (see Table 19-17).

Lifting Magnets These devices operate in either a continuous or a cyclic manner. Continuous devices usually have a belt which moves over the lifting magnetic poles to carry the magnetized particles into a region of low or zero magnetic field, where they are released. Depending on the design of the poles, these units can be either high- or low-intensity devices. Figure 19-41 shows the in-line and cross-belt methods for installing a lifting magnet above a conveyor belt.

Cross-belt magnetic separators are based on the same principle as lifting magnets. Although these units have relatively low capacities, the same unit can produce selective separations with different products by using different pole gaps and field strengths. (See Fig. 19-42.)

The magnet designs shown in Figs. 19-41 and 19-42 are used for tramp-iron removal. Suspended magnets are positioned from 5 to 10 cm above the highest point of the material on the conveyor and may be designed to be self-cleaning. Sizes for devices of this type range up to 2.8/1.6 m. The installation shown in Fig. 19-41a is often preferred because it requires a less powerful magnet and can clean material from a higher-speed conveyor belt (over 1.75 m/s). For self-cleaning units, the belt is run at up to 2.5 m/s.

Drum and Pulley Magnets Since Thomas Edison invented and developed the magnetic pulley for the concentration of nickel ore, drums and pulleys have become the most common types of magnetic separators. These devices can be built with either a permanent magnet or an electromagnet, and the drum separator can operate with

TABLE 19-17 Tramp-Iron Removal with Plate Magnet, 0.6 m from Top*

Particle size	Relative capacity, percent		
	Chute angle, 35°	Chute angle, 45°	Chute angle, 60°
Over 30 g (1 oz)	125	100	75
Over 8 mesh (2.38 mm) to 30 g	100	75	45
Under 8 mesh (2.38 mm)	33	25	10

*Courtesy of Eriez Magnetics.

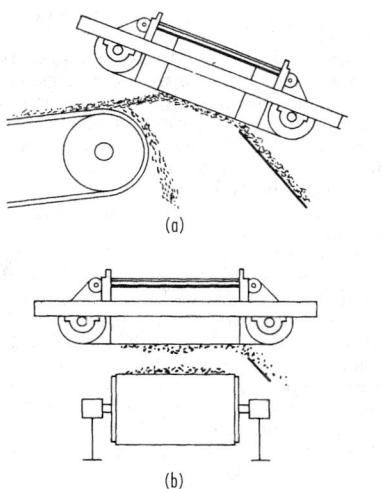

FIG. 19-41 Types of lifting magnets. (*a*) In-line lifting magnet. (*b*) Cross-belt lifting magnet. (*Courtesy of Eriez Magnetics.*)

either dry or wet feeds. Figure 19-43 is a schematic for mounting a magnetic pulley.

Dry magnetic drums can be designed to perform as lifting magnets or pulleys. Magnetic drum devices have stationary magnets; pulley drums rotate. Other schematics of possible arrangements are presented in Fig. 19-44.

In the drum-separator category, several specialized devices are worthy of mention.

Alternating-polarity drum separator. This device is used for the treatment of coarse material (minus 40 mm, plus 0.15 mm) containing strongly magnetic particles when a high-grade concentrate is required. The capacity of this device varies with feed-particle size, up to 100 t/(h·m).

Unigap drum separator. This device is used for materials finer than 6 mm at feed rates of up to 10 t/(h·m).

High-speed, low-intensity drum magnetic separator. This device is designed to handle very fine material (minus 0.15 mm and finer) to produce a high-grade magnetic concentrate.

Depending on the required results—high recovery of magnetics or high-grade concentrates (clean magnetics)—wet drum separators are designed to work in concurrent, countercurrent, or counterrotating fashion by using one or more drums in any possible combina-

tion. Figure 19-45 presents schematics of these wet drum magnetic separators.

Magnetic pulleys. These vary in size from 0.203 to 1.219 m in diameter and from 2.03 to 1.526 m in width. The acceptable depth of the material on the conveyor belt depends on the diameter of the pulley and the linear velocity of the belt (see Table 19-18). Table 19-19 indicates the maximum capacity for such units. Depending on the application, the correction factors given in Table 19-20 should be applied. For sizing and maximum efficiency, multiply the actual volume of material to be handled by the correction factor shown and select the magnetic pulley having a capacity equal to or greater than the resultant volume.

Wet Drum Magnetic Separators These devices are used for the concentration of strongly magnetic coarse particles. The size of the separator is influenced by several variables: slurry volume, percent solids in the slurry, percent magnetics in the slurry, required recovery of magnetic particles, and required concentration of magnetic product. This type of separator is built by several manufacturers; drum sizes range from 0.023 to 1.2 m in diameter, with widths up to 3.0 m. The concurrent type can process slurries with 20 percent solids by weight for single-drum separators and with 35 to 45 percent for units with two drums. Recommended maximum particle size is 6 mm (¼ in), but with special tanks these devices can handle even coarser material. Countercurrent-type separators can handle particles finer than 0.8 mm (20 mesh) and obtain optimum results with slurries containing about 30 percent solids. This design has the advantage of being able to handle wide fluctuations in throughput. The counterrotating separator is recommended for applications in which recovery is more important than grade. This unit can handle particles up to 3 or 4 mm (⅛ in) in size, but with less satisfactory results for particles finer than 0.5 mm and slurries containing 30 to 40 percent solids by weight. Figure 19-46 shows a gauss (tesla) chart for a 1.2-m-diameter wet drum separator. The influence of drum diameter in separation is shown in Table 19-21 and in Fig. 19-47. These data result from the processing of a partially martised magnetite ground to 75 percent minus 0.044 mm (325 mesh). The influence of drum diameter and grind in separator capability is summarized in Table 19-22. Figure 19-48 shows the influence of drum diameter on investment. The **installed cost** of a wet single-drum magnetic separator can vary between $25,000 and $75,000 per meter of magnet width, depending on the design, dimensions, and manufacturer. Multiple-drum costs increase in about direct proportion to the number of drums required. **Maintenance costs** per year vary between 3 and 5 percent of the initial investment.

Induced-Roll Separators These devices, which have been in commercial use since 1890, handle only dry, granulated material. They are similar to drum separators, with the difference that the cylinder rotates in the gap of an electromagnet. Magnetic-field gradients are obtained by creating sharply edged ridges on the surface of the cylinder or by constructing a cylinder of alternate magnetic and non-

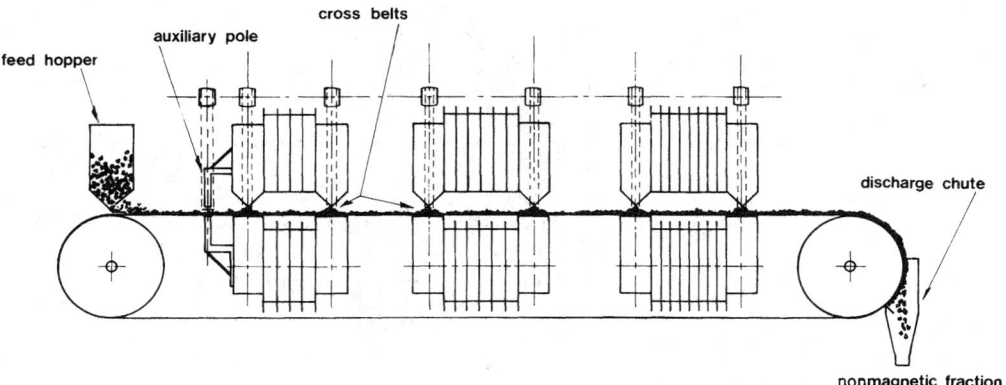

FIG. 19-42 Six-pole, seven-cross-belt magnetic separator. (*Courtesy of Readings, Inc.*)

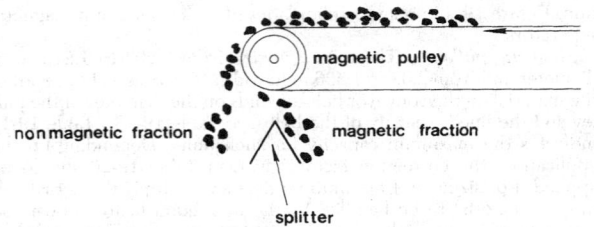

FIG. 19-43   Magnetic pulley.

magnetic disks. A schematic of an induced-roll separator is shown in Fig. 19-49. The best particle-size distribution for separation is minus 2 mm, plus 0.074 mm (minus 10 mesh, plus 200 mesh). Industrial devices are built with multiple rolls, which operate either in a series or in parallel, and can be used as concentrators or as purifiers (see Fig. 19-50). Standard widths for the rolls are 0.25, 0.5, and 0.75 m. Capacities vary between 1.5 and 18 t/(h·m). Induced-roll separators are used

only to process weakly magnetic materials. **Capital costs** for this type of device are relatively low compared with those of other high-intensity magnetic separators, but **total process costs** are high owing to moisture-free feed requirements. A wet-process induced-roll separator was developed in the U.S.S.R. during the early 1960s and is reported to have a capacity of up to 100 t/h. In 1964 an Australian manufacturer introduced a wet-type induced-roll separator designed with a laminated, grooved rotor that rotates around a vertical axis (pole). These devices are built with up to 10 poles and are used principally to concentrate ilmenite sands. Capacity is approximately 0.8 t/h per magnetic pole.

Separations similar to those obtained with dry induced-roll devices can be obtained with cross-belt separators (Fig. 19-42). These units are built with up to eight poles, each of which can operate at different magnetic-field intensities to allow simultaneous production of different concentrates. However, capacity is low, and installed costs per ton capacity are high compared with induced-roll units.

Induced-Pole Separators   In devices of this type, magnetic-field gradients are produced by the application of background magnetic field to a ferromagnetic matrix, thereby inducing magnetic poles around matrix edges. The correlation of edge and field directions

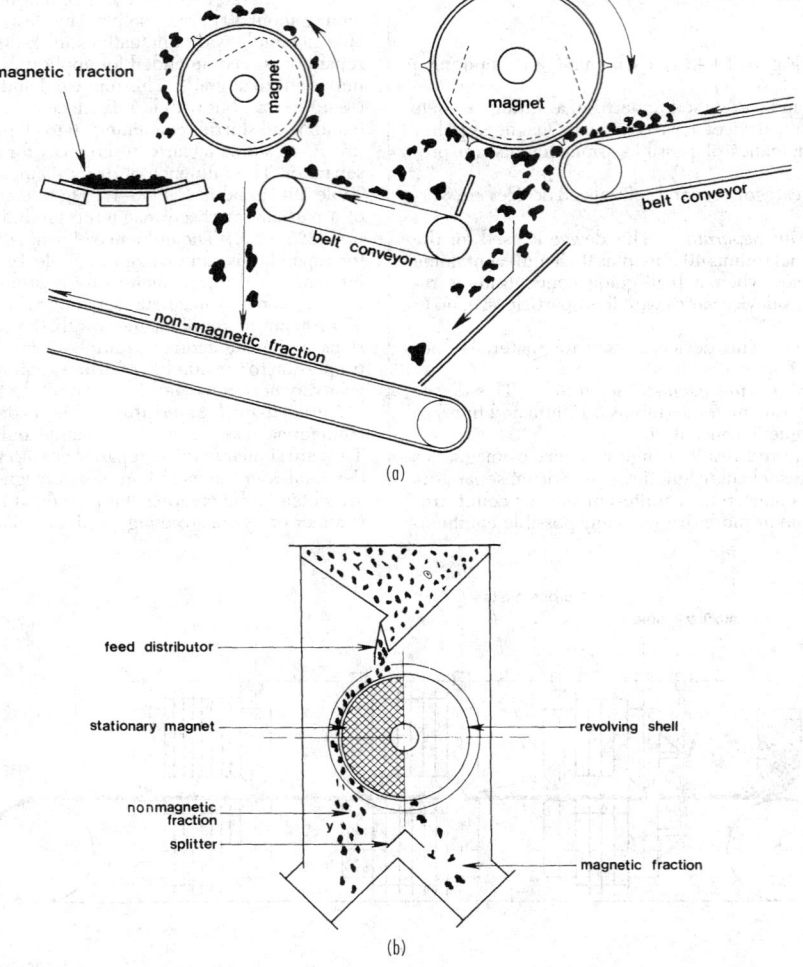

FIG. 19-44   Arrangement of magnetic drum separators. (*a*) Magnetic drum operating as a lifting magnet. (*b*) Magnetic drum operating as a pulley. (*Adapted from design courtesy of Eriez Magnetics.*)

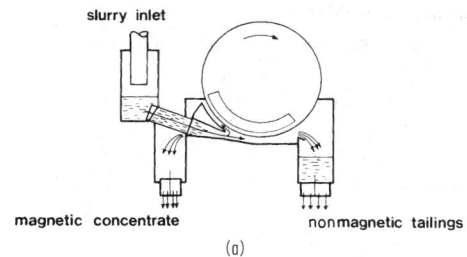

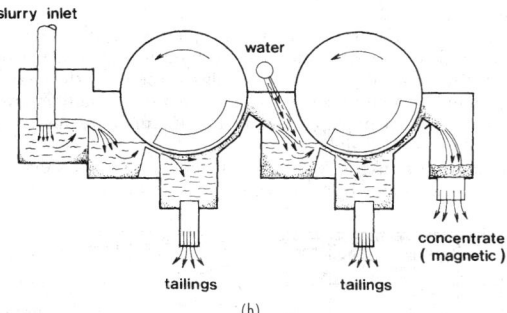

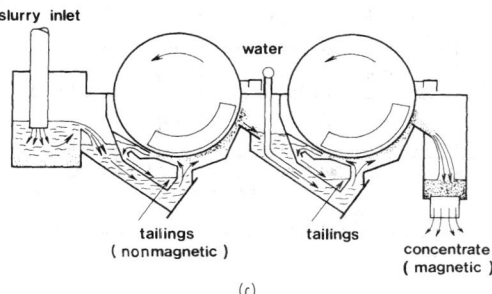

FIG. 19-45 Wet-drum-magnetic-separator arrangements. (*a*) Counterrotation-type wet magnetic drum separator. (*Courtesy of Sala International, Inc.*) (*b*) Concurrent-type wet magnetic double-drum separator. (*c*) Countercurrent-type wet magnetic double-drum separator.

determines whether the separator is a parallel field-to-flow unit or a perpendicular field-to-flow unit. In this category there are only two practical types of separator constructions, C-frame and solenoid. Uniformity of background magnetic field depends on design.

1. *C-frame magnets.* As shown in Fig. 19-38*b*, these magnets employ a ferromagnetic matrix placed between the poles of an electromagnet. With this design, however, the background magnetic field

TABLE 19-18   Maximum Depth of Material for Separator by Magnetic Pulley Based on Pulley Diameter and Linear-Velocity Belt

Diameter of pulley, mm	Belt linear velocity, m/s	Depth of mterial, mm
203	0.584	38
305	0.890	70
380	1.017	89
508	1.271	121
610	1.448	140
762	1.678	165
914	1.855	191
1067	2.033	210
1219	2.211	235

TABLE 19-19   Maximum Capacity for Magnetic Pulley Separator*

Pulley diameter, mm	Belt width, mm	Belt velocity, m/s	Capacity, m³/h
203	203	0.585	12.2
	406		24.9
	610		62.3
	914		133.0
381	305	1.017	38.0
	406		50.0
	610		113.0
	914		255.0
	1219		515.0
457	305	1.143	47.0
	406		59.0
	610		130.0
	914		300.0
	1219		623.0
610	406	1.448	88.0
	610		170.0
	914		374.0
	1219		755.0
	1524		1133.0
914	457	1.855	153.0
	610		218.0
	914		481.0
	1219		935.0
	1529		1500.0

*Courtesy of Eriez Magnetics.

TABLE 19-20   Correction Factors for Magnetic-Pulley Capacities*

Type of application	Type of tramp iron to be removed	Correction factor
Crusher and primary-mill protection	Large and medium, over 30 g (1 oz)	1.0
Secondary-mill, pulverizer, and general separation	Large, over 30 g	1.0
	Medium, 30 to 240 g	1.3
	Small, 8 mesh (2.38 mm) to 30 g	2.0
Product purification	Fine ferrous contamination; finer than 8 mesh (2.38 mm)	4.0

*Courtesy of Eriez Magnetics.

TABLE 19-21   Influence of Magnetic-Drum Diameter on Separation*

Separator diameter, m	No. of stages	Composition of concentrates		Fe recovery, %	Feed rate, t/(h·m)
		SiO₂, %	Fe, %		
0.600	6	1.1–1.3	70.0	92	10–12
0.916	6	0.9–1.0	70.0	98	28–33
1.200	4	0.9–1.0	70.0	98	62–85

*Courtesy of Sala International Inc.

is not uniform. Also, high-magnetic-fringe fields are usually noticed in the flush region of these separators, and these can cause possible matrix clogging when even relatively small amounts of ferromagnetic particles are present in the slurry. The ferromagnetic material used to transfer the magnetic-flux lines from pole to pole occupies between 40 and 80 percent of the magnetized volume.

2. *Solenoid magnets.* These devices can be designed for wet or dry feeds. Depending on design, they can have a relatively uniform background magnetic field. It can be noted from Fig. 19-51 that the use of the return frame is important for generating a uniform mag-

TABLE 19-22 Influence of Drum Diameter and Grind on Separator Capability*

Feed			Recommended capacities, t/(h·m)		
Description	Percent of feed minus 74 μm	Separator arrangement	Diameter of drum, m		
			0.60	0.90	1.20
Coarse	15–25	Concurrent	15–25	70–90	120–160
Medium	50	Concurrent or full countercurrent	10–15	35–50	60–90
Fine	75–95	Semicountercurrent	6–10	30–50	60–90

*Courtesy of Sala International Inc.

netic field. Filamentary-type matrices, which occupy less than 10 percent of the magnetized volume yet still provide very high field gradients, can be used with these types of magnets.

The most familiar of the C-frame, matrix-type industrial magnetic separators are the Carpco, Eriez, Readings, and Jones devices. The Carpco separator employs steel balls as a matrix, Eriez uses a combination of expanded metal matrices, and the Readings and Jones separators have grooved-plate matrices. Capacities for this type of unit are reported to up to 180 t/h (in the case of Brazilian-hematite processing).

Solenoid magnetic separators are designed for batch-type, cyclic, and continuous operation. Devices which can use matrices of expanded metal, grooved plates, steel balls, or filamentary metals have been designed. Continuous separators with capacities to 600 t/h for iron ores (similar to the Brazilian hematite) are commercially available (Sala International Inc.). Selection of the method of operation is application-dependent, being based on variables such as temperature, pressure, volume of magnetics in the feed, etc.

A familiar type of cyclically operated solenoid electromagnet is the Franz separator, a well-known continuous type of solenoid separator manufactured by Krupp-Sol. An enclosed flux return-frame solenoid design for cyclic and continuous use is built by Sala International Inc.

A schematic of a continuous Sala high-gradient magnetic separator is shown in Fig. 19-52.

Depending on the type of matrix used, induced-pole magnetic separators can be classified as either high-intensity magnetic separators, which utilize grooved plates or steel balls as the matrix material, or as high-gradient magnetic separators, which use filamentary matrices such as steel wool or expanded metal. Filamentary matrices have proved to be more advantageous.

The maximum magnetic field produced by a C-type device is 2 T. For solenoids, conventional designs produce magnetic-field intensi-

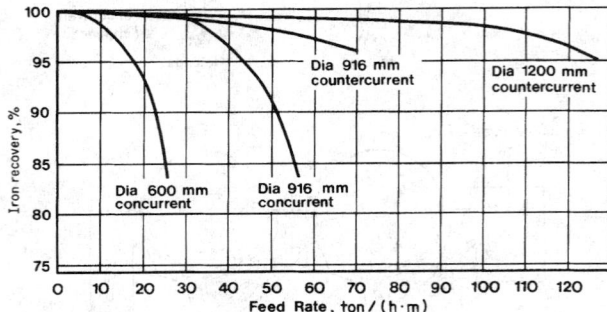

FIG. 19-47 Influence of drum diameter on separation. (*Courtesy of Sala International, Inc.*)

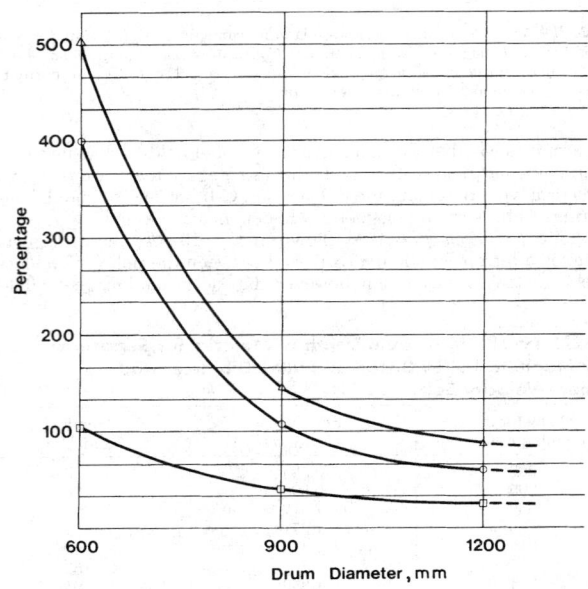

FIG. 19-48 Influence of drum diameter on separation cost. △, Two stages coarse separation; ○, three stages fine separation; □, sum of coarse and fine separations. (*Courtesy of Sala International, Inc.*)

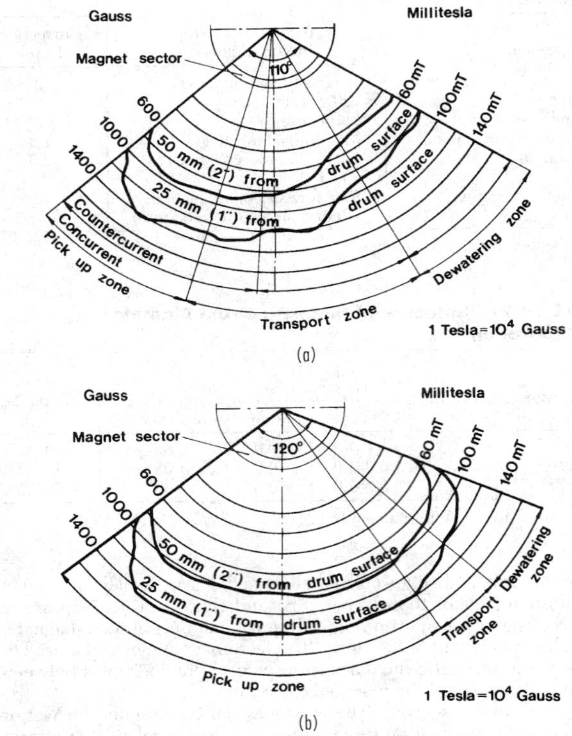

FIG. 19-46 Magnetic-field distribution charts. (*a*) Concurrent and countercurrent wet drum magnetic separator, 1.2-m diameter. (*b*) Counterrotation wet drum magnetic separator, 1.2-m diameter. (*Courtesy of Sala International, Inc.*)

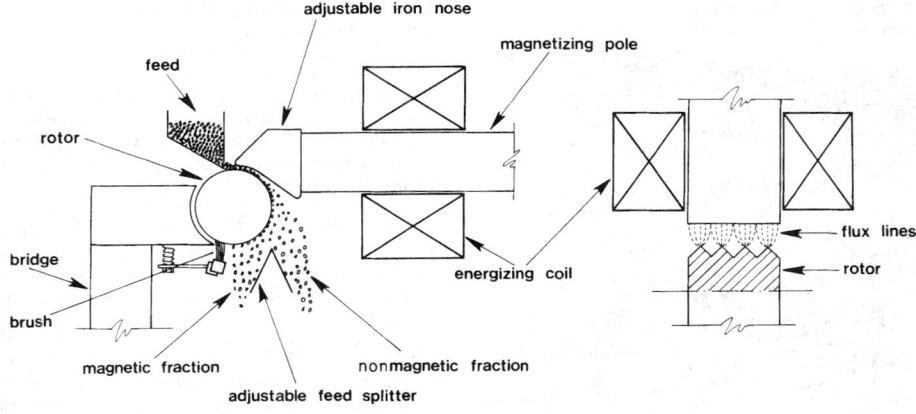

FIG. 19-49 Schematic diagram of an induced-roll separator.

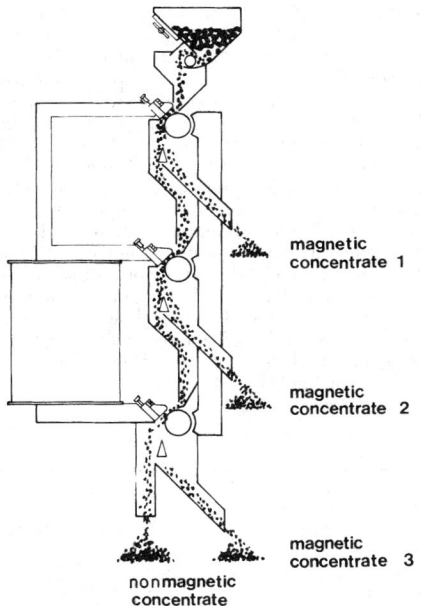

FIG. 19-50 Multiple induced-roll magnetic separator.

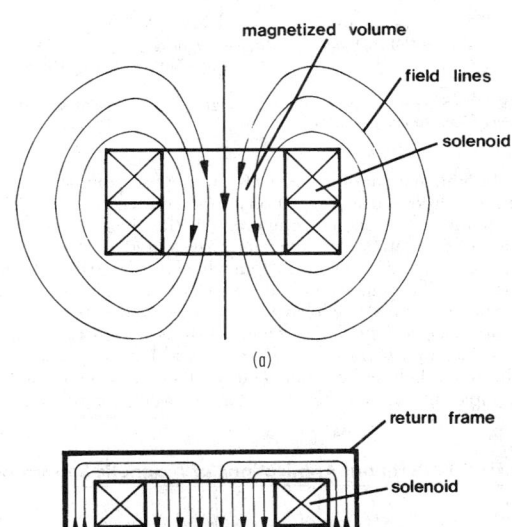

FIG. 19-51 Solenoid-type magnets. (*a*) Nonuniform field. (*b*) Uniform field by use of return frame.

ties up to 2 T, while superconducting units can be constructed with ratings up to 8 T.

In all induced-pole devices, the more magnetic particles are retained on the matrix while the less magnetic fraction is carried away in the slurry.

Dynamic (or Deflecting) Devices The oldest type of dynamic separation device is the Franz Isodynamic separator. This laboratory device has a dipole configuration with the poles shaped so that the value of $H\,dB/dz$ is constant throughout the working separation volume. Material to be separated can be fed through a vibratory chute or dropped between the poles, producing a separation based only on relative magnetic susceptibility.

There are a variety of new developments in magnetic separation which are of possible interest, but their commercial applicability is still not yet assured. These include quadripole separators and a spiral-flow device.

Table 19-23 lists potential applications for all types of magnetic separators.

SUPERCONDUCTING MAGNETIC SEPARATION

Superconducting Magnets In a superconducting magnet the magnetic field is generated in exactly the same way as on a normal electrical solenoid, coil, or winding. The only real difference is that the conductor is made from superconducting alloy which has to be maintained at a suitably low temperature to maintain the superconducting state. There are a number of possible material compositions which can be used for the superconducting winding but for industrial applications where economics and reliability play a key role an alloy of niobium and titanium is presently the preferred choice.

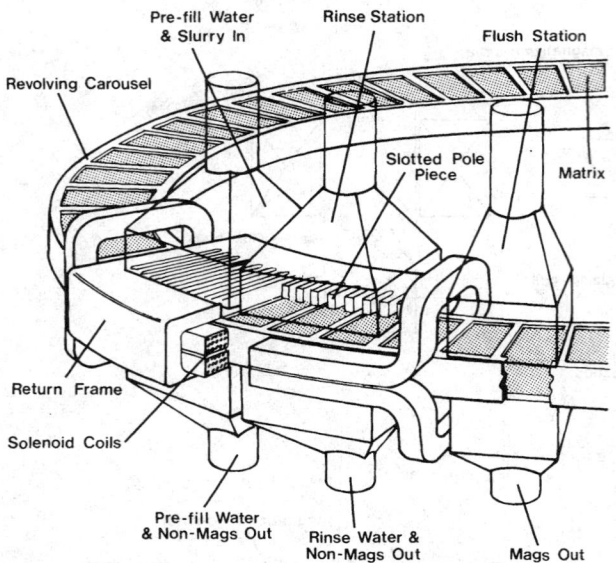

FIG. 19-52 Schematic of continuous high-gradient magnetic separator. (*Courtesy of Sala International, Inc.*)

Superconducting windings of N6Ti are now so commonplace that they can be considered as conventional superconductors. What is more relevant to the construction of a superconducting magnetic separator is the choice of refrigeration or cryogenic system used to cool the magnet winding, and it is the cryogenic system which to a large extent dictates the economics and practicality of these machines. There are presently three cryogenic routes which have been successfully applied.

1. *Closed-cycle liquefier systems.* In this design the superconductor resides in a bath of liquid helium and boil-off gas is recirculated through a helium liquefier. The installation of such a system is quite complex but these installations have proved good reliability pro-viding there are no long-term interruptions to the supply of electrical power and cooling water.

2. *Low-loss system.* In a low-loss system the winding also resides in a reservoir of liquid helium but a very efficient insulation system enables the magnet to operate for long periods, typically 1 year or more, between liquid helium refills. An important feature of these systems is that they are relatively immune to short-term electrical power failures, which has enabled complete reliability even in extremely difficult environments.

3. *Indirect cooling.* The advent of heat engines based on the Gifford McMahon cycle to generate temperatures of 4 kelvin or less has made it possible to cool superconducting windings without the need for liquid helium. This technique offers great potential for small-scale systems where the economics of helium supply or the cost of a liquefier cannot be justified. The only drawback is that a constant supply of electrical power is essential for reliable operation.

The key benefits offered by superconducting magnets are (1) very low power consumption resulting from zero resistance of the magnet winding and (2) much higher magnetic fields which can be generated.

Superconducting magnets are presently being used in two distinct types of devices: *high-gradient magnetic separators* (HGMS) and *open-gradient magnetic separators* (OGMS). We shall consider these in turn.

Superconducting HGMS The HGMS principle relies on the capture of magnetic particles on a magnetized ferromagnetic matrix, as described in previous sections. In this type of device where an increase in magnetic-field induction enables capture of weaker magnetics and power consumption of large-scale systems is an important economic factor it is not surprising that superconducting magnets have made a significant impact. Indeed, for large-scale HGMS, superconducting magnets are by far the preferred choice.

A key feature of the HGMS process is that periodically the matrix must be demagnetized to flush out the captured magnetics. For superconducting magnets this demagnetization can be achieved by either de-energizing the magnet (switched-mode HGMS) or by moving the matrix canister (referred to as reciprocating canister HGMS) out of the magnetic field. The reciprocating matrix canister method is unique to superconducting HGMS and is shown schematically in Fig. 19-53. The differences resulting from the superconducting and resistive electromagnet HGMS systems are quite evident from the engineering data shown in Table 19-24.

TABLE 19-23 Potential Applications of Magnetic Separators

Device type	Type of construction	Maximum background magnetic field, Oe	Type of matrix which can be used	Maximum field gradient obtainable, G/cm	Required magnetic susceptibility for particulates	Particle size to be treated, mm	Materials which can be treated; fields of use
Grate	Permanent magnet	500	Rods	500	Ferro	<12	Tramp and fine iron
Pulley	Permanent magnet and electromagnet	100–200	—	100–1000	Ferro, strongly	<50	Ferro and strongly magnetic
Belt	Electromagnet	100–1000	—	100–1000	Strongly	0.15–30	Strongly magnetic
Drum	Permanent magnet and electromagnet	500–1000	—	500–1000	Strongly	0.02–20	Magnetite processing
Franz Isodynamic	Electromagnet	10,000	—	2000	Strongly, weakly	>0.01	Only for laboratory
Solenoid; Franz ferrofilter	Electromagnet	20,000	Steel ribbons, balls	200,000	Strongly, weakly	>0.01	Tramp and fine iron, ceramic slurries, industrial minerals, chemical industry
Induced rolls	Electromagnet	20,000	—	200,000	Strongly	0.03–3	Dry, dedusted, weakly magnetic particles
C-frame type; Jones	Electromagnet	20,000	Grooved plates	200,000	Strongly, weakly	0.01–2	Iron ores, industrial minerals
Carpco	Electromagnet	20,000	Steel balls	45,000	Weakly	0.01–1	Iron ores, industrial minerals
Marston Sala high-gradient magnetic separator	Electromagnet, superconducting	20,000 50,000	Steel wool, expanded metal, steel balls	25×10^6	Strongly to very weakly	0.0001–2	Iron ores, industrial minerals, coal, liquefied coal, wastewaters, purifiers, catalyst recovery, chemical industry

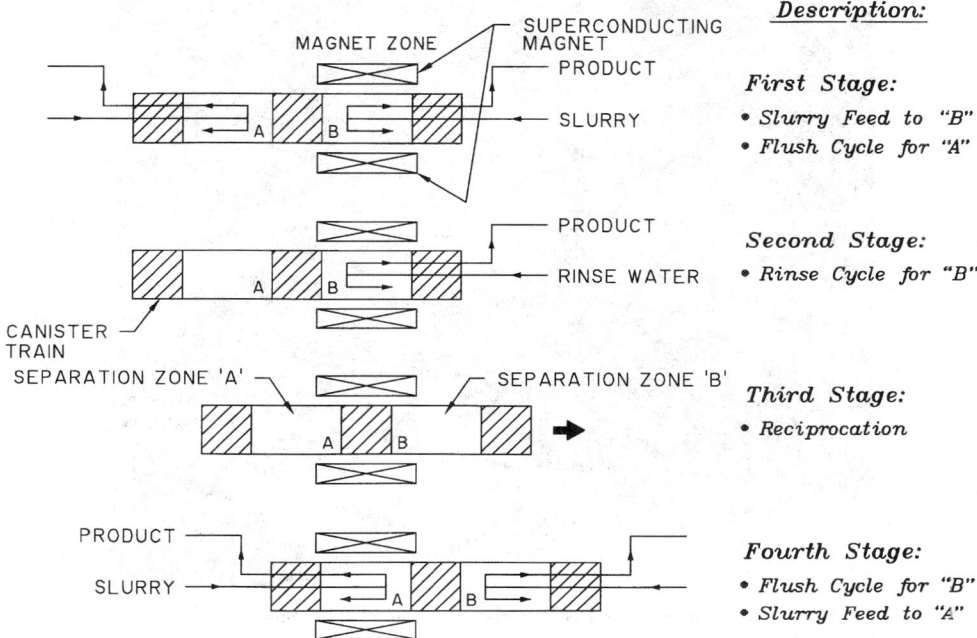

First Stage:
- *Slurry Feed to "B"*
- *Flush Cycle for "A"*

Second Stage:
- *Rinse Cycle for "B"*

Third Stage:
- *Reciprocation*

Fourth Stage:
- *Flush Cycle for "B"*
- *Slurry Feed to "A"*

FIG. 19-53   Basic process cycle for a reciprocating canister superconducting magnetic separator. (*Courtesy Carpco, Inc.*)

One might expect that the perceived complexities of superconducting magnets would restrict their use to highly developed and industrialized areas. It is therefore noteworthy that the simplicity and reliability of the combination of low-loss cryogen technology coupled with the reciprocating canister principle has enabled a number of these HGMS units to operate with total reliability in areas as remote as the Amazon rain forest areas of Munguba and Rio Capim. Figure 19-54 shows the installation of a typical large-scale reciprocating canister HGMS.

A new development which shows promise of imminent industrial application is superconducting HGMS designed for treating dry feeds. One such unit employing a vibrating matrix in a reciprocating canister design has been evaluated with promising results. Once again the advantages of low power consumption and significantly improved levels of beneficiation are the key factors in driving this technology into industry.

Superconducting OGMS   In open-gradient magnetic separators (OGMS) the magnet structure is arranged to provide a region in open space with a highly divergent field. Thus, the magnet geometry provides both the magnetic field and field gradient. Any paramagnetic material passing through this region will experience a force directly proportional to the field intensity and magnitude of field gradient. There are many conventional devices ranging from lift magnets to rare earth-drum and roll-type separators that operate on this principle. Superconducting OGMS has the benefit of offering not just higher magnetic force profiles, but a significantly greater depth of reach (i.e., larger separation volume) than permanent magnet and electromagnet devices. All superconducting OGMS units at the present operate with dry feeds usually with particle size >75 μm.

The earliest industrial application of superconducting OGMS was based on a drum separator design referred to as the "desces" separator. The drum was 1 meter in diameter and generated a peak field of 3 tesla and field gradient of 40 tesla/meter. A helium re-liquefier was required to provide adequate cryogenic capacity which significantly affected the capital cost, nevertheless, the unit has been operated successfully for many years in the beneficiation of magnesite processing normally 150 mm material at feed rates of up to 100 TPH.

A somewhat simpler and more compact OGMS device referred to as the "Cryostream" has recently been introduced. This unit operates on the inclined-fall process in which the magnet is held at an incline and material is simply allowed to fall through the magnetic region. The treatment size for the Cryofilter is 25 × 0.5 mm. The winding structure and overall geometry of this system make it an ideal candidate for indirect cooling by providing both overall simplicity and economic benefits.

TABLE 19-24   Comparative Data for 50 TPH HGMS Plant

	Superconducting Cryofilter 5T/460	Switched-mode superconducting HGMS	Conventional electromagnet
Magnetic induction	5 tesla	2 tesla	2 tesla
Separator weight	45 tons	250 tons	300–400 tons
Operator power consumption	10 kW	40–80 kW	280–400 kW
Overhead service crane for matrix canister exchange	2 tonne rail mounted	15 tonne gantry	15 tonne gantry
External cryogenic requirements	None required	1. Helium liquefier 2. Helium compressor 3. Liquid helium storage tank 4. Liquid nitrogen storage tank 5. Helium gas ballast tank	Not applicable

FIG. 19-54   Superconducting magnetic separator operating in a kaolin treatment plant. (*Courtesy Carpco, Inc.*)

The Cryostream operates with a peak field of 4 tesla and a magnetic force of 250 T²/m allows separation of minerals with magnetic susceptibilities in the order of 10^6 emu/g. In essence the Cryostream is an industrial-scale version of the well known laboratory Frantz Isodynamic Separator.

ELECTROSTATIC SEPARATION

GENERAL REFERENCES: *SME Mineral Processing Handbook,* Society of Mining Engineers of the AIMMPE, NY, 1985 Edition. Moore, *Electrostatics and Its Applications,* Wiley-Interscience, New York, 1973. Knoll, Taylor, *Advances in Electrostatic Separation,* Minerals & Metallurgical Processing, 1985. E. Tondu et al., *Commercial Separation of Unburned Carbon from Fly Ash,* Mining Engineering, June 1996, p. 47–50.

General Principles   Electrostatic separation (of particles), also commonly known as high-tension separation, is a method of separation based on the differential attraction or repulsion of charged particles under the influence of an electrical field. Applying an electrostatic charge to the particles is a necessary step before particle separation can be accomplished. Various techniques can be used for charging. These include contact electrification, conductive induction, and ion bombardment.

Regardless of the method of charging, the amount of charge that can be accumulated on a particle is limited by the maximum achievable charge density and the surface area of the particle. Electrostatic separation of mixed particles is possible when the electrostatic force acting on some particles is great enough to overcome gravity or inertial forces. Because the surface area of a solid varies as the square of a linear dimension whereas the mass varies as the cube of that dimension, gravity and inertial forces acting on solid particles increase faster with particle size than do electrostatic forces for charged particles in electric fields. Thus, there are upper size limits beyond which electrostatic separation of particles of a given shape is not feasible. For granular materials, this upper size limit is about 4 mm; for thin pieces of large cross-sectional area and for long pieces of small cross-sectional area, the limit can be greater than 25 mm.

The motion of fine particles immersed in a moving fluid is more greatly affected by fluid drag forces than that for similar large particles. For very small particles in a fluid, particle motion approximates the motion of the enveloping fluid. Industrial electrostatic separation of solid particles, which is universally conducted in air (or other easily ionizable gas), is difficult at particle sizes less than about 0.074 mm (200 mesh).

Charging Mechanisms

Contact Electrification (Fig. 19-55a)   When dissimilar materials touch each other, there is an opportunity for the transfer of electric charges. The extent of charge transfer can be such that a significant surface charge of opposite sign is developed when the materials are

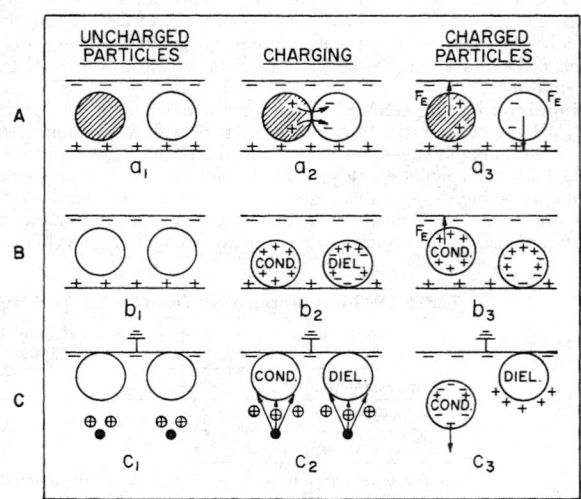

FIG. 19-55   Schematic representation of charging mechanisms. (*A*) Contact electrification. (*B*) Conductive induction. (*C*) Ion bombardment. Cond. = conductor particle; diel. = dielectric particle; ● = high-voltage dc electrode; ⊕ = ions from corona discharge at high-voltage electrode.

later separated. High temperatures and low humidity favor the development of high surface charges through the mechanism of contact electrification. Rubbing the materials together to increase the area of effective contact can also lead to high surface charges.

Particles carrying charges of opposite polarity due to contact electrification will be attracted to opposite electrodes when passing through an electric field and thus can be separated from each other.

Conductive Induction (**Fig. 19-55b**) The term *conductive induction* describes the process by which an initially uncharged particle that comes into contact with a charged surface assumes the polarity and, eventually, the potential of the surface. A particle that is an electrical conductor will assume the polarity and potential of the charged surface very rapidly. However, a nonconducting particle will become polarized so that the side of the particle away from the charged surface develops the same polarity as the surface. Particles of intermediate conductivity may be initially polarized but approach the potential of the charged surface at a rate depending on their conductivity.

If a conductor particle and a nonconductor particle are just separated from contact with a charged plate, the conductor particle will be repelled by the charged plate and the nonconducting particle will be neither repelled nor attracted by it.

The charged plate must be balanced by other oppositely charged (or earthed) bodies to maintain overall neutrality. In electrostatic separation, this is usually accomplished by means of a single electrode of charge opposite in sign to that of the charged plate. The conductor particle is then in the electrical field between the two electrodes and experiences a net electrostatic force in the direction of the second electrode. The nonconducting particle, having no net charge, experiences no electrostatic force in a uniform electric field. Electrostatic separation of the conductor and nonconductor particles can be accomplished by movement of the conductors in the electric field.

Ion Bombardment (**Fig. 19-55c**) The most positive and strongest method of charging particles for electrostatic separation is ion bombardment. Use of ion bombardment in charging materials of dissimilar properties may be visualized by considering conductor and nonconductor particles touching the grounded conducting surface of Fig. 19-55c. Both particles are bombarded by ions of atmospheric gases generated by an electrical corona discharge from a high-voltage electrode (usually a fine tungsten-alloy wire at 20 to 30 kV with respect to ground and several centimeters away from the particles). When ion bombardment ceases, the conductor particle loses its acquired charge to ground very rapidly and experiences an opposite electrostatic force tending to repel it from the conducting surface. The nonconducting particle, however, being coated on its side away from the conducting surface with ions of charge opposite in electrical polarity to that of the surface, experiences an electrostatic force tending to hold it to the surface. If the electrostatic force is larger than the force of gravity or other forces tending to separate the nonconducting particle from the conducting surface, the particle is held in contact with the surface and is said to be "pinned."

Electrostatic-Separation Machines The first electrostatic machines to be used commercially employed the principle of contact electrification. These were free-fall devices incorporating large vertical plates between which an electrostatic field was maintained. Triboelectric separation (contact charging) has experienced an increase in applications due to advances in mechanical self-cleaning and electrical design as well as the development of efficient precharging techniques.

Triboelectric Separators There are currently two industrial forms of triboelectric separators installed commercially to treat minerals and recycled plastics:

Tube-type. These separators are typically divided into two sections: (1) precharging and (2) separation. The precharging section is designed to create or enhance the charge difference between particles to be separated (typically by some form of contact mechanism or external pretreatment to render one constituent positive or negative in comparison to the other materials present. The separation section consists of two vertical walls of tubes opposing each other. Each tube

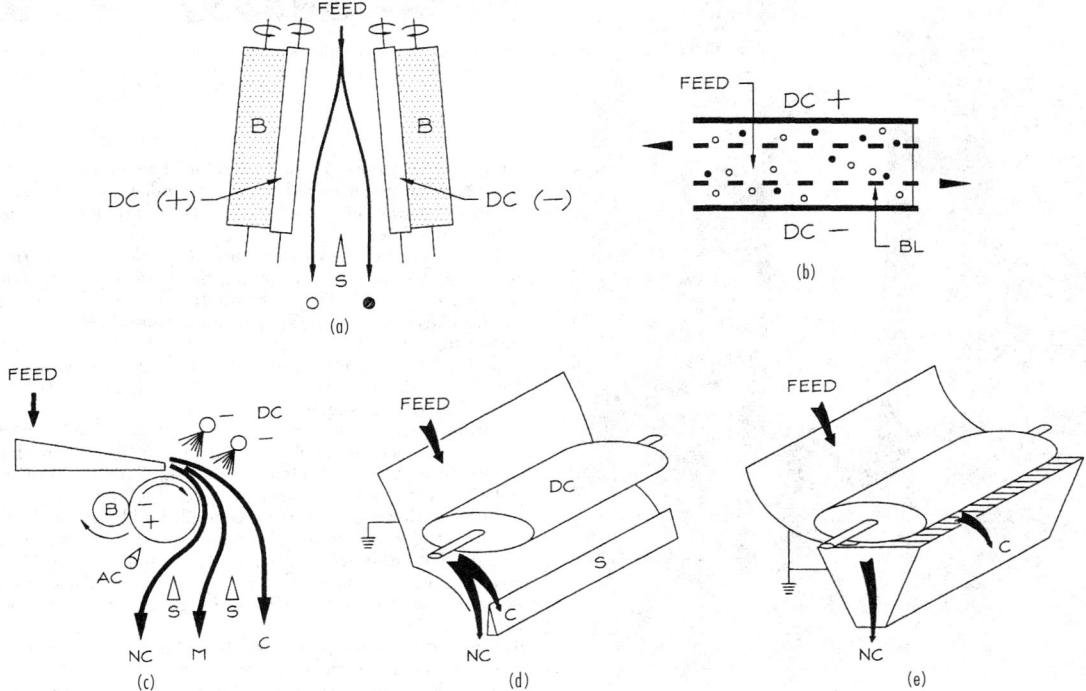

FIG. 19-56 Operating principles of electrostatic separators. *C* = conductors; *NC* = nonconductors; *M* = middling; *DC* = high-voltage dc electrodes; *AC* = high-voltage ac wiper (electrodes); *B* = brush; *S* = splitter; ○ = negatively charged particles; ● = positively charged particles; *BL* = belt.

"wall" is electrified with the opposite potential, and product splitters running parallel to the electrode walls at the base of unit separate materials attracted to the oppositely charged tube electrodes. This arrangement is illustrated in Fig. 19-57.

Belt-type. Figure 19-58 illustrates a horizontal belt-type separator equipped with fast-moving belts that travel in opposite directions adjacent to suitably placed plate electrodes of the opposite polarity. Material is fed into a thin gap between two parallel electrodes. The particles are swept upward by a moving open-mesh belt and conveyed in opposite directions, thus facilitating particle charging by contact with other particles. The electric field attracts particles up or down depending on their charge. The moving belts transport the particles adjacent to each electrode toward opposite ends of the separator.

The common types of other industrial electrostatic separators employ charging by conductive induction and/or ion bombardment. Figure 19-56 illustrates the principles of application.

Conductive-Induction Machines Electrostatic separators exploiting the principle of conductive induction will generally use the following electrode designs:

Plate separators. These separators introduce material typically by gravity onto a grounded-metal slide in front of which is placed a static electrode of large surface area. The elaborate contact electrification used in earlier free-fall electrostatic separators was avoided in later devices by incorporating the slide principle. Separation occurs by particles selectively acquiring an induced charge from the grounded plate and then being attracted in the direction of the charged electrode. Refer to Fig. 19-59.

Screen-plate separators. These include a metal slide at ground potential that is extended with a conducting screen of suitable screen-opening size to allow easy passage of the largest grains being treated. A stationary electrode is placed above the slide and screen sections as

FIG. 19-58 Triboelectric separators. Belt-type electrostatic separator for separation of carbon from fly ash. (*Courtesy of Separation Technologies, Inc.*)

FIG. 19-57 Triboelectric separators. V-Stat electrostatic separator for silica removal from industrial minerals. (*Courtesy of Carpco, Inc.*)

shown in Fig. 19-56e. Particles capable of assuming an induced charge from the slide are attracted by the electrode and prevented from passing through the screen grid; the other particles pass through the screen unaffected.

Ion-Bombardment Machines *Conductive roll (drum) separators.* These separators are, by far, the most widely employed industrial-machine type. The electrostatic elements of these machines consist of a conductive rotating drum at ground potential coupled with one or more high-voltage ionizing electrodes. Suitably placed nondischarging (static) electrodes are often used in conjunction with an ionizing electrode to create a static field which aids centrifugal force in removing conductive particles from the drum surface.

Drum construction is typically of carbon or stainless steel when treating granular materials minus 1 mm in size. Drum diameter has ranged from 0.150 to 0.360 m, while drum length varies from 0.460 to 3.050 m in industrial ion-bombardment (high-tension) machines.

Feeding these separators is accomplished by vibratory, belt, rotary spline, or gravity methods, depending on the particle size being treated. Vibratory and belt feeding techniques are preferred for coarser sizes, and rotary spline and gravity methods are normally used for finer materials. Exceptions to this generalization can be observed in plant practice. The ionizing electrodes employed in these machines vary considerably in appearance, but all produce a corona.

An alternating-current electrode system referred to in the industry as a "wiper" is often installed in the nonconductor product-collection section behind each drum. The function of the wiper is to use an ac corona to neutralize the charge on the nonconductor particles pinned

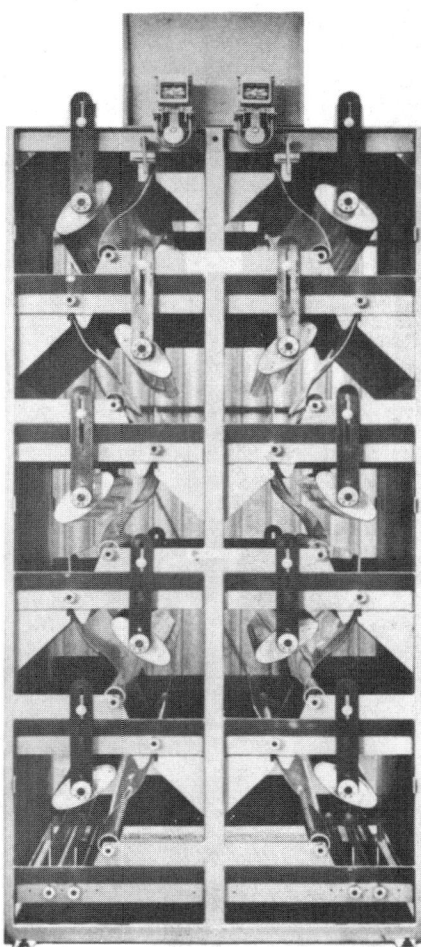

FIG. 19-59 Conductive-induction plate-type electrostatic separator. (*Courtesy of Mineral Technology, Ltd.*)

FIG. 19-60 Internal configuration of roll-type electrostatic separator, ionizing mode. (*Courtesy Carpco, Inc.*)

High-voltage controllers which regulate primary input voltage to the rectifier and wiper transformer and house primary current-limiting protection, meters, and instrumentation are designed for local or remote operation.

Machine Capacities Table 19-25 presents machine-capacity information for electrostatic separators.

Applications of Electrostatic Separation

Mineral Beneficiation Electrostatic methods are widely used in the processing of ores with mineral concentrates. Generally, electrostatic separation is used as a part of an overall flow sheet comprising various combinations of physical separation procedures. It is particularly well established in the processing of heavy-mineral beach sands from which are recovered ilmenite, rutile, zircon, monazite, silicates, and quartz. High-grade specular hematite concentrates have been recovered at rates of 1000 tons/h in Labrador. Applications also include processing tin ores to separate cassiterite from columbite and ilmenite. Refer to Fig. 19-61a.

Charging by ion bombardment is the technique used in most mineral separations. The conductive-induction (nonionizing) plate types of separators have also been used. Applications of this device in the minerals industry include its use as a final cleaning step when concentrating rutile and zircon.

Generally, separators of the conductive-induction type have a lower capacity per unit length of electrode than the ion-bombardment (ionizing) type of apparatus, and multipass operation is typically required. This disadvantage is offset by the ability of these separators to (1) produce high-grade concentrates from ore materials that are otherwise difficult to process and (2) process a coarser material than competitive

to the surface of the drum and thereby reduce the workload for mechanically operated brushing systems.

Ion-bombardment machines are available in horizontal and vertical (stacked) configurations. Horizontal units are preferred for large-tonnage applications in which machines are arranged in rows for ease of maintenance and operation. Stacked units up to four rolls high have been used to reduce material-handling costs in multipass treatment schemes when the material is capable of being passed vertically from one roll to another by gravity. Internal details of industrial roll-type separators are shown in Fig. 19-60.

Power Supplies High-voltage ac and dc power supplies for electrostatic separators are usually of solid-state construction and feature variable outputs ranging from 0 to 30,000 V for ac wiper transformers to 0 to 60,000 for the dc supply. The maximum current requirement is approximately 1.0 to 1.5 mA/m of electrode length. Power supplies for industrial separators are typically oil-insulated, but smaller dry-epoxy-insulated supplies are also available.

Features common to most high-voltage dc power supplies include reversible polarity, short-circuit and current-limiting protection, and automatic residual-charge dissipation to ground.

TABLE 19-25 Machine Capacities of Electrostatic Separators for Mineral Applications

Type	Capacity, t/h
A) Trioelectric separators	
• Belt type (1 pass)	10
• Tube type (1 pass)	20
B) Conductive-induction electrostatic separators	
Plate separator: basis, 5 pass × 2 start; 1.42 m in length per start; capacity, 900 kg/(h·m)	2.0
C) Ion-bombardment electrostatic separators	
2-m units-basis:	
• 1 pass × 6 starts × 2 m × 2000 kg/(h·m)	24
• 3 pass × 2 start × 2 m	8

NOTE: Capacity information is based on the treatment of industrial minerals having a specific gravity of 2.6 to 4.0.

(a)

(b)

FIG. 19-61   (*a*) Mineral separation: high-capacity 4-roll (250 mm dia × 2000 mm long) separator featuring all-start × 2-pass separation. Siz-roll units also available. (*b*) Recycling separator for nonferrous metals and plastics: 4-roll (350 mm dia × 1500 mm long) separator featuring 2-start × 2-pass or 1-start × 4-pass separation. (*Courtesy Carpco, Inc.*)

processes such as froth flotation. These units are used as a final cleaning stage for rutile and zircon.

Electrostatic-type separation is being tested as an alternative to the presently used process of flotation of pebble phosphates for coarser-size fractions. Advantages sought include reduced reagent costs, a lower water requirement, and fewer tailings-disposal problems when a part of the flotation circuit is eliminated. The largest application of triboelectric separation is in the salt industry where sodium and potassium salts are separated after preconditioning.

Plastic and Metals Recycling   Electrostatic separation has been increasingly applied to recover nonferrous metals from industrial plastics (telephone and communication scrap). It also is an important step in the recycling of beverage bottles to reject any remaining nonferrous metals. Both of these recycling applications make use of roll-type ion-bombardment separators (Fig. 19-61b).

A new application of triboelectric separation involves the separation of PVC from PET and other plastics. Recent developments in pre-charging technology permit PVC to assume a strong negative charge and be removed efficiently from properly protected mixed plastic feedstocks (Fig. 19-62).

Other Applications   Electrostatic separators have been used to separate a number of different types of materials not only on the basis of differences in dielectric properties but also in combination with differences in surface conductivity and shape factors. Among these operations are seed sorting, cleaning of spices, separating of pill coatings from base materials, removal of textile from reclaimed plastics, and separation of paper and plastic. Electrostatic separation has also been adapted for use in classification and sizing when elongated particles or extremely fine sizes cause difficulty in conventional dry-screening applications.

Typical Operating Conditions   Table 19-26 presents some typical values of important operating conditions for the separation of several different types of feed materials. In considering candidate processes for a given separation job, the table can sometimes be helpful in showing that materials of similar properties and/or economic value can be treated by electrostatic separation.

FIG. 19-62   New Triboelectric separator for separation of PVC from other plastics. (*Courtesy Carpco, Inc.*)

TABLE 19-26   Typical Operating Conditions for Electrostatic Separations

Type of particle charging	Feed	Separation	Type of separator	Feed temperature, °C	Feed size, mm	Feed rate, metric tons per hour per start*	No. of stages of separation
Triboelectric	Silica from limestone	Reduction of quartz by 80–90%	Tube type	80–100	−1.0 + 0.015	20	1
	Florida pebble-phosphate flotation conc.	Residual silica from pebble phosphate	Tube type	70–90	−1.0 + 0.10	10–15	1
Conductive induction	Zircon or rutile concentrate (eastern Australia)	Residual conductor minerals from rutile and zircon; upgrading of concentrate from 98.95 to 99.35% zircon at 92% recovery	Plate	50–80	−0.21 + 10.074	0.6–0.7	5–10
Ion bombardment	Heavy-mineral concentrate	Conductor minerals (ilmenite, rutile) from non-conductor minerals (zircon, monazite, aluminum silicates, quartz and others)	Roll	120	−1.0 + 0.04	2.5	3–6
	Iron ore	Iron oxides from quartz and silicates	Roll	120	−1.0	6–7	2–4
	Tungsten concentrate	Scheelite from iron oxides and other conductor minerals	Roll	150	−0.6	1.0–1.5	3
	Chrome ore	Chromite from silica and silicates	Roll	120	−0.85		
	Chopped wire	Metal from plastic insulation	Roll	Ambient	−12.5	1.5–2.0	2–4
	Metal powder	Removal of nonmetallic impurities	Roll	Ambient to 120	−0.20	1–2	2–4

*To convert metric tons per hour per start to kilograms per second per start, multiply by 0.2778.

FLOTATION

GENERAL REFERENCES: "Flotation," *SME Mineral Processing Handbook,* section 5, Society for Mining, Metallurgy and Exploration, Littleton, CO, 1985. "Flotation Operating Practices and Fundamentals," *Proceedings of XIX International Mineral Processing Congress,* vol. 3, Society for Mining, Metallurgy and Exploration, Littleton, CO, 1995. R. D. Crozier, *Flotation: Theory, Reagents and Ore Testing,* Pergamon Press, 1992. J. A. Finch and G. S. Dobby, *Column Flotation,* Pergamon Press, 1990. D. W. Fuerstenau (ed.), *Froth Flotation,* American Institute of Mining, Metallurgical, and Petroleum Engineers, New York, 1962. D. W. Fuerstenau, "Fine Particle Flotation," *Fine Particles Processing,* P. Somasundaran (ed.), American Institute of Mining, Metallurgical, and Petroleum Engineers, New York, 1980. D. W. Fuerstenau, "Where We Are in Flotation Chemistry after 70 Years of Research," *Proceedings of XIX International Mineral Processing Congress,* vol. 3, Society for Mining, Metallurgy and Exploration, Littleton, CO, 1995. M. C. Fuerstenau (ed.), *Flotation,* American Institute of Mining, Metallurgical, and Petroleum Engineers, New York, 1976. Kenneth J. Ives (ed.), *The Scientific Basis of Flotation,* NATO ASI Series E, Applied Sciences, no. 75, The Hague, Boston, Martinus Nijhoff, 1984. K. J. Ives and H. J. Bernhardt (eds.), "Flotation Processes in Water and Sludge Treatment," *Water Science and Technology,* vol. 31, nos. 3–4, 1995. S. G. Malghan, "Typical Flotation Circuit Configurations": *Design and Installation of Concentration and Dewatering Circuits,* A. L. Mular and M. A. Anderson (eds.), Society of Mining Engineers, Littleton, CO, 1986. K. A. Matis (ed.), *Flotation Science and Engineering,* Marcel Dekker, 1995. P. Mavros and K. A. Matis (eds.), *Innovations in Flotation Technology,* NATO ASI Series E, Applied Sciences, vol. 208, Kluwer Academic Publishers, 1992. K. S. Moon and L. L. Sirois, *Theory and Application of Column Flotation,* CANMET Report No. 87-7E, Energy, Mines and Resources Canada, 1988. M. W. Ranney (ed.), *Flotation Agents and Processes,* Noyes Data Corporation, 1980. J. B. Rubinstein, *Column Flotation: Processes, Designs and Practices,* Gordon Breach Science Publishers, 1994. K. V. S. Sastry (ed.), *Column Flotation '88,* Society of Mining Engineers, Littleton, CO, 1988. K. V. S. Sastry and M. C. Fuerstenau (eds.), *Challenges in Mineral Processing,* Society of Mining Engineers, Littleton, CO, 1989. H. J. Schulze, *Physico-Chemical Elementary Processes in Flotation,* Elsevier, 1984. F. Sebba, *Ion Flotation,* Elsevier, 1962. G. Barbery, "Engineering Aspects of Flotation in the Minerals Industry: Flotation Machines, Circuits and their Simulation," in *The Scientific Basis of Flotation,* Kenneth J. Ives (ed.), NATO ASI Series E, no. 75, Martinus Nijhoff, 1984, p. 289. R. Clayton, G. J. Jameson, and E. V. Manlapig, "The Development and Application of the Jameson Cell," *Minerals Engineering,* **4,** 1991, p. 925. V. A. Glembotskii, V. I. Klassen, and I. N. Plaksin, *Flotation,* Primary Sources, New York, 1972. C. C. Harris, "Flotation Machines" in *Flotation,* M. C. Fuerstenau (ed.), American Institute of Mining, Metallurgical, and Petroleum Engineers, New York, 1976, p. 753. P. Mavros, "Mixing and Hydrodynamics in Flotation Cells" in *Innovations in Flotation Technology,* P. Mavros and K. A. Matis (eds.), NATO ASI Series E, **208,** Kluwer Academic Publishers, 1992, p. 211. J. D. Miller et al., "Design and Operating Variables in Flotation Separation with the Air-Sparged Hydrocyclone," *Proc. XVI International Mineral Processing Congress,* K. S. E. Forssberg (ed.), Elsevier, 1988, p. 499. D. C. Yang, "A New Packed Column Flotation System" in *Column Flotation '88,* K. V. S. Sastry (ed.), Society of Mining Engineers, Littleton, 1988, p. 257. P. Young, "Flotation Machines," *Mining Magazine,* **146,** 1982, p. 35. A. J. Zouboulis, K. A. Matis, and G. A. Stalidis, "Flotation Techniques in Waste Water Treatment" in *Innovations in Flotation Technology,* P. Mavros and K. A. Matis (eds.), NATO ASI Series E, vol. 208, Kluwer Academic Publishers, 1992, p. 475.

INTRODUCTION

Mixed liberated particles can be separated from each other by flotation if there are sufficient differences in their wettability. The flotation process operates by preparing a water suspension of a mixture of relatively fine-sized particles (smaller than 150 micrometers) and by contacting the suspension with a swarm of air bubbles of air in a suitably designed process vessel. Particles that are readily wetted by water (hydrophilic) tend to remain in suspension, and those particles not wetted by water (hydrophobic) tend to be attached to air bubbles, levitate (float) to the top of the process vessel, and collect in a froth layer. Thus, differences in the surface chemical properties of the solids are the basis for separation by flotation.

Surfaces that do not have strong surface chemical bonds that were broken tend to be nonpolar and are not readily wetted. Substances such as graphite and talc are examples that can be broken along weakly bonded layer planes without rupturing strong chemical bonds. These solids are naturally floatable. Also, polymeric particles possess

nonpolar surfaces and are naturally hydrophobic. By contrast, most naturally occurring materials are polar and exhibit high free energy at the polar surface. The polar surfaces react strongly with water and render those particles naturally hydrophilic. The relative wettability of the solids in a mixture can be enhanced by the addition of various surface chemical agents that are adsorbed selectively on the particle surface.

Mineral Applications. The flotation process is most widely used in the mineral process industry to concentrate mineral values in the ores.

A U.S. Bureau of Mines survey covering 202 froth flotation plants in the United States showed that 198 million tons of material were treated by flotation in 1960 to recover 20 million tons of concentrates which contained approximately $1 billion in recoverable products. Most of the world's copper, lead, zinc, molybdenum, and nickel are produced from ores that are concentrated first by flotation. In addition, flotation is commonly used for the recovery of fine coal and for the concentration of a wide range of mineral commodities including fluorspar, barite, glass sand, iron oxide, pyrite, manganese ore, clay, feldspar, mica, sponumene, bastnaesite, calcite, garnet, kyanite, and talc.

Other Applications. In addition to the minerals industry, flotation is finding a variety of new applications in other fields. The next largest application is for wastewater treatment to remove particulate, organic, and biological contaminants. Other applications include extraction of metallic values or removal of heavy metal compounds from hydrometallurgical streams by precipitate flotation, recovery of bitumen from tar sands, deinking of waste paper, recovery of solids from white water in paper making, recovery of glass sands from industrial wastes, removal of impurities from peas, removal of ergot from rye, separation of proteins from milk, and clarification of fruit juices. Ion flotation and foam fractionation are the slight modifications in the basic flotation process and are sometimes referred to as "adsorptive bubble separation." These methods are used for the extraction of soluble species.

GENERAL ASPECTS

Unit operation of flotation is based on two major steps: (1) *conditioning* and (2) *separation,* as is schematically depicted in Fig. 19-63. During the first step, the *slurry* or the *pulp,* consisting of particles to be separated, the particle size of which is already properly adjusted, is fed to the conditioning unit, to which the necessary flotation reagents are added. The main purpose of the conditioning step is to create physical-chemical conditions for achieving appropriate selectivity between particle species that are to be separated. The second step is then intended to generate and introduce air bubbles into the process vessel for contacting them with particulate species so as to affect their separation by flotation. Particles attached to the air bubbles are in most applications removed from the process vessel as froth. Accordingly,

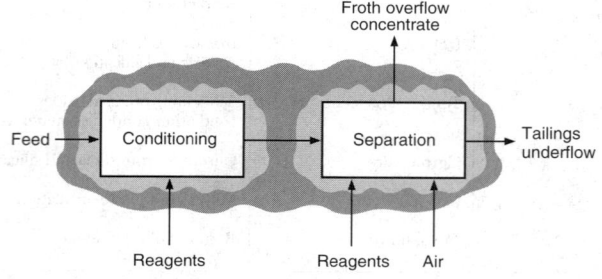

FIG. 19-63 Basic steps in a flotation system.

the unit operation of flotation is often referred to as *froth flotation.* The froth overflow stream is called a *concentrate* in the minerals industry, and the slurry underflow is termed *tailings.* Depending on the application, these two steps may be carried out in two distinctly different process units or in one combined unit.

Flotation Reagents.   Three types of chemical reagents are used during the froth flotation process: collectors, frothers, and modifiers.

Collectors.   These are surface-active agents that are added to the flotation pulp, where they adsorb selectively on the surface of the particles and render them hydrophobic. A convenient classification of the commonly used collectors is shown in Fig. 19-64. The nonionizing collectors (fuel oils and kerosene) are practically insoluble in water and cause the particles to become hydrophobic by covering them with a thin film. The ionizing collectors dissociate into ions in water and are made up of complex heteropolar molecules in that the molecule contains both a nonpolar hydrocarbon group with pronounced hydrophobic properties and a polar group with hydrophilic properties. The ionizing collectors adsorb either physically or chemically on the particle surface and can further be classified into anionic or cationic collectors depending on the nature of the nonpolar hydrocarbon group. Common examples of the ionizing collectors include fatty acids, long-chain sulfates, sulfonates and amines, xanthates, and dithiophosphates. Dosage requirements for collectors depend on the mechanisms by which they interact with the particle surface, but just enough is needed to form a monomolecular layer. As a rule, high dosages are required for nonionizing collectors and physisorbing ionizing collectors (in the order of 0.1 to 1 g of reagent per kg of solids) and low dosages for chemisorbing ionizing collectors (0.01 to 0.1 g of reagent per kg of solid). Addition of excess quantities of a collector is not desirable because it results in reducing the selectivity and increasing the cost.

Frothers.   These are also surface-active agents added to the flotation pulp primarily to stabilize the air bubbles for effective particle-bubble attachment, carryover of particle-laden bubbles to the froth, and removal of the froth. The frother action is similar to the ionizing collectors except that they concentrate primarily at the air-liquid interface. Commonly used frothers are pine oil, cresylic acid, polypropylene glycol, short-chain alcohols, and 5- to 8-carbon aliphatic alcohols. Quantities of frothers required are usually 0.01 to 0.1 g per kg of solids.

Modifiers.   Flotation modifiers include several classes of chemicals.

1. *Activators.*   These are used to make a mineral surface amenable to collector coating. Copper ion is used, for example, to activate sphalerite (ZnS), rendering the sphalerite surface capable of absorbing a xanthate or dithiophosphate collector. Sodium sulfide is used to coat oxidized copper and lead minerals so that they can be floated by a sulfide mineral collector.

2. *pH regulators.*   Regulators such as lime, caustic soda, soda ash, and sulfuric acid are used to control or adjust pH, a very critical factor in many flotation separations.

3. *Depressants.*   Depressants assist in selectivity (sharpness of separation) or stop unwanted minerals from floating. Typical are sodium or calcium cyanide to depress pyrite (Fe_2S_2) while floating galena (PbS), sphalerite (ZnS), or copper sulfides; zinc sulfate to depress ZnS while floating PbS; sodium ferrocyanide to depress copper sulfides while floating molybdenite (MoS_2); lime to depress pyrite; sodium silicate to depress quartz; quebracho to depress calcite ($CaCO_3$) during fluorite (CaF_2) flotation; and lignin sulfonates and dextrins to depress graphite and talc during sulfide flotation.

4. *Dispersants and flocculants.*   These are important for the control of slimes that sometimes interfere with the selectivity and increase reagent consumption. For example, soda ash, lime sodium silicate, and lignin sulfonates are used as dispersants, and starch and polyacrylamide are used as flocculants.

Quantities of modifying agents used vary widely, ranging from as little as 0.01 to 0.1 g/kg to as high as 1 to 2 g/kg of solids, depending upon the reagent and the metallurgical problem.

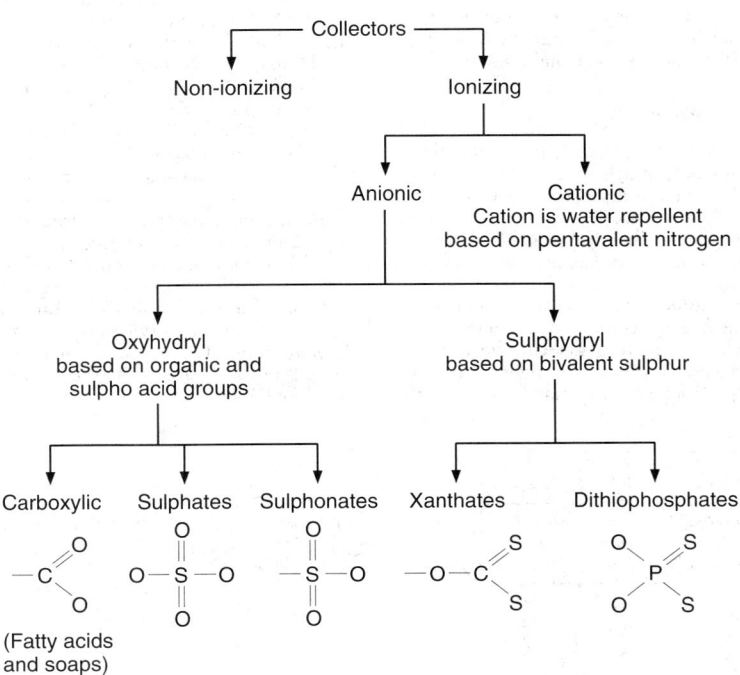

FIG. 19-64   Classification of collectors (after Glembotakii et al., 1972).

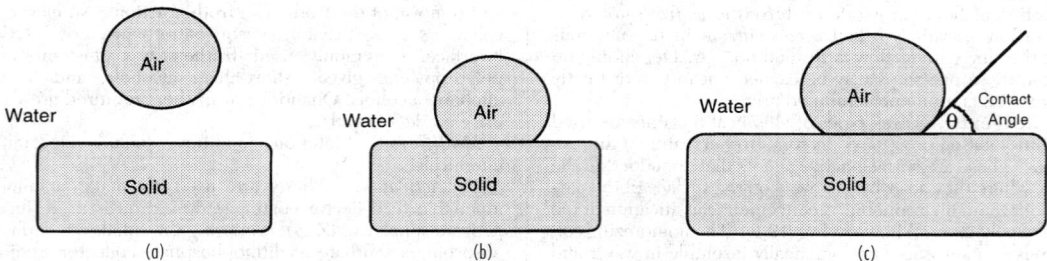

FIG. 19-65 Schematic representation of air bubble-water-solid particle system: (a) before, (b) after particle-bubble attachment, and (c) equilibrium force balance.

FUNDAMENTALS

Flotation is a physical process involving relative interaction of three phases: solid, water, and air. An understanding of the wettability of the solid surface, physical surface, and chemical phenomena by which the flotation reagents act and the mechanical factors that determine particle-bubble attachment and removal of particle-laden bubbles, is helpful in designing and operating flotation systems successfully.

Thermodynamics of Wetting. The fundamental objective of flotation is to contact solid particles suspended in water with air bubbles (Fig. 19-65a) and cause a stable bubble-particle attachment (Fig. 19-65b). It is seen that attachment of the particle to an air bubble destroys the solid-water and air-water interfaces and creates air-solid interface. The free energy change, on a unit area basis, is given by

$$\Delta G = \gamma_{AS} - (\gamma_{SW} + \gamma_{AW}) \qquad (19\text{-}21)$$

where γ terms are the interfacial tensions of the air-solid (AS), solid-water (SW), and air-water (AW) interfaces, respectively. A force balance for the air-water-solid particle system (Fig. 19-65c) yields the familiar Young's Equation

$$\gamma_{AS} = \gamma_{SW} + \gamma_{AW} \cos \theta \qquad (19\text{-}22)$$

where θ is the contact angle (measured through the water phase). It must be seen that the contact angle is an equilibrium measure of the interfacial energy of the air-water-solid system. Combining the above two equations, one obtains

$$\Delta G = \gamma_{AW}(\cos \theta - 1) \qquad (19\text{-}23)$$

Thus, for any finite value of the contact angle, the free energy change becomes negative and particle-bubble attachment can take place. As mentioned above, polar solids have high surface energy and are wet by water. Therefore, the contact angle is zero. The wettability of solids can be controlled through adsorption of chemical reagents, which can change the interfacial tensions, so that the contact angle becomes finite and flotation can take place.

Physical-Chemical Phenomena. Several physical-chemical phenomena occur when chemical reagents are added to an air-water solid system due to the interaction of the reagents with the air-water, water-solid, and air-solid interfaces. This causes changes in the solution chemistry in which the particles are suspended. Some of the important phenomena that occur due to the addition of reagents include: solubility and dissociation of reagents in water, change of pH of the suspension, change of air-water surface tension, physical and chemical adsorption of the dissolved species on the solid surfaces due to hydrogen bond formation, electrostatic interactions, hydrophobic bonding, chemical bond formation, and fixation of reagent species in the solid lattice. All these phenomena in essence result in affecting the contact angle and flotation nature of solid particles and their attachment to air bubbles. A simplified pictorial representation of collector adsorption on particle surface, action of frother on the air bubble formation, and particle-bubble contact is shown in Fig. 19-66. An adequate understanding of the role played by the reagents and their proper choice to create the desired conditions is paramount to successful flotation.

Particle-Bubble Attachment. In the above, principles leading to creation of desired hydrophobicity/hydrophilicity of the particles has been discussed. The next step is to create conditions for particle-bubble contact, attachment, and their removal, which is simply described as a combination of three stochastic events with which are associated the probability of particle-bubble collision, probability of attachment, and probability of retention of attachment. The first term is controlled by the hydrodynamic conditions prevailing in the flotation unit. The second is determined by the surface forces. The third is dependent on the survival of the laden bubble by liquid turbulence and impacts by the other suspended particles. A detailed description of the hydrodynamic and other physical aspects of flotation is found in the monograph by Schulze (1984).

Process Variables. There are a number of variables that govern the flotation process. These include particle characteristics (size, shape, and chemical and mineralogical composition), chemical variables (type and amount of flotation reagents added), flotation machine variables (equipment size, internal geometry of the device, speed of operation, etc.) and operating variables (slurry feed rate and percent solids). A combined effect of all these variables can be represented by two independent variables—specific flotation rate (representing the rate of flotation of particles per unit time) and residence time of the pulp in the flotation device—and two dependent variables of grade (composition of the desired component) and recovery (ratio of the weight of the desired component in the froth product to that in the feed) of the froth product.

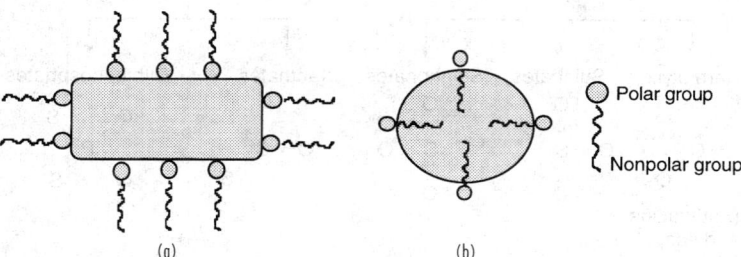

FIG. 19-66 Schematics of (a) collector adsorption at the particle-water interface and (b) action of the frother.

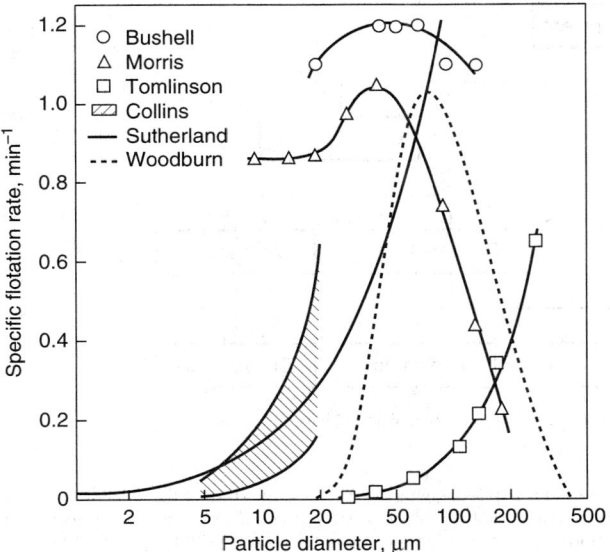

FIG. 19-67   Effect of particle diameter on specific flotation rate (Fuerstenau, 1980).

Several laboratory procedures are available to investigate the flotation response of any solid-solid system and in generating basic data for the selection and sizing of the flotation units and circuits. Also available are various process models for flotation with varying degrees of sophistication and representation. These process models can form a quantitative basis during all stages of engineering flotation systems.

Finally, numerous types of flotation reagents and flotation equipment with different design details are currently available, and their proper choice has to be made depending on the kind of separation task one has on hand. Monographs listed in the general references provide a good starting point towards an understanding of these aspects.

Effect of Particle Size.   Particle size is the most significant variable of flotation separation. The effect of particle size on the flotation rate is shown in Fig. 19-67. Particles in the size range of 20 to 60 μm have the highest flotation rate. Larger particles, being heavy, cannot be easily levitated and recovered, even though proper thermodynamic conditions might exist. In contrast, as particles become small, they become lighter and their surface-to-volume ratio becomes large. There are several factors that enter into making the flotation of small particles quite inefficient (Fig. 19-68).

EQUIPMENT

Various types of flotation machine designs can be classified into different categories based on the methods used for the generation and introduction of air bubbles into the equipment (Fig. 19-69). Each of the techniques of air bubble generation and particle-bubble contact along with the special features associated with different kinds of equipment has its own advantages and limitations. These must be considered carefully in selecting the equipment for a specific application. Individual manufacturers can provide basic help in selecting the equipment.

Electrolytic Flotation Units.   Electrolytic or electroflotation is based on the generation of hydrogen and oxygen bubbles in a dilute aqueous solution by passing direct current between two electrodes. Choice of electrode materials include aluminum, platinized titanium, titanium coated with lead dioxide, and stainless steel of varying grades. Figure 19-70 illustrates the basic arrangement of an electrolytic flotation unit.

Electrical power to the electrodes is supplied at a low voltage potential of 5 to 10 volts. The power consumption is in the range of 0.5 to 0.7 kW/m^2 of flotation tank surface area depending on the conductiv-

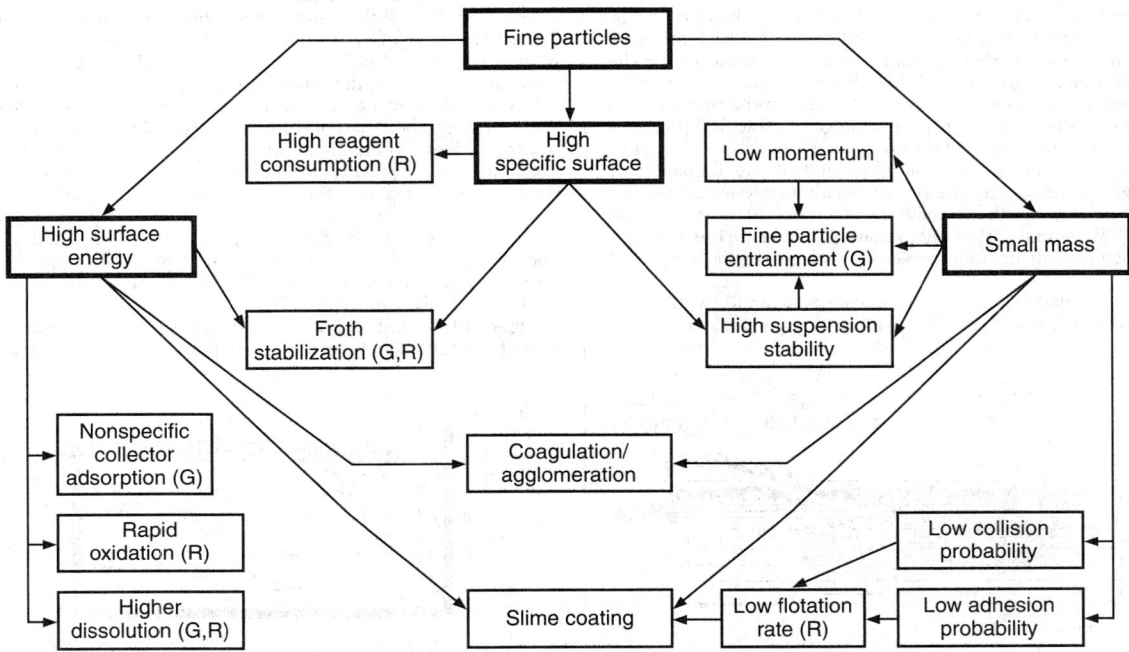

FIG. 19-68   The schematic diagram showing the relationship between the physical and chemical properties of fine particles and their behavior in flotation. (G) and (R) refer to whether the phenomena affects grade and/or recovery. The arrows indicate the various factors contributing to a particular phenomena observed in flotation of fine particles (Fuerstenau, 1980).

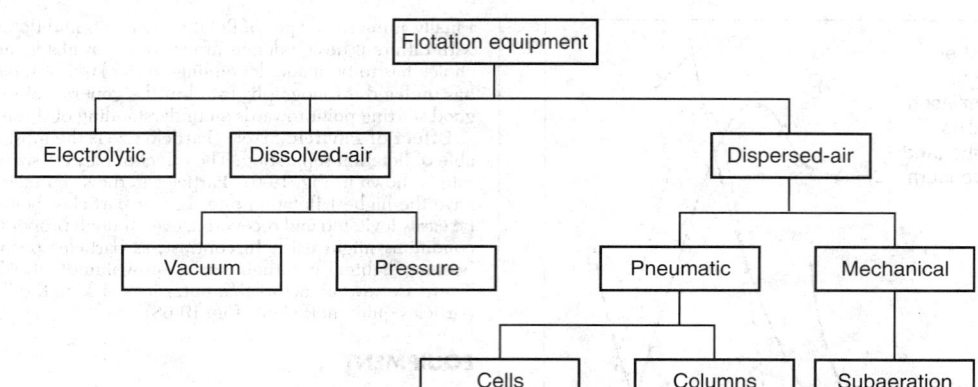

FIG. 19-69   Classification of flotation equipment based on the generation and introduction of air bubbles.

ity of the liquid and the distance between the electrodes. Such a unit produces approximately 50 to 1 of gas/h/m² of tank area. The main drawback of the electroflotation units is associated with the electrodes in terms of their fouling requiring mechanical cleaning devices and their consumption needing replacement at frequent intervals.

The bubble size in these cells tends to be the smallest (10 to 50 µm) as compared to the dissolved-air and dispersed-air flotation systems. Also, very little turbulence is created by the bubble formation. Accordingly, this method is attractive for the separation of small particles and fragile flocs. To date, electroflotation has been applied to effluent treatment and sludge thickening. However, because of their bubble generation capacity, these units are found to be economically attractive for small installations in the flow-rate range of 10 to 20 m³/h. Electroflotation is not expected to be suitable for potable water treatment because of the possible heavy metal contamination that can arise due to the dissolution of the electrodes.

Dissolved-Air Flotation Units.   Dissolved-air flotation entails saturating the process stream with air and generating air bubbles by releasing the pressure. Particle-bubble contact is achieved by the direct nucleation and growth of air bubbles on the particles, and very little mechanical agitation is employed. The dissolved-air precipitates in the form of fine bubbles in the size range of 20 to 100 µm. This method of air bubble generation does not require the addition of frother-type chemical reagents and often limits the total quantity of aeration possible. As such, dissolve-air flotation systems are used to treat process streams with low solids concentration (0.01 to 2 percent by volume). Vacuum flotation and pressure flotation are the two main types of dissolved-air flotation processes, with the latter being most widely used.

In vacuum flotation, the process stream is saturated with air at atmospheric pressure and introduced to the flotation tank on which a vacuum is applied, giving rise to the generation of the air bubbles. The process can be run only as a batch process and requires sophisticated equipment to produce and maintain the vacuum. By and large, the amount of air released during flotation is limited by the vacuum achievable.

In contrast to vacuum flotation, dissolved-air flotation units can be operated on a continuous basis by the application of pressure. This consists of pressurizing and aerating the process stream and introducing it into the flotation vessel that is maintained at the atmospheric pressure. The reduction of pressure results in the formation of fine air bubbles and the collection of fine particulates to be floated and removed as sludge.

Pressurization could be carried out on the entire feed stream (full-flow pressure flotation) or a fraction of the feed stream while the remainder is introduced directly without aeration into the flotation tank (split-flow pressure flotation). The split-flow system offers a cost saving over the full-flow units, since only a portion of the influent needs to be pressurized. In both cases, however, if the solid particles in the feed stream are flocculated before introducing to the flotation tank, the high shear during pressurization, aeration, and pressure release can destroy the flocs. Also, if the particle loading in the feed stream is high, both systems are susceptible to blockage of the air release devices. To minimize these problems, recycle-flow pressure flotation is often practiced (Fig. 19-71). In this process, the feed stream, flocculated or otherwise, is introduced directly into the process vessel, and part of the clarified effluent is pressurized, aerated, and recycled to the flotation tank in which it is mixed with the flocculated feed. The air bubbles are released as they attach to the flocs and float to the tank surface. The recycle-flow devices are found to offer the highest unit capacities.

Figure 19-72 illustrates a dissolved-air flotation plant flowsheet for water treatment. The flowsheet shows that the incoming raw water is

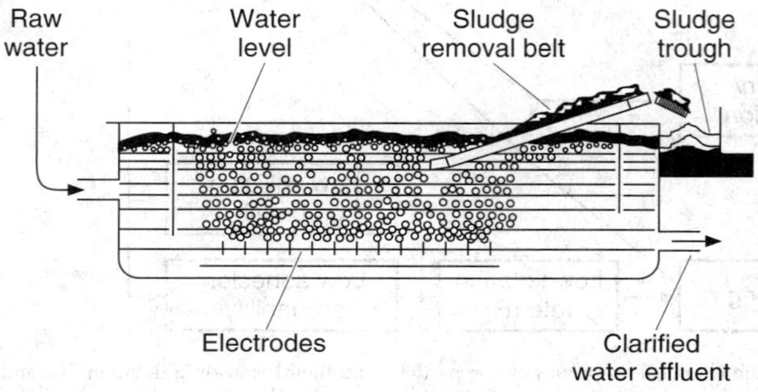

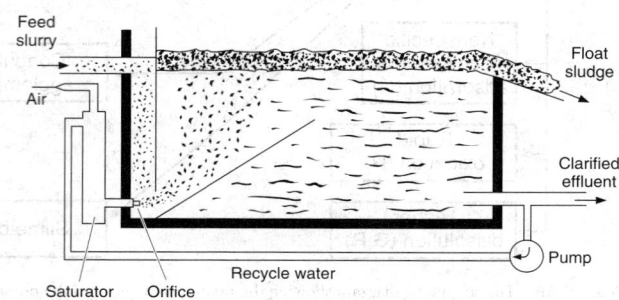

FIG. 19-70   Schematic diagram of an electrolytic flotation plant.

FIG. 19-71   Schematic diagram of a dissolved air flotation plant.

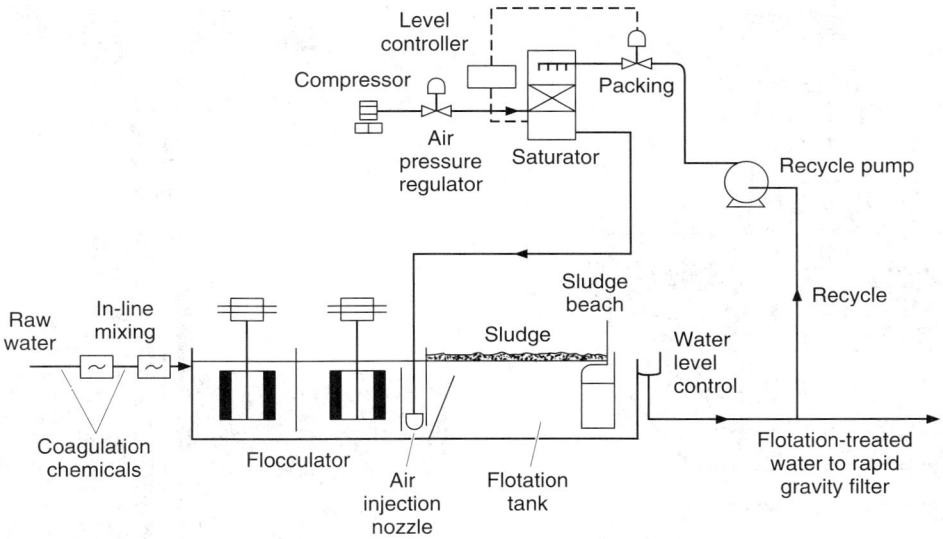

FIG. 19-72 Schematic diagram of a recycle dissolved-air flotation plant for water treatment.

conditioned with the addition of coagulation chemicals in a flocculator. This device pressurizes and aerates part of the treated water and recycles it to the flotation unit.

The dissolved-air flotation process is most commonly used for sewage and potable water treatment. It is also gaining popularity for the treatment of slaughterhouse, poultry processing, seafood processing, soap, and food processing wastes (Zoubulis et. al., 1991).

Dispersed-Air Flotation Units.   Dispersed-air flotation involves the generation of air bubbles, either pneumatically or by mechanical means. In both cases, relatively large air bubbles (at least 1 mm in size) are generated. In order to control the size and stability of air bubbles, frothers are added to the flotation devices. These devices represent the workhorses of the minerals industry in beneficiating metallic and nonmetallic ore bodies and cleaning of high-ash and high-sulfur coals in which feed streams contain relatively high percent solids (5 to 50 percent by volume), and high throughputs are maintained (in excess of 4000 t/h). Handling of large quantities of solids in these flotation devices requires such special design considerations as maintaining the solids in suspension, promoting particle-bubble collisions leading to attachment, providing a quiescent pulp region below the froth to minimize pulp entrainment, and finally providing sufficient froth depth to permit washing and drainage of hydrophilic solids entering the froth region.

Mechanical flotation machines are most commonly used in the mineral industry, while pneumatic column-type units are gaining popularity in recent years. Surveys by Harris (1976), Young (1982), Barbery (1982), and Mavros (1991) provide a detailed overview of the process-engineering aspects of mineral flotation devices in particular and systems in general.

Mechanical Cells.   Figure 19-73 presents a schematic representation of a typical mechanical device commonly known as a *flotation cell.* It is characterized by a cubic or cylindrical shape, equipped with an impeller surrounded by baffles with provisions for introduction of the feed slurry and removal of froth overflow and tailings underflow. The machines receive the supply of air through a concentric pipe surrounding the impeller shaft, either by self-aeration due to the pressure drop created by the rotating impeller or by air injection by means of an external blower. In a typical installation, a number of flotation cells are connected in series such that each cell outputs froth into a launder and the underflow from one cell goes to the next one. The cell design may be such that the flow of slurry from one cell to another can either be "restricted" by weirs or unrestricted.

The mechanical cells that are most widely used today in sulfide, coal, and nonmetallic flotation operations in the western hemisphere are made by Fagergren (by WEMCO Division of Envirotech Corporation), D-R Denver (by Denver Equipment Corporation of Sala International), Agitair (supplied by Galligher Ash Company), and Outokumpy (by Outokumpu Oy).

These machines provide mechanical agitation and aeration by means of a rotation impeller on an upright shaft. In addition, the Agitair and Denver cells also utilize air from a blower to help aerate the pulp.

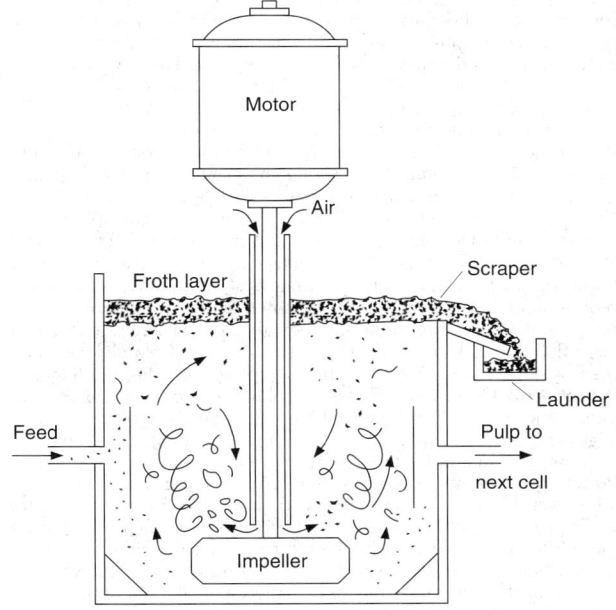

FIG. 19-73 Schematic of a mechanical flotation cell.

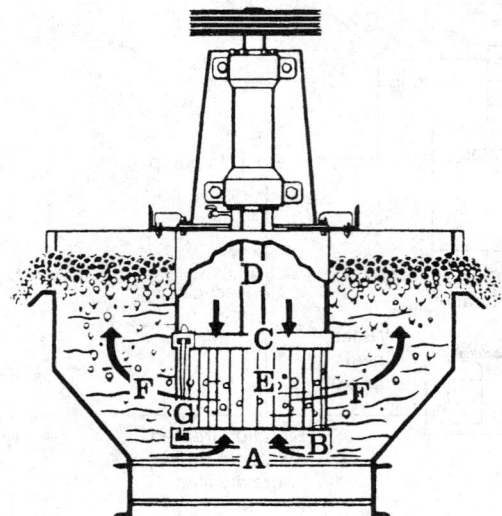

FIG. 19-74 Fagergren flotation machine.

In the Fagergren machine (Fig. 19-74), pulp is drawn upward into the rotor A by the rotor's lower portion B. Simultaneously the rotor's upper end C draws air down the standpipe D for thorough mixing with the pulp inside the rotor E. The aerated pulp is then expelled by a strong centrifugal force F. The shearing action of the stator G, a stationary cage fitting closely around the rotor, breaks the air into minute bubbles. This action uniformly distributes a large volume of air in the form of minute bubbles in all parts of the cell.

In the D-R Denver machine (Fig. 19-75), the pulp enters the top of the recirculation well A, while the low-pressure air enters through the air passage B. Pulp and air are intimately mixed and thrown outward by the rotating impeller C through the stationary diffuser D. The collector-coated mineral particles adhere to be removed in the froth product.

In the Agitair flotation machine (Fig. 19-76), the impeller is a flat rubber-covered disk with steel fingers extending downward from the periphery. A rubber-covered stabilizer eliminates dead spots in the agitation zone and improves bubble-ore contact. The degree of aeration is controlled by regulating air volume on each cell with an individual air valve. Air is supplied at 10×10^3 Pa (1.5 lbf/in²).

Modern mineral-processing plants are being designed with capacities on the order of 500 to 1000 kg/s (2000 to 4000 tons/h). The unit capacities of flotation machines now being manufactured are 10 times greater than those in common use 15 to 20 years ago (Fig. 19-76). Examples of large flotation cells that are currently available on the market include Denver Equipment (36.1 m³), Agitair (42.5 m³), and Wemco (85 m³). Larger-scale flotation machines offer advantages of lower installed cost, lower operating cost, and lower floor-space requirements. However, it should be noted that large flotation cells do not permit a reduction in the number of cells in a series. The use of large flotation cells does enable a fewer number of parallel rows and thereby permits a reduction in pumps, piping, and other auxiliaries.

Flotation Columns.   Flotation columns belong to the class of pneumatic devices in that air-bubble generation is accomplished by a gas-sparging system and no mechanical agitation is employed. Columns are built of long tubes of either circular or square cross sections that are commonly fitted with internal baffling. They are usually 10 or even 15 m high with a cross sectional area of 5 to 10 m². Figure 19-78 presents a schematic of a typical flotation column unit. Inputs to the column include preconditioned slurry feed and air and washwater spray, which are introduced at about two-thirds of the height from the bottom, in the bottom region, and at the top of the column, respectively. The outputs are froth overflow, consisting of hydrophobic particles from the top, and underflow from the bottom of the column, carrying the nonfloatable hydrophilic particles. Flotation columns

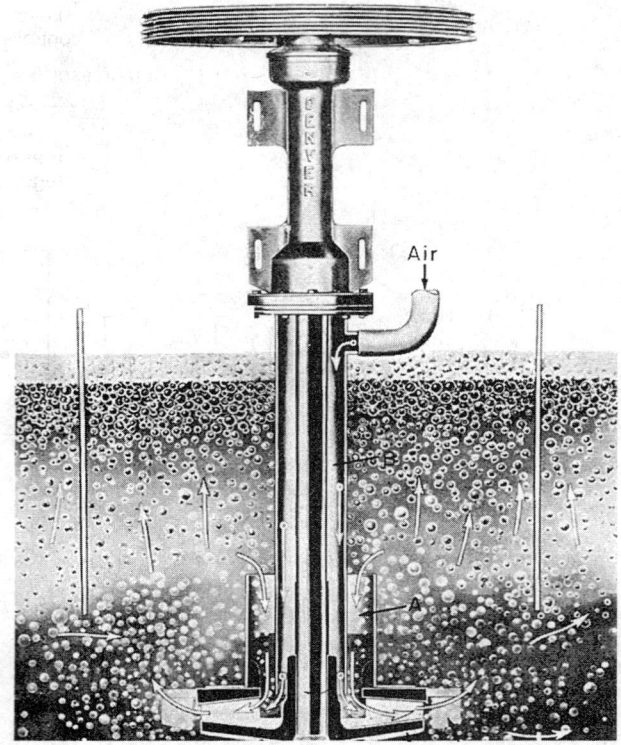

FIG. 19-75   D-R Denver flotation machine.

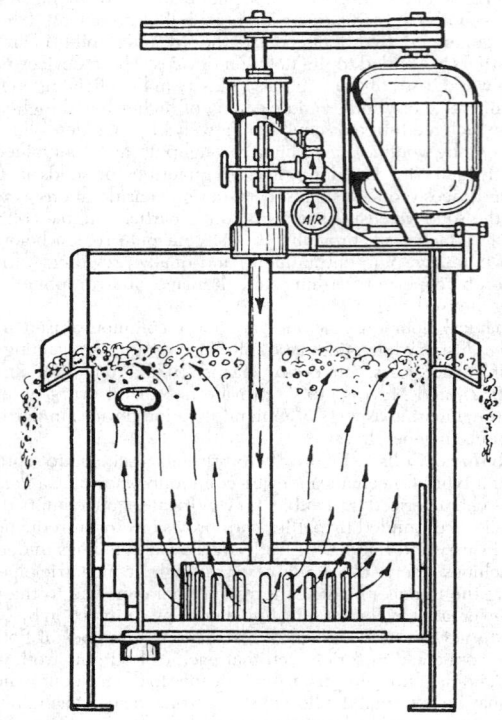

FIG. 19-76   Agitair flotation machine.

FIG. 19-77   Large flotation cell No. 165 AX 1500 Agitair, 42.5 m³ (1500 ft³). (*Courtesy of Caligher Ash Company.*)

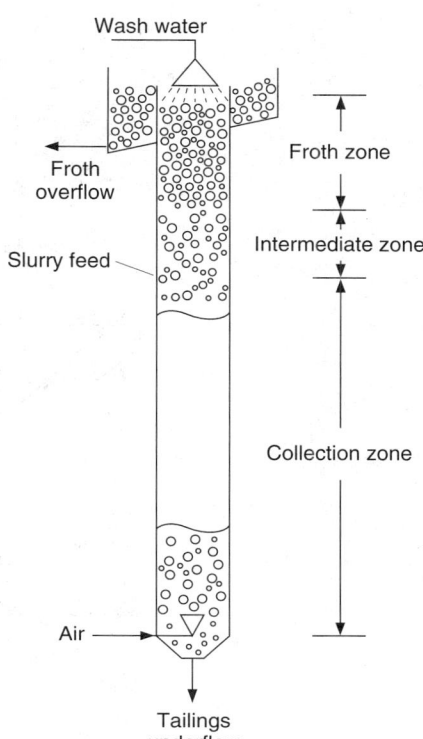

FIG. 19-78   Schematic of a flotation column.

make use of the countercurrent flow principle in that the swarm of air bubbles rises through the downward-flowing slurry during which time transfer of hydrophobic particles occurs between the slurry and bubble phases. The particle transfer process occurs in three distinct zones known as collection, intermediate, and froth zones. Properly designed baffles reduce short circuiting and promote better bubble-particle contact. Recovery of hydrophobic particles by the air bubbles takes place in the collection zone. Underflow removal rate and washwater addition rate are regulated such that there exists downward flow of slurry throughout the height of the column, thus ensuring that there is no bypass of the feed slurry in the upward direction. Further, the downward pattern of the flow of liquid helps in minimizing the entrainment of hydrophilic particles with the uprising air bubbles in the collection zone and in stripping the hydrophilic particles attached to the air bubbles in all three regions. All in all, the performance of the columns in terms of the recovery of hydrophobic particles and the grade of the froth concentrate is determined primarily by the slurry feed rate, air flow rate, and the surface area of the air bubbles.

Because of their inherently simple design, it is fairly common for flotation columns to be constructed in-house except for using the patented air-sparging systems. Several sparger designs are available that include simple porous plugs made from glass, stainless steel, and rubber, or more sophisticated venturi or in-line mixer configurations (Finch and Dobby, 1990). Some of the advantages claimed with flotation columns include improved separation performance, particularly for fine materials; low capital and operating costs; low plant floorspace requirements; and easy adaptability to automatic control. Flotation columns are being used in iron ore, copper, lead, zinc, and coal flotation applications and are expected to become even more popular because of their simplicity in construction and flexibility of operation.

Several modifications to the basic column design have become available over the years. Figure 19-79 shows three such designs. The first design variation is a packed column (Fig. 19-79a), which represents a minor variation to the basic column design in that it provides for corrugated plate-type packing. The packing feature enables uni-

form bubble size throughout the height, intimate particle-bubble contact, increased residence time of the slurry and bubble phases in the column, and a deeper froth zone (Yang, 1988). The Jameson cell (Fig. 19-79b), by contrast, is a combination column-cell design. It includes a vertical downcomer column in which the air and pulp are dispersed into a dense foam of fine bubbles creating a favorable environment for particle-bubble contact. The bubbly mixture is then discharged into a cell, which allows the separation of particle-laden bubbles from the pulp (Clayton et. al., 1991). Air-sparged hydrocyclone (Figure 19-79c) is a distinctly different design consisting of two concentric tubes with a conventional cyclone header at the top, providing for a tangential entry of the feed. As the feed slurry swirls down the inner porous tube through which air is sparged, collision between centrifuged particles and air bubbles takes place, leading to the recovery of hydrophobic particles. The bubble-hydrophobic particle aggregates are transported into the overflow stream as froth, while the nonfloating particles are removed with the underflow (Miller et al., 1988). Each of these and other design variations to the basic pneumatic flotation column concept is found to offer process improvements to specific applications.

Monographs by Sastry (1988), Finch and Dobby (1990), and Rubinstein (1994) provide an overview of the design and operational aspects of flotation columns.

FLOTATION PLANT OPERATION

Ores must be ground to a point of complete or nearly complete liberation. Even though this might possibly be accomplished by coarse crushing, grinding to finer than 10 mesh in all cases and finer than 48 mesh in most cases is necessary prior to flotation. Grinding is done in closed circuit with classifiers.

In many instances, superior flotation results are obtained by conditioning the ore with the reagents before the flotation step. Oily-type collectors are sometimes added to the grinding circuit to ensure dis-

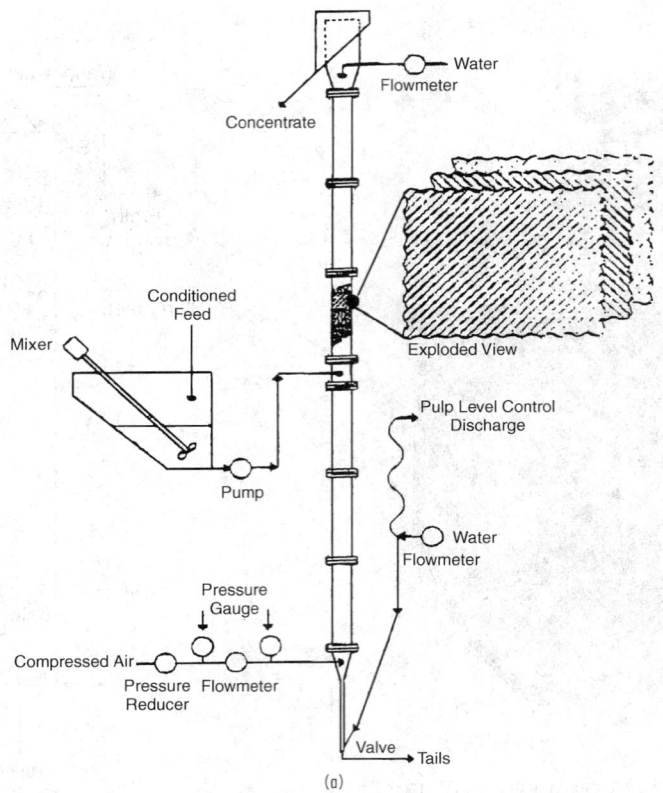

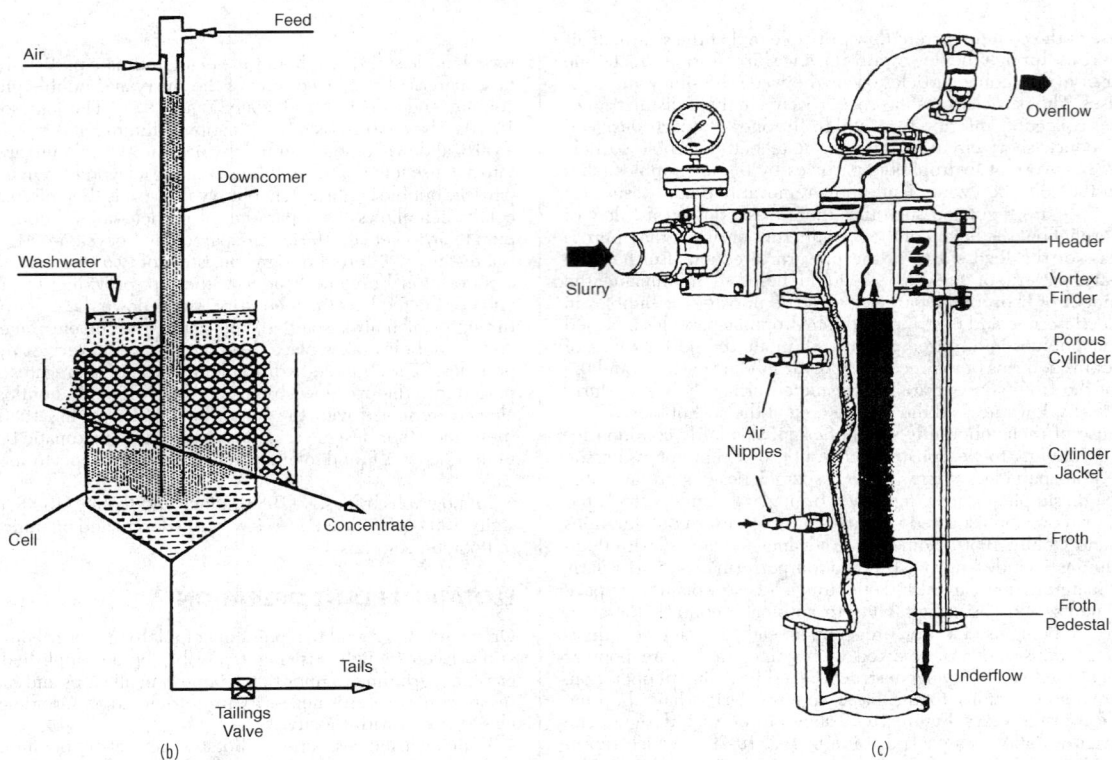

FIG. 19-79 Variations in the basic column design: (*a*) packed column, (*b*) Jameson cell, and (*c*) air-sparged hydrocyclone.

MODELING AND SIMULATION OF GRANULATION PROCESSES

Nomenclature and Units for Size Enlargement and Practice

Symbol	Definition	SI units	U.S. customary units	Symbol	Definition	SI units	U.S. customary units
A	Parameter in Eq. (20-47)			k	Coalescence rate constant	1/s	1/s
A	Apparent area of indentor contact	cm²	in²	K	Agglomerate deformability		
A	Attrition rate	cm³/s	in³/s	K_c	Fracture toughness	MPa·m$^{1/2}$	MPa·m$^{1/2}$
A_i	Spouted-bed inlet orifice area	cm²	in²	l	Wear displacement of indentor	cm	in
B	Nucleation rate	cm³/s	in³/s	L	Roll loading	dyn	lbf
B_f	Fragmentation rate	g/s	lb/s	$(\Delta L/L)_c$	Critical agglomerate deformation strain		
B_f	Wear rate	g/s	lb/s	N_t	Granules per unit volume	1/cm³	1/ft³
c	Crack length	cm	in	n	Feed droplet size	cm	in
δ_c	Effective increase in crack length due to process zone	cm		$n(v,t)$	Number frequency size distribution by size volume	1/cm⁶	1/ft⁶
c	Unloaded shear strength of powder	kg/cm²	psf	N_c	Critical drum or disc speed	rev/s	rev/s
d	Harmonic average granule diameter	cm	in	P	Applied load	dyn	lbf
d	Primary particle diameter	cm	in	P	Pressure in powder	kg/cm²	psf
d	Impeller diameter	cm	in	Q	Maximum compressive force	kg/cm²	psf
d	Roll press pocket depth	cm	in	Q	Granulator flow rate	cm³/s	ft³/s
d_i	Indentor diameter	cm	in	r_p	Process zone radius	cm	in
d_p	Average feed particle size	cm	in	R	Capillary radius	cm	in
D	Die diameter	cm	in	S	Volumetric spray rate	cm³/s	ft³/s
D	Disc or drum diameter	cm	in	St	Stokes number, Eq. (20-48)		
D	Roll diameter	cm	in	St°	Critical Stokes number representing energy required for rebound		
D_c	Critical limit of granule size	cm	in	St$_0$	Stokes number based on initial nuclei diameter		
e_r	Coefficient of restitution			t	Time	s	s
E	Strain energy stored in particle	J	J	u,v	Granule volumes	cm³	in³
$E°$	Reduced elastic modulus	kg/cm²	psf	u_0	Relative granule collisional velocity	cm/s	in/s
f_c	Unconfined yield stress of powder	kg/cm²	psf	U	Fluidization gas velocity	cm/s	ft/s
g	Acceleration due to gravity	cm/s2	ft/s2	U_{mf}	Minimum fluidization gas velocity	cm/s	ft/s
G_c	Critical strain energy release rate	J/m²	J/m²	U_i	Spouted-bed inlet gas velocity	cm/s	ft/s
F	Indentation force	dyn	lbf	V	Volumetric wear rate	cm³/s	in³/s
F	Roll separating force	dyn	lbf	V_R	Mixer swept volume ratio of impeller	cm³/s	ft³/s
G	Layering rate	cm³/s	in³/s	V	Volume of granulator	cm³	ft³
h	Height of liquid capillary rise	cm	in	w	Weight fraction liquid		
h	Roll press gap distance	cm	in	w	Granule volume	cm³	in³
h	Binder liquid layer thickness	cm	in	$w°$	Critical average granule volume	cm³	in³
h_b	Fluid-bed height	cm	in	W	Roll width	cm	in
h_a	Height of surface asperities	cm	in	x	Granule or particle size	cm	in
h_e	Maximum height of liquid capillary rise	cm	in	y	Liquid loading		
H	Individual bond strength	dyn	lbf	Y	Calibration factor		
H	Hardness of agglomerate or compact	kg/cm²	psf				

Greek symbols							
$\beta(u, v)$	Coalescence rate constant for collisions between granules of volumes u and v	1/sec	1/sec	$\Delta\rho$	Relative fluid density with respect to displaced gas or liquid	gm/cm³	
ε	Porosity of packed powder			ρ	Apparent agglomerate or granule density	gm/cm³	lb/ft³
ε_b	Interagglomerate bed voidage			ρ_a	Apparent agglomerate or granule density	gm/cm³	lb/ft³
ε_g	Intraagglomerate granule porosity			ρ_b	Bulk density	gm/cm³	lb/ft³
κ	Compressibility of powder			ρ_g	Apparent agglomerate or granule density	gm/cm³	lb/ft³
ϕ	Disc angle to horizontal	deg	deg	ρ_l	Liquid density	gm/cm³	lb/ft³
ϕ	Internal angle of friction	deg	deg	ρ_s	True skeletal solids density	gm/cm³	lb/ft³
ϕ_e	Effective angle of friction	deg	deg	σ_0	Applied axial stress	kg/cm²	psf
ϕ_w	Wall angle of friction	deg	deg	σ_z	Resulting axial stress in powder	kg/cm²	psf
ϕ_w	Roll friction angle	deg	deg	σ	Powder normal stress during shear	kg/cm²	psf
$\varphi(\eta)$	Relative size distribution			σ_c	Powder compaction normal stress	kg/cm²	psf
γ^{lv}	Liquid-vapor interfacial energy	dyn/cm	dyn/cm	σ_f	Fracture stress under three-point bend loading	kg/cm²	psf
γ^{sl}	Solid-liquid interfacial energy	dyn/cm	dyn/cm	σ_T	Granule tensile strength	kg/cm²	psf
γ^{sv}	Solid-vapor interfacial energy	dyn/cm	dyn/cm	σ_y	Granule yield strength	kg/cm²	psf
μ	Binder or fluid viscosity	poise		τ	Powder shear stress	kg/cm²	psf
μ	Coefficient of internal friction			θ	Contact angle	°	°
ω	Impeller rotational speed	rad/s	rad/s	ς	Parameter in Eq. (20-47)		
				η	Parameter in Eq. (20-47)		

Nomenclature and Units

Symbol	Definition	SI units	U.S. customary units	Symbol	Definition	SI units	U.S. customary units
A	Coefficient in double Schumann equation			q_f	Fine-fraction mass flow rate	g/s	lb/s
a	Constant			q_o	Feed mass flow rate	g/s	lb/s
$a_{k,k}$	Coefficient in mill equations			q_p	Mass flow rate of classifier product	g/s	lb/s
$a_{k,n}$	Coefficient in mill equations			q_R	Mass flow rate of classifier tailings	g/s	lb/s
$\mathbf{B}$	Matrix of breakage function			q_R	Recycle mass flow rate to a mill	g/s	lb/s
$\Delta B_{k,u}$	Breakage function			R	Recycle		
b	Constant			R	Reid solution		
C	Constant			r	Dimensionless parameter in size-distribution equations		
C_s	Impact-crushing resistance	kWh/cm	(ft·lb)/in	S	Rate function	S^{-1}	S^{-1}
D	Diffusivity	m²/s	ft²/s	S	Corrected rate function	S^{-1}	S^{-1}
D	Mill diameter	m	ft	$\mathbf{S'}$	Matrix of rate function	Mg/kWh	ton/(hp·h)
D_b	Ball or rod diameter	cm	in	$S_G(X)$	Grindability function	S^{-1}	S^{-1}
D_{mill}	Diameter of mill	m	ft	S_u	Grinding-rate function		
d	Differential			s	Parameter in size-distribution equations		
d	Distance between rolls of crusher	cm	in	s	Peripheral speed of rolls	cm/min	in/min
E	Work done in size reduction	kWh	hp·h	t	Time	s	s
E	Energy input to mill	kW	hp	u	Settling velocity of particles	cm/s	ft/s
E_i	Bond work index	kWh/Mg	(hp·h)/ton	$\mathbf{W}$	Vector of differential size distribution of a stream		
E_i	Work index of mill feed			w_k	Weight fraction retained on each screen		
E_2	Net power input to laboratory mill	kW	hp	w_u	Weight fraction of upper-size particles		
erf	Normal probability function			w_t	Material holdup in mill	g	lb
F	As subscript, referring to feed stream			X	Particle size or sieve size	cm	in
F	Bonding force	kg/kg	lb/lb	X'	Parameter in size-distribution equations	cm	in
g	Acceleration due to gravity	cm/s²	ft/s²	ΔX_i	Particle-size interval	cm	in
$\mathbf{I}$	Unit matrix in mill equations			X_i	Midpoint of particle-size interval, ΔX_i	cm	in
i	Tensile strength of agglomerates	kg/cm²	lb/in²	X_0	Constant, for classifier design		
K	Constant			X_f	Feed-particle size	cm	in
k	Parameter in size-distribution equations	cm	in	X_m	Mean size of increment in size-distribution equations	cm	in
k	As subscript, referring to size of particles in mill and classifier parameters			X_p	Product-particle size	cm	in
L	As subscript, referring to discharge from a mill or classifier			X_p	Size of coarser feed to mill	cm	in
L	Length of rolls	cm	in	X_{25}	Particle size corresponding to 25 percent classifier-selectivity value	cm	in
L	Inside length of tumbling mill	m	ft	X_{50}	Particle size corresponding to 50 percent classifier-selectivity value	cm	in
$\mathbf{M}$	Mill matrix in mill equations			X_{75}	Particle size corresponding to 75 percent classifier-selectivity value	cm	in
m	Dimensionless parameter in size-distribution equations			ΔX_k	Difference between opening of successive screens	cm	in
N	Mean-coordination number			x	Weight fraction of liquid		
N_c	Critical speed of mill	r/min	r/min	Y	Cumulative fraction by weight undersize in size-distribution equations		
ΔN	Incremental number of particles in size-distribution equation			Y	Cumulative fraction by weight undersize or oversize in classifier equations		
n	Dimensionless parameter in size-distribution equations			ΔY	Fraction of particles between two sieve sizes		
n	Constant, general			ΔY	Incremental weight of particles in size-distribution equations	g	lb
n_r	Percent critical speed of mill			ΔY_{ci}	Cumulative size-distribution intervals of coarse fractions	cm	in
O	As subscript, referring to inlet stream			ΔY_{fi}	Cumulative size-distribution intervals of fine fractions	cm	in
P	As subscript, referring to product stream			$\mathbf{Z}$	Matrix of exponentials		
P_k	Fraction of particles coarser than a given sieve opening						
p	Number of short-time intervals in mill equations						
Q	Capacity of roll crusher	cm³/min	ft³/min				
q	Total mass throughput of a mill	g/s	lb/s				
q_c	Coarse-fraction mass rate	g/s	lb/s				
q_F	Mass flow rate of fresh material to mill	g/s	lb/s				

Greek symbols							
β	Sharpness index of a classifier			ρ_ℓ	Density of liquid	g/cm³	lb/in³
δ	Angle of contact	rad	0	ρ_s	Density of solid	g/cm³	lb/in³
ε	Volume fraction of void space			Σ	Summation		
Z	Residence time in the mill	s	s	σ	Standard deviation		
η_x	Size-selectivity parameter			σ	Surface tension	N/cm	dyn/cm
μ	Viscosity of fluid	(N·S)/m²	P	υ	Volumetric abundance ratio of gangue to mineral		
ρ_f	Density of fluid	g/cm³	lb/in³				

PARTICLE-SIZE ANALYSIS

GENERAL REFERENCES: Allen, *Particle Size Measurement*, Chapman and Hall, 4th ed. 1990. Barth and Sun, *Particle Size Analysis Review,* Anal. Chem., 57, 151R, 1985. Miller and Lines, *Critical Reviews in Analytical Chemistry,* **20**(2), 75–116, 1988. Herdan, *Small Particle Statistics,* Butterworths, London. Orr and DalleValle, *Fine Particle Measurement,* Macmillan, New York, 2d ed., 1960. Kaye, *Direct Characterization of Fine Particles,* Wiley, New York, 1981. Van de Hulst, *Light Scattering by Small Particles,* Wiley, New York, 1957.

PARTICLE-SIZE DISTRIBUTION

Specification for Particulates Feed, recycle, and product from size reduction operations are defined in terms of the sizes involved. It is also important to have an understanding of the degree of aggregation or agglomeration that exists in the measured distribution.

The fullest description of a powder is given by its **particle-size distribution.** This can be presented in tabular or graphical form. The simplest presentation is in linear form with equal size intervals (Table 20-1). The significance of the distribution is more easily grasped when the data are presented pictorially, the simplest form of which is the histogram. More usually the plot is of cumulative percentage oversize or undersize against particle diameters, or percentage frequency against particle diameters. It is common to use a weight basis for percentage but surface or number may, in some cases, be more relevant. The basis of percentage; weight, surface, or volume should be specified, together with the basis of diameter; sieve, Stokes, or otherwise. The measuring procedure should also be noted.

Figure 20-1 presents the data from Table 20-1 in both cumulative and frequency format. In order to smooth out experimental errors it is best to generate the frequency curve from the slope of the cumulative curve, to use wide-size intervals or a data-smoothing computer program. The advantage of this method of presenting frequency data is that the area under the frequency curve equals 100 percent, hence, it is easy to visually compare similar data. A typical title for such a presentation would be: *Relative and cumulative mass distributions of quartz powder by pipet sedimentation.*

An alternative presentation of the same data is given in Fig. 20-2. In this case the sizes on the abscissa are in a logarithmic progression [log (x)] and the frequency is [dP/d ln (x)] so that the area under the frequency curve is, again, 100. This form of presentation is useful for wide-size distributions: Many instrument software programs generate data in a logarithmic-size interval and information is compressed in the finer-size intervals if an arithmetic-size progression is used.

It is always preferable to plot data so that the area under the frequency curve is normalized to 100 percent since this facilitates data comparison.

Particle-Size Equations It is common practice to plot size-distribution data in such a way that a straight line results, with all the advantages that follow from such a reduction. This can be done if the curve fits a standard law such as the normal probability law. According to the normal law, differences of equal amounts in excess or deficit from a mean value are equally likely. In order to maintain a symmetrical bell-shaped curve for the frequency distribution it is necessary to plot the population density (e.g., percentage per micron) against size.

With the **log-normal probability law,** it is ratios of equal amounts which are equally likely. In order to obtain a symmetrical bell-shaped frequency curve it is therefore necessary to plot the population density per log (micron) against log (size) [Hatch and Choate, *J. Franklin Inst., 207,* 369 (1929)]:

$$Y = \text{erf}\left(\frac{\ln{(X/X')}}{\sigma}\right) \tag{20-1}$$

Other equations in general use include the Gates-Gaudin-Schumann [Schumann, *Am. Inst. Min. Metall. Pet. Eng.,* Tech. Paper 1189, Min. Tech. (1940)]:

$$Y = (X/k)^m \tag{20-2}$$

The Rosin-Rammler-Bennett [Rosin and Rammler, *J. Inst. Fuel,* **7,** 29–36 (1933); Bennett, ibid., **10,** 22–29 (1936)]:

$$Y = 1 - [\exp\{-(X/X')^n\}] \tag{20-3}$$

The Gaudin-Meloy [Gaudin and Meloy, *Trans. Am. Inst. Min. Metall. Pet. Eng.,* **223,** 40–50 (1962)]:

$$Y = 1 - \left[1 - \left(\frac{X}{X'}\right)\right]^r \tag{20-4}$$

where Y = cumulative fraction by weight undersize; x = size; k, X' = parameters with dimensions of size, m, n, r = dimensionless exponents; erf = normal probability function; and σ = standard deviation parameter.

The Rosin-Rammler is useful for monitoring grinding operations for highly skewed distributions, but should be used with caution since the device of taking logs always reduces scatter, hence taking logs twice is not to be recommended. The Gates-Gaudin-Schumann has the advantage of simplicity and the Gaudin-Meloy can be fitted to a variety of distributions found in practice. The log-normal distribution

TABLE 20-1 Tabular Presentation of Particle Size Data

Particle size in microns (x)	Percentage undersize (P)	Percentage per micron $\left(\dfrac{dP}{dx}\right)$	Percentage per log (micron) $\dfrac{dP}{d \log (x)}$	Particle size in microns (x)	Percentage undersize (P)	Percentage per micron $\left(\dfrac{dP}{dx}\right)$	Percentage per log (micron) $\dfrac{dP}{d \log (x)}$
2.5	0.00	0.000	0.00	52.5	97.41	0.425	13.21
7.5	0.29	0.058	0.12	57.5	98.59	0.236	8.16
12.5	5.06	0.953	3.21	62.5	99.24	0.130	4.92
17.5	19.44	2.875	15.54	67.5	99.59	0.071	2.92
22.5	39.93	4.098	34.68	72.5	99.79	0.039	1.71
27.5	59.68	3.951	47.97	77.5	99.89	0.021	1.00
32.5	74.91	3.047	48.92	82.5	99.95	0.012	0.58
37.5	85.18	2.054	41.04	87.5	99.98	0.006	0.34
42.5	91.55	1.273	30.31	92.5	100.00	0.004	0.20
47.5	95.28	0.748	20.58			0.000	0.12

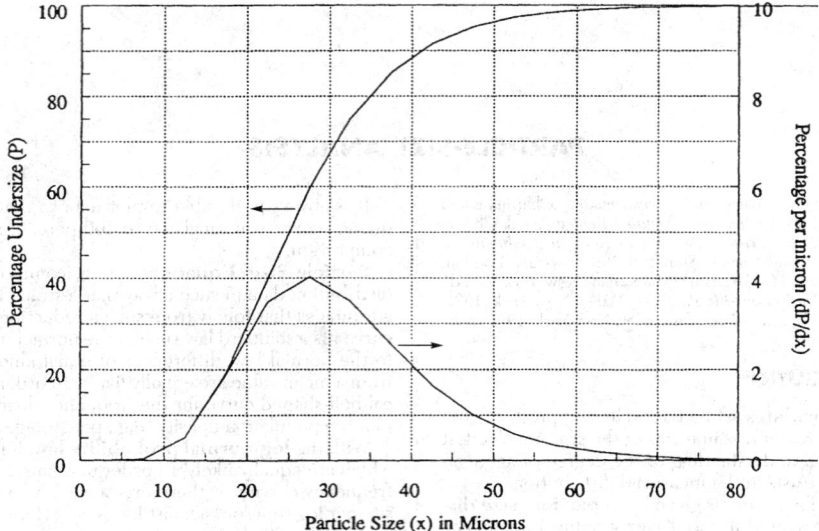

FIG. 20-1   Particle-size distribution curve plotted on linear axes.

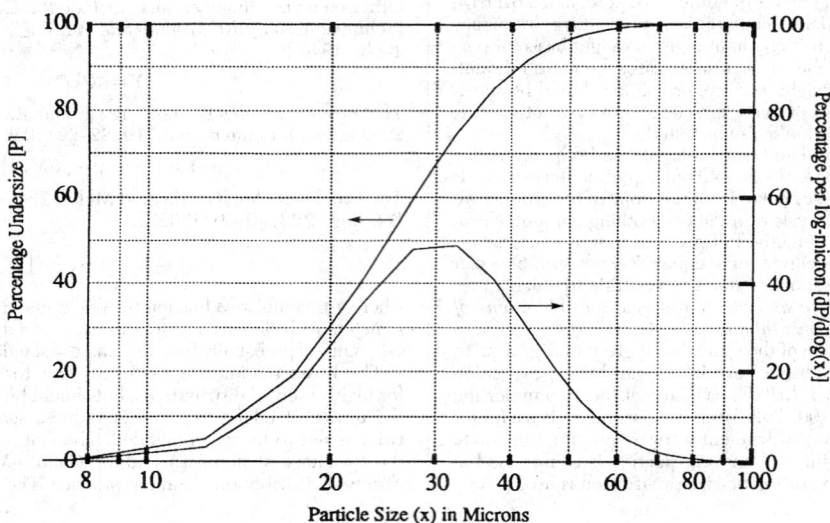

FIG. 20-2   Particle-size distribution curve plotted using a logarithmic scale for the abscissa.

has an advantage in that transformations between distributions is simple; that is, if the number distribution is log-normal, the surface and volume distributions are also log-normal with the same slope (σ).

Average Particle Size   A powder has many average sizes; hence it is essential that they be well specified. The median is the 50 percent size; half the distribution is coarser and half finer. The mode is a high-density region; if there is more than one peak in the frequency curve, the distribution is said to be multimodal. The mean is the center of gravity of the distribution. The center of gravity of a mass (volume) distribution is defined by: $X_{VM} = \sum X dV / \sum dV$ where $dV = X^3 dN$: dV is the volume of dN particles of size X. This is defined as the *volume-moment mean diameter* and differs from the mean for a number or surface distribution.

Specific Surface   This can be calculated from size distribution data. For example, the Gates diagram employs a plot of cumulative

percent by weight undersize versus reciprocal diameter; the area beneath the curve represents surface. Likewise the area under the Roller diagram of weight percent per micron against log of diameter represents surface (Work and Whitby, op.cit., p 477).

Sampling of Powders   An important prerequisite to accurate particle size analysis is proper powder sampling. Powders may be classified as nonsegregating (cohesive) or segregating (free-flowing). Representative samples are more easily taken from cohesive powders *provided* they have previously been mixed. It is difficult to mix free-flowing powders; hence it is advisable to sample them in motion. (1) A powder should always be sampled when in motion. (2) The whole of the stream of powder should be taken for many short increments of time in preference to part of the stream being taken for the whole of the time. The estimated maximum sample errors on a 60:40 blend of free-flowing sand using different sampling techniques are given in Table 20-2.

TABLE 20-2 Reliability of Selected Sampling Methods

Method	Estimated maximum sample error
Cone and quartering	22.7%
Scoop sampling	17.1%
Table sampling	7.0%
Chute splitting	3.4%
Spinning riffling	0.42%

The spinning riffler obeys the "Golden Rules of Sampling" given above and, therefore, generates the most representative samples [Allen and Khan, *The Chemical Engineer*, 103–112 (May 1970)]. In this device (Fig. 20-3) a ring of containers rotates under the powder feed. If the powder flows for a long time compared with the period of rotation (a ratio of at least 30:1), the sample in each container will be made up of many small portions drawn from all parts of the bulk. Several different configurations of this device are available for both cohesive and free-flowing powders. In the one illustrated, a single pass divides the bulk into 16 parts, two passes increases this subdivision to 256:1, and so on.

PARTICLE-SIZE MEASUREMENT

There are many techniques available for measuring the particle-size distribution of powders. The wide size range covered, from nanometers to millimeters, cannot be analyzed using a single measurement principle. Added to this are the usual constraints of capital costs versus running costs, speed of operation, degree of skill required, and, most important, the end-use requirement.

If the particle-size distribution of a powder composed of hard, smooth spheres is measured by any of the techniques, the measured values should be identical. However, there are many different size distributions that can be defined for any powder made up of nonspherical particles. For example, if a rod-shaped particle is placed on a sieve, its diameter, not its length, determines the size of aperture through which it will pass. If, however, the particle is allowed to settle in a viscous fluid, the calculated diameter of a sphere of the same substance that would have the same falling speed in the same fluid (i.e., the Stokes diameter) is taken as the appropriate size parameter of the particle.

Since the Stokes diameter for the rod-shaped particle will obviously differ from the rod diameter, this difference represents added information concerning particle shape. The ratio of the diameters measured by two different techniques is called a **shape factor.**

Heywood [Heywood, *Symposium on Particle Size Analysis*, Inst. Chem. Engrs. (1947), Suppl. 25, 14] recognized that the word "shape" refers to two distinct characteristics of a particle—form and proportion. The first defines the degree to which the particle approaches a definite form such as cube, tetrahedron, or sphere, and the second by the relative proportions of the particle which distinguish one cuboid, tetrahedron, or spheroid from another in the same class. He replaced historical qualitative definitions of shape by numerical shape coefficients.

Gravitational Sedimentation Methods In gravitational sedimentation methods, particle size is determined from settling velocity and undersize fraction by changes of concentration in a settling suspension. The equation relating particle size to settling velocity is known as Stokes law:

$$d_{St} = \sqrt{\frac{18\eta u}{(\rho_s - \rho_f)g}} \qquad (20\text{-}5)$$

where d_{St} is the Stokes diameter, η is viscosity, u is particle settling velocity under gravity, ρ_s is the particle density, ρ_f is the fluid density, and g is the acceleration due to gravity.

Stokes diameter is defined as the diameter of a sphere having the same density and the same velocity as the particle in a fluid of the same density and viscosity settling under laminar flow conditions. Correction for deviation from Stokes law may be necessary at the large end of the size range. Sedimentation methods are limited to sizes above a µm due to the onset of thermal diffusion (Brownian motion) at smaller sizes.

An experimental problem is to obtain adequate dispersion of the particles before sedimentation analysis. For powders that are difficult to disperse the addition of dispersing agents is necessary, together with ultrasonic probing. It is essential to examine a sample of the dispersion under a microscope to ensure that the sample is fully dispersed.

Equations to calculate size distributions from sedimentation data are based on the assumption that the particles fall freely in the suspension. In order to ensure that particle-particle interaction does not prevent free fall, an upper-volume concentration limit of around 0.2 percent is recommended.

There are various procedures available for determining the changing solids concentration of a sedimenting suspension:

In the *pipet method*, concentration changes are monitored by extracting samples from a sedimenting suspension at known depths of fall and predetermined times. The method is best known as the Andreasen modification [Andreasen, *Kolloid-Z.*, **49**, 253 (1929)] shown in Fig. 20-4. Two 10-mL samples are withdrawn from a fully dispersed, agitated suspension at zero time to corroborate the 100 percent concentration given by the known weight of powder and volume of liquid making up the suspension. The suspension is then allowed to settle, and 10-mL samples are taken at time intervals in a geometric 2:1 time progression starting at 1 minute (i.e., 1, 2, 4, 8, 16, 32, 64 minutes); if longer time intervals than this are used it is necessary to enclose the pipet in a temperature-controlled environment. The amounts of powder in the extracted samples are determined by

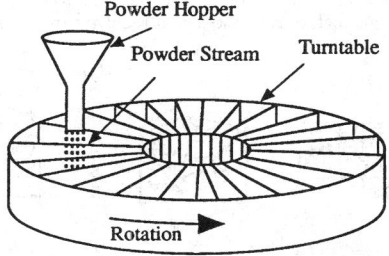

FIG. 20-3 Spinning riffler sampling device.

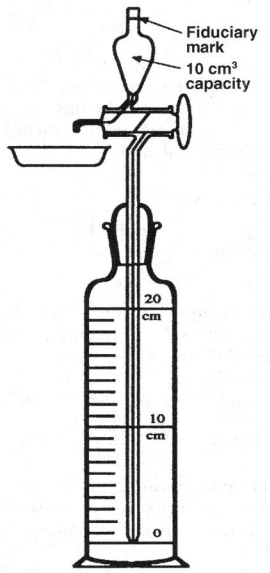

FIG. 20-4 Equipment used in the pipet method of size analysis.

drying, cooling in a dessicator, and weighing. Stokes diameters are determined from the predetermined times and the depths, with correction for the changes in depth due to the extractions. The cumulative, mass, undersize distribution comprises a plot of the normalized concentration against the Stokes diameter. A reproducibility of ±2 percent is possible using this apparatus. The technique is versatile in that it is possible to analyze most powders which are dispersible in liquids; its disadvantages are that it is a labor-intensive procedure and a high level of skill is needed.

The *hydrometer method* is simpler in that the density of the suspension, which is related to the concentration, is read directly from the stem of the hydrometer while the depth is determined by the distance of the hydrometer bulb from the surface (ASTM Spec. Pub. 234, 1959). The method has low resolution but is widely used in soil science studies.

In *gravitational photosedimentation methods* the changing concentration with time and depth of fall is monitored using a light beam. These methods give a continuous record of changing optical density with time and depth and have the added advantage that the beam can be scanned to the surface to reduce the measurement time. A correction needs to be applied to compensate for the breakdown in the laws of geometric optics (due to diffraction effects the particles cut off more light than geometric optics predicts). The normalized measurement is a cumulative surface undersize.

In *gravitational X-ray sedimentation methods* the changing concentration with time and depth of fall is monitored using an X-ray beam. These methods give a continuous record of changing X-ray density with time and depth and have the added advantage that the beam can be scanned to the surface to reduce the measurement time. The method is limited to materials having a high atomic mass (i.e., X-ray opaque material) and gives a mass undersize distribution directly.

Sedimentation Balance Methods   In sedimentation balances the weight of sediment is measured as it accumulates on a balance pan suspended in an initially homogeneous suspension. The technique is slow because of the time required for the smallest fine particle to settle out over a given column height. The relationship between settled weight (P), weight undersize (W) and time (t) is given by the following equation.

$$P = W - \frac{dP}{d \ln (t)} \qquad (20\text{-}6)$$

Centrifugal Sedimentation Methods   These methods extend sedimentation methods into the submicron size range. Sizes are calculated from a modified version of Stokes equation:

$$d_{St} = \sqrt{\frac{18\eta u}{(\rho_s - \rho_f)\omega^2 r}} \qquad (20\text{-}7)$$

where r is the measurement radius and ω is the radial velocity of the centrifuge. The concentration calculations are complicated due to radial dilution effects (i.e., particles do not travel in parallel paths as in gravitational sedimentation but move away from each other as they settle radially outwards).

Particle velocities are given by:

$$u = \frac{\ln \left(\dfrac{r}{s}\right)}{t} \qquad (20\text{-}8)$$

where both r, the measurement radius, and s, the surface radius can be varying; the former varies if the system is a scanning system, and the latter if the surface falls due to the extraction of samples.

Concentration undersize D_m is determined using Kamack's equation [Kamack, *Br. J. Appl. Phys.*, **5**, 1962–68 (1972)]:

$$Q(D_m) = \int_0^{D_m} \left(\frac{r_i}{s_i}\right)^2 f(D)dD \qquad (20\text{-}9)$$

where r_i is the measurement radius and s_i is the surface radius, either or both of which may vary during the analysis. $f(D)dD = F(D)$ is the fraction of particles in the narrow-size range dD. $(r_i/s_i)^2$ is the radial dilution correction factor.

The disc centrifuge, developed by Slater and Cohen and modified by Allen and Svarovsky [Allen and Svarovsky, *Dechema Monogram*,

Nuremberg, Numbers 1589–1615, 279–292 (1975)], is essentially a centrifugal pipet device. Size distributions are calculated from the measured solids concentrations of a series of samples withdrawn through a central drainage pillar at various time intervals.

In the centrifugal disc photosedimentometer concentration changes are monitored by a light beam. In one high-resolution mode of operation, the suspension under test is injected into clear fluid in the spinning disc through an entry port, and a layer of suspension is formed over the free surface of the liquid (the line-start technique). The analysis can also be carried out using a homogeneous suspension. Very low concentrations are used, but the light-scattering properties of small particles make it difficult to interpret the measured data.

Several centrifugal cuvet photocentrifuges are commercially available. These instruments use the same theory as the disc photocentrifuges but are limited in operation to the homogeneous mode of operation.

The X-ray disc centrifuge is a centrifugal version of the gravitational instruments and extends the measuring technique well into the sub-μm-size range.

Microscope Methods   In microscope methods of size analysis, direct measurements are made on enlarged images of the particles. In the simplest technique, linear measurements of particles are made by using a calibrated scale on top of the particle image. Alternatively, the projected areas of the particles can be compared to areas of circles. **Feret's diameter** (Fig. 20-5) is the perpendicular projection, in a fixed direction, of the tangents to the extremities of the particle profile. **Martin's diameter** is a line, parallel to a fixed direction, which divides the particle profile into two equal areas. Since the magnitude of these statistical diameters varies with particle orientation, these diameters have meaning only when a sufficient number of measurements are averaged.

Quantitative image microscopy has revolutionized microscopic methods of size and shape analysis. The sizes of large numbers of particles can be rapidly determined and the data manipulated. The speed and sophistication of such devices make it possible to devise new methods for characterizing the shape of fine particles. In **Fourier techniques** the shape characteristic is transformed into a signature waveform. Beddow and coworkers (Beddow, *Particulate Science and Technology*, Chemical Publishing, New York, 1980) take the particle centroid as a reference point. A vector is then rotated about this centroid with the tip of the vector touching the periphery. A plot of the magnitude of the vector against its angular position is a wave-type function. This wave form is then subjected to Fourier analysis. The lower frequency harmonics constituting the complex wave correspond to the gross external morphology, whereas the higher frequencies correspond to the texture of the fine particle. **Fractal logic** was introduced into fine-particle science by Kaye and coworkers [Kaye, op. cit. (1981)], who show that the non-Euclidean logic of Mandelbrot can be applied to describe the ruggedness of a particle profile. A combination of fractal dimension, and geometric-shape factors such as aspect ratio, can be used to describe a population of fine particles of various shapes, and these can be related to the functional properties of the particle.

Stream Scanning Methods   In these techniques, the particles to be measured are examined individually in a stream of fluid. As the particles pass through a sensing zone they are counted and measured through their interaction with the sensor. It is essential to use very low particle concentrations since the signals received from two particles is indistinguishable from the signal received from a single larger particle.

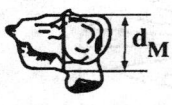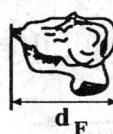

FIG. 20-5   Statistical (Martin's and Feret's) and projected area diameters for an irregular particle.

In the **electrical sensing zone method** a dilute well-dispersed suspension in an electrolyte is caused to flow through a small aperture [Kubitschek, *Research,* **13,** 129 (1960)]. The changes in the resistivity between two electrodes on either side of the aperture, as the particles pass through, are directly related to the volumes of the particles. The pulses are fed to a pulse-height analyzer where they are counted and scaled. The method is limited by the pulse-height analyzer which can resolve pulses in the range 16,000:1 (i.e., a volume diameter range of about 25:1) and the need to suspend the particles in an electrolyte.

In **light blockage methods** the size of the particle is determined from the amount of light blocked off by the particle as it passes through a sensing zone. In **light scattering methods** the particle size is determined either from the light scattered in the forward direction or at some angle, usually 90° from the direction of the incident beam. The light source can be incandescent or laser and the detecting mechanism ranges from simple photodetectors to parabolic mirrors. The particles can be suspended in a liquid or gas.

In the Lasentec instruments a chord-length distribution is generated, from a rotating infrared beam, and this is converted to a size distribution. Since highly concentrated systems can be interrogated this system can be used for on-line size analysis.

Field Scanning Methods   In these techniques, the particles to be measured are examined collectively, and the signal from the assembly of particles is deconvoluted to generate a size distribution.

Light Diffraction Methods   These comprise one of several field scanning procedures in which an assembly of particles is irradiated with a laser beam. The forward light-scattered flux contains information on the size distribution of the particles. Several assumptions are made in the transformation of the diffraction pattern into particle-size data, and the various companies manufacturing these instruments offer different interpretive programs. The dozen or so instruments currently marketed tend to disagree with each other and also generate a wider distribution than other methods. Their advantages are ease of use and high reproducibility. Although low-angle laser light scattering is only applicable down to around 0.7 μm, the lower limit can be extended using secondary measurements of 90 degree scattering of white light, polarization ratio, and so on.

The principle of ultrasonic attenuation is that plane sound waves moving through a slurry are attenuated according to the size and concentration of the particles in the slurry. Two instruments have been described, one for sizing in the plus-one μm up to mm sizes (Reibel and Loffler, *European Symposium on Particle Characterization,* Nuremberg, Germany, publ. Nuremberg Messe, 1989) and the other for sub-μm sizing (Alba, US Patent 5121629, June 16, 1992). Although in its infancy, the technique shows high promise for on-line size analysis at high-volume (+50%) concentrations.

Photon Correlation Spectroscopy (PCS)   The size distribution of particles ranging in size from a few nanometers to a few μm can be determined from their random motion due to molecular bombardment. This technique involves passing a laser beam into a suspension and measuring the Doppler shift of the frequency of light scattered at an angle (usually 90 degrees) with respect to the incident beam. The Doppler shift is related to particle velocity which, in turn, is inversely related to their size. Multiangle instruments are also available to generate the angular variation of scattered light intensity for derivation of molecular weight, radius of gyration, translational and rotational diffusion coefficients, and other molecular properties.

Through dynamic light scattering in the controlled reference method, a laser beam is fed into an agitated measuring cell or flowing suspension using an optical-wave guide. Particles within 50 microns of the tip of the wave guide (a fiber-optic probe) scatter light some of which is reflected back into the fiber and transmitted back through the guide. The reflected light from the interface between the guide tip and the suspension is also transmitted back. If these two components are coherent they will interfere with each other and result in a component of signal which has the difference or "beat" frequency between the reflected and scattered components. The difference frequencies are the same as the desired Doppler shifts. The received signal resembles random noise at the output of the silicon photodiode as a result of the mixing of the Doppler shifts from all the particles scattering the laser light. The photodiode output is digitized and the

power spectrum of the signal is determined using Fast Fourier Transform techniques. The spectrum is then analyzed to determine the particle-size distribution. Two instruments based on this phenomenon are available, the Microtrac UPA (Fig. 20-6) [Trainer, Freud, and Weiss, *Pittsburgh Conference, Analytical and Applied Spectroscopy* Symp. Particle Size Analysis (March 1990)] and the Malvern Hi-C which operates in a similar manner to cover the size range 0.015 μm to 1 μm at concentrations from 0.01 to 50 percent solids.

Sieving Methods and Classification   Sieving is probably the most frequently used and abused method of analysis because the equipment, analytical procedure, and basic concepts are deceptively simple. In sieving, the particles are presented to equal-size apertures that constitute a series of go-no-go gauges. Sieve analysis presents three major difficulties: (1) with woven-wire sieves, the weaving process produces three-dimensional apertures with considerable tolerances, particularly for fine-woven mesh; (2) the mesh is easily damaged in use; (3) the particles must be efficiently presented to the sieve apertures.

Sieves are often referred to by their mesh size, which is the number of wires per linear unit. The U.S. Standard Sieve Series as described by the American Society of Testing and Materials (ASTM) document E-11-87 *Standard Specification for Wire-cloth Sieves for Testing Purposes* addresses sieve opening sizes from 20 μm (635 mesh) to 125 mm (5.00 in). Electroformed sieves with square or round apertures and tolerances of ±2 μm, are also available.

For coarse separation **dry sieving** is used, but other procedures are necessary as the powder becomes finer and more cohesive. Machine sieving is performed by stacking sieves in ascending order of aperture size and placing the powder on the top sieve. The most aggressive action is performed with Pascal Inclyno and Tyler Ro-tap sieves which combine a gyratory and jolting movement, although a simple vibratory action may be suitable in many cases. With the Air-Jet sieve a rotating jet below the sieving surface cleans the apertures and helps the passage of fines through the apertures. The sonic sifter combines two actions, a vertically oscillating column of air, and a repetitive mechanical pulse. Wet sieving is frequently used with cohesive powders.

Elutriation Methods and Classification   In gravity elutriation the particles are classified, in a column, by a rising fluid current. In centrifugal elutriation the fluid moves inward against the centrifugal force. A *cyclone* is a centrifugal elutriator, though not usually so regarded (see Sec. 17: "Dust Collection Equipment"). The cyclosizer is a series of inverted cyclones with added apex chambers through which water flows. Suspension is fed into the largest cyclone and particles are separated into different size ranges.

Surface Area Determination   The surface-to-volume ratio is an important powder property since it governs the rate at which a powder interacts with its surroundings. Surface area may be determined from size-distribution data or measured directly by flow through a powder bed or the adsorption of gas molecules on the powder surface. Other methods such as gas diffusion, dye adsorption from solution, and heats of adsorption have also been used. It is emphasized that a powder does not have a unique surface, unless the surface is considered to be absolutely smooth, and the magnitude of the measured surface depends upon the level of scrutiny (e.g., the smaller the gas molecules used for gas adsorption measurement, the larger the measured surface).

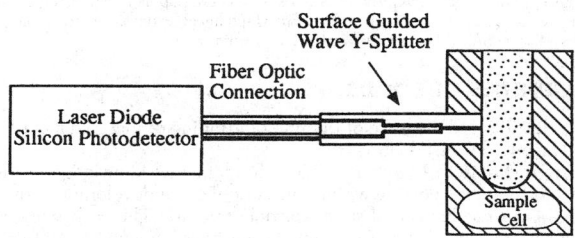

FIG. 20-6   Diagram of the Leeds and Northrup Ultrafine Particle Analyzer (UPA).

Gas adsorption is the preferred method of surface-area determination. An isotherm is generated of the amount of gas adsorbed against gas pressure, and the amount of gas required to form a monolayer is determined. The surface area can then be calculated using the cross-sectional area of the gas molecule. Outgassing of the powder before analysis should be conducted very carefully to ensure reproducibility. Commonly, nitrogen at liquid nitrogen vapor pressure is used but, for low surface-area powders, the adsorbed amounts of krypton or xenon are more accurately found. Many theories of gas adsorption have been advanced, but measurements are usually interpreted by using the BET theory [Brunauer, Emmett, and Teller, *J. Am. Chem. Soc.*, **60**, 309 (1938)].

In the **static method** the powder is isolated under high vacuum and surface gases driven off by heating the container. The container is next immersed in liquid nitrogen and known amounts of nitrogen vapor are admitted into the container at measured increasing pressures in the relative pressure range 0.05 to 0.35.

In the **dynamic method** the powder is flushed with an inert gas during degassing, nitrogen is then adsorbed on the powder in a carrier of helium gas at known relative pressure while the powder is in a container surrounded by liquid nitrogen. The changing concentration of nitrogen is measured by a calibrated conductivity cell so that the amount adsorbed can be determined.

Permeametry   The flow of fluid through a packed bed of powder can be related to the surface area of the powder using the Carman-Arnel Equation [Carman and Arnell, *Can. J. Res.*, **26**, 128 (1948)]. Flow takes place through two mechanisms, viscous and diffusional flow. The latter term is often neglected leading to erroneous results [British Standard BS 4359: Part 2: (1982) *Determination of specific surface of powders. Recommended air permeability methods*]. Essentially, the pressure drop across the bed is directly related to the flow rate through it, and the constant of proportionality includes surface area. The assumptions made in deriving the equation are so sweeping that the derived value is better considered as a surface-related parameter.

On-line Procedures   The growing trend toward automation in industry has resulted in many studies of rapid procedures for generating size information so that feedback loops can be instituted as an integral part of a process. Many of these techniques are modifications of more traditional methods. The problems associated with on-line methods include: allocation and preparation of a representative sample; analysis of the sample; evaluation of the results. The interface between the measuring apparatus and the process has the potential of high complexity, and consequently, high costs [Leschonski, *Particle Characterization*, **1**, 1 (July 1984)].

PRINCIPLES OF SIZE REDUCTION

GENERAL REFERENCES: Annual reviews of size reduction, *Ind. Eng. Chem.*, October or November issues, by Work from 1934 to 1965, by Work and Snow in 1966 and 1967, and by Snow in 1968, 1969, and 1970; and in *Powder Technol.*, **5**, 351 (1972), and **7** (1973); Snow and Luckie, **10**, 129 (1973), **13**, 33 (1976), **23**(1), 31 (1979). *Chemical Engineering Catalog*, Reinhold, New York, annually. Cremer-Davies, *Chemical Engineering Practice*, vol. 3: *Solid Systems*, Butterworth, London, and Academic, New York, 1957. *Crushing and Grinding: A Bibliography*, Chemical Publishing, New York, 1960. *European Symposia on Size Reduction:* 1st, Frankfurt, 1962, publ. 1962, Rumpf (ed.), Verlag Chemie, Düsseldorf; 2d, Amsterdam, 1966, publ. 1967, Rumpf and Pietsch (eds.), *DECHEMA-Monogr.*, **57**; 3d, Cannes, 1971, publ. 1972, Rumpf and Schönert (eds.), *DECHEMA-Monogr.*, **69**. Gaudin, *Principles of Mineral Dressing*, McGraw-Hill, New York, 1939. International Mineral Processing Congresses: *Recent Developments in Mineral Dressing*, London, 1952, publ. 1953, Institution of Mining and Metallurgy; *Progress in Mineral Dressing*, Stockholm, 1957, publ. London, 1960, Institution of Mining and Metallurgy; 6th, Cannes, 1962, publ. 1965, Roberts (ed.), Pergamon, New York, 1964, publ. 1965, Arbiter (ed.), vol. 1: *Technical Papers*, vol. 2: *Milling Methods in the Americas*, Gordon and Breach, New York; 8th, Leningrad, 1968; 9th, Prague, 1970; 10th, London, 1973; 11th, Cagliari, 1975; 12th, São Paulo, 1977. Lowrison, *Crushing and Grinding*, CRC Press, Cleveland, 1974. *Pit and Quarry Handbook*, Pit & Quarry Publishing, Chicago, 1968. Richards and Locke, *Text Book of Ore Dressing*, 3d ed., McGraw-Hill, New York, 1940. Rose and Sullivan, *Ball, Tube and Rod Mills*, Chemical Publishing, New York, 1958. Snow, *Bibliography of Size Reduction*, vols. 1 to 9 (an update of the previous bibliography to 1973, including abstracts and index), U.S. Bur. Mines Rep. SO122069, available IIT Research Institute, Chicago, Ill. 60616. Stern, "Guide to Crushing and Grinding Practice," *Chem. Eng.*, **69**(25), 129 (1962). Taggart, *Elements of Ore Dressing*, McGraw-Hill, New York, 1951.

Since a large part of the literature is in the German language, availability of English translations is important. Translation numbers cited in this section refer to translations available through the National Translation Center, Library of Congress, Washington, DC. Also, volumes of selected papers in English translation are available from the Institute for Mechanical Processing Technology, Karlsruhe Technical University, Karlsruhe, Germany.

PROPERTIES OF SOLIDS

Grindability is a measure of the rate of grinding of material in a particular mill (discussed later).

Single-Particle Fracture   More fundamental knowledge of the breaking action occurring within mills depends on developing knowledge of the mechanism of single-particle fracture. The early workers [Smekal, *Z. Ver. Dtsch. Ing. Beh. Verfahrenstech.*, no. 6, 159–165 (1938), NTC translation 70-14798; and Smekal Z. *Ver. Dtsch. Ing.*, **81**(46), 1321–1326 (1937), NTC translation 70-14799] investigated the breakage of cubes. This gives misleading results when cubes are crushed between platens because surface irregularities concentrate the load and give nonuniform load distribution. More meaningful measurements can be made with spheres, which approximate the shapes of particles broken in mills.

The force required to crush a single particle that is spherical near the contact regions is given by the equation of Hertz (Timoschenko and Goodier, *Theory of Elasticity*, 2d ed., McGraw-Hill, New York, 1951).

In an experimental and theoretical study on glass spheres Frank and Lawn [*Proc. R. Soc. (London)*, **A299**(1458), 291 (1967)] observed the repeated formation of ring cracks as increasing load was applied, causing the circle of contact to widen. Eventually a load is reached at which the crack deepens to form a cone crack, and at a sufficient load this propagates across the sphere to cause breakage into fragments. The authors' photographs show how the size of flaws that happen to be encountered at the edge of the circle of contact can result in a distribution of breakage strengths. Thus the mean value of breakage strength depends partly on intrinsic strength and partly on the extent of flaws present. From the measured breaking load and the Hertz theory one can calculate the apparent tensile strength σ_0, which is the maximum stress under the circle of contact normal to the direction of crack propagation. This tensile strength is the most appropriate one to use for breakage in mills, although the crushing strength of cubes still is often used as a rule of thumb. The propagation of cracks across spheres and disks has been recorded by high-speed spark cinematographs by Rumpf et al. (*Second European Symposium on Size Reduction*, op. cit., 1966, p. 57). They attempt to extend the Hertz theory deeper into the sphere although it is not valid far from the point of load application. The stress at points within the sphere far from the point of load application is given by the Bousinesque theory [Sternberg and Rosenthal, *J. Appl. Mech.*, **12**, 413 (1952); and Hiramatsu and Oka, *Int. J. Rock Mech. Min. Sci.*, **3**, 89 (1966)].

Snow and Paulding (Heywood Memorial Conference, Loughborough University, England, September 1973) observed that when breakage occurs, the finest fragments arise near the circle of contact where the stored internal stress is highest. They postulated that the fragment-size distribution could be calculated by assuming that the local fragment size is correlated with the locally stored stress energy just before fracture occurs. Calculated fragment-size distributions are roughly similar to those that they measured for glass spheres and various hard minerals as well as to distributions measured by Bergstrom and Sollenberger [*Trans. Am. Inst. Min. Metall. Pet. Eng.*, **220**, 373–

379 (1961)]. From this it can be concluded that the wide distribution of fragment sizes from milling is inherent in the breakage process itself and that attempts to improve grinding efficiency by weakening the particles will result in coarser fragments which may require a further break to reach the desired size.

Different mills are designed to apply the force in different ways [Rumpf, *Chem. Ing. Tech.*, **31**, 323–327 (1959), NTC translation 61-12395]. The detailed prediction of grinding rates and product-size distribution from mills awaits the development of a simulation model based on the physics of fracture. An initial attempt is that of Buss and Shubert (*Third European Symposium on Size Reduction*, op. cit., 1972, p. 233), who assume that mill performance is given by the sum of breakage events which are similar to *single-particle breakage experiments* in the laboratory. A paper by Schönert [*Trans. Am. Inst. Min. Metall. Pet. Eng.*, **252**(1), 21 (1972)] summarizes single-particle-breakage data from numerous publications from the Technical University of Karlsruhe, Germany. Hildinger [*Freiberg. Forschungsh.*, **A480**, 19 (1970)] and Steier and Schönert (*Third European Symposium on Size Reduction*, op. cit., 1972, p. 135) report more experimental results on the probability of breakage of single particles by drop-weight experiments.

Grindability Grindability is the amount of product from a particular mill meeting a particular specification in a unit of grinding time e.g., tons per hour passing 200 mesh. The chief purpose of a study of grindability is to evaluate the size and type of mill needed to produce a specified tonnage and the power requirement for grinding. So many variables affect grindability that this concept can be used only as a rough guide to mill sizing; it says nothing about product-size distribution or type or size of mill. If a particular energy law is assumed, then the grinding behavior in various mills can be expressed as an energy coefficient or **work index** (discussed later). This more precise concept is limited by the inadequacies of these laws but often provides the only available information.

The technology based on grindability and energy considerations is being supplanted by computer simulation of milling circuits (see subsection "Simulation of Milling Circuits"), in which the gross concept of grindability is replaced by the **rate of breakage function** (sometimes called the selection function), which is the grindability of each particle size referred to the fraction of that size present.

Factors of hardness, elasticity, toughness, and cleavage are important in determining grindability. Grindability is related to modulus of elasticity and speed of sound in the material [Dahlhoff, *Chem. Ing. Tech.*, **39**(19), 1112–1116 (1967)].

The **hardness** of a mineral as measured by the **Mohs scale** is a criterion of its resistance to crushing [Fahrenwald, *Trans. Am. Inst. Min. Metall. Pet. Eng.*, **112**, 88 (1934)]. It is a fairly good indication of the abrasive character of the mineral, a factor that determines the wear on the grinding media. Arranged in increasing order or hardness, the Mohs scale is as follows: 1, talc; 2, gypsum; 3, calcite; 4, fluoride; 5, apatite; 6, feldspar; 7, quartz; 8, topaz; 9, corundum; and 10, diamond.

Materials of hardness 1 to 3 inclusive may be classed as soft; 4 to 7, as intermediate; and the others, as hard. Examples are:

Soft Materials (1) Talc, dried filter-press cakes, soapstone, waxes, aggregated salt crystals; (2) gypsum, rock salt, crystalline salts in general, soft coal; (3) calcite, marble, soft limestone, barites, chalk, brimstone.

Intermediate Hardness (4) Fluorite, soft phosphate, magnesite, limestone; (5) apatite, hard phosphate, hard limestone, chromite, bauxite; (6) feldspar, ilmenite, orthoclase, hornblendes.

Hard Materials (7) Quartz, granite; (8) topaz; (9) corundum, sapphire, emery; (10) diamond.

A hardness classification of stone based on the **compressive strength** of 1-in cubes is as follows, for loadings in pounds-force per square inch: very soft, 10,000; soft, 15,000; medium, 20,000; hard, 25,000; very hard, 30,000.

Grindability Methods Laboratory experiments on single particles have been used to correlate grindability. In the past it has usually been assumed that the total energy applied could be related to the grindability whether the energy is applied in a single blow or by repeated dropping of a weight on the sample [Gross and Zimmerly, *Trans. Am. Inst. Min. Metall. Pet. Eng.*, **87**, 27, 35 (1930)]. In fact, the

results depend on the way in which the force is applied (Axelson, Ph.D. thesis, University of Minnesota, 1949). In spite of this, the results of large mill tests can often be correlated within 25 to 50 percent by a simple test, such as the number of drops of a particular weight needed to reduce a given amount of feed to below a certain mesh size.

Two methods having particular application for coal are known as the ball-mill and Hardgrove methods. In the ball-mill method, the relative amounts of energy necessary to pulverize different coals are determined by placing a weighed sample of coal in a ball mill of a specified size and counting the number of revolutions required to grind the sample so that 80 percent of it will pass through a No. 200 sieve. The grindability index in percent is equal to the quotient of 50,000 divided by the average of the number of revolutions required by two tests (ASTM designation D-408).

In the **Hardgrove method** a prepared sample receives a definite amount of grinding energy in a miniature ball-ring pulverizer. The unknown sample is compared with a coal chosen as having 100 grindability. The Hardgrove grindability index = $13 + 6.93W$, where W is the weight of material passing the No. 200 sieve (see ASTM designation D-409).

Chandler [*Bull. Br. Coal Util. Res. Assoc.*, **29**(10), 333; (11), 371 (1965)] finds no good correlation of grindability measured on 11 coals with roll crushing and attrition, and so these methods should be used with caution. The Bond grindability method is described in the subsection "Capacity and Power Consumption."

Manufacturers of various types of mills maintain laboratories in which grindability tests are made to determine the suitability of their machines. When grindability comparisons are made on small equipment of the manufacturers' own class, there is a basis for scale-up to commercial equipment. This is better than relying on a grindability index obtained in a ball mill to estimate the size and capacity of different types such as hammer or jet mills.

OPERATIONS

Mill Wear Wear of mill components costs nearly as much as the energy required for comminution, hundreds of millions of dollars a year. The finer stages of comminution result in the most wear, because the grinding effort is greatest, as measured by the energy input per unit of feed. Parameters that affect wear fall under three categories: (1) the ore, including hardness, presence of corrosive minerals, and particle size; (2) the mill, including composition, microstructure and mechanical properties of the material of construction, size of mill, and mill speed; and (3) the environment, including water chemistry and pH, oxygen potential, slurry solids content, and temperature [Moore et al., *International J. Mineral Processing*, **22**, 313–343 (1988)].

An abrasion index in terms of kilowatthour input per pound of metal lost furnishes a useful indication. Rough values are quoted in Table 20-3.

The use of hard-surfacing techniques by welding and by inserts has contributed greatly to better maintenance and lower downtime [Lutes and Reid, *Chem. Eng.*, **63**(6), 243 (1956)].

In *wet grinding* a synergy between mechanical wear and corrosion results in higher metal loss than with either mechanism alone [Iwasaki, *International J. Mineral Processing*, **22**, 345–360 (1988)]. This is due to removal of protective oxide films by abrasion, and by increased corrosion of stressed metal around gouge marks (Moore, loc. cit.). Wear rate is higher at lower solids content, since ball coating at high solids protects the balls from wear. This indicates that the mechanism is different from dry grinding. The rate without corrosion can be measured with an inert atmosphere such as nitrogen in the mill. Insertion of marked balls into a ball mill best measures the wear rate at conditions in industrial mills, so long as there is not a galvanic effect due to a different composition of the balls. The mill must be cleared of dissimilar balls before a new composition is tested. Sulfide ores promote corrosion due to galvanic coupling by a chemical reaction with oxygen present. Increasing the pH generally reduces corrosion.

The use of harder materials enhances wear resistance, but this conflicts with achieving adequate ductility to avoid catastrophic brittle failure, so these two effects must be balanced. Wear-resistant materials can be divided into three groups: (1) abrasion-resistant steels, (2)

TABLE 20-3   Abrasion Index Test Results*

Material	Number tested	Product diameter, μm	Work index E_i	Average abrasion index† A_i
Alnico	1			0.3850
Alumina	6	15,500		0.6447
Asbestos cement pipe	1	13,330		0.0073
Cement clinker	2	12,100	10.9	0.0409
Cement raw material	4		10.5	0.0372
Chrome ore	1	10,200	9.6	0.1200
Coke	1		20.7	0.3095
Copper ore	12	12,900	11.2	0.0950
Coral rock	1			0.0061
Diorite	1		19.4	0.2303
Dolomite	5		11.3	0.0160
Gneiss	1		20.1	0.5360
Gold ore	2		14.8	0.2000
Granite	11	15,200	14.4	0.3937
Gravel	2		19.0	0.3051
Hematite	3		8.6	0.0952
Iron ore (misc.)	4		5.4	0.0770
Lead zinc ore	3		8.3	0.1520
Limestone	19	13,000	12.1	0.0256
Magnesite	3	14,400	16.8	0.0750
Magnetite	2		10.2	0.2517
Manganese ore	1		17.2	0.1133
Nickel ore	2		11.9	0.0215
Perlite	2			0.0452
Pumice	1		11.9	0.1187
Quartz	7		12.8	0.1831
Quartzite	3		12.2	0.6905
Rare earths	1			0.0288
Rhyolite	2	13,200		0.4993
Schist-biotite	1		23.5	0.1116
Shale	2	11,200	11.2	0.0060
Slag	1		15.8	0.0179
Slate	1		13.8	0.1423
Sulfur	1		11.5	0.0001
Taconite	7		16.2	0.6837
Trap rock	11	14,900	19.9	0.3860
Average		13,250	13.8	0.228

*Allis-Chalmers Corporation.

†Abrasion index is the fraction of a gram weight lost by the standard steel paddle in 1 h of beating 1600 g of ¾- by ½-in particles. The product averages 80 percent passing 13,250 μm.

alloyed cast irons, and (3) nonmetallics [Durman, *International J. Mineral Processing*, **22**, 381–399 (1988)].

Manganese steel, containing 12 to 14 percent manganese and 1 to 1.5 percent carbon, is characterized by its exceptional toughness combined with adequate wear resistance, and enhanced by the austenitic microstructure having the ability to work harden. Although the relatively low yield strength can lead to problems of spreading in service, this can be compensated for by the addition of chromium or molybdenum; or by the use of a lower manganese grade (6–7 percent) also alloyed with up to 1 percent molybdenum. Applications are generally those involving the highest levels of stress, particularly by high-impact loading, such as jaw-crusher elements, gyratory cone-crusher mantles, and primary hammer-mill parts.

For medium- to high-impact applications a wide range of low-alloy steels are produced containing some chromium, molybdenum, phosphorus, and silicon. Economy of manufacture is a benefit for selection of these steels.

Low-alloy steel production falls into two metallurgical types. The traditional approach is heat treatment to produce a pearlitic microstructure. The other is to add sufficient alloying constituents to permit thermal processing to produce a martensitic structure. Typical applications are ball mill liners and grinding balls. Many of these components are consumed in high tonnages, and this combines with metallurgical characteristics to favor production by forging. Low-alloy steel liners, grates, and balls are also produced as castings.

Alloyed white cast irons fall into the second category. This group includes the nickel-chromium grades known as Ni-hards (Durman,

loc. cit.). These contain sufficient chromium to ensure solidification as a white cast iron (at least 2 percent), and sufficient nickel (normally at least 4 percent) to induce hardenability and prevent transformation to soft pearlitic iron. Grade 2 Ni-hard contains massive areas of ledeburite carbide within a matrix of austenite and martensite. Ni-hard 2 is a good general-purpose wear-resistant alloy, but is limited due to an inherent low level of toughness attributed to the presence of the carbide phase. It is suited to small-section components involving low-stress abrasion, such as secondary mill liners. Grade 4 Ni-hard typically contains 5 percent nickel and 8 percent chromium. This alloy is heat-treated to a martensitic matrix and has a modified carbide structure which improves toughness. It is more suited to thicker section castings where heat treatment can consistently control structure.

High-chromium cast irons contain between 12 and 30 percent chromium, 1.5 to 3.5 percent carbon, and frequently contain molybdenum and nickel as secondary constituents. They have become standard in most secondary and tertiary dry-grinding applications (Durman, loc. cit.). These alloys form a metastable austenite structure on casting. Subsequent thermal processing forms secondary chromium-carbide particles dispersed through the matrix. This depletes the austenite of alloy content and facilitates transformation to martensite on quenching. The chromium carbide results in a slightly higher level of toughness than Ni-hard, and higher wear resistance because of greater hardness of chromium carbide. Molybdenum may be added to increase hardenability in heavy sections. Elimination of austenite in the structure can improve resistance to spalling, although spalling limits the range of uses. Hardness can range from 52 to 65 Rockwell. For ball mill balls the dry wear rate is often ⅒ that of cast or forged steel. The cost is 2–3 times as great, so there is an economic advantage. In wet grinding, however, the wear rate of chrome alloys is greater, so the cost may not be competitive. For ring-roll mills, high-chromium molybdenum parts have improved wear costs over use of Ni-hard, and also reduced labor costs for maintenance.

Recent nonmetallic developments include natural rubber, polyurethane, and ceramics. Rubber, due to its high resilience, is extremely wear-resistant in low-impact abrasion. It is inert to corrosive wear in mill liners, pipe linings, and screens. It is susceptible to cutting abrasion, so that wear increases in the presence of heavy particles which penetrate, rather than rebound from the wear surface. Rubber can also swell and soften in solvents. Advantages are its low density leading to energy savings, ease of installation, and sound-proofing qualities. Polyurethane has similar resilient characteristics. Its fluidity at the formation stage makes it suitable for the production of the wearing surface of screens, diaphragms, grates, classifiers, and pump and flotation impellers. The low heat tolerance of elastomers limits their use in dry processing where heat may build up.

Ceramics fill a specialized niche in comminution where metallic contamination cannot be tolerated. Therefore ceramics are used for milling cements and pigments. Ceramic tiles have been used for lining roller mills and chutes and cyclones, where there is a minimum of impact.

Safety   The explosion hazard of such nonmetallic materials as sulfur, starch, wood flour, cereal dust, dextrin, coal, pitch, hard rubber, and plastics is often not appreciated (Hartmann and Nagy, U.S. Bur. Mines Rep. Invest. 3751, 1944). Explosions and fires may be initiated by discharges of static electricity, sparks from flames, hot surfaces, and spontaneous combustion. Metal powders present a hazard because of their **flammability**. Their combustion is favored during grinding operations in which ball, hammer, or ring-roller mills are employed and during which a high grinding temperature may be reached.

Many finely divided metal powders in suspension in air are potential **explosion hazards**, and causes for ignition of such dust clouds are numerous [Hartmann and Greenwald, *Min. Metall.*, **26**, 331 (1945)]. Concentration of the dust in air and its particle size are important factors that determine explosibility. Below a lower limit of concentration, no explosion can result because the heat of combustion is insufficient to propagate it. Above a maximum limiting concentration, an explosion cannot be produced because insufficient oxygen is available. The finer the particles, the more easily is ignition accomplished and the more rapid is the rate of combustion. This is illustrated in Fig. 20-7.

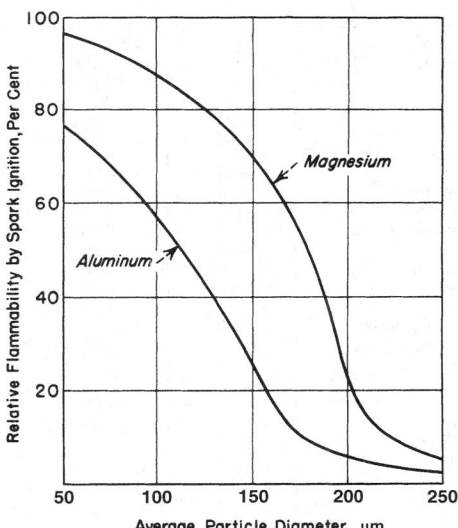

FIG. 20-7 Effect of fineness on the flammability of metal powders. (*Hartmann, Nagy, and Brown, U.S. Bur. Mines Rep. Invest. 3722, 1943.*)

Isolation of the mills, use of nonsparking materials of construction, and magnetic separators to remove foreign magnetic material from the feed are useful **precautions** (Hartman, Nagy, and Brown, U.S. Bur. Mines Rep. Invest. **3722**, 1943). Stainless steel has less sparking tendency than ordinary steel or forgings.

Reduction of the oxygen content of air present in grinding systems is a means for preventing dust explosions in equipment (Brown, U.S. Dep. Agri. Tech. Bull. **74**, 1928). Maintenance of oxygen content below 12 percent should be safe for most materials, but 8 percent is recommended for sulfur grinding. The use of **inert gas** has particular adaptation to pulverizers equipped with air classification; flue gas can be used for this purpose, and it is mixed with the air normally present in a system (see subsection "Chemicals and Soaps" for sulfur grinding). Despite the protection afforded by the use of inert gas, equipment should be provided with explosion vents, and structures should be designed with venting in mind [Brown and Hanson, *Chem. Metall. Eng.*, **40**, 116 (1933)].

Hard rubber presents a fire hazard when reduced on steam-heated rolls (see subsection "Organic Polymers"). Its dust is explosive [Twiss and McGowan, *India Rubber J.*, **107**, 292 (1944)].

An annual publication, *National Fire Codes for the Prevention of Dust Explosions*, is available from the National Fire Protection Association, Quincy, Massachusetts, and should be of interest to those handling hazardous powders.

ATTAINABLE PRODUCT SIZE AND ENERGY REQUIRED

The fineness to which a material is ground has a marked effect on its production rate. Figure 20-8 is an example showing how the capacity decreases and the specific energy and cost increase as the product is ground finer.

Concern about the rising cost of energy has led to publication of a report (National Materials Advisory Board, *Comminution and Energy Consumption*, Publ. NMAB-364, National Academy Press, Washington, 1981; available National Technical Information Service, Springfield, Va. 22151). This has shown that United States industries use approximately 32 billion kWh of electrical energy per annum in size-reduction operations. More than half of this energy is consumed in the crushing and grinding of minerals, one-quarter in the production of cement, one-eighth in coal, and one-eighth in agricultural products. The report recommends that five areas be considered to save energy: classification-device design, mill design, control, addi-

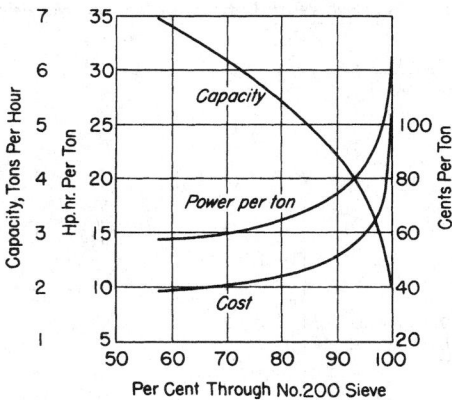

FIG. 20-8 Variation in capacity, power, and cost of grinding relative to fineness of product.

tives, and materials to resist wear. It reviews these areas with an extensive bibliography.

Energy Laws Several laws have been proposed to relate size reduction to a single variable, the energy input to the mill. These laws are encompassed in a general differential equation (Walker, Lewis, McAdams, and Gilliland, *Principles of Chemical Engineering*, 3d ed., McGraw-Hill, New York, 1937):

$$dE = -C\, dX/X^n \tag{20-10}$$

where E is the work done, X is the particle size, and C and n are constants. For $n = 1$ the solution is *Kick's law* (Kick, *Das Gasetz der proportionalen Widerstande und seine Anwendung*, Leipzig, 1885). The law can be written

$$E = C \log (X_F/X_P) \tag{20-11}$$

where X_F is the feed-particle size, X_P is the product size, and X_F/X_P is the reduction ratio. For $n > 1$ the solution is

$$E = \left(\frac{C}{n-1}\right)\left(\frac{1}{X_P^{n-1}} - \frac{1}{X_F^{n-1}}\right) \tag{20-12}$$

For $n = 2$ this becomes *Rittinger's law*, which states that the energy is proportional to the new surface produced (Rittinger, *Lehrbuch der Aufbereitungskunde*, Ernst and Korn, Berlin, 1867).

The *Bond law* corresponds to the case in which $n = 1.5$ [Bond, *Trans. Am. Inst. Min. Metall. Pet. Eng.*, **193**, 484 (1952)]:

$$E = 100E_i\left(\frac{1}{\sqrt{X_P}} - \frac{1}{\sqrt{X_F}}\right) \tag{20-13}$$

where E_i is the **Bond work index,** or work required to reduce a unit weight from a theoretical infinite size to 80 percent passing 100 µm. Extensive data on the work index have made this law useful for rough mill sizing. Summary data are given in Table 20-4.

The work index may be found experimentally from laboratory crushing and grinding tests or from commercial mill operations. Some rules of thumb for extrapolating the work index to conditions different from those measured are that for dry grinding the index must be increased by a factor of 1.34 over that measured in wet grinding; for open-circuit operations another factor of 1.34 is required over that measured in closed circuit; if the product size X_p is extrapolated below 70 µm, an additional correction factor is $(10.3 + X_p)/1.145X_p$. Also for a jaw or gyratory crusher the work index may be estimated from

$$E_i = 2.59C_s/\rho_s \tag{20-14}$$

where C_s = impact crushing resistance, (ft · lb)/in of thickness required to break; ρ_s = specific gravity; and E_i is expressed in kWh/ton.

None of the energy laws apply well in practice, and they have failed to yield a starting point for further development of understanding of milling. They are mainly of historical interest. Most of the early papers

TABLE 20-4   Average Work Indices for Various Materials*

Material	No. of tests	Specific gravity	Work index†	Material	No. of tests	Specific gravity	Work index†
All materials tested	2088	—	13.81	Taconite	66	3.52	14.87
Andesite	6	2.84	22.13	Kyanite	4	3.23	18.87
Barite	11	4.28	6.24	Lead ore	22	3.44	11.40
Basalt	10	2.89	20.41	Lead-zinc ore	27	3.37	11.35
Bauxite	11	2.38	9.45	Limestone	119	2.69	11.61
Cement clinker	60	3.09	13.49	Limestone for cement	62	2.68	10.18
Cement raw material	87	2.67	10.57	Manganese ore	15	3.74	12.46
Chrome ore	4	4.06	9.60	Magnesite, dead burned	1	5.22	16.80
Clay	9	2.23	7.10	Mica	2	2.89	134.50
Clay, calcined	7	2.32	1.43	Molybdenum	6	2.70	12.97
Coal	10	1.63	11.37	Nickel ore	11	3.32	11.88
Coke	12	1.51	20.70	Oil shale	9	1.76	18.10
Coke, fluid petroleum	2	1.63	38.60	Phosphate fertilizer	3	2.65	13.03
Coke, petroleum	2	1.78	73.80	Phosphate rock	27	2.66	10.13
Copper ore	308	3.02	13.13	Potash ore	8	2.37	8.88
Coral	5	2.70	10.16	Potash salt	3	2.18	8.23
Diorite	6	2.78	19.40	Pumice	4	1.96	11.93
Dolomite	18	2.82	11.31	Pyrite ore	4	3.48	8.90
Emery	4	3.48	58.18	Pyrrhotite ore	3	4.04	9.57
Feldspar	8	2.59	11.67	Quartzite	16	2.71	12.18
Ferrochrome	18	6.75	8.87	Quartz	17	2.64	12.77
Ferromanganese	10	5.91	7.77	Rutile ore	5	2.84	12.12
Ferrosilicon	15	4.91	12.83	Sandstone	8	2.68	11.53
Flint	5	2.65	26.16	Shale	13	2.58	16.40
Fluorspar	8	2.98	9.76	Silica	7	2.71	13.53
Gabbro	4	2.83	18.45	Silica sand	17	2.65	16.46
Galena	7	5.39	10.19	Silicon carbide	7	2.73	26.17
Garnet	3	3.30	12.37	Silver ore	6	2.72	17.30
Glass	5	2.58	3.08	Sinter	9	3.00	8.77
Gneiss	3	2.71	20.13	Slag	12	2.93	15.76
Gold ore	209	2.86	14.83	Slag, iron blast furnace	6	2.39	12.16
Granite	74	2.68	14.39	Slate	5	2.48	13.83
Graphite	6	1.75	45.03	Sodium silicate	3	2.10	13.00
Gravel	42	2.70	25.17	Spodumene ore	7	2.75	13.70
Gypsum rock	5	2.69	8.16	Syenite	3	2.73	14.90
Ilmenite	7	4.27	13.11	Tile	3	2.59	15.53
Iron ore	8	3.96	15.44	Tin ore	9	3.94	10.81
Hematite	79	3.76	12.68	Titanium ore	16	4.23	11.88
Hematite—specular	74	3.29	15.40	Trap rock	49	2.86	21.10
Oolitic	6	3.32	11.33	Uranium ore	20	2.70	17.93
Limanite	2	2.53	8.45	Zinc ore	10	3.68	12.42
Magnetite	83	3.88	10.21				

*Allis-Chalmers Corporation.
†Caution should be used in applying the average work index values listed here to specific installations, since individual variations between materials in any classification may be quite large.

supporting one law or another were based on extrapolations of size distributions to finer sizes on the assumption of one or another size-distribution law. With present particle-size-analysis techniques applicable to the finest sizes, such confusion is no longer necessary. The relation of energy expenditure to the size distribution produced has been thoroughly examined [Arbiter and Bhrany, *Trans. Am. Inst. Min. Metall. Pet. Eng.,* **217,** 245–252 (1960); Harris, *Inst. Min. Metall. Trans.,* **75**(3), C37 (1966); Holmes, *Trans. Inst. Chem. Eng. (London),* **35,** 125–141 (1957); and Kelleher, *Br. Chem. Eng.,* **4,** 467–477 (1959); **5,** 773–783 (1960)].

Grinding Efficiency   The energy efficiency of a grinding operation is defined as the energy consumed compared with some ideal energy requirement.

The theoretical energy efficiency of grinding operations is 0.06 to 1 percent, based on values of the surface energy of quartz [Martin, *Trans. Inst. Chem. Eng. (London),* **4,** 42 (1926); Gaudin, *Trans. Am. Inst. Min. Metall. Pet. Eng.,* **73,** 253 (1926)]. Uncertainty in these results is due to uncertainty in the theoretical surface energy.

A definitive monograph (Kuznetsov, *Surface Energy of Solids,* English translation, H. M. Stationery Office, London, 1957) established that most laboratory methods of measuring surface energy introduce large errors, but the cleavage method of Obreimov [Gilman, *J. Appl. Phys.,* **31,** 2208 (1960)] gave results for sodium chloride that agree with theoretical lattice calculations. Later studies by Raasch [*Int. J. Frac. Mech.,* **7**(9), 289 (1971)] and by Burns [*Philos.*

Mag., **25**(1), 131 (1972)] conclude that these measurements are valid when 50 percent corrections are added for the bending energy of the crystal. Kuznetsov ranks other materials by a relative wear test. His results substantiate the efficiencies given earlier. Attempts to measure efficiency of the grinding process by calorimetry involve errors that exceed the theoretical surface energy of the material being ground.

Practical energy efficiency is defined as the efficiency of technical grinding compared with that of laboratory crushing experiments. Practical efficiencies of 25 to 60 percent have been shown [Wilson, *Min. Technol.,* Tech. Publ. 810, 1937; and Bond and Maxson, *Trans. Am. Inst. Min. Metall. Pet. Eng.,* **134,** 296 (1939)].

An **energy coefficient** is sometimes based on Rittinger's law, i.e., new surface produced per unit of energy input. Usually time of grinding is the experimental variable, which is expressed indirectly as energy. The energy coefficient may also be expressed as tons per horsepower-hour passing a certain size. The value of this coefficient is between about 0.02 and 0.1 for wet ball-mill pulverizing hard to medium-hard minerals to No. 200 sieve size (74 μm).

The curves in Fig. 20-9 show decreasing production rate with increasing **moisture content.** (Occasionally, a small amount of water may be beneficial over complete dryness.) All three materials were being ground to 99.9 percent through a No. 200 sieve.

Fine Size Limit   (See also Single-Particle Fracture above.)

It has long been thought that a **limiting size** is attainable. New technologies such as pressed ceramics and Xerox toners require finer sizes,

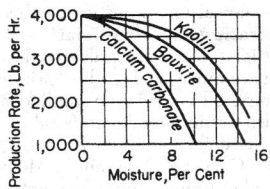

FIG. 20-9 Effect of moisture on the production rate of a pulverizer. [*Work*, *Chem. Metall. Eng.*, **40**, 306 (1933).]

and this again questions the existence of a limit. There are three theories for such a limit. Bradshaw [*J. Chem. Phys.*, **19**, 1057–1059 (1951)] thought that *reagglomeration* is responsible, especially in ball mills. Schönert and Steier [*Chem. Ing. Tech.*, **43**(13), 773 (1971)] suggest two other causes: *plastic deformation* and the difficulty of *stressing* fine particles to their breaking point. The latter stems from the Griffith crack theory, which requires that the particle have enough stored stress energy to allow a crack to propagate. A 10-μm glass particle requires 140 kPa/mm² tensile stress. Although both of these mechanisms can be limiting, recent experimental evidence indicates that plastic deformation can increase the resistance of even the most brittle materials on a fine scale. Rumpf and Schönert (*Third European Symposium on Size Reduction*, op. cit., 1972, p. 27) observed plastic deformation in crushing fine glass spheres. Schönert and Steier (loc. cit.) in electron-microscope photographs observed plastic deformation in crushing limestone particles as large as 3 to 4 μm and quartz particles of 2 to 3 μm. This deformation spreads a stress that would otherwise produce brittle fracture. Gane [*Philos. Mag.*, **25**(1), 25 (1972)] observed plastic deformation in magnesium oxide crystals 0.2 to 0.4 μm in size. The strengths average 180 kg/mm², which is 15 times the strength of large MgO crystals but one-tenth of the theoretical strength. Further proof is given by Weichert and Schönert [*J. Mech. Phys. Solids*, **22**, 127 (1974)], who analyze and measure the temperature rise at a propagating crack tip. They estimate that irreversible deformation occurs in a zone of radius about 30 A running along at the tip. The energy release causes temperatures as high as 1500 K above ambient temperature at the tip. This temperature explains plastic flow and even emitted light in some cases. Therefore, it is proved that plastic deformation can limit the grinding size attainable. Other means than size reduction must be found if particles much finer than 0.5 μm are wanted.

Dry versus Wet Grinding (See under Ball Mills, and Wear.) In practice it is found that finer size can be achieved by **wet grinding** than by **dry grinding.** In wet grinding by ball mills or vibratory mills with suitable surfactants, product sizes of 0.5 μm are attainable. In dry grinding the size is generally limited by ball coating (Bond and Agthe, *Min. Technol.*, AIME Tech. Publ. 1160, 1940) to about 15 μm. In dry grinding with hammer mills or ring-roller mills the limiting size is about 10 to 20 μm. Jet mills are generally limited to a product mean size of 15 μm, although dense particles can be ground to 5 μm because of the greater ratio of inertia to aerodynamic drag.

Dispersing Agents and Grinding Aids There is no doubt that grinding aids are helpful under some conditions. For example, surfactants make it possible to ball-mill magnesium in kerosine to 0.5-μm size [Fochtman, Bitten, and Katz, *Ind. Eng. Chem. Prod. Res. Dev.*, **2**, 212–216 (1963)]. Without surfactants the size attainable was 3 μm, and of course the rate of grinding was very slow at sizes below this. Also, the water in wet grinding may be considered to act as an additive.

Chemical agents that increase the rate of grinding are an attractive prospect since their cost is low. However, despite a voluminous literature on the subject, there is no accepted scientific method to choose such aids; there is not even agreement on the mechanisms by which they work. The subject has been recently reviewed [Fuerstenau, *KONA Powder and Particle*, **13**, 5–17 (1995)].

In wet grinding there are several theories, which have been reviewed [Somasundaran and Lin, *Ind. Eng. Chem. Process Des. Dev.*, **11**(3), 321 (1972); Snow, annual reviews, op. cit., 1970–1974. See also Rose, *Ball and Tube Milling*, Constable, London, 1958, pp. 245–249]. The *Rehbinder theory* (Rehbinder, Schreiner, and Zhi-

gach, *Hardness Reducers in Rock Drilling*, Moscow Academy of Science, 1944, transl. Council for Scientific and Industrial Research, Melbourne, Australia, 1948).

Additives can alter the rate of wet ball milling by changing the slurry viscosity or by altering the location of particles with respect to the balls. These effects are discussed under "Tumbling Mills." In conclusion, there is still no theoretical way to select the most effective additive. Empirical investigation, guided by the principles discussed earlier, is the only recourse. There are a number of commercially available grinding aids that may be tried. Also, a kit of 450 surfactants that can be used for systematic trials (Model SU-450, Chem Service Inc., West Chester, PA 19380) is available.

Numerous experimental studies lead to the conclusion that dry grinding is limited by ball coating and that additives function by reducing the tendency to coat (Bond and Agthe, op. cit.). Most materials coat if they are ground fine enough, and softer materials coat at larger sizes than hard materials. The presence of more than a few percent of soft gypsum promotes ball coating in cement-clinker grinding. The presence of a considerable amount of coarse particles above 35 mesh inhibits coating. Balls coat more readily as they become scratched. Small amounts of moisture may increase or decrease ball coating, and dry materials also coat.

Materials used as grinding aids include solids such as graphite, oleoresinous liquid materials, volatile solids, and vapors. The complex effects of vapors have been extensively studied [Goette and Ziegler, *Z. Ver. Dtsch. Ing.*, **98**, 373–376 (1956); and Locher and von Seebach, *Ind. Eng. Chem. Process Des. Dev.*, **11**(2), 190 (1972)], but water is the only vapor used in practice.

The most effective additive for dry grinding is fumed silica that has been treated with methyl silazane [Tulis, *J. Hazard. Mater.*, **4**, 3 (1980)].

SIZE REDUCTION COMBINED WITH OTHER OPERATIONS

Practically every solid material undergoes size reduction at some point in its processing cycle. Some of the reasons for size reduction are: (1) to liberate a desired component for subsequent separation, as in separating ores from gangue; (2) to prepare the material for subsequent chemical reaction, i.e., by enlarging the specific surface as in cement manufacture; (3) to subdivide the material so that it can be intimately blended with other components; (4) to meet a size requirement for the quality of the end product, as in fillers or pigments for paints, plastics, agricultural chemicals etc.; (5) to prepare wastes for recycling.

Systems Involving Size Reduction Industrial applications usually involve a number of processing steps combined with size reduction [Hixon, *Chemical Engineering Progress*, **87**, 36–44 (May 1991)]. The most common of these is **size classification.** Often only a particular range of product sizes is wanted for a given application. Since the particle breakage process always yields a spectrum of sizes, the product size can not be directly controlled; however, mill operation can sometimes be varied to produce less fines at the expense of producing more coarse particles. By recycling the classified coarse fraction and regrinding it, production of the wanted size range is optimized. Such an arrangement of classifier and mill is called a **mill circuit,** and is dealt with further below.

More complex systems may include several unit operations such as mixing (Sec. 18), drying (Sec. 12), and agglomerating (see Size Enlargement, this section). Inlet and outlet silencers are helpful to reduce noise from high-speed mills. Chillers, air coolers, and explosion proofing may be added to meet requirements. Weighing and packaging facilities complete the system.

Batch ball mills with low ball charges can be used in dry mixing or standardizing of dyes, pigments, colors, and insecticides to incorporate wetting agents and inert extenders (see also Sec. 21). Disk mills, hammer mills, and other high-speed disintegration equipment are useful for intensive blending of insecticide compositions, earth colors, cosmetic powders, and a variety of other finely divided materials that tend to agglomerate in ribbon and conical blenders. Liquid sprays or gases may be injected into the mill or air stream, for mixing with the material being pulverized to effect chemical reaction or surface treatment.

The **drying** of materials while they are being pulverized or disintegrated is known variously as "flash" or "dispersion" drying; a generic term is "pneumatic conveying" drying. Data for the grinding and drying of bauxite in a ring-roller mill are given in Table 20-5. A drying system is shown under "Clays and Kaolins," Fig. 20-58.

Milling Problems   Materials with low-softening temperatures, such as chocolate, are amenable to pulverizing if proper temperature control is exercised. Compositions containing fats and waxes are pulverized and blended readily if refrigerated air is introduced into their grinding systems (U.S. Patents 1,739,761 and 2,098,798; see also subsection "Organic Polymers" and Hixon loc. cit. for flow sheets). Some materials, such as salt, are very hygroscopic; they pick up water from air and deposit on mill surfaces, forming a hard cake. Mills with air-classification units may be equipped so that the circulating air can be conditioned by mixing with hot or cold air or gases introduced into the mill or by dehumidification to prepare the air for the grinding of hygroscopic materials. Flow sheets including air dryers are also described by Hixon. All organic materials and most metals can form **flammable** or **explosive** mixtures with air (see Safety above). The feed material may be toxic, bioreactive or radioactive, requiring isolation. This may best be accomplished by batch operation. Or the material may be corrosive to mill components. Iron contamination from wear or corrosion of the mill is often a problem. Jet milling is used to produce ultrapure materials for semiconductor manufacture.

Continuous Operation   Advantages, which can be very important (Hixon, loc. cit.), are: (1) There is increased economical operation. Less time is needed for start-up and shutdown, and often less maintenance is needed. Because the operating conditions are constant, less operator attention may be needed, and automatic control is more readily applied. (2) It may be more feasible to increase the scale of a continuous system, thus improving unit economy. (3) There is better quality control. This may be the major benefit. Once the operating parameters are properly set, the continuous system will provide a more consistent product. (4) The improvement of the competitive position. A clever arrangement of a system may give an advantage either in product quality or cost. The disadvantages of a continuous system are (1) increased complexity of equipment over batch processing, and (2) the need for more thorough planning, usually requiring pilot-scale testing.

Some special requirements of continuous systems are: (1) Metering the feed. A continuous system must be fed at a precise, uniform rate. (See Sec. 21.) (2) Dust collection. This is a necessary part of most dry-processing systems. Filters are available that can effectively remove dust down to 10 mg/m³ or less, and operate automatically. (Dust collection is covered in Sec. 17.) (3) On-line analysis. For more precise operation, on-line analysis of product particle size and composition may be desirable. (4) Computer control. Simulation can aid in optimizing system design and computer control.

Beneficiation   Ball and pebble mills, batch or continuous, offer considerable opportunity for combining a number of **processing steps** that include grinding [Underwood, *Ind. Eng. Chem.*, **30**, 905 (1938)]. Mills followed by air classifiers can serve to **separate components of mixtures** because of differences in specific gravity and particle size. The removal of impurities by this means is known as **cleaning, concentrating,** or **beneficiating.** Screens are used to separate coarse particles, not easily pulverized, from fine particles of the component that are pulverized readily. Grinding followed by *froth flotation* has become the beneficiation method most widely used for metallic ores and also for nonmetallic minerals such as feldspar. Magnetic separation is the chief means used for upgrading taconite iron ore (see subsection "Ores and Minerals"). Magnetic separators frequently are employed to remove tramp magnetic solids from the feed to high-speed hammer and disk mills.

Liberation   Most ores are heterogeneous, and the objective of grinding is to release the valuable mineral component so that it can be separated. Calculations based on a random-breakage model assuming no preferential breakage [Wiegel and Li, *Trans. Am. Inst. Min. Metall. Pet. Eng.*, **238**, 179–191 (1967)] agreed at least in general trends with plant data on the efficiency of release of mineral grains. Figure 20-10 shows that the desired mineral B can be liberated by coarse grinding when the grade is high so that mineral A becomes a small fraction and mineral B a large fraction of the total volume; mineral B can be liberated only by fine grinding below the grain size, when the grade is low so that there is a small proportion of grains of B. Similar curves, somewhat displaced in size, resulted from a more detailed integral geometry analysis by Barbery [*Minerals Engineering*, **5**(2), 123–141 (1992)]. There is at present no way to measure grain size on-line, and thus to control liberation. The current status of liberation modeling is given by Mehta et al. [*Powder Technology*, **58**(3), 195–209 (1989)].

Many authors have assumed that breakage occurs preferentially along grain boundaries, but there is scant evidence for this. On the contrary, Gorski [*Bull. Acad. Pol. Sci. Ser. Sci. Tech.*, **20**(12), 929 (1972); CA 79, 20828k], from analysis of microscope sections, finds an intercrystalline character of comminution of dolomite regardless of the type of crusher used.

The liberation of a valuable constituent does not necessarily translate directly into recovery in downstream processes. For example, flotation tends to be more efficient in intermediate sizes than at coarse or fine sizes [McIvor and Finch, *Minerals Engineering*, **4**(1), 9–23 (1991)]. For coarser sizes, failure to liberate may be the limitation; finer sizes that are liberated may still be carried through by the water flow. A conclusion is that overgrinding should be avoided by judicious use of size classifiers with recycle grinding.

Size Reduction Combined with Size Classification   Grinding systems are batch or continuous in operation (Fig. 20-11). Most large-scale operations are continuous; batch ball or pebble mills are used

TABLE 20-5   Operating Data for Grinding and Drying of Bauxite in a Ring-Roller Mill

Initial moisture, %	9.75
Final moisture, %	0.75
Feed, lb./hr	12,560
Product, lb./hr	11,420
Moisture evaporated, lb	1,140
Temperature of gases entering mill, °F	700
Temperature of gases leaving mill, °F	170
Temperature of feed, °F	70
Temperature of material leaving mill, °F	150
Oil consumed, gal	14.3
Heating value of oil, B.t.u./gal	142,000
Thermal efficiency, %	68.5
Total power for drying and pulverizing, hp	105
Power for drying, hp	10
Final product, % through No. 100 sieve	90

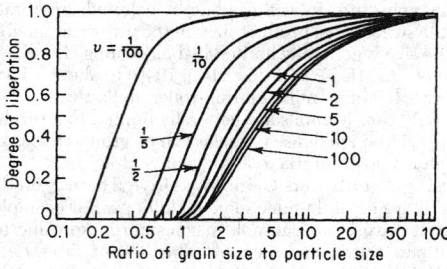

FIG. 20-10   Fraction of mineral B that is liberated as a function of volumetric abundance ratio υ of gangue to mineral B (1/grade), and ratio of grain size to particle size of broken fragments (1/fineness). [*Wiegel and Li*, Trans. Soc. Min. Eng.-Am. Inst. Min. Metall. Pet. Eng., **238**, 179 (1967).]

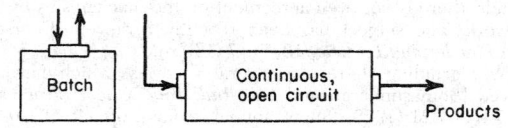

FIG. 20-11   Batch and continuous grinding systems.

only when small quantities are to be processed. Batch operation involves a high labor cost for charging and discharging the mill.

Continuous operation is accomplished in open or closed circuit, as illustrated in Figs. 20-11 and 20-12. **Operating economy** is the object of closed-circuit grinding with size classifiers. The idea is to remove the material from the mill before all of it is ground, separate the fine product in a classifier, and return the coarse for regrinding with the new feed to the mill. A mill with the fines removed in this way performs much more efficiently. Coarse material returned to a mill by a classifier is known as the **circulating load;** its rate may be from 1 to 10 times the production rate. The ability of the mill to transport material may limit the recycle rate; tube mills for use in such circuits may be designed with a smaller length-to-diameter ratio and hence a larger hydraulic gradient for more flow or with compartments separated by diaphragms with lifters.

Internal size classification plays an essential role in the functioning of machines for dry grinding in the fine-size range; particles are retained in the grinding zone until they are as small as required in the finished product; then they are allowed to discharge.

By closed-circuit operation the product size distribution is narrower and will have a larger proportion of particles of the desired size. On the other hand, making a *product size within narrow limits* (such as between 20 and 40 μm) is often requested but usually is not possible regardless of the grinding circuit used. The reason is that particle breakage is a random process, both as to the probability of breakage of particles and as to the sizes of fragments produced from each breakage event. The narrowest size distribution ideally attainable is one that has a slope of 1.0 when plotted on Gates-Gaudin-Schumann coordinates [Eq. (20-2) and Fig. 20-13]. This can be demonstrated by examining the Gaudin-Meloy size distribution [Eq. (20-4)]. This is the distribution produced in a mill when particles are cut into pieces of random size, with r cuts per event. The case in which r is large corresponds to a breakage event producing many fines. The case in which r is 1 corresponds to an ideal case such as a knife cutter, in which each particle is cut once per event and the fragments are removed immediately by the classifier. The Meloy distribution with $r = 1$ reduces to the Schumann distribution with a slope of 1.0. Therefore, no practical grinding operation can have a slope greater than 1.0. Slopes typically range from 0.5 to 0.7. The specified product may still be made, but the finer fraction may have to be disposed of in some way. Within these limits, the size distribution of the classifier product depends both on the recycle ratio and on the sharpness of cut of the classifier used.

Characteristics of Size Classifiers (See Sec. 19: "Screening" on screening equipment and "Wet Classification" on wet classifiers.) Types of classifiers and commercially available equipment are de-

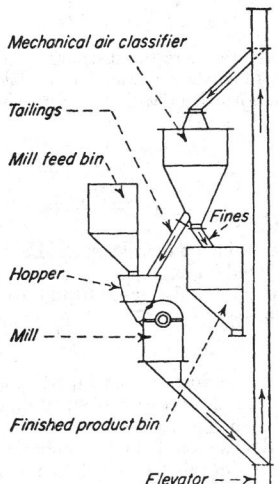

FIG. 20-12 Hammer mill in closed circuit with an air classifier.

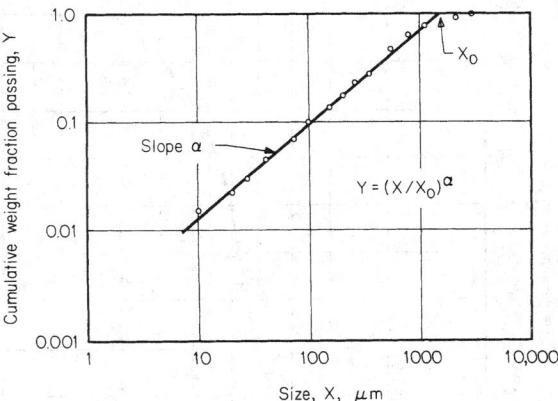

FIG. 20-13 Example of a Gates-Gaudin-Schumann plot of mill-product-size distribution.

scribed in the subsection "Particle-Size Classifiers Used with Grinding Mills." The American Institute of Chemical Engineers Equipment Testing Procedures Committee has published a procedure for particle-size classifiers (*Particle-Size Classifiers—A Guide to Performance Evaluation,* 2d ed. American Institute of Chemical Engineers, New York, 1994), including definitions which are followed here.

Three parameters define the performance of a classifier. These are *cut size, sharpness* of cut, and *capacity.* Cut size, X_{50}, is the size at which 50 percent of the material goes into the coarse product and 50 percent into the fine. (This should not be confused with the "cutoff size," a name sometimes given to the top size of the fine product.)

Size selectivity is the most thorough method of expressing classifier performance under a given set of operating conditions. Cut size and sharpness can be calculated from size-selectivity data. Size selectivity is defined by

$$\eta_X = \frac{\text{quantity of size } X \text{ entering coarse fraction}}{\text{quantity of size } X \text{ in feed}} \quad (20\text{-}15)$$

An equivalent mathematical expression is, on a *mass* basis,

$$\eta_X = \frac{q_c dY_c}{q_0 dY_0} = \frac{q_c dY_c}{q_c dY_c + q_f dY_f} \quad (20\text{-}16)$$

where Y_c is the cumulative percent by mass of coarse fraction less than particle size X, Y_f is the cumulative percent by mass of fine fraction less than particle size X, Y_0 is the cumulative percent by mass of feed less than particle size X, q_c is the coarse-fraction mass flow rate, q_f is the fine-fraction mass flow rate, and q_0 is the feed mass flow rate.

For purposes of calculating size selectivity from cumulative particle size distribution data, Eq. (20-16) can be expressed in incremental form as follows:

$$\eta_{X_i} = \frac{q_c \Delta Y_{ci}}{q_c \Delta Y_{ci} + q_f \Delta Y_{fi}} \quad (20\text{-}17)$$

where ΔY_{ci} and ΔY_{fi} are the cumulative size-distribution intervals of coarse and fine fractions associated with the size interval ΔX_i respectively. An interval representative size X_i is arbitrarily taken as the midpoint of ΔX_i.

See the American Institute of Chemical Engineers classifier test procedure for a sample calculation of classifier selectivity. This example is plotted in Fig. 20-14.

There are many ways in which sharpness can be expressed. One index that has been widely used is the ratio

$$\beta = X_{25}/X_{75} \quad (20\text{-}18)$$

where β is the sharpness index, X_{75} is the particle size corresponding to the 75 percent classifier selectivity value, and X_{25} is the particle size corresponding to the 25 percent value. For perfect classification, β has a value of unit; the smaller β, the poorer the sharpness of classification.

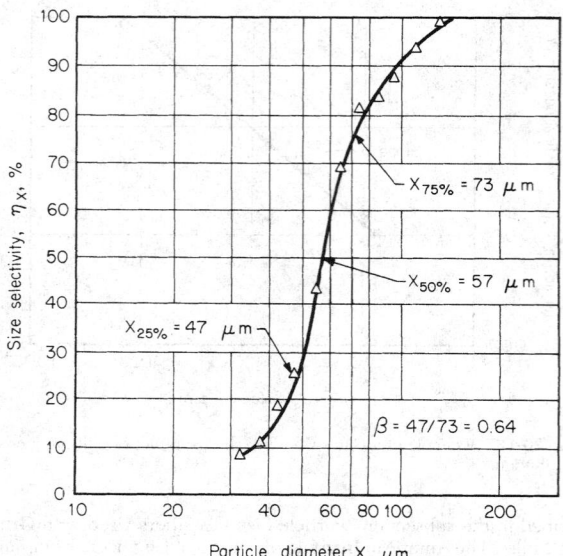

FIG. 20-14 Size-selectivity example.

Several empirical formulas for classifier selectivity have been proposed. Such a formula is needed for computer simulation of mill circuits. The following formula has been found to fit data from several field installations for classifiers of many types, including vibrating screens (Vaillant, AIME Tech. Pap. 67B26, 1967).

$$C_x = 1 - (1-a) \exp b \left(1 - \frac{X}{X_0}\right) \qquad \text{for } X > X_0$$

$$= a \qquad \text{for } X \leq X_0 \qquad (20\text{-}19)$$

where a, b, and X_0 are constants and X is the particle size. The agreement is especially good for wet classifying systems. For wet cyclones the factor a and Y_{50} are related to the ratio of overflow to underflow rates [Draper and Lynch, *Proc. Australas. Inst. Min. Metall.*, **209**, 109 (1964); Mizrahi and Cohen, *Trans. Inst. Min. Metall.*, C318–329 (December 1966); and Lynch and Rao, *Indian J. Tech.*, **6**, 106–114 (April 1968)]. An equation developed by stochastic reasoning for cyclones involves a similar exponential form [Molerus, *Chem. Ing. Tech.*, **39**(13), 792–796 (1967)].

It has been suggested that the circulating load can be calculated by a material balance from size analyses of the feed, fine product, and coarse product of the classifier in a closed-circuit grinding system [Bond, *Rock Prod.*, **41**, 64 (January 1938)]. However, since size analyses are subject to error, it is better to use this information to check the size analyses (Vaillant, op. cit.). The appropriate equation is (Dahl, *Classifier Test Manual*, Portland Cem. Assoc. Bull. MRB-53, 1954)

$$\frac{q_R}{q_R + q_P} = R = \frac{Y_L(X) - Y_P(X)}{Y_R(X) - Y_P(X)} = \text{a constant for all } X \qquad (20\text{-}20)$$

where q_R is tailings, q_P is classifier product, L is mill discharge, R is recycle, and Y is either the fraction of particles in a stream between two sieve sizes or the cumulative fraction retained or passing a sieve of size X.

SIMULATION OF MILLING CIRCUITS

The energy laws of Bond, Kick, and Rittinger relate to grinding from some average feed size to some product size but do not take into account the behavior of different sizes of particles in the mill. Computer simulation, based on population-balance models [Bass, *Z. Angew. Math. Phys.*, **5**(4), 283 (1954)], traces the breakage of each size of particle as a function of grinding time. Furthermore, the simu-

lation models separate the breakage process into two aspects: a breakage rate and a mean fragment-size distribution. These are both functions of the size of particle being broken. They usually are not derived from knowledge of the physics of fracture but are empirical functions fitted to milling data. The following formulation is given in terms of a discrete representation of size distribution; there are comparable equations in integrodifferential form.

Batch Grinding Let w_k = the weight fraction of material retained on each screen of a nest of n screens; w_k is related to P_k, the fraction coarser than size X_k, by

$$w_k = (\partial P_k / \partial X_k) \Delta X_k \qquad (20\text{-}21)$$

where ΔX_k is the difference between the openings of screens k and $k+1$. The **grinding-rate function** S_u is the rate at which the material of upper size u is selected for breakage in an increment of time, relative to the amount of that size present:

$$dw_u/dt = -S_u w_u \qquad (20\text{-}22)$$

The **breakage function** $\Delta B_{k,u}$ gives the size distribution of product breakage of size u into all smaller sizes k. Since some fragments from size u are large enough to remain in the range of size u, the term $\Delta B_{u,u}$ is not zero, and

$$\sum_{k=n}^{u} \Delta B_{k,u} = 1 \qquad (20\text{-}23)$$

The differential equation of batch grinding is deduced from a balance on the material in the size range k. The rate of accumulation of material of size k equals the rate of production from all larger sizes minus the rate of breakage of material of size k:

$$\frac{dw_k}{dt} = \sum_{u=1}^{k} [w_u S_u(t) \Delta B_{k,u}] - S_k(t) w_k \qquad (20\text{-}24)$$

In general, S_u is a function of all the milling variables. $\Delta B_{k,u}$ is also a function of breakage conditions. If it is assumed that these functions are constant, then relatively simple solutions of the grinding equation are possible, including an analytical solution [Reid, *Chem. Eng. Sci.*, **20**(11), 953–963 (1965)] and matrix solutions [Broadbent and Callcott, *J. Inst. Fuel*, **29**, 524–539 (1956); **30**, 18–25 (1967); and Meloy and Bergstrom, *7th Int. Min. Proc. Congr. Tech. Pap.*, 1964, pp. 19–31].

Solution of Batch-Mill Equations In general, the grinding equation can be solved by numerical methods—for example, the Euler technique (Austin and Gardner, *1st European Symposium on Size Reduction*, 1962) or the Runge-Kutta technique. The matrix method is a particularly convenient formulation of the Euler technique.

Reid's **analytical solution** is useful for calculating the product as a function of time t for a constant feed composition. It is

$$w_{L,k} = \sum_{n=1}^{k} a_{k,n} \exp(-\overline{S}_n \Delta t) \qquad (20\text{-}25)$$

where the subscript L refers to the discharge of the mill, zero to the entrance, and $\overline{S}_n = 1$ "corrected" rate function defined by $\overline{S}_n = (1 - \Delta B_{n,n})$ and B is then normalized with $B_{n,n} = 0$. The coefficients are

$$a_{k,k} = w_{0k} - \sum_{n=1}^{k-1} a_{k,n} \qquad (20\text{-}26)$$

and

$$a_{k,n} = \sum_{u=n}^{k-1} \frac{S_u \Delta B_{k,u} a_{n,u}}{\overline{S}_k - \overline{S}_n} \qquad (20\text{-}27)$$

The coefficients are evaluated in order since they depend on the coefficients already obtained for larger sizes.

The basic idea behind the **Euler method** is to set the change in w per increment of time as

$$\Delta w_k = (dw_k/dt) \Delta t \qquad (20\text{-}28)$$

where the derivative is evaluated from Eq. (20-24). Equation (20-28) is applied repeatedly for a succession of small time intervals until the desired duration of milling is reached.

In the matrix method a modified rate function is defined, $S'_k = S_k \Delta t$ as the amount of grinding that occurs in some small time Δt. The result is

$$\mathbf{w}_L = (\mathbf{I} + \mathbf{S'B} - \mathbf{S'})\mathbf{w}_F = \mathbf{M}\mathbf{w}_F \qquad (20\text{-}29)$$

where the quantities **w** are vectors, **S′** and **B** are the matrices of rate and breakage functions, and **I** is the unit matrix. This follows because the result obtained by multiplying these matrices is just the sum of products obtained from the Euler method. Equation (20-29) has a physical meaning. The unit matrix times $\mathbf{w}_F$ is simply the amount of feed that is not broken. $\mathbf{S'Bw}_F$ is the amount of feed that is selected and broken into the vector of products. $\mathbf{S'w}_F$ is the amount of material that is broken out of its size range and hence must be subtracted from this element of the product. The entire term in parentheses can be considered as a mill matrix **M**. Thus the milling operation transforms the feed vector into the product vector. Meloy and Bergstrom (op. cit.) pointed out that when Eq. (20-29) is applied over a series of p short-time intervals, the result is

$$\mathbf{w}_L = \mathbf{M}^p \mathbf{w}_F \qquad (20\text{-}30)$$

Matrix multiplication happens to be cummutative in this special case. It is easy to raise a matrix to a power on a computer since three multiplications give the eighth power, etc. Therefore the matrix formulation is well adapted to computer use.

Continuous-Mill Simulation   Batch-grinding experiments are the simplest type of experiments to produce data on grinding coefficients. But scale-up from batch to continuous mills must take into account the **residence-time distribution** in a continuous mill. This distribution is apparent if a tracer experiment is carried out. For this purpose background ore is fed continuously, and a pulse of tagged feed is introduced at time t_0. This tagged material appears in the effluent distributed over a period of time, as shown by a typical curve in Fig. 20-15. Because of this distribution some portions are exposed to grinding for longer times than others. Levenspiel (*Chemical Reaction Engineering*, Wiley, New York, 1962) shows several types of residence-time distribution that can be observed. Data on large mills indicate that a curve like that of Fig. 20-15 is typical (Keienberg et al., *3d European Symposium on Size Reduction*, op. cit., 1972, p. 629). This curve can be accurately expressed as a series of arbitrary functions (Merz and Molerus, *3d European Symposium on Size Reduction*, op. cit., 1972, p. 607). A good fit is more easily obtained if we choose a function that has the right shape since then only the first two moments are needed. The log-normal probability curve fits most available mill data, as was demonstrated by Mori [*Chem. Eng. (Japan)*, **2**(2), 173 (1964)]. Two examples are shown in Fig. 20-16. The log-normal plot fails only when the mill acts nearly as a perfect mixer.

To measure a residence-time distribution, a pulse of tagged feed is inserted into a continuous mill and the effluent is sampled on a schedule. If it is a dry mill, a soluble tracer such as salt or dye may be used and the samples analyzed conductimetrically or colorimetrically. If it is a wet mill, the tracer must be a solid of similar density to the ore. Materials like copper concentrate, chrome brick, or barites have been used as tracers and analyzed by X-ray fluorescence. To plot results in log-normal coordinates, the concentration data must first be normalized from the form of Fig. 20-15 to the form of cumulative percent discharged, as in Fig. 20-16. For this, one must either know the total amount of pulse fed or determine it by a simple numerical integration

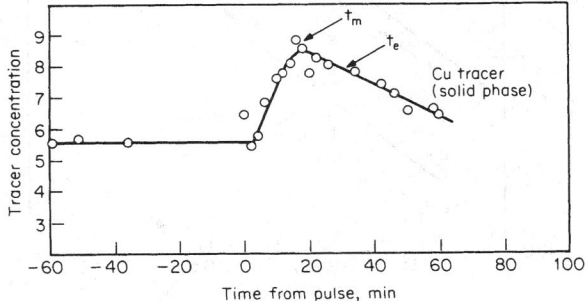

FIG. 20-15   Ore transit through a ball mill. Feed rate is 500 lb h. (*Courtesy Phelps Dodge Corporation.*)

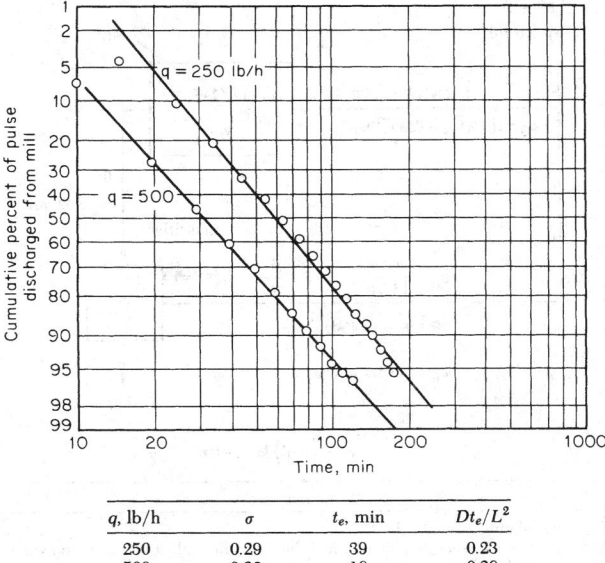

q, lb/h	σ	t_e, min	Dt_e/L^2
250	0.29	39	0.23
500	0.32	18	0.28

FIG. 20-16   Log-normal plot of residence-time distribution in Phelps Dodge mill.

by using a computer. The data are then plotted as in Fig. 20-16, and the coefficients in the log-normal formula of Mori can be read directly from the graph. Here $t_e = t_{50}$ is the time when 50 percent of the pulse has emerged. The standard deviation σ is the time between t_{16} and t_{50} or between t_{50} and t_{84}. Knowing t_e and σ, one can reconstruct the straight line in log-normal coordinates. One can also calculate the vessel dispersion number, Dt_e/L^2, which is a measure of the sharpness of the pulse (Levenspiel, *Chemical Reactor Omnibook*, p. 100.6, Oregon State University Bookstores Inc., 1979). This number has erroneously been called by some the Peclet number). Here D is the particle diffusivity. A few available data are summarized [Snow, *International Conference on Particle Technology*, IIT Research Institute, Chicago, Ill. 60616, 1973, p. 28) for wet mills. Other experiments are presented for dry mills [Hogg et al., *Trans. Am. Inst. Min. Metall. Pet. Eng.*, **258**, 194 (1975)]. The most important variables affecting the vessel dispersion number are L/diameter of the mill, ball size, mill speed, scale expressed either as diameter or as throughput, degree of ball filling, and degree of material filling.

Solution for Continuous Mill   In the method of Mori (op. cit.) the residence-time distribution is broken up into a number of segments, and the batch-grinding equation is applied to each of them. The resulting size distribution at the mill discharge is

$$\mathbf{w}(L) = \mathbf{w}(t)\,\Delta\varphi \qquad (20\text{-}31)$$

where $\mathbf{w}(t)$ is a matrix of solutions of the batch equation for the series of times t, with corresponding segments of the cumulative residence-time curve.

Using the Reid solution, Eq. (20-25), this becomes

$$\mathbf{w}(L) = \mathbf{RZ}\,\Delta\varphi \qquad (20\text{-}32)$$

since the Reid solution [Eq. (20-25)] can be separated into a matrix **Z** of exponentials exp $(-\bar{S}t)$ and another factor R involving only particle sizes. Austin, Klimpel, and Luckie [*Process Engineering of Size Reduction: Ball Milling*, Society of Mining Engineers of AIME, (1984)] incorporated into this form a tanks-in-series model for the residence time distribution.

Closed-Circuit Milling   In closed-circuit milling the tailings from a classifier are mixed with fresh feed and recycled to the mill. Calculations can be based on a material balance and an explicit solu-

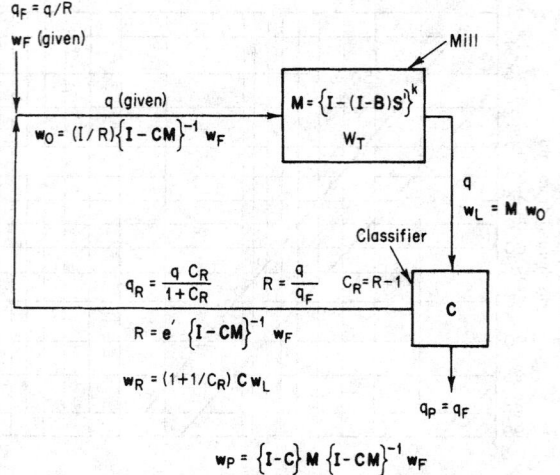

Nomenclature

C_R = circulating load, $R - 1$
C = classifier selectivity matrix, which has classifier selectivity-function values η on diagonal zeros elsewhere
I = identity matrix, which has ones on diagonal, zeros elsewhere
$\mathbf{M}$ = mill matrix, which transforms mill-feed-size distribution into mill-product-size distribution
q = flow rate of a material stream
R = recycle ratio q/q_F
w = vector of differential size distribution of a material stream
W_T = holdup, total mass of material in mill

Subscripts:
0 = inlet to mill
F = feed stream
L = mill-discharge stream
P = product stream
R = recycle stream, classifier tailings

FIG. 20-17 Normal closed-circuit continuous grinding system with stream flows and composition matrices, obtained by solving material-balance equations. [*Callcott, Trans. Inst. Min. Metall.,* **76**(1), C1-11 (1967).]

tion such as Eq. (20-30). Material balances for the normal circuit arrangement (Fig. 20-17) give

$$q = q_F + q_R \qquad (20\text{-}33)$$

where q = total mill throughput, q_F = rate of feed of new material, and q_R = recycle rate. A material balance on each size gives

$$w_{0,k} = \frac{q_F w_{Fk} + \dfrac{q_R}{R}\,\eta_k w_{L,k}}{q} \qquad (20\text{-}34)$$

where $w_{0,k}$ = fraction of size k in the mixed feed streams, R = the recycle ratio, and η_k = classifier selectivity for size k. With these conditions a calculation of the transient behavior of the mill can be performed by using any method of solving the milling equation and iterating over intervals of time τ = residence time in the mill. This information is important for evaluating mill-circuit-control stability and strategies. If the throughput q is controlled to be a constant, as is often the case, then τ is constant, and a closed-form matrix solution can be found for the steady state [Callcott, *Trans. Inst. Min. Metall.,* **76**(1), C1-11 (1967)]. The resulting flow rates and composition vectors are given in Fig. 20-17. Equations for the reverse-circuit case, in which the feed is classified before it enters the mill, are given by Calcott (loc. cit.). These results can be used to investigate the effects of changes in feed composition on the product. Separate calculations can be made to find the effects of classifier selectivity, mill throughput or recycle, and

grindability (rate function) to determine optimum mill-classifier combinations [Lynch, Whiten, and Draper, *Trans. Inst. Min. Metall.,* **76**, C169, 179 (1967)]. Equations such as these form the basis for computer codes that are available for modeling mill circuits [Austin, Klimpel, and Luckie (loc. cit.)].

Data on Behavior of Grinding Functions   Although several breakage functions were early suggested [Gardner and Austin, *1st European Symposium on Size Reduction,* op. cit., 1962, p. 217; Broadbent and Calcott, *J. Inst. Fuel,* **29**, 524 (1956); 528 (1956); 18 (1957); **30**, 21 (1957)], the simple Gates-Gaudin-Schumann equation [Eq. (20-2) and Fig. 20-13] has been most widely used to fit ball-mill data. For example, this form was assumed by Herbst and Fuerstenau [*Trans. Am. Inst. Min. Metall. Pet. Eng.,* **241**(4), 538 (1968)] and Kelsall et al. [*Powder Technol.,* **1**(5), 291 (1968); **2**(3), 162 (1968); **3**(3), 170 (1970)]. More recently it has been observed that when the Schumann equation is used, the amount of coarse fragments cannot be made to agree with the mill-product distribution regardless of the choice of rate function. This points to the need for a breakage function that has more coarse fragments, such as the function used by Reid and Stewart (Chemica meeting, 1970) and Stewart and Restarick [*Proc. Australas. Inst. Min. Metall.,* **239**, 81 (1971)] and shown in Fig. 20-18. This graph can be fitted by a *double Schumann equation*

$$B(X) = A\left(\frac{X}{X_0}\right)^s + (1 - A)\left(\frac{X}{X_0}\right)^r \qquad (20\text{-}35)$$

where A is a coefficient less than 1.

In the investigations mentioned earlier the breakage function was assumed to be normalizable; i.e., the shape was independent of X_0. Austin and Luckie [*Powder Technol.,* **5**(5), 267 (1972)] allowed the coefficient A to vary with the size of particle breaking when grinding soft feeds.

Grinding-Rate Functions   These were determined by tracer experiments in laboratory mills by Kelsall et al. (op. cit.) as shown in Fig. 20-19 and in similar work by Szantho and Fuhrmann [*Aufbereit. Tech.,* **9**(5), 222 (1968)]. These curves can be fitted by the following equation:

$$\frac{S}{S_{\max}} = \left(\frac{X}{X_{\max}}\right)^\alpha \exp\left(-\frac{X}{X_{\max}}\right) \qquad (20\text{-}36)$$

That a maximum must exist should be apparent from the observation of Coghill and Devaney (U.S. Bur. Mines Tech. Pap., 1937, p. 581) that there is an optimum ball size for each feed size. Figure 20-19

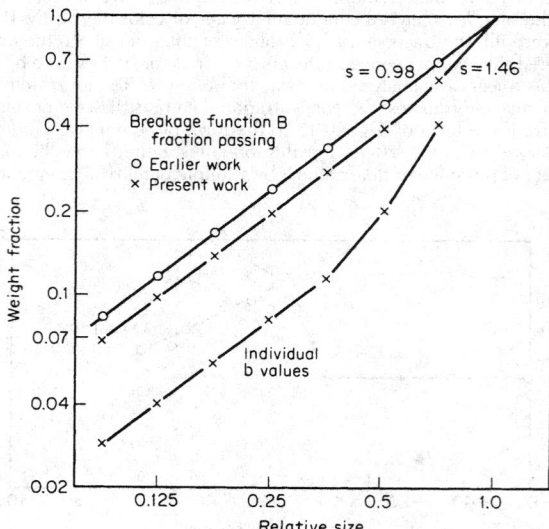

FIG. 20-18 Experimental breakage functions. (*Reid and Stewart, Chemica meeting, 1970.*)

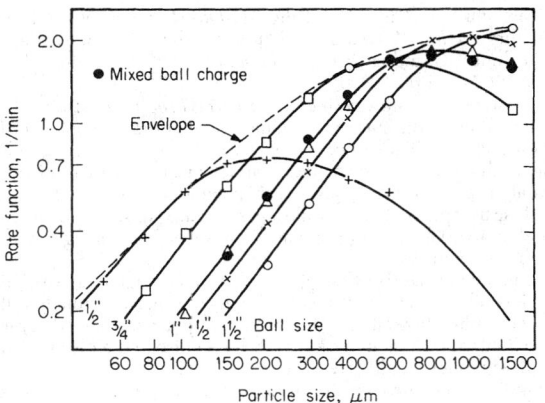

FIG. 20-19 Variation of rate function with size of feed particles and size of balls in a ball mill. [*Kelsall, Reid, and Restarick,* Powder Technol., *1(5), 291 (1968).*]

shows that the position of this maximum depends on the ball size. In fact, the feed size for which S is a maximum can be estimated by inverting the formula for optimum ball size given by Coghill and Devaney under "Tumbling Mills."

Scale-Up Based on Energy Since large mills are usually sized on the basis of power draft (see subsection "Energy Laws"), it is appropriate to scale up or convert from batch to continuous data by

$$S(X)_{\text{cont}} = S(X)_{\text{batch}} \frac{(W_T/KW)_{\text{batch}}}{(W_T/KW)_{\text{cont}}} \qquad (20\text{-}37)$$

W_T is usually not known for continuous mills, but it can be determined from $W_T = t_e Q$, where t_e is determined by a tracer measurement. Eq. (20-37) will be valid if the holdup W_T is geometrically similar in the two mills or if operating conditions are in the range in which total production is independent of holdup. From studies of the kinetics of milling [Patat and Mempel, *Chem. Ing. Tech.,* **37**(9), 933; (11), 1146; (12), 1259 (1965)] there is a range of holdup in which this is true. More generally, Austin, Luckie, and Klimpel (loc. cit.) developed empirical relations to predict S as holdup varies. In particular, they observe a slowing of grinding rate when mill filling exceeds ball void volume due to cushioning.

Parameters for Scale-Up Before simulation equations can be used, the parameter matrices **S** and **B** must be back-calculated from experimental data, which turns out to be difficult. One reason is that **S** and **B** occur as a product, so they are to some extent indeterminate; errors in one tend to be compensated by the other. Also, the number of parameters is larger than the number of data values from a single size-distribution measurement; but this is overcome by using data from grinding tests at a series of grinding times. This should be done anyway, since the empirical parameters should be determined to be valid over the experimental range of grinding times.

It may be easier to fit the parameters by forcing them to follow specified functional forms. In earliest attempts it was assumed that the forms should be normalizable (have the same shape whatever the size being broken). With complex ores containing minerals of different friability, the grinding functions **S** and **B** exhibit complex behavior near the grain size (Choi et al., *Particulate and Multiphase Processes Conference Proceedings,* **1**, 903–916.) **B** is not normalizable with respect to feed size and **S** does not follow a simple power law.

There are also experimental problems: When a feed size distribution is ground for a short time, there is not enough change in the size distribution in the mill to distinguish between particles being broken into and out of intermediate sizes, unless individual feed-size ranges are tagged. Feeding narrow-size fractions alone solves the problem, but changes the milling environment; the presence of fines affects the grinding of coarser sizes. Gupta et al. [*Powder Technology,* **28**(1), 97–106 (1981)] ground narrow fractions separately, but subtracted out the effect of the first 3 min of grinding, after which the behavior had

become steady. Another experimental difficulty arises from the recycle of fines in a closed circuit, which soon "contaminates" the size distribution in the mill; it is better to conduct experiments in open circuit, or in batch mills on a laboratory scale.

There are few data demonstrating scale-up of the grinding-rate functions **S** and **B** from pilot- to industrial-scale mills. Weller et al. [*International J. Mineral Processing,* **22**, 119–147 (1988)] ground chalcopyrite in pilot and plant mills and compared predicted parameters with laboratory data of Kelsall [*Electrical Engineering Transactions,* Institution of Engineers Australia, **EE5**(1), 155–169 (1969)] and Austin Klimpel & Luckie (*Process Engineering of Size Reduction, Ball Milling,* Society of Mining Engineers, NY, 1984) for quartz. **S** has a maximum for a particle size that depends on ball size according to Fig. 20-19, which can be expressed as

$$X_s/X_t = (d_s/d_t)^{2.4}$$

where s = scaled-up mill, t = test mill, d = ball size, X = particle size of maximum rate. Changing ball size also changes the rates according to $\mathbf{S}_s/\mathbf{S}_t = (d_s/d_t)^{0.55}$. These relations shift one rate curve onto another and allow scale-up to a different ball size. Mill diameter also affects rate by a factor $(D_s/D_t)^{0.5}$. Lynch [*Mineral crushing and grinding circuits, their simulation optimization design and control,* Elsevier Scientific Publishing Co., Amsterdam, New York (1977)], and Austin, Klimpel, and Luckie (loc. cit.) developed scale-up factors for ball load, mill filling, and mill speed. In addition, slurry solids content is known to affect the rate, through its effect on slurry rheology. Austin, Klimpel, and Luckie (loc. cit.) present more complete simulation examples and compare them with experimental data to study scale-up and optimization of open and closed circuits, including classifiers such as hydrocyclones and screen bends. Differences in the classifier will affect the rates in a closed circuit. For these reasons scale-up is likely to be uncertain unless conditions in the large mill are as close as possible to those in the test mill.

Control of Grinding Circuits Mineral processing plants require constant supervision and intervention by operators or controllers. Conditions such as feed hardness, grade, grain size, etc., change substantially as ore is delivered from different locations in the mine. Typically the objective of control is to maximize the production of the valuable component per unit time, or to maximize throughput while maintaining a constant grind size. Also, it is necessary to control individual process units so that they run smoothly and in harmony.

Measuring process parameters on full-scale plants is notoriously difficult, but is needed for control. Usually few of the important variables are accessible to measurement. Recycle of material makes it difficult to isolate the effects of changes to individual process units in the circuit. Newer plants have more instrumentation, including on-line viscosimeters [Kawatra and Eisele, *International J. Mineral Processing,* **22**, 251–259 (1988)], mineral composition by on-line X-ray fluorescence, belt feeder weighers, etc., but the information is always incomplete. Therefore it is helpful to have models to predict quantities that cannot be measured while measuring those that can.

Some plants have been using computer control for 20 years. Control systems in industrial use typically consist of individual feedback and feedforward loops. Horst and Enochs [*Engineering & Mining J.,* **181**(6), 69–171 (1980)] reported that installation of single-variable automatic controls improved performance of 20 mineral processing plants by 2 to 10 percent. But interactions among the processes make it difficult for independent controllers to control the circuit optimally.

Optimal control refers first to controlling the circuit dynamically so that it operates close to its optimum state. The *state* is the combination of variables that define the operation of the circuit. The optimum state varies as conditions change. Second, optimal control refers to a mathematical process to find an optimum path to move from a given state to the optimum state, based on minimizing an objective function [as used in control theory [Herbst et al., *International J. Mineral Processing,* **22**, 275–296 (1988)]. Other mathematical approaches have been published as well (Hulbert et al., Automation in Mining, Mineral and Metal Processing: Proceedings of 3rd IFAC Symposium, 311–322, 1980; Romberg, First IFAC Symposium on Automation for Mineral Resource Development, 289, 1985)].

The difficulty with individual PID (proportional-integral-differential) controllers is that each controller only manipulates one variable to achieve a desired effect, whereas the grinding circuit is *multivariate*. For example [Rajamani and Herbst, *Chem. Engr. Science*, **46**(3), 861–879 (1991)], to control the fineness of a ball-mill circuit one can vary either the ore-feed rate to the circuit or the water addition to the sump. Adding water dilutes the slurry going to the hydrocyclone, causing it to separate at a finer size. But then the cyclone sends more recycle back to the mill, loading it more and resulting in a coarser grind. If instead one controls the feed rate, the mill grinds finer, and indirectly the cyclone separates out more fines. Thus, both ore-feed rate and water addition influence more than one variable. As a result PID controllers cause the output to oscillate, as Rajamani and Herbst showed experimentally. The circuit tends to be unstable, and long time delays exist (Metzner and MacLeod, 7th IFAC Symposium on Intelligent Tuning and Adaptive Control, 163–169, 1991). Another difficulty with PID controllers is that their tuning changes depending on the operating conditions. They can be tuned for a rapid response when the ore is soft, but when it is harder the response is more sensitive and the gains have to be reduced to prevent overshoot.

Developing a multivariate control model is difficult because the process is complex. One approach is simplification; meaningful control results can be obtained with as few as two particle sizes in the model (Rajamani and Herbst, loc. cit.). Another approach is to use more powerful inexpensive computers. Complex calculations that previously seemed only of academic interest are now or will soon become practical to perform on-line. The mathematical complexity is also an impediment to understanding, but the commercial availability of packaged software for on-line control will overcome this problem. Software packages that enable grinding-circuit analysis, scale-up design, and flow-sheet optimization have been developed and are widely applied (Herbst et al., *MODSIM User's Manual*, Univ of Utah, Salt Lake City, 1986; Herbst et al., *ESTIMILL User's Manual*, Univ

of Utah, Salt Lake City, 1977; King, *MODSIM, Report No. 9*, Dept. of Metallurgy, Univ. Witwatersrand, Johannesburg, 1983; Jamsa, Acta Polytechnica Scandinavica, *Mathematics and Computer Science Series No. 57*, 32 pp., 1990).

Herbst et al. [*International J. Mineral Processing*, **22**, 273–296 (1988)] describe the software modules in an optimum controller for a grinding circuit. The *process model* can be an empirical model as some authors have used. A phenomenological model can give more accurate predictions, and can be extrapolated, for example from pilot- to full-scale application, if scale-up rules are known. Normally the model is a variant of the population balance equations given in the previous section.

Rajamani and Herbst (loc. cit.) compared control of an experimental pilot-mill circuit using feedback and optimal control. Feedback control resulted in oscillatory behavior. Optimal control settled rapidly to the final value, although there was more noise in the results. A more complete model would give even better results.

If individual controllers are used instead of optimal computer control, several strategies are possible. In one strategy (Lynch and Elber, *3rd IFAC Symposium on Automation of Mining, Mineral and Metal Processing*, 25–32, 1980) the slurry-pump rate is controlled to maintain sump-level constant, which results in smooth cyclone operation. The water-feed rate is ratioed to the ore-feed rate, which keeps the circulating load from oscillating. The ore-feed rate is then controlled to maintain product-particle size.

Improved instrumentation can improve control by measuring more directly the variables governing the internal behavior of the mill. By installing an electrical conductivity probe in the wall of the mill, Moys and Montini [*CIM Bulletin*, **80**(907), 52–6 (1987)] were able to detect the position of the ball mass during dynamic operation. This together with on-line measurement of slurry viscosity (see rheological properties, p. 20–32d) made it possible to control the mill at the desired operating point.

CRUSHING AND GRINDING EQUIPMENT

CLASSIFICATION AND SELECTION OF EQUIPMENT

A wide variety of size-reduction equipment is available. The chief reasons for lack of standardization are the variety of products to be ground and product qualities demanded, the limited amount of useful grinding theory, and the requirements by different industries in the economic balance between investment cost and operating cost. Some differences exist for the sake of difference; sometimes similarities are advertised as differences [Rumpf, *Chem. Ing. Tech.*, **37**(3), 187–202 (1965)].

Equipment may be classified according to the way in which forces are applied, as follows (Rumpf, loc. cit.):
1. Between two solid surfaces (Fig. 20-20*a*, crushing or attrition; Fig. 20-20*b*, shearing; Fig. 20-20*c*, crushing in a particle bed)
2. Impact at one solid surface (Fig. 20-20*d*), or between particles (Fig. 20-20*e*)
3. By shear action of the surrounding medium (Fig. 20-20*f*, colloid mill)
4. Nonmechanical introduction of energy (Thermal shock, explosive shattering, electrohydraulic)

A practical **classification of crushing and grinding equipment** is given in Table 20-6.

A guide to the *selection* of equipment may be based on feed *size* and *hardness* (see subsection "Grindability") as shown in Table 20-7. It should be emphasized that Table 20-7 is merely a guide and that exceptions can be found in practice.

A number of general principles govern the **selection of crushers** [Riley, *Chem. Process Eng.* (January 1953)]. When the rock contains a predominant amount of material that has a tendency to be cohesive when moist, such as clay, any form of repeated pressure crusher will show a tendency for the fines to pack in the outlet of the crushing zone and prevent free discharge at fine settings. Impact breakers are then

suitable, provided that the rock is not harder and more abrasive than limestone with 5 percent silica.

When the rock is not hard but cohesive, toothed rolls give satisfactory performance. With harder rocks, jaw and gyratory crushers are required, and the jaw crusher is less prone to clogging than the gyratory. In crushing throughputs of a few hundred tons per hour, a jaw or impact crusher may be satisfactory, but for the largest capacities the gyratory is unsurpassed. For secondary crushing the high-speed conical-head gyratory is unsurpassed except when sticky material precludes its use. For very hard ores a rod mill may compete effectively. If a wide size distribution is to be avoided, a compression-type crusher is best; if the product requires fragments of compact shape,

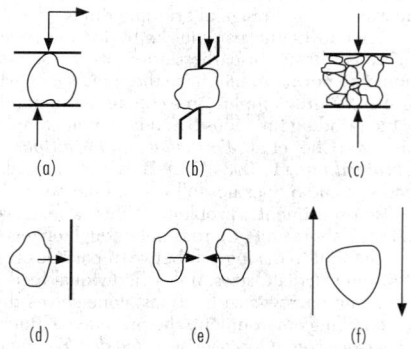

FIG. 20-20   Stressing mechanisms to cause size reduction. [*Rumpf*, Chem. Eng. Tech., *37(3), 187–202 (1965).*]

TABLE 20-6   Types of Size-Reduction Equipment

A. Jaw crushers:
   1. Blake
   2. Overhead eccentric
B. Gyratory crushers:
   1. Primary
   2. Secondary
   3. Cone
C. Heavy-duty impact mills:
   1. Rotor breakers
   2. Hammer mills
   3. Cage impactors
D. Roll crushers:
   1. Smooth rolls (double)
   2. Toothed rolls (single and double)
   3. Roll press
E. Dry pans and chaser mills
F. Shredders:
   1. Toothed shredders
   2. Cage disintegrators
   3. Disk mills
G. Rotary cutters and dicers
H. Media mills:
   1. Ball, pebble, rod, and compartment mills:
      a. Batch
      b. Continuous
   2. Autogenous tumbling mills
   3. Stirred ball and bead mills
   4. Vibratory mills
I. Medium peripheral-speed mills:
   1. Ring-roll and bowl mills
   2. Roll mills, cereal type
   3. Roll mills, paint and rubber types
   4. Buhrstones
J. High-peripheral-speed mills:
   1. Fine-grinding hammer mills
   2. Pin mills
   3. Colloid mills
   4. Wood-pulp beaters
K. Fluid-energy superfine mills:
   1. Centrifugal jet
   2. Opposed jet
   3. Jet with anvil
   4. Fluidized-bed jet

TABLE 20-7   Guide to Selection of Crushing and Grinding Equipment

Size-reduction operation	Hardness of material	Size* Range of feeds, in.† Max.	Size* Range of feeds, in.† Min.	Size* Range of products, in.† Max.	Size* Range of products, in.† Min.	Reduction ratio‡	Types of equipment
Crushing:							
Primary	Hard	60	12	20	4	3 to 1	A to B
		20	4	5	1	4 to 1	
Secondary	Hard	5	1	1	0.2	5 to 1	A to E
		1.5	0.25	0.185 (4)	0.033 (20)	7 to 1	
	Soft	60	4	2	0.4	10 to 1	C to G
Grinding:							
Pulverizing:							
Coarse	Hard	0.185 (4)	0.033 (20)	0.023 (28)	0.003 (200)	10 to 1	D to I
Fine	Hard	0.046 (14)	0.0058 (100)	0.003 (200)	0.00039 (1250)	15 to 1	H to K
Disintegration:							
Coarse	Soft	0.5	0.065	0.023 (28)	0.003	20 to 1	F, I
Fine	Soft	0.156 (5)	0.0195 (32)	0.003 (200)	0.00039 (1250)	50 to 1	I to K

*85% by weight smaller than the size given.
†Sieve number in parentheses, mesh per inch
‡Higher reduction ratios for closed-circuit operations.
NOTE: To convert inches to millimeters, multiply by 25.4.

an impact crusher or a gyratory is best. Further information is given under each type of mill, and also in the subsection on crushed stone and aggregate.

JAW CRUSHERS

Design and Operation   These crushers may be divided into two main groups (Fig. 20-21), the *Blake,* with a movable jaw pivoted at the top, giving greatest movement to the smallest lumps; and the *overhead eccentric,* which is also hinged at the top, but through an *eccentric-driven* shaft which imparts an elliptical motion to the jaw. Both types have a removable crushing plate, usually corrugated, fixed in a vertical position at the front end of a hollow rectangular frame. A similar plate is attached to the swinging movable jaw. The Blake jaw (Fig. 20-22) is moved through a knuckle action by the rising and falling of a second lever (pitman) carried by an eccentric shaft. The vertical movement is communicated horizontally to the jaw by double toggle plates. Because the jaw is pivoted at the top, the throw is greatest at the discharge, preventing choking.

Crushing angles in standard Allis-Chalmers-Svedala Blake-type machines generally are near 0.47 rad (27°) (see Fig. 20-21). The reduction ratios at minimum recommended settings and with straight jaw plates average about 8:1. Curved (or concave) jaw plates are designed to minimize choking.

The **overhead eccentric jaw crusher** (*Nordberg, Telsmith Inc., and Cedarapids*) falls into the second type. These are single-toggle machines. The lower end of the jaw is pulled back against the toggle by a tension rod and spring.

The choice between the two types of jaw crushers is generally dictated by the feed characteristics, tonnage, and product requirements (Pryon, *Mineral Processing*, Mining Publications, London, 1960; Wills, *Mineral Processing Technology*, Pergamon, Oxford, 1979.) Greater wear caused by the elliptical motion of the overhead eccentric and direct transmittal of shocks of this type to the bearing limit use of this type to readily breakable material, although Cedarapids has reduced the elliptical wear effect by moving the bearing over the crushing chamber. Overhead eccentric crushers are generally preferred for crushing rocks with a hardness equal to or lower than that of limestone. Operating costs of the overhead eccentric are higher for the crushing of hard rocks, but its large reduction ratio is useful for simplified low-tonnage circuits with fewer grinding steps. Double-toggle type crushers cost about 50 percent more than similar overhead-eccentric-type crushers.

Comparison of Crushers   The jaw crusher can accommodate the same size rocks as a gyratory, with lower capacity and also lower capital and maintenance costs, but similar installation costs. Therefore they are preferred when the crusher gape is more important than the throughput. If required throughput in tons/h is less than the square of the gape in inches, a jaw crusher is more economical (Taggart, *Handbook of Mineral Dressing*, Wiley, New York, 1945). Relining the gyratory requires more effort than for the jaw, and also more space above and below the crusher. Improved alloys have reduced the need for relining, however. Gyratories are preferred for larger throughputs. In metal mines continuity of operation favors gyratories over jaws because of their low maintenance. Quarries on the other hand can use hammer or other impact crushers, while maintenance is done on second shift.

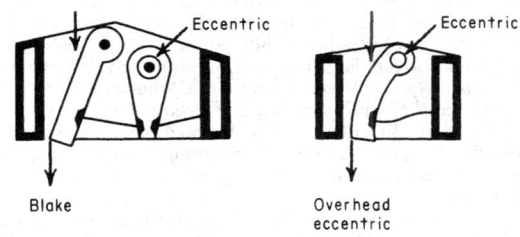

FIG. 20-21   Jaw-crusher designs.

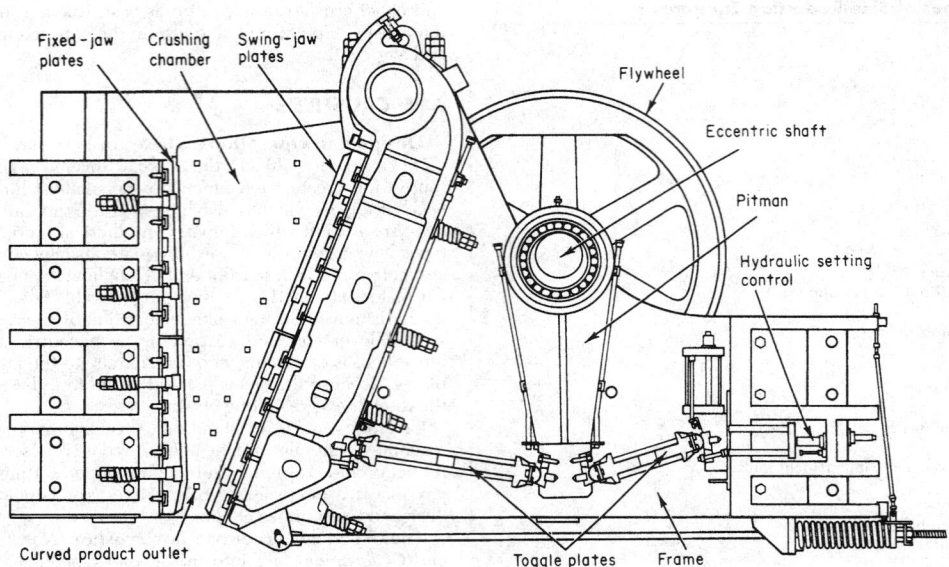

FIG. 20-22 Blake jaw crusher. (*Allis Mineral Systems Grinding Div., Svedala Industries, Inc.*)

Performance Jaw crushers are applied to the primary crushing of hard materials and are usually followed by other types of crushers. In smaller sizes they are used as single-stage machines. Jaw crushers are usually rated by the dimensions of their feed area. This depends on the width of the crushing jaws and the *gape*, which is the maximum distance between the fixed and movable jaws at the feed opening. The *setting* of a jaw crusher is the *closed* (or close) or the *wide opening* between the moving jaws at the outlet end, usually measured between tips of the corrugations. The reciprocating motion of the jaws causes the opening to vary between closed and wide, and the difference is the *throw.* Specifications are usually based on the closed settings. The setting is adjustable.

The throw of Blake jaw crushers is determined by the hardness of the ore as well as the size of the machine [Mollick et al., *Engineering & Mining J.*, **181**(6), 69–171 (1980)]. It may vary from ¼ in for hard but friable ores to 3 in for resilient material. The *Big Bite* jaw crusher (*Kue-Ken Div, Process Technology Inc.*) is a Blake double-toggle type in which the fulcrum pinion is positioned well over the center of the grinding chamber [Anon., *Quarry Management*, **18**(1), 25–7 (1991)]. This increases the throw at the top of the jaw opening, allowing it to better crush large rocks. It can be shipped in sections. Capacities of Kue-Ken jaw crushers (Blake type) are given in Table 20-8, including both standard and Big Bite types. Performance data of overhead eccentric crushers with straight jaw plates are given in Table 20-9.

Dragon single-toggle primary jaw crushers are supplied (by *Fives-Cail Babcock*) in eight sizes. The smallest is the MR53 with feed opening 550 × 350 mm and the largest is the MR200 with feed opening 2000 × 1600 mm, the corresponding capacities being from 15 to 800 ton/h. These units can be supplied fully mobile or on skid frames. The BCS series are extra heavy-duty crushers. High flywheel inertia enables them to accept large lumps of limestone and achieve throughputs up to 6000 ton/h. Pegson Telsmith in UK has a new single-toggle primary jaw crusher with roller bearings sized at 1000 × 800 mm. It is expected to be the first of a new line that incorporates the latest manufacturing techniques. Universal Engineering Corp. offers 27 sizes of overhead eccentric crusher, with long jaws and small nip angle to reduce rebound.

Smaller jaw crushers are available from Sturtevant Inc, with capacities of 1 to 10 hp (0.7 to 7 kW). Jaw settings range from ⅛ to 2.5 in (3 to 65 mm).

GYRATORY CRUSHERS

The development of improved supports and drive mechanisms has allowed gyratory crushers to take over most large hard-ore and mineral-crushing applications. The largest expense of these units is in relining them. Operation is intermittent; so power demand is high, but the total power cost is not great.

Design and Operation The gyratory crusher (see Fig. 20-23) consists of a cone-shaped pestle oscillating within a larger cone-shaped mortar or bowl. The angles of the cones are such that the width of the passage decreases toward the bottom of the working faces. The pestle consists of a mantle which is free to turn on its spindle. The spindle is oscillated from an eccentric bearing below. Differential motion causing attrition can occur only when pieces are caught simultaneously at the top and bottom of the passage owing to different radii at these points.

The circular geometry of the crusher gives a favorably small **nip angle** in the horizontal direction. The nip angle in the vertical direction is less favorable and limits feed acceptance. The vertical nip angle

TABLE 20-8 Performance Data for Blake Jaw Crushers

Model	Feed opening, in.	hp	Closed setting, in.	Capacity, ton/h	Weight, 1000 lb.	Max. r/min.
Standard 22	3 × 12	10	¼	2–5	3	400
			2	19–32		
54	8 × 24	15–25	¾	15–26	7	400
			3	59–88		
81	12 × 36	25–40	1	32–50	17	390
			3	95–135		
95	24 × 36	30–50	2	64–95	27	390
			6	195–270		
Big Bite 110	25 × 42	50–60	2	85–120	40	390
			5	205–275		
160	42 × 48	125–150	4	215–290	117	290
			10	520–770		
440	66 × 84	350–400	7	835–1310	455	210
			22	2625–4120		

To convert inches to millimeters, multiply by 2.54; to convert pounds to kilograms, multiply by 0.4535; to convert horsepower to kilowatts, multiply by 0.746; and to convert tons per hour to kilograms per hour, multiply by 907.

TABLE 20-9 Performance Data for Overhead Eccentric Jaw Crushers*

Crusher Size, in.	r/min.	hp	Capacity, tons/h — Setting, in.								
			¾	1	1½	2	3	4	6	8	12
10 × 16	300	20–30	15	20	30	40	60				
12 × 36	275	60–75			65	85	130	175			
24 × 36	250	125–150					130	175	265		
30 × 42	250	125–175						200	300	400	610
54 × 60	200	350–450						290	440	580	880

*Cedar Rapids, CO. Div. of Raytheon Co., Pocket Reference Book, 13 ed., pp. 8–12, 1993.
Capacity can vary depending on breaking characteristics and compression strength of each: installed horsepower, size of feed, rate of feed, type of fall, and proper operating conditions.

is determined by the shape of the mantle and bowl liner; it is similar to that of a jaw crusher.

Primary crushers have a steep cone angle and a small reduction ratio. **Secondary** crushers have a wider cone angle; this allows the finer product to be spread over a larger passage area and also spreads the wear over a wider area. Wear occurs to the greatest extent in the lower, fine-crushing zone. These features are further extended in cone crushers; therefore secondary gyratories are much less popular than secondary cone crushers, but they can be used as primaries when quarrying produces suitable feed sizes.

The three general types of gyratory crusher are the **suspended-spindle,** the **supported-spindle,** and the **fixed-spindle** types. Primary gyratories are designated by the size of feed opening, and

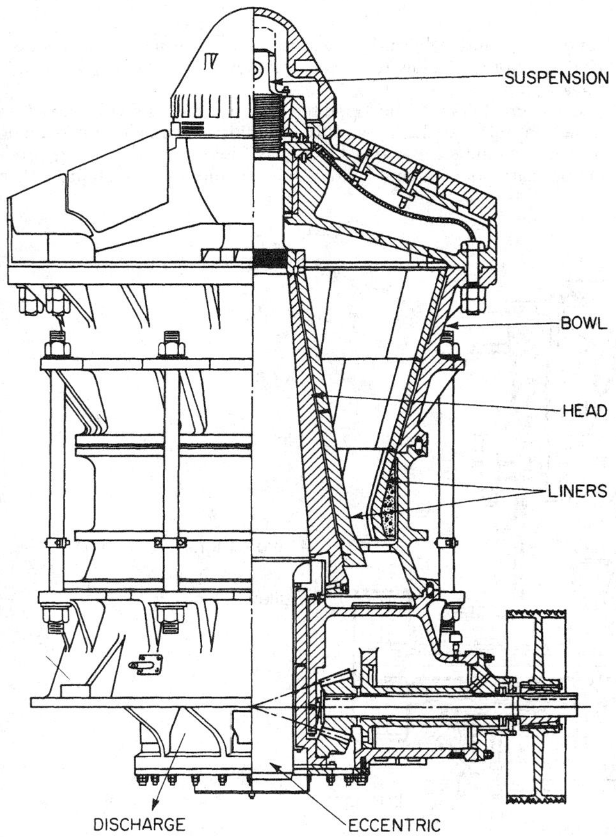

FIG. 20-23 Primary gyratory crusher with spider suspension. (*Nordberg, Inc.*)

secondary or reduction crushers by the diameter of the head in feet and inches. There is a close opening and a wide opening as the mantle gyrates with respect to the concave ring at the outlet end. The close opening is known as the close setting or the close-side setting or the closed-side setting, while the wide opening is known as the wide-side or open-side setting. Specifications usually are based on closed settings. The setting is adjustable by raising or lowering the mantle.

The length of the crushing stroke greatly affects the capacity and the screen analysis of the crushed product. A very short stroke will give a very evenly crushed product but will not give the greatest capacity. A very long stroke will give the greatest capacity, but the product will contain a wider product-size distribution.

Performance Crushing occurs through the full cycle in a gyratory crusher, and this produces a higher crushing capacity than a similar-sized jaw crusher, which crushes only in the shutting half of the cycle. Gyratory crushers also tend to be easier to operate. They operate most efficiently when they are fully charged, with the main shaft fully buried in charge. Power consumption for gyratory crushers is also lower than that of jaw crushers. These are preferred over jaw crushers when capacities of 800 Mg/h (900 tons/h) or higher are required.

Gyratories make a product with open-side settings of 5 to 10 in at discharge rates from 600 to 6000 ton/h, depending on size. Most manufacturers offer a throw from ¼ to 2 in. The throughput and power draw depend on the throw and the hardness of the ore, and on the amount of undersized material in the feed. Removal of undersize (which can amount to ⅓ of the feed) by a stationary grizzly can reduce power draw.

Crusher Product Sizes Table 20-10 relates product size to the discharge setting of the crusher in terms of the percent smaller than that size in the product. Size-distribution curves differ for various types of materials crushed, and a general set of curves is not valid.

Primary gyratories will accept feed directly from truck or railcar. Most manufacturers make both mechanical and hydraulically supported types. Figure 20-23 shows a Nordberg primary gyratory crusher with spider suspension. It is available in 1- to 1.5-m (42-, 48-, 54-, and 60-in) feed sizes. Table 20-11 gives capacity data for the Superior gyratory crusher (*Allis-Chalmers*).

Gyratory crushers that feature wide-cone angles are called **cone crushers.** These are suitable for secondary crushing, because crushing of fines requires more work and causes more wear; the cone shape provides more working area than primary or jaw crushers for grinding of the finer product. Crusher performance is harmed by sticky material in the feed, more than 10 percent fines in the feed smaller than the crusher setting, excessive feed moisture, feed-size segregation, uneven distribution of feed around the circumference, uneven feed control, insufficient capacity of conveyors and closed-circuit screens, extremely hard or tough feed material, and operation at less than recommended speed. Rod mills are sometimes substituted for crushing of tough ore, since they provide more easily replaceable metal for wear.

HP Series Cone Crushers (*Nordberg Inc.*) (Fig. 20-24) are available in four sizes. The previous standard- and short-head versions have been combined into one machine, with replaceable liner shapes corresponding to either version, adapted respectively to relatively coarse and fine grinding. In addition, the throw has been increased so as to

TABLE 20-10   Relation of Product Size to Discharge Setting of Crusher*

Setting measured on open side	Kind of feed	% of product passing a square opening equal to discharge setting of crusher			
		Limestone	Granite	Trap rock	Ores
Primary service:					
Jaw crusher†	Quarry-run	85–90	70–75	65–70	85–90
	Prescalped	80–85	65–70	60–65	80–85
Gyratory crusher	Quarry-run	85–90	75–80	65–70	85–90
	Prescalped	80–85	70–75	60–65	80–85
Secondary service:					
Gyratory crusher‡	Screened	85–90	80–85	75–80	85–90

*From "Crushing Theory and Practice," Allis-Chalmers Mfg. Co.
†Blake type, or crushers with equivalent speeds and throws: opening measured from tip of corrugations on one jaw plate to bottom of corrugations in opposing plate.
‡For standard, or reduction, types with nonchoking concaves. Single-toggle jaw product, on screened feed, will approximate that of gyratory-type secondary crushers with nonchoking concaves.

TABLE 20-11   Performance Data for Primary Gyratory Crushers*
Capacities for crushing limestone, tons/hr.

Crusher size	Approx. feed opening, in.	Gyrations per min.	Pinion r.p.m.	Max. hp.	Eccentric throw, in.	Open-side setting of discharge opening, in																				
						2½	3	3½	4	4½	5	5½	6	6½	7	7½	8	8½	9	9½	10	10½	11	11½	12
30–55	30 × 78	175	585	150	⅝	150	205	270	335	390	450	510													
					1¼				605	675	735	800													
42–65	42 × 108	150	497	265	1					540	660	790	920	1040	1170									
					1½								1040	1260	1490									
54–74	54 × 132	135	497	300	1							960	1040	1100	1160	1240	1330							
					1⅜										1950	2070	2210							
60–109	60 × 150	100	400	1000	1½													3250	3500	3750	4000	4250	4500	4750	5000

*Allis-Chalmers Corporation. To convert inches to millimeters, multiply by 2.514; to convert horsepower to watts, multiply by 746.

replace the previous Gyradisc models, designed to reduce wear by facilitating interparticle breakage with increased particle-bed depth. A built-in hydraulic motor unscrews the head for easy change-out of the liner. The feed opening ranges from 0.8 to 14.3 in depending on the mill size and the liner chosen. The drive shaft turns an eccentric, which causes the head to oscillate in the mantle, while a head ball resists vertical thrust. Bronze bearings withstand severe operating conditions, and labyrinth seals prevent bearing contamination. Hydraulic pistons allow remote or automated control of settings and release tramp iron.

Cone crushers can be operated in open circuit with capacities shown in Table 20-12, or in closed circuit in parallel with scalping screens for multistage size reduction, as Table 20-13 shows (the product sizes are not comparable). The feed should not contain more than

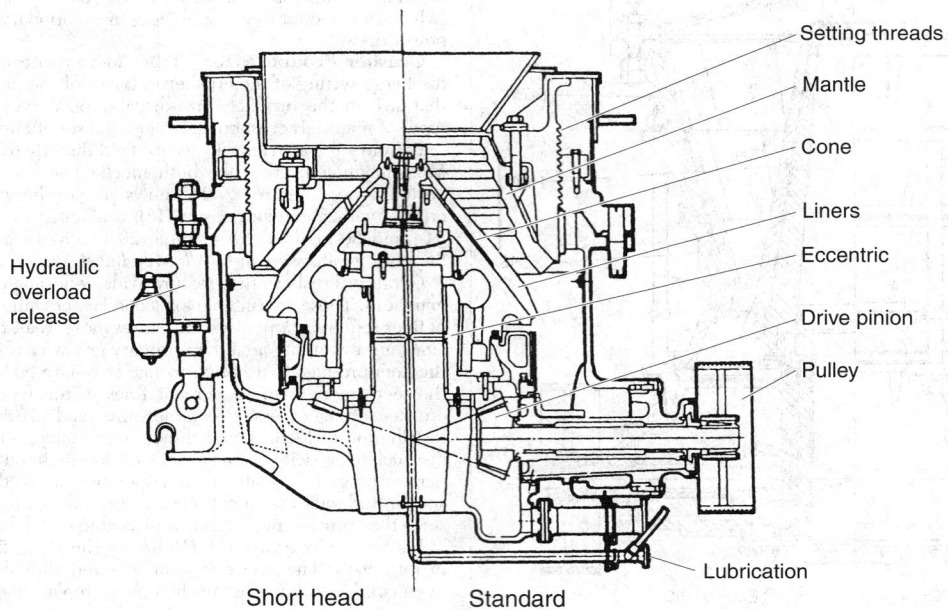

FIG. 20-24   HP700 cone crusher. (*Nordberg, Inc.*)

10 percent fines less than the discharge setting, else scalping screens should be used. A series of Waterflush Crushers is also available from Nordberg, which flush out fines and increase capacity with some feeds, and allow its use in closed circuit with wet ball and autogenous mills (qv). It can give products with 80 percent passing as small as 0.15 in.

The Eljay Rollercone crusher (*Cedarapids Eljay Div.*) uses a different drive principle: a wedge-shaped plate rotates on a ring of roller bearings under the crushing cone, causing it to oscillate. Low wear of these bearings maintains close tolerance of settings and hence of product size, and their overdesign results in long bearing life. A hydraulic-relief system passes larger pieces of tramp iron from spring-relief systems, while a hydraulic hold-down system allows quick changing of crusher-setting shims. The massive base frame directs compression forces to the product, reducing energy losses due to structural deflection of the members. The crusher is available in sizes from 36- to 66-in diameter. With standard head the capacity ranges from 36 to 580 ton/h at closed-side settings of ⅜ to 2 in; with a fine-crushing head the capacity ranges from 37 to 368 ton/h with closed-side setting of 1.4 to 1 in and recirculating rates from 18 to 30 percent.

Control of Crushers   Lower-grade raw materials, higher energy costs, larger-scale operations, and more complex, capital-intensive plants make automatic control of size-reduction equipment more important (Suominen, 21st International Symposium—Applications of Computers and Operations Research in the Mineral Industry, 1011–1018). Benefits are: increased productivity, process stability and safety, improved recovery of mineral values, and reduced costs [Horst and Enochs, *Engineering & Mining J.*, **181**(6), 69–171 (1980)].

Improved sensors allow computer monitoring of the system for safety and protection of the equipment from damage. Sensors include lubrication-flow monitors and alarms, bearing-temperature sensors, belt scales, rotation sensors, and proximity sensors to detect ore level under the crusher. The latter prevent jamming of the output with too high an ore level, and protect the conveyor from impact of lumps with too low an ore level. Motion detectors assure that the conveyor is moving. Control applied to crusher systems including conveyors can facilitate use of mobile crushers in quarries and mines, since these can be controlled remotely by computer with reduced labor.

The objective of crusher control is usually to maximize crusher throughput at some specified product size, without overloading the crusher. Usually only three variables can be adjusted: feed rate, crusher opening, and feed size in the case of a secondary crusher. Four modes of control for a crusher are: (1) Setting overload control, where the gape setting is fixed except that it opens when overload occurs. A hardness change during high throughput can cause a power overload on the crusher, which control should protect against. (2) Constant power setting control, which maximizes throughput. (3) Pressure control, which provides settings that give maximum crusher force, and hence, also throughput. (4) Feeding-rate control, for smooth operation. Setting control influences mainly product size and quality, while feed control determines capacity. Flow must also be synchronized with the feed requirements of downstream processes such as ball mills, and improved crusher efficiency can reduce the load on the more costly downstream grinding.

ROLL CRUSHERS

Once popular for coarse crushing, these devices long ago lost favor to gyratory and jaw crushers because of their poorer wear characteristics with hard rocks. Roll crushers are still commonly used for both primary and secondary crushing of coal and other friable rocks such as oil shale and phosphate. The roll surface is smooth, corrugated, or toothed, depending on the application. *Smooth rolls* tend to wear ring-shaped corrugations that interfere with particle nipping, although some designs provide a mechanism to move one roll from side to side to spread the wear. *Corrugated rolls* give a better bite to the feed, but wear is still a problem. *Toothed rolls* are still practical for rocks of not too high silica content, since the teeth can be regularly resurfaced with hard steel by electric-arc welding. Toothed rolls are frequently used for crushing coal and chemicals. For further details, see edition 6 of this handbook.

The **capacity** of roll crushers is calculated from the ribbon theory, according to the following formula:

$$Q = dLs/2.96 \qquad (20\text{-}38)$$

where Q = capacity, cm³/min; d = distance between rolls, cm; L = length of rolls, cm; and s = peripheral speed, cm/min. The denominator becomes 1728 in engineering units for Q in cubic feet per minute, d and L in inches, and s in inches per minute. This gives the theoretical capacity and is based on the rolls' discharging a continuous, solid uniform ribbon of material. The actual capacity of the crusher depends on roll diameter, feed irregularities, and hardness and varies between 25 and 75 percent of theoretical capacity.

Sturtevant two-roll crushers are available in capacities from 2 to 30 hp (1.5 to 22 kW). The gap is controlled by a hand-wheel. Rolls are hard-surfaced for wear resistance.

TABLE 20-12   Cone Crusher Capacity in Open Circuit, ton/h*

Mill	Feed opening +, in.	Motor hp	r/min	¼	⅜	½	¾	1	1.5	2
				Closed-Side Discharge Setting, in						
HP200	¾–7	150	1200	85	110	150	190	220	200	
HP500	¾–12	450	900	180	230	290	390	445	540	605
HP700	1½–14	600	850	300	370	425	600	675	845	975
MP1000	5–18	1000		600	750	900	1100	1200	1800	2400

*Nordberg Inc.—assumes feed weighing 100 lb/ft³ and with a work index of 13.
Range covered by short-head and standard versions.

TABLE 20-13   Cone Crusher Capacity in Closed-Circuit, ton/h*

Mill	¼		⅜		½		¾		1		1½	
	Top Size of Product from Screen (98% passing), in											
	A	B	A	B	A	B	A	B	A	B	A	B
HP200	65	90	80	115	115	165	140	200	160	230	180	265
HP500	135	195	175	250	220	315	280	405	335	465	380	545
HP700	215	310	275	390	215	445	425	610	480	680	590	850
MP1000	320	480	980	720	630	800	840	1100	1000	1200		

*Nordberg Inc. Assumes feed weighing 100 lb/ft³ and with a work index of 13.
A—Net finished product from screen.
B—Recirculating load.

ROLL PRESS

A novel comminution device, the **roll press,** has achieved commercial success [Schoenert, *International J. Mineral Processing,* **22,** 401–412 (1988)]. It is used for fine crushing, replacing the function of a coarse-ball mill or of tertiary crushers. Unlike ordinary roll crushers which crush individual particles, the roll press is choke fed and acts on a thick stream or ribbon of feed. Particles are crushed mostly against other particles, so wear is very low. Energy efficiency is also greater than in ball mills.

The product is in the form of agglomerated slabs. These are broken up either in a ball mill or an impact or hammer mill running at a speed too slow to break individual particles. Some materials may even deagglomerate from the handling that occurs in conveyors. Product-size distributions are shown in Fig. 20-25. Note that the press can handle a hard rock like quartz. Pressure in Fig. 20-25 is the calculated force-per-unit contact area in the nip. A large proportion of fines is produced, but a fraction of coarse material survives. This makes recycle necessary. From experiments to grind cement clinker −80 µm, as compression is increased from 100 to 300 MPa, the required recycle ratio decreases from 4 to 2.8. The energy required per ton of throughput increases from 2.5 to 3.5 kWh/ton. These data are for a 200-mm diameter pilot-roll press.

Status of 150 installations in the cement industry are reviewed [Strasser et al., *Rock Products,* **92**(5), 60–72 (1989)]. In cement-clinker milling, wear is usually from 0.1 to 0.8 g/ton, and for cement raw materials it is between 0.2 and 1.2 g/ton, whereas it may be 20- to 40-in ball mills. New surface of 100 to 130 cm²/g is produced with cement. The size of the largest feed particles should not exceed 0.04 × roll diameter (*D*) according to Schoenert (loc. cit.). However, it has been found [Wuestner et al., *Zement-Kalk-Gips,* **41**(7), 345–353 (1987); English edition, 207–212] that particles as large as 3–4 times the roll gap may be fed to an industrial press. Particles receive preliminary breakage at the top of the nip.

The hybrid circuit in Fig. 20-26 has proven most versatile, and can increase capacity of an existing ball-mill circuit by 30 to 100 percent. Recycling the rock as in the hybrid flow sheet reduces the need for coarse-ball milling.

Roll presses are now manufactured by most cement-equipment manufacturers, for example Krupp-Polysius AG. For drives up to 1200 kW they are equipped with V-belt drives, allowing speed to be changed. Rolls in one particular crusher are 48 inches (1200 mm) in diameter and 30 inches (700-mm) wide and each weighs 90 tons. Machines with up to 2500 kW installed power and 1000 ton/h (900

tonne/h) capacity have been installed. The largest presses can supply feed for four or five ball mills.

Operating experience (Wuestner et al., loc. cit.) has shown that roll diameters about 1 m are preferred, as a compromise between production rate and stress on the equipment. The press must be operated choke fed, with a substantial depth of feed in the hopper; otherwise it will act like an ordinary roll crusher.

The advent of the roll press has greatly improved the economy of cement milling. More information is given later under cement.

IMPACT BREAKERS

Impact breakers include heavy-duty hammer crushers and rotor impact breakers. Fine hammer mills are described in a subsequent subsection.

Not all rocks shatter well by impact. Impact breaking is best suited for the reduction of relatively nonabrasive and low-silica-content materials such as limestone, dolomite, anhydrite, shale, and cement rock, the most popular application being on limestone.

Hammer Crusher   (Fig. 20-27) Pivoted hammers are mounted on a horizontal shaft, and crushing takes place by impact between the hammers and breaker plates. Heavy-duty hammer crushers are frequently used in the quarrying industry, for processing municipal solid waste, and scrap automobiles.

These crushers are of two types, with and without grates or screens. The bottom of the **Pennsylvania reversible impactor** (Fig. 20-27) is open, and the sized material passes through almost instantaneously. Particles acquire high velocities, and this leads to little control on particle size and a much higher proportion of fines than slow-speed crushers. A product-size distribution curve (*Handbook of Crushing,* Pennsylvania Crusher Corp, 1994) shows the following for crushing of minus 10-in subbituminous coal from Wyoming: 92 percent minus 2 in, 48 percent minus 0.53 in, 15 percent minus 10 mesh/in.

In the second type of mill (Fig. 20-28), a cylindrical grating is provided beneath the rotor for product discharge. Some hammer crushers are symmetrically designed so that the direction of rotation can be reversed to distribute wear evenly on the hammer and breaker plates. When such a **Pennsylvania nonreversible hammermill** is used for reduction, material is broken first by impact against hammers and then by rubbing action (attrition) against screen bars. Performance data of Pennsylvania reversible hammer mills are shown in Table 20-14.

Sturtevant Inc. produces three sizes of screen-type hammermills with 5 to 50 hp (4 to 37 kW) capacity, suitable for recycle of glass and

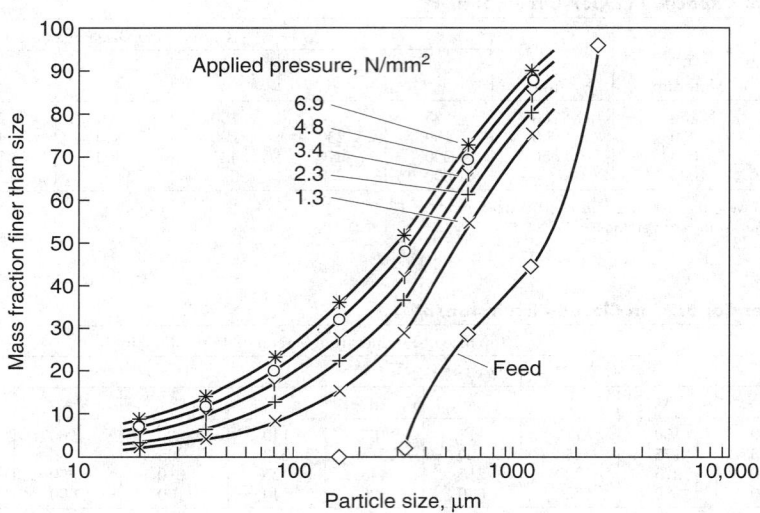

FIG. 20-25   Size distribution of roll-press products. Feed is quartz, 300–2500 µm, dry. [*Schonert, International J. Mineral Processing,* **22,** *401–412 (1988).*]

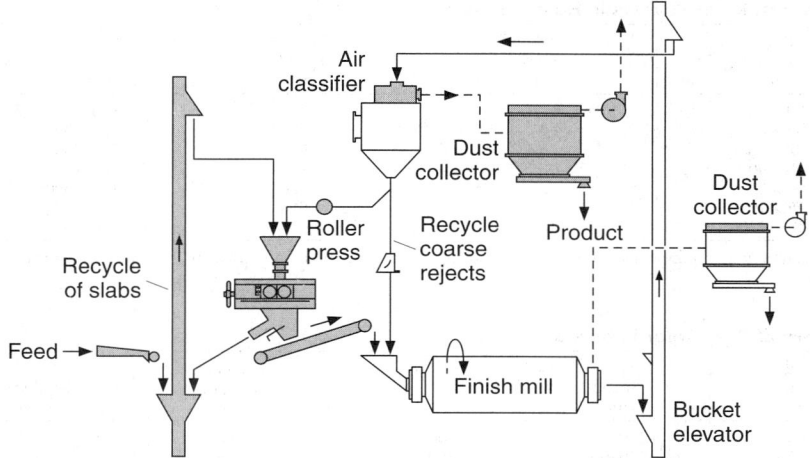

FIG. 20-26 Hybrid flow sheet to combine a roll press and ball mill with a classifier. (*Conroy, 31st IEEE Cement Industry Technical Conference, 509–542, 1988. Copyright IEEE.*)

wastes, and for mineral processing. **Jeffrey hammermills** (*Jeffrey Div, Indresco Inc.*) are available with screen grates having capacities from 50 to 140 ton/h (135 to 350 Mg/h) for a product with top size 1.75 in (45 mm). The screenless, reversible type is also available.

Each hammer may weigh several hundred kilograms. Speed varies between 500 to 1800 r/min, depending on the size of the machine.

Rotor Impactors   The rotor of these machines is a cylinder to which is affixed a tough steel bar. Breakage can occur against this bar or on rebound from the walls of the device. Free impact breaking is the principle of the rotor breaker, and it does not rely on pinch crushing or attrition grinding between rotor hammers and breaker plates.

The result is a high reduction ratio and elimination of secondary and tertiary crushing stages. The investment cost may be a third of that for a two-stage jaw and gyratory crusher plant producing 180 Mg/h (200 tons/h) and half for a plant crushing 540 Mg/h (600 tons/h) [Godfrey, *Quarry Managers' J.*, 405–416 (October 1964)].

By adding a screen on a portable mounting, a complete, compact mobile crushing plant of high capacity and efficiency for use in any location is provided.

The peripheral speed of the rotors manufactured by *KHD Humboldt Wedag* is 34 m/sec; the 52-ton rotor can handle a lump as large as 8 ton. Controllers stop the feeder if rotor speed drops below 75 percent of normal [Schaefer and Gallus, *Zement-Kalk-Gips*, **41**(10), 486–492 (1988); English ed., 277–280].

Pennsylvania Crusher Corp. offers a variety of impactor types: granulators, Bradford breakers, roll crushers, and jaw crushers. (See Table 20-15.)

The **ring-type granulator** (*Pennsylvania Crusher Corp.*) features a rotor assembly with loose crushing rings, held outwardly by centrifugal force, which chop the feed. It is suitable for highly friable materials which may give excessive fines in an impact mill. For example, bituminous coal is ground to a product below 2 cm (¾ in). They have also been successfully used to grind abrasive quartz to sand size, due to the ease of replacement of the ring impact elements.

Cage Mills   The **Stedman disintegrator** (*Stedman Machine Co.*), commonly referred to as a cage mill, is used for crushing quarry rock, phosphate rock, and fertilizer and for disintegrating clays, colors, press cake, and bones. Cages of one, two, three, four, six, and eight rows, with bars of special alloy steel, revolving in opposite directions, produce a powerful impact action that pulverizes many materials. (See Fig. 20-29.)

The life of a cage may be a few months and may produce 9000 Mg (10,000 tons) of quarry rock. A gray-iron cage is used for alumina grinding, with metal particles removed magnetically. The advantage of

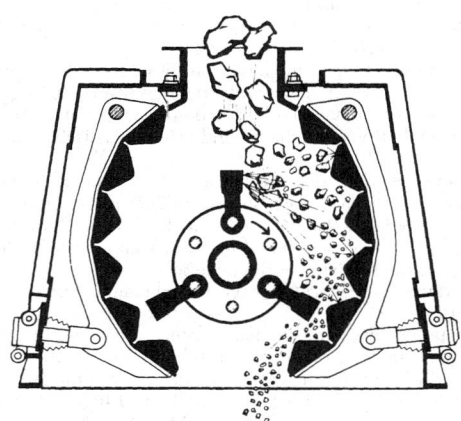

FIG. 20-27   Reversible impactor. (*Pennsylvania Crusher Corp.*)

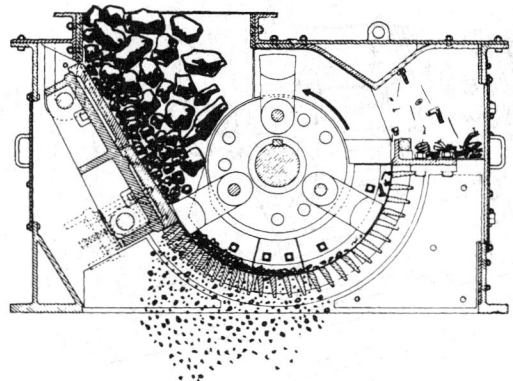

FIG. 20-28   Nonreversible hammermill (*Pennsylvania Crusher Corp.*)

TABLE 20-14 Performance Data for Reversible Hammer Mills

Model no.	Rotor dimensions, in	Maximum feed size, in	Maximum speed, r/min	hp	Capacity, tons/h
505	30 × 30	2½	1200	100–200	40–60
605	36 × 30	4	1200	200–300	80–100
708	42 × 48	8	900	300–550	140–180
815	48 × 90	10	900	900–1200	330–400
1014	60 × 84	12	720	1100–1500	450–500
1217	72 × 102	14	600	1550–2000	620–685
1221	72 × 126	14	600	1900–2500	760–850

NOTE: To convert inches to centimeters, multiply by 2.54; to convert horsepower to kilowatts, multiply by 0.746; and to convert tons per hour to megagrams per hour, multiply by 0.907.

TABLE 20-15 Performance of Dual-Rotor Impact Breakers

Model no.	Feed opening, in	Speed, r/min	hp	Hammer weight, lb	Product size, in	Capacity, tons/h
3648	36 × 48	550–990	250–300	300	2	300
4850	48 × 50	550–990	300–400	400	3	500
5462	54 × 62	480–750	400–500	500	4–5	700
6072	60 × 72	300–600	500–600	600	5–6	1200

To convert pounds to kilograms, multiply by 0.4535; to convert inches to centimeters, multiply by 2.54; to convert horsepower to kilowatts, multiply by 0.746; and to convert tons per hour to megagrams per hour, multiply by 0.907.

multiple-row cages is the achievement of a greater reduction ratio in a single pass.

These features and the low cost of the mills make them suitable for medium-scale operations where complicated circuits cannot be justified. The maximum feed size is 20 cm (8 in), and the product size may be as fine as 325 mesh.

Prebreakers   Aside from the normal problems of grinding, there are special procedures and equipment for breaking large masses of feed to smaller sizes for further grinding. There is the breaking or shredding of bales, as with rubber, cotton, or hay, in which the compacted mass does not readily come apart. There also is often caking in bags of plastic or hygroscopic materials which were originally fine. Although crushers are sometimes used, the desired size-reduction ratio often is not obtainable. Furthermore, a lower capital investment may result through choosing a less rugged device which progressively attacks the large mass to remove only small amounts at a time. In structure such a device comprises a toothed rotating shaft in a casing. *Prater Industries, Inc.,* offers a double roll with heavy-duty teeth as a precrusher feeder. The **Mikro roll crusher** (*Hosokawa Micron*

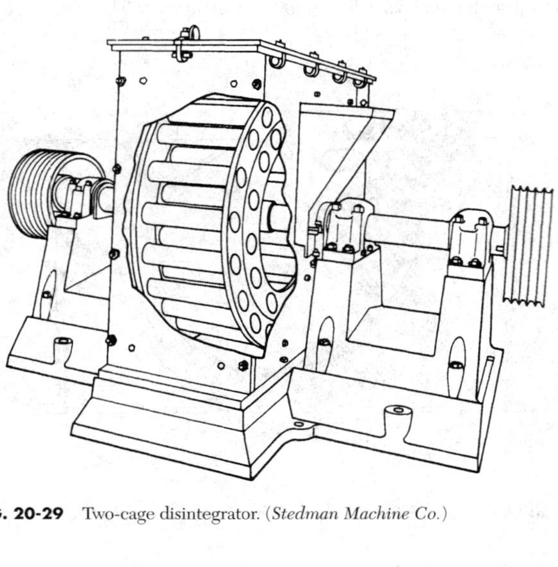

FIG. 20-29   Two-cage disintegrator. (*Stedman Machine Co.*)

Powder Systems) serves as grinder or prebreaker and is of similar type. The **Rietz prebreaker** differs somewhat from these.

Precision Cutters and Slitters   Often it becomes desirable to reduce the size of a solid mass to regular smaller sizes.

Precision knife cutters differ from random cutting mills in that a feeder is synchronized with the knives. This ensures the exact size, whether it be slit widths in a sheet, fiber length from a strand, or both width and length from a sheet, as in dicing. In the **Giant dicing cutter,** the sheet stock is first slit lengthwise with opposing sets of circular knives. The slit strands then pass between pressure rolls to a rotary cutter which operates against an adjustable bed knife. Capacities range up to 18 Mg/h (20 tons/h), with sheet stock up to 60 cm (24 in) wide.

PAN CRUSHERS

Design and Operation   The pan crusher (Fig. 20-30) consists of one or more grinding wheels or mullers revolving in a pan; the pan may remain stationary and the mullers be driven, or the pan may be driven while the mullers revolve by friction. The mullers are made of tough alloys such as Ni-Hard. Iron scrapers or plows at a proper angle feed the material under the mullers.

The **Chambers dry pan** (*Bonnot Co.*) uses air cylinders to regulate the grinding pressure under each of the muller tires from 33,000 to 90,000 N (7500 to 20,000 lb).

The pan bottom rotates and has a central solid crushing ring as well as an outer ring of screen plates with openings from 0.16 to 1.3 cm (1/16 to ½ in). In some instances a solid pan bottom is used in place of a perforated screen bottom, and the ground material is discharged through a slot in the rim.

Performance   The dry pan is useful for crushing medium-hard and soft materials such as clays, shales, cinders, and soft minerals such as barites. Materials fed should normally be 7.5 cm (3 in) or smaller, and a product able to pass No. 4 to No. 16 sieves can be delivered, depending on the hardness of the material.

High reduction ratios with low power and maintenance are features of pan crushers.

The Chambers dry pan is available from 1.8 to 3 m (6 to 10 ft) pan diameter with mullers ranging from 0.71 to 1.6 m (28 to 62 in) in diameter with 13- to 46-cm (5- to 18-in) face. Power ranges from 11 to 75 kW (15 to 100 hp) or from 0.8 to 4 kW/Mg (1 to 5 hp/ton) of product. Production rate varies from 1 to 54 Mg/h (1 to 60 tons/h) according to pan size and hardness of material as well as fineness of feed and product.

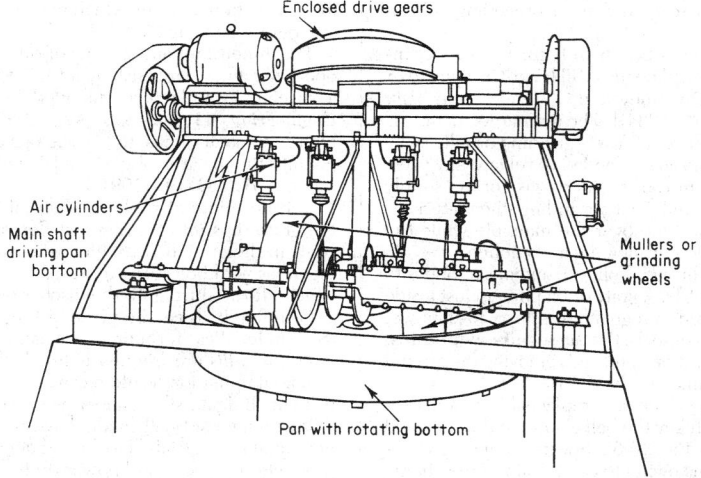

FIG. 20-30 Chambers 10-ft, 100-hp dry pan. (*Bonnot Co.*)

The **wet pan** is used for developing plasticity or molding qualities in ceramic feed materials. The abrasive and kneading actions of the mullers blend finer particles with the coarser particles as they are crushed (Greaves-Walker, *Am. Refract. Inst. Tech. Bull.* **64,** 1937), and this is necessary so that a high packing density can be achieved to result in strength.

TUMBLING MILLS

Ball, pebble, rod, tube, and compartment mills have a cylindrical or conical shell, rotating on a horizontal axis, and are charged with a grinding medium such as balls of steel, flint, or porcelain or with steel rods. The **ball mill** differs from the tube mill by being short in length; its length, as a rule, is not far from its diameter (Fig. 20-31). Feed to ball mills can be as large as 2.5 to 4 cm (1 to 1½ in) for very fragile materials, although the top size is generally 1 cm (½ in). Most ball mills operate with a reduction ratio of 20 to 200:1. The largest balls are typically 13 cm (5 in) in diameter.

The tube mill is generally long in comparison with its diameter, uses smaller balls, and produces a finer product. The compartment mill consists of a cylinder divided into two or more sections by perforated partitions; preliminary grinding takes place at one end and finish grinding at the discharge end. These mills have a length-to-diameter ratio in excess of 2 and operate with a reduction ratio of up to 600:1. Rod mills deliver a more uniform granular product than other revolving mills while minimizing the percentage of fines, which are sometimes detrimental.

The **pebble mill** is a tube mill with flint or ceramic pebbles as the grinding medium and may be lined with ceramic or other nonmetallic liners. The **rock-pebble mill** is an autogenous mill in which the

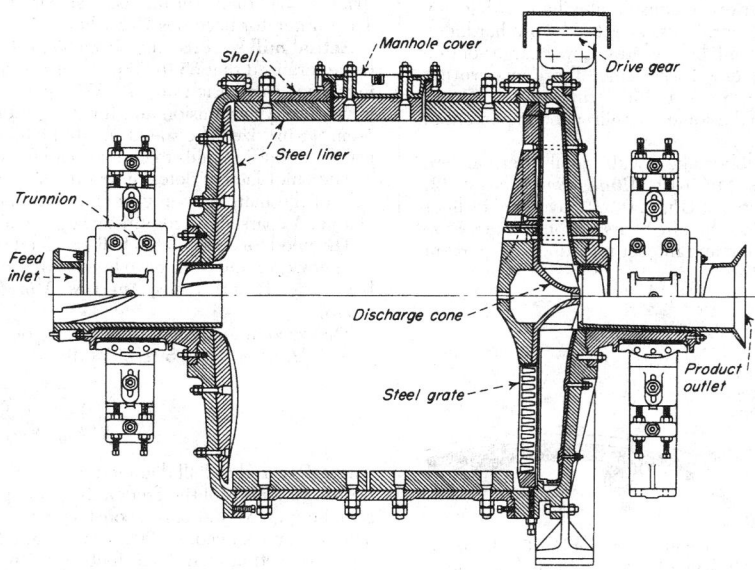

FIG. 20-31 Marcy grate-type continuous ball mill. (*Allis Mineral Systems, Svedala Inc.*)

medium consists of larger lumps scalped from a preceding step in the grinding flow sheet.

Design   The conventional type of **batch mill** consists of a cylindrical steel shell with flat steel-flanged heads. Mill length is equal to or less than the diameter [Coghill, De Vaney, and O'Meara, *Trans. Am. Inst. Min. Metall. Pet. Eng.*, **112**, 79 (1934)]. The discharge opening is often opposite the loading manhole and for wet grinding usually is fitted with a valve. One or more vents are provided to release any pressure developed in the mill, to introduce inert gas, or to supply pressure to assist discharge of the mill. In dry grinding, the material is discharged into a hood through a grate over the manhole while the mill rotates. Jackets can be provided for heating and cooling.

Material is fed and discharged through hollow trunnions at opposite ends of **continuous mills** (Fig. 20-31). A grate or diaphragm just inside the discharge end may be employed to regulate the slurry level in wet grinding and thus control retention time. In the case of **air-swept mills,** provision is made for blowing air in at one end and removing the ground material in air suspension at the same or other end.

Ball mills usually have **liners** which are replaceable when they wear. Optimum liner shapes which key the ball charge to the shell and prevent slippage are illustrated in Fig. 20-32. Special operating problems occur with smooth-lined mills owing to erratic slip of the charge against the wall. At low speeds the charge may **surge** from side to side without actually tumbling; at higher speeds tumbling with **oscillation** occurs. The use of lifters prevents this [Rose, *Proc. Inst. Mech. Eng.* (*London*), **170**(23), 773–780 (1956)]. Power consumption in a smooth mill depends in a complex way on operating conditions such as feed viscosity, whereas it is more predictable in a mill with lifters [Kitchener and Clarke, *Br. Chem. Eng.*, **13**(7), 991 (1968)].

Plant data on an 8-ft diameter rod mill showed that when smooth liners were replaced with lifter bars spaced 300 mm (12 in) apart, liner wear decreased by 13-fold, production increased 50 percent, and power per ton decreased by 34 percent [Howat and Vermeulen, *Journal South African Institute Mining and Metallurgy*, **86**, 251–259 (July 1986)]. When a smooth-walled ball mill was fitted with lifter bars the production of −75 μm material increased 8 percent and the power/ton decreased 10 percent. Plant tests confirmed previous results [Meaders and McPherson, *Mining Engineering*, **16**, 81–84 (Sept. 1964)] that optimum lifter bar spacing is from 3 to 5 times bar height.

Liner wear increases with the size, hardness, and sharpness of feed more than with ball size. The hardness of manganese steel corresponds to softer types of ore, while Nihard is about the same as magnetite [Moore et al., *International J. of Mineral Processing*, **22**, 313–343 (1983)]. Quartz and pyrite are considerably harder than any metals used. Rubber, being resilient, is less affected by ore hardness, and therefore has the advantage with harder ores. Low-charge volume below 35 percent results in increased wear since the liners are not protected by a covering of ore. Several studies indicate that wear increases at least proportional to the square of mill speed in percent of critical.

Both all-rubber liners, and rubber liners with metal lifter bars are currently used in large ball mills [McTavish, *Mining Engineering*, **42**, 1249–1251 (Nov. 1990)]. Both types (Fig. 20-32) have rubber-liner plates, separated and held in place by lifter bars. Rubber liners have the following advantages: Abrasion and impact resistance, 15 percent

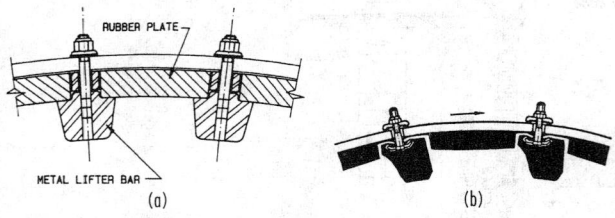

FIG. 20-32   Types of ball-mill liners: (*a*) combination liner [McTavish, *Mining Engineering*, **42**, 11, 1249–1251 (1990) Exton Industries, Inc.]; (*b*) all-rubber liner [Moller and Brough, *Mining Engineering*, **41**, 8, 849–853 (1989) Skega Ltd.].

lower weight than steel, which makes replacement easier, lower noise, and tight sealing against the shell.

The dimensions and spacing of rubber lifter bars are critical to good operation and wear resistance. Figure 20-32 shows rubber lifters with a spacing formula recommended by Moller and Brough [*Mining Engineering*, **41**, 849–853 (Aug. 1989)]. Lifters must be at least as high as the ball radius, to key the ball charge and assure that the balls fall into the toe area of the mill [Powell, *International J. Mineral Processing*, **31**, 163–193 (1991)].

The flexibility of rubber is helpful in discharge grates. Each time the grate dips into the slurry it flexes, tending to work out particles stuck in the grate holes. Metal frames support the rubber while leaving areas with holes free to flex. The grate is backed up by lifter bars which lift the slurry into the discharge trommel.

Grinding balls can be made of forged steel, cast steel, or cast iron. (See under Wear for more information on alloys.)

Pebble mills are frequently lined with nonmetallic materials when iron contamination would harm a product such as a white pigment or cement. Belgian silex (silica) or porcelain block are popular linings. Silica linings and ball media have proved to wear better than other nonmetallic materials. The higher density of silica media increases the production capacity and power draft of a given mill.

Capacities of pebble mills are generally 30 to 50 percent of the capacity of the same size of ball mill with steel grinding media and liners; this depends directly on the density of the media.

Smaller mills, up to about 0.19-m³ (50-gal) capacity, are made in one piece with a cover. U.S. Stoneware Co. makes these in wear-resistant Burundum-fortified ceramic and also makes larger three-piece units, in a metal protective case, up to 0.8-m³ (210-gal) capacity. A handbook on pebble milling is available from Paul O. Abbe, Inc., Little Falls, NJ.

Operation   **Cascading** and **cataracting** are the terms applied to the motion of grinding media. The former applies to the rolling of balls or pebbles from top to bottom of the heap, and the latter refers to the throwing of the balls through the air to the toe of the heap.

The criterion by which the ball action in mills of various sizes may be compared is the concept of **critical speed.** It is the theoretical speed at which the centrifugal force on a ball in contact with the mill shell at the height of its path equals the force on it due to gravity:

$$N_c = 42.3/\sqrt{D} \qquad (20\text{-}39)$$

where N_c is the critical speed, r/min, and D is diameter of the mill, m (ft). For a ball diameter that is small with respect to the mill diameter. The numerator becomes 76.6 when D is expressed in feet.

Actual mill speeds range from 65 to 80 percent of critical. It might be generalized that 65 to 70 percent is required for fine wet grinding in viscous suspension and 70 to 75 percent for fine wet grinding in low-viscosity suspension and for dry grinding of large particles up to 1-cm (½-in) size. The speeds might be increased by 5 percent of critical for unbaffled mills to compensate for slip.

The chief factors determining the size of **grinding balls** are fineness of the material being ground and maintenance cost for the ball charge. A coarse feed requires a larger ball than a fine feed.

The need for a calculated ball-size feed distribution is open to question; however, methods have been proposed for calculating a rationed ball charge [Bond, *Trans. Am. Inst. Min. Metall. Pet. Eng.*, **153**, 373 (1943)].

The recommended optimum size of makeup rods and balls is [Bond, *Min. Eng.*, **10**, 592–595 (1958)]

$$D_b = \sqrt{\frac{X_p E_i}{K n_r}} \sqrt{\frac{\rho_s}{\sqrt{D}}} \qquad (20\text{-}40)$$

where D_b = rod or ball diameter, cm (in); D = mill diameter, m (ft); E_i is the work index of the feed; n_r is speed, percent of critical; ρ_s is feed specific gravity; and K is a constant = 214 for rods and 143 for balls. The constant K becomes 300 for rods and 200 for balls when D_b and D are expressed in inches and feet respectively. This formula gives reasonable results for production-sized mills but not for laboratory mills. The ratio between the recommended ball and rod sizes is 1.23.

A graded charge of rods results from wear in a rod mill. **Rod diameter** may range from 10 to 2.5 cm (4 to 1 in), for example. A new rod load usually is patterned after a used one found to give good results. The maximum length of a rod mill appears to be 20 ft, because longer rods tend to twist and bend.

Tumbling-Mill Circuits Tumbling mills may be operated in **normal closed circuit,** as shown in Fig. 20-45 or 20-59, or in **reverse arrangement** in which the feed passes through the classifier before entering the mill (see secondary mill in Fig. 20-44 or 20-59).

Material and Ball Charges The load of a grinding medium can be expressed in terms of the percentage of the volume of the mill that it occupies; i.e., a bulk volume of balls half filling a mill is a 50 percent ball charge. The void space in a static bulk volume of balls is approximately 41 percent. Since the medium expands as the mill is rotated, the actual running volume is unknown.

Simple relationships govern the amount of balls and voids in the mill. The weight of balls = $\rho_b \epsilon_b V_m$, where ρ_b = bulk density of balls, g/cm^3 (lb/ft^3); ϵ_b = apparent ball filling fraction; and V_m = volume of mill = $\pi D^2 L/4$. Steel balls have a bulk density of approximately 4.8 g/cm^3 (300 lb/ft^3), stone pebbles, 1.68 g/cm^3 (100 lb/ft^3); and alumina balls, 2.4 g/cm^3 (150 lb/ft^3).

The amount of material in a mill can be expressed conveniently as the ratio of its volume to that of the voids in the ball load. This is known as the **material-to-void ratio.** If the solid material and its suspending medium (water, air, etc.) just fill the ball voids, the M/V ratio is 1, for example. Grinding-media loads vary from 20 to 50 percent in practice, and M/V ratios are usually near 1.

The material charge of continuous mills called the **holdup** cannot be set directly but is indirectly determined by operating conditions. There is a maximum throughput rate that depends on the shape of the mill, the flow characteristics of the feed, the speed of the mill, and the type of feed and discharge arrangement. Above this rate the holdup increases unstably.

The holdup of material in a continuous mill determines the mean residence time, and thus the extent of grinding. Gupta et al. [*International J. Mineral Processing,* **8,** 345–358 (Oct. 1981)] analyzed published experimental data on a 40 × 40-cm grate discharge laboratory mill, and determined that holdup was represented by H_w = (4.020 − 0.176 WI)F_w + (0.040 + 0.01237 WI)S_w − (4.970 + 0.395 WI), where WI is Bond work index based on 100 percent passing a 200-mesh sieve, F_w is the solids feed rate in kg/min, and S_w is weight percent solids in the feed. This represents experimental data for limestone, feldspar, sulfide ore, and quartz. The influence of WI is believed to be due to its effect on amount of fines present in the mill. Parameters that did not affect H_w are specific gravity of feed material, and feed size over the narrow range studied.

Sufficient data was not available to develop a correlation for overflow mills, but the data indicated a linear variation of H_w with F as well.

The mean residence time τ (defined as H_w/F) is the most important parameter, since it determines the time over which particles are exposed to grinding. Measurements on several industrial mills (Weller, *Automation in Mining Mineral and Metal Processing,* 3d IFAC Symposium, 303–309, 1980) (measured on the water, not the ore) showed that the maximum mill filling was about 40 percent, and the maximum flow velocity through the mill is 40 m/h.

Swaroop et al. [*Powder Technology,* **28,** 253–260 (Mar.–Apr. 1981)] found that the material holdup is higher and the vessel dispersion number $D\tau/L^2$ (see subsection on Continuous Mill Simulation) is lower in the rod mill than in the ball mill under identical dimensionless conditions. This indicates that the known narrow-product-size distribution from rod mills is partly due to less mixing in the rod mill, in addition to different breakage kinetics.

The holdup in grate-discharge mills depends on the grate openings. Kraft et al. [*Zement-Kalk-Gips International,* **42**(7), 353–9 (1989); English edition, 237–9] measured the effect of various hole designs in wet milling. They found that slots tangential to the circumference gave higher throughput and therefore lower holdup in the mill. Total hole area had little effect until the feed rate was raised to a critical value (30 m/h in a mill 0.26 m diam × 0.6 m long); above this the larger area led to lower holdup. The open area is normally specified between 3 and 15 percent, depending on the number of grinding chambers and

other conditions. The slots should be 1.5 to 16 mm wide, tapered toward the discharge side by a factor of 1.5 to 2 to prevent blockage by particles.

Dry versus Wet Grinding The choice between wet and dry grinding is generally dictated by the end use of product. If the presence of liquid with the finished product is not objectionable or the feed is moist or wet, wet grinding generally is preferable to dry grinding, but power consumption, liner wear, and capital costs determine the choice. Other factors that influence choice are the performance of subsequent dry or wet classification steps, the cost of drying, and the capability of subsequent processing steps for handling a wet product.

The net production in wet grinding in the Bond grindability test varies from 145 to 200 percent of that in dry grinding depending on mesh [Maxson, Cadena, and Bond, *Trans. Am. Int. Min. Metall. Pet. Eng.,* **112,** 130–145, 161 (1934)]. Ball mills have a large field of application for wet grinding in closed circuit with size classifiers, which also perform advantageously wet.

In fine dry grinding, surface forces come into action to cause cushioning and ball coating with a less efficient use of energy. Grinding-media and liner-wear consumption per ton of ground product is lower for a dry-grinding system. However, power consumption for dry grinding is about 30 percent larger than for wet grinding. Dry grinding requires the use of dust-collecting equipment.

In **wet ball milling** the grinding rate increases with solids content up to 70 wt % (35 vol %), as Fig. 20-33 shows, due to pulp rheology. Examination of gouge marks indicated that most breakage was by impact of balls on particles rather than by abrasion.

The **rheological properties** of the slurry affect the grinding behavior in ball mills. Rheology depends on solids content, particle size, and mineral chemical properties [Kawatra and Eisele, *International J. Mineral Processing,* **22,** 251–259 (1988)]. Up to 50 percent solids by volume a mineral slurry exhibits dilatent behavior, i.e. shear stress increases more than proportionate to shear rate. Above 50 percent it becomes pseudoplastic, i.e., it exhibits a yield value (Austin, Klimpel, and Luckie, *Process Engineering of Size Reduction: Ball Milling,* AIME, 1984). Above the yield value the grinding rate decreases, and this is believed to be due to adhesion of grinding media to the mill wall causing centrifuging [Tangsatitkulchai and Austin, *Powder Technology,* **59**(4), 285–293 (1989)]. Maximum power draw and fines production are achieved when the solids content is just below that which produces the critical yield.

The solids concentration in a pebble-mill slurry should be high enough to give a slurry viscosity of at least 0.2 Pa·s (200 cP) for best grinding efficiency [Creyke and Webb, *Trans. Br. Ceram. Soc.,* **40,** 55

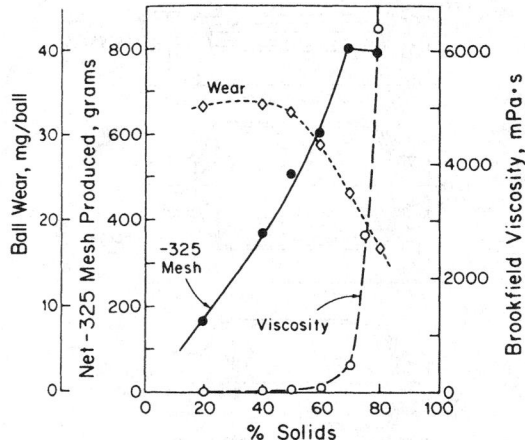

FIG. 20-33 Effect of percent solids on the wear of mild steel balls, net weight of 325 mesh material produced, and pulp viscosity after grinding magnetic taconite for 60 min [*Iwasaki et al., International J. Mineral Processing,* **22,** 345–360 (1988).]

(1941)], but this may have been required to key the charge to the walls of the smooth mill used.

Since viscosity increases with amount of fines present, mill performance can often be improved by closed-circuit operation to remove fines. Chemicals such as surfactants allow the solids content to be increased without increasing the yield value of the pseudoplastic slurry, allowing a higher throughput. They may cause foaming problems downstream, however. Increasing temperature lowers the viscosity of water, which controls the viscosity of the slurry under high-shear conditions such as those encountered in the cyclone, but does not greatly affect chemical forces. Slurry viscosity can be most directly controlled by controlling solids content. To control viscosity it is necessary to measure the viscosity on line. Kawatra and Eisele (loc. cit.) have described viscometers that can measure viscosity of a flowing slurry stream with high-solids content. A satisfactory spindle type operates in an overflow reservoir of the slurry (Napier-Munn et al., DeBeers Diamond Research Laboratory, Johannesburg, 1985, in Kawatra and Eisele, loc. cit.). Vibrating and ultrasonic types can also be used.

A simple device [Montini and Moys, *J. South African Inst. Mining & Metallurgy*, **88**(6), 199–206 (1988)] measures slurry viscosity and hence percent solids within the mill. It consists of a bolt that passes through insulating washers in the shell. A measurement of electrical conduction is taken at a moment when the ball charge has just fallen away from the rotating shell. This measures the thickness of the slurry adhering to the vicinity of the bolt, which is related to the slurry viscosity. The device can distinguish among slurry concentrations of 60, 65, and 70 percent solids, and should be useful for computer control of viscosity in the mill.

Feed and Discharge   Feed and discharge arrangements for ball and rod mills depend on their mode of operation. Various feed and discharge mechanisms are shown in Fig. 20-34.

Mill feeders attached to the feed trunnion of the conical mill and used to pass the feed into the mill without backspill are of several types. A feed chute is generally used for dry grinding, this consisting of an inclined chute which is sealed at the outer edge of the trunnion and down which the material slides to pass through the trunnion and into the mill. A screw feeder, consisting of a short section of screw conveyor which extends partway into the opening in the feed trunnion and conveys the material into the mill, may also be used when dry

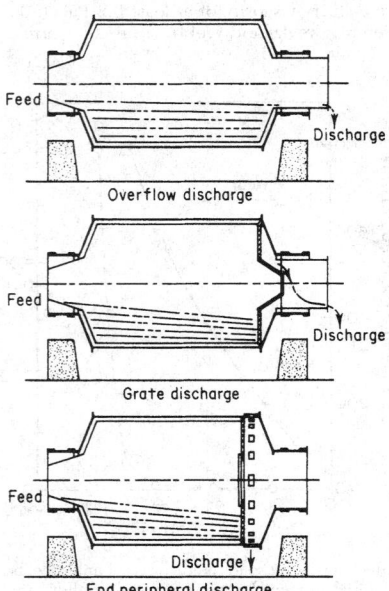

FIG. 20-34   Continuous ball-mill discharge arrangements for wet grinding.

Feed

Discharge

Overflow discharge

Feed

Discharge

Grate discharge

Feed

Discharge

End peripheral discharge

grinding. For wet grinding, several different types of feeders are available: the scoop feeder, which is attached to and rotates with the mill trunnion and which dips into a stationary box to pick up the material and pass it into the mill; a drum feeder attached to and rotating with the feed trunnion, having a central opening into which the material is fed and an internal deflector or lifter to pass the material through the trunnion into the mill; or a combination drum and scoop feeder, in which the new feed to the mill is fed through the central opening of the drum while the scoop picks up the oversize being returned from a classifier to a scoop box well below the centerline of the mill.

Control of pulp level to obtain high-circulating load is accomplished by use of **grate-discharge mills.** In one case the grates allowed passage of sufficient pulp to maintain the circulating load at 400 percent (Duggan, *Min. Technol.,* Tech. Publ. 1456, March 1942).

Mill Efficiencies   The controlling factors conceded to govern the ore-grinding efficiency of cylindrical mills are as follows:

1.   Speed of mill affects capacity, also liner and ball wear, in direct proportion up to 65–75 percent of critical speed.

2.   Ball charge equal to 35–50 percent of the mill volume gives the maximum capacity.

3.   Minimum-size balls capable of grinding the feed give maximum efficiency.

4.   Bar-type lifters are essential for smooth operation.

5.   Material filling equal to ball-void volume is optimum.

6.   Higher-circulating loads tend to increase production and decrease the amount of unwanted fine material.

7.   Low-level or grate discharge with recycle from a classifier increases grinding capacity over the center or overflow discharge; but liner, grate, and media wear is higher.

8.   Ratio of solids to liquids in the mill must be considered on the basis of slurry rheology.

Experimental evidence presented in a classic paper by Coghill and De Vaney ("Ball Mill Grinding," U.S. Bur. Mines Tech. Publ. 581, 1937) indicated that the efficiency of battered reject balls was about 11 percent less than that of new spherical balls. These and other results have been graphically presented (Rose and Sullivan, *Ball Rod and Tube Mills,* Chemical Publishing Co., NY, 1958).

Selection of Mill   The selection of a ball- or rod-mill grinding unit is based on pilot-mill experiments, scaled up on the basis that production is proportional to energy input. When pilot experiments cannot be undertaken, performance is based on published data for similar types of materials, expressed in terms of either grindability or an energy requirement (see subsections "Grindability" and "Energy Laws"). Newer methods of sizing mills and determining operating conditions for optimum circuit performance are based on computer solutions of the grinding equations with values of rate and breakage functions determined from pilot or full-scale tests (see subsection "Simulation of Milling Circuits").

Capacity and Power Consumption   One of the methods of mill sizing is based on the observation that the amount of grinding depends on the amount of energy expended, if one assumes comparable good practice of operation in each case. The energy applied to a ball mill is primarily determined by the size of mill and load of balls. Theoretical considerations show the net power to drive a ball mill to be proportional to $D^{2.5}$, but this exponent may be used without modification in comparing two mills only when operating conditions are identical [Gow, Guggenehim, Campbell, and Coghill, *Trans. Am. Inst. Min. Metall. Pet. Eng.,* **112**, 24 (1934)]. The net power to drive a ball mill was found to be

$$E = [(1.64L - 1)K + 1][(1.64D)^{2.5}E_2] \qquad (20\text{-}41)$$

where L is the inside length of the mill, m (ft); D is the mean inside diameter of the mill, m (ft); E_2 is the net power used by a 0.6- by 0.6-m (2- by 2-ft) laboratory mill under similar operating conditions; and K is 0.9 for mills less than 1.5 m (5 ft) long and 0.85 for mills over 1.5 m long. This formula may be used to scale up pilot-milling experiments in which the diameter and length of the mill are changed but the size of balls and the ball loading as a fraction of mill volume are unchanged. More accurate computer models are now available.

Morrell [*Trans. Instn. Min. Metall.,* Sect. C, **101**, 25–32 (1992)] established equations to predict power draft based on a model of the shape of the rotating ball mass. Photographic observations from laboratory and plant-sized mills, including autogenous, semiautogenous and ball mills showed that the shape of the material charge could roughly be represented by angles that gave the position of the toe and shoulder of the charge. The power is determined by the angular speed and the torque to lift the balls. The resulting equations show that power increases rapidly with mill filling up to 35 percent, then varies little between 35 and 50 percent. Also, net power is related to mill diameter to an exponent less than 2.5. This agrees with Bond [*Brit. Chem. Engr.*, 378–385 (1960)] who stated from plant experience that power increases with diameter to the 2.3 exponent or more for larger mills. Power input increases faster than volume, which varies with diameter squared. The equations can be used to estimate holdup for control of autogenous mills.

The gross power drawn by the mill is the net power plus the power drawn by the empty mill due to friction of its weight on the bearings.

Motor and Drive   The *ring motor* is a low-frequency synchronous motor with its windings mounted directly on the mill shell. Hamdani and Zarif (*Proc. IEEE Cement Industry Tech. Conference*, 32d, 57–80, 1990) compared ring-motor cement mills with geared mills. Although the ring motor has higher initial cost, the operating cost is lower. In particular the downtime for maintenance is lower; the availability based on 20 years experience is >99 percent, while it is 89 percent with geared drives. Ring motors were said to become economically favorable over gear drives at sizes above 8500 hp [Englund, *Power*, **125**(9), 111–2 (1981)].

Recent innovations [Knecht, *IEEE Transactions on Industry Applications*, **28**(4), 962–969 (1992)] have improved the cost effectiveness and reliability of geared-mill designs as well. Slide-shoe bearings support the ends of the mill body and eliminate alignment problems caused by strain on the ends, and allow more compact end design. The bearing pads can be adjusted to compensate for deviations of the slide ring or settlement of the foundation, better than trunnion bearings.

One major requirement of gears is checking and adjusting to close tolerance the alignment of the girth gear and pinion which drives it. To avoid much of this effort, a self-adjusting pinion has been developed which is mounted in tilting bearings. Another improvement is the mounting of tandem-pinion gears in a common housing with the bearing system (Combiflex drive, *Krupp Polysius AG*), which further reduces maintenance problems. Maintenance is also reduced by covering the ring gear and using a circulating oil-lubrication system, instead of manually-applied grease.

Girth-gear drives can cause the concrete foundations to fail if not designed to resist harmonic vibrations (Saxer and Van der Heuvel, 31st IEEE Cement Industry Conference, 1989).

Performance of Proprietary Equipment

Joy Denver Equipment Div   Manufactures a variety of grinding mills. Designs incorporate cast steel heads, heavy rolled steel shell, replaceable cast steel trunnions, hydraulic starting lubricator, inter-nally lubricated trunnion bearings, hard iron trunnion linings, heavy-duty spherical self-aligning anti-friction roller bearings, reversible forged and machined drive pinions, and split construction, reversible mill gears. Sole plates are adjustable. Drive types include V-belt, gear reducer with air clutch or direct, and drum, scoop, scoop-drum or spout feeders. Grinding media may be rod with overflow, end peripheral or center peripheral discharge, or balls with overflow or grate discharge.

Marcy Ball Mill   (Fig. 20-31; *Mine and Smelter Div., Allis Mineral Processing Systems.*) This is traditionally a grate-discharge mill, used to give a high throughput rate for a high circulating load in the wet and dry grinding of ores. The data in Table 20-16 must not be used for design but for orientation only. Mill design must be based on pilot experiments or other techniques previously discussed.

Krupp Polysius AG Mineral Processing Systems   Provides rod- and tube-ball mills from 0.6 to 6 m diameter and up to 15.7 m long. Structural analyses are by computer methods. The patented slide-shoe design compensates for journal deformation, and hardened-steel spherical pads allow the shoes to pivot. Lubrication oil is provided at high pressure for startup/shutdown, and low pressure for continuous running. Drive can be by single pinion to 5,200 kW, twin pinion to 4,500 kW or higher, or gearless from 6,700 kW. Mill linings may be cast Cr-Mo steel alloy, or alloys with Ni for abrasion resistance. Rubber linings are fitted for many wet-grinding processes.

Multicompartmented mills feature grinding of coarse feed to finished product in a single operation, wet or dry. The primary grinding compartment carries large grinding balls or rods; one or more secondary compartments carry smaller media for finer grinding.

STIRRED MEDIA MILLS

Applications overlap with those of vibratory mills, described in the next group. Vibrational equipment is generally used for hard-grinding operations ($ZrSiO_4$, SiO_2, TiO_2, Al_2O_3, etc.), while stirred grinders are mainly used for dispersion and soft grinding (dyes, clays, $CaCO_3$, biological cells, etc). Stirred mills are also called **sand mills** when Ottawa sand is used as media. Contamination and grinder-body wear may be minimized in both types by the use of resilient coatings. Stirred mills use media 6 mm (¼ in.) in size or smaller, whereas vibratory mills use larger media for the same power input. Vibratory mills may grind dry, but most stirred mills are restricted to wet milling. Solids vary from 25 to 70 percent, depending on the feed size and rheology. Unlike in rotary-ball mills, some sedimentation may occur. The media filling ranges from 60 to 90 percent of apparent filling of the mill volume.

Design   In stirred mills, a central paddle wheel or disced armature stirs the media at speeds from 100 to 1500 r/min. The media oscillate in one or more planes and commonly rotate very slowly.

In the **Attritor** (*Union Process Inc.*) a single vertical armature rotates several long radial arms. These are available in batch, continuous, and circulation types. Morehouse-Cowles media mills comprise a

TABLE 20-16   Illustrative Performance of Marcy Ball Mills

Size, ft.	Ball charge, tons	Hp to run	Mill speed, r.p.m.	Capacity, tons/24 hr. (based on medium-hard ore)								
				No. 8 sieve* 20% −200	No. 20 sieve 35% −200	No. 35 sieve 50% −200	No. 48 sieve 60% −200	No. 65 sieve 70% −200	No. 80 sieve 80% −200	No. 100 sieve 85% −200	No. 150 sieve 93% −200	No. 200 sieve 97% −200
3 × 2	0.85	5–7	35	19	15	12	10	8	6½	5	4	3
4 × 3	2.73	20–24	30	80	64	53	45	36	28	22	18	14
5 × 4	5.25	44–50	27	180	145	120	102	82	63	51	41	32
6 × 4½	8.90	85–95	24	375	300	250	210	170	135	105	85	66
7 × 5	13.10	135–150	22½	640	510	425	360	290	225	180	145	113
8 × 6	20.2	220–245	21	1100	885	735	625	500	390	310	250	195
9 × 7	30.0	345–380	20	1800	1450	1200	1020	815	635	505	410	315
10 × 10	56.50	700–750	18	3680	2960	2450	2100	1700	1325	1050	850	655
12 × 12	90.5	1260–1345	16.4	7125	5725	4750	4070	3290	2570	2035	1650	1275

*Sieve through which substantially all the material can pass.

NOTE: To convert horsepower to kilowatts, multiply by 0.746; to convert tons to megagrams, multiply by 0.907; and to convert tons per 24 hours to megagrams per day, multiply by 0.907.

vertical tubular chamber with coaxial rotating disks, and an integral variable-flow diaphragm pump. Models are available from 5 to 100 hp for aqueous and solvent slurries. The Netzsch LME4 mill can be operated with a feed rate up to 100 L/hr [Kula and Schuette, *Biotechnology Progress*, **3**(1), 31–42 (1987)].

Figure 20-35 illustrates the Drais continuous stirred-media mill. The media are stirred by discs mounted on a central shaft. The advantage of horizontal machines is the elimination of gravity segregation of the feed. The feed slurry is pumped in at one end and discharged at the other, where the media are retained by a screen or array of closely spaced flat discs. The latter is useful with slurries having viscosity up to 50 Pa·s (50,000 cP), while screens are useful up to 1 Pa·s. Hydrodynamically shaped screen cartridges can accommodate media as fine as 0.2 mm. German manufacturers [Stadler et al., *Chemie-Ingenieur-Technik*, **62**(11), 907–915 (1990)] have produced mills of various shapes, primarily to aid separation of beads from product. When the mill body rotates with the screen at the axis, centrifugal force aids this separation.

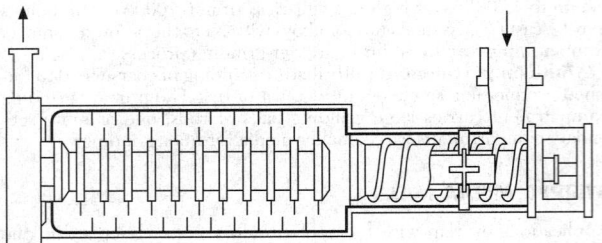

FIG. 20-35 Drais wet-grinding and dispersing system (U.S. patent 3,957,210) Draiswerke Gmbh. [*Stehr*, International J. Mineral Processing, **22**(1–4), 431–444 (1988).]

Agitator discs are available is several forms: smooth, perforated, eccentric, and pinned. Effect of disc design has received limited study, but pinned discs are usually reserved for highly viscous materials.

Cooling water is circulated through a jacket and sometimes through the central shaft. The working speed of disc tips ranges from 5 to 15 m/sec, regardless of mill size.

The continuously stirred mill can be fed by screw feeders from several material bins simultaneously, thus blending uniform compositions, without incurring problems of transporting imperfectly blended or agglomerated mixtures. A series of mills may be used with decreasing media size and increasing rotary speed to achieve desired fine-particle size. No additional feed pumps are needed.

The **annular gap mill** shown in Fig. 20-36 is a variation of the bead mill. It has a high-energy input as shown in Fig. 20-37. It may be lined with polyurethane and operated in multipass mode to narrow the residence-time distribution and to aid cooling.

Performance of Bead Mills Materials processed in stirred-media mills are listed in Table 20-17. Variables affecting the milling process are listed below.

Stirred bead-mill process variables:
Agitator speed
Feed rate
Size of beads
Bead charge, percent of mill volume
Cell concentration in feed
Density of beads
Temperature
Design of blades
Shape of mill chamber
Residence time

The availability of more powerful, continuous machines has extended the possible applications to both lower- and higher-size ranges, from 5 to 200 μm product size, and to feed size as large as 5 mm. The energy density may be 50 times larger than in tumbling-ball mills, so that a smaller mill is required (see Fig. 20-37). Mills range in size from 1 to 1000 L, with installed power up to 320 kW. Specific

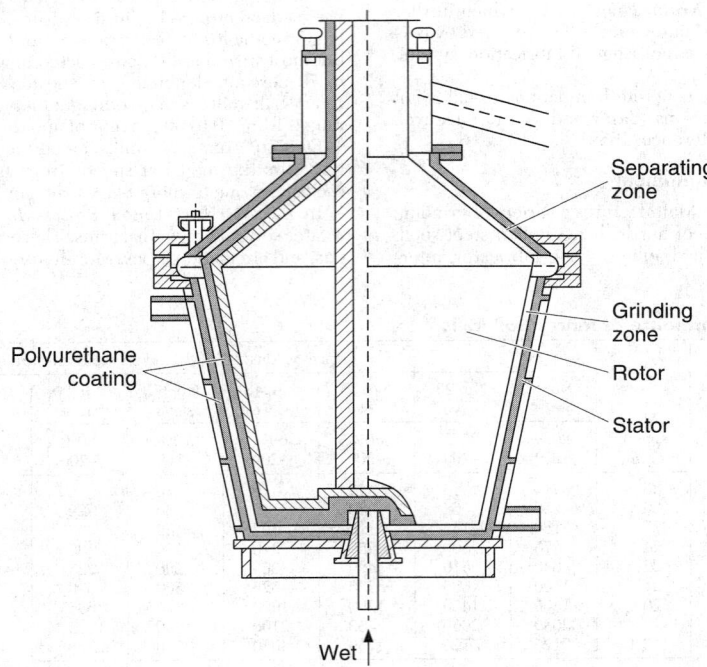

FIG. 20-36 Annular gap-bead mill (*Welte Mahltechnik, Gmbh.*) [*Kolb*, Ceramic Forum International, **70**(5), 212–6 (1993)].

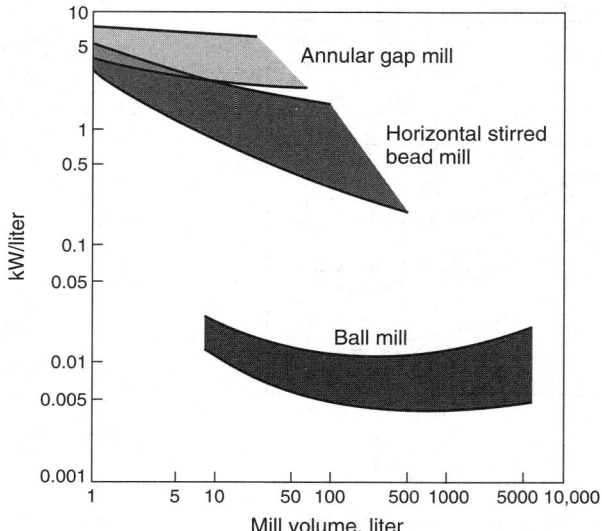

FIG. 20-37 Specific power of bead and ball mills [*Kolb*, Ceramic Forum International, **70**(5), 212–6 (1993)].

TABLE 20-17 Materials Processed in Stirred-Media Mills (Stehr, *International J. Mineral Processing*, 22 1–4, 431–4, 1988).

Industry	Product
Paint & Lacquer	Primer coatings
	Lacquer
	Dispersion paints
Ink	Printing inks
	Dyestuffs
	Textile inks
	Pigment crudes activating
Chemical and pharmaceutical	Various
Electronics	High-grade ceramics
	Oxides for electric components
	Ferrites for permanent magnets
	Audio & Video coatings
Minerals	Limestone
	Fillers
	Paper coatings
	Flue gas desulfurization
	Kaolin
	Gypsum
	Alumina
	Precious metals liberation
Agrochemical	Pesticides
	Insecticides
	Herbicides
	Fungicides
Foods	Cocoa nibs
	Milk chocolate
	Peanuts
	Sesame
Biotechnology	Cell disruption
	Yeast disruption for enzyme extraction
Rubber	Dissolving polymers in solvents
Coal, energy	Coal-oil mixtures
	Coal-water slurry
	Gas turbine coal fuel
	Diesel coal fuel
	Fuel beneficiation, desulfurization

power ranges from 10 to 200 or even 2000 kWh/tonne, with feed rates usually less than 1 tonne/h. Required energy can be scaled-up using dimensionless numbers, as shown in Fig. 20-38.

The nature of the media can influence the grinding rate. Table 20-18 lists some currently used media. Steel balls were more effective in the fine grinding of coal than glass balls [Mankosa et al., *Powder Technology*, **49**(1), 75–82 (1986)], presumably because they have a higher density and greater impact energy than glass. There is also an optimum size of media, for coal about 20 times the size of the particle being ground. For large particles this appears to be related to the coefficient of friction which determines the nip angle. For small particles the capture zone is limited by the curvature of the bead, so that it is not feasible to reduce the size more than one order of magnitude with a given bead size. (For further examples see under Micas and Alumina.)

In vertical disc-stirred mills the media should be in a fluidized condition (White, *Media Milling*, Premier Mill Co., 1991). Particles can pack in the bottom if there is not enough stirring action or feed flow; or in the top if flow is too high. These conditions are usually detected by experiment.

A study of bead milling [Gao and Forssberg, *International J. Mineral Processing*, **32**(1–2), 45–59 (1993)] was done in a continuous Drais mill of 6 L capacity having seven 120 × 10 mm horizontal discs. Twenty-seven tests were done with variables at three levels. Dolomite was fed with 2 m²/g surface area in a slurry ranging from 65 to 75 percent solids by weight, or 39.5 to 51.3 percent by volume. Surface area produced was found to increase linearly with grinding time or specific-energy consumption. The variables studied strongly affected milling rate; two extremes differed by a factor of 10. An optimum bead density for this feed material was 3.7. Evidently the discs of the chosen design could not effectively stir the more-dense beads. Higher slurry concentration above 70 wt % solids reduced the surface production per unit energy. The power input increased more than proportional to speed.

Residence Time Distribution Commercially available bead mills have a diameter-to-length ratio ranging from 1:2.5 to 1:3.5. The ratio is expected to affect the *residence time distribution* (RTD). A wide distribution results in overgrinding some feed and undergrinding another. Data from Kula and Schuette [*Biotechnology Progress*, **3**(1), 31–42 (1987)] shows that in a Netzsch LME20 mill RTD extends from 0.2 to 2.5 times the nominal time, indicating extensive stirring. (See "Cell Disruption" under Applications.) The RTD is even more important when the objective is to reduce the top size of the product as Stadler et al. [*Chemie-Ingenieur-Technik*, **62**(11), 907–915 (1990)] showed, because much of the feed received less than half the nominal residence time. A narrow RTD could be achieved by rapidly flowing material through the mill for as many as 10 passes.

VIBRATORY MILLS

The dominant form of industrial vibratory mill is the type with two horizontal tubes, called the **horizontal tube mill.** These tubes are mounted on springs and given a circular vibration by rotation of a counterweight, shown in Fig. 20-39. Many feed flow arrangements are possible, adapting to various applications. Variations include polymer lining to prevent iron contamination, blending of several components, milling under inert gas and at high and low temperatures. An example is the Palla vibratory mill (*ABB Raymond Div, Combustion Engineering Inc.*).

The Vibro-Energy (*Sweco, Inc.*), a **vertical vibratory mill** (Fig. 20-40) has a mill body in a ring-shaped trough form but uses horizontal vibrations at a frequency of about 20 Hz of the contained media, usually alumina spheres or cylinders. Other characteristics appear in Table 20-19.

The vertical vibratory mill has good wear values and a low-noise output. It has an unfavorable residence-time distribution, since in continuous operation it behaves like a well-stirred vessel. Tube mills are better for continuous operation. The mill volume of the vertical mill cannot be arbitrarily scaled up because the static load of the upper media, especially with steel beads, prevents thorough energy introduction into the lower layers. Larger throughputs can therefore only be obtained by using more mill troughs, as in tube mills.

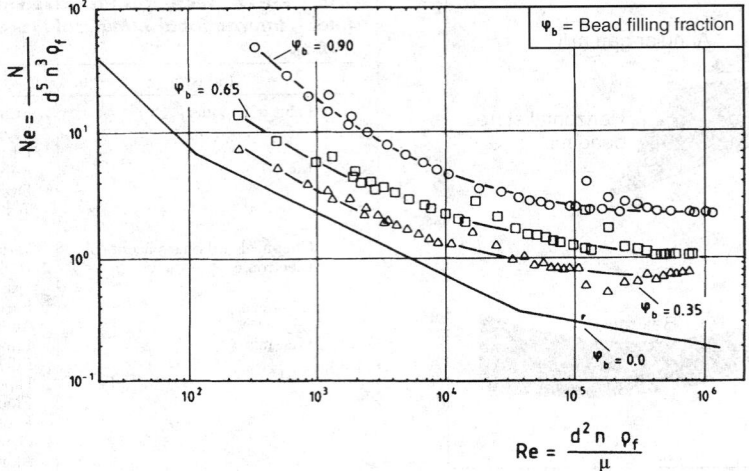

FIG. 20-38 Newton number as a function of Reynolds number for a horizontal stirred bead mill, with fluid alone and with various filling fractions of 1-mm glass beads [*Weit and Schwedes, Chemical Engineering and Technology,* **10**(6), 398–404 (1987)]. (N = power input, W; d = stirrer disk diameter, m; n = stirring speed, 1/s; μ = liquid viscosity, Pa·s; Q_f = feed rate, m³/s.)

The primary applications of vibratory mills are in fine milling of medium to hard minerals primarily in dry form, producing particle sizes of 1 μm and finer. Throughputs are typically 10 to 20 t/hr. Grinding increases with residence time, active mill volume, the energy density and the vibration frequency, and media filling and feed charge. The amount of energy that can be applied limits the tube size to 600 mm, although one design reaches 1000 mm. Larger vibratory amplitudes are more favorable for comminution than higher frequency. The development of larger vibratory mills is unlikely in the near future because of excitation problems. This has led to the use of mills with as many as six grinding tubes.

Performance The grinding-media diameter should preferably be 10 times that of the feed and should not exceed 100 times the feed diameter. To obtain improved efficiency when reducing size by several orders of magnitude, several stages should be used with different media diameters. As fine grinding proceeds, rheological factors alter the charge ratio, and power requirements may increase.

A variety of grinding media are available, as shown in Table 20-18. Size availability varies, ranging from 1.3 cm (½ in) down to 325 mesh (44 μm).

Although there are no definitive data on media shape and grinding, ball-milling data indicate that spheres are the most effective shape. [Norris, *Trans. Inst. Min. Metall.,* **63**(567), 197–209 (1954)].

Although each machine has its peculiar characteristics and time requirements for various types of grinding, Fig. 20-40 illustrates some typical results obtained under optimum conditions for several materials.

Advantages of vibratory mills are (1) simple construction and low-capital cost, (2) very fine product size attainable with large reduction

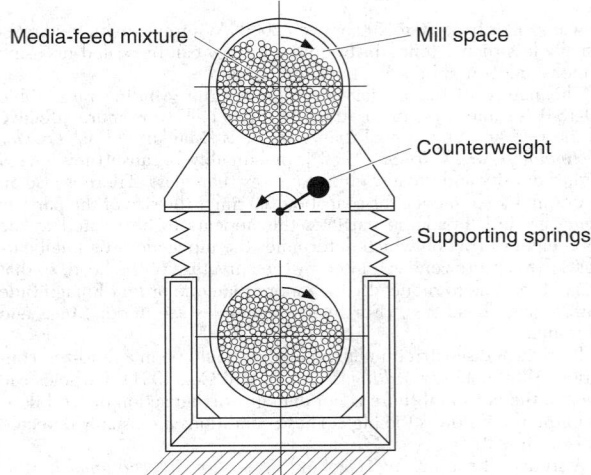

FIG. 20-39 Two-tube vibratory mill. (*Hoeffl,* Freiberger Forschungshefte A **750**, 125 pp., 1988.)

TABLE 20-18 Media for Stirred and Vibratory Mills
(* S = spheres, C = cylinders, I = irregular shapes)

Material	Common name	Density, g/cm³	Diameter, mm	Forms available*
Perfluoroethylene	Teflon	1.	—	C
Silicon dioxide	Ottawa sand	2.8	0.4–3.0	I
Annealed glass	Hard glass, unleaded	2.5	0.3–5.0	S
Annealed glass	Hard glass, leaded	2.9	0.3–5.0	S
Aluminum oxide	Alumina, Corundum	3.4	0.5–15.0	S,C,I
Zirconium silicate	Zircon	3.8	0.3–15.0	S,C,I
Zirconium oxide	Zirconia, Zircoa	5.4	0.5–3.0	S,C,I
Steel	Steel shot	7.6	0.2–15.0	S
Steel	Chrome steel	7.85	1.0–12.0	S

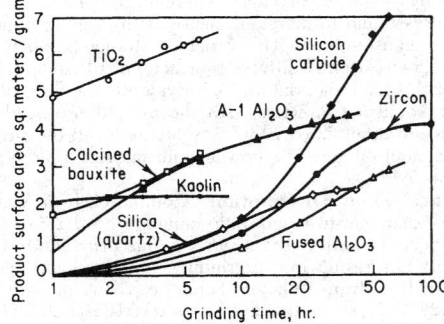

FIG. 20-40 Vibratory-mill typical performance. (*Sweco, Inc.*)

TABLE 20-19   Characteristics of Sweco Vibratory Mills

Designation	Capacity	Typical sample charge, lb	Motor	Mill diameter, in
M-18	2.6 gal	5–20	¼	18
M-45	20 gal	50–200	5	45
M-60	70 gal	200–1000	10	60
M-80	182 gal	500–2000	40	80
DM-1	0.125 ft³	3–5	⅓	24
DM-3	0.5 ft³	20–60	1¼	30
DM-10	3 ft³	100–400	5	45
DM-20	65 ft³	200–800	10	60
DM-70	23 ft³	900–3000	40	95

NOTE: To convert gallons to cubic meters, multiply by 3.785×10^{-3}; to convert pounds to kilograms, multiply by 0.4535; to convert horsepower to kilowatts, multiply by 0.746; and to convert inches to centimeters, multiply by 2.54.

ratio in a single pass, (3) good adaptation to many uses (4) Small space and weight requirements, (5) ease and low cost of maintenance. Disadvantages are (1) limited mill size and throughput, (2) vibration of the support and foundation, (3) high-noise output, especially when run dry.

Residence Time Distribution   Hoeffl (Freiberger *Forschungshefte A*, **750**, 119 pp., 1988) carried out the first investigations of residence time distribution and grinding on vibratory mills, and derived differential equations describing the motion. In vibratory horizontal-tube mills the mean axial transport velocity increases with increasing vibrational velocity, defined as the product $r_s \Omega$, where r_s = amplitude and Ω = frequency. Apparently the media act as a filter for the feed particles, and are opened by vibrations. Nevertheless, good uniformity of transport is obtained, indicated by *vessel dispersion numbers* $D\tau/L^2$ (see "Simulation of Milling Circuits" above) in the range 0.06 to 0.08 measured in limestone grinding under conditions where both throughput and vibrational acceleration are optimum.

The vibratory-tube mill is also suited to wet milling. In fine wet milling this narrow residence time distribution lends itself to a simple open circuit with a small throughput. But for tasks of grinding to colloid-size range, the stirred media mill has the advantage.

NOVEL MEDIA MILLS

Planetary Ball Milling   This is a method of increasing the gravitational force acting on balls in a ball mill. For example, refractory metals and carbides can be ground to 1 to 2.6 µm in 5 to 20 min in an apparatus capable of applying a centrifugal force of 10 to 50 G. [Dobrovol'skii, *Poroshk. Metall.*, **7**(6), 1–7 (1967)].

Pulverit planetary mills are available from Geoscience Inc. High-speed planetary-ball mills can be used to perform rapid tests to simulate ball milling of materials [Vock, *DECHEMA-Monogr.*, **69**, III-8 (1972)]. The size of high-speed mills will be much smaller than the size of same-capacity ball mills [Bradley, *S. Afr. Mech. Eng.*, **22**, 129 (1972)].

PARTICLE-SIZE CLASSIFIERS USED WITH GRINDING MILLS

Ball mills or tube mills can be operated in closed circuit with external air classifiers with or without air sweeping being employed. If air sweeping is employed, a cyclone separator may be placed between mill and classifier. (The principles of size reduction combined with size classification are discussed under "Characteristics of Size Classifiers.") Likewise other types of grinding mill can be operated in closed circuit with external size classifiers (Fig. 20-12), as will be described at appropriate places on succeeding pages. However, many types of grinders are air-swept and are so closely coupled with their classifiers that the latter are termed internal classifiers.

Dry Classifiers   Dry **screens** are used primarily in crusher circuits, since they are most effective down to 4 mesh. They can sometimes be used to 35 mesh. Examples are *Hummer* screens (*W. S. Tyler, Inc.*), *Rotex* screens (*Rotex Inc.*), and the *Vibro-Energy separator* (*Sweco, Inc.*).

Most dry-milling circuits use **air classifiers.** There are a number of types, but all use the principles of air drag and particle inertia, which

depend on particle size. The simplest type of air classifier is an elutriator. A countercurrent multielement elutriator is the Zig-Zag classifier (*Hosokawa Micron Powder Systems Div.*). The sharpness of separation increases with the number of elements. These devices are effective in the 30- to 80-mesh range.

Another type of classifier directs an air stream across a stream of the particles to be classified. An example is the radial-flow classifier (*Kennedy Van Saun Corp.*), which features adjustable elements to control the flow and classification. A further development on this principle is the Vari-Mesh classifier (*Kennedy Van Saun Corp.*), which controls classification by adjustable flow baffles. A change in direction of air flow is the operating principle of the reverse-flow Superfine classifier (*Hosokawa Mineral Processing Systems*).

Rotating blades are the main elements of several types of classifiers. The blades set up a centrifugal motion that tends to throw coarser particles outward. An example is the Mikro-ACM Pulverizer, in which an external fan forces air inwardly through the blades, carrying with it the fines. Centrifugal motion returns coarse particles to the hammers. The whizzer blades shown on the Raymond vertical mill (Fig. 20-49) have a similar centrifugal effect, throwing coarse particles to the wall of the chamber, where the lower boundary-layer air velocity allows them to fall back to the grinding zone.

Rotor blades also form an element of several external classifiers that are used in closed-circuit dry milling. These are generally called mechanical air separators or classifiers. Examples are the Whirlwind classifier (*Sturtevant Inc.*), the Gayco centrifugal separator (*Universal Road Machinery Co.* see Fig. 20-41), and the whizzer separator (*Raymond Division of Combustion Engineering Inc.*).

Some mechanical air classifiers are designed so that the fine product must pass radially inward through rotor blades instead of spirally moving across them as with whizzer blades. Examples are the Mikron separator (*Hosokawa Micron Powder Systems Div.*), Sturtevant Side Draft separator, and the Majac classifier shown attached to the Majac jet mill (Fig. 20-55).

There are several mechanical air classifiers designed to operate in the superfine 10- to 90-µm range. Two of these are the Mikroplex spiral air classifier MPVI (*Hosokawa Micron Powder Systems Div.*) and the classifier which is an intregal part of the Hurricane pulverizer-classifier (*ABB Raymond Div, Combustion Engineering Inc.*) described under "Hammer Mills." Others are the Majac classifier (*Hosokawa Micron Powder Systems Div.*), the Sturtevant Superfine Air Separator, and the Bradley RMC classifier. These also use a vaned rotor, but operate at higher speed with higher power input and lower throughput.

Performance   Deflector-type classifiers without rotating elements may give a product 85 percent through a 250-µm sieve, although more typically they give a product 95 percent below 74 µm. Mechanical air classifiers with rotating elements can give a product from 85 percent through 250 µm to as fine as 99.9 percent below 44 µm. The single whizzer is designed for operation where finenesses range to about 95 percent passing 74 µm, whereas the double-whizzer separator is intended for use where higher-fineness products, in the range of 99.9 percent or better passing 44 µm, are required. Sizes of mechanical air classifiers range from 1 to 7 m (3 to 24 ft) in diameter, with power requirements from 2 to 450 kW (3 to 600 hp). Superfine types can give a product 98 percent through 10 µm.

Typical separation efficiency curves of an air classifier versus particle size are given in Fig. 20-14. The amount of top size in the fines may be very low, but there is typically 10 to 30 percent fines in the coarse product; that is, the low end of the curve tends to flatten out at 10 to 30 percent. In addition, the separation at the cut size is typically a gradual curve. Data of this sort, which are needed to evaluate closed-circuit mill performance, are seldom available. See subsection on characteristics of size classifiers for a testing method.

The Sturtevant air separator (*Sturtevant Inc.*) incorporates a hydraulic mechanism to adjust the width of the ring baffle over the spinner blades, which allows adjusting the separation curve within limits, at some cost in production rate. For example, the residue on 74-µm screen can be varied from 1 to 20 percent, with a production rate shown in Table 20-20 for cement rock.

An improved version of the Raymond classifier sends all the air to the product-collector cyclone, returning only solids. This improves

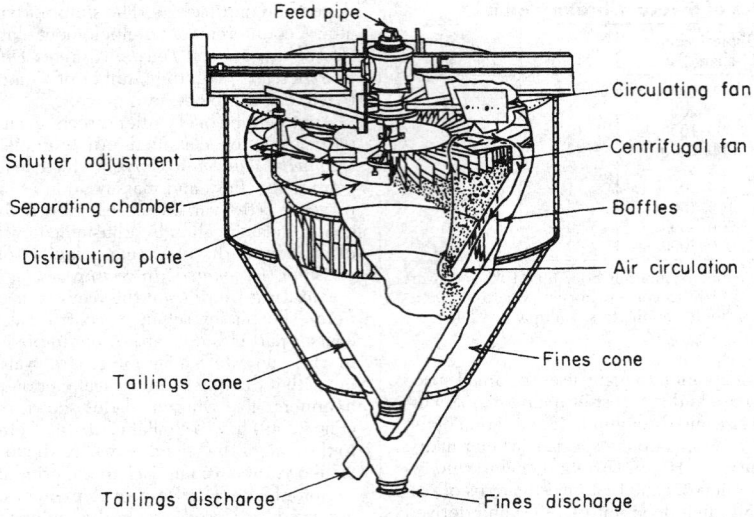

FIG. 20-41 Typical centrifugal separator.

fines production, since air returned with coarse material otherwise carries fines.

Wet Classifiers   Closed-circuit wet milling is the rule in large-scale operations because of its greater production and economy. The simplest wet classifier is a **settling basin** arranged so that the fines do not have time to settle but are drawn off while the thickened coarse product is raked to a central discharge. Examples are the Hardinge Hydro-Classifier and the Dorr thickener. For classification near micrometer size a **continuous centrifuge** such as the Sharpless super-centrifuge or the Bird centrifuge is effective. The separation is not sharp in settlers, and the large space requirement is a detraction. **Rake classifiers** and **screw classifiers** are described in Sec. 21. The action is countercurrent, so separation of coarse grains is more effective. Examples are the Hardinge countercurrent (screw) classifier (Fig. 20-42) and the Dorr-Oliver rake classifier (Fig. 20-43). Typical circuits used with these classifiers in cement- and ore-processing plants are shown in Figs. 20-43 and 20-44. **Hydrocyclones** have become the most popular wet classifiers in ore operations owing to their compact design and economy of operation. Control is effected by feeding at a constant rate from a sump, in which the liquid level is maintained by varying water addition as the slurry feed rate varies (see Fig. 20-45).

In the 1930s there were attempts to use **screens** for wet closed-circuit milling, but operating cost was prohibitive. Recently screens have been developed that are practical for mill circuits. The first of these was the Dutch State Mines screen, which has the screen cloth on a curved incline, with the dilute slurry flowing over and through it. Because of the angle of attack, the effective opening is smaller than the physical opening, so blinding does not occur, but the separation is less sharp.

The use of rubber screen cloths also solves problems of blinding [Wessel, *Aufbereit. Tech.*, **8**(2), 53–62; (5), 167–80; (8), 417–428 (1967);

TABLE 20-20   Production Rate of Sturtevant Air Separator at 74 μm

Separator diameter		Fines production	
m	ft	Mg/h	ton/h
0.9	3	0.4–1.0	0.4–1.1
2.5	8	3–9	3–10
4.3	14	15–40	17–45
6.1	20	77–210	85–230
8.0	26	180–450	200–500

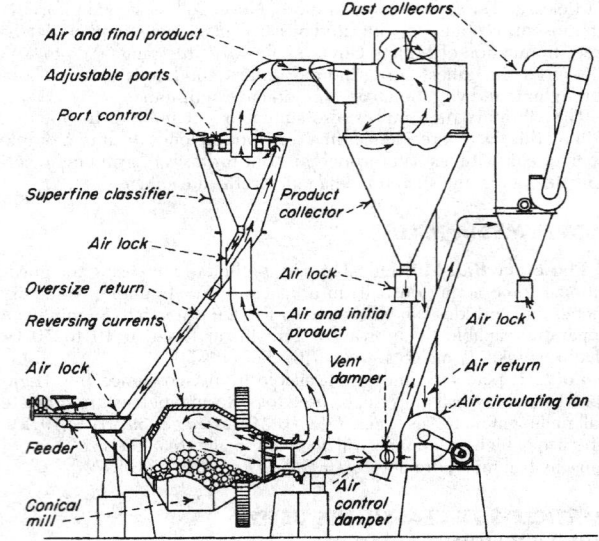

FIG. 20-42   Hardinge conical mill with reversed-current air classifier.

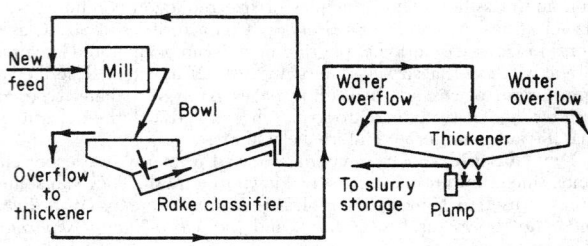

FIG. 20-43   Single-stage closed-circuit wet-grinding system. [*Tonry*, Pit Quarry (*February–March 1959*).]

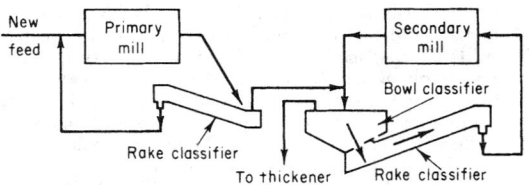

FIG. 20-44 Two-stage closed-circuit wet-grinding system. [*Tonry, Pit Quarry (February–March 1959).*]

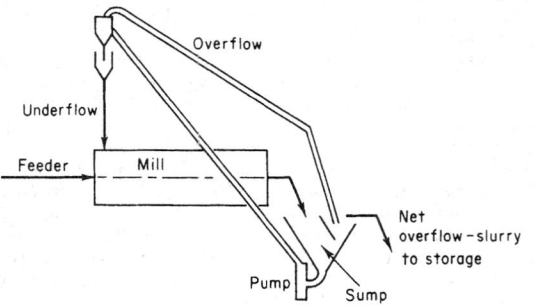

FIG. 20-45 Closed-circuit wet grinding with liquid-solid cyclone. [*Tonry, Pit Quarry (February–March 1959).*]

Michel, *Min. Mag. (London)*, annual review issue (5), 189–193, 207 (1968)]. An upper layer of rubber is perforated with fine slots for particle sizes from 0.2 to 2.5 mm., and this is supported by a lower layer with coarse holes. The vibration rate is 2500 to 3000 cycles/min. The advantage of screens over other classifiers is that a considerably sharper separation can be effected, and less fines are returned to the mill. Screen separation is considerably less than perfect, although there are few published data. The selectivity of a screen depends on the screen opening size and the feed rate, as well as other variables. A semiempirical model for an inclined vibratory screen [Karra, *CIM Bulletin*, 167–171 (Apr. 1979); *Proc. 14th International Mining Congress*, Toronto, **III**(6), 1–614; King, *International J. Mineral Processing*, **29**(3–4), 249–265 (1990)] includes factors for these effects.

HAMMER MILLS

Hammer mills for fine pulverizing and disintegration are operated at high speeds. The rotor shaft may be vertical or horizontal, generally the latter. The shaft carries hammers, sometimes called beaters. The hammers may be T-shaped elements, stirrups, bars, or rings fixed or pivoted to the shaft or to disks fixed to the shaft. The grinding action results from **impact** and **attrition** between lumps or particles of the material being ground, the housing, and the grinding elements. A cylindrical screen or grating usually encloses all or part of the rotor. The fineness of product can be regulated by changing rotor speed, feed rate, or clearance between hammers and grinding plates, as well as by changing the number and type of hammers used and the size of discharge openings.

The **screen** or **grating discharge** for a hammer mill serves as an internal classifier, but its limited area does not permit effective usage when small apertures are required. A larger external screen may then be required.

The feed must be nonabrasive with a hardness of 1.5 or less. The mill is capable of taking 2-cm (¾-in) feed material, depending on the size of the feed throat, and reducing it to a product substantially all able to pass a No. 200 sieve. For producing materials in the fine-size range, it may be operated in conjunction with external air classifiers. Such an arrangement is shown in Figs. 20-12 or 20-42. A number of machines have internal air classifiers.

Hammer Mills without Internal Air Classifiers   The **Mikro-Pulverizer** (Fig. 20-46) (*Hosokawa Micron Powder Systems Div.*) is a close-clearance, screen discharge, high-speed, controlled sealed-feed hammer mill used for a wide range of nonabrasive materials, the major applications being sugar, carbon black, chemicals, pharmaceuticals, plastics, dyestuffs, dry colors, and cosmetics. For performance see Table 20-21. Speeds, types of hammers, feed devices, housing variations, and perforations of screens are all varied to fit applications, with the result that finenesses and character of grind cover a wide range. Some of the grinds are as fine as 99.9 percent through a 325-mesh screen. Feed material should usually be down to 4 cm (1½ in) or finer. If feed is larger, an auxiliary crusher may be required, preferably as a separate unit, because synchronization is difficult since the crusher has larger capacities than the pulverizer. Tie-in is possible with careful regulation of relative speeds of crusher and feed screw or screws.

A replaceable liner for the mill housing cover is made with multiple serrations. Hammer tips can be provided with tungsten carbide inserts for greatest wear resistance or with Hastellite tipping. An air-injector feeder can be supplied to project the feed particles directly in front of the hammer tips, to provide a more direct blow and thus increase mill efficiency. Cinematographic studies show that otherwise the particles receive a glancing blow from hammers. Wet feed can be charged with feed screws or pumps for wet grinding. Mikro-Pulverizers are made in five sizes shown in Table 20-21, plus a 150-hp size. The smallest size is the Bantam, which is widely used in laboratories for development and pilot work. Results will indicate in a qualitative way what may be expected of full-scale production units.

The **Imp pulverizer** (*Raymond Division, Combustion Engineering Inc.*) is an air-swept hammer mill. This machine is made in many sizes from the smallest, using 18 kW (25 hp), to the largest size, with six rows of hammers and requiring 700 kW (1000 hp) to drive it. The machines are equipped with a hopper below which is the star feeder, actuated by a pawl-and-ratchet mechanism. The Imp mill is generally used as part of an air-swept classifying circuit such as shown in Fig. 20-42. Any type of classifier can be used, depending on the application. Any solid material with a top size of 1 in. that is softer than 2 on the Moh scale can be processed to a fineness ranging from 1000 to 45 μm.

The **Blue Streak dual-screen pulverizer** (*Prater Industries, Inc.*) is used for the grinding of resins, chemical salts, plastic scrap, food products, and similar materials to a granular uniform powder of No. 30 or No. 40 sieve fineness. Feed enters opposite ends of the rotor and undergoes three stages of size reduction by hammers of decreasing size. Two perforated screens cover more than 70 percent of the area of the final sizing drum through which the product passes.

The **Atrita pulverizer** (*Riley Stoker Corp.*) is available in several single- and duplex types. Capacities vary from 3400 to 25,000 kg/h (7500 to 54,000 lb/h). The coal is carried in a pin mill where most of the pulverizing is done, followed by a recycling classifier.

Hot air can be introduced into the machine for drying the coal. Air at 150°C dries coal with 8 percent moisture down to about 1 percent.

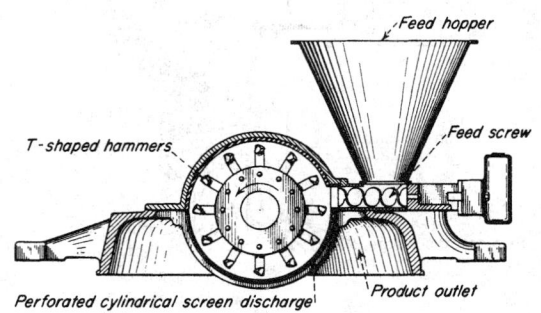

FIG. 20-46 Mikro-Pulverizer hammer mill. (*Hosokawa Micron Powder Systems Div.*)

TABLE 20-21 **Mikro-Pulverizer Performance**

Size	Rotor diam., in.	Max. r.p.m.	Hp.	Avg. capacities, lb./hr.		
				6X sugar	Clay-graphite water slurry	Pigments and colors (dry)
Bantam	5	16,000	¾–1	75–100	75–100	70–90
1	8	9,600	3–5	350–550	550	300–500
2	12	6,900	7½–15	800–1500	750–1600	800–2000
3	18	4,600	20–40	2000–4500	4500	2500–4500
4	24	3,450	40–100	4000–9000	7000	4500–7000

NOTE: To convert inches to centimeters, multiply by 2.54; to convert horse-power to kilowatts, multiply by 0.746; and to convert pounds per hour to kilograms per hour, multiply by 0.4535.

The **Aero pulverizer** (*Foster Wheeler Corp.*) is used for coal, pitch, and coke, blowing the ground material directly into the furnace. The housing is divided into two or three short cylindrical pulverizing chambers.

Hot gases can be introduced to dry the fuel being pulverized. Refractory material such as tramp iron is removed in the first pulverizing chamber and eliminated through a tramp-iron pocket.

Disintegrator The Rietz machine (Fig. 20-47) consists of a rotor running inside a 360° screen enclosure. The rotor includes a number of hammers designed to run at fairly close clearance relative to the inside of the cylindrical screen enclosing the disintegration chamber. The hammers are normally fixed rigidly to the shaft by keyways, pins, or welding, but swing hammers are used when indicated.

Rietz disintegrators are supplied in three types. *In-line disintegrators* (RI series) are designed for in-line installation, in which they function without impeding the process flow. Their primary applications are the mixing, delumping, and dissolving of fluids, slurries, and pastes and the grinding and separation of high-fiber solids.

Angle disintegrators (RA and RP series) are used for the fine pulping of many food products and for fine dispersion and homogenizing in the food and chemical industries. *Vertical disintegrators* (RD series) are used for dry pulverizing, wet grinding to produce slurries or pastes, shredding, defibrizing, and the fine pulping of soft fruits and vegetables.

Rietz disintegrators are normally supplied in rotor diameters from 10 to 60 cm (4 to 24 in), with rotational speeds to produce hammer tip speeds in ranges of 300 to 6700 m/min (1000 to 22,000 ft/min) and power ranges from 0.4 to 150 kW (½ to 200 hp). Higher speeds and higher power are available. AC variable-frequency drives can eliminate belts and provide easier variation of speed. Models are available

in various materials of construction and in highly sanitary, easy-cleaning models or heavy-duty industrial construction (Table 20-22).

Fitz mills (*Fitzpatrick Co.*) consist of several series of hammer mills in configurations adapted to a variety of uses in food processing. There are high-speed screen hammer mills with flat hammers for impact, and narrow hammers or sharp hammers for tough plastic or fibrous materials. There are long, small-diameter rotating mills for processing pastes and two-shaft toothed masticators. There are also single-roll toothed choppers and shredders with fixed knives.

Prater Industries, Inc., manufactures narrow swing-hammer screened mills for oilseeds and fibrous materials.

Turbo pulverizers and **turbo mills** (*Pallmann Pulverizer Co.*) combine the action of hammer and attrition mills, finding special application for grinding plastic materials that would be softened under high-energy warm-mill conditions.

Pin Mills In contrast to peripheral hammers of the rigid or swing types, there is a class of high-speed mills having pin breakers in the grinding circuit. These may be on a rotor with stator pins between circular rows of pins on the rotor disk, or they may be on rotors operating in opposite directions, thereby securing an increased differential of speed. See also the Mikro-ACM pulverizer described later.

Fine impact mills (*Hosokawa Micron Powder Systems Div.*) are high-speed impact mills with one stationary and one rotating stud disk. The mills are operated without a sieve and hence can be used with materials that tend to block (see Fig. 20-48). The **Contraplex wide chamber** is a similar mill with both disks rotating. It is suitable for grinding materials that tend to form deposits or for greasy, heat-sensitive products. These mills are used in the grinding of food, pesticides, pigments, and soft minerals, the wet grinding of PVC suspensions, and the crushing of cacao beans, etc. A laboratory model is also available.

Entoleter impact mills (*Entoleter, Inc.*) and the **Simpactor impact Mill** (*Sturtevant Inc.*) are a class of vertical-shaft devices in which feed at the shaft is caused to move rotationally and is thrown outward from the rotor to impact on an outer ring. Pin-type structures have been found effective; and in these the pins on the rotor do primary breakage, while the outer ring of pins gives further reduction. A wide range of speeds is employed, the higher ones being used for fine pulverizing. These mills grind a great variety of free-flowing or semi-free-flowing substances. Among these are plastics, rubber, grain and flour, coal, clay, slag, and salts. In some cases, external classification is required to remove oversize for return to the mill, or the mill may be combined with a classifier that contains no driven elements. Plastic materials are embrittled by liquid nitrogen or other suitable refrigerants to reduce their elasticity. For the highest speeds, the stator pins are mounted on a ring which is moving in reverse rotation to the central rotor.

Hammer Mills with Internal Air Classifiers The rotating components of the Raymond **vertical mill** are carried on its vertical shaft. They are the grinding element, double-whizzer classifier, and fan, as shown in Fig. 20-49. This mill has a hammer-tip speed of 7600 m/min (25,000 ft/min), so that it is effective for finer grinding than the Imp mill, which has a tip speed of 6400 m/min (21,000 ft/min).

The fine product is carried in the air stream through the fan and discharge port and separated by a cyclone collector into a suitable container.

Machines are available with rotor diameters of 45.7 and 88.9 cm (18 and 35 in), driven by 15- and 110-kW (20- and 150-hp) motors respectively. The larger mill is directly connected to a vertical motor. Normal rotor speed for the 45.7-cm (18-in) Raymond vertical mill is 6000 r/min and 3600 r/min for the 88.9-cm (35-in) machine.

The field of application of the Raymond vertical mill is the production of soft materials that range in size from those having 99 percent passing a 44-μm sieve to those having 99 percent smaller than 15 μm, depending on the state of aggregation of the feed. A production rate of 227 kg/h (500 lb/h) is achieved with a chemical in a 45.7-cm (18-in) machine, consuming 13.4 kW (18 hp) when the product is substantially smaller than 15 μm. In a talc operation on an 88.9-cm (35-in) machine requiring 110 kW (150 hp), a production rate of 320 kg/h (700 lb/h) is obtained. At a production rate of 2250 kg/h (5000 lb/h), a sample of the product leaves only a trace of talc on a No. 325 testing sieve.

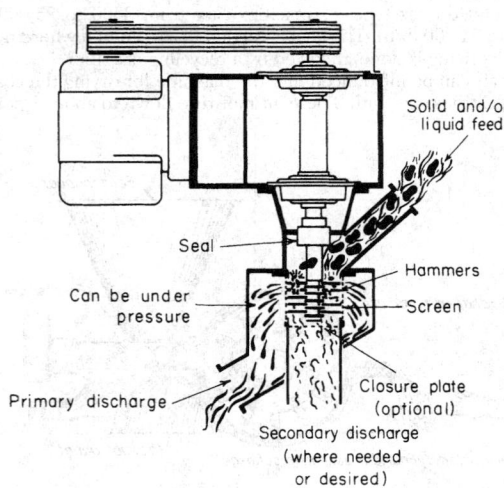

FIG. 20-47 Rietz disintegrator. (*Rietz Div. Hosokawa Bepex Corp.*)

TABLE 20-22   Performance of Rietz Disintegrators

Model	Rotor diam., in.	Max. r.p.m.	Hp. range	Screen perforation, in.	Typical applications	
					Material	Capacity
RA-1	4	16,000	½–5	¹⁄₃₂–¼	General lab use	1–10 lb./min.
RP-6	6	3,600	1–20	³⁄₁₆	Horseradish	300 lb./hr.
RI-2	6 or 8	5,000	3–20	¹⁄₁₆	Detergent delumping	100 gal./min.
RD-8	8	8,400	3–20	⅛	Color coat	3600 lb./hr.
RA-2	8 or 12	8,400	3–20	¹⁄₃₂	Meat, cooked	3000–5000 lb./hr.
RP-8	8	3,600	10–60	¼	Blood declotting	20 gal./min.
RD-12	12	7,200	15–50	¼–¾	Polystyrene	3000–10,000 lb./hr.
RA-3	12 or 18	6,500	10–75	³⁄₆₄	Corn, heated	350 lb./min.
RP-12	12	3,600	20–75	⅜	Asbestos-cement slurry	200 gal./min.
RD-18	18	3,600	30–150	⅜	Chemical-fertilizer delumping	15 tons/hr.
RP-18	18	3,600	25–100	¼	Animal fat (90°F.)	15,000 lb./hr.
RD-24	24	3,600	75–400	1	Wood-chip shredding	30 tons/hr.
RI-4	24	3,600	50–200	¼	Bagasse depithing	30 tons/hr. (dry)

Maximum horsepower depends upon maximum speed.
RA and RP models are normally supplied with stainless-steel contact parts.
Some disintegrators are available for operation under pressure.
Screens are available in various sizes and types of perforations down to 0.006 in.
NOTE: To convert inches to centimeters, multiply by 2.54; to convert horsepower to kilowatts, multiply by 0.746; and to convert pounds per hour to kilograms per hour, multiply by 0.4535.

The **Mikro-ACM pulverizer** is a pin mill with the feed being carried through the rotating pins and recycled through an attached vane classifier. The classifier rotor is separately driven through a speed control which may be adjusted independently of the pin-rotor speed. Oversize particles are carried downward by the internal circulating air stream and are returned to the pin rotor for further reduction. The constant flow of air through the ACM maintains a reasonable low temperature which makes it ideal for handling heat-sensitive materials. Typical capacities are given in Table 20-23. The mill is built in eight sizes: model 2 to model 400 with drive motor from 5 to 400 hp and production rates from 0.2 to 36 times the rate in Table 20-26.

The **Pulvocron** (*Hosokawa Bepex Corp.*) employs one or more beater plates, around the periphery of which are attached rigid hammers of hard metal. It is driven within a casing at clearances of small fractions of an inch, the periphery of which is generally V-cut. The grinding ring has provision for cooling with liquid in direct contact with its periphery. Feed enters around the driving shaft and is broken by breaker plates. It then travels to a classifying chamber, in which is a separately driven and controlled rotor with vanes. See Table 20-24.

RING-ROLLER MILLS

Ring-roller mills (Fig. 20-50) are equipped with rollers that operate against grinding rings. Pressure may be applied with heavy springs or by centrifugal force of the rollers against the ring. Either the ring or the rollers may be stationary. The grinding ring may be in a vertical

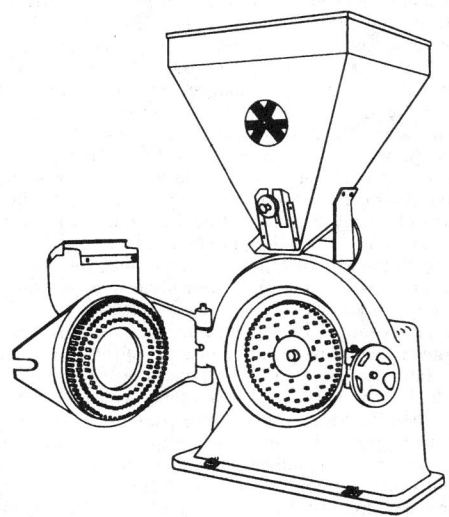

FIG. 20-48   Fine-impact mill. (*Hosokawa Micron Powder Systems Div.*)

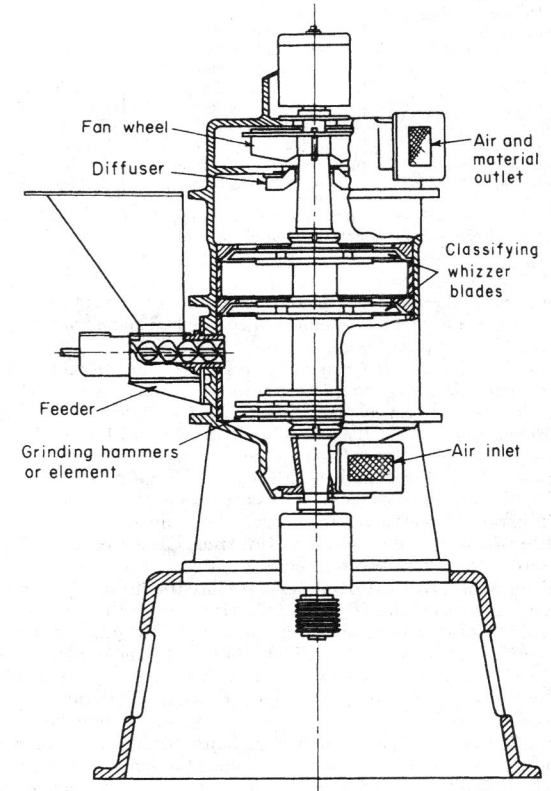

FIG. 20-49   Raymond vertical mill. (*Raymond Division, Combustion Engineering Inc.*)

TABLE 20-23   Test Results on Model 10 Mikro-ACM Pulverizer

Material	Fineness, d97, μm		Output, lb/h
	Feed	Product	
Alumina	45	11	519
Calcined coke	3000	79	315
Clay	850	45	750
Cocoa/sugar	300	44	100
Corn starch	105	62	300
Dextrose	1200	75	700
Diatomaceous earth	70	35	800
Graphite	150	45	250
Herbicide	150	45	790
Kaolin	200	9.6	600
Limestone	6000	45	360
Phenolic resin	9000	75	1140
Starch	25000	180	400
Sugar	450	75	1000
Sugar	450	30	850
Talc	43	12	80
Titanium dioxide	6000	30	1000
Wood flour	2000	125	40
Xanthan gum	6000	180	70

NOTE: 100 mesh = 150 μm; 200 mesh = 75 μm; 325 mesh = 45 μm

TABLE 20-24   Performance of the 20-in Pulvocron

Material	Particle analysis, by weight	Capacity, lb./hr.	Hp.
Sucrose	97.5% minus 325 mesh	1800	60
Sodium chloride	99.4% minus 100 mesh	3600	50
	99.95% minus 325 mesh	160	45
Urea-formaldehyde and melamine molding compounds	99.2% minus 80 mesh	1600	45
Paraformaldehyde	99.7% minus 325 mesh	1300	40
Casein	99% minus 80 mesh	650	50
Corn flour	88% minus 200 mesh	800	35
Soy flakes	95% minus 200 mesh	2000	60
Sterols	100% minus 5 μ	700	60
Lactose	98.5% minus 200 mesh	1200	40
Alumina, hydrated	99% minus 325 mesh	700	30
Cinnamon	99.7% minus 60 mesh	1000	50

NOTE: To convert pounds per hour to kilograms per hour, multiply by 0.4535; to convert horsepower to kilowatts, multiply by 0.746.

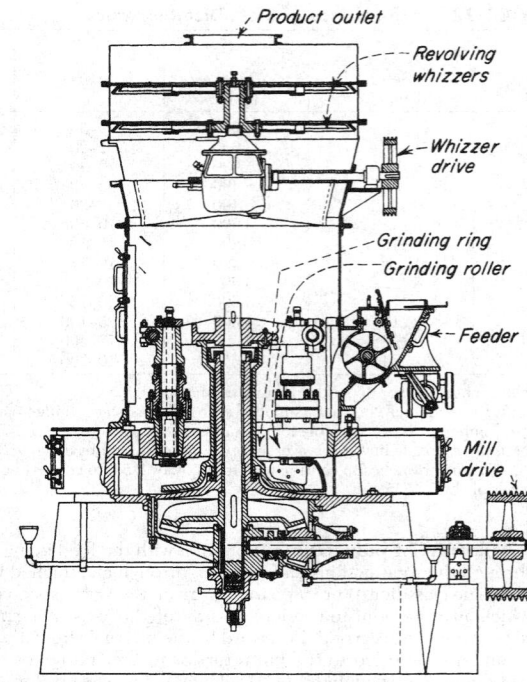

FIG. 20-50   Raymond high-side mill with an internal whizzer classifier. (*ABB Raymond Div., Combustion Engineering Inc.*)

or a horizontal position. Ring-roller mills also are referred to as ring-roll mills or roller mills or medium-speed mills. The ball-and-ring and bowl mills are types of ring-roller mill.

Ring-roller mills are more energy efficient than ball mills or hammer mills. The energy to grind coal to 80 percent passing 200 mesh was determined (Luckie and Austin, *Coal Grinding Technology—A Manual for Process Engineers*) as: ball mill—13 hp/ton; hammer mill—22 hp/ton; roller mill—9 hp/ton.

Ring-roller mills should be distinguished from roller mills. Paint roll mills are described under "Disk Attrition Mills," and flour roll mills under "Cereals and Other Vegetable Products."

Ring-Roller Mills without Internal Classification   Known examples are no longer produced.

Ring Mills with Internal Screen Classification   The grinding action of the Hercules (*Bradley Pulverizer Co.*) is that of three rolls that are revolved around and against a die to create grinding pressures of approximately 100 MPa (15,000 lbf/in²). It can produce minus-20-mesh agricultural limestone or phosphate rock from minus-5-cm (−2-in) feed. The material is discharged from the grinding chamber through a surrounding screen. Capacity is relatively high, being 23 to 45 Mg/h (25 to 50 tons/h) of average-hardness dry limestone. Other product sizes may be obtained by changing the screen aperture.

The B & W pulverizer, Type E, consists of a single row of balls operating between a rotating bottom ring and a stationary top ring. Externally adjusted springs apply pressure to the top ring to give the required loading for proper pulverizing. In operation wet raw coal is admitted inside the ball row and is fed through the grinding elements by centrifugal force. The Type E pulverizer is particularly suited to the direct firing of rotary kilns and industrial furnaces when close temperature control is required and long periods of continuous operation are essential. It is built in 17 sizes with capacities up to 12.6 Mg/h (14 tons/h) or more.

The Raymond ring-roller mill (Fig. 20-50) is of the internal air-classification type. The base of the mill carries the grinding ring, rigidly fixed in the base and lying in the horizontal plane. Underneath the grinding ring are tangential air ports through which the air enters the grinding chamber. A vertical shaft driven from below carries the roller journals. Centrifugal force urges the pivoted rollers against the ring. The raw material from the feeder drops between the rolls and ring and is crushed. Both centrifugal air motion and plows move the coarse feed to the nips. The air entrains fines and conveys them up from the grinding zone, providing some classification at this point. An air classifier is also mounted above the grinding zone to return oversize.

The method of classification used with Raymond mills depends on the fineness desired. If a medium-fine product is required (up to 85 or 90 percent through a No. 100 sieve), a single-cone air classifier is used. This consists of a housing surrounding the grinding elements with an outlet on top through which the finished product is discharged. This is known as the low-side mill. For a finer product and when frequent changes in fineness are required, the whizzer-type classifier is used. This type of mill is known as the *high-side mill.*

The Raymond ring-roll mill with internal air classification is used for the large-capacity fine grinding of most of the softer nonmetallic minerals. Materials with a Mohs-scale hardness up to and including 5 are handled economically on these units. Typical natural materials handled include barites, bauxite, clay, gypsum, magnesite, phosphate rock, iron oxide pigments, sulfur, talc, graphite, and a host of similar materials. Many of the manufactured pigments and a variety of chemicals are pulverized to high fineness on such units. Included are such materials as calcium phosphates, sodium phosphates, organic insecticides, powdered cornstarch, and many similar materials. When properly operated under suction, these mills are entirely dust-free and automatic. They are available in nine basic sizes. Connected power ranges from 10 to 400 kW (15 to 600 hp). Capacities range from 0.5 to 450 Mg/h (0.5 to 50 tons/h), depending upon nature of material and exact fineness of grind.

The **Bradley pneumatic** (air-swept type) **Hercules mills** (*Bradley Pulverizer Co.*) are centrifugal ring-roll-type pulverizing mills which can be fitted with either two or three rolls. These mills are suitable for the pulverization of many materials to produce as coarse as 98 percent minus-20 mesh to as fine as 99.5 percent minus-325 mesh. The size of the pulverized product can be varied by adjusting the fineness selector mounted on top of the mill. Capacities range from 225 kg/h (500 lb/h) to 90 Mg/h (100 tons/h). When combined with the superfine *Bradley RMC classifier,* this mill can make a product finer than 11 μm.

The Williams ring-roller mill (*Williams Patent Crusher & Pulverizer Co.*) is an air-swept mill with integral classifier of the rotating-blade type (the Spinner air classifier) or a double-cone type. The fluid-bed roller mill system has jets to introduce hot air into the bed of coal in the mill to dry it.

Of similar design is the **Raymond VR mill,** which is designed to run with hot gases to simultaneously dry and grind. It accepts feed as large as 3 in.

The **MBF pulverizer** (*Foster Wheeler Corp.*) for coal grinding also has three grinding rollers pivoted off the grinding housing. These pulverizers are commonly used in the utility industry, and capacities of up to 80 Mg/h (90 tons/h) are available.

Bowl Mills In the Raymond bowl mill the journals that carry the grinding rollers are stationary while the grinding ring rotates. The grinding pressure is produced by means of springs, which may be adjusted to give the required pressure, and the distance between the rollers and the ring may be set to a predetermined clearance. The rollers do not touch the ring, there being no metal-to-metal contact.

DISK ATTRITION MILLS

The disk or attrition mill is a modern counterpart of the early buhrstone mill. Stones are replaced by steel disks mounting interchangeable metal or abrasive grinding plates rotating at higher speeds, thus permitting a much broader range of application. They have a place in the grinding of tough organic materials, such as wood pulp and corn grits. Grinding takes place between the plates, which may operate in a vertical or a horizontal plane. One or both disks may be rotated; if both, then in opposite directions. The assembly, comprising a shaft, disk, and grinding plate, is called a **runner.** Feed material enters a chute near the axis, passes between the grinding plates, and is discharged at the periphery of the disks. The grinding plates are bolted to the disks; the distance between them is adjustable.

The **Andritz-Sprout-Bauer attrition mill** (Fig. 20-51) is available in single- and double-runner models with 30- to 91-cm- (12- to 36-in.-) diameter disks and with power ranging up to 750 kW (1000 hp). By the use of a variety of plates and shell constructions these units are represented in such applications as coarse granulating, pulverizing, and shredding.

In general, single-runner mills are used for the same purposes as double-runner mills, excepting that they will accept a coarser feedstock, their range of reduction for a given material is more limited, and they offer correspondingly higher outputs at lower power. While spike-tooth plates can be used in certain applications to simulate hammer-mill action, they are more generally applied to specialized tasks involving tearing, shredding, or controlled shattering, as in dehulling, and in wet corn-milling, where the germ must be separated from the starchy part and the hulls. The performance data presented in Table 20-25 typify the applications of the attrition mill.

Buhrstone mills are attrition mills with hard circular stones serving as grinding media, generally French, American, Esopus buhrstones, or rock emery. Buhrstone mills are still employed for grinding special cereals and grains. Feed enters the mill through a center hole in one of the stones. It is distributed between the stone faces and ground while working its way to the periphery.

DISPERSION AND COLLOID MILLS

When the problem is to disrupt lightly bonded clusters or agglomerates, a new aspect of fine grinding enters. This may be illustrated by the breakdown of pigments to incorporate them in liquid vehicles in the making of paints, and the disruption of biological cells to release soluble products. Purees, food pastes, pulps, and the like are processed by this type of mill. Dispersion is also associated with the formation of emulsions which are basically two-fluid systems. Syrups, sauces, milk, ointments, creams, lotions, and asphalt and water-paint emulsions are in this category.

Mills employed for dispersion and colloidal operations operate on the principle of high-speed fluid shear. They produce dispersed droplets of fine size, around 3 to 5 μm.

Paint-grinding roller mills (Fig. 20-52) consist of two to five smooth rollers (sometimes called rolls) operating at differential speeds. A paste is fed between the first two, or low-speed, rollers and is discharged from the final, or high-speed, roller by a scraping blade. The paste passes from the surface of one roller to that of the next because of the differential speed, which also applies shear stress to the film of

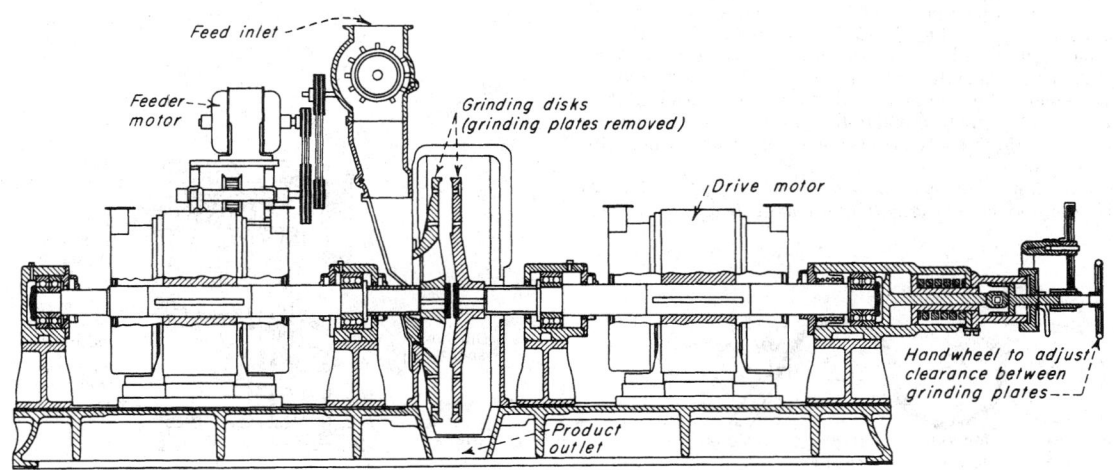

FIG. 20-51 Double-runner attrition mill. (*Andritz-Sprout-Baner Inc.*)

TABLE 20-25 Performance of Disk Attrition Mills

Material	Size-reduction details	Unit°	Capacity lb./hr.	Hp.
Alkali cellulose	Shredding for xanthation	B	4,860	5
Asbestos	Fluffing and shredding	C	1,500	50
Bagasse	Shredding	B	1,826	5
Bronze chips	⅛ in. to No. 100 sieve size	A	50	10
Carnauba wax	No. 4 sieve to 65% < No. 60 sieve	D	1,800	20
Cast-iron borings	¼ in. to No. 100 sieve	A	100	10
Cast-iron turnings	¼ in. to No. 100 sieve	E	500	50
Coconut shells	2 × 2 × ¼ in. to 5/100 sieve	B	1,560	17
	5/100 sieve to 43% < No. 200 sieve	D	337	20
Cork	2/20† sieve to 20/120 < No. 200 sieve	D	145	15
Corn cobs	1 in. to No. 10 sieve	F	1,500	150
Cotton seed oil and solvent	Oil release from 10/200 sieve product	B	2,400	30
Mica	4 × 4 × ¼ in. to 3/60 sieve	B	2,800	6
	8/60 to 75% < 60/200 sieve	D	510	7.5
Oil-seed cakes (hydraulic)	1½ in. to No. 16 sieve	F	15,000	100
Oil-seed residue (screw press)	1 in. to No. 16 sieve size	F	25,000	100
Oil-seed residue (solvent)	¼ in. to No. 16 sieve	F	35,000	100
Rags	Shredding for paper stock	B	1,440	11
Ramie	Shredding	B	820	10
Sodium sulfate	35/200 sieve to 80/325 sieve	B	11,880	10
Sulfite pulp sheet	Fluffing for acetylation, etc.	C	1,500	50
Wood flour	10/50 sieve to 35% < 100 sieve	D	130	15
Wood rosin	4 in. max. to 45% < 100 sieve	B	7,200	15

°A—8 in. single-runner mill   D—20 in. double-runner mill
  B—24 in. single-runner mill   E—24 in. double-runner mill
  C—36 in. single-runner mill   F—36 in. double-runner mill
†2/20, or smaller than No. 2 and larger than No. 20 sieve size.

NOTE: To convert inches to centimeters, multiply by 2.54; to convert pounds per hour to kilograms per hour, multiply by 0.4535; and to convert horsepower to kilowatts, multiply by 0.746.

material passing between the rollers. Roller-mill technique and action have been studied by Hummel [*J. Oil Colour Chem. Assoc.*, 270–277 (June 1950)], and the breakup of agglomerates in this mill has been discussed by Krekel [*Chem. Ing. Tech.*, **38**(3), 229 (1966)].

Colloid mills which are employed for dispersion or for emulsification fall into four main groups: the hammer or turbine, the smooth-surface disk, the rough-surface type, and valve or orifice devices.

A mathematical analysis of the action in **Kady** and other colloid mills checks well with experimental performance [Turner and McCarthy, *Am. Inst. Chem. Eng. J.*, **12**(4), 784 (1966)]. Various models of the Kady mill have been described, and capacities and costs given by Zimmerman and Lavine [*Cost Eng.*, **12**(1), 4–8 (1967)]. Energy requirements differ so much with the materials involved that other devices are often used to obtain the same end. These include high-speed stirrers, turbine mixers, bead mills, and vibratory mills. In some cases, sonic devices are effective.

The concentration of energy in mills of this class is high, and there is a considerable amount of heating. This is materially reduced by use

of a cooling-water jacket. In other cases, as when emulsions are made hot, the jacket is employed for heating.

The **Morehouse mill** (*Morehouse Industries, Inc.*) is a high-speed disk type of mill (Fig. 20-53). The undispersed phase is fed at the top and passes between converging disks, being thrown outward at the periphery.

In the **Premier Mill** the rotor is shaped like the frustrum of a cone, similar to that in Fig. 20-53. Surfaces are smooth, and adjustment of the clearance can be made from 25 μm (0.001 in) upward. A small impeller helps to feed material into the rotor gap. The mill is jacketed for temperature control. Direct-connected liquid-type mills are available with 15- to 38-cm (6- to 15-in) rotors. These mills operate at 3600 r/min at capacities up to 2 m³/h (500 gal/h). They are powered with up to 28 kW (40 hp). Working parts are made of Invar alloy, which does not expand enough to change the grinding gap if heating occurs. The rotor is faced with Stellite or silicon carbide for wear resistance. For pilot-plant operations, the Premier Mill is available with 7.5- and 10-cm (3- and 4-in) rotors. These mills are belt-driven and operate at 7200 to 17,000 r/min with capacities of 0.02 to 2 m³/h (5 to 50 gal/h).

The **Charlotte mill** (*Chemicolloid Corp.*) also employs high speed of rotation with the fluid flowing between a grooved conical rotor and a corresponding grooved conical stator. Clearance between them is regulated by an external calibrated adjustment device. Laboratory model W-10 operates at 0.75 kW (1 hp) with a capacity of 4 to 190 L/h (1 to 50 gal/h). Sanitary models are available for processing foodstuffs.

The **APV Gaulin colloid mill** has a smooth rotor, shaped like a discus.

A **high-pressure valve homogenizer** such as the Gaulin and Rannie (APV Gaulin Group) forces the suspension through a narrow orifice. The equipment has two parts: a high-pressure piston pump and a homogenizer valve [Kula and Schuette, *Biotechnology Progress*, **3**(1), 31–42 (1987)]. The pump in production machines may have up to 6 pistons. The valve, illustrated in Fig. 20-54, opens at a preset or adjustable value, and the suspension is released at high velocity (300 m/sec) and impinges on an impact ring. The flow changes direction twice by 90 degrees, resulting in turbulence. Machines studied so far compress up to 60 Mpa, but higher pressures are becoming available. For machines with feed rates greater than 2000 L/h and pressures up to 100 Mpa tungsten carbide or special ceramics are used for the valve components to reduce erosion. There is also a 2-stage valve, but it has been shown better to expend all the pressure across a single stage.

The temperature of the suspension increases about 2.5°C per 10 Mpa pressure drop. Therefore intermediate cooling is required for

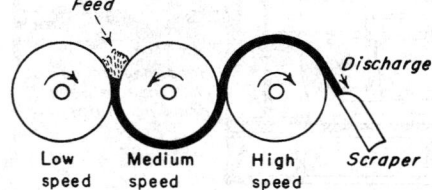

FIG. 20-52 Roller mill for paint grinding.

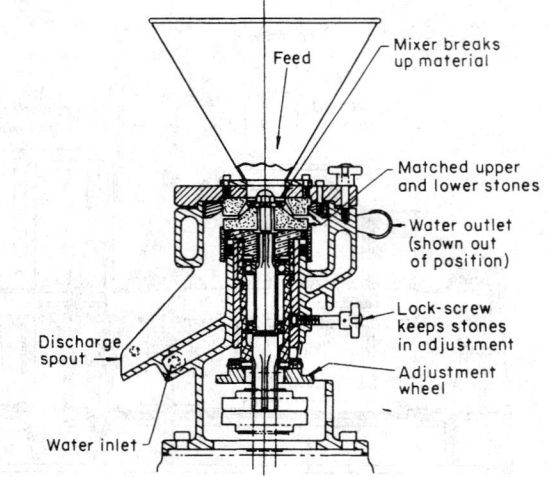

FIG. 20-53 Model M colloid mill. (*Morehouse Industries, Inc.*)

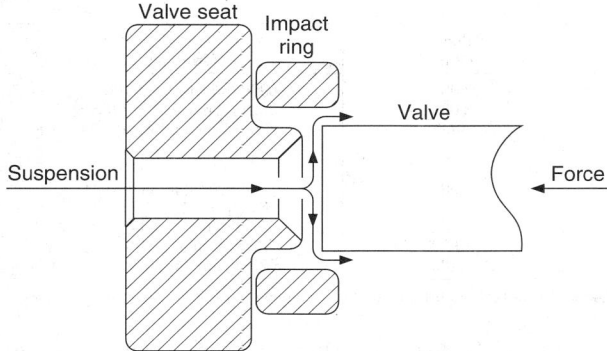

FIG. 20-54 Details of valve seat of the Gaulin high-pressure homogenizer, type CD.

TABLE 20-26 Micronizer Performance*

Material	Product average size, μm	Feed Size sieve no.	Feed Rate, lb/h	Fluid consumption, g fluid/g solid Air	Fluid consumption, g fluid/g solid Steam
Ceylon graphite	2	3	200	—	8.5
Cryolite	3	60	900	—	4.0
Limestone	3.5	80	1000	—	4.0
Hard talc	3.5	20	1000	—	4.0
Silica gel	5.5	8	500	—	3.5
Soft talc	6.5	20	1800	—	2.5
Barite	3.5	40	1800	—	2.2
Bituminous coal	2	10	1300	—	1.2
Copal resin	5	2	600	7.5	
Wolframite ore	5.5	10	800	5.6	
Sulfur	3.5	3	1300	3.5	

*Ind. Eng. Chem., **38**, 672 (1946). To convert pounds per hour to kilograms per hour, multiply by 0.4535.

multiple passes. In the Gaulin M3 homogenizer operated with a feed rate of 400 L/h and a pressure of 55 Mpa around 6000 kcal/hr have to be removed by a cooling system.

These units have been used to disrupt bacterial cells for release of enzymes. (See "Cell Disruption.")

FLUID-ENERGY OR JET MILLS

A detailed description of mills of this type has been presented by Gossett [Chem. Process. (Chicago), **29**(7), 29 (1966)]. **Fluid-energy mills** may be classified in terms of the nature of the mill action. In one class of mills, the fluid energy is admitted in fine high-velocity streams at an angle around a portion or all of the periphery of a grinding and classifying chamber. In this class are the Micronizer, jet-verizer, Reductionizer, Jet-O-Mizer, and others of somewhat similar structure. In the other class the fluid streams convey the particles at high velocity into a chamber where two streams impact upon each other. The **Majac** and the **Fluidized-Bed Jet** mills are in this class. Whether the particles are conveyed with the jet or are intercepted by jets, there is a high-energy release and a high order of turbulence which causes the particles to grind upon themselves and to be ruptured. Not all the particles are fully ground; so it is necessary to carry out a classifying operation and to return the oversize for further grinding. Most of these mills utilize the energy of the flowing-fluid stream to effect a centrifugal classification. The Majac mill differs, using a mechanical air classifier.

The **Micronizer** (*Sturtevant Inc.*) consists of a shallow circular grinding chamber in which the material to be pulverized is acted upon by a number of gaseous fluid jets issuing through orifices spaced around the periphery of the chamber. The rotating gas must discharge at the center, carrying the fines with it, while the coarse particles are thrown toward the wall, where they are subjected to further reduction by impact from particles entrained by incoming jets.

The action in these mills has been studied photographically and mathematically [Rumpf, Chem. Ing. Tech., **32**(3), 129–135; (5), 335–342 (1960); ATS translations, 668GJ, 844GJ].

Micronizer mills are constructed in nine standard sizes from 5 to 107 cm (2 to 42 in) in diameter, with capacities from 250 g/h to 1.8 Mg/h (½ lb/h to 2 tons/h). The feed size should be smaller than 1 cm (¼ in). Production rate, fluid consumption, and fineness figures are shown in Table 20-26.

The **jet pulverizer** (*Jet Pulverizer Co.*) is another mill of the shallow-pan, angle-jet, and radial-inward-classification type, like the Micronizer.

The **Jet-O-Mizer** (*Fluid Energy Processing & Equipment Co.*) is one of a group employing a hollow elongated torus which is placed vertically. The operating principle is similar to that of the Micronizer, with the feed entering tangentially to the whirling fluid stream and the fines leaving centrally.

Trost air mills from *Colt Industries* are available in five sizes. The smallest mill (Gem T) is a research unit and can be used for fine-grinding studies. Capacities of 1 to 2300 kg/h (2 to 5000 lb/h) are available. Air-flow rates vary from 0.2 to 28 m³/min (6.5 to 1000 ft³/min).

The **Majac jet pulverizer** (*Hosokawa Micron Powder Systems Div.*) is an opposed-jet type with a mechanical classifier (Fig. 20-55). Fineness is controlled primarily by the classifier speed and the amount of fan air delivered to the classifier, but other effects can be achieved by variation of nozzle pressure, distance between the muzzles of the gun barrels, and position of the classifier disk. These pulverizers are available in 30 sizes, operated on quantities of compressed air ranging from approximately 0.6 to 13.0 m³/min (20 to 4500 ft³/min). In most applications, the economics of the use of this type of jet pulverizer becomes attractive in the range of 98 percent through 200 mesh or finer.

Materials illustrated in Table 20-27 are shown because of their wide range and type.

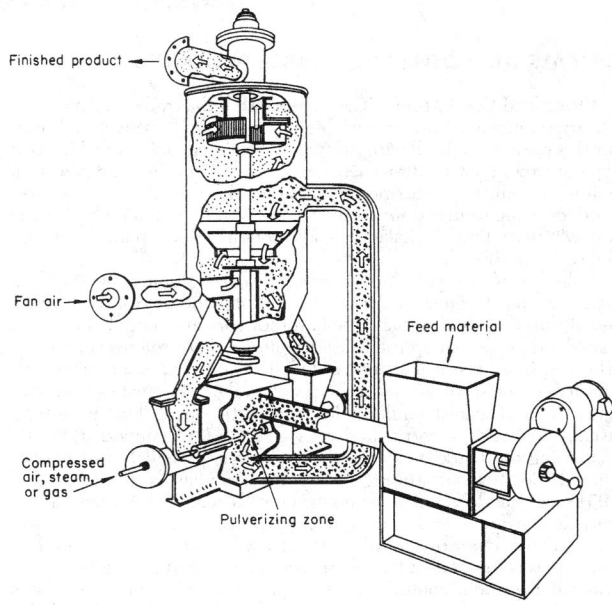

FIG. 20-55 Majac jet pulverizer. (*Hosokawa Micron Powder Systems Div.*)

TABLE 20-27 *Majac Jet Pulverizer Capacities**

Material	Finished particle size	Mill size	Production, lb h	Grinding fluid used
Alumina	−325 mesh, 3 μm average	15	12,000	6300 lb/h steam at 100 psig, 750°F.
Coal, bituminous	90%, −325 mesh	20	8,000	3000 ft³/min air at 100 psig, 70°F.
Diphenyl phthalate	−325 mesh, 20–30 μm maximum, 4.2 μm average	2–6	435	300 ft³/min air at 100 psig, 70°F.
Feldspar, silica	99%, −200 mesh	15	8,500	1350 ft³/min air at 100 psig, 800°F.
Graphite	90%, −10 μm	8.5–2.5	50	75 ft³/min air at 100 psig, 70°F.
Mica	95%, −325 mesh	8	1,600	720 ft³/min air at 100 psig, 800°F.
Rare-earth ore	60%, −1 μm	8–15	400	720 ft³/min air at 100 psig, 800°F.

*°C = (°F − 32) × ⅝. To convert pounds per hour to kilograms per hour, multiply by 0.4535; to convert pounds per square inch gauge to kilopascals, multiply by 7.0.

Fluidized-bed opposed-jet mills (*Hosokawa Micron Powder Systems Div.*) differ from the Majac mill in that powder is not fed into the jets, but the jets impinge into a chamber which contains suspended powder. The powder is entrained into the jets. This eliminates wear on the nozzles, and reduces contamination. Otherwise, construction and applications are similar to the Majac mill. The fluidized-bed level is maintained a few inches above the jets. The Fluidized-bed mill is available in 13 sizes with air volumes ranging from 50 to 11,000 m³/h. One application is for toner grinding.

NOVEL METHODS

Only once in 15 years does a truly novel method of size reduction become successful. The roll press is a truly successful example of a novel mill, and stirred-bead mills are older mills that have reached a new state of development. Many more methods are proposed and studied; some of these are described below. The information may be useful to judge other novel methods that may be proposed.

Avoiding Size Reduction Since size reduction is a difficult and inefficient operation, it is sometimes better to avoid it and use another approach. Thus rather than make large crystals and then grind them, one may be able to precipitate or crystallize material in the desired fine size. It may even be possible to control the process to give a more narrow size distribution than would be possible by size reduction.

In the case of lactose manufacture, crystals of uniform size are produced by first grinding part of a previous batch and taking a quantity with the required number of particles, then introducing these as seed crystals into a solution that is gradually cooled with gentle stirring. Variation in size of the seed crystals does not affect the size of the product crystals.

Some materials that are prepared in the molten state are converted advantageously to flake form by cooling a thin layer continuously on the surface of a rotating drum. Another way is to spray cool from the melt, using a spray dryer with cold air. Thus, massive cooling and subsequent pulverizing are avoided. See the Index for details of these other methods.

Ultrafine powders can be prepared in high-temperature plasmas. Particles below 1 μm and larger particles with unusual surface structures are formed according to Waldie [*Trans. Inst. Chem. Eng.*, **48**(3), T90 (1970)]. Energy costs are discussed.

Bond [*Min. Eng. (London)*, **60**(1), 63–64 (1968)] reviewed attempts to induce breakage without wastefully applying pressure and concluded that inherent practical limitations have been found for the following methods: spinning particles, resonant vibration, electrohydraulic crushing, induction heating, sudden release of gas pressure, and chisel-effect breakers. For a review of more recent efforts, see edition 6 of this handbook.

CRUSHING AND GRINDING PRACTICE

CEREALS AND OTHER VEGETABLE PRODUCTS

Flour and Feed Meal The **roller mill** is the traditional machine for grinding wheat and rye into high-grade flour. A typical mill used for this purpose is fitted with two pairs of rolls, capable of making two separate reductions. After each reduction the product is taken to a bolting machine or classifier to separate the fine flour, the coarse product being returned for further reduction. Feed is supplied at the top, where a vibratory shaker spreads it out in a thin stream across the full width of the rolls.

Rolls are made with various types of corrugation. Two standard types are most generally used: the dull and the sharp, the former mainly on wheat and rye, and the latter for corn and feed. Under ordinary conditions, a sharp roll is used against a sharp roll for very tough wheat, a sharp fast roll against a dull slow roll for moderately tough wheat, a dull fast roll against a sharp slow roll for slightly brittle wheat, and a dull roll against a dull roll for very brittle wheat. The speed ratio usually is 2½:1 for corrugated rolls and 1¼:1 for smooth rolls. By examining the marks made on the grain fragments it has been concluded (Scott, *Flour Milling Processes*, Chapman & Hall, London, 1951) that the differential action of the rolls actually can open up the berry and strip the endosperm from the hulls.

High-speed **hammer** or **pin mills** result in some *selective grinding*. Such mills combined with air classification can produce fractions with controlled protein content. An example of such a combination is a Bauer hurricane hammer mill combined with the Alpine Mikroplex superfine classifier. Flour with different protein content is needed for

the baking of breads and cakes, and these types of flour were formerly available only by selection of the type of wheat, which is limited by growing conditions prevailing in particular locations [Wichser, *Milling*, **3**(5), 123–125 (1958)].

Soybeans, Soybean Cake, and Other Pressed Cakes After granulation on rolls the granules are generally treated in presses or solvent extracted to remove the oil. The product from the presses goes to attrition mills or flour rolls and then to bolters, depending upon whether the finished product is to be a feed meal or a flour.

The method used for grinding pressed cakes depends upon the nature of the cake, its purity, residual oil, and moisture content. If the whole cake is to be pulverized without removal of fibrous particles, it may be ground in a hammer mill with or without air classification. A 15-kW (20-hp) hammer mill with an air classifier, grinding pressed cake, had a capacity of 136 kg/h (300 lb/h), 90 percent through No. 200 sieve; a 15-kW (20-hp) screen-hammer mill grinding to 0.16-cm (¹⁄₁₆-in) screen produced 453 kg/h (1000 lb/h). In many cases the hammer mill is used merely as a preliminary disintegrator, followed by an attrition mill. Typical performance of the attrition mill is given in Table 20-25. A finer product may be obtained in a hammer mill in closed circuit with an external screen or classifier.

High-speed hammer mills are extensively used for the grinding of soya flour. For example, the Raymond Imp mill with an air classifier is used, primarily with solvent-extracted soya.

Starch and Other Flours Grinding of starch is not particularly difficult, but precautions must be taken against explosions; starches must not come in contact with hot surfaces, sparks, or flame when suspended in air. See "Properties of Solids: Safety" for safety precautions. When a product of medium fineness is required, a hammer mill of the screen type is employed. Potato flour, tapioca, banana, and similar flours are handled in this manner. For finer products a high-speed impact mill such as the Entoleter pin mill is used in closed circuit with bolting cloth, an internal air classifier, or vibrating screens.

ORES AND MINERALS

Metalliferous Ores   The most extensive grinding operations are done in the ore-processing and cement industries. Grinding is one of the major problems in milling practice and one of the main items of expense. Mill manufacturers, operators, and engineers find it necessary to compare grinding practice in one plant with that of another, attempting to evaluate circuits and practices (Arbiter, *Milling in the Americas,* 7th International Mineral Processing Congress, Gordon and Breach, New York, 1964). Direct-shipping ores are high in metal assay, and require only preliminary crushing before being fed to a blast furnace or smelter. As these high-grade ores have been depleted, it has become necessary to concentrate ores of lower mineral value. The native copper ores of Michigan have given way to porphyry copper ores of the southwest. Initially the deposits containing 2 to 4 percent copper were worked, but now ores of 0.40 percent must be processed by grinding and flotation or leaching. The effectiveness of closed-circuit milling with wet classifiers reopened the Iron Range of Minnesota by permitting economic beneficiation of taconite iron ores, which contain up to two-thirds hard cherty gangue. By grinding, magnetic separation or froth flotation and pelletizing, a blast-furnace feed is produced that is more uniform and gives a higher iron yield than the direct-shipping ores.

Three types of milling circuits are used in large ore-processing plants [Allis Chalmers, *Engineering & Mining J.,* **181**(6), 69–171 (1980)]. (1) Three stages of gyratory crushers, followed by a wet rod mill followed by a ball mill (Fig. 20-56). This combination has high-power efficiency and low steel consumption, but higher-investment cost because rod mills are limited in length to 20 ft by potential tangling of the rods. (2) Similar crusher equipment followed by one or two stages of large ball mills. One stage may suffice if product size can be as coarse as 65 mesh. This circuit has lower capital cost but higher energy and wear costs. (3) One stage of gyratory crusher followed by large-diameter semiautogenous ball mills followed by a second stage of autogenous or ball mills (Fig. 20-57). The advantage of autogenous mills is reduction of ballwear costs, but power costs are at least 25 percent greater, because

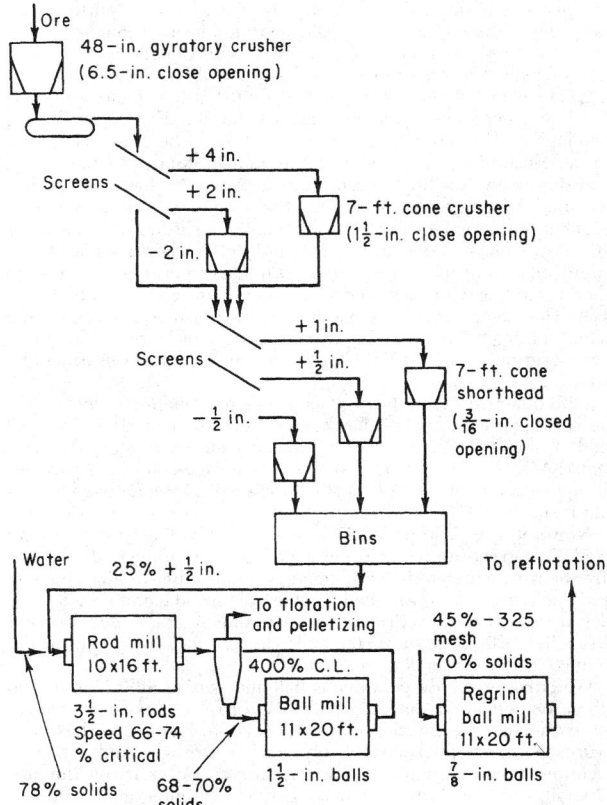

FIG. 20-56   Ball- and rod-mill circuit. Simplified flow sheet of the Cleveland-Cliffs Iron Co. Republic mine iron-ore concentrator. To convert inches to centimeters, multiply by 2.54; to convert feet to centimeters, multiply by 30.5. (*Johnson and Bjorne,* Milling in the Americas, *Gordon and Breach, New York, 1964.*)

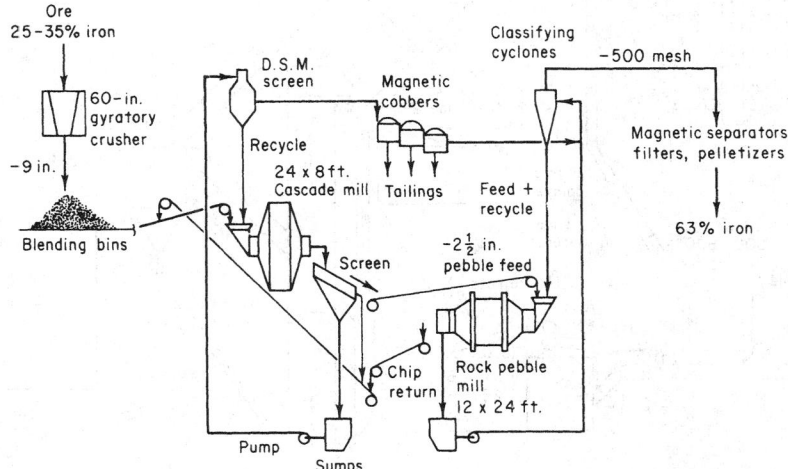

FIG. 20-57   Autogenous mill circuit. Simplified flow diagram of the Cleveland-Cliffs Iron Co. Empire iron-mine concentrator with two autogenous wet-grinding stages. To convert inches to centimeters, multiply by 2.54; to convert feet to centimeters, multiply by 30.5.

irregular-shaped media are less effective than balls. A fourth circuit using the roll press has been widely accepted in the cement industry (see "Roll Press" and "Cement Industry"), and could be used in other mineral plants. It could replace the last stage of crushers and the first stage of ball or rod mills, at substantially reduced power and wear.

A flow sheet for one iron ore process is shown in Fig. 20-56. For the grinding of softer copper ore the rod mill might be eliminated, both coarse-crushing and ball-milling ranges being extended to fill the gap.

Autogenous milling of iron and copper ores has been widely accepted. When successful, this method results in economies due to elimination of media wear. Probably another reason for efficiency is the use of higher circulating loads and better classification. These improvements resulted from the need to use larger-diameter mills to obtain grinding with rock media that have a lower density than steel balls. The major difficulty is in arranging the crushing circuits and the actual mining so as to assure a steady supply of large ore lumps to serve as grinding media. With rocks that are too friable this cannot be achieved.

With other ores there has been a problem of buildup of intermediate-sized particles, but this has been solved either by adding a small load of steel balls, thus converting to a semiautogenous grinding system (SAG), or by sending the scalped intermediate-sized particles through a cone crusher. A flow sheet for a typical wet autogenous circuit is shown in Fig. 20-57.

Nonmetallic Minerals Dry and wet grinding processes are used. **Dry grinding** is less expensive than wet grinding in the coarser sizes because a dry product is obtained without a final drying step. Dry grinding is carried out with ball, pebble, roller, and hammer mills with closed-circuit air classification. The product may be 99.8 percent through a 325-mesh sieve. Jet mills can produce a product in the range 5 to 15 µm, but at greater cost.

Wet processes use continuous ball and pebble mills, stirred and vibratory media mills, and pug mills. Wet processes take advantage of more effective classification in water, using bowl and cone classifiers, hydroseparators, thickeners, continuous centrifuges and cyclones, vacuum filters, and rotary, tray, or tunnel dryers. After drying, the cake generally has to be broken up in some type of disintegrator or pulverizer unless it was spray-dried. The objective of a process may be to obtain many grades of the same material by tying classifiers and screens into the grinding system to remove various-sized products. Choice of equipment generally depends on (1) hardness and (2) contaminations. Capacity of any system decreases rapidly with increasing fineness of the material; this applies particularly to nonmetallics, for which extreme fineness is usually required.

Clays and Kaolins Because of declining quality of available clay deposits, beneficiation is becoming more required [Uhlig, *Ceramics Forum International*, **67**(7–8), 299–304 (1990), English and German text]. Beneficiation normally begins with a size-reduction step, not to break particles but to dislodge adhering clay from coarser impurities. In **dry processes** this is done with low-energy impact mills.

Mined clay with 22 percent moisture is broken up into pieces of less than 5 cm (2 in) in a rotary impact mill without screen, and fed to a rotary gas-fired kiln for drying (see Fig. 20-58). The moisture content is then 8 to 10 percent, and this material is fed to a mill, usually a Raymond ring-roll mill with an internal whizzer classifier or a pan mill. Hot gases introduced to the mill complete the drying while the material is being pulverized to the required fineness.

To grind 3.2 Mg/h (3½ tons/h) of a raw clay, power consumption will be about 75 kW (100 hp), and it takes about 31 m³ (1100 ft³) of natural gas 3.7 MJ/m³ (1000 Btu/ft³) to dry the clay from 10 percent moisture down to about 1 percent. The product is used in paint pigments and rubber fillers.

Larger amounts of clay are being produced in dry powdered form [Anon., *Ceramics Forum International*, **67**(7–8), 330–4 (1990)]. After grinding, the clay is agglomerated to a flowable powder with water mist in a balling drum.

In the **wet process,** the clay is masticated in a pug mill to break up lumps and dispersed with a dispersing aid and water to make a 40 percent solids slip of low viscosity. A high-speed agitator such as a Cowles dissolver is used for this purpose. Sands are settled out, and then the clay is classified into two size fractions in either a Hydrosettler or a continuous Sharples or Bird centrifuge. The fine fraction, with sizes of less than 1 µm, is used as a pigment and for paper coating, while the coarser fraction is used as a paper filler.

A process for upgrading kaolin by grinding in a stirred bead mill has been reported (Stanczyk and Feld, *U.S. Bur. Mines Rep. Invest.* 6327 and 6694, 1965). By this means the clay particles are delaminated, and the resulting platelets give a much improved surface on coated paper.

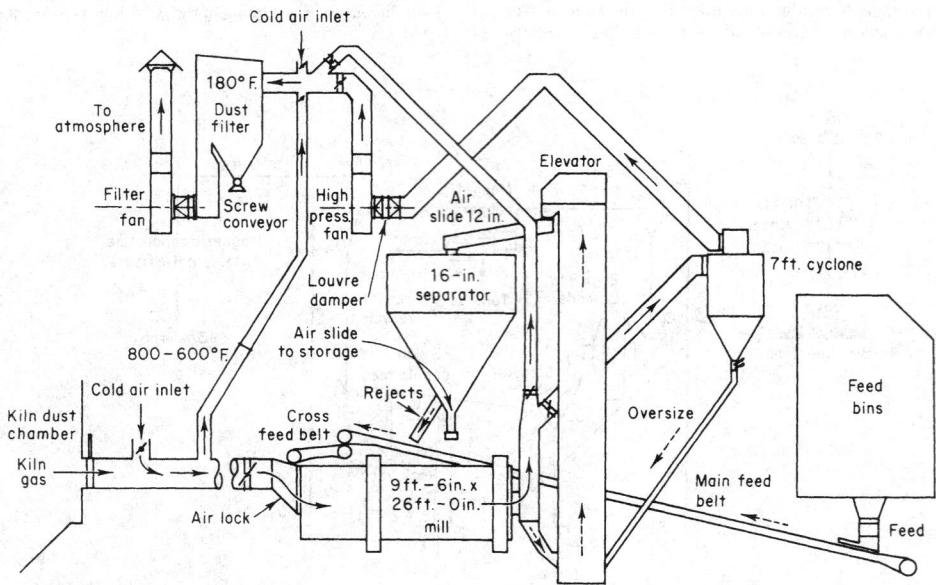

FIG. 20-58 Combined drying-grinding system using ball-mill and hot kiln exhaust gases. To convert inches to centimeters, multiply by 2.54; to convert feet to centimeters, multiply by 30.5; °C = (°F − 32) × 5/9. [*Tonry*, Pit Quarry (*February-March* 1959).]

Talc and Soapstone Generally these are easily pulverized. Certain fibrous and foliated talcs may offer greater resistance to reduction to impalpable powder, but these are no longer produced because of their asbestos content.

Talc milling is largely a grinding operation accompanied by air separation. Most of the industrial talcs are dry-ground. Dryers are commonly employed to predry ahead of the milling operation because of the wet material reduces mill capacity by as much as 30 percent.

Conventionally, in talc milling, rock taken from the mines is crushed in primary and then in secondary crushers to at least 1.25 cm (½ in) and frequently as fine as 0.16 cm (1⁄16 in).

Ring-roll mills with internal air separation are widely used for the large-capacity fine grinding of the softer talcs. High-speed hammer mills with internal air separation are also an outstanding success on some of the softer high-purity talcs for very fine fineness.

The mills in the western United States, in which generally are ground softer talcs than those of New York, have simple flow sheets. Single-stage crushing is employed, and the talc is merely ground in Raymond roller mills in closed circuit with air separators.

For a ring-roller mill receiving 2.54-cm (1-in) feed, production rates range from 1360 to 2720 kg/h (3000 to 6000 lb/h) for 60-kW (80-hp) grinding to 99 to 99.5 percent able to pass a No. 200 sieve.

Talcs of extreme fineness and high surface area are rapidly attaining industrial importance and are used for various purposes in the paint, paper, plastics, and rubber industries.

Carbonates and Sulfates Carbonates include limestone, calcite, marble, marls, chalk, dolomite, and magnesite; the most important sulfates are barite, celestite, anhydrite, and gypsum; these are used as fillers in paint, paper, and rubber. (Gypsum and anhydrite are discussed below as part of the cement, lime, and gypsum industries.)

A Raymond 5057 ring-roller mill pulverizing fluorspar produces 4500 kg (10,000 lb) of product per hour 95 percent minus 200 mesh with 104 kW (140-hp) operating power, or 26 kWh/Mg (23.4 kWh/ton). Fluorspar is also ground in continuous-tube mills with classification.

Silica and Feldspar These are ground in silex-lined mills with flint balls (see Table 20-28). At a mine near Cairo, Illinois, silica is successfully crushed prior to ball-milling in American rotary impact mills having loose crushing rings made of hard alloy steel. The rings are easily replaced as they wear.

Feldspar for the ceramic and chemical industries is ground finer than for the glass industry. A feldspar mill is described in U.S. Bur. Mines Cir. 6488, 1931. It uses pebble mills with a Gayco air classifier.

Table 20-28 gives the results obtained with Hardinge pebble mills, grinding several siliceous refractory materials.

Asbestos and Mica Asbestos is no longer mined in the United States because of the severe health hazard, but it is still mined and processed in Canada. See previous editions of this handbook for process descriptions.

The micas, as a class, are difficult to grind to a fine powder; one exception is disintegrated schist, in which the mica occurs in minute flakes. For dry grinding, hammer mills equipped with an air-transport system are generally used. Maintenance is often high. It has been established that the method of milling has a definite effect on the particle characteristics of the final product. Dry grinding of mica is customary for the coarser sizes down to 100 mesh. Micronized mica, produced by high-pressure steam jets, is considered to consist of highly delaminated particles.

Conditions for grinding micas and kaolin in a Drais stirred-bead mill were investigated [Sivamohan and Vachot, *Powder Technology*, **61**(2), 119–129 (1990)]. Muscovite with a high aspect ratio of 50:1 imparts strength and electrical breakdown resistance to molded-plastic parts. A feed material with d_{90} of 180 μm and aspect ratio 3:1 was ground in 1 to 10 passes with 4 min residence time. It gave aspect ratio as high as 27:1 and d_{50} of 8 μm with the longest grinding times. Glass beads of 0.3 and 0.8 mm diameter had the same effect. Pulp density of 12 wt % gave better grinding rate than 25 percent. Low-dispersant concentration of 0.35 wt % was best.

Wollastonite with an aspect ratio of 15:1 is useful as a replacement for asbestos and as a high-strength filler for plastics. The feed material with d_{90} of 45 μm was similarly ground. Beads of 0.3 mm gave faster grinding than 0.8 mm beads, and these corresponded to a bead-particle-size ratio of 19, confirming other results.

Refractories Refractory bricks are made from fireclay, alumina, magnesite, chrome, forsterite, and silica ores. These materials are crushed and ground, wetted, pressed into shape, and fired. To obtain the maximum brick density, furnishes of several sizes are prepared and mixed. Thus a magnesia brick may be made from 40 percent coarse, 40 percent middling, and 20 percent fines. Theorems have been proposed for calculating weight ratios of sizes to produce maximum packing density of powder mixtures [Lewis and Goldman, *J. Am. Ceram. Soc.*, **49**(6), 323 (1966)]. Preliminary crushing is done in jaw or gyratories, intermediate crushing in pan mills or ring rolls, and fine grinding in open-circuit ball mills. Since refractory plants must make a variety of products in the same equipment, pan mills and ring rolls are preferred over ball mills because the former are more easily cleaned.

Sixty percent of refractory magnesite is made synthetically from Michigan brines. When calcined, this material is one of the hardest refractories to grind. Gyratory crushers, jaw crushers, pan mills, and ball mills are used.

Alumina produced by the Bayer process is precipitated and then calcined [Krawczyk, *Ceramic Forum International*, **67**(7–8), 342–8 (1990)]. Aggregates are typically 20 to 70 μm, and have to be reduced. The standard product is typically made in continuous dry ball or vibratory mills to give a product d_{50} size of 3–7 μm, 98 percent finer than 45 μm. The mills are lined with wear-resistant alumina blocks, and balls or cylinders are used with an alumina content of 80–92 percent. The products containing up to 96 percent Al_2O_3 are used for bricks, kiln furniture, grinding balls and liners, high voltage insulators, catalyst carriers, etc.

Ultrafine grinding is carried out batchwise in vibratory or ball mills, either dry or wet. The purpose of batch operation is to avoid the residence time distribution which would pass less-ground material through a continuous mill. The energy input is 20–30 times greater than for standard grinding, with inputs of 1300–1600 kWh/ton compared to 40–60. Jet milling is also used, followed by air classification, which can reduce top size below 8 μm.

TABLE 20-28 Grinding Refractory Siliceous Materials in Pebble Mills

	Feldspar	Silica sand	Enamel frit	Grog
Size of mill	8′ × 60″	8′ × 48″	4½′ × 16″	5′ × 22″
Feed size	2″	20 mesh	⅛″	1½″
Size of product	99% through No. 200 sieve	98% through No. 325 sieve	97% through No. 100 sieve	95% through No. 10 sieve
Capacity, tons/hr	1.75	1.25	0.225	5
Power for mill, hp	68	58	8.5	28
Power for auxiliaries, hp	21	20		
Pebble load, lb	10,000	12,000	2000	2800
Speed of mill, r.p.m.	22	18	30	30
Moisture, %	1	1	0	1
Type of classifier	Hardinge	Air	Trommel screen on mill	
Lining and grinding mediums		Flint blocks and flint pebbles	Steel	

NOTE: To convert feet to centimeters, multiply by 30.5; to convert inches to centimeters, multiply by 2.54; to convert tons per hour to megagrams per hour, multiply by 0.907; and to convert horsepower to kilowatts, multiply by 0.746.

Among new mill developments, annular-gap bead mills and stirred bead mills are being used. These have a high cost, but result in a steep particle-size distribution when used in multipass mode [Kolb, *Ceramic Forum International,* **70**(5), 212–216 (1993)]. Costs for fine grinding typically exceed the cost of raw materials. Products are used for high-performance ceramics.

Silicon carbide grains were reduced from 100–200 mesh to 80 percent below 1 μm in a version of stirred bead mill, using 20–30 mesh silicon carbide as media (Hoyer, Report Investigations U.S. Bureau Mines 9097, 9 pp., 1987).

Crushed Stone and Aggregate   In-pit crushing is increasingly being used to reduce the rock to a size that can be handled by a conveyor system. In quarries with a long, steep haul, conveyors may be more economic than trucks. The primary crusher is located near the quarry face, where it can be supplied by shovels, front-end loaders or trucks. The crusher may be fully mobile or semimobile. It can be of any type listed below. The choices depend on individual quarry economics, and are described by Faulkner [*Quarry Management and Products,* **7**(6), 159–168 (1980)].

Primary crushers used are jaw, gyratory, impact, and toothed roll crushers. Impact mills are limited to limestone and softer stone. With rocks containing more than 5 percent quartz, maintenance of hammers may become prohibitive. Gyratory and cone crushers dominate the field for secondary crushing of hard and tough stone. Rod mills have been employed to manufacture stone sand when natural sands are not available.

Crushed stone for road building must be relatively strong and inert, and must meet specifications regarding size distribution and shape. Both size and shape are determined by the crushing operation. Table 20-29 lists specifications for a few size ranges.

The purpose of these specifications is to produce a mixture where the fines fill the voids in the coarser fractions, thus to increase load-bearing capacity. (See "Refractories" above.)

Sometimes a product that does not meet these requirements must be adjusted by adding a specially crushed fraction. No crushing device available will give any arbitrary size distribution, and so crushing with a small reduction ratio and recycle of oversize is practiced when necessary.

The claim of various crushers to produce a cubical product is exaggerated. However, there are differences. If an impact mill is designed to apply an excess of energy at each blow, then production of slivers can be avoided but a larger amount of fines is produced.

A survey of research on product shape and size distribution (**grading**) was given by Rösslein [translation by Shergold, *Quarry Managers J.,* 207–222 (October 1946)]. The main study involved tests on 30 jaw crushers, with shapes of over 1 million particles being measured. Concerning **particle shape,** there is a tendency for hard rocks to produce more numerous flaky chippings than soft rocks. The size of feed to a crusher does not affect the shape of products. With jaw crushers, the largest and finest sizes contain the highest proportion of flaky pieces, but even the intermediate sizes are irregular. An increase in the reduction ratio of jaw crushers increases the flakiness of the product.

Smooth jaws produce more numerous flaky pieces than corrugated jaws. Curved jaws produce less fines but more numerous flaky particles. The presence of material too small to be crushed has a deleterious effect on the shape of products. Secondary crushers with a small reduction ratio can improve the shape of primary crushed material, but secondary crushers are not inherently different from primary crushers. Slotted screens can remove flaky particles from the product. Impact crushers produce fewer flaky particles than any other type.

Grading of the product (i.e., size distribution) depends on the discharge opening, which is difficult to measure and adjust owing to wear of the plates in a jaw crusher. Size of feed and wear of plates do not affect grading significantly. Table 20-10 relates sizes of product to crusher settings.

FERTILIZERS AND PHOSPHATES

Many of the materials used in the fertilizer industry are pulverized, such as those serving as sources for calcium, phosphorus, potassium, and nitrogen. The most commonly used for their lime content are limestone, oyster shells, marls, lime, and, to a small extent, gypsum. Limestone is generally ground in hammer mills, ring-roller mills, and ball mills. Fineness required varies greatly from No. 10 sieve to 75 percent through No. 100 sieve.

Oyster Shells and Lime Rock   Operating characteristics for hammer mills grinding oyster shells and burned lime for agricultural purposes are given in Table 20-30.

Phosphates   Phosphate rock is generally ground for one of two major purposes: for direct application to the soil or for acidulation with mineral acids in the manufacture of fertilizers. Because of larger capacities and fewer operating-personnel requirements, plant installations involving production rates over 900 Mg/h (100 tons/h) have used ball-mill grinding systems. Ring-roll mills are used in smaller applications. Rock for direct use as fertilizer is usually ground to various specifications, ranging from 40 percent minus 200 mesh to 70 percent minus 200 mesh. For manufacture of normal and concentrated superphosphates the fineness of grind ranges from 65 percent minus 200 mesh to 85 percent minus 200 mesh.

Grindability of phosphate rocks from different areas varies widely; in Table 20-31 typical work-index data are shown.

Grinding-media wear in phosphate ball-mill grinding systems ranges from 5 to 25 g/Mg (0.05 to 0.20 lb/ton) ground; ball-mill liners show an average consumption of 2.5 to 100 g/Mg (0.01 to 0.05 lb/ton) ground.

Inorganic salts often do not require fine pulverizing, but they frequently become lumpy. In such cases, they are passed through a double-cage mill or some type of hammer mill.

Basic slag is often used as a source of phosphorus. Its grinding resistance depends largely upon the way in which it has been cooled, slowly cooled slag generally being more easily pulverized. The most common method for grinding basic slag is in a ball mill, followed by a tube mill or a compartment mill. Both systems may be in closed circuit with an air classifier. A 2.1- by 1.5-m (7- by 5-ft) mill, requiring

TABLE 20-29   Selected Standard Sizes of Coarse Aggregate*

Size number	Nominal size, square openings	4 in	3½ in	3 in	2½ in	2 in	1½ in	1 in	¾ in	½ in	⅜ in	No. 4 (4760 μm)	No. 8 (2380 μm)	No. 16 (1190 μm)	No. 50 (297 μm)	No. 100 (149 μm)
						Amounts finer than each laboratory sieve (square openings), percent by weight										
1	3½ to 1½ in.	100	90 to 100	—	25 to 60	—	0 to 15	—	0 to 5							
2	2½ to 1½ in.	—	—	100	90 to 100	35 to 70	0 to 15	—	0 to 5							
3	2 to 1 in.	—	—	—	100	90 to 100	35 to 70	0 to 15	—	0 to 5						
4	1½ to ¾ in.	—	—	—	—	100	90 to 100	20 to 55	0 to 15	—	0 to 5					
5	1 to ½ in.	—	—	—	—	—	100	90 to 100	20 to 55	0 to 10	0 to 5					
6	¾ to ⅜ in.	—	—	—	—	—	—	100	90 to 100	20 to 55	0 to 15	0 to 5				
7	½ in. to No. 4	—	—	—	—	—	—	—	100	90 to 100	40 to 70	0 to 15	0 to 5			
8	⅜ in. to No. 8	—	—	—	—	—	—	—	—	100	85 to 100	10 to 30	0 to 10	0 to 5		
9	No. 4 to No. 16	—	—	—	—	—	—	—	—	—	100	85 to 100	10 to 40	0 to 10	0 to 5	

*ASTM *Standards on Mineral Aggregates and Concrete,* March 1956.
NOTE: To convert inches to centimeters, multiply by 2.54.

TABLE 20-30 Operating Data for Grinding Oyster Shells and Burned Lime in Hammer Mills

Type of mill	Material	Size, in.	Capacity, tons/hr.	Hp.
Jeffrey	Oyster shells	15 × 8	0.5–0.75	8
		20 × 12	1–1.5	12
		24 × 18	2–3	20
		30 × 24	4–5	30
		36 × 24	8–10	40
Stedman	Burned lime	12 × 9	1.5	8
		20 × 12	4	20
		24 × 20	8	40
		30 × 30	12	60
		36 × 36	20	100

NOTE: To convert inches to centimeters, multiply by 2.54; to convert tons per hour to megagrams per hour, multiply by 0.907; and to convert horsepower to kilowatts, multiply by 0.746.

TABLE 20-31 Ball-Mill Grindability of Various Phosphate Rocks

Rock type	Calcium phosphate content, %	Work index, kw.-hr./ton
Central Florida, pebble	68	14.5–16.5
North Florida, pebble	72	18.0–21.0
North Carolina, calcined concentrate	62–67	14.2–18.0
Central Florida, concentrate	72–74	16.0–23.0
Morocco	80–82	10.1–23.6
Idaho		19.0–24.7

NOTE: To convert kilowatthours per ton to kilowatthours per megagram, multiply by 1.1.

94 kW (125 hp), operating with a 4.2-m (14-ft) 22.5-kW (30-hp) classifier, gave a capacity of 4.5 Mg/h (5 tons/h) from the classifier, 95 percent through a No. 200 sieve. Mill product was 68 percent through a No. 200 sieve, and circulating load 100 percent.

CEMENT, LIME, AND GYPSUM

Portland-cement manufacture requires grinding on a very large scale and entails a large use of electric power. Raw materials consist of sources of lime, alumina, and silica and range widely in properties, from crystalline limestone with silica inclusions to wet clay. Therefore a variety of crushers are needed to handle these materials. Typically a crushability test is conducted by measuring the product size from a laboratory impact mill on core samples [Schaefer and Gallus, *Zement-Kalk-Gips*, **41**(10), 486–492 (1988); English ed., 277–280]. Abrasiveness is measured by weight loss of the hammers. The presence of 5 to 10 percent silica can result in an abrasive rock, but only if the silica grain size exceeds 50 μm. Silica inclusions can also occur in soft rocks. The presence of sticky clay will usually result in handling problems, but other rocks can be handled even if moisture reaches 20 percent.

If the rock is abrasive, the first stage of crushing may use gyratory or jaw crushers, otherwise a rotor-impact mill. Their reduction ratio is only 1:12 to 1:18, so they often must be followed by a hammer mill, or they can feed a roll press. Rotor crushers have become the dominant primary crusher for cement plants because of the characteristics. All of these types of crushers may be installed in moveable crusher plants.

In the grinding of raw materials, two processes are used: the dry process in which the materials are dried to less than 1 percent moisture and then ground to a fine powder, and the wet process in which the grinding takes place with addition of water to the mills to produce a slurry. The two processes are used about equally in the United States.

Dry-Process Cement After crushing, the feed may be ground from a size of 5 to 6 cm (2 to 2½ in) to a powder of 75 to 90 percent passing a 200-mesh sieve in one or several stages.

The first stage, reducing the material size to approximately 20 mesh, may be done in vertical, roller, ball-race, or ball mills. The last-

named rotate from 15 to 18 r/min and are charged with grinding balls 5 to 13 cm (2 to 5 in) in diameter. The second stage is done in tube mills charged with grinding balls of 2 to 5 cm (¾ to 2 in).

Frequently ball and tube mills are combined into a single machine consisting of two or three compartments, separated by perforated-steel diaphragms and charged with grinding media of different size. Rod mills are hardly ever used in cement plants. The compartments of a tube mill may be combined in various circuit arrangements with classifiers, as shown in Fig. 20-59.

A dry-process plant has been described by Bergstrom [*Rock Prod.*, 59–62 (August 1968)].

Wet-Process Cement Ball, tube, and compartment mills of essentially the same construction as for the dry process are used for grinding. Water or clay slip is added at the feed end of the initial grinder, together with the roughly proportioned amounts of limestone and other components.

In modern installations wet grinding is sometimes accomplished in ball mills alone, operating with excess water in closed circuit with classifiers and hydroseparators.

Figures 20-43 to 20-44 illustrate single-stage and two-stage closed-circuit wet-grinding systems. The circuits of Fig. 20-59 may also be used as a closed-circuit wet-grinding system incorporating a liquid-solid cyclone as the classifier.

A wet-process plant making cement from shale and limestone has been described by Bergstrom [*Rock Prod.*, 64–71 (June 1967)]. There are separate facilities for grinding each type of stone. The ball mill operates in closed circuit with a battery of Dutch State Mines screens. Material passing the screens is 85 percent minus 200 mesh. The entire process is extensively instrumented and controlled by computer. Automatic devices sample crushed rock, slurries, and finished product for chemical analysis by X-ray fluorescence. Mill circuit feed rates and water additions are governed by conventional controllers.

Finish-Grinding of Cement Clinker Typically the hot clinker is first cooled and then ground in a compartment mill in closed circuit with an air classifier. To crush the clinkers, balls as large as 5 in may be needed in the first compartment. A roll press added before the ball mill can reduce clinkers to a fine size, and thus reduce the load on the ball mills. The main reasons for adding a roll press has been to increase capacity of the plant, and lower cost.

Installation of roll presses in several cement plants is described (31st IEEE Cement Industry Technical Conference, 1989). Considerable modification of the installation was required because of the characteristics of the press. A roll press is a constant throughput machine, and the feed rate cannot easily be reduced to match the rate accepted by the ball mill that follows it. Several mills attempted to control the rate by increasing the recycle of coarse rejects from the air classifier, but the addition of such fine material was found to reduce the pull-in capacity of the rolls, for example from 180 to 250 t/hr. With the resulting high recycle ratio of 5:1 the roll operation became unstable, and power peaks occurred. Deaeration of fines occurs in the nip, and this also interferes with feeding fines to the rolls. In some plants these problems were overcome by recirculating slabs of product directly from the roll discharge (Fig. 20-26). In other cases the rolls were equipped with variable speed drives to allow more versatile operation when producing several different grades/finenesses of cement. The roll press was found to be 2.5 times as efficient as the ball mill, in terms of new surface per unit energy.

Tests showed that the slab from pressing of clinker at 120 bar and 20 percent recycle contained 97 percent finer than 2.8 mm, and 39

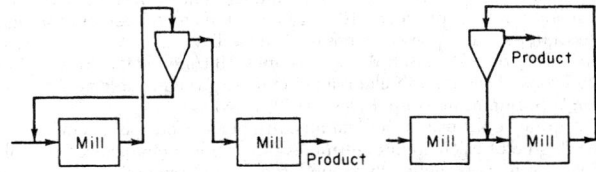

FIG. 20-59 Two cement-milling circuits. [*For others, see Tonry, Pit Quarry (February-March 1959).*]

percent finer than 48 μm. Current operation is at 160 bar. The wear was small; after 4000 h operation and 1.5 million tons of throughput the wear rate was less than 0.1 g per ton, or 0.215 g/ton of finished cement. There is some wear of the working parts of the press, requiring occasional maintenance. The press is controlled by four control loops. The main control adjusts the gates that control slab recycle. Since this adjustment is sensitive, the level in the feed bin is controlled by adjusting the clinker-feed rate to assure choke-feed conditions. Hydraulic pressure is also controlled. Separator-reject rate is fixed. The investment cost was only $42,000 per ton of increased capacity. Energy savings was 15 kWh/ton. This together with off-peak power rates results in energy cost savings of $500,000/yr.

Particle-Size Control The strength of cements varies with fineness to which it is ground, and also with size distribution, which affects particle packing. Controlling the particle-size distribution of cements to achieve a higher packing density can give higher strengths, (Helmuth, 23rd IEEE Cement Industry Technical Conference, 1981, 33 pp.). See subsection "Refractories" for further discussion of particle packing.

There have been efforts to take advantage of the greater energy efficiency and lower cost of the roll press for product grinding [Wuestner et al., *Zement-Kalk-Gips*, **41**(7), 345–353 (1987); English ed., 207–212]. However, the much steeper size distribution [Tamashige, *32d IEEE Cement Industry Technical Conference*, 319–340 (1990)], requires a fine size that results in greatly increased water requirement and very rapid setting, which are unacceptable. Also, insufficient heat is generated to dehydrate the gypsum component [Rosemann et al., *Zement-Kalk-Gips*, **42**(4), 165–9 (1987); English ed., 141–3]. It has also been found that the impact deagglomerators tried so far did not completely break up agglomerates as well as a ball mill. Thus there is room for further improvements in developing ways to optimize cement particle-size distribution.

Lime Lime used for agricultural purposes generally is ground in hammer mills. It includes burnt, hydrated, and raw limestone. When a fine product is desired, as in the building trade and for chemical manufacture, ring-roller mills, ball mills, and certain types of hammer mills are used.

Gypsum When gypsum is calcined in rotary kilns, it is first crushed and screened. After calcining it is pulverized. Tube mills are usually used. These impart plasticity and workability. Occasionally such calcined gypsum is passed through ring-roller mills ahead of the tube mills.

COAL, COKE, AND OTHER CARBON PRODUCTS

Bituminous Coal The grinding characteristics of bituminous coal are affected by impurities contained, such as inherent ash, slate, gravel, sand, and sulfur balls. The grindability of coal is determined by grinding it in a standard laboratory mill and comparing the results with the results obtained under identical conditions on a coal selected as a standard. This standard coal is a low-volatile coal from Jerome Mines, Upper Kittanning bed, Somerset County, Pennsylvania, and is assumed to have a grindability of 100. Thus a coal with a grindability of 125 could be pulverized more easily than the standard, while a coal with a grindability of 70 would be more difficult to grind. (Grindability and grindability methods are discussed under "Properties of Solids.")

Anthracite Anthracite is harder to reduce than bituminous coal. It is pulverized for foundry-facing mixtures in ball mills or hammer mills followed by air classifiers. Only to a lesser extent is it used for fuel in powdered form.

A 3- by 1.65-m (10-ft by 66-in) Hardinge mill in closed circuit with an air classifier as shown in Fig. 20-42, grinding 4-mesh anthracite with 3.5 percent moisture, produced 10.8 Mg/h (12 tons/h), 82 percent through No. 200 sieve. The power required for the mill was 278 kW (370 hp); for auxiliaries, 52.5 kW (70 hp); speed of mill, 19 r/min; ball load, 25.7 Mg (28.5 tons). Data for a similar pilot circuit are given in Table 20-32 (Sanner, U.S. Bur. Mines Rep. Invest. 7170, 1968).

Anthracite for use in the manufacture of electrodes is calcined, and the degree of calcination determines the grinding characteristics. Calcined anthracite is generally ground in ball and tube mills or ring-roller mills equipped with air classification. A Raymond high-side ring-roller mill grinding calcined anthracite for electrode manufacture has a capac-

TABLE 20-32 Closed-Circuit Continuous Grinding of Anthracite in an Air-Swept Ball Mill

Production rate, lb/h	Mean product size, μm	Circulating load, %	Recirculated material, −37 μm, %	Energy, kWh ton
19.8	7.3	277	42	330
23.0	6.6	283	59	280
27.0	6.1	757	31	246

NOTE: To convert pounds per hour to kilograms per hour, multiply by 0.454; to convert kilowatthours per ton to kilowatthours per megagram, multiply by 1.1.

ity of 2.1 Mg/h (4600 lb/h) for a product fineness of 76 percent passing a No. 200 sieve and 52.5-kW (70-hp) power requirement.

Coke The grinding characteristics of coke vary widely. By-product coke is hard and abrasive, while certain foundry and retort coke is extremely hard to grind. For certain purposes it may be necessary to produce a uniform granule with minimum fines. This is best accomplished in rod or ball mills in closed circuit with screens, operating on by-product-coke breeze, was 8.1 Mg (9 tons), 100 percent through No. 10 sieve, and 73 percent on No. 200 sieve; power requirement, 30 kW (40 hp).

Petroleum coke is generally pulverized for the manufacture of electrodes; ring-roller mills with air classification and tube mills are generally used. A No. 5057 Raymond ring-roller mill gave an hourly output of 3.8 tons, 78.5 percent through No. 200 sieve, with 67 kW (90 hp).

Other Carbon Products **Pitch** may be pulverized as a fuel or for other commercial purposes; in the former case the unit system of burning is generally employed, and the same equipment is used as described for coal. Grinding characteristics vary with the melting point, which may be anywhere from 50 to 175°C.

Natural graphite may be divided into three grades in respect to grinding characteristics: flake, crystalline, and amorphous. Flake is generally the most difficult to reduce to fine powder, and the crystalline variety is the most abrasive. Graphite is ground in ball mills, tube mills, ring-roller mills, and jet mills with or without air classification. Beneficiation by flotation is an essential part of most current procedures.

Majac jet-pulverizer performance on natural graphite is given in Table 20-27. Graphite for pencils has 47, 83, 91, and 94 percent by weight smaller than 4, 9, 18, and 31 μm respectively.

Artificial graphite has been ground in ball mills in closed circuit with air classifiers. For lubricants the graphite is ground wet in a paste in which water is eventually replaced by oil. The colloid mill is used for production of graphite paint.

Mineral black, a shale sometimes erroneously called "rotten stone," contains a large amount of carbon and is used as a filler for paints and other chemical operations. It is pulverized and classified with the same equipment as shale, limestone, and barite.

Bone black is sometimes ground very fine for paint, ink, or chemical uses. A tube mill often is used, the mill discharging to a fan which blows the material to a series of cyclone collectors in tandem.

Decolorizing carbons of vegetable origin should not be ground too fine. Standard fineness varies from 100 percent through No. 30 sieve to 100 percent through No. 50, with 50 to 70 percent on No. 200 sieve as the upper limit. Ball mills, hammer mills, and rolls, followed by screens, are used. When the material is used for filtering, a product of uniform size must be used.

Charcoal usually is ground in hammer mills with screen or air classification. For absorption of gases it is usually crushed and graded to about No. 16 sieve size. Care should be taken to prevent it from igniting during grinding.

Gilsonite sometimes is used in place of asphalt or pitch. It is easily pulverized and is generally reduced on hammer mills with air classification.

CHEMICALS, PIGMENTS, AND SOAPS

Colors and Pigments **Dry colors** and **dyestuffs** generally are pulverized in hammer mills (see Table 20-21 or 20-23). The jar mill or a large pebble mill is often used for small lots. There is a special problem with some dyes which are coarsely crystalline. These are ground to the desired fineness with hammer or jet mills using air classification to limit the size. Synthetic mineral pigments are usually fine agglomerates. They may be disintegrated with hammer or jet mills without elaborate pregrinding. A 1.5- by 0.4-m (4.5-ft by 16-in) Hardinge conical mill in closed circuit with classifier, grinding 50-mesh iron oxide with 33 percent moisture for the paint trade, showed a capacity of 22.5 Mg/day (25 tons/day), 100 percent through No. 200 sieve. Power consumption was 15 kW (20 hp), mill speed 30 r/min, ball load 1800 kg (4000 lb). The conditions necessary for dispersing pigments in paint by means of steel-ball mills have been investigated [Fischer, *Ind. Eng. Chem.*, **33**(12), 1465–1472 (1941)]. Other examples for iron pigments are given in Table 20-33.

Easily dispersible colors are not ordinarily ground fine, since they are subsequently processed in a liquid medium in pebble mills, rolls, or colloid mills. There is, however, a tendency to grind them wet with a dispersing agent, then drying and pulverizing, after which they are easily dispersed in the vehicle in which they are used.

White pigments are the basic commodities processed in large quantities. Titanium dioxide is the most important. The problem of cleaning the mill between batches does not exist as with different colors. These pigments are finish-ground to sell as dry pigments using mills with air classification. For the denser, low-oil-absorption grades, roller and pebble mills are employed. For looser, fluffier products, hammer and jet mills are used. Often a combination of the two mill actions is used to set the finished quality. Jet-mill performance for a number of extenders is given in Table 20-26.

Lead Oxides Leady litharge containing 25 to 30 percent free lead is required for storage-battery plates. It is processed on Raymond Imp mills. They have the ability to produce litharge that has a desired low density of 1.1 to 1.3 g/cm³ (18 to 22 g/in³). A 56-kW (75-hp) unit produces 860 kg/h (1900 lb/h) of material having this density.

The processing of **diatomite** is unique, since its particle-size control is effected by calcination treatments and air classification.

Chemicals The fineness obtainable with a hammer mill on rock salt and chemicals is given in Tables 20-23 and 20-24.

Sulfur The ring-roller mill can be used for the fine grinding of sulfur. Inert gases are supplied instead of hot air (see "Properties of Solids: Safety" for use of inert gas). Performance of a Raymond No. 5057 ring-roller mill is given in Table 20-34. The total cost might be 3 to 4 times the power cost and include labor, inert gas, maintenance, and fixed charges.

Soaps Soaps in a finely divided form may be classified as soap powder, powdered soap, and chips or flakes. The term "soap powder" is applied to a granular product, No. 12 to No. 16 sieve size with a certain amount of fines, which is produced in hammer mills with perforated or slotted screens.

The oleates and erucates are best pulverized by multicage mills; laurates and palmitates, in cage mills and also in hammer mills if particu-

TABLE 20-34 Grinding Sulfur

Fineness, % through No. 325 sieve	Capacity, tons/hr.	Power, kw.-hr./ton
90	6.0	13.7
95	5.0	16.4
99	3.5	23.4
99.9	2.5	32.7

NOTE: To convert tons per hour to megagrams per hour, multiply by 0.907; to convert kilowatthours per ton to kilowatthours per megagram, multiply by 1.1.

larly fine division is not required; stearates may generally be pulverized in multicage mills, screen mills, and air-classification hammer mills.

ORGANIC POLYMERS

The grinding characteristics of various resins, gums, waxes, hard rubbers, and molding powders depend greatly upon their softening temperatures. When a finely divided product is required, it is often necessary to use a water-jacketed mill or a pulverizer with an air classifier in which cooled air is introduced into the system. Hammer and cage mills are used for this purpose. Some low-softening-temperature resins can be ground by mixing with 15 to 50 percent by weight of dry ice before grinding. Refrigerated air sometimes is introduced into the hammer mill to prevent softening and agglomeration [Dorris, *Chem. Metall. Eng.*, **51**, 114 (July 1944)].

Most **gums and resins,** natural or artificial, when used in the paint, varnish, or plastic industries, are not ground very fine, and hammer or cage mills will produce a suitable product. Typical performance of the disk attrition mill is given in Table 20-25. Roll crushers will often give a sufficiently fine product.

The Raymond ring-roll mill with its internal air separation is widely used to pulverize **phenolformaldehyde resins.** The usual fineness of grind is finer than 99 percent minus 200 mesh. Air at 4°C (40°F) is usually introduced into the mill to limit temperature rise. A typical 3036 Raymond mill using 34 kW (45 hp) will produce better than 900 kg/h (2000 lb/h) at 99 percent minus 200 mesh.

Hard rubber is one of the few combustible materials which is generally ground on heavy steam-heated rollers. The raw material passes to a series of rolls in closed circuit with screens and air classifiers. Farrel-Birmingham rolls are used extensively for this work. There is a differential in the roll diameters. The motor should be separated from the grinder by a fire wall.

Specifications for **molding powders** vary widely, from a No. 8 to a No. 60 sieve product; generally the coarser products are No. 12, 14, or 20 sieve material. Specifications usually prescribe a minimum of fines (below No. 100 and No. 200 sieve). Molding powders are produced with hammer mills, either of the screen type or equipped with air classifiers.

The following materials may be ground at ordinary temperatures if only the regular commercial fineness is required: amber, arabac, tragacanth, rosin, olibanum, gum benzoin, myrrh, guaiacum, and montan wax. If a finer product is required, hammer mills or attrition mills in closed circuit, with screens or air classifiers, are used.

PROCESSING WASTE

In flowsheets for processing *municipal solid waste* (MSW), the objective is to separate the waste into useful materials, such as scrap metals, plastics, and *refuse-derived fuels* (RDF). Usually size reduction is the first step, followed by separations with screens or air classifiers, which attempt to recover concentrated fractions [Savage and Diaz, *Proceedings ASME National Waste Processing Conference*, Denver, Colo, 361–373 (1986)]. Many installed circuits proved to be ineffective or not cost-effective, however. Begnaud and Noyon [*Biocycle*, **30**(3), 40–41 (1989)] concluded from a study of French operations that milling could not grind selectively enough to separate different materials.

Size reduction uses either hammer mills or blade cutters (shredders). Hammer mills are likely to break glass into finer sizes making it hard to separate. Better results may be obtained in a flowsheet where size reduction follows separation (Savage, Seminar on the Application

TABLE 20-33 Grinding Iron Oxides in a Ring-Roller Mill

Material	Fineness	Capacity, lb./hr.	Total hp.-hr./ton
Raw sienna	99% through No. 200 sieve	5950	23.5
Burnt sienna	99.5% through No. 200 sieve	5800	22.1
Raw umber	99% through No. 200 sieve	5200	26.9
Burnt umber	99.5% through No. 200 sieve	5400	25.9
Natural ocher	99.9% through No. 200 sieve	4500	31.0
Iron oxide (ore)	99% through No. 325 sieve	3100	45.0
Iron oxide (precipitated)	99.9% through No. 325 sieve	1800	72.5

NOTE: To convert pounds per hour to kilograms per hour, multiply by 0.4535; to convert horsepower-hours per ton to kilowatthours per megagram, multiply by 0.82.

of U.S. Water and Air Pollution Control Technology to Korea, Korea, May 1989).

The energy requirement for reducing MSW to 90 percent passing 10 cm is typically 6 kWh/ton, or 50 kWh/ton for passing 1 cm. Wear is also a major cost, and wear rates are shown in Fig. 20-60. The maximum capacity of commercially available hammer mills is about 100 ton/h.

CRYOGENIC GRINDING

This is a process used for recovering recycled materials [Biddulph, *Chemical Engineering*, **87**(3), 93–96 (1980)]. This process permits efficient separation of materials in cases where some materials in a mixture become brittle, while others do not, and it also reduces crusher energy requirements. Examples are rubber crumb recovered in a pure state from scrap tires, which has found uses including road and sports surfaces and polymer fillers. Stripping of polymeric insulation from copper wires occurs when the chilled material is passed through crushing rolls. The Inchscrap process shreds chilled baled automobiles, magnetically separating pure steel and other components. Biddulph (loc. cit.) presents graphs for design of a heat-transfer tunnel using liquid nitrogen. Countercurrent flow of the gas also extracts the latent heat of the nitrogen. The process can be economic

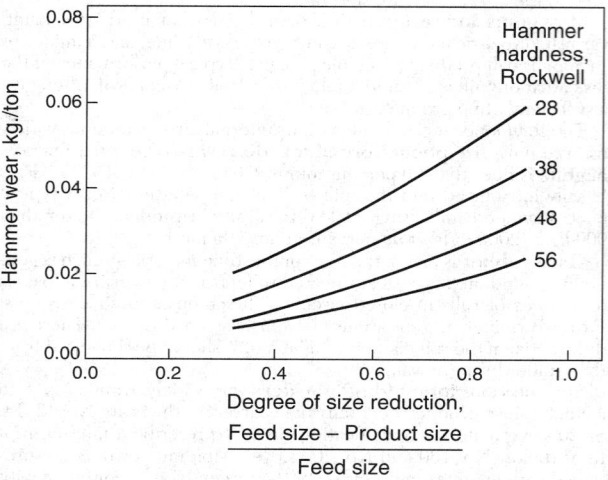

FIG. 20-60 Hammer wear as a consequence of shredding municipal solid waste. (*Savage and Diaz, Proceedings ASME National Waste Processing Conference, Denver, CO, 361–373, 1986.*)

where the added value of the recovered product ranges from a few cents to 15 cents/kg.

Results for cryogrinding of polymers in an opposed-jet laboratory mill are given by [Haesse, *Kunststoffe-German Plastics*, **70**(12), 9–10 (1980)].

CELL DISRUPTION

Mechanical disruption is the most practical first step in the release and isolation of proteins and enzymes from microorganisms on a commercial scale. The size-reduction method must be gently tuned to the strength of the organisms to minimize formation of fine fragments that interfere with subsequent clarification by centrifugation or filtration. Typically, fragments as fine as 0.3 μm are produced. High-speed stirred-bead mills and high-pressure homogenizers have been applied for cell disruption [Kula and Schuette, *Biotechnology Progress*, **3**(1), 31–42 (1987)].

There are two limiting cases in operation of bead mills for disruption of bacterial cells. When the energy imparted by collision of beads is insufficient to break all cells, the rate of breakage is proportional to the specific energy imparted [Bunge et al., *Chemical Engineering Science*, **47**(1), 225–232 (1992)]. On the other hand, when the energy is high due to higher speed above 8 m/sec, larger beads above 1 mm, and low concentrations of 10 percent, each bead impact has more than enough energy to break any cells that are captured, which causes problems during subsequent separations.

The strength of cell walls differs among bacteria, yeasts, and molds. The strength also varies with the species and the growth conditions, and must be determined experimentally. Beads of 0.5 mm are typically used for yeast and bacteria. Recommended bead charge is 85 percent for 0.5 mm beads, and 80 percent for 1 mm beads [Schuette et al., *Enzyme Microbial Technology*, **5**, 143 (1983)].

Residence-time distribution is important in continuous mills. Further data are given in the above references.

While the above discussion centered on the rate of disruption, the objective is usually to attain at least 90 percent release of the valuable protein from the cells. Cell disruption with protein solubilization is considered to be first order in amount of protein remaining [Currie et al., *Biotechnol. Bioeng.*, **14**, 725 (1972)]:

$$(R_m - R)/R_m = \exp(-kt)$$

where R is residual protein, and R_m is maximum protein removable.

Valve homogenizers have been used to disrupt cells and release soluble components (Pandolf, *Cell Disruption by Homogenization*, APV Gaulin, 1993). The cell disruption is believed due to the sudden pressure drop, although impact may also be a cause [Brookman, *Biotechnol. Bioeng.*, **16**, 371 (1974); Engler and Robinson, *Biotechnol. Bioeng.*, **23**, 765 (1981)]. The release of glycose-6-phosphate dehydrogenase from Saccharomyces cerevisiae is linear with pressure beginning at 200 bar and reaches 40 percent at 550 bar in the Gaulin M3 unit.

PRINCIPLES OF SIZE ENLARGEMENT

GENERAL REFERENCES: Ennis & Litster, *The Science & Engineering of Granulation Processes*, Blackie Academic Ltd., 1997. Ennis (ed.), *Proceedings of the First International Particle Technology Forum, Vol. 1*, AIChE, Denver, 1994. Ennis, *Powder Technology*, **88**, 203 (1996). Kapur, *Adv. Chem. Eng.*, **10**, 55, 1978. Kristensen, *Acta Pharm. Suec.*, **25**, 187, 1988. Pietsch, *Size Enlargement by Agglomeration*, John Wiley & Sons Ltd., Chichester, 1992. Randolph & Larson, *Theory of Particulate Processes*, Academic Press, Inc., San Diego, 1988. Stanley-Wood, *Enlargement and Compaction of Particulate Solids*, Butterworth & Co. Ltd., 1983. Ball et al., *Agglomeration of Iron Ores*, Heinemann, London, 1973. Capes, *Particle Size Enlargement*, Elsevier, New York, 1980. King, "Tablets, Capsules and Pills," in *Remington's Pharmaceutical Sciences*, Mack Pub. Co., Easton, Pa., 1970. Knepper (ed.), *Agglomeration*, Interscience, New York, 1962. Mead (ed.), *Encyclopedia of Chemical Process Equipment*, Reinhold, New York, 1964. Pietsch, *Roll Pressing*, Heyden, London, 1976. Sastry (ed.), *Agglomeration 77*, AIME, New York, 1977. Sauchelli (ed.), *Chemistry and Technology of Fertilizers*, Reinhold, New York, 1960. Sherrington and Oliver, *Granulation*, Heyden, London, 1981.

SCOPE AND APPLICATIONS

Size enlargement is any process whereby small particles are gathered into larger, relatively permanent masses in which the original particles can still be distinguished. The term encompasses a variety of unit-operations or processing techniques dedicated to particle agglomeration. **Agglomeration** is the formation of aggregates through the sticking together of feed and/or recycle material. These processes can be loosely broken down into **agitation** and **compression** methods. Although terminology is industry specific, agglomeration by agitation will be referred to as **granulation.** Here, a particulate feed is introduced to a process vessel and is agglomerated, either batchwise and continuously, to form a granulated product. Agitative processes include fluid-bed, pan (or disc), drum, and mixer granulators. The feed typically consists of a mixture of solid ingredients, referred to as a **formulation,** which includes an active or key ingredient, binders, diluents, flow aids, surfactants, wetting agents, lubricants, fillers, or end-use aids (e.g., sintering aids, colors or dyes, taste modifiers). The agglomeration can be induced in several ways. A solvent or slurry can be atomized onto the bed of particles which either coats the particle or granule surfaces promoting agglomeration, or the spray drops can form small nuclei in the case of a powder feed which subsequently can agglomerate. The solvent or slurry may contain a binder, or solid binder may be present as one component of the feed. Alternatively, the solvent may induce dissolution and recrystallization in the case of soluble particles. Slurries often contain the same particulate matter as the dry feed, and granules may be formed, either completely or partially, as the droplets solidify in flight prior to reaching the particle bed. **Spray-drying** is an extreme case with no further, intended agglomeration taking place after granule formation. Agglomeration may also be induced by heat, which either leads to controlled **sintering** of the particle bed or induces sintering or partial melting of a binder component of the feed, e.g., a polymer.

An alternative approach to size enlargement is by **compression agglomeration,** where the mixture of particulate matter is fed to a compression device which promotes agglomeration due to pressure. Either continuous sheets of solid material are produced or some solid form such as a **briquette** or **tablet.** Heat or cooling may be applied, and reaction may be induced as for example with sintering. Carrier fluids may be present, either added or induced by melting, in which case the product is **wet extruded.** Continuous compaction processes include roll presses, briquetting machines, and extrusion, whereas batch-like processes include tableting. Some processes operate in a semicontinuous fashion, such as ram extrusion.

At the level of a manufacturing plant, the size-enlargement process involves several peripheral, unit operations such as milling, blending, drying or cooling, and classification, referred to generically as an **agglomeration circuit** (Fig. 20-61). In addition, more than one

agglomeration step may be present as in the case of a pharmaceutical process which often involves both an agitative-granulation technique followed by the compressive technique of tableting.

Numerous benefits result from size-enlargement processes, as will be appreciated from Table 20-35. A wide variety of size-enlargement methods are available; a classification of these is given in Table 20-36.

APPROACHING THE DESIGN OF SIZE-ENLARGEMENT PROCESSES

Agglomeration Kinetics A change in particle size of a particulate material due to agglomeration is akin to a change in chemical species, and so analogies exist between agglomeration and chemical kinetics and the unit operations of size enlargement and chemical reaction. The performance of a **granulator** or **compactor** may be described by the **extent of agglomeration** of a species, typically represented by a loss in number of particles. Let $(x_1, x_2, \ldots, x_n)$ represent a list of attributes such as average particle size, porosity, strength, surface properties, and any generic quality metric and associated variances. Alternatively, $(x_1, x_2, \ldots, x_n)$ might represent the concentrations or numbers of certain size or density classes, just as in the case of chemical reactors. The proper design of a chemical reactor or an **agglomerator** then relies on understanding and controlling the evolution (both time and spatial) of the feed vector **X** to the desired product vector **Y.** Inevitably, the reactor or granulator is contained within a larger plant-scale process chain, or **manufacturing circuit,** with overall plant performance being dictated by the interactions between individual unit operations. For successful plant design and operation, there are four natural levels of scrutiny (Fig. 20-62). Conceptually, the design of chemical reactors and agglomeration processes differ in that the former deals with chemical transformations whereas the latter deals primarily with **physical transformations** with the mechanisms or rate processes of agglomeration controlled by a set of key **physicochemical interactions.**

Granulation Rate Processes Granulation is controlled by four key rate processes. These include wetting, coalescence or growth, consolidation, and breakage (Fig. 20-63). Initial **wetting** of the particles by the binding fluid is strongly influenced by spray or fluid distri-

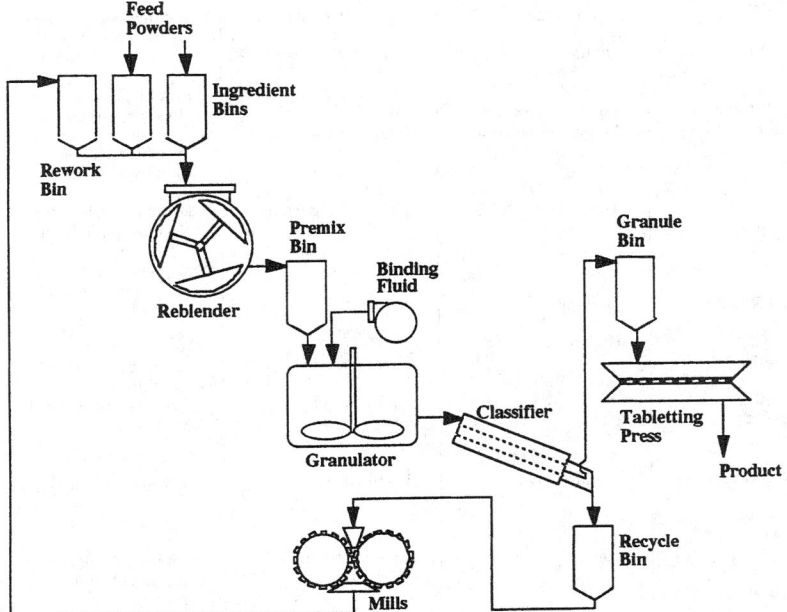

FIG. 20-61 A typical agglomeration circuit utilized in the processing of pharmaceutical or agricultural chemicals involving both granulation and compression techniques. Reprinted from *Granulation and Coating Technologies for High-Value-Added Industries,* Ennis and Litster (1996) with permission of E & G Associates. All rights reserved.

TABLE 20-35 Objectives of Size Enlargement

Production of useful structural forms, as in pressing of intricate shapes in powder metallurgy.
Provision of a defined quantity for dispensing and metering, as in agricultural chemical granules or pharmaceutical tablets.
Elimination of dust handling hazards or losses, as in briquetting of waste fines.
Improved product appearance, or product renewal.
Reduced caking and lump formation, as in granulation of fertilizer.
Improved flow properties, as in granulation of pharmaceuticals for tabletting or ceramics for pressing.
Increased bulk density for storage.
Creation of non-segregating blends of powder ingredients, as in sintering of fines for steel or agricultural chemical granules.
Control of solubility, as in instant food products.
Control of porosity and surface-to-volume ratio, as with catalyst supports.
Improvement of heat transfer characteristics, as in ores or glass for furnace feed.
Removal of particles from liquid, as with polymer additives which induce clay flocculation.

bution as well as feed-formulation properties. Often wetting agents such as surfactants are carefully chosen to enhance poorly wetting feeds. In the **coalescence** or **growth** stage, partially wetted primary particles and larger nuclei coalesce to form granules composed of several particles. The term **nucleation** is typically applied to the initial coalescence of primary particles in the immediate vicinity of the larger-wetting drop, whereas the more general term of **coalescence** refers to the successful collision of two granules to form a new larger granule. Nucleation is promoted from some initial distribution of moisture, such as a drop or from the homogenization of a fluid feed to the bed, as with high-shear mixing. The nucleation process is strongly linked with the wetting stage. As granules grow, they become consolidated by the compaction forces of the bed due to agitation. This **consolidation** stage strongly influences final granule porosity, and therefore end-use properties such as granule strength or dispersability. Formed granules may be particularly susceptible to **breakage** if they are inherently weak or if flaws develop during drying. The final size distribution and other end-use properties of the product are determined by the complex interaction of all these rate processes acting simultaneously.

Compaction Rate Processes The performance of compaction techniques is controlled by the ability of the particulate phase to uniformly transmit stress and the relationship between applied stress and the compaction and strength characteristics of the final compacted particulate phase. The general area of study relating compaction and stress transmission is referred to as **powder mechanics** (Brown & Richards, *Principles of Powder Mechanics*, Pergamon Press Ltd., Oxford, 1970).

Process versus Formulation Design The end-use properties of the agglomerated material are controlled by agglomerate size and porosity. Granule structure may also influence properties. To achieve a desired **product quality** as defined by metrics of end-use properties, size and porosity may be manipulated by changes in either process **operating** or product **material variables** (Fig. 20-63). The first approach is the realm of traditional **process engineering,** whereas the second is **product engineering.** Both approaches are

TABLE 20-36 Size-Enlargement Methods and Application

Method	Product size (mm)	Granule density	Scale of operation	Additional comments	Typical applications
Tumbling granulators Drums Discs	0.5 to 20	Moderate	0.5–800 ton/hr	Very spherical granules	Fertilizers, iron ore, non-ferrous ore, agricultural chemicals
Mixer granulators Continuous high shear (e.g. Shugi mixer)	0.1 to 2	Low to high	Up to 50 ton/hr	Handles very cohesive materials well, both batch and continuous	Chemicals, detergents, clays, carbon black
Batch high shear (e.g. paddle mixer)	0.1 to 2	High	Up to 500 kg batch		Pharmaceuticals, ceramics
Fluidized granulators Fluidized beds Spouted beds Wurster coaters	0.1 to 2	Low (agglomerated) Moderate (layered)	100–900 kg batch 50 ton/hr continuous	Flexible, relatively easy to scale, difficult for cohesive powders, good for coating applications	Continuous: fertilizers, inorganic salts, detergents Batch: pharmaceuticals, agricultural chemicals, nuclear wastes
Centrifugal granulators	0.3 to 3	Moderate to high	Up to 200 kg batch	Powder layering and coating applications	Pharmaceuticals, agricultural chemicals
Spray methods Spray drying	0.05 to 0.5	Low		Morphology of spray dried powders can vary widely	Instant foods, dyes, detergents, ceramics
Prilling	0.7 to 2	Moderate			Urea, ammonium nitrate
Pressure compaction Extrusion Roll press Tablet press Molding press Pellet mill	 >0.5 >1 10	High to very high	 Up to 5 ton/hr Up to 50 ton/hr Up to 1 ton/hr	Very narrow size distributions, very sensitive to powder flow and mechanical properties	Pharmaceuticals, catalysts, inorganic chemicals, organic chemicals, plastic preforms, metal parts, ceramics, clays, minerals, animal feeds
Thermal processes Sintering	2 to 50	High to very high	Up to 100 ton/hr	Strongest bonding	Ferrous & non-ferrous ores, cement clinker, minerals, ceramics
Liquid systems Immiscible wetting in mixers Sol-gel processes Pellet flocculation	<0.3	Low	Up to 10 ton/hr	Wet processing based on flocculation properties of particulate feed	Coal fines, soot and oil removal from water Metal dicarbide, silica hydrogels Waste sludges and slurries

Level of scrutiny   Area of analysis

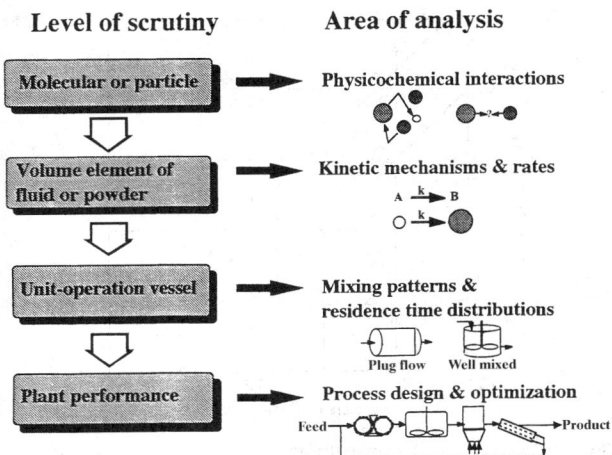

FIG. 20-62 Comparisons of levels of analysis of chemical reaction and size-enlargement processes. Reprinted from *Granulation and Coating Technologies for High-Value-Added Industries,* Ennis and Litster (1996) with permission of E & G Associates. All rights reserved.

critical and must be integrated to achieve a desired end point in product quality. Operating variables are defined by the chosen agglomeration technique and peripheral processing equipment. In addition, the choice of agglomeration technique dictates the mixing pattern of the vessel. Material variables include parameters such as binder viscosity, surface tension, particle-size distribution and friction, and the adhesive properties of the solidified binder. Material variables are specified by the choice of ingredients, or **product formulation.** Both operating and material variables together define the kinetic mechanisms and

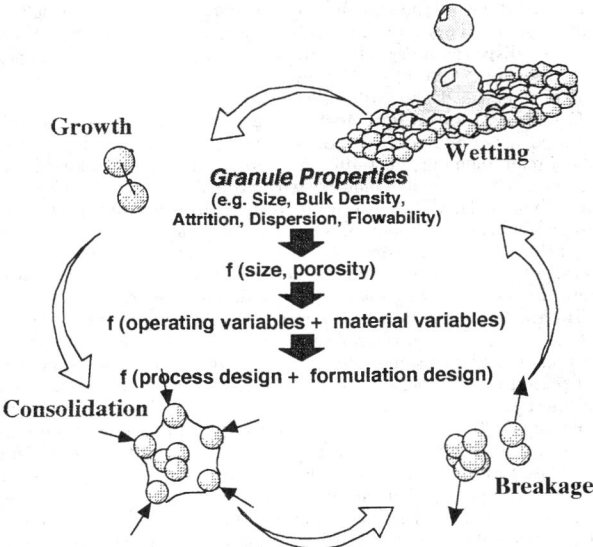

FIG. 20-63 The rate processes of agitative agglomeration, or granulation, which include powder wetting, granule growth, granule consolidation, and granule attrition. These processes combined control granule size and porosity, and they may be influenced by formulation or process-design changes. Reprinted from *Granulation and Coating Technologies for High-Value-Added Industries,* Ennis and Litster (1996) with permission of E & G Associates. All rights reserved.

rate constants of wetting, growth, consolidation, attrition, and powder flow. Overcoming a given size-enlargement problem often requires changes in both processing conditions and in product formulation.

PRODUCT CHARACTERIZATION

Powders are agglomerated to modify physical or physicochemical properties. Effective measurement of agglomerate properties is vital. However, many tests are industry specific and take the form of empirical indices based on standardized protocols. Such tests as described below are useful for quality control if used with care. However, since they often reflect an end use rather than a specifically defined agglomerate property, they often are of little developmental utility for recommending process or formulation changes. Significant improvements have been made in the ability to measure real agglomerate properties. Key agglomerate properties are **size, porosity,** and **strength** and their associated distributions. These properties directly affect end-use attributes of the product such as attrition resistance, flowability, bulk-solid permeability, wettability and dispersibility, appearance, or active-agent release rate.

Size Agglomerate **mean size** and **size distribution** are both important properties. (See subsection "Particle-Size Measurement".) For granular materials, sieve analysis is the most common sizing technique. Care is needed in sizing wet granules. Handling during sampling and sieving can cause changes to the size distribution through coalescence or breakage. Sieves are also easily blinded. Snap freezing the granules with liquid nitrogen prior to sizing overcomes these problems [Hall, *Chem. Eng. Sci.,* **41,** 187 (1986)]. On-line or in-line measurement of granules as large as 9 mm is now available by laser diffraction techniques, making improved granulation control schemes possible [Ogunnaike et al., *I.E.C. Fund.,* 1997].

Porosity and Density There are three important densities of granular or agglomerated materials: **bulk density** ρ_b (related to the volume occupied by the bulk solid), the apparent or **agglomerate density** ρ_g (related to the volume occupied by the agglomerate including internal porosity) and the true or **skeletal-solids density** ρ_s. These densities are related to the each other and the interagglomerate **voidage** ε_b and the intra-agglomerate **porosity** ε_g:

$$\varepsilon_b = 1 - \frac{\rho_b}{\rho_a}; \ \varepsilon_g = 1 - \frac{\rho_a}{\rho_s} \qquad (20\text{-}42)$$

Bulk density is easily measured from the volume occupied by the bulk solid and is a strong function of sample preparation. True density is measured by standard techniques using liquid or gas picnometry. Apparent (agglomerate) density is difficult to measure directly. Hinkley et al. [*Int. J. Min. Proc.,* **41,** 53–69 (1994)] describe a method for measuring the apparent density of wet granules by kerosene displacement. Agglomerate density may also be inferred from direct measurement of true density and porosity using Eq. (20-42).

Agglomerate porosity can be measured by gas adsorption or mercury porosimetry. However, any breakage or compression of the granules under high pressure during porosimetry can invalidate the results.

Strength of Agglomerates Agglomerate **bonding mechanisms** may be divided into five major groups [Rumpf, in Knepper (ed.), *Agglomeration,* op. cit., p. 379]. More than one mechanism may apply during a given size-enlargement operation. (In addition, see Krupp [*Adv. Colloid. Int. Sci.,* **1,** 111 (1967)] for a review of adhesion mechanisms.)

Solid bridges can form between particles by the sintering of ores, the crystallization of dissolved substances during drying as in the granulation of fertilizers, and the hardening of bonding agents such as glue and resins.

Mobile liquid binding produces cohesion through interfacial forces and capillary suction. Three states can be distinguished in an assembly of particles held together by a mobile liquid (Fig. 20-64). Small amounts of liquid are held as discrete lens-shaped rings at the points of contact of the particles; this is the **pendular state.** As the liquid content increases, the rings coalesce and there is a continuous network of liquid interspersed with air; this is the **funicular state.** When all the pore spaces in the agglomerate are completely filled, the **capillary state** has been reached. When a mobile liquid bridge fails, it con-

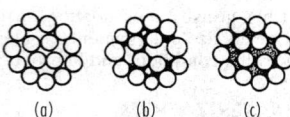

FIG. 20-64 Three states of liquid content for an assembly of spherical particles. (*a*) Pendular state. (*b*) Funicular state. (*c*) Capillary state. [*Newitt and Conway-Jones, Trans. Inst. Chem. Eng. (London),* **36**, *422 (1958).*]

stricts and divides without fully exploiting the adhesion and cohesive forces in the bridge in the absence of viscous effects. Binder viscosity markedly increases the strength of the pendular bridge due to dynamic lubrication forces, and aids the transmission of adhesion. For many systems, viscous forces outweigh interfacial capillary effects, as demonstrated by Ennis et al. [*Chem. Eng. Sci.,* **45**, 3071 (1990)].

In the limit of high viscosity, **immobile liquid bridges** formed from materials such as asphalt or pitch fail by tearing apart the weakest bond. Then adhesion and/or cohesion forces are fully exploited, and binding ability is much larger.

Intermolecular and **electrostatic forces** bond very fine particles without the presence of material bridges. Such bonding is responsible for the tendency of particles less than about 1 μm diameter to form agglomerates spontaneously under agitation. With larger particles, however, these short-range forces are insufficient to counterbalance the weight of the particle, and adhesion does not occur without applied pressure. High compaction pressures act to plastically flatten interparticle contacts and substantially enhance short-range forces.

Mechanical interlocking of particles may occur during the agitation or compression of, for example, fibrous particles, but it is probably only a minor contributor to agglomerate strength in most cases.

For an agglomerate of equal-sized spherical particles (Rumpf, loc. cit., 379), the tensile strength is

$$\sigma_T = \frac{9}{8} \frac{1 - \varepsilon_g}{\varepsilon_g} \frac{H}{d^2} \qquad (20\text{-}43)$$

where H is the individual bonding force per point of contact and d is the size of particles making up the granule. For other bonding mechanisms, Fig. 20-65 indicates values of tensile strength to be expected in various size enlargement processes. In particular, it should be noted that viscous mechanisms of binding (e.g., adhesives) can exceed capillary effects in determining agglomerate strength.

Strength Testing Methods Compressed agglomerates often fail in **tension** along their diameter. This is the basis of the commonly used measurement of **crushing strength** of a agglomerate as a method to assess tensile strength. However, the brittle failure of a granule depends on the flaw distribution as well as the inherent tensile strength of bonds as given by the Griffith crack theory. [Lawn, *Fracture of Brittle Solids,* 2d ed., Cambridge University Press, 1975] Therefore, it is more appropriate to characterize granule strength by **fracture toughness** K_c [Kendall, *Nature,* **272**, 710 (1978); See also subsections "Size Reduction: Properties of Solids" and "Breakage"].

Several strength-related indices are measured in different industries which give some measure of resistance to attrition. These tests do not measure strength or toughness directly, but rather the size distribution of fragments after handling the agglomerates in a defined way. The handling could be repeated drops, tumbling in a drum, fluidizing, circulating in a pneumatic conveying loop, etc. These indices should only be used for quality control if the test procedure simulates the actual handling of the agglomerates during processing and transportation.

Flow Property Tests Flowability of the product granules can be characterized by **unconfined yield stress** and **angle of friction** by

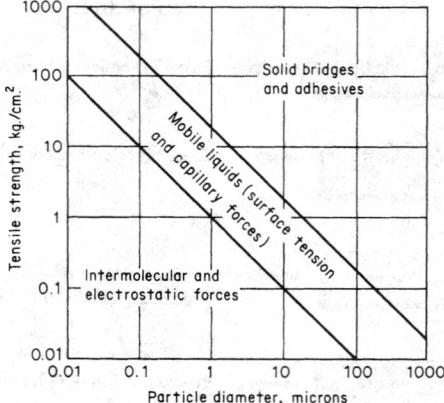

FIG. 20-65 Theoretical tensile strength of agglomerates. [*Adapted from Rumpf, "Strength of Granules and Agglomerates," in Knepper (ed.),* Agglomeration, *Wiley, New York, 1962.*]

shear-cell tests as used generally for bulk solids (see subsection "Powder Mechanics and Powder Compaction" and also Section 21, Solids Transport and Storage). **Caking** refers to deterioration in the flow properties of the granules due to chemical reaction or hydroscopic effects. Caking tests as used for fertilizer granules consist of two parts (Bookey and Raistrick, in Sauchelli (ed.), *Chemistry and Technology of Fertilizers,* p. 454). A cake of the granules is first formed in a compression chamber under controlled conditions of humidity, temperature, etc. The crushing strength of the cake is then measured to determine the degree of caking.

Redispersion Tests Agglomerated products are often **redispersed** in a fluid by a customer. Examples include the dispersion of fertilizer granules in spray-tank solutions or of tablets within the gastrointestinal tract of the human body. The mechanisms comprising this redispersion process of product **wetting**, agglomerate **disintegration**, and final **dispersion** are related to interfacial properties (For details, see subsection "Wetting"). There are a wide range of industry-specific empirical indices dealing with redispersion assessment.

Disintegration height tests consist of measuring the length required for complete agglomerate disintegration in a long, narrow tube. Small fragments may still remain after initial agglomerate disintegration. The residual of material which remains undispersed is measured by a related test, or **long-tube sedimentation test.** The residual undispersed material is reported by the level in the bottom tip of the tube. A variation of this test is the **wet screen test,** which measures the residual remaining on a fine mesh screen (e.g., 350 mesh) following pouring the beaker solution through the screen.

Tablet-disintegration tests consist of cyclical immersion in a suitable dissolving fluid of pharmaceutical tablets contained in a basket. Acceptable tablets disintegrate completely by the end of the specified test period (*United States Pharmacopeia,* 17th rev., Mack Pub. Co., Easton, Pa., 1965, p. 919).

Permeability **Bulk solid permeability** is important in the iron and steel industry where gas-solid reactions occur in the sinter plant and blast furnace. It also strongly influences compaction processes where entrapped gas can impede compaction, and solids-handling equipment where restricted gas flow can impede product flowability. The permeability of a granular bed is inferred from measured pressure drop under controlled gas-flow conditions.

AGGLOMERATION RATE PROCESSES

GENERAL REFERENCES: Adetayo *et al.*, *Powder Tech.*, **82**, 37 (1995). Brown & Richards, *Principles of Powder Mechanics*, Pergamon Press (1970). Carson & Marinelli, *Chemical Eng.*, April (1994). Ennis & Litster, *The Science & Engineering of Granulation Processes*, Blackie Academic Ltd., 1997. Ennis *et al.*, *Powder Tech.*, **65**, 257 (1991). Ennis & Sunshine, *Tribology International*, **26**, 319 (1993). Ennis (ed.), *Proceedings of the First International Particle Technology Forum, Vol. 1*, AIChE, Denver, 1994. Ennis, *Powder Technology*, **88**, 203 (1996). Kristensen, *Acta Pharm. Suec.*, **25**, 187, 1988. Lawn, *Fracture of Brittle Solids*, 2nd ed., Cambridge University Press, 1975. Owens & Wendt, *J. Appl. Polym. Sci.*, **13**, 1741 (1969). Pan *et al.*, *Dynamic Properties of Interfaces & Association Structure*, American Oil Chem. Soc. Press (1995). Parfitt (ed.), *Dispersion of Powders in Liquids*, Elsevier Applied Science Publishers Ltd., 1986.

WETTING

The initial distribution of binding fluid can have a pronounced influence on the size distribution of seed granules, or **nuclei**, which are formed from fine powder. Both the final **extent** of and the **rate** at which the fluid wets the particulate phase are important. For a poor wetting fluid, a substantial portion of the powder remains ungranulated. When the size of a particulate feed material is larger than drop size, wetting dynamics controls the distribution of coating material which has a strong influence on the later stages of growth. Wetting phenomena also influence redistribution of individual ingredients within a granule, drying processes, and redispersion of granules in a fluid phase (e.g., agricultural chemicals and pharmaceutical products).

Methods of Measurement   Methods of characterizing the rate process of wetting include four approaches as illustrated in Table 20-37. The first considers the ability of a drop to spread across the powder. This approach involves the measurement of a **contact angle** of a drop on a powder compact. The contact angle is a measure of the affinity of the fluid for the solid as given by the Young-Dupré equation, or

$$\gamma^{sv} - \gamma^{sl} = \gamma^{lv} \cos \theta \qquad (20\text{-}44)$$

TABLE 20-37   Methods of Characterizing Wetting Dynamics of Particulate Systems*

Mechanism of wetting	Characterization method

Spreading of drops on powder surface 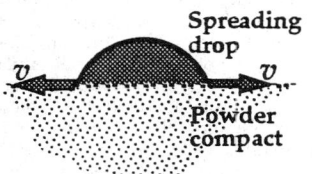	**Contact angle goniometer**
	  Contact angle
	  Drop height or volume
	  Spreading velocity
	References: Kossen & Heertjes, *Chem. Eng. Sci.*, **20**, 593 (1965). Pan et al., *Dynamic Properties of Interfaces & Association Structure*, American Oil Chem. Soc. Press (1995)

Penetration of drops into powder bed 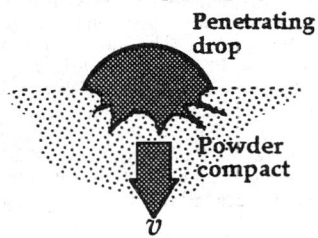	**Washburn test**
	  Rate of penetration by height or volume
	Bartell cell
	  Capillary pressure difference
	References: Parfitt (ed.), *Dispersion of Powders in Liquids*, Elsevier Applied Science Publishers Ltd., 1986. Washburn, *Phys. Rev.*, **17**, 273 (1921). Bartell & Osterhof, *Ind. Eng. Chem.*, **19**, 1277 (1927).

Penetration of particles into fluid 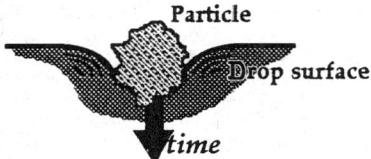	**Flotation tests**
	  Penetration time
	  Sediment height
	  Critical solid surface energy distribution
	References: R. Ayala, Ph.D. Thesis, Chem. Eng., Carnegie Mellon Univ. (1985). Fuerstaneau et al., *Colloids & Surfaces*, **60**, 127 (1991). Vargha-Butler, et al., *Colloids & Surfaces*, **24**, 315 (1987).

Chemical probing of powder 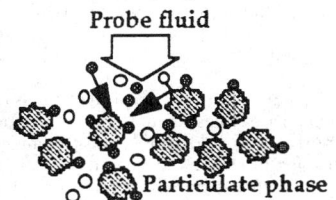	**Inverse gas chromatography**
	  Preferential adsorption with probe gases
	Electrokinetics
	  Zeta potential and charge
	Surfactant adsorption
	  Preferential adsorption with probe surfactants
	References: Lloyd et al. (eds.), *ACS Symposium Series* **391**, ACS, Washington D.C. (1989). Aveyard & Haydon, *An Introduction to the Principles of Surface Chemistry*, Cambridge Univ. Press (1973). Shaw, *Introduction to Colloid & Surface Chemistry*, Butterworths & Co. Ltd. (1983).

* Reprinted from *Granulation and Coating Technologies for High-Value-Added Industries*, Ennis and Litster (1996) with permission of E & G Associates. All rights reserved.

where γ^{sv}, γ^{sl}, γ^{lv} are the solid-vapor, solid-liquid, and liquid-vapor interfacial energies, respectively, and θ is the contact angle measured through the liquid as illustrated in Fig. 20-66. When the solid-vapor interfacial energy exceeds the solid-liquid energy, the fluid wets the solid with a contact angle less than 90°. In the limit of $\gamma^{sv} - \gamma^{sl} \geq \gamma^{lv}$, the contact angle equals 0° and the fluid **spreads** on the solid. The extent of wetting is controlled by the group $\gamma^{lv} \cos \theta$ which is referred to as **adhesion tension.** Sessile drop studies of contact angle can be performed on powder compacts in the same way as on planar surfaces. Methods involve (1) direct measurement of the contact angle from the tangent to the air-binder interface, (2) solution of the Laplace-Young equation involving the contact angle as a boundary condition, or (3) indirect calculations of the contact angle from measurements of e.g., drop height (Ennis & Litster, *The Science & Engineering of Granulation Processes*, Blackie Academic Ltd., 1997). The compact can either be saturated with the fluid for static measurements, or dynamic measurements may be made through a computer-imaging goniometer [Pan et al., *Dynamic Properties of Interfaces & Association Structure*, American Oil Chem. Soc. Press (1995)].

For granulation processes, the dynamics of wetting are often crucial, requiring that powders be compared on the basis of a short time scale, **dynamic contact angle.** Important factors are the physical nature of the powder surface (particle size, pore size, porosity, environment, roughness, pretreatment). Powders which are formulated for granulation often are composed of a combination of ingredients. The dynamic wetting process is therefore influenced by the rates of ingredient dissolution and surfactant adsorption and desorption kinetics. [Pan et al., loc. cit.].

The second approach to characterize wetting considers the ability of the fluid to penetrate a powder bed. It involves the measurement of the extent and rate of fluid rise by capillary suction into a column of powder, better known as the **Washburn test.** Considering the powder to consist of capillaries of radius R, the equilibrium height of rise h_e is determined by equating capillary and gravimetric pressures, or

$$h_e = \frac{2\gamma^{lv} \cos \theta}{\Delta \rho g R} \qquad (20\text{-}45)$$

where $\Delta \rho$ is the fluid density with respect to air, g is gravity, and $\gamma^{lv} \cos \theta$ is the adhesion tension as before. In addition to the equilibrium height of rise, the dynamics of penetration can be equally important. Ignoring gravity and equating viscous losses with the capillary pressure, the rate (dh/dt) and dynamic height of rise h are given by

$$\frac{dh}{dt} = \frac{R\gamma^{lv} \cos \theta}{4\mu h}, \quad \text{or} \quad h = \sqrt{\left[\frac{R\gamma^{lv} \cos \theta}{2\mu}\right]t} \qquad (20\text{-}46)$$

where t is time and μ is binder-fluid viscosity. [Parfitt (ed.), *Dispersion*

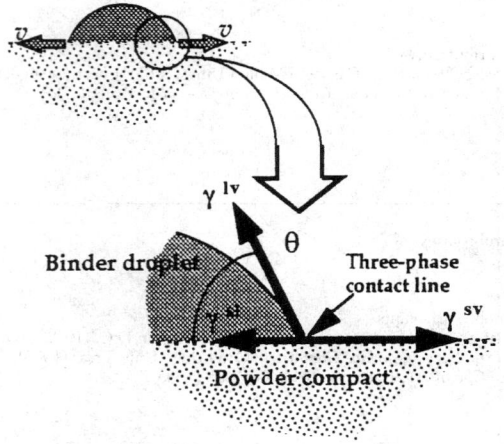

FIG. 20-66 Contact angle on a powder surface, where γ^{sv}, γ^{sl}, γ^{lv} are the solid-vapor, solid-liquid, and liquid-vapor interfacial energies, and θ is the contact angle measured through the liquid.

of Powders in Liquids, Elsevier Applied Science Publishers Ltd., 10 (1986).] The grouping of terms in brackets involves the material properties which control the dynamics of fluid penetration, namely average pore radius, or **tortuosity** R (related to particle size and void distribution of the powder), adhesion tension, and binder viscosity.

The contact angle or adhesion tension of a binder solution with respect to a powder can be determined from the slope of the penetration profile. Washburn tests can also be used to investigate the influence of powder preparation on penetration rates. The **Bartell cell** is related to the Washburn test except that adhesion tension is determined by a variable gas pressure which opposes penetration. [Bartell & Osterhof, *Ind. Eng. Chem.*, **19**, 1277 (1927).]

The contact angle of a binder-particle system is not itself a primary thermodynamic quantity, but rather a reflection of individual interfacial energies [Eq. (20-44)] which are a function of the molecular interactions of each phase with respect to one another. An interfacial energy may be broken down into its **dispersion** and **polar** components. These components reflect the chemical character of the interface, with the polar component due to hydrogen bonding and other polar interactions and the dispersion component due to van der Waals interactions. These components may be determined by the wetting tests described here, where a variety of solvents are chosen as the wetting fluids to probe specific molecular interactions as described by Zisman [Contact Angle, Wettability, & Adhesion, *Advances in Chemistry Series*, ACS, **43**, 1 (1964)]. These components of interfacial energy are strongly influenced by trace impurities which often arise in crystallization of the active ingredient, or other forms of processing such as grinding, and they may be modified by judicious selection of **surfactants.** [R. Ayala, Ph.D. Thesis, *Chem. Eng.*, Carnegie Mellon Univ. (1985).] Charges may also exist at interfaces. In the case of solid-fluid interfaces, these may be characterized by **electrokinetic** studies [Shaw, *Introduction to Colloid & Surface Chemistry*, Butterworths & Co. Ltd. (1983)].

The total solid-fluid interfacial energy (i.e., both dispersion and polar components) is also referred to as the **critical solid-surface energy** of the particulate phase. It is equal to the surface tension of a fluid which *just* wets the solid with zero-contact angle. This property of the particle feed may be determined by a third approach to characterize wetting, involving the penetration of particles into a series of fluids of varying surface tension. [R. Ayala, Ph.D. Thesis, *Chem. Eng.*, Carnegie Mellon Univ. (1985).]; Fuerstaneau et al., *Colloids & Surfaces*, **60**, 127 (1991).] The critical surface energy may also be determined from the variation of sediment height with the surface tension of the solvent. [Vargha-Butler, et al., *Colloids & Surfaces*, **24**, 315 (1987).] Distributions in surface energy and its components often exist in practice, and these may be determined by the wetting measurements described here.

Examples of the Impact of Wetting   Wetting dynamics has a pronounced influence on the initial nuclei distribution formed from fine powder. The influence of powder-contact angle on the average size of nuclei formed in fluid-bed granulation is illustrated in Fig. 20-67 where contact angle of the powder with respect to water was varied by changing the weight ratios of the ingredients of lactose and salicylic acid, which are hydrophilic and hydrophobic, respectively. [Aulton & Banks, *Proceedings of Powder Technology in Pharmacy Conference*, Powder Advisory Centre, Basel, Switzerland (1979).] Aulton et al. [*J. Pharm. Pharmacol.*, **29**, 59P (1977)] also demonstrated the influence of surfactant concentration on shifting nuclei size due to changes in adhesion tension.

The effect of fluid-penetration rate and the extent of penetration on granule-size distribution for drum granulation was shown by Gluba et al. [*Powder Hand. & Proc.*, **2**, 323 (1990).] Increasing penetration rate increased granule size and decreased asymmetry of the granule-size distribution.

Controlling Wetting   Table 20-38 summarizes typical changes in material & operating variables which are necessary to improve wetting uniformity. Also listed are appropriate routes to achieve these changes in a given variable through changes in either the formulation or in processing. Improved wetting uniformity generally implies a tighter granule-size distribution and improved-product quality due to a better-controlled manufacturing process. Eqs. (20-44) to (20-46), shown previously, provide the basic trends of the effect of material variables on both wetting dynamics and wetting extent.

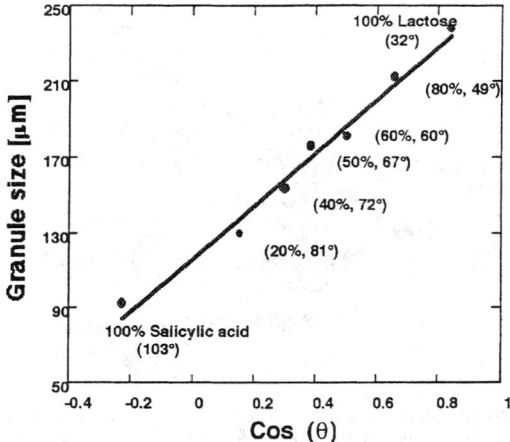

FIG. 20-67 The influence of contact angle on nuclei size formed in fluid-bed granulation of lactose/salicylic acid mixtures. Formulations ranged from hydrophobic (100% salicylic acid) to hydrophilic (100% lactose). Powder contact angle θ was determined by gonimetry and percent lactose of each formulation is given in parentheses. [*Aulton & Banks*, Proceedings of Powder Technology in Pharmacy Conference, *Powder Advisory Centre, Basel, Switzerland (1979).*]

Since drying occurs simultaneously with wetting, the effect of drying can substantially modify the expected impact of a given process variable and this should not be overlooked. In addition, simultaneously drying often implies that the *dynamics* of wetting are far more important than the *extent*.

Adhesion tension should be maximized to increase the rate and extent of both binder spreading and binder penetration. Maximizing adhesion tension is achieved by minimizing contact angle and maximizing surface tension of the binding solution. These two aspects work against one another as surfactant is added to a binding fluid, and in general, there is an optimum surfactant concentration which must be determined for the formulation (R. Ayala, loc. cit.). In addition, surfactant type influences adsorption and desorption kinetics at the three-phase contact line. An inappropriate choice of surfactant can lead to Marangoni interfacial stresses which slow the dynamics of wetting [Pan et al., loc. cit.]. Additional variables which influence adhesion tension include (1) impurity profile and particle habit/morphology typically controlled in the particle-formation stage such as crystallization, (2) temperature of granulation, and (3) technique of grinding, which is an additional source of impurity as well.

Decreases in binder viscosity enhance the rate of both binder spreading and binder penetration. The prime control over the viscosity of the binding solution is through binder concentration. Therefore, liquid loading and drying conditions strongly influence binder viscosity. For processes *without* simultaneous drying, binder viscosity generally decreases with increasing temperature. For processes *with* simultaneous drying, however, the dominantly observed effect is that lowering temperature lowers binder viscosity and enhances wetting due to decreased rates of drying and increased liquid loading.

Changes in particle-size distribution affect the pore distribution of the powder. Large pores between particles enhance the *rate* of binder penetration, whereas they decrease the final *extent*. In addition, the particle-size distribution affects the ability of the particles to pack within the drop as well as the final degree of saturation [Waldie, *Chem. Engin. Sci.,* **46,** 2781 (1991)].

The drop distribution and spray rate of binder fluid have a major influence on wetting. Generally, finer drops will enhance wetting as well as the distribution of binding fluid. The more important question, however, is how large may the drops be or how high of a spray rate is

TABLE 20-38   Controlling Wetting In Granulation Processes*

Typical changes in material or operating variables which improve wetting uniformity	Appropriate routes to alter variable through formulation changes	Appropriate routes to alter variable through process changes
Increase adhesion tension. Maximize surface tension. Minimize contact angle.	Alter surfactant concentration or type to maximize adhesion tension and minimize Marangoni effects. Precoat powder with wettable monolayers, e.g., coatings or steam.	Control impurity levels in particle formation. Alter crystal habit in particle formation. Minimize surface roughness in milling.
Decrease binder viscosity.	Lower binder concentration. Change binder. Decrease any diluents and polymers which act as thickeners.	Raise temperature for processes without simultaneous drying. Lower temperature for processes with simultaneous drying since binder concentration will decrease due to increased liquid loading. This effect generally offsets inverse relationship between viscosity and temperature.
Increase pore size to increase rate of fluid penetration. Decrease pore size to increase extent of fluid penetration.	Modify particle size distribution of feed ingredients.	Alter milling, classification or formation conditions of feed if appropriate to modify particle size distribution.
Improve spray distribution.	Improve atomization by lowering binder fluid viscosity.	Increase wetted area of the bed per unit mass per unit time by increasing the number of spray nozzles, lowering spray rate, increase air pressure or flow rate of two-fluid nozzles.
Increase solids mixing.	Improve powder flowability of feed.	Increase agitation intensity (e.g., impeller speed, fluidization gas velocity, or rotation speed).
Minimize moisture buildup & losses.	Avoid formulations which exhibit adhesive characteristics with respect to process walls.	Maintain spray nozzles to avoid caking and nozzle drip. Avoid spray entrainment in process air streams, and spraying process walls.

* Reprinted from *Granulation and Coating Technologies for High-Value-Added Industries,* Ennis and Litster (1996) with permission of E & G Associates.

possible. The answer depends on the wetting propensity of the feed. If the liquid loading for a given spray rate exceeds the ability of the fluid to penetrate and spread on the powder, maldistribution in binding fluid will develop in the bed. This maldistribution increases with increasing spray rate, increasing drop size, and decreasing spray area (due to, e.g., bringing the nozzle closer to the bed or switching to fewer nozzles). The maldistribution will lead to large granules on the one hand and fine ungranulated powder on the other. In general, the width of the granule-size distribution will increase and generally the average size will decrease. Improved spray distribution can be aided by increases in agitation intensity (e.g., mixer impeller or chopper speed, drum rotation rate, or fluidization gas velocity) and by minimizing moisture losses due to spray entrainment, dripping nozzles, or powder caking on process walls.

GROWTH AND CONSOLIDATION

The evolution of the granule-size distribution of a particulate feed in a granulation process is controlled by several mechanisms, as illustrated in Table 20-39. These include the **nucleation** of fine powder to form initial primary granules, the **coalescence** of existing granules and the **layering** of raw material onto previously formed nuclei or granules. Granules may simultaneously be compacted by **consolidation** and reduced in size by **breakage.** There are strong interactions between these rate processes. Dominant mechanisms of growth and consolidation are dictated by the relationship between critical particle properties and operating variables as well as by mixing, size distribution, and the choice of processing.

Growth Physics and Contact Mechanics   In order for two colliding granules to coalesce rather than break up, the **collisional kinetic energy** must first be dissipated to prevent rebound as illustrated in Fig. 20-68. In addition, the strength of the bond must resist any subsequent breakup forces due to **bed agitation intensity.** The ability of the granules to **deform** during processing increases the bonding or **contact area** thereby dissipating breakup forces and has a large effect on growth rate. From a balance of binding and separating forces and torque acting within the area of granule contact, Ouchiyama & Tanaka [*I & EC Proc. Des. & Dev.,* **21,** 29 (1982)] derived a **critical limit of size** above which coalescence becomes impossible, given by

TABLE 20-39   Growth and Breakage Mechanisms in Granulation Processes

Granule growth		Granule breakage	
Nucleation	$jp_1 \rightarrow P_j$	Shatter	$P_j \rightarrow jp_1$
Coalescence	$P_i + P_j \rightarrow P_{i+j}$	Fragmentation	$P_j \rightarrow P_j + P_{i-j}$
Layering	$P_i + jp_1 \rightarrow P_{i+j}$	Wear	$P_i \rightarrow P_{i-j} + jp_1$
Abrasion transfer	$P_i + P_j \Big\{ \begin{array}{l} \rightarrow P_{i+1} + P_{j-1} \\ \rightarrow P_{i-1} + P_{j+1} \end{array}$ or		
Free fines P_1		Working unit P_i	

*After Sastry and Fuerstenau in *Agglomeration '77,* Sastry (ed.), *AIME,* New York, 381 (1977).

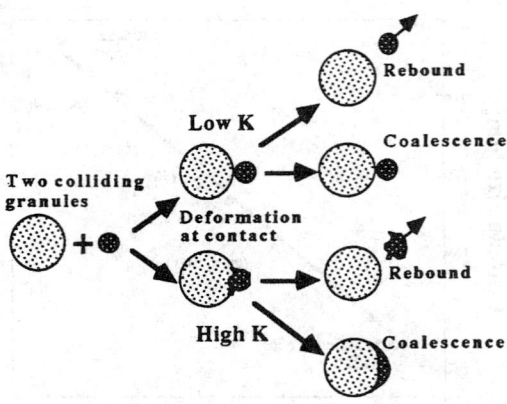

FIG. 20-68 Mechanisms of granule coalescence for low- and high-deformability systems. Rebound occurs for average granule sizes greater than the critical granule size D_c. K = deformability.

$$D_c = (AQ^{3\zeta/2}K^{3/2}\sigma_T)^{1/[4-(3/2)\eta]} \qquad (20\text{-}47)$$

K is **deformability,** a proportionality constant relating the maximum compressive force Q to the deformed-contact area, A is a constant with units of (L^3/F) which relates granule volume to impact compression force and σ_T is the **tensile strength** of the granule bond [see Eq. (20-43)]. The parameters ζ and η depend on the deformation mechanism acting within the contact area, with their values bounded by the cases of complete plastic or complete elastic deformation. For plastic deformation, $\zeta = 1$, $\eta = 0$, and $K \propto 1/H$ where H is hardness. For elastic, Hertzian deformation, $\zeta = \frac{2}{3}$, $\eta = \frac{2}{3}$, and $K \propto (1/E^*)^{2/3}$ where E^* is reduced elastic modulus. Granule deformation is generally dominated by **inelastic** behavior of the contacts during collision, with such deformation treated by the area of inelastic **contact mechanics.** [Johnson, *Contact Mechanics,* Cambridge University Press (1985).]

The value D_c represents a harmonic **average** granule size. Therefore, it is possible for the collision of two large granules to be unsuccessful, their average being beyond this critical size, whereas the collision of a large and small granule leads to successful coalescence. D_c is a strong function of moisture, as illustrated in Fig. 20-69 by the marked increase in average granule size with moisture which is related to deformability.

Granules are compacted as they collide due to bed agitation intensity. This process of **granule consolidation** expels pore fluid to the granule surface, thereby increasing local liquid saturation in the contact area of colliding granules. This surface fluid (1) increases the tensile strength of the liquid bond σ_T, and (2) increases surface plasticity and deformability K. Equation (20-47) shows that these factors generally increase the extent of granule growth, particularly for systems of low deformability. In such **low-deformability systems,** growth is largely controlled by the extent of this fluid layer. For **deformable systems,** however, the combined effect of these increases is more complex. Deformability K is related to both the **yield strength** of the material σ_y, i.e., the ability of the material to resist stresses, and the ability of the surface to be strained without degradation or rupture of the granule, with this maximum allowable **critical deformation strain** denoted by $(\Delta L/L)_c$. Figure 20-70 illustrates the stress-strain behavior of cylindrical compact agglomerates during compression as a function of liquid saturation, with **strain** denoted by $\Delta L/L$. In general, high deformability K requires low yield strength σ_y and high critical strain $(\Delta L/L)_c$. For this formulation, increasing moisture increases deformability by lowering interparticle frictional resistance which also increases mean granule size (Fig. 20-69).

Effect of Equipment Mechanical Variables   The importance of deformability to the growth process depends on **bed agitation intensity** in comparison to the strength of the formed granules for agitative granulation processes. **Low agitation intensity** processes include fluid-bed, drum, and pan granulators. **High agitation intensity** processes include pin and plow shear-type mixers, and high-shear

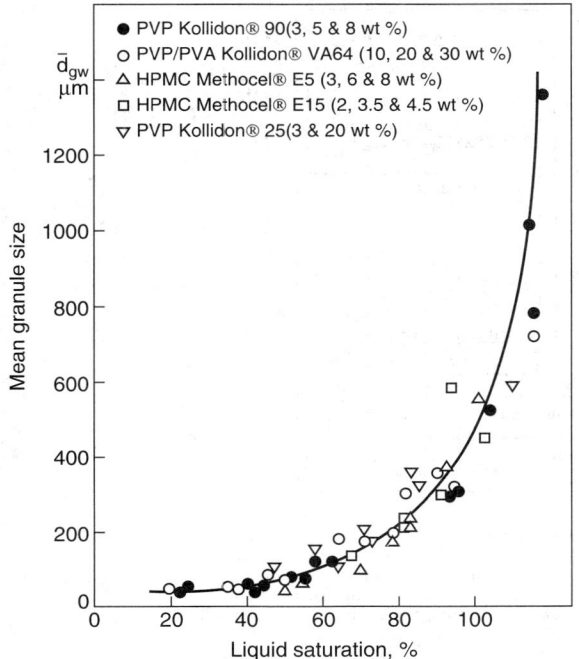

FIG. 20-69 Effect of moisture as a percentage of granule saturation on mean granule diameter, indicating the marked increase in granule deformability with increased moisture. Mean granule diameter is a measure of the critical limit of size D_c. Granulation of calcium hydrogen phosphate with aqueous binder solutions in a Fielder PMAT 25 VG high-shear mixer. [*Ritala et al., Drug Dev. & Ind. Pharm., 14(8), 1041 (1988).*]

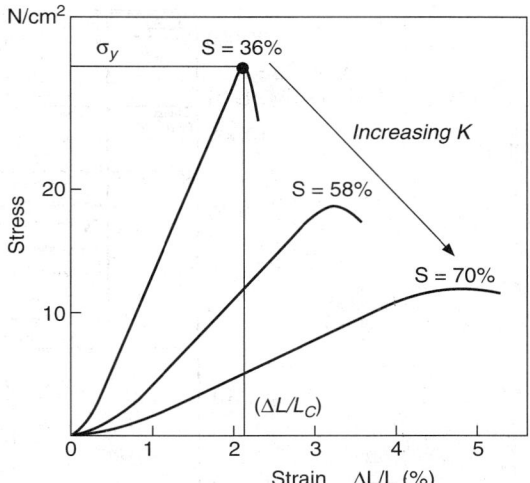

FIG. 20-70 The influence of moisture as a percentage of sample saturation S on granule deformability. Here, deformation strain ($\Delta L/L$) is measured as a function of applied stress, with the peak stress and strain denoted by tensile strength σ_y and critical strain ($\Delta L/L)_c$ of the material. Dicalcium phosphate with a 15 wt % binding solution of PVP/PVA Kollidon® VA64. [*Holm et al.,* Powder Tech., **43**, 213 (1985).] With kind permission from Elsevier Science SA, Lausanne, Switzerland.

mixers involving choppers. Bed agitation intensity and compaction pressure are controlled by **mechanical variables** of the process such as fluid-bed excess gas velocity or mixer-impeller speed. Agitation intensity controls the relative **collisional velocities** of granules within the process and therefore growth, breakage, consolidation, and final-product density. Figure 20-71 summarizes typical characteristic velocities, agitation intensities and compaction pressures, and product-relative densities achieved for a variety of size-enlargement processes.

Low-Agitation Intensity Growth   For low-agitation processes, consolidation of the granules occurs at a much slower rate than growth, and granule deformation can be ignored to a first approximation. The growth process can be modeled by the collision of two rigid granules each coated by a liquid layer of thickness h. The probability of successful coalescence is governed by a dimensionless **Stokes number St** given by

$$St = \frac{4\rho u_0 d}{9\mu} \qquad (20\text{-}48)$$

where u_0 is the relative collisional velocity of the granules, ρ is granule density, d is the harmonic average of granule diameter, and μ is the solution phase binder viscosity. The Stokes number represents the ratio of initial collisional kinetic energy to the energy dissipated by viscous lubrication forces, and it is one measure of normalized bed agitation energy. (The granule Stokes number is akin to the particle Reynolds number in two-phase systems.) Successful growth by coalescence or layering requires that

$$St < St^\circ \quad \text{where} \quad St^\circ = \left(1 + \frac{1}{e_r}\right) \ln(h/h_a) \qquad (20\text{-}49)$$

where St° is a critical Stokes number representing the energy required for rebound. The binder layer thickness h is related to liquid loading, e_r is the coefficient of restitution of the granules and h_a is a

measure of surface roughness or asperities. The critical condition given by Eq. (20-49) controls the growth of low-deformability systems (Fig. 20-68). [Ennis et al., *Powder Tech.*, **65**, 257 (1991).]

Both the **binder solution viscosity** μ and the granule density are largely properties of the feed. Binder viscosity is also a function of local temperature, collisional strain rate (for non-Newtonian binders) and binder concentration. It can be controlled as discussed above through judicious selection of binding and surfactant agents and measured by standard rheological techniques. [Bird et al., *Dynamics of Polymeric Liquids*, vol. 1, John Wiley & Sons, Inc. (1977).] The **collisional velocity** is a function of process design and operating variables, and is related to bed agitation intensity and mixing. Possible choices of u_0 are summarized in Fig. 20-71. Note that u_0 is an interparticle collisional velocity, which is not necessarily the local average granular flow velocity.

Three regimes of granule growth may be identified for low-agitation intensity processes [Ennis et al., *Powder Tech.*, **65**, 257 (1991)]. For small granules or high binder viscosity lying within a **noninertial regime** of granulation, all values of St will lie below the critical value St° and therefore all granule collisions result in successful growth *provided* binder is present. Growth rate is independent of granule kinetic energy, particle size, and binder viscosity (provided other rate processes are constant). Distribution of binding fluid then controls growth, and this is strongly coupled with the rate process of wetting. (See subsection "Wetting.") As granules grow in size, their collisional momentum increases, leading to localized regions in the process where St exceeds the critical value St°. In this **inertial regime** of granulation, granule size, binder viscosity, and collision velocity determine the proportion of the bed in which granule rebound is possible. Increases in binder viscosity and decreases in agitation intensity increase the extent of granule growth. When the spatial average of St exceeds St°, growth is balanced by granule disruption or breakup, leading to the **coating regime** of granulation. Growth continues by coating of granules by binding fluid alone. Transitions between granulation regimes depends on bed hydrodynamics. As demonstrated by Ennis et al. [*Powder Tech.*, **65**, 257 (1991)], granulation of an initially fine powder may exhibit characteristics of all three granulation regimes as time progresses, since St increases with increasing granule size. Implications and additional examples regarding the regime analysis are highlighted by Ennis [*Powder Tech.*, **88**, 203 (1996).]

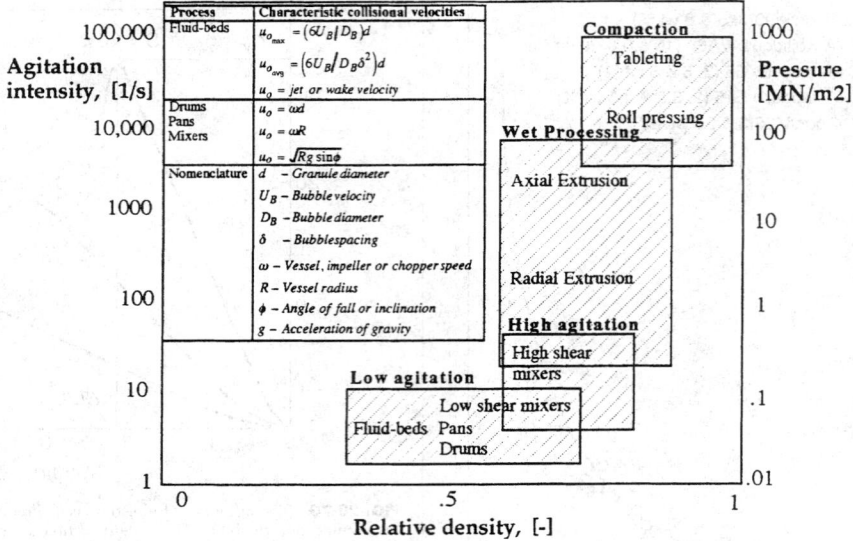

FIG. 20-71 Classification of agglomeration processes by agitation intensity and compaction pressure. Relative density is with respect to primary particle density and equals $(1 - \varepsilon)$ where ε is the solid volume fraction. Reprinted from *Granulation and Coating Technologies for High-Value-Added Industries*, Ennis and Litster (1996) with permission of E & G Associates. All rights reserved.

High-Agitation Intensity Growth For **high-agitation** processes involving **high-shear mixing** granule deformability and plastic deformation can no longer be neglected as they occur at the same rate as granule growth. Typical growth profiles for high-shear mixers are illustrated in Fig. 20-72. Two stages of growth are evident, which reveal the possible effects of binder viscosity and impeller speed. The *initial, nonequilibrium stage of growth* is controlled by granule deformability, and is of most practical significance in manufacturing. Increases in St due to lower viscosity or higher impeller speed increase the rate of growth as shown in Fig. 20-72, since the system becomes more deformable and easier to **knead** into larger-granule structures. These effects are contrary to what is predicted from the Stokes analysis based on rigid, elastic granules.

Growth continues until disruptive and growth forces are balanced in the process. This last *equilibrium stage of growth* represents a balance between dissipation and collisional kinetic energy, and so increases in St decrease the final granule size, as expected from the Stokes analysis. Note that the equilibrium-granule diameter decreases with the inverse square root of the impeller speed, as it should based on $St = St^\circ$ with $u_0 = d \cdot (du/dx) = \omega d$. The Stokes analysis is used to determine the effect of operating variables and binder viscosity on *equilibrium* growth, where disruptive and growth forces are balanced.

In the early stages of growth for high-shear mixers, the Stokes analysis in its present form is inapplicable. Freshly formed, uncompacted granules are easily deformed, and as growth proceeds and consolidation of granules occur, they will surface harden and become more resistant to deformation. This increases the importance of the elasticity of the granule assembly. Therefore, in later stages of growth, older granules approach the ideal Stokes model of rigid, elastic collisions. For these reasons, the Stokes approach has had reasonable success in providing an overall framework with which to compare a wide variety of granulating materials [Ennis, *Powder Tech.*, **88**, 203 (1996)]. In addition, the Stokes number controls in part the degree of deformation occurring during a collision since it represents the importance of collision kinetic energy in relation to viscous dissipation, although the exact dependence of deformation on St is presently unknown.

Extent of Noninertial Growth Growth by coalescence in granulation processes may be modeled by the population balance. (See the

Modeling and Simulation subsection.) It is necessary to determine both the mechanism and kernels which describe growth. For fine powders within the noninertial regime of growth, all collisions result in successful coalescence *provided* binder is present. Coalescence occurs via a random, size-independent kernel which is only a function of liquid loading y, or

$$\beta(u,v) = k = k^\circ f(y) \tag{20-50}$$

The dependence of growth on liquid loading $f(y)$ strongly depends on wetting properties. For random growth, it may be shown that the average granule size increases exponentially with time, or $d = d_0 e^{kt}$, and that the **maximum extent** of granulation $(kt)_{max}$ occurring within the noninertial regime is given by

$$(kt)_{max} = 6 \ln\left(\frac{St^\circ}{St_0}\right) \tag{20-51}$$

$$(kt)_{max} \propto \ln\left(\frac{\mu}{\rho u_0 d_0}\right) \tag{20-52}$$

St_0 is the Stokes number based on initial nuclei diameter d_0 [Adetayo et al., *Powder Tech.*, **82**, 37 (1995)]. Extent $(kt)_{max}$ depends logarithmically on binder viscosity and inversely on agitation velocity. Maximum granule size depends linearly on these variables. Also, $(kt)_{max}$ has been observed to depend linearly on liquid loading y. Therefore, the maximum granule size depends exponentially on liquid loading. Fig. 20-73 illustrates this normalization of extent $(kt)_{max}$ for the drum granulation of limestone and fertilizers.

Determination of St° The extent of growth is controlled by some limit of granule size, either reflected by the critical Stokes number St° or by the critical limit of granule size D_c. There are three possible methods to determine this critical limit. The first involves measuring the critical rotation speed for the survival of a series of liquid-binder drops during drum granulation [Ennis, Ph.D. Thesis (1990)]. A second refined version involves measuring the survival of granules in a couette-fluidized shear device [Tardos & Khan, AIChE Annual Meeting, Miami (1995)]. Both the onset of granule deformation and complete granule rupture are determined from the dependence of granule shape and the number of surviving granules,

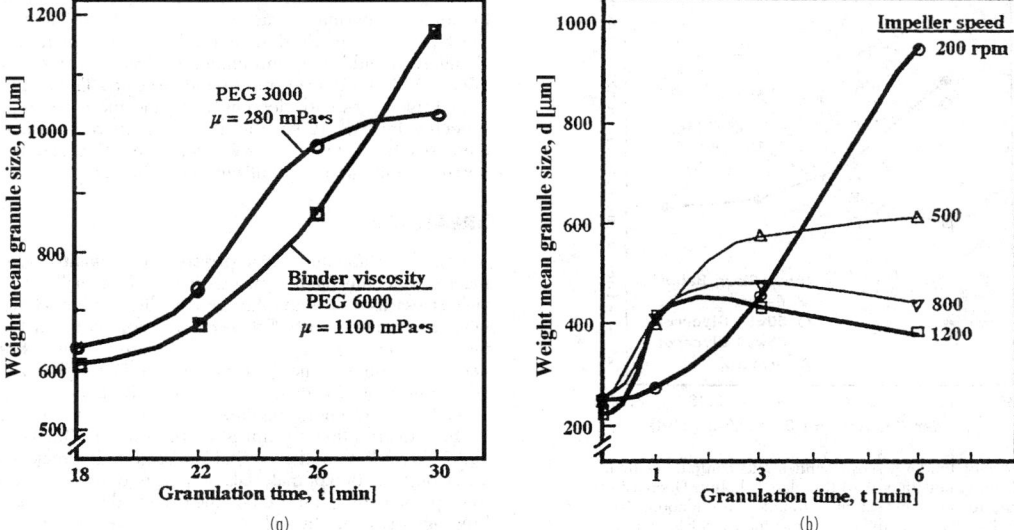

FIG. 20-72   Granule diameter as a function of time for high-shear mixer granulation, illustrating the influence of deformability on growth behavior. (*a*) 10-liter vertical high-shear melt granulation of lactose with liquid loading of 15 wt % binder and impeller speed of 1400 rpm for two different viscosity grades of polyethylene glycol binders. [*Schaefer et al.*, Drug Dev. & Ind. Pharm., *16(8), 1249 (1990)*]. (*b*) 10-liter vertical high-shear mixer granulation of dicalcium phosphate with 15 wt % binder solution of PVP/PVA Kollidon® VA64, liquid loading of 16.8 wt % and chopper speed of 1000 rpm for varying impeller speed. [*Schaefer et al.*, Pharm. Ind., *52(9), 1147 (1990)*.]

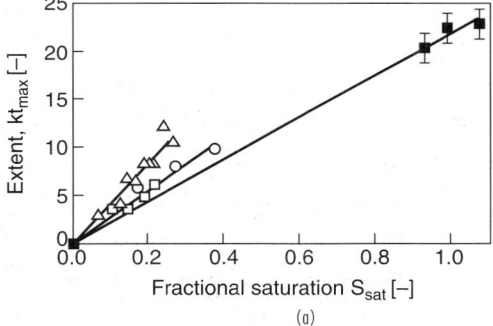

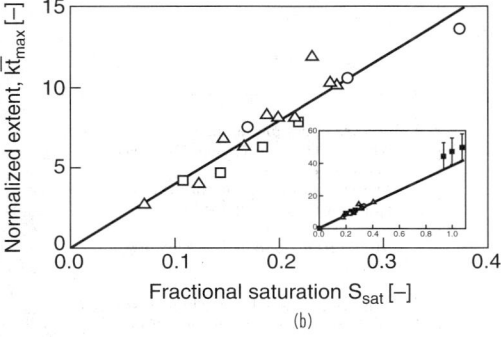

FIG. 20-73   (*a*) Maximum extent of noninertial growth $(kt)_{max}$ as a function of fractional saturation of the powder feed S_{sat} for drum granulation, and (*b*) maximum extent normalized for differences in St_0. Feed powders: ammonium sulfate (O); monoammonium phosphate (□), diammonium phosphate (△), limestone (■). [*Adetayo et al.*, Powder Tech., **82**, 37 (1995).] With kind permission from Elsevier Science SA, Lausanne, Switzerland.

respectively, on shear rate. Granule breakdown and deformation are controlled by a generalization of St, or a yield number Y given by

$$Y = \frac{\rho u_0^2}{\tau_y} = \frac{\rho (du_0/dx)^2 d^2}{\tau_y} \qquad (20\text{-}53)$$

Viscosity has been replaced by a generalized form of plastic deformation controlled by a yield stress τ_y, which may be determined by compression experiments. Compare with Eq. (20-48). The critical shear rate describing complete granule rupture defines St°, whereas the onset of deformation and the beginning of granule breakdown defines an additional critical value St^y.

The last approach is to measure the deviation in the growth-rate curve from random exponential growth [Adetayo & Ennis, *AIChE J.*, (1997)]. The deviation from random growth indicates a value of w°, or the **critical granule diameter** at which noninertial growth ends. This value is related to D_c. (See the Modeling and Simulation subsection for further discussion.)

Consolidation   Consolidation of granules determines **granule porosity,** and hence **granule density.** Granules may consolidate over extended times and achieve high densities if there is no simultaneous drying to stop the consolidation process. The extent and rate of consolidation are determined by the balance between the collision energy and the granule resistance to deformation. Decreasing feed-particle size decreases the rate of consolidation due to the high specific surface area and low permeability of fine powders. The effects of binder viscosity and liquid content are complex and interrelated. For low-viscosity binders, consolidation *increases* with liquid content, but for high-viscosity binders consolidation *decreases* with increasing liquid content (see Fig. 20-74). The exact effect of liquid content and binder viscosity is determined by the balance between viscous dissipation and particle frictional losses and is difficult to predict [Iveson et al., *Powder Tech.*, (1996)]. Increasing agitation intensity increases the degree of consolidation by increasing the energy of collision and compaction.

Controlling Growth and Consolidation   Table 20-40 summarizes typical changes in material & operating variables which maximize granule growth and consolidation. Also listed are appropriate routes to achieve these changes in a given variable through changes in either the formulation or in processing. Growth and consolidation of

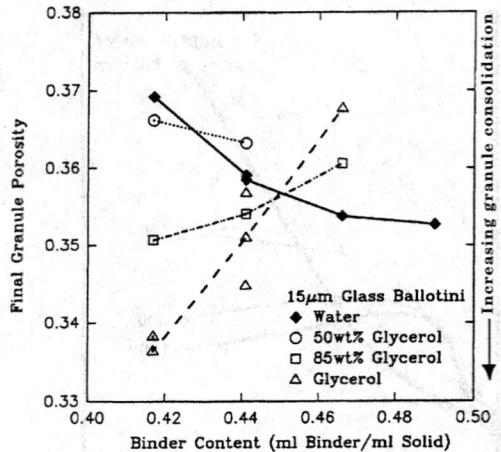

FIG. 20-74 Effect of binder viscosity and liquid content on final granule porosity for the drum granulation of 15 μm glass ballotini. Decreasing granule porosity corresponds to increasing extent of granule consolidation. [*Iveson et al.,* Powder Tech., **88,** 15 *(1996).*] With kind permission from Elsevier Science SA, Lausanne, Switzerland.

granules are strongly influenced by St. Increasing St increases energy with respect to dissipation during deformation of granules. Therefore, the rate of growth for deformable systems (e.g., deformable formulation or high-shear mixing) and the rate of consolidation of granules generally increases with increasing St. St may be increased by de-

creasing binder viscosity or increasing agitation intensity. Changes in binder viscosity may be accomplished by formulation changes (e.g., the type or concentration of binder) or by operating temperature changes. In addition, simultaneous drying strongly influences the effective binder concentration and viscosity. The *maximum* extent of growth increases with decreasing St and increased liquid loading, as reflected by Eqs. (20-50) and (20-51). Increasing particle size also increases the rate of consolidation, and this can be modified by upstream milling or crystallization conditions.

BREAKAGE

A granule is a nonuniform **physical composite** possessing certain macroscopic mechanical properties, such as a generally anisotropic yield stress, as well as an inherent flaw distribution. Hard materials may fail in tension (see brittle fracture under size reduction), with the breaking strength being much less than the inherent tensile strength of bonds because of the existence of flaws. [Lawn, *Fracture of Brittle Solids,* 2d ed., Cambridge University Press (1975).] Bulk breakage tests of granule strength measure both the inherent bond strength of granule as well as its flaw distribution. [Bemros & Bridgwater, *Powder Tech.,* **49,** 97 (1987).] In addition, the mechanism of granule breakage (Table 20-39) is a strong function of materials properties of the granule itself as well as the type of loading imposed by the test conditions. Ranking of product-breakage resistance by ad hoc tests may be test specific, and in the worst case differ from actual process conditions. Standardized mechanical tests should be employed instead to measure material properties which *minimize* the effect of flaws and loading conditions under well-defined geometries of internal stress, as described below.

Fracture Properties   Fracture toughness defines the stress distribution in the body just before fracture and is given by

$$K_c = Y\sigma_f \sqrt{\pi c} \qquad (20\text{-}54)$$

TABLE 20-40   Controlling Growth and Consolidation in Granulation Processes*

Typical changes in material or operating variables which maximize growth and consolidation	Appropriate routes to alter variable through formulation changes	Appropriate routes to alter variable through process changes
Rate of growth (low deformability): Increase rate of nuclei formation	Improve wetting properties. (See "Wetting" subsection.) Increase binder distribution.	Increase spray rate and number of drops.
Increase collision frequency		Increase mixer impeller or drum rotation speed or fluid-bed gas velocity.
Increase residence time		Increase batch time or lower feed rate.
Rate of growth (high deformability): Decrease binder viscosity	Decrease binder concentration or change binder. Decrease any diluents and polymers which act as thickeners.	Decrease operating temperature for systems with simultaneous drying. Otherwise increase temperature.
Increase agitation intensity		Increase mixer impeller or drum rotation speed or fluid-bed gas velocity.
Increase particle density Increase rate of nuclei formation, collision frequency & residence time, as above for low deformability systems.		
Extent of growth: Increase binder viscosity	Increase binder concentration, change binder, or add diluents and polymers as thickeners.	Increase operating temperature for systems with simultaneous drying. Otherwise decrease temperature.
Decrease agitation intensity		Decrease mixer impeller or drum rotation speed or fluid-bed gas velocity.
Decrease particle density Increase liquid loading		Extent observed to increase linearly with moisture.
Rate of consolidation: Decrease binder viscosity Increase agitation intensity Increase particle density Increase particle size	As above for high deformability systems. Particle size and friction strongly interact with binder viscosity to control consolidation. Feed particle size may be increased and fine tail of distribution removed.	As above for high deformability systems. In addition, increase compaction forces by increasing bed weight, or altering mixer impeller or fluid-bed distributor plate design. Size is controlled in milling and particle formation.

where σ_f is the applied fracture stress, c is the length of the crack in the body, and Y is a calibration factor introduced to account for different body geometries. (Lawn, loc. cit.)

The elastic stress cannot exceed the **yield stress** of the material, implying a region of local yielding at the crack tip. Nevertheless, to apply the simple framework of linear elastic fracture mechanics, Irwin [*J. Applied Mechanics*, **24**, 361 (1957)] proposed that this **process zone size** r_p be treated as an *effective* increase in crack length δc. Fracture toughness is then given by

$$K_c = Y\sigma_f\sqrt{\pi(c + \delta c)} \quad \text{with} \quad \delta c \sim r_p \qquad (20\text{-}55)$$

The process zone is a measure of the yield stress or plasticity of the material in comparison to its brittleness. Yielding within the process zone may take place either plastically or by diffuse microcracking, depending on the brittleness of the material. For plastic yielding, r_p is also referred to as the **plastic zone size**.

The **critical strain energy release rate** G_c is the energy equivalent to fracture toughness, first proposed by Griffith [*Phil. Trans. Royal Soc.*, **A221**, 163 (1920)]. They are related by

$$G_c = K_c^2/E \qquad (20\text{-}56)$$

Fracture Measurements In order to ascertain fracture properties in any reproducible fashion, very specific test geometry must be used since it is necessary to know the stress distribution at predefined, induced cracks of known length. Three traditional methods are (1) the **three-point bend test**, (2) **indentation fracture testing,** and (3) **Hertzian contact compression** between two spheres of the material (See fracture under size reduction). In the case of the three-point bend test, toughness is determined from the variance of fracture stress on induced crack length, as given by Eq. (20-55) where δc is initially taken as zero and determined in addition to toughness. (Ennis & Sunshine, *Tribology International*, **26**, 319 (1993).) In the case of indentation fracture, one determines **hardness** H from the area of the residual plastic impression and fracture toughness from the lengths of cracks propagating from the indent as a function of indentation load F. [Johnsson & Ennis, *Proceedings of the First International Particle Technology Forum*, vol. 2, AIChE, Denver, 178 (1994).] Hardness is a measure of the yield strength of the material. Toughness and hardness in the case of indentation are given by

$$K_c = \beta\sqrt{\frac{E}{H}}\frac{F}{c^{3/2}} \quad \text{and} \quad H \sim \frac{F}{A} \qquad (20\text{-}57)$$

Table 20-41 compares typical fracture properties of agglomerated materials. Fracture toughness K_c is seen to range from 0.01 to 0.06 MPa$\cdot$m$^{1/2}$, less than that typical for polymers and ceramics, presumably due to the high agglomerate voidage. Critical strain energy release rates G_c from 1 to 200 J/m^2, typical for ceramics but less than that for polymers. Process zone sizes δc are seen to be large and of the order of 0.1–1 mm, values typical for polymers. Ceramics on the other hand typically have process zone sizes less than 1 μm. Critical displacements required for fracture may be estimated by the ratio G_c/E, which is an indication of the **brittleness** of the material. This value was of the order of 10^{-7}–10^{-8} mm for polymer-glass agglomerates, similar to polymers, and of the order of 10^{-9} mm for herbicide bars, similar to ceram-

ics. In summary, granulated materials behave similar to brittle ceramics which have small critical displacements and yield strains but also similar to ductile polymers which have large process or plastic zone sizes.

Mechanisms of Breakage The process zone plays a large role in determining the mechanism of granule breakage (Table 20-39). [Ennis & Sunshine, loc. cit.] Agglomerates with process zones small in comparison to granule-size break by a brittle-fracture mechanism into smaller fragments, or **fragmentation** or **fracture**. On the other hand for agglomerates with process zones of the order of their size, there is insufficient volume of agglomerate to concentrate enough elastic energy to propagate gross fracture during a collision. The mechanism of breakage for these materials is one of **wear, erosion,** or **attrition** brought about by diffuse microcracking. In the limit of very weak bonds, agglomerates may also **shatter** into small fragments or primary particles.

Each mechanism of breakage implies a different functional dependence of breakage rate on material properties. For the case of **abrasive wear** of ceramics due to surface scratching by loaded indentors, Evans & Wilshaw [*Acta Metallurgica*, **24**, 939 (1976)] determined a volumetric wear rate V of

$$V = \frac{d_i^{1/2}}{A^{1/4}K_c^{3/4}H^{1/2}}P^{5/4}l \qquad (20\text{-}58)$$

where d_i is indentor diameter, P is applied load, l is wear displacement of the indentor and A is apparent area of contact of the indentor with the surface. Therefore, wear rate depends inversely on fracture toughness. For the case of fragmentation, Yuregir et al. [*Chem. Eng. Sci.*, **42**, 843 (1987)] have shown that the fragmentation rate of organic and inorganic crystals is given by

$$V \sim \frac{H}{K_c^2}\rho u^2 a \qquad (20\text{-}59)$$

where a is crystal length, ρ is crystal density, and u is impact velocity. Note that hardness plays an opposite role for fragmentation than for wear, since it acts to concentrate stress for fracture. Fragmentation rate is a stronger function of toughness as well.

Drawing on analogies with this work, the breakage rates by wear B_w and fragmentation B_f for fluid-bed processing should be of the forms:

$$B_w = \frac{d_0^{1/2}}{K_c^{3/4}H^{1/2}}h_b^{5/4}(U - U_{mf}) \qquad (20\text{-}60)$$

$$B_f \sim \frac{H}{K_c^2}\rho(U - U_{mf})^2 a \qquad (20\text{-}61)$$

where d is granule diameter, d_0 is primary particle diameter, $(U - U_{mf})$ is fluid-bed excess gas velocity, and h_b is bed height. Fig. 20-75 illustrates the dependence of erosion rate on material properties for granules undergoing a wear mechanism of breakage, as governed by Eq. (20-60).

Controlling Breakage Table 20-42 summarizes typical changes in material & operating variables which are necessary to minimize breakage. Also listed are appropriate routes to achieve these changes in a given variable through changes in either the formulation or in processing. Both fracture toughness and hardness are strongly influenced by the compatibility of the binder with the primary particles, as well as the elastic/plastic properties of the binder. In addition, hardness and toughness increase with decreasing voidage and are influenced by

TABLE 20-41 Fracture Properties of Agglomerated Materials*

Material	K_c (MPa$\cdot$m$^{1/2}$)	G_c (J/m^2)	δc (μm)	E (MPa)	G_c/E (m)
Bladex 60™†	0.070	3.0	340	567	5.29e–09
Bladex 90™†	0.014	0.96	82.7	191	5.00e–09
Glean™†	0.035	2.9	787	261	1.10e–08
Glean Aged™†	0.045	3.2	3510	465	6.98e–09
CMC-Na (M)‡	0.157	117.0	641	266	4.39e–07
Klucel GF‡	0.106	59.6	703	441	1.35e–07
PVP 360K‡	0.585	199.0	1450	1201	1.66e–07
CMC 2% 1kN‡	0.097	16.8	1360	410	4.10e–08
CMC 2% 5kN‡	0.087	21.1	1260	399	5.28e–08
CMC 5% 1kN‡	0.068	15.9	231	317	5.02e–08

*Ennis & Sunshine, *Tribology International*, **26**, 319 (1993)
†DuPont corn herbicides
‡50 μm glass beads with polymer binder

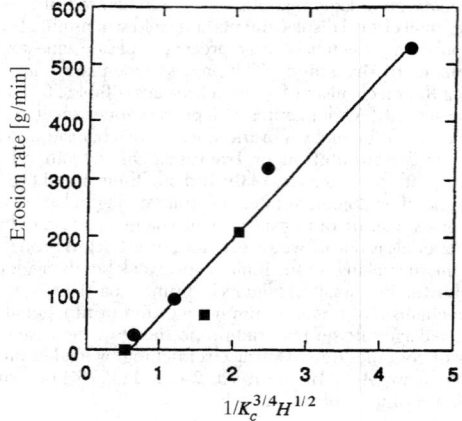

FIG. 20-75 Fluid-bed erosion or wear rate as a function of granule material properties. K_C is fracture toughness and H is hardness as measured by three-point bend tests. [*Ennis & Sunshine, Tribology International,* **26**, *319 (1993).*]

previous consolidation of the granules (see subsection of Growth and Consolidation). While the direct effect of increasing gas velocity and bed height is to increase breakage of dried granules, increases in these variables may also act to increase consolidation of wet granules, lower voidage, and therefore lower the final-breakage rate. Granule structure also influences breakage rate, e.g., a layered structure is less prone to breakage than a raspberry-shaped agglomerate. However, it

may be impossible to compensate for extremely low toughness by changes in structure. Measurements of fracture properties help define expected breakage rates for a product and aid product development of formulations.

POWDER MECHANICS & POWDER COMPACTION

The ability of powders to freely flow, easily compact, and maintain strength during stress unloading determines the success of compaction techniques of agglomeration. These attributes are strongly influenced by mechanical properties of the feed. [Brown & Richards, *Principles of Powder Mechanics,* Pergamon Press (1970).] The flow properties summarized below are also relevant to the design of bulk powder handling systems such as feeders and hoppers [see Section 21 & Carson & Marinelli, *Chemical Eng.,* April (1994)].

Powder Mechanics Measurements   As opposed to fluids, powders may withstand applied shear stress similar to a bulk solid due to **interparticle friction.** As the applied shear stress is increased, the powder will reach a maximum sustainable shear stress τ, at which point it yields or flows. This limit of shear stress τ increases with increasing applied normal load σ, with the functional relationship being referred to as a **yield locus.** A well-known example is the Mohr-Coulomb yield locus, or

$$\tau = c + \mu\sigma = c + \sigma \tan \phi \qquad (20\text{-}62)$$

Here, μ is the coefficient of internal friction, ϕ is the internal angle of friction, and c is the shear strength of the powder in the absence of any applied normal load. The yield locus of a powder may be determined from a **shear cell,** which typically consists of a cell composed of an upper and lower ring. The normal load is applied to the powder vertically while shear stresses are measured while the lower half of the cell is either translated or rotated [Carson & Marinelli, loc. cit.]. Over-

TABLE 20-42   Controlling Breakage in Granulation Processes*

Typical changes in material or operating variables which minimize breakage	Appropriate routes to alter variable through formulation changes	Appropriate routes to alter variable through process changes
Increase fracture toughness Maximize overall bond strength Minimize agglomerate voidage	Increase binder concentration or change binder. Bond strength strongly influenced by formulation and compatibility of binder with primary particles.	Decrease binder viscosity to increase agglomerate consolidation by altering process temperatures (usually decrease for systems with simultaneous drying). Increase bed-agitation intensity (e.g., increase impeller speed, increase bed height) to increase agglomerate consolidation. Increase granulation-residence time to increase agglomerate consolidation, but minimize drying time.
Increase hardness to reduce wear Minimize binder plasticity Minimize agglomerate voidage	Increase binder concentration or change binder. Binder plasticity strongly influenced by binder type.	See above effects which decrease agglomerate voidage.
Decrease hardness to reduce fragmentation Maximize binder plasticity Maximize agglomerate voidage	Change binder. Binder plasticity strongly influenced by binder type. Apply coating to alter surface hardness.	Reverse the above effects to increase agglomerate voidage.
Decrease load to reduce wear	Lower-formulation density.	Decrease bed-agitation and compaction forces (e.g., mixer impeller speed, fluid-bed height, bed weight, fluid-bed excess gas velocity, drum rotation speed).
Decrease contact displacement to reduce wear		Decrease contacting by lowering mixing and collision frequency (e.g., mixer impeller speed, fluid-bed excess gas velocity, drum rotation speed).
Decrease impact velocity to reduce fragmentation	Lower-formulation density.	Decrease bed-agitation intensity (e.g., mixer impeller speed, fluid-bed excess gas velocity, drum rotation speed). Also strongly influenced by distributor-plate design in fluid-beds, or impeller and chopper design in mixers.

* Reprinted from *Granulation and Coating Technologies for High-Value-Added Industries,* Ennis and Litster (1996) with permission of E & G Associates. All rights reserved.

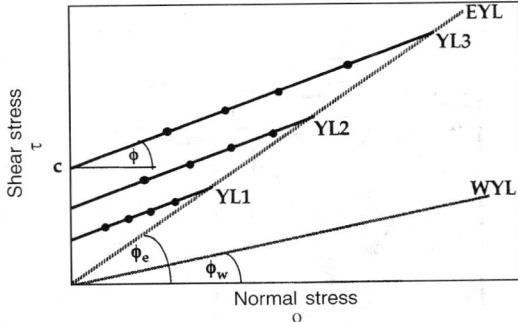

FIG. 20-76 The yield loci of a powder, reflecting the increased shear stress required for flow as a function of applied normal load. YL1 through YL3 represent yield loci for increasing previous compaction stress. EYL and WYL are the effective and wall yield loci, respectively.

compacted powders dilate when sheared, and the ability of powders to change volume with shear results in the powder's shear strength τ being a strong function of previous compaction. There are therefore a series of *yield loci* (YL) as illustrated in Fig. 20-76 for increasing previous **compaction stress.** The individual yield loci terminate at a critical state of stress, which when joined together form the *effective yield locus* (EYL) which typically passes through the stress-strain origin, or

$$\tau = \mu_e \sigma = \sigma \tan \phi_e \qquad (20\text{-}63)$$

This line represents the critical shear stress that a powder can withstand which has not been over or underconsolidated, i.e., the stress typically experienced by a powder which is in a constant state of shear. When sheared powders also experience friction along a wall, this relationship is described by the wall yield locus, or

$$\tau = \mu_w \sigma = \sigma \tan \phi_w \qquad (20\text{-}64)$$

The angles ϕ_e and ϕ_w are the **effective angle** and **wall angle of friction,** respectively.

A powder's strength increases significantly with increasing previous compaction. The relationship between the unconfined yield stress f_c, or a powder's strength, and compaction pressure σ_c is described by the powder's **flow function** *FF*. The flow function is the paramount characterization of powder strength and flow properties, and it is calculated from the yield loci determined from shear cell measurements. [Jenike, *Storage and Flow of Solids*, Univ. of Utah, *Eng. Exp. Station Bulletin*, no. 123, November (1964). See also Sec. 21 on storage bins, silos, and hoppers.]

Compact Density Compact strength depends on the number and strength of interparticle bonds (Eq. 20-43) created during consolidation, and both generally increase with increasing compact density. Compact density is in turn a function of the maximum pressure achieved during compaction. The **mechanisms of compaction** have been discussed by Cooper & Eaton [*J. Am. Ceramic Soc.*, **45**, 97 (1962)] in terms of two largely independent, probabilistic processes. The first is the filling of large holes with particles from the original size distribution. The second is the filling of holes smaller than the original particles by **plastic flow** or **fragmentation.** Additional, possible mechanisms include the low pressure elimination of arches and cavities created during die filling due to wall effects, and the final high-pressure consolidation of the particle phase itself. As these mechanisms manifest themselves over different pressure ranges, four stages of compression are generally observed in the compressibility diagram when density is measured over a wide pressure range (Fig. 20-77). The slope of the intermediate- and high-pressure regions is defined as $1/\kappa$ where κ is the **compressibility** of the powder. The density at an arbitrary pressure σ is given by a compaction equation of the form

$$\rho = \rho_0 \left[\frac{\sigma}{\sigma_0} \right]^{1/\kappa} \qquad (20\text{-}65)$$

where ρ_0 is the density at an arbitrary pressure or stress σ_0. For a complete review of compaction equations, see Kawakita & Lüdde [*Powder*

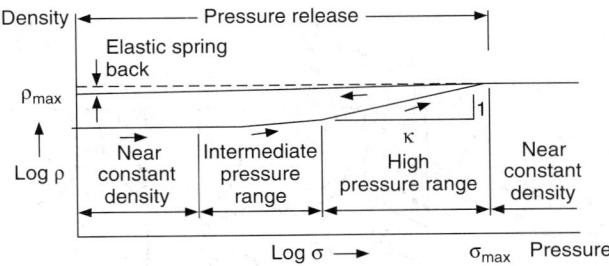

FIG. 20-77 Compressibility diagram of a typical powder illustrating four stages of compaction.

Tech., **4**, 61 (1970/71)] and Hersey et al. [*Proceedings of First International Conf. of Compaction & Consolidation of Particulate Matter,* Brighton, 165 (1972)].

Transmission of Forces As pressure is applied to a powder in a die or roll press, various zones in the compact are subjected to differing intensities of pressure and shear. Compaction stress decreases with axial distance from the applied pressure [Strijbos *et al., Powder Tech.*, **18**, 187 & 209 (1977)] due to **frictional properties** of the powder and die wall. For example, the axial pressure experienced within a cylindrical die with an applied axial load σ_0 may be estimated to a first approximation by

$$\sigma_z = \sigma_0 e^{-(4\mu_w \lambda_\phi/D)z} \qquad (20\text{-}66)$$

where D is die diameter, z is axial distance from the applied load, λ is the ratio of radial to axial pressure given by $\lambda \approx K_\phi = (1 - \sin \phi_e)/(1 + \sin \phi_e)$, and μ_w and ϕ_e are the wall friction coefficient and effective angle of friction defined above (Eqs. 20-63 and 64). Typical pressure and density distributions for uniaxial compaction are shown in Fig. 20-78. High- and low-density annuli are apparent along the die corners, with a dense axial core in the lower part of the compact and a low-density core just below the moving upper punch. These density variations are due to the formation of a dense conical wedge acting along the top punch (A) with a resultant force directed toward the center of the compact (B). The wedge is densified to the greatest extent by the shearing forces developed by the axial motion of the upper punch along the stationary wall, whereas the corners along the bottom stationary die are densified the least (C). The lower axial core (B) is densified by the wedge, whereas the upper low-density region (D) is shielded by the wedge from the full axial compressive force. These variations in compact density lead to local variations in strength as well as differential zones of expansion upon compact unloading, which in turn can lead to flaws in the compact.

Compact Strength Both **particle size** and **bond strength** control final compact strength (Eq. 20-43). Although particle surface energy and elastic deformation play a role, **plastic deformation** at particle contacts is likely the major mechanism contributing to large *permanent* bond formation and successful compaction in practice. Figure 20-79 illustrates the strength of mineral compacts of varying hardness and size cut. To obtain significant strength, Benbow [*Enlargement & Compaction of Particulate Solids*, Stanley-Wood (ed.), Butterworths, 169 (1983)] found that a **critical yield pressure** must be exceeded which was independent of size but increased linearly with particle hardness. Strength also increases linearly with compaction pressure, with the slope inversely related to particle size. Similar results were obtained by others for ferrous powder, sucrose, sodium chloride, and coal [Hardman & Lilly, *Proc. Royal Soc. A., 333,* 183 (1973)]. Particle hardness and elasticity may be characterized directly by nanoindentation [Johnson & Ennis, *Proceedings of the First International Particle Technology Forum*, vol. 2, AIChE, Denver, 178 (1994)].

The development of flaws and the loss of interparticle bonding during decompression substantially weaken compacts (see breakage subsection). **Delamination** during load removal involves the fracture of the compact into layers, and it is induced by strain recovery in excess of the elastic limit of the material which cannot be accommodated by

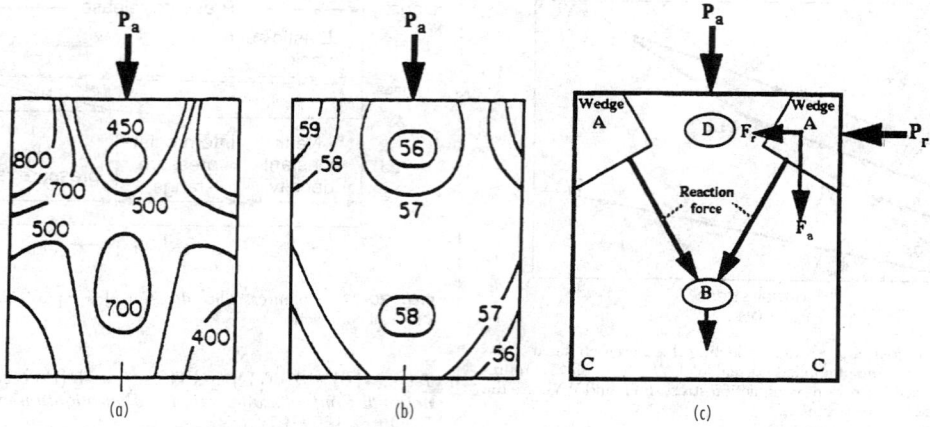

FIG. 20-78 Reaction in compacts of magnesium carbonate when pressed ($P_a = 671$ kg/cm²). (*a*) Stress contour levels in kilograms per square centimeter. (*b*) Density contours in percent solids. (*c*) Reaction force developed at wedge responsible for stress and density patterns. [*Train, Trans. Inst. Chem. Eng. (London), 35, 258 (1957).*]

plastic flow. Delamination also occurs during compact ejection, where the part of the compact which is clear of the die elastically recovers in the radial direction while the lower part remains confined. This differential strain sets up shear stresses causing fracture along the top of the compact referred to as **capping.**

Hiestand Tableting Indices Likelihood of failure during decompression depends on the ability of the material to relieve elastic stress by plastic deformation without undergoing brittle fracture, and this is time dependent. Those which relieve stress rapidly are less

likely to cap or delaminate. Hiestand & Smith [*Powder Tech.*, **38**, 145 (1984)] developed three pharmaceutical **tableting indices,** which are applicable for general characterization of powder compactability. The **strain index** (SI) is a measure of the elastic recovery following plastic deformation, the **bonding index** (BI) is a measure of plastic deformation at contacts and bond survival, and the **brittle fracture index** (BFI) is a measure of compact brittleness.

Compaction Cycles Insight into compaction performance is gained from direct analysis of pressure/density data over the cycle of

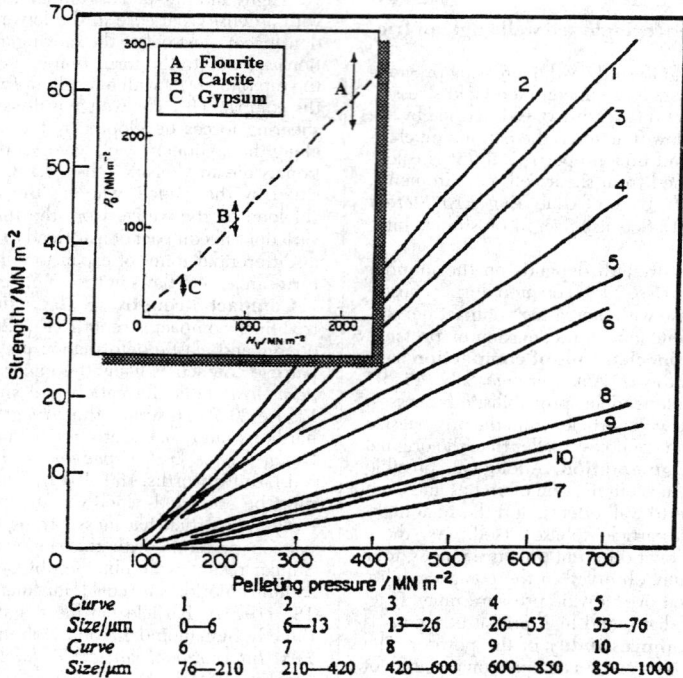

Curve	1	2	3	4	5
Size/μm	0—6	6—13	13—26	26—53	53—76
Curve	6	7	8	9	10
Size/μm	76—210	210—420	420—600	600—850	850—1000

FIG. 20-79 Effect of pelleting pressure on axial crushing strength of compacted calcite particles of different sizes demonstrating existence of a critical yield pressure. Inset shows the effect of hardness on critical yield pressure. [*Benbow, Enlargement and Compaction of Particulate Solids, Stanley-Wood (ed.), Butterworths, 169 (1983).*]

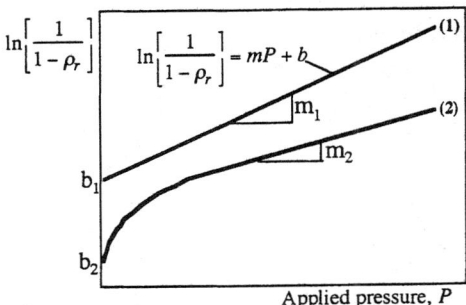

FIG. 20-80 Heckel profiles of the unloaded relative compact density for (1) a material densifying by pure plastic deformation, and (2) a material densifying with contributions from brittle fragmentation and particle rearrangement.

axial compact compression and decompression. Figure 20-80 illustrates typical **Heckel profiles** for plastic and brittle deforming materials which are determined from density measurements of *unloading* compacts. The slope of the curves gives an indication of the yield pressure of the particles. The contribution of fragmentation and rearrangement to densification is indicated by the low-pressure deviation from linearity. In addition, elastic recovery contributes to the degree of hysterisis which occurs in the *at-pressure* density curve during compression followed by decompression. [Doelker, *Powder Technology & Pharmaceutical Processes*, Chulia et al. (eds.), Elsevier, 403 (1994).]

Powder Feeding Bulk density control of feed materials and reproducible powder feeding are crucial to the smooth operation of compaction techniques. In the case of pharmaceutical tableting, reproducible tablet weights with variations of less than 2 percent are required. Feeding problems are most acute for direct powder-filling of compression devices, as opposed to granular feeds. **Flowability data** developed from shear cell measurements are invaluable in designing machine hoppers for device filling. In addition, complex-force feeding systems have been developed to aid filling and ensure uniform bulk density. Lubricants or glidants such as fumed silica and talc are also added in small amounts to improve flow properties. These additives act to modify coefficients of friction of the particulate feed.

Controlling Powder Compaction Compaction properties of powders are generally improved by altering flow properties. In particular, stress transmission improves with either low wall-friction angle or raising the internal angle of friction of the powder. **Internal lubricants** may be mixed with the feed material to be compacted. They may aid stress transmission by reducing the wall angle of friction of the material, but may also weaken bonding properties and the unconfined yield stress of the powder. **External lubricants** are applied to the die surface. They reduce sticking to the die and aid stress transmission by reducing the wall angle of friction of the material.

Binders improve the strength of compacts through increased plastic deformation or chemical bonding. They may be classified as **matrix type, film type,** and **chemical.** Komarek [*Chem. Eng.,* **74**(25), 154 (1967)] provides a classification of binders and lubricants used in the tableting of various materials.

Particle properties such as size, shape, elastic/plastic properties, and surface properties are equally important. Their direct effect, however, is difficult to ascertain without thorough particle characterization. Generally, friction coefficients decrease with decreasing particle hardness. Increasing particle size and decreasing the spread of the size distribution lowers powder friction, thereby aiding stress transmission. However, increasing particle size may also lower compact strength, as described by Eq. (20-43).

SIZE ENLARGEMENT EQUIPMENT AND PRACTICE

GENERAL REFERENCES: Ball et al., *Agglomeration of Iron Ores,* Heinemann, London, 1973. Capes, *Particle Size Enlargement,* Elsevier, New York, 1980. Ennis & Litster, *The Science & Engineering of Granulation Processes,* Blackie Academic Ltd., 1997. Knepper (ed.) *Agglomeration,* Interscience, New York, 1962. Kristensen, *Acta Pharm. Suec.,* **25,** 187, 1988. Pietsch, *Size Enlargement by Agglomeration,* John Wiley & Sons Ltd., Chichester, 1992. Pietsch, *Roll Pressing,* Heyden, London, 1976. Sastry (ed.), *Agglomeration 77,* AIME, New York, 1977. Sherrington & Oliver, *Granulation,* Heyden, London, 1981. Stanley-Wood, *Enlargement and Compaction of Particulate Solids,* Butterworth & Co. Ltd., 1983.

Particle size enlargement equipment can be classified into several groups, with advantages, disadvantages, and applications summarized in Table 20-36. Comparisons of bed-agitation intensity, compaction pressures, and product bulk density for selected agglomeration processes are highlighted above in Fig. 20-71.

Particle size terminology is industry specific. In the following discussion, particle size enlargement in tumbling, mixer and fluidized-bed granulators is referred to as **granulation.** Granulation includes **pelletization** or **balling** as used in the iron-ore industry, but does not include the breakdown of compacts as used in some tableting industries. The term pelleting or pelletization will be used for extrusion processes only.

TUMBLING GRANULATORS

In **tumbling granulators,** particles are set in motion by the tumbling action caused by the balance between gravity and centrifugal forces. The most common types of tumbling granulators are **disc** and **drum** granulators. Their use is widespread, including the iron-ore industry (where the process is sometimes called **balling** or **wet pelletization**), fertilizer manufacture, agricultural chemicals and pharmaceuticals.

Tumbling granulators generally produce granules in the size range 1 to 20 mm and are not suitable for making granules smaller than 1 mm. Granule density generally falls between that of fluidized-bed and mixer granulators (Fig. 20-71), and it is difficult to produce highly porous agglomerates in tumbling granulators. Tumbling equipment is also suitable for coating large particles, but it is difficult to coat small particles, since growth by coalescence of the seed particles is hard to control.

Drum and disc granulators generally operate in **continuous feed** mode. A key advantage to these systems is the ability to run at large scale. Drums with diameters up to four meters and throughputs up to 100 ton/hr are widely used in the mineral industry.

Disc Granulators Figure 20-81 shows the elements of a **disc granulator.** It is also referred to as a **pelletizer** in the iron-ore industry or a **pan granulator** in the agricultural chemical industry. The equipment consists of a rotating, tilted disc or pan with a rim. Solids and wetting agents are continuously added to the disc. A coating of the feed material builds up on the disc and the thickness of this layer is controlled by scrapers or a plow, which may oscillate mechanically. The surface of the pan may also be lined with expanded metal or an abrasive coating to promote proper lifting and cascading of the particulate bed, although this is generally unnecessary for fine materials. Solids are typically fed by either volumetric or gravimetric feeders. Gravimetric feeding generally improves granulation performance due to smaller fluctuations in feed rates which act to disrupt rolling action in the pan. Wetting fluids which promote growth are generally applied by a series of single-fluid spray nozzles distributed across the face of the bed. Solids feed and spray nozzle locations have a pronounced effect on granulation performance and granule structure.

Variations of the simple pan shape include (1) an outer reroll ring which allows granules to be simultaneously coated or densified without further growth, (2) multistepped sidewalls, and (3) a pan in the form of a truncated cone (Capes, *Particle Size Enlargement,* Elsevier, 1980).

The required disc-rotation speed is given in terms of the **critical speed,** i.e., the speed at which a single particle is held stationary on

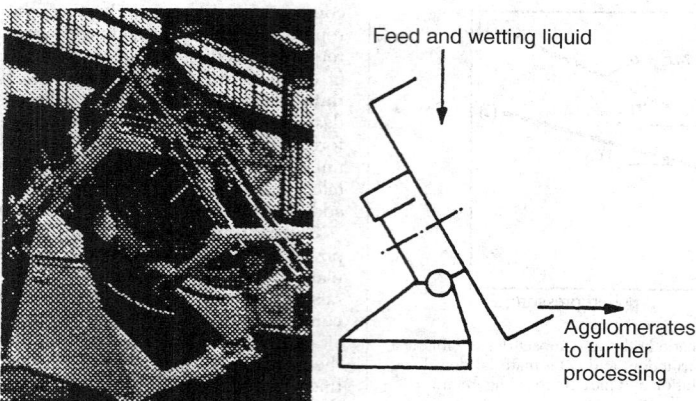

FIG. 20-81   A typical disc granulator (*Capes*, Particle Size Enlargement, *Elsevier, 1980*).

the rim of the disc due to centripetal forces. The critical speed N_c is given by:

$$N_c = \sqrt{\frac{g \sin \phi}{2\pi^2 D}} \qquad (20\text{-}67)$$

where g is the gravitational acceleration, ϕ is the angle of the disc to the horizontal, and D is the disc diameter. The typical operating range for discs is 50 to 75 percent of critical speed, with angles ϕ of 45–55°. This range ensures a good tumbling action. If the speed is too low, sliding will occur. If the speed is too high, particles will be thrown off the disc or openings develop in the bed, allowing spray blowthrough and uneven buildup on the disc bottom. Proper speed is influenced by flow properties of the feed materials in addition to granulation performance.

Discs range in size from laboratory models 30 cm in diameter up to production units of 10 meters in diameter with throughputs of 100 ton/hr. Figure 20-82 shows throughput capacities for discs of varying diameter for different applications and formulation feed densities. When scaling up from laboratory or pilot tests it is usual to keep the

same disc angle and fraction of critical speed. Power consumption and throughput are approximately proportional to the square of disc diameter, and disc height is typically 10-20 percent of diameter. It should be emphasized that these relationships are best used as a guide and in combination with actual experimental data on the system in question in order to indicate the approximate effect of scale-up.

A key feature of disc operation is the inherent **size classification** (Fig. 20-83). Product overflows from the **eye** of the tumbling granules where the large granules are segregated. The overflow size distribution is narrow compared to drum granulators, and discs typically operate with little or no pellet recycle. Due to this segregation, positioning of the feed and spray nozzles is key in controlling the balance of granulation-rate processes.

Total holdup and granule residence time distribution vary with changes to operating parameters which affect granule motion on the disc. Total holdup (mean residence time) increases with decreasing pan angle, increasing speed and increasing moisture content. The residence time distribution for a disc lies between the mixing extremes of plug flow and completely mixed. Increasing the disc angle narrows the residence time distribution. Several mixing models for disc granulators have been proposed (see Table 20-58 in modeling and simulation subsection). One to two-minute residence times are common.

Drum Granulators   Granulation drums are common in the metallurgical and fertilizer industries and are primarily used for very large throughput applications (see Table 20-43). In contrast to discs, there is no output size classification and high recycle rates of off-size product are common. As a first approximation, granules can be considered to flow through the drum in plug flow, although backmixing to some extent is common.

A granulation drum consists of an inclined cylinder which may be either open-ended or fitted with annular retaining rings. Feeds may

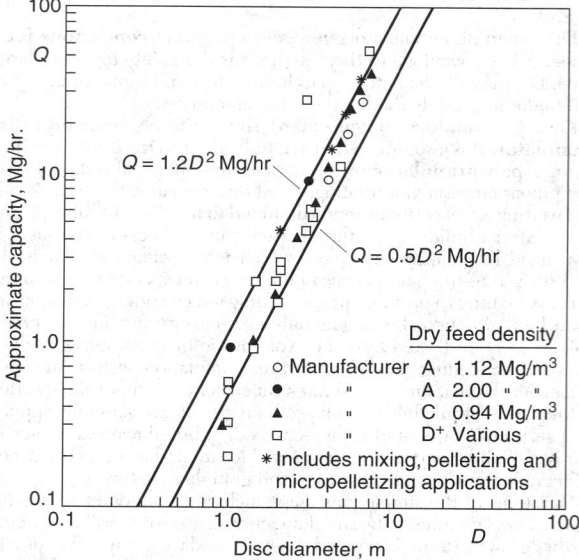

FIG. 20-82   Capacity of inclined disc granulators of varying diameter and formulation feed densities. (*Capes*, Particle Size Enlargement, *Elsevier, 1980.*)

In the figure:
- $Q = 1.2D^2$ Mg/hr
- $Q = 0.5D^2$ Mg/hr
- Axes: Approximate capacity, Mg/hr (Q) vs. Disc diameter, m (D)

Dry feed density

○	Manufacturer	A	1.12 Mg/m³
●	"	A	2.00 " "
▲	"	C	0.94 Mg/m³
□	"	D⁺	Various
*	Includes mixing, pelletizing and micropelletizing applications		

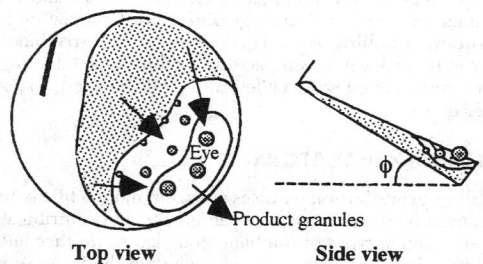

FIG. 20-83   Granule segregation on a disc granulator, illustrating a size-classified granular bed sitting on ungranulated feed powder. Reprinted from *Granulation and Coating Technologies for High-Value-Added Industries*, Ennis and Litster (1996) with permission of E & G Associates. All rights reserved.

TABLE 20-43   Characteristics of Large-Scale Granulation Drums*

Diameter, (ft)	Length, (ft)	Power, (hp)	Speed, (rpm)	Approximate capacity, (ton/hr)†
Fertilizer granulation				
5	10	15	10–17	7.5
6	12	25	9–16	10
7	14	30	9–15	20
8	14	60	20–14	25
8	16	75	20–14	40
10	20	150	7–12	50
Iron-ore balling				
9	31	60	12–14	54
10	31	60	12–14	65
12	33	75		98

*Capes, *Particle Size Enlargement*, Elsevier (1980)
†Capacity excludes recycle. Actual drum throughput may be much higher.
NOTE: To convert feet to centimeters, multiply by 30.48; to convert tons/hr to megagrams per hour, multiply by 0.907; and to convert horsepower to kilowatts, multiply by 0.746.

be either premoistened by mixers to form granule nuclei, or liquid may be sprayed onto the tumbling bed via nozzles or distributor-pipe systems. Drums are usually tilted longitudinally a few degrees from the horizontal (0–10°) to assist flow of granules through the drum. The critical speed for the drum is calculated from Eq. 20-67 with $\phi = 80$–$90°$. To achieve a cascading, tumbling motion of the load, drums operate at lower fractions of critical speed than discs, typically 30–50% of N_c. Scrapers of various designs are often employed to control buildup of the drum wall. Holdup in the drum is between 10 and 20% of the drum volume. Drum length ranges from 2–5 times diameter, and power and capacity scale with drum volume. Holdup and mean residence time are controlled by drum length, with difficult systems requiring longer residence times than those that agglomerate readily. One to five-minute residence times are common.

A variation of the basic cylindrical shape is the multicone drum which contains a series of compartments formed by annular baffles [Stirling, in Knepper (ed.), *Agglomeration*, Interscience, New York, 1962].

Drum granulation plants often have significant recycle of undersize, and sometimes crushed oversize granules. Recycle ratios between 2:1 and 5:1 are common in iron-ore balling and fertilizer granulation circuits. This large recycle stream has a major effect on circuit operation, stability, and control. A surge of material in the recycle stream affects both the moisture content and the size distribution in the drum. Surging and limit-cycle behavior are common. There are several possible reasons for this, including:

1. A shift in controlling mechanism from coalescence to layering when the ratio of recycled pellets to new feed changes [Sastry and Fuerstenau, *Trans. Soc. Mining Eng.*, AIME, **258**, 335–340 (1975)].

2. Significant changes in the moisture content in the drum due to recycle fluctuations (recycle of dry granules in fertilizer granulation) [Zhang et al., *Control of Particulate Processes IV* (1995)].

In many cases, plants simply live with these problems. However, use of modern model-based control schemes in conjunction with improved methods for on-line moisture and particle size analysis can help overcome these effects [Ennis (ed.), *Powder Tech.*, **82** (1995); Zhang et al., *Control of Particulate Processes IV* (1995)].

Granulation Rate Processes and Effect of Operating Variables   Granulation rate processes have been discussed in detail above (see "Agglomeration Rate Processes"). Nucleation, coalescence, consolidation, and layering are all important processes in tumbling granulation. In tumbling granulators, the growth processes are complex for a number of reasons:

1. Granules remain wet and can deform and consolidate. The behavior of a granule is therefore a function of its history.

2. Different granulation behavior is observed for broad- and narrow-size distributions.

3. There is often complex competition between growth mechanisms.

Consolidation of the granules in tumbling granulators directly determines granule density and porosity. Since there is typically no in situ drying to stop the consolidation process, granules consolidate over extended times. Increasing size, speed, and angle of drums and discs will increase the rate of consolidation. Increasing residence time through lower feed rates will increase the extent of consolidation. With disc granulators, residence time can also be increased by increasing bed depth (controlled by bottom inserts), raising disc speed, or lowering disc angle. With drum granulators, residence time can be influenced by internal baffling.

Growth rate is very sensitive to liquid content for narrow initial-size distributions, with increases in liquid content for fine powders leading to an approximate exponential increase in granule size. For low-viscosity liquids, granulation occurs when very close to the saturation of the granule. This leads to the following equation to estimate moisture requirements [Capes, *Particle Size Enlargement*, Elsevier (1980)]:

$$w = \frac{\varepsilon\rho_l}{\varepsilon\rho_l + (1-\varepsilon)\rho_s}$$

$$w = \frac{1}{1 + 1.85(\rho_s/\rho_l)}; \quad d_p < 30 \ \mu m \qquad (20\text{-}68)$$

$$w = \frac{1}{1 + 2.17(\rho_s/\rho_l)}; \quad d_p > 30 \ \mu m$$

where w is the weight fraction of the liquid, ε is the porosity of the close-packed material, ρ_s is true particle density, ρ_l is liquid density, and d_p is the average size of the feed material. Equation (20-68) is suitable for preliminary mass-balance requirements for liquid binders with similar properties to water. If possible, however, the liquid requirements should be measured in a balling test on the material in question, since unusual packing and wetting effects, particle internal porosity, and solubility, air inclusions, etc. may cause error. Approximate moisture requirements for balling several materials are given in Table 20-44. In addition, for materials containing soluble constituents, such as fertilizer formulations, the total solution-phase ratio controls growth, and not simply the amount of binding fluid used.

When fines are recycled, as in iron ore, sinter feed, or fertilizer drum granulation, they are rapidly granulated and removed from the

TABLE 20-44   Moisture Requirements for Granulating Various Materials

Raw material	Approximate size of raw material, less than indicated mesh	Moisture content of balled product, wt % H_2O
Precipitated calcium carbonate	200	29.5–32.1
Hydrated lime	325	25.7–26.6
Pulverized coal	48	20.8–22.1
Calcined ammonium metavanadate	200	20.9–21.8
Lead-zinc concentrate	20	6.9–7.2
Iron pyrite calcine	100	12.2–12.8
Specular hematite concentrate	150	8.0–10.0
Taconite concentrate	150	8.7–10.1
Magnetic concentrate	325	9.8–10.2
Direct-shipping open-pit iron ores	10	10.3–10.9
Underground iron ore	¼ in.	10.4–10.7
Basic oxygen converter fume	1 μ	9.2–9.6
Raw cement meal	150	13.0–13.5
Fly ash	150	24.9–25.8
Fly ash-sewage sludge composite	150	25.7–27.1
Fly ash-clay slurry composite	150	22.4–24.9
Coal-limestone composite	100	21.3–22.8
Coal-iron ore composite	48	12.8–13.9
Iron ore-limestone composite	100	9.7–10.9
Coal-iron ore-limestone composite	14	13.3–14.8

*Dravo Corp.

distribution up to some critical size, which is a function of both moisture content and binder viscosity. Changing the initial-size distribution changes the granule porosity and hence moisture requirements [Adetayo et al., *Chem. Eng. Sci.*, **48**, 3951 (1993)]. Since recycle rates in drum systems are high, differences in size distribution between feed and recycle streams are one source of the limit-cycle behavior observed in practice.

Growth by layering is important for the addition of fine powder feed to recycled well-formed granules in drum granulation circuits and for disc granulators. In each case, layering will compete with nuclei formation and coalescence as growth mechanisms. Layered growth leads to a smaller number of larger, denser granules with a narrower-size distribution than growth by coalescence. Layering is favored by a high ratio of pellets to new feed, low moisture and positioning powder feed to fall onto tumbling granules.

Granulator-Dryers for Layering and Coating Some designs of tumbling granulators also act as dryers specifically to encourage layered growth or coating and discourage coalescence or agglomeration, e.g., the fluidized-drum granulator [Anon, *Nitrogen*, **196**, 3–6 (1992)]. These systems have drum internals designed to produce a falling curtain of granules past an atomized feed solution or slurry. Layered granules are dried by a stream of warm air before circulating through the coating zone again. Applications are in fertilizer and industrial chemicals manufacture. Analysis of these systems is similar to fluidized-bed granulator dryers.

In the pharmaceutical industry, pan granulators are still widely used for coating application. Pans are suitable for coating only relatively large granules or tablets. For smaller particles, the probability of coalescence is too high.

Relative Merits of Disc versus Drum Granulators The principal difference between disc and drum granulators is the classifying action of the disc, resulting in disc granulators having narrower exit-granule-size distributions than drums. This alleviates the need for product screening and recycle for disc granulators in some industries with only moderate-size specifications. For industries with tighter specifications, however, recycle rates are rarely more than 1:2 compared to drum recycle rates often as high as 5:1. The classified mixing action of the disc affects product bulk density, growth mechanisms, and granule structure as well. Generally, drum granulators produce denser granules than discs. Control of growth mechanisms on discs is complex, since regions of growth overlap and mechanisms compete. Both layered and partially agglomerated structures are therefore possible in disc granulators.

Other advantages claimed for the disc granulator include low equipment cost, sensitivity to operating controls, and easy observation of the granulation/classification action, all of which lend versatility in agglomerating many different materials. Dusty materials and chemical reactions such as the ammoniation of fertilizer are handled less readily in the disc granulator than in the drum.

Advantages claimed for the drum granulator over the disc are greater capacity, longer retention time for materials difficult to agglomerate or of poor flow properties, and less sensitivity to upsets in the system due to the damping effect of a large, recirculated load. Disadvantages are high recycle rates which can promote limit-cycle behavior or degradation of properties of the product.

MIXER GRANULATORS

Mixer granulators contain an agitator to mix particles and liquid to cause granulation. In fact, mixing any wet solid will cause some granulation, even if unintentionally. Mixer granulators have a wide range of applications including pharmaceutical, agrochemical, and detergent (Table 20-36), and they have the following advantages:
- They can process plastic, sticky materials.
- They can spread viscous binders.
- They are less sensitive to operating conditions than tumbling granulators.
- They can produce small (<2 mm) high-density granules.

Power and maintenance costs are higher than for tumbling granulators. Outside of high-intensity continuous systems (e.g., the Schugi in-line mixer), mixers are not feasible for very large throughput applications if substantial growth is required. Granules produced in mixer granulators may not be as spherical as those produced in tumbling granulators, and are generally denser due to higher-agitation intensity (see Fig. 20-71). Control of the amount of liquid phase and the intensity and duration of mixing determine agglomerate size and density. Due to greater compaction and kneading action, generally less liquid phase is required in mixers than in tumbling granulation.

Low-Speed Mixers **Pug mills** and **paddle mills** are used for both batch and continuous applications. These devices have horizontal troughs in which rotate central shafts with attached mixing blades of bar, rod, paddle, and other designs. (See Sec. 19.) The vessel may be of single- or double-trough design. The rotating blades throw material forward and to the center to achieve a kneading, mixing action. Characteristics of a range of pug mills available for fertilizer granulation are given in Table 20-45. These mills have largely been replaced by tumbling granulators in metallurgical and fertilizer applications, but they are still used as a premixing step for blending very different raw materials e.g., filter cake with dry powder.

Batch planetary mixers are used extensively in the pharmaceutical industry for powder granulation. A typical batch size of 100 to 200 kg has a power input of 10 to 20 kW. Mixing times in these granulators are quite long (20 to 40 min).

High-Speed Mixers High-speed mixers include continuous-shaft mixers and batch high-speed mixers. **Continuous-shaft mixers** have blades or pins rotating at high speed on a central shaft. Both horizontal and vertical shaft designs are available. Examples include the vertical **Schugi mixer** (Fig. 20-84) and the horizontal **pin** or **peg**

TABLE 20-45 Characteristics of Pug Mixers for Fertilizer Granulation*

Model	Material bulk density, lb/ft³	Approximate capacity, tons/h	Size (width × length), ft	Plate thickness, in	Shaft diameter, in	Speed, r/min	Drive, hp
A	25	8	2 × 8	¼	3	56	15
	50	15	2 × 8	¼	3	56	20
	75	22	2 × 8	¼	3	56	25
	100	30	2 × 8	¼	3	56	30
B	25	30	4 × 8	⅜	4	56	30
	50	60	4 × 8	⅜	4	56	50
	75	90	4 × 8	⅜	4	56	75
	100	120	4 × 8	⅜	5	56	100
C	25	30	4 × 12	⅜	5	56	50
	50	60	4 × 12	⅜	5	56	100
	75	90	4 × 12	⅜	6	56	150
	100	120	4 × 12	⅜	6	56	200
	125	180	4 × 12	⅜	7	56	300

*Feeco International, Inc. To convert pounds per cubic foot to kilograms per cubic meter, multiply by 16; to convert tons per hour to megagrams per hour, multiply by 0.907; to convert feet to centimeters, multiply by 30.5; to convert inches to centimeters, multiply by 2.54; and to convert horsepower to kilowatts, multiply by 0.746.

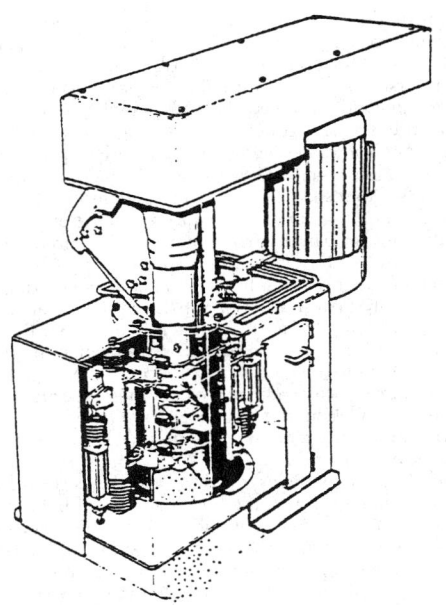

FIG. 20-84 The Schugi Flexomix® vertical high-shear continuous granulator. Shaft speeds range from 100–3500 rpm, capacities from 0.1 to 200 megagrams per hour, and power requirements from 1–100 kW. (*Courtesy of Hosokawa Micron Australia Pty. Ltd.*)

mixers. These mixers operate at high speed (200 to 3500 rpm) to produce granules of size 0.5 to 1.5 mm with a residence time of a few seconds. Schugi capacities may range up to 50 tons/hr. Typical plant capacities of lower-shear peg mixers are 10-20 tons/hr [Capes, *Particle Size Enlargement*, 1980]. Examples of applications include detergents, agricultural chemicals, clays, ceramics, and carbon black.

Batch high-shear mixer granulators are used extensively in the pharmaceutical industry. Plow-shaped mixers rotate on a horizontal shaft at 60 to 800 rpm. Separate high-speed cutters or choppers rotate at much higher speed (500 to 3500 rpm) and are used to limit the maximum granule size. Granulation times of the order 5–10 mins are common, which includes both wet massing and granulation stages operating at low and high impeller speed, respectively. Several designs with both vertical and horizontal shafts are available (Fig. 20-85). Popular designs include horizontal Lödige and vertical Diosna, Fielder, and Gral granulators [Schaefer, *Acta. Pharm. Sci.*, **25**, 205 (1988)].

Granulation-Rate Processes and Effect of Operating Variables Granule deformation is important due to the higher agitation intensity existing in high-shear mixers as compared with tumbling granulators (see Fig. 20-71 and "Growth and Consolidation: High Agitation Intensity Growth"). As deformability is linked to granule saturation and interparticle, frictional forces, consolidation and growth are highly coupled as illustrated in Fig. 20-86 where continued growth is associated with continued compaction and decreases in granule porosity. Impeller shaft power intensity has been used both as a rheological tool to characterize formulation deformability as well as a control technique to judge granulation end-point. [See Kristensen et al., *Acta. Pharm. Sci.*, **25**, 187 (1988) and Holm et al., *Powder Tech.*, **43**, 225 (1985).]

Scale-Up and Operation Scale-up of pharmaceutical mixer granulators is difficult because geometric similarity is often not preserved. Kristensen recommends constant relative swept volume ratio $\dot{V}_R$ as a scale-up parameter defined as:

$$\dot{V}_R = \frac{\dot{V}_{\text{imp}}}{V_{\text{tot}}} \tag{20-69}$$

where $\dot{V}_{\text{imp}}$ is the volume rate swept by the impeller and V_{tot} is the total volume of the granulator. Depending on mixer design, relative swept volume may decrease significantly with scale [Schaefer, loc. cit.] requiring increases in impeller speed with scale-up. In general, scale-up leads to poorer liquid distribution, higher-porosity granules and wider granule-size distributions. Required granulation time may increase with scale, though this depends on the importance of consolidation kinetics as discussed above.

Power dissipation can lead to temperature increases of up to 40°C in the mass. Note that evaporation of liquid as a result of this increase needs to be accounted for in determining liquid requirements for granulation. Liquid should be added through an atomizing nozzle to aid uniform liquid distribution in many cases. In addition, power intensity (kW/kg) has been used with some success to judge granulation end point and for scale-up, primarily due to its relationship to granule deformation [Holm loc. cit.]. Swept volume ratio is a preliminary estimate of expected power intensity.

FLUIDIZED-BED AND RELATED GRANULATORS

In fluidized granulators (**fluidized beds** and **spouted beds**), particles are set in motion by air, rather than by mechanical agitation. Applications include fertilizers, industrial chemicals, agricultural chemicals, pharmaceutical granulation, and a range of coating processes. Fluidized granulators produce either high-porosity granules due to the agglomeration of powder feeds or high-strength layered granules due to coating of seed particles or granules by liquid feeds.

Figure 20-87 shows a typical production-size **batch** fluid-bed granulator. The air-handling unit dehumidifies and heats the inlet air. Heated fluidization air enters the processing zone through a distribu-

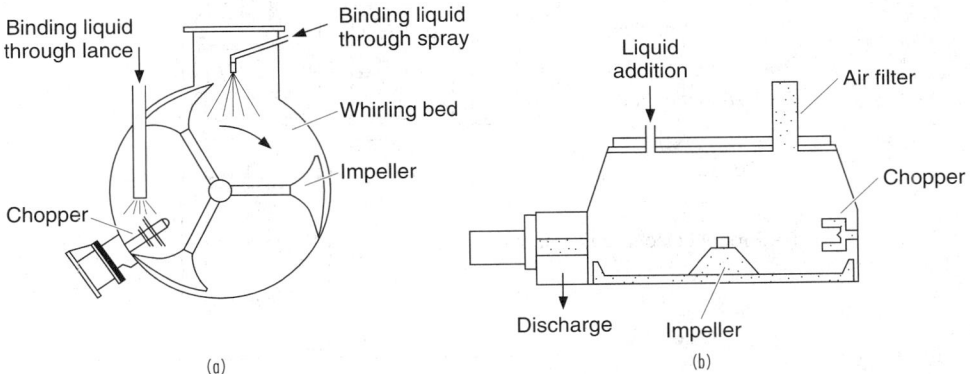

FIG. 20-85 (*a*) Horizontal and (*b*) vertical high-shear mixer granulators for pharmaceutical granule preparation for subsequent tableting.

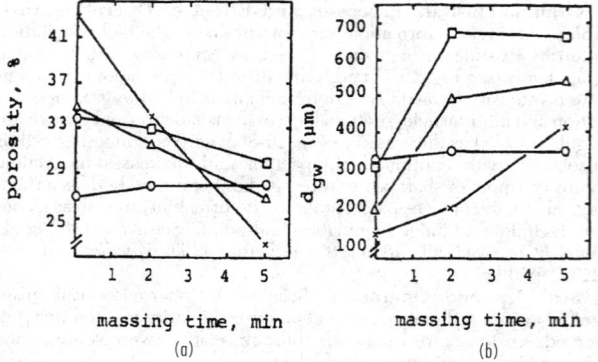

FIG. 20-86 Relationship between consolidation and growth kinetics in high-shear mixer granulators. Effect of wet-massing time on intragranular porosity (*a*) and granule size (*b*) in a high-shear mixer, Fielder PMAT 25 VG. Impeller speed: 250 rpm. Chopper speed: 3000 rpm. Starting materials: Lactose, o; Dicalcium phosphate, x. Dicalcium phosphate/cornstarch 85/15 w/wt %: △. Dicalcium phosphate/cornstarch 55/45 w/wt %: □ [*Kristensen et al., Acta Pharm. Sci.*, **25**, *187 (1988)*.]

tor which also supports the particle bed. Liquid binder is sprayed through an air-atomizing nozzle located above, in, or below the bed. Bag filters or cyclones are needed to remove dust from the exit air. Other fluidization gases such as nitrogen are also used in place of or in combination with air to avoid potential explosion hazards due to fine powders.

Continuous fluid-bed granulators are used in the fertilizer and detergent industries. For fertilizer applications, near-size granules are recycled to control the granule size distribution. Dust is not recycled directly, but first remelted or slurried in the liquid feed.

Advantages of fluidized-beds over other granulation systems include high-volumetric intensity, simultaneous drying and granulation, high heat and mass-transfer rates, and robustness with respect to operating variables on product quality. Disadvantages include the effect of high operating costs with respect to air handling and dust containment, and the potential of defluidization due to uncontrolled growth.

Hydrodynamics The **hydrodynamics** of fluidized beds is covered in detail in Sec. 17. Only aspects specifically related to particle size enlargement are discussed here. Granular product from fluidized beds are generally **group B** or **group D** particles under Geldart's powder classification. However, for batch granulation, the bed may initially consist of a **group A** powder. For granulation, fluidized beds typically operate in the range $1.5U_{mf} < U < 5U_{mf}$ where U_{mf} is the minimum fluidization velocity and U is the operating superficial gas velocity. For batch granulation, the gas velocity may need to be increased significantly during operation to maintain the velocity in this range as the bed particle size increases.

For group B and D particles, nearly all the excess gas velocity ($U - U_{mf}$) flows as bubbles through the bed. The flow of bubbles controls particle mixing, attrition, and elutriation. Therefore, elutriation and attrition rates are proportional to excess gas velocity. Readers should refer to Sec. 17 for important information and correlations on Geldart's powder classification, minimum fluidization velocity, bubble growth and bed expansion, and elutriation.

Mass and Energy Balances Due to the good mixing and heat-transfer properties of fluidized beds, the exit-gas temperature is assumed to be the same as the bed temperature. Fluidized bed gran-

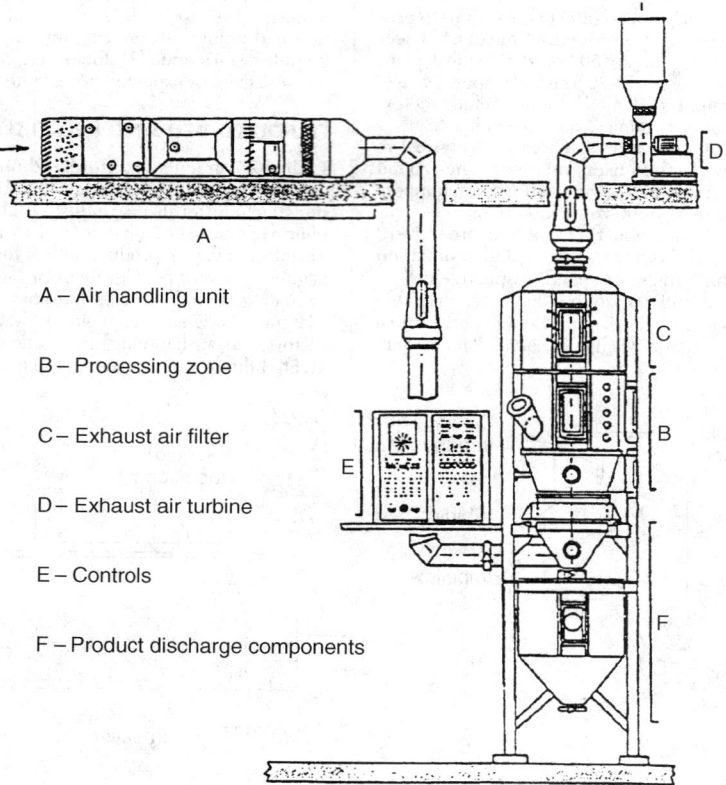

A – Air handling unit

B – Processing zone

C – Exhaust air filter

D – Exhaust air turbine

E – Controls

F – Product discharge components

FIG. 20-87 Fluid-bed granulator for batch processing of powder feeds. [*Ghebre-Selassie (ed.)*, Pharmaceutical Pelletization Technology, *Marcel Dekker (1989).*]

ulators also act simultaneously as dryers and therefore are subject to the same mass and energy-balance limits as driers, namely:

1. The concentration of solvent of the atomized binding fluid in the exit air cannot exceed the saturation value for the solvent in the fluidizing gas at the bed temperature.

2. The supplied energy in the inlet air must be sufficient to evaporate the solvent and maintain the bed at the desired temperature.

Both these limits restrict the maximum rate of liquid feed or binder addition for given inlet gas velocity and temperature. The liquid feed rate may be further restricted to avoid excess coalescence or quenching.

Granulation Rate Processes and Effect of Operating Variables Table 20-46 summarizes the typical effect of feed properties (material variables) and operating variables on fluidized-bed granulation. Due to the range of mechanisms operating simultaneously, the combined effect of these variables can be complex. "Rules of thumb" for operation are very dependent on knowing the dominant granulation mechanisms or rate processes. Understanding individual rate processes allows at least semiquantitative analysis to be used in design and operation. See Tables 20-38, 20-40, and 20-42 on controlling the individual granulation rate processes of wetting, coalescence and consolidation, and breakage, respectively.

Competing mechanisms of growth include layering which results in dense, strong granules with a very tight size distribution and coalescence which results in raspberrylike agglomerates of higher voidage. Growth rates range from 10–100 μm hr^{-1} to 100–2000 μm hr^{-1} for growth by layering and coalescence, respectively.

Air atomizing nozzles are commonly used to control the droplet-size distribution independently of the liquid feed rate and to minimize the chances of **defluidization** due to uncontrolled growth or large droplets.

Equipment Operation **Spray nozzles** suffer from caking on the outside and clogging on the inside. When the nozzle is below the bed surface, fast capture of the liquid drops by bed particles, as well as scouring of the nozzle by particles, prevents caking. Blockages inside the nozzle are also common, particularly for slurries. The nozzle design should be as simple as possible and provision for in situ cleaning or easy removal is essential.

The formation of large, wet agglomerates that dry slowly is called **wet quenching.** Large agglomerates defluidize, causing channeling and poor mixing, ultimately leading to shutdown. Sources of wet quenching include high liquid-spray rates, large spray droplets, or dripping nozzles. **Dry quenching** (uncontrolled coalescence) is the formation and defluidization of large, stable, dry agglomerates, which also may ultimately lead to shutdown. Early detection of quenching is important. The initial stages of defluidization are detected by monitoring the bed temperature just above the distributor. A sudden increase (dry quenching) or decrease (wet quenching) indicates the onset of bed defluidization. Wet quenching is avoided by reducing the liquid feed rate and improving the nozzle operation. In situ jet grinding is sometimes used to limit the maximum stable size of dry agglomerates.

Control is accomplished by monitoring bed temperature, as well as granule size and density of samples. Temperature is controlled best by adjusting the liquid feed rate. For batch granulation, the fluidizing air velocity must be increased during the batch. **Bed pressure fluctuations** can be used to monitor the quality of fluidization and to indicate when gas velocity increases are required. In addition, intermittent sampling systems may be employed with **on-line size analyzers** to monitor granule size. There is no simple heuristic to control the final-granule-size distribution (batch) or exit-granule-size distribution (continuous). A good knowledge of the granulation processes combined with model-based control schemes can be used to fix the batch time (batch), or adjust the seed-granule recycle (continuous) to maintain product quality.

Scale-up of fluid-bed granulators relies heavily on pilot-scale tests. The pilot-plant fluid bed should be at least 0.3 meters in diameter so that bubbling rather than slugging fluidization behavior occurs. Key in scale-up is the increase in agitation intensity with increasing bed height. In particular, granule density and attrition resistance increase linearly with operating bed height whereas the rate of granule dispersion decreases.

Draft Tube Designs and Spouted Beds A **draft tube** is often employed to regulate particle circulation patterns. The most common design is the **Wurster** draft tube fluid bed employed extensively in the pharmaceutical industry, usually for coating and layered growth applications. The **Wurster coater** uses a bottom positioned spray, but other variations are available (Table 20-47).

The **spouted-bed granulator** consists of a central high-velocity **spout** surrounded by a **moving bed annular region.** (See Sec. 17.) All air enters through the orifice at the base of the spout. Particles entrained in the spout are carried to the bed surface and rain down on the annulus as a fountain. Bottom-sprayed designs are the most common. Due to the very high gas velocity in the spout, granules grow by layering only. Therefore, spouted beds are good for coating applications. However, attrition rates are also high, so the technique is not suited to weak granules. Spouted beds are well suited to group D particles and are more tolerant of nonspherical particles than a fluid bed. Particle circulation is better controlled than in a fluidized bed, unless a draft tube design is employed. Spouted beds are difficult to scale past two meters in diameter.

The liquid spray rate to a spouted bed may be limited by agglomerate formation in the spray zone causing spout collapse [Liu & Litster, *Powder Tech.*, **74**, 259 (1993)]. The maximum liquid spray rate increases with increasing gas velocity, increasing bed temperature, and decreasing binder viscosity (see Fig. 20-88). The maximum liquid flow rate is typically between 20 and 90 percent of that required to saturate the exit air, depending on operating conditions. Elutriation of fines from spouted-bed granulators is due mostly to the attrition of newly layered material, rather than spray drying. The elutriation rate is proportional to the kinetic energy in the inlet air (see Eq. 20-71).

CENTRIFUGAL GRANULATORS

In the pharmaceutical industry, a range of centrifugal granulator designs are used. In each of these, a horizontal disc rotates at high speed causing the feed to form a rotating **rope** at the walls of the

TABLE 20-46 Effect of Variables on Fluidized-Bed Granulation*

Operating or material variable	Effect of increasing variable
Liquid feed or spray rate	Increase size and spread of granule-size distribution
	Increase granule density and strength
	Increase chance of defluidization due to quenching
Liquid droplet size	Increase size and spread of granule-size distribution
Gas velocity	Increase attrition and elutriation rates (major effect)
	Decrease coalescence for inertial growth
	No effect on coalescence for noninertial growth
	Increase granule consolidation and density
Bed height	Increase granule density and strength
Bed temperature	Decrease granule density and strength
Binder viscosity	Increase coalescence for inertial growth
	No effect on coalescence for noninertial growth
	Decrease granule density
Particle or granule size	Decrease chance of coalescence
	Increase required gas velocity to maintain fluidization

TABLE 20-47 Sizes & Capacities of Wurster Coaters*

Bed diameter, inches	Batch size, kg
7	3–5
9	7–10
12	12–20
18	35–55
24	95–125
32	200–275
46	400–575

*Ghebre-Sellasie (ed.), *Pharmaceutical Pelletization Technology,* Marcel Dekker (1989).

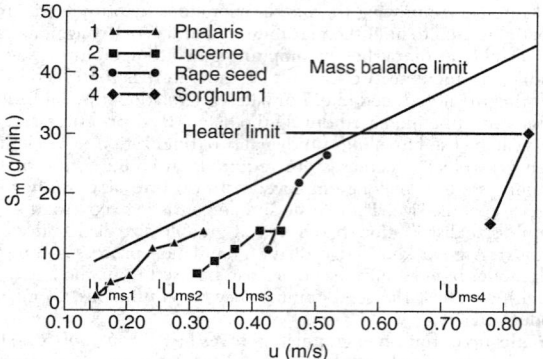

FIG. 20-88 Effect of gas velocity on maximum liquid rate for a spouted-bed seed coater. [*Liu & Litster*, Powder Tech., *74, 259 (1993).*] With kind permission from Elsevier Science SA, Lausanne, Switzerland.

vessel (see Fig. 20-89). There is usually an allowance for drying air to enter around the edge of the spinning disc. Applications of such granulators include spheronization of extruded pellets, dry-powder layering of granules or sugar spheres, and coating of pellets or granules by liquid feeds.

Centrifugal Designs Centrifugal granulators tend to give denser granules or powder layers than fluidized beds and more spherical granules than mixer granulators. Operating costs are reasonable but capital cost is generally high compared to other options. Several types are available including the **CF granulator** (Fig. 20-89) and **rotary fluidized-bed** designs which allow high gas volumes and therefore significant drying rates (Table 20-48). CF granulator capacities range from 3–80 kg with rotor diameters of .36–1.3 meters and rotor speeds of 45–360 rpm [Ghebre-Selassie (ed.), *Pharmaceutical Pelletization Technology,* Marcel Dekker, (1989)].

Particle Motion and Scale-Up Very little fundamental information is published on centrifugal granulators. Qualitatively, good operation relies on maintaining a smoothly rotating **stable rope** of tumbling

particles. Operating variables which affect the particle motion are disc speed, peripheral air velocity, and the presence of baffles.

For a given design, good rope formation is only possible for a small range of disc speeds. If the speed is too low, a rope does not form. If the speed is too high, very high attrition rates can occur. Scale-up on the basis of either **constant peripheral speed** (DN = const.), or **constant Froude number** (DN^2 = const.) is possible. Increasing peripheral air velocity and baffles helps to increase the rate of rope turnover. In designs with tangential powder or liquid feed tubes, additional baffles are usually not necessary. The motion of particles in the equipment is also a function of the frictional properties of the feed, so the optimum operating conditions are feed specific.

Granulation Rate Processes Possible granulation processes occurring in centrifugal granulators are extrudate breakage, consolidation, rounding (spheronization), coalescence, powder layering and coating, and attrition. Very little information is available about these processes as they occur in centrifugal granulators, however, similar principles from tumbling and fluid-bed granulators will apply.

SPRAY PROCESSES

Spray processes include **spray driers, prilling towers,** and **flash driers.** Feed solids in a fluid state (solution, gel, paste, emulsion, slurry, or melt) are dispersed in a gas and converted to granular solid products by heat and/or mass transfer. In spray processes, the size distribution of the particulate product is largely set by the drop size distribution, i.e., **nucleation** is the dominant granulation process. Exceptions are where fines are recycled to coalesce with new spray droplets and where spray-dried powders are rewet in a second tower to encourage agglomeration. For spray drying, a large amount of solvent must be evaporated whereas prilling is a spray-cooling process. Fluidized or spouted bed must be used to capture nucleated fines as hybrid granulator designs, e.g., the **fluid-bed spray dryer.**

Product diameter is small and bulk density is low in most cases, except prilling. Feed liquids must be pumpable and capable of atomization or dispersion. Attrition is usually high, requiring fines recycle or recovery. Given the importance of the droplet-size distribution, nozzle design and an understanding of the fluid mechanics of drop formation are critical. In addition, heat and mass-transfer rates during

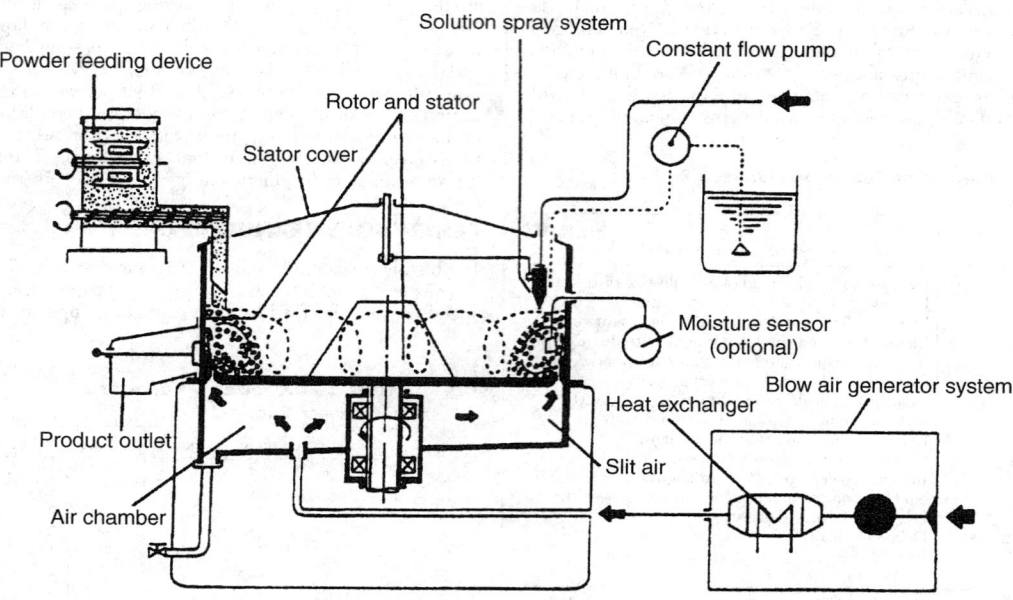

FIG. 20-89 Schematic of a CF granulator. [*Ghebre-Selassie (ed.),* Pharmaceutical Pelletization Technology, *Marcel Dekker (1989).*]

TABLE 20-48   Specifications of Glatt Rotary Fluid-Bed Granulators*

Parameter	15	60	200	500
Volume, liters	45	220	670	1560
Fan				
Power, kW	11	22	37	55
Capacity, m³/hr	1500	4500	8000	12000
Heating capacity, kW	37	107	212	370
Diameter, m	1.7	2.5	3.45	4.0

*Glatt Company, in Ghebre-Selassie (ed.), *Pharmaceutical Pelletization Technology,* Marcel Dekker (1989).

drying can strongly effect the particle morphology, of which a wide range of characteristics are possible.

Spray Drying   Detailed descriptions of spray dispersion dryers, together with application, design, and cost information, are given in Sec. 17. Product quality is determined by a number of properties such as particle form, size, flavor, color, and heat stability. Particle size and size distribution, of course, are of greatest interest from the point of view of size enlargement.

In general, particle size is a function of atomizer-operating conditions and also of the solids content, liquid viscosity, liquid density, and feed rate. Coarser, more granular products can be made by increasing viscosity (through greater solids content, lower temperature, etc.), by increasing feed rate, and by the presence of binders to produce more agglomeration of semidry droplets. Less-intense atomization and spray-air contact also increase particle size, as does a lower exit temperature, which yields a moister (and hence a more coherent) product. This latter type of spray-drying agglomeration system has been described by Masters and Stoltze [*Food Eng.,* 64 (February 1973)] for the production of instant skim-milk powders in which the completion of drying and cooling takes place in vibrating conveyors (see Sec. 17) downstream of the spray dryer.

Prilling   The prilling process is similar to spray drying and consists of spraying droplets of liquid into the top of a tower and allowing these to fall against a countercurrent stream of air. During their fall the droplets are solidified into approximately spherical particles or prills which are up to about 3 mm in diameter, or larger than those formed in spray drying. The process also differs from spray drying since the droplets are formed from a melt which solidifies primarily by cooling with little, if any, contribution from drying. Traditionally, ammonium nitrate, urea, and other materials of low viscosity and melting point and high surface tension have been treated in this way. Improvements in the process now allow viscous and high-melting point materials and slurries containing undissolved solids to be treated as well.

The design of a prilling unit first must take into account the properties of the material and its sprayability before the tower design can proceed. By using data on the melting point, viscosity, surface tension, etc., of the material, together with laboratory-scale spraying tests, it is possible to specify optimum temperature, pressure, and orifice size for the required prill size and quality. Tower sizing basically consists of specifying the cross-sectional area and the height of fall. The former is determined primarily by the number of spray nozzles necessary for the desired production rate. Tower height must be sufficient to accomplish solidification and is dependent on the heat-transfer characteristics of the prills and the operating conditions (e.g., air temperature). Because of relatively large prill size, narrow but very tall towers are used to ensure that the prills are sufficiently solid when they reach the bottom. Table 20-49 describes the principal characteristics of a typical prilling tower.

Theoretical calculations are possible to determine tower height with reasonable accuracy. Simple parallel streamline flow of both droplets and air is a reasonable assumption in the case of prilling towers compared with the more complex rotational flows produced in spray dryers. For velocity of fall, see, for example, Becker [*Can. J. Chem.,* **37,** 85 (1959)]. For heat transfer, see, for example, Kramers [*Physica,* **12,** 61 (1946)]. Specific design procedures for prilling towers are available in the *Proceedings of the Fertilizer Society (England);* see Berg and Hallie, no. 59, 1960; and Carter and Roberts, no. 110, 1969.

Recent developments in nozzle design have led to drastic reductions

TABLE 20-49   Some Characteristics of a Typical Prilling Operation*

Tower size		
Prill tube height, ft	130	
Rectangular cross		
section, ft	11 by 21.4	
Cooling air		
Rate, lb/h	360,000	
Inlet temperature	Ambient	
Temperature rise, °F	15	
Melt		Ammonium nitrate
Type	Urea	
Rate, lb/h	35,200 (190 lb H₂O)	43,720 (90 lb H₂O)
Inlet temperature, °F	275	365
Prills		
Outlet temperature, °F	120	225
Size, mm	Approximately 1 to 3	

*HPD Incorporated. To convert feet to centimeters, multiply by 30.5; to convert pounds per hour to kilograms per hour, multiply by 0.4535; °C = (°F − 32) × 5⁄9.

in the required height of prilling towers. However, such nozzle designs are largely proprietary and little information is openly available.

Flash Drying   Special designs of pneumatic-conveyor dryers, described in Sec. 17 can handle filter and centrifuge cakes and other sticky or pasty feeds to yield granular size-enlarged products. The dry product is recycled and mixed with fresh, cohesive feed, followed by disintegration and dispersion of the mixed feed in the drying-air stream.

PRESSURE COMPACTION PROCESSES

The success of compression agglomeration depends on the effective utilization and transmission of the applied external force and on the ability of the material to form and maintain interparticle bonds during pressure compaction (or consolidation) and decompression. Both these aspects are controlled in turn by the geometry of the confined space, the nature of the applied loads and the physical properties of the particulate material and of the confining walls. (See the section on Powder Mechanics and Powder Compaction.)

Pressure compaction is carried out in two classes of equipment. These are **confined-pressure devices** (molding, piston, tableting, and roll presses), in which material is directly consolidated in closed molds or between two opposing surfaces; and **extrusion devices** (pellet mills, screw extruders), in which material undergoes considerable shear and mixing as it is consolidated while being pressed through a die. See Table 20-36 for examples of uses. Product densities and applied pressures are substantially higher than agitative agglomeration techniques, as shown in Fig. 20-71.

Piston and Molding Presses   Piston or molding presses are used to create uniform and sometimes intricate compacts, especially in powder metallurgy and plastics forming. Equipment comprises a mechanically or hydraulically operated press and, attached to the platens of the press, a two-part mold consisting of top (male) and bottom (female) portions. The action of pressure and heat on the particulate charge causes it to flow and take the shape of the cavity of the mold. Compacts of metal powders are then sintered to develop metallic properties, whereas compacts of plastics are essentially finished products on discharge from the molding machine.

Tableting Presses   Tableting presses are employed in applications having strict specifications for weight, thickness, hardness, density, and appearance in the agglomerated product. They produce simpler shapes at higher production rates than do molding presses. A single-punch press is one that will take one station of tools consisting of an upper punch, a lower punch, and a die. A rotary press employs a rotating round die table with multiple stations of punches and dies. Older rotary machines are single-sided; that is, there is one fill station and one compression station to produce one tablet per station at every revolution of the rotary head. Modern high-speed rotary presses are double-sided; that is, there are two feed and compression stations to

TABLE 20-50 Characteristics of Tableting Presses*

	Single-punch	Rotary
Tablets per minute	8–140	72–6000
Tablet diameter, in.	⅛–4	⅝–2½
Pressure, tons	1½–100	4–100
Horsepower	¼–15	1½–50

*Browning, *Chem. Eng.,* **74**(25), 147 (1967).

NOTE: To convert inches to centimeters, multiply by 2.54; to convert tons to megagrams, multiply by 0.907; and to convert horsepower to kilowatts, multiply by 0.746.

produce two tablets per station at every revolution of the rotary head. Some characteristics of tableting presses are shown in Table 20-50.

For successful tableting, a material must have suitable flow properties to allow it to be fed to the tableting machine. Wet or dry granulation is used to improve the flow properties of materials. In the case of **wet granulation,** agitative granulation techniques such as fluidized beds or mixer granulators as discussed above are often employed.

In **dry granulation,** the blended dry ingredients are first densified in a heavy-duty rotary tableting press which produces "slugs" 1.9 to 2.5 cm (¾ to 1 in) in diameter. These are subsequently crushed into particles of the size required for tableting. Predensification can also be accomplished by using a rotary compactor-granulator system. A third technique, direct compaction, uses sophisticated devices to feed the blended dry ingredients to a high-speed rotary press.

Excellent accounts of tableting in the pharmaceutical industry have been given by Kibbe [*Chem. Eng. Prog.,* **62**(8), 112 (1966)], Carstensen [*Handbook of Powder Science & Technology,* Fayed & Otten (eds.), Van Nostrand Reinhold Inc., **252** (1983)], Stanley-Wood (ed.) (loc. cit.), and Doelker (loc. cit.).

Roll Presses Roll presses compact raw material as it is carried into the gap between two rolls rotating at equal speeds (Fig. 20-90). The size and shape of the agglomerates are determined by the geometry of the roll surfaces. Pockets or indentations in the roll surfaces form briquettes the shape of eggs, pillows, teardrops, or similar forms from a few grams up to 2 kg (5 lb) or more in weight. Smooth or corrugated rolls produce a solid sheet which can be granulated or broken down into the desired particle size on conventional grinding equipment.

Roll presses can produce large quantities of materials at low cost, but the product is less uniform than that from molding or tableting presses. The introduction of the proper quantity of material into each of the rapidly rotating pockets in the rolls is the most difficult problem in the briquetting operation. Various types of feeders have helped to overcome much of this difficulty.

The impacting rolls can be either solid or divided into segments. Segmented rolls are preferred for hot briquetting, as the thermal expansion of the equipment can be controlled more easily.

Roll presses provide a **mechanical advantage** in amplifying the feed pressure P_0 to some maximum value P_m. This maximum pressure P_m and the roll compaction time control compact density. Generally speaking, as compaction time decreases (e.g., by increasing roll speed), the minimum necessary pressure for quality compacts increases. There may be an upper limit of pressure as well for friable materials or elastic materials prone to delamination.

Pressure amplification occurs in two regions of the press (Fig. 20-90). Above the **angle of nip,** sliding occurs between the material and roll surface as material is forced into the rolls, with intermediate pressure ranging from 1–10 psi. Energy is dissipated primarily through overcoming particle friction and cohesion. Below the angle of nip, no slip occurs as the powder is compressed into a compact and pressure may increase up to several thousand psi. Both these intermediate and high-pressure regions of densification are indicated in compressibility diagram of Fig. 20-77.

The overall performance of the press and its mechanical advantage (P_m/P_0) depend on the mechanical and frictional properties of the powder. (See "Powder Mechanics and Powder Compaction" section.) For design procedures, see Johanson [*Proc. Inst. Briquet. Agglom. Bien. Conf.,* **9,** 17 (1965).] Nip angle α generally increases with decreasing compressibility κ, or with increasing roll friction angle ϕ_w and effective angle of friction ϕ_e. Powders which compress easily and have high-friction grip high in the rolls. The mechanical advantage pressure ratio (P_m/P_0) increases and the time of compaction decreases with decreasing nip angle since the pressure is focused over a smaller roll area. In addition, the mechanical advantage generally increases with increasing compressibility and roll friction.

The most important factor that must be determined in a given application is the pressing force required for the production of acceptable compacts. Roll loadings (i.e., roll separating force divided by roll width) in commercial installations vary from 4.4 MN/m to more than 440 MN/m (1000 lb/in to more than 100,000 lb/in). Roll sizes up to 91 cm (36 in) in diameter by 61 cm (24 in) wide are in use.

The roll loading L is related to the maximum developed pressure and roll diameter by

$$L = \frac{F}{W} = \frac{1}{2} f P_m D \propto P_m D^{1/2} (h + d)^{1/2} \qquad (20\text{-}70)$$

where F is the **roll-separating force,** D and W are the roll diameter and width, f is a **roll-force factor** dependent on compressibility κ and gap thickness as given in Fig. 20-91, h is the gap thickness and $d/2$ is the pocket depth for briquette rolls. [Pietsch, *Size Enlargement by Agglomeration,* John Wiley & Sons Ltd., Chichester, 1992.] The maximum pressure P_m is established on the basis of required compact density and quality, and it is a strong function of roll gap distance and powder properties as discussed above, particularly compressibility.

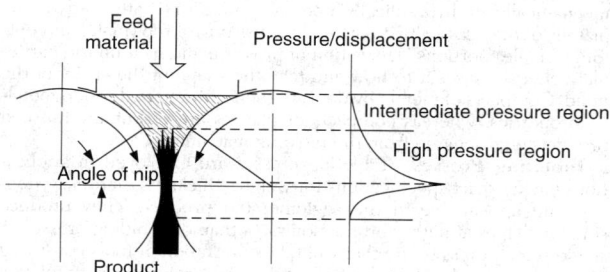

FIG. 20-90 Regions of compression in roll presses. Slippage and particle rearrangement occurs above the angle of nip, and powder compaction at high pressure occurs in the nonslip region below the angle of nip. Reprinted from *Granulation and Coating Technologies for High-Value-Added Industries,* Ennis and Litster (1996) with permission of E & G Associates. All rights reserved.

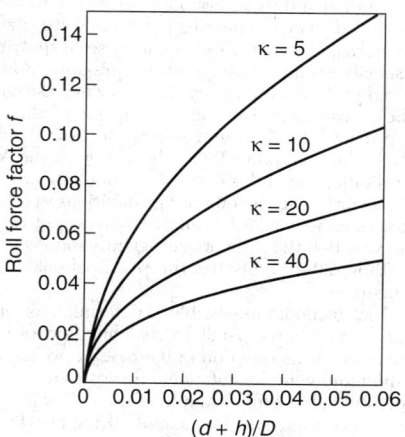

FIG. 20-91 Roll-force factor as a function of compressibility κ and dimensionless gap distance $(d + h)/D$. [Pietsch (ed.), Roll Pressing, *Powder Advisory Centre, London (1987).*]

Small variations in feed properties can have a pronounced effect on maximum pressure P_m and press performance. Roll presses are scaled on the basis of constant maximum pressure. The required roll loading increases approximately with the square root of increasing roll diameter or gap width.

The allowable roll width is inversely related to the required pressing force because of mechanical-design considerations. The throughput of a roll press at constant roll speed decreases as pressing force increases since the allowable roll width is less. Machines with capacities up to 45 Mg/h (50 tons/h) are available. Some average figures for the pressing force and energy necessary to compress a number of materials on roll-type briquette machines are given in Table 20-51. Typical capacities are given in Table 20-52.

During compression in the slip region, escaping air may induce fluidization or erratic pulsating of the feed. This effect, which is controlled by the permeability of the powder, limits the allowable roll speed of the press, and may also enduce compact delamination. Increases in roll speed or decreases in permeability require larger feed pressures.

Pellet Mills Pellet mills operate on the principle shown in Fig. 20-92. Moist, plastic feed is pushed through holes in dies of various shapes. The friction of the material in the die holes supplies the resistance necessary for compaction. Adjustable knives shear the rodlike extrudates into pellets of the desired length. Although several designs are in use, the most commonly used pellet mills operate by applying power to the die and rotating it around a freely turning roller with fixed horizontal or vertical axis.

Pellet quality and capacity vary with properties of the feed such as

TABLE 20-51 Pressure and Energy Requirements to Briquette Various Materials*

Pressure range, lb./sq. in.	Approximate energy required, kw.-hr./ton	Type of material being briquetted or compacted		
		Without binder	With binder	Hot
Low 500–20,000	2–4	Mixed fertilizers, phosphate ores, shales, urea	Coal, charcoal, coke, lignite, animal feed, candy	Phosphate ores, urea
Medium 20,000–50,000	4–8	Acrylic resins, plastics, PVC, ammonium chloride, DMT, copper compounds, lead	Ferroalloys, fluorspar, nickel	Iron, potash, glass-making mixtures
High 50,000–80,000	8–16	Aluminum, copper, zinc, vanadium, calcined dolomite, lime, magnesia, magnesium carbonates, sodium chloride, sodium and potassium compounds	Flue dust, natural and reduced iron ores	Flue dust, iron oxide, natural and reduced iron ores, scrap metals
Very high >80,000	>16	Metal powders, titanium	—	Metal chips

*Courtesy Bepex Corporation. To convert pounds per square inch to newtons per square meter, multiply by 6895; to convert kilowatthours per ton to kilowatthours per megagram, multiply by 1.1.

TABLE 20-52 Some Typical Capacities (tons/h) for a Range of Roll Presses*

	10	16	12	10.3	13	20.5	28	36
Roll diameter, in	10	16	12	10.3	13	20.5	28	36
Maximum roll-face width, in	3.25	6	4	6	8	13.5	27	10
Roll-separating force, tons	25	50	40	50	75	150	300	360
Carbon								
Coal, coke		2	1		3	6	25	
Charcoal			8			13		
Activated					3	7		
Metal and ores								
Alumina					5	10	28	
Aluminum				2	4	8	20	
Brass, copper	0.5			1.5	3	6	16	
Steel-mill waste					5	10		
Iron					3	15	40	
Nickel powder					2.5	5.0		
Nickel ore				2	5	20	40	
Stainless steel				2	5	10		25
Steel								
Bauxite		1.5				10	20	
Ferrometals						10		
Chemicals								
Copper sulfate	0.5	1.5		1	3	6	15	
Potassium hydroxide				1	4	8		
Soda ash	0.5				3	6	15	
Urea	0.25					10		
DMT	0.25				2	6		
Minerals								
Potash						20	80	
Salt				2	5	9		
Lime				4	8		15	
Calcium sulfate							13	40
Fluorspar					5	10		28
Magnesium oxide						1.5	5	
Asbestos						1.5	3	
Cement						5	12	
Glass batch						5		

*Courtesy Bepex Corporation. To convert inches to centimeters, multiply by 2.54; to convert tons to megagrams, multiply by 0.907; and to convert tons per hour to megagrams per hour, multiply by 0.907.

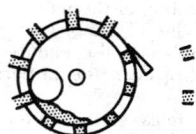

FIG. 20-92 Operating principle of a pellet mill.

moisture, lubricating characteristics, particle size, and abrasiveness, as well as die characteristics and speed. A readily pelleted material will yield about 122 kg/kWh [200 lb/(hp·h)] by using a die with 0.6-cm (¼-in) holes. Some characteristics of pellet mills are given in Table 20-53.

Screw Extruders Screw extruders employ a screw to force material in a plastic state continuously through a die. If the die hole is round, a compact in the form of a rod is formed, whereas if the hole is a thin slit, a film or sheet is formed. Many other forms are also possible.

Basic types of extruders include axial end plate, radial screen, rotary cylinder or gear, and ram or piston. For a review, see Newton [*Powder Technology & Pharmaceutical Processes*, Chulia et al. (eds.), Elsevier, 391 (1994).]

Both wet and dry extrusion techniques are available, and both are strongly influenced by the frictional properties of the particulate phase and wall. In the case of wet extrusion, rheological properties of the liquid phase are equally important. See Pietsch [*Size Enlargement by Agglomeration*, John Wiley & Sons Ltd., Chichester, 346 (1992)] and Benbow et al. [*Chem. Eng. Sci.*, **422**, 2151 (1987)] for a review of design procedures for dry and wet extrusion, respectively.

A common use of screw extruders is in the forming and compounding of plastics. Table 20-54 shows typical outputs that can be expected per horsepower for various plastics and the characteristics of several popular extruder sizes.

Deairing pug-mill extruders which combine mixing, densification, and extrusion in one operation are available for agglomerating clays, catalysts, fertilizers, etc. Table 20-55 gives data on screw extruders for the production of catalyst pellets.

THERMAL PROCESSES

Bonding and agglomeration by temperature elevation or reduction are applied either in conjunction with other size-enlargement processes or as a separate process. Agglomeration occurs through one or more of the following mechanisms:
1. Drying of a concentrated slurry or wet mass of fines
2. Fusion
3. High-temperature chemical reaction
4. Solidification and/or crystallization of a melt or concentrated slurry during cooling

Sintering and Heat Hardening In powder metallurgy compacts are sintered with or without the addition of binders. In ore processing the agglomerated mixture is either sintered or indurated. Sintering refers to a process in which fuel is mixed with the ore and burned on a grate. The product is a porous cake. Induration, or heat hardening, is accomplished by combustion of gases passed through

TABLE 20-53 Characteristics of Pellet Mills

Horsepower range	10–250
Capacity, lb/(hp·h)	75–300
Die characteristics	
Size	Up to 26 in inside diameter × approximately 8 in wide
Speed range	75–500 r/min
Hole-size range	¹⁄₁₆–1¼ in inside diameter
Rollers	As many as three rolls; up to 10-in diameter

NOTE: To convert horsepower to kilowatts, multiply by 0.746; to convert pounds per horsepower-hour to kilograms per kilowatthour, multiply by 0.6; and to convert inches to centimeters, multiply by 2.54.

TABLE 20-54 Characteristics of Plastics Extruders*

Efficiencies	lb/(hp·h)
Rigid PVC	7–10
Plasticized PVC	10–13
Impact polystyrene	8–12
ABS polymers	5–9
Low-density polyethylene	7–10
High-density polyethylene	4–8
Polypropylene	5–10
Nylon	8–12

	Relation of size, power, and output		
	Diameter		
hp	in	mm	Output, lb/h, low-density polyethylene
15	2	45	Up to 125
25	2½	60	Up to 250
50	3½	90	Up to 450
100	4½	120	Up to 800

*The Encyclopedia of Plastics Equipment, Simonds (ed.), Reinhold, New York, 1964.

NOTE: To convert inches to centimeters, multiply by 2.54; to convert horsepower to kilowatts, multiply by 0.746; to convert pounds per hour to kilograms per hour, multiply by 0.4535; and to convert pounds per horsepower-hour to kilograms per kilowatthour, multiply by 0.6.

TABLE 20-55 Characteristics of Pelletizing Screw Extruders for Catalysts*

Screw diameter, in	Drive hp	Typical capacity, lb/h
2.25		60
4	7.5–15	200–600
6	Up to 60	600–1500
8	75–100	Up to 2000

*Courtesy The Bonnot Co. To convert inches to centimeters, multiply by 2.54; to convert horsepower to kilowatts, multiply by 0.746; and to convert pounds per hour to kilograms per hour, multiply by 0.4535.

NOTE:
1. Typical feeds are high alumina, kaolin carriers, molecular sieves, and gels.
2. Water-cooled worm and barrel, variable-speed drive.
3. Die orifices as small as ¹⁄₁₆ in.
4. Vacuum-deairing option available.

the bed. The aim is to harden the pellets without fusing them together, as is done in the sintering process.

Ceramic bond formation and grain growth by diffusion are the two prominent reactions for bonding at the high temperature (1100 to 1370°C, or 2000 to 2500°F, for iron ore) employed. The minimum temperature required for sintering may be measured by modern **dilatometry** techniques, as well as by differential scanning calorimetry. See Compo et al. [*Powder Tech.*, **51**(1), 87 (1987); *Particle Characterization*, **1**, 171 (1984)] for reviews.

In addition to agglomeration, other useful processes may occur during sintering and heat hardening. For example, carbonates and sulfates may be decomposed, or sulfur may be eliminated. Although the major application is in ore beneficiation, other applications, such as the preparation of lightweight aggregate from fly ash and the formation of clinker from cement raw meal, are also possible. Nonferrous sinter is produced from oxides and sulfides of manganese, zinc, lead, and nickel. An excellent account of the many possible applications is given by Ban et al. [Knepper (ed.), *Agglomeration*, op. cit. p. 511] and Ball et al. [Agglomeration of Iron Ores (1973)]. The highest tonnage use at present is in the beneficiation of iron ore.

The machine most commonly used for sintering iron ores is a traveling grate, which is a modification of the Dwight-Lloyd continuous sintering machine formerly used only in the lead and zinc industries. Modern sintering machines may be 4 m (13 ft) wide by 60 m (200 ft) long and have capacities of 7200 Mg/day (8000 tons/day).

The productive capacity of a sintering strand is related directly to the rate at which the burning zone moves downward through the bed. This rate, which is of the order of 2.5 cm/min (1 in/min), is controlled by the air rate through the bed, with the air functioning as the heat-transfer medium.

Heat hardening of green iron-ore pellets is accomplished in a vertical shaft furnace, a traveling-grate machine, or a grate-plus-kiln combination (see Ball et al., op. cit.).

Drying and Solidification Granular free-flowing solid products are often an important result of the drying of concentrated slurries and pastes and the cooling of melts. Size enlargement of originally finely divided solids results. Pressure agglomeration including extrusion, pelleting, and briquetting is used to preform wet material into forms suitable for drying in through-circulation and other types of dryers. Details are given in Sec. 12 in the account of solids-drying equipment.

Rotating-drum-type and belt-type heat-transfer equipment forms granular products directly from fluid pastes and melts without intermediate preforms. These processes are described in Sec. 5 as examples of indirect heat transfer to and from the solid phase. When solidification results from melt freezing, the operation is known as *flaking*. If evaporation occurs, solidification is by drying.

MODELING AND SIMULATION OF GRANULATION PROCESSES

For granulation processes, granule size distribution is an important if not the most important property. The evolution of the granule size distribution within the process can be followed using population balance modeling techniques. This approach is also used for other size-change processes including crushing and grinding (Sec. 20: "Principles of Size Reduction") and crystallization (Sec. 18). The use of the **population balance (PB)** is outlined briefly below. For more in-depth analysis see Randolph & Larson (*Theory of Particulate Processes, 2d ed.*, Academic Press, 1991), Ennis & Litster (*The Science and Engineering of Granulation Processes*, Chapman-Hall, 1997), and Sastry & Loftus [*Proc. 5th Int. Symp. Agglom.*, IChemE, 623 (1989)].

The key uses of PB modeling of granulation processes are:
• Critical evaluation of data to determine controlling granulation mechanisms
• In design, to predict the mean size and size distribution of product granules
• Sensitivity analysis: to analyze quantitatively the effect of changes to operating conditions and feed variables on product quality
• Circuit simulation, optimization, and process control

The use of PB modeling by practitioners has been limited for two reasons. First, in many cases the kinetic parameters for the models have been difficult to predict and are very sensitive to operating conditions. Second, the PB equations are complex and difficult to solve. However, recent advances in understanding of granulation micromechanics, as well as better numerical solution techniques and faster computers, means that the use of PB models by practitioners should expand.

THE POPULATION BALANCE

The PB is a statement of continuity for particulate systems. It includes a **kinetic expression** for each mechanism which changes a particle property. Consider a section of a granulator as illustrated in Fig. 20-93. The PB follows the change in the granule size distribution as granules are born, die, grow, and enter or leave the control volume. As discussed in detail previously ("Agglomeration Rate Processes"), the granulation mechanisms which cause these changes are nucleation, layering, coalescence, and attrition (Tables 20-39 and 20-56). The number of particles-per-unit volume of granulator between size volume v and $v + dv$ is $n(v)dv$, where $n(v)$ is the **number frequency size distribution** by size volume, having dimensions of number per-unit-granulator and volume per-unit-size volume. For constant granulator volume, the macroscopic PB for the granulator in terms of $n(v)$ is:

$$\frac{\partial n(v,t)}{\partial t} = \frac{Q_{\text{in}}}{V} n_{\text{in}}(v) - \frac{Q_{\text{ex}}}{V} n_{\text{ex}}(v) - \frac{\partial (G^\circ - A^\circ) n(v,t)}{\partial v}$$

$$+ B_{\text{nuc}}(v) + \frac{1}{2N_t} \int_0^y \beta(u, v-u, t) n(u,t) n(v-u, t) du$$

$$- \frac{1}{N_t} \int_0^\infty \beta(u, v, t) n(u,t) n(v,t) du \quad (20\text{-}71)$$

where V is the volume of the granulator; Q_{in} and Q_{ex} are the inlet and exit flow rates from the granulator; $G(v)$, $A(v)$, and $B_{\text{nuc}}(v)$ are the layering, attrition, and nucleation rates, respectively; $\beta(u,v,t)$ is the coalescence kernel and N_t is the total number of granules-per-unit volume of granulator. The left-hand side of Eq. 20-71 is the accumulation of particles of a given size volume. The terms on the right-hand side are in turn: the bulk flow into and out of the control volume, the convective flux along the size axis due to layering and attrition, the birth of new particles due to nucleation, and birth and death of granules due to coalescence. Equation 20-71 is written in terms of granule volume v, but could also be written in terms of granule size x or could also be expanded to follow changes in other granule properties, e.g., changes in granule density or porosity due to consolidation.

MODELING INDIVIDUAL GROWTH MECHANISMS

The **granule size distribution (GSD)** is a strong function of the balance between different mechanisms for size change shown in Table 20-53—layering, attrition, nucleation, and coalescence. For example, Fig. 20-94 shows the difference in the GSD for a doubling in mean granule size due to (1) layering only, or (2) coalescence only for batch, plug-flow, and well-mixed granulators. Table 20-56 describes how four key rate mechanisms effect the GSD.

Nucleation Nucleation increases both the mass and number of the granules. For the case where new granules are produced by liquid

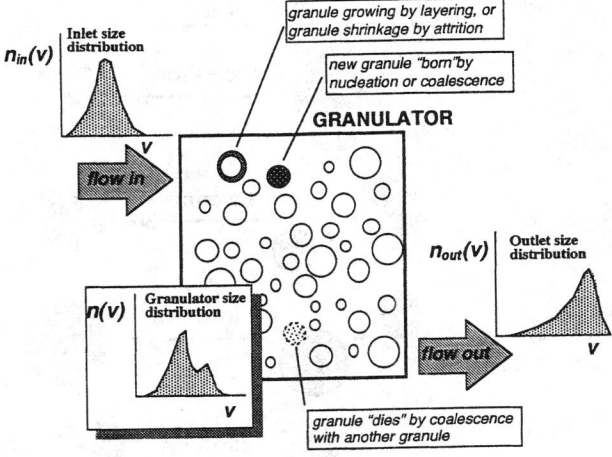

FIG. 20-93 Changes to the granule size distribution due to granulation-rate processes as particles move through the granulator. Reprinted from *Granulation and Coating Technologies for High-Value-Added Industries*, Ennis and Litster (1996) with permission of E & G Associates. All rights reserved.

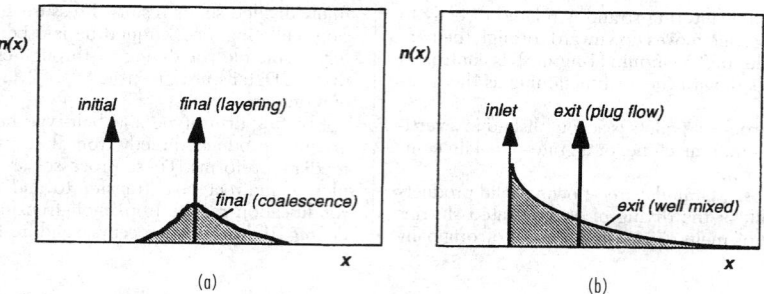

FIG. 20-94   The effect of growth mechanism and mixing on product granule size distribution for (*a*) batch growth by layering or coalescence, and (*b*) layered growth in well-mixed or plug-flow granulators. Reprinted from *Granulation and Coating Technologies for High-Value-Added Industries,* Ennis and Litster (1996) with permission of E & G Associates. All rights reserved.

feed which dries or solidifies, the nucleation rate is given by the new feed droplet size n_s and the volumetric spray rate S:

$$B(v)_{nuc} = Sn_S(v) \qquad (20\text{-}72)$$

In processes where new powder feed has a much smaller particle size than the smallest granular product, the feed powder can be considered as a **continuous phase** which can nucleate to form new granules [Sastry & Fuerstenau, *Powder Tech.,* **7**, 97 (1975)]. The size of the nuclei is then related to nucleation mechanism. In the case of nucleation by spray, the size of the nuclei is of the order of the droplet size and proportional to $\cos\theta$, where θ is binder fluid-particle contact angle (see Fig. 20-67 of "Wetting" section).

Layering   Layering increases granule size and mass by the progressive coating of new material onto existing granules, but it does not alter the number of granules in the system. As with nucleation, the new feed may be in liquid form (where there is simultaneous drying or cooling) or may be present as a fine powder. Where the feed is a powder, the process is sometimes called **pseudolayering** or **snowballing.** It is often reasonable to assume a linear-growth rate $G(x)$ which is independent of granule size. For batch and plug-flow granulators, this causes the initial feed distribution to shift forward in time

with the shape of the GSD remaining unaltered and governed by a traveling-wave equation (Table 20-59). As an example, Fig. 20-95 illustrates size-independent growth of limestone pellets by snowballing in a batch drum. Size-independent linear growth rate implies that the volumetric growth rate $G^\circ(v)$ is proportional to projected granule surface area, or $G^\circ(v) \propto v^{2/3} \propto x^2$. This assumption is true only if all granules receive the same exposure to new feed. Any form of segregation will invalidate this assumption [Liu and Litster, *Powder Tech.,* **74**, 259 (1993)]. The growth rate $G^\circ(v)$ by layering only can be calculated directly from the mass balance:

$$\dot{V}_{feed} = (1 - \varepsilon) \int_0^\infty G^\circ(v)n(v)dv \qquad (20\text{-}73)$$

where $\dot{V}_{feed}$ is the volumetric flow rate of new feed and ε is the granule porosity.

Coalescence   Coalescence is the most difficult mechanism to model. It is easiest to write the population balance (Eq. 20-71) in terms of number distribution by volume $n(v)$ because granule volume is conserved in a coalescence event. The key parameter is the **coalescence kernel** or rate constant $\beta(u,v)$. The kernel dictates the overall rate of coalescence, as well as the effect of granule size on coalescence

TABLE 20-56   Impact of Granulation Mechanisms on Size Distribution

Mechanism	Changes number of granules?	Changes mass of granules?	Discrete or differential?
nucleation	yes	yes	discrete
layering	no	yes	differential
coalescence	yes	no	discrete
attrition	no	yes	differential

Reprinted from *Granulation and Coating Technologies for High-Value-Added Industries,* Ennis and Litster (1996) with permission of E & G Associates. All rights reserved.

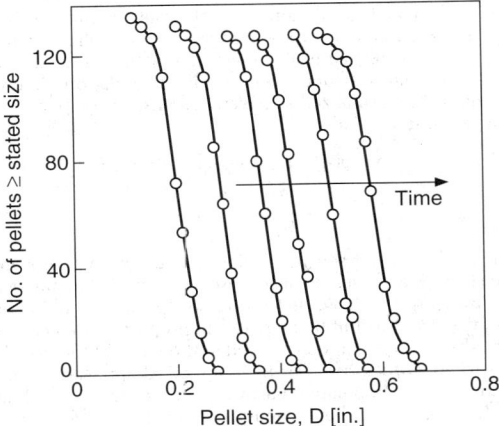

FIG. 20-95 Batch drum growth of limestone pellets by layering with a size-independent linear growth rate [*Capes*, Chem. Eng., **45**, CE78 (1967).]

rate. The **order** of the kernel has a major effect on the shape and evolution of the granule size distribution. [See Adetayo & Ennis, *AIChE J.* 1997 (In press).] Several empirical kernels have been proposed and used (Table 20-57).

All the kernels are empirical, or semiempirical and must be fitted to plant or laboratory data. The kernel proposed by Adetayo and Ennis is consistent with the granulation regime analysis described above (see section on growth) and is therefore recommended:

$$\beta(u,v) = \begin{cases} k, w < w^\circ \\ 0, w > w^\circ \end{cases} \quad w = \frac{(uv)^a}{(u+v)^b} \qquad (20\text{-}74)$$

where w° is the **critical average granule volume** in a collision corresponding to $St = St^\circ$, and it is related to the critical cutoff diameter defined above (Eq. 20-47). For fine powders in the noninertial regime (See section on growth) where $St \ll St^\circ$, this kernel collapses to the simple **random** or size-independent kernel $\beta = k$ for which the mean granule size increases exponentially with time (Eq. 20-51). Where deformation is unimportant, coalescence occurs only in the noninertial regime and stops abruptly when $St_v = St_v^\circ$. Based on the granulation regime analysis, the effects of feed characteristics and operating variables on granulation extent has been predicted [Adetayo et al., *Powder Tech.*, **82**, 47–59 (1995)]. (See "Extent of Noninertial Growth" subsection.)

Modeling growth where deformation is significant is more difficult. It can be assumed that a critical cutoff size exists w° which determines which combination of granule sizes are capable of coalescence, based on their inertia. When the harmonic average of sizes of two colliding granules w is less than this critical cut-off size w°, coalescence is successful, or

$$w = \frac{(uv)^b}{(u+v)^a} = w^\circ = \frac{\pi}{6}\left(\frac{16\mu}{\rho u_0} St^\circ\right)^3 \qquad (20\text{-}75)$$

where a and b are model parameters expected to vary with granule deformability, and u and v are granule volumes. To be dimensionally consistent, $2b - a = 1$. w° and w involving the parameters a and b represent a generalization of the Stokes analysis for non-deforming systems, for which case $a = b = 1$. For deformable systems, the kernel is then represented by Eq. 20-74. Figure 20-96 illustrates the evolution of the granule size distribution as predicted by this cutoff-based kernel which accounts for deformability. The cutoff kernel is seen to clearly track the experimental average granule size over the life of the granulation, illustrating that multiple kernels are not necessary to describe the various stages of granule growth, including the initial stage of random noninertial coalescence and the final stage of nonrandom preferential inertial growth by balling or crushing and layering (see Table 20-39).

Attrition The wearing away of granule surface material by attrition is the direct opposite of layering. It is an important mechanism when drying occurs simultaneously with granulation and granule velocities are high, e.g., fluidized beds and spouted beds. In a fluid bed [Ennis & Sunshine, *Tribology International*, **26**, 319 (1993)], attrition rate is proportional to excess gas velocity $(U - U_{mf})$ and approximately inversely proportional to granule-fracture toughness K_c, or $A \propto (U - U_{mf})/K_c$. For spouted beds, most attrition occurs in the spout and the attrition rate may be expressed as:

$$A \propto \frac{A_i U_i^3}{K_c} \qquad (20\text{-}76)$$

where A_i and U_i are the inlet orifice area and gas velocity, respectively. Attrition rate also increases with increasing slurry feed rate [Liu and Litster, *Powder Tech.*, **74**, 259 (1993)]. Granule breakage by fragmentation is also possible, with its rate being described by an **on function** which plays a similar role as the coalescence kernel does for growth. (See "Principles of Size Reduction" and "Breakage" sections for additional details.)

SOLUTION OF THE POPULATION BALANCE

Effects of Mixing As with chemical reactors, the **degree of mixing** within the granulator has an important effect on the final granule size distribution because of its influence on the **residence time distribution.** Fig. 20-94 shows the difference in exit size distribution for a plug-flow and well-mixed granulator for growth by layering only. In general, the exit size distribution is broadened and the extent of growth (for constant rate constants) is diminished for an increased degree of mixing in the granulator. With layering and attrition rates playing the role of generalized velocities, coalescence, and fragmentation rates, the role of reaction rate constants, methodologies of traditional reaction engineering may be employed to design granu-

TABLE 20-57   Coalescence Kernels for Granulation

Kernel	Reference and comments
$\beta = \beta_o$	Kapur & Fuerstenau [*I&EC Proc. Des. & Dev.*, **8**(1), 56 (1969)], size-independent kernel.
$\beta = \beta_o \dfrac{(u+v)^a}{(uv)^b}$	Kapur [*Chem. Eng. Sci.*, **27**, 1863 (1972)], preferential coalescence of limestone.
$\beta = \beta_o \dfrac{(u^{2/3} + v^{2/3})}{1/u + 1/v}$	Sastry [*Int. J. Min. Proc.*, **2**, 187 (1975)], preferential balling of iron ore and limestone.
$\beta(u,v) = \begin{cases} k, & w < w^\circ \\ 0, & w > w^\circ \end{cases} \quad w = \dfrac{(uv)^a}{(u+v)^b}$	Adetayo & Ennis [*AIChE J.*, (1997) In press], based on granulation regime analysis.

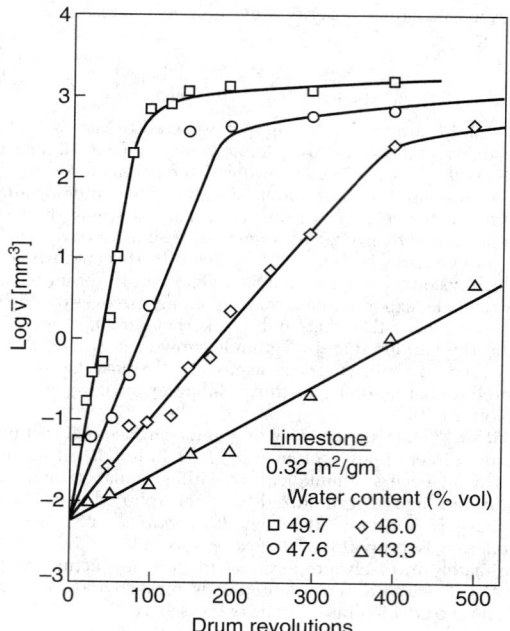

FIG. 20-96 Batch drum growth of limestone by coalescence. Note granule size increases exponentially with time in the first stage of noninertial growth. Experimental data of Kapur [*Adv. Chem. Eng.*, *10*, *56 (1978)*] compared with single deformable granulation kernel (Eqs. 20-74, 20-75). [*Adetayo & Ennis, AIChEJ. (In press.)*] Reproduced with permission of the American Institute of Chemical Engineers. Copyright AIChE. All rights reserved.

lation systems or optimize the granule size distribution. [For the related example of crystallization, see Randolph & Larson (*Theory of Particulate Processes,* 2d ed., Academic Press, 1991).] Table 20-58 lists some mixing models that have been used for several types of granulators.

Analytical Solutions   Solution of the population balance is not trivial. Analytical solutions are available for only a limited number of special cases, of which some of examples of practical importance are summarized in Table 20-59. For other analytical solutions, see general references on population balances given above.

In general, analytical solutions are only available for specific initial or inlet size distributions. However, for batch granulation where the only growth mechanism is coalescence, at long times the size distribution may become **self-preserving.** The size distribution is self-preserving if the normalized size distributions $\varphi = \varphi(\eta)$ at long time are independent of mean size $\bar{v}$, or

$$\varphi = \varphi(\eta) \quad \text{only where} \quad \eta = v/\bar{v}$$

$$\bar{v} = \int_0^\infty v \cdot n(v,t)\, dv \qquad (20\text{-}77)$$

Analytical solutions for self-preserving growth do exist for some coalescence kernels and such behavior is sometimes seen in practice (Fig. 20-97). Roughly speaking, self-preserving growth implies that the width of the size distribution increases in proportion to mean granule size, i.e., the width is uniquely related to the mean of the distribution.

Numerical Solutions   For many practical applications, numerical solutions to the population balance are necessary. Several numerical solution techniques have been proposed. It is usual to break the size range into discrete intervals and then solve the resulting series of ordinary differential equations. A geometric discretization reduces the number of size intervals (and equations) that are required. Litster et al. [*AIChE J.*, (1995)] give a general discretized PB for nucleation, growth, and coalescence with a geometric discretization of $v_j = 2^{1/q} v_{j-1}$ where q is an integer. Accuracy is increased (at the expense of

TABLE 20-58   Mixing Models for Continuous Granulators

Granulator	Mixing Model	Reference
Fluid bed	Well-mixed	See Sec. 17
Spouted bed	Well-mixed	Liu and Litster, *Powder Tech.*, **74**, 259 (1993)
	Two-zone model	Litster, et al. [*Proc. 6th Int. Symp. Agglom.*, Soc. Powder Tech., Japan, **123** (1993)]
Drum	Plug-flow	Adetayo et al., *Powder Tech.*, **82**, 47–59 (1995)
Disc	Two well-mixed tanks in series with classified exit	Sastry & Loftus [*Proc. 5th Int. Symp. Agglom.*, IChemE, 623 (1989)]
	Well-mixed tank and plug-flow in series with fines bypass	Ennis, Personal communication (1986)

TABLE 20-59   Some Analytical Solutions to the Population Balance*

Mixing state	Mechanisms operating	Initial or inlet size distribution	Final or exit size distribution
Batch	Layering only: $G(x) = constant$	Any initial size distribution, $n_0(x)$	$n(x) = n_0(x - \Delta x)$ where $\Delta x = Gt$
Continuous & well-mixed	Layering only: $G(x) = constant$	$n_{in}(x) = N_{in}\, \delta(x - x_{in})$	$n(x) = \dfrac{N_0 G}{\tau} \exp\left(-\dfrac{\tau(x - x_{in})}{G}\right)$
Batch	Coalescence only, size independent: $\beta(u, v) = \beta_o$	$n_o(v) = N_0\delta(v - v_o)$	$n(v) = \dfrac{N_0}{\bar{v}} \exp\left(-\dfrac{v}{\bar{v}}\right)$ where $\bar{v} = v_0 \exp\left(\dfrac{\beta_o t}{6}\right)$
Batch	Coalescence only, size independent: $\beta(u, v) = \beta_o$	$n_o(v) = \dfrac{N_0}{v_0} \exp\left(-\dfrac{v}{v_o}\right)$	$n(v) = \dfrac{4 N_0}{v_0(N_0\beta_o t + 2)^2} \exp\left[\dfrac{-2v/v_0}{N_0\beta_o t + 2}\right]$

*Randolph and Larson, *Theory of Particulate Processes,* 2d ed., Academic Press, New York (1988); Gelbart and Seinfeld, *J. Computational Physics,* **28**, 357 (1978).

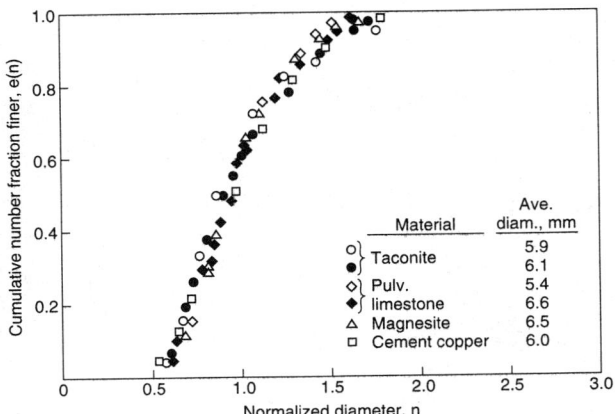

FIG. 20-97 Self-preserving size distributions for batch coalescence in drum granulation. [*Sastry,* Int. J. Min. Proc., **2,** *187 (1975).*] With kind permission of Elsevier Science -NL, 1055 KV Amsterdam, the Netherlands.

computational time) by increasing the value of q. Their discretized PB is recommended for general use.

SIMULATION OF GRANULATION CIRCUITS WITH RECYCLE

When granulation circuits include recycle streams, both steady-state and dynamic responses can be important. Computer simulation packages are now widely used to design and optimize many process flow sheets, e.g., comminution circuits, but simulation of granulation circuits is much less common. Commercial packages do not contain library models for granulators. Some researchers have developed simulations and used these for optimization and control studies [Sastry, *Proc. 3rd Int. Symp. Agglom.* (1981); Adetayo et al., *Computers Chem. Eng.,* **19,** 383 (1995); Zhang et al., *Control of Part. Processes IV* (1995)]. For these simulations, dynamic population-balance models have been used for the granulator. Standard literature models are used for auxiliary equipment such as screens, dryers, and crushers. These simulations are valuable tools for optimization studies and development of control strategies in granulation circuits, and may be employed to investigate the effects of transient upsets in operating variables, particularly moisture level and recycle ratio, on circuit performance.

Handling of Bulk Solids and Packaging of Solids and Liquids

Grantges J. Raymus, M.E., M.S., *President, Raymus Associates, Incorporated, Packaging Consultants; Adjunct Professor and Program Coordinator, Center for Packaging Science and Engineering, College of Engineering, Rutgers, The State University of New Jersey; formerly Manager of Packaging Engineering, Union Carbide Corporation; Registered Professional Engineer, California; Member, Institute of Packaging Professionals, ASME.*

INTRODUCTION

During the period between this edition and the sixth, notable changes have taken place which influence packaging and material handling in the chemical industry. These are: Rapid change in the technologies involved; the impact of governmental regulations on both packages and equipment; and the global nature of the chemical industry. These are all interrelated and must be taken into account by the chemical engineer who is planning and choosing packages for chemical products and the packaging machinery and material handling equipment which will be used. Packaging and material handling technology—which had largely been an art, blending science and engineering—has seen dramatic changes. Influencing these are: The advent of special purpose and personal computers; availability of software which allows the rapid calculation and presentation of data and control of processes; and the unusual degree of development of sensing devices.

With these technologies, literally anything in the operation of a packaging or material handling system can be sensed and then via computer, using most often commercially available software, incredible levels of calculations can be performed. These can be used to control the process producing the data and can present operators, engineers, and managers with data and calculations describing virtually any facet of the packaging or material handling systems operation. Closely akin to these is the advent of robots which can carry out many human tasks. Robots are controlled by computers, under the direction of software which relies on sensors and computers. A further outgrowth of this technical revolution occurs in the management and control of inventories—whether they be in the raw material stage, in processing, or in a finished goods sales warehouse. Automatic storage and retrieval equipment under the direction of a special-purpose computer allows unbelievable degrees of identifying all items in the inventory: Where they are, their age, and which inventory to choose to optimize the profit plan of the corporation which owns them. Underlying the control of this sophisticated inventory-management system is an emerging technology known as **bar coding.** This is a logical outgrowth of the development of the personal computer and the software for its operation. The placing of lasers into printing devices has created the ability to produce labels with bar codes. Production of labels has become a real-time packaging-line-type operation rather than an operation in which the labels have to be ordered and inventoried, as was the case before the dramatic technical revolution just described.

The weighing and proportioning of liquids and solids has also benefited from this technical revolution. Sensing devices and special-purpose computers give a level of precision and speed not possible in the era of electromechanical devices. The net result is that packaging and material handling systems now have the sophistication of chemical processes.

All of this gives the chemical engineer new levels of ability to design, operate, and manage complex material handling and packaging systems. Since it is still incumbent on the engineer to define production and performance requirements, some definitions of productivity are useful.

PHYSICAL-DISTRIBUTION CONCEPT

Systems Approach   **Physical distribution** is a term applied to a systems concept that comprises the entire spectrum of materials movement. The system begins with the storage and handling of raw materials and follows right on through the packaging and disposition of the finished product. The aim is the attainment of the **lowest over-all cost** for the system as a whole, comprising the expenses borne by the manufacturer, transport carrier, warehouser, distributor, and customer. Even the manner in which the customer will handle the product is often taken into account.

Two main benefits accrue from a systems approach to materials handling and packaging. First, a trade-off of investment and operating costs is made possible; higher costs in some parts of a system become permissible in return for much lower costs in other parts. The net result is usually the lowest overall cost. If this is not the case, the reasons for incurring the higher costs can be identified and justified. The second benefit is that customers are not offended by ill-conceived packages, delivery vehicles, or product characteristics.

Mathematical modeling, using digital computers, aids in performing a systems-type analysis for either the entire system or parts of it. By means of integer or linear-programming techniques, optimum systems can be identified. The dynamic performance of these can then be determined by simulation techniques.

Determining the capacities of material-handling and packaging equipment is a primary consideration. Many **interacting variables** often are involved, such as an ever-changing or intermittent material-delivery rate, the capacity of intermediate storage and receiver bins, random stoppage or failure of equipment in the system, and the setup and cleanup time between product grades or blends. Variables frequently interact in such complex ways that conventional capacity analyses are impossible, especially if the interaction varies with time. Under such conditions the question of whether the system will deliver the required output can be answered only by simulation techniques.

Even when a total system analysis is unnecessary, the methodology of mathematical modeling is useful, because by considering each component of a system as a block of a flow sheet, the interrelationships become much clearer. Additional alternatives often become apparent, as does the need for more equipment-performance data.

Capacity Definitions   In any analysis, the capacity per unit time of dynamic equipment (such as conveyors and bagging machines), as well as the rates at which they actually perform, must be defined more precisely and realistically than by a mere statement of kilograms or pounds per hour. Some useful definitions employed by the equipment industry are the following:

Instantaneous Rate   This is a short-term rate when the equipment operates at the design rate or faster. Typical is the average weight handled over a short period of time, not exceeding 5 min.

Hourly Rate   This intermediate rate takes into account equipment stoppages due principally to mechanical downtime rather than the equipment's idle time while it waits for action by other parts of the system.

Shift Rate   A long-term rate, this reflects all causes of downtime, including idle time. Thus, the average per shift will vary, but by examining its range the practical capacity can be determined. Production time lost due to scheduling of the equipment affects the shift rate. On certain days the equipment has a shift rate close to the hourly rate, while on other days this rate is only half of the hourly rate. Examination of the reasons for this difference often identifies scheduled events as being responsible: the equipment was shut down for cleanup between product grades, product was unavailable for packaging because a bulk order had to be filled, or the product scheduled had a production rate which was half of the products normally made.

These capacity definitions are used to define responsibilities of both vendors and buyers. For instance, often a vendor is called in to examine a piece of equipment that does not perform at the "guaranteed" rate. Records of shift production are offered as proof. Yet the vendor then makes a test and shows that the guaranteed rate is met over a short interval. Who is correct? By defining rates the engineer responsible for the installation not only can avoid these situations but can obtain a better appreciation of potential plant situations.

CONVEYING OF BULK SOLIDS

CONVEYOR SELECTION

Selection of the correct conveyor for a specific bulk material in a specific situation is complicated by the large number of interrelated factors that must be considered. First, the alternatives among basic types must be weighed, and then the correct model and size must be chosen. Workability is the first criterion, but the degree of performance perfection that can be afforded must be established.

Because **standardized equipment designs** and complete engineering data are available for many common types of conveyors, their performance can be accurately predicted when they are used with materials having well-known conveying characteristics. However, even the best conveyors can perform disappointingly if material characteristics are unfavorable. It is often true that conveyor engineering is more of an art than a science; problems involving unusual materials or equipment should be approached with caution.

Many preengineered conveyor components can be purchased off the shelf; they are economical and easy to assemble, and they perform well on conventional applications (for which they are designed). However, it is advisable to check with the manufacturer to be sure that the application is proper.

Capacity requirement is a prime factor in conveyor selection. Belt conveyors, which can be manufactured in relatively large sizes to operate at high speeds, deliver large tonnages economically. On the other hand, screw conveyors become extremely cumbersome as they get larger and cannot be operated at high speeds without creating serious abrasion problems.

Length of travel is definitely limited for certain types of conveyors. With high-tensile-strength belting, the length limit on belt conveyors can be a matter of miles. Air conveyors are limited to 305 m (1000 ft); vibrating conveyors, to hundreds of meters or feet. In general, as length of travel increases, the choice among alternatives becomes narrower.

Lift can usually be handled most economically by vertical or inclined bucket elevators, but when lift and horizontal travel are combined, other conveyors should be considered. Conveyors that combine several directions of travel in a single unit are generally more expensive, but since they require only a single drive, this feature often compensates for the added base cost.

Material characteristics, both chemical and physical, should be considered, especially flowability. Abrasiveness, friability, and lump size are also important. Chemical effects (e.g., the effect of oil on rubber or of acids on metal) may dictate the structural materials out of which conveyor components are fabricated. Moisture or oxidation effects from exposure to the atmosphere may be harmful to the material being conveyed and require total enclosure of the conveyor or even an artificial atmosphere. Obviously, certain types of conveyors lend themselves to such special requirements better than others.

Processing requirements can be met by some conveyors with little or no change in design. For example, a continuous-flow conveyor may provide a desired cooling of the solids simply because it puts the conveyed material into direct contact with heat-conducting metals. Screen decks can be readily attached to vibrating conveyors for simple sizing and scalping operations, and special flights or casings on screw conveyors are available for a wide variety of processing operations such as mixing, dewatering, heating, and cooling.

Initial cost of a conveyor system is usually related to **life expectancy** as well as to the flow rate chosen. There is a great temptation to overdesign, which should be resisted. The first really long-distance belt conveyor was designed and fabricated to extremely high standards of quality. After 35 years it was still in operation with almost all its original components. Had this operation been planned for only a 10-year life, the conveyor system would have represented a bad case of overdesign. While there is a market for used conveyor equipment, it is extremely limited. Thus, it is important to choose conveyor quality for expected life of project.

Comparative costs for conveyor systems can be based only on studies of specific problems. For example, belt-conveyor idlers are available in a range of qualities that may make the best unit cost three times as much as the cheapest. Bearing quality, steel thickness, and diameter of rolls all affect cost, as does design for easy maintenance and repair. Therefore, it is necessary to make cost comparisons on the basis of a specific study for each conveyor application.

As a general guide to conveyor selection, Table 21-1 indicates **conveyor choices** on the basis of some common functions. Table 21-2 is designed to aid in **feeder selection** on the basis of the physical characteristics of the material to be handled. Table 21-3 is a coded listing of **material characteristics** to be used with Table 21-4, which describes the conveying qualities of some common materials. While these tables may serve as valuable guides, conveyor selection must be based on the **as-conveyed characteristics** of a material. For instance, if packing or aerating can occur in the conveyor, the machine's performance will not meet expectations if calculations are based on an average weight per cubic meter. Storage conditions, variations in ambient temperature and humidity, and discharge methods may all affect conveying characteristics. Such factors should be carefully considered before making a final conveyor selection.

To obtain a reliable **measurement of bulk density,** any wide-mouthed vessel with a capacity of 1 ft³ or more may be used. When such a determination must be made often, it is worthwhile to con-

TABLE 21-1   Conveyors for Bulk Materials*

Function	Conveyor type
Conveying materials horizontally	Apron, belt, continuous flow, drag flight, screw, vibrating, bucket, pivoted bucket, air
Conveying materials up or down an incline	Apron, belt, continuous flow, flight, screw, skip hoist, air
Elevating materials	Bucket elevator, continuous flow, skip hoist, air
Handling materials over a combination horizontal and vertical path	Continuous flow, gravity-discharge bucket, pivoted bucket, air
Distributing materials to or collecting materials from bins, bunkers, etc.	Belt, flight, screw, continuous flow, gravity-discharge bucket, pivoted bucket, air
Removing materials from rail cars, trucks, etc.	Car dumper, grain-car unloader, car shaker, power shovel, air

*From FMC Corporation, Material Handling Systems Division.

TABLE 21-2   Feeders for Bulk Materials*

Material characteristics	Feeder type
Fine, free-flowing materials	Bar flight, belt, oscillating or vibrating, rotary vane, screw
Nonabrasive and granular materials, materials with some lumps	Apron, bar flight, belt, oscillating or vibrating, reciprocating, rotary plate, screw
Materials difficult to handle because of being hot, abrasive, lumpy, or stringy	Apron, bar flight, belt, oscillating or vibrating, reciprocating
Heavy, lumpy, or abrasive materials similar to pit-run, stone, and ore	Apron, oscillating or vibrating, reciprocating

*From FMC Corporation, Material Handling Systems Division.

TABLE 21-3 Classification System for Bulk Solids*

	Material characteristics	Class
Size	Very fine—< 149 μm (100 mesh)	A
	Fine—149 μm to 3.18 mm (100 mesh to ⅛ in)	B
	Granular—3.18 to 12.7 mm (⅛ in to ½ in)	C
	Lumpy—containing lumps > 12.7 mm (½ in)	D
	Irregular—being fibrous, stringy, or the like	H
Flowability	Very free-flowing—angle of repose up to 30°	1
	Free-flowing—angle of repose 30 to 45°	2
	Sluggish—angle of repose 45° and up	3
Abrasiveness	Nonabrasive	6
	Mildly abrasive	7
	Very abrasive	8
Special characteristics	Contaminable, affecting use or salability	K
	Hygroscopic	L
	Highly corrosive	N
	Mildly corrosive	P
	Gives off dust or fumes harmful to life	R
	Contains explosive dust	S
	Degradable, affecting use or salability	T
	Very light and fluffy	W
	Interlocks or mats to resist digging	X
	Aerates and becomes fluid	Y
	Packs under pressure	Z

*From FMC Corporation, Material Handling Systems Division.

Example: A material which is granular, very free-flowing, mildly abrasive, and mildly corrosive would fall in classes C, 1, 7, and P, making its classification C17P.

struct a test box from wood or light metal having a dimension of exactly 1 ft for length, width, and depth. The material to be weighed is poured into the test box so as to overfill it slightly. After the material has been leveled, the test box and its contents are weighed, and after adjustment has been made for the tare weight, the weight obtained is equivalent to the bulk density in the loose or flowing condition. If a loose density is to be determined, care should be exercised in filling the test box so as not to rap or vibrate the material. When a settled density is needed, the filling portion of the procedure is accompanied by rapping the walls of the box until no more material can be added. The density value obtained by this experiment (pounds per cubic foot) can be directly converted to SI units by multiplying by 16.02, giving density in kilograms per cubic meter.

Conveyor Drives   Conveyor drives may account for from 10 to 30 percent of the total cost of the conveyor system, depending on specific job requirements. They may be either fixed-speed or adjustable-speed type. **Fixed-speed drives** are used when the initially chosen conveyor speed does not require change during the course of normal operation. Simple sheave or sprocket changes suffice should minor speed alterations be needed. However, for major adjustments motor or speed-reducer changes are required. In any event, the conveyor must be shut down while the speed change is made. **Adjustable-speed drives** are designed for changing speed either manually or automatically while the conveyor is in operation, to meet variations in processing requirements.

The number of **speed reductions** is another way to classify conveyor drives. Most common of the speed-reduction methods is the two-step system, in which the motor is coupled to a speed reducer and the slow-speed shaft of the reducer is connected to the conveyor-drive shaft by a V belt or a roller chain. The second reduction not only permits the use of a simpler speed reducer but also allows a more flexible layout of the motor and reducer mounting plate. On many installations this eliminates the need for a specially designed drive mount.

Since it is good practice to maintain a selected inventory of spare parts for drives, economy can be achieved by **standardizing conveyor drives** throughout the plant. For example, intermediate speed reduction by means of V belts, sheaves or chains, and sprockets can frequently permit using the same speed-reducer size for several drives. Thus, it may be necessary to keep only one repair-stock speed reducer for a number of conveyors.

Conveyor Motors   Motors for conveyor drives are generally three-phase, 60-Hz, 220-V units; 220/440-V; 550-V; four-wire, 208-V. Also common are 240- and 480-V ratings. Although many adjustable-

speed drives use alternating-current induction motors, powered by ac alternators or ac-driven eddy-current clutches, there is a strong preference for direct-current motors when speed adjustments are required over a wide range at extremely accurate settings.

The **silicon-controlled rectifier** with a dc motor has become predominant in adjustable-speed drives for almost all commonly used conveyors when speed adjustment to process conditions is necessary. The low cost of this control device has influenced its use when speed synchronization among conveyors is required. This can also be done, of course, by changing sheave or sprocket ratios.

The **squirrel-cage motor** is most commonly used with belt conveyors and with drives up to 7.457 kW (10 hp); across-the-line starting is generally specified. Between 7.457 and 37.285 kW (10 and 50 hp), squirrel-cage motors are usually started by means of a manual reduced-voltage starter or a magnetic primary-resistance starter. Normal-torque motors are generally specified, with the assumption that if power is sufficient to drive the belt, sufficient starting torque can be developed. Motor selection for large conveyors should be based on a careful study, with particular emphasis on starting conditions.

Auxiliary Equipment   Elevating conveyors must be equipped with some form of **holdback** or **brake** to prevent reversal of travel and subsequent jamming when power is unexpectedly cut off. Ratchet and wedge roller-type holdbacks are commonly used. Solenoid brakes and spring clutches may also be employed.

Another problem with most conveyors is to cut out the driving force when a conveyor jams. **Torque-limiting devices** are often used, as are electrical controls which cut power to the drive motor. However, because of the high inertia of the motor rotor, it is sometimes desirable to eliminate the torque surge which may occur when the conveyor jams. A shear-pin hub is generally used in these cases, power being transmitted through a set of pins which are designed to shear at a fixed maximum torque. While equipment remains down until the pins can be replaced, there is an immediate disconnect between motor and conveyor which may prevent serious equipment damage. Special clutches are also used.

Unless a material discharges freely, **cleaners** are required on belt conveyors and may be helpful on others. Common types use a rotating brush, powered from the conveyor head-pulley shaft or independently, or a spring-mounted blade. The latter is applicable only at some point where the belt conveyor lies reasonably flat. Whenever cleaners are used, provision should be made for catching and chuting the material back into the main discharge stream or to a collecting container which can be periodically emptied.

Control of Conveyors   Control has been enhanced considerably with the introduction of process-control computers and programmable controllers, which can be used to maintain rated capacities to close tolerances. This ability is especially useful if feed to the conveyor tends to be erratic. Through variable-speed drives, outputs can be adjusted automatically for changes in processing conditions. When the control devices are used in conjunction with strain-gauge or load-cell weight-sensing devices, actual discharge rates can be measured and employed in process calculations made by these devices, and output adjustments can be made automatically and accurately.

SCREW CONVEYORS

The screw conveyor is one of the oldest and most versatile conveyor types. It consists of a helicoid flight (helix rolled from flat steel bar) or a sectional flight (individual sections blanked and formed into a helix from flat plate), mounted on a pipe or shaft and turning in a trough. Power to convey must be transmitted through the pipe or shaft and is limited by the allowable size of this member. Screw-conveyor capacities are generally limited to around 4.72 m³/min (10,000 ft³/h).

In addition to their conveying ability, screw conveyors can be adapted to a wide variety of **processing operations**. Almost any degree of mixing can be achieved with screw-conveyor flights cut, cut and folded, or replaced by a series of paddles. Use of ribbon flights allows sticky materials to be handled. Variable-pitch, tapered-flight, or stepped-flight units can give excellent control for feeder applications

TABLE 21-4   Material Classes and Bulk Densities*

Material	Average bulk density, lb/ft³†	Class‡	Material	Average bulk density, lb/ft³†	Class‡
Alum, lumpy	50–60	D26§	Lime, ground, ⅛ in and under	60	B36Z
Alum, fine	45–50	B26§	Lime, hydrated, ⅛ in and under	40	B26YZ
Alumina	60	B28	Lime, hydrated, pulverized	32–40	A26YZ
Alumina gel	45	B27	Lime, pebble	53–56	D36
Aluminum hydrate	18	C26	Limestone, agricultural, ⅛ in and under	68	B27§
Ammonium chloride, crystalline	52	B26	Limestone, crushed	85–90	D27§
Ammonium sulfate	45–58	§	Limestone dust	75	A37Y§
Antimony powder		B27	Magnesium chloride	33	C36
Asbestos shred	20–25	H37WZ	Manganese sulfate	70	C28
Ashes, coal, dry, 3 in. and under	35–40	D37	Marl	80	D27§
Asphalt, crushed, ½ in and under	45	C26	Mica, flakes	17–22	B17WY
Bagasse	7–10	H36WXZ	Mica, ground	13–15	B27
Baking powder	41	A26	Mica, pulverized	13–15	A27Y
Bark, wood, refuse	10–20	H37X§	Muriate of potash	77	B28
Bauxite, crushed, 3 in and under	75–85	D28§	Naphthalene flakes	45	§
Bentonite, 100 mesh and under	50–60	A27Y§	Oxalic acid crystals	60	B36L
Bicarbonate of soda	41	A26	Oyster shells, ground, ½ in and under	53	C27
Boneblack, 100 mesh and under	20–25	A27§	Oyster shells, whole		D27X
Bonechar, ⅛ in and under	27–40	B27	Phenol-formaldehyde molding powder	30–40	A36
Bonemeal	55–60	B27	Phosphate rock	75–85	D27§
Borate of lime		A26§	Phosphate sand	90–100	B28
Borax, fine	53	B26	Phthalic anhydride flakes	30–35	C36XZ
Boric acid, fine	55	B26	Polyethylene pellets, high-density	35–45	C16K
Calcium carbide	70–80	D27	Polyethylene pellets, low-density	28–40	C16K
Carbon black, pelletized	20–25	B16TZ§	Polypropylene pellets	35–50	C16K
Carbon black, powder	4–6	§	Polystyrene cubes	35–40	C16K
Casein	36	B27§	Polyvinyl chloride pellets, compounds	35–55	C16K
Cast-iron chips	130–200	C37	Polyvinyl chloride resin, dispersion-type	12–18	A36KPY
Cement, Portland	65–85	A27Y	Polyvinyl chloride resin, solvent, non-solvent, suspension types	20–35	A26KY
Cement clinker	75–80	D28§	Potassium nitrate	76	C17P
Chalk, lumpy	85–90	D37Z	Pumice, ⅛ in and under	42–45	B38§
Chalk, 100 mesh and under	70–75	A37YZ	Salt, common dry, coarse	45–50	C37PL§
Charcoal	18–25	D37T	Salt, common dry, fine	70–80	B27PL§
Cinders, coal	40	D28§	Salt cake, dry, coarse	85	D27
Clay (see bentonite, fuller's earth, kaolin, and marl)			Salt cake, dry, pulverized	65–85	B27
Coal, anthracite	60	C27P	Saltpeter	80	B26S
Coal, bituminous, mined, 50 mesh and under	50	B36P	Sand, bank, dry	90–110	B28
Coal, bituminous, mined, sized	50	D26PT	Sand, silica, dry	90–100	B18
Coal, bituminous, mined, slack, ½ in and under	50	C36P	Sawdust	10–13	§
Coke, loose	23–32	D38TX§	Shale, crushed	85–90	C27
Coke, petroleum, calcined	35–45	D28X	Shellac, powdered or granulated	31	B26K§
Coke breeze, ¼ in and under	25–35	C38	Silica gel	45	B28
Copper sulfate		D26	Slag, furnace, granulated	60–65	C28
Cork, fine ground	12–15	B36WY	Slate, crushed, ½ in and under	80–90	C27
Cork, granulated	12–15	C36	Slate, ground, ⅛ in and under	82	B27
Cryolite	110	D27	Soap beads or granules		B26T
Cullet	80–120	D28§	Soap chips	15–25	C26T§
Dicalcium phosphate	43	A36	Soap flakes	5–15	B26T§
Dolomite, lumpy	90–100	D27§	Soap powder	20–35	B26§
Ebonite, crushed, ½ in and under	63–70	C26	Soapstone talc, fine	40–50	A37Z
Epsom salts	40–50	B26	Soda ash, heavy	55–65	B27
Feldspar, ground, ⅛ in and under	65–70	B27	Soda ash, light	20–35	A27W
Ferrous sulfate	50–75	C27	Sodium nitrate	70–80	§
Flour, wheat	35–40	A36K§	Sodium sulfate (see salt cake)		
Fluorspar	82	C37	Starch	25–50	§
Fly ash, dry	35–45	A18Y§	Steel chips, crushed	100–150	D38
Fuller's earth, oil filter, burned	40	B28	Sugar, granulated	50–55	B26KT
Fuller's earth, oil filter, raw	35–40	B27	Sugar, raw, cane, or beet	55–65	B36Z§
Fuller's earth, oil filter, spent	60–65	§	Sugar-beet pulp, dry	12–15	§
Glass batch	90–100	D28§	Sugar-beet pulp, wet	25–45	§
Glue, ground, ⅛ in and under	40	B27	Sulfur, crushed, ½ in and under	50–60	C26S§
Graphite, flake	40	C26	Sulfur, lumpy, 3 in and under	80–85	D26S§
Graphite, flour	28	A16Y	Sulfur, powdered	50–60	B26SY§
Gypsum, calcined, ½ in and under	55–60	C27	Talcum powder	40–60	A27Y
Gypsum, calcined, powdered	60–80	A37	Trisodium phosphate	60	B27
Gypsum, raw, 1 in and under	90–100	D27	Vermiculite, expanded	16	C37W
Ice, crushed	35–45	D16	Vermiculite ore	80	D27
Ilmenite	140	B28	Wood chips	10–30	H36WX§
Kaolin clay, 3 in and under	163	D27	Wood flour	16–36	§
Lead arsenate	72	B36R	Zinc oxide, heavy	30–35	A36Z§
Lignite, air dried	45–55	D26	Zinc oxide, light	10–15	A36WZ§

*Data supplied mostly by FMC Corporation, Material Handling Systems Division. To convert pounds per cubic foot to kilograms per cubic meter, multiply by 16.02.
†Weights of material, loose or slightly agitated. Weights are usually different when materials are settled or packed as in bins or containers.
‡These classes represent observations under general conditions. Specific conditions may vary because of manufacturing processes and handling.
§Class may vary considerably because of conditions.

or on conveyors when precise control of the transport rate is required. Short-pitch screws are used for inclined and vertical conveying applications, and double-flight short-pitch units effectively deter flushing action. In addition to a wide variety of designs for components, screw conveyors may be fabricated in materials ranging from cast iron to stainless steel.

Use of hollow screws and pipes for circulating hot or cold fluids allows the screw conveyor to be used for heating, cooling, and drying operations. Jacketed casings may be used for the same purpose. It is relatively easy to seal a screw conveyor from the outside atmosphere so that it can operate outdoors without special protection. In fact, the conveyor can be completely sealed to operate in its own atmosphere at positive or negative pressure, and the casing can be insulated to maintain internal temperatures in areas of high or low ambient temperature. A further advantage is the fact that the casing can be designed with a drop bottom for easy cleaning to avoid contamination when different materials are to be run through the same system.

Since screw conveyors are usually made up of standard sections coupled together, special attention should be given to bending stresses in the couplings. Hanger bearings supporting the flights obstruct the flow of material when the trough is loaded above their level. Thus, with difficult materials, the load in the trough must be kept below this level, or special hanger bearings which minimize obstruction should be selected. Since screw conveyors operate at relatively low rotational speeds, the fact that the outer edge of the flight may be moving at a relatively high linear speed is often neglected. This may create a wear problem; if wear is too severe, it can be reduced by the use of hard-surfaced edges, detachable hardened flight segments, rubber covering, or high-carbon steels.

Power calculations for screw conveyors are well standardized. However, each manufacturer has grouped numerical constants in a different fashion and assigned slightly different values on the basis of individual design variations. Thus, in comparing screw-conveyor power requirements it is advisable to use a specific formula for specific equipment.

Required power is made up of two components, that necessary to drive the screw empty and that necessary to move the material. The first component is a function of conveyor length, speed of rotation, and friction in the conveyor bearings. The second is a function of the total weight of material conveyed per unit of time, conveyed length, and depth to which the trough is loaded. The latter power item is in turn a function of the internal friction and friction on metal of the conveyed material.

Table 21-5 indicates **screw-conveyor performance** on the basis of material classifications as listed in Table 21-4 and defined in Table 21-3. Table 21-6 gives a wide range of **capacities** and **power requirements** for various sizes of screws handling 801 kg/m³ (50 lb/ft³) of material of average conveyability. Within reasonable limits, values from Tables 21-5 and 21-6 can be interpolated for preliminary estimates and designs.

Typical **feed arrangements** are shown in Fig. 21-1. Plain spouts (Fig. 21-1a) may be used when the feed rate is fairly uniform and controlled by preceding equipment. The capacity of the conveyor should be well above the maximum rate of feed from either single or multiple feed points. The rotary cutoff valve (Fig. 21-1b) is an enclosed dust-tight quick-acting valve for free-flowing materials. The rotary-vane feeder (Fig. 21-1c) delivers a uniform predetermined volume of material and may be driven from the screw or independently by constant- or variable-speed drive. Rack-and-pinion gates (Fig. 21-1d) are well suited to free-flowing materials in bins, hoppers, tanks, or silos and are also used as side inlet gates (Fig. 21-1e) for heavy or lumpy materials.

Typical **discharge arrangements** are shown in Fig. 21-2. Plain discharge openings (Fig. 21-2a) equipped with a discharge spout (Fig. 21-2b) are most common, although the open-end trough (Fig. 21-2c) is frequently used, as is the discharge-trough end (Fig. 21-2e). Open-bottom troughs (Fig. 21-2g) are often used for spreading material uni-

TABLE 21-5 Screw-Conveyor Capacities and Loading Conditions*

| Material class† | Screw diam., in | Max. lump size, in | | Capacity, cu ft/hr‡ | | Approx. area occupied by material¶ |
		25% lumps	100% lumps	At 1 rpm.	At max. rpm.§	
A, B, C, D, and H 16, 26, 36	6	¾	½	2.27	375	
	9	1½	¾	8.0	1,200	
	12	2	1	19.3	2,700	
	14	2½	1¼	30.8	4,000	
	16	3	1½	46.6	5,600	
	18	3	2	66.1	7,600	
	20	3½	2	95.0	10,000	45%
A, B, C, D, and H 17, 27, 37	6	¾	½	1.5	75	
	9	1½	¾	5.6	280	
	12	2	1	13.3	665	
	14	2½	1¼	21.1	1,055	
	16	3	1½	31.4	1,570	
	18	3	2	45.4	2,270	
	20	3½	2	62.1	3,105	30%
A, B, C, D, and H 18, 28, 38	6	¾	½	0.75	25	
	9	1½	¾	2.8	90	
	12	2	1	6.7	200	
	14	2½	1¼	10.5	300	
	16	3	1½	15.7	425	
	18	3	2	22.7	590	
	20	3½	2	31.1	780	15%

*FMC Corporation, Material Handling Systems Division. To convert cubic feet per hour to cubic meters per hour, multiply by 0.02832; to convert screw diameter in inches to the nearest screw size in centimeters, multiply by 2.5. See elsewhere for conversion of particle sizes from one measurement system to another.

†These classifications cover a broad list of materials that generally can be handled in a screw conveyor. Special consideration must be given to applications handling materials with the following characteristics:

Highly corrosive, Class N
Degradable, affecting use or salability, Class T
Interlocks or mats, Class X
Highly aerated or of fluid nature, Class Y

‡Capacity for horizontal conveyor uniformly fed. Volumetric capacity is based on material slightly agitated or fluffed. Material highly fluffed or aerated will decrease in weight and increase in volume.

§Maximum capacity for economical service.

¶Percentages higher than those indicated will result in excessive wear on hanger bearings and couplings.

TABLE 21-6   Screw-Conveyor Data for 50-lb/ft³ Material and Pipe-Mounted Sectional Spiral Flights*

Capacity†		Diam. of flights, in	Diam. of pipe, in‡	Diam. of shafts, in	Hanger centers, ft	Max. size of lumps			Speed, r/min	Max. torque capacity, in·lb	Feed section diam., in	hp at motor§					Max. hp capacity at speed listed
Tons/h	ft³/h					All lumps	Lumps 20 to 25%	Lumps 10% or less				15-ft max. length	30-ft max. length	45-ft max. length	60-ft max. length	75-ft max. length	
5	200	9	2½	2	10	¾	1½	2¼	40	7,600	6	0.43	0.85	1.27	1.69	2.11	4.8
10	400	10	2½	2	10	¾	1½	2½	55	7,600	9	0.85	1.69	2.25	3.00	3.75	6.6
15	600	10	2½	2	10	¾	1½	2½	80	7,600	9	1.27	2.25	3.38	3.94	4.93	9.6
		12	2½	2	12	1	2	3	45	7,600	10	1.27	2.25	3.38	3.94	4.93	5.4
		12	3½	3						16,400		1.27	2.25	3.38	3.94	4.93	11.7
20	800	12	2½	2	12	1	2	3	60	7,600	10	1.69	3.00	3.94	4.87	5.63	7.2
			3½	3						16,400		1.69	3.00	3.94	4.87	5.63	15.6
25	1000	12	2½	2	12	1	2	3	75	7,600	10	2.12	3.75	4.93	5.63	6.55	9.0
			3½	3						16,400		2.12	3.75	4.93	5.63	6.55	9.0
		14	3½	3		1¼	2½	3½	45	16,400	12	2.12	3.75	4.93	5.63	6.55	11.7
30	1200	14	3½	3	12	1¼	2½	3½	55	16,400	12	2.25	3.94	5.05	6.75	7.50	14.3
35	1400	14	3½	3	12	1¼	2½	3½	65	16,400	12	2.62	4.58	5.90	7.00	8.75	16.9
40	1600	16	3½	3	12	1½	3	4	50	16,400	14	3.00	4.50	6.75	8.00	10.00	13.0

*Fairfield Engineering Co. data in U.S. customary system. To convert cubic feet per hour to cubic meters per hour, multiply by 0.02832; to convert tons per hour to metric tons per hour, multiply by 0.9078; and to convert screw size in inches to the nearest screw size in centimeters, multiply by 2.5.
†Capacities are based on screws carrying 31 percent of their cross section and, in the case of feed sections with half-pitch flights, based on 100 percent of their cross section.
‡Pipe sizes given are for ¼-in (6.35-mm) flights.
§Horsepowers listed are calculated for average conditions and are of the proper motor size with factors for length of conveyor, momentary overloads, etc., taken into consideration.

formly over a storage area. Flat-bottomed rack-and-pinion gates (Fig. 21-2f) allow selective discharge, as do hand slide gates (Fig. 21-2d). However, for perishable materials, the curved slide gate (Fig. 21-2h) eliminates the dead-storage pocket. Enclosed rack-and-pinion gates (Fig. 21-2j) give dust-tight operation, and rotary cutoff valves (Fig. 21-2i) allow quick shutoff and are readily adaptable to remote control. Air-cylinder-actuated gates have become more and more prominent because of the low investment required and the ease of connecting to automatic process-control centers.

BELT CONVEYORS

The belt conveyor is almost universal in application. It can travel for miles at speeds up to 5.08 m/s (1000 ft/min) and handle up to 4539 metric tons/h (5000 tons/h). It can also operate over short distances at speeds slow enough for manual picking, with a capacity of only a few kilograms per hour. However, it is not normally applicable to processing operations, except under unusual conditions.

Belt-conveyor slopes are limited to a maximum of about 30°, with those in the 18 to 20° range more common. Direction changes can occur only in the vertical plane of the belt path and must be carefully designed as vertical curves or relatively flat bends. Belt conveyors inside the plant may have higher initial cost than some other types of conveyors and, depending on idler design, may or may not require more maintenance. However, a belt conveyor given good routine maintenance can be expected to outlast almost any other type of conveyor. Thus, in terms of cost per ton handled, outstanding economy records have been established by belt conveyors.

Belt-conveyor design begins with a study of the material to be handled. Since weight per cubic meter or foot is an important factor, it should be accurately determined with the material in an as-handled condition. It is not wise to rely solely on published tables of weight per cubic meter or foot for various materials, since many processing operations will affect this by fluffing or compacting the material. Lump size is important, too. For a 600-mm (24-in) belt, uniform lump size can range up to about 102 mm (4 in). For each 152-mm (6-in) increase in belt width, lump size can increase by about 51 mm (2 in). If material contains around 90 percent fines, lump size can be increased by around 50 percent. However, care should be taken to maintain uniform flow of material, with fine material reaching the belt first to protect it from impact damage. The larger the lump, the more danger of

its falling off the belt or rolling back on inclines. With the belt running horizontally or sloping only slightly at the feed point, the problem of lumps falling off is minimized, especially if particular care is taken with feed-chute design.

Temperature and chemical activity of the conveyed material play important roles in **belt selection.** For example, natural rubber should be avoided with oily materials even when the material does not present an obviously oily surface. Special rubber, cotton, and asbestos-fiber belts are available to meet varying degrees of material temperature, and they should be used whenever high temperatures exist. Belts can be seriously and quickly damaged by high temperature, and the investment in what at first glance seems to be an extremely high-priced belt may prove most economical in the long run. There are many superperformance elastomers available for belt construction. These include neoprene, Teflon, Buna N rubber, and vinyls. Manufacturers are able to test products to be handled and often recommend several elastomer grades that will perform satisfactorily, each grade having a different first-cost–operating-life relation.

Moisture may create poor discharge conditions because of material sticking to the belt and to chutes, or it may even reduce capacity if it is present in enough quantity to give the material fluid properties. Even though abrasion may create problems with belt conveyors, these are easier to solve with properly designed belt systems than with most other conveyors.

In establishing belt-conveyor **tonnage requirements** it is important to work with peak rather than average loads. Only occasionally, because of intentional or accidental variations in production rates, are these two figures identical. The belt that runs empty half the time must carry twice the average load when it is working.

When a belt conveyor must **change direction,** it is often easier to use more than one conveyor. However, vertical curves can be designed and upward changes of direction accomplished with a pair of snub pulleys. If the belt pull is downward on the idlers, a simple flat pulley can be used for minor directional changes. In any case, using a single continuous belt eliminates the need for more than one drive. With a pair of snub pulleys, the carrying face of the belt is brought in contact with the pulley; hence special care must be taken to get a good discharge. When bending the belt over a flat pulley, belt speed must be slow enough to keep material from flying off the belt. In many situations the smooth curve, either concave or convex, is preferable. For a 61-cm (24-in) belt the minimum curve radius is about 61 m (200 ft),

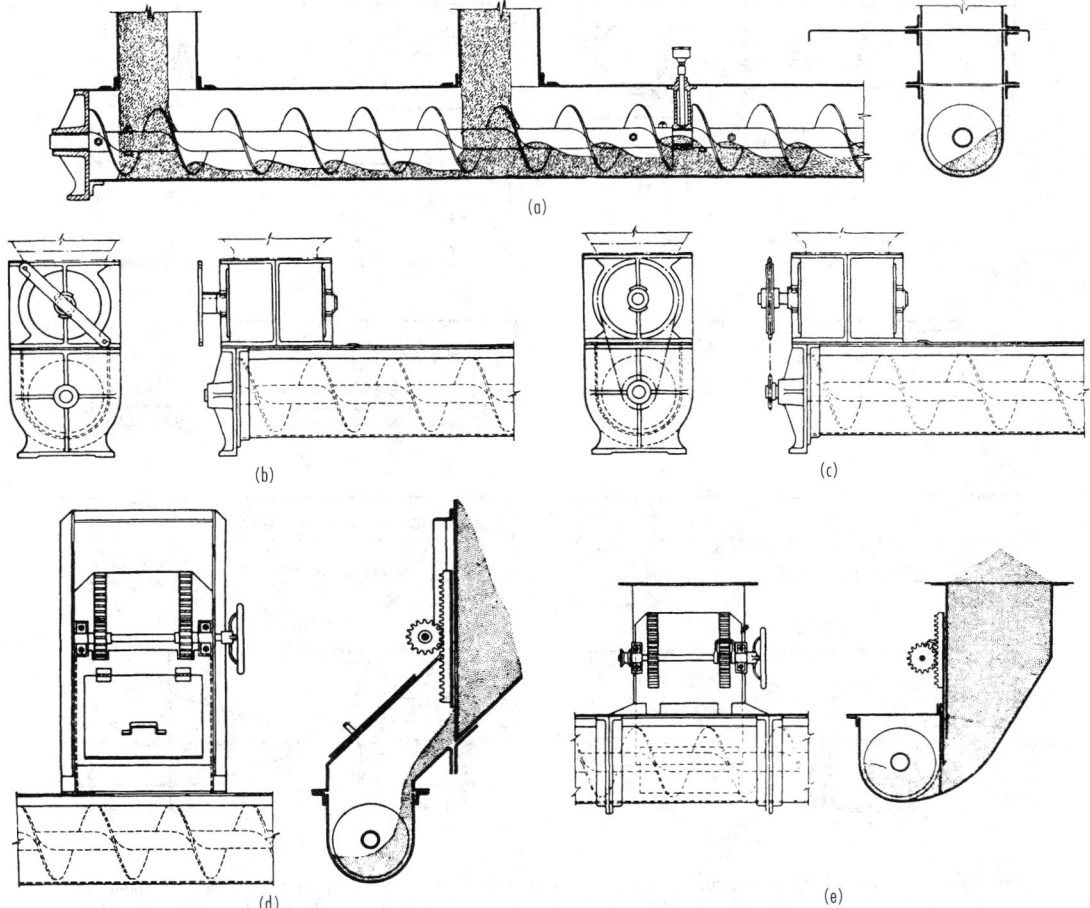

FIG. 21-1 Typical feed arrangements for screw conveyors. (*a*) Plain spouts of chutes. (*b*) Rotary cutoff valve. (*c*) Rotary-vane feeder. (*d*) Bin gate. (*e*) Side inlet gate. (*FMC Corporation, Material Handling Systems Division.*)

but for best operating conditions the curve should be carefully designed.

Operating conditions which affect belt-conveyor design include climate, surroundings, and hours of continuous service. Temperature and humidity extremes may dictate total enclosure of the belt; surroundings which involve such conditions as high temperature or corrosive atmosphere can affect belt, machinery, and structure; and continuous service may require extremely high-quality components and even specially designed equipment for servicing while the belt is in operation. For example, idlers may be obtained with tilting stands which allow them to be tipped out of the way for service while the belt is running.

Belt width and **speed** are functions of bulk density of the material and lump size. Lowest first cost can often be obtained by using the narrowest possible belt for a given lump size and operating it at maximum speed. However, speed often may be limited by dusting, and sometimes it may be better economy to use a wider belt with fewer plies to combine the necessary tensile strength with good belt-troughing characteristics. Abrasiveness of the material can strongly affect speed and also lump size, for at higher speeds abrasive wear is increased and there is greater danger of lumps rolling off the belt. Ideally a belt should run with lump size, slope, and load of less than recommended maximums and with uniform feed introduced to the belt centrally as nearly as possible in the direction and speed of belt travel.

Power to drive a belt conveyor is made up of five components: power to drive the empty belt, to move the load against friction of the rotating parts, to raise or lower the load, to overcome inertia in putting material into motion, and to operate a belt-driven tripper if required. As with most other conveyor problems, it is advisable to work with formulas and constants from a specific manufacturer in making these calculations. For estimating purposes, typical data are given in Table 21-7.

Belt selection depends on power and development of the required tensile strength. Knowing drive-shaft power, belt tension can be calculated and a belt selected. However, since various combinations of width and ply thickness will develop the required strength, final selection is influenced by lump size, troughability of the belt, and ability of the belt to support the load between idlers. Thus it is necessary to use an empirical approach to arrive at a belt selection which meets all requirements.

Once final belt selection has been made, **idlers** and **return rolls** can also be selected. Figure 21-3 indicates the wide variety of belt supports for bulk-handling applications. Figure 21-3*a* and *b* consists of flat-belt arrangements of rollers or plate which allow material to be discharged by simple V-shaped plows. The flat plate-supported belt allows sidewalls to be erected to prevent dribble or to build up larger loads on the flat belt. As in Fig. 21-3*f*, larger capacity can also be achieved by troughing the plate. The 20° troughing idler with equal-length rolls (Fig. 21-3*c*) is the most common, with lighter materials adaptable to 45° idlers with short or long side rolls (Fig. 21-3*d* and *e*).

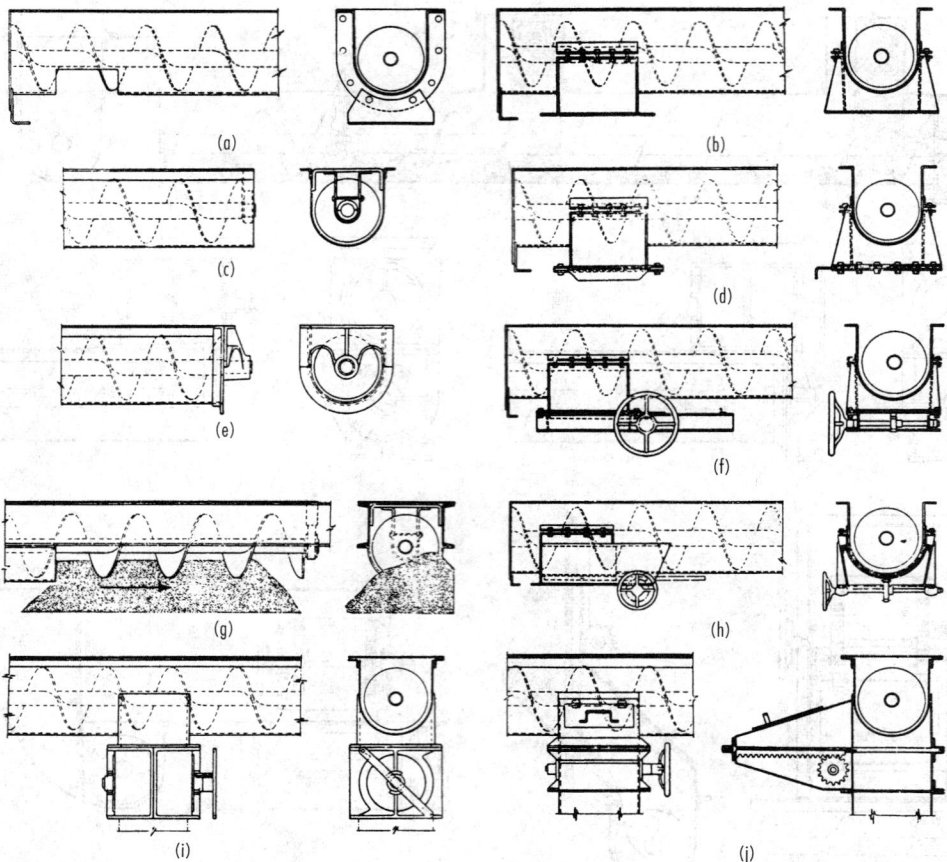

FIG. 21-2 Typical discharge arrangements for screw conveyors. (*a*) Plain discharge opening. (*b*) Discharge spout. (*c*) Open-end trough. (*d*) Hand slide gate. (*e*) Discharge trough end. (*f*) Rack-and-pinion flat side gate. (*g*) Open-bottom trough. (*h*) Rack-and-pinion curved slide gate. (*i*) Rotary cutoff valve. (*j*) Enclosed rack-and-pinion gate. (*FMC Corporation, Material Handling Systems Division.*)

Since the lighter materials do not require stiff belts for tensile strength, there is usually no problem with troughing.

With the proper idlers selected for size and service conditions, the most important step is to locate them properly. For long belts the tension varies considerably, and idlers should be spaced to hold belt sag to reasonable limits along the full length of travel. Too much belt sag can cause a significant power loss, but for most belts of ordinary length it is usually satisfactory to space idlers fairly closely at the feed point and then farther apart and uniformly for the rest of the conveyor length.

Loading and discharge points on belt conveyors need to accommodate several factors. Figure 21-4*a* shows details for one type of rubber seal on a metal skirt plate. It is particularly important that material be loaded onto the belt in its center and in the direction of its travel, preferably with lumps falling on a layer of fine material. Fines can be delivered to the belt first by notching the feed chute or installing a screen section or grizzly bars. Figure 21-4*b* shows a heavy-duty loading-section design using not only rubber idler rolls but an additional short pad belt. Mass-flow bins and/or bin-flow-assisting devices are often used to minimize segregation of fines and to assure a uniform feed from a hopper onto a conveyor belt.

A clean discharge is vital to good belt life. On the return run the carrying side of the belt is in contact with the return rollers, and any material adhering to it is ground in or deposited on the roller. Extremely sticky material may require a belt-cleaning device in the form of a revolving brush, spring-mounted steel scrapers, rubber scraper blades, or sometimes a taut wire. When these devices are used, care should be taken that the dribble does not fall on the belt. Refer to the subsection "Storage and Weighing of Solids in Bulk," which deals with the criteria for bin design. For non-free-flowing materials, the combination of correct bin-discharge design and feeder-loading design is often a critical relation in which a slight error in either may produce a system in which the material will not flow at all.

BUCKET ELEVATORS

Bucket elevators are the simplest and most dependable units for making vertical lifts. They are available in a wide range of capacities and may operate entirely in the open or be totally enclosed. The trend is toward highly standardized units, but for special materials and high capacities it is wise to use specially engineered equipment. Main variations in quality are in casing thickness, bucket thickness, belt or chain quality, and drive equipment.

Spaced-Bucket Centrifugal-Discharge Elevators These elevators (Fig. 21-5*a*) are the most common. They are usually equipped with the style 1 or 2 buckets shown in Fig. 21-5*h*. Mounted on a belt or a chain, the buckets are spaced to prevent interference in loading or discharging. This type of elevator will handle almost any free-flowing fine or small-lump material such as grain, coal, or dry chemicals. Buckets are loaded partly by material flowing directly into them

TABLE 21-7 Belt-Conveyor Data for Troughed Antifriction Idlers*

Belt width in (cm)	Cross-sectional area of load ft² (m²)	Belt speed, ft/min (m/min) Normal	Belt speed, ft/min (m/min) Maximum	Belt plies Minimum	Belt plies Maximum	Max lump size Sized material, 80% under in (mm)	Max lump size Unsized material, not over 20% in (mm)	Belt speed, ft/min (m/min)	Capacity tons/h (metric tons/h)	hp/10-ft (3.05-m) lift	hp/100-ft (30.48-m) centers	Add for tripper hp†
14 (35)	0.11 (.010)	200 (61)	300 (91)	3	5	2.0 (51)	3.0 (76)	100 (30.5)	32 (29)	0.34	0.44	2.0
								200 (61.0)	64 (58)	0.68	0.68	
								300 (91.5)	96 (87)	1.04	1.32	
16 (40)	0.14 (.013)	200 (61)	300 (91)	3	5	2.5 (64)	4.0 (102)	100 (30.5)	44 (40)	0.46	0.56	2.5
								200 (61.0)	88 (80)	0.90	1.12	
								300 (91.5)	132 (120)	1.36	1.68	
18 (45)	0.18 (.017)	250 (76)	350 (107)	4	6	3.0 (76)	5.0 (127)	100 (30.5)	54 (49)	0.58	0.70	3.0
								250 (76.2)	134 (122)	1.42	1.76	
								350 (106.7)	190 (172)	2.00	2.42	
20 (50)	0.22 (.020)	250 (76)	350 (107)	4	6	3.5 (89)	6.0 (152)	100 (30.5)	66 (60)	0.70	0.84	3.20
								250 (76.2)	164 (148)	1.72	2.06	
								350 (106.7)	230 (209)	2.44	2.90	
24 (60)	0.33 (.030)	300 (91)	400 (122)	4	7	4.5 (114)	8.0 (203)	100 (30.5)	98 (89)	1.02	1.02	3.5
								300 (91.5)	294 (267)	3.06	3.04	
								400 (121.9)	392 (356)	4.08	4.04	
30 (75)	0.53 (.049)	300 (91)	450 (137)	4	8	7.0 (178)	12.0 (305)	100 (30.5)	158 (143)	1.60	1.50	5.0
								300 (91.5)	474 (430)	4.80	4.50	
								450 (137.2)	710 (645)	7.20	6.74	
36 (90)	0.78 (.072)	400 (122)	600 (183)	4	9	8.0 (203)	15.0 (381)	100 (30.5)	230 (209)	2.44	1.59	7.0
								400 (121.9)	920 (835)	9.74	6.36	
								600 (182.9)	1380 (1253)	14.60	9.52	
42 (105)	1.09 (.101)	400 (122)	600 (183)	4	10	10.0 (254)	18.0 (457)	100 (30.5)	330 (300)	3.50	2.28	9.5
								400 (121.9)	1320 (1198)	14.00	9.12	
								600 (182.9)	1980 (1797)	23.20	13.68	
48 (120)	1.46 (.136)	400 (122)	600 (183)	4	12	12.0 (305)	21.0 (533)	100 (30.5)	440 (399)	4.66	3.04	12.8
								400 (121.9)	1760 (1598)	18.70	12.14	
								600 (182.9)	2640 (2397)	28.00	18.20	
54 (135)	1.90 (.177)	450 (137)	600 (183)	6	14	14.0 (356)	24.0 (610)	100 (30.5)	570 (517)	6.04	3.94	20.0
								450 (137.2)	2564 (2328)	27.20	17.70	
								600 (182.9)	3420 (3105)	36.20	23.60	
60 (150)	2.40 (.223)	450 (137)	600 (183)	6	16	16.0 (406)	28.0 (711)	100 (30.5)	720 (654)	7.64	4.98	23
								450 (137.2)	3240 (2941)	34.40	22.40	
								600 (182.9)	4320 (3921)	45.80	29.90	

*Fairfield Engineering Co. data in U.S. customary system. Metric conversion is rounded off. For inclined conveyors, add lift horsepower to center horsepower for total horsepower. For terminals multiply horsepower by the following factors: 0–50 ft (15.2 m), 1.20; 51–100 ft (30.5 m), 1.10; 101–150 ft (45.7 m), 1.05. For countershaft drives, multiply horsepower by 1.05 for each reduction (cut gears).

†Tripper horsepower is based on material bulk density of 100 lb/ft³ (1602 kg/m³) and a belt speed of 300 ft/min (91.4 m/min).

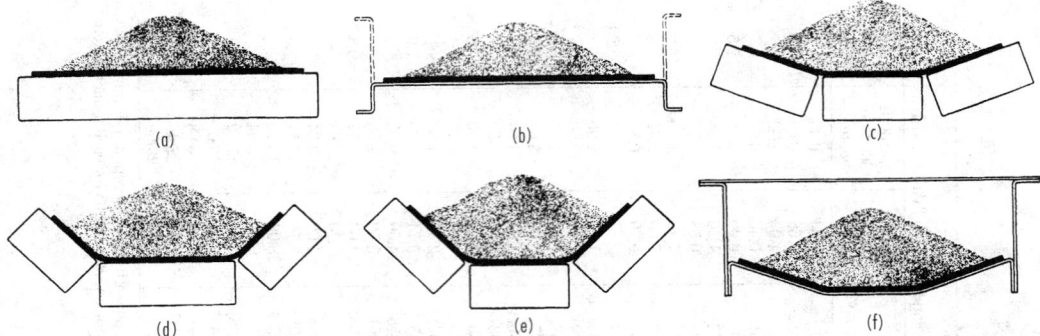

FIG. 21-3   Typical belt-conveyor idler and plate-support arrangements. (*a*) Flat belt on flat-belt idlers. (*b*) Flat belt on continuous plate. (*c*) Troughed belt on 20° idlers. (*d*) Troughed belt on 45° idlers with rolls of unequal length. (*e*) Troughed belt on 45° idlers with rolls of equal length. (*f*) Troughed belt on continuous plate. (*FMC Corporation, Material Handling Systems Division.*)

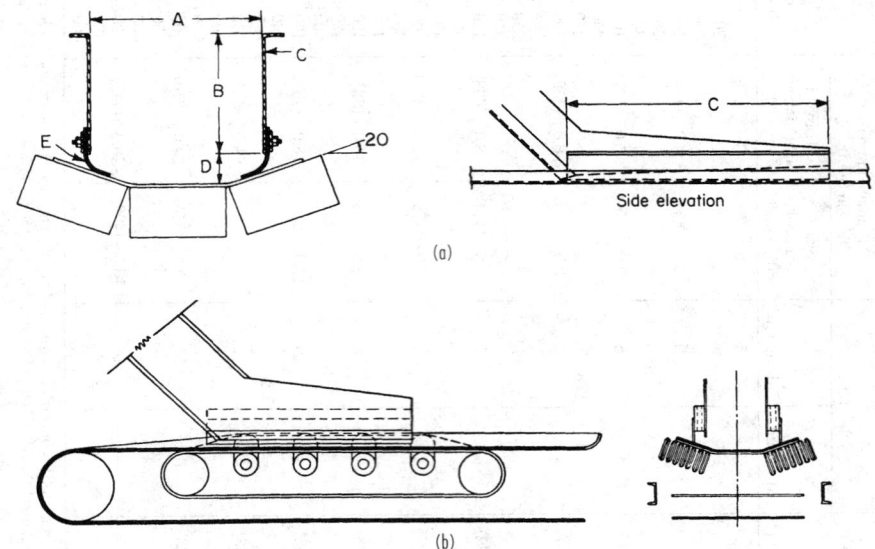

Skirt Plate and Seal Dimensions

Symbol	Name	Dimensions
A	Trough width	See table for (*a*).
B	Skirt depth	Minimum: 6 in (150 mm); maximum: 12 in (300 mm)
C	Skirt length	Minimum: 6 ft (1.8 m); maximum: 8 ft (2.4 m)
D	Skirt-to-belt clearance	See table for (*a*).
E	Seal specification	0.25-in- (6-mm-) thick live rubber, 6 in (150 mm) × C

Belt width, in (cm)	14 (35)	16 (40)	18 (45)	20 (50)	24 (60)	30 (75)	36 (90)	42 (110)	48 (125)	54 (140)	60 (155)
Trough width A, in (cm)	9 (23)	11 (28)	12 (30)	13 (33)	16 (41)	20 (51)	24 (61)	28 (71)	32 (81)	36 (91)	40 (102)
Skirt seal D, in (cm)	2.0 (5.1)	2.25 (5.7)	2.25 (5.7)	2.88 (7.3)	2.88 (7.3)	3.13 (8.0)	3.63 (9.2)	4.0 (10.1)	4.38 (11.1)	4.75 (12.1)	5.25 (13.3)

FIG. 21-4   Belt-conveyor loading details. (*a*) Typical skirt-plate design and dimensions. (*b*) Pad belt and special roller-bearing idlers for heavy-duty loading. (*Stephens-Adamson Division, Allis-Chalmers Corporation.*)

and partly by scooping material from the boot as shown in Fig. 21-5e. Speeds can be relatively high for fairly dense materials but must be lowered considerably for aerated or low-bulk-density materials [under 641 kg/m³ (40 lb/ft³)] to prevent fanning action.

Spaced-Bucket Positive-Discharge Elevators Elevators of this type (Fig. 21-5b) are essentially the same as centrifugal-discharge units except that the buckets are mounted on two strands of chain and are snubbed back under the head sprocket to invert them for positive discharge. These units are designed especially for materials which are

sticky or tend to pack, and the slight impact of the chain seating on the snub sprocket combined with complete bucket inversion is generally sufficient to empty the buckets completely. In extreme cases, knockers may be used to hit the buckets at the discharge point to help free material. The speed of these units is relatively slow, and buckets must be larger or more closely spaced to reach capacity levels of the centrifugal style.

Continuous-Bucket Elevators These elevators (Fig. 21-5c) are generally used for larger-lump materials or for materials too difficult

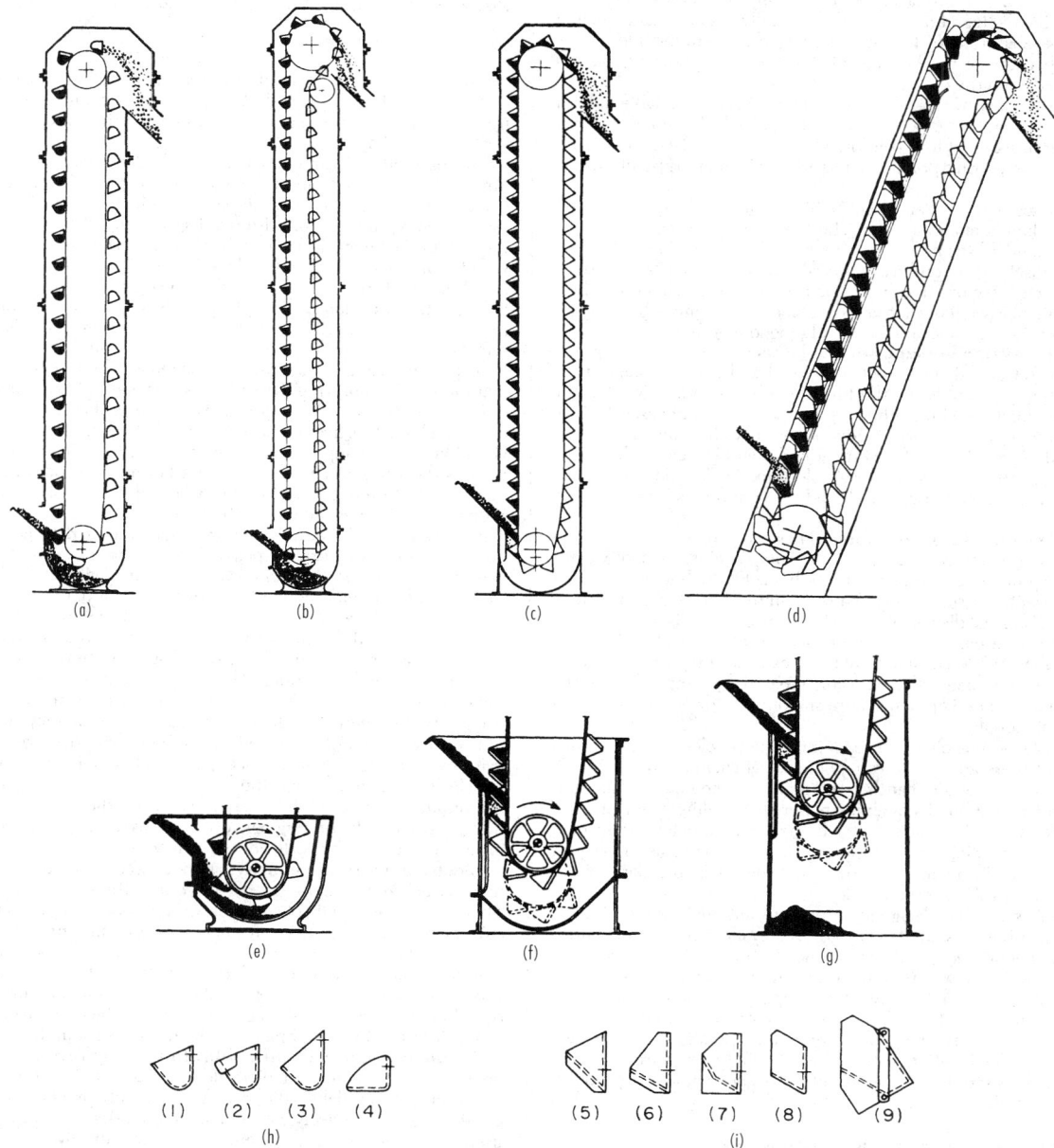

FIG. 21-5 Bucket-elevator types and bucket details. (a) Centrifugal-discharge spaced buckets. (b) Positive-discharge spaced buckets. (c) Continuous bucket. (d) Supercapacity continuous bucket. (e) Spaced buckets receive part of lead direct and part by scooping from bottom. (f) Continuous: buckets are filled as they pass through loading leg, with feed spout above tail wheel. (g) Continuous: buckets in bottomless boot, with cleanout door. (h) Malleable-iron spaced buckets for centrifugal discharge. (i) Steel buckets for continuous-bucket elevators. (Stephens-Adamson Division, Allis-Chalmers Corporation.)

to handle with centrifugal-discharge units. Buckets are closely spaced, with the back of the preceding bucket serving as a discharge chute for the bucket which is dumping as it rounds the head pulley. Close bucket spacing reduces the speed at which the elevator must run to maintain capacities comparable with the spaced-bucket elevator. Gentle discharge prevents excessive degradation and makes this type of elevator effective for handling finely pulverized or aerated materials. Two boot styles and typical loading conditions are illustrated in Fig. 21-5f and g.

Supercapacity Continuous-Bucket Elevators   Elevators of this type (Fig. 21-5d) are designed for high lifts and large-lump material. They handle high tonnages and are usually operated at an incline to improve loading and discharge conditions. Operating speeds are low, and because of the heavy loads the bucket-supporting chain is usually guided on the elevating and return runs.

Buckets for spaced-type elevators (Fig. 21-5h) are available in both malleable iron and steel in a variety of styles. Style 1 is standard, with style 2 identical except for a reinforced lip. Styles 3 and 4 are low-front designs for wet, stringy, or sticky materials which are difficult to discharge.

Continuous-type buckets (Fig. 21-5i) are generally back-mounted to chain or belt at close intervals. They are usually fabricated of steel. Style 5 is standard for normal materials, with style 6 a low-front type for better discharge of difficult materials. Style 7 buckets are used for additional capacity or large lumps, and style 8 for inclined crusher-type elevators. Style 9 buckets are designed for extremely high capacities and are usually side-mounted and hinged together.

Bucket-elevator horsepower can be calculated quite easily. For spaced buckets and digging boots it is equal to the desired capacity in tons per hour multiplied by the lift in feet and divided by 500. For continuous buckets with loading leg, the divisor is increased to 550. Both formulas include normal drive losses as well as loading pickup losses and are applicable for vertical and slightly inclined lifts. For estimating purposes, general bucket-elevator specifications are given for centrifugal units in Table 21-8 and for continuous units in Table 21-9.

V-Bucket Elevator-Conveyors   These are still used for handling heavy materials, for coal, and, in light-duty designs, for lightweight free-flowing materials. Similar to the V-bucket type, but with buckets swinging freely on supporting shafts mounted between two strands of roller chain, is the pivoted-bucket conveyor. This type can be equipped with a fixed or movable tripper to dump buckets by overturning them. While considerably more expensive than the V-bucket conveyor, it eliminates the abrasion created by dragging material along in a trough and operates more smoothly at lower power per ton for heavy materials.

The most common chain conveyor is the bucket elevator already discussed, but there are a wide variety of special chain conveyors which are used so infrequently that they should be selected only on the specific recommendation of a qualified materials-handling engineer.

Skip Hoists   These hoists, which operate on a batch rather than continuous principle, are not so widely used as in the past. However, for high lifts and extremely lumpy or hot materials, the skip hoist is still an economical and practical device.

Skip hoists may be designed to operate automatically or from a manual push-button station. They are usually classified as uncounterweighted, counterweighted, or balanced. Both the latter systems reduce operating-power requirements, and the balanced unit, using two buckets, can operate at twice the capacity of the others. Figure 21-6 illustrates these types as well as some of the common paths of travel which skip hoists may follow. Speed of operation is also a basis for skip-hoist classification, with multispeed motors required on high-speed operations to slow down bucket travel speed at loading and discharge points.

VIBRATING OR OSCILLATING CONVEYORS

Most vibrating conveyors are essentially directional-throw units which consist of a spring-supported horizontal pan vibrated by a direct-connected eccentric arm, rotating eccentric weights, an electromag-

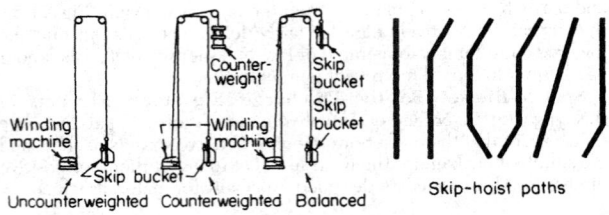

FIG. 21-6 Types of skip hoists and skip-hoist paths. (*Fairfield Engineering Co.*)

net, or a pneumatic or hydraulic cylinder. The motion imparted to the material particles may vary, but its purpose is to throw the material upward and forward so that it will travel along the conveyor path in a series of short hops.

The **capacity** of directional-throw vibrating conveyors is determined by the magnitude of trough displacement, frequency of this displacement, angle of throw, slope of trough, and ability of the material to receive and transmit through its mass the directional throw of the trough. The material itself is the most important factor. To be conveyed properly it should have a high friction factor on steel as well as a high internal friction factor so that conveying action is transmitted through its entire depth. Thus deep loads tend to move more slowly than thin ones. Material must also be dense enough to minimize the effect of air resistance on its trajectory, and it should not aerate. Tests have shown that granular materials handle better than pulverized materials and flat or irregular shapes better than spherical ones.

Classification of vibrating conveyors can probably best be based on drive characteristics as shown in Fig. 21-7. All these types transmit vibration to their supporting structures, but the direct, or positive, drive is the worst offender and should be mounted on a heavy supporting structure if it is not counterbalanced. Semipositive- and nonpositive-drive types reduce vibration effects because thrust is transmitted over the entire support length rather than at a specific point. Regardless of drive type, care should be taken to mount the conveyor properly so that supporting structures will not be damaged. The frequency of vibration of the conveyor should in no case be at or near the natural frequency of the supporting structure.

Mechanical vibrating conveyors are designed to operate at specific frequencies and do not perform well at other frequencies without carefully designed alterations. Thus they are not adapted to frequent capacity changes except by varying the depth of material fed to the trough. Positive eccentric drives maintain their frequency and magnitude of stroke regardless of load, and serious drive damage can result from overloading. Rotating eccentric weights can also provide the motive force, and although they maintain a constant frequency, the magnitude of stroke is definitely affected by the load. Directional-throw mechanical vibrating conveyors are used primarily for conveying and do not usually perform well as feeders.

Electrical vibrating conveyors are characterized by the fact that there is no contact between the drive and the conveying medium. They operate on a pull-release cycle or a pull-push cycle, using direct current and pulsating electromagnets or alternating current combined with electromagnets and permanent magnets. While most electrical vibrating units are used as feeders, they also work well as conveyors. Most of them offer the advantage of capacity regulation through control of the electric current magnitude via rheostats. Figure 21-8 gives capacities as a function of pan size and power consumption.

Pneumatic and **hydraulic vibrating conveyors** have as their greatest asset elimination of explosion hazards. If pressurized air, water, or oil is available, they can be extremely practical since their drive design is relatively simple and pressure-control valves can be used to vary capacity either manually or automatically.

The **capacity** of vibrating conveyors is extremely broad, ranging from thousands of tons down to grams or ounces. Since so many variables affect their ability to convey, there is no simple formula for figuring capacity and power. Available data are generally the results of

TABLE 21-8 Bucket-Elevator Specifications for Centrifugal-Discharge Buckets on Belt, Malleable-Iron, or Steel Buckets*

Size of bucket, in (mm), and bucket spacing, in (mm)†	Elevator centers, ft‡	Capacity, tons/h (metric tons/h)§	Size of lumps handled, in (mm)¶	Bucket Bucket speed, ft/min (m/min)	r/min, head shaft	hp required at head shaft	Additional hp/ft for intermediate lengths	Head	Tail	Head	Tail	Belt width, in
6 × 4 × 4¼ – 12	25	14 (12.7)	¾ (19.0)	225 (68.6)	43	1.0	0.02	1$\frac{15}{16}$	1$\frac{11}{16}$	20	14	7
	50	14 (12.7)	¾ (19.0)	225 (68.6)	43	1.6	0.02	1$\frac{15}{16}$	1$\frac{11}{16}$	20	14	7
(152 × 102 × 108) – (305)	75	14 (12.7)	¾ (19.0)	225 (68.6)	43	2.1	0.02	1$\frac{15}{16}$	1$\frac{11}{16}$	20	14	7
8 × 5 × 5½ – 14	25	27 (24.5)	1 (25.4)	225 (68.6)	43	1.6	0.04	1$\frac{15}{16}$	1$\frac{11}{16}$	20	14	9
	50	30 (27.2)	1 (25.4)	260 (79.2)	41	3.5	0.05	1$\frac{15}{16}$	1$\frac{11}{16}$	24	14	9
(203 × 127 × 140) – (356)	75	30 (27.2)	1 (25.4)	260 (79.2)	41	4.8	0.05	2$\frac{7}{16}$	1$\frac{11}{16}$	24	14	9
10 × 6 × 6¼ – 16	25	45 (40.8)	1¼ (32.0)	225 (68.6)	43	3.0	0.063	1$\frac{15}{16}$	1$\frac{15}{16}$	20	16	11
	50	52 (47.2)	1¼ (32.0)	260 (79.2)	41	5.2	0.07	2$\frac{7}{16}$	1$\frac{15}{16}$	24	16	11
(254 × 152 × 159) – (406)	75	52 (47.2)	1¼ (32.0)	260 (79.2)	41	7.2	0.07	2$\frac{15}{16}$	1$\frac{15}{16}$	24	16	11
12 × 7 × 7¼ – 18	25	75 (68.1)	1½ (38.1)	260 (79.2)	41	4.7	0.1	2$\frac{7}{16}$	1$\frac{15}{16}$	24	18	13
	50	84 (76.3)	1½ (38.1)	300 (91.4)	38	8.9	0.115	2$\frac{15}{16}$	1$\frac{15}{16}$	30	18	13
(305 × 178 × 184) – (457)	75	84 (76.3)	1½ (38.1)	300 (91.4)	38	11.7	0.115	3$\frac{7}{16}$	2$\frac{7}{16}$	30	18	13
14 × 7 × 7¼ – 18	25	100 (90.8)	1¾ (44.5)	300 (91.4)	38	7.3	0.14	2$\frac{15}{16}$	2$\frac{7}{16}$	30	18	15
	50	100 (90.8)	1¾ (44.5)	300 (91.4)	38	11.0	0.14	3$\frac{7}{16}$	2$\frac{7}{16}$	30	18	15
(355 × 179 × 184) – (457)	75	100 (90.8)	1¾ (44.5)	300 (91.4)	38	14.3	0.14	3$\frac{7}{16}$	2$\frac{7}{16}$	30	18	15
16 × 8 × 8½ – 18	25	150 (136.2)	2 (50.8)	300 (91.4)	38	8.5	0.165	2$\frac{15}{16}$	2$\frac{7}{16}$	30	20	18
	50	150 (136.2)	2 (50.8)	300 (91.4)	38	12.6	0.165	3$\frac{7}{16}$	2$\frac{7}{16}$	30	20	18
(406 × 203 × 216) – (457)	75	150 (136.2)	2 (50.8)	400 (121.9)	38	16.7	0.165	3$\frac{15}{16}$	2$\frac{7}{16}$	30	20	18

*From Stephens-Adamson Division, Allis-Chalmers Corporation.

†Bucket size given: width × projection × depth. Assumed bucket linear speed is 150 ft/min (45.7 m/min).

‡Elevator centers to nearest SI equivalent are 25 ft ≃ 8 m, 50 ft ≃ 15 m, and 75 ft ≃ 23 m.

§Capacities and horsepowers are given for materials having bulk densities of 100 lb/ft^3 (1602 kg/m^3). For other densities these will vary in direct proportion: a 50-lb/ft^3 material will reduce the capacity and horsepower required by 50 percent.

¶If the amount of lump product is less than 15 percent of the total, lump size may be twice that given.

TABLE 21-9 Bucket-Elevator Specifications for Continuous Buckets on Chain*

Size of bucket and bucket spacing, in (mm)†	Elevator centers, ft‡	Capacity, tons/h (metric tons/h)§		Size of lumps handled, in (mm)¶		r/min, head shaft	hp required at head shaft	Additional hp/ft for intermediate lengths	Head	Tail	Head	Tail
8 × 5½ × 7¾ − 8	25	35	(31.7)	1	(25.4)	28	1.8	0.06	1¹⁵⁄₁₆	1¹¹⁄₁₆	20½	14
(203 × 140 × 197) − (203)	50	35	(31.7)	1	(25.4)	28	3.4	0.06	2⁷⁄₁₆	1¹¹⁄₁₆	20½	14
	75	35	(31.7)	1	(25.4)	28	5.0	0.06	2¹⁵⁄₁₆	1¹¹⁄₁₆	20½	14
10 × 7 × 11¾ − 12	25	60	(54.5)	1½	(38.1)	23	3.0	0.10	2⁷⁄₁₆	1¹⁵⁄₁₆	25	17½
(254 × 178 × 298 − (305)	50	60	(54.5)	1½	(38.1)	23	5.5	0.10	2⁷⁄₁₆	1¹⁵⁄₁₆	25	17½
	75	60	(54.5)	1½	(38.1)	23	8.0	0.10	2¹⁵⁄₁₆	1¹⁵⁄₁₆	25	17½
12 × 7 × 11¾ − 12	25	70	(63.5)	1½	(38.1)	23	3.5	0.12	2⁷⁄₁₆	1¹⁵⁄₁₆	25	17½
(305 × 178 × 298) − (305)	50	70	(63.5)	1½	(38.1)	23	6.5	0.12	2¹⁵⁄₁₆	1¹⁵⁄₁₆	25	17½
	75	70	(63.5)	1½	(38.1)	23	9.5	0.12	3⁷⁄₁₆	2⁷⁄₁₆	25	17½
14 × 7 × 11¾ − 12	25	80	(72.6)	1¾	(44.5)	23	4.0	0.14	2⁷⁄₁₆	2⁷⁄₁₆	25	17½
(356 × 178 × 298) − (305)	50	80	(72.6)	1¾	(44.5)	20	7.5	0.14	2¹⁵⁄₁₆	2⁷⁄₁₆	29	17½
	75	80	(72.6)	1¾	(44.5)	20	11	0.14	3⁷⁄₁₆	2⁷⁄₁₆	29	17½
14 × 8 × 11¾ − 12	25	100	(90.8)	2	(50.8)	20	5.0	0.17	2¹⁵⁄₁₆	2⁷⁄₁₆	29	17½
(356 × 203 × 298) − (305)	50	100	(90.8)	2	(50.8)	20	9.3	0.17	3⁷⁄₁₆	2⁷⁄₁₆	29	17½
	75	100	(90.8)	2	(50.8)	20	13.3	0.17	3¹⁵⁄₁₆	2⁷⁄₁₆	29	17½
16 × 8 × 11¾ − 12	25	115	(104.4)	2	(50.8)	20	6.0	0.20	2¹⁵⁄₁₆	2⁷⁄₁₆	29	17½
(406 × 203 × 298) − (305)	50	115	(104.4)	2	(50.8)	20	11	0.20	3¹⁵⁄₁₆	2⁷⁄₁₆	29	17½
	75	115	(104.4)	2	(50.8)	20	16	0.20	4⁷⁄₁₆	2⁷⁄₁₆	29	17½
18 × 8 × 11¾ − 12	25	130	(118.0)	2	(50.8)	20	7	0.22	2¹⁵⁄₁₆	2⁷⁄₁₆	29	17½
(406 × 203 × 298) − (305)	50	130	(118.0)	2	(50.8)	20	13	0.22	3¹⁵⁄₁₆	2⁷⁄₁₆	29	17½
	75	130	(118.0)	2	(50.8)	20	20	0.22	4⁷⁄₁₆	2⁷⁄₁₆	29	17½

*From Stephens-Adamson Division, Allis-Chalmers Corporation.
†Bucket size given: width × projection × depth. Assumed bucket linear speed is 150 ft/min (45.7 m/min).
‡Elevator centers to nearest SI equivalent are 25 ft ≈8 m, 50 ft ≈15 m, and 75 ft ≈23 m.
§Capacities and horsepowers are given for materials having bulk densities of 100 lb/ft³ (1602 kg/m³). For other densities these will vary in direct proportion: a 50-lb/ft³ material will reduce the capacity and horsepower required by 50 percent.
¶If the total amount of lump product is less than 15 percent of the total, lump size may be twice that given.

experiments and empirical equations, with most manufacturers providing selection charts for specific types of conveyors and materials. A typical leaf-spring unit is shown in Fig. 21-8, along with the graphical information required to select a standard unit. Conveyor lengths are limited to about 61 m (200 ft) with multiple drives and about 30.5 m (100 ft) with a single drive. There are many exceptions to these general limitations, and they should not preclude study of a specific problem when vibrating conveyors seem desirable.

Processing operations of many types can be carried out in vibrating conveyors because their simple conveying troughs can be modified quite easily. While tube and flat-pan troughs are most common, troughs can be provided in a wide variety of shapes and materials.

Although conveying action is usually so gentle that abrasion problems do not arise, such problems can be easily solved when they do occur by the use of special materials or liners. Troughs are easily sealed to prevent contamination or for operation under positive or negative pressure. With screen or perforated deck plates, vibrating conveyors can dewater, rough-screen, scalp, or dry. Heating and cooling can also be handled by the use of air streams blowing over or through the material, infrared panels, resistance-heating panels, or contact with air- or water-cooled or heated trough casings. Special vibrating-conveyor designs are available for elevating at relatively steep slopes or up a spiral trough. There is probably no other conveyor so readily adaptable to the solution of processing problems.

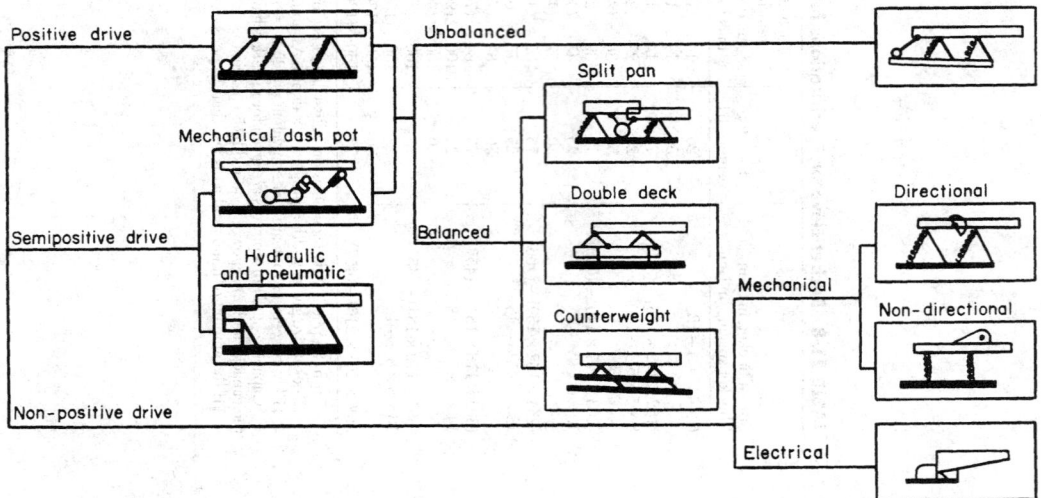

FIG. 21-7 Vibration-conveyor classification. (*Modern Materials Handling.*)

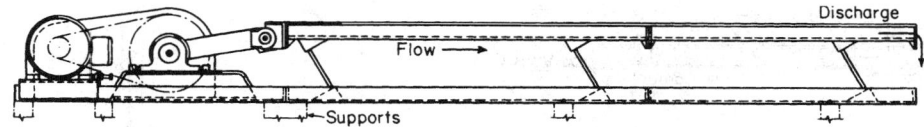

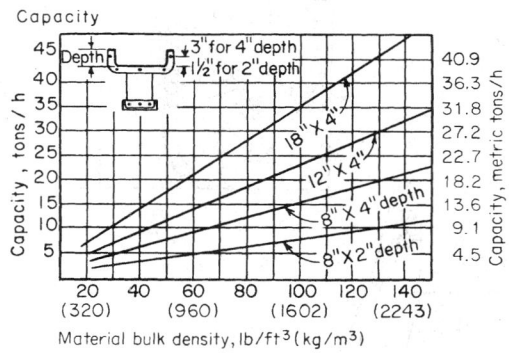

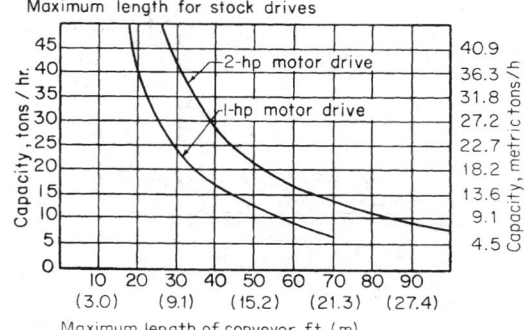

Pan width and stream depth: nearest metric equivalents

in	mm	in	mm
18 x 4	450 x 100	8 x 4	200 x 100
12 x 4	300 x 100	8 x 2	200 x 50

FIG. 21-8 Standardized leaf-spring mechanical oscillating conveyor with selection charts. Multiply pounds per cubic foot by 16.02 to get kilograms per cubic meter; multiply feet by 0.3048 to get meters. (*FMC Corporation, Materials Handling Systems Division.*)

CONTINUOUS-FLOW CONVEYORS

The principle of the continuous-flow conveyor is that when a surface is pulled transversely through a mass of granular, powdered, or small-lump material, it will pull along with it a cross section of material which is greater than the area of the surface itself. The conveying action of various designs of continuous-flow conveyors varies with the type of conveying flight but theoretically is not comparable with the action in a flight or drag conveyor. Flights vary from solid surfaces to skeleton designs, as shown in Fig. 21-9.

The continuous-flow conveyor is a totally enclosed unit which has a relatively high capacity per unit of cross-sectional area and can follow an irregular path in a single plane. These features make it extremely versatile. Figure 21-10 shows some typical arrangements and applica-tions possible with these conveyors. Included is an example of the unit acting as a dewatering device (Fig. 21-10c).

These conveyors employ a chain-supported conveying element (some are cast integrally with the chain, which is designed with easily detachable knuckle joints). Thus the connecting element runs along the outside of the casing so that head and tail sections do not become excessively large because of projecting conveying elements. This means that the material feeding into the conveyor must fall past the chain element and travel in a reverse direction before passing into the actual conveying leg (see Fig. 21-10c). Since this affects the lump size that the conveyor can conveniently handle, the loop design (Fig. 21-10c) is sometimes used for better feeding conditions, or separate carrying runs and return runs are provided with inclined loading chutes to the lower carrying run. In any event, lump size and abrasive

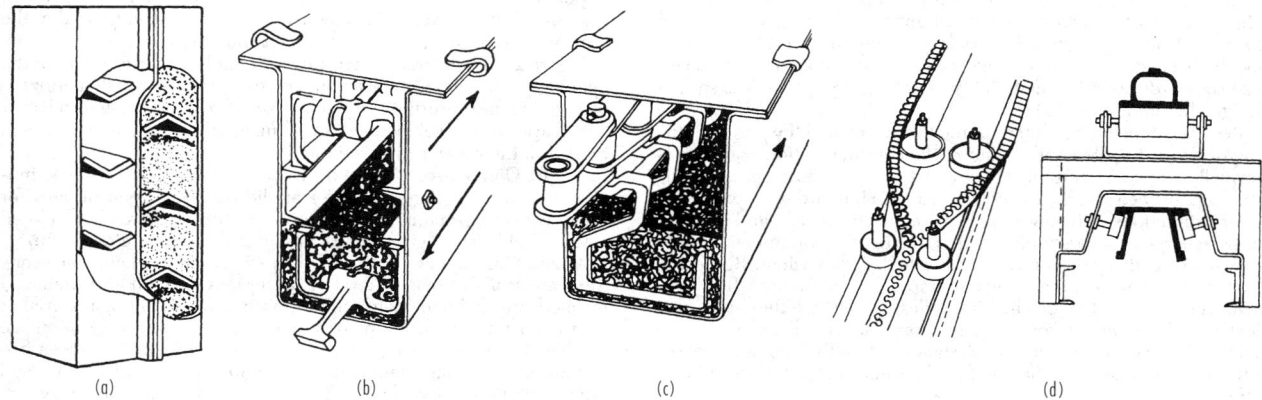

(a)   (b)   (c)   (d)

FIG. 21-9 Closed and open flights for continuous-flow conveyors. (*a*) and (*b*) Conveyor-elevator. (*c*) Horizontal conveyor with side-pull chain. (*d*) Detail of closed-belt conveyor; opening and closing rollers mesh and unmesh teeth in the same manner as a conventional clothing fastener. (*FMC Corporation, Material Handling Systems Division; Stephens-Adamson Division, Allis-Chalmers Corporation.*)

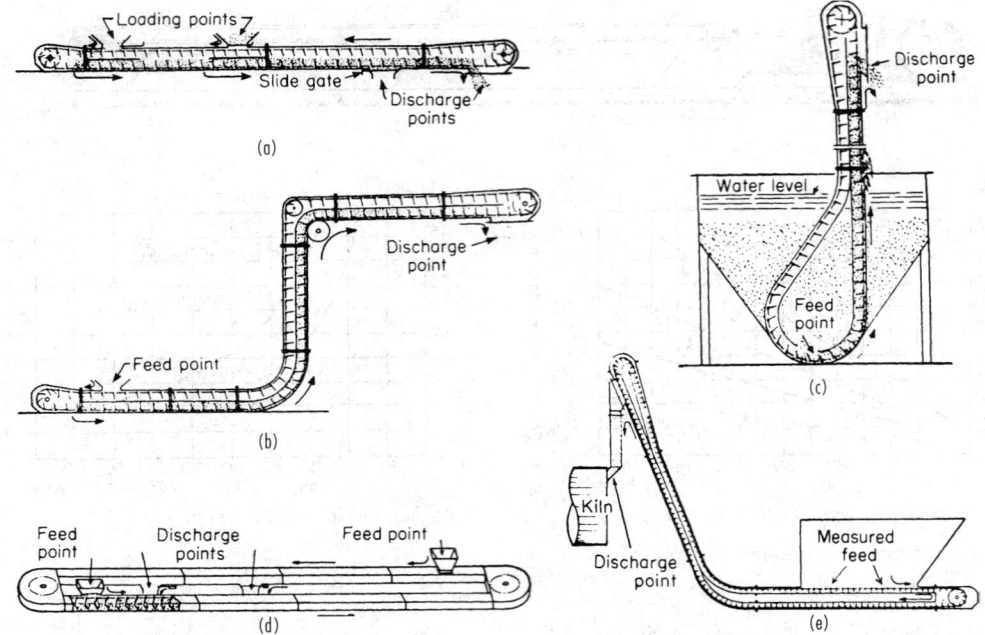

FIG. 21-10 Typical arrangements and applications for continuous-flow conveyors. (*a*) Horizontal conveyor. (*b*) Z-type conveyor-elevator. (*c*) Loop-feed elevator used for dewatering. (*d*) Side-pull horizontal recirculating conveyor. (*e*) Horizontal inclined conveyor-elevator. (*Stephens-Adamson Division, Allis-Chalmers Corporation.*)

characteristics of material are important considerations in the selection of continuous-flow conveyors.

The **side-pull** continuous-flow conveyor can follow a variety of paths in a horizontal plane, picking up and discharging material at many different points. Figure 21-9*c* is a detailed illustration of one type of conveying element, and Fig. 21-10*d* shows a typical arrangement with 180° turns. Triangular arrangements and rectangular layouts with 90° corners are also available.

The **capacity** of the continuous-flow conveyor is dependent on the particular design being considered. Limiting speeds are subject to considerable controversy. It is advisable to follow the manufacturer's recommendations closely for best conveyor service. **Power** calculations depend on a number of experimentally determined constants which vary for different conveyor designs. One factor contributing to total power requirements is the power required on bend corners where flights assume a radial position and tend to compress material which was fed between them when they were running in a parallel position. Noncompressible materials may require special clearances and feed conditions. Thus, while conveyor components have been well standardized, many materials will not convey well unless special design alterations are made.

Because of the fabrication required for casings and the precision fitting of conveying elements within it, the continuous-flow conveyor is normally an expensive unit. However, it occupies little space, needs little support because the casing forms a rigid box girder, may travel in several directions with only a single drive, is self-feeding, and can feed and discharge at several points. These factors may often compensate for what sometimes appears as a rather high cost per foot. Because it is adaptable to many processing operations, the continuous-flow conveyor is widely used in the chemical industry, in which there is a great deal of rehandling or requirements for many feed and discharge points. The conveyors can be designed for self-cleaning to allow different materials to be handled in the same unit without contamination.

Closed-Belt Conveyor This device, with zipperlike teeth which mesh to form a closed tube, is particularly adaptable to the problem of handling fragile materials which cannot be subjected to degradation.

Since the belt is wrapped snugly around the material, it moves with the belt and is not subject to any form of internal movement except at feed and discharge. In addition, the belt can operate in many planes, with twists and turns to meet almost any layout condition within the fixed limit of curvature placed on the loaded belt. It can convey and elevate with only a single drive; multiple feed and discharge points are relatively easy to arrange.

The closed-belt conveyor is not readily adaptable to the handling of sticky materials, and special designs may be required for materials which are highly susceptible to aeration. Initial cost per foot is relatively high because of belting cost, but power requirements are low and with proper installation and maintenance belt life is good.

Since this type of conveyor is available in only one standard size, its capacity is determined by the belt speed and the fixed cross-sectional area. Tons-per-hour capacity is figured by multiplying the bulk density in pounds per cubic foot by the speed in feet per minute and a constant of 0.0021. Power requirements are quite low and figured in the same way as those for conventional belt conveyors.

Figure 21-9*d* illustrates a typical closed-belt-conveyor detail of the opening or closing mechanism and a cross section through a horizontal carrying-and-return run. Designs using two conventional conveyor belts have been developed to elevate material by pressing it between them, but their application is limited.

Flight Conveyors These devices are available in an almost infinite variety. Most flight-conveyor applications are open designs for rough conveying operations, but some are built with totally enclosed casings. Table 21-10 gives typical design and capacity information.

Apron Conveyors Probably the most common chain conveyors, these are available in a wide variety of designs for both horizontal and inclined travel. Their main application is the feeding of material at controlled rates, with lump sizes that are large enough to minimize dribble. The typical design is a series of pans mounted between two strands of roller chain, with pans overlapping to eliminate dribble, and often equipped with end plates for deeper loads. Pan design may vary according to material requirements. Figure 21-11 illustrates a typical apron-conveyor design, and Table 21-11 gives capacities for units with and without skirt plates. Apron-feeder applications range from fairly

TABLE 21-10   Flight-Conveyor Capacities*

Flight size and no. of strands, in (mm)	Maximum size of lumps				Capacity, tons/h (metric tons/h)† for various flight spacings, conveyor, horizontal, in (mm)						Design type‡
	All lumps, in (mm)		10% lumps, in (mm)		18 (460)		24 (610)		36 (915)		
10 × 4 (255 × 100)—1	1½	(38)	3	(76)	32	(29)	25	(23)	16	(15)	1
12 × 5 (305 × 130)—1	1¾	(45)	3½	(89)	46	(42)	35	(32)	23	(21)	1
15 × 5 (380 × 130)—1	2	(51)	4	(102)	66	(60)	50	(45)	33	(30)	1
15 × 6 (380 × 155)—2	3½	(89)	7	(178)	87	(79)	67	(61)	44	(40)	2
16 × 8 (405 × 205)—2	4	(102)	8	(203)	110	(99)	82	(74)	55	(50)	2
18 × 8 (460 × 205)—2	5	(127)	9	(229)	124	(113)	93	(84)	62	(56)	2
20 × 10 (510 × 255)—2	6	(152)	10	(254)	—	—	141	(128)	94	(85)	2
24 × 10 (610 × 255)—2	8	(203)	13	(305)	—	—	176	(160)	116	(105)	2
30 × 10 (765 × 255)—2	10	(254)	14	(355)	—	—	—	—	250	(227)	2
12 × 5 (305 × 130)—1	1¾	(45)	3½	(89)	56	(51)	42	(38)	28	(25)	3
15 × 7 (380 × 180)—1	2½	(64)	4½	(114)	78	(71)	58	(53)	39	(35)	3
18 × 8 (460 × 205)—1	3	(76)	5	(127)	124	(113)	93	(84)	62	(56)	3
12 × 5 (305 × 130)—2	2	(51)	4	(102)	56	(51)	4	(38)	28	(25)	4
15 × 6 (380 × 155)—2	3	(76)	5	(127)	76	(69)	57	(52)	38	(34)	4
18 × 7 (460 × 180)—2	4	(102)	8	(203)	96	(87)	72	(65)	48	(44)	4
24 × 8 (610 × 205)—2	8	(203)	12	(305)	—	—	124	(113)	83	(75)	4

*Data from Fairfield Engineering Co.
†Basis: 30-lb/ft³ (480-kg/m³) bulk density and conveyor velocity of 100 ft/min (30.5 m/min). For inclined conveyors capacities are reduced by factors given:

Slope off horizontal	Factor
15°	0.80
30°	0.55
45°	0.33

‡Type 1: malleable-iron conveyor flights; type 2: steel flights on roller chain; type 3: steel flights with wear shoes or rollers; type 4: steel flights on plain chain.

light-duty applications with light-gauge steel pans up to extremely heavy-duty applications requiring reinforced manganese steel pans with center supports. Table 21-11 values may be used in calculating capacities of other sizes, since this is a function of width of carrying surface, height of sides, speed, and bulk density. Apron-conveyor speeds are typically 0.25 to 0.38 m/s (50 to 75 ft/min). When these conveyors are used as feeders, velocities are kept in the 0.05- to 0.15-m/s (10- to 30-ft/min) range.

PNEUMATIC CONVEYORS

One of the most important material-handling techniques in the chemical industry is the movement of material suspended in a stream of air over horizontal and vertical distances ranging from a few to several hundred feet. Materials ranging from fine powders through 6.35-mm (¼-in) pellets and bulk densities of 16 to more than 3200 kg/m³ (1 to more than 200 lb/ft³) can be handled. A large, capable manufacturing

TABLE 21-11   Apron-Conveyor Capacities*

Apron width, in (mm)	Capacity without skirts for various speeds and bulk densities for material depth of 4 in (102 mm) on pans											
	50 ft/min (15.2 m/min)						100 ft/min (30.5 m/min)					
			tons/h		(metric tons/h)				ton/h		(metric tons/h)	
	ft³/h	(m³/h)	50 lb/ft³ (801 kg/m³)		100 lb/ft³ (1602 kg/m³)		ft³/h	(m³/h)	50 lb/ft³ (801 kg/m³)		100 lb/ft³ (1602 kg/m³)	
18 (460)	1125	(31.9)	28	(25)	56	(51)	2250	(63.7)	56	(51)	112	(102)
24 (610)	1500	(42.5)	38	(34)	75	(68)	3000	(85.0)	75	(68)	150	(136)
30 (765)	1875	(53.2)	47	(43)	94	(85)	3750	(106.2)	94	(85)	188	(171)
36 (915)	2250	(63.7)	56	(51)	113	(102)	4500	(127.4)	113	(102)	226	(205)
42 (1070)	2625	(74.3)	66	(60)	131	(119)	5250	(148.7)	131	(119)	262	(238)
48 (1220)	3000	(85.0)	75	(68)	150	(136)	6000	(170.0)	150	(136)	300	(272)
54 (1370)	3375	(95.6)	85	(77)	169	(153)	6750	(191.2)	169	(153)	338	(307)
60 (1525)	3750	(106.2)	94	(85)	188	(171)	7500	(212.4)	188	(171)	376	(341)

Pan width, in (mm)	Width between skirts, in (mm)		Max. lump size, in (mm)		Capacities with skirts for various material depths, 0.75 loaded cross section and 10-ft/min (3-m/min) velocity											
					Capacity, tons/h (metric tons/h); 50-lb/ft³ (801-kg/m³) bulk-density material; material depth on pans, in (mm)											
					4 (105)		8 (205)		12 (305)		18 (460)		21 (535)		24 (610)	
18 (460)	16	(410)	3	(76)	5.0	(4.5)	10.0	(9.1)	15.0	(13.6)	22.5	(20.4)	26.3	(23.9)	30.0	(27.2)
24 (610)	22	(560)	4	(102)	6.9	(6.3)	13.7	(12.4)	20.6	(18.7)	31.0	(28.1)	36.1	(32.8)	41.2	(37.4)
30 (765)	28	(715)	6	(152)	8.8	(8.0)	17.5	(15.9)	26.2	(23.8)	39.3	(35.7)	45.9	(41.7)	52.5	(47.7)
36 (915)	34	(865)	8	(203)	10.7	(9.7)	21.3	(19.3)	32.0	(29.1)	48.0	(43.6)	56.0	(51.8)	64.0	(58.1)
42 (1070)	40	(1020)	10	(254)	12.5	(11.3)	25.0	(22.7)	37.5	(34.0)	56.3	(51.1)	65.7	(59.6)	75.0	(68.1)
48 (1220)	46	(1170)	12	(305)	14.4	(13.1)	28.8	(26.1)	43.2	(39.2)	64.8	(58.8)	75.6	(68.6)	86.3	(78.3)

*Data from Fairfield Engineering Co.

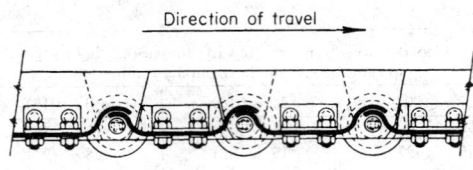
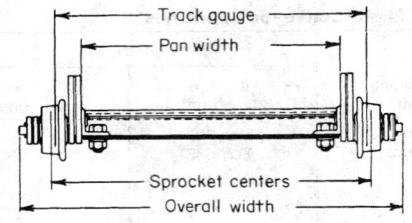

FIG. 21-11 Apron conveyors. (*Fairfield Engineering Co.*)

industry supplies complete systems as well as components that users can incorporate into their own designs. Much engineering information is available from this industry in the form of brochures, data sheets, and nomographs.

The **capacity** of a pneumatic-conveying system depends on (1) product bulk density (and particle size and shape to some extent), (2) energy content of the conveying air over the entire system, (3) diameter of conveying line, and (4) equivalent length of conveying line.

Minimum capacity is achieved when the energy of the conveying air is just sufficient to move the product through the line without stoppage. To prevent such stoppage, it is good practice to provide an additional increment of air energy so that a factor of safety exists that allows for minor changes in product characteristics. An **optimum system** is one that repays, through operating economies, all design features above the minimum required, within the return-on-investment criteria set by the owner.

While successful and economical system designs can be devised by experienced process engineers, the competent technical aid available from equipment suppliers has led to a growing trend toward the purchase of complete systems, even on small jobs, rather than in-plant assembly from components on the basis of in-house designs. An idea of the change in **capital investment** for typical pneumatic-conveyor systems as a function of increasing transfer rates is given in Table 21-12.

Conveyor installations may be permanent or a combination of permanent and portable. The latter kind is often mounted on a bulk-delivery vehicle, which permits fast unloading into the customer's silo by the carrier without effort or equipment from the customer. Controls range from simple motor starters and hand-connected hoses to sophisticated, microprocessor-electropneumatic control systems.

Types of Systems Generally, pneumatic conveyors are classified according to five basic types: pressure, vacuum, combination pressure and vacuum, fluidizing, and the blow tank.

In **pressure systems** (Fig. 21-12a), material is dropped into an air stream (at above atmospheric pressure) by a rotary air-lock feeder. The velocity of the stream maintains the bulk material in suspension until it reaches the receiving vessel, where it is separated from the air by means of an air filter or cyclone separator.

Pressure systems are used for free-flowing materials of almost any particle size, up to 6.35-mm (1/4-in) pellets, where flow rates over

151 kg/min (20,000 lb/h) are needed and where pressure loss through the system is about 305 mmHg (12 inHg). These systems are favored when one source must supply several receivers. Conveying air is usually supplied by positive-displacement blowers.

Vacuum systems (Fig. 21-12b) are characterized by material moving in an air stream of pressure less than ambient. The advantages of this type are that all the pumping energy is used to move the product and that material can be sucked into the conveyor line without the need of a rotary feeder or similar seal between the storage vessel and the conveyor. Material remains suspended in the air stream until it reaches a receiver. Here, a cyclone separator or filter (Fig. 21-12c) separates the material from the air, the air passing through the separator and into the suction side of the positive-displacement blower or some other power source.

Vacuum systems are typically used when flows do not exceed 6800 kg/h (15,000 lb/h), the equivalent conveyor length is less than 305 m (1000 ft), and several points are to be supplied from one source. They are widely used for finely divided materials. Of special interest are vacuum systems designed for flows under 7.6 kg/min (1000 lb/h), used to transfer materials short distances from storage bins or bulk containers to process units. This type of conveyor is widely used in plastics and other processing operations where the variety of conditions requires flexibility in choosing pickup devices, power sources, and receivers. Capital investment can be kept low, often in the range of $2000 to $7000.

Pressure-vacuum systems (Fig. 21-12c) combine the best of both the pressure and the vacuum methods. A vacuum is used to induce material into the conveyor and move it a short distance to a separator. Air passes through a filter and into the suction side of a positive-displacement blower. Material then is fed by a rotary feeder into the conveyor positive-pressure air stream, which comes from the blower discharge. Application can be very flexible, ranging from a central control station, with all interconnection activities electrically controlled and sequenced, to one in which activities are handled by manually changing conveyor connections. The most typical application is the combined bulk vehicle unloading and transferring to product storage (Fig. 21-12d).

Fluidizing systems generally convey prefluidized, finely divided, non-free-flowing materials over short distances, such as from storage bins or transportation vehicles to the entrance of a main conveying

TABLE 21-12 Approximate Pneumatic-Conveyor Costs*

Flow rate, lb/h (kg/h)		Conveyor pipe, inside diameter, in (mm)		Power required, hp	Range of investment, $†	
					Manual‡	Automatic§
10,000	(4,536)	4	(100)	25	83,000	46,000
25,000	(11,340)	6	(155)	60	135,000	89,000
50,000	(22,680)	6	(155)	125	200,000	155,000
100,000	(45,360)	8	(205)	200	356,000	312,000

*Product: Plastic pellets, 1/8-in (3.2-mm) cubes, 30-lb/ft³ (481-kg/m³) bulk density; equivalent length of system, 600 ft (183 m).
†1995 costs. Equipment includes motor and blower package, cyclone receivers, railcar-unloading connections, high-level interlocks for stopping the motor and blower combination when the silos reach a full level, and all necessary piping. Installation is not included.
‡System includes a minimum control package, with most activities person-actuated, including the changing of feed lines to storage silos.
§System includes automatic actuation of most activities, with changing of feed lines to silos accomplished by diverter valves controlled automatically by process control computer.

system. A particular advantage in storage-bin applications is that the bottom of the bin is permitted to be nearly horizontal. Fluidizing is accomplished by means of a chamber in which air is passed through a porous membrane that forms the bottom of the conveyor, upon which the material to be conveyed rests. As air passes through the membrane, each particle is surrounded by a film of air (Fig. 21-12e). At the point of incipient fluidization the material takes on the characteristics of free flow. It can then be passed into a conveyor air stream by a rotary feeder.

Prefluidizing has the advantage of reducing the volume of conveying air needed; consequently, less power is required. The characteristics of the rest of this system are similar to those of regular pressure- or vacuum-type conveyors. Of special concern is the tendency of material to stick to and build up on surfaces of the system compo-

nents. The most common application of this type of conveyor is the well-known railroad Airslide covered hopper car.

An early application of pneumatic conveying was the **blow tank.** This device functions by introducing pressurized air on top of a head of material contained in a pressure vessel. If the material is free-flowing, it will flow through a valve at the bottom of the chamber and move through a short conveying line, usually limited to a maximum of 16 m (50 ft), depending on the product, although systems as long as 457 m (1500 ft) are in use. Of special concern when using this system are the surges of air caused either by the tank emptying or by the air breaking through the product.

The blow-tank principle can be used to feed regular pneumatic conveyors. Use of an Airslide or other fluidizing device at the bottom of the blow tank permits handling non-free-flowing materials. This

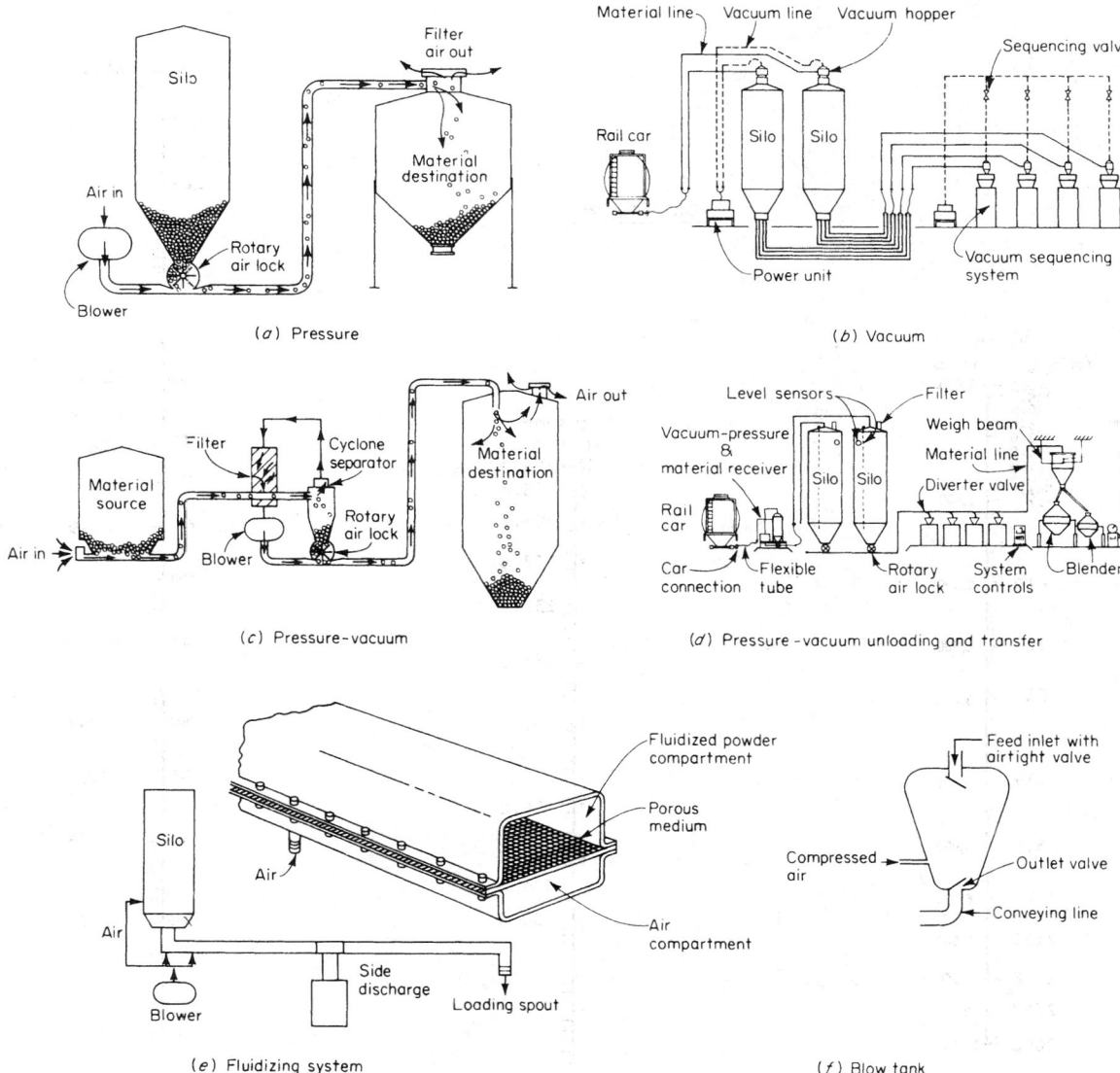

FIG. 21-12 Types of air-conveying systems. (*a*) Pressure. (*b*) Vacuum. (*c*) Pressure-vacuum. (*d*) Pressure vacuum unloading and transfer. (*Whitlock, Inc.*) (*e*) Fluidizing system. (*Fuller Co.*) (*f*) Blow tank.

principle is used extensively in pressure-fluidizing-type valve-bag-packing machines.

Nomographs for Preliminary Design A useful set of nomographs[°] for determining conveyor-design parameters is given in Fig. 21-13. With these charts, conservative approximations of conveyor

[°] Nomographs prepared from data supplied by Flotronics Division, Allied Industries.

size and power for given product bulk density, conveyor equivalent length, and required capacity can be obtained. Because pneumatic conveyors and their components are subject to continual improvements by a fast-changing supplier industry, manufacturers should be invited to submit alternative designs to that resulting from the use of the nomograph. Some large users of pneumatic conveyors have found it expedient to write computer programs for calculating system parameters.

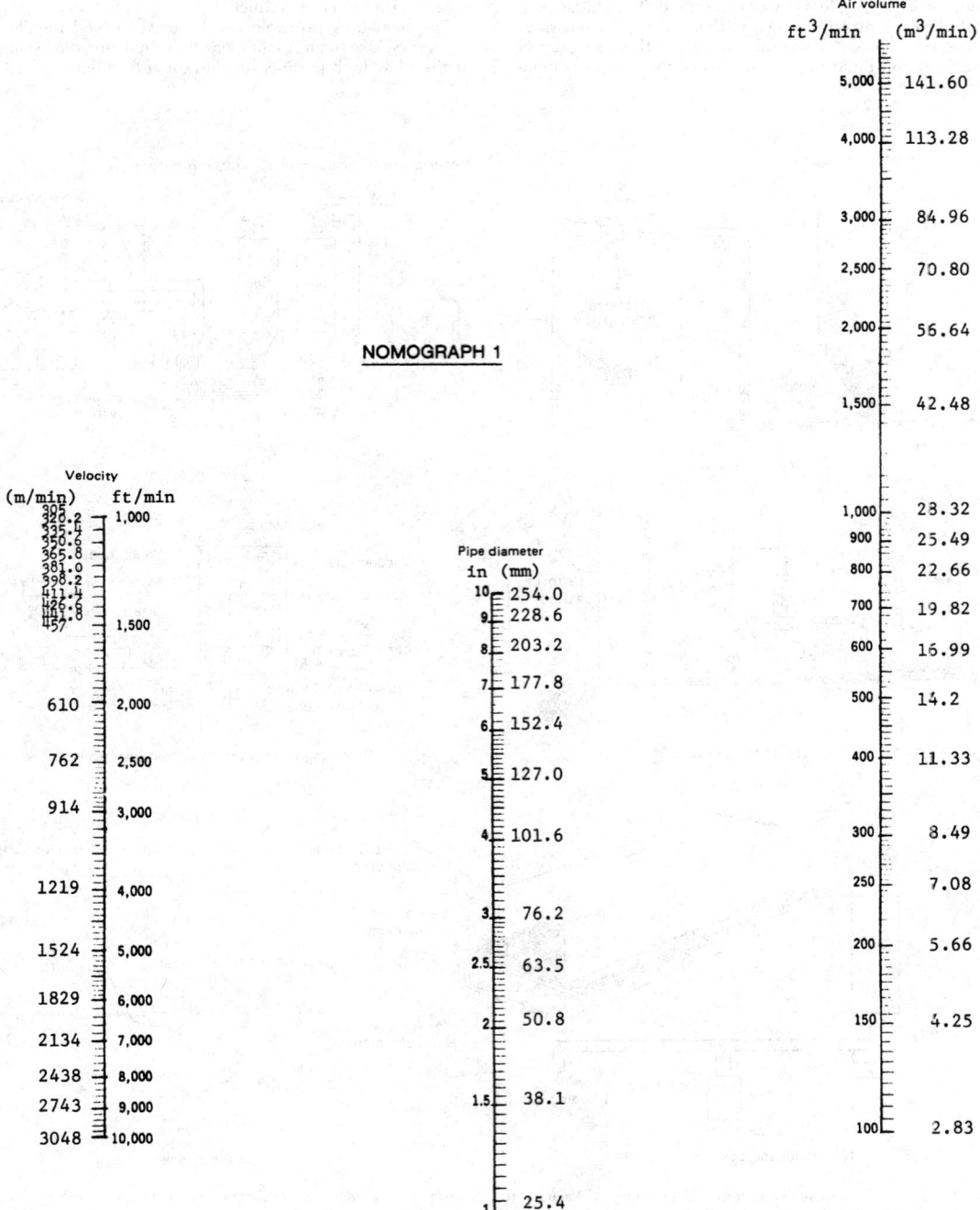

FIG. 21-13 Nomographs for determining conveyor-design parameters.

NOMOGRAPH 2

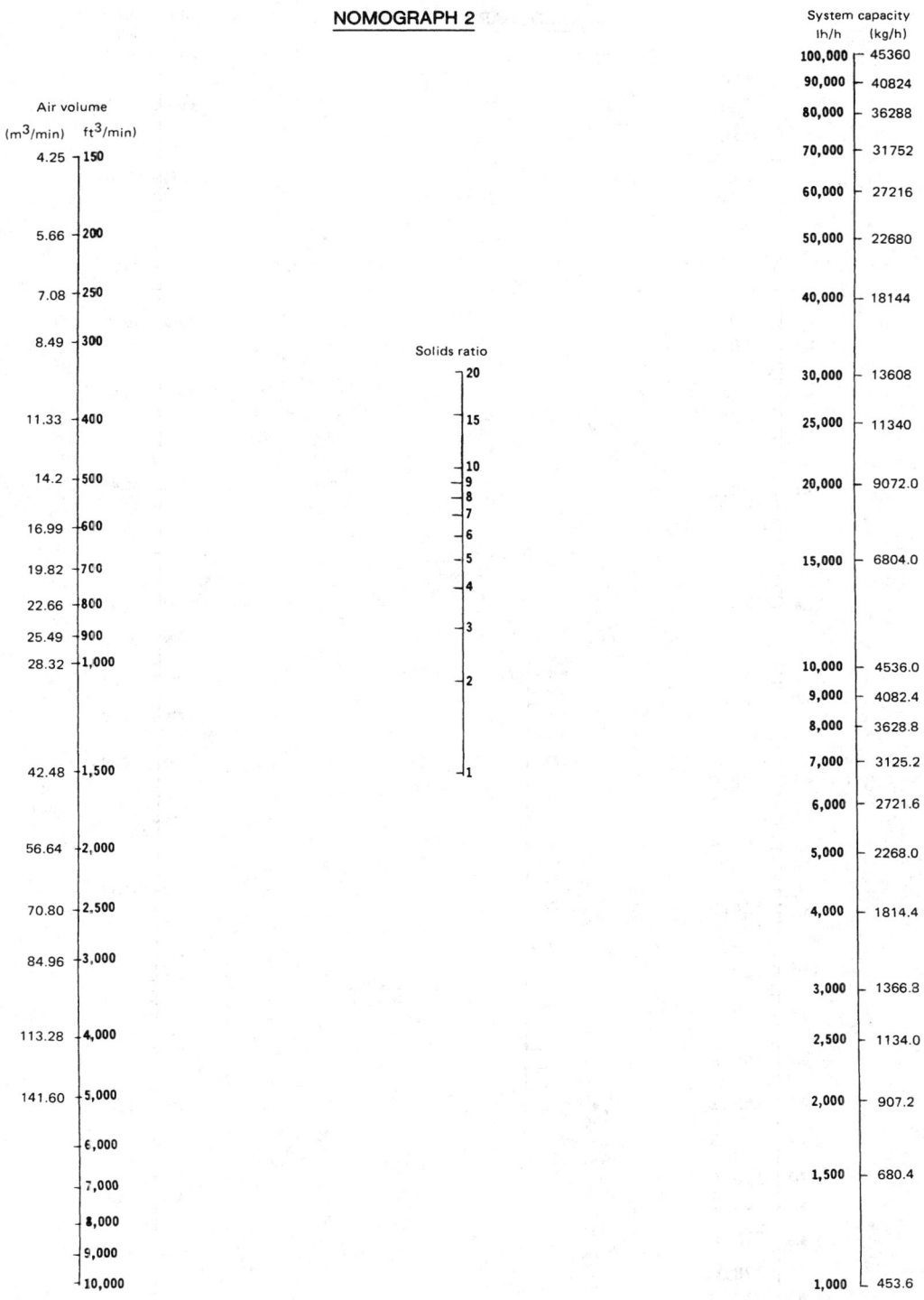

FIG. 21-13   *(Continued)*

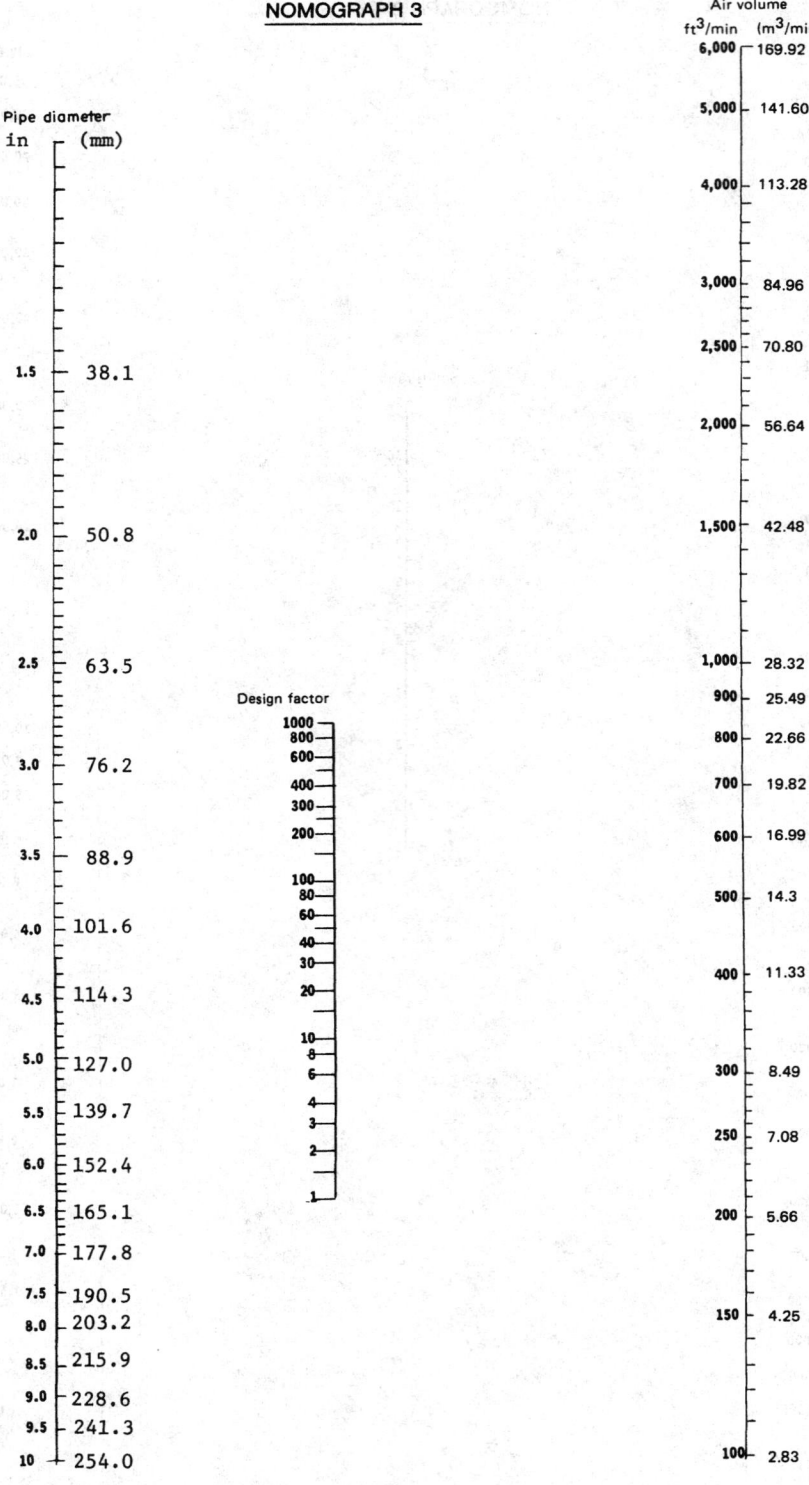

FIG. 21-13 *(Continued)*

NOMOGRAPH 4

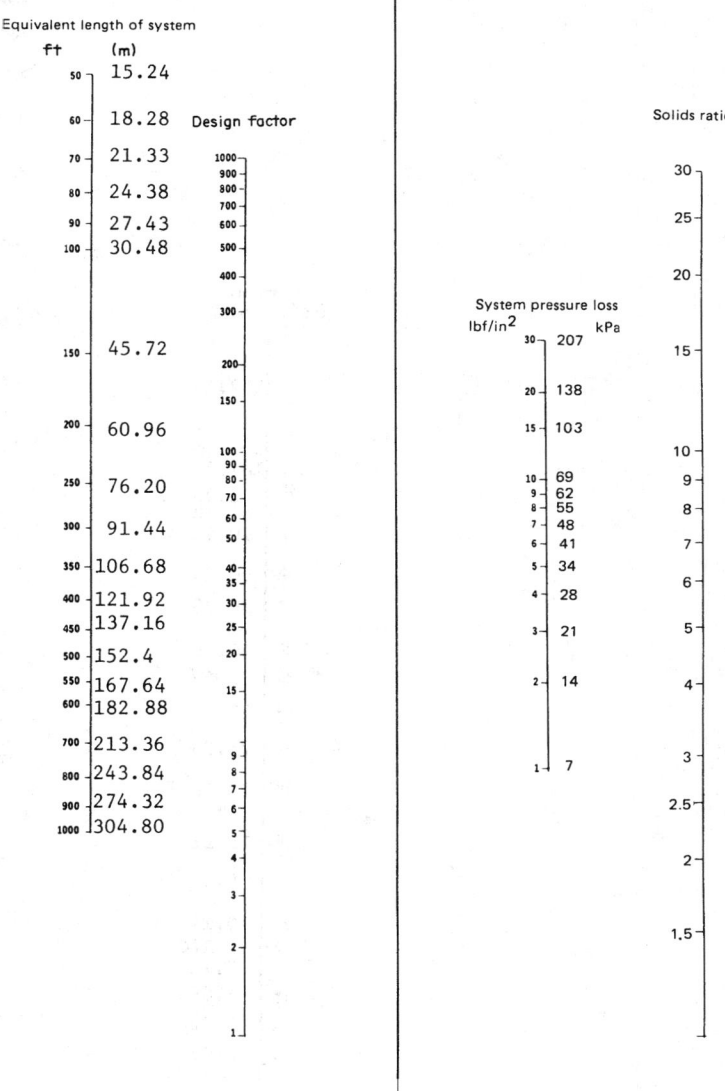

FIG. 21-13   *(Continued)*

To begin preliminary calculations, first determine the equivalent length of the system being considered. This length is the sum of the vertical and horizontal distances, plus an allowance for the pipe fittings used. Most common of these fittings are the long-radius 90° elbow pipe [equivalent length = 25 ft (7.6 m)] and the 45° elbow [equivalent length = 15 ft (4.6 m)].

The second step consists of choosing from Table 21-13 an initial air velocity that will move the product. An iterative procedure then begins by assuming a pipe diameter for the required capacity of the system.

Referring now to Nomograph 1, draw a straight line between the air-velocity and the pipe-diameter scales so that when the line is extended it will intersect the air-volume scale at a certain point.

Turn now to Nomograph 2 and locate in their respective scales the air volume and the calculated system capacity. A straight line between these two points intersects the scale in between them, thus providing at the intersection point the value of the solids ratio. If the solids ratio exceeds 15, assume a larger line size.

Locate in Nomograph 3 the pipe diameter and the air volume found in Nomograph 1. A line between these two points yields the

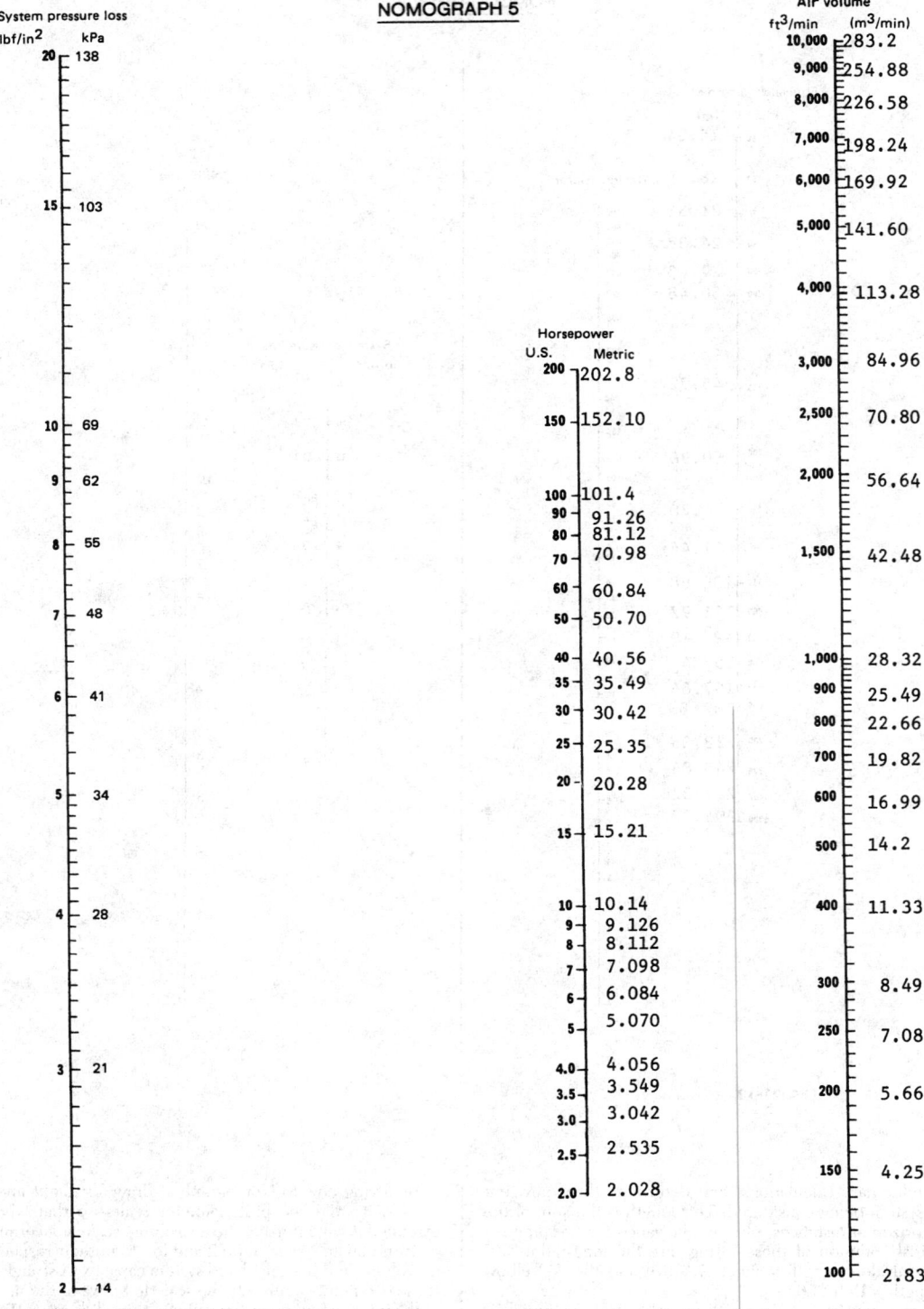

FIG. 21-13   (*Continued*)

TABLE 21-13   Air Velocities Needed to Convey Solids of Various Bulk Densities*

Bulk density		Air velocity		Bulk density		Air velocity	
lb/ft³	kg/m³	ft/min	m/min	lb/ft³	kg/m³	ft/min	m/min
10	160	2900	884	70	1120	7700	2347
15	240	3590	1094	75	1200	8000	2438
20	320	4120	1256	80	1280	8250	2515
25	400	4600	1402	85	1360	8500	2591
30	480	5050	1539	90	1440	8700	2652
35	560	5500	1676	95	1520	9000	2743
40	640	5840	1780	100	1600	9200	2804
45	720	6175	1882	105	1680	9450	2880
50	800	6500	1981	110	1760	9700	2957
55	880	6800	2072	115	1840	9900	3118
60	960	7150	2179	120	1920	10500	3200
65	1040	7450	2270				

*Courtesy of Flotronics Division, Allied industries.

design factor, or P 100 (30.5), the pressure drop per 100 ft (30.5 m), at the intersection of the center scale.

Locating now in their respective scales on Nomograph 4 the design factor (from Nomograph 3) and the calculated equivalent length, draw an extended straight line to intersect the pivot line in the center. Now connect this point in the pivot line with the solids-ratio scale (from Nomograph 2), and read the system pressure loss.

If the value of this loss exceeds 10 lb/in² (70 kPa), assume a larger pipe diameter and repeat all these steps, beginning with Nomograph 1. After a pressure drop of 10 lb/in² (70 kPa) or less is found, turn to Nomograph 5 and locate this pressure loss, as well as the corresponding air volume (from Nomograph 2), and draw a straight line between the two points. The intersection of the horsepower scale will provide the value of the power required. From this, the system cost can now be approximated by consulting Table 21-12.

STORAGE AND WEIGHING OF SOLIDS IN BULK

STORAGE PILES

Discharge Arrangements Open-yard storage is probably best handled by belt conveyor when tonnages are large. Figure 21-14 shows some of the many discharge arrangements possible for single, multiple, or moving-tripper discharge from belt conveyors. Also shown is a tilting-plow arrangement for discharging flat belts. Most of these discharge methods are equally applicable for indoor storage. Large traveling stackers may also be used for outdoor storage. They may move along the length of a belt, forming a pile on one or both sides of the belt, or pivot about a fixed axis to form a circular pile.

Reclaiming Underground-tunnel belts fed by special gates (Fig. 21-15) are often used for reclaiming, as is mobile shovel equipment. Cable-drag scrapers are also used for large outside storage areas and sometimes on inside storage when large, flat areas are used. A drag-scraper system may follow a single fixed cable line, or back posts may be provided to allow relocation of the cable line to cover almost any storage-space shape.

One development for handling large tonnages of bulk materials from storage is the **bucket-wheel reclaimer,** which consists of a series of buckets placed about the periphery of a large wheel that is carried by a fixed propulsion unit. The buckets empty onto a removal conveyor, usually of the belt type, which takes the product to further processing or handling. Bucket-wheel reclaimers capable of handling as little as 150 tons/h to as much as 20,000 tons/h (see Fig. 21-16) have been built.

Mobile equipment is often preferred to fixed types. Front-end loaders, scrapers, and bulldozers are used with increasing frequency, especially on projects of short duration or when capital investment must be limited. Front-end loaders are especially advantageous because of their ability to carry material as well as to plow or bull-doze it.

Angle of repose is the angle at which a material will rest on a pile. It is useful for determining the capacity of a bin or a pile. The angle of the cone that develops at the top of the pile when a bin is being filled will be somewhat flatter than the angle of repose because of the effect of impact.

STORAGE BINS, SILOS, AND HOPPERS

Probably no section of the materials-handling and -storage art advanced as far in a decade (the 1960s) as did that of bin storage of bulk materials. Prior to this time, **storage-bin design** was a hit-or-miss empirical affair, in which success was assured only if the product was free-flowing. This was changed radically as a result of research led by Andrew W. Jenike. This work, which resulted in identifying the cri-

teria that affect material flow in storage vessels, was first reported in Jenike's paper "Gravity Flow of Bulk Solids" (Bull. 108, University of Utah Engineering Experiment Station, October 1961). This paper set forth the equations defining bulk flow and the coefficients affecting flow.

Continuing experimentation verified these criteria, and in Bulletin 123 (November 1964) the subject was further defined by providing flow factors for a number of bin-hopper designs as well as specifications for determining experimentally the characteristics of bulk material affecting flow and storage. Along with the theory, Jenike produced a method of applying it, which includes equations and the physical measurement of material characteristics.

In what follows, a storage vessel is considered as consisting of a bin and a hopper. A bin is the upper section of the vessel and has vertical sides. The hopper, which has at least one sloping side, is the section between the bin and the outlet of the vessel.

Material-Flow Characteristics Two important definitions of the flow characteristics of a storage vessel are **mass flow,** which means that all the material in the vessel moves whenever any is withdrawn (Fig. 21-17), and **funnel flow,** which occurs when only a portion of the material flows (usually in a channel or rathole in the center of the system) when any material is withdrawn (Fig. 21-18). Some typical mass-flow designs are shown in Fig. 21-19.

Mass-flow bins feature the most sought-after characteristics of a storage vessel: unassisted flow whenever the bottom gate is opened. A funnel-flow bin may or may not flow but probably can be made to flow by some means.

Until Jenike developed the rationale for storage-vessel design, a common criterion was to measure the angle of repose, use this value as the hopper angle, and then fit the bin to whatever space was available. Too often, bins were designed from an architectural or structural-engineering viewpoint rather than from the role they were to play in a process. Economy of space is certainly one valid criterion in bin design, but others must be considered equally as well. Table 21-14 compares the principal characteristics of mass-flow and funnel-flow bins.

Although a mass-flow bin is obviously preferable to a funnel-flow vessel, the additional investment generally required must be justified. Often, this can be done by the reduced operating costs. But when installation space is limited, a compromise must be made, such as providing a special hopper design and sometimes even a feeder. Certainly, with mass-flow bins the feeder is not required for flow, but it might still be used for other reasons, such as conveying the material to the next process step.

Design Criteria Jenike's criteria permit an **engineering-economic analysis of storage** with about the same confidence level

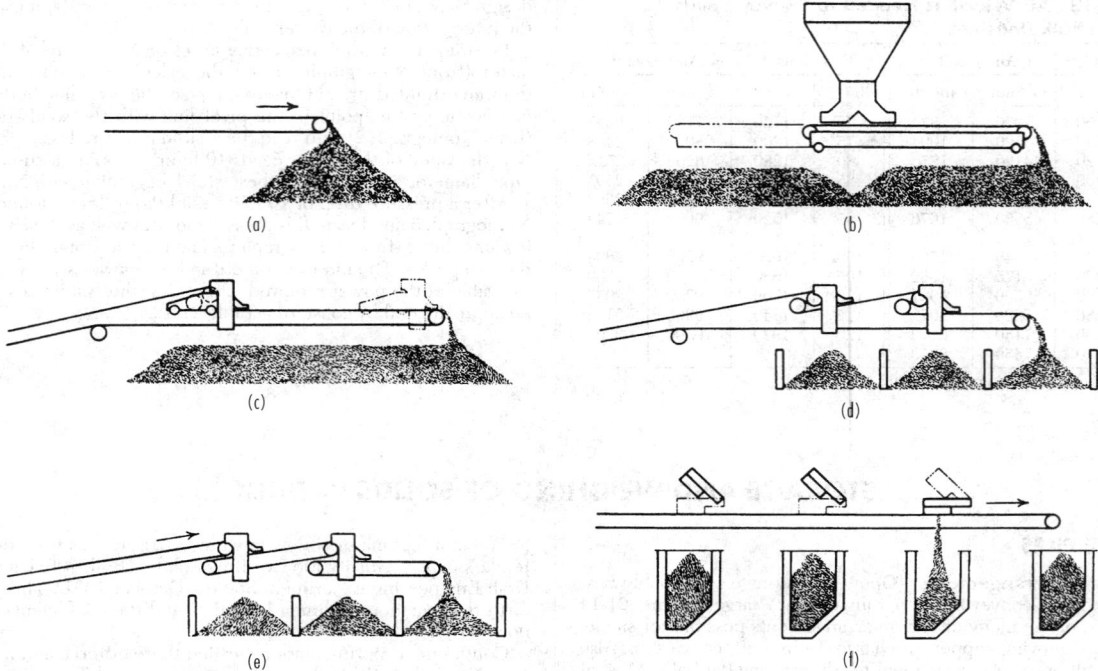

FIG. 21-14   Belt-conveyor discharge arrangements. (*a*) Discharge over an end pulley forms a conical pile at the end of the belt. (*b*) Discharge over either end pulley to distribute lengthwise by a reversible-shuttle conveyor. (*c*) Discharge through a traveling tripper, with or without a cross conveyor, to distribute material to one or both sides of the conveyor for the entire distance of tripper travel. Trippers can be propelled by a conveyor belt or by a separate motor. Motor-propelled trippers can also be automatically reversing to distribute material evenly or can be manually controlled to discharge at any desired point. (*d*) Discharge through fixed trippers, with or without a cross conveyor to one or both sides of the belt, to fixed bin openings or pile locations. This can also be done with multiple conveyors as shown in (*e*) or by stopping traveling trippers in the desired position. (*e*) Discharge from multiple conveyors through fixed discharge chutes, with or without a cross conveyor to one or both sides of the belt, to fixed bin openings or pile locations. (*f*) Discharge by hinged plows to one or more fixed locations along one or both sides of the conveyor. Plows may be adjusted to divide the discharge in several places simultaneously in the proportion desired. (*FMC Corporation, Material Handling Systems Division.*)

as in the rest of the process plant. His quantitative methods may be used to determine (1) whether the vessel will function with mass or funnel flow and (2) the outlet dimensions of the hopper so that product will flow. His methods also provide criteria for making engineering trade-offs between mass flow and funnel flow when product characteristics, space limitations, etc., dictate against design for mass flow.

The **relation between mass and funnel flows** for conical bins is shown in Fig. 21-20. The angle of kinematic friction ϕ', which is a measure of the friction coefficient between the solid and the material of construction used for the conical-shaped hopper, is measured with the "flow-factor tester." The degree of finish of the metal surface can have a large effect in determining whether the vessel will function in mass or funnel flow. Finer degrees of finish are being used more fre-

quently, mostly because intuition has recommended this course. The kinematic angle of friction is also related to the degree of compression that the product undergoes in storage.

Once a decision for mass or funnel flow has been made or a compromise made by including an expanded-flow bin, the hopper outlet and the type of feeder must be considered. Jenike's teaching on the flow through the bin opening is that materials that can be compacted (as opposed to being free-flowing) will be compacted because of storage-vessel shape and the packing characteristics of the product. When this happens, the material forms an arch that is capable of withstanding considerable stress.

Since the arch transfers the load to the hopper walls and in doing so applies so much pressure to them, the kinematic coefficient of friction

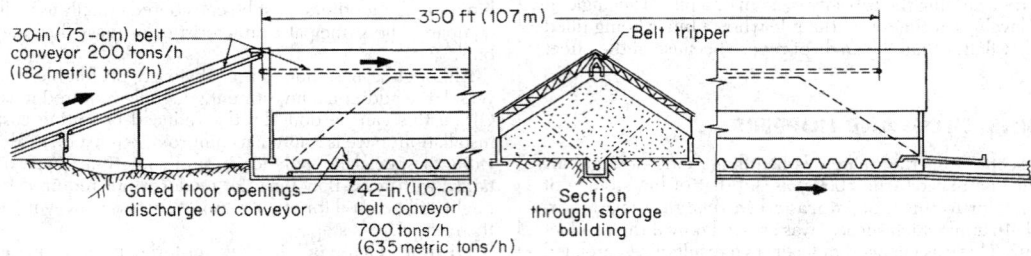

FIG. 21-15   Belt-conveyor storage and reclaiming in a flat-floor building. (*Stephens-Adamson Division, Allis-Chalmers Corporation.*)

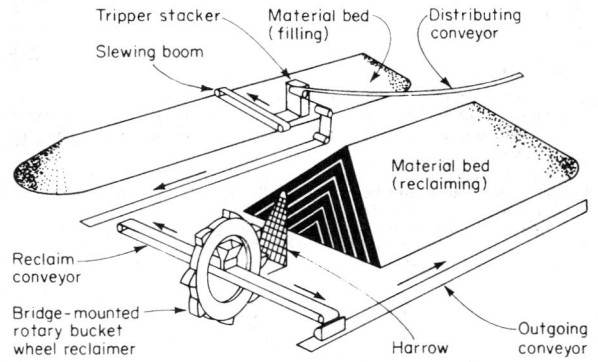

FIG. 21-16   Bucket-wheel reclaimer. Digging buckets mounted on wheel discharge on a belt conveyor for material transfer. (*Courtesy of* Mechanical Engineering.)

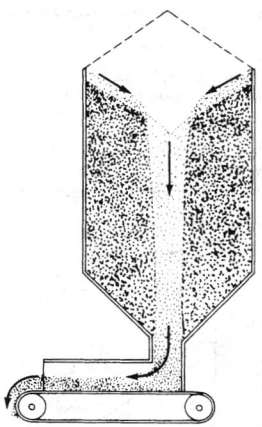

FIG. 21-18   Funnel-flow bin. The material segregates and develops ratholes. (*Courtesy of* Chemical Engineering.)

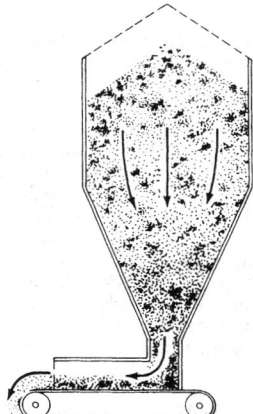

FIG. 21-17   Mass-flow bin. The material does not channel on discharge. (*Courtesy of* Chemical Engineering.)

ϕ' becomes great. The net result is that the "dome" or "bridge" that forms prevents any flow from the vessel. Force must then be applied to the arch so that it will collapse and flow will begin, even if erratically.

According to Jenike, when the strength of the arch f is exceeded by the internal stress s generated by a force applied above the dome, flow takes place. Summarizing:

When $f < s$, flow occurs.

When $f > s$, there is no flow.

When $f = s$, the critical point is reached.

To make a **flow analysis** when $f < s$, an element of material is observed as it moves through a storage vessel (Fig. 21-21). The pressure p on the element increases from zero at the entrance to a maximum value at the transition from the bin to the hopper. The pressure then decreases to zero linearly at the vertex of the hopper cone. The resultant strength f follows a similar pattern, though usually it has some value greater than zero. The stresses induced in the material in the hopper bottom by the weight of material above it are constant but decrease linearly to zero at the cone vertex. The f and s curves intersect at a point corresponding to the critical dimensions of the **bin opening** B.

Reducing this analysis to a technique for determining B, Jenike's method provides a practical way to measure and interpret the strength

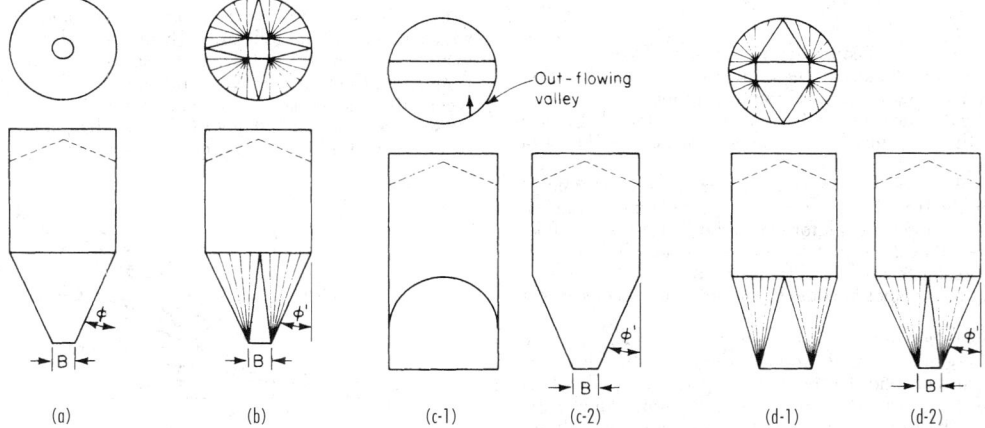

FIG. 21-19   Types of mass-flow bins. Type c is simple but has a valley. Although more difficult to make, type d has no valleys and is usually recommended. (*Courtesy of* Mechanical Engineering.)

TABLE 21-14 **Principal Characteristics of Mass-Flow and Funnel-Flow Bins**

Mass-flow bins	Funnel-flow bins
1. Particles segregate, but remit on discharge	1. Particles segregate and remain segregated
2. Powders deaerate and do not flood when the system discharges	2. First portion in is last one out
3. Flow is uniform	3. Product can remain in dead zones until complete cleanout of the system
4. Density of flow is constant	4. Product tends to bridge or arch, and then to rat-hole when discharging
5. Level indicators work reliably	5. Flow is erratic
6. Product does not remain in dead zones, where degradation can occur	6. Density can vary
7. Bin can be designed to yield non-segregating storage, or to function as a blender	7. Level indicators must be placed in critical positions so they will work properly
	8. Bins perform satisfactorily with free-flowing, large-particle solids

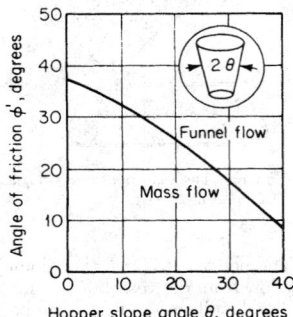

FIG. 21-20 Relation between mass and funnel flows for conical bins. (*Courtesy of* Chemical Engineering.)

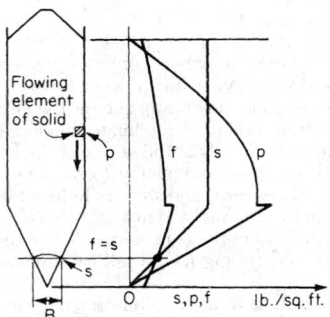

FIG. 21-21 Flow analysis is made by observing an element of material as it moves through the bin. (*Courtesy of* Chemical Engineering.)

of a bulk solid as a function of consolidating pressure. To develop this relation, Jenike developed a **shear tester** that gives a flow function *FF*, which is a curve through a locus of points resulting from values of *f* and *p* obtained by the shear tester. This *FF* curve is plotted against a flow factor *ff* for the particular hopper being designed, as shown in Fig. 21-22.

The method makes use of the principle that a constant ratio of induced stress *s* in the stored contents to the consolidating pressure *p* exists. Thus, for any hopper design for which the *ff* curve is available, the shear-tester results can be plotted, and the point where *f* = *s* is located. Since the distance at which this occurs above the hopper vertex is also known, these values become the hopper dimensions at that point.

A useful approximation of *B* for a conical hopper is $B = 22f/\alpha$, where α is the bulk density of the stored product. The apparatus for determining the properties of solids has been developed and is offered for sale by the consulting firm of Jenike and Johansen, Winchester, Massachusetts, which also performs these tests on a contract basis. The flow-factor *FF* tester, a constant-rate-of-strain, direct-shear-type machine, gives the locus of points for the *FF* curve as well as ϕ', the

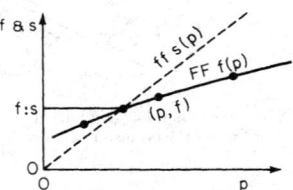

FIG. 21-22 Flow takes place only where *FF* lies below *ff*. (*Courtesy of* Chemical Engineering.)

kinematic coefficient of friction. A consolidating bench is used to prepare samples having different degrees of compaction for the flow-factor tester. This can be supplemented with the same bench enclosed in a controlled-temperature cabinet.

It is possible for some materials to produce an *FF* plot having no intersection with the *ff* curve. This indicates that a different hopper-bin design is needed or that the material cannot be made to flow. Figure 21-23 shows *FF* curves for several materials.

Jenike's method allows the chemical engineer to design bulk-storage vessels and to weigh cost versus performance with a high level of confidence that if the conditions in the real storage system are the same as those prevailing during the tests, the product will flow. It is up to the engineer, however, to establish the bounds of conditions that the product will encounter and to make appropriate tests. A product may not flow if its characteristics change, if radical temperature changes are encountered at the plant, or if moisture is left from an underdesigned dryer.

A further use of the Jenike method is its extension to the critical **structural design** of storage vessels. Because pressures can be calculated, it is possible to design for actual conditions rather than estimates. Also, flow-corrective devices may be designed by using his theory.

Specifying Bulk Materials for Best Flow Many flow problems can be eliminated at the source by rigid, accurate, and sensible specification of the physical characteristics of the material.

Particle size is one of the most common and controllable factors which affect the flowability of a given material. In general, it may be assumed that the larger the particle size and the freer the material is from fines, the more easily the material will flow. Specifications can dictate the desired particle size and uniformity of particle size for purchased raw materials. Stage grinding in the plant can reduce waste and improve flowability by producing a ground material with a mini-

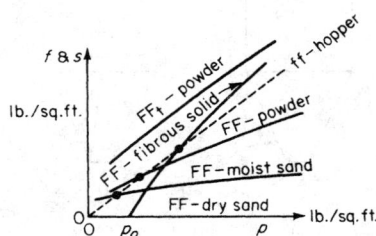

FIG. 21-23 *FF* curves for various materials. Multiply pounds-force per square foot by 0.0479 to get kilopascals. (*Courtesy of* Chemical Engineering.)

mum of fines, but this involves extra operations which may not be economically defensible.

Handling ease is often enhanced by **pelleting** the raw materials. The large particle size, uniformity of particle size, and hard, smooth surface of pellets all contribute to good flow.

Moisture content is another common and controllable flow factor. Most materials can safely absorb moisture up to a certain point; further addition of moisture can cause significant flow problems. Specifications can control the amount of moisture content present in purchased raw materials. Moisture content can be lowered in the plant by including a drying operation in the process line. The costs incurred in drying may be offset by more efficient flow, lower shipping cost, and control of deterioration losses.

Moisture control can also be effected by replacing the air in the material container or bin with a dry, stable gas—nitrogen, for example. This technique is also used to protect the material from certain types of deterioration, such as vitamin loss from food materials.

High temperatures can cause serious flow problems in some materials which contain glutens, sugars, or other soluble or low-melting-point components. These materials become sticky at high temperatures, and it may be necessary to install cooling equipment. As with drying equipment, a study should be made to determine if the additional cost of cooling can be offset by the savings effected by improved flow. Other possible advantages, such as the keeping qualities of the product at lower temperatures, should of course be considered.

Age appears to improve the flowability of certain materials. This is probably the result of particle-surface oxidation, more even moisture distribution, and the rounding of particle corners caused by handling.

Oil content does not materially decrease flowability. For example, the addition of oils and fats to animal-feed ingredients improves the quality of pellets made from these materials, making the pellet surfaces harder and enabling the pellets to resist attrition.

Gates (Fig. 21-24) are used to control flow from bins, hoppers, and processing equipment to feeders or directly to conveyors. They are available in a wide range of styles, from the simple hand slide gate (which can frequently be very difficult to operate by hand) to the precision rack-and-pinion design, which is usually tightly sealed against dust and dribble. The rack-and-pinion gate operates manually with a minimum of effort and is easily adapted to electric, pneumatic, or hydraulic operation. The lever-operated quadrant gate is most often used when a quick-opening gate is desired. It is not designed to control the flow of material but rather to allow the free discharge of lumpy materials. There are hundreds of gate styles to select from, and when properly applied they can often eliminate the need for a more expensive feeder.

Solids-level controls are important for determining the level of materials in bins and hoppers and can also protect conveyors from damage due to jamming if placed in transfer and discharge chutes. They may simply activate an audio or visual warning signal, or they may be electrically tied into the conveying system to start or stop conveyors automatically. Many designs are available, based on principles such as ultrasonics, lasers, radar, and switches operated by diaphragms or paddles. The two designs shown in Fig. 21-25 depend on limit switches, with activation from a pendant cone on one and from a stainless-steel diaphragm on the other. In either case, the presence of material resting against the cone or diaphragm opens or closes the switch, activating a warning signal in the latter case and turning off power to the conveyor in the former.

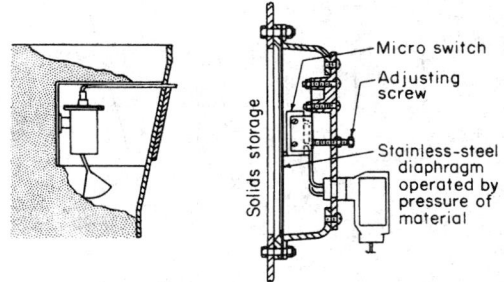

FIG. 21-25 Bin-level-control units.

FLOW-ASSISTING DEVICES AND FEEDERS

Often there are situations in which mass-flow bins cannot be installed for reasons such as space limitations and capacity requirements. Also, sometimes the product to be stored has an *FF* flow function that lies below the flow factor *ff*, bridging takes place, and unassisted mass flow is not possible. To handle these situations, a number of **flow assisters** are available, the most desirable of which use a feeder and a short mass-flow hopper to enlarge the flow channel of a funnel-flow bin. The choice of feeder or flow assister should always be made as part of the storage-vessel analysis. The resulting systems are then usually as effective as the mass-flow types.

Vibrating hoppers are one of the most important and versatile flow assisters. They are used to enlarge the storage-bin opening and to cause flow by breaking up material bridges. Figure 21-26 shows this type of feeder. Two basic types of vibrating hoppers are common: the gyrating kind, in which vibration is applied perpendicularly to the flow channel, and the whirlpool type, which by providing a combined twist and lift to the material, causes bridging to break. One version of this type of flow assister is a bin that vibrates or oscillates in its entirety. Such bins are usually limited to a capacity of about 2.8 m³ (100 ft³).

Screw feeders are also used to assist in bin unloading and in producing uniform feed. Of importance here is the need for a variable-pitch screw to produce a uniform draw of material across the entire hopper opening (Fig. 21-27). For uniform flow to occur, the screw-feeder opening-to-diameter ratio should not exceed 6.

Belt or apron feeders can also be used to give uniform feed from a bin, but care must be taken that dead spots are not produced in the

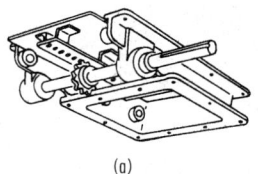

(a)

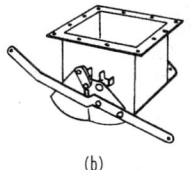

(b)

FIG. 21-24 (a) Rack-and-pinion gate. (b) Double-quadrant gate.

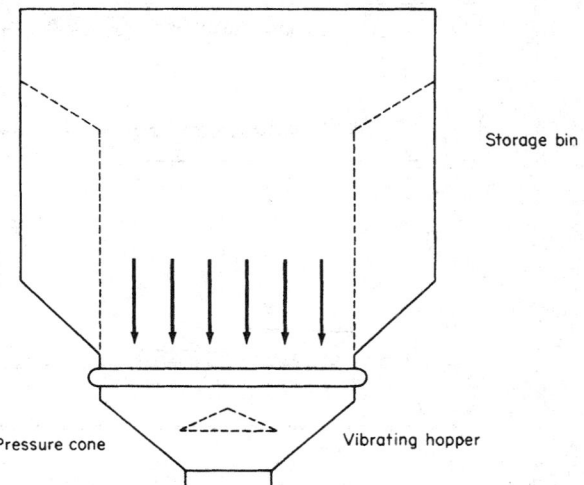

FIG. 21-26 Vibrating hopper. It enlarges the storage-bin opening and breaks up material bridging. (*Courtesy of* Mechanical Engineering.)

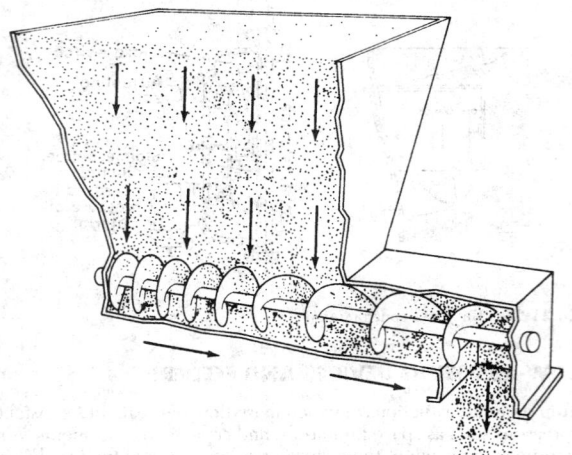

FIG. 21-27 Screw feeder. It needs a variable-pitch screw to produce a uniform draw of material. (*Courtesy of* Chemical Engineering.)

flow channel above the feeder belt (Fig. 21-28). The capacities of these feeders can be increased by tapering the outlet in the horizontal and vertical planes. To ensure the flow of non-free-flowing solids along the front bin wall, a sloping striker plate at the front of the hopper is necessary. Taper may be in one direction only. An apron feeder for large rock, for instance, would have bin skirts tight against the pan to prevent the rock from wedging between the hopper and feeder, and the taper would be in the horizontal plane only. For long slots, however, increasing slot width to provide taper becomes impractical.

Belts have been used successfully under slot openings as long as 30 m (100 ft), with a constant slot width of 205 mm (8 in). Provisions should be made for field adjustment of the space between the skirt and the belt to provide uniform flow along the entire length. Since the minimum distance between the skirt and the belt should allow the largest particle to pass under, very long belt feeders are limited to the finer solids.

The same principles apply to **table feeders.** The skirt is raised above the table in a spiral pattern to provide increased capacity in the

direction of rotation (Fig. 21-29). The plow, located outside the bin, plows only the material that flows from under the skirt.

Vibratory feeders also provide uniform flow along a slot opening of limited length (Fig. 21-30). Here also, the distance between the feeder pan and the hopper is increased in the feed direction. Slot length is limited by the motion of the feeder. Because in long slots the upward component of motion is not relieved by the front opening, solids tend to pack. This can cause flow problems with sticky solids as well as a large demand of power for free-flowing materials. To circumvent these difficulties, vibratory feeders and reciprocating-plate feeders are designed to feed across the slot. Although this kind of feeder may require several drives to accommodate extreme width, the drives are small because of the feeder's short length.

Star feeders with a collecting-screw conveyor (Fig. 21-31) provide highly uniform withdrawal along a slot opening. A vertical section of at least one outlet width should be added above the feeder to ensure uniform withdrawal across the opening.

Other methods of aiding bin unloading are rotating-arm units and air fluidizing pads.

WEIGHING OF BULK SOLIDS

Automatic weighing has largely replaced manual weighing in the chemical-process industries because of the advent of larger-capacity processes and the need to economize on labor. Also, the dependability of weighing equipment has increased markedly, and investment cost has decreased. Both batch and continuous weighing are used.

Batch Weighing In batch weighing, a given unit of weight is measured, and then the desired total weight is obtained through multiples of the given unit. Batching scales find use when small weighings are carried out either singly or a few in sequence.

Most batch scales involve a vessel mounted on a weigh beam, which is counterbalanced by a set of weights approximately equal to the desired weighing. A feed source mounted over the weigh vessel is activated or stopped by a signal generated by motion of the scale beam. Straight mechanical scale-control systems have largely been replaced by those having air or hydraulic-cylinder control of the feed source and weigh-vessel discharge. These are activated by electrical controls.

The **principle of operation** of batch-type scales is based on the concept that a flowing stream of material has constant density. If this is true, then if at some point in advance of the desired batch weight the stream is cut off, the amount of material flowing will remain con-

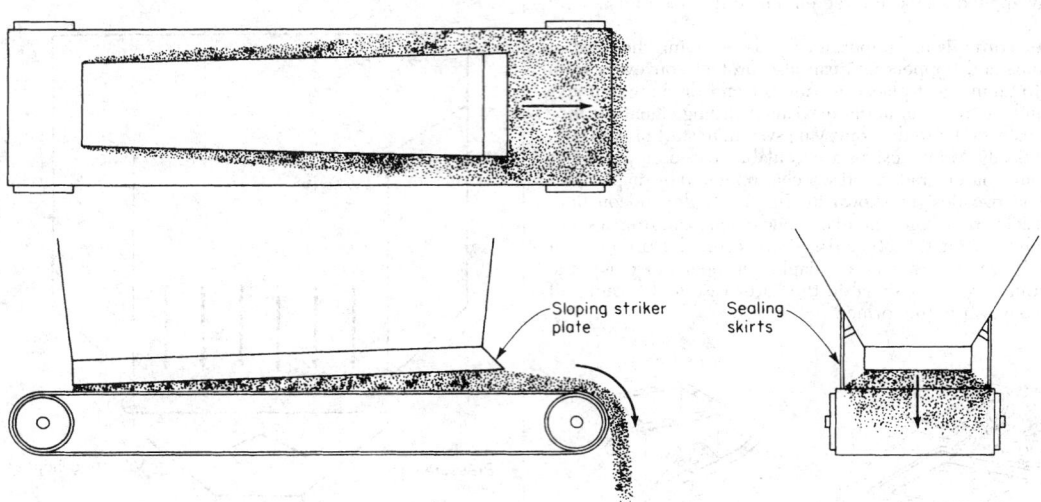

FIG. 21-28 Sloping striker plate in the belt of an apron feeder ensures the flow of non-free-flowing solids. (*Courtesy of* Chemical Engineering.)

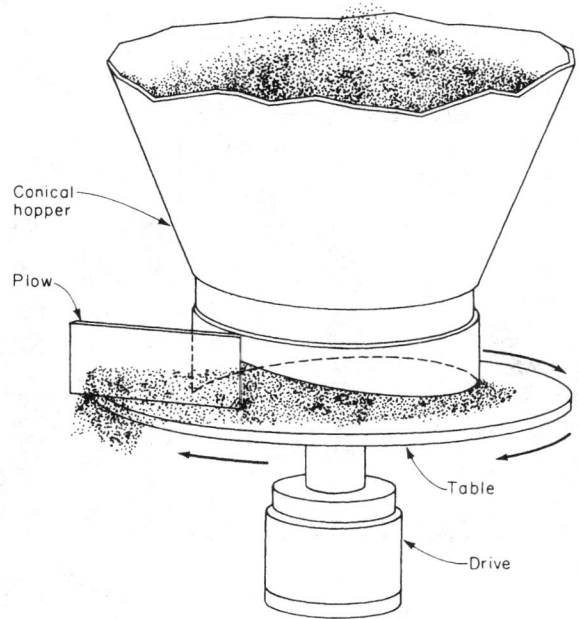

FIG. 21-29 Table feeder. The skirt is raised in a spiral pattern for increased capacity in the direction of rotation. (*Courtesy of* Chemical Engineering.)

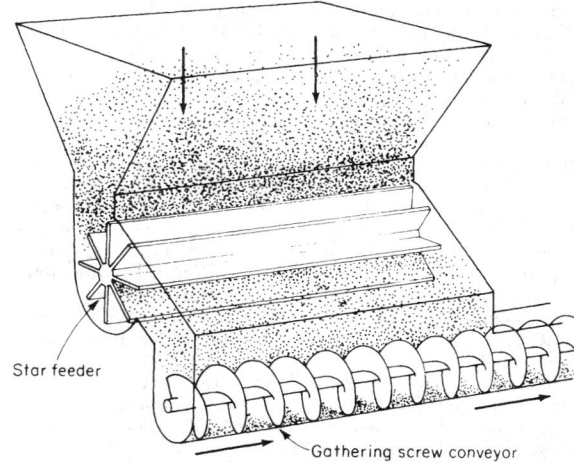

FIG. 21-31 Star feeder. The collecting screw ensures uniform withdrawal. (*Courtesy of* Chemical Engineering.)

stant between the time when the weight is sensed and the time when the flow is stopped. The total weight in the weigh vessel is the sum of the charge due to flow and the amount that flows during the cutoff period. For this reason, feed conditions to the scale are important. **Uniform flow** is essential for accurate batch weighings.

If the material is free-flowing, a mass-flow hopper (Fig. 21-32) can be used. If it is not free-flowing, an appropriate feeder such as a screw, belt, or vibratory feeder should be used. These feeders are described in the subsection "Flow-Assisting Devices and Feeders."

Of special interest in scale-control systems is the type in which the motion of the scale beam is sensed by a differential transformer or a group of load cells. The output of such devices is proportional to the displacement of the scale beam, which in turn is proportional to the amount of material in the weigh bucket. Many designs use load-sensing devices such as strain gauges or transducers. These eliminate the need for a scale-beam mechanism. The weigh vessel is mounted directly on the load-sensing devices. This provides many benefits in

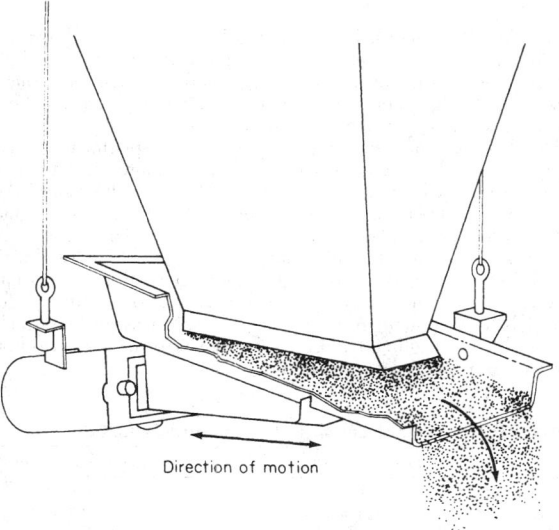

FIG. 21-30 Vibratory feeder. The distance between the feeder pan and the hopper is increased in the direction of feed. (*Courtesy of* Chemical Engineering.)

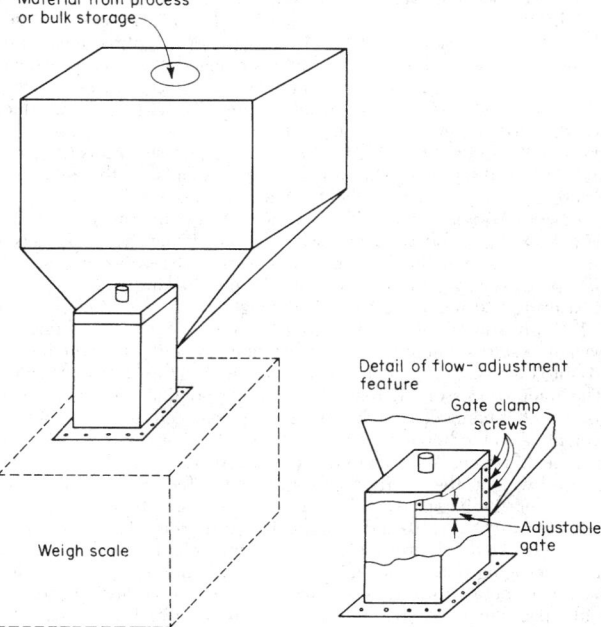

FIG. 21-32 Mass-flow hopper for free-flowing products, used with simultaneous fill-and-weigh and preweigh scales.

addition to accurate weights. One of the most notable is the ability to use the output to indicate actual weight in the weigh vessels or, through a different calibration, to read variations from the desired weight.

Use of microprocessors allows this signal to be employed in a variety of useful ways: controlling weight, adjusting the scale to accommodate the slight changes in bulk density inherent in flowing bulk materials, and activating recording devices, printing heads, and label printers. These features were not possible with straight mechanical scales. Because the microprocessor can perform arithmetic operations and be programmed by using algebraic logic, many products previously considered difficult or impossible to weigh accurately can now be weighed with accuracies equal to those for free-flowing materials. This permits weighing to be done by **addition** or by **subtraction**. With the **additive**-type scale, material is **added** to the weigh vessel, which is being sensed by the scale-control system. With a **loss-of-weight scale,** material flows out of a vessel which is being continuously weighed.

When extreme accuracy of weighings is required, the feed to the weigh vessel is divided into two successive portions: a large **bulk charge,** followed by a final short **dribble feed,** which should have a flow rate of about 0.01 percent of the bulk rate.

A typical application of batch scales is the weighing of charges for packaging machines. Another is the weighing of given amounts of raw material and then dropping these into the next process unit, such as a mixer or an autoclave. Batch weigh scales are capable of a weighing accuracy of within ±0.1 percent when they are equipped with bulk and dribble controls. When also equipped with high-sensitivity weight-sensing devices and microprocessor control, with the latter continuously plotting actual versus desired weight, batch weigh scales are capable of ±0.001 percent accuracy within 3 sigma limits.

Additive Weigh Scale   The sequence of operations involved in weighing a charge of material (Fig. 21-33) is as follows. A free-flowing product is available in the scale feed hopper (1). On depressing the manual start switch (19), the bulk gate (5) and the dribble gate (6) open. The product flows into the scale weigh bucket (2). Weight is sensed by strain gauges (13), (14), whose analog output is converted to a digital output by a circuit in the microprocessor (18), which reads the weight X times each second, depending on the sensivity needed. When a preset bulk weight (approximately 98 percent of the desired weight) is reached in the scale bucket (2), the microprocessor closes the bulk gate (5) and opens the dribble gate (6). Dribble feed commences, and when the desired weight is reached, the microprocessor causes the dribble gate to close, completing the weight measurement. The scale bucket gate automatically opens, discharging the product weighed to the next process stage. The microprocessor then displays the actual weight of the charge (20) and records, lists, prints (21), and signals any discrepancy in weight if this is outside the tolerance desired. It also can print a label for the batch if desired.

Loss-of-Weight Scale   The sequence of operations in this scale (Fig. 21-34) is as follows: Depressing the initializing switch (1) causes the feeder (2) to fill the weigh hopper (3) until the level-control switch (5) opens, stopping the flow, closing the interlocking switch (5), and measuring and recording the initial weight W_0 in the weigh hopper (3). Depressing the start button (6) causes the feeder (4) and the bag packer to start simultaneously. The product is conveyed by the feeder (4) into the packer (7) and by the bag packer into the bag (not shown). The microprocessor (8) reads the analog-to-digital-converter signal (8), which is connected to the strain-gauge load cell (9), and subtracts weight W_i in the hopper (3) at time t_i from the initial weight W_0. The weight difference W_j is summed and recorded. When $W_j = W_s$, the desired weight, the microprocessor stops the feeder (4) and, X seconds later, the bag packer (7). The microprocessor then displays the value of weight W_f (10) and records, lists, and prints (11) any discrepancy between the desired weight and W_f, the weight actually obtained. The packaging system shown is designed for handling products which have very poor or erratic flow characteristics. In addition to its use with packaging equipment, the **loss-of-weight scale** can be employed for a wide variety of process applications.

Continuous Weighing   This procedure involves a device that is sensitive both to the total amount of material flowing and to changes

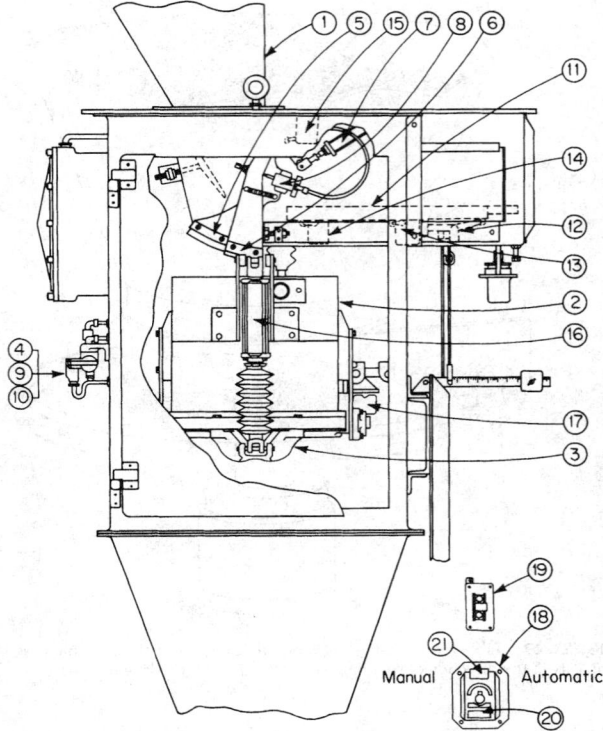

FIG. 21-33   Typical batch-type additive automatic scale. Components: (1) Bin. (2) Scale bucket. (3) Bucket gate. (4) Solenoid valve. (5) Bulk gate. (6) Dribble gate. (7) and (8) Air cylinders powered by solenoid-operated valves. (9) and (10) Solenoid-operated valves. (11) Scale beam. (12) Booster device. (13) and (14) Load cells. (15) Microswitches. (16) Air cylinder. (17) Proximity switch. (18) Microprocessor. (19) Manual start switch. (20) Weight display of the most recent weighing. (21) Average weight display for all weighments of the current product batch.

in the flow. The material is continuously brought over the weight-sensing elements of the continuous-weigh scale, which is capable of keeping track of the flow and its changes and eventually accounts for these when totaling them. Continuous-weighing scales use a section of a belt conveyor, over which the material to be weighed passes.

The belt is mounted on a weight-sensitive platform, typically equipped with load cells, which can detect minute changes in the weight of material passing over the belt. The load-cell output (which is usually a change in resistance proportional to weight) is integrated over short intervals and the condition of flow given. This may be a rate of flow or, at the end of a weight measurement, the total weight. Figure 21-35 shows a continuous weigher, sometimes referred to as a proportioner. Continuous-weighing scales are used mostly to feed materials to continuous processes at uniform, measured rates. They are capable of weighing within ±1 percent error or even within 0.1 percent error under certain conditions.

TABLE 21-15   Weight Sensing Devices and Sensitivity

Device	Sensitivity (one part in)
Beam-Microswitch	1,000
Beam-Differential Transformer	10,000
Strain Gauge Type Load Cell	20,000
Magnetic Force Restoration Transducer	500,000
Variable Capacitance Transducer	1,000,000

Data courtesy of Kg Systems, Inc., Bloomfield, NJ.

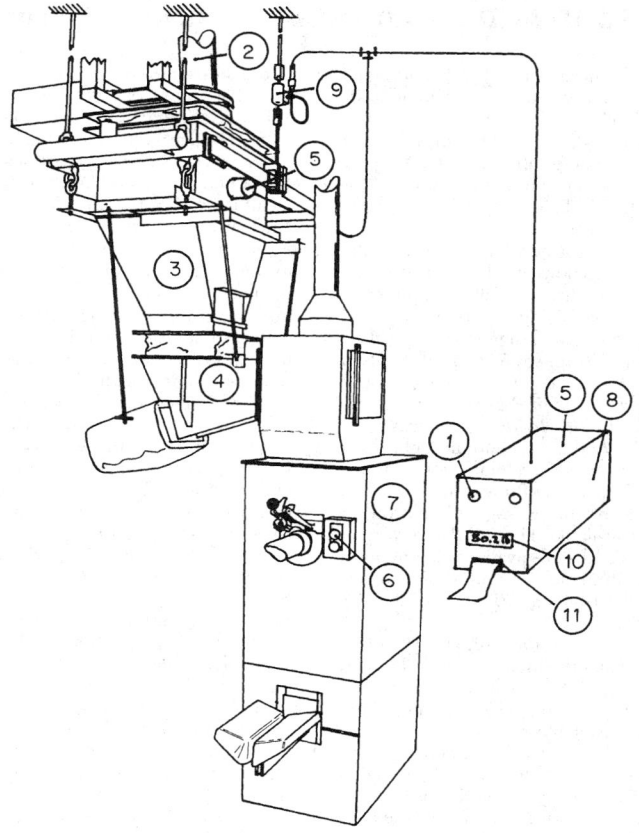

FIG. 21-34 Loss-of-weight-type scale used with a bag-packaging machine for non-free-flowing products. (*Courtesy of H. F. Henderson Industries, Inc., Caldwell, NJ.*)

All scales require continuous monitoring to assure that the desired set weight is maintained and does not drift off because of changes in product bulk-density or flow characteristics. Microprocessors can perform this task automatically.

Weight Sensing These devices have been the subject of intensive research, development, and applications. Increased sensitivity and reliability have been the result of this effort, which has been driven by the increased availability of special purpose computers and the data processing capability of low-cost personal computers. Table 21-15 lists those commonly available and gives their sensitivity. As a result of this, custom-designed weighing equipment has become an important alternative to standardized or "off the shelf" designs. This is especially true when there is a necessity to modify a standard design, which often is more expensive than a custom design.

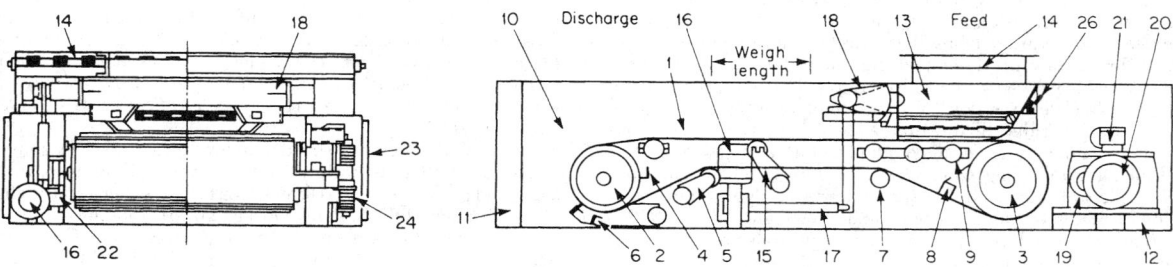

FIG. 21-35 Bulk continuous weigher. (1) Conveyor belt. (2) Head pulley. (3) Tail pulley. (4) Pulley scraper. (5) Spring-loaded take-up assembly. (6) Outer belt scraper (spring-loaded). (7) Belt tracking control. (8) Belt scraper (inner). (9) Pulley scraper. (10) Side channel. (11) Cross channel. (12) Cross channel. (13) Transition chute (optional). (14) Feed-cutoff gates. (15) Weigh idlers. (16) Gate-screw drive motor. (17) Gate screw. (18) Manual gate-adjustment screw. (19) Tachogenerator (optional). (20) Variable-speed drive. (21) Speed adjustment for motor drive (20). (22) Coupling. (23) Side cover. (24) Weight-sensing elements. (25) Hopper storage (optional). (26) Adjustable heel plate (optional). (*Howe Richardson Company.*)

PACKAGING OF SOLID AND LIQUID PRODUCTS AND HANDLING OF PACKAGES

Packaging is often defined by the chemical industry as including all packages or containers which will hold 2 metric tons (4400 lb) of product or less. These containers include bags, cartons, drums and pails, cans, bottles, bulk bags, and metal tanks. Materials of construction can be paper, plastic, metal, glass, wood, or composites of these. A representative list showing the wide variety of containers available for chemical products is given in Table 21-16, which includes typical specifications and representative 1995 costs.

Packaging information has assumed a new importance as the chemical industry in the United States has downsized and the position of the professional packaging engineer has virtually disappeared. Often, the chemical engineer without packaging experience is faced with a project in which the package must be designed and specified and the packaging machinery chosen. Useful sources of information are texts which deal specifically with the subject of packaging. Prominent among these are: *The Wiley Handbook of Packaging Engineering*, published by John Wiley & Sons; *The Handbook of Packaging Engineering* by Joseph Hanlon, published by Technomic Publishing Co., Inc.; and *Fundamentals of Packaging Technology* by Walter Soroka, published by the Institute of Packaging Professionals (IOPP). These and other useful texts are sold by the packaging bookstore of IOPP which has an extensive catalog available. The *IOPP Directory of Packaging Consultants* gives services and skills available in chemical packaging and labeling. Address: Institute of Packaging Professionals, 481 Carlisle Drive, Herndon, VA 22070-4823, 703-318-8970, FAX 703-318-0310.

The Gottscho Packaging Information Center at Rutgers, The State University of New Jersey, College of Engineering, Center for Packaging Science and Engineering, Building 3529, Busch Campus, Piscataway, NJ 08855 is a unique library devoted to packaging and related subjects. A literature search service is available.

Ecology concerns have resulted from an environmentally conscious world. In many countries, laws regulate the type of packaging that can be used and the manner in which it must be disposed of. In Europe, the German packaging law has become a model for the control of packaging and its disposal. It has been copied by many nations. While consumer packaging is the principal aim of this law, industrial packaging is also controlled. Arrangements must be made for the safe and approved disposal of used packaging. The effect of this has caused the chemical industry to reexamine whether disposable packaging should be used in view of the cost of disposal. By way of example, a wood pallet, used to contain one metric ton of product in multiwall paper bags, might have a purchase price in the United States of $12.00. The disposal costs for such a pallet could range between $30 and $65. This gives rise to the use of returnable packaging components which would also include steel drums and pails, plastic drums and pails, pallets, bulk corrugated boxes, corrugated shipping containers, and woven mesh bulk bags. Because this is such a rapidly changing field it is incumbent on those involved with package design, development, and logistics to know the local regulations where the package will be sent. Logistics is an important consideration. If returnable packages are to be used, their return must be planned and the costs for such return developed. An option for package disposal where returnable packages are involved is to sell them for reuse in the country they are destined for. Further complicating the disposal of packages is any residue of the product which they originally contained. For example, it is not uncommon to find a 55 gal (208 liter) steel drum after unloading, containing a "heel" of 1 to 4 liters of product. The collection and disposal of such materials is highly regulated in the United States by the Environmental Protection Agency (EPA) CFR 40 and state and local laws. Similar regulations are in effect in most of the developed nations.

Regulation of packaging during the past decade has changed significantly. There are more governmental units involved, innumerable regulations which must be complied with, and substantial penalties for failure to comply. This applies to packaging regulations especially in the United States but also in the rest of the world. The United Nations (UN) acts as a worldwide regulatory agency. Most industrialized nations have their own governmental units which regulate packaging based on UN regulations.

In the United States, the Department of Transportation (U.S. DOT) regulates the packaging, handling, and transport of all materials which are regarded as hazardous or dangerous. In addition the Environmental Protection Agency (EPA), the Occupational Safety and Health Act (OSHA), the Food and Drug Administration (FDA), the Nuclear Regulatory Commission (NRC), and the U.S. Department of Agriculture (USDA) all exercise a regulatory influence over packaging for materials or products which by law are mandated to them. Those regulatory agencies with whom the practicing chemical engineer involved with packaging must contend are principally the U.S. DOT and to some extent, the EPA and OSHA. In the pharmaceutical and food industries, the FDA is the primary regulator. In the alcoholic beverages industry, the U.S. Department of the Treasury's Bureau of Alcohol, Tobacco, and Firearms (ATF) branch regulates the packaging used for these products. Most notably, however, is the impact of the regulations of the U.S. DOT. During the past two decades the U.S. DOT has changed the approach to packaging regulations from one of strict packaging material specifications to that of requiring specific package performance. Under the former, construction features of a package required for a specific hazardous material was spelled out in minute detail. Under the performance-packaging approach the same size and type container must now be capable of handling performance tests without failure for the particular packing group that is required for the hazardous nature of the product.

The specific regulations of the U.S. DOT are found in the Code of Federal Regulations (CFR) Title 49, Parts 100–199. A key part of CFR 49 is Part 172.101 a portion which is illustrated as Table 21-17.

Where a product is not deemed as hazardous and thus not regulated by any of the above governmental bodies, in the United States, acceptable packaging is defined by industry associations for the different transportation modes which might transport the product. Their interest is in protecting the product from damage and, consequently, claims against the transportation companies for improper handling and for loss or damage to the product. The National Motor Freight Classification (NMFC) of the American Trucking Associations, Inc. and the Uniform Freight Classification of the Association of American Railroads are two such regulatory bodies. Failure to comply with their regulations does not carry the penalties of law, but it does allow the transportation company to disallow payment of claims for loss or damage to freight while in their care. In addition to these, other agencies are involved in the regulation of packaging. Table 21-18 summarizes these.

The CFRs can be obtained from the Superintendent of Documents, U.S. Government Printing Office, Washington, DC 20402, telephone 202-783-3238. The National Motor Freight Commission booklet 6000 may be obtained from the American Trucking Associations, Inc., 7200 Mill Road, Alexandria, VA 22314. The Uniform Freight classification, ratings, rules, and regulations for railroad shipment can be obtained from the Uniform Freight Classification Commission, Tariff Publishing Officer, Suite 1160, 222 South Riverside Place, Chicago, IL 60606. UN regulations are contained in the publication *Recommendations on the Transport of Dangerous Goods*, which is available from the UN Publications Section, UN Plaza, New York, NY 10017.

The United States has moved to follow the approach established by the United Nations to performance-based specifications for the packaging and packaging materials used for hazardous materials. The U.S. DOT, through Rule Making Docket HM 181 defined the United States position for performance-oriented packaging. Through many hearings and requests for industry comment, the DOT has now published in CFR 49 the changes necessary to bring U.S. packaging specification for hazardous materials in compliance with the United Nations specifications.

The nature of the change to performance-tested packaging is best illustrated with an example of a shipment of ethyl alcohol in a 55-gallon (208 liter) steel drum. Under the old rules one could look up

TABLE 21-16 Typical 1996 Cost of Containers for Chemical Products

Container size and description	Unit cost,[i] US$	Usable volume ft³	Usable volume m³
Metal drums[a,b,c]			
55 gal (208 L), steel, tight head, 18-20-18 gauge,[d] DOT-P.G. II	$50.15	7.35	0.208
55 gal (208 L), steel, tight head, all 16 gauge, DOT P.G. I	75.00	7.35	0.208
55 gal (208 L), steel, open head, 18 gauge, Rule 40[e]	55.71	7.35	0.208
55 gal (208 L), steel, open head, 16-18-18 gauge, DOT-P.G. III	70.15	7.35	0.208
55 gal (208 L), steel, tight head or open head, 18 or 18-20-18 gauge, used, reconditioned	22.80	7.35	0.208
55 gal (208 L), Type 304 stainless steel, tight head, 16 gauge, DOT P.G. I	985.00	7.35	0.208
30 gal (113 L), steel, tight head, 20 gauge, DOT	41.28	4.00	0.113
30 gal (113 L), steel, open head, 20 gauge, Rule 40[e]	33.23	4.00	0.113
16 gal (61 L), steel, open head, lug cover, without fittings, 22 gauge	21.71	2.14	0.061
55 gal (208 L), steel, dip-galvanized, tight head, 18 gauge, DOT	100.00	7.35	0.208
56.1 gal (204 L), steel, with 40-mil PE insert, external fittings, 18-20-18 gauge, 55-gal usable volume, DOT, open head[e]	89.00	7.20	0.204
56.1 gal (204 L), but tight head	90.75	7.20	0.204
55 gal (208 L), of blow-molded high-density PE,	45.15	7.20	0.204
Cans and pails[b]			
Pail, 5 gal (19 L), steel, tight head, 26 gauge, black steel, PE pour spout, unlined	$10.91	0.67	0.019
Pail, 5 gal (19 L), 26 gauge, black steel, open head, unlined, lug cover, wire-bail handle.	8.15	0.67	0.019
Can, 1 gal (3.8 L), friction-wedge lid, handle (paint can)	1.90	0.1335	0.004
Can, 1 qt (0.95 L), friction-wedge lid (paint can)	0.79	0.034	0.001
Can, 1 gal (3.8 L), oblong F style, handle, 1¼-in (32-mm) screw cap	2.90	0.1335	0.004
Can, 1 pt (0.48 L), oblong F style, 1-in (25-mm) screw cap	1.05	0.0167	0.0005
Square can, 5 gal (20 L), blow-molded PE with 2¾-in cap (often called "5-gal squares", or Jug)	7.43	0.67	0.019
Can, 1 pt (0.48 L), round-cone-top style, with 1-in (25-mm) cap	1.15	0.0167	0.0005
Fiber drums[b,f]			
61 gal (231 L), 9 ply, 400-lb (181-kg) load limit, dry products only, Rule 403	$22.50	8.15	0.231
55 gal (208 L), 9 ply, 400-lb (181-kg) load limit, dry products only, Rule 403	20.67	7.35	0.208
47 gal (178 L), 9 ply, 400-lb (181-kg) load limit, dry products only Rule 403	19.89	6.28	0.179
41 gal (155 L), 9 ply, 400-lb (181-kg) load limit, dry products only, Rule 403	18.24	5.48	0.155
30 gal (113 L), 9 ply, 400-lb (181-kg) load limit, dry products only, Rule 403	16.20	4.00	0.113
30 gal (113 L), 7 ply, 225-lb (102-kg) load limit, dry products only, Rule 403	14.48	4.00	0.113
15 gal (56.8 L), 6 ply, 150-lb (68-kg) load limit, dry products only Rule 403	10.75	2.00	0.057
55 gal (208 L), 9 ply, PE barrier, 400-lb (181-kg) load limit, Rule 403	21.05	7.35	0.208
55 gal (208 L), 9 ply, PE-aluminum foil liner, 400-lb (181-kg) load limit, Rule 403	31.32	7.35	0.208
55 gal (208 L), 10 ply, blow-molded, 15-mil PE liquidtight liner, tight head, steel cover with 2-in and ¾-in NPT1 openings, 600-lb (272-kg) load limit,	47.31	7.35	0.208
30 gal (113 L), 9 ply, same as preceding except 450-lb (204-kg) load limit	36.51	4.00	0.113
30 gal (113 L), 8 ply, 300-lb (136-kg) load limit, removable fiber cover, no barrier	17.28	4.00	0.113
15 gal (56.8 L), 6 ply, same as preceding except 150-lb (68-kg) load limit	11.41	2.00	0.057
1 gal (3.8 L), 5 ply, same as preceding except 150-lb (68-kg) load limit	5.01	0.1335	0.004
55 gal (208 L), 9 ply, 400-lb (181-kg) load limit, semisquare removable fiber cover, Ro-Con style	20.21	7.35	0.208
45 gal (170 L)	12.21	6.01	0.170

Remarks

Black steel drums lined at extra cost:

Lining	No. coats	Cost/drum, $
Baked resin,	1	7.00
pigmented	2	9.60
Epoxy phenolic composite	2	11.14

Prices shown include ¾-in and 2-in fittings and are for unlined drums; add $1.25 per drum for delivery.

Approximate steel drums per truckload

55-gal size = 360 drums
30-gal size = 592 drums
16-gal size = 1225 drums

U.S. standard gauge equivalents

Gauge	in	mm
16	0.0598	0.0152
18	0.0478	0.0121
20	0.0359	0.0091
22	0.0299	0.0076
24	0.0239	0.0061

Approximate fiber drums per truckload

61-gal size = 300 drums
55-gal size = 318 drums
47-gal size = 424 drums
41-gal size = 552 drums
30-gal size = 592 drums
15-gal size = 1272 drums
1-gal size = 17,365 drums

Fiber drum type	Outside dimensions: Diameter, in	Diameter, cm	Height, in	Height, cm
55-gal lever top	21	53.3	40¾	103.5
55-gal lever top	23½	59.7	30¾	78.1
55 gal lever top	22	55.9	34¾	88.2
41-gal lever top	20½	45.1	30¼	76.8
30-gal lever top	19	48.3	26¼	66.7
6.28-ft³ rectangular	17⅝°	44.8	37½	95.3
55-gal liquid	22	55.9	37½	95.3
30-gal liquid	19	48.3	28	71.1
55-gal fiber	20⅜	51.8	40¾	103.5
30-gal fiber	17⅜	44.1	30¾	78.1

°Side dimension, square.

TABLE 21-16 Typical 1996 Cost of Containers for Chemical Products (*Continued*)

Container size and description	Unit cost, US$	Usable volume ft³	m³
Bags: multiwall paper and polyethylene film[g]			
Pasted valve bag, 20½- × 22-in face, 5½-in top and bottom (520 × 560 × 140 mm) with 1-mil free film, 2/50, 1/60 kraft, PE internal sleeve	$0.51	1.33	0.038
Sewn valve bag, 15 × 5½ × 30¼ in, 5½-in internal sleeve (380 × 140 × 770 × 140 mm) with 1-mil free film, 2/50, 1/60 kraft	0.61	1.33	0.038
Pasted valve bag, 18½ × 22¾ in, 3½-in top and bottom (470 × 580 × 90 mm), PE internal sleeve, 3/50 kraft	0.36	0.84	0.024
Sewn open-mouth bag, 20 × 4 × 30¾ in (510 × 100 × 780 mm), 3/50, 1/60 kraft	0.49	2.00	0.057
Sewn valve bag, 19 × 5 × 33½ in, 5½-in tuck-in sleeve (480 × 130 × 850 × 140 mm), 3/50, 1/60 kraft	0.65	2.00	0.057
Pasted valve bag, 24 × 25¼ in, 8-in top and bottom (610 × 640 × 200 mm), tuck-in sleeve, 3/50, 1/60 kraft	0.61	2.00	0.057
Pasted open-mouth baler bags, 22 × 24 in, 6-in bottom (560 × 610 × 150 mm), 1/130 kraft (or 2/70)	0.36	1.33	0.038
Flat tube open-mouth bag, 10-mil PE film, plain, 20½ × 34¼ in (520 × 870 mm)	0.73	1.33	0.038
Square-end valve bag, 20½- × 22-in face, 5½-in top and bottom (520 × 560 × 140 mm), 8-mil PE film, plain	0.49	1.33	0.038
Pinch-style open-mouth bag, 20 × 4 × 30¾ (510 × 100 × 780 mm), 1/10 PE 50, 2/50, 1/60 kraft, plain, no printing	0.65	2.00	0.057
Small bags, pouches, and folding boxes[h]			
Pouch, 8¾ × 16¾ in (220 × 425 mm), 2-ply PE film, 2-mil- (0.05-mm) thickness per ply	$0.13	0.12	0.0034
Bag, sugar-packet style, 6 × 2¾ × 16¾ in (150 × 70 × 425 mm), 2/40-lb basis weight, natural kraft paper	0.11	0.12	0.0034
Bag, pinch style, 8¾ × 3 × 21 in (220 × 75 × 530 mm), 2/40-lb basis weight, natural kraft paper	0.11	0.12	0.0034
Folding box, 5 × 1 × 8 in (125 × 25 × 200 mm), reverse-tuck design, 12-point kraft board with bleached white exterior	0.23	0.028	0.0008
Folding box, 9½ × 4½ × 15 in (240 × 115 × 380 mm), full-overlap top and bottom, 30-point chipboard with bleached white exterior	0.47	0.37	0.0105

For tuck-in sleeve, add $0.05/bag. Unit cost is for unprinted bag. For printing add the following up charges. U.S. dollars per 1000 bags:

1 side, 1 color, $13.50
1 side, 2 colors, $16.85
1 side, 3 colors, $22.15
2 sides, 1 color, $16.85
2 sides, 2 colors, $22.15
2 sides, 3 colors, $28.50

Polyethylene-film gauges

mil	Actual mm	Nearest mm
0.5	0.0127	0.01
1.0	0.0254	0.03
1.5	0.0381	0.04
1.75	0.0445	0.04
2.0	0.0508	0.05
8.0	0.2032	0.20

Multiwall kraft-paper basis-weight equivalents

U.S. customary, lb/3000 ft² ream	SI, g/m²
40	65
50	81
60	97

Permeability of common packaging films*

Type of film	Water-vapor transmission†	Gas permeability‡ O₂	N₂	CO₂	Water absorption
Cellophane, nitrocellulose-coated	0.3	1	1	13	High
Nylon	19	25	160	160	Medium
Polycarbonate	11	300	50	1000	Medium
Polyester, oriented	1.7	4	1	16	Low
Polyethylene, low-density	1.3	550	180	2900	Low
Polyethylene, high-density	0.3	600	70	4500	Low
Polypropylene	0.7	240	60	800	Low
Saran	0.2	14	12	4	Low

*From J. R. Hanlon, *Handbook of Package Engineering*, Technomic Publishing Co., Lancaster, PA 17604, 1992 ed.
†g loss, 24 h/(100 in²·mil), at 95°F, 90 percent relative humidity.
‡cc, 24 h/(100²·mil), at 77°F, 50 percent relative humidity; ASTM D1434.

TABLE 21-16 Typical 1996 Cost of Containers for Chemical Products (*Concluded*)

Container size and description	Unit cost, US$	Usable volume ft³	Usable volume m³
Corrugated cartons and bulk boxes[b]			
Regular slotted carton (RSC), 24 × 16 × 6 in (610 × 405 × 150 mm), 275-lb-test double wall, stapled (stitched) joint	1.25	1.33	0.038
RSC, 16 × 6 × 24 in (405 × 150 × 610 mm), 275-lb-test double wall, stitched joint, end-opening style	0.90	1.33	0.038
Bag in box, RSC, 15 × 15 × 22 in (380 × 380 × 560 mm), 275-lb-test double wall, stitched liner, 600-lb-test double wall, 6-mil (0.15-mm) PE liner	3.80	2.86	0.081
Bulk box, 600/600 (test in lb for both pieces), laminated inner lining, approximately 41 × 34 × 36 in (1040 × 865 × 915 mm); includes special wood pallet and 8-mil (0.2-mm) blown low-density PE liner	35.00	5.00	0.142
Carboys, plastic drums, jars, and bottles[b]			
Carboy, 13½ gal (51 L), PE, blow-molded	39.90	1.35	0.038
Drum, PE, 15 gal (57 L), blow-molded,	41.00	2.00	0.057
Carboy, 15 gal (57 L), glass, nitric acid service, wooden crate	128.00	2.00	0.057
Jug, 1 gal (3.78 L), glass, with finger handle, plastic 38-mm cap, with corrugated reshipper carton	3.10	0.1335	0.004
Bottle, 1 qt (0.95 L), glass, Boston round, plastic 28-mm cap	1.50	0.034	0.001
Jar, 1 qt (0.95 L), glass, wide mouth, plastic 89-mm cap	1.60	0.034	0.001
Jar, 1 qt (0.95 L), glass, plastic 63-mm cap	1.42	0.034	0.001
Jar, 1 gal (3.78 L), PE, wide mouth, plastic 100-mm cap	1.47	0.1335	0.004
Bottle, 1 gal (3.78 L), round, PE, narrow neck, plastic 38-mm cap	1.85	0.1335	0.004
Bottle, 1 qt (0.95 L), PE, narrow neck, plastic 28-mm cap	0.94	0.034	0.001
Jar, 1 pt (0.47 L), PE, wide mouth, plastic 53-mm cap	0.71	0.017	0.0005

Cost US$[h]

	Expendable grade		Warehouse reusable grade	
	9-block type	Stringer type	9-block type	Stringer type
Pallets[b]				
40 × 48 in (1015 × 1220 mm)	11.76	10.90	17.78	16.32
35 × 42 in (890 × 1065 mm)	10.17	10.17	16.49	15.17
42 × 48 in (1065 × 1220 mm)	17.40	17.40	20.10	18.51
48 × 48 in (1220 × 1220 mm)	18.81	18.81	21.69	19.97
44 × 50 in (1115 × 1270 mm)	21.45	21.45	24.53	22.60

Wrap materials	US$/lb
Film, PE, Grade ADL, blown type	1.05
Film, PE, Grade ASF (shrinkable)	1.25
Film, polypropylene, shrinkable, yield before shrinkage = 31,100 in²/(lb·mil)	3.37
Paper, kraft, wrapping quality, 50-lb/ream basis-weight yield = 3000 ft²/ream	0.50
Film, PE, stretchable type for pallet wrap, 1.5 mil × 20 in (0.04 × 510 mm) wide	1.15

[a] Drum has 2-in and ¾-in national-pipe-thread (NPT) openings in head.
[b] Truckload quantity price, FOB east-coast manufacturer's plant.
[c] DOT = U.S. Department of Transportation. Also UN (United Nations).
[d] Sequence of top, body, and bottom gauges. For example, 18-20-18 = 18-gauge top, 20-gauge body, and 18-gauge bottom.
[e] Removable head secured with bolted ring with screw draw-up.
[f] Drums are of plain fiber, have steel cover and bottom, and have lever-operated closing ring.
[g] Truckload-quantity price, FOB buyer's plant.
[h] Truckload-quantity price, FOB east-coast buyer's plant.
[i] Prices given are adequate for comparing alternatives. For budget purposes, actual, recent vendor quotation must be used due to marketplace fluctuations in prices.

Explanation of U.S. DOT term:
PACKING GROUP I, II, III
P.G. I Great danger
P.G. II Medium danger
P.G. III Minor danger

Polyethylene-film* yield table

Thickness in	Thickness mm	Yield in²/lb	Yield m²/kg
0.001	0.025	30,000	19.4
0.0015	0.040	24,000	15.5
0.002	0.050	15,000	9.7
0.003	0.075	10,000	6.5
0.004	0.100	7,500	4.8
0.008	0.200	3,750	2.4
0.010	0.250	3,000	1.9

*Flat sheeting.

TABLE 21-17 Abstract of Part 172-101 of Code of Federal Regulations (CFR) Title 49 to Illustrate the Ethyl Alcohol Example in the Text. (U.S. Government Printing Office)

Symbols (1)	Hazardous materials descriptions and proper shipping names (2)	Hazard class or Division (3)	Identification Numbers (4)	Packing group (5)	Label(s) required (if not excepted) (6)	Special provisions (7)	(8) Packaging authorizations (§ 173.***)			(9) Quantity limitations		(10) Vessel stowage requirements	
							Exceptions (8A)	Non-bulk packaging (8B)	Bulk packaging (8C)	Passenger aircraft or railcar (9A)	Cargo aircraft only (9B)	Vessel stowage (10A)	Other stowage provisions (10B)
—	Ethanol or Ethyl alcohol or Ethanol solutions or Ethyl alcohol solutions.	den 3	UN1170	II	FLAMMABLE LIQUID.	T1	150	202	242	5 L	60 L	A	
		—	—	III	FLAMMABLE LIQUID.	B1, T1	150	203	242	60 L	220 L	A	

TABLE 21-18 Agency and Administrative Law

Title	Symbol	Regulate or affect
International		
United Nations	UN	Packages and labeling for products moving among member nations.
Intergovernmental Maritime Consultative Organization	IMCO	
Federal		
Department of the Treasury Bureau of Alcohol, Tobacco and Firearms, Title 27	ATF	Packages and labeling for alcohols, tobacco, firearms, and explosives.
Department of Transportation Transportation Safety Act, Title 49	DOT	Packaging and labeling for all hazardous materials shipped in interstate commerce.
U.S. Coast Guard, Title 46	USCG	Set packaging, labeling, blocking, and bracing for all freight moving by United States–registry ships on lakes, rivers, or oceans.
Department of Labor Occupational Safety and Health Act, Title 29	OSHA	Package-filling and -handling machinery, workplace design, warehouse practice, and acceptability of packages from workplace and warehouse viewpoint.
Food and Drug Administration Federal Food, Drug, and Cosmetic Act, Title 21	FDA	Packages, packaging machinery, and workplace from viewpoint of their effect on food and drug purity; package labeling and marking.
Environmental Protection Agency, Title 40	EPA	Packaging facilities, packaging and labeling, package disposal, workplace refuse disposal, cleanup and disposal of spills.
Clean Air Act	CAA	
Clean Water Act	CWA	
Resource Conservation and Recovery Act	RCRA	
Federal Insecticide, Fungicide, and Rodenticide Act	FIFRA	
Toxic Substances Control Act	TSCA	
Nuclear Regulatory Commission Title 10	NRC	Packaging and labeling for nuclear materials and wastes
State		
Department of Transportation Labor Environmental Protection Agriculture Others	Example: New Jersey Department of Transportation	Packaging, labeling, workplace, packaging machinery, etc., in *intrastate* commerce. Regulations generally parallel those of federal departments but frequently have important differences and additional requirements.
City		
Departments of Health Labor Fire Protection	City fire department; example: NYFD	Packages, packaging machinery, packaging facilities, and materials which are transported, stored, and handled in the city. These local laws are in addition to the requirements of state and federal law.
Industry associations		
Air Line Pilots Association	APA	Materials which can be carried on commercial aircraft piloted by Air Line Pilots Association members.
International Air Transport Association	IATA	Materials which can be carried on members' aircraft and packaging and labeling requirements for them.
International Civil Aviation Organization	ICAO	Materials which can be carried and the packaging and labeling for them.
Association of American Railroads Bureau of Explosives	AAR B of E	Packages and loading, blocking, and bracing for all hazardous products shipped by rail in the United States. Bureau standards are generally accepted by all railroads and are the basis for R. M. Graziano's Tariff *Hazardous Materials Regulations of the Department of Transportation by Air, Rail, Highway, and Water,* latest edition.
Uniform Freight Classification Committee, Rules 40 and 41	UFC	Set packaging standards for all freight moving by *rail.*
American Bureau of Shipping National Cargo Bureau	ABS	Packaging, labeling, loading, blocking, and bracing for all freight moving by United States–registry ships on lakes, rivers, or oceans.
National Motor Freight Traffic Association National Motor Freight Classification	NMFC	Set packaging standards for all freight moving by highway.
National Fire Protection Association	NFPA	Packages, packaging facilities, and warehouse designs and operation.
Special carriers		
United Parcel Service	UPS	Packaging, labeling, and size and weight of small packages carrying hazardous materials. Requirements meet DOT standards. Quantities generally do not exceed 1 gal (3.785 L).
U.S. Postal Service	USPS	Materials which may be shipped and packaging, labeling, and size and weight of small packages handled by parcel post.
Federal Express Corp.	FEDEX	Packaging and labeling for hazardous and nonhazardous materials and products which they will carry.

TABLE 21-19   Performance Testing of Steel Drums—Type 'A' for Packaging Group II

Test	CFR 49 reference	Criteria summary
Drop test	178.603(e); (ii)	1.2 Meters (3.0 ft.)—no leakage
Leak proofness test	178.604(e); (2)	20 kPa (3.9 PSIG)—no leakage
Hydrostatic test	178.605(d); (1)	100 kPA (15 PSIG)—no leakage
Stacking test	178.606(c); (2), (ii)	2000 lbs. (907 kg)—no leakage
(By compression machine)		
Vibration	178.608(b), (3)	1 in. (2.54 cm) Amplitude at resonant frequency for 1 hour—no leakage

This information is taken from CFR 49, issue of October 1, 1992. The table is intended to be illustrative of the use of CFR 49. It should not be used as the basis for choosing a drum or other type package. Each package and product requires specific analysis to identify the container which meets customer needs and complies fully with U.S. DOT and UN regulations.

ethyl alcohol in part 172.101 and find that the required packaging was described in parts 173.125 and 173.119. According to Part 173.119, a 55-gallon (208 liter) steel drum was authorized for ethyl alcohol and its specifications could be found in Part 178.116. The type 17E drum was completely specified in this section. Under the new performance-oriented specification approach, the packaging requirements are specified in Part 172.101 which states that ethyl alcohol falls into the hazard described under Packing Group II, and packaging authorizations are found in Part 173.202. This shows that a Type 1A1, 55-gallon (208 liter) drum which meets standards given in Part 178.504 and performance specified in 178 subpart M can be used. Table 21-19 summarizes the test criteria. Any drum which can meet these criteria is authorized and may be used for this product. The manufacturer of the drum and the seller of the drum must warrant that the drum does meet these criteria and that either they have tested this design on a routine basis to verify that it meets these criteria, or that a third party has done so. They must be able to produce records that these test requirements have been complied with.

Under the old regulations, the marking on the bottom of the 17E drum was as shown in Fig. 21-36. Under the new regulations the marking on the bottom of the drum is as shown in Fig. 21-37. After October 1, 1994 all packages must be marked only with the new marking. Those new packages which are in the distribution system and contain the old mark may be used until October 1, 1996. After this date only packages marked with the approved UN marking are authorized by the U.S. DOT.

Also of note is a growing trend among packagers of hazardous materials to determine quantitatively the degree to which the package they will use exceeds that minimum performance requirement as specified by the U.S. DOT or UN. Using programmable shock and vibration machines and the damage-boundary-curve method, it is possible to develop the fragility of any package which then permits comparison of one alternative design with another, on a quantitative and economic basis. Rutgers, The State University of New Jersey, Center for Packaging Science and Engineering has been conducting research in this field.

Competent advice on the correct packaging to use for hazardous materials or other products is obtainable from consultants in the field of packaging who are members of the Institute of Packaging Professionals, consultants counsel. A brochure listing the qualifications of member consultants is obtainable from the IOPP.

Whatever hazardous materials are involved, whether they be new products, an existing product in a new package type, hazardous waste, or any other hazard category, the proposed packaging and all conditions which are expected to be incident to its use should be reviewed by a competent attorney who specializes in distribution and packaging matters. In the larger corporations, the law department usually has such a specialist. If not, local bar associations can provide the names of attorneys with this specialty.

It cannot be overemphasized that in the packaging of hazardous materials, extra steps and precautions must be taken to provide absolute compliance with established regulations. The penalties for failure to do this are exceedingly high, but the potential for serious injury to people and damage to the environment are of greater significance. A publication which addresses changes and issues in the packaging and shipping of hazardous materials is *HAZMAT PACKAGER AND SHIPPER* published by Packaging Research International, Inc., PO Box 3144, West Chester, PA 19381-3144, phone 610-436-8292, FAX 610-436-9422. Changes in CFR49 occur often. Before being published, they appear in the *FEDERAL REGISTER* which is published five times per week. CFR 49 is published on October 1, of each year and incorporates changes of the previous year.

Once the package alternatives permitted by government or transportation companies have been determined and marketing and production considerations are known, performance and economic evaluation must be made. This evaluation should consider packaging as part of a system. Not only must the package itself be considered, but so must factors which affect the package or are affected by it. If a choice of shipping in bulk form or in packages exists, cost comparisons must be made (Table 21-20).

Metric-system dimensions for packaging are not used extensively in the United States, but initial steps are being taken to permit use of these dimensions. The subject is under intensive study by both the packaging-supply and the packaging-using industries. SI equivalents are usually available from package suppliers, but at present all ordering in the United States is done in the U.S. customary system. Table 21-21 gives the degree of expected metric conversion. In the United States suppliers are using millimeters as the principal metric measure. When a soft conversion is made, increments of 5 mm are used. A converted package dimension is rounded up or down to the nearest multiple of 5 mm. For example, a bag-face width of 16 in equals 406.4 mm, which would be rounded down to 405 mm.

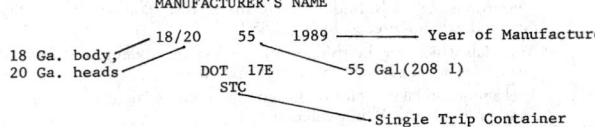

FIG. 21-36   Marking authorized by the U.S. DOT prior to October 1, 1994 prior to the change to performance-oriented specifications. Marking is found on the bottom head of the drum.

FIG. 21-37   Marking which complies with U.S. DOT and UN regulations and applies to a specific open-head steel drum. (*Courtesy of the Steel Shipping Container Institute, Union, NJ.*)

TABLE 21-20 Container Comparative Data[a]

Product: Thermoplastic pellets of 33-lb/ft³ (528.7-kg/m³) bulk density

Parameter	Domestic paper bag	Bulk corrugated-paper box	Intermediate bulk container	Bulk hopper truck	Ship container	Railroad hopper car	Intermediate bulk container
Construction	Multiwall pasted valve bag, 4-ply construction, with inner ply having an extrusion coating of low-density PE	Corrugated box, one-half RSC design, made of 600-lb burst-strength-test double wallboard, A-C flute laminated; includes cap, inner, and wood pallet	Flexible bag made of woven polypropylene of 2000 denier, 12 × 12 weave, with PE liner and nylon support straps	Welded-aluminum tank with pneumatic unloading pump; undercarriage equipped with pneumatic tires	Welded- or riveted-aluminum construction, with International Organization for Standardization (ISO) end castings for lifting by standard spreader	Welded-steel construction with plastic-coated interior, equipped with 100-ton trucks for 4-ft, 8½-in-gauge tracks	Rigid container made of welded-aluminum construction with butterfly discharge valve and fill port
Size	16 × 25 × 6.5 in (405 × 635 × 165 mm)	41 × 34 × 36 in (1040 × 865 × 915 mm)	53 in diameter × 53 in high (1350 × 1350 mm)	8 ft wide × 30 ft long (2.4 × 9.1 m)	8 × 8 × 35 ft (2.4 × 2.4 × 10.7 m)	5700 ft³ (161.4 m³)[b]	42 × 48 × 84 in (1070 × 1220 × 2135 mm)
Capacity, lb (kg)	55.1 (25)	900 (408)	2205 (1000)	42,000 (19,051)	50,000 (22,680)	180,000 (81,648)	2205 (1000)
Tare weight, lb (kg)	0.7 (0.32)	50 (22.7)	20 (9.1)	20,000 (9072)	4200 (1905)	100,000 (45,360)	255 (115.7)
Unit cost, US$	$0.55	$27.00	$59.00	$75,000	$16,000	$100,000	$2500
Lease cost, US$/month				$920	—	$685	$40[d]
Useful life	1 trip	1 trip	10 trips	15 years	5 years	25 years	10 years
Practical shipping radius, mi (km)	Any	Any	Any	250 (402) maximum	Any	300 (483) minimum	Any
Cost of typical system, US$/100 lb (45 kg)[e]	$6.81	$7.76	$5.08	$0.80	$0.80	$0.32	$4.63
Plant investment, US$ × 1000	$90–180[b]	$22–238	$27–252	$22–315	$45–315	$36–315	$27–252

[a] Based on 1996 prices.
[b] See Fig. 21-56 for typical plant layout.
[c] Mileage credit is paid to owners or lessees of this type of hopper car. To gain mileage credit car must be loaded. Rate is negotiable. A typical rate is $0.43 per loaded mile.
[d] Lease cost is based on a period equal to the useful life.
[e] Includes cost of container, filling, handling, and storage for minimum volume of 1000 tons (908 metric tons per month) but does not include freight or amortization of filling equipment.
[f] Represents typical investment in filling and handling equipment. Actual amount depends on plant layout and nature of existing facilities.

TABLE 21-21 Expected Metric Conversion of Packages for Chemical Products in the United States

Package type	Degree of conversion*
Bags (paper, plastic)	Hard
Boxes (paper)	Hard
Drums (fiber)	Hard
Drums (steel)	Soft
Pails (steel, plastic)	Soft
Cans (steel, fiber)	Hard
Bottles (glass, plastic)	Hard
Paper (kraft)	Hard

*Hard conversion: resized package dimensions to hold the nearest acceptable metric unit. Example: 50-lb (22.7-kg) multiwall paper bag changed to hold 25 kg (55.1 lb). Size can be limited by the maximum-size package available.

Soft conversion: the volume of the package is unchanged, and its metric equivalent is stated. Example: 55-gal drum equals 208 L.

For kraft paper, conversion will eventually be a hard conversion to two basis weights of 75 and 90 g/m², which will replace the 40-, 50-, and 60-lb basis weight currently used.

Package development involves the design, specification, and testing of packages. Sophistication of chemical and pharmaceutical industry products sold to industry and consumers is increasing. Package design takes into account four principal considerations. First is the preference of the customer for a particular type of package and whose preference is often opposed to that of the producer of the product. Second is compliance with applicable regulations. Third is the effect which a package choice will have on production-plant-operating conditions. A fourth consideration is the effect of atmospheric gases on the product itself, especially when the product is packaged in other than metal or glass. Of concern are water vapor, oxygen, and carbon dioxide, which can permeate most flexible package materials. There are thermoplastic films available that offer barrier properties which restrict atmospheric gases from entering a package. Additionally, the thermoplastic films can be combined with aluminum foil to produce impermeance. Aluminum foil has the propensity to stress crack when used in flexible packages. To overcome this problem the film industry has devised a method of depositing vaporized aluminum directly onto a thermoplastic film. This notably improves its permeation properties. Table 21-16 lists some of the common thermoplastic films and their barrier properties. Where pharmaceutical products are involved, and this would also include intermediate chemicals used in their formulation, the FDA requires that any reactions between the packaging material and the product itself must be known and often this must be reported in parts per million or finer. CFR 29, Part 1200 gives this for many types of packaging materials and containers.

Package testing is mandatory for chemical products which are deemed as hazardous, as set forth in CFR 49, Part 172.101. A package for hazardous material being offered by a supplier must be tested by them or by a qualified third party. The tester must certify that the package complies with the DOT regulations and the design has been tested satisfactorily for a given packing group. The purchaser does not have to be concerned with the testing itself but only that it has been done and that the supplier can certify that the package has passed the required tests.

A second type of tests are those which shippers require when using a new package never before used for a given product, or an improved existing one. This type of testing is done in a simulated distribution environment using the actual package and often the product which is involved. The expected shock, vibration, compression, and impact shock which might be encountered in a distribution environment can be simulated in the laboratory, and a fair assessment made of the ability of the package to withstand or exceed expected conditions. One test protocol which accomplishes this is that published by the American Society for Testing Materials (ASTM) No. D4169. This protocol allows the user to define the shipping environment in minute detail, to carry out tests which simulate shock, vibration, compression, etc., and then to appraise the results of these tests. Copies of this test protocol may be obtained from ASTM, 1916 Race Street, Philadelphia, PA 19102.

Another test protocol is that specified by the International Safe Transit Association (ISTA), East Lansing, Michigan. This also simulates transportation and handling conditions. Transportation companies—truck, rail—accept ISTA test results and will transport packages bearing the ISTA marks.

Competent testing laboratories which can provide tests according to the ASTM D1469, ISTA, DOT, UN, and other protocols are located strategically throughout the United States and in most of the developed nations. The yellow pages of the telephone directory are useful to find such laboratories.

LIQUID PACKAGING

Containers Containers for liquids consist principally of drums, pails, and cans made of steel or plastic and of bottles and vials made of plastic or glass. The chemical industry is often involved with all these containers, but the most frequently used packages for industrial chemicals are steel drums and pails. For exotic products, stainless-steel drums and pails are available. The most common types used are 208-L (55-gal) drums and 19-L (5-gal) pails.

Once the appropriate package has been determined by consulting governmental and carrier regulations, the type of material compatible with the product needs to be determined. A wide variety of coatings is available for lining carbon steel drums and pails. Suppliers are often able on the basis of experience to assist in determining a lining which will be compatible with the product. When prior information is unavailable, laboratory tests can determine compatibility. Laboratory tests are often desirable before field trials. In some instances, a product may not be compatible with metal. This circumstance has led to an important container, the **all-plastic 208-L (55-gal) drum.** Made from blow-molded high-density polyethylene, this container is especially useful for products which might react with carbon steel or whose value does not warrant stainless steel. Special treatments are available to make the inner surface of the plastic drum impervious to penetration of many products. Sulfonation and fluorination are prominent among these processes.

Two basic designs of steel drums and pails exist: the tight-head and the open-head. Tight-head drums have both top and bottom members permanently fastened to the drum body. Open-head drums have only the bottom permanently attached. As the term "open-head" implies, the top of the drum does not have a permanently fixed cover; rather, a removable head is used. This head is designed so that a locking ring secures it to the drum body. Open-head drums and pails are usually employed for viscous products or for mixtures and slurries which are difficult to pump through lines 50 mm (2 in) or smaller. Tight-head drums and pails are used for low-viscosity products. No set rule can be given for the viscosities above which an open-head drum or pail must be used.

Reconditioned and remanufactured drums are authorized by the U.S. DOT for certain hazardous materials. CFR 49 paragraph 173.28 provides details of the reconditioning process. This consists essentially of rinsing the inside of the drum, removing any dents which deform the chime, grit blasting the exterior to remove all previous labels and paint, and then recoating with a new outer finish paint. The reconditioner must put their company mark on the drum. Reconditioned drums costs 50 to 70 percent of the price of a new drum and as such are useful for packaging marginally profitable products. The regulations set forth by the U.S. DOT must be complied with in complete detail.

Remanufactured drums are also permitted. Remanufacturing involves the dismantling of the drum, usually involving removal of the top and bottom head. All dents, rust, and other corrosion are removed by the grit blasting. New heads are then flanged onto the drum. An interior coating is given the drum when required, and the outside is painted. The drum remanufacturer must put their company mark on the drum. Remanufactured drums must withstand the same tests as a new drum. Details of DOT requirements for remanufactured drums are contained in CFR 49 paragraph 178.16. For products which are not classified as hazardous, the DOT and UN regulations do not apply but they are useful in setting a minimum performance standard for the packaging of any nonhazardous material.

Closures for drums and pails need to be determined together with the gasket material to be used. Consideration must be given to compatibility with the product and to the vibration which the container will encounter during transportation. The torque required to produce closure integrity is thus a significant factor. The typical closure sizes used in United States practice for tight-head drums are a 2-in national pipe thread (NPT) and a ¾-in NPT in the top head. For open-head drums, market considerations determine whether or not these fittings are used.

Steel drums are an ideal package because they can be stored out-of-doors and are generally impervious to weather conditions. Because of the drum's top head being recessed into the drum body, water can accumulate following a rain storm. While rusting of the drum head is undesirable a greater problem is the obliteration of whatever printing and labeling is on the drum head. To overcome this problem a patented wicking device "Drumwic" is available from Lee Technology Inc., Huntington, West Virginia. It is a low-cost way for quick removal of any accumulated water. The device is reusable. An illustration is contained in Fig. 21-40(o).

Tamper-evident seals and closures are commonly added to all packages. For certain products, child resistant closures are also added. There are several types of tamper-evident closures of which three are most common to liquid chemical products. First is a metal enclosure which covers the drum bung and which is crimped to it. The pulling of a tab breaks the tampered-evident enclosure which cannot be used again. It is customary to have the tamper-evident seal on both drum bungs. This holds also for pails. For small glass or plastic packages a seal is fitted into the cap which is then heat sealed to the bottle by means of an induction sealer. The construction of this seal consists of an outer layer of a thermoplastic material such as polyethylene, followed by aluminum foil, followed by a bleached kraft paper. The induction sealer induces eddy currents in the aluminum foil which raises its temperature to above the melting point of the polyethylene. The polyethylene melt then fuses the seal to the bottle. Another type also used for plastic and glass bottles is an external sleeve of shrinkable PVC (polyvinylchloride). This is usually applied by machine as the bottles move down the packaging line after being capped. They pass through a heated tunnel which raises the temperature of the seal to where it shrinks tightly around the closure, thereby providing tamper

evidence. Child-resistant closures are those which are designed to be sufficiently complicated that the hand-eye coordination of the child is inadequate to open the package.

Filling Line Among filling-line considerations are filling and weighing equipment, mechanical handling of empty and filled drums, loading of filled drums onto transportation vehicles, workstation design for the safe and efficient use of personnel, and conformance to Occupational Safety and Health Administration (OSHA) and other codes. A typical drum-filling line, capable of handling two drums per minute, is shown in Fig. 21-38.

Filling and Weighing of Drums This procedure is divided into two parts: delivery of the liquid to the drum and weighing out the desired amount. Pumping the liquid product through a series of delivery pipes to the drum-filling point should follow good practice whereby reasonable velocities and pressure losses are maintained. The terminal point of the filling line is a control valve which is activated by a signal from a weight-sensing unit or scale. Valves may be pneumatically, hydraulically, or electrically operated, their operation being actuated either by an electric or pneumatic system or manually. The filling nozzle may be either top-fill or bottom-fill. Top filling is usually employed for most products, especially viscous materials or slurries. Bottom filling is used for low-viscosity products, for those having flash points under 37.8°C (100°F), or for places where static electricity is a concern. Also, products which tend to foam are bottom-filled. With a bottom-fill installation sufficient headroom is needed to permit the filling nozzle to be withdrawn from the drum. With both types of filling nozzles, a provision must be made for collecting product which dribbles from the end of the nozzle after filling is complete.

Weighing The weighing apparatus can be as simple as a platform scale in which the operator shuts off the filling nozzle when the desired weight has been reached. Automatic weighing can employ a load-cell system activating the flow-cutoff mechanism through a microprocessor. The same principles of filling and weighing as were described in the subsection "Weighing of Bulk Solids" hold for liquids. The advent of the microprocessor, image recognition, and stepper motor controls together with precision weight sensing has led to a custom-made system which can fill any drum or pail from 5 gal (19 L) to 55 gal (208 L) in sequence without the operator having to insert the fill nozzle into each container. (See Fig. 21-39.)

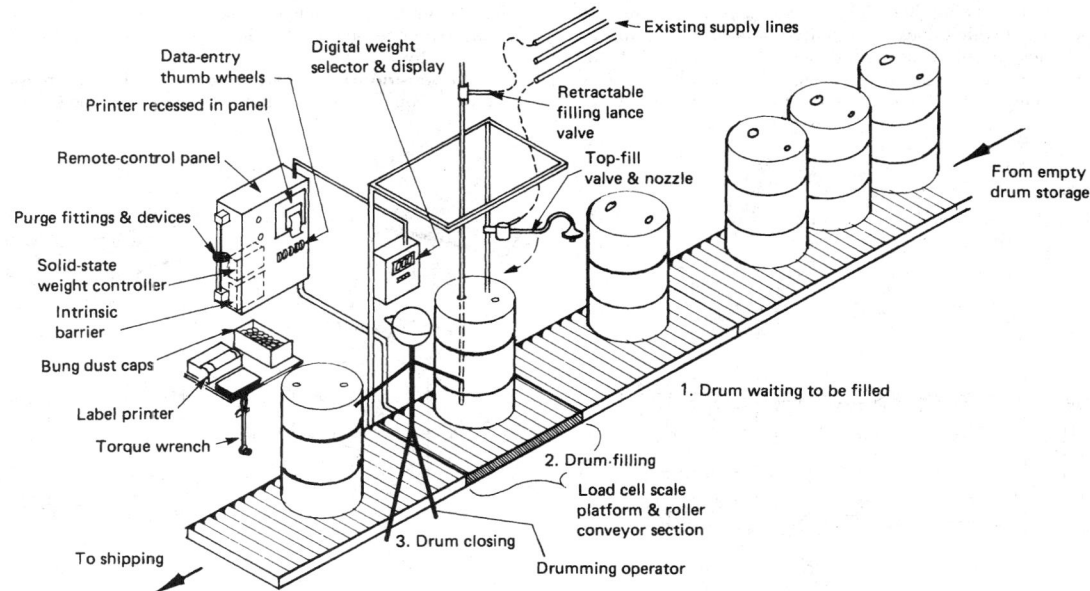

FIG. 21-38 Typical high-precision liquid filling and weighing system for packaging 208-L (55-gal) steel drums and similar smaller containers. (*Courtesy of H. F. Henderson Industries, Inc., Caldwell, NJ.*)

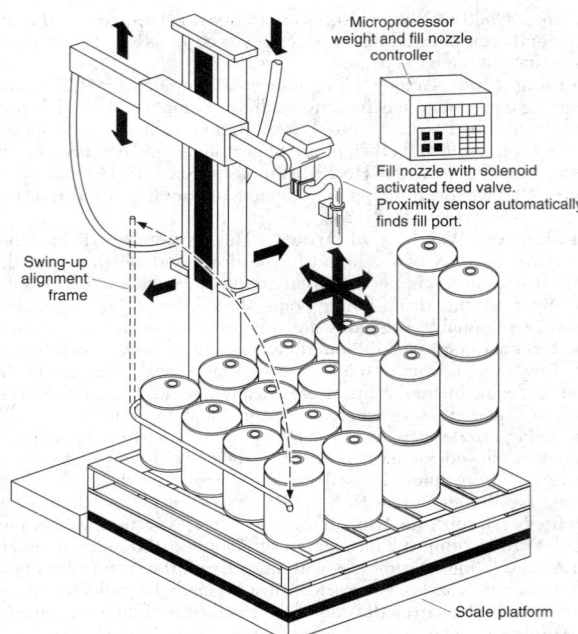

FIG. 21-39 Microprocessor-controlled drum and pail filling and weighing machine. Will fill drums or pails on one tier without operator intervention. (*Courtesy of Kg Systems, Inc., Bloomfield, NJ.*)

Work-Station Design Critical consideration must be given to the design of work stations so that filling operators work in a safe environment and are used productively. Methods design, time studies, and predetermined work-element data are helpful in determining the amount of work involved and the proper sequence of operations to permit good productivity. Of special importance (and often overlooked) is having the operator perform service functions on a drum while another drum is being filled. This is especially significant with an automatic system which does not require operator attention. The following activities can often be undertaken while a drum is being filled: removing closure plugs from the empty drum which will be filled next; replacing and tightening the closures in the drum which

preceded the drum being filled; labeling and marking code numbers on the drum; and starting the filled, sealed, and labeled drum down the handling system leading to storage or transportation.

Safety Regulations Consideration must be given to safety regulations for the electrical grounding of the drum during filling, the handling of product vapors, the handling of possible inadvertent spills and splashes of product, and the design of the work station to conform to OSHA and state and local codes. Operators must be protected from contact with the product, and their physical movements must not be such as could cause potential injury. Work-station design benefits from consultation with governmental bodies and with equipment vendors and consultants.

Small Liquid Packages The packaging of small packages of liquids is a specialized field. High-speed bottle and can fillers typically are of volumetric rather than weigh design. Up to a size of 3.8 L (1 gal) volumetric fillers are used almost universally when the filling rate exceeds 10 containers per minute. Below this rate, filling is controlled by weight or even volumetrically by an operator activating manual controls.

SOLIDS PACKAGING

Containers for solids include bags, bulk boxes, cartons, and drums. While the intermediate flexible bulk container (IBC) has become an important package of world commerce, the most used package remains the multiwall paper bag, supplemented by bags of similar design made of plastic film or plastic woven mesh.

Multiwall Paper Bags These bags (Fig. 21-40), made from plies of kraft paper or from combinations of kraft and special-purpose papers and plastics, are the most common packages for almost any pelleted or powdered material as well as for briquettes or bats of such solids as synthetic rubber, waxes, and insulation.

Empty bags are ordinarily shipped compressed (to obtain high load density) and on pallets, the most common of which measure 1220 by 1065 mm (48 by 42 in), 1220 by 1015 mm (48 by 40 in), and 1270 by 1115 mm (50 by 44 in). The number of empty bags per pallet varies with size, 1500 to 2000 being common. A typical filled pallet weighs about 907 kg (2000 lb). Pallet loads are often triple-tiered in warehouses.

Two bag designs are common: the **valve** and the **open-mouth** types. The **valve** bag has both ends closed during fabrication, filling being accomplished through a small opening (valve) in one corner of the bag. The open-mouth bag has one end closed at the factory and the other after filling.

Most **open-mouth bags** are closed by sewing, whereas adhesive is used for the pinch type. The pinch bag has been the subject of inten-

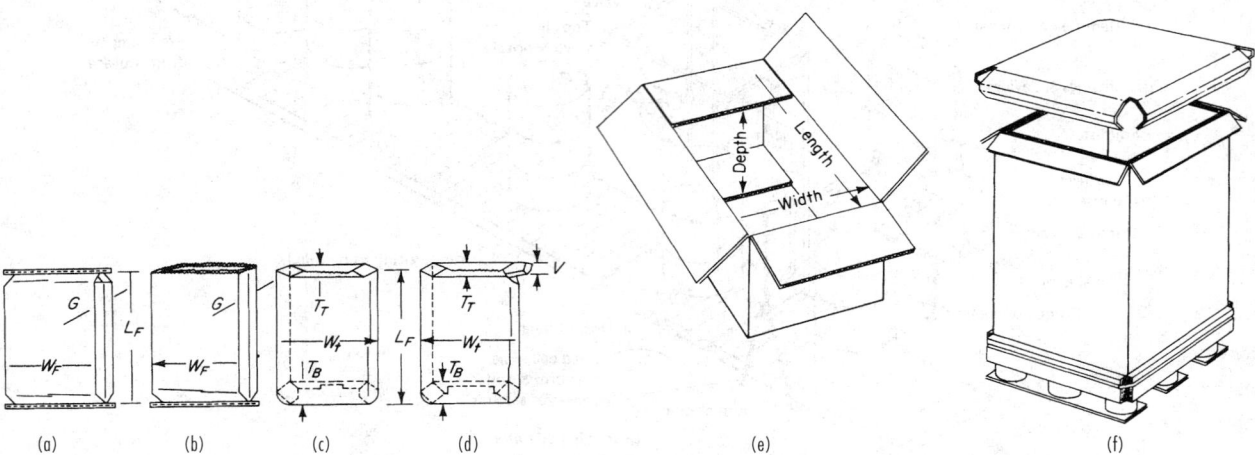

FIG. 21-40 Typical packages used for chemical products. (*a*) Sewn valve bag. (*b*) Sewn open-mouth bag; pinch-bottom-type open-mouth bag. (*c*) Pasted valve bag. (*d*) Pasted valve tuck-in-sleeve bag. (*e*) Principal (inside) dimensions of a regular slotted carton (RSC). (*f*) Bulk box of corrugated fiberboard for product weighing 450 kg (990 lb).

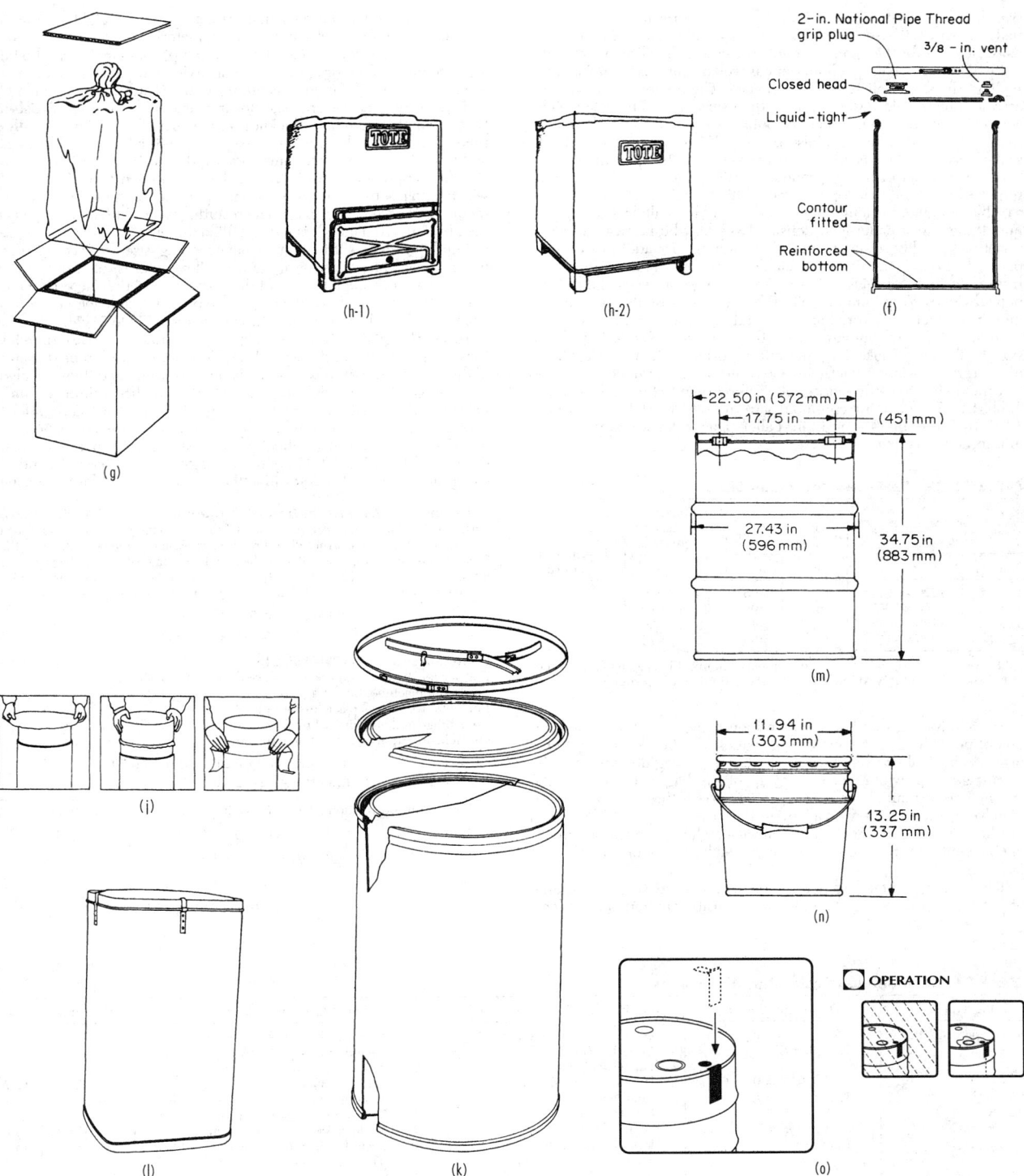

FIG. 21-40 *(Continued)* Typical packages used for chemical products. (*g*) Bag in box. (*h*) Tote-bin rigid intermediate bulk container: 1—for solids, 2—for liquids. (*Courtesy of Hoover Universal, Inc., Materials Handling Division.*) (*i*) Liquid-type polyethylene drum with fiber overpack. (*j*) All-fiber drum with removal top. (*k*) Lever-locking fiber drum. (*l*) Ro-Con rectangular fiber drum; clip-mounted top. (*Courtesy of Greif Bros. Corp.*) (*m*) 208-L (55-gal) DOT Packing Group II closed-head steel drum. (*n*) 19-L (5-gal) DOT Packing Group III open-head universal steel pail with lug-type cover. (*o*) DRUMWIC™ device for removal of water from drum heads stored out-of-doors. U.S. Patent No. 5,373,962. (*From Lee Technologies, Inc. Huntington, WV.*)

sive development by the bag industry, and as a result it has substantially displaced the sewn open-mouth bag and to some extent, valve bags. Reasons for this change include the ease, reliability, and repeatability of closing (sealing) equipment as well as the close control of the sealing adhesive applied by the bagmaker. The preapplied adhesive is activated by the closing machine in the user's plant. The positive closure of the pinch bag produces a container which completely seals in the product. Valve bags can also be sealed if they are provided with sealable sleeves. An advantage of valve bags over the open-mouth type is the availability of highly productive filling machines that not only require less labor than open-mouth filling equipment but also are capable of higher packing rates. In addition, the sealed valve bag permits the greatest pallet-load density. Bags may be fabricated from a variety of readily available flexible materials. In addition to kraft paper, barrier materials are available to prevent moisture or gases from entering or leaving the bag. These range in permeability from polyethylenes to aluminum foil. Table 21-16 presents some of the more prominent moisture barriers and their typical properties.

In the interest of standardization, the Institute of Packaging Professionals Chemical Packaging Committee (formerly the Chemical Manufacturers Association Packaging Committee) recommends four sizes of expendable four-way-entry pallets for bagged chemicals given in Table 21-22. The most common pattern consists of five bags on a 1220- by 1065-mm (48- by 42-in) pallet. This pallet size permits maximum trailer loading.

TABLE 21-22 Preferred Bag-Pallet Sizes*

Pallet size		Filled-bag face width		Filled-bag face length		Bags per tier (pattern)
in	mm	in	mm	in	mm	
48 × 40	1220 × 1015†	16	405	24	610	5
48 × 42	1220 × 1065	16	405	24	610	5
52 × 44	1320 × 1115	17.25	440	26	660	5
52 × 36	1320 × 915	16	405	36	915	3

*From Institute of Packaging Professionals Chemical Packaging Committee.
†Also GMA (Grocery Manufacturers Association) and ISO standard.

Sewn valve bags and sewn open-mouth bags are less important, except possibly for products with densities over 960 kg/m³ (60 lb/ft³) or for individual or small-lot shipments. These bag designs have the advantage of providing an easy grasp of the bag at the end of the sewing line without allowing fine powders to sift through the closure.

Valve bags usually rely on a labyrinth of paper or plastic film to seal off the valve. The automatic internal valve, while adequately protecting the contents of the bag, does allow a small amount of sifting of fine powders.

The starting point in bag-size determination is the weight or volume of product to be packaged and its bulk density (aerated and settled).

Also to be considered are particle size, shape, and weight; degree of aeration at time of packaging; flow characteristics; temperature and relative humidity; type of handling system up to and including the filling machine; bag-closing method; bag style; and pallet size and pattern. Three sets of dimensions are needed: (1) **tube**—outside length and width of tube before bag closures are fabricated; (2) **finished face**—length, width, and thickness of bag after fabrication; (3) **filled face**—length, width, and thickness of bag after filling. Table 21-23 and Fig. 21-40 show these dimensions and their interrelations.

A first approximation of size can be determined from Fig. 21-41, which applies to sewn valve, sewn open-mouth, pinch-type open-mouth, and pasted valve bags. The resulting tube width and length can then be converted to finished and filled dimensions, and bag samples ordered for field verification. Changing bag size to accommodate different weights, density variations, pallet patterns, etc., becomes a simple matter through the use of the graph. Correction factors for particular situations such as special plies, type of filling machine, storage system, and product characteristics are given in Table 21-24.

To use the graph, given the weight of material to be packed, follow these steps: (1) obtain the settled and loose (or aerated) bulk densities of the product (a 1-ft³ box serves this purpose well), and then calculate the average of the two densities; (2) calculate the bag-volume requirement from the relation $V_b = [W$ lb (weight to be packed)·1728] d lb/ft³ (average density); (3) multiply V_b by the product of the correction factors (Table 21-24), which reflect product, storage, and packaging conditions; (4) from Fig. 21-41 obtain the bag-tube equivalent T_e; and (5) using the corrected V_b, determine the bag size needed for palletizing.

Example 1: Determination of Proper Bag Size 55.1 lb (25 kg) of plastic pellets having a bulk density of 38.5 lb/ft³ (615 kg/m³) are to be packaged in pasted valve bags constructed of three kraft plies and a free polyethylene (PE) 2-mil (0.05-mm) liner. Bags will be palletized in a 5-bag pattern, 40 bags per pallet, on 48- by 40-in (1220- by 1016-mm) pallets. Filled bags are permitted to overhang the pallet by 0.5 in (15 mm). Determine the proper bag size.

$$\text{Density} = 38.5 \text{ lb/ft}^3$$
$$V_b = (50/35) \times 1728 = 2470$$

Correction factor (from Table 21-24):
For barrier sheet of 2-mil polyethylene film   1.05
For filling machine, fluidizing type   1.02
For ⅛-in- (3.2-mm-) particle-size pellets   1.00
For storage and handling 24 h   1.00
Overall correction factor (product of above)   1.07

$$\text{Corrected } V_b = 2470 \times 1.07 = 2650$$
$$T_e \text{ (from Fig. 21-41)} = 640$$

For first approximation let $T_T = T_B = 6$ in, and $L_f = 24 - 1 = 23$ in.

Since
$$L_f = L_t - (T_T + T_B)/2 - 1$$
$$L_t = 23 + 6 + 1 = 30 \text{ in}$$
and
$$T_e = W_t L_t = 640$$
$$W_t = 640/L_t = 640/30 = 21.3 \text{ in}$$

TABLE 21-23 Multiwall-Paper-Bag Dimensions

Bag type	Tube dimensions	Finished-face dimensions	Filled-face dimensions*	Valve dimensions
Sewn open-mouth	Width = $W_t = W_f + G_f$ Length = $L_t = L_f$	Width = $W_f = W_t - G_f$ Length = $L_f = L_t$ Gusset = G_f	Width = $W_F = W_f + ½$ in. Length = $L_F = L_f - 0.67 G_f$ Thickness = $G_F = G_f + ½$ in.	
Sewn valve	Width = $W_t = W_f + G_f$ Length = $L_t = L_f$	Width = $W_f = W_t - G_f$ Length = $L_f = L_t$ Gusset = G_f	Width = $W_F = W_f + 1$ in. Length = $L_F = L_f - 0.67 G_f$ Thickness = $G_F = G_f + 1$ in.	Width = $V = G_f ± ½$ in. †
Pasted valve	Width = $W_t = W_f$ Length = L_t	Width = $W_f = W_t$ Length = $L_f = L_t - (T_T + T_B)/2 - 1$ Thickness at top = T_T‡ Thickness at bottom = T_B‡	Width = $W_F = W_f - T_T + 1$ in. Length = $L_F = L_f - T_T + 1$ in. Thickness = $T_F = T_T + ½$ in.	Width = $V = T_T \begin{cases} +0 \text{ in.} \\ -1 \text{ in.} \end{cases}$ §

Meaning of subscripts: B = bottom; f = finished-face; F = filled-face; t = tube; T = top.
*Formulas are based on conditions of bags after mechanical flattening.
†Valve dimension is flat width, which must not exceed ±½ in. + G to maintain good closure. Circumference of valve = twice the width.
‡T_T and T_B are usually equal; if they differ, use average. T = thickness.
§Valve dimension is flat width. Valve width can be made less than top width without affecting closure properties.

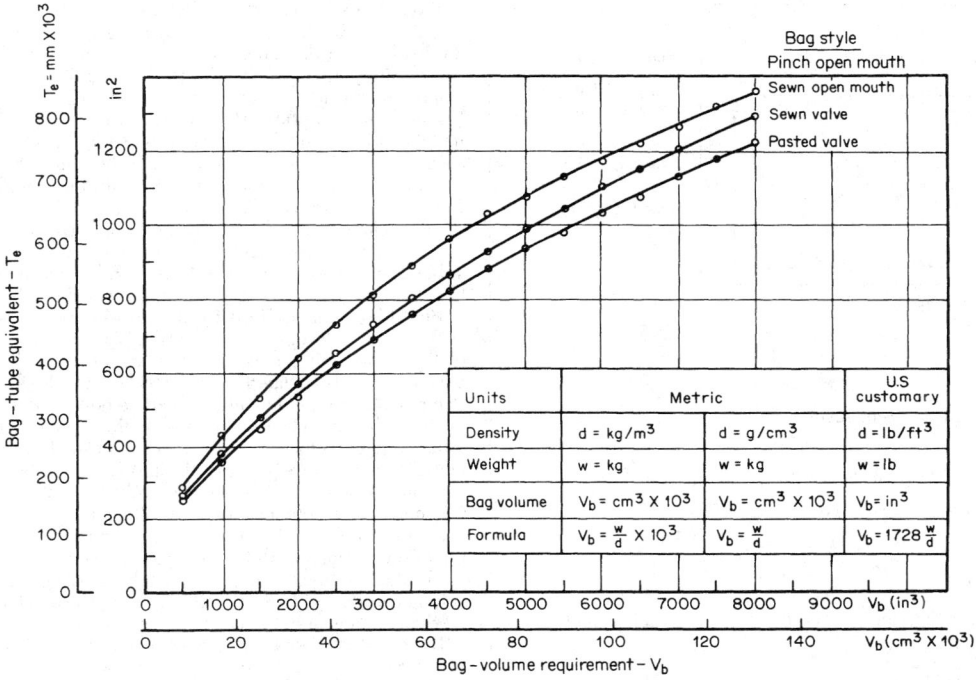

FIG. 21-41 Multiwall-bag-sizing graph. (*Raymus Associates, Inc.*)

TABLE 21-24 Bag-Volume-Equivalent (BVE) Correction Factors *f* for Specific Conditions*

Barrier and type of bag construction material		Filling-machine characteristic		Conveying, handling, and storage conditions					
						Mechanical conveying		Pneumatic conveying	
				Material particle characteristic		Off-product stream	From storage	Off-product stream	From storage
Type of material†	*f*	Type	*f*	Size, in.	Shape	*f*	*f*	*f*	*f*
Asphalt-kraft laminates	1.03	Auger, gross weigh	1.00	1/16	Pellets, round	1.00	1.00	1.00	1.00
				1/8, 3/16	Pellets, round	1.00	1.00	1.00	1.00
Polyethylene extrusion-coated on kraft	1.05	Auger, net weigh	1.03	1/8, 1/4	Cubes	1.01	1.01	1.01	1.01
Polyethylene extrusion-coated on kraft	1.01	Belt	1.05	+200 mesh	Granules, sharp edges	1.02	1.01	1.03	1.02
with random partial perforations°		Preweigh-belt	1.01	+200 mesh	Granules, smooth edges	1.02	1.02	1.03	1.03
Polyethylene-aluminum foil-kraft laminate	1.15	Fluidizing	1.02	−50, +200 mesh	Platelets (tiny flakes)	1.03	1.02	1.05	1.03
Wax-coated kraft	1.09	Gravity	1.15	+1/4, −3/8	Flakes	1.02		1.03	
Glassine	1.04	Preweigh open-mouth	1.00						
Free polyethylene film 1 to 4 mils thickness	1.03	Gross-weigh open-mouth	1.10	−30 to +200 mesh mix	Granules, sharp	1.01	1.01	1.03	1.02
Polyethylene-film bag (extruded tubular film with heat-sealed ends)	1.10 to 1.15			−30 to +200 mesh mix	Granules, round	1.02	1.02	1.05	1.03
				−325 mesh	Granules, platelets	1.04	1.03	1.07	1.03

NOTE: Refer to Table 21-6 for metric particle sizes.
*Approximate factors are based on many observations of each material or condition stated.
†Applies to all commercially available materials of designated type, unless noted otherwise.

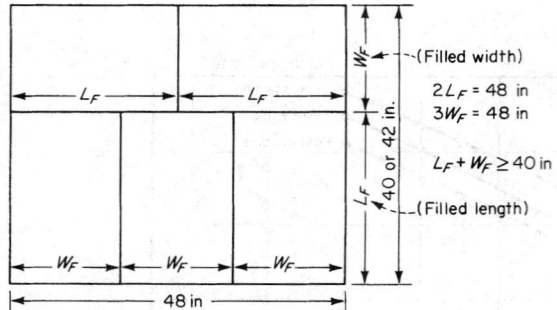

Since
$$W_F = W_f - T_f + 1, \text{ and } W_t = W_f$$

then
$$W_F = 21.3 - 6 + 1 = 16.3 \text{ in}$$

Checking for pallet-length conformity,

$$2L_F = 48 \text{ in, or } 2(24) = 48 \text{ in}$$

$$3W_F = 48 \text{ in, or } 3(16.3) = 48.9 \text{ in}$$

Checking for pallet-width conformity,

$$W_F + L_F \geq 40 \text{ in}$$

or
$$16.3 + 24 = 40.3 \text{ in}$$

Summary:
Bag size: 21½ in (face width) × 23 in (face length) × 6 in top and bottom (545 × 585 × 150 mm).
Pallet size: Use 48 in × 40 in (1220 × 1015 mm).
Overall filled pallet dimensions: 48.9 × 40.3 in (1242 × 1024 mm).
This example can also be carried out in the SI system by using Fig. 21-41.

Liners   At the time of setup of filling, many containers for bulk solids are lined with a polyethylene (PE)-film bag, the purpose being to prevent sifting of fine particles, retard moisture pickup or release, or prevent product contamination by the construction material of the container. Liner length should be sufficient to permit the top to be closed by heat sealing or wire tying. The film gauge (thickness) needed depends on the weight, bulk density, and particle roughness of the contents. Gauges of 0.05 to 0.25 mm (2 to 10 mils) are common. For ease of placing the liner in the carton, the gusseted type is preferred; and for convenience in handling, liners are usually made as continuous tubes, with heat seals and tear-off perforations at intervals equal to one bag length (see Table 21-17 for costs).

Form-Fill-Seal, Small Bags and Pouches, and Baler Bags   Product weights from a few grams to 11 kg (25 lb) are often placed in packages made of plastic film, paper, or combinations of these. Groups of packages are then shipped in cartons or baler bags (see Table 21-17 for costs).

Form-fill-seal is a machine process that forms a tube from plastic-coated paper stock, heat-seals one end, fills the resultant bag, heat-seals the other end, and then cuts off the filled bag. This method has the advantage over filling small bags or pouches in that the cost of package fabrication is avoided until the package is actually needed; packaging labor is reduced to one attendant who can service a number of machines. Also, order lead time is shortened because standard, merchant plastic film or paper can be bought from local stock, often avoiding waits of 4 to 8 weeks for fabricated bags. Offsetting this is the higher investment in equipment and the service and maintenance problems associated with automatic equipment.

Small bags and pouches are made from one or more plies of paper or plastic film. The two main types of paper bags are the satchel-bottom and the pinch-bottom. Both types usually have a gusset that helps form a rectangular cross section (a useful trait when packing in cartons or baler bags). Although order lead time is longer than for form-fill-seal and operating labor is greater, capital and maintenance costs are smaller (Table 21-17), and equipment reliability is greater. These small packages require a master shipping container. Corrugated cartons are used extensively, as is the flexible baler bag.

Baler bags are pasted open-mouth bags with one or more plies and of either satchel-bottom or self-opening (gusseted) design. Pouches are loaded into the baler with their long axis parallel to that of the baler. Since the pouches must be tightly packed, mechanical-compression loading equipment is mandatory.

Rigid Intermediate Bulk Containers   Rigid IBCs are made of metal or plastic suitable for the product and service intended. Sizes available range from 0.17 m³ (6 ft³) to 2.83 m³ (100 ft³). This type of container is intended for reuse and a useful life of up to 20 years. Important economic considerations are the cost of returning the empty container to the filling location and the cleaning, handling, and storage of it. Figure 21-40 (h-1, h-2) illustrates a metal container. Table 21-20 gives economic information comparing this type of container with other containers of larger or smaller volume.

Flexible Intermediate Bulk Containers   These containers are an important development of the 1970s. Made from woven poly-olefins or other materials, flexible IBCs are available in a wide variety of volumes and can handle up to 1800 kg (4000 lb), depending on construction. This type of container can be equipped with a thermoplastic liner when it is necessary to protect the product against moisture or other contamination. Handling is accomplished by forklift truck or by hoist. Filling and weighing of flexible IBCs can be accomplished on specially designed weigh scales or volumetrically if the container is weighed at a remote location after filling and that weight is used as a basis for invoicing. Filling is carried out through a flexible port at the top of the container, while unloading is accomplished through a similar flexible member at the bottom. Table 21-25 gives dimensional and volumetric data. Figure 21-42 shows typical container designs and types of loading and discharge spouts.

Boxes   Bulk boxes (Fig. 21-40) of corrugated kraft paper for dry bulk products fall into two broad categories: large, for 0.5- to 2-ton loads, and small, for loads of 23 to 68 kg (50 to 150 lb). Large boxes are used extensively for resin shipment; small ones, for certain regulated materials (such as caustic soda) and for low-bulk-density products that are assessed excessive freight rates if packed in drums.

A bulk box, sometimes called *bag in box*, consists of a box within a box plus other elements such as end pads, PE bag liners, and closing materials (tape, glue, staples). The double-wall corrugated kraft board consists of an outside liner, a corrugating medium, a center liner, another corrugating medium, and an inside liner; the single-wall board

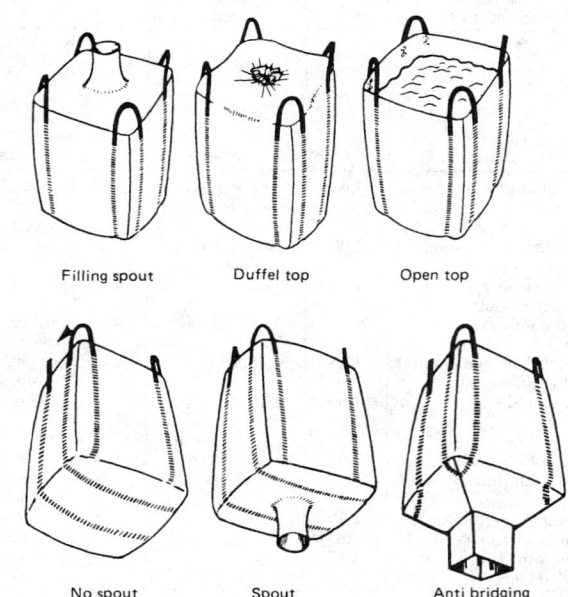

FIG. 21-42   Typical flexible-bulk-container designs and loading- and unloading-spout designs. (*Courtesy of Bonar Co., Ltd.*)

TABLE 21-25 Flexible-Type Intermediate Bulk Containers: Dimension and Capacity Data (Variable Data)*

Height		Usable volume		Maximum bulk density	
mm	in	m³	ft³	kg/m³	lb/ft³
800	31.5	0.7	24.6	1445	90
915	36.0	0.8	28.1	1250	78
1040	41.0	0.9	31.7	1120	70
1145	45.0	1.0	35.3	995	62
1270	50.0	1.1	38.3	930	58
1385	54.5	1.2	42.4	835	52
1500	59.0	1.3	45.9	770	48
1600	63.0	1.4	49.4	720	45

*From Bonar Co., Ltd.

NOTE: Maximum weight, 1 metric ton (2205 lb); cross-section dimensions, 890 by 890 mm (35 by 35 in); tare weight, 3 kg (7 lb); material of construction, woven polypropylene body and polyester lifting straps.

consists of an inner and outer liner with a corrugating medium center. The specifications for each depend on service requirements; 4100 kPa (600 lbf/in²) burst strength is common for 454-kg (1000-lb) loads; 1900 kPa (275 lbf/in²), for 68-kg (150-lb) loads. Materials of construction that resist high humidity and wetting with water are available.

Advantages of this container are its reclosing feature and its efficient use of storage and shipping space. Disadvantages are the space required to store box components before assembly and the limited reuse market. The lead time for ordering made-to-order boxes ranges from 3 to 6 weeks. Filling equipment is similar to that for drums. Setting up the box can require two persons because of the unwieldiness of the components. Table 21-26 gives an idea of filling speeds for several types of filling arrangements and box styles.

Wire-bound wood boxes (typical loads, 1 to 2 tons) have limited use for chemical products. The box body, consisting of thin wooden slats held in place by steel wire twisted around each slat, is fastened to a solid-deck wood pallet. The top also consists of wire-bound wooden pieces. A PE liner protects the product and prevents it from falling through the slats. Disadvantages of the container are the labor needed for setup and the space required for knocked-down boxes. Since manufacturers are usually near sources of hardwood, shipping costs to users may be high; lead times of 3 to 4 weeks are common.

Folding boxes made from chipboard are used for consumer-size units [from a few grams or ounces to about 11 kg (25 lb)] of such products as insecticides, snow-melting compounds, salt, and food additives. PE liners are often included to protect the product from moisture or to prevent it from sifting through minute openings in the top and bottom folds. Order lead time is 6 to 8 weeks. Knocked-down folding boxes are dense and store efficiently when palletized. A typical pallet measures 760 by 915 mm (30 by 36 in) with the load 1220 mm (48 in) high. Filling equipment, which can also be used for filling pouches, small bags, and glass jars, ranges from small, manually operated units to high-production, fully automatic units. Most common is the manually operated gross-weigher type.

Shipping cartons for liquids in cans and bottles, bulk solids in jars, pouches, and folding boxes, and briquetted items with or without individual packaging are usually made of corrugated kraft paper. The most common styles are the regular slotted carton (RSC), the end-opening RSC, and the center special-full overlap slotted container (SFF). End joints may be stapled, stitched, glued, or taped.

Specifications include dimensions of length, width, and depth, in that order (Fig. 21-40e). When boxes are set up and closed by automatic equipment, dimensional tolerances become critical. Cartons are shipped knocked down to the user from plants located in all industrial centers. Because order lead time is 4 to 6 weeks, inventories of empty boxes require considerable space. A useful booklet describing all aspects of corrugated box designs and materials is the *Fiber Box Handbook* available from The Fiber Box Association, 2850 Gulf Road, Rolling Meadows, IL 60008.

Often, the size of items packaged in corrugated cartons either does not permit interlocking of layers of cartons, or leaves considerable void space between them. Since calculating by hand the best size of

carton for maximum palletizing density requires considerable effort, computer software is available. Examples are CAPE by CAPE Systems, Inc., Plano, Texas and TOPS by TOPS Engineering Corp., Plano, Texas. An analysis of pallet and trailer loading using CAPE is given in Fig. 21-63a,b.

Drums Drums (Fig. 21-42), made of either **steel** or **fiber,** rank next in importance after the multiwall paper bag. For dry solids or slurries, fiber drums predominate; for liquids, the steel drum. Steel drums of the open-head design are used for dry products when the product is hazardous, is to be stored outdoors, or is of a density that will cause reasonable weights to exceed the limits for fiber drums. Although only a few sizes are common, fiber drums can be made to order in almost any size and diameter-length combinations for volumes of 2 to 285 L (0.75 to 75 gal) and for weights ranging from 25 to 250 kg (60 to 550 lb; Table 21-16).

Advantages of the drum are protection of contents, ease of reclosure, and appreciable reuse-resale value. A serious limitation is the inefficient use of space because of the cylindrical shape, which results in high storage and transportation costs. To overcome this, a fiber drum with a square cross section (Ro-Con drum) and the bulk corrugated bag in box have been developed.

Fiber drums decorated with advertising adds $.80 to $3.50 to the basic cost of the drum. The most common type is the multiple-ply kraft-paper body with a steel bottom and a reinforcing top hoop crimpled to the drum. A steel lid, secured by a locking ring tightened by a lever system, fits over the body. For **vapor protection,** barriers are incorporated among the plies, or liners are used as the first ply in contact with the product. Among barrier and liner materials are PE, aluminum and steel foil, polyesters, and silicones. When liquids are to be contained, blow-molded PE liners are used. Free-film PE liners inserted by the user yield a combination of barrier and liner properties at less cost than having the liners as part of the drum body.

Fiber drums are also made with a removable fiber top and a fiber bottom that is either removable or permanently fastened to the body. This drum has limited reuse, but it costs less than the lever-locked metal-top type. Filling equipment consists most commonly of an operator-controlled spout connected to a supply bin resting on a platform scale. Table 21-26 shows the labor productivity of several systems.

Steel drums are made from cold-rolled-steel sheet formed into a cylinder. The longitudinal seam is made by electric resistance welding. The rolling hoops are expanded into the body wall by a special hydraulic fixture, and the ends are crimped to the body to form a leakproof joint. Sealing compounds are often used to assure leakproof joints. The top head has openings to allow installation of the closures or bungs. These closures have U.S. standard-pipe-thread fittings, usually one 2-in and one ¾-in fitting, to allow connection to the loading and unloading equipment.

PACKAGING OPERATIONS

Dry-bulk-packaging operations are divided into two categories: weighing and filling a package that is itself the shipping container and weighing and filling small packages that are in turn placed in outer packages for shipment. The choice of equipment and the way in which it is combined into a system depend on such factors as the product and its chemical, physical, and rheological properties; the type of package to be filled; the total packaging output required; the instantaneous and average rates of filling; cost, attitude, and availability of labor; space available for equipment; storage, shipping, and transportation conditions; cost and availability of capital; seasonality of packaging activity; expected duration of the venture; sanitary, safety, packaging, and working conditions imposed by regulatory bodies; maintainability and reliability of equipment; changes expected in the product and in the demand for it; and nature of the product market (i.e., industrial, consumer, agricultural, or government).

Weighing and Proportioning These are terms used in the packaging industry to describe methods of measuring out an amount of product—into the packaged weight unit which is offered for sale. It could also be done as a process step where a given amount of material must be added to a process on a continuous or a regular basis. With

TABLE 21-26   Performance Data for Packaging Systems

Type of filling and weighing machine	No. of filling spouts	Type	Package detail				Product detail			
			Size, in	Size, mm	Construction	Closure	Material	Bulk density lb/ft³	kg/m³	Particle size, U.S. standard§
Fluidizing, SFW‡	4	Pasted valve bag	20 × 25 5 top width	510 × 635 125	4-170 (= 4-ply, 170 lb)	Inner sleeve	PVC‡	38	609	−60 mesh
PWS, open-mouth filler	1	Pinch bag	16 × 5 × 30	405 × 125 × 760	4-170 PE barrier	Adhesive	PE	30	481	⅛-in pellets
Fluidizing, SFW‡	2	Pasted valve bag	20 × 25—5¼ top	510 × 635 135	4-170	Tuck-in sleeve	PVC	36	577	−60 mesh
Fluidizing, SFW	3	Pasted valve bag	21 × 25	535 × 635	4-170	Tuck-in sleeve	PE	30	481	⅛-in pellets
Impeller, SFW	4	Pasted valve bag	18½ × 27½ (face)	470 × 700	3-170	Insert sleeve	Portland cement	94	1506	−325 mesh
Auger, SFW, net-weigh	1	Sewn valve bag	16 × 5 × 28	405 × 125 × 710	4-190	Tuck-in sleeve	PE	32	513	⁵⁄₃₂-in cubes
Centrifugal belt, SFW	2	Sewn valve bag	15 × 5 × 36	380 × 125 × 915	3-150 PE barrier	Insert sleeve	Fertilizer	55	881	⅛-in pellets
PWS, open-mouth filler,	1	Pinch bag	17 × 4 × 36	431 × 100 × 915	3-150 PE barrier	Adhesive	Fertilizer	55	881	⅛-in pellets
Fluidizing, SFW	4	Pasted valve bag	18½ × 26 5¼ top	470 × 660 135	3-150 PE barrier	PE inner sleeve	Fertilizer	55	881	⅛-in pellets
PWS, open-mouth, heat sealer	1	PE flat-tube bag	16 × 30½	405 × 775	10-mil PE	Heat-sealed	Fertilizer	55	881	⅛-in pellets
PWS, form-fill-seal	1	F/F/S gusseted bag	16 × 5 × 30	405 × 125 × 760	6 mil PE	Heat-sealed	LDPE	30	480	⅛ in
SFW, liquid fill and weigh	1	Steel drum U.S. DOT 1A1 PG Y	55 gal (208 L)— 23.5 in dia. × 34.75 in high	596 × 883	18-ga (0.0428-in) ends, 20-ga (0.0324-in) body	2-in, ¾-in NPT bungs	Lacquer solvent		0.839	sp. gr.
Gravity, SFW	1	Sewn valve bag	16 × 5 × 28	405 × 125 × 710	4-190	Tuck-in sleeve	Polystyrene	32	513	⁵⁄₃₂-in cubes
Platform scale, autofill cutoff, SFW	1	Drum	55 gal	208L	6-ply fiber (300 lb)	Lever-locked steel cover	PE master batch	30	481	⅛-in pellets
Platform scale, manual cutoff	1	Drum	55 gal	208L	6-ply fiber (300 lb)	Lever-locked steel cover	Cleaning compound	45	721	−20 to +80 mesh
Platform scale, autofill cutoff, SFW	1	Bulk box, 3-mil PE liner	15 × 15 × 24	380 × 380 × 610	Outer 275-lb test DW‡ liner 600-lb test, DW	Staples	Insecticide, technical grade	40	640	−200 mesh
Platform scale, autofill cutoff, SFW, automatic staple closer	1	Bulk box	41 × 34 × 36	1040 × 860 × 915	Inner, outer boxes: 600-lb test, DW kraft board	Staples	PE	30	481	⅛-in pellets
Vertical auger, SFW	1	Small bag Pouch	10 × 4 × 25 14 × 27	255 × 100 × 635 355 × 685	3-120 paper, 2- to 4-mil PE	Glued, heat-sealed	Insecticide powder	20	320	−325 mesh
Vertical auger, SFW	1	Folding box	6½ × 3½ × 9	165 × 90 × 230	12-point reprocessed board with 2-mil PE liner	Glued, tied PE liner	Sprayable insecticide powder	20	320	−10 μm
Form-fill pouch maker, PWS	2	Pouch	8½ × 15	215 × 380	1- to 3-mil PE film	Heat-seal, hot-wire cutoff	Detergent, spray-dried	39	625	−30 to +60 mesh
Baler, manual package in feed, mechanized closing	—	Baler bag	23 × 30	585 × 760	2-140	Glued	12, 5-lb (2.3 kg) bags herbicide	45	721	−325 mesh
Corrugated case, manual package in feed, mechanized closing	—	Regular slotted carton	24 × 16 × 7	610 × 405 × 180	275 DW	Glued	12, 5-lb (2.3 kg) bags herbicide	45	721	−325 mesh
Carousel liquid filler	18	Round jug	4.8d × 9.7h	96d × 223h	Plastic 100 gr.	38 mm cap plastic	Laundry bleach	—	—	1.1 sp. gr
In line liquid filler	6	'Boston Round' bottle	3.8d × 8.7h	123 × 245	Glass	33 mm cap plastic	Isopropyl alcohol USP	—	—	0.9 sp. gr

*Fractions indicate the portion of a person's time required to perform activity; these are additive to compute the number of people needed.

†Includes equipment and installation but not building or services needed.

‡Definition of abbreviations: SFW = simultaneous fill-and-weigh; PWS = preweigh scale; SMC = sewing-machine closer; DW = double wall; SOM = sewn open-mouth; PE = polyethylene; PVC = polyvinyl chloride.

§Metric equivalent of particle sizes given elsewhere.

[1]For existing equipment, remanufactured to new machine standards and guarantees, multiply the above investment values by 0.5.

[2]Where ² is shown after the system investment, add $181,000 for an automated inspection system comprising an X-ray metal detection machine, and 3 machine vision units to verify closure in place and label and bar code are correct and in place.

[3]Investment data courtesy of In Plant Packaging Systems, Inc., Metuchen, NJ.

[4]The above data is useful in comparing alternative systems and for order of magnitude investment values. However, for capital and other budgets, recent actual quotations from manufacturers should always be used.

Weight of contents		Packaging rate, packages/min		Weight variation from average		Packaging personnel needed°				Package handling		Approximate 1996 investment ($ × 1000)	
lb	kg	Avg	Instant	oz	gr	Package setup, supply	Filling-machine operators	Package closers	Palletizers, loaders, attendants	Package conveyorized	Automatic palletizing	Filling machine[1]	System[2]
50	22.7	12	17	4	114	1	1	0	2	Yes	No	110	437
50	22.7	8	12	0.5	14	1	1	1	2	Yes	No	45	200[2]
50	22.7	6	8	4	114	1	1	0	1	Yes	No	78	330[2]
50	22.7	16	24	3	85	1	1	0	1	Yes	Yes	96	550[2]
94	42.7	22	28	8	227	1.5	1	0	0.5	Yes	Yes	110	655
50	22.7	1	2	3	85	0.25	0.25	0	0.5	No	No	27	52
80	36.4	12	16	8	227	1	1	0	2	Yes	No	66	330
80	36.4	16	22	4	114	1	1	2	2	Yes	No	44	240
80	36.4	18	24	16	455	1	1	0	2	Yes	No	119	350
50	22.7	18	24	4	114	1	1	1	2	Yes	No	66	285
50	25	8	12	1	28	1	0	0	0	Yes	Yes	1,075	3,300[2]
385	175	2	3	6	170	0.25	0.5	0.5	1.75	Yes	No	116	178
50	22.7	0.2	0.4	16	455	0.25	0.5	0	0.25	No	No	7	13
250	113.6	1	4	2	57	1	0.5	0.5	1	Yes	No	27	110
300	136.4	0.5	1	0.5	14	0.25	0.25	0.25	0.25	Yes	No	7	28
100	45.5	0.5	1	4	114	1	1	1	1	No	No	45	87
900	409.1	0.33	0.50	8	227	2	0.75	0.25	—	Yes		45	218
10	4.5	5	10	1	28	1	1	1	3	Yes	No	21	110
1.5	0.682	8	12	0.5	14	1	1	1	3	Yes	No	21	45
2.5	1.136	10	12	0.5	14	1	—	—	2	Yes	No	110	153
60	27.3	1.5	3	—	—	0.5	0.5	0.5	0.5	Yes	No	45	66
60	27.3	1	2	—	—	0.5	0.5	0	1	Yes	No	30	45
64 fl.oz.	1.81	40	50	±0.1 fl.oz.	3cc	0.5	0.5	0.5	0.5	Yes	Yes	175	750
32 fl.oz.	0.941	14	18	±0.1 fl.oz.	3cc	0.5	0.5	0.5	0.5	Yes	No	55	256

each, certain degrees of precision are required. There are several terms which are used by the scale industry. A **net weigher** (Fig. 21-43) is a device with a scale system for weighing bulk solid materials. The analog of the weight being measured is sensed by a mechanical or an electrical sensor system. Sensor output is interpreted by a control system, either electrically or mechanically or by a combination of both, which controls the flow of product into the scale and hence the weight. Net weighers are rarely used for liquids. The term **gross weigher** applies to a type of device which is becoming obsolete in the packaging industry. This type of device has a relatively large equipment mass holding the package into which the product is being weighed, with the result that high weight accuracy is not possible. The term **proportioning** is used to describe a system where a given volume of material is moved in a given period of time into a weigh vessel—often the package itself—without the weight actually being sensed. Proportioners operate under the assumption that product density is constant. For liquids, this is usually true at any given temperature. For solids this is rarely true. As a result, proportioning devices rely heavily on either having constant density or relatively uniform density which varies only slightly over time. An example of a volumetric filler is given in Fig. 21-44. This device is a vertical auger designed for filling powdered materials into glass jars. The assumption of constant density is applied. The machine is set for the required time to fill the desired weight into the package. This type of device has benefited from the microprocessor era in that downstream check weighers are used to determine the net weight packaged in each container. Using a software to calculate trends, the check weigher sends a signal to the proportioning device to increase or decrease the filling time because the weight being filled now shows a trend to drift as a result of density change. As the density increases, for the same constant period of time the filled weight will increase. The converse is true when the density decreases.

There are two principal types of package-weighing and -filling equipment: **simultaneous fill-and-weigh,** with which the material is weighed as it is poured into the container; and **preweigh,** with which the material is weighed prior to being poured into the package. The former applies mainly to valve bags, pouches, bulk boxes, and bags in boxes; the latter, to open-mouth bags, small bags, and cartons, to form-fill-seal, and, at times, to valve bags.

There is a further distinction between net weighers and gross weighers. **Net weighers** are defined by the ratio (0.3 to 0.5) of weight of charged material to weight of weighing vessel and associated parts. Preweigh scales are examples of net weighers. With **gross weighers,** of which simultaneous fill-and-weigh is an example, the ratio is usually greater than unity. Net weighers are accurate within ±0.125 to ±0.25 percent; gross weighers, from ±0.5 to 1.0 percent. Maintaining certain scale-feed conditions is critical in obtaining accuracy and sustaining a given production rate; appropriate feeding devices and surge bins are of great importance. If desired, weight accuracy can be increased, at greater cost, by special modifications and accessories such as load cells and microprocessor controls, bulk and dribble devices, and feeders and bulk density tracking software.

The **weight accuracy** of a dynamic weighing device is expressed as a plus or minus percentage deviation from a given *set weight,* which can only approximate the desired *actual weight.* The dynamic nature of weighing requires that the scale respond to changing static conditions as well as to a series of constant dynamic conditions. Minor variations in product density can cause the set weight to drift, the result being unacceptable packaged weight. Scale sensitivity is often suspected to be at fault, when in fact it is the set weight that has drifted. This is easily verified by check-weighing a series of weighings and determining their standard deviation.

Check Weighing   Because of drifting set weight and the influence of federal and state legislation on allowable deviation from

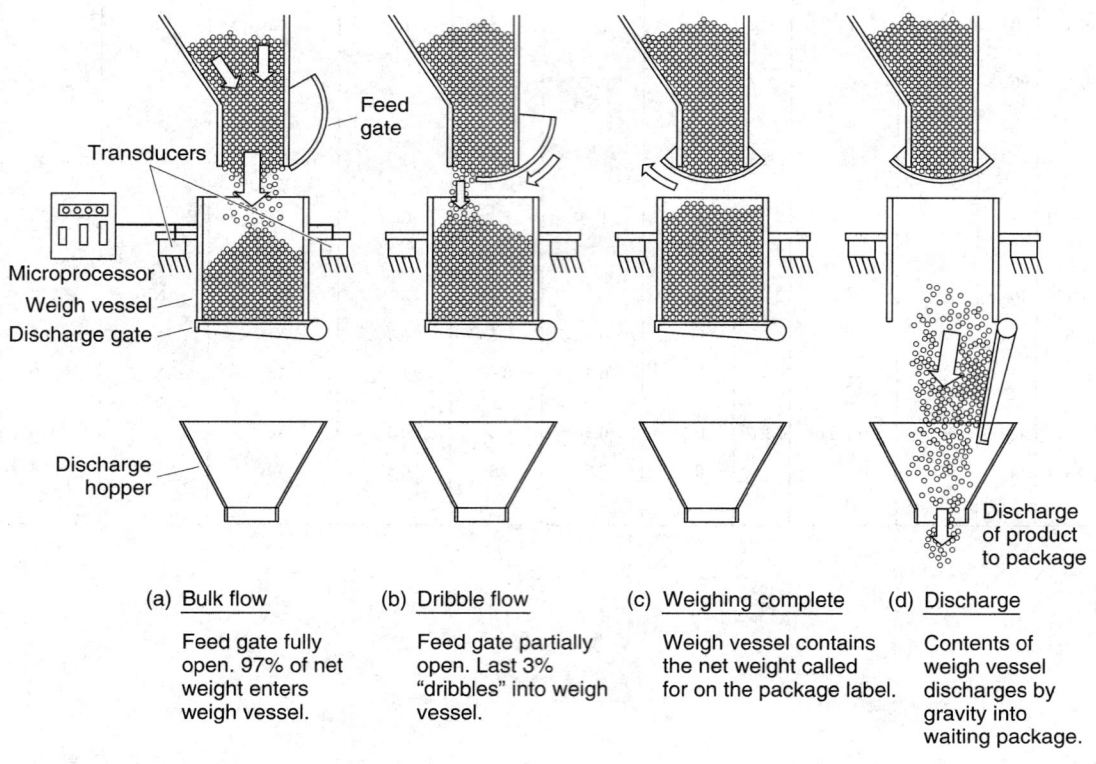

(a) Bulk flow	(b) Dribble flow	(c) Weighing complete	(d) Discharge
Feed gate fully open. 97% of net weight enters weigh vessel.	Feed gate partially open. Last 3% "dribbles" into weigh vessel.	Weigh vessel contains the net weight called for on the package label.	Contents of weigh vessel discharges by gravity into waiting package.

FIG. 21-43   Netweigh scale operating concept.

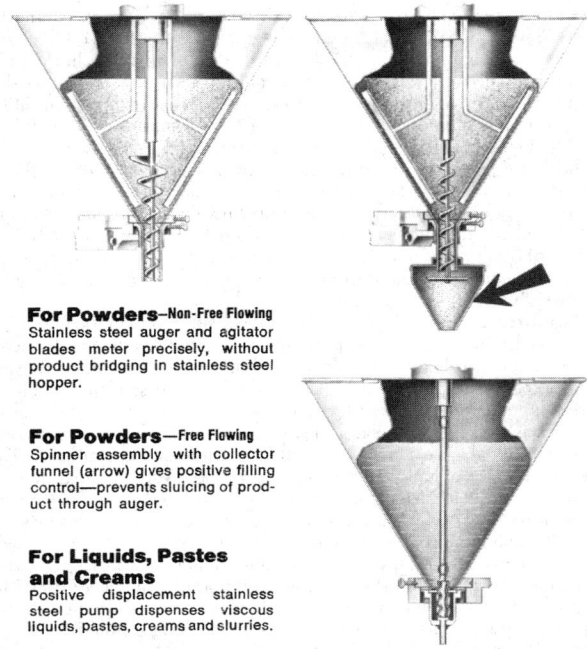

For Powders—Non-Free Flowing
Stainless steel auger and agitator blades meter precisely, without product bridging in stainless steel hopper.

For Powders—Free Flowing
Spinner assembly with collector funnel (arrow) gives positive filling control—prevents sluicing of product through auger.

For Liquids, Pastes and Creams
Positive displacement stainless steel pump dispenses viscous liquids, pastes, creams and slurries.

FIG. 21-44 Volumetric proportioner designs for various product types and consistencies. (*Courtesy of Mateer-Burt Co., Inc., Wayne, PA 19087.*)

advertised weights, a major new phase of package filling and weighing is that of check weighing. This can be done manually with a platform scale and then following a simple statistical procedure and control chart. There are devices applicable to preweigh scales which perform and record a static weighing just prior to discharging to the filling machine. There are in-line check weighers that weigh each package and pass or reject it depending on its weight, and keep a log of the results. One development permits continuous automatic readjustment of the scale set weight by means of a microprocessor that records each weighing and then, from a series of these, computes whether or not the set weight is drifting. An automatic adjustment is then made. The use of high-accuracy transducers aids this process.

Filling and Weighing Equipment   Of special interest in the selection of filling equipment from the wide variety available (Table 21-26) and combining it into the total system is the equipment's relation to instantaneous output, average output, and personnel. Methodizing, subdividing into work elements, and prediction of the time required for each job function by means of standardized data such as methods-time measurement (MTM) and general-purpose data (GPD) permit accurate identification of jobs and work content. Actual average output can thus be calculated and, from this, the instantaneous output.

Instantaneous rate, which is the rate that equipment manufactuers imply in their guarantees of performance, is defined as the number of packages produced per minute with the equipment operating under steady-state conditions. **Average rate,** the measure that the user needs to plan production and output commitments, can be defined as the arithmetic average (packages per minute) produced over a production shift (usually 8 h). Equipment reliability must be taken into consideration in determining average rates because malfunction downtime can have a significant effect on rate values. Also, the effect of production scheduling on equipment idle time and changeover from one product to another needs to be considered.

Valve-Bag-Filling Equipment   Although multiwall paper bags and plastic bags can be filled by a wide variety of equipment, the simultaneous fill-and-weigh (gross-weigher) type predominates. Net-weigher-type equipment using a preweigh scale which discharges into

a valve-bag filler is finding increased favor when greater weight accuracy is required. The most widely used category is the gross weigher of the pressure-fluidizing type, for which these parameters and ranges hold:

Parameter	Capability range[*]
Particle size	⅜-in (9.5-mm) pellets to submicrometer
Bulk density	0.5 to 200 lb/ft³ (8 to 3200 kg/m³)
Filling spout	1 to 4
Bagged weight	20 to 150 lb (10 to 70 kg)
Bagged volume	1 to 6 ft³ (0.03 to 0.17 m³)
Material of construction in contact with product	Carbon steel, stainless steel, plastic-coated steel, aluminum
Output capability	1 to 30 bags per minute
Bag-valve size	3 to 5½ in (75 to 140 mm)
Weight error: simultaneous fill-and-weigh scale	+2 to ±4 oz (60 to 120 g)
Weight error: preweigh scale	±1 oz (30 g)

[*]SI equivalents are rounded off.

Fluidizing Bag Fillers   These fillers can meet any production requirements, ranging from pilot-plant scale through heavy-duty, conveyorized high-tonnage installations. A chamber is provided with an air pad at the bottom, adjacent to a filling spout. A column of product over this section, which is what causes flow, may be opened to the atmosphere or enclosed and pressurized. When the desired bag weight is reached, a system is activated by the integral weigh scale to close the valve through which material flows to the bag. Fluidizing and pressurizing air is best provided by a positive-displacement blower at 1.5 kW (3 hp) per filling spout.

Of special interest on multiple-spout conveyor-equipped fluidizers and on certain types of screw and belt filling machines is a combination operator's seat, bag rest, and tuck-in-sleeve work aid. This device places the operator in an optimum work position after filling, to allow easy and positive tucking of the sleeve. Extensive use of the polyethylene film internal sleeve, however, has reduced the significance of the tuck-in-sleeve feature. Several types of heat-sealable valve-bag sleeves are available, as is equipment for closing them automatically. These are used when even slight leakage of product from an internal sleeve bag is unacceptable.

Bags can be automatically placed on valve-bag packers by means of an **automatic bag-placing device,** which consists of a magazine holding approximately 100 empty pasted valve bags and of a mechanism for removing the bag from the magazine, opening the valve and placing it on the filling spout, and initiating the filling-discharge cycle. The device's installed cost can be recovered in about 1 year's operation, based on typical wage rates paid in the United States for packaging-line labor.

Auger or Screw-Type Bag Fillers   These fillers are usually applied to tuck-in-sleeve-type valve bags, for which production rates of one to two bags per minute and weight error limits of ±1 percent are required. Single-screw filling-spout designs (ordinarily of the net-weigh type) with simultaneous fill-and-weigh features are most common.

Gross-weight fillers need a feeding device such as a screw, vibrator, or belt, depending on the product. Particle size from 12.7-mm (½-in) pellets to 44-micrometer (325-mesh) powders can be handled, as can bulk densities ranging from 80 to 3200 kg/m³ (5 to 200 lb/ft³). Power requirements range from 373 W to 5.6 kW (0.5 to 7.5 hp). Weight accuracy is obtained by braking the motor to a rapid stop once the correct weight has been reached and the scale system has actuated the electrical or mechanical control system. Although fluidizing packers have diminished the importance of the screw type, the latter will always find application when space is a problem and investment must be low.

Centrifugal Belt-Type Packers   This packer is used to a limited extent for granular or pelleted products whose bulk densities range from 400 to 1600 kg/m³ (25 to 100 lb/ft³). Single-spout, simultaneous fill-and-weigh fillers, which consist basically of a short-belt conveyor, handle one to three bags per minute at weight accuracies within ±1 percent; the two-spout design is most common in high-speed conveyor-equipped installations, with which preweigh scales are used. Up to 30 bags per minute can be handled, with weight accuracy within ±0.1 percent or better.

Impeller-Type Fillers Used extensively for finely divided materials such as portland cement, plaster, lime, and talc, these fillers contain an impeller that turns in a casing (similar to a centrifugal pump) to move the product into the bag. Most impeller machines are installed with conveyors, although single-spout machines have been used when bag handling is done manually. Bulk densities are limited to 800 kg/m³ (50 lb/ft³) and higher. Portland-cement filling rates of up to thirty 43-kg (94-lb) bags per minute are possible with weight accuracies within about ±2 percent. Power requirements range from 3.7 to 7.4 kW (5 to 10 hp) per filling spout. Impeller fillers are being superseded by the fluidizing type because of the latter's better weight accuracy, cleanliness, and reduced investment and operating cost.

Gravity-Type Fillers These fillers are available in either the gross-weigher type or the net-weight type using a preweigh scale. Gross-weigher types are used in marginal operations for which investment must be limited and performance is not critical. Packing rates of 0.5 bag per minute and weight accuracies within ±5 percent are possible. Only free-flowing pellets and granules can be handled practically. The net-weight type utilizes a highly accurate preweigh scale which is placed 3 to 5 m (10 to 15 ft) over the bag-filling spout. Gravitational energy of the falling charge of product is used to force the product into the bag. Rates of up to six 25-kg (50-lb) bags per minute per scale-fill spout unit are possible. This type of equipment requires a free-flowing material and can handle the range of 250-micrometer through 4.8-mm (60-mesh through 4-mesh) pellets. Bulk densities as low as 400 kg/m³ (25 lb/ft³) can be handled.

Open-Mouth-Bag-Filling Equipment Two considerations in choosing this type of equipment (Table 21-26) and in deciding between open-mouth and valve-bag systems are the labor required for a given output and the capacity limitation of the closing system. With open-mouth bags, weighing and filling are usually done by a net-weight preweigh scale; gross weighers are sometimes used on low outputs. Operating principles and installation practice for automatic scales have been described earlier in this subsection.

Preweigh scales discharge to a chute system to which a bag is attached. The kinetic energy of the charge as it reaches the bottom permits the bag to stand without lateral support on a closing-machine conveyor. The filled bag is then dropped to a short-belt conveyor that passes the bag through a closing machine. Empty bags are held onto the chute system by hand or by a bag-clamp arrangement. These scales handle from 8 to 35 charges per minute. Weight accuracies are commensurate with product value and weight laws.

Bag Closures Conventional multiwall paper open-mouth bags are closed by sewing; the pinch-bottom type, by hot-melt adhesive. Three styles of sewn closure are used. The simplest and fastest consists of sewing with cotton or polyester thread, with needle and looper threads entwined in a chain-fashion stitch. This is adequate for low-cost products, for which sifting through the sewing is not objectionable. An improved method consists of adding a flat tape over the open mouth and sewing through it with the needle and looper threads. An additional thread, called filter cord, can be added between the needle thread and the tape to increase siftproofness, but this reduces closing rates.

Complete **siftproofness** can be had by the "tape-over-sewn" procedure, whereby the tape is glued onto the finished sewn closure by a device downstream from the sewing head. For siftproofness at high production rates, the pinch-style glued closure is used. The pinch-bag closure has the adhesive preapplied to the open end by the bagmaker. After the bag has been filled, the closing machine reactivates the adhesive by heat prior to sealing.

Polyethylene film bags are closed by heat-sealing together the face and back of the bag. The closing unit consists of a pair of belts that support the top of the bag and guide it through a heated section that fuses the face and back. This is followed by a cooling section.

Drum and Bulk-Box Filling This process consists of three operations: setting up, filling and weighing, and closing. Because setting up bulk boxes is cumbersome, a well-methodized workplace, equipped with work aids, is recommended. Weighing and filling can

be done manually or automatically. There is enough similarity between the two ways for manual systems to be mechanized later.

The most common installation consists of a conveyor line with a platform scale at a central location. This scale may be a simple dial type, which the operator watches to stop flow. The first mechanization step is to add a cutoff switch to the scale. Filling rates of 5 to 10 kg/s (10 to 20 lb/s), with weight accuracies within ±1 percent, are possible. Check weighing is easily accomplished by observing the net weight on the dial. A skilled worker can operate a manual system to within a few grams or ounces of the desired weight. Preweigh scales are occasionally used for free-flowing products, when the net weight is 100 kg (200 lb) or less or is a multiple of a weight that can be set on the scale and repeated to get the desired total weight. The main advantage of preweighing is higher accuracy.

Maintenance This is an important consideration in the operation of a packaging line. It is especially true with the advent of microprocessor control and the sophisticated devices for sensing of packaging-process variables. Two requirements for successful maintenance are needed. First is skilled technicians capable of handling the electronics as well as the mechanical parts of the packaging line. The community colleges are a source of skilled maintenance technicians. The second is the availability of repair parts. In the present industrial environment these are usually minimized and confined to those which are known to fail or which are prone to premature wear. Such parts are usually maintained in inventory. However, other parts are usually available on an overnight basis from equipment suppliers regardless of location. This is true whether the equipment is made outside the United States or the country of use. The overnight-shipping services such as Federal Express make minimum stocking a practical reality. With the use of microprocessors the trend is to have duplicates of all of the circuit boards in the plant inventory. Should there be a failure, the board can be replaced immediately and the old board returned to the maker for remanufacturing.

Small Packages These are of importance in the consumer chemical and the reagent chemical businesses and for shipping samples of products. Bulk solids, liquids, and gel-type products are involved. For bulk solids such as powders, granules, and pellets, the most often used packages are folding cartons, multiwall paper pockets (bags), or form-fill-seal plastic bags. For liquids, glass or plastic bottles and jars are used. Weight units are typical up to 20 lbs, or 10 kg. Volume units are typical up to 1 gallon or 4 liters. Where the product is a gel or a thixotrope, jars are used, with weight unit to 1 lb, or 0.5 kg.

Low-speed operations for these packages use in-line-type equipment in which transfer in, filling, and transfer out are sequential. Rates of production are typically 10 to 20 packages per minute. For high-speed packaging, the in-line equipment is usually too slow or requires a high investment for multiple lines. High-speed filling is typically 40 to 100 packages per minute, but it is not unusual to find certain products packaged at over 100 units per minute. To obtain such production rates, the packaging machinery industry has developed a carousel type of equipment wherein the filling operation is continuous. The package enters at one end of the carousel, is filled as it moves around the carousel, and is discharged having the required weight or volume of product. From that point it moves to a capping or closing machine. Figure 21-45 shows a carousel unit for liquids in bottles or jars. An example of an in-line liquid filler is shown in Fig. 21-46. Figure 21-47a shows a carousel-filling unit for bulk solids. Small-package operations for bulk solids involve two main procedures: filling and closing.

Weighing and filling may involve either preweighing scales, proportioners, or simultaneous fill-and-weigh. Preweigh scales are preferred when high weight accuracy is required. With appropriate package-handling equipment, these weighing devices can be used to fill cartons, bags, jars, or bottles. Preweigh scales of the multiple or "gang" arrangement are used to obtain both high accuracy and rates of production on dry products. One design uses 17 separate preweigh scales. Each weighs a charge equal to one quarter of the desired weight. A process-control computer monitors all weigh vessels and selects the four whose sum of weights is nearest to the desired weight and directs the weigh vessels to discharge to the waiting package. Accuracies of 1 gram in 1 kg are typical, and production rates of 150 weighings per

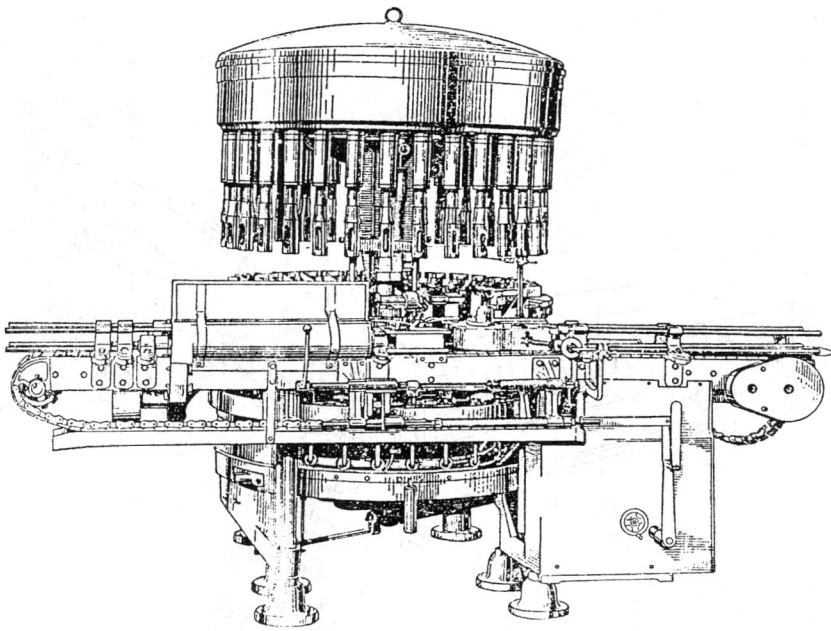

FIG. 21-45   HORIX 32 station carousel-type liquid filler for glass and plastic bottles and metal cans.

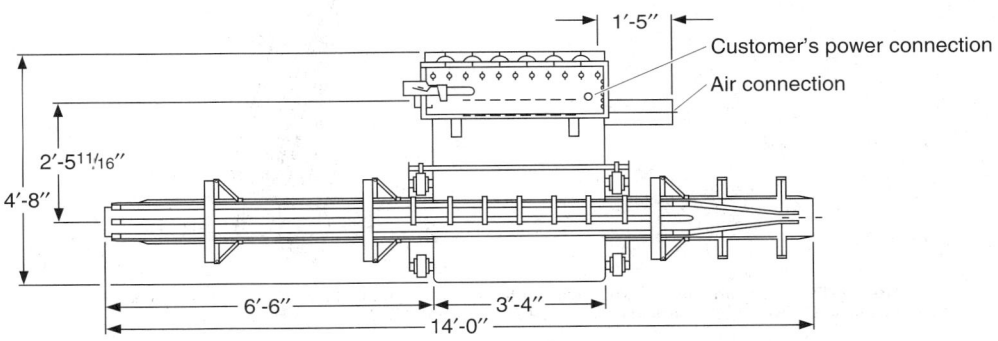

FIG. 21-46   In-line, 6-head, 2-lane filler for liquids in glass and plastic bottles and jars. (*Courtesy of National Controls, Inc., Baltimore, MD.*)

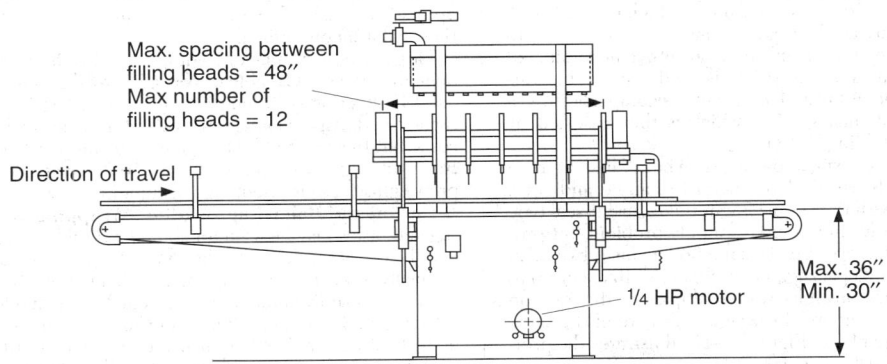

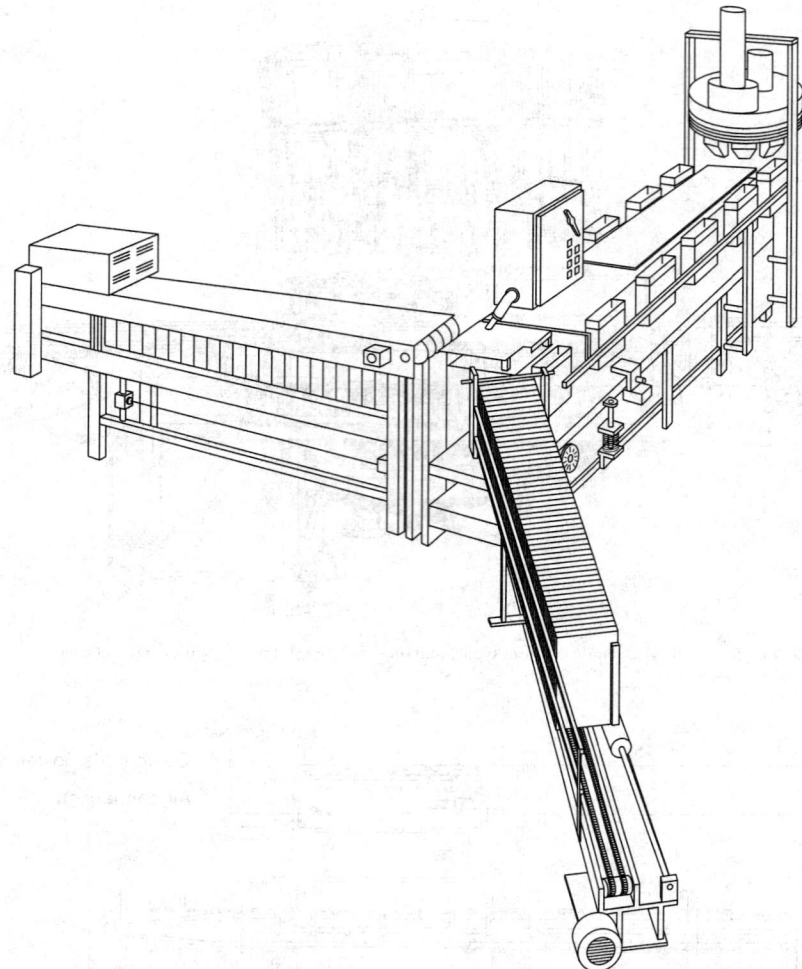

FIG. 21-47a Clybourn carousel-type folding carton set-filler-closer for bulk solids such as powders and granules. (*Courtesy of Clybourn Machine Co., A Division of PAXALL, Inc., Skokie, IL 60076.*)

minute are possible. Figure 21-47b illustrates a "gang"-type scale system. Figure 21-47c shows how such a weighing system can be integrated into a business record-keeping system.

Where folding cartons are to contain a product which is sensitive to moisture pickup, barriers are available which can be incorporated into the carton construction. An example of this is polyethylene laminated to paperboard. In some cases an insert bag is used instead, but this requires an additional operation and is usually accomplished by a form-fill-seal machine that makes a bag which is then inserted by another machine into the folding carton.

Form-Fill-Seal Premade small packages, when filled at high rates, present a problem because of the need to handle and store empty packages. To cope with this requirement the form-fill-seal type of packaging, which not only simplifies the supply problem but produces a superior package, has evolved. This method involves two main functions: a weigh cycle and a package make-fill cycle. Rates of up to 50 packages per minute are possible, with multiples of this rate on machines with two or more stations. Preweigh scales are of the same type used for small-package filling. Figure 21-47d illustrates the principles of a form-fill-seal (F/F/S) machine.

At present, form-fill-seal is limited to products having reasonably free-flowing particles with low dust concentrations. Because heat seal-

ing to form the pouch has been largely responsible for the success of this system, thermoplastic films or other plastics and papers with a thermoplastic coating are required. The choice of form-fill-seal versus premade packages depends on economics but usually applies to materials that are nonseasonal.

Large-size form-fill-seal equipment for industrial packages has been introduced. Capable of packaging 25-kg (50-lb) bags, such units use PE sheeting or tubing in roll form. A bag is made just prior to being filled. An advantage of this system is lower labor and material costs, but this is offset by the increased complexity of the equipment. Rates of eight to twenty 25-kg (50-lb) bags are possible. Since preweigh scales are used, high accuracy can be attained.

Carton and Baler-Bag Loading, Wrapping, and Sealing Corrugated boxes may be used for shipping flexible or rigid small containers; baler bags, for flexible ones. Corrugated boxes are loaded manually, semiautomatically (manual loading of the carton set up by machine), or fully automatically. Manual setup and loading are practical for up to 3 cases per minute, semiautomatic up to 10, and automatic up to 40. Associated with each are conveyors that bring packages to the carton loader and remove filled cartons from the sealer.

Carton sealing is carried out automatically by adhesives, tape, or staples or manually by tape or staples. Carton closer-sealers have

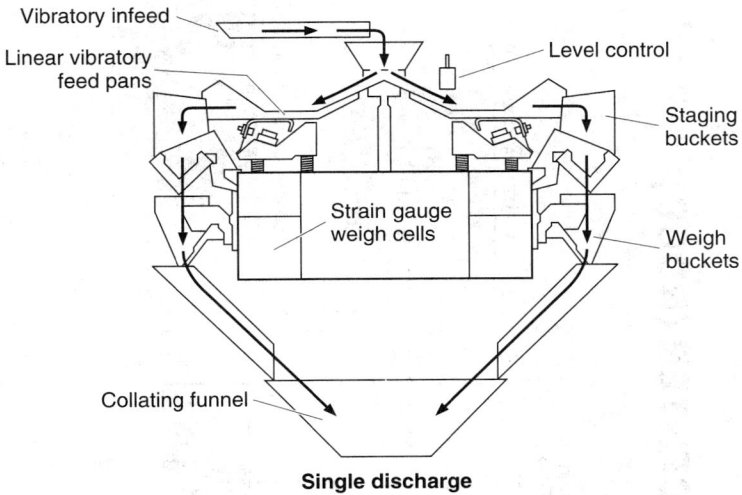

Single discharge

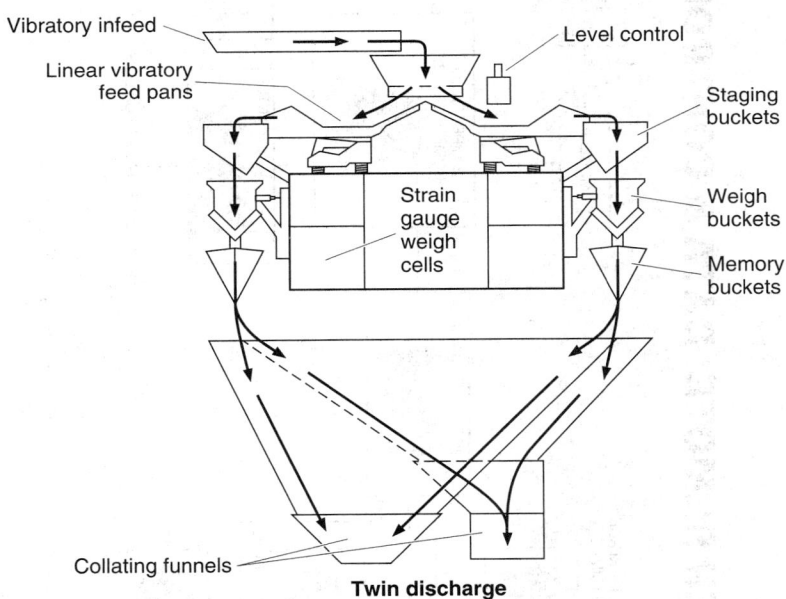

Twin discharge

FIG. 21-47b   Hayssen Yamato "Dataweigh" gang-type scale system. (*Courtesy of Hayssen Mfg. Co., Duncan, SC 29334.*)

become so attractively priced that small operations can justify their use even when the balance of the line is manually operated. Case closer-sealers that take different package sizes in random order are available.

Baler bags can be manually loaded, but the preferred practice involves specifically designed compression units. Their use permits making an integral load in which all package parts share the forces imposed by shipping. A manually loaded compression unit handles 2 to 4 balers per minute; semiautomatic units, with a mechanical package feed and a manual baler-bag application unit, can handle 15 to 20. Baler bags are automatically closed with tape or adhesive, the latter being preferred especially for automated operations.

Wrapping, Bundling, and Shrink Packaging   These techniques have limited applications for chemical products. Wrapping and

bundling are substitutes for cartons and baler bags, their advantage being that the package is made from roll stock.

Shrink packaging is a significant development. The most important application in the chemical industry is in unitizing packages for palletized shipment. A cover of shrinkable PE film serves to bind a pallet load and permit it to absorb considerably higher transportation forces than it would if packed by any other method. Palletizing, adhesives, and strapping are eliminated, which offsets the cost of the shrink wrap. But it is the reduced damage in shipment that makes shrink wrap so economically attractive.

The hand-applied shroud of shrinkable PE usually consists of a premade bag large enough to envelop the load. Equipment to shrink the wrap ranges from small propane-fired hand-held units, which take

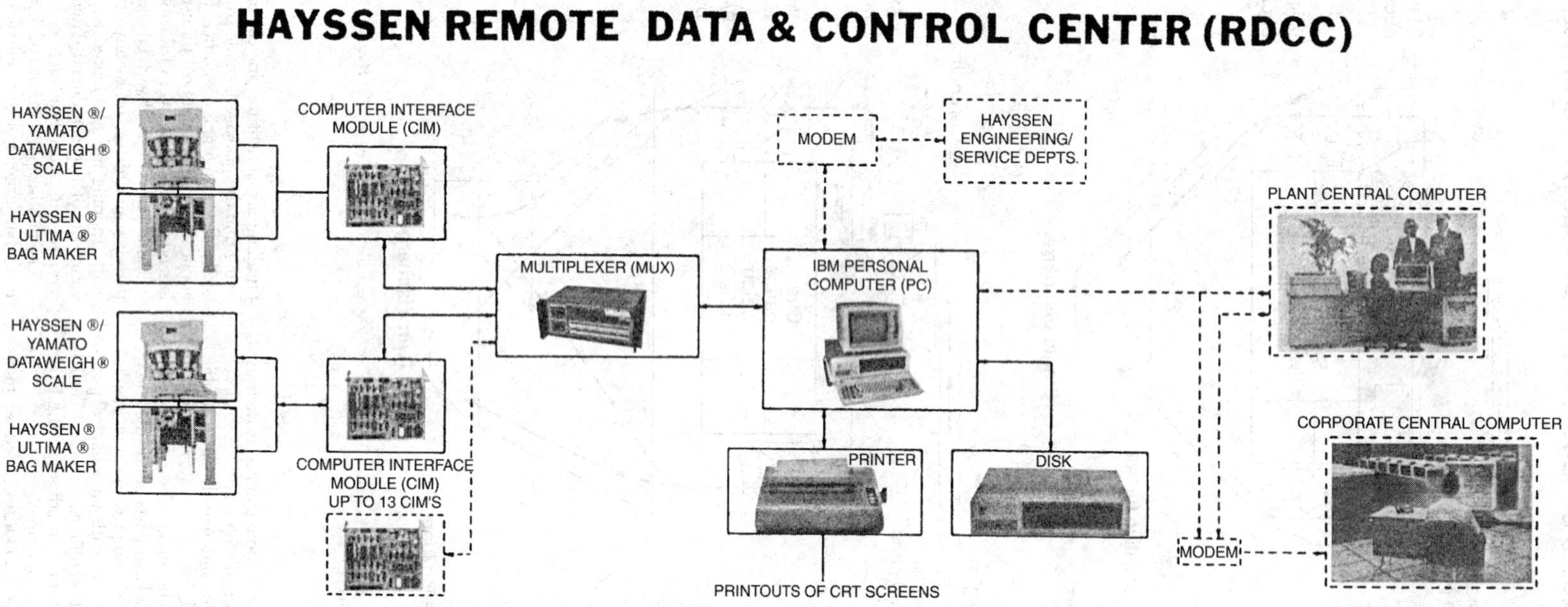

HAYSSEN REMOTE DATA & CONTROL CENTER (RDCC)

PLANT CENTRAL COMPUTER

CORPORATE CENTRAL COMPUTER

MODEM

HAYSSEN ENGINEERING/ SERVICE DEPTS.

MODEM

IBM PERSONAL COMPUTER (PC)

DISK

PRINTER

PRINTOUTS OF CRT SCREENS

MULTIPLEXER (MUX)

COMPUTER INTERFACE MODULE (CIM)

COMPUTER INTERFACE MODULE (CIM) UP TO 13 CIM'S

HAYSSEN ®/ YAMATO DATAWEIGH® SCALE

HAYSSEN® ULTIMA® BAG MAKER

HAYSSEN ®/ YAMATO DATAWEIGH® SCALE

HAYSSEN® ULTIMA® BAG MAKER

FIG. 21-47c Business data and report system using output of Hayssen Yamato "Dataweigh" scale as input to computers. *(Courtesy of Hayssen Mfg. Co., Duncan, SC 29334.)*

ULTIMA® II OPERATIONAL DRAWING

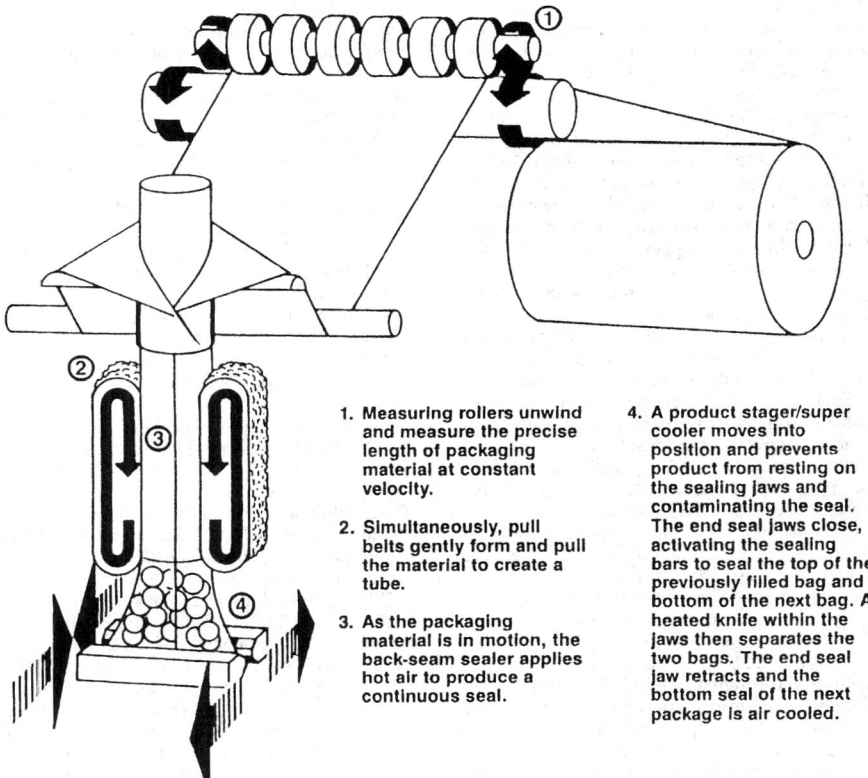

1. **Measuring rollers unwind and measure the precise length of packaging material at constant velocity.**

2. **Simultaneously, pull belts gently form and pull the material to create a tube.**

3. **As the packaging material is in motion, the back-seam sealer applies hot air to produce a continuous seal.**

4. **A product stager/super cooler moves into position and prevents product from resting on the sealing jaws and contaminating the seal. The end seal jaws close, activating the sealing bars to seal the top of the previously filled bag and bottom of the next bag. A heated knife within the jaws then separates the two bags. The end seal jaw retracts and the bottom seal of the next package is air cooled.**

FIG. 21-47d Hayssen patent no. 4,288,965 form/fill/seal bag-making system. (*Courtesy of Hayssen Mfg. Co., Duncan, SC 29334, 864-486-4000.*)

about 5 min to shrink a 1-ton load, to fully automatic conveyor lines handling up to 30 pallets per minute.

Stretch-Wrap Packaging   This type of packaging is an alternative to shrink packaging. It consists of wrapping a pallet load of product with a thermoplastic film which is applied under tension and which envelops the sides of the entire load. Special machinery accomplishes this procedure correctly. An advantage of stretch wrap is that it does not require the precise fit and shrinkage properties of a shrink-wrap bag to ensure tight binding of the load to the pallet. As long as the stretch film meets tensile and elongation properties and the stretch-wrap machine is in proper adjustment, a satisfactory result will be obtained. Material cost for stretch wrap is 75 to 100 percent of that of shrink wrap. An economic advantage is that stretch wrap is less energy-intensive than shrink wrap by a factor of 100 or more. Labor requirements are also less, in that one operator can start the stretch-wrap process by attaching the film to the pallet and the machine completes the operation automatically. With shrink-wrap, two operators are usually needed because of the unwieldiness of the shrink bag.

Batch Inclusion Packaging   This has been the result of the environmental consciousness of the chemical industry worldwide. The meaning of this term is that the product contained in the package and the package itself are added to a process and the packaging material actually becomes part of the product being made. By way of example, an antioxidant in a batch inclusion package is added to a polyolefin compounding operation. Not only is the antioxidant incorporated in the product but so is the package. The package material is chosen such that it is compatible with the end product and usually represents a minute fraction of that end product. A second advantage is that many

companies offering this type package will offer a specific weight requested by the customer, which simplifies their formulating operations. The development of strain gauge-microprocessor-controlled weighing equipment allows packagers to offer custom package weights at little additional costs over offering a single size. A marketing advantage can come from this type of customer service. A wide variety of film-type materials are available to packagers. These include polyethylene, polypropylene, nylon, polycarbonate, partially polymerized rubber, fluoro-chloro-ethylene polymers and others. Predictions have been made that in a matter of a few years most industrial chemicals will be packaged in batch inclusion packages. The reason is there is no costly disposal of empty packages which, depending on the contained material, may be subject to government disposal regulations.

Fine-Particle Packaging   This term is applied to powders and similar finely divided materials. These present significant flow problems in material handling and packaging. Surface chemistry, particle shape, and electrostatic forces have a great effect on flowability. Weighing and filling of fine powders in plastic packages is a difficult and sometimes impossible task because of these characteristics. While much work has been done in the characterization of fine particles the development of packaging machinery to handle such difficult products remains more art than science and engineering. A notable effort in packaging such materials is that by William J. Runo of Allentown, Pennsylvania, who has developed a process which uses compressive shock to orient particles and drive them into the package so that they occupy minimum volume and consequently have maximum possible density. This process is proprietary and is licensed by Dr. Runo. An example of the ability to increase bulk density is a product having an average particle size of eight microns and a loose bulk density of 8 pounds per cubic foot (0.128 grams per cc) which was increased to 24 pounds per cubic foot (0.385 grams per cc). The principle of operation of the Runo system is that it is independent of the package type and can be used on all types of packages used for chemical products—drums, bags, batch inclusion bags, bulk bags, and others—by having the appropriate holder for the package.

Chopped-Fiber Packaging This type of packaging is complicated by a geometry of the fiber. The ratio of fiber length to diameter can be several thousand with the result that it is impossible to obtain any degree of uniform flow of the fiber into packaging machinery, into the package itself, or when the fiber is being handled, as during incorporation into a product such as a composite. A proprietary process for accomplishing uniform feeding of chopped or milled fibers is described by U.S. Patents numbers 4,669,887 and 4,953,135. This technology can be licensed and apparatus purchased from Lee Technologies, Inc., Huntington, West Virginia.

Labeling This has become a very complex issue. Labeling requirements for the products of the chemical industry are regulated by the U.S. Department of Transportation, the Occupational Safety and Health Agency, the Environmental Protection Agency, and other government organizations. Customers may also require special labeling. Requirements mandated by government must be strictly adhered to. In many states, the right-to-know laws impose additional requirements. Products of the pharmaceutical industry have additional requirements imposed by the Food and Drug Administration (FDA).

While the foregoing refers principally to the information the label must contain, there is the further consideration of how the label is to be printed, the inks which are to be used, the method of adhering the label to the package, and the method of applying the label. Application methods can range from hand to automatic machinery applications. Influencing label production and strongly benefiting it is the advent of the personal computer. Labeling systems are available in which the information is entered through a terminal, it is displayed on a CRT, and when complete, the microprocessor directs a laser type or a dot matrix printer which prints the complete label which must be attached to the package. Figure 21-48 illustrates such a label. Another system actually sprays information onto the package as it passes by the labeler which is adjacent to the package conveyor. Figure 21-49 illustrates this type of noncontact system.

Label information may be divided into two classes, that fixed for each container and that which varies from package to package or from batch to batch. Examples of fixed information are name and address, net weight, name of product, and warnings about product hazards. Variable information includes batch, blend, or lot number, consecutive package number, coded information, and possibly date of manufacture. Export packages require outside-package dimensions and the gross, tare, and net weights in U.S. customary and SI units. When there is uncertainty as to label requirements, experienced legal advice should be sought. Fixed information is usually printed by the package maker. Variable information can be applied manually with rubber stamps or stencils or by automatic in-line marking equipment.

H4504LL/F

```
F1035557%M0363LL/R

SATO M-8400
4MB RAM
2MB ROM
5"/sec
5"MAX. WIDTH
203/152 DPI
150 CUSTOM FONTS
RS232C/RS422

SATO AMERICA, INC.
2761-A MARINE WAY
MOUNTAIN VIEW CA 94043
TEL : (415)964-5445
```

FIG. 21-48 Label example printed by a high density dot matrix printer. (*Courtesy of SATO AMERICA, Inc., 2761-A Marine Way, Mountain View, CA 94043.*)

Bar Coding Bar coding represents a major advance in being able to track identities and inventories of products on a real-time basis. The impetus for the development of bar coding was the Universal Product Code (UPC) developed under the auspices of the American grocery industry for the purpose of allowing automatic reading of product and price information at the checkout counters of supermarkets and mass merchandisers. It has since grown into a widely used system to track an inventory item or group of items. It has facilitated the use of automatic storage and retrieval systems (ASRS) described later in this section. There are currently five types of bar codes in use worldwide, illustrated in Fig. 21-50, (a)–(e).

Universal Product Code (UPC) See Fig. 21-50(a).
- Incorporates numeric characters only.
- Usually includes 12 digits and allows bidirectional scanning.
- Zero-suppressed version is printed using seven digits.
- Check digit is incorporated into code.
- Quiet zone is nine times the narrow bar width on both the left and the right.
- At 100 percent magnification, required size for a 12-digit UPC with the quiet zone is approximately 1.5″ horizontally and 1.0″ vertically.
- EAN variations are used in Europe.
- Common applications include retail, packaging, counting, and data processing.

When scanned, the UPC will be decoded as a 12-digit number. These 12 digits represent the following: digit 1 is the number system character; digits 2, 3, 4, 5, and 6 make up the manufacturer's ID number; digits 7, 8, 9, 10, and 11 are the vendor's item number(s); digit 12 is the check digit.

When scanned, the UPC zero-suppressed will be decoded as a 12-digit number. These 12 digits represent the following: digit 1 is the number system character, which is always zero when printing zero-suppressed UPCs; digits 2, 3, 4, 5, and 6 make up the manufacturer's ID number; digits 7, 8, 9, 10, and 11 are the vendor's item number(s); digit 12 is the check digit. However, only seven human-readable numbers appear when printing zero-suppressed UPCs.

Interleaved 2-of-5 (I 2 of 5) See Fig. 21-50(b).
- Incorporates numeric characters only.
- Can be of variable length, but must have an even number of characters.
- Common applications include warehousing, product/container identification, general industrial, and automotive.
- Often used in UPC Shipping Container Code formats.
- Quiet zone is ten times the width of the narrow bar.

The Interleaved 2-of-5 bar code is a bidirectional, continuous, self-checking numeric bar code. It uses a series of wide and narrow bars or spaces to represent each character, and each symbol employs unique Start and Stop elements.

The symbology requires an even number of characters to be interleaved together. The bars represent data characters occupying the odd positions, and the spaces represent characters in the even positions. Additionally, each data character must be composed of five elements, two wide and three narrow. Character pairing begins with the most significant digit (left-most digit) and continues two at a time until all characters are used. The Start element consist of two narrow bars while the Stop element combines a wide and narrow bar.

Code 128 See Fig. 21-50(c).
- Employs alphanumeric characters.
- Can be of variable length.
- Common applications include general industrial, inventory control, and retail container marking.
- Often used in UCC/EAN Serial Shipping Container Code formats.
- Quiet zone is ten times the width of the narrow bar.

This code has 128 characters. Like Code 39, Code 128 offers variable-length symbols. But at the same time, Code 128 is more compact.

Code 128 allows the user to encode any character found on a CRT keyboard, including the control characters. This gives the user more encoding versatility than previously possible in an industrial bar code.

Code 39 (3 of 9) See Fig. 21-50(d).
- Incorporates alphanumeric characters.
- Can be of variable length.

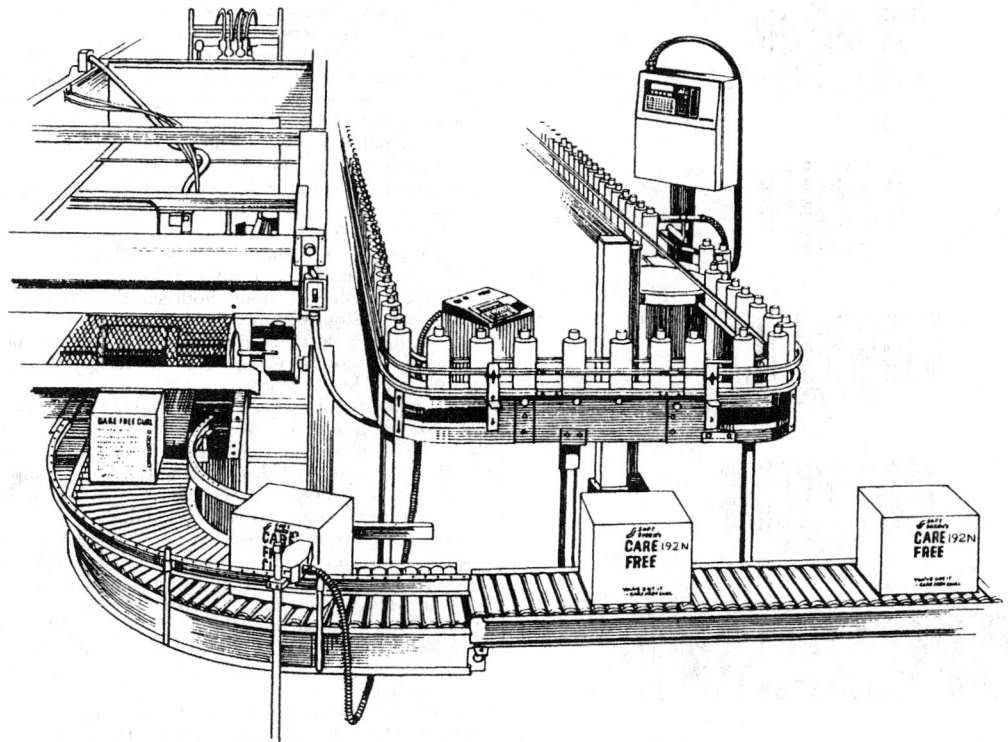

FIG. 21-49 Application of a noncontact ink jet printing system used for coding plastic bottles and corrugated fiberboard cartons. (*VIDEOJET™ shown, courtesy of VIDEOJET Systems International, Inc., 1500 Mittal Blvd., Wood Dale, IL 60191-1073.*)

• Check digit is optional but normally not used.
• Common applications include LOGMARS (Department of Defense), GSA, AIAG (automotive), general industrial, and HIBCC (health industry).
• Quiet zone is 10 times the width of the narrow bar.

The 3 of 9 bar code is a variable-length, bidirectional, discrete, self-checking, alphanumeric bar code. Its data character set contains 43 characters: 0–9, A–Z, -, ., $, /, +, %, and space. Three of the nine elements are wide and six are narrow. A common character (°) is used exclusively for both a Start and Stop character. The Start/Stop characters must be included in every bar code. It's the Start/Stop pattern that allows symbols to be scanned bidirectionally.

Code 39's flexibility to encode both text and numbers has contributed to its widespread use.

PDF417 See Fig. 21-50(e)

• Self-checking, two-dimensional bar code.
• Encodes up to 810,900 different character sets and/or interpretations, plus 256 international characters and binary data.
• Allows for bidirectional scanning.
• Symbology includes a Start/Stop pattern, left/right row indicators, and data codewords.
• Quiet zones are two times the X-dimension.

PDF417 is a multirow, continuous symbology capable of encoding large quantities of information. It's just what its name suggests—a Portable Data File.

Being one of the first two-dimensional bar codes, the symbology has not yet been standardized by any industry. However, it is being considered for coding shipping manifest information.

The symbology can vary in height and width because any number of rows of information (from 3–90) can be stacked vertically, plus a varying amount of data codewords (from 1–30) can make up the length.

Each PDF417 bar code also incorporates two parity-check codewords, which act as the symbol's error-correction code. The codewords carry out the same functions as check digits in other bar codes.

PDF417 is able to condense so much information into such a small space that it could soon prove to be one of the most flexible bar code symbologies around.

Where products are sold at retail to consumers it is necessary to have a Universal Product Code (UPC) printed on its label. The Universal Product Codes are assigned by the Uniform Code Counsel, 8163 Old Yankee Road, Suite J, Dayton, Ohio 45458. With the bar coding system, the information which is most meaningful to its user can be represented. This can include numeric and alpha (word) representations.

The economic incentive behind the development of the universal product code to allow for automated checkout, has resulted in the development of a whole industry infrastructure which supports bar coding. Automatic reading of bar codes can be done with fully automated systems with the reader as part of a conveyor line, where it reads the bar code from packages as they pass by on a conveyor. Simple systems consist of a handheld "wand" which a person points at the bar code and the code is read. Once read, the bar code analog is then processed by computer which then translates the bar code into the desired information. Such information can include product identity, package size, date of manufacture, manufacturing site, manufacturing process unit, and any special information that applies to the product, such as whether it needs specialized storage conditions, whether it is a hazardous material, and whether it has a shelf life. In Fig. 21-50, (e) is a schematic for a bar code system which shows both automatic and manual reading, the translation into useful information, the printing of reports, and the activation of sorting machinery. Two texts having detailed information are: *A Guide to Bar Coding* published by Bar-

FIG. 21-50 Bar code designs. (*Courtesy of Weber Marking Systems, Inc.*)

coding Systems, Inc., and *Handbook of Bar Coding Systems* by Harry E. Burke. Both are available from the Packaging Book-Store, Institute of Packaging Professional, Herndon, VA 22070. A very good summary of bar codes is the booklet "How to Stay on Top of Bar Codes," by Weber Marking Systems, Inc., 711 W. Algonquin Rd., Arlington Heights, IL 60005-4457, 708-364-8500.

Package Inspection   A more critical part of packaging operations is package inspection. In the past this was largely carried out by people, under the heading of "quality control." With the increases in output of typical packaging lines the inspection task becomes greater than the human person is able to accomplish. Several important electronic techniques have been developed which allow for graphic inspection of a number of packaging variables and the rapid rejection of those which do not meet an established standard, and with the passing of those which do. One notable variable is the accuracy of weight or fill volume. Automatic check-weight systems handle all of these. In the case of a product sold by volume a machine-vision system can determine whether the liquid level, in a bottle for example, is at the proper level. Labeling is another variable which has assumed critical importance, especially for products which are regulated by the U.S. DOT or the FDA. Again, machine-vision systems are able to scan each label to be sure that it is correctly applied and that the text is correct for the product being packaged. Metal detection in a product and/or package can be accomplished with several techniques with an x-ray able to detect particles as small as 0.01 mm at high line outputs. Leaking packages can be detected at high speed with helium leak detectors. Figure 21-51a shows the installation of machine-vision inspection systems on a typical integrated-packaging line packing tableted products in glass bottles. Figure 21-51b shows an X-ray inspection system designed specifically for packaging line use.

Packaging-Line Integration   The matching and balancing of all components in a given line so that the line will perform as designed is known as **packaging-line integration.** Computer simulation aids in this integration by allowing test cases of "what if" to run. By defining all conditions under which a line may be expected to function, simulation allows rapid determination of those conditions it can handle, those it cannot, and the quantitative degree to which it can. An example of an integrated line for packaging a tableted pharmaceutical is shown in Fig. 21-51c. This line includes inspection by machine vision at critical points and is computer directed and controlled.

Robotics   The introduction of **robotics** has given a new dimension to packaging in that it is now possible to do repetitive tasks with speed and accuracy at notably lower cost than if done by people. The manufacture of robots is well established with corporations of substantial resources providing a quality product with continuity of service, supply, and software support. There is also a specialty industry which is available to supply both accessory hardware and software which are custom designed to handle specific user situations. Economic analysis needs to be done before making the decision as to whether to automate using robots, fixed automation, or the labor of people aided by work aids.

There are two principal classes of robots. One type involves a **fixed position** for a central control and manipulator unit, illustrated by Fig. 21-52. This type of device is particularly useful where a repetitive motion is required, such as taking a package component from one position and then rapidly and accurately placing it in another position. The value of the robot increases when there are more than one downstream positions and when sorting must be done. The capability of this type of robot can be further expanded by having a manipulator at its pickup point which also can function in an X-Y-Z axis basis. This permits the device to perform relatively crude tasks such as picking up a component, orienting it, and then moving it to the desired place and precisely positioning it in the X, Y, or Z planes. The term **package components** can mean any part of the package itself or the product which is to be packaged.

A second type, generally regarded as being more versatile than the fixed-position robot, is the **gantry robot.** This device also offers capability of the X, Y, and Z directions. Programming is usually more simple for the gantry than for the fixed-position robot. The gantry robot can also use a manipulator at its pickup and discharge points. This often is as simple as a clamp or a device that has its own X-Y-Z degrees of freedom. Figure 21-53 gives an example of the gantry-type robot that is used for a wide variety of packaging activities, such as palletizing.

A robot often can be economically justified when the task of doing a certain packaging operation is analyzed in detail. For example, the palletizing of the fiber drums can be accomplished by human labor but work aids would be necessary in order to have acceptable production rates, reasonable operator fatigue, and a safe working environment. When the work aids are considered and their cost determined, the additional cost for providing robot capability is often of a small magnitude, which justifies its use to replace human labor.

Another reason for considering the use of robots is the availability of people who are willing to do the hard, manual, repetitive tasks which go with much of the packaging in the chemical industry. In the United States, the cost of such labor—if at all available—can often justify the use of robotic equipment. There are many examples where a single operator controls an entire production and packaging operation where robotics do all of the manual tasks. The robots are under the direction of their software. The operator is often a person who has at least an associate in science degree from a county college. Programming language used for robots is becoming more standardized. This allows robotic equipment to be reused many times after the original operation has been abandoned. It is not unusual to see either a gantry or a fixed-position robot reprogrammed and reequipped with new pickup members, doing an entirely different task than the one it was originally purchased for. One example is a system of a tabletop gantry-type robot used originally in an assembly operation for placing spots of adhesive at precise places on a matrix. After that project was completed the robot was reprogrammed and reused on a packaging line where it places large capsules in blister packs. The investment required for reuse of this robotic system was approximately 25 percent of the cost of a new system. There is much to recommend considering the use of robotics for the packaging activities of the chemical and pharmaceutical industries.

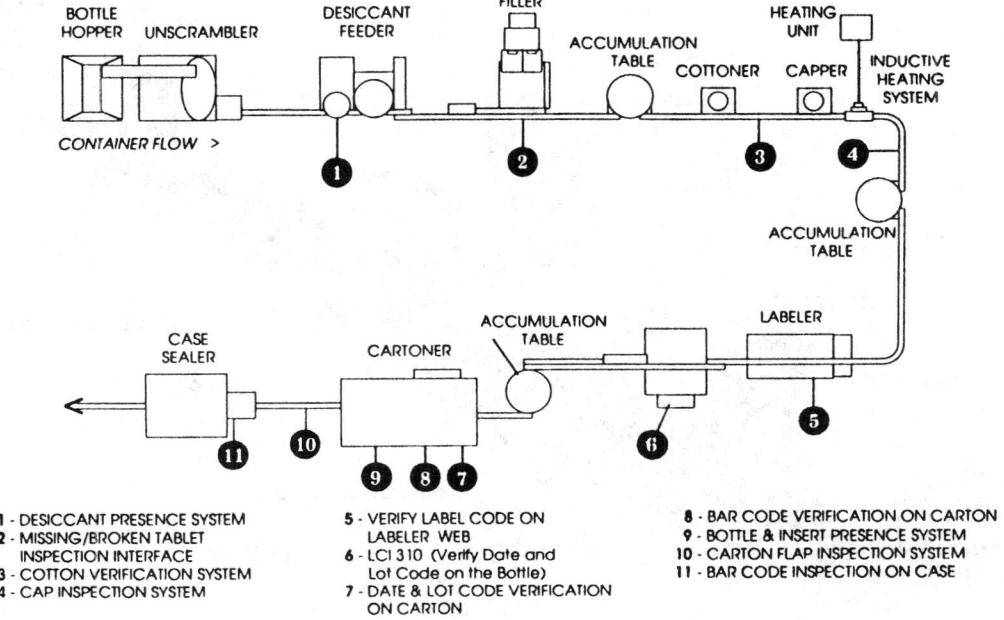

1 - DESICCANT PRESENCE SYSTEM
2 - MISSING/BROKEN TABLET
    INSPECTION INTERFACE
3 - COTTON VERIFICATION SYSTEM
4 - CAP INSPECTION SYSTEM

5 - VERIFY LABEL CODE ON
    LABELER WEB
6 - LCI 310 (Verify Date and
    Lot Code on the Bottle)
7 - DATE & LOT CODE VERIFICATION
    ON CARTON

8 - BAR CODE VERIFICATION ON CARTON
9 - BOTTLE & INSERT PRESENCE SYSTEM
10 - CARTON FLAP INSPECTION SYSTEM
11 - BAR CODE INSPECTION ON CASE

FIG. 21-51*a*   Inspection system for an integrated-packaging line packing tablets into glass bottles. Machine vision, bar code technology, and sensor technology are linked together by a supervisory system. (*Courtesy of AGR International, Inc., Butler, PA 16003.*)

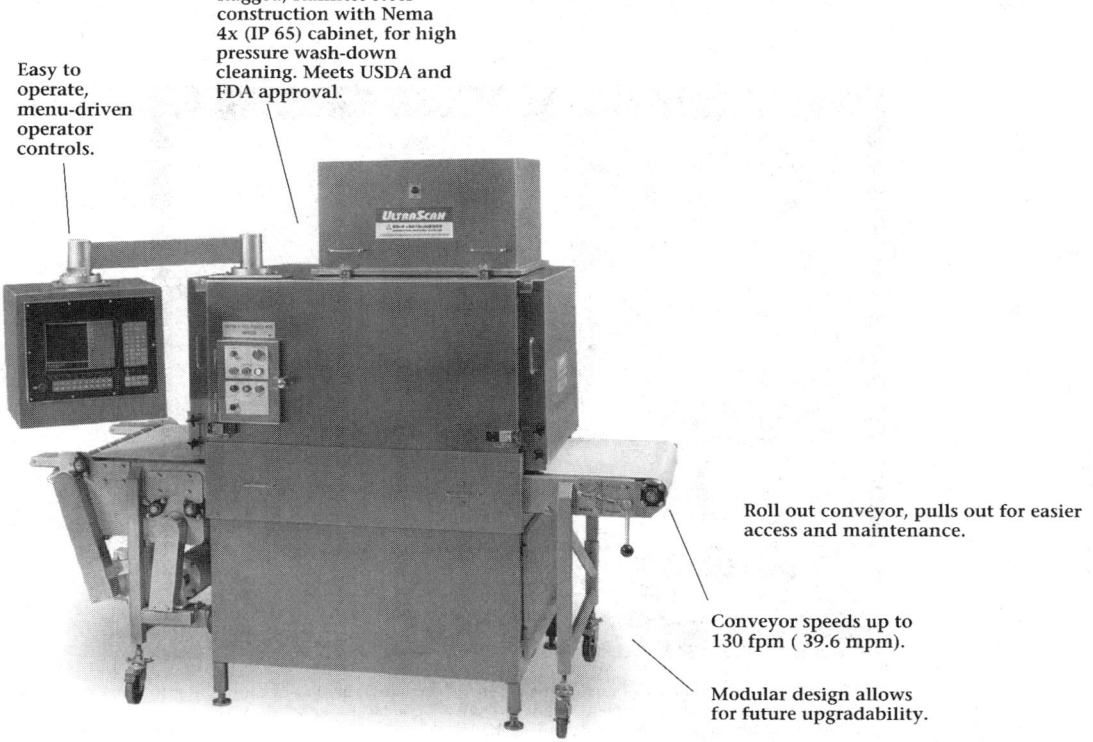

Easy to operate, menu-driven operator controls.

Rugged, stainless steel construction with Nema 4x (IP 65) cabinet, for high pressure wash-down cleaning. Meets USDA and FDA approval.

Roll out conveyor, pulls out for easier access and maintenance.

Conveyor speeds up to 130 fpm (39.6 mpm).

Modular design allows for future upgradability.

FIG. 21-51*b*   An X-ray system for detecting foreign matter in packages. (*Courtesy of EG&G Instruments, Inspection System Division, Oak Ridge, TN 37830.*)

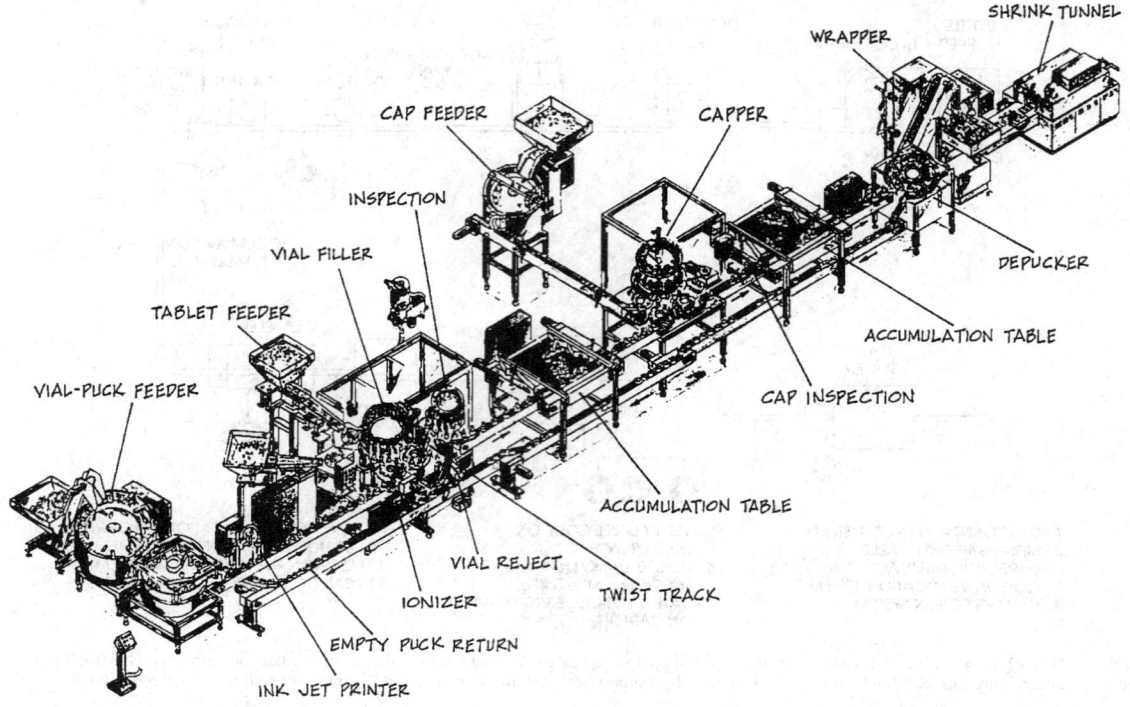

FIG. 21-51c An integrated-packaging line for a tableted product in glass vials (bottles). Machine vision inspection systems check key variables. (*Courtesy of Pharmaceutical & Medical Packaging News, Paoli, PA 19301.*)

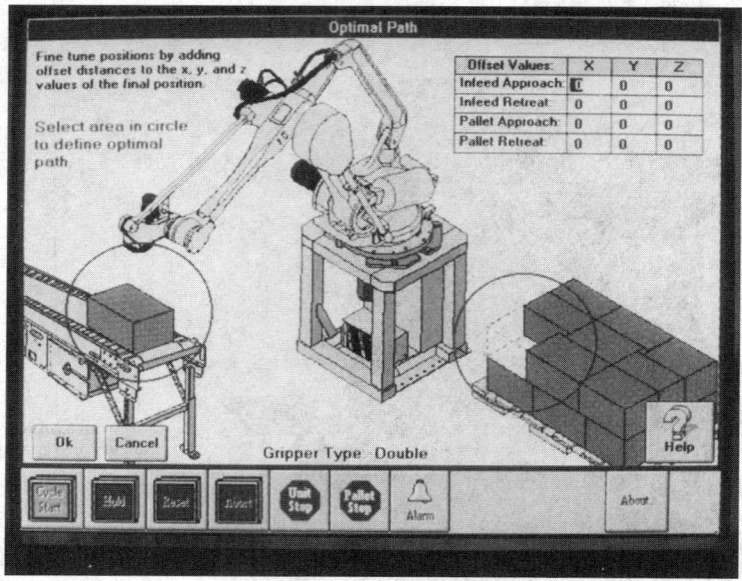

FIG. 21-52 Fixed-position robot used for palletizing of corrugated fiberboard cartons. Robot is integrated with a personal computer using WINDOWS™ graphical-user interface. (*Courtesy of FANUC Robotics North America, Auburn Hill, MI.*)

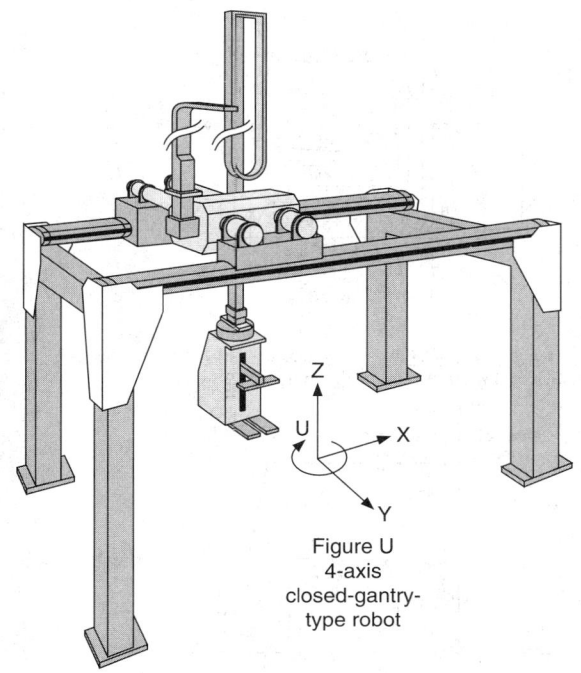

FIG. 21-53   Example of gantry-type robot having 4 axes which would be used for palletizing cartons or bags. (*Courtesy of C & D ROBOTICS, A Division of Ohmstead, Inc., Beaumont, TX 77701.*)

PACKAGE HANDLING AND STORAGE

Warehouse Requirements   Finished packages of the chemical industry are usually bags, drums, pails, or cartons (the last-named containing smaller units). Equipment for package handling and storage may be grouped into three main performance categories: (1) from packaging to pallet-unit loading, (2) from pallet-unit loading to storage or shipping, and (3) from storage to shipping.

The trend has been to the use of decentralized warehouses, with less and less finished product being stored at the producing plant. A typical plant inventory consists of 2 to 3 days of production, but at stocking points the inventory may amount to as much as 15 to 30 days. Although high inventory turnover is desirable, the variety of product grades and specifications often leads to longer storage times. Because of this, storage equipment and related conveyors are available to permit either a high velocity of product movement in and out of the plant warehouse or a virtually static one.

Mechanical handling of products in warehouses began with the forklift truck and pallet combination. Since warehouses had no moving equipment, pallet loads were set on the floor or placed on top of one another. Although this procedure is still practiced, loads now move into a storage-rack system, which permits storing pallet loads in vertical columns that make fuller use of the volume or height of the warehouse. Conveyors are also used to carry the pallet loads to storage, retrieve them, and send them to shipping. Forklift trucks are usually involved in all these movements.

Package-Handling Systems   The **control of package-handling systems** may depend on simple motor starters, on interlocked relays with photocell control, or on computers. Solid-state controls are finding much application in the last two systems.

A second type of control required is that of the package or pallet itself as it is handled by conveyors and other equipment. This handling may consist of right-angle transfers in a vertical lift or of a set of restrainers on the sides of a belt conveyor.

System Analysis   The choice of a specific handling system must take into consideration trade-offs that can be made among different types of equipment and between people-operated and -controlled equipment and automation. A disadvantage of automation is the high cost of specialized maintenance required, which can cost annually between 5 and 10 percent of the original equipment cost and sometimes more. New technical skills are also often necessary.

Factors that enter into any economic analysis of handling-warehousing systems are (1) expected mechanical and economic life of the system; (2) annual maintenance cost; (3) capital requirements and expected return on investment; (4) building-construction cost and land value; (5) detailed analysis of each work position (to determine trade-offs of labor and equipment; expected future costs and availability of labor are important); (6) relation of system control and personnel used in system (trade-offs of people versus mechanical control); (7) type of information system (computerized or manual); and (8) expected changed in product, container, unit pallet loads, and customer preferences during the life of the system.

Forklift Trucks   The backbone of most in-plant handling systems in the chemical industry is the forklift truck. Available in capacities ranging from 1 to 50 tons, the most commonly used are 1-, 1.5-, and 2-ton vehicles, with the 3-ton unit occasionally being used (Fig. 21-54). The trucks are usually powered by internal-combustion engines that consume liquefied petroleum gas (LPG) or by electricity by means of storage batteries.

With internal-combustion engines, automatic transmissions are frequently used; these are easily justified when vehicles must make many moves during the day. Smooth as is the control afforded by automatic transmissions, it is nevertheless inferior to that provided by electric trucks, especially those with solid-state controls. Gasoline and diesel power are also used, but mostly for outdoor equipment and very-heavy-duty units.

The lift-truck industry is competitive, with innovations being introduced frequently. Competent sales and service are available at low cost from most manufacturers or their dealers. Application sales engineering (a very worthwhile service) is generally supplied at no cost.

The many **options available** for lift trucks fall into two classes: vehicle specialties, which include controls, transmissions, guards, etc.; and accessories, which are devices that handle specific types of loads (Fig. 21-55). Included in this second category are high-lift masts, up to 7 m (24 ft); handling attachments for circular products, such as drums and roll goods; attachments such as carton clamps; and the fork side-to-side shifting mechanism.

Worthy of particular notice among accessories is the **side shifter** that is used to move trucks horizontally, about 100 mm (4 in) from side to side. The modest cost of this feature is returned in a few months' operation through reduced handling time, maintenance, and product damage. The driver first positions the truck approximately in front of where the load is to be set down and then makes the final horizontal adjustment by means of the side shifter. Without this mechanism, two or three maneuverings of the truck are necessary, with the load never quite being placed in the ideal spot. Correct positioning is important for pallet loads, which should be placed as tightly together as possible.

Lift trucks are available to meet a variety of **clearance restrictions.** Noteworthy is narrow-aisle equipment. Another accessory worthy of consideration is the multilift mast, which permits lifting loads over 3.7 m (12 ft). Of special importance in specifying any mast is that it will clear the various door openings it must enter, which includes those of trucks, railcars, and buildings. To meet most conditions, the collapsed height of the mast must be 2235 mm (88 in). An ideal lift truck for chemical-plant distribution warehouses would have 2000-kg (4000-lb) capacity; electric (battery) propulsion; solid-state controls; power steering; Trilift mast, up to 4.9 m (16 ft) [2235 mm (88 in) collapsed]; side shifter; operator guard; solid tires (except for outside use); and adjustable forks.

Exceptions to the preceding requirements would apply where explosionproof equipment is needed; building ceiling heights are such that the standard 3.7-m (12-ft) lift is all that will ever be needed; and loads will never exceed 1 to 1.5 tons. Safety requirements for lift trucks are mandated by OSHA, by NIOSH (National Institute of Occupational Safety and Health), by State Depts. of Labor, and often by individual company standards. Among these requirements are backup-movement signals, seat belts, overhead framework for pro-

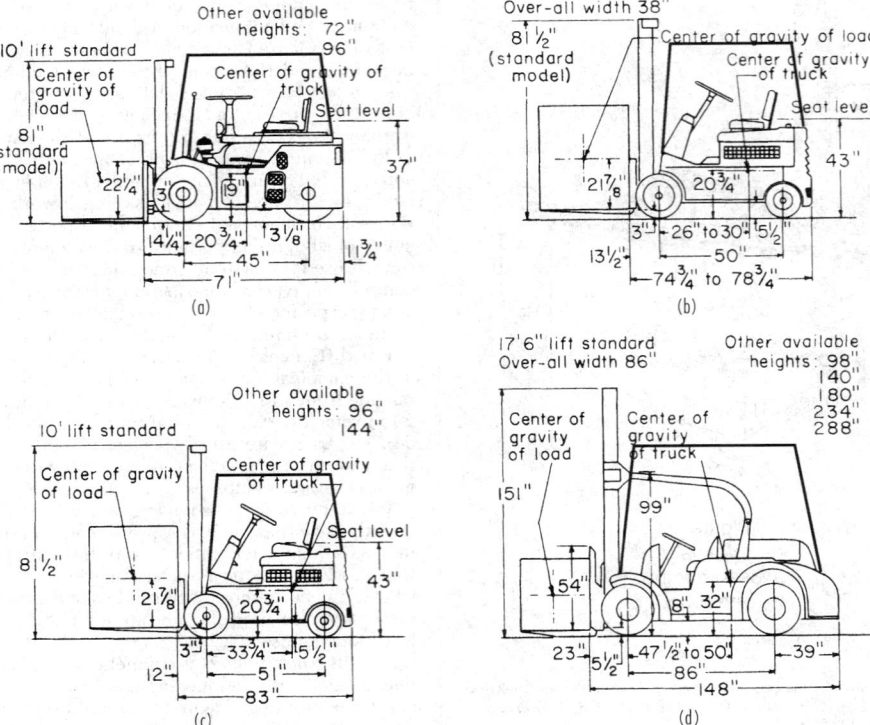

FIG. 21-54 Dimensions of representative forklift trucks. (*a*) 1000- to 2000-lb capacity. (*b*) 3000- to 4000-lb capacity. (*c*) 5000-lb capacity. (*d*) 10,000- to 12,000-lb capacity. Multiply pounds by 0.4536 to get kilograms; multiply inches by 0.0254 to get meters. (*Hyster Co.*)

tecting the operator from falling items, flashing lights, and two-way radio transmitters. Driver training is also mandated. A prominent training program is that of the National Safety Council. It includes video instruction, home workbooks, written and oral tests, and a demonstration of competency in operating an actual lift truck.

Capital investments in forklift equipment vary with specifications. Table 21-27 compares the cost of the electric-propulsion truck just described with an LPG-operated alternative. The operating cost is primarily for energy, with electric consumption being cheaper than liquid fuels.

Maintenance on gas trucks is also higher than with electric vehicles. About 5 percent annually of the initial cost applies to internal-combustion equipment, and about 2 percent annually to electric. A special feature on electric trucks with solid-state controls is the use of modules or circuit boards, which can be replaced as units and rebuilt at the factory. Typical maintenance costs for trucks operating five 8-h shifts per week are in the order of $3.15 per hour for gas vehicles and $1.78 per hour for electric ones. Under these conditions, energy costs are typically 9.3 cents per hour for gas trucks and 5.1 cents per hour for the electric units.

The **straddle truck,** designed for lifting bolsters with heavy loads or materials such as structural steel, is also finding application in handling van-type containers of packaged goods. For example, it can straddle a flatcar, pick off a van container, and deposit it directly on a truck-trailer rig. It can also be used for loading railroad flatcars and even oceangoing vessels. It is just one of many special pieces of mobile equipment available for special handling problems.

Slide Conveyors   Simple gravity slides and spiral chutes, while not technically conveyors, are widely used with conveyor systems or as separate units for lowering materials from one floor to another. They are low in cost and require little floor space if slopes are held at fairly steep angles. However, they must be used only after a careful study

of possible damage to containers from bumping either together or against the sides of the chutes or slides. Enclosed units are available for outside operation, and fire doors can be provided to meet requirements of local building codes. Multiple-blade chutes may be used for service to several floors, with separate inlet and outlet points. Blades may be lapped and riveted to eliminate the possibility of containers hanging up on exposed edges. Flight sections may also be flanged and bolted together.

Speed of containers sliding down a spiral may be controlled by the pitch of the spiral or by banking the outer or inner edge of the blade. Banking tends to throw the container to one side of the blade, thus varying its total travel distance. While usually fabricated of steel, blades may be specified in different materials, as required by specific applications.

Because of the steep pitch required, **slides** are limited in application. They are most commonly used to bridge the gap between roller-conveyor systems on two floors, because the roller conveyor can take the container off the slide rapidly and eliminate or reduce the chance for collisions. Slides may also be used when containers can be chuted from an upper floor to a manually loaded carrier. The use of several rollers at the feed point is recommended for easy delivery to the sloping section. If the drop is short and containers light, a roller cleanout will prevent backup of containers on the slide. The slope of gravity slides is a function of container weight, size, and friction characteristics and should be selected with care to be sure that containers do not move either too swiftly or not at all. Slides usually use flat steel sheet.

Gravity Wheel Conveyors   These can be used as pusher units set horizontally or inclined for gravity flow. They are highly standardized and are usually sold in 1.5- or 3-m (5- or 10-ft) sections; special lengths are available at extra charge. Since wheel conveyors give what is essentially "point" support to containers, it is generally recommended that at least six wheels be located under the load at all times.

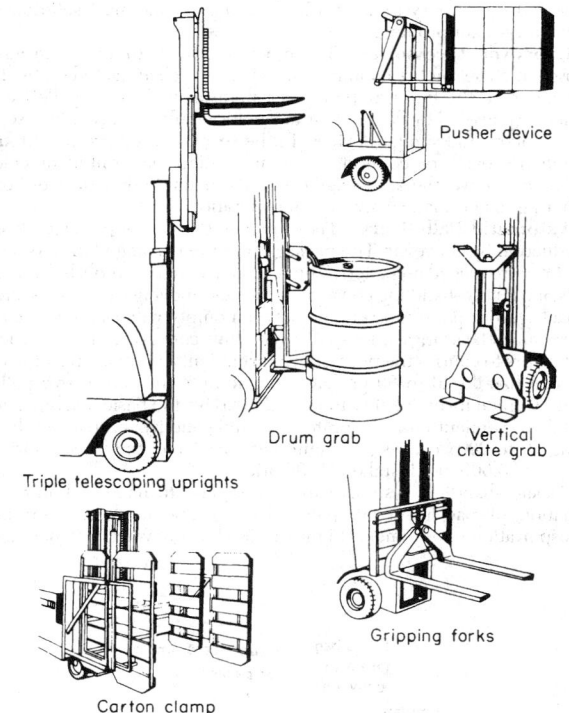

Pusher device

Triple telescoping uprights

Drum grab

Vertical crate grab

Gripping forks

Carton clamp

FIG. 21-55 Various types of fork-truck attachments.

Thus wheel arrangement is dictated by the smallest container that the line will handle. Only flat-bottomed containers can be handled on wheel conveyors, with the exception of fairly stiff-walled bags, which handle satisfactorily. This is due to the fact that the separate roller supports tend to pull the bag wall taut and flatten it out. Roller conveyors, on the contrary, tend to ripple the bag surface and prevent its movement. Wheel conveyors may also be specially designed for handling smooth-walled cylindrical shapes.

Wheels are available in a number of different designs, including variations in contour and material in contact with the container. Rubber or plastic tires are not uncommon. Through shafts may be used, with several wheels mounted on each shaft; stub bolts with a single wheel may be mounted to the side frame, or short shafts supported by bent bars may also be used. Wheel conveyors are generally used on lighter loads, and although manufacturers may offer widths up to 915 mm (36 in) or more, the smaller widths [up to about 457 mm (18 in)] are generally standard. Load ratings are generally given as the total uniform load which a standard section will support.

Since wheel units are relatively light, they have relatively low inertia, and loads may be started and stopped quite easily. In addition, wheel bearings are designed with loose tolerance to reduce starting friction. Metal plates or projecting hardwood slats are commonly used as stops on conveyor lines. Special hinged sections for passage of personnel through the conveyor line are available, and standard supports from floor or ceiling are recommended. Wheel-conveyor units are widely used for live storage, and special telescoping units are available for extension and retraction to meet variable conditions. Wheel conveyors are sometimes powered by a pressure belt or other methods but are most widely used as pusher or gravity lines. They are adaptable only for end discharge or side discharge by lifting, since the individual rollers tend to grip the container and prevent its sliding off the line at right angles to direction of travel.

Roller Conveyors Gravity rollers are considerably heavier than the wheels on wheel conveyors, and the weight is concentrated at a

greater distance from the shaft centerline. Hence, roller conveyors have a greater inertia; they are harder to start and harder to stop, require more slope than wheel units, and on long runs tend to speed up containers at an accelerating rate.

Spiral-roller units are usually equipped with tapered rollers to compensate for the difference in distance traveled by the inner and outer edges of the container. Tapered rollers are also used on curved sections of ordinary roller-conveyor lines.

Rollers are available in a wide **variety of constructions,** with tube ends either bored or formed to take the bearing insert. Bearings may be plain, with nylon rapidly becoming the most popular material for this type. Ball bearings are probably most common and are available with a variety of seals, or the bearing may be left unprotected. Lubrication fittings may be provided on a drilled shaft, or bearings may be prelubricated and sealed for life. Roller shafts are usually nonrotating and may be cut from hexagonal stock to fit a similar opening in the side frame, or they may be round with ends milled flat to prevent turning. Rollers may be mounted in side frames in a variety of ways, above the side frames when containers are to be slid off the line or below when there is danger of the containers falling off.

Gravity roller conveyors can handle containers with protruding edges, i.e., steel drums, which is one of their advantages over wheel conveyors. However, they are not generally suitable for bags since the sides tend to sag between supports and prevent forward motion.

As with gravity wheel conveyors, roller units are highly standardized and auxiliary equipment is available for supporting the line from ceiling or floor. Many special rollers are available for retarding containers if speed becomes too great for safe handling. Switches, brakes, hinged sections, spurs, and frogs are also available.

Roller conveyors are quite frequently **powered,** the simplest method being use of a pressure belt in contact with the lower surface of the rolls. A special ripple belt with raised pads is capable of starting up the load but does not build up excessive blocked pressure if the line fills up. Other similar drives are available, with varying degrees of control over the applied power. Most expensive of the powered roller units are those in which each roll is equipped with V-belt or chain drives. Pusher bars suspended from overhead chain conveyors may also be used to move containers along a roller line.

One of the most important control devices on roller-conveyor lines is the escapement mechanism which allows containers to be released from a line individually. Powered escapement mechanisms are commonly available on highly mechanized systems. Their main function is to space out the containers so that they can be handled as discrete units.

Flat-Belt Conveyors These powered conveyors can lift containers up inclines. With the aid of special belt surfacing, grades can be quite steep. Belts also keep containers spaced out in exactly the way in which they are placed on the conveyor. However, because of the relatively high friction containers cannot be slid off belts by pushing devices.

Belt-conveyor designs use both roller and slider bed supports for the flat belt. The variety of designs available allows proper selection of flat belts for heavy or light loads and for various applications such as carton filling or emptying.

TABLE 21-27 Initial Capital Investment Comparison between Liquefied Petroleum Gas and Electric Forklift Trucks of 2-Ton Capacity*

Item	Liquefied petroleum gas	Electric
Basic truck	$40,000	$46,000
Automatic transmission	$ 1,300	
Solid-state controls (standard)		
Trilift mast, to 4800 mm (189 in)	$ 4,500	$ 4,500
Side shifter	$ 4,500	$ 4,500
Power steering (standard)		
Solid tires (standard)		
Storage battery and charger		$16,000
Total	$50,300	$71,000

*Based on 1995 prices (U.S. dollars). Initial inventory of repair parts is not included in these prices.

Chain Conveyors   These devices for handling containers are available in either roller-chain designs or less costly types. There is a variety of **slat conveyors** that use both single and double strands of roller chain, as well as a slider type using cheaper chain. In general, slat chain conveyors are used only on loads which are too heavy for economical handling by belt, roller, or wheel units or which have odd shapes not suitable for roller or wheel units. They are particularly adaptable to pallet handling, as are simple open strands of chain with flat-surfaced attachments.

The most commonly used warehouse chain conveyor is the **tow chain.** Chain may be mounted overhead or in the floor, and trucks being towed can be designed for automatic detachment at a specific point. While the overhead chain is often used and is usually easy to support from structural members in the ceiling, the in-floor chain is probably most common. Automatic disengagement is possible should trucks encounter an obstruction or accidentally strike warehouse personnel. The two-chain conveyor is, of course, most economical when large tonnages are moved over a fixed path.

Chain-type **elevators,** such as arm and tray units, are commonly used for drums and barrels. Slight gravity runs at feed and discharge allow these units to roll on and off the conveyor easily and without special equipment.

Elevators   Cable-type elevators are usually selected for heavy loads such as full pallets or large containers. They can be made fully automatic and are able to serve many floor levels. The use of properly designed elevator systems is often the only economical solution to multistory-plant problems.

Conveyor Accessories   These may be divided into two groups, those which act on the container and those which are acted on by the container. In the first group are such items as deflectors, palletizers, pushers (powered by fluid, air, or mechanical linkage), upenders, sealers, staplers, and similar devices. In the second group are such items as electric eyes for counting or identification via printed or color codes, check weighers, mechanical counters, and other devices contributing to automatic conveyor-line operation.

Automatic Palletizers   These machines receive packages from production by conveyor. The packages are then arranged in tiers, and the tiers are placed on pallets. The mechanism to accomplish this consists of package-handling conveyors, package-moving stops, rams, etc.; a package-tier-pattern assembly plate; an empty-pallet-handling conveyor and elevator; a filled-pallet-handling conveyor; and electrical regulators to control the tier-pattern formation. Automatic pallet loaders can handle 40 to 80 packages per minute, or one to two pallet loads. Capital investment is about $225,000 for the basic machine, not installed. Semiautomatic operator-directed palletizers capable of handling 10 to 20 packages per minute are available; they cost approximately $95,000, not installed (1995 prices).

Package-handling systems can be designed to handle almost any situation of package type, packaging machine, and warehousing-transportation requirement. Figure 21-56 shows a typical bag-handling

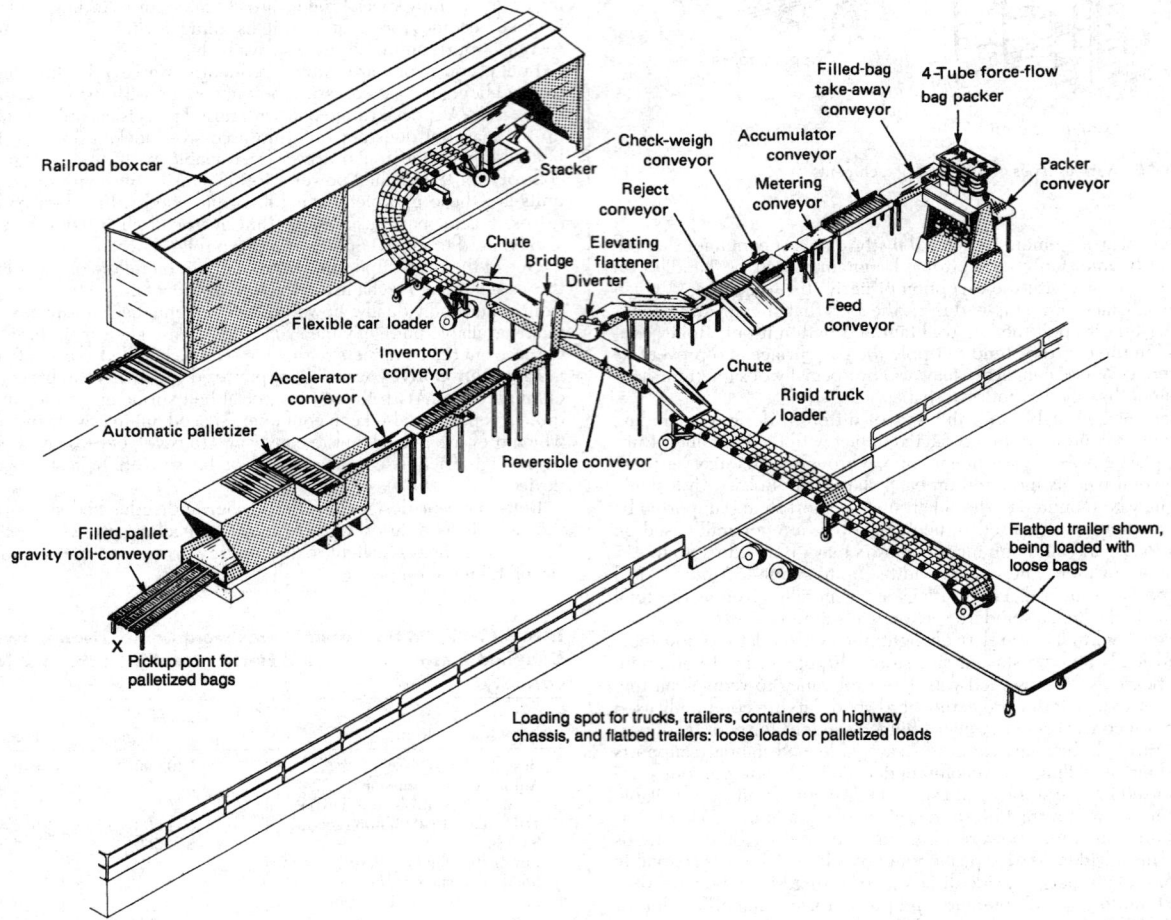

FIG. 21-56   Typical four-tube force-flow valve-bag packer with automatic palletizing and truck- and railcar-loading facilities. (*Courtesy of Stone Container Corp.*)

system in which both palletized loads and loose bags are handled. A system of this type can handle thirty 25-kg (50-lb) bags per minute. Figure 21-57 shows a system using the pinch-bottom-type open-mouth bag. Such a system can handle ten 25-kg bags per minute. Performance data for both examples can be found in Table 21-26.

Storage of Packaged Items   The inventory needed to support a given sales level is increasing in quantity as well as in the number of places where inventories are maintained to provide better service. Since a major portion of the chemical-process industries is located in urban centers where space is extremely valuable, efficient ways of storing packaged inventory have become very important. A similar situation exists when inventories are maintained at production plants. Here, space may be more readily available than in urban locations, but there is the question of whether to use the space for storage or for processing. These situations have led to the development of the storage-rack concept.

Storage racks permit storing pallet loads of packages vertically as well as horizontally. Most pallet loads can be tiered two or three pallets high, with one resting on top of another (provided the packages are able to withstand the weight of the pallets above). Because the racks bear the pallet weight, stacks six to eight and even more pallets high are possible. Forklift trucks and stacker cranes are used to place and remove the pallets.

From an inventory-turnover point of view, four major rack-storage systems are possible: drive-in, drive-through, flow, and aisle.

Drive-in racks, which are practical up to a height of 10 m (30 ft), are serviced by forklift trucks. The inventory system required is last-in–first-out (LIFO), which many consider inefficient. Capital invest-

ment (installed) for a 5000-pallet rack system is about $90.00 per stored pallet, lift truck not included. Drive-in racks make good use of floor space, having a higher ratio of storage to aisle space than aisle racks.

A typical drive-in rack consists of a steel structure to support palletized goods at the pallet edge, with the center of the pallet unsupported. The space between pallet support members is sufficient to permit a lift truck to drive in to place or retrieve a load. These racks are usually made to accommodate 12 pallets, which are positioned from the service aisle to the end of the rack. Because of the rack, each pallet position has the ability to hold 6 to 8 pallets vertically.

In operation, the lift truck takes the first pallet load and drives to the end of the rack to set down the pallet. With the second pallet, the truck enters the rack with the pallet elevated to permit clearing the support member. This procedure is repeated until the rack is filled. Lift-truck productivity is low because the driver must possess agility and skill to manipulate pallets extended on the truck.

Drive-through racks are similar to the drive-in kind, differing mainly in having lift-truck access at both ends. The main advantage of drive-through racks is that they allow a first-in–first-out (FIFO) type of inventory management. Capital investment, installed, is about $100 per pallet for a 5000-pallet rack structure. In operation, the rack is loaded in the same way as a drive-in rack. The unloading is different, in that removal of pallets begins at the opposite end from the loading point.

Flow racks are similar to drive-through racks in that they are loaded from one end and unloaded from the opposite end. However, the truck does not enter the rack. Rather, each lane in the rack is equipped with a conveyor (roller, wheel, or belt, depending on pallet characteristics) which both supports the pallet and transports it (by

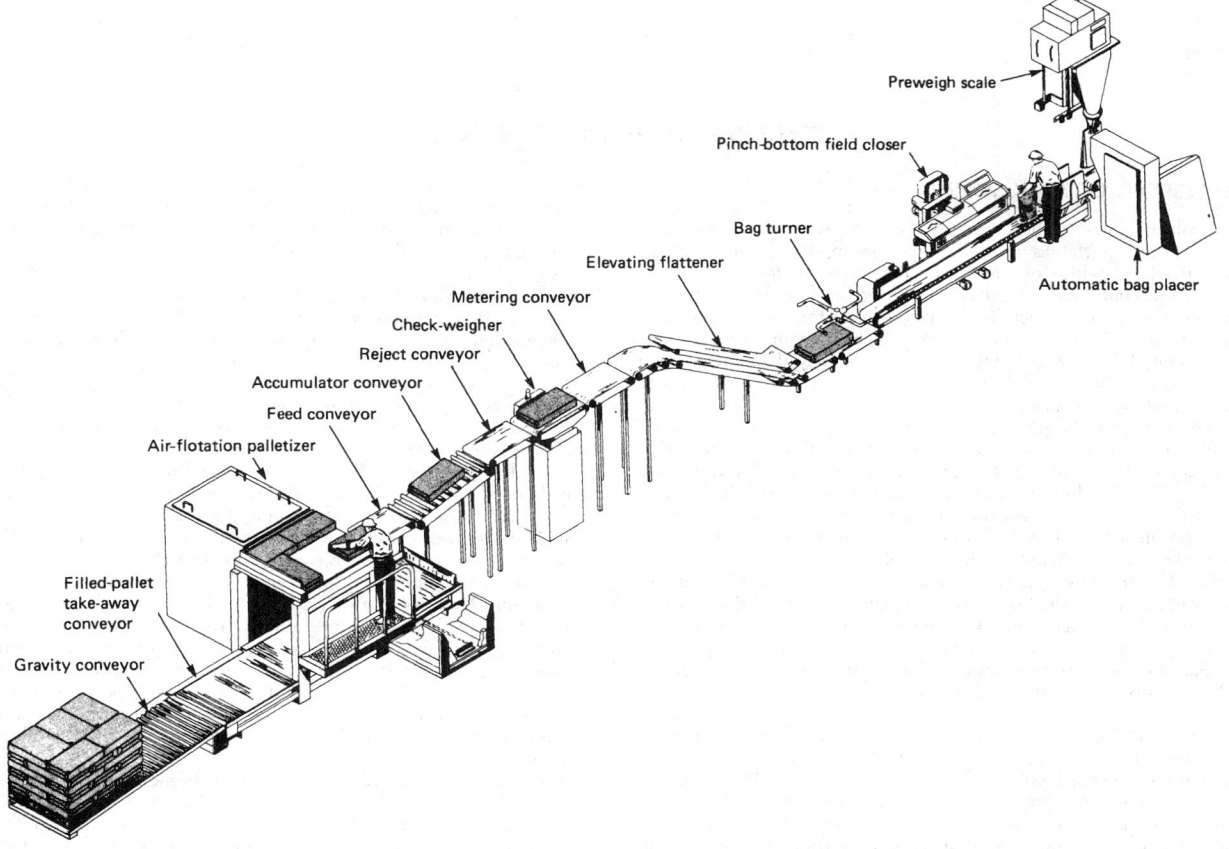

FIG. 21-57   Typical pinch-bottom system with automatic bag hanging and semiautomatic palletizing. Filled bags are handled on pallets only. (*Courtesy of Stone Container Corp.*)

gravity) from the entry point to the discharge end or to the nearest pallet. As a pallet is removed, the remaining ones flow to the removal point. This is a FIFO system of inventory.

The installed capital investment is about $375 per pallet for a 5000-pallet system. A characteristic of drive-in, drive-through, and flow racks is that, at any one point in time, only one product can occupy a given storage lane. Products are not mixed because of the complications that this practice presents in inventory management. In any event, there is seldom any need to mix products in the chemical industry because products are made in lots, blends, etc., and a storage lane is ordinarily designed to accommodate either a complete lot or some fraction of a lot. The result is that the total storage space available rarely is completely used. This is a problem that aisle racks overcome.

Aisle racks, used when there is a rapid turnover of inventory, permit a storage depth of only one or two pallet loads but offer the advantage of instant access to the stored item. Although this requires only a minimum of lift-truck time to store or retrieve a pallet, a high percentage of floor space must be devoted to aisles.

The inventory system needed is FIFO, which is desirable when inventories are subject to obsolescence or deterioration or when they consist of raw materials that fluctuate widely in value. The capital investment for a typical aisle-rack storage system having a 5000-pallet capacity (1 deep) is $65 per stored pallet (this does not include lift-truck investment).

Lift-truck operation is very simple and productive, even at 7-m (20-ft) elevations. Aisle racks can be as high as 30 m (100 ft), but above 7 m stacker cranes are favored over lift trucks because cranes allow servicing high storage at high rates. The installed investment in aisle racks, including a stacker crane, is about $500 per pallet.

Automatic Storage and Retrieval Systems   These systems are of increasing importance in the warehousing of chemical products whether they are for consumer or industrial markets. They are comprised of warehouse racks which can be several stories high. These racks are serviced by stacker cranes equipped for palletized handling. Each rack is divided into storage modules. Each module is capable of holding one pallet load of product and has an address which is stored in a process control computer memory. Stacker cranes are under the direction of the process-control computer. Through use of bar codes marked on packages and pallets, products entering the warehouse are identified by such variables as nomenclature, weight, package type, and any expiration date. The entering pallet load is identified by its bar code through use of a laser scanner which picks up the information from the bar code and transmits it to the process-control computer. The computer then directs the stacker crane to an empty storage module which is available and makes a record of the product and its storage location. The stacker crane places the pallet load in the module. The computer verifies this record before the stacker crane leaves the storage module. Stacker cranes often are equipped with an enclosure for a person who can ride the crane and inspect any storage module. This is also used where an operator does order picking, as in the case when single cartons of product are ordered. The term *paperless warehouse* is a very apt description of automatic storage and retrieval systems since they rely entirely on bar code, scanning, and computers, with a minimum of personnel. Pallets for such systems must be accurately made to required dimensions and have sufficient mechanical strength to withstand the repeated handling. Often, metal pallets are used to support wood pallets and their loads. The metal pallets never leave the warehouse, and assure trouble-free operation.

TRANSPORTATION OF SOLIDS

TRANSPORT OF BULK SOLIDS

Originally confined to the shipment of crude raw materials and fuels, the term "transportation of bulk solids" now applies also to manufactured products, which often become raw materials for other industries. In recent years, increasing tonnages of highly processed, finished chemical products have moved to customers in large bulk units. A useful definition of a bulk shipment is any unit greater than 2000 kg (4000 lb) or 2 m^3 (70 ft^3). The containers available range from small portable hoppers of 2-m^3 (70-ft^3) capacity to railroad cars of 255-m^3 (9000-ft^3) capacity.

The choice of shipping in package or bulk depends on market requirements and economics. Products from different sources that tend to have the same characteristics (appearance, quality, price) are usually offered in bulk form. Those tending to be specialties, while sometimes offered in small bulk units, usually are sold in packages. Many products are sold in both ways. A comparison of the costs of typical package and bulk units is given in Table 21-17.

Bulk Containers   These containers may be either open or closed. Generally, it is the effect of the weather on the product that governs the choice. High-value materials, such as certain ores, may be shipped in open containers, while relatively low-cost items, such as portland cement, require closed containers. Further influencing the choice of bulk containers is whether deliveries are made by truck, railroad, or water.

When customers maintain small inventories, **truck** delivery is often used, provided the location of the supply point is nearby, usually 550 km (300 mi) or less, and deliveries are frequent. If, however, a user maintains large inventories, deliveries are ordinarily made by **rail.** Other parameters influencing choice are transportation cost; operating costs of supplier loading facilities; customer receiving and unloading facilities; turnaround time for the container and the number of trips made per year (hence, investment write-off per trip); and container-operating cost, exclusive of transportation.

In planning for railroad-car loading or unloading facilities, many dimensional and weight factors must be dealt with. The common carriers that are to serve the facility are usually able to provide technical assistance as to clearances and weights to be handled.

An interesting new concept in planning for finished goods and bulk storage (when rail is used principally for customer delivery) is the use of hopper cars instead of fixed storage bins. Since products are eventually to be loaded into cars, there is much to be saved by avoiding double handling and capital investment. A systemwide analysis often will show this to be the least costly method, especially if there is a policy of minimum finished-goods inventory.

The most important bulk containers are railroad hopper cars, highway hopper trucks, portable bulk bins, van-type (ship) containers, barges, and ships. Factors determining the suitability of any of these containers (after establishing whether open or closed containers are to be used) depend on product physical properties, the most important of which are ease of flow, corrosiveness, and sensitivity to contamination.

Railroad Hopper Cars   Hopper cars follow three basic designs: (1) covered, with bottom unloading ports; (2) open, with bottom unloading ports; and (3) open, without unloading ports. Three types of unloading systems are used: gravity, pressure-differential, and fluidized. For the open-type car without unloading ports, clamshell buckets are often used. The car is loaded through ports located on the top of the car. Figure 21-58 shows a common type of covered hopper car.

Table 21-28 gives dimensions of hopper cars and other cars typically used in the chemical industry. Vacuum-pressure systems are used most frequently for unloading covered hopper cars. For certain free-flowing materials, in both covered and open-top hopper cars, shake-out devices are useful.

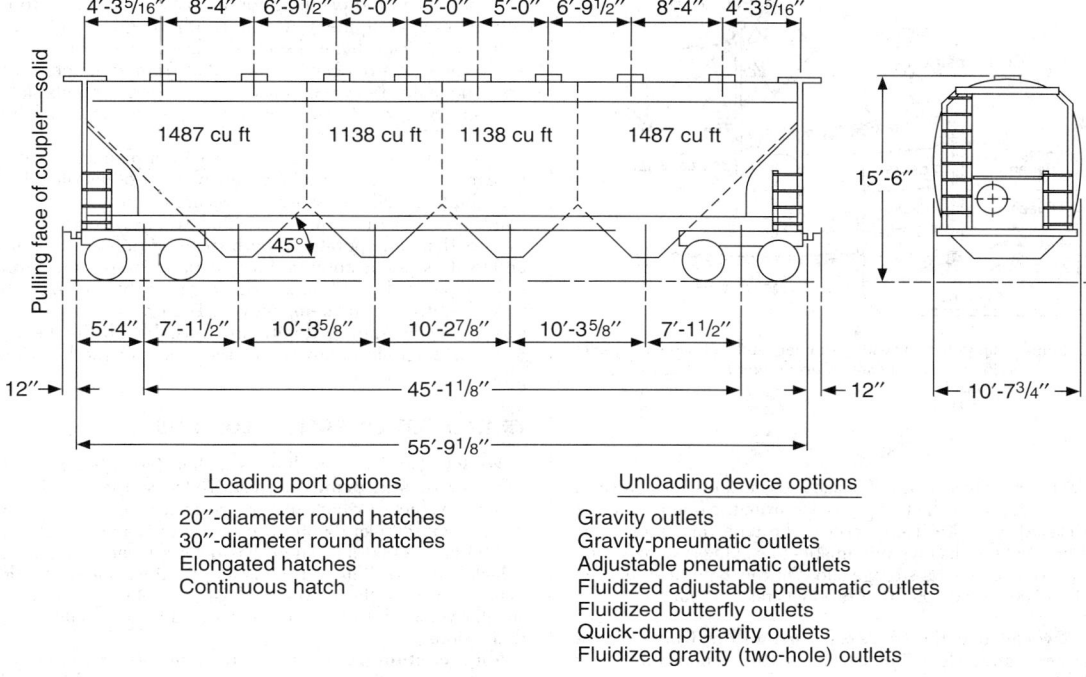

Loading port options
20″-diameter round hatches
30″-diameter round hatches
Elongated hatches
Continuous hatch

Unloading device options
Gravity outlets
Gravity-pneumatic outlets
Adjustable pneumatic outlets
Fluidized adjustable pneumatic outlets
Fluidized butterfly outlets
Quick-dump gravity outlets
Fluidized gravity (two-hole) outlets

FIG. 21-58 Typical railroad covered hopper car, 5250 ft³ (148.7 m³) ACF Centerflow designed for ladings having a bulk density of 42 lb/ft³ (672 kg/m³). Load limit 200,000 lb (90,720 kg). Unloaded weight with 20 in (508 mm) loading ports and gravity pneumatic unloading devices is 66,920 lb (33,355 kg). (*Courtesy of ACF Shippers Car Line Division.*) Change dimensions to inches and multiply by 25.4 to get mm. Multiply ft³ by .02832 to obtain m³.

Because of the railroad-car shortage that has persisted for many years, boxcars are often used for bulk materials. Lined with suitable materials to prevent contamination and with special bulkheads at each door, these cars are acceptable substitutes for covered hopper cars even though unloading is more difficult. Vacuum conveying wands are used to pick up the material, as are front-end-loader-type vehicles.

Loading of hopper cars and trucks can be done with most types of conveyors: air, belt, screw, etc. When an extremely full loading is required, centrifugal trimmers are frequently used. Available in a range of capacities, they can be engineered for any size of unit, up to a shiphold cargo (Fig. 21-59).

Track hoppers are needed for some boxcar and bottom-dump-car shipments. Since boxcars discharge to one side, fairly light construction can be used for the hoppers, which are located to one side of the tracks. However, for bottom-dump cars, the hoppers must be located on the centerline of the tracks. This requires heavy track girders over a hopper and feeder conveyor pit, but hopper depth must be set to give sufficient angle for material to flow well. Belts or reciprocating-plate feeders commonly carry the material to the bucket elevator.

TABLE 21-28   Typical Railroad-Car Dimensions and Capacities*

Type of car	AAR° class	Nominal inside dimensions,			Nominal outside dimensions,			Cargo (lading) volume, cu. ft. × 100	Cargo (lading) weight, lb. × 1000
		Length	Width	Height	Length	Width	Height		
ACF center-flow hopper car	LO				39 ft. 8 in.	10 ft. 8 in.	14 ft. 10 in.	29.7	207
ACF center-flow hopper car	LO				54 ft. 8 in.	10 ft. 9 in.	15 ft. 1 in.	47.0	200
ACF center-flow hopper car	LO				59 ft. 2 in.	10 ft. 9 in.	15 ft. 1 in.	52.5	200
GATX	LO				42 ft. 0 in.	10 ft. 8 in.	14 ft. 4 in.	26.0	140
Airslide hopper car	LO				54 ft. 6 in.	10 ft. 7 in.	14 ft. 6 in.	41.8	192
Hopper car	HT	42 ft. 10 in.	9 ft. 8 in.		43 ft. 10 in.	10 ft. 6 in.	10 ft. 8 in.	27.5	157
Gondola car	GB	41 ft. 6 in.	9 ft. 4 in.	2 ft. 5 in.	42 ft. 9 in.	10 ft. 2 in.	6 ft. 2 in.	9.6	100
Boxcar	XM	50 ft. 7 in.	9 ft. 6 in.	10 ft. 8 in.	55 ft. 2 in.	10 ft. 6 in.	15 ft. 9 in.	51.2	100
Boxcar with DF† equipment	XL	50 ft. 6 in.	9 ft. 5 in.	10 ft. 6 in.	57 ft. 7 in.	10 ft. 6 in.	14 ft. 10 in.	49.5	100

*From Association of American Railroads. Data are given for United States railroads, which do not use SI dimensions. To convert to SI dimensions (millimeters), change dimensions shown in inches and multiply by 25.4. To convert volume to cubic meters, multiply by 0.02832. For weight, multiply by 0.4536 to obtain kilograms.
†Damage-free.

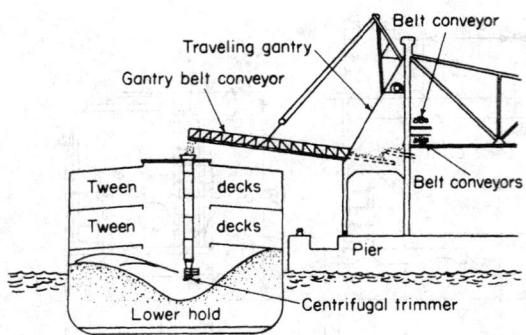

FIG. 21-59 Ship-loading system with trimmer and telescoping chute. (*Stephens-Adamson Division, Allis-Chalmers Corporation.*)

Hopper Trucks   These trucks are used to transport by highway a wide variety of materials. Vehicle types range from the open-dumping kind to the closed type. Most common is the type that unloads by pressure differential into its own pneumatic-conveying system, which is temporarily connected to a storage silo. On this type of truck, the unloading of 18,100 kg (40,000 lb) of products takes about 1 h, sometimes less.

The actual weight that the truck can carry in the United States depends on state-highway load limits, which in turn depend on the net vehicle weight and the number of axles on the truck (and tractor, when a trailer arrangement is used). The accepted maximum combined total weight of vehicle and cargo is 36,200 kg (80,000 lb). In some states, this is reduced slightly, while in others it is exceeded.

Of significance is the rapidly developing containerization system used for package cargo. SeaLand Corp. has developed a patented liner device which can be used to convert a cargo container into a bulk carrier. Figure 21-60a provides dimensions of typical bulk hopper-truck equipment.

Important in the planning for an installation that is to handle rail and highway equipment are the width, length, height, and turning radius of vehicles that will serve the facility. These dimensions can be easily obtained from carriers as well as from equipment manufacturers. Adequate clearances must be provided for railroad and other work crews. The clearances are often specified in state labor-practice codes.

Movement of railcars and trucks within plants is frequently done by carrier crews. Since, however, plant production schedules and availability of railroad switch crews are often not compatible, many plants provide their own switching service. Specially built prime movers that can operate on both roads or rails are available. Front-end loaders can be equipped with couplers to permit car movement. Cable-operated car pullers are now generally in disfavor because of lack of control of cars being moved. Trailers are often moved by tractors especially equipped with an adjustable "fifth-wheel" coupling, which will couple to any trailer regardless of the height of its coupling.

TRANSPORT OF PACKAGED ITEMS

Vehicle Choice   Small units such as **bags, boxes, cartons, carboys, cans, and drums** are usually transported in closed van-type highway vehicles, which may range from small pickup and delivery vehicles of 1400-kg (3000-lb) capacity to trailers capable of holding 23,600 kg (52,000 lb). There has been a trend to higher and wider vehicles, but loading and unloading facilities should be designed to handle not only the newest and largest vehicles but also the older, smaller versions. Figure 21-60b shows a typical trailer with principal dimensions.

Ship containers are now predominant for ocean transport of freight in specially designed containerships. They fall into three categories: package-freight containers, tank containers for liquids, and open containers for handling unwieldy items of large size, such as chemical-processing machinery, which are mounted on wood skids. The dominant container sizes are the 20, 40, and 35 foot length. Table 21-29 gives the principal inside dimensions for a variety of package-freight-type containers.

The use of closed railroad cars has declined somewhat in the handling of packaged chemical products in favor of trailers hauled pig-

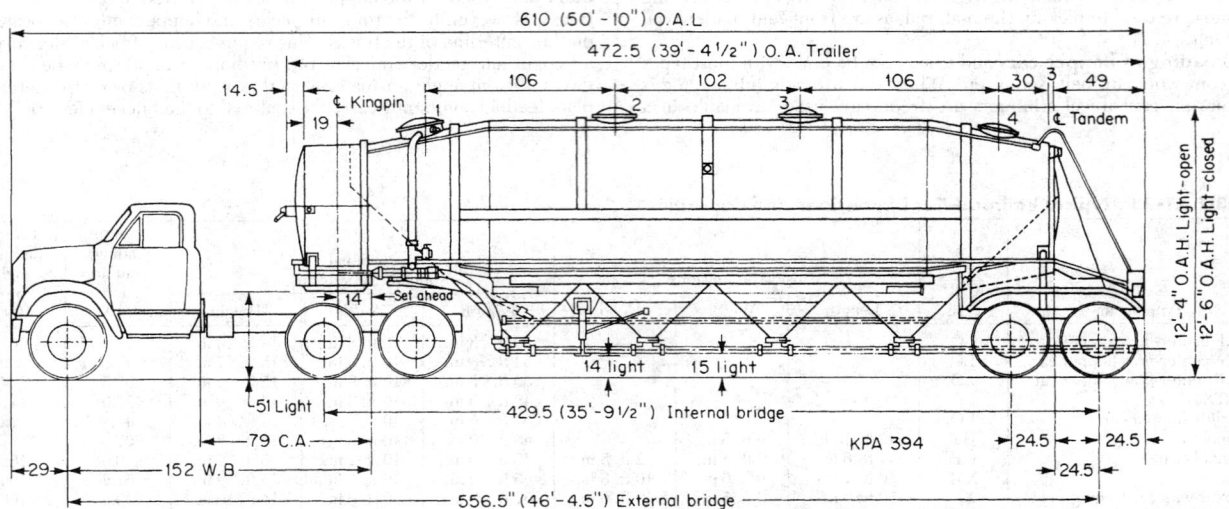

FIG. 21-60a Bulk hopper truck. Tractor trailer used for plastics. (*Butler Mfg. Co.*). To convert data to the SI system, change the dimensions shown to inches and multiply by 25.4. To convert volume to cubic meters, multiply cubic feet by 0.02832.

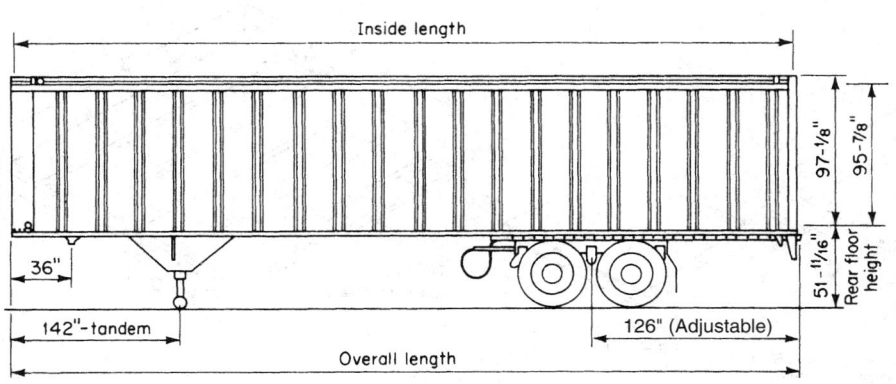
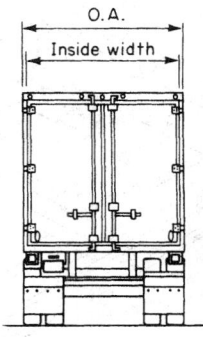

Typical inside dimensions for trailers used in North America.

Width: 92½ in, 2350 mm
Height: 96 in, 2438 mm min; 111 in, 2820 mm
Lengths: 28 ft, 8534 mm
         45 ft, 13716 mm
         55 ft, 16764 mm

FIG. 21-60b Typical inside dimensions for trailers used in North America. Dimensions shown are for an inside height of 96 in. (2438 mm).

gyback fashion by the railroads. This trailer-on-flatcar approach combines the convenience and flexibility of trucks with the low cost and high speed offered by the railroads. Covered railroad cars for hauling packaged freight include not only standard boxcars but much special equipment offering heating, insulation, refrigeration, high volume for low-density products, and special protection for fragile items. Table 21-28 shows principal dimensions for some of this equipment. Of special note is the so-called damage-free (DF) equipment that provides bulkheads with the car. These form modules within the car to keep the freight from shifting during car movement, thus reducing damage.

Pallets These portable platforms, on which packaged materials can be handled and stored (Fig. 21-61 shows several designs), can be

had in a variety of standard sizes and in almost any custom-made size. The dimensions, however, tend to be set by the transportation vehicle in which they will move. The older and most common 2235-mm (88-in) truck width and the 2743-mm (108-in) boxcar width have resulted in a "standard" pallet size of 1065 by 1220 mm (42 by 48 in), which fits two across in a truck (the 1065-mm side) and two across in a boxcar (the 1220-mm side), with adequate clearance for maneuvering the lift truck handling them.

There are several variations of this basic size, including the well-used Grocery Manufacturers of America size of 1220 by 1015 mm (48 by 40 in). The choice of the exact size depends on the truck and boxcar width normally available, the size of the package load, and the customer's receiving and handling facilities. Ideally, the sum of the

TABLE 21-29 Typical-Ship-Container Data*

Type	Length		Width		Height		Volume		Capacity			
									Maximum load†		Tare weight	
	ft	mm	ft	mm	ft	mm	ft³	m³	lb	kg	lb	kg
20-ft standard‡												
Out	20	6,096	8	2438	8	2438	1,123	31.8	52,913	24,000	4410	2000
In	19.479	5,935	7.771	2370	7.406	2258						
20-ft high§												
Out	20	6,096	8	2438	8.5	2591	1,197	33.9	52,913	24,000	4585	2080
In	19.479	5,935	7.771	2370	7.813	2383						
40-ft standard												
Out	40	12,192	8	2438	8.5	2591	2,430	68.8	59,500	26,990	7700	3490
In	39.594	12,069	7.781	2373	7.896	2405						
40-ft high												
Out	40	12,192	8	2438	9.5	2895	2,684	76.0	60,400	27,400	6800	3080
In	39.563	12,059	7.688	2344	8.823	2689						

*From Hapag Lloyd.
†This is the maximum load that the container will safely carry. The actual load will usually be less because of road weight limits, which vary from country to country and among states and other political subdivisions. In planning, these limits need to be determined with governmental authorities.
‡Liquid tank containers having these outside dimensions and holding 5055 U.S. gal (19,140 L) are available.
§This type of container is also available for dry bulk cargo.

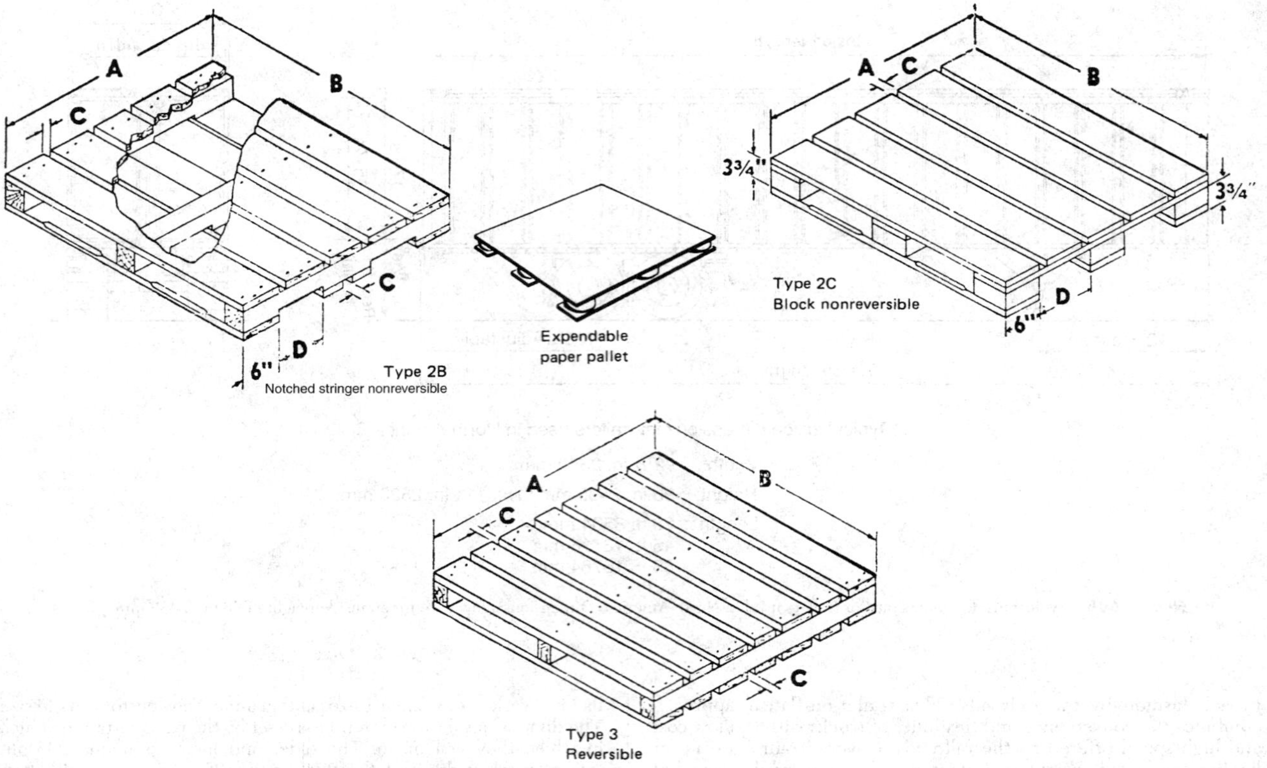

FIG. 21-61 Types of pallets. Designations are standard designs based on nomenclature of the National Wooden Pallet and Container Association, Washington, D.C. Types 2B and 2C are used for bags and corrugated cartons. Type 3 is used for drums and pails.

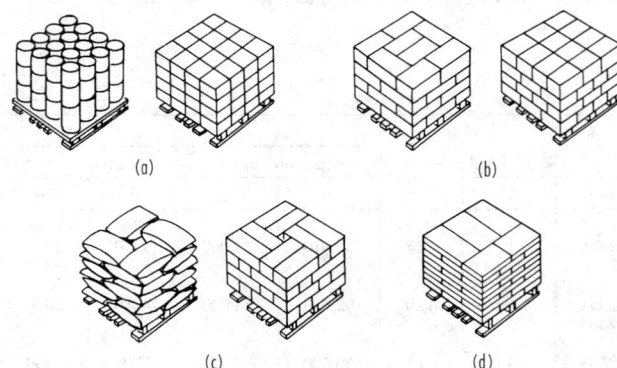

FIG. 21-62 Typical pallet patterns. (*a*) Block pattern is commonly used, although it is often unstable. It may be made more secure by encircling the top tier of containers with wire or strapping. (*b*) Brick pattern is the most commonly used. Containers are interlocked to make a relatively stable load by placing alternate tiers at a 90° position to each other. (*c*) Pinwheel pattern is used when the brick pattern is found to be unstable. Alternate tiers can be interlocked. (*d*) Three-to-two interlocking pattern is used extensively for bagged products. All pallet loads benefit from load-securing systems such as stretch wrap, shrink wrap, or palletizing adhesive.

package dimensions should exactly fit the pallet, but in practice this is virtually impossible. The following rules of thumb are helpful:

For bags: exact pallet dimensions, or up to 13-mm (½-in) overhang on each side

For cartons: pallet dimensions or underhang by 13 mm on each dimension

For drums, cylinders, etc.: pallet dimensions or underhang by as much as 25 mm (1 in)

Pallet patterns to achieve these conditions are numerous. Figure 21-62 shows common patterns used in the chemical-process industries.

The traditional material for pallet construction has been hardwood such as oak, ash, and maple. Yellow pine is also often used. Nails and adhesives are used to join component pieces.

The growing shortage of hardwood has increased the cost of wooden pallets to a point at which plastic pallets and composites of wood, paper, and plastics are economically feasible. Much development work is being done on plastic-pallet design to handle typical loadings. Because of the cost of disposing of expendable pallets, returnable ones are often justified.

Blocking and Bracing of Packaged Loads All transportation vehicles impart significant forces to the packages they contain. Forces of up to 2 G are regularly encountered in rail and ocean shipments, and of up to 1 G in trucks. Some forces are caused by vibration of the vehicle in the vertical plane; vibration frequencies are in the 20- to 40-Hz range. Longitudinal forces caused by starting or braking are of similar magnitude under normal conditions, but

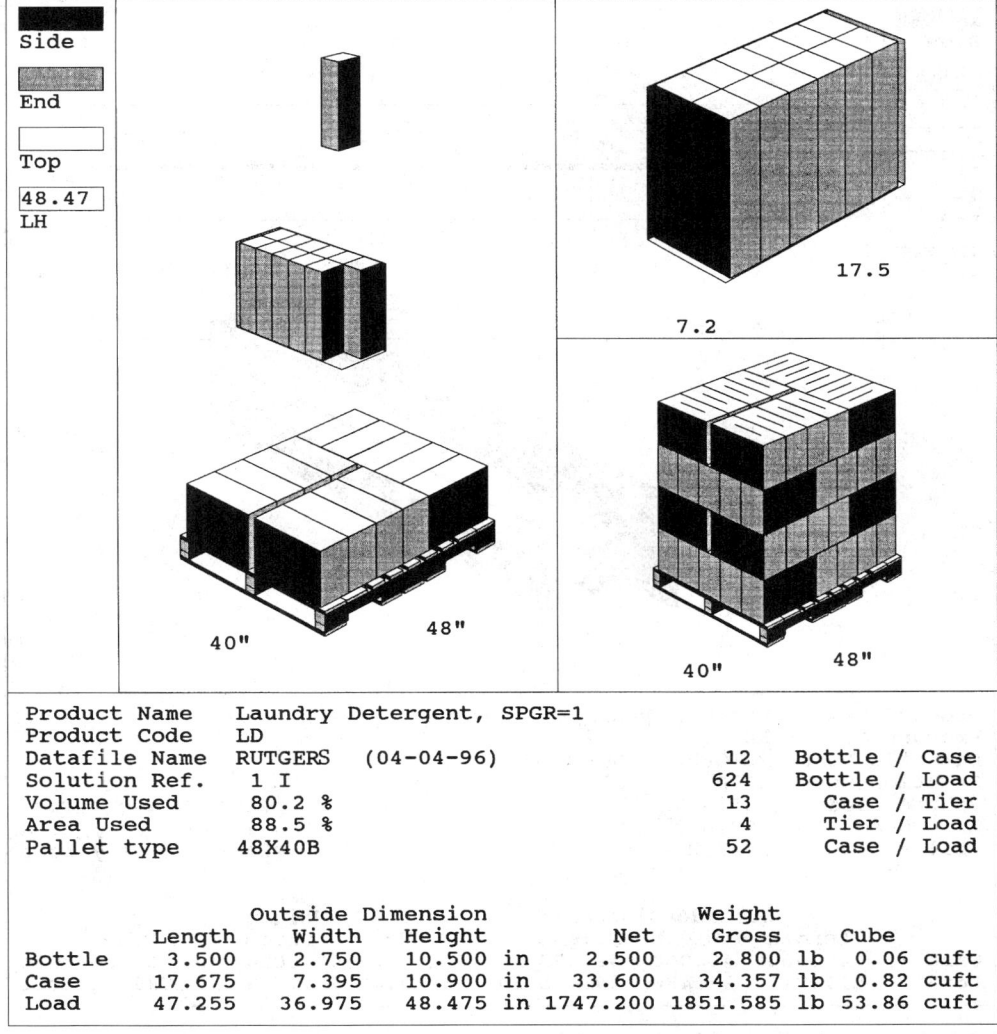

Product Name	Laundry Detergent, SPGR=1					
Product Code	LD					
Datafile Name	RUTGERS (04-04-96)			12	Bottle / Case	
Solution Ref.	1 I			624	Bottle / Load	
Volume Used	80.2 %			13	Case / Tier	
Area Used	88.5 %			4	Tier / Load	
Pallet type	48X40B			52	Case / Load	

		Outside Dimension			Weight		
	Length	Width	Height	Net	Gross	Cube	
Bottle	3.500	2.750	10.500 in	2.500	2.800 lb	0.06 cuft
Case	17.675	7.395	10.900 in	33.600	34.357 lb	0.82 cuft
Load	47.255	36.975	48.475 in	1747.200	1851.585 lb	53.86 cuft

FIG. 21-63a Use of CAPE software system for determining carton, shipping case, and pallet load dimensions and patterns for defined conditions. Shipping cases must fit within the pallet dimensions. (*Courtesy of CAPE Systems, Inc., Plano, TX, 800-229-3434.*)

severe coupling action or the starting of a long train, through slack action, can cause these forces to reach 6 to 8 G. To protect packaged freight from damage restraining systems have been developed. There are **energy-absorbing blocking and bracing systems** which can absorb these forces with little damage to the packages. Readily available at low cost, these systems are economically justified by reducing or eliminating repackaging cost and by lowering product loss due to package failure. This subject is treated in detail by J. J. Dempsey in *Methods for Loading, Bracing and Blocking of Packaged Goods in Transportation Equipment* (E. I. du Pont de Nemours & Co., Applied Technology Division, Wilmington, DE 19898), which may be purchased from Du Pont at nominal cost.
Simulation of transportation systems on a laboratory scale is used to predict the effectiveness of freight-restraining systems before actual tryout. The effect of a system on controlling damage at various levels of impact and vibration can be determined quickly and at low cost. As a result, the risk of substantial product loss during ini-

tial trials of a new system is reduced significantly. This service is offered by several firms and by the Center for Packaging Science and Engineering, Shock and Vibration Laboratory, Rutgers, The State University of New Jersey, Piscataway, New Jersey.

Distribution packaging design applies primarily to small packages such as found in the household-chemical industry. The competitive nature of household chemicals sold in supermarkets and by mass merchandisers involves frequent change in the design of the primary package—that which is on the supermarket shelf—with a consequent redesign of the secondary protective package in which the point-of-sale packages are shipped. This further influences pallet pattern and loading. Redesign using manual methods can be time consuming, and finding an optimum package is difficult. The era of the personal computer has changed all this. Software which can be run on most 486 personal computers can rapidly evaluate alternative designs and a near optimum established. The CAPE software mentioned previously provides a real-time ability to not only rapidly design the primary and

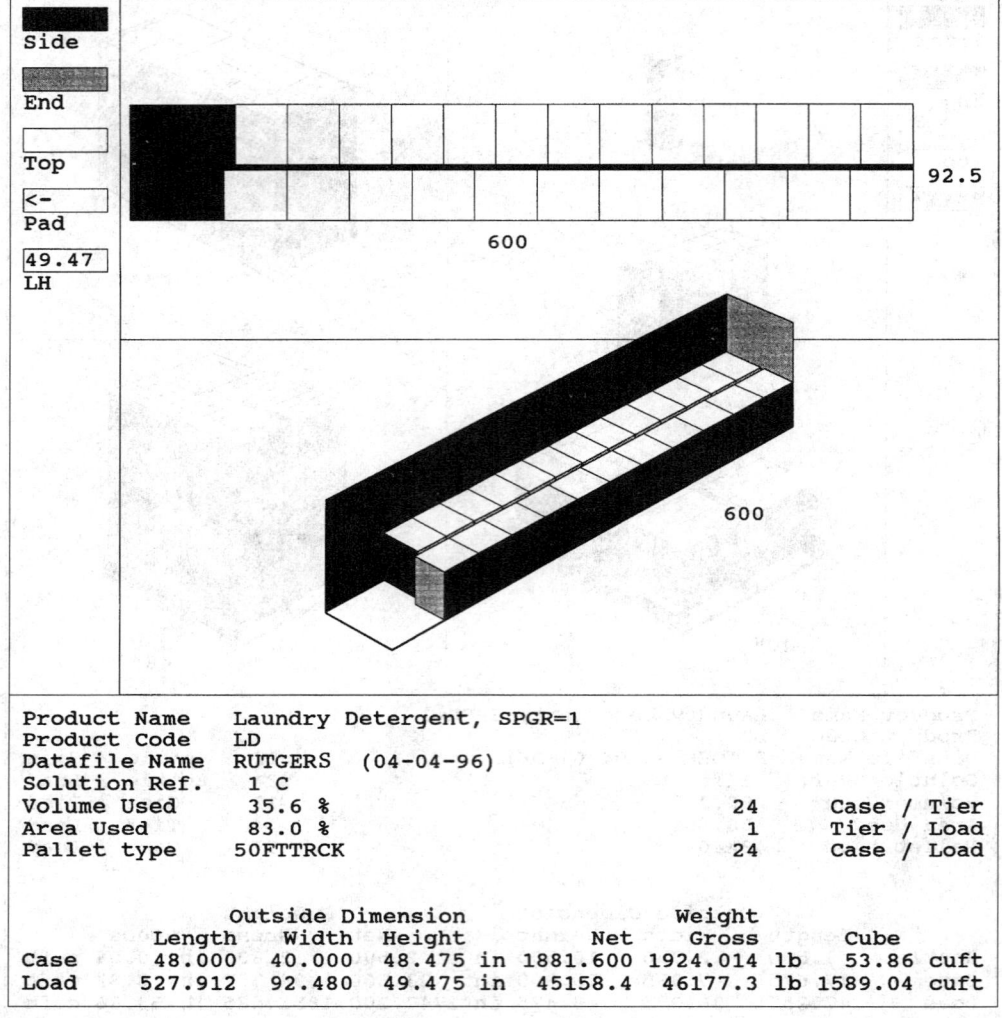

FIG. 21-63b   Use of CAPE for determining trailer loading for conditions of Fig. 21-63a, where the maximum cargo weight cannot exceed 46,800 lbs (21,228 kg). Note: CAPE can also give SI units. For Fig. 21-63a and b, multiply in × 25.4 for mm; cu ft × .02832 for m³, and lbs × 0.4538 for kg. (*Courtesy of CAPE Systems, Inc., Plano, TX, 800-229-3434.*)

secondary packages and pallet loading, but also determine truck-loading patterns based on highway and road-limit restrictions. Figures 21-63, *a* and *b*, show the design for a one-liter bottle of laundry detergent whose basic dimensions of width, thickness, and height were determined by the industrial designer responsible for the artistic design of the bottle. The CAPE software rapidly evaluates this and in the example given has the restrictions that there can be no overhang of shipping containers over the pallet edges, and that the number of point-of-sale packages in a shipping container is twelve but it can be in any pattern or combination so long as there are twelve bottles in a shipping container. The highway-load restriction is 46,800 pounds. The operation of the CAPE program including loading it into the PC, setting up the case study given, and obtaining results took less than five minutes.

Alternative Separation Processes

Joseph D. Henry, Jr., Ph.D., P.E., *Senior Fellow, Department of Engineering and Public Policy, Carnegie Mellon University; Member, American Institute of Chemical Engineers, American Society for Engineering Education. (Section Editor, Alternative Solid/Liquid Separations, Crystallization from the Melt)*

Michael E. Prudich, Ph.D., *Professor and Chair of Chemical Engineering, Ohio University; Member, American Institute of Chemical Engineers, American Chemical Society, Society of Mining Engineers, American Society for Engineering Education. (Section Editor, Alternative Solid/Liquid Separations)*

William Eykamp, Ph.D., *Adjunct Professor of Chemical Engineering, Tufts University; Formerly President, Koch Membrane Systems; Member, American Institute of Chemical Engineers, American Chemical Society, American Association for the Advancement of Science, North American Membrane Society, European Society of Membrane Science and Technology. (Membrane Separation Processes)*

T. Alan Hatton, Ph.D., *Ralph Landau Professor and Director of the David H. Koch School of Chemical Engineering Practice, Massachusetts Institute of Technology; Founding Fellow, American Institute of Medical and Biological Engineering; Member, American Institute of Chemical Engineers, American Chemical Society, International Association of Colloid and Interface Scientists, American Association for the Advancement of Science, Neutron Scattering Society of America. (Selection of Biochemical Separation Processes)*

Keith P. Johnston, Ph.D., P.E., *Professor of Chemical Engineering, University of Texas (Austin); Member, American Institute of Chemical Engineers, American Chemical Society, Univ. of Texas Separations Research Program. (Supercritical Fluid Separation Processes)*

Richard M. Lemert, Ph.D., P.E., *Assistant Professor of Chemical Engineering, University of Toledo; Member, American Institute of Chemical Engineers, Society of Mining Engineers, American Society for Engineering Education, American Chemical Society. (Supercritical Fluid Separation Processes)*

Robert Lemlich, Ph.D., P.E., *Professor of Chemical Engineering Emeritus, University of Cincinnati; Fellow, American Institute of Chemical Engineers; Member, American Chemical Society, American Society for Engineering Education; Fellow, American Association for the Advancement of Science. (Adsorptive-Bubble Separation Methods)*

Charles G. Moyers, Jr., Ph.D., P.E., *Principal Engineer, Union Carbide Corporation, Fellow, American Institute of Chemical Engineers. (Crystallization from the Melt)*

John Newman, Ph.D., *Professor of Chemical Engineering, University of California, Berkeley; Principal Investigator, Inorganic Materials Research Division, Lawrence Berkeley Laboratory. (Separation Processes Based Primarily on Action in an Electric Field: Theory of Electrical Separations)*

Herbert A. Pohl, Ph.D., *(deceased), Professor of Physics, Oklahoma State University. (Separation Processes Based Primarily on Action in an Electric Field: Dielectrophoresis)*

Kent Pollock, Ph.D., *Member of Technical Staff, Group 91, Space Surveillance Techniques, MIT Lincoln Laboratory. (Separation Processes Based Primarily on Action in an Electric Field: Dielectrophoresis)*

Michael P. Thien, Sc.D., *Senior Research Fellow, Merck & Company, Inc.; Member, American Institute of Chemical Engineers, American Chemical Society, International Society for Pharmaceutical Engineering. (Selection of Biochemical Separation Processes)*

CRYSTALLIZATION FROM THE MELT

GENERAL REFERENCES: Mullin, *Crystallization, 3d ed.,* Butterworth-Heinemann, 1993. Myerson, *Handbook of Industrial Crystallization,* Butterworth-Heinemann, 1993. Pfann, *Zone Melting, 2d ed.,* Wiley, New York, 1966. U.S. Patents 3,621,664 and 3,796,060. Zief and Wilcox, *Fractional Solidification,* Marcel Dekker, New York, 1967.

INTRODUCTION

Purification of a chemical species by solidification from a liquid mixture can be termed either **solution** crystallization or crystallization from the **melt**. The distinction between these two operations is somewhat subtle. The term **melt crystallization** has been defined as the separation of components of a binary mixture without addition of solvent, but this definition is somewhat restrictive. In **solution crystallization** a diluent solvent is added to the mixture; the solution is then directly or indirectly cooled, and/or solvent is evaporated to effect crystallization. The solid phase is formed and maintained somewhat below its pure-component freezing-point temperature. In melt crystallization no diluent solvent is added to the reaction mixture, and the solid phase is formed by cooling of the melt. Product is frequently maintained near or above its pure-component freezing point in the refining section of the apparatus.

A large number of techniques are available for carrying out crystallization from the melt. An abbreviated list includes partial freezing and solids recovery in cooling crystallizer-centrifuge systems, partial melting (e.g., sweating), staircase freezing, normal freezing, zone melting, and column crystallization. A description of all these methods is not within the scope of this discussion. Zief and Wilcox (op. cit.) and Myerson (op. cit.) describe many of these processes. Three of the more common methods—progressive freezing from a falling film, zone melting, and melt crystallization from the bulk—are discussed here to illustrate the techniques used for practicing crystallization from the melt.

High or ultrahigh product purity is obtained with many of the melt-purification processes. Table 22-1 compares the product quality and product form that are produced from several of these operations. Zone refining can produce very pure material when operated in a batch mode; however, other melt crystallization techniques also provide high purity and become attractive if continuous high-capacity processing is desired. Comparison of the features of melt crystallization and distillation are shown on Table 22-2.

A brief discussion of solid-liquid phase equilibrium is presented prior to discussing specific crystallization methods. Figures 22-1 and 22-2 illustrate the phase diagrams for binary solid-solution and eutec-

TABLE 22-1 Comparison of Processes Involving Crystallization from the Melt

Processes	Approximate upper melting point, °C	Materials tested	Minimum purity level obtained, ppm, weight	Product form
Progressive freezing	1500	All types	1	Ingot
Zone melting				
Batch	3500	All types	0.01	Ingot
Continuous	500	SiI$_4$	100	Melt
Melt crystallization				
Continuous	300	Organic	10	Melt
Cyclic	300	Organic	10	Melt

Abbreviated from Zief and Wilcox, *Fractional Solidification,* Marcel Dekker, New York, 1967, p. 7.

TABLE 22-2 Comparison of Features of Melt Crystallization and Distillation

Distillation	Melt crystallization
Phase equilibria	
Both liquid and vapor phases are totally miscible.	Liquid phases are totally miscible; solid phases are not.
Conventional vapor/liquid equilibrium.	Eutectic system.
Neither phase is pure.	Solid phase is pure, except at eutectic point.
Separation factors are moderate and decrease as purity increases.	Partition coefficients are very high (theoretically, they can be infinite).
Ultrahigh purity is difficult to achieve.	Ultrahigh purity is easy to achieve.
No theoretical limit on recovery.	Recovery is limited by eutectic composition.
Mass-transfer kinetics	
High mass-transfer rates in both vapor and liquid phases.	Only moderate mass-transfer rate in liquid phase, zero in solid.
Close approach to equilibrium.	Slow approach to equilibrium; achieved in brief contact time. Included impurities cannot diffuse out of solid.
Adiabatic contact assures phase equilibrium.	Solid phase must be remelted and refrozen to allow phase equilibrium.
Phase separability	
Phase densities differ by a factor of 100–10,000:1.	Phase densities differ by only about 10%.
Viscosity in both phases is low.	Liquid phase viscosity moderate, solid phase rigid.
Phase separation is rapid and complete.	Phase separation is slow; surface-tension effects prevent completion.
Countercurrent contacting is quick and efficient.	Countercurrent contacting is slow and imperfect.

Wynn, *Chem. Eng. Prog.,* **88,** 55 (1992). Reprinted with permission of the American Institute of Chemical Engineers.

tic systems respectively. In the case of binary solid-solution systems, illustrated in Fig. 22-1, the liquid and solid phases contain equilibrium quantities of both components in a manner similar to vapor-liquid phase behavior. This type of behavior causes separation difficulties since multiple stages are required. In principle, however, high purity and yields of both components can be achieved since no eutectic is present.

If the impurity or minor component is completely or partially soluble in the solid phase of the component being purified, it is convenient to define a distribution coefficient k, defined by Eq. (22-1):

$$k = C_s/C_\ell \qquad (22\text{-}1)$$

C_s is the concentration of impurity or minor component in the solid phase, and C_ℓ is the impurity concentration in the liquid phase. The distribution coefficient generally varies with composition. The value of k is greater than 1 when the solute raises the melting point and less than 1 when the melting point is depressed. In the regions near pure A or B the liquidus and solidus lines become linear; i.e., the distribution coefficient becomes constant. This is the basis for the common assumption of constant k in many mathematical treatments of fractional solidification in which ultrapure materials are obtained.

In the case of a simple eutectic system shown in Fig. 22-2, a pure solid phase is obtained by cooling if the composition of the feed mix-

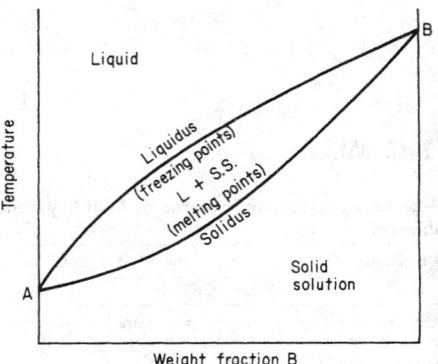

FIG. 22-1 Phase diagram for components exhibiting complete solid solution. (*Zief and Wilcox,* Fractional Solidification, *vol. 1, Marcel Dekker, New York, 1967, p. 31.*)

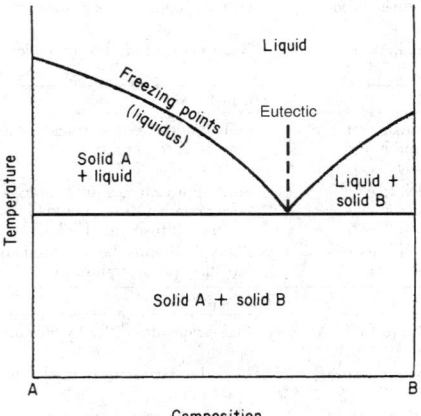

FIG. 22-2 Simple eutectic-phase diagram at constant pressure. (*Zief and Wilcox,* Fractional Solidification, *vol. 1, Marcel Dekker, New York, 1967, p. 24.*)

ture is not at the eutectic composition. If liquid composition is eutectic, then separate crystals of both species will form. In practice it is difficult to attain perfect separation of one component by crystallization of a eutectic mixture. The solid phase will always contain trace amounts of impurity because of incomplete solid-liquid separation, slight solubility of the impurity in the solid phase, or volumetric inclusions. It is difficult to generalize on which of these mechanisms is the major cause of contamination because of analytical difficulties in the ultrahigh-purity range.

The distribution-coefficient concept is commonly applied to fractional solidification of eutectic systems in the ultrapure portion of the phase diagram. If the quantity of impurity entrapped in the solid phase for whatever reason is proportional to that contained in the melt, then assumption of a constant k is valid. It should be noted that the theoretical yield of a component exhibiting binary eutectic behavior is fixed by the feed composition and position of the eutectic. Also, in contrast to the case of a solid solution, only one component can be obtained in a pure form.

There are many types of phase diagrams in addition to the two cases presented here; these are summarized in detail by Zief and Wilcox (op. cit., p. 21). Solid-liquid phase equilibria must be determined experimentally for most binary and multicomponent systems. Predictive methods are based mostly on ideal phase behavior and have limited accuracy near eutectics. A predictive technique based on extracting liquid-phase activity coefficients from *vapor-liquid* equilib-

ria that is useful for estimating nonideal binary or multicomponent *solid-liquid* phase behavior has been reported by Muir (Pap. 71f, 73d ann. meet., AIChE, Chicago, 1980).

PROGRESSIVE FREEZING

Progressive freezing, sometimes called *normal* freezing, is the slow, directional solidification of a melt. Basically, this involves slow solidification at the bottom or sides of a vessel or tube by indirect cooling. The impurity is rejected into the liquid phase by the advancing solid interface. This technique can be employed to concentrate an impurity or, by repeated solidifications and liquid rejections, to produce a very pure ingot. Figure 22-3 illustrates a progressive freezing apparatus. The solidification rate and interface position are controlled by the rate of movement of the tube and the temperature of the cooling medium. There are many variations of the apparatus; e.g., the residual-liquid portion can be agitated and the directional freezing can be carried out vertically as shown in Fig. 22-3 or horizontally (see Richman et al., in Zief and Wilcox, op. cit., p. 259). In general, there is a solute redistribution when a mixture of two or more components is directionally frozen.

Component Separation by Progressive Freezing When the distribution coefficient is less than 1, the first solid which crystallizes contains less solute than the liquid from which it was formed. As the fraction which is frozen increases, the concentration of the impurity in the remaining liquid is increased and hence the concentration of impurity in the solid phase increases (for $k < 1$). The concentration gradient is reversed for $k > 1$. Consequently, in the absence of diffusion in the solid phase a concentration gradient is established in the frozen ingot.

One extreme of progressive freezing is equilibrium freezing. In this case the freezing rate must be slow enough to permit diffusion in the solid phase to eliminate the concentration gradient. When this occurs, there is no separation if the entire tube is solidified. Separation can be achieved, however, by terminating the freezing before all the liquid has been solidified. Equilibrium freezing is rarely achieved in practice because the diffusion rates in the solid phase are usually negligible (Pfann, op. cit., p. 10).

If the bulk-liquid phase is well mixed and no diffusion occurs in the solid phase, a simple expression relating the solid-phase composition to the fraction frozen can be obtained for the case in which the distribution coefficient is independent of composition and fraction frozen [Pfann, *Trans. Am. Inst. Mech. Eng.,* **194**, 747 (1952)].

$$C_s = kC_0(1 - X)^{k-1} \tag{22-2}$$

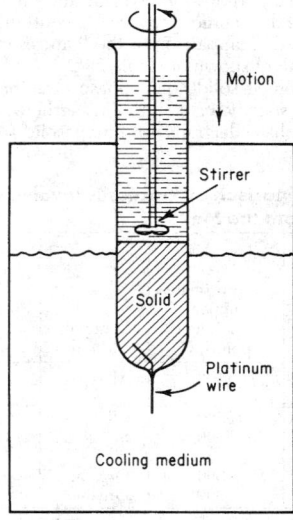

FIG. 22-3 Progressive freezing apparatus.

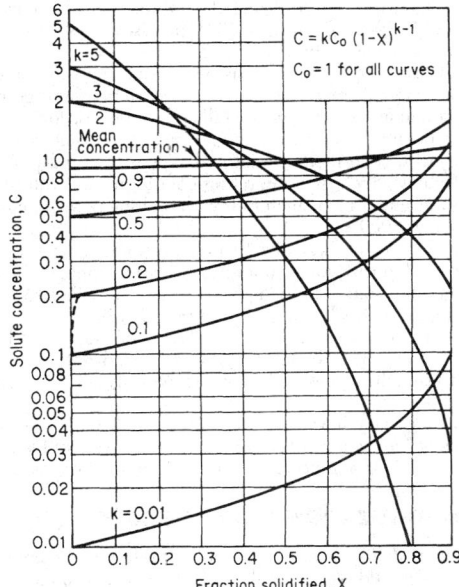

FIG. 22-4 Curves for progressive freezing, showing solute concentration C in the solid versus fraction-solidified X. (*Pfann, Zone Melting, 2d ed., Wiley, New York, 1966, p. 12.*)

C_0 is the solution concentration of the initial charge, and X is the fraction frozen. Figure 22-4 illustrates the solute redistribution predicted by Eq. (22-2) for various values of the distribution coefficient.

There have been many modifications of this idealized model to account for variables such as the freezing rate and the degree of mixing in the liquid phase. For example, Burton et al. [*J. Chem. Phys.*, **21**, 1987 (1953)] reasoned that the solid rejects solute faster than it can diffuse into the bulk liquid. They proposed that the effect of the freezing rate and stirring could be explained by the diffusion of solute through a stagnant film next to the solid interface. Their theory resulted in an expression for an effective distribution coefficient k_{eff} which could be used in Eq. (22-2) instead of k.

$$k_{\text{eff}} = \frac{1}{1 + (1/k - 1)e^{-f\delta/D}} \qquad (22\text{-}3)$$

where f = crystal growth rate, cm/s; δ = stagnant film thickness, cm; and D = diffusivity, cm²/s. No further attempt is made here to summarize the various refinements of Eq. (22-2). Zief and Wilcox (op. cit., p. 69) have summarized several of these models.

Pertinent Variables in Progressive Freezing The dominant variables which affect solute redistribution are the degree of mixing in the liquid phase and the rate of solidification. It is important to attain sufficient mixing to facilitate diffusion of the solute away from the solid-liquid interface to the bulk liquid. The film thickness δ decreases as the level of agitation increases. Cases have been reported in which essentially no separation occurred when the liquid was not stirred. The freezing rate which is controlled largely by the lowering rate of the tube (see Fig. 22-3) has a pronounced effect on the separation achieved. The separation is diminished as the freezing rate is increased. Also fluctuations in the freezing rate caused by mechanical vibrations and variations in the temperature of the cooling medium can decrease the separation.

Applications Progressive freezing has been applied to both solid solution and eutectic systems. As Fig. 22-4 illustrates, large separation factors can be attained when the distribution coefficient is favorable. Relatively pure materials can be obtained by removing the desired portion of the ingot. Also in some cases progressive freezing provides a convenient method of concentrating the impurities; e.g., in the case of $k < 1$ the last portion of the liquid that is frozen is enriched in the distributing solute.

Progressive freezing has been applied on the commercial scale. For example, aluminum has been purified by continuous progressive freezing [Dewey, *J. Metals*, **17**, 940 (1965)]. The Proabd refiner described by Molinari (Zief and Wilcox, op. cit., p. 393) is also a commercial example of progressive freezing. In this apparatus the mixture is directionally solidified on cooling tubes. Purification is achieved because the impure fraction melts first; this process is called sweating. This technique has been applied to the purification of naphthalene and *p*-dichlorobenzene and commercial equipment is available from BEFS PROKEM, Houston, Tx.

ZONE MELTING

Zone melting also relies on the distribution of solute between the liquid and solid phases to effect a separation. In this case, however, one or more liquid zones are passed through the ingot. This extremely versatile technique, which was invented by W. G. Pfann, has been used to purify hundreds of materials. Zone melting in its simplest form is illustrated in Fig. 22-5. A molten zone can be passed through an ingot from one end to the other by either a moving heater or by slowly drawing the material to be purified through a stationary heating zone.

Progressive freezing can be viewed as a special case of zone melting. If the zone length were equal to the ingot length and if only one pass were used, the operation would become progressive freezing. In general, however, when the zone length is only a fraction of the ingot length, zone melting possesses the advantage that a portion of the ingot does not have to be discarded after each solidification. The last portion of the ingot which is frozen in progressive freezing must be discarded before a second freezing.

Component Separation by Zone Melting The degree of solute redistribution achieved by zone melting is determined by the zone length l, ingot length L, number of passes n, the degree of mixing in the liquid zone, and the distribution coefficient of the materials being purified. The distribution of solute after one pass can be obtained by material-balance considerations. This is a two-domain problem; i.e., in the major portion of the ingot of length $L - l$ zone melting occurs in the conventional sense. The trailing end of the ingot of length l undergoes progressive freezing. For the case of constant-distribution coefficient, perfect mixing in the liquid phase, and negligible diffusion in the solid phase, the solute distribution for a single pass is given by Eq. (22-4) [Pfann, *Trans. Am. Inst. Mech. Eng.*, **194**, 747 (1952)].

$$C_s = C_0[1 - (1 - k)e^{-kx/l}] \qquad (22\text{-}4)$$

The position of the zone x is measured from the leading edge of the ingot. The distribution for multiple passes can also be calculated from a material balance, but in this case the leading edge of the zone encounters solid corresponding to the composition at the point in question for the previous pass. The multiple-pass distribution has been numerically calculated (Pfann, *Zone Melting*, 2d ed., Wiley, New York, 1966, p. 285) for many combinations of k, L/l, and n. Typical solute-composition profiles are shown in Fig. 22-6 for various numbers of passes.

The ultimate distribution after an infinite number of passes is also shown in Fig. 22-6 and can be calculated for $x < (L - l)$ from the following equation (Pfann, op. cit., p. 42):

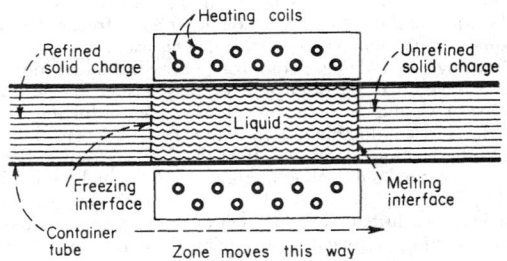

FIG. 22-5 Diagram of zone refining.

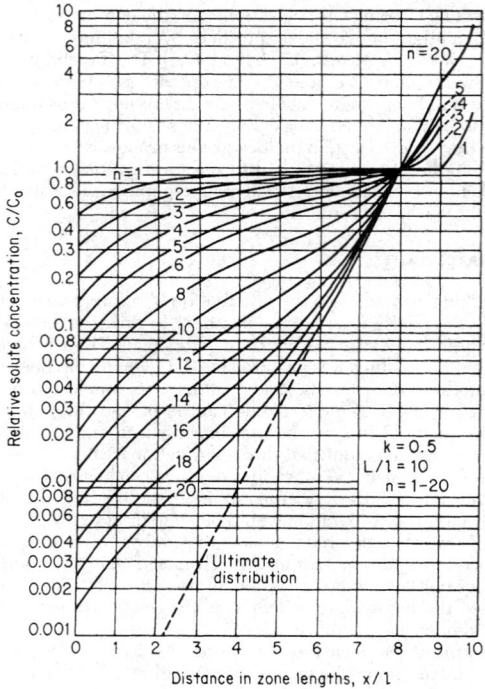

FIG. 22-6 Relative solute concentration C/C_o (logarithmic scale) versus distance in zone lengths x/ℓ from beginning of charge, for various numbers of passes n. L denotes charge length. (*Pfann, Zone Melting, 2d ed., Wiley, New York, 1966, p. 290.*)

$$C_s = Ae^{BX} \qquad (22\text{-}5)$$

where A and B can be determined from the following relations:

$$k = B\ell/(e^{B\ell} - 1) \qquad (22\text{-}6)$$

$$A = C_0BL/(e^{BL} - 1) \qquad (22\text{-}7)$$

The ultimate distribution represents the maximum separation that can be attained without cropping the ingot. Equation (22-5) is approximate because it does not include the effect of progressive freezing in the last zone length.

As in progressive freezing, many refinements of these models have been developed. Corrections for partial liquid mixing and a variable distribution coefficient have been summarized in detail (Zief and Wilcox, op. cit., p. 47).

Pertinent Variables in Zone Melting The dominant variables in zone melting are the number of passes, ingot-length–zone-length ratio, freezing rate, and degree of mixing in the liquid phase. Figure 22-6 illustrates the increased solute redistribution that occurs as the number of passes increases. Ingot-length–zone-length ratios of 4 to 10 are commonly used (Zief and Wilcox, op. cit., p. 624). An exception is encountered when one pass is used. In this case the zone length should be equal to the ingot length; i.e., progressive freezing provides the maximum separation when only one pass is used.

The freezing rate and degree of mixing have effects in solute redistribution similar to those discussed for progressive freezing. Zone travel rates of 1 cm/h for organic systems, 2.5 cm/h for metals, and 20 cm/h for semiconductors are common. In addition to the zone-travel rate the heating conditions affect the freezing rate. A detailed summary of heating and cooling methods for zone melting has been outlined by Zief and Wilcox (op. cit., p. 192). Direct mixing of the liquid region is more difficult for zone melting than progressive freezing. Mechanical stirring complicates the apparatus and increases the probability of contamination from an outside source. Some mixing occurs because of natural convection. Methods have been developed to stir

the zone magnetically by utilizing the interaction of a current and magnetic field (Pfann, op. cit., p. 104) for cases in which the charge material is a reasonably good conductor.

Applications Zone melting has been used to purify hundreds of inorganic and organic materials. Many classes of inorganic compounds including semiconductors, intermetallic compounds, ionic salts, and oxides have been purified by zone melting. Organic materials of many types have been zone-melted. Zief and Wilcox (op. cit., p. 624) have compiled tables which give operating conditions and references for both inorganic and organic materials with melting points ranging from $-115°C$ to over $3000°C$.

Some materials are so reactive that they cannot be zone-melted to a high degree of purity in a container. Floating-zone techniques in which the molten zone is held in place by its own surface tension have been developed by Keck et al. [*Phys. Rev.*, **89**, 1297 (1953)].

Continuous-zone-melting apparatus has been described by Pfann (op. cit., p. 171). This technique offers the advantage of a close approach to the ultimate distribution, which is usually impractical for batch operation.

Performance data have been reported by Kennedy et al. (*The Purification of Inorganic and Organic Materials*, Marcel Dekker, New York, 1969, p. 261) for continuous-zone refining of benzoic acid.

MELT CRYSTALLIZATION FROM THE BULK

Conducting crystallization inside a vertical or horizontal column with a countercurrent flow of crystals and liquid can produce a higher product purity than conventional crystallization or distillation. The working concept is to form a crystal phase from the bulk liquid, either internally or externally, and then transport the solids through a countercurrent stream of enriched reflux liquid obtained from melted product. The problem in practicing this technology is the difficulty of controlling solid-phase movement. Unlike distillation, which exploits the specific-gravity differences between liquid and vapor phases, melt crystallization involves the contacting of liquid and solid phases that have nearly identical physical properties. Phase densities are frequently very close, and gravitational settling of the solid phase may be slow and ineffective. The challenge of designing equipment to accomplish crystallization in a column has resulted in a myriad of configurations to achieve reliable solid-phase movement, high product yield and purity, and efficient heat addition and removal.

Investigations Crystallization conducted inside a column is categorized as either **end-fed** or **center-fed** depending on whether the feed location is upstream or downstream of the crystal forming section. Figure 22-7 depicts the features of an end-fed commercial column described by McKay et al. [*Chem. Eng. Prog. Symp. Ser.*, no. 25, 55, 163 (1969)] for the separation of xylenes. Crystals of *p*-xylene are formed by indirect cooling of the melt in scraped-surface heat exchangers, and the resultant slurry is introduced into the column at

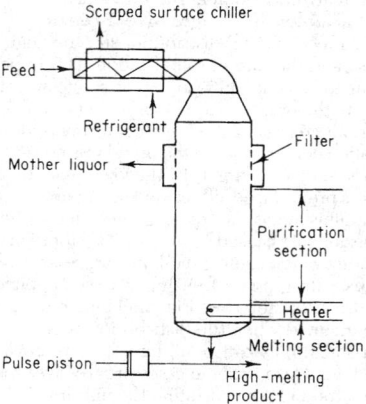

FIG. 22-7 End-fed column crystallizer. (*Phillips Petroleum Co.*)

the top. This type of column has no mechanical internals to transport solids and instead relies upon an imposed hydraulic gradient to force the solids through the column into the melting zone. Residue liquid is removed through a filter directly above the melter. A pulse piston in the product discharge improves washing efficiency and column reliability.

Figure 22-8 shows the features of a horizontal center-fed column [Brodie, *Aust. Mech. Chem. Eng. Trans.*, **37** (May 1979)] which has been commercialized for continuous purification of naphthalene and p-dichlorobenzene. Liquid feed enters the column *between* the hot purifying section and the cold freezing or recovery zone. Crystals are formed internally by indirect cooling of the melt through the walls of the refining and recovery zones. Residue liquid that has been depleted of product exits from the coldest section of the column. A spiral conveyor controls the transport of solids through the unit.

Another center-fed design that has only been used on a preparative scale is the vertical spiral conveyor column reported by Schildknecht [*Angew. Chem.*, **73**, 612 (1961)]. In this device, a version of which is shown on Fig. 22-9, the dispersed-crystal phase is formed in the freezing section and conveyed downward in a controlled manner by a rotating spiral with or without a vertical oscillation.

Differences have been observed in the performance of end- and center-fed column configurations. Consequently, discussions of center- and end-fed column crystallizers are presented separately. The design and operation of both columns are reviewed by Powers (Zief and Wilcox, op. cit., p. 343). A comparison of these devices is shown on Table 22-3.

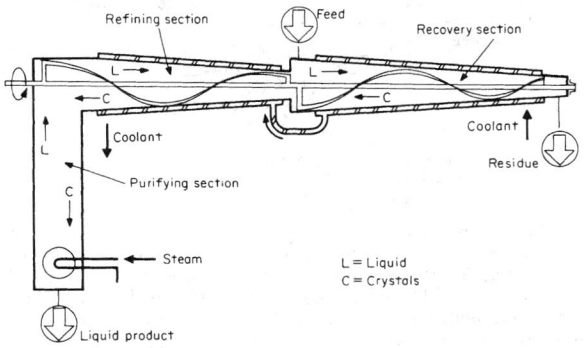

FIG. 22-8 Horizontal center-fed column crystallizer. (*The C. W. Nofsinger Co.*)

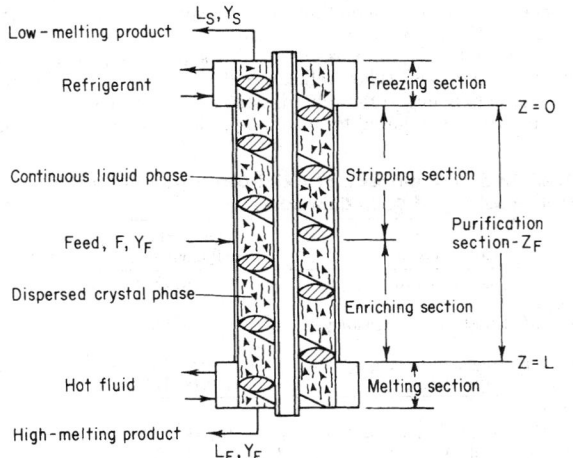

FIG. 22-9 Center-fed column crystallizer with a spiral-type conveyor.

TABLE 22-3 Comparison of Melt-Crystallizer Performance

Center-fed column	End-fed column
Solid phase is formed internally; thus, only liquid streams enter and exit the column.	Solid phase is formed in external equipment and fed as slurry into the purifier.
Internal reflux can be controlled without affecting product yield.	The maximum internal liquid reflux is fixed by the thermodynamic state of the feed relative to the product stream. Excessive reflux will diminish product yield.
Operation can be continuous or batchwise at total reflux.	Total reflux operation is not feasible.
Center-fed columns can be adapted for both eutectic and solid-solution systems.	End-fed columns are inefficient for separation of solid-solution systems.
Either low- or high-porosity-solids-phase concentrations can be formed in the purification and melting zones.	End-fed units are characterized by low-porosity-solids packing in the purification and melting zones.
Scale-up depends on the mechanical complexity of the crystal-transport system and techniques for removing heat. Vertical oscillating spiral columns are likely limited to about 0.2 m in diameter, whereas horizontal columns of several meters are possible.	Scale-up is limited by design of melter and/or crystal-washing section. Vertical or horizontal columns of several meters in diameter are possible.

Center-Fed Column Crystallizers Two types of center-fed column crystallizers are illustrated on Figs. 22-8 and 22-9. As in a simple distillation column, these devices are composed of three distinct sections: a freezing or recovery section, where solute is frozen from the impure liquor; the purification zone, where countercurrent contacting of solids and liquid occurs; and the crystal-melting and -refluxing section. Feed position separates the refining and recovery portions of the purification zone. The section between feed location and melter is referred to as the refining or enrichment section, whereas the section between feed addition and freezing is called the recovery section. The refining section may have provisions for sidewall cooling. The published literature on column crystallizers connotes stripping and refining in a reverse sense to distillation terminology, since refined product from a melt crystallizer exits at the hot section of the column rather than at the cold end as in a distillation column.

Rate processes that describe the purification mechanisms in a column crystallizer are highly complex since phase transition and heat- and mass-transfer processes occur simultaneously. Nucleation and growth of a crystalline solid phase along with crystal washing and crystal melting are occurring in various zones of the apparatus. Column hydrodynamics are also difficult to describe. Liquid- and solid-phase mixing patterns are influenced by factors such as solids-transport mechanism, column orientation, and, particularly for dilute slurries, the settling characteristics of the solids.

Most investigators have focused their attention on a differential segment of the zone between the feed injection and the crystal melter. Analysis of crystal formation and growth in the recovery section has received scant attention. Table 22-4 summarizes the scope of the literature treatment for center-fed columns for both solid-solution and eutectic forming systems.

The dominant mechanism of purification for column crystallization of solid-solution systems is recrystallization. The rate of mass transfer resulting from recrystallization is related to the concentrations of the solid phase and free liquid which are in intimate contact. A model based on height-of-transfer-unit (HTU) concepts representing the composition profile in the purification section for the high-melting component of a binary solid-solution system has been reported by Powers et al. (in Zief and Wilcox, op. cit., p. 363) for total-reflux operation. Typical data for the purification of a solid-solution system, azobenzene-stilbene, are shown in Fig. 22-10. The column crystallizer was operated

TABLE 22-4 Column-Crystallizer Investigations

	Treatments	
	Theoretical	Experimental
Solid solutions		
Total reflux—steady state	1, 2, 4, 6	1, 4, 6
Total reflux—dynamic	2	
Continuous—steady state	1, 4	4, 8, 9
Continuous—dynamic		
Eutectic systems		
Total reflux—steady state	1, 3, 4, 7	1, 3, 6
Total reflux—dynamic		
Continuous—steady state	1, 5, 10, 11, 12	5, 8, 9, 10, 11, 13
Continuous—dynamic		

1. Powers, *Symposium on Zone Melting and Column Crystallization,* Karls-ruhe, 1963.
2. Anikin, *Dokl. Akad. Nauk SSSR,* **151,** 1139 (1969).
3. Albertins et al., *Am. Inst. Chem. Eng. J.,* **15,** 554 (1969).
4. Gates et al., *Am. Inst. Chem. Eng. J.,* **16,** 648 (1970).
5. Henry et al., *Am. Inst. Chem. Eng. J.,* **16,** 1055 (1970).
6. Schildknecht et al., *Angew. Chem.,* **73,** 612 (1961).
7. Arkenbout et al., *Sep. Sci.,* **3,** 501 (1968).
8. Betts et al., *Appl. Chem.,* **17,** 180 (1968).
9. McKay et al., *Chem. Eng. Prog. Symp. Ser.,* no. 25, 55, 163 (1959).
10. Bolsaitis, *Chem. Eng. Sci.,* **24,** 1813 (1969).
11. Moyers et al., *Am. Inst. Chem. Eng. J.,* **20,** 1119 (1974).
12. Griffin, M.S. thesis in chemical engineering, University of Delaware, 1975.
13. Brodie, *Aust. Mech. Chem. Eng. Trans.,* 37 (1971).

at total reflux. The solid line through the data was computed by Powers et al. (op. cit., p. 364) by using an experimental HTU value of 3.3 cm.

Most of the analytical treatments of center-fed columns describe the purification mechanism in an adiabatic oscillating spiral column (Fig. 22-9). However, the analyses by Moyers (op. cit.) and Griffin (op. cit.) are for a nonadiabatic dense-bed column. Differential treatment of the horizontal-purifier (Fig. 22-8) performance has not been reported; however, overall material and enthalpy balances have been described by Brodie (op. cit.) and apply equally well to other designs.

A dense-bed center-fed column (Fig. 22-11) having provision for internal crystal formation and variable reflux was tested by Moyers et al. (op. cit.). In the theoretical development (ibid.) a nonadiabatic, plug-flow axial-dispersion model was employed to describe the performance of the entire column. Terms describing interphase transport of impurity between adhering and free liquid are not considered.

A comparison of the axial-dispersion coefficients obtained in oscillating-spiral and dense-bed crystallizers is given in Table 22-5. The dense-bed column approaches axial-dispersion coefficients similar to those of densely packed ice-washing columns.

The concept of minimum reflux as related to column-crystallizer

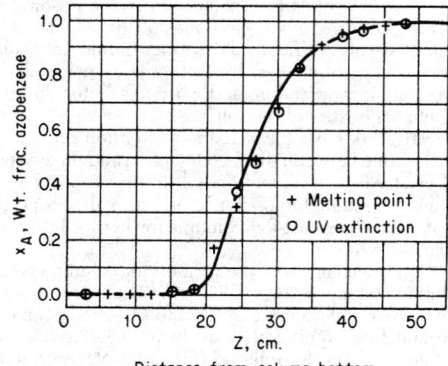

FIG. 22-10 Steady-state separation of azobenzene and stilbene in a center-fed column crystallizer with total-reflux operation. To convert centimeters to inches, multiply by 0.3937. (*Zief and Wilcox,* Fractional Solidification, *vol. 1, Marcel Dekker, New York, 1967, p. 356.*)

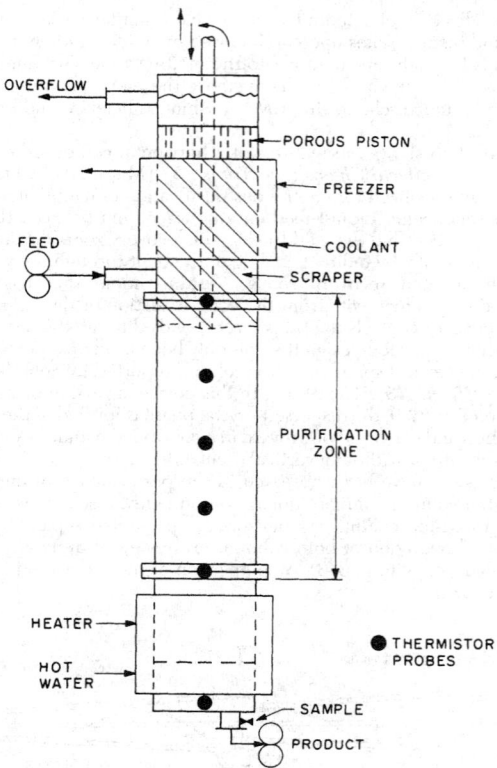

FIG. 22-11 Dense-bed center-fed column crystallizer. [*Moyers et al.,* Am. Inst. Chem. Eng. J., **20,** *1121 (1974).*]

operation is presented by Brodie (op. cit.) and is applicable to all types of column crystallizers, including end-fed units. In order to stabilize column operation the sensible heat of subcooled solids entering the melting zone should be balanced or exceeded by the heat of fusion of the refluxed melt. The relationship in Eq. (22-8) describes the minimum reflux requirement for proper column operation.

$$R = (T_P - T_F)\,C_P/\lambda \qquad (22\text{-}8)$$

R = reflux ratio, g reflux/g product; T_P = product temperature, °C; T_F = saturated-feed temperature, °C; C_P = specific heat of solid crystals, cal/(g·°C); and λ = heat of fusion, cal/g.

All refluxed melt will refreeze if reflux supplied equals that computed by Eq. (22-8). When reflux supplied is greater than the minimum, jacket cooling in the refining zone or additional cooling in the recovery zone is required to maintain product recovery. Since high-

TABLE 22-5 Comparison of Axial-Dispersion Coefficients for Several Liquid-Solid Contactors

Column type	Dispersion coefficient, cm²/s	Reference
Center-fed crystallizer (oscillating spiral)	1.6–3.5	1
Center-fed crystallizer (oscillating spiral)	1.3–1.7	2
Countercurrent ice-washing column	0.025–0.17	3
Center-fed crystallizer	0.12–0.30	4

References:
1. Albertins et al., *Am. Inst. Chem. Eng. J.,* **15,** 554 (1969).
2. Gates et al., *Am. Inst. Chem. Eng. J.,* **16,** 648 (1970).
3. Ritter, Ph.D. thesis, Massachusetts Institute of Technology, 1969.
4. Moyers et al., *Am. Inst. Chem. Eng. J.,* **20,** 1119 (1974).

purity melts are fed near their pure-component freezing temperatures, little refreezing takes place unless jacket cooling is added.

To utilize a column-crystallizer design or rating model, a large number of parameters must be identified. Many of these are empirical in nature and must be determined experimentally in equipment identical to the specific device being evaluated. Hence macroscopic evaluation of systems by large-scale piloting is the rule rather than the exception. Included in this rather long list of critical parameters are factors such as impurity level trapped in the solid phase, product quality as a function of reflux ratio, degree of liquid and solids axial mixing in the equipment as a function of solids-conveyor design, size and shape of crystals produced, and ease of solids handling in the column. Heat is normally removed through metal surfaces; thus, the stability of the solution to subcooling can also be a major factor in design.

End-Fed Column Crystallizer End-fed columns were developed and successfully commercialized by the Phillips Petroleum Company in the 1950s. The sections of a typical end-fed column, often referred to as a Phillips column, are shown on Fig. 22-7. Impure liquor is removed through filters located between the product-freezing zone and the melter rather than at the end of the freezing zone, as occurs in center-fed units. The purification mechanism for end-fed units is basically the same as for center-fed devices. However, there are reflux restrictions in an end-fed column, and a high degree of solids compaction exists near the melter of an end-fed device. It has been observed that the free-liquid composition and the fraction of solids are relatively constant throughout most of the purification section but exhibit a sharp discontinuity near the melting section [McKay et al., *Ind. Eng. Chem.*, **52**, 197 (1969)]. Investigators of end-fed column behavior are listed in Table 22-6. Note that end-fed columns are adaptable only for eutectic-system purification and cannot be operated at total reflux.

Performance information for the purification of *p*-xylene indicates that nearly 100 percent of the crystals in the feed stream are removed as product. This suggests that the liquid which is refluxed from the melting section is effectively refrozen by the countercurrent stream of subcooled crystals. A high-melting product of 99.0 to 99.8 weight percent *p*-xylene has been obtained from a 65 weight percent *p*-xylene feed. The major impurity was *m*-xylene. Figure 22-12 illustrates the column-cross-section-area–capacity relationship for various product purities.

Column crystallizers of the end-fed type can be used for purification of many eutectic-type systems and for aqueous as well as organic systems (McKay, loc. cit.). Column crystallizers have been used for xylene isomer separation, but recently other separation technologies including more efficient melt crystallization equipment have tended to supplant the Phillips style crystallizer.

Commercial Equipment and Applications In the last two decades the practice of melt crystallization techniques for purification of certain organic materials has made significant commercial progress. The concept of refining certain products by countercurrent staging of crystallization in a column has completed the transition from laboratory and pilot equipment to large-scale industrial configurations. Chemicals which have been purified by suspension crystallization-purifier column techniques are listed on Table 22-7. The practice of crystal formation and growth from the bulk liquid (as is practiced in suspension crystallization techniques described in Sec. 18 of this handbook) and subsequent crystal melting and refluxing in a purifier column has evolved into two slightly different concepts: (1) the hori-

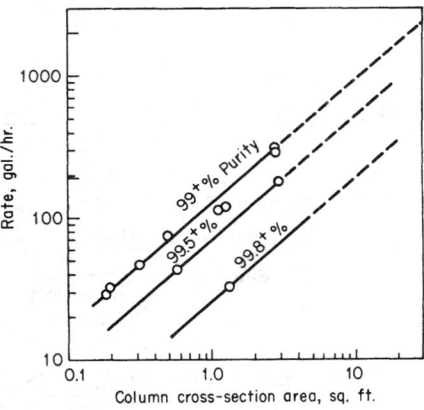

FIG. 22-12 Pulsed-column capacity versus column size for 65 percent *p*-xylene feed. To convert gallons per hour to cubic meters per hour, multiply by 0.9396; to convert square feet to square meters, multiply by 0.0929. (*McKay et al., prepr., 59th nat. meet. AIChE, East Columbus, Ohio.*)

zontal continuous crystallization technique with vertical purifier invented by Brodie (op. cit.) and (2) the continuous multistage or stepwise system with vertical purifier developed by Tsukishima Kikai Co., Ltd. (TSK). A recent description of these processes has been published by Meyer [*Chem. Proc.*, **53**, 50 (1990)].

The horizontal continuous Brodie melt crystallizer is basically an indirectly cooled crystallizer with an internal ribbon conveyor to transport crystals countercurrent to the liquid and a vertical purifier for final refining. Figure 22-8 describes the operation of a single tube unit and Fig. 22-13 depicts a multitube unit. The multitube design has been successfully commercialized for a number of organic chemicals. The Brodie purifier configuration requires careful control of process and equipment temperature differences to eliminate internal encrustations and is limited by the inherent equipment geometry to capacities of less than 15,000 tons per year per module.

In the multistage process described on Fig. 22-14 feed enters one of several crystallizers installed in series. Crystals formed in each crystallizer are transferred to a hotter stage and the liquid collected in the clarified zone of the crystallizer is transferred to a colder stage and eventually discharged as residue. At the hot end, crystals are transferred to a vertical purifier where countercurrent washing is performed by pure, hot-product reflux. TSK refers to this multistage process as the countercurrent cooling crystallization (CCCC) process. In principle any suitable type of crystallizer can be used in the stages

TABLE 22-6 End-Fed-Crystallizer Investigations

Eutectic systems	Treatments	
	Theoretical	Experimental
Continuous—steady state	1, 2, 4	1, 4
Batch	3	3

1. McKay et al., *Ind. Eng. Chem. Process Des. Dev.*, **6**, 16 (1967).
2. Player, *Ind. Eng. Chem. Process Des. Dev.*, **8**, 210 (1969).
3. Yagi et al., *Kagaku Kogaku*, **72**, 415 (1963).
4. Shen and Meyer, Prepr. 19F, AIChE Symp., Chicago, 1970.

TABLE 22-7 Chemicals Purified by TSK CCCC Process (The C. W. Nofsinger Co.)

Acetic acid	Isophthaloyl chloride
Acrylic acid	Isopregol
Adipic acid	Lutidine
Benzene	Maleic anhydride
Biphenyl	Naphthalene
Bisphenol-A	*p*-Nitrochloro benzene
Caprolactam	*p*-Nitrotoluene
Chloroacetic acid	Phenol
p-Chloro toluene	*b*-Picoline
p-Cresol	*g*-Picoline
Combat (proprietary)	Pyridine
Dibutyl hydroxy toluene (BHT)	Stilbene
p-Dichloro benzene	Terephthaloyl chloride
2,5 Dichlorophenol	Tertiary butyl phenol
Dicumyl peroxide	Toluene diisocyanate
Diene	Trioxane
Heliotropin	*p*-Xylene
Hexachloro cyclo butene	3,4 Xylidine
Hexamethylene diamine	

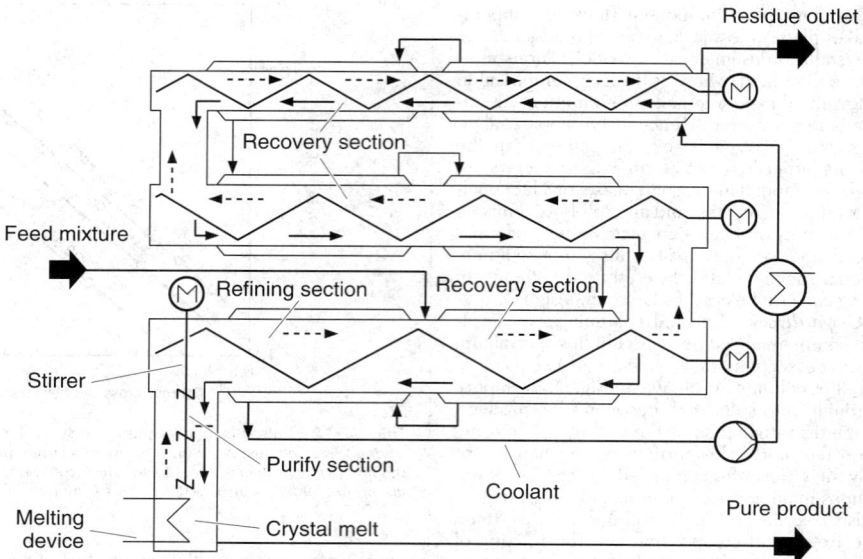

FIG. 22-13   Horizontal continuous Brodie melt crystallizer—multitube unit. (*The C. W. Nofsinger Co.*)

TABLE 22-8   Commercial TSK Crystallization Operating Plants

	Capacity MM lbs/yr	Company	Date & location
Countercurrent Cooling Crystallization (CCCC) Process			
Nofsinger license			
Nofsinger design & construct			
p-Dichlorobenzene	Confidential	Monsanto Co.	1989—Sauget, IL
TSK license			
TSK design & construct			
Confidential[1]	Confidential	Confidential	1988—Japan
p-Xylene	137	MGC	1986—Mizushima, Japan
Confidential[2]	Confidential	Confidential	1985—Japan
p-Xylene	132	MGC	1983—Mizushima, Japan
	expanded to 160		1984—Japan
p-Xylene	26.5	MGC	1981—Mizushima, Japan
Brodie			
TSK license			
TSK design & construct			
Naphthalene	8	SHSM	1985—China
p-DCB	13	Hodogaya	1981—Japan
p-DCB	5.5	Sumitomo	1978—Japan
UCAL license & design			
TSK hardware			
Naphthalene	16	Nippon Steel	1974—Japan
p-DCB	13	Hodogaya	1974—Japan
UCAL license & design			
Naphthalene	10	British Tar	1972—U.K.
p-DCB	3	UCAL	1969—Australia

ABBREVIATIONS:
  Nofsinger   The C. W. Nofsinger Company
  TSK   Tsukishima Kikai Co., Ltd.
  SHSM   Shanghai Hozan Steel Mill
  UCAL   Union Carbide Australia, Ltd.
  MGC   Mitsubishi Gas Chemical Co., co-developer with TSK of the application for *p*-Xylene
  1. Commercial scale plant started up in the spring of 1988 purifying a bulk chemical. This is the first application of the CCCC process on this bulk chemical.
  2. This small unit is operating in Japan with an 800 mm crystallizer and 300 mm purifier. Because of confidentiality, we cannot disclose the company, capacity, or product.

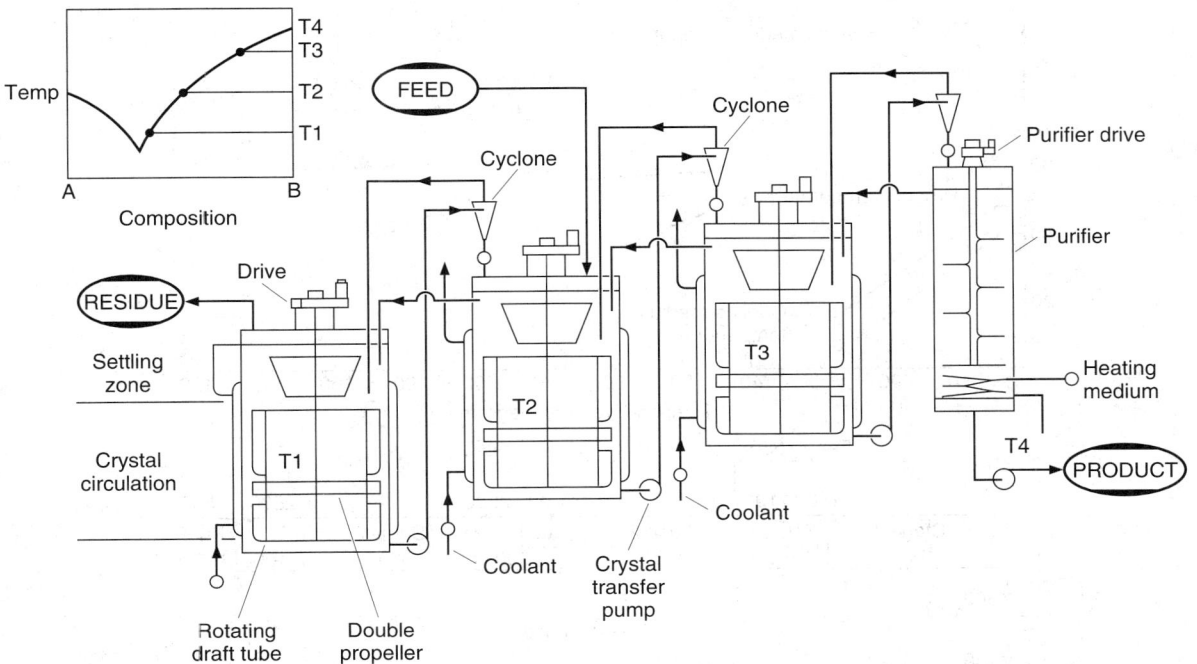

FIG. 22-14   Crystallizer—multistage process. (*The C. W. Nofsinger Co.*)

as long as the crystals formed can be separated from the crystallizer liquid and settled and melted in the purifier.

Commercial applications for both the Brodie and CCCC process are indicated on Table 22-8. Both the Brodie Purifier and the CCCC processes are available from The C. W. Nofsinger Company, PO Box 419173, Kansas City, MO 64141-0173.

FALLING-FILM CRYSTALLIZATION

Falling-film crystallization utilizes progressive freezing principles to purify melts and solutions. The technique established to practice the process is inherently cyclic. Figure 22-15 depicts the basic working concept. First a crystalline layer is formed by subcooling a liquid film

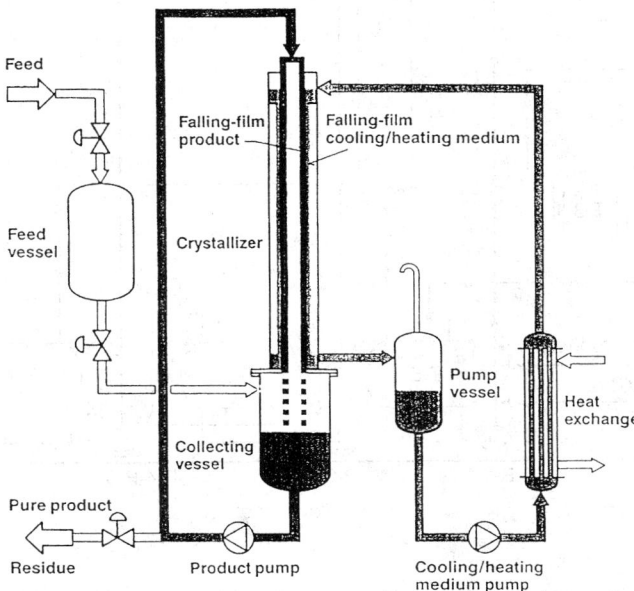

FIG. 22-15   Dynamic crystallization system. (*Sulzer Chemtech*)

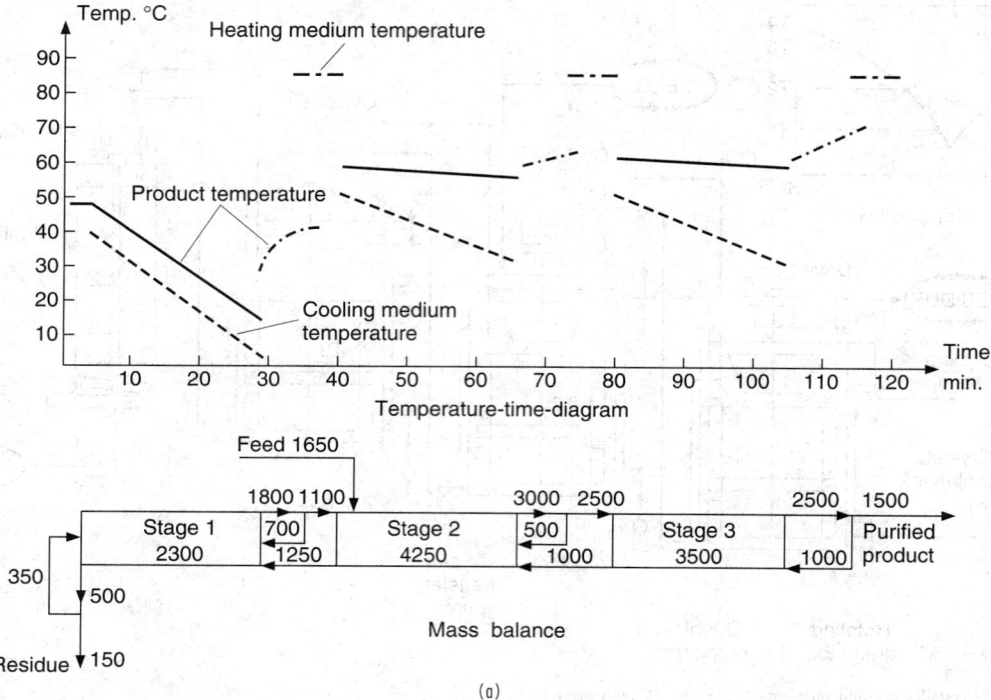

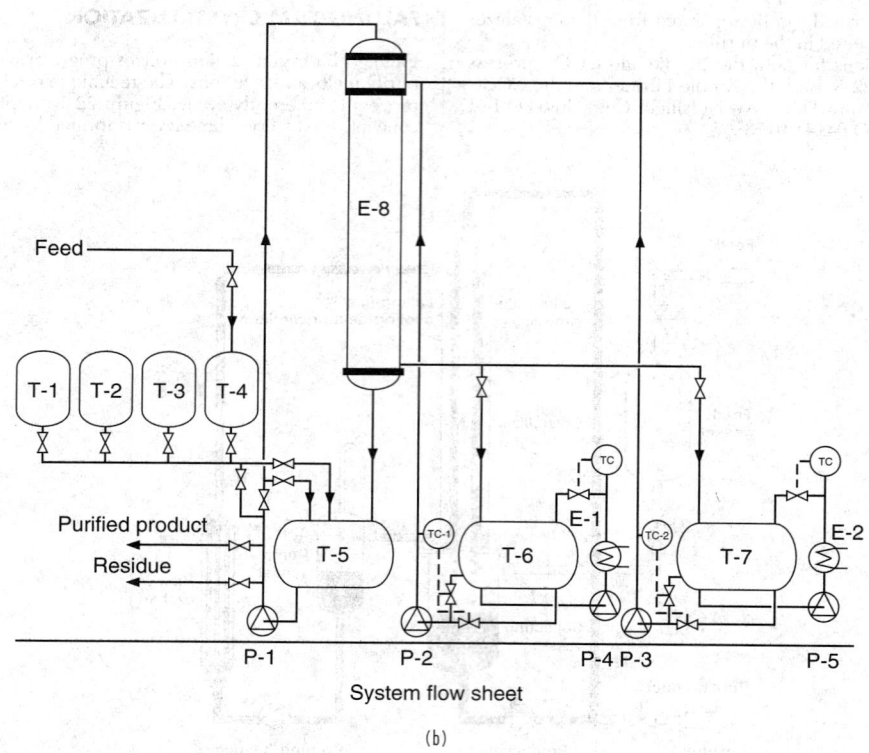

FIG. 22-16   Sulzer MWB-crystallization process. (*a*) Stepwise operation of the process. (*b*) System flow sheet. (*Sulzer Chemtech*)

on a vertical surface inside a tube. This coating is then grown by extracting heat from a falling film of melt (or solution) through a heat transfer surface. Impure liquid is then drained from the crystal layer and the product is reclaimed by melting. Variants of this technique have been perfected and are used commercially for many types of organic materials. Both static and falling-film techniques have been described by Wynn [*Chem. Eng. Progr.*, (1992)]. Mathematical models for both static and dynamic operations have been presented by Gilbert [*AIChE J.*, **37**, 1205 (1991)].

Principles of Operation Figure 22-16 describes a typical three-stage falling-film crystallization process for purification of MCA (monochloro acetic acid). Crystallizer E-8 consists of a number of vertical tubular elements working in parallel enclosed in a shell. Normal tube length is 12 meters with a 50- to 75-millimeter tube inside diameter. Feed enters stage two of the sequential operation, is added to the kettle (T-5), and is then circulated to the top of the crystallizer and distributed as a falling film inside the tubes. Nucleation is induced at the inside walls and a crystal layer starts to grow. Temperature of the coolant is progressively lowered to compensate for reduced heat

transfer and lower melt freezing point until the thickness inside the tube is between 5 and 20 millimeters depending on the product. Kettle liquid is evacuated to the first-stage holding tank (T-3) for eventual recrystallization at a lower temperature to maximize product yield and to strip product from the final liquid residue. Semirefined product frozen to the inside of the tube during operation of stage two is first heated above its melting point and slightly melted (sweated). This semipurified melted material (sweat) is removed from the crystallizer kettle, stored in a stage tank (T-4), and then added to the next batch of fresh feed. The remaining material inside the crystallizer is then melted, mixed with product sweat from stage three, recrystallized, and sweated to upgrade the purity even further (stage 3).

Commercial Equipment and Applications The falling-film crystallization process was invented by the MWB company in Switzerland. The process is now marketed by Sulzer Chemtech. Products successfully processed in the falling-film crystallizer are listed on Table 22-9. The falling-film crystallization process is available from the Chemtech Div. of Sulzer Canada Inc., 60 Worcester Rd., Rexdale, Ontario N9W 5X2 Canada.

TABLE 22-9 Fractional Crystallization Reference List

Product	Main characteristics	Capacity, tons/year	Purity	Type of plant	Country	Client
Acrylic acid	Very low aldehyde content, no undesired polymerization in the plant	Undisclosed Undisclosed	99.95% 99.9%	Falling film Falling film	Undisclosed Undisclosed	Undisclosed Undisclosed
Benzoic acid	Pharmaceutical grade, odor- and color free	4,500	99.97%	Falling film	Italy	Chimica del Friuli
Bisphenol A	Polycarbonate grade, no solvent required	150,000	Undisclosed	Falling film	USA	General Electric
Carbonic acid		1,200	Undisclosed	Falling film	Germany	Undisclosed
Fatty acid	Separation of tallow fatty acid into saturated and unsaturated fractions	20,000	Stearic acid: Iodine no. 2 Oleic acid: Cloud pt 5°C	Falling film	Japan	Undisclosed
Fine chemicals		<1,000 <1,000 <1,000 <1,000 <1,000 <1,000 <1,000	Undisclosed Undisclosed Undisclosed Undisclosed Undisclosed Undisclosed Undisclosed	Falling film Static Falling film Falling film Falling film Falling film Falling film	GUS Switzerland Switzerland Switzerland USA Germany Japan	Undisclosed Undisclosed Undisclosed Undisclosed Undisclosed Undisclosed Undisclosed
Hydrazine	Satellite grade	3	>99.9%	Falling film	Germany	ESA
Monochloro acetic acid (MCA)	Low DCA content	6,000	>99.2%	Falling film	USA	Undisclosed
Multipurpose	Separation or purification of two or more chemicals, alternatively	1,000 1,000	Various grades Undisclosed	Falling film Falling film	Belgium Belgium	UCB Reibelco
Naphthalene	Color free and color stable with low thionaphthene content	60,000 20,000 10,000 12,000	99.5% 99.5% 99.8% Various grades	Falling film Falling film/static Falling film/static Falling film	Germany P.R. China P.R. China The Netherlands	Rütgers-Werke Anshan Jining Cindu Chemicals
p-Dichlorobenzene	No solvent washing required	40,000 5,000 4,000 3,000 3,000	99.95% 99.98% 99.8% 99.95% >97%	Falling film Falling film Falling film/distillation Falling film Static	USA Japan Brazil P.R. China P.R. China	Standard Chlorine Toa Gosei Nitroclor Fuyang Shandong
p-Nitrochlorobenzene		18,000 10,000	99.3% 99.5%	Falling film/distillation Static	P.R. China India	Jilin Chemical Mardia
Toluene diisocyanate (TDI)	Separation of TDI 80 into TDI 100 & TDI 65	22,000	Undisclosed	Falling film	Undisclosed	Undisclosed
Trioxan		<1,000	99.97%	Falling film	Undisclosed	Undisclosed

SUPERCRITICAL FLUID SEPARATION PROCESSES

GENERAL REFERENCES: Bruno and Ely, *Supercritical Fluid Technology: Reviews in Modern Theory and Applications*, CRC Press, Boca Raton, 1991. Brunner, *Gas Extraction: An Introduction to Fundamentals of Supercritical Fluids and the Application to Separation Processes*, Springer, New York, 1994. Gloyna and Li, *Waste Management*, **13**, 379 (1993). Johnston and Penninger, *Supercritical Fluid Science and Technology*, Am. Chem. Soc. Symp. Series, **406**, 1989. Kiran and Brennecke, *Am. Chem. Soc. Symp. Ser.*, **512**, 1993. Kiran and Sengers, *Supercritical Fluids: Fundamentals for Application;* NATO Adv. Study Inst. Series E, 1994, Kluwer, Boston, 1994, p. 273. McHugh and Krukonis, *Supercritical Fluid Extraction: Principles and Practice, 2d ed.*, Butterworth-Heinemann, Boston, 1994. Paulaitis, Krukonis, Kurnik, and Reid, *Rev. Chem. Eng.*, **1**, 179 (1983). Savage, Gopalan, Mizan, Martino, and Brock, *AIChE J.*, **41**, 1723 (1995). Shaw, Brill, Clifford, Eckert, and Franck, *Chem. Eng. News*, **69**, 26 (1991). Stahl, Quirin, and Gerard, *Dense Gases for Extraction and Refining*, Springer-Verlag, Berlin, 1988. Tester, Holgate, Armellini, Webley, Killilea, Hong, and Barnes, *Am. Chem. Soc. Symp. Ser.*, **518**, 1993, p. 35.

INTRODUCTION

Fluids above their critical temperatures and pressures, called supercritical fluids (SCFs), exhibit properties intermediate between those of gases and liquids. Consequently, each of these two boundary conditions shed insight into the nature of these fluids. Unlike gases, SCFs possess a considerable solvent strength, transport properties are more favorable (e.g., lower viscosities and higher diffusion coefficients, than in liquid solvents). In regions where a SCF is highly compressible, its density and hence its solvent strength may be adjusted over a wide range with modest variations in temperature and pressure. This tunability may be used to control phase behavior, separation processes (e.g., SCF extraction), rates and selectivities of chemical reactions, and morphologies in materials processing. A variety of advantages of SCF separation processes are given in Table 22-10. In some cases these advantages compensate for the disadvantage of the need for elevated pressure. Despite the diversity of SCF separation processes (see Table 22-11), an attempt will be made to identify unifying themes.

TABLE 22-10   Advantages of Supercritical Fluid Separations

Adjustable solvent strength to tailor selectivities and yields.
Higher diffusion coefficients and lower viscosities compared with liquids.
Rapid diffusion of CO_2 through condensed phases, e.g. polymers.
Solvent recovery is fast and complete, with minimal residue in product.
Properties of CO_2 as a solvent:
   Environmentally acceptable solvent for waste minimization, nontoxic,
      nonflammable, inexpensive, usable at mild temperatures.
Properties of water as a solvent:
   Nontoxic, nonflammable substitute for organic solvents.
   Extremely wide variation in solvent strength with temperature and pressure.
Collapse of structure due to capillary forces is prevented during solvent removal.

TABLE 22-11   Commercial Applications of Supercritical Fluid Separations Technology

Extraction of foods and pharmaceuticals
   Coffee and tea decaffeination
   Flavors from hops
   Cholesterol and fat from eggs
   Nicotine from tobacco
   Acetone from antibiotics
Extraction of organics from water
Extraction of volatile substances from substrates
   Drying and aerogel formation
   Cleaning, e.g. quartz rods for light guide fibers
   Removal of monomers, oligomers, and solvent from polymers
Fractionation
   Residuum oil supercritical extraction-petroleum deasphalting
   Polymer fractionation
   Edible oils fractionation
Analytical SCF extraction and chromatography
Reactive separations
   Extraction of sec-butanol from isobutene
   Hydrothermal oxidation of organic wastes in water

The two fluids most often studied in supercritical fluid technology, carbon dioxide and water, are the two least expensive of all solvents. Carbon dioxide is nontoxic, nonflammable, and has a near-ambient critical temperature of 31.1°C. CO_2 is an environmentally friendly substitute for organic solvents including chlorocarbons and chlorofluorocarbons. Supercritical water ($T_c = 374$°C) is of interest as a substitute for organic solvents to minimize waste in extraction and reaction processes. Additionally, it is used for hydrothermal oxidation of hazardous organic wastes (also called supercritical water oxidation) and hydrothermal synthesis.

PHYSICAL PROPERTIES OF PURE SUPERCRITICAL FLUIDS

Thermodynamic Properties   The variation in solvent strength of a supercritical fluid from gaslike to liquidlike values may be described qualitatively in terms of the density, ρ, or the solubility parameter, δ (square root of the cohesive energy density). It is shown for gaseous, liquid, and SCF CO_2 as a function of pressure in Fig. 22-17 according to the rigorous thermodynamic definition:

$$\delta = \left(\frac{u^{ig} - u}{v}\right)^{1/2} = \left(\frac{h^{ig} - RT - h + Pv}{v}\right)^{1/2} \quad (22\text{-}9)$$

where u is the internal energy, v is the molar volume, h is the enthalpy, and the superscript ig refers to the ideal gas. Similar characteristics are observed for a plot of other density-dependent variables versus pressure, e.g., density, enthalpy, entropy, viscosity, and diffusion coefficient. However, unlike δ, some of these properties decrease with density. The δ for gaseous carbon dioxide is essentially zero; whereas, the value for liquid carbon dioxide is like that of a hydrocarbon. At −30°C there is a large increase in δ upon condensation from vapor to liquid. Above the critical temperature, it is possible to tune the solubility parameter continuously over a wide range with either a small isothermal pressure change or a small isobaric temperature change. This ability to tune the solvent strength of a supercritical fluid is its

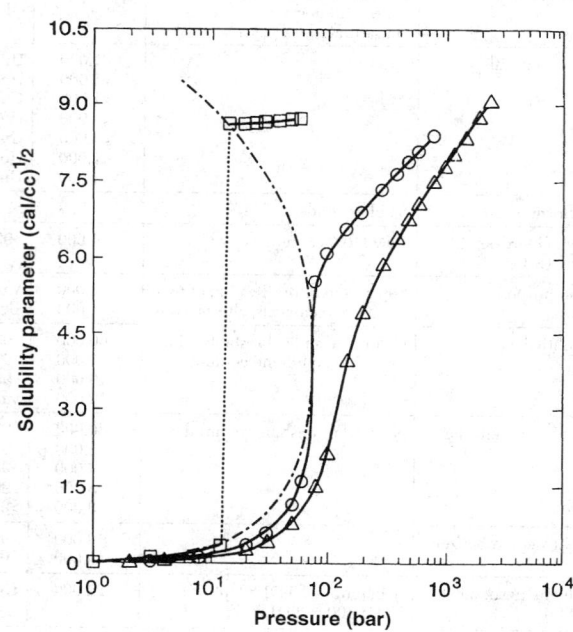

FIG. 22-17   Solubility parameter of CO_2 as a function of pressure in the gas, liquid, and supercritical states ($\cdots$: −30°C; O: 31°C; Δ: 70°C).

unique feature, and it can be used to extract and then recover selected products. Note that density and δ are more direct measures of the solvent strength of a SCF than pressure.

Although density (either mass or molar) is a good indicator of solvent strength for a single SCF, it is not a useful indicator for comparing different fluids. For example, CF_3Cl at 40°C and 1300 bar has a mass density of 1.95 g/cm^3, yet it is a weaker solvent than the much less dense fluid SCF CO_2 or liquid hexane. The same argument applies for SF_6. A better indicator of the van der Waals forces contributed by a SCF is obtained by multiplying ρ by the molecular polarizability, α, which is a constant for a given molecule. The solubility parameter δ of CO_2 can be misleading. It is larger than ethane's even though ethane has a larger value of $\alpha\rho$. However about 20 percent of δ for CO_2 may be attributed to its large quadrupole moment. For nonpolar solutes, where this quadrupole moment is unimportant, CO_2 is a much weaker solvent than n-hexane, and is more like fluorocarbons, which also have small values of $\alpha\rho$.

Water, a key SCF, undergoes profound changes upon heating to the critical point. It expands by a factor of 3 destroying about $\frac{2}{3}$ of the hydrogen bonds, and the dielectric constant drops from 80 to 5 (Shaw et al., op. cit.). (See Fig. 22-18.) *Supercritical water* (SCW) therefore behaves like a "nonaqueous" solvent, and it dissolves many organics and even gases such as O_2. At 400°C and 350 bar, the density of water is 0.47 g/mL, the dielectric constant, ε, is 10, and the ion product, K_w, is 7×10^{-14} compared with 10^{-14} at room temperature. Here, water behaves as a dense fluid which can dissolve electrolytes, with high diffusion coefficients and ion mobilities. At 500°C and the same pressure, the density of water is only 0.144 g/mL, ε is 2, and K_w is 2×10^{-20}. At these conditions, water is a high-temperature gas which does not solvate ions significantly.

Transport Properties   Although the densities of supercritical fluids approach those of conventional liquids, their transport properties are closer to those of gases, as shown for a typical SCF such as CO_2 in Table 22-12. For example, the viscosity is several orders of magnitude lower than at liquidlike conditions. The self-diffusion coefficient ranges between 10^{-3} and 10^{-5} cm^2/s, and binary-diffusion coefficients are similar [Liong, Wells, and Foster, *J. Supercritical Fluids* **4**, 91 (1991); Catchpole and King, *Ind. Eng. Chem. Research*, **33**,

1828 (1994)]. These values are as much as one hundred times larger than those typically observed in conventional liquids. The improved transport rates in SCFs versus liquid solvents are important in practical applications including supercritical extraction. Furthermore, carbon dioxide diffuses through condensed-liquid phases (e.g., adsorbents and polymers) faster than do typical solvents which have larger molecular sizes.

PROCESS CONCEPTS IN SUPERCRITICAL FLUID EXTRACTION

Figure 22-19 shows a one-stage extraction process that utilizes the adjustability of the solvent strength with pressure in a separation process. The solvent flows through the extraction chamber at a relatively high pressure to extract the components of interest from the feed. The products are then recovered in the separator by depressurization, and the solvent is recompressed and recycled. The products can also be precipitated from the extract phase by raising the temperature after the extraction to lower the solvent density. In the increasing pressure profiling approach, conditions are set so that only the lightest components in the feed are extracted in the first fraction. The recovery vessel is then replaced, and the pressure is increased to collect the next heavier fraction. In the multistage isothermal decreasing pressure profiling process, all but the heaviest fraction are extracted in the first vessel. The extract then passes through a series of recovery vessels held at successively lower pressures, each of which precipitates the next lower molecular-weight fraction in the raffinate. A new process, critical isobaric temperature-rising elution fractionation, is a supercritical variation on temperature-rising elution fractionation in a liquid solvent (McHugh and Krukonis, op. cit.).

Solids may be processed continuously or semicontinuously by pumping slurries or by using lock hoppers. An example is the separation of insoluble polymers by floatation with a variable-density SCF. For liquid feeds, multistage separation may be achieved by continuous counter-current extraction, much like conventional liquid-liquid extraction. The final products may be recovered from the extract phase by a depressurization, a temperature change, or by conventional distillation.

PHASE EQUILIBRIA

Liquid-Fluid Equilibria   Nearly all binary liquid-fluid phase diagrams can be conveniently placed in one of six classes (Fig. 22-20). Two-phase regions are represented by an area and three-phase regions by a line. In Class I, the two components are completely miscible, and a single critical mixture curve connects their critical points. Class II behavior is similar, except that a region of liquid-liquid immiscibility is found at lower temperatures. As the two components become increasingly dissimilar, the *upper critical solution temperature* (UCST) line merges with the branch of the critical mixture curve that begins at the heavier component's critical point, and Class III

TABLE 22-12   Density and Transport Properties of a Gas, Supercritical Fluid, and a Liquid

State	ρ (g/cm³)	μ (g/cm·s)	D (cm²/s)
Gas, 1 bar	10^{-3}	10^{-4}	0.2
SCF (T_c, P_c)	0.3	10^{-4}	10^{-3}
Liquid	1	10^{-2}	10^{-5}

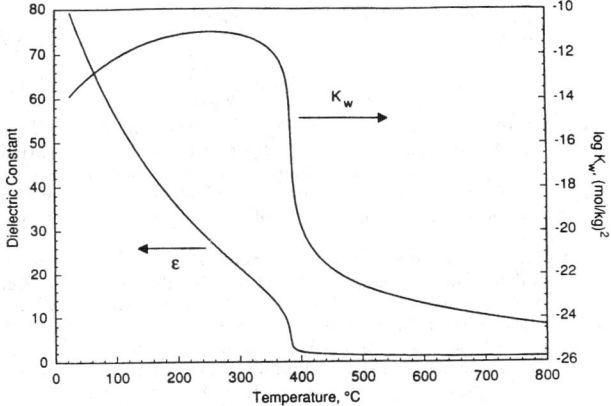

FIG. 22-18   Dielectric constant and dissociation constant, K_w, of water at 250 bar (*Tester et al., op. cit.*).

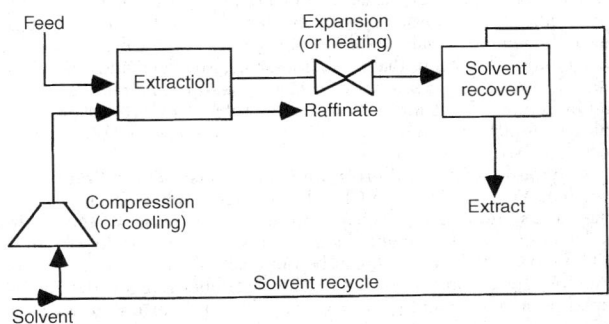

FIG. 22-19   Schematic diagram of a typical supercritical fluid-extraction process.

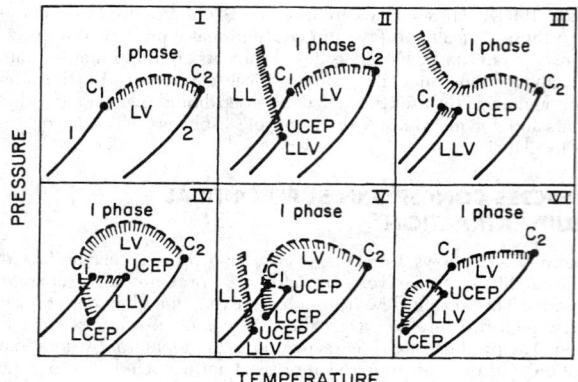

FIG. 22-20   Six classes of binary liquid-fluid phase diagrams (*Prausnitz et al., Molecular Thermodynamics of Fluid-Phase Equilibria*, © 1986. *Reprinted by permission of Prentice-Hall, Inc.*).

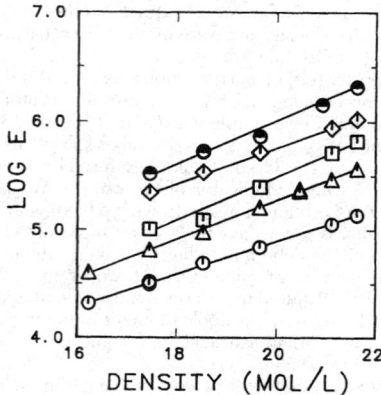

FIG. 22-21   Enhancement factor for solids with a variety of polar functionalities in CO_2 at 35°C (from bottom to top: hexamethylbenzene, 2-naphthol, phthalic anhydride, anthracene, acridine).

behavior is observed. In Class IV behavior, the mixture critical curve bends down to low pressures and intercepts the three-phase *liquid-liquid-vapor* (LLV) line at the lower critical end point. Class V resembles Class IV except it includes an additional LL critical curve. Class VI has features of Class II except that the critical curve intersects the LLV line twice.

For a ternary system, the phase diagram appears much like that in conventional liquid-liquid equilibrium. However, because a SCF solvent is compressible, the slopes of the tie lines (distribution coefficients) and the size of the two-phase region can vary significantly with pressure as well as temperature. Furthermore, at lower pressures, LLV tie-triangles appear upon the ternary diagrams and can become quite large.

Solid-Fluid Equilibria   The phase diagrams of binary mixtures in which the heavier component (the solute) is normally a solid at the critical temperature of the light component (the solvent) include *solid-liquid-vapor* (SLV) curves which may or may not intersect the LV critical curve. The solubility of the solid is very sensitive to pressure and temperature in compressible regions where the solvent's density and solubility parameter are highly variable. In contrast, plots of the log of the solubility versus density at constant temperature exhibit fairly simple linear behavior.

To understand the role of solute-solvent interactions on solubilities and selectivities, it is instructive to define an enhancement factor, E, as the actual solubility, y_2, divided by the solubility in an ideal gas, so that $E = y_2 P/P_2^{sat}$, where P_2^{sat} is the vapor pressure. This factor is a normalized solubility because it removes the effect of the vapor pressure, providing a means to focus on interactions in the SCF phase. For a given fluid at a particular temperature and pressure, enhancement factors do not vary much for many types of organic solids of similar molecular weight. As shown in Fig. 22-21, Es fall within a range of only about 1.5 orders of magnitude for substances with a variety of polar functional groups, even though the actual solubilities (not shown) vary by many orders of magnitude. This means that solubilities, and also selectivities, in carbon dioxide are governed primarily by vapor pressures and only secondarily by solute-solvent interactions in the SCF phase. However, fluid-phase interactions can be especially important if cosolvents are added which are strong Lewis acids or bases.

Polymer-Fluid Equilibria and the Glass Transition   Most polymer systems fall in the Class III or Class V phase diagrams, and the same system can often change from one class into the other as the polymer's molecular weight changes. Most polymers are insoluble in CO_2 below 100°C, yet CO_2 can be quite soluble in the polymer. For example, the sorption of CO_2 into silicone rubber is highly dependent upon temperature and pressure, since these properties have a large influence on the density and activity of CO_2.

For glassy polymers, sorption isotherms are more complex and hysteresis between the pressurization and depressurization steps may

appear. CO_2 adds free volume to the polymer that can relax very slowly. Furthermore, CO_2 can act as a plasticizer and depress the glass transition temperature by 100°C or even more. Not only do the mechanical properties change as the polymer is plasticized, but the diffusion coefficient of CO_2 and other solutes can increase by orders of magnitude. In PMMA, for instance, carbon dioxide's diffusion coefficient increases by as much as two orders-of-magnitude as the pressure is increased by 75 bar at 35°C.

Cosolvents and Surfactants   Many nonvolatile polar substances cannot be dissolved at moderate temperatures in nonpolar fluids such as CO_2. Cosolvents (also called entrainers, modifiers, moderators) such as alcohols and acetone have been added to fluids to raise the solvent strength. The addition of only 2 mol % of the complexing agent tri-*n*-butyl phosphate (TBP) to CO_2 increases the solubility of hydroquinone by a factor of 250 due to Lewis acid-base interactions. Very recently, surfactants have been used to form reverse micelles, microemulsions, and polymeric latexes in SCFs including CO_2. These organized molecular assemblies can dissolve hydrophilic solutes and ionic species such as amino acids and even proteins. Examples of surfactant tails which interact favorably with CO_2 include fluoroethers, fluoroacrylates, fluoroalkanes, propylene oxides, and siloxanes.

Phase Equilibria Models   Two approaches are available for modeling the fugacity of a solute, f_i, in a supercritical fluid solution. The compressed gas approach is the most common where:

$$f_i^G = y_i \phi_i P \qquad (22\text{-}10)$$

and ϕ_i is the fugacity coefficient of component i. The "expanded liquid" approach is given as:

$$f_i^L = x_i \gamma_i(P^0, x_i) f_i^{0L}(P^0) \exp\left\{ \int_{P^0}^{P} \frac{\bar{v}_i}{RT} \, dP \right\} \qquad (22\text{-}11)$$

where x_i is the mole fraction, γ_i is the activity coefficient, P^0 and f_i^0 are the reference pressure and fugacity, respectively, and $\bar{v}_i$ is the partial molar volume of component i. In principle this approach has an advantage in that γ_i can be chosen to give exact results at a pressure in the near-critical region, but the use of γ_i introduces an additional parameter.

A variety of equations-of-state have been applied to supercritical fluids, ranging from simple cubic equations like the Peng-Robinson equation-of-state to the Statistical Associating Fluid Theory. All are able to model nonpolar systems fairly successfully, but most are increasingly challenged as the polarity of the components increases. The key is to calculate the solute-fluid molecular interaction parameter from the pure-component properties. Often the standard approach (i.e. corresponding states based on critical properties) is of limited accuracy due to the vastly different critical temperatures of the solutes (if known) and the solvents; other properties of the solute are more appropriate [Johnston et al., *Ind. Eng. Chem. Research.*, **28**, 1115 (1989)].

MASS TRANSFER

Experimental gas-solid mass-transfer data have been obtained for naphthalene in CO_2 to develop correlations for mass-transfer coefficients [Lim et al., *Am. Chem. Soc. Symp. Ser.*, **406**, 379 (1989)]. The data were correlated over a wide range of conditions with the following equation for combined natural and forced convection:

$$Sh/(Sc \cdot Gr)^{1/4} = e(Re/Gr^{1/2})^f \qquad (22\text{-}12)$$

where Sh, Sc, Gr, and Re are the Sherwood, Schmidt, Grashof, and Reynolds numbers, respectively, and e and f are constants. The mass-transfer coefficient increases dramatically near the critical point, goes through a maximum and then decreases gradually. The strong natural convection at SCF conditions leads to higher mass-transfer rates than in liquid solvents.

A comprehensive mass-transfer model has been developed for SCF extraction from an aqueous phase to CO_2 in countercurrent sieve tray and packed columns [Seibert and Moosberg, *Sep. Sci. Technol.*, **23**, 2049 (1988)]. Both the hydraulics and mass-transfer coefficients were obtained from models developed for conventional liquid extraction, and the results were in good agreement with experiment for a 10-cm diameter column either with sieve trays or packing. If interfacial tensions are comparable, mass-transfer rates for extraction of organics from aqueous solutions are higher for CO_2 than hydrocarbon solvents. For this type of extraction, it was found that CO_2 preferentially wets ceramic and metal packings; consequently, trays are more efficient than packings.

APPLICATIONS

Food and Pharmaceutical Applications These applications are driven by the environmental acceptability of CO_2, as well as by the ability to tailor the extraction with the adjustable solvent strength. The General Foods coffee decaffeination plant in Houston, Texas is designed to process between 15,000 and 30,000 pounds of coffee beans per hour (McHugh and Krukonis, op. cit.). See Fig. 22-22. The moist, green coffee beans are charged to an extraction vessel approximately 7 ft diameter by 70 ft high, and carbon dioxide is used to extract the caffeine from the beans. Various methods have been proposed for recovery of the caffeine including washing with water and adsorption. Often the recovery of a particular component of an extract is the key challenge in SCF extraction. Thus, SCF extraction is frequently combined with another process such as distillation, absorption, or adsorption.

Temperature-Controlled Residuum Oil Supercritical Extraction (ROSE) The Kerr-McGee ROSE process has been licensed by over a dozen companies worldwide. The extraction step uses a liquid solvent, and the solvent is recovered at supercritical conditions to save energy as shown in Fig. 22-23. The residuum is contacted with butane or pentane to precipitate the heavy asphaltene fraction. The extract is then passed through a series of heaters, where it goes from the liquid state to a lower-density supercritical fluid state. Because the entire process is carried out at conditions near the critical point, a relatively small temperature change is required to produce a fairly large density change. After the light oils have been removed, the solvent is cooled back to the liquid state and recycled.

Extraction from Aqueous Solutions Critical Fluid Technologies, Inc. has developed a continuous countercurrent extraction process based on a 0.5- by 10-m column to extract residual organic solvents such as trichloroethylene, methylene chloride, benzene, and chloroform from industrial wastewater streams. Typical solvents include supercritical CO_2 and near-critical propane. The economics of these processes are largely driven by the hydrophilicity of the product, which has a large influence on the distribution coefficient. For example, at 16°C, the partition coefficient between liquid CO_2 and water is 0.4 for methanol, 1.8 for *n*-butanol, and 31 for *n*-heptanol.

Adsorption and Desorption Adsorbents may be used to recover solutes from supercritical fluid extracts; for example, activated carbon and polymeric sorbents may be used to recover caffeine from CO_2. This approach may be used to improve the selectivity of a supercritical fluid extraction process. SCF extraction may be used to regenerate adsorbents such as activated carbon and to remove contaminants from soil. In many cases the chemisorption is sufficiently strong that regeneration with CO_2 is limited, even if the pure solute is quite soluble in CO_2. In some cases a cosolvent can be added to the SCF to displace the sorbate from the sorbent. Another approach is to use water at elevated or even supercritical temperatures to facilitate desorption. Many of the principles for desorption are also relevant to extraction of substances from other substrates such as natural products and polymers.

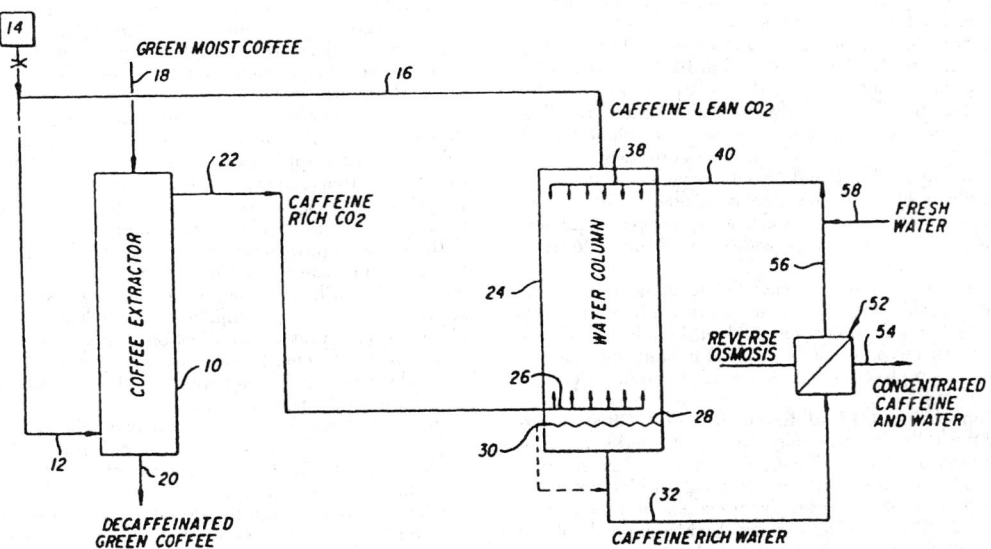

FIG. 22-22 Schematic diagram of the Kraft process for producing decaffeinated coffee using supercritical carbon dioxide (*McHugh and Krukonis, op. cit.*).

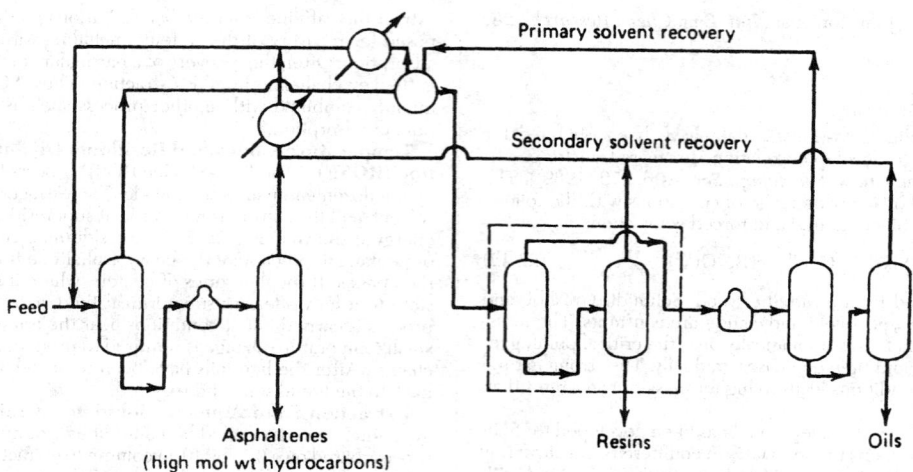

FIG. 22-23 Schematic diagram of the Kerr-McGee ROSE process.

Polymer Devolatilization and Fractionation Supercritical fluids may be used to extract solvent, monomers, and oligomers from polymers. After extraction the pressure is reduced to atmospheric leaving little residue in the substrate; furthermore, the extracted impurities are easily recovered from the SCF. To aid process design, partition coefficients of various solutes between polymers and CO_2 have been measured with static and dynamic techniques such as inverse supercritical fluid chromatography. The swelling and lowering of the glass transition temperature of the polymer by the SCF can increase mass-transfer rates markedly. The polymer may or may not return to its original dimensions, depending upon factors such as the glass transition properties and crystallinity.

Supercritical fluids may be used to fractionate polymers on the basis of molecular weight and/or composition. The most common techniques are isothermal increasing-pressure profiling and isothermal decreasing-pressure profiling as discussed in the above section on process concepts. The critical isobaric temperature rising elution fractionation process can be used to fractionate polymers as a function of crystallinity (e.g., due to branching), based on the melting points in the presence of the fluid (McHugh and Krukonis, op. cit.).

Drying and Aerogel Formation One of the oldest practical applications of supercritical fluids, developed in 1932, is supercritical fluid drying. Here the solvent is extracted from a porous solid with a SCF fluid, and then the fluid is depressurized. Because the fluid expands from the solid without crossing a liquid-vapor phase boundary, capillary forces are not present which would otherwise collapse the structure. Using supercritical fluid drying, aerogels have been prepared with densities so low that they essentially float in air and look like a cloud of smoke. The process is used in a commercial instrument to dry samples for electron microscopy without perturbing the structure.

Cleaning Supercritical fluids such as CO_2 are being used to clean and degrease quartz rods used to produce optical fibers, products used in the fabrication of printed circuit boards, oily chips from machining operations, and precision bearings in military applications, and so on. Here, CO_2 replaces conventional chlorocarbon or chlorofluorocarbon solvents.

Analytical Supercritical Fluid Extraction and Chromatography Supercritical fluids, especially CO_2, are used widely to extract a wide variety of solid and liquid matrices to obtain samples for analysis. Benefits compared with conventional Soxhlet extraction include minimization of solvent waste, faster extraction, tunability of solvent strength, and simple solvent removal with minimal solvent contamination in the sample. Compared with high-performance liquid chromatography, the number of theoretical stages is higher in

SCF chromatography due to the more favorable transport rates. A limitation in each of these applications is the low solvent strength of CO_2; often cosolvents are required.

Precipitation with a Compressed Fluid Antisolvent (PCA) Because fluids such as CO_2 are weak solvents, they are often more effective as antisolvents. In this process, the antisolvent may be a compressed gas, pressurized liquid, or a supercritical fluid. Mixing of a solution with the antisolvent leads to a precipitated product. There are two primary process configurations for this mixing: (1) a pure, compressed fluid may be added to a liquid solution or (2) a liquid solution may be sprayed through a nozzle into a pure, compressed fluid. Gaseous CO_2 is quite soluble in a number of organic solvents such as methanol, toluene, dimethylformamide, and tetrahydrofuran, at pressures from 10 to 100 bar. As CO_2 mixes with the liquid phase, it decreases the cohesive energy density (solvent strength) substantially, leading to precipitation of dissolved solutes (e.g., crystals of progesterone). It has been demonstrated that the rate of addition of a fluid antisolvent or the liquid solvent may be programmed to control crystal morphology, size, and size distribution over a wide range from 1 to 100 μm. The high-diffusion rates of the organic solvent into CO_2 and vice-versa can lead to rapid-phase separation. This process may be used to precipitate a solute from a solvent or for separation of solutes.

Crystallization Solutes may be crystallized from supercritical fluids by temperature and/or pressure changes, and by the PCA process described above. In the *rapid expansion from supercritical solution* (RESS) process, a SCF containing a dissolved solute is expanded through a nozzle or orifice in less than 1 ms to form small particles or fibers. A variety of inorganic crystals have been formed naturally and synthetically in SCF water.

Reactive Separations Reactions may be integrated with SCF separation processes to achieve a large degree of control for producing a highly purified product. Reaction products may be recovered by volatilization into, or precipitation from, a SCF phase. A classic example is the high-pressure production of polyethylene in the reacting solvent SCF ethylene. The molecular-weight distribution may be controlled by choosing the temperature and pressure for precipitating the polymer from the SCF phase.

In the last few years, Idemitsu commercialized a 5000 metric ton/year integrated reaction and separation process in SCF isobutene, as shown in Fig. 22-24. The reaction of isobutene and water takes place in the water phase and is acid catalyzed. The product, *sec*-butanol, is extracted into the isobutene phase to drive the reversible reaction to the right. The *sec*-butanol is then recovered from the isobutene by depressurizing the SCF phase, and the isobutene is recompressed and recycled.

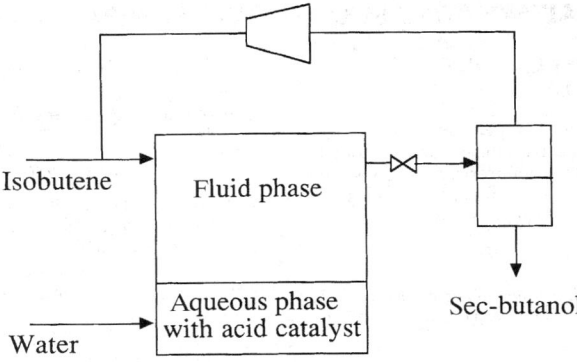

FIG. 22-24 Process for the integrated reaction and separation of *sec*-butanol from isobutene.

Supercritical fluid solvents have been tested for reactive extractions of liquid and gaseous fuels from heavy oils, coal, oil shale, and biomass. In some cases the solvent participates in the reactions, as in the hydrolysis of coal and heavy oils with water. Related applications include conversion of cellulose to glucose in water, delignification of wood with ammonia, and liquefaction of lignin in water.

Hydrothermal oxidation (HO) (also called supercritical water oxidation) is a reactive process to separate aqueous wastes into water, CO_2, nitrogen, salts, and other byproducts. It is an enclosed and complete water-treatment process making it more desirable to the public than incineration (Fig. 22-25) (Tester et al., op. cit.; Gloyna and Li,

op. cit.; Shaw et al., op. cit.). As mentioned above, organics and oxygen mix in a single phase in SCW due the low dielectric constant. Oxidation is rapid and efficient in this one-phase solution, so that wastewater containing 1 to 20 wt % organics may be oxidized rapidly in SCW with higher energy efficiency and much less air pollution than in conventional incineration. Temperatures range from about 375 to 650°C and pressures from 3000 to about 5000 psia. Conversions can be greater than 99.99 percent for reactor residence times of a minute or less. Organics are oxidized to CO_2, H_2O, and molecular nitrogen with little NO_x. A commercial plant designed by Eco-Waste Technology appeared in Austin, Texas in 1994.

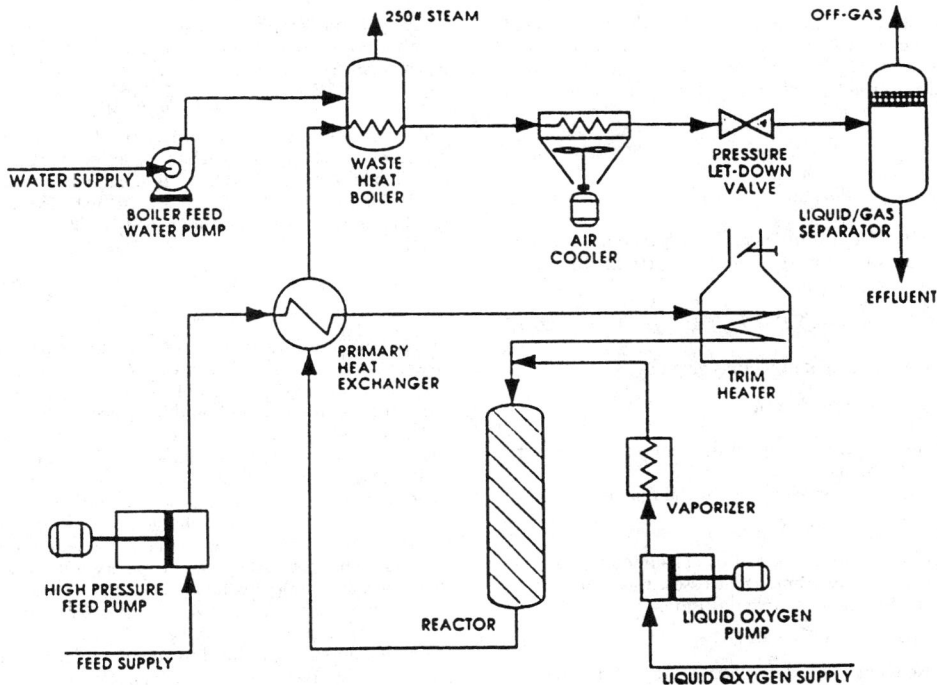

FIG. 22-25 Hydrothermal-oxidation process (also called supercritical water oxidation) for wastewater purification. (*Courtesy Eco-Waste Technologies.*)

ALTERNATIVE SOLID/LIQUID SEPARATIONS

SEPARATION PROCESSES BASED PRIMARILY ON ACTION IN AN ELECTRIC FIELD

Differences in mobilities of ions, molecules, or particles in an electric field can be exploited to perform useful separations. Primary emphasis is placed on electrophoresis and dielectrophoresis. Analogous separation processes involving magnetic and centrifugal force fields are widely applied in the process industry (see Secs. 18 and 19).

Theory of Electrical Separations

GENERAL REFERENCES: Newman, *Adv. Electrochem. Electrochem. Eng.,* **5,** 87 (1967); *Ind. Eng. Chem.,* **60**(4), 12 (1968). Ptasinski and Kerkhof, *Sep. Sci. Technol.,* **27,** 995 (1992).

For electrolytic solutions, migration of charged species in an electric field constitutes an additional mechanism of mass transfer. Thus the flux of an ionic species N_i in $(\text{g·mol})/(\text{cm}^2\text{·s})$ in dilute solutions can be expressed as

$$N_i = -z_i u_i \mathscr{F} c_i \nabla E - D_i \nabla c_i + c_i v \qquad (22\text{-}13)$$

The ionic mobility u_i is the average velocity imparted to the species under the action of a unit force (per mole). v is the stream velocity, cm/s. In the present case, the electrical force is given by the product of the electric field ∇E in V/cm and the charge $z_i \mathscr{F}$ per mole, where $\mathscr{F}$ is the Faraday constant in C/g equivalent and z_i is the valence of the ith species. Multiplication of this force by the mobility and the concentration c_i $[(\text{g·mol})/\text{cm}^3]$ yields the contribution of migration to the flux of the ith species.

The diffusive and convective terms in Eq. (22-13) are the same as in nonelectrolytic mass transfer. The ionic mobility u_i, $(\text{g·mol·cm}^2)/(\text{J·s})$, can be related to the ionic-diffusion coefficient D_i, cm²/s, and the ionic conductance of the ith species λ_i, cm²/(Ω·g equivalent):

$$u_i = D_i/RT = \lambda_i / |z_i| \mathscr{F}^2 \qquad (22\text{-}14)$$

where T is the absolute temperature, K; and R is the gas constant, 8.3143 J/(K·mol). Ionic conductances are tabulated in the literature (Robinson and Stokes, *Electrolyte Solutions,* Academic, New York, 1959). For practical purposes, a bulk electrolytic solution is electrically neutral.

$$\sum_i z_i c_i = 0 \qquad (22\text{-}15)$$

since the forces required to effect an appreciable separation of charge are prohibitively large.

The **current density** (A/cm²) produced by movement of charged species is described by summing the terms in Eq. (22-16) for all species:

$$i = \mathscr{F} \sum_i z_i \mathbf{N}_i = -\kappa \nabla E - \mathscr{F} \sum_i z_i D_i \nabla c_i \qquad (22\text{-}16)$$

where the electrical conductivity κ in S/cm is given by

$$\kappa = \mathscr{F}^2 \sum_i z_i^2 u_i c_i \qquad (22\text{-}17)$$

In solutions of uniform composition, the diffusional terms vanish and Eq. (22-16) reduces to Ohm's law.

Conservation of each species is expressed by the relation

$$\partial c_i / \partial t = -\nabla \cdot \mathbf{N}_i \qquad (22\text{-}18)$$

provided that the species is not produced or consumed in homogeneous chemical reactions. In two important cases, this conservation law reduces to the equation of convective diffusion:

$$(\partial c_i / \partial t) + \mathbf{v}\nabla \cdot c_i = D \nabla^2 c_i \qquad (22\text{-}19)$$

First, when a large excess of inert electrolyte is present, the electric field will be small and migration can be neglected for minor ionic components; Eq. (22-19) then applies to these minor components, where D is the ionic-diffusion coefficient. Second, Eq. (22-19) applies when the solution contains only one cationic and one anionic species.

The electric field can be eliminated by means of the electroneutrality relation.

In the latter case the diffusion coefficient D of the electrolyte is given by

$$D = (z_+ u_+ D_- - z_- u_- D_-)/(z_+ u_+ - z_- u_-) \qquad (22\text{-}20)$$

which represents a compromise between the diffusion coefficients of the two ions. When Eq. (22-19) applies, many solutions can be obtained by analogy with heat transfer and nonelectrolytic mass transfer.

Because the solution is electrically neutral, conservation of charge is expressed by differentiating Eq. (22-16):

$$\nabla \cdot \mathbf{i} = 0 = -\kappa \nabla^2 E - \mathscr{F} \sum_i z_i D_i \nabla^2 c_i \qquad (22\text{-}21)$$

For solutions of uniform composition, Eq. (22-21) reduces to Laplace's equation for the potential:

$$\nabla^2 E = 0 \qquad (22\text{-}22)$$

This equation is the starting point for determination of the current-density distributions in many electrochemical cells.

Near an interface or at solution junctions, the solution departs from electroneutrality. Charges of one sign may be preferentially adsorbed at the interface, or the interface may be charged. In either case, the charge at the interface is counterbalanced by an equal and opposite charge composed of ions in the solution. Thermal motion prevents this countercharge from lying immediately adjacent to the interface, and the result is a "diffuse-charge layer" whose thickness is on the order of 10 to 100 Å.

A tangential electric field ∇E_t acting on these charges produces a relative motion between the interface and the solution just outside the diffuse layer. In view of the thinness of the diffuse layer, a balance of the tangential viscous and electrical forces can be written

$$\mu(\partial^2 v_t / \partial y^2) + \rho_e \nabla E_t = 0 \qquad (22\text{-}23)$$

where μ is the viscosity and ρ_e is the electric-charge density, C/cm³. Furthermore, the variation of potential with the normal distance satisfies Poisson's equation:

$$\partial^2 E / \partial y^2 = -(\rho_e / \varepsilon) \qquad (22\text{-}24)$$

with ε defined as the **permittivity** of the solution. [The relative dielectric constant is $\varepsilon/\varepsilon_0$, where ε_0 is the permittivity of free space; $\varepsilon_0 = 8.8542 \times 10^{-14}$ C/(V·cm).] Elimination of the electric-charge density between Eqs. (22-23) and (22-24) with two integrations, gives a relation between ∇E_t and the velocity v_0 of the bulk solution relative to the interface.

$$\mu[v_t(\infty) - v_t(0)] = \varepsilon \nabla E_t[E(\infty) - E(0)] \qquad (22\text{-}25)$$

or

$$v_0 = -(\varepsilon \nabla E_t \zeta / \mu) \qquad (22\text{-}26)$$

The potential difference across the mobile part of the diffuse-charge layer is frequently called the **zeta potential,** $\zeta = E(0) - E(\infty)$. Its value depends on the composition of the electrolytic solution as well as on the nature of the particle-liquid interface.

There are four related electrokinetic phenomena which are generally defined as follows: *electrophoresis*—the movement of a charged surface (i.e., suspended particle) relative to a stationary liquid induced by an applied electrical field, *sedimentation potential*—the electric field which is crested when charged particles move relative to a stationary liquid, *electroosmosis*—the movement of a liquid relative to a stationary charged surface (i.e., capillary wall), and *streaming potential*—the electric field which is created when liquid is made to flow relative to a stationary charged surface. The effects summarized by Eq. (22-26) form the basis of these electrokinetic phenomena.

For many particles, the diffuse-charge layer can be characterized adequately by the value of the zeta potential. For a spherical particle of radius r_0 which is large compared with the thickness of the diffuse-charge layer, an electric field uniform at a distance from the particle will produce a tangential electric field which varies with position on the particle. Laplace's equation [Eq. (22-22)] governs the distribution of potential outside the diffuse-charge layer; also, the Navier-Stokes equation for a creeping-flow regime can be applied to the velocity distribution. On account of the thinness of the diffuse-charge layer, Eq. (22-26) can be used as a local boundary condition, accounting for the effect of this charge in leading to movement of the particle relative to the solution. The result of this computation gives the velocity of the particle as

$$v = \varepsilon \zeta \, \nabla E / \mu \tag{22-27}$$

and it may be convenient to tabulate the mobility of the particle

$$U = v / \nabla E = \varepsilon \zeta / \mu \tag{22-28}$$

rather than its zeta potential. Note that this mobility gives the velocity of the particle for unit electric field rather than for unit force on the particle. Related equations can be developed for the velocity of electroosmotic flow. The subsections presented below ("Electrophoresis," "Electrofiltration," and "Cross-Flow–Electrofiltration") represent both established and emerging commercial applications of electrokinetic phenomena.

Electrophoresis

GENERAL REFERENCE: Wankat, *Rate-Controlled Separations*, Elsevier, London, 1990.

Electrophoretic Mobility Macromolecules move at speeds measured in tenths of micrometers per second in a field (gradient) of 1 V/cm. Larger particles such as bubbles or bacteria move up to 10 times as fast because U is usually higher. To achieve useful separations, therefore, voltage gradients of 10 to 100 V/cm are required. High voltage gradients are achieved only at the expense of power dissipation within the fluid, and the resulting heat tends to cause undesirable convection currents.

Several devices are available commercially to measure mobility. One of these (Zeta-Meter Inc., New York) allows direct microscopic measurement of individual particles. Another allows measurement in more concentrated suspensions (Numinco Instrument Corp., Monroeville, Pa.). The state of the charge can also be measured by a streaming-current detector (Waters Associates, Inc., Framingham, Mass.). For macromolecules, more elaborate devices such as the Tiselius moving-boundary apparatus are used.

Mobility is affected by the dielectric constant and viscosity of the suspending fluid, as indicated in Eq. (22-28). The ionic strength of the fluid has a strong effect on the thickness of the double layer and hence on ζ. As a rule, mobility varies inversely as the square root of ionic strength [Overbeek, *Adv. Colloid Sci.*, **3**, 97 (1950)].

Modes of Operation There is a close analogy between sedimentation of particles or macromolecules in a gravitational field and their electrophoretic movement in an electric field. Both types of separation have proved valuable not only for analysis of colloids but also for preparative work, at least in the laboratory. Electrophoresis is applicable also for separating mixtures of simple cations or anions in certain cases in which other separating methods are ineffectual.

Electrodecantation or electroconvection is one of several operations in which one mobile component (or several) is to be separated out from less mobile or immobile ones. The mixture is introduced between two vertical semipermeable membranes; for separating cations, anion membranes are used, and vice versa. When an electric field is applied, the charged component migrates to one or another of the membranes; but since it cannot penetrate the membrane, it accumulates at the surface to form a dense concentrated layer of particles which will sink toward the bottom of the apparatus. Near the top of the apparatus immobile components will be relatively pure. Murphy [*J. Electrochem. Soc.*, **97**(11), 405 (1950)] has used silver-silver chloride electrodes in place of membranes. Frilette [*J. Phys. Chem.*, **61**, 168 (1957)], using anion membranes, partially separated H+ and Na+, K+ and Li+, and K+ and Na+. Unfortunately no simple electrodecantation apparatus is available for bench-scale testing. A rather complex device described by Polson and Largier [in Alexander and Block (eds.), *Analytical Methods of Protein Chemistry*, vol. I, Pergamon, New York, 1960] is available commercially (Quickfit Reeve Angel, Inc., Clifton, NJ).

Countercurrent electrophoresis can be used to split a mixture of mobile species into two fractions by the electrical analog of elutriation. In such countercurrent electrophoresis, sometimes termed an ion still, a flow of the suspending fluid is maintained parallel to the direction of the voltage gradient. Species which do not migrate fast enough in the applied electric field will be physically swept out of the apparatus. An apparatus based mainly on this principle but using also natural convection currents has been developed (Bier, *Electrophoresis*, vol. II, Academic, New York, 1967).

Membrane electrophoresis which is based upon differences in ion mobility, has been studied by Glueckauf and Kitt [*J. Appl. Chem.*, **6**, 511 (1956)]. Partial exclusion of coions by membranes results in large differences in coion mobilities. Superposing a cation and an anion membrane gives high transference numbers (about 0.5) for both cations and anions while retaining the selectivity of mobilities. Large voltages are required, and flow rates are low.

In continuous-flow zone electrophoresis the "solute" mixture to be separated is injected continuously as a narrow source within a body of carrier fluid flowing between two electrodes. As the "solute" mixture passes through the transverse field, individual components migrate sideways to produce zones which can then be taken off separately downstream as purified fractions.

Resolution depends upon differences in mobilities of the species. Background electrolyte of low ionic strength is advantageous, not only to increase electrophoretic (solute) mobilities, but also to achieve low electrical conductivity and thereby to reduce the thermal-convection current for any given field [Finn, in Schoen (ed.), *New Chemical Engineering Separation Techniques*, Interscience, New York, 1962].

The need to limit the maximum temperature rise has resulted in two main types of apparatus, illustrated in Fig. 22-26. The first consists of multicomponent ribbon separation units—apparatus capable of separating small quantities of mixtures which may contain few or many species. In general, such units operate with high voltages, low currents, a large transverse dimension, and a narrow thickness between cooling faces. Numerous units developed for analytic chemistry, generally with filter-paper curtains but sometimes with granular "anticonvectant" packing, are of this type. The second type consists of block separation units—apparatus designed to separate larger quantities of a mixture into two (or at most three) species or fractions. Such units generally use low to moderate voltages and high currents, with cooling by circulation of cold electrolyte through the electrode compartments. Scale-up can readily be accomplished by extending the thickness dimension w.

Both types of units have generally been operated in trace mode; that is, "background" or "elutant" electrolyte is fed to the unit along with the mixture to be separated. A desirable and possible means of

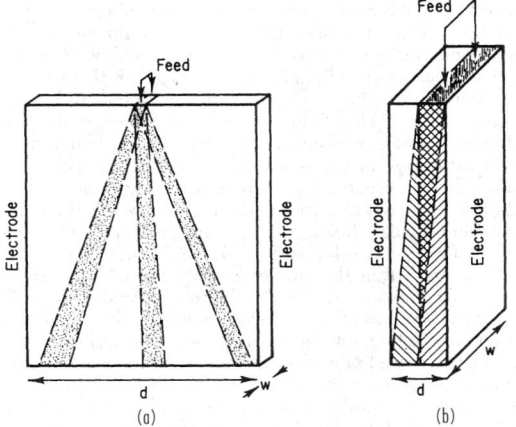

FIG. 22-26 Types of arrangement for zone electrophoresis or electrochromatography. (*a*) Ribbon unit, with $d > w$; cooling at side faces. (*b*) Block unit, with $w > d$; cooling at electrodes.

operation for preparative applications is in bulk mode, in which one separated component follows the other without background electrolyte being present, except that other ions may be required to bracket the separated zones. Overlap regions between components should be recycled, and pure components collected as products.

For block units, the need to stabilize flow has given rise to a number of distinct techniques.

Free flow. Dobry and Finn [*Chem. Eng. Prog.,* **54,** 59 (1958)] used upward flow, stabilized by adding methyl cellulose, polyvinyl alcohol, or dextran to the background solution. Upward flow was also used in the electrode compartments, with cooling efficiency sufficient to keep the main solution within 1°C of entering temperature.

Density gradients to stabilize flow have been employed by Philpot [*Trans. Faraday Soc.,* **36,** 38 (1940)] and Mel [*J. Phys. Chem.,* **31,** 559 (1959)]. Mel's Staflo apparatus [*J. Phys. Chem.,* **31,** 559 (1959)] has liquid flow in the horizontal direction, with layers of increasing density downward produced by sucrose concentrations increasing to 7.5 percent. The solute mixture to be separated is introduced in one such layer. Operation at low electrolyte concentrations, low voltage gradients, and low flow rates presents no cooling problem.

Packed beds. A packed cylindrical electrochromatograph 9 in (23 cm) in diameter and 48 in (1.2 m) high, with operating voltages in the 25- to 100-V range, has been developed by Hybarger, Vermeulen, and coworkers [*Ind. Eng. Chem. Process Des. Dev.,* **10,** 91 (1971)]. The annular bed is separated from inner and outer electrodes by porous ceramic diaphragms. The unit is cooled by rapid circulation of cooled electrolyte between the diaphragms and the electrodes.

An interesting modification of zone electrophoresis resolves mixtures of ampholytes on the basis of **differing isoelectric points** rather than differing mobilities. Such **isoelectric spectra** develop when a pH gradient is established parallel to the electric field. Each species then migrates until it arrives at the region of pH where it possesses no net surface charge. A strong focusing effect is thereby achieved [Kolin, in Glick (ed.), *Methods of Biochemical Analysis,* vol. VI, Interscience, New York, 1958].

Electrofiltration

GENERAL REFERENCE: P. Krishnaswamy and P. Klinkowski, "Electrokinetics and Electrofiltration," in *Advances in Solid-Liquid Separation,* H. S. Muralidhara (ed.), Battelle Press, Columbus, OH, 1986.

Process Concept The application of a direct electric field of appropriate polarity when filtering should cause a net charged-particle migration relative to the filter medium (electrophoresis). The same direct electric field can also be used to cause a net fluid flow relative to the pores in a fixed filter cake or filter medium (electroosmosis). The exploitation of one or both of these phenomena form the basis of conventional electrofiltration.

In conventional filtration, often the object is to form a high-solids-content filter cake. At a single-filter surface, a uniform electric field can be exploited in one of two ways. The first method of exploitation occurs when the electric field is of a polarity such that the charged-particle migration occurs toward the filter medium. In this case, the application of the electric field increases the velocity of the solid particles toward the filter surface (electrosedimentation), thereby hastening the clarification of the feed suspension and, at the same time, increasing the compaction of the filter cake collected on the filter surface. In this first case, electroosmotic flow occurs in a direction away from the filter media. The magnitude of the pressure-driven fluid flow toward the filter surface far exceeds the magnitude of the electroosmotic flow away from the surface so that the electroosmotic flow results in only a minor reduction of the rate of production of filtrate. The primary benefits of the applied electric field in this case are increased compaction, and hence increased dewatering, of the filter cake and an increased rate of sedimentation or movement of the particles in bulk suspension toward the filter surface.

The second method of exploitation occurs when the electric field is of a polarity such that the charged-particle migration occurs away from the filter medium. The contribution to the net-particle velocity of the electrophoretically induced flow away from the filter medium is generally orders of magnitude less than the contribution to the net-

particle velocity of the flow induced by drag due to the pressure-induced flow of the bulk liquid toward the filter media. (In conventional or cake filtration, the velocity of liquid in dead-end flow toward the filter is almost always sufficient to overcome any electrophoretic migration of particles away from the filter media so that the prevention of the formation of filter cake is not an option. This will not necessarily be the case for cross-flow electrofiltration.) The primary enhancement to filtration caused by the application of an electrical field in this manner is the increase in the filtrate flux due to electroosmotic flow through the filter cake. This electroosmotic flow is especially beneficial during the latter stages of filtration when the final filter-cake thickness has been achieved. At this stage, electroosmosis can be exploited to draw filtrate out from the pore structure of the filter cake. This type of drying of the filter cake is sometimes called **electroosmotic dewatering.**

Commercial Applications Krishnaswamy and Klinkowski, op. cit., describe the Dorr-Oliver EAVF®. The EAVF® combines vacuum filtration with electrophoresis and electroosmosis and has been described as a series of parallel platelike electrode assemblies suspended in a tank containing the slurry to be separated. When using the EAVF®, solids are collected at both electrodes, one collecting a compacted cake simply by electrophoretic attraction and the second collecting a compacted cake though vacuum filtration coupled with electroosmotic dewatering. Upon the completion of a collection cycle, the entire electrode assembly is withdrawn from the slurry bath and the cake is removed. The EAVF® is quoted as being best suited for the dewatering of ultrafine slurries (particle sizes typically less than 10 μm).

Cross-Flow–Electrofiltration

GENERAL REFERENCES: Henry, Lawler, and Kuo, *Am. Inst. Chem. Eng. J.,* **23**(6), 851 (1977). Kuo, Ph.D. dissertation, West Virginia University, 1978.

Process Concept The application of a direct electric field of appropriate polarity when filtering should cause a net charged-particle migration away from the filter medium. This electrophoretic migration will prevent filter-cake formation and the subsequent reduction of filter performance. An additional benefit derived from the imposed electric field is an electroosmotic flux. The presence of this flux in the membrane and in any particulate accumulation may further enhance the filtration rate.

Cross-flow–electrofiltration (CF-EF) is the multifunctional separation process which combines the electrophoretic migration present in electrofiltration with the particle diffusion and radial-migration forces present in cross-flow filtration (CFF) (microfiltration includes cross-flow filtration as one mode of operation in "Membrane Separation Processes" which appears later in this section) in order to reduce further the formation of filter cake. Cross-flow–electrofiltration can even eliminate the formation of filter cake entirely. This process should find application in the filtration of suspensions when there are charged particles as well as a relatively low conductivity in the continuous phase. Low conductivity in the continuous phase is necessary in order to minimize the amount of electrical power necessary to sustain the electric field. Low-ionic-strength aqueous media and nonaqueous suspending media fulfill this requirement.

Cross-flow–electrofiltration has been investigated for both aqueous and nonaqueous suspending media by using both rectangular- and tubular-channel processing configurations (Fig. 22-27). Henry, Lawler, and Kuo (op. cit.), using a rectangular-channel system with a 0.6-μ-pore-size polycarbonate Nuclepore filtration membrane, investigated CF-EF for 2.5-μm kaolin-water and 0.5- to 2-μm oil-in-water emulsion systems. Kuo (op. cit.), using similar equipment, studied 5-μm kaolin-water, ~100-μm Cr_2O_3-water, and ~6-μm Al_2O_3-methanol and/or -butanol systems. For both studies electrical fields of 0 to 60 V/cm were used for aqueous systems, and to 5000 V/cm were used for nonaqueous systems. The studies covered a wide range of processing variables in order to gain a better understanding of CF-EF fundamentals. Lee, Gidaspow, and Wasan [*Ind. Eng. Chem. Fundam.,* **19**(2), 166 (1980)] studied CF-EF by using a porous stainless-steel tube (pore size = 5 μm) as the filtration medium. A platinum wire running down the center of the tube acted as one electrode, while the

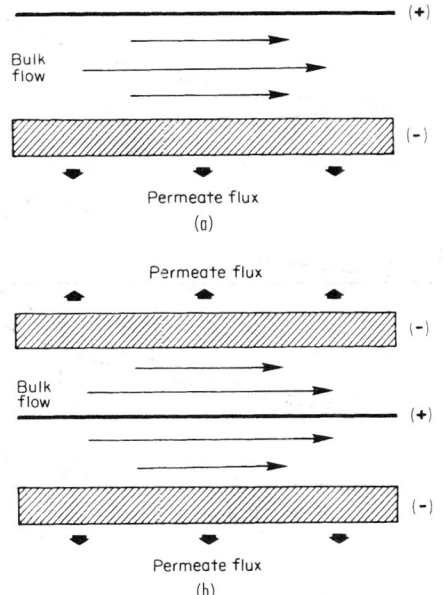

FIG. 22-27 Alternative electrode configurations for cross-flow–electrofiltration.

porous steel tube itself acted as the other electrode. Nonaqueous suspensions of 0.3- to 2-μm Al_2O_3-tetralin and a coal-derived liquid diluted with xylene and tetralin were studied. By operating with applied electric fields (1000 to 10,000 V/cm) above the critical voltage, clear particle-free filtrates were produced. It should be noted that the pore size of the stainless-steel filter medium (5 μm) was greater than the particle size of the suspended Al_2O_3 solids (0.3 to 2 μm). Crossflow—electrofiltration has also been applied to biological systems. Brors, Kroner, and Deckwer [ECB6: Proc. 6th Eur. Cong. Biotech., 511 (1994)] separated malate dehydrogenase from the cellular debris of Baker's yeast using CF-EF. A two- to fivefold increase in the specific enzyme transport rate was reported when electric field strengths of 20 to 40 V/cm were used.

Theory   Cross-flow–electrofiltration can theoretically be treated as if it were cross-flow filtration with superimposed electrical effects. These electrical effects include electroosmosis in the filter medium and cake and electrophoresis of the particles in the slurry. The addition of the applied electric field can, however, result in some qualitative differences in permeate-flux-parameter dependences.

The membrane resistance for CF-EF can be defined by specifying two permeate fluxes as

$$J_{om} = \Delta P / R_{om} \tag{22-29}$$

$$J_m = \Delta P / R_m \tag{22-30}$$

where J_{om} is the flux through the membrane in the absence of an electric field and any other resistance, m/s; J_m is the same flux in the presence of an electric field; and R_{om} is the membrane resistance in the absence of an electric field, $(N \cdot s)/m^3$. When electroosmotic effects do occur,

$$J_m = J_{om} + K_m E \tag{22-31}$$

where K_m is the electroosmotic coefficient of the membrane, $m^2/(V \cdot s)$; and E is the applied-electric-field strength, V/m. Equations (22-29), (22-30), and (22-31) can be combined and rearranged to give Eq. (22-32), the membrane resistance in the presence of an electric field.

$$R_m = \frac{R_{om}}{1 + \left(\dfrac{K_m E}{J_{om}}\right)} \tag{22-32}$$

Similarly, cake resistance can be represented as

$$R_c = \frac{R_{oc}}{1 + \left(\dfrac{K_c E}{J_{oc}}\right)} \tag{22-33}$$

where J_{oc} is the flux through the cake in the absence of an electric field or any other resistance, R_{oc} is the cake resistance in the absence of an electric field, and K_c is the electroosmotic coefficient of the cake. The cake resistance is not a constant but is dependent upon the cake thickness, which is in turn a function of the transmembrane pressure drop and electrical-field strength.

Particulate systems require the addition of the term $\mu_e E$ in order to account for the electrophoretic migration of the particle. The constant μ_e is the electrophoretic mobility of the particle, $m^2/(V \cdot s)$. For the case of the CF-EF, the film resistance R_f can be represented as

$$R_f = \frac{\Delta P}{k \ln\left(\dfrac{C_s}{C_b}\right) + U_r + \mu_e E} \tag{22-34}$$

The resistances, when incorporated into equations descriptive of cross-flow filtration, yield the general expression for the permeate flux for particulate suspensions in cross-flow-electrofiltration systems.

There are three distinct regimes of operation in CF-EF. These regimes (Fig. 22-28) are defined by the magnitude of the applied electric field with respect to the critical voltage E_c. The critical voltage is defined as the voltage at which the net particle migration velocity toward the filtration medium is zero. At the critical voltage, there is a balance between the electrical-migration and radial-migration velocities away from the filter and the velocity at which the particles are swept toward the filter by bulk flow. There is no diffusive transport at $E = E_c$ (Fig. 22-28b) because there is no gradient in the particle concentration normal to the filter surface. At field strengths below the critical voltage (Fig. 22-28a), all migration velocities occur in the same direction as in the cross-flow-filtration systems discussed earlier. At values of applied voltage above the critical voltage (Fig. 22-28c) qualitative differences are observed. In this case, the electrophoretic-migration velocity away from the filter medium is greater than the velocity caused by bulk flow toward the filtration medium. Particles concentrate away from the filter medium. This implies that particle concentration is lowest next to the filter medium (in actuality, a clear boundary layer has been observed). The influence of fluid shear still improves the transfer of particles down the concentration gradient, but in this case it is toward the filtration medium. When the particles are small and diffusive transport dominates radial migration, increasing the circulation velocity will decrease the permeate flux rate in this regime. When the particles are large and radial migration dominates, the increase in circulation velocity will still improve the filtration rate. These effects are illustrated qualitatively in Fig. 22-29a. The solid lines represent systems in which the particle diffusive effect dominates the radial-migration effect, while the dashed lines represent the inverse. Figure 22-29b illustrates the increase in filtration rate with increasing electric field strength. For field strengths $E > E_c$, increases in permeate flux rate are due only to electroosmosis in the filtration medium.

One potential difficulty with CF-EF is the electrodeposition of the particles at the electrode away from the filtration medium. This phenomenon, if allowed to persist, will result in performance decay of CF-EF with respect to maintenance of the electric field. Several approaches such as momentary reverses in polarity, protection of the electrode with a porous membrane or filter medium, and/or utilization of a high fluid shear rate can minimize electrodeposition.

Dielectrophoresis

GENERAL REFERENCES: Pohl, in Moore (ed.), *Electrostatics and Its Applications*, Wiley, New York, 1973, chap. 14 and chap. 15 (with Crane). Pohl, in Catsimpoolas (ed.), *Methods of Cell Separation*, vol. I, Plenum Press, New York, 1977, chap. 3. Pohl, *Dielectrophoresis: The Behavior of Matter in Nonuniform Electric Fields*, Cambridge, New York, 1978.

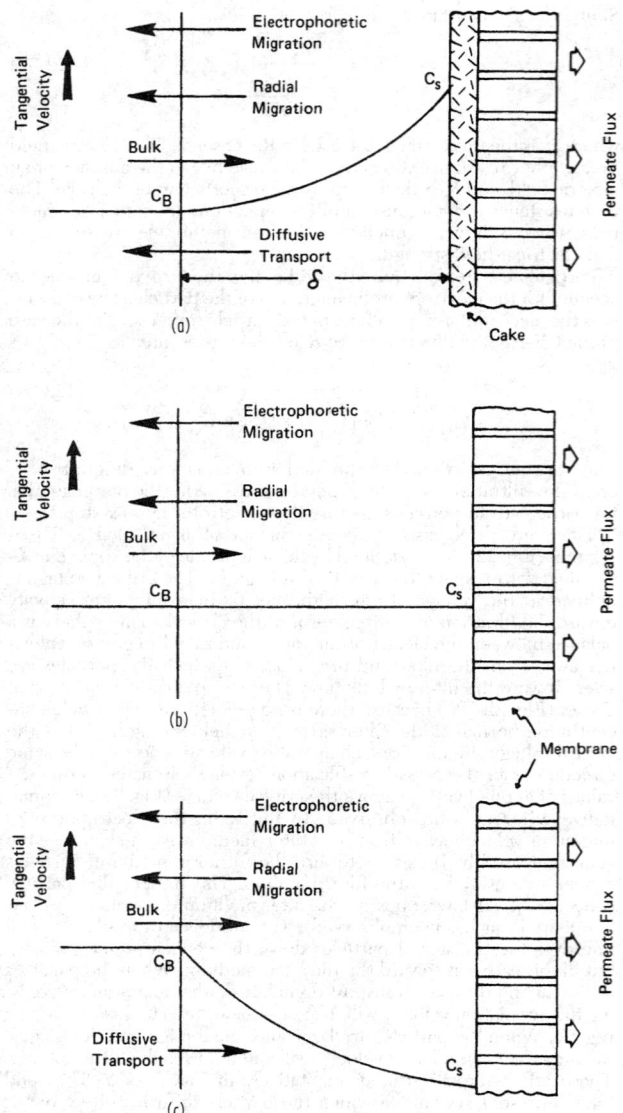

FIG. 22-28 Regimes of operation of cross-flow–electrofiltration: (*a*) voltage less than critical, (*b*) voltage equal to the critical voltage, (*c*) voltage greater than critical.

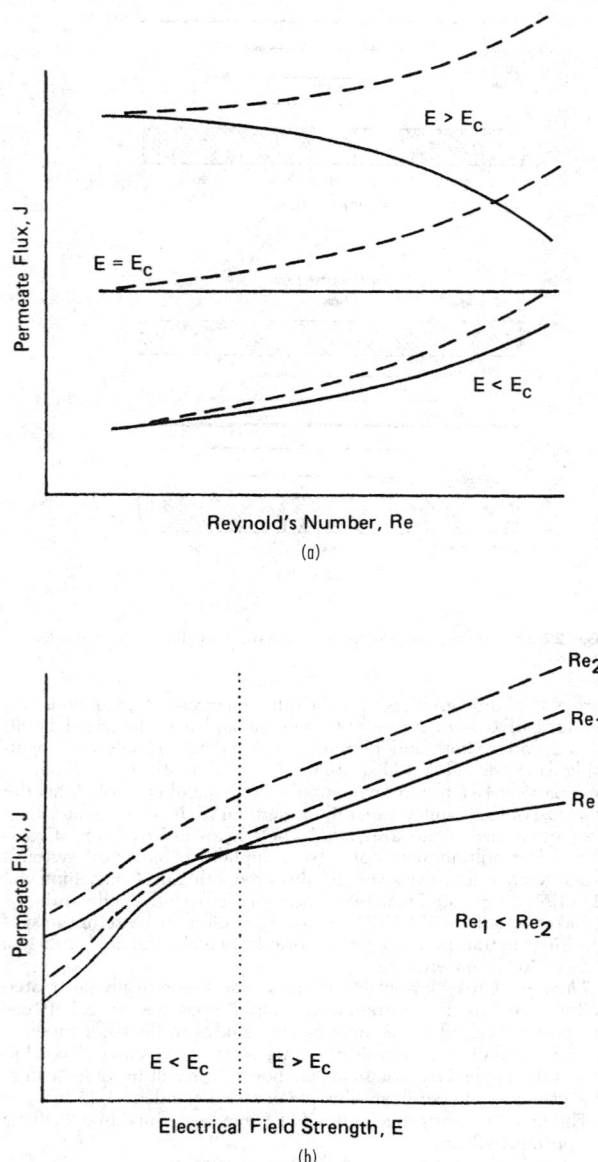

FIG. 22-29 Qualitative effects of Reynolds number and applied-electric-field strength on the filtration permeate flux *J*. Dashed lines indicate large particles (radial migration dominates); solid lines, small particles (particle diffusion dominates).

Introduction   Dielectrophoresis (DEP) is defined as the motion of neutral, polarizable matter produced by a nonuniform electric (ac or dc) field. DEP should be distinguished from electrophoresis, which is the motion of charged particles in a uniform electric field (Fig. 22-30).

The DEP of numerous particle types has been studied, and many applications have been developed. Particles studied have included aerosols, glass, minerals, polymer molecules, living cells, and cell organelles. Applications developed include filtration, orientation, sorting or separation, characterization, and levitation and materials handling. Effects of DEP are easily exhibited, especially by large particles, and can be applied in many useful and desirable ways. DEP effects can, however, be observed on particles ranging in size even down to the molecular level in special cases. Since thermal effects tend to disrupt DEP with molecular-sized particles, they can be controlled only under special conditions such as in molecular beams.

Principle   The principle of particle and cell separation, control, or characterization by the action of DEP lies in the fact that a net force can arise upon even neutral particles situated in a nonuniform electric field. The force can be thought of as rising from the imaginary two-step process of (1) induction or alignment of an electric dipole in a particle placed in an electric field followed by (2) unequal forces on the ends of that dipole. This arises from the fact that the force of an electric field upon a charge is equal to the amount of the charge and to the *local* field strength at that charge. Since the two (equal) charges of the (induced or oriented) dipole of the particle lie in unequal field strengths of the diverging field, a net force arises. If the particle is suspended in a fluid, then the polarizability of that medium enters, too. If, for example, the particle is more polarizable than the fluid, then the

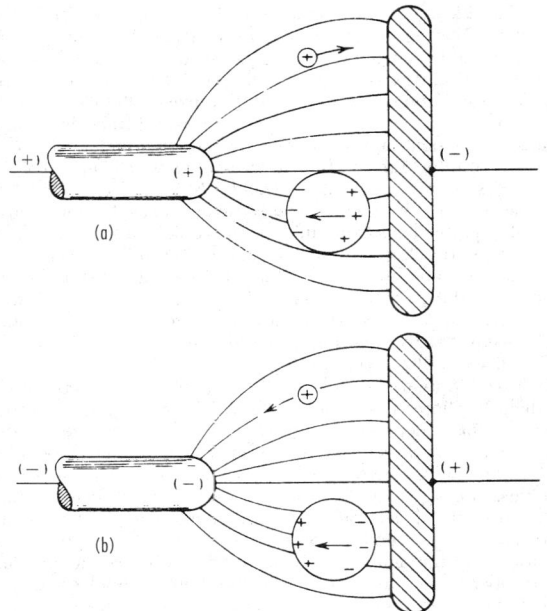

FIG. 22-30 Comparison of behaviors of neutral-charged bodies in an alternating nonuniform electric field. (*a*) Positively charged body moves toward negative electrode. Neutral body is polarized, then is attracted toward point where field is strongest. Since the two charge regions on the neutral body are equal in amount of charge but the force is proportional to the local field, a net force toward the region of more intense field results. (*b*) Positively charged body moves toward the negative electrode. Again, the neutral body is polarized, but it does not reverse direction although the field is reversed. It still moves toward the region of highest field intensity.

net force is such as to impel the particle to regions of greater field strength. Note that this statement implies that the effect is independent of the absolute sign of the field direction. This is found to be the case. Even rapidly alternating (ac) fields can be used to provide *unidirectional* motion of the suspended particles.

Formal Theory A small neutral particle at equilibrium in a static electric field experiences a net force due to DEP that can be written as $\mathbf{F} = (\mathbf{p} \cdot \boldsymbol{\nabla})\mathbf{E}$, where $\mathbf{p}$ is the dipole moment vector and $\mathbf{E}$ is the external electric field. If the particle is a simple dielectric and is isotropically, linearly, and homogeneously polarizable, then the dipole moment can be written as $\mathbf{p} = \alpha v \mathbf{E}$, where α is the (scalar) polarizability, v is the volume of the particle, and $\mathbf{E}$ is the external field. The force can then be written as:

$$\mathbf{F} = \alpha v (\mathbf{E} \cdot \boldsymbol{\nabla})\mathbf{E} = \tfrac{1}{2}\alpha v \boldsymbol{\nabla}|\mathbf{E}|^2 \qquad (22\text{-}35)$$

This force equation can now be used to find the force in model systems such as that of an ideal dielectric sphere (relative dielectric constant K_2) in an ideal perfectly insulating dielectric fluid (relative dielectric constant K_1). The force can now be written as

$$\mathbf{F} = 2\pi a^3 \varepsilon_0 K_1 \left(\frac{K_2 - K_1}{K_2 + 2K_1}\right) \boldsymbol{\nabla}|\mathbf{E}|^2 \qquad (22\text{-}36)$$

(ideal dielectric sphere in ideal fluid).

Heuristic Explanation As we can see from Fig. 22-31, the DEP response of *real* (as opposed to perfect insulator) particles with frequency can be rather complicated. We use a simple illustration to account for such a response. The force is proportional to the difference between the dielectric permittivities of the particle and the surrounding medium. Since a part of the polarization in real systems is thermally activated, there is a delayed response which shows as a phase lag between $\mathbf{D}$, the dielectric displacement, and $\mathbf{E}$, the electric-field intensity. To take this into account we may replace the simple (absolute) dielectric constant ε by the complex (absolute) dielectric

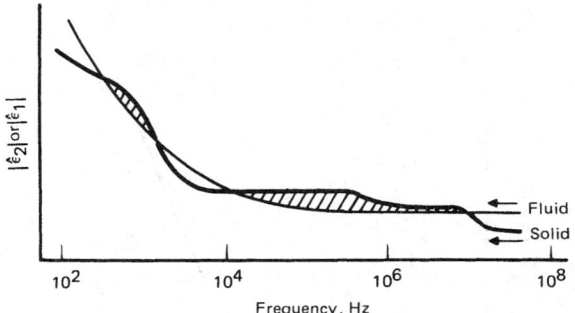

FIG. 22-31 A heuristic explanation of the dielectrophoretic-collection-rate (DCR)-frequency spectrum. The curves for the absolute values of the complex permittivities of the fluid medium and of the suspended particles are shown lying nearly, but not entirely, coincident over the frequency range of the applied electric field. When the permittivity (dielectric constant) of the particles exceeds that of the suspending medium, the collection, or "positive dielectrophoresis," occurs. In the frequency ranges in which the permittivity of the particles is less than that of the suspending medium no collection at the regions of higher field intensity occurs. Instead there is "negative dielectrophoresis," i.e., movement of the particles into regions of lower field intensity.

constant $\hat{\varepsilon} = \varepsilon' - i\varepsilon'' = \varepsilon' - i\sigma/w$, where ω is the angular frequency of the applied field. For treating spherical objects, for example, the replacement

$$F \propto \frac{\varepsilon_1(\varepsilon_2 - \varepsilon_1)}{\varepsilon_2 + 2\varepsilon_1} \rightarrow \mathrm{Re}\left\{\frac{\hat{\varepsilon}_1^{\circ}(\hat{\varepsilon}_2 - \hat{\varepsilon}_1)}{\hat{\varepsilon}_2 + 2\hat{\varepsilon}_1}\right\} \qquad (22\text{-}37)$$

can be made, where $\hat{\varepsilon}^{\circ}$ is the complex conjugate of $\hat{\varepsilon}$.

With this force expression for real dielectrics, we can now explain the complicated DEP response with the help of Fig. 22-31.

A particle, such as a living cell, can be imagined as having a number of different frequency-dependent polarization mechanisms contributing to the total effective polarization of the particle $|\hat{\varepsilon}_2|$. The heavy curve in Fig. 22-31 shows that the various mechanisms in the particle drop out stepwise as the frequency increases. The light curve in Fig. 22-31 shows the polarization for a simple homogeneous liquid that forms the surrounding medium. This curve is a smooth function which becomes constant at high frequency. As the curves cross each other (and hence $|\hat{\varepsilon}_2| = |\hat{\varepsilon}_1|$), various responses occur. The particle can thus be attracted to the strongest field region, be repelled from that region, or experience no force depending on the frequency.

Limitations It is desirable to have an estimate for the smallest particle size that can be effectively influenced by DEP. To do this, we consider the force on a particle due to DEP and also due to the osmotic pressure. This latter diffusional force will randomize the particles and tend to destroy the control by DEP. Figure 22-32 shows a plot of these two forces, calculated for practical and representative conditions, as a function of particle radius. As we can see, the smallest particles that can be effectively handled by DEP appear to be in range of 0.01 to 0.1 μm (100 to 1000 Å).

Another limitation to be considered is the volume that the DEP force can affect. This factor can be controlled by the design of electrodes. As an example, consider electrodes of cylindrical geometry. A practical example of this would be a cylinder with a wire running down the middle to provide the two electrodes. The field in such a system is proportional to $1/r$. The DEP force is then $\mathbf{F}_{\mathrm{DEP}} \propto \nabla|E^2| \propto 1/r^3$, so that any differences in particle polarization might well be masked merely by positional differences in the force. At the outer cylinder the DEP force may even be too small to affect the particles appreciably. The most desirable electrode shape is one in which the force is independent of position within the nonuniform field. This "isomotive" electrode system is shown in Fig. 22-33.

Applications of Dielectrophoresis Over the past 20 years the use of DEP has grown rapidly to a point at which it is in use for biological, colloidal, and mineral materials studies and handling. The effects of nonuniform electric fields are used for handling particulate matter far more often than is usually recognized. This includes the

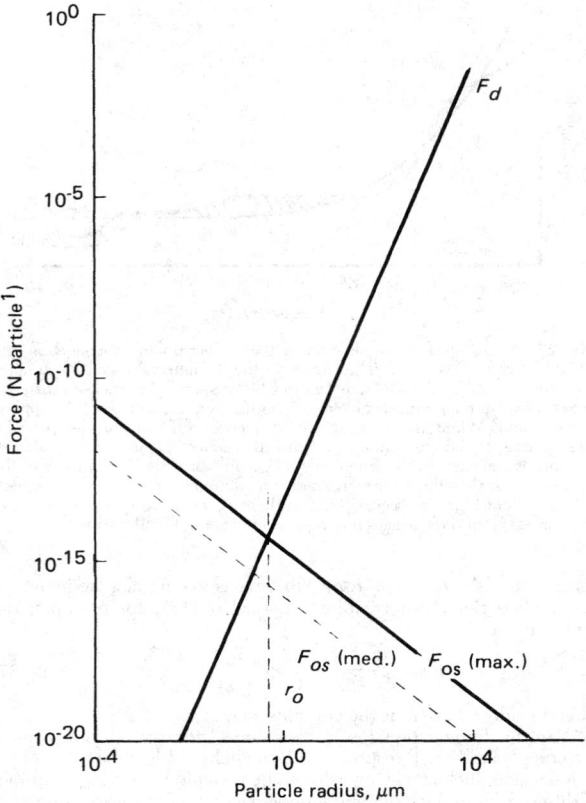

FIG. 22-32 Comparison of the dielectrophoretic (F_d) and osmotic (F_{os}) forces as functions of the particle size.

removal of particulate matter by "electrofiltration," the sorting of mixtures, or its converse, the act of mixing, as well as the coalescence of suspensions. In addition to these effects involving the translational motions of particles, some systems apply the orientational or torsional

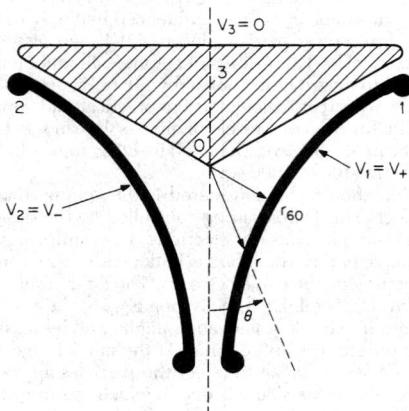

FIG. 22-33 A practical isomotive field geometry, showing r_{60}, the critical radius characterizing the isomotive electrodes. Electrode 3 is at ground potential, while electrodes 1 and 2 are at $V_1 = V_+$ and $V_2 = V_- = -V_+$ respectively. The inner faces of electrodes 1 and 2 follow $r = r_0 [\sin(3\theta/2)]^{-2/3}$, while electrode 3 forms an angle of 120° about the midline.

forces available in nonuniform fields. One well-known example of the latter is the placing of "tip-up" grit on emery papers commercially. Xerography and many other imaging processes are examples of multi-billion-dollar industries which depend upon DEP for their success.

A clear distinction between **electrophoresis** (field action on an object carrying excess free charges) and **dielectrophoresis** (field gradient action on neutral objects) must be borne in mind at all times.

A dielectrofilter [Lin and Benguigui, *Sep. Purif. Methods*, **10**(1), 53 (1981); Sisson et al., *Sep. Sci. Technol.*, **30**(7–9), 1421 (1995)] is a device which uses the action of an electric field to aid the filtration and removal of particulates from fluid media. A dielectrofilter can have a very obvious advantage over a mechanical filter in that it can remove particles which are much *smaller* than the flow channels in the filter. In contrast, the ideal mechanical filter must have all its passages smaller than the particles to be removed. The resultant flow resistance can be use-restrictive and energy-consuming unless a phenomenon such as dielectrofiltration is used.

Dielectrofiltration can (and often does) employ both electrophoresis and dielectrophoresis in its application. The precise physical process which dominates depends on a number of physical parameters of the system. Factors such as field intensity and frequency and the electrical conductivity and dielectric constants of the materials present determine this. Although these factors need constant attention for optimum operation of the dielectrofilter, this additional complication is often more than compensated for by the advantages of dielectrofiltration such as greater throughput and lesser sensitivity to viscosity problems, etc. To operate the dielectrofilter in the dominantly electrophoretic mode requires that excess free charges of one sign or the other reside on the particulate matter. The necessary charges can be those naturally present, as upon a charged sol; or they may need to be artificially implanted such as by passing the particles through a corona discharge. Dielectrofiltration by the corona-charging, electrophoresis-dominated Cottrell technique is now widely used.

To operate the dielectrofilter (dominantly dielectrophoretic mode), on the other hand, one must avoid the presence of free charge on the particles. If the particles can become charged during the operation, a cycle of alternate charging and discharging in which the particles dash to and from the electrodes can occur. This is most likely to occur if static or very low frequency fields are used. For this reason, corona and like effects may be troublesome and need often to be minimized. To be sure, the DEP force is proportional to the field applied [actually to $\nabla(E)^2$], but fields which are too intense can produce such troublesome charge injection. A compromise for optimal operation is necessary between having $\nabla(E)^2$ so low that DEP forces are insufficient for dependable operation, on the one hand, and having **E** so high that troublesome discharges (e.g., coronalike) interfere with dependable operation of the dielectrofilter. In insulative media such as air or hydrocarbon liquids, for example, one might prefer to operate with fields in the range of, say, 10 to 10,000 V/cm. In more conductive media such as water, acetone, or alcohol, for example, one would usually prefer rather lower fields in the range of 0.01 to 100 V/cm. The higher field ranges cited might become unsuitable if conductive sharp asperities are present.

Another factor of importance in dielectrofiltration is the need to have the DEP effect firmly operative upon *all* portions of the fluid passing through. Oversight of this factor is a most common cause of incomplete dielectrofiltration. Good dielectrofilter design will emphasize this crucial point. To put this numerically, let us consider the essential field factor for DEP force, namely $\nabla(\mathbf{E}_0)^2$. Near sharp points, e.g., **E**, the electric field varies with the radial distance r as $E \propto r^{-2}$; hence our DEP force factor will vary as $\nabla(E)^2 \propto r^{-5}$. In the neighborhood of sharp "line" sources such as at the edge of electrode plates, $E \propto r^{-1}$, hence, $\nabla(E)^2 \propto r^{-3}$. If, for instance, the distance is varied by a factor of 4 from the effective field source in these cases, the DEP force can be expected to weaken by a factor of 1024 or 64 respectively for the point source and the line source. The matter is even more keenly at issue when field-warping dielectrics (defined later) are used to effect maximal filtration. In this case the field-warping material is made to produce dipole fields as induced by the applied electric field. If we ask how the crucial factor, $\nabla(E)^2$, varies with distance away from such a dipole, we find that since the field E_d about a dipole varies approximately as r^{-3}, then $\nabla(E)^2$ can be expected to vary as r^{-7}. It then becomes critically important that the particles to be removed from the passing fluid do, indeed, pass very close to the surface of the field-warping material, or it will not be effectively handled. Clearly, it would be difficult to maintain successfully uniform dielectrofiltration treatment of fluid passing through such wildly variant regions. The problems can be minimized by ensuring that all the elements of the passing fluid go closely by such field sources in the dielectrofilter. In practice this is done by constructing the dielectrofilter from an assembly of highly comminuted electrodes or else by a set of relatively simple and widely spaced metallic electrodes between which is set an assembly of more or less finely divided solid dielectric material having a complex permittivity different from that of the fluid to be treated. The solid dielectric (fibers, spheres, chunks) serves to produce field nonuniformities or **field warpings** to which the particles to be filtered are to be attracted. In treating fluids of low dielectric constant such as air or hydrocarbon fluids, one sees field-warping materials such as sintered ceramic balls, glass-wool matting, open-mesh polyurethane foam, alumina, chunks, or $BaTiO_3$ particles.

An example of a practical dielectrofilter which uses both of the features described, namely, sharp electrodes *and* dielectric field-warping filler materials, is that described in Fig. 22-34 [H. J. Hall and R. F. Brown, *Lubric. Eng.*, **22**, 488 (1966)]. It is intended for use with hydraulic fluids, fuel oils, lubricating oils, transformer oils, lubricants, and various refinery streams. Performance data are cited in Fig. 22-35. It must be remarked that in the opinion of Hall and Brown the action of the dielectrofilter was "electrostatic" and due to free charge on the particles dispersed in the liquids. It is the present authors' opinion, however, that both electrophoresis and dielectrophoresis are operative here but that the dominant mechanism is that of DEP, in which neutral particles are polarized and attracted to the regions of highest field intensity.

A second commercial example of dielectric filtration is the Gulftronic® separator [G. R. Fritsche, *Oil & Gas J.*, **75**, 73 (1977)] which was commercialized in the late 1970s by Gulf Science and Technology Company. Instead of using needle-point electrodes as shown in Fig. 22-34, the Gulftronic® separator relied on the use of a bed of glass beads to produce the field nonuniformities required for dielectric filtration. Either ac or dc electric fields could be used in this separator. The Gulftronic® separator has been used primarily to remove catalyst fines from FCC decant oils and has been reported to exhibit removal efficiencies in excess of 80 percent for this fine-particle separation problem.

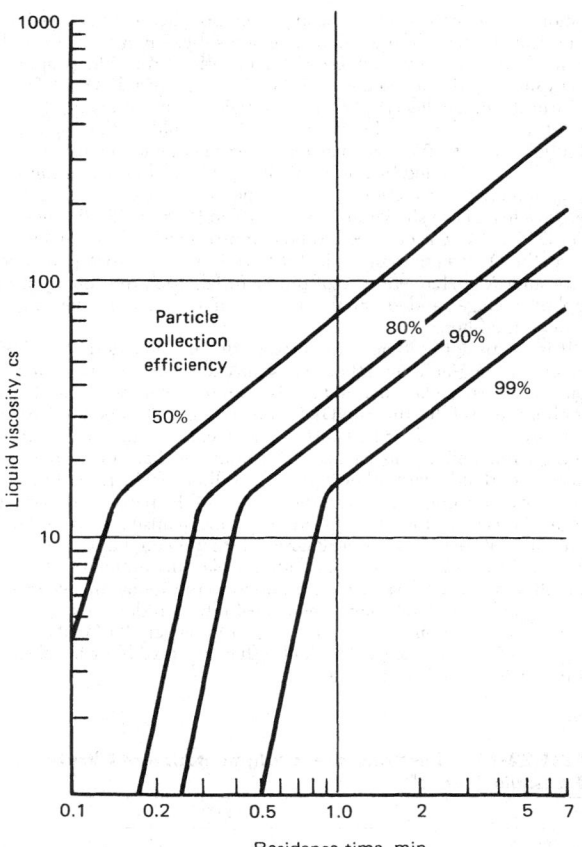

FIG. 22-35 Performance data for a typical high-efficiency electrostatic liquid cleaner.

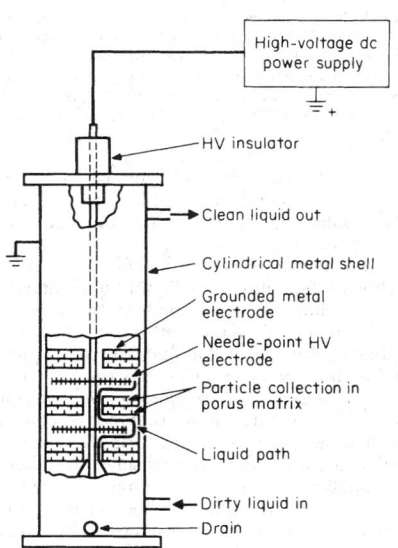

FIG. 22-34 Diagram illustrating the function of an electrostatic liquid cleaner.

Another example of the commercial use of DEP is in polymer clarification [A. N. Wennerberg, U.S. Patent 2,914,453, 1959; assignor to Standard Oil Co. (Indiana)]. Here, either ac or dc potentials were used while passing suspensions to be clarified through regions with an area-to-electrode-area ratio of 10:1 or 100:1 and with fields in the order of 10 kV/cm. Field warping by the presence of various solid dielectrics was observed to enhance filtration considerably, as expected for DEP. The filtration of molten or dissolved polymers to free them of objectionable quantities of catalyst residues, for example, was more effective if a solid dielectric material such as Attapulgus clay, silica gel, fuller's earth, alumina, or bauxite was present in the region between the electrodes. The effectiveness of percolation through such absorptive solids for removing color bodies is remarkably enhanced by the presence of an applied field. A given amount of clay is reported to remove from 4 to 10 times as much color as would be removed in the absence of DEP. Similar results are reported by Lin et al. [Lin, Yaniv, and Zimmels, *Proc. XIIIth Int. Miner. Process. Congr.*, Wrocław, Poland, 83–105 (1979)].

The instances cited were examples of the use of DEP to filter liquids. We now turn to the use of DEP to aid in dielectrofiltration of gases. Fielding et al. observe that the effectiveness of high-quality fiberglass air filters is dramatically improved by a factor of 10 or more by incorporating DEP in the operation. Extremely little current or power is required, and no detectable amounts of ozone or corona need result. The DEP force, once it has gathered the particles, continues to act on the particles already sitting on the filter medium, thereby improving adhesion and minimizing blowoff.

The degree by which the DEP increases the effectiveness of gas filtration, or the dielectrophoretic augmentation factor (DAF), is definable. It is the ratio of the volumes of aerosol-laden gas which can be cleaned effectively by the filter with and without the voltage applied. For example, the application of 11 kV/cm gave a DAF of 30 for 1.0-μm-diameter dioctyl phthalate particles in air, implying that the penetration of the glass filter is reduced thirtyfold by the application of a field of 1100 kV/m. Similar results were obtained by using "standard" fly ash supplied by the Air Pollution Control Office of the U.S. Environmental Protection Agency. The data obtained for several aerosols tested are shown in Table 22-13 and in Fig. 22-36. The relation DAF = kV²/v is observed to hold approximately for each aerosol. Here, the DEP augmentation factor DAF is observed to depend upon a constant K, a characteristic of the material, upon the square of the applied voltage, and upon the inverse of the volume flow rate v through the filter.

It is worth noting that in the case of the air filter described DEP serves as an augmenting rather than as an exclusive mechanism for the removal of particulate material. It is a unique feature of the dielectrophoretic gas filter that the DEP force is maximal when the particulates are at or on the fiber surface. This causes the deposits to be strongly retained by this particular filtration mechanism. It thus contrasts importantly with other types of gas filter in which the filtration mechanism no longer acts after the capture of the particle. In particular, in the case of the older electrostatic mechanisms involving only coulombic attraction, a simple charge alternation on the particle, such as caused by normal conduction, often evokes disruption of the filter operation because of particle repulsion from the contacting electrode. On the other hand, ordinary mechanical filtration depends upon the action of adventitious particle trapping or upon van der Waals forces, etc., to hold the particles. The high efficiency possible with electrofilters suggests their wider use.

TABLE 22-13   Dielectrophoretic Augmentation of Filtration of a Liquid Aerosol*

Air speed, cm/s	DAF at			
	2 kV	3.5 kV	5 kV	7 kV
0.3-μm-diameter dioctyl phthalate aerosol				
3	8	19	95	330
6	3	13	39	120
9	3	11	28	100
15	2	6	13	42
20	2	5	9	27
28	2	4	6	14
39	2	3	4	9
50	1	2	3	6
1.0-μm-diameter dioctyl phthalate aerosol				
3	30	110	300	1100
6	6	3	95	360
9	4	18	50	170
15	3	10	20	50
20	2	6	13	35
28	2	4	8	18
39	2	3	5	11
50	1	2	3	7
Fly-ash aerosol				
6	10	30	80	
10	8	30	80	
14	5	20	40	
20	4	10	30	70
35	3	7	10	20
45	1	2	6	
53	1	2	7	10

*Experimentally measured dielectrophoretic augmentation factor DAF as a function of air speed and applied voltage for a glass-fiber filter (HP-100, Farr Co.). Cf. Fielding, Thompson, Bogardus, and Clark, *Dielectrophoretic Filtration of Solid and Liquid Aerosol Particulates*, Prepr. 75-32.2, 68th ann. meet., Air Pollut. Control Assoc., Boston, June 1975.

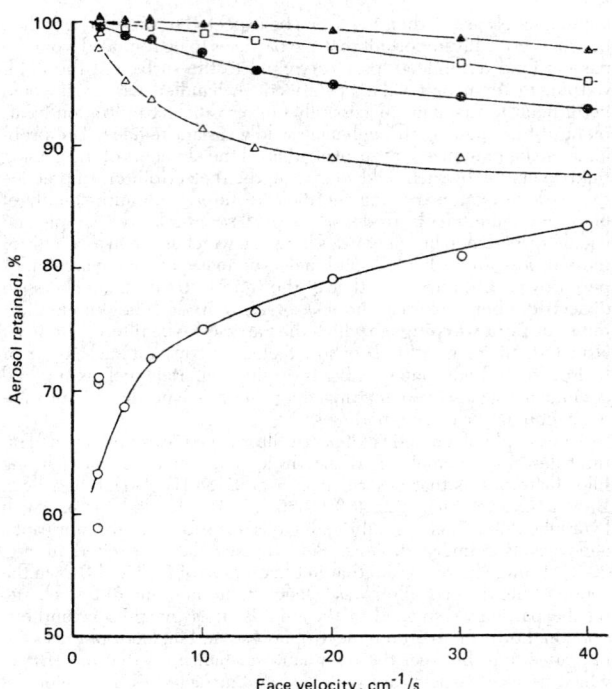

FIG. 22-36   Efficiency of an electrofilter as a function of gas flow rate at 5 different voltages. Experimental materials: 1-μm aerosol of dioctyl phthalate; glass-fiber filter. Symbols: ○, no voltage applied; △, 2 kV; ●, 3.5 kV; □, 5 kV; ▲, 7 kV. (*After Fielting et al.*, Dielectrophoretic Filtration of Solid and Liquid Aerosol Particulates, *Prepr. 75-32.2, 68th ann. meet., Air Pollut. Control Assoc., Boston, June 1975.*)

SURFACE-BASED SOLID-LIQUID SEPARATIONS INVOLVING A SECOND LIQUID PHASE

GENERAL REFERENCES: Fuerstenau, "Fine Particle Flotation," in Somasundaran (ed.), *Fine Particles Processing*, vol. 1, American Institute of Mining, Metallurgical, and Petroleum Engineers, New York, 1980. Henry, Prudich, and Lau, *Colloids Surf.*, **1**, 335 (1980). Henry, Prudich, and Vaidyanathan, *Sep. Purif. Methods*, **8**(2), 31 (1979). Jacques, Hovarongkura, and Henry, *Am. Inst. Chem. Eng. J.*, **25**(1), 160 (1979). Stratton-Crawley, "Oil Flotation: Two Liquid Flotation Techniques," in Somasundaran and Arbiter (eds.), *Beneficiation of Mineral Fines*, American Institute of Mining, Metallurgical, and Petroleum Engineers, New York, 1979.

Process Concept   Three potential surface-based regimes of separation exist when a second, immiscible liquid phase is added to another, solids-containing liquid in order to effect the removal of solids. These regimes (Fig. 22-37) are:

1. Distribution of the solids into the bulk second liquid phase
2. Collection of the solids at the liquid-liquid interface
3. Bridging or clumping of the solids by the added fluid in order to form an agglomerate followed by settling or filtration

These separation techniques should find particular application in systems containing fine particles. The surface chemical differences involved among these separation regimes are only a matter of degree; i.e., all three regimes require the wetting of the solid by the second liquid phase. The addition of a surface-active agent is sometimes needed in order to achieve the required solids wettability. In spite of this similarity, applied processing (equipment configuration, operating conditions, etc.) can vary widely. Collection at the interface would normally be treated as a flotation process (see also Sec. 22: "Adsorptive-Bubble Separation Methods"; and Sec. 19: "Flotation"), distribu-

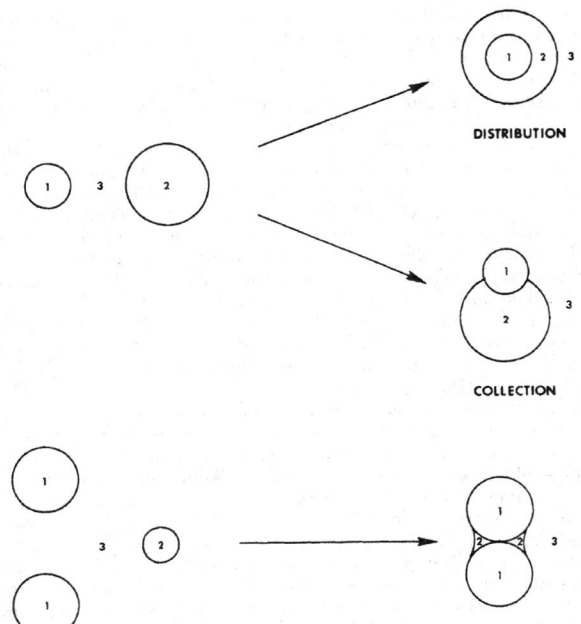

FIG. 22-37 Regimes of separation in a liquid-solid-liquid system. Phase 1 = particle; phase 2 = liquid (dispersed); phase 3 = liquid (continuous).

tion to the bulk liquid as a liquid-liquid extraction analog, and particle bridging as a settling (sedimentation) or filtration process.

Even though surface-property-based liquid-solid-liquid separation techniques have yet to be widely used in significant industrial applications, several studies which demonstrate their effectiveness have appeared in literature.

Albertsson (*Partition of Cell Particles and Macromolecules*, 3d ed., Wiley, New York, 1986) has extensively used particle distribution to fractionate mixtures of biological products. In order to demonstrate the versatility of particle distribution, he has cited the example shown in Table 22-14. The feed mixture consisted of polystyrene particles, red blood cells, starch, and cellulose. Liquid-liquid particle distribution has also been studied by using mineral-matter particles (average diameter = 5.5 μm) extracted from a coal liquid as the solid in a xylene-water system [Prudich and Henry, *Am. Inst. Chem. Eng. J.*, **24**(5), 788 (1978)]. By using surface-active agents in order to enhance the water wettability of the solid particles, recoveries of better than 95 percent of the particles to the water phase were observed. All particles remained in the xylene when no surfactant was added.

Particle collection at a liquid-liquid interface is a particularly favor-

able separation process when applied to fine-particle systems. Advantages of this type of processing include:
- Decreased liquid-liquid interfacial tension (when compared with a gas-liquid system) results in higher liquid-liquid interfacial areas, which favor solid-particle droplet collisions.
- Liquid-solid interactions due to long-range intermolecular forces are much larger than are gas-solid interactions. This means that it is easier to collect fine particles at a liquid-liquid interface than at a gas-liquid interface.
- The increased momentum of liquid droplets (when compared with gas) should favor solid-particle collection.

Fuerstenau [Lai and Fuerstenau, *Trans. Am. Inst. Min. Metall. Pet. Eng.*, **241**, 549 (1968); Raghavan and Fuerstenau, *Am. Inst. Chem. Eng. Symp. Ser.*, **71**(150), 59 (1975)] has studied this process with respect to the removal of alumina particles (0.1 μm) and hematite particles (0.2 μm) from an aqueous solution by using isooctane. The use of isooctane as the collecting phase for the hematite particles resulted in an increase in particle recovery of about 50 percent over that measured when air was used as the collecting phase under the same conditions. The effect of the wettability of the solid particles (as measured by the three-phase contact angle) on the recovery of hematite in the water-isooctane system is shown in Fig. 22-38. This behavior is typical of particle collection. Particle collection at an oil-water interface has also been studied with respect to particle removal from a coal liquid. Particle removals averaging about 80 percent have been observed when water is used as the collecting phase (Lau, master's thesis, West Virginia University, 1979). Surfactant addition was necessary in order to control the wettability of the solids.

Particle bridging has been chiefly investigated with respect to spherical agglomeration. Spherical agglomeration involves the collecting or transferring of the fine particles from suspension in a liquid phase into spherical aggregates held together by a second liquid phase. The aggregates are then removed from the slurry by filtration or settling. Like the other liquid-solid-liquid separation techniques, the solid must be wet by the second liquid phase. The spherical agglomeration process has resulted in the development of a pilot unit called the Shell Pelletizing Separator [Zuiderweg and Van Lookeren Campagne, *Chem. Eng. (London)*, **220**, CE223 (1968)]. A detailed discussion of spherical agglomeration can be found in Sec. 20: "Size Enlargement."

The ability to determine in advance which of the separation regimes is most advantageous for a given liquid-solid-liquid system would be desirable. No set of criteria with which to make this determination presently exists. Work has been done with respect to the identification of system parameters which make these processes technically feasible. The results of these studies can be used to guide the selection of the second liquid phase as well as to suggest approximate operating conditions (dispersed-liquid droplet size, degree and type of mixing, surface-active-chemical addition, etc.).

TABLE 22-14   Separations of Particles between Two Phases

System	Top phase	Bottom phase
Polyethylene glycol / salt	Polystyrene	All others
PEG / Dextran; 20,000 MW	Algae	All others
PEG / Dextran; 200,000 MW	Red cells	All others
Methyl cellulose / Dextran	Cellulose particles	Starch

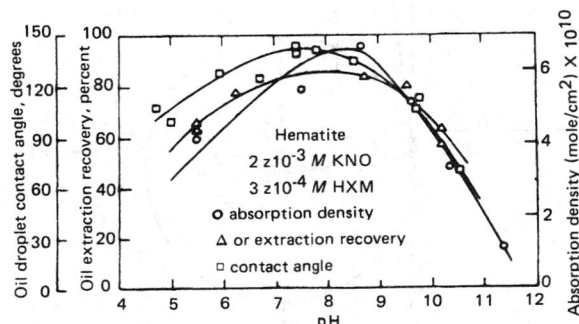

FIG. 22-38 The variation of adsorption density, oil-droplet contact angle, and oil-extraction recovery of hematite as a function of pH. To convert gram-moles per square centimeter to pound-moles per square foot, multiply by 2.048. [*From Raghavan and Fuerstenau, Am. Inst. Chem. Eng. Symp. Ser.*, **71**(150), 59 (1975).]

Theory   Theoretical analyses of spherical particles suspended in a planar liquid-liquid interface have appeared in literature for some time, the most commonly presented forms being those of a free energy and/or force balance made in the absence of all external body forces. These analyses are generally used to define the boundary criteria for the shift between the collection and distribution regimes, the bridging regime not being considered. This type of analysis shows that for a spherical particle possessing a three-phase contact angle between 0 and 180°, as measured through the receiving or collecting phase, collection at the interface is favored over residence in either bulk phase. These equations are summarized, using a derivation of Young's equation, as

$$\frac{\gamma_{s2} - \gamma_{s1}}{\gamma_{12}} > 1 \text{ particles wet to phase 1} \qquad (22\text{-}38)$$

$$\frac{\gamma_{s2} - \gamma_{s1}}{\gamma_{12}} < -1 \text{ particles wet to phase 2} \qquad (22\text{-}39)$$

$$\left| \frac{\gamma_{s2} - \gamma_{s1}}{\gamma_{12}} \right| \leq 1 \text{ particle at interface} \qquad (22\text{-}40)$$

where γ_{ij} is the surface tension between phases i and j, N/m (dyn/cm); s indicates the solid phase; and subscripts 1 and 2 indicate the two liquid phases.

Several additional studies [Winitzer, *Sep. Sci.*, **8**(1), 45 (1973); ibid., **8**(6), 647 (1973); Maru, Wasan, and Kintner, *Chem. Eng. Sci.*, **26**, 1615 (1971); and Rapacchietta and Neumann, *J. Colloid Interface Sci.*, **59**(3), 555 (1977)] which include body forces such as gravitational acceleration and buoyancy have been made. A typical example of a force balance describing such a system (Fig. 22-39) is summarized in Eq. (22-41).

$$[(\gamma_{s1} - \gamma_{s2}) \cos \delta + \gamma_{12} \cos B]L = g[V_{total}\rho_s - V_1\rho_1 - V_2\rho_2] \qquad (22\text{-}41)$$

where V_1 is the volume of the particle in fluid phase 1, V_2 is the volume in fluid phase 2, L is the particle circumference at the interface between the two liquid phases, ρ_i is the density of phase i, and g is the gravitational constant. The left-hand side of the equation represents the surface forces acting on the solid particle, while the right-hand side includes the gravitational and buoyancy forces. This example illustrates the fact that body forces can have a significant effect on system behavior. The solid-particle size as well as the densities of the solid and both liquid phases are introduced as important system parameters.

A study has also been performed for particle distribution for cases

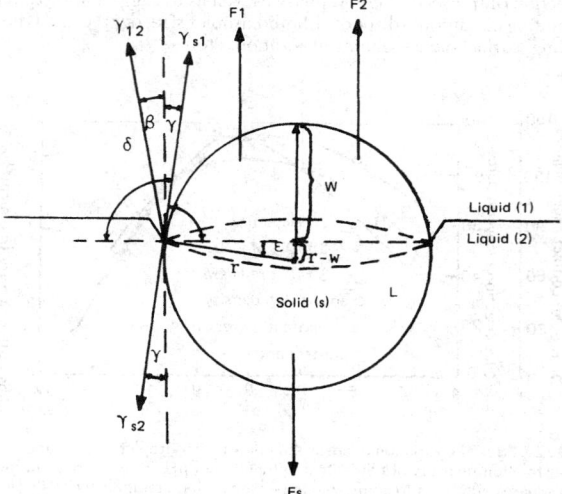

FIG. 22-39   Solid sphere suspended at the liquid-liquid interface. F_1 and F_2 are buoyancy forces; F_s is gravity. [*From Winitzer, Sep. Sci., 8(1), 45 (1973).*]

in which the radii of curvature of the solid and the liquid-liquid interface are of the same order of magnitude [Jacques, Hovarongkura, and Henry, *Am. Inst. Chem. Eng. J.*, **25**(1), 160 (1979)]. Differences between the final and initial surface free energies are used to analyze this system. Body forces are neglected. Results (Fig. 22-40) demonstrate that n, the ratio of the particle radius to the liquid-liquid-interface radius, is an important system parameter. Distribution of the particle from one phase to the other is favored over continued residence in the original phase when the free-energy difference is negative. For a solid particle of a given size, these results show that as the second-phase droplet size decreases, the contact angle required in order to effect distribution decreases (the required wettability of the solid by the second phase increases). The case of particle collection at a curved liquid-liquid interface has also been studied in a similar manner [Smith and Van de Ven, *Colloids Surf.*, **2**, 387 (1981)]. This study shows that collection is preferred over distribution for any n in systems without external body forces when the contact angle lies between 0 and 180°.

While thermodynamic-stability studies can be valuable in evaluating the technical feasibility of a process, they are presently inadequate in determining which separation regime will dominate a particular liquid-solid-liquid system. These analyses ignore important processing phenomena such as the mechanism of encounter of the dispersed-phase liquid with the solid particles, the strength of particle attachment, and the mixing-energy input necessary to effect the separation. No models of good predictive value which take all these variables into account have yet been offered. Until the effects of these and other system variables can be adequately understood, quantified, and combined into such a predictive model, no a priori method of performance prediction will be possible.

ADSORPTIVE-BUBBLE SEPARATION METHODS

GENERAL REFERENCES: Lemlich (ed.), *Adsorptive Bubble Separation Techniques*, Academic, New York, 1972. Carlson, "Adsorptive Bubble Separation Processes" in Scamehorn and Harwell (eds.), *Surfactant-Based Separation Processes*, Marcel Dekker, New York, 1989.

Principle   The adsorptive-bubble separation methods, or adsubble methods for short [Lemlich, *Chem. Eng.* **73**(21), 7 (1966)], are based on the selective adsorption or attachment of material on the surfaces of gas bubbles passing through a solution or suspension. In most of the methods, the bubbles rise to form a foam or froth which carries the material off overhead. Thus the material (desirable or undesirable) is removed from the liquid, and not vice versa as in, say, filtration. Accordingly, the foaming methods appear to be particularly (although not exclusively) suited to the removal of small amounts of material from large volumes of liquid.

For any adsubble method, if the material to be removed (termed the **colligend**) is not itself surface-active, a suitable surfactant (termed the **collector**) may be added to unite with it and attach or adsorb it to the bubble surface so that it may be removed (Sebba, *Ion Flotation*, Elsevier, New York, 1962). The union between colligend and collector may be by chelation or other complex formation. Alternatively, a charged colligend may be removed through its attraction toward a collector of opposite charge.

Definitions and Classification   Figure 22-41 outlines the most widely accepted classification of the various adsubble methods [Karger, Grieves, Lemlich, Rubin, and Sebba, *Sep. Sci.*, **2**, 401 (1967)]. It is based largely on actual usage of the terms by various workers, and so the definitions include some unavoidable inconsistencies and overlap.

Among the methods of foam separation, **foam fractionation** usually implies the removal of dissolved (or sometimes colloidal) material. The overflowing foam, after collapse, is called the *foamate*. The solid lines of Fig. 22-42 illustrate simple continuous foam fractionation. (Batch operation would be represented by omitting the feed and bottoms streams.)

On the other hand, **flotation** usually implies the removal of solid particulate material. Most important under the latter category is **ore flotation**, which is covered separately in Sec. 19.

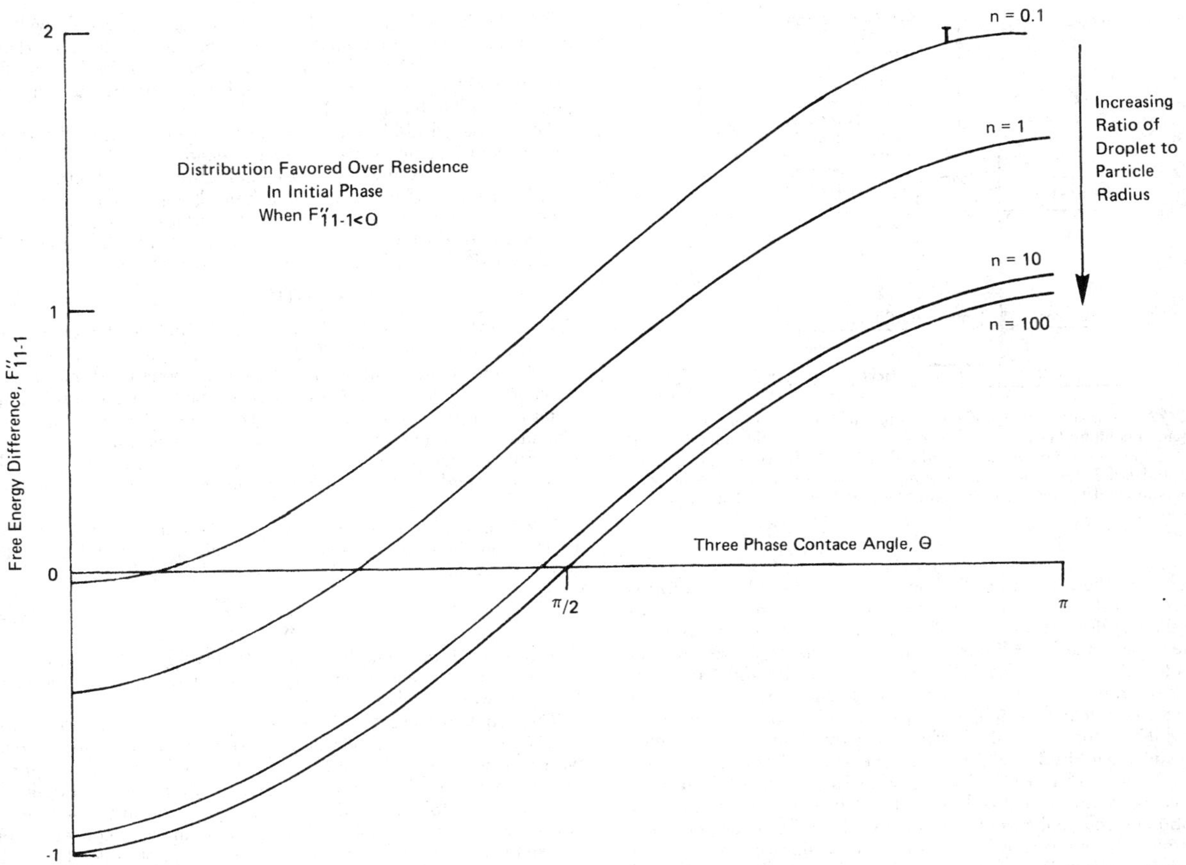

n = 0.1

n = 1

n = 10

n = 100

Increasing
Ratio of
Droplet to
Particle
Radius

Distribution Favored Over Residence
In Initial Phase
When $F''_{11-1} < 0$

Three Phase Contace Angle, Θ

$\pi/2$

π

Free Energy Difference, F''_{11-1}

FIG. 22-40 Normalized free-energy difference between distributed (II) and nondistributed (I) states of the solid particles versus three-phase contact angle (collection at the interface is not considered). A negative free-energy difference implies that the distributed state is preferred over the nondistributed state. Note especially the significant effect of n, the ratio of the liquid droplet to solid-particle radius. [*From Jacques, Hovarongkura, and Henry*, Am. Inst. Chem. Eng. J., **25**(*1*), *160* (*1979*).]

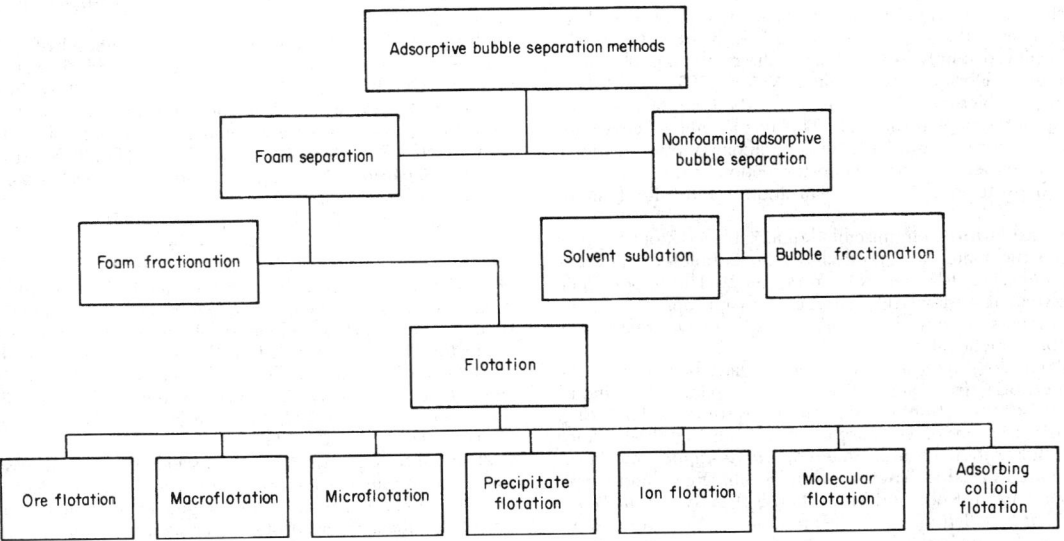

FIG. 22-41 Classification for the adsorptive-bubble separation methods.

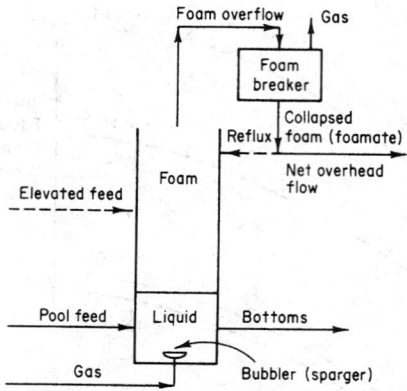

FIG. 22-42 Four alternative modes of continuous-flow operation with a foam-fractionation column: (1) The simple mode is illustrated by the solid lines. (2) Enriching operation employs the dashed reflux line. (3) In stripping operation, the elevated dashed feed line to the foam replaces the solid feed line to the pool. (4) For combined operation, reflux and elevated feed to the foam are both employed.

Also under the category of flotation are to be found **macroflotation,** which is the removal of macroscopic particles; **microflotation** (also called **colloid flotation**), which is the removal of microscopic particles, particularly colloids or microorganisms [Dognon and Dumontet, *Comptes Rendus,* **135,** 884 (1941)]; **molecular flotation,** which is the removal of surface-inactive molecules through the use of a collector (surfactant) which yields an insoluble product; **ion flotation,** which is the removal of surface-inactive ions via a collector which yields an insoluble product, especially a removable scum [Sebba, *Nature,* **184,** 1062 (1959)]; **adsorbing colloid flotation,** which is the removal of dissolved material in piggyback fashion by adsorption on colloidal particles; and **precipitate flotation,** in which a precipitate is removed by a collector which is not the precipitating agent [Baarson and Ray, "Precipitate Flotation," in Wadsworth and Davis (eds.), *Unit Processes in Hydrometallurgy,* Gordon and Breach, New York, 1964, p. 656]. The last definition has been narrowed to precipitate flotation of the first kind, the second kind requiring no separate collector at all [Mahne and Pinfold, *J. Appl. Chem.,* **18,** 52 (1968)].

A separation can sometimes be obtained even in the absence of any foam (or any floated floc or other surrogate). In **bubble fractionation** this is achieved simply by lengthening the bubbled pool to form a vertical column [Dorman and Lemlich, *Nature,* **207,** 145 (1965)]. The ascending bubbles then deposit their adsorbed or attached material at the top of the pool as they exit. This results in a concentration gradient which can serve as a basis for separation. Bubble fractionation can operate either alone or as a booster section below a foam fractionator, perhaps to raise the concentration up to the foaming threshold.

In **solvent sublation** an immiscible liquid is placed atop the main liquid to trap the material deposited by the bubbles as they exit [Sebba, *Ion Flotation,* Elsevier, New York, 1962]. The upper liquid should dissolve or at least wet the material. With appropriate selectivity, the separation so achieved can sometimes be much greater than that with bubble fractionation alone.

The droplet analogs to the adsubble methods have been termed the **adsoplet methods** (from adsorptive droplet separation methods) [Lemlich, "Adsorptive Bubble Separation Methods," *Ind. Eng. Chem.,* **60**(10), 16 (1968)]. They are omitted from Fig. 22-41, since they involve adsorption or attachment at *liquid-liquid* interfaces. Among them are **emulsion fractionation** [Eldib, "Foam and Emulsion Fractionation," in Kobe and McKetta (eds.), *Advances in Petroleum Chemistry and Refining,* vol. 7, Interscience, New York, 1963, p. 66], which is the analog of foam fractionation; and **droplet fractionation** [Lemlich, loc. cit.; and Strain, *J. Phys. Chem.,* **57,** 638

(1953)], which is the analog of bubble fractionation. Similarly, the old beneficiation operation called bulk oil flotation (Gaudin, *Flotation,* 2d ed., McGraw-Hill, New York, 1957) is the analog of modern ore flotation. By and large, the adsoplet methods have not attracted the attention accorded to the adsubble methods.

Of all the adsubble methods, foam fractionation is the one for which chemical engineering theory is the most advanced. Fortunately, some of this theory also applies to other adsubble methods.

Adsorption   The separation achieved depends in part on the selectivity of adsorption at the bubble surface. At equilibrium, the adsorption of dissolved material follows the Gibbs equation (Gibbs, *Collected Works,* Longmans Green, New York, 1928).

$$d\gamma = -\mathbf{R}T\Sigma\Gamma_i d \ln a_i \qquad (22\text{-}42)$$

Γ_i is the surface excess (Davies and Rideal, *Interfacial Phenomena,* 2d ed., Academic, New York, 1963). For most purposes, it is sufficient to view Γ_i as the concentration of adsorbed component i at the surface in units of, say, (g·mol)/cm². $\mathbf{R}$ is the gas constant, T is the absolute temperature, γ is the surface tension, and a_i is the activity of component i. The minus sign shows that material which concentrates at the surface generally lowers the surface tension, and vice versa. This can sometimes be a guide in determining preliminarily what materials can be separated.

When applied to a nonionic surfactant in pure water at concentrations below the critical micelle concentration, Eq. (22-42) simplifies into Eq. (22-43)

$$\Gamma_s = -\frac{1}{\mathbf{R}T}\frac{d\gamma}{d \ln C_s} \qquad (22\text{-}43)$$

C is the concentration in the bulk, and subscript s refers to the surfactant. Under some conditions, Eq. (22-43) may apply to an ionic surfactant as well (Lemlich, loc. cit.).

The major surfactant in the foam may usually be considered to be present at the bubble surfaces in the form of an adsorbed monolayer with a substantially constant Γ_s, often of the order of 3×10^{-10} (g·mol)/cm², for a molecular weight of several hundred. On the other hand, trace materials follow the linear-adsorption isotherm $\Gamma_i = K_iC_i$ if their concentration is low enough. For a wider range of concentration a Langmuir or other type of isotherm may be applicable (Davies and Rideal, loc. cit.).

Factors Affecting Adsorption K_i for a colligend can be adversely affected (reduced) through an insufficiency of collector. It can also be reduced through an excess of collector, which competes for the available surface against the collector-colligend complex [Schnepf, Gaden, Mirocznik, and Schonfeld, *Chem. Eng. Prog.,* **55**(5), 42 (1959)].

Excess collector can also reduce the separation by forming micelles in the bulk which adsorb some of the colligend, thus keeping it from the surface. This effect of the micelles on K_i for the colligend is given theoretically [Lemlich, "Principles of Foam Fractionation," in Perry (ed.), *Progress in Separation and Purification,* vol. 1, Interscience, New York, 1968, chap. 1] by Eq. (22-44) [Lemlich (ed.), *Adsorptive Bubble Separation Techniques,* Academic, New York, 1972] if Γ_s is constant when $C_s > C_{sc}$:

$$\frac{1}{K_2} = \frac{1}{K_1} + \frac{C_s - C_{sc}}{\Gamma_s E} \qquad (22\text{-}44)$$

K_1 is K_i just below the collector's critical micelle concentration, C_{sc}. K_2 is K_i at some higher collector concentration, C_s. E is the relative effectiveness, in adsorbing colligend, of surface collector versus micellar collector. Generally, $E > 1$. Γ_s is the surface excess of collector. More about each K is available [Lemlich, "Adsubble Methods," in Li (ed.), *Recent Developments in Separation Science,* vol. 1, CRC Press, Cleveland, 1972, pp. 113–127; Jashnani and Lemlich, *Ind. Eng. Chem. Process Des. Dev.,* **12,** 312 (1973)].

The controlling effect of various ions can be expressed in terms of thermodynamic equilibria [Karger and DeVivo, *Sep. Sci.,* **3,** 393 1968]. Similarities with ion exchange have been noted. The selectivity of counterionic adsorption increases with ionic charge and decreases with hydration number [Jorne and Rubin, *Sep. Sci.,* **4,** 313 (1969); and Kato and Nakamori, *J. Chem. Eng. Japan,* **9,** 378 (1976)].

By analogy with other separation processes, the relative distribution in multicomponent systems can be analyzed in terms of a selectivity coefficient $\alpha_{mn} = \Gamma_m C_n / \Gamma_n C_m$ [Rubin and Jorne, *Ind. Eng. Chem. Fundam.*, **8**, 474 (1969); *J. Colloid Interface Sci.*, **33**, 208 (1970)].

Operation in the Simple Mode If there is no concentration gradient within the liquid pool and if there is no coalescence within the rising foam, then the operation shown by the solid lines of Fig. 22-42 is truly in the simple mode, i.e., a single theoretical stage of separation. Equations (22-45) and (22-46) will then apply to the steady-flow operation.

$$C_Q = C_W + (GS\Gamma_W/Q) \tag{22-45}$$

$$C_W = C_F - (GS\Gamma_W/F) \tag{22-46}$$

C_F, C_W, and C_Q are the concentrations of the substance in question (which may be a colligend or a surfactant) in the feed stream, bottoms stream, and foamate (collapsed foam) respectively. G, F, and Q are the volumetric flow rates of gas, feed, and foamate respectively. Γ_W is the surface excess in equilibrium with C_W. S is the surface-to-volume ratio for a bubble. For a spherical bubble, $S = 6/d$, where d is the bubble diameter. For variation in bubble sizes, d should be taken as $\Sigma n_i d_i^3 / \Sigma n_i d_i^2$, where n_i is the number of bubbles with diameter d_i in a representative region of foam.

Finding Γ Either Eq. (22-45) or Eq. (22-46) can be used to find the surface excess indirectly from experimental measurements. To assure a close approach to operation as a single theoretical stage, coalescence in the rising foam should be minimized by maintaining a proper gas rate and a low foam height [Brunner and Lemlich, *Ind. Eng. Chem. Fundam.* **2**, 297 (1963)]. These precautions apply particularly with Eq. (22-45).

For laboratory purposes it is sometimes convenient to recycle the foamate directly to the pool in a manner analogous to an equilibrium still. This eliminates the feed and bottoms streams and makes for a more reliable approach to steady-state operation. However, this recycling may not be advisable for colligend measurements in the presence of slowly dissociating collector micelles.

To avoid spurious effects in the laboratory, it is advisable to employ a prehumidified chemically inert gas.

Bubble Sizes Subject to certain errors (de Vries, *Foam Stability*, Rubber-Stichting, Delft, 1957), foam bubble diameters can be measured photographically. Some of these errors can be minimized by taking pains to generate bubbles of fairly uniform size, say, by using a bubbler with identical orifices or by just using a bubbler with a single orifice (gas rate permitting). Otherwise, a correction for planar statistical sampling bias in the bubble diameters should be incorporated with actual diameters [de Vries, op. cit.] or truncated diameters [Lemlich, *Chem. Eng. Commun.* **16**, 153 (1982)]. Also, size segregation can reduce mean mural bubble diameter by roughly half the standard deviation [Cheng and Lemlich, *Ind. Eng. Chem. Fundam.* **22**, 105 (1983)]. Bubble diameters can also be measured in the liquid pool, either photographically or indirectly via measurement of the gas flow rate and stroboscopic determination of bubble frequency [Leonard and Lemlich, *Am. Inst. Chem. Eng. J.*, **11**, 25 (1965)].

Bubble sizes at formation generally increase with surface tension and orifice diameter. Prediction of sizes in swarms from multiple orifices is difficult. In aqueous solutions of low surface tension, bubble diameters of the order of 1 mm are common. Bubbles produced by the more complicated techniques of pressure flotation or vacuum flotation are usually smaller, with diameters of the order of 0.1 mm or less.

Enriching and Stripping Unlike truly simple foam fractionation without significant changes in bubble diameter, coalescence in a foam column destroys some bubble surface and so releases adsorbed material to trickle down through the rising foam. This downflow constitutes internal reflux, which enriches the rising foam by countercurrent action. The result is a richer foamate, i.e., higher C_Q than that obtainable from the single theoretical stage of the corresponding simple mode. Significant coalescence is often present in rising foam, but the effect on bubble diameter and enrichment is frequently overlooked.

External reflux can be furnished by returning some of the externally

broken foam to the top of the column. The concentrating effect of reflux, even for a substance which saturates the surface, has been verified [Lemlich and Lavi, *Science*, **134**, 191 (1961)].

Introducing the feed into the foam some distance above the pool makes for stripping operation. The resulting countercurrent flow in the foam further purifies the bottoms, i.e., lowers C_W.

Enriching, stripping, and combined operations are shown in Fig. 22-42.

Foam-Column Theory The counterflowing streams within the foam are viewed as consisting *effectively* of a descending stream of interstitial liquid (equal to zero for the simple mode) and an ascending stream of interstitial liquid plus bubble surface. (By considering this ascending surface as analogous to a vapor, the overall operation becomes analogous in a way to distillation *with entrainment*.)

An effective concentration $[\bar{C}]$ in the ascending stream at any level in the column is defined by Eq. (22-47):

$$\bar{C} = C + (GS\Gamma/U) \tag{22-47}$$

where U is the volumetric rate of interstitial liquid upflow, C is the concentration in this ascending liquid at that level, and Γ is the surface excess in equilibrium with C. Any effect of micelles should be included.

For simplicity, U can usually be equated to Q. An effective equilibrium curve can now be plotted from Eq. (22-47) in terms of $\bar{C}$ (or rather $\bar{C}°$) versus C.

Operating lines can be found in the usual way from material balances. The slope of each such line is $\Delta \bar{C}/\Delta C = L/U$, where L is the downflow rate in the particular column section and C is now the concentration in the descending stream.

The number of theoretical stages can then be found in one of the usual ways. Figure 22-43 illustrates a graphical calculation for a stripper.

Alternatively, the number of transfer units (NTU) in the foam based on, say, the ascending stream can be found from Eq. (22-48):

$$\text{NTU} = \int_{\bar{C}_W}^{C_Q} \frac{d\bar{C}}{\bar{C}° - \bar{C}} \tag{22-48}$$

$\bar{C}°$ is related to C by the effective equilibrium curve, and $\bar{C}_W°$ is similarly related to C_W. $\bar{C}$ is related to C by the operating line.

To illustrate this integration analytically, Eq. (22-48) becomes Eq. (22-49) for the case of a stripping column removing a colligend which is subject to the linear-equilibrium isotherm $\Gamma = KC$.

$$\text{NTU} = \frac{F}{GSK - W} \ln \frac{FW + F(GSK - W)C_F/C_W}{GSK(GSK + F - W)} \tag{22-49}$$

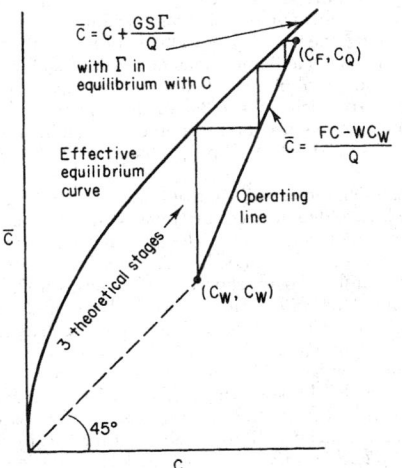

FIG. 22-43 Graphical determination of theoretical stages for a foam-fractionation stripping column.

As another illustration, Eq. (22-48) becomes Eq. (22-50) for an enriching column which is concentrating a surfactant with a constant Γ:

$$NTU = R \ln \frac{RGS\Gamma(F - D)}{(R + 1)GS\Gamma(F - D) - (R + 1)FD(C_D - C_F)} \quad (22\text{-}50)$$

Unless the liquid pool is purposely lengthened vertically in order to give additional separation via bubble fractionation, it is usually taken to represent one theoretical stage. A bubbler submergence of 30 cm or so is usually ample for a solute with a molecular weight that does not exceed several hundred.

In a colligend stripper, it may be necessary to add some collector to the pool as well as the feed because the collector is also stripped off.

Limiting Equations If the height of a foam-fractionation column is increased sufficiently, a concentration pinch will develop between the counterflowing interstitial streams (Brunner and Lemlich, loc. cit.). For an enricher, the separation attained will then approach the predictions of Eq. (22-51) and, interestingly enough, Eq. (22-46).

$$C_D = C_W + (GS\Gamma_W/D) \quad (22\text{-}51)$$

D is the volumetric rate at which net foamate (net overhead liquid product) is withdrawn. $D = Q/(R + 1)$. The concentration in the net foamate is C_D. In the usual case of total foam breakage (no dephlegmation), $C_D = C_Q$.

If the tall column is a stripper, the separation will approach that of Eqs. (22-52) and (22-53):

$$C_Q = C_F + (GS\Gamma_F/Q) \quad (22\text{-}52)$$

$$C_W = C_F - (GS\Gamma_F/W) \quad (22\text{-}53)$$

For a sufficiently tall combined column, the separation will approach that of Eqs. (22-54) and (22-53):

$$C_D = C_F + (GS\Gamma_F/D) \quad (22\text{-}54)$$

The formation of micelles in the foam breaker does not affect the limiting equations because of the theoretically unlimited opportunity in a sufficiently tall column for their transfer from the reflux to the ascending stream [Lemlich, "Principles of Foam Fractionation," in Perry (ed.), *Progress in Separation and Purification*, vol. 1, Interscience, New York, 1968, chap. 1].

In practice, the performance of a well-operated foam column several feet tall may actually approximate the limiting equations, provided there is little channeling in the foam and provided that reflux is either absent or is present at a low ratio.

Column Operation To assure intimate contact between the counterflowing interstitial streams, the volume fraction of liquid in the foam should be kept below about 10 percent—and the lower the better. Also, rather uniform bubble sizes are desirable. The foam bubbles will thus pack together as blunted polyhedra rather than as spheres, and the suction in the capillaries (Plateau borders) so formed will promote good liquid distribution and contact. To allow for this desirable deviation from sphericity, $S = 6.3/d$ in the equations for enriching, stripping, and combined column operation [Lemlich, *Chem. Eng.*, **75**(27), 95 (1968); **76**(6), 5 (1969)]. Diameter d still refers to the sphere.

Visible channeling or significant deviations from plug flow of the foam should be avoided, if necessary by widening the column or lowering the gas and/or liquid rates. The superficial gas velocity should probably not exceed 1 or 2 cm/s. Under proper conditions, HTU values of several cm have been reported [Hastings, Ph.D. dissertation, Michigan State University, East Lansing, 1967; and Jashnani and Lemlich, *Ind. Eng. Chem. Process Des. Dev.*, **12**, 312 (1973)]. The foam column height equals NTU × HTU.

For columns that are wider than several centimeters, reflux and feed distributors should be used, particularly for wet foam [Haas and Johnson, *Am. Inst. Chem. Eng. J.*, **11**, 319 (1965)]. Liquid content within the foam can be monitored conductometrically [Chang and Lemlich, *J. Colloid Interface Sci.*, **73**, 224 (1980)]. See Fig. 22-44. Theoretically, as the limit $\mathcal{D} = K = 0$ is very closely approached, $\mathcal{D} = 3K$ [Lemlich, *J. Colloid Interface Sci.*, **64**, 107 (1978)].

Wet foam can be handled in a bubble-cap column (Wace and Ban-

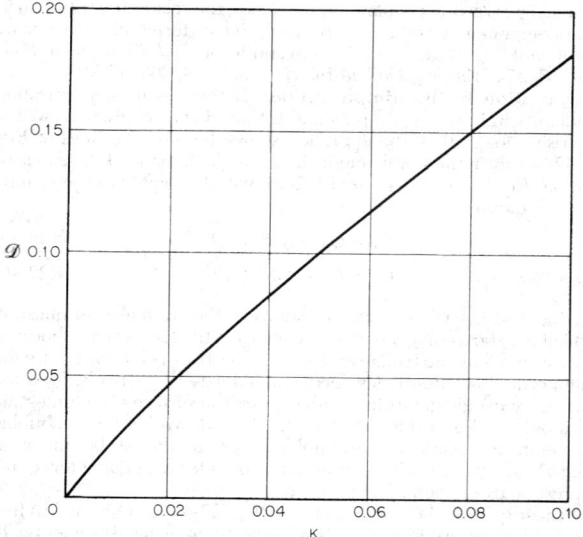

FIG. 22-44 Empirical relationship between $\mathcal{D}$, the volumetric fraction of liquid in common polydisperse foam, and K, the electrical conductivity of the foam divided by the electrical conductivity of the liquid. [*Chang and Lemlich*, J. Colloid Interface Sci., **73**, 224 (1980).]

field, *Chem. Process Eng.*, **47**(10), 70 (1966)] or in a sieve plate column [Aguayo and Lemlich, *Ind. Eng. Chem. Process Des. Dev.*, **13**, 153 (1974)]. Alternatively, individual short columns can be connected in countercurrent array [Banfield, Newson, and Alder, *Am. Inst. Chem. Eng. Symp. Ser.*, **1**, 3 (1965); Leonard and Blacyki, *Ind. Eng. Chem. Process Des. Dev.*, **17**, 358 (1978)].

A high gas rate can be used to achieve maximum throughput in the simple mode (Wace, Alder, and Banfield, AERE-R5920, U.K. Atomic Energy Authority, 1968) because channeling is not a factor in that mode. A horizontal drainage section can be used overhead [Haas and Johnson, *Ind. Eng. Chem. Fundam.*, **6**, 225 (1967)]. The highly mobile dispersion produced by a very high gas rate is not a true foam but is rather a so-called **gas emulsion** [Bikerman, *Ind. Eng. Chem.*, **57**(1), 56 (1965)].

A very low gas rate in a column several feet tall with internal reflux can sometimes be used to effect difficult multicomponent separations in batch operation [Lemlich, "Principles of Foam Fractionation," in Perry (ed.), *Progress in Separation and Purification*, vol. 1, Interscience, New York, 1968, chap. 1].

The same end may be achieved by continuous operation at total external reflux with a small U bend in the reflux line for foamate holdup [Rubin and Melech, *Can. J. Chem. Eng.*, **50**, 748 (1972)].

The slowly rising foam in a tall column can be employed as the sorbent for continuous chromatographic separations [Talman and Rubin, *Sep. Sci.*, **11**, 509 (1976)]. Low gas rates are also employed in short columns to produce the scumlike froth of batch-operated ion flotation, microflotation, and precipitate flotation.

Foam Drainage and Overflow The rate of foam overflow on a gas-free basis (i.e., the total volumetric foamate rate Q) can be estimated from a detailed theory for foam drainage [Leonard and Lemlich, *Am. Inst. Chem. Eng. J.*, **11**, 18 (1965)]. From the resulting relationship for overflow [Fanlo and Lemlich, *Am. Inst. Chem. Eng. Symp. Ser.*, **9**, 75, 85 (1965)], Eq. (22-55) can be employed as a convenient approximation to the theory so as to avoid trial and error over the usual range of interest for foam of low liquid content ascending in plug flow:

$$\frac{Q}{G} = 22 \left(\frac{v_G^3 \mu \mu_s^2}{g^3 \rho^3 d^8} \right)^{1/4} \quad (22\text{-}55)$$

The superficial gas velocity v_g is G/A, where A is the horizontal cross-sectional area of the empty vertical foam column. Also, g is the acceleration of gravity, ρ is the liquid density, μ is the ordinary liquid viscosity, and μ_s is the effective surface viscosity.

To account for inhomogeneity in bubble sizes, d in Eq. (22-55) should be taken as $\sqrt{\Sigma n_i d_i^3 / \Sigma n_i d_i}$ and evaluated at the top of the vertical column if coalescence is significant in the rising foam. Note that this average d for overflow differs from that employed earlier for S. Also, see "Bubble Sizes" regarding the correction for planar statistical sampling bias and the presence of size segregation at a wall.

For theoretical reasons, Q determined from Eq. (22-55) should be multiplied by the factor $(1 + 3Q/G)$ to give a final Q. However, for foam of sufficiently low liquid content this multiplication can be omitted with little error.

The effective surface viscosity is best found by experiment with the system in question, followed by back calculation through Eq. (22-55). From the precursors to Eq. (22-55), such experiments have yielded values of μ_s on the order of 10^{-4} (dyn·s)/cm for common surfactants in water at room temperature, which agrees with independent measurements [Lemlich, *Chem. Eng. Sci.*, **23**, 932 (1968); and Shih and Lemlich, *Am. Inst. Chem. Eng. J.*, **13**, 751 (1967)]. However, the expected high μ_s for aqueous solutions of such skin-forming substances as saponin and albumin was not attained, perhaps because of their non-newtonian surface behavior [Shih and Lemlich, *Ind. Eng. Chem. Fundam.*, **10**, 254 (1971); and Jashnani and Lemlich, *J. Colloid Interface Sci.*, **46**, 13 (1974)].

The drainage theory breaks down for columns with tortuous cross section, large slugs of gas, or heavy coalescence in the rising foam.

Foam Coalescence   Coalescence is of two types. The first is the growth of the larger foam bubbles at the expense of the smaller bubbles due to interbubble gas diffusion, which results from the smaller bubbles having somewhat higher internal pressures (Adamson, *The Physical Chemistry of Surfaces*, 4th ed., Wiley, New York, 1982). Small bubbles can even disappear entirely. In principle, the rate at which this type of coalescence proceeds can be estimated [Ranadive and Lemlich, *J. Colloid Interface Sci.*, **70**, 392 (1979)].

The second type of coalescence arises from the rupture of films between adjacent bubbles [Vrij and Overbeek, *J. Am. Chem. Soc.*, **90**, 3074 (1968)]. Its rate appears to follow first-order reaction kinetics with respect to the number of bubbles [New, *Proc. 4th Int. Congr. Surf. Active Substances*, Brussels, 1964, **2**, 1167 (1967)] and to decrease with film thickness [Steiner, Hunkeler, and Hartland, *Trans. Inst. Chem. Eng.*, **55**, 153 (1977)]. Many factors are involved [Bikerman, *Foams*, Springer-Verlag, New York, 1973; and Akers (ed.), *Foams*, Academic, New York, 1976].

Both types of coalescence can be important in the foam separations characterized by low gas flow rate, such as batchwise ion flotation producing a scum-bearing froth of comparatively long residence time. On the other hand, with the relatively higher gas flow rate of foam fractionation, the residence time may be too short for the first type to be important, and *if* the foam is sufficiently stable, even the second type of coalescence may be unimportant.

Unlike the case for Eq. (22-55), when coalescence is significant, it is better to find S from d evaluated at the feed level for Eqs. (22-52) to (22-54) and at the pool surface for Eqs. (22-46) and (22-51).

Foam Breaking   It is usually desirable to collapse the overflowing foam. This can be accomplished by chemical means (Bikerman, op. cit.) if external reflux is not employed or by thermal means [Kishimoto, *Kolloid Z.*, **192**, 66 (1963)] if degradation of the overhead product is not a factor.

Foam can also be broken with a rotating perforated basket [Lemlich, "Principles of Foam Fractionation," in Perry (ed.), *Progress in Separation and Purification*, vol. 1, Interscience, New York, 1968, chap. 1]. If the foamate is aqueous (as it usually is), the operation can be improved by discharging onto Teflon instead of glass [Haas and Johnson, *Am. Inst. Chem. Eng. J.*, **11**, 319 (1965)]. A turbine can be used to break foam [Ng, Mueller, and Walden, *Can. J. Chem. Eng.*, **55**, 439 (1977)]. Foam which is not overly stable has been broken by running foamate onto it [Brunner and Stephan, *Ind. Eng. Chem.*, **57**(5), 40 (1965)]. Foam can also be broken by sound or ultrasound, a rotating disk, and other means [Ohkawa, Sakagama, Sakai, Futai, and Takahara, *J. Ferment. Technol.*, **56**, 428, 532 (1978)].

If desired, dephlegmation (partial collapse of the foam to give reflux) can be accomplished by simply widening the top of the column, provided the foam is not too stable. Otherwise, one of the more positive methods of foam breaking can be employed to achieve dephlegmation.

Bubble Fractionation   Figure 22-45 shows continuous bubble fractionation. This operation can be analyzed in a simplified way in terms of the adsorbed carry-up, which furthers the concentration gradient, and the dispersion in the liquid, which reduces the gradient [Lemlich, *Am. Inst. Chem. Eng. J.*, **12**, 802 (1966); **13**, 1017 (1967)].

To illustrate, consider the limiting case in which the feed stream and the two liquid takeoff streams of Fig. 22-45 are each zero, thus resulting in batch operation. At steady state the rate of adsorbed carry-up will equal the rate of downward dispersion, or $af\Gamma = DAdC/dh$. Here a is the surface area of a bubble, f is the frequency of bubble formation. D is the dispersion (effective diffusion) coefficient based on the column cross-sectional area A, and C is the concentration at height h within the column.

There are several possible alternative relationships for Γ (Lemlich, op. cit.). For simplicity, consider $\Gamma = K'C$, where K' is not necessarily the same as the equilibrium constant K. Substituting and integrating from the boundary condition of $C = C_B$ at $h = 0$ yield

$$C/C_B = \exp(Jh) \qquad (22\text{-}56)$$

C_B is the concentration at the bottom of the column, and parameter $J = K'af/DA$. Combining Eq. (22-56) with a material balance against the solute in the initial charge of liquid gives

$$\frac{C}{C_i} = \frac{JH \exp(Jh)}{\exp(JH) - 1} \qquad (22\text{-}57)$$

C_i is the concentration in the initial charge, and H is the total height of the column.

The foregoing approach has been extended to steady continuous flow as illustrated in Fig. 22-45 [Cannon and Lemlich, *Chem. Eng. Prog. Symp. Ser.*, **68**(124), 180 (1972); Bruin, Hudson, and Morgan, *Ind. Eng. Chem. Fundam.*, **11**, 175 (1972); and Wang, Granstrom, and Kown, *Environ. Lett.*, **3**, 251 (1972), **4**, 233 (1973), **5**, 71 (1973)]. The extension includes a rough method for estimating the optimum feed location as well as a very detailed analysis of column performance which takes into account the various local phenomena around each rising bubble (Cannon and Lemlich, op. cit.).

Uraizee and Narsimhan [*Sep. Sci. Technol.*, **30**(6), 847 (1995)] have provided a model for the continuous separation of proteins from dilute solutions. Although their work is focused on protein separation, the model should find general application to other separations.

In agreement with experiment [Shah and Lemlich, *Ind. Eng. Chem. Fundam.*, **9**, 350 (1970); and Garmendia, Perez, and Katz,

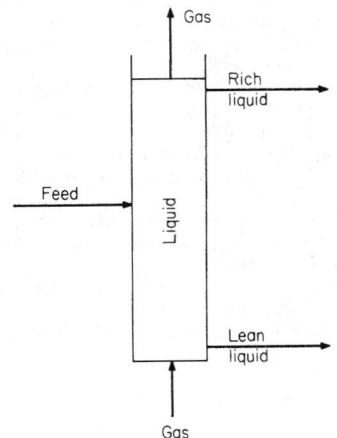

FIG. 22-45   Continuous bubble fractionation.

J. Chem. Educ., **50,** 864 (1973)], theory shows that the degree of separation that is obtained increases as the liquid column is made taller. But unfortunately it decreases as the column is made wider. In simple terms, the latter effect can be attributed to the increase in the dispersion coefficient as the column is widened.

In this last connection it is important that the column be aligned precisely vertically (Valdes-Krieg, King, and Sephton, *Am. Inst. Chem. Eng. J.,* **21,** 400 (1975)]. Otherwise, the bubbles with their dragged liquid will tend to rise up one side of the column, thus causing liquid to flow down the other side, and in this way largely destroy the concentration gradient. A vertical foam-fractionation column should also be carefully aligned to be plumb.

The escaping bubbles from the top of a bubble-fractionation column can carry off an appreciable quantity of adsorbed material in an aerosol of very fine film drops [various papers, *J. Geophys. Res., Oceans Atmos.,* **77**(27), (1972)]. If the residual solute is thus appreciably depleted, C_i in Eq. (22-57) should be replaced with the average residual concentration.

This carry-off of film drops, which may also occur with breaking foam, in certain cases can partially convert water pollution into air pollution. Such is the case, it may be desirable to recirculate the gas. Such recirculation is also indicated if hydrocarbon vapors or other volatiles are incorporated in the gas stream to improve adsorptive selectivity [Maas, *Sep. Sci.,* **4,** 457 (1969)].

A small amount of collector (surfactant) or other appropriate additive in the liquid may greatly increase adsorption (Shah and Lemlich, op. cit.). Column performance can also be improved by skimming the surface of the liquid pool or, when possible, by removing adsorbed solute in even a tenuous foam overflow. Alternatively, an immiscible liquid can be floated on top. Then the concentration gradient in the tall pool of main liquid, plus the trapping action of the immiscible layer above it, will yield a combination of bubble fractionation and solvent sublation.

Systems Separated   Some of the various separations reported in the literature are listed in Rubin and Gaden, "Foam Separation," in Schoen (ed.), *New Chemical Engineering Separation Techniques,* Interscience, New York, 1962, chap. 5; Lemlich, *Ind. Eng. Chem.,* **60**(10), 16 (1968); Pushkarev, Egorov, and Khrustalev, *Clarification and Deactivation of Waste Waters by Frothing Flotation,* in Russian, Atomizdat, Moscow, 1969; Kuskin and Golman, *Flotation of Ions and Molecules,* in Russian, Nedra, Moscow, 1971; Lemlich (ed.), *Adsorptive Bubble Separation Techniques,* Academic, New York, 1972; Lemlich, "Adsubble Methods," in Li (ed.), *Recent Developments in Separation Science,* vol. 1, CRC Press, Cleveland, 1972, chap. 5; Grieves, *Chem. Eng. J.,* **9,** 93 (1975); Valdes-Krieg King, and Sephton, *Sep. Purif. Methods,* **6,** 221 (1977); Clarke and Wilson, *Foam Flotation,* Marcel Dekker, New York, 1983; and Wilson and Clarke, "Bubble and Foam Separations in Waste Treatment," in Rousseau (ed.), *Handbook of Separation Processes,* Wiley, New York, 1987.

Of the numerous separations reported, only a few can be listed here. Except for minerals beneficiation [ore flotation] which is covered in Sec. 21, the most important industrial applications are usually in the area of pollution control.

A pilot-sized foaming unit reduced the alkyl benzene sulfonate concentration of 500,000 gal of sewage per day to nearly 1 mg/L, using a G/F of 5 and producing a Q/F of no more than 0.03 [Brunner and Stephan, *Ind. Eng. Chem.,* **57**(5), 40 (1965); and Stephan, *Civ. Eng.,* **35**(9), 46 (1965)]. A full-scale unit handling over 45,420 m³/day (12 million gal/day) performed nearly as well. The foam also carried off some other pollutants. However, with the widespread advent of biodegradable detergents, large-scale foam fractionation of municipal sewage has been discontinued.

Other plant-scale applications to pollution control include the flotation of suspended sewage particles by depressurizing so as to release dissolved air [Jenkins, Scherfig, and Eckhoff, "Applications of Adsorp-

tive Bubble Separation Techniques to Wastewater Treatment," in Lemlich (ed.), *Adsorptive Bubble Separation Techniques,* Academic, New York, 1972, chap. 14; and Richter, *Internat. Chem. Eng.,* **16,** 614 (1976)]. Dissolved-air flotation is also employed in treating wastewater from pulp and paper mills [Coertze, *Prog. Water Technol.,* **10,** 449 (1978); and Severeid, *TAPPI* **62**(2), 61, 1979]. In addition, there is the flotation, with electrolytically released bubbles [Chambers and Cottrell, *Chem. Eng.,* **83**(16), 95 (1976)], of oily iron dust [Ellwood, *Chem. Eng.,* **75**(16), 82 (1968)] and of a variety of wastes from surface-treatment processes at the maintenance and overhaul base of an airline [Roth and Ferguson, *Desalination,* **23,** 49 (1977)].

Fats and, through the use of lignosulfonic acid, proteins can be floated from the wastewaters of slaughterhouses and other food-processing installations [Hopwood, *Inst. Chem. Eng. Symp. Ser.,* **41,** M1 (1975)]. After further treatment, the floated sludge has been fed to swine.

A report of the recovery of protein from potato-juice wastewater by foaming [Weijenberg, Mulder, Drinkenberg, and Stemerding, *Ind. Eng. Chem. Process Des. Dev.,* **17,** 209 (1978)] is reminiscent of the classical recovery of protein from potato and sugar-beet juices [Ostwald and Siehr, *Kolloid Z.,* **79,** 11 (1937)]. The isoelectric pH is often a good choice for the foam fractionation of protein (Rubin and Gaden, loc. cit.). Adding a salt to lower solubility may also help. Additional applications of foam fractionation to the separation of protein have been reported by Uraizee and Narsimhan [*Enzyme Microb. Technol.* **12,** 232 (1990)].

With the addition of appropriate additives as needed, the flotation of refinery wastewaters reduced their oil content to less than 10 mg/L in pilot-plant operation [Steiner, Bennett, Mohler, and Clere, *Chem. Eng. Prog.,* **74**(12), 39 (1978)] and full-scale operation (Simonsen, *Hydrocarb. Process. Pet. Refiner,* **41**(5), 145, 1962). Experiments with a cationic collector to remove oils reportedly confirmed theory [Angelidon, Keskavarz, Richardson, and Jameson, *Ind. Eng. Chem. Process Des. Dev.,* **16,** 436 (1977)].

Pilot-plant [Hyde, Miller, Packham, and Richards, *J. Am. Water Works Assoc.,* **69,** 369 (1977)] and full-scale [Ward, *Water Serv.,* **81,** 499 (1977)] flotation in the preparation of potable water is described.

Overflow at the rate of 2700 m³ (713,000 gal) per day from a zinc-concentrate thickener is treated by ion flotation, precipitate flotation, and ultrafine-particle flotation [Nagahama, *Can. Min. Metall. Bull.,* **67,** 79 (1974)]. In precipitate flotation only the surface of the particles need be coated with collector. Therefore, in principle less collector is required than for the equivalent removal of ions by foam fractionation or ion flotation.

By using an anionic collector and external reflux in a combined (enriching and stripping) column of 3.8-cm (1.5-in) diameter with a feed rate of 1.63 m/h [40 gal/(h·ft²)] based on column cross section, D/F was reduced to 0.00027 with C_w/C_F for Sr^{2+} below 0.001 [Shonfeld and Kibbey, *Nucl. Appl.,* **3,** 353 (1967)]. Reports of the adsubble separation of 29 heavy metals, radioactive and otherwise, have been tabulated [Lemlich, "The Adsorptive Bubble Separation Techniques," in Sabadell (ed.), *Proc. Conf. Traces Heavy Met. Water,* 211–223, Princeton University, 1973, EPA 902/9-74-001, U.S. EPA, Reg. II, 1974). Some separation of ^{15}N from ^{14}N by foam fractionation has been reported [Hitchcock, Ph.D. dissertation, University of Missouri, Rolla, 1982].

The numerous separations reported in the literature include surfactants, inorganic ions, enzymes, other proteins, other organics, biological cells, and various other particles and substances. The scale of the systems ranges from the simple Crits test for the presence of surfactants in water, which has been shown to operate by virtue of transient foam fractionation [Lemlich, *J. Colloid Interface Sci.,* **37,** 497 (1971)], to the natural adsubble processes that occur on a grand scale in the ocean [Wallace and Duce, *Deep Sea Res.,* **25,** 827 (1978)]. For further information see the reviews cited earlier.

MEMBRANE SEPARATION PROCESSES

GENERAL REFERENCES: Noble and Stern (eds.), *Membrane Separations Technology*, Elsevier, 1995. Howell, Sanchez, and Field, *Membranes in Bioprocessing*, Chapman & Hall, 1993. Ho and Sirkar (eds.), *Membrane Handbook*, Van Nostrand Reinhold, 1992. Mulder, *Basic Principles of Membrane Technology*, Kluwer Academic Publishers, 1992. Bhave, *Inorganic Membranes: Synthesis, Characteristics, and Applications*, Chapman & Hall, 1991. Baker, Cussler, Eykamp, Koros, Riley, and Strathmann, *Membrane Separation Systems*, U.S. Department of Energy, DOE/ER/30133-H1, 1990. Porter (ed.), *Handbook of Industrial Membrane Technology*, Noyes, 1990. Wankat, *Rate-Controlled Separations*, chapters 12 and 13, Elsevier, 1990. Rautenbach and Albrecht, *Membrane Processes*, Wiley, 1989 Mohr, Leeper, Engelgau, and Charboneau, *Membrane Applications and Research in Food Processing*, Noyes, 1989. Li and Strathmann, *Separation Technology*, United Engineering Trustees, 1988. Cheryan, *Ultrafiltration Handbook*, Technomic Publishing, Lancaster, PA, 1986. Speigler and Laird, *Principles of Desalination, 2d ed.*, Academic Press, 1980. Many membrane research papers are published in *J. Membrane Sci.*, Elsevier.

Topics Omitted from This Section In order to concentrate on the membrane processes of widest industrial interest, several are left out.

Dialysis and Hemodialysis Historically, dialysis has found some industrial use. Today, much of that is supplanted by ultrafiltration. Donan dialysis is treated briefly under electrodialysis. Hemodialysis is a huge application for membranes, and it dominates the membrane field in area produced and in monetary value. This medical application is omitted here.

An excellent description of the engineering side of both topics is provided by Kessler and Klein [in Ho and Sirkar (eds.), op. cit., pp. 163–216]. A comprehensive treatment of diffusion appears in: Von Halle and Shachter, "Diffusional Separation Methods," in *Encyclopedia of Chemical Technology*, pp. 149–203, Wiley, 1993.

Facilitated Transport Transport by a reactive phase through a membrane is promising but problematic. Way and Noble [in Ho and Sirkar (eds.), op. cit., pp. 833–866] have a description and a complete bibliography.

Liquid Membranes Several types of liquid membranes exist: molten salt, emulsion, immobilized/supported, and hollow-fiber-contained liquid membranes. Araki and Tsukube (*Liquid Membranes: Chemical Applications*, CRC Press, 1990) and Sec. IX and Chap. 42 in Ho and Sirkar (eds.) (op. cit., pp. 724, 764–808) contain detailed information and extensive bibliographies.

Catalytic Membranes Falconer, Noble, and Sperry (Chap. 14—"Catalytic Membrane Reactors" in Noble and Stern, op. cit., p. 669–712) give a detailed review and an extensive bibliography. Additional information can be found in a work by Tsotsis et al. ["Catalytic Membrane Reactors," pp. 471–551, in Becker and Pereira (eds.), *Computer-Aided Design of Catalysts*, Dekker, 1993].

BACKGROUND AND DEFINITIONS

This section describes the use of separation processes which utilize membranes. Placement in this chapter is in recognition of the recent ascendency of industrial-scale membrane-based separations, but it also reflects the view that within a decade, many of these separation processes will be mainstream unit operations. Some approach that status already. Figure 22-46 shows the relative size of things important in membrane separations.

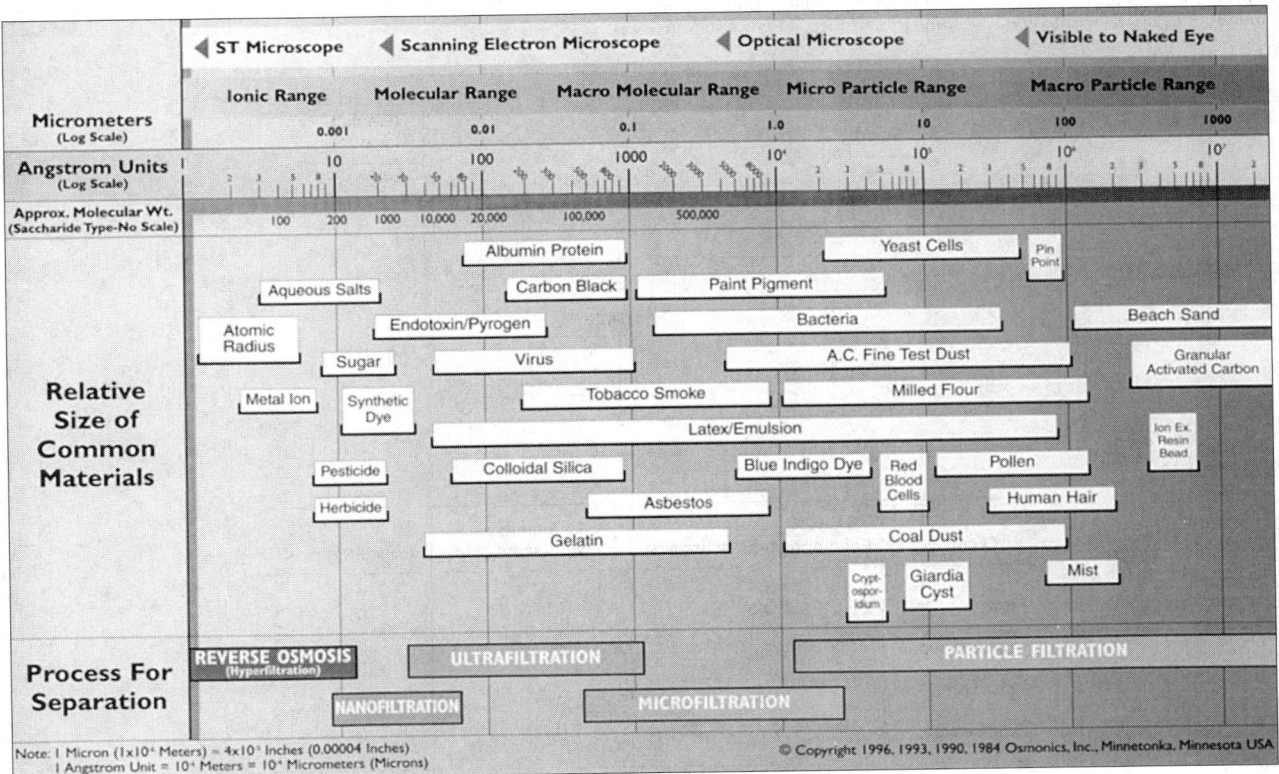

FIG. 22-46 The filtration spectrum. (*Copyright © 1996. Reprinted by permission of Osmonics, Minnetonka, MN.*)

Advantages to Membrane Separation This subsection covers the commercially important membrane applications. All except electrodialysis are pressure driven. All except pervaporation involve no phase change. All tend to be inherently low-energy consumers in theory if not in practice. They operate by a different mechanism than do other separation methods, so they have a unique profile of strengths and weaknesses. In some cases they provide unusual sharpness of separation, but in most cases they perform a separation at lower cost, provide more valuable products, and do so with fewer undesirable side effects than older separations methods. The membrane interposes a new phase between feed and product. It controls the transfer of mass between feed and product. It is a kinetic, not an equilibrium process. In a separation, a membrane will be selective because it passes some components much more rapidly than others. Many membranes are very selective. Membrane separations are often simpler than the alternatives.

General Examples No artificial membrane yet compares to the ones surrounding every cell in nature, but many are very sophisticated in what they do. Thousands of tons of drinking water are produced every day in the Middle East, on islands, and in certain coastal areas by passing pressurized seawater across a very thin membrane that permeates water with practically none of the dissolved salts. Huge quantities of nitrogen are purified from air by membranes operating from the output of a simple, single stage air compressor. Tons of water are purified to an exquisite degree for use in making microchips, most of the purification having been accomplished by passing the water through membranes. Municipalities depend on electrodialysis to remove salt from brackish aquifers to provide high-quality potable water.

Basic Equations All of the processes described in this section depend to some extent on the following background theory. Substances move through membranes by several mechanisms. For porous membranes, such as are used in microfiltration, viscous flow dominates the process. For electrodialytic membranes, the mass transfer is caused by an electrical potential resulting in ionic conduction. For all membranes, Fickian diffusion is of some importance, and it is of dom-

inant importance in gas permeation and reverse osmosis. In the following, almost every step is accompanied by a new assumption. Generally, they are reasonable assumptions, but when using the result, it is important to remember the many assumptions used to reach the convenient form of the equation. The driving force for Fickian transport of a substance is a gradient in chemical potential:

$$N_i = -D_i C \frac{d(\mu_i/RT)}{dx} \tag{22-58}$$

where N_i is the mass of component i transported, kmol/m²·s, D_i is diffusivity of component i, m²/s, C is concentration, kmol/m³, μ_i is chemical potential of the substance diffusing, J/kmol·K, and x is distance, m.

In most cases, activity coefficients are close to one, and Fick's first law is written as:

$$N_i = -D_i \frac{dC_i}{dx} \tag{22-59}$$

Assuming D_i is constant, and in particular that it is independent of C_i, and that the concentrations in the fluid phases are in equilibrium with the membrane. Fick's law may be written:

$$N_i = D_i \frac{(C_f - C_p)}{z} = D_i \frac{\Delta C}{z} \tag{22-60}$$

where z is the thickness of the membrane active layer, and C_f and C_p are concentrations in the feed and the permeate, respectively. $\overline{D}_i$ is the diffusivity in the membrane. The concentration of a component in the membrane phase will be quite different than its concentration in the fluid phase even though they are in equilibrium. The diffusivity in the membrane phase will always be much different than it is in the fluid phases, which must be remembered when applying Eq. (22-60). More complete nomenclature is shown in Fig. 22-47.

If Henry's law applies, the concentrations in fluid phases and the membrane are related by:

$$C_i = S \cdot p_i \tag{22-61}$$

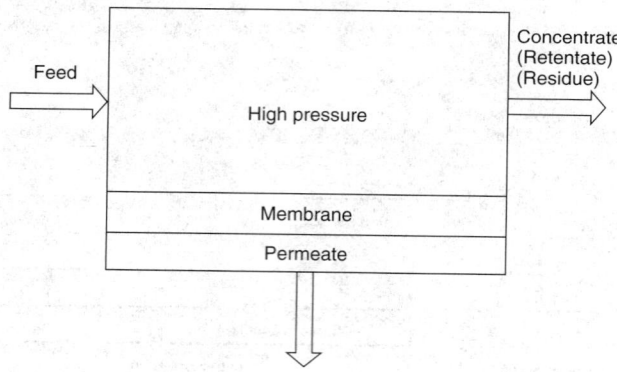

Quantity	Feed or high-pressure side	Permeate or low-pressure side	Concentrate, etc.
Flow	$Q_f N$	(J)(A),N	Q_r
Concentration	C_f	C_p	C_c
Partial pressure	p_{feed}	$p_{permeate}$	$p_{residue}$
Gas feed	L	V	R

FIG. 22-47 Schematic of pressure-driven processes showing nomenclature.

where S is a proportionality factor, kmol/m³·Pa, specific for a membrane and a penetrant and p is the partial pressure of component i, Pa.

If the membrane and its immediate surroundings are isothermal (generally except for pervaporation) and if S is a function only of temperature, then:

$$N_i = \left(\frac{S \cdot \overline{D}_i}{z}\right)(p_f - p_p) \qquad (22\text{-}62)$$

In membrane separations, the product $S \cdot \overline{D}_i$ is referred to as the permeability, ρ_i (kmol/m·s·Pa). The rate of passage of material through a membrane is referred to as flux, with symbol J_i. J_i is equal to N_i in the equations given above. Generally J_i has the dimensions of velocity, m/s (more conveniently, μm/s), or conventionally as ℓ/m²·hr, gal/ft²·day, or ft³/ft²·day. For most applications, throughput is expressed in volumes instead of moles or mass.

An important consideration is that the "goodness" of the separation is almost independent of the membrane thickness, z, and the *rate* of the process is *inversely* proportional to z, giving rise to a major emphasis on making the separating layer of a membrane very thin. It is rare that z is known for a commercial membrane, and J_i is stated without regard to z.

Basic Concepts

Membrane Porosity   Separation membranes run a gamut of porosity (see Fig. 22-48). Polymeric and metallic gas separation membranes, electrodialysis membranes, pervaporation membranes, and reverse osmosis membranes are nonporous, although there is lingering controversy over the nonporosity of the latter. Porous membranes are used for microfiltration and ultrafiltration. Nanofiltration membranes are probably charged porous structures.

Solution-Diffusion   Mass passes through nonporous membranes either by ionic conduction (electrodialysis) or by dissolving in the membrane [Eq. (22-61)] then diffusing through the membrane in response to a chemical potential gradient, which may be a change in concentration vapor pressure or an electric potential. Materials have differing Henry's law constants, which was an early emphasis in membrane material selection. Currently, especially in gas membranes, the structural differences leading to enhanced diffusivity selection is an important research area.

Pores   Even porous membranes can give very high selectivity. Molecular sieve membranes exist that give excellent separation factors for gases. Their commercial scale preparation is a formidable obstacle. At the other extreme, UF_6 separations use Knudsen flow barriers, with a very low separation factor. Microfiltration (MF) and ultrafiltration (UF) membranes are clearly porous, their pores ranging in size from 3 nm to 3 μm. Nanofiltration (NF) membranes have smaller pores.

TABLE 22-15   Liquid Flux Conversion Factors

TABLE 22-15   Liquid Flux Conversion Factors

To convert	Into	Multiply by	Inverse
U.S. gallons/ft²·day (gfd)	Liters/m²·hr (ℓmh)	1.6976	0.5891
gfd	μm/s	0.4715	2.121
ℓmh	μm/s	0.2778	3.600

Separation Factor   The separation factor, α, is defined consistent with other separation methods. It is important to recall that in membranes, α is the result of differing rates, and that it has no equilibrium implications. The convention in membrane separations is to define the separation so that $\alpha > 1$.

$$\alpha = \frac{(C_i/C_j)_p}{(C_i/C_j)_f} \qquad (22\text{-}63)$$

α is widely used in gas separations, and is used occasionally in other separations.

Retention, Rejection, and Reflection   **Retention** and **rejection** are used almost interchangeably. A third term, **reflection,** includes a measure of solute-solvent coupling, and is the term used in irreversible thermodynamic descriptions of membrane separations. It is important in only a few practical cases. Rejection is the term of trade in reverse osmosis (RO) and NF, and retention is usually used in UF and MF.

$$R = 1 - \left(\frac{C_p}{C_f}\right) \qquad (22\text{-}64)$$

where C is the concentration of the material being retained/rejected.

By convention, C_f is measured in the bulk feed, not at the membrane. Clearly, the concentration at the membrane is the important one, but the convention is well established and it simplifies calculations on yield and material balance. Concentration at the membrane, C_{wall} may be calculated by the method shown in Eq. (22-91).

Cross Flow   Most membrane processes are operated in cross flow, and only a few have the option to operate in the more conventional dead-end flow. In cross flow, the retentate passes parallel to the separating membrane, often at a velocity an order of magnitude higher than the velocity of the stream passing through the membrane. Microfiltration is the major membrane process in which a significant number if applications may be run with dead-end flow.

Staging   Membranes are rarely staged. Except for gas separations, it is unusual for the product to pass through more than one membrane. If the membrane does not make the required separation in one pass, other means of separation will normally be employed. Exceptions are noted for specific applications.

Flux is the term used to describe how fast a product passes through a membrane. A velocity, almost always reported as volume/area-time, it does not take membrane thickness into account. For most users,

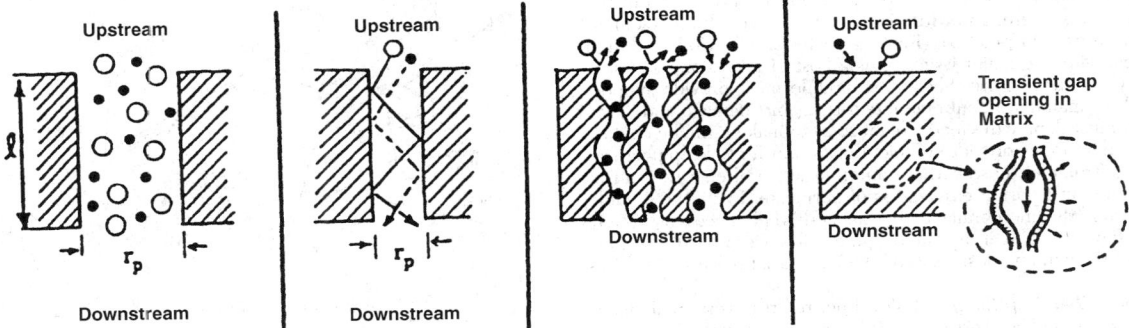

FIG. 22-48 Transport mechanisms for separation membranes: (*a*) Viscous flow, used in UF and MF. No separation achieved in RO, NF, ED, GAS, or PV. (*b*) Knudsen flow used in some gas membranes. Pore diameter < mean free path. (*c*) Ultramicroporous membrane—precise pore diameter used in gas separation. (*d*) Solution-diffusion used in gas, RO, PV. Molecule dissolves in the membrane and diffuses through. Not shown: Electrodialysis membranes and metallic membranes for hydrogen.

membrane thickness is unknown. Flux is specific for the membrane, for the application, for the operating conditions, and usually for time. Some generalization is possible. The rate-limiting step is very different among the operations, and membrane operations have quite different equations for flux.

Fouling   Flux declines with time in most membrane operations. A decline to 80 percent of initial output can take minutes or months. The principal cause is fouling, defined as an irreversible decline in output resulting from interaction with components in the feed. "Irreversible" is not synonymous with permanent, and describes a decline that can't be recovered by merely restoring a previous set of process conditions. Cleaning is a common way of restoring much or all of a fouled membrane's former output. Some forms of fouling are permanent, such as those that change membrane structure. Compaction, resulting from polymer creep, is permanent but is not fouling.

Membrane Types   A detailed taxonomy of membranes is beyond the scope of this handbook. Membranes may be made from physical solids (metal, ceramic, etc.), homogeneous films (polymer, metal, etc.), heterogeneous solids (polymer mixes, mixed glasses, etc.), solutions (usually polymer), asymmetric structures, and liquids.

Ceramic membranes are quite important since microporous ceramics are the principal barrier in UF_6 separation. Similar devices are used for microfiltration membranes and to a lesser extent for ultrafiltration. Homogeneous films are transformed into microporous devices by irradiation followed by selective leaching of the radiation damaged tracks, by stretching (Gortex® is one well-known example), or by electrochemical attack on aluminum. A few membranes are made by selective leaching of one component from a solid, as in membranes derived from glass or by selective extraction of polymer blends.

Liquid membranes are a specialty, either adsorbed in capillaries or emulsified. They are much studied, but little commercial application is found.

Polymeric membranes dominate the membrane separation field because they are well developed and quite competitive in separation performance and economics. Their usual final form is as **hollow fibers** or **capillaries** or as **flat sheet,** either of which is incorporated into a large module. Fiber-type membranes are solution or melt spun, and undergo some sort of transformation into a membrane shortly after the spinning head. **Thermal inversion** involves quenching a solution to a temperature regime where the polymer precipitates. **Solvent spinning** involves quenching in a nonsolvent, such as water, to produce a highly heterogeneous structure. Flat sheet membrane is made by preparing a casting dope of polymer in solvent, then casting it into a uniform film, then removing the solvent or introducing a nonsolvent (usually both) in such a way as to produce a membrane with a thin, active, separating layer backed by a porous, but mechanically robust sublayer.

An important variant is the composite membrane in which a relatively porous membrane (which often has its own skin) is coated by an even more selective layer, applied by a technique resulting in a very thin separating layer.

Module Types   The term "module" is universally used, but the definitions vary. Here, a **module** is the simplest membrane element that can be used in practice. (Figs. 22-48 to 22-53). Module design must deal with five major issues, plus a host of minor ones. First is economy of manufacture. Second, a module must provide support and seals to maintain membrane integrity against damage and leaks. Third, it must deploy the feed stream so as to make intimate contact with the membrane, provide sufficient mass transfer to keep polarization in control, and do so with a minimum waste of energy. Fourth, the module must permit easy egress of permeate. Fifth, the module must permit the membrane to be cleaned when necessary. Many module types have been invented, quite a few were used commercially, but the winning designs as of 1996 are variations on a few simple themes.

Hollow Fiber-Capillary   Hollow fiber refers to very small diameter membranes. The most successful one has an outer diameter of only 93 μm and is used for reverse osmosis. Capillary membranes are larger-diameter membranes used for liquid separations. The distinction between them has blurred to the point where there is a virtual

continuum of membrane diameters for gases and liquids from the smallest all the way up to 25-mm tubes.

Gas separation membranes have diameters as small as 135 μm outer diameter by 95 μm inner diameter. For low-pressure applications such as air, they may run with tube-side feed. Gas membranes operating at high pressure (above 1.5 MPa) are almost always run with shell-side feed. The outer diameter for gas membranes may be as high as 500 μm.

Self-supported cylindrical membranes for liquid separations are made from 250 μm up to 6 mm, but there is no obvious limit to future offerings. Membrane devices for liquids are almost always tube-side feed, with two major exceptions at the extremes of porosity. High-pressure RO is almost always shell-side feed, and one supplier of very low-pressure MF also runs with shell-side feed.

All types of membrane in this configuration are fashioned into modules by potting the ends with a curable liquid (Figs. 22-49 and 22-50).

Tubular   Tubular membranes (Fig. 22-51) are supported by a pressure vessel, usually perforated or porous. It can be as simple as a wrapped nonwoven fabric, or as robust as a stainless-steel tube. All run with tube-side feed. They are mainly used for UF, with some RO applications, particularly for food and dairy. The primary diameters available are 12 and 25 mm. Tubes are often connected in series; parallel bundles, gasketed or potted, are also common.

Monolith   Ceramic membranes are usually monoliths of tubular capillaries (Fig. 22-52), although one supplier has square passages. Channel sizes are in the millimeter range. By strict definition, a monolith becomes a module by attaching end fittings and a means of permeate collection. In practice, many monoliths are usually incorporated into one modular housing.

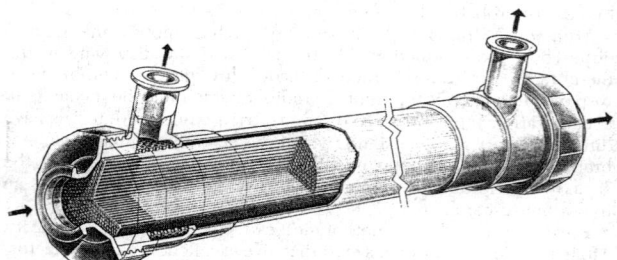

FIG. 22-49   Cutaway view of low-pressure capillary-membrane module. (*Courtesy Pall Corporation.*)

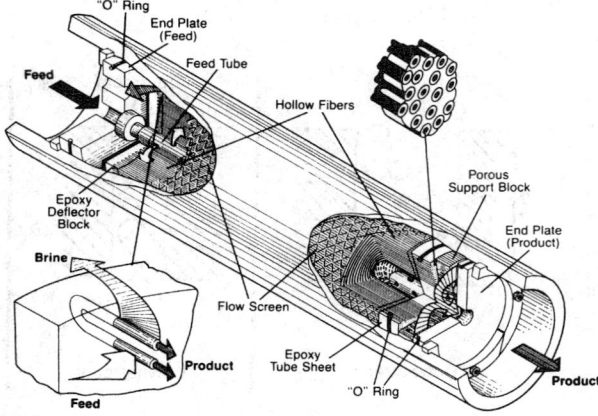

FIG. 22-50   Cutaway view of high-pressure hollow-fine-fiber reverse-osmosis module. (*Courtesy DuPont.*)

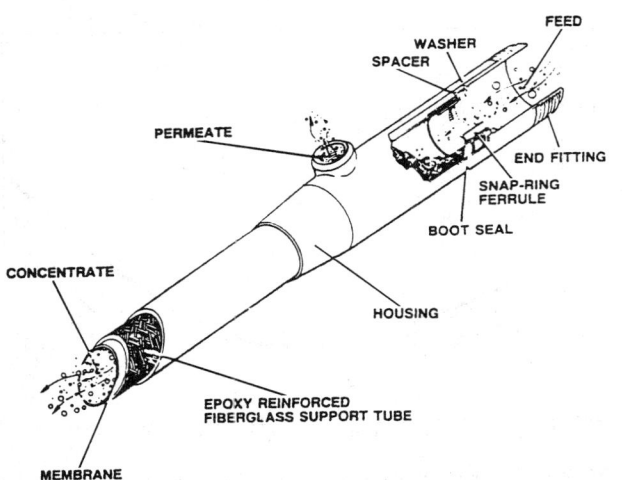

FIG. 22-51 25-mm diameter tubular-membrane assembly. (*Courtesy Koch Membrane Systems.*)

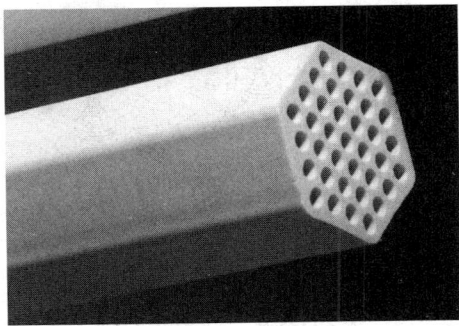

FIG. 22-52 Ceramic element cutaway. Membrane is inside 3 mm circular channels. (*Courtesy SCT/U.S. Filter.*)

Spiral Flat-sheet membrane may be fashioned into an inexpensive and compact module by spiral winding (Fig. 22-53). Membrane is laminated with a feed spacer separating two sheets of membrane. The permeate side of the membrane contacts a fluid-conductive fabric, in turn connected to a perforated central pipe. The edges are glued to make a complete seal between the feed and permeate sides of the device, and the finished round module is fitted into a pressure vessel where feed enters through one face of the cylinder and leaves through the other. Permeate is collected from the central tube. Multiple leaves are used because the pressure drop in the permeate-conducting fabric becomes limiting at leaf lengths much over 1 m. The spiral-wound

membrane module is now a highly evolved device, made as large as 16 inches (400 mm) in diameter, using many leaves. It has slowly increased its market share and is now the dominant design (meaning the "base case" against which other module types are measured) in RO and UF. Spiral-wound modules are occasionally used in gas separation and pervaporation.

Plate-and-Frame Conceptually the simplest, it is very much like a filter press. Once found in RO, UF, and MF, it is still the only module commonly used in electrodialysis (ED). A few applications in pressure-driven membrane separation remain (see Sec. 18 for a description of a plate-and-frame filter press).

Other The cassette (Fig. 22-54), a modification of a plate-and-frame device that is favored because of the ease of scale-up from laboratory to small plants is widely used in pharmaceutical microfiltration and ultrafiltration. An entirely different module also called a **cassette** is used in the MF of water. There are a host of other clever module designs in use, and new ones appear frequently.

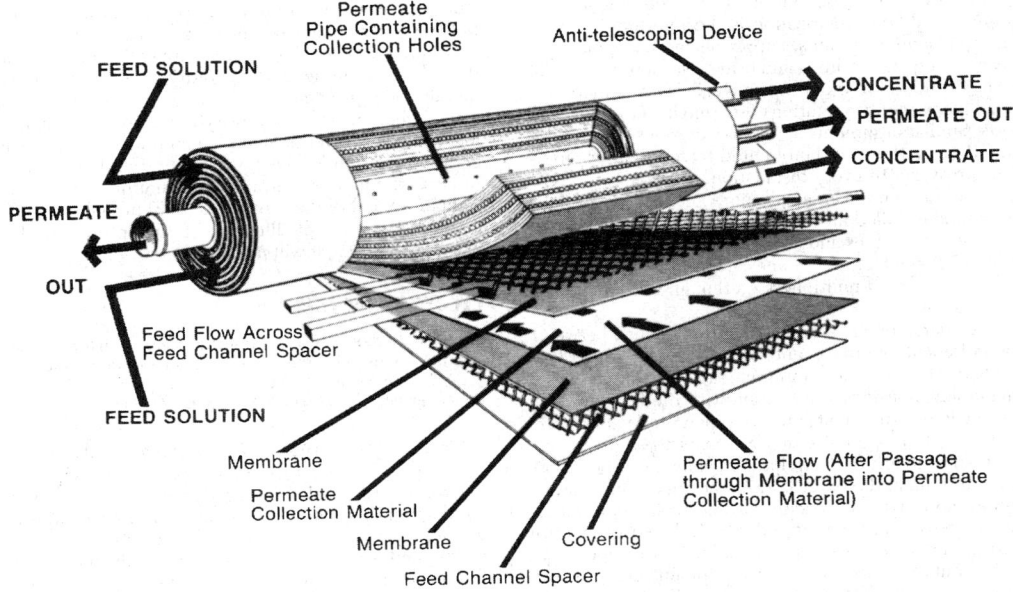

FIG. 22-53 Spiral-wound module used in many membrane processes. Permeate collection material is wound on a perforated permeate pipe. A membrane "sandwich" is constructed over the permeate carrier using glue seams as seals. Membrane "sandwiches" are separated by feed-channel spacers, through which the feed stream is passed. (*Courtesy Koch Membrane Systems.*)

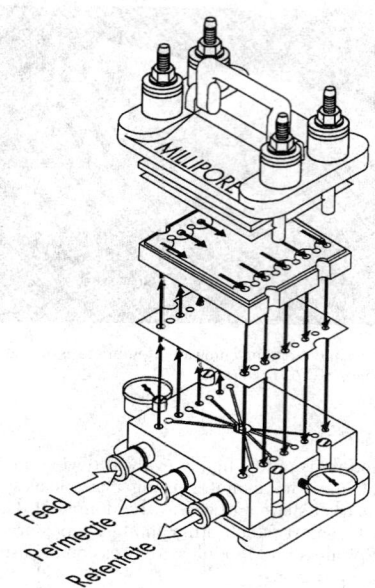

FIG. 22-54   Exploded view of cassette membrane assembly. (*Courtesy Millipore Corporation.*)

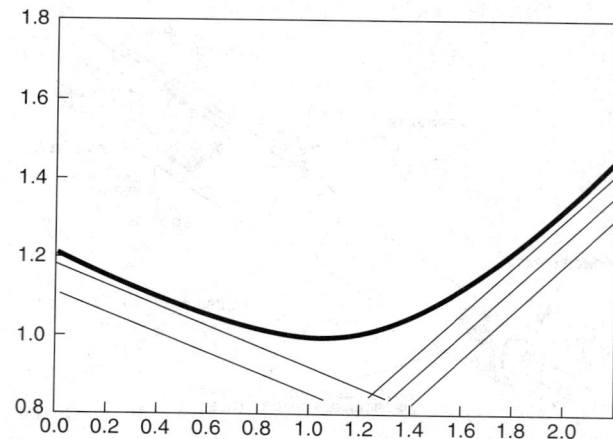

FIG. 22-55   Typical capital-cost schematic for membrane equipment showing trade-off for membrane area and mechanical equipment. Lines shown are from families for parallel lines showing limiting costs for membrane and for ancillary equipment. Abscissa: Relative membrane area installed in a typical membrane process. Minimum capital cost is at 1.0. Ordinate: Relative cost. Line with positive slope is total membrane cost. Line with negative slope is total ancillary equipment cost. Curve is total capital cost. Minimum cost is at 1.0.

Economics   The economics of each membrane process are distinct, but there is an underlying concept that is almost universal. What all membrane processes have in common is a membrane device per se, having a characteristic economy of scale cost = $r(area)^a$, and equipment surrounding and supporting the membrane having an economy of scale cost = $s(size)^b$. r and s are empirical constants. For all membrane equipment, $a > b$. For almost any membrane separation, there is a trade-off between the amount of membrane area required and the supporting equipment. Where polarization is the rate-limiting step, the designer can put in more pumps, supplying more depolarizing energy, thereby achieving higher fluxes and using less membrane. The result is lower investment for membranes, higher investment for pumps and pipes, higher energy operating costs, and lower membrane replacement costs. Similar arguments apply to operations that are not polarization limited, since they are then limited by some other physical factor such as pressure. In every installation, there is a trade-off between adding more membrane area and more ancillary equipment to make the membrane installed more productive. The values r and s vary with the *kind* of membranes and ancillary equipment used. Systems based on different *kinds* of membranes, ceramic versus organic, for example, will have different multipliers r although the exponent a should be about the same.

Figure 22-55 is a characteristic schematic of the balance between membrane area and ancillary equipment for a particular plant. For actual cases studied, the cost of membrane and its immediately related costs (manifolds, housings, etc.) varied with the 0.9 power of area. The cost of ancillary equipment (pumps, pipes, etc.) varied with the 0.4 power of size. Total capital cost represents the sum of the membrane related and the ancillary equipment, shown in Fig. 22-55 as a curve with a minimum at the point 1, 1. As the curve moves away from the minimum, costs rise approaching a line with slope 0.9 as membrane area increases, representing in the limit the cost of membrane. Moving to the left, the costs approach a line with slope −0.4, representing in the limit the costs of ancillary equipment. It is obvious that every plant needs some of both. For both membrane and ancillary equipment, there will be a series of parallel lines depending on the values of r and s for the different *kinds* of equipment chosen for the separation. For example, substituting a ceramic membrane for an organic membrane does not imply that the *kind* of pumping equipment employed will change. Assuming the ceramic membrane is costlier, using a membrane cost cure shifted upward will change the optimum-cost point, shifting it to the left.

An important caveat: The lines are shown as continuous functions, a considerable oversimplification. Pumps, pipes, valves, and even membrane assemblies come in discrete sizes and capacities, sometimes giving a project cost with a sharper minimum and one displaced from the ideal minimum. Every process has different characteristics, but the general shape of a broad economic minimum is characteristic.

Another similar curve can be drawn for total operating costs. The three biggest elements in operating costs are usually capital charges, membrane replacement, and energy. For most industrial installations, capital charges dominate everything. For municipal and a few other installations, power cost and membrane replacement are often of greater magnitude. Depending on the local economics and the process, the total operating cost minimum may be shifted a bit to the right or the left of the capital cost minimum, but usually not by a great amount. If membrane life is short, the optimum will move left, while if energy is costly, it will move right.

ELECTRODIALYSIS

GENERAL REFERENCES: This section is based on three publications by Heiner Strathmann, to which the interested reader is referred for greater detail [chap. 6, pp. 213–281, in Noble and Stern (eds.), op. cit.; sec. V, pp. 217–262, in Ho and Sirkar, op. cit.; chap. 8, pp. 8-1–8-53, Baker et al., op. cit.].

Process Description   *Electrodialysis* (ED) is a membrane separation process in which ionic species are separated from water, macrosolutes, and all uncharged solutes. Ions are induced to move by an electrical potential, and separation is facilitated by ion-exchange membranes. Membranes are highly selective, passing either anions or cations and very little else. The principle of ED is shown in Fig. 22-56.

The feed solution containing ions enters a compartment whose walls are a cation-exchange and an anion-exchange membrane. If the anion-exchange membrane is in the direction of the anode, as shown for the middle feed compartment, anions may pass through that membrane in response to an electrical potential. The cations can likewise

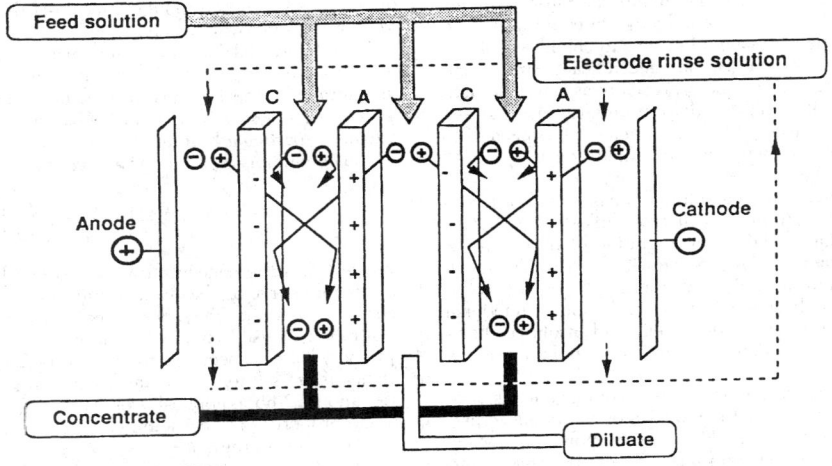

C: Cation transfer membrane
A: Anion transfer membrane

FIG. 22-56 Schematic diagram of electrodialysis. Solution containing electrolyte is alternately depleted or concentrated in response to the electrical field. Feed rates to the concentrate and diluate cells need not be equal. In practice, there would be many cells between electrodes.

move toward the cathode. When the ions arrive in the adjacent compartments, however, their further progress toward the electrodes is prevented by a membrane having the same electrical charge as the ion. The two feed compartments to the left and right of the central compartment are concentrate compartments. Ions entering these two compartments, either in the feed or by passing through a membrane, are retained, either by a same-charged membrane, or by the EMF driving the operation. The figure shows two cells (four membranes) between anode and cathode. In an industrial application, a membrane stack can be composed of hundreds of cells, where mobile ions are alternately being depleted and concentrated.

Many related processes use charged membranes and/or EMF. Electrodialytic water dissociation (water splitting), diffusion dialysis, Donnan dialysis, and electrolysis are related processes. **Electrolysis** (chlorine-caustic) is a process of enormous importance much of which is processed through very special membranes.

Leading Examples   Electrodialysis has its greatest use in removing salts from brackish water, where feed salinity is around 0.05–0.5 percent. For producing high-purity water, ED can economically reduce solute levels to extremely low levels as a hybrid process in combination with an ion-exchange bed. ED is not economical for the production of potable water from seawater. Paradoxically, it is also used for the concentration of seawater from 3.5 to 20 percent salt. The concentration of monovalent ions and selective removal of divalent ions from seawater uses special membranes. This process is unique to Japan, where by law it is used to produce essentially all of its domestic table salt. ED is very widely used for deashing whey, where the desalted product is a useful food additive, especially for baby food.

Many ED-related processes are practiced on a small scale, or in unique applications. Electrodialysis may be said to do these things well: separate electrolytes from nonelectrolytes and concentrate electrolytes to high levels. It can do this even when the pH is very low. ED does not do well at: removing the last traces of salt (although the hybrid process, electrodeionization, is an exception), running at high pH, tolerating surfactants, or running under conditions where solubility limits may be exceeded. Hydroxyl ion and especially hydrogen ion easily permeate both types of ED membrane. Thus, processes that generate a pH gradient across a membrane are limited in their scope.

Water splitting, a closely related process, is useful for reconstituting an acid and a base out of a salt. It is used to reclaim salts produced during neutralization.

Membranes   Ion-exchange membranes are highly swollen gels containing polymers with a fixed ionic charge. In the interstices of the polymer are mobile counterions. A schematic diagram of a cation-exchange membrane is depicted in Fig. 22-57.

Figure 22-57 is a schematic diagram of a cation-exchange membrane. The parallel, curved lines represent the polymer matrix composed of an ionic, crosslinked polymer. Shown in the polymer matrix are the fixed negative charges on the polymer, usually from sulfonate groups. The spaces between the polymer matrix are the water-swollen interstices. Positive ions are mobile in this phase, but negative ions are repelled by the negative charge from the fixed charges on the polymer.

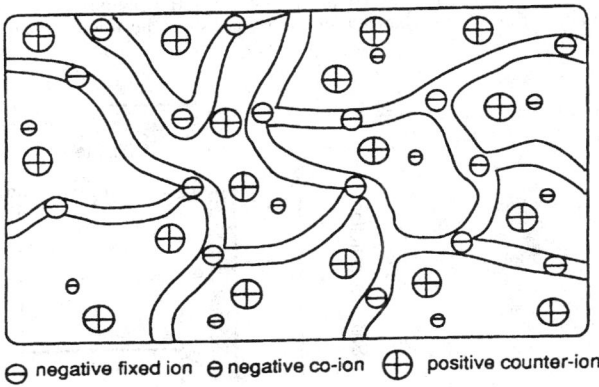

⊖ negative fixed ion   ⊖ negative co-ion   ⊕ positive counter-ion

FIG. 22-57 Schematic diagram of a cation-exchange membrane showing the polymer matrix with fixed negative charges and mobile positive counterions. The density of fixed negative charges is sufficient to prevent the passage (exchange) of anions.

In addition to high permselectivity, the membrane must have low-electrical resistance. That means it is conductive to counterions and does not unduly restrict their passage. Physical and chemical stability are also required. Membranes must be mechanically strong and robust, they must not swell or shrink appreciably as ionic strength changes, and they must not wrinkle or deform under thermal stress. In the course of normal use, membranes may be expected to encounter the gamut of pH, so they should be stable from $0 < pH < 14$ and in the presence of oxidants.

Optimization of an ion-exchange membrane involves major trade-offs. Mechanical properties improve with cross-link density, but so does high electrical resistance. High concentration of fixed charges favors low electrical resistance and high selectivity, but it leads to high swelling, thus poor mechanical stability. Membrane developers try to combine stable polymeric backbones with stable ionic functional groups. The polymers are usually hydrophobic and insoluble. Polystyrene is the major polymer used, with polyethylene and polysulfone finding limited application.

Most commercial ion-exchange membranes are homogeneous, produced either by polymerization of functional monomers and cross-linking agents, or by chemical modification of polymers. Many heterogeneous membranes have been prepared both by melting and pressing a mixture of ion-exchange resin and nonfunctional polymer, or by dissolving or dispersing both functional and support resins in a solvent and casting into a membrane. Microheterogeneous membranes have been made by block and graft polymerization.

No membrane and no set of membrane properties has universal applicability. Manufacturers who service multiple applications have a variety of commercial membranes. One firm lists twenty different membranes having a broad spectrum of properties.

Cation-Exchange Membranes Polystyrene copolymerized with divinylbenzene, then sulfonated, is the major building block for cation-exchange membranes. These membranes have reasonable stability and versatility and are highly ionized over most of the pH range. Other chemistries mentioned in the literature include carboxylic acid membranes based on acrylic acid, PO_3^{2-}, HPO_2^-, AsO_3^{2-}, and SeO_3^-. Many specialty membranes have been produced for electrodialysis applications. A notable example is a membrane selective to monovalent cations made by placing a thin coating of positive charge on the cation-exchange membrane. Charge repulsion for polyvalent ions is much higher than that for monovalent ions, but the resistance of the membrane is also higher.

Anion-Exchange Membranes Quaternary amines are the major charged groups in anion-exchange membranes. Polystyrene-divinyl-benzene polymers are common carriers for the quaternary amines. The literature mentions other positive groups based on N, P, and S. Anion-exchange membranes are problematic, for the best cations are less robust chemically than their cation exchange counterparts. Since most natural foulants are colloidal polyanions, they adhere preferentially to the anion-exchange membrane, and since the anion-exchange

membrane is exposed to higher local pH there is a greater likelihood that precipitates will form there.

Membrane Efficiency The permselectivity of an ion-exchange membrane is the ratio of the transport of electric charge through the membrane by specific ions to the total transport of electrons. Membranes are not strictly semipermeable, for coions are not completely excluded, particularly at higher feed concentrations. For example, the Donnan equilibrium for a univalent salt in dilute solution is:

$$C_{Co}^M = \left(\frac{(C_{Co}^F)^2}{C_R^M} \right) \left(\frac{\gamma_{\pm}^F}{\gamma_{\pm}^M} \right)^2 \tag{22-65}$$

where C_{Co}^M is the concentration of coions (the ions having the same electrical charge as the fixed charges on the membrane); C_{Co}^F is the concentration of such coions in the ambient solution; C_R^M is the concentration of fixed charges in the gel water of the membrane; and $\gamma_{\pm}^F$ and $\gamma_{\pm}^M$ are respectively the geometric mean of the activity coefficients of the salt ions in the ambient solution and in the membrane. Equation (22-65) is applicable only when $C_{Co}^F \ll C_R^M$. Since the membrane properties are constant, coion transport rises roughly with the square of concentration.

Process Description Figure 22-58 gives a schematic view of an ED cell pair, showing the salt-concentration profile. In the solution compartment on the left, labeled "Diluate," anions are being attracted to the right by the anode. The high density of fixed cations in membrane "A" is balanced by a high mobile anion concentration within it. Cations move toward the cathode at the left, and there is a similar high mobile cation concentration in the fixed anion membrane "C" separating the depleting compartment from the concentrating compartment further left (not shown). For the anion permeable membrane "A," two boundary layers are shown. They represent depletion in the boundary layer on the left, and an excess in the boundary layer on the right, both due to the concentration polarization effects common to all membrane processes—ions must diffuse through a boundary layer whether they are entering or leaving a membrane. That step must proceed down a concentration gradient.

With every change in ion concentration, there is an electrical effect generated by an electrochemical cell. The anion membrane shown in the middle has three cells associated with it, two caused by the concentration differences in the boundary layers, and one resulting from the concentration difference across the membrane. In addition, there are ohmic resistances for each step, resulting from the E/I resistance through the solution, boundary layers, and the membrane. In solution, current is carried by ions, and their movement produces a friction effect manifested as a resistance. In practical applications, I^2R losses are more important than the power required to move ions to a compartment with a higher concentration.

Transfer of Ions Mass transfer of ions in ED is described by many electrochemical equations. The equations used in practice are empirical. If temperature, the flux of individual components, elec-

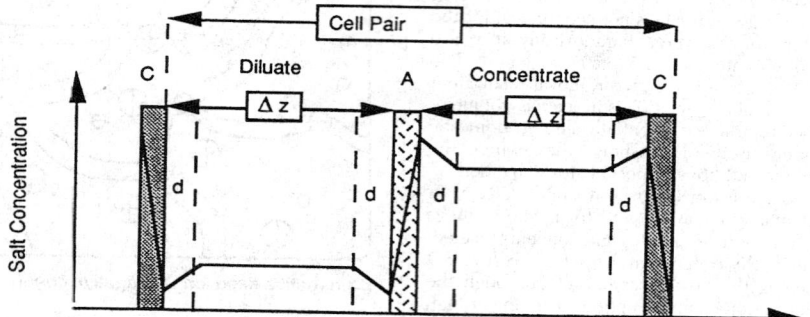

FIG. 22-58 Concentration profile of electrolyte across an operating ED cell. Ion passage through the membrane is much faster than in solution, so ions are enriched or depleted at the cell-solution interface. "d" is the concentration boundary layer. The cell gap, Δz should be small. The ion concentration in the membrane proper will be much higher than shown. (*Courtesy Elsevier.*)

troosmotic effects, streaming potential and other indirect effects are minor, an equation good to a reasonable level of approximation is:

$$J_n = C_{mn} U_{mn}\, \Delta\varphi/\Delta x \qquad (22\text{-}66)$$

where J is the component flux through the membrane kmol/m²·s, C_{mn} is the concentration in phase m of component n, kmol/m³, U_{mn} is the ion mobility of n in m, m²/v·s, φ is the electrical potential, volts, and x is distance, m.

Equation (22-66) assumes that all mass transport is caused by an electrical potential difference acting only on cations and anions. Assuming the transfer of electrical charges is due to the transfer of ions.

$$i = F \sum_n |z_n J_n| \qquad (22\text{-}67)$$

where i is the current density, amperes/m², F is the Faraday constant and z is the valence. The transport number, T, is the ratio of the current carried by an ion to the current carried by all ions.

$$T_n = \frac{J_n z_n}{\sum_n J_n z_n} \qquad (22\text{-}68)$$

The transport number is a measure of the permselectivity of a membrane. If, for example, a membrane is devoid of coions, then all current through the membrane is carried by the counterion, and the transport number = 1. The transport numbers for the membrane and the solution are different in practical ED applications.

Concentration Polarization   As is shown in the flow schematic, ions are depleted on one side of the membrane and enriched on the other. The ions leaving a membrane diffuse through a boundary layer into the concentrate, so the concentration of ions will be higher at the membrane surface, while the ions entering a membrane diffuse through a boundary layer from the diluate, so the bulk concentration in the diluate must be higher than it is at the membrane. These effects occur because the transport number of counterions in the membrane is always very much higher than their transport number in the solution. Were the transport numbers the same, the boundary layer effects would vanish. This concentration polarization is similar to that experienced in reverse osmosis, except that it has both depletion and enrichment components. The equations governing concentration polarization and depolarization of a membrane are given in the section describing ultrafiltration. The depolarizing strategies used for ED are similar to those employed in other membrane processes, as they involve induced flow past the membrane.

Two basic flow schemes are used: tortuous path flow and sheet flow (Fig. 22-59). Tortuous path spacers are cut to provide a long path between inlet and outlet, providing a relatively long residence time and high velocity past the membrane. The flow channel is open. Sheet flow units have a net spacer separating the membranes. Mass transfer is enhanced either by the spacer or by higher velocity.

Enhanced depolarization requires capital equipment and energy, but it achieves savings in overall capital costs (permits the use of a smaller stack) and energy (permits lower voltage.) The designer's task is to achieve an optimum balance in these requirements. Sheet-flow units have lower capital and operating costs in general, yet both sheet-flow and tortuous-flow units remain competitive. The fact that most ED reversal units (see below) are tortuous flow, and that ED reversal is the dominant technology for water and many waste treatment applications may explain the paradox.

Limiting Current Density   As the concentration in the diluate becomes ever smaller, or as the current driving the process is increased, eventually a situation arises in which the concentration of ions at the membranes surrounding the diluate compartment approaches zero. When that occurs, there are insufficient ions to carry additional current, and the cell has reached the limiting current. Forcing the voltage higher results in the dissociation of water at the membrane, giving rise to a dramatic increase in pH due to OH⁻ ions emerging from the anion-exchange membrane. The high pH promotes precipitation of metal hydroxides and CaCO₃ on the membrane surface leading to flow restrictions, poor mass transfer, and subsequent membrane damage. Once a precipitate forms, its presence ini-

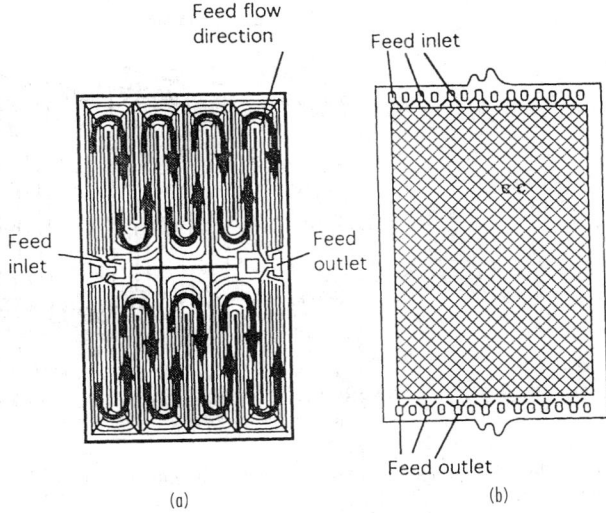

FIG. 22-59   Schematic of two ways to pass solution across an ED membrane. Tortuous flow (left) uses a special spacer to force the solution through a narrow, winding path, raising its velocity, mass transfer, and pressure drop. Sheet feed (right) passes the solution across the plate uniformly, with lower pressure drop and mass transfer. (*Courtesy Elsevier.*)

tiates a vicious cycle fostering the formation of more precipitates. The general expression for the limiting current density is $i_{\lim} \sim (C_{\text{bulk diluate}})$ (diluate velocity)^m, where m is an experimental constant (often 0.6). Thus, the concentration at the membrane is limiting, and at constant current that is proportional to the bulk concentration and the mass-transfer coefficient. Flow in the compartment is laminar but mass transfer is enhanced by the spacer. A normal operating practice is to operate a stack at around 75 percent of the limiting current.

Process Configuration   Figure 22-56 shows a basic cell pair. A stack is an assembly of many cell pairs, electrodes, gaskets, and manifolds needed to supply them. An exploded schematic of a portion of a sheet-flow stack is shown in Fig. 22-60.

Gaskets are very important, since they not only keep the streams separated and prevent leaks from the cell, they have the manifolds to conduct feeds, both concentrate and diluate, built into them. No other practical means of feeding the stack is used in the very cramped space required by the need to keep cells thin because the diluate has very low conductivity. The manifolds are formed by aligning holes in membrane and gasket.

The membranes are supported and kept apart by feed spacers. A typical cell gap is 0.5–2 mm. The spacer also helps control solution distribution and enhances mass transfer to the membrane. Given that an industrial stack may have up to 500 cell pairs, assuring uniform flow distribution is a major design requirement.

Since electrodialysis membranes are subject to fouling, it is sometimes necessary to disassemble a stack for cleaning. Ease of reassembly is a feature of ED.

Process Flow   The schematic in Fig. 22-56 may imply that the feed rates to the concentrate and diluate compartments are equal. If they are, and the diluate is essentially desalted, the concentrate would leave the process with twice the salt concentration of the feed. A higher ratio is usually desired, so the flow rates of feed for concentrate and feed for diluate can be independently controlled. Since sharply differing flow rates lead to pressure imbalances within the stack, the usual procedure is to recirculate the brine stream using a feed-and-bleed technique This is usually true for ED reversal plants. Some nonreversal plants use slow flow on the brine side avoiding the recirculating pumps.. Diluate production rates are often 10× brine-production rates.

Electrodes   No matter how many cells are put in series, there will be electrodes. The more cells, the less the relative importance of the

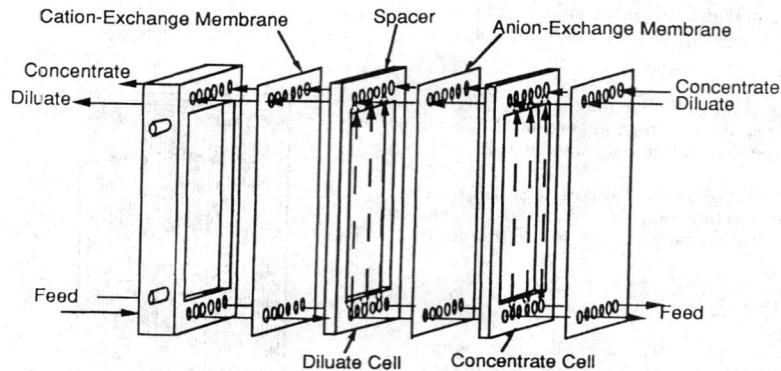

FIG. 22-60 Exploded view of a sheet-feed ED stack. Manifolds are built into the membranes and spacers as the practical way to maintain a narrow cell gap. (*Courtesy Elsevier.*)

electrodes. The cathode reactions are relatively mild, and depending on the pH:

$$2H^+ + 2e^- \rightarrow H_2$$

$$2H_2O + 2e^- \rightarrow 2OH^- + H_2$$

Anode reactions can be problematic. The anode may dissolve or be oxidized. Or, depending on the pH and the chloride concentration:

$$2H_2O \rightarrow O_2 + 4H^+ + 4e^-$$

$$4OH^- \rightarrow O_2 + 2H_2O + 4e^-$$

$$2Cl^- \rightarrow Cl_2 + 2e^-$$

Dissolution of metal is avoided by selecting a resistant material such as Pt, Pt coated on Ti, or Pt on Nb. Base metals are sometimes used, as are graphite electrodes.

Electrode isolation is practiced to minimize chlorine production and to reduce fouling. A flush solution free of chlorides or with reduced pH is used to bathe the electrodes in some plants. Further information on electrodes may be found in a work by David ["Electrodialysis," pp. 496–499, in Porter (ed.), op. cit.].

Peripheral Components In addition to the stack, a power supply, pumps for diluate and concentrate, instrumentation, tanks for cleaning, and other peripherals are required. Safety devices are mandatory given the dangers posed by electricity, hydrogen, and chlorine.

Pretreatment Feed water is pretreated to remove gross objects that could plug the stack. Additives that inhibit the formation of scale, frequently acid, may be introduced into the feed.

Electrodialysis Reversal Two basic operating modes for ED are used in large-scale installations. **Unidirectional operation** is the mode described above in the general explanation of the process. The electrodes maintain their polarity and the ions always move in a constant direction. **ED reversal** is an intermittent process in which the polarity in the stack is reversed periodically. The interval may be from several minutes to several hours. When the polarity is reversed, the identity of compartments is also reversed, and diluate compartments become concentrate compartments and vice versa. The scheme requires instruments and valves to redirect flows appropriately after a reversal. The advantages that often justify the cost are a major reduction in membrane scaling and fouling, a reduction in feed additives required to prevent scaling, and less frequent stack-cleaning requirements.

Water Splitting A modified electrodialysis arrangement is used as a means of regenerating an acid and a base from a corresponding salt. For instance, NaCl may be used to produce NaOH and HCl. Water splitting is a viable alternative to disposal where a salt is produced by neutralization of an acid or base. Other potential applications include the recovery of organic acids from their salts and the treating of effluents from stack gas scrubbers. The new component required is a bipolar membrane, a membrane that splits water into H^+ and OH^-. At its simplest, a bipolar membrane may be prepared by

laminating a cation and an anion membrane. In the absence of mobile ions, water sorbed in the membrane splits into its components when a sufficient electrical gradient is applied. The intimate contact of the two membranes minimizes the problem of the low ionic conductivity of ion-depleted water. As the water is split, replacement water readily diffuses from the surrounding solution. Properly configured, the process is energy efficient.

A schematic of the production of acid and base by electrodialytic water dissociation is shown in Fig. 22-61. The bipolar membrane is inserted in the ED stack as shown. Salt is fed into the center compartment, and base and acid are produced in the adjacent compartments. The bipolar membrane is placed so that the cations are paired with OH^- ions and the anions are paired with H^+. Neither salt ion penetrates the bipolar membrane. As is true with conventional electrodialysis, many cells may be stacked between the anode and the cathode.

If recovery of both acid and base is unnecessary, one membrane is left out. For example, in the recovery of a weak acid from its salt, the anion-exchange membrane may be omitted. The process limitations relate to the efficiency of the membranes, and to the propensity for H^+ and OH^- to migrate through membranes of like fixed charge, limiting the attainable concentrations of acid and base to 3–5 N. The problem is at its worst for HCl and least troublesome for organic acids. Ion leakage limits the quality of the products, and the regenerated acids and bases are not of high enough quality to use in regenerating a mixed-bed ion-exchange resin.

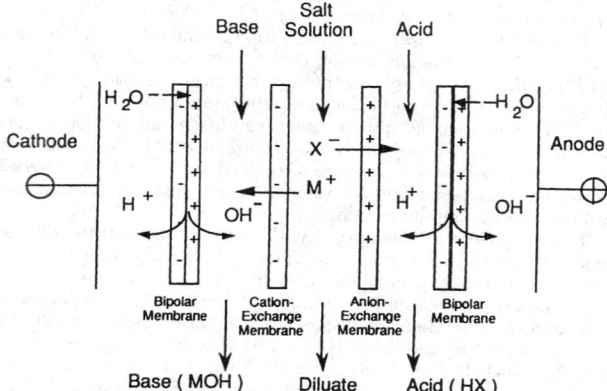

FIG. 22-61 Electrodialysis water dissociation (water splitting) membrane inserted into an ED stack. Starting with a salt, the device generates the corresponding acid and base by supplying H^+ and OH^- from the dissociation of water in a bipolar membrane. (*Courtesy Elsevier.*)

Diffusion Dialysis   The propensity of H^+ and OH^- to penetrate membranes is useful in diffusion dialysis. An anion-exchange membrane will block the passage of metal cations while passing hydrogen ions. This process uses special ion-exchange membranes, but does not employ an applied electric current.

As an example, in the aircraft industry heavy-aluminum sections are shaped as airfoils, then masked. The areas where the metal is not required to be strong are then unmasked and exposed to NaOH to etch away unneeded metal for weight reduction. Sodium aluminate is generated, a potential waste problem. Cation-exchange membranes leak OH^- by a poorly understood mechanism that is not simply the transport of OH^- with its waters of hydration. The aluminate anion is retained in the feed stream while the caustic values pass through. NaOH recovery is high, because all the Na^+ participates in the driving force. There is considerable passage of water due to the osmotic pressure difference as well. This scheme operates efficiently only because aluminum hydroxide forms highly supersaturated solutions. Hydroxide precipitation within the apparatus is reported to be a minor problem. $Al(OH)_3$ is precipitated in a downstream crystallizer, and is reported to be of high quality.

Donnan Dialysis   Another nonelectrical process using ED membranes is used to exchange ions between two solutions. The common application is to use H^+ to drive a cation from a dilute compartment to a concentrated one. A schematic is shown in Fig. 22-62. In the right compartment, the pH is 0, thus the H^+ concentration is 10^7 higher than in the pH 7 compartment on the left. H^+ diffuses leftward, creating an electrical imbalance that can only be satisfied by a cation diffusing rightward through the cation-selective membrane. By this scheme, Cu^{++} can be "pumped" from left to right against a significant concentration difference.

Electrodialysis-Moderated Ion Exchange   The production of ultrapure water is facilitated by incorporating a mixed-bed ion-exchange resin between the membranes of an ion-exchange stack. Already pure water is passed through the bed, while an electric current is passed through the stack. Provided the ion-exchange beads are in contact with each other and with the membranes, an electrical current can pass through the bed even though the conductivity of the very pure water is quite low. In passing, the current conducts any ions present into adjacent compartments, simultaneously and continuously regenerating the resin in situ.

Energy Requirements   The thermodynamic limit on energy is the ideal energy needed to move water from a saline solution to a pure phase. The theoretical minimum energy is given by:

$$\Delta G = RT \ln (a/a_s) \qquad (22\text{-}69)$$

where ΔG is the Gibbs free energy required to move one mole of

water from a solution, a is the activity of pure water ($\equiv 1$), and a_s is the activity of water in the salt solution. In a solution, the activity of water is approximately equal to the molar fraction of water in the solution. So that approximate activity is:

$$a_s = \frac{n_s}{n_s + \nu \cdot s_s} \qquad \frac{1}{a_s} = 1 + \frac{\nu \cdot s_s}{n_s} \qquad (22\text{-}70)$$

where n_s is the number of moles of water in the salt solution, ν is the number of atoms in the salt molecule (2 for NaCl, 3 for $CaCl_2$) and s_s is the number of moles of salt in the salt solution. The ratio of moles of salt in the salt solution to the number of moles of water in the salt solution is a very small number for a dilute solution. This permits using the approximation $\ln (1 + x) = x$, when x is of the magnitude 0.01, making this an applicable approximation for saline water. That permits rewriting Eq. (22-69) as:

$$\Delta G = \nu RT (s_s/n_s) \qquad (22\text{-}71)$$

where ΔG is still the free energy required to move one mole of water from the saline solution to the pure water compartment.

The conditions utilized in the above development of minimum energy are not sufficient to describe electrodialysis. In addition to the desalination of water, salt is moved from a saline feed to a more concentrated compartment. That free-energy change must be added to the free energy given in Eq. (22-71), which describes the movement of water from salt solution, the reverse of the actions in the diluate compartment (but having equal free energy). *Schaffer & Mintz* develop that change, and after solving the appropriate material balances, they arrive at a practical simplified equation for a monovalent ion salt, where activities may be approximated by concentrations:

$$\Delta G = RT(C_f - C_d)\left(\frac{\ln C_{fc}}{C_{fc} - 1} - \frac{\ln C_{fd}}{C_{fd} - 1} \right); C_{fc} = \frac{C_f}{C_c}; C_{fd} = \frac{C_f}{C_d} \quad (22\text{-}72)$$

where C_f is the concentration of ions in the feed, C_d is the concentration in the diluate, and C_c is the concentration in the concentrate, all in $kmol/m^3$. When $C_{fc} \to 1$ and $C_{fd} \to$ infinity, the operation is one approximating the movement of salt from an initial concentration into an unlimited reservoir of concentrate, while the diluate becomes pure. This implies that the concentrate remains at a constant salt concentration. In that case, Eq. (22-72) reduces to $RT(C_f - C_d)$. As a numerical example of Eq. (22-72) consider the desalting of a feed with initial concentration 0.05 M to 0.005 M, roughly approximating the production of drinking water from a saline feed. If 10 ℓ of product are produced for every 3 ℓ of concentrate, the concentrate leaves the process at 0.2 M. The energy calculated from Eq. (22-72) is 0.067 kWh/m^3 at 25°C. If the concentrate flow is infinite, $C_c = 0.05$ M, and the energy decreases to 0.031 kWh/m^3.

This minimum energy is that required to move ions only, and that energy will be proportional to the ionic concentration in the feed. It assumes that all resistances are zero, and that there is no polarization. In a real stack, there are several other important energy dissipaters. One is overcoming the electrical resistances in the many components. Another is the energy needed to pump solution through the stack to reduce polarization and to remove products. Either pumping or desalting energy may be dominant in a working stack.

Energy Not Transporting Ions   Not all current flowing in an electrodialysis stack is the result of the transport of the intended ions. Current paths that may be insignificant, minor, or significant include electrical leakage through the brine manifolds and gaskets, and transport of co-ions through a membrane. A related indirect loss of current is water transport through a membrane either by osmosis or with solvated ions, representing a loss of product, thus requiring increased current.

Pump Energy Requirements   If there is no forced convection within the cells, the polarization limits the current density to a very uneconomic level. Conversely, if the circulation rate is too high, the energy inputs to the pumps will dominate the energy consumption of the process. Furthermore, supplying mechanical energy to the cells raises the pressure in the cells, and raises the pressure imbalance between portions of the stack, thus the requirements of the confining gear and the gaskets. Also, cell plumbing is a design problem made more difficult by high circulation rates.

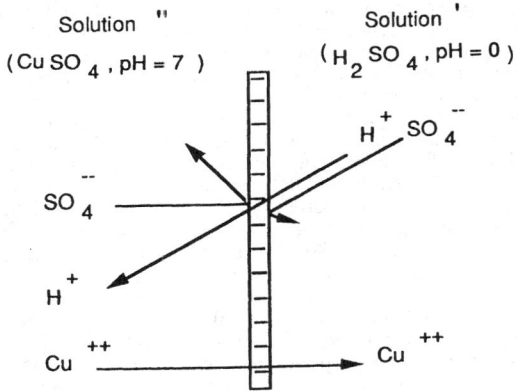

Solution "   Solution '

(Cu SO₄ , pH = 7)   (H₂ SO₄ , pH = 0)

FIG. 22-62   Schematic of Donnan dialysis using a cation-exchange membrane. Cu^{2+} is "pumped" from the lower concentration on the left to a higher concentration on the right maintaining electrical neutrality accompanying the diffusion of H^+ from a low pH on the right to a higher pH on the left. The membrane's fixed negative charges prevent mobile anions from participating in the process. (*Courtesy Elsevier.*)

A rule of thumb for a modern ED stack is that the pumping energy is roughly 0.5 kWh/m³, about the same as is required to remove 1700 mg/ℓ dissolved salts.

Equipment and Economics A very large electrodialysis plant would produce 500 ℓ/s of desalted water. A rather typical plant was built in 1993 to process 4700 m³/day (54.4 ℓ/s). Capital costs for this plant, running on low-salinity brackish feed were $1,210,000 for all the process equipment, including pumps, membranes, instrumentation, and so on. Building and site preparation cost an additional $600,000. The building footprint is 300 m². For plants above a threshold level of about 40 m³/day, process-equipment costs usually scale at around the 0.7 power, not too different from other process equipment. On this basis, process equipment (excluding the building) for a 2000 m³/day plant would have a 1993 predicted cost of $665,000.

The greatest operating-cost component, and the most highly variable, is the charge to amortize the capital. Many industrial firms use capital charges in excess of 30 percent. Some municipalities assign long amortization periods and low-interest rates, reflecting their cost of capital. Including buildings and site preparation, the range of capital charges assignable to 1000 m³ of product is $90 to $350.

On the basis of 1000 m³ of product water, the operating cost elements (as shown in Table 22-16) are anticipated to be:

TABLE 22-16 Electrodialysis Operating Costs

$ 66	Membrane-replacement cost (assuming seven-year life)
32	Plant power
16	Filters and pretreatment chemicals
11	Labor
8	Maintenance
$133	Total

These items are highly site specific. Power cost is low because the salinity removed by the selected plant is low. The quality of the feed water, its salinity, turbidity, and concentration of problematic ionic and fouling solutes, is a major variable in pretreatment and in conversion.

REVERSE OSMOSIS AND NANOFILTRATION

Process Description *Reverse osmosis* (RO) and *nanofiltration* (NF) processes utilize a membrane that selectively restricts flow of solutes while permitting flow of the solvent. The processes are closely related, and NF is sometimes called "loose RO." They are kinetic processes, not equilibrium processes. The solvent is almost always water.

Leading Examples

Potable Water RO and NF both play a major role in providing potable water, defined either by the WHO criterion of <1000 ppm *total dissolved solids* (TDS) or the U.S. EPA limit of 500 ppm TDS. RO is most prominent in the Middle East and on islands where potable-water demand has outstripped natural supply. A plant awaiting startup at Al Jubail, Saudi Arabia produces over 1 m³/s of fresh water (see Table 22-17). Small units are found on ships and boats. Seawater RO competes with *multistage flash distillation* (MSF) and *multieffect distillation* (MED) (see Sec. 13: "Distillation"). It is too expensive to compete with conventional civil supply (canals, pipelines, wells) in most locations. Low-pressure RO and NF compete with electrodialysis for the desalination of brackish water. The processes overlap economically, but they are sufficiently different so that the requirements of the application often favor one over the others.

TABLE 22-17 Water Volume Conversion Factors

To convert	Into	Multiply by	Inverse
1000 U.S. gallons	Cubic meters	3.785	0.2642
Acre-feet	Cubic meters	1233	8.11×10^{-4}
100 cubic feet	Cubic meters	2.831	0.3532
k-gal/day	m³/s	4.381×10^{-5}	22,827
MGD	m³/s	4.381×10^{-2}	22,827

Ocean water has an osmotic pressure of about 2.6 MPa, with some locations as high as 3.5 MPa. Recovery (r) is normally around 45 percent, occasionally higher. Osmotic pressure in the concentrate rises as $1/(1 - r)$ and significant overpressure (at least 1 MPa) is required to maintain good-quality permeate. Normal operating pressures are 6–8 MPa.

Brackish water has lower TDS than seawater. It ranges from diluted seawater to natural sources containing various salts. Some of the sources are quite large, and they may provide an attractive supplemental source of potable water. Disposal of the concentrate (brine) can be a problem for inland aquifers.

Where the TDS in water supplies is above the taste threshold, or where there is concern about the safety of the water supply, simple RO systems operating on line pressure have made a major impact. These compact units are usually kitchen installed, and are simple, small, and cheap. For typical line pressure of 400 kPa, 95 percent reduction of TDS is feasible if the inlet concentration is below 2000 mg/ℓ.

Process Water Purification Boiler feed water is a major process application of RO. Scalants and colloids are particularly well rejected by membranes, and TDS is reduced to a level that makes ion exchange or continuous deionization for the residual ions very economical. Even the extremely high quality water required for nuclear power plants can be made from seawater. The ultra-high quality water required for production of electronic microcircuits is usually processed starting with two RO systems operating in series, followed by many other steps.

Process Dewatering Applications RO is useful in many small applications where there is a volume of water containing a small amount of contaminant. RO is often able to recover most of the water at a purity high enough for reuse. The waste is concentrated making its disposal less costly, which generally pays for the recovery process.

Food and Beverage Applications RO achieved modest success in dewatering and concentrating food streams. The first food application pursued by early workers was the dewatering of orange juice prior to freezing. Cheryan [in Noble and Stern (eds.), op. cit., p. 452] describes a proprietary process that achieves high quality and high concentration by removing the flavor-laden phase first, concentrating the sugar stream with tight RO, then at high sugar concentrations using looser membranes to reduce $\Delta\Pi$ (concentrate-permeate). The high sugar permeate from the final stage is recycled through a lower-concentration stage, from which permeate is rejected. Cheryan provides a process schematic on p. 456. Apple-juice concentration has received much attention, but commercial success has been elusive. Other membrane processes are widely used in the juice industry, but for concentration applications, high osmotic pressure and flavor leakage through the membrane are barriers to wider adoption. After many years only a few plants are in operation.

Whey concentration, both of whole whey and ultrafiltration permeate, is practiced successfully, but the solubility of lactose limits the practical concentration of whey to about 20 percent total solids, about a 4× concentration factor. (Membranes do not tolerate solids forming on their surface.) Nanofiltration is used to soften water and clean up streams where complete removal of monovalent ions is either unnecessary or undesirable. Because of the ionic character of most NF membranes, they reject polyvalent ions much more readily than monovalent ions. NF is used to treat "salt whey," the whey expressed after NaCl is added to curd. Nanofiltration permits the NaCl to permeate while retaining the other whey components, which may then be blended with ordinary whey. NF is also used to deacidify whey produced by the addition of HCl to milk in the production of casein.

Basic Principles of Operation RO and NF are pressure-driven processes where the solvent is forced through the membrane by pressure, and the undesired coproducts frequently pass through the membrane by diffusion. The major processes are rate processes, and the relative rates of solvent and solute passage determine the quality of the product. The general consensus is that the solution-diffusion mechanism describes the fundamental mechanism of RO membranes, but a minority disagrees. Fortunately, the equations presented below describe the observed phenomena and predict experimental outcomes regardless of mechanism.

Driving Force For RO and NF, Eq. (22-62) becomes:

$$J = \frac{\rho_w}{z}(P_f - P_p) - (\Pi_f - \Pi_p) = \frac{\rho_w}{z}(\Delta P - \Delta \Pi) \qquad (22\text{-}73)$$

where ρ_w is the water permeability of the membrane, m²/Pa·s, and the subscripts f and p refer to feed and permeate. Π is the osmotic pressure, Pa. Since the thickness of the active layer z is almost never known, Eq. (22-73) is usually modified to the form

$$J = \left(\frac{1}{R_m + R_n \cdots}\right)(\Delta P - \Delta \Pi) \qquad (22\text{-}74)$$

where R_m is the membrane resistance, Pa·s/m. Other resistance terms $(R_n \cdots)$ may be added, such as terms for fouling or compaction. Normally, the important terms are the inherent membrane resistance, the driving pressure P, and the osmotic pressure in the feed, Π. For a high rejection RO membrane, the back-pressure and pressure terms for the permeate are insignificant. For most work, the van't Hoff approximation for osmotic pressure gives an adequate estimate:

$$\Pi = \nu n_s RT \qquad (22\text{-}75)$$

where νn_s is the total concentration of ions, kmol/m³ [Eq. (22-71)] and $R = 8.313$ kPa·m³/kmol·K. This equation should not be used for any unusually high concentration operation, or where accuracy is important.

Salt flux across a membrane is due to effects coupled to water transport, usually negligible, and diffusion across the membrane. Eq. (22-60) describes the basic diffusion equation for solute passage. It is independent of pressure, so as $\Delta P - \Delta \Pi \to 0$, rejection $\to 0$. This important factor is due to the kinetic nature of the separation. Salt passage through the membrane is concentration dependent. Water passage is dependent on $P - \Pi$. Therefore, when the membrane is operating near the osmotic pressure of the feed, the salt passage is not diluted by much permeate water.

The flux equation assumes constant temperature. As T rises, Π rises slowly, but around 25°C the viscosity of water drops enough to produce about a 3 percent rise in flux per °C.

Effects of Operating Variables Figure 22-63 shows trends in effects as various operating variables change in RO. Similar effects apply to NF.

RO and NF Membranes Nanofiltration membranes, sometimes called "loose RO," are more open than RO, and lie between reverse osmosis and ultrafiltration. This definition is not precise. NF membranes are usually negatively charged composite membranes. Donnan exclusion is an important rejection mechanism for charged membranes, and charged NF membranes reject polyvalent anions to a much greater degree than monovalent ions. They are usually tight enough to reject uncharged solutes heavier than a few hundred daltons and can, under ideal conditions, be highly retentive to lactose while passing most monovalent salts in a whey stream. Degrees of ion passage are strongly concentration dependent, with all rejections of charged ions dropping rapidly with increasing ionic strength. MgSO₄ rejection may remain high while NaCl rejection drops at around a few thousand ppm.

Methods of Production Modern RO and NF membranes are of two basic types. The more traditional is the asymmetric (skinned) membrane formed by the addition of a nonsolvent to a thin coating of homogeneous polymer solution. As solvent leaves the polymer solution and nonsolvent enters it, the surface polymer precipitates which alters the mechanism of solvent-nonsolvent interchange in the layers below. Thus the top layer, or skin, has a fundamentally different structure. Researchers have learned techniques for preparing membranes with widely differing properties, including some with excellent salt rejection properties. Membranes may be made either in flat sheet or fiber form using this technique. Strathmann [in Porter (ed.), op. cit., pp. 1–60] gives a complete summary.

The second major membrane type is a composite. Starting with a loose asymmetric membrane, usually a UF membrane, a coating is applied which is polymerized in situ to become the salt rejecting membrane. This process is used for most high-performance flat-sheet RO membranes, as well as for many commercial nanofiltration membranes. The chemistry of the leading RO membranes is known, but details on NF membranes are mostly proprietary. It may be inferred

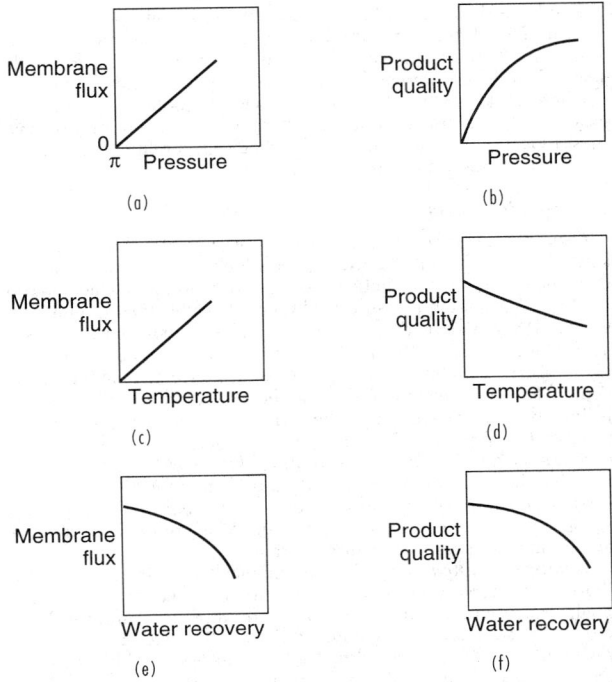

FIG. 22-63 Effects of operating variables on the performance of an RO membrane. (*a*) Water passage increases with pressure, assuming Π is constant. (*b*) Solute rejection rises with pressure, since solvent flux increases and solute diffusion does not. (*c*) Flux rises with temperature because viscosity declines. (*d*) Solute rejection declines with a temperature rise because of the osmotic pressure increase with temperature and because solute passage has a higher activation energy than does solvent passage. (*e*) Flux declines with increasing solute concentration, an osmotic-pressure effect. (*f*) At low-feed velocity past the membrane, solute is polarized at the membrane, while as velocity increases, mass transfer redisperses more of the polarized solute, lowering the effective solute concentration at the membrane. (Rejection is measured between permeate and bulk-feed concentration.) Bulk concentration of salt rises with recovery.

from published properties that various chemical approaches are utilized. Further information on composites may be found in Petersen [*J. Membrane Sci.*, **83**, 81–150 (1993)].

Membrane Characterization Membranes are always rated for flux and rejection. NaCl is always used as one measure of rejection, and for a very good RO membrane, it will be 99.7 percent or more. Nanofiltration membranes are also tested on a larger solute, commonly MgSO₄. Test results are very much a function of how the test is run, and membrane suppliers are usually specific on the test conditions. Salt concentration will be specified as some average of feed and exit concentration, but both are bulk values. Salt concentration at the membrane governs performance. Flux, pressure, membrane geometry, and cross-flow velocity all influence polarization and the other variables shown in Fig. 22-63.

Membrane Limitations Chemical attack, fouling, and compaction are prominent problems with RO and NF membranes. Compaction is the most straightforward. It is the result of creep, slow cold flow of the polymer resulting in a loss of water permeability. It is measured by the slope of log flux versus log time in seconds. It is independent of the flux units used and is reported as a slope, sometimes with the minus sign omitted. A slope of −0.001, typical for noncellulosic membranes, means that for every threefold increase in log(time), 10³ seconds, a membrane looses 10 percent of its flux. Since membranes are rated assuming that the dramatic early decline in permeability has already occurred, the further decline after the first few weeks is very slow. Compaction is specific to pressure, temperature, and environment.

Fouling is defined in "Background and Definitions" and is a significant problem in most process applications, and somewhat of a problem in most water applications. RO membranes may be fouled by sparingly soluble scalants which supersaturate at the membrane.

Chemical attack is often a result either of fouling prevention or cleaning in response to fouling. Chlorine and hypochlorite damage most RO and NF membranes, as do oxidants generally (see discussion of chlorine tolerance below).

Process Limitations

Osmotic Pressure   In RO, and to a great extent NF, osmotic pressure is a critically important design consideration. A proper thermodynamic treatment of osmotic pressure may be found in Cheryan (op. cit., p. 13). Osmotic pressure is always calculated for the bulk-feed stream. It varies along the membrane train as salt concentration rises. The osmotic pressure that really matters is the one at the membrane, higher by the amount polarization raises the concentration there. As a general rule for a new membrane application, the inlet concentration is limited to about 0.5 N, for which $\Pi \approx 2.5$ MPa, giving a final concentrate Π of 5 MPa for 50 percent conversion. A few systems may be designed at much higher pressure, notably one of Du Pont's hollow fiber bundles, which is rated at 8 MPa. It is rated for 65 percent conversion on ocean water, and can concentrate sucrose to 60 percent using a special technique and membrane. Much of the appeal of NF membranes is their low-pressure operation.

Membrane Chemistry   Three chemical families dominate the RO-NF membrane industry. Many other products are made on a small scale, and the field continues to attract significant R&D resources. But three types command most of the market.

Cellulose esters are the oldest. *Cellulose acetate* (CA) blend, a mixture of cellulose acetate (40.1 percent acetyl) and triacetate (43.2 percent acetyl) is the major polymer used; some cellulose triacetate and cellulose acetate-butyrate are used as well. Outstanding features of CA are its known behavior in many applications over many years, its slower fouling rate in some applications, its relative tolerance to chlorine, and foremost, its low cost. Where CA works well, there is little incentive to replace it by other materials. It has outstanding weaknesses including chemical and pH susceptibility, possible biodegradation, high compaction especially at elevated temperature, and poor rejection for organic solutes (due to their high solubility in CA). It is usually made as flat sheet, and occasionally as fibers.

Aromatic polyamide (aramid) membranes are a copolymer of 1-3 diaminobenzene with 1-3 and 1-4 benzenedicarboxylic acid chlorides. They are usually made into fine hollow fibers, 93 μm outer diameter by 43 μm inner diameter. Some flat sheet is made for spirals. These membranes are widely used for seawater desalination and to some extent for other process applications. The hollow fibers are capable of very high-pressure operation and have considerably greater hydrolytic resistance than does CA. Their packing density in hollow-fiber form makes them very susceptible to colloidal fouling (a permeator 8 inches in diameter contains 3 M fibers), and they have essentially no resistance to chlorine.

Cross-linked aromatic polyamides are the third major membrane type. The best known and most successful form has been an interfacially polymerized polyamide formed from 1-3 diaminobenzene and 1,3,5 benzenetricarboxylic acid chloride. This membrane is noted for excellent salt rejection and some degree of chemical resistance, including some resistance to chlorine, but it will not tolerate continuous exposure. The residual negative charge resulting from incomplete reaction and subsequent hydrolysis and ionization of the third acid-chloride group seems to help somewhat in improving its resistance to colloidal fouling. This membrane is made as flat sheet, and can be made in tubular form (13 mm inner diameter).

Chlorine Tolerance   Most of the best RO membranes are attacked by oxidants, and they are particularly susceptible to chlorine. A particularly sensitive locus for attack is the amidic hydrogen. Cellulosic membranes are generally less sensitive, and pass the chlorine into the permeate giving downstream biocidal activity, very useful for under-the-sink RO. These factors are largely responsible for CA's survival in RO membranes. Chlorine, whatever its vices, has the virtue of being a known, effective, residual bactericide and a good inhibitor of bacterial growth on membranes (see fouling). It is also very useful in

membrane cleaning, because not only does it kill bacteria, it breaks down some membrane deposits.

Chlorine is desirable as a bulk pretreatment biocide for inlet water, but its subsequent removal upstream of the membrane is absolutely necessary and difficult. $NaHSO_3$ is a common additive to dechlorinate before membranes. It is customarily added at 3–5 mg/l, an excess over the stoichiometric requirement. NH_3 is sometimes added to convert the chlorine to chloramine, a much less damaging biocide. Heavy metals present in seawater seem to amplify the damaging effects of chlorine and other oxidants.

Membranes are commonly rated for their chlorine tolerance in "ppm-hours," simply the product of the concentration and the contact time. Tolerance is temperature dependent.

Concentration Polarization   Concentration polarization is a function of both flux, which increases the mass rate of material stranded at the membrane and cross-flow velocity, which reduces polarization by enhancing feed-side mass transfer. Polarization is far less of a problem in reverse osmosis and nanofiltration than it is in ultrafiltration or microfiltration, but it cannot be ignored. If cross-flow velocity is insufficient, rejected species concentrate near the membrane to an unacceptable level. The resulting increase in osmotic pressure and the precipitation of sparingly soluble species (scaling) are concerns. Scale inhibitors are normally added to water when they are appropriate and, for these feeds, careful consideration of cross-flow velocity is required. Hollow-fiber modules operate at low flux and at low cross-flow velocity so diffusion is better able to reduce polarization; spirals have much better redispersion rates, but can be overdriven if operated at fluxes above the design values. The equations for polarization are given below in the section describing ultrafiltration.

Rejection   Rejection is defined in "Background and Definitions." The highest-rejection membranes are those designed for single-pass production of potable water from the sea. The generally accepted criterion is 99.4 percent rejection of NaCl. Some membranes, notably cellulose triacetate fibers are rated even higher. A whole range of membranes is available as rejection requirements ease, and membranes with excellent chlorine resistance and hydrolytic stability can be made with salt rejection over 90 percent.

Plugging   Silt carried into a membrane module may deposit and plug it. Membrane configurations differ markedly in their ability to tolerate suspended solids, with fine hollow fibers fed shell-side being least tolerant, and large bore tubes fed tube-side being most tolerant. Membrane manufacturers use the *silt density index* (SDI) to show the tolerance of their wares for suspended solids. The SDI is an arbitrary test, used to determine the plugging propensity of an RO feed. A 0.45 μm microfiltration membrane, 47 mm in diameter, is used for the test. The feed to be tested is passed through the filter at 30 psi (206 kPa).

The time required for the first 100 ml to pass through the filter is recorded. That is defined as t_0. The flow is continued for 15 minutes, then the time required for an additional 100 ml to pass through the filter is recorded. That is defined as t_{15}. If the flow has continued throughout the test, that is, the stream has not been reduced to dropwise flow, the SDI is defined as:

$$\text{SDI} = \frac{100(t_{15} - t_0)}{15(t_{15})} = 6.67 \frac{(t_{15} - t_0)}{(t_{15})} \qquad (22\text{-}76)$$

If the flow becomes dropwise during the 15 minutes, the formula is instead:

$$\text{SDI} = \frac{100}{\text{(minutes until flow becomes dropwise)}} \qquad (22\text{-}77)$$

Examples: (1) The first 100 ml required 11 seconds, and after 15 minutes of flow, an additional 100 ml required 95 seconds. The SDI would be 5.9 (2) The first 100 ml took 37 seconds, but the flow became dropwise after 4 minutes. The SDI would be 25.

Fouling   Fouling is as inevitable as death and taxes. All membranes foul. Prevention and remediation of fouling are major economic and operating concerns in the design of a membrane facility. Scaling results from the precipitation of sparingly soluble species. Colloids deposit on the membrane in spite of cross-flow, and biofouling is a problem for most feeds. Much can be done to increase the interval at which a membrane unit must be shut down and cleaned. The silt density index is a reasonable, if qualitative, guide to the degree to which colloidal fouling becomes a dominating problem. Biofouling is highly dependent on the feed-biota level and on nutrient levels. Reduction of biological load through pretreatment with chlorine or another biocide is a common practice, and on a pilot scale, microfiltration is being tried. Limited success has come from attempts to maintain anaerobic conditions in the feed. Research in biofouling prevention and remediation is an active area.

Pretreatment   For most membrane applications, particularly for RO and NF, pretreatment of the feed is essential. If pretreatment is inadequate, success will be transient. For most applications, pretreatment is location specific. Well water is easier to treat than surface water and that is particularly true for sea wells. A reducing (anaerobic) environment is preferred. If heavy metals are present in the feed even in small amounts, they may catalyze membrane degradation. If surface sources are treated, chlorination followed by thorough dechlorination is required for high-performance membranes [Riley in Baker et al., op. cit., p. 5–29]. It is normal to adjust pH and add antiscalants to prevent deposition of carbonates and sulfates on the membrane. Iron can be a major problem, and equipment selection to avoid iron contamination is required. Freshly precipitated iron oxide fouls membranes and requires an expensive cleaning procedure to remove. Humic acid is another foulant, and if it is present, conventional flocculation and filtration are normally used to remove it. The same treatment is appropriate for other colloidal materials. Ultrafiltration or microfiltration are excellent pretreatments, but in general they are uneconomic.

Process Configuration

Osmotic Pinch Effect   Feed is pumped into the membrane train, and as it flows through the membrane array, sensible pressure is lost due to friction effects. Simultaneously, as water permeates, leaving salts behind, osmotic pressure increases. There is no known practical alternative to having the lowest pressure and the highest salt concentration occur simultaneously at the exit of the train, the point where $\Delta P - \Delta \Pi$ is minimized. This point is known as the "osmotic pinch," and it is the point backward from which hydraulic design takes place. A corollary factor is that the permeate produced at the pinch is of the lowest quality anywhere in the array. Commonly, this permeate is below the required quality, so the usual practice is to design around average-permeate quality, not incremental quality. A 1 MPa overpressure at the pinch is preferred, but the minimum brine pressure tolerable is 1.1 times Π.

Brine Staging   Velocity past the membrane is important. If too low, polarization is excessive, local Π rises, and rejection declines. Fouling occurs faster. If too high, pressure losses are higher than they need be, and the osmotic pinch is premature. Since the volume of feed declines continuously, the hydraulic design needs periodic rearrangement. This is commonly done as shown in Fig. 22-64, sometimes known as a Christmas tree. This design is commonly used where the fluid is pumped once, as in RO, NF, and gas-separation systems, but not where recirculation is practiced, as in ultrafiltration.

Economics   The largest application for RO and NF is water treatment. Brackish water desalination for drinking water is the largest, and seawater desalination is perhaps the best known. Electrodialysis competes with pressure-driven membranes for brackish water applications. Seawater desalination is dominated by evaporation, with membranes an active competitor for midsize plants. There are numerous smaller process applications. The economic examples given are to illustrate the considerable difference between a small process plant and a large seawater RO plant. For the process plant, the cost given is for a skid-mounted self-contained unit for which considerable on-site additions would be required (tanks, piping, utilities, etc.). The seawater economics assume a leveled site but include everything necessary to deliver the product to the site boundary. Brackish water plants are considerably cheaper than seawater plants.

Process Applications   The diversity of application for process RO permits only a few generalities. Sugar concentration, wastewater recovery, and beverage uses are a few of the currently popular applications. Taking a fairly standard 100 gpm (6.3 ℓ/s) (based on water permeated) system, a package unit (uninstalled) would cost approximately $90,000 (1995). Standard plant design uses 8-inch multileaf spiral modules. About 25 to 30 percent of the capital cost is for the membranes and housings. This is in line with costs for other membrane processes. Operating costs are predominantly for membrane replacement and power, each in the order of $80/1000 m³ permeated. Approximately $25/1000 m³ will be consumed for pretreatment, maintenance, and cleaning. Concentrate disposal costs are highly variable, and capital charges must be added. Spirals are the standard design for process plants, but tubes, plate and frame, and hollow fibers with boreside feed have market niches.

Seawater Desalination   Seawater plants use spirals and hollow-fiber modules. Competition between these types is keen, with fibers having the advantage in high-salinity waters: polyamide fibers have higher pressure limits and cellulose triacetate fibers possess very high rejection. Spirals have an edge on cost, robustness, and fouling resistance. Seawater plant economics reflect the costs of high operating pressure, the need to handle and pretreat large quantities of seawater (2–2.5× rated capacity), and so on.

Capital Costs   A typical medium-scale RO seawater plant might produce 0.25 m³/s (6 MGD). For a plant with an open sea intake, seawater salinity of 38 g/l, and conversion of 45 percent, the overall cost would be $26.5 million (1996). A capital breakdown is given in Table 22-18. Capital charges are site specific, and are sensitive to the salinity of the feed. A plant of this size would likely contain six trains. For seawater RO, the best estimate for the slopes of the family of lines in Fig. 22-55 is −0.6 for the equipment and 0.95 for the membranes. Capital charges, shown in Table 22-19, usually dominate the overall economics; the numbers presented are only an example. Seawater economics are based on Shields and Moch, *Am. Desalination Assn. Conf. Monterey CA* (1996).

Operating Costs   Annual operating costs for the example in Table 22-18 are shown in Table 22-19. For this 0.25 m³/s plant, a service factor of 90 percent is assumed.

Energy   For the plant described in Table 22-18, the energy consumption is given in Table 22-20. For any simple membrane plant, the power consumption is simply the feed pressure divided by the yield (P/Y) (Pa). Energy consumption is expressed as kWh/m³, measured as net power fed to the plant divided by net permeate leaving the plant. Energy recovery can be very important in seawater plants, as shown here. Significant losses occur in motors, couplings, pumps, pipes, and manifolds.

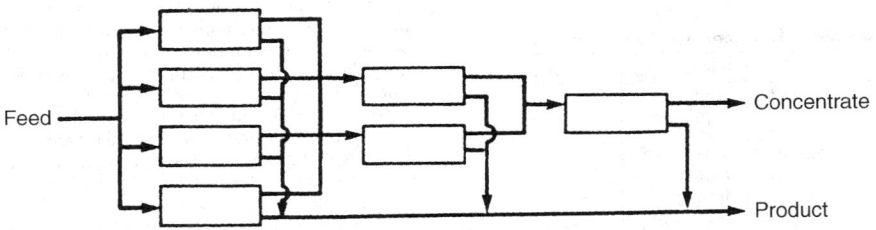

FIG. 22-64   Cascade arrangement for membrane processing to maintain cross-flow velocity as permeate is removed.

TABLE 22-18   RO Seawater Plant, Capital Costs

Item	$000	% of category	% of total capital
Membranes and housings, installed	3,600	21	14
Process equipment	13,700	79	52
Total installed equipment	17,300	100	65
Site development	500	21	2
Intake and outfall structures	1,850	79	7
Subtotal for site costs	2,350	100	9
Construction interest	983	14	4
Contingency	1,572	23	6
A&E fees	3,341	49	13
Working capital	983	14	4
Total indirect capital	6,879	100	26
Total capital employed	26,529		100

TABLE 22-19   RO Seawater Plant, Operating Costs

Item	$000	% of operating cost	% operating + capital	$/m³ water
Electrical power @ $0.08/kWh	3,163	70	44	0.40
Consumables and chemicals	187	4	3	0.02
Maintenance and parts	482	11	7	0.06
Supervision and labor	265	6	4	0.03
Membrane replacement	390	9	5	0.05
Total	4,487	100	63	0.56
Amortization, 20 years, 8 percent	2,664		37	0.34
Total operating plus capital charges	7,151		100	0.90

ULTRAFILTRATION

Process Description   *Ultrafiltration* (UF) is the membrane process which will retain soluble macromolecules and everything larger while passing solvent, ions, and other small soluble species. It is almost always operated with some means of forced convection near the membrane. Conventional dead-end filtration is rarely an option for UF, since most applications behave as if they have a "cake compressibility" factor of 1 in the filtration equation. Cross-flow filtration is practically universal for UF. An illustrative example of UF is its use for whey processing. Whey production exceeds 4×10^7 tons/year worldwide. It is a byproduct of cheese manufacture. Whey is composed of roughly 0.6 percent true proteins, 0.2 percent nonprotein nitrogen, 5 percent lactose, 1 percent salts, some lactic acid, and the balance water at a pH between 3.5 and 6. It contains trace amounts of casein fines and butterfat globules, and a large population of bacteria. UF retains the two principle proteins, alpha-lactalbumin (molar mass 17,000 daltons) and beta lactogloublin (molar mass 36,000 daltons), along with the large casein and butterfat particles and the bacteria. UF passes water, lactose, salts, and nonprotein nitrogen through the membrane into the permeate.

UF is widely used to concentrate oil-in-water emulsions, the byproduct of many metal-working applications, because the membrane retains stable emulsified oil while the water and the very low concentration of dissolved oil and free surfactant pass through it. UF is likewise useful for the concentration of dilute latex. Special membranes

TABLE 22-20   RO Seawater Plant, Energy Consumption

Item	kWh/m³	% of total
Low-pressure pumps	0.70	14
High-pressure pumps	5.79	116
Energy recovery turbine	(1.94)	−39
Degasification	0.02	0
Product pump	0.16	3
Plant services	0.28	6
Total energy	5	100

are used to remove virus from solution to help achieve the 12-log reduction required for vaccine manufacture. UF is useful for protein recovery in many fermentation operations, from the commodity-scale production of enzymes to small-scale specialty pharmaceutical manufacture.

Ultrafiltration may be distinguished from other membrane operations by example: When reverse osmosis is used to process whey, it passes only the water and some of the lactic acid (due to the solubility of lactic acid in RO membranes). Nanofiltration used on whey will pass most of the sodium salts while retaining the calcium salts and most of the lactose. Microfiltration will pass everything except the particulates and the bacteria.

UF Membranes   Design of UF membranes prizes high retention, hydrolytic stability, and good process flux. Since fouling is the principal impediment to flux, and membranes which are hydrophilic generally foul less rapidly, there is competition between the truly stable hydrophobic membranes and the less-fouling-prone hydrophilic ones.

Cellulosic Membranes   The first commercial UF membranes were made from *cellulose acetate* (CA), with an acetyl content of about 37 percent. They are prized for their low level of interaction with proteins and are still used in other applications where long life is not critical.

Polymeric Membranes   Economically important applications required membranes that could operate at higher pH than could CA, for which the optimum is around pH = 5. Many polymeric membranes are now available, most of which have excellent hydrolytic stability. Particularly prominent are polysulfone, polyvinylidene fluoride, polyethersulfone, polyvinyl alcohol-polyethylene copolymers, and acrylic copolymers.

Ceramic Membranes   Alumina-based microfiltration membranes and porous carbon substrates are tightened for use as UF membranes usually by depositing a layer of zirconium oxide on the surface.

UF Membranes as a Substrate for RO   An important use of UF membranes is as a substrate for composite reverse-osmosis membranes. After the UF membrane (usually polysulfone) is prepared, it is coated with an aqueous solution of an amine, then dipped in an organic solution of an acid chloride to produce an interfacially polymerized membrane coating.

Membrane Characterization   The two important characteristics of a UF membrane are its permeability and its retention characteristics. Ultrafiltration membranes contain pores too small to be tested by bubble point. Direct microscopic observation of the surface is difficult and unreliable. The pores, especially the smaller ones, usually close when samples are dried for the electron microscope. **Critical-point drying** of a membrane (replacing the water with a fluid which can be removed at its critical point) is utilized; even though this procedure has complications of its own it has been used to produce a few good pictures.

Water Flux   The permeability of a UF membrane is determined by pore size, pore density, and the thickness of the membrane active layer. Water flux is measured in the absence of solute, generally on a newly made or freshly cleaned sample. The test is simple, and involves passing water through the membrane generally in dead-end flow under carefully controlled conditions. In a water flux test, the membrane behaves as a porous medium with the flow described by Darcy's law. Adjustments for viscosity and pressure are made to correct the results to standard conditions, typically the viscosity of water at 25°C and the pressure to 50 psi (343 kPa). The water flux will be many multiples of the process flux when the membrane is being used for a separation. Virgin membrane has a standard water flux of over 1 mm/sec. By the time the membrane is incorporated into a device and used in an application, that flux drops to perhaps 100 μm/s. Process fluxes are much lower.

Molecular Weight Cutoff   The best-known method for characterizing UF membranes is molecular weight cutoff. Unfortunately, it is widely misunderstood and has been the source of much error. The concept of *molecular weight cutoff* (MWCO) is powerful and deceptively simple. Ultrafilters retain soluble molecules, so their retention is measured by seeing which molecules will pass through them. The definition, generally *but not universally* followed is: molecular weight cutoff is the molar mass of the globular protein which is 90 percent retained by the membrane.

There are many complications with interpreting MWCO data. First, UF membranes have a distribution of pore sizes. In spite of decades of effort to narrow the distribution, most commercial membranes are not notably "sharp." What little is known about pore-size distribution in commercial UF membranes fits the Poisson distribution or log-normal distribution. Some pore-size distributions may be polydisperse.

Second, most membrane materials adsorb proteins. Worse, the adsorption is membrane-material specific and is dependent on concentration, pH, ionic strength, temperature, and so on. Adsorption has two consequences: it changes the membrane pore size because solutes are adsorbed near and in membrane pores; and it removes protein from the permeate by adsorption in addition to that removed by **sieving.** Porter (op. cit., p. 160) gives an illustrative table for adsorption of Cytochrome C on materials used for UF membranes, with values ranging from 1 to 25 percent. Because of the adsorption effects, membranes are characterized only when clean. Fouling has a dramatic effect on membrane retention, as is explained in its own section below.

Third, picking the point on the curve of retention versus molar mass where "90 percent" falls is inexact. The retention curve usually bends in a way that makes picking the "90 percent point" somewhat arbitrary.

Fourth, selection of the marker molecule can effect the MWCO measured. Markers for UF membranes are usually protein, but always polymeric. Polymers of the same molar mass can have very different molecular size, and MWCO is more a measure of size than anything else. To further complicate the picture, molecular shape can change in the vicinity of a membrane. One well-known example [Porter (ed.), op. cit., pp. 156–160] is Dextran 250, a branched polysaccharide with molar mass 250 kilodaltons which passes through a 50 kD MWCO membrane. Linear molecules, such as polyacrylic acid, with a given molecular mass passes easily through a membrane that retains a globular protein of the same molecular mass. The definition requires globular proteins, for which many of these effects are mitigated.

When testing a membrane using protein, in keeping with the definition of MWCO, it is necessary to keep the concentration in the feed and the flux very low to minimize polarization effects. Any polarization of the marker at the membrane will alter the measured value, and significant accumulation will result in **autofiltration.** The result is a measurement of the boundary layer rather than the membrane itself. What is needed are conditions in which J/k [see Eq. (22-91)] is close to 1. Reducing the marker concentration to minimize these problems raises the probability that adsorption will become important in reducing the concentration of marker in the permeate. A lack of reproducibility between laboratories is one manifestation of the intractability of the MWCO problem.

Membrane manufacturers require a standard test to maintain batch-to-batch quality. Few use proteins. Materials selected are ones for which the complications are minimized, the probe is simple, fast, and cheap to detect, does not readily biodegrade, and gives results, whatever they are, which are reproducible. There is no standardization of these tests within the industry.

Misunderstandings arise when membrane users assume that MWCO means what it seems to say. The definition implies that a 50 kD membrane will separate a 25 kD material from a 75 kD material. The rule of thumb is that the molecular mass must differ by a factor of ten for a good separation. Special techniques are used to permit the separation of proteins with much smaller mass ratio.

In an ideal world, membranes would contain a very high density of fully uniform, cylindrical pores. It is perhaps natural to envision a membrane as a uniform plane featuring cylindrical holes, challenged by rigid, spherical particles. None of these preconceptions is true. Membrane surfaces may be relatively rough, openings are neither uniform nor cylindrical, and are randomly spaced. A tiny minority of retained species are spherical and rigid; the vast majority is neither. It is perhaps instructive that in spite of the very creative effort invested in "sharp" membranes, their share of the overall membrane market is very small. Practically speaking, ordinary membranes have proven to be adequate for most separations. Very few membranes are used under conditions remotely approaching a test for MWCO. They are

usually highly polarized with retention determined more by autofiltration than by inherent properties.

Autofiltration   The retention of any material at the surface of the membrane gives rise to the possibility of a secondary or a dynamic membrane being formed. This is a significant problem for fractionation by ultrafiltration because microsolutes are partially retained by almost all retained macrosolutes. The degree of retention is quite case-specific. As a rule of thumb, higher pressure and more polarization results in more autofiltration. Autofiltration is particularly problematic in attempts to fractionate macromolecules.

Process Limitations

Concentration Polarization   Throughput data from countless ultrafiltration experiments are shown in two characteristic curves. Figure 22-65 shows flux as a function pressure. Figure 22-66 shows flux as a function of the log of retentate concentration. Figure 22-65 is for a fixed solute concentration. At low transmembrane pressure, Region I, flux is governed by the rate at which solvent passes a porous material—Darcy's law. The magnitude of flux will approximate the water flux if viscosity is the same. Increase the pressure, and at first the flux responds. Soon, however, there is no response to pressure at all (Region III), and in some extreme cases there are reports of a negative response to pressure. This counterintuitive response to an increase in driving force has received considerable attention, and what is going on can be described by looking at the concentration of retained solute stranded at the membrane. It is a given that cross-filtration membrane processes operate at steady state. It isn't unusual for a process to operate for weeks or months at a flux essentially the same as that measured a few minutes after startup. It is therefore apparent that there is no long-term buildup of retained material at the membrane. Therefore, the redistribution of retained material must equal its rate of transport toward the membrane. Flux in the pressure-independent portion of Fig. 22-65 is quantitatively described by making a material balance on the retained solute, and solving the mass-transfer equations for its redispersal. These same equations describe the phenomenon shown in Fig. 22-66, flux at a constant stirring rate but with concentration as a variable.

Because this mass-transfer step is so vital, conventional dead-end operation of ultrafilters is very rare. There are many ways to depolarize a membrane. Cross-flow is by far the most common. Turbulent flow is more common than laminar flow.

The mass-transfer coefficient, k, is contained in the Sherwood number:

$$\text{Sh} = \frac{kd_h}{D} \qquad (22\text{-}78)$$

For turbulent flow, a correlation attributed to Chilton-Colburn is:

$$\text{Sh} = 0.023\text{Re}^{0.8}\text{Sc}^{0.33} \qquad (22\text{-}79)$$

$$\text{Sc} = \frac{\nu}{D} \qquad (22\text{-}80)$$

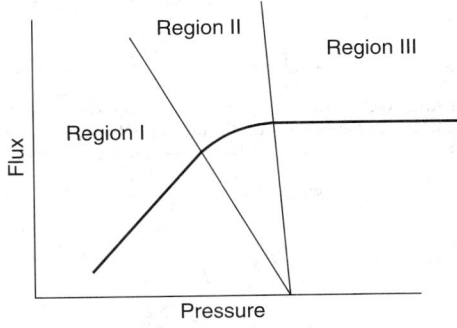

FIG. 22-65   Characteristic curve for flux as a function of pressure for cross-flow membrane processes limited by mass transfer at the membrane.

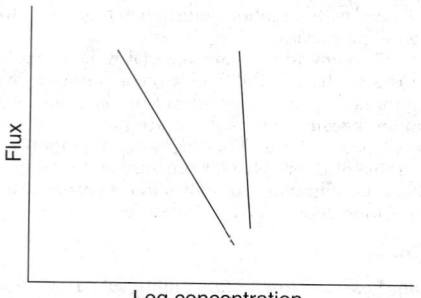

FIG. 22-66 Characteristic curve for flux as a function of feed composition for cross-flow membrane. Right curve is for a higher Sherwood number than the left curve.

$$Re = \frac{Vd}{\nu} \tag{22-81}$$

D is diffusivity, m²/s, d_h is hydraulic diameter, (4)(cross sectional area)/(wetted perimeter), m, k is the mass-transfer coefficient, m/s, V is the velocity, m/s, and ν is the kinematic viscosity, m²/s.

For cylindrical flow channels, $d_h = d$, and Re expressed in terms of volumetric flow rate, Q, is:

$$Re = 4Q/\pi\nu d \tag{22-82}$$

where Q is in m³/s. Defining J as acting along the x axis, passing from the feed through the membrane into the permeate, and recalling the steady-state stipulation, from Eq. (22-59), the rate of redispersion of retained material from the membrane is:

$$J_{solute} \approx -N_{solute} = D\frac{dC}{dx} \tag{22-83}$$

where x is the distance normal to the membrane.

A material balance on solute, ignoring the effects parallel to the membrane, is:

$$J \cdot C_x - J \cdot C_{perm} - D\frac{dC}{dx} = 0 \tag{22-84}$$

Calling the thickness of the concentration boundary layer, δ, Eq. (22-84) can be integrated to give:

$$J_{solute} \sim \left(\frac{D}{\delta}\right) \ln\left(\frac{C_{wall} - C_{perm}}{C_{bulk} - C_{perm}}\right) = k \ln\left(\frac{C_{wall} - C_{perm}}{C_{bulk} - C_{perm}}\right) \tag{22-85}$$

Since D/δ is the mass-transfer coefficient. For the portion of the operating curve in which flux is invariant, the wall concentration is apparently invariant. The mechanism governing why and how that occurs is the subject of a continuing debate in the literature.

For the usual case when $R = 1$ (total retention of the solute), $C_{perm} = 0$ and combining these equations gives a general expression for flux in a turbulent-flow membrane system. For any given solute concentration:

$$J \sim J_{solute} \sim k \sim \frac{Q^{0.8}D^{0.7}}{d^{1.8}\nu^{0.5}} \tag{22-86}$$

Flux is a function of solute concentration as is shown in Fig. 22-66. The exponent on Q is not always found to be 0.8 experimentally.

For laminar flow in a circular tube, the Leveque relationship is:

$$Sh = 1.62\left(ReSc\frac{d}{l}\right)^{0.33} \tag{22-87}$$

$$k = 1.62\left(\frac{VD^2}{ld}\right)^{0.33} \tag{22-88}$$

$$J \sim J_{solute} \sim k \sim \left(\frac{QD^2}{ld^3}\right)^{0.33} \tag{22-89}$$

where l is the distance from the channel entrance.

Equations (22-86) and (22-89) are the turbulent- and laminar-flow flux equations for the pressure-independent portion of the ultrafiltration operating curve. They assume complete retention of solute. Appropriate values of diffusivity and kinematic viscosity are rarely known, so an a priori solution of the equations isn't usually possible. Interpolation, extrapolation, even prediction of an operating curve may be done from limited data. For turbulent flow over an unfouled membrane of a solution containing no particulates, the exponent on Q is usually 0.8. Fouling reduces the exponent and particulates can increase the exponent to a value as high as 2. These equations also apply to some cases of reverse osmosis and microfiltration. In the former, the constancy of C_{wall} may not be assumed, and in the latter, D is usually enhanced very significantly by the action of materials not in true solution.

Usually, diffusivity and kinematic viscosity are given properties of the feed. Geometry in an experiment is fixed, thus d and averaged l are constant. Even if values vary somewhat, their presence in the equations as factors with fractional exponents dampens their numerical change. For a continuous steady-state experiment, and even for a batch experiment over a short time, a very useful equation comes from taking the logarithm of either Eq. (22-86) or (22-89) then the partial derivative:

$$\left(\frac{\partial \log J}{\partial \log Q}\right)_{\nu,d,D} = m \tag{22-90}$$

Equation (22-90) is the basis for the ubiquitous plots of log J versus log Q. Such plots are powerful tools for analyzing experimental data. The known range of observed values of m in well-developed turbulent flow is $0.8 < m < 2.0$. For laminar flow, $m = 0.33$ for true solutions, with values up to around 0.8 for systems with particulates. It is important to determine the experimental value, both for design optimization and for prediction of long-term effects. It is important that all values of flux be taken in the pressure-independent region of operation (see Fig. 22-65). High values of m are usually found in systems containing large, dense particles. Polyvinyl chloride latex with particles over 0.5 μm diameter is a classic high-slope example.

The slope of the flux-flow line is an indicator of fouling. In turbulent flow, $m < 0.8$ indicates fouling. A decline in the value of m with time is the most sensitive indicator of fouling. While the slope is difficult to obtain unconfounded by changes in pressure, a well-designed J versus Q experiment yields results with good predictive value for fouling. As a special precaution when using spiral-wound modules Da Costa, Fane, and Wiley [*J. Membrane Sci.*, **87**, 79–98 (1994)] found that while the pressure-drop data in a spiral module behave as if the flow is turbulent, the mass-transfer data are consistent with laminar flow.

Prediction of C_{wall} Equation (22-85) shows a semilog dependency of wall concentration on flux. Experimentally, the dependence of flux on concentration usually deviates significantly from linearity well before the zero-flux intercept extrapolated from data in Fig. 22-66. Experimental data at very low flux are difficult to gather, but usually the flux is much higher than the values extrapolated assuming linearity. The mass-transfer equations predict a mass-transfer coefficient, k, without reference to flux, as they were formulated for nonmembrane systems. This k is used by assumption to predict the wall concentration up to the point at which flux becomes independent of pressure.

$$C_{wall} - C_{perm} = (C_{bulk} - C_{perm})\exp(J/k) \tag{22-91}$$

When flux becomes independent of pressure, C_{wall} becomes **constant.** The assumptions underlying the equations, such as constancy of D and especially of ν are unlikely to hold, and other means are needed to determine the true value of C_{wall}. Field [in Howell, Sanchez, and Field, op. cit., pp. 87–95] gives a detailed treatment of this regime.

Osmotic Pressure, Gel Effect, Etc. The reason for the apparent constancy of C_{wall} under usual operating conditions is still tentative. For many years, it was thought that the macro solute forms a new phase near the membrane—that of a gel or gel-like layer. The model provided good correlations of experimental data and has been widely used. It does not fit known experimental facts. An explanation that fits the known data well is based on osmotic pressure. The van't Hoff equation [Eq. (22-75)] is hopelessly inadequate to predict the osmotic pressure of a macromolecular solution. Using the empirical expression

$$\Pi(C) = \sum_{n=1}^{n=3} a_i C^n \qquad (22\text{-}92)$$

where a_i are experimental constants it is possible to correlate experimental osmotic-pressure data for macro solutes and to predict the concentration at the membrane. Data now confirmed show osmotic pressures high enough to counter the transmembrane driving force [Jonsson, *Desalination*, **51**, 61–77 (1984)]. Other theories of the boundary layer exist, but they have not attracted adherents. In terms of predictive power, both the gel theory and the osmotic-pressure theory provide a framework for correlating data.

Fouling Everything fouls (Fig. 22-67), and in process UF fouling is a major concern. Because of its importance, industrial suppliers of membrane equipment place a major emphasis on understanding, controlling, and preventing fouling. Equipment and process conditions can be specified with confidence for most applications because of the extensive knowledge base accumulated in response to the problem.

Fouling is the term used to describe the loss of throughput of a membrane device as it becomes chemically or physically changed by the process fluid (often by a minor component or a contaminant). A manifestation of fouling in cross-flow UF is that the membrane becomes unresponsive to the hydrodynamic mass transfer which is rate-controlling for most UF. Fouling is different from concentration polarization. Both reduce output, and their resistances are additive. Raising the flow rate in a cross-flow UF will increase flux, as in Eq. (22-90). If the system is badly fouled, $m \sim 0$, and increasing or decreasing flow at constant pressure has little effect on flux. However, raising the pressure may raise flux. For an unfouled system in laminar flow $0.33 < m < 0.8$; for turbulent flow, $0.8 < m < 2$.

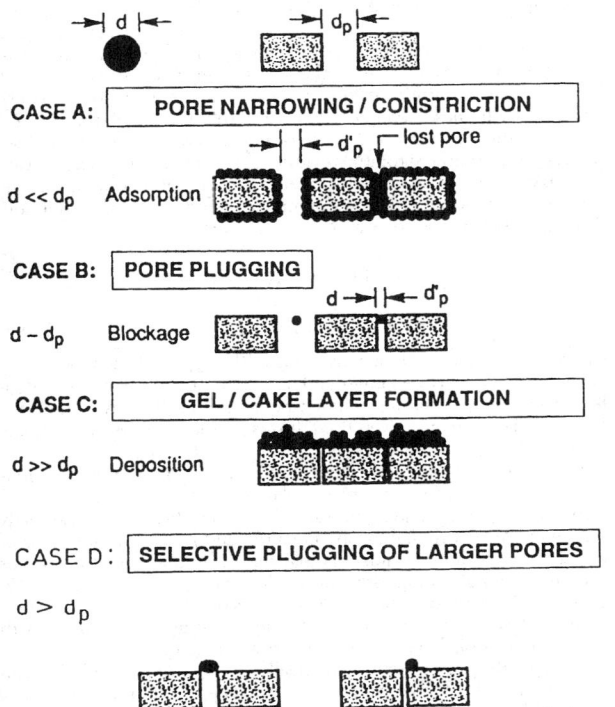

CASE A: **PORE NARROWING / CONSTRICTION**

$d \ll d_p$ Adsorption

CASE B: **PORE PLUGGING**

$d - d_p$ Blockage

CASE C: **GEL / CAKE LAYER FORMATION**

$d \gg d_p$ Deposition

CASE D : **SELECTIVE PLUGGING OF LARGER PORES**

$d > d_p$

FIG. 22-67 Fouling schematics. Case A—Particles plug narrow pores and narrow larger ones. Case B—Particles plug narrow pores. Case C—Particles form a layer on the membrane. Case D—Particles or debris plug the largest pores. [*Courtesy Elsevier (modified)*.]

Fouling affects flux dramatically. The pure water flux through a new UF membrane is commonly tenfold greater than the water flux after the membrane has been exposed to protein. Processing fluxes commonly decline roughly as $J = J_o t^{-n}$ where t is an arbitrary dimensionless time and n is small. J_o is the flux when $t = 1$. Thorough cleaning is required to restore $J = J_o$ as incompletely cleaned membranes foul faster than completely cleaned membranes. Fouling prevention is an important part of process design. Proper selection of membrane, operating conditions, feed pretreatment, startup techniques, and cleaning type and frequency can make a major difference in fouling, thus throughput, thus cost. There can be a startling difference between a well-designed process and a haphazard combination of membrane and process stream. Fouling also strongly influences retention.

Since flow through a porous membrane is always laminar, the volumetric flow through an individual pore is proportional to the fourth power of diameter, as known from the Pouiselle equation. Pore plugging, as in Case D, will dramatically lower flux and significantly increase retention, while Case B will have far less of an effect, lowering flux marginally and probably lowering retention. Even a slight reduction in pore diameter from an adsorption phenomenon (Case A) will have dramatic results. Some commercial membranes are designed with the inevitability of fouling in mind, and their behavior in the first minutes is inferior to their steady-state performance.

Cleaning membranes to restore their efficiency is normal in UF. Food and dairy systems require daily cleaning in any event for hygiene; more frequent cleaning is economically intolerable. A few industrial systems operate for six months between cleanings. Cleaning shortens membrane life, and it is often the major determinant of membrane-replacement frequency.

Among techniques to prevent fouling, pretreatment is widely practiced. Free-oil phases must be removed or stabilized, and whey processing benefits from holding the whey at a mildly elevated temperature for some minutes. Membranes operated at high transmembrane pressure foul much faster. The optimum operating point on Fig. 22-65 is Region II; however it is very difficult to design economical equipment to operate there. Pretreatment is often stream- or site-specific, and it has received little attention in the literature.

Some fouling occurs simply by contact, almost certainly due to adsorption. Some occurs slowly as material is processed, some of that due to trace components in the feed and some due to slow accumulation and rearrangement processes.

Process Configurations Ultrafiltration membranes are produced in four basic forms: tubular, hollow fiber/capillary, flat sheet, and ceramic monolith. Commercial diameter of tubes is 5 to 25 mm. Fibers range from a few mm down to 250 μm. Flat-sheet membrane is most common, made into spiral-wound modules, cassettes, and plate-and-frame devices. Spiral-wound flat sheet is by far the dominant commercial configuration, followed by capillaries in the millimeter range, and then 13 and 25 mm tubes. Spirals have the economic edge, and where they will work they are used. Tubes have the advantage of being tolerant of high solids loadings and of being extremely forgiving of process upsets. The finest fibers are only used for very clean feeds such as protein solutions and water.

The simplest ultrafiltration is the stirred cell, a batch operation. The most complex is a continuous stages-in-series operation incorporating diafiltration. Industrial practice incorporates the full gamut of complexity.

Process Objective UF is used for three principle objectives. First, to **fractionate,** to pass selectively one component through the membrane with the solvent. Second, to **concentrate,** to pass the solvent. These two, while different, are related and it is common to purify and concentrate a component simultaneously. The third objective, quite different, is to **produce a solvent stream** as a product. An example is the operation of an ultrafilter for producing low-cost permeate. An important application of UF is in the automotive industry where UF is used to remove water and microsolutes from huge electrophoretic paint tanks for use in rinsing excess paint (dragout) from newly primed auto bodies. Permeate and recovered paint are returned to the paint tank. It is undesirable to concentrate the paint, so the increase in paint solids within the UF loop is maintained below 10 percent.

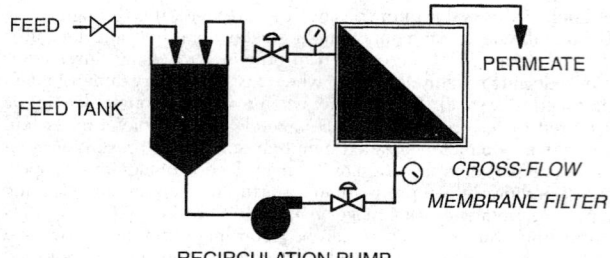

FIG. 22-68 Flow schematic for batch (feed valve closed) or semibatch (feed valve open) operation. (*Courtesy Koch Membrane Systems.*)

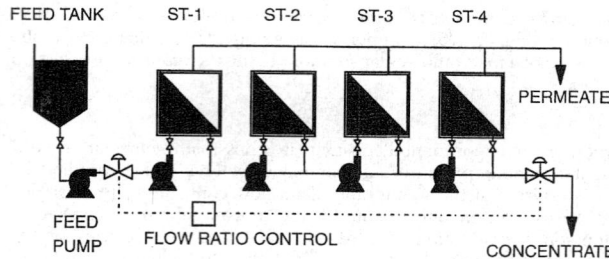

FIG. 22-69 Flow schematic for stages-in-series. In operation, the block valves are open. All pump inlets are connected, but in such a way as to prevent feed bypassing a stage. (*Courtesy Koch Membrane Systems.*)

Batch and Semibatch Straight batch is the least common industrial process, but semibatch is quite common (see Fig. 22-68). In **batch,** the entire quantity to be processed is put in a tank, and a feed from it is pumped across a UF membrane, with the concentrate returned to the tank. Conversion per pass is very low, but with recirculation any desired level of concentration may be obtained. It is the industrial version of the stirred cell. **Semibatch** uses a smaller tank to which fresh feed is introduced continuously. As the process proceeds, the contents of the batch tank become more concentrated with time, and at some point the feed is shut off and the tank contents are concentrated to the desired end point. Batch operation has the advantage of requiring the minimum membrane area, and the disadvantage of requiring the largest tank. Pumping costs are high because it is rarely practical to pressurize the feed tank or locate it at a height, so the circulating loop pressure is lost continuously.

For batch concentration, the yield equation is:

$$Y = \left(\frac{V_o}{V_f}\right)^{R-1} \qquad (22\text{-}93)$$

Y is fractional yield of retained species, and V_o and V_f are the volume of process fluid at the beginning and end of a batch run, respectively. This equation is valid only when R, retention, is constant.

Diafiltration If a batch process is run so that the permeate is replaced by an equal volume of fresh solvent, unretained solutes are flushed through the system more efficiently. A major use of UF is fractionation, where a solvent, a retained solute and an unretained solute are present. An example is whey, containing water, protein, and lactose. If the retention of protein is 1 and the retention of lactose is 0, the concentration of protein in the retentate rises during UF. The ratio of protein to lactose rises, but the feed concentration of lactose is unchanged in retentate and permeate. Diafiltration dilutes the feed, and permits the concentration of lactose to be reduced. Diafiltration is used to produce high-purity products, and is used to fractionate high-value products. R is always <1 and >0 for every component.

For diafiltration, the yield equation is:

$$Y = \exp\left[\left(\frac{V_D}{V_o}\right)(R - 1)\right] \qquad (22\text{-}94)$$

where V_D is the volume of diafiltration solvent (water) added, equal to the volume of permeate removed.

The combination of diafiltration and batch concentration can be used to fractionate two macrosolutes whose retentions differ by as little as 0.2. It is possible *in principle* to achieve separations that are competitive with chromatography. When tanks and other equipment are considered, as well as the floor space they occupy, the economics of membrane separation of proteins may be attractive [R. van Reis, U.S. Patent 5,256,294 (1993)].

Stages in Series Large-scale UF normally operates in a mode called **stages-in-series** (see Fig. 22-69). Feed is pumped from a feed tank, which in principle can be quite small, to a first recirculating stage. A pipe connects the low-pressure side of this stage to the low-pressure side of another stage. When more than five stages are linked in this manner, the membrane area required drops to within 20 percent of the minimum batch membrane area requirement. The system

is self-adjusting and stable. The preferred means of control is to use a volume-ratio controller to regulate the product-exit valve on the last stage. The economic advantages from modular fabrication, reduced tankage, and continuous operation make this the scheme of choice for most large fractionation and concentration operations.

Economic Yield Both in a high-value protein separation and in a low-value commodity concentration, economic yield is vital. Economic yield is defined as the fraction of useful product entering the process that leaves it in salable form. The yield equations used in the industry focus on retention, so they deal only with direct losses through the membrane. These losses result both in direct (product not sold) and indirect costs from a waste stream whose disposal or subsequent use may be more expensive when it is contaminated by macrosolute. There are additional indirect losses, mainly product left in the equipment, particularly that left adhering to the membrane. Costs of cleaning and disposal of this indirect loss, while hard to measure, are usually higher than the cost of product lost through the membrane.

Decoupled Driving Force and Depolarization Needs for improved fractionation motivate designers to reduce autofiltration. Using fluid velocity for depolarization means that hydrodynamic pressure drop will be additive to the transmembrane pressure driving force. Schemes to limit this effect confront a harsh economic reality. Two novel schemes decouple the driving from the depolarizing force.

When fluid flows around a curve in a duct, or when fluid is confined between differentially rotating cylinders, secondary flows called **Taylor vortices** are generated. If a membrane is mounted on a rotating cylinder, these secondary flows minimize polarization independent of driving pressure. The relevant equations are:

$$Ta = \frac{\omega R g}{v}\sqrt{\frac{g}{R}} \qquad (22\text{-}95)$$

$$Sh = c\, Ta^{0.5} Sc^{0.33} \qquad (22\text{-}96)$$

where c is an experimental constant, R is the radius of the inner cylinder, g is the gap between inner and outer cylinders, and ω is the angular velocity of the rotating cylinder, radian/s. Ta is the Taylor number. For a fixed device on a given fluid, flux is predicted to be proportional to $\omega^{1/2}$.

By mounting a plate-and-frame membrane assembly atop a torsion-bar spring, membranes may be depolarized by vibrating the stack at a resonant frequency. The membrane moves and the fluid is essentially stationary. At first thought to be ideal for, and limited to, solutions of very high viscosity or solids loading, the devices are now viewed as another economic competitor for a broader range of applications. No adequate theory is available to explain mass transfer in vibrating membrane systems. Some data correlate depolarizing mass transfer with the first power of shear rate over a narrow range of conditions.

Energy Requirements Practically all the energy needed to run an ultrafilter is depolarization energy. The thermodynamic work required for any UF separation is approximately 0.01 kWh/m³ of permeate. Energy requirements vary widely by application and economics. Design is a classic trade-off between membrane area and pump related equipment. The range of energy requirements from the easiest to the most difficult for modern designs runs in the magnitude of 0.5 to 5 kWh/m³ of permeate.

Design   UF equipment has considerable variety of design, but the trend is toward more compact, energy efficient, and lower-cost designs. Much of the robustness characteristic of older designs is now available in less costly versions.

Hydraulic Design   Looking over a wide spectrum of UF applications, the hydraulic energy delivered to the membrane falls within a characteristic range. At the high extreme, a large-diameter tubular plant operating at very high velocity in order to retard fouling on a stream where long operating cycles between cleanings are valued, power dissipation at the membrane may exceed 150 watts/m². A well-designed, relatively large, hydraulically efficient plant will deliver power from electrical source to membrane at 64 percent efficiency. If the flux in the installation is 50 ℓmh, the power consumption of the process is 150/(0.64)(50) = 4.7 kWh/m³. A large plant designed using less-efficient sanitary standard pumps and spiral-wound membranes could deliver 20 watts/m² to the membrane with an energy conversion of 50 percent and a flux of 30 ℓmh. That plant would consume 1.3 kWh/m³. Very efficient plants for benign feed streams may consume half as much energy.

Module Types Favored   Because of their low cost, spiral-wound membranes are the first choice for industrial ultrafiltration. Over the years, the number of applications for which spirals are a good choice has gone from practically none to over 50 percent, owing to very intensive process development and a related extensive modification of the spiral module as it was known in the RO field. When the spiral is not appropriate—examples include feeds where fibers, debris, certain types and loadings of suspended solids, most emulsified oils, etc. are present and too expensive to remove by pretreatment—other module designs are used. Capillary membranes are the usual next choice, and they are in fact preferred for some applications. Then comes open tubes, known for being expensive and practically indestructible. Tubes also have an edge in a few applications, such as apple juice, because they are able to recover more juice before they plug up with pomace. Cassettes are used because of direct scale-up from bench to plant in applications where equipment and operating cost are not paramount. Plate-and-frame modules are still found on occasion.

Economics   The general examples section found under "Background and Definitions" is directly applicable to the following.

Capital Costs   Package UF units are sold for many applications. Prices vary widely by application, with equipment designed for food and pharmaceutical applications priced higher than general industrial equipment. All package units would include membranes, one or more pumps, a cleaning system, piping, instrumentation, an electrical control panel, and perhaps a process tank, all designed for rapid field installation. A 1996 budget price for a typical industrial spiral or capillary unit containing 100 m² of active membrane, with a process output of the magnitude of 1.0–1.5 ℓ/s is $250–500/m². The replaceable membrane component of that cost is $30–40/m².

UF/MF applications with plant cost of $10⁶ are considered large. In the decade ending in 1995, fewer than ten large plants were sold worldwide in any year. During that decade, the process-membrane industry matured. More vendors for equipment and membranes broadened the applications and lowered the costs. The industry-cost picture is changing fast enough that only broad guidelines are given here. Fresh information from vendors and users is needed if accuracy is required.

UF and MF use energy to depolarize membranes so as to increase flux. As is shown in Fig. 22-55, membranes and mechanical equipment are traded off to achieve an overall economic minimum. Three things can drive a design toward the use of more membranes and less mechanical equipment: cheaper membranes, very high flux, and very low flux. The availability of lower-cost membranes is easiest to understand. In the five years ending in 1995, the cost of both membrane area and membrane housings was driven down by competition. Pumps, pipes, and other peripheral equipment also declined, but not as much. By the principles of Fig. 22-55, this pushed the optimum design point to the right. Membrane costs maintained their *slope* but came down in *position*. Neither high-flux applications (potable water

for example, average design flux for UF is 125 ℓmh, and for MF it is 185 ℓmh) or very low flux (polyvinylalcohol, average design flux 5.8 ℓmh) are very responsive to mechanical energy applied at the membrane. In the case of water, there is little to depolarize, and most systems operate in Region I of Fig. 22-13. Polyvinylalcohol is so viscous that it is, or becomes, laminar in a spiral module. The pressure drop through a spiral limits the velocity and forces the economics toward plants with high-area and moderate-pumping packages.

In the case of whey, paint, and other midflux process fluids, mechanical energy at the membrane surface produces a larger dividend. For these applications, pumping for depolarization is much more important economically, but the trend toward lower-cost membranes has nonetheless shifted systems toward more membrane area.

In 1996, a $1M plant would process: 6 ℓ/s of polyvinylalcohol (UF), 17 ℓ/s of whey (UF), 35 ℓ/s dextrose (MF), or 108 ℓ/s water (MF or UF).

During 1990–1995, capital costs for large UF/MF plants broke down into the ranges shown in Table 22-21.

TABLE 22-21   Capital-Cost Distribution for Components in Large UF/MF Plants

Cost distribution	% of total, range
Membranes and membrane housings	17–40
Pumps, motors, etc.	15–9
Pipes, valves, and framework	35–31
Cleaning system	18–10
Control panel	15–10

Operating Costs   Operating expenses again span a considerable range, but there are fairly consistent operating norms (Table 22-22). A few very high flux MF applications with fluxes of up to 500 ℓmh are just starting to become commercial. These applications use large pumps to maintain output. It is not yet know whether these cost estimates will apply to those plants.

TABLE 22-22   Operating-Cost Range for Large UF/MF Plants

Expense item	Range of variable commonly encountered
Energy consumption	0.5–5 kWh/m³ permeated
Cleaning chemicals and lost product	$10–100/m² membrane installed-year
Membrane replacement	1–5 years at $20–40 m²; 10–20 years at $200/m²
Operating, cleaning, and maintenance labor	2–3% of installed capital
Maintenance materials	0.6–0.06% of installed capital

MICROFILTRATION

Process Description   Microfiltration (MF) separates particles from true solutions, be they liquid or gas phase. Alone among the membrane processes, microfiltration may be accomplished without the use of a membrane. The usual materials retained by a microfiltration membrane range in size from several μm down to 0.2 μm. At the low end of this spectrum, very large soluble macromolecules are retained by a microfilter. Bacteria and other microorganisms are a particularly important class of particles retained by MF membranes. Among membrane processes, dead-end filtration is uniquely common to MF, but cross-flow configurations are often used.

Brief Examples   Microfiltration is the oldest and largest membrane field. It was important economically when other disciplines were struggling for acceptance, yet because of its incredible diversity and lack of large applications, it is the most difficult to categorize. Nonetheless, it has had greater membrane sales than all other membrane applications combined throughout most of its history. The early success of microfiltration was linked to an ability to separate microorganisms from water, both as a way to detect their presence, and as a means to remove them. Both of these applications remain important.

Laboratory   Microfiltration membranes have countless laboratory uses, such as recovering biomass, measuring particulates in water, clarifying and sterilizing protein solutions, and so on. There are countless examples for both general chemistry and biology, especially for analytical procedures. Most of these applications are run in dead-end flow, with the membrane replacing a more conventional medium such as filter paper.

Medical MF membranes provide a convenient, reliable means to sterilize fluids without heat. Membranes are used to filter injectable fluids during manufacture. Sometimes they are inserted into the tube leading to a patient's vein.

Process Membrane microfiltration competes with conventional filtration, particularly with diatomaceous earth filtration in general-process applications. A significant advantage for membrane MF is the absence of a diatomaceous earth residue for disposal. Membranes have captured most of the final filtration of wine (displacing asbestos), are gaining market share in the filtration of gelatin and corn syrup (displacing diatomaceous earth), are employed for some of the cold pasteurization of beer, and have begun to be used in the pasteurization of milk. Wine and beer filtration operate dead-end; gelatin, corn syrup, and milk are cross-flow operations. MF is used to filter all fluid reactants in the manufacture of microcircuits to ensure the absence of particulates, with point-of-use filters particularly common.

Gas Phase Microfiltration plays an important and unique role in filtering gases and vapors. One important example is maintaining sterility in tank vents, where incoming air passes through a microfilter tight enough to retain any microorganisms, spores, or viruses. A related application is the containment of biological activity in purge gases from fermentation. An unrelated application is the filtration of gases, even highly reactive ones, in microelectronics fabrication to prevent particulates from contaminating a chip.

Downstream Processing Microfiltration plays a significant role in downstream processing of fermentation products in the pharmaceutical and bioprocessing industry. Examples are clarification of fermentation broths, sterile filtration, cell recycle in continuous fermentation, harvesting mammalian cells, cell washing, mycelia recovery, lysate recovery, enzyme purification, vaccines, and so forth.

MF Membranes Microfiltration is a mature field that has proliferated and subdivided. The scope and variety of MF membranes far exceeds that in any other field. A good overview is given by Strathmann [in Porter (ed.), op. cit., pp. 1–78]. MF membranes may be classified into those with tortuous pores or those with capillary pores. Tortuous-pore membranes are far more common, and are spongelike structures. The pore openings in MF are much larger than those in any other membrane. Surface pores may be observed by electron microscope, but tortuous pores are much more difficult to observe directly. Membranes may be tested by bubble-point techniques. Many materials not yet useful for tighter membranes are made into excellent MF membranes. Retention is the primary attribute of an MF membrane, but important as well are permeability, chemical and temperature resistance, dirt capacity (for dead-end filters), FDA-USP approval, inherent strength, adsorption properties, wetting behavior, and service life.

Membrane-production techniques listed below are applicable primarily or only to MF membranes. In addition, the Loeb-Sourirjain process, used extensively for reverse osmosis and ultrafiltration membranes, is used for some MF membranes.

Membranes from Solids Membranes may be made from microparticles by sintering or agglomeration. The pores are formed from the interstices between the solid particles. The simplest of this class of membrane is formed by sintering metal, metal oxide, graphite, ceramic, or polymer. Silver, tungsten, stainless steel, glass, several ceramics, and other materials are made into commercial membranes. Sintered metal may be coated by TiO_2 or zirconium oxide to produce MF and UF membranes. Membranes may be made by the careful winding of microfibers or wires.

Ceramic Ceramic membranes are made generally by the sol-gel process, the successive deposition of ever smaller ceramic precursor spheres, followed by firing to form multitube monoliths. The diameter of the individual channels is commonly about 2 to 6 mm. Monoliths come in a variety of shapes and sizes. A 19-channel design is common. One manufacturer makes large monoliths with square channels.

Track-Etched Track-etched membranes (Fig. 22-70) are now made by exposing a thin polymer film to a collimated beam of radiation strong enough to break chemical bonds in the polymer chains. The film is then etched in a bath which selectively attacks the damaged polymer. The technique produces a film with photogenic pores, whose diameter may be varied by the intensity of the etching step. Commercially available membranes have a narrow pore size distribution and are reportedly resistant to plugging. The membranes have low flux, because it is impossible to achieve high pore density without sacrificing uniformity of diameter.

Chemical Phase Inversion Symmetrical phase-inversion membranes (Fig. 22-71) remain the most important commercial MF membranes produced. The process produces tortuous-flow membranes. It involves preparing a concentrated solution of a polymer in a solvent. The solution is spread into a thin film, then precipitated through the slow addition of a nonsolvent, usually water, sometimes from the vapor phase. The technique is impressively versatile, capable of producing fairly uniform membranes whose pore size may be varied within broad limits.

Thermal Phase Inversion Thermal phase inversion is a technique which may be used to produce large quantities of MF membrane economically. A solution of polymer in poor solvent is prepared at an elevated temperature. After being formed into its final shape, a sudden drop in solution temperature causes the polymer to precipitate. The solvent is then washed out. Membranes may be spun at high rates using this technique.

Stretched Polymers MF membranes may be made by stretching (Fig. 22-72). Semicrystalline polymers, if stretched perpendicular to the axis of crystallite orientation, may fracture in such a way as to make reproducible microchannels. Best known are Goretex® produced from Teflon®, and Celgard® produced from polyolefin. Stretched polymers have unusually large fractions of open space, giving them

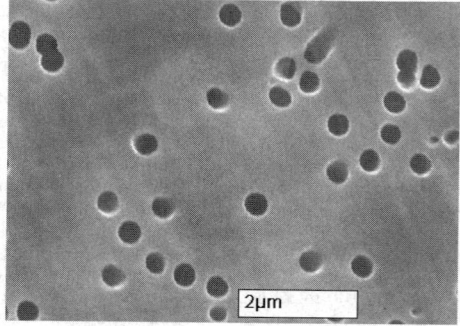

FIG. 22-70 Track-etched 0.4 μm polycarbonate membrane. (*Courtesy Millipore Corporation.*)

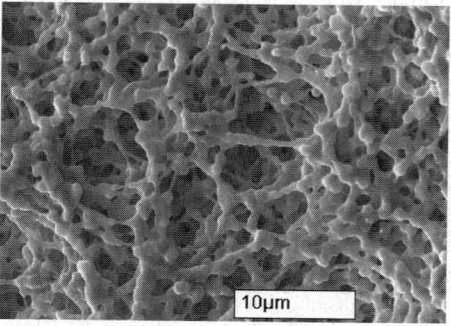

FIG. 22-71 Chemical phase inversion 0.45 μm polyvinylidene fluoride membrane. (*Courtesy Millipore Corporation.*)

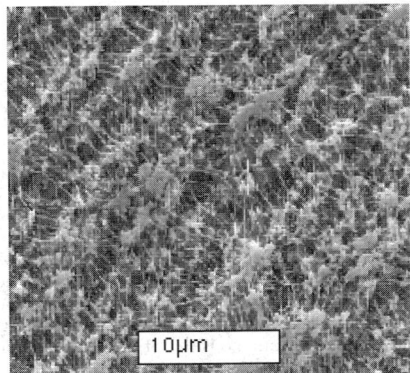

FIG. 22-72 Stretched polyetrafluoroethylene membrane. (*Courtesy Millipore Corporation.*)

very high fluxes in the microfiltration of gases, for example. Most such materials are very hydrophobic.

Membrane Characterization   MF membranes are rated by flux and pore size. Microfiltration membranes are uniquely testable by direct examination, but since the number of pores that may be observed directly by microscope is so small, microscopic pore size determination is mainly useful for membrane research and verification of other pore-size-determining methods. Furthermore, the most critical dimension may not be observable from the surface. Few MF membranes have neat, cylindrical pores. Indirect means of measurement are generally superior. Accurate characterization of MF membranes is a continuing research topic for which interested parties should consult the current literature.

Bubble Point   Large areas of microfiltration membrane can be tested and verified by a bubble test. Pores of the membrane are filled with liquid, then a gas is forced against the face of the membrane. The Young-Laplace equation. $\Delta P = (4\gamma \cos \Theta)/d$, relates the pressure required to force a bubble through a pore to its radius, and the interfacial surface tension between the penetrating gas and the liquid in the membrane pore. γ is the surface tension (N/m), d is the pore diameter (m), and P is transmembrane pressure (Pa). Θ is the liquid-solid contact angle. For a fluid wetting the membrane perfectly, $\cos \Theta = 1$.

By raising the gas pressure on a wet membrane until the first bubble appears, the largest pore may be identified, and its size computed. This is a good test to run on a membrane apparatus used to sterilize a fluid, since bacteria larger than the identified largest pore (or leak) cannot readily penetrate the assembly. Pore-size distribution may also be run by bubble point. Bubble-point testing is particularly useful in assembled microfilters, since the membrane and all seals may be verified. Periodic testing insures that the assembly retains its integrity. Diffusional flow of gas is a complication in large MF assemblies. It results from gas dissolving in pore liquid at the high-pressure side, and desorbing at the low-pressure side If the number of pores and the average pore length are known, the effect can be computed. Special protocols are used when this method is used for critical applications. Detail is provided in ASTM F316-86, "Standard test method for pore size characteristics of membrane filters by bubble point and mean flow pore test." The bubble-point test may also be run using two liquids. Because interfacial surface tensions of liquids can be quite low, this technique permits measurements on pores as small as 10 nm.

Charged Membranes   The use of tortuous-flow membranes containing a positive electrical charge may reduce the quantity of negatively charged particles passing even when the pore size is much larger than the particle. The technique is useful for making prefilters or layered membranes that withstand much higher solids loadings before becoming plugged.

Bacteria Challenge   Membranes are further tested by challenge with microorganisms of known size: their ability to retain all of the organisms is taken as proof that all pores are smaller than the organism. The best-known microorganism for pore-size determination is *Pseudomonas diminuta*, an asporogenous gram-negative rod with a mean diameter of 0.3 μm. Membranes with pore size smaller than that are used to ensure sterility in many applications. Leahy and Sullivan [*Pharmaceutical Technology*, **2**(11), 65 (1978)] provide details of this validation procedure.

Membrane thickness is a factor in microbial retention. Tortuous-pore membranes rated at 0.22 μm typically have surface openings as large as 1 μm (Fig. 22-71). Narrower restrictions are found beneath the surface. In challenge tests, *P. diminuta* organisms are found well beneath the surface of an 0.2 μm membrane, but not in the permeate.

Latex   Latex particles of known size are available as standards. They are useful to challenge MF membranes.

Process Configuration   As befits a field with a vast number of important applications and a history of innovation, there are countless variations on how an MF process is run.

Dead-end versus Cross-flow   Conventional filtration is usually run dead-end, and is facilitated by amendments that capture the particulates being removed. Membranes have very low dirt capacity, so only applications with very low solids to be removed are run in conventional dead-end flow. A rough upper limit to solids content is about 0.5 percent; streams containing <0.1 percent are almost always processed by dead-end devices. Since dead-end membrane equipment is much less expensive than cross-flow, great ingenuity is applied to protecting the critical membrane pores by structured prefilters to remove larger particles and debris. The feed may also be pretreated. It is common practice to dispose of the spent membrane rather than clean it. The membrane may be run inverted. A review of dead-end membrane filtration is given by Davis and Grant [in Ho and Sirkar (eds.), op. cit., pp. 461–479].

Cross-flow is the usual case where cake compressibility is a problem. Cross-flow microfiltration is much the same as cross-flow ultrafiltration in principle. In practice, the devices are often different. As with UF, spiral-wound membranes provide the most economical configuration for many large-scale installations. However, capillary devices and cassettes are widely employed, especially at smaller scale. A detailed description of cross-flow microfiltration had been given by Murkes and Carlsson [*Crossflow Filtration*, Wiley, NY (1988)].

Membrane Inverted   Most membranes have larger openings on one face than on the other. Common practice is to run the tightest face against the feed in order to avoid plugging of the backing by particles. The rationale is that anything that makes it past the "skin" will have relatively unimpeded passage into the backing and out with the permeate. For very low solids this convention is reversed, the rationale being that the porous backing provides a trap for particulates, rather like filter aid.

If the complete passage of soluble macromolecules is required, a highly polarized membrane is an advantage. The upside-down membrane hinders the back diffusion of macrosolutes. Countering the tendency of the retained particulates to "autofilter" soluble macrosolutes, the inhibition of back diffusion raises the polarization and thus the passage of macrosolutes such as proteins. The particulates are physically retained by the membrane. Blinding and plugging can be controlled if the membrane is backwashed frequently. This technique has been demonstrated at high solids loadings in an application where high passage of soluble material is critical, the microfiltration of beer [Wenten, Rasmussen, and Jonsson, *North Am. Membrane Soc. Sixth Annual Meeting*, Breckenridge, CO (1994)].

Liquid Backpulse   Solid membranes are backwashed by forcing permeate backward through the membrane. Frequent pulsing seems to be the key.

Air Backflush   A configuration unique to microfiltration feeds the process stream on the shell side of a capillary module with the permeate exiting the tube side. The device is run as an intermittent dead-end filter. Every few minutes, the permeate side is pressurized with air. First displacing the liquid permeate, a blast of air pushed backward through the membrane pushes off the layer of accumulated solids. The membrane skin contacts the process stream, and while being backwashed, the air simultaneously expands the capillary and membrane pores slightly. This momentary expansion facilitates the removal of imbedded particles.

Process Limitations The same sorts of process limitations affecting UF apply to MF. The following section will concentrate on the differences.

Concentration Polarization The equations governing cross-flow mass transfer are developed in the section describing ultrafiltration. The velocity, viscosity, density, and channel-height values are all similar to UF, but the diffusivity of large particles (MF) is orders-of-magnitude lower than the diffusivity of macromolecules (UF). It is thus quite surprising to find the fluxes of cross-flow MF processes to be similar to, and often higher than, UF fluxes. Two primary theories for the enhanced diffusion of particles in a shear field, the inertial-lift theory and the shear-induced theory, are explained by Davis [in Ho and Sirkar (eds.), op. cit., pp. 480–505], and Belfort, Davis, and Zydney [*J. Membrane. Sci.,* **96,** 1–58 (1994)]. While not clear-cut, shear-induced diffusion is quite large compared to Brownian diffusion except for those cases with very small particles or very low cross-flow velocity. The enhancement of mass transfer in turbulent-flow microfiltration, a major effect, remains completely empirical.

Fouling Fouling affects MF as it affects all membrane processes. One difference is that the fouling effect caused by deposition of a foulant in the pores or on the surface of the membrane can be confounded by a rearrangement or compression of the solids cake which may form on the membrane surface. Also, the high, open space found in tortuous-pore membranes makes them slower to foul and harder to clean.

Equipment Configuration Since the early days when membrane was available only in flat-sheet form, the variety of offerings of various geometry and fabricated filter component types has grown geometrically. An entire catalog is devoted just to list the devices incorporating membranes whose area ranges from less than 1 cm^2 up to 3 m^2. Microfiltration has grown to maturity selling these relatively small devices. Replacement rather than reuse has long been the custom in MF, and only with later growth of very large applications, such as water, sewage, and corn sweeteners, has long membrane life become an economic necessity on a large scale.

Conventional Designs Designs familiar from other unit operations are also used in microfiltration. Cartridge-filter housings may be fitted with pleated MF membrane making a high-area dead-end membrane filter. Plate-and-frame type devices are furnished with MF sheet stock, and are common in some applications. Capillary bundles with tube-side feed are used for cross-flow applications, and are occasionally used in dead-end flow. A few tubular membranes are used. Spiral-wound modules are becoming increasingly important for process applications where economics are paramount. Belt filters have been made using MF membrane.

Ceramics Ceramic microfilters for commercial applications are almost always employed as tube-side feed multitube monoliths. They are also available as flat sheet, single tubes, discs, and other forms primarily suited to lab use. They are used for a few high-temperature applications, in contact with solvents, and particularly at very high pH.

Cassettes *Cassette* is a term used to describe two different cross-flow membrane devices. The less-common design is a usually large stack of membrane separated by a spacer, with flow moving in parallel across the membrane sheets. This variant is sometimes referred to as a **flat spiral,** since there is some similarity in the way feed and permeate are handled. The more common cassette has long been popular in the pharmaceutical and biotechnical field. It too is a stack of flat-sheet membranes, but the membrane is usually connected so that the feed flows across the membrane elements in series to achieve higher conversion per pass. Their popularity stems from easy direct scale-up from laboratory to plant-scale equipment. Their limitation is that fluid management is inherently very limited and inefficient. Both types of cassette are very compact and capable of automated manufacture.

Representative Process Applications

Pharmaceutical Removal of suspended matter is a frequent application for MF. Processes may be either clarification, in which the main product is a clarified liquid, or solids recovery. Separating cells or their fragments from broth is the most common application. Clarification of the broth in preparation for product recovery is the usual objective, but the primary goal may be recovery of cells. Cross-flow microfiltration competes well with centrifugation, conventional filtration by rotary vacuum filter or filter press and decantation. MF delivers a cleaner permeate, an uncontaminated, concentrated cell product which may be washed in the process, and generally gives high yields. There is no filter-aid disposal problem. Microfiltration has higher capital costs than the other processes, although total cost may be lower. The recovery of penicillin is an example of a process for which cross-flow microfiltration is generally accepted.

Water and Wastewater Microfiltration is beginning to be applied to large-scale potable-water treatment. Its major advantage is positive removal of cryptosporidium and giardia cysts, and its major disadvantage is cost. MF is used in a few large sewage-treatment facilities, where its primary advantage is that it permits a major reduction in the physical size of the facility.

Chemical MF is used in several applications to recover caustic values from cleaning or processing streams. An example is the caustic solution used to clean dairy evaporators, which may be cleaned for reuse by passing it through a microfilter. Significant savings in caustic purchase and disposal costs provide the incentive. Acids are also recovered and reused. Ceramic microfilters are most commonly used in these applications.

Food and Dairy Microfiltration has many applications in the food and dairy industries. An innovative dairy application uses MF membranes to remove bacteria as a nonthermal means of disinfection for milk. A special flow apparatus maintains a carefully controlled transmembrane pressure as the milk flows across the membrane. The concentrate contains the bacteria and spores, as well as any fat. The concentrate may be heat sterilized and recombined with the sterile permeate. In another milk application, some success is reported in separating fat from milk or other dairy streams by cross-flow microfiltration instead of centrifugation. Transmembrane pressure must be kept very low to prevent fat penetration into the membrane. In the food industry, MF membranes are replacing diatomaceous earth filtration in the processing of gelatin. The gelatin is passed with the permeate, but the haze producing components are retained. UF may be used downstream to concentrate the gelatin.

Flow Schemes The outline of batch, semibatch, and stages-in-series is given in the section describing ultrafiltration. Diafiltration is also described there. All these techniques are common in MF, except for stages-in-series, used rarely. MF features uses of special techniques to control transmembrane pressure in some applications. An example is one vendor's device for the microfiltration of milk. In most devices the permeate simply leaves by the nearest exit, but for this application the permeate is pumped through the device in such a way as to duplicate the pressure drop in the concentrate side, thus maintaining a constant transmembrane-pressure driving force. In spite of the low-pressure driving force, the flux is extremely high.

Limitations Some applications which seem ideal for MF, for example the clarification of apple juice, are done with UF instead. The reason is the presence of deformable solids which easily plug and blind an MF membrane. The pores of an ultrafiltration membrane are so small that this plugging does not occur, and high fluxes are maintained. UF can be used because there is no soluble macromolecule in the juice that is desired in the filtrate. There are a few other significant applications where MF seems obvious, but is not used because of deformable particle plugging.

Economics Microfiltration may be the triumph of the Lilliputians; nonetheless, there are a few large-industrial applications. Dextrose plants are very large, and as membrane filtration displaces the precoat filters now standard in the industry, very large membrane microfiltration equipment will be built.

Site Size Most MF processes require a smaller footprint than competing processes. Reduction in total-area requirements are sometimes a decisive economic advantage for MF. It may be apparent that the floor-space costs in a pharmaceutical facility are high, but municipal facilities for water and sewage treatment are often located on expensive real estate, giving MF an opportunity despite its higher costs otherwise.

Large Plants The economics of microfiltration units costing about 10^6 is treated under ultrafiltration. When ceramic membranes are used, the cost optimum may shift energy consumption upward to as much as 10 kWh/m³.

Disposables For smaller MF applications, short membrane life is a traditional characteristic. In these applications, costs are dominated by the disposables, and an important characteristic of equipment design is the ease, economy, and safety of membrane replacement.

Hygiene and Regulation Almost unique to MF is the influence of regulatory concerns in selection and implementation of a suitable microfilter. Since MF is heavily involved with industries regulated by the Food and Drug Administration, concerns about process stability, consistency of manufacture, virus reduction, pathogen control, and material safety loom far larger than is usually found in other membrane separations.

GAS-SEPARATION MEMBRANES

Process Description Gas-separation membranes separate gases from other gases. Some gas filters, which remove liquids or solids from gases, are microfiltration membranes. Gas membranes generally work because individual gases differ in their solubility and diffusivity through nonporous polymers. A few membranes operate by sieving, Knudsen flow, or chemical complexation.

Selective gas permeation has been known for generations, and the early use of palladium silver-alloy membranes achieved sporadic industrial use. Gas separation on a massive scale was used to separate U^{235} from U^{238} using porous (Knudsen flow) membranes. An upgrade of the membranes at Oak Ridge cost $1.5 billion. Polymeric membranes became economically viable about 1980, introducing the modern era of gas-separation membranes. H_2 recovery was the first major application, followed quickly by acid gas separation (CO_2/CH_4) and the production of N_2 from air.

Three basic mechanisms can be used for membranes in gas separation. They are types (b), (c), and (d) in Fig. 22-47. Membranes of type (d) are by far the dominant type.

The more permeable component is called the *fast gas*, so it is the one enriched in the permeate stream. Permeability through polymers is the product of solubility and diffusivity. The diffusivity of a gas in a membrane is inversely proportional to its kinetic diameter, a value determined from zeolite cage exclusion data (see Table 22-23 after Breck, *Zeolite Molecular Sieves*, Wiley, NY, 1974, p. 636).

Leading Examples These applications are commercial, some on a very large scale. They illustrate the range of application for gas-separation membranes. Unless otherwise specified, all use polymeric membranes.

Hydrogen Hydrogen recovery was the first large commercial membrane gas separation. Polysulfone fiber membranes became available in 1980 at a time when H_2 needs were rising, and these novel membranes quickly came to dominate the market. Applications include recovery of H_2 from ammonia purge gas, and extraction of H_2 from petroleum cracking streams. Hydrogen once diverted to low-quality fuel use is now recovered to become ammonia, or is used to desulfurize fuel, etc. H_2 is the fast gas.

Carbon Dioxide-Methane Much of the natural gas produced in the world is coproduced with an acid gas, most commonly CO_2 and/or H_2S. While there are many successful processes for separating the gases, membrane separation is a commercially successful competitor, especially for small installations. The economics work best for feeds with very high or very low CH_4 content. Methane is a slow gas; CO_2, H_2S, and H_2O are fast gases.

Oxygen-Nitrogen Because of higher solubility, in many polymers, O_2 is faster than N_2 by a factor of 5. Water is much faster still. Since simple industrial single-stage air compressors provide sufficient pressure to drive an air-separating membrane, moderate purity N_2 (95–99.5%) may be produced in low to moderate quantities quite economically by membrane separation. (Argon is counted as part of the nitrogen.) An O_2-enriched stream is a coproduct, but it is rarely of economic value. The membrane process to produce O_2 as a primary product has a limited market.

Helium Helium is a very fast gas, and may be recovered from natural gas through the use of membranes. More commonly, membranes are used to recover He after it has been used and become diluted.

Gas Dehydration Water is extremely permeable in polymer membranes. Dehydration of air and other gases is a growing membrane application.

Vapor Recovery Organic vapors are recovered from gas streams using highly permeable rubbery polymer membranes which are generally unsuitable for permanent gas separations because of poor selectivity. The high sorption of vapors in these materials makes them ideal for stripping and recovering vapors from gases.

Competing Processes Membranes are not the only way to make these separations, neither are they generally the dominant way. In many applications, membranes compete with cryogenic distillation and with pressure-swing adsorption; in others, physical absorption is the dominant method. The growth rate for membrane capacity is higher than that for any competitor.

Basic Principles of Operation Gas-separation literature often uses nomenclature derived from distillation, a practice that will generally be followed here. L is the molar feed rate, V is the molar permeate rate, R is the molar residue ($L - V$). Mole fractions of components i, j, in the feed-residue phase will be x_i, x_j . . . and in the per-

TABLE 22-23 Kinetic Diameters for Important Gases

Penetrant	He	H_2	NO	CO_2	Ar	O_2	N_2	CO	CH_4	C_2H_4	Xe	C_3H_8
Kinetic dia, nm	0.26	0.289	0.317	0.33	0.34	0.346	0.364	0.376	0.38	0.39	0.396	0.43

TABLE 22-24 Gas-Permeation Units

Quantity	Engineering units	Literature units	SI units
Permeation rate	Standard cubic feet/minute		kmol/s
Permeation flux	ft³/ft²·day	cm/sec (STP)	kmol/m²·s
Permeability	ft³·ft/ft²·day·psi	Barrers	kmol/m·s·Pa
Permeance	ft³/ft²·day·psi	Barrers/cm	kmol/m²·s·Pa

TABLE 22-25 Barrer Conversion Factors

Quantity	Multiply	By	To get
Permeability	Barrers	3.348×10^{-19}	kmol/m·s·Pa
Permeability	Barrers	4.810×10^{-8}	ft³(STP)/ft·psi·day
Permeance	Barrers/cm	3.348×10^{-17}	kmol/m²·s·Pa
Permeance	Barrers/cm	1.466×10^{-6}	ft³(STP)/ft²·psi·day

TABLE 22-26 Industry-Specific Gas Measures

Industry-unit	How measured	Cubic feet per pound mole	kmol per mscf
STP, Mscf	1000 ft³ at 32°F	359.3	1.262
Gas industry, Mscf	1000 ft³ at 60°F	379.8	1.194
Air industry, Mnsf	1000 ft³ at 70°F	387.1	1.172

meate phase $y_i, y_j, \ldots$. Stage cut, Θ, is permeate volume/feed volume, or V/L.

Basic Equations   In "Background and Definitions," the basic equation for gas permeation was derived with the major assumptions noted. Equation (22-62) may be restated as:

$$J_i \sim N_i = (\rho_i/z)(p_{i,\text{feed}} - p_{i,\text{permeate}}) \qquad (22\text{-}97)$$

where ρ_i is the permeability of component i through the membrane, J_i is the flux of component i through the membrane for the *partial pressure difference* (Δp) of component i. z is the effective thickness of the membrane. By choosing units appropriately, $J = \rho\Delta p$.

A similar equation may be written for a second component, j, and any additional number of components, employing partial pressures:

$$J_j = (\rho_j/z)(p_{j,\text{feed}} - p_{j,\text{permeate}}) \qquad (22\text{-}98)$$

The total pressure is the sum of the partial pressures:

$$P_{\text{feed}} = \sum (p_{\text{feed}})_{i,j,\ldots} \qquad (22\text{-}99)$$

$$P_{\text{permeate}} = \sum (p_{\text{permeate}})_{i,j\ldots} \qquad (22\text{-}100)$$

For simplicity, consider a two component system. The volume fraction of a component is

$$x_i = \frac{p_{\text{feed}}}{P_{\text{feed}}} \qquad (22\text{-}101)$$

$$y_i = \frac{p_{\text{permeate}}}{P_{\text{permeate}}} \qquad (22\text{-}102)$$

When two species are permeating through a membrane, the ratio of their fluxes can be written following Eqs. (22-97) and (22-98) as:

$$\frac{J_i}{J_j} = \frac{\dfrac{\rho_i}{z}(p_{i,\text{feed}} - p_{i,\text{permeate}})}{\dfrac{\rho_i}{z}(p_{j,\text{feed}} - p_{j,\text{permeate}})} \qquad (22\text{-}103)$$

Recalling Eq. (22-63), and restating it in the nomenclature for gas membranes:

$$\alpha = \frac{y_i/y_j}{x_i/x_j} \qquad (22\text{-}104)$$

Defining the pressure ratio $\Phi = P_{\text{feed}}/P_{\text{permeate}}$ and applying Eqs. (22-99) through (22-102) gives:

$$\frac{J_i}{J_j} = \alpha\left[\frac{x_i - (y_i/\Phi)}{x_j - (y_j/\Phi)}\right] \qquad (22\text{-}105)$$

Combining these equations and rearranging, the permeate composition may be solved explicitly:

$$y_i = \left(\frac{\Phi}{2}\right)\left[x_i + \frac{1}{\Phi} + \frac{1}{\alpha-1} - \sqrt{\left(x_i + \frac{1}{\Phi} + \frac{1}{\alpha-1}\right)^2 - \frac{4\alpha x_i}{(\alpha-1)\Phi}}\right]$$

$$(22\text{-}106)$$

Equation (22-106) gives a permeate concentration as a function of the feed concentration at a stage cut, $\Theta = 0$. To calculate permeate composition as a function of Θ, the equation may be used iteratively if the permeate is unmixed, such as apply in a test cell. The calculation for real devices must take into account the fact that the driving force is variable due to changes on both sides of the membrane, as partial pressure is a point function, nowhere constant. Using the same caveat, permeation rates may be calculated component by component using Eq. (22-98) and permeance values. For any real device, both concentration and permeation require iterative calculations dependent on module geometry.

Driving Force   Gas moves across a membrane in response to a difference in chemical potential. Partial pressure is sufficiently proportional to be used as the variable for design calculations for most gases of interest, but fugacity must be used for CO_2 and usually for H_2 at high pressure. Gas composition changes as a gas passes along a membrane. As the fast gas passes through the membrane, x_i drops. Total pressure on the upstream side of the membrane drops because of frictional losses in the device. Frictional losses on the permeate side

will affect the permeate pressure. The partial pressure of a component in the permeate may thus rise rapidly. Permeation rate is a point function, dependent on the difference in partial pressures at a point on the membrane. Many variables affect point partial pressures, among them are membrane structure, module design, and permeate gas-sweep rates. Juxtaposition of feed and permeate is a function of permeator design, and a rapid decline in driving force may result when it is not expected (see "Membrane System Design Features"). An additional complication may arise in a few cases from the Joule-Thompson effect during expansion of a gas through a membrane changing the temperature. High-pressure CO_2 is an example.

Plasticization   Gas solubility in the membrane is one of the factors governing its permeation, but the other factor, diffusivity, is not always independent of solubility. If the solubility of a gas in a polymer is too high, plasticization and swelling result, and the critical structure that controls diffusion selectivity is disrupted. These effects are particularly troublesome with condensable gases, and are most often noticed when the partial pressure of CO_2 or H_2S is high. H_2 and He do not show this effect. This problem is well known, but its manifestation is not always immediate.

Limiting Cases   Equation (22-106) has two limiting cases for a binary system. First, when $\alpha \gg \phi$. In this case, selectivity is no longer very important.

$$y_i \cong x_i \frac{P_{\text{feed}}}{P_{\text{permeate}}} = x_i\Phi \qquad (22\text{-}107)$$

Module design is very important for this case, as the high α may result in high permeate partial pressure. An example is the separation of H_2O from air.

Conversely, when $\Phi \gg \alpha$, pressure ratio loses its importance, and permeate composition is:

$$y_i \cong \frac{x_i\alpha}{[1 + x_i(\alpha - 1)]} \qquad (22\text{-}108)$$

Module design for this case is of lesser importance.

Selectivity and Permeability

State of the Art   A desirable gas membrane has high separating power (α) and high permeability to the fast gas, in addition to critical requirements discussed below. The search for an ideal membrane produced copious data on many polymers, neatly summarized by Robeson [*J. Membrane Sci.*, **62**, 165 (1991)]. Plotting log permeability versus log selectivity (α), an "upper bound" is found (see Fig. 22-73) which all the many hundreds of data points fit. The data were taken between 20–50°C, generally at 25 or 35°C.

The lower line in Fig. 22-73 shows the upper bound in 1980. Although no breakthrough polymers have been reported in the past few years, it would be surprising if these lines remain the state of the art forever.

The upper-bound line connects discontinuous points, but polymers exist near the bound for separations of interest. Whether these will be available as membranes is a different matter. A useful membrane requires a polymer which can be fabricated into a device having an active layer around 50 nm thick. At this thickness, membrane properties may vary significantly from bulk properties, although not by a factor of 2.

The data reported are *permeabilities*, not *fluxes*. Flux is proportional to permeability/thickness. The separations designer must deal with real membranes, for which thickness is determined by factors outside the designer's control. It is vital that flux data are used in design.

TABLE 22-27   High-Performance Polymers for O₂/N₂*

Polymer	α (O_2/N_2)	ρ (O_2)—Barrers
Poly (trimethylsilylpropyne)	2.0	4000
Tetrabromo Bis A polycarbonate	7.47	1.36
Poly (tert-butyl acetylene)	3.0	300
Vectra polyester	15.3	0.00046
Poly (triazole)	9.0	1.2
Polypyrrolone	6.5	7.9

*Polymers that are near the upper bound and their characteristics.

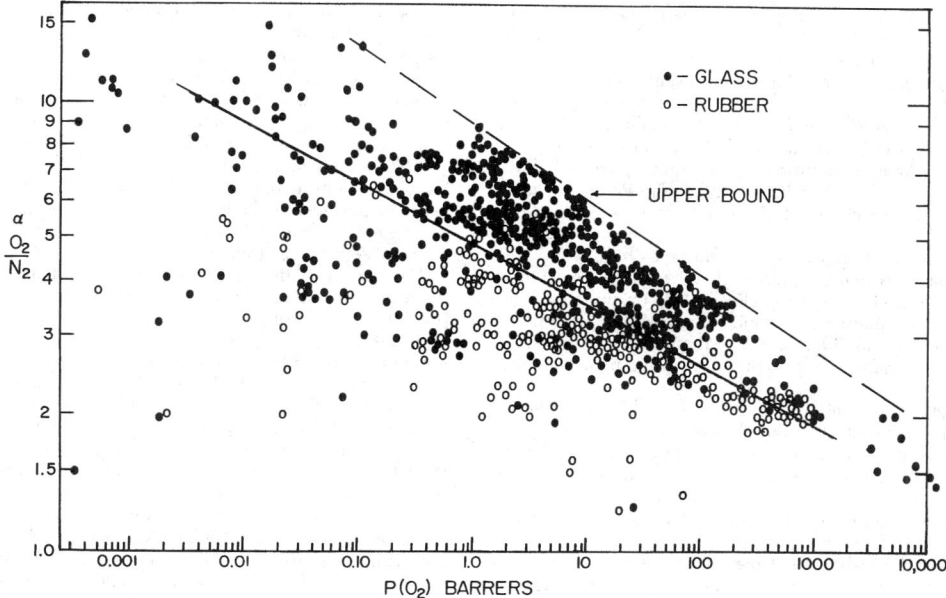

FIG. 22-73   Plot of separation factor versus permeability for many polymers, O_2/N_2. Abscissa—"Fast Gas Permeability, $\rho(O_2)$ Barrers." Ordinate—"Selectivity, α (O_2/N_2)."

Glassy polymers are significantly overrepresented in the high-selectivity region near the upper bound, and rubbery polymers are overrepresented at the high-permeability end, although the highest-permeability polymer discovered, poly(trimethylsilylpropyne) is glassy. Most of the polymers with interesting properties are noncrystalline. Current membrane-materials research is strongly focused on glassy materials and on attempts to improve diffusivity, as it seems more promising than attempts to increase solubility [Koros, *North Am. Membrane Soc. Sixth Annual Meeting*, Breckenridge, CO (1994)].

Robeson [*J. Membrane Sci.*, **62**, 165 (1991); *Polymer*, **35**, 4970 (1994)] has determined upper-bound lines for many permeant pairs in hundreds of polymers. These lines may be drawn from Eq. (22-109) and the data included in Table 22-28. These values will give ρ_i in Barrers; α is dimensionless. Robeson [op.cit., (1991); op. cit., (1994)] lists high-performance polymers for most of these gas pairs, like Table 22-28.

$$\log \rho_i = \log k - m \log \alpha_{ij} \qquad (22\text{-}109)$$

Temperature Effects   A temperature increase in a polymer membrane permits larger segmental motions in the polymer, producing a dramatic increase in diffusivity. Countering this is a decrease in solubility. It increases the size of the gaps in the polymer matrix, decreasing diffusivity selectivity. The net result is that for a glassy

TABLE 22-28   Upper-Bound Coordinates for Gas Pairs

Gas pair	$\log k$	m
He/N_2	4.0969	1.0242
H_2/N_2	4.7236	1.5275
He/CH_4	3.6991	0.7857
H_2/CH_4	4.2672	1.2112
O_2/N_2	5.5902	5.800
He/O_2	3.6628	1.295
H_2/O_2	4.5534	2.277
CO_2/CH_4	6.0309	2.6264
He/H_2	2.9823	4.9535
He/CO_2	2.8482	1.220
H_2/CO_2	3.0792	1.9363

polymer, permeability rises while selectivity declines. For organic permeants in rubbery polymers, this trend is often reversed.

Plasticization and Other Time Effects   Most data from the literature, including those presented above are taken from experiments where one gas at a time is tested, with α calculated as a ratio of the two permeabilities. If either gas permeates because of a high-sorption coefficient rather than a high diffusivity, there may be an increase in the permeability of all gases in contact with the membrane. Thus, the α actually found in a real separation may be much lower than that calculated by the simple ratio of permeabilities. The data in the literature do not reliably include the plasticization effect. If present, it results in the sometimes slow relaxation of polymer structure giving a rise in permeability and a dramatic decline in selectivity.

Other Caveats   Transport in glassy polymers is different from transport in rubbery ones. In glassy polymers, there are two sites in which sorption may occur, and the literature dealing with dual-mode sorption is voluminous. The simplest case describes behavior when the downstream pressure is zero. It is of great help in understanding the theory but of limited value in practice. There are concerns about permeation of mixtures in glassy polymers with reports of crowding out and competitive sorption. Practical devices are built and operated for many streams, and the complications are often minor. But taking data independently determined for two pure gases and dividing them to obtain α in the absence of other facts is risky.

Gas-Separation Membranes

Organic   Organic polymer membranes are the basis for almost all commercial gas-separation activity. Early membranes were cellulose esters and polysulfone. These membranes have a large installed base. New installations are dominated by specialty polymers designed for the purpose, including some polyimides and halogenated polycarbonates. In addition to skinned membranes, composites are made from "designer" polymers, requiring as little as 2 g/1000 m². The rapid rise of N_2/O_2 membranes in particular is the result of stunning improvements in product uniformity and quality. A few broken fibers in a 100 m² module results in the module's being scrapped.

Caulked   Membrane manufacturing defects are unavoidable, and pinholes are particularly deleterious in gas-separation membranes. A very effective remedy is to **caulk** the membrane by applying a highly permeable, very thin topcoat over the finished membrane. While the coating will have poor selectivity, it will plug up the gross leaks while impairing the fast gas permeance only slightly. Unless the as-cast membrane is almost perfect, caulking dramatically improves membrane performance.

Metallic   Palladium films pass H_2 readily, especially above 300°C. α for this separation is extremely high, and H_2 produced by purification through certain Pd alloy membranes is uniquely pure. Pd alloys are used to overcome the crystalline instability of pure Pd during heating-cooling cycles. Economics limit this membrane to high-purity applications.

Advanced Materials   Experimental membranes have shown remarkable separations between gas pairs such as O_2/N_2 whose kinetic diameters (see Table 22-23) are quite close. Most prominent is the carbon molecular sieve membrane, which operates by ultramicroporous molecular sieving (see Fig. 22-48c). Preparation of large-scale permeators based on ultramicroporous membranes has proven to be a major challenge.

Catalytic   A catalytic-membrane reactor is a combination heterogeneous catalyst and permselective membrane that promotes a reaction, allowing one component to permeate. Many of the reactions studied involve H_2. Membranes are metal (Pd, Ag), nonporous metal oxides, and porous structures of ceramic and glass. Falconer, Noble, and Sperry [in Noble and Stern (eds.), op. cit., pp. 669–709] review status and potential developments.

Membrane System Design Features   For the rate process of permeation to occur, there must be a driving force. For gas separations, that force is partial pressure (or fugacity). Since the ratio of the component fluxes determines the separation, the partial pressure of each component at each point is important. There are three ways of driving the process: Either high partial pressure on the feed side (achieved by high total pressure), or low partial pressure on the permeate side, which may be achieved either by vacuum or by introduction of a sweep gas. Both of the permeate options have negative economic implications, and they are less commonly used.

Figure 22-74 shows three of the principal operating modes for gas membranes. A critical issue is the actual partial pressure of permeant

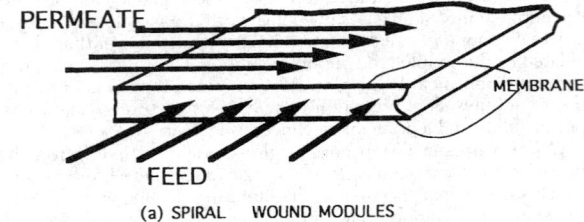

(a) SPIRAL   WOUND MODULES

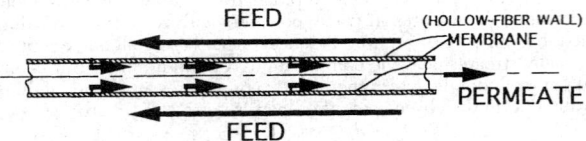

(b) HOLLOW FIBER MODULES WITH COUNTERCURRENT FLOW

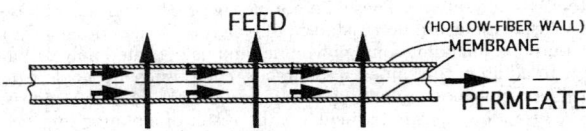

(c) HOLLOW-FIBER MODULES WITH CROSS FLOW

FIG. 22-74   Flow paths in gas permeators. (*Courtesy Elsevier.*)

at a point on the membrane. Flow arrangements for the permeate are very important in determining the efficiency of the separation, in rough analogy to the importance of arrangements in heat exchangers.

Spiral membranes are the usual way to form flat sheet into modules. They have the characteristic that the feed and the permeate move at right angles. Since the membrane is always cast on a porous support, point-permeate values are influenced by the substrate.

Hollow-fiber membranes may be run with shell-side or tube-side feed, cocurrent, countercurrent or in the case of shell-side feed and two end permeate collection, co- and countercurrent. Not shown is the scheme for feed inside the fiber, common practice in lower-pressure separations such as air.

The design of the membrane device will influence whether the membrane is operating near its theoretical limit. Sengupta and Sirkar [in Noble and Stern (eds.), op. cit., pp. 499–552] treat module design thoroughly (including numerical examples for most module configurations) and provide an extensive bibliography.

For a hollow-fiber device running with shell-side feed with the membrane on the outside, Giglia et al. [*Ind. Eng. Chem Research*, **29**, 1239–1248 (1991)] analyzed the effect of membrane-backing porosity on separation efficiency. The application is production of N_2 from air, the desired result being the lowest possible O_2 content in the retentate at a given stage cut. The modules used were operated cocurrent and countercurrent. If the porous-membrane backing prevented the permeate from mixing with the gas adjacent to the membrane, a result approximating cross-flow is expected. For the particular membrane structure used, the experimental result for cocurrent flow was quite close to the calculated cocurrent value, while for countercurrent flow, the experimental data were between the values calculated for the crosscurrent model and the countercurrent model. For a membrane to be commercially useful in this application, the mass transfer on the permeate side must exceed the crosscurrent model.

Air is commonly run with tube-side feed. The permeate is run countercurrent with the separating skin in contact with the permeate. (The feed gas is in contact with the macroporous back side of the membrane.) This configuration has proven to be superior, since the permeate-side mass-transfer problem is reduced to a minimum, and the feed-side mass-transfer problem is not limiting.

Partial Pressure Pinch   An example of the limitations of the partial pressure pinch is the dehumidification of air by membrane. While O_2 is the fast gas in air separation, in this application H_2O is faster still. Special dehydration membranes exhibit $\alpha = 20,000$. As gas passes down the membrane, the partial pressure of H_2O drops rapidly in the feed. Since the H_2O in the permeate is diluted only by the O_2 and N_2 permeating simultaneously, p_{H2O} rises rapidly in the permeate. Soon there is no driving force. The commercial solution is to take some of the dry air product and introduce it into the permeate side as a countercurrent sweep gas, to dilute the permeate and lower the H_2O partial pressure. It is in effect the introduction of a leak into the membrane, but it is a controlled leak and it is introduced at the optimum position.

Fouling   Industrial streams may contain condensable or reactive components which may coat, solvate, fill the free volume, or react with the membrane. Gases compressed by an oil-lubricated compressor may contain oil, or may be at the water dew point. Materials that will coat or harm the membrane must be removed before the gas is treated. Most membranes require removal of compressor oil. The extremely permeable poly(trimethylsilylpropyne) may not become a practical membrane because it loses its permeability rapidly. Part of the problem is pore collapse, but it seems extremely sensitive to contamination even by diffusion pump oil and gaskets [Robeson, op. cit., (1994)].

A foulinglike problem may occur when condensable vapors are left in the residue. Condensation may result which in the best case results in blinding of the membrane, and in the usual case, destruction of the membrane module. Dew-point considerations must be part of any gas-membrane design exercise.

Modules and Housings   Modern gas membranes are packaged either as hollow-fiber bundles or as spiral-wound modules. The former uses extruded hollow fibers. Tube-side feed is preferable, but it is limited to about 1.5 MPa. Higher-pressure applications are usually fed

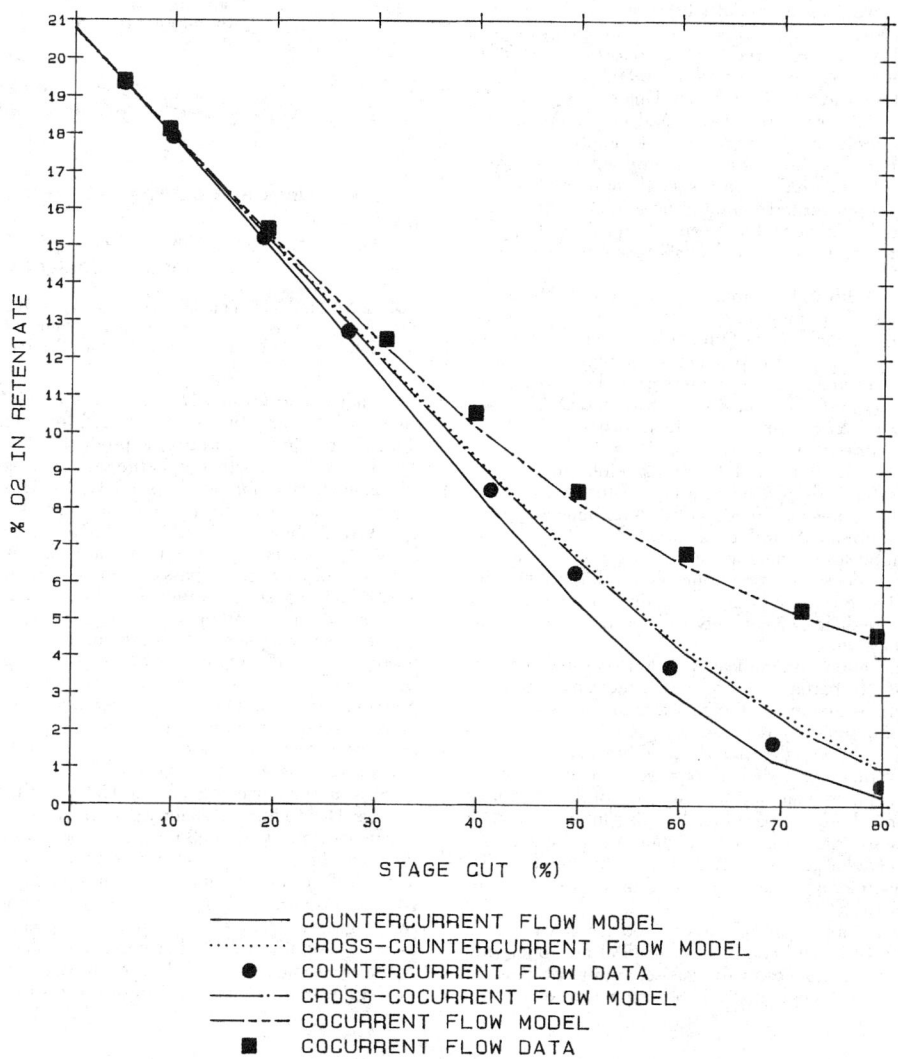

O2 RETENTATE CONCENTRATION vs STAGE CUT
FLOW PATTERN COMPARISON

————— COUNTERCURRENT FLOW MODEL
·········· CROSS-COUNTERCURRENT FLOW MODEL
● COUNTERCURRENT FLOW DATA
—·—· CROSS-COCURRENT FLOW MODEL
— — — COCURRENT FLOW MODEL
■ COCURRENT FLOW DATA

FIG. 22-75 Air fractionation by membrane. O_2 in retentate as a function of feed fraction passed through the membrane (stage cut) showing the different result with changing process paths. Process has shell-side feed at 690 kPa (abs) and 298 K. Module comprised of hollow fibers, diameter 370 μm od × 145 μm id × 1500 mm long. Membrane properties $\alpha = 5.7$ (O_2/N_2), permeance for $O_2 = 3.75 \times 10^{-6}$ Barrer/cm. (*Courtesy Innovative Membrane Systems/ Praxair.*)

on the shell side. A large industrial permeator contains fibers 400 μm by 200 μm i.d. in a 6-inch shell ten feet long. Flat-sheet membrane is wound into spirals, with an 8- by 36-inch permeator containing 25 m² of membrane. Both types of module are similar to those illustrated in "Background and Definitions." Spiral modules are useful when feed flows are very high and especially in vapor-permeation applications. Otherwise, fiber modules have a large and growing share of the market.

Energy Requirements The thermodynamic minimum energy requirement to separate a metric ton of N_2 from air and compress it to atmospheric pressure at 25°C is independent of separation method, 20.8 MJ or 5.8 kWh. In practice, a cryogenic distillation plant requires twice this energy, and it produces a very pure product as a matter of

course. The membrane process requires somewhat more energy than distillation at low purity and much more energy at high purity. Membranes for O_2-reduced air are economical and are a rapidly growing application, but not because of energy efficiency. Figure 22-75 may be used to estimate membrane energy requirements. From the required purity, locate the stage cut on either of the countercurrent flow curves. The compressor work required is calculated using the pressure and the term (product flow rate)/$(1 - \Theta)$. The N_2-rich product is produced at pressure, and the O_2-enriched permeate is vented. For example, if gas is required containing 97 percent inerts, assume the product composition would be 96 percent N_2, 2 percent O_2, 1 percent Ar and 3 percent O_2, giving a calculated molar mass of 28.2. One tonne would thus contain 35.4 kmol. For 98 percent inerts, Fig. 22-75 shows a stage cut of 67 percent when operating at 690 kPa (abs) and 25°C. Therefore, $35.5/(1 - 0.67) = 108$ kmol of air are required as feed. The adiabatic compression of 108 kmol of air from atmospheric pressure and 300K (it would subsequently be cooled to $273 + 25 = 298$K) requires 192 kWh. Assuming 75 percent overall compressor + driver efficiency, 256 kWh are required. For comparison, a very efficient, large N_2 distillation plant would produce 99.99 percent N_2 at 690 kPa for 113 kWh/metric ton. The thermodynamic minimum to separate ($N_2 + Ar$) and deliver it at the given P and T is 72 kWh.

Economics It is ironic that a great virtue of membranes, their versatility, makes economic optimization of a membrane process very difficult. Designs can be tailored to very specific applications, but each design requires a sophisticated computer program to optimize its costs. Spillman [in Noble and Stern (eds)., op. cit., pp. 589–667] provides an overall review and numerous specific examples including circa 1989 economics.

Rules of Thumb With a few notable exceptions such as H_2 through Pd membranes, membrane separations are not favored when a component is required at high purity. Often, membranes serve these needs by providing a moderate purity product which may be inexpensively upgraded by a subsequent process. Increasing the purity of N_2 by the introduction of H_2 or CH_4 to react with unwanted O_2 is a good example. Unless permeates are recycled, high product purity is accompanied by lower product recovery.

Pressure ratio (Φ) is quite important, but transmembrane ΔP matters as well. Consider the case of a vacuum permeate ($\Phi = \infty$): The membrane area will be an inverse function of P. Φ influences separation and area, transmembrane ΔP influences area.

In a binary separation, the highest purity of integrated permeate occurs at $\Theta = 0$. Purity decreases monotonically until it reaches the feed purity at $\Theta = 1$. In a ternary system, the residue concentration of the gas with the intermediate permeability will reach a maximum at some intermediate stage cut.

Concentration polarization is a significant problem only in vapor separation. There, because the partial pressure of the penetrant is normally low and its solubility in the membrane is high, there can be depletion in the gas phase at the membrane. In other applications it is usually safe to assume bulk gas concentration right up to the membrane.

Another factor to remember is that for $\alpha = 1$, or for $\Phi = 1$, or for $(1/\Theta) = 1$ there is no separation at all. Increasing any of the quantities as defined make for a better separation, but the improvement is diminishing in all cases as the value moves higher. An example of the economic tradeoff between permeability and α is illustrated in Fig. 22-76 where the economics are clearly improved by sacrificing selectivity for flux.

Compression If compression of either feed or permeate is required, it is highly likely that compression capital and operating costs will dominate the economics of the gas-separation process. In some applications, pressure is essentially "free," such as when remov-

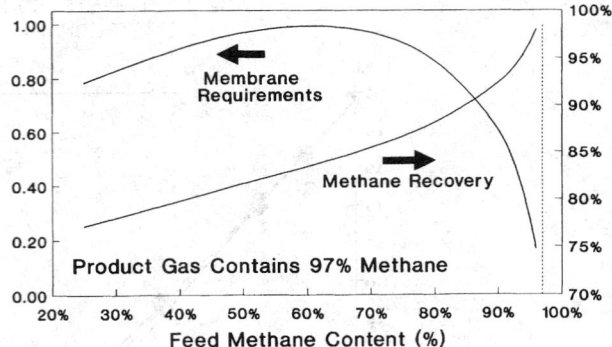

FIG. 22-77 Influence of feed purity on total membrane area when the residue gas at fixed purity is the product. Feed-gas volume is constant. CO_2/CH_4 cellulose-acetate membrane, $\alpha = 21$. (*Courtesy W. R. Grace.*)

ing small quantities of CO_2 from natural gas. The gas is often produced at pressure, but is compressed for transmission anyway, and since the residue constitutes the product, it continues downstream at pressure. H_2 frequently enters the separation process at pressure, an advantage for membranes. Unlike CH_4, the H_2 is in the permeate, and recompression may be a significant cost. A relative area cure for CO_2/CH_4 is shown in Fig. 22-77. When the permeate is the product (H_2, CO_2) the increasing membrane area shown in Fig. 22-78 is largely the effect of more gas to pass through the membrane, since the curve is based on a constant volume of feed gas, not a constant output of H_2. The facts of life in compressor economics are in painful opposition to the desires of the membrane designer. Pressure ratios higher than six become expensive; vacuum is very expensive, and scale is important. Because of compressor economics, staging membranes with recompression is unusual. Designers can assume that a flow sheet that mixes unlike streams or reduces pressure through a throttling valve will increase cost in most cases.

Product Losses Account must be taken of the product loss, the slow gas in the permeate (such as CH_4), or the fast gas in the residue (such as H_2). Figure 22-79 illustrates the issue for a membrane used to purify natural gas from 93 percent to pipeline quality, 98 percent. In the upper figure, the gas is run through a permeator bank operating as a single stage. For the membrane and module chosen, the permeate contains 63 percent CH_4. By dividing the same membrane area into two stages, two permeates (or more) may be produced, one of which may have significantly higher economic value than the single mixed permeate. In fact, where CH_4 is involved, another design parameter is the local economic value of various waste streams as fuel.

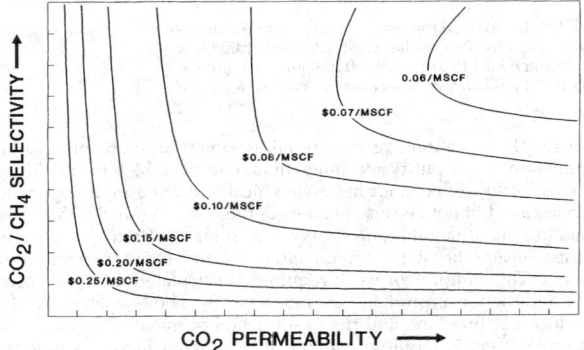

FIG. 22-76 Constant-cost lines as a function of permeability and selectivity for CO_2/CH_4. Cellulose-acetate membrane "mscf" is one thousand standard cubic feet. (*Courtesy W. R. Grace.*)

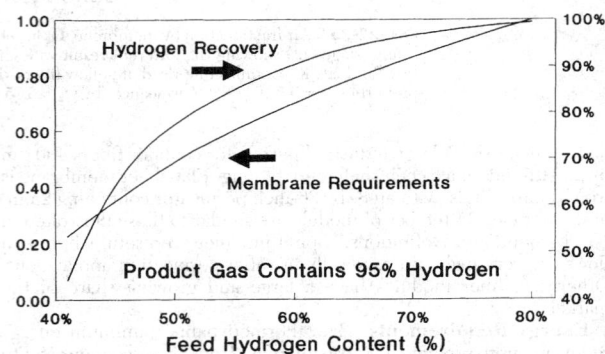

FIG. 22-78 Influence of feed purity on total membrane area when the permeate gas at fixed purity is the product. Feed-gas volume is constant. H_2/CH_4 cellulose-acetate membrane, $\alpha = 45$. (*Courtesy W. R. Grace.*)

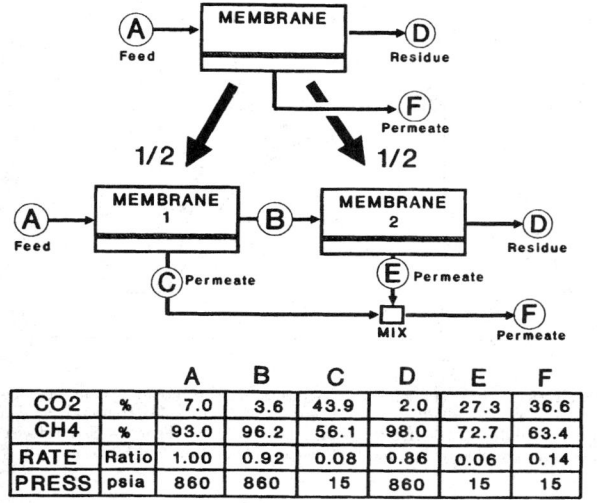

		A	B	C	D	E	F
CO2	%	7.0	3.6	43.9	2.0	27.3	36.6
CH4	%	93.0	96.2	56.1	98.0	72.7	63.4
RATE	Ratio	1.00	0.92	0.08	0.86	0.06	0.14
PRESS	psia	860	860	15	860	15	15

FIG. 22-79   Effect on permeate of dividing a one-stage separation into two equal stages having the same total membrane area. Compositions of A, D, and F are equal for both cases. (*Courtesy W. R. Grace.*)

Membrane Replacement   Membrane replacement is a significant cost factor, but membrane life and reliability are now reasonable. Membranes are more susceptible to operating upsets than more traditional equipment, but their field-reliability record in properly engineered, properly maintained installations is good to excellent. In N_2 separations, membrane life is very long.

Competing Technologies   The determination of which separation technique is best for a specific application is a dynamic function of advances in membranes and several other technologies. At this writing, very small quantities of pure components are best obtained by purchase of gas in cylinders. For N_2, membranes become competitive at moderate flow rates where purity required is <99.5 percent. At higher flow rates and higher purity, *pressure swing adsorption* (PSA) is better (see Sec. 16: "Adsorption and Ion Exchange"). At still higher volumes, delivered liquid N_2, pipeline N_2, or on-site distillation will be superior. For O_2, membranes have little economic importance. The problem is the economic cost of using vacuum on the permeate side, or the equally unattractive prospect of compressing the feed and operating at low stage cut. For H_2, membranes are dominant when the feed is at high pressure, except for high purity (excepting Pd noted above) and very large volume. Higher purity or lower feed pressure

favor PSA. Very high volume favors cryogenic separation. For CH_4, membranes compete at low purity and near pipeline (98 percent) purity, but distillation and absorption are very competitive at large scale and for intermediate CO_2 contamination levels.

Membranes are found as adjuncts to most conventional processes, since their use can improve overall economics in cases where membrane strength coincides with conventional process weakness.

PERVAPORATION

Process Description   **Pervaporation** is a separation process in which a liquid mixture contacts a nonporous permselective membrane. One component is transported through the membrane preferentially. It evaporates on the downstream side of the membrane leaving as a vapor. The name is a contraction of permeation and evaporation. Permeation is induced by lowering partial pressure of the permeating component, usually by vacuum or occasionally with a sweep gas. The permeate is then condensed or recovered. Thus, three steps are necessary: Sorption of the permeating components into the membrane, diffusive transport across the nonporous membrane, then desorption into the permeate space, with a heat effect. Pervaporation membranes are chosen for high selectivity, and the permeate is often highly purified.

In the flow schematic (Fig. 22-80), the condenser controls the vapor pressure of the permeating component. The vacuum pump, as shown, pumps both liquid and vapor phases from the condenser. Its major duty is the removal of noncondensibles. Early work in pervaporation focused on organic-organic separations. Many have been demonstrated; few if any have been commercialized. Still, there are prospects for some difficult organic separations.

An important characteristic of pervaporation that distinguishes it from distillation is that it is a rate process, not an equilibrium process. The more permeable component may be the less-volatile component. Pervaporation has its greatest utility in the resolution of azeotropes, as an adjunct to distillation. Selecting a membrane permeable to the minor component is important, since the membrane area required is roughly proportional to the mass of permeate. Thus pervaporation devices for the purification of the ethanol-water azeotrope (95 percent ethanol) are always based on a hydrophilic membrane.

Pervaporation membranes are of two general types. **Hydrophilic membranes** are used to remove water from organic solutions, often from azeotropes. **Hydrophobic membranes** are used to remove organic compounds from water. The important operating characteristics of hydrophobic and hydrophilic membranes differ. Hydrophobic membranes are usually used where the solvent concentration is about 6 percent or less—above this value, other separations methods are usually cheaper unless the flow rate is small. At low-solvent levels the usual membrane (silicone rubber) is not swollen appreciably, and movement of solvent into the membrane makes depletion of solvent

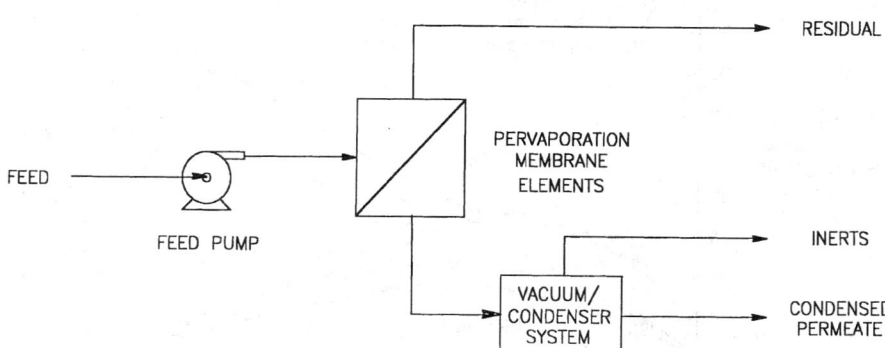

FIG. 22-80   Simplified flow schematic for a pervaporation system. Heated feed enters from left through a feed pump. Heaters in a recirculating feed loop may be required (not shown). Stripped liquid exits at the top of the pervaporation membrane. Vapor exits at the bottom to a condenser. Liquid and noncondensibles are removed under vacuum. (*Courtesy Hoechst Celanese.*)

in the boundary layer (concentration polarization) an important design problem. Hydrophilic pervaporation membranes operate such that the upstream portion is usually swollen with water, while the downstream face is low in water concentration because it is being depleted by vaporization. Fluxes are low enough (<5 kg/m²·hr) that boundary layer depletion (liquid side) is not limiting.

The simplifying assumptions that make Fick's law useful for other processes are not valid for pervaporation. The activity gradient across the membrane is far more important than the pressure gradient. Equation (22-110) is generally used to describe the pervaporation process:

$$J_i = -D_i C \frac{d \ln a_i}{dz} \qquad (22\text{-}110)$$

where a_i is the activity coefficient of component i.

This equation is not particularly useful in practice, since it is difficult to quantify the relationship between concentration and activity. The Flory-Huggins theory does not work well with the cross-linked semi-crystalline polymers that comprise an important class of pervaporation membranes. Neel (in Noble and Stern, op. cit., pp. 169–176) reviews modifications of the Stefan-Maxwell approach and other equations of state appropriate for the process.

A typical permeant-concentration profile in a pervaporation membrane is shown in Fig. 22-81. The concentration gradient at the permeate (vapor side) of the hydrophilic membrane is usually rate controlling. Therefore the downstream pressure (usually controlled by condenser temperature) is very important for flux and selectivity. Since selectivity is the ratio of fluxes of the components in the feed, as downstream pressure increases, membrane swelling at the permeate interface increases, and the concentration gradient at the permeate interface decreases. The permeate flux drops, and the more swollen membrane is less selective. A rise in permeate pressure may result in a drastic drop in membrane selectivity. This effect is diminished at low water concentrations, where membrane swelling is no longer dominant. In fact, when the water concentration drops far enough, permeate backpressure looses its significance. (See Fig. 22-82.)

For rubbery membranes (hydrophobic), the degree of swelling has less effect on selectivity. Thus the permeate pressure is less critical to the separation, but it is critical to the driving force, thus flux, since the vapor pressure of the organic will be high compared to that of water.

Definitions   Following the practice presented under "Gas-Separation Membranes," distillation notation is used. Literature articles often use mass fraction instead of mole fraction, but the substitution of one to the other is easily made.

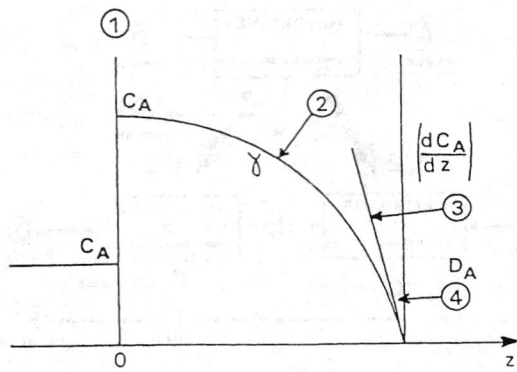

FIG. 22-81   Permeant-concentration profile in a pervaporation membrane. 1—Upstream side (swollen). 2—Convex curvature due to concentration-dependent permeant diffusivity. 3—Downstream concentration gradient. 4—Exit surface of membrane, depleted of permeant, thus unswollen. (*Courtesy Elsevier.*)

$$\beta \equiv \frac{y_i}{x_i} \qquad (22\text{-}111)$$

where β is the enrichment factor. β is related to α, [Equation (22-104)] the separation factor by

$$\beta = \frac{\alpha}{1 + (\alpha - 1)x_i} \qquad \alpha = \frac{\beta(1 - x_i)}{1 - x_i\beta} \qquad (22\text{-}112)$$

α is larger than β, and conveys more meaning when the membrane approaches the ideal. β is preferred in pervaporation because it is easier to use in formulations for cost, yield, and capacity. In fact, note that neither factor is constant, and that both generally change with x_i and temperature.

Operational Factors   In industrial use, pervaporation is a continuous-flow single-stage process. Multistage cascade devices are unusual. Pervaporation is usually an adjunct separation, occasionally a principal one. It is used either to break an azeotrope or to concentrate a minor component. Large stand-alone uses may develop in areas where distillation is at a disadvantage, such as with closely boiling components, but these developments, have not yet been made. Notable exceptions occur when a stream is already fairly pure, such as a contaminated water source, or isopropanol from microelectronics fabrication washing containing perhaps 15 percent water.

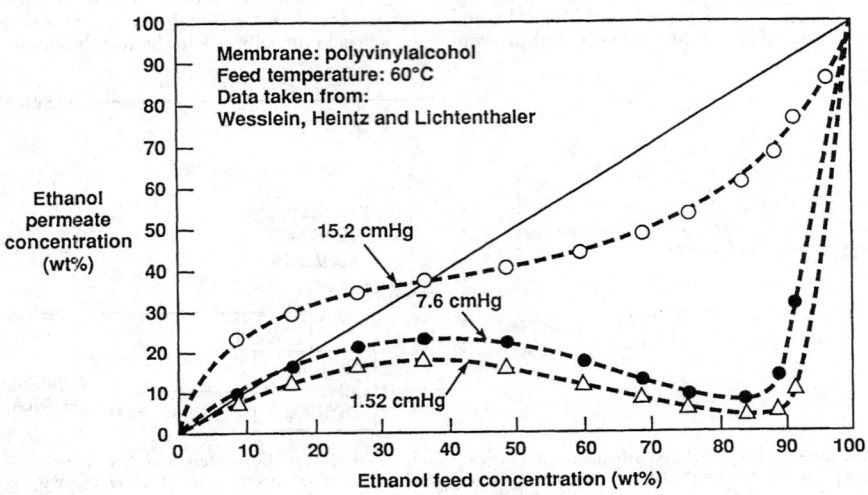

FIG. 22-82   Pervaporation schematic for ethanol-water. The illustration shows the complex behavior for a simple system at three pressures. Only the region above 90 percent is of commercial interest. (*Courtesy Elsevier.*)

In continuous flow, the feed will begin with concentration ($x_{i,\text{feed}}$) and be stripped until it reaches some desired ($x_{i,\text{residue}}$). If significant mass is removed from the feed, it will cool, since the general rule for liquids is that the latent heat exceeds the specific heat per °C by two orders of magnitude (more for water). So the membrane has a concentration gradient and a temperature gradient both across it and along it. Pervaporation is a complex multigradient process.

The term β is not constant for some important separations. Even worse, it can exhibit maxima that make analytic treatment difficult. Operating diagrams are often used for preliminary design rather than equations. Because of the very complex behavior in the membrane as concentrations change, all design begins with an experiment. For water removal applications, design equations often mispredict the rate constants as the water content of the feed approaches zero. The estimates tend to be lower than the experimental values, which would lead to overdesign. Therefore it is necessary to obtain experimental data over the entire range of water concentrations encountered in the separation. Once the kinetic data are available, heat transfer and heat capacity are the problem. It is general practice to pilot the separation on a prototype module to measure the changes due to thermal effects. This is particularly true for water as the permeant, given its high latent heat.

For removal of an organic component from water, swelling of the organophillic membrane would result in a higher flux and lower α. At organic levels below about 10 percent, that has not been a major problem. In most applications, boundary-layer mass-transfer limitations become limiting. Pilot data must therefore be taken with hydrodynamic similarity to the module that will be used and the actual organic permeation rate may become limited by the boundary layer. In organic-organic pervaporation, membrane swelling is a major concern.

Vapor Feed A variant on pervaporation is to use vapor, rather than liquid, as a feed. While the resulting process could be classified along with gas-separation membranes, it is customarily regarded as pervaporation.

The residue at the top of a distillation column is a vapor, so there is logic in using it as the feed to a membrane separator. Weighed against the obvious advantage are disadvantages of vapor handling (compressors cost more than pumps) and equipment size to handle the larger vapor volume. When the more volatile component permeates, heat must be added to maintain superheat. The vapor feed technique has been used in a few large installations where the advantages outweigh the disadvantages.

Leading Examples

Dehydration The growing use of isopropanol as a clean-rinse fluid in microelectronics produces significant quantities of an 85–90 percent isopropanol waste. Removing the water and trace contaminants is required before the alcohol can be reused. Pervaporation produces a 99.99 percent alcohol product in one step. It is subsequently polished to remove metals and organics. In Europe, dehydration of ethanol is the largest pervaporation application. For the very large ethanol plants typical of the United States, pervaporation is not competitive with thermally integrated distillation.

Organic from Water An area where pervaporation may become important is in flavors, fragrances, and essential oils. Here, high-value materials with unique properties are recovered from aqueous or alcohol solutions.

Pollution Control Pervaporation is used to reduce the organic loading of a waste stream, thus effecting product recovery and a reduction in waste-treatment costs. An illustration is a waste stream containing 11 percent (wt) n-propanol. The residue is stripped to 0.5 percent and 96 percent of the alcohol is recovered in the permeate as a 45 percent solution. This application uses a hydrophobic, rubbery membrane. The residue is sent to a conventional waste-treatment plant.

Pervaporation Membranes Pervaporation has a long history, and many materials have found use in pervaporation experiments. Cellulosic-based materials have given way to polyvinyl alcohol and blends of polyvinyl alcohol and acrylics in commercial water-removing membranes. These membranes are typically solution cast (from water) on ultrafiltration membrane substrates. It is important to have enough cross linking in the final polymer to avoid dissolution of the membrane in use. A very thin membrane with little penetration into the UF substrate is required. The substrate can be a problem, as it provides significant pressure drop to the vapor passing through it which, as mentioned above, has serious ramifications for flux and separation efficiency. Many new membranes are under development. Ion-exchange membranes, and polymers deposited by or polymerized by plasma, are frequently mentioned in the literature.

Modules Every module design used in other membrane operations has been tried in pervaporation. One unique requirement is for low hydraulic resistance on the permeate side, since permeate pressure is very low (0.1–1 Pa). The rule for near-vacuum operation is the bigger the channel, the better the transport. Another unique need is for heat input. The heat of evaporation comes from the liquid, and intermediate heating is usually necessary. Of course economy is always a factor. Plate-and-frame construction was the first to be used in large installations, and it continues to be quite important. Some smaller plants use spiral-wound modules, and some membranes can be made as capillary bundles. The capillary device with the feed on the inside of the tube has many advantages in principle, such as good vapor-side mass transfer and economical construction, but it is still limited by the availability of membrane in capillary form.

SELECTION OF BIOCHEMICAL SEPARATION PROCESSES

GENERAL REFERENCES: Albertsson, *Partition of Cell Particles and Macromolecules,* 3d ed., John Wiley & Sons, New York, 1986. Belter, Cussler, and Hu, *Bioseparations,* Wiley Interscience, New York, 1988. Cooney and Humphrey (eds.), *Comprehensive Biotechnology,* vol. 2, Pergamon, Oxford, 1985. Janson and Ryden (eds.), *Protein Purification,* VCH, New York, 1989. Scopes, *Protein Purification,* 2d ed., Springer-Verlag, New York, 1987. Stephanopoulos (ed.), *Biotechnology,* 2d ed., vol. 3, VCH, Weinheim, 1993. Walter, Brooks, and Fisher (eds.), *Partitioning in Aqueous Two-Phase Systems,* Academic Press, Orlando, FL, 1985.

GENERAL BACKGROUND

The biochemical industry derives its products from two primary sources. Natural products are yielded by plants, animal tissue, and fluids, and obtained via fermentation from bacteria, molds and fungi, and from mammalian cells. Products can also be obtained by recombinant methods through the insertion of foreign DNA directly into the hosting microorganism to allow overproduction of the product in this unnatural environment. The range of bioproducts is enormous, and the media in which they are produced are generally complex and ill-defined, containing many unwanted materials in addition to the desired product. The product is invariably at low concentration. The goals of downstream processing operations include removal of these unwanted impurities, bulk-volume reduction with concomitant concentration of the desired product, and, for protein products, transfer of the protein to an environment where it will be stable and active, ready for its intended application. This requires on average three to six separate processing steps. Some of the purification methods in general use are shown in Fig. 22-83 [Bonnerjea et al., *Bio/Technology,* **4,** 954–958 (1986)] which indicates the average stage at which different methods are used in the downstream processing train. A general strategy for downstream processing of biological materials and the types of operations that may be used in the different steps is shown in Fig. 22-84 [see also Ho, in M. R. Ladisch et al. (eds.), *Protein Purification from Molecular Mechanisms to Large-Scale Processes,* ACS Symp. ser. 427, ACS, Washington DC, 1990, pp. 14–34]. Section 24 in this handbook provides a general discussion of biochemical engineering.

Low-molecular-weight products, generally secondary metabolites such as alcohols, carboxylic and amino acids, antibiotics, and vitamins,

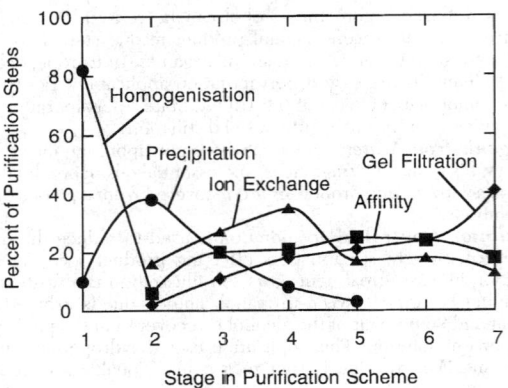

FIG. 22-83 Frequency of use of purification methods at different stages in purification schemes published in the literature. (*Adapted from Bonnerjea et al., op. cit.*)

can be recovered using many of the standard operations such as liquid-liquid extraction, adsorption and ion-exchange, described elsewhere in this handbook. Proteins require special attention, however, as they are sufficiently more complex, their function depending on the integrity of a delicate three-dimensional tertiary structure that can be disrupted if the protein is not handled correctly. For this reason, this section focuses primarily on protein separations. Cell separations, as a necessary part of the downstream processing sequence, are also covered.

Techniques used in bioseparations depend on the nature of the product (i.e., the unique properties and characteristics which provide a "handle" for the separation), and on its state (i.e., whether soluble or insoluble, intra- or extracellular, etc.). All early isolation and recovery steps remove whole cells, cellular debris, suspended solids, and colloidal particles, concentrate the product, and, in many cases, achieve some degree of purification, all the while maintaining high yield. For intracellular compounds, the initial harvesting of the cells is important for their concentration prior to release of the product. Following this phase, a range of purification steps is employed to remove the remaining impurities and enhance the product purity; this purification phase, in turn, is followed by polishing steps to remove the last traces of con-

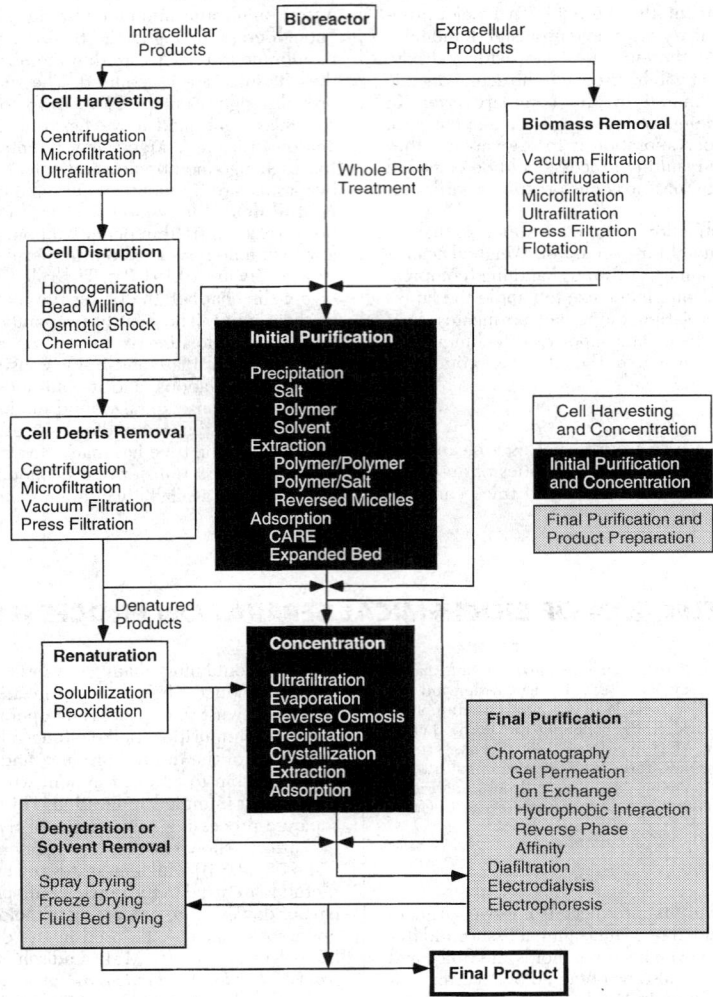

FIG. 22-84 General stages in downstream processing for protein production indicating representative types of operations used at each stage.

taminating components and process-related additions (e.g., buffer salts, detergents, etc.), and to prepare the product for storage and/or distribution. The prevention and/or avoidance of contamination is another important goal of downstream processing.

Even for good yields of 80 to 95 percent per step, the overall yield can be poor for any process that requires a large number of steps. Thus, careful consideration must be given to optimization of the process in terms of both the unit operations themselves, and their sequencing. It is usually desirable to reduce the process volume early in the downstream processing, and to remove any components that can be removed fairly easily (particulates, small solutes, large aggregates, nucleic acids, etc.) so as not to overly burden the more refined separation processes downstream. Possible shear and temperature damage, and deactivation by endogenous proteases must be considered in the selection of separation processes.

The purification of proteins to be used for therapeutic purposes presents more than just the technical problems associated with the separation process. Owing to the complex nature and intricate three-dimensional structure, the routine determination of protein structure as a quality-control tool, particularly in its final medium for use, is not well established. In addition, the complex nature of the human immune system allows for even minor quantities of impurities and contaminants to be biologically active. Thus, regulation of biologics production has resulted in the concept of the process defining the product since even small and inadvertent changes in the process may affect the safety and efficacy of the product. Indeed, it is generally acknowledged that even trace amounts of contaminants introduced from other processes, or contaminants resulting from improper equipment cleaning can compromise the product. From a regulatory perspective, then, operations should be chosen for more than just efficiency. The consistency of the unit operation, particularly in the face of potentially variable feed from the culture/fermentation process, is the cornerstone of the process definition. Operations that lack robustness or are subject to significant variation should not be considered. Another aspect of process definition is the ability to quantify the operation's performance. Finally, the ease with which the equipment can be cleaned in a verifiable manner should play a role in unit-operation selection.

In the development of new products, optimization of the fermentation medium for titer only often ignores the consequences of the medium properties on subsequent downstream processing steps such as filtration and chromatography. It is imperative, therefore, that there be effective communication and understanding between workers on the upstream and downstream phases of the product development if rational trade-offs are to be made to ensure overall optimality of the process. One example is to make the conscious decision, in collaboration with those responsible for the downstream operations, whether to produce a protein in an unfolded form or in its native folded form; the purification of the aggregated unfolded proteins is simpler than that of the native protein, but the refolding process itself to obtain the product in its final form may lack scalability.

INITIAL PRODUCT HARVEST AND CONCENTRATION

The initial processing steps are determined to a large extent by the location of the product species, and generally consist of cell/broth separation and/or cell-debris removal. For products retained within the biomass during production, it is first necessary to concentrate the cell suspension before homogenization or chemical treatment to release the product. Clarification to remove the suspended solids is the process goal at this stage.

Regardless of the location of the protein and its state, cell separation needs to be inexpensive, simple, and reliable, as large amounts of fermentation-broth dilute in the desired product may be handled. The objectives are to obtain a well-clarified supernatant and solids of maximum dryness, avoiding contamination by using a contained operation. Mechanical methods, almost exclusively centrifugation and cross-flow filtration, are preferred for cell separations [Datar and Rosen, in Stephanopoulos (ed.), op. cit., pp. 369–503].

Intracellular products can be present either as folded, soluble proteins, or as dense masses of unfolded protein (inclusion bodies). For these products, it is first necessary to concentrate the cell suspension before effecting release of the product. Filtration can result in a suspension of cells that can be of any desired concentration up to 15–17 percent, and that can be diafiltered into the desired buffer system. In contrast, the cell slurry that results from centrifugation will either be that of a dry mass (requiring resuspension but substantially free of residual broth, i.e., from a tubular bowl centrifuge) or a wet slurry (containing measurable residual broth and requiring additional resuspension). During the separation, conditions which result in cell lysis (such as extremes in temperature) must be avoided. In addition, while soluble protein is generally protected from shear and external proteolysis, these proteins are still subject to thermal denaturation.

For extracellular products, which are invariably water-soluble, the goal is the removal of whole cells (clarification) and, in the case of typical protein products, the removal of dissolved low-molecular-weight compounds. This must be done under relatively gentle conditions to avoid undesired denaturation of the product. Again, either filtration or centrifugation can be applied. Filtration results in a cell-free supernatant with dilution associated with the diafiltration of the final cell slurry. Centrifugation, regardless of the mode, will result in a small amount of cells in the centrate, but there is no dilution of the supernatant. During the process development careful studies should be conducted to examine the effects of pH and ionic strength on the yield, as cells and cell debris may retain the product through charge interactions. If the broth or cell morphology does not allow for filtration or if dry cell mass is required, tubular-bowl centrifugation is typically utilized. It should be noted that plant and animal cells cannot sustain the same degree of applied shear as microbial cells, and thus cross-flow filtration or classical centrifugation may not be applicable. Alternatives using low-shear equipment under gentle conditions are often employed in these situations.

For whole broths the range of densities and viscosities encountered affect the concentration factor that can be attained in the process, and can also render cross-flow filtration uneconomical because of the high pumping costs, and so on. Often, the separation characteristics of the broth can be improved by broth conditioning using physiocochemical or biological techniques, usually of a proprietary nature. The important characteristics of the broth are rheology and conditioning.

Centrifugation Centrifugation relies on the enhanced sedimentation of particles of density different from that of the surrounding medium when subjected to a centrifugal-force field [Axelsson, in Cooney and Humphrey (eds.), op. cit., pp. 325–346] (see also Sec. 18: "Liquid-Solid Operations and Equipment"). Advantages of centrifugal separations are that they can be carried out continuously and have short retention times, from a fraction of a second to seconds, which limit the exposure time of sensitive biologicals to shear stresses. Yields are high, provided that temperature and other process conditions are adequately controlled. They have small space requirements, and an adjustable separation efficiency makes them a versatile unit operation. They can be completely closed to avoid contamination, and, in contrast to filtration, no chemical external aids are required that can contaminate the final product. The ability now to contain the aerosols typically generated by centrifuges adds to their operability and safety.

Sedimentation rates must be sufficiently high to permit separation, and can be enhanced by modifying solution conditions to promote the aggregation of proteins or impurities. An increase in precipitation of the contaminating species can often be accomplished by a reduction in pH, or an elevation in temperature. Flocculating agents, which include polyelectrolytes, polyvalent cations, and inorganic salts, can cause a 2000-fold increase in sedimentation rates. Some examples are polyethylene imine, EDTA, and calcium salts. Cationic bioprocessing aids (cellulosic or polymeric) reduce pyrogen, nucleic acid, and acidic protein loads which can foul chromatography columns. The removal of these additives both during centrifugation and subsequent processing must be clearly demonstrated.

There are many different types of centrifuges, classified according to the way in which the transport of the sediment is handled. The selection of a particular centrifuge type is determined by its capacity for handling sludge; the advantages and disadvantages of various separator types are discussed by Axelsson [in Cooney and Humphrey (eds.), op. cit., pp. 325–346]. Solids-retaining centrifuges are operated in a semibatch mode, as they must be shut down periodically to remove the accumulated solids; they are primarily used when solids concentrations are low, and have found application during the clarification and simultaneous separation of two liquids. In solids-ejecting centrifuges, the solids are removed intermittently either through radial slots or axially while the machine is running at full speed. These versatile machines can be used to handle a variety of feeds, including yeast, bacteria, mycelia, antibiotics, enzymes, and so on. Solids-discharging nozzle centrifuges have a large capacity, and can accommodate up to 30 percent solids loading. Decanter centrifuges consist of a drum, partly cylindrical and partly conical, and an internal screw conveyor for transport of the solids, which are discharged at the conical end; liquids are discharged at the cylindrical end. Levels within the drum are set by means of external nozzles.

Continuous-flow units, the scroll decanter and disk-stack centrifuges, are easiest to use from an operational perspective; shutdown of the centrifuge during the processing of a batch is not expected. While the disk-stack centrifuge enjoys popularity as a process instrument within the pharmaceutical and biotechnology industries, the precise timing of solids ejection and continuous high-speed nature of the device make for complex equipment and frequent maintenance. It is often used to harvest cells, since the solids generated are substantially wet and could lead to measurable yield losses in extracellular product systems. For intracellular product processing, the wet cell sludge is easily resuspended for use in subsequent processing.

The tubular bowl, in contrast, is a semibatch processing unit owing to the limited solids capacity of the bowl. The use of this unit requires shutdown of the centrifuge during the processing of the batch. The semibatch nature of these centrifuges can thus greatly increase processing-cycle times. The introduction of disposable sheets to act as bowl liners has significantly impacted turnaround times during processing. The dry nature of the solids generated makes the tubular-bowl centrifuge well-suited for extracellular protein processing, since losses to the cell sludge are minimal. In contrast, the dry, compact nature of the sludge can make the cells difficult to resuspend. This can be problematic for intracellular protein processing where cells are homogenized in easily clogged, mechanical disrupters.

Filtration   Cross-flow filtration (microfiltration includes cross-flow filtration as one mode of operation in "Membrane Separation Processes" which appears earlier in this section) relies on the retention of particles by a membrane. The driving force for separation is pressure across a semipermeable membrane, while a tangential flow of the feed stream parallel to the membrane surface inhibits solids settling on and within the membrane matrix (Datar and Rosen, loc. cit.).

Microfiltration is used for the removal of suspended particles, recovery of cells from fermentation broth, and clarification of homogenates containing cell debris. Particles removed by microfiltration typically average greater than 500,000 nominal molecular weight [Tutunjian, in Cooney and Humphrey (eds.), op. cit., pp. 367–381; Gobler, in Cooney and Humphrey (eds.), op. cit., pp. 351–366]. Ultrafiltration focuses on the removal of low-molecular-weight solutes and proteins of various sizes, and operates in the less than 100,000 nominal-molecular-weight cutoff range [Le and Howell, in Cooney and Humphrey (eds.), op. cit., pp. 383–409]. Both operations consist of a concentration segment (of the larger particles) followed by diafiltration of the retentate [Tutunjian, in Cooney and Humphrey (eds.), op. cit., pp. 411–437].

Generally, the effectiveness of the separation is determined not by the membrane itself, but rather by the formation of a secondary or dynamic membrane caused by interactions of the solutes and particles with the membrane. The buildup of a gel layer on the surface of an ultrafiltration membrane owing to rejection of macromolecules can provide the primary separation characteristics of the membrane. Similarly, with colloidal suspensions, pore blocking and bridging of pore entries can modify the membrane performance, while molecules of size similar to the membrane pores can adsorb on the pore walls, thereby restricting passage of water and smaller solutes. Media containing poorly defined ingredients may contain suspended solids, colloidal particles, and gel-like materials that prevent effective microfiltration. In contrast to centrifugation, specific interactions can play a significant role in membrane-separation processes.

The factors to consider in the selection of cross-flow filtration include the cross-flow velocity, the driving pressure, the separation characteristics of the membrane (permeability and pore size), size of particulates relative to the membrane pore dimensions, and the hydrodynamic conditions within the flow module. Again, since particle-particle and particle-membrane interactions are key, broth conditioning (ionic strength, pH, etc.) may be necessary to optimize performance.

Selection of Cell-Separation Unit Operation   The unit operation selected for cell separations can depend on the subsequent separation steps in the train. In particular, when the operation following cell separation requires cell-free feed (e.g. chromatography), filtration is used, since centrifugation is not absolute in terms of cell separation. In addition, if cells are to be stored (i.e., they contain the desired product) because later processing is more convenient (e.g., only two-shift operation, facility campaigns equipment with other products, batch is too big for single pass in equipment), it is generally better to store the cells as a frozen concentrate rather than a paste, since the concentrate thaws more completely, avoiding small granules of unfrozen cell solids that can foul homogenizers, columns, and filters. Here the retentate from filtration is desired, although the wet cell mass from a disc stack-type centrifuge may be used.

Centrifugation is generally necessary for complex media used to make natural products, for, while the media components may be sifted prior to use, they can still contain small solids that can easily foul filters. The medium to be used should be tested on a filter first to determine the fouling potential. Some types of organisms, such as filamentous organisms may sediment too slowly owing to their larger cross sections, and are better treated by filtration (mycelia have the potential to easily foul tangential-flow units; vacuum-drum filtration using a filter aid, e.g., diatomaceous earth, should also be considered). Often the separation characteristics of the broth can be improved by broth conditioning using physicochemical or biological techniques, usually of a proprietary nature.

Regardless of the machine device, centrifuges are typically maintenance-intensive. Filters can be cheaper in terms of capital and maintenance and should be considered first unless centrifugal equipment already exists. Small facilities (<1000 liters) use filtration, since centrifugation scale-down is constrained by equipment availability. Comparative economics of the two classes of operations are discussed by Datar and Rosen (loc. cit.).

Cell Disruption   Intracellular protein products are present as either soluble, folded proteins or inclusion bodies. Release of folded proteins must be carefully considered. Active proteins are subject to deactivation and denaturation, and thus require the use of "gentle" conditions. In addition, due consideration must be given to the suspending medium; lysis buffers are often optimized to promote protein stability and protect the protein from proteolysis and deactivation. Inclusion bodies, in contrast, are protected by virtue of the protein agglomeration. More stressful conditions are typically employed for their release, which includes going to higher temperatures if necessary. For "native" proteins, gentler methods and temperature control are required.

The release of intracellular protein product is achieved through rupture of the cell walls, and release of the protein product to the surrounding medium, through either mechanical or nonmechanical means, or through chemical, physical, or enzymatic lysis [Engler, in Cooney and Humphrey (eds.), op. cit., pp. 305–324; Schutte and Kula, in Stephanopoulos (ed.), op. cit., pp. 505–526]. Mechanical methods use pressure, as in the Manton/APV-Gaulin/French Press, or the Microfluidizer, or mechanical grinding, as in ball mills, the latter being used typically for flocs and usually only for natural products. Nonmechanical means include use of desiccants or solvents, while cell lysis can also be achieved through physical means (osmotic shock, freeze/thaw cycles), chemical (detergents, chaotropes) or enzymatic (lysozyme, phages).

In pressure-based homogenizers, cells suspended in an aqueous medium are forced at high velocity through a narrow, adjustable gap between a valve and its seat at pressures in excess of 50 MPa. Product release, which generally follows first-order kinetics, occurs through impingement of the high-velocity cell-suspension jet on the stationary surfaces, and possibly also by the high-shear forces generated during the acceleration of the liquid through the gap. While sufficiently high pressures can be attained using commercially available equipment to ensure good release in a single pass, the associated adiabatic temperature increases (~1.8°C/1000 psig) may cause unacceptable activity losses for heat-labile proteins. Further denaturation can occur on exposure to the lysis medium. Thus, multiple passes may be preferred, with rapid chilling of the processed-cell suspension between passes. The number of passes and the heat removal ability should be carefully optimized. The efficiency of the process depends on the homogenizing pressure and the choice of the valve unit, for which there are many designs available. Materials of construction are important to minimize erosion of the valve, to provide surface resistance to aggressive cleaning agents and disinfectants, and to permit steam cleaning and sanitization.

The release of inclusion bodies, in contrast, may follow a different strategy. Since inclusion bodies are typically recovered by centrifugation, it is often advantageous to send the lysate through the homogenizer with multiple passes to decrease the particle size of the cell debris. Since the inclusion bodies are much denser than the cell debris, the debris, now much reduced in size, can be easily separated from the inclusion bodies by centrifugation at low speeds. The inclusion bodies may be resuspended and centrifuged multiple times (often in the presence of low concentrations of denaturants) to clean up these aggregates. Since the inclusion bodies are already denatured, temperature control is not as important as in the case of native proteins.

In high-speed-agitation ball mills, cells suspended with beads are agitated by disks rotating at high speed. Ball mills have longer residence times than the pressure homogenizers, and are susceptible to channeling and shedding of the ball material.

Chemical lysis, or solubilization of the cell wall, is typically carried out using detergents such as Triton X-100, or the chaotropes urea, and guanidine hydrochloride. This approach does have the disadvantage that it can lead to some denaturation or degradation of the product. While favored for laboratory cell disruption, these methods are not typically used at the larger scales. Enzymatic destruction of the cell walls is also possible, and as more economical routes to the development of appropriate enzymes are developed, this approach could find industrial application. Again, the removal of these additives is an issue.

Physical methods such as osmotic shock, in which the cells are exposed to high salt concentrations to generate an osmotic pressure difference across the membrane, can lead to cell-wall disruption. Similar disruption can be obtained by subjecting the cells to freeze/thaw cycles, or by pressurizing the cells with an inert gas (e.g., nitrogen) followed by a rapid depressurization. These methods are not typically used for large-scale operations.

On homogenization, the lysate may drastically increase in viscosity due to DNA release. This can be ameliorated to some extent using multiple passes to reduce the viscosity. Alternatively, precipitants or nucleic acid digesting enzymes can be used to remove these viscosity-enhancing contaminants.

For postlysis processing, careful optimization must be carried out with respect to pH and ionic strength. Often it is necessary to do a buffer exchange. Cell debris can act as an ion exchanger and bind proteins ionically, thus not allowing them to pass through a filtration device or causing them to be spun out in a centrifuge. Once optimal conditions are found, these conditions can be incorporated in the lysis buffer by either direct addition (if starting from cell paste) or diafiltration (if starting from a cell concentrate).

Protein Refolding The products of recombinant DNA technology are frequently not produced in their native, biologically active form, because the foreign hosts in which they are produced lack the appropriate apparatus for the folding of the proteins. Thus, the overproduced proteins are generally recovered as refractile or inclusion bodies, or aggregates, typically 1–3 μm in size, and all cysteine residues are fully reduced. It is necessary at some stage in the processing to dissolve the aggregates and then refold them to obtain the desired biologically active product [Cleland and Wang, in Stephanopoulos (ed.), op. cit., pp. 527–555].

Advantages of inclusion bodies in the production stage are their ease of separation by centrifugation following cell disruption because of their size and density, and their provision of excellent initial-purification possibilities, as long as impurities are not co-purified to any significant extent with the inclusion bodies. They also provide a high expression level and prevent endogenous proteolysis. There can be, however, significant product loss during protein refolding to the active form.

Following cell disruption, and washing, the pellet is solubilized through the disruption of hydrogen and ionic bonds, and hydrophobic interactions by the addition of chaotropes such as guanidine hydrochloride (4–9 M), urea (7–8 M), sodium thiocyanate (4–9 M), or detergents such as Triton X-100 or sodium dodecyl sulfate. This step may also require the breaking of all incorrectly formed intramolecular disulfide bonds through the addition of appropriate reducing agents (e.g., beta mercaptoethanol). To permit proper refolding of the protein, it is necessary to reduce the chaotrope concentration either by dilution, or by solvent exchange using dialysis against a buffer (poor scale-up potential), diafiltration using the desired buffers, or electrodialysis (good scale-up potential). High-protein concentrations can lead to aggregate formation, while if the concentration is too low, the volumes to be processed become inhibitive. Subsequently, oxidation of the cysteine residues is needed to allow for correct disulfide bond formation in the native protein.

INITIAL PURIFICATION

In initial purification steps the goal is to obtain concentration with partial purification of the product, which is recovered either as a precipitate (precipitation), a solution in a second phase (liquid-liquid extraction), or adsorbed to solids (adsorption, chromatography).

Precipitation Precipitation of products, impurities or contaminants can be induced by the addition of solvents, salts, or polymers to the solution, by increasing temperature, or by adjusting the solution pH (Scopes, op. cit., pp. 41–71; Ersson et al., in Janson and Ryden, op. cit., pp. 3–32). This operation is used most often in the early stages of the separation sequence, particularly following centrifugation, filtration, and/or homogenization steps. Precipitation is often carried out in two stages, the first to remove bulk impurities and the second to precipitate and concentrate the target protein. Generally, amorphous precipitates are formed, owing to occlusion of salts or solvents, or to the presence of impurities.

Salts can be used to precipitate proteins by "salting out" effects. The effectiveness of various salts is determined by the Hofmeister series, with anions being effective in the order citrate > $PO_4^=$ > $SO_4^=$ > CH_3COO^- > Cl^- > NO_3^-, and cations according to NH_4^+ > K^+ > Na^+ (Ersson et al., op. cit., p. 10; Belter et al., op. cit., pp. 221–236). Salts should be inexpensive owing to the large quantities used in precipitation operations. Ammonium sulfate is the most commonly used precipitant. Drawbacks to this approach include low selectivity, high sensitivity to operating conditions, and downstream complications associated with salt removal and disposal of the high-nitrogen-content stream. Generally, aggregates formed on precipitation with ammonium sulfate are fragile, and are easily disrupted by shear. Thus, these precipitation operations are, following addition of salt, often aged without stirring before being fed to a centrifuge by gravity feed or using low-shear pumps (e.g., diaphragm pumps).

The organic solvents most commonly used for protein precipitation are acetone and ethanol (Ersson et al., op. cit.). These solvents can cause some denaturation of the protein product. Temperatures below 0°C can be used, since the organic solvents depress the freezing point of the water. The precipitate formed is often an extremely fine powder that is difficult to centrifuge and handle. With organic solvents, in-line mixers are preferred, as they minimize solvent-concentration gradients, and regions of high-solvent concentrations, which can lead to significant denaturation and local precipitation of undesired components typically left in the mother liquors. In general, precipitation with organic solvents at lower temperature increases yield and reduces denaturation. It is best carried out at ionic strengths of 0.05–0.2 M.

Water-soluble polymers and polyelectrolytes (e.g., polyethylene glycol, polyethylene imine polyacrylic acid) have been used successfully in protein precipitations, and there has been some success in affinity precipitations wherein appropriate ligands attached to polymers can couple with the target proteins to enhance their aggregation. Protein precipitation can also be achieved using pH adjustment, since proteins generally exhibit their lowest solubility at their isoelectric point. Temperature variations at constant salt concentration allow for fractional precipitation of proteins.

Precipitation is typically carried out in standard cylindrical tanks with low-shear impellers. If in-line mixing of the precipitating agent is to be used, this mixing is employed just prior to the material entering the aging tank. Owing to their typically poor filterability, precipitates are normally collected using a centrifugal device.

Extraction Partitioning of the desired protein to a second phase in liquid-liquid extraction operations can be achieved using two-phase aqueous polymer systems (Albertsson, op. cit.), reversed micellar systems [Kelley and Hatton, in Stephanopoulos (ed.), op. cit., pp. 593–616], or phase-separated micellar solutions [Pryde and Phillips, *Biochem. J.*, **233**, 525–533 (1986)]. Organic solvents more typically identified with solvent-extraction operations cannot be used here because of protein solubility constraints, and because they lead to protein denaturation and degradation. Aqueous two-phase polymer systems are good for unclarified broths since particles tend to collect at the interface between the two phases, making their removal very efficient. They can also be used early on in the processing train for initial bulk-volume reduction and partial purification. These processes have not been applied widely as yet, owing both to a general lack of experience with these phase systems, and the need to remove phase-forming reagents (polymers, salts, detergents) from the products.

The basis for the separation is that when two polymers, or a polymer and certain salts, are mixed together in water, they are incompatible, leading to the formation of two immiscible but predominantly aqueous phases, each rich in only one of the two components [Albertsson, op. cit.; Kula, in Cooney and Humphrey (eds.), op. cit., pp. 451–471]. A phase diagram for a polyethylene glycol (PEG)-Dextran, two-phase system is shown in Fig. 22-85. Proteins are known to distribute unevenly between these phases. This uneven distribution can be used for the selective concentration and partial purification of the products. Partitioning between the two phases is controlled by the polymer molecular weight and concentration, protein net charge and size, and hydrophobic and electrostatic interactions. Affinity ligands covalently bonded to one of the phase-forming polymers have been found to be effective in enhancing dramatically the selectivity and partitioning behavior.

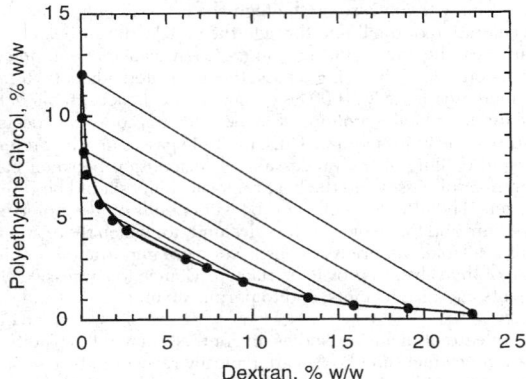

FIG. 22-85 Phase diagram for a PEG/Dextran, biphasic, aqueous-polymer system used in liquid-liquid extraction operations for protein separations. (*Albertsson, Partition of Cell Particles and Macromolecules, 3d ed., Copyright © 1986. Reprinted by permission of John Wiley & Sons, Inc.*)

Product recovery from these systems can be accomplished by either changes in temperature or system composition. Composition changes can be affected by dilution, back extraction and micro- and ultrafiltration. As the value of the product decreases, recovery of the polymer may take on added significance. A flow diagram showing one possible configuration for the extraction and product and polymer recovery operations is shown in Fig. 22-86 [Greve and Kula, *J. Chem. Tech. Biotechnol.*, **50**, 27–42 (1991)]. The phase-forming polymer and salt are added directly to the fermentation broth. The cells or cell debris and contaminating proteins report to the salt-rich phase and are discarded. Following pH adjustment of the polymer-rich phase, more salt is added to induce formation of a new two-phase system in which the product is recovered in the salt phase, and the polymer can be recycled. In this example, disk-stack centrifuges are used to enhance the phase-separation rates. Other polymer recycling options include extraction with a solvent or supercritical fluid, precipitation, or diafiltration. Electrodialysis can be used for salt recovery and recycling.

Reversed micellar solutions can also be used for the selective extraction of proteins (Kelley and Hatton, op. cit.). In these systems,

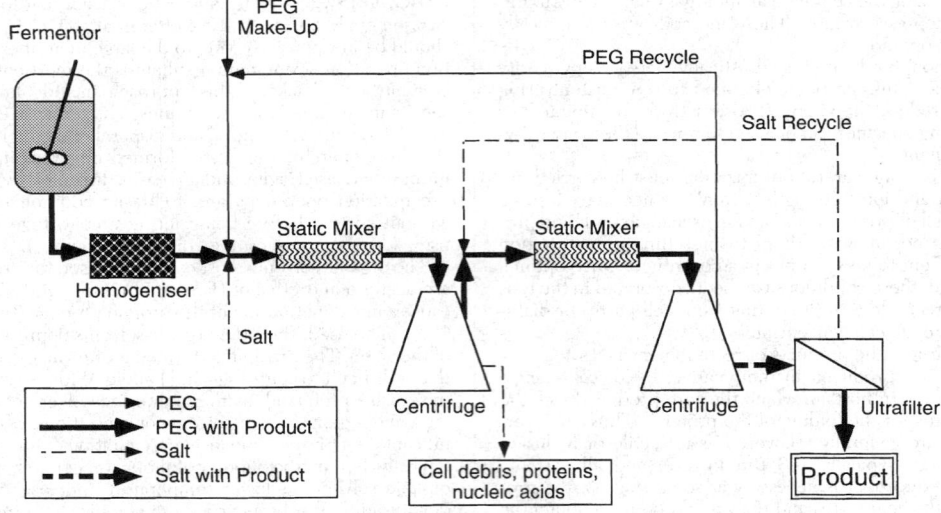

FIG. 22-86 Process scheme for protein extraction in aqueous two-phase systems for the downstream processing of intracellular proteins, incorporating PEG and salt recycling. [*Reprinted from Kelly and Hatton in Stephanopoulos (ed.), op. cit.; adapted from Greve and Kula, op. cit.*]

detergents soluble in an oil-phase aggregate to stabilize small water droplets having dimensions similar to those of the proteins to be separated. These droplets can host hydrophilic species such as proteins in an otherwise inhospitable organic solvent, thus enabling these organic phases to be used as protein extractants. Factors affecting the solubilization effectiveness of the solvents include charge effects, such as the net charge determined by the pH relative to the protein isoelectric point, charge distribution and asymmetry on the protein surface, and the type (anionic or cationic) of the surfactant used in the reversed micellar phase. Ionic strength and salt type affect the electrostatic interactions between the proteins and the surfactants, and also affect the sizes of the reversed micelles. Attachment of affinity ligands to the surfactants has been demonstrated to lead to enhancements in extraction efficiency and selectivity [Kelley et al., *Biotech. Bioeng.*, **42**, 1199–1208 (1993)].

Product recovery from reversed micellar solutions can often be attained by simple back extraction, by contacting with an aqueous solution having salt concentration and pH that disfavors protein solubilization, but this is not always a reliable method. Addition of cosolvents such as ethyl acetate or alcohols can lead to a disruption of the micelles and expulsion of the protein species, but this may also lead to protein denaturation. These additives must be removed by distillation, for example, to enable reconstitution of the micellar phase. Temperature increases can similarly lead to product release as a concentrated aqueous solution. Removal of the water from the reversed micelles by molecular sieves or silica gel has also been found to cause a precipitation of the protein from the organic phase.

Aqueous-detergent solutions of appropriate concentration and temperature can phase separate to form two phases, one rich in detergents, possibly in the form of micelles, and the other depleted of the detergent (Pryde and Phillips, op. cit.). Proteins distribute between the two phases, hydrophobic (e.g., membrane) proteins reporting to the detergent-rich phase and hydrophilic proteins to the detergent-free phase. Indications are that the size-exclusion properties of these systems can also be exploited for viral separations. These systems would be handled in the same way as the aqueous two-phase systems.

On occasion, for extracellular products, cell separation can be combined with an initial volume reduction and purification step using liquid-liquid extraction. This is particularly true for low-molecular-weight products, and has been used effectively for antibiotic and vitamin recovery. Often scroll decanters can be used for this separation. The solids are generally kept in suspension (which requires that the solids be denser than heavy phase), while the organic phase, which must be lighter than water (cells typically sink in water) is removed. Experience shows that scrolls are good for handling the variability seen in fermentation feedstock. Podbielniak rotating-drum extraction units have been used often, but only when solids are not sticky, gummy or flocculated, as they can get stuck in perforations of the concentric drums, but will actually give stages to the extraction in short-residence time (temperature-sensitive product). The Karr reciprocating-plate column can handle large volumes of whole-broth materials efficiently, and is amenable to ready scale-up from small laboratory-scale systems to large plant-scale equipment.

Adsorption Adsorption (see also Sec. 16: "Adsorption and Ion Exchange") can be used for the removal of pigments and nucleic acids, for example, or can be used for direct adsorption of the desired proteins. Stirred-batch or expanded-bed operations allow for presence of particulate matter, but fixed beds are not recommended for unclarified broths owing to fouling problems. These separations can be effected through charge, hydrophobic, or affinity interactions between the species and the adsorbent particles, as in the chromatographic steps outlined below. The adsorption processes described here are different from those traditionally ascribed to chromatography in that they do not rely on packed-bed operations.

In *continuous affinity recycle extraction* (CARE) operations, the adsorbent beads are added directly to the cell homogenate and the mixture is fed to a microfiltration unit. The beads loaded with the desired solute are retained by the membrane, and the product is recovered in a second stage by changing the buffer conditions to disfavor binding.

Stable expanded-bed operations promise the ability to handle whole broths efficiently, all the while maintaining plug-flow characteristics. Magnetically stabilized fluidized beds have been shown to work effectively for bioproduct separations, but are not yet used commercially. A commercially available process uses well-designed beads of appropriate densities and sizes to enable bed fluidization and stable operation without appreciable recirculation.

Membrane Processes Membrane processes are also used; diafiltration is convenient for the removal of small contaminating species such as salts and smaller proteins, and can be combined with subsequent steps to concentrate the protein. Provided that proper membrane materials have been selected to avoid protein-membrane interactions, diafiltration using ultrafiltration membranes is typically straightforward, high-yielding and capital-sparing. These operations can often tolerate the concentration of the desired protein to its solubility limit, maximizing process efficiency.

FINAL PURIFICATION AND PRODUCT FORMULATION

The final purification steps are responsible for the removal of the last traces of impurities. The volume reduction in the earlier stages of the separation train are necessary to ensure that these high-resolution operations are not overloaded. Generally, chromatography is used in these final stages. Electrophoresis can also be used, but since it is rarely found in process-scale operations, it is not addressed here. The final product preparation may require removal of solvent and drying, or lyophilization, of the product.

Chromatography Chromatography is the most widely used downstream processing operation because of its versatility, high selectivity and efficiency, in addition to its adequate scale-up potential based on wide experience in the biochemical processing industries. As familiarity is gained with other techniques such as liquid-liquid extraction they will begin to find more favor in the early stages of the separation train, but are unlikely to replace chromatography in the final stages where high purities are needed.

Chromatography is a fixed-bed adsorption operation, in which a column filled with chromatographic packing materials is fed with the mixture of components to be separated. Apart from gel-permeation chromatography, in the most commonly practiced industrial processes the solutes are adsorbed strongly to the packing materials until the bed capacity has been reached. The column may then be washed to remove impurities in the interstitial regions of the bed prior to elution of the solutes. This latter step is accomplished by using buffers or solvents which weaken the binding interaction of the proteins with the packings, permitting their recovery in the mobile phase. Different elution strategies (isocratic, gradient elution) can be used to ensure adequate separation of the species to be resolved. In gel-permeation chromatography (discussed below), and in high-performance liquid chromatography, the principle of operation is different in that it is the differential migration of the various components owing to their different affinities for the packing materials that allows the species to be separated.

Advantages of chromatography for protein separations include the large number of possible chemical interactions resulting from variations in the frequency and distribution of the amino-acid side chains on the surfaces of the proteins, and the availability of a wide array of different adsorption media. Chromatography has high efficiency and selectivity, and adequate scale-up potential.

Types of Chromatography Practiced Separation of proteins using chromatography can exploit a range of different physical and chemical properties of the proteins and the chromatography adsorption media [Janson and Ryden, op. cit.; Scopes, op. cit.; Egerer, in Finn and Prave (eds.), *Biotechnology Focus 1*, Hanser Publishers, Munich, 1988, pp. 95–151]. Parameters that must be considered in the selection of a chromatographic method include composition of the reaction mixture, the chemical structure and stability of the components, the electric charge at a defined pH value and the isoelectric point of the proteins, the hydrophilicity and hydrophobicity of the components, and molecular size. The different types of interactions are illustrated schematically in Fig. 22-87.

Ion-exchange chromatography relies on the coulombic attraction between the ionized functional groups of proteins and oppositely

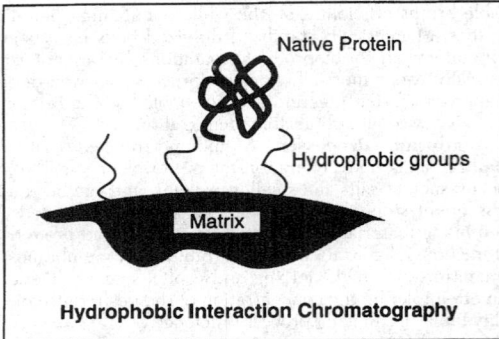

Hydrophobic Interaction Chromatography

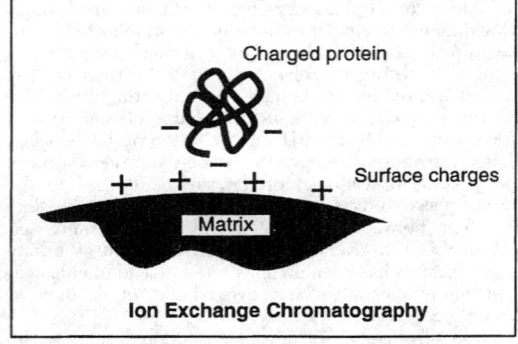

Ion Exchange Chromatography

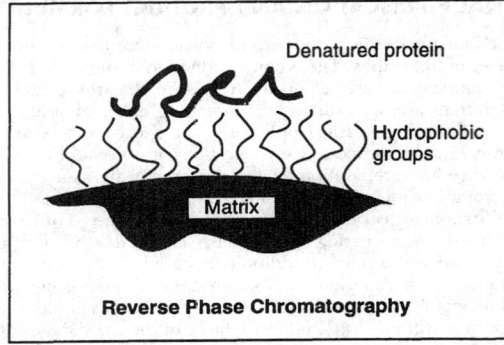

Reverse Phase Chromatography

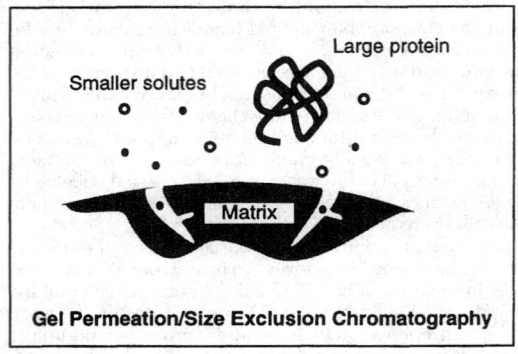

Gel Permeation/Size Exclusion Chromatography

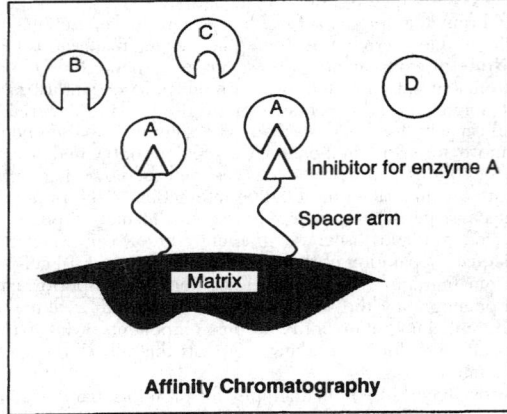

Affinity Chromatography

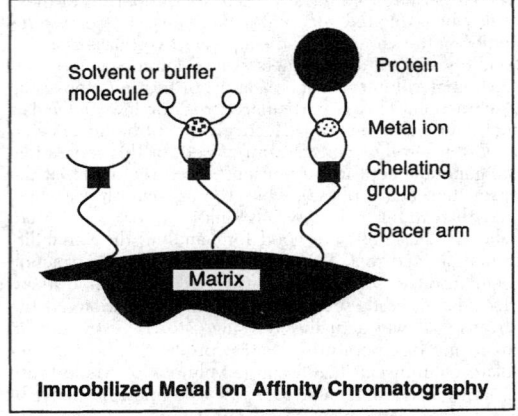

Immobilized Metal Ion Affinity Chromatography

FIG. 22-87   Schematic illustration of the chromatographic methods most commonly used in downstream processing for protein recovery.

charged functional groups on the chromatographic support. It is used to separate the product from contaminating species having different charge characteristics under well-defined eluting conditions, and for concentration of the product, owing to the high-adsorptive capacity of most ion-exchange resins, and the resolution attainable. It is used effectively at the front end of a downstream processing train for early volume reduction and purification.

The differences in sizes and locations of hydrophobic pockets or patches on proteins can be exploited in *hydrophobic interaction chromatography* (HIC) and *reversed-phase chromatography* (RPC); discrimination is based on interactions between the exposed hydrophobic residues and hydrophobic ligands which are distributed evenly throughout an hydrophilic porous matrix. As such, the binding characteristics complement those of other chromatographic methods, such as ion-exchange chromatography.

In HIC, the hydrophobic interactions are relatively weak, often driven by salts in moderate concentration (1 to 2 M), and depend primarily on the exposed residues on or near the protein surface; preservation of the native, biologically active state of the protein is an important feature of HIC. Elution can be achieved differentially by decreasing salt concentration or increasing the concentration of polarity perturbants (e.g., ethylene glycol) in the eluent.

Reversed-phase chromatography relies on significantly stronger hydrophobic interactions than in HIC, which can result in unfolding and exposure of the interior hydrophobic residues, i.e., leads to protein denaturation and irreversible inactivation; as such, RPC depends on total hydrophobic-residue content. Elution is effected by organic solvents applied under gradient conditions.

Ligands for both HIC and RPC are straight chain alkanes or simple aromatic compounds. Increasing the carbon number and the graft density of the ligands on the support surface leads to increasing strength of interaction and passing from HIC to RPC mode of operation. Raising the temperature increases the hydrophobic interactions at the temperatures commonly encountered in biological processing.

HIC is most effective during the early stages of a purification strategy and has the advantage that sample pretreatment such as dialysis or desalting after salt precipitation is not usually required. It is also finding increased use as the last high-resolution step to replace gel filtration. It is a group-separation method, and generally 50 percent or more of extraneous impurities are removed. This method is characterized by high-adsorption capacity, good selectivity, and satisfactory yield of active material.

Despite the intrinsically nonspecific nature of ion-exchange and reversed-phase/hydrophobic interactions, it is often found that chromatographic techniques based on these interactions can exhibit remarkable resolution; this is attributed to the dynamics of multisite interactions being different for proteins having differing surface distributions of hydrophobic and/or ionizable groups.

Gel-permeation chromatography separates proteins nominally on the basis of size only, where the effective size of the protein is determined by its geometry and solvation characteristics. Smaller proteins are able to penetrate the pore volumes of the beads, and are therefore retained relative to the larger proteins, which remain in the fast-flowing fluid in the interstitial regions of the bed. Such ideal behavior is rarely observed, however, as proteins can interact adsorptively with the gel matrix, thus affecting their relative elution behaviors. Industrially, gel-permeation chromatography is used for desalting, removing low-molecular-weight impurities, and removal of desired product oligomers. It is used in the latter stages of the separation sequence, often as a final "polishing" step.

Protein affinity chromatography can be used for the separation of an individual compound, or a group of structurally similar compounds from crude-reaction mixtures, fermentation broths, or cell lysates by exploiting very specific and well-defined molecular interactions between the protein and affinity groups immobilized on the packing-support material. Examples of affinity interactions include antibody-antigen, hormone-receptor, enzyme-substrate/analog/inhibitor, metal ion-ligand, and dye-ligand pairs. Monoclonal antibodies are particularly effective as biospecific ligands for the purification of pharmaceutical proteins. Affinity chromatography may be used for the isolation of a pure product directly from crude fermentation mixtures in a single chromatographic step. Immunosorbents should not be subjected to crude extracts, however, as they are particularly susceptible to fouling and inactivation. Affinity chromatography does not find wide use on the process scale because of its high cost, if a protein ligand such as A or G is to be used.

Immobilized metal ion affinity chromatography (IMAC) relies on the interaction of certain amino-acid residues, particularly histidine, cysteine, and tryptophan, on the surface of the protein with metal ions fixed to the support by chelation with appropriate chelating compounds, invariably derivatives of iminodiacetic acid. Commonly-used metal ions are Cu^{2+}, Zn^{2+}, Ni^{2+} and Co^{2+}. Despite its relative complexity in terms of the number of factors which influence the process, IMAC is beginning to find industrial applications. The choice of chelating group, metal ion, pH, and buffer constituents will determine the adsorption and desorption characteristics. Elution can be effected by several methods, including pH gradient, competitive ligands, organic solvents, and chelating agents.

Following removal of unbound materials in the column by washing, the bound substances are recovered by changing conditions to favor desorption. A gradient or stepwise reduction in pH is often suitable. Otherwise, one can use competitive elution with a gradient of increasing concentration. IMAC eluting agents include ammonium chloride, glycine, histamine, histidine, or imidazol. Inclusion of a chelating agent such as EDTA in the eluent will allow all proteins to be eluted indiscriminately along with the metal ion.

Chromatographic Development   The basic concepts of chromatographic adsorption separations are described elsewhere in this handbook. Proteins differ from small solutes in that the large number of charged and/or hydrophobic residues on the protein surface provide multiple binding sites, which ensure stronger binding of the proteins to the adsorbents, as well as some discrimination based on the surface distribution of amino-acid residues. The proteins are recovered by elution with a buffer that reduces the strength of this binding and permits the proteins to be swept out of the column with the buffer solution. In isocratic elution, the buffer concentration is maintained constant during the elution period. Since the different proteins may have significantly different adsorption isotherms, the recovery may not be complete, or it may take excessive processing time to recover all proteins from the column. In gradient elution operations, the composition of the mobile phase is changed during the process to decrease the binding strength of the proteins successively, the more loosely bound proteins being removed first before the eluent is strengthened to enable recovery of the more strongly adsorbed species. The change in eluent composition can be gradual and continuous, or it can be stepwise. Industrially, in large-scale columns it is difficult to maintain a continuous gradient owing to difficulties in fluid distribution, and thus stepwise changes are almost universally used. In some adsorption modes, the protein can be recovered by the successive addition of competing compounds to displace the adsorbed proteins. In all cases, the product is eluted as a Gaussian (to a first approximation) peak, with some possible overlap between adjacent product peaks.

Displacement chromatography relies on a different mode of elution. Here a displacer which is more strongly adsorbed than any of the proteins is introduced with the mobile phase. As the displacer concentration front develops, it pushes the proteins ahead of itself. The more strongly adsorbed proteins then act as displacers for the less strongly bound proteins, and so on. This leads to the development of a displacer train in which the different molecules are eluted from the column in abutting rectangular peaks in the order of their interaction strength with the adsorption sites of the column. Despite its apparent process advantages, displacement chromatography has not yet become an accepted industrial operation, primarily because of lack of suitable displacers, and possible contamination of the protein product with these displacers.

For efficient adsorption it is advisable to equilibrate both the column and the sample with the optimum buffer for binding. Prior to this, the column must be cleaned to remove tightly bound impurities by increasing the salt concentration beyond that used in the product-elution stages. At the finish of cleaning operation the column should be washed with several volumes of the starting buffer to remove remaining adsorbed material. In desorption, it is necessary to drive the favored binding equilibrium for the adsorbed substance from the stationary to the mobile phase. Ligand-protein interactions are generally a combination of electrostatic, hydrophobic, and hydrogen bonds, and the relative importance of each of these and the degree of stability of the bound protein must be considered in selecting appropriate elution conditions; frequently compromises must be made. Gradient elution often gives good results.

Changes in pH or ionic strength are generally nonspecific in elution performance; ionic-strength increases are effective when the protein binding is predominantly electrostatic, as in IEC. Polarity changes are effective when hydrophobic interactions play the primary role in protein binding. By reducing the polarity of the eluting mobile phase, this phase becomes a more thermodynamically favorable environment for the protein than adsorption to the packing support. A chaotropic salt (KSCN, KCNO, KI in range 1–3 M) or denaturing agent (urea, guanidine HCl; 3–4 M) in the buffer can also lead to enhanced desorption. For the most hydrophobic proteins (e.g., membrane proteins) one can use detergents just below their critical micelle concentrations to solubilize the proteins and strip them from the packing surface.

Specific elution requires more selective eluents. Proteins can be desorbed from ligands by competitive binding of the eluting agent (low concentration 5–100 mM) either to the ligand or to the protein. Specific eluents are most frequently used with group-specific adsorbents since selectivity is greatly increased in the elution step. The effectiveness of the elution step can be tailored using a single eluent, pulses of different eluents, or eluent gradients. These systems are generally characterized by mild desorption conditions. If the eluting agent is bound to the protein, it can be dissociated by desalting on a gel-filtration column or by diafiltration.

Column Packings   The quality of the separation obtained in chromatographic separations will depend on the capacity, selectivity, and hydraulic properties of the stationary phase, which usually consists of porous beads of hydrophilic polymers filled with the solvent. The xerogels (e.g., cross-linked dextran) shrink and swell depending on solvent conditions, while aerogels have sizes independent of solution conditions. A range of materials is used for the manufacture of gel beads, classified according to whether they are inorganic, synthetic, or polysaccharides. The most widely used materials are based on neutral polysaccharides and polyacrylamide. Cellulose gels, such as cross-linked dextran, are generally used as gel-filtration media, but can also be used as a matrix for ion exchangers. The primary use of these gels is for desalting and buffer exchange of protein solutions, as nowadays, fractionation by gel filtration is performed largely with composite gel matrices. Agarose, a low-charge fraction of the seaweed polysaccharide agar, is a widely used packing material.

Microporous gels made by point cross-linking dextran or polyacrylamides are used for molecular-sieve separations such as size-exclusion chromatography and gel filtration, but are generally too soft at the porosities required for efficient protein chromatography. Macroporous gels are most often obtained from aggregated and physically cross-linked polymers. Examples include agarose, macroreticular polyacrylamide, silica, and synthetic polymers. These gels are good for ion-exchange and affinity chromatography, as well as for other adsorption chromatography techniques. Composite gels, in which the microporous gel is introduced into the pores of macroreticular gels, combine the advantages of both types.

High matrix rigidity is offered by porous silica, which can be derivatized to enhance its compatibility with proteins, but it is unstable at alkaline pH. Hydroxyapatite particles have high selectivity for a wide range of proteins and nucleic acids.

Sequencing of Chromatography Steps   The sequence of chromatographic steps used in a protein purification train should be designed such that the more robust techniques are used first, to obtain some volume reduction (concentration effect) and to remove major impurities that might foul subsequent units; these robust units should have high chemical and physical resistance to enable efficient regeneration and cleaning, and should be of low material cost. These steps should be followed by the more sensitive and selective operations, sequenced such that buffer changes and concentration steps between applications to chromatographic columns are avoided. Frequently, ion-exchange chromatography is used as the first step. The elution peaks from such columns can be applied directly to hydrophobic-interaction chromatographic columns or to a gel-filtration unit, without the need for desalting of the solution between applications. These columns can also be used as desalting operations, and the buffers used to elute the columns can be selected to permit direct application of the eluted peaks to the next chromatographic step.

Factors to be considered in making the selection of chromatography processing steps are cost, sample volume, protein concentration and sample viscosity, degree of purity of protein product, presence of nucleic acids, pyrogens, and proteolytic enzymes. Ease with which different types of adsorbents can be washed free from adsorbed contaminants and denatured proteins must also be considered.

Lyophilization and Drying   After the last high-performance purification steps it is usually necessary to prepare the finished product for special applications. For instance, final enzyme products are often required in the form of a dry powder to provide for stability and ease of handling, while pharmaceutical preparations also require high purity, stability during formulation, absence of microbial load, and extended shelf life. This product-formulation step may involve drying of the final products by freeze drying, spray drying, fluidized-bed drying, or crystallization (Golker, in Stephanopoulos, op. cit., pp. 695–714).

Freeze drying, or lyophilization, is normally reserved for temperature-sensitive materials such as vaccines, enzymes, microorganisms, and therapeutic proteins, as it can account for a significant portion of total production cost. This process is characterized by three distinct steps, beginning with freezing of the product solution, followed by water removal by sublimation in a primary drying step, and ending with secondary drying by heating to remove residual moisture.

Freezing is carried out on cooled plates in trays or with the product distributed as small particles on a drum cooler; by dropping the product solution in liquid nitrogen or some other cooling liquid; by cospraying with liquid CO_2 or liquid nitrogen; or by freezing with circulating cold air. The properties of the freeze-dried product, such as texture and ease of rehydration, depend on the size and shape of the ice crystals formed, which in turn depend on the degree of undercooling. It is customary to cool below the lowest equilibrium eutectic temperature of the mixture, although many multicomponent mixtures do not exhibit eutectic points. Freezing should be rapid to avoid effects from local concentration gradients. Removal of water from solution by the formation of ice crystals leads to changes in salt concentration and pH, as well as enhanced concentration of the product, in the remaining solution; this in turn can enhance reaction rates, and even reaction order can change, resulting in cold denaturation of the product. With a high initial protein concentration the freeze concentration factor and the amount of ice formed will be reduced, resulting in greater product stability. For aseptic processing, direct freezing in the freeze-drying plant ensures easier loading of the solution after filtration than if transferred separately from remote freezers.

In the primary drying step, heat of sublimation is supplied by contact, conduction, or radiation to the sublimation front. It is important to avoid partial melting of the ice layer. Many pharmaceutical preparations dried in ampoules are placed on heated shelves. The drying time depends on the quality of ice crystals, indicating the importance of controlling the freezing process; smaller crystals offer higher interfacial areas for heat and mass transfer, but larger crystals provide pores for diffusion of vapor away from the sublimation front.

A high percentage of water remains after the sublimation process, present as adsorbed water, water of hydration or dissolved in the dry amorphous solid; this is difficult to remove. Usually, shelf-temperature is increased to 25 to 40°C and chamber pressure is lowered as far as possible. This still does not result in complete drying, however, which can be achieved only by using even higher temperatures, at which point thermally induced product degradation can occur.

Excipients can be used to improve stability and prevent deterioration and inactivation of biomolecules through structural changes such as dissociation from multimeric states into subunits, decrease in α-helical content accompanied by an increase in β-sheet structure, or complete unfolding of helical structure. These are added prior to the freeze-drying process. Examples of these protective agents include sugars, sugar derivatives, and various amino acids, as well as polymers such as dextran, polyvinyl pyrrolidone, hydroxyethyl starch, and polyethylene glycol. Some excipients, the **lyoprotectants,** provide protection during freezing, drying, and storage, while others, the **cryoprotectants,** offer protection only during the freezing process.

Spray drying can use up to 50 percent less energy than freeze-drying operations and finds application in the production of enzymes used as industrial catalysts, as additives for washing detergents, and as the last step in the production of single-cell protein. The product is usually fed to the dryer as a solution, a suspension, or a free-flowing wet substance. Spray drying is an adiabatic process, the energy being provided by hot gas (usually hot air) at temperatures between 120 and 400°C. Product stability is assured by a very short drying time in the spray-drying equipment, typically in the subsecond to second range, which limits exposure to the elevated temperatures in the dryer. Protection can be offered by addition of additives (e.g., galactomannan, polyvinyl pyrrolidone, methyl cellulose, cellulose).

The spray-drying process requires dispersion of the feed as small droplets to provide a large heat and mass-transfer area. The dispersion of liquid is attained using rotating disks, different types of nozzles or ultrasound, and is affected by interfacial tension, density, and dynamic viscosity of the feed solution, as well as the temperature and relative velocities of the liquid and air in the mixing zone. Rotating-disk atomizers operate at 4000 to 50,000 rpm to generate the centrifugal forces needed for dispersion of the liquid phase; typical droplet sizes of 25 to 950 μm are obtained. These atomizers are specially suitable for dispersing suspensions that would tend to clog nozzles.

For processing under aseptic conditions, the spray drier must be connected to a filling line that allows aseptic handling of the product.

INTEGRATION OF FERMENTATION AND DOWNSTREAM PROCESSING OPERATIONS

Traditionally, the upstream fermentation and cell culture processes have been viewed as being distinct from the subsequent downstream processing and purification steps, and the two different sets of processes have been optimized individually. In some instances, careful consideration of the conditions used in the fermentation process, or manipulation of the genetic makeup of the host, can simplify and even eliminate some unit operations in the downstream processing sequence [Kelley and Hatton, *Bioseparation*, **1**, 303–349 (1991)]. Some of the advances made in this area are the engineering of strains of *Escherichia coli* to allow the inducible expression of lytic enzymes capable of disrupting the wall from within for the release of intracel-lular protein products, the use of secretion vectors for the expression of proteins in bacterial production systems, and protein synthesis to include a peptide or protein fusion to confer unique properties to the product to facilitate subsequent downstream processing. The cell culture medium can be selected to avoid components that can hinder subsequent purification procedures. Integration of the fermentation and initial separation/purification steps in a single operation can also lead to enhanced productivity, particularly when the product can be removed as it is formed to prevent its proteolytic destruction by the proteases which are frequently the by-product of fermentation processes. The introduction of a solvent directly to the fermentation medium (e.g., phase-forming polymers), the continuous removal of products using ultrafiltration membranes, or the use of continuous fluidized-bed operations are examples of this integration.

Chemical Reactors

Stanley M. Walas, Ph.D., *Professor Emeritus, Department of Chemical and Petroleum Engineering, University of Kansas; Fellow, American Institute of Chemical Engineers*

Nomenclature and Units

Following is a listing of typical nomenclature and units expressed in SI and U.S. customary. Specific definitions and units are stated at the place of application in the section.

Symbol	Definition	SI units	U.S. customary units
A, B, C, . . .	Names of substances, or their concentrations		
A^0	Free radical, as CHj^0		
C_a	Concentration of substance A	kg mol/m³	lb mol/ft³
C^0	Initial mean concentration in vessel	kg mol/m³	lb mol/ft³
C_p	Heat capacity	kJ/(kg·K)	Btu/(lbm·°F)
CSTR	Continuous stirred tank reactor		
D, D_e, D_x	Dispersion coefficient	m²/s	ft²/s
D_{eff}	Effective diffusivity	m²/s	ft²/s
D_K	Knudsen diffusivity	m²/s	ft²/s
$E(t)$	Residence time distribution		
$E(t_r)$	Normalized residence time distribution		
f_a	C_a/C_{a0} or n_a/n_{a0}, fraction of A remaining unconverted		
$F(t)$	Age function of tracer		
ΔG	Gibbs energy change	kJ	Btu
Ha	Hatta number		
ΔH_r	Heat of reaction	kJ/kg mol	Btu/lb mol
K, K_c, K_y, K_ϕ	Chemical equilibrium constant		
k, k_c, k_p	Specific rate of reaction	Variable	Variable
L	Length of path in reactor	m	ft
n	Parameter of Erlang or gamma distribution, or number of stages in a CSTR battery		
n_a	Number of mols of A present		
n_a'	Number of mols flowing per unit time, the prime (') may be omitted when context is clear		
n_t	Total number of mols		
p_a	Partial pressure of substance A	kPa	psi
Pe	Peclet number for dispersion		
PFR	Plug flow reactor		
Q	Heat transfer	kJ	Btu
r	Radial position	m	ft
r_a	Rate of reaction of A per unit volume	Variable	Variable
R	Radius of cylindrical vessel	m	ft
Re	Reynolds number		
Sc	Schmidt number		

Symbol	Definition	SI units	U.S. customary units
t	Time	s	s
$\bar{t}$	Mean residence time	s	s
t_r	$t/\bar{t}$, reduced time		
TFR	Tubular flow reactor		
u	Linear velocity	m/s	ft/s
u(t)	Unit step input		
V	Volume of reactor contents	m³	ft³
V'	Volumetric flow rate	m³/s	ft³/s
V_r	Volume of reactor	m³	ft³
x	Axial position in a reactor	m	ft
x_a	$1 - f_a = 1 - C_a/C_{a0}$ or $1 - n_a/n_{a0}$, fraction of A converted		
z	x/L, normalized axial position		

Greek letters

Symbol	Definition	SI units	U.S. customary units
β	r/R, normalized radial position		
$\gamma^3(t)$	Skewness of distribution		
$\delta(t)$	Unit impulse input, Dirac function		
ε	Fraction void space in a packed bed		
ϑ	$t/\bar{t}$, reduced time		
η	Effectiveness of porous catalyst		
$\Lambda(t)$	Intensity function		
μ	Viscosity	Pa·s	lbm/(ft·s)
ν	v/ρ, kinematic viscosity	m²/s	ft²/s
π	Total pressure	Pa	psi
ρ	Density	kg/m³	lbm/ft³
ρ	r/R, normalized radial position in a pore		
$\sigma^2(t)$	Variance		
$\sigma^2(t_r)$	Normalized variance		
τ	$t/\bar{t}$, reduced time		
τ	Tortuosity		
φ	Thiele modulus		
ϕ_m	Modified Thiele modulus		

Subscripts

Symbol	Definition	SI units	U.S. customary units
0	Subscript designating initial or inlet conditions, as in $C_{a0}, n_{a0}, V_0', \ldots$		

MODELING CHEMICAL REACTORS

GENERAL REFERENCES: The General References listed in Sec. 7 are applicable for Sec. 23. References to specific topics are made throughout this section.

An industrial chemical reactor is a complex device in which heat transfer, mass transfer, diffusion, and friction may occur along with chemical reaction, and it must be safe and controllable. In large vessels, questions of mixing of reactants, flow distribution, residence time distribution, and efficient utilization of the surface of porous catalysts also arise. A particular process can be dominated by one of these factors or by several of them; for example, a reactor may on occasion be predominantly a heat exchanger or a mass-transfer device. A successful commercial unit is an economic balance of all these factors.

Some modes of heat transfer to stirred tank reactors are shown in Fig. 23-1 and to packed bed reactors in Fig. 23-2. Temperature and composition profiles of some processes are shown in Fig. 23-3. Operating data, catalysts, and reaction times are stated for a number of industrial reaction processes in Table 23-1.

Many successful types of reactors are illustrated throughout this section. Additional sketches may be found in other books on this topic, particularly in Walas (*Chemical Process Equipment Selection and Design*, Butterworths, 1990) and Ullmann (*Encyclopedia of Chemical Technology* (in German), vol. 3, Verlag Chemie, 1973, pp. 321–518).

The general characteristics of the main types of reactors—batch and continuous—are clear. Batch processes are suited to small production rates, to long reaction times, or to reactions where they may have superior selectivity, as in some polymerizations. They are conducted in tanks with stirring of the contents by internal impellers, gas bubbles, or pumparound. Temperature control is with internal surfaces or jackets, reflux condensers, or pumparound through an exchanger.

Large daily production rates are mostly conducted in continuous equipment, either in a series of stirred tanks or in units in which some degree of plug flow is attained. Many different equipment configurations are illustrated throughout this section for reactions of liquids, gases, and solids, singly or in combinations. By showing how something has been done previously, this picture gallery may suggest how a similar new process could be implemented.

Continuous stirred tank reactors (CSTRs) are frequently employed multiply and in series. Reactants are continuously fed to the first vessel; they overflow through the others in succession, while being thoroughly mixed in each vessel. Ideally, the composition is uniform in individual vessels, but a stepped concentration gradient exists in the system as a whole. For some cases, a series of five or six vessels approximates the performance of a plug flow reactor. Instead of being in distinct vessels, the several stages of a CSTR battery can be put in a single shell. If horizontal, the multistage reactor is compartmented by vertical weirs of different heights, over which the reacting mixture cascades. When the reactants are of limited miscibilities and have a sufficient density difference, the vertical staged reactor lends itself to countercurrent operation, a real advantage with reversible reactions. A small fluidized bed is essentially completely mixed. A large commercial fluidized bed reactor is of nearly uniform temperature, but the flow patterns consist of mixed and plug flow and in-between zones.

Tubular flow reactors (TFRs) are characterized by continuous gradients of concentration in the direction of flow that approach plug flow, in contrast to the stepped gradient characteristic of the CSTR battery. They may have several pipes or tubes in parallel. The reactants are charged continuously at one end and products are removed at the other end. Normally a steady state is attained, a fact of importance for automatic control and for laboratory work. Both horizontal and vertical orientations are common. When heat transfer is needed, individual tubes are jacketed or shell-and-tube construction is used. In the latter case the reactants may be on either the shell or the tube side. The reactant side may be filled with solid particles, either catalytic (if required) or inert, to improve heat transfer by increased turbulence or to improve interphase contact in heterogeneous reactions. Large-diameter vessels with packing or trays may approach plug flow behavior and are widely employed. Some of the configurations in use are axial flow, radial flow, multiple shell, with built-in heat exchangers, horizontal, vertical and so on. *Quasi-plug flow reactors* have continuous gradients but are not quite in plug flow.

Semiflow or batch flow operations may employ a single stirred tank or a series of them. Some of the reactants are loaded into the reactors as a single charge and the remaining ones are then fed gradually. This mode of operation is especially favored when large heat effects occur and heat-transfer capability is limited, since exothermic reactions can be slowed down and endothermic rates maintained by limiting the concentrations of some of the reactants. Other situations making this sort of operation desirable occur when high concentrations may result in the formation of undesirable side products, or when one of the reactants is a gas of limited solubility so that it can be fed only at the dissolution rate.

Relative advantages and fields of application of continuous stirred and plug flow reactors may be indicated briefly. A reaction battery is a highly flexible device, although both mechanically and operationally more expensive and complex than tubular units. Relatively slow reactions are best conducted in a CSTR battery, which is usually cheaper than a single reactor for moderate production rates. The tubular reactor is especially suited to cases needing considerable heat transfer, where high pressures and very high or very low temperatures occur, and when relatively short reaction times suffice.

MATHEMATICAL MODELS

A model of a reaction process is a set of data and equations that is believed to represent the performance of a specific vessel configuration (mixed, plug flow, laminar, dispersed, and so on). The equations include the stoichiometric relations, rate equations, heat and material balances, and auxiliary relations such as those of mass transfer, pressure variation, contacting efficiency, residence time distribution, and so on. The data describe physical and thermodynamic properties and, in the ultimate analysis, economic factors.

Correlations of heat and mass-transfer rates are fairly well developed and can be incorporated in models of a reaction process, but the chemical rate data must be determined individually. The most useful rate data are at constant temperature, under conditions where external mass transfer resistance has been avoided, and with small particles

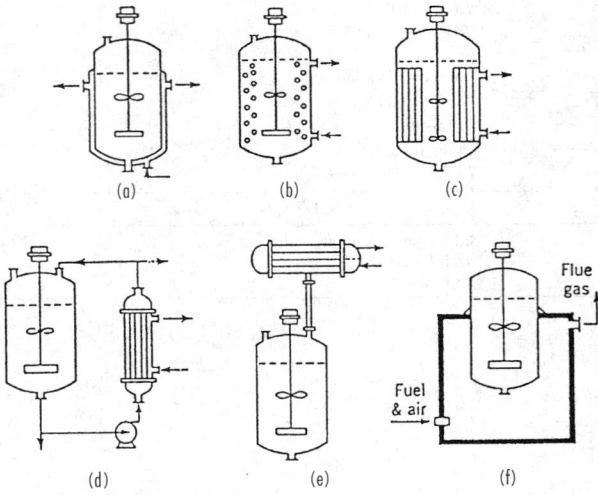

FIG. 23-1   Heat transfer to stirred tank reactors. (*a*) Jacket. (*b*) Internal coils. (*c*) Internal tubes. (*d*) External heat exchanger. (*e*) External reflux condensor. (*f*) Fired heater. (*Walas*, Reaction Kinetics for Chemical Engineers, *McGraw-Hill, 1959*).

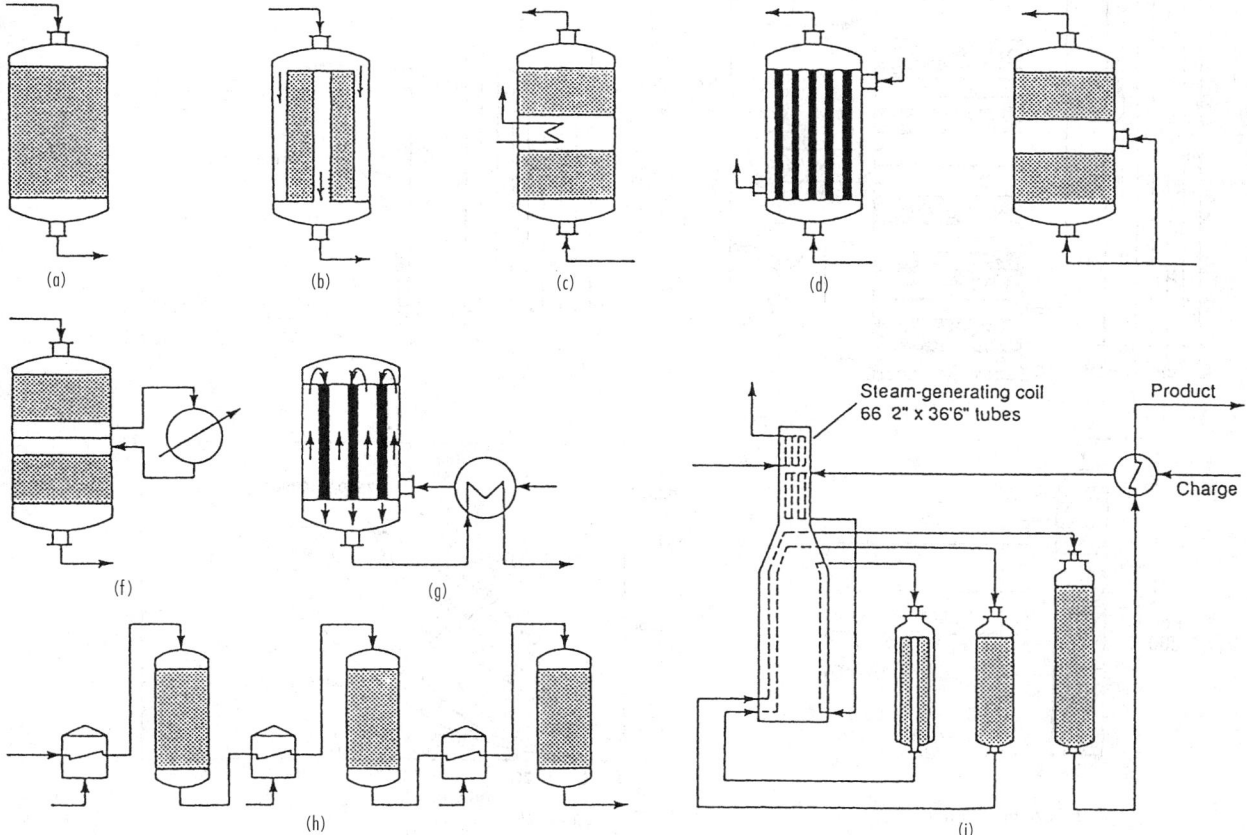

FIG. 23-2   Heat exchange in packed reactors. (*a*) Adiabatic downflow. (*b*) Adiabatic radial flow, low ΔP. (*c*) Built-in interbed exchanger. (*d*) Shell and tube. (*e*) Interbed cold-shot injection. (*f*) External interbed exchanger. (*g*) Autothermal shell, outside influent/effluent heat exchanger. (*h*) Multibed adiabatic reactors with interstage heaters. (*i*) Platinum catalyst, fixed bed reformer for 5,000 BPSD charge rate; reactors 1 and 2 are 5.5 by 9.5 ft and reactor 3 is 6.5 by 12.0 ft; temperatures $502 \Rightarrow 433$, $502 \Rightarrow 471$, $502 \Rightarrow 496°C$. To convert ft to m, multiply by 0.3048; BPSD to m³/h, multiply by 0.00662.

with complete catalyst effectiveness. Equipment for obtaining such data is now widely used, especially for reactions with solid catalysts, and it is virtually essential for serious work. Simpler equipment gives data that are more difficult to interpret but may be adequate for exploratory work.

Once fundamental data have been obtained, the goal is to develop a mathematical model of the process and to utilize it to explore such possibilities as product selectivity, start-up and shut-down behavior, vessel configuration, temperature, pressure, and conversion profiles, and so on.

How complete a model has to be is an open question. Very elaborate ones are justifiable and have been developed only for certain widely practiced and large-scale processes, or for processes where operating conditions are especially critical. The only policy to follow is to balance the cost of development of the model against any safety factors that otherwise must be applied to a final process design. Engineers constantly use heat and material balances without perhaps realizing, like Molière's Bourgeois, that they are modeling mathematically. The simplest models may be adequate for broad discrimination between alternatives.

MODELING PRINCIPLES

How a differential equation is formulated for some kinds of ideal reactors is described briefly in Sec. 7 of this Handbook and at greater length with many examples in Walas (*Modeling with Differential Equations in Chemical Engineering*, Butterworth-Heineman, 1991).

First, a mechanism is assumed: whether completely mixed, plug flow, laminar, with dispersion, with bypass or recycle or dead space, steady or unsteady, and so on. Then, for a differential element of space and/or time the elements of a conservation law,

$$\text{Inputs} + \text{Sources} = \text{Outputs} + \text{Sinks} + \text{Accumulations}$$

are formulated and put together. Any transport properties are introduced through known correlations together with the parameters of specified rate equations. The model can be used to find the performance under various conditions, or its parameters can be evaluated from experimental data. The two comprehensive examples in this section evaluate performance when all the parameters are known. There are other examples in these sections where the parameters of rate equations, residence time distributions, or effectiveness are to be found from rate or tracer data.

CHEMICAL KINETIC LAWS

The two basic laws of kinetics are the *law of mass action* for the rate of a reaction and the *Arrhenius equation* for its dependence on temperature. Both of these are strictly empirical. They depend on the structures of the molecules, but at present the constants of the equations cannot be derived from the structures of reacting molecules. For a reaction, $a\text{A} + b\text{B} \Rightarrow \text{Products}$, the combined law is

$$r_a = -\frac{1}{V_r}\frac{dn_a}{dt} = \exp\left(\gamma + \frac{\delta}{T}\right) C_a^{\alpha} C_b^{\beta} \tag{23-1}$$

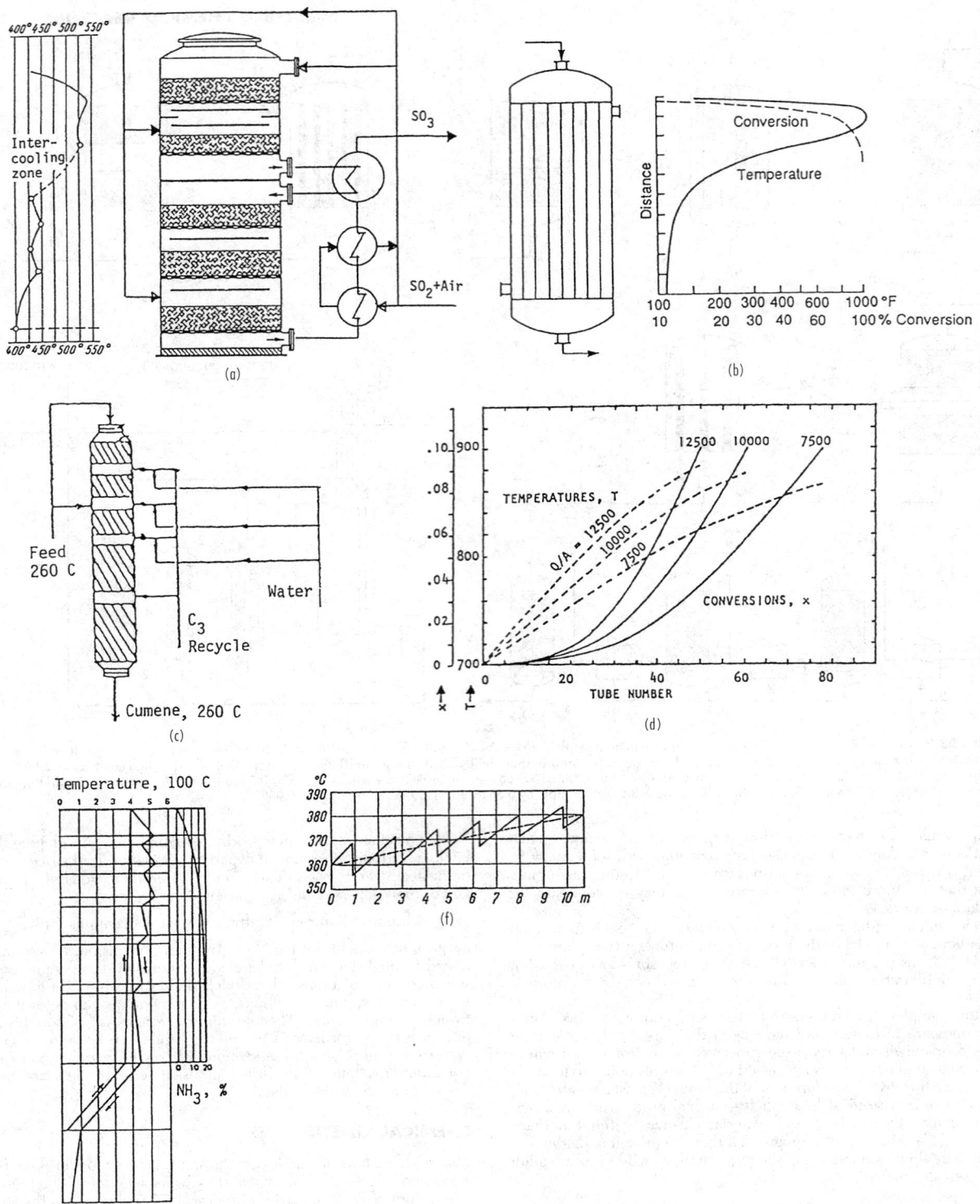

FIG. 23-3 Temperature and composition profiles. (*a*) Oxidation of SO₂ with intercooling and two cold shots. (*b*) Phosgene from CO and Cl₂, activated carbon in 2-in tubes, water cooled. (*c*) Cumene from benzene and propylene, phosphoric acid on quartz, with four quench zones, 260°C. (*d*) Mild thermal cracking of a heavy oil in a tubular furnace, back pressure of 250 psig and several heat fluxes, Btu/(ft²·h), *T* in °F. (*e*) Vertical ammonia synthesizer at 300 atm, with five cold shots and an internal exchanger. (*f*) Vertical methanol synthesizer at 300 atm, Cr₂O₃-ZnO catalyst, with six cold shots totaling 10 to 20 percent of the fresh feed. To convert psi to kPa, multiply by 6.895; atm to kPa, multiply by 101.3.

TABLE 23-1 Residence Times and/or Space Velocities in Industrial Chemical Reactors*

Product (raw materials)	Type	Reactor phase	Catalyst	T, °C	P, atm	Residence time or space velocity	Source and page†
Acetaldehyde (ethylene, air)	FB	L	Cu and Pd chlorides	50–100	8	6–40 min	[2] 1, [7] 3
Acetic anhydride (acetic acid)	TO	L	Triethylphosphate	700–800	0.3	0.25–5 s	[2]
Acetone (i-propanol)	MT	LG	Ni	300	1	2.5 h	[1] 1 314
Acrolein (formaldehyde, acetaldehyde)	FL	G	MnO, silica gel	280–320	1	0.6 s	[1] 1 384, [7] 33
Acrylonitrile (air, propylene, ammonia)	FL	G	Bi phosphomolybdate	400	1	4.3 s	[3] 684, [2] 47
Adipic acid (nitration of cyclohexanol)	TO	L	Co naphthenate	125–160	4–20	2 h	[2] 51, [7] 49
Adiponitrile (adipic acid)	FB	G	H_3BO_3 H_3PO_4	370–410	1	3.5–5 s 350–500 GHSV	[1] 2 152, [7] 52
Alkylate (i-C_4, butenes)	CST	L	H_2SO_4	5–10	2–3	5–40 min	[4] 223
Alkylate (i-C_4, butenes)	CST	L	HF	25–38	8–11	5–25 min	[4] 223
Allyl chloride (propylene, Cl_2)	TO	G	NA	500	3	0.3–1.5 s	[1] 2 416, [7] 67
Ammonia (H_2, N_2)	FB	G	Fe	450	150	28 s 7,800 GHSV	[6] 61
Ammonia (H_2, N_2)	FB	G	Fe	450	225	33 s 10,000 GHSV	[6] 61
Ammonia oxidation	Flame	G	Pt gauze	900	8	0.0026 s	[6] 115
Aniline (nitrobenzene, H_2)	B	L	$FeCl_2$ in H_2O	95–100	1	8 h	[1] 3 289
Aniline (nitrobenzene, H_2)	FB	G	Cu on silica	250–300	1	0.5–100 s	[7] 82
Aspirin (salicylic acid, acetic anhydride)	B	L	None	90	1	>1 h	[7] 89
Benzene (toluene)	TU	G	None	740	38	48 s 815 GHSV	[6] 36, [9] 109
Benzene (toluene)	TU	G	None	650	35	128 s	[1] 4 183, [7] 98
Benzoic acid (toluene, air)	SCST	LG	None	125–175	9–13	0.2–2 h	[7] 101
Butadiene (butane)	FB	G	Cr_2O_3, Al_2O_3	750	1	0.1–1 s	[7] 118
Butadiene (1-butene)	FB	G	None	600	0.25	0.001 s 34,000 GHSV	[3] 572
Butadiene sulfone (butadiene, SO_2)	CST	L	t-Butyl catechol	34	12	0.2 LHSV	[1] 5 192
i-Butane (n-butane)	FB	L	$AlCl_3$ on bauxite	40–120	18–36	0.5–1 LHSV	[4] 239, [7] 683
i-Butane (n-butane)	FB	L	Ni	370–500	20–50	1–6 WHSV	[4] 239
Butanols (propylene hydroformylation)	FB	L	PH_3-modified Co carbonyls	150–200	1,000	100 g L·h	[1] 5 373
Butanols (propylene hydroformylation)	FB	L	Fe pentacarbonyl	110	10	1 h	[7] 125
Calcium stearate	B	L	None	180	5	1–2 h	[7] 135
Caprolactam (cyclohexane oxime)	CST	L	Polyphosphoric acid	80–110	1	0.25–2 h	[1] 6 73, [7] 139
Carbon disulfide (methane, sulfur)	Furn.	G	None	500–700	1	1.0 s	[1] 6 322, [7] 144
Carbon monoxide oxidation (shift)	TU	G	Cu-Zn or Fe_2O_3	390–220	26	4.5 s 7,000 GHSV	[6] 44
Portland cement	Kiln	S		1,400–1,700	1	10 h	[11]
Chloral (Cl_2, acetaldehyde)	CST	LG	None	20–90	1	140 h	[7] 158
Chlorobenzenes (benzene, Cl_2)	SCST	LG	Fe	40	1	24 h	[1] 8 122
Coking, delayed (heater)	TU	LG	None	490–500	15–4	250 s	[1] 10 8
Coking, delayed (drum, 100 ft max height)	B	LG	None	500–440	4	0.3–0.5 ft/s vapor	[1] 10 8
Cracking, fluid catalytic	Riser	G	Zeolite	520–540	2–3	2–4 s	(14) 353
Cracking, hydro (gas oils)	FB	LG	Ni, SiO_2, Al_2O_3	350–420	100–150	1–2 LHSV	[11]
Cracking (visbreaking residual oils)	TU	LG	None	470–495	10–30	450 s, 8 LHSV	[11]
Cumene (benzene, propylene)	FB	G	H_3PO_4	260	35	23 LHSV	[11]
Cumene hydroperoxide (cumene, air)	CST	L	Metal porphyrins	95–120	2–15	1–3 h	[7] 191
Cyclohexane (benzene, H_2)	FB	G	Ni on Al_2O_3	150–250	25–55	0.75–2 LHSV	[7] 201
Cyclohexanol (cyclohexane, air)	SCST	LG	None	185–200	48	2–10 min	[7] 203
Cyclohexanone (cyclohexanol)	CST	L	N.A.	107	1	0.75 h	[8] (1963)
Cyclohexanone (cyclohexanol)	MT	G	Cu on pumice	250–350	1	4–12 s	[8] (1963)
Cyclopentadiene (dicyclopentadiene)	TJ	G	None	220–300	1–2	0.1–0.5 LHSV	[7] 212
DDT (chloral, chlorobenzene)	B	L	Oleum	0–15	1	8 h	[7] 233
Dextrose (starch)	CST	L	H_2SO_4	165	1	20 min	[8] (1951)
Dextrose (starch)	CST	L	Enzyme	60	1	100 min	[7] 217
Dibutylphthalate (phthalic anhydride, butanol)	B	L	H_2SO_4	150–200	1	1–3 h	[7] 227
Diethylketone (ethylene, CO)	TO	L	Co oleate	150–300	200–500	0.1–10 h	[7] 243
Dimethylsulfide (methanol, CS_2)	FB	G	Al_2O_3	375–535	5	150 GHSV	[7] 266
Diphenyl (benzene)	MT	G	None	730	2	0.6 s 3.3 LHSV	[7] 275, [8] (1938)
Dodecylbenzene (benzene, propylene tetramer)	CST	L	$AlCl_3$	15–20	1	1–30 min	[7] 283
Ethanol (ethylene, H_2O)	FB	G	H_3PO_4	300	82	1,800 GHSV	[2] 356, [7] 297
Ethyl acetate (ethanol, acetic acid)	TU, CST	L	H_2SO_4	100	1	0.5–0.8 LHSV	[10] 45, 52, 58
Ethyl chloride (ethylene, HCl)	TO	G	$ZnCl_2$	150–250	6–20	2 s	[7] 305
Ethylene (ethane)	TU	G	None	860	2	1.03 s 1,880 GHSV	[3] 411, [6] 13
Ethylene (naphtha)	TU	G	None	550–750	2–7	0.5–3 s	[7] 254
Ethylene, propylene chlorohydrins (Cl_2, H_2O)	CST	LG	None	30–40	3–10	0.5–5 min	[7] 310, 580
Ethylene glycol (ethylene oxide, H_2O)	TO	LG	1% H_2SO_4	50–70	1	30 min	[2] 398
Ethylene glycol (ethylene oxide, H_2O)	TO	LG	None	195	13	1 h	[2] 398
Ethylene oxide (ethylene, air)	FL	G	Ag	270–290	1	1 s	[2] 409, [7] 322
Ethyl ether (ethanol)	FB	G	WO_3	120–375	2–100	30 min	[7] 326
Fatty alcohols (coconut oil)	B	L	Na, solvent	142	1	2 h	[8] (1953)

TABLE 23-1   Residence Times and/or Space Velocities in Industrial Chemical Reactors (Concluded)

Product (raw materials)	Type	Reactor phase	Catalyst	T, °C	P, atm	Residence time or space velocity	Source and page†
Formaldehyde (methanol, air)	FB	G	Ag gauze	450–600	1	0.01 s	[2] 423
Glycerol (allyl alcohol, H_2O_2)	CST	L	H_2WO_4	40–60	1	3 h	[7] 347
Hydrogen (methane, steam)	MT	G	Ni	790	13	5.4 s 3,000 GHSV	[6] 133
Hydrodesulfurization of naphtha	TO	LG	Co-MO	315–500	20–70	1.5–8 LHSV 125 WHSV	[4] 285, [6] 179, [9] 201
Hydrogenation of cottonseed oil	SCST	LG	Ni	130	5	6 h	[6] 161
Isoprene (i-butene, formaldehyde)	FB	G	HCl, silica gel	250–350	1	1 h	[7] 389
Maleic anhydride (butenes, air)	FL	G	V_2O_5	300–450	2–10	0.1–5 s	[7] 406
Melamine (urea)	B	L	None	340–400	40–150	5–60 min	[7] 410
Methanol (CO, H_2)	FB	G	ZnO, Cr_2O_3	350–400	340	5,000 GHSV	[7] 421
Methanol (CO, H_2)	FB	G	ZnO, Cr_2O_3	350–400	254	28,000 GHSV	[3] 562
o-Methyl benzoic acid (xylene, air)	CST	L	None	160	14	0.32 h 3.1 LHSV	[3] 732
Methyl chloride (methanol, Cl_2)	FB	G	Al_2O_3 gel	340–350	1	275 GHSV	[2] 533
Methyl ethyl ketone (2-butanol)	FB	G	ZnO	425–475	2–4	0.5–10 min	[7] 437
Methyl ethyl ketone (2-butanol)	FB	G	Brass spheres	450	5	2.1 s 13 LHSV	[10] 284
Nitrobenzene (benzene, HNO_3)	CST	L	H_2SO_4	45–95	1	3–40 min	[7] 468
Nitromethane (methane, HNO_3)	TO	G	None	450–700	5–40	0.07–0.35 s	[7] 474
Nylon-6 (caprolactam)	TU	L	Na	260	1	12 h	[7] 480
Phenol (cumene hydroperoxide)	CST	L	SO_2	45–65	2–3	15 min	[7] 520
Phenol (chlorobenzene, steam)	FB	G	Cu, Ca phosphate	430–450	1–2	2 WHSV	[7] 522
Phosgene (CO, Cl_2)	MT	G	Activated carbon	50	5–10	16 s 900 GHSV	[11]
Phthalic anhydride (o-xylene, air)	MT	G	V_2O_5	350	1	1.5 s	[3] 482, 539, [7] 529
Phthalic anhydride (naphthalene, air)	FL	G	V_2O_5	350	1	5 s	[9] 136, [10] 335
Polycarbonate resin (bisphenol-A, phosgene)	B	L	Benzyltriethylammonium chloride	30–40	1	0.25–4 h	[7] 452
Polyethylene	TU	L	Organic peroxides	180–200	1,000–1,700	0.5–50 min	[7] 547
Polyethylene	TU	L	Cr_2O_3, Al_2O_3, SiO_2	70–200	20–50	0.1–1,000 s	[7] 549
Polypropylene	TO	L	R_3AlCl, $TiCl_4$	15–65	10–20	15–100 min	[7] 559
Polyvinyl chloride	B	L	Organic peroxides	60	10	5.3–10 h	[6] 139
i-Propanol (propylene, H_2O)	TO	L	H_2SO_4	70–110	2–14	0.5–4 h	[7] 393
Propionitrile (propylene, NH_3)	TU	G	CoO	350–425	70–200	0.3–2 LHSV	[7] 578
Reforming of naphtha (H_2/hydrocarbon = 6)	FB	G	Pt	490	30–35	3 LHSV 8,000 GHSV	[6] 99
Starch (corn, H_2O)	B	L	SO_2	25–60	1	18–72 h	[7] 607
Styrene (ethylbenzene)	MT	G	Metal oxides	600–650	1	0.2 s 7,500 GHSV	[5] 424
Sulfur dioxide oxidation	FB	G	V_2O_5	475	1	2.4 s 700 GHSV	[6] 86
t-Butyl methacrylate (methacrylic acid, i-butene)	CST	L	H_2SO_4	25	3	0.3 LHSV	[1] 5 328
Thiophene (butane, S)	TU	G	None	600–700	1	0.01–1 s	[7] 652
Toluene diisocyanate (toluene diamine, phosgene)	B	LG	None	200–210	1	7 h	[7] 657
Toluene diamine (dinitrotoluene, H_2)	B	LG	Pd	80	6	10 h	[7] 656
Tricresyl phosphate (cresyl, $POCl_3$)	TO	L	$MgCl_2$	150–300	1	0.5–2.5 h	[2] 850, [7] 673
Vinyl chloride (ethylene, Cl_2)	FL	G	None	450–550	2–10	0.5–5 s	[7] 699
Aldehydes (diisobutene, CO)	CST	LG	Co Carbonyl	150	200	1.7 h	(12) 173
Allyl alcohol (propylene oxide)	FB	G	Li phosphate	250	1	1.0 LHSV	(15) 23
Automobile exhaust	FB	G	Pt-Pd: 1–2 g/unit	400–600+	1		
Gasoline (methanol)	FB	G	Zeolite	400	20	2 WHSV	(13)3 383
Hydrogen cyanide (NH_3, CH_4)	FB	G	Pt-Rh	1150	1	0.005 s	(15) 211
Isoprene, polymer	B	L	$Al(i-Bu)_3 \cdot TiCl_4$	20–50	1–5	1.5–4 h	(15) 82
NO_x pollutant (with NH_3)	FB	G	$V_2O_5 \cdot TiO_2$	300–400	1–10		(14) 332
Vinyl acetate (ethylene + CO)	TO	LG	Cu-Pd	130	30	1 h L, 10 s G	(12) 140

*Abbreviations: reactors: batch (B), continuous stirred tank (CST), fixed bed of catalyst (FB), fluidized bed of catalyst (FL), furnace (Furn.), multitubular (MT), semicontinuous stirred tank (SCST), tower (TO), tubular (TU). Phases: liquid (L), gas (G), both (LG). Space velocities (hourly): gas (GHSV), liquid (LHSV), weight (WHSV). Not available, NA. To convert atm to kPa, multiply by 101.3.

†1. J. J. McKetta, ed., *Encyclopedia of Chemical Processing and Design*, Marcel Dekker, 1976 to date (referenced by volume).
2. W. L. Faith, D. B. Keyes, and R. L. Clark, *Industrial Chemicals*, revised by F. A. Lowenstein and M. K. Moran, John Wiley & Sons, 1975.
3. G. F. Froment and K. B. Bischoff, *Chemical Reactor Analysis and Design*, John Wiley & Sons, 1979.
4. R. J. Hengstebeck, *Petroleum Processing*, McGraw-Hill, New York, 1959.
5. V. G. Jenson and G. V. Jeffreys, *Mathematical Methods in Chemical Engineering*, 2d ed., Academic Press, 1977.
6. H. F. Rase, *Chemical Reactor Design for Process Plants*, Vol. 2: Case Studies, John Wiley & Sons, 1977.
7. M. Sittig, *Organic Chemical Process Encyclopedia*, Noyes, 1969 (patent literature exclusively).
8. Student Contest Problems, published annually by AIChE, New York (referenced by year).
9. M. O. Tarhan, *Catalytic Reactor Design*, McGraw-Hill, 1983.
10. K. R. Westerterp, W. P. M. van Swaaij, and A. A. C. M. Beenackers, *Chemical Reactor Design and Operation*, John Wiley & Sons, 1984.
11. Personal communication (Walas, 1985).
12. B. C. Gates, J. R. Katzer, and G. C. A. Schuit, *Chemistry of Catalytic Processes*, McGraw-Hill, 1979.
13. B. E. Leach, ed., *Applied Industrial Catalysts*, 3 vols., Academic Press, 1983.
14. C. N. Satterfield, *Heterogeneous Catalysis in Industrial Practice*, McGraw-Hill, 1991.
15. C. L. Thomas, *Catalytic Processes and Proven Catalysts*, Academic Press, 1970.

When the density is constant, replace $n_a/V_r = C_a$. How the constants α, β, γ, and δ are found from experimental conversion data is explained in Sec. 7.

BASIC REACTOR ELEMENTS

Reactions are carried out as batches or with continuous streams through a vessel. Flow reactors are distinguished by the degree of mixing of successive inputs. The limiting cases are: (1) with complete mixing, called an *ideal continuous stirred tank reactor* (CSTR), and (2) with no axial mixing, called a *plug flow reactor* (PFR).

Real reactors deviate more or less from these ideal behaviors. Deviations may be detected with *residence time distributions* (RTD) obtained with the aid of tracer tests. In other cases a mechanism may be postulated and its parameters checked against test data. The commonest models are combinations of CSTRs and PFRs in series and/or parallel. Thus, a stirred tank may be assumed completely mixed in the vicinity of the impeller and in plug flow near the outlet.

The combination of reactor elements is facilitated by the concept of *transfer functions*. By this means the Laplace transform can be found for the overall model, and the residence time distribution can be found after inversion. Finally, the chemical conversion in the model can be developed with the *segregation* and *maximum mixed* models.

Simple combinations of reactor elements can be solved directly. Figure 23-8, for instance, shows two CSTRs in series and with recycle through a PFR. The material balances with an n-order reaction $r = kC^n$ are

$$\frac{C_3}{C_2} = \frac{1}{1 + \dfrac{kV_r C_2}{RV'}}$$

$$V'C_f + RC_3 = (V' + R)C_1 + kV_{r1}C_1^n$$

$$(V' + R)C_1 = (V' + R)C_2 + kV_{r2}C_2^n \qquad (23\text{-}2)$$

Elimination of C_1 and C_3 from these equations will result in the desired relation between inlet C_f and outlet C_2 concentrations, although not in an explicit form except for zero or first-order reactions. Alternatively, the Laplace transform could be found, inverted and used to evaluate segregated or max mixed conversions that are defined later. Inversion of a transform like that of Fig. 23-8 is facilitated after replacing the exponential by some ratio of polynomials, a Padé approximation, as explained in books on linear control theory. Numerical inversion is always possible.

A stirred tank sometimes can be modeled as having a fraction α in bypass and a fraction β of the reactor volume stagnant. The material balance then is made up of

$$C = \alpha C_0 + (1 - \alpha)C_1 \qquad (23\text{-}3)$$

$$(1 - \alpha)V'C_0 = (1 - \alpha)V'C_1 + (1 - \beta)kV_rC_1^n \qquad (23\text{-}4)$$

where C_1 is the concentration leaving the active zone of the tank. Elimination of C_1 will relate the input and overall output concentrations. For a first-order reaction,

$$\frac{C_0}{C} = 1 + \frac{kV_r(1 - \beta)}{V'(1 - \alpha)} \qquad (23\text{-}5)$$

The two parameters α and β may be expected to depend on the amount of agitation.

A flow reactor with some deviation from plug flow, a quasi-PFR, may be modeled as a CSTR battery with a characteristic number n of stages, or as a dispersion model with a characteristic value of the dispersion coefficient or Peclet number. These models are described later.

MATERIAL BALANCES

Material and energy balances of common types of reactors are summarized in several tables of Sec. 7. For review purposes some material balances are restated here. For the nth stage of a CSTR battery,

$$V'_{n-1}C_{n-1} = V''_n C_n + V_r r_n \qquad (23\text{-}6)$$

or at constant density and a power law rate,

$$C_{n-1} = C_n + k\bar{t}_n C_n^\alpha \qquad (23\text{-}7)$$

The concentrations of all stages are found in succession when C_0 is known. For a PFR,

$$-V' \, dn_a = r_a \, dV_r = k\left(\frac{n_a}{V'}\right)^\alpha dV_r$$

$$kV_r = \int_{n_a}^{n_{a0}} V'\left(\frac{V'}{n_a}\right)^\alpha dn_a \qquad (23\text{-}8)$$

For operation with an inert tracer, the material balances are conveniently handled as Laplace transforms. For a stirred tank, the differential equation

$$C_0 = C + \bar{t}\,\frac{dC}{dt}, \qquad C = 0 \quad \text{when} \quad t = 0 \qquad (23\text{-}9)$$

becomes

$$\frac{\overline{C}}{\overline{C}_0} = \frac{1}{1 + \bar{t}s} \qquad (23\text{-}10)$$

and the partial differential equation of a plug flow vessel becomes

$$\frac{\overline{C}}{\overline{C}_0} = \exp(-\bar{t}s) \qquad (23\text{-}11)$$

The terms on the right are the transfer functions.

With the two units in series,

$$\frac{\overline{C}_2}{\overline{C}_0} = \left(\frac{\overline{C}_2}{\overline{C}_1}\right)\left(\frac{\overline{C}_1}{\overline{C}_0}\right) = \frac{\exp(-\bar{t}_2 s)}{1 + \bar{t}_1 s} \qquad (23\text{-}12)$$

and in parallel with a fraction β going to the mixed unit

$$\frac{\overline{C}}{\overline{C}_0} = \frac{\beta}{1 + \bar{t}_1 s} + (1 - \beta)\exp(-\bar{t}_2 s) \qquad (23\text{-}13)$$

where $\quad \bar{t}_1 = \dfrac{V_{r1}}{\beta V'}, \qquad \bar{t}_2 = \dfrac{V_{r2}}{(1 - \beta)V'}$

HEAT TRANSFER AND MASS TRANSFER

Temperature affects rates of reaction, degradation of catalysts, and equilibrium conversion. Some of the modes of heat transfer applied in reactors are indicated in Figs. 23-1 and 23-2. Profiles of some temperatures and compositions in reactors are in Figs. 23-3 to 23-6, 23-22, and 23-40. Many reactors with fixed beds of catalyst pellets have divided beds, with heat transfer between the individual sections. Such units can take advantage of initial high rates at high temperatures and higher equilibrium conversions at lower temperatures. Data for two such cases are shown in Table 23-2. For SO_2 the conversion attained in the fourth bed is 97.5 percent, compared with an adiabatic single bed value of 74.8 percent. With the three-bed ammonia reactor, final ammonia concentration is 18.0 percent, compared with the one-stage adiabatic value of 15.4 percent. Some catalysts deteriorate at much above 500°C, another reason for limiting temperatures.

Since reactors come in a variety of configurations, use a variety of operating modes, and may handle mixed phases, the design of provisions for temperature control draws on a large body of heat transfer theory and data. These extensive topics are treated in other sections of this Handbook and in other references. Some of the high points pertinent to reactors are covered by Rase (*Chemical Reactor Design for Process Plants*, Wiley, 1977). Two encyclopedic references are *Heat Exchanger Design Handbook* (5 vols., Hemisphere, 1983–date), and Cheremisinoff, ed. (*Handbook of Heat and Mass Transfer*, 4 vols., Gulf, 1986–1990), which has several articles addressed specifically to reactors.

References to mass transfer are made throughout this section wherever multiple phases are discussed.

CASE STUDIES

Exploration for an acceptable or optimum design of a new reaction process may need to consider reactor types, several catalysts, specifications of feed and product, operating conditions, and economic evaluations. Modifications to an existing process likewise may need to consider many cases. These efforts can be eased by commercial kinetics services. A typical one can handle up to 20 reactions in CSTRs or

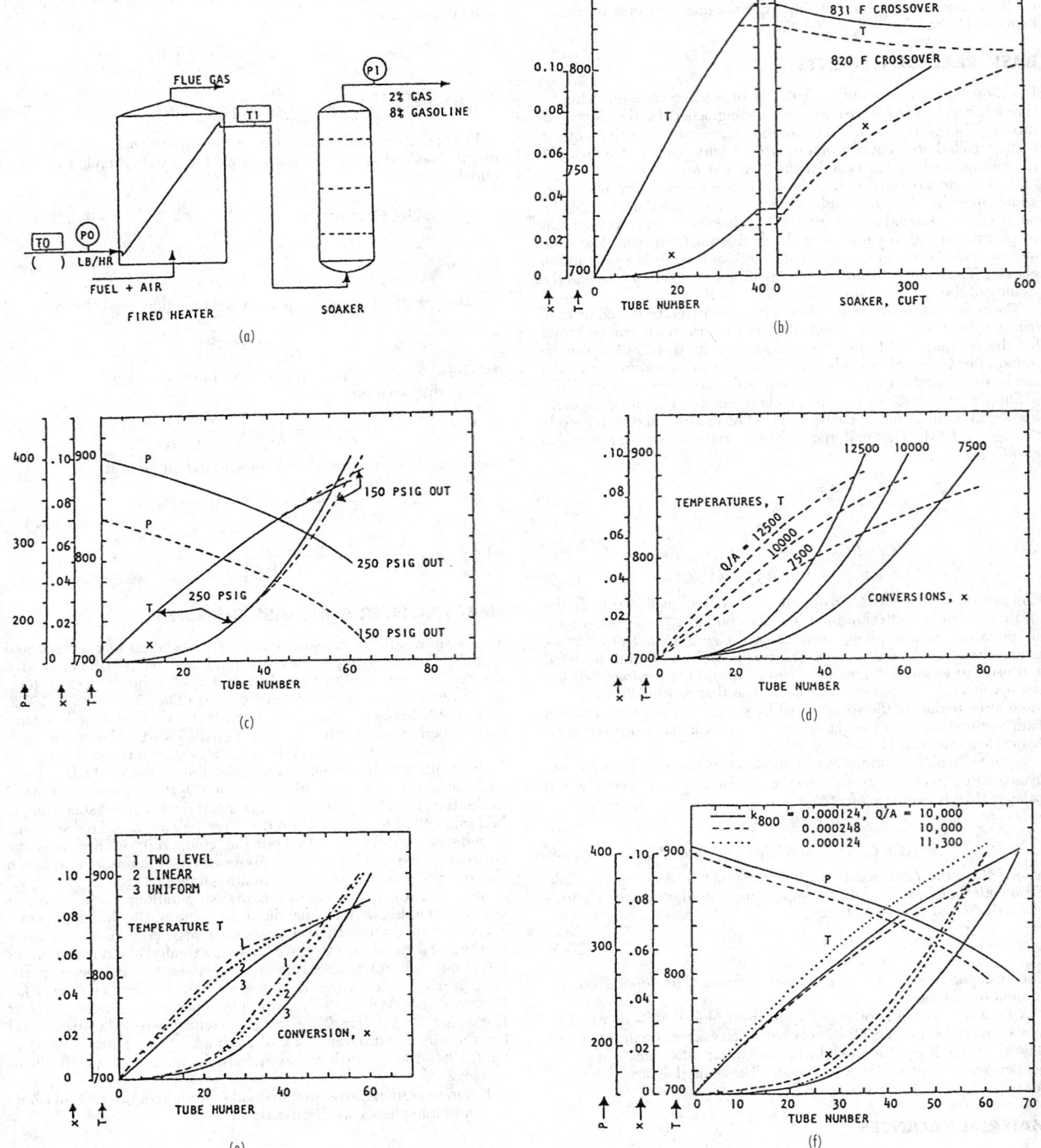

FIG. 23-4   (a) Visbreaking flowsketch, feed 160,000 lbm/h, $k_{800} = 0.000248$/s, tubes 5.05 in ID by 40 ft; (b) Q/A = 10,000 Btu/(ft²·h), P_{out} = 250 psig; (c) Q/A = 10,000 Btu/(ft²·h), P_{out} = 150 or 250 psig; (d) three different heat fluxes, P_{out} = 250 psig; (e) variation of heat flux, average 10,000 Btu/(ft²·h), P_{out} = 250 psig; (f) halving the specific rate. T in °F. To convert psi to kPa, multiply by 6.895; ft to m, multiply by 0.3048; in to cm, multiply by 2.54.

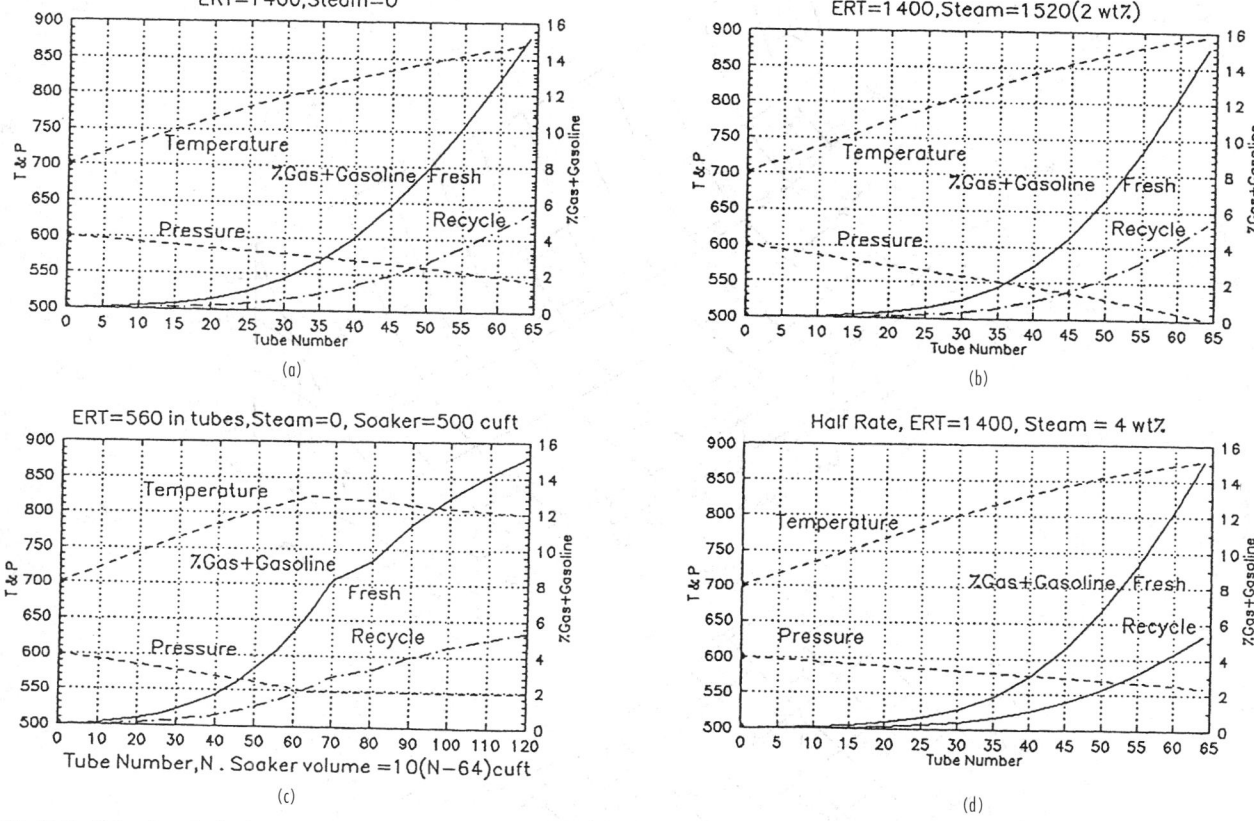

FIG. 23-5 Visbreaking fresh oil at 38,000 lbm/h, $k_{800} = 0.00012/s$, plus heavy gas oil at 38,000 lbm/h, $k_{800} = 0.00004/s$, 2% steam to put flow in the turbulent range. Tubes 4.25 in ID by 69 ft. (*a*) Heat flux 2,300 without steam. (*b*) Heat flux 2,600 with steam. (*c*) Soaker operation cuts the peak temperature by 50 ft and the heat flux by 30%. (*d*) When the feed rate turndown is 50%, steam rate is increased to 4% to keep flow turbulent. To convert ft to m, multiply by 0.3048; lb to kg/h, multiply by 0.454; $\Delta t°F$ to $\Delta t°C$, multiply by 0.556.

PFRs, under isothermal, adiabatic, or heat transfer conditions in one or two phases. Outputs can provide profiles of composition, pressure, and temperature as well as vessel size.

When the kinetics are unknown, still-useful information can be obtained by finding equilibrium compositions at fixed temperature or adiabatically, or at some specified approach to the adiabatic temperature, say within 25°C (45°F) of it. Such calculations require only an input of the components of the feed and products and their thermodynamic properties, not their stoichiometric relations, and are based on Gibbs energy minimization. Computer programs appear, for instance, in Smith and Missen (*Chemical Reaction Equilibrium Analysis Theory and Algorithms*, Wiley, 1982), but the problem often is laborious enough to warrant use of one of the several available commercial services and their data banks. Several simpler cases with specified stoichiometries are solved by Walas (*Phase Equilibria in Chemical Engineering*, Butterworths, 1985).

For some widely practiced processes, especially in the petroleum industry, reliable and convenient computerized models are available from a number of vendors or, by license, from proprietary sources. Included are reactor-regenerator of fluid catalytic cracking, hydrotreating, hydrocracking, alkylation with HF or H_2SO_4, reforming with Pt or Pt-Re catalysts, tubular steam cracking of hydrocarbon fractions, noncatalytic pyrolysis to ethylene, ammonia synthesis, and other processes by suppliers of catalysts. Vendors of some process simulations are listed in the *CEP Software Directory* (AIChE, 1994).

Several excellent case studies that appear in the literature are listed, following.

Rase (*Case Studies and Design Data*, vol. 2 of *Chemical Reactor Design for Process Plants*, Wiley, 1977) has these items:
Styrene polymerization
Cracking of ethane to ethylene
Quench cooling in the ethylene process
Toluene dealkylation
Shift conversion
Ammonia synthesis
Sulfur dioxide oxidation
Catalytic reforming
Ammonia oxidation
Phthalic anhydride production
Steam reforming
Vinyl chloride polymerization
Batch hydrogenation of cottonseed oil
Hydrodesulfurization
Rase (*Fixed Bed Reactor Design and Diagnostics*, Butterworths, 1990) has a general computer program for reactor design and these case studies:
Methane-steam reaction
Hydrogenation of benzene to cyclohexane
Dehydrogenation of ethylbenzene to styrene
Tarhan (*Catalytic Reactor Design*, McGraw-Hill, 1983) has computer programs and results for these cases:
Toluene hydrodealkylation to benzene and methane
Phthalic anhydride by air oxidation of naphthalene
Trickle bed reactor for hydrosulfurization

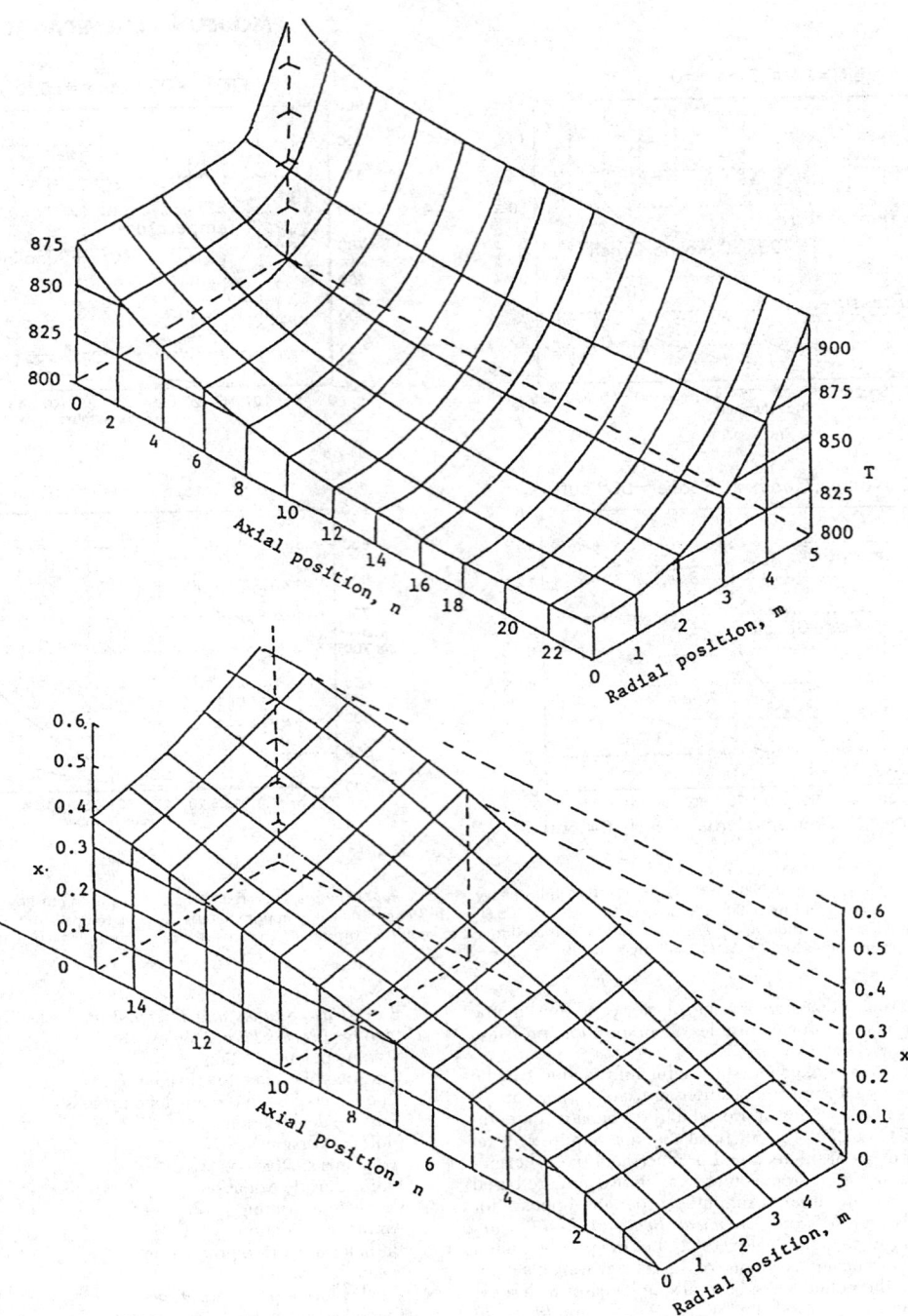

FIG. 23-6 Temperature in K and composition profiles in the styrene reactor. (a) Temperature profiles with $T' = 1000$, $T_{5.0} = 936.5$, and $hU/k_E = 0.5$. (b) Composition profiles with $T' = 1000$, $hU/k_E = 1000$. h = radial increment, U = heat-transfer coefficient at the wall, k_E = thermal conductivity, T in K. To convert K to R, multiply by 1.8.

T_0	T'	hU/k_E	$T_{5.0}$	n at 50% conversion
873	1000	0	873	—°
873	1000	0.5	936.5	25.3
873	1000	0.5	873	26.1
873	1000	1000	1000	15.6
873	1050	0.5	962.5	21.6
873	873	0.5	873	51.2

°The adiabatic reaction temperature reaches essentially the steady conditions, $x = 0.426$ and $T = 784$, after about 70 axial increments.

TABLE 23-2 Multibed Reactors, Adiabatic Temperature Rises and Approaches to Equilibrium*

Oxidation of SO_2 at atmospheric pressure in a four-bed reactor. Feed 6.26% SO_2, 8.3% O_2, 5.74% CO_2, and 79.7% N_2.

°C		Conversion, %	
In	Out	Plant	Equilibrium
463.9	592.8	68.7	74.8
455.0	495.0	91.8	93.4
458.9	465.0	96.0	96.1
435.0	437.2	97.5	97.7

Ammonia synthesis in a three bed reactor at 225 atm. Feed 22% N_2, 66% H_2, 12% inerts.

°C		Ammonia, %	
In	Out	Calculated	Equilibrium
399	518.9	13.0	15.4
427	488.9	16.0	19.0
427	470.0	18.0	21.7

*To convert atm to kPa multiply by 101.3.
SOURCE: Plant data and calculated design values from Rase, *Chemical Reactor Design for Process Plants*, Wiley, 1977.

Ramage, Graziani, Schipper, Krambeck, and Choi (*Advances in Chemical Engineering*, vol. 13, Academic Press, 1987, pp. 193–266):
  Mobil's Kinetic Reforming Model
  Dente and Ranzi (in Albright et al., eds., *Pyrolysis Theory and Industrial Practice*, Academic Press, 1983, pp. 133–175):
  Mathematical modeling of hydrocarbon pyrolysis reactions
  Shah and Sharma (in Carberry and Varma, eds., *Chemical Reaction and Reaction Engineering Handbook*, Dekker, 1987, pp. 713–721):
  Hydroxylamine phosphate manufacture in a slurry reactor
  Some aspects of a kinetic model of methanol synthesis are described in the first example, which is followed by a second example that describes coping with the multiplicity of reactants and reactions of some petroleum conversion processes. Then two somewhat simplified industrial examples are worked out in detail: mild thermal cracking and production of styrene. Even these calculations are impractical without a computer. The basic data and mathematics and some of the results are presented.

Example 1: Kinetics and Equilibria of Methanol Synthesis
Although methanol from synthesis gas has been a large-scale industrial chemical for 70 years, the scientific basis of the manufacture apparently can stand some improvement, which was undertaken by Beenackers, Graaf, and Stamhuis (in Cheremisinoff, ed., *Handbook of Heat and Mass Transfer*, vol. 3, Gulf, 1989, pp. 671–699). The process occurs at 50 to 100 atm with catalyst of oxides of Cu-Zn-Al and a feed stream of H_2, CO, and CO_2. Three reactions were taken for the process:

$$CO + 3H_2 \Leftrightarrow CH_3OH$$
$$CO_2 + H_2 \Leftrightarrow CO + H_2O \quad \text{(shift reaction)}$$
$$CO_2 + 3H_2 \Leftrightarrow CH_3OH$$

In some earlier work the shift reaction was assumed always at equilibrium. Fugacities were calculated with the SRK and Peng-Robinson equations of state, and correlations were made of the equilibrium constants.
  Various Langmuir-Hinshelwood mechanisms were assumed. CO and CO_2 were assumed to adsorb on one kind of active site, s1, and H_2 and H_2O on another kind, s2. The H_2 adsorbed with dissociation and all participants were assumed to be in adsorptive equilibrium. Some 48 possible controlling mechanisms were examined, each with 7 empirical constants. Variance analysis of the experimental data reduced the number to three possibilities. The rate equations of the three reactions are stated for the mechanisms finally adopted, with the constants correlated by the Arrhenius equation.
  Kinetic studies were made with a spinning basket reactor using catalyst HaldorTopsoe MK 101, at three pressures and three temperatures.
  Effectiveness factors of the porous catalyst were found by comparison with crushed particles 0.15 to 0.2 mm in diameter. For methanol formation the range of effectiveness was from approximately 0.9 at 490 K to 0.5 at 560 K, and for water formation 0.8 at 490 K and 0.5 at 560 K.

  Simpler, mostly power law rate equations for the production of mixed alcohols from synthesis gas are cited by Forzatti, Tronconi, and

Villa (in *Handbook of Heat & Mass Transfer*, vol. 4, Gulf, 1990, pp. 289–311). Agreements between their correlations and some data were deemed satisfactory. Rate equations are not necessarily the same on different catalysts.

Example 2: Coping with Multiple Reactants and Reactions
  (a) Thermal cracking of ethane and propane is done to make primarily olefins but other products also are formed. A number of simplified reaction models are in use. One of these (Sundaram and Froment, *Chem. Eng. Sci.*, **32**, 601–617 [1977]; *Ind. Eng. Chem. Fund.*, **17**, 174–182 [1978]) takes these reactions to represent the cracking processes:

Ethane cracking

$$C_2H_6 \xrightleftharpoons{k_1} C_2H_4 + H_2$$
$$2C_2H_6 \xrightarrow{k_2} C_3H_8 + CH_4$$
$$C_3H_8 \xrightarrow{k_3} C_3H_6 + H_2$$
$$C_3H_8 \xrightarrow{k_4} C_2H_4 + CH_4$$
$$C_3H_6 \xrightleftharpoons{k_5} C_2H_2 + CH_4$$
$$C_2H_2 + C_2H_4 \xrightarrow{k_6} C_4H_6$$

Propane cracking

$$C_3H_8 \longrightarrow C_2H_4 + CH_4$$
$$C_3H_8 \rightleftharpoons C_3H_6 + H_2$$
$$C_3H_8 + C_2H_4 \longrightarrow C_2H_6 + C_3H_6$$
$$2C_3H_6 \longrightarrow 3C_2H_4$$
$$2C_3H_6 \longrightarrow 0.5C_6 + 3CH_4$$
$$C_3H_6 \rightleftharpoons C_2H_2 + CH_4$$
$$C_3H_6 + C_2H_6 \longrightarrow C_4H_8 + CH_4$$
$$C_2H_6 \rightleftharpoons C_2H_4 + H_2$$
$$C_2H_4 + C_2H_2 \longrightarrow C_4H_6$$

In the second paper the models were amplified: for ethane, 49 reactions with 11 molecular species and 9 free radicals; for propane, 80 reactions with 11 molecular species and 11 free radicals. The second paper has a list of 133 reactions involving light hydrocarbons and their first- or second-order specific rates.
  (b) The feed to a typical fluidized catalytic cracking unit consists of liquid C_5 to C_{40} and contains thousands of individual species. This stream is made amenable to kinetic modeling by a process of lumping. A Mobil Corporation model (Weekman, "Lumps Models and Kinetics in Practice," *AIChE Monograph Series* No. 11, 1979) employs 10 lumps and 20 reactions. There are four lumps each in the boiling range, 222 to 342°C, of paraffins, naphthenes, aromatics, and aromatic substituents and also four above 342°C. The other two lumps are total C_1 to C_4 plus coke, and C_5 to 222°C. The specific rates are presumably proprietary data. In 1991, a seventh generation of this model was said to be in use.
  (c) Commercial catalytic reformers upgrade naphthas in the C_5 to C_{12} range to high-octane gasoline with coproducts. Some 300 participants have been identified. In a Mobil Corporation model (Ramage et al., *Advances in Chemical Engineering*, vol. 13, Academic Press, 1987, pp. 193–266; Sapre, in Sapre and Krambeck, eds., *Chemical Reactions in Complex Mixtures*, Van Nostrand Reinhold, 1991, pp. 222–253) these have been lumped into pseudocomponents identified by carbon number and chemical nature. These lumps vary with the age of the catalyst. The table shows the 13 lumps adopted for the "start of the kinetic cycle."

Carbon number	Six-carbon-ring naphthenes, N_6	Five-carbon-ring naphthenes, N_5	Paraffins, P	Aromatics, A
C_8	C_8 cyclohexanes (1)	C_8 cyclopentanes (2)	C_8 paraffins (3)	C_8 aromatics (4)
C_7	Methylcyclohexane (5)	Cyclopentanes (6)	Heptanes (7)	Toluene (8)
C_6	Cyclohexane (9)	Methylcyclopentane (10)	Hexanes (11)	Benzene (12)
C_5				C_5 hydrocarbons (13)

The reaction network is shown in the paper. The kinetic characteristics of the lumps are proprietary. Originally, the model required 30 person-years of effort on paper and in the laboratory, and it is kept up to date.

Example 3: Thermal Cracking of Heavy Oils (Visbreaking)
Mild thermal cracking is conducted in the tubes of a fired heater, sometimes fol-

lowed by an adiabatic holding drum. The cases to be studied are for making a product with reduced viscosity and with 8 percent gasoline plus 2 percent gas. A tube diameter is first selected with the rule that the cold oil velocity be 5 to 6 ft/s; this diameter is subject to change during the course of the calculation. For a specified heat flux, Q/A Btu/(h·ft^2 of tube surface), a number of parameters can be explored:

- Number of tubes of a given length, with and without a soaking drum
- Profiles of T, P, and fractional conversion x

A key consideration is that coking becomes likely in the vicinity of 900°F.

For the process of the flowsketch two different cases will be examined:
(a) Oil feed 160,000 lbm/h, $k_{800} = 0.000248$/s, tubes 5.05 in ID, 40 ft long.
(b) 38,000 lbm/h of fresh oil, $k_{800} = 0.00012$/s, + 38,000 lbm/h heavy gas oil, $k_{800} = 0.00004$/s, + 1,520 lbm/h steam; 64 tubes in series, 4.25 in ID, each 69 ft long.

In view of the necessity to be able to design for charge stocks of varied and not-well-characterized properties, some approximations are made. The main properties and operating variables are listed. n is the increment number of the integration and x is the fraction decomposed to gas + gasoline.
1. The components are original oil M, lbm/h, 10°API; cracked oil, 30°API; steam S, lbm/h; gasoline product 114 mol wt.; gas product 28 mol wt.
2. The gas produced is a fraction $y = 0.2$ of the gas + gasoline made.
3. The specific rate at temperature T_n is

$$k = k_{800} \exp\left[50,000\left(\frac{1}{1,260} - \frac{1}{T_n + 460}\right)\right]$$

4. The mean density is figured on the assumption that the gas and gasoline are in the vapor phase and the cracked oil is in the liquid phase:

$$\rho_n = f(T_n, P_n, x_n) \tag{A}$$

5. The viscosity is that of the 30°API stock. $\nu_n = f(T_n)$, cSt.
6. Heat of reaction $\Delta H_r = 332$ Btu/(lbm gas + gasoline)
7. $Q_t A_t/M$ = heat input per increment of reactor, Btu/(h·lbm feed).
8. Each component of the feed has the enthalpy equation $H_i = A_i + B_i(T - 800)$.
9. The enthalpies of the mixtures leaving increments n and $n + 1$ of the reaction are expressed per unit mass of fresh oil,

$$H_n = f(T_n, x_n), \qquad H_{n+1} = f(T_{n+1}, x_{n+1})$$

10. The enthalpy balance over an increment of tubes is

$$H_n + \frac{Q_t A_t}{M} = H_{n+1} + \Delta H_r(x_{n+1} - x_n)$$

11. The temperature is derived from the enthalpy balance and is

$$T_{n+1} = T_0 + f(T_n, x_n, x_{n+1}) \tag{B}$$

where $T_0 = 800$ is the reference temperature for enthalpy.
12. The Reynolds number is

$$\text{Re}_n = \frac{6.316\,(M + S)}{D\rho_n \nu_n}$$

D, in; ρ, lbm/ft^3; ν, cSt
13. The friction factor is given by Round's equation with a roughness factor $\varepsilon = 0.00015$ ft,

$$f_n = \frac{1.6434}{\ln(0.000243/D + 6.5/\text{Re}_n)}$$

14. The pressure relation, lbm/in^2 is

$$P_{n+1} = P_n + \frac{3.356(10^{-6})L_t}{D^5}(M + S)^2\left(\frac{f_n}{\rho_n} + \frac{f_{n+1}}{\rho_{n+1}}\right) \tag{C}$$

The differential material balance of the first-order flow reaction is

$$M\,dx = r\,dV_r = kC\,dV_r = \frac{kM(1 - x)}{V'}dV_r = \frac{kM(1 - x)}{(M + S)/\rho}dV_r$$

$$= k\rho\left(\frac{M}{M + S}\right)(1 - x)\,V_t\,dN$$

The integral is rearranged to

$$x_{n+1} = 1 - (1 - x_n)\exp\left[-\frac{V_t}{M + S}\int_N^{N + \Delta N} k\rho\,dN\right]$$

Integration is by successive approximation, using essentially Euler's method. For the first trial,

$$x_{n+1}^{(1)} = 1 - (1 - x_n)\exp\left(-\frac{3600V_t}{M + S}(k\rho)_n \Delta N\right) \tag{D}$$

and for subsequent trials,

$$x_{n+1}^{(i+1)} = 1 - (1 - x_n)\exp\left\{-\frac{1800V_t}{M + S}\left[(k\rho)_n + (k\rho)_{n+1}^{(i)}\right]\right\} \tag{E}$$

Here,
k_n is a function of T_n
ρ_n is a function of T_n, P_n and x_n
V_t is the volume of one tube
ΔN is the number of tubes per reactor increment
3,600 compensates for the fact that k_n is per s and $M + S$ is per h
Calculation procedure
(a) In the preheat zone of the reactor, before cracking is appreciable, usually below 800°F, the pressure drop is found adequately enough by taking average densities and Reynolds numbers over this zone. The conditions at the inlet to the reaction zone are designated x_0, T_0, and P_0.
(b) Evaluate k_n, H_n, ρ_n, and ν_n at the inlet, where $n = 0$.
(c) Apply Eq. (D) to find the first approximation $x_{n+1}^{(1)}$ at the end of the first reactor increment.
(d) Find the next approximation $x_{n+1}^{(2)}$ with Eq. (E).
(e) If the condition

$$\frac{|x_{n+1}^{(2)} - x_{n+1}^{(1)}|}{x_{n+1}^{(2)} + x_{n+1}^{(1)}} \leq 0.01$$

is not satisfied, repeat the process; otherwise, continue with the next increment until the specified conversion is attained or some other specification is met or violated.

Equivalent residence time (ERT) can be found after the temperature profile has been established:

$$\text{ERT} = \frac{V_r}{V'\rho_0}\int_0^1\left(\frac{k}{k_{800}}\right)\rho\,dz \tag{F}$$

where z is the fractional distance along the total tube length. Engineers in the industry have a feel for what value of ERT is desirable with particular stocks and their conversion. The number usually is in the range of 500 to 1,500.

Results The results of computer calculations are summarized for the two operations by Figs. 23-4 and 23-5. In the second design, the equations had to be modified to take account of two feeds with different specific rates.

Example 4: Styrene from Ethylbenzene
The principal reaction in the dehydrogenation of ethylbenzene is to styrene and hydrogen,

$$C_6H_5C_2H_5 \Leftrightarrow C_6H_5C_2H_3 + H_2$$

For a catalyst packed reactor, a rate equation was found by Wenner and Dybdal (*Chem. Eng. Prog.*, **44**, 275–286 [1948]):

$$R_c = k_r\left(P_E - \frac{(P_S P_H)^2}{K_e}\right) \text{ kg mol/(h·kg catalyst)} \tag{A}$$

$$k_r = \exp\left(9.44 - \frac{11,000}{T}\right) \tag{B}$$

$$K_e = \exp\left(15.69 - \frac{15,000}{T}\right) \tag{C}$$

where the pressures are in bar and temperatures in K.

The material balance with bulk flow in the axial direction z and diffusion in the radial direction r with diffusivity D gives rise to the equation

$$\frac{\partial(uC)}{\partial z} - \frac{D}{u}\left[\frac{\partial^2(uC)}{\partial r^2} + \frac{1}{r}\frac{\partial(uC)}{\partial r}\right] + \rho R_c = 0 \tag{D}$$

In terms of the fraction converted, the material balance becomes

$$\frac{\partial x}{\partial z} - \frac{D}{u}\left[\frac{\partial^2 x}{\partial r^2} + \frac{1}{r}\frac{\partial x}{\partial r}\right] - \frac{\rho}{u_0 C_0}R_c = 0 \tag{E}$$

Similarly, the heat balance gives rise to

$$\frac{\partial T}{\partial z} - \frac{k_c}{GC_p}\left[\frac{\partial^2 T}{\partial r^2} + \frac{1}{r}\frac{\partial T}{\partial r}\right] + \frac{\Delta H_r \rho}{GC_p}R_c = 0 \tag{F}$$

where G is the mass flow rate (2,500 kg/h as developed later).

Operating conditions The reactor is 10 cm ID, input of ethylbenzene is 0.069 kg mol/h, input of steam is 0.69 kgmol/h, total of 2,500 kg/h. Pressure is 1.2 bar, inlet temperature is 600°C. Heat is supplied at some constant temperature in a jacket. Performance is to be found with several values of heat transfer coefficient at the wall, including the adiabatic case.

Data

Bulk density of catalyst	$\rho = 1,440$ kg/m^3
Heat of reaction	$\Delta H_r = 140,000$ kJ/kg mol
Thermal conductivity	$k_c = 0.45$ W/m K
Ratio of diffusivity to linear velocity	$D/u = 0.000427$ m
Specific heat of the mixture	$C_p = 2.18$ kJ/kg K

In terms of the fractional conversion and the ideal gas law, the rate equation becomes

$$R_c = 1.2\exp\left(9.44 - \frac{11,000}{T}\right)\left[\frac{1 - x}{11 + x} - \frac{1.2}{K_e}\left(\frac{x}{11 + x}\right)^2\right] \tag{G}$$

Substitution of the data transforms the heat and material balances into Eqs. (H) and (I):

$$\frac{\partial T}{\partial z} = a\left(\frac{\partial^2 T}{\partial r^2} + \frac{1}{r}\frac{\partial T}{\partial r}\right) - bR_c, \qquad a = 0.000297, \qquad b = 37{,}000 \tag{H}$$

$$\frac{\partial x}{\partial z} = c\left(\frac{\partial^2 x}{\partial r^2} + \frac{1}{r}\frac{\partial x}{\partial r}\right) + dR_c, \qquad c = 0.000427, \qquad d = 164 \tag{I}$$

At the center, Eqs. (H) and (I) become, after application of l'Hopital's rule,

$$\frac{\partial T}{\partial z} = 2a\frac{\partial^2 T}{\partial r^2} - bR_c \tag{J}$$

$$\frac{\partial x}{\partial r} = 2c\frac{\partial^2 T}{\partial r^2} + dR_c \tag{K}$$

At the wall,

$$-k_e\frac{\partial T}{\partial r} = U(T - T') \tag{L}$$

Eqs. (H) through (L) will be solved by the explicit finite difference method. Substitute

$$\frac{\partial T}{\partial z} = \frac{T_{m,\,a+1} - T_{m,\,n}}{k}, \qquad k = \Delta z \tag{M}$$

$$\frac{\partial T}{\partial r} = \frac{T_{m,\,a+1} - T_{m,\,n-1}}{2h}, \qquad h = \Delta r \tag{N}$$

$$\frac{\partial^2 T}{\partial r^2} = \frac{T_{m+1,\,n} - 2T_{m,\,n} + T_{m-1,\,n}}{h^2} \tag{O}$$

and similarly for variable x.

The substituted finite difference equations will contain the terms

$$M' = \frac{0.000427k}{h^2} = 0.250, \qquad \text{for stability} \tag{P}$$

$$M = \frac{0.000297k}{h^2} = 0.174$$

The choice $M' = 0.250$ is to prevent negative coefficients and thus an unstable iterative process.

The finite difference equations will be formulated with five radial increments $m = 0, 1, 2, 3, 4, 5$, and for as many axial increments as necessary to obtain 50 percent conversion. Accordingly,

$$h = \Delta r = 1 \text{ cm}$$

$$k = \Delta z = \frac{0.250}{0.000427} = 585 \text{ cm}$$

At the center, $m = 0$, the finite difference equations become

$$T_{0,\,n+1} = 4MT_{1,\,n} + (1 - 4M)T_{0,\,n} - kbR_{c0,\,n} \tag{Q}$$

$$x_{0,\,n+1} = 4M'x_{1,\,n} + (1 - 4M')x_{0,\,n} + kdR_{c0,\,n} \tag{R}$$

When $m = 1, 2, 3,$ or 4,

$$T_{m,\,n+1} = M\left(1 + \frac{1}{2m}\right)T_{m+1,\,n} + (1 - 2M)\,T_{m,\,n}$$

$$+ M\left(1 - \frac{1}{2m}\right)T_{m-1,\,n} - kbR_{cm,\,n} \tag{S}$$

$$x_{m,\,n+1} = M\left(1 + \frac{1}{2m}\right)x_{m+1,\,n} + (1 - 2M)\,x_{m,\,n}$$

$$+ M\left(1 - \frac{1}{2m}\right)x_{m-1,\,n} + kdR_{cm,\,n} \tag{T}$$

At the wall, $m = 5$, the heat-transfer coefficient is U and the heating gas has temperature T'_n,

$$T_{5,\,n} = \frac{T_{4,\,n} + (hU/k_e)T'_n}{1 + hU/k_e} \tag{U}$$

$$x_{5,\,n} = x_{4,\,n} \tag{V}$$

The rate equation becomes

$$R_{cm,\,n} = 1.2\,k_r\left[\frac{1 - x_{m,\,n}}{11 + x_{m,\,n}} - \frac{1.2}{K_e}\left(\frac{x_{m,\,n}}{11 + x_{m,\,n}}\right)^2\right] \tag{W}$$

$$k_r = \exp\left(9.44 - \frac{11{,}000}{T_{m,\,n}}\right) \tag{X}$$

$$K_e = \exp\left(15.69 - \frac{15{,}000}{T_{m,\,n}}\right) \tag{Y}$$

At this point the computer takes over. Cases with several values of jacket temperature and several values of heat-transfer coefficient, or hU/k_e, are examined, and also several assumptions about the temperature at the wall at the inlet. Eq. (U) with $n = 0$ could be used. The number of axial increments are found for several cases of 50% conversion. Two of the profiles of temperature or conversion are shown in Fig. 23-6.

The Crank-Nicholson implicit method and the method of lines for numerical solution of these equations do not restrict the radial and axial increments as Eq. (P) does. They are more involved procedures, but the burden is placed on the computer in all cases.

A version of this problem is solved by Jenson and Jeffreys (*Mathematical Methods in Chemical Engineering*, Academic Press, 1977).

More up-to-date data of this process are employed in a study by Rase (*Fixed Bed Reactor Design and Diagnostics*, Butterworths, 1990, pp. 275–286). In order to keep the pressure drop low, radial flow reactors are used, two units in series with reheating between them. Simultaneous formation of benzene, toluene, and minor products is taken into account. An economic comparison is made of two different catalysts under a variety of operating conditions. Some of the computer printouts are shown there.

RESIDENCE TIME DISTRIBUTION (RTD) AND REACTOR EFFICIENCY

The distribution of residence times of reactants or tracers in a flow vessel, the RTD, is a key datum for determining reactor performance, either the expected conversion or the range in which the conversion must fall. In this section it is shown how tracer tests may be used to establish how nearly a particular vessel approaches some standard ideal behavior, or what its efficiency is. The most useful comparisons are with complete mixing and with plug flow. A glossary of special terms is given in Table 23-3, and major relations of tracer response functions are shown in Table 23-4.

TRACERS

Nonreactive substances that can be used in small concentrations and that can easily be detected by analysis are the most useful tracers. When making a test, tracer is injected at the inlet of the vessel along with the normal charge of process or carrier fluid, according to some definite time sequence. The progress of both the inlet and outlet concentrations with time is noted. Those data are converted to a residence time distribution (RTD) that tells how much time each fraction of the charge spends in the vessel.

An RTD, however, does not represent the mixing behavior in a vessel uniquely, because several arrangements of the internals of a vessel may give the same tracer response, for example, any series arrangements of reactor elements such as plug flow or complete mixing. This is a consequence of the fact that tracer behavior is represented by linear differential equations. The lack of uniqueness limits direct application of tracer studies to first-order reactions with constant specific rates. For other reactions, the tracer curve may determine the upper and lower limits of reactor performance. When this range is not too broad, the result can be useful. Tracer data also may be taken at several representative positions in the vessel in order to develop a realistic model of the reactor.

REACTOR EFFICIENCY

One quantitative measure of reactor efficiency at a conversion level x is the ratio of the mean residence time or the reactor volume in a plug flow reactor to that of the reactor in question,

$$\eta_x = \left(\frac{t_{pf}}{t}\right)_x = (V_{pf}/V)_x \tag{23-14}$$

TABLE 23-3 Glossary of RTD Terms

Closed end vessel One in which the inlet and outlet streams are completely mixed and dispersion occurs only between the terminals. At the inlet where $z = 0$, $uC_0 = [uC - D_e(\partial C/\partial z)]_{z=0}$; at the outlet where $x = L$, $(\partial C/\partial z)_{z=L} = 0$. These are called *Danckwerts' boundary conditions*.

Concentration The main special kinds are: C_δ, that of the effluent from a vessel with impulse input of tracer; $C^0 = m/V_r$, the initial mean concentration resulting from impulse input of magnitude m; C_u, that of the effluent from a vessel with a step input of magnitude C_f.

Dispersion The movement of aggregates of molecules under the influence of a gradient of concentration, temperature, and so on. The effect is represented by Fick's law with a dispersion coefficient substituted for molecular diffusivity. Thus, rate of transfer $= -D_e(\partial C/\partial z)$.

Impulse An amount of tracer injected instantaneously into a vessel at time zero. The symbol $m\delta(t - a)$ represents an impulse of magnitude m injected at time $t = a$. The effluent concentration resulting from an impulse input is designated C_δ.

Maximum mixedness Exists when any molecule that enters a vessel immediately becomes associated with those molecules with which it will eventually leave the vessel; that is, with those molecules that have the same life expectation. A state of MM is associated with every RTD.

Mixing, ideal or complete A state of complete uniformity of composition and temperature in a vessel. In flow, the residence time varies exponentially, from zero to infinity.

Peclet number for dispersion $Pe = uL/D_e$ where u is a linear velocity, L is a linear dimension, and D_e is the dispersion coefficient. In packed beds, $Pe = ud_p/D_e$, where u is the interstitial velocity and d_p is the pellet diameter.

Plug flow A condition in which all effluent molecules have had the same residence time.

Residence time distribution (RTD) In the case of elutriation of tracer from a vessel that contained an initial average concentration C^0, the area under a plot of $E(t) = C_{\text{effluent}}/C^0$ between the ordinates at t_1 and t_2 is the fraction of the molecules that have residence times in this range. In the case of constant input of concentration C_f to a vessel with zero initial concentration, the ratio $F(t) = C_{\text{effluent}}/C_f$ at t_1 is the fraction of molecules with residence times less than t_1.

Residence time, mean The average time spent by the molecules in a vessel. Mathematically, it is the first moment of the effluent concentration from a vessel with impulse input, or

$$\bar{t} = \frac{\int_0^\infty tC_\delta\, dt}{\int_0^\infty C_\delta\, dt}$$

Segregated flow Occurs when all molecules that enter together also leave together. A state of aggregation is associated with every RTD. Each aggregate of molecules reacts independently of every other aggregate; thus, as an individual batch reactor.

Skewness The third moment of a residence time distribution:

$$\gamma^3(t) = \int_0^\infty (t - \bar{t})^3 E(t)\, dt.$$

It is a measure of asymmetry.

Step An input in which the concentration of tracer is changed to some constant value C_f at time zero and maintained at this level indefinitely. The symbol $C_f u(t - a)$ represents a step of magnitude C_f beginning at $t = a$. The resulting effluent concentration is designated C_u.

Variance The second moment of the RTD. There are two forms: one in terms of the absolute time, designated $\sigma^2(t)$; and the other in terms of reduced time, $t_r = t/\bar{t}$, designated $\sigma^2(t_r)$.

$$\sigma^2(t) = \frac{\int_0^\infty (t - \bar{t})^2 C_\delta\, dt}{\int_0^\infty C_\delta\, dt} = \int_0^\infty (t - \bar{t})^2 E(t)\, dt$$

$$\sigma^2(t_r) = \frac{\sigma^2(t)}{\bar{t}^2} = \int_0^\infty (t_r - 1)^2 E(t_r)\, dt_r$$

TABLE 23-4 Tracer Response Functions

Mean residence time:

$$\bar{t} = \frac{\int_0^\infty tC_\delta\, dt}{\int_0^\infty C_\delta\, dt} = \frac{\int_0^{C_{u\infty}} t\, dC_u}{C_{u\infty}}$$

Initial mean concentration with impulse input,

$$C^0 = \frac{m}{V_r} = \left(\frac{V'}{V_r}\right)\int_0^\infty C_\delta\, dt = \frac{\int_0^\infty C_\delta\, dt}{\bar{t}}$$

Reduced time:

$$t_r = \frac{t}{\bar{t}}$$

Residence time distribution:

$$E(t) = \frac{C_\delta}{\int_0^\infty C_\delta\, dt} = \frac{E(t_r)}{\bar{t}} = \frac{dF(t)}{dt}$$

Residence time distribution, normalized,

$$E(t_r) = \frac{\text{impulse output}}{\text{initial mean concentration}}$$

$$= \frac{C_\delta}{C^0} = \frac{\bar{t}C_\delta}{\int_0^\infty C_\delta\, dt} = \bar{t}\, E(t) = \frac{dF(t)}{dt}$$

Age:

$$F(t) = \frac{\text{step output}}{\text{step input}}$$

$$= \frac{C_u}{C_f} = \frac{\int_0^t C_\delta\, dt}{\int_0^\infty C_\delta\, dt} = F(t_r)$$

Internal age:

$$I(t) = 1 - F(t)$$

Intensity:

$$\Lambda(t) = \frac{E(t)}{1 - F(t)} = \frac{E(t)}{I(t)}$$

Variance:

$$\sigma^2(t) = \int_0^\infty (t - \bar{t})^2 E(t)\, dt = -\bar{t}^2 + \frac{\int_0^\infty t^2 C_\delta\, dt}{\int_0^\infty C_\delta\, dt}$$

Variance, normalized:

$$\sigma^2(t_r) = \frac{\sigma^2(t)}{\bar{t}^2} = -1 + \frac{\int_0^\infty t^2 C_\delta\, dt}{\int_0^\infty C_\delta\, dt}$$

$$= \int_0^1 (t_r - 1)^2\, dF(t_r)$$

Skewness, third moment:

$$\gamma^3(t_r) = \int_0^\infty (t_r - 1)^3 E(t_r)\, dt_r$$

The conversion level sometimes is taken at 95 percent of equilibrium, but there is no universal standard.

Other measures of efficiency are derived from the experimental RTD, which is characterized at least approximately by the variance $\sigma^2(t_r)$. This quantity is zero for plug flow and unity for complete mixing, and thus affords natural bounds to an efficiency equated to the variance. It is possible, however, for the variance to fall out of the range (0,1) when stagnancy or bypassing occurs.

A related measure of efficiency is the equivalent number of stages n_{erlang} of a CSTR battery with the same variance as the measured RTD. Practically, in some cases 5 or 6 stages may be taken to approximate plug flow. The dispersion coefficient D_e also is a measure of deviation from plug flow and has the merit that some limited correlations in terms of operating conditions have been obtained.

No correlations of $\sigma^2(t_r)$ or n_{erlang} have been achieved in terms of operating variables. At present, the chief value of RTD studies is for the diagnosis of the performance of existing equipment; for instance, maldistribution of catalyst in a packed reactor, or bypassing or stagnancy in stirred tanks. Reactor models made up of series and/or parallel elements also can be handled theoretically by these methods.

TRACER RESPONSE

The unsteady material balances of tracer tests are represented by linear differential equations with constant coefficients that relate an input function $C_f(t)$ to a response function of the form

$$\sum_0^n a_n \frac{dC^n}{dt^n} = C_f \tag{23-15}$$

The general form of the material balance is the familiar one,

$$\text{Inputs} + \text{Sources} = \text{Outputs} + \text{Sinks} + \text{Accumulations}$$

as described in the "Modeling Chemical Reactors" section. For the special case of initial equilibrium or dead state (all derivatives = 0 at time 0), the transformed function of the preceding equation is

$$\overline{C} = \frac{\overline{C_f}}{\sum a_n s^n} \qquad (23\text{-}16)$$

and $C(t)$ is found by inversion with a table of Laplace transform pairs. The individual transform is defined as

$$\overline{C} = C(s) = \int_0^\infty \exp{(-st)}C(t)\,dt \qquad (23\text{-}17)$$

The ratio of transforms,

$$G(s) = \frac{C(s)}{C_f(s)} \qquad (23\text{-}18)$$

is called a *transfer function.* The concept is useful in the representation of systems consisting of several elements in series and parallel.

A particularly useful property of linear differential equations may be explained by comparing an equation and its derivative in operator form,

$$f(D)y = g(t) \qquad \text{and} \qquad f(D)z = \frac{dg(t)}{dt} \qquad (23\text{-}19)$$

where the RHS of the second equation is the derivative of the RHS of the first. The property in question is that $z = dy/dt$. The chief use of this property in this area is with the step and impulse functions, the impulse being the derivative of the step. Often problems are easier to visualize and formulate in terms of the step input, but the desired solution may be for the impulse input which gives the RTD directly.

Kinds of Inputs Since a tracer material balance is represented by a linear differential equation, the response to any one kind of input is derivable from some other known input, either analytically or numerically. Although in practice some arbitrary variation of input concentration with time may be employed, five mathematically simple input signals supply most needs. *Impulse* and *step* are defined in the Glossary (Table 23-3). *Square pulse* is changed at time a, kept constant for an interval, then reduced to the original value. *Ramp* is changed at a constant rate for a period of interest. A *sinusoid* is a signal that varies sinusoidally with time. Sinusoidal concentrations are not easy to achieve, but such variations of flow rate and temperature are treated in the vast literature of automatic control and may have potential in tracer studies.

RESPONSE FUNCTIONS

The chief quantities based on tracer tests are summarized in Table 23-4. Effluent concentrations resulting from impulse and step inputs are designated C_δ and C_u, respectively. The initial mean concentration resulting from an impulse of magnitude m into a vessel of volume V_r is $C^0 = m/V_r$. The mean residence time is the ratio of the vessel volume to the volumetric flow rate, $\bar{t} = V_r/V'$ or $\bar{t} = \int_0^\infty tC_\delta\,dt/\int_0^\infty C_\delta\,dt$. The reduced time is $t_r = t/\bar{t}$.

Residence time distributions are used in two forms: Normalized, $E(t_r) = C_\delta/C^0$; or plain, $E(t) = C_\delta/\int_0^\infty C_\delta\,dt$. The relation between them is $E(t_r) = \bar{t}E(t)$. On time plots, the area under either RTD is unity: $\int_0^\infty E(t_r)\,dt_r = \int_0^\infty E(t)\,dt = 1$. Moreover, the area between the ordinates at t_1 and t_2 is the fraction of the total effluent that has spent the period between those times in the vessel.

The *age function* is defined in terms of the step input as

$$F(t) = \frac{C_u}{C_f} = \int_0^t E(t)\,dt$$

Relations to other functions are in Table 23-3.

The intensity function $\Lambda(t) = E(t)/[1 - F(t)]$ occurs in the maximum mixing concept and is of value, for instance, in detecting maldistributions of catalyst and channeling in a packed vessel.

The Erlang number n_{erlang} and the variances $\sigma^2(t_r)$ and $\sigma^2(t)$ are single parameter characterizations of RTD curves. The skewness $\gamma^3(t)$, and higher moments can be used to represent RTD curves more closely if the data are accurate enough.

ELEMENTARY MODELS

Real reactors may conform to some sort of ideal mixing patterns, or their performance may be simulated by combinations of ideal models. The commonest ideal models are the following:

- *Plug flow reactor (PFR)*, in which all portions of the charge have the same residence time. The concentration varies with time and position, according to the equation,

$$\frac{\partial C}{\partial t} + V'\frac{\partial C}{\partial V_r} = 0$$

- *Continuous stirred tank reactor (CSTR)*, with the effluent concentration the same as the uniform vessel concentration. With a mean residence time $\bar{t} = V_r/V'$, the material balance is

$$\bar{t}\frac{dC}{dt} + C = \text{Input concentration}$$

- *Dispersion model* is based on Fick's diffusion law with an empirical dispersion coefficient D_e substituted for the diffusion coefficient. The material balance is

$$\frac{\partial C}{\partial t} + V'\frac{\partial C}{\partial V_r} - D_e\frac{\partial^2 C}{\partial V_r^2} = 0$$

- *Laminar* or power law velocity distribution in which the linear velocity varies with radial position in a cylindrical vessel. Plug flow exists along any streamline and the mean concentration is found by integration over the cross section.

- *Distribution models* are curvefits of empirical RTDs. The Gaussian distribution is a one-parameter function based on the statistical rule with that name. The Erlang and gamma models are based on the concept of the multistage CSTR. RTD curves often can be well fitted by ratios of polynomials of the time.

Figure 23-7 illustrates the responses of CSTRs and PFRs to impulse or step inputs of tracers.

REAL BEHAVIOR

An empty tubular reactor often can be simulated as a PFR or by a dispersion model with a small value of the dispersion coefficient. Stirred tank performance often is nearly completely mixed (CSTR), or the model may be modified by taking account of bypass zones, stagnant zones, or other parameters associated with the geometry and operation of the vessel. The additional parameters contribute to the mathematical complexity of the model. Sometimes the vessel can be visualized as a zone of complete mixing in the vicinity of impellers followed by plug flow zones elsewhere; thus, CSTRs followed by PFRs. Packed beds usually deviate substantially from plug flow. The dispersion model and some combinations of PFRs and CSTRs or multiple CSTRs in series may approximate their behavior. Fluidized beds in small sizes approximate CSTR behavior, but large ones exhibit bypassing, stagnancy, nonhomogeneous regions, and several varieties of contact between particles and fluid.

The concept of transfer functions facilitates the combination of linear elements. The rule is:

Output transform = (Transfer function) (Input transform)

Figure 23-8 develops the overall transform of a process with a PFR in parallel with two CSTRs in series. $C(t)$ is found from $C(s)$ by inversion of the output transform.

TRACER EQUATIONS

Differential equations and their solutions will be stated for the elementary models with the main kinds of inputs. Since the ODEs are linear, solutions by Laplace transforms are feasible.

Ideal CSTR With a step input of magnitude C_f the unsteady material balance is

$$V_r\frac{dC}{dt} + V'C = V'C_f \qquad (23\text{-}20)$$

whose integral is

$$\frac{C}{C_f} = F(t_r) = 1 - \exp{(-t_r)} \qquad (23\text{-}21)$$

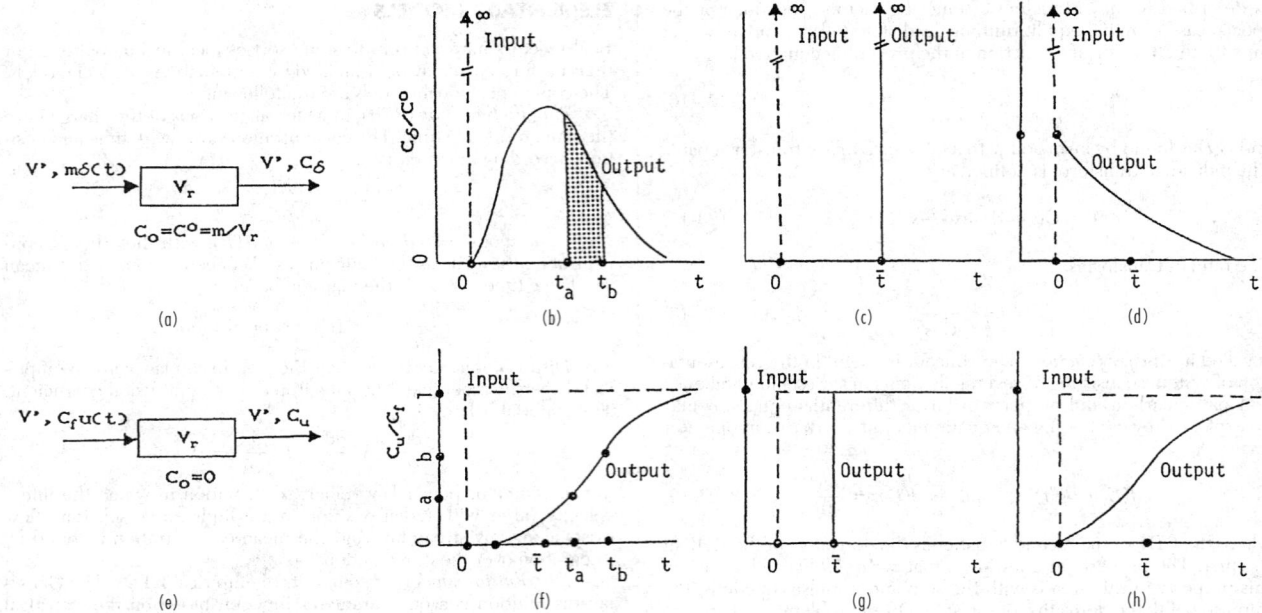

FIG. 23-7   Impulse and step inputs and responses. Typical, PFR and CSTR. (*a*) Experiment with impulse input of tracer. (*b*) Typical behavior; area between ordinates at t_a and t_b equals the fraction of the tracer with residence time in that range. (*c*) Plug flow behavior; all molecules have the same residence time. (*d*) Completely mixed vessel; residence times range between zero and infinity. (*e*) Experiment with step input of tracer; initial concentration zero. (*f*) Typical behavior; fraction with ages between t_a and t_b equals the difference between the ordinates, $b - a$. (*g*) Plug flow behavior; zero response until $t = \bar{t}$ has elapsed, then constant concentration C_f (*h*) Completely mixed behavior; response begins at once, and ultimately reaches feed concentration.

With an impulse input of magnitude m or an initial mean concentration $C^0 = m/V_r$, the material balance is

$$\frac{dC}{dt_r} + C = 0 \qquad \text{when} \qquad C = C^0, t = 0 \qquad (23\text{-}22)$$

and

$$\frac{C}{C^0} = E(t_r) = \exp(-t_r) \qquad (23\text{-}23)$$

From these results it is clear that $E(t_r) = dF(t_r)/dt_r$.

Plug Flow Reactor (PFR)   The material balance over a differential volume dV_r is

$$V'C = V'(C + dC) + dV_r \frac{\partial C}{\partial t}$$

or

$$\frac{\partial C}{\partial t} + V' \frac{\partial C}{V_r} = 0 \qquad (23\text{-}24)$$

With step input the boundary conditions are

$$C(0,t) = C_f u(t) \qquad C(V_r,0) = 0 \qquad (23\text{-}25)$$

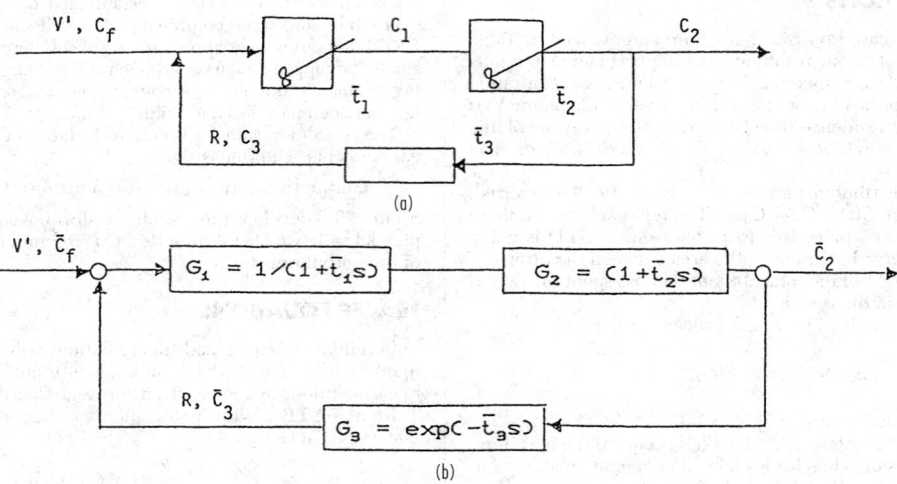

FIG. 23-8   Diagrams of a composite process: (*a*) process flow diagram, (*b*) transfer function diagram. The overall transfer function is:

$$\frac{\overline{C}_2}{\overline{C}_f} = \frac{1}{(1 + R/V')(1 + t_1 s)(1 + t_2 s) - (R/V') \exp(-t_3 s)}$$

By Laplace transform the solution is

$$\frac{C}{C_f} = F(t) = u(t - \bar{t}) = 0 \qquad \text{when} \qquad t \leq \bar{t}$$

$$= 1 \qquad \text{when} \qquad t \geq \bar{t} \qquad (23\text{-}26)$$

The response to impulse input is found by differentiation,

$$\frac{C}{C^0} = E(t_r) = \bar{t}\delta(t - \bar{t}) = \bar{t}\delta[\bar{t}(t_r - 1)] = \delta(t_r - 1) \qquad (23\text{-}27)$$

that is, the effluent is an impulse that is delayed from the input impulse by $t_r = 1$ or $t = \bar{t}$.

Multistage CSTR This model has a particular importance because its RTD curve is bell-shaped like those of many experimental RTDs of packed beds and some empty tubes. The RTD is found by induction by solving the equations of one stage, two stages, and so on, with the result,

$$E(t_r) = \frac{C_n}{C^0} = \frac{n^n}{(n-1)!} t_r^{n-1} \exp(-nt_r) \qquad (23\text{-}28)$$

Step response is found by integration, thus

$$F(t_r) = \int_0^{t_r} E(t_r) \, dt_r = 1 - \exp(-nt_r) \sum_{j=0}^{n-1} \frac{(nt_r)^j}{j!} \qquad (23\text{-}29)$$

Plots of $E(t_r)$ and $f(t_r)$ for various values of n appear in Fig. 23-9. The bell shapes of $E(t_r)$ are more distinctive. The experimental curves of Figs. 23-10 and 23-11 clearly are of that family.

The peak of the E curve is reached at $t_r = (n-1)/n$ and has a magnitude

$$E(t_r)_{\text{max}} = \frac{n(n-1)^{n-1}}{(n-1)!} \exp(1-n) \qquad (23\text{-}30)$$

from which n can be found when $E(t_r)_{\text{max}}$ has been measured.

Another significant characteristic of the E curve is the variance or the second moment which is

$$\sigma^2(t_r) = \int_0^\infty (t_r - 1)^2 E(t_r) \, dt_r = \frac{1}{n} \qquad (23\text{-}31)$$

The E equation can be rearranged into a linear form,

$$\ln[t_r E(t_r)] = \ln\left[\frac{n^n}{(n-1)!}\right] + n \ln[t_r \exp(-nt_r)] \qquad (23\text{-}32)$$

With appropriate coordinates the slope of a loglog plot is n.

Combined Models, Transfer Functions The transfer function relation is

$$\overline{C}_{\text{output}} = (\text{Transfer function}) \, \overline{C}_{\text{input}}$$

$$= G(s)\overline{C}_{\text{input}} \qquad (23\text{-}33)$$

Some common transfer functions are

Element	Transfer function, $G(s)$
Ideal CSTR	$\dfrac{1}{(1 + \bar{t}s)}$
PFR	$\exp(-\bar{t}s)$
n-stage CSTR (Erlang)	$\dfrac{1}{(1 + \bar{t}s)^n}$
Erlang with time delay	$\dfrac{\exp(-\bar{t}_1 s)}{(1 + \bar{t}_2 s)^n}$

The last item is of a PFR and an n-stage CSTR in series.

Although a transfer function relation may not be always invertible analytically, it has value in that the moments of the RTD may be derived from it, and it is thus able to represent an RTD curve. For instance, if G_0' and G_0'' are the limits of the first and second derivatives of the transfer function $G(s)$ as $s \Rightarrow 0$, the variance is

$$\sigma^2(t) = G_0'' - (G_0')^2$$

Characterization of RTD Curves An empirical RTD curve can be represented by equations of more than one algebraic form. The characteristic bell shape of many RTDs is evident in the real examples of Figs. 23-10 and 23-11. Such shapes invite comparison with some well-known statistical distributions and representation of the RTDs by their equations. Many of the standard statistical distributions are described by Hahn and Shapiro (*Statistical Models in Engineering*, Wiley, 1967) with their applicabilities. The most useful models in the present area are the gamma (or Erlang) and the Gaussian together with its Gram-Charlier extension. These distributions are representable by only a few parameters that define the asymmetry, the peak, and the shape in the vicinity of the peak. The moments—variance, skewness, and kurtosis—are such parameters.

Gamma or Erlang Distribution For nonintegral values, the factorial of the function $E(t_r)$ of Eq. (23-28) replaced as $(n-1)! = \Gamma(n)$. The result is called the gamma distribution,

$$E(t_r)_{\text{gamma}} = \frac{n^n}{\Gamma(n)} t_r^{n-1} \exp(-nt_r) \qquad (23\text{-}34)$$

The value of n is the only parameter in this equation. Several procedures can be used to find its value when the RTD is known by experiment or calculation: from the variance, as in $n = 1/\sigma^2(t_r) = 1/\bar{t}^2\sigma^2(t)$, or from a suitable loglog plot or the peak of the curve as explained for the CSTR battery model. The Peclet number for dispersion is also related to n, and may be obtainable from correlations of operating variables.

Gaussian Distribution The best-known statistical distribution is the normal, or Gaussian, whose equation is

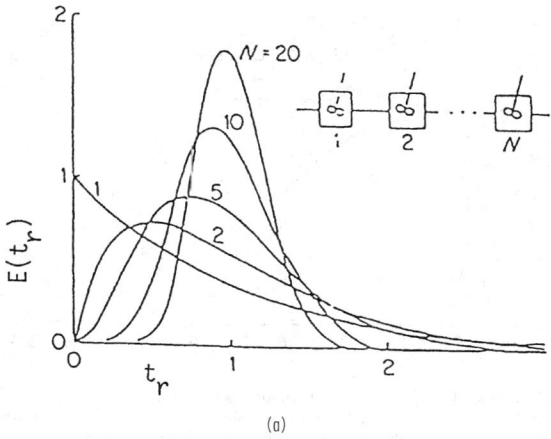

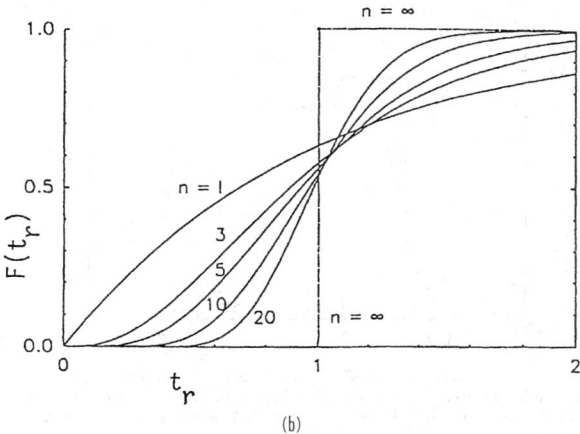

FIG. 23-9 Tracer responses to n-stage continuous stirred tank batteries; the Erlang model: (*a*) impulse inputs, (*b*) step input.

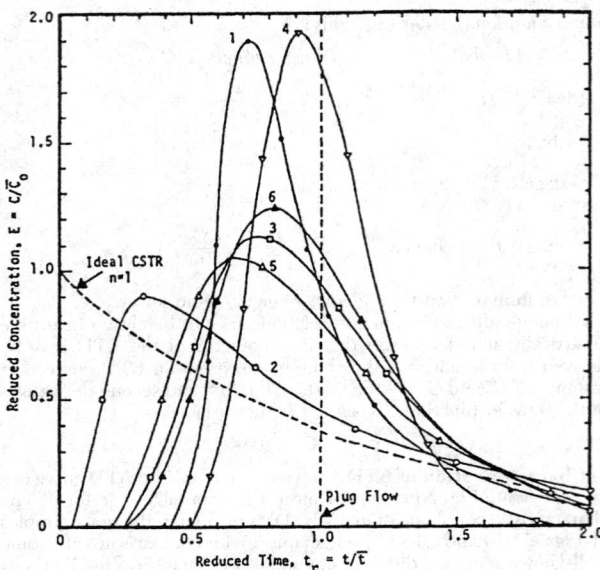

FIG. 23-10 Residence time distributions of pilot and commercial reactors. σ^2 = variance of the residence time distribution, n = number of stirred tanks with the same variance, Pe = Peclet number.

No.	Code	Process	σ^2	n	Pe
1	○	Aldolization of butyraldehyde	0.050	20.0	39.0
2	●	Olefin oxonation pilot plant	0.663	1.5	1.4
3	□	Hydrodesulfurization pilot plant	0.181	5.5	9.9
4	▽	Low-temp hydroisomerization pilot	0.046	21.6	42.2
5	△	Commercial hydrofiner	0.251	4.0	6.8
6	▲	Pilot plant hydrofiner	0.140	7.2	13.2

(*Walas*, Chemical Process Equipment, *Butterworths, 1990.*)

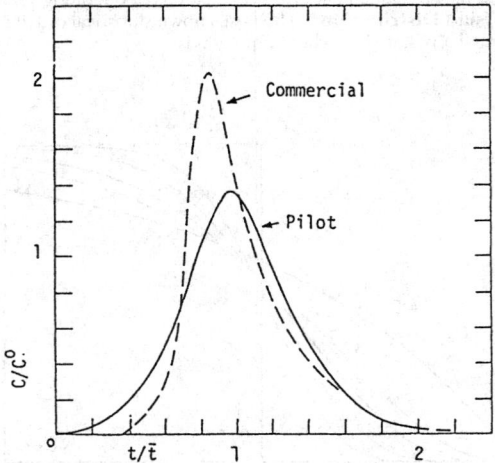

FIG. 23-11 Tracer tests on hydrosulfurizers with 10-mm catalyst pellets. Commercial, 3 ft by 30 ft, $\bar{t} = 23$ s, $n = 12.9$. Pilot, 4 in by 9 ft, $\bar{t} = 14$ s, $n = 9.3$. To convert ft to m, multiply by 0.3048. (*Sherwood*, A Course in Process Design, *MIT Press, 1963.*)

$$C(t_r) = \frac{1}{\sigma\sqrt{2\pi}} \exp\left[-\frac{(t_r-1)^2}{2\sigma^2}\right], \qquad -\infty \le t_r \le +\infty \qquad (23\text{-}35)$$

Since only positive values of t_r are of concern in RTD work, this function is normalized by dividing by the integral from 0 to ∞ with the result

$$E(t_r)_{\text{gauss}} = f(\sigma) \exp\left[-\frac{(t_r-1)^2}{2\sigma^2}\right] \qquad (23\text{-}36)$$

where

$$f(\sigma) = \frac{\sqrt{2/\pi\sigma^2}}{1 + \text{erf}(1/\sigma\sqrt{2})} \qquad (23\text{-}37)$$

Gram-Charlier Series This is an infinite series whose coefficients involve the Gaussian distribution and its derivatives (Kendall, *Advanced Theory of Statistics,* vol. 1, Griffin, 1958). The derivatives, in turn, are expressed in terms of the moments. The series truncated at the coefficient involving the fourth moment is

$$E(t_r)_{\text{GC}} = E(t_r)_{\text{gauss}}\left[\frac{1 - m_3(3z - z^3)}{6} + \frac{(m_4 - 3)(z^4 - 6z^2 + 3)}{24}\right] \qquad (23\text{-}38)$$

where $\quad z = \frac{(t_r - 1)}{\sigma}$

$$m_3 = \left(\frac{\gamma}{\sigma}\right)^3 = \int_0^\infty \left(\frac{t_r - 1}{\sigma}\right)^3 E(t_r)\, dt_r$$

$$m_4 = \left(\frac{\delta}{\sigma}\right)^4 = \int_0^\infty \left(\frac{t_r - 1}{\sigma}\right)^4 E(t_r)\, dt_r$$

More terms of the series are usually not justifiable because the higher moments cannot be evaluated with sufficient accuracy from experimental data. A comparison of the fourth-order GC with other distributions is shown in Fig. 23-12, along with calculated segregated conversions of a first-order reaction. In this case, the GC is the best fit to the original. At large variances the finite value of the ordinate at $t_r = 0$ appears to be a fatal objection to both the Gaussian and GC distributions. On the whole, the gamma distribution is perhaps the best representation of experimental RTDs.

Empirical Equations Tabular (C,t) data are easier to use when put in the form of an algebraic equation. Then necessary integrals and derivatives can be formed most readily and accurately. The calculation of chemical conversions by such mechanisms as segregation, maximum mixedness, or dispersion also is easier with data in the form of equations.

Procedures for curve fitting by polynomials are widely available. Bell-shaped curves, however, are fitted better and with fewer constants by ratios of polynomials. For figuring chemical conversions, the

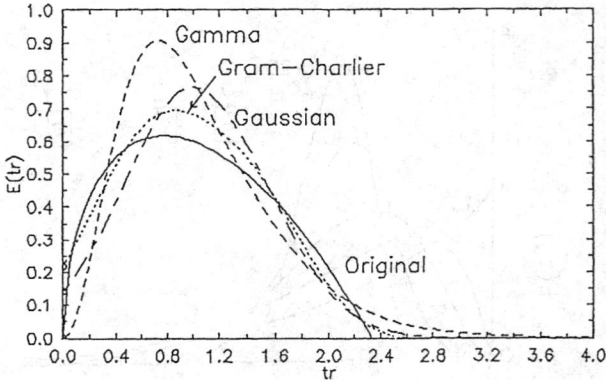

FIG. 23-12 Comparison of RTD models, all with the same variance and skewness. Values of C/C_0 of segregated conversion of a first-order reaction with $kt = 3$: original, 0.1408; gamma, 0.1158; Gauss, 0.1148; GC, 0.1418.

curve fit need not be accurate at small values of $E(t)$, since those regions do not affect the overall conversion significantly.

CHEMICAL CONVERSION

A distinction is to be drawn between situations in which (1) the flow pattern is known in detail, and (2) only the residence time distribution is known or can be calculated from tracer response data. Different networks of reactor elements can have similar RTDs, but fixing the network also fixes the RTD. Accordingly, reaction conversions in a known network will be unique for any form of rate equation, whereas conversions figured when only the RTD is known proceed uniquely only for linear kinetics, although they can be bracketed in the general case.

When the flow pattern is known, conversion in a known network and flow pattern is evaluated from appropriate material and energy balances. For first-order irreversible isothermal reactions, the conversion equation can be obtained from the transfer function by replacing s with the specific rate k. Thus, if $G(s) = \overline{C}/\overline{C}_0 = 1/(1 + \bar{t}s)$, then $C/C_0 = 1/(1 + kt)$. Complete knowledge of a network enables incorporation of energy balances into the solution, whereas the RTD approach cannot do that.

Segregated Flow   In segregated flow the molecules travel as distinct groups, in which all molecules that enter the vessel together leave together. The groups are small enough so that the RTD of the whole system is represented by a smooth curve. For reaction orders above one, with a given RTD, the conversion is a maximum in segregated flow and a minimum under *maximum mixedness* conditions. This point is discussed in detail later.

Each group of molecules reacts independently of any other group, that is, as a batch reactor. Batch conversion equations for power law rate equations are:

$$\left(\frac{C}{C_0}\right)_{\text{batch}} = \exp(-kt) = \exp(-k\bar{t}t_r), \qquad \text{first order} \qquad (23\text{-}39)$$

$$\left[\frac{1}{1 + (q-1)kC^{q-1}\bar{t}t_r}\right]^{1/(q-1)}, \qquad \text{order } q \qquad (23\text{-}40)$$

For other rate equations a numerical solution may be needed.

The mean conversion of all the groups is the sum of the products of the individual conversions and their volume fractions of the total flow. Since the groups are small, the sum is replaced by the integal. Thus,

$$\left(\frac{C}{C_0}\right)_{\text{segregated}} = \int_0^\infty \left(\frac{C}{C_0}\right)_{\text{batch}} E(t)\,dt \qquad (23\text{-}41)$$

$$= \int_0^\infty \left(\frac{C}{C_0}\right)_{\text{batch}} E(t_r)\,dt_r \qquad (23\text{-}42)$$

Conversion in segregated flow is less than in plug flow and somewhat greater than in a CSTR battery with the same variance of the RTD.

When a conversion and an RTD are known, the specific rate can be found by trial: Values of k are estimated until one is found that makes the segregated integral equal to the known value. Moreover, if a series of conversions are known at several residence times, the order of the reaction can be found by trying different orders and noting which give a constant series of specific rates. A catch here, however, is that the RTD depends on the hydrodynamics of the process and may change with the residence time.

Maximum Mixedness   The flow pattern for maximum mixedness is compared with segregated flow in Fig. 23-13. Segregated flow is in a vessel with multiple side *outlets* that result in a particular RTD. Maximum mixedness occurs in a plug flow vessel with multiple side *inlets* whose flow pattern is given by the same RTD. The main flow in the vessel is plug flow, but at each inlet the incoming material is completely mixed across the cross section with the axial flow. This means that each portion of fresh material is mixed with all the material that has the same life expectation, regardless of the actual residence time in the vessel up to the time of mixing. The life expectation under plug flow conditions is measured by the distance remaining to be traveled before leaving the vessel.

In contrast to segregated flow, in which the mixing occurs only after each sidestream leaves the vessel, under maximum mixedness mixing of all molecules having a certain period remaining in the vessel (the life expectation) occurs at the time of introduction of fresh material. These two mixing extremes—as late as possible and as soon as possible, both consistent with the same RTD—correspond to performance extremes of the vessel as a chemical reactor.

The differential equation of maximum mixedness was obtained by Zwietering (*Chem. Eng. Sci.,* **11,** 1 [1959]). It is:

$$\frac{dC}{dt} = R_c - \frac{E(t)}{1 - F(t)}(C_0 - C) \qquad (23\text{-}43)$$

where R_c is the chemical rate equation; for an order q, $R_c = kC^q$. In the normalized units $f = C/C_0$ and $t_r = t/\bar{t}$,

$$\frac{df}{dt_r} = k\bar{t}C_0^{q-1}f^q - \frac{E(t_r)}{1 - F(t_r)}(1 - f) \qquad (23\text{-}44)$$

The boundary condition is

$$\frac{df}{dt_r} = 0 \qquad \text{for} \qquad t_r \Rightarrow \infty \qquad (23\text{-}45)$$

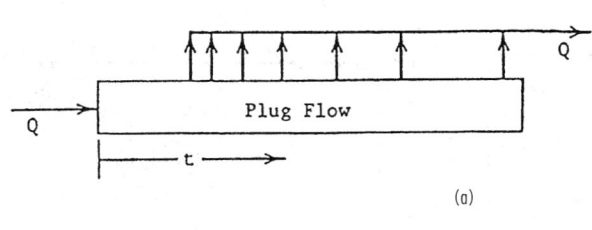
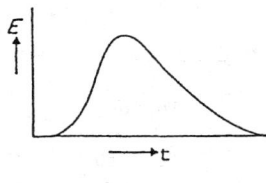

(a)

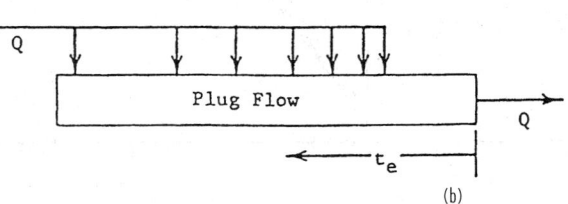
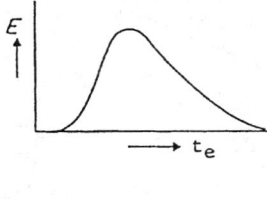

(b)

FIG. 23-13   The two limiting flow patterns with the same RTD. (*a*) Segregated flow, in which all molecules of any exit stream have the same residence time. (*b*) Maximum mixed flow, in which all molecules of an external stream with a certain life expectation are mixed with all molecules of the internal stream that have the same life expectation.

which makes

$$k\bar{t}C_0^{q-1}f_\infty^q - \frac{E(\infty)}{1-F(\infty)}(1-f_\infty) = 0 \qquad (23\text{-}46)$$

The conversion achieved in the vessel is obtained by the solution of the differential equation at the exit of the vessel where the life expectation is $t = 0$. The starting point for the integration is (f_∞, t_∞). When integrating numerically, however, the RTD becomes essentially 0 by the time t_r becomes 3 or 4, and the value of the integral beyond that point becomes nil. Accordingly, the integration interval is from $(f_\infty, t_r \leq 3 \text{ or } 4)$ to $(f_{\text{effluent}}, t_r = 0)$. f_∞ is found from Eq. (23-46).

Numerical solutions of the maximum mixedness and segregated flow equations for the Erlang model have been obtained by Novosad and Thyn (*Coll. Czech. Chem. Comm.*, **31**, 3,710–3,720 [1966]). A few comparisons are made in Fig. 23-14. In some ranges of the parameters n or R_c the differences in conversion or reactor sizes for the same conversions are substantial. On the basis of only an RTD for the flow pattern, perhaps only an average of the two calculated extreme performances is justifiable.

Experimental confirmations of these mixing mechanisms are scarce. One study was with a 50-gal stirred tank reactor (Worrell and Eagleton, *Can. J. Chem. Eng.*, 254–258 [Dec. 1964]). They found seg-

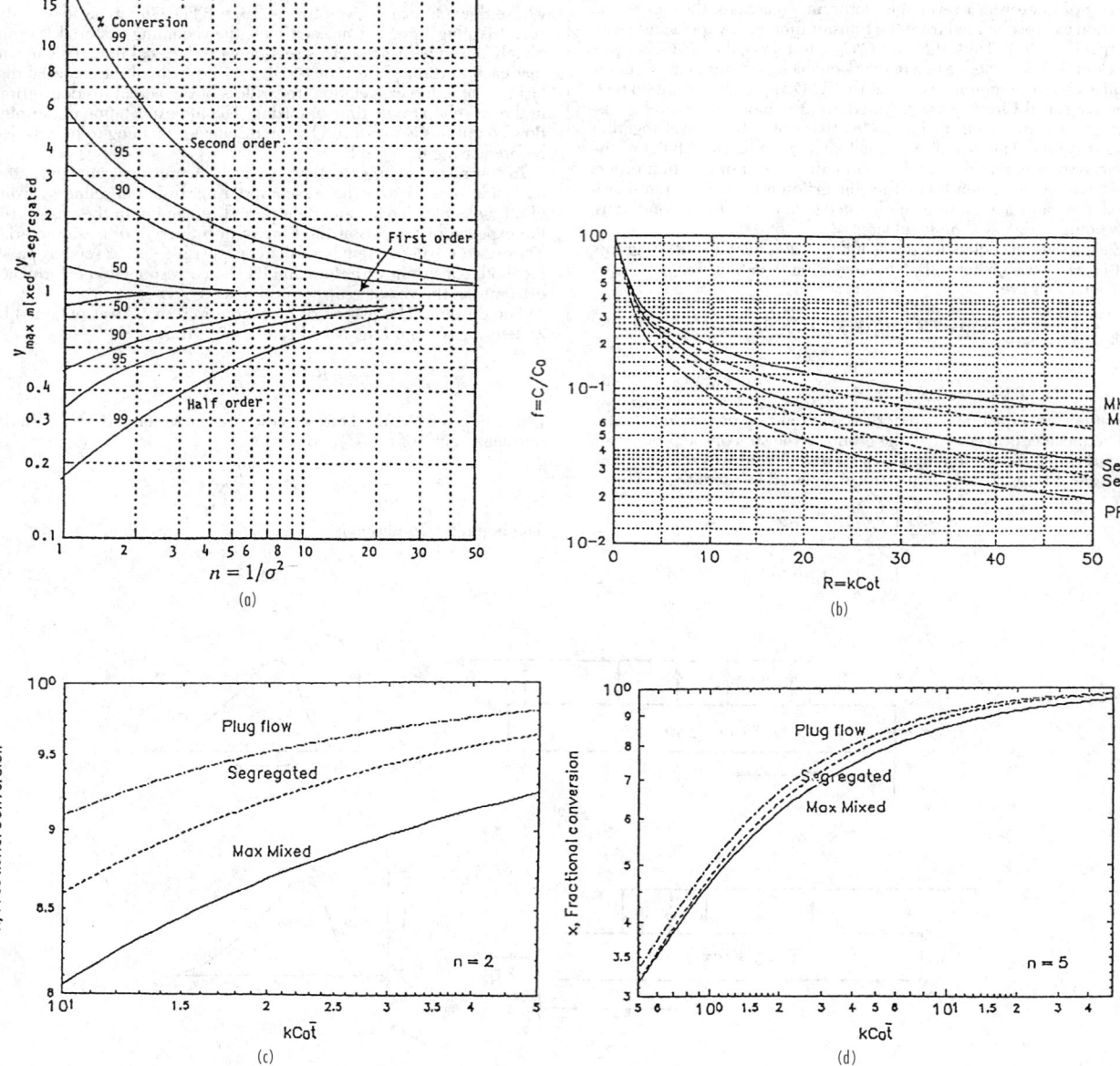

FIG. 23-14   Comparison of maximum mixed, segregated, and plug flows. (*a*) Relative volumes as functions of variance or n, for several reaction orders. (*b*) Second-order reaction with $n = 2$ or 3. (*c*) Second-order, $n = 2$. (*d*) Second-order, $n = 5$.

regation at low agitation, and were able to correlate complete mixing and maximum mixedness in terms of the power input and recirculation within the vessel.

Dispersion Model   An impulse input to a stream flowing through a vessel may spread axially because of a combination of molecular diffusion and eddy currents that together are called *dispersion*. Mathematically, the process can be represented by Fick's equation with a dispersion coefficient replacing the diffusion coefficient. The dispersion coefficient D_e is associated with a linear dimension L and a linear velocity in the Peclet number, $Pe = uL/D_e$. In plug flow, $D_e = 0$ and $Pe \Rightarrow \infty$; and in a CSTR, $D_e \Rightarrow \infty$ and $Pe = 0$.

The dispersion coefficient is orders of magnitude larger than the molecular diffusion coefficient. Some rough correlations of the Peclet number are proposed by Wen (in Petho and Noble, eds., *Residence Time Distribution Theory in Chemical Engineering*, Verlag Chemie, 1982), including some for fluidized beds. Those for axial dispersion are:

1. Axial dispersion in empty tubes,

$$\frac{1}{Pe} = \frac{1}{(Re)(Sc)} + \frac{(Re)(Sc)}{192},$$

$$1 \le Re \le 2000, \qquad 0.2 \le Sc \le 1000 \quad (23\text{-}47)$$

$$\frac{1}{Pe} = \frac{3(10^7)}{(Re)^{2.1}} + \frac{1.35}{(Re)^{0.125}}, \qquad Re \ge 2000 \quad (23\text{-}48)$$

2. Axial dispersion of gases in packed tubes,

$$\frac{1}{Pe} = \frac{0.3}{(Re)(Sc)} + \frac{0.5}{1 + \dfrac{3.8}{(Re)(Sc)}},$$

$$0.008 \le Re \le 400, \qquad 0.28 \le Sc \le 2.2 \quad (23\text{-}49)$$

where Pe = Peclet number, $d_p u_0 / \varepsilon D$
Re = Reynolds number, $d_p s_p u_0 / \nu$
Sc = Schmidt number, ν/D
D = axial dispersion coefficient
d_p = Diameter of particle or empty tube
ε = Fraction voids in packed bed
u_0 = Superficial velocity in the vessel.

In a vessel with axial dispersion, the steady-state equation for a reaction of order q is

$$\frac{d^2 f}{dz^2} = Pe\left(\frac{df}{dz} - R_c\right) = Pe\left(\frac{df}{dz} - k\bar{t}C_0^{q-1}f^q\right) \quad (23\text{-}50)$$

where the normalized variables are $f = C/C_0$ and $z = x/L$. For tracer flow, $R_c = 0$ and the time derivative appears,

$$\frac{\partial f}{\partial t_r} = \frac{1}{Pe}\frac{\partial^2 f}{\partial z^2} - \frac{\partial f}{\partial z} \quad (23\text{-}51)$$

The solution of this partial differential equation is recorded in the literature (Otake and Kunigata, *Kagaku Kogaku*, **22**, 144 [1958]). The plots of $E(t_r)$ against t_r are bell-shaped, resembling the corresponding Erlang plots. A relation is cited later between the Peclet number, n_{erlang}, and $\sigma^2(t_r)$.

Boundary Conditions   In normal operation with "closed ends," reactant is brought in by bulk flow and carried away by both bulk and dispersion flow. At the inlet where $L = 0$ or $z = 0$,

$$uC_0 = \left(uC - D\frac{\partial C}{\partial L}\right)_{L=0} \quad (23\text{-}52)$$

or

$$f_0 = \left(f - \frac{1}{Pe}\frac{\partial f}{\partial z}\right)_{z=0} \quad (23\text{-}53)$$

At the exit where $z = 1$,

$$\left(\frac{\partial f}{\partial z}\right)_{z=1} = 0 \quad (23\text{-}54)$$

With these two-point boundary conditions the dispersion equation, Eq. (23-50), may be integrated by the *shooting method*. Numerical solutions for first- and second-order reactions are plotted in Fig. 23-15.

The discontinuity of concentration at the inlet is commonly observed with CSTR operations, where $Pe = 0$. At other values of Pe, the effects are shown in Fig. 23-16.

Comparison of Models   Only scattered and inconclusive results have been obtained by calculation of the relative performances of the different models as converters. Both the RTD and the dispersion coefficient require tracer tests for their accurate determination, so neither method can be said to be easier to apply. The exception is when one of the cited correlations of Peclet numbers in terms of other groups can be used, although they are rough. The tanks-in-series model, however, provides a mechanism that is readily visualized and is therefore popular.

The Erlang (or gamma) and dispersion models can be related by equating the variances of their respective $E(t_r)$ functions. The result for the "closed-ends" condition is

$$\sigma^2(t_r) = \frac{2[Pe - 1 + \exp(-Pe)]}{Pe^2} \quad (23\text{-}55)$$

$$= \frac{1}{n_{erlang}} \quad (23\text{-}56)$$

For both large and small values of Pe,

$$n_{erlang} \Rightarrow \frac{Pe}{2} \quad (23\text{-}57)$$

MULTIPLICITY AND STABILITY

Normally, when a small change is made in the condition of a reactor, only a comparatively small change in the response occurs. Such a system is uniquely stable. In some cases, a small positive perturbation can result in an abrupt change to one steady state, and a small negative perturbation to a different steady condition. Such multiplicities occur most commonly in variable temperature CSTRs. Also, there are cases where a process occurring in a porous catalyst may have more than one effectiveness at the same Thiele number and thermal balance. Some isothermal systems likewise can have multiplicities, for instance, CSTRs with rate equations that have a maximum, as in Example (*d*) following.

Conditions at steady state are determined by heat and material balances. Such balances for a CSTR can be put in the form,

Heat generated by reaction = Sensible heat pickup

or

$$-\Delta H_r V_r r_c = V' \rho C_p (T - T_f) \quad (23\text{-}58)$$

Four examples of multiplicity and stability follow.

Example (a)   For a first-order reaction in a CSTR, the rate of reaction is:

$$r_c = kC = \frac{kC_f}{1 + k\bar{t}} = \frac{C_f \exp(a + b/T)}{1 + \bar{t}\exp(a + b/T)}$$

When this is substituted into the previous equation, both sides become functions of T and may be plotted against each other. As Fig. 23-17 of a typical case shows, as many as three steady states are possible. When generation is greater than removal (as at points A− and B+), the temperature will rise to the next higher steady state; when generation is less than removal (as at points A+ and B−), it will fall to the next steady state. Point B is an unsteady state, while A and C are steady.

Example (b)   In terms of fractional conversion, $f = 1 - C/C_f$, the material and energy balances for a first-order CSTR are:

$$f = \frac{k\bar{t}}{1 + k\bar{t}} = \frac{50k}{1 + 50k} = \frac{\rho C_p(T - T_f)}{-\Delta H_r C_f} = \frac{1.2(0.9)(T - T_f)}{46,000C_f}$$

The two equations are plotted for several combinations of C_f and T_f. The number of steady states can be one, two, or three if zero or complete conversion are considered possibilities.

Example (c)   For the reactions $A \xrightarrow{1} B \xrightarrow{2} C$ the concentrations are:

$$A = \frac{A_f}{1 + k_1\bar{t}}, \qquad B = \frac{k_1\bar{t}A_f}{(1 + k_1\bar{t}(1 + k_2\bar{t})}$$

The heat balance is:

$$-\Delta H_a(A_f - A) - \Delta H_b(B - B_f) = \rho C_p(T - T_f)$$

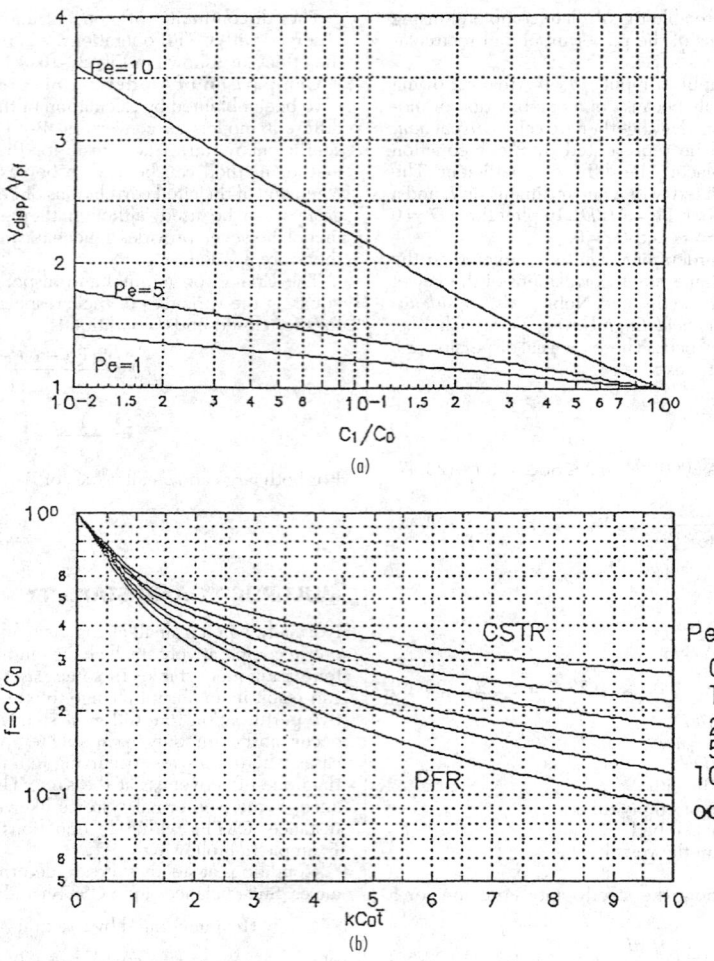

FIG. 23-15 Chemical conversion by the dispersion model. (*a*) First-order reaction, volume relative to plug flow against residual concentration ratio. (*b*) Second-order reaction, residual concentration ratio against $kC_0\bar{t}$.

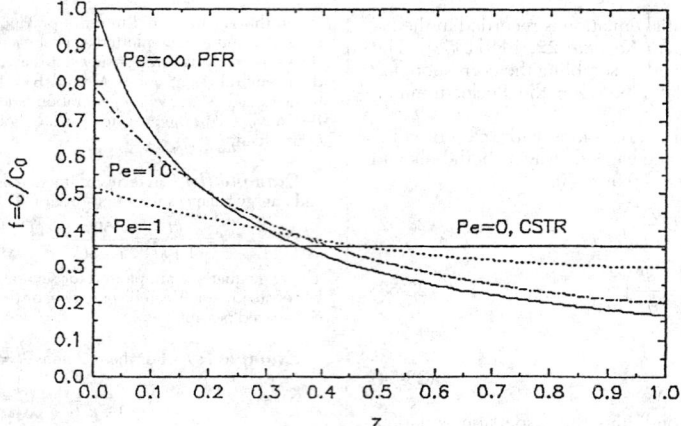

FIG. 23-16 Concentration jump at the inlet of a "closed ends" vessel with dispersion. Second-order reaction with $kC_0\bar{t} = 5$.

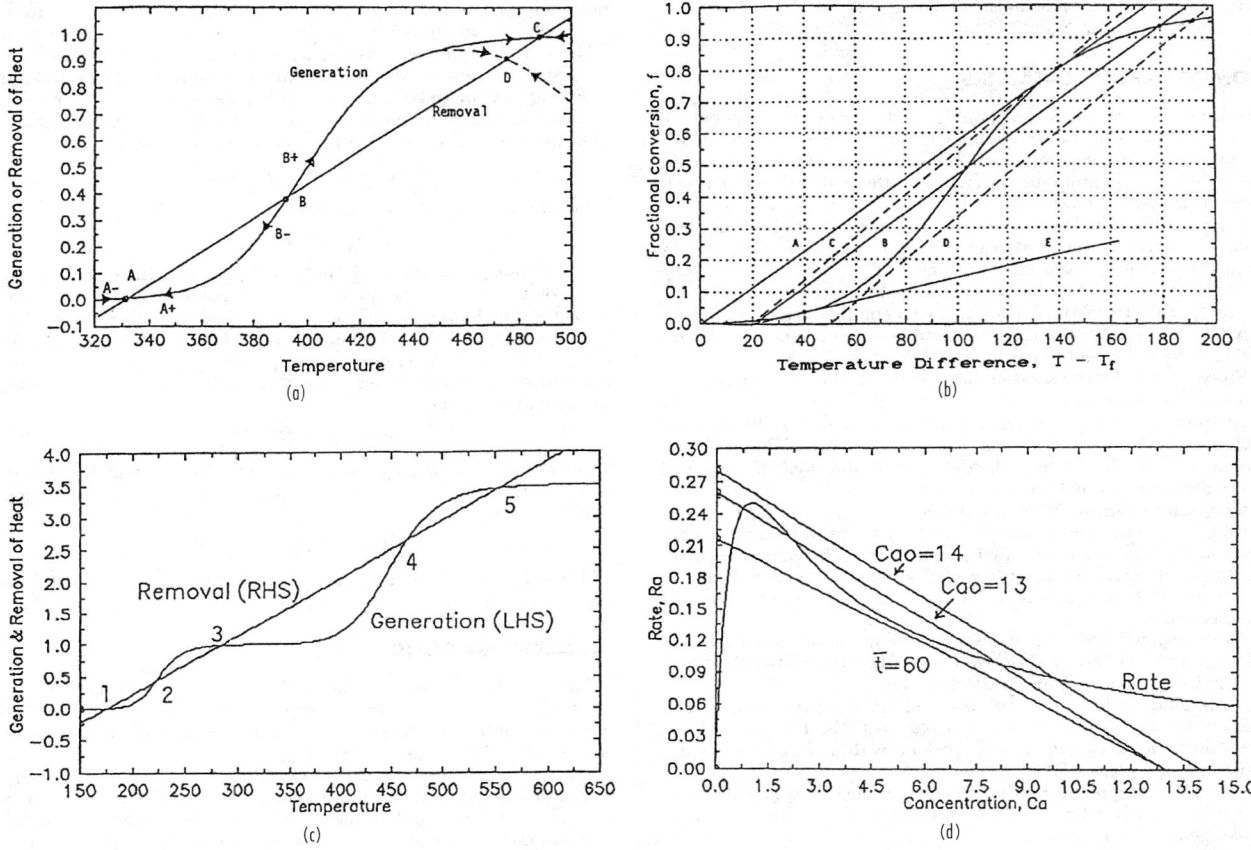

FIG. 23-17 Multiple steady states of CSTRs, stable and unstable, adiabatic except the last item. (a) First-order reaction, A and C stable, B unstable, A is no good for a reactor, the dashed line is of a reversible reaction. (b) One, two, or three steady states depending on the combination (C_f, T_f). (c) The reactions $A \Rightarrow B \Rightarrow C$, with five steady states, points 1, 3, and 5 stable. (d) Isothermal operation with the rate equation $r_a = C_a/(1 + C_a)^2 = (C_{a0} - C_a)/\bar{t}$.

In a specific case this becomes

$$110\left(\frac{k_1\bar{t}}{1+k_1\bar{t}}\right)\left(\frac{1+2.5k_2\bar{t}}{1+k_2\bar{t}}\right) = T - 175$$

with $k_1\bar{t} = \exp(20 - 4{,}500/T)$ and $k_2\bar{t} = \exp(20 - 9{,}000/T)$. The plot shows 5 steady states, of which 1, 3, and 5 are stable. T is in K.

Example (d) The rate equation and the CSTR material balance of this process are:

$$r_a = \frac{C_a}{(1 + C_a)^2} = \frac{C_{a0} - C_a}{\bar{t}}$$

The plots of these equations with $C_{a0} = 13$ and $\bar{t} = 50$ show three intersections, and only one with $C_{a0} = 14$, but the conversion is poor. Keeping $C_{a0} = 13$ but making $\bar{t} = 60$ changes the stable operation to $C_a = 0.6$, or 95 percent conversion.

Instances of multiplicities in CSTR batteries and in PFRs also can be developed.

Only a very few experimental studies have been made for detection of multiplicities of steady states to check on theoretical predictions. The studies of multiplicities and of oscillations of concentrations have similar mathematical bases. Comprehensive reviews of these topics are by Schmitz (*Adv. Chem. Ser.,* **148,** 156, ACS [1975]), Razon and Schmitz (*Chem. Eng. Sci.,* **42,** 1,005–1,047 [1987]), Morbidelli, Varma, and Aris (in Carberry and Varma, eds., *Chemical Reaction and Reactor Engineering,* Dekker, 1987, pp. 975–1,054).

CATALYSIS

A catalyst is a substance that increases the rate of a reaction by participating chemically in intermediate stages of reaction and is ultimately liberated in a chemically unchanged form. Over a period of time, however, permanent changes in the catalyst—deactivation—may occur. *Inhibitors* are substances that slow down rates of reaction. The turnover ratio, the number of molecules converted per molecule of catalyst, can be in the millions. Many catalysts have specific actions in that they influence only one reaction or group of definite reactions. The outstanding example is the living cell in which there are several hundred catalysts, called *enzymes,* each one favoring a specific chemical process. When a reaction can proceed by more than one path, a particular catalyst may favor one path over another and thus lead to a product distribution different from an uncatalyzed reaction. A catalytic reaction requires a lower energy of activation, thus permitting a reduction of temperature at which the reaction can proceed favorably. The equilibrium condition is not changed since both forward and reverse rates are accelerated equally. For example, a good hydrogenation catalyst also is a suitable dehydrogenation accelerator but possibly with a different most-favorable temperature.

Catalytic processes may be homogeneous in the liquid or gas phase (for instance, nitrogen oxides in the Chamber process for sulfuric acid), but industrial examples are most often heterogeneous with a solid catalyst and fluid reactants. Some large industrial reactors of this type are illustrated in Fig. 23-18.

HOMOGENEOUS CATALYSIS

The most numerous cases of homogeneous catalysis are by certain ions or metal coordination compounds in aqueous solution and in biochemistry, where enzymes function catalytically. Many ionic effects are known. The hydronium ion H_3O^+ and the hydroxyl ion OH^- catalyze hydrolyses such as those of esters; ferrous ion catalyzes the decomposition of hydrogen peroxide; decomposition of nitramide is catalyzed by acetate ion. Other instances are inversion of sucrose by HCl, halogenation of acetone by H^+ and OH^-, hydration of isobutene by acids, hydrolysis of esters by acids, and others.

The specific action of a particular metal complex can be altered by varying the ligands or coordination number of the complex or the oxidation state of the central metal atom.

Probable mechanisms often have been deduced: The reactant forms a short-lived intermediate with the catalyst that subsequently decomposes into the product and regenerated catalyst. In fluid phases such intermediates can be detected spectroscopically. This is in contrast to solid catalysis, where the detection of intermediates is much more difficult and is not often accomplished.

Significant characteristics of homogeneous catalysis are that they are highly specific and proceed under relatively mild conditions—again in contrast to solid catalysis, which is less discriminating as to reaction and may require extremes of temperature and pressure. A problem with homogeneous operation is the difficulty of separating product and catalyst.

A review of industrial processes that employ homogeneous catalysts is by Jennings, ed. (*Selected Developments in Catalysis,* Blackwell Scientific, 1985). Some of those processes are:

- Alkylation of isobutane with C_3-C_4 olefins in the presence of HF at 25 to 35°C or 90 to 98% sulfuric acid at 10°C (50°F).
- Alkylation of isobutane and ethylene with a complex of liquid hydrocarbon + $AlCl_3$ + HCl.
- The Wacker process for the oxidation of ethylene to acetaldehyde with $PdCl_2/CuCl_2$ at 100°C (212°F) with 95 percent yield and 95 to 99 percent conversion per pass.
- The OXO process for higher alcohols: $CO + H_2 + C_3H_6 \Rightarrow$ *n*-butanal $\Rightarrow$ further processing. Catalyst is rhodium triphenylphosphine coordination compound, 100°C (212°F), 30 atm (441 psi).
- Acetic acid from methanol by the Monsanto process, $CH_3OH + CO \Rightarrow CH_3COOH$, rhodium iodide catalyst, 3 atm (44 psi), 150°C (302°F), 99 percent selectivity of methanol.
- Ammonia is a cyclic reagent that is recovered by the end of the Solvay process for sodium carbonate from lime and salt. Although there is nothing obscure about the intermediate reactions, ammonia definitely participates in a catalytic sequence.

Immobilized or Polymer Bound, or Heterogenized Catalysts The specificity of homogeneous and separability of solid catalysts are realized by attaching the catalyst to a solid support. This is commonly done for enzymes. In use as carriers are organic polymers such as polystyrene and inorganic polymers such as zeolites, silica, or alumina. A catalyst metal atom, for instance, is anchored to the polymer through a group that is chemically bound to the polymer with a coordinating site such as $-P(C_6H_5)_2$ or $-C_5H_4$ (cyclopentadienyl). Immobilized catalysts have applications in hydrogenation, hydroformylation, and polymerization reactions (Lieto and Gates, *CHEMTECH*, 46–53 [Jan. 1983]).

Phase-Transfer Catalysis In phase-transfer catalysis (PTC), reactions between reactants located in different phases (typically organic and aqueous liquids) are brought about or accelerated by a catalyst that migrates between the phases. Ultimately, the reaction takes place in the organic phase. Although no one mechanism explains everything, the catalyst may combine with a reactant in one phase, migrate to the other phase in which it is also miscible, and react there. Ultimately, the reaction is homogeneous. The catalysts used most extensively are quaternary ammonium or phosphonium salts and crown ethers and cryptates, for instance benzyltriethylammonium chloride and the cheaper methyltrioctylammonium chloride. Reac-

tions helped by PTC include making higher esters, ethers, and alkylates, polymerization of butyl acrylate, and oxidation of olefins with $KMnO_4$. In most of these instances the reaction rate is nil without the catalyst (Dehmlow, *Phase Transfer Catalysis,* Verlag Chemie, 1992).

Effect of Concentration All catalytic reactions appear to involve the formation of intermediate compounds with the catalyst. For a reactant A, product B, and catalyst the corresponding equation is

$$A + C \Rightarrow AC \Rightarrow B + C$$

When the first step is rate controling,

$$r_a = k[A][C]^\alpha$$

In some cases, the exponent is unity. In other cases, the simple power law is only an approximation for an actual sequence of reactions. For instance, the chlorination of toluene catalyzed by acids was found to have $\alpha = 1.15$ at 6°C (43°F) and 1.57 at 32°C (90°F), indicating some complex mechanism sensitive to temperature. A particular reaction may proceed in the absence of catalyst but at a reduced rate. Then the rate equation may be

$$r_a = (k_1 + k_2[C]^\alpha)[A]$$

An instance of *autocatalysis* is the hydrolysis of methyl acetate, which is catalyzed by product acetic acid, $A \Rightarrow C$. The rate equation may be

$$r_a = k[A][C] = k[A][A_0 - A]$$

which will have a maximum value that is characteristic of autocatalytic reactions in general.

CATALYSIS BY SOLIDS

Solid catalysts are widely employed because they are usually cheap, are easily separated from the reaction medium, and are adaptable to either flow or nonflow reactors. Their drawbacks are lack of specificity and possibly high temperatures and pressures.

Usually they are employed as porous pellets in a packed bed. Some exceptions are platinum for the oxidation of ammonia, which is in the form of several layers of fine-mesh wire gauze, and catalysts deposited on membranes. Pore surfaces can be several hundred m^2/g and pore diameters of the order of 100 Å. The entire structure may be of catalytic material (silica or alumina, for instance, sometimes exert catalytic properties) or an active ingredient may be deposited on a porous refractory carrier as a thin film. In such cases the mass of expensive catalytic material, such as Pt or Pd, may be only a fraction of 1 percent.

The principal components of most solid catalysts are three in number:

1. A catalytically active substance or mixture.
2. A carrier of more or less large specific surface, often refractory to withstand high temperatures. A carrier may have some promotion action; for example, silica carrier helps chromia catalyst.
3. Promoters, usually present in small amount, which enhance activity or retard degradation; for instance, rhenium slows coking of platinum reforming, and KCl retards vaporization of $CuCl_2$ in oxychlorination for vinyl chloride.

Selection of Catalysts A basic catalyst often can be selected by using general principles, but the subsequent fine tuning of a commercially attractive recipe must be done experimentally. A start for catalyst design usually is by analogy to what is known to be effective in chemically similar problems, although a scientific basis is being developed. This involves a study in detail of the main possible intermediate reactions that could occur and of the proton and electron receptivity of the catalyst and possible promoters, as well as reactant bond lengths and crystal lattice dimensions. Several designs are made from this point of view by Trimm (*Design of Industrial Catalysts,* Elsevier, 1980). Some of this scientific basis is treated by Hegedus et al. (*Catalyst Design Progress and Perspectives,* Wiley, 1987). Preparation of specific catalysts is described by Stiles (*Catalyst Manufacture,* Dekker, 1983). A thorough coverage of catalytic reactions and catalysts arranged according to the periodic table is in a series by Roiter ed. (*Handbook of Catalytic Properties of Substances,* in Russian, 1968). Industrial catalyst practice is summarized by Thomas (*Cat-*

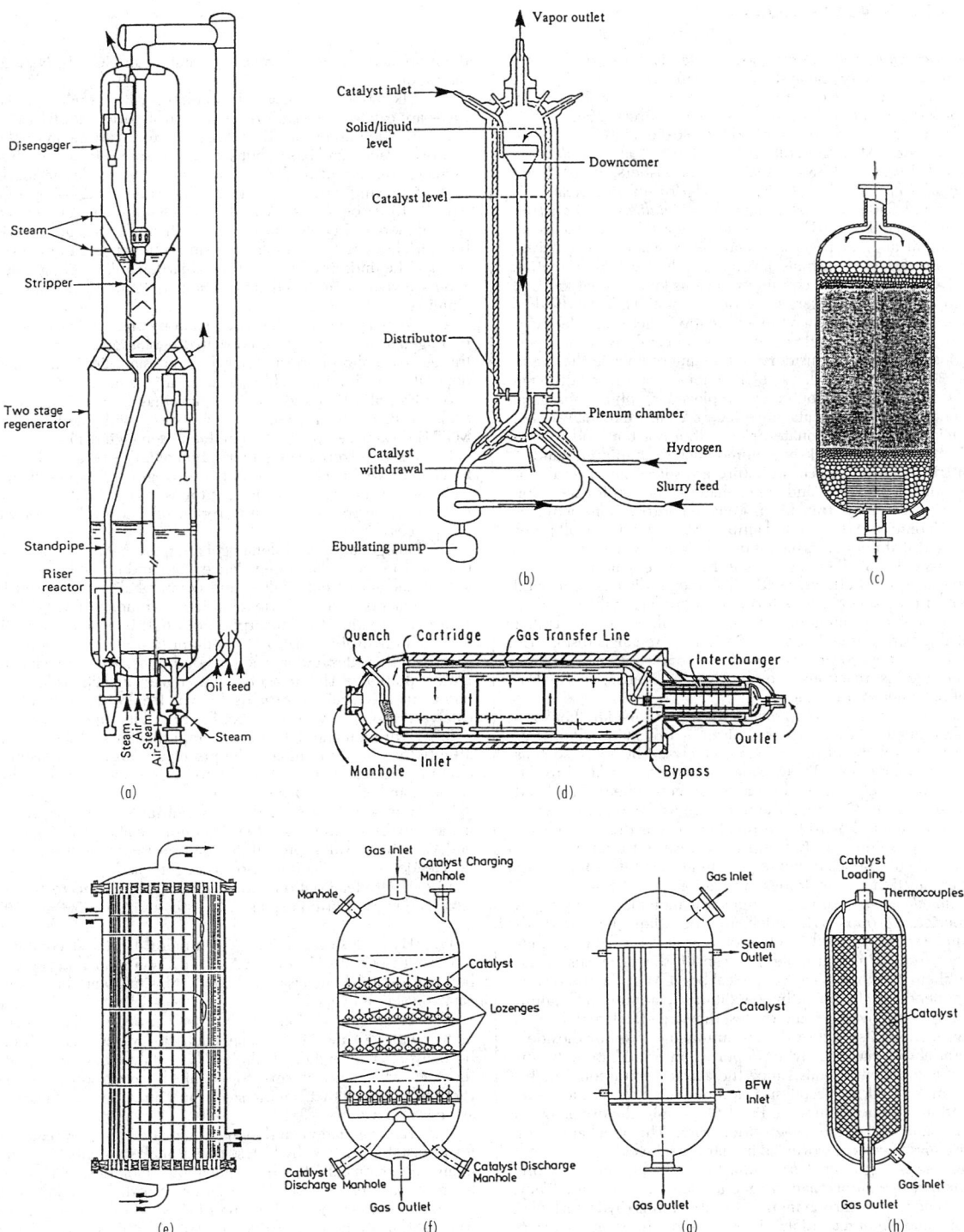

FIG. 23-18 Reactors with solid catalysts. (*a*) Riser cracker with fluidized zeolite catalyst, 540°C; circulation by density difference, 34 to 84 kg/m³ in riser, 420 to 560 kg/m² in regenerator. (*b*) Ebullating fluidized bed for conversion of heavy stocks to gas and light oils. (*c*) Fixed bed unit with support and hold-down zones of larger spheres. (*d*) Horizontal ammonia synthesizer, 26 m long without the exchanger, 2,000 ton/d (*M W Kellogg Co.*). (*e*) Shell-and-tube vessel for hydrogenation of crotonaldehyde has 4,000 packed tubes, 30 mm ID, 10.7 m long, aldehyde feed 209 kg mol/h, hydrogen feed 2,090 kg mol/h. (*After Berty, in Leach*, ed., *Applied Industrial Catalysis, vol. 1, Academic Press, 1983, p. 51*). (*f*), (*g*), (*h*) Methanol synthesizers, 50 to 100 atm, 230 to 300°C, Cu catalyst; ICI quench type, Lurgi tubular, Haldor Topsoe radial flow (*Marschner and Moeller, in Leach, loc. cit.*). To convert ton/d to kg/h, multiply by 907; atm to kPa, multiply by 101.3.

alytic Processes and Proven Catalysts, Academic Press, 1970) who names manufacturers of catalysts for specific processes. Specific processes and general aspects of catalysts are covered by Leach, ed. (*Applied Industrial Catalysis*, 3 vols., Academic Press, 1983–1984). Many industrial processes are described by Gates et al. (*Chemistry of Catalytic Processes*, McGraw-Hill, 1979), Matar et al. (*Catalysis in Petrochemical Processes*, Kluwer Academic Publishers, 1989), Pines (*Chemistry of Catalytic Conversions of Hydrocarbons*, Academic Press, 1981) and Satterfield (*Heterogeneous Catalysis in Industrial Practice*, McGraw-Hill, 1991). Books and encyclopedia articles on particular chemicals and processes may be consulted for catalytic data. There are also eight journals devoted largely to catalytic kinetics.

Kinds of Catalysts To a certain extent it is known what kinds of reactions are speeded up by certain classes of catalysts, but individual members of the same class may differ greatly in activity, selectivity, resistance to deactivation, and cost. Since solid catalysts are not particularly selective, there is considerable crossing of lines in the classification of catalysts and the kinds of reactions they favor. Although some trade secrets are undoubtedly employed to obtain marginal improvements, the principal catalytic effects are known in many cases.

Strong acids are able to donate protons to a reactant and to take them back. Into this class fall the common acids, aluminum halides, and boron trifluoride. Also acid in nature are silica, alumina, aluminosilicates, metal sulfates and phosphates, and sulfonated ion exchange resins. They can transfer protons to hydrocarbons acting as weak bases. Zeolites are dehydrated aluminosilicates with small pores of narrow size distribution, to which is due their highly selective action since only molecules small enough to enter the pores can react.

Base catalysis is most effective with alkali metals dispersed on solid supports or, in the homogeneous form, as aldoxides, amides, and so on. Small amounts of promoters form organoalkali comnpounds that really contribute the catalytic power. Basic ion exchange resins also are useful. Base-catalyzed processes include isomerization and oligomerization of olefins, reactions of olefins with aromatics, and hydrogenation of polynuclear aromatics.

Metal oxides, sulfides, and hydrides form a transition between acid/base and metal catalysts. They catalyze hydrogenation/dehydrogenation as well as many of the reactions catalyzed by acids, such as cracking and isomerization. Their oxidation activity is related to the possibility of two valence states which allow oxygen to be released and reabsorbed alternately. Common examples are oxides of cobalt, iron, zinc, and chromium and hydrides of precious metals that can release hydrogen readily. Sulfide catalysts are more resistant than metals to the formation of coke deposits and to poisoning by sulfur compounds; their main application is to hydrodesulfurization.

Metals and alloys, the principal industrial metallic catalysts, are found in periodic group VIII, which are transition elements with almost-completed 3*d*, 4*d*, and 5*d* electronic orbits. According to theory, electrons from adsorbed molecules can fill the vacancies in the incomplete shells and thus make a chemical bond. What happens subsequently depends on the operating conditions. Platinum, palladium, and nickel form both hydrides and oxides; they are effective in hydrogenation (vegetable oils) and oxidation (ammonia or sulfur dioxide). Alloys do not always have catalytic properties intermediate between those of the component metals, since the surface condition may be different from the bulk and catalysis is a function of the surface condition. Addition of some rhenium to Pt/Al_2O_3 permits the use of lower temperatures and slows the deactivation rate. The mechanism of catalysis by alloys is still controversial in many instances.

Transition-metal organometallic catalysts in solution are more effective for hydrogenation than are metals such as platinum. They are used for reactions of carbon monoxide with olefins (hydroformylation) and for some oligomerizations. They are sometimes immobilized on polymer supports with phosphine groups.

Kinds of Catalyzed Organic Reactions A fundamental classification of organic reactions is possible on the basis of the kinds of bonds that are formed or destroyed and the natures of eliminations, substitutions, and additions of groups. Here a more pragmatic list of 20 commercially important kinds or classes of reactions will be discussed. In all instances of solid-catalyzed reactions, chemisorption is a primary step. Often molecules are dissociated on chemisorption into the more active atomic forms, for instance, of CO, H_2, N_2, and O_2 on platinum.

1. Alkylations—for example, of olefins with aromatics or isoparaffins—are catalyzed by sulfuric acid, hydrofluoric acid, BF_3 and $AlCl_3$.

2. Condensations of aldehydes and ketones are catalyzed homogeneously by acids and bases, but solid bases are preferred, such as anion exchange resins and alkali or alkaline earth hydroxides or phosphates.

3. Cracking, a rupturing of carbon-carbon bonds—for example, of gas oils to gasoline—is favored by silica-alumina, zeolites, and acid types generally. Zeolites have pores with small and narrow size distribution. They crack only molecules small enough to enter the pores. To restrain the undesirable formation of carbon and C_3-C_4 hydrocarbons, zeolite activity is reduced by dilution to 10 to 15 percent in silica-alumina.

4. Dehydration and dehydrogenation combined utilizes dehydration agents together with mild dehydrogenation agents. Included in this class are phosphoric acid, silica-magnesia, silica-alumina, alumina derived from aluminum chloride, and various metal oxides.

5. Esterification and etherification may be catalyzed by mineral acids or BF_3. The reaction of isobutylene with methanol to make MTBE is catalyzed by a sulfonated ion exchange resin.

6. Fischer-Tropsch oligomerization of CO + H_2 to make hydrocarbons and oxygenated compounds was originally catalyzed by cobalt, which forms the active carbonyl, but now iron promoted by potassium is favored. Dissociative chemisorption of CO has been observed in this process.

7. Halogenation and dehalogenation are catalyzed by substances that exist in more than one valence state and are able to donate and accept halogens freely. Silver and copper halides are used for gas-phase reactions, and ferric chloride commonly for liquid phase. Hydrochlorination (the absorption of HCl) is promoted by $BiCl_3$ or $SbCl_3$ and hydrofluorination by sodium fluoride or chromia catalysts that form fluorides under reaction conditions. Mercuric chloride promotes addition of HCl to acetylene to make vinyl chloride. Oxychlorination in the Stauffer process for vinyl chloride from ethylene is catalyzed by $CuCl_2$ with some KCl to retard its vaporization.

8. Hydration and dehydration employ catalysts that have a strong affinity for water. Alumina is the principal catalyst, but also used are aluminosilicates, metal salts and phosphoric acid or its metal salts on carriers, and cation exchange resins.

9. Hydrocracking is catalyzed by substances that promote cracking and hydrogenation together. In commercial use are Ni, Co, Cr, W, and V or their oxides, presulfided before use, on acid supports. Zeolites loaded with palladium also have been used.

10. Hydrodealkylation—for instance, of toluene to benzene—is catalyzed by supported oxides of Cr, Mo, and Co at 500 to 650°C and 50 atm.

11. Hydrodesulfurization. A commercial catalyst contains about 4 percent CoO and 12 percent MoO_3 on γ-alumina and is presulfided before use. Molybdena is a weak catalyst by itself and the cobalt has no catalytic action by itself.

12. Hydroformylation, or the OXO process, is the reaction of olefins with CO and H_2 to make aldehydes, which may subsequently be converted to higher alcohols. The catalyst base is cobalt naphthenate, which transforms to cobalt hydrocarbonyl in place. A rhodium complex that is more stable and functions at a lower temperature is also used.

13. Hydrogenation and dehydrogenation employ catalysts that form unstable surface hydrides. Raney nickel or cobalt are used for many reductions. In the hydrogenation of olefins, both olefin and hydrogen are chemisorbed, the latter with dissociation. Transition-group and bordering metals such as Fe, Ni, Co, and Pt are suitable, as well as transition-group oxides or sulfides. This class of reactions includes the important examples of ammonia and methanol syntheses, the Fischer-Tropsch, OXO, and SYNTHOL processes, and the production of alcohols, aldehydes, ketones, amines, and edible oils. In the sequence for making nylon, phenol is hydrogenated to cyclohexanol with nickel catalyst at 150°C (302°F) and 15 atm (221 psi); the product is hydrogenated to cyclohexanone at 400°C (752°F) using a zinc or copper catalyst; that product is hydrogenated to cyclohexane carboxylic acid at 150°C (302°F) and 15 atm (221 psi) with palladium on

charcoal; this last product is a precursor for ε-caprolactam, which goes on to make nylon.

14.   Hydrolysis of esters is speeded up by both acids and bases. Soluble alkylaryl sulfonic acids or sulfonated ion exchange resins are suitable.

15.   Isomerization is promoted by either acids or bases. Higher alkylbenzenes are isomerized in the presence of $AlCl_3/HCl$ or BF_3/HF; olefins with most mineral acids, acid salts and silica-alumina; saturated hydrocarbons with $AlCl_3$ or $AlBr_3$ promoted by 0.1 percent of olefins.

16.   Metathesis is the rupture and reformation of carbon-carbon bonds—for example, of propylene into ethylene plus butene. Catalysts are oxides, carbonyls, or sulfides of Mo, W, or Re.

17.   Oxidation catalysts are either metals that chemisorb oxygen readily, such as platinum or silver, or transition metal oxides that are able to give and take oxygen by reason of their having several possible oxidation states. Ethylene oxide is formed with silver, ammonia is oxidized with platinum, and silver or copper in the form of metal screens catalyze the oxidation of methanol to formaldehyde. Cobalt catalysis is used in the following oxidations: butane to acetic acid and to butyl-hydroperoxide, cyclohexane to cyclohexylperoxide, acetaldehyde to acetic acid and toluene to benzoic acid. $PdCl_2$-$CuCl_2$ is used for many liquid-phase oxidations and V_2O_5 combinations for many vapor-phase oxidations.

18.   Polymerization of olefins such as styrene is promoted by acid or base or sodium catalysts, and polyethylene is made with homogeneous peroxides. Condensation polymerization is catalyzed by acid-type catalysts such as metal oxides and sulfonic acids. Addition polymerization is used mainly for olefins, diolefins, and some carbonyl compounds. For these processes, initiators are coordination compounds such as Ziegler-type catalysts, of which halides of transition metals Ti, V, Mo, and W are important examples.

19.   Reforming is the conversion primarily of naphthenes and alkanes to aromatics, but other reactions also occur under commercial conditions. Platinum or platinum/rhenium are the hydrogenation/dehydrogenation component of the catalyst and alumina is the acid component responsible for skeletal rearrangements.

20.   Steam reforming is the reaction of steam with hydrocarbons to make town gas or hydrogen. The first stage is at 700 to 830°C (1,292 to 1,532°F) and 15–40 atm (221 to 588 psi). A representative catalyst composition contains 13 percent Ni supported on α-alumina with 0.3 percent potassium oxide to minimize carbon formation. The catalyst is poisoned by sulfur. A subsequent *shift reaction* converts CO to CO_2 and more H_2, at 190 to 260°C (374 to 500°F) with copper metal on a support of zinc oxide which protects the catalyst from poisoning by traces of sulfur.

Physical Characteristics   With a few exceptions, solid catalysts are employed as porous pellets in a fixed or fluidized bed. Their physical characteristics of major importance are as follows.

• *Pellet size* is a major consideration. Shapes are primarily spherical, short cylindrical, or irregular. Special shapes like those used in mass-transfer equipment may be used to minimize pressure drop. In gas fluidized beds the diameters average less than 0.1 mm (0.0039 in); smaller sizes impose too severe loading on the entrainment recovery equipment. In slurry beds the diameters can be about 1.0 mm (0.039 in). In fixed beds the range is 2 to 5 mm (0.079 to 0.197 in). The competing factors are the pressure drop and accessibility of the internal surface, which vary in opposite directions with changes in diameter. With poor thermal conductivity, severe temperature gradients or peaks arise with large pellets that may lead to poor control of the reaction and undesirable side reactions like coking.

• *Specific surface* of solid spheres of 0.1 mm (0.0039 in) dia is 0.06 m^2/ml (18,300 ft^2/ft^3) and a porous activated alumina pellet has about 600 m^2/ml (1.83×10^8 ft^2/ft^3). Other considerations aside, a large surface is desirable because the rate of reaction is proportional to the accessible surface. On the other hand, large specific surface means pores of small diameter.

• *Pore diameter and distribution* are important factors. Small pores limit the accessibility of internal surface because of increased resistance to diffusion of reactants inwards. Diffusion of products outwards also is slowed, and degradation of those products may result.

When the catalyst is expensive, the inaccessible internal surface is a liability, and in every case it makes for a larger reactor size. A more or less uniform pore diameter is desirable, but this is practically realizable only with molecular sieves. Those pellets that are extrudates of compacted masses of smaller particles have bimodal pore size distributions, between the particles and inside them. Micropores have diameters of 10 to 100 Å, macropores of 1,000 to 10,000 Å. The macropores provide rapid mass transfer into the interstices that lead to the micropores where the reaction takes place.

• *Diffusivity and tortuosity* affect resistance to diffusion caused by collision with other molecules (bulk diffusion) or by collision with the walls of the pore (Knudsen diffusion). Actual diffusivity in common porous catalysts is intermediate between the two types. Measurements and correlations of diffusivities of both types are known. Diffusion is expressed per unit cross section and unit thickness of the pellet. Diffusion rate through the pellet then depends on the porosity ϑ and a tortuosity factor τ that accounts for increased resistance of crooked and varied-diameter pores. Effective diffusion coefficient is $D_{eff} = D_{theo}\vartheta/\tau$. Empirical porosities range from 0.3 to 0.7, tortuosities from 2 to 7. In the absence of other information, Satterfield (*Heterogeneous Catalysis in Practice*, McGraw-Hill, 1991) recommends taking $\vartheta = 0.5$ and $\tau = 4$. In this area, clearly, precision is not a feature.

Rate of Reaction   Rate equations of fluid reactions catalyzed by solids are of two main types:

1.   Power law type, based directly on the law of mass action, say,

$$r_a = kP_a^\alpha P_b^\beta P_c^\gamma P_d^\delta \ldots \tag{23-59}$$

with a term for every reactant or product. The exponents are empirical and may be positive or negative, integral or fractional.

2.   Hyperbolic, based on the Langmuir adsorption principle; for instance,

$$r_a = \frac{k(P_a P_b - P_c P_d/K_e)}{(1 + k_a P_a + K_b P_b + K_c P_c + K_d P_d)^2} \tag{23-60}$$

The latter kind of formulation is described at length in Sec. 7. The assumed mechanism is comprised of adsorption and desorption rates of the several participants and of the reaction rates of adsorbed species. In order to minimize the complexity of the resulting rate equation, one of the several rates in series may be assumed controlling. With several controlling steps the rate equation usually is not explicit but can be used with some extra effort.

Two quite successful rate equations of catalytic industrial processes are cited by Rase (*Chemical Reactor Design for Process Plants*, vol. 2, Wiley, 1977):

1.   For the synthesis of ammonia,

$$r_{NH_3} = k(P_{N_2} P_{H_2} P_{NH_3}^{-1.5} P_{NH_3} P_{H_2}^{-1.5}) \tag{23-61}$$

2.   For the oxidation of sulfur dioxide,

$$r_{SO_2} = \frac{k_1 P_{SO_2}}{(P_{SO_3}^{0.5} + k_2 P_{SO_2})^2} \left(P_{O_2} - \frac{P_{SO_3}}{K_e P_{SO_2}} \right)^2 \tag{23-62}$$

The first is a power law type modified for reversibility, and the second is a modified hyperbolic.

The partial pressures in the rate equations are those in the vicinity of the catalyst surface. In the presence of diffusional resistance, in the steady state the rate of diffusion through the stagnant film equals the rate of chemical reaction. For the reaction $A + B \Rightarrow C + \ldots$, with rate of diffusion of A limited,

$$r = r_d = r_s = k_1(P_{ag} - P_{as}) = \frac{k_2 P_{as} P_b}{(1 + K_a P_{as} + P_b + K_c P_c + \cdots)^2} \tag{23-63}$$

The unknown partial pressure at the external surface can be eliminated as $P_{as} = P_{ag} - r/k_1$), which results in a cubic equation for r.

Another complication arises when not all of the internal surface of a porous catalyst is accessed. Then a factor called the effectiveness η is applied, making the power law equation, for instance,

$$r = k\eta P_a^\alpha P_b^\beta P_c^\gamma P_d^\delta \tag{23-64}$$

The effectiveness is known from experiment for important industrial catalysts and is correlated, in general, in terms of pore characteristics, concentrations, and specific rate equations.

From a statistical viewpoint, there is often little to choose between power law and hyperbolic equations as representations of data over an experimental range. The fact, however, that a particular hyperbolic equation is based on some kind of possible mechanism may lead to a belief that such an equation may be extrapolated more safely outside the experimental range, although there may be no guarantee that the controlling mechanism will remain the same in the extrapolated region.

Effectiveness As a reactant diffuses into a pore, it undergoes a falling concentration gradient and a falling rate of reaction. The concentration depends on the radial position in the pores of a spherical pellet according to

$$\frac{d^2C}{dr^2} + \frac{2}{r}\frac{dC}{dr} = \frac{R_c}{D} \qquad (23\text{-}65)$$

where R_C is the rate of reaction per unit volume; for a reaction of order n, $R_c = kC^n$. In terms of the normalized variables $f = C/C_s$ and $\rho = r/R$, and the Thiele modulus for a sphere,

$$\phi_s = R\sqrt{\frac{kC_s^{n-1}}{D}} \qquad (23\text{-}66)$$

this becomes

$$\frac{d^2f}{d\rho^2} + \frac{2}{\rho}\frac{df}{d\rho} = \phi_s^2 f^n \qquad (23\text{-}67)$$

At the inlet to the pore, $\rho = 1$ and $f = f_s$. At the center, $\rho = 0$ and $df/d\rho = 0$.

Although the point values of the rate diminish with ρ, in the steady state the rate of reaction equals the rate of diffusion at the mouth of the pores. The effectiveness of the catalyst is a ratio

$$\eta = \frac{r_{\text{actual}}}{r_{\text{ideal}}} = \frac{D(dC/dr)_{r=R}}{\dfrac{4\pi R^3}{3} kC_s^n} \qquad (23\text{-}68)$$

where r_{actual} = rate of diffusion at the mouth of the pore
r_{ideal} = rate on the assumption that all of the pore surface is exposed to the concentration at the external surface
C_s = concentration at the external surface

Numerical and some analytical solutions of the diffusion/reaction equations are represented closely by an empirical curve/fit,

$$\eta = \frac{1.0357 + 0.3173\phi_m + 0.000427\phi_m^2}{1 + 0.4172\phi_m + 0.1390\phi_m^2} \qquad (23\text{-}69)$$

where the modified Thiele modulus for an nth order reaction, $R_c = kC^n$ per unit volume, is

$$\phi_m = 3\left(\frac{V_p}{A_p}\right)\left(\frac{3}{n+1}\right)^{1/2}\left(\frac{kC_s^{n-1}}{D_{\text{eff}}}\right)^{1/2} \qquad (23\text{-}70)$$

where V_p/A_p = (volume of pellet)/(external surface of pellet)
= $R/3$ for spheres of radius R
= L for a slab with one permeable face
= $R/2$ for a cylinder with sealed flat ends

The analytical result for a first-order reaction in a spherical pellet is:

$$\eta = \frac{3}{\phi^2}\left(\frac{\phi}{\tanh\phi} - 1\right), \qquad \phi = R\sqrt{\frac{k}{D_{\text{eff}}}} \qquad (23\text{-}71)$$

The effectiveness of a given size of pellet can be found experimentally by running tests of reaction conversion with a series of diminishing sizes of pellets until a limiting rate is found. Then η will be the ratio of the rate with the pellet size in question to the limiting value.

Since theoretical calculation of effectiveness is based on a hardly realistic model of a system of equal-sized cylindrical pores and a shaky assumption for the tortuosity factor, in some industrially important cases the effectiveness has been measured directly. For ammonia synthesis by Dyson and Simon (*Ind. Eng. Chem. Fundam.*, **7**, 605 [1968]) and for SO_2 oxidation by Kadlec et al. (*Coll. Czech. Chem. Commun.*, **33**, 2388, 2526 [1968]).

When account is taken of the effectiveness, the rate of reaction becomes

$$R_c = k\eta C_s^n$$

where η depends on C_s except for first-order reactions.

Example 5: Application of Effectiveness For a second-order reaction in a plug flow reactor the Thiele modulus is $\phi = 8\sqrt{C_s}$, and inlet concentration is $C_{s0} = 1.0$. The equation will be integrated for 80 percent conversion with Simpson's rule. Values of η are

$$(C_s, \eta) = (1.0, 0.272), (0.6, 0.338), (0.2, 0.510)$$

The material balance and the integral become

$$-V'dC_s = k\eta C_s^2 dV_r$$

$$kV_r/V' = \int_{0.2}^{1.0}\left(\frac{1}{\eta C_s^2}\right)dC = \left(\frac{0.4}{3}\right)[3.68 + 4\,(8.22) + 49.02] = 11.4$$

Adiabatic Reactions Aside from the Thiele modulus, two other parameters are necessary in this case:

$$\beta = -\frac{\Delta H_r DC_s}{\lambda T_s}$$

$$\gamma = \frac{E}{RT_s}$$

where ΔH_r = heat of reaction
λ = thermal conductivity
E = energy of activation
R = gas constant, 1.987 cal/g mol

Figure 23-19 is one of several by Weisz and Hicks (*Chem. Eng. Sci.*, **17**, 263 [1962]). Although this predicts some very large values of η in some ranges of the parameters, these values are mostly not realized in practice, as Table 23-5 shows. The modified Lewis number is $Lw' = \lambda_s/\rho_s C_{ps} D_{\text{eff}}$.

Deactivation in Process The active surface of a catalyst can be degraded by chemical, thermal, or mechanical factors. Poisons and

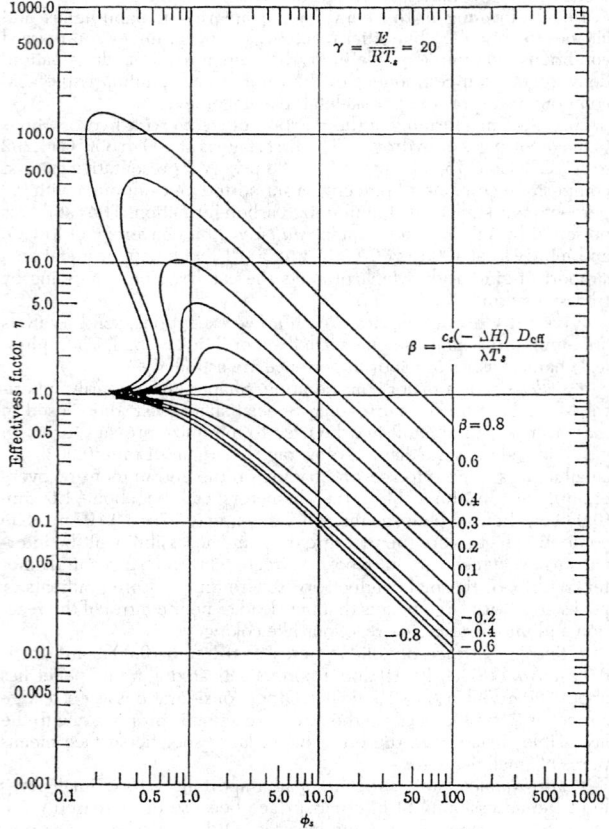

FIG. 23-19 Effectiveness of first-order reactions in spheres under adiabatic conditions (*Weisz and Hicks*, Chem. Eng. Sci., **17**, 265 [1962]).

TABLE 23-5 Parameters of Some Exothermic Catalytic Reactions

Reaction	β	γ	$\gamma\beta$	Lw'	ϕ
NH$_3$ synthesis	0.000061	29.4	0.0018	0.00026	1.2
Synthesis of higher alcohols from CO and H$_2$	0.00085	28.4	0.024	0.00020	—
Oxidation of CH$_3$OH to CH$_2$O	0.0109	16.0	0.175	0.0015	1.1
Synthesis of vinyl chloride from acetylene and HCl	0.25	6.5	1.65	0.1	0.27
Hydrogenation of ethylene	0.066	23–27	2.7–1	0.11	0.2–2.8
Oxidation of H$_2$	0.10	6.75–7.52	0.21–2.3	0.036	0.8–2.0
Oxidation of ethylene to ethylenoxide	0.13	13.4	1.76	0.065	0.08
Dissociation of N$_2$O	0.64	22.0	1.0–2.0	—	1–5
Hydrogenation of benzene	0.12	14–16	1.7–2.0	0.006	0.05–1.9
Oxidation of SO$_2$	0.012	14.8	0.175	0.0415	0.9

SOURCE: After Hlavacek, Kubicek, and Marek, *J. Catal.*, **15**, 17, 31 (1969).

inhibitors are known in specific cases. For instance, Thomas (*Catalytic Processes and Proven Catalysts,* Academic Press, 1970) often cites the poisons for the processes he describes. Potent poisons are compounds of P, S, As, Te, and Bi that have free electron pairs. In some cases a reduced life is simply accepted, as in the case of slow accumulation of trace metals from feed to catalytic cracking, but in other cases the deactivation is too rapid. Sulfur and water are removed from feed to ammonia synthesis, sulfur from the feed to platinum reforming, and arsenic from feed to SO$_2$ oxidation with platinum but not necessarily with vanadium. The catalyst also can be modified by additives; for instance, chromia to nickel to prevent sintering, rhenium to platinum to reduce coking, and so on. Reactivation sometimes is done in place; for instance, coke is burned off cracking catalyst or off nickel and nickel-molybdenum catalysts in a fluidized reactor/regenerator system. Platinum-alumina catalyst is regenerated in place by chlorine treatment. Much work has been done in this general field (Butt and Petersen, *Activation, Deactivation and Poisoning of Catalysts,* Academic Press, 1988). A list of 18 important industrial processes with catalyst lives and factors influencing them is in Delmon and Froment (*Catalyst Deactivation,* Elsevier, 1980). The lives range from a few days to several years.

Dependence of activity α may be simply on time onstream. One index is the ratio of the rate at time t to the rate with fresh catalyst,

$$\alpha = \frac{r_c @ t}{r_c @ t = 0} \quad (23\text{-}72)$$

The rate of destruction of active sites and pore structure can be expressed as a mass-transfer relation; for instance, as a second-order reaction

$$-\frac{d\alpha}{dt} = k_d \alpha^2 \quad (23\text{-}73)$$

and the corresponding integral

$$\alpha = \frac{1}{1 + k_d t} \quad (23\text{-}74)$$

The specific rate is expected to have an Arrhenius dependence on temperature. Deactivation by coke deposition in cracking processes apparently has this kind of correlation.

Assumption of a first-order rate law gives rise to

$$\alpha = \exp(k_1 - k_2 t) \quad (23\text{-}75)$$

Another relation that is sometimes successful is

$$\alpha = \frac{1}{1 + k_1 t^{k_2}} \quad (23\text{-}76)$$

When the feedstock contains constant proportions of reactive impurities, the rate of decline also may depend on the concentration of the main reactant, thus:

$$-\frac{d\alpha}{dt} = k_d \, \alpha^p \, C^q \quad (23\text{-}77)$$

Such a differential equation together with a rate equation for the main reactant constitutes a pair that must be solved simultaneously. Take the example of a CSTR for which the unsteady material balance is

$$C_0 = C + \bar{k}\bar{t}\alpha C^q + \bar{t}\,\frac{dC}{dt} \quad (23\text{-}78)$$

With most values of the constants of these two equations a numerical solution will be needed.

The constants of the various time dependencies of activity are found by methods like those for finding constants of any rate equation, given suitable (α, t) data.

Uniform deactivation is one of the two limiting cases of the behavior of catalyst poisoning that are recognized. In one, the poison is distributed uniformly throughout the pellet and degrades it gradually. In the other, the poison is so effective that it kills completely as it enters the pore and is simultaneously removed from the stream. Complete deactivation begins at the mouth and moves gradually inward.

When uniform poisoning occurs the specific rate declines by a factor $1 - \beta$ where β is the fractional poisoning. Then a power law rate equation becomes

$$r_c = k_v(1 - \beta)\,\eta C_s^q \quad (23\text{-}79)$$

The effectiveness also depends on β through the Thiele modulus,

$$\phi = L\sqrt{\frac{k_v(1-\beta)C_s^{q-1}}{D}} \quad (23\text{-}80)$$

To find the effectiveness under poisoned conditions, this form of the modulus is substituted into the appropriate relation for effectiveness. For first-order reaction in slab geometry, for instance,

$$\eta = \frac{1}{L}\sqrt{\frac{D}{k_v(1-\beta)}}\tanh\left[L\sqrt{\frac{k_v(1-\beta)}{D}}\right] \quad (23\text{-}81)$$

Pore mouth (or shell) poisoning occurs when the poisoning of a pore surface begins at the mouth and moves gradually inward. In this case the reactant must diffuse through the dead zone before it starts to react. β is the fraction of the pore that is deactivated, C_1 is the concentration at the end of the inactive region, and $x = (1 - \beta)L$ is the coordinate there.

The rate of diffusion into the pore equals the rate of diffusion through the dead zone

$$D\left(\frac{dC}{dx}\right)_{x=(1-\beta)L} = D\,\frac{\Delta C}{\Delta x} = d\,\frac{C_s - C_1}{\beta L} \quad (23\text{-}82)$$

The concentration profile in a porous slab is represented by

$$\frac{d^2C}{dx^2} = \frac{k}{D}C^q \quad (23\text{-}83)$$

At a sealed face or at the center of a slab with two permeable faces, the condition is:

$$\left(\frac{dC}{dx}\right)_{x=0} = 0 \quad (23\text{-}84)$$

The three preceding equations may be solved simultaneously by the shooting method. A result for a first-order reaction is shown in Fig. 23-20, together with the case of uniform poisoning.

Distribution of Catalyst in Pores Because of the practical requirements of manufacturing, commercial impregnated catalysts usually have a higher concentration of active ingredient near the outside than near the tip of the pores. This may not be harmful, because it seems that effectiveness sometimes is better with some kind of nonuniform distribution of a given mass of catalyst. Such effects may be present in cases where the rate exhibits a maximum as a function of

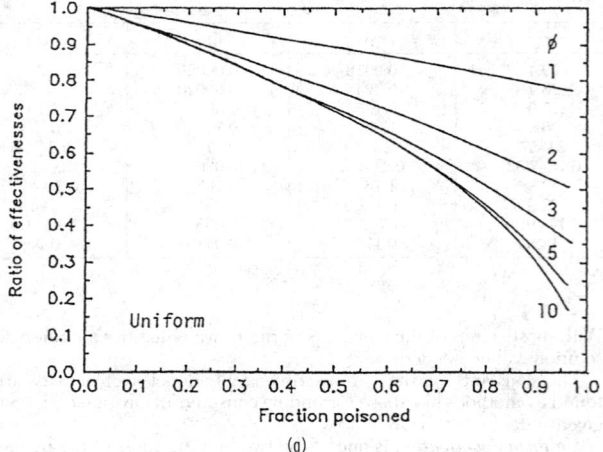

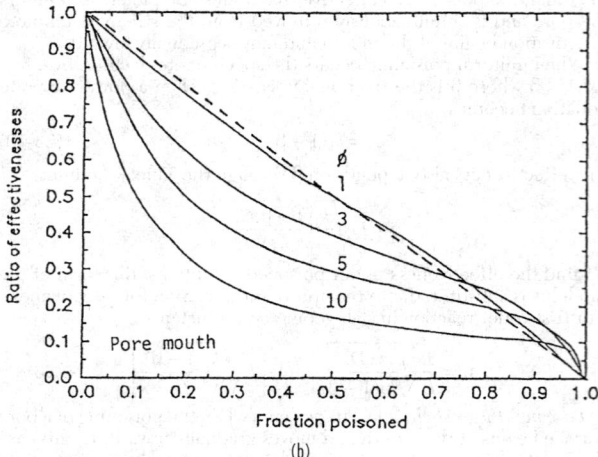

FIG. 23-20 Poisoning of a first-order reaction: (*a*) uniform poisoning, (*b*) pore mouth poisoning.

concentration of reactant, which is the case, for instance, with bimolecular Langmuir-Hinshelwood kinetics. Even under isothermal conditions, a catalyst's effectiveness can be several times unity at an optimum location of catalyst in the pore. Theoretical studies have recognized three zones: inner, middle, and outer (or egg yolk, egg white, and eggshell). Rather more theoretical studies have been made than experimental ones.

When a carrier is impregnated with a solution, where the catalyst deposits will depend on the rate of diffusion and the rate of adsorption on the carrier. Many studies have been made of Pt deposition from chloroplatinic acid (H_2PtCl_6) with a variety of acids and salts as coimpregnants. HCl results in uniform deposition of Pt. Citric or oxalic acid drive the Pt to the interior. HF coimpregnant produces an egg white profile. Photographs show such varied distributions in a single pellet.

Some studies of potential commercial significance have been made. For instance, deposition of catalyst some distance away from the pore mouth extends the catalyst's life when pore mouth deactivation occurs. Oxidation of CO in automobile exhausts is sensitive to the catalyst profile. For oxidation of propane the activity is eggshell > uniform > egg white. Nonuniform distributions have been found superior for hydrodemetallation of petroleum and hydrodesulfurization with molybdenum and cobalt sulfides. Whether any commercial processes with programmed pore distribution of catalysts are actually in use is not mentioned in the recent extensive review of Gavrillidis et al. (in Becker and Pereira, eds., *Computer-Aided Design of Catalysts*, Dekker, 1993, pp. 137–198), with the exception of monolithic automobile exhaust cleanup where the catalyst may be deposited some distance from the mouth of the pore and where perhaps a 25-percent longer life thereby may be attained.

Catalytic Membrane Reactors Membrane reactors combine reaction and separation in a single vessel. By removing one of the products of reaction, the membrane reactor can make conversion beyond thermodynamic equilibrium in the absence of separation.

For these studies, laboratory reactors have been of two main types: (1) a bed of catalyst pellets in series with a membrane, and (2) a membrane with catalyst deposited on the pore surface. Industrial membranes must be sturdy, temperature resistant, and affordable. Palladium alloys have high hydrogen permselectivity but are not commercially feasible because of their high cost. Microporous ceramic membranes in use thus far are able to separate gases only in accordance with the Knudsen diffusion law; that is, permeability is inversely proportional to the square root of the molecular weight. Efforts are being made for the development of membranes with molecular sieving properties, including zeolite, carbon, and polyphosphazene membranes.

Dehydrogenation processes in particular have been studied, with conversions in most cases well beyond thermodynamic equilibrium: Ethane to ethylene, propane to propylene, water-gas shift reaction $CO + H_2O \Leftrightarrow CO_2 + H_2$, ethylbenzene to styrene, cyclohexane to benzene, and others. Some hydrogenations and oxidations also show improvement in yields in the presence of catalytic membranes, although it is not obvious why the yields should be better since no separation is involved: hydrogenation of nitrobenzene to aniline, of cyclopentadiene to cyclopentene, of furfural to furfuryl alcohol, and so on; oxidation of ethylene to acetaldehyde, of methanol to formaldehyde, and so on.

At present, according to the review of Tsotsis et al. (in Becker and Pereira, eds., *Computer-Aided Design of Catalysts*, Dekker, 1993, pp. 471–551), there is no record of industrial implementation of reactors with catalytic membranes.

HOMOGENEOUS REACTIONS

Much of the basic theory of reaction kinetics presented in Sec. 7 of this Handbook deals with homogeneous reactions in batch and continuous equipment, and that material will not be repeated here. Material and energy balances and sizing procedures are developed for batch operations in ideal stirred tanks—during startup, continuation, and shutdown—and for continuous operation in ideal stirred tank batteries and plug flow tubulars and towers.

LIQUID PHASE

Batch reactions of single or miscible liquids are almost invariably done in stirred or pumparound tanks. The agitation is needed to mix multiple feeds at the start and to enhance heat exchange with cooling or heating media during the process.

Topics that acquire special importance on the industrial scale are the quality of mixing in tanks and the residence time distribution in vessels where plug flow may be the goal. The information about agitation in tanks described for gas/liquid and slurry reactions is largely applicable here. The relation between heat transfer and agitation also is discussed elsewhere in this Handbook. Residence time distribution is covered at length under "Reactor Efficiency." A special case is that of laminar and related flow distributions characteristic of non-Newtonian fluids, which often occurs in polymerization reactors.

Laminar Flow A mathematically simple deviation from uniform flow across a cross section is that of power law fluids whose linear velocity in a tube depends on the radial position $\beta = r/R$, according to the equation

$$u = \bar{u}\left(\frac{2n+1}{n+1}\right)(1 - \beta^{(n+1)/n}) \quad (23\text{-}85)$$

where $\bar{u}$ is the average velocity. For normal fluids $n = 0$, for laminar ones $n = 1$, and other values apply to pseudoplastic and dilatant fluids. Along any particular radius, all molecules have the same residence time; that is, plug flow is achieved on that streamline. The average over the cross section is the value of primary interest.

For laminar flow, the velocity at the centerline is $u_0 = 2\bar{u}$. For a power law rate equation $r_c = kC^q$, the differential material balance on a streamline is

$$-u_0(1 - \beta^2)dC = kC^q dL \quad (23\text{-}86)$$

when $q = 1$,

$$\frac{C}{C_0} = \exp(-kt) = \exp\left[\frac{kL}{u_0(1-\beta^2)}\right] \quad (23\text{-}87)$$

and when $q = 2$,

$$\frac{1}{C} - \frac{1}{C_0} = kt = \frac{kL}{u_0(1-\beta^2)} \quad (23\text{-}88)$$

The average concentration is found by integration over the cross section. After some algebraic manipulation, the result is

$$\frac{\overline{C}}{C_0} = 2t_0^2 \int_{t_0}^{\infty} C\, \frac{dt}{t^3} \quad (23\text{-}89)$$

where $t_0 = 0.5\bar{t}$ is the residence time along the centerline and C is to be substituted from the previous equations. Figure 23-21 compares the performances of plug flow and laminar reactors of first- and second-order processes.

When it is deleterious, laminar flow can be avoided by mixing over the cross section. For this purpose static mixers in line can be provided. For very viscous materials and pastes, screws of the type used for pumping and extrusion are used as reactors.

Nonisothermal Operation Some degree of temperature control of a reaction may be necessary. Figures 23-1 and 23-2 show some of the ways that may be applicable to homogeneous liquids. More complex modes of temperature control employ internal surfaces, recycles, split flows, cold shots, and so on. Each of these, of course, requires an individual design effort.

Heat transfer through a vessel wall is often satisfactory:

1. In tubular reactors of only a few cm in diameter, the temperature is substantially uniform over the cross section so only an axial gradient occurs in the heat balance.

2. In towers with inert packing, both radial and axial gradients occur, although conduction in the axial direction often is neglected in view of the preponderant transfer of sensible enthalpy in a flow system.

3. In an ideal CSTR, there are no gradients of temperature or composition, only the overall changes.

The simultaneous equations of heat and material balance and rate equations for these three cases are stated in several tables of Sec. 7.

The profiles of temperature and composition shown in Fig. 23-3 are not of homogeneous liquid reactions, but are perhaps representative of all kinds of reactions. Only in stirred tanks and some fluidized beds are nearly isothermal conditions practically attainable.

GAS PHASE

Although they are termed homogeneous, most industrial gas-phase reactions take place in contact with solids, either the vessel wall or particles as heat carriers or catalysts. With catalysts, mass diffusional resistances are present; with inert solids, the only complication is with heat transfer. A few of the reactions in Table 23-1 are gas-phase type, mostly catalytic. Usually a system of industrial interest is liquefied to take advantage of the higher rates of liquid reactions, or to utilize liquid homogeneous catalysts, or simply to keep equipment size down. In this section, some important noncatalytic gas reactions are described.

Mixing of feed gases and temperature control are major process requirements. Gases are usually mixed by injecting one of the streams from a high-speed nozzle into the rest of the gases, as in the flame reactor shown in Fig. 23-22d. Different modes of heat transfer are described, along with some processes that utilize each particular mode, following.

1. Heat is supplied from combustion gases through tubes in fired heaters. Olefins are made this way from light hydrocarbons and naphthas at 800°C (1472°F) and enough above atmospheric pressure to overcome friction. Superheated steam is injected to bring the final temperature up quickly and to retard carbon deposits. Contact times are 0.5 to 3.0 s, followed by rapid quenching. The total tube length of an industrial furnace may be more than 1,000 m. Other important gas-phase cracking processes are: Toluene $\Rightarrow$ benzene, diphenyl $\Rightarrow$ benzene, dicyclopentadiene $\Rightarrow$ CPD, and butene-1 $\Rightarrow$ butadiene. Figure 23-22a shows a cracking furnace.

2. Heat is transferred by direct contact with solids that have been preheated by combustion gases. The process is a cycle of alternate heating and reacting periods. The Wulf process for acetylene by pyrolysis of natural gas utilizes a heated brick checkerwork on a 4-min cycle of heating and reacting. The temperature play is 15°C (59°F), peak temperature is 1,200°C (2,192°F), residence time is 0.1 s of which 0.03 s is near the peak (Faith, Keyes, and Clark, *Industrial Chemicals*, vol. 27, Wiley, 1975).

3. The pebble heater recirculates refractory pebbles continuously through heating and reaction zones. The Wisconsin process for the fixation of nitrogen from air operates at 2,200°C (3,992°F), followed by extremely rapid quenching to freeze the small equilibrium content of nitrogen oxide that is made (Ermenc, *Chem. Eng. Prog.*, **52**, 149 [1956]). Such moving-bed units have been proposed for cracking to olefins but have been obsolesced like most moving-bed reactors.

4. The heat-carrying solids are particles of fluidized sand that circulate between the heating and reaction zones. The reaction section for light hydrocarbons is at 720 to 850°C (1,328 to 1,562°F), the regenerated sand returns at 50 to 100°C (122 to 212°F) above the reactor temperature. The heat comes mostly from the burning of carbon deposited on the sand. This equipment is perhaps competitively suited to cracking heavy stocks that coke readily.

5. Inert combustion gases are injected directly into the reacting stream in flame reactors. Figures 23-22a and 23-22d show two such devices used for making acetylene from light hydrocarbons and naphthas; Fig. 23-22e shows a temperature profile, reaction times in ms.

6. Burning a portion of a combustible reactant with an additive of air or oxygen. Such oxidative pyrolysis of light hydrocarbons to acetylene is done in a special burner, at 0.001 to 0.01 s reaction time, peak at 1,400°C (2,552°F), followed by rapid quenching with oil or water.

7. Exothermic processes, with cooling through heat transfer surfaces or cold shots. In use are shell-and-tube reactors with small-diameter tubes, or towers with internal recirculation of gases, or multiple stages with intercooling. Chlorination of methane and other hydrocarbons results in a mixture of products whose relative amounts

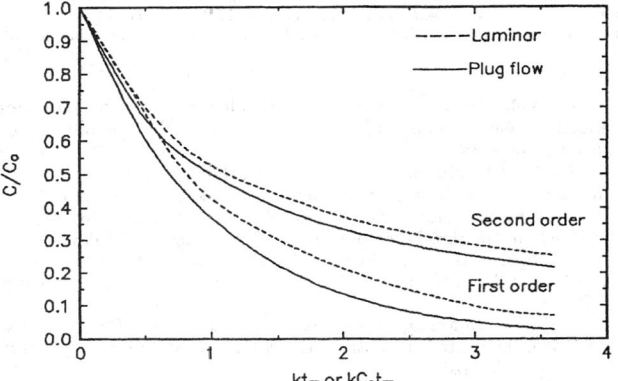

FIG. 23-21 Laminar compared with plug flow of first- and second-order reactions.

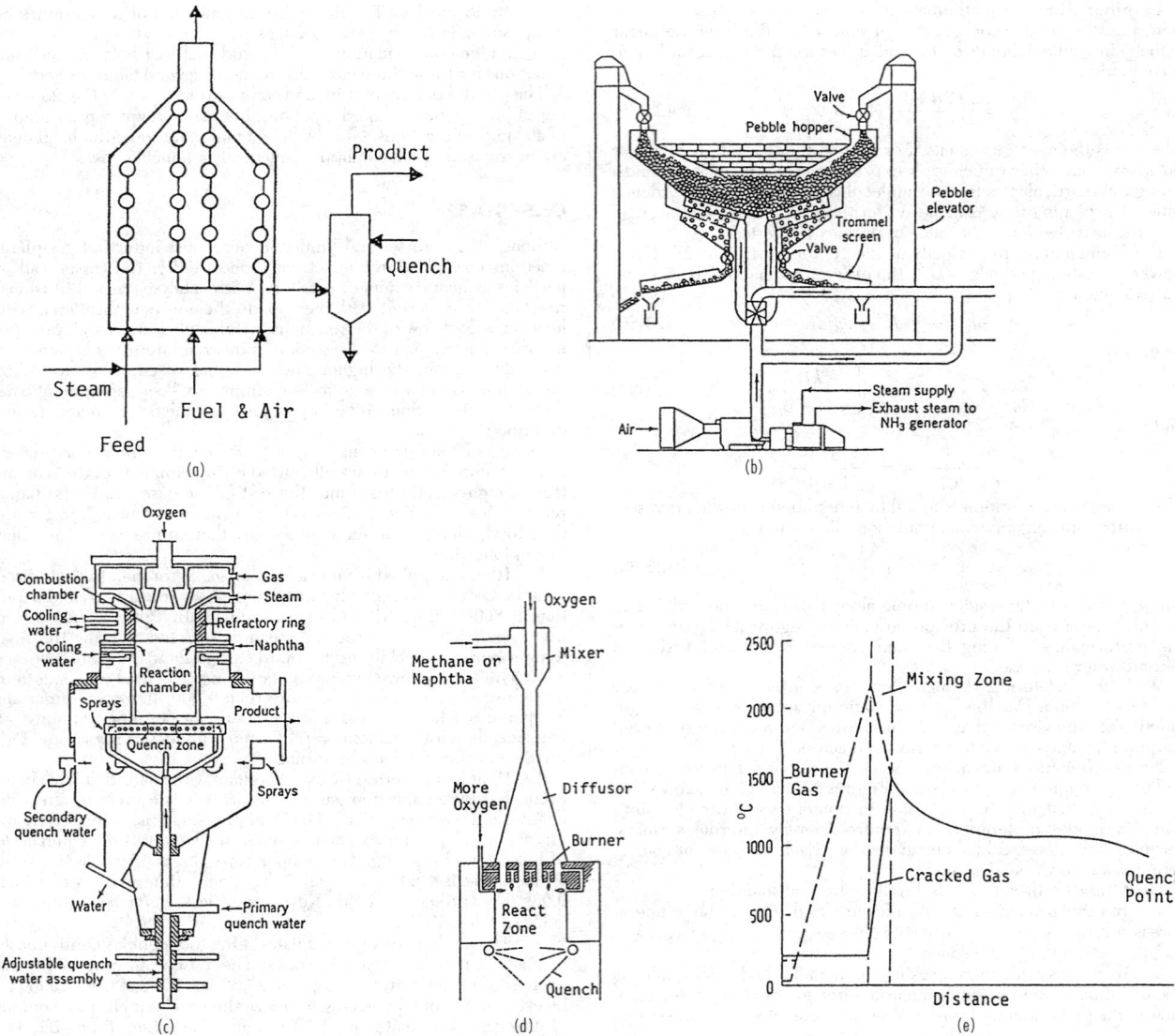

FIG. 23-22 Noncatalytic gas phase reactions. (*a*) Steam cracking of light hydrocarbons in a tubular fired heater. (*b*) Pebble heater for the fixation of nitrogen from air. (*c*) Flame reactor for the production of acetylene from hydrocarbon gases or naphthas (*Patton, Grubb, and Stephenson, Pet. Ref. **37**(11), 180 [1958]*). (*d*) Flame reactor for acetylene from light hydrocarbons (*BASF*). (*e*) Temperature profiles in a flame reactor for acetylene (*Ullmann Encyclopadie der Technischen Chemie, vol. 3, Verlag Chemie, 1973, p. 335*).

can be controlled by varying the Cl/HC ratio and recycling unwanted derivatives; for instance, recycling the mono and di derivatives when only the tri and tetra derivatives are of value, or keeping the chlorine ratio low when emphasizing the lower derivatives. Chlorination temperatures are normally kept in the range 230 to 400°C (446 to 752°F) to limit carbon formation, but may be raised to 500°C (932°F) when favoring CCl$_4$.

SUPERCRITICAL CONDITIONS

The critical properties of water are 374°C (705°F) and 218 atm (3,205 psi). Above this condition a heterogeneous mixture of water, organic compounds, and oxygen may become homogeneous. Then the rate of oxidation may be considerably accelerated because of (1) elimination of diffusional resistances, (2) increase of oxygen concentration by rea-

son of greater density of the mixture, (3) enhanced solubility of oxygen and the organic compound, and (4) increase of the specific rate of reaction by pressure.

That the specific rate is affected by extremes of pressure—sometimes upward, sometimes downward—is well known. A review of this subject is by Kohnstam ("The Kinetic Effects of Pressure," in *Progress in Reaction Kinetics,* Pergamon, 1970). Three examples follow:

1. Thermal decomposition of Di-*t*-butyl peroxide in toluene at 120°C (248°F): $10^6 k$, 1/s, = 13.4 @ 1 atm (14.7 psi), = 5.7 @ 53 atm (779 psi)

2. Isomerization of cyclopropane at 491°C (916°F): $10^4 k$, 1/s, = 0.303 @ 0.067 atm (0.98 psi), = 1.30 @ 1.37 atm (20.1 psi), = 2.98 @ 84.1 atm (1,236 psi)

3. Fading of bromphenol blue at 25°C (77°F), $10^4 k$, L/m·s, = 9.3 @ 1 atm (14.7 psi), = 17.9 @ 1088 atm (16,000 psi)

Solubilities also may be changed greatly in the supercritical region. For naphthalene in ethylene at 35°C (95°F), the mol fraction of naphthalene goes from 0.004 @ 20 atm (294 psi) to 0.02 at 100 atm (1,470 psi) and 0.05 at 300 atm (4,410 psi).

High destructive efficiencies (above 99.99 percent) of complex organic compounds in water can be achieved with residence times under 5 min. Although there are some disagreements, the rate appears to be first order in the organic compound and first or zero order in oxygen.

Recent reviews of research in this area are: Bruno and Ely, eds., *Supercritical Fluid Technology*, CRC Press, 1991; Kiran and Brennecke, eds., *Supercritical Engineering Science*, ACS, 1992.

There is no mention in these reviews of any industrial implementation of supercritical kinetics. Two areas of interest are wastewater treatment—for instance, removal of phenol—and reduction of coking on catalysts by keeping heavy oil decomposition products in solution.

POLYMERIZATION

Polymers that form from the liquid phase may remain dissolved in the remaining monomer or solvent, or they may precipitate. Sometimes beads are formed and remain in suspension; sometimes emulsions form. In some processes solid polymers precipitate from a fluidized gas phase.

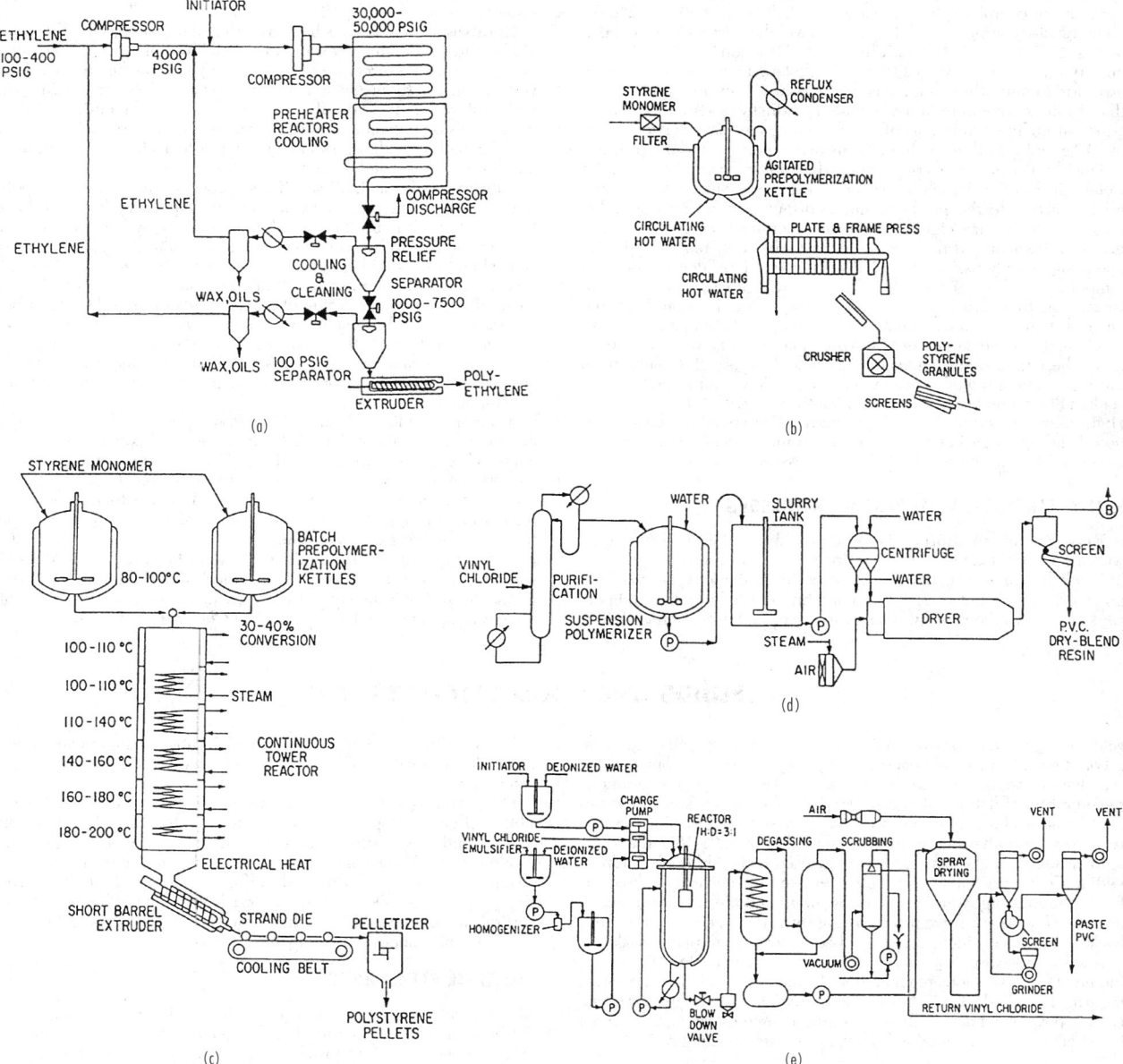

FIG. 23-23 Batch and continuous polymerizations. (*a*) Polyethylene in a tubular flow reactor, up to 2 km long by 6.4 cm ID. (*b*) Batch process for polystyrene. (*c*) Batch-continuous process for polystyrene. (*d*) Suspension (bead) process for polyvinylchloride. (*e*) Emulsion process for polyvinylchloride. (*Ray and Laurence, in Lapidus and Amundson, eds., Chemical Reactor Theory Review, Prentice-Hall, 1977.*)

Stirred batch and continuous reactors are widely used because of their flexibility, but a variety of reactor configurations are in use for particular cases. A selection may be made on rational grounds, or for historical reasons, or due simply to individual taste or a sense of proprietorship. For this complex area, in a given space, it is hardly possible to improve on the guide to selection of polymerization reactors by Gerrens (*German Chemical Engineering*, **4**, 1–13 [1981]; *ChemTech*, 380–383, 434–443 [1982]). Interested parties should go there. A general reference is Rodriguez (*Principles of Polymer Systems*, McGraw-Hill, 1989).

Polymerization processes are characterized by extremes. Industrial products are mixtures with molecular weights of 10^4 to 10^7. In a particular polymerization of styrene the viscosity increased by a factor of 10^6 as conversion went from 0 to 60 percent. The adiabatic reaction temperature for complete polymerization of ethylene is 1,800 K (3,240 R). Heat transfer coefficients in stirred tanks with high viscosities can be as low as 25 W/(m²·°C) (16.2 Btu/[h·ft²·°F]). Reaction times for butadiene-styrene rubbers are 8 to 12 h; polyethylene molecules continue to grow for 30 min; whereas ethyl acrylate in 20% emulsion reacts in less than 1 min, so monomer must be added gradually to keep the temperature within limits. Initiators of the chain reactions have concentration of 10^{-8} g mol/L so they are highly sensitive to poisons and impurities.

The physical properties of polymers depend largely on the molecular weight distribution, which can cover a wide range. Since it is impractical to fractionate the products and reformulate them into desirable ranges of molecular weights, immediate attainment of desired properties must be achieved through the correct choice of reactor type and operating conditions, notably of distributions of residence time and temperature. Those factors are influenced by high viscosities. In tubular reactors there are strong gradients in the radial direction. In stirred tanks ideal mixing is not attainable; wide variations in temperatures may result, and stagnant zones and bypassing may exist. Devices that counteract these unfavorable characteristics include inserts that cause radial mixing, scraping impellers, screw feeders, hollow-shaft impellers with coolant flow through them, recirculation through draft tubes, and so on. High viscosities of bulk and melt polymerizations are avoided with solution, bead, or emulsion operations. Then more nearly normal RTDs exist in CSTR batteries and tubular flow vessels.

KINDS OF POLYMERIZATION PROCESSES

Bulk Polymerization The monomer and initiators are reacted without or with mixing; without mixing to make useful shapes directly, like bakelite products. Because of viscosity limitations, stirred bulk polymerization is not carried to completion but only to 30 to 60 percent or so, with the remaining monomer stripped out and recycled. A

variety of processes is in use for polystyrene, two of which are represented in Figs. 23-23b and 23-23c. A twin-screw extruder is used for polymerization of trioxane and of polyamide.

Bead Polymerization Bulk reaction proceeds in independent droplets of 10 to 1,000 μm diameter suspended in water or other medium and insulated from each other by some colloid. A typical suspending agent is polyvinyl alcohol dissolved in water. The polymerization can be done to high conversion. Temperature control is easy because of the moderating thermal effect of the water and its low viscosity. The suspensions sometimes are unstable and agitation may be critical. Only batch reactors appear to be in industrial use: polyvinyl acetate in methanol, copolymers of acrylates and methacrylates, polyacrylonitrile in aqueous $ZnCl_2$ solution, and others. Bead polymerization of styrene takes 8 to 12 h.

Emulsions Emulsions have particles of 0.05 to 5.0 μm diameter. The product is a stable latex, rather than a filterable suspension. Some latexes are usable directly, as in paints, or they may be coagulated by various means to produce massive polymers. Figures 23-23d and 23-23e show bead and emulsion processes for vinyl chloride. Continuous emulsion polymerization of butadiene-styrene rubber is done in a CSTR battery with a residence time of 8 to 12 h. Batch treating of emulsions also is widely used.

Solution Polymerization These processes may retain the polymer in solution or precipitate it. Polyethylene is made in a tubular flow reactor at supercritical conditions so the polymer stays in solution. In the Phillips process, however, after about 22 percent conversion when the desirable properties have been attained, the polymer is recovered and the monomer is flashed off and recyled (Fig. 23-23a). In another process, a solution of ethylene in a saturated hydrocarbon is passed over a chromia-alumina catalyst, then the solvent is separated and recyled. Another example of precipitation polymerization is the copolymerization of styrene and acrylonitrile in methanol. Also, an aqueous solution of acrylonitrile makes a precipitate of polyacrylonitrile on heating to 80°C (176°F).

A factor in addition to the RTD and temperature distribution that affects the molecular weight distribution (MWD) is the nature of the chemical reaction. If the period during which the molecule is growing is short compared with the residence time in the reactor, the MWD in a batch reactor is broader than in a CSTR. This situation holds for many free radical and ionic polymerization processes where the reaction intermediates are very short lived. In cases where the growth period is the same as the residence time in the reactor, the MWD is narrower in batch than in CSTR. Polymerizations that have no termination step—for instance, polycondensations—are of this type. This topic is treated by Denbigh (*J. Applied Chem.*, **1**, 227 [1951]).

FLUIDS AND SOLID CATALYSTS

In the design of reactors for fluids in the presence of granular catalysts, account must be taken of heat transfer, pressure drop and contacting of the phases, and, in many cases, of provision for periodic or continuous regeneration of deteriorated catalyst. Several different kinds of vessel configurations for continuous processing are in commercial use. Some reactors with solid catalysts are represented in Figs. 23-18 and 23-24.

Most solid catalytic processes employ fixed beds. Although fluidized beds have the merit of nearly uniform temperature and can be designed for continuous regeneration, they cost more and are more difficult to operate, require extensive provisions for dust recovery, and suffer from backmixing. Accordingly, they have been adopted on a large scale for only a few processes. Ways have been found in some instances to avoid the need for continuous regeneration. In the case of platinum reforming with fixed beds, for instance, a large recycle of hydrogen prevents coke deposition while a high temperature compensates for the retarding effect of hydrogen on this essentially dehydrogenating process.

SINGLE FIXED BEDS

These are used for adiabatic processing or when it is practical to embed heat-transfer surface in the bed. Usually, heat transfer is more

effective with the catalyst inside small tubes than outside them. Hydrodesulfurization of petroleum fractions is one large-scale application of single-bed reactors.

During filling, the catalyst is distributed uniformly to avoid the possibility of channeling that could lead to poor heat transfer, poor conversion, and harm to the catalyst because of hot spots. During startup, sudden surges of flow may disturb the bed and are to be avoided. For instance, in a study of a hydrodesulfurizer by Murphree et al. (*Ind. Eng. Chem. Proc. Des. & Dev.*, **3**, 381 [1964]) the efficiency of conversion in a commercial size unit varied between 47 and 80 percent with different modes of loading and startup.

MULTIPLE FIXED BEDS

These enable temperature control with built-in exchangers between the beds or with pumparound exchangers. Converters for ammonia, SO_3, cumene, and other processes may employ as many as five or six beds in series. The Sohio process for vapor-phase oxidation of propylene to acrylic acid uses two beds of bismuth molybdate at 20 to 30 atm (294 to 441 psi) and 290 to 400°C (554 to 752°F). Oxidation of ethylene to ethylene oxide also is done in two stages with supported

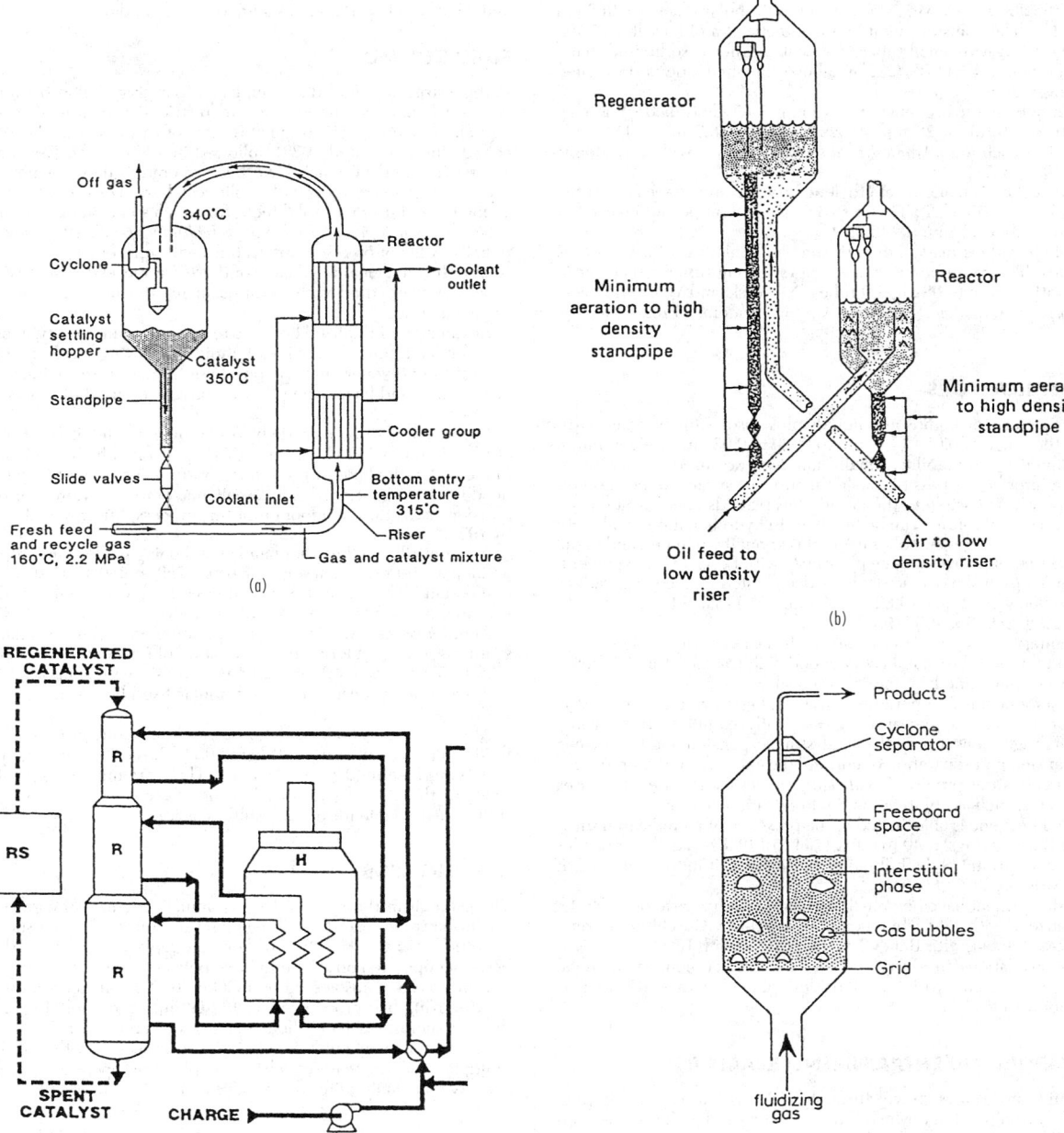

FIG. 23-24   Reactors with moving catalysts. (*a*) Transport fluidized type for the Sasol Fischer-Tropsch process, nonregenerating. (*b*) Esso type of stable fluidized bed reactor/regenerator for cracking petroleum oils. (*c*) UOP reformer with moving bed of platinum catalyst and continuous regeneration of a controlled quantity of catalyst. (*d*) Flow distribution in a fluidized bed; the catalyst rains through the bubbles.

silver catalyst, the first stage to 30 percent conversion, the second to 76 percent, with a total of 1.0 s contact time.

MULTITUBULAR REACTORS

These reactors are of shell-and-tube configuration and mostly have the catalyst in the tubes, although some ammonia converters have the catalyst on the shell side. Hundreds of tubes of a few cm diameter may be required. Their diameters may be approximately 8 times the diameters of the pellets and lengths limited by allowable pressure drop. Catalyst pellet sizes usually are in the range of 0.3 to 0.5 cm (0.76 to 1.27 in). Uniform loading is ensured by using special equipment that charges the same amount of catalyst to each tube at a definite rate. After filling, each tube is checked for pressure drop.

Maleic anhydride is made by oxidation of benzene with air above 350°C (662°F) with V-Mo catalyst in a multitubular reactor with 2-cm tubes. The heat-transfer medium is a eutectic of molten salt at 375°C (707°F). Even with small tubes, the heat transfer is so limited that a peak temperature 100°C (212°F) above the shell side is developed and moves along the tubes.

Butanol by the hydrogenation of crotonaldehyde is made in a reactor with 4,000 tubes, 28 mm (0.029 ft) ID by 10.7 m (35.1 ft) long (Berty, in Leach, ed., *Applied Industrial Catalysis,* vol. 1, Academic Press, 1983, p. 51).

Vinyl acetate is made from ethylene, oxygen, and acetic acid in the vapor phase at 150 to 175°C (302 to 347°F) with supported Pd catalyst in packed tubes, 25 mm (0.082 ft) ID.

Vinyl chloride is made from ethylene and chlorine with Cu and K chlorides. The Stauffer process employs 3 multitubular reactors in series with 25 mm (0.082 ft) ID tubes (Naworski and Velez, in Leach, ed., *Applied Industrial Catalysis,* vol. 1, Academic Press, 1983, p. 251).

SLURRY REACTORS

These reactors for liquids and liquids plus gases employ small particles in the range of 0.05 to 1.0 mm (0.0020 to 0.039 in), the minimum size limited by filterability. Small diameters are used to provide as large an interface as possible since the internal surface of porous pellets is poorly accessible to the liquid phase. Solids concentrations up to 10 percent by volume can be handled. In hydrogenation of oils with Ni catalyst, however, the solids content is about 0.5 percent, and in the manufacture of hydroxylamine phosphate with Pd-C it is 0.05 percent. Fischer-Tropsch slurry reactors have been tested with concentrations of 10 to 950 g catalyst/L (0.624 to 59.3 lbm/ft³) (Satterfield and Huff, *Chem. Eng. Sci.,* **35,** 195 [1980]).

Advantages of slurry reactors are high heat capacity, which makes for good temperature stability, and good heat transfer. Catalyst activity can be maintained by partial removal of degraded material and replenishment during operation. Disadvantages are a lower conversion for a given size because of essentially complete backmixing, power for agitation to keep the catalyst in suspension and to enhance heat transfer, and separation of entrained catalyst from the product.

Most industrial processes with slurry reactors are used for gases with liquids, such as chlorination, hydrogenation, and oxidation.

Liquid benzene is chlorinated in the presence of metallic iron turnings or Raschig rings at 40 to 60°C (104 to 140°F). Carbon tetrachloride is made from CS_2 by bubbling chlorine into it in the presence of iron powder at 30°C (86°F).

Substances that have been hydrogenated in slurry reactors include: nitrobenzene with Pd-C, butynediol with Pd-CaCO₃, chlorobenzene with Pt-C, toluene with Raney Ni, and acetone with Raney Ni.

Some oxidations in slurry reactors include: cumene with metal oxides, cyclohexene with metal oxides, phenol with CuO, and n-propanol with Pt.

TRANSPORT (OR ENTRAINMENT) REACTORS

The fluid and catalyst travel through the vessel in essentially plug flow and are separated downstream by settling and with cyclones and filters. The main publicized applications are for cracking to gasoline-range hydrocarbons with highly active zeolite catalysts and in the Sasol Fischer-Tropsch process. A considerable body of experience with transport of solids by entrainment with gases (pneumatic conveying) and with pneumatic drying has been accumulated; nevertheless, five years elapsed before the Sasol reactor was made to function satisfactorily. A principal advantage of transport reactors is that they approach plug flow, whereas stable fluidized beds have much backmixing. Figure 23-24a shows an in-line heat exchanger in the Sasol unit. The catalyst is promoted iron. It circulates through the 1.0-m

(3.28-ft) ID riser at 72,600 kg/h (160,000 lbm/h) at 340°C (644°F) and 23 atm (338 psi) and has a life of about 50 days.

FLUIDIZED BEDS

Particle sizes in fluidized bed applications average below 0.1 mm, but very small particles impose severe restrictions on the recovery of entrained material. The original reactor of this type was the Winkler coal gasifier (patented 1922), followed in 1940 by the Esso cracker, several hundred of which have been operated; they are now being replaced by riser reactors with zeolite catalysts. The other large application is combustion of solid fuels, for which some 30 installations are listed in *Encyclopedia of Chemical Technology,* vol. 10, Wiley, 1980, p. 550). A list of 55 other applications with references is in the same source. It is not clear how many of these are successful because many processes were tried with enthusiasm for the new technology and found wanting.

Advantages of fluidized beds are temperature uniformity, good heat transfer, and transportability of rapidly decaying catalyst between reacting and regenerating sections. Disadvantages are attrition, recovery of fines, and backmixing. Baffles have been used to reduce backmixing.

Phthalic anhydride is made by oxidation of naphthalene at temperatures of 340 to 380°C (644 to 716°F) controlled by heat exchangers immersed in the bed. At these temperatures the catalyst is stable and need not be regenerated. The excellence of temperature control was a major factor for the adoption of this process, but it was obsolesced by 1972.

Acrylonitrile, on the other hand, is still being made from propylene, ammonia, and oxygen at 400 to 510°C (752 to 950°F) in this kind of equipment. The good temperature control with embedded heat exchangers permits catalyst life of several years.

Another process where good temperature control is essential is the synthesis of vinyl chloride by chlorination of ethylene at 200 to 300°C (392 to 572°F), 2 to 10 atm (29.4 to 147 psi), with supported cupric chloride, but a process with multitubular fixed beds is a strong competitor.

Although it is not a catalytic process, the roasting of iron sulfide in fluidized beds at 650 to 1,100°C (1,202 to 2,012°F) is analogous. The pellets are 10-mm (0.39-in) diameter. There are numerous plants, but they are threatened with obsolescence because cheaper sources of sulfur are available for making sulfuric acid.

MOVING BEDS

The catalyst, in the form of large granules, circulates by gravity and gas lift between reaction and regeneration zones. The first successful operation was the Houdry cracker that replaced a plant with fixed beds that operated on a 10-min cycle between reaction and regeneration. It was soon obsolesced by FCC units. The only currently publicized moving bed process is a UOP platinum reformer (Fig. 23-24c) that regenerates a controlled quantity of catalyst on a continuous basis. It is not known how competitive this process is with units having multiple reactors that regenerate in place or operate at such low severity that catalyst life of several years is obtained.

THIN BEDS AND WIRE GAUZES

Fast catalytic reactions that must be quenched rapidly are done in contact with wire screens or thin layers of fine granules. Ammonia in a 10% concentration in air is oxidized by flow through a fine gauze catalyst made of 2 to 10% Rh in Pt, 10 to 30 layers, 0.075-mm (0.0030-in) diameter wire. Contact time is 0.0003 s at 750°C (1,382°F) and 7 atm (103 psi) followed by rapid quenching. Methanol is oxidized to formaldehyde in a thin layer of finely divided silver or a multilayer screen, with a contact time of 0.01 s at 450 to 600°C (842 to 1,112°F).

GAS/LIQUID REACTIONS

Industrial gas/liquid reaction processes are of four main categories:

1.   Gas purification or the removal of relatively small amounts of impurities such as CO_2, CO, COS, SO_2, H_2S, NO, and others from air, natural gas, hydrogen for ammonia synthesis, and others

2.   Liquid phase processes, such as hydrogenation, halogenation, oxidation, nitration, alkylation, and so on

3.   Manufacture of pure products, such as sulfuric acid, nitric acid, nitrates, phosphates, adipic acid, and so on

4.   Biochemical processes, such as fermentation, oxidation of sludges, production of proteins, biochemical oxidations, and so on

Reaction between an absorbed solute and a reagent reduces the equilibrium partial pressure of the solute, thus increasing the rate of mass transfer. The mass-transfer coefficient likewise is enhanced, which contributes further to increased absorption rates. Extensive theoretical analyses of these effects have been made, but rather less experimental work and design guidelines.

For reaction between a gas and a liquid, three modes of contacting are possible: (1) the gas is dispersed as bubbles in the liquid, (2)

the liquid is dispersed as droplets in the gas, and (3) the liquid and gas are brought together as thin films over a packing or wall. The choice between these modes is an important problem. Some considerations are the magnitude and distribution of the residence times of the phases, the power requirements, the scale of the operation, the opportunity for heat transfer, and so on. Industrial equipment featuring particular factors is available. The main types of apparatus appear in Fig. 23-25, but many variations are practiced. Specific liquid/gas processes are represented in Fig. 23-26. Equipment is selected and designed by a combination of theory, pilot plant work, and experience, with rather less reliance on theory than in some other areas of reactor design.

MASS TRANSFER COEFFICIENTS

The resistance to transfer of mass between a gas and a liquid is assumed confined to that of fluid films between the phases. Let

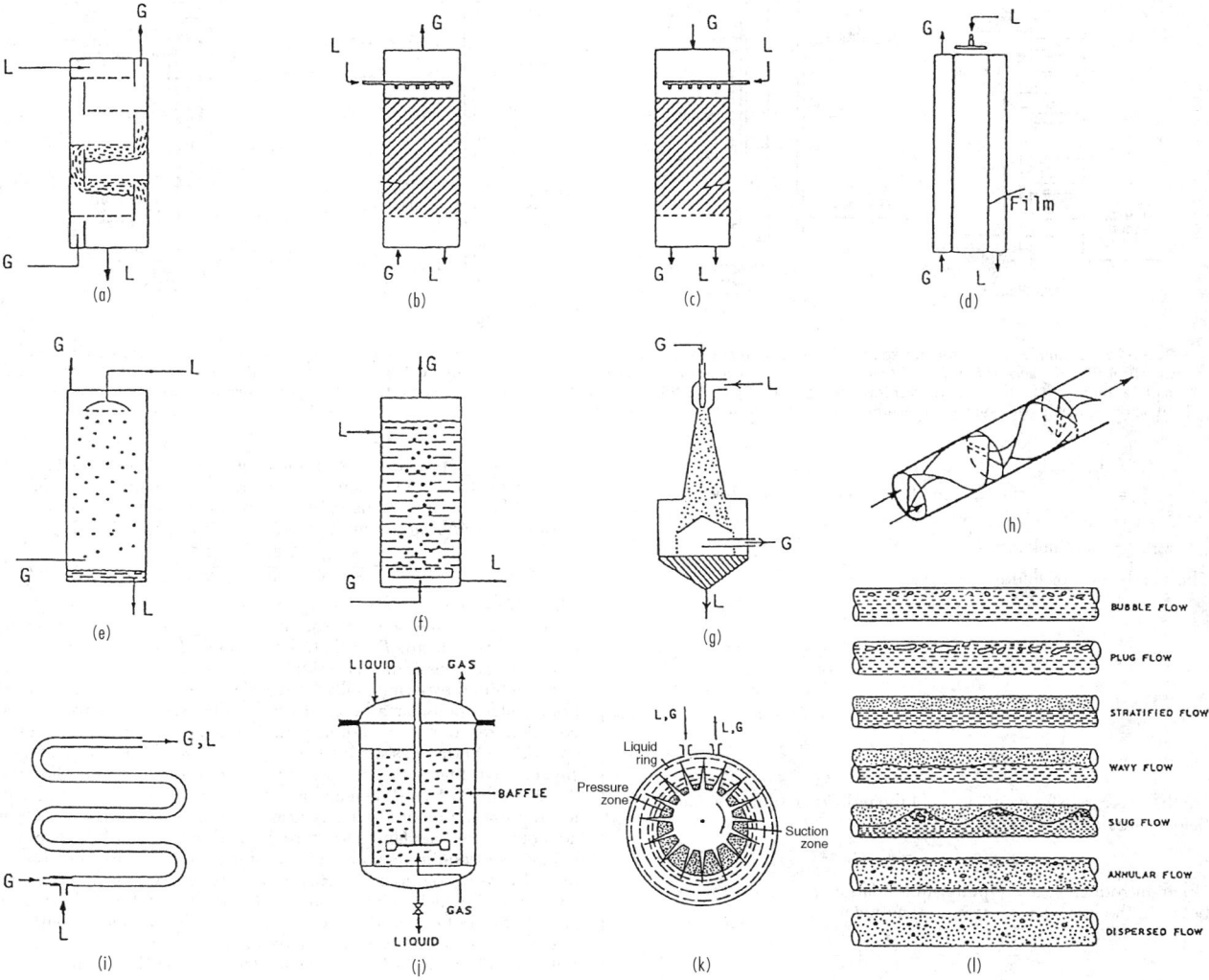

FIG. 23-25   Types of industrial gas/liquid reactors. (*a*) Tray tower. (*b*) Packed, counter current. (*c*) Packed, parallel current. (*d*) Falling liquid film. (*e*) Spray tower. (*f*) Bubble tower. (*g*) Venturi mixer. (*h*) Static in line mixer. (*i*) Tubular flow. (*j*) Stirred tank. (*k*) Centrifugal pump. (*l*) Two-phase flow in horizontal tubes.

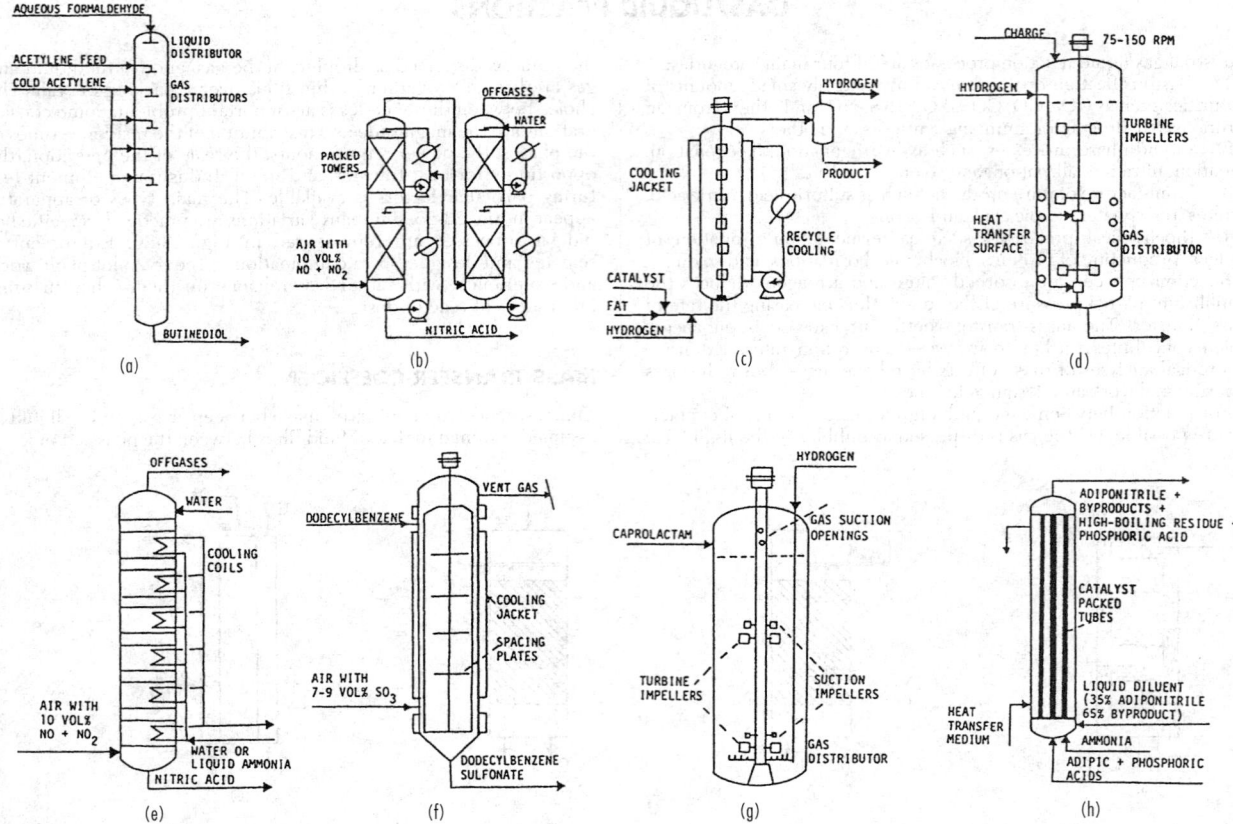

FIG. 23-26 Examples of reactors for specific liquid/gas processes. (a) Trickle reactor for synthesis of butinediol, 1.5 m diameter by 18 m high. (b) Nitrogen oxide absorption in packed columns. (c) Continuous hydrogenation of fats. (d) Stirred tank reactor for batch hydrogenation of fats. (e) Nitrogen oxide absorption in a plate column. (f) A thin-film reactor for making dodecylbenzene sulfonate with SO_3. (g) Stirred tank reactor for the hydrogenation of caprolactam. (h) Tubular reactor for making adiponitrile from adipic acid in the presence of phosphoric acid.

D = diffusivity

$p_i = f(C_i)$ or $p_i = HC_i$, equilibrium relation at the interface

a = interfacial area/unit volume

z_g, z_L = film thicknesses

The steady rates of solute transfer are

$$r = k_g a(p_g - p_i)$$
$$= k_L a(C_i - C_L)$$

where

$$k_g = \frac{D}{z_g}$$

$$k_L = \frac{D}{z_L}$$

are the mass-transfer coefficients of the individual films. Overall coefficients are defined by

$$r = K_g a(p_g - p_L) = K_L a(C_g - C_L)$$

Upon introducing the equilibrium relation $p = HC$, the relation between the various mass-transfer coefficients is

$$\frac{1}{K_g a} = \frac{H}{K_L a} = \frac{1}{k_g a} + \frac{H}{k_L a} \qquad (23\text{-}90)$$

When the solubility is low, H is large and $k_L \Rightarrow K_L$; when the solubility is high, H is small and $k_g \Rightarrow K_g$.

For purely physical absorption, the mass-transfer coefficients depend on the hydrodynamics and the physical properties of the phases. Many correlations exist; for example, that of Dwivedi and Upadhyay (*Ind. Eng. Chem. Proc. Des. & Dev.*, **16**, 157 [1977]),

$$k = \frac{u'}{Sc^{2/3}}\left(\frac{0.765}{Re^{0.82}} + \frac{0.365}{Re^{0.386}}\right) \qquad (23\text{-}91)$$

where $Re = \rho u' d_p / \mu$.

With a reactive solvent, the mass-transfer coefficient may be enhanced by a factor E so that, for instance, K_g is replaced by EK_g. Like specific rates of ordinary chemical reactions, such enhancements must be found experimentally. There are no generalized correlations. Some calculations have been made for idealized situations, such as complete reaction in the liquid film. Tables 23-6 and 23-7 show a few spot data. On that basis, a tower for absorption of SO_2 with NaOH is smaller than that with pure water by a factor of roughly $0.317/7.0 = 0.045$. Table 23-8 lists the main factors that are needed for mathematical representation of $K_g a$ in a typical case of the absorption of CO_2 by aqueous monethanolamine. Figure 23-27 shows some of the complex behaviors of equilibria and mass-transfer coefficients for the absorption of CO_2 in solutions of potassium carbonate. Other than Henry's law, $p = HC$, which holds for some fairly dilute solutions, there is no general form of equilibrium relation. A typically complex equation is that for CO_2 in contact with sodium carbonate solutions (Harte, Baker, and Purcell, *Ind. Eng. Chem.*, **25**, 528 [1933]), which is

$$p_{CO_2} = \frac{137f^2 N^{1.29}}{S(1-f)(365-T)}, \qquad \text{Torr} \qquad (23\text{-}92)$$

TABLE 23-6   Typical Values of K_Ga for Absorption in Towers Packed with 1.5-in Intalox Saddles at 25% Completion of Reaction*

Absorbed gas	Absorbent	K_Ga, lb mol/(h·ft³·atm)
Cl_2	$H_2O\cdot NaOH$	20.0
HCl	H_2O	16.0
NH_3	H_2O	13.0
H_2S	$H_2O\cdot MEA$	8.0
SO_2	$H_2O\cdot NaOH$	7.0
H_2S	$H_2O\cdot DEA$	5.0
CO_2	$H_2O\cdot KOH$	3.10
CO_2	$H_2O\cdot MEA$	2.50
CO_2	$H_2O\cdot NaOH$	2.25
H_2S	H_2O	0.400
SO_2	H_2O	0.317
Cl_2	H_2O	0.138
CO_2	H_2O	0.072
O_2	H_2O	0.0072

*To convert in to cm, multiply by 2.54; lb mol/(h·ft³·atm) to kg mol/(h·m³·kPa), multiply by 0.1581.
SOURCE: From Eckert, et al., *Ind. Eng. Chem.*, **59**, 41 (1967).

TABLE 23-7   Selected Absorption Coefficients for CO_2 in Various Solvents in Towers Packed with Raschig Rings*

Solvent	K_ga, lb mol/(h·ft³·atm)
Water	0.05
1-N sodium carbonate, 20% Na as bicarbonate	0.03
3-N diethanolamine, 50% converted to carbonate	0.4
2-N sodium hydroxide, 15% Na as carbonate	2.3
2-N potassium hydroxide, 15% K as carbonate	3.8
Hypothetical perfect solvent having no liquid-phase resistance and having infinite chemical reactivity	24.0

*Basis: $L = 2,500$ lb/(h·ft²); $G = 300$ lb/(h·ft²); $T = 77°F$; pressure, 1.0 atm. To convert lb mol/(h·ft³·atm) to kg mol/(h·m³·kPa) multiply by 0.1581.
SOURCE: From Sherwood, Pigford, and Wilke, *Mass Transfer*, McGraw-Hill, 1975, p. 305.

TABLE 23-8   Correlation of K_Ga for Absorption of CO_2 by Aqueous Solutions of Monoethanolamine in Packed Towers*

$$K_Ga = F\left(\frac{L}{\mu}\right)^{2/3}[1 + 5.7(C_e - C)M\, e^{0.0067T - 3.4p}]$$

where K_Ga = overall gas-film coefficient, lb mol/(h·ft³·atm)
   μ = viscosity, centipoises
   C = concentration of CO_2 in the solution, mol/mol monoethanolamine
   M = amine concentration of solution (molarity, g mol/L)
   T = temperature, °F
   p = partial pressure, atm
   L = liquid-flow rate, lb/(h·ft²)
   C_e = equilibrium concentration of CO_2 in solution, mol/mol monoethanolamine
   F = factor to correct for size and type of packing

Packing	F	Basis for calculation of F
5- to 6-mm glass rings	7.1×10^{-3}	Shneerson and Leibush data, 1-in column, atmospheric pressure
⅜-in ceramic rings	3.0×10^{-3}	Unpublished data for 4-in column, atmospheric pressure
¾- by 2-in polyethylene Tellerettes	3.0×10^{-3}	Teller and Ford data, 8-in column, atmospheric pressure
1-in steel rings		
1-in ceramic saddles	2.1×10^{-3}	
1½- and 2-in ceramic rings	$0.4–0.6 \times 10^{-3}$	Gregory and Scharmann and unpublished data for two commercial plants, pressures 30 to 300 psig

*To convert in to cm multiply by 2.54.
SOURCE: From Kohl and Riesenfeld, *Gas Purification*, Gulf, 1985.

where f = fraction of total base present as bicarbonate
   N = normality, 0.5 to 2.0
   S = solubility of CO_2 in pure water at 1 atm, g mol/L
   T = temperature, 65 to 150°F

COUNTERCURRENT ABSORPTION TOWERS

Consider mass transfer in a countercurrent tower, packed or spray or bubble. Let

$$G_m = \frac{\text{mol inert gas}}{(\text{unit time})(\text{unit cross section})}$$

$$L_m = \frac{\text{mol solute-free liquid}}{(\text{unit time})(\text{unit cross section})}$$

$$Y = \frac{y}{(1-y)} = \frac{\text{mol solute, gas phase}}{\text{mol inert gas, vapor phase}}$$

$$X = \frac{x}{(1-x)} = \frac{\text{mol solute in the liquid}}{\text{mol inert solvent in the liquid}}$$

Z = height of active tower section.

The material balance over a differential height is

$$G_m dY = L_m dX$$

In terms of gas film conditions,

$$G_m dY = k_g a(p_g - p_i)dZ$$

The partial pressure of the gas is related to the total pressure π by

$$p_g = y\pi = \frac{Y}{1+Y}\,\pi, \qquad p_i = \frac{Y_i}{1+Y_i}\,\pi$$

Substitution and rearrangement leads to the equation for the tower height,

$$Z = \frac{G_m}{\pi}\int_{Y_2}^{Y_1} \frac{(1+Y)(1+Y_i)}{k_g a(Y - Y_i)}\,dY \qquad (23\text{-}93)$$

Similarly, in terms of the liquid-phase condition,

$$C = \frac{X}{1+X}\,C_t$$

$$C_t = \frac{\text{mol (solute + solvent)}}{\text{volume of liquid}}$$

$$L_m dX = k_L a(C_i - C)dZ$$

$$Z = \frac{L_m}{C_t}\int_{x_2}^{x_1} \frac{(1+X)(1+X_i)}{k_L a(X_i - X)}\,dX \qquad (23\text{-}94)$$

The balance around one end of the tower can be written:

$$X = X_1 + \frac{G_m}{L_m}(Y - Y_1) \qquad (23\text{-}95)$$

If the equilibrium relation is

$$y_i = mx \qquad \text{or} \qquad \frac{Y_i}{1+Y_i} = m\frac{X}{1+X} \qquad (23\text{-}96)$$

or any other functional form, substitution of Eqs. (23-95) and (23-96) into (23-93) will enable the last equation to be solved numerically. Graphical methods of solution are explained in the chapter on physical absorption.

For physical absorption, values of the mass-transfer coefficients may not vary greatly, so a mean value could be adequate and could be taken outside the integral sign, but for reactive absorption the variation usually is too great.

Note that the tower height is inversely proportional to the enhanced mass-transfer coefficient, or to the enhancement factor itself.

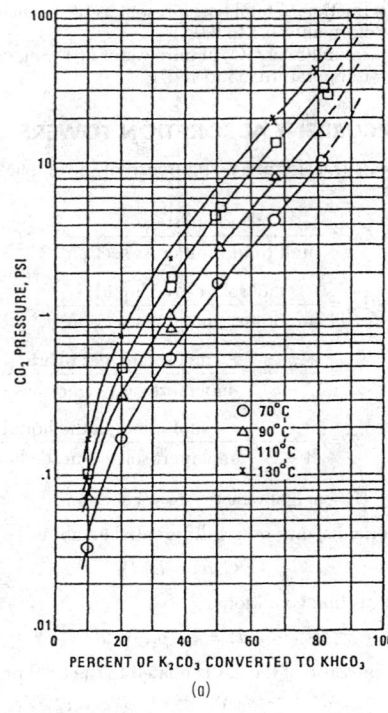

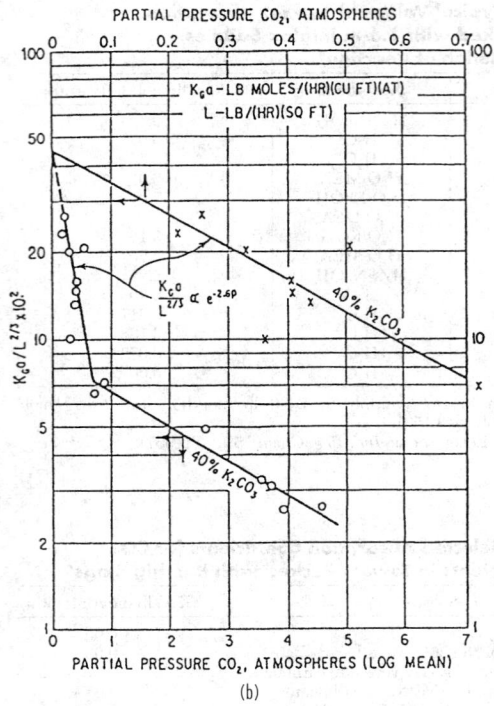

FIG. 23-27 CO_2 in potassium carbonate solutions: (*a*) equilibrium in 20% solution, (*b*) mass-transfer coefficients in 40% solutions. (*Data cited by Kohl and Riesenfeld,* Gas Purification, *Gulf Publishing, 1985.*)

FIRST-ORDER OR PSEUDO-FIRST-ORDER REACTION IN A LIQUID FILM

A reactant A diffuses into a stagnant liquid film where the concentration of excess reactant B remains essentially constant at C_{b0}. At the inlet face the concentration is C_{ai}. Making the material balance over a differential dz of the distance leads to the second-order diffusional equation,

$$\frac{d^2C_a}{dz^2} = \frac{k_cC_{b0}}{D}C_a = \alpha^2C_a$$

The boundary conditions are $C_a = C_{ai}$ where $z = 0$, and $C_a = C_{aL}$ where $z = z_L$. The integral is

$$C_a = \frac{C_{aL}\sinh(\alpha z) + C_{ai}\sinh[\alpha(z_L - z)]}{\sinh(\alpha z_L)}$$

Since this is a steady condition, the rate of reaction in the film equals the rate of input to the film,

$$r = -D\left(\frac{dC_a}{dz}\right)_{C_a = C_{ai}} = \frac{\alpha D[\cosh(\alpha z_L) + C_{aL}]}{\sinh(\alpha z_L)} \quad (23\text{-}97)$$

An important special case is that of complete reaction in the film; that is, for $C_{aL} = 0$ where $z = z_L$. Then,

$$r = \alpha D\coth(\alpha z_L)(C_{ai} - 0)$$
$$= \alpha z_L\left(\frac{D}{z_L}\right)\coth(\alpha z_L)(C_{ai} - 0)$$
$$= k_L\beta\coth(\beta)\,\Delta C_a = k_L E\,\Delta C_a \quad (23\text{-}98)$$

where the enhancement factor is

$$E = \beta\coth(\beta)$$

and a parameter called the *Hatta number* is

$$\beta = \alpha z_L = z_L\sqrt{\frac{k_cC_{b0}}{D}} = \frac{\sqrt{k_cDC_{b0}}}{k_L} \quad (23\text{-}99)$$

Note that this parameter has the same form as the Thiele number which occurs in the theory of diffusion/reaction in catalyst pores.

SECOND-ORDER REACTION IN A LIQUID FILM

A pure gas A diffuses into a liquid film where it reacts with B from the liquid phase. Material balances on the two participants are:

$$D_a\frac{d^2C_a}{dz^2} = k_cC_aC_b \quad (23\text{-}100)$$

$$D_b\frac{d^2C_b}{dz^2} = k_cC_aC_b \quad (23\text{-}101)$$

At the gas/liquid interface, $z = 0$, $C_a = C_{ai}$, $dC_b/dz = 0$. On the liquid side of the film, $z = z_L$, $C_b = C_{bL}$. The volume of the bulk liquid per unit of interfacial area

$$V_L = \text{total volume} - \text{film volume}$$
$$= \frac{\varepsilon}{a - z_L} \quad (23\text{-}102)$$

where ε is fractional holdup of liquid and a is interfacial area per unit volume of liquid. The remaining boundary condition at z_L is

$$-D_a\left(\frac{dC_a}{dz}\right) = k_cC_{aL}C_{bL}\left(\frac{\varepsilon}{a - z_L}\right) \quad (23\text{-}103)$$

The numerical solution of these equations is shown in Fig. 23-28. This is a plot of the enhancement factor E against the Hatta number, with several other parameters. The factor E represents an enhancement of the rate of transfer of A caused by the reaction compared with physical absorption with zero concentration of A in the liquid. The uppermost line on the upper right represents the pseudo-first-order reaction, for which $E = \beta\coth\beta$.

Three regions are identified with different requirements of ε and a, and for which particular kinds of contacting equipment may be best:

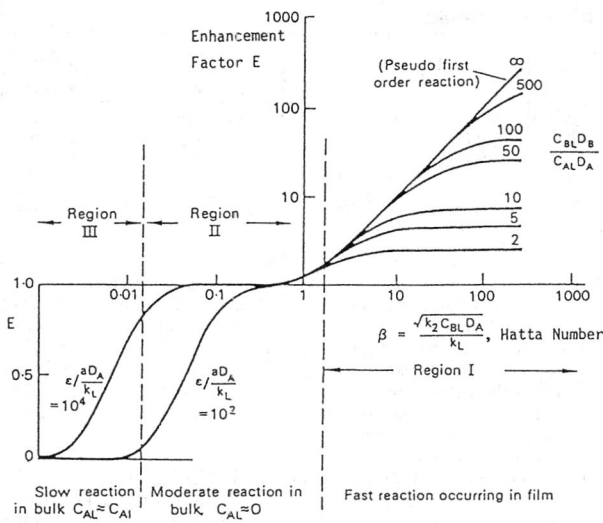

FIG. 23-28 Enhancement factor E and Hatta number of first- and second-order gas/liquid reactions, numerical solutions by several hands.

$$\text{Hatta number} = \beta = \frac{\sqrt{k_2 C_{B2} D_A}}{k_L}$$

(*Coulson and Richardson, Chemical Engineering, vol. 3, Pergamon, 1971, p. 80.*)

Region I, $\beta > 2$. Reaction is fast and occurs mainly in the liquid film so $C_{aL} \Rightarrow 0$. The rate of reaction $r_a = k_L a E C_{ai}$ will be large when a is large, but liquid holdup is not important. Packed towers or stirred tanks will be suitable.

Region II, $0.02 < \beta < 2$. Most of the reaction occurs in the bulk of the liquid. Both interfacial area and holdup of liquid should be high. Stirred tanks or bubble columns will be suitable.

Region III, $\beta < 0.02$. Reaction is slow and occurs in the bulk liquid. Interfacial area and liquid holdup should be high, especially the latter. Bubble columns will be suitable.

SCALE-UP FROM LABORATORY DATA

Three criteria for scale-up are that the laboratory and industrial units have the same mass-transfer coefficients k_g and Ek_L and the same ratio of the specific interfacial surface and liquid holdup a/ε_L. Tables 23-9 and 23-10 give order-of-magnitude values of some parameters that may be expected in common types of liquid/gas contactors.

Auxiliary data are the sizes of bubbles and droplets. These data and the holdups of the two phases are measured by a variety of standard techniques. Interfacial area measurements utilize techniques of transmission or reflection of light. Data on and methods for finding solubilities of gases or the relation between partial pressure and concentration in liquid are also well established.

Hatta Number A film-conversion parameter is defined as

$$M = \frac{\text{maximum possible conversion in the film}}{\text{maximum diffusional transport through the film}}$$

$$= \frac{k_c C_a C_{b0} z_L}{Da(C_a - 0)/z_L} = \frac{k_c C_{b0} z_L^2}{D_a} = \frac{D_a k_c C_{b0}}{k_L^2} = \overset{2}{\mathrm{Ha}}$$

(23-104)

When $\overset{2}{\mathrm{Ha}} \gg 1$, all of the reaction occurs in the film and the amount of interfacial area is controlling, necessitating equipment that has a large interfacial area. When $\mathrm{Ha} \ll 1$, no reaction occurs in the film and the bulk volume controls. The Hatta criteria are often applied in the following form.

$\mathrm{Ha} < 0.3$	Reaction needs large bulk liquid volume
$0.3 < \mathrm{Ha} < 3.0$	Reaction needs large interfacial area and large bulk liquid volume
$\mathrm{Ha} > 3.0$	Reaction needs large interfacial area

Of the parameters making up the Hatta number, liquid diffusivity data and measurement methods are well reviewed in the literature.

Specific Rate k_c For the liquid-phase reaction without the complications of diffusional resistances, the specific rate can be determined after dissolving the gas solute and liquid reactant separately in the same solvent, mixing the two liquids quickly and thoroughly and following the progress of the liquid-phase reaction at the elevated pressure. Unless the reaction is very fast, the mixing time may be ignored. There may be an advantage in employing supercritical conditions at which gas solubility may be appreciably enhanced.

A number of successful devices have been in use for finding mass-transfer coefficients, some of which are sketched in Fig. 23-29, and all of which have known or adjustable interfacial areas. Such laboratory testing is reviewed, for example, by Danckwerts (*Gas-Liquid Reactions*, McGraw-Hill, 1970) and Charpentier (in Ginetto and Silveston, eds., *Multiphase Chemical Reactor Theory, Design, Scaleup*, Hemisphere, 1986).

Gas-Film Coefficient k_g Since the gas film is not affected by the liquid-phase reaction, one of the many available correlations for physical absorption may be applicable. The coefficient also may be found directly after elimination of the liquid-film coefficient by employing a solution that reacts instantaneously and irreversibly with the dissolved gas, thus canceling out any backpressure. Examples of such systems are SO_2 in NaOH and NH_3 in H_2SO_4.

Liquid-Film Coefficients k_L (Physical) and Ek_L (Reactive) The gas-side resistance can be eliminated by employing a pure gas, thus leaving the liquid film as the only resistance. Alternatively, after the gas-film resistance has been found experimentally or from corre-

TABLE 23-9 Mass-Transfer Coefficients, Interfacial Areas and Liquid Holdup in Gas/Liquid Reactions

Type of reactor	ε_L, %	k_G, gm mol/(cm^2·s·atm) s·atm) $\times 10^4$	k_L cm/s $\times 10^2$	a, cm^2/cm^3 reactor	$k_L a$, $s^{-1} \times 10^2$
Packed columns					
Countercurrent	2–25	0.03–2	0.4–2	0.1–3.5	0.04–7
Cocurrent	2–95	0.1–3	0.4–6	0.1–17	0.04–102
Plate columns					
Bubble cap	10–95	0.5–2	1–5	1–4	1–20
Sieve plates	10–95	0.5–6	1–20	1–2	1–40
Bubble columns	60–98	0.5–2	1–4	0.5–6	0.5–24
Packed bubble columns	60–98	0.5–2	1–4	0.5–3	0.5–12
Tube reactors					
Horizontal and coiled	5–95	0.5–4	1–10	0.5–7	0.5–70
Vertical	5–95	0.5–8	2–5	1–20	2–100
Spray columns	2–20	0.5–2	0.7–1.5	0.1–1	0.07–1.5
Mechanically agitated bubble reactors	20–95	—	0.3–4	1–20	0.3–80
Submerged and plunging jet	94–99	—	0.15–0.5	0.2–1.2	0.03–0.6
Hydrocyclone	70–93	—	10–30	0.2–0.5	2–15
Ejector reactor	—	—	—	1–20	—
Venturi	5–30	2–10	5–10	1.6–25	8–25

SOURCE: From Charpentier, *Advances In Chemical Engineering*, vol. 11, Academic Press, 1981, pp. 2–135).

TABLE 23-10 Order-of-Magnitude Data of Equipment for Contacting Gases and Liquids

Device	$k_L a$, s^{-1}	V, m^3	$k_L a V$, m^3/s (duty)	a, m^{-1}	ε_L	Liquid mixing	Gas mixing	Power per unit volume, kW/m^3
Baffled agitated tank	0.02–0.2	0.002–100	10^{-4}–20	~200	0.9	~Backmixed	Intermediate	0.5–10
Bubble column	0.05–0.01	0.002–300	10^{-5}–3	~20	0.95	~Plug	Plug	0.01–1
Packed tower	0.005–0.02	0.005–300	10^{-5}–6	~200	0.05	Plug	~Plug	0.01–0.2
Plate tower	0.01–0.05	0.005–300	10^{-5}–15	~150	0.15	Intermediate	~Plug	0.01–0.2
Static mixer (bubble flow)	0.1–2	Up to 10	1–20	~1000	0.5	~Plug	Plug	10–500

SOURCE: From J. C. Middleton, in Harnby, Edwards, and Nienow, *Mixing in the Process Industries,* Butterworth, 1985.

lations, the liquid-film coefficient can be calculated from a measured overall liquid-film coefficient with the relation

$$\frac{1}{K_L} = \frac{1}{Hk_g} + \frac{1}{k_L} \qquad (23\text{-}105)$$

In order to allow integration of countercurrent relations like Eq. (23-93), point values of the mass-transfer coefficients and equilibrium data are needed, over ranges of partial pressure and liquid-phase compositions. The same data are needed for the design of stirred tank performance. Then the conditions vary with time instead of position. Because of limited solubility, gas/liquid reactions in stirred tanks usually are operated in semibatch fashion, with the liquid phase charged at once, then the gas phase introduced gradually over a period of time. CSTR operation rarely is feasible with such systems.

INDUSTRIAL GAS/LIQUID REACTION PROCESSES

Two lists of gas/liquid reactions of industrial importance have been compiled recently. The literature survey by Danckwerts (*Gas-Liquid Reactions,* McGraw-Hill, 1970) cites 40 different systems. A supplementary list by Doraiswamy and Sharma (*Heterogeneous Reactions: Fluid-Fluid-Solid Reactions,* Wiley, 1984) cites another 50 items, and indicates the most suitable kind of reactor to be used for each. Estimates of values of parameters that may be expected of some types of gas/liquid reactors are in Tables 23-9 and 23-10.

Examples are given of common operations such as absorption of ammonia to make fertilizers and of carbon dioxide to make soda ash. Also of recovery of phosphine from offgases of phosphorous plants; recovery of HF; oxidation, halogenation, and hydrogenation of various organics; hydration of olefins to alcohols; oxo reaction for higher aldehydes and alcohols; ozonolysis of oleic acid; absorption of carbon monoxide to make sodium formate; alkylation of acetic acid with isobutylene to make *tert*-butyl acetate, absorption of olefins to make various products; HCl and HBr plus higher alcohols to make alkyl halides; and so on.

By far the greatest number of installations is for the removal or recovery of mostly small concentrations of acidic and other components from air, hydrocarbons, and hydrogen. Hundreds of such plants are in operation, many of them of great size. They mostly employ either packed or tray towers. Power requirements for such equipment are small. When the presence of solid impurities could clog the equipment or when the pressure drop must be low, spray towers are used in spite of their much larger size for a given capacity and scrubbing efficiency.

Removal of CO_2 and H_2S from Inert Gases, Packed and Tray Towers The principal reactive solvents for the removal of acidic constituents from gas streams are aqueous solutions of monethanolamine (MEA), diethanolamine (DEA) and K_2CO_3. These are all regenerable. Absorption proceeds at a lower temperature or higher pressure and regeneration in a subsequent vessel at higher temperature or lower pressure, usually with some assistance from stripping steam. CO_2 is discharged to the atmosphere or recovered to make dry ice. H_2S is treated for recovery of the sulfur. Any COS in the offgases is destroyed by catalytic hydrogenation.

Some performance data of plants with DEA are shown in Table 23-11. Both the absorbers and strippers have trays or packing. Vessel diameters and allowable gas and liquid flow rates are established by the same correlations as for physical absorptions. The calculation of tower heights utilizes data of equilibria and enhanced mass-transfer coeffi-

cients like those of Fig. 23-27 but for DEA solutions. Such calculations are complex enough to warrant the use of the professional methods of tower design that are available from a number of service companies.

Partly because of their low cost, aqueous solutions of sodium or potassium carbonate also are used for CO_2 and H_2S. Potassium bicarbonate has the higher solubility so the potassium salt is preferred. In view of the many competitive amine and carbonate plants that are in operation, fairly close figuring apparently is required to find an economic superiority, but other intangibles may be involved.

Both the equilibria and the enhancement of the coefficients can be improved by additives, of which sodium arsenite is the major one in use, but sodium hypochlorite and small amounts of amines also are effective. *Sterically hindered amines* as promoters are claimed by Say et al. (*Chem. Eng. Prog.,* **80**(10), 72–77 [1984]) to result in 50 percent more capacity than ordinary amine promoters of carbonate solutions.

Many operating data for carbonate plants are cited by Kohl and Riesenfeld (*Gas Purification,* Gulf, 1985) but not including tower heights. Pilot plant tests, however, are reported on 0.10- and 0.15-m (4- and 6-in) columns packed to depths of 9.14 m (30 ft) of Raschig rings by Benson et al. (*Chem. Eng. Prog.,* **50**, 356 [1954]).

Sulfur Dioxide, Spray Towers Flue gases and offgases from sulfuric acid plants contain less than 0.5 percent SO_2; smelter gases like those from ore processing plants may contain 8 percent. The high-concentration streams are suitable for the manufacture of sulfuric acid. The low concentrations usually are regarded as contaminants to be destroyed or recovered as elemental sulfur by, for example, the Claus process.

Of the removal processes that have attained commercial status, the current favorite employs a slurry of lime or limestone. The activity of the reagent is promoted by the addition of small amounts of carboxylic acids such as adipic acid. The gas and the slurry are contacted in a spray tower. The calcium salt is discarded. A process that employs aqueous sodium citrate, however, is suited for the recovery of elemental sulfur. The citrate solution is regenerated and recycled. (Kohl and Riesenfeld, *Gas Purification,* Gulf, 1985, p. 356.)

Limestone is pulverized to 80 to 90 percent through 200 mesh. Slurry concentrations of 5 to 40% have been checked in pilot plants. Liquid to gas ratios are 0.2 to 0.3 gal/MSCF. Flue gas enters at 149°C (300°F) at a velocity of 2.44 m/s (8 ft/s). Utilization of 80 percent of the solid reagent is approached. Flow is in parallel downward. Residence times are 10 to 12 s. At the outlet the particles are made just dry enough to keep from sticking to the wall, and the gas is within 11 to 28°C (20 to 50°F) of saturation. The fine powder is recovered with fabric filters.

Rotary wheel atomizers require 0.8 to 1.0 kWh/1,000 L. The lateral throw of a spray wheel requires a large diameter to prevent accumulation on the wall; length to diameter ratios of 0.5 to 1.0 are in use in such cases. The downward throw of spray nozzles permits smaller diameters but greater depths; L/D ratios of 4 to 5 or more are used. Spray vessel diameters of 15 m (50 ft) or more are known. The technology of spray drying is applicable.

In one test facility, a gas with 4,000 ppm SO_2 had 95 percent removal with lime and 75 percent removal with limestone.

Stirred Vessels Gases may be dispersed in liquids by spargers or nozzles and redispersed by packing or trays. More intensive dispersion and redispersion is obtained by mechanical agitation. At the same time, the agitation will improve heat transfer and will keep catalyst particles in suspension if necessary. Power inputs of 0.6 to 2.0 kW/m^3 (3.05 to 10.15 hp/1,000 gal) are suitable.

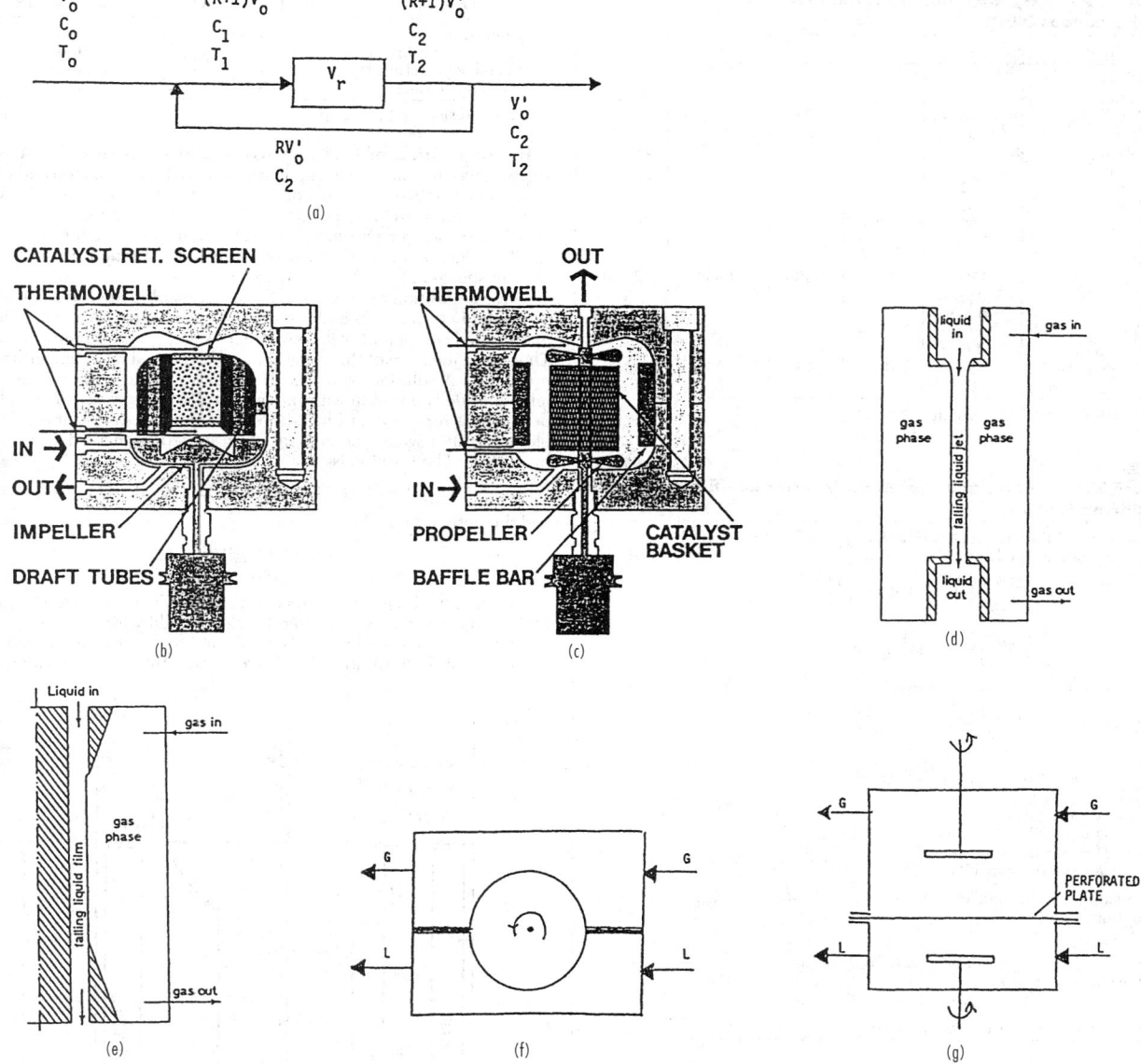

FIG. 23-29 Laboratory reactors for gas/liquid and fluid/solid processes. (*a*) Plug flow with external recycle, isothermal approach, $T_1 - T_2 = (T_0 - T_2)/(R + 1) \Rightarrow 0$. (*b*) Internal recycle, fixed basket. (*Berty, Autoclave Engineers Inc.*) (*c*) Rotating basket. (*Carberry, Autoclave Engineers Inc.*) (*d*) Falling liquid jet, known interfacial area, $t = 0.001–0.1$ s. (*e*) Falling film, $t = 0.1–0.25$ s. (*f*) Rotating drum, $t = 0.01–0.25$ s. (*Danckwerts, Gas-Liquid Reactions, McGraw-Hill, 1970.*) (*g*) L and G stirred, gradientless, $t = 1–10$ s. (*Levenspiel and Godfrey, Chem. Eng. Sci., **29**, 1723 [1974].*)

Bubble sizes tend to a minimum regardless of power input because coalescence eventually sets in. Pure liquids are coalescing type; solutions with electrolytes are noncoalescing but their bubbles also tend to a minimum. Agitated bubble size in air/water is about 0.5 mm (0.020 in), holdup fractions are about 0.10 coalescing and 0.25 noncoalescing, but more elaborate correlations have been made.

Mass-transfer coefficients seem to vary as the 0.7 exponent on power input per unit volume, with the dimensions of the vessel and impeller and the superficial gas velocity as additional factors. A survey of such correlations is made by van't Riet (*Ind. Eng. Chem. Proc. Des. Dev.*, **18**, 357 [1979]). Table 23-12 shows some of the results.

A basic stirred tank design is shown in Fig. 23-30. Height to diameter ratio is $H/D = 2$ to 3. Heat transfer may be provided through a jacket or internal coils. Baffles prevent movement of the mass as a whole. A draft tube enhances vertical circulation. The vapor space is about 20 percent of the total volume. A hollow shaft and impeller increase gas circulation (as in Fig. 23-31). A splasher can be attached to the shaft at the liquid surface to improve entrainment of gas. A variety of impellers is in use. The pitched propeller moves the liquid axially, the flat blade moves it radially, and inclined blades move it both axially and radially. The anchor and some other designs are suited to viscous liquids.

TABLE 23-11   Hydrocarbon Gas Treatments with Aqueous DEA*

Item	A	B	C	D	E
Absorber					
P, atm	68	14	7	15	12
T, °C			19	60	52
Trays	30	23	16		
Packing				26 ft, 3 in	30 ft, ¾
Stripper					
P, atm	2	2			
T, °C	133	118			
Trays	20	20			
Input					
CO_2	10%	0.35%			
H_2S gr/100 scf	15%	170	3,196	1,490	2,500
COS	300 ppm				
CS_2	600 ppm				
Output					
CO_2 gr/100 scf	1.6				
H_2S gr/100 scf	0.28	0.3	15	26	15
COS gr/100 scf	0				

*1 gr/100 scf = 0.0229 g/std m³

TABLE 23-12   Correlations of Mass-Transfer Coefficients in Stirred Tanks

Correlations of Koetsier, et al. (*Chem. Eng. Journal*, **5**, 61, 71 [1973])

For coalescing liquids:

$$k_L a = 0.05\left(\frac{N^2 D_i^2}{D_t^{1.5}}\right)^{1.95} = 0.05\, E^{0.65}\left(\frac{D_i}{D_t}\right)^{0.65} D_t^{-0.33}, \qquad 1/s$$

$$k_L = 0.002\left(\frac{N D_i^2}{D_t^{1.6}} - 0.45\right), \qquad m/s$$

For noncoalescing liquids (electrolytes):

$$k_L a = 0.11\, E^{0.7}\left(\frac{D_i}{D_t}\right)^{0.7} D_t^{-0.35}$$

$$k_L = 0.000325\, E^{0.3}\left(\frac{D_i}{D_t}\right)^{0.7} D_t^{-0.35}$$

where N = revolutions/s
D_i = impeller diameter, m, 6-blade turbine
D_t = tank diameter, m
E = power input, kW/m³

Correlation cited by Middleton (in Harnby, et al., *Mixing in the Process Industries*, Butterworth, 1985)

Coalescing:

$$k_L a = 1.2\, E^{0.7}\, u_s^{0.6}, \qquad 1/s$$

Noncoalescing:

$$k_L a = 2.3\, E^{0.7}\, u_s^{0.6}, \qquad 1/s$$

where u_s = superficial gas velocity, m/sec

For gas dispersion the six-bladed turbine is preferred. When the ratio of liquid height to diameter is $H/D \le 1$ a single impeller suffices, and in the range $1 \le H/D \le 1.8$ two are needed. The oil hydrogenator of Fig. 23-32, which is to scale, uses three impellers. The greater depth there will give longer contact time, which is desirable for slow reactions. The best position for inlet of gas is below and at the center of the impeller, or at the bottom of the draft tube. An open pipe is in common use, but a sparger may be helpful. A two-speed motor is desirable to prevent overloading, the lower speed to cut in when the gas supply is cut off but agitation is to continue, since gassed power requirement is significantly less than ungassed.

In tanks of 5.7 to 18.9 m³ (1,500 to 5,000 gal) rotation speeds are from 50 to 200 rpm and power requirements are 2 to 75 hp; both depend on superficial velocities of gas and liquid (Hicks and Gates, *Chem. Eng.*, 141–148 [July 1976]). As a rough guide, power requirements and impeller tip speeds are as follows.

Operation	hp/1,000 gal*	Tip speed, ft/s
Homogeneous reaction	0.5–1.5	7.5–10
With heat transfer	1.5–5	10–15
Liquid/liquid mixing	5	15–20
Gas/liquid mixing	5–10	15–20

*1 hp/1,000 gal = 0.197 kw/m³.

Hydrogenation of Oils in Stirred Tanks   Large-scale uses of stirred tanks include the gas/liquid reactions of hydrogenation and fermentation. For hydrogenation of vegetable and animal oils, semibatch operations often are preferred to continuous ones because of the variety of feedstocks or product specifications or long reaction times or small production rates. Sketches of batch and continuous hydrogenators are shown in Figs. 23-32 and 23-33.

The composition of an oil and the progress of its hydrogenation is expressed in terms of its iodine value (IV). Edible oils are mixtures of unsaturated compounds with molecular weights in the vicinity of 300. The IV is a measure of this unsaturation. It is found by a standardized procedure. A solution of ICl in a mixture of acetic acid and carbon tetrachloride is mixed in with the oil and allowed to react to completion, usually for less than 1 h. Halogen addition takes place at the double bond, after which the amount of unreacted iodine is determined by analysis. The reaction is

$$ICl + RCH{=}CHR_1 \Rightarrow RCHI{-}CHClR_1$$

and the definition is

$$IV = \frac{I\ absorbed}{100\ g\ oil}$$

To start a hydrogenation process, the oil and catalyst are charged first, then the vessel is evacuated for safety and hydrogen is supplied continuously from storage and kept at some fixed pressure, usually in the range of 1 to 10 atm (14.7 to 147 psi). Internal circulation of

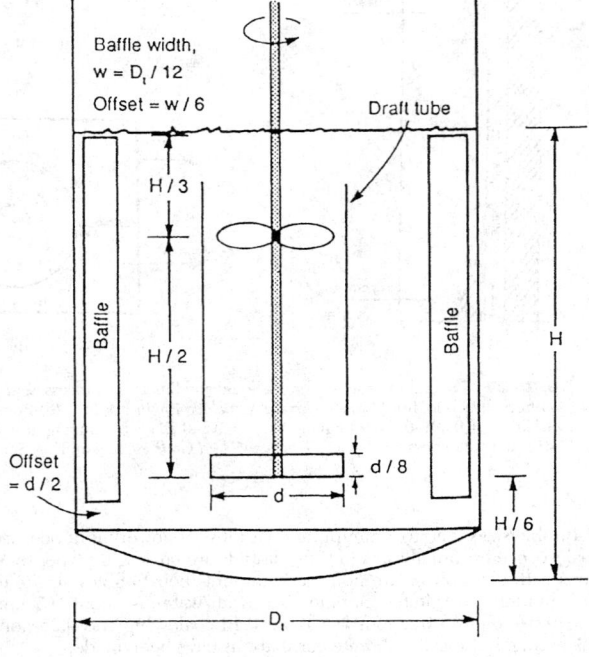

Baffle width,
w = D_t / 12
Offset = w / 6

Draft tube

H / 3
H / 2
Baffle
Offset = d / 2
d / 8
d
H
H / 6
D_t

FIG. 23-30a   A basic stirred tank design, not to scale, showing a lower radial impeller and an upper axial impeller housed in a draft tube. Four equally spaced baffles are standard. H = height of liquid level, D_t = tank diameter, d = impeller diameter. For radial impellers, $0.3 \le d/D_t \le 0.6$.

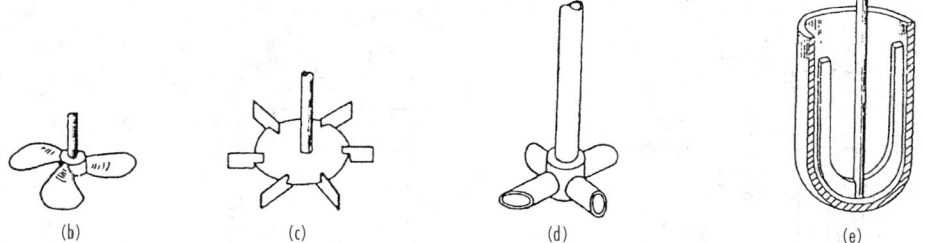

FIG. 23-30 Basic stirred tank design and selected kinds of impellers. (*b*) Propeller. (*c*) Turbine. (*d*) Hollow. (*e*) Anchor.

hydrogen is provided by axial and radial impellers or with a hollow impeller that throws the gas out centrifugally and sucks gas in from the vapor space through the hollow shaft. Some plants have external gas circulators. Reaction times are 1 to 4 h.

For edible oils the temperature is kept at about 180°C (356°F). Consumption of hydrogen per unit change of IV is

$$\frac{0.0795 \text{ kg } H_2}{1,000 \text{ kg oil}}$$

$$\frac{883.3 \text{ L STP } H_2}{1,000 \text{ kg oil}}$$

Solubility of hydrogen depends on the temperature and pressure but only slightly on the natures of the oils that are usually processed.

$$S = (0.04704 + 0.000294T)P, \quad \frac{\text{L STP}}{\text{kg oil}} \quad (23\text{-}106)$$

with T in °C and P in atm.

Heat evolution is 0.94 to 1.10 kcal/(kg oil)(unit drop of IV) (1.69 to 1.98 Btu/[lbm oil][unit drop of IV]). Because space for heat-transfer coils in the vessel is limited, the process is organized to give a maximum IV drop of about 2.0/min. The rate of reaction, of course, drops off rapidly as the reaction proceeds, so a process may take several hours. The end point of a hydrogenation is a specified IV of the prod-

uct, but the progress of a reaction before the end can be followed by measuring hardness or refractive index.

Saturation of the oil with hydrogen is maintained by agitation. The rate of reaction depends on agitation and catalyst concentration. Beyond a certain agitation rate, resistance to mass transfer is eliminated and the rate becomes independent of pressure. The effect of catalyst concentration also reaches limiting values. The effects of pressure and temperature on the rate are indicated by Fig. 23-34 and of catalyst concentration by Fig. 23-35. Reaction time is related to temperature, catalyst concentration, and IV in Table 23-13.

Nickel is the most used catalyst, 20 to 25 percent Ni on a porous siliceous support in the form of flakes that are readily filterable. The pores allow access of the reactants to the extended pore surface, which is in the range of 200 to 600 m²/g (977 × 10³ to 2,931 × 10³ ft²/lbm) of

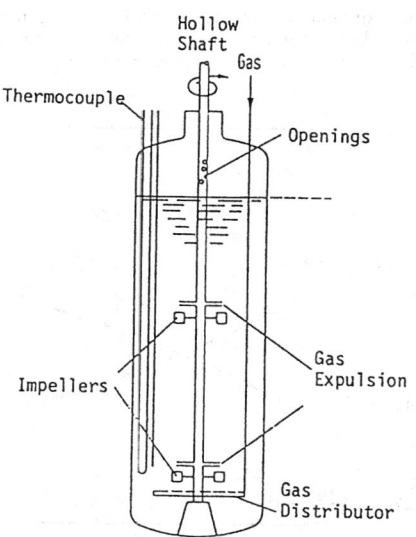

FIG. 23-31 Stirred tank with hollow shaft for hydrogenation of nitrocaprolactam. (*Dierendonk et al.*, 5th European Symp. Chem. React. Eng., *Pergamon,* 1972, pp. B6–45.)

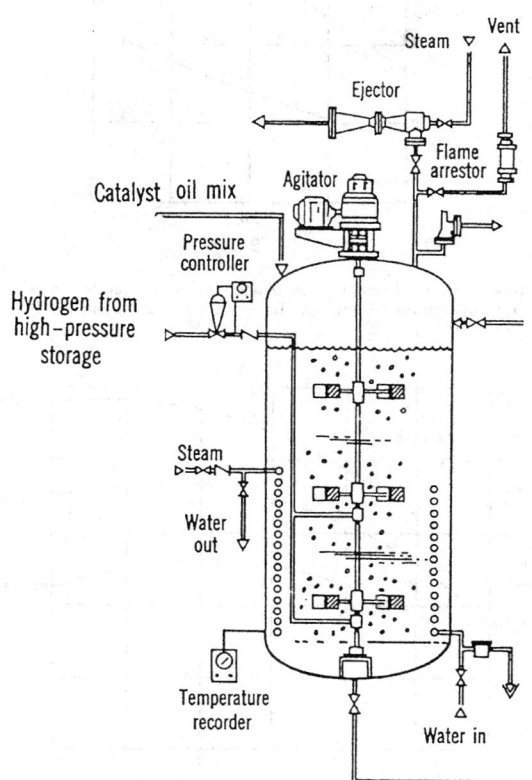

FIG. 23-32 Stirred tank hydrogenator for edible oils. (*Votator Division, Chemetron Corporation.*)

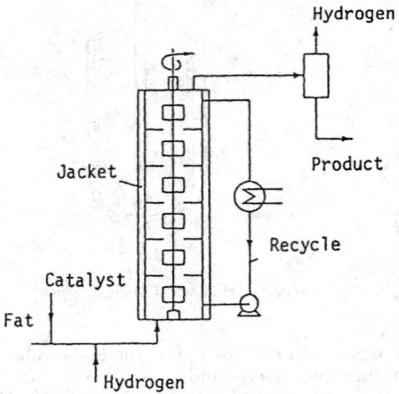

FIG. 23-33 Continuous hydrogenation of fats (*Albright*, Chem. Eng., **74**, *249* [9 Oct. 1967].)

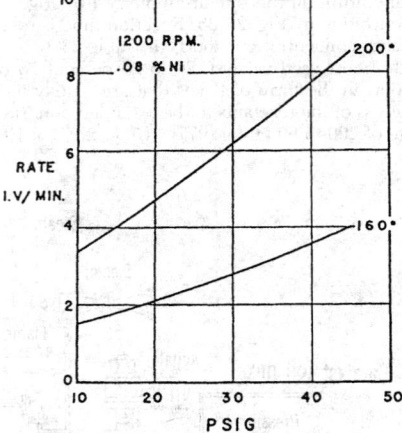

FIG. 23-34 Effect of reaction pressure and temperature on the rate of hydrogenation of soybean oil. (*Swern, ed.,* Bailey's Industrial Oil and Fat Products, *vol. 2, Wiley, 1979.*)

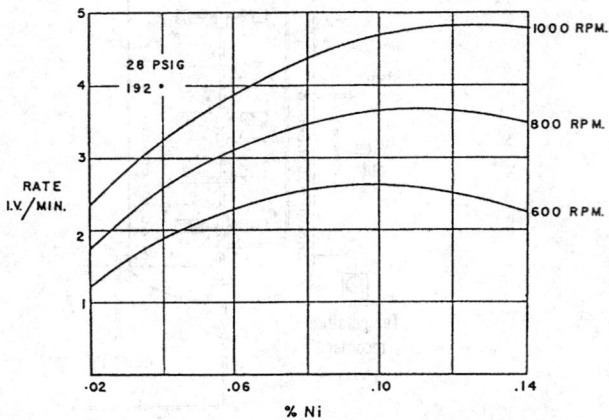

FIG. 23-35 Effect of catalyst concentration and stirring rate on hydrogenation of soybean oil. (*Swern, ed.,* Bailey's Industrial Fat and Oil Products, *vol. 2, Wiley, 1979.*)

TABLE 23-13   Time, Temperature, and Iodine Value of Tallow Hydrogenation*

t, min	0.03% Ni, P = 20 atm		0.04% Ni, P = 25 atm		0.08% Ni, P = 3 atm	
	T, °C	IV	T, °C	IV	T, °C	IV
0	140	42.3	140	44.1	160	42.3
5	145	39.7			165	35.0
10	150	37.3	147	38.0	170	27.8
30	160	27.1	160	26.6	180	8.4
60	180	14.5	180	13.4	200	1.7
90	190	5.4	180	5.6	200	0.3
120	200	1.0	180	0.5	200	0.25
180	200	0.3			200	0.1

*To convert atm to kPa, multiply by 101.3.
SOURCE: From Patterson, *Hydrogenation of Fats and Oils,* Applied Science Publishers, 1983.

which 20 to 30 percent is catalytically active. The concentration of catalyst in the slurry can vary over a wide range but is usually under 0.1% Ni. Catalysts are subject to degradation and poisoning, particularly by sulfur compounds. Accordingly, about 10 to 20 percent of the recovered catalyst is replaced by fresh before recycling. Other catalysts are applied in special cases. Expensive palladium has about 100 times the activity of nickel and is effective at lower temperatures.

While the liquid is saturated with hydrogen the reaction is pseudo-first-order. One sequence of reactions of acids that has been investigated (Swern, ed., *Bailey's Industrial Oil and Fat Products,* vol. 2, Wiley, 1979, p. 12) is

$$\text{Linolenic} \overset{1}{\rightleftharpoons} \text{Linoleic} \overset{2}{\rightleftharpoons} \text{Oleic} \overset{3}{\rightleftharpoons} \text{Stearic}$$

At 175°C (347°F), 0.02% Ni, 1 atm, 600 rpm the specific rates are: $k_1 = 0.367$, $k_2 = 0.159$, $k_3 = 0.013$/min.

Figure 23-36 shows a computer calculation with these specific rates, but which does not agree quantitatively with the figure shown by Swern. The time scales appear to be different, but both predict a peak in the amount of oleic acid and rapid disappearance of the first two acids.

A case study of the hydrogenation of cottonseed oil is made by Rase (*Chemical Reactor Design for Process Plants,* vol. 2, Wiley, 1977, pp. 161–178).

Semibatch hydrogenation of edible oils has a long history and a well-established body of practice by manufacturers and catalyst suppliers. Problems of new oils, new specifications, new catalyst poisons,

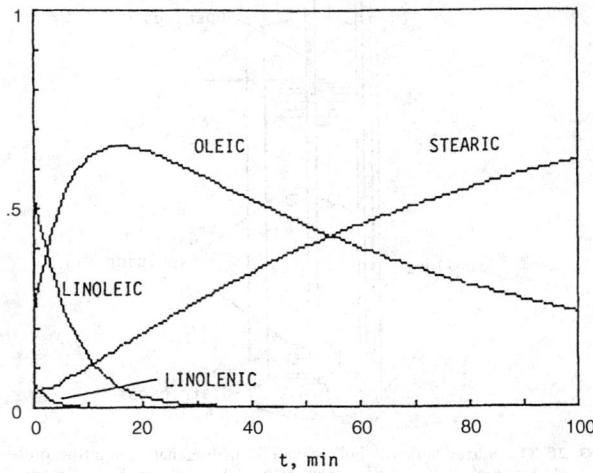

FIG. 23-36 Product compositions of the first-order sequence of fatty acids: Linolenic ⇒ Linoleic ⇒ Oleic ⇒ Stearic.

and even new scales of operation can probably be handled with a minimum amount of laboratory or pilot plant work.

Aerobic Fermentation The classic example of large-scale aerobic fermentation is the production of penicillin by the growth of a specific mold. Commercial vessel sizes are 40,000 to 200,000 L (1,400 to 7,000 ft^3). The operation is semibatch in that the lactose or glucose nutrient and air are charged at controlled rates to a precharged batch of liquid nutrients and cell mass. Reaction time is 5 to 6 days.

The broth is limited to 7 to 8 percent sugars, which is all the mold will tolerate. Solubility of oxygen is so limited that air must be supplied over a long period as it is used up. The pH is controlled at about 6.5 and the temperature at 24°C (75°F). The air is essential to the growth. Dissolved oxygen must be kept at a high level for the organism to survive. Air also serves to agitate the mixture and to sweep out the CO_2 and any noxious byproducts that are formed. Air supply is in the range of 0.5 to 1.5 volumes/(volume of liquid)(min). For organisms grown on glucose the oxygen requirement is 0.4 g/g dry weight; on methanol it is 1.2 g/g.

The heat of reaction requires cooling water at the rate of 10 to 40 L/(1,000 L holdup)(h). Vessels under about 500 L (17.6 ft^3) are provided with jackets, larger ones with coils. For a 55,000-L vessel, 50 to 70 m^2 may be taken as average.

Mechanical agitation is needed to break up the gas bubbles but must avoid rupturing the cells. The disk turbine with radial action is most suitable. It can tolerate a superficial gas velocity up to 120 m/h (394 ft/h) without flooding, whereas the propeller is limited to about 20 m/h (66 ft/h). When flooding occurs, the impeller is working in a gas phase and cannot assist the transfer of gas to the liquid phase. Power input by agitation and air sparger is 1 to 4 W/L (97 to 387 Btu/[ft^3·h]) of liquid.

Bubble Reactors In bubble columns the gas is dispersed by nozzles or spargers without mechanical agitation. In order to improve the operation, redispersion at intervals may be effected by static mixers, such as perforated plates. The liquid may be clear or be a slurry.

Because of their large volume fraction of liquid, bubble reactors are suited to slow reactions; that is, those whose rate of reaction is smaller than their rate of diffusion. Major advantages are the absence of moving parts, the ability to handle solid particles without erosion or plugging, good heat transfer at the wall or coils, high interfacial area, and high mass-transfer coefficients. A disadvantage is the occurrence of backmixing of the liquid phase and some of the gas phase, which may result in poor selectivity with some complex reactions. High pressure drop because of the static head of the liquid also is harmful. At high height/diameter ratios (>15), the effective interfacial area decreases rapidly. Generally, the tower height is greater than for tray or packed towers.

Two complementary reviews of this subject are by Shah et al. (*AIChE Journal*, **28**, 353–379 [1982]) and Deckwer (in de Lasa, ed., *Chemical Reactor Design and Technology*, Martinus Nijhoff, 1985, pp. 411–461). Useful comments are made by Doraiswamy and Sharma (*Heterogeneous Reactions*, Wiley, 1984). Charpentier (in Gianetto and Silveston, eds., *Multiphase Chemical Reactors*, Hemisphere, 1986, pp. 104–151) emphasizes parameters of trickle bed and stirred tank reactors. Recommendations based on the literature are made for several design parameters: namely, bubble diameter and velocity of rise, gas holdup, interfacial area, mass-transfer coefficients k_La and k_L but not k_g, axial liquid-phase dispersion coefficient, and heat-transfer coefficient to the wall. The effect of vessel diameter on these parameters is insignificant when $D \geq 0.15$ m (0.49 ft), except for the dispersion coefficient. Application of these correlations is to: (1) chlorination of toluene in the presence of $FeCl_3$ catalyst, (2) absorption of SO_2 in aqueous potassium carbonate with arsenite catalyst, and (3) reaction of butene with sulfuric acid to butanol.

Some qualitative observations can be made. Increase of the superficial gas velocity increases the holdup of gas, the interfacial area, and the overall mass-transfer coefficient. The ratio of height to diameter is not important in the range of 4 to 10. Increase of viscosity and decrease of surface tension increase the interfacial area. Electrolyte solutions have smaller bubbles, higher gas holdup, and higher interfacial area. Sparger design is unimportant for superficial gas velocities > 5 to 10 cm/s (0.16 to 0.32 ft/s). Gas conversion falls off at higher superficial velocities, so values under 10 cm/s (0.32 ft/s) are advisable.

The gas approximates plug flow except in wide columns, but the liquid undergoes considerable backmixing. The latter effect can be reduced with packing or perforated plates. The effect on selectivity may become important. In the oxidation of liquid *n*-butane, for instance, the ratio of methyl ethyl ketone to acetic acid is much higher in plug flow than in mixed. Similarly, in the air oxidation of isobutane to *tert*-butyl hydroperoxide, where *tert*-butanol also is obtained, plug flow is more desirable.

Bubble action provides agitation about equivalent to that of mechanical stirrers, and thus about the same heat-transfer coefficients.

In cases of high liquid velocities (>30 cm/s [0.98 ft/s]) and low gas velocities (1 to 3 cm/s [0.033 to 0.098 ft/s]) it is advantageous to have concurrent downward flow.

Pulsations can improve performance. According to Baird and Garstang (*Chem. Eng. Sci.*, **22**, 1663 [1967]; **27**, 823 [1972]) at low gas velocity (0.8 to 2.4 cm/s [0.026 to 0.079 ft/s]) the value of k_La can be increased by as much as three-fold by pulsations.

Packed bubble columns operate with flooded packing, in contrast with normal packed columns which usually operate below 70 percent of the flooding point. With packing, liquid backmixing is reduced and interfacial area is increased 15 to 80 percent, but the true mass-transfer coefficient remains the same. The installation of perforated plates or grids at intervals also reduces liquid backmixing. At relatively high superficial gas velocities (10 to 15 cm/s [0.33 to 0.49 ft/s]) mixing between zones is small so the vessel performs as a CSTR battery. Radial baffles (also called disk-and-doughnut baffles) are also helpful. One set of rules is that the hole should be about 0.7 times the vessel diameter and the spacing should be 0.8 times the diameter.

Liquid Dispersion Spray columns are used with slurries or when the reaction product is a solid. The absorption of SO_2 by a lime slurry is an example. In the treatment of phosphate rock with sulfuric acid, off-gases contain HF and SiF_4. In a spray column with water, solid particles of fluorosilic acid are formed but do not harm the spray operation. The coefficient k_L in spray columns is about the same as in packed columns, but the spray interfacial area is much lower. Considerable backmixing of the gas also takes place, which helps to make the spray volumetrically inefficient. Deentrainment at the outlet usually is needed.

In Venturi scrubbers the gas is the motive fluid. This equipment is of simple design and is able to handle slurries and large volumes of gas, but the gas pressure drop may be high. When the reaction is slow, further holdup in a spray chamber is necessary.

In liquid ejectors or aspirators, the liquid is the motive fluid, so the gas pressure drop is low. Flow of slurries in the nozzle may be erosive. Otherwise, the design is as simple as that of the Venturi.

The application of liquid dispersion reactors to the absorption of fluorine gases is described by Kohl and Riesenfeld (*Gas Purification*, Gulf, 1985, pp. 268–288).

Tubular Reactors In a tubular reactor with concurrent flow of gas and liquid, the variety of flow patterns ranges from a small quantity of bubbles in the liquid to small quantities of droplets in the gas, depending on the rates of the two streams. Figure 23-25 shows the patterns in horizontal flow; those in vertical flow are a little different. This equipment has good heat transfer, accommodates wide ranges of T and P, and is primarily in plug flow, and the high velocities prevent settling of slurries or accumulations on the walls. Mixing of the phases can be improved by helical static mixing inserts like those made by Kenics Corporation and others.

The reasons why a tubular reactor was selected for the production of adipic acid nitrile from adipic acid and ammonia are discussed by Weikard (in Ullmann, *Enzyklopaedie*, 4th ed., vol. 3, Verlag Chemie, 1973, p. 381).

1. The process has a large Hatta number; that is, the rate of reaction is much greater than the rate of diffusion, so a large interfacial area is desirable for carrying out the reaction in the film.

2. With normal excess ammonia the gas/liquid ratio is about 3,500 m^3/m^3. At this high ratio there is danger of fouling the surface with tarry reaction products. The ratio is brought down to a more satisfactory value of 1,000 to 1,500 by recycle of some of the effluent.

3. High selectivity of the nitrile is favored by short contact time.

4. The reaction is highly endothermic so heat input must be at a high rate.

Points 2 and 4 are the main ones governing the choice of reactor type. The high gas/liquid ratio restricts the choice to types *d, e, i,* and *k* of Fig. 23-25, but because of the high rate of heat transfer that is needed the choice falls to the falling film or tubular types.

The final selection was a tubular reactor with upward concurrent flow, with liquid holdup of 20 to 30 percent, and with residence times of 1.0 s for gas and 3 to 5 min for liquid.

Reaction in a Centrifugal Pump In the reaction between acetic acid and gaseous ketene to make acetic anhydride, the pressure must be kept low (0.2 atm) to prevent polymerization of ketene. A packed tower with low pressure drop could be used but the required volume is very large because of the low pressure. Spes (*Chem. Ing. Tech.,* **38,** 963–966 [1966]) selected a centrifugal pump reactor where

compression to atmospheric pressure does take place but the contact time is too short for polymerization yet long enough for the reaction to occur. The reaction product is cooled in an external unit and partly recycled for temperature control.

Falling Film Reactor Dodecylbenzene sulfonic acid is made by reacting 5 to 10% SO$_3$ in air with dodecylbenzene. The gas/liquid ratio is about 1,000 m^3/m^3, and the product is highly viscous. Process requirements are a short contact time, a high rate of heat removal (170 kJ/g mol) and minimum backmixing of the liquid phase to avoid byproducts. To satisfy these requirements, a falling film reactor was selected by Ujidhy, Babos, and Farady (*Chemische Technik,* **18,** 652–654 [1966]). The gas velocity was made 12 to 15 m/s (39 to 49 ft/s) through a narrowed passage counter to the liquid flow.

LIQUID/LIQUID REACTIONS

Liquid/liquid reactions of industrial importance are fairly numerous. A list of 26 classes of reactions with 61 references has been compiled by Doraiswamy and Sharma (*Heterogeneous Reactions,* Wiley, 1984). They also indicate the kind of reactor normally used in each case. The reactions range from such prosaic examples as making soap with alkali, nitration of aromatics to make explosives, and alkylation of C$_4$s with sulfuric acid to make improved gasoline, to some much less familiar operations.

EQUIPMENT

Equipment suitable for reactions between liquids is represented in Fig. 23-37. Almost invariably, one of the phases is aqueous with reactants distributed between phases; for instance, NaOH in water at the start and an ester in the organic phase. Such reactions can be carried out in any kind of equipment that is suitable for physical extraction, including mixer-settlers and towers of various kinds: empty or packed, still or agitated, either phase dispersed, provided that adequate heat transfer can be incorporated. Mechanically agitated tanks are favored because the interfacial area can be made large, as much as 100 times that of spray towers, for instance. Power requirements for L/L mixing are normally about 5 hp/1,000 gal and tip speeds of turbine-type impellers are 4.6 to 6.1 m/s (15 to 20 ft/s).

Table 23-14 gives data for common types of L/L contactors. Since the given range of $k_L a$ is more than 100/1, this information is not of direct value for sizing of equipment. The efficiencies of various kinds of small liquid/liquid contactors are summarized in Fig. 23-38. Larger units may have efficiencies of less than half these values.

MECHANISMS

Few mechanisms of liquid/liquid reactions have been established, although some related work such as on droplet sizes and power input has been done. Small contents of surface-active and other impurities in reactants of commercial quality can distort a reactor's predicted performance. Diffusivities in liquids are comparatively low, a factor of 10^5 less than in gases, so it is probable in most industrial examples that they are diffusion controlled. One consequence is that L/L reactions may not be as temperature sensitive as ordinary chemical reactions, although the effect of temperature rise on viscosity and droplet size can result in substantial rate increases. L/L reactions will exhibit behavior of homogeneous reactions only when they are very slow, nonionic reactions being the most likely ones. On the whole, in the present state of the art, the design of L/L reactors must depend on scale-up from laboratory or pilot plant work.

Particular reactions can occur in either or both phases or near the interface. Nitration of aromatics with HNO$_3$-H$_2$SO$_4$ occurs in the aqueous phase (Albright and Hanson, eds., *Industrial and Laboratory Nitrations,* ACS Symposium Series **22** [1975]). An industrial example of reaction in both phases is the oximation of cyclohexanone, a step in the manufacture of caprolactam for nylon (Rod, *Proc. 4th International/6th European Symposium on Chemical Reactions,* Heidelberg, Pergamon, 1976, p. 275). The reaction between butene and isobutane

to form isooctane in the presence of sulfuric acid is judged to occur at the acid/hydrocarbon interface, although side reactions to form higher hydrocarbons may occur primarily in the acid phase (Albright, in Albright and Goldsby, eds., *Industrial and Laboratory Nitrations,* ACS Symposium Series **55,** 145 [1977]). The formation of dioxane from isobutene in a hydrocarbon phase and aqueous formaldehyde occurs preponderantly in the aqueous phase where the rate equation is first-order in formaldehyde, although the specific rate is also proportional to the concentration of isobutene in the organic phase (Hellin et al., *Genie. Chim.,* **91,** 101 [1964]).

Reactions involving ions can be favored to occur in the organic phase by use of phase-transfer catalysts. Thus the conversion of 1-chlorooctane to 1-cyanooctane with aqueous NaCN is vastly accelerated in the organic phase by 1.3 percent of tributyl (hexadecyl) phosphonium bromide in the aqueous phase. (Starks and Owens, *J. Am. Chem. Soc.,* **95,** 3613 [1973]). A large class of such promotions is known.

There are instances where an extractive solvent is employed to force completion of a reversible homogeneous reaction by removing the reaction product. In the production of KNO$_3$ from KCl and HNO$_3$, for instance, the HCl can be removed continuously from the aqueous phase by contact with amyl alcohol, thus forcing completion (Baniel and Blumberg, *Chim. Ind.,* **4,** 27 [1957]).

OPERATING DATA

Not many operating data of large-scale liquid/liquid reactions are published. One study was made of the hydrolysis of fats with water at 230 to 260°C (446 to 500°F) and 41 to 48 atm (600 to 705 psi) in a continuous commercial spray tower. A small amount of water dissolved in the fat and reacted to form an acid and glycerine. Then most of the glycerine migrated to the water phase. The tower was operated at about 18 percent of flooding, at which condition the HETS was found to be about 9 m (30 ft) compared with an expected 6 m (20 ft) for purely physical extraction (Jeffreys, Jenson, and Miles, *Trans. Inst. Chem. Eng.,* **39,** 389–396 [1961]). A similar mathematical treatment of a batch hydrolysis is made by Jenson and Jeffreys (*Inst. Chem. Engrs. Symp. Ser.,* No. 23 [1967]).

LABORATORY STUDIES

For many laboratory studies, a suitable reactor is a cell with independent agitation of each phase and an undisturbed interface of known area, like the item shown in Fig. 23-29*d.* Whether a rate process is controlled by a mass-transfer rate or a chemical reaction rate sometimes can be identified by simple parameters. When agitation is sufficient to produce a homogeneous dispersion and the rate varies with further increases of agitation, mass-transfer rates are likely to be significant. The effect of change in temperature is a major criterion: a rise of 10°C (18°F) normally raises the rate of a chemical reaction by a factor of 2 to 3, but the mass-transfer rate by much less. There may be instances, however, where the combined effect on chemical equilibrium, diffusivity, viscosity, and surface tension also may give a comparable enhancement.

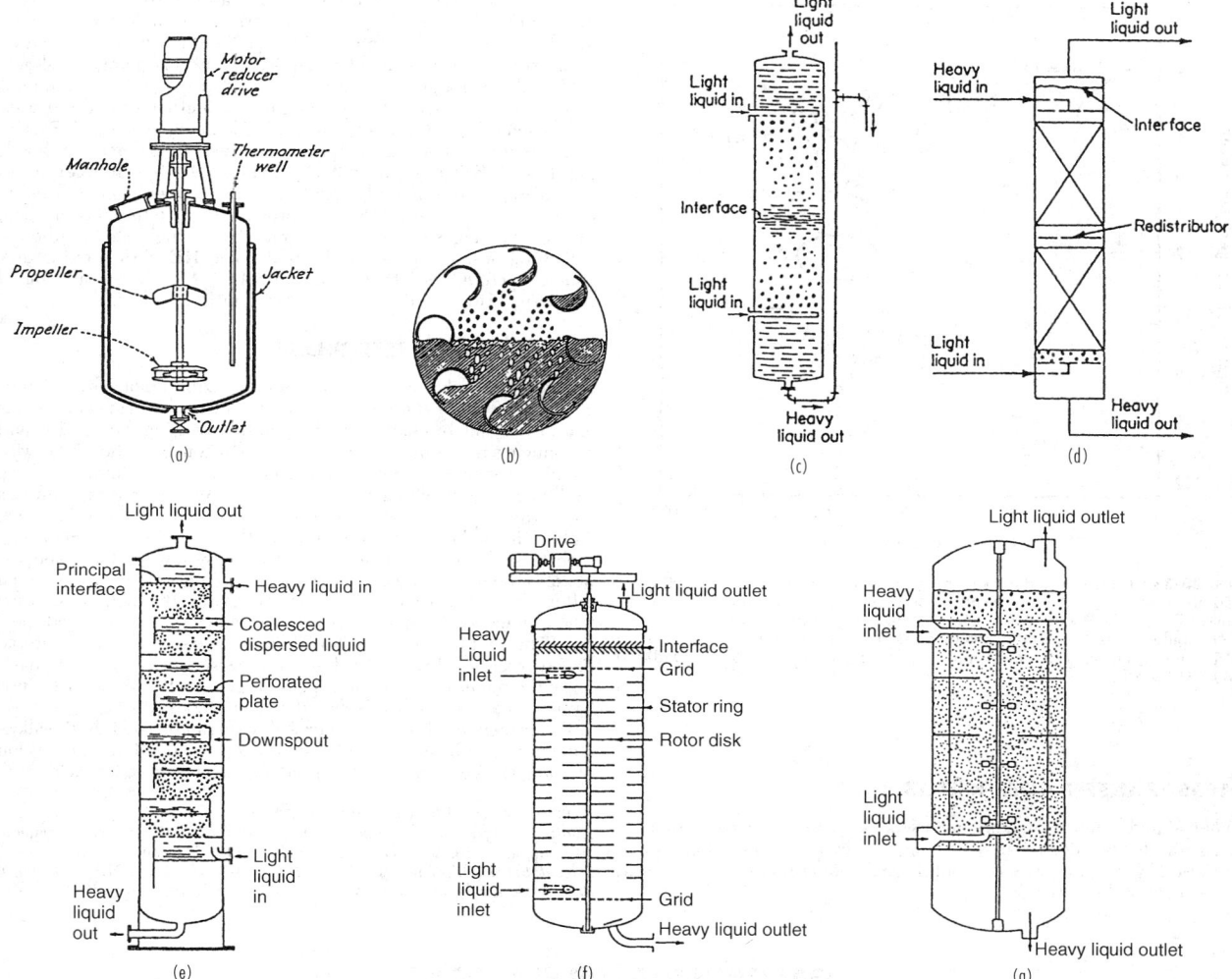

FIG. 23-37   Equipment for liquid/liquid reactions. (*a*) Batch stirred sulfonator. (*b*) Raining bucket (*RTL S A, London*). (*c*) Spray tower with both phases dispersed. (*d*) Two-section packed tower with light phase dispersed. (*e*) Sieve tray tower with light phase dispersed. (*f*) Rotating disk contactor (RDC) (*Escher B V, Holland*). (*g*) Oldshue-Rushton extractor (*Mixing Equipment Co.*).

For a chemically controlled process, conversion depends only on the residence time and not on which phase is dispersed, whereas the interfacial area and, consequently, the rate of mass transfer will change when the relative volumes of the phases are changed. If a reaction is known to occur in a particular phase, and the conversion is found to depend on the residence time in that phase, chemical reaction is controlling.

Laboratory investigations may possibly establish reaction mechanisms, but quantitative data for design purposes require pilot plant work with equipment of the type expected to be used in the plant.

TABLE 23-14   Continuous-Phase Mass-Transfer Coefficients and Interfacial Areas in Liquid/Liquid Contactors*

Type of equipment	Dispersed phase	Continuous phase	ε_D	τ_D	$k_L \times 10^2$, cm/s	a, cm²/cm³	$k_L a \times 10^2$, s⁻¹
Spray columns	P	M	0.05–0.1	Limited	0.1–1	1–10	0.1–10
Packed columns	P	P	0.05–0.1	Limited	0.3–1	1–10	0.3–10
Mechanically agitated contactors	PM	M	0.05–0.4	Can be varied over a wide range	0.3–1	1–800	0.3–800
Air-agitated liquid/liquid contactors	PM	M	0.05–0.3	Can be varied over a wide range	0.1–0.3	10–100	1.0–30
Two-phase cocurrent (horizontal) contactors	P	P	0.05–0.2	Limited	0.1–1.0	1–25	0.1–25

*°P = plug flow, M = mixed flow, ε_D = fractional dispersed phase holdup, τ_D = residence time of the dispersed phase.
SOURCE: From Doraiswamy and Sharma, *Heterogeneous Reactions*, Wiley, 1984.

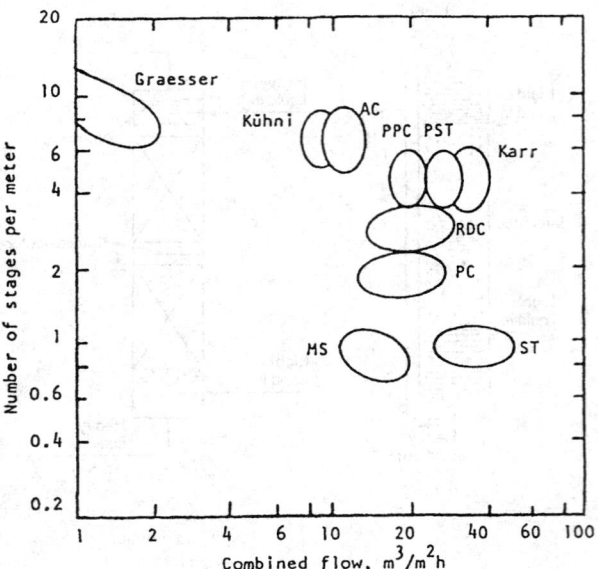

FIG. 23-38 Efficiency and capacity range of small-diameter extractors, 50 to 150 mm diameter. Acetone extracted from water with toluene as the disperse phase, $V_d/V_c = 1.5$. Code: AC = agitated cell; PPC = pulsed packed column; PST = pulsed sieve tray; RDC = rotating disk contactor; PC = packed column; MS = mixer-settler; ST = sieve tray. (*Stichlmair*, Chem. Ing. Tech. **52**(3), 253–255 [1980]).

MASS-TRANSFER COEFFICIENTS

When liquid/liquid contactors are used as reactors, values of their mass-transfer coefficients may be enhanced by reaction, analogously to those of gas/liquid processes, but there do not seem to be any published data of this nature.

Mass-transfer coefficients and other characteristics of some types of liquid/liquid contactors are summarized in Table 23-14, which may be compared with Tables 23-9 and 23-10 for gas-liquid contactors. Efficiencies of several kinds of small-scale extractors are shown in Fig. 23-38. Larger-diameter equipment may have less than one-half these efficiencies. Spray columns are inefficient and are used only when other kinds of equipment could become clogged. Packed columns as liquid/liquid reactors are operated at 20 percent of flooding. Their HETS range from 0.6 to 1.2 m (1.99 to 3.94 ft). Sieve trays minimize backmixing and provide repeated coalescence and redispersion. Mixer-settlers provide approximately one theoretical stage, but several stages can be incorporated in a single shell, although with some loss of operating flexibility. The HETS of rotating disk contactor (RDC) is 1 to 2 m (3.2 to 6.4 ft). More elaborate staged extractors bring this down to 0.35 to 1.0 m (1.1 to 3.3 ft).

CHOICE OF DISPERSED PHASE

It is difficult to disperse a liquid when it occupies more than 75 percent of the volume. Otherwise, either liquid can be made continuous in a stirred tank by charging that liquid first, starting the agitator, and introducing the liquid to be dispersed. Customarily, the phase with the higher volumetric rate is dispersed since a larger interfacial area results in this way with a given droplet size. When a reactant diffuses away from a phase, that phase should be dispersed since the travel path then will be lower. In equipment that is subject to backmixing, such as spray or packed towers but not tray towers, the dispersed phase is the one with the smaller volumetric rate. When a substantial difference is known to exist for the two phases, the high phase resistance should be compensated for with increased surface by dispersion. The continuous phase should be the one that wets the material of construction. Usually, it is best to disperse a highly viscous phase. Since the holdup of continuous phase is greater, the phase that is less hazardous or less expensive should be continuous.

Experimentally, both modes probably should be tried. In the alkylation of C_4s with sulfuric acid, for instance, the continuous emulsion of acid produces a much better product and consumes less acid.

REFERENCES FOR LIQUID/LIQUID REACTORS
Hanson, C., ed., *Recent Advances in Liquid-Liquid Extraction*, Pergamon, 1971, pp. 429–453. Lo, T. C., M. H. I. Baird, and C. Hanson, eds., *Handbook of Solvent Extraction*, Wiley, 1983, pp. 37–52, 615–618. Rase, H. F., *Chemical Reactor Design for Process Plants*, vol. 1, Wiley, 1977, pp. 715–733.

GAS/LIQUID/SOLID REACTIONS

In many important cases of reactions involving gas, liquid, and solid phases, the solid phase is a porous catalyst. It may be in a fixed bed or it may be suspended in the fluid mixture. In general, the reaction occurs either in the liquid phase or at the liquid/solid interface. In fixed-bed reactors the particles have diameters of about 3 mm (0.12 in) and occupy about 50 percent of the vessel volume. Diameters of suspended particles are limited to 0.1 to 0.2 mm (0.004 to 0.008 in) minimum by requirements of filterability and occupy 1 to 10 percent of the volume in stirred vessels.

A list of 74 GLS reactions with literature references has been compiled by Shah (*Gas-Liquid-Solid Reactions*, McGraw-Hill, 1979), classified into groups where the solid is a reactant, or a catalyst, or inert. A list of 75 reactions made by Ramachandran and Chaudhari (*Three-Phase Chemical Reactors*, Gordon and Breach, 1983) identifies reactor types, catalysts, temperature, and pressure. They classify the processes according to hydrogenation of fatty oils, hydrodesulfurization, Fischer-Tropsch reactions, and miscellaneous hydrogenations and oxidations.

Some contrasting characteristics of the main kinds of three-phase reactors are summarized in Table 23-15. In *trickle bed reactors* both phases usually flow down, the liquid as a film over the packing. In *flooded reactors*, the gas and liquid flow upward through a fixed bed. *Slurry reactors* keep the solids in suspension mechanically; the overflow may be a clear liquid or a slurry, and the gas disengages from the vessel. The fluidized three-phase mixture is pumped through an *entrained solids reactor* and the effluent is separated into its phases in downstream equipment. In petroleum cracking technology this kind of equipment is called a *transfer line reactor*. In *fluidized bed reactors*, a stable bed of solids is maintained in the vessel and only the fluid phases flow through, except for entrained very fine particles. Most of the concern in this section is with trickle bed reactors, but some superior features of the other types are cited.

OVERALL RATE EQUATIONS WITH DIFFUSIONAL RESISTANCES

Say the concentration of dissolved gas A is $A°$. The series rates involved are from the gas to the interface where the concentration is A_i and from the interface to the surface of catalyst where the concentration is A_s and where the reaction rate is $\eta w k_m A_s^m$. At steady state,

$$r_A = k_L a_B (A° - A_i) = k_s a_p (A_i - A_s) = \eta w k_m A_s^m \quad (23\text{-}107)$$

For a first-order reaction, $m = 1$, the catalyst effectiveness η is independent of A_s, so that after elimination of A_i and A_s the explicit solution for the rate is

$$r_A = A° \left[\frac{1}{k_L a_B} + \frac{1}{k_s a_p} + \frac{1}{\eta w k_1} \right]^{-1} \quad (23\text{-}108)$$

TABLE 23-15   Characteristics of Gas/Liquid/Solid Reactors

Property	Trickle bed	Flooded	Stirred tank	Entrained solids	Fluidized bed
Gas holdup	0.25–0.45	Small	0.2–0.3		
Liquid holdup	0.05–0.25	High	0.7–0.8		
Solid holdup	0.5–0.7		0.01–0.10		0.5–0.7
Liquid distribution	Good only at high liquid rate		Good	Good	Good
RTD, liquid phase	Narrow	Narrower than for entrained solids reactor	Wide	Wide	Narrow
RTD, gas phase	Nearly plug flow		Backmixed	Backmixed	Narrow
Interfacial area	20–50% of geometrical	Like trickle bed reactor	100–1,500 m^2/m^3	100–400 m^2/m^3	Less than for entrained solids reactor
MTC, gas/liquid	High		Intermediate		
MTC, liquid/solid	High		High		
Radial heat transfer	Slow		Fast	Fast	Fast
Pressure drop	High with small d_p		Hydrostatic head		

Analytical solutions also are possible when η is constant and $m = 0$, ½, or 2. More complex chemical rate equations will require numerical solutions. Such rate equations are applied to the sizing of plug flow, CSTR, and dispersion reactor models by Ramachandran and Chaudhari (*Three-Phase Chemical Reactors*, Gordon and Breach, 1983).

TRICKLE BEDS

The catalyst is a fixed bed. Flows of gas and liquid are cocurrent downwards. Liquid feed is at such a low rate that it is distributed over the packing as a thin film and flows by gravity, helped along by the drag of the gas. This mode is suited to reactions that need only short reaction times, measured in seconds, short enough to forestall undesirable side reactions such as carbon formation. In the simplest arrangement the liquid distributor is a perforated plate with about 10 openings/dm² (10 openings/15.5 in²) and the gas enters through several risers about 15 cm (5.9 in) high. More elaborate distributor caps also are used. Thicknesses of liquid films have been estimated to vary between 0.01 and 0.2 mm (0.004–0.008 in).

Liquid holdup is made up of a dynamic fraction, 0.03 to 0.25, and a stagnant fraction, 0.01 to 0.05. The high end of the stagnant fraction includes the liquid that partially fills the pores of the catalyst. The effective gas/liquid interface is 20 to 50 percent of the geometric surface of the particles, but it can approach 100 percent at high liquid loads with a consequent increase of reaction rate as the amount of wetted surface changes.

Both phases are substantially in plug flow. Dispersion measurements of the liquid phase usually report Peclet numbers, $u_L d_p/D$, less than 0.2. With the usual small particles, the wall effect is negligible in commercial vessels of a meter or so in diameter, but may be appreciable in lab units of 50 mm (1.97 in) diameter. Laboratory and commercial units usually are operated at the same space velocity, LHSV, but for practical reasons the lengths of lab units may be only 0.1 those of commercial units.

Countercurrent gas flow is preferred in pollution control when removal of gaseous impurities is desired.

Trickle Bed Hydrodesulfurization   The first large-scale application of trickle bed reactors was to the hydrodesulfurization of petroleum oils in 1955. The temperature is elevated to enhance the specific rate and the pressure is elevated to improve the solubility of the

hydrogen. A large commercial reactor may have 20 to 25 m (66 to 82 ft) total depth of catalyst, and may be up to 3 m (9.8 ft) diameter in several beds of 3 to 6 m (9.8 to 19.7 ft), limited by the crushing strength of the catalyst and the need for cold shots. Each bed is adiabatic, but the rise in temperature usually is limited to 30°C (86°F) by injection of cold hydrogen between beds. Conditions depend on the boiling range of the oil. Pressures are 34 to 102 atm (500 to 1,500 psi), temperatures 345 to 425°C (653 to 797°F). Catalyst granules are 1.5 to 3.0 mm (0.06 to 0.12 in), sometimes a little more. Catalysts are 10 to 20 percent Co and Mo on alumina.

Limiting flow rates are listed in Table 23-16. The residence times of the combined fluids are figured for 50 atm (735 psi), 400°C (752°F), and a fraction free volume between particles of 0.4. In a 20-m (66-ft) depth, accordingly, the contact times range from 6.9 to 960 s in commercial units. In pilot units the packing depth is reduced to make the contact times about the same.

An apparent first-order specific rate increases with liquid rate as the fraction of wetted surface improves. Catalyst effectiveness of particles 3 to 5 mm (0.12 to 0.20 in) diameter has been found to be about 40 to 60 percent.

A case study has been made (Rase, *Chemical Reactor Design for Process Plants*, vol. 2, Wiley, 1977, pp. 179–182) for removing 50 percent of the 1.9 percent sulfur from a 0.92 SG oil at the rate of 24,000 bbl/d with 2,300 SCF H_2/bbl at 375°C (707°F) and 50 atm (735 psi). For a particular catalyst, the bed height was 8.75 m (29 ft) and the diameter 2.77 m (9.09 ft).

Figure 23-39 is a sketch of a unit to handle 20,000 bbl/d of a light cracker oil with a gas stream containing 75% H_2. Liquid rate was 115,000 kg/h (253,000 lbm/h), gas rate 12,700 kg/hr (28,000 lbm/h), catalyst charge 40,000 kg (88,000 lbm) or 45 m³ (1600 ft³), LHSV ≅ 3. Operating conditions were 370°C (698°F) and 27 atm (397 psi). Vessel dimensions were not revealed, but with an $H/D = 5$, the catalyst bed will have the dimensions 2.25 × 11.25 m (7.38 × 36.9 ft).

FLOODED FIXED BED REACTORS

When the gas and liquid flows are cocurrent upward, a screen is needed at the top to retain the catalyst particles. Such a unit has been used for the hydrogenation of nitro and double bond compounds and

TABLE 23-16   Hydrodesulfurization Feed Rates of Gas and Liquid and Residence Times of the Mixture

	Superficial liquid velocity		Superficial gas velocity				Residence time s/m	
			ft/h (STP)		kg/m²·s			
	ft/h	kg/m²·s	A°	B°	A°	B°	A	B
Pilot plant†	1–30	0.08–2.5	180–5,400	890–27,000	0.0013–0.040	0.0066–0.20	480–16	97–3.5
Commercial reactor	10–300	0.8–25	1,800–54,000	8900–270,000	0.013–0.40	0.066–2.0	48–1.6	9.7–0.35

°The values of gas velocity are shown for (A) 1,000 and (B) 5,000 std ft³ of H_2 per barrel, assuming that all the oil is in the liquid phase. To convert ft³/bbl to m³/m³, multiply by 0.178.
†The length of the pilot-plant reactor was assumed to be one-tenth the length of the commercial reactor.
SOURCE: Partly after Satterfield, *AIChE Journal*, **21**, 209–228 (1975).

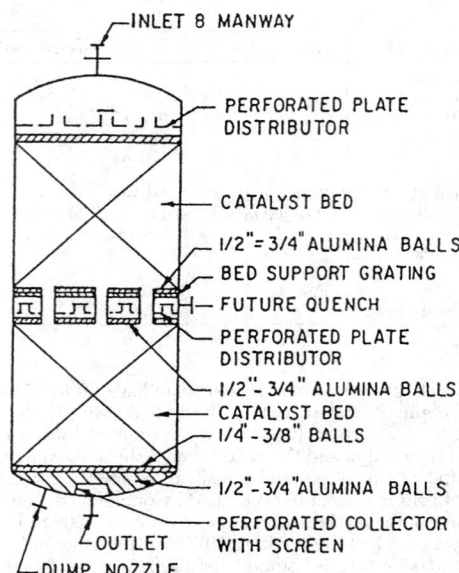

FIG. 23-39 Trickle bed reactor for hydrotreating 20,000 bbl/d of light catalytic cracker oil at 370°C and 27 atm. To convert atm to kPa, multiply by 101.3. (*Gianetto and Silveston,* Multiphase Chemical Reactors, *Hemisphere, 1986, pp. 533–563.*)

nitriles (Ovcinnikov et al., *Brit. Chem. Eng.,* **13,** 1,367 [1968]). High gas rates can cause movement and attrition of the particles. Accordingly, such equipment is restricted to low gas flow rates; for instance, where a hydrogen atmosphere is necessary but the consumption of hydrogen is slight. Backmixing is substantial in commercial-size columns, but less than in bubble columns. Liquid distribution is not a problem, and heat transfer is much better than in the trickle vessel. Liquid holdup and residence times are greater under flooding conditions, which may encourage side reactions.

Downward flow of both fluids imposes no restriction on the gas rate, except that the pressure drop will be high. On the whole, the trickle bed is preferred to the flooded bed.

SUSPENDED CATALYST BEDS

There are three main types of three-phase (GLS) reactors in which the catalyst particles move about in the fluid, as follows.

Slurry Reactors with Mechanical Agitation The catalyst may be retained in the vessel or it may flow out with the fluid and be separated from the fluid downstream. In comparison with trickle beds, high heat transfer is feasible, and the residence time can be made very great. Pressure drop is due to sparger friction and hydrostatic head. Filtering cost is a major item.

Entrained Solids Bubble Columns with the Solid Fluidized by Bubble Action The three-phase mixture flows through the vessel and is separated downstream. Used in preference to fluidized beds when catalyst particles are very fine or subject to disintegration in process.

GLS Fluidized with a Stable Level of Catalyst Only the fluid mixture leaves the vessel. Gas and liquid enter at the bottom. Liquid is continuous, gas is dispersed. Particles are larger than in bubble columns, 0.2 to 1.0 mm (0.008 to 0.04 in). Bed expansion is small. Bed temperatures are uniform within 2°C (3.6°F) in medium-size beds, and heat transfer to embedded surfaces is excellent. Catalyst may be bled off and replenished continuously, or reactivated continuously. Figure 23-40 shows such a unit.

In the reactor shown in Fig. 23-41, a stable fluidized bed is maintained by recirculation of the mixed fluid through the bed and a draft tube. An external pump sometimes is used instead of the built-in

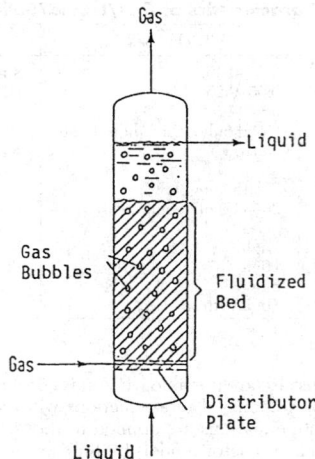

FIG. 23-40 Three-phase fluidized bed reactor.

impeller shown. Such units were developed for the liquefaction of coal and are called *ebullating beds.*

Three-phase fluidized bed reactors are used for the treatment of heavy petroleum fractions at 350 to 600°C (662 to 1,112°F) and 200 atm (2,940 psi). A biological treatment process (Dorr-Oliver Hy-Flo) employs a vertical column filled with sand on which bacterial growth takes place while waste liquid and air are charged. A large interfacial area for reaction is provided, about 33 cm²/cm³ (84 in²/in³), so that an 85 to 90 percent BOD removal in 15 min is claimed compared with 6 to 8 h in conventional units.

TRICKLE BED PARAMETERS

Numerous studies have been made of the hydrodynamics and other aspects of the behavior of gas/liquid/solid systems, in particular of trickle beds, and including absorption and extraction in packed beds. Some of the literature is reviewed in the references at the end of this subsection.

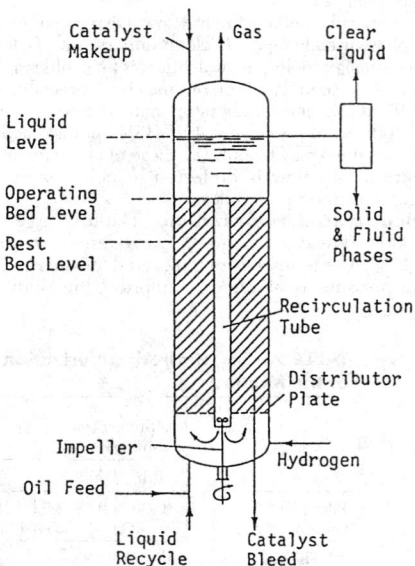

FIG. 23-41 Gas/liquid fluidized (ebullating) bed reactor for hydroliquefaction of coal. (*Kampiner, in Winnacker-Keuchler,* Chemische Technologie, *vol. 3, Hanser, 1972, p. 252.*)

Although many processes have progressed satisfactorily through the laboratory, pilot plant, and commercial scales, design from first principles and scale-up procedures have proved elusive. The various pieces of this puzzle that will be cited in this section, however, do tell us some of what is going on and should aid in the appraisal of the effects of changes in capacity and operating conditions of existing units.

Pressure Drop   Some models regard trickle bed flow as analogous to gas/liquid flow in pipe lines. Various flow regimes may exist like those typified in Fig. 23-25l but in a vertical direction. The two-phase ΔP_{GL} is related to the pressure drops of the individual phases on the assumptions that they are flowing alone. The relation proposed by Larkin et al. (*AIChE Journal*, **7**, 231 [1961]) is

$$\ln \frac{\Delta P_{GL}}{\Delta P_L + \Delta P_G} = \frac{5.0784}{3.531 + (\ln X)^2}$$

$$X = \sqrt{\frac{\Delta P_L}{\Delta P_G}}, \qquad 0.05 \leq X \leq 30 \qquad (23\text{-}109)$$

Several other correlations are cited in the literature, some of which agree with the one quoted here. Pressure drop usually is not a major factor in the design of a trickle bed.

Example 6: Conditions of a Trickle Bed   A system has the properties of air/water at room conditions. Liquid is at 0.5 cm/s, gas at 10 cm/s, free volume fraction $\varepsilon_B = 0.4$, particle diameter $d_p = 0.5$ cm. The pressure drops by one of the many correlations for packed bed flow are $\Delta P_L = 217$, $\Delta P_G = 100$, so that $X = 1.47$ and $\Delta P_{GL} = 1,270$ N/m^3, Newtons/m^3 or 0.0124 atm/(m^2·m).

Liquid Holdup   The major factor influencing this property is the liquid flow rate, but the shape, size, and wetting characteristics of the particles and the gas rate and the initial distribution of liquid also enter in. One of the simpler correlations is that of Midoux et al. (*J. Chem. Eng. Japan*, **9**, 350 [1976]).

$$\frac{\varepsilon_L}{\varepsilon_B} = \frac{0.66X^{0.81}}{1 + 0.66X^{0.81}}, \qquad 0.1 \leq X \leq 80 \qquad (23\text{-}110)$$

where ε_B is the bed porosity. For Example 6 with $X = 1.47$ and $\varepsilon_B = 0.4$,

$$\varepsilon_L = 0.4(0.474) = 0.190$$

A correlation due to Sato (*J. Chem. Eng. Japan*, **6**, 147 [1973]) is

$$\frac{\varepsilon_L}{\varepsilon_B} = 0.185 \left[\frac{6(1 - \varepsilon_B)}{d_p} \right]^{1/3} X^{0.22} = 0.3659 \qquad (23\text{-}111)$$

for the same example. There are also correlations for the static holdup; that is, when the flow rate is zero after wetting. For non-porous catalysts, usually $\varepsilon_{\text{static}} < 0.05$.

Gas/Liquid Mass Transfer   This topic has been widely investigated for gas absorption in packed beds, usually countercurrent. One correlation for cocurrent flow in catalyst beds is by Sato et al. (*First Pacific Chemical Engineering Congress*, Pergamon, 1972, p. 187):

$$k_L a_B = 6.185(10^{-3})d_p^{-0.5}u_L^{0.8}u_G^{0.8}, \qquad 1/s$$

For Example 6, this gives $k_L a_B = 8.8(10^{-3})$/s.

Gas/Liquid Interfacial Area   This has been evaluated by measuring absorption rates like those of CO_2 in NaOH. A correlation by Charpentier (*Chem. Eng Journal*, **11**, 161 [1976]) is

$$\frac{a_E}{a_p(1 - \varepsilon_B)} = 0.05 \left[\frac{\Delta P_{GL}\varepsilon_B}{a_p(1 - \varepsilon_B)} \right]^{1.2} \qquad (23\text{-}112)$$

With $a_p = 6/d_p = 1,200$/m, $\varepsilon_B = 0.4$, and $\Delta P_{GL} = 1,260$ N/m^3,

$$a_B = 0.05(1,200)(0.6) \left[\frac{1,260(2/3)}{1,200} \right]^{1.2} = 23.5 \text{ m}^2/\text{m}^3$$

For comparison, the geometrical area of the particles is

$$a_{\text{geom}} = \frac{6(1 - \varepsilon_B)}{d_p} = \frac{6(0.6)}{0.005} = 72 \text{ m}^2/\text{m}^3$$

Liquid/Solid Mass Transfer   The dissolved gas and the solvent react in contact with the surface of the catalyst. For studying the rate of transfer to the surface, an often-used system was benzoic acid or naphthalene in contact with water. A correlation of Dharwadkar and Sylvester (*AIChE Journal*, **23**, 376 [1977]) that agrees well with some others is

$$k_s = 1.637u_L\text{Re}_L^{-0.331} \left(\frac{\rho_L D}{\mu_L} \right)^{2/3} \qquad (23\text{-}113)$$

For the example with diffusivity of oxygen $D = 2.1(10^{-5})$ cm^2/s,

$$k_s = 1.637(0.1) \left[\frac{0.5(0.1)(1)}{0.01} \right]^{-0.331} \left[\frac{1(2.1)(10^{-5})}{0.01} \right]^{2/3}$$

$$= 1.7(10^{-3}) \text{ cm/s}$$

Axial Dispersion and the Peclet Number   Peclet numbers are measures of deviation from plug flow. They may be calculated from residence time distributions found by tracer tests. Their values in trickle beds are ⅓ to ⅙ those of flow of liquid alone at the same Reynolds numbers. A correlation by Michell and Furzer (*Chem. Eng. J.*, **4**, 53 [1972]) is

$$\text{Pe}_L = \frac{u_L d_p}{D_{EL}} = \left(\frac{\text{Re}_L}{\varepsilon_L} \right)^{0.7} \left(\frac{\mu_L^2}{d_p^3 g \rho_L^2} \right)^{0.32} \qquad (23\text{-}114)$$

where D_{EL} is the axial dispersion coefficient of the liquid, cm^2/s. In the range of Re = 10 to 100, the Peclet number is in the range 0.2 to 0.6 cm^2/s (0.03 to 0.09 in^2/s). It is insensitive to the kind of packing and to the gas flow. Gas-phase dispersions also have been measured and found to be one or two orders of magnitude less than in single-phase gas flows.

Plug flow is approached at low values of the dispersion coefficient or high values of Peclet number. A criterion developed by Mears (*Chem. Eng. Sci.*, **26**, 1361 [1971]) is that conversion will be within 5 percent of that predicted by plug flow when

$$\text{Pe} = \frac{u_L L}{D} > 20 \, n \ln \frac{1}{1 - x_B} \qquad (23\text{-}115)$$

where n is the order of the reaction with respect to B and x_B is the fractional conversion of B. For instance, when $L = 100$ cm, $n = 1$, $u_L = 0.8$ cm/s, and $x_B = 0.96$, then $D = 1.25$ cm^2/s. Note that the Pe numbers of the last two equations do not have the same linear term.

REFERENCES FOR GAS/LIQUID/SOLID REACTIONS
de Lasa, H. I., *Chemical Reactor Design and Technology*, Martinus Nijhoff, 1986. Gianetto, A., and P. L. Silveston, eds., *Multiphase Chemical Reactors*, Hemisphere, 1986. Ramachandran, P. A., and R. V. Chaudhari, *Three-Phase Chemical Reactors*, Gordon & Breach, 1983. Rodrigues, A. E. et al., eds., *Multiphase Chemical Reactors*, vol. 2, Sijthoff & Noordhoff, 1981. Satterfield, C. N., "Trickle Bed Reactors," *AIChE Journal*, **21**, 209–228 (1975). Shah, Y. T., *Gas-Liquid-Solid Reactor Design*, McGraw-Hill, 1979.

REACTIONS OF SOLIDS

Many reactions of solids are industrially feasible only at elevated temperatures which are often obtained by contact with combustion gases, particularly when the reaction is done on a large scale. A product of reaction also is often a gas that must diffuse away from a remaining solid, sometimes through a solid product. Thus, thermal and mass-transfer resistances are major factors in the performance of solid reactions.

There are a number of commercial operations where the object is to make useful products with solid reactions. Design and practice, however, do not appear to rely generally on sophisticated kinetics, and they are rarely divulged completely. The desirable information is about temperatures, configuration, quality of mixing, and residence times or space velocities. Most of the information about current practice of reactions of solids is proprietary. Some of the sparse published data can be cited. Scientific data for solid reactions are abundant but not very coherent. The same data can be fitted, equally badly, by several models. Nevertheless, quotation of some of those kinetic and mechanistic conclusions from the literature may throw some light on these processes. Some data for reactions of solids are shown in Fig. 23-42.

THERMAL DECOMPOSITIONS

All substances are unstable above certain temperatures. The main theory of the rates of decomposition of solids is that they begin at positions of strain on the surface, called *nuclei* or *active sites;* as the reaction progresses, the number of nuclei and their sizes grow. In accordance with this theory the conversion varies as some power of the time, that is, as t^n, where n commonly assumes a value of 3 to 6 but the range is from about 1 to 8. Other rate expressions also have been fitted to certain results. Activation energies represented by the Arrhenius equation sometimes change during the course of a reaction, indicating a change of mechanism.

Decompositions may be exothermic or endothermic. Solids that decompose without melting upon heating are mostly such that can give rise to gaseous products. When a gas is made, the rate can be affected by the diffusional resistance of the product zone. Particle size is a factor. Aging of a solid can result in crystallization of the surface that has been found to affect the rate of reaction. Annealing reduces strains and slows any decomposition rates. The decompositions of some fine powders follow a first-order law. In other cases, the decomposed fraction x is in accordance with the Avrami-Erofeyev equation (cited by Galwey, *Chemistry of Solids*, Chapman Hall, 1967)

$$-\ln(1-x) = kt^n \qquad (23\text{-}116)$$

with $n = 3.5$ to 4. Another rule of simple form also cited by Galwey that sometimes applies is

$$x = k(t - t_1)^n \qquad (23\text{-}117)$$

with n as great as 6. The equation of Prout and Tompkins is of autocatalytic form,

$$\frac{dx}{dt} = kx(1-x), \qquad \ln\frac{x}{1-x} = kt + C \qquad (23\text{-}118)$$

It states that the rate is proportional to the fraction x that has decomposed (which is dominant early in the reaction) and to the fraction not decomposed (which is dominant in latter stages of reaction). The decomposition of potassium permanganate and some other solids is in accordance with this equation. The shape of the plot of x against t is sigmoid in many cases, with slow reactions at the beginning and end, but no theory has been proposed that explains everything.

Organic Solids A few organic compounds decompose before melting, mostly nitrogen compounds: azides, diazo compounds, and nitramines. The processes are exothermic, classed as explosions, and may follow an autocatalytic law. Temperature ranges of decomposition are mostly 100 to 200°C (212 to 392°F). Only spotty results have been obtained, with no coherent pattern. The decomposition of malonic acid has been measured for both the solid and the supercooled liquid. The first-order specific rates at 126.3°C (259.3°F) were 0.00025/min for solid and 0.00207 for liquid, a ratio of 8; at 110.8°C (231.4°F), the values were 0.000021 and 0.00047, a ratio of 39. The decomposition of oxalic acid (m.p. 189°C) obeyed a zero-order law at 130 to 170°C (266 to 338°F).

Exothermic Decompositions These decompositions are nearly always irreversible. Solids with such behavior include oxygen-containing salts and such nitrogen compounds as azides and metal styphnates. When several gaseous products are formed, reversal would require an unlikely complex of reactions. Commercial interest in such materials is more in their storage properties than as a source of desirable products, although ammonium nitrate is an important explosive. A few typical examples will be cited to indicate the ranges of reaction conditions. They are taken from the review by Brown et al. ("Reactions in the Solid State," in Bamford and Tipper, *Comprehensive Chemical Kinetics*, vol. 22, Elsevier, 1980).

Silver oxalate decomposes smoothly and completely in the range 100 to 160°C (212 to 320°F). One investigation showed $x = k(t - t_1)^n$

with $n = 3.5$ to 4, and an activation energy of 27 kcal/g mol (48,600 Btu/lb mol).

Ammonium chromates and some other solids exhibit aging effects. Material that has been stored for months or years follows a cubic law with respect to time, but fresh materials are about fifth-order. At 199°C (390°F), the Prout-Tompkins law was followed.

Ammonium nitrate decomposes into nitrous oxide and water. In the solid phase, decomposition begins at about 150°C (302°F) but becomes extensive only above the melting point (170°C) (338°F). The reaction is first-order, with activation energy about 40 kcal/g mol (72,000 Btu/lb mol). Traces of moisture and Cl^- lower the decomposition temperature; thoroughly dried material has been kept at 300°C (572°F). All oxides of nitrogen, as well as oxygen and nitrogen, have been detected in decompositions of nitrates.

Styphnic acid is a nitrogen compound. Lead styphnate monohydrate was found to detonate at 229°C (444°F), but the course of decomposition could be followed at 228°C and below.

Many investigations are reported on azides of barium, calcium, strontium, lead, copper, and silver in the range 100 to 200°C (212 to 392°F). Time exponents were 6 to 8 and activation energies of 30 to 50 kcal/g mol (54,000 to 90,000 Btu/lb mol) or so. Some difficulties with reproducibility were encountered with these hazardous materials.

Endothermic Decompositions These decompositions are mostly reversible. The most investigated substances have been hydrates and hydroxides, which give off water, and carbonates, which give off CO_2. Dehydration is analogous to evaporation, and its rate depends on the water content of the gas. Activation energies are nearly the same as reaction enthalpies. As the reaction proceeds in the particle, the rate of reaction is impeded by resistance to diffusion of the water through the already formed product. A particular substance may have several hydrates. Which one is present will depend on the partial pressure of water vapor in contact. $FeCl_2$, for instance, combines with 4, 5, 7, or 12 molecules of water with melting points ranging from about 75 to 40°C (167 to 104°F).

Dehydration of $CuSO_4$ pentahydrate at 53 to 63°C (127 to 145°F) and of the trihydrate at 70 to 86°C (158 to 187°F) obey the Avrami-Erofeyev equation. The rate of water loss from $Mg(OH)_2$ at lower temperatures is sensitive to partial pressure of water. Its decomposition above 297°C (567°F) yields appreciable amounts of hydrogen and is not reversible.

Carbonates decompose at relatively high temperatures, 660 to 740°C (1,220 to 1,364°F) for $CaCO_3$. When large samples are used the rate of decomposition can be controlled by the rate of heat transfer or the rate of CO_2 removal.

Some ammonium salts decompose reversibly with release of ammonia, for example,

$$(NH_4)_2SO_4 \Leftrightarrow NH_4HSO_4 + NH_3$$

at 250°C (482°F). Further heating can release SO_3 irreversibly.

The decomposition of silver oxide was one of the earliest solid reactions studied. It is smoothly reversible below 200°C (392°F) with equation for partial pressure of oxygen,

$$\frac{dp}{dt} = k\left(\frac{1-p}{p_{\text{equilib}}}\right)$$

The reaction is sensitive to the presence of metallic silver at the start, indicating autocatalysis, and to the presence of silver carbonate, which was accidentally present in some investigations.

SOLID REACTION EXAMPLES

Diffusion of ions or molecules in solids is preliminary to reaction. It takes place through the normal crystal lattices of reactants and products as well as in channels and fissures of imperfect crystals. It is slow in comparison with that in fluids even at the elevated temperatures at which such reactions have to be conducted. In cement manufacture, for instance, reaction times are 2 to 3 h at 1,200 to 1,500°C (2,192 to 2,732°F) even with 200-mesh particles.

Large contact areas between solid phases are essential. For experimental purposes they are enhanced by forming and mixing fine powders and compressing them. The practices of powder metallurgy are

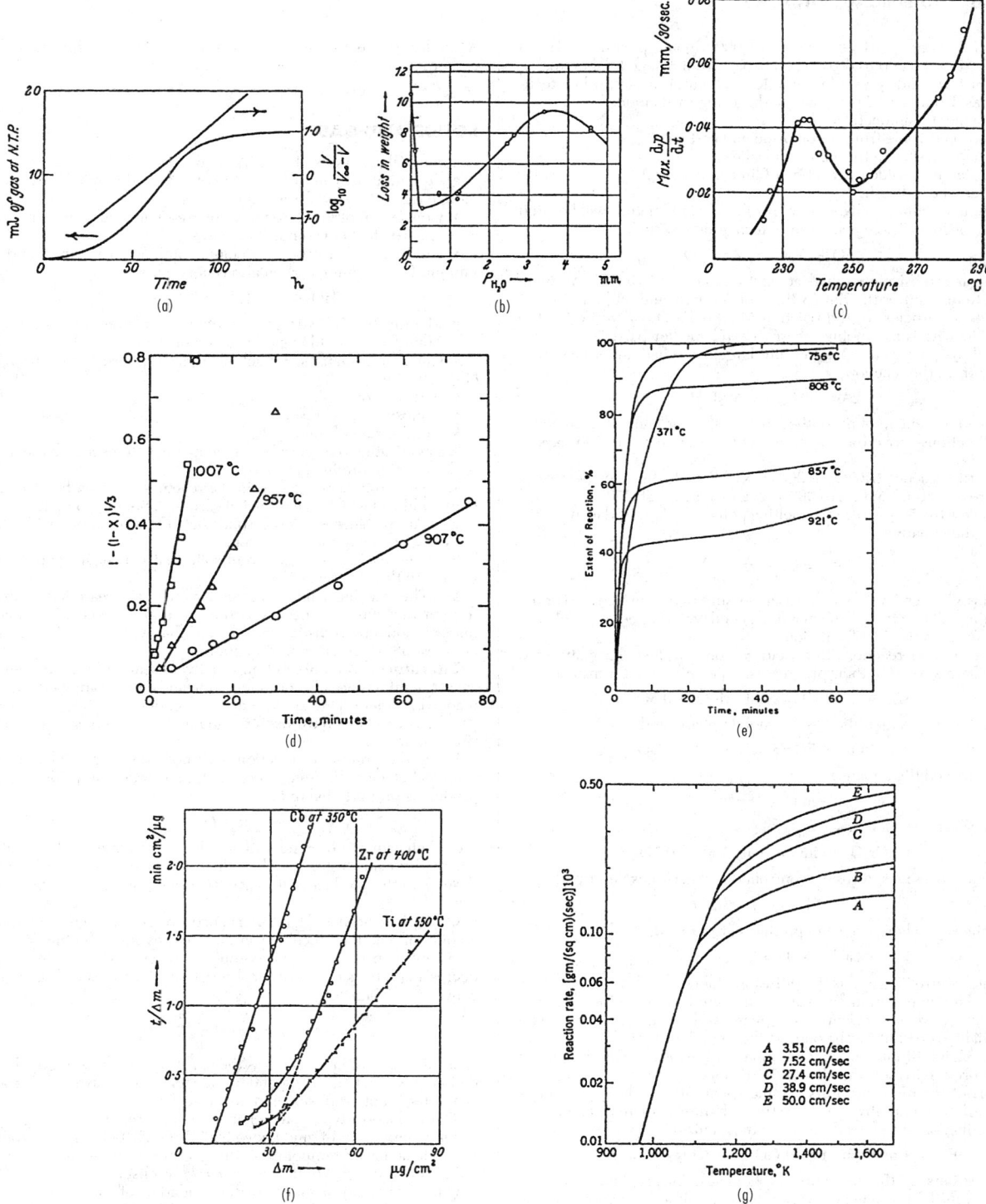

FIG. 23-42 Data for reactions of solids. (*a*) Thermal decomposition of 2-nitrobenzene-4-diazo-1 oxide at 99.1°C (*Vaughan and Williams*, J. Chem. Soc., *p. 1,560 [1946]*). (*b*) Decomposition of CuSO₄·5H₂O; the first four molecules are removed easily, the last with difficulty (*Kohlschutter and Nitschmann, 1931*). (*c*) Decomposition of ammonium perchlorate (*Bircumshaw and Newman*, Proc. Roy. Soc., *A227, 228 [1955]*). (*d*) Reduction of hematite by graphite in the presence of lithium oxide catalyst (*Rao*, Met. Trans., *2, 1439 [1971]*). (*e*) Reduction of nickel oxide by hydrogen; sintering begins just above 750°C (*Hashimoto and Silveston*, AIChE Journal, *19, 268 [1973]*). (*f*) Oxidation of metals, examples of parabolic law, with rate inversely proportional to the weight gain (*Gulbransen and Andrew*, Trans. Electrochem. Soc., *96, 364 [1949]*). (*g*) Effect of air velocity on combustion of carbon (*Tu et al.*, Ind. Eng. Chem., *26, 749 [1934]*).

examples, where particles are 0.1 to 1,000 μm and pressures are 138 to 827 MPa (20,000 to 60,000 psi). Apart from research laboratories, reactions of solids occur in ceramic, metallurgical, and other technologies. Commercial processes of this category include:
- Cement manufacturing
- Boron carbide from boron oxide and carbon
- Calcium silicate from lime and silica
- Calcium carbide by reaction of lime and carbon
- Leblanc soda ash

A limited number of laboratory results of varied nature will be cited for perspective. The mechanism of formation of zinc ferrite,

$$ZnO + Fe_2O_3 \Rightarrow ZnFe_2O_4$$

has been extensively studied at up to 1,200°C (2,192°F). At some lower temperatures the ZnO is the mobile phase and migrates to coat the Fe_2O_3 particles. In the reaction $MgO + Fe_2O_3 \Rightarrow MgFe_2O_4$, the MgO likewise is the mobile phase. There, smaller particles (<1 μm) obeyed the power law $x = k \ln t$, but larger ones had a more complex behavior. In the reaction

$$2AgI + HgI_2 \Rightarrow Ag_2HgI_4$$

nearly equivalent amounts of the ions Ag^+ and Hg^{2+} migrating in opposite directions arrived at their respective interfaces after 66 days at 65°C (149°F).

For the reaction $KClO_4 + 2C \Rightarrow KCl + CO_2$, fine powders were compressed to 69 MPa (10,000 psi) and reacted at 350°C (662°F), well below the 500°C (932°F) melting point. The kinetic data were fitted by the equation

$$\frac{dx}{dt} = k(a-x)^{2/3}x^{-1/3}$$

The term $(a-x)^{2/3}$ was included as a measure of the surface area of the oxidizing salt and the $x^{-1/3}$ term is associated with the reduction of contact area from product formation.

Several other reactions that yield gaseous products have attracted investigators because their progress is easily followed, for instance,

$$MnO_3 + 2MoO_3 \Rightarrow 2MnMoO_4 + 0.5O_2$$

where MoO_3 was identified as the mobile phase, and

$$Ca_3(PO_4)_2 + 5C \Rightarrow 3CaO + P_2 + 5CO$$

which obeyed the equation

$$-\ln(1-x) = kt$$

The reaction

$$CuCr_2O_4 + CuO \Rightarrow Cu_2Cr_2O_4 + 0.5O_2$$

eventually becomes diffusion-controlled and becomes described by

$$[1-(1-x)^{1/3}]^2 = k \ln t$$

In the case where two solid products are formed,

$$CsCl + NaI \Rightarrow CsI + NaCl$$

the rate-controlling step is the diffusion of iodide ion in CsCl.

The kinetic equation can vary with a number of factors. For the reaction between tricalcium phosphate and urea, relatively coarse material (−180+200 mesh) obeyed the law $x^2 = kt$ with $E = 18$ kcal/g mol (32,400 Btu/lb mol) and finer material (−300+320 mesh) obeyed a first-order equation with $E = 28$ kcal/g mol.

Carbothermic Reactions Some apparently solid/solid reactions with carbon apparently take place through intermediate CO and CO_2. The reduction of iron oxides has the mechanism

$$Fe_xO_y + yCO \Rightarrow xFe + yCO_2, \qquad CO_2 + C \Rightarrow 2CO$$

Some results of the reduction of hematite by graphite at 907 to 1,007°C in the presence of lithium oxide catalyst were correlated by the equation $1-(1-x)^{1/3} = kt$. The reaction of solids ilmenite and carbon has the mechanism

$$FeTiO_3 + CO \Rightarrow Fe + TiO_2 + CO_2, \qquad CO_2 + C \Rightarrow 2CO$$

and that between chromium oxide and chromium carbide,

$$\frac{1}{6}Cr_{23}O_6 + CO_2 \Rightarrow \frac{23}{6}Cr + 2CO, \qquad \frac{1}{3}Cr_2O_3 + CO \Rightarrow \frac{2}{3}Cr + CO_2$$

A similar case is the preparation of metal carbides from metal and carbon,

$$C + 2H_2 \Rightarrow CH_4, \qquad Me + CH_4 \Rightarrow MeC + 2H_2$$

SOLIDS AND GASES

Commercial processes of solid/gas reactions include:
- Oxidation with air of sulfides in ores to oxides or sulfates that are more easily processed for metal recovery.
- Conversion of Fe_2O_3 to magnetic Fe_3O_4 in contact with reducing atmosphere of CO in combustion gases.
- Chlorination of ores of uranium, titanium, zirconium and aluminum. For titanium, carbon also is needed:

$$TiO_2 + C + 2Cl_2 \Rightarrow TiCl_4 + CO_2$$

- Manufacture of hydrogen by action of steam on iron.
- Manufacture of blue gas by action of steam on carbon.
- Calcium cyanamide by action of atmospheric nitrogen on calcium carbide.
- Nitriding of steel.
- Atmospheric corrosion.
- Combustion of solid fuels.

A classification of processes can be made with respect to the nature of the reaction product, as follows:
1. The product is a gas, as in the reaction between finely divided nickel and CO which makes nickel carbonyl, boiling point 42°C.
2. The product is a loose solid that offers no appreciable diffusional resistance to the gas reactant.
3. The product is an adherent solid that may or may not be permeable to the gas reactant.
4. The reacting solid is in granular form. Decrease in the area of the reaction interface occurs as the reaction proceeds. The mathematical modeling is distinguished from that with flat surfaces, which are most often used in experimentation.

Literature A number of informative researches can be cited, but again the difficulties of experimentation and complicating factors have made the kinetic patterns difficult to generalize. The most investigated gas reactants have been oxygen and hydrogen and some chlorine systems.

When the product of reaction does not prove a barrier to further chemical change, the rate is constant, zero-order, and the weight of product is proportional to time,

$$w = kt$$

Such behavior is observed with alkali and alkaline earth oxidations where the oxide volume is less than the metal volume and cracks develop in the product coat, permitting ready access for further reaction.

Oxide films of Al and Cr, for instance, are protective beyond a small amount of reaction, since they are nonporous and adherent.

When the product layer is porous the reaction will continue but at decreasing rate as the diffusional resistance increases with increasing conversion. Then,

$$\frac{dw}{dt} = \frac{k}{w}$$

This rate expression is known as the *parabolic law*. It is obeyed by oxidation of Ni, Ti, Cu, and Cr and by halogenation of silver. The product coat retards both diffusion and heat transfer.

The behavior type may change with temperature range. For instance, oxidation of zinc above 350°C (662°F) obeys the parabolic law, but at lower temperature the product coat seems to develop cracks and the *logarithmic law*, $\ln w = kt$, is observed.

One mathematical model of the oxidation of nickel spheres was confirmed when it took into account the decrease in the reaction surface as the reaction proceeded.

EQUIPMENT AND PROCESSES

Reactions of solids of commercial interest almost always involve a gas as a reagent and/or a heat source. Some of the equipment in use is represented in Fig. 23-43. Temperatures are usually high so the

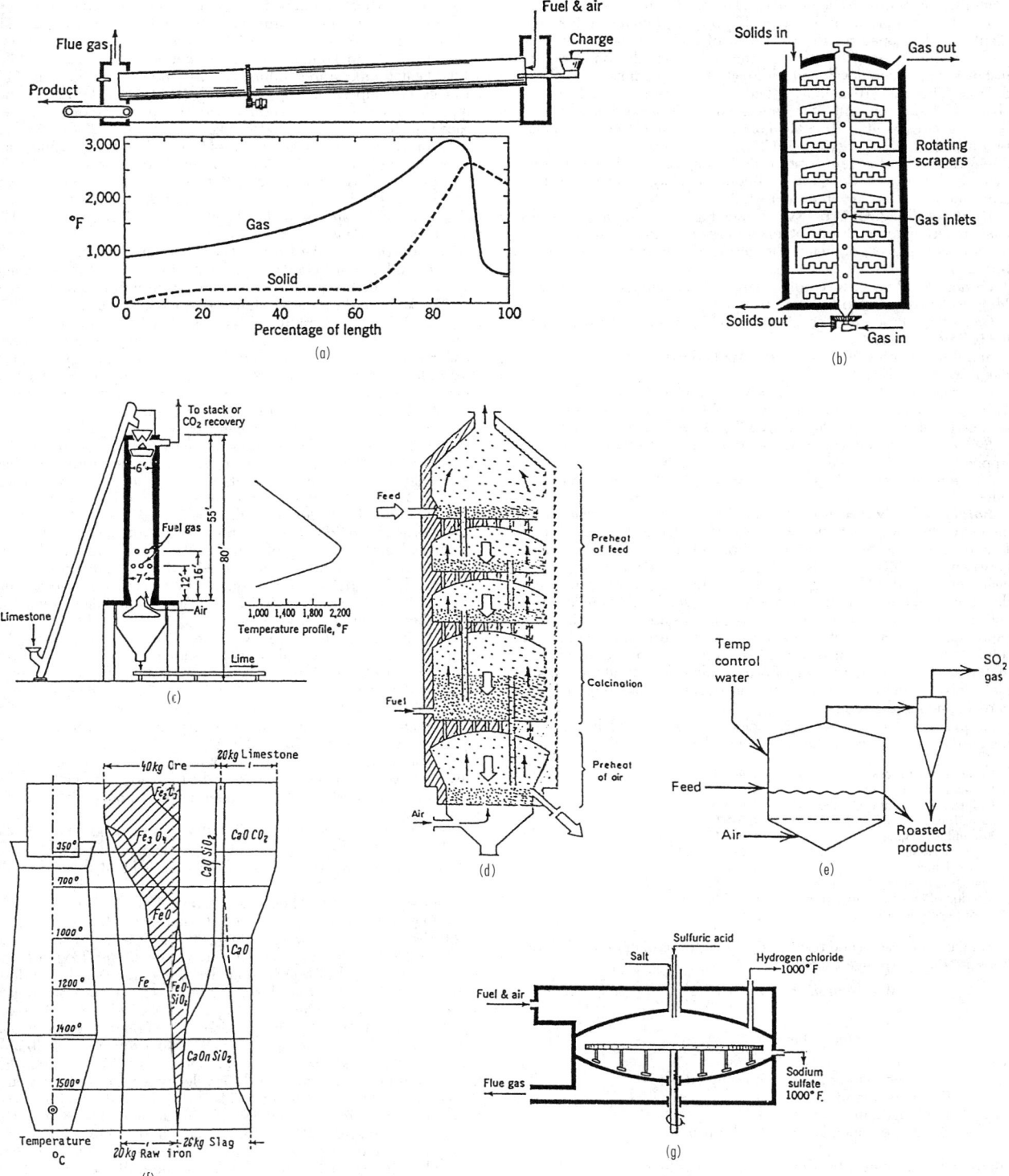

FIG. 23-43   Reactors for solids. (*a*) Temperature profiles in a rotary cement kiln. (*b*) A multiple hearth reactor. (*c*) Vertical kiln for lime burning, 55 ton/d. (*d*) Five-stage fluidized bed lime burner, 4 by 14 m, 100 ton/d. (*e*) A fluidized bed for roasting iron sulfides. (*f*) Conditions in a vertical moving bed (blast furnace) for reduction of iron oxides. (*g*) A mechanical salt cake furnace. To convert ton/d to kg/h, multiply by 907.

equipment is mostly refractory-lined. The solid is in granular form, at most a few mm or cm in diameter. Historically, much of the equipment was developed for the treatment of ores and the recovery of metals. More than 50 processes for reduction of iron ores, for instance, have been developed, although the clear winner is the blast furnace. This is a vertical moving-bed device; iron oxides and coal are charged at the top and flow countercurrently to combustion and reducing gases. Units of 1,080 to 4,500 m³ (38,000 to 159,000 ft³) may produce up to 9×10^6 kg (20×10^6 lbm) of molten iron per day. Figure 23-43f identifies the temperature and composition profiles (Blom, Stahl, and Eisen, **47**, 955 [1927]). Reduction is with CO and H_2 that are made from coal, air, and water within the reactor.

Pyrometallurgical Processes   Such high temperature processes convert certain minerals into others for easier separation from gangue or for easier recovery of metal. They are accomplished in kilns, hearth furnaces or fluidized bed reactors.

Oxidation   Proceeds according to the stoichiometry $MeS + 1.5O_2 \Rightarrow MeO + CO_2$. Applied to Fe, Pb, Cu, and Ni.

Calcining   Proceeds according to $MeCO_3 \Rightarrow MeO + CO_2$. Applied to Ca, Mg, and Ba.

Sulfating $MeS + 2O_2 \Rightarrow MeSO_4$. Applied to Cu, of which the sulfate is water-soluble.

Chlorination $MeO + Cl_2 + C \Rightarrow MeCl_2 + CO$. Applied to Mg, Be, Ti, and Zr, whose chlorides are water-soluble. The chlorine can be supplied indirectly, as in $Cu_2S + 2NaCl + O_2 \Rightarrow 2CuCl + Na_2SO_4$.

Reduction $MeO + H_2 \Rightarrow Me + H_2O$, $MeO + CO \Rightarrow Me + CO_2$. Applied to Fe, W, Mo, Ge, and Zn.

Magnetic Roasting $Fe_2O_3 + CO \Rightarrow Fe_3O_4$. Recovered magnetically.

Rotary Kiln Equipment   A rotary kiln is a long, narrow cylinder inclined 2 to 5 degrees to the horizontal and rotated at 0.25 to 5 rpm. It is used for the decomposition of individual solids, for reactions between finely divided solids, and for reactions of solids with gases or even with liquids. The length/diameter ratio ranges from 10 to 35, depending on the reaction time needed. The solid is in granular form and may have solid fuel mixed in. The granules are scooped into the vapor space and are heated as they cascade downwards. Holdup of solids is 8 to 15 percent of the cross section. For most free-falling materials, the solids pattern approaches plug flow axially and complete mixing laterally. Rotary kilns can tolerate some softening and partial fusion of the stock.

Approximate ranges of space velocities in rotary kilns, metric tons/(m³·d), of selected processes are as follows:

Cement, dry process	0.4–1.1
Cement, wet process	0.4–0.8
Cement, with heat exchange	0.6–1.8
Lime burning	0.5–0.9
Dolomite burning	0.4–0.6
Pyrite roasting	0.2–0.35
Clay calcination	0.5–0.8
Magnetic roasting	1.5–2.0
Ignition of inorganic pigments	0.15–2.0
Barium sulfide preparation	0.35–0.8

Formulas for capacity and residence time in terms of operating conditions are due to Heiligenstaedt:

$$W = 148n\phi D^3 \tan \vartheta, \qquad t/(m^3 \cdot d) \qquad (23\text{-}119)$$

$$\tau = \frac{L}{60\pi nD \tan \vartheta}, \qquad h \qquad (23\text{-}120)$$

where n = rpm
   ϕ = fraction of cross section occupied
   D = diameter, m
   L = length, m
   ϑ = degrees inclination to the horizontal

High-temperature heat transfer from the gas is by radiation and convection.

Vertical Kilns   These kilns are used primarily where no fusion or softening occurs, as in the burning of limestone or dolomite, although rotary kilns also are used for these operations. The blast furnace, Fig.

23-43f, is a vertical kiln in which fusion takes place in the lower section. A cross section of a continuous lime kiln is shown in Fig. 23-43c which is for 50,000 kg/day (110,000 lbm/d). The diameter range of these kilns are 2.4 to 4.5 m (7.9 to 14.8 ft) and height 15 to 24 m (49 to 79 ft). Peak temperatures in lime calcination are 1,200°C (2,192°F), although decomposition proceeds freely at 1,000°C (1,832°F). Fuel supply may be coke mixed and fed with the limestone or other fuel. Space velocity of the kiln is 14 to 25 kg CaO/(m³·h) (0.87 to 1.56 lbm/ [ft²·h]) or 215 to 485 kg CaO/(m²·h) (44 to 99 lbm/[ft²·h]), depending on the size and modernity of the kiln, the method of firing, and the lump size, which is in the range of 10 to 25 cm (3.9 to 9.8 in). A five-stage fluidized bed calciner is sketched in Fig. 23-43d. Such a unit 4 m (13 ft) in diameter and 14 m (46 ft) high has a production of 91,000 kg CaO/d (200,000 lbm/d).

Cement Manufacture Kilns   These kilns are up to 6 m (17 ft) in diameter and 200 m (656 ft) long. Inclination is 3 to 4 degrees, and rotation is 1.2 to 2.0 rpm. Typical temperature profiles are shown in Fig. 23-43a. Near the flame the temperature is 1,800 to 2,000°C (3272 to 3632°F). The temperature of the solid reaches 1,350 to 1,500°C (2,462 to 2,732°F) which is necessary for clinker formation. In one smaller kiln, a length of 23 m (75 ft) was allowed for drying, 34 m (112 ft) for preheating, 19 m (62 ft) for calcining, and 15 m (49 ft) for clinkering. Total residence time is 40 min to 5 h, depending on the type of kiln. The time near the clinkering temperature of 1,500°C (2,732°F) is 10 to 20 min. Subsequent cooling is as rapid as possible.

A kiln 6 m (20 ft) in diameter by 200 m (656 ft) can produce 2.7×10^6 kg/d (6×10^6 lbm/d) of cement. For production rates less than 270,000 kg/d (600,000 lbm/d), shaft kilns are used. These are vertical cylinders 2 to 3 m (6.5–10 ft) by 8 to 10 m (26–33 ft) high, fed with raw meal pellets and finely ground coal.

Roasting of Sulfide Ores   In this process iron sulfide (pyrite) is burned with air for recovery of sulfur and to make the iron oxide from which the metal is more easily recovered. Sulfides of other metals also are roasted. The original kind of equipment was a multiple hearth furnace, as shown in Fig. 23-43b. In some designs the plates rotate, in others the scraper arms rotate, and in still others the arms oscillate and discharge the material to lower plates at each stroke. Material is charged at the top, moves along as the rotation proceeds, and drops onto successively lower plates while combustion gases or gaseous reactants flow upward. A reactor with 9 trays 5 m (16 ft) in diameter and 12 m (39 ft) high can roast about 600 kg/h (1,300 lbm/h) of pyrite. Another unit with 11 trays 2 m (7 ft) in diameter is said to have a capacity of 114,000 kg/d (250,000 lbm/d).

A major portion of the reaction is found to occur in the vapor space between trays. A unit in which most of the trays are replaced by empty space is called a *flash roaster;* its mode of operation is like that of a spray dryer.

Molybdenum sulfide is roasted at the rate of 5500 kg/d (12,000 lbm/d) in a unit with 9 stages, 5 m (16 ft) diameter, at 630 ± 15°C (1,166 ± 27°F), sulfur going from 35.7 percent to 0.04 to 0.006 percent.

A Dorr-Oliver fluidized bed roaster 5.5 m (18 ft) in diameter, 7.6 m (25 ft) high, with a bed height of 1.2 to 1.5 m (3.9 to 4.9 ft) has a capacity of 154,000 to 200,000 kg/d (340,000–440,000 lbm/d) at 650 to 700°C (1,200 to 1,300°F) (Kunii and Levenspiel, *Fluidization Engineering*, Butterworth, 1991). Two modes of operation can be used for a fluidized bed unit like that shown in Fig. 23-43e. In one operation, a stable bed level is maintained at a superficial gas velocity of 0.48 m/s (1.6 ft/s); a unit 4.8 m (16 ft) in diameter, 1.5 m (4.9 ft) bed depth, 3 m (9.8 ft) freeboard, has a capacity of 82,000 kg/d (180,000 lbm/d) pyrrhotite, 200 mesh, 53 percent entrained solids, 875°C (1,600°F). In the other mode, the superficial gas velocity is 1.1 m/s (3.6 ft/s) and 100 percent entrainment occurs. This is called *transfer line* or *pneumatic transport reaction;* a unit 6.6 m (22 ft) diameter by 1.8 m (5.9 ft) handles 545,000 kg/d (1.2×10^6 lbm/d), 200 mesh, 780°C (1,436°F).

Magnetic Roasting   In this process ores containing Fe_2O_3 are reduced with CO to Fe_3O_4, which is magnetically separable from gangue. Rotary kilns are used, temperatures 700 to 800°C (1,292 to 1,472°F). Higher temperatures form FeO. A reducing-gas atmosphere is created by firing a deficiency of air with carbonaceous fuels. Data for two installations are as follows.

Measurements	Charge, t/d	Product, t/d
3.6 by 44 m	900–960, 21% Fe	350–368, 41–42% Fe
3.6 by 50 m	1,300	—

A unit for 2.3×10^6 kg/d (5×10^6 lbm/d) has a power consumption of 0.0033 to 0.0044 kwh/kg (3 to 4 kwh/ton) and a heat requirement of 180,000 to 250,000 kcal/ton (714,000 to 991,000 Btu/ton). The magnetic concentrate can be agglomerated for further treatment by pelletizing or sintering.

Sodium Sulfate   A single-hearth furnace is used, like that shown in Fig. 23-40g. Sodium chloride and sulfuric acid are charged continuously to the center of the pan and the rotating scrapers gradually work the reacting mass to the periphery, where the sodium sulfate is discharged at 540°C (1,000°F). Pans are 3.3 to 5.5 m (11 to 18 ft) in diameter and can handle 5,500 to 9,000 kg/d (12,000 to 20,000 lbm/d) of salt. Rotary kilns also are used for this purpose. Such a unit 1.5 m (4.9 ft) in diameter by 6.7 m (22 ft) has a capacity of 22,000 kg/d (48,000 lbm/d) of salt cake. A pan furnace also is used, for instance, in the Leblanc soda ash process and for making sodium sulfide from sodium sulfate and coal.

Desulfurization with Dry Lime   Limestone or lime or dolomite ($CaCO_3 \cdot MgCO_3$) in a fluidized bed coal combustor reacts with SO_2 in the gas to make $CaSO_4$. The most favorable conditions with lime are 1 atm (14.7 psi) and 800 to 850°C (1,472 to 1,562°F). At higher temperatures the sulfate tends to decomposition, according to $CaSO_4 + CO \Rightarrow CaO + CO_2 + SO_2$. Dolomite works better at 950°C (1,742°F) and 3 to 10 atm (44 to 147 psi). Although $MgSO_4$ is not formed, the pore structure with dolomite is more open to diffusion of SO_2. The pores of the CaO tend to be clogged by $CaSO_4$, but the attrition in a fluidized bed helps the extent of utilization of the lime which, at best, is only partial. A mol ratio Ca/S of 3 to 5 is needed to get as much as 70 to 80 percent removal of the sulfur. Reactivities of limestones are variable, depending on the content of impurities. Ground particle sizes are 1.5 to 5 mm (0.059 to 0.20 in), bed depths 1.2 to 3 m (3.9 to 9.8 ft), residence times of lime 0.5 to 2.5 s. The reaction seems to be first-order in SO_2. The solid $CaSO_4$ is a waste product. It is technically, although not economically, feasible to use it for making sulfuric acid.

REFERENCES FOR REACTIONS OF SOLIDS: Brown, W. E., D. Dollimore, and A. K. Galwey, "Reactions in the Solid State," in Bamford and Tipper, eds., *Comprehensive Chemical Kinetics,* vol. 22, Elsevier, 1980. Galwey, A. K., *Chemistry of Solids,* Chapman and Hall, 1967. Sohn, H. Y., and W. E. Wadsworth, eds., *Rate Processes of Extractive Metallurgy,* Plenum Press, 1979. Szekely, J., J. W. Evans, and H. Y. Sohn, *Gas-Solid Reactions,* Academic Press, 1976. Ullmann, ed., *Enzyklopaedie der technischen Chemie,* "Uncatalyzed Reactions with Solids," vol. 3, 4th ed., Verlag Chemie, 1973, pp. 395–464.

Biochemical Engineering

Henry R. Bungay, P.E., Ph.D., *Professor of Chemical and Environmental Engineering, Rensselaer Polytechnic Institute; Member, American Institute of Chemical Engineers, American Chemical Society, American Society for Microbiology, American Society for Engineering Education, Society for General Microbiology. (Section Editor)*

Arthur E. Humphrey, Ph.D., *Retired, Professor of Chemical Engineering, Pennsylvania State University; Member, U.S. National Academy of Engineering, American Institute of Chemical Engineers, American Chemical Society, American Society for Microbiology.*

George T. Tsao, Ph.D., *Director, Laboratory for Renewable Resource Engineering, Purdue University; Member, American Institute of Chemical Engineers, American Chemical Society, American Society for Microbiology.*

Assisted by David T. Tsao.

Nomenclature and Units

Symbol	Definition	SI units	U.S. customary units
A	Empirical constant	Dimensionless	Dimensionless
C	Concentration (mass)	kg/m^3	lb/ft^3
C	Concentration	mol/m^3	$(lb \cdot mol)/ft^3$
D	Diameter	m	ft
D	Effective diffusivity	m^2/s	ft^2/h
D	F/V	s^{-1}	h^{-1}
DRT	Decimal reduction time for sterilization	s	h
E	Activation energy	cal/mol	$Btu/(lb \cdot mol)$
F	Flow or feed rate	m^3/s	ft^3/h
H	Concentration of host organisms	kg/m^3	lb/ft^3
K	Rate coefficient	Units dependent on order of reaction	Units dependent on order of reaction
K_M	Michaelis constant	kg/m^3	lb/ft^3
$K_l a$	Lumped mass-transfer coefficient	s^{-1}	h^{-1}
K_d	Death-rate coefficient	s^{-1}	h^{-1}
K_s	Monod coefficient	kg/m^3	lb/ft^3
k	Kinetic constants	Dependent on reaction order	Dependent on reaction order
M	Coefficient for maintenance energy	Dimensionless	Dimensionless
N	Numbers of organisms or spores	Dimensionless	Dimensionless
P	Product concentration	kg/m^3	lb/ft^3
Q_{O_2}	Specific-respiration-rate coefficient	$kg\ O_2/(kg\ organism \cdot s)$	$lb\ O_2/(lb\ organism \cdot h)$
R	Universal-gas-law constant	$8314\ J/(mol \cdot K)$	$0.7299\ (ft^3)(atm)/(lb \cdot mol \cdot R)$
r	Radial position	m	ft
S	Substrate concentration	kg/m^3	lb/ft^3
S	Shear	N/m^2	lbf/ft^2
S_o	Substrate concentration in feed	kg/m^3	lb/ft^3
T	Temperature	K	°F
t	Time	s	h
V	Velocity of reaction	mol/s	$(lb \cdot mol)/h$
V_m	Maximum velocity of reaction	mol/s	$(lb \cdot mol)/h$
V	Air velocity	m/s	ft/h
V	Fermenter volume	m^3	ft^3
VVM	Volume of air/volume of fermentation broth per minute	Dimensionless	Dimensionless
X	Organism concentration	kg/m_3	lb/ft^3
Y	Yield coefficient	kg/kg	lb/lb
		Greek symbols	
β	Dimensionless Michaelis constant	Dimensionless	Dimensionless
μ	Specific-growth-rate coefficient	s^{-1}	h^{-1}
$\hat{\mu}$ or μ_{max}	Maximum-specific-growth-rate coefficient	s^{-1}	h^{-1}
ω	Recycle ratio	Dimensionless	Dimensionless
ϕ	Thiele modulus	Dimensionless	Dimensionless

GENERAL REFERENCES

1. Aiba, S., A. E. Humphrey, and N. F. Millis, Biochemical Engineering, 2d ed., University of Tokyo Press, 1973.
2. Atkinson, B., and F. Mavituna, *Biochemical Engineering and Biotechnology Handbook,* 2d ed., Stockton, New York, 1991.
3. Bailey, J. E., and D. F. Ollis, *Biochemical Engineering Fundamentals,* 2d ed., McGraw-Hill, 1986.
4. Baltz, R. H., and G. D. Hegeman (eds.), *Industrial Microorganisms,* ASM Press, Washington, DC, 1993.
5. Bu'Lock, J. D., and B. Kristiansen (eds.), *Basic Biotechnology,* Academic Press, London, 1987.
6. Bungay, H. R., *Energy: The Biomass Options,* Wiley, 1981.
7. Bungay, H. R., *BASIC Biochemical Engineering,* BiLine Assoc., Troy, New York, 1993.
8. Bungay, H. R., *BASIC Environmental Engineering,* BiLine Assoc., Troy, New York, 1992.
9. Coombs, J., *Dictionary of Biotechnology,* Stockton Press, New York, 1992.
10. Demain, A., and N. Solomon, *Biology of Industrial Microorganisms,* Butterworth/Heinemann, Stoneham, Massachusetts, 1985.
11. Dibner, M. D., *Biotechnology Guide U.S.A.: Companies, Data, and Analysis,* Stockton Press, New York, 1991.
12. Dunn, I. J., E. Heinzele, J. Ingham, and J. E. Prenosil, *Biological Reactor Engineering—Principles, Applications, and Modelling with PC Simulation,* VCH—Weinheim, New York, 1992.
13. Fiechter, A., H. Okada, and R. D. Tanner (eds.), *Bioproducts and Bioprocesses,* Springer-Verlag, Berlin, 1989.
14. Finkelstein, D. B., and C. Ball, *Biotechnology of Filamentous Fungi,* Butterworth/Heinemann, Stoneham, Massachusetts, 1992.
15. Fleschar, M. H., and K. R. Nill, *Glossary of Biotechnology Terms,* Technomic Pub. Co., Lancaster, PA, 1993.
16. Glick, B. R., and J. J. Pasternak, *Molecular Biotechnology,* ASM Press, Washington, DC, 1994.
17. Ho, C. S., and D. I. C. Wang, *Animal Cell Bioreactors,* Butterworth/Heinemann, Stoneham, Massachusetts, 1991.
18. Jackson, A. T., *Process Engineering in Biotechnology,* Prentice Hall, Englewood Cliffs, New Jersey, 1991.
19. Lancini, G., and R. Lorenzetti, *Biotechnology of Antibiotics and Other Bioactive Microbial Metabolites,* Plenum, New York, 1993.
20. Laskin, A., *Enzymes and Immobilized Cells in Biotechnology,* Butterworth/Heinemann, Stoneham, Massachusetts, 1985.
21. Lee, J. M., *Biochemical Engineering,* Prentice Hall, Englewood Cliffs, New Jersey, 1991.
22. Lydersen, B., N. A. D'Elia, and K. L. Nelson, *Bioprocess Engineering,* Wiley, New York, 1994.
23. McDuffie, N. G., *Bioreactor Design Fundamentals,* Butterworth/Heinemann, Stoneham, Massachusetts, 1991.
24. Moo-Young, M. (ed.), *Comprehensive Biotechnology: The Principles, Applications and Regulations of Biotechnology in Industry, Agriculture and Medicine,* Pergamon Press, Oxford, 1985.
25. Murooka, Y., and T. Imanaka, *Recombinant Microbes for Industrial and Agricultural Applications,* M. Dekker, New York, 1993.
26. Nielsen, J., and J. Villadsen, *Bioreaction Engineering Principles,* Plenum, New York, 1994.
27. Pons, M.-N. (ed.), *Bioprocess Monitoring and Control,* Hanser, 1991.
28. Richardson, J. F., and D. G. Peacock, *Chemical Engineering,* vol. 3, Pergamon/Elsevier, Oxford, 1994.
29. Solomons, G. L., *Materials and Methods in Fermentation,* Academic Press, 1969.
30. Shuler, M. L. (ed.), *Chemical Engineering Problems in Biotechnology,* Am. Inst. Chem. Engr., New York, 1989.
31. Thilly, W., *Mammalian Cell Technology,* Butterworth/Heinemann, Stoneham, Massachusetts, 1986.
32. Twork, J. V., and A. M. Yacynych, *Sensors in Bioprocess Control,* Dekker, New York, 1993.
33. Vanek, Z., and Z. Hostalek, *Overproduction of Microbial Metabolites,* Butterworth/Heinemann, Stoneham, Massachusetts, 1986.
34. Vieth, W. R., *Bioprocess Engineering: Kinetics, Mass Transport, Reactors, and Gene Expression,* Wiley, New York, 1994.
35. Vogel, H. C., *Fermentation and Biochemical Engineering Handbook: Principles, Process Design, and Equipment,* Noyes Publications, Park Ridge, New Jersey, 1983.
36. Volesky, B., and J. Votruba, *Modeling and Optimization of Fermentation Processes,* Elsevier, 1992.

INTRODUCTION TO BIOCHEMICAL ENGINEERING

The differences between biochemical engineering and chemical engineering lie not in the principles of unit operations and unit processes but in the nature of living systems. The commercial exploitation of cells or enzymes taken from cells is restricted to conditions at which these systems can function. Most plant and animal cells live at moderate temperatures and do not tolerate extremes of pH. The vast majority of microorganisms also prefer mild conditions, but some thrive at temperatures above the boiling point of water or at pH values far from neutrality. Some can endure concentrations of chemicals that most other cells find highly toxic. Commercial operations depend on having the correct organisms or enzymes and preventing inactivation or the entry of foreign organisms that could harm the process.

The pH, temperature, redox potential, and nutrient medium may favor certain organisms and discourage the growth of others. For example, pickles are produced in vats by lactobacilli well-suited to the acid conditions and with small probability of contamination by other organisms. In mixed culture systems, especially those for biological waste treatment, there is an ever shifting interplay between microbial populations and their environments that influences performance and control. Although open systems may be suitable for hardy organisms or for processes in which the conditions select the appropriate culture, many bioprocesses are closed and have elaborate precautions to prevent contamination. The optimization of the complicated biochemical activities of isolated strains, of aggregated cells, of mixed populations, and of cell-free enzymes or components presents engineering challenges that are sophisticated and difficult. Performance of a bioprocess can suffer from changes in any of the many biochemical steps functioning in concert, and genetic controls are subject to mutation. Offspring of specialized mutants that yield high concentrations of product tend to revert during propagation to less productive strains—a phenomenon called *rundown.*

This section emphasizes cell cultures and microbial and enzymatic processes and excludes medical, animal, and agricultural engineering systems. Engineering aspects of biological waste treatment are covered in Sec. 25.

Biotechnology has a long history—fermented beverages have been produced for several thousand years. But biochemical engineering is not yet fully mature. Developments such as immobilized enzymes and cells have been exploited partially, and many exciting advances should be forthcoming. Genetic manipulations through recombinant DNA techniques are leading to practical processes for molecules that could previously be found only in trace quantities in plants or animals. Biotechnology is now viewed as a highly profitable route to relatively valuable products. In the near future, costs of environmental protection may force more companies to switch from chemical processing that generates wastes that are costly to treat to biochemical methods with wastes that are easily broken down by biological waste treatment processes and that present much less danger to the environment. Some commercial bioprocesses could have municipal and industrial wastes as feedstocks, and the credits for accepting them should improve the economic prospects. When petroleum runs out and the prices soar for petrochemicals, there will be large profits for fermentations that produce equivalent compounds.

BIOLOGICAL CONCEPTS

Cells The cell is the unit of life. Cells in multicellular organisms function in association with other specialized cells, but many organisms are free-living single cells. Although differing in size, shape, and functions, there are basic common features in all cells. Every cell contains cytoplasm, a colloidal system of large biochemicals in a complex solution of smaller organic molecules and inorganic salts. The cytoplasm is bounded by a semielastic, selectively permeable cell membrane that controls the transport of molecules into and out of the cell. There are biochemical transport mechanisms that spend energy to bring substances into the cell despite unfavorable concentration gradients across the membrane. Cells are protected by rigid cell walls external to the cell membranes. Certain bacteria, algae, and protozoa have gelatinous sheaths of inorganic materials such as silica.

Sequences of genes along a threadlike chromosome encode information that controls cellular activity. As units of heredity, genes determine the cellular characteristics passed from one generation to the next. In most cells, the chromosomes are surrounded by a membrane to form a conspicuous nucleus. Cells with organized nuclei are described as eukaryotic. Other intracellular structures serve as specialized sites for cellular activities. For example, photosynthesis is carried out by organelles called chloroplasts. In bacteria and cyanobacteria (formerly called blue-green algae), the chromosomes are not surrounded by a membrane, and there is little apparent subcellular organization. Lacking a discrete nucleus, these organisms are said to be prokaryotic.

Microorganisms of special concern to biochemical engineering include yeasts, bacteria, algae, and molds. The protozoa can feed on smaller organisms in natural waters and in waste-treatment processes but are not useful in producing materials of commercial value. Certain viruses called phages are also important in that they can infect microorganisms and may destroy a culture. A beneficial feature of microbial viruses is the ability to convey genetic materials from other sources into an organism. This is called transduction. Each species of microorganisms grows best within certain pH and temperature ranges, commonly between 20 and 40°C (68–104°F) and not too far from neutral pH.

Bacteria The bacteria are tiny single-cell organisms ranging from 0.5–20 μm in size, although some may be smaller, and a few exceed 100 μm in length. The cell wall imparts a characteristic round or ovoid, rod, or spiral shape to the cell. Some bacteria can vary in shape, depending on culture conditions; this is termed *pleomorphism*. Certain species are further characterized by the arrangement of cells in clusters, chains, or discrete packets. Some cells produce various pigments that impart a characteristic color to bacterial colonies. The cytoplasm of bacteria may also contain numerous granules of storage materials such as carbohydrates and lipids. Bacteria can contain plasmids that are pieces of genetic material existing outside the main genome. Plasmids can be used as vectors for introducing foreign genes into the bacteria that can impart new synthetic capabilities to an otherwise "wild" bacterial strain. Many bacteria exhibit motility by means of one or more hairlike appendages called flagella. Bacteria reproduce by dividing into equal parts, a process termed *binary fission*.

Under adverse conditions, certain microorganisms produce spores that germinate upon return to a favorable environment. Spores are a particularly stable form or state of bacteria that may survive dryness and temperature extremes. Some microorganisms form spores at a stage in their normal life cycle.

Many species may, under appropriate circumstances, become surrounded by gelatinous material that provides a means of attachment and some protection from other organisms. If many cells share the same gelatinous covering, it is called a slime; otherwise each is said to have a capsule.

Algae Algae are a very diverse group of photosynthetic organisms that range from microscopic size to giant kelp that may reach lengths of 20 m (66 ft). Some commercial biochemicals come from algal seaweeds, and algae supply oxygen and consume nutrients in some processes used for biological waste treatment. Although their rapid growth rates relative to other green plants offer great potential for producing biomass for energy or a chemical feedstock, there is little industrial use of algae. One proposed process uses Dunaliella, a species that grows in high salinity and accumulates glycerol internally to counter the high external osmotic pressure. Outdoor ponds are most suitable for growing algae because vast surfaces and high illumination are needed.

Fungi As a group, fungi are characterized by simple vegetative bodies from which reproductive structures are elaborated. All fungal cells possess distinct nuclei and produce spores in specialized fruiting bodies at some stage in their life cycles. The fungi contain no chlorophyll and therefore require sources of complex organic molecules for growth. Many species grow on dead organic material; others live as parasites.

Yeasts are one kind of fungi. They are unicellular organisms surrounded by a cell wall and possessing a distinct nucleus. With very few exceptions, yeasts reproduce by a process known as budding, where a small new cell is pinched off the parent cell. Under certain conditions, an individual yeast cell may become a fruiting body, producing spores.

Isolated Plant and Animal Cells Biotechnology includes recovery of biochemicals from intact animals and plants, but the care and feeding of them is beyond the scope of this section. Processes with their isolated cells have much in common with processes based on microorganisms. The cells tend to be much more fragile than microbial cells, and allowable ranges of pH and temperature are quite narrow. These cells occur in aggregates and usually require enzymes to free them. There is a strong tendency for the cells to attach to something, and cell cultures often exploit attachment to surfaces.

Plant and animal cells have numerous chromosomes. Growth rates are relatively slow. A typical nutrient medium will contain a large number of vitamins and growth factors in addition to complex nitrogen sources, because other specialized cells in the original structures supply these needs. A plant or animal cell is not like a microbial cell in its ability to function independently.

Viruses Viruses are particles of a size below the resolution of the light microscope and are composed mainly of nucleic acid, either DNA or RNA, surrounded by a protein sheath. Lacking metabolic machinery, viruses exist only as intracellular highly host-specific parasites. Many bacteria and certain molds are subject to invasion by virus particles. Those that attack bacteria are called bacteriophages. They may be either virulent or temperate (lysogenic). Virulent bacteriophages divert the cellular resources to the manufacture of phage particles; new phage particles are released to the medium as the host cell dies and lyses. Temperate bacteriophages have no immediate effect upon the host cell; they become attached to the bacterial chromosome. They may be carried through many generations before being triggered to virulence by some physical or chemical event.

Biochemistry All organisms require sources of carbon, oxygen, nitrogen, sulfur, phosphorus, water, and trace elements. Some have specific vitamin requirements as well. Green plants need only carbon dioxide, nitrate or ammonium ions, dissolved minerals, and water to manufacture all of their cellular components. Photosynthetic bacteria require specific sources of hydrogen ions, and the chemosynthetic bacteria must have an oxidizable substrate. Some microorganisms have the ability to "fix" atmospheric nitrogen by reduction. Organisms that use only simple inorganic compounds as nutrients are said to be *autotrophic* (self-nourishing).

Organisms that require compounds manufactured by other organisms are called *heterotrophs* (other-nourishing). Many heterotrophs secrete enzymes (exoenzymes) that hydrolyze large molecules such as starch and cellulose to smaller units that can readily enter the cell.

Proteins are macromolecules that play many roles such as serving as enzymes or components of cell membranes and muscle. The antibodies that protect against invasion by foreign substances are themselves proteins. There are twenty-odd amino acids found regularly in most naturally occurring proteins. Because of the great length of protein chains and the various sequences of amino acids, the theoretical number of possible proteins is astronomical. The amino acid sequence is referred to as the primary structure of a protein. The polypeptide chain is usually coiled or folded to provide secondary structure to the molecule, and linkages through other functional groups (mainly disulfide bonds) form the tertiary structure. For some protein molecules, there may be spatial arrangement forming defined aggregates, known as the quarternary structure of proteins. For a polypeptide polymer to have biological activity a certain molecular arrangement is necessary. This requires not only the primary and secondary but also tertiary and sometimes quarternary structure. Such a strict structural requirement explains the high specificity of proteins. In the presence of certain chemical reagents, excessive heat, radiation, unfavorable pH, and so on, the protein structure may become disorganized. This is called *denaturation* and may be reversible if not too severe.

A special class of proteins, the enzymes, are biological catalysts that expedite reactions by lowering the amount of activation energy required for the reactions to go. An enzyme has an active site that may be thought of as an atomic vise that orients a portion of a molecule for its reaction. The rest of the enzyme is not just an inert glob. Regions that are recognized by antibodies enable living systems to identify and inactivate foreign proteins. Immunological reactions involving antibodies are a defense against such foreign proteins. Enzymes function in conjunction with another special class of compounds known as coenzymes. Coenzymes are not proteins; many of the known coenzymes include vitamins, such as niacin and riboflavin, as part of their molecular structure. Coenzymes carry reactant groups or electrons between substrate molecules in the course of a reaction. As coenzymes serve merely as carriers and are constantly recycled, only small amounts are needed to produce large amounts of biochemical product.

Hundreds of metabolic reactions take place simultaneously in cells. There are branched and parallel pathways, and a single biochemical may participate in several distinct reactions. Through mass action, concentration changes caused by one reaction may effect the kinetics and equilibrium concentrations of another. In order to prevent accumulation of too much of a biochemical, the product or an intermediate in the pathway may slow the production of an enzyme or may inhibit the activation of enzymes regulating the pathway. This is termed feedback control and is shown in Fig. 24-1. More complicated examples are known where two biochemicals act in concert to inhibit an enzyme. As accumulation of excessive amounts of a certain biochemical may be the key to economic success, creating mutant cultures with defective metabolic controls has great value to the production of a given product.

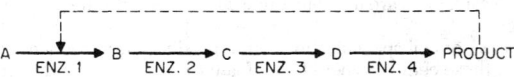

FIG. 24-1   Feedback control. Product inhibits the first enzyme.

Cell efficiency is improved by inhibiting or regulating the synthesis of unneeded enzymes, so there are two classes of enzymes—those that are constitutive and always produced and those that are inducible, i.e., synthesized when needed in response to an inducer, usually the initial substrate in a pathway. Enzymes that are induced in one organism may be constitutive in another.

Microorganisms exhibit nutritional preferences. The enzymes for common substrates such as glucose are usually constitutive, as are the enzymes for common or essential metabolic pathways. Furthermore, the synthesis of enzymes for attack on less common substrates such as lactose is repressed by the presence of appreciable amounts of common substrates or metabolites. This is logical for cells to conserve their resources for enzyme synthesis as long as their usual substrates are readily available. If presented with mixed substrates, those that are in the main metabolic pathways are consumed first, while the other substrates are consumed later after the common substrates are depleted. This results in diauxic behavior. A diauxic growth curve exhibits an intermediate growth plateau while the enzymes needed for the uncommon substrates are synthesized (see Fig. 24-2). There may also be preferences for the less common substrates such that a mixture shows a sequence of each being exhausted before the start of metabolism of the next.

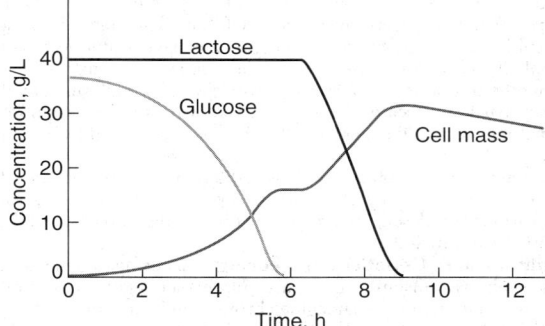

FIG. 24-2   Computer simulation of typical diauxic behavior.

Energy   Many metabolic reactions, once activated, proceed spontaneously with a net release of energy. Hydrolysis and molecular rearrangements are examples of spontaneous reactions. The hydrolytic splitting of starch to glucose, for instance, results in a net release of energy. But a great many biochemical reactions are not spontaneous and therefore require an energy input. In living systems this requirement is met by coupling an energy-requiring reaction with an energy-releasing reaction. If a sufficient amount of energy is produced by a metabolic reaction, it may be used to synthesize a high-energy compound such as adenosine triphosphate (ATP). When the terminal phosphate linkage is broken, adenosine diphosphate (ADP) and inorganic phosphate are formed, and energy is provided. When sufficient energy becomes available, ATP is reformed from ADP.

In biological systems, the most frequent mechanism of oxidation is the removal of hydrogen, and conversely, the addition of hydrogen is the common method of reduction. Nicotinamide-adenine dinucleotide (NAD) and nicotinamide-adenine dinucleotide phosphate (NADP) are two coenzymes that assist in oxidation and reduction. These cofactors can shuttle between biochemical reactions so that one drives another, or their oxidation can be coupled to the formation of ATP. However, stepwise release or consumption of energy requires driving forces and losses at each step such that overall efficiency suffers.

Overall redox potential of a system determines the amount of energy that cells can derive from their nutrients. When oxygen is present to be the ultimate acceptor of electrons, complete oxidation of organic molecules yields maximum energy and usually results in the production of H_2O and CO_2. However, inside animals, in polluted waters, in the benthos (bottom region) of natural waters, and elsewhere, there is little or no free oxygen. In these environments, organisms develop that can partially oxidize substrates or can derive a small amount of energy from reactions where some products are oxidized while others are reduced. The pathways for complete oxidation may be absent and the presence of oxygen can disrupt the mechanisms for anaerobic metabolism so that the cell is quickly killed. The differences in efficiency are striking: Aerobic metabolism of one molecule of glucose can generate bond energy as much as 33 molecules of ATP, while anaerobic metabolism can yield as little as two molecules of ATP. Natural anaerobic processes accumulate compounds such as ethanol, acetoin, acetone, butanol, lactate, and malate. Products of natural aerobic metabolism are water and carbon dioxide, cell mass, and secondary metabolic products such as antibiotics.

Photosynthesis   All living cells synthesize ATP, but only green plants and a few photosynthetic (or phototrophic) microorganisms can drive biochemical reactions to form ATP with radiant energy through the process of photosynthesis. All photosynthetic organisms contain one or more of the group of green pigments called chlorophylls. In plants, these are contained in organelles called chloroplasts. The number per cell of membrane-surrounded chloroplasts varies with species and environmental conditions. In higher plants, numerous chloroplasts are found in each cell of the mesophyll tissue of leaves, while an algal cell may contain a single chloroplast. A chloroplast has a sandwich of many layers alternating between pigments and enzymatic proteins such that electromagnetic excitation from light becomes chemical bond energy. Prokaryotic organisms have a unique type of chlorophyll and do not possess chloroplasts organelles. Instead, their photosynthetic systems are associated with the cell membrane or with lamellar structures located in organelles known as chromatophores. Chromatophores, unlike chloroplasts, are not surrounded by a membrane.

The net result of photosynthesis is reduction of carbon dioxide to form carbohydrates. A key intermediate is phosphoglyceric acid, from which various simple sugars are produced and disproportionated to form other carbohydrates.

Mutation and Genetic Engineering   Exposing organisms to agents such as mustard chemicals, ultraviolet light, and x-rays increases mutation rate by damaging chromosomes. In strain development through mutagenesis, the idea is to limit the mutagen exposure to kill about 99 percent of the organisms. The few survivors of this intense treatment are usually mutants. Most of the mutations are harmful to the cell, but a very small number may have economic importance in that impaired cellular control may result in better yields of product. The key is to have a procedure for selecting out the useful mutants. Screening of many strains to find the very few worthy of further study is tedious and expensive. Such screening that was so very important to biotechnology a few decades ago is becoming obsolete because of genetic improvements based on recombinant DNA technology.

Whereas mutagenic agents delete or scramble genes, recombinant DNA techniques add desirable genetic material from very different cells. The genes may come from plant, animal, or microbial cells, or in a few instances they may be synthesized in the laboratory from known nucleic acid sequences in natural genes. Opening a chromosome and splicing in foreign DNA is simple in concept, but there are complications. Genes in fragments of DNA must have control signals from other nucleic acid sequences in order to function. Both the gene and its controls must be spliced into the chromosomes of the receiving culture. Bacterial chromosomes (circular DNA molecules) are cut open with enzymes, mixed with the new fragments to be incorporated, and closed enzymatically. The organism will acquire new traits. This technique is referred to as recombinant technology.

There are many tricks and some art in genetic engineering. Examples would be using bacteriophage infection to introduce a gene for producing a new enzyme in a cell. Certain strains of *E. coli, B. subtilis*, yeast, and streptomyces are the usual working organisms (cloning vectors) to which genes are added. The reason for this is that the genetics of these organisms is well understood and the methodology has become fairly routine.

ADDITIONAL REFERENCES: Murooka, Y. and T. Imanka (ed.), *Recombinant Microbes for Industrial and Agricultural Applications*, Dekker, NY, 1993. Glick, B. R. and J. J. Pasternak, *Molecular Biotechnology: Principles and Applications of Recombinant DNA*, ASM Press, Herndon, VA, 1994. Bajpai, Rakesh K., and Ales Prokop, eds. *Recombinant DNA Technology II, Annals of the New York Academy of Sciences*, vol. 721, 1993.

CELL AND TISSUE CULTURES

Mammalian Cells   Unlike microbial cells, mammalian cells do not continue to reproduce forever. Cancerous cells have lost this natural timing that leads to death after a few dozen generations and continue to multiply indefinitely. Hybridoma cells from the fusion of two mammalian lymphoid cells, one cancerous and the other normal, are important for mammalian cell culture. They produce monoclonal antibodies for research, for affinity methods for biological separations, and for analyses used in the diagnosis and treatment of some diseases. However, the frequency of fusion is low. If the unfused cells are not killed, the myelomas will overgrow the hybrid cells. The myelomas can be isolated when there is a defect in their production of enzymes involved in nucleotide synthesis. Mammalian cells can produce the necessary enzymes and thus so can the fused cells. When the cells are placed in a medium in which the enzymes are necessary for survival, the myelomas will not survive. The unfused normal cells will die because of their limited life span. Thus, after a period of time, the hybridomas will be the only cells left alive.

A hybridoma can live indefinitely in a growth medium that includes salts, glucose, glutamine, certain amino acids, and bovine serum that provides essential components that have not been identified. Serum is expensive, and its cost largely determines the economic feasibility of a particular culture system. Only recently have substitutes or partial replacements for serum been found. Antibiotics are often included to prevent infection of the culture. The pH, temperature and dissolved oxygen, and carbon dioxide concentration must be closely controlled. The salt determines the osmotic pressure to preserve the integrity of the fragile cell.

Most glucose is metabolized to lactate because glycolysis is usually much faster than uptake rate of glycolytic intermediates. Glutamine acts as the primary source of nitrogen as well as providing additional carbon and energy. After glutamine is partially oxidized to glutamate, it can enter the TCA cycle and emerge as pyruvate. It has been estimated that between 30 and 65 percent of the cell energy requirement is derived from glutamine metabolism when both glucose and glutamine are available. Ammonia is produced in the deamination of glutamine to form glutamate and in the formation of alpha-ketoglutarate.

Plant Cells and Tissues   It is estimated that today some 75 percent of all pharmaceuticals originate in plants. Typically, these compounds are derived from the secondary metabolic pathways of the cells. When plant or animal cells are cultured, concepts from microbiology come into play. Only specialized cells are used, and these can be improved with mutation, selection, and recombinant DNA techniques. One very major difference between cell and tissue cultures and most microbiological processes is very high susceptibility to contamination by foreign organisms. Most microorganisms grow rapidly and compete well; some are aided by their own changes to the environment. When a microbial process changes the pH to be far from neutrality or when the product such as ethanol is inhibitory to other organisms, growth of contaminants is discouraged. Cell and tissue cultures require rich media and are characterized by slow growth rates. There is seldom any protection by the products of the process. Optimum conditions for production of the secondary metabolites are not likely to be the same as for growth. Economics may hinge on a good balance of growing sufficient cells and favoring product formation.

Only a few biochemicals derived from plant cell and tissue cultures have high volume/low value products, but some have sizeable markets as specialty chemicals such as dyes, fragrances, insecticides, and pesticides. These differ from the low volume/very high value compounds that typify life-saving drugs and pharmaceuticals. Examples for both of these categories are listed in Table 24-1 along with the plant species of origin.

Because of cell specialization, some products are produced in cultures of those cellular types. Three main classifications of the types of plant cell and tissue cultures are:

Undifferentiated cell cultures.   Aggregate clumps of cells on solid media (callus) or in liquid media (suspension)

Protoplast cultures.   Cellular tissues devoid of cell wall material in culture

Organ cultures.   Differentiated tissues of shoots, roots, anthers, ovaries, or other plant organs in culture

Primary Growth Requirements   Primary growth is defined as the processes in a plant that are essential for the growth of the meristematic regions such as the shoot apex, root tip, and axillary meristems. Plant cell and tissue cultures have specific optima for their primary growth in terms of lighting, temperature, aeration, a nutrient medium that must supply a carbon source, vitamins, hormones, and inorganic constituents and with pH typically between 5.5 and 6.5. Aeration can be critical depending upon the species.

TABLE 24-1   Typical Products Derived from Plants

Compound	Application	Volume/ value	Source
Shikonin	Dye	High/low	*Lithospermum erthrorhizon*
Warfarin	Pesticide	High/low	Sweet clover
Gossypol	Pesticide/anti-fertility	Low/high	Cotton
Scopolamine	Antispasmodic	Low/high	*Daturastramonium*
Ajmalicine	Circulatory agent	Low/high	*Catharanthusroseus*
Taxol	Anticancer	Low/high	California yew tree

Although a few exceptions do exist where glucose, fructose, or galactose is preferred, the majority of the plant cultures use sucrose. Usual trace requirements are thiamine, niacin, riboflavin, pyridoxine, choline, ascorbic acid, and inositol. Hormones such as auxins and cytokinins promote an undifferentiated state or trigger differentiation into specific plant tissues. As with the whole plant, cellular groups, either differentiated or undifferentiated, require a set of inorganic elements such as nitrogen, phosphorus, potassium, magnesium, calcium, sulfur, iron, chlorine, boron, manganese, and zinc. The exact compositions and concentrations of these inorganic elements that are optimum for a particular plant species can be highly variable. However, prepackaged formulations of these salts that can even include the carbon source, vitamins, hormones, and pH buffers are commercially available.

Secondary Metabolic Requirements A difference between the growth and secondary metabolic phase is that the latter gains importance when approaching the reproductive stages. For example, many of the pigments of flowers are secondary metabolites (e.g., shikonin). Secondary mechanisms are typical responses to stress, such as change in pH (e.g., alkaloid production in *Hyocyamus muticus* cell cultures is optimum at pH 3.5, while growth is best at 5.0). Similarly, carbon-source concentrations affect *Morinda citrifolia* cell cultures that grow best at 5 percent sucrose but produce the anthraquinone secondary metabolites optimally at 7 percent. Temperature changes can cause flowering; several plants require a cold treatment to induce flowering. This is called *vernalization*. Secondary metabolic pathways in plant cell and tissue cultures seem to be highly controlled by the hormone level in the medium. Another method of eliciting secondary metabolites employs the natural defense mechanisms of the plants that have developed through evolution. For example, gossypol produced by *Gossypium hirsutum* (cotton) cells is a natural response of the plant when subjected to the infections of the wilt-producing fungus *Verticillium dahliae*.

ADDITIONAL REFERENCES: Lambert, K. J. and J. R. Birch, "Cell Growth Media" in *Animal Cell Biology*, vol. 1, 1985, pp. 85–122. van Wezel, A. L., C. A. M. van der Velden-de Groot, H. H. de Haan, N. van der Heuvel, and R. Schasfoort, "Large-Scale Animal Cell Cultivation for Production of Cellular Biologicals," *Dev. Bio. Stand.*, **60**, 229–236 (1985). Altman, D. W., R. D. Stipanovic, D. M. Mitten, and P. F. Heinstein, *In Vitro Cell. Dev. Biol.*, **21**, 659 (1985). Toivonen, L., M. Ojala, and V. Kauppinen, *Biotechnol. Bioeng.*, **37**, 673 (1991). Calcott, P. H., *Continuous Cultures of Cells*, vols. 1 and 2, CRC Press, 1981. Maramorosch, K. and A. H. McIntosh, *Insect Cell Biotechnology*, CRC Press, Boca Raton, 1994. Endress, R., *Plant Cell Biotechnology*, Springer-Verlag, Berlin, New York 1994. Morgan, S. J. and D. C. Darling, *Animal Cell Culture: Introduction to Biotechniques*, BIOS Scientific Pub, 1993. Goosen, M., A. Daugulis, and P. Faukner (eds.), *Insect Cell Culture Engineering*, M. Dekker, New York, 1993.

RECENT EMPHASES

Commercial use of cell and tissue culture continues to expand. Improvement of organisms through recombinant nucleic acid techniques has become commonplace. Formerly, a few laboratories were well ahead of most others, but now the methods have been perfected for routine use. Another technique that is widely practiced is culturing of cells that excrete high concentrations of just one antibody protein. The specificity of antibodies and antigens is exploited in medical testing procedures using these pure monoclonal antibodies.

Environmental issues are driving several aspects of biotechnology. Sites contaminated by toxic wastes can be cleaned by several alternative methods, but all are expensive. The most certain way to remove toxic materials from soil is to excavate it for incineration, but this requires much labor, energy, and money. Bioremediation in situ tends to be much less expensive on one hand but is slow and uncertain on the other. Microbial growth rates approach zero as nutrient levels fall to the low concentrations required for approval of the toxic site remediation. This means that rates tend to be unacceptable when striving for complete removal. Many toxic materials do not support growth of microorganisms. However, they may be degraded as the microorganisms grow on other nutrients. This is termed *cometabolism*.

Materials that are easily biodegraded could substitute for plastics and other organic chemicals that damage the environment. There has been some progress with natural surfactants produced by microorganisms; these would be used in detergents if properties were acceptable and costs were competitive. Biodegradable polymers such as poly-beta-hydroxybutyrate or its derivatives should eventually substitute for polyethylene and polypropylene, but costs are still too high.

BIOLOGICAL REACTORS

FERMENTERS

The term *fermentation* formerly distinguished processes from which air was absent, but the term has now been extended to aerobic processes. Bioprocessing is usually aseptic (free of unwanted organisms) in vessels held under positive pressure of sterile air to resist entry of contaminating microorganisms. A few processes such as the production of pathogenic organisms for medical purposes or for biological warfare operate below atmospheric pressure because safety of the plant operators is more important than the integrity of the product. Older processes such as manufacture of pickles had no special measures against contamination, but many of these have been converted to aseptic operations to prevent impairment of product quality by foreign organisms. Biological waste treatment employs elective cultures of microorganisms in relatively crude, open equipment.

Activities associated with bioreactors include gas/liquid contacting, on-line sensing of concentrations, mixing, heat transfer, foam control, and feed of nutrients or reagents such as those for pH control. The workhorse of the fermentation industry is the conventional batch fermenter shown in Fig. 24-3. Not shown are ladder rungs inside the vessel, antifoam probe, antifoam system, and sensors (pH, dissolved oxygen, temperature, and the like). Note that coils may lie between baffles and the tank wall or connect to the top to minimize openings below the water level, and bottom-entering mixers are used frequently. There is extensive process piping, and copper or brass fittings are taboo for some processes because of highly deleterious effects of copper (more than 50 percent reduction in yield has been noted in penicillin fermentations when a bronze valve was in a feed line). Cooling coils must be used for larger tanks because the heat-transfer area of a jacket is inadequate for cooling from sterilization temperature to operating temperature in a reasonable time. Some features of interest for a conventional fermenter are that (1) a bypass valve in the air system allows diversion of air so that foaming is not excessive and the redox potential is not too high during the early stage of fermentation when the inoculum is becoming established; (2) antifoam is added when excessive foam reaches a conductive or capacitive electronic probe; (3) all piping is sterilized by the use of steam and is protected by steam until put into use; (4) the level of liquid when filling the vessel is determined by reference to a calibration chart based on points in the tank such as a rung on the ladder; (5) the weight of the tank contents can be determined by the hydrostatic balance against air bubbled slowly through the sparger; and (6) pumps are very uncommon because it is so easy to force fluid from a pressurized vessel.

The need for highly cost-efficient oxygen transfer in fermentations such as those with hydrocarbon feedstocks has led to air-lift fermenters as shown in Fig. 24-4. The world's largest industrial fermenter was

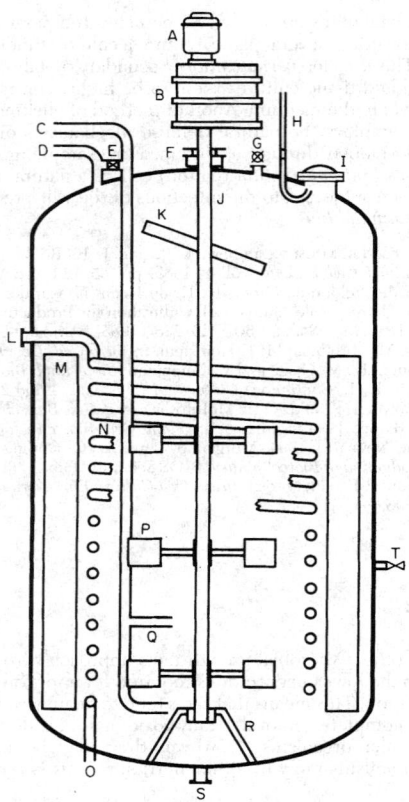

FIG. 24-3 Conventional batch fermenter. A = agitator motor; B = speed-reduction unit; C = air inlet; D = air outlet; E = air bypass valve; F = shaft seal; G = sight glass with light; H = sight-glass clean-off line; I = manhole with sight glass; J = agitator shaft; K = paddle to break foam; L = cooling-water outlet; M = baffle; N = cooling coils; O = cooling-water inlet; P = mixer; Q = sparger; R = shaft bearing and bracket; S = outlet (steam seal not shown); T = sample valve (steam seal not shown).

designed for producing single-cell protein from hydrocarbons at Billingham, England, U.K. Its dimensions are 100 m (328 ft) in height and 10 m (33 ft) in diameter.

A few variations on the standard fermenter have been attempted, but none has become popular. An obsolete design in which the fermenter was rotated to aerate the medium is shown in Fig. 24-5. Performance was unsatisfactory, and the units were turned on end, with spargers and agitation added. One of the largest fermenters used for antibiotics is a horizontal cylinder with several agitators, as in Fig. 24-6. Multiple agitator motors and shafts have also been used with vertical cylindrical vessels.

Another innovative design is the toroidal fermenter shown in Fig. 24-7. Motion in an axial direction allows intimate mixing with air. A special case of the air-lift fermenter is shown in Fig. 24-8. The feed enters at moderate pressure and is drawn downward to regions of very high hydrostatic pressure that provides a great driving force for gas transfer. In the other leg, lowering pressure allows gases to expand to induce circulation. Experimental units have been built in elevator shafts.

Ethanol fermentation is a particularly good example of product accumulation inhibiting the microbial culture. Most strains of yeast have a much slower alcohol production rate when ethanol reaches about ten percent, and the wine or saki strains that achieve over 20 percent by volume of ethanol are very, very slow. A system known as the Vacuferm for removal of alcohol by distillation as it is formed is

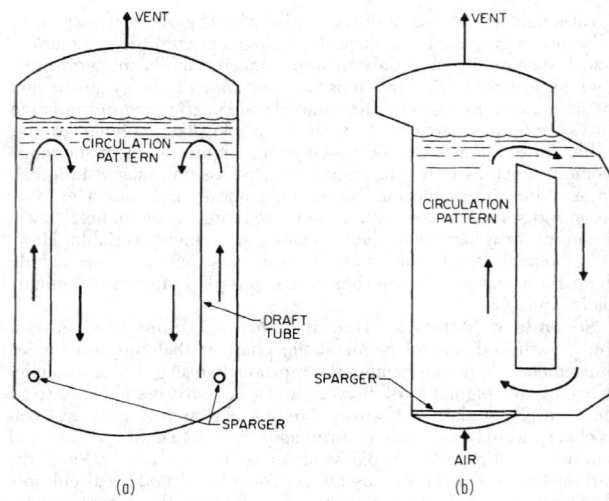

FIG. 24-4 Air-lift fermenters: (*a*) Concentric cylinder; (*b*) external recycle.

shown in Fig. 24-9. The vacuum is adjusted to the vapor pressure of the alcohol-water solution at the fermentation temperature, 30 to 40°C (86 to 104°F). Volumetric productivity is far better than that of a conventional fermenter, but there is a killing disadvantage of having to recompress large volumes of vapor so that alcohol can be condensed with normal cooling water instead of expensive cold brine. Furthermore, the large amounts of carbon dioxide generated by fermentation are evacuated and recompressed along with the alcohol and water vapors. A far better design is shown in Fig. 24-10, where the fermenter operates at normal pressure so that carbon dioxide escapes, and broth is circulated through the flash pot for vaporization of the ethanol. Although this system may seem to have attractive energy economy because metabolic heat is removed as the vapor flashes,

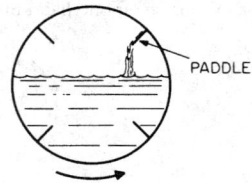

FIG. 24-5 Rotating fermenter.

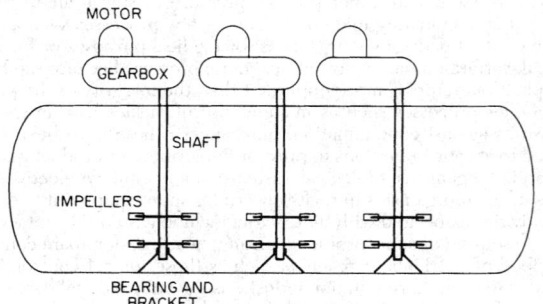

FIG. 24-6 Horizontal fermenter.

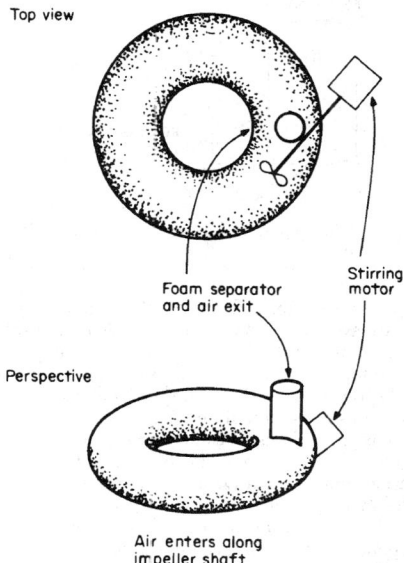

FIG. 24-7   Toroidal fermenter.

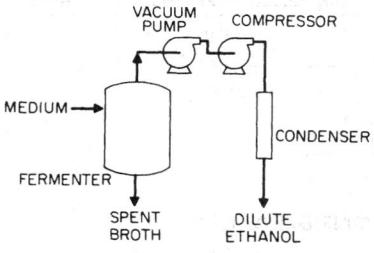

FIG. 24-9   Vacuferm.

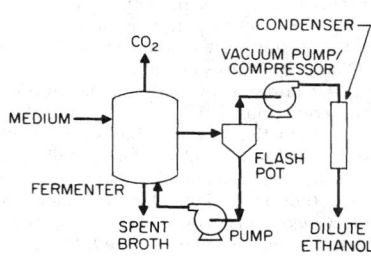

FIG. 24-10   Flash-pot fermenter.

initial investment and the cost of pumping the vapors are high. Operating the ethanol fermentation at higher temperatures with thermophillic organisms has better economics in terms of milder vacuum and less recompression because of the higher vapor pressure, but evolving carbon dioxide can strip out product. Other alternatives for overcoming inhibition by the product are extraction from the fermentation broth with an immiscible solvent and/or operating with very dense cultures so that low productivity per cell is compensated by having many more cells.

Elevated cell concentrations can be achieved by separating cells from the effluent and recycling them to the fermenter. This has been

standard practice for many years in biological waste treatment where dilute feed streams result in slow growth of the culture. Producing cell flocs that are collected easily by sedimentation is aided by recycling those cells that do settle. This is a selective advantage that may allow them to dominate. Industrial fermentations can afford more expense than can waste treatment, and centrifuges for collecting cells are not uncommon. Recycle of yeast cells can lower the fermentation time for ethanol significantly. Heavily coagulated cells can be retained in the fermenter; the tower fermenter shown in Fig. 24-11 uses this principle with yeast strains that flocculate naturally. Cells can also be retained in the reactor by attachment to a support. Vinegar is sometimes produced in a generator filled with wood shavings to which bacteria attach. Rocks or plastic support materials are used in trickling filters for waste treatment (see Sec. 25). Chemical agents can link cells to the support materials when simple adsorption does not hold them tightly enough. Gel entrapment can provide extremely high cell concentrations because the cells continue to multiply within the gel. Comparisons of results with fermentation of ethanol are in Table 24-2.

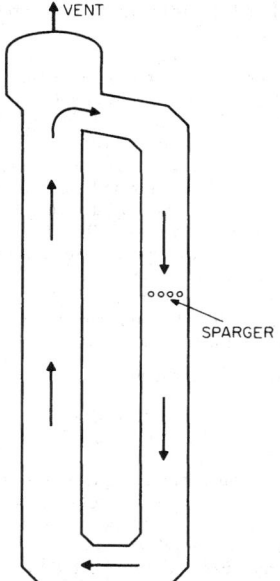

FIG. 24-8   Deep-shaft fermenter.

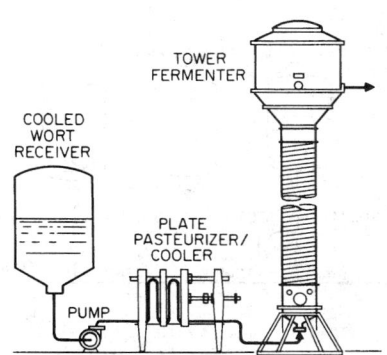

FIG. 24-11   Tower fermenter. (*Compliments of APV Corp.*)

TABLE 24-2 Comparison of Ethanol Fermenters

System	Typical time (h)	Typical ethanol concentration (%)
Conventional	72	10
Cell recycle	12	8
Tower fermenter	3	8
Gel immobilization	1	10

PROCESS CONSIDERATIONS

Fermentation can be combined with other operations. For example, feedback inhibition of enzymatic hydrolysis of cellulose can be relieved by removal of the product glucose by fermentation as it forms. This is termed *simultaneous-saccharification-fermentation* (SSF).

Valves and pumps that have a potential path for contaminating organisms are taboo for aseptic operations. Rising stem valves could bring organisms to the sterile side by the in and out motion as the valve operates. Diaphragm valves are still commonly used, but heating, cooling, and the abrasion by solids in the nutrient media are somewhat severe conditions leading to occasional rupture of a diaphragm and contamination of a run. Ball valves or plug valves do not have an absolute seal to the outside, but the direction of motion does not tend to bring organisms in. Contamination is seldom attributed to these valves; they are designed for easy maintenance in place, and there is the very nice human advantage that a glance at the handle tells easily whether the valve is open or closed. Many runs have been spoiled or impaired because a manual valve was left in the wrong position. For plant operations, pumps with diaphragms are satisfactory. In the lab or pilot plant, peristaltic pumps (also known as tubing squeezers) predominate.

Transfer of fluid in a fermentation plant usually makes use of air pressure differences. One or more manifold headers may interconnect many vessels. As transfers may have to be aseptic, headers are pressurized with steam until needed. A typical arrangement of steam seals is shown in Fig. 24-12.

Sample lines commonly have steam seals too. A typical layout is shown in Fig. 24-13. In the closed position, steam provides an absolute barrier to contamination. To take a sample, the steam line and the trap line are closed, and fermentation medium is flowed to waste until the pipes are cool to the touch so that sensitive products do not give false assays because of thermal destruction. Cooling takes up to 5 liters of medium if not done carefully, and bad practices can waste considerably more. Pilot-sized tanks have less massive fittings that are easier to cool; less medium is wasted, but oversampling to the point where the fermenter volume is low can be a problem. For this reason, alternate sampling methods have been devised. For example, a sterile syringe and needle may be used to sample through a rubber diaphragm in the wall of the tank. Although such methods appear reliable, there is a tendency to scrap all innovations and to return to tried and true steam seals when the factory encounters any period of contamination.

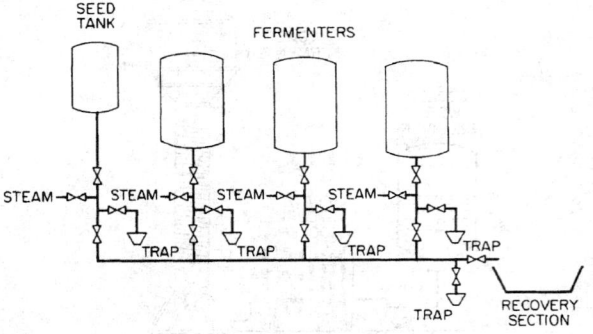

FIG. 24-12 Inoculation and harvest header.

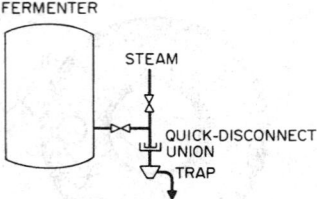

FIG. 24-13 Sample-line piping. (A valve to the sewer allows bypass of the trap while cooling the line.)

All piping to a fermenter is flushed with steam during the sterilization period. A clever means for sight-glass cleaning uses steam condensate that is naturally sterile in a dead leg. Steam pressure behind the condensate forces this water to the sight glass. Without cleaning, splashing and spray can quickly cover the sight glass with a thick coating of microorganisms and medium.

While it is easy to add materials to a fermentation, removal is difficult. Membrane devices have been placed in the fermenter or in external recycle loops to dialyze away a soluble component. Cells release wastes or metabolites that can be inhibitory; these are sometimes referred to as *staling factors*. Their removal by dialysis has allowed cell concentrations to reach ten to one hundred times that of control cultures.

Solid substrates such as pulverized wood cannot be stirred when slurry concentration exceeds about 5 percent. For saccharification prior to ethanol fermentation, keeping sugar concentration high can avoid an evaporation step. In a batch reactor, mixing limitations with the wood results in a dilute sugar solution. This has been circumvented by placing the wood in a column and percolating the solution through. As wood dissolves, more is added. Simultaneous fermentation of the sugars formed in the column is possible.

Oxygen Transfer Supplying sufficient oxygen can be a very challenging engineering problem for some aerobic fermentations. Oxygen is sparingly soluble in water; saturation with pressurized air at room temperature provides only 6 or 7 milligrams per liter of oxygen. A vigorous process can deplete the dissolved oxygen in several seconds when aeration is stopped. Mass transfer of gases to liquids is covered in Sec. 5. Emphasis is somewhat different for biological systems that commonly have bubble aeration. Because the number and size of bubbles is very difficult to estimate, transfer area is usually lumped with the mass-transfer coefficient as a $K_l a$ term. The "l" subscript in K_l signifies that liquid film resistance should greatly predominate for a sparingly soluble gas such as oxygen. The relationship between oxygen concentration and growth is of a Michaelis-Menten type (see Fig. 24-20). When a process is rate-limited by oxygen, the specific respiration rate (Q_{O2}) also increases steeply with dissolved oxygen concentration until a plateau is reached. The concentration below which respiration is severely limited is termed the critical oxygen concentration, which typically ranges from 0.5 to 2.0 ppm for well-dispersed bacteria, yeast, and fungi growing at 20 to 30°C (68 to 86°F). Above this critical concentration, the specific oxygen uptake increases only slightly with increasing oxygen concentrations.

A plot of the specific respiration rate Q_{O2} versus the specific growth rate coefficient μ is linear, with the intercept on the ordinate equal to the oxygen uptake rate for cell maintenance. A formulation of this is:

$$\text{Uptake rate} = \text{uptake for maintenance} + \text{uptake for growth}$$

or

$$Q_{O2}X = \frac{(Q_{O2})_M X + \mu X}{Y_g} \qquad (24\text{-}1)$$

where X is organism concentration, the subscript M denotes maintenance, Y_g is yield of cell mass per mass of oxygen, and the Q terms signify oxygen uptake rates in mass O_2 per mass of organisms.

This type of correlation applies to almost any substrate involved in cellular energy metabolism and is supported by experimental data and energetic considerations. However, it is based on assumptions true at or near the steady-state equilibrium conditions and may not be valid during transient states. The oxygen-uptake equation should be modified when other cellular activities requiring oxygen can be identified. For example, use of oxygen for product formation would be represented by:

Uptake rate = maintenance uptake + growth uptake + product uptake

or
$$Q_{O2}X = (Q_{O2})_M X + \frac{dX}{dt}\frac{1}{Y_G} + \frac{dP}{dt}\frac{1}{Y_P} \qquad (24\text{-}2)$$

where P is the product concentration and Y is the yield of product per unit weight of limiting nutrient. Oxygen uptake is distributed between that for growth and that for cellular activities dependent on cell concentration.

As the oxygen transfer rate under steady-state conditions must equal oxygen uptake, $K_l a$ may be calculated:

$$K_l a = \frac{\text{overall oxygen uptake rate}}{(C° - C)_{\text{mean}}} \qquad (24\text{-}3)$$

where $C°$ = concentration of oxygen in the liquid that would be in equilibrium with the gas-bubble concentration and $K_l a$ = the volumetric oxygen transfer rate.

A convenient method for measuring oxygen transfer rates in microbial systems depends on dissolved oxygen electrodes with relatively fast response times. Quite inexpensive oxygen electrodes are available for use with open systems, and steam-sterilizable electrodes are available for aseptic systems. There are two basic types: One develops a voltage from an electrochemical cell based on oxygen, and the other is a polarographic cell whose current depends on the rate at which oxygen arrives. See Fig. 24-14. Measurement of oxygen transfer properties requires only a brief interruption of oxygen supply. A mass balance for oxygen is:

$$\frac{dO}{dt} = \text{rate of supply} - \text{uptake rate} \qquad (24\text{-}4)$$

A tracing of the electrode signal during a cycle of turning aeration off and on is shown in Fig. 24-15. The rate of supply is zero (after bubbles have escaped) in the first portion of the response curve; thus, the slope equals the uptake rate by the organisms. When aeration is resumed, both the supply rate and uptake rate terms apply. The values for $C° - C$ can be calculated from the data, the slope of the response curve at a given point is measured to get dC/dt, and the equation can be solved for $K_l a$ because all the other values are known.

Measurements of the rate of change in concentration of oxidizable chemicals in aerated vessels have questionable value for assessing rates with biological systems. Not only are flow patterns and bubble sizes different for biological systems, but surface active agents and

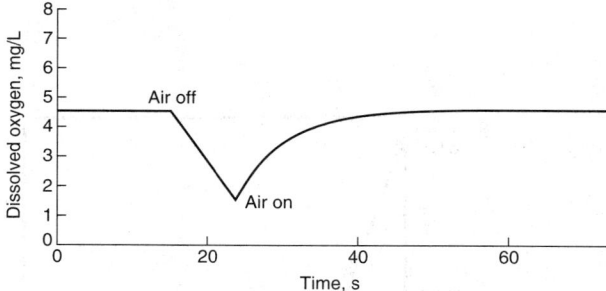

FIG. 24-15 Computer simulation of response for dynamic measurement of $K_l a$.

suspended particles can seriously impair gas transfer. Fig. 24-16 shows the effects of various particles. In general, spherical particles have a small effect, elongated particles have more effect, and entangled particles markedly impair transfer. As many mold cultures are intertwined and lipids and proteins are present with strong surface activity, oxygen transfer to a fermentation can be much slower than that to simple aqueous solutions.

Except as an index of respiration, carbon dioxide is seldom considered in fermentations but plays important roles. Its participation in carbonate equilibria affects pH; removal of carbon dioxide by photosynthesis can force the pH above 10 in dense, well-illuminated algal cultures. Several biochemical reactions involve carbon dioxide, so their kinetics and equilibrium concentrations are dependent on gas concentrations, and metabolic rates of associated reactions may also change. Attempts to increase oxygen transfer rates by elevating pressure to get more driving force sometimes encounter poor process performance that might be attributed to excessive dissolved carbon dioxide.

Sparger Systems Gas distributors in tanks are shown in Sec. 6. Large openings are desirable for spargers in industrial fermentations to avoid clogging by microbial growth, but the diameter is usually designed for the acoustic velocity that insures small bubbles. Relatively small holes or diffusers are used in activated sludge units for biological waste treatment, but there is commonly a means for swinging a section of the aerator out of the vessel for cleaning. Newer designs for fermenters were conceived as answers to the problems of oxygen transfer. The air lift fermenter (Fig. 24-4) creates intimate mixing of air and medium while using the buoyancy of the gas to mix the fluid.

Surface active substances also lead to foaming that can be so bad that most of the contents of the fermenter are lost. Mechanical antifoam devices are helpful but cannot function alone except when there is little propensity for foaming. The mechanical foam breakers rupture the large, weak bubbles while allowing tiny, rugged bubbles to accumulate. Surface active antifoam agents tend to reduce elasticity of the bubbles so that mechanical shocks are easily transmitted to encourage rupture. Several antifoam delivery systems are shown in Fig. 24-17. Some lipids used as antifoams are metabolized by the culture and must be replaced. The nutrition supplied by these oils may be beneficial, but they are much more expensive than their equivalents in carbohydrate nutrients. Furthermore, it is troublesome to have nutrition coupled to foam control. Several synthetic antifoam agents are not nutrients and tend to persist. Their tendency to be lost by coating solid surfaces in the fermenter means that more must be added occasionally. These synthetic antifoam agents are toxic to some organisms, but one of the many types is usually satisfactory.

Scale-Up Fermenters ranging from about two to over 100 liters (0.07–3.5 ft³) have been used for research and development, but the smaller sizes provide too little volume for sampling and are difficult to replicate, while large vessels are expensive and use too much medium. Autoclavable small fermenters that are placed in a water bath for temperature control are less expensive than vessels with jackets or coils, but much labor is required for handling them. Pressure vessels that

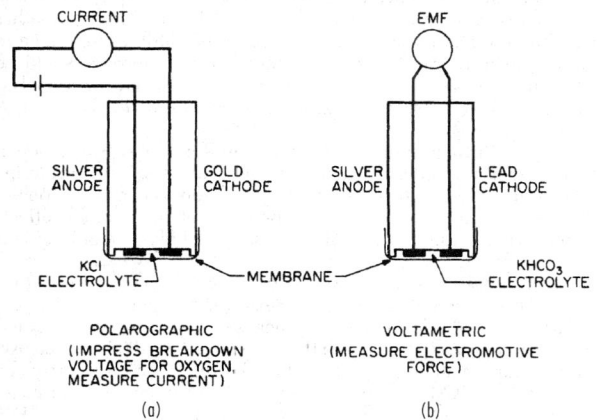

FIG. 24-14 Dissolved-oxygen electrodes: (a) polarographic (impress breakdown voltage for oxygen; measure current); (b) voltametric (measure electromotive force).

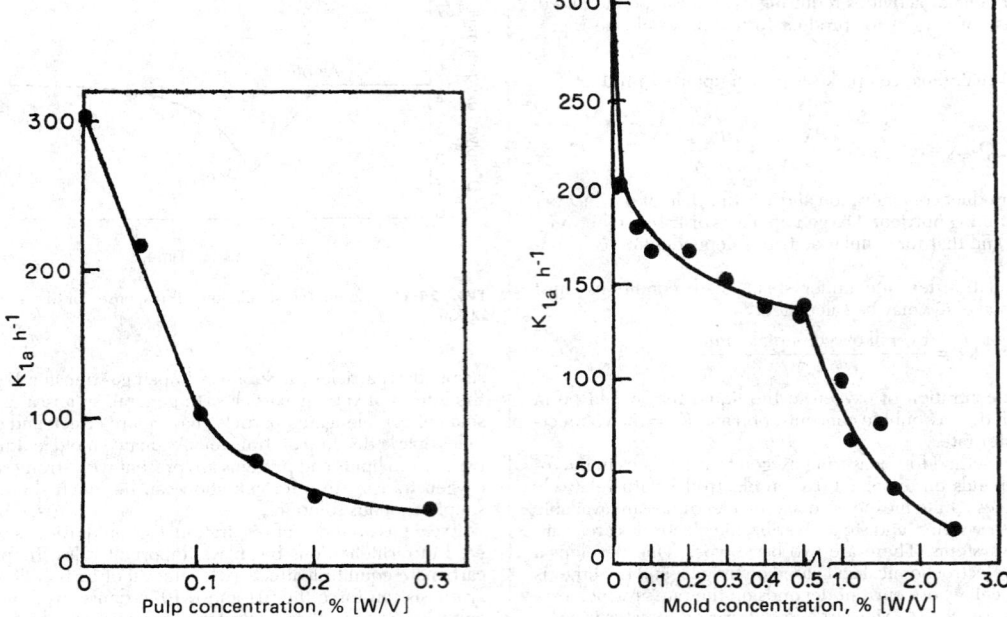

FIG. 24-16 Effect of solids on $K_l a$. Operating conditions: agitator speed, 800 r/min; air flow, 2.5 min. [*M. R. Brierly and R. Steel, Appl. Microbiol.*, **7**, 57 (1959). *Courtesy of American Society for Microbiology.*]

are sterilized in place are more convenient, but initial investment is high. Judgement is needed to select the most economical equipment and to plan for cost-effective experimentation.

A suitable means of scale-up for aerobic processes is to measure the dissolved oxygen level that is adequate in small equipment and to adjust conditions in the plant until this level of dissolved oxygen is reached. However, some antibiotic fermentations and the production of fodder yeast from hydrocarbon substrates have very severe requirements, and designers are hard-pressed to supply enough oxygen.

Older methods of fermentation scale up insisted on geometric similarity based on proportional physical dimensions. It was thought that applying the same power per unit volume as in the pilot equipment would give an equivalent process performance in large fermenters. Antibiotic fermentations aim for mixer power in the range of 0.2 to 4 Kw/m (0.1 to 2 HP/100 gal). As mixing devices have areas (dimensions squared) to supply a volume (dimensions cubed), methods based on dimensional similarity are fundamentally unsound. Scale-up based on equivalent oxygen transfer coefficient $K_l a$ has been reasonably successful.

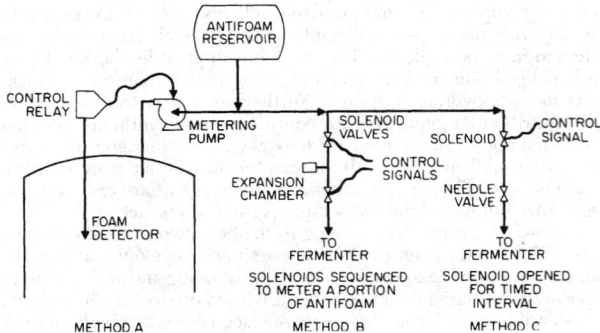

FIG. 24-17 Antifoam systems.

Impeller Reynolds number and equations for mixing power for particle suspensions are in Sec. 5. Dispersion of gasses into liquids is in Sec. 14. Usually, an increase in mechanical agitation is more effective than is an increase in aeration rate for improving mass transfer.

Other scale-up factors are shear, mixing time, Reynolds number, momentum, and the mixing provided by rising bubbles. Shear is maximum at the tip of the impeller and may be estimated from Eq. (24-5), where the subscripts s and l stand for *small* and *large* and Di is impeller diameter [R. Steel and W. D. Maxon, *Biotechnol. Bioengr.*, **4**, 231 (1962)].

$$S_s = S_l \left(\frac{Di_s}{Di_l} \right)^{1/3} \tag{24-5}$$

Some mycelial fermentations exhibit early sporulation, breakup of mycelium, and low yields if the shear is excessive. A tip speed of 250 to 500 cm/s (8 to 16 ft/s) is considered permissible. Mixing time has been proposed as a scale-up consideration, but little can be done to improve it in a large fermenter because gigantic motors would be required to get rapid mixing. Culturing cells from plants or animals is beset by mixing problems because these cell are easily damaged by shear.

Constant Reynolds number is not used for fermentation scale-up; it is only one factor in the aeration task. This is also true for considering the impeller as a pump and attempting scale-up by constant momentum. As mechanical mixing tends to predominate over bubble effects in improving aeration, scale-up equations including bubble effects have had little use.

Fermentation biomass productivities usually range from 2 to 5 g/(l·h). This represents an oxygen demand in the range of 1.5 to 4 g O/(l·h). In a 500-m fermenter, this means achievement of a volumetric oxygen transfer coefficient in the range of 250 to 400 h⁻¹. Such oxygen-transfer capabilities can be achieved with aeration rates of the order of 0.5 VVM (volume of air at STP/volume of broth) and mechanical agitation power inputs of 2.4 to 3.2 Kw/m (1.2 to 1.6 HP/100 gal).

Often heat removal causes design problems for scale-up. Mechanical agitation coupled with a metabolic heat from the growing biomass overwhelms cooling capacity of a large fermenter with only a jacket. External circulation through a heat exchanger or extensive coils inside the fermenter must be used. In highly viscous fermentations, internal cooling coils are usually not desirable because of interference with mixing patterns. Numerous schemes exist for heat removal in large fermenters such as half-coil baffles (plate coils) and draft tubes. Heat removal limits the size of packed-cell bioreactors. There is also a serious problem of gas evolution for bioreactors when the cells are immobilized with a membrane or by retention in a gel because the gas can rupture the structure.

Evaporation of medium provides a little cooling. The inlet air to particulate filters must not be near saturation because condensation of moisture on the filter medium mobilizes contaminating microorganisms so that their chances of penetration are greatly increased. Sometimes humidified air is used, and the filter unit is heated to prevent condensation. However, this is common only for small equipment where the extra operations are relatively easy to install and maintain.

Once a plant is built, the conditions of agitation, aeration, oxygen transfer, and heat transfer are more or less set, and sterilization cycles are defined. Those environmental conditions achievable in plant-scale equipment should be scaled down to the pilot plant and laboratory equipment (shaken flasks) to insure that results can be translated.

Sterilization Some old, traditional fermentations such as those for alcohol and pickles are conducted by organisms that are hardy and help their own cause by creating conditions that are unfavorable for competitors. Yeasts, for example, lower pH by producing acids from sugars, and tolerance to alcohol is another powerful advantage that allows their domination. Nevertheless, modern factories use aseptic techniques or extreme care to minimize contamination that can jeopardize product quality by affecting taste, texture, aroma, or appearance. Bioprocesses such as tissue culture to produce vaccines are very easily contaminated because there is an abundance of nutrients and no inherent protection against foreign organisms. Thus, bioprocesses range from relatively good self-protection to practically none; the value of the product and the need for quality control determine the extent to which precautions must be taken. Contamination in the practical sense is statistical in that foreign organisms can be present but may not propagate rapidly enough to damage the run. Good defense against contamination is relatively inexpensive, while absolute protection is impossible and attempts to achieve it can be inordinately costly. The production of agents for biological warfare takes extreme pains to keep the organisms away from the workers, yet people operating the fermentations are occasionally killed. The most expensive and best protected industrial fermentations are sometimes contaminated. Sterilization and aseptic techniques to keep bioprocesses uncontaminated can be crucial.

Common ways of sterilization are removal of microorganisms by filtration or killing them with heat or chemicals. Sterilization by filtration follows a standard unit operation that is covered in Sec. 18. The differences for biochemical engineering are: (1) the filter medium and the downstream lines are steamed at a pressure where the temperature kills all organisms and spores; (2) the sizes of the particles being removed are in the micrometer range; and (3) the filter medium should be reusable and not degraded by repeated heating.

Sterilization by Filtration Air is almost always sterilized by filtration. The alternative of heating to sterilize has been successful for small installations, but large equipment for heating an air stream to sterilizing temperatures has not been sufficiently reliable. While it seems simple enough to maintain a section of the air-supply pipeline at high temperature, automatic control is needed to adjust for varying heat transfer as flow rate changes, and an air cooling section is needed to prevent excessive heat load on the fermenter. Furthermore, energy costs are now much higher than when heating seemed a promising alternative to filtration. For exit gases from a fermentation that has hazardous organisms, heating is a reasonable precaution, and cost is not the key factor. A clever heating method failed in full-scale testing. The air was compressed to raise its temperature by the Joule-Thompson effect, but the method was abandoned because several batches became contaminated. Any flow system for heat sterilization is crippled by process upsets because a slug of material can have inadequate temperature or exposure time for killing.

Sterilization of liquids by filtration has performed very well since the advent of membrane filters of small pore size. When heat can damage the ingredients, filtration is an ideal choice. However, the extra handling and equipment mitigate against filtration because heat sterilization of a batch is easy and relatively inexpensive. A tank must be steam-sterilized anyway, so it is convenient to fill it first and sterilize the contents as well. When some constituents must not be subjected to heat, it is customary to sterilize the rest of the medium with heat and to filter concentrated solutions of the delicate ingredients. Very large vessels can have insufficient heat-transfer surface; the tank is sterilized empty, and the medium is sterilized by flow through a continuous sterilizer.

The magnitude of the air sterilization problem is seen from the usual needs of a highly aerobic fermentation where roughly 1 volume of air per volume of medium per minute may be used. For a factory with 20 fermenters of 100,000 l (3500 ft^3) each, 2 million l/m (70,000 ft^3/m) of air is handled. Very large compressors are used, and at least two are required so that one can be down for maintenance.

Fibrous or particulate filters are not important anymore because membrane filters are relatively compact and perform very well. For filtration by straining, there is an intermediate air velocity at which filtration efficiency is a minimum because different collection mechanisms predominate at different ranges of velocity. At low velocities, diffusional and electrostatic forces on the particle are important, and increased velocity shortens the time for them to operate. At high velocities, inertial forces that increase with air velocity come into play; below a certain air velocity, their effect on collection is zero. Surges or brief power failures could change velocity and collection efficiency.

Membrane filters for air sterilization are reliable and relatively small. Membranes are quite efficient for filtering air and tend to capture particles larger than the pore size. To ensure safety, a pore size of 0.2 to 0.3 micrometers is recommended. Hydrophilic membranes should not be used because moisture is held tightly in their pores and not dislodged unless quite high pressure drops are created across the membrane. Moisture tends to drain from hydrophobic membranes and collect in a sump. Sizing of a membrane unit for air filtration is based on the number of cartridges needed. Only 60 percent of the available pressure drop should be used in the calculation to allow for increased resistance as particles collect on the membrane. Figure 24-18 shows membrane cartridges in parallel in a housing.

Sterilization of Media First-order kinetics may be assumed for heat destruction of living matter, and this leads to a linear relationship when logarithm of the fraction surviving is plotted against time. However, nonlogarithmic kinetics of death are quite often found for bacterial spores. One model for such behavior assumes inactivation of spores via a sensitive intermediate state by the mechanism:

$$C_r \rightarrow C_s \rightarrow C_d \tag{24-6}$$

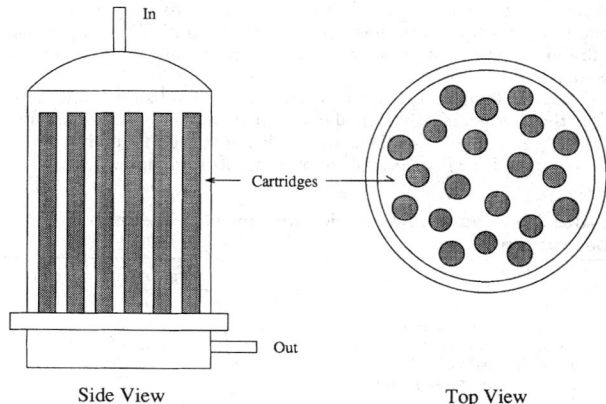

FIG. 24-18 Sketch of housing and membrane cartridges for air filtration. Typical cartridges are 76 cm long and 7.36 cm in diameter of polyvinylidene difluoride with 0.22-μm pores.

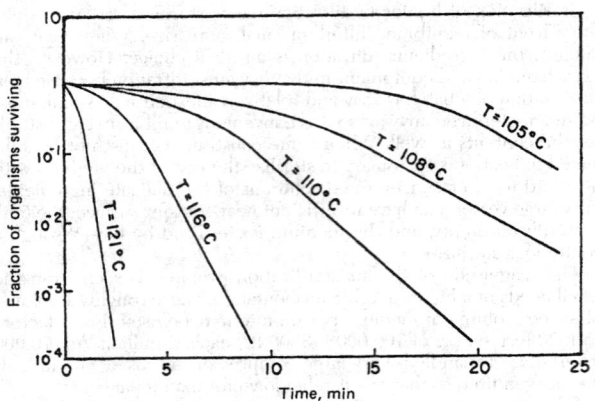

FIG. 24-19 Typical death-rate data for bacterial spores (*B. stearothermophilus*). To convert °C to °F, multiply by 1.8 and add 32 (*Wang et al., Fermentation and Enzyme Technology, Wiley-Interscience, New York, 1979, p. 140.*)

where C_r, C_s, and C_d are concentrations of resistant, sensitive, and dead spores, respectively. Typical plots are in Fig. 24-19.

Relative thermal resistance for the different types of microorganisms encountered in typical environments associated with fermentation broths is shown in Table 24-3. Bacterial spores are far more resistant to moist heat than are any other type of microbial contaminants; thus, a sterilization cycle based on the destruction of bacterial spores should destroy all life.

TABLE 24-3   Relative Resistance to Killing by Moist Heat

	Relative resistance
Vegetative bacteria or yeast	1.0
Bacterial spores	3,000,000
Mold spores	2 to 10
Virus and bacteriophage	1 to 5

As predicted by the Arrhenius equation (Sec. 4), a plot of microbial death rate versus the reciprocal of the temperature is usually linear with a slope that is a measure of the susceptibility of microorganisms to heat. Correlations other than the Arrhenius equation are used, particularly in the food processing industry. A common temperature relationship of the thermal resistance is *decimal reduction time* (DRT), defined as the time required to reduce the microbial population by one-tenth. Over short temperature intervals (e.g., 5.5°C) DRT is useful, but extrapolation over a wide temperature interval gives serious errors.

The activation energy (E) associated with microbial death is larger than the thermal inactivation of chemical compounds in fermentation broths (see Table 24-4). Thus by sterilizing at high temperatures for short times (HTST), overcooking of nutrients is minimized.

TABLE 24-4   Various Activation Energies for Thermal Destruction

	Activation energy, cal/mol
Folic acid	16,800
d-Panthothenyl alcohol	21,000
Cyanocobalamin	23,100
Thiamine hydrochloride	22,000
Bacillus stearothermophilus	67,700
Bacillus subtilis	76,000
Clostridium botulinum	82,000
Putrefactive anaerobe NCA 3679	72,400

To convert calories per mole to British Thermal Units, multiply by 1.8.

Batch Sterilization   Assuming that the presence of one single contaminating organism could cause ultimate failure of the desired fermentation, it is necessary to assign some probability of success. For example, if one contaminated fermentation per thousand can be tolerated, the design calculation for the sterilization cycle should use 0.001 organisms per fermentation. However, in batch sterilization, the heating, holding, and cooling portions of the cycle all contribute toward the reduction of the microbial contaminants. Furthermore, the specific death-rate constant varies because the temperature of the medium changes. Therefore, the design criterion total is composed of

$$\nabla \text{ total} = \nabla \text{ heating} + \nabla \text{ holding} + \nabla \text{ cooling}$$

$$\nabla \text{ heating} = \ln \frac{N_0}{N_1} = A \int_0^{t_1} \exp\left(-\Delta E/RT\right) dt \qquad (24\text{-}7)$$

$$\nabla \text{ holding} = \ln \frac{N_1}{N_2} = A \int_0^{t_2} \exp\left(-\Delta E/RT\right) dt \qquad (24\text{-}8)$$

$$\nabla \text{ cooling} = \ln \frac{N_2}{N} = A \int_0^{t_3} \exp\left(-\Delta E/RT\right) dt \qquad (24\text{-}9)$$

Other parameters affecting the temperature profile include the viscosity of the medium and amount of suspended solids or insoluble materials such as vegetable oils that can foul heat-transfer surfaces. Releasing part of the pressure after the holding period gives flash cooling. The temperature-time profile during the cooling portion of the sterilization cycle includes effects of cooling coils or the fermenter jacket and evaporative cooling if sterile air is injected to remove heat. Most of the sterilization is derived from the holding portion of the cycle; the cooling cycle contributes little to the overall process. An extremely long heating cycle should be avoided because its contribution towards microbial destruction is far outweighed by its detrimental biochemical effects.

Continuous Sterilization   Continuous sterilization permits short detention times at high temperature to avoid overcooking and has potential for improvement in yield. Tubular or plate-and-frame heat exchangers in a system for continuous sterilization provide economical heat exchange between process streams. However, direct injection of steam offers almost instantaneous heating to the sterilization temperature but wastes energy because heating and cooling are not integrated. Steam injection has no heating surface to foul.

In the holding section of a continuous sterilizer, correct exposure time and temperature must be maintained. Because of the distribution of residence times, the actual reduction of microbial contaminants in the holding section is significantly lower than that predicted from plug flow assumption. The difference between actual and predicted reduction in viable microorganisms can be several orders of magnitude; therefore, a design based on ideal flow conditions may fail.

Cell Culture   Single plant and animal cells are much larger than microbial cells and are easily damaged by shear due to intense agitation. Standard stirred tank bioreactor designs tend not to work well for undifferentiated plant and animal cells. Attempts to substitute paddle and spiral agitators has proven successful in some cases, but shear can still cause a morphological change in the structure of the cells. Usually, single cells and cellular aggregates tend to be smaller and produce significantly less secondary metabolites with improper agitation. Airlift reactors may work well, and cultured cells tend not to have severe oxygen requirements. There has been a thrust for new systems that can maintain absolute sterility without depending on antibiotics that can mask the slow development of a low-grade infection in a culture and can affect the metabolism of cultured cells in subtle ways that cannot always be predicted.

Cells that must attach to a surface can be grown on spongy polymers, arrays of thin tubing or hollow fibers, stacks of thin plates, or microscopically small beads called microcarriers. Damage from shear can be a serious problem; thus, aeration systems cannot employ the vigorous mixing that aids mass transfer in microbial processes. Airlift bioreactors have been developed that are tapered instead of being cylindrical with a central draft tube. Above a zone of intense aeration are the suspended cells circulating with little more mixing than would result from Brownian motion.

Improvements in fermentation include microcarriers that not only provide support for anchorage-dependent cells but also aid in harvesting at the end of a run. While microcarriers based on dextran, polystyrene and polyacrylamide beads have been widely used in the past, new materials for microcarriers make separating cells from their growth medium easier. Some are made of collagen that is detached from the cells by immersion in dilute collagenase. However, since some collagen can be left in solution, downstream processing can be made difficult. An alternative is to use plastic beads coated with collagen. The plastic can be easily separated after the cells are released enzymatically. Another alternative is glass-coated particles that induce attachment of the cells' long slender filopodia. At the end of a process, a brief incubation in dilute trypsin gently removes the cells from the beads.

Supplying sufficient oxygen can be difficult when dealing with differentiated plant tissues such as root cultures that can reach lengths of several decimeters and can be highly branched and complex struc-turally. In this case, mechanical agitation is impractical because the mass of roots can occupy approximately 50 percent of the reactor volume. Alternatives are bubble columns, rotating drums, and trickle bed reactors where the medium is recycled and sprayed over the column of roots.

Because of the differences in primary and secondary metabolism, a reactor may have a dual-stage fed-batch system. In other words, fed-batch operation optimizes growth with little or no product formation. When sufficient biomass has accumulated, a different fed-batch protocol comes into play.

ADDITIONAL REFERENCES: Asenjo, J. A., and J. C. Merchuck, *Bioreactor System Design*, Dekker, New York, 1994. Rehm, H.-J., and G. Reed, *Biotechnology*, vol. 6b, VCH Verlagsgesellschaft, 1988. Chang, H. N., "Membrane Bioreactors, Engineering Aspects," *Biotechnol. Adv.*, **5**, 129–145 (1987). Cheryan, M. and M. A. Mehaia, "Membrane Bioreactors" in McGregor, W. C. (ed.), *Membrane Separations: Biotechnology*, Marcel Dekker, New York, 1989. Heath, C. A. and G. Belfort, "Membranes and Bioreactors," *Int. J. Biochem.*, **22**(8), 823–835 (1990).

PRODUCT RECOVERY

Although most of the purification equipment in a large biotechnological factory is the same as that used throughout the chemical process industries, there are fewer separations in which the product reaches elevated temperatures. Most biochemicals are destroyed if heated. Recovery of products from the bioprocess fluid can be more difficult and expensive than all of the previous steps. The ratio of recovery costs to cost of creating the product can range from about one to more than ten because the investment for the recovery facilities may be several times that for the fermenter vessels and their auxiliary equipment. As much as 60 percent of the fixed costs of fermentation plants for organic acids or amino acids is attributable to the recovery section. The costs for recovery of proteins based on recombinant DNA techniques are particularly high.

Research and development for better recovery of existing products may provide diminishing returns as time passes, and some companies focus on new products. Government regulations are different for drugs that are sold in very nearly pure state and for biologicals that may be ill-defined. Little paperwork is required for a process improvement for a pure drug. For biologicals intended for humans, securing government approval makes it unwise to modify the recovery process except when the potential savings are very great. The extensive and expensive testing and validation of such therapeutic agents is keyed to the processes for making and purifying them. Only trivial changes are permitted; otherwise the testing must be repeated. Market forces make it important to have a new product tested and ready for sale as quickly as possible, but this usually means that the process has not been optimized. Whereas competition and process improvements resulted in remarkable lowering of prices in the past, the regulations now discourage investment in process development other than at the early stages prior to submission of the documents for government approval.

Certain products (e.g., inclusion bodies) are contained inside the cells and are not released or only partially released to the medium. It may be possible to flush impurities from the cells before breaking them to get the product. Cell disruption is a unit operation peculiar to biochemical engineering. The equipment, however, may be borrowed from other industries. Colloid mills and shear devices used for manufacturing paint and other products effectively rupture walls of many types of cells. Cells with high resistance to shear can be passed through the unit several times, but heat generation from the process can cause loss of product. Shear alone can denature sensitive proteins. Ultrasonic energy is commonly used on a small scale for cell disintegration but is impractical for large batches. Grinding with sand or beads, high-pressure pumping through a tiny orifice, freezing and thawing, dessication, adding lytic enzymes, inducing autolysis with a chemical such as chloroform, and various means of creating shear are alternative or synergistic means of rupturing cells. There are encour-aging results with special mutants of cells with impaired ability to form cell walls at temperatures slightly above their normal growth temperatures. When shifted to the elevated temperature, cell division gives damaged walls and lets the cell contents leak out. The bioprocess operates at its optimum temperature until the temperature is raised shortly before harvest. Cells that form at the elevated temperature release their contents easily.

Whenever possible, the fermentation fluid goes directly to ion exchange, solvent extraction, or some other step. However, prior removal of biological cells and other solids is usually necessary. Centrifugation can be considered, but rotary drum filters with string discharge are commonly used. In the past, large amounts of filter aid were added because many fermentation broths are slimy and hard to filter. Present practice employs polymeric bridging agents to agglomerate the solids. This allows good filtration with only small amounts of filter aid.

The most popular steps for recovery are ion exchange or solvent extraction because selectivity is good, costs are reasonable, and large scale is feasible. Unfortunately, some biochemicals neither exchange ions nor extract well. These and other purification steps can be affected by modifications in fermentation. Adding excess lipids or antifoam oils to a fermentation can aggravate emulsion problems for solvent extraction or impair ion exchange by coating the resin. Stability of biochemical products can be troublesome. For example, penicillin fermentation broth is acidified just prior to contact with the extracting solvent because low pH causes very rapid destruction of penicillin in water. The most popular immiscible solvent is methylisobutylketone (MIBK); halogenated hydrocarbons are avoided because of their hazard for humans. Penicillin is extracted back into an aqueous phase at pH 7.5 to 8 using bicarbonate as a buffer because harsher agents are difficult to control. The Podbelniak design of centrifuge (see Sec. 15) is widely used in the United States for extraction of fermentation broths, while the Westphalia design is common in Europe. Countercurrent flow through a centrifuge should result in more than one equilibrium contact, but emulsions may carry product into the phase being wasted. Lead-and-trail operation of the centrifuges can improve yields. Extraction back into water is seldom troubled by emulsions, and DeLavalle separators work well.

Some products are precipitated from the fermentation broth. The insoluble calcium salts of some organic acids precipitate and are collected, and adding sulfuric acid regenerates the acid while forming gypsum (calcium sulfate) that constitutes a disposal problem. An early process for recovering the antibiotic cycloserine added silver nitrate to the fermentation broth to precipitate an insoluble silver salt. This process was soon obsolete because of poor economics and because the silver salt, when dry, exploded easily.

Of the various types of purification steps based on sorption to a solid phase, ion exchange is the most straightforward. See Sec. 22 for a discussion of techniques. Ion-exchange resins actually adsorb Vitamin B_{12} well instead of exchanging it. Carbon adsorption has considerable importance to biochemical engineering, primarily for the removal of traces of colored impurities. When neither solvent extraction nor ion exchange can be used as the primary concentration/purification step because of the chemical properties of the desired product, an alternative may be adsorption on carbon. Although carbon adsorption may be unselective and low in capacity, it can provide a roughing step to get the purification scheme started.

Isolation procedures for many biochemicals are based on chromatography. Practically any substance can be selected from a crude mixture and eluted at relatively high purity from a chromatographic column with the right combination of adsorbent, conditions, and eluant. For bench scale or for a small pilot plant, such chromatography has rendered alternate procedures such as electrophoresis nearly obsolete. Unfortunately, as size increases, dispersion in the column ruins resolution. To produce small amounts or up to tens of kilograms per year, chromatography is an excellent choice. When the scale-up problem is solved, these procedures should displace some of the conventional steps in the chemical process industries.

Affinity chromatography uses ligands with high specificity for certain compounds. There are several types of affinity that can be employed: antigen-antibody, enzyme-substrate, enzyme-cofactor, chelation with metal ions, or special biochemical attractions such as the protein avidin for the vitamin biotin. Numerous purifications have been devised wherein affinity chromatography is able to isolate quite pure product from a very crude mixture. Expensive affinity agents are regenerated and reused many times. In some cases, the attraction is so strong that the adsorbent can be added batchwise. This scales up well but is less convenient than column operations in terms of collection, elution, and regeneration of the affinity agent.

Conventional elution chromatography has the serious disadvantage of dilution, and usually a concentration step must follow. The technique of displacement chromatography circumvents dilution and may even result in an eluant more concentrated than the feed. A displacer compound breaks the desired product from the chromatographic material sharply, and a column heavily loaded with several biochemicals will release them one at a time depending on their adsorption equilibria. However, the displacers tend to be expensive and can be troublesome to remove from the product.

A number of water-soluble polymers will cause phase separation when present together at concentrations of a few percent. The most widely used polymers are polyethylene glycol (PEG) and dextran. Proteins, other macromolecules, and cell components such as mitochondria distribute in the phases or collect at the interface. Proteins are destabilized at organic solvent/water interfaces, but when each solvent is water, the interfacial tension is negligible. Some salts such as potassium phosphate will also induce phase separation when a polymer is present, but the salt concentration must be high. Two-phase aqueous systems provide a mild method for purification of proteins, and scale-up to large volumes presents no engineering problems. The polymers can have functional groups that improve distribution coefficients of the biochemical products, but the costs for these polymers are high. Although highly promising, two-phase aqueous methods are used only for valuable products because the cost of the polymers is too high and they are not easily recovered for reuse. Another drawback is distribution coefficients not far from 1 for most proteins; several extraction stages are needed to get acceptable yields when the distribution coefficients are unfavorable.

Surface-active agents and liquids immiscible in water can form tiny dispersed units called reverse micelles. These can extract biochemicals from water or permit complexing or reacting in ways not possible in simple aqueous systems.

Crystallization is the preferred method of forming many final products because very high purification is possible. High purity antibiotic crystals can be produced from colored, rather impure solutions if the filter cake is uniform and amenable to good washing to remove the mother liquor. When a sterile pharmaceutical product is desired, crystals are formed from liquid streams that have been sterilized by filtration.

ADDITIONAL REFERENCES: Belter, P. A., E. L. Cussler, and W.-S. Hu, *Bioseparations: Downstream Processing for Biotechnology*, Wiley, New York, 1988. Li, N. N., and J. M. Calo (ed.), *Separation and Purification Technology*, Dekker, New York, 1992. Harrison, R. G. (ed.), *Protein Purification Process Engineering*, Dekker, New York, 1993. Zaslavsky, B. Y., *Aqueous Two-Phase Partitioning: Physical Chemistry and Bioanalytical Applications*, Dekker, New York, 1994. Belfort, G., *Synthetic Membrane Processes: Fundamentals and Water Applications*, Academic Press, New York, 1984. Belfort, G. "Membranes and Bioreactors: A Technical Challenge in Biotechnology," *Biotechnol. Bioeng.*, **33**, 1047–1066, 1989. Brandt, S., R. A. Goffe, S. B. Kessler, J. L. O'Connor, and S. E. Zale, Membrane-Based Affinity Technology for Commercial Scale Purifications, *Bio/Technology*, **6**, 779, 1988. Hanisch, W., "Cell Harvesting" in McGregor, W. C. (ed.), *Membrane Separations in Biotechnology*, Marcel Dekker, New York, 1986. Heath, C. A., and G. Belfort, "Synthetic Membranes in Biotechnology: Realities and Possibilities," *Advances in Biochem. Engr. and Biotechnol.*, **47**, 45–88, 1992. Klein, E., *Affinity Membranes*, John Wiley & Sons, New York, 1991. Matson, S. L., and J. A. Quinn, "Membrane Reactors in Bioprocessing" in *Biochemical Engineering IV*, vol. 49, New York Academy of Sciences, New York, 1986. Mattiasson, G., and W. Ramstorp, "Ultrafiltration Affinity Purification" in *Biochemical Engineering III: Annals of the New York Academy of Sciences*, vol. 413, 1983. Crespo, Jaoa, and Karl Boddeker (eds.), "Membrane Processes" in *Separation and Purification*, Kluwer Academic Publishers, The Netherlands, 1994. Michaels, A. S., "Membranes, Membrane Processes and Their Applications: Needs, Unsolved Problems and Challenges of the 1990s," *Desalination*, **77**, 5–34, 1990. Schugerl, K., *Solvent Extraction in Biotechnology: Recovery of Primary and Secondary Metabolites*, Springer-Verlag, Berlin, New York, 1994. Mattiasson, B., and O. Holst (eds.), *Extractive Bioconversions*, Marcel Dekker, New York, 1991. Asenjo, J. A. (ed.), *Separation Processes in Biotechnology*, Marcel Dekker, New York, 1990. Ladisch, M. R. (ed.), *Protein Purification: From Molecular Mechanisms to Large-Scale Processes*, Am. Chem. Soc. Div. of Biochemical Technol., Washington, DC, 1990. Dechow, F. J., *Separation and Purification Techniques in Biotechnology*, Noyes Publications, Park Ridge, New Jersey, 1989. Verrall, M. S., and M. J. Hudson (eds.), *Separations for Biotechnology*, Ellis Harwood, Wiley, New York, 1987. McGregor, W. C. (ed.), *Membrane Separations in Biotechnology*, Dekker, New York, 1986. Ataai, M. M., and S. K. Sikdar (eds.), *New Developments in Bioseparation*, AIChE Symposium Series, vol. 88, New York, 1993.

PROCESS MODELING

It is generally assumed that properties of very large numbers of cells can be treated as continuous functions having average properties because there are so many cell divisions occurring that the overall rates follow smooth curves. There is an exception in which the cells can all be induced to divide at the same time because events such as illumination or temperature changes slow or halt a step in division. The cells can be triggered to proceed together from that point with overall numbers that are stepwise with time. This is termed a *synchronous culture;* the steps are seldom distinct for more than a few generations unless the triggering event continues to be applied periodically.

Mass balances for common, unsynchronized batch culture give:

$$\frac{dX}{dt} = \mu X - K_d X \tag{24-10}$$

$$\frac{dS}{dt} = -\frac{\mu X}{Y} \tag{24-11}$$

$$\mu = f(S) \tag{24-12}$$

Various functional relationships between μ and S have been proposed, but the Monod equation is used almost exclusively:

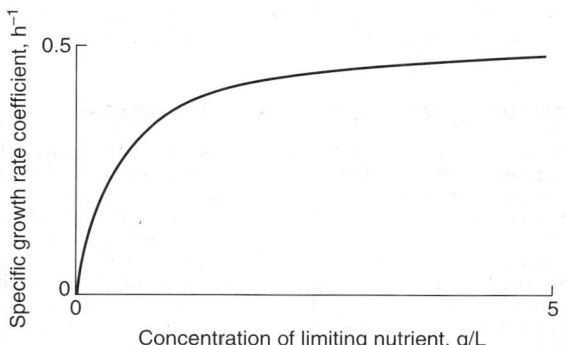

FIG. 24-20   Plot of the Monoc equation.

$$\mu = \mu_{max} \times \frac{S}{K_s + S} \qquad (24\text{-}13)$$

A graph of the Monod equation is shown as Fig. 24-20.

The death rate coefficient is usually relatively small unless inhibitory substances accumulate, so Eq. (24-10) shows an exponential rise until S becomes depleted to reduce μ. This explains the usual growth curve (Fig. 24-21) with its lag phase, logarithmic phase, resting phase, and declining phase as the effect of K_d takes over.

Structured Models   Meaningful detail can be added to culture models in several ways. Cells can be compartmentalized according to biochemical functions, and the components can interact. For example, there can be a group of equations for carbohydrate metabolism, a group for protein synthesis, another for nucleic acid synthesis, and so on. This permits a much more intricate description of cell activities but at the expense of having so many rate constants that assigning values to them may end up as guesswork. For cells with distinct life cycles, a structured model may have compartments corresponding to each stage in the cycle. In addition, each compartment may be subdivided into the biochemical functions mentioned above. Such complicated models have had limited practical use but have great value for directing research toward areas where information is lacking.

Continuous Culture   Continuous culture has been a goal of bioengineers for several decades because batch culture has inherent down time for cleaning and sterilization and long lags before the organisms enter a brief period of high productivity. Continuous runs can last many weeks, but there must be stoppages for cleaning and maintenance. Bacteria may foul surfaces to a small extent, but molds tend to form thick coatings on the shaft, coils, and any protuberances in the fermenter after several weeks of continuous cultivation, that seriously impair mixing and mass transfer.

The nutrition and the product mix can be advantageously manipulated as functions of dilution rate. A serious problem, however, is

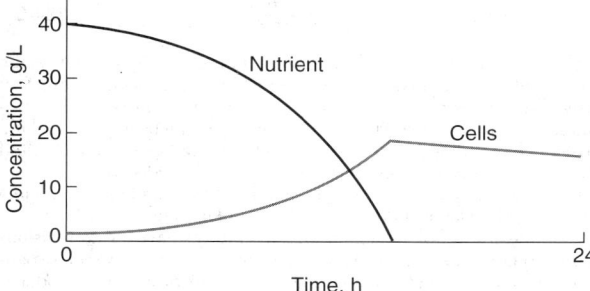

FIG. 24-21   Microbial growth curve. $dx/dt = \mu X - K_d X$; $ds/dt = -\mu X/Y$; $\mu = \mu_{max} S/(K_s + S)$; $\mu_{max} = 0.35$; $K_d = 0.025$; $K_s = 12.0$ mg/L; $Y = 0.48$.

instability of the culture itself. There is a tendency to revert to less productive strains that quickly replace the finely tuned mutants that achieve high titers of product. The main successes with continuous fermentation have been with rugged strains that are producing either cell mass for cattle feed or a simple enzyme or metabolite. When a single stage is used for a product that is elaborated from cells that are not growing, it is difficult to optimize simultaneously cell growth, product production, and efficient use of substrates.

In view of its few industrial applications, continuous culture gets a disproportionate amount of attention from academicians. As a research tool, batch culture suffers from changing concentrations of products and reactants; varying pH and redox potential; and a complicated mix of growing, dying, and dead cells. Data from continuous cultures are much easier to interpret because steady states are achieved or there are repeatable excursions from steady state. The usual explanations for limited use of continuous culture in industry are: culture instability, difficulty of maintaining asepsis, insufficient knowledge of microbial behavior, and reluctance to convert existing factories. Overall cost savings can be relatively small for continuous cultivation because productivity of the bioreator is not very important compared to high product concentration. Another factor is the cost of each research station. Rapid progress in research and development requires multiple vessels for screening many variables, but there are usually only one or two continuous fermenters in the lab or pilot plant because the cost of pumps, reservoirs, sterilizers, and controls is relatively high.

Conventional means for continuous culturing are the chemostat in which nutrient is fed to a reactor at constant rate and the turbidostat that employs feedback control of pumping rate to maintain a fixed turbidity of the culture. Another alternative with feedback control of a nutrient or product concentration has been termed *auxostat, nustat,* or *nutristat.* Proportional control of the pumping rate is desirable because continuous cultures can have oscillatory responses induced by turning the feed pump on or off. A chemostat tends to be unstable and erratic at dilution rates that approach the maximum specific growth rate of the organisms. This is explained by the adjustment of growth rate to nutrient concentration in the region where a small change in dilution rate equates to a big change in nutrient concentration. An auxostat has little advantage over a chemostat at moderate dilution rates but is stable at the high dilution rates at which the chemostat is unreliable.

MATHEMATICAL ANALYSIS

The concept of a limiting nutrient is essential to the theory of continuous culture. There will only be exact stoichiometric balance of all the ingredients going into the cells when a very deliberate and time-consuming effort has been made to determine the details of cell nutrition. Even then, there may be a different balance if the growth rate is changed or kinetic rather than stoichiometric limitations may apply. The ingredient in short supply relative to the other ingredients will be exhausted first and thus limit cellular growth or product synthesis. The other ingredients may exhibit toxicity or influence cellular activities, but there will not be acute shortage as in the case of the limiting nutrient.

Mass balances for one vessel in a series of continuous fermenters give:

Rate of change = rate in − rate out + rate of production

or
$$\frac{V dX_n}{dt} = F X_{n-1} - F X + V \mu_v X_n \qquad (24\text{-}14)$$

Dividing through by V and substituting D = F/V:

$$\frac{dX_n}{dt} = D(X_{n-1} - X_n) + \mu_n X_n \qquad (24\text{-}15)$$

and
$$\frac{V dS_n}{dt} = F S_{n-1} - F S_n - \frac{V \mu_n X_n}{Y} - V M X_n \qquad (24\text{-}16)$$

$$\frac{dS_n}{dt} = D(S_{n-1} - S_n) - \frac{\mu_n X_n}{Y} - M X_n \qquad (24\text{-}17)$$

For a single vessel with sterile feed, this reduces to:

$$\frac{dX}{dt} = \mu X - DX \tag{24-18}$$

$$\frac{dS}{dt} = D(S_o - S) - \frac{\mu X}{Y} - MX \tag{24-19}$$

This is an old, familiar analysis that applies to any continuous culture with a single growth-limiting nutrient that meets the assumptions of perfect mixing and constant volume. The fundamental mass balance equations are used with the Monod equation, which has no time dependency and should be applied with caution to transient states where there may be a time lag as μ responds to changing S. At steady state, the rates of change become zero, and $\mu = D$. Substituting:

$$D = \mu_{max} \frac{S}{K_s + S} \tag{24-20}$$

and

$$S = \frac{DK_s}{\mu_{max} - D} \tag{24-21}$$

Solving for X gives:

$$X = DY \times \frac{S - S_o}{D + MY} = Y(S_o - S) \tag{24-22}$$

From these equations, the behavior of X and S as functions of dilution rate can be plotted as in Fig. 24-22.

The interesting features are: (1) X goes to zero and S reaches S_o as D approaches μ_{max}; (2) S is not a function of S_o when D is less than μ_{max}; (3) the maintenance coefficient is very important at low dilution rate but has little effect afterwards; and (4) S_o is never so high that μ_{max} can be reached, thus washout always occurs before μ_{max} and is a function of S_o.

Mixing has been shown to be critical at low dilution rates because uptake of substrate is extremely rapid for cells in a starved condition. Vigorous agitation is required in small vessels to insure homogeneous distribution of the feed; such intense agitation is probably impractical in large vessels, but good dispersion has been achieved by distributing the feed from many fine openings throughout the vessel.

It is easy to postulate advantages for multistage continuous culture but very difficult to conduct all of the research and development of the many parameters that should be optimized. Each stage could have its feed streams, control of pH and other conditions, and recycle of cells or fluids from other steps in the process. Not only are there many parameters to study for each stage, but changes in one stage can markedly affect other stages. It can be quite troublesome to get representative conditions and cultures in a given stage to begin research because of the complicated interactions with other stages. Time delays in lines and separators for recycling plus complexities from nonideal flow regimes cause a theoretical analysis to be faulty. An optimized multistage continuous fermentation system with recycle and control is a most difficult engineering feat, and the dynamics of microbial responses to upsets are poorly understood. Nevertheless, a three-stage continuous fermentation process is used in industry for the manufacture of a vitamin.

COMPUTER AIDS FOR ANALYSIS AND DESIGN

Specialized programs make layout easy in the form of diagrams of the individual components and their interconnections. Often there are databases for thermodynamic properties as well as routines that will calculate approximate values for properties of the compounds handled in the factory. The programs save the engineer from many tedious calculations, and it is practical to investigate options for equipment sizes and operating conditions. Advice on selecting modeling software is available [Chan, W. K., J. F. Boston, and L. B. Evans, "Select the Right Software for Modeling Separation Processes," *Chem. Engr. Prog.*, **87**, 63–69 (1991)]. A partial list of programs is:

PRO/II from Simulation Sciences Inc., Fullerton CA

HYSIM from Hyprotech, Ltd., Calgary, Alberta; Houston TX; Whittier CA

CHEMCAD from Coade Engr. Software, Houston TX

DesignPFD from ChemShare, Houston TX

Aspen/SP from JSD Simulation Service Co., Denver CO

ELECTROSIM (processes dealing with dissociation and chemical reactions), from Real Time Simulation

POWERTRAN-PC, from Bigelow Systems

DataLogiX, from DataLogiX Formula Systems

G2, a bioprocess expert program with simulation and control from GENSYM, Cambridge, MA

BioPro Designer from Intelligen, Inc., Scotch Plains, NJ

PLANT CELL AND TISSUE CULTURES

Monod kinetics where the growth is limited by a single substrate can apply but only for specific stages of growth and not for the entire growth cycle of plant cells. An example of this is cells grown in a 14 l (0.5 ft^3) fermenter where the latter stages of growth were adequately described by the Monod equation but the initial stages were not. It is understandable that differentiated plant tissues with their distinct cellular characteristics may not be suited to a simple growth-rate expression. For example, root tissues have a meristematic stage where cells are actively dividing. This is followed by cells expanding and maturing to transport nutrients. Older cells may sequester certain nutrients until required by the younger cells and thus alter the availability of substrates.

A kinetic model originally derived by Nyholm is distinguished from Monod's model by the fate of a limiting substrate. Instead of immediate metabolism, the substrate in Nyholm's model is sequestered. The governing equations are:

$$\frac{dC_i}{dt} = v - \mu C_i \tag{24-23}$$

$$\mu = f(C_i) \tag{24-24}$$

$$v = \frac{1}{x} \frac{dS_i}{dt} \tag{24-25}$$

In these equations, μ = specific growth rate coefficient, v = specific rate of substrate uptake, t = time, x = biomass concentration, S_i = intracellular substrate, and C_i = concentration of intracellular substrate. Several examples where these equations can be applied include nitrogen limitations in *M. citrifolia* cultures and phosphate limited growth in *C. roseus, N. tabacum,* and *Papaver somniferum.*

The production of secondary metabolites has often been characterized using the classical equations of Leudeking and Piret. However, the complexities of plant cell and tissue cultures have led to revisions to this equation to include fresh cell weight and viability, cell expansion, and culture death phase. Therefore, the production model is written as the following:

$$\frac{dP}{dt} = \alpha V \frac{dX_d}{dt} + \beta V X_d + \frac{P}{V} \frac{dV}{dt} \tag{24-26}$$

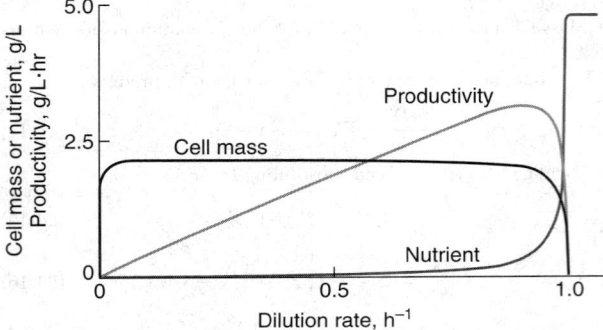

FIG. 24-22   Computer analysis of steady states in continuous culture; $\mu = 2.38$ h^{-1}; $S_o = 1100$ mg/l; $Y = 0.45$; $K_s = 35$ mg/l; $M = 0.05$.

where P = total intracellular product concentration, t = time, V = fraction of viable cells, and X_d = total dry cell weight. Furthermore, a and b represent the usual growth and nongrowth associated production constants, respectively. This type of model has been successfully applied to the production of ajmalicine and serpentine from *C. roseus* cell cultures. Similarly, the production of gossypol by *G. hirsutum* cell cultures used this type of production model but took into account the further metabolism of the phytoalexin due to the growth of new cells.

Because of the differences in primary and secondary metabolism, a reactor may have a dual-stage fed-batch system. In other words, fed-batch operation optimizes growth with little or no product formation. When sufficient biomass has accumulated, a different fed-batch protocol comes into play.

When the production of the secondary metabolites coincides with the death and general lysis of the cells, the recovery of the product is simply a matter of separation from the spent production solution downstream of the reactor. An example of this type of operation was initially used in Japan during the production of shikonin. However, if the secondary metabolites are stored in the vacuole of the cells and the cells remain viable but dormant during the production phase, then a permeabilizing agent such as dimethylsulfoxide (DMSO), detergents, proteins, and antibiotics may be employed in some cases in concentrations that make the cells leak product out but maintain cell viability. Success for this type of product recovery has been reported in *C. roseus, Datura innoxia,* and *Daucus carota* cell cultures.

ADDITIONAL REFERENCES: Nyholm, N., *Biotechnol. Bioeng.* **18,** 1043 (1976). Bailey, C. M. and H. Nicholson, *Biotechnol. Bioeng.* **34,** 1331 (1989). Cazzulino, D. L., H. Pedersen, C. K. Chin, and D. Styer, *Biotechnol. Bioeng.* **35,** 781 (1990). Staba, E. J., *Plant Tissue Cultures as a Source of Biochemicals,* CRC Press, 1980. Thorpe, T. A., *Plant Tissue Culture: Methods and Applications in Agriculture,* Academic Press, 1981. Payne, G., V. Bringi, C. Prince, and M. Shuler, *Plant Cell and Tissue Culture in Liquid Systems,* Hanser Publishers, 1992. Brodelius, P. and K. Nilson, *Eur. J. Appl. Microbiol. Biotechnol.,* **17,** 275 (1983). Rehm, H.-J. and G. Reed, *Biotechnology,* vol. 6b, VCH Verlagsgesellschaft, 1988.

Recycle   Separation and recycle of cells results in much longer residence times for the cells than for the fluid and permits relatively high cell concentrations. In waste treatment, the dilute feed leads to slow growth rates, so more rapid processing is attained through cell recycle to establish a higher population. A higher percentage of the cells may be dead in a recycle system because all are in a starved state. High rates of production are also important in industrial fermentations with cell recycle, and there is the added advantage of reusing cells instead of diverting expensive substrate to producing more cells.

In waste treatment with activated sludge units, organisms not associated with flocs are not collected and tend to leave the system as recycle increases the proportions of flocculating types. Recycle of collectible algae to outdoor ponds has profound influence on the population, but seasonal changes can develop small algae despite retention of large algae; thus, recycle fails as the few large algae die or escape collection.

Recycle of fermentation fluids has quite different objectives than those for cell recyle that aims for population control or greater productivity. Spent broths have leftover nutrients, so the recycling process can save on costs of nutrients and make up water while greatly reducing the volumes sent to waste treatment. Of course, total recycle is bad because undesirable materials build up in concentration and can poison the fermentation. This buildup determines the amount of recycle, but there may be purification steps to remove toxic substances. For example, glycerol is a by-product of alcohol fermentation, but its low volatility means that much is in the stillage from alcohol recovery. Its removal by a physical or chemical step would prevent accumulation in the fermenter by recycling. Alternatively, it could be metabolized by a special strain or mixed culture. Glycerol recovery from the recycle stream in the bioproduction of ethanol now competes with older processes for recovery in the oil and fat industries.

Figure 24-23 is a sketch of continuous culture with recycle. The symbols for flow rates and organism concentrations are F and X, respectively. Assuming perfect mixing and steady state so that the derivatives can be set to zero, mass balances lead to:

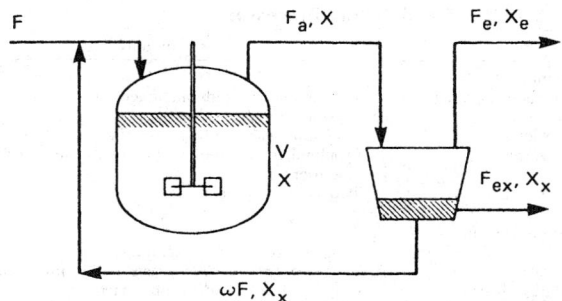

FIG. 24-23   Continuous culture with recycle. (*A. E. Humphrey, "Biochemical Engineering"* in Encyclopedia of Chemical Processing and Design, *vol. 4, July 1977, pp. 359–394.*)

$$\mu = D\left[1 + \omega\left(1 - \frac{1+\omega}{1+\omega-\dfrac{F_e}{F}}\right)\right] \qquad (24\text{-}27)$$

Without recycle, washout occurs when D is greater than μ_{max}, but recycle permits operation with D far greater than μ_{max}. A family of curves is shown in Fig. 24-24 for concentrations of cell mass and nutrient at different recycle ratios. The distinct differences from Fig. 24-22 with no recycle are obvious.

Mixed Cultures   Mixtures of microorganisms characterize the processes shown in Table 24-5.

Processes for biological waste treatment have elective cultures, and the proportions of different species can shift dramatically in response to changing nutrition or physiological conditions. There is an interesting area of research on defined mixtures of microorganisms, but there has been little practical application of the results. The definitions of various interactions are in Table 24-6. These definitions are difficult to apply to real systems where there is highly complicated interplay among organisms that play various roles with respect to each other.

Two types of interaction, competition, and predation are so important that worthwhile insight comes from considering mathematical formulations. Assuming that specific growth-rate coefficients are different, no steady state can be reached in a well-mixed continuous culture with both types present because, if one were at steady state with $\mu = D$, the other would have μ unequal to D and a rate of change unequal to zero. The net effect is that the faster-growing type takes over while the other declines to zero. In real systems—even those that approximate well-mixed continuous cultures—there may be profound

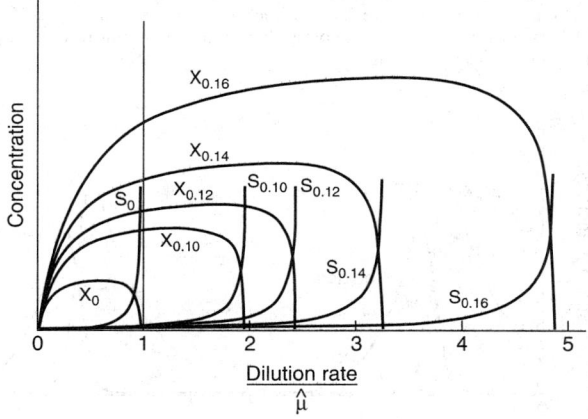

FIG. 24-24   Effect of recycle on steady-state concentrations of cell mass and limiting nutrient. 5-fold increase in cell concentration in separator. Subscripts denote fraction of cell concentrate recycled.

TABLE 24-5   Mixed Culture Processes

Process	Types of organisms
Commercial	
Alcoholic beverages	Various yeasts, molds, and bacteria
Sauerkraut	L. plantarum plus other bacteria
Pickles	L. plantarum plus other bacteria
Cheeses	Propionibacteria, molds, and possibly many other microorganisms
Lactic acid	Two lactobacillus species
Waste treatment	
Trickling filters	Zoogloea, protozoa, algae, fungi
Activated sludge	Zoogloea, Sphaerotilus, yeasts, molds, protozoa
Sludge digestion	Cellulolytic and acid-forming bacteria, methanogenic bacteria
Sewage lagoons	Many types from most microbial families

TABLE 24-6   Some Definitions of Microbial Interactions

Competition	A race for nutrients and space
Predation	One feeds on another
Commensalism	One lives off another with negligible help or harm
Mutualism	Each benefits the other
Synergism	Combination has cooperative metabolism
Antibiosis	One excretes a factor harmful to the other

changes in relative numbers of the various organisms present, but complete takeover by one type is extremely uncommon. Survival of a broad range of species is highly advantageous in natural systems because a needed type will be present should an uncommon nutrient (pollutant?) be added or the conditions change.

Prey-predator or host-parasite systems can be analyzed by mass balance equations:

$$\frac{dH}{dt} = \mu_H H - DH - KHP \qquad (24\text{-}28)$$

$$\frac{dP}{dt} = \mu_P P - DP \qquad (24\text{-}29)$$

$$\frac{dS}{dt} = D(S_o - S) - \frac{\mu_H H}{Y} \qquad (24\text{-}30)$$

where H = the concentration of hosts (prey)
P = the concentration of predators
S = substrate concentration (food for prey)
K = a coefficient for killing

and μ_H and μ_P are Monod functions of S and H respectively.

Computer simulation of these equations is shown in Fig. 24-25. Real systems do have this type of oscillating behavior, but frequencies and amplitudes are erratic.

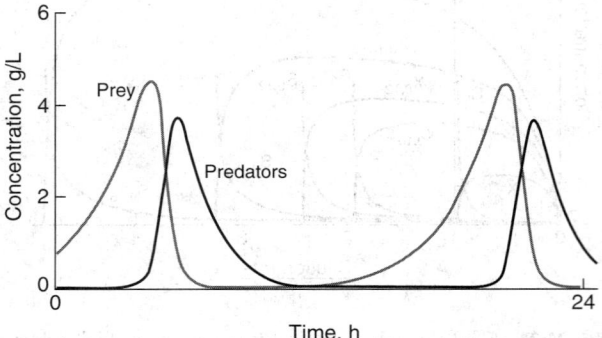

FIG. 24-25   Computer simulation of prey-predator kinetics.

Another interaction with grave consequences is attack on a species by a phage (microbial virus) that is usually highly specific. Infection of a cell by a virulent phage results in the production of 10 to several hundred new phage particles as phage nucleic acid takes over control of cellular activities. The cell disintegrates and releases phage that infect other cells to reach high phage titers quickly. A few cells of the host species may be resistant to phage; such resistance can be acquired through mutation. These cells have fewer competitors and may thrive. However, mutations also occur in phage, so highly complicated behavior occurs as the hosts mutate and mutate further as the phage mutates to counter host resistance.

Commercial fermentation groups usually maintain different strains of cultures suitable for production so that phage attacks can be thwarted by substituting a nonsusceptible culture. After a period of time for the phage to dissipate, it may be possible to return the most desirable production strain.

Bioprocess Control   An industrial fermenter is a fairly sophisticated device with control of temperature, aeration rate, and perhaps pH, concentration of dissolved oxygen, or some nutrient concentration. There has been a strong trend to automated data collection and analysis. Analog control is still very common, but when a computer is available for on-line data collection, it makes sense to use it for control as well. More elaborate measurements are performed with research bioreactors, but each new electrode or assay adds more work, additional costs, and potential headaches. Most of the functional relationships in biotechnology are nonlinear, but this may not hinder control when bioprocess operate over a narrow range of conditions. Furthermore, process control is far advanced beyond the days when the main tools for designing control systems were intended for linear systems.

Many of the sensor problems such as those with steam-sterilizable pH electrodes and dissolved oxygen probes have been solved. Perhaps the most important factor for bioprocessing is the concentration of organisms, but there is no practical method for continuous measurement. Samples of the process fluid must be filtered and dried to get the mass concentration of cells. Numbers can be obtained by direct counting with a microscope or by counting the colonies that form when samples are cultured with nutrient medium in Petri dishes. In lieu of direct measurement, many other ways to estimate cell concentration are tried. Turbidity of the culture fluid can be correlated with cell concentration, but properties and calibration change during the process, and the optical surfaces of the sensors tend to become fouled. Alternatives such as measuring electrical conductivity or capacitance of the fluid sometimes are useful but often are suited only to specific cases. Indirect methods such as measuring protein produced by cells or monitoring nucleic acids are reported, but their proportionality to cell mass may vary during the fermentation. An important advance was made by developing computer models that can interpret measured variables to calculate cell mass or product concentration that may be difficult or impractical to measure on-line.

Mounting electrodes in a bioreactor is costly, and there is an additional contamination risk for sensitive cell cultures. Some other sensors of practical importance are those for dissolved oxygen and for dissolved carbon dioxide. The analysis of gas exiting from a bioreactor with an infrared unit that detects carbon dioxide or a paramagnetic unit that detects oxygen (after carbon dioxide removal) has been replaced by mass spectrophotometry. Gas chromatographic procedures coupled with a mass spectrophotometer will detect all the volatile components.

A useful index of process performance is the oxygen uptake rate, OUR, that is calculated from the difference in oxygen concentration of the inlet air and the exiting gas. Also important is the respiration ratio defined as the carbon dioxide evolved divided by the oxygen consumed.

Although dynamic responses of microbial systems are poorly understood, models with some basic features and some empirical features have been found to correlate with actual data fairly well. Real fermentations take days to run, but many variables can be tried in a few minutes using computer simulation. Optimization of fermentation with models and real-time dynamic control is in its early infancy; however, bases for such work are advancing steadily. The foundations for all such studies are accurate material balances.

The common indices of the physical environment are: temperature, pressure, shaft power input, impeller speed, foam level, gas flow rate, liquid feed rates, broth viscosity, turbidity, pH, oxidation-reduction potential, dissolved oxygen, and exit gas concentrations. A wide variety of chemical assays can be performed; product concentration, nutrient concentration, and product precursor concentration are important. Indices of respiration were mentioned with regard to oxygen transfer and are particularly useful in tracking fermentation behavior. Computer control schemes for fermentation can focus on high productiv-

ity, high product titer, or minimum cost. Computer systems may perform on-line optimization of fermentation. Progress has been slow by empirical methods because there is a multiplicity of variables and because statistical techniques suffer from the relatively poor reproducibility of fermentations. Careful attention to preparation of inoculum, time/temperature factors of sterilization, and the timing of inoculation and feeding can greatly reduce variability of bioprocess performance.

ENZYME ENGINEERING

ENZYMATIC REACTION KINETICS

Enzymes are excellent catalysts for two reasons: great specificity and high turnover rates. With but few exceptions, all reactions in biological systems are catalyzed by enzymes, and each enzyme usually catalyzes only one reaction. For most of the important enzymes and other proteins, the amino-acid sequences and three-dimensional structures have been determined. When the molecular structure of an enzyme is known, a precise molecular weight could be used to state concentration in molar units. However, the amount is usually expressed in terms of catalytic activity because some of the enzyme may be denatured or otherwise inactive. An international unit (IU) of an enzyme is defined as the amount capable of producing one micromole of its reaction product in one minute under its optimal (or some defined) reaction conditions. Specific activity, the activity per unit mass, is an index of enzyme purity.

Although the mechanisms may be complicated and varied, some simple equations can often describe the reaction kinetics of common enzymatic reactions quite well. Each enzyme molecule is considered to have an active site that must first encounter the substrate (reactant) to form a complex so that the enzyme can function. Accordingly, the following reaction scheme is written:

$$E + S \underset{2}{\overset{1}{\rightleftharpoons}} ES \overset{3}{\rightarrow} P + E \qquad (24\text{-}31)$$

where E = enzyme, S = substrate, ES = enzyme-substrate complex, and P = product.

Reactions 1 and 2 may be assumed to be in equilibrium soon after the enzyme is exposed to its substrate. Rate equations for these reactions are:

$$\frac{d(S)}{dt} = k_2(ES) - k_1(E)(S) \qquad (24\text{-}32)$$

$$\frac{d(ES)}{dt} = k_1(E)(S) - (k_1 + k_3)(ES) \qquad (24\text{-}33)$$

$$\frac{d(P)}{dt} = k_3(ES) \qquad (24\text{-}34)$$

where k_1, k_2, k_3 = kinetic constants shown with the arrows in Eq. (24-31). Analysis leads to the Michaelis-Menten equation:

$$\frac{d(P)}{dt} = \frac{V_{max}(S)}{K_M + (S)} \qquad (24\text{-}35)$$

where K_M = Michaelis Constant and V_{max} = maximum rate of reaction.

This equation successfully describes the kinetic behavior of a surprisingly large number of reactions of different enzymes. Taking reciprocals of both sides gives:

$$\frac{dt}{d(P)} = \frac{K_M}{(S) V_{max}} + \frac{1}{V_{max}} \qquad (24\text{-}36)$$

A linear plot of the reciprocal of the reaction rate versus $1/(S)$ will allow the determination of K_M and V_{max} from experimental data.

Kinetic behavior becomes complicated when there are two chemical species that can both complex with the enzyme molecules. One of the species might behave as an inhibitor of the enzyme reaction with

the other as the substrate. Depending upon the nature of the complex, different inhibition patterns will yield different kinetic equations. For example:

$$E + S \rightleftharpoons ES \rightarrow P + E \qquad (24\text{-}37)$$
$$E + I \rightleftharpoons EI \qquad (24\text{-}38)$$

Since the EI complex does not yield product P, and I competes with S for E, there is a state of *competitive inhibition*. By analogy to the Michaelis-Menten equation:

$$\frac{dt}{d[P]} = \frac{K_M}{V_{max}} \frac{1}{S} \left(1 + \frac{[I]}{K_i}\right) + \frac{1}{V_{max}} \qquad (24\text{-}39)$$

where I = concentration of the competitive inhibitor and K_i = inhibition constant.

Enzyme reactions are also sensitive to pH and temperature changes. In characterizing an enzyme, its optimal pH and optimal temperature are conditions at which the enzyme has its highest catalytic activity.

For a somewhat more extensive exposure to enzyme reaction kinetics, consult standard biochemistry texts and also Dixon, M. and E. C. Webb, *Enzymes*, 2d ed., Academic Press, 1964; Segal, I. H., *Enzyme Kinetics*, Wiley, 1975; Gacesa, P. and J. Hubble, *Enzyme Technology*, Open University Press, England, 1987.

Immobilized Enzymes   One factor that usually impedes the development of wide industrial application of enzymes is high cost. Immobilization is a technique to retain enzyme molecules for repeated use. The method of immobilization can be adsorption, covalent bonding, or entrapment. Semipermeable membranes in the form of flat sheets or hollow fibers are one way to restrain the enzyme while allowing smaller molecules to pass. Polyacrylamide gel, silica gel, and other similar materials have been used for entrapment of biologically active materials including enzymes. Encapsulation is another means of capture by coating liquid droplets containing enzymes with some semipermeable materials formed in situ. Generally speaking, entrapment does not involve a chemical or physical/chemical reaction directly with the enzyme molecules; and the enzyme molecules are not altered. Physical adsorption on active carbon particles and ionic adsorption on ion-exchange resins are important for enzyme immobilization. A method with a myriad of possible variations is covalent bonding of the enzyme to a selected carrier. Materials such as glass particles, cellulose, silica, and so on, have been used as carriers for immobilization. Enzymes immobilized by entrapment and adsorption may be subject to loss due to leakage or desorption. On the other hand, the chemical treatment in forming the covalent bond between an enzyme and its carrier may permanently damage some enzyme molecules. In enzyme immobilization, two efficiency terms are often used. Immobilization yield can be used to describe the percent of enzyme activity that is immobilized,

$$\% \text{ yield} = 100 \times \frac{\text{activity immobilized}}{\text{starting activity}}$$

Immobilization efficiency describes the percent of enzyme activity that is observed:

$$\text{percent efficiency} = 100 \times \frac{\text{observed activity}}{\text{activity immobilized}}$$

When an enzyme molecule is attached to a carrier, its active site might be sterically blocked and thus its activity becomes unobservable (inactivated).

One of the most important parameters of an immobilized-carrier complex is stability of its activity. Catalytic activity of the complex diminishes with time because of leakage, desorption, deactivation, and the like. The half-life of the complex is often used to describe the activity stability. Even though there may be frequent exceptions, linear decay is often assumed in treating the kinetics of activity decay of an immobilized complex.

Immobilization by adsorption or by covalent bonding often helps to stabilize the molecular configurations of an enzyme against alternations including those that may cause thermal deactivation. Immobilized enzymes tend to be less sensitive to pH changes than are free enzymes. Although careful choice of the immobilization chemistry can result in stabilized activity, there are some enzymes that are much less stable after immobilization. Most carriers are designed to have high porosity and large internal surface areas so that a relatively large amount of enzymes can be immobilized onto a given volume or given weight of the carrier. Therefore, in an immobilized enzyme-carrier complex, the enzyme molecules are subject to the effect of the microenvironment in the pores of the complex. Surface charges and other microenvironmental effects can create a shift up or down of optimal pH of the enzyme activity.

An immobilized enzyme-carrier complex is a special case that can employ the methodology developed for evaluation of a heterogeneous catalytic system. The enzyme complex also has external diffusional effects, pore diffusional effects, and an effectiveness factor. When carried out in aqueous solutions, heat transfer is usually good, and it is safe to assume that isothermal conditions prevail for an immobilized enzyme complex.

The Michaelis-Menten equation and other similar nonlinear expressions characterize immobilized enzyme kinetics. Therefore, for a spherical porous carrier particle with enzyme molecules immobilized on its external as well as internal surfaces, material balance of the substrate will result in the following:

$$2\,\frac{D_e}{r}\,\frac{dS}{dr} + D_e\,\frac{d^2S}{dr^2} = \frac{V_{max}S}{K_M + S} \qquad (24\text{-}40)$$

with also the usual boundary conditions, at $r = R$, $S = S$ and at $r = 0$, $dS/dr = 0$ where R = radius of the sphere, r = distance from sphere center, S = substrate concentration, and D_e = effective diffusivity. Normalizing results in:

$$\frac{d^2y}{dx^2} + \frac{2}{x}\,\frac{dy}{dx} - \phi^2\beta\left(\frac{y}{\beta} + y\right) = 0 \qquad (24\text{-}41)$$

where y is dimensionless concentration, x is dimensionless distance, and ϕ and β are dimensionless constants; ϕ is sometimes referred to as the Thiele modulus of the immobilized enzyme complex. The boundary conditions are $x = 1$, $y = 1$ and at $x = 0$, $dy/dx = 0$. Graphical solutions are available in standard tests. Two meaningful asymptotic conditions have analytical solutions. In one extreme, $\beta \rightarrow 0$, meaning $S \gg K_m$, and accordingly the Michaelis-Menten equation reduces to a zero-order reaction with $V = V_{max}$. This is the condition of saturation (i.e., the substrate supply is high and saturates all of the active sites of the enzyme molecules). In the other extreme, $\beta \rightarrow \infty$, meaning $K_m \gg S$, and accordingly the Michaelis-Menten equation approaches that of a first-order reaction with $V = V_{max}S/K_m$. This is the condition of a complete substrate control.

Enzymatic Reactors   Adding free enzyme to a batch reactor is practical only when the value of the enzyme is relatively low. With expensive enzymes, reuse by retaining the enzyme with some type of support makes great economic sense. As some activity is usually lost in tethering the enzyme and the additional operations cost money, stability is very important. However, many enzymes are stabilized by immobilization; thus, many reuses may be possible.

Methods of immobilization have already been discussed, and various reactor configurations are possible. An enzyme immobilized on beads of a support material or captured in a gel droplet is essentially a catalytic particle. Mounted in a packed column, there may be upflow or downflow of the feed solution, and a fluidized bed may be feasible except that particle collision often endangers stability of the enzyme. A serious problem is growth of microorganisms on the particles because enzymes are proteins that are nutritious. As immobilized enzymes often have more thermal stability than do free enzymes, the columns can be run at elevated temperatures (50 to 65°C, or 122 to 149°F) to improve reaction rate and to inhibit most but not all contaminating organisms. Sterile feed solutions and aseptic technique can minimize contamination, but, more commonly, antiseptics are added to the feed, and there is occasional treatment with a toxic chemical to wash organisms from the column. Particles with immobilized enzymes are sometimes added to a reactor and recovered later by filtration or by some trick such as using magnets to collect enzymes attached to iron.

Cellulose is hydrolyzed by a complex of several enzymes. The mix of enzyme activities produced by mold cultures can have insufficient amounts of the enzyme beta-glucosidase to maintain a commercially acceptable hydrolysis rate. This enzyme can be produced with a different microbial culture and used to supplement the original enzyme mix, but the cost is high. It is logical to immobilize the beta-glucosidase for multiple use. Handling is minimized by circulating fluid from the main reactor through an external packed column of immobilized enzyme.

Enzymes can be immobilized in sheets. One design had discs of enzymes fastened to a rotating shaft to improve mass transfer, and an alternate design had the feed stream flowing back and forth through sandwiches of sheets with enzyme. However, volumetric efficiency of such reactors is low because sheets with finite spacing offer less area than that of packed particles.

It is possible to add free enzyme and recover it by ultrafiltration, but sufficient membrane surface to get good rates and the required auxiliary equipment are expensive. A hollow fiber device packs a vast amount of membrane area into a small volume. Enzyme may be immobilized inside or outside of the fiber, and it is easy to flush and replace the enzyme. Drawbacks to this design are: (1) The stability of the enzyme is not affected—activity that survives the immobilization step can have enhanced long-term stability; (2) there are two mass-transfer steps, as the substrate must diffuse through the fiber to reach the enzyme and the product must diffuse back; and (3) diffusion is poor on the outside of packed fibers. There is a scheme with the enzyme immobilized on or in the membrane to provide excellent contact as the feed is forced through. Although not yet commercialized, this method appears quite attractive. Recent Russian research with quick freezing has produced gels of enzymes that have high activity, good stability, and a temperature range up to the point where the gel collapses.

ADDITIONAL REFERENCES: Baldwin, T. O., F. M. Raushel, and A. I. Scott, *Chemical Aspects of Enzyme Technology—Fundamentals*, Plenum Press, New York, 1990. Various eds., *Enzyme Engineering*, vols. 2–5, Plenum Press, New York, 1974–1980. Wingard, L. B., I. V. Berezin, and A. A. Klyosov (eds.), *Enzyme Engineering—Future Directions*, Plenum Press, 1980.

ENZYME IMMOBILIZATION: Zaborsky, O. R., *Immobilized Enzymes*, CRC Press, 1973. Lee, Y. Y. and G. T. Tsao, "Engineering Problems of Immobilized Enzymes," *J. Food Technol.*, **39**, 667 (1974). Messing, R. A., *Immobilized Enzymes for Industrial Reactors*, Academic Press, 1975. Torry, S., *Enzyme Technology*, Noyes Data Corp., Park Ridge, New Jersey, 1983.

ENGINEERING ASPECTS: Trevan, M. D., *Immobilized Enzymes: An Introduction and Applications in Biotechnology*, Wiley, 1980. Moo-Young, M., *Bioreactors: Immobilized Enzymes and Cells: Fundamentals and Applications*, Elsevier, London, 1988.

REVIEW OF HETEROGENEOUS CATALYSIS: Satterfield, C. N., *Mass Transfer in Heterogeneous Catalysis*, M.I.T. Press, 1970. Sherwood, T. K., R. L. Pigford, and C. R. Wilke, *Mass Transfer*, McGraw-Hill, 1975.

Waste Management

Louis Theodore, Sc.D., *Professor of Chemical Engineering, Manhattan College; Member, Air and Waste Management Association. (Section Coeditor; Pollution Prevention)*

Anthony J. Buonicore, M.CH.E, P.E., *Diplomate AAEE, CEO, Environmental Data Resources, Inc.; Member, American Institute of Chemical Engineers, Air and Waste Management Association. (Section Coeditor; Introduction and Regulatory Overview)*

John D. McKenna, Ph.D., *President and Chairman, ETS International, Inc., Member, American Institute of Chemical Engineers, Air and Waste Management Association. (Air Pollution Management of Stationary Sources)*

Irwin J. Kugelman, Sc.D., *Professor of Civil Engineering, Lehigh University; Member, American Society of Civil Engineering, Water Environment Federation. (Wastewater Management)*

John S. Jeris, Sc.D., P.E., *Professor of Environmental Engineering, Manhattan College Environmental Consultant; Member, American Water Works Association, Water Environment Federation Section Director. (Wastewater Management)*

Joseph J. Santoleri, P.E., *Senior Consultant, RMT/Four Nines; Member, American Institute of Chemical Engineers, American Society of Mechanical Engineers (Research Committee on Industrial and Municipal Waste), Air and Waste Management Association. (Solid Waste Management)*

Thomas F. McGowan, P.E., *Senior Consultant, RMT/Four Nines; Member, American Institute of Chemical Engineers, American Society of Mechanical Engineers, Air and Waste Management Association. (Solid Waste Management)*

The contributions of Dr. Ross E. McKinney and Dr. George Tchobanoglous to material from the sixth edition of this handbook are acknowledged.

List of Abbreviations

Abbreviation	Definition
3P	Pollution prevention pays
ABS	Alkyl benzene sulfonate
ACC	Annualized capital costs
BACT	Best available control technology
BAT	Best available technology
BCOD	Biodegradable chemical oxygen demand
BCT	Best conventional technology
BOD	Biochemical oxygen demand
BSRT	Biomass solids retention time
BTEX	Benzene, toluene, xylene
CAA	Clean Air Act
CAAA	Clean Air Act Amended
CCP	Comprehensive costing procedures
CFR	Code of federal regulations
COD	Chemical oxygen demand
CPI	Chemical process industries
CRF	Capital recovery factor
CTDMPLUS	Complex terrain dispersion model plus algorithms for unstable situations
CWRT	Center for Waste Reduction Technologies
DCF	Direct installation cost factor
DO	Dissolved oxygen
DRE	Destruction and removal efficiency
EBCT	Empty bed contact time
EPA	Environmental Protection Agency
FML	Flexible membrane liner
GAX	Granular activated carbon
HAPS	Hazardous air pollutants
HAZWOPER	Hazardous waste operators
HCS	Hauled-container systems
HRT	Reactor hydraulic retention time
HSWA	Hazardous and Solid Waste Act
I-TEF	International toxic equivalency factor
ICF	Indirect installation cost factor
ISO	International Organization for Standardization
LCA	Life cycle assessment
LCC	Life cycle costing
LOX	Liquid oxygen

Abbreviation	Definition
MACT	Maximum achievable control technology
MSDA	Material safety data sheets
MSW	Municipal solid waste
MWC	Municipal waste combustors
MWI	Medical waste incinerators
NBOD	Nitrogenous biochemical oxygen demand
NIMBY	Not in my back yard
NPDES	National pollutant discharge elimination system
NSPS	New source performance standards
PCB	Polychlorinated biphenyl
PIES	Pollution prevention information exchange systems
PM	Particulate matter
POTW	Publicly owned treatment work
PPIC	Pollution prevention information clearinghouse
PSD	Prevention of significant deterioration
RCRA	Resource Conservation and Recovery Act
RDF	Refuse-derived fuel
SARA	Superfund Amendments and Reauthorization Act
SCR	Selective catalytic reduction
SCS	Stationary-container systems
SE	Strength of the treated waste
SMART	Save money and reduce toxics
SO	Strength of the untreated waste
SS	Suspended solids
TCC	Total capital cost
TCP	Traditional costing procedures
TGNMO	Total gas nonmethane organics
TOC	Total organic carbon
TSA	Total systems approach
TSCA	Toxic Substances Control Act
TSD	Treatment, storage, and disposal
UASB	Upflow anerobic sludge blanket
VOC	Volatile organic compound
VOST	Volatile organic sampling train
VSS	Volatile suspended solids
WRAP	Waste reduction always pays
WTE	Waste-to-energy (systems)

GENERAL REFERENCES
1. United States EPA, *Pollution Prevention Fact Sheet*, Washington, DC, March 1991.
2. Keoleian, G. and D. Menerey, "Sustainable Development by Design: Review of Life Cycle Design and Related Approaches", *Air & Waste*, **44**, May, 1994.
3. Theodore, L. Personal notes.
4. Theodore, L. and Y. McGuinn, *Pollution Prevention*, New York: Van Nostrand Reinhold, 1992.
5. World Wildlife Fund, *Getting at the Source*, 1991, p. 7.
6. United States EPA, *1987 National Biennial RCRA Hazardous Waste Report—Executive Summary*, Washington, DC, GPO, 1991, p. 10.
7. ASTM, Philadelphia, PA.
8. Theodore, L. and R. Allen, *Pollution Prevention: An ETS Theodore Tutorial*, Roanoke, VA, ETS International, Inc., 1993.
9. United States EPA, *The EPA Manual for Waste Minimization Opportunity Assessments*, Cincinnati, OH, August 1988.
10. Theodore, L. and J. R. Reynolds, *Introduction to Hazardous Waste Incineration*, New York, Wiley-Interscience, 1989.
11. ICF Technology Incorporated, *New York State Waste Reduction Guidance Manual*, Alexandria, VA, 1989.
12. Details available from Louis Theodore.
13. Neveril, R. B., *Capital and Operating Costs of Selected Air Pollution Control Systems*, EPA Report 450/5-80-002, Gard, Inc., Niles, IL, December 1978.
14. Vatavuk, W. M., and R. B. Neveril, "Factors for Estimating Capital and Operating Costs," *Chemical Engineering*, November 3, 1980, pp. 157–162.
15. Vogel, G. A. and E. J. Martin, "Hazardous Waste Incineration," *Chemical Engineering*, September 5, 1983, pp. 143–146 (part 1).
16. Vogel, G. A. and E. J. Martin, "Hazardous Waste Incineration," *Chemical Engineering*, October 17, 1983, pp. 75–78 (part 2).
17. Vogel, G. A. and E. J. Martin, "Estimating Capital Costs of Facility Components," *Chemical Engineering*, (November 28, 1983): pp. 87–90.
18. Ulrich, G. D., *A Guide to Chemical Engineering Process Design and Economics*, New York, Wiley-Interscience, 1984.
19. California Department of Health Services, *Economic Implications of Waste Reduction, Recycling, Treatment, and Disposal of Hazardous Wastes: The Fourth Biennial Report*, California, 1988, p. 110.
20. Zornberg, R. and B. Wainwright, *Waste Minimization: Applications in Industry and Hazardous Waste Site Remediation*, Riverdale, New York, 1992/1993, p. 37.
21. Theodore, L., *A Citizen's Guide to Pollution Prevention*, East Williston, NY, 1993.
22. Varga, A., *On Being Human*, Paulist Press, New York, 1978.
23. Theodore, L., "Dissolve the USEPA . . . Now," *Environmental Manager* (AWA publication), vol. 1, Nov., 1995.
24. Yang, Yonghua and Eric R. Allen, "Biofiltration Control of Hydrogen Sulfide. 2. Kinetics, Biofilter Performance, and Maintenance," *JAWA*, vol. 44, p. 1315.
25. Mycock, J. and J. McKenna, *Handbook of Air Pollution Control and Technology*, ETS, Inc., chap. 21.
26. Ottengraf, S. P. P., "Biological Systems for Waste Gas Elimination," 1987.
27. Hubert, Fleming L., "Consider Membrane Pervaporation," *Chemical Engineering Progress*, July 1992, p. 46.
28. Caruana, Claudia M., "Membranes Vie for Pollution Cleanup Role," *Chemical Engineering Progress*, October 1993, p. 11.
29. Winston, W. S. and Kamalesh K. Sirkar, *Membrane Handbook*, Van Nostrand Reinhold, NY, 1992, p. 78.
30. Tabak, et al., "Biodegradability Studies with Organic Priority Pollutant Compounds," USEPA, MERL, Cincinnati, Ohio, April 1980.
31. Levin, M. A. and M. A. Gealt, *Biotreatment of Industrial and Hazardous Waste*, McGraw-Hill Inc., 1993.
32. Sutton, P. M. and P. N. Mishra, "Biological Fluidized Beds for Water and Wastewater Treatment: A State-of-the-Art Review," WPCF Conference, October 1990.
33. Envirex equipment bulletin FB. 200-R2 and private communication, Waukesha, WI, 1994.
34. Donavan, E. J. Jr., "Evaluation of Three Anaerobic Biological Systems Using Paper Mill Foul Condensate," HydroQual, Inc., EPA, IERL contract 68-03-3074.
35. Mueller, J. A., K. Subburama, and E. J. Donavan, *Proc. 39th Ind. Waste Conf.*, 599, Ann Arbor, 1984.

INTRODUCTION TO WASTE MANAGEMENT

In this section, a number of references are made to laws and procedures that have been formulated in the United States with respect to waste management. An engineer handling waste-management problems in another country would well be advised to know the specific laws and regulations of that country. Nevertheless, the treatment given here is believed to be useful as a general guide.

Multimedia Approach to Environmental Regulations in the United States Among the most complex problems to be faced by industry during the 1990s is the proper control and use of the natural environment. In the 1970s the engineering profession became acutely aware of its responsibility to society, particularly for the protection of public health and welfare. The decade saw the formation and rapid growth of the U.S. Environmental Protection Agency (EPA) and the passage of federal and state laws governing virtually every aspect of the environment. The end of the decade, however, brought a realization that only the more simplistic problems had been addressed. A limited number of large sources had removed substantial percentages of a few readily definable air pollutants from their emissions. The incremental costs to improve the removal percentages would be significant and would involve increasing numbers of smaller sources, and the health hazards of a host of additional toxic pollutants remained to be quantified and control techniques developed.

Moreover, in the 1970s, air, water, and waste were treated as separate problem areas to be governed by their own statutes and regulations. Toward the latter part of the decade, however, it became obvious that environmental problems were closely interwoven and should be treated in concert. The traditional type of regulation—command and control—had severely restricted compliance options.

The 1980s began with EPA efforts redirected to take advantage of the case-specific knowledge, technical expertise, and imagination of those being regulated. Providing plant engineers with an incentive to find more efficient ways of abating pollution would greatly stimulate innovation in control technology. This is a principal objective, for example, of EPA's "controlled trading" air pollution program, established in the Offsets Policy Interpretative Ruling issued by the EPA in 1976, with statutory foundation given by the Clean Air Act Amendments of 1977. The Clean Air Act Amendments of 1990 expanded the program even more to the control of sulfur oxides under Title IV. In effect, a commodities market on "clean air" was developed.

The rapidly expanding body of federal regulation presents an awesome challenge to traditional practices of corporate decision-making, management, and long-range planning. Those responsible for new plants must take stock of the emerging requirements and construct a fresh approach.

The full impact of the Clean Air Act Amendments of 1990, the Clean Water Act, the Safe Drinking Water Act, the Resource Conservation and Recovery Act, the Comprehensive Environmental Responsibility, Compensation and Liability (Superfund) Act, and the Toxic Substances Control Act is still not generally appreciated. The combination of all these requirements, sometimes imposing conflicting demands or establishing differing time schedules, makes the task of obtaining all regulatory approvals extremely complex.

One of the dominant impacts of environmental regulations is that the lead time required for the planning and construction of new plants is substantially increased. When new plants generate major environmental complexities, the implications can be profound. Of course, the exact extent of additions to lead time will vary widely from one case to another, depending on which permit requirements apply and on what difficulties are encountered. For major expansions in any field of heavy industry, however, the delay resulting from federal requirements could conceivably add 2 to 3 years to total lead time. Moreover, there is always the possibility that regulatory approval will be denied. So, contingency plans for fulfilling production needs must be developed.

Any company planning a major expansion must concentrate on environmental factors from the outset. Since many environmental approvals require a public hearing, the views of local elected officials and the community at large are extremely important. To an unprecedented degree, the political acceptability of a project can now be crucial.

Plant Strategies At the plant level, a number of things can be done to minimize the impact of environmental quality requirements. These include:

1. Maintaining an accurate source-emission inventory
2. Continually evaluating process operations to identify potential modifications that might reduce or eliminate environmental impacts
3. Ensuring that good housekeeping and strong preventive-maintenance programs exist and are followed
4. Investigating available and emerging pollution-control technologies
5. Keeping well informed of the regulations and the directions in which they are moving
6. Working closely with the appropriate regulatory agencies and maintaining open communications to discuss the effects that new regulations may have
7. Keeping the public informed through a good public-relations program.

It is unrealistic to expect that at any point in the foreseeable future Congress will reverse direction, reduce the effect of regulatory controls, or reestablish the preexisting legal situation in which private companies are free to construct major industrial facilities with little or no restraint by federal regulation.

Corporate Strategic Planning Contingency planning represents an essential component of sound environmental planning for a new plant. The environmental uncertainties surrounding a large capital project should be specified and related to other contingencies (such as marketing, competitive reactions, politics, foreign trade, etc.) and mapped out in the overall corporate strategy.

Environmental factors should also be incorporated into a company's technical or research and development program. Since the planning horizons for new projects may now extend to 5 to 10 years, R&D programs can be designed for specific projects. These may include new process modifications or end-of-pipe control technologies.

Another clear need is to integrate environmental factors into financial planning for major projects. It must be recognized that strategic environmental planning is as important to the long-range goal of the corporation as is financial planning. Trade-off decisions regarding financing may have to change as the project goes through successive stages of environmental planning and permit negotiations. For example, requirements for the use of more expensive pollution control technology may significantly increase total project costs; or a change from end-of-pipe to process modification technology may preclude the use of industrial revenue bond financing under Internal Revenue Service (IRS) rules. Regulatory delays can affect assumptions as to both the rate of expenditure and inflation factors. Investment, production, environmental, and legal factors are all interrelated and can have a major impact on corporate cash flow.

Most companies must learn to deal more creatively with local officials and public opinion. The social responsibility of companies can become an extremely important issue. Companies should apply thoughtfulness and skill to the timing and conduct of public hearings. Management must recognize that local officials have views and constituencies that go beyond attracting new jobs.

From all these factors, it is clear that the approval and construction of major new industrial plants or expansions is a far more complicated operation than it has been in the past, even the recent past. Stringent environmental restrictions are likely to preclude construction of certain facilities at locations where they otherwise might have been built. In other cases, acquisition of required approvals may generate a heated technical and political debate that can drag out the regulatory process for several years.

In many instances, new requirements may be imposed while a company is seeking approval for a proposed new plant. Thus, companies intending to expand their basic production facilities should anticipate their needs far in advance, begin preparation to meet the regulatory challenge they will eventually confront, and select sites with careful consideration of environmental attributes. It is the objective of this section to assist the engineer in meeting this environmental regulatory challenge.

UNITED STATES AIR QUALITY LEGISLATION AND REGULATIONS

Although considerable federal legislation dealing with air pollution has been enacted since the 1950s, the basic statutory framework now in effect was established by the Clean Air Act of 1970; amended in 1974 to deal with energy-related issues; amended in 1977, when a number of amendments containing particularly important provisions associated with the approval of new industrial plants were adopted; and amended in 1990 to address toxic air pollutants and ozone nonattainment areas.

Clean Air Act of 1970 The Clean Air Act of 1970 was founded on the concept of attaining National Ambient Air Quality Standards (NAAQS). Data were accumulated and analyzed to establish the quality of the air, identify sources of pollution, determine how pollutants disperse and interact in the ambient air, and define reductions and controls necessary to achieve air-quality objectives.

EPA promulgated the basic set of current ambient air-quality standards in April 1971. The specific regulated pollutants were particulates, sulfur dioxide, photochemical oxidants, hydrocarbons, carbon monoxide, and nitrogen oxides. In 1978, lead was added. Table 25-1 enumerates the present standards.

To provide basic geographic units for the air-pollution control program, the United States was divided into 247 air quality control regions (AQCRs). By a standard rollback approach, the total quantity of pollution in a region was estimated, the quantity of pollution that could be tolerated without exceeding standards was then calculated, and the degree of reduction called for was determined. States were required by EPA to develop state implementation plans (SIPs) to achieve compliance.

The act also directed EPA to set new source performance standards (NSPS) for specific industrial categories. New plants were required to use the best system of emission reduction available. EPA gradually issued these standards, which now cover a number of basic industrial categories (as listed in Table 25-2). The 1977 amendments to the Clean Air Act directed EPA to accelerate the NSPS program and included a regulatory program to prevent significant deterioration in those areas of the country where the NAAQS were being attained.

Finally, Sec. 112 of the Clean Air Act required that EPA promulgate National Emission Standards for Hazardous Air Pollutants (NESHAPs). Between 1970 and 1989, standards were promulgated for asbestos, beryllium, mercury, vinyl chloride, benzene, arsenic, radionuclides, and coke-oven emissions.

Prevention of Significant Deterioration (PSD) Of all the federal laws placing environmental controls on industry (and, in particular, on new plants), perhaps the most confusing and restrictive are the limits imposed for the prevention of significant deterioration (PSD) of air quality. These limits apply to areas of the country that are already cleaner than required by ambient air-quality standards. This regulatory framework evolved from judicial and administrative action under the 1970 Clean Air Act and subsequently was given full statutory foundation by the 1977 Clean Air Act Amendments.

EPA established an area classification scheme to be applied in all such regions. The basic idea was to allow a moderate amount of industrial development but not enough to degrade air quality to a point at which it barely complied with standards. In addition, states were to designate certain areas where pristine air quality was especially desirable. All air-quality areas were categorized as Class I, Class II, or Class III. Class I areas were pristine areas subject to the tightest control. Permanently designated Class I areas included international parks, national wilderness areas, memorial parks exceeding 5000 acres, and national parks exceeding 6000 acres. Although the nature of these areas is such that industrial projects would not be located within them, their Class I status could affect projects in neighboring areas where meteorological conditions might result in the transport of emissions into them. Class II areas were areas of moderate industrial growth. Class III areas were areas of major industrialization. Under EPA regulations promulgated in December 1974, all areas were initially categorized as Class II. States were authorized to reclassify specified areas as Class I or Class III.

The EPA regulations also established another critical concept

TABLE 25-1   National Primary and Secondary Ambient Air Quality Standards

Pollutant	Averaging time	Primary standards	Secondary standards
Sulfur oxides	Annual arithmetic mean 24 h 3 h	80 $\mu g/m^3$ (0.03 ppm) 365 $\mu g/m^3$ (0.14 ppm)	1,300 $\mu g/m^3$ (0.5 ppm)
Particulate matter (PM_{10}, particulates with aerodynamic diameter less than or equal to 10 microns)	Annual geometric mean 24 h	75 $\mu g/m^3$ 260 $\mu g/m^3$	60 $\mu g/m^3$ 150 $\mu g/m^3$
Ozone	1 h	240 $\mu g/m^3$ (0.12 ppm)	Same as primary standard
Carbon monoxide	8 h 1 h	10 mg/m^3 40 mg/m^3 (35 ppm)	Same as primary standard
Nitrogen oxides	Annual arithmetic mean	100 $\mu g/m^3$ (0.05 ppm)	Same as primary standard
Lead	3 months	1.5 $\mu g/m^3$	Same as primary standard

NOTE: National standards, other than those based on annual arithmetic means or annual geometric means, are not to be exceeded more than once a year.

TABLE 25-2   Source Categories for Which New Source Performance Standards Have Been Set as of 1991

Fossil-Fuel-Fired Steam Generators for Which Construction Commenced after August 17, 1971

Electric Utility Steam Generating Units for Which Construction Commenced after September 18, 1978

Industrial-Commercial-Institutional Steam Generating Units

Incinerators

Portland Cement Plants

Nitric Acid Plants

Sulfuric Acid Plants

Asphalt Concrete Plants

Petroleum Refineries

Storage Vessels for Petroleum Liquids for Which Construction, Reconstruction, or Modification Commenced after June 11, 1973, and Prior to May 19, 1978

Storage Vessels for Petroleum Liquids for Which Construction, Reconstruction, or Modification Commenced after May 18, 1978, and Prior to July 23, 1984

Volatile Organic Liquid Storage Vessels (Including Petroleum Liquid Storage Vessels) for Which Construction, Reconstruction, or Modification Commenced after July 23, 1984

Secondary Lead Smelters

Secondary Brass and Bronze Production Plants

Primary Emissions from Basic Oxygen Process Furnaces for Which Construction Commenced after June 11, 1973

Secondary Emissions from Basic Oxygen Process Steelmaking Facilities for Which Construction Commenced after January 20, 1983

Sewage Treatment Plants

Primary Copper Smelters

Primary Zinc Smelters

Primary Lead Smelters

Primary Aluminum Reduction Plants

Phosphate Fertilizer Industry: Wet-Process Phosphoric Acid Plants

Phosphate Fertilizer Industry: Superphosphoric Acid Plants

Phosphate Fertilizer Industry: Diammonium Phosphate Plants

Phosphate Fertilizer Industry: Triple Superphosphate Plants

Phosphate Fertilizer Industry: Granular Triple Superphosphate Storage Facilities

Coal Preparation Plants

Ferroalloy Production Facilities

Steel Plants: Electric Arc Furnaces Constructed after October 21, 1974, and on or before August 17, 1983

Steel Plants: Electric Arc Furnaces and Argon-Oxygen Decarburization Vessels Constructed after August 7, 1983

Kraft Pulp Mills

Glass Manufacturing Plants

Grain Elevators

Surface Coating of Metal Furniture

Stationary Gas Turbines

Lime Manufacturing Plants

Lead-Acid Battery Manufacturing Plants

Metallic Mineral Processing Plants

Automobile and Light-Duty Truck Surface Coating Operations

Phosphate Rock Plants

Ammonium Sulfate Manufacture

Graphic Arts Industry: Publication Rotogravure Printing

Pressure-Sensitive Tape and Label Surface Coating Operations

Metal Coil Surface Coating

Asphalt Processing and Asphalt Roofing Manufacture

Equipment Leaks of VOC in the Synthetic Organic Chemicals Manufacturing Industry

Beverage Can Surface Coating Industry

Bulk Gasoline Terminals

New Residential Wood Heaters

Rubber Tire Manufacturing Industry

Flexible Vinyl and Urethane Coating and Printing

Equipment Leaks of VOC in Petroleum Refineries

Synthetic Fiber Production Facilities

Petroleum Dry Cleaners

Equipment Leaks of VOC from Onshore Natural Gas Processing Plants

Onshore Natural Gas Processing; SO_2 Emissions

Nonmetallic Mineral Processing Plants

Wool Fiberglass Insulation Manufacturing Plants

VOC Emissions from Petroleum Refinery Wastewater Systems

Magnetic Tape Coating Facilities

Industrial Surface Coating: Surface Coating of Plastic Parts for Business Machines

Volatile Organic Compound Emissions from Synthetic Organic Chemical Manufacturing Industry (SOCMI) Air Oxidation Unit Processes

Volatile Organic Compound Emissions from Synthetic Organic Chemical Manufacturing Industry Distillation Operations

Polymeric Coating of Supporting Substrates Facilities

known as the *increment*. This was the numerical definition of the amount of additional pollution that may be allowed through the combined effects of all new growth in a particular locality (see Table 25-3). To assure that the increments would not be used up hastily, EPA specified that each major new plant must install best available control technology (BACT) to limit emissions. This reinforced the same policy underlying the NSPS; and where an NSPS had been promulgated, it would control determinations of BACT. Where such standards had not been promulgated, an ad hoc determination was called for in each case.

To implement these controls, EPA requires that every new source undergo preconstruction review. The regulations prohibited a company from commencing construction on a new source until the review had been completed and provided that, as part of the review procedure, public notice should be given and an opportunity provided for a public hearing on any disputed questions.

Sources Subject to Prevention of Significant Deterioration (PSD)   Sources subject to PSD regulations (40 CFR, Sec. 52.21, Aug. 7, 1980) are major stationary sources and major modifications located in attainment areas and unclassified areas. A major stationary source was defined as any source listed in Table 25-4 with the potential to emit 100 tons per year or more of any pollutant regulated under the Clean Air Act (CAA) or any other source with the potential to emit 250 tons per year or more of any CAA pollutant. The "potential to emit" is defined as the maximum capacity to emit the pollutant under applicable emission standards and permit conditions (after application of any air pollution control equipment) excluding secondary emissions. A "major modification" is defined as any physical or operational change of a major stationary source producing a "significant net emissions increase" of any CAA pollutant (see Table 25-5).

Ambient monitoring is required of all CAA pollutants with emissions greater than or equal to Table 25-5 values for which there are

TABLE 25-3 Prevention of Significant Deterioration (PSD) Air-Quality Increments

	Maximum allowable increase over baseline air quality, $\mu g/m^3$			Primary ambient air-quality standard, $\mu g/m^3$
	Class I	Class II	Class III	
Particulate matter				
Annual geometric mean	5	19	37	75
24-h maximum	10	37	75	260
SO_2				
Annual arithmetic mean	2	20	40	80
24-h maximum	5	91	182	365
3-h maximum	25	512	700	1300°

°Secondary standard rather than primary standard.

TABLE 25-4 Sources Subject to PSD Regulation if Their Potential to Emit Equals or Exceeds 100 Tons per Year

Fossil-fuel-fired steam electric plants of more than 250 million Btu/h heat input
Coal-cleaning plants (with thermal dryers)
Kraft-pulp mills
Portland-cement plants
Primary zinc smelters
Iron and steel mill plants
Primary aluminum-ore-reduction plants
Primary copper smelters
Municipal incinerators capable of charging more than 250 tons of refuse per day
Hydrofluoric, sulfuric, and nitric acid plants
Petroleum refineries
Lime plants
Phosphate-rock-processing plants
Coke-oven batteries
Sulfur-recovery plants
Carbon-black plants (furnace process)
Primary lead smelters
Fuel-conversion plants
Sintering plants
Secondary metal-production plants
Chemical-process plants
Fossil-fuel boilers (or combinations thereof) totaling more than 250 million Btu/h heat input
Petroleum-storage and -transfer units with total storage capacity exceeding 300,000 bbl
Taconite-ore-processing plants
Glass-fiber-processing plants
Charcoal-production plants

TABLE 25-5 Significant Net Emissions Increase

Pollutant	Tons/year
CO	100
NO_x (as NO_2)	40
SO_2	40
Particulate matter	25
Ozone	40 of volatile organic compounds
Lead	0.6
Asbestos	0.007
Beryllium	0.0004
Mercury	0.1
Vinyl chloride	1
Fluorides	3
Sulfuric acid mist	7
Hydrogen sulfide	10
Total reduced sulfur	10
Reduced-sulfur compounds	10
Other CAA pollutants	>0

NAAQS. Continuous monitoring is also required for other CAA pollutants for which the EPA or the state determines that monitoring is necessary. The EPA or the state may exempt any CAA pollutant from these monitoring requirements if the maximum air-quality impact of the emissions increase is less than the values in Table 25-6 or if present concentrations of the pollutant in the area that the new source would affect are less than the Table 25-6 values. The EPA or the state

may accept representative existing monitoring data collected within 3 years of the permit application to satisfy monitoring requirements.

EPA regulations provide exemption from BACT and ambient air-impact analysis if the modification that would increase emissions is accompanied by other changes within the plant that would net a zero increase in total emissions. This exemption is referred to as the "bubble" or "no net increase" exemption.

A full PSD review would include a case-by-case determination of the controls required by BACT, an ambient air-impact analysis to determine whether the source might violate applicable increments or air-quality standards; an assessment of the effect on visibility, soils, and vegetation; submission of monitoring data; and full public review.

EPA regulations exempted smaller sources from the major elements of PSD review and, in particular, relieved those sources from compliance with BACT (though they still had to comply with applicable NSPS as well as with SIP requirements under the SIP program). Smaller sources were also exempted from conducting ambient air-impact analysis and submitting data supporting an ambient air-quality analysis. Smaller sources, however, were not exempted from the program altogether. They remained subject to the statutory requirements to obtain preconstruction approval, including procedures for public review, and they still might be required, at EPA request, to submit data supporting their applications. Also, if emissions from a smaller source would affect a Class I area or if an applicable increment were already being violated, the full PSD requirements for ambient air-impact analysis would apply.

Nonattainment (NA) Those areas of the United States failing to attain compliance with ambient air-quality standards were considered nonattainment areas. New plants could be constructed in nonattainment areas only if stringent conditions were met. Emissions had to be controlled to the greatest degree possible, and more than equivalent offsetting emission reductions had to be obtained from other sources to assure progress toward achievement of the ambient air-quality standards. Specifically, (1) the new source must be equipped with pollution controls to assure lowest achievable emission rate (LAER), which in no case can be less stringent than any applicable NSPS; (2) all existing sources owned by an applicant in the same region must be in compliance with applicable state implementation plan requirements or be under an approved schedule or an enforcement order to achieve such compliance; (3) the applicant must have sufficient offsets to more than

TABLE 25-6 Concentration Impacts below Which Ambient Monitoring May Not Be Required

	$\mu g/m^3$	Average time
CO	575	8-h maximum
NO_2	14	24-h maximum
TSP	10	24-h maximum
SO_2	13	24-h maximum
Lead	0.1	24-h maximum
Mercury	0.25	24-h maximum
Beryllium	0.0005	24-h maximum
Fluorides	0.25	24-h maximum
Vinyl chloride	15	24-h maximum
Total reduced sulfur	10	1-h maximum
H_2S	0.04	1-h maximum
Reduced-sulfur compounds	10	1-h maximum

make up for the emissions to be generated by the new source (after application of LAER); and (4) the emission offsets must provide "a positive net air quality benefit in the affected area."

LAER was deliberately a technology-forcing standard of control. The statute stated that LAER must reflect (1) the most stringent emission limitation contained in the implementation plan of any state for such category of sources unless the applicant can demonstrate that such a limitation is not achievable, or (2) the most stringent limitation achievable in practice within the industrial category, whichever is more stringent. In no event could LAER be less stringent than any applicable NSPS. While the statutory language defining BACT directed that "energy, environmental, and economic impacts and other costs" be taken into account, the comparable provision on LAER provided no instruction that economics be considered.

For existing sources emitting pollutants for which the area is nonattainment, reasonable available control technology (RACT) would be required. EPA defines RACT by industrial category.

Controlled-Trading Program   The legislation enacted under the Clean Air Act Amendments of 1977 provided the foundation for EPA's controlled-trading program, the essential elements of which include:

- Bubble policy (or bubble exemption under PSD)
- Offsets policy (under nonattainment)
- Banking and brokerage (under nonattainment)

While these different policies vary broadly in form, their objective is essentially the same: to substitute flexible economic-incentive systems for the current rigid, technology-based regulations that specify exactly how companies must comply. These market mechanisms have made regulating easier for EPA and less burdensome and costly for industry.

Bubble Policy   The bubble concept introduced under PSD provisions of the Clean Air Act Amendments of 1977 was formally proposed as EPA policy on Jan. 18, 1979, the final policy statement being issued on Dec. 11, 1979. The bubble policy allows a company to find the most efficient way to control a plant's emissions as a whole rather than by meeting individual point-source requirements. If it is found less expensive to tighten control of a pollutant at one point and relax controls at another, this would be possible as long as the total pollution from the plant would not exceed the sum of the current limits on individual point sources of pollution in the plant. Properly applied, this approach would promote greater economic efficiency and increased technological innovation.

There are some restrictions, however, in applying the bubble concept:

1. The bubble may be only used for pollutants in an area where the state implementation plan has an approved schedule to meet air-quality standards for that pollutant.

2. The alternatives used must ensure that air-quality standards will be met.

3. Emissions must be quantifiable, and trades among them must be even. Each emission point must have a specific emission limit, and that limit must be tied to enforceable testing techniques.

4. Only pollutants of the same type may be traded; that is, particulates for particulates, hydrocarbons for hydrocarbons, etc.

5. Control of hazardous pollutants cannot be relaxed through trades with less toxic pollutants.

6. Development of the bubble plan cannot delay enforcement of federal and state requirements.

Some additional considerations must be noted:

1. The bubble may cover more than one plant within the same area.

2. In some circumstances, states may consider trading open dust emissions for particulates (although EPA warns that this type of trading will be difficult).

3. EPA may approve compliance date extensions in special cases. For example, a source may obtain a delay in a compliance schedule to install a scrubber if such a delay would have been permissible without the bubble.

EPA will closely examine particulate size distribution in particulate emission trades because finer particulates disperse more widely, remain in the air longer, and frequently are associated with more adverse health effects.

It will be the responsibility of industry to suggest alternative control approaches and demonstrate satisfactorily that the proposal is equivalent in pollution reduction, enforceability, and environmental impact to existing individual process standards.

Offsets Policy   Offsets were EPA's first application of the concept that one source could meet its environmental protection obligations by getting another source to assume additional control actions. In nonattainment areas, pollution from a proposed new source, even one that controls its emissions to the lowest possible level, would aggravate existing violations of ambient air-quality standards and trigger the statutory prohibition. The offsets policy provided these new sources with an alternative. The source could proceed with construction plans, provided that:

1. The source would control emissions to the lowest achievable level.

2. Other sources owned by the applicant were in compliance or on an approved compliance schedule.

3. Existing sources were persuaded to reduce emissions by an amount at least equal to the pollution that the new source would add.

Banking and Brokerage Policy   EPA's banking policy is aimed at providing companies with incentives to find more offsets. Under the original offset policy, a firm shutting down or modifying a facility could apply the reduction in emissions to new construction elsewhere in the region only if the changes were made simultaneously. However, with banking a company can "deposit" the reduction for later use or sale. Such a policy will clearly establish that clean air (or the right to use it) has direct economic value.

Clean Air Act of 1990   In November, 1990, Congress adopted the Clean Air Act Amendments of 1990, providing substantial changes to many aspects of the existing CAAA. The concepts of NAAQS, NSPS, and PSD remain virtually unchanged. However, significant changes have occurred in several areas that directly affect industrial facilities and electric utilities and air-pollution control at these facilities. These include changes and additions in the following major areas:

Title I	Nonattainment areas
Title III	Hazardous air pollutants
Title IV	Acid deposition control
Title V	Operating permits

Title I: Nonattainment Areas   The existing regulations for nonattainment areas have been made more stringent in several areas. The CAAA of 1990 requires the development of comprehensive emission inventory-tracking for all nonattainment areas and establishes a classification scheme that defines nonattainment areas into levels of severity. For example, ozone nonattainment areas are designated as marginal, moderate, serious, severe (two levels), and extreme, with compliance deadlines of 3, 6, 9, 15–17, and 20 years, respectively, with each classification having more stringent requirements regarding strategies for compliance (see Table 25-7). Volatile organic compound (VOC) emissions reductions of 15 percent are required in moderate areas by 1996 and 3 percent a year thereafter for severe or extreme areas until compliance is achieved. In addition, the definition of a major source of ozone precursors (previously 100 tons per year of NO_x, CO, or VOC emissions) was redefined to as little as 10 tons per year in the extreme classification, with increased offset requirements of 1.5 to 1 for new and modified sources. These requirements place major constraints on affected industries in these nonattainment areas. A similar approach is being taken in PM_{10} and CO nonattainment areas.

Title III: Hazardous Air Pollutants   The Title III provisions on hazardous air pollutants (HAPs) represent a major departure from the previous approach of developing NESHAPs. While only eight HAPs were designated in the 20 years since enactment of the CAAA of 1970, the new CAAA of 1990 designated 189 pollutants as HAPs requiring regulation. These are summarized in Table VIII and will affect over 300 major source categories. Major sources are defined as any source (new or existing) that emits (after control) 10 tons a year or more of any regulated HAP or 25 tons a year or more of any combination of HAPs. The deadlines for promulgation of the source categories and appropriate emission standards are as follows:

First 40 source categories	November 15, 1992
Coke oven batteries	December 31, 1992
25% of all listed categories	November 15, 1994

TABLE 25-7 Ozone Nonattainment Area Classifications and Associated Requirements

Nonattainment area classification	One-hour ozone concentration design value, ppm	Attainment date	Major source threshold level, tons VOCs/yr	Offset ratio for new/modified sources
Marginal	0.121–0.138	Nov. 15, 1993	100	1.1 to 1
Moderate	0.138–0.160	Nov. 15, 1996	100	1.15 to 1
Serious	0.160–0.180	Nov. 15, 1999	50	1.2 to 1
Severe	0.180–0.190	Nov. 15, 2005	25	1.3 to 1
	0.190–0.280	Nov. 15, 2007	25	1.3 to 1
Extreme	0.280 and up	Nov. 15, 2010	10	1.5 to 1

Publicly owned treatment works — November 15, 1995
50% of all listed categories — November 15, 1997
100% of all categories — November 15, 2000

Each source will be required to meet maximum achievement control technology (MACT) requirements. For existing sources, MACT is defined as a stringency equivalent to the average of the best 12 percent of the sources in the category. For new sources, MACT is defined as the best controlled system. New sources are required to meet MACT immediately, while existing sources have three years from the date of promulgation of the appropriate MACT standard. As an early incentive, existing sources that undergo at least a 90 percent reduction in emissions of a HAP (or 95 percent for a hazardous particulate) prior to the promulgation of the MACT standard will be issued a six-year extension on the deadline for final compliance.

Title IV: Acid Deposition Control The Acid Deposition Control Program is designed to reduce emissions of SO_2 in the United States by 10 million tons per year, resulting in a net yearly emission of 8.9 million tons by the year 2000. Phase I of the program requires 111 existing uncontrolled coal-fired power plants ($\geq$100 MW) to reduce emissions to 2.5 pounds of SO_2 per 10^6 Btu by 1995 (1997 if scrubbers are used to reduce emissions by at least 90 percent). The reduction is to be accomplished by issuing all affected units emission "allowances" equivalent to what their annual average SO_2 emissions would have been in the years 1985–1987 based on 2.5 pounds SO_2 per 10^6 Btu coal. The regulations represent a significant departure from previous regulations where specified SO_2 removal efficiencies were mandated; rather, the utilities will be allowed the flexibility of choosing which strategies will be used (e.g., coal washing, low-sulfur coal, flue gas desulfurization, etc.) and which units will be controlled, as long as the overall "allowances" are not exceeded. Any excess reduction in SO_2 by a utility will create "banked" emissions that can be sold or used at another unit.

Phase II of Title IV limits the majority of plants $\geq$20 MW and all plants $\geq$75 MW to maximum emissions of 1.2 pounds of SO_2 per 10^6 Btu after the year 2000. In general, new plants would have to acquire banked emission allowances in order to be built. Emission allowances will be traded through a combination of sell/purchase with other utilities, EPA auctions and direct sales.

Control of NO_x under the CAAA of 1990 will be accomplished through the issuance of a revised NSPS in 1994, with the objective of reducing emissions by 2 million tons a year from 1980 emission levels. The technology being considered is the use of low-NO_x burners (LNBs). The new emission standards will not apply to cyclone and wet bottom boilers, unless alternative technologies are found, as these cannot be retrofitted with existing LNB technologies.

Title V: Operating Permits Title V of the 1990 Clean Air Act Amendments established a new operating permit program to be administered by state agencies in accordance with federal guidelines. An approved state program will basically require a source to obtain a permit that covers each and every requirement applicable to the source under the Clean Air Act. A major impact of the Title V program is that sources are required to implement measures to demonstrate routinely that they are operating in compliance with permit terms. This represents a dramatic shift from previous practices where, to bring an enforcement action, regulatory authorities were required to demonstrate that a source was in noncompliance.

Initially, all "major" sources of air pollution are required to obtain an operating permit. However, any state permitting authority may extend the applicability of the operating permit to minor sources as well. Once a source is subject to the permit program as a major source for any one pollutant, emissions of every regulated air pollutant must be addressed in the permit application.

The operating permit must outline specifically how and when a source will be allowed to operate over the five-year term of the permit. The permit the state develops from an application becomes the principal mechanism for enforcement of all air-quality regulations. As such, it is critically important to submit an application that allows maximum operating flexibility.

Sources must also include in their permit applications monitoring protocols sufficient to document compliance with each permit term and condition.

The Title V operating permit requires the submission of at least five types of reports:

1. The initial compliance report
2. The annual compliance certification
3. Monitoring reports submitted at least every six months
4. Progress reports for sources not in compliance when the application is submitted
5. Prompt reports on any deviations from the permit terms

Permit fees are a mandatory element of the Title V program. In most cases, fees will be assessed for emissions on a dollars-per-ton-emitted basis.

Regulatory Direction The current direction of regulations and air-pollution control efforts is clearly toward significantly reducing the emissions to the environment of a broad range of compounds, including:

1. Volatile organic compounds and other ozone precursors (CO and NO_x)
2. Hazardous air pollutants, including carcinogenic organic emissions and heavy metal emissions
3. Acid rain precursors, including SO_x and NO_x

In addition, the PM_{10} NAAQS will continue to place emphasis on quantifying and reducing particulate emissions in the less than 10-μm particle-size range. Particle size-specific emission factors have been developed for many sources, and size-specific emission standards have been developed in a number of states. These standards are addressing concerns related to HAP emissions of heavy metals, which are generally associated with the submicron particles.

Although it is not possible to predict the future, it is possible to prepare for it and influence it. It is highly recommended that maximum flexibility be designed into new air-pollution control systems to allow for increasingly more stringent emission standards for both particulates and gases. Further, it is everyone's responsibility to provide a thorough review of existing and proposed new processes and to make every attempt to identify economical process modifications and/or material substitutions that reduce or, in some cases, eliminate both the emissions to the environment and the overdependency on retrofitted or new end-of-pipe control systems.

UNITED STATES WATER QUALITY LEGISLATION AND REGULATIONS

Federal Water Pollution Control Act In 1948, the original Federal Water Pollution Control Act (FWPCA) was passed. This act and its various amendments are often referred to as the Clean Water Act (CWA). It provided loans for treatment plant construction and temporary authority for federal control of interstate water pollution. The enforcement powers were so heavily dependent on the states as to make the act almost unworkable. In 1956, several amendments to the FWPCA were passed that made federal enforcement procedures less cumbersome. The provision for state consent was removed by amendments passed in 1961, which also extended federal authority to include navigable waters in the United States.

In 1965 the Water Quality Act established a new trend in water pollution control. It provided that the states set water quality standards in accordance with federal guidelines. If the states failed to do so, the standards would be set by the federal government subject to a review hearing. In 1966, the Clean Water Restoration Act transferred the Federal Water Pollution Control Administration from the Department of Health, Education and Welfare to the Department of the Interior. It also gave the Interior Department the responsibility for the Oil Pollution Act.

After the creation of EPA in 1970, the EPA was given the responsibility previously held by the Department of the Interior with respect to water pollution control. In subsequent amendments to the FWPCA in 1973, 1974, 1975, 1976, and 1977, additional Federal programs were established. The goals of these programs were to make waterways of the United States fishable and swimmable by 1983 and to achieve zero discharge of pollutants by 1985. The National Pollutant Discharge Elimination System (NPDES) was established as the basic regulatory mechanism for water pollution control. Under this program, the states were given the authority to issue permits to "point-source" dischargers provided the dischargers gave assurance that the following standards would be met:

1. Source-specific effluent limitations (including New Source Performance Standards)
2. Toxic pollutant regulations (for specific substances regardless of source)
3. Regulations applicable to oil and hazardous substance liability

In order to achieve that stated water-quality goal of fishable and swimmable waters by 1983, each state was required by EPA to adopt water-quality standards that met or exceeded the Federal water quality criteria. After each state submitted its own water-quality standards, which were subsequently approved by EPA, the Federal criteria were removed from the Code of Federal Regulations. The state water-quality standards are used as the basis for establishing both point-source-based effluent limitations and toxic pollutant limitations used in issuing NPDES permits to point-source discharges.

Source-Based Effluent Limitations   Under the FWPCA, EPA was responsible for establishing point-source effluent limitations for municipal dischargers, industrial dischargers, industrial users of municipal treatment works, and effluent limitations for toxic substances (applicable to all dischargers).

Standards promulgated or proposed by EPA under 40 CFR, Parts 402 through 699, prescribe effluent limitation guidelines for existing sources, standards of performance for new sources, and pretreatment standards for new and existing sources. Effluent limitations and new source performance standards apply to discharges made directly into receiving bodies of water. The new standards require best available technology (BAT) and are to be used by the states when issuing NPDES permits for all sources 18 months after they are made final by EPA. Pretreatment standards apply to waste streams from industrial sources that are sent to publicly owned treatment works (POTW) for final treatment. These regulations are meant to protect the POTW from any materials that would either harm the treatment facility or pass through untreated. They are to be enforced primarily by the local POTW. These standards are applicable to particular classes of point-sources and pertain to discharges into navigable waters without regard to the quality of the receiving water. Standards are specific for numerous subcategories under each point-source category.

Limitations based upon application of the best practicable control technology currently available (BPT) apply to existing point-sources and should have been achieved by July 1, 1977. Limitations based upon application of the BATEA (Best Available Technology Economically Achievable) that will result in reasonable further progress toward elimination of discharges had to be achieved by July 1, 1984.

Clean Water Act of 1977   The 1977 Clean Water Act directed EPA to review all BAT guidelines for conventional pollutants in those industries not already covered.

On August 23, 1978 (43 FR 37570), the EPA proposed a new approach to the control of conventional pollutants by effluent guideline limitations. The new guidelines were known as best conventional pollutants control technology (BCT). These guidelines replaced the existing BAT limitations, which were determined to be unreasonable for certain categories of pollutants.

In order to determine if BCT limitations would be necessary, the cost effectiveness of conventional pollutant reduction to BAT levels beyond BPT levels had to be determined and compared to the cost of removal of this same amount of pollutant by a publicly owned treatment works of similar capacity. If it was equally cost-effective for the industry to achieve the reduction required for meeting the BAT limitations as the POTW, then the BCT limit was made equal to the BAT level. When this test was applied, the BAT limitation set for certain categories were found to be unreasonable. In these subcategories EPA proposed to remove the BAT limitations and revert to the BPT limitations until BCT control levels could be formulated.

Control of Toxic Pollutants   Since the early 1980s, EPA's water-quality standards guidance placed increasing importance on toxic pollutant control. The Agency urged states to adopt criteria into their standards for the priority toxic pollutants, particularly those for which EPA had published criteria guidance. EPA also provided guidance to help and support state adoption of toxic pollutant standards with the *Water Quality Standards Handbook* (1983) and the *Technical Support Document for Water Quality Toxics Control* (1985 and 1991).

Despite EPA's urging and guidance, state response was disappointing. A few states adopted large numbers of numeric toxic pollutant criteria, primarily for the protection of aquatic life. Most other states adopted few or no water-quality criteria for priority toxic pollutants. Some relied on "free from toxicity" criteria and so-called "action levels" for toxic pollutants or occasionally calculated site-specific criteria. Few states addressed the protection of human health by adopting numeric human health criteria.

State development of case-by-case effluent limits using procedures that did not rely on the statewide adoption of numeric criteria for the priority toxic pollutants frustrated Congress. Congress perceived that states were failing to aggressively address toxics and that EPA was not using its oversight role to push the states to move more quickly and comprehensively. Many in Congress believed that these delays undermined the effectiveness of the Act's framework.

1987 CWA Amendments   In 1987, Congress, unwilling to tolerate further delays, added Section 303 (c) (2) (B) to the CWA. The Section provided that, whenever a state reviews water-quality standards or revises or adopts new standards, the state had to adopt criteria for all toxic pollutants listed pursuant to Section 307 (a) (1) of the Act for which criteria have been published under Section 304 (a), the discharge or presence of which in the affected waters could reasonably be expected to interfere with those designated uses adopted by the state, as necessary to support such designated uses. Such criteria had to be specific numerical criteria for such toxic pollutants. When such numerical criteria are not available, wherever a state reviews water-quality standards, or revises or adopts new standards, the state has to adopt criteria based on biological monitoring or assessment methods consistent with information published pursuant to Section 304 (a) (8). Nothing in this Section was to be construed to limit or delay the use of effluent limitations or other permit conditions based on or involving biological monitoring or assessment methods or previously adopted numerical criteria.

In response to this new Congressional mandate, EPA redoubled its efforts to promote and assist state adoption of numerical water-quality standards for priority toxic pollutants. EPA's efforts included the development and issuance of guidance to the states on acceptable implementation procedures. EPA attempted to provide the maximum flexibility in its options that complied not only with the express statutory language but also with the ultimate congressional objective: prompt adoption of numeric toxic pollutant criteria. The Agency believed that flexibility was important so that each state could comply with Section 303 (c) (2) (B) within its resource constraints. EPA distributed final guidance on December 12, 1988. This guidance was similar to earlier drafts available for review by the states. The availability of the guidance was published in the *Federal Register* on January 5, 1989 (54 FR 346).

The structure of Section 303 (c) is to require states to review their water-quality standards at least once each three-year period. Section 303 (c) (2) (B) instructs states to include reviews for toxics criteria whenever they initiate a triennial review. EPA initially looked at February 4, 1990, the 3-year anniversary of the 1987 CWA amendments, as a convenient point to index state compliance. The April 17, 1990 *Federal Register* Notice (55 *FR* 14350) used this index point for the preliminary assessment of state compliance. However, some states were very nearly completing their state administrative processes for ongoing reviews when the 1987 amendments were enacted and could not legally amend those proceedings to address additional toxics criteria. Therefore, in the interest of fairness, and to provide such states a full 3-year review period, EPA's FY 1990 Agency Operating Guidance provided that states should complete adoption of the numeric criteria to meet Section 303 (c) (2) (B) by September 30, 1990.

Section 303 (c) does not provide penalties for states that do not complete timely water-quality standard reviews. In no previous case had an EPA Administrator found that state failure to complete a review within three years jeopardized the public health or welfare to such an extent that promulgation of Federal standards pursuant to Section 303 (c) (4) (B) was justified. However, the pre-1987 CWA never mandated state adoption of priority toxic pollutants or other specific criteria. EPA relied on its water-quality standards regulation (40 CFR 131.11) and its criteria and program guidance to the states on appropriate parametric coverage in state water quality standards, including toxic pollutants. With Congressional concern exhibited in the legislative history for the 1987 Amendments regarding undue delays by states and EPA, and because states have been explicitly required to adopt numeric criteria for appropriate priority toxic pollutants since 1983, the Agency is proceeding to promulgate Federal standards pursuant to Section 303 (c) (4) (B) of the CWA and 40 CFR 131.22 (b).

States have made substantial recent progress in the adoption, and EPA approval, of toxic pollutant water-quality standards. Furthermore, virtually all states have at least proposed new toxics criteria for priority toxic pollutants since Section 303 (c) (2) (B) was added to the CWA in February of 1987. Unfortunately, not all such state proposals address, in a comprehensive manner, the requirements of Section 303 (c) (2) (B). For example, some states have proposed to adopt criteria to protect aquatic life, but not human health; other states have proposed human health criteria that do not address major exposure pathways (such as the combination of both fish consumption and drinking water). In addition, in some cases final adoption of proposed state toxics criteria that would be approved by EPA has been substantially delayed due to controversial and difficult issues associated with the toxic pollutant criteria adoption process.

Biological Criteria   While the overall mandate of the Clean Water Act may now be more clearly stated and understood, the tools needed are still under development, and their full application is being worked out. The direction is towards a more comprehensive approach to water quality protection, which might be more appropriately termed "water resource protection" to encompass the living resources and their habitat along with the water itself.

In 1991, EPA directed states to adopt biological criteria into their water-quality standards by September 30, 1993. To assist the states, EPA issued its "Policy on the Use of Biological Assessments and Criteria in the Water-Quality Program," which sets forth the key policy directions governing the shift to the use of biological criteria:

• "Biological surveys shall be fully integrated with toxicity and chemical-specific assessment methods in state water-quality programs."

• "Biological surveys should be used together with whole-effluent and ambient toxicity testing, and chemical-specific analyses to assess the attainment/nonattainment of designated aquatic life uses in state water-quality standards."

• "If any one of the three assessment methods demonstrate that water-quality standards are not attained, it is EPA's policy that appropriate action should be taken to achieve attainment, including use of regulatory authority" (the independent applicability policy).

• "States should designate aquatic life uses that appropriately address biological integrity and adopt biological criteria necessary to protect those uses."

These policy statements are founded on the existing language and authorities in Clean Water Act Sections 303 (c) (2) (A) and (B). EPA defined biological criteria as "numerical values or narrative expressions used to describe the expected structure and function of the aquatic community."

Most states currently conduct biological surveys. These consist of the collection and analysis of resident aquatic-community data and a subsequent determination of the aquatic community's structure and function. A limiting factor in the number of surveys conducted is simply the funding to support field sampling. Most states also take their survey data through the next level of analysis, which is a biological assessment, in which the biological condition of the water body is evaluated. A state needs to conduct a biological assessment to justify designated waters as having uses less than the Clean Water Act goal of "fishable/swimmable" [Sec. 101 (a) (2)].

Implementing biological criteria requires that the state establish what the expected condition of a biological community should be. This expected condition (or reference condition) is based on measurements at either unimpacted sites or sites that are minimally impacted by human activities. A key element of developing and adopting biological criteria is developing a set of metrics (biological measurements) and indices (summations of those metrics) that are sensitive to reflecting the community changes that occur as a result of human perturbation. Metrics may include measures of species richness, percentages of species of a particular feeding type, and measures of species abundance and condition. The development of metrics and indices have been major technical advances that have effectively brought the field to the point of implementing biological integrity measures.

Aside from the need for additional monitoring and data as a basis for biological criteria, the largest impediment is currently related to the question of how or whether biological criteria should be translated directly into NPDES permit requirements. A strict reading of the NPDES regulations leads some individuals to conclude that every water-quality standard potentially affected by a discharge must have a specific effluent limit. In the case of biological criteria, the direct translation isn't always possible. The preferred approach is to use biological criteria as a primary means to identify impaired waters that will then need additional study. Appropriate discharge controls can then be applied.

The other area of contention in adopting biocriteria is the EPA policy of "independent applicability." This policy recognizes that chemical-specific criteria, toxicity testing (WET tests), and biological criteria each have unique as well as overlapping attributes, limitations, and program applications. No single approach is superior to another; rather, the approaches are complementary. When any one of these approaches indicates that water quality is impaired, appropriate regulatory action is needed. Critics of this approach most frequently cite biological criteria as the superior, or determining, measure and would use it to override an exceedance of chemical-specific criteria.

EPA and many states are now moving towards a watershed or basin management approach to water management. This is a holistic approach that looks at the multiple stresses and activities that affect a basin and evaluates these effects in the context of the survival of ecological systems. Biological criteria are recognized by states as a necessary tool for detecting impacts to important aquatic resources that would otherwise be missed, particularly those caused by nonpoint source activities, and for defining the desired endpoint of environmental restoration activities.

To assist states, EPA has issued programmatic guidance (*Biological Criteria: National Program Guidance for Surface Waters*, April 1990; *Procedures for Initiating Narrative Biological Criteria*, October 1992). In addition, EPA will be issuing over the next couple of years technical guidance specific to developing biological criteria for streams and small rivers, lakes and reservoirs, estuaries, and wetlands.

Metal Bioavailability and Toxicity   Another area of policy and regulatory change that bears directly on questions of biological integrity is the application of toxic-metal water-quality criteria. EPA is in the midst of reconsidering its approach to implementing toxic metals criteria. This was prompted in part by the difficulty that some dischargers are having in meeting the state ambient water-quality criteria for metals, which generally are expressed as total recoverable metals (a measurement that includes the metal dissolved in water plus metal that becomes dissolved when the sample is treated with acid). At issue is whether the criteria can be expressed in a form that more accurately addresses the toxicity of the metal to aquatic life, thereby providing the desired level of aquatic life protection, while allowing more leeway in the total amount of metal discharged. This is a technically complex issue that will require long-term research even if short-term solutions are implemented.

The bioavailability, and hence the toxicity, of metal depends on the physical and chemical form of the metal, which in turn depends on the chemical characteristics of the surrounding water. The dissolved form of the metal is generally viewed as more bioavailable and therefore more toxic than the particulate form. Particulate matter and dissolved organic matter can bind the metal, making it less bioavailable. What is not well known or documented is the various chemical transformations that occur both within the effluent stream and when the effluent reaches and mixes with the receiving water. Metal that is not bioavailable in the effluent may become bioavailable under ambient chemical conditions.

The NPDES regulations (40 CFR 122.45) require effluent limits to be expressed as total recoverable metal. This requirement makes sense as a means to monitor and regulate both the total metal loading and also the effectiveness of wastewater treatment that involves chemical precipitation of the metal.

From the perspective of ecological integrity called for in the Clean Water Act, any adjustment to the implementation of toxic metals criteria needs to be integrated with both sediment criteria and biological criteria to provide ecosystem protection envisioned by the Act.

Regulatory Direction   EPA and states are directed by the Clean Water Act to develop programs to meet the Act's stated objective: "to restore and maintain the chemical, physical, and biological integrity of the Nation's waters" [Sec. 101 (a)]. Efforts to date have emphasized "clean water" quite literally, by focusing on the chemical makeup of discharges and their compliance with chemical water-quality standards established for surface water bodies. These programs have successfully addressed many water-pollution problems, but they are not sufficient to identify and address all of them. A large gap in the current regulatory scheme is the absence of a direct measure of the condition of the biological resources that we are intending to protect (the biological integrity of the water body). Without such a measure it will be difficult to determine whether our water-management approaches are successful in meeting the intent of the Act. Taken together, chemical, physical, and biological integrity are equivalent to the "ecological integrity" of a water body. It is highly likely the future will see interpretation of biological integrity and ecological integrity, and the management of ecological systems, assuming a much more prominent role in water-quality management.

UNITED STATES SOLID WASTE LEGISLATION AND REGULATIONS

Much of the current activity in the field of solid waste management, especially with respect to hazardous wastes and resources recovery, is a direct consequence of recent legislation. Therefore, it is important to review the principal legislation that has affected the entire field of solid-waste management.

What follows is a brief review of existing legislation that affects the management of solid wastes. The actual legislation must be consulted for specific detail. Implementation of the legislation is accomplished through regulations adopted by federal, states, and local agencies. Because these regulations are revised continuously, they must be monitored continuously, especially when design and construction work is to be undertaken.

Rivers and Harbors Act, 1899   Passed in 1899, the Rivers and Harbor Act directed the U.S. Army Corps of Engineers to regulate the dumping of debris in navigable waters and adjacent lands.

Solid Waste Disposal Act, 1965   Modern solid-waste legislation dates from 1965, when the Solid Waste Disposal Act, Title II of Public Law 88-272, was enacted by Congress. The principal intent of this act was to promote the demonstration, construction, and application

of solid waste management and resource-recovery systems that preserve and enhance the quality of air, water, and land resources.

National Environmental Policy Act, 1969   The National Environmental Policy Act (NEPA) of 1969 was the first federal act that required coordination of federal projects and their impacts with the nation's resources. The act specified the creation of Council on Environmental Quality in the Executive Office of the President. This body has the authority to force every federal agency to submit to the council an environmental impact statement on every activity or project which it may sponsor or over which it has jurisdiction.

Resource Recovery Act, 1970   The Solid Waste Disposal Act of 1965 was amended by Public Law 95-512, the Resources Recovery Act of 1970. This act directed that the emphasis of the national solid-waste-management program should be shifted from disposal as its primary objective to that of recycling and reuse of recoverable materials in solid wastes or the conversion of wastes to energy.

Resource Conservation and Recovery Act, 1976   RCRA is the primary statute governing the regulation of solid and hazardous waste. It completely replaced the Solid Waste Disposal Act of 1965 and supplemented the Resource Recovery Act of 1970; RCRA itself was substantially amended by the Hazardous and Solid Waste Amendments of 1984 (HSWA). The principal objectives of RCRA as amended are to:

- Promote the protection of human health and the environment from potential adverse effects of improper solid and hazardous waste management
- Conserve material and energy resources through waste recycling and recovery
- Reduce or eliminate the generation of hazardous waste as expeditiously as possible

To achieve these objectives, RCRA authorized EPA to regulate the generation, treatment, storage, transportation, and disposal of hazardous wastes. The structure of the national hazardous waste regulatory program envisioned by Congress is laid out in Subtitle C of RCRA (Sections 3001 through 3019), which authorized EPA to:

- Promulgate standards governing hazardous waste generation and management
- Promulgate standards for permitting hazardous waste treatment, storage, and disposal facilities
- Inspect hazardous waste management facilities
- Enforce RCRA standards
- Authorize states to manage the RCRA Subtitle C program, in whole or in part, within their respective borders, subject to EPA oversight

Federal RCRA hazardous waste regulations are set forth in 40 CFR Parts 260 through 272. The core of the RCRA regulations establishes the "cradle to grave" hazardous waste regulatory program through seven major sets of regulations:

- Identification and listing of regulated hazardous wastes (Part 261)
- Standards for generators of hazardous waste (Part 262)
- Standards for transporters of hazardous waste (Part 263)
- Standards for owners/operators of hazardous waste treatment, storage, and disposal facilities (Parts 264, 265, and 267)
- Standards for the management of specific hazardous wastes and specific types of hazardous waste management facilities (Part 266)
- Land disposal restriction standards (Part 268)
- Requirements for the issuance of permits to hazardous waste facilities (Part 270)
- Standards and procedures for authorizing state hazardous waste programs to be operated in lieu of the federal program (Part 271)

EPA, under Section 3006 of RCRA, may authorize a state to administer and enforce a state hazardous waste program in lieu of the federal Subtitle C program. To receive authorization, a state program must:

- Be equivalent to the federal Subtitle C program
- Be consistent with, and no less stringent than, the federal program and other authorized state programs
- Provide adequate enforcement of compliance with Subtitle C requirements

Toxic Substances Control Act, 1976   The two major goals of the Toxic Substances Control Act (TSCA), passed by Congress in

1976, are (1) the acquisition of sufficient information to identify and evaluate potential hazards from chemical substances and (2) the regulation of the manufacture, processing, distribution, use, and disposal of any substance that presents an unreasonable risk of injury to health of the environment.

Under TSCA, the EPA has issued a ban on the manufacture, processing, and distribution of products containing PCBs. Exporting of PCB has also been banned. TSCA also required that PCB mixtures containing more than 50 ppm PCBs must be disposed of in an acceptable incinerator or chemical waste landfill. All PCB containers or products containing PCBs had to be clearly marked and records maintained by the operator of each facility handling at least 45 kilograms of

PCB. These records include PCBs in use in transformers and capacitors, PCBs in transformers and capacitors removed from service, PCBs stored for disposal, and a report on the ultimate disposal of the PCBs.

TSCA also placed restrictions on the use of chlorofluorocarbons, asbestos, and fully halogenated chlorofluoroalkanes such as aerosol propellants.

Regulatory Direction   There is no doubt that pollution prevention continues to be the regulatory direction. The development of new and more efficient processes and waste minimization technologies will be essential to support this effort.

POLLUTION PREVENTION

This subsection is drawn in part from "Pollution Prevention Overview," prepared by B. Wainwright and L. Theodore, copyrighted, 1993.

FURTHER READING: American Society of Testing and Materials, *Standard Guide for Industrial Source Reduction*, draft copy dated June 16, 1992. American Society of Testing and Materials, *Pollution Prevention, Reuse, Recycling and Environmental Efficiency*, June, 1992. California Department of Health Services, *Economic Implications of Waste Reduction, Recycling, Treatment, and Disposal of Hazardous Wastes: The Fourth Biennial Report*, California, 1988. Citizen's Clearinghouse for Hazardous Waste, *Reduction of Hazardous Waste: The Only Serious Management Option*, Falls Church, CCHW, 1986. Congress of the United States. Office of Technology Assessment, *Serious Reduction of Hazardous Waste: For Pollution Prevention and Industrial Efficiency*, Washington, D.C., GPO, 1986. Friedlander, S., "Pollution Prevention—Implications for Engineering Design, Research, and Education," *Environment*, May 1989, p. 10. Theodore, L. and Y. McGuinn, *Pollution Prevention*, New York, Van Nostrand Reinhold, 1992. Theodore, L. and R. Allen, *Pollution Prevention: An ETS Theodore Tutorial*, Roanoke, VA, ETS International, Inc., 1994. Theodore, L., *A Citizen's Guide to Pollution Prevention*, East Williston, NY, 1993. United States EPA, *Facility Pollution Prevention Guide*. (EPA/600/R-92/088), Washington, D.C., May 1992. United States EPA, "Pollution Prevention Fact Sheets," Washington, D.C., GPO, 1991. United States EPA, *1987 National Biennial RCRA Hazardous Waste Report—Executive Summary*, Washington, D.C., GPO, 1991. United States EPA, Office of Pollution Prevention, *Report on the U.S. Environmental Protection Agency's Pollution Prevention Program*, Washington, D.C., GPO, 1991. World Wildlife Fund, *Getting at the Source—Executive Summary*, 1991.

INTRODUCTION

The amount of waste generated in the United States has reached staggering proportions; according to the United States Environmental Protection Agency (EPA), 250 million tons of solid waste alone are generated annually. Although both the Resource Conservation and Recovery Act (RCRA) and the Hazardous and Solid Waste Act (HSWA) encourage businesses to minimize the wastes they generate, the majority of the environmental protection efforts are still centered around treatment and pollution clean-up.

The passage of the Pollution Prevention Act of 1990 has redirected industry's approach to environmental management; pollution prevention has now become the environmental option of this decade and the 21st century. Whereas typical waste-management strategies concentrate on "end-of-pipe" pollution control, pollution prevention attempts to handle waste at the source (i.e., source reduction). As waste handling and disposal costs increase, the application of pollution prevention measures is becoming more attractive than ever before. Industry is currently exploring the advantages of multimedia waste reduction and developing agendas to *strengthen* environmental design while *lessening* production costs.

There are profound opportunities for both industry and the individual to prevent the generation of waste; indeed, pollution prevention is today primarily stimulated by economics, legislation, liability concerns, and the enhanced environmental benefit of managing waste at the source. The Pollution Prevention Act of 1990 has established pollution prevention as a national policy, declaring that "waste should be prevented or reduced at the source wherever feasible, while pollu-

tion that cannot be prevented should be recycled in an environmentally safe manner" (Ref. 1). The EPA's policy establishes the following hierarchy of waste management:
1. Source reduction
2. Recycling/reuse
3. Treatment
4. Ultimate disposal

The hierarchy's categories are prioritized so as to promote the examination of each individual alternative prior to the investigation of subsequent options (i.e., the most preferable alternative should be thoroughly evaluated before consideration is given to a less accepted option). Practices that decrease, avoid, or eliminate the generation of waste are considered source reduction and can include the implementation of procedures as simple and economical as good housekeeping. Recycling is the use, reuse, or reclamation of wastes and/or materials that may involve the incorporation of waste recovery techniques (e.g., distillation, filtration). Recycling can be performed at the facility (i.e., on-site) or at an off-site reclamation facility. Treatment involves the destruction or detoxification of wastes into nontoxic or less toxic materials by chemical, biological, or physical methods, or any combination of these control methods. Disposal has been included in the hierarchy because it is recognized that residual wastes will exist; the EPA's so-called "ultimate disposal" options include landfilling, land farming, ocean dumping, and deep-well injection. However, the term *ultimate disposal* is a misnomer, but is included here because of its adaptation by the EPA.

Table 25-8 provides a rough timetable demonstrating the United States' approach to waste management. Note how waste management has begun to shift from pollution *control*-driven activities to pollution *prevention* activities.

The application of waste-management practices in the United States has recently moved toward securing a new pollution prevention ethic. The performance of pollution prevention assessments and their subsequent implementation will encourage increased activity into methods that will further aid in the reduction of hazardous wastes. One of the most important and propitious consequences of the pollution-prevention movement will be the development of life-cycle design and standardized life-cycle cost-accounting procedures. These two consequences are briefly discussed in the two paragraphs that follow. Additional information is provided in a later subsection.

TABLE 25-8 Waste Management Timetable

Prior to 1945	No control
1945–1960	Little control
1960–1970	Some control
1970–1975	Greater control (EPA is founded)
1975–1980	More sophisticated control
1980–1985	Beginning of waste-reduction management
1985–1990	Waste-reduction management
1990–1995	Formal pollution prevention programs (Pollution Prevention Act)
1995–2000	Widespread acceptance of pollution prevention
After 2000	???

The key element of life-cycle design is Life-Cycle Assessment (LCA). LCA is generally envisioned as a process to evaluate the environmental burdens associated with the cradle-to-grave life cycle of a product, process, or activity. A product's life cycle can be roughly described in terms of the following stages:

1. Raw material
2. Bulk material processing
3. Production
4. Manufacturing and assembly
5. Use and service
6. Retirement
7. Disposal

Maintaining an objective process while spanning this life cycle can be difficult given the varying perspective of groups affected by different parts of that cycle. LCA typically does not include any direct or indirect monetary costs or impacts to individual companies or consumers.

Another fundamental goal of life-cycle design is to promote sustainable development at the global, regional, and local levels. There is significant evidence that suggests that current patterns of human and industrial activity on a global scale are not following a sustainable path. Changes to achieve a more sustainable system will require that environmental issues be more effectively addressed in the future. Principles for achieving sustainable development should include (Ref. 2):

1. *Sustainable resource use (conserving resources, minimizing depletion of nonrenewable resources, using sustainable practices for managing renewable resources).* There can be no product development or economic activity of any kind without available resources. Except for solar energy, the supply of resources is finite. Efficient designs conserve resources while also reducing impacts caused by material extraction and related activities. Depletion of nonrenewable resources and overuse of otherwise renewable resources limits their availability to future generations.

2. *Maintenance of ecosystem structure and function.* This is a principal element of sustainability. Because it is difficult to imagine how human health can be maintained in a degraded, unhealthy natural world, the issue of ecosystem health should be a more fundamental concern. Sustainability requires that the health of all diverse species as well as their interrelated ecological functions be maintained. As only one species in a complex web of ecological interactions, humans cannot separate their success from that of the total system.

3. *Environmental justice.* The issue of environmental justice has come to mean different things to different people. Theodore (Ref. 3) has indicated that the subject of environmental justice contains four key elements that are interrelated: environmental racism, environmental health, environmental equity, and environmental politics. (Unlike many environmentalists, Theodore has contended that only the last issue, politics, is a factor in environmental justice.) A major challenge in sustainable development is achieving both intergenerational and intersocietal environmental justice. Overconsuming resources and polluting the planet in such a way that it enjoins future generations from access to reasonable comforts irresponsibly transfer problems to the future in exchange for short-term gain. Beyond this intergenerational conflict, enormous inequities in the distribution of resources continue to exist between developed and less developed countries. Inequities also occur within national boundaries.

Life cycle is a perspective that considers the true costs of product production and/or services provided and utilized by analyzing the price associated with potential environmental degradation and energy consumption, as well as more customary costs like capital expenditure and operating expenses. Unfortunately, a host of economic and economic-related terms have appeared in the literature. Some of these include total cost assessment, life-cycle costing, and full-cost accounting. Unfortunately, these terms have come to mean different things to different people at different times. In an attempt to remove this ambiguity, the following three economic terms are defined below. The reader is also referred to Fig. 25-1 for additional details.

1. *Traditional costing procedure (TCP).* This accounting procedure *only* takes into account capital and operating (including environmental) costs.

2. *Comprehensive costing procedure (CCP).* The economic procedure includes not only the traditional capital and operating costs but

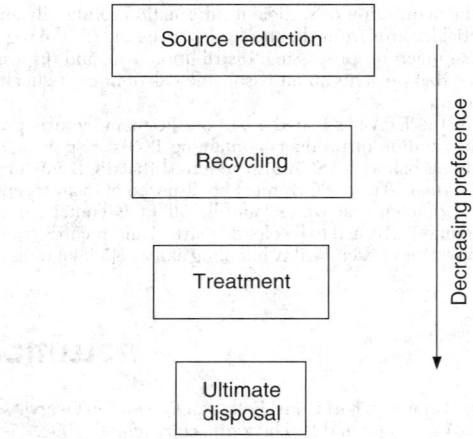

FIG. 25-1 Pollution prevention hierarchy.

also peripheral costs such as liability, regulatory related expensive, borrowing power, and social considerations.

3. *Life-cycle costing (LCC).* This type of analysis requires that all the traditional costs of project or product system, from raw-material acquisition to end-result product disposal, be considered.

The TCP approach is relatively simple and can be easily applied to studies involving comparisons of different equipments, different processes, or even parts of processes. CCP has now emerged as the most realistic approach that can be employed in economic project analyses. It is the recommended procedure for pollution-prevention studies. The LCC approach is usually applied to the life-cycle analysis (LCA) of a product or service. It has found occasional application in project analysis.

The remainder of this subsection on pollution prevention will be concerned with providing the reader with the necessary background to understand the meaning of pollution prevention and its useful implementation. Assessment procedures and the economic benefits derived from managing pollution at the source are discussed along with methods of cost accounting for pollution prevention. Additionally, regulatory and nonregulatory methods to promote pollution prevention and overcome barriers are examined, and ethical considerations are presented. By eliminating waste at the source, all can participate in the protection of the environment by reducing the amount of waste material that would otherwise need to be treated or ultimately disposed; accordingly, attention is also given to pollution prevention in both the domestic and business office environments.

POLLUTION-PREVENTION HIERARCHY

As discussed in the introduction, the hierarchy set forth by the USEPA in the Pollution Prevention Act establishes an order to which waste-management activities should be employed to reduce the quantity of waste generated. The preferred method is source reduction, as indicated in Fig. 25-1. This approach actually precedes traditional waste management by addressing the source of the problem prior to its occurrence.

Although the EPA's policy does not consider recycling or treatment as actual pollution prevention methods per se, these methods present an opportunity to reduce the amount of waste that might otherwise be discharged into the environment. Clearly, the definition of pollution prevention and its synonyms (e.g., waste minimization) must be understood to fully appreciate and apply these techniques.

Waste minimization generally considers all of the methods in the EPA hierarchy (except for disposal) appropriate to reduce the volume or quantity of waste requiring disposal (i.e., source reduction). The definition of *source reduction* as applied in the Pollution Prevention Act, however, is "any practice that reduces the amount of any hazardous substance, pollutant, or contaminant entering any waste stream or otherwise released into the environment . . . prior to recycling, treatment or disposal" (Ref. 1). Source reduction reduces the amount of waste generated; it is therefore considered true pollution prevention and has the highest priority in the EPA hierarchy.

Recycling (or reuse) refers to the use (or reuse) of materials that would otherwise be disposed of or treated as a waste product. A good example is a rechargeable battery. Wastes that cannot be directly reused may often be recovered on-site through methods such as distillation. When on-site recovery or reuse is not feasible due to quality specifications or the inability to perform recovery on-site, off-site recovery at a permitted commercial recovery facility is often a possibility. Such management techniques are considered secondary to source reduction and should only be used when pollution cannot be prevented.

The treatment of waste is the third element of the hierarchy and should be utilized only in the absence of feasible source reduction or recycling opportunities. Waste treatment involves the use of chemical, biological, or physical processes to reduce or eliminate waste material. The incineration of wastes is included in this category and is considered "preferable to other treatment methods (i.e., chemical, biological, and physical) because incineration can permanently destroy the hazardous components in waste materials" (Ref. 4). It can also be employed to reduce the volume of waste to be treated.

Of course, several of these pollution-prevention elements are used by industry in combination to achieve the greatest waste reduction. Residual wastes that cannot be prevented or otherwise managed are then disposed of only as a last resort.

Figure 25-2 provides a more detailed schematic representation of the two preferred pollution prevention techniques (i.e., source reduction and recycling).

MULTIMEDIA ANALYSIS AND LIFE-CYCLE ANALYSIS

Multimedia Analysis   In order to properly design and then implement a pollution prevention program, sources of all wastes must be fully understood and evaluated. A multimedia analysis involves a multifaceted approach. It must not only consider one waste stream but all potentially contaminant media (e.g., air, water, land). Past waste-management practices have been concerned primarily with treatment. All too often, such methods solve one waste problem by transferring a contaminant from one medium to another (e.g., air stripping); such waste shifting is *not* pollution prevention or waste reduction.

Pollution prevention techniques must be evaluated through a thorough consideration of all media, hence the term multimedia. This approach is a clear departure from previous pollution treatment or control techniques where it was acceptable to transfer a pollutant from one source to another in order to solve a waste problem. Such strategies merely provide short-term solutions to an ever increasing problem. As an example, air pollution control equipment prevents or reduces the discharge of waste into the air but at the same time can produce a solid (hazardous) waste problem.

Life-Cycle Analysis   The aforementioned multimedia approach to evaluating a product's waste stream(s) aims to ensure that the treatment of one waste stream does not result in the generation or increase in an additional waste output. Clearly, impacts resulting during the production of a product or service must be evaluated over its entire history or life cycle. This life-cycle analysis or total systems approach (Ref. 3) is crucial to identifying opportunities for improvement. As described earlier, this type of evaluation identifies "energy use, material inputs, and wastes generated during a product's life: from extraction and processing of raw materials to manufacture and transport of a product to the marketplace and finally to use and dispose of the product" (Ref. 5).

During a forum convened by the World Wildlife Fund and the Conservation Foundation in May 1990, various steering committees recommended that a three-part life-cycle model be adopted. This model consists of the following:

1. An inventory of materials and energy used, and environmental releases from all stages in the life of a product or process

2. An analysis of potential environmental effects related to energy use and material resources and environmental releases

3. An analysis of the changes needed to bring about environmental improvements for the product or process under evaluation

Traditional cost analysis often fails to include factors relevant to future damage claims resulting from litigation, the depletion of nat-

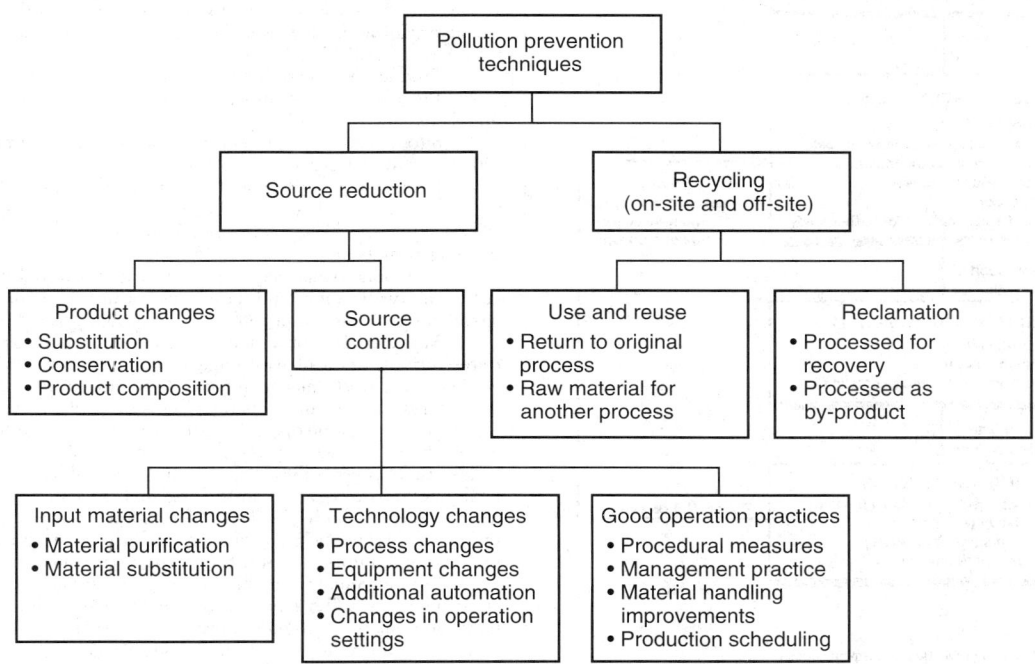

FIG. 25-2   Pollution prevention techniques.

ural resources, the effects of energy use, and the like. As such, waste-management options such as treatment and disposal may appear preferential if an overall life-cycle cost analysis is not performed. It is evident that environmental costs from cradle to grave have to be evaluated together with more conventional production costs to accurately ascertain genuine production costs. In the future, a total systems approach will most likely involve a more careful evaluation of pollution, energy, and safety issues. For example, if one was to compare the benefits of coal versus oil as a fuel source for an electric power plant, the use of coal might be considered economically favorable. In addition to the cost issues, however, one must be concerned with the environmental effects of coal mining (e.g., transportation and storage prior to use as a fuel). Society often has a tendency to overlook the fact that there are serious health and safety matters (e.g., miner exposure) that must be considered along with the effects of fugitive emissions. When these effects are weighed alongside standard economic factors, the full cost benefits of coal usage may be eclipsed by environmental costs. Thus, many of the economic benefits associated with pollution prevention are often unrecognized due to inappropriate cost-accounting methods. For this reason, economic considerations are detailed later.

POLLUTION-PREVENTION ASSESSMENT PROCEDURES

The first step in establishing a pollution prevention program is the obtainment of management commitment. This is necessary given the inherent need for project structure and control. Management will determine the amount of funding allotted for the program as well as specific program goals. The data collected during the actual evaluation is then used to develop options for reducing the types and amounts of waste generated. Figure 25-3 depicts a systematic approach that can be

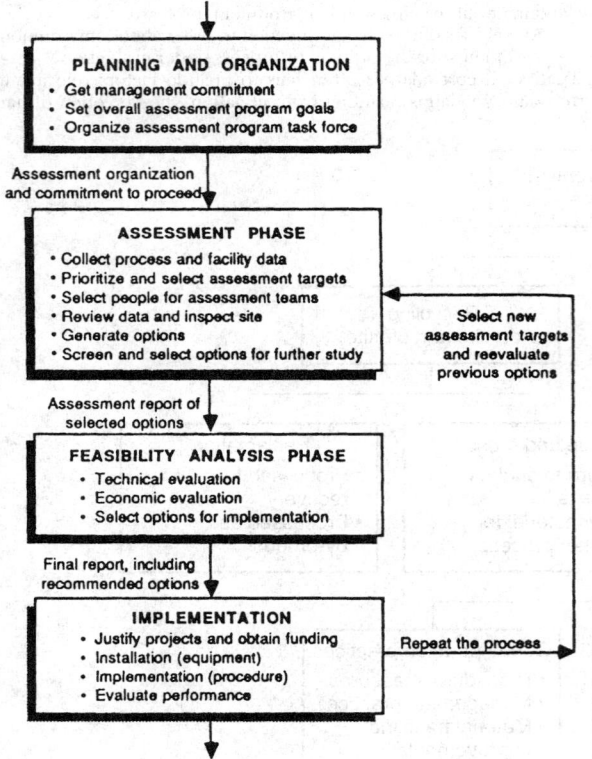

Pollution prevention assessment procedure.

FIG. 25-3 Pollution prevention assessment procedure.

used during the procedure. After a particular waste stream or area of concern is identified, feasibility studies are performed involving both economic and technical considerations. Finally, preferred alternatives are implemented. The four phases of the assessment (i.e., planning and organization, assessment, feasibility, and implementation) are introduced in the following subsections. Sources of additional information as well as information on industrial programs is also provided in this section.

Planning and Organization The purpose of this phase is to obtain management commitment, define and develop program goals, and assemble a project team. Proper planning and organization are crucial to the successful performance of the pollution-prevention assessment. Both managers and facility staff play important roles in the assessment procedure by providing the necessary commitment and familiarity with the facility, its processes, and current waste-management operations. It is the benefits of the program, including economic advantages, liability reduction, regulatory compliance, and improved public image that often leads to management support.

Once management has made a commitment to the program and goals have been set, a program task force is established. The selection of a team leader will be dependent upon many factors including their ability to effectively interface with both the assessment team and management staff.

The task force must be capable of identifying pollution reduction alternatives as well as be cognizant of inherent obstacles to the process. Barriers frequently arise from the anxiety associated with the belief that the program will negatively affect product quality or result in production losses. According to an EPA survey, 30 percent of industry comments responded that they were concerned that product quality would decline if waste minimization techniques were implemented (Ref. 6). As such, the assessment team, and the team leader in particular, must be ready to react to these and other concerns (Ref. 2).

Assessment Phase The assessment phase aims to collect data needed to identify and analyze pollution-prevention opportunities. Assessment of the facility's waste-reduction needs includes the examination of hazardous waste streams, process operations, and the identification of techniques that often promise the reduction of waste generation. Information is often derived from observations made during a facility walk-through, interviews with employees (e.g., operators, line workers), and review of site or regulatory records. One professional organization suggests the following information sources be reviewed, as available (Ref. 7):

1. Product design criteria
2. Process flow diagrams for all solid waste, wastewater, and air emissions sources
3. Site maps showing the location of all pertinent units (e.g., pollution-control devices, points of discharge)
4. Environmental documentation, including: Material Safety Data Sheets (MSDS), military specification data, permits (e.g., NPDES, POTW, RCRA), SARA Title III reports, waste manifests, and any pending permits or application information
5. Economic data, including: cost of raw material management; cost of air, wastewater, and hazardous waste treatment; waste management operating and maintenance costs; and waste disposal costs
6. Managerial information: environmental policies and procedures; prioritization of waste-management concerns; automated or computerized waste-management systems; inventory and distribution procedures; maintenance scheduling practices; planned modifications or revisions to existing operations that would impact waste-generation activities; and the basis of source reduction decisions and policies

The use of process flow diagrams and material balances are worthwhile methods to quantify losses or emissions and provide essential data to estimate the size and cost of additional equipment, other data to evaluate economic performance, and a baseline for tracking the progress of minimization efforts (Ref. 3). Material balances should be applied to individual waste streams or processes and then utilized to construct an overall balance for the facility. Details on these calculations are available in the literature (Ref. 8). In addition, an introduction to this subject is provided in the next section.

The data collected is then used to prioritize waste streams and operations for assessment. Each waste stream is assigned a priority based on corporate pollution-prevention goals and objectives. Once waste origins are identified and ranked, and potential methods to reduce the waste stream are evaluated. The identification of alternatives is generally based on discussions with the facility staff; review of technical literature; and contacts with suppliers, trade organizations, and regulatory agencies.

Alternatives identified during this phase of the assessment are evaluated using screening procedures so as to reduce the number of alternatives requiring further exploration during the feasibility analysis phase. The criteria used during this screening procedure include: cost-effectiveness; implementation time; economic, compliance, safety and liability concerns; waste-reduction potential; and whether the technology is proven (Refs. 4, 8). Options that meet established criteria are then examined further during the feasibility analysis.

Feasibility Analysis   The selection procedure is performed by an evaluation of technical and economic considerations. The technical evaluation determines whether a given option will work as planned. Some typical considerations follow:
1. Safety concerns
2. Product quality impacts or production delays during implementation
3. Labor and/or training requirements
4. Creation of new environmental concerns
5. Waste reduction potential
6. Utility and budget requirements
7. Space and compatibility concerns

If an option proves to be technically ineffective or inappropriate, it is deleted from the list of potential alternatives. Either following or concurrent with the technical evaluation, an economic study is performed, weighing standard measures of profitability such as payback period, investment returns, and net present value. Many of these costs (or, more appropriately, cost savings) may be substantial yet are difficult to quantify. (Refer to *Economic Considerations Associated with Pollution Prevention.*)

Implementation   The findings of the overall assessment are used to demonstrate the technical and economic worthiness of program implementation. Once appropriate funding is obtained, the program is implemented not unlike any other project requiring new procedures or equipment. When preferred waste-pollution-prevention techniques are identified, they are implemented and should become part of the facility's day-to-day management and operation. Subsequent to the program's execution, its performance should be evaluated in order to demonstrate effectiveness, generate data to further refine and augment waste-reduction procedures, and maintain management support.

It should be noted that waste reduction, energy conservation, and safety issues are interrelated and often complementary to each other. For example, the reduction in the amount of energy a facility consumes usually results in reduced emissions associated with the generation of power. Energy expenditures associated with the treatment and transport of waste are similarly reduced when the amount of waste generated is lessened; at the same time, worker safety is elevated due to reduced exposure to hazardous materials. However, this not always the case. Addition of air-pollution control systems at power plants decreases net power output due to the power consumed by the equipment. This in turn requires more fuel to be combusted for the same power exported to the grid. This additional fuel increases pollution from coal mining, transport, ash disposal, and the like. In extreme cases, very high recovery efficiencies for some pollutants can raise, not lower, total emissions. Seventy percent removal might produce a 2 percent loss in power output; 90 percent recovery could lead to a 3 percent loss; 95 percent, a 5 percent loss; 99 percent, a 10 percent loss, and so on. The point to be made is that pollution control is generally not a "free lunch."

Sources of Information   The successful development and implementation of any pollution prevention program is not only dependent on a thorough understanding of the facility's operations but also requires an intimate knowledge of current opportunities and advances in the field. In fact, 32 percent of industry respondents to an EPA survey identified the lack of technical information as a major factor delaying or preventing the implementation of a waste-

minimization program (Ref. 6). One of EPA's positive contributions has been the development of a national Pollution Prevention Information Clearinghouse (PPIC) and the Pollution Prevention Information Exchange System (PIES) to facilitate the exchange of information needed to promote pollution prevention through efficient information transfer (Ref. 2).

PPIC is operated by the EPA's Office of Research and Development and the Office of Pollution Prevention. The clearinghouse is comprised of four elements:
1.   Repository, including a hard copy reference library and collection center and an on-line information retrieval and ordering system.
2.   PIES, a computerized conduit to databases and document ordering, accessible via modem and personal computer: (703) 506-1025.
3.   PPIC uses the RCRA/Superfund and Small Business Ombudsman Hotlines as well as a PPIC technical assistance line to answer pollution-prevention questions, access information in the PPIC, and assist in document ordering and searches. To access PPIC by telephone, call:
   RCRA/Superfund Hotline, (800) 242-9346
   Small Business Ombudsman Hotline, (800) 368-5888
   PPIC Technical Assistance, (703) 821-4800
4.   PPIC compiles and disseminates information packets and bulletins and initiates networking efforts with other national and international organizations.

Additionally, the EPA publishes a newsletter entitled *Pollution Prevention News* that contains information including EPA news, technologies, program updates, and case studies. The EPA's Risk Reduction Engineering Laboratory and the Center for Environmental Research Information has published several guidance documents, developed in cooperation with the California Department of Health Services. The manuals supplement generic waste reduction information presented in the EPA's *Waste Minimization Opportunity Assessment Manual* (Ref. 9).

Pollution prevention or waste minimization programs have been established at the State level and as such are good sources of information. Both Federal and State agencies are working with universities and research centers and may also provide assistance. For example, the American Institute of Chemical Engineers has established the Center for Waste Reduction Technologies (CWRT), a program based on targeted research, technology transfer, and enhanced education.

Industry Programs   A significant pollution-prevention resource may very well be found with the "competition." Several large companies have established well-known programs that have successfully incorporated pollution-prevention practices into their manufacturing processes. These include, but are not limited to: 3M—Pollution Prevention Pays (3P); Dow Chemical—Waste Reduction Always Pays (WRAP); Chevron—Save Money and Reduce Toxics (SMART); and, the General Dynamics Zero Discharge Program.

Smaller companies can benefit by the assistance offered by these larger corporations. It is clear that access to information is of major importance when implementing efficient pollution-prevention programs. By adopting such programs, industry is affirming pollution prevention's application as a good business practice and not simply a "noble" effort.

ASSESSMENT PHASE MATERIAL BALANCE CALCULATIONS

(The reader is directed to Refs. 4 and 10 for further information.)

One of the key elements of the assessment phase of a pollution prevention program involves mass balance equations. These calculations are often referred to as material balances; the calculations are performed via the conservation law for mass. The details of this often-used law are described below.

The conservation law for mass can be applied to any process or system. The general form of the law follows:

$$mass\ in - mass\ out + mass\ generated = mass\ accumulated$$

This equation can be applied to the total mass involved in a process or to a particular species, on either a mole or mass basis. The conservation law for mass can be applied to steady-state or unsteady-state processes and to batch or continuous systems. A steady-state system is one in which there is no change in conditions (e.g., temperature, pressure) or rates of flow with time at any given point in the system; the accumulation term then becomes zero. If there is no chemical reaction, the generation term is zero. All other processes are classified as unsteady state.

To isolate a system for study, the system is separated from the surroundings by a boundary or envelope that may either be real (e.g., a reactor vessel) or imaginary. Mass crossing the boundary and entering the system is part of the mass-in term. The equation may be used for any compound whose quantity does not change by chemical reaction or for any chemical element, regardless of whether it has participated in a chemical reaction. Furthermore, it may be written for one piece of equipment, several pieces of equipment, or around an entire process (i.e., a total material balance).

The conservation of mass law finds a major application during the performance of pollution-prevention assessments. As described earlier, a pollution-prevention assessment is a systematic, planned procedure with the objective of identifying methods to reduce or eliminate waste. The assessment process should characterize the selected waste streams and processes (Ref. 11)—a necessary ingredient if a material balance is to be performed. Some of the data required for the material balance calculation may be collected during the first review of site-specific data; however, in some instances, the information may not be collected until an actual site walk-through is performed.

Simplified mass balances should be developed for each of the important waste-generating operations to identify sources and gain a better understanding of the origins of each waste stream. Since a mass balance is essentially a check to make sure that what goes into a process (i.e., the total mass of all raw materials), which leaves the process (i.e., the total mass of the products and by-products), the material balance should be made individually for all components that enter and leave the process. When chemical reactions take place in a system, there is an advantage to doing "elemental balances" for specific chemical elements in a system. Material balances can assist in determining concentrations of waste constituents where analytical test data are limited. They are particularly useful when there are points in the production process where it is difficult or uneconomical to collect analytical data.

Mass-balance calculations are particularly useful for quantifying fugitive emissions such as evaporative losses. Waste stream data and mass balances will enable one to track flow and characteristics of the waste streams over time. Since in most cases the accumulation equals zero (steady-state operation), it can then be assumed that any buildup is actually leaving the process through fugitive emissions or other means. This will be useful in identifying trends in waste/pollutant generation and will also be critical in the task of measuring the performance of implemented pollution prevention options. The result of these activities is a catalog of waste streams that provides a description of each waste, including quantities, frequency of discharge, composition, and other important information useful for material balance. Of course, some assumptions or educated estimates will be needed when it is impossible to obtain specific information.

By performing a material balance in conjunction with a pollution prevention assessment, the amount of waste generated becomes known. The success of the pollution prevention program can therefore be measured by using this information on baseline generation rates (i.e., that rate at which waste is generated without pollution prevention considerations).

BARRIERS AND INCENTIVES TO POLLUTION PREVENTION

As discussed previously, industry is beginning to realize that there are profound benefits associated with pollution prevention including cost effectiveness, reduced liability, enhanced public image, and regulatory compliance. Nevertheless, there are barriers or disincentives identified with pollution prevention. This section will briefly outline both barriers and incentives that may need to be confronted or considered during the evaluation of a pollution prevention program.

Barriers to Pollution Prevention ("The Dirty Dozen")   There are numerous reasons why more businesses are not reducing the wastes they generate. The following "dirty dozen" are common disincentives:

1. *Technical limitations.*   Given the complexity of present manufacturing processes, waste streams exist that cannot be reduced with current technology. The need for continued research and development is evident.

2. *Lack of information.*   In some instances, the information needed to make a pollution-prevention decision may be confidential or is difficult to obtain. In addition, many decision makers are simply unaware of the potential opportunities available regarding information to aid in the implementation of a pollution-prevention program.

3. *Consumer preference obstacles.*   Consumer preference strongly affects the manner in which a product is produced, packaged, and marketed. If the implementation of a pollution-prevention program results in the increase in the cost of a product or decreased convenience or availability, consumers might be reluctant to use it.

4. *Concern over product quality decline.*   The use of a less hazardous material in a product's manufacturing process may result in decreased life, durability, or competitiveness.

5. *Economic concerns.*   Many companies are unaware of the economic advantages associated with pollution prevention. Legitimate concerns may include decreased profit margins or the lack of funds required for the initial capital investment.

6. *Resistance to change.*   The unwillingness of many businesses to change is rooted in their reluctance to try technologies that may be unproven or based on a combination of the barriers discussed in this section.

7. *Regulatory barriers.*   Existing regulations that have created incentives for the control and containment of wastes are at the same time discouraging the exploration of pollution-prevention alternatives. Moreover, since regulatory enforcement is often intermittent, current legislation can weaken waste-reduction incentives.

8. *Lack of markets.*   The implementation of pollution-prevention processes and the production of environmentally friendly products will be of no avail if markets do not exist for such goods. As an example, the recycling of newspaper in the United States has resulted in an overabundance of waste-paper without markets prepared to take advantage of this raw material.

9. *Management apathy.*   Many managers capable of making decisions to begin pollution-prevention activities, do not realize the potential benefits of pollution prevention.

10. *Institutional barriers.*   In an organization without a strong infrastructure to support pollution-prevention plans, waste-reduction programs will be difficult to implement. Similarly, if there is no mechanism in place to hold individuals accountable for their actions, the successful implementation of a pollution-prevention program will be limited.

11. *Lack of awareness of pollution prevention advantages.*   As mentioned in reason no. 5, decision makers may merely be uninformed of the benefits associated with pollution reduction.

12. *Concern over the dissemination of confidential product information.*   If a pollution-prevention assessment reveals confidential data pertinent to a company's product, fear may exist that the organization will lose a competitive edge with other businesses in the industry.

Pollution-Prevention Incentives ("A Baker's Dozen")   Various means exist to encourage pollution prevention through regulatory measures, economic incentives, and technical assistance programs. Since the benefits of pollution prevention can surpass prevention barriers, a "baker's dozen" incentives is presented below:

1. *Economic benefits.*   The most obvious economic benefits associated with pollution prevention are the savings that result from the elimination of waste storage, treatment, handling, transport, and disposal. Additionally, less tangible economic benefits are realized in terms of decreased liability, regulatory compliance costs (e.g., permits), legal and insurance costs, and improved process efficiency. Pollution prevention almost always pays for itself, particularly when the time required to comply with regulatory standards is considered. Several of these economic benefits are discussed separately below.

2. *Regulatory compliance.* Quite simply, when wastes are not generated, compliance issues are not a concern. Waste-management costs associated with record keeping, reporting, and laboratory analysis are reduced or eliminated. Pollution prevention's proactive approach to waste management will better prepare industry for the future regulation of many hazardous substances and wastes that are currently unregulated. Regulations have, and will continue to be, a moving target.

3. *Liability reduction.* Facilities are responsible for their wastes from "cradle-to-grave." By eliminating or reducing waste generation, future liabilities can also be decreased. Additionally, the need for expensive pollution liability insurance requirements may be abated.

4. *Enhanced public image.* Consumers are interested in purchasing goods that are safer for the environment, and this demand, depending on how they respond, can mean success or failure for many companies. Business should therefore be sensitive to consumer demands and use pollution-prevention efforts to their utmost advantage by producing goods that are environmentally friendly.

5. *Federal and state grants.* Federal and state grant programs have been developed to strengthen pollution-prevention programs initiated by states and private entities. The EPA's "Pollution Prevention by and for Small Business" grant program awards grants to small businesses to assist their development and demonstration of new pollution-prevention technologies.

6. *Market incentives.* Public demand for environmentally preferred products has generated a market for recycled goods and related products; products can be designed with these environmental characteristics in mind, offering a competitive advantage. In addition, many private and public agencies are beginning to stimulate the market for recycled goods by writing contracts and specifications that call for the use of recycled materials.

7. *Reduced waste-treatment costs.* As discussed in reason no. 5 of the dirty dozen, the increasing costs of traditional end-of-pipe waste-management practices are avoided or reduced through the implementation of pollution-prevention programs.

8. *Potential tax incentives.* In an effort to promote pollution prevention, taxes may eventually need to be levied to encourage waste generators to consider reduction programs. Conversely, tax breaks could be developed for corporations that utilize pollution-prevention methods to foster pollution prevention.

9. *Decreased worker exposure.* By reducing or eliminating chemical exposures, businesses benefit by lessening the potential for chronic workplace exposure and serious accidents and emergencies. The burden of medical monitoring programs, personal exposure monitoring, and potential damage claims are also reduced.

10. *Decreased energy consumption.* As mentioned previously, methods of energy conservation are often interrelated and complementary to each other. Energy expenditures associated with the treatment and transport of waste are usually but not always reduced when the amount of waste generated is lessened, while at the same time the pollution associated with energy consumed by these activities is abated.

11. *Increased operating efficiencies.* A potential beneficial side effect of pollution-prevention activities is a concurrent increase in operating efficiency. Through a pollution-prevention assessment, the assessment team can identify sources of waste that result in hazardous waste generation *and* loss in process performance. The implementation of a reduction program will often rectify such problems through modernization, innovation, and the implementation of good operating practices.

12. *Competitive advantages.* By taking advantage of the many benefits associated with pollution prevention, businesses can gain a competitive edge.

13. *Reduced negative environmental impacts.* Through an evaluation of pollution-prevention alternatives, which consider a total systems approach, consideration is given to the negative impact of environmental damage to natural resources and species that occur during raw-material procurement and waste disposal. The performance of pollution-prevention endeavors will therefore result in enhanced environmental protection.

ECONOMIC CONSIDERATIONS ASSOCIATED WITH POLLUTION-PREVENTION PROGRAMS

The purpose of this subsection is to outline the basic elements of a pollution-prevention cost-accounting system that incorporates both traditional and less tangible economic variables. The intent is not to present a detailed discussion of economic analysis but to help identify the more important elements that must be considered to properly quantify pollution-prevention options.

The greatest driving force behind any pollution-prevention plan is the promise of economic opportunities and cost savings over the long term. Pollution prevention is now recognized as one of the lowest-cost options for waste/pollutant management. Hence, an understanding of the economics involved in pollution prevention programs/options is quite important in making decisions at both the engineering and management levels. Every engineer should be able to execute an economic evaluation of a proposed project. If the project cannot be justified economically after *all* factors—and these will be discussed in more detail below—have been taken into account, it should obviously not be pursued. The earlier such a project is identified, the fewer resources will be wasted.

Before the true cost or profit of a pollution-prevention program can be evaluated, the factors contributing to the economics must be recognized. There are two traditional contributing factors (capital costs and operating costs), but there are also other important costs and benefits associated with pollution prevention that need to be quantified if a meaningful economic analysis is going to be performed. Table 25-9 demonstrates the evolution of various cost-accounting methods. Although Tables 25-8 (see introduction) and 25-9 are not directly related, the reader is left with the option of comparing some of the similarities between the two.

The Total Systems Approach (TSA) referenced in Table 25-9 aims to quantify not only the economic aspects of pollution prevention but also the social costs associated with the production of a product or service from cradle to grave (i.e., life cycle). The TSA attempts to quantify less tangible benefits such as the reduced risk derived from not using a hazardous substance. The future is certain to see more emphasis placed on the TSA approach in any pollution-prevention program. As described earlier, a utility considering the option of converting from a gas-fired boiler to coal-firing is usually not concerned with the environmental effects and implications associated with such activities as mining, transporting, and storing the coal prior to its usage as an energy feedstock. Pollution-prevention approaches in the mid-to-late 1990s will become more aware of this need.

The economic evaluation referred to above is usually carried out using standard measures of profitability. Each company and organization has its own economic criteria for selecting projects for implementation. (For example, a project can be judged on its payback period. For some companies, if the payback period is more than 3 years, it is a dead issue.) In performing an economic evaluation, various costs and savings must be considered. The economic analysis presented in this subsection represents a preliminary, rather than a detailed, analysis. For smaller facilities with only a few (and perhaps simple) processes, the entire pollution-prevention assessment procedure will tend to be much less formal. In this situation, several obvious pollution-prevention options such as the installation of flow controls and good operating practices may be implemented with little

TABLE 25-9 Economic Analysis Timetable

Prior to 1945	Capital costs only
1945–1960	Capital and some operating costs
1960–1970	Capital and operating costs
1970–1975	Capital, operating, and some environmental control costs
1975–1980	Capital, operating, and environmental control costs
1980–1985	Capital, operating, and more sophisticated environmental control costs
1985–1990	Capital, operating, and environmental controls, and some life-cycle analysis (Total Systems Approach)
1990–1995	Capital, operating, and environmental control costs and life-cycle analysis (Total Systems Approach)
1995–2000	Widespread acceptance of Total Systems Approach
After 2000	???

or no economic evaluation. In these instances, no complicated analyses are necessary to demonstrate the advantages of adopting the selected pollution-prevention option. A proper perspective must also be maintained between the magnitude of savings that a potential option may offer and the amount of manpower required to do the technical and economic feasibility analyses. A short description of the various economic factors—including capital and operating costs and other considerations—follows.

Once identified, the costs and/or savings are placed into their appropriate categories and quantified for subsequent analysis. Equipment cost is a function of many variables, one of the most significant of which is capacity. Other important variables include operating temperature and/or pressure conditions, and degree of equipment sophistication. Preliminary estimates are often made using simple cost-capacity relationships that are valid when the other variables are confined to a narrow range of values.

The usual technique for determining the capital costs (i.e., total capital costs, which include equipment design, purchase, and installation) for the facility is based on the factored method of establishing direct and indirect installation costs as a function of the known equipment costs. This is basically a modified Lang method, whereby cost factors are applied to known equipment costs (Refs. 13 and 14). The first step is to obtain from vendors the purchase prices of the primary and auxiliary equipment. The total base price, designated by X, which should include instrumentation, control, taxes, freight costs, and so on, serves as the basis for estimating the direct and indirect installation costs. These costs are obtained by multiplying X by the cost factors, which are available in the literature (see Refs. 13–20).

The second step is to estimate the direct installation costs by summing all the cost factors involved in the direct installation costs, which include piping, insulation, foundation and supports, and so on. The sum of these factors is designated as the DCF (direct installation cost factor). The direct installation costs are then the product of the DCF and X. The third step consists of estimating the indirect installation costs; that is, all the cost factors for the indirect installation costs (engineering and supervision, startup, construction fees, and so on) are added; the sum is designated by ICF (indirect installation cost factor). The indirect installation costs are then the product of ICF and X. Once the direct and indirect installation costs have been calculated, the total capital cost (TCC) may be evaluated as follows:

$$TCC = X + (DCF)(X) + (ICF)(X)$$

This is then converted to annualized capital costs (ACC) with the use of the capital recovery factor (CRF), which can be calculated from the following equation:

$$CRF = \frac{(i)(1+i)^n}{(1+i)^n - 1}$$

where n = projected lifetime of the project, years
i = annual interest rate, expressed as a fraction

The annualized capital cost (ACC) is the product of the CRF and TCC and represents the total installed equipment cost distributed over the lifetime of the project. The ACC reflects the cost associated with the initial capital outlay over the depreciable life of the system. Although investment and operating costs can be accounted for in other ways such as present-worth analysis, the capital recovery method is preferred because of its simplicity and versatility. This is especially true when comparing somewhat similar systems having different depreciable lives. In such decisions, there are usually other considerations besides economic, but if all other factors are equal, the alternative with the lowest total annualized cost should be the most viable.

Operating costs can vary from site to site since these costs reflect local conditions (e.g., staffing practices, labor, utility costs). Operating costs, like capital costs, may be separated into two categories: direct and indirect costs. Direct costs are those that cover material and labor and are directly involved in operating the facility. These include labor, materials, maintenance and maintenance supplies, replacement parts, wastes, disposal fees, utilities, and laboratory costs. Indirect costs are those operating costs associated with, but not directly involved in, operating the facility; costs such as overhead (e.g., building/land leas-

ing and office supplies), administrative fees, property taxes, and insurance fees) fall into this category. However, the major direct operating costs are usually those associated with labor and materials.

The main problem with the traditional type of economic analysis is that it is difficult—nay, in some cases impossible—to quantify some of the not-so-obvious economic merits of a pollution-prevention program. Several considerations have just recently surfaced as factors that need to be taken into account in any meaningful economic analysis of a pollution-prevention effort. What follows is a summary listing of these considerations, most which have been detailed earlier.
1. Decreased long-term liabilities
2. Regulatory compliance
3. Regulatory recordkeeping
4. Dealings with the EPA
5. Dealings with state and local regulatory bodies
6. Elimination or reduction of fines and penalties
7. Potential tax benefits
8. Customer relations
9. Stockholder support (corporate image)
10. Improved public image
11. Reduced technical support
12. Potential insurance costs and claims
13. Effect on borrowing power
14. Improved mental and physical well-being of employees
15. Reduced health-maintenance costs
16. Employee morale
17. Other process benefits
18. Improved worker safety
19. Avoidance of rising costs of waste treatment and/or disposal
20. Reduced training costs
22. Reduced emergency response planning

Many proposed pollution-prevention programs have been squelched in their early stages because a comprehensive economic analysis was not performed. Until the effects described above are included, the true merits of a pollution-prevention program may be clouded by incorrect and/or incomplete economic data. Can something be done by industry to remedy this problem? One approach is to use a modified version of the standard Delphi panel. In order to estimate these other economic benefits of pollution prevention, several knowledgeable individuals within and perhaps outside the organization are asked to independently provide estimates, with explanatory details, on these economic benefits. Each individual in the panel is then allowed to independently review all responses. The cycle is then repeated until the group's responses approach convergence.

Finally, pollution-prevention measures can provide a company with the opportunity of looking their neighbors in the eye and truthfully saying that all that can reasonably be done to prevent pollution is being done. In effect, the company is doing right by the environment. Is there an economic advantage to this? It is not only a difficult question to answer quantitatively but also a difficult one to answer qualitatively. The reader is left to ponder the answer to this question.

POLLUTION PREVENTION AT THE DOMESTIC AND OFFICE LEVELS

Concurrent with the United States' growth as an international economic superpower during the years following World War II, a new paradigm was established whereby society became accustomed to the convenience and ease with which goods could be discarded after a relatively short useful life. Individuals have come to expect these everyday comforts with what may be considered an unconscious ignorance towards the ultimate effect of the throwaway lifestyle. In fact, many, while fearful of environmental degradation, are not aware of the ill effect these actions have on the world as a whole. Many individuals who abide by the "not in my back yard" (NIMBY) mind-set also feel pollution prevention does not have to occur "in my house."

The past two decades have seen an increased social awareness of the impact of modern-day lifestyles on the environment. Public environmental concerns include issues such as waste disposal; hazardous-material regulations; depletion of natural resources; and air, water, and land pollution. Nevertheless, roughly one-half of the total quantity of waste generated each year can be attributed to domestic sources!

More recently, concern about the environment has begun to stimulate environmentally correct behavior. After all, the choices made today affect the environment of tomorrow. Simple decisions can be made at work and at home that conserve natural resources and lessen the burden placed on a waste-management system. By eliminating waste at the source, society is participating in the protection of the environment by reducing the amount of waste that would otherwise need to be treated or ultimately disposed.

There are numerous areas of environmental concern that can be directly influenced by the consumer's actions. The first issue, which is described above, is that of waste generation. Second, energy conservation has significantly affected Americans and has resulted in cost-saving measures that have directly reduced pollution. As mentioned previously, energy conservation is directly related to pollution prevention since a reduction in energy use usually corresponds to less energy production and, consequently, less pollution output. A third area of concern is that of accident and emergency planning. Relatively recent accidents like Chernobyl and Bhopal have increased public awareness and helped stimulate regulatory policies concerned with emergency planning. Specifically, Title III of the Superfund Amendments and Reauthorization Act (SARA) of 1986 established the Emergency Planning and Community Right-to-Know Act and forever changed the concept of environmental management. This law attempts to avert potential emergencies through careful planning and the development of contingency plans. Planning for emergency situations can help protect human health and the environment (Ref. 21).

Based on the above three areas of potential concern, a plethora of waste reduction activities can be performed at home or in an office environment. One can easily note the similarities between these activities and those pollution-prevention activities performed in industry. The following few examples are identified by category according to this notation (Ref. 21):

- ●   Waste reduction
- ★   Accidents, health, and safety
- ■   Energy conservation

At home
- ●   Purchase products with the least amount of packaging
- ●   Borrow items used infrequently
- ★   Handle materials to avoid spills (and slips, trips)
- ★   Keep hazardous materials out-of-reach of children
- ■   Use energy-efficient lighting (e.g., florescent)
- ■   Install water-flow restriction devices on sink faucets and showerheads

At the office
- ●   Pass on verbal memos when written correspondence isn't required
- ●   Reuse paper before recycling it
- ★   Know building evacuation procedures
- ★   Adhere to company medical policies (e.g., annual physicals)
- ■   Don't waste utilities simply because you are not paying for them
- ■   Take public transportation to the office

The use of pollution-prevention principles on the home front clearly does not involve the use of high-technology equipment or major lifestyle changes; success is only dependent upon active and willing public participation. All can help to make a difference.

Before pollution prevention becomes a fully accepted way of life, considerable effort still needs to be expended to change the way one looks at waste management. A desire to use "green" products and services will be of no avail in a market where these goods are not available. Participation in pollution-prevention programs will increase through continued education, community efforts, and lobbying for change. Market incentives can be created and strengthened by tax policies, price preferences, and packaging regulations created at the federal, state, and local levels. Each individual should take part by communicating with industry and expressing concerns. For example,

citizens should feel free to write letters to decision makers at both business organizations and government institutions regarding specific products or legislation. Letters can be positive, demonstrating personal endorsement of a green product, or disapproving, expressing discontent and reluctance to use a particular product because of its negative environmental effect (Ref. 21).

By starting now, society will learn through experience to better manage its waste while providing a safer and cleaner environment for future generations.

ETHICAL CONSIDERATIONS

Given the evolutionary nature of pollution prevention, it is evident that as technology changes and continued progress is achieved, society's opinion of both what is possible and desirable will also change. Government officials, scientists, and engineers will face new challenges to fulfill society's needs while concurrently meeting the requirements of changing environmental regulations. It is now apparent that attention should also be given to ethical considerations and their application to pollution prevention policy. How one makes decisions on the basis of ethical beliefs is clearly a personal issue but one that should be addressed. In order to examine this issue, the meaning of ethics must be known, although the intent here is not to provide a detailed discussion regarding the philosophy of ethics or morality.

Ethics can simply be defined as the analysis of the rightness and wrongness of an act or actions. According to Dr. Andrew Varga, director of the Philosophical Resources Program at Fordham University, in order to discern the morality of an act, it is customary to look at the act on the basis of four separate elements: its object, motive, circumstances, and consequences (Ref. 22). Rooted in this analysis is the belief that if one part of the act is bad, then the overall act itself cannot be considered good. Of course, there are instances where the good effects outweigh the bad, and therefore there may be a reason to permit the evil. The application of this principle is not cut and dry and requires decisions to be made on a case-by-case basis. Each must make well-judged decisions rooted in an understanding of the interaction between technology and the environment. After all, the decisions made today will have an impact not only on this generation but on many generations to come. If one chooses today not to implement a waste-reduction program in order to meet a short-term goal of increased productivity, this might be considered a good decision since it benefits the company and its employees. However, should a major release occur that results in the contamination of a local sole-source of drinking water, what is the good?

As an additional example, toxicological studies have indicated that test animals exposed to small quantities of toxic chemicals had better health than control groups that were not exposed. A theory has been developed that says that a low-level exposure to the toxic chemical results in a challenge to the animal to maintain homeostasis; this challenge increases the animal's vigor and, correspondingly, its health. However, larger doses seem to cause an inability to adjust, resulting in negative health effects. Based on this theory, some individuals would believe that absolute pollution reduction might not be necessary.

FUTURE TRENDS

The reader should keep an open mind when dealing with the types of issues discussed above and facing challenges perhaps not yet imagined. Clearly, there is no simple solution or answer to many of the questions. The EPA is currently *attempting* to develop a partnership with government, industry, and educators to produce and distribute pollution-prevention educational materials. Given EPA's past history and performance, there are understandable doubts as to whether this program will succeed (Ref. 23).

Finally, no discussion on pollution prevention would be complete without reference to the activities of the 49-year-old International Organization for Standardization (ISO). ISO recently created Technical Committee 207 (TC 207) to begin work on new standards for environmental management systems (EMS). The ramifications, especially to the chemical industry, which has become heavily involved in the development of these standards, will be great. TC 207's activities are being scrutinized closely and will continue to be into 1997 when the first environmental standards are expected to be published. This new environmental management set of standards will be entitled ISO 14000. Once this new standard is in place, it is expected that customers will require their product suppliers to be certified by ISO 14000. In addition, some service suppliers who have an impact on the environment will also probably be expected to obtain an ISO 14000 certification. Certification will imply to the customer that, when the product or service was prepared, the environment was not significantly damaged in the process. This will effectively require that the life-cycle design mentality discussed earlier be applied to all processes. Implementing the life-cycle design framework will require significant organizational and operational changes in business. To effectively promote the goals of sustainable development, life-cycle designs will have to successfully address cost, performance, and cultural and legal factors.

The impact of all of the above on both the industry and the consumer will be significant. All have heard the expression "I've met the enemy . . . and we're it." After all, it is the consumer who can and will ultimately have the final say. Once the consumer refuses to buy and/or accept products and/or services that damage the environment, that industry is either out of business or must change its operations to environmentally acceptable alternatives. With the new standards, customers (in addition to other organizations) of certified organizations will be assured that the products or services they purchase have been produced in accordance with universally accepted standards of environmental management. Organizational claims, which today can be misleading or erroneous, will, under the standards, be backed up by comprehensive and detailed environmental management systems that must withstand the scrutiny of intense audits.

Can all of the above be achieved in the near future? If it can, then society need only key its environmental efforts on educating the consumer.

AIR-POLLUTION MANAGEMENT OF STATIONARY SOURCES

FURTHER READING: Billings and Wilder, *Fabric Filter Handbook*, U.S. EPA, NTIS Publ. PB 200-648 (vol. 1), PB 200-649 (vol. 2), PB 200-651 (vol. 3), and PB 200-650 (vol. 4), 1970. Buonicore, "Air Pollution Control," *Chem. Eng.*, **87**(13), 81 (June 30, 1980). Buonicore and Theodore, *Industrial Control Equipment for Gaseous Pollutants*, vols. I and II, CRC Press, Boca Raton, Florida, 1975. Calvert, *Scrubber Handbook*, U.S. EPA, NTIS Publ. PB 213-016 (vol. 1) and PB 213-017 (vol. 2), 1972. Danielson, *Air Pollution Engineering Manual*, EPA Publ. AP-40, 1973. Davis, *Air Filtration*, Academic, New York, 1973. Kleet and Galeski, *Flare Systems Study*, EPA-600/2-76-079 (NTIS), 1976. Lund, *Industrial Pollution Control Handbook*, McGraw-Hill, New York, 1971. Oglesby and Nichols, *Manual of Electrostatic Precipitator Technology*, U.S. EPA, NTIS Publ. PB 196-380 (vol. 1), PB 196-381 (vol. 2), PB 196-370 (vol. 3), and PB 198-150 (vol. 4), 1970. *Package Sorption System Study*, EPA-R2-73-202 (NTIS), 1973. Rolke et al., *Afterburner Systems Study*, U.S. EPA, NTIS Publ. PB 212-500, 1972. Slade, *Meteorology and Atomic Energy*, AEC (TID-24190), Oak Ridge, Tennessee, 1969. Stern, *Air Pollution*, Academic, New York, 1974. Strauss, *Industrial Gas Cleaning*, Pergamon, New York, 1966. Buonicore and Theodore, *Industrial Control Equipment for Gaseous Pollutants*, vol. 1, 2d ed., CRC Press, Boca Raton, Florida, 1992. Theodore and Buonicore, *Air Pollution Control Equipment: Selection, Design, Operation, and Maintenance*, Prentice Hall, Englewood Cliffs, New Jersey, 1982. Treybal, *Mass Transfer Operations*, 3d ed., McGraw-Hill, New York, 1980. Turner, *Workbook of Atmospheric Dispersion Estimates*, U.S. EPA Publ. AP-26, 1970. White, *Industrial Electrostatic Precipitation*, Addison-Wesley, Reading, Massachusetts, 1963. McKenna, Mycock, and Theodore, *Handbook of Air Pollution Control Engineering and Technology*, CRC Press, Boca Raton, Florida, 1995. Theodore and Allen, *Air Pollution Control Equipment*, ETSI Prof. Training Inst., Roanoke, VA, 1993. *Air Pollution Engineering Manual*, Van Nostrand Reinhold, New York, 1992. *Guideline on Air Quality Models (GAQM)*, rev. ed., EPA-450/2-78-027R, July 1986. *Compilation of Air Pollution Emission Factors (AP-42)*, 4th ed., U.S. EPA, Research Triangle Park, North Carolina, September, 1985. McKenna and Turner, *Fabric Filter—Baghouses I—Theory, Design, and Selection*, ETSI Prof. Training Inst., Roanoke, Virginia, 1993. Greiner, *Fabric Filter—Baghouses II—Operation, Maintenance, and Troubleshooting*, ETS Prof. Training Inst., Roanoke, Virginia, 1993. Yang, Yonghua, and Allen, "Biofiltration Control of Hydrogen Sulfide. 2. Kinetics Biofilter Performance and Maintenance," *JAWA*, **44**, p. 1315.

INTRODUCTION

Air pollutants may be classified into two broad categories: (1) natural and (2) human-made. Natural sources of air pollutants include:

- Windblown dust
- Volcanic ash and gases
- Ozone from lightning and the ozone layer
- Esters and terpenes from vegetation
- Smoke, gases, and fly ash from forest fires
- Pollens and other aeroallergens
- Gases and odors from biologic activities
- Natural radioactivity

Such sources constitute background pollution and that portion of the pollution problem over which control activities can have little, if any, effect.

Human-made sources cover a wide spectrum of chemical and physical activities and are the major contributors to urban air pollution. Air pollutants in the United States pour out from over 10 million vehicles, the refuse of over 250 million people, the generation of billions of kilowatts of electricity, and the production of innumerable products demanded by everyday living. Hundreds of millions of tons of air pollutants are generated annually in the United States alone. The five main classes of pollutants are particulates, sulfur dioxide, nitrogen oxides, volatile organic compounds, and carbon monoxide. Total emissions in the United States are summarized by source category for the year 1993 in Table 25-10.

Air pollutants may also be classified as to the origin and state of matter:

1. Origin
   a. Primary. Emitted to the atmosphere from a process
   b. Secondary. Formed in the atmosphere as a result of a chemical reaction
2. State of matter
   a. Gaseous. True gases such as sulfur dioxide, nitrogen oxide, ozone, carbon monoxide, etc.; vapors such as gasoline, paint solvent, dry cleaning agents, etc.
   b. Particulate. Finely divided solids or liquids; solids such as dust, fumes, and smokes; and liquids such as droplets, mists, fogs, and aerosols

Gaseous Pollutants   Gaseous pollutants may be classified as inorganic or organic. Inorganic pollutants consist of:

1. *Sulfur gases.*   Sulfur dioxide, sulfur trioxide, hydrogen sulfide
2. *Oxides of carbon.*   Carbon monoxide, carbon dioxide
3. *Nitrogen gases.*   Nitrous oxide, nitric oxide, nitrogen dioxide, other nitrous oxides
4. *Halogens, halides.*   Hydrogen fluoride, hydrogen chloride, chlorine, fluorine, silicon tetrafluoride
5. *Photochemical products.*   Ozone, oxidants
6. *Cyanides.*   Hydrogen cyanide
7. *Ammonium compounds.*   Ammonia
8. *Chlorofluorocarbons.*   1,1,1-trichloro-2,2,2-trifluoroethane; trichlorofluoromethane, dichlorodifluoromethane; chlorodifluoromethane; 1,2-dichloro-1,1,2,2-tetrafluoroethane; chloropentafluoroethane

Organic pollutants consist of:

1. Hydrocarbons
   a. Paraffins. Methane, ethane, octane
   b. Acetylene
   c. Olefins. Ethylene, butadiene
   d. Aromatics. Benzene, toluene, benzpyrene, xylene, styrene
2. Aliphatic oxygenated compounds
   a. Aldehydes. Formaldehyde
   b. Ketones. Acetone, methylethylketone

TABLE 25-10 1993 National Emissions by Source Category in short tons ×1000

Source category	Particulate (PM-10)	Sulfur dioxide	Nitrogen oxides	Volatile organic compounds	Carbon monoxide
Fuel combustion					
Electric utility	270	15836	7782	36	322
Industrial	219	2830	3176	271	667
Other	723	600	732	341	4444
Chemical and allied product manufacturing	75	450	414	1811	1998
Metals processing	141	580	82	74	2091
Petroleum and related industries	26	409	95	720	398
Other industrial processes	311	413	314	486	732
Solvent utilization	2	1	3	6249	2
Storage and transport	55	5	3	1861	56
Waste disposal and recycling	248	37	84	2271	1732
Highway vehicles	197	438	7437	6094	59999
Off-highway	395	278	2986	2207	15272
Natural sources—wind erosion	628				
Miscellaneous	42200	11	296	893	9506
Fugitive dust	41801				
Nonfugitive dust	399				
Total	45490	21888	23404	23314	97209

SOURCE: EPA-454/R-94-027, National Air Pollutant Emission Trends, 1900–1993.

    c. Organic acids
    d. Alcohols. Methanol, ethanol, isopropanol
    e. Organic halides. Cyanogen chloride bromobenzyl cyanide
    f. Organic sulfides. Dimethyl sulfide
    g. Organic hydroperoxides. Peroxyacetyl nitrite or nitrate (PAN)

The most common gaseous pollutants and their major sources and significance are presented in Table 25-11.

Particulate Pollutants Particulates may be defined as solid or liquid matter whose effective diameter is larger than a molecule but smaller than approximately 100 μm. Particulates dispersed in a gaseous medium are collectively termed an *aerosol*. The terms *smoke, fog, haze,* and *dust* are commonly used to describe particular types of aerosols, depending on the size, shape, and characteristic behavior of the dispersed particles. Aerosols are rather difficult to classify on a scientific basis in terms of their fundamental properties such as settling rate under the influence of external forces, optical activity, ability to absorb an electrical charge, particle size and structure, surface-to-volume ratio, reaction activity, physiological action, and so on. In general, particle size and settling rate have been the most characteristic properties for many purposes. On the other hand, particles on the order of 1 μm or less settle so slowly that, for all practical purposes, they are regarded as permanent suspensions. Despite possible advantages of scientific classification schemes, the use of popular descriptive terms such as smoke, dust, and mist, which are essentially based on the mode of formation, appears to be a satisfactory and convenient method of classification. In addition, this approach is so well established and understood that it undoubtedly would be difficult to change.

Dust is typically formed by the pulverization or mechanical disintegration of solid matter into particles of smaller size by processes such as grinding, crushing, and drilling. Particle sizes of dust range from a lower limit of about 1 μm up to about 100 or 200 μm and larger. Dust particles are usually irregular in shape, and particle size refers to some average dimension for any given particle. Common examples include fly ash, rock dusts, and ordinary flour. *Smoke* implies a certain degree of optical density and is typically derived from the burning of organic materials such as wood, coal, and tobacco. Smoke particles are very fine, ranging in size from less than 0.01 μm up to 1 μm. They are usually spherical in shape if of liquid or tarry composition and irregular in shape if of solid composition. Owing to their very fine particle size, smokes can remain in suspension for long periods of time and exhibit lively brownian motion.

Fumes are typically formed by processes such as sublimation, condensation, or combustion, generally at relatively high temperatures. They range in particle size from less than 0.1 μm to 1 μm. Similar to smokes, they settle very slowly and exhibit strong brownian motion.

Mists or fogs are typically formed either by the condensation of water or other vapors on suitable nuclei, giving a suspension of small liquid droplets, or by the atomization of liquids. Particle sizes of natural fogs and mists lie between 2 and 200 μm. Droplets larger than 200 μm are more properly classified as drizzle or rain. Many of the important properties of aerosols that depend on particle size are presented in Sec. 17, Fig. 17-34.

When a liquid or solid substance is emitted to the air as particulate matter, its properties and effects may be changed. As a substance is broken up into smaller and smaller particles, more of its surface area is exposed to the air. Under these circumstances, the substance, whatever its chemical composition, tends to combine physically or chemically with other particles or gases in the atmosphere. The resulting combinations are frequently unpredictable. Very small aerosol particles (from 0.001 to 0.1 μm) can act as condensation nuclei to facilitate the condensation of water vapor, thus promoting the formation of fog and ground mist. Particles less than 2 or 3 μm in size (about half by weight of the particles suspended in urban air) can penetrate the mucous membrane and attract and convey harmful chemicals such as sulfur dioxide. In order to address the special concerns related to the effects of very fine, inhalable particulates, EPA replaced its ambient air standards for total suspended particulates (TSP) with standards for particulate matter less than 10 μm in size (PM_{10}).

By virtue of the increased surface area of the small aerosol particles and as a result of the adsorption of gas molecules or other such properties that are able to facilitate chemical reactions, aerosols tend to exhibit greatly enhanced surface activity. Many substances that oxidize slowly in their massive state will oxidize extremely fast or possibly even explode when dispersed as fine particles in the air. Dust explosions, for example, are often caused by the unstable burning or oxidation of combustible particles, brought about by their relatively large specific surfaces. Adsorption and catalytic phenomena can also be extremely important in analyzing and understanding the problems of particulate pollution. The conversion of sulfur dioxide to corrosive sulfuric acid assisted by the catalytic action of iron oxide particles, for example, demonstrates the catalytic nature of certain types of particles in the atmosphere. Finally, aerosols can absorb radiant energy and rapidly conduct heat to the surrounding gases of the atmosphere. These are gases that ordinarily would be incapable of absorbing radiant energy by themselves. As a result, the air in contact with the aerosols can become much warmer.

Estimating Emissions from Sources Knowledge of the types and rates of emissions is fundamental to evaluation of any air pollution problem. A comprehensive material balance on the process can often assist in this assessment. Estimates of the rates at which pollutants are discharged from various processes can also be obtained by utilizing published emission factors. See *Compilation of Air Pollution Emission Factors (AP-42)*, 4th ed., U.S. EPA, Research Triangle Park, North Carolina, September, 1985, with all succeeding supplements and the EPA Technology Transfer Network's CHIEF. The emission factor is a statistical average of the rate at which pollutants are emitted from the

TABLE 25-11 Typical Gaseous Pollutants and Their Principal Sources and Significance

Air pollutants	From manufacturing sources such as these	In typical industries	Cause these damaging effects
Alcohols	Used as a solvent in coatings	Surface coatings, printing	Sensory and respiratory irritation
Aldehydes	Results from thermal decomposition of fats, oil, or glycerol; used in some glues and binders	Food processing, light process, wood furniture, chip board	An irritating odor, suffocating, pungent, choking; not immediately dangerous to life; can become intolerable in a very short time
Ammonia	Used in refrigeration, chemical processes such as dye making, explosives, lacquer, fertilizer	Textiles, chemicals	Corrosive to copper, brass, aluminum, and zinc; high concentration producing chemical burns on wet skin
Aromatics	Used as a solvent in coatings	Surface coatings, printing	Irritation of mucous membranes, narcotic effects; some are carcinogens
Arsine	Any soldering, pickling, etching, or plating process involving metals or acids containing arsenic	Chemical processing, smelting	Breakdown of red cells in blood
Carbon dioxide	Fuel combustion; calcining	Industrial boilers, cement and lime production	Greenhouse gas
Carbon monoxide	Fuming of metallic oxides, gas-operated fork trucks	Primary metals; steel and aluminum	Reduction in oxygen-carrying capacity of blood
Chlorine	Manufactured by electrolysis, bleaching cotton and flour; by-product of organic chemicals	Textiles, chemicals	Attacks entire respiratory tract and mucous membrane of eye
Chlorofluoro-carbons	Used in refrigeration and production of porous foams; degreasing agent	Refrigeration, plastic foam production, metal fabricating	Attacks stratospheric ozone layer; greenhouse gas
Hydrochloric acid	Combustion of coal or wastes containing chlorinated plastics	Coal-fired boilers, incinerators	Irritant to eyes and respiratory system
Hydrogen cyanide	From metal plating, blast furnaces, dyestuff works	Metal fabricating, primary metals, textiles	Capable of affecting nerve cells
Hydrogen fluoride	Catalyst in some petroleum refining, etching glass, silicate extraction; by-product in electrolytic production of aluminum	Petroleum, primary metals, aluminum	Strong irritant and corrosive action on all body tissue; damage to citrus plants, effect on teeth and bones of cattle from eating plants
Hydrogen sulfide	Refinery gases, crude oil, sulfur recovery, various chemical industries using sulfur compounds	Petroleum and chemicals; Kraft pulping process	Foul odor of rotten eggs; irritating to eyes and respiratory tract; darkening exterior paint
Ketones	Used as a solvent in coatings	Surface coatings, printing	Sensory and respiratory irritation
Lead	Incineration, smelting and casting, transportation.	Copper and lead smelting, MSWs	Neurological impairments; kidney, liver, and heart damage.
Nitrogen oxides	High-temperature combustion: metal cleaning, fertilizer, explosives, nitric acid; carbon-arc combustion; manufacture of H_2SO_4	Metal fabrication, heavy chemicals	Irritating gas affecting lungs; vegetation damage
Odors	Slaughtering and rendering animals, tanning animal hides, canning, smoking meats, roasting coffee, brewing beer, processing toiletries	Food processing, allied industries	Objectionable odors
Ozone	Reaction product of VOC and nitrogen oxides	Not produced directly	Irritant to eyes and respiratory system
Phosgenes	Thermal decomposition of chlorinated hydrocarbons, degreasing, manufacture of dyestuffs, pharmaceuticals, organic chemicals	Metal fabrication, heavy chemicals	Damage capable of leading to pulmonary edema, often delayed
Sulfur dioxide	Fuel combustion (coal, oil), smelting and casting, manufacture of paper by sulfite process	Primary metals (ferrous and nonferrous); pulp and paper	Sensory and respiratory irritation, vegetation damage, corrosion, possible adverse effect on health

burning or processing of a given quantity of material or on the basis of some other meaningful parameter. Emission factors are affected by the techniques employed in the processing, handling, or burning operations, by the quality of the material used, and by the efficiency of the air-pollution control. Since the combination of these factors tends to be unique to a source, emission factors appropriate for one source may not be satisfactory for another. Hence, care and good judgment must be exercised in identifying appropriate emission factors. If appropriate emission factors cannot be found or if air-pollution control equipment is to be designed, specific source sampling should be conducted. The major industrial sources of pollutants, the air con-

taminants emitted, and typical control techniques are summarized in Table 25-12.

Effects of Air Pollutants

Materials The damage that air pollutants can do to some materials is well known: ozone in photochemical smog cracks rubber, weakens fabrics, and fades dyes; hydrogen sulfide tarnishes silver; smoke dirties laundry; acid aerosols ruin nylon hose. Among the most important effects are discoloration, corrosion, the soiling of goods, and impairment of visibility.

TABLE 25-12 Control Techniques Applicable to Unit Processes at Important Emission Sources

Industry	Process of operation	Air contaminants emitted	Control techniques
Aluminum reduction plants	Materials handling: Buckets and belt Conveyor or pneumatic conveyor	Particulates (dust)	Exhaust systems and baghouse
	Anode and cathode electrode preparation: Cathode (baking) Anode (grinding and blending)	Hydrocarbon emissions from binder Particulates (dust)	Exhaust systems and mechanical collectors
	Baking	Particulates (dust), CO, SO$_2$, hydrocarbons, and fluorides	High-efficiency cyclone, electrostatic precipitators, scrubbers, catalytic combustion or incinerators, flares, baghouse
	Pot charging	Particulates (dust), CO, HF, SO$_2$, CF$_4$, and hydrocarbons	High-efficiency cyclone, baghouse, spray towers, floating-bed scrubber, electrostatic precipitators, chemisorption, wet electrostatic precipitators
	Metal casting	Cl$_2$, HCl, CO, and particulates (dust)	Exhaust systems and scrubbers
Asphalt plants	Materials handling, storage and classifiers: elevators, chutes, vibrating screens	Particulates (dust)	Wetting; exhaust systems with a scrubber or baghouse
	Drying: rotary oil- or gas-fired	Particulates, SO$_2$, NO$_x$, VOC, CO, and smoke	Proper combustion controls, fuel-oil preheating where required; local exhaust system, cyclone and a scrubber or baghouse
	Truck traffic	Dust	Paving, wetting down truck routes
Cement plants	Quarrying: primary crusher, secondary crusher, conveying, storage	Particulates (dust)	Wetting; exhaust systems with fabric filters
	Dry processes: materials handling, air separator (hot-air furnace)	Particulates (dust)	Local exhaust system with mechanical collectors and baghouse
	Grinding	Particulates (dust)	Local exhaust system with cyclones and baghouse
	Pneumatic, conveying and storage	Particulates (dust)	
	Wet process: materials handling, grinding, storage	Wet materials, no dust	
	Kiln operations: rotary kiln	Particulates (dust), CO, SO$_x$, NO$_x$, hydrocarbons, aldehydes, ketones	Electrostatic precipitators, acoustic horns and baghouses, scrubber
	Clinker cooling: materials handling	Particulates (dust)	Local exhaust system and electrostatic precipitators or fabric filters
	Grinding and packing, air separator, grinding, pneumatic conveying, materials handling, packaging	Particulates (dust)	Local exhaust system and fabric filters
Coal-preparation plants	Materials handling: conveyors, elevators, chutes	Particulates (dust)	Local exhaust system and cyclones
	Sizing: crushing, screening, classifying	Particulates (dust)	Local exhaust system and cyclones
	Dedusting	Particulates (dust)	Local exhaust system, cyclone precleaners, and baghouse
	Storing coal in piles	Blowing particulates (dust)	Wetting, plastic-spray covering
	Refuse piles	H$_2$S, particulates, and smoke from burning storage piles	Digging out fire, pumping water onto fire area, blanketing with incombustible material
	Coal drying: rotary, screen, suspension, fluid-bed, cascade	Dust, smoke, particulates, sulfur oxides, H$_2$S	Exhaust systems with cyclones and fabric filters
Coke plants	By-product-ovens charging	Smoke, particulates (dust)	Pipe-line charging, careful charging techniques, portable hooding and scrubber or baghouses
	Pushing	Smoke, particulates (dust), SO$_2$	Minimizing green-coke pushing, scrubbers and baghouses
	Quenching	Smoke, particulates (dust and mists), phenols, and ammonia	Baffles and spray tower
	By-product processing	CO, H$_2$S, methane, ammonia, H$_2$, phenols, hydrogen cyanide, N$_2$, benzene, xylene, etc.	Electrostatic precipitator, scrubber, flaring
	Material storage (coal and coke)	Particulates (dust)	Wetting, plastic spray, fire-prevention techniques
Fertilizer industry (chemical)	Phosphate fertilizers: crushing, grinding, and calcining	Particulates (dust)	Exhaust system, scrubber, cyclone, baghouse
	Hydrolysis of P$_2$O$_5$	PH$_3$, P$_2$O$_5$PO$_4$ mist	Scrubbers, flare
	Acidulation and curing	HF, SiF$_4$	Scrubbers
	Granulation	Particulates (dust)(product recovery)	Exhaust system, scrubber, or baghouse
	Ammoniation	NH$_3$, NH$_4$Cl, SiF$_4$, HF	Cyclone, electrostatic precipitator, baghouse, high-energy scrubber
	Nitric acid acidulation	NO$_x$, gaseous fluoride compounds	Scrubber, addition of urea
	Superphosphate storage and shipping	Particulates (dust)	Exhaust system, cyclone or baghouse
	Ammonium nitrate reactor	NH$_3$, NO$_x$	Scrubber
	Prilling tower	NH$_4$, NO$_3$	Proper operation control, scrubbers

TABLE 25-12 Control Techniques Applicable to Unit Processes at Important Emission Sources (*Continued*)

Industry	Process of operation	Air contaminants emitted	Control techniques
Foundries: Iron	Melting (cupola:) Charging Melting Pouring Bottom drop	Smoke and particulates Smoke and particulates, fume Oil, mist, CO Smoke and particulates	Closed top with exhaust system, CO afterburner, gas-cooling device and scrubbers, baghouse or electrostatic precipitator, wetting to extinguish fire
Brass and bronze	Melting: Charging Melting Pouring	Smoke particulates, oil mist Zinc oxide fume, particulates, smoke Zinc oxide fume, lead oxide fume	Low-zinc-content red brass: use of good combustion controls and slag cover; high-zinc-content brass: use of good combustion controls, local exhaust system, and baghouse or scrubber
Aluminum	Melting: charging, melting, pouring	Smoke and particulates	Charging clean material (no paint or grease); proper operation required; no air-pollution-control equipment if no fluxes are used and degassing is not required; dirty charge requiring exhaust system with scrubbers and baghouses
Zinc	Melting: Charging Melting Pouring Sand-handling shakeout Magnetic pulley, conveyors, and elevators, rotary cooler, screening, crusher-mixer Coke-making ovens	Smoke and particulates Zinc oxide fume Oil mist and hydrocarbons from diecasting machines Particulates (dust), smoke, organic vapors Particulates (dust) Organic acids, aldehydes, smoke, hydrocarbons	Exhaust system with cyclone and baghouse, charging clean material (no paint or grease) Careful skimming of dross Use of low-smoking die-casting lubricants Exhaust system, cyclone, and baghouse Use of binders that will allow ovens to operate at less than 204°C (400°F) or exhaust systems and afterburners
Galvanizing operations	Hot-dip-galvanizing-tank kettle: dipping material into the molten zinc; dusting flux onto the surface of the molten zinc	Fumes, particulates (liquid), vapors: NH_4Cl, ZnO, $ZnCl_2$, Zn, NH_3, oil, and carbon	Close-fitting hoods with high in-draft velocities (in some cases, the hood may not be able to be close to the kettle, so the in-draft velocity must be very high), baghouses, electrostatic precipitators
Kraft pulp mills	Digesters: batch and continuous Multiple-effect evaporators Recovery furnace Weak and strong black-liquor oxidation Smelt tanks Lime kiln	Mercaptans, methanol (odors) H_2S, other odors H_2S, mercaptans, organic sulfides, and disulfides H_2S Particulates (mist or dust) Particulates (dust), H_2S	Condensers and use of lime kiln, boiler, or furnaces as afterburners Caustic scrubbing and thermal oxidation of noncondensables Proper combustion controls for fluctuating load and unrestricted primary and secondary air flow to furnace and dry-bottom electrostatic precipitator; noncontact evaporator Packed tower and cyclone Demisters, venturi, packed tower, or impingement-type scrubbers Venturi scrubbers
Municipal and industrial incinerators	Single-chamber incinerators Multiple-chamber incinerators (retort, inline): Flue-fed Wood waste Municipal incinerators (50 tons and up per day):	Particulates, smoke, volatiles, CO, SO_x, ammonia, organic acids, aldehydes, NO_x, dioxins hydrocarbons, odors, HCl, furans Particulates, smoke, and combustion contaminants Particulates, smoke, and combustion contaminants Particulates, smoke, and combustion contaminants Particulates, smoke, volatiles, CO, ammonia, organic acids, aldehydes, NO_x, furans, hydrocarbons, SO_x, hydrogen chloride, dioxins and odors	Afterburner, combustion controls Operating at rated capacity, using auxiliary fuel as specified, and good maintenance, including timely cleanout of ash Use of charging gates and automatic controls for draft; afterburner Continuous-feed systems; operation at design load and excess air; cyclones Preparation of materials, including weighing, grinding, shredding; control of tipping area, furnace design with proper automatic controls; proper start-up techniques; maintenance of design operating temperatures; use of electrostatic precipitators, scrubbers, and baghouses; proper ash cleanout

TABLE 25-12 Control Techniques Applicable to Unit Processes at Important Emission Sources (*Continued*)

Industry	Process of operation	Air contaminants emitted	Control techniques
Municipal and industrial incinerators	Pathological incinerators	Odors, hydrocarbons, HCl, dioxins, furans	Proper charging, acid gas scrubber, baghouse
	Industrial waste	Particulates, smoke, and combustion contaminants	Modified fuel feed, auxiliary fuel and dryer systems, cyclones, scrubbers
Nonferrous smelters, primary: Copper	Roasting	SO_2, particulates, fume	Exhaust system, settling chambers, cyclones or scrubbers and electrostatic precipitators for dust and fumes and sulfuric acid plant for SO_2
	Reverberatory furnace	Smoke, particulates, metal oxide fumes, SO_2	Exhaust system, settling chambers, cyclones or scrubbers and electrostatic precipitators for dust and fumes and sulfuric acid plant for SO_2
	Converters: charging, slag skim, pouring, air or oxygen blow	Smoke, fume, SO_2	Exhaust system, settling chambers, cyclones or scrubbers and electrostatic precipitators for dust and fumes and sulfuric acid plant for SO_2
Lead	Sintering	SO_2, particulates, smoke	Exhaust system, cyclones and baghouse or precipitators for dust and fumes, sulfuric acid plant for SO_2
	Blast furnace	SO_2, CO, particulates, lead oxide, zinc oxide	Exhaust system, settling chambers, afterburner and cooling device, cyclone, and baghouse
	Dross reverberatory furnace	SO_2, particulates, fume	Exhaust system, settling chambers, cyclone and cooling device, baghouse
	Refining kettles	SO_2, particulates	Local exhaust system, cooling device, baghouse or precipitator
Cadmium	Roasters, slag, fuming furnaces, deleading kilns	Particulates	Local exhaust system, baghouse or precipitator
Zinc	Roasting	Particulates (dust) and SO_2	Exhaust system, humidifier, cyclone, scrubber, electrostatic precipitator, and acid plant
	Sintering	Particulates (dust) and SO_2	Exhaust system, humidifier, electrostatic precipitator, and acid plant
	Calcining	Zinc oxide fume, particulates, SO_2, CO	Exhaust system, baghouse, scrubber or acid plant
	Retorts: electric arc		
Nonferrous smelters, secondary	Blast furnaces and cupolas-recovery of metal from scrap and slag	Dust, fumes, particulates, oil vapor, smoke, CO	Exhaust systems, cooling devices, CO burners and baghouses or precipitators
	Reverberatory furnaces	Dust, fumes, particulates, smoke, gaseous fluxing materials	Exhaust systems, and baghouses or precipitators, or venturi scrubbers
	Sweat furnaces	Smoke, particulates, fumes	Precleaning metal and exhaust systems with afterburner and baghouse
	Wire reclamation and autobody burning	Smoke, particulates	Scrubbers and afterburners
Paint and varnish manufacturing	Resin manufacturing: closed reaction vessel	Acrolein, other aldehydes and fatty acids (odors), phthalic anhydride (sublimed)	Exhaust systems with scrubbers and fume burners
	Varnish: cooking-open or closed vessels	Ketones, fatty acids, formic acids, acetic acid, glycerine, acrolein, other aldehydes, phenols and terpenes; from tall oils, hydrogen sulfide, alkyl sulfide, butyl mercaptan, and thiofen (odors)	Exhaust system with scrubbers and fume burners; close-fitting hoods required for open kettles
	Solvent thinning	Olefins, branched-chain aromatics and ketones (odors), solvents	Exhaust system with fume burners
Rendering plants	Feedstock storage and housekeeping	Odors	Quick processing, washdown of all concrete surfaces, paving of dirt roads, proper sewer maintenance, enclosure, packed towers
	Cookers and percolators	SO_2, mercaptans, ammonia, odors	Exhaust system, condenser, scrubber, or incinerator
	Grinding	Particulates (dust)	Exhaust system and scrubber
Roofing plants (asphalt saturators)	Felt or paper saturators: spray section, asphalt tank, wet looper	Asphalt vapors and particulates (liquid)	Exhaust system with high inlet velocity at hoods (3658 m/s [>200 ft/min]) with either scrubbers, baghouses, or two-stage low-voltage electrostatic precipitators
	Crushed rock or other minerals handling	Particulates (dust)	Local exhaust system, cyclone or multiple cyclones

TABLE 25-12   Control Techniques Applicable to Unit Processes at Important Emission Sources (Concluded)

Industry	Process of operation	Air contaminants emitted	Control techniques
Steel mills	Blast furnaces: charging, pouring	CO, fumes, smoke, particulates (dust)	Good maintenance, seal leaks; use of higher ratio of pelletized or sintered ore; CO burned in waste-heat boilers, stoves, or coke ovens; cyclone, scrubber, and baghouse
	Electric steel furnaces: charging, pouring, oxygen blow	Fumes, smoke, particulates (dust), CO	Segregating dirty scrap; proper hooding, baghouses or electrostatic precipitator
	Open-hearth furnaces: oxygen blow, pouring	Fumes, smoke, SO_x, particulates (dust), CO, NO_x	Proper hooding, settling chambers, waste-heat boiler, baghouse, electrostatic precipitator, and wet scrubber
	Basic oxygen furnaces: oxygen blowing	Fumes, smoke, CO, particulates (dust)	Proper hooding (capturing of emissions and dilute CO), scrubbers, or electrostatic precipitator
	Raw material storage	Particulates (dust)	Wetting or application of plastic spray
	Pelletizing	Particulates (dust)	Proper hooding, cyclone, baghouse
	Sintering	Smoke, particulates (dust), SO_2, NO_x	Proper hooding, cyclones, wet scrubbers, baghouse, or precipitator

1. *Discoloration.*   Many air pollutants accumulate on and discolor buildings. Not only does sooty material blacken buildings, but it can accumulate and become encrusted. This can hide lines and decorations and thereby disfigure structures and reduce their aesthetic appeal. Another common effect is the discoloration of paint by certain acid gases. A good example is the blackening of white paint with a lead base by hydrogen sulfide.

2. *Corrosion.*   A more serious effect and one of great economic importance is the corrosive action of acid gases on building materials. Such acids can cause stone surfaces to blister and peel; mortar can be reduced to powder. Metals are also damaged by the corrosive action of some pollutants. Another common effect is the deterioration of tires and other rubber goods. Cracking and apparent "drying" occur when these goods are exposed to ozone and other oxidants.

3. *Soiling of goods.*   Clothes, real estate, automobiles, and household goods can easily be soiled by air contaminants, and the more frequent cleaning thus required can become expensive. Also, more frequent cleaning often leads to a shorter life span for materials and to the need to purchase goods more often.

4. *Impairment of visibility.*   The impairment of atmospheric visibility (i.e., decreased visual range through a polluted atmosphere) is caused by the scattering of sunlight by particles suspended in the air. It is not a result of sunlight being obscured by materials in the air. Since light scattering, and not obscuration, is the main cause of the reduction in visibility, reduced visibility due to the presence of air pollutants occurs primarily on bright days. On cloudy days or at night there may be no noticeable effect, although the same particulate concentration may exist at these times as on sunny days. Reduction in visibility creates several problems. The most significant are the adverse effects on aircraft, highway, and harbor operations. Reduced visibility can reduce quality of life and also cause adverse aesthetic impressions that can seriously affect tourism and restrict the growth and development of any area. Extreme conditions such as dust storms or sandstorms can actually cause physical damage by themselves.

Vegetation   Vegetation is more sensitive than animals to many air contaminants, and methods have been developed that use plant response to measure and identify contaminants. The effects of air pollution on vegetation can appear as death, stunted growth, reduced crop yield, and degradation of color. It is interesting to note that in some cases of color damage such as the silvering of leafy vegetables by oxidants, the plant may still be used as food without any danger to the consumer; however, the consumer usually will not buy such vegetables on aesthetic grounds, so the grower still sustains a loss. Among the pollutants that can harm plants are sulfur dioxide, hydrogen fluoride, and ethylene. Plant damage caused by constituents of photochemical smog has been studied extensively. Damage has been attributed to ozone and peroxyacetyl nitrites, higher aldehydes, and products of the reaction of ozone with olefins. However, none of the cases precisely duplicates all features of the damage observed in the field, and the question remains open to some debate and further study.

Animals   Considerable work continues to be performed on the effects of pollutants on animals, including, for a few species, experiments involving mixed pollutants and mixed gas-aerosol systems. In general, such work has shown that mixed pollutants may act in several different ways. They may produce an effect that is additive, amounting to the sum of the effects of each contaminant acting alone; they may produce an effect that is greater than the simply additive (synergistic) or less than the simply additive (antagonistic); or they may produce an effect that differs in some other way from the simply additive.

The mechanism by which an animal can become poisoned in many instances is completely different from that by which humans are affected. As in humans, inhalation is an important route of entry in acute air-pollution exposures such as the Meuse Valley and Donora incidents (see the paragraph on **humans** below). However, probably the most common exposure for herbivorous animals grazing within a zone of pollution will be the ingestion of feed contaminated by air pollutants. In this case, inhalation is of secondary importance.

Air pollutants that present a hazard to livestock, therefore, are those that are taken up by vegetation or deposited on the plants. Only a few pollutants have been observed to cause harm to animals. These include arsenic, fluorides, lead, mercury, and molybdenum.

Humans   There seems to be little question that, during many of the more serious episodes, air pollution can have a significant effect on health, especially upon the young, elderly, or people already in ill health. Hundreds of excess deaths have been attributed to incidents in London in 1952, 1956, 1957, and 1962; in Donora, Pennsylvania, in 1948; in New York City in 1953, 1963, and 1966; and in Bhopal, India in 1989. Many of the people affected were in failing health, and they were generally suffering from lung conditions. In addition, hundreds of thousands of persons have suffered from serious discomfort and inconvenience, including eye irritation and chest pains, during these and other such incidents. Such acute problems are actually the lesser of the health problems. There is considerable evidence of a chronic threat to human health from air pollution. This evidence ranges from the rapid rise of emphysema as a major health problem, through identification of carcinogenic compounds in smog, to statistical evidence that people exposed to polluted atmospheres over extended periods of time suffer from a number of ailments and a reduction in their life span. There may even be a significant indirect exposure to air pollution. As noted above, air pollutants may be deposited onto vegetation or into bodies of water, where they enter the food chain. The impact of such indirect exposures is still under review.

Sufficient evidence is available to indicate that atmospheric pollution in varying degrees does affect health adversely. [Amdur, Melvin, and Drinker, "Effect of Inhalation of Sulfur Dioxide by Man," *Lancet,* **2,** 758 (1953); Barton, Corn, Gee, Vassallo, and Thomas, "Response of Healthy Men to Inhaled Low Concentrations of Gas-Aerosol Mixtures," *Arch. Environ. Health,* **18,** 681 (1969); Bates, Bell, Burnham, Hazucha, and Mantha, "Problems in Studies of Human Exposure to Air Pollutants," *Can. Med. Assoc. J.,* **103,** 833 (1970); Ciocco and

Thompson, "A Follow-Up of Donora Ten Years After: Methodology and Findings," *Am. J. Public Health*, **51**, 155 (1961); Daly, "Air Pollution and Causes of Death," *Br. J. Soc. Med.*, **13**, 14 (1959); Jaffe, "The Biological Effect of Photochemical Air Pollutants on Man and Animals," *Am. J. Public Health*, New York, **57**, 1269 (1967); New York Academy of Medicine, Committee on Public Health, "Air Pollution and Health," *Bull. N.Y. Acad. Med.*, **42**, 588 (1966); Pemberton and Goldberg, "Air Pollution and Bronchitis," *Br. Med. J.*, London, **2**, 567 (1954); Snell and Luchsinger, "Effect of Sulfur Dioxide on Expiratory Flowrates and Total Respiratory Resistance in Normal Human Subjects," *Arch. Environ. Health*, Chicago, **18**, 693 (1969); Speizer and Frank, "A Comparison of Changes in Pulmonary Flow Resistance in Healthy Volunteers Acutely Exposed to SO_2 by Mouth and by Nose," *J. Ind. Med.*, **23** 75 (1966); Stocks, "Cancer and Bronchitis Mortality in Relation to Atmospheric Deposit and Smoke," *Br. Med. J. London*, **1**, 74 (1959); Toyama, "Air Pollution and Its Health Effects in Japan," *Arch. Environ. Health*, Chicago, **8**, 153 (1963); U.K. Ministry of Health, "Mortality and Morbidity During London Fog of December 1952," Report on Public Health and Medical Subjects No. 95, London, 1954; U.S. Public Health Service, "Air Pollution in Donora, Pa.: Preliminary Report," *Public Health Bull.* 306.] It contributes to excesses of death, increased morbidity, and earlier onset of chronic respiratory diseases. There is evidence of a relationship between the intensity of the pollution and the severity of attributable health effects and a consistency of the relationship between these environmental stresses and diseases of the target organs. Air pollutants can both initiate and aggravate a variety of respiratory diseases including asthma. In fact, the clinical presentation of asthma may be considered an air pollution host-defense disorder brought on by specific airborne irritants: pollens, infectious agents, and gaseous and particulate chemicals. The bronchopulmonary response to these foreign irritants is bronchospasm and hypersecretion; the airways are intermittently and reversibly obstructed.

Air-pollutant effects on neural and sensory functions in humans vary widely. Odorous pollutants cause only minor annoyance; yet, if persistent, they can lead to irritation, emotional upset, anorexia, and mental depression. Carbon monoxide can cause death secondary to the depression of the respiratory centers of the central nervous system. Short of death, repeated and prolonged exposure to carbon monoxide can alter sensory protection, temporal perception, and higher mental functions. Lipid-soluble aerosols can enter the body and be absorbed in the lipids of the central nervous system. Once there, their effects may persist long after the initial contact has been removed. Examples of agents of long-term chronic effects are organic phosphate pesticides and aerosols carrying the metals lead, mercury, and cadmium.

The acute toxicological effects of most air contaminants are reasonably well understood, but the effects of exposure to heterogenous mixtures of gases and particulates at very low concentrations are only beginning to be comprehended. Two general approaches can be used to study the effects of air contaminants on humans: epidemiology, which attempts to associate the effect in large populations with the cause, and laboratory research, which begins with the cause and attempts to determine the effects. Ideally, the two methods should complement each other.

Epidemiology, the more costly of the two, requires great care in planning and often suffers from incomplete data and lack of controls. One great advantage, however, is that moral barriers do not limit its application to humans as they do with some kinds of laboratory research. The method is therefore highly useful and has produced considerable information. Laboratory research is less costly than epidemiology, and its results can be checked against controls and verified by experimental repetition.

A SOURCE-CONTROL-PROBLEM STRATEGY

Strategy   Control technology is self-defeating if it creates undesirable side effects in meeting objectives. Air pollution control must be considered in terms of regulatory requirements, total technological systems (equipment and processes), and ecological consequences, such as the problems of treatment and disposal of collected pollutants.

It should be noted that the 1990 Clean Air Act Amendments (CAAA) have impacted on the control approach in a significant manner. In particular, the CAAA have placed an increased emphasis on control technology by requiring Best Available Control Technology (BACT) on new major sources and modifications, and by requiring Maximum Achievable Control Technology (MACT) on new and *existing* major sources of Hazardous Air Pollutants (HAPs).

The control strategy for environmental-impact assessment often focuses on five alternatives whose purpose would be the reduction and/or elimination of pollutant emissions:

1. Elimination of the operation entirely or in part
2. Modification of the operation
3. Relocation of the operation
4. Application of appropriate control technology
5. Combinations thereof

In light of the relatively high costs often associated with pollution-control systems, engineers are directing considerable effort toward process modification to eliminate as much of the pollution problem as possible at the source. This includes evaluating alternative manufacturing and production techniques, substituting raw materials, and improving process-control methods. Unfortunately, if there is no alternative, the application of the correct pollution-control equipment is essential. The equipment must be designed to comply with regulatory emission limitations on a continual basis, interruptions being subject to severe penalty depending upon the circumstances. The requirement for design performance on a continual basis places very heavy emphasis on operation and maintenance practices. The escalating costs of energy, labor, and materials can make operation and maintenance considerations even more important than the original capital cost.

Factors in Control-Equipment Selection   In order to solve an air-pollution problem, the problem must be defined in detail. A number of factors must be considered prior to selecting a particular piece of air-pollution-control equipment. In general, these factors can be grouped into three categories: environmental, engineering, and economic.

Environmental Factors   These include (1) equipment location, (2) available space, (3) ambient conditions, (4) availability of adequate utilities (i.e., power, water, etc.) and ancillary-system facilities (i.e., waste treatment and disposal, etc.), (5) maximum allowable emission (air pollution codes), (6) aesthetic considerations (i.e., visible steam or water-vapor plume, etc.), (7) contributions of the air-pollution-control system to wastewater and land pollution, and (8) contribution of the air-pollution-control system to plant noise levels.

Engineering Factors   These include:

1. Contaminant characteristics (e.g., physical and chemical properties, concentration, particulate shape and size distribution [in the case of particulates], chemical reactivity, corrosivity, abrasiveness, and toxicity)
2. Gas-stream characteristics (e.g., volume flow rate, temperature, pressure, humidity, composition, viscosity, density, reactivity, combustibility, corrosivity, and toxicity)
3. Design and performance characteristics of the particular control system (i.e., size and weight, fractional efficiency curves [in the case of particulates]), mass-transfer and/or contaminant-destruction capability (in the case of gases or vapors), pressure drop, reliability, turndown capability, power requirements, utility requirements, temperature limitations, maintenance requirements, operating cycles (including startup and shutdown) and flexibility toward complying with more stringent air-pollution codes.

Economic Factors   These include capital cost (equipment, installation, engineering, etc.), operating cost (utilities, maintenance, etc.), emissions fees, and life-cycle cost over the expected equipment lifetime.

Comparing Control-Equipment Alternatives   The final choice in equipment selection is usually dictated by the equipment capable of achieving compliance with regulatory codes at the lowest uniform annual cost (amortized capital investment plus operation and maintenance costs). To compare specific control-equipment alternatives, knowledge of the particular application and site is essential. A preliminary screening, however, may be performed by reviewing the advantages and disadvantages of each type of air-pollution-control equipment. General advantages and disadvantages of the most popular types of air-pollution equipment for gases and particulates are presented in Tables 25-13 through 25-25. Other activities that must be accomplished before final compliance is achieved are presented in Table 25-26.

In addition to using annualized cost comparisons in evaluating an air-pollution-control (APC) equipment installation, the impact of the 1990 Clean Air Act Amendments (CAAA) and resulting regulations also must be included in the evaluation. The CAAA prescribes specific pollution-control requirements for particular industries and locations. As an example, the CAAA requires that any major stationary source or

TABLE 25-13   Advantages and Disadvantages of Cyclone Collectors

Advantages
1. Low cost of construction
2. Relatively simple equipment with few maintenance problems
3. Relatively low operating pressure drops (for degree of particulate removal obtained) in the range of approximately 2- to 6-in water column
4. Temperature and pressure limitations imposed only by the materials of construction used
5. Collected material recovered dry for subsequent processing or disposal
6. Relatively small space requirements

Disadvantages
1. Relatively low overall particulate collection efficiencies, especially on particulates below 10 μm in size
2. Inability to handle tacky materials

TABLE 25-14   Advantages and Disadvantages of Wet Scrubbers

Advantages
1. No secondary dust sources
2. Relatively small space requirements
3. Ability to collect gases as well as particulates (especially "sticky" ones)
4. Ability to handle high-temperature, high-humidity gas streams
5. Capital cost low (if wastewater treatment system not required)
6. For some processes, gas stream already at high pressures (so pressure-drop considerations may not be significant)
7. Ability to achieve high collection efficiencies on fine particulates (however, at the expense of pressure drop)
8. Ability to handle gas streams containing flammable or explosive materials

Disadvantages
1. Possible creation of water-disposal problem
2. Product collected wet
3. Corrosion problems more severe than with dry systems
4. Steam plume opacity and/or droplet entrainment possibly objectionable
5. Pressure-drop and horsepower requirements possibly high
6. Solids buildup at the wet-dry interface possibly a problem
7. Relatively high maintenance costs
8. Must be protected from freezing
9. Low exit gas temperature reduces exhaust plume dispersion
10. Moist exhaust gas precludes use of most additional controls

TABLE 25-15   Advantages and Disadvantages of Dry Scrubbers

Advantages
1. No wet sludge to dispose of
2. Relatively small space requirements
3. Ability to collect acid gases at high efficiencies
4. Ability to handle high-temperature gas streams
5. Dry exhaust allows addition of fabric filter to control particulate

Disadvantages
1. Acid gas control efficiency not as high as with wet scrubber
2. No particulate collection—dry scrubber generates particulate
3. Corrosion problems more severe than with dry systems
4. Solids buildup at the wet-dry interface possibly a problem
5. Relatively high maintenance costs
6. Must be protected from freezing
7. Low exit gas temperature reduces exhaust plume dispersion

TABLE 25-16   Advantages and Disadvantages of Electrostatic Precipitators

Advantages
1. Extremely high particulate (coarse and fine) collection efficiencies attainable (at a relatively low expenditure of energy)
2. Collected material recovered dry for subsequent processing or disposal
3. Low pressure drop
4. Designed for continuous operation with minimum maintenance requirements
5. Relatively low operating costs
6. Capable of operation under high pressure (to 150 lbf/in^2) or vacuum conditions
7. Capable of operation at high temperatures [to 704°C(1300°F)]
8. Relatively large gas flow rates capable of effective handling

Disadvantages
1. High capital cost
2. Very sensitive to fluctuations in gas-stream conditions (in particular, flows, temperature, particulate and gas composition, and particulate loadings)
3. Certain particulates difficult to collect owing to extremely high- or low-resistivity characteristics
4. Relatively large space requirements required for installation
5. Explosion hazard when treating combustible gases and/or collecting combustible particulates
6. Special precautions required to safeguard personnel from the high voltage
7. Ozone produced by the negatively charged discharge electrode during gas ionization
8. Relatively sophisticated maintenance personnel required
9. Gas ionization may cause dissociation of gas stream constituents and result in creation of toxic byproducts
10. Sticky particulates may be difficult to remove from plates
11. Not effective in capturing some contaminants that exist as vapors at high temperatures (e.g., heavy metals, dioxins)

TABLE 25-17   Advantages and Disadvantages of Fabric-Filter Systems

Advantages
1. Extremely high collection efficiency on both coarse and fine (sub-micrometer) particles
2. Relatively insensitive to gas-stream fluctuation; efficiency and pressure drop relatively unaffected by large changes in inlet dust loadings for continuously cleaned filters
3. Filter outlet air capable of being recirculated within the plant in many cases (for energy conservation)
4. Collected material recovered dry for subsequent processing or disposal
5. No problems with liquid-waste disposal, water pollution, or liquid freezing
6. Corrosion and rusting of components usually not problems
7. No hazard of high voltage, simplifying maintenance and repair and permitting collection of flammable dusts
8. Use of selected fibrous or granular filter aids (precoating), permitting the high-efficiency collection of submicrometer smokes and gaseous contaminants
9. Filter collectors available in a large number of configurations, resulting in a range of dimensions and inlet and outlet flange locations to suit installment requirements
10. Relatively simple operation

Disadvantages
1. Temperatures much in excess of 288°C (550°F) requiring special refractory mineral or metallic fabrics that are still in the developmental stage and can be very expensive
2. Certain dusts possibly requiring fabric treatments to reduce dust seeping or, in other cases, assist in the removal of the collected dust
3. Concentrations of some dusts in the collector (~50 g/m^3) forming a possible fire or explosion hazard if a spark or flame is admitted by accident; possibility of fabrics burning if readily oxidizable dust is being collected
4. Relatively high maintenance requirements (bag replacement, etc.)
5. Fabric life possibly shortened at elevated temperatures and in the presence of acid or alkaline particulate or gas constituents
6. Hygroscopic materials, condensation of moisture, or tarry adhesive components possibly causing crusty caking or plugging of the fabric or requiring special additives
7. Replacement of fabric, possibly requiring respiratory protection for maintenance personnel
8. Medium pressure-drop requirements, typically in the range 4- to 10-in water column

TABLE 25-18 Advantages and Disadvantages of Absorption Systems (Packed and Plate Columns)

Advantages
1. Relatively low pressure drop
2. Standardization in fiberglass-reinforced plastic (FRP) construction permitting operation in highly corrosive atmospheres
3. Capable of achieving relatively high mass-transfer efficiencies
4. Increasing the height and/or type of packing or number of plates capable of improving mass transfer without purchasing a new piece of equipment
5. Relatively low capital cost
6. Relatively small space requirements
7. Ability to collect particulates as well as gases
8. Collected substances may be recovered by distillation

Disadvantages
1. Possibility of creating water (or liquid) disposal problem
2. Product collected wet
3. Particulates deposition possibly causing plugging of the bed or plates
4. When FRP construction is used, sensitive to temperature
5. Relatively high maintenance costs
6. Must be protected from freezing

TABLE 25-19 Comparison of Plate and Packed Columns

Packed column
1. Lower pressure drop
2. Simpler and cheaper to construct
3. Preferable for liquids with high-foaming tendencies

Plate column
1. Less susceptible to plugging
2. Less weight
3. Less of a problem with channeling
4. Temperature surge resulting in less damage

TABLE 25-20 Advantages and Disadvantages of Adsorption Systems

Advantages
1. Possibility of product recovery
2. Excellent control and response to process changes
3. No chemical-disposal problem when pollutant (product) recovered and returned to process
4. Capability of systems for fully automatic, unattended operation
5. Capability to remove gaseous or vapor contaminants from process streams to extremely low levels

Disadvantages
1. Product recovery possibly requiring an exotic, expensive distillation (or extraction) scheme
2. Adsorbent progressively deteriorating in capacity as the number of cycles increases
3. Adsorbent regeneration requiring a steam or vacuum source
4. Relatively high capital cost
5. Prefiltering of gas stream possibly required to remove any particulate capable of plugging the adsorbent bed
6. Cooling of gas stream possibly required to get to the usual range of operation (less than 49°C [120°F])
7. Relatively high steam requirements to desorb high-molecular-weight hydrocarbons
8. Spent adsorbent may be considered a hazardous waste
9. Some contaminants may undergo a violent exothermic reaction with the adsorbent

TABLE 25-21 Advantages and Disadvantages of Combustion Systems

Advantages
1. Simplicity of operation
2. Capability of steam generation or heat recovery in other forms
3. Capability for virtually complete destruction of organic contaminants

Disadvantages
1. Relatively high operating costs (particularly associated with fuel requirements)
2. Potential for flashback and subsequent explosion hazard
3. Catalyst poisoning (in the case of catalytic incineration)
4. Incomplete combustion, possibly creating potentially worse pollution problems
5. Even complete combustion may produce SO_2, NO_x, and CO_2
6. High temperature components and exhaust may be hazardous to maintenance personnel and birds
7. High maintenance requirements—especially if operation is cyclic

TABLE 25-22 Advantages and Disadvantages of Condensers

Advantages
1. Pure product recovery (in the case of indirect-contact condensers)
2. Water used as the coolant in an indirect-contact condenser (i.e., shell-and-tube heat exchanger), not in contact with contaminated gas stream, and is reusable after cooling
3. May be used to produce vacuum to remove contaminants from process

Disadvantages
1. Relatively low removal efficiency for gaseous contaminants (at concentrations typical of pollution-control applications)
2. Coolant requirements possibly extremely expensive
3. Direct-contact condenser may produce water discharge problems

TABLE 25-23 Advantages and Disadvantages of Biofiltration

Advantages
1. Uses natural biological processes and materials
2. Relatively simple and economical
3. High destruction efficiencies for oxygen-rich, low-contaminant concentration gas streams
4. Waste products are CO_2 and water

Disadvantages
1. Raw gas must not be lethal to microorganisms
2. Gas stream must be maintained at proper temperature and humidity
3. Heavy particulate loadings can damage pore structure of filter bed

TABLE 25-24 Advantages and Disadvantages of Membrane Filtration

Advantages
1. Low energy utilization
2. Modular design
3. Low capital costs
4. Low maintenance
5. Superior separation ability

Disadvantages
1. Potential particulate fouling problems if not properly designed

TABLE 25-25 Advantages and Disadvantages of Selective Catalytic Reduction of Nitrogen Oxides

Advantages
1. Capable of 90% NO_x removal
2. Uses readily available urea reagent
3. Exhaust products are N_2 and water

Disadvantages
1. Spent catalyst may be considered hazardous waste
2. Gas stream must be maintained at proper temperature and humidity
3. Heavy particulate loadings can damage pore structure of filter bed

major modification plan that is subject to Prevention of Significant Deterioration (PSD) requirements must under go a Best Available Control Technology (BACT) analysis. The BACT analysis is done on a case-by-case basis. The process involves evaluating the possible types of air-pollution-control equipment that could be used for technology and energy and in terms of the environment and economy. The analysis uses a top-down approach that lists all available control technologies in descending order of effectiveness. The most stringent technology is chosen unless it can be demonstrated that, due to tech-

TABLE 25-26 Compliance Activity and Schedule Chart

Milestones

Milestones

1. Date of submittal of final control plan to appropriate agency
2. Date of award of control-device contract
3. Date of initiation of on-site construction or installation of emission-control equipment
4. Date by which on-site construction or installation of emission-control equipment is completed
5. Date by which final compliance is achieved

Activities

Designation	Activity	Designation	Activity
A-C	Preliminary investigation.	K-L	Review and approval of assembly drawings.
A-B	Source tests, if necessary.	L-M	Vendor prepares fabrication drawings.
C-D	Evaluate control alternatives.	M-N	Fabricate control device.
D-E	Commit funds for total program.	L-O	Prepare engineering drawings.
E-F	Prepare preliminary control plan and compliance schedule for agency.	O-P	Procure construction bids.
		P-Q	Evaluate construction bids.
F-G	Agency review and approval.	Q-3	Award construction contract.
G-I	Finalize plans and specifications.	3-N	On-site construction.
I-H	Procure control-device bids.	N-R	Install control device.
H-J	Evaluate control-device bids.	R-4	Complete construction (system tie-in).
J-2	Award control-device contract.	4-5	Startup, shakedown, source test.
2-K	Vendor prepares assembly drawings.		

nical, energy, environmental, or economic considerations, this type of APC technology is not feasible.

The 1990 CAAA introduced a new level of control for hazardous (toxic) air pollutants (HAPs). As a result, EPA has identified 189 HAPs for regulation. Rather than rely upon ambient air quality standards to set acceptable exposures to HAPs, the CAAA requires that EPA promulgate through the end of the decade Maximum Achievable Control Technology (MACT) standards for controlling HAPs emitted from specified industries. These standards are based on the level of control established by the best performing 12 percent of industries in each of the categories identified by EPA.

DISPERSION FROM STACKS

Stacks discharging to the atmosphere have long been the most common industrial method of disposing of waste gases. The concentrations to which humans, plants, animals, and structures are exposed at ground level can be reduced significantly by emitting the waste gases from a process at great heights. Although tall stacks may be effective in lowering the ground-level concentration of pollutants, they do not in themselves reduce the amount of pollutants released into the atmosphere. However, in certain situations, their use can be the most practical and economical way of dealing with an air-pollution problem.

Preliminary Design Considerations To determine the acceptability of a stack as a means of disposing of waste gases, the acceptable ground-level concentration (GLC) of the pollutant or pollutants must be determined. The topography of the area must also be considered so that the stack can be properly located with respect to buildings and hills that might introduce a factor of air turbulence into the operation of the stack. Awareness of the meteorological conditions prevalent in the area, such as the prevailing winds, humidity, and rainfall, is also essential. Finally, an accurate knowledge of the constituents of the waste gas and their physical and chemical properties is paramount.

Wind Direction and Speed Wind direction is measured at the height at which the pollutant is released, and the mean direction will indicate the direction of travel of the pollutants. In meteorology, it is conventional to consider the wind direction as the direction from which the wind blows; therefore, a northwest wind will move pollutants to the southeast of the source.

The effect of wind speed is twofold: (1) Wind speed will determine the travel time from a source to a given receptor; and (2) wind speed

will affect dilution in the downwind direction. Generally, the concentration of air pollutants downwind from a source is inversely proportional to wind speed.

Wind speed has velocity components in all directions so that there are vertical motions as well as horizontal ones. These random motions of widely different scales and periods are essentially responsible for the movement and diffusion of pollutants about the mean downwind path. These motions can be considered atmospheric turbulence. If the scale of a turbulent motion (i.e., the size of an eddy) is larger than the size of the pollutant plume in its vicinity, the eddy will move that portion of the plume. If an eddy is smaller than the plume, its effect will be to diffuse or spread out the plume. This diffusion caused by eddy motion is widely variable in the atmosphere, but even when the effect of this diffusion is least, it is in the vicinity of three orders of magnitude greater than diffusion by molecular action alone.

Mechanical turbulence is the induced-eddy structure of the atmosphere due to the roughness of the surface over which the air is passing. Therefore, the existence of trees, shrubs, buildings, and terrain features will cause mechanical turbulence. The height and spacing of the elements causing the roughness will affect the turbulence. In general, the higher the roughness elements, the greater the mechanical turbulence. In addition, mechanical turbulence increases as wind speed increases.

Thermal turbulence is turbulence induced by the stability of the atmosphere. When the Earth's surface is heated by the sun's radiation, the lower layer of the atmosphere tends to rise and thermal turbulence becomes greater, especially under conditions of light wind. On clear nights with wind, heat is radiated from the Earth's surface, resulting in the cooling of the ground and the air adjacent to it. This results in extreme stability of the atmosphere near the Earth's surface. Under these conditions, turbulence is at a minimum. Attempts to relate different measures of turbulence of the wind (or stability of the atmosphere) to atmospheric diffusion have been made for some time. The measurement of atmospheric stability by temperature-difference measurements on a tower is frequently utilized as an indirect measure of turbulence, particularly when climatological estimates of turbulence are desired.

Lapse Rate and Atmospheric Stability Apart from mechanical interference with the steady flow of air caused by buildings and other obstacles, the most important factor that influences the degree of turbulence and hence the speed of diffusion in the lower air is the variation of temperature with height above the ground, referred to as the "lapse rate." The dry-adiabatic lapse rate (DALR) is the temperature change for a rising parcel of dry air. The dry-adiabatic lapse rate can be approximated as $-1°$ C per 100 m, or $dT/dz = -10^{-2}$ °C/m or $-5.4°$ F/1000 ft. If the rising air contains water vapor, the cooling due to adiabatic expansion will result in the relative humidity being increased, and saturation may be reached. Further ascent would then lead to condensation of water vapor, and the latent heat thus released would reduce the rate of cooling of the rising air. The buoyancy force on a warm-air parcel is caused by the difference between its density and that of the surrounding air. The perfect-gas law shows that, at a fixed pressure (altitude), the temperature and density of an air parcel are inversely related; temperature is normally used to determine buoyancy because it is easier to measure than density. If the temperature gradient (lapse rate) of the atmosphere is the same as the adiabatic lapse rate, a parcel of air displaced from its original position will expand or contract in such a manner that its density and temperature remain the same as its surroundings. In this case there will be no buoyancy forces on the displaced parcel, and the atmosphere is termed "neutrally stable."

If the atmospheric temperature decreases faster with increasing altitude than the adiabatic lapse rate (superadiabatic), a parcel of air displaced upward will have a higher temperature than the surrounding air. Its density will be lower, giving it a net upward buoyancy force. The opposite situation exists if the parcel of air is displaced downward, and the parcel experiences a downward buoyancy force. Once a parcel of air has started moving up or down, it will continue to do so, causing unstable atmospheric conditions. If the temperature decreases more slowly with increasing altitude than the adiabatic lapse rate, a displaced parcel of air experiences a net restoring force. The buoyancy forces then cause stable atmospheric conditions (see Fig. 25-4).

Strongly stable lapse rates are commonly referred to as *inversions*. The strong stability inhibits mixing across the inversion layer. Normally these conditions of strong stability extend for only several hundred meters vertically. The vertical extent of the inversion is referred to as the *inversion depth*. Two distinct types are observed: the ground-level inversion, caused by radiative cooling of the ground at night, and inversions aloft, occurring between 500 and several thousand meters above the ground (see Fig. 25-5). Some of the more common lapse-rate profiles with the corresponding effect on stackplumes are presented in Fig. 25-6.

From the viewpoint of air pollution, both stable surface layers and low-level inversions are undesirable because they minimize the rate of dilution of contaminants in the atmosphere. Even though the surface layer may be unstable, a low-level inversion will act as a barrier to vertical mixing, and contaminants will accumulate in the surface layer below the inversion. Stable atmospheric conditions tend to be more frequent and longest in persistence in the autumn, but inversions and stable lapse rates are prevalent at all seasons of the year.

Design Calculations For a given stack height, the calculational sequence begins by first estimating the effective height of the emission, employing an applicable plume-rise equation. The maximum GLC may then be determined by using an appropriate atmospheric-diffusion equation. A simple comparison of the calculated GLC for the particular pollutant with the maximum GLC permitted by the local air-pollution codes dictates whether the stack is operating satisfactorily.

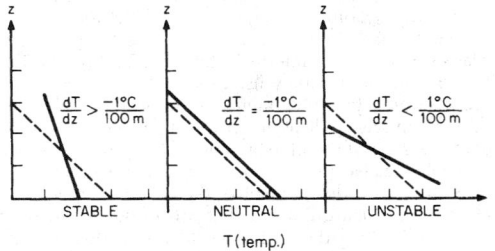

FIG. 25-4 Stability criteria with measured lapse rate.

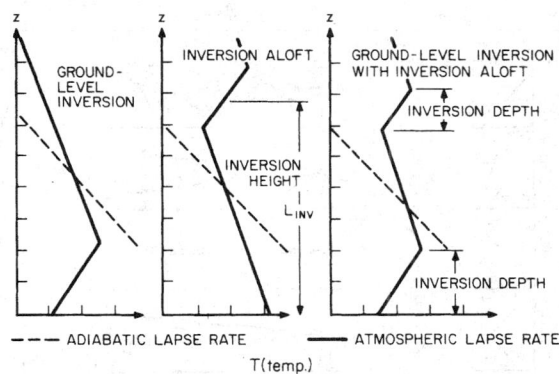

FIG. 25-5 Characteristic lapse rates under inversion conditions.

Conversely, with knowledge of the maximum acceptable GLC standards, a stack that will satisfy these standards can be properly designed.

Effective Height of an Emission The effective height of an emission rarely corresponds to the physical height of the stack. If the plume is caught in the turbulent wake of the stack or of buildings in the vicinity of the stack, the effluent will be mixed rapidly downward toward the ground. If the plume is emitted free of these turbulent zones, a number of source emission characteristics and meteorological factors influence the rise of the plume. The source emission characteristics include the gas flow rate and the temperature of the effluent at the top of the stack and the diameter of the stack opening. The meteorological factors influencing plume rise include wind speed, air temperature, shear of the wind speed with height, and atmospheric stability. No theory on plume rise presently takes into account all of these variables. Most of the equations that have been formulated for computing the effective height of an emission are semi-empirical. When considering any of these plume-rise equations, it is important to evaluate each in terms of the assumptions made and the circumstances existing at the time that the particular correlation was formulated. The formulas generally are not applicable to tall stacks (above 305 m [1000 ft] effective height).

The effective stack height (equivalent to the effective height of the emission) is the sum of the actual stack height, the plume rise due to the exhaust velocity (momentum) of the issuing gases, and the buoyancy rise, which is a function of the temperature of the gases being emitted and the atmospheric conditions.

Some of the more common plume-rise equations have been summarized by Buonicore and Theodore (*Industrial Control Equipment for Gaseous Pollutants*, vol. 2, CRC Press, Boca Raton, Florida, 1975) and include:

- ASME, *Recommended Guide for the Prediction of the Dispersion of Airborne Effluents*, ASME, New York, 1968.
- Bosanquet-Carey-Halton, *Proc. Inst. Mech. Eng.* (London), **162**, 355 (1950).
- Briggs, *Plume Rise*, AEC Critical Review ser., U.S. Atomic Energy Commission, Div. Tech. Inf.
- Brummage et al., *The Calculation of Atmospheric Dispersion from a Stack*, CONCAWE, The Hague, 1966.
- Carson and Moses, *J. Air Pollut. Control Assoc.*, **18**, 454 (1968) and **19**, 862 (1969).
- Csnaday, *Int. J. Air Water Pollut.*, **4**, 47 (1961).
- Davison-Bryant, *Trans. Conf. Ind. Wastes*, 14th Ann. Meet. Ind. Hyg. Found. Am., **39**, 1949.
- Holland, *Workbook of Atmospheric Dispersion Estimates*, U.S. EPA Publ. AP-26, 1970.
- Lucas, Moore, and Spurr, *Int. J. Air Water Pollut.*, **7**, 473 (1963).
- Montgomery et al., *J. Air Pollut. Control Assoc.*, **22**(10), 779, TVA (1972).
- Stone and Clarke, *British Experience with Tall Stacks for Air Pollution Control on Large Fossil-Fueled Power Plants*, American Power Conference, Chicago, 1967.

Temperature gradient	Observation	Description

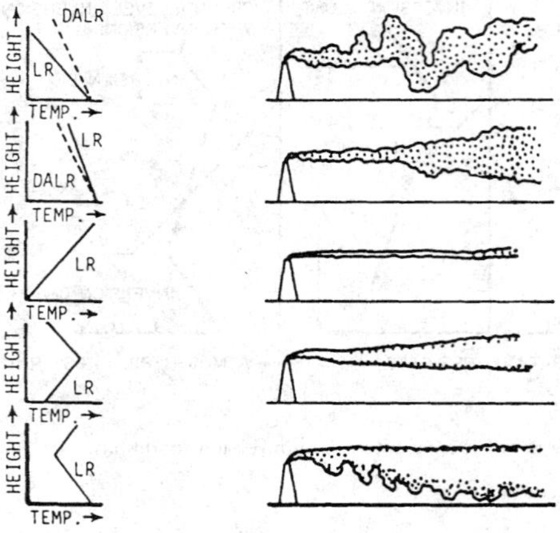

Strong lapse—looping (unstable)
This is usually a fair-weather daytime condition, since strong solar heating of the ground is required. Looping is not favored by cloudiness, snow cover, or strong winds.
Weak lapse—coning (slightly unstable or neutral)
This is usually favored by cloudy and windy conditions and may occur day or night. In dry climates it may occur infrequently, and in cloudy climates it may be the most frequent type observed.
Inversion—fanning (stable)
This is principally a nighttime condition. It is favored by light winds, clear skies, and snow cover. The condition may persist in some climates for several days at a time during winter, especially in the higher latitudes.
Inversion below, lapse aloft—lofting (transition from unstable to stable)
This condition occurs during transition from lapse to inversion and should be observed most frequently near sunset; it may be very transitory or persist for several hours. The shaded zone of strong effluent concentration is caused by trapping by the inversion of effluent carried into the stable layer by turbulent eddies that penetrate the layer for a short distance.
Lapse below, inversion aloft—fumigation (transition from stable to unstable)
This occurs when the nocturnal inversion is dissipated by heat from the morning sun. The lapse layer usually starts at the ground and works its way upward (less rapidly in winter than in summer). Fumigation may also occur in sea-breeze circulations during late morning or early afternoon. The shaded zone of strong concentration is that portion of the plume which has not yet been mixed downward.

FIG. 25-6   Lapse-rate characteristics of atmospheric-diffusion transport of stack emissions.

- Stumke, *Staub,* **23,** 549 (1963).
See also:
- Briggs, G. A., *Plume Rise Predications: Lectures on Air Pollution and Environmental Impact Analyses,* Workshop Proceedings, American Meteorological Society, Boston, Massachusetts, 1975, pp. 59–111.
- Randerson, Darryl (ed.), *Plume Rise and Bouyancy Effects, Atmospheric Science and Power Production,* DOE Report DOE/TIC-27601, 1981.

Maximum Ground-Level Concentrations   The effective height of an emission having been determined, the next step is to study its path downward by using the appropriate atmospheric-dispersion formula. Some of the more popular atmospheric-dispersion calculational procedures have been summarized by Buonicore and Theodore (op. cit.) and include:

- Bosanquet-Pearson model [Pasquill, *Meteorol. Mag.,* **90,** 33, 1063 (1961), and Gifford, *Nucl. Saf.,* **2,** 4, 47 (1961).]
- Sutton model [*Q.J.R. Meteorol. Soc.,* **73,** 257 (1947).]
- TVA Model [Carpenter et al., *J. Air Pollut. Control Assoc.,* **21**(8), (1971), and Montgomery et al., *J. Air Pollut. Control Assoc.,* **23**(5), 388 (1973).]
See also:
- Bornstein et al., *Simulation of Urban Barrier Effects on Polluted Urban Boundary Layers Using the Three-Dimensional URBMET\ TVM Model with Urban Topography,* Air Pollution Proceedings, 1993.
- EPA, *Guidance on the Application of Refined Dispersion Models for Air Toxins Releases,* EPA-450/4-91-007.
- Zannetti, Paolo, *Air Pollution Modeling: Theories, Computational Methods, and Available Software,* Van Nostrand, Reinhold, New York, 1990.
- Zannetti, Paolo, *Numerical Simulation Modeling of Air Pollution: An Overview,* Ecological Physical Chemistry, 2d International Workshop, May 1992.

Miscellaneous Effects

Evaporative Cooling   When effluent gases are washed to absorb certain constituents prior to emission, the gases are cooled and become saturated with water vapor. Upon release of the gases, further cooling due to contact with cold surfaces of ductwork or stack is likely. This cooling causes water droplets to condense in the gas stream. Upon release of the gases from the stack, the water droplets evaporate, withdrawing the latent heat of vaporization from the air and cooling the plume. The resulting negative buoyancy reduces the effective stack height. The result may be a plume (with a greater density than that of the ambient atmosphere) that will fall to the ground. If any pol-

lutant remains after scrubbing, its full effect will be felt on the ground in the vicinity of the stack.

Aerodynamic Downwash   Should the stack exit velocity be too low as compared with the speed of the crosswind, some of the effluent can be pulled downward by the low pressure on the lee side of the stack. This phenomenon, known as "stack-tip downwash," can be minimized by keeping the exit velocity greater than the mean wind speed (i.e., typically twice the mean wind speed). Another way to minimize stack-tip downwash is to fit the top of the stack with a flat disc that extends for at least one stack diameter outward from the stack.

If it becomes necessary to increase the stack-gas exit velocity to avoid downwash, it may be necessary to remodel the stack exit. A venturi-nozzle design has been found to be the most effective. This design also keeps pressure losses to a minimum.

Building Downwash   A review must be conducted for each stack to determine if building downwash effects need to be considered. Atmospheric flow is disrupted by aerodynamic forces in the immediate vicinity of structures or terrain obstacles. The disrupted flow near either building structures or terrain obstacles can both enhance the vertical dispersion of emissions from the source and reduce the effective height of the emissions from the source, resulting in an increase in the maximum GLC.

EPA Air Dispersion Models   EPA addresses modeling techniques in the "Guideline on Air Quality Models" (GAQM-Revised. EPA-450/2-78-027R, July 1986). Several computational models are available for performing air quality analyses. However, there is no one model capable of properly addressing all conceivable modeling situations. The EPA recommends a case-by-case approach to selection of appropriate models. All models incorporate assumptions that are designed to predict conservative results. Since the general intention of modeling is to determine if the maximum GLC is below regulatory limits, modeling is performed beginning with the simpler models with the most conservative assumptions and proceeding to more complex models with more sophisticated input data. Input data includes source characteristics, topography, and meteorological factors.

The air quality modeling procedures can be categorized into four generic classes: Gaussian, numerical, statistical, and physical. Gaussian models are the most widely used techniques for estimating the impact from nonreactive pollutants. Numerical models are more appropriate for urban applications involving reactive pollutants. In cases where the scientific understanding of the physical or chemical processes involved is lacking, a statistical modeling approach may be necessary. Statistical modeling involves the collection of a large number of site-specific measurements that are evaluated to develop a predictive modeling tool that is limited to use at that site. Physical modeling is performed using a wind tunnel or other fluid modeling facility. Physical modeling may be useful for complex flow situations, such as building, terrain, or stack downwash conditions, plume impact on elevated terrain, diffusion in an urban environment, or diffusion in complex terrain. It is generally limited to use within a few square kilometers surrounding the source.

Most modeling can be performed using Gaussian models. The simplest Gaussian model is the EPA SCREEN2 model. This is an interactive computational model that incorporates algorithms for calculating maximum GLC in simple terrain (below stack top) or complex terrain (above stack top) considering downwash and cavity effects. The SCREEN2 model is designed for single source modeling. A set of conservative meteorological conditions is contained within the model.

The Industrial Source Complex (ISC2) model allows for modeling to be performed in simple terrain for multiple sources and allows for use of real meteorological data. Complex I is a screening level model commonly used in complex terrain modeling. This model may use real meteorological data and allows for modeling multiple sources. The Complex Terrain Dispersion Model Plus Algorithms for Unstable Situations (CTDMPLUS) is a refined point source Gaussian air quality model for use in all stability conditions for complex terrain applications. Its use of meteorological and terrain data is different from other EPA models. CTDMPLUS requires the parameterization of individual hill shapes using a terrain preprocessor. CTSCREEN is a screening version of CTDMPLUS. It contains the same algorithms but is run with synthetic meteorological data.

SOURCE CONTROL OF GASEOUS EMISSIONS

There are four chemical-engineering unit operations commonly used for the control of gaseous emissions:
1. *Absorption.* Sec. 4, "Thermodynamics"; and Sec. 18, "Liquid Gas Systems." For plate columns, see Sec. 18, "Gas-Liquid Contacting: Plate Columns." For packed columns, see Sec. 18, "Gas-Liquid Contacting: Packed Columns."
2. *Adsorption.* See Sec. 16, "Adsorption and Ion Exchange."
3. *Combustion.* See Sec. 9, "Heat Generation" and "Fired Process Equipment." For incinerators, see material under these subsections.
4. *Condensation.* See Sec. 10, "Heat Transmission"; Sec. 11, "Heat-Transfer Equipment"; and Sec. 12, "Evaporative Cooling." For direct-contact condensers, See Sec. 11, "Evaporator Accessories"; and Sec. 12, "Evaporative Cooling." For indirect-contact condensers, see Sec. 10, "Heat Transfer with Change of Phase" and "Thermal Design of Heat-Transfer Equipment"; and Sec. 11, "Shell and Tube Heat Exchangers" and "Other Heat Exchangers for Liquids and Gases."

These operations, which are routine chemical engineering operations, have been treated extensively in other sections of this handbook.

There are three additional chemical engineering unit operations that are increasing in use in recent years. They are:
1. *Biofiltration*
2. *Membrane filtration*
3. *Selective catalytic reduction*
and are discussed further in this section.

Absorption The engineering design of gas absorption equipment must be based on a sound application of the principles of diffusion, equilibrium, and mass transfer as developed in Secs. 5 and 14 of the handbook. The main requirement in equipment design is to bring the gas into intimate contact with the liquid; that is, to provide a large interfacial area and a high intensity of interface renewal and to minimize resistance and maximize driving force. This contacting of the phases can be achieved in many different types of equipment, the most important being packed and plate columns. The final choice between them rests with the various criteria that must be met. For example, if the pressure drop through the column is large enough that compression costs become significant, a packed column may be preferable to a plate-type column because of the lower pressure drop.

In most processes involving the absorption of a gaseous pollutant from an effluent gas stream, the gas stream is the processed fluid; hence, its inlet condition (flow rate, composition, and temperature) are usually known. The temperature and composition of the inlet liq-

uid and the composition of the outlet gas are usually specified. The main objectives in the design of an absorption column, then, are the determination of the solvent flow rate and the calculation of the principal dimensions of the equipment (column diameter and height to accomplish the operation). These objectives can be obtained by evaluating, for a selected solvent at a given flow rate, the number of theoretical separation units (stage or plates) and converting them into practical units column heights or number of actual plates by means of existing correlations.

The general design procedure consists of a number of steps to be taken into consideration. These include:
1. Solvent selection
2. Equilibrium-data evaluation
3. Estimation of operating data (usually consisting of a mass and energy in which the energy balance decides whether the absorption balance can be considered isothermal or adiabatic)
4. Column selection (should the column selection not be obvious or specified, calculations must be carried out for the different types of columns and the final based on economic considerations)
5. Calculation of column diameter (for packed columns, this is usually based on flooding conditions, and, for plate columns, on the optimum gas velocity or the liquid-handling capacity of the plate)
6. Estimation of column height or number of plates (for packed columns, column height is obtained by multiplying the number of transfer units, obtained from a knowledge of equilibrium and operating data, by the height of a transfer unit; for plate columns, the number of theoretical plates determined from the plot of equilibrium and operating lines is divided by the estimated overall plate efficiency to give the number of actual plates, which in turn allows the column height to be estimated from the plate spacing)
7. Determination of pressure drop through the column (for packed columns, correlations dependent of packing type, column-operating data, and physical properties of the constituents involved are available to estimate the pressure drop through the packing; for plate columns, the pressure drop per plate is obtained and multiplied by the number of plates)

Solvent Selection The choice of a particular solvent is most important. Frequently, water is used, as it is very inexpensive and plentiful, but the following properties must also be considered:
1. *Gas solubility.* A high gas solubility is desired, since this increases the absorption rate and minimizes the quantity of solvent necessary. Generally, solvents of a chemical nature similar to that of the solute to be absorbed will provide good solubility.
2. *Volatility.* A low solvent vapor pressure is desired, since the gas leaving the absorption unit is ordinarily saturated with the solvent, and much may thereby be lost.
3. *Corrosiveness.*
4. *Cost.*
5. *Viscosity.* Low viscosity is preferred for reasons of rapid absorption rates, improved flooding characteristics, lower pressure drops, and good heat-transfer characteristics.
6. *Chemical stability.* The solvent should be chemically stable and, if possible, nonflammable.
7. *Toxicity.*
8. *Low freezing point.* If possible, a low freezing point is favored, since any solidification of the solvent in the column makes the column inoperable.

Equipment The principal types of gas-absorption equipment may be classified as follows:
1. Packed columns (continuous operation)
2. Plate columns (staged operations)
3. Miscellaneous

Of the three categories, the packed column is by far the most commonly used for the absorption of gaseous pollutants. Miscellaneous gas-absorption equipment could include acid gas scrubbers that are commonly classified as either "wet" or "dry." In wet scrubber systems, the absorption tower uses a lime-based sorbent liquor that reacts with the acid gases to form a wet/solid by-product. Dry scrubbers can be grouped into three categories: (1) spray dryers; (2) circulating spray dryers; and (3) dry injection. Each of these systems yields a dry product that can be captured with a fabric filter baghouse downstream and thus avoids costly wastewater treatment systems. The baghouse is highly efficient in capturing the particulate emissions, and a portion of the overall acid gas removal has been found to occur within the baghouse. Additional information may be found by referring to the appropriate sections in this *handbook* and the many excellent texts available, such as McCabe and Smith, *Unit Operations of Chemical Engineering,* 3d ed., McGraw-Hill, New York, 1976; Sherwood and Pigford, *Absorption and Extraction,* 2d ed., McGraw-Hill, New York, 1952; Smith, *Design of Equilibrium Stage Processes,* McGraw-Hill, New York, 1963; Treybal, *Mass Transfer Operations,* 3d ed., McGraw-Hill, New York, 1980; McKenna, Mycock, and Theodore, *Handbook of Air Pollution Control Engineering and Technology,* CRC Press, Boca Raton, Florida, 1995; Theodore and Allen, *Air Pollution Control Equipment,* ETSI Prof. Training Inst., Roanoke, Virginia, 1993.

Adsorption   The design of gas-adsorption equipment is in many ways analogous to the design of gas-absorption equipment, with a solid adsorbent replacing the liquid solvent (see Secs. 16 and 19). Similarity is evident in the material- and energy-balance equations as well as in the methods employed to determine the column height. The final choice, as one would expect, rests with the overall process economics.

Selection of Adsorbent   Industrial adsorbents are usually capable of adsorbing both organic and inorganic gases and vapors. However, their preferential adsorption characteristics and other physical properties make each of them more or less specific for a particular application. General experience has shown that, for the adsorption of vapors of an organic nature, activated carbon has superior properties, having hydrocarbon-selective properties and high adsorption capacity for such materials. Inorganic adsorbents, such as activated alumina or silica gel, can also be used to adsorb organic materials, but difficulties can arise during regeneration. Activated alumina, silica gel, and molecular sieves will also preferentially adsorb any water vapor with the organic contaminant. At times this may be a considerable drawback in the application of these adsorbents for organic-contaminant removal.

The normal method of regeneration of adsorbents is by use of steam, inert gas (i.e., nitrogen), or other gas streams, and in the majority of cases this can cause at least slight decomposition of the organic compound on the adsorbent. Two difficulties arise: (1) incomplete recovery of the adsorbate, although this may be unimportant; and (2) progressive deterioration in capacity of the adsorbent as the number of cycles increases owing to blocking of the pores from carbon formed by hydrocarbon decomposition. With activated carbon, a steaming process is used, and the difficulties of regeneration are thereby overcome. This is not feasible with silica gel or activated alumina because of the risk of breakdown of these materials when in contact with liquid water.

In some cases, none of the adsorbents has sufficient retaining capacity for a particular contaminant. In these applications, a large-surface-area adsorbent can be impregnated with an inorganic compound or, in rare cases, with a high-molecular-weight organic compound that can react chemically with the particular contaminant. For example, iodine-impregnated carbons are used for removal of mercury vapor and bromine-impregnated carbons for ethylene or propylene removal. The action of these impregnants is either catalytic conversion or reaction to a nonobjectionable compound or to a more easily adsorbed compound. For this case, general adsorption theory no longer applies to the overall effects of the process. For example, mercury removal by an iodine-impregnated carbon proceeds faster at a higher temperature, and a better overall efficiency can be obtained than in a low-temperature system.

Since adsorption takes place at the interphase boundary, the adsorption surface area becomes an important consideration. Generally, the higher the adsorption surface area, the greater its adsorption capacity. However, the surface area has to be "available" in a particular pore size within the adsorbent. At low partial pressure (or concentration) a surface area in the smallest pores in which the adsorbate can enter is the most efficient. At higher pressures the larger pores become more important; at very high concentrations, capillary condensation will take place within the pores, and the total micropore volume becomes the limiting factor.

The action of molecular sieves is slightly different from that of other adsorbents in that selectivity is determined more by the pore-size limitations of the particular sieve. In selecting molecular sieves, it is important that the contaminant to be removed be smaller than the available pore size. Hence, it is important that the particular adsorbent not only have an affinity for the contaminant in question but also have sufficient surface area available for adsorption.

Design Data   The adsorbent having been selected, the next step is to calculate the quantity of adsorbent required and eventually consider other factors such as the temperature rise of the gas stream due to adsorption and the useful life of the adsorbent under operating conditions. The sizing and overall design of the adsorption system depend on the properties and characteristics of both the feed gas to be treated and the adsorbent. The following information should be known or available for design purposes:

1. Gas stream
   a. Adsorbate concentration.
   b. Temperature.
   c. Temperature rise during adsorption.
   d. Pressure.
   e. Flow rate.
   f. Presence of adsorbent contaminant material.
2. Adsorbent
   a. Adsorption capacity as used on stream.
   b. Temperature rise during adsorption.
   c. Isothermal or adiabatic operation.
   d. Life, if presence of contaminant material is unavoidable.
   e. Possibility of catalytic effects causing an adverse chemical reaction in the gas stream or the formation of solid polymerizates on the adsorbent bed, with consequent deterioration.
   f. Bulk density.
   g. Particle size, usually reported as a mean equivalent particle diameter. The dimensions and shape of particles affect both the pressure drops through the adsorbent bed and the diffusion rate into the particles. All things being equal, adsorbent beds consisting of smaller particles, although causing a higher pressure drop, will be more efficient.
   h. Pore data, which are important because they may permit elimination from consideration of adsorbents whose pore diameter will not admit the desired adsorbate molecule.
   i. Hardness, which indicates the care that must be taken in handling adsorbents to prevent the formation of undesirable fines.
   j. Regeneration information.

The design techniques used include both stagewise and continuous-contacting methods and can be applied to batch, continuous, and semicontinuous operations.

Adsorption Phenomena   The adsorption process involves three necessary steps. The fluid must first come in contact with the adsorbent, at which time the adsorbate is preferentially or selectively adsorbed on the adsorbent. Next the fluid must be separated from the adsorbent-adsorbate, and, finally, the adsorbent must be regenerated by removing the adsorbate or by discarding used adsorbent and replacing it with fresh material. Regeneration is performed in a variety of ways, depending on the nature of the adsorbate. Gases or vapors are usually desorbed by either raising the temperature (thermal cycle) or reducing the pressure (pressure cycle). The more popular thermal cycle is accomplished by passing hot gas through the adsorption bed in the direction opposite to the flow during the adsorption cycle. This ensures that the gas passing through the unit during the adsorption cycle always meets the most active adsorbent last and that the adsorbate concentration in the adsorbent at the outlet end of the unit is always maintained at a minimum.

In the first step, in which the molecules of the fluid come in contact with the adsorbent, an equilibrium is established between the adsorbed fluid and the fluid remaining in the fluid phase. Figures 25-7 through 25-9 show several experimental equilibrium adsorption isotherms for a number of components adsorbed on various adsorbents. Consider Fig. 25-7, in which the concentration of adsorbed gas on the solid is plotted against the equilibrium partial pressure p^0 of the vapor or gas at constant temperature. At 40° C, for example, pure propane vapor at a pressure of 550 mm Hg is in equilibrium with an adsorbate concentration at point P of 0.04 lb adsorbed propane per pound of silica gel. Increasing the pressure of the propane will cause

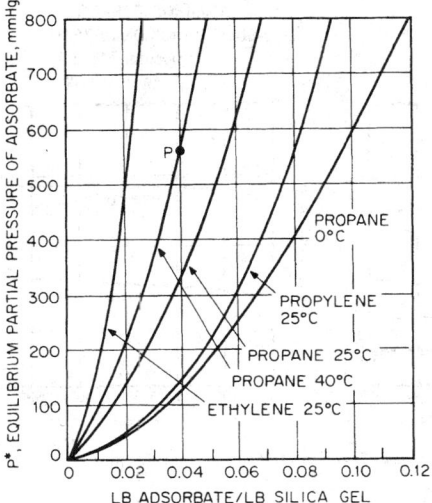

FIG. 25-7 Equilibrium partial pressures for certain organics on silica gel.

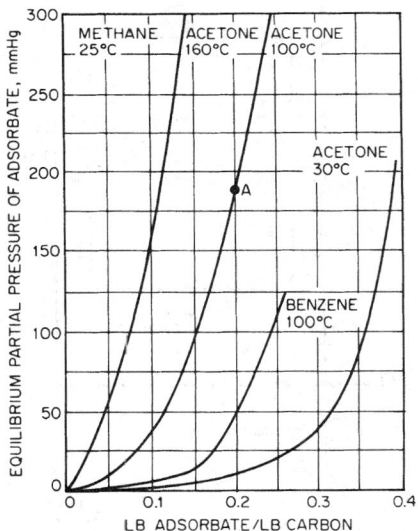

FIG. 25-8 Equilibrium partial pressures for certain organics on carbon.

more propane to be adsorbed, while decreasing the pressure of the system at *P* will cause propane to be desorbed from the carbon.

The adsorptive capacity of activated carbon for some common solvent vapors is shown in Table 25-27.

Adsorption-Control Equipment   If a gas stream must be treated for a short period, usually only one adsorption unit is necessary, provided, of course, that a sufficient time interval is available between adsorption cycles to permit regeneration. However, this is usually not the case. Since an uninterrupted flow of treated gas is often required, it is necessary to employ one or more units capable of operating in this fashion. The units are designed to handle gas flows without interruption and are characterized by their mode of contact, either staged or continuous. By far the most common type of adsorption system used to remove an objectionable pollutant from a gas stream consists of a number of fixed-bed units operating in such a sequence that the gas flow remains uninterrupted. A two- or three-bed system is usually

employed, with one or two beds bypassed for regeneration while one is adsorbing. A typical two-bed system is shown in Fig. 25-10, while a typical three-bed system is shown in Fig. 25-11. The type of system best suited for a particular job is determined from several factors, including the amount and rate of material being adsorbed, the time between cycles, the time required for regeneration, and the cooling time, if required.

Typical of continuous-contact operation for gaseous-pollutant adsorption is the use of a fluidized bed. During steady-state staged-contact operation, the gas flows up through a series of successive fluidized-bed stages, permitting maximum gas-solid contact on each stage. A typical arrangement of this type is shown in Fig. 25-12 for multistage countercurrent adsorption with regeneration. In the upper part of the tower, the particles are contacted countercurrently on perforated trays in relatively shallow beds with the gas stream containing the pollutant, the adsorbent solids moving form tray to tray through downspouts. In the lower part of the tower, the adsorbent is regenerated by similar contact with hot gas, which desorbs and carries off the pollutant. The regenerated adsorbent is then recirculated by an airlift to the top of the tower.

Although the continuous-countercurrent type of operation has found limited application in the removal of gaseous pollutants from process streams (for example, the removal of carbon dioxide and sulfur compounds such as hydrogen sulfide and carbonyl sulfide), by far the most common type of operation presently in use is the fixed-bed adsorber. The relatively high cost of continuously transporting solid particles as required in steady-state operations makes fixed-bed adsorption an attractive, economical alternative. If intermittent or batch operation is practical, a simple one-bed system, cycling alternately between the adsorption and regeneration phases, will suffice.

Additional information may be found by referring to the appropriate sections of this *handbook*. A comprehensive treatment of adsorber design principles is given in Buonicore and Theodore, *Industrial Control Equipment for Gaseous Pollutants*, vol. 1, 2d ed., CRC press, Boca Raton, Florida, 1992.

Combustion   Many organic compounds released from manufacturing operations can be converted to innocuous carbon dioxide and water by rapid oxidation (chemical reaction): combustion. However, combustion of gases containing halides may require the addition of acid gas treatment to the combustor exhaust.

Three rapid oxidation methods are typically used to destroy combustible contaminants: (1) flares (direct-flame-combustion), (2) thermal combustors, and (3) catalytic combustors. The thermal and flare methods are characterized by the presence of a flame during combustion. The combustion process is also commonly referred to as "afterburning" or "incineration."

To achieve complete combustion (i.e., the combination of the combustible elements and compounds of a fuel with all the oxygen that they can utilize), sufficient space, time, and turbulence and a temperature high enough to ignite the constituents must be provided.

The three T's of combustion—time, temperature, and turbulence—govern the speed and completeness of the combustion reaction. For complete combustion, the oxygen must come into intimate contact with the combustible molecule at sufficient temperature and for a sufficient length of time for the reaction to be completed. Incomplete reactions may result in the generation of aldehydes, organic acids, carbon, and carbon monoxide.

Combustion-Control Equipment   Combustion-control equipment can be divided into three types: (1) flares, (2) thermal incinerators, (3) catalytic incinerators.

Flares   In many industrial operations and particularly in chemical plants and petroleum refineries, large volumes of combustible waste gases are produced. These gases result from undetected leaks in the operating equipment, from upset conditions in the normal operation of a plant in which gases must be vented to avoid dangerously high pressures in operating equipment, from plant startups, and from emergency shutdowns. Large quantities of gases may also result from off-specification product or from excess product that cannot be sold. Flows are typically intermittent, with flow rates during major upsets of up to several million cubic feet per hour.

The preferred control method for excess gases and vapors is to

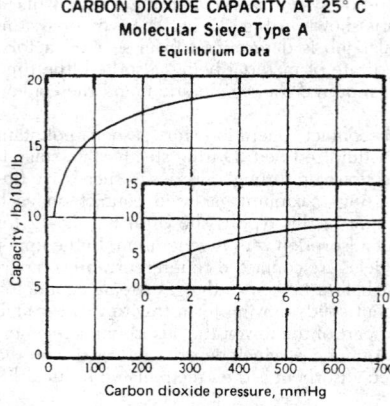

CARBON DIOXIDE CAPACITY AT 25° C
Molecular Sieve Type A
Equilibrium Data

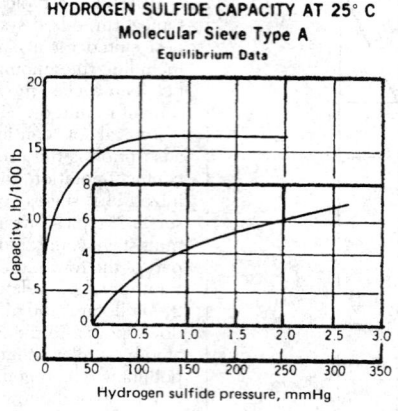

HYDROGEN SULFIDE CAPACITY AT 25° C
Molecular Sieve Type A
Equilibrium Data

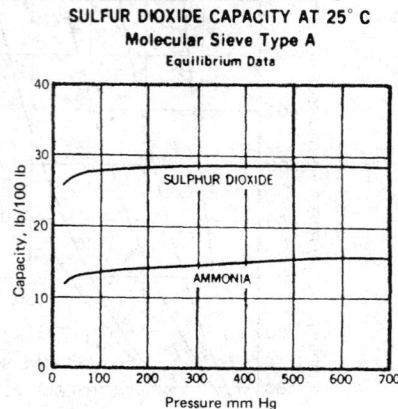

AMMONIA CAPACITY AT 25° C
SULFUR DIOXIDE CAPACITY AT 25° C
Molecular Sieve Type A
Equilibrium Data

FIG. 25-9   Equilibrium partial pressures for certain gases on molecular sieves. (A. J. Buonicore and L. Theodore, Industrial Control Equipment for Gaseous Pollutants, vol. I, CRC Press, Boca Raton, Fla., 1975.)

recover them in a blowdown recovery system. However, large quantities of gas, especially those during upset and emergency conditions, are difficult to contain and reprocess. In the past all waste gases were vented directly to the atmosphere. However, widespread venting caused safety and environmental problems, and, in practice, it is now customary to collect such gases in a closed flare system and to burn them as they are discharged.

Although flares can be used to dispose of excess waste gases, such systems can present additional safety problems. These include the explosion potential, thermal-radiation hazards from the flame, and the problem of toxic asphyxiation during flameout. Aside from these safety aspects, there are several other problems associated with flaring that must be dealt with during the design and operation of a flare system. These problems include the formation of smoke, the luminosity of the flame, noise during flame, and the possible emission of by-product air pollutants during flaring.

The heat content of the waste stream to be disposed is another important consideration. The heat content of the waste gas falls into two classes. The gases can either support their own combustion or not. In general, a waste gas with a heating value greater than 7443 kJ/m³ (200 Btu/ft³) can be flared successfully. The heating value is based on the lower heating value of the waste gas at the flare. Below 7443 kJ/m³, enriching the waste gas by injecting another gas with a higher heating value may be necessary. The addition of such a rich gas is called "endothermic flaring." Gases with a heating value as low as 2233 kJ/m³ (60 Btu/ft³) have been flared but at a significant fuel demand. It is usually not feasible to flare a gas with a heating value below 3721 kJ/m³ (100 Btu/ft³). If the flow of low-Btu gas is continuous, thermal or catalytic incineration can be used to dispose of the gas. For intermittent flows, however, endothermic flaring may be the only possibility.

TABLE 25-27   Adsorptive Capacity of Common Solvents on Activated Carbons*

Solvent	Carbon bed weight, %†
Acetone	8
Heptane	6
Isopropyl alcohol	8
Methylene chloride	10
Perchloroethylene	20
Stoddard solvent	2–7
1,1,1-Trichloroethane	12
Trichloroethylene	15
Trichlorotrifluoroethane	8
VM&P naphtha	7

*Assuming steam desorption at 5 to 10 psig.
†For example, 8 lb of acetone adsorbed on 100 lb of activated carbon.

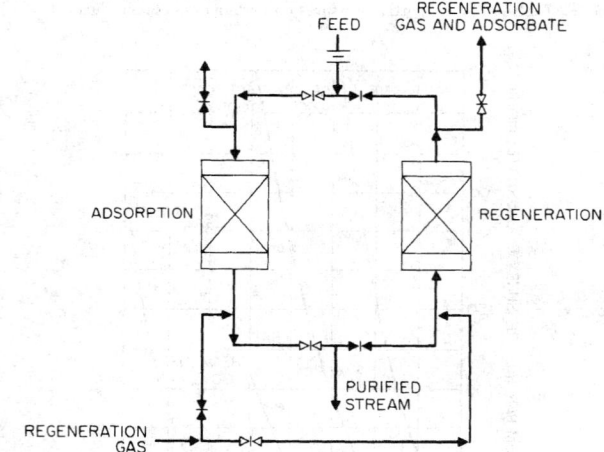

FIG. 25-10   Typical two-bed adsorption system.

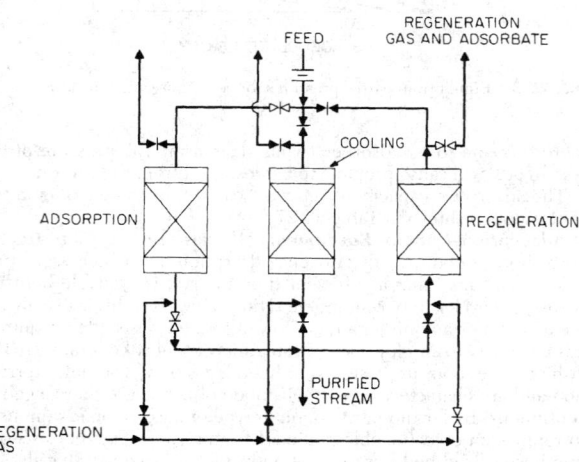

FIG. 25-11   Typical three-bed adsorption system.

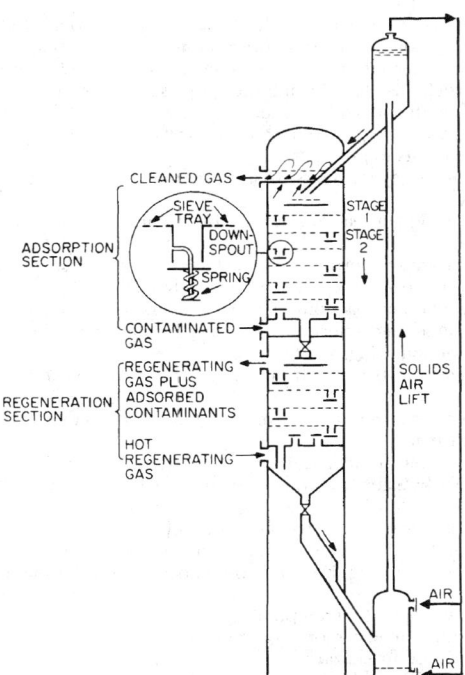

FIG. 25-12 Multistage countercurrent adsorption with regeneration.

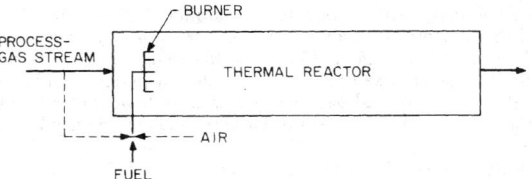

FIG. 25-13 Thermal-combustion device.

enter the reaction zone of the unit. The pollutants in the process-gas stream are then reacted at the elevated temperature. Thermal incinerators generally require operating temperatures in the range of 650 to 980° C (1200 to 1800° F) for combustion of most organic pollutants (see Table 25-28). A residence time of 0.2 to 1.0 is often recommended, but this factor is dictated primarily by complex kinetic considerations. The kinetics of hydrocarbon (HC) combustion in the presence of excess oxygen can be simplified into the following first-order rate equation:

$$\frac{d(HCl)}{dt} = -k[HCl] \qquad (25\text{-}1)$$

where k = pseudo-first-order rate constant (s^{-1}). If the initial concentration is $C_{A,o}$, the solution of Eq. (25-1) is:

$$\ln\left(\frac{C_A}{C_{A,o}}\right) = -kt \qquad (25\text{-}2)$$

Equation (25-2) is frequently used for a kinetic modeling of a burner using mole fractions in the range of 0.15 and 0.001 for oxygen and HC, respectively. The rate constant is generally of the following Arrhenius form:

$$k = Ae^{-E/RT} \qquad (25\text{-}3)$$

where: A = pre-exponential factor, s- (see Table 2, *Air Pollution Engineering Manual*, Van Nostrand Reinhold, New York, 1992, p. 62)
E = activation energy, cal/gmol (see Table 2, *Air Pollution Engineering Manual*, Van Nostrand Reinhold, New York, 1992, p. 62)
R = universal gas constant, 1.987 cal/gmol °K
T = absolute temperature, °K

The referenced table for A and E is a summary of first-order HC combustion reactions.

Although most flares are used to dispose of intermittent waste gases, some continuous flares are in use, but generally only for relatively small volumes of gases. The heating value of large-volume continuous-flow waste gases is usually too valuable to lose in a flare. Vapor recovery or the use of the vapor as a fuel in a process heater is preferred over flaring. Since auxiliary fuel must be added to the gas in order to flare, large continuous flows of a low-heating-value gas are usually more efficient to burn in a thermal incinerator than in the flame of a flare.

Flares are mostly used for the disposal of hydrocarbons. Waste gases composed of natural gas, propane, ethylene, propylene, butadiene, and butane probably constitute over 95 percent of the material flared. Flares have been used successfully to control malodorous gases such as mercaptans and amines, but care must be taken when flaring these gases. Unless the flare is very efficient and gives good combustion, obnoxious fumes can escape unburned and cause a nuisance.

Flaring of hydrogen sulfide should be avoided because of its toxicity and low odor threshold. In addition, burning relatively small amounts of hydrogen sulfide can create enough sulfur dioxide to cause crop damage or a local nuisance. For gases whose combustion products may cause problems, such as those containing hydrogen sulfide or chlorinated hydrocarbons, flaring is not recommended.

Thermal Incinerators   Thermal incinerators or afterburners can be used over a fairly wide but low range of organic vapor concentration. The concentration of the organics in air must be substantially below the lower flammable level (lower explosive limit). As a rule, a factor of four is employed for safety precautions. Reactions are conducted at elevated temperatures to ensure high chemical-reaction rates for the organics. To achieve this temperature, it is necessary to preheat the feed stream using auxiliary energy. Along with the contaminant-laden gas stream, air and fuel are continuously delivered to the incinerator (see Fig. 25-13). The fuel and contaminants are combusted with air in a firing unit (burner). The burner may utilize the air in the process-waste stream as the combustion air for the auxiliary fuel, or it may use a separate source of outside air. The products of combustion and the unreacted feed stream are intensely mixed and

TABLE 25-28 Thermal Afterburners: Conditions Required for Satisfactory Performance in Various Abatement Applications

Abatement category	Afterburner residence time, s	Temperature, °F
Hydrocarbon emissions: 90 + % destruction of HC	0.3–0.5	1100–1250°
Hydrocarbons + CO: 90 + % destruction of HC + CO	0.3–0.5	1250–1500
Odor		
50–90% destruction	0.3–0.5	1000–1200
90–99% destruction	0.3–0.5	1100–1300
99 + % destruction	0.3–0.5	1200–1500
Smokes and plumes		
White smoke (liquid mist)		
Plume abatement	0.3–0.5	800–1000†
90 + % destruction of HC + CO	0.3–0.5	1250–1500
Black smoke (soot and combustible particulates)	0.7–1.0	1400–2000

°Temperatures of 1400 to 1500°F (760 to 816°C) may be required if the hydrocarbon has a significant content of any of the following: methane, cellosolve, and substituted aromatics (e.g., toluene and xylenes).

†Operation for plume abatement only is not recommended, since this merely converts a visible hydrocarbon emission into an invisible one and frequently creates a new odor problem because of partial oxidation in the afterburner.

The end combustion products are continuously emitted at the outlet of the reactor. The average gas velocity can range from as low as 3 m/s (10 ft/s) to as high as 15 m/s (50 ft/s). These high velocities are required to prevent settling of particulates (if present) and to minimize the dangers of flashback and fire hazards. Space velocity calculations are given in the "Incinerator Design and Performance Equation" section.

The fuel is usually natural gas. The energy liberated by reaction may be directly recovered in the process or indirectly recovered by suitable external heat exchange (see Fig. 25-14).

Because of the high operating temperatures, the unit must be constructed of metals capable of withstanding this condition. Combustion devices are usually constructed with an outer steel shell that is lined with refractory material. Refractory-wall thickness is usually in the 0.05- to 0.23-m (2- to 9-in) range, depending upon temperature considerations.

Some of the advantages of the thermal incinerators are:
1. Removal of organic gases
2. Removal of submicrometer organic particles
3. Simplicity of construction
4. Small space requirements

Some of the disadvantages are:
1. High operating costs
2. Fire hazards
3. Flashback possibilities

Catalytic Incinerators Catalytic incinerators are an alternative to thermal incinerators. For simple reactions, the effect of the presence of a catalyst is to (1) increase the rate of the reaction, (2) permit the reaction to occur at a lower temperature, and (3) reduce the reactor volume.

In a typical catalytic incinerator for the combustion of organic vapors, the gas stream is delivered to the reactor continuously by a fan at a velocity in the range of 3 to 15 m/s (10 to 30 ft/s), but at a lower temperature, usually in the range of 350 to 425° C (650 to 800° F), than the thermal unit. (Design and performance equations used for calculating space velocities are given in the next section). The gases, which may or may not be preheated, pass through the catalyst bed, where the combustion reaction occurs. The combustion products, which again are made up of water vapor, carbon dioxide, inerts and unreacted vapors, are continuously discharged from the outlet at a higher temperature. Energy savings can again be effected by heat recovery from the exit stream.

Metals in the platinum family are recognized for their ability to promote combustion at low temperatures. Other catalysts include various oxides of copper, chromium, vanadium, nickel, and cobalt. These catalysts are subject to poisoning, particularly from halogens, halogen and sulfur compounds, zinc, arsenic, lead, mercury, and particulates. It is therefore important that catalyst surfaces be clean and active to ensure optimum performance.

Catalysts may be porous pellets, usually cylindrical or spherical in shape, ranging from 0.16 to 1.27 cm (1/16 to 1/2 in) in diameter. Small sizes are recommended, but the pressure drop through the reactor increases. Among other shapes are honeycombs, ribbons, and wire mesh. Since catalysis is a surface phenomenon, a physical property of these particles is that the internal pore surface is nearly infinitely greater than the outside surface.

The following sequence of steps is involved in the catalytic conversion of reactants to products:
1. Transfer of reactants to and products from the outer catalyst surface
2. Diffusion of reactants and products within the pores of the catalyst
3. Activated adsorption of reactants and the desorption of the products on the active centers of the catalyst
4. Reaction or reactions on active centers on the catalyst surface

At the same time, energy effects arising from chemical reaction can result in the following:
1. Heat transfer to or from active centers to the catalyst-particle surface
2. Heat transfer to and from reactants and products within the catalyst particle
3. Heat transfer to and from moving streams in the reactor
4. Heat transfer from one catalyst particle to another within the reactor
5. Heat transfer to or from the walls of the reactor

Some of the advantages of catalytic incinerators are:
1. Lower fuel requirements as compared with thermal incinerators
2. Lower operating temperatures
3. Minimum insulation requirements
4. Reduced fire hazards
5. Reduced flashback problems

The disadvantages include:
1. Higher initial cost than thermal incinerators
2. Catalyst poisoning
3. Necessity of first removing large particulates
4. Catalyst-regeneration problems
5. Catalyst disposal

Incinerator Design and Performance Equations The key incinerator design and performance calculations are the required fuel usage and physical dimensions of the unit. The following is a general calculation procedure to use in solving for these two parameters, assuming that the process gas stream flow, inlet temperature, combustion temperature, and required residence time are known. The combustion temperature and residence time for thermal incinerators can be estimated using Table 25-28 and Eqs. (25-1) through (25-3).

1. The heat load needed to heat the inlet process gas stream to the incinerator operating temperature is:

$$Q = \Delta H \tag{25-4}$$

2. Correct the heat load for radiant heat losses, RL:

$$Q = (1 + RL)(\Delta H); RL = \text{fractional basis} \tag{25-5}$$

3. Assuming that natural gas is used to fire the burner with a known heating value of HV_G, calculate the available heat at the operating temperature. A shortcut method usually used for most engineering purposes is:

$$HA_T = (HV_G)\left(\frac{HA_T}{HV_G}\right)_{\text{ref}} \tag{25-6}$$

where the subscript "ref" refers to a reference fuel. For natural gas with a reference HV_G of 1059 Btu/scf, the heat from the combustion using no excess air would be given by:

$$(HA_T)_{\text{ref}} = -0.237(t) + 981; T = F \tag{25-7}$$

4. The amount of natural gas needed as fuel (NG) is given by:

$$NG = \frac{Q}{HA}; \text{consistent units} \tag{25-8}$$

5. The resulting volumetric flow rate is the sum of the combustion of the natural gas q and the process gas stream p at the operating temperature:

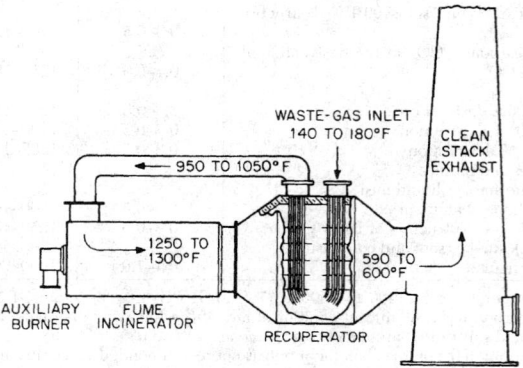

WASTE-GAS INLET
140 TO 180°F
CLEAN STACK EXHAUST
950 TO 1050°F
1250 TO 1300°F
590 TO 600°F
AUXILIARY BURNER
FUME INCINERATOR
RECUPERATOR

FIG. 25-14 Thermal combustion with energy (heat) recovery.

$$q_T = q_p + q_c \qquad (25\text{-}9)$$

A good estimate for q_c is:

$$q_c = (11.5)(NG) \qquad (25\text{-}10)$$

6. The diameter of the combustion device is given by:

$$S = \frac{q_T}{v_t} \qquad (25\text{-}11)$$

where v_t is defined as the velocity of the gas stream at the incinerator operating temperature.

Condensation Frequently in air-pollution-control practice, it becomes necessary to treat an effluent stream consisting of a condensable pollutant vapor and a noncondensable gas. One control method to remove such pollutants from process-gas streams that is often overlooked is condensation. Condensers can be used to collect condensable emissions discharged to the atmosphere, particularly when the vapor concentration is high. This is usually accomplished by lowering the temperature of the gaseous stream, although an increase in pressure will produce the same result. The former approach is usually employed by industry, since pressure changes (even small ones) on large volumetric gas-flow rates are often economically prohibitive.

Condensation Equipment There are two basic types of condensers used for control: contact and surface. In contact condensers, the gaseous stream is brought into direct contact with a cooling medium so that the vapors condense and mix with the coolant (see Fig. 25-15). The more widely used system, however, is the surface condenser (or heat exchanger), in which the vapor and the cooling medium are separated by a wall (see Fig. 25-16). Since high removal efficiencies cannot be obtained with low-condensable vapor concentrations, condensers are typically used for pretreatment prior to some other more efficient control device such as an incinerator, absorber, or adsorber.

Contact Condensers Spray condensers, jet condensers, and barometric condensers all utilize water or some other liquid in direct contact with the vapor to be condensed. The temperature approach between the liquid and the vapor is very small, so the efficiency of the condenser is high, but large volumes of the liquid are necessary. If the vapor is soluble in the liquid, the system is essentially an absorptive one. If the vapor is not soluble, the system is a true condenser, in which case the temperature of the vapor must be below the dew point. Direct-contact condensers are seldom used for the removal of organic solvent vapors because the condensate will contain an organic-water mixture that must be separated or treated before disposal. They are, however, the most effective method of removing heat from hot gas streams when the recovery of organics is not a consideration.

In a direct-contact condenser, a stream of water or other cooling liquid is brought into direct contact with the vapor to be condensed. The liquid stream leaving the chamber contains the original cooling liquid plus the condensed substances. The gaseous stream leaving the chamber contains the noncondensable gases and such condensable vapor as did not condense; it is reasonable to assume that the vapors in the exit gas stream are saturated. It is then the temperature of the exit gas stream that determines the collection efficiency of the condenser.

The advantages of contact condensers are that (1) they can be used to produce a vacuum, thereby creating a draft to remove odorous vapors and also reduce boiling points in cookers and vats; (2) they usually are simpler and less expensive than the surface type; and (3) they usually have considerable odor-removing capacity because of the greater condensate dilution (13 lb of 60° F water is required to condense 1 lb of steam at 212° F and cool the condensate to 140° F). The principal disadvantage is the large water requirement. Depending on the nature of the condensate, odor in the wastewater can be offset by using treatment chemicals.

Direct-contact condensers involve the simultaneous transfer of heat and mass. Design procedures available for absorption, humidification, cooling towers, and the like may be applied with some modifications.

Surface Condensers Surface condensers (indirect-contact condensers) are used extensively in the chemical-process industry. They are employed in the air-pollution-equipment industry for recovery, control, and/or removal of trace impurities or contaminants. In the surface type, coolant does not contact the vapor condensate. There are various types of surface condensers including the shell-and-tube, fin-fan, finned-hairpin, finned-tube-section, and tubular. The use of surface condensers has several advantages. Salable condensate can be recovered. If water is used for coolant, it can be reused, or the condenser may be air-cooled when water is not available. Also, surface condensers require less water and produce 10 to 20 times less condensate. Their disadvantage is that they are usually more expensive and require more maintenance than the contact type.

Biofilters Biofilters are an APC technology that uses microorganisms, generally bacteria, to treat odorous off-gas emissions in an environmentally safe and economic manner. The biofilters consist of a porous filter media through which a waste gas stream is distributed. Microorganisms that feed on the waste gas are attached to this porous substrate. The biofiltration process is related to conventional activated sludge treatment in that, in both instances, the microorganisms are used to completely oxidize organic compounds into CO_2 and water. Biofilters are used to control the off-gas emissions from composting operations, rendering plants, food and tobacco processing, chemical manufacturing, iron and steel foundries, and other industrial facilities. Biofilters are in widespread use in Europe and Japan and are slowly becoming more acceptable as an APC technology in the United States.

General Process Description Biofilters are fixed film bioreactors that use microorganisms attached to substrate materials such as compost, peat, bark, soil, or inert materials to convert organic and inorganic waste products into CO_2 and water. The substrate provides structural support and elemental nutrients for the microbes. Its porous structure should provide adequate surface area at a reasonably low gas pressure drop. As waste gases are passed through the reactor, the target pollutants diffuse into the biofilm. The pollutants are then decomposed through the natural aerobic biodegradation process.

Biofilters are most economic when applied to low-concentration gas streams (<1000 ppm) that are also oxygen rich. Greater than 90 percent destruction efficiencies can be obtained for water-soluble organics such as alcohols, aldehydes, and amines. Water-soluble inor-

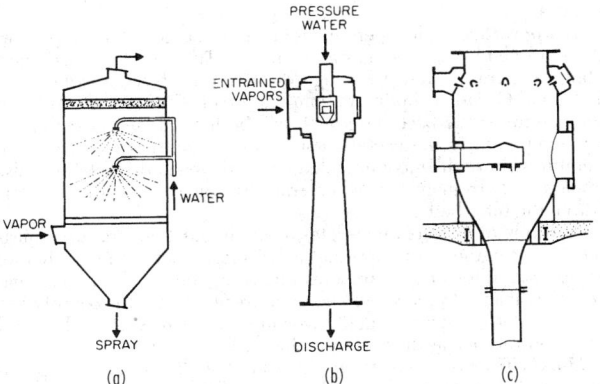

FIG. 25-15 Typical direct-contact condensers. (a) Spray chamber. (b) Jet. (c) Barometric.

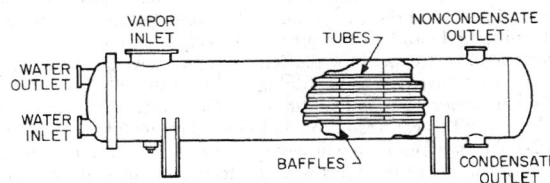

FIG. 25-16 Typical surface condenser (shell-and-tube).

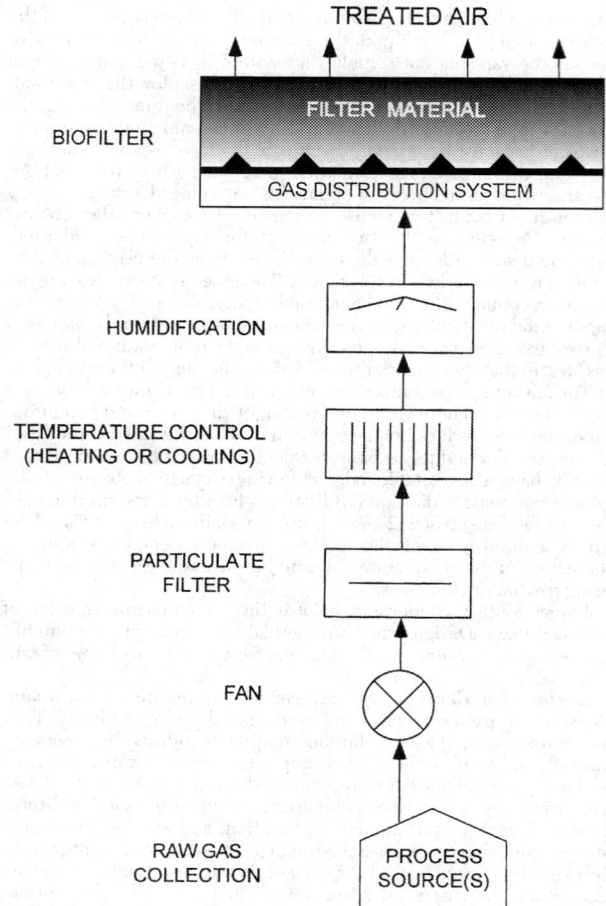

TREATED AIR

FILTER MATERIAL

BIOFILTER

GAS DISTRIBUTION SYSTEM

HUMIDIFICATION

TEMPERATURE CONTROL
(HEATING OR COOLING)

PARTICULATE
FILTER

FAN

RAW GAS
COLLECTION

PROCESS
SOURCE(S)

FIG. 25-17 Schematic flowsheet illustrating the individual elements of an open, single-layer biofilter system. Particulate filtration and/or temperature adjustment is often combined with the equipment to adjust gas humidity content.

TABLE 25-29 Microorganisms Frequently Identified in Biofilters

Bacteria	Fungi
Actinomyces globisporus	*Penicillium* spp.
Micrococcus albus	*Cephalosporium* spp.
Micromonospora vulgarus	*Mucor* spp.
Bacillus cereus	*Circinella* spp.
Streptomyces spp.	*Cephalotecium* spp.
	Ovularia spp.
	Stemphilium spp.

SOURCE: Data compiled from Ottengraf, S. P. P., "Biological Systems for Waste Gas Elimination," 1987, Table 3.

space, target removal efficiency, gas species, and gas loading. Proper design of the main biofilter components is very important to ensure a feasible, cost-effective operation.

Raw Gas Composition The suitability of the raw gas stream must first be determined. The raw gas stream must contain the following in order to ensure both reasonable removal efficiencies and microorganism life expectancy:

1. An oxygen concentration equal to ambient level.
2. Gas concentrations below lethal levels for microorganisms used.
3. Lethal gas species must be absent from the raw gas stream.

Raw Gas Transport The gas is collected from the processing area and is transported by ductwork and ID fans to the preconditioning equipment.

Raw Gas Preconditioning In order to ensure the required destruction efficiency and continued biofilter life, the raw gas stream must be adjusted to a preselected range of particulate loading, temperature, and humidification before the gas can be safely introduced into the biofilter.

Preconditioning for Particulates Heavy particulate loading of the inlet gas with dust, grease, oils, or other aerosols can be very damaging to the pore structure of the filter bed, resulting in an eventual pressure-drop increase. Oils and heavy metals that are deposited on the filter bed can be poisonous to the microorganisms that live within the biofilm. Particulate APC equipment such as fabric filters and venturi scrubbers are generally adequate for this level of particulate removal.

Temperature The operating temperature of a biofilter is primarily controlled by the inlet gas temperature. The recommended operating temperature range for high destruction efficiency is between 20° to 40° C, with an optimum temperature of 37° C (98° F). At lower temperatures, the bacteria growth will be limited, and at extremely low temperatures the bacteria could possibly be destroyed. At temperatures above the recommended range, the bacteria's activity is also impaired. Extremely high temperatures will destroy the bacteria within the filter bed.

In terms of an economic determination, gas temperature adjustment is often the most important cost factor in determining whether to use a biofilter or a more conventional system. If the process gas stream is at an extremely high temperature (+100° C), the cost of cooling the inlet gas stream might favor more conventional methods for odor control such as thermal oxidation.

Humidification The microorganisms that digest the target pollutants live in a thin water layer called the biofilm that surrounds the filter substrate. Without the biofilm, the microorganisms would die; therefore, maintaining a wetted surface within the filter bed is crucial. Insufficient moisture can also lead to shrinking and cracking of the filter media, resulting in reduced active surface area and gas by-passing.

Humidification of the gas stream is the preferred method of keeping the filter bed moist. Gas moisture is usually added to the incoming gas stream downstream of the particulate removal APC equipment by either water sprays or steam. Adding moisture directly to the top of the bed in order to maintain filter media moisture is not recommended since this can result in: (1) localized drying of the substrate, and (2) cold water addition will reduce the activity of the microorganisms until the water becomes warmed to the steady-state filter-bed temperature.

Gas Distribution System The function of the gas distribution

ganics, such as H_2S and NH_3, can also readily undergo aerobic decomposition.

The basic biofilter process steps involved are shown in Fig. 25-17 and are as follows:

1. Collection and transportation of raw waste gases from the processing/manufacturing area to a gas pretreatment area via ductwork and blowers.
2. Pretreatment of the raw waste gas to remove particulates, adjust temperature and humidify to saturation.
3. The pretreated gas is evenly distributed throughout the biofilter.

Microorganisms The naturally occurring microorganisms that are commonly used in biofilters are the same bacteria and fungi that are currently used in activated-sludge wastewater treatment and landfills. Genetically engineered microbes have been created to digest manmade chemical species, such as the organic aromatics xylene and styrene. On-going genetic engineering research is expected to increase the number of chemical species that can be biodegraded. This will also contribute to lowering the cost and size of current filter beds by decreasing the required digestion time. The more commonly used microorganisms are listed in Table 25-29.

Design and Construction The capacity and efficiency of biofilter operation are a direct function of active surface area, filter void

system is to ensure even flow of the preconditioned gas stream to all areas of the filter bed. In upflow biofilter designs, the gas distribution system also provides the following:

1. A means of drainage, collection, and transportation of excess water within the filter bed
2. Prevents the potential contamination of surrounding soil by leaking filter leachate
3. A structural base for the filter bed media

The gas distribution system can be composed of a network of perforated pipe, slotted or vented concrete block, or metal grating. When there are no space limitations, single-level filters are used. In regions where footprint space is limited, like Japan, multiple-deck filter beds have become commonplace. If inorganic compounds are being treated, corrosion-resistant materials of construction are used due to the acidic by-products of the bioreaction.

Filter Matrix The most common filter substrates in use today are soils or compost produced from leaves, bark, wood chips, activated sludge, paper, or other organic materials. In selecting a proper filter substrate for a specific use, the following should be considered:

1. The particle size and porosity of the filter media, since operating efficiency is directly related to the available biofilm surface area.
2. The filter media must be a source of inorganic nutrients for the microbes. In cases of long-term operation, inorganic nutrients can be periodically added to the bed.
3. Compaction of the filter bed over time will result in gas channeling and pressure-drop increases. This can be avoided by adding large, rigid particles such as plastic spheres, ceramics, or wood/bark chips to provide additional support to the filter substrate.
4. Good bed drainage characteristics are necessary to ensure that reaction products are easily transported out of the filter media. The leachate is generally recycled through the humidification process to reduce the wastewater stream.
5. The filter media should have buffering capacity in order to maintain a pH of at least 3. This is especially a concern when inorganic compounds are targeted for reduction by the biofilter.
6. The filter media should be composed of materials that have a nonobjectional odor.

Kinetics The capacity and efficiency of biofilter operation is a function of active surface area, filter void space, target removal efficiency, gas species, gas concentration, and gas flow rate. A simplified theoretical model described by S.P.P. Ottengraf et al. is schematically represented by in Fig. 25-18. The mass balance made around the liquid-phase biolayer can be described as follows:

$$D \times \left(\frac{d^2 C_1}{dx^2}\right) - R = 0 \qquad (25\text{-}12)$$

where: D = the mass-transfer coefficient, L^2/T
C_1 = liquid phase concentration, M/L^3
x = distance through biolayer, L
R = substrate utilization rate, $M/L^3/T$ (biodegradation rate)

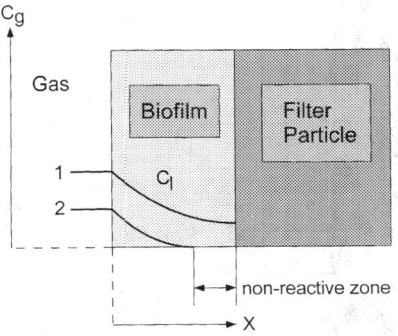

FIG. 25-18 Biophysical model for the biolayer. C_g is the concentration in the gas phase. The two concentration profiles shown in the biolayer (C_l) refer to (1) elimination reaction rate limited, and (2) diffusion limited. (SOURCE: Redrawn from Ref. 26.)

The Ottengraf biolayer mass balance assumes the following:

1. Monod kinetic model applies.
2. Biodegradation occurs in the biofilm liquid phase.
3. The biofilm thickness is small compared to the diameter of the substrate; therefore, the biofilm can be regarded as a flat surface.
4. Plug flow of the gas through the filter media.
5. Gas and liquid phase concentrations follow Henry's law.
6. Ideal gas law applies; i.e., no gas species interactions.
7. Steady state.

The biodegradation rate R is characterized by the Monod (or Michaelis-Menten) following relationship:

$$R = R_{max} \times \frac{C_1}{(C_1 + K_m)} \qquad (25\text{-}13)$$

where: K_m = the Monod (Michaelis-Menten) constant, M/L^3
R_{max} = the maximum substrate utilization rate, $M/L^3/T$

R_{max} is a function of the active microorganism concentration in the biofilm and is defined as:

$$R_{max} = X \times \frac{\mu_m}{y_i} \qquad (25\text{-}14)$$

where: X = cell concentration of the active microbes, M/L^3;
μ_m = the maximum growth rate of microbe species, $1/T$;
y_i = cell yield coefficient of microbe species, dimensionless

Two rate-limiting cases exist with the above mass balance:

1. *Reaction rate limited (zero-order kinetics).* In this case, the biofilm concentration has no effect on reaction rate, and the biodegradation breakthrough curve is linear.
2. *Diffusion rate limited (first-order kinetics).* In this case, the reaction rate is controlled by the rate of diffusion of the pollutant species into the biofilm.

The concentration profiles that result from the above two rate-controlling mechanisms are shown in Fig. 25-18.

One of the listed assumptions for Ottengraf's kinetic model was that no gas-phase interactions occur between different chemical species (the ideal-gas assumption). Under actual operating conditions, gas-phase interactions can either have a negative or positive impact on biofilter operation. These interactions include:

1. Cometabolism, which increases the biodegradation rate of the multiple targeted compounds.
2. Cross-inhibition, which decreases the biodegradation rate of the multiple targeted compounds.
3. Vertical stratification, where the most easily degraded compounds are metabolized first upon entering the filter bed. The more difficult-to-metabolize compounds pass through the lower region of the bed and are metabolized in the upper levels.

Based upon the above-mentioned species interactions, pilot-scale testing is generally recommended to accurately size a biofilter bed for a multicomponent waste gas stream.

Membrane Filtration Membrane systems have been used for several decades to separate colloidal and molecular slurries by the chemical process industries (CPI). Membrane filtration was not viewed as a commercially viable pollution control technology until recently. This conventional wisdom was due to fouling problems exhibited when handling process streams with high solid content. This changed with the advent of membranes composed of high-flux cellulose acetate. In Europe and Asia, membrane filtration systems have been used for several years, primarily for ethanol dewatering for synthetic fuel plants. Membrane systems were selected in these cases for their (1) low energy utilization; (2) modular design; (3) low capital costs; (4) low maintenance; and (5) superior separations. In the United States, as EPA regulations become increasingly stringent, a renewed interest in advanced membrane filtration systems has occurred. In this case, the driving factors are the membrane system's ability to operate in a pollution-free, closed-loop manner with minimum wastewater output. Low capital and maintenance costs also result from the small amount of moving parts within the membrane systems.

Process Descriptions Selectively permeable membranes have an increasingly wide range of uses and configurations as the need for more advanced pollution control systems are required. There are four major types of membrane systems: (1) pervaporation; (2) reverse osmosis (RO); (3) gas absorption; and (4) gas adsorption. Only membrane pervaporation is currently commercialized.

Membrane Pervaporation   Since 1987, membrane pervaporation has become widely accepted in the CPI as an effective means of separation and recovery of liquid-phase process streams. It is most commonly used to dehydrate liquid hydrocarbons to yield a high-purity ethanol, isopropanol, and ethylene glycol product. The method basically consists of a selectively-permeable membrane layer separating a liquid feed stream and a gas phase permeate stream as shown in Fig. 25-19. The permeation rate and selectivity is governed by the physicochemical composition of the membrane. Pervaporation differs from reverse osmosis systems in that the permeate rate is not a function of osmotic pressure, since the permeate is maintained at saturation pressure (Ref. 24).

Three general process groups are commonly used when describing pervaporation: (1) water removal from organics; (2) organic removal from water (solvent recovery); and (3) organic/organic separation. Organic/organic separations are very uncommon and therefore will not be discussed further.

Ethanol Dehydration   The membrane-pervaporation process for dehydrating ethanol was first developed by GFT in West Germany in the mid 1970s, with the first commercial units being installed in Brazil and the Philippines. At both sites, the pervaporation unit was coupled to a continuous sugarcane fermentation process that produced ethanol at concentrations up to 96 percent after vacuum pervaporation. The key advantages of the GFT process are (1) no additive chemicals are required; (2) the process is skid-mounted (low capital costs and small footprint); (3) and there is a low energy demand. The low energy requirement is achieved because only a small fraction of the water is actually vaporized and that the required permeation driving force is provided by only a small vacuum pump. The basic ethanol dehydration process schematic is shown in Fig. 25-20. A key advantage of all pervaporation processes is that vapor-liquid equilibria and possible resulting azeotropic effects are irrelevant (see Fig. 25-21 and Ref. 24).

Solvent Recovery   The largest current industrial use of pervaporation is the treatment of mixed organic process streams that have become contaminated with small (10 percent) quantities of water. Pervaporation becomes very attractive when dehydrating streams down to less than 1 percent water. The advantages result from the small operating costs relative to distillation and adsorption. Also, distillation is often impossible, since azeotropes commonly form in multicomponent organic/water mixtures.

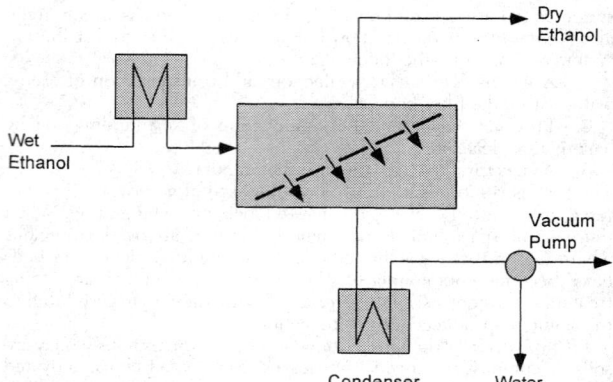

FIG. 25-20   Ethanol dehydration using pervaporation membrane. (SOURCE: Redrawn from Ref. 24.)

Pervaporation occurs in three basic steps:
1. Preferential sorption of chemical species
2. Diffusion of chemical species through the membrane
3. Desorption of chemical species from the membrane

Steps 1 and 2 are controlled by the specific polymer chemistry and its designed interaction with the liquid phase. The last step, consisting of evaporation of the chemical species, is considered to be a fast, nonselective process. Step 2 is the rate-limiting step. The development by the membrane manufacturers of highly selective, highly permeable composite membranes that resist fouling from solids has subsequently been the key to commercialization of pervaporation systems. The membrane composition and structure are designed in layers, with each layer fulfilling a specific requirement. Using membrane dehydration as an example, a membrane filter would be composed of a support layer of nonwoven porous polyester below a layer of polyacrylonitrile (PAN) or polysulfone ultrafiltration membrane and a layer of 0.1-μm-thick crosslinked polyacrylate, polyvinyl alcohol (PVA). Other separations membranes generally use the same two sublayers. The top layer is interchanged according to the selectivity desired.

Emerging Membrane Control Technologies   The recent improvements in membrane technology have spawned several potentially commercial membrane filtration uses.

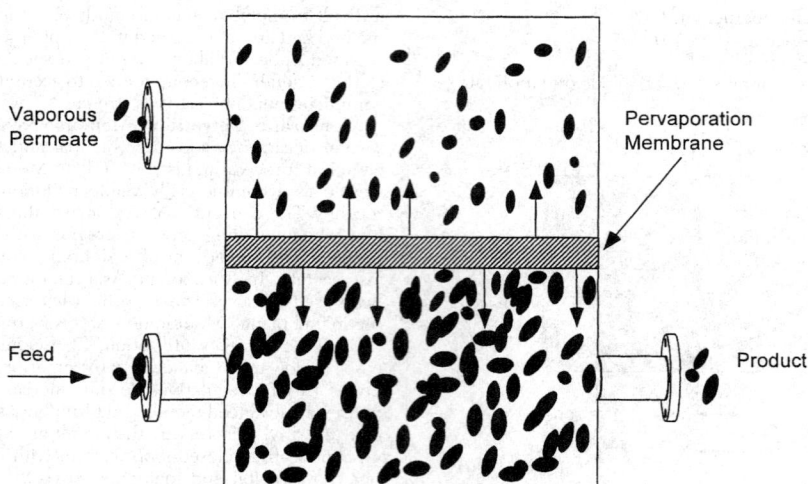

FIG. 25-19   Pervaporation of gas from liquid feed across membrane to vaporous permeate. (SOURCE: Redrawn from Ref. 24.)

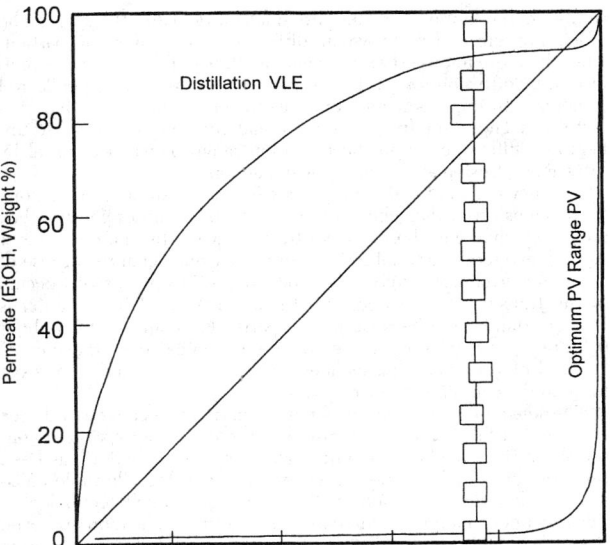

FIG. 25-21   Comparison of two types of pervaporation membranes to distillation of ethanol-water mixtures. (SOURCE: Redrawn from Ref. 24.)

Reverse Osmosis (RO) Membranes   A type of membrane system for treating oily wastewater is currently undergoing commercialization by Bend Research, Inc. The system uses a tube-side feed module that yields high fluxes while being able to handle high-solids-content waste streams (Ref. 25). Another type of reverse osmosis technique is being designed to yield ultrapurified HF recovered from spent etching solutions. It is estimated that 20,000 tons of spent solution is annually generated in the United States and that using membrane RO could save 1 million bbl/yr of oil.

In Situ Filter Membranes   In situ membranes are being fitted into incinerator flue-gas stacks in an attempt to reduce hydrocarbon emissions. Two types of commercially available gas separation membranes are being studied: (1) flat cellulose acetate sheets; and (2) hollow-tube fiber modules made of polyamides.

Vibratory Shear-Enhanced Membranes   The vibratory shear-enhancing process (VSEP) is just starting commercialization by Logic International, Emeryville, CA. It employs the use of intense sinusiodal shear waves to ensure that the membrane surfaces remain active and clean of solid matter. The application of this technology would be in the purification of wastewater (Ref. 2).

Vapor Permeation   Vapor permeation is similar to vapor pervaporation except that the feed stream for permeation is a gas. The future commercial viability of this process is based upon energy and capital costs savings derived from the feed already being in the vapor-phase, as in fractional distillation, so no additional heat input would be required. Its foreseen application areas would be the organics recovery from solvent-laden vapors and pollution treatment. One commercial unit was installed in Germany in 1989 (Ref. 26).

Selective Catalytic Reduction of Nitrogen Oxides   The traditional approach to reducing ambient ozone concentrations has been to reduce VOC emissions, an ozone precurssor. In many areas, it has now been recognized that elimination of persistent exceedances of the National Ambient Air Quality Standard for ozone may require more attention to reductions in the other ingredients in ozone formation, nitrogen oxides (NO_x). In such areas, ozone concentrations are controlled by NO_x rather than VOC emissions.

Selective catalytic reduction (SCR) has been used to control NO_x emissions from utility boilers in Europe and Japan for over a decade. Applications of SCR to control process NO_x emissions in the chemical industry are becoming increasingly common. A typical SCR system is shown in Fig. 25-22.

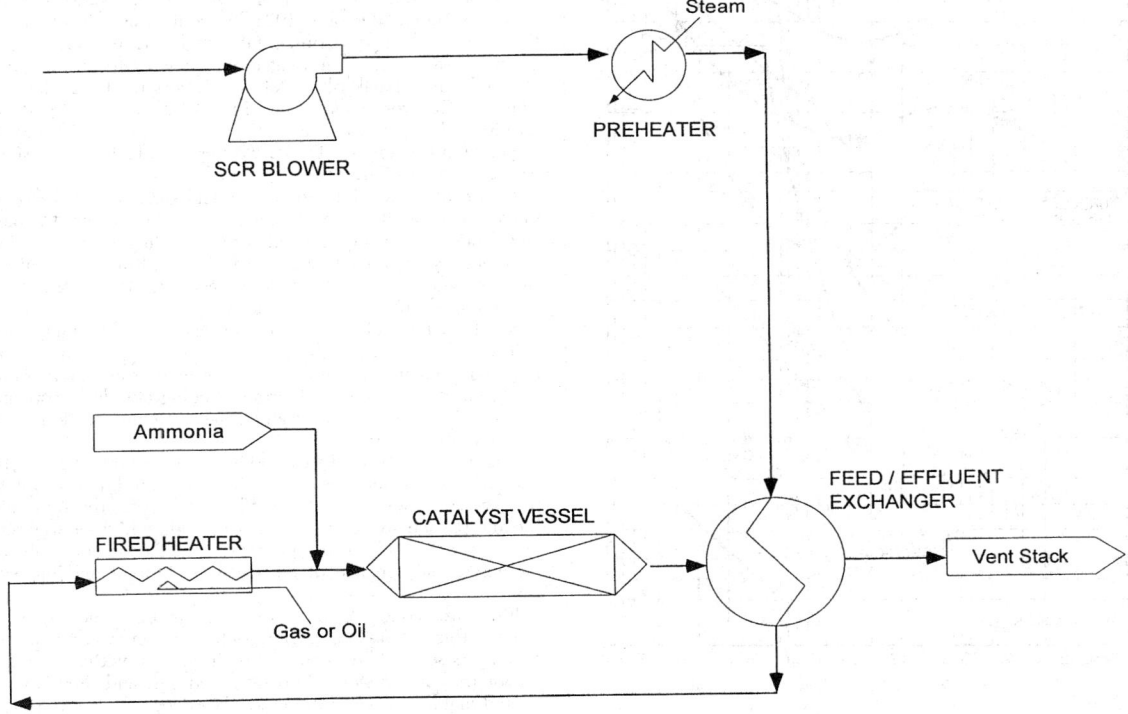

FIG. 25-22   Selective catalytic reduction of nitrogen oxides.

NO_x-laden fumes are preheated by effluent from the catalyst vessel in the feed/effluent heat exchanger and then heated by a gas- or oil-fired heater to over 600° F. A controlled quantity of ammonia is injected into the gas stream before it is passed through a metal oxide, zeolite, or promoted zeolite catalyst bed. The NO_x is reduced to nitrogen and water in the presence of ammonia in accordance with the following exothermic reactions:

$$6NO + 4NH_3 \rightarrow 5N_2 + 6H_2O$$

$$4NO + 4NH_3 + O_2 \rightarrow 4N_2 + 6H_2O$$

$$2NO_2 + 4NH_3 + O_2 \rightarrow 3N_2 + 6H_2O$$

$$NO + NO_2 + 2NH_3 \rightarrow 2N_2 + 3H_2O$$

NO_x analyzers at the preheater inlet and catalyst vessel outlet monitor NO_x concentrations and control the ammonia feed rate. The effluent gives up much of its heat to the incoming gas in the feed/effluent exchanger. The vent gas is discharged at about 350° F.

SOURCE CONTROL OF PARTICULATE EMISSIONS

There are four conventional types of equipment used for the control of particulate emissions:
1. Mechanical collectors
2. Wet scrubbers
3. Electrostatic precipitators
4. Fabric filters

Each is discussed in Sec. 17 of this handbook under "Gas-Solids Separations." The effectiveness of conventional air-pollution-control equipment for particulate removal is compared in Fig. 25-23. These fractional efficiency curves indicate that the equipment is least efficient in removing particulates in the 0.1- to 1.0-μm range. For wet

scrubbers and fabric filters, the very small particulates (0.1 μm) can be efficiently removed by brownian diffusion. The smaller the particulates, the more intense their brownian motion and the easier their collection by diffusion forces. Larger particulates (>1 μm) are collected principally by impaction, and removal efficiency increases with particulate size. The minimum in the fractional efficiency curve for scrubbers and filters occurs in the transition range between removal by brownian diffusion and removal by impaction.

A somewhat similar situation exists for electrostatic precipitators. Particulates larger than about 1 μm have high mobilities because they are highly charged. Those smaller than a few tenths of a micrometer can achieve moderate mobilities with even a small charge because of aerodynamic slip. A minimum in collection efficiency usually occurs in the transition range between 0.1 and 1.0 μm. The situation is further complicated because not all particulates smaller than about 0.1 μm acquire charges in an ion field. Hence, the efficiency of removal of very small particulates decreases after reaching a maximum in the submicrometer range.

The selection of the optimum type of particulate collection device (i.e., ESP or fabric filter baghouse) is often not obvious without conducting a site-specific economic evaluation. This situation has been brought about by both the recent reductions in the allowable emissions levels and advancements with fabric filter and ESP technologies. Such technoeconomic evaluations can result in application and even site-specific differences in the final optimum choice (see *Precip Newsletter,* 220, June, 1994 and *Fabric Filter Newsletter,* 223, June, 1994).

Improvements in existing control technology for fine particulates and the development of advanced techniques are top-priority research goals. Conventional control devices have certain limitations. Precipitators, for example, are limited by the magnitude of charge on the particulate, the electric field, and dust reentrainment. Also, the resistivity of the particulate material may adversely affect both charge and electric field. Advances are needed to overcome resistivity and extend the performance of precipitators not limited by resistivity (see Buonicore and Theodore, "Control Technology for Fine Particulate Emissions," DOE Rep. ANL/ECT-5, Argonne National Laboratory, Argonne, Illinois, October 1978). Recent design developments with the potential to improve precipitator performance include pulse energization, electron beam ionization, wide plate spacing, and precharged units [Balakrishnan et al., "Emerging Technologies for Air Pollution Control," *Pollut. Eng.,* **11,** 28–32 (Nov. 1979); "Pulse Energization," *Environ. Sci. Technol.* **13**(9), 1044 (1974); and Midkaff, "Change in Precipitator Design Expected to Help Plants Meet Clean Air Laws," *Power,* **126**(10), 79 (1979)].

Fabric filters are limited by physical size and bag-life considerations. Some sacrifices in efficiency might be tolerated if higher air-cloth ratios could be achieved without reducing bag life (improved pulse-jet systems). Improvements in fabric filtration may also be possible by enhancing electrostatic effects that may contribute to rapid formation of a filter cake after cleaning.

Scrubber technology is limited by scaling and fouling, overall reliability, and energy consumption. The use of supplementary forces acting on particulates to cause them to grow or otherwise be more easily collected at lower pressure drops is being closely investigated. The development of electrostatic and flux-force-condensation scrubbers is a step in this direction.

The electrostatic effect can be incorporated into wet scrubbing by charging the particulates and/or the scrubbing-liquor droplets. Electrostatic scrubbers may be capable of achieving the same efficiency for fine-particulate removal as is achieved by high-energy scrubbers, but at substantially lower power input. The major drawbacks are increased maintenance of electrical equipment and higher capital cost.

Flux-force-condensation scrubbers combine the effects of flux force (diffusiophoresis and thermophoresis) and water-vapor condensation. These scrubbers contact hot, humid gas with subcooled liquid, and/or they inject steam into saturated gas, and they have demonstrated that a number of these novel devices can remove fine particulates (see Fig. 25-24). Although limited in terms of commercialization, these systems may find application in many industries.

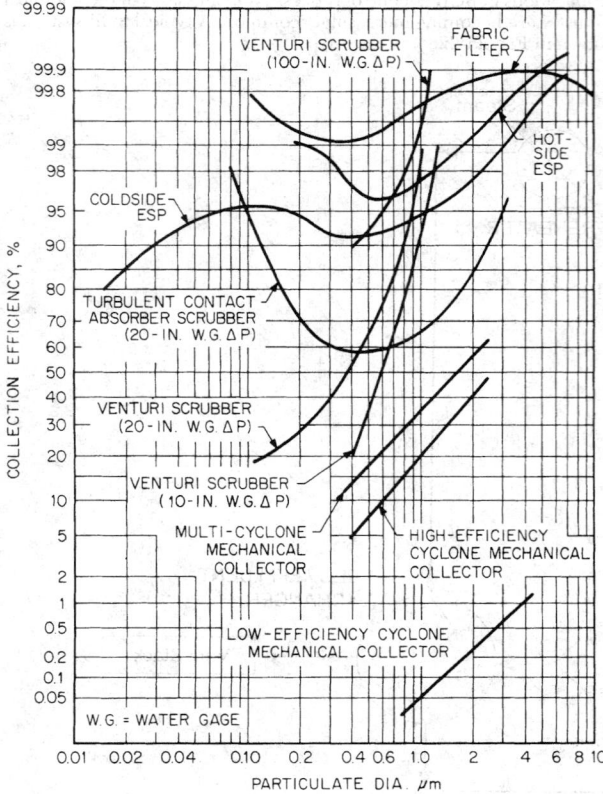

FIG. 25-23 Fractional efficiency curves for conventional air-pollution-control devices. [Chem. Eng., *87*(13), 83 (June 30, 1980).]

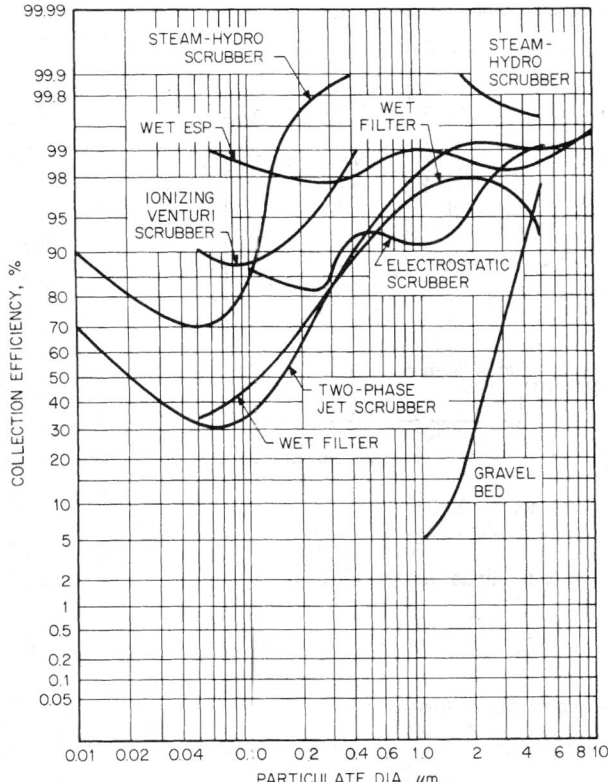

FIG. 25-24 Fractional efficiency curves for novel air-pollution-control devices. [*Chem. Eng.*, **87**(13), 85 (*June 30, 1980*).]

EMISSIONS MEASUREMENT

Introduction An accurate quantitative analysis of the discharge of pollutants from a process must be determined prior to the design and/or selection of control equipment. If the unit is properly engineered by utilizing the emission data as input to the control device and the code requirements as maximum-effluent limitations, most pollutants can be successfully controlled.

Sampling is the keystone of source analysis. Sampling methods and tools vary in their complexity according to the specific task; therefore, a degree of both technical knowledge and common sense is needed to design a sampling function. Sampling is done to measure quantities or concentrations of pollutants in effluent gas streams, to measure the efficiency of a pollution-abatement device, to guide the designer of pollution-control equipment and facilities, and/or to appraise contamination from a process or a source. A complete measurement requires a determination of the concentration and contaminant characteristics as well as the associated gas flow. Most statutory limitations require mass rates of emissions; both concentration and volumetric-flow-rate data are therefore required.

The selection of a sampling site and the number of sampling points required are based on attempts to get representative samples. To accomplish this, the sampling site should be at least eight stack or duct diameters downstream and two diameters upstream from any flow disturbance, such as a bend, expansion, contraction, valve, fitting, or visible flame.

Once the sampling location has been decided on, the flue cross section is laid out in a number of equal areas, the center of each being the point where the measurement is to be taken. For rectangular stacks, the cross section is divided into equal areas of the same shape, and the traverse points are located at the center of each equal area, as shown

in Fig. 25-25. The ratio of length to width of each elemental area should be selected. For circular stacks, the cross section is divided into equal annular areas, and the traverse points are located at the centroid of each area. The location of the traverse points as a percentage of diameter from the inside wall to the traverse point for circular-stack sampling is given in Table 25-30. The number of traverse points necessary on each of two perpendiculars for a particular stack may be estimated from Fig. 25-26.

Once these traverse points have been determined, velocity measurements are made to determine gas flow. The stack-gas velocity is usually determined by means of a pitot tube and differential-pressure gauge. When velocities are very low (less than 3 m/s [10 ft/s]) and when great accuracy is not required, an anemometer may be used. For gases moving in small pipes at relatively high velocities or pressures, orifice-disk meters or venturi meters may be used. These are valuable as continuous or permanent measuring devices.

Once a flow profile has been established, sampling strategy can be considered. Since sampling collection can be simplified and greatly reduced depending on flow characteristics, it is best to complete the flow-profile measurement before sampling or measuring pollutant concentrations.

Sampling Methodology The following subsections review the methods specified for sampling commonly regulated pollutants as well as sampling for more exotic volatile and semivolatile organic compounds. In all sampling procedures, the main concern is to obtain a representative sample; the U.S. EPA has published reference sampling methods for measuring emissions of specific pollutants so that uniform procedures can be applied in testing to obtain a representative sample. Table 25-31 provides an overview of the regulatory citation for selected test methods. The test methods reviewed in the following subsections address measuring the emissions of the following pollutants: particulate matter, sulfur dioxide, nitrogen oxides, carbon monoxide, fluorides, hydrogen chloride, total gaseous organics, multiple metals, volatile organic compounds, and semivolatile organic compounds.

Each sampling method requires the use of complex sampling equipment that must be calibrated and operated in accordance with specified reference methods. Additionally, the process or source that is being tested must be operated in a specific manner, usually at rated capacity, under normal procedures.

Velocity and Volumetric Flow Rate The U.S. EPA has published Method 2 as a reference method for determining stack-gas velocity and volumetric flow rate. At several designated sampling points, which represent equal portions of the stack volume (areas in the stack), the velocity and temperature are measured with instrumentation shown in Fig. 25-27.

Measurements to determine volumetric flow rate usually require approximately 30 min. Since sampling rates depend on stack-gas velocity, a preliminary velocity check is usually made prior to testing for pollutants to aid in selecting the proper equipment and in determining the approximate sampling rate for the test.

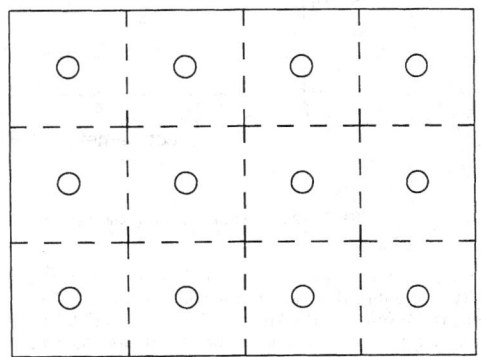

FIG. 25-25 Example showing rectangular stack cross section divided into 12 equal areas, with a traverse point at centroid of each area.

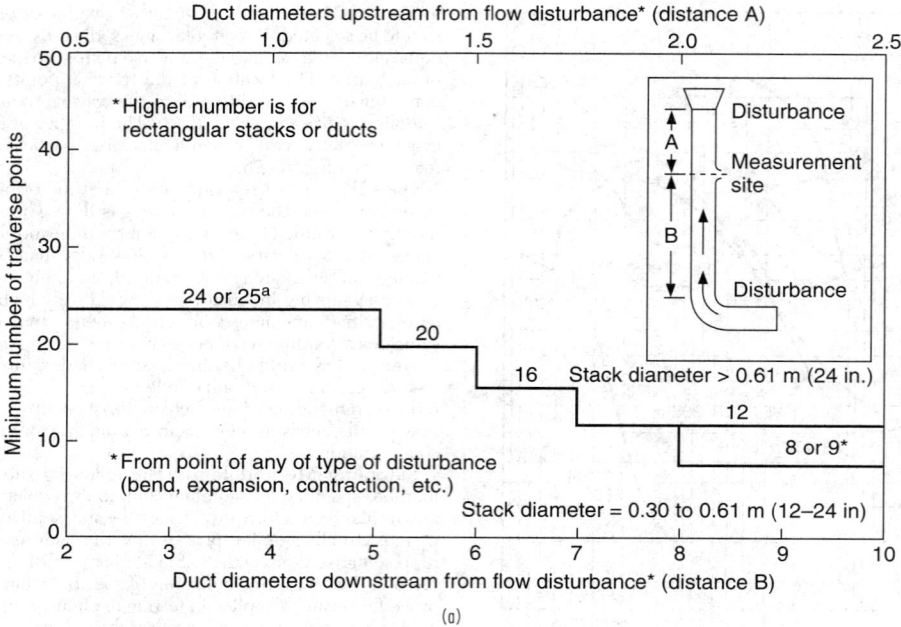

FIG. 25-26a   Minimum number of traverse points for particulate traverses.

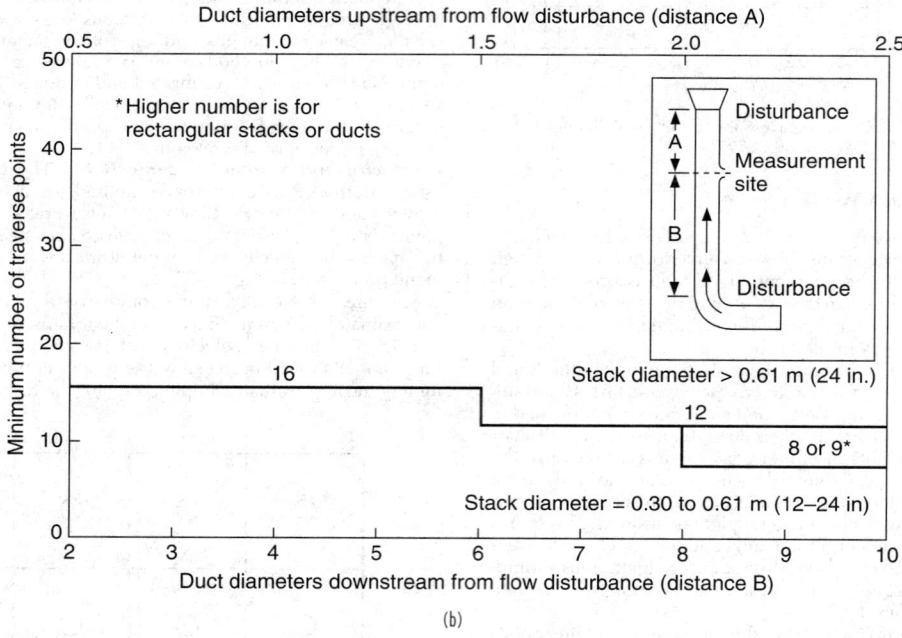

FIG. 25-26b   Minimum number of traverse points for velocity (nonparticulate) traverse.

The volumetric flow rate determined by this method is usually within ±10 percent of the true volumetric flow rate.

Molecular Weight   EPA Method 3 is used to determine carbon dioxide and oxygen concentrations and dry molecular weight of the stack-gas stream. Depending on the intended use of the data, these values can be obtained with an integrated sample (see Fig. 25-28) or a grab sample (see Fig. 25-29). In addition, the instrumental analyzer

method, EPA Method 3A, is used for gas compositional analyses (O_2 and CO_2) in determining the sample gas molecular weight.

With the grab sampling technique, a sampling probe is placed at the center of the stack, and a sample is drawn directly into an Orsat analyzer or a Fyrite-type combustion-gas analyzer. The sample is then analyzed for carbon dioxide and oxygen content. With these data, the dry molecular weight of the gas stream can then be calculated.

TABLE 25-30 Location of Traverse Points in Circular Stacks
(Percent of stack diameter from inside wall to traverse point)

Traverse point number on a diameter	Number of traverse points on a diameter										
	2	4	6	8	10	12	14	18	20	22	24
1	14.6	6.7	4.4	3.2	2.6	2.1	1.8	1.4	1.3	1.1	1.1
2	85.4	25.0	14.6	10.5	8.2	6.7	5.7	4.4	3.9	3.5	3.2
3		75.0	29.6	19.4	14.6	11.8	9.9	7.5	6.7	6.0	5.5
4		93.3	70.4	32.3	22.6	17.7	14.6	10.9	9.7	8.7	7.9
5			85.4	67.7	34.2	25.0	20.1	14.6	12.9	11.6	10.5
6			95.6	80.6	65.8	35.6	26.9	18.8	16.5	14.6	13.2
7				89.5	77.4	64.4	36.6	23.6	20.4	18.0	16.1
8				96.8	85.4	75.0	63.4	29.6	25.0	21.8	19.4
9					91.8	82.3	73.1	38.2	30.6	26.2	23.0
10					97.4	88.2	79.9	61.8	38.8	31.5	27.2
11						93.3	85.4	70.4	61.2	39.3	32.3
12						97.9	90.1	76.4	69.4	60.7	39.8
13							94.3	81.2	75.0	68.5	60.2
14							98.2	85.4	79.6	73.8	67.7
15								89.1	83.5	78.5	72.8
16								92.5	87.1	82.0	77.0
17								95.6	90.3	85.4	80.8
18								98.6	93.3	88.4	83.9
19									96.1	91.3	86.8
20									96.7	94.0	89.5
21										96.5	92.1
22										98.9	94.5
23											96.8
24											96.9

TABLE 25-31 Regulatory Citations for Selected Test Methods

Citation	Description	Selected methods included	Test parameter(s)
40 CFR Part 60 (Appendix A)	New source performance standards	Methods 1–4	Test location, volumetric flow rate, gas composition, moisture content
		Method 5	Particulate matter
		Method 6/6C	Sulfur dioxide
		Method 7/7E	Nitrogen oxides
		Method 9	Opacity of visible emissions
		Method 10	Carbon monoxide
		Method 13A or 13B	Total fluoride
		Method 23	Dioxins and furans
		Method 25/25A	Total gaseous non-methane organics (VOCs)
		Method 26/26A	Halogens, halides (primarily HCl, HF, Cl_2)
40 CFR Part 51 (Appendix M)	State implementation plans	Method 201A	Particulate matter of less than or equal to 10 µg (PM10)
40 CFR Part 266 (Appendix IX)	Boiler and industrial furnace (BIF) regulations	"Multiple metals"°	Cr, Cd, As, Ni, Mn, Be, Cu, Zn, Pb, Se, P, Tl, Ag, Sb, Ba, Hg
		Method Cr+6	Hexavalent chromium
		Method 0050	Isokinetic HCl/Cl_2
40 CFR Part 61 (Appendix B)	NESHAP regulations	Method 101A	Mercury
		Method 104	Beryllium
		Method 108	Arsenic
SW-846†		Method 0010	Semivolatile organics
		Method 0030	Volatile organics

NOTES:
°The multiple metals method is also currently published as Draft EPA Method 29 for inclusion in 40 CFR 60.
†Full citation is: *Test Methods for Evaluating Solid Waste: Physical/Chemical Methods*, SW-846, 3d ed., July 1992.

The instrumental analyzer procedure, EPA Method 3A, is commonly used for the determination of oxygen and carbon dioxide concentrations in emissions from stationary sources. An integrated continuous gas sample is extracted from the test location and a portion of the sample is conveyed to one or more instrumental analyzers for determination of O_2 and CO_2 gas concentrations (see Fig. 25-30). The sample gas is conditioned prior to introduction to the gas analyzer by removing particulate matter and moisture. Sampling is conducted at a constant rate for the entire test run. Performance specifications and test procedures are provided in the method to ensure reliable data.

Moisture Content   EPA Method 4 is the reference method for determining the moisture content of the stack gas. A value for moisture content is needed in some of the calculations for determining pollution-emission rates.

A sample is taken at several designated points in the stack, which represent equal areas. The sampling probe is placed at each sampling point, and the apparatus is adjusted to take a sample at a constant rate. As the gas passes through the apparatus, a filter collects the particulate matter, the moisture is removed, and the sample volume is measured. The collected moisture is then measured, and the moisture content of the gas stream is calculated. A schematic of the sampling used in this reference method is shown in Fig. 25-31.

Particulates   Procedures for testing a particulate source are more detailed than those used for sampling gases. Because particulates exhibit inertial effects and are not uniformly distributed within a stack, sampling to obtain a representative sample is more complex than for gaseous pollutants. EPA Method 5 (as shown in Fig. 25-32) is the most widely used procedure for determination of particulate emissions from a stationary source. In-stack sampling guidelines are presented in EPA Method 17.

According to Method 5 (except as applied to fossil-fuel-fired steam generators), a particulate is defined as any material collected at

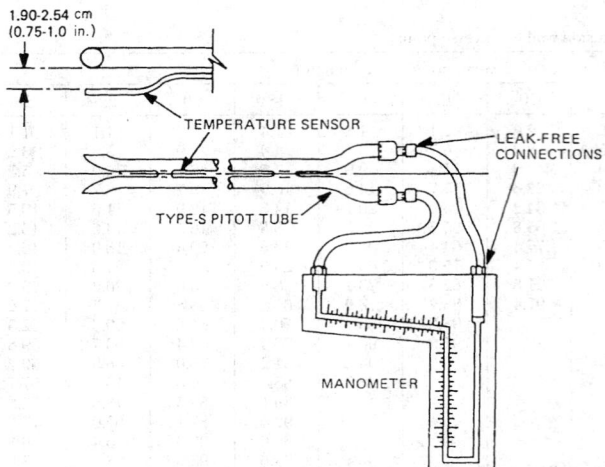

FIG. 25-27   Velocity-measurement system.

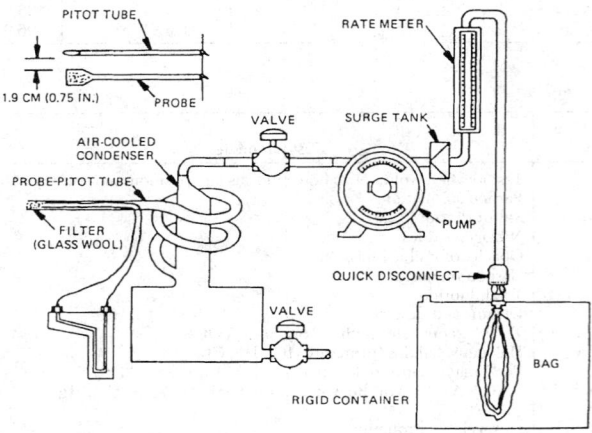

FIG. 25-28   Integrated-sample setup for molecular-weight determination.

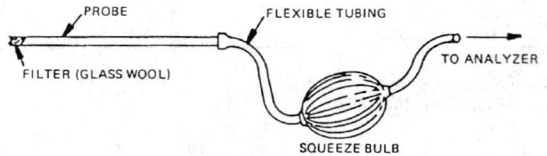

FIG. 25-29   Grab-sample setup for molecular-weight determination.

121° C (250° F) on a filtering medium. The sampling apparatus, however, may have to be modified to conform with the state's definition of a particulate. For example, a state may define *particulate* as any material collectible at stack conditions, a definition that would allow the filtering medium to be located in the stack.

In performing a particulate source test, samples are taken at several designated sampling points in the stack, which represent equal areas. At each sampling point, the velocity, temperature, molecular weight, and static pressure of the particulate-laden gas stream are measured. The sampling probe is placed at the first sampling point, and the sampling apparatus is adjusted to take a sample at the conditions measured at this point, and the process is repeated continuously until a sample has been taken from each designated sampling point. To achieve valid results in a particulate-source test, the sample must be taken isokinetically. Measurement of stack conditions allows adjustment of the sampling rate to meet this requirement.

As the gas stream proceeds through the sampling apparatus, the particulate matter is trapped on a filter, the moisture is removed, and the volume of the sample is measured. Upon completion of sampling, the collected material is recovered and sent to a laboratory for a gravimetric determination or analysis.

Sulfur Dioxide   EPA Method 6 is the reference method for determining emissions of sulfur dioxide (SO_2) from stationary sources. As the gas goes through the sampling apparatus (see Fig. 25-33), the sulfuric acid mist and sulfur trioxide are removed, the SO_2 is removed by a chemical reaction with a hydrogen peroxide solution, and, finally, the sample gas volume is measured. Upon completion of the run, the sulfuric acid mist and sulfur trioxide are discarded, and the collected material containing the SO_2 is recovered for analysis at the laboratory. The concentration of SO_2 in the sample is determined by a titration method.

For determination of the total mass-emission rate of SO_2, the moisture content and the volumetric flow rate of the exhaust gas stream must also be measured.

The minimum detectable limit has been determined to be 3.4 mg of SO_2 per cubic meter of gas (2.1×10^{-7} lb of SO_2 per cubic foot of gas). Although no upper limit has been established, the theoretical upper concentration limit in a 20-liter sample is about 93,300 mg/m^3.

EPA Method 6C is the instrumental analyzer procedure used to determine sulfur dioxide emissions from stationary sources (see Fig. 25-30). An integrated continuous gas sample is extracted from the test location, and a portion of the sample is conveyed to an instrumental analyzer for determination of SO_2 gas concentration using an ultraviolet (UV), nondispersive infrared (NDIR), or fluorescence analyzer. The sample gas is conditioned prior to introduction to the gas analyzer by removing particulate matter and moisture. Sampling is conducted at a constant rate for the entire test run.

Quality control elements required by the instrumental analyzer method include: analyzer calibration error (± 2 percent of instrument span allowed); verifying the absence of bias introduced by the sampling system (less than ± 5 percent of span for zero and upscale calibration gases); and verification of zero and calibration drift over the test period (less than ± 3 percent of span of the period of each run).

The analytical range is determined by the instrumental design. For this method, a portion of the analytical range is selected by choosing the span of the monitoring system. The span of the monitoring system is selected such that the pollutant gas concentration equivalent to the emission standard is not less than 30 percent of the span. If at any time during a run the measured gas concentration exceeds the span, the run is considered invalid.

The minimum detectable limit depends on the analytical range, span, and signal-to-noise ratio of the measurement system. For a well-designed system, the minimum detectable limit should be less than 2 percent of the span.

Nitrogen Oxides (NO_x)   EPA Method 7 is the reference method for determining emissions of nitrogen oxides from stationary sources. Sampling for NO_x by this method is relatively simple with the proper equipment.

A sampling probe is placed at any location in the stack, and a grab sample is collected in an evacuated flask. This flask contains a solution of sulfuric acid and hydrogen peroxide, which reacts with the NO_x. The volume and moisture content of the exhaust-gas stream must be determined for calculation of the total mass-emission rate. The sample is sent to a laboratory, where the concentration of nitrogen oxides, except nitrous oxide, is determined colorimetrically.

Each grab sample is obtained fairly rapidly (15 to 30 s), and four grab samples constitute one run; a total of 12 grab samples is required for a complete series of three runs. An interval of 15 min between grab samples is required. The range of this method has been determined to be 2 to 400 mg of NO_x (as NO_2) per dry standard cubic meter (without dilution). Figure 25-34 shows a schematic of the sampling apparatus for an NO_x source test.

EPA Method 7E is the instrumental analyzer procedure used to

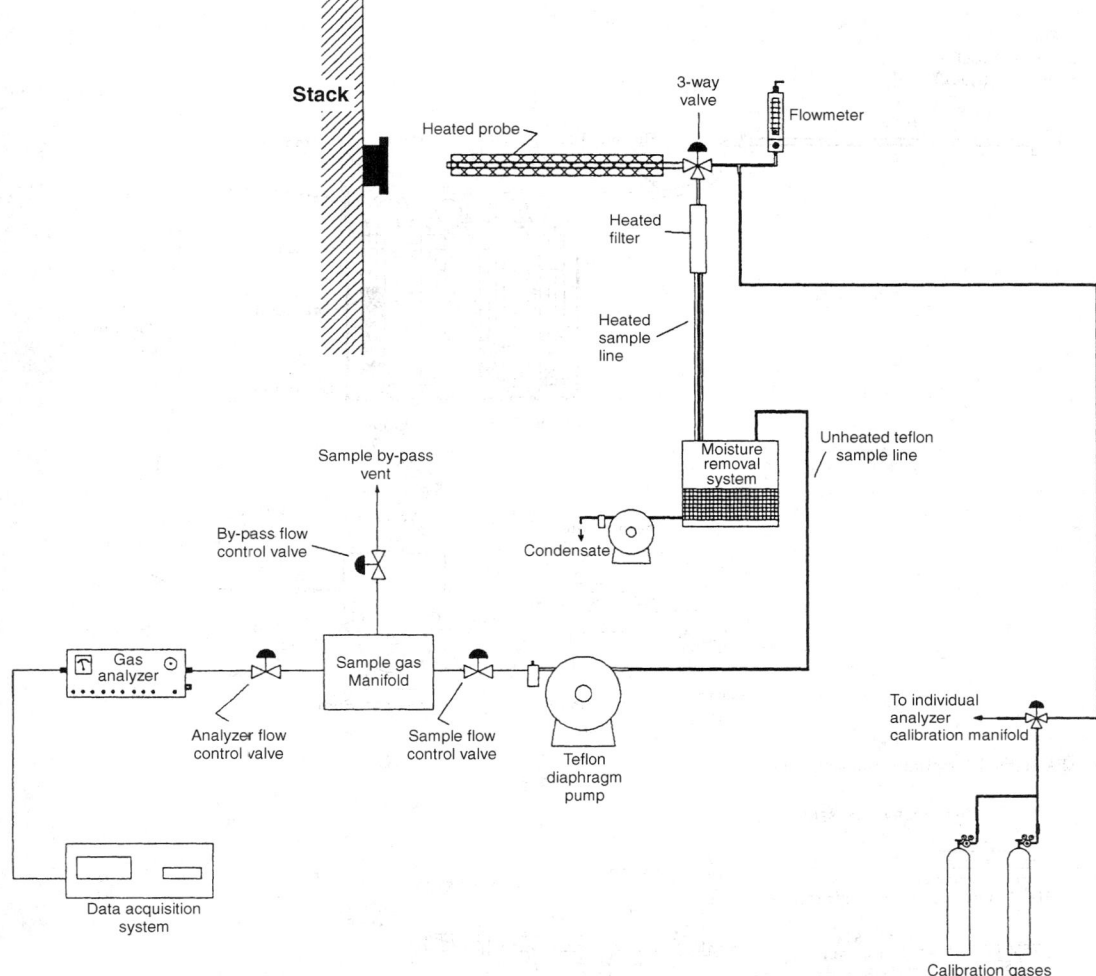

FIG. 25-30   Measurement system schematic for continuous emissions sampling (EPA Method 3A, 6C, or 7E).

determine NO_x emissions from stationary sources (see Fig. 26-28). An integrated continuous gas sample is extracted from the test location, and a portion of the sample is conveyed to an instrumental chemiluminescent analyzer for determination of NO_x gas concentration. The sample gas is conditioned prior to introduction to the gas analyzer by removing particulate matter and moisture. Sampling is conducted at a constant rate for the entire test run.

The NO_x analyzer is based on the principles of chemiluminescence to determine continuously the NO_x concentration in the sample gas stream. The analyzer should contain a NO_2-to-NO converter, which converts the nitrogen dioxide (NO_2) in the sample gas to nitrogen oxide (NO). An NO_2-to-NO converter is not necessary if data are presented to demonstrate that the NO_2 portion of the exhaust gas is less than 5 percent of the total NO_2 concentration.

Quality control elements required by the instrumental analyzer method include: analyzer calibration error (± 2 percent of instrument span allowed); verifying the absence of bias introduced by the sampling system (less than ± 5 percent of span for zero and upscale calibration gases); verification of zero and calibration drift over the test period (less than ± 3 percent of span of the period of each run).

Carbon Monoxide (CO)   EPA Method 10 is the reference method for determining emissions of carbon monoxide from station-

ary sources. An integrated or a continuous gas sample may be required, depending on operating conditions.

When the operating conditions are uniform and steady (there are no fluctuations in flow rate or in concentration of CO in the gas stream), the continuous sampling method can be used. A sampling probe is placed in the stack at any location, preferably near the center. The sample is extracted at a constant sampling rate. As the gas stream passes through the sampling apparatus, any moisture or carbon dioxide in the sample gas stream is removed. The CO concentration is then measured by a nondispersive infrared analyzer, which gives direct readouts of CO concentrations.

Figure 25-35 is a schematic of an assembled sampling apparatus used to determine CO concentrations by the continuous sampling method.

For an integrated sample, the sampling probe is located at any point near the center of the stack, and the sampling rate is adjusted proportionately to the stack-gas velocity. As the stack gas passes through the sampling apparatus, moisture is removed and the sample gas is collected in a flexible bag. Analysis of the sample is then performed in a laboratory with a nondispersive infrared analyzer. Figure 25-36 is a schematic of an assembled apparatus for the integrated sampling of CO.

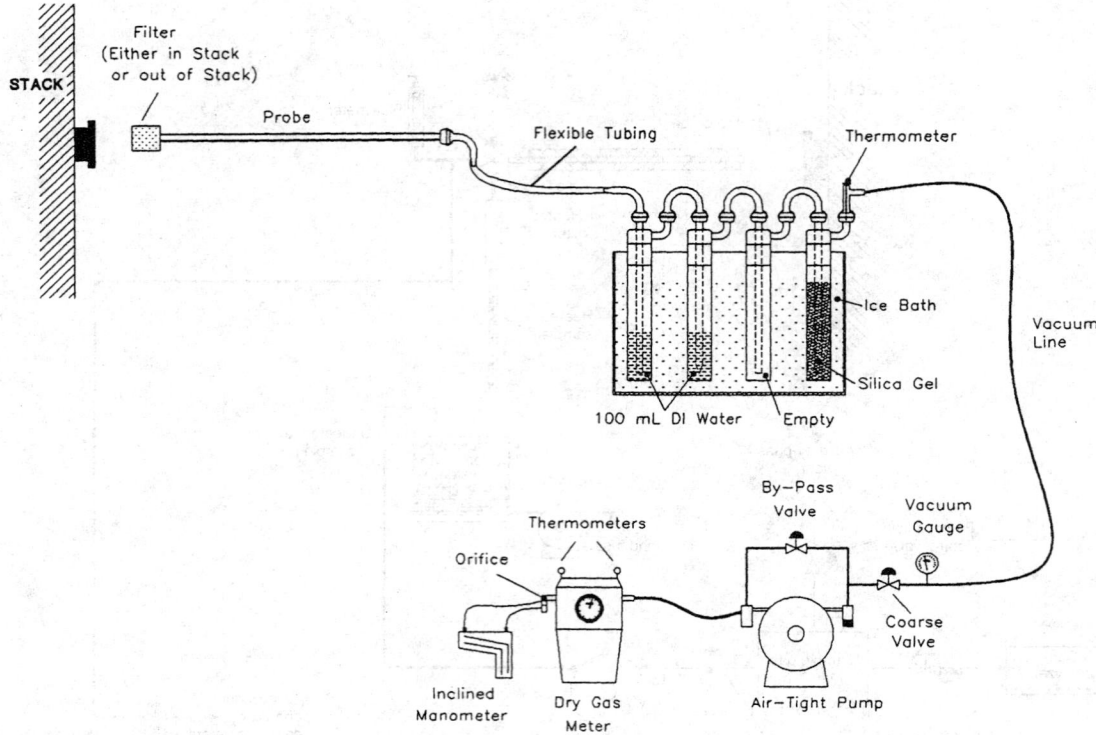

FIG. 25-31   EPA Method 4: moisture sampling train.

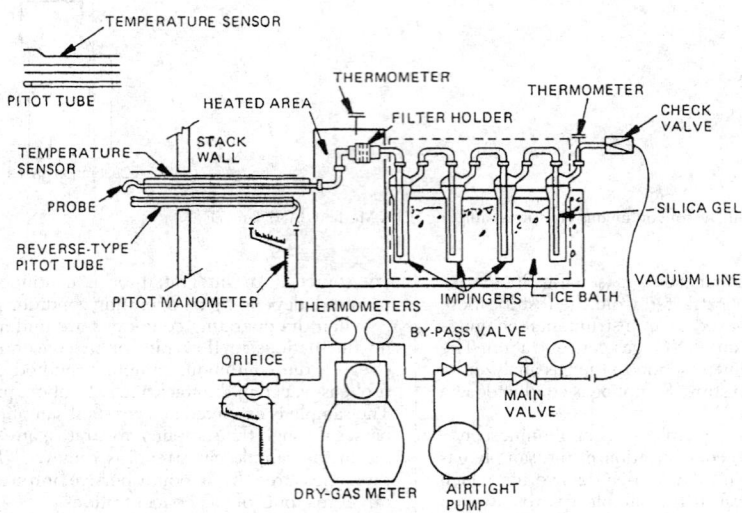

FIG. 25-32   EPA Method 5 particulate-sample apparatus.

A one-hour sampling period is generally required for this method. Sampling periods are specified by the applicable standard; e.g., standards for petroleum refineries require sampling for one hour or more.

For Method 10, the minimum detectable concentration of CO has been determined to be 20 ppm in a range of 1 to 1000 ppm.

Fluorides   Two EPA reference methods, Method 13A and Method 13B, can be used to determine total fluoride emissions from a stationary source. The difference in the two methods is the analyti-

cal procedure for determining total fluorides. Fluorides can occur as particulates or as gaseous fluorides; the particulates are captured on a filter, and the gaseous fluorides are captured in a chemical reaction with water.

Samples for either Method 13A or Method 13B are obtained by the procedures outlined in Method 5 for particulates. As the gas stream passes through the sampling apparatus, the gaseous fluorides are removed by a chemical reaction with water, the particulate fluorides

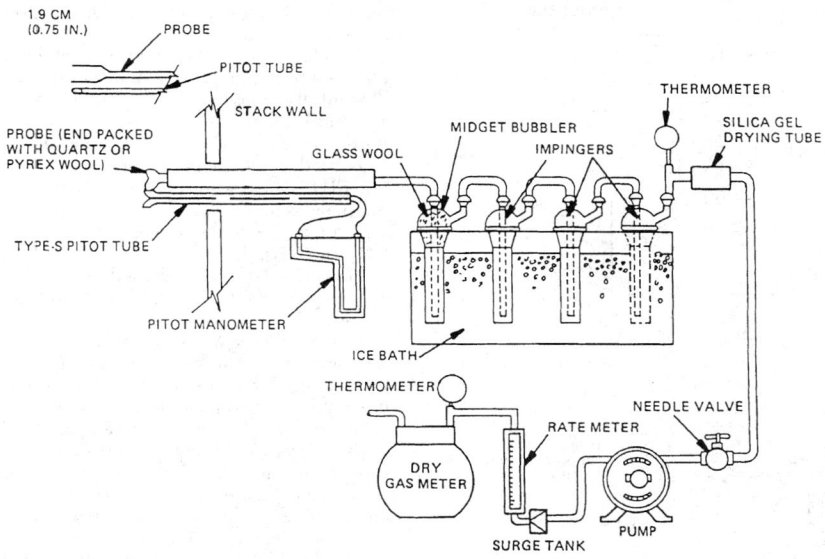

FIG. 25-33 EPA Method 6 sulfur dioxide sample train.

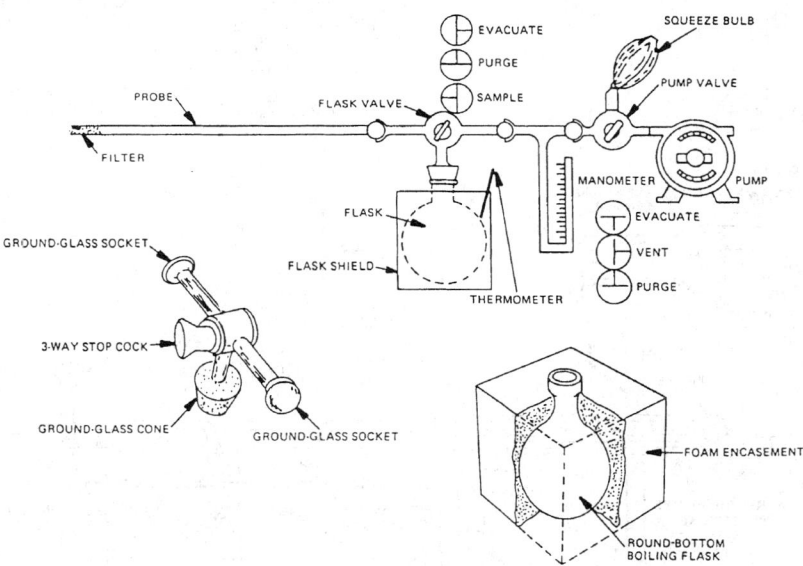

FIG. 25-34 EPA Method 7 nitrogen oxide sample train.

are captured on a filter, and the sample volume is measured. The sample is recovered and sent to the laboratory for analysis. Procedures of Methods 13A and 13B are complex and should be performed by an experienced chemist. Method 13A is a colorimetric method, and Method 13B utilizes a specific ion electrode.

A one-hour sampling period is generally required for both methods. Sampling periods are specified by the applicable standard; e.g., standards applicable to triple-superphosphate plants require sampling of one hour or more. The standard may also specify a minimum sample volume that will dictate the minimum length of the sampling period.

The determination range of Method 13A is 0 to 1.4 μg of fluoride per milliliter; the range of Method 13B is 0.02 to 2000 μg of fluo-

ride per milliliter. Figure 25-37 is a schematic of an assembled fluoride sampling apparatus used in Methods 13A and 13B.

Total Gaseous Organics Concentration The U.S. EPA has published two reference-source testing methods for determination of total gaseous organics: EPA Method 25 for the determination of total gaseous nonmethane organics (TGNMO) and EPA Method 25A as the instrumental analyzer method for determination of total gaseous organics.

Method 25 applies to the measurement of volatile organic compounds (VOC) as nonmethane organics (TGNMO), reported as carbon. Organic particulate matter will interfere with the analysis, and, therefore, in some cases, an in-stack particulate filter will be required. The method requires an emission sample to be withdrawn at a con-

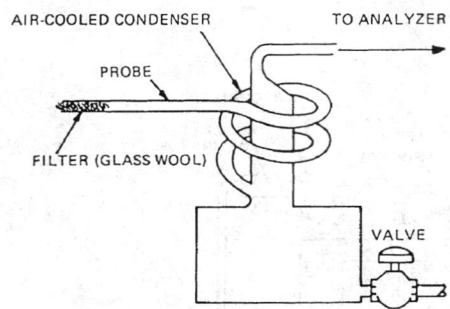

FIG. 25-35   Continuous sample train for CO.

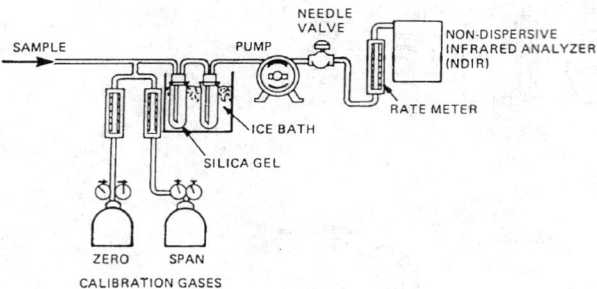

FIG. 25-36   Sampling apparatus for CO.

stant rate through a chilled-condensate trap by means of an evacuated-sample tank.

TGNMO are determined by combining the analytical results obtained from independent analyses of the condensate-trap and sample-tank fractions. After sampling has been completed, the organic contents of the condensate trap are oxidized to carbon dioxide, which is quantitatively collected in an evacuate vessel; then a portion of the CO_2 is reduced to methane and measured by a flame ionization detector (FID). The organic content of the sample fraction collected in the sampling tank is measured by a chromatographic column to achieve separation of the nonmethane organics from carbon monoxide (CO), CO_2, and CH_4; the nonmethane organics (NMO) are oxidized to CO_2, reduced to CH_4, and measured by an FID. In this manner, the variable response of the FID associated with different types of organics is eliminated.

The sampling system consists of a condensate trap, flow-control system, and sample tank (Fig. 25-38). The analytical system consists of two major subsystems: an oxidation system for the recovery and conditioning of the condensate-trap contents and an NMO analyzer. The NMO analyzer is a gas chromatograph with backflush capability for NMO analysis and is equipped with an oxidation catalyst, a reduction catalyst, and an FID. The system for the recovery and conditioning of the organics captured in the condensate trap consists of a heat source, an oxidation catalyst, a nondispersive infrared (NDIR) analyzer, and an intermediate collection vessel.

EPA Method 25A is the instrumental analyzer method for determination of total gaseous organic concentration using a flame ionization analyzer. The method applies to the measurement of total gaseous organic concentration of vapors consisting primarily of alkanes, alkenes, and/or arenes (aromatic hydrocarbons). The concentration is expressed in terms of propane (or other appropriate organic calibration gas) or in terms of carbon.

A gas sample is extracted from the source through a heated sample line, if necessary, and glass fiber filter to a flame ionization analyzer

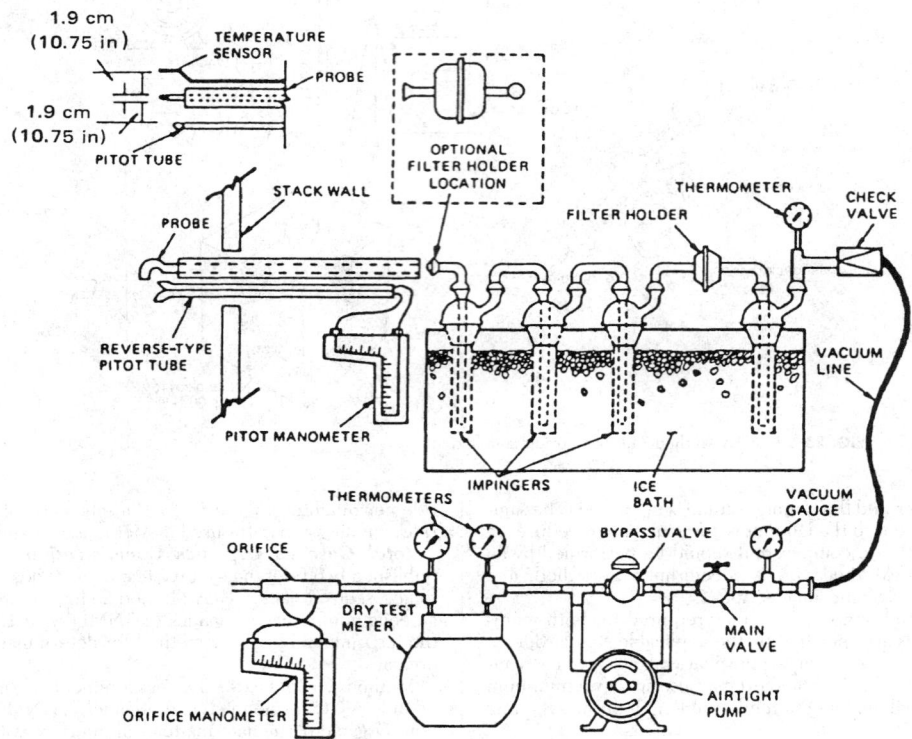

FIG. 25-37   Sampling apparatus for fluoride.

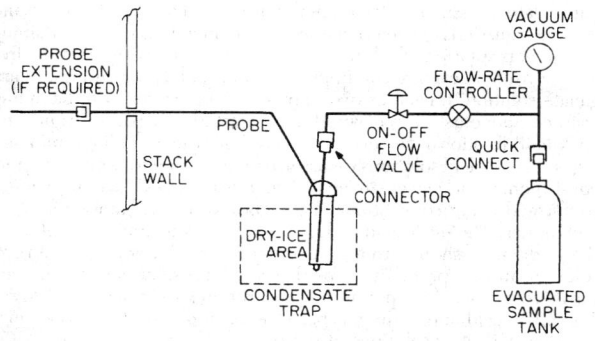

FIG. 25-38 EPA Method 25 sampling system.

(FIA). Figure 25-39 presents a schematic of the sampling system. Results are reported as volume concentration equivalents of the calibration gas or as carbon equivalents.

The upper limit of a gas concentration measurement range is usually 1.5 to 2.5 times the applicable emission limit. If no span value is provided, a span value equivalent to 1.5 to 2.5 times the expected concentration is used. For convenience, the span value should correspond to 100 percent of the recorder scale.

Hydrogen Chloride (HCl) EPA Method 26 is the reference method used to measure hydrogen chloride emissions from stationary sources. The method is applicable for determining emissions of hydrogen halides (HX) such as hydrogen chloride (HCl), hydrogen

bromide (HBr), and hydrogen fluoride (HF), and halogens (X_2) like chlorine (Cl_2) and bromine (Br_2), from stationary sources. Sources, such as those controlled by wet scrubbers, that emit acid particulate matter must be sampled using Method 26A.

An integrated sample is extracted from the source and passed through a prepurged heated probe and filter into dilute sulfuric acid and dilute sodium hydroxide solutions that collect the gaseous hydrogen halides and halogens, respectively. The filter collects other particulate matter, including halide salts. The hydrogen halides are solubilized in the acidic solution and form chloride (Cl), bromide (Br^-), and fluoride (F^-) ions. The halogens have a very low solubility in the acidic solution and pass through to the alkaline solution, where they are hydrolyzed to form a proton (H+), the halide ion, and the hypohalous acid (HClO or HBrO). Sodium thiosulfate is added in excess to the alkaline solution to assure reaction with the hypohalous acid to form a second halide ion such that two halide ions are formed for each molecule of halogen gas. The halide ions in the separate solutions are measured by ion chromatography (IC).

Volatile materials, such as chlorine dioxide (ClO_2) and ammonium chloride (NH_4Cl), which produce halide ions upon dissolution during sampling, are potential interferents. Interferents for the halide measurements are the halogen gases, which disproportionate to a hydrogen halide and a hydrohalous acid upon dissolution in water. However, the use of acidic rather than neutral or basic solutions for collection of the hydrogen halides greatly reduces the dissolution of any halogens passing through this solution. The simultaneous presence of HBr and Cl_2 may cause a positive bias in the HCl result with a corresponding negative bias in the Cl_2 result as well as affecting the HBr/Br_2 split. High concentrations of nitrogen oxides (NO_x) may produce sufficient nitrate (NO_3^-) to interfere with measurements of very low Br^- levels.

The collected Cl^- samples can be stored for up to 4 weeks. The ana-

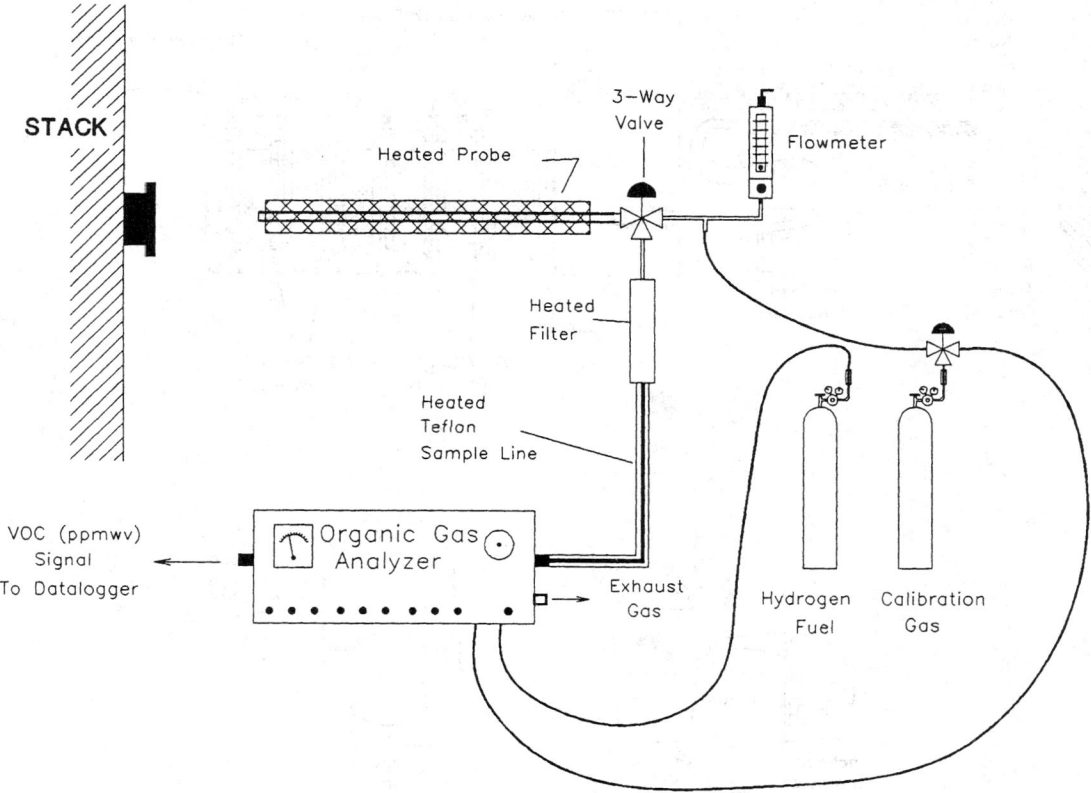

FIG. 25-39 Measurement system schematic for continuous emissions sampling for total gaseous organics.

lytical detection limit for Cl⁻ is 0.1 μg/ml. Detection limits for the other analyses should be similar.

The previous discussion of test methods has reviewed selected reference methods contained in Title 40 the Code of Federal Regulations, Part 60 (40 CFR 60), Appendix A. The following subsections review test methods contained in other regulations and publications. These additional test methods are employed at a greater frequency as emissions of hazardous air pollutants have become increasingly regulated. The regulatory citation for each method is provided in the narrative for each test procedure reviewed. In addition, Table 26-22 provides a reference listing of selected test methods.

Multiple Metals Testing The sampling method commonly used to measure emissions of metals from stationary sources is contained in 40 CFR 266, Appendix IX. The procedure is titled "Methodology for the Determination of Metals Emissions in Exhaust Gases from Hazardous Waste Incineration and Similar Combustion Processes." It is also currently published as Draft EPA Method 29 for inclusion in 40 CFR 60.

This method is used for the determination of total chromium (Cr), cadmium (Cd), arsenic (As), nickel (Ni), manganese (Mn), beryllium (Be), copper (Cu), zinc (Zn), lead (Pb), selenium (Se), phosphorus (P), thallium (Tl), silver (Ag), antimony (Sb), barium (Ba), and mercury (Hg) stack emissions from stationary sources. This method may also be used for the determination of particulate emissions following the procedures and precautions described. However, modifications to the sample recovery and analysis procedures described in the method for the purpose of determining particulate emissions may potentially impact the front-half mercury determination.

The stack sample is withdrawn isokinetically from the source, with particulate emissions collected in the probe and on a heated filter and gaseous emissions collected in a series of chilled impingers containing an aqueous solution of dilute nitric acid combined with dilute hydrogen peroxide in each of two impingers, and acidic potassium permanganate solution in each of two impingers (see Fig. 25-40). Sampling train components are recovered and digested in separate front- and back-half fractions. Materials collected in the sampling train are digested with acid solutions to dissolve organics and to remove organic constituents that may create analytical interferences. Acid digestion is performed using conventional Parr® Bomb or microwave digestion techniques. The nitric acid and hydrogen peroxide impinger solution, the acidic potassium permanganate impinger solution, the HCl rinse solution, and the probe rinse and digested filter solutions are analyzed for mercury by cold-vapor atomic absorption spectroscopy (CVAAS). The nitric acid and hydrogen peroxide solution and the probe rinse and digested filter solutions of the train catches are analyzed for Cr, Cd, Ni, Mn, Be, Cu, Zn, Pb, Se, P, Tl, Ag, Sb, Ba, and As by inductively coupled argon plasma emission spectroscopy (ICAP) or atomic absorption spectroscopy (AAS). Graphite furnace atomic absorption spectroscopy (GFAAS) is used for analysis of antimony, arsenic, cadmium, lead, selenium, and thallium, if these elements require greater analytical sensitivity than can be obtained by ICAP.

Volatile Organics Sampling EPA Method 0030 as contained in *Test Methods for Evaluating Solid Waste*, 3d ed., Report No. SW-846, is used to measure emissions of volatile organic compounds with boiling points less than 100° C. Method 0030 is designed to determine destruction and removal efficiency (DRE) of volatile principal organic hazardous constituents (POHCs) from stack gas effluents of hazardous waste incinerators. The lower boiling point POHCs (less than

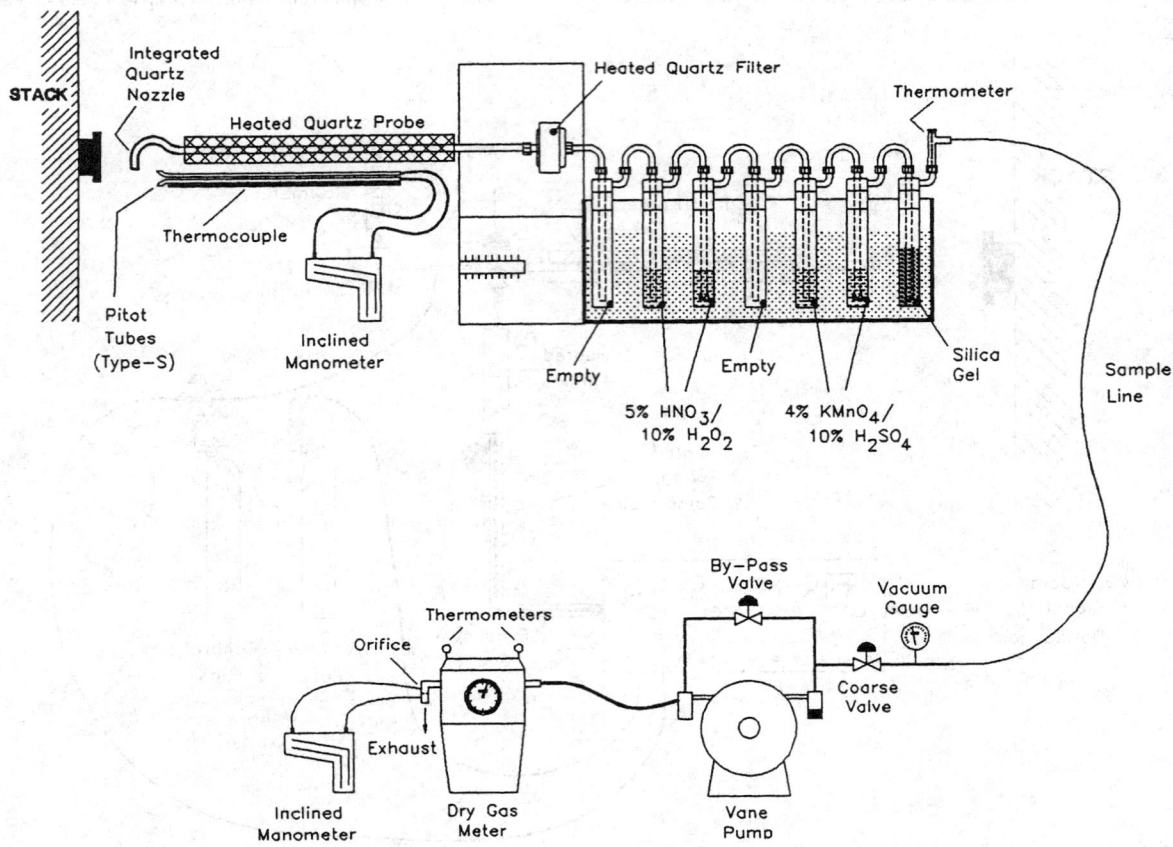

FIG. 25-40 Multiple-metals sampling train.

30° C) may break through the sorbent under the conditions of the sample collection procedure.

The procedure involves withdrawing a 20-L sample of effluent gas from the source at a flow rate of 1 L/min, using a glass-lined probe and a volatile organic sampling train (VOST). Figure 25-41 provides a schematic of the VOST set-up. The gas sample is cooled to 20° C by passage through a water-cooled condenser and volatile POHCs are collected on a pair of sorbent resin traps. Liquid condensate is collected in an impinger placed between the two resin traps. The first resin trap (front trap) contains approximately 1.6 g Tenax® resin and the second trap (back trap) contains approximately 1 g each of Tenax® and petroleum-based charcoal, 3:1 by volume. The first trap retains most of the higher boiling analytes. Lower boiling analytes and the portion of the higher boiling analytes that break through the first cartridge are retained on the second trap. Analytes that collect in the condensate trap are purged into the second trap and condenser units. The VOST sorbent cartridges are thermally desorped and analyzed by gas chromatography with mass spectroscopy (GC/MS).

Semivolatile Organics Sampling EPA Method 0010 as contained in *Test Methods for Evaluating Solid Waste*, 3d ed., Report No.

SW-846, is used to measure emissions of semivolatile principal organic constituents. Method 0010 is designed to determine destruction and removal efficiency (DRE) of POHCs from incineration systems. The method involves a modification of the EPA Method 5 sampling train and may be used to determine particulate emission rates from stationary sources. The method is applied to semivolatile compounds, including polychlorinated biphenyls (PCBs), chlorinated dibenzodioxins and dibenzofurans, polycyclic organic matter, and other semivolatile organic compounds.

Gaseous and particulate pollutants are withdrawn isokinetically from an emission source and collected in a multicomponent sampling train. Principal components of the train include a high-efficiency glass- or quartz-fiber filter and a packed bed of porous polymeric adsorbent resin (typically XAD-2® or polyurethane foam for PCBs). The filter is used to collect organic-laden particulate materials and the porous polymeric resin to adsorb semivolatile organic species (compounds with a boiling point above 100° C). Figure 25-42 presents an illustration of the Method 0010 sampling train. Comprehensive chemical analyses, using a variety of applicable analytical methodologies, are conducted to determine the identity and concentration of the organic materials.

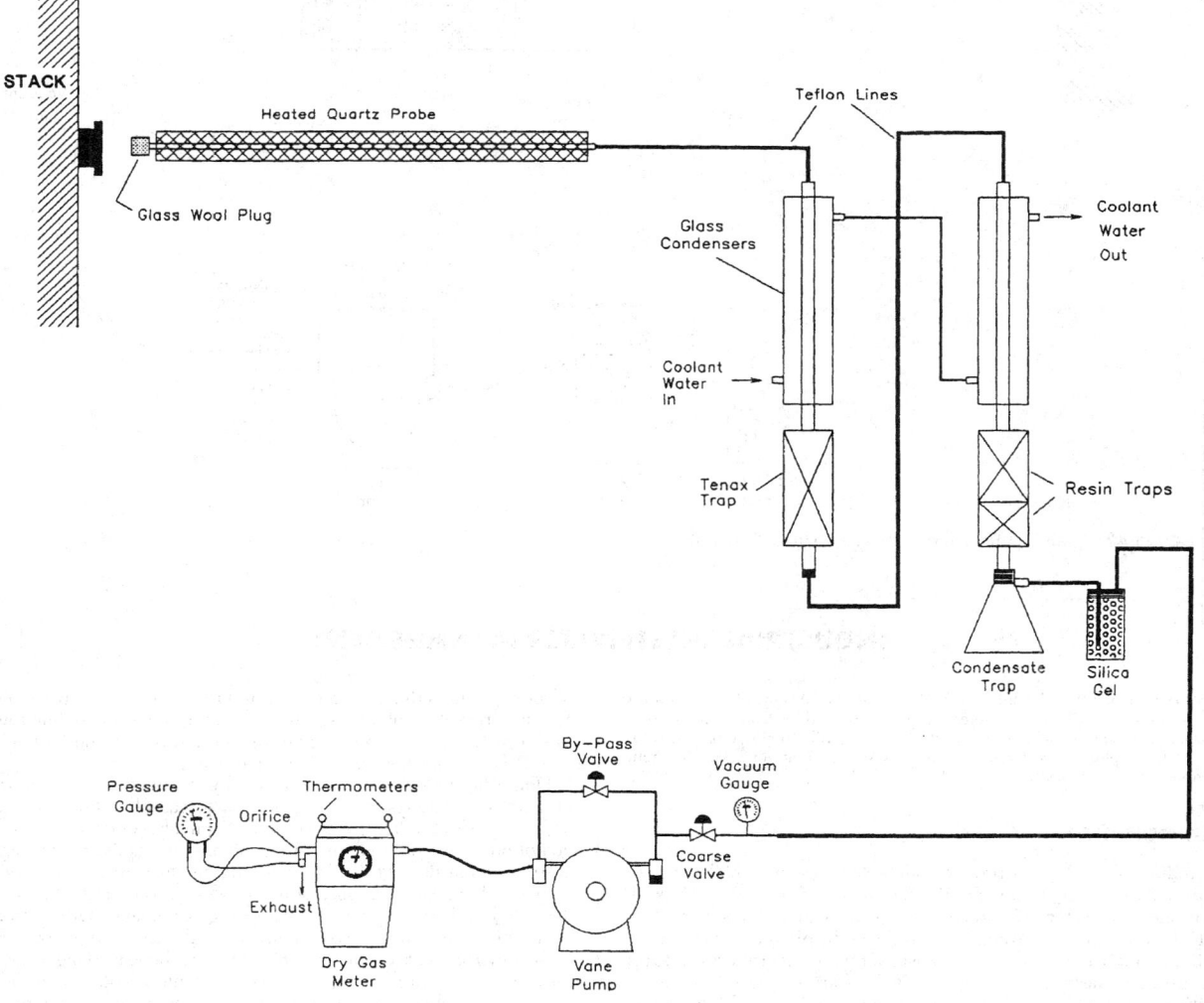

FIG. 25-41 Volatile organic sampling train (Method 0030).

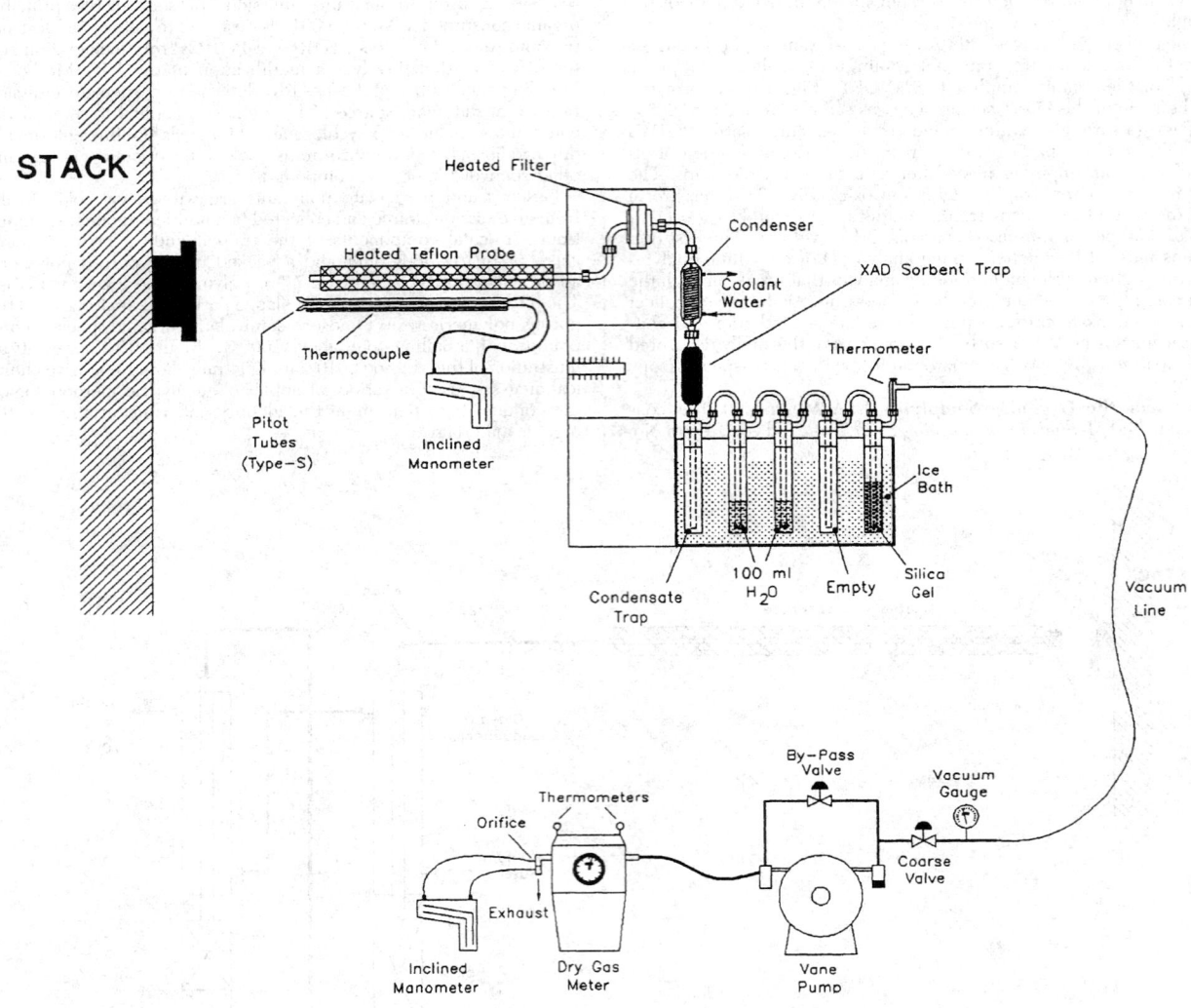

FIG. 25-42   Method 0010: sampling train for semivolatile organics.

INDUSTRIAL WASTEWATER MANAGEMENT

FURTHER READING: Eckenfelder, W. W., *Industrial Water Pollution Control,* 2d ed., McGraw-Hill, 1989. Metcalf & Eddy, Inc., *Wastewater Engineering, Treatment, Disposal and Reuse,* 3d ed., McGraw-Hill, 1990. Nemerow, N. L. and A. Dasgupta, Industrial and Hazardous Waste Treatment, Van Nostrand Reinhold, New York, 1991.

INTRODUCTION

All industrial operations produce some wastewaters which must be returned to the environment. Wastewaters can be classified as (1) domestic wastewaters, (2) process wastewaters, and (3) cooling wastewaters. Domestic wastewaters are produced by plant workers, shower facilities, and cafeterias. Process wastewaters result from spills, leaks, and product washing. Cooling wastewaters are the result of various cooling processes and can be once-pass systems or multiple-recycle cooling systems. Once-pass cooling systems employ large volumes of cooling waters that are used once and returned to the environment. Multiple-recycle cooling systems have various types of cooling towers to return excess heat to the environment and require periodic blow-down to prevent excess buildup of salts.

Domestic wastewaters are generally handled by the normal sanitary-sewerage system to prevent the spread of pathogenic micro-organisms which might cause disease. Normally, process wastewaters do not pose the potential for pathogenic microorganisms, but they do pose potential damage to the environment through either direct or indirect chemical reactions. Some process wastes are readily bio-degraded and create an immediate oxygen demand. Other process wastes are toxic and represent a direct health hazard to biological life in the environment. Cooling wastewaters are the least dangerous, but they can contain process wastewaters as a result of leaks in the cooling systems. Recycle cooling systems tend to concentrate both inorganic and organic contaminants to a point at which damage can be created.

Recently, concern for subtle aspects of environmental damage has started to take precedence over damage referred to above. It has been realized that the presence of substances in water at concentrations far below those that will produce overt toxicity or excessive reduction of dissolved oxygen levels can have a major impact by altering the predomination of organisms in the aquatic ecosystem. This is beginning to have an effect on water-quality standards and, consequently, allowable discharges. Unfortunately, the extent of knowledge is not sufficient upon which to base definitive standards for these subtle effects. Over the next decade, it is anticipated that accumulation of knowledge will make it possible to delineate defensible standards. Another recent concern is that of contaminated storm water from industrial sites. Federal guidelines to control this wastewater are presently being developed. Most concern is for product spills on plant property but outside the production facility and from rain contact with spoil piles.

UNITED STATES LEGISLATION, REGULATIONS, AND GOVERNMENTAL AGENCIES

Federal Legislation   Public Law 92-500 promulgated in 1972 created the primary framework for management of water pollution in the United States. This act has been amended many times since 1972. Major amendments occurred in 1977 (Public Laws 95-217 and 95-576), 1981 (Public Laws 97-117 and 97-164), and 1987 (Public Law 100-4). This law and its various amendments are referred to as the Clean Water Act. The Clean Water Act addresses a large number of issues of water pollution management. A general review of these issues has been presented in an earlier chapter in this handbook. Primary with respect to control of industrial wastewater is the National Pollutant Discharge Elimination System (NPDES), originally established by PL 92-500. Any municipality or industry that discharges wastewater to the navigable waters of the United States must obtain a discharge permit under the regulations set forth by the NPDES. Under this system, there are three classes of pollutants (conventional pollutants, priority pollutants, and nonconventional/nonpriority pollutants). Conventional pollutants are substances such as biochemical oxygen demand (BOD), suspended solids (SS), pH, oil and grease, and coliforms. Priority pollutants are a list of 129 substances originally set forth in a consent decree between the Environmental Protection Agency and several environmental organizations. This list was incorporated into the 1977 amendments. Most of the substances on this list are organics, but it does include most of the heavy metals. These substances are generally considered to be toxic. However, the toxicity is not absolute; it depends on the concentration. In addition, many of the organics on the list are under appropriate conditions biodegradable. In reality, substances for inclusion on the priority pollutant list were chosen on the basis of a risk assessment rather than only a hazard assessment. The third class of pollutants could include any pollutant not in the first two categories. Examples of substances that are presently regulated in the third category are nitrogen, phosphorous, sodium, and chlorine residual. The overall goals of the Clean Water Act are to restore and maintain water quality in the navigable waters of the United States. The initial standard was to ensure that these waters would be clean enough for recreation (swimmable) and ecologically secure (fishable). Initially this was to be achieved by curtailment of the discharge of pollutants. Eventually the regulatory system was essentially to phase out discharge of any pollutant into water. Obviously, if no discharge of pollutants occurs, the waters will be maintained in close to a pristine condition. However, the early reduction of pollutant discharge to zero was realized as impractical. Thus, at present, the NPDES has prescribed the limits on discharges as a function of the type of pollutant, the type of industry discharging, and the desired water quality. The specific requirements for each individual discharger are established in the NPDES permit issued to the discharger. These permits are reviewed every five years and are subject to change at the time of review.

A major tactic that was adopted in the Clean Water Act was to establish uniform technology standards, by class of pollutant and specific industry type, which applied nationwide to all dischargers. Thus, a kraft mill in Oregon would have to meet essentially the same discharge standards as a kraft mill in New York. In establishing these

standards, the EPA took into account the state of the art of waste treatment in each particular industry as well as cost and ecological effectiveness. These discharge standards have been published in the *Federal Register* for more than thirty industrial categories and several hundred subcategories (the Commerce Department Industrial Classification System was used to establish these categories) in the *Federal Register*. These have been promulgated over an extensive period of time. The reader is advised to consult the index to this document to ascertain regulations that apply to a particular industry. Table 25-32 presents the major industrial categories.

As indicated above, not only are the discharge standards organized on an industry-by-industry basis, but they are different depending upon which of the three classes of pollutants is being regulated. The standards for conventional pollutants are referred to as best conventional technology (BCT). The standards for priority pollutants are referred to as best available technology (BAT), as are those for nonconventional, nonpriority pollutants. These standards envision that the technology will not be limited to treatment but may include revision in the industrial processing and/or reuse of effluents (pollution prevention). They are usually presented as mass of pollutant discharged per unit of product produced.

In some situations it is anticipated that application of BCT and BAT may not be sufficient to ensure that water-quality standards are achieved in a stream segment. Studies have indicated that approximately 10 percent of the stream segments in the United States will have their water-quality standards violated, even if all the dischargers to that stream segment meet BCT and BAT regulations. These segments are referred to as water quality limited segments. The NPDES permit for those who discharge into water quality limited segments must provide for pollutant removal in excess of that required by BCT and BAT so that water-quality standards (which are jointly established by each state and EPA) are achieved. The stream segments in which water-quality standards will be met by application of BCT and BAT are referred to as "effluent quality limited segments."

For industrial discharges that enter municipal sewers and thus eventually municipal treatment plants, NPDES regulations are nominally the same as for industries that discharge directly to navigable

TABLE 25-32   Industry Categories

1. Adhesives and sealants
2. Aluminum forming
3. Asbestos manufacturing
4. Auto and other laundries
5. Battery manufacturing
6. Coal mining
7. Coil coating
8. Copper forming
9. Electric and electronic components
10. Electroplating
11. Explosives manufacturing
12. Ferroalloys
13. Foundries
14. Gum and wood chemicals
15. Inorganic chemicals manufacturing
16. Iron and steel manufacturing
17. Leather tanning and finishing
18. Mechanical products manufacturing
19. Nonferrous metals manufacturing
20. Ore mining
21. Organic chemicals manufacturing
22. Pesticides
23. Petroleum refining
24. Pharmaceutical preparations
25. Photographic equipment and supplies
26. Plastic and synthetic materials manufacturing
27. Plastic processing
28. Porcelain enamelling
29. Printing and publishing
30. Pulp and paperboard mills
31. Soap and detergent manufacturing
32. Steam electric power plants
33. Textile mills
34. Timber products processing

waters; i.e., they must meet BCT and BAT standards. These are referred to as "categorical industrial pretreatment standards." However if it can be demonstrated that the municipal treatment plant can remove a pollutant in the industrial waste, a "removal credit" can be assigned to the permit, thus lowering the requirement on the industrial discharger. The monetary charge to the industry by the municipality for this service is negotiable but must be within parameters established by EPA for industrial user charges if the municipality received a federal grant for construction of the treatment plant. A removal credit cannot be assigned if removal of the industrial pollutant results in difficulty in ultimate disposal of sludge from the municipal plant. In addition, an industry cannot discharge any substance that can result in physical damage to the municipal sewer and/or interfere with any aspect of treatment plant performance even, if the substance is not covered by BCT or BAT regulations.

Environmental Protection Agency   President Nixon created the EPA in 1970 to coordinate all environmental pollution-control activities at the federal level. The EPA was placed directly under the Office of the President so that it could be more responsive to the political process. In the succeeding decade, the EPA produced a series of federal regulations increasing federal control over all wastewater-pollution-control activities. In January 1981, President Reagan reversed the trend of greater federal regulation and began to decrease the role of the federal EPA. However, during the 1980s, an equilibrium was achieved between those who wish less regulation and those who wish more. In general, industry and the EPA reached agreement on the optimum level of regulation.

State Water-Pollution-Control Offices   Every state has its own water-pollution-control office. Some states have reorganized along the lines of the federal EPA with state EPA offices, while others have kept their water-pollution-control offices within state health departments. Prior to 1965, each state controlled its own water-pollution-control programs. Conflicts between state and uneven enforcement of state regulations resulted in the federal government's assuming the leadership role. Unfortunately, conflicts between states shifted to being conflicts between the states and the federal EPA. By 1980, the state water-pollution-control offices were primarily concerned with handling most of the detailed permit and paperwork for the EPA and in furnishing technical assistance to industries at the local level. In general, most of the details of regulation are carried out by state water-pollution-control agencies with oversight by EPA.

WASTEWATER CHARACTERISTICS

Wastewater characteristics vary widely from industry to industry. Obviously, the specific characteristics will affect the treatment techniques chosen for use in meeting discharge requirements. Some general characteristics that should be considered in planning are given in Table 25-33. Because of the large number of pollutant substances, wastewater characteristics are not usually considered on a substance-by-substance basis. Rather, substances of similar pollution effects are grouped together into classes of pollutants or characteristics as indicated below.

Priority Pollutants   Recently, greatest concern has been for this class of substances for the reasons given previously. These materials are treated on an individual-substance basis for regulatory control. Thus, each industry could receive a discharge permit that lists an acceptable level for each priority pollutant. Table 25-34 presents a list of these substances; most are organic, but some inorganics are included. All are considered toxic, but, as indicated previously, there is wide variation in their toxicity. Most of the organics are biologically degradable despite their toxicity (Refs. 30 and 31). USEPA has collected data on the occurrence of these substances in various industrial wastes and their treatability. A recent trend has been to avoid their use in industrial processing.

Organics   The organic composition of industrial wastes varies widely, primarily due to the different raw materials used by each specific industry. These organics include proteins, carbohydrates, fats and oils, petrochemicals, solvents, pharmaceutical, small and large molecules, solids and liquids. Another complication is that a typical industry produces many diverse wastestreams. Good practice is to conduct a material balance throughout an entire production facility. This survey should include a flow diagram, location and sizes of piping, tanks and flow volumes, as well as an analysis of each stream. Results of an industrial waste survey for an industry are given in Table 25-35. Noteworthy is the range in waste sources, including organic soap, toilet articles, ABS (alkyl benzene sulfonate), and the relatively clean but hot condenser water that makes up half the plant flow, while the strongest wastes—spent caustic and fly ash—have the lowest flows. See Tables 25-35 and 25-36 for information on the average characteristics of wastes from specific industries.

An important measure of the waste organic strength is the 5-day biochemical oxygen demand (BOD_5). As this test measures the demand for oxygen in the water environment caused by organics released by industry and municipalities, it has been the primary parameter in determining the strength and effects of a pollutant. This test determines the oxygen demand of a waste exposed to biological organisms (controlled seed) for an incubation period of five days. Usually this demand is caused by degradation of organics according to the following simplified equation, but reduced inorganics in some industries may also cause demand (i.e., Fe^{2+}, S^{2-}, and SO_3^{2-}).

$$\text{Organic waste} + O_2 \text{ (D.O.)} \xrightarrow[\text{microbes}]{\text{seed}} CO_2 + H_2O$$

This wet lab test measures the decrease in dissolved oxygen (D.O.) concentration in 5 days, which is then related to the sample strength. If the test is extended over 20 days, the BOD_{20} (ultimate BOD) is obtained and corresponds more closely to the Chemical Oxygen Demand (COD) test. The COD test uses strong chemical oxidizing agents with catalysts and heat to oxidize the wastewater and obtain a value that is almost always larger than the 5- and 20-day BOD values. Some organic compounds (like pyridene, a ring structure containing nitrogen) resists chemical oxidation giving a low COD. A major advantage of the COD test is the completion time of less than 3 hours, versus 5 days for the BOD_5 test. Unfortunately, state and federal regulations generally require BOD_5 values, but approximate correlations can be made to allow computation of BOD from COD. A more rapid measure of the organic content of a waste is the instrumental test for total organic carbon (TOC), which takes a few minutes and may be correlated to both COD and BOD for specific wastes. Unfortunately, BOD_5 results are subject to wide statistical variations and require close scrutiny and experience. For municipal wastewaters, BOD_5 is about 67 percent of the ultimate BOD and 40–45 percent of the COD, indicating a large amount of nonbiodegradable COD and the continuing need to run BOD as well as COD and TOC. An example of BOD, COD, and TOC relationships for chemical industry wastewater is given in Table 25-36. The concentrations of the wastewaters vary by two orders of magnitude, and the BOD/COD, COD/TOC, and BOD/TOC ratios vary less than twofold. The table indicates that correlation/codification is possible, but care and continual scrutiny must be exercised.

TABLE 25-33   Wastewater Characteristics

Property	Characteristic	Example	Size or concentration
Solubility	Soluble	Sugar	>100 gm/L
	Insoluble	PCB	<1 mg/L
Stability, biological	Degradable	Sugar	
	Refractory	DDT, metals	
Solids	Dissolved	NaCl	$<10^{-9}$ m
	Colloidal	Carbon	$>10^{-6}$–$<10^{-9}$ m
	Suspended	Bacterium	$>10^{-6}$ m
Organic	Carbon	Alcohol	
Inorganic	Inorganic	Cu^{2+}	
pH	Acidic	HNO_3	
	Neutral	Salt (NaCl)	1–12
	Basic	NaOH	
Temperature	High–low	Cooling	>5°
		Heat exchange	>30°
Toxicity	Biological effect	Heavy metals	Varies
		Priority compounds	
Nutrients	N	NH_3	Varies
	P	PO_4^{3-}	

TABLE 25-34 List of Priority Chemicals*

Compound name	Compound name	Compound name
1. Acenaphthene†	37. 1,2-Diphenylhydrazine†	83. Indeno (1,2,3-cd) pyrene (2,3-o-phenylenepyrene)
2. Acrolein†	38. Ethylbenzene†	
3. Acrylonitrile†	39. Fluoranthene†	84. Pyrene
4. Benzene†		85. Tetrachloroethylene†
5. Benzidine†	Haloethers† (other than those listed elsewhere)	86. Toluene†
6. Carbon tetrachloride† (tetrachloromethane)	40. 4-Chlorophenyl phenyl ether	87. Trichloroethylene†
	41. 4-Bromophenyl phenyl ether	88. Vinyl chloride† (chloroethylene)
Chlorinated benzenes (other than dichloroben-zenes)	42. Bis(2-chloroisopropyl) ether	
	43. Bis(2-chloroethoxy) methane	Pesticides and metabolites†
7. Chlorobenzene		89. Aldrin†
8. 1,2,4-Trichlorobenzene		90. Dieldrin†
9. Hexachlorobenzene	Halomethanes† (other than those listed else-where)	91. Chlordane† (technical mixture and metabo-lites)
	44. Methylene chloride (dichloromethane)	
Chlorinated ethanes† (including 1,2-dichloroethane, 1,1,1-trichloroethane, and hexachloroethane)	45. Methyl chloride (chloromethane)	DDT and metabolites†
	46. Methyl bromide (bromomethane)	92. 4-4'-DDT
10. 1,2-Dichloroethane	47. Bromoform (tribromomethane)	93. 4,4'-DDE (p,p'-DDX)
11. 1,1,1-Trichloroethane	48. Dichlorobromomethane	94. 4,4'-DDD (p,p'-TDE)
12. Hexachloroethane	49. Trichlorofluoromethane	
13. 1,1-Dichloroethane	50. Dichlorodifluoromethane	Endosulfan and metabolites†
14. 1,1,2-Trichloroethane	51. Chlorodibromomethane	95. α-Endosulfan-alpha
15. 1,1,2,2-Tetrachloroethane	52. Hexachlorobutadiene†	96. β-Endosulfan-beta
16. Chloroethane (ethyl chloride)	53. Hexachlorocyclopentadiene†	97. Endosulfan sulfate
	54. Isophorone†	
Chloroalkyl ethers† (chloromethyl, chloroethyl, and mixed ethers)	55. Naphthalene†	Endrin and metabolites†
	56. Nitrobenzene†	98. Endrin
17. Bis(chloromethyl) ether		99. Endrin aldehyde
18. Bis(2-chloroethyl) ether	Nitrophenols† (including 2,4-dinitrophenol and dinitrocresol)	
19. 2-Chloroethyl vinyl ether (mixed)		Heptachlor and metabolites†
	57. 2-Nitrophenol	100. Heptachlor
Chlorinated napthalene†	58. 4-Nitrophenol	101. Heptachlor epoxide
20. 2-Chloronapthalene	59. 2,4-Dinitrophenol†	
	60. 4,6-Dinitro-o-cresol	Hexachlorocyclohexane (all isomers)†
Chlorinated phenols† (other than those listed elsewhere; includes trichlorophenols and chlorinated cresols)		102. α-BHC-alpha
	Nitrosamines†	103. β-BHC-beta
21. 2,4,6-Trichlorophenol	61. N-Nitrosodimethylamine	104. γ-BHC (lindane)-gamma
22. para-Chloro-meta-cresol	62. N-Nitrosodiphenylamine	105. δ-BHC-delta
23. Chloroform (trichloromethane)†	63. N-Nitrosodi-n-propylamine	
24. 2-Chlorophenol†	64. Pentachlorophenol†	Polychlorinated biphenyls (PCB)†
	65. Phenol†	106. PCB-1242 (Arochlor 1242)
Dichlorobenzenes†		107. PCB-1254 (Arochlor 1254)
25. 1,2-Dichlorobenzene		108. PCB-1221 (Arochlor 1221)
26. 1,3-Dichlorobenzene	Phthalate esters†	109. PCB-1232 (Arochlor 1232)
27. 1,4-Dichlorobenzene	66. Bis(2-ethylhexyl) phthalate	110. PCB-1248 (Arochlor 1248)
	67. Butyl benzyl phthalate	111. PCB-1260 (Arochlor 1260)
Dichlorobenzidine†	68. Di-n-butyl phthalate	112. PCB-1016 (Arochlor 1016)
28. 3,3'-Dichlorobenzidine	69. Di-n-octyl phthalate	113. Toxaphene†
	70. Diethyl phthalate	114. antimony (total)
Dichloroethylenes† (1,1-dichloroethylene and 1,2-dichloroethylene)	71. Dimethyl phthalate	115. arsenic (total)
		116. asbestos (fibrous)
29. 1,1-Dichloroethylene	Polynuclear aromatic hydrocarbons (PAH)†	117. beryllium (total)
30. 1,2-trans-Dichloroethylene	72. Benzo(a)anthracene (1,2-benzanthracene)	118. cadmium (total)
31. 2,4-Dichlorophenol†	73. Benzo(a)pyrene (3,4-benzopyrene)	119. chromium (total)
	74. 3,4-Benzofluoranthene	120. copper (total)
Dichloropropane and dichloropropene†	75. Benzo(k)fluoranthene (11,12-benzofluoranthene)	121. cyanide (total)
32. 1,2-Dichloropropane		122. lead (total)
33. 1,2-Dichloropropylene (1,2-dichloropropene)	76. Chrysene	123. mercury (total)
	77. Acenaphthylene	124. nickel (total)
34. 2,4-Dimethylphenol†	78. Anthracene	125. selenium (total)
	79. Benzo(ghl)perylene (1,12-benzoperylene)	126. silver (total)
Dinitrotoluene†	80. Fluorene	127. thallium (total)
35. 2,4-Dinitrotoluene	81. Phenanthrene	128. zinc (total)
36. 2,6-Dinitrotoluene	82. Dibenzo(a,h)anthracene (1,2,5,6-dibenzanthracene)	129. 2,3,7,8-Tetrachlorodibenzo-p-dioxin (TCDD)

*Adapted from Eckenfelder, W. W. Jr., *Industrial Water Pollution Control*, 2d ed., McGraw-Hill, New York, 1989.
†Specific compounds and chemical classes as listed in the consent degree.

Another technique for organics measurement that overcomes the long period required for the BOD test is the use of continuous respirometry. Here the waste (full-strength rather than diluted as in the standard BOD test) is contacted with biomass in an apparatus that continuously measures the dissolved oxygen consumption. This test determines the ultimate BOD in a few hours if a high level of biomass is used. The test can also yield information on toxicity, the need to develop an acclimated biomass, and required rates of oxygen supply.

In general, low-molecular-weight water-soluble organics are biodegraded readily. As organic complexity increases, solubility and biodegradability decrease. Soluble organics are metabolized more easily than insoluble organics. Complex carbohydrates, proteins, and fats and oils must be hydrolyzed to simple sugars, aminos, and other organic acids prior to metabolism. Petrochemicals, pulp and paper,

TABLE 25-35 Industrial Waste Components of a Soap, Detergents, and Toilet Articles Plant

Waste source	Sampling station	COD, mg/l	BOD, mg/l	SS, mg/l	ABS, mg/l	Flow, gal/min
Liquid soap	D	1,100	565	195	28	300
Toilet articles	E	2,650	1,540	810	69	50
Soap production	R	29	16	39	2	30
ABS production	S	1,440	380	309	600	110
Powerhouse	P	66	10	50	0	550
Condenser	C	59	21	24	0	1100
Spent caustic	B	30,000	10,000	563	5	2
Tank bottoms	A	120,000	150,000	426	20	1.5
Fly ash	F			6750		10
Main sewer		450	260	120	37	2150

NOTE: gal/min = 3.78×10^{-3} m³/min.
SOURCE: Eckenfelder, W. W., *Industrial Water Pollution Control*, 2d ed., McGraw-Hill, New York, 1989.

TABLE 25-36 BOD, COD, and TOC Relationships

Type of waste	BOD$_5$, mg/L	COD, mg/L	TOC, mg/L	BOD$_5$/COD	COD/TOC	BOD$_5$/TOC
Chemical	700	1,400	450	0.50	3.12	1.55
Chemical	850	1,900	580	0.45	3.28	1.47
Chemical	8,000	17,500	5,800	0.46	3.02	1.38
Chemical	9,700	15,000	5,500	0.65	2.72	1.76
Chemical	24,000	41,300	9,500	0.58	4.35	2.53
Chemical	60,700	78,000	26,000	0.78	3.00	2.34
Chemical	62,000	143,000	48,140	0.43	2.96	1.28

Adapted from Eckenfelder, W.W. and D.L. Ford, *Water Pollution Control*, Pemberton Press, Austin and New York, 1970.

slaughterhouse, brewery, and numerous other industrial wastes containing complex organics have been satisfactorily treated biologically, but proper testing and evaluation is necessary.

Inorganics The inorganics in most industrial wastes are the direct result of inorganic compounds in the carriage water. Soft-water sources will have lower inorganics than hard-water or saltwater sources. However, some industrial wastewaters can contain significant quantities of inorganics which result from chemical additions during plant operation. Many food processing wastewaters are high in sodium. While domestic wastewaters have a balance in organics and inorganics, many process wastewaters from industry are deficient in specific inorganic compounds. Biodegradation of organic compounds requires adequate nitrogen, phosphorus, iron, and trace salts. Ammonium salts or nitrate salts can provide the nitrogen, while phosphates supply the phosphorus. Either ferrous or ferric salts or even normal steel corrosion can supply the needed iron. Other trace elements needed for biodegradation are potassium, calcium, magnesium, cobalt, molybdenum, chloride, and sulfur. Carriage water or demineralizer wastewaters or corrosion products can supply the needed trace elements for good metabolism. Occasionally, it is necessary to add specific trace elements or nutrient elements.

pH and Alkalinity Wastewaters should have pH values between 6 and 9 for minimum impact on the environment. Wastewaters with pH values less than 6 will tend to be corrosive as a result of the excess hydrogen ions. On the other hand, raising the pH above 9 will cause some of the metal ions to precipitate as carbonates or as hydroxides at higher pH levels. Alkalinity is important in keeping pH values at the right levels. Bicarbonate alkalinity is the primary buffer in wastewaters. It is important to have adequate alkalinity to neutralize the acid waste components as well as those formed by partial metabolism of organics. Many neutral organics such as carbohydrates, aldehydes, ketones, and alcohols are biodegraded through organic acids which must be neutralized by the available alkalinity. If alkalinity is inadequate, sodium carbonate is a better form to add than lime. Lime tends to be hard to control accurately and results in high pH levels and precipitation of the calcium which forms part of the alkalinity. In a few instances, sodium bicarbonate may be the best source of alkalinity.

Temperature Most industrial wastes tend to be on the warm side. For the most part, temperature is not a critical issue below 37° C

if wastewaters are to receive biological treatment. It is possible to operate thermophilic biological wastewater-treatment systems up to 65° C with acclimated microbes. Low-temperature operations in northern climates can result in very low winter temperatures and slow reaction rates for both biological treatment systems and chemical treatment systems. Increased viscosity of wastewaters at low temperatures makes solid separation more difficult. Efforts are generally made to keep operating temperatures between 10 and 30° C if possible.

Dissolved Oxygen Oxygen is a critical environmental resource in receiving streams and lakes. Aquatic life requires reasonable dissolved-oxygen (DO) levels. EPA has set minimum stream DO levels at 5 mg/L during summer operations, when the rate of biological metabolism is a maximum. It is important that wastewaters have maximum DO levels when they are discharged and have a minimum of oxygen-demanding components so that DO remains above 5 mg/L. DO is a poorly soluble gas in water, having a solubility around 9.1 mg/L at 20° C and 101.3-kPa (1-atm) air pressure. As the temperature increases and the pressure decreases with higher elevations above sea level, the solubility of oxygen decreases. Thus, DO is a minimum when BOD rates are a maximum. Lowering the temperature yields higher levels of DO saturation, but the biological metabolism rate decreases. Warm-wastewater discharges tend to aggravate the DO situation in receiving waters.

Solids Total solids is the residue remaining from a wastewater dried at 103–105° C. It includes the fractions shown in Fig. 25-43. The first separation is the portion that passes through a 2-μm filter (dissolved) and those solids captured on the filter (suspended). Combustion at 500° C further separates the solids into volatile and ash (fixed) solids. Although ash and volatile solids do not distinguish inorganic from organic solids exactly, due to loss of inorganics on combustion, the volatile fraction is often used as an approximate representation of the organics present. Another type of solids, settleable solids, refers to solids that settle in an Imhoff cone in one hour. Industrial wastes vary substantially in these types of solids and require individual wastewater treatment process analysis. An example of possible variation is given in Table 25-37.

Nutrients and Eutrophication Nitrogen and phosphorus cause significant problems in the environment and require special attention in industrial wastes. Nitrogen, phosphorus, or both may cause aquatic biological productivity to increase, resulting in low dissolved oxygen and eutrophication of lakes, rivers, estuaries, and marine waters. Table 25-38 gives the primary nutrient forms causing problems, while the following equation shows the biological oxidation or oxygen-consuming potential of the most common nitrogen forms.

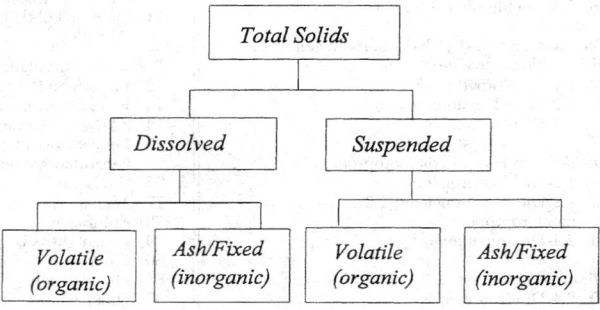

FIG. 25-43 Solids identification. Abbreviations: TS, total solids; SS, suspended solid; D, dissolved; V, volatile.

TABLE 25-37 Solids Variation in Industrial Wastewater

Type of solids	Plating	Pulp and paper
Total	High	High
Dissolved	High	High
Suspended	Low	High
Organic	Low	High
Inorganic	High	High

TABLE 25-38 Nutrient Forms

Parameter	Example
Organic N	protein
Ammonia	NH_3
Ammonium	NH_4^+
Nitrite	NO_2^-
Nitrate	NO_3^-
Organic P	malathion
Ortho-P	PO_4^{3-}
Poly-P	$(PO_4^{3-})_x$

When organics containing reduced nitrogen are degraded, they usually produce ammonium, which is in equilibrium with ammonia. As the pK for $NH_3 \leftrightarrow NH_4^+$ is 9.3, the ammonium ion is the primary form present in virtually all biological treatment systems, as they operate at pH < 8.5 and usually in the pH range of 6.5–7.5. In aerobic reactions, ammonium is oxidized by nitrifying bacteria (nitrosomonas) to nitrite and each mg of NH_4^+—N oxidized will require 3.43 mg D.O. Further oxidation of nitrite by nitrobacter yields nitrate and uses an additional 1.14 mg of D.O. for a total D.O. consumption of 4.57 mg. Thus, organic and ammonium nitrogen can exert significant biochemical oxygen demand in the water environment. This nitrogen demand is referred to as nitrogenous or NBOD, whereas organic BOD is CBOD (carbonaceous). In treatment of wastewaters with organics and ammonium, the total oxygen demand (TOD) may have to be satisfied in accordance with the approximate formula:

$$TOD \cong 1.5\ BOD_5 + 4.5\ (TKN)$$

where TKN (total Kjeldahl nitrogen) = Organic N + NH_4^+—N.

Phosphorus is not oxidized or reduced biologically, but ortho-P may be formed from organic and poly-P. Ortho-P may be removed by chemical precipitation or biologically with sludges and will be covered in a later section.

While many industrial wastes are so low in nitrogen and phosphorus that these must be added if biologically based treatment is to be used, others contain very high levels of these nutrients. For example, paint-production wastes are high in nitrogen, and detergent production wastes are high in phosphorus. Treatment for removal of these nutrients is required in areas where eutrophication is a problem.

Oil and Grease These substances are found in many industrial wastes (i.e., meat packing, petrochemical, and soap production). They tend to float on the water surface, blocking oxygen transfer, interfering with recreation, and producing an aesthetically poor appearance in the water. Measurement is by a solvent extraction procedure. Many somewhat different substances will register as oil and grease in this test. Often oil and grease interfere with other treatment operations, so they must be removed as part of the initial stages of treatment.

WASTEWATER TREATMENT

As indicated above, industrial wastewater contains a vast array of pollutants in soluble, colloidal, and particulate forms, both inorganic and organic. In addition, the required effluent standards are also diverse, varying with the industrial and pollutant class. Consequently, there can be no standard design for industrial water-pollution control. Rather, each site requires a customized design to achieve optimum performance. However, each of the many proven processes for industrial waste treatment is able to remove more than one type of pollutant and is in general applicable to more than one industry. In the sections that follow, waste-treatment processes are discussed more from the broad-based generalized perspective than with narrow specificity. Generally, a combination of several processes is utilized to achieve the degree of treatment required at the least cost.

Much of the experience and data from wastewater treatment has been gained from municipal treatment plants. Industrial liquid wastes are similar to wastewater but differ in significant ways. Thus, typical design parameters and standards developed for municipal wastewater operations must not be blindly utilized for industrial wastewater. It is best to run laboratory and small pilot tests with the specific industrial wastewater as part of the design process. It is most important to understand the temporal variations in industrial wastewater strength, flow, and waste components and their effect on the performance of various treatment processes. Industry personnel in an effort to reduce cost often neglect laboratory and pilot studies and depend on waste characteristics from similar plants. This strategy often results in failure, delay, and increased costs. Careful studies on the actual waste at a plant site cannot be overemphasized.

PRETREATMENT

Many industrial-wastewater streams should be pretreated prior to discharge to municipal sewerage systems or even to a central industrial sewerage system. Pretreatment of individual streams should be considered whenever these streams might have an adverse effect on the total treatment system.

Equalization Equalization is one of the most important pretreatment devices. The batch discharge of concentrated wastes is best suited for equalization. It may be important to equalize wastewater flows, wastewater concentrations, or both. Periodic wastewater discharges tend to overload treatment units. Flow equalization tends to level out the hydraulic loads on treatment units. It may or may not level out concentration variations, depending upon the extent of mixing within the equalization basin. Mechanical mixing may be adequate if the wastes are purely chemical in their reactivity. Biodegradable wastes normally require aeration mixing so that the microbes are kept aerobic and nuisance odors are prevented. Diffused aeration systems offer better mixing under variable load conditions than mechanical surface aeration equipment. Mixing and oxygen transfer are both important with biodegradable wastewaters. Operation on regular cycles determines the size of the equalization basin. There is no advantage in making the equalization basin any larger than necessary to level out wastewater variations. Industrial operation on a 5-day, 40-h week will normally make a 2-day equalization basin as large as needed for continuous operation of the wastewater-treatment system under uniform conditions.

Neutralization Acidic or basic wastewaters must be neutralized prior to discharge. If an industry produces both acidic and basic wastes, these wastes may be mixed together at the proper rates to obtain neutral pH levels. Equalization basins can be used as neutralization basins. When separate chemical neutralization is required, sodium hydroxide is the easiest base material to handle in a liquid form and can be used at various concentrations for in-line neutralization with a minimum of equipment. Yet, lime remains the most widely used base for acid neutralization. Limestone is used when reaction rates are slow and considerable time is available for reaction. Sulfuric acid is the primary acid used to neutralize high-pH wastewaters unless calcium sulfate might be precipitated as a result of the neutralization reaction. Hydrochloric acid can be used for neutralization of basic wastes if sulfuric acid is not acceptable. For very weak basic wastewaters carbon dioxide can be adequate for neutralization.

Grease and Oil Removal Grease and oils tend to form insoluble layers with water as a result of their hydrophobic characteristics. These hydrophobic materials can be easily separated from the water phase by gravity and simple skimming, provided they are not too well mixed with the water prior to separation. If the oils and greases form emulsions with water as a result of turbulent mixing, the emulsions are difficult to break. Separation of oil and grease should be carried out near the point of their mixing with water. In a few instances, air bubbles can be added to the oil and grease mixtures to separate the hydrophobic materials from the water phase by flotation. Chemicals have also been added to help break the emulsions. American Petroleum Institute (API) separators have been used extensively by the petroleum industry to remove oils from wastewaters. The food industries use grease traps to collect the grease prior to its discharge. Unfortunately, grease traps are designed for regular cleaning of the trapped grease. Too often they are allowed to fill up and discharge the excess grease into the sewer or are flushed with hot water and steam to fluidize the grease for easy discharge to the sewer. A grease trap should be designed for a specific volume of grease to be collected over specific time periods. Care should be taken to design the trap so that the grease can easily be removed and properly handled. Neglected or poorly designed grease traps are worse than no grease traps at all.

Toxic Substances   Recent federal legislation has made it illegal for industries to discharge toxic materials in wastewaters. Each industry is responsible for determining if any of its wastewater components are toxic to the environment and to remove them prior to the wastewater discharge. The EPA has identified a number of priority pollutants which must be removed and kept under proper control from their origin to their point of ultimate disposal. Major emphasis has recently been placed on heavy metals and on complex organics that have been implicated in possible cancer production. Pretreatment is essential to reduce heavy metals below toxic levels and to prevent discharge of any toxic organics. Fortunately, toxic organics can ultimately be destroyed by various chemical oxidation systems. Incineration appears to be the most economical method for destroying toxic organics. To make incineration economical, the organics must be kept separated from the dilute wastewaters and treated in their concentrated form. If the heavy metals cannot be reused, they must be concentrated and placed into insoluble materials which will not leach the heavy metals. Toxic substances currently pose the greatest challenge to industries since very little attention has been paid to these materials in the past.

PRIMARY TREATMENT

Wastewater treatment is directed toward removal of pollutants with the least effort. Suspended solids are removed by either physical or chemical separation techniques and handled as concentrated solids.

Screens   Fine screens such as hydroscreens are used to remove moderate-size particles that are not easily compressed under fluid flow. Fine screens are normally used when the quantities of screened particles are large enough to justify the additional units. Mechanically cleaned fine screens have been used for separating large particles. A few industries have used large bar screens to catch large solids that could clog or damage pumps or equipment following the screens.

Grit Chambers   Industries with sand or hard, inert particles in their wastewaters have found aerated grit chambers useful for the rapid separation of these inert particles. Aerated grit chambers are relatively small, with total volume based on 3-min retention at maximum flow. Diffused air is normally used to create the mixing pattern shown in Fig. 25-44, with the heavy, inert particles removed by centrifugal action and friction against the tank walls. The air flow rate is adjusted for the specific particles to be removed. Floatable solids are removed in the aerated grit chamber. It is important to provide for

regular removal of floatable solids from the surface of the grit chamber; otherwise, nuisance conditions will be created. The settled grit is normally removed with a continuous screw and buried in a landfill.

Gravity Sedimentation   Slowly settling particles are removed with gravity sedimentation tanks. For the most part, these tanks are designed on the basis of retention time, surface overflow rate, and minimum depth. A sedimentation tank can be rectangular or circular. The important factor affecting its removal efficiency is the hydraulic flow pattern through the tank. The energy contained in the incoming-wastewater flow must be dissipated before the solids can settle. The wastewater flow must be distributed properly through the sedimentation volume for maximum settling efficiency. After the solids have settled, the settled effluent can be collected without creating serious hydraulic currents that could adversely affect the sedimentation process. Effluent weirs are placed at the end of rectangular sedimentation tanks and around the periphery of circular sedimentation tanks to ensure uniform flow out of the tanks. Once the solids have settled, they must be removed from the sedimentation-tank floor by scraping and hydraulic flow. Conventional sedimentation tanks have sludge hoppers to collect the concentrated sludge and to prevent removal of excess volumes of water with the settled solids. Cross-sectional diagrams of conventional sedimentation tanks are shown in Figs. 25-45 and 25-46.

Design criteria for gravity sedimentation tanks normally provide for 2-h retention based on average flow, with longer retention periods used for light solids or inert solids that do not change during their retention in the tank. Care should be taken that sedimentation time is not too long; otherwise, the solids will compact too densely and affect solids collection and removal. Organic solids generally will not compact to more than 5 to 10 percent. Inorganic solids will compact up to 20 or 30 percent. Centrifugal sludge pumps can handle solids up to 5 or 6 percent, while positive-displacement sludge pumps can handle solids up to 10 percent. With solids above 10 percent the sludge tends to lose fluid properties and must be handled as a semi-solid rather than a fluid. Circular sedimentation tanks have steel truss boxes with angled sludge scrapers on the lower side. As the sludge scrapers rotate, the solids are pushed toward the sludge hopper for removal on a continuous or semicontinuous basis. The rectangular sedimentation tanks employ chain-and-flight sludge collectors or rail-mounted sludge collectors. When floating solids can occur in primary sedimentation tanks, surface skimmers are mounted on the sludge scrapers so that the surface solids are removed at regular intervals.

The surface overflow rate (SOR) for primary sedimentation is normally held close to 40.74 $m^3/(m^2 \cdot day)$ [1000 gal/($ft^2 \cdot day$)] for average flow rates, depending upon the solids characteristics. Lowering the SOR below 40.74 $m^3/(m^2 \cdot day)$ does not produce improved effluent

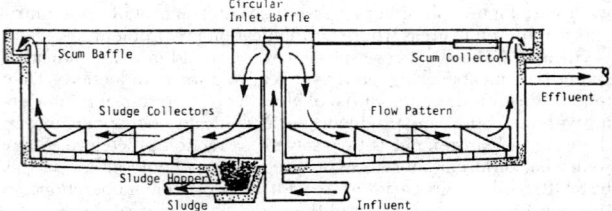

FIG. 25-45   Schematic diagram of a circular sedimentation tank.

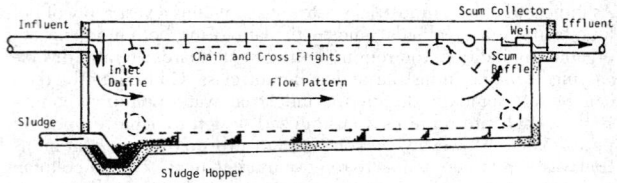

FIG. 25-46   Schematic diagram of a rectangular sedimentation tank.

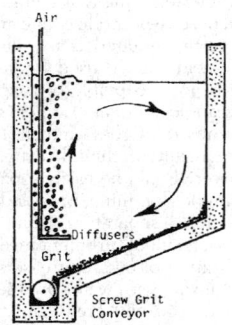

FIG. 25-44   Schematic diagram of an aerated grit chamber.

quality in proportion to the reduction in SOR. Generally, the minimum depth of sedimentation tanks is 3.0 m (10 ft), with circular sedimentation tanks having a minimum diameter of 6.0 m (20 ft) and rectangular sedimentation tanks having length-to-width ratios of 5:1. Chain-and-flight limitations generally keep the width of rectangular sedimentation tanks to increments of 6.0 m (20 ft) or less. While hydraulic overflow rates have been limited on the effluent weirs, operating experience has indicated that the recommended limit of 186 m³/(m·day) [15,000 gal/(ft·day)] is lower than necessary for good operation. A circular sedimentation tank with a single-edge weir provides adequate weir length and is easier to adjust than one with a double-sided weir. More problems appear to be created from improper adjustment of the effluent weirs than from improper length.

Chemical Precipitation   Lightweight suspended solids and colloidal solids can be removed by chemical precipitation and gravity sedimentation. In effect, the chemical precipitate is used to agglomerate the tiny particles into large particles that settle rapidly in normal sedimentation tanks. Aluminum sulfate, ferric chloride, ferrous sulfate, lime, and polyelectrolytes have been used as coagulants. The choice of coagulant depends upon the chemical characteristics of the particles being removed, the pH of the wastewaters, and the cost and availability of the precipitants. While the precipitation reaction results in removal of the suspended solids, it increases the amount of sludge to be handled. The chemical sludge must be considered along with the characteristics of the original suspended solids in evaluating sludge-processing systems.

Normally, chemical precipitation requires a rapid mixing system and a flocculation system ahead of the sedimentation tank. With a rectangular sedimentation tank, the rapid-mixer and flocculation units are added ahead of the tank. With a circular sedimentation tank the rapid-mixer and flocculation units are built into the tank. Schematic diagrams of chemical treatment systems are shown in Figs. 25-47 and 25-48. Rapid mixers are designed to provide 30-s retention at average flow with sufficient turbulence to mix the chemicals with the incoming wastewaters. The flocculation units are designed for slow mixing at 20-min retention. These units are designed to cause the particles to collide and increase in size without excessive shearing. Care must be taken to move the flocculated mixture from the flocculation unit to the sedimentation unit without disrupting the large floc particles.

The parameter used to design rapid mix and flocculation systems is the root mean square velocity gradient G, which is defined by equation

$$G = \left(\frac{P}{VU}\right)^{1/2} \left(\frac{1}{\text{sec}}\right)$$

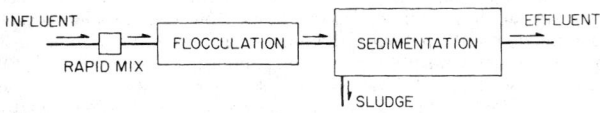

FIG. 25-47 Schematic diagram of a chemical precipitation system for rectangular sedimentation tanks.

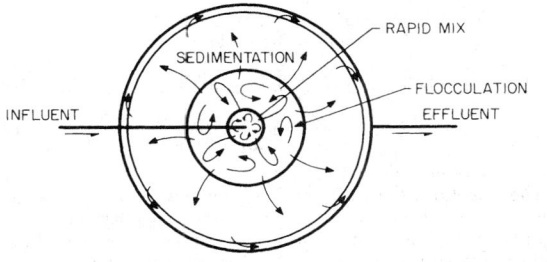

FIG. 25-48 Schematic diagram of a chemical precipitation system for circular sedimentation tanks.

where: P = Power input to the water (ft·lb/sec)
V = Mixer or flocculator volume (ft)³
U = Absolute viscosity of water (lb·sec/ft²)

Optimum mixing usually requires a G value of greater than 1000 inverse seconds. Optimum flocculation occurs when G is in the range 10–100 inverse seconds.

Chemical precipitation can remove 95 percent of the suspended solids, up to 50 percent of the soluble organics and the bulk of the heavy metals in a wastewater. Removal of soluble organics is a function of the coagulant chemical, with iron salts yielding best results and lime the poorest. Metal removal is primarily a function of pH and the ionic state of the metal. Guidance is available from solubility product data.

SECONDARY TREATMENT

Secondary treatment utilizes processes in which microorganisms, primarily bacteria, stabilize waste components. The mixture of microorganisms is usually referred to as biomass. A portion of the waste is oxidized, releasing energy, the remainder is utilized as building blocks of protoplasm. The energy released by biomass metabolism is utilized to produce the new units of protoplasm. Thus, the incentive for the biomass to stabilize waste is that it provides the energy and basic chemical components required for reproduction. The process of biological waste conversion is illustrated by Eq. (25-15).

$$\text{Waste (electron donor)} + \text{Biomass} + \text{Electron acceptor} \xrightarrow[\substack{\text{Proper environmental conditions}}]{} \text{More biomass} + \substack{\text{End products: Oxidized electron donor Reduced electron acceptor}}$$

$$(25-15)$$

As this equation indicates, the waste generally serves as an electron donor, necessitating that an electron acceptor be supplied. A variety of substances can be utilized as electron acceptors, including molecular oxygen, carbon dioxide, oxidized forms of nitrogen, sulfur, and organic substances. The characteristics of the end products of the reaction are determined by the electron acceptor. Table 25-39 is a list of typical end products as a function of the electron acceptor. In general, the end products of this reaction are at a much lower energy level than the waste components, thus resulting in the release of energy referred to above. Although this process is usually utilized for the stabilization of organic substances, it can also be utilized for oxidation of inorganics. For example, biomass-mediated oxidation of iron, nitrogen, and sulfur is known to occur in nature and in anthropogenic processes.

Equation (25-15) describes the biomass-mediated reaction and indicates that proper environmental conditions are required for the reaction to take place. These conditions are required by the biomass, not the electron donor or acceptor. The environmental conditions include pH, temperature, nutrients, ionic balance, and so on. In general, biomass can function over a wide pH range generally from 5 to 9. However, some microbes require a much narrower pH range; i.e., effective methane fermentation requires a pH in the range of 6.5–7.5. It is just as important to maintain a relatively constant pH in the process as it is to stay within the range given above. Microorganisms can function effectively at the extremes of their pH range provided they are given the opportunity to acclimate to these conditions. Continual changes in pH are detrimental, even if the organisms are on the average near the middle of their effective pH range. A similar situa-

TABLE 25-39 Electron Acceptors and End Products for Biological Reactions

Electron acceptors	End product
Molecular oxygen	Water, CO_2, oxidized nitrogen
Oxidized nitrogen	N_2, N_2O, NO, CO_2, H_2O
Oxidized sulfur	H_2S, S, CO_2, H_2O
CO_2, acetic acid, formic acid	CH_4, CO_2, H_2
Complex organics	H_2, simple organics, CO_2, H_2O

tion prevails for temperature. Most organisms can function well over a broad range of temperature but do not adjust well to frequent fluctuations of even a few degrees. There are three major temperature ranges in which microorganisms function. The psychrophilic range (5° C to 20° C), the mesophilic range (20° C to 45° C), and the thermophilic range (45° C to 70° C). In general the microbes that function in one of these temperature ranges cannot function efficiently in the other ranges. As it is generally uneconomical to adjust the temperature of a waste, most processes are operated in the mesophilic range. If the normal temperature of the waste is above or below the mesophilic range, the process will be operated in the psychrophilic or thermophilic range as appropriate. However, occasionally the temperature of the waste is altered to improve performance. For example, some anaerobic treatment processes are operated under thermophilic conditions, even though the waste must be heated to achieve this temperature range. This is carried out in order to speed up the degradation of complex organics and/or to achieve kill of mesophilic pathogens. It should be noted that any time the biological operation of a process moves away from its optimum or most effective range, be it pH, temperature, nutrients, or what have you, the rate of biological processing is reduced.

All microorganisms require varying amounts of a large number of nutrients. These are required because they are necessary components of bacterial protoplasm. The nutrients can be divided into three groups: macro, minor, and micro. The macronutrients are those that comprise most of the biomass. These are given by the commonly accepted formula for biomass ($C_{60}H_{87}O_{23}N_{12}P$). The carbon, hydrogen, and oxygen are normally supplied by the waste and water, but the nitrogen and phosphorous must often be added to industrial wastes to ensure that a sufficient amount is present. A good rule is that the mass of nitrogen should be at least 5 percent of the BOD, and the mass of phosphorous should be at least 20 percent of the mass of nitrogen. One of the major operational expenses is the purchase of nitrogen and phosphorous for addition to biologically based treatment processes. The quantities of nitrogen and phosphorous referred to above as required are actually in excess of the minimum amounts needed. The actual amount required depends upon the quantity of excess biomass wasted from the system and the amount of N and P available in the waste. This will be expanded upon later in this section. The minor nutrients include the typical inorganic components of water. These are given in Table 25-40. The range of concentrations required in the wastewater for the minor nutrients is 1–100 mg/L. The micronutrients include the substances that we normally refer to as trace metals and vitamins. It is interesting to note that the trace metals include virtually all of the toxic heavy metals. This reinforces the statement made above that toxicity is a function of concentration and not an absolute parameter. Whether or not the substances referred to as vitamins will be required depends upon the type of microorganisms required to stabilize the waste materials. Many microorganisms have the ability to make their own vitamins from the waste components; thus, a supplement is not needed. However, occasionally the addition of an external source of vitamins is essential to the success of a biologically based waste-treatment system. In general, the trace nutrients must be present in a waste at a level of a few micrograms per liter.

One aspect of the basic equation describing biological treatment of waste that has not been referred to previously is that biomass appears on both sides of the equation. As was indicated above, the only reason that microorganisms function in waste-treatment systems is because it enables them to reproduce. Thus, the quantity of biomass in a waste-treatment system is higher after the treatment process than before it.

TABLE 25-40   Minor and Micro Nutrients Required for Biologically Mediated Reactors

Minor 1–100 mg/L
Sodium, potassium, calcium, magnesium, iron, chloride, sulfate
Micro 1–100 µg/L
Copper, cobalt, nickel, manganese, boron, vanadium, zinc, lead, molybdenum, various organic vitamins, various amino acids

This is favorable in that there is a continual production of the organisms required to stabilize the waste. Thus, one of the major reactants is, in effect, available free of charge. However, there is an unfavorable side in that unless some organisms are wasted from the system, an excess level will build up, and the process could choke on organisms. The wasted organisms are referred to as sludge. A major cost component of all biologically based processes is the need to provide for the ultimate disposal of this sludge.

Biologically based treatment processes probably account for the majority of the treatment systems used for industrial waste management because of their low cost and because most substances are amenable to biological breakdown. However, some substances are difficult to degrade biologically. Unfortunately, it is not possible at present to predict a priori the biodegradability of a specific organic compound; rather, we must depend upon experience and testing. The collective experience of the field has been put into compendia by EPA in a variety of documents. However, these data are primarily qualitative. There have been some attempts to develop a system of prediction of biodegradability based on a number of compound parameters such as solubility, presence or absence of certain functional groups, compound polarity, and so on. Unfortunately, none of these systems has advanced to the point where reliable quantitative predictions are possible. Another complication is that some organics that are easily biodegradable at low concentration exert a toxic effect at high concentration. Thus, literature data can be confusing. Phenol is a typical compound that shows ease of biodegradation when the concentration is below 500 mg/L but poor biodegradation at higher concentrations. Another factor affecting both biodegradation and toxicity is whether or not a substance is in solution. In general, if a substance is not in solution, it is not available to affect the biomass. Thus, the presence of a waste in substances that can precipitate, complex, or absorb other waste components can have a significant effect on reports of biodegradability and/or toxicity. A quantitative estimate of toxicity can be obtained in terms of the change in kinetic parameters of a system. These kinetic parameters are discussed below.

Design of Biological Treatment Systems   In the past, the design of biologically based waste-treatment systems has been derived from rules of thumb. During the past two decades, however, a more fundamental system has been developed and is presently widely used to design such systems. This system is based upon a fundamental understanding of the kinetics and stoichiometry of biological reactions. The system is codified in terms of equations in which four pseudo constants appear. These pseudo constants are k_m, the maximum substrate utilization rate (1/time); K_s, the half maximal velocity concentration (mg/L); Y, the yield coefficient; and b, the endogenous respiration rate (1/time). These are referred to as pseudo constants because, in the mathematical manipulation of the equations in which they appear, they are treated as constants. However, the value of each is a function of the nature of the microbes, the pH, the temperature, and the components of the waste. It is important to remember that if any of these change, the value of the pseudo constants may change as well. The kinetics of biological reactions are described by Eq. (25-16).

$$\frac{dS}{dt} = \frac{k_m S X}{K_s + S} \qquad (25\text{-}16)$$

where t = time (days)
   S = waste concentration (mg/L)
   X = biomass concentration (mg/L)
   k_m = (mg/L substrate ÷ mg/L biomass) – time

The accumulation or growth of biosolids is given by Eq. (25-17).

$$\frac{dX}{dt} = Y \frac{dS}{dt} - bX \qquad (25\text{-}17)$$

where X = biomass level (mg/L).

The equations that have been developed for design using these pseudo constants are based on steady-state mass balances of the biomass and the waste components around both the reactor of the system and the device used to separate and recycle microorganisms. Thus, the equations that can be derived will be dependent upon the characteristics of the reactor and the separator. It is impossible here to present equations for all the different types of systems. As an illustration, the equations for a common system (a complete mix – stirred tank reactor with recycle) are presented below.

$$S_e = \frac{K_s[1 + \text{BSRT}(b)]}{\text{BSRT}(YK_m - b) - 1} \qquad (25\text{-}18)$$

$$X = \frac{(\text{BSRT})(Y)(S_o - S_e)}{\text{HRT}[1 + (b)\text{BSRT}]} \qquad (25\text{-}19)$$

where: S_e = influent waste concentration (mg/L)
         S_o = treated waste concentration (mg/L)
         HRT = reactor hydraulic retention time

Note that these equations predict some unexpected results. The strength of the untreated waste (S_o) has no effect on the strength of the treated waste (S_e). Neither does the size of the reactor (HRT). Rather, a parameter referred to as the biomass solids retention time (BSRT) is the key parameter determining the system performance. This is illustrated in Fig. 25-49. As the BSRT increases, the concentration of untreated waste in the effluent decreases irrespective of the reactor size or the waste strength. However, the reactor size and waste strength have a significant effect on the level of biomass (X) that is maintained in the system at steady state. Since the development of an excess level of biomass in the system can lead to system upset, it is important to take cognizance of the waste strength and the reactor size. But treatment performance with respect to removal of the waste components is again a function only of BSRT and the value of the pseudo constants referred to above. The BSRT also has an effect on the level of biomass in the system and the quantity of excess biomass produced (Fig. 25-50). The latter will determine the quantity of waste sludge which must be dealt with as well as the N and P requirement. The N and P in the biomass removed from the system each day must be replaced. From the formula given previously for biomass, N is 12 percent and P is 2.3 percent by weight in biomass grown under ideal conditions (see Fig. 25-50). Also, note that m, the minimum BSRT, is the minimum time required for the microorganisms to double in mass. Below this minimum, washout occurs and substrate removal approaches zero.

Although identical results are not obtained for other reactor configurations, the design equations yield similar patterns. The dominant parameters in determining system performance are again the BSRT and the pseudo constants. The latter are not under the control of the design engineer as they are functions of the waste and the microorganisms that developed in the system. The BSRT is the major design parameter under the control of the design engineer. This parameter has been defined as the ratio of the biomass in the reactor to the biomass produced from the waste each day. At steady state, the level of biomass in the system is constant; thus, the biomass produced must equal the biomass wasted. The minimum BSRT that can be utilized is that which will produce the degree of treatment required. Generally,

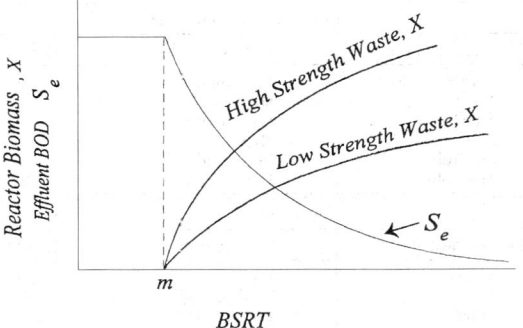

FIG. 25-49   Effect of BSRT on biological treatment process performance. m = minimum BSRT.

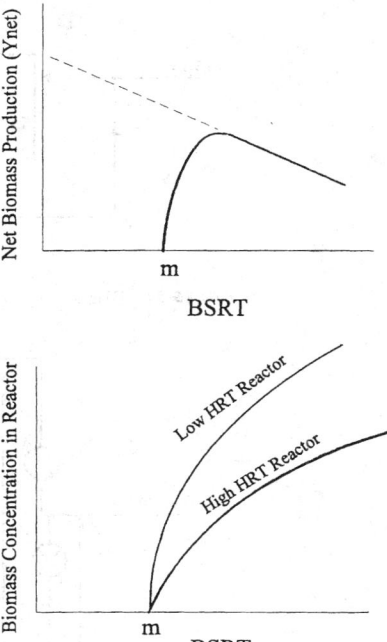

FIG. 25-50   Biomass production.

this minimum value is less than the range of BSRT used in design and operation. This provides not only a safety factor, but, in addition, it is necessary in order to foster the growth of certain favorable groups of microorganisms in the system. Generally, BSRT values in the range of 3–15 days are utilized in most systems, although BSRT values of less than 1 day for high rate systems are usually adequate to ensure greater than 95 percent destruction of waste components in all but anaerobic systems. For the latter, a minimum BSRT of 8 days is needed for 95 percent destruction. BSRT is controlled by the concentration of biomass (X) in the system and the quantity wasted each day; thus, the treatment plant operator can alter BSRT by altering the rate of biomass wasting.

Reactor Concepts   A large number of reactor concepts seem to be used in biologically based treatment systems. However, this diversity is more apparent than real. In reality, there are only two major reactor types that are used. One is referred to as a suspended growth reactor, the other is referred to as a fixed film reactor. In the former, the waste and the microorganisms move through the reactor, with the microorganisms constantly suspended in the flow. After exiting the reactor, the suspension flows through a separator, which separates the organisms from the liquid. Some of the organisms are wasted as sludge, while the remainder are returned to the reactor. The supernatant is discharged either to the environment or to other treatment units (Fig. 25-51). The functioning of the separator is very important, as poor performance of the separator will result in high solids in the effluent and reduction of organisms in the recycle and eventually also in the reactor. As indicated above, the level of the BSRT used in design of suspended growth systems is set to produce a biomass of the proper level in the system; i.e., a level at which waste degradation is rapid but not so high that excess loading on the separator will occur. BSRT also influences the ability of biomass to self flocculate and thus be removed in the separator (usually a clarifier).

In the fixed-film reactor, the organisms grow on an inert surface that is maintained in the reactor. The inert surface can be granular material, proprietary plastic packing, rotating discs, wood slats, mass-transfer packing, or even a sponge-type material. The reactor can be flooded or have a mixed gas-liquid space (Fig. 25-52). The biomass level on the

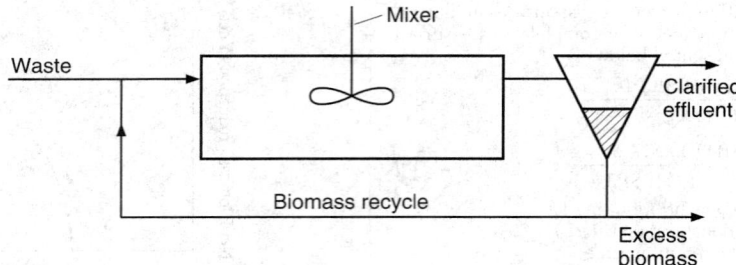

FIG. 25-51   Diagram of a suspended growth system.

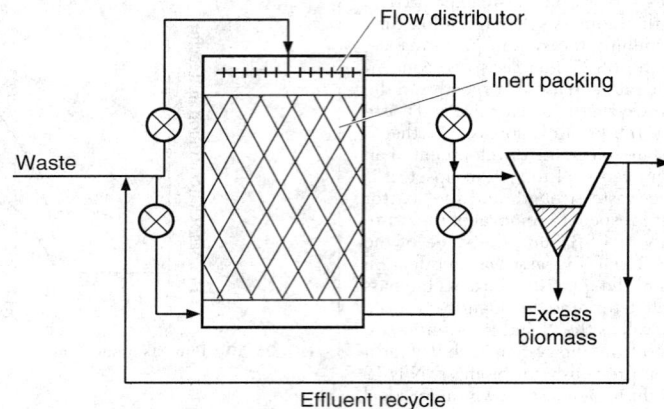

FIG. 25-52   Diagram of a fixed-film system.

packing is controlled by hydraulic scour produced as the waste liquid flows through the reactor. There is no need of a separator to ensure that the biomass level in the reactor is maintained. The specific surface area of the packing is the design parameter used to ensure an adequate level of biomass. However, a separator is usually supplied to capture biomass washed from the packing surface by the flow of the waste, thus providing a clarified effluent. As with the suspended growth system, if this biomass escapes the system, it will result in a return to the environment of organic laden material, thus negating the effectiveness of the waste treatment. As indicated above, biomass level in a fixed film reactor is maintained by a balance between the rate of growth on the packing (which is a function of the strength of the waste, the yield coefficient and the BSRT), and the rate of hydraulic flushing by the waste flow. Recycle of treated wastewater is used to control the degree of hydraulic flushing. Thus, typical design parameters for fixed film reactors include both an organic loading and a hydraulic loading. Details on the typical levels for these will be given in a later section.

The reactor concepts described above can be utilized with any of the electron acceptor systems previously discussed, although some reactor types perform better with specific electron acceptors. For example, suspended growth systems are generally superior to fixed film systems when molecular oxygen is the electron acceptor because it is easier to supply oxygen to suspended growth systems.

On the other hand, fixed film systems are superior for systems in which nitrate is the electron acceptor because the nitrogen bubbles produced in suspended growth systems tend to float sludge to the top of the clarifiers used as biomass separators. Another factor affecting reactor type and electron acceptor selection is waste strength. The higher the waste strength, the higher the level of biomass in the reactor and vice versa. When waste strength is low, fixed film reactors are favored as it is easier to maintain an adequate biomass level. When waste strength is high, it is better to use anaerobic electron acceptors than aerobic because oxygen supply is not limiting for anaerobic sys-

tems. In addition, the yield coefficient Y is much lower for anaerobic systems, so the probability of excessive biomass development in the system is much less. Table 25-41 summarizes the advantages and disadvantages of various combinations of reactor type, electron acceptor, and waste strength.

Recently, a new concept in fixed film reactors that uses an expanded or fluidized bed of particles as the biomass support medium has been introduced. This reactor type can easily handle both low- and high-strength wastes with most electron acceptors. It will be discussed in detail in a later section.

TABLE 25-41   Favorable (F) and Unfavorable (U) Combinations of Electron Acceptor, Waste Strength, and Reactor Type

Suspended growth reactor		
Electron acceptor	Waste strength	Condition
Aerobic	Low–modest	F
Aerobic	High	U
Anoxic	Low–modest	F
Anoxic	High	U
Anaerobic	Low–modest	U
Anaerobic	High	F

Fixed film reactor		
Electron acceptor	Waste strength	Condition
Aerobic	Low	F
Aerobic	Modest–high	U
Anoxic	Low–modest	F
Anoxic	High	U
Anaerobic	Low–modest	F
Anaerobic	High	F

Determination of Kinetic and Stoichiometric Pseudo Constants As indicated above, these parameters are most important for predicting the performance of biologically based treatment systems. It would be ideal if tabulations of these were available for various industrial wastes as a function of pH temperature and nutrient levels. Unfortunately, little reliable data has been codified. Only certain trends have been established, and these are primarily the result of studies on municipal wastewater. For example, the yield coefficient Y has been shown to be much higher for systems that are aerobic (molecular oxygen as the electron acceptor) than for anaerobic systems (sulfate or carbon dioxide as the electron acceptors). Systems where oxidized nitrogen is the electron acceptor (termed *anoxic*) exhibit yield values intermediate between aerobic and anaerobic systems. The endogenous respiration rate is higher for aerobic and anoxic systems than anaerobic systems. However, no trends have been established for values of the maximum specific growth rate or the half maximal velocity concentration. Thus, the values applicable to a specific waste must be determined from laboratory studies.

The laboratory studies utilized small-scale (1–5-L) reactors. These are satisfactory because the reaction rates observed are independent of reactor size. Several reactors are operated in parallel on the waste, each at a different BSRT. When steady state is reached after several weeks, data on the biomass level (X) in the system and the untreated waste level in the effluent (usually in terms of BOD or COD) are collected. These data can be plotted for equation forms that will yield linear plots on rectangular coordinates. From the intercepts and the slope of the lines, it is possible to determine values of the four pseudo constants. Table 25-42 presents some available data from the literature on these pseudo constants. Figure 25-53 illustrates the procedure for their determination from the laboratory studies discussed previously.

Activated Sludge This treatment process is the most widely used aerobic suspended growth reactor system. It will consistently produce a high-quality effluent (BOD$_5$ and SS of 20–30 mg/L). Operational costs are higher than for other secondary treatment processes primarily because of the need to supply molecular oxygen using energy-intensive mechanical aerator- or sparger-type equipment. Removal of soluble organics, colloidal, particulates, and inorganics are achieved in this system through a combination of biological metabolism, adsorption, and entrapment in the biological floc. Indeed, many pollutants that are not biologically degradable are removed during activated sludge treatment by adsorption or entrapment by the floc. For example, most heavy metals form hydroxide or carbonate precipitates under the pH conditions maintained in activated sludge, and most organics are easily adsorbed to the surface of the biological floc. A qualitative guide to the latter is provided by the octanol-water partition coefficient of a compound.

All activated sludge systems include a suspended growth reactor in which the wastewater, recycled sludge, and molecular oxygen are mixed. The latter must be dissolved in the water; thus the need for an energy-intensive pure oxygen or air supply system. Usually, air is the source of the molecular oxygen rather than pure oxygen. Energy for mixing of the reactor contents is supplied by the aeration equipment. All systems include a separator and pump station for sludge recycle and sludge wasting. The separator is usually a sedimentation tank that is designed to function as both a clarifier and a thickener. Many modifications of the activated sludge process have been developed over

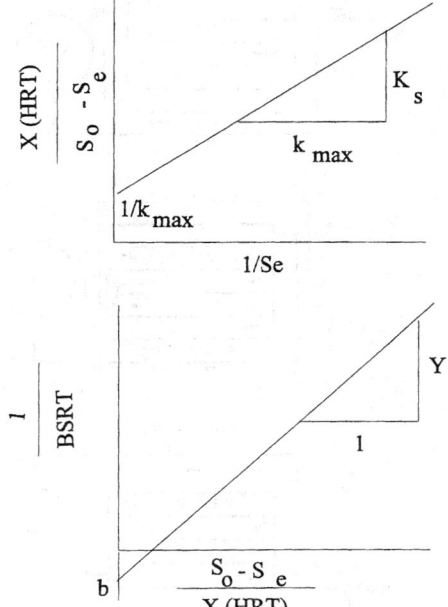

FIG. 25-53 Determination of pseudo-kinetic constants.

the years and are described below. Most of these involve differences in the way the reactor is compartmentalized with respect to introduction of waste, recycle, and/or oxygen supply.

Modifications The modifications of activated sludge systems offer considerable choice in processes. Some of the most popular modifications of the activated sludge process are illustrated in Fig. 25-54. Conventional activated sludge uses a long narrow reactor with air supplied along the length of the reactor. The recycle sludge and waste are introduced at the head end of the reactor producing a zone of high waste to biomass concentration and high oxygen demand. This modification is used for relatively dilute wastes such as municipal wastewater. Step aeration systems distribute the waste along the length of the aerator, thus reducing the oxygen demand at the head end of the reactor and spreading the oxygen demand more uniformly over the whole reactor. In the complete mix system, the waste and sludge recycle are uniformly distributed over the whole reactor, resulting in the waste load and oxygen demand being uniform in the entire reactor. Complete-mixing activated sludge is the most popular system for industrial wastes because of its ability to absorb shock loads better than other modifications. Contact stabilization is a modification of activated sludge that is best suited to wastewaters having high suspended solids and low soluble organics. Contact stabilization employs a short-term mixing tank to adsorb the suspended solids and metabolize the soluble organics, a sedimentation tank for solids separation,

TABLE 25-42 Typical of Values for Pseudo Constants

Biomass type	Substrate	k_{max}, $\dfrac{\text{mg substrate}}{\text{mg biomass·day}}$	Y, $\dfrac{\text{mg biomass}}{\text{mg substrate}}$	K_s, $\dfrac{\text{mg substrate}}{\text{liter}}$	b, $\dfrac{1}{\text{day}}$	Remarks
Mixed culture	Sewage (COD)	5–10	0.5	50	0.05	Aerobic 20°C
Mixed culture	Glucose (COD)	7.5	0.6	10	0.07–.1	Aerobic 20°C
Mixed culture	Skim milk (COD)	5	—	100	0.05	Aerobic 20°C
Mixed culture	Soybean waste (COD)	12	—	355	0.144	Aerobic 20°C
Methane bacteria	Acetic acid (COD)	8.7	.04	165	0.035	35°C
Anaerobic mixed	Propanoic acid (COD)	7.7	.04	60	0.035	35°C
Anaerobic mixed	Sewage sludge (COD)	6.7	.04	1,800	0.03	35°C
Anoxic	NO$_3$ as N	0.375	0.8	0.1	0.04	Methanol feed
Aerobic	NH$_4^+$ as N	5	0.2	1.4	.05	20°C

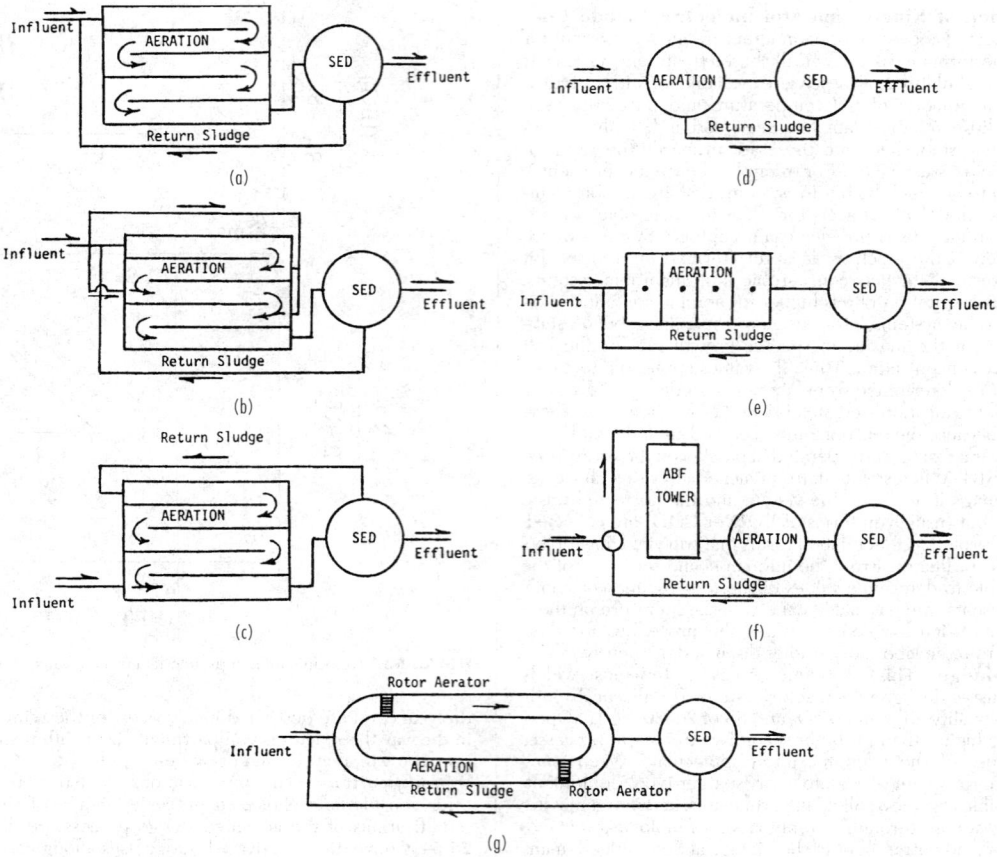

FIG. 25-54 Schematic diagrams of various modifications of the activated-sludge process. (*a*) Conventional activated sludge. (*b*) Step aeration. (*c*) Contact stabilization. (*d*) Complete mixing. (*e*) Pure oxygen. (*f*) Activated biofiltration (ABF). (*g*) Oxidation ditch.

and a reaeration tank for stabilization of the suspended organics. Extended-aeration systems are actually long-term-aeration, completely mixed activated-sludge systems. They employ 24- to 48-h aeration periods and high mixed-liquor suspended solids to provide complete stabilization of the organics and aerobic digestion of the activated sludge in the same aeration tank. The oxidation ditch is a popular form of the extended-aeration system employing mechanical aeration. Pure-oxygen systems are designed to treat strong industrial wastes in a series of completely mixed units having relatively short contact periods. One of the latest modifications of activated sludge employs powdered activated carbon to adsorb complex organics and assist in solids separation. Another modification employs a redwood-medium trickling filter ahead of a short-term aeration tank with mixed liquor recycled over the redwood-medium tower to provide heavy microbial growth on the redwood as well as in the aeration tank.

As indicated previously, success with the activated sludge process requires that the biomass have good self-flocculating properties. Significant research effort has been expended to determine the conditions that favor the development of good settling biomass cultures. These have indicated that nutritional deficiency, levels of dissolved oxygen between 0 to 0.5 mg/L, and pH values below 6.0 will favor the predomination of filamentous biomass. Filamentous organisms settle and compact poorly and thus are difficult to separate from liquid. By avoiding the above conditions, and with the application of selector technology, the predomination of filaments in the biomass can be eliminated. A selector is often a short contact (15–30 min) reactor set ahead of the main activated sludge reactor. All of the recycle sludge

and all of the waste are routed to the selector. In the selector, either a high rate of aeration (aerobic selector) is used to keep the dissolved oxygen above 2 mg/L, or no aeration occurs (anaerobic selector) so that the dissolved oxygen is zero. Because filamentous organisms are microaerophilic, they cannot predominate when the dissolved oxygen level is zero or high. The anaerobic selector not only selects in favor of nonfilamentous biomass but also fosters luxury uptake of phosphorous and is used in systems where phosphorous removal is desired.

Aeration Systems   These systems control the design of aeration tanks. Aeration equipment has two major functions: mixing and oxygen transfer. Diffused-aeration equipment employs either a fixed-speed positive-displacement blower or a high-speed turbine blower for readily adjustable air volumes. Air diffusers can be located along one side of the aeration tank or spread over the entire bottom of the tank. They can be either fine-bubble or coarse-bubble diffusers. Fine-bubble diffusers are more efficient in oxygen transfer but require more extensive air-cleaning equipment to prevent them from clogging as a result of dirty air. Mechanical-surface-aeration equipment is more efficient than diffused-aeration equipment but is not as flexible. Economics has dictated the use of large-power aerators, but tank configuration has tended to favor the use of greater numbers of lower-power aerators. Oxidation ditches use horizontal rotor-type aerators. Mixing is a critical problem with mechanical-surface aerators since they are a point-source pump of limited capacity. Experience has indicated that bearings are a serious problem with mechanical-aeration equipment. Wave action generated within the aeration tank tends to produce lateral stresses on the bearings and has resulted in failures and increased maintenance costs. Slow-speed mechanical-surface-aeration units present fewer problems than the high-speed mechanical-surface-aeration units. Deep tanks, greater than 3.0 m (10 ft), require draft tubes to ensure proper hydraulic flow through the aeration tank. Short-circuiting is one of the major problems associated with mechanical aeration equipment. Combined mechanical- and diffused-aeration systems have enjoyed some popularity for industrial-waste systems that treat variable organic loads. The mechanical mixers provide the fluid mixing with the diffused aeration varied for different oxygen-transfer rates.

Diffused-aeration systems transfer from 20 to 40 mg/(L O₂·h). Combined mechanical- and diffused-aeration systems can transfer up to 65 mg/(L O₂·h), while mechanical-surface aerators can provide up to 90 mg/(L O₂·h). Pure-oxygen systems can provide the highest oxygen-transfer rate, up to 150 mg/(L O₂·h). Aeration equipment must provide sufficient oxygen to meet the peak oxygen demand; otherwise, the system will fail to provide proper treatment. For this reason, the peak oxygen demand and the rate of transfer for the desired equipment determine the size of the aeration tank in terms of retention time. Economics dictates a balance between the size of the aeration tank and the size of the aeration equipment. As the cost of power increases, economics will favor constructing a larger aeration tank and smaller aerators. It is equally important to examine the hydraulic flow pattern around each aerator to ensure maximum efficiency of oxygen transfer. Improper spacing of aeration equipment can waste energy.

There is no standard aeration-tank shape or size. Aeration tanks can be round, square, or rectangular. Shallow aeration tanks are more difficult to mix than deeper tanks. Yet aeration-tank depths have ranged from 0.6 m (2 ft) to 18 m (60 ft). The oxidation-ditch systems tend to be shallow, while some high-rate diffused-aeration systems have used very deep tanks to provide more efficient oxygen transfer.

Regardless of the aeration equipment employed, oxygen-transfer rates must provide from 0.6 to 1.4 kg of oxygen/kg BOD₅ (0.6 to 1.4 lb oxygen/lb BOD₅) stabilized in the aeration tank for carbonaceous-oxygen demand. Nitrogen oxidation can increase oxygen demand at the rate of 4.3 kg (4.3 lb) of oxygen/kg (lb) of ammonia nitrogen oxidized. At low oxygen-transfer rates more excess activated sludge must be removed from the system than at high oxygen-transfer rates. Here again the economics of sludge handling must be balanced against the cost of oxygen transfer. The quantity of waste activated sludge will depend upon wastewater characteristics. The inert suspended solids entering the treatment system must be removed with the excess activated sludge. The soluble organics are stabilized by converting a portion of the organics into suspended solids, producing from 0.3 to 0.8 kg (0.3 to 0.8 lb) of volatile suspended solids/kg (lb) of BOD₅ stabilized. Biodegradable suspended solids in the wastewaters will result in destruction of the original suspended solids and their conversion to a new form. Depending upon the chemical characteristics of the biodegradable suspended solids, the conversion factor will range from 0.7 to 1.2 kg (0.7 to 1.2 lb) of microbial solids produced/kg (lb) of suspended solids destroyed. If the suspended solids produced by metabolism are not wasted from the system, they will eventually be discharged in the effluent. While considerable efforts have been directed toward developing activated-sludge systems which totally consume the excess solids, no such system has proved to be practical. The concept of total oxidation of excess sludge is fundamentally unsound and should be recognized as such.

A definitive determination of the waste sludge production and the oxygen requirement can be obtained using the pseudo constants referred to previously. The ultimate BOD in the waste will be accounted for by the sum of the oxidation and sludge synthesis (Eq. 25-20).

Thus:

Waste BOD$_u$ = Oxygen required + BOD$_u$ of sludge produced   (25-20)

where: BOD$_U$ of sludge = $(Y_{net})(1.42)$(Waste BOD$_U$)

$$Y_{net} = \frac{Y}{1 + b(\text{BSRT})}$$

1.42 = Factor converting biomass to oxygen units; i.e.,
BOD$_u$ ≈ COD

Sedimentation Tanks These tanks are an integral part of any activated-sludge system. It is essential to separate the suspended solids from the treated liquid if a high-quality effluent is to be produced. Circular sedimentation tanks with various types of hydraulic sludge collectors have become the standard secondary sedimentation system. Square tanks have been used with common-wall construction for compact design with multiple tanks. Most secondary sedimentation tanks use center-feed inlets and peripheral-weir outlets. Recently, efforts have been made to employ peripheral inlets with submerged-orifice flow controllers and either center-weir outlets or peripheral-weir outlets adjacent to the peripheral-inlet channel.

Aside from flow control, basic design considerations have centered on surface overflow rates, retention time, and weir overflow rate. Surface overflow rates have been slowly reduced from 33 m³/(m²·day) [800 gal/(ft²·day)] to 24 m³/(m²·day) to 16 m³/(m²·day) [600 gal/(ft²·day) to 400 gal/(ft²·day)] and even to 12 m³ (m²·day) [300 gal/(ft²·day)] in some instances, based on average raw-waste flows. Operational results have not demonstrated that lower surface overflow rates improve effluent quality, making 33 m³/(m²·day) [800 gal/(ft²·day)] the design choice in most systems. Retention time has been found to be an important design factor, averaging 2 h on the basis of raw-waste flows. Longer retention periods tend to produce rising sludge problems, while shorter retention periods do not provide for good solids separation with high-return sludge flow rates. Effluent-weir overflow rates have been limited to 186 m³/(m·day) [15,000 gal/(ft·day)] with a tendency to reduce the rate to 124 m³/(m·day) [10,000 gal/(ft·day)]. Lower effluent-weir overflow rates are obtained by using dual-sided effluent weirs cantilevered from the periphery of the tank. Unfortunately, proper adjustment of dual-side effluent weirs has created more hydraulic problems than the weir overflow rate. Field data have shown that effluent quality is not really affected by weir overflow rates up to 990 m³/(m·day) [80,000 gal/(ft·day)] or even 1240 m³/(m·day) [100,000 gal/(ft·day)] in a properly designed sedimentation tank. A single peripheral weir, being easy to adjust and keep clean, appears to be optimal for secondary sedimentation tanks from an operational point of view.

Depth tends to be determined from the retention time and the surface overflow rate. As surface overflow rates were reduced, the depth of sedimentation tanks was reduced to keep retention time from being excessive. It was recognized that depth was a valid design parameter and was more critical in some systems than retention time. As mixed-liquor suspended-solids (MLSS) concentrations increase, the depth should also be increased. Minimum sedimentation-tank depths for variable operations should be 3.0 m (10 ft) with depths to 4.5 m (15 ft) if 3000 mg/L MLSS concentrations are to be maintained under variable hydraulic conditions. With MLSS concentrations above 4000 mg/L, the depth of the sedimentation tank should be increased to 6.0 m (20 ft). The key is to keep a definite freeboard over the settled-sludge blanket so that variable hydraulic flows do not lift the solids over the effluent weir.

Scum baffles around the periphery of the sedimentation tank and radial scum collectors are standard equipment to ensure that rising solids or other scum materials are removed as quickly as they form. Hydraulic sludge-collection tubes have replaced the center sludge well, but they have caused a new set of operational problems. These tubes were designed to remove the settled sludge at a faster rate than conventional sludge scrapers. To obtain good hydraulic distribution in the sludge-collection tubes, it was necessary to increase the rate of return sludge flow and decrease the concentration of return sludge. The higher total inflow to the sedimentation tank created increased forces that lifted the settled-solids blanket at the wall, causing loss of excessive suspended solids and lower effluent quality. Operating data tend to favor conventional secondary sedimentation tanks over hydraulic sludge-collection systems. Return-sludge rates normally range from 25 to 50 percent for MLSS concentrations up to 3300 mg/L. Most return-sludge pumps are centrifugal pumps with capacities up to 100 percent raw-waste flow.

Gravity settling can concentrate activated sludge to 10,000 mg/L, but hydraulic sludge-collecting tubes tend to operate best below 8,000 mg/L. The excess activated sludge can be wasted either from the return sludge or from a separate waste-sludge hopper near the center of the tank. The low solids concentrations result in large volumes of waste activated sludge in comparison with primary sludge. Unfortunately, the physical characteristics of waste activated sludge prevent significant concentration without the expenditure of considerable energy. Gravity thickening can produce 2 percent solids, while air flotation can produce 4 percent solids concentration. Centrifuges are able to concentrate activated sludge from 10 to 15 percent solids, but the capture is limited. Vacuum filters can equal the performance of centrifuges if the sludge is chemically conditioned. Filter presses and belt-press filters can produce cakes with 15 to 25 percent solids. It is very important that the excess activated sludge formed in the aeration tanks be wasted on a regular basis; otherwise, effluent quality will deteriorate. Care should be taken to ensure that sludge-thickening systems do not control activated-sludge operations. Alternative sludge-handling provisions should be available during maintenance on sludge-thickening equipment. At no time should final sedimentation tanks be used for the storage of sludge beyond that required by daily operational variations.

Anaerobic/Anoxic Activated Sludge   The activated sludge concept (i.e., suspended growth reactor) can be used for anaerobic or anoxic systems in which no oxygen or air is added to the reactor. An anoxic activated sludge is used for systems in which removal of nitrate is a goal or where nitrate is used as the electron acceptor. These systems (denitrification) will be successful if the nitrate is reduced to low levels in the reactor so that nitrate reduction to nitrogen gas does not take place in the clarifier-thickener. Nitrogen gas production in the clarifier will result in escape of biological solids with the effluent as nitrogen bubbles floating sludge to the clarifier surface. For nitrogen reduction, a source of organics (electron donor) is required. Any inexpensive carbohydrate can be effectively used for nitrate removal. Many systems utilize methanol as the donor because it is rapidly metabolized; others use the organics in sewage in order to reduce chemical costs. Nitrate reduction is invariably used as part of a system in which organics and nitrogen removal are goals. In such systems, the nitrogen in the waste is first oxidized to nitrate and then reduced to nitrogen gas. Figure 25-55 presents some flow sheets for such

TABLE 25-43   Design Parameter for Nitrogen Removal

System	BSRT, d	HRT, h	X, mg/L	pH
Carbon removal†	2–5	1–3	1000–2000	6.5–8.0
Nitrification†	10–20	0.5–3	1000–2000	7.4–8.6
Denitrification†	1–5	0.2–2	1000–2000	6.5–7.0
Internal recycle°	10–40	8–20	2000–4000	6.5–8.0

°Total for all reactors
†Separate reactors

systems. Table 25-43 presents some design parameters for nitrate-reduction-activated sludge systems. Anaerobic activated sludge has been used for strong industrial wastes high in degradable organic solids. In these systems, a high rate of gasification takes place in the sludge separator so that a highly clarified effluent is usually not obtained. A vacuum degasifier is incorporated in such systems to reduce solids loss. Such systems have been used primarily with meat packing wastes that are warm, high in BOD and yield a high level of bicarbonate buffer as a result of ammonia release from protein breakdown. All of these conditions favor anaerobic processing. The use of this process scheme provides a high BSRT (15–30 days), usually required for anaerobic treatment, at a low HRT (1–2 days).

Lagoons   Lagoons are low-cost, easy-to-operate wastewater-treatment systems capable of producing satisfactory effluents. Nominally, a lagoon is a suspended-growth no-recycle reactor with a variable degree of mixing. In lagoons in which mechanical or diffused aeration is used, mixing may be sufficient to approach complete mixing (i.e., solids maintained in suspension). In other types of lagoons, most solids settle and remain on the lagoon bottom, but some mixing is achieved as a result of gas production from bacterial metabolism and wind action. Lagoons are categorized as aerobic, facultative, or anaerobic on the basis of degree of aeration. Aerobic lagoons primarily depend on mechanical or diffused air supply. Significant oxygen supply is also realized through natural surface aeration. A facultative lagoon is dependent primarily on natural surface aeration and oxygen

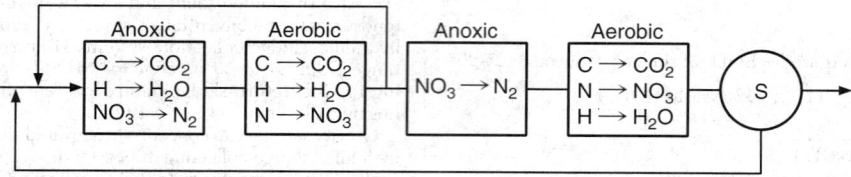

FIG. 25-55   Nitrogen removal systems.

generated by algal cells. These two lagoon types are relatively shallow to encourage surface aeration and provide for maximum algal activity. The third type of lagoon is maintained under anaerobic conditions to foster methane fermentation. This system is often covered with floating polystyrene panels to block surface aeration and help prevent a drop in temperature. Anaerobic lagoons are several meters deeper than the other two types. Lagoon flow schemes can be complex, employing lagoons in series and recycle from downstream to upstream lagoons. The major effect of recycle is to maintain control of the solids. If solids escape the lagoon system, a poor effluent is produced. Periodically controlled solids removal must take place or solids will escape.

Lagoons are, in effect, inexpensive reactors. They are shallow basins either cut below grade or formed by dikes built above grade or a combination of a cut and dike. The bottom must be lined with an impermeable barrier and the sides protected from wind erosion. These systems are best used where large areas of inexpensive land are available.

Facultative Lagoons These lagoons have been designed to use both aerobic and anaerobic reactions. Normally, facultative lagoons consist of two or more cells in series. The settleable solids tend to settle out in the first cell and undergo anaerobic metabolism with the production of organic acids and methane gas, which bubbles out to the atmosphere. Algae at the surface of the lagoon utilize sunlight for their energy in converting carbon dioxide, water, and ammonium ions into algal protoplasm with the release of oxygen as a waste product. Aerobic bacteria utilize the oxygen released by the algae to stabilize the soluble and colloidal organics. Thus, the bacteria and algae form a symbiotic relationship as shown in Fig. 25-56. The interesting aspect of facultative lagoons is that the organic matter in the incoming wastewaters is not stabilized but rather is converted to microbial protoplasm, which has a slower rate of oxygen demand. In fact, in some facultative lagoons inorganic compounds in the wastewaters are converted to organic compounds with a total increase in organics within the lagoon system.

Facultative lagoons are designed on the basis of organic load in relationship to the potential sunlight availability. In the northern part of the United States facultative lagoons are designed on the basis of 2.2 g/(m²·day) [20 lb BOD₅/(acre·day)]. In the middle part of the United States the organic load can be increased to 3.4 to 4.5 g/(m²·day) [30 to 40 lb BOD₅/(acre·day)], while in the southern part the organic load can be increased to 6.7 g/(m²·day) [60 lb BOD₅/(acre·day)]. The depth of lagoons is normally maintained between 1.0 and 1.7 m (3 and 5 ft). A depth less than 1.0 m (3 ft) encourages the growth of aquatic weeds and permits mosquito breeding. In dry areas the maximum depth may be increased above 1.7 m (5 ft) depending upon evaporation. Most facultative lagoons depend upon natural wind action for mixing and should not be placed in screened areas where wind action is blocked.

Effluent quality from facultative lagoons is related primarily to the suspended solids created by living and dead microbes. The long retention period in the lagoons allows the microbes to die off, leaving a small particle that settles slowly. The release of nutrients from the dead microbes permits the algae to survive by recycling the nutrients.

Thus, the algae determine the ultimate effluent quality. Use of series ponds with well-designed transfer structures between ponds permits maximum retention of algae within the ponds and the best-quality effluent. Normally the soluble BOD₅ is under 5 or 10 mg/L with a total effluent BOD₅ under 30 mg/L. The effluent suspended solids will vary widely during the different seasons of the year, being a maximum of 70 to 100 mg/L in the summer months and a minimum of 10 to 20 mg/L in the winter months. If suspended-solids removal is essential, chemical precipitation is the best method available at the present time. Slow sand filters and rock filters have been studied for suspended-solids removal; they work well as long as the effluent suspended solids are relatively low, 40 to 70 mg/L.

Aerated Lagoons These lagoons originated from efforts to control overloaded facultative lagoons. Since the lagoons were deficient in oxygen, additional oxygen was supplied by either mechanical surface aerators or diffused aerators. Mechanical surface aerators were quickly accepted as the primary aerators because they could be quickly added to existing ponds and moved to strategic locations. Unfortunately, the high-speed, floating surface aeration units were not efficient, and large numbers were required for existing lagoons. The problem was simply one of poor mixing in a very shallow lagoon.

Eventually, diffused aeration equipment was added to relatively deep lagoons [3.0 to 6.0 m (10 to 20 ft)]. Mixing became the most significant parameter for good oxygen transfer in aerated lagoons. From an economical point of view, it was found that a completely mixed aerated lagoon with 24-h retention provided the best balance between mixing and oxygen transfer. As the organic load increased, the fluid-retention time also increased. Short-term aeration permitted metabolism of the soluble organics by the bacteria, but time did not permit metabolism of the suspended solids. The suspended solids were combined with the microbial solids produced from metabolism and discharged from the aerated lagoon to a solids-separation pond. Data from the short-term aerated lagoon indicated that 50 percent BOD₅ stabilization occurred, with conversion of the soluble organics to microbial cells. The problem was separation and stabilization of the microbial cells. Short-term sedimentation ponds permitted separation of the solids without significant algae growths but required cleaning at frequent intervals to keep them from filling with solids and flowing into the effluent. Long-term lagoons permitted solids separation and stabilization but also permitted algae to grow and affect effluent quality.

Aerated lagoons were simply dispersed microbial reactors which permitted conversion of the organic components in the wastewaters to microbial solids without stabilization. The residual organics in solution were very low, less than 5 mg/L BOD₅. By adding oxygen and improving mixing, the microbial metabolism reaction was speeded up, but the stabilization of the microbial solids has remained a problem to be solved.

Anaerobic Lagoons These lagoons were developed when a major fraction of the organic contaminants consisted of suspended solids that could be removed easily by gravity sedimentation. The anaerobic lagoons are relatively deep [8.0 to 6.0 m (10 to 20 ft)], with a short fluid-retention time (3 to 5 days) and a high BOD₅ loading rate, up to 3.2 kg/(m³·day) [200 lb/(1000 ft³·day)]. Microbial metabolism in the settled-solids layer produces methane and carbon dioxide, which quickly rise to the surface, carrying some of the suspended solids. A scum layer that retards oxygen transfer and release of obnoxious gases is quickly produced in anaerobic lagoons. Mixing with a grinder pump can provide a better environment for metabolism of the suspended solids. The key for anaerobic lagoons is adequate buffer to keep the pH between 6.5 and 8.0. Protein wastes have proved to be the best pollutants to be treated by anaerobic lagoons, with the ammonium ions reacting with carbon dioxide and water to form ammonium bicarbonate as the primary buffer. High-carbohydrate wastes are poor in anaerobic lagoons since they produce organic acids without adequate buffer, making it difficult to maintain a suitable pH for good microbial growth.

Anaerobic lagoons do not produce a high-quality effluent but are able to reduce the BOD load by 80 to 90 percent with a minimum of effort. Since anaerobic lagoons work best on strong organic wastes, their effluent must be treated by either aerated lagoons or facultative

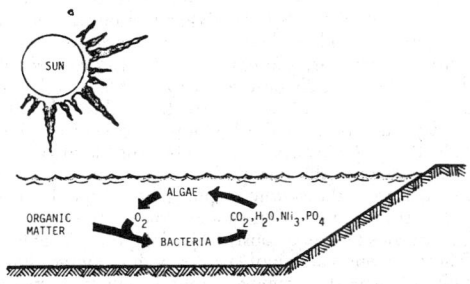

FIG. 25-56 Schematic diagram of oxidation-pond operations.

lagoons. An anaerobic lagoon is simply the first stage in the treatment of strong organic wastewaters.

Fixed Film Reactor Systems   A major advantage of fixed film systems is that a flocculent-type biomass is not necessary as the biomass remains in the reactor attached to inert packing. Biomass does periodically slough off or break away from the packing, usually in large chunks that can be easily removed in a clarifier. On the other hand, the time of contact between the biomass and the waste is much shorter than in suspended growth systems, making it difficult to achieve the same degree of treatment especially in aerobic systems. Aerobic, anoxic, and anaerobic fixed film systems are utilized for waste treatment.

Aerobic systems including trickling filters and rotating biological contactors (RBC) are operated in a nonflooded mode to ensure adequate oxygen supply. Other aerobic, anoxic, and anaerobic systems employ flooded reactors. The most common systems are packed beds (anaerobic trickling filter) and fluidized or expanded bed systems.

Trickling Filters   For years trickling filters were the mainstay of biological wastewater treatment systems because of their simplicity of design and operation. Trickling filters were displaced as the primary biological treatment system by activated sludge because of better effluent quality. Trickling filters are simply fixed-medium biological reactors with the wastewaters being spread over the surface of a solid medium where the microbes are growing. The microbes remove the organics from the wastewaters flowing over the fixed medium. Oxygen from the air permits aerobic reactions to occur at the surface of the microbial layer, but anaerobic metabolism occurs at the bottom of the microbial layer where oxygen does not penetrate.

Originally, the medium in trickling filters was rock, but rock has largely been replaced by plastic, which provides greater void space per unit of surface area and occupies less volume within the filter. A plastic medium permitted trickling filters to be increased from a medium depth of 1.8 m (6 ft) to one of 4.2 m (14 ft) and even 6.0 m (20 ft). The wastewaters are normally applied by a rotary distributor or a fixed-spray nozzle. The spraying or discharging of wastewaters above the trickling-filter medium permits better distribution over the medium and oxygen transfer before reaching the medium. The effluent from the trickling-filter medium is captured in a clay-tile underdrain system or in a tank below the plastic medium. It is important that the bottom of the trickling filter be open for air to move quickly through the filter and bring adequate oxygen for the microbial reactions.

If a high-quality effluent is required, trickling filters must be operated at a low hydraulic-loading rate and a low organic-loading rate. Low-rate trickling filters are operated at hydraulic loadings of 2.2 × 10^{-5} to 4.3 × 10^{-5} m³/(m²·s) [2 million to 4 million gal/(acre·day)]. High-rate trickling filters are designed for 10.8 × 10^{-5} to 40.3 × 10^{-5} m³/(m²·s) [10 million to 40 million gal/(acre·day)] hydraulic loadings and organic loadings up to 1.4 kg/(m³·day) [90 lb BOD_5/(1000 ft³·day)]. Plastic-medium trickling filters have been designed to operate up to 108 × 10^{-5} m³/(m²·s) [100 million gal/(acre·day)] or even higher, with organic loadings up to 4.8 kg/(m³·day) [300 lb BOD_5/ (1000 ft³·day)]. Low-rate trickling filters will produce better than 90 percent BOD_5 and suspended-solids reductions, while high-rate trickling filters will produce from 65 to 75 percent BOD_5 reduction. Plastic-medium trickling filters will produce from 59 to 85 percent BOD_5 reduction depending upon the organic-loading rate. It is important to recognize that concentrated industrial wastes will require considerable hydraulic recirculation around the trickling filter to obtain the proper hydraulic-loading rate without excessive organic loads. With high recirculation rates the organic load is distributed over the entire volume of the trickling filter for maximum organic removal. The short fluid-retention time within the trickling filter is the primary reason for the low treatment efficiency.

Rotating Biological Contactors (RBC)   The newest form of trickling filter is the rotating biological contactor with a series of circular plastic disks, 3.0 to 3.6 m (10 to 12 ft) in diameter, immersed to approximately 40 percent diameter in a shaped contact tank. The RBC disks rotate at 2 to 5 r/min. As the disks travel through the wastewaters, a small layer adheres to them. As the disks travel into the air, the microbes on the disk surface oxidize the organics. Thus, only a small amount of energy is required to supply the required oxygen for

wastewater treatment. As the microbes build up on the plastic disks, the shearing velocity that is created by the movement of the disks through the water causes the excess microbes to be removed from the disks and discharged to the final sedimentation tank.

Rotating biological contactors have been very popular in treating industrial wastes because of their relatively small size and their low energy requirements. Unfortunately, there have occurred a number of problems which should be recognized prior to using RBCs. Strong industrial wastes tend to create excessive microbial growths which are not easily sheared off and which create high oxygen-demand rates with the production of hydrogen sulfide and other obnoxious odors. The heavy microbial growths have damaged some of the disks and caused some shaft failures. The disks are currently being covered with plastic shells to prevent nuisance odors from occurring. Air must be forced through the covered RBC systems and be chemically treated before being discharged back into the environment. Recirculation of wastewater flow around the RBC units can distribute the load over all the units and reduce the heavy initial microbial growths. RBC units also work best under uniform organic loads, requiring surge tanks for many industrial wastes. The net result has been for the cost of RBC units to approach that of other treatment units in terms of organic matter stabilized.

RBCs should be designed on both a hydraulic-loading rate and an organic-loading rate. Normally, hydraulic-loading rates of up to 0.16 m³/(m²·day) [4 gal/(ft²·day)] of surface area are used with organic loading rates up to 44 kg/(m²·day) [9 lb BOD_5/(ft²·day)]. Treatment efficiency is primarily a function of the fluid-retention time and the organic-loading rate. At low organic-loading rates the RBC units will produce nitrification in the same way as low-rate trickling filters.

Packed-Bed Fixed-Film Systems   These systems were originally termed anaerobic trickling filters because the first systems were submerged columns filled with stones run under anaerobic conditions (Fig. 25-57). A wide variety of packed media is now used ranging in size from granules 40 mesh to 7.5-cm (3–in) stones. Many systems use open structure plastic packing similar to that used in aerobic trickling filters.

The systems using granular media packing are used for anoxic denitrification. They are usually downflow, thus serving the dual function of filtration and denitrification. Contact times are short (EBCT < 15 min), but excellent removal is achieved due to the high level of biomass retained in the reactor. Pacing the methanol dose to the varying feed nitrate concentration is crucial. Frequent, short-duration backwash (usually several times per day) is required or the nitrogen bubbles formed will bind the system, causing poor results. Extended backwash every two to three days is required or the system will clog on the biomass growth. Thus, several units in parallel or a large holding tank are needed to compensate for the down time during backwash. Backwash does not remove all the biomass; a thin film remains coating the packing. Thus, denitrification begins immediately when the flow is restored.

The systems using the larger packing are used in the treatment of relatively strong, low-suspended-solids industrial waste. These systems are closed columns usually run in an upflow mode with a gas space at the top. These are operated under anaerobic conditions with waste conversion to methane and carbon dioxide as the goal. Effluent recycle is often used to help maintain the pH in the inlet zone in the correct range 6.5–7.5 for the methane bacteria. Some wastes require the addition of alkaline material to prevent a pH drop. Sodium bicarbonate is often recommended for pH control because it is easier to handle than lime or sodium hydroxide, and because an overdose of bicarbonate will only raise the pH modestly. An overdose of lime or sodium hydroxide can easily raise the pH above 8.0. Table 25-44 gives some performance data with systems treating industrial wastes. HRTs of 1 to 2 days are used, as the buildup of growth on the packing ensures a BSRT of 20–50 days. It should be possible to lower the HRT further, but in practice this has not been successful because biomass starts to escape from the system or plugging occurs. Some escape is due to high gasification rates, and some is due to the fact that anaerobic sludge attaches less tenaciously to packing than aerobic or anoxic sludge. These systems can handle wastes with moderate solids levels. Periodically, solids must be removed from the reactor to prevent plugging of the packing or loss of solids in the effluent.

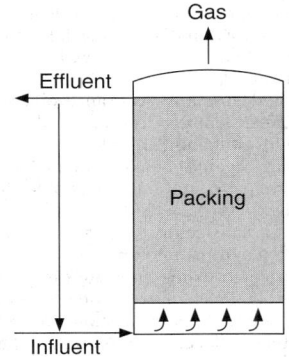
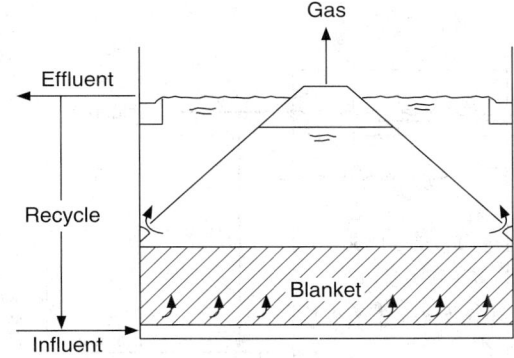

FIG. 25-57 Anaerobic processes.

Biological Fluidized Beds This high-rate process has been used successfully for aerobic, anoxic, and anaerobic treatment of municipal and industrial wastewaters. Numerous small- and large-scale applications for hazardous waste, contaminated groundwater, nontoxic industrial waste and municipal wastewater have been reported (Refs. 32 and 33). The basic element of the process is a bed of solid carrier particles, such as sand or granular activated carbon, placed in a reactor through which wastewater is passed upflow with sufficient velocity to impart motion or fluidize the carrier. An active growth of biological organisms grow as firmly attached mass surrounding each of the carrier particles. As the wastewater contaminants pass by the biologically covered carrier, they are removed from the wastewater through biological and adsorptive mechanisms. Figure 25-58 is a schematic of the process.

The influent wastewater enters the reactor through a pipe manifold and is introduced downflow through nozzles that distribute the flow uniformly at the base of the reactor. Reversing direction at the bottom, the flow fluidizes the carrier when the fluid drag overcomes the buoyant weight of the carrier and its attached biomass layer. During startup (before much biomass has accumulated), the flow velocity required to achieve fluidization is higher than after the biomass attaches. Recycle of treated effluent is adjusted to achieve the desired degree of fluidization. As biomass accumulates, the particles of coated biomass will separate to a greater extent at constant flow velocity. Thus, as the system ages and more biomass accumulates, the extent of bed expansion increases (the volume of voids increases). This phenomenon is advantageous because it prevents clogging of the bed with biomass. Consequently, higher levels of biomass attachment are possible than in other types of fixed film systems. However, eventually, the degree of bed expansion may become excessive. Reduction of recycle will reduce expansion but may not be feasible because recycle has several purposes (i.e., supply of nutrients, alkalinity, and dilution of waste strength). Control of the expanded bed surface level is automatically accomplished using a sensor that activates a biomass growth control system at a prescribed level and maintains the bed at the proper depth. A pump removes a portion of the attached biomass, separates the biomass and inert carrier by abrasion, and pumps the mixture into a separator. Here the heavy carrier settles back into the fluidized bed and the abraded biomass, which is less dense, is removed from the system by gravity or a second pump. Other growth control designs are also used. Effluent is withdrawn from the supernatant layer above the fluidized bed. The reactor is usually not covered unless it is operating under anaerobic conditions and methane, odorous gases, or other safety precautions are mandated.

When aerobic treatment is to be provided to high concentrations of organics, pure oxygen or hydrogen peroxide may be injected into the wastewater prior to entering the reactor. Liquid oxygen (LOX) or pressure swing absorption (PSA) systems have been used to supply oxygen. Air may be used at low D.O. demands.

In full-scale applications, this process has been found to operate at significantly higher volumetric loading rates for wastewater treatment than other processes. The primary reasons for the very high rates of contaminant removal is the high biologically active surface area available (approximately 1000 ft²/ft³ of reactor) and the high concentration of reactor biological solids (8,000–40,000 mg/L) that can be maintained (Table 25-45). Because of these atypical high values, designs usually indicate a 200–500 percent reduction in reactor volume when compared to other fixed film and suspended growth treatment processes. In Table 25-46, is a list of full-scale commercial applications of the process operated at high wastewater concentrations including

TABLE 25-44 Anaerobic Process Performance on Industrial Wastewater UASB, Submerged filter (SF), FBR

Process	Wastewater	Reactor size, MG	COD, g/l	OVL, kg/m³d	%CODr	HRT-d	°C
SF	Rum slops	3.5	80–105	15	71	7.8	35
SF	Modified guar	0.27	9.1	7.5	60	1.0	37
SF	Chemical	1.5	14	11	90	0.7	H
SF	Milk	0.2	3	7.5	60	0.5	32
SF	PMFC	10⁻⁵	13.7	23	72	0.6	36
UASB	Potato	0.58	2.5	3	85	0.7	35
UASB	Sugar beet	0.21	3	16	88	—	—
UASB	Brewery	1.16	1.6–2.2S	4.4	83	0.4	30
UASB	Brewery	1.16	2.0–2.4S	8.7	78	0.2	30
UASB	PMFC	10⁻⁵	13.7	4–5	87	2.9	36
FBR	PMFC	10⁻⁵	13.7	35–48	88	0.4	36
FBR	Soft drink	0.04	3.0	6–7	75	0.5	35
FBR	Chemical	0.04	35	14	95	2.5	35

OVL = organic volumetric load (COD)
S = settled effluent
H = heated
PMFC = paper mill foul condensate (5,6)
NOTE: MG = 3785 m³.

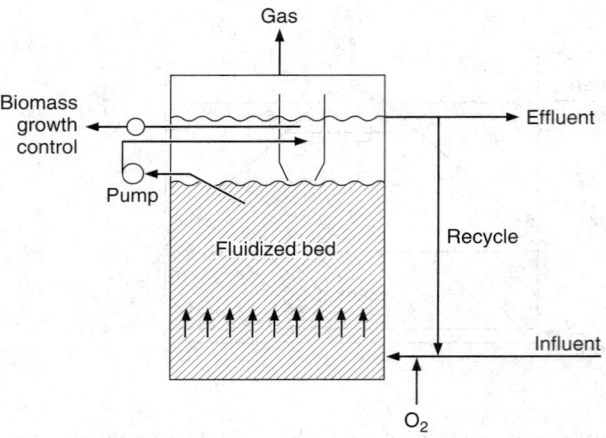

FIG. 25-58 Schematic of fluidized bed process.

aerobic oxidation of organics, anoxic denitrification and anaerobic treatment systems.

Of special note is the enhancement to the process when granular activated carbon (GAC) is used as the carrier. Because GAC has adsorptive properties, organic compounds present in potable waters and wastewater at low concentrations, often less than ten mg/L, are removed by adsorption and subsequently consumed by the biological organisms that grow in the fluidized bed. The BTEX compounds, methylene chloride, chlorobenzene, plastics industry toxic effluent, and many others are removed in this manner. BTEX contamination of groundwater from leaking gasoline storage tanks is a major problem, and sixteen full-scale fluidized bed process applications have been made. Contaminated groundwater is pumped to the ground surface for treatment using the fluidized bed in the aerobic mode. Often the level of BTEX is 1–10 mg/L and about 99 percent is removed in less than ten minutes detention time. Installations in operation range in size from 30 to 3000 gpm. The smaller installations are often skid-

TABLE 25-45 Process Comparisons

Biological process	MLVSS, mg/L	Surface area, ft²/ft³
Activated sludge	1500–3000	—
Pure oxygen activated sludge	2000–5000	—
Suspended growth nitrification	1000–2000	—
Suspended growth denitrification	1500–3000	—
Fluidized bed-CBOD removal	12,000–20,000	800–1200
Fluidized bed-nitrification	8,000–12,000	800–1200
Fluidized bed-denitrification	25,000–40,000	800–1200
Trickling filter-CBOD removal	—	12–50
RBC-CBOD removal	—	30–40

TABLE 25-46 Full-Scale Commercial Applications of Fluidized Bed Process

Application	Type	Reactor volume, m³
CBOD—paint—General Motors	Aerobic	108
C,NBOD—sanitary and automotive—GM	Aerobic	165
Chemical—Grindsted Products Denmark	Aerobic	730
Nitrification—fish hatchery—Idaho	Aerobic	820
BTEX, groundwater—Ohio	Aerobic	250
Chemical, Toxicity reduction—Texas	Aerobic	225
Denitrification, Nuclear Fuel, DOE, Fornald, OH	Anoxic	53
CBOD, Denitrification—municipal, Nevada	Anoxic	1450
Petrochemical—Reliance Ind. India	Anaerobic	850
Soft drink, Delaware	Anaerobic	166
Brewery—El Aguila, Spain	Anaerobic	1570

mounted and may be moved from location to location at a given site. A major advantage of this process over stripping towers and vacuum systems for treating volatile organics (VOCs) is the elimination of effluent gas treatment.

Also, pilot plant and laboratory scale anaerobic studies have demonstrated successful treatment of wastewaters of 5,000 to 50,000 mg/L COD from corn chips containing soluble and colloidal corn starch and protein, cheese whey, organic chemicals, food, bakery, brewery, paper mill foul condensate, paint, and numerous other hazardous and non-hazardous materials.

Design criteria for satisfactory biological fluidized bed treatment systems include the major parameters given in Table 25-47.

Earlier the anaerobic trickling filter was discussed and in the above section the fluidized bed reactor was reviewed. A third type of reactor system with some similarity is the upflow anaerobic sludge blanket (UASB) system. Here, a flocculent biomass is retained in the reactor with no recycle of sludge necessary. The sludge is maintained in the system by the use of a relatively low flow rate. The sludge formed seems to be granular in nature and has a relatively high specific gravity. It is thought that the presence of small particles of $CaCO_3$ and/or clay in the waste may contribute to the formation of the dense sludge. A diagram of this system is shown in Fig. 25-56. The supplier of this system provides a startup seed of this granular-type biomass.

The anaerobic filter, UASB, and fluidized bed reactors have all been used for anaerobic treatment of industrial wastes, as each is especially suited for use in anaerobic treatment. Table 25-44 presents results from these applications.

PHYSICAL-CHEMICAL TREATMENT

Processes and/or unit operations that fall under this classification include adsorption, ion exchange, stripping, chemical oxidation, and membrane separations. All of these are more expensive than biological treatment but are used for removal of pollutants that are not easily removed by biomass. Often these are utilized in series with biological treatment; but sometimes they are used as stand-alone processes.

Adsorption This is the most widely used of the physical-chemical treatment processes. It is used primarily for the removal of soluble organics with activated carbon serving as the adsorbent. Most liquid-phase-activated carbon adsorption reactions follow a Freundlich Isotherm [Eq. (25-21)].

$$y = kc^{1/n} \qquad (25-21)$$

where: y = adsorbent capacity, mass pollutant/mass carbon
c = concentration of pollutant in waste, mass/volume
k and n = empirical constants

EPA has compiled significant data on values of k and n for environmentally significant pollutants with typical activated carbons. Assuming equilibrium is reached, the isotherm provides the dose of carbon required for treatment. In a concurrent contacting process, the capacity is set by the required effluent concentration. In a countercurrent process, the capacity of the carbon is set by the untreated waste pollutant concentration. Thus countercurrent contacting is preferred.

Activated carbon is available in powdered form (200–400 mesh) and granular form (10–40 mesh). The latter is more expensive but is easier to regenerate and easier to utilize in a countercurrent contactor. Powdered carbon is applied in well-mixed slurry-type contactors for

TABLE 25-47 Typical Design Parameters for Fluidized Beds

Waste	Mode	Influent concentration, mg/L	Volumetric load, lb/ft³·d	Hydraulic detention time, hr	Volatile solids (biosolids), g/L
Organic COD	Aerobic	<2000	0.1–0.6	0.5–2.5	12–20
Organic COD	Aerobic	<10	0.02–0.1	0.1–0.5	4–8
NH_3-N	Aerobic	<25	0.05–0.3	0.1–0.5	7–14
NO_3^--N	Anoxic	<5,000	0.4–2	0.1–2.4	20–40
Organic COD	Anaerobic	>4000	0.5–4	2–24	20–40

NOTE: lb/ft³·d = 16 kg/m³·d.

detention times of several hours after which separation from the flow occurs by sedimentation. Often coagulation, flocculation, and filtration are required in addition to sedimentation. As it is difficult to regenerate, powdered carbon is usually discarded after use. Granular carbon is used in column contactors with EBCT of 30 minutes to 1 hour. Often several contactors are used in series, providing for full countercurrent contact. A single contactor will provide only partial countercurrent contact. When a contactor is exhausted, the carbon is regenerated either by a thermal method or by passing a solvent through the contactor. For waste-treatment applications where a large number of pollutants must be removed but the quantity of each pollutant is small, thermal regeneration is favored. In situations where a single pollutant in large quantity is removed by the carbon, solvent extraction regeneration can be used, especially where the pollutant can be recovered from the solvent and reused. Thermal regeneration is a complex operation. It requires removal of the carbon from the contactor, drainage of free water, transport to a furnace, heating under controlled conditions of temperature, oxygen, time, water vapor partial pressure, quenching, transport back to the reactor, and reloading of the column. Five to ten percent of the carbon is lost in this regeneration process due to burning and attrition during each regeneration cycle. Multiple hearth, rotary kiln, and fluidized bed furnaces have all been successfully used for carbon regeneration.

Pretreatment prior to carbon adsorption is usually for removal of suspended solids. Often this process is used as tertiary treatment after primary and biological treatment. In either situation, the carbon columns must be designed to provide for backwash. Some solids will escape pretreatment, and biological growth will occur on the carbon, even with extensive pretreatment. Originally carbon treatment was viewed only as applicable for removal of toxic organics or those that are difficult to degrade biologically. Present practice applies carbon adsorption as a procedure for removal of all types of organics. It is realized that some biological activity will occur in virtually any activated carbon unit, so the design must be adjusted accordingly.

Ion Exchange This process has been employed for many years for treatment of industrial water supplies but not for wastes. However, some new ion-exchange materials have recently been developed that can be used for removal of specific pollutants. These new resins are primarily useful for selective removal of heavy metals, even though the target metals are present at low concentration in a wastewater containing many other inorganics. An ion-exchange process is usually operated with the waste being run downflow through a series of columns containing the appropriate ion-exchange resins. EBCT contact times of 30 min to 1 hour are used. Pretreatment for suspended solids and organics removal is practiced as well as pH adjustment.

The capacity of an ion-exchange resin is a function of the type and concentration of regenerant. Because ion-exchange resins are so selective for the target compound, a significant excess of the regenerant must be used. Up to a point, the more regenerant, the greater the capacity of the resin. Unfortunately, this results in waste of most of the regenerant. In fact, the bulk of the operational cost for this process is for regenerant purchase and its disposal; thus, regeneration must be optimized. Regenerants used (selection depends on the ion exchange resin) include sodium chloride, sodium carbonate, sulfuric acid, sodium hydroxide, and ammonia. Table 25-48 gives information on some of the new highly selective resins. These resins may provide the only practical method for reduction of heavy metals to the very low levels required by recent EPA regulations.

Stripping Air stripping is applied for the removal of volatile substances from water. Henry's law is the key relationship for use in design of stripping systems. The minimum gas-to-liquid ratio required for stripping is given by:

$$\frac{G}{L} = \frac{S_{in} - X_{out}}{(H)X_{out}}$$

for a concurrent process and

$$\frac{G}{L} = \frac{X_{in} - X_{out}}{(H)X_{in}}$$

for a countercurrent process where:
  G = air flow rate, mass/time
  L = waste flow rate, volume/time
  X = concentration of pollutant in waste, mass/volume
  H = Henry's constant for the pollutant in water, volume/mass

Higher ratios of gas to liquid flow are required than those computed above because mass-transfer limitations must be overcome. Stripping can occur by sparging air into a tank containing the waste. Indeed, stripping of organics from activated sludge tanks is of concern because of the possibility that the public will be exposed to airborne pollutants including odors. A much more efficient stripping procedure is to use counterflow or crossflow contact towers. Procedures for design of these and mass-transfer characteristics of various packings are available elsewhere in this handbook.

Stripping has been successfully and economically employed for removal of halogenated organics from water and wastes with dispersion of the effluent gas to the atmosphere. However, recent EPA regulations have curtailed this practice. Now removal of these toxic organics from the gas stream is also required. Systems employing activated carbon (prepared for use with gas streams) are employed as well as systems to oxidize the organics in the gas stream. However, the cost of cleaning up the gas stream often exceeds the cost of stripping these organics from the water.

Chemical Oxidation This process has not been widely utilized because of its high cost. Only where the concentration of the target compound is very low will the quantity of oxidant required be low enough to justify treatment by chemical oxidation. The efficiency of this process is also low, as many side reactions can occur that will consume the oxidant. In addition, complete oxidation of organics to carbon dioxide and water often will not occur unless a significant overdose is used. However, renewed interest has recently occurred for two reasons.

1. It has been found that partial oxidation is satisfactory as a pretreatment to biological or carbon adsorption treatment. Partial oxidation often seems to make recalcitrant organics easier to degrade biologically and easier to adsorb.

2. A combination of ozone oxidation with simultaneous exposure to ultraviolet light seems to produce a self-renewing chain reaction that can significantly reduce the dose of ozone needed to accomplish oxidation.

Oxidants commonly used include ozone, permanganate, chlorine, chlorine dioxide, and ferrate, often in combination with catalysts. Standard-type mixed reactors are used with contact times of several minutes to an hour. Special reactors for use with ultraviolet light have been developed.

Membrane Processes These processes use a selectively permeable membrane to separate pollutants from water. Most of the mem-

TABLE 25-48 Selective Ion-Exchange Resins

Application	Exchanger type	Composition	Regenerant
Softening	Cation	Polystyrene matrix Sulfonic acid functional groups	NaCl
Heavy metals	Cation	Polystyrene matrix Chelating functional groups	Mineral acids
Chromate	Anion	Polystyrene matrix Tertiary or quaternary ammonium functional groups	Sodium carbonate or alkaline NaCl
Nitrate	Anion	Polystyrene matrix Tributyl ammonium functional group	NaCl

branes are formulated from complex organics that polymerize during membrane preparation. This allows the membrane to be tailored to discriminate by molecular size or by degree of hydrogen bonding potential. Ultrafiltration membranes discriminate by molecular size or weight, while reverse osmosis membranes discriminate by hydrogen-bonding characteristics. The permeability of these membranes is low: from 0.38–3.8 m/day (10–100 gal/d/ft²). The apparatus in which they are used must provide a high surface area per unit volume. The membranes are also fragile and are often subjected to several hundred psi of pressure. Thus the apparatus must provide a rugged support structure and efficient seals to prevent leakage. Membrane support configurations include tubular, spiral-wound, and hollow fine fiber systems. An operational problem is the buildup of pollutant on the side of the membrane where rejection occurs. This leads to fouling of the membrane surface which initially reduces water flux. Eventually, deterioration and membrane failure occur. Membrane life is in the range 1–2 years. New membrane materials with improved properties are being sought. Membrane separation-based systems require extensive pretreatment and continual maintenance to reduce fouling. These systems have significant potential, but much developmental work is needed on membrane characteristics and the support apparatus.

SLUDGE PROCESSING

Objectives   Sludges consist primarily of the solids removed from liquid wastes during their processing. Thus, sludges could contain a wide variety of pollutants and residuals from the application of treatment chemicals; i.e., large organic solids, colloidal organic solids, metal sulfides, heavy-metal hydroxides and carbonates, heavy-metal organic complexes, calcium and magnesium hydroxides, calcium carbonate, precipitated soaps and detergents, and biomass and precipitated phosphates. As sludge even after extensive concentration and dewatering is still greater than 50 percent by weight water, it can also contain soluble pollutants such as ammonia, priority pollutants, and nonbiologically degradable COD.

The general treatment or management of sludge involves stabilization of biodegradable organics, concentration and dewatering, and ultimate disposal of the stabilized and dewatered residue. A large number of individual unit processes and unit operations are used in a sludge-management scheme. Those most frequently used are discussed below. Occasionally, only one of these is needed, but usually several are used in a series arrangement.

Because of the wide variability in sludge characteristics and the variation in acceptability of treated sludges for ultimate disposal (this is a function of the location and characteristics of the ultimate disposal site), it is impossible to prescribe any particular sludge-management plan. In the sections below, general performance of individual sludge-treatment processes and operations is presented.

Concentration: Thickening and Flotation   Generated sludges are often dilute (1–2 percent solids by weight). In order to reduce the volumetric loading on other processes, the first step in sludge processing is often concentration. The most popular process is gravity thickening which is carried out in treatment units similar to circular clarifiers. Organic sludges from primary treatment can usually be concentrated to 5–8 percent solids. Sludges from secondary treatment can be thickened to 2 percent solids. The potential concentration with completely inorganic sludges is higher (greater than 10 percent solids) except for sludges high in metal hydroxides. Polymers are often used to speed up thickening and increase concentration. Thickening is enhanced by long retention of the solids in the thickening apparatus. However, when biodegradable organics are present, solids retention time must be at a level that will not foster biological activity, lest odors, gas generation, and solids hydrolysis occur. Loading rates on thickeners range from 50–122 kg/m²/d (10–25 lb/ft²/d) for primary sludge to 12–45 kg/m²/d (2.5–9 lb/ft²/d). Solids detention time is 0.5 days in summer to several days in winter.

Flotation   Air flotation has proved to be successful in concentrating secondary sludges to about 4 percent solids. The incoming solids are normally saturated with air at 275 to 350 kPa (40 to 50 psig) prior to being released in the flotation tank. As the air comes out of solution, the fine bubbles are trapped under the suspended solids and carry them to the surface of the tank. The air bubbles compact the solids as a floating mass. Normally, the air-to-solids ratio is about 0.01–0.05 L/g (0.16–0.8 ft³/lb). The thickened solids are scraped off the surface, while the effluent is drawn off the middle of the tank and returned to the treatment system. In large flotation tanks with high flows the effluent rather than the incoming solids is pressurized and recycled to the influent. The size of the flotation tanks is determined primarily by the solids-loading rate, directly or indirectly. A solids loading of 25 to 97 kg/(m²·day) [5 to 20 lb/(ft²·day)] has been found to be adequate. On a flow basis this translates into 0.14 to 2.7 L/(m²·day) [0.2 to 4 gal/(ft²·min)] surface area.

As with thickening, air flotation is enhanced by the addition of polymers. Flotation has been successfully used with wholly inorganic metal hydroxide sludges. Polymers and surfactants are used as additives. Engineering details on air flotation equipment has been developed by and is available from various equipment manufacturing companies. Liquid removed during thickening and flotation is usually returned to the head end of the plant.

Stabilization (Anaerobic Digestion, Aerobic Digestion, High Lime Treatment)   Sludges high in organics can be stabilized by subjecting them to biological treatment. The most popular system is anaerobic digestion.

Anaerobic Digestion   Anaerobic digesters are large covered tanks with detention times of 30 days, based on the volume of sludge added daily. Digesters are usually heated with an external heat exchanger to 35–37° C to speed the rate of reaction. Mixing is essential to provide good contact between the microbes and the incoming organic solids. Gas mixing and mechanical mixers have been used to provide mixing in the anaerobic digester. Following digestion, the sludge enters a holding tank, which is basically a solids-separation unit and is not normally equipped for either heating or mixing. The supernatant is recycled back to the treatment plant, while the settled sludge is allowed to concentrate to 3–6 percent solids before being further processed.

Anaerobic digestion results in the conversion of the biodegradable organics to methane, carbon dioxide, and microbial cells. Because of the energy in the methane, the production of microbial mass is quite low, less than 0.1 kg/kg (0.1 lb volatile suspended solids (VSS)/lb) BCOD metabolized except for carbohydrate wastes. The production of methane is 0.35 m³/kg (5.6 ft³/lb) BCOD destroyed. Digester gases range from 50 to 80 percent methane and 20 to 50 percent carbon dioxide, depending on the chemical characteristics of the waste organics being digested. The methane is often used on site for heat and power generation.

There are three major groups of bacteria that function in anaerobic digestion. The first group hydrolyzes large soluble and nonsoluble organic compounds such as proteins, fats and oils (grease), and carbohydrates, producing smaller water-soluble compounds. These are then degraded by acid-forming bacteria, producing simple volatile organic acids (primarily acetic acid) and hydrogen. The last group (the methane bacteria) split acetic acid to methane and carbon dioxide and produce methane from carbon dioxide and hydrogen. Good operation requires the destruction of the volatile acids as quickly as they are produced. If this does not occur, the volatile acids will build up and depress the pH, which will eventually inhibit the methane bacteria. To prevent this from occurring, feed of organics to the digester should be as uniform as possible.

If continuous addition of solids is not possible, additions should be made at as short intervals as possible. Alkalinity levels are normally maintained at about 3000 to 5000 mg/L to keep the pH in the range 6.5–7.5 as a buffer against variable organic-acid production with varying organic loads. Proteins will produce an adequate buffer, but carbohydrates will require the addition of alkalinity to provide a sufficient buffer. Sodium bicarbonate should be used to supply the buffer.

An anaerobic digester is a no-recycle complete mix reactor. Thus, its performance is independent of organic loading but is controlled by hydraulic retention time (HRT). Based on kinetic theory and values of the pseudo constants for methane bacteria, a minimum HRT of 3 to 4 days is required. To provide a safety factor and compensate for load variation as indicated earlier, HRT is kept in the range 10 to 30 days. Thickening of feed sludge is used to reduce the tank volume required to achieve the long HRT values. When the sludge is high in protein, the alkalinity can increase to greater than 5000 mg/L, and the pH will rise past 7.5. This can result in free ammonia toxicity. To avoid this situation, the pH should be reduced below 7.5 with hydrochloric acid. Use of nitric or sulfuric acid will result in significant operational problems.

Aerobic Digestion Waste activated sludge can be treated more easily in aerobic treatment systems than in anaerobic systems. The sludge has already been partially aerobically digested in the aeration tank. For the most part, only about 25 to 35 percent of the waste activated sludge can be digested. An additional aeration period of 15 to 20 days should be adequate to reduce the residual biodegradable mass to a satisfactory level for dewatering and return to the environment. One of the problems in aerobic digestion is the inability to concentrate the solids to levels greater than 2 percent. A second problem is nitrification. The high protein concentration in the biodegradable solids results in the release of ammonia, which can be oxidized during the long retention period in the aerobic digester. Limiting oxygen supply to the aerobic digester appears to be the best method to handle nitrification and the resulting low pH.

A new concept is to use an on/off air supply cycle. During aeration, nitrates are produced. When the air is shut off, nitrates are reduced to nitrogen gas. This prevents acid buildup and removes nitrogen from the sludge. High power cost for aerobic digestion restricts the applicability of this process.

High Lime Treatment Uses doses of lime sufficient to raise the pH of sludge to 12 or above. As long as the pH is maintained at this level, biological breakdown will not occur. In this sense, the sludge is stable. However, any reduction of pH as a result of contact with CO_2 in the air will allow biological breakdown to begin. Thus, this technique should only be used as temporary treatment until further processing can occur. It is not permanent stabilization as is provided by anaerobic or aerobic digestion.

Sludge Dewatering Dewatering is different from concentration in that the latter still leaves a substance with the properties of a liquid. The former produces a product which is essentially a friable solid. When the water content of sludge is reduced to <70–80 percent, it forms a porous solid called sludge cake. There is no free water in the cake, as the water is chemically combined with the solids or tightly adsorbed on the internal pores. The operations below which are used to dewater sludge can be applied at any stage of the sludge management process, but often they follow concentration and/or biological stabilization. Chemical conditioning is almost always used to aid dewatering.

Lime, alum, and various ferric salts have been used to condition sludge prior to dewatering. Lime reacts to form calcium carbonate crystals, which act as a solid matrix to hold the sludge particles apart and allow the water to escape during dewatering. Alum and iron salts help displace some of the bound water from hydrophilic organics and form part of the inorganic matrix. Chemical conditioning increases the mass of sludge to be ultimately handled from 10 to 25 percent, depending upon the characteristics of the individual sludge. Chemical conditioning can also help remove some of the fine particles by incorporating them into insoluble chemical precipitates. The water (supernatant) removed from the sludge during dewatering is often high in suspended solids and organics. Addition of polymers prior to, during, or after dewatering will often reduce the level of pollutants in the supernatant.

Centrifugation Both basket and solid-bowl centrifuges have been used to concentrate waste sludges. Field data have shown that it is possible to obtain 10 to 20 percent solids with waste activated sludge, 15 to 30 percent solids with a mixture of primary and waste activated sludge, and up to 30 to 35 percent solids with primary sludge alone. Centrifuges result in 85 to 90 percent solids capture with good operation. The problem is that the centrate contains the fine solids not easily removed. The centrate is normally returned to the treatment process, where it may or may not be removed. Economics do not favor centrifuges unless the sludge cake produced is at least 25 to 30 percent solids. For the most part, centrifuges are designed by equipment manufacturers from field experience. With varying sludge characteristics centrifuge characteristics will also vary widely.

Vacuum Filtration Vacuum filtration has been the most common method employed in dewatering sludges. Vacuum filters consist of a rotary drum covered with a cloth-filter medium. Various plastic fibers as well as wool have been used for the filter cloth. The filter operates by drawing a vacuum as the drum rotates into chemically conditioned sludge. The vacuum holds a thin layer of sludge, which is dewatered as the drum rotates through the air after leaving the vat. When the drum rotates the cloth to the opposite side of the apparatus, air-pressure jets replace the vacuum, causing the sludge cake to separate from the cloth medium as the cloth moves away from the drum. The cloth travels over a series of rollers, with the sludge being separated by a knife edge and dropping onto a conveyor belt by gravity. The dewatered sludge is moved on the conveyor belt to the next concentration point, while the filter cloth is spray-washed and returned to the drum prior to entering the sludge vat. Vacuum filters yield the poorest results on waste activated sludge and the best results on primary sludge. Waste activated sludge will concentrate to between 12 and 18 percent solids at a rate of 4.9 to 9.8 kg dry cake/$(m^2 \cdot h)$ [1 to 2 lb/$(ft^2 \cdot h)$]. Primary sludge can be dewatered to 25 to 30 percent solids at a rate of 49 kg dry cake/$(m^2 \cdot h)$ [10 lb/$(ft^2 \cdot h)$].

Pressure Filtration Pressure filtration has been used increasingly since the early 1970s because of its ability to produce a drier sludge cake. The pressure filters consist of a series of plates and frames separated by a cloth medium. Sludge is forced into the filter under pressure, while the filtrate is drawn off. When maximum pressure is reached, the influent-sludge flow is stopped and the pressure filter is allowed to discharge the residual filtrate prior to opening the filter and allowing the filter cake to drop by gravity to a conveyor belt below the filter press. The pressure filter operates at a pressure between 689 and 1380 kPa (100 and 200 psig) and takes 1.5 to 4 h for the pressure cycle. Normally, 20 to 30 min is required to remove the filter cake. The sludge cakes will vary from 20 to 25 percent for waste activated sludge to 50 percent for primary sludge. Chemical conditioning is necessary to obtain good dewatering of the sludges.

Belt-Press Filters The newest filter for handling waste activated sludge is the belt-press filter. The belt press utilizes a continuous cloth-filter belt. Waste activated sludge is spread over the filter medium, and water is removed initially by gravity. The open belt with the sludge moves into contact with a second moving belt, which squeezes the sludge layer between rollers with ever-increasing pressure. The sludge cake is removed at the end of the filter press by a knife blade, with the sludge dropping by gravity to a conveyor belt. Belt-press filters can produce sludge with 20–30 percent solids.

Sand Beds Sand filter beds can be used to dewater either anaerobically or aerobically digested sludges. They work best on relatively small treatment systems located in relatively dry areas. The sand bed consists of coarse gravel graded to fine sand in a series of layers to a depth of 0.45 to 0.6 m (1.5 to 2 ft). The digested sludge is placed over the entire filter surface to a depth of 0.3 m (12 in) and allowed to sit until dry. Free water will drain through the sand bed to an open pipe underdrain system and be removed from the filter. Air drying will slowly remove the remaining water. The sludge must be cleaned from the bed by hand prior to adding a second layer of sludge. The sludge layer will drop from an initial thickness of 3 m (12 in) to about 0.006 m (¼ in). An open sand bed can generally handle 49 to 122 kg dry solids/$(m^2 \cdot year)$ [10 to 25 lb/$(ft^2 \cdot year)$]. Covered sand beds have been used in wet climates as well as in cold climates, but economics does not favor their use.

SLUDGE DISPOSAL

Incineration Incineration has been used to reduce the volume of sludge after dewatering. The organic fractions in sludges lend themselves to incineration if the sludge does not have an excessive water content. Multiple-hearth and fluid-bed incinerators have been extensively used for sludge combustion.

A multiple-hearth incinerator consists of several hearths in a vertical cylindrical furnace. The dewatered sludge is added to the top hearth and is slowly pushed through the incinerator, dropping by gravity to the next lower layer until it finally reaches the bottom layer. The top layer is used for drying the sludge with the hot gases from the lower layers. As the temperature of the furnace increases, the organics begin to degrade and undergo combustion. Air is used to add the necessary oxygen and to control the temperature during combustion. It is very important to keep temperatures above 600° C to ensure complete oxidation of the volatile organics. One of the problems with the multiple-hearth incinerator is volatilization of odorous organics during the drying phase before the temperature reaches combustion levels. Even afterburners on the exhaust-gas line may not be adequate for complete oxidation. Air-pollution-control devices are required on all incinerators to remove fly ash and corrosive gases. The ash from the incinerator must be cooled, collected, and conveyed back to the environment, normally to a sanitary landfill for burial. The residual ash will weigh from 10 to 30 percent of the original dry weight of the sludge. Supplemental fuels are needed to start the incinerator and to ensure adequate temperatures with sludges containing excessive moisture, such as activated sludge. Heat recovery from wastes is being given more consideration. It is possible to combine the sludges with other wastes to provide a better fuel for the incinerator.

A fluid-bed incinerator uses hot sand as a heat reservoir for dewatering the sludge and combusting the organics. The turbulence created by the incoming air and the sand suspension requires the effluent gases to be treated in a wet scrubber prior to final discharge. The ash is removed from the scrubber water by a cyclone separator. The scrubber water is normally returned to the treatment process and diluted with the total plant effluent. The ash is normally buried.

Sanitary Landfills Dewatered sludge, either raw or digested, is often buried in a sanitary landfill to minimize the environmental impact. Increased concern over sanitary landfills has made it more difficult simply to bury dewatered sludge. Sanitary landfills must be made secure from leachate and be monitored regularly to ensure that no environmental damage occurs. The moisture content of most sludges makes them a problem at sanitary landfills designed for solid wastes, requiring separate burial even at the same landfill.

Land Spreading The nutrient content of most sludges makes them useful as fertilizers or as soil conditioners if properly mixed with the surface soil. Land spreading has gained in popularity in agricultural areas. Normally, the rate of application of sludge to land is controlled by the nitrogen content of the sludge. Since nitrogen uptake varies with different crops, nitrogen application is limited to approximately twice the annual uptake of nitrogen by the proposed crop. Approximately one-half of the nitrogen is readily available in sludge. Nutrient release with sludge is slower than with chemical fertilizers, allowing the nutrients to become available as the crop needs it. Activated sludge appears to be an excellent soil conditioner because the humus material in the sludge provides a good matrix for root growth, while the nutrient elements are released in approximately the right combination for optimal plant growth. There is a growing concern over heavy metals in some sludge, and care should be taken to minimize heavy-metal concentrations in sludges placed on the land. Since heavy metals cannot be easily removed from sludges, it is important to prevent them from entering the wastewater-treatment system. Greater concern will be placed on other potentially toxic or hazardous materials, including some organic compounds such as pesticides and PCBs. Land spreading of sludge requires careful application of the sludge at the surface and its mixing with the soil. Soil microbes will assist in further stabilization of any biodegradable organics remaining. Land spreading of sludge will become more popular as energy and nutrients become scarcer.

MANAGEMENT OF SOLID WASTES

INTRODUCTION

"Solid wastes" are all the wastes arising from human and animal activities that are normally solid and that are discarded as useless or unwanted. The term as used in this subsection is all-inclusive, and it encompasses the heterogeneous mass of throwaways. The three R's should be applied to Solid Wastes: Reuse, Recycle, and Reduce. When these have been implemented, management of residual solid waste can be addressed.

Functional Elements The activities associated with the management of solid wastes from the point of generation to final disposal have been grouped into the functional elements identified in Fig. 25-59. By considering each fundamental element separately, it is possible to (1) identify the fundamental element and (2) develop, when possible, quantifiable relationships for the purpose of making engineering comparisons, analyses, and evaluations.

Waste Reduction Processes can be redesigned to reduce the amount of waste generated. For example, transfer lines between processes can be blown clear pneumatically to drive liquid into the batch mix tank.

Waste Generation Waste generation encompasses those activities in which materials are identified as no longer being of value and are either thrown away or gathered together for disposal. From the standpoint of economics, the best place to sort waste materials for recovery is at the source of generation.

Reuse Waste may be diverted to reuse. For example, containers may be cleaned and reused or the waste from one process may become the feedstock for another.

On-Site Handling, Storage, and Processing This functional element encompasses those activities associated with the handling, storage, and processing of solid wastes at or near the point of generation. On-site storage is of primary importance because of the aesthetic considerations, public health, public safety, and economics involved.

Collection The functional element of collection includes the gathering of solid wastes and the hauling of wastes after collection to the location where the collection vehicle is emptied. As shown in Fig. 25-59, this location may be a transfer station, a processing station, or a landfill disposal site.

Transfer and Transport The functional element of transfer and transport involves two steps: (1) the transfer of wastes from the smaller collection vehicle to the larger transport equipment and (2) the subsequent transport of the wastes, usually over long distances, to the disposal site.

Processing and Recovery The functional element of processing and recovery includes all the techniques, equipment, and facilities used both to improve the efficiency of the other functional elements and to recover usable materials, conversion products, or energy from solid wastes. Materials that can be recycled are exported to facilities equipped to do so. Residues go to disposal.

Disposal The final functional element in the solid-waste-management system is disposal. Disposal is the ultimate fate of all solid wastes, whether they are wastes collected and transported directly to a landfill site, semisolid wastes (sludge) from industrial treatment plants and air-pollution-control devices, incinerator residue, compost, or other substances from various solid-waste processing plants that are of no further use.

Solid-Waste-Management Systems Practical aspects associated with solid-waste-management systems not covered in the presentation include financing, operations, equipment management, personnel, reporting, cost accounting and budgeting, contract administration, ordinances and guidelines, and public communications.

UNITED STATES LEGISLATION, REGULATIONS, AND GOVERNMENTAL AGENCIES

Much of the current activity in the field of solid-waste management, especially with respect to hazardous wastes and resources recovery, is a direct consequence of legislation. It is imperative to have a working knowledge of waste regulations, including RCRA (for EPA hazardous waste); TSCA (Toxic Substances Control Act) for PCBs and toxic waste; Solid Waste Disposal Act; the Clean Air Act; and PSD (prevention of

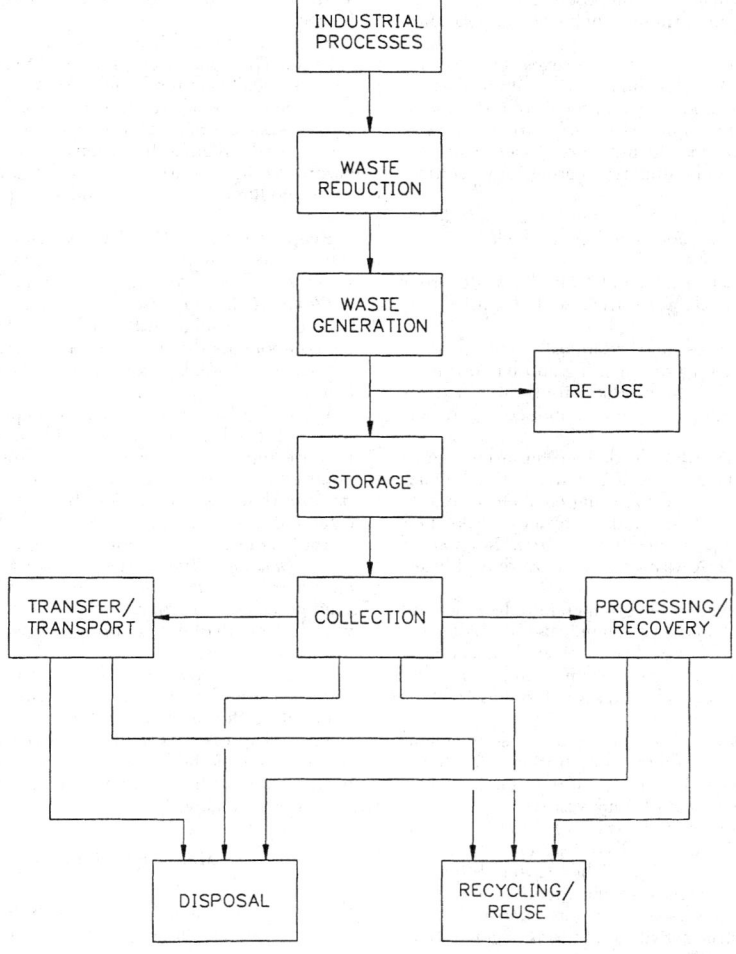

FIG. 25-59 Functional elements in a solid-waste management system. (*Updated from G. Tchobanoglous, H. Theisen, and R. Eliassen,* Solid Wastes: Engineering and Management Issues, *McGraw-Hill, New York, 1977.*)

significant deterioration) regulations. The Clean Air Act has far-reaching implications for any waste-related operation (e.g., incineration or landfill) that emits particulates or gaseous emissions. Primary elements of the act are Title I, dealing with fugitive emissions monitoring, Title III with reduction of organic hazardous air pollutants, and Title V, operating permits. The 1990 amendments have added a significant permitting burden for industrial plants and require additional controls for HAPs (hazardous air pollutants). In addition, state and local regulations apply to waste operations. Regulations are explored in detail in "Introduction to Waste Management" elsewhere in this section.

Regulations have been adopted by federal, state, and local agencies to implement the legislation. Before any design and construction work is undertaken, these regulations must be monitored due to continuous revisions.

GENERATION OF SOLID WASTES

Solid wastes, as noted previously, include all solid or semisolid materials that are no longer considered of sufficient value to be retained in a given setting. The types and sources of solid wastes, the physical and chemical composition of solid wastes, and typical solid-waste generation rates are considered in this subsection.

Types of Solid Wastes   The term *solid wastes* is all-inclusive and encompasses all sources, types of classifications, compositions, and properties. As a basis for subsequent discussions, it will be helpful to define the various types of solid wastes that are generated. It is important to note that the definitions of solid-waste terms and the classifications vary greatly in practice and in literature. Consequently, the use of published data requires considerable care, judgment, and common sense. The following definitions are intended to serve as a guide.

1. *Food wastes.*   Food wastes are the animal, fruit, or vegetable residues (also called *garbage*) resulting from the handling, preparation, cooking, and eating of foods. The most important characteristic of these wastes is that they are putrescible and will decompose rapidly, especially in warm weather.

2. *Rubbish.*   Rubbish consists of combustible and noncombustible solid wastes, excluding food wastes or other putrescible materials. Typically, combustible rubbish consists of materials such as paper, cardboard, plastics, textiles, rubber, leather, wood, furniture, and garden trimmings. Noncombustible rubbish consists of items such as glass, crockery, tin cans, aluminum cans, ferrous and other nonferrous metals, dirt, and construction wastes.

3. *Ashes and residues.*   These are the materials remaining from the burning of wood, coal, coke, and other combustible wastes.

Residues from power plants normally are composed of fine, powdery materials, cinders, clinkers, and small amounts of burned and partially burned materials.

4. *Demolition and construction wastes.* Wastes from razed building and other structures are classified as demolition wastes. Wastes from the construction, remodeling, and repair of commercial and industrial buildings and other similar structures are classified as construction wastes. These wastes may include dirt, stones, concrete, bricks, plaster, lumber, shingles, and plumbing, heating, and electrical parts.

5. *Special wastes.* Wastes such as street sweepings, roadside litter, catch-basin debris, dead animals, and abandoned vehicles are classified as special wastes.

6. *Treatment-plant wastes.* The solid and semisolid wastes from water, wastewater, and industrial waste-treatment facilities are included in this classification.

7. *Agricultural wastes.* Wastes and residues resulting from diverse agricultural activities, such as the planting and harvesting of row, field, and tree and vine crops, the production of milk, the production of animals for slaughter, and the operation of feedlots are collectively called agricultural wastes.

Hazardous Wastes The U.S. EPA has defined hazardous waste in RCRA regulations, CFR Parts 260 and 261. A waste may be hazardous if it exhibits one or more of the following characteristics: (1) ignitability, (2) corrosivity, (3) reactivity, and (4) toxicity. A detailed definition of these terms was first published in the *Federal Register* on May 19, 1980, pages 33, 121–122. A waste may be hazardous if listed in Appendix VIII.

In the past, hazardous wastes were often grouped into the following categories: (1) radioactive substances, (2) chemicals, (3) biological wastes, (4) flammable wastes, and (5) explosives. The chemical category included wastes that were corrosive, reactive, and toxic. The principal sources of hazardous biological wastes are hospitals and biological-research facilities.

Sources of Industrial Wastes Knowledge of the sources and types of solid wastes, along with data on the composition and rates of generation, is basic to the design and operation of the functional elements associated with the management of solid wastes.

Conventional Wastes Sources and types of industrial solid wastes generated by Standard Industrial Classification (SIC) group classification are reported in Table 25-49. The expected specific wastes in the table are those that are most readily identifiable.

Hazardous Wastes Hazardous wastes are generated in limited amounts throughout most industrial activities. In terms of generation, concern is with the identification of amounts and types of hazardous wastes developed at each source, with emphasis on those sources where significant waste quantities are generated.

The USEPA released "Biennial RCRA Hazardous Waste Report" in November 1994. The six-volume report, which represents comprehensive data on the generation and management of hazardous waste in the United States during 1991, noted that 306 million tons of hazardous waste generated that year represented an increase of 108 million tons compared with the 198 million tons generated in 1989. The number of large-quantity generators increased by about 3000 from 20,426 reporting in 1989. The number of waste treatment, storage, and disposal (TSD) facilities increased from 3000 to 3862 during the same period. Less than 5 percent—13 million tons—of the 306 million tons generated during 1991 were shipped offsite to commercial TSD facilities.

The majority—76 percent—of the 306 million tons of hazardous waste generated during 1991 was managed in aqueous waste-treatment units. Land disposal accounted for another 9 percent of the total, divided as follows: 23 million tons injected into underground wells; 1.7 million tons sent to landfills; 240,00 tons placed in surface impoundments; and 52,000 tons managed by land farming. Resource-recovery operations managed 2 percent of the 306 million tons, with solvent-recovery units managing 3.6 million tons; fuel-blending units, 1.4 million tons; metals recovery, 1 million tons; and other methods such as acid regeneration and waste oil recovery accounting for 480,000 tons. Thermal treatment systems burned 1.1 percent of the hazardous wastes generated in 1991: 1.9 million tons were incinerated, and 1.5 million tons were used as fuel in boilers and industrial furnaces.

The generation of hazardous wastes by spillage must also be considered. The quantities of hazardous wastes that are involved in spillage usually are not known. After a spill, the wastes requiring collection and disposal are often significantly greater than the amount of spilled wastes, especially when an absorbing material, such as straw, is used to soak up liquid hazardous wastes or when the soil into which a hazardous liquid waste has percolated must be excavated. Both the straw and liquid and the soil and the liquid are classified as hazardous wastes.

Properties of Solid Wastes Information on the properties of solid wastes is important in evaluating alternative equipment needs, systems, and management programs and plans.

Physical Composition Information and data on the physical composition of solid wastes including (1) identification of the individual components that make up industrial and municipal solid wastes, (2) density of solid wastes, and (3) moisture content are presented below.

1. *Individual components.* Components that typically make up most industrial and municipal solid wastes and their relative distribution are reported in Table 25-50. Although any number of components could be selected, those listed in the table have been chosen because they are readily identifiable, are consistent with component categories reported in the literature, and are adequate for the characterization of solid wastes for most applications.

2. *Density.* Typical densities for various wastes as found in containers are reported by source in Table 25-50. Because the densities of solid wastes vary markedly with geographical location, season of the year, and length of time in storage, great care should be used in selecting typical values.

3. *Moisture content.* The moisture content of solid wastes usually is expressed as the mass of moisture per unit mass of wet or dry material. In the wet-mass method of measurement, the moisture in a sample is expressed as a percentage of the wet mass of the material; in the dry-mass method, it is expressed as a percentage of the dry mass of the material. In equation form, the wet-mass moisture content is expressed as follows:

$$\text{Moisture content (\%)} = \left(\frac{a - b}{a}\right) \times 100 \qquad (25\text{-}22)$$

where: a = initial mass of sample as delivered
 b = mass of sample after drying

Typical data on the moisture content for the solid-waste components are given in Table 25-51. For most industrial solid wastes, the moisture content will vary from 10 to 25 percent.

Chemical Composition Information on the chemical composition of solid wastes is important in evaluating alternative processing and recovery options. If solid wastes are to be used as fuel, the four most important properties to be known are:

1. Proximate analysis
 a. Moisture (loss at 105° C for 1 h)
 b. Volatile matter (additional loss on heating to 950° C)
 c. Ash (residue after burning)
 d. Fixed carbon (remainder)
2. Fusion point of ash
3. Ultimate analysis, percent of C (carbon), H (hydrogen), O (oxygen), N (nitrogen), S (sulfur), and ash
4. Heating value
5. Organic chlorine

Typical proximate-analysis data for the combustible components of industrial and municipal solid wastes are presented in Table 25-52.

Typical data on the inert residue and energy values for solid wastes may be converted to a dry basis by using Eq. (25-23).

$$\frac{\text{kJ}}{\text{kg (dry basis)}} = \frac{\text{kJ}}{\text{kg (as discarded)}} \left(\frac{100}{100 - \% \text{ moisture}}\right) \qquad (25\text{-}23)$$

The corresponding equation on an ash-free dry basis is

TABLE 25-49 Sources and Types of Industrial Wastes*

Code	SIC group classification	Waste-generating processes	Expected specific wastes
19	Ordnance and accessories	Manufacturing, assembling	Metals, plastic, rubber, paper, wood, cloth, chemical residues
20	Food and kindred products	Processing, packaging, shipping	Meats, fats, oils, bones, offal, vegetables, fruits, nuts and shells, cereals
22	Textile mill products	Weaving, processing, dyeing, shipping	Cloth and filter residues
23	Apparel and other finished products	Cutting, sewing, sizing, pressing	Cloth, fibers, metals, plastics, rubber
24	Lumber and wood products	Sawmills, millwork plants, wooden containers, miscellaneous wood products, manufacturing	Scrap wood, shavings, sawdust; in some instances, metals, plastics, fibers, glues, sealers, paints, solvents
25a	Furniture, wood	Manufacture of household and office furniture, partitions, office and store fixtures, mattresses	Those listed under Code 24; in addition, cloth and padding residues
25b	Furniture, metal	Manufacture of household and office furniture, lockers, springs, frames	Metals, plastics, resins, glass, wood, rubber, adhesives, cloth, paper
26	Paper and allied products	Paper manufacture, conversion of paper and paperboard, manufacture of paperboard boxes and containers	Paper and fiber residues, chemicals, paper coatings and filters, inks, glues, fasteners
27	Printing and publishing	Newspaper publishing, printing, lithography, engraving, bookbinding	Paper, newsprint, cardboard, metals, chemicals, cloth, inks, glues
28	Chemicals and related products	Manufacture and preparation of inorganic chemicals (ranging form drugs and soaps to paints and varnishes and explosives)	Organic and inorganic chemicals, metals, plastics, rubber, glass, oils, paints, solvents, pigments
29	Petroleum refining and related industries	Manufacture of paving and roofing materials	Asphalt and tars, felts, paper, cloth, fiber
30	Rubber and miscellaneous plastic products	Manufacture of fabricated rubber and plastic products	Scrap rubber and plastics, lampblack, curing compounds, dyes
31	Leather and leather products	Leather tanning and finishing, manufacture of leather belting and packing	Scrap leather, thread, dyes, oils, processing and curing compounds
32	Stone, clay, and glass products	Manufacture of flat glass, fabrication or forming of glass,; manufacture of concrete, gypsum, and plaster products; forming and processing of stone products, abrasives, asbestos, and miscellaneous non-mineral products	Glass, cement, clay, ceramics, gypsum, asbestos, stone, paper, abrasives
33	Primary metal industries	Melting, casting, forging, drawing, rolling, forming, extruding operations	Ferrous and non-ferrous metals scrap, slag, sand, cores, patterns, bonding agents
34	Fabricated metal products	Manufacture of metal cans, hand tools, general hardware, non-electrical heating apparatus, plumbing fixtures, fabricated structural products, wire, farm machinery and equipment, coating and engraving of metal	Metals, ceramics, sand, slag, scale, coatings, solvents, lubricants, pickling liquors
35	Machinery (except electrical)	Manufacture of equipment for construction, elevators, moving stairways, conveyors, industrial trucks, trailers, stackers, machine tools, etc.	Slag, sand, cores, metal scrap, wood, plastics, resins, rubber, cloth, paints, solvents, petroleum products
36	Electrical	Manufacture of electrical equipment, appliances and communication apparatus, machining, drawing, forming, welding, stamping, winding, painting, plating, baking, firing operations	Metal scrap, carbon, glass, exotic metals, rubber, plastics, resins, fibers, cloth residues, PCBs
37	Transportation equipment	Manufacture of motor vehicles, truck and bus bodies, motor-vehicle parts and accessories, aircraft and parts, ship and boat building, repairing motorcycles and bicycles and parts, etc.	Metal scrap, glass, fiber, wood, rubber, plastics, cloth, paints, solvents, petroleum products
38	Professional scientific controlling instruments	Manufacture of engineering, laboratory, and research instruments and associated equipment	Metals, plastics, resins, glass, wood, rubber, fibers, abrasives
39	Miscellaneous manufacturing	Manufacture of jewelry, silverware, plated ware, toys, amusement, sporting and athletic goods, costume novelties, buttons, brooms, brushes, signs, advertising displays	Metals, glass, plastics, resin, leather, rubber, composition, bone, cloth, straw, adhesives, paints, solvents

*From C.L. Mantell (ed.), *Solid Wastes: Origin, Collection, Processing and Disposal*, Wiley-Interscience, New York, 1975. PCBs added to item 36.

$$\frac{kJ}{kg \text{ (ash-free dry basis)}} = \frac{kJ}{kg \text{ (as discarded)}}$$

$$\times \left(\frac{100}{100 - \% \text{ ash} - \% \text{ moisture}} \right) \quad (25\text{-}24)$$

Representative data on the ultimate analysis of typical industrial and municipal-waste components are presented in Table 25-53. If energy values are not available, approximate values can be determined by using Eq. (25-25), known as the modified Dulong formula, and the data in Table 25-53.

$$\frac{kJ}{kg} = 337C + 1428 \left(H - \frac{1}{8}O \right) + 95S \quad (25\text{-}25)$$

where: C = carbon, percent
H = hydrogen, percent
O = oxygen, percent
S = sulfur, percent

Quantities of Solid Wastes Representative data on the quantities of solid wastes and factors affecting the generation rates are considered briefly in the following paragraphs.

TABLE 25-50   Typical Data on Distribution of Industrial Wastes Generated by Major Industries and Municipalities*

SIC code	Food wastes†	Paper	Wood	Leather	Rubber	Plastics	Metals	Glass	Textiles	Miscellaneous
20 Food and kindred products	15–20	50–60	5–10	0–2	0–2	0–5	5–10	4–10	0–2	5–15
22 Textile mill products	0–2	40–50	0–2	0–2	0–2	3–10	0–2	0–2	20–40	0–5
23 Apparel and other finished products	0–2	40–60	0–2	0–2	0–2	0–2	0–2	0–2	30–50	0–5
24 Lumber and wood products	0–2	10–20	60–80	0–2	0–2	0–2	0–2	0–2	0–2	5–10
25a Furniture, wood	0–2	20–30	30–50	0–2	0–2	0–2	0–2	0–2	0–5	0–5
25b Furniture, metal	0–2	20–40	10–20	0–2	0–2	0–2	20–40	0–2	0–5	0–10
26 Paper and allied products	0–2	40–60	10–15	0–2	0–2	0–2	5–15	0–2	0–2	10–20
27 Printing and publishing	0–2	60–90	5–10	0–2	0–2	0–2	0–2	0–2	0–2	0–5
28 Chemicals and related products	0–2	40–60	2–10	0–2	0–2	5–15	5–10	0–5	0–2	15–25
29 Petroleum refining and related industries	0–2	60–80	5–15	0–2	0–2	10–20	2–10	0–12	0–2	2–10
30 Rubber and miscellaneous plastic products	0–2	40–60	2–10	0–2	5–20	10–20	0–2	0–2	0–2	0–5
31 Leather and leather products	0–2	5–10	5–10	40–60	0–2	0–2	10–20	0–2	0–2	0–5
32 Stone, clay, and glass products	0–2	20–40	2–10	0–2	0–2	0–2	5–10	10–20	0–2	30–50
33 Primary metal industries	0–2	30–50	5–15	0–2	0–2	2–10	2–10	0–5	0–2	20–40
34 Fabricated metal products	0–2	30–50	5–15	0–2	0–2	0–2	15–30	0–2	0–2	5–15
35 Machinery (except electrical)	0–2	30–50	5–15	0–2	0–2	1–5	15–30	0–2	0–2	0–5
36 Electrical	0–2	60–80	5–15	0–2	0–2	2–5	2–5	0–2	0–2	0–5
37 Transportation equipment	0–2	40–60	5–15	0–2	0–2	2–5	0–2	0–2	0–2	15–30
38 Professional scientific controlling instruments	0–2	30–50	2–10	0–2	0–2	5–10	5–15	0–2	0–2	0–5
39 Miscellaneous manufacturing	0–2	40–60	10–20	0–2	0–2	5–15	2–10	0–2	0–2	5–15
Municipal	10–20	40–60	1–4	0–2	0–2	2–10	3–15	4–16	0–4	5–30

*Adapted in part from D. G. Wilson (ed.), *Handbook of Solid Waste Management*, van Nostrand Reinhold, New York, 1977.
†With the exception of food and kindred products, food wastes are from company cafeterias, canteens, etc.

Typical Generation Rates   Typical unit waste-generation rates for selected industrial sources are reported in Table 25-54. Because waste-generation practices are changing so rapidly, the presentation of "typical" waste-generation data may not be reliable.

Factors That Affect Generation Rates   Factors that influence the quantity of industrial wastes generated include (1) the extent of salvage and recycle operations, (2) company attitudes, and (3) legislation. The existence of salvage and recycling operations within an industry definitely affects the quantities of wastes collected. Whether such operations affect the quantities generated is another matter. The section entitled "Pollution Prevention" by Louis Theodore points out activities by industrial plants as a result of the 1990 Pollution Prevention Act. Significant reductions in the quantities of solid wastes that are generated will occur when and if companies are willing to change—on their own volition—to conserve national resources, handle waste at its source rather than "end of pipe" pollution control, and to reduce the economic burdens associated with the management of solid wastes. Perhaps the most important factor affecting the generation of certain types of wastes is the existence of local, state, and federal regulations concerning the use and disposal of specific material. In general, the more regulated the waste, the higher the cost for treat-

ment and disposal and the greater the incentive to reduce generation of the waste.

ON-SITE HANDLING, STORAGE, AND PROCESSING

The handling, storage, and processing of solid wastes at the source before they are collected is the second of the six functional elements in the solid-waste-management system.

On-Site Handling   On-site handling refers to the activities associated with the handling of solid wastes until they are placed in the containers used for storage before collection. Depending on the type of collection service, handling may also be required to move loaded containers to the collection point and to return the empty containers to the point where they are stored between collections.

Conventional Solid Wastes   In most office, commercial, and industrial buildings, solid wastes that accumulate in individual offices or work locations usually are collected in relatively large containers mounted on casters. Once filled, these containers are removed by means of the service elevator, if there is one, and emptied into (1) large storage containers, (2) compactors used in conjunction with the storage containers, (3) stationary compactors that can compress the

TABLE 25-51   Typical Density and Moisture-Content Data for Domestic, Commercial, and Industrial Solid Waste

Item	Density, kg/m³ Range	Density, kg/m³ Typical	Moisture content, % by mass Range	Moisture content, % by mass Typical
Residential (Uncompacted)				
Food wastes (mixed)	130–480	290	50–80	70
Paper	40–130	85	4–10	6
Cardboard	40–80	50	4–8	6
Plastics	40–130	65	1–4	2
Textiles	40–100	65	6–15	10
Rubber	100–200	130	1–4	2
Leather	100–260	160	8–12	10
Garden trimmings	60–225	100	30–80	60
Wood	130–320	240	15–40	20
Glass	160–480	195	1–4	2
Tin cans	50–160	90	2–4	3
Nonferrous metals	65–240	160	2–4	2
Ferrous metals	130–1150	320	2–4	2
Dirt, ashes, etc.	320–1000	480	6–12	8
Ashes	650–830	745	6–12	6
Rubbish (mixed)	90–180	130	5–20	15
Residential (compacted)				
In compactor truck	180–450	300	15–40	20
In landfill (normally compacted)	360–500	450	15–40	30°
In landfill (well-compacted)	590–740	600	15–40	30°
Commercial				
Food wastes (wet)	475–950	535	50–85	75
Appliances	150–200	180	0–5	
Wooden crates	110–160	110	10–30	20
Tree trimmings	100–180	150	20–80	50
Rubbish (combustible)	50–180	120	5–25	15
Rubbish (non-combustible)	180–360	300	5–15	10
Rubbish (mixed)	140–180	160	5–20	12
Construction; demolition				
Mixed demolition (non-combustible)	1000–1600	1420	2–10	4
Mixed demolition (combustible)	300–400	360	4–15	8
Mixed construction (combustible)	180–360	260	4–15	8
Broken concrete	1200–1800	1540	0–5	
Industrial wastes				
Chemical sludges (wet)	800–1100	1000	75–99	80
Fly ash	700–900	800	2–10	4
Leather scraps	100–250	160	6–15	10
Metal scraps (heavy)	1500–2000	1780	0–5	
Metal scrap (light)	500–900	740	0–5	
Metal scrap (mixed)	700–1500	900	0–5	
Oils, tars, asphalt	800–1000	950	0–5	2
Sawdust	100–350	290	10–40	15
Textile wastes	100–220	180	6–15	10
Wood (mixed)	400–675	500	10–40	20
Agricultural wastes				
Agricultural (mixed)	400–750	560	40–80	50
Fruit wastes (mixed)	250–750	360	60–90	75
Manure (wet)	900–1050	1000	75–96	94
Vegetable wastes (mixed)	200–700	360	50–80	65

°Depends on degree of surface-water infiltration.

material into bales or into specially designed containers, or (4) other processing equipment.

Hazardous Wastes   When hazardous wastes are generated, special containers are usually provided, and trained personnel (OSHA 1910.120 required such workers to have HAZWOPER training) are responsible (or should be) for the handling of these wastes. Hazardous wastes include solids, sludges, and liquids; hence, container requirements vary with the form of waste.

On-Site Storage   Factors that must be considered in the on-site storage of solid wastes include (1) the type of container to be used, (2) the container location, (3) public health and aesthetics, (4) the collection methods to be used, and (5) future transport method.

Containers   To a large extent, the types and capacities of the containers used depend on the characteristics of the solid wastes to be collected, the collection frequency, and the space available for the placement of containers.

1. *Containers for conventional wastes.*   The types and capacities of containers now commonly used for on-site storage of solid wastes

are summarized in Table 25-55. The small containers are used in individual offices and work stations. The medium-size and large containers are used at locations where large volumes are generated.

2. *Containers for hazardous wastes.*   On-site storage practices are a function of the types and amounts of hazardous wastes generated and the time period over which waste generation occurs. Usually, when large quantities are generated, special facilities that have sufficient capacity to hold wastes accumulated over a period of several days are used. When only small amounts of hazardous wastes are generated on an intermittent basis, they may be containerized, and limited quantities may be stored for periods covering months or years. General information on the storage containers used for hazardous wastes and the conditions of their use is presented in Table 25-56.

Container Location.   The location of containers at existing commercial and industrial facilities depends on both the location of available space and service-access conditions. In newer facilities, specific service areas have been included for this purpose. Often, because the containers are not owned by the commercial or industrial activity, the

TABLE 25-52 Typical Proximate-Analysis and Energy-Content Data for Components in Domestic, Commercial, and Industrial Solid Waste*

Component	Proximate analysis, % by mass				Energy content, kJ/kg		
	Moisture	Volatile matter	Fixed carbon	Non-combustible	As collected	Dry	Moisture- and ash-free
Food and food products							
Fats	2.0	95.3	2.5	0.2	37,530	38,296	38,374
Food wastes (mixed)	70.0	21.4	3.6	5.0	4,175	13,917	16,700
Fruit wastes	78.7	16.6	4.0	0.7	3,970	18,638	19,271
Meat wastes	38.8	56.4	1.8	3.1	17,730	28,970	30,516
Paper products							
Cardboard	5.2	77.5	12.3	5.0	16,380	17,278	18,240
Magazines	4.1	66.4	7.0	22.5	12,220	12,742	16,648
Newsprint	6.0	81.1	11.5	1.4	18,550	19,734	20,032
Paper (mixed)	10.2	75.9	8.4	5.4	15,815	17,611	18,738
Waxed cartons	3.4	90.9	4.5	1.2	26,345	27,272	27,615
Plastics							
Plastics (mixed)	0.2	95.8	2.0	2.0	32,000	32,064	32,720
Polyethylene	0.2	98.5	<0.1	1.2	43,465	43,552	44,082
Polystyrene	0.2	98.7	0.7	0.5	38,190	38,266	38,216
Polyurethane	0.2	87.1	8.3	4.4	26,060	26,112	27,316
Polyvinyl chloride	0.2	86.9	10.8	2.1	22,690	22,735	23,224
Wood, trees, etc.							
Garden trimmings	60.0	30	9.5	0.5	6,050	15,125	15,316
Green wood	50.0	42.3	7.3	0.4	4,885	9,770	9,848
Hardwood	12.0	75.1	12.4	0.5	17,100	19,432	19,542
Wood (mixed)	20.0	67.9	11.3	0.8	15,444	19,344	19,500
Leather, rubber, textiles, etc.							
Leather (mixed)	10	68.5	12.5	9.0	18,515	20,572	22,858
Rubber (mixed)	1.2	83.9	4.9	9.9	25,330	25,638	28,493
Textiles (mixed)	10	66.0	17.5	6.5	17,445	19,383	20,892
Glass, metals, etc.							
Glass and mineral	2	—	—	96–99+	196†	200	200
Metal, tin cans	5	—	—	94–99+	1,425†	1,500	1,500
Metals, ferrous	2	—	—	96–99+	—	—	—
Metals, nonferrous	2	—	—	94–99+	—	—	—
Miscellaneous							
Office sweepings	3.2	20.5	6.3	70	8,535	8,817	31,847
Multiple wastes	20 (15–40)	53 (30–60)	7 (5–15)	20 (9–30)	10,470	13,090	17,450
Industrial wastes	15 (10–30)	58 (30–60)	7 (5–15)	20 (10–30)	11,630	13,682	17,892

*Adapted in part from D. G. Wilson (ed.), *Handbook of Solid Waste Management*, Van Nostrand Reinhold, New York, 1977, and G. Tchobanoglous, H. Theisen, and R. Eliassen, *Solid Wastes: Engineering Principles and Management Issues*, McGraw-Hill, New York, 1977.
†Energy content is from coatings, labels, and attached materials.

locations and types of containers to be used for on-site storage must be worked out jointly between the industry and the public or private collection agency.

On-Site Processing of Solid Wastes On-site-processing methods are used to (1) recover usable materials from solid wastes, (2) reduce the volume, or (3) alter the physical form. The most common on-site-processing operations as applied to large commercial and industrial sources include manual sorting, compaction, and incineration. These and other processing operations are considered in the portion of this section dealing with processing and resource recovery. Factors that should be considered in the selection of on-site-processing equipment are summarized in Table 25-57.

COLLECTION OF SOLID WASTES

Information on collection, one of the most costly functional elements, is presented in four parts dealing with (1) the types of collection services, (2) the types of collection systems, (3) an analysis of collection systems, and (4) the general methodology involved in setting up collection routes.

Collection Services The various types of collection services now used for commercial-industrial sources are described in this subsection.

Commercial-Industrial The collection service provided to large commercial and industrial activities typically is centered in the use of large moveable and stationary containers and large stationary com-

pactors. Compactors are of the type that can be used to compress material directly into large containers (see Fig. 25-60) or to form bales that are then placed in large containers.

Hazardous Wastes Hazardous Wastes for delivery to a treatment or disposal facility normally are collected by the waste producer or a licensed, specialized hauler. Typically, the loading of collection vehicles is completed in one of two ways: (1) wastes stored in large-capacity tanks are either drained or pumped into collection vehicles, and (2) wastes stored in sealed drums or other sealed containers are loaded by hand or by mechanical equipment onto flatbed trucks. To avoid accidents and possible loss of life, two collectors should always be assigned when hazardous wastes are to be collected.

Types of Collection Systems On the basis of their mode of operation, collection systems are classified into two categories: (1) hauled-container systems and (2) stationary-container systems.

Hauled-Container Systems (HCS) Collection systems in which the containers used for storage of wastes are hauled to the processing, transfer, or disposal site, emptied, and returned to either their original location or some other location are defined as "hauled-container systems." In most hauled-container systems, a single collector is used. The collector is responsible for driving the vehicle, loading full containers and unloading empty containers, and emptying the contents of the container at the disposal site. In some cases, for safety reasons, both a driver and helper are used.

There are three main types of hauled-container systems: (1) hoist

TABLE 25-53 Typical Ultimate-Analysis Data for Components in Domestic, Commercial and Industrial Solid Waste*

Components	Percent by mass (dry basis)					
	Carbon	Hydrogen	Oxygen	Nitrogen	Sulfur	Ash
Foods and food products						
Fats	73.0	11.5	14.8	0.4	0.1	0.2
Food wastes (mixed)	48.0	6.4	37.6	2.6	0.4	5.0
Fruit wastes	48.5	6.2	39.5	1.4	0.2	4.2
Meat wastes	59.6	9.4	24.7	1.2	0.2	4.9
Paper products	45.4	6.1	42.1	0.3	0.1	6.0
Cardboard	43.0	5.9	44.8	0.3	0.2	5.0
Magazines	32.9	5.0	38.6	0.1	0.1	23.3
Newsprint	49.1	6.1	43.0	<0.1	0.2	23.3
Paper (mixed)	43.4	5.8	44.3	0.3	0.2	6.0
Waxed cartons	59.2	9.3	30.1	0.1	0.1	1.2
Plastics						
Plastics (mixed)	60.0	7.2	22.8	—	—	10.0
Polyethylene	85.2	14.2	—	<0.1	<0.1	0.4
Polystyrene	87.1	8.4	4.0	0.2	—	0.3
Polyurethane†	63.3	6.3	17.6	6.0	<0.1	4.3
Polyvinyl chloride†	45.2	5.6	1.6	0.1	0.1	2.0
Wood, trees, etc.						
Garden trimmings	46.0	6.0	38.0	3.4	0.3	6.3
Green timber	50.1	6.4	42.3	0.1	0.1	1.0
Hardwood	49.6	6.1	43.2	0.1	<0.1	0.9
Wood (mixed)	49.5	6.0	42.7	0.2	<0.1	1.5
Wood chips (mixed)	48.1	5.8	45.5	0.1	<0.1	0.4
Glass, metals, etc.						
Glass and mineral‡	0.5	0.1	0.4	<0.1	—	98.9
Metals (mixed)	4.5	0.6	4.3	<0.1	—	90.5
Leather, rubber, textiles						
Leather (mixed)	60.0	8.0	11.6	10.0	0.4	10.0
Rubber (mixed)	69.7	8.7	—	—	1.6	20.0
Textiles (mixed)	48.0	6.4	40.0	2.2	0.2	3.2
Miscellaneous						
Office sweepings	24.3	3.0	4.0	0.5	0.2	68.0
Oils, paints	66.9	9.6	5.2	2.0	—	16.3
Refuse-derived fuel (RAF)	44.7	6.2	38.4	0.7	<0.1	9.9

*Adapted in part from D. G. Wilson (ed.), *Handbook of Solid Waste Management*, Van Nostrand Reinhold, New York, 1977, and G. Tchobanoglous, H. Theisen, and R. Eliassen, *Solid Wastes: Engineering Principles and Management Issues*, McGraw-Hill, New York, 1977.
†Remainder is chlorine.
‡Organic content is from coatings, labels, and other attached materials.

truck, (2) tilt-frame container, and (3) tractor trailer. Typical data on the containers and collection vehicles used with these systems are reported in Tables 25-58 and 25-59, respectively.

1. *Container carrier systems.* Container carrier systems are being used in a limited number of cases, the most important of which are (1) the collection of wastes from only a few pickup points at which a considerable amount of waste is generated and (2) the collection of bulky items and industrial rubbish not suitable for collection with compaction vehicles, and (3) small, heavy loads such as scrap metal.

2. *Roll-off container systems.* Systems that use tilt-frame-loaded vehicles and large containers, often called "roll-off boxes," are ideally suited for the collection of all typed of solid waste and rubbish from locations where the generation rate warrants the use of large containers. Open-top containers are used routinely at warehouses and construction sites. Large containers used in conjunction with stationary compactors are common at commercial and industrial services and transfer stations. Because of the large volume that can be hauled, the use of tilt-frame hauled-container systems has become widespread, especially among private collectors servicing industrial accounts. Roll-off boxes now predominate; however, permanently-attached, open-top, self-dumping containers are still in use. Self-contained compactors of 14–30 m³ capacity are used in conjunction with hoist trucks. This type has an integral ram and cylinder permanently built into the roll-off container. Other compactors are of the external type, with a detachable roll-off container. Compactors are excellent for volume reduction of bulky material such as cardboard boxes. In many areas, cardboard-recycling facilities will provide the compacting equipment, and the generator experiences a reduction in waste disposal costs.

3. *Tractor-trailer systems.* The application of tractor trailers is similar to that for tilt-frame-container systems. The use of a separate tractor increases the number of axles and the net weight of waste that can be hauled. Tractor trailers are better for the collection of especially heavy rubbish, such as sand, timber, and metal scrap, and they are often are used for the collection of demolition wastes at construction sites.

Stationary-Container Systems (SCS) Collection systems in which the containers used for the storage of wastes remain at the point of waste generation, except for occasional short trips to the collection vehicle, are defined as stationary-container systems. Labor requirements for mechanically loaded stationary-container systems are essentially the same as for hauled-container systems. There are two main types of stationary-container systems: (1) those in which self-loading compactors are used and (2) those in which manually loaded vehicles are used.

1. *Systems with self-loading compactors.* Container size and utilization are not as critical in stationary-container systems using self-loading collection vehicles equipped with a compaction mechanism (see Fig. 25-61 and Table 25-59) as they are in hauled-container systems. Trips to the disposal site, transfer station, or processing station are made after the contents of a number of containers have been collected and compacted and the collection vehicle is full. Because a variety of container sizes and types are available, these systems may be used for the collection of all types of wastes. Container sizes vary from relatively small sizes (0.6 m³) to sizes comparable to those handled with a hoist truck (see Table 25-58).

2. *Systems with manually loaded vehicles.* The major application of manual transfer and loading methods is in the collection of res-

TABLE 25-54 Unit Solid-Waste-Generation Rates for Selected Industrial Sources

Source	Unit	Range
Canned and frozen foods	Metric tons/metric tons of raw product	0.04–0.06
Printing and publishing	Metric tons/metric tons of raw paper	0.08–0.10
Automotive	Metric tons/vehicle produced	0.6–0.8
Petroleum refining	Metric tons/(employee day)	0.04–0.05
Rubber	Metric tons/metric tons of raw rubber	0.01–0.3

TABLE 25-55 Data on the Types and Sizes of Containers Used for On-Site Storage of Solid Wastes

Container type	Capacity			Dimensions°	
	Unit	Range	Typical	Unit	Typical
Small capacity					
Plastic or metal (office type)	L	16–40	28	mm	(180×300) B $\times (260 \times 380)$ T $\times 380$ H
Plastic or galvanized metal	L	75–150	120	mm	510 D $\times$ 660 H
Barrel, plastic, aluminum, or fiber barrel	L	20–250	120	mm	510 D $\times$ 660 H
Disposable paper bags (standard, leak-resistant, and leakproof)	L	75–210	120	mm	380 W $\times$ 300 d $\times$ 1100 H
Disposable plastic bag	L	20–200	170	mm	460 W $\times$ 380 d $\times$ 1000 H
Medium capacity					
Side or top loading	m^3	0.75–9	3	mm	1830 W $\times$ 1070 d $\times$ 1650 H
Bulk bags	m^3	0.3–2	1	mm	1000 W $\times$ 1000 d $\times$ 1000 H
Large capacity					
Open-top, roll-off (also called debris boxes)	m^3	9–38	27	mm	2440 W $\times$ 1830 H $\times$ 6100 L
Used with stationary compactor	m^3	15–30	23	mm	2440 W $\times$ 1830 H $\times$ 5490 L
Equipped with self-contained compaction mechanism	m^3	15–30	23	mm	2440 W $\times$ 2440 H $\times$ 6710 L
Trailer-mounted					
Open-top	m^3	15–38	27	mm	2440 W $\times$ 3660 H $\times$ 6100 L
Enclosed, equipped with self-contained compaction mechanism	m^3	15–30	27	mm	2440 W $\times$ 3660 H $\times$ 7320 L

°B = bottom, T = top, D = diameter, H = height, L = length, W = width, and d = depth.

TABLE 25-56 Typical Data on Containers Used for Storage and Transport of Hazardous Wastes

Waste category	Container		Auxiliary equipment and conditions of use
	Type	Capacity	
Radioactive substances	Lead encased in concrete	Varies with waste	Isolated storage buildings; high-capacity hoists and lighting
	Lined metal drums	210 L	equipment; special container markings
Corrosive, reactive, and toxic chemicals	Metal drums	210 L	Washing facilities for empty containers; special blending
	Plastic drums	210 L, up to 500 L	precautions to prevent hazardous reactions; incompatible
	Lined metal drums	210 L	wastes stored separately
	Lined and unlined storage tanks	Up to 20 m^3	
Liquids and sludges	Drums	142–3,400 L	Drum hand-trucks, pallets, forklifts
	Trucks	11–30 m^3	Transfer piping, hoses, pumps
	Vacuum tankers	11–15 m^3	Transfer piping, hoses, pumps
Biological wastes	Sealed plastic bags	120 L	Heat sterilization prior to bagging; special heavy-duty bags
	Lined boxes, Lined metal drums	57 L	with hazard warning printed on sides
Flammable wastes	Metal drums	210 L	Fume ventilation; temperature control
	Storage tanks	Up to 20 m^3	
Explosives	Shock-absorbing containers	Varies	Temperature control; special container markings

idential wastes and litter. For industrial and hazardous waste, manual loading is the exception, not the rule. It is used for the collection of industrial wastes when pickup points are inaccessible to the collection vehicle.

Equipment for Hazardous-Wastes Collection The equipment used for collection varies with the characteristics of the wastes. For short-haul distances, drum storage and collection with an enclosed trailer are often preferred methods. Full-size drums are usually shipped four to a pallet. Smaller sizes are stacked and/or wrapped up to 10 per pallet. As hauling distances increase, larger tank trucks, trailers, and railroad tank cars are used.

Equipment for Superfund Waste Shipment RCRA hazardous waste that has been spilled, improperly landfilled, or dredged from defunct lagoons is a CERCLA waste, more commonly referred to as a Superfund waste. For clean-ups where offsite treatment is the chosen solution, soil is excavated and placed in 15-m^3 roll-off box or dump body truck. The trucks may be lined with polyethylene to reduce

decontamination and eliminate sticking of the load to the steel body during freezing weather. The waste is shipped to a landfill, incinerator, or other treatment facility. If the waste is a sludge (as found in a lagoon), dewatering to 50 percent solids is normally performed on site via plate and frame filter press. This may be followed by mixing with bulking reagents to eliminate free water. The alternative is shipping the waste in a tanker or other container that retains free liquid.

Occasionally, a small clean-up of a few tons may occur. These are usually handled via placing the waste in open-topped drums that are sealed with a lid and ring closure. Consultation with the ultimate treatment facility is important to select the correct drum size and material and eliminate the need for repackaging the waste.

Transportation Costs Most waste trucking is done by commercial and hazardous waste firms. Costs are quoted per load, based on cost for transport (charged per load mile) from point of generation to its destination at a landfill or TSD facility. A strategy which permits shipment of full truckloads minimizes the transportation cost per ton.

TABLE 25-57 Factors That Should Be Considered in Evaluating On-Site Processing Equipment

Factor	Evaluation
Capabilities	What will the device or mechanism do? Will its use be an improvement over conventional practices?
Reliability	Will the equipment perform its designated functions with little attention beyond preventive maintenance? Has the effectiveness of the equipment been demonstrated in use over a reasonable period of time or merely predicted?
Service	Will servicing capabilities beyond those of the local building maintenance staff be required occasionally? Are properly trained service personnel available through the equipment manufacturer or the local distributor?
Safety of operation	Is the proposed equipment reasonably foolproof so that it may be operated by tenants or building personnel with limited mechanical knowledge or abilities? Does it have adequate safeguards to discourage careless use?
Ease of operation	Is the equipment easy to operate by a tenant or by building personnel? Unless functions and actual operations of equipment can be carried out easily, they may be ignored or bypassed by paid personnel and most often by "paying" tenants.
Efficiency	Does the equipment perform efficiently and with a minimum of attention? Under most conditions, equipment that completes an operational cycle each time that it is used should be selected.
Environmental effects	Does the equipment pollute or contaminate the environment? When possible, equipment should reduce environmental pollution presently associated with conventional functions.
Health hazards	Does the device, mechanism, or equipment create or amplify health hazards?
Aesthetics	Do the equipment and its arrangement offend the senses? Every effort should be made to reduce or eliminate offending sights, odors, and noises.
Economics	What are the economics involved? Both first and annual costs must be considered. Future operation and maintenance costs must be assessed carefully. All factors being equal, equipment produced by well-established companies, having a proven history of satisfactory operation, should be given appropriate consideration.
Flexibility	Will the equipment or its placement allow future changes to the process to handle wastes with differing characteristics?

*From G. Tchobanoglous, H. Theisen, and R. Eliassen, *Solid Wastes: Engineering Principles and Management Issues,* McGraw-Hill, New York, 1977.

FIG. 25-60 Small compactor used in conjunction with a detachable roll-off container. Note the wheeled container used to collect wastes within the warehouse.

TRANSFER AND TRANSPORT

The functional element of transfer and transport refers to the means, facilities, and appurtenances used to effect the transfer of wastes from relatively small collection vehicles to larger vehicles and to transport them over extended distances to either processing centers or disposal sites. Transfer and transport operations become a necessity when haul distances to available disposal sites or processing centers increase to a point at which direct hauling is no longer economically feasible.

Transfer Stations Important factors that must be considered in the design of transfer stations include (1) the type of transfer operation to be used, (2) capacity requirements, (3) equipment and accessory requirements, and (4) environmental requirements.

Types of Transfer Stations Depending on the method used to load the transport vehicles, transfer stations may be classified into three types: (1) direct-discharge, (2) storage-discharge, and (3) combined direct- and storage-discharge.

1. *Direct discharge.* In a direct-discharge transfer station, wastes from the collection vehicles are usually emptied directly into the vehicle to be used to transport them to a place of final disposition. To accomplish this, these transfer stations are usually constructed in a two-level arrangement. Either the unloading dock or platform from which wastes from collection vehicles are discharged into the transport trailers is elevated, or the transport trailers are located in a depressed ramp (see Fig. 25-62). Direct-discharge transfer stations employing stationary compactors also are popular.

TABLE 25-58 Typical Data on Container Capacities for Use with Various Collection Systems for Solid Waste

Vehicle	Collection	Typical range of container capacities,* m³
	Container type	
Hauled-container systems		
Hoist-truck	Open and covered, also used with stationary compactor	2–9
Tilt-frame/roll-off container	Open-top, roll-off containers	8–40
	Can be used in conjunction with stationary compactor	10–30
	Equipped with self-contained compaction mechanism	15–30
Tractor trailer	Open-top trash trailers	10–30
	Enclosed trailer-mounted containers equipped with self-contained compaction mechanism	15–30
	Long-body dump trailer	15–30
Stationary-container systems		
Compactor, mechanically loaded	Open-top and enclosed top- and front- or rear-loading	0.6–6
Compactor, manually loaded	Small plastic or galvanized metal containers, disposable paper and plastic bags	75–200 L†

*See Table 25-55 for typical dimensions.
†Loaded mass of container should not exceed 30 kg.

TABLE 25-59   Typical Data on Vehicles Used for Collection of Solid Wastes

| Collection vehicle | | | Typical overall collection-vehicle dimensions | | | | |
Type	Available container or truck body capacities,* m³	Number of axles	With indicated container or truck body capacity,* m³	Width, mm	Height, mm	Length,* mm	Unloading method
Hauled-container systems							
Hoist-truck	2–9	2	7.6	2440	2030–2540	2800–3800	Gravity, bottom-opening
Tilt-frame	8–40	3	22.9	2440	2030–2290	5590–7620	Gravity, inclined-tipping
Truck-tractor trash trailer	10–30	3	30.6	2440	2290–3800	5590–11,430	Gravity, inclined-tipping
Roll-off container	8–30	3–4	37	2440	12400–2700	9100–10,000	Gravity, inclined tipping
Stationary-container systems							
Compactor (mechanically loaded)							
Front-loading	15–35	3	22.9	2440	3560–3800	6100–7370	Hydraulic ejector panel
Side-loading	6–28	3	22.9	2440	3350–3000	5590–6600	Hydraulic ejector panel
Rear-loading	6–24	2	15.3	2440	3175–3430	5330–5824	Hydraulic ejector panel
Compactor (manually loaded)							
Side-loading	6–28	3	28.3	2440	3350–3800	6100–7620	Hydraulic ejector panel
Rear-loading	6–24	2	15.3	2440	3175–3430	5330–5840	Hydraulic ejector panel

*From front of truck to rear of container or truck body.

2. *Storage discharge.*   In the storage-discharge transfer station, wastes are emptied either into a storage pit or onto a platform from which they are loaded into transport vehicles by various types of auxiliary equipment. In a storage-discharge transfer station, storage volume varies from about one-half to 2 days' volume of wastes.

3. *Combined direct and storage discharge.*   In some transfer stations, both direct-discharge and storage-discharge methods are used. Usually, these are multipurpose facilities designed to service a broader range of users than a single-purpose facility. In addition to serving a broader range of users, a multipurpose transfer station may house a materials-salvage operation recovery facility.

Capacity Requirements   The operating capacity of a transfer station must be such that collection vehicles do not have to wait long to unload. In most cases, it will not be cost-effective to design the station to handle the ultimate peak number of hourly loads. An economic trade-off analysis should be made between the annual cost for the time spent by the collection vehicles waiting to unload against the incremental annual cost of a larger transfer station and/or the use of more transport equipment. Because of the increased cost of transport equipment, a trade-off analysis must also be made between the capacity of the transfer station and the cost of the transport operation, including both equipment and labor components.

Equipment and Accessory Requirements   The types and amounts of equipment required vary with the capacity of the station and its function in the waste-management system. Specifically, truck scales should be provided at all medium-size and large transfer stations, both to monitor the operation and to develop meaningful management and engineering data.

Environmental Requirements   Most large modern transfer stations are enclosed and are constructed of materials that can be maintained and cleaned easily. For direct-discharge transfer stations with open loading areas, special attention must be given to the problem of blowing papers. Windscreens or other barriers are commonly used. Regardless of type, the station should be designed and constructed so that all accessible areas where rubbish or paper can accumulate are eliminated.

Transfer Means and Methods   Motor vehicles, railroads, and barges are the principal means used to transport solid wastes. Pneumatic and hydraulic systems have also been used.

Motor-Vehicle Transport   Motor vehicles used to transport solid wastes on highways should satisfy the following requirements: (1) Wastes must be transported at minimum cost. (2) Wastes must be covered during the haul operation. (3) Vehicles must be designed for highway traffic. (4) Vehicle capacity must be such that allowable weight limits are not exceeded. (5) Methods used for unloading must be simple and dependable. The maximum volume that can be hauled in highway transport vehicles depends on the regulations in force in the state in which these vehicles are operated.

1. *Trailers and semitrailers.*   In recent years, because of their simplicity and dependability, open-top trailers and semitrailers have found wide acceptance (see Table 25-60). Some trailers are equipped with sumps to collect any liquids that accumulate from the solid wastes. The sumps are equipped with drains so that they can be emptied at the disposal site.

FIG. 25-61 Typical front-loading compactor used in a stationary-container collection system.

FIG. 25-62 Direct-discharge transfer station with open-top trailers located in a depressed ramp under the loading hoppers.

TABLE 25-60 Typical Data on Haul Vehicles Used at Transfer Stations

Type	Capacity per trailer		Dimensions of single trailer			Length of tractor and trailer units, m
	m³	Metric tons	Width, m	Length, m	Approximate height, empty, m	
Tractor-tandem-trailer	54	11.4	2.44	8.25	4.12	19.8
Tractor-trailer	54	10.0	2.44	8.50	4.12	12.0
	74	17.3	2.44	12.2	4.12	16.0
Tractor-compactor-trailer	58	18.0	2.44	10.2	4.12	14.0

Methods used to unload the transport trailers may be classified as (1) self-emptying and (2) requiring the aid of auxiliary equipment. Self-emptying transport trailers are equipped with mechanisms such as pusher rams, dump body, and moving floors that are part of the vehicle. An advantage of the moving-floor trailer is the rapid turnaround time (typically 6 to 10 min) achieved at the disposal site without the need for auxiliary equipment. Unloading systems that require auxiliary equipment are usually of the pull-off type, in which the wastes are pulled out of the truck by either a moveable bulkhead or wire-cable slings placed forward of the load. Walking floor trailers are self-unloading.

Another auxiliary unloading system that has proven to be very effective and efficient involves the use of moveable, hydraulically operated truck dumps located at the disposal site. Operationally, the trailer is backed up onto one of the tipping ramps, with or without its tractor. The back of the trailer is opened, and the unit is then tilted upward until the wastes fall out by gravity. The time required for the entire unloading operation typically is about 5 min per trip.

2. *Compactors.* Large-capacity containers and container-trailers are used in conjunction with stationary compactors at transfer stations. In some cases, the compaction mechanism is an integral part of the container. When containers are equipped with a self-contained compaction mechanism, the moveable bulkhead used to compress the wastes is also used to discharge the compacted wastes.

Railroad Transport Renewed interest is developing in the use of railroads for hauling solid wastes, especially in heavily populated areas where landfill space is scarce and tipping fees are high. Containerized waste on "piggyback" rail cars are used. Full-size bulk material cars are also used for shipping solid wastes.

Water Transport Barges, scows, and special boats have been used in the past to transport solid wastes to processing locations and to seaside and ocean disposal sites, but ocean disposal is no longer practiced by the United States. Although some self-propelled vessels (such as U.S. Navy garbage scows and other special boats) have been used, most common practice is to use vessels towed by tugs or other special boats.

Pneumatic Transport Both low-pressure air-vacuum conduit transport systems have been used to transport solid wastes. The most common application is the transport of wastes from high-density apartments or commercial activities to a central location for processing or for loading into transport vehicles. The largest pneumatic system in use in the United States is at the Walt Disney World amusement park in Orlando, Florida.

Location of Transfer Stations Whenever possible, transfer stations should be located (1) as near as possible to the weighted center of the individual solid-waste-production ares to be served, (2) within easy access of major arterial highways as well as near secondary or supplemental means of transportation, (3) where there will be a minimum of public and environmental objection to the transfer operations, and (4) where construction and operation will be most economical. Additionally, if the transfer-station site is to be used for processing operations involving material recovery and/or energy production, the requirements for those operations must be considered.

Transfer and Transport of Hazardous Wastes The facilities of a hazardous-waste transfer station are quite different from those of an industrial or municipal solid-waste transfer station. Typically, hazardous wastes are not compacted (mechanical volume reduction),

discharged at differential levels, or delivered by numerous collection companies. Instead, liquid hazardous wastes are generally pumped from collection vehicles, and sludges or solids are reloaded without removal from the collection containers for transport to processing and disposal facilities. Repacking may occur at a TSD (transport, storage, and disposal facility) to put the waste into standard size and weight containers for reshipment to final treatment facilities. Most often, this involves subdividing larger containers into smaller containers and/or use of compatible container material. Examples are the use of fiber drums rather than steel barrels and prepacking to limit the heat content or volatility of waste destined for hazardous waste incineration.

It is unusual to find a hazardous-waste-transfer facility at which wastes are simply transferred to larger transport vehicles. Some processing and storage facilities are often part of the materials-handling sequence at a transfer section. For example, neutralization of corrosive wastes will result in the use of lower-cost holding tanks on transport vehicles.

PROCESSING AND RESOURCE RECOVERY

The purpose of this subsection is to introduce the reader to the techniques and methods used to recover materials, conversion products, and energy from solid wastes. Topics to be considered include (1) processing techniques for solid waste, (2) processing techniques for hazardous wastes, (3) materials-recovery systems, (4) recovery of biological conversion products, (5) thermal processes, and (6) waste-to-energy systems.

Because many of the techniques, especially those associated with the recovery of materials and energy and the processing of solid hazardous wastes, are in a state of flux with respect to application and design criteria, the objective here is only to introduce them to the reader. If these techniques are to be considered in the development of waste-management systems, current engineering design and performance data must be obtained from consultants, operating records, field tests, equipment manufacturers, and available literature.

Processing Techniques for Solid Wastes Processing techniques are used in solid-waste-management systems to (1) improve the efficiency of the systems, (2) to recover resources (usable materials), and (3) to prepare materials for recovery of conversion products and energy. The more important techniques used for processing solid wastes are summarized in Tables 25-61 and 25-62.

Manual Component Separation The manual separation of solid-waste components can be accomplished at the source where solid wastes are generated, at a transfer station, at a centralized processing station, or at the disposal site. Manual sorting at the source of generation is the most positive way to achieve the recovery and reuse of materials. The number and types of components salvaged or sorted (e.g., cardboard and high-quality paper, metals, and wood) depend on the location, the opportunities for recycling, and the resale market. There has been an evolution in the solid waste industry to combine manual and automatic separation techniques to reduce overall costs and produce a cleaner product, especially for recyclable materials.

Storage and Transfer When solid wastes are to be processed for material recovery, storage and transfer facilities should be considered an essential part of the processing operation. Important factors in the design of such facilities include (1) the size of the material before and after processing, (2) the density of the material, (3) the angle of repose before and after processing, (4) the abrasive characteristics of the material, and (5) the moisture content.

Mechanical Volume Reduction Mechanical volume reduction is perhaps the most important factor in the development and operation of solid-waste-management systems. Vehicles equipped with compaction mechanisms are used for the collection of most industrial solid wastes. To increase the useful life of landfills, wastes are compacted. Paper for recycling is baled for shipping to processing centers. When compacting industrial solid wastes, it has been found that the final density (typically about 1,100 kg/m³) is essentially the same regardless of the starting density and applied pressure. This fact is important in evaluating manufacturers' claims.

Chemical Volume Reduction Incineration has been the method commonly used to reduce the volume of wastes chemically. One of the most attractive features of the incineration process is that it can be used to reduce the original volume of combustible solid wastes by 80 to 90 percent. The technology of incineration has advanced since 1960 with many mass burn facilities now having two or more combustors with capacities of 1000 tons per day of refuse per unit. However, regulations of metal and dioxin emissions have resulted in higher costs and operating complexity.

Mechanical Size Alteration The objective of size reduction is to obtain a final product that is reasonably uniform and considerably reduced in size in comparison to its original form. It is important to note that size reduction does not necessarily imply volume reduction. In some situations, the total volume of the material after size reduction may be greater than the original volume. Shredding is most often used for size reduction. Use of two-stage (coarse, fine) low-speed shredders or a single-stage shredder with screen and recycle of oversize are the most common systems. The gain in ease of material handling must be weighed against the substantial operating costs for shredding equipment.

Mechanical Component Separation Component separation is a necessary operation in the recovery of resources from solid wastes and in instances when energy and conversion products are to be recovered from processed wastes. Mechanical separation techniques that have been used are reported in Table 25-61.

Magnetic and Electromechanical Separation Magnetic separation of ferrous materials is a well-established technique. More recently, a variety of electromechanical techniques have been developed for the removal of several nonferrous materials (see Table 25-62).

Drying and Dewatering In many solid-waste energy-recovery and incineration systems, the shredded light fraction is predried to decrease weight. Although energy requirements for drying wastes vary with the application, the required energy input can be established by using a value of about 4300 kJ/kg of water evaporated. Drying can frequently increase waste throughput in many treatment systems and for incinerators; it produces more stable combustion and better ash quality.

Processing of Hazardous Wastes As with conventional solid wastes, the processing of hazardous wastes is undertaken for three purposes: (1) to recover useful materials, (2) to reduce the amount of wastes that must be disposed in landfills, and (3) to prepare the wastes for ultimate disposal.

Processing Techniques The processing of hazardous wastes on a batch basis can be accomplished by physical, chemical, thermal, and biological means. The various individual processes in each category are reported in Table 25-63. Clearly, the number of possible treatment-process combinations is staggering. In practice, the physical, chemical, and thermal treatment operations and processes are the ones most commonly used.

Identification of Waste Constituents In any processing (and disposal) scheme, the key item is knowledge of the characteristics of the wastes to be handled. Without this information, effective processing or treatment is impossible. For this reason, the characteristics of the wastes must be known before they are accepted and hauled to a treatment or disposal site. In most states, proper identification of the constituents of the waste is the responsibility of the waste generator.

Materials-Recovery Systems Paper, rubber, plastics, textiles, glass, metals, and organic and inorganic materials are the principal recoverable materials contained in industrial solid wastes.

Once a decision has been made to recover materials and/or energy, process flow sheets must be developed for the removal of the desired components, subject to predetermined materials specifications. A typical flow sheet for the recovery of specific components and the preparation of combustible materials for use as a fuel source is presented in Fig. 25-63. The light combustible materials are often identified as refuse-derived fuel (RDF).

The design and layout of the physical facilities that make up the processing-plant flow sheet are an important aspect in the implementation and successful operation of such systems. Important factors that must be considered in the design and layout of such systems include (1) process performance efficiency, (2) reliability and flexibility, (3) ease and economy of operation, (4) aesthetics, and (5) environmental controls.

Recovery of Biological Conversion Products Biological conversion products that can be derived from solid wastes include compost, methane, various proteins and alcohols, and a variety of other intermediate organic compounds. The principal processes that have been used are reported in Table 25-64. Composting and anaerobic digestion, the two most highly developed processes, are considered further. The recovery of gas from landfills is discussed in the portion of this section dealing with ultimate disposal.

TABLE 25-61 Mechanical Methods for Separating Solid-Waste Components

Method	Function	Equipment and/or facilities and applications	Method	Function
Screening	Used to separate solid waste components by size	Trommels and horizontal and vibrating screens for unprocessed and processed wastes; disk screens with processed wastes	Optical sorting	Used to separate plastics
Air separation	Used to separate light (organic) materials from heavy (inorganic) materials in solid waste	Zig-zag-air, vibrating-air, rotary-air, and air-knife classifiers used with processed wastes	Sink-float, flotation, Inertial, Inclined-table, shaking-table	Used to separate light and heavy materials in solid wastes
Jig separation	Used to separate light and heavy materials in solid waste by means of density separation			
Pneumatic separation (stoners)	Used to separate light and heavy materials in solid waste			

TABLE 25-62 Summary of Techniques Used for Processing Solid Wastes

Processing technique	Function	Representative equipment and/or facilities and applications
Manual component separation	Separation of recoverable materials, usually at point of generation	Visual inspection and removal via conveyor belt picking stations
Storage and transfer	Storage and transfer of wastes to be processed	Open storage pits for unprocessed wastes, storage bins and silos for processed wastes; transfer equipment including front-end loaders, metal and rubber belt conveyors, vibratory conveyors with unprocessed wastes, pneumatic conveyors, and screw conveyors with processed wastes
Mechanical volume reduction	Reduction of solid-waste volume; alteration of shape of solid-waste components; all modern collection vehicles essentially equipped with compaction equipment	Hydraulic piston-type compactors for collection vehicles, on-site compactors, and transfer-station compactors; roll crushers used to fracture brittle materials and to crush tin and aluminum cans and other ductile materials
Chemical volume reduction	Reduction of volume of solid wastes through burning (incineration)	Mass-fired incinerators, with and without heat recovery, for unprocessed wastes; rotary kilns for hazardous/containerized and bulk solid/sludge waste
Mechanical size and shape alteration	Alteration of size and shape of solid-waste components	Equipment used to reduce the size of solid waste including hammer mills, shredders, roll crushers, grinders, chippers, jaw crushers, rasp mills, and hydropulpers; briquettes
Mechanical component separation	Separation of recoverable materials, usually at a processing facility	(See Table 26-40)
Magnetic and electro-mechanical separation	Separation of ferrous and nonferrous materials from processed solid wastes	Magnetic separation for ferrous materials; eddy-current separation for aluminum; electrostatic separation for glass from wastes free of ferrous and aluminum scrap; magnetic fluid separation for nonferrous materials from processed wastes
Drying and dewatering	Removal of moisture from solid wastes	Convection, conduction, and radiation dryers used for solid wastes and sludge; centrifuge and filtration used to dewater treatment-plant sludge

Composting   If the organic materials, excluding plastics, rubber, and leather, are separated from municipal solid wastes and subjected to bacterial decomposition, the end product remaining after dissimilatory and assimilatory bacterial activity is called *compost* or *humus*. The entire process involving both separation and bacterial conversion of the organic solid wastes is known as *composting*. Decomposition of the organic solid wastes may be accomplished either aerobically or anaerobically, depending on the availability of oxygen.

Most composting operations involve three basic steps: (1) preparation of solid wastes, (2) decomposition of the solid wastes, and (3) product preparation and marketing. Receiving, sorting, separation, size reduction, and moisture and nutrient addition are part of the preparation step. Several techniques have been developed to accomplish the decomposition step. Once the solid wastes have been converted to a humus, they are ready for the third step of product preparation and marketing. This step may include fine grinding, blending with various additives, granulation, bagging, storage, shipping, and, in some cases, direct marketing. The principal design considerations associated with the biological decomposition of prepared solid wastes are presented in Table 25-65.

Anaerobic Digestion   Anaerobic digestion or anaerobic fermentation, as it is often called, is the process used for the production of methane from solid wastes. In most processes in which methane is to be produced from solid wastes by anaerobic digestion, three basic steps are involved. The first step involves preparation of the organic fraction of the solid wastes for anaerobic digestion and usually includes receiving, sorting, separation, and size reduction. The second step involves the addition of moisture and nutrients, blending, pH adjustment to about 6.7, heating of the slurry to between 327 and 333 K (130 and 140° F), and anaerobic digestion in a reactor with continuous flow, in which the contents are well mixed for a period of time varying from 8 to 15 days. The third step involves capture, storage, and, if necessary, separation of the gas components evolved during the digestion process. The fourth step is the disposal of the digested sludge is an additional task that must be accomplished. Some important design considerations are reported in Table 25-66. Because of the variability of the results reported in the literature, it is recommended that pilot-plant studies be conducted if the digestion process is to be used for the conversion of solid wastes.

Thermal Processes   Conversion products that can be derived from solid wastes include heat, gases, a variety of oils, and various related organic compounds. The principal thermal processes that have been used for the recovery of usable conversion products from solid wastes are reported in Table 25-64.

Incineration with Heat Recovery   Heat contained in the gases produced from the incineration of solid wastes can be recovered as steam. The low-level heat remaining in the gases after heat recovery can also be used to preheat the combustion air, boiler makeup water, or solid-waste fuel.

1. *In existing incinerators.*   With existing incinerators, waste-heat boilers can be installed to extract heat from the combustion gases without introducing excess amounts of air or moisture. Typically, incinerator gases will be cooled from a range of 1250 to 1375 K (1800 to 2000° F) to a range from 500 to 800 K (600 to 1000° F) before being discharged to the air pollution control system. Apart from the production of steam, the use of a boiler system is beneficial in reducing the volume of gas to be processed in the air-pollution-control equipment. The compounds in the waste stream will generate products of combustion and ash that may create serious corrosion and fouling problems in waste-heat boilers.

2. *In water-wall incinerators.*   The internal walls of the combustion chamber are lined with boiler tubes that are arranged vertically and welded together in continuous sections. When water walls are employed in place of refractory materials, they are not only useful for the recovery of steam but also extremely effective in controlling furnace temperature without introducing excess air; however, they are subject to corrosion by the hydrochloric acid produced from the burning of some plastic compounds and the molten ash containing salts (chlorides and sulfates) that attach to the tubes.

Combustion   Combustion of industrial and municipal waste is an attractive waste management option because it reduces the volume of waste by 70 to 90 percent. In the face of shrinking landfill availability, municipal waste combustion capacity in the United States has grown at an astonishing rate, significantly faster than the growth rate for municipal refuse generation.

Types of Combustors   The three main classes of facilities used to combust municipal refuse are mass burn, modular, and RDF-fired facilities. Mass-burn combustors are field erected and generally range in size from 50 to 1000 tons/day of refuse feed per unit (Fig. 25-64). Modular combustors burn waste with little more pre-

TABLE 25-63   Treatment Operations and Processes for Hazardous Wastes[a]

Operation or process	Functions performed[b]	Types of wastes[c]	Forms of waste[d]
Physical treatment			
Adsorption[e]	Se	1, 2, 3,	L
Aeration	Se	1, 2, 3, 4, 5	L
Ammonia stripping	VR, Se	1, 2, 3, 4	L
Carbon sorption	VR, Se	1, 2, 3, 4, 5	L, G
Centrifugation	VR, Se	1, 2, 3, 4, 5	L
Dialysis	VR, Se	1, 2, 3, 4	L
Distillation[e]	VR, Se	1, 2, 3, 4, 5	L
Electrodialysis	VR, Se	1, 2, 3, 4, 6	L
Encapsulation	St	1, 2, 3, 4, 6	L, S
Evaporation	VR, Se	1, 2, 5	L
Filtration[e]	VR, Se	1, 2, 3, 4, 5	L, G
Flocculation or setting	VR, Se	1, 2, 3, 4, 5	L
Flotation[e]	Se	1, 2, 3, 4	L
Reverse osmosis	VR, Se	1, 2, 4, 6	L
Screening	Se	1, 2, 3, 4, 5	L
Sedimentation[e]	VR, Se	1, 2, 3, 4, 5	L
Shredding	VR, Se	1, 2, 3, 4	S
Solar evaporation[e]	VR, Se	1, 2, 5	L
Solvent extraction	Se	1, 2, 3, 4, 5	L
Thickening	Se	1, 2, 3, 4	L
Ultrafiltration	Se	1, 2, 3, 4, 5	L
Vapor scrubbing	VR, Se	1, 2, 3, 4	L
Chemical treatment			
Calcination	VR	1, 2, 5	L
Chemical dechlorination	De	1, 3	L
Ion exchange	VR, Se, De	1, 2, 3, 4, 5	L
Neutralization[e]	De	1, 2, 3, 4	L
Oxidation	De	1, 2, 3, 4	L
Precipitation[e]	VR, Se	1, 2, 3, 4, 5	L
Reduction	De	1, 2	L
Sorption	De	1, 2, 3, 4	L
Stabilization or solidification[e]	De	1, 2, 3, 4	L
Thermal treatment			
Desorption	VR, De	1, 2, 3, 4	S
Incineration[e]	VR, De	3, 5, 6, 7, 8	S, L, G
Pyrolysis	VR, De	3, 4, 6	S, L, G
Biological treatment			
Activated sludge[e]	De	3	L
Aerated lagoons	De	3	L
Anaerobic digestion	De	3	L
Anaerobic filters	De	3	L
Trickling filters	De	3	L
Waste-stabilization ponds[e]	De	3	L

[a]Adapted from *Report to Congress: Disposal of Hazardous Wastes*, U.S. EPA Publ. SW-115, 1974.

[b]Functions: VR, volume reduction; Se, separation; De, detoxification; St, storage.

[c]Waste types: 1, inorganic chemical without heavy metals; 2, inorganic chemical with heavy metals; 3, organic chemical without heavy metals; 4, organic chemical with heavy metals; 5, radiological; 6, biological; 7, flammable; 8, explosive.

[d]Waste forms: S, solid; L, liquid; G, gas.

[e]Most widely used technologies for hazardous-waste management.

processing than do the mass-burn units. These range in size from 5 to 100 tons/day. The third major class of municipal waste combustor burns RDF. The types of waste-to-energy boilers used to combust RDF can include suspension, stoker, and fluidized bed designs. RDF fuels can also be fired directly in large industrial boilers that are now used for the production of power with pulverized or stoker coal, oil, and gas. Although the process is not well established with coal, it appears that about 15 to 20 percent of the heat input can be from RDF. With oil as the fuel, about 10 percent of the heat input can be from RDF. Depending on the degree of processing, suspension, spreader-stoker, and double-vortex firing systems have been used.

Gasification   The gasification process involves the partial combustion of a carbonaceous or hydrocarbon fuel to generate a combustible fuel gas rich in carbon monoxide and hydrogen. A gasifier is basically an incinerator operating under reducing conditions. Heat to sustain the process is derived from exothermic reactions, while the combustible components of the low-energy gas are primarily generated by endothermic reactions. The reaction kinetics of the gasification process are quite complex and still the subject of considerable debate.

When a gasifier is operated at atmospheric pressure with air as the oxidant, the end products of the gasification process are a low-energy gas typically containing (by volume) 10 percent CO_2, 20 percent CO, 15 percent H_2, and 2 percent CH_4, with the balance being N_2, and a carbon-rich ash. Because of the diluting effect of the nitrogen in the input air, the low-energy gas has an energy content in the range of the 5.2 to 6.0 MJ/m³ (140 to 160 Btu/ft³). When pure oxygen is used as the oxidant, a medium-energy gas with an energy content in the range of 12.9 to 13.8 MJ/m³ (345 to 370 Btu/ft³) is produced. Gasifiers were in widespread use on coal and wood until natural gas displaced them in the 1930s–1950s. Some large coal gasifiers are in use today in the United States and worldwide. While the process can work on solid waste, incinerators (which gasify and burn in one chamber) are favored over gasifiers.

Pyrolysis   Of the many alternative chemical conversion processes that have been investigated, pyrolysis has received the most attention. Pyrolysis has been tested in countless pilot plants, and many full-scale demonstration systems have been operated. Few attained any long-term commercial use. Major issues were lack of market for the unstable and acidic pyrolytic oils and the char.

Depending on the type of reactor used, the physical form of solid wastes to be pyrolyzed can vary from unshredded raw wastes to the finely ground portion of the wastes remaining after two stages of shredding and air classification. Upon heating in an oxygen-free atmosphere, most organic substances can be split via thermal cracking and condensation reactions into gaseous, liquid, and solid fractions. Pyrolysis is the term used to describe the process. In contrast to the combustion process, which is highly exothermic, the pyrolytic process is highly endothermic. For this reason, the term *destructive distillation* is often used as an alternative for *pyrolysis*.

The characteristics of the three major component fractions resulting from the pyrolysis are (1) a gas stream containing primarily hydrogen, methane, carbon monoxide, carbon dioxide, and various other gases, depending on the organic characteristics of the material being pyrolyzed; (2) a fraction that consists of a tar and/or oil stream that is liquid at room temperatures and has been found to contain hundreds of chemicals such as acetic acid, acetone, methanol, and phenols; and (3) a char consisting of almost pure carbon plus any inert material that may have entered the process. It has been found that distribution of the product fractions varies with the temperature at which the pyrolysis is carried out. Under conditions of maximum gasification, the energy content of the resulting gas is about 26.1 MJ/kg (700 Btu/ft³). The energy content of pyrolytic oils has been estimated to be about 23.2 MJ/kg (10,000 Btu/lb).

Waste-to-Energy Systems   The preceding section ended at production of steam. WTE (waste-to-energy) systems take over at this point, using high-pressure/high-temperature steam to drive turbines and produce shaft horsepower for prime movers at industrial plants or to generate electricity.

The fuel may be solid waste or gas or oil from a gasifier or pyrolysis system.

Typical flow sheets for alternative energy-recovery systems are shown in Fig. 25-65. Perhaps the most common flow sheet for the production of electric energy involves the use of a steam turbine-generator combination (see Fig. 25-65). As shown, when solid wastes are used as the basic fuel source, four operating modes are possible. A flow sheet using a gas-turbine-generator combination is shown in Fig. 25-65. The low-energy gas is compressed under high pressure so that it can be used more effectively in the gas turbine. Use of low- or medium-Btu gas for gas turbines has been attempted, and success requires good design and operation of gas cleaning equipment prior to introduction into the combustor of the gas turbine.

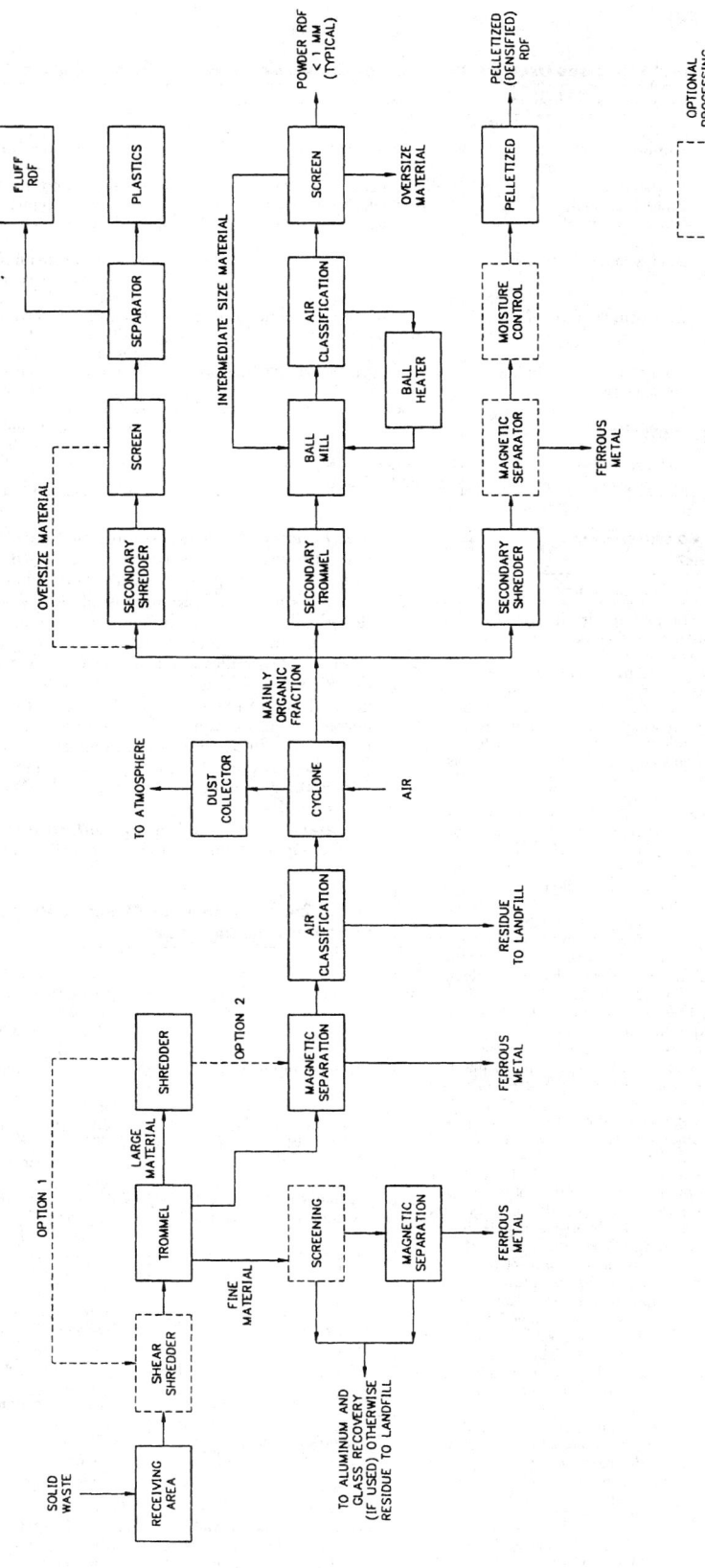

FIG. 25-63 Typical flow sheet for the recovery of materials and production of refuse-derived fuels (RDF). [*Adapted in part from D. C. Wilson (ed.), Waste Management: Planning, Evaluation, Technologies, Oxford University Press, Oxford, 1981.*]

TABLE 25-64   Biological and Thermal Processes Used for Recovery of Conversion Products from Solid Waste

Process	Conversion product	Preprocessing required	Comments
Biological			
Composting	Humuslike material	Shredding, air separation	Lack of markets primary shortcoming; technically proven in full-scale application
Anaerobic digestion	Methane gas	Shredding, air separation	Technology on laboratory scale only
Biological conversion to protein	Protein, alcohol	Shredding, air separation	Technology on pilot scale only
Biological fermentation	Glucose, furfural	Shredding, air separation	Used in conjunction with the hydrolytic process
Thermal			
Incineration with heat recovery	Energy in the form of steam	None	Markets for steam required; proven in numerous full-scale applications; air-quality regulations possibly prohibiting use
Supplementary fuel firing in boilers	Energy in the form of steam	Shredding, air separation, magnetic separation	If least capital investment desired, existing boiler required to be capable of modification; air-quality regulations possibly prohibiting use
Gasification	Energy in the form of low energy gas	Shredding, air separation, magnetic separation	Gasification also capable of being used for codisposal for industrial sludges
Pyrolysis	Energy in the form of gas or oil	Shredding, magnetic separation	Technology proven only in pilot applications; even though pollution is minimized, air-quality regulations possibly prohibiting use
Hydrolysis	Glucose, furfural	Shredding, air separation	Technology on pilot scale only
Chemical conversion	Oil, gas, cellulose acetate	Shredding, air separation	Technology on pilot scale only

TABLE 25-65   Important Design Considerations for Aerobic-Composting Processes*

Item	Comment
Particle size	For optimum results the size of solid wastes should be between 25 and 75 mm (1 and 3 in).
Seeding and mixing	Composting time can be reduced by seeding with partially decomposed solid wastes to the extent of about 1 to 5 percent by weight. Sewage sludge can also be added to prepared solid wastes. When sludge is added, the final moisture content is the controlling variable.
Mixing or turning	To prevent drying, caking, and air channeling, material in the process of being composted should be mixed or turned on a regular schedule or as required. Frequency of mixing or turning will depend on the type of composting operation.
Air requirements	Air with at least 50 percent of the initial oxygen concentration remaining should reach all parts of the composting material for optimum results, especially in mechanical systems.
Total oxygen requirements	The theoretical quantity of oxygen required can be estimated.
Moisture content	Moisture content should be in the range between 50 and 60 percent during the composting process. The optimum value appears to be about 55 percent.
Temperature	For best results temperature should be maintained between 322 and 327 K (130 and 140°F) for the first few days and between 327 and 333 K (130 and 140°F) for the remainder of the active composting period. If temperature goes beyond 339 K (150°F), biological activity is reduced significantly.
Carbon-nitrogen ratio	Initial carbon-nitrogen ratios (by mass) between 35 and 50 are optimum for aerobic composting. At lower ratios ammonia is given off. Biological activity is also impeded at lower ratios. At higher ratios nitrogen may be a limiting nutrient.
pH	To minimize the loss of nitrogen in the form of ammonia gas, pH should not rise above about 8.5.
Control of pathogens	If the process is properly conducted, it is possible to kill all the pathogens, weeds, and seeds during the composting process. To do this, the temperature must be maintained between 333 and 344 K (140 and 160°F) for 24 h.

*Adapted from G. Tchobanoglous, H. Theisen, R. Eliassen, *Solid Wastes: Engineering Principles and Management Issues*, McGraw-Hill, New York, 1977.

Efficiency Factors   Representative efficiency data for boilers, pyrolytic reactors, gas turbines, steam-turbine-generator combinations, electric generators, and related plant use and loss factors are given in Table 25-67. In any installation in which energy is being produced, allowance must be made for the power needs of the station or process and for unaccounted-for-process-heat losses. Typically, the auxiliary power allowance varies from 4 to 8 percent of the power produced. Process-heat losses usually will vary from 2 to 8 percent. In general, steam pressures of 600 psig and temperatures of 650° F are considered minimum for economical power generation. Industrial plants may choose a cogeneration topping cycle, with steam exhaust from the turbine at the plant's process steam pressure, typically in the 125–250-psig range. For commercial WTE plants, condensing turbines are the norm.

Determination of Energy Output and Efficiency for Energy-Recovery Systems   An analysis of the amount of energy produced

TABLE 25-66   Important Design Considerations for Anaerobic Digestion*

Item	Comment
Size of material shredded	Wastes to be digested should be shredded to a size that will not interfere with the efficient functioning of pumping and mixing operations.
Mixing equipment	To achieve optimum results and to avoid scum buildup, mechanical mixing is recommended.
Percentage of solid wastes mixed with sludge	Although amounts of waste varying from 50 to 90+ percent have been used, 60 percent appears to be a reasonable compromise.
Hydraulic and mean cell residence time, $\Theta_h = \Theta_c$	Washout time is in the range of 3 to 4 days. Use 8 to 15 days for design or base design on results of pilot-plant studies.
Loading rate	0.6 to 1.6 kg/(m³·day) [0.04 to 0.10 lb/(ft³·day)]. Not well defined at present time. Significantly higher rates have been reported.
Temperature	Between 327 and 333 K (130 and 140°F).
Destruction of volatile solid wastes	Varies from about 60 to 80 percent; 70 percent can be used for estimating purposes.
Total solids destroyed	Varies from 40 to 60 percent, depending on amount of inert material present originally.
Gas production	0.5 to 0.75 m³/kg (8 to 12 ft³/lb) of volatile solids destroyed ($CH_4 = 60$ percent; $CO_2 = 40$ percent).

*From G. Tchobanoglous, H. Theisen, and R. Eliassen, *Solid Wastes: Engineering Principles and Management Issues*, McGraw-Hill, New York, 1977.

†Actual removal rates for volatile solids may be less, depending on the amount of material diverted to the scum layer.

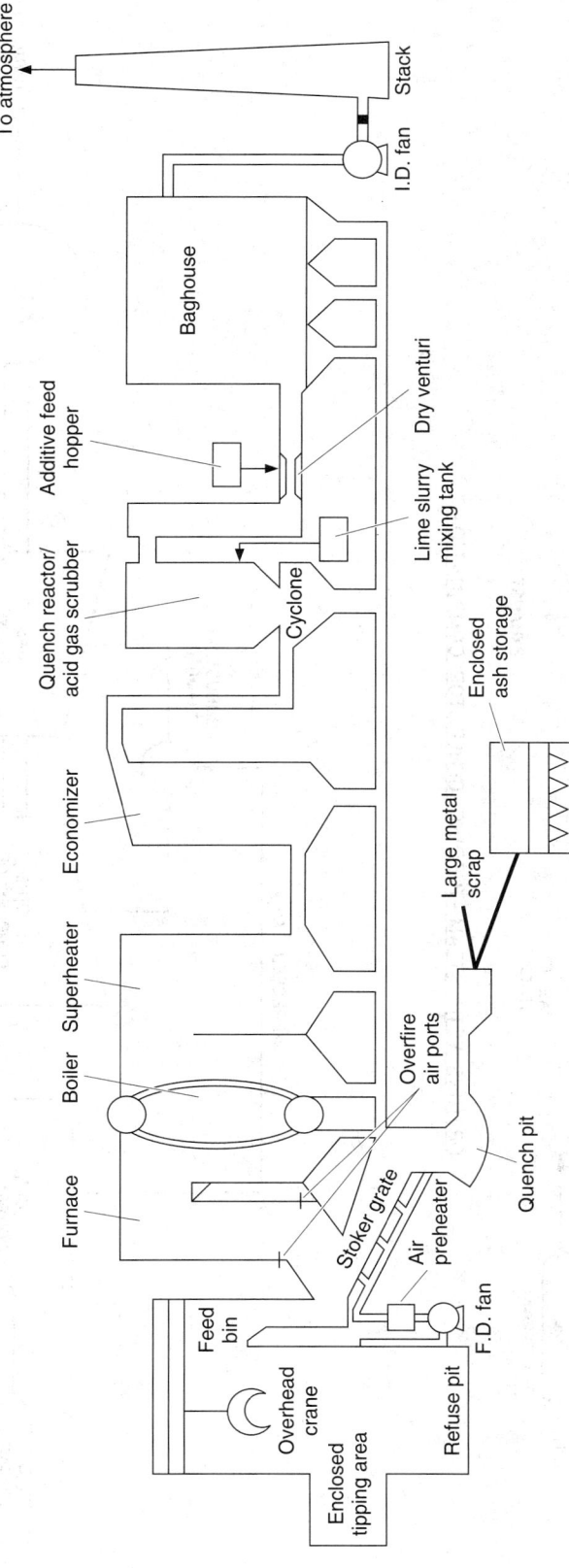

FIG. 25-64 Diagram of a modern mass-burn facility. (*From Municipal Waste Combustion Study Report to Congress, June 1982, PB87-206074.*)

25-97

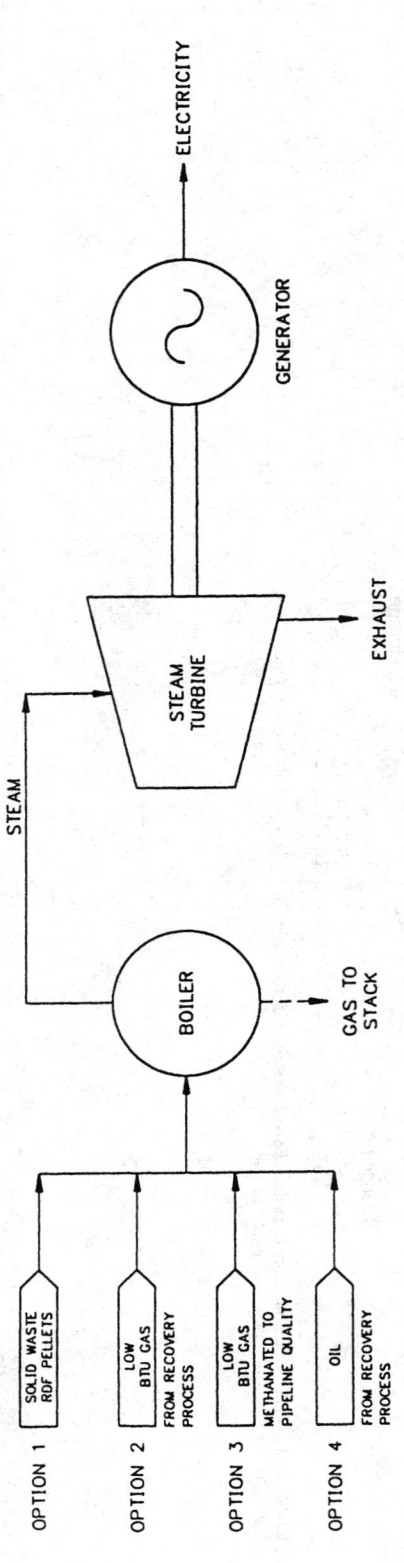

OPTIONS WITH STEAM–TURBINE–GENERATOR COMBINATION

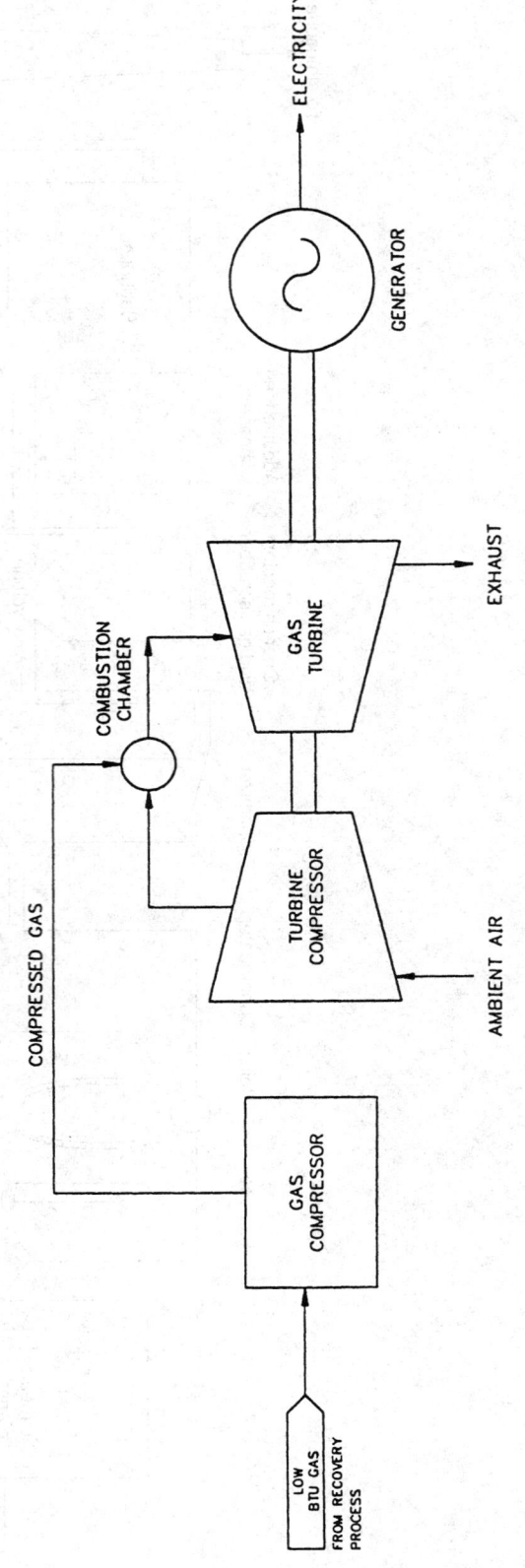

OPTIONS WITH GAS–COMPRESSOR–GAS–TURBINE–GENERATOR COMBINATION

FIG. 25-65 Flow sheet—alternative energy recovery systems.

TABLE 25-67 Typical Thermal Efficiency and Plant Use and Loss Factors for Individual Components and Processes Used for Recovery of Energy from Solid Wastes

Component	Efficiency° Range	Efficiency° Typical	Comments
Incinerator-boiler	40–68	63	Mass-fired
Boiler			
Solid fuel	60–75	72	Processed solid wastes (RDF)
Low-Btu gas	60–80	75	Necessity to modify burners
Oil-fired	65–85	80	Oils produced from solid wastes possibly required to be blended to reduce corrosiveness
Gasifier	60–70	70	
Pyrolysis reactor	65–75	70	
Turbines			
Combustion gas			
Simple cycle	8–12	10	
Regenerative	20–26	24	Including necessary appurtenances
Expansion gas	30–50	40	
Steam-turbine-generator system			
Less than 10 MW	24–40	29†‡	Including condenser, heaters, and all other necessary appurtenances but not boiler
Over 10 MW	28–32	31.6†‡	
Electric generator			
Less than 10 MW	88–92	90	
Over 10 MW	94–98	96	
Plant use and loss factors			
Station-service allowance			
Steam-turbine-generator plant	4–8	6	
Unaccounted heat losses	2–8	5	

°Theoretical value for mechanical equivalent of heat = 3600 kJ/kWh.
†Efficiency varies with exhaust pressure. Typical value given is based on an exhaust pressure in the range of 50 to 100 mmHg.
‡Heat rate = 11,395 kJ/kWh = (3600 kJ/kWh)/0.316.

TABLE 25-68 Energy Output and Efficiency for 1000 Metric Ton of Waste/Day Steam-Boiler Turbine-Generator Energy-Recovery Plant Using Unprocessed Industrial Solid Wastes with Energy Content of 12,000 kJ/kg

Item	Value
Energy available in solid wastes, million kJ/h [(1000 metric tons/day × 1000 kg/metric ton × 12,000 kJ/kg)/(24h/day × 10⁶ kJ/million kJ)]	500
Steam energy available, million kJ/h (500 million kJ/h × 0.7)	350
Electric power generation, kW (350 million kJ/h)/(11,395 kJ/kWh)°	30,715
Station-service allowance, kW [30,715 (0.06)]	−1,843
Unaccounted heat losses, kW [30,715 (0.05)]	−1,536
Net electric power for export, kW	27,336
Overall efficiency, percent {(27,336 kW)/[(500,000,000 kJ/h)/(3,600 kJ/kWh)]}(100)	19.7

°11,395 kJ/kWh = (3600 kJ/kWh)/0.316.

from a solid-waste energy conversion system using an incinerator-boiler-steam-turbine-electric-generator combination with a capacity of 1000 metric tons per day of waste is presented in Table 25-68. If it is assumed that 10 percent of the power generated is used for the front-end processing system (typical values vary from 8 to 14 percent), then the net power for export is 24,604 kW and the overall efficiency is 17.5 percent.

Concentration of WTE Incinerators The total number of municipal waste incinerator facilities as listed in the *Solid Waste Digest*, vol. 4, no. 9 September 1994 (a publication of Chartwell Information Publishers of Alexandria, VA) is 62. See Table 25-69, which covers over 200 existing units. The wastes burned in these facilities totals 8.44 percent of total municipal wastes managed in landfills, incinerators, and transfer stations. This amounts to 88,470 tons per day combusted municipal waste.

One notes that the heavily populated areas of the country also have the highest number of WTE facilities as well as the highest intake of municipal waste into incinerators. This is also due to the lack of open space for landfills compared to the midwest and western states. The amount of waste combusted in the northeastern states is 20.7 percent of the total generated compared to 8.44 percent of all municipal wastes combusted nationwide. The cost per ton averages $77 in New Jersey and New York compared to $57 for the nation. Incinerator costs are similar to landfill costs in the northeastern states. However, landfill costs are far lower than WTE tipping fees. For example, in states like Idaho (landfill costs, $12/ton) and Texas (landfill costs, $10/ton), WTE incinerator plants cannot compete. The NIMBY (Not in my Backyard) syndrome concerning incineration systems has also hurt the siting and permitting of many of these facilities. Figure 25-66 shows the range of costs for incineration in September 1994. Figure 25-67 shows the range costs for all solid waste management in September 1994. The National index was $37.93/ton in 1994, up from $33.64/ton in 1992. This is an average increase of 6.5% per year which is above the inflation rate for this two-year period.

REGULATIONS APPLICABLE TO MUNICIPAL WASTE COMBUSTORS

New Source Performance Standards (NSPS) were promulgated under Sections 111(b) and 129 of the CAA Amendments of 1990. The NSPS applies to new municipal solid-waste combustors (MWCs) with

TABLE 25-69 Solid Waste Price Index, WTE Incinerator Intake TPD-Tip Fee, September 1994

Region	Number of facilities	Intake, TPD	% of daily intake	Tip fee, $/ton
Northeastern states				
Connecticut	5	6410	60.63	66.35
Delaware	0	0	0	n/a
Massachusetts	2	2460	13.33	60.52
Maryland	3	3562	19.78	64.65
Maine	3	2407	77.94	61.55
New Hampshire	1	997	19.76	52.70
New Jersey	4	6575	22.64	80.40
New York	8	14,150	29.22	74.22
Pennsylvania	6	7580	10.32	62.57
Rhode Island	0	0	0	n/a
Vermont	0	0	0	n/a
Total northeastern states	32	44,141	20.7	66.56
Southern states				
Alabama	1	690	6.39	39.00
Arkansas	0	0	0	n/a
Florida	10	16,633	26.08	58.20
Georgia	0	0	0	n/a
Kentucky	0	0	0	n/a
Louisiana	0	0	0	n/a
Mississippi	0	0	0	n/a
North Carolina	0	0	0	n/a
South Carolina	1	600	4.44	51.50
Tennessee	2	1970	9.81	28.64
Virginia	3	4975	11.12	45.04
West Virginia	0	0	0	n/a
Total southern states	17	24,868	10.32	52.53
Midwestern states				
Iowa	0	0	0	n/a
Illinois	0	0	0	n/a
Indiana	1	2362	6.45	25.00
Kansas	0	0	0	n/a
Michigan	1	1630	5.22	52.66
Minnesota	3	5187	35.4	61.56
Missouri	0	0	0	n/a
North Dakota	0	0	0	n/a
Nebraska	0	0	0	n/a
Ohio	4	5437	9.16	31.32
South Dakota	0	0	0	n/a
Wisconsin	0	0	0	n/a
Total midwestern states	9	14,616	5.89	43.41
Western states				
Arizona	0	0	0	n/a
Colorado	0	0	0	n/a
Idaho	0	0	0	n/a
Montana	0	0	0	n/a
New Mexico	0	0	0	n/a
Nevada	0	0	0	n/a
Oklahoma	1	1230	15.1	41.91
Texas	0	0	0	n/a
Utah	0	0	0	n/a
Wyoming	0	0	0	41.91
Total western states	1	1230	0.76	41.91
Pacific states				
Alaska	0	0	0	n/a
California	2	2673	1.79	27.50
Hawaii	0	0	0	n/a
Oregon	0	0	0	n/a
Washington	1	944	5	87.61
Total Pacific states	3	3617	1.95	43.91
National total	62	88,471	8.44	57.49

capacity to combust more than 250 tons per day of municipal solid waste that commenced construction after December 20, 1989. The proposed standards and guidelines were published in the *Federal Register* on December 20, 1989.

Section 129 of the CAAA of 1990 applies to a range of solid waste incinerators including MWCs, medical waste incinerators (MWIs) and industrial waste incinerators. Incinerators for hazardous solid and liquid wastes are covered under RCRA regulations (40 CFR Parts 260 through 272) and TSCA, Toxic Substances Control Act, 1976 (40 CFR Parts #700–766).

Regulated Pollutants The NSPS regulates MWC emissions, and nitrogen oxides (NO_x) emissions from individual MWC units larger

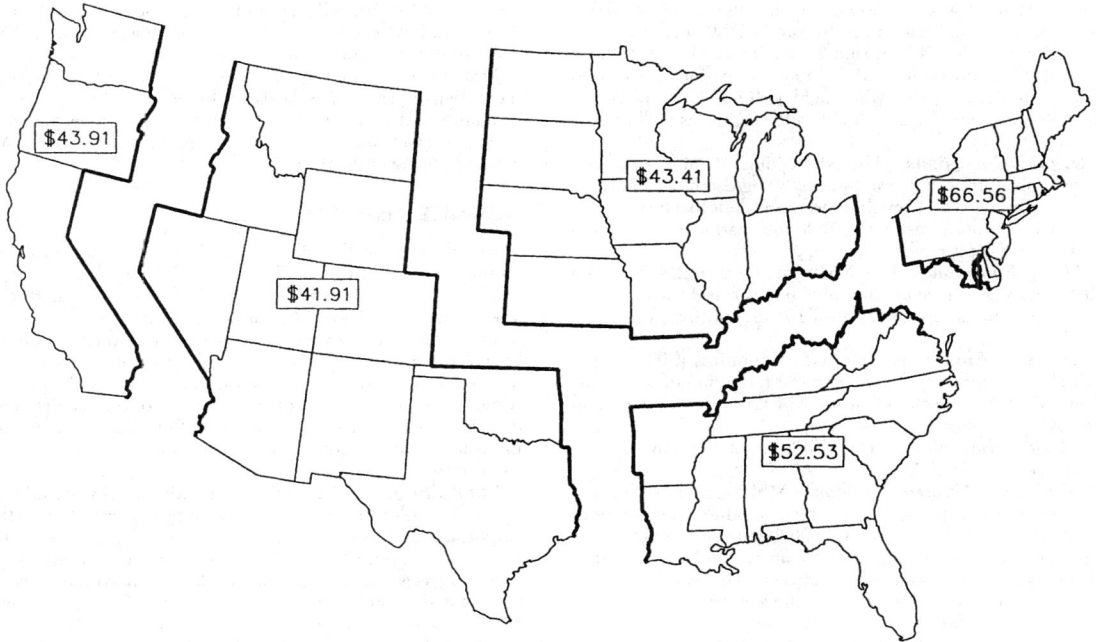

FIG. 25-66 Waste-to-energy price index. National index, September 1994, $57.49/ton. (*Data from* Solid Waste Digest, *vol. 4, no. 9, Sept. 1994. Published by Chartwell Information Publishers, Alexandria, VA.*)

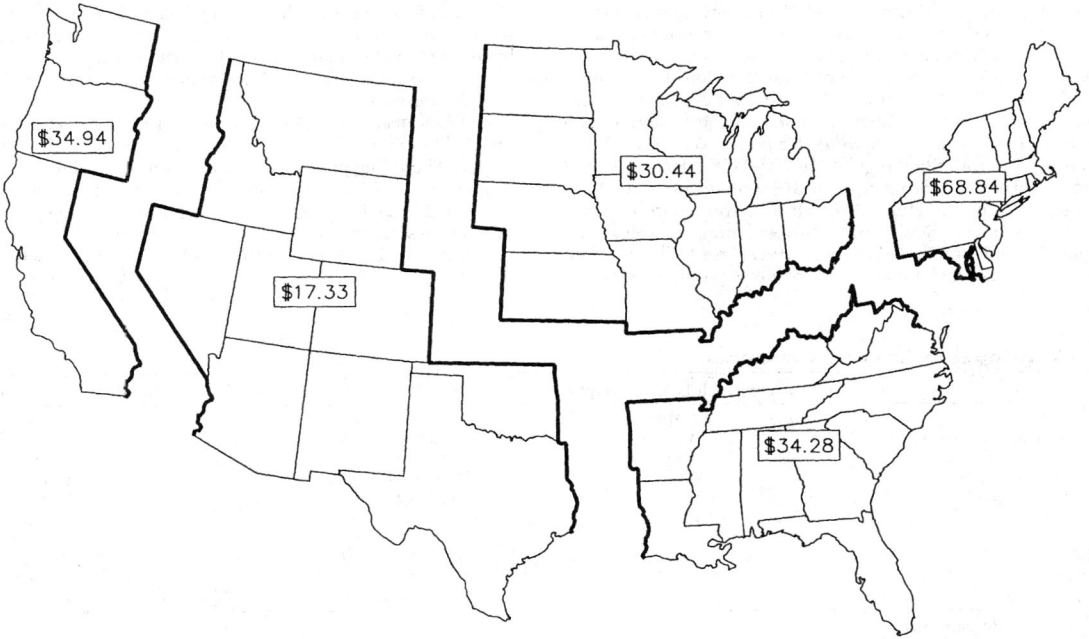

FIG. 25-67 Solid Waste Price Index™. National index, September 1994, $37.93/ton; September 1992, $33.64/ton. (*Data from* Solid Waste Digest, *vol. 4, no. 9, Sept. 1994. Published by Chartwell Information Publishers, Alexandria, VA.*)

than 250 tpd capacity. MWC emissions are subcategorized as MWC metal emissions, MWC organic emissions, and MWC acid gas emissions. The NSPS establishes emission limits for organic emissions (measured as dioxins and furans), MWC metal emissions (measured as particulate matter [PM]), and MWC acid gas emissions (measured as sulfur dioxide [SO_2] and hydrogen chloride [HCl]), as well as NO_x emission limits.

MWCs Organic Emissions   The NSPS limits organic emissions to a total dioxin plus furan emission limit of 30 ng/dscm (at 7 percent O_2 dry volume). This level is approximately equivalent to a toxic equivalent (TEQ) of 1.0 ng/dscm, using the 1990 international toxic equivalency factor (I-TEF) approach.

MWCs Metal Emissions   The NSPS includes a PM emission limit of 0.015 grains per dry standard cubic feet (gr/dscf) at 7 percent oxygen dry/volume and an opacity limit of 10 percent (6 minute average).

MWCs Acid Gas Emissions   The NSPS requires a 95 percent reduction of HCl emissions and an 80 percent reduction of SO_2 emissions for new MWCs or an emission limit of 25 ppmv for HCl and 30 ppmv for SO_2 (at 7 percent O_2 dv).

Nitrogen Oxides Emissions   The NSPS limits NO_x emissions to 180 ppmv (at 7 percent O_2 dv).

MSC Air-Pollution-Control Systems   MSCs generate flue gas that contains particulates, acid gases and trace amounts of organic and volatile metals. Particulates have traditionally been removed by use of cyclone separators and electrostatic precipitators. Acid gases require neutralization and removal from the gas stream. This can be accomplished by adding solutions or chemicals to the gas stream and removing the products of the chemical reaction when these materials are mixed together. Two major types of APCs are employed:

• Dry systems, where the gas stream is humidified and chemicals are added to the system

• Wet systems, where large quantities of water containing chemicals wash the gas stream

Incinerator Performance   Environmental organizations have had concerns regarding performance of existing incineration plants. Major concerns focus on dioxin, mercury, and ash. Groups opposing a proposed WTE facility in New York State set emission standards for the facility. Table 25-70 compares the recommendations for emission levels made by the Natural Resources Defense Council, Environmental Defense Fund, INFORM, Environmental Action Coalition, and Scenic Hudson in *A Solid Waste Blue Print for New York State*, (March, 1988), New York State regulations (GNYCRR Part 219), and actual tests performed at Hempstead in the summer of 1990. Note that the results from the Hempstead stack testing performed in August, 1990 exceed the New York State regulations as well as the environmentalist's recommendation in every category. This was not necessarily true in the case of many MSC plants without proper com-

bustion controls or APC systems. With the standards established in Section 129 of the CAAA of 1990, all facilities must retrofit to conform to these standards or be shut down.

MSC Facilities are required to meet some of the toughest environmental air emission standards in the country. Complying with these standards makes modern waste combustors among the cleanest producers of electricity—and may even provide a means of improving a community's overall air quality.

ULTIMATE DISPOSAL

Disposal on or in the earth's mantle is, at present, the only viable method for the long-term handling of (1) solid wastes that are collected and are of no further use, (2) the residual matter remaining after solid wastes have been processed, and (3) the residual matter remaining after the recovery of conversion products and/or energy has been accomplished. The three land disposal methods used most commonly are (1) landfilling, (2) landfarming, and (3) deepwell injection. Although incineration is being used more often as a disposal method, it is, in reality, a processing method. Recently, the concept of using muds in the ocean floor as a waste-storage location also has received some attention.

Landfilling of Solid Wastes   Landfilling involves the controlled disposal of solid wastes on or in the upper layer of the earth's mantle. Important aspects in the implementation of sanitary landfills include (1) site selection, (2) landfilling methods and operations, (3) occurrence of gases and leachate in landfills, (4) movement and control of landfill gases and leachate, and (5) landfill design. The landfilling of hazardous wastes is considered separately.

Site Selection   Factors that must be considered in evaluating potential solid-waste-disposal sites are summarized in Table 25-71. Final selection of a disposal site usually is based on the results of a preliminary site survey, results of engineering design and cost studies, and an environmental-impact assessment.

Landfilling Methods and Operations   To use the available area at a landfill site effectively, a plan of operation for the placement of solid wastes must be prepared. Various operational methods have been developed primarily on the basis of field experience. The principal methods used for landfilling dry areas may be classified as (1) area, and (2) depression.

1. *Area method.*   The area method is used when the terrain is unsuitable for the excavation of trenches in which to place the solid wastes. The filling operation usually is started by building an earthen levee against which wastes are placed in thin layers and compacted (see Fig. 25-68). Each layer is compacted as the filling progresses until the thickness of the compacted wastes reaches a height varying from 2 to 3 m (6 to 10 ft). At that time and at the end of each day's operation, a 150- to 300-mm (6- to 12-in) layer of cover material is placed

TABLE 25-70   Hempstead Emissions Comparison

Parameter°	Environmentalists recommendations	New York State regulations	Actual test results from Hempstead
Particulates (gr/dscf)	0.010	0.010	0.00053
Opacity (clarity of air from the stack)	5%	10%	<5%
Dioxin† (ng/dscm)	0.1	0.2 (goal)	0.0155
Hydrogen chloride (ppmv)	50	50 ppmv or 90% removal	41.7
Sulfur dioxide‡ (ppmv)	50	Use HCl	22.9
Lead (lb/ton)	0.0005	Risk assessment	0.000011
Arsenic (lb/ton)	0.00001	Risk assessment	<0.000003
Mercury (lb/ton)	0.0015	Risk assessment	0.0006
Cadmium (lb/ton)	0.00002	Risk assessment	<0.000003
Carbon monoxide§ (ppmv)	50	Use combustion efficiency	46.2

gr/dscf = grains per dry standard cubic foot
ng/dscm = nanograms per dry standard cubic meter
ppmv = parts per million by volume
lb/ton = pounds per ton of refuse processed
°Concentrations are corrected to 12 percent CO_2 basis.
†PCDD/PCDF toxic equivalents as defined by NYSDEC.
‡8-hour average.
§4-hour average.
Data provided by Taconic Resources, Inc., Jan. 12, 1993. Inform Publication, *A Solid Waste Blueprint for New York State,* March 1988.

TABLE 25-71   Important Factors in Preliminary Selection of Landfill Sites

Factor	Remarks
Available land area	In selecting potential land disposal sites, it is important to ensure that sufficient land area is available. Sufficient area to operate for at least 1 year at a given site is needed to minimize costs.
Impact of processing and resource recovery	It is important to project the extent of resource-recovery-processing activities that are likely to occur in the future and determine their impact on the quantity and condition of the residual materials to be disposed of.
Haul distance	Although minimum haul distances are desirable, other factors must also be considered. These include collection-route location, types of wastes to be hauled, local traffic patterns, and characteristics of the routes to and from the disposal site (condition of the routes, traffic patterns, and access conditions).
Soil conditions and topography	Because it is necessary to provide material for each day's landfill and a final layer of cover after the filling has been completed, data on the amounts and characteristics of the soils in the area must be obtained. Local topography will affect the type of landfill operation to be used, equipment requirements, and the extent of work necessary to make the site usable.
Climatological conditions	Local weather conditions must also be considered in the evaluation of potential sites. Under winter conditions where freezing is severe, landfill cover material must be available in stockpiles when excavation is impractical. Wind and wind patterns must also be considered carefully. To avoid blowing or flying papers, windbreaks must be established.
Surface-water hydrology	The local surface-water hydrology of the area is important in establishing the existing natural drainage and runoff characteristics that must be considered. Other conditions of flooding must also be identified.
Geologic and hydrogeologic conditions	Geologic and hydrogeologic conditions are perhaps the most important factors in establishing the environmental suitability of the area for a landfill site. Data on these factors are required to assess the pollution potential of the proposed site and to establish what must be done to the site to control the movement of leachate or gases from the landfill.
Local environmental conditions	The proximity of both residential and industrial developments is extremely important. Great care must be taken in their operation if they are to be environmentally sound with respect to noise, odor, dust, flying paper, and vector control.
Ultimate uses	Because the ultimate use affects the design and operation of the landfill, this issue must be resolved before the layout and design of the landfill are started.

over the completed fill. The cover material must be hauled in by truck or earth-moving equipment from adjacent land or from borrow-pit areas. In some newer landfill operations, the daily cover material is omitted. A completed lift, including the cover material is called a "cell" (see Fig. 25-69). Successive lifts are placed on top of one another until the final grade called for in the ultimate development plan is reached. A final layer of cover material is used when the fill reaches the final design height.

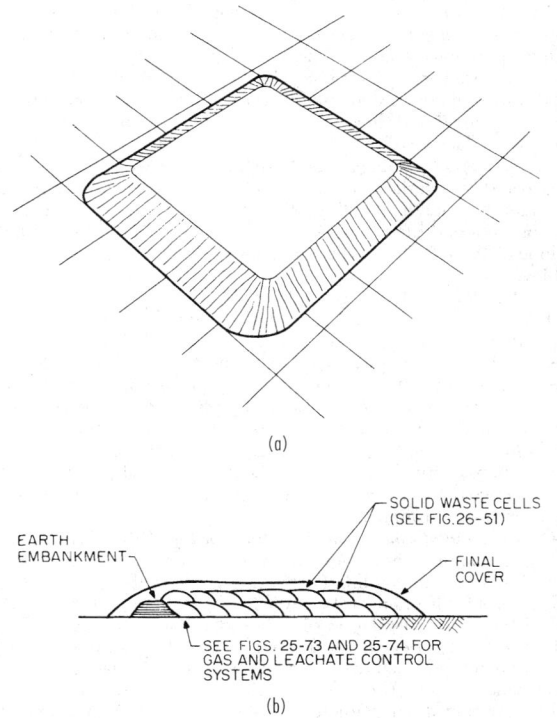

FIG. 25-68   Area method for landfilling solid wastes. (*a*) Pictorial view of completed landfill. (*b*) Section through landfill.

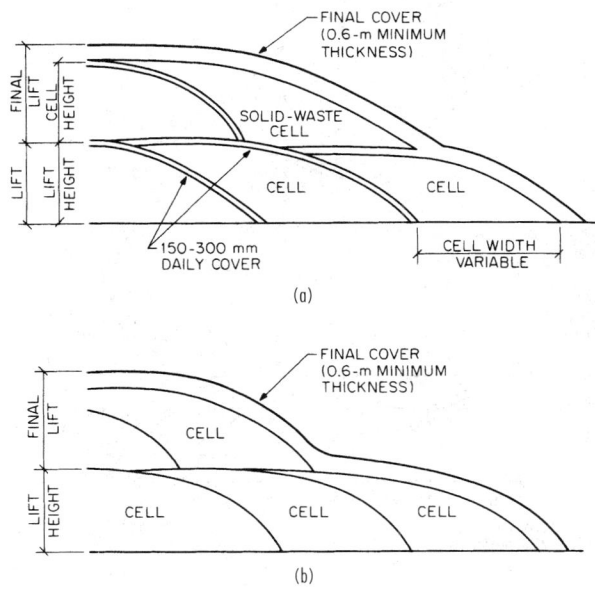

FIG. 25-69   Typical section through a landfill. (*a*) With daily or intermediate cover. (*b*) Without daily or intermediate cover.

2. *Depression method.*   At locations where natural or artificial depressions exist, it is often possible to use them effectively for land-filling operations. Canyons, ravines, dry borrow pits, and quarries have been used for this purpose. The techniques to place and compact solid wastes in depression landfills vary with the geometry of the site, the characteristics of the cover material, the hydrology and geology of the site, and access of the site.

In a canyon site, filling starts at the head end of the canyon (see Fig. 25-70) and ends at the mouth. The practice prevents the accumulation of water behind the landfill. Wastes usually are deposited on the canyon floor and from there are pushed up against the canyon face at a slope of about 2 to 1. In this way, a high degree of compaction can be achieved.

3. *Landfills in wet areas.*   Because of the problems associated with contamination of local groundwaters, the development of odors, and structural stability, landfills must be avoided in wetlands. If wet areas such as ponds, pits, or quarries must be used as landfill sites, special provisions must be made to contain or eliminate the movement of leachate and gases from completed cells. Usually this is accomplished by first draining the site and then lining the bottom with a clay liner or other appropriate sealants. If a clay liner is used, it is important to continue operation of the drainage facility until the site is filled to avoid the creation of uplift pressures that could cause the liner to rupture from heaving.

Occurrence of Gases and Leachate in Landfills   The following biological, physical, and chemical events occur when solid wastes are placed in a sanitary landfill: (1) biological decay of organic materials, either aerobically or anaerobically, with the evolution of gases and liquids; (2) chemical oxidation of waste materials; (3) escape of gases from the fill; (4) movement of liquids caused by differential heads; (5) dissolving and leaching of organic and inorganic materials by water and leachate moving through the fill; (6) movement of dissolved material by concentration gradients and osmosis; and (7) uneven settlement caused by consolidation of material into voids.

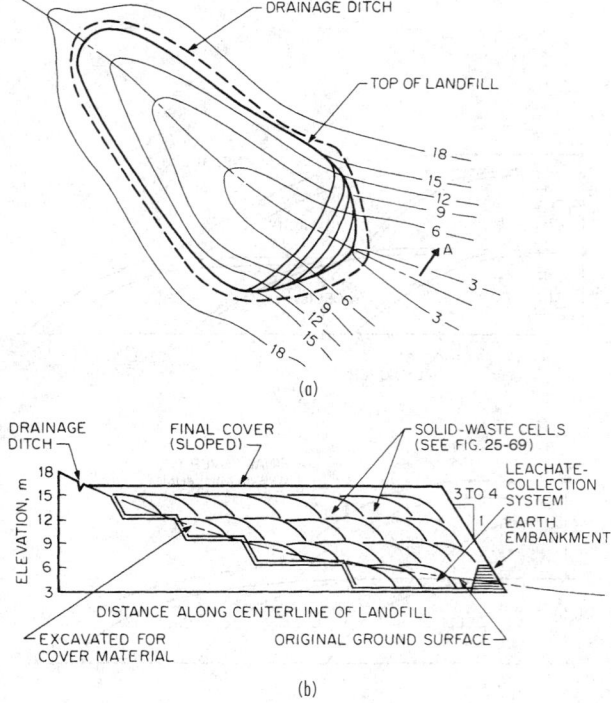

FIG. 25-70 Depression method for landfilling solid wastes. (*a*) Plan view: canyon-site landfill. (*b*) Section through landfill.

With respect to item 1, bacterial decomposition initially occurs under aerobic conditions because a certain amount of air is trapped within the landfill. However, the oxygen in the trapped air is exhausted within days, and long-term decomposition occurs under anaerobic conditions.

1. *Gases in landfills.*   Gases found in landfills include air, ammonia, carbon dioxide, carbon monoxide, hydrogen, hydrogen sulfide, methane, nitrogen, and oxygen. Data on the molecular weight and density of these gases are presented in Sec. 2. Carbon dioxide and methane are the principal gases produced from the anaerobic decomposition of the organic solid-waste components.

The anaerobic conversion of organic compounds is thought to occur in three steps: The first involves the enzyme-mediated transformation (hydrolysis) of higher-weight molecular compounds into compounds suitable for use as a source of energy and cell carbon; the second is associated with the bacterial conversion of the compounds resulting from the first step into identifiable lower-molecular-weight intermediate compounds; and the third step involves the bacterial conversion of the intermediate compounds into simpler end products, such as carbon dioxide (CO_2) and methane (CH_4). The overall anaerobic conversion of organic industrial wastes can be represented with the following equation:

$$C_aH_bO_cN_d \rightarrow nC_wH_xO_yN_z + mCH_4 + sCO_2 + rH_2O + (d - nz)NH_3$$

$$(25\text{-}26)$$

where $s = a - nw - m$
$r = c - ny - 2s$

The terms $C_aH_bO_cN_d$ and $C_wH_xO_yN_z$ are used to represent on a molar basis the composition of the material present at the start of the process. If it is assumed that the organic wastes are stabilized completely, the corresponding expression is

$$C_aH_bO_cN_d + \left(\frac{4a - b - 2c + 3d}{4}\right)H_2O \rightarrow \left(\frac{4a + b - 2c - 3d}{8}\right)CH_4$$

$$+ \left(\frac{4a - b + 2c + 3d}{8}\right)CO_2 + dNH_3 \quad (25\text{-}27)$$

The rate of decomposition in unmanaged landfills, as measured by gas production, reaches a peak within the first 2 years and then slowly tapers off, continuing in many cases for periods up to 25 years or more. The total volume of the gases released during anaerobic decomposition can be estimated in a number of ways. If all the organic constituents in the wastes (with the exception of plastics, rubber, and leather) are represented with a generalized formula of the form $C_aH_bO_cN_d$, the total volume of gas can be estimated by using Eq. (25-27) with the assumption of completed conversion to carbon dioxide and methane.

2. *Leachate in landfills.*   Leachate may be defined as liquid that has percolated through solid waste and has extracted dissolved or suspended materials from it. In most landfills, the liquid portion of the leachate is composed of the liquid produced from the decomposition of the wastes and liquid that has entered the landfill from external sources, such as surface drainage, rainfall, groundwater, and water form underground springs. Representative data on chemical characteristics of leachate are reported in Table 25-72.

Gas and Leachate Movement and Control   Under ideal conditions, the gases generated from a landfill should be either vented to the atmosphere or, in larger landfills, collected for the production of energy. Landfills with >2.5 million cubic meters of waste or >50 Mg/y NMOC (nonmethane organic compounds) emissions may require landfill-gas collection and flare systems, per EPA support WWW, CFR 60 Regulations. The leachate should be either contained within the landfill or removed for treatment.

1. *Gas movement.*   In most cases, over 90 percent of the gas volume produced from the decomposition of solid wastes consists of methane and carbon dioxide. Although most of the methane escapes to the atmosphere, both methane and carbon dioxide have been found in concentrations of up to 40 percent at lateral distances of up to 120 m (400 ft) from the edges of landfills. Methane can accumulate below buildings or in other enclosed spaces on or close to a sanitary landfill. With proper venting, methane should not pose a problem.

TABLE 25-72   Typical Leachate Quality of Municipal Waste

S1 Number	Parameter	Overall range (mg/liter except as indicated)
1	TDS	584–55,000
2	Specific conductance	480–72,500 u.mho/cm
3	Total suspended solids	2–140,900
4	BOD	ND–195,000
5	COD	6.6–99,000
6	TOC	ND–40,000
7	pH	3.7–8.9 units
8	Total alkalinity	ND–15,050
9	Hardness	0.1–225,000
10	Chloride	2–11,375
11	Calcium	3.0–2,500
12	Sodium	12–6,010
13	Total Kjeldahl nitrogen	2–3,320
14	Iron	ND–4,000
15	Potassium	ND–3,200
16	Magnesium	4.0–780
17	Ammonia-nitrogen	ND–1,200
18	Sulfate	ND–1,850
19	Aluminum	ND–85
20	Zinc	ND–731
21	Manganese	ND–400
22	Total phosphorus	ND–234
23	Boron	0.87–13
24	Barium	ND–12.5
25	Nickel	ND–7.5
26	Nitrate-nitrogen	ND–250
27	Lead	ND–14.2
28	Chromium	ND–5.6
29	Antimony	ND–3.19
30	Copper	ND–9.0
31	Thallium	ND–0.78
32	Cyanide	ND–6
33	Arsenic	ND–70.2
34	Molybdenum	0.01–1.43
35	Tin	ND–0.16
36	Nitrite-nitrogen	ND–1.46
37	Selenium	ND–1.85
38	Cadmium	ND–0.4
39	Silver	ND–1.96
40	Beryllium	ND–0.36
41	Mercury	ND–3.0
42	Turbidity	40–500 Jackson units

From McGinely, P. M. and Kmet, P., *Formation Characteristics, Treatment, and Disposal of Leachate from Municipal Solid Waste Landfills*, Bureau of Solid Waste Management, Wisconsin Department of Natural Resources, Madison, 1984.

Because carbon dioxide is about 1.5 times as dense as air and 2.8 times as dense as methane, it tends to move toward the bottom of the landfill. As a result, the concentration of carbon dioxide in the lower portions of landfill may be high for years. Ultimately, because of its density, carbon dioxide will also move downward through the underlying formation until it reaches the groundwater. Because carbon dioxide is readily soluble in water, it usually lowers the pH, which in turn can increase the hardness and mineral content of the groundwater through the solubilization of calcium and magnesium carbonates.

2. *Control of gas movement.* The lateral movement of gases produced in a landfill can be controlled by installing vents made of materials that are more permeable than the surrounding soil. Typically, as shown in Fig. 25-71a, gas vents are constructed of gravel. The spacing of cell vents depends on the width of the waste cell but usually varies from 18 to 60 m (60 to 200 ft). The thickness of the gravel layer should be such that it will remain continuous even though there may be differential settling; 300 to 450 mm (12 to 18 in) is recommended. Barrier vents (see Fig. 25-71b) also can be used to control the lateral movement of gases. Well vents are often used in conjunction with lateral surface vents buried below grade in a gravel trench (see Fig. 25-71c). Details of a gas vent are shown in Fig. 25-72. Control of the downward movement of gases can be accomplished by installing perforated pipes in the gravel layer at the bottom of the landfill. If the gases cannot be vented laterally, it may be necessary to install gas wells

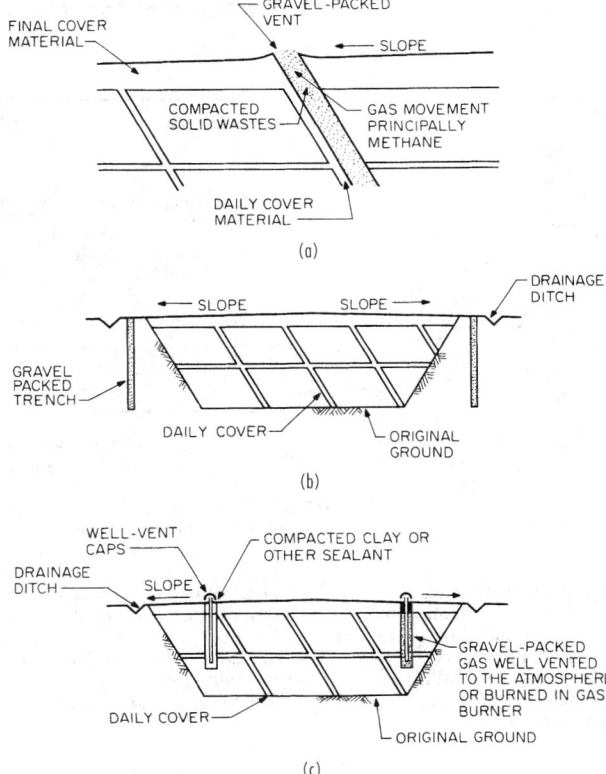

FIG. 25-71   Vents used to control the lateral movement of gases in landfills. (*a*) Cell. (*b*) Barrier. (*c*) Well. (*From G. Tchobanoglous, H. Theisen, and R. Eliassen, Solid Wastes: Engineering Principles and Management Issues, McGraw-Hill, New York, 1977.*)

and vent the gas to the atmosphere. This is considered a passive venting system. See Fig. 25-73.

The movement of landfill gases through adjacent soil formations can be controlled by constructing barriers of materials that are more impermeable than the soil (see Fig. 25-74a). Some of the landfill

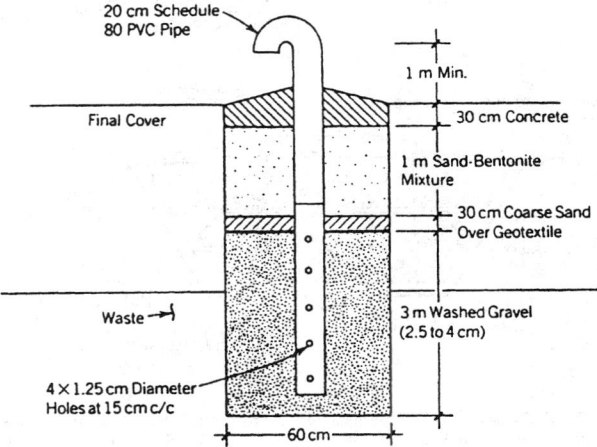

FIG. 25-72   Typical detail of an isolated gas vent. (*From Bagchi, A., Design, Construction, and Monitoring of Sanitary Landfill, Wiley, 1990.*)

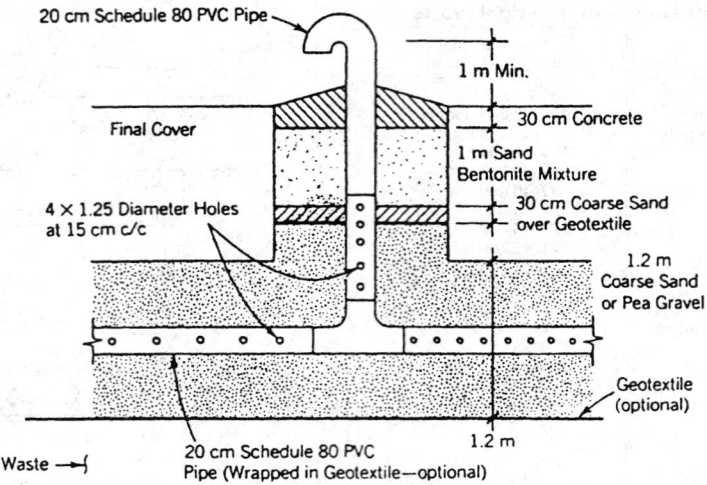

FIG. 25-73 Typical detail of a passive gas venting system with a header pipe. (*From Bagchi, A., Design, Construction, and Monitoring of Sanitary Landfill, Wiley, 1990.*)

sealants that are available for this use are identified in Table 25-73. Of these, the use of compacted clays is the most common. The thickness will vary depending on the type of clay and the degree of control required; thickness ranging from 0.15 to 1.25 m (6 to 48 in) have been used. Covers of landfills are also typically multi-layer-foundation layer, clay layer, membrane, drainage layer (synthetic or natural) root zone layer, or soil.

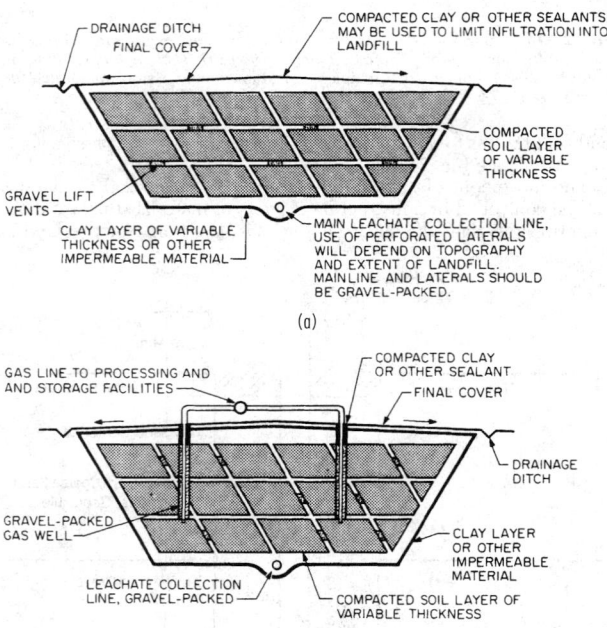

(a)

(b)

FIG. 25-74 Use of an impermeable liner to control the movement of gases and leachate in landfills. (*a*) Without gas recovery. (*b*) With gas recovery. (*From G. Tchobanoglous, H. Theisen, and R. Eliassen, Solid Wastes: Engineering Principles and Management Issues, McGraw-Hill, New York, 1977.*)

3. *Control of gas movement by recovery.* The movement of gases in landfills can also be controlled by installing gas-recovery wells in completed landfills (see Fig. 25-74*b*). This is considered an active venting system. Clay and other liners are used when landfill gas is to be recovered. In some gas-recovery systems, leachate is collected and recycled to the top of the landfill and reinjected through perforated lines located in drainage trenches. Typically, the rate of gas production is greater in leachate-recirculation systems.

Gas-recovery systems have been installed in some large municipal landfills. The economics of such operations must be reviewed for each

TABLE 25-73   Landfill Sealants for the Control of Gas and Leachate Movement

Sealant		Remarks
Classification	Representative types	
Compacted soil		Should contain some clay or fine silt.
Compacted clay	Bentonites, illites, kaolinites	Most commonly used sealant for landfills; layer thickness varies from 0.15–1.25 m; layer must be continuous and not be allowed to dry out and crack.
Inorganic chemicals	Sodium carbonate, silicate, or pyrophosphate	Use depends on local soil characteristics.
Synthetic chemicals	Polymers, rubber latex	Experimental; use not well established.
Synthetic membrane liners	Polyvinyl chloride, butyl rubber, Hypalon, polyethylene, nylon-reinforced liners	Expensive; may be justified where gas is to be recovered.
Asphalt	Modified asphalt, asphalt-covered polypropylene fabric, asphalt concrete	Layer must be thick enough to maintain continuity under differential settling conditions.
Others	Gunite concrete, soil cement, plastic soil cement	

*From G. Tchobanoglous, H. Theisen, and R. Eliassen, *Solid Wastes: Engineering Principles and Management Issues,* McGraw-Hill, New York, 1977.

site. The end use of gas will affect the overall economics. The cost of gas-cleanup and -processing equipment will limit the recovery of landfill gases, especially from small landfills.

4. *Leachate movement.* Under normal conditions, leachate is found in the bottom of landfills. From there it moves through the underlying strata, although some lateral movement may also occur, depending on the characteristics of the surrounding material. The rate of seepage of leachate from the bottom of a landfill can be estimated by Darcy's law by assuming that the material below the landfill to the top of the water table is saturated and that a small layer of leachate exists at the bottom of the fill. Under these conditions, the leachate discharge rate per unit area is equal to the value of the coefficient of permeability K, expressed in meters per day. The computed value represents the maximum amount of seepage that would be expected, and this value should be used for design purposes. Under normal conditions, the actual rate would be less than this value because the soil column below the landfill would not be saturated. Models have been developed to aid in the estimation of leachate quantity. Bagchi has covered details of these models in the *Design, Construction, and Monitoring of Sanitary Landfill.*

5. *Control of leachate movement.* As Leachate percolates through the underlying strata, many of the chemical and biological constituents originally contained in it will be removed by the filtering and adsorptive action of the material composing the strata. In general, the extent of this action depends on the characteristics of the soil, especially the clay content. Because of the potential risk involved in allowing leachate to percolate to the groundwater, best practice calls for its elimination or containment. Ultimately, it will be necessary to collect and treat the leachate.

The use of clay has been the favored method of reducing or eliminating the percolation of leachate (see Fig. 25-74 and Table 25-73). Membrane liners are used most often today but require care so that they will not be damaged during the filling operations. Equally important in controlling the movement of leachate is the elimination of surface-water infiltration, which is the major contributor to the total volume of leachate. With the use of an impermeable clay layer, membrane liners, an appropriate surface slope (1 to 2 percent), and adequate drainage, surface infiltration can be controlled effectively.

6. *Settlement and structural characteristics of landfills.* The settlement of landfills depends on the initial compaction, characteristics of wastes, degree of decomposition, and effects of consolidation when the leachate and gases are formed in the landfill. The height of the completed fill will also influence the initial compaction and degree of consolidation.

Design and Operation of Landfills   Important design considerations in the design and operation of landfills include (1) land requirements, (2) types of wastes that must be handled, (3) evaluation of seepage potential, (4) design of drainage and seepage-control facilities, (5) development of a general operation plan, (6) design of solid-waste-filling plan, and (7) determination of equipment requirements. The more important individual factors that must be considered in the design of a landfill are reported in Table 25-74. The last three items are considered further in the following discussion.

1. *Landfill-operation plan.* The layout of the site and the development of a workable operating schedule are the main features of a landfill-operation plan. In planning the layout of a landfill site, the location of the following must be determined: (1) access roads; (2) equipment shelters; (3) scales, if used; (4) storage sites for special wastes; (5) topsoil-stockpile sites; (6) landfill areas; and (7) plantings.

2. *Solid-waste-filling plan.* The specific method of filling will depend on the characteristics of the site, such as the amount of available cover material, the topography, and local hydrology and geology. To assess future development plans, it will be necessary to prepare a detailed plan for the layout of the individual solid-waste cells. On the basis of the characteristics of the site or the method of operation (e.g., gas recovery), it may be necessary to incorporate special features for the control of the movement of gases and leachate from the landfill.

3. *Equipment requirements.* The types of equipment that have been used at sanitary landfills include both crawler and rubber-tired tractors, scrapers, compactors, draglines, and graders. The size and amount of equipment required will depend primarily on local site

TABLE 25-74   Important Factors That Must Be Considered in Design and Operation of Solid-Waste Landfills

Factor	Remarks	Factor	Remarks
Design		Landfilling method	Selection of method will vary with terrain and available cover.
Access	Paved all-weather access roads to landfill site; temporary roads to unloading areas.	Litter control	Use movable fences at unloading areas; crews should pick up litter at least once per month or as required.
Cell design and construction	Will vary depending on terrain, landfilling method, and whether gas is to be recovered.	Operation plan	With or without the codisposal of treatment-plant sludges and the recovery of gas.
Cover material	Maximize use of on-site earth materials; approximately 1 m³ of cover material will be required for every 4 to 6 m³ of solid wastes; mix with sealants to control surface infiltration. In some designs, intermediate cover is not used.	Spread and compaction	Spread and compact waste in 0.6-m (2-ft) layers.
		Unloading area	Keep small, generally under 30 m (100 ft).
Drainage	Install drainage ditches to divert surface-water runoff; maintain 1 to 2 percent grade on finished fill to prevent ponding. TCLP Tests	Operation	
		Communications	Telephone for emergencies.
Equipment requirements	Vary with size of landfills.	Days and hours of operation	Usual practice is 5 to 6 days/week and 8 to 10 h/day.
Fire prevention	Water on site; if nonpotable, outlets must be marked clearly; proper cell separation prevents continuous burn-through if combustion occurs.	Employee facilities	Rest rooms and drinking water should be provided.
		Equipment maintenance	A covered shed should be provided for field maintenance of equipment.
Groundwater protection	Divert any underground springs; if required, install sealants for leachate control; install wells for gas and groundwater monitoring.	Operational records	Tonnage, transactions, and billing if a disposal fee is charged.
		Salvage	No scavenging; salvage should occur away from the unloading area; no salvage storage on site.
Land area	Area should be large enough to hold all wastes for a minimum of 5 years but preferably for 25 to 30 years.	Scales	Essential for record keeping.

conditions, the size of the landfill operation, and the method of operation.

Landfilling of Hazardous Wastes   In many states, the only disposal option available for most hazardous wastes is landfilling. The basis for the management of hazardous-wastes landfills is set forth in the Resource Conservation and Recovery Act of 1976. In general, disposal sites for hazardous wastes should be separate from sites for municipal solid wastes. If separate sites are not possible, great care must be taken to ensure that separate disposal operations are maintained.

Requirements   The requirements for a hazardous-waste landfill are detailed in RCRA and the regulations developed to implement the act. From a design standpoint, two of the most important requirements are (1) complete leachate containment, and (2) control of the surface water on and around the site.

Site Selection   Factors that must be considered in evaluating potential sites for the disposal of hazardous waste are covered in state and federal regulations. In California, landfills where hazardous wastes can be received are referred to as Class I disposal sites. To qualify as a Class I site, it must be shown that:

1.   Geological conditions are naturally capable of preventing vertical hydraulic continuity between liquids and gases emanating from the waste in the site and usable surface or groundwaters.

2.   Geological conditions are naturally capable of preventing lateral hydraulic continuity between liquids and gases emanating from wastes in the site and usable surface or groundwaters, or the disposal area has been modified to achieve such capability.

3.   Underlying geological formations that contain rock fractures or fissures of questionable permeability must be permanently sealed to provide a competent barrier to the movement of liquids or gases from the disposal site to usable water.

4.   Inundation of disposal areas shall not occur until the site is closed in accordance with requirements of the regional board.

5.   Disposal areas shall not be subject to washout.

6.   Leachate and subsurface flow into the disposal areas shall be contained within the site unless other disposition is made in accordance with requirements of the regional board.

7.   Site shall not be located over zones of active faulting or where other forms of geological change would impair the competence of natural features or artificial barriers which prevent continuity with usable waters.

8.   Sites made suitable for use by human-made physical barriers shall not be located where improper operations or maintenance of such structures could permit the waste, leachate, or gases to contact usable groundwater of surface water.

9.   Sites that comply with the above-noted clauses but would be subject to inundation by a tide or a flood of greater than 100-year frequency may be considered by the regional board as limited Class I disposal sites.

Landfilling Methods and Operations   Operation of a landfill for hazardous wastes is quite different from that of a conventional landfill.

Many but not all hazardous wastes can be disposed of on land in properly designed landfills. To minimize potentially adverse environmental effects from wastes deposited at hazardous-waste landfill sites, the U.S. Environmental Protection Agency (EPA) has developed specific regulations regarding the characteristics of wastes suitable for landfilling. These regulations (40 CFR 265) include a prohibition on the placement of:

• Noncontainerized hazardous wastes containing free liquids, whether or not absorbents have been added.

• Containers holding free liquids unless all freestanding liquid has been removed by decanting or other methods or has been mixed with absorbent or solidified so that freestanding liquid is no longer observed. Such containers must be at least 90 percent full or, if empty, reduced in size as much as possible via crushing or shredding prior to disposal.

The following containers are exempt from the above regulations:

• Very small containers, such as ampules, and containers holding liquids for use other than storage, such as batteries, which may be disposed directly in a hazardous-waste landfill.

• Small lab-pack containers of hazardous waste if they are first placed in nonleaking, larger containers. These containers must be filled to capacity and surrounded by enough absorbent material to contain the liquid contents of the lab pack. The resultant container must then be placed in a larger container packed with absorbent material which will not react with, become decomposed by, or ignited by the contents of the inside containers. Incompatible wastes may not be packed and disposed of together in this manner.

Design of Hazardous-Waste Landfills   Most of the regulations governing the design of hazardous-waste landfills have been resolved. Although specific requirements will vary, the factors identified in Table 25-74 can be used as a design guide. Some special precautions that can be taken to prevent contamination of underlying strata are shown in Figs. 25-75 to 25-77. Figures 25-76 and 25-77 illustrate a conceptual design for a typical control-cell grid system for a hazardous waste landfill.

Landfarming   Landfarming is a waste-disposal method in which the biological, chemical, and physical processes that occur in the surface of the soil are used to treat biodegradable industrial wastes. Wastes to be treated are either applied on top of the land which has

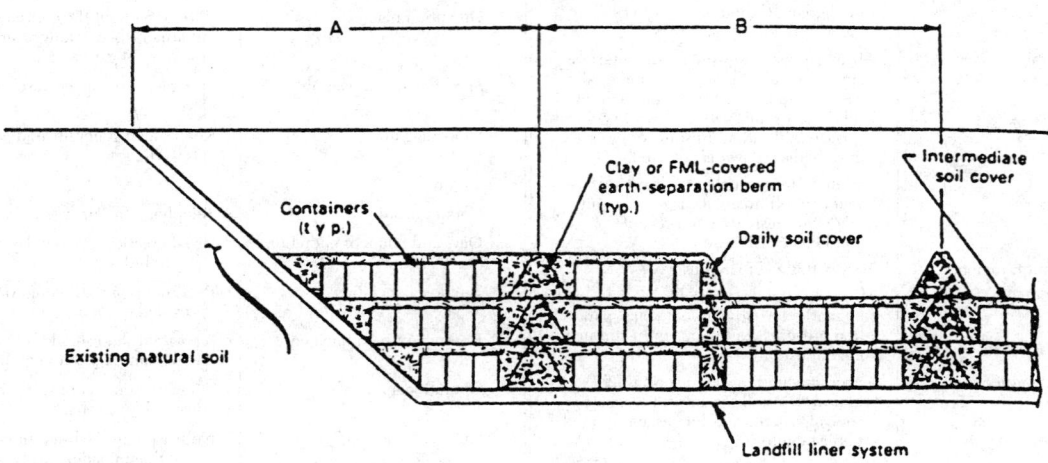

FIG. 25-75   Conceptual design for control cells for hazardous-waste disposal (section view). FML = flexible-membrane liner. (*From Freeman, H. M., Standard Handbook of Hazardous Waste Treatment and Disposal, McGraw-Hill, 1988.*)

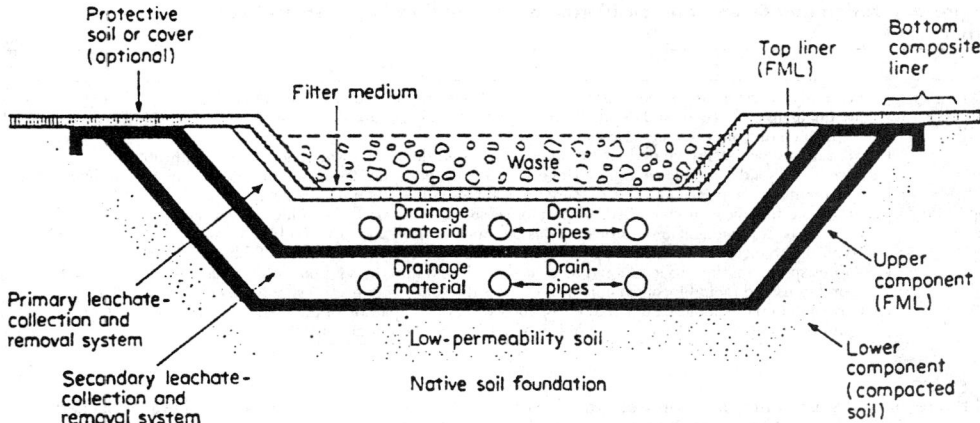

FIG. 25-76   Schematic of an FML plus compacted-soil double-liner system for a landfill. (Drawing not to scale.) (*U.S. EPA, EPA/530/SW-85-012. Washington, D.C., 1985. From Freeman, H. M., Standard Handbook of Hazardous Waste Treatment and Disposal, McGraw-Hill, 1988.*)

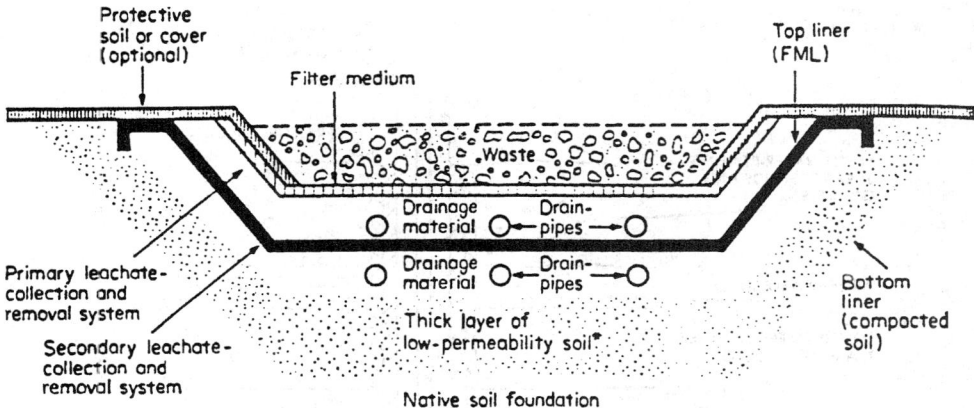

*Thickness to be determined by breakthrough time.

FIG. 25-77   Schematic of an FML plus composite double-liner system for a landfill. (Drawing not to scale.) (*U.S. EPA, EPA/530/SW-85-012, Washington, D.C., 1985. From Freeman, H. M., Standard Handbook of Hazardous Waste Treatment and Disposal, McGraw-Hill, 1988.*)

been prepared to receive the wastes or injected below the surface of the soil.

Process Description   When organic wastes are added to the soil, they are subjected simultaneously to the following processes: (1) bacterial and chemical decomposition, (2) leachating of water-soluble components in the original wastes and from the decomposition products, and (3) volatilization of selected components in the original wastes and from the products of decomposition.

Factors that must be considered in evaluating the biodegradability of organic wastes in a landfilling application include (1) composition of the waste; (2) compatibility of wastes and soil microflora; (3) environmental requirements including oxygen, temperature, pH, and inorganic nutrients; and (4) moisture content of soil-waste mixture.

Although most of the volatile components are released to the atmosphere, a small fraction is dissolved and/or carried away with the water in the soil matrix. Leached waters are carried with the water as it percolates through the underlying soil strata. Most of the organic constituents contained in the leachate receive additional treatment as they pass through the soil column. Leached wastes can also be lost in surface runoff.

Applications   Landfarming is suitable for wastes that contain organic constituents that are biodegradable and are not subject to significant leaching while the bioconversion process is occurring. For example, petroleum oily wastes and oily sludges are ideally suited for disposal by landfarming. A variety of other organic wastes with similar characteristics are also suitable. Properly managed landfarming sites can be reused at frequent intervals with no adverse effects.

Design and Operation   Important consideration in the design and operation of landfarming systems include (1) site selection, (2) site preparation, (3) waste characteristics, (4) method of waste application, (5) waste-application rate, (6) site management, and (7) monitoring. Important factors related to these design and operation considerations are reported in Table 25-75.

Deep-Well Injection   Deep-well injection for the disposal of liquid wastes involves injecting the wastes deep in the ground into permeable rock formation (typically limestone or dolomite) or underground caverns.

Process Description   The installation of deep wells for the injection of wastes closely follows the practices used for the drilling and completion of oil and gas wells.

TABLE 25-75 Important Design and Operation Considerations for Landfarming Systems Used for Waste Treatment

Item	Remarks
Site selection location	Proximity to critical areas specified in government regulations, accessibility, site geology and hydrology.
Site selection: soil characteristics	Adequate area soil cover and depth to groundwater usually greater than 1.5 m (4 ft). Slope should not exceed 5 to 8 percent. Soil type, including ion-exchange capacity.
Site preparation	Area should be fenced, graded for runoff control, and disked or plowed before waste application.
Waste characterization	Suspended solids, organic content, nitrogen (all forms), phosphorus pH, and inorganic metals including arsenic, barium, cadmium, chromium, copper, lead, mercury, selenium, silver, sodium, and zinc.
Method of waste application	Ridge and furrow, sprinkling (fixed or portable systems), tank-truck spreading, subsurface injection.
Waste-application rates	For petroleum crude oil and lubricating oils the range is from 250 to 1250 bbl/(ha · year) with a value of 400 bbl/(ha · year) being typical. A typical value for general refinery oils and wastes would be about 150 bbl/(ha · year).
Site management	Wastes spread on the surface should be disked or plowed into the soil soon after application (1 to 7 days). To promote aerobic conditions and rapid bioconversion of the wastes the soil-waste mixture should be cultivated periodically.
Monitoring	Periodic samples should be taken to assess the extent of completion of the bioconversion process. Core samples should be taken annually to monitor the movement of leached wastes in the underlying strata.

Examination of the records of wastewater injection wells that have been constructed in the United States shows that almost all the wells constructed thus far have been completed by one of three methods or close variations of them. The methods are:

1. Open-hole completion in competent formations

2. Screened or screened and gravel-packed in incompetent sands and gravels

3. Fully cased and cemented with the casing perforated in either competent or somewhat incompetent formations

Most wastewater injection wells will be constructed with injection

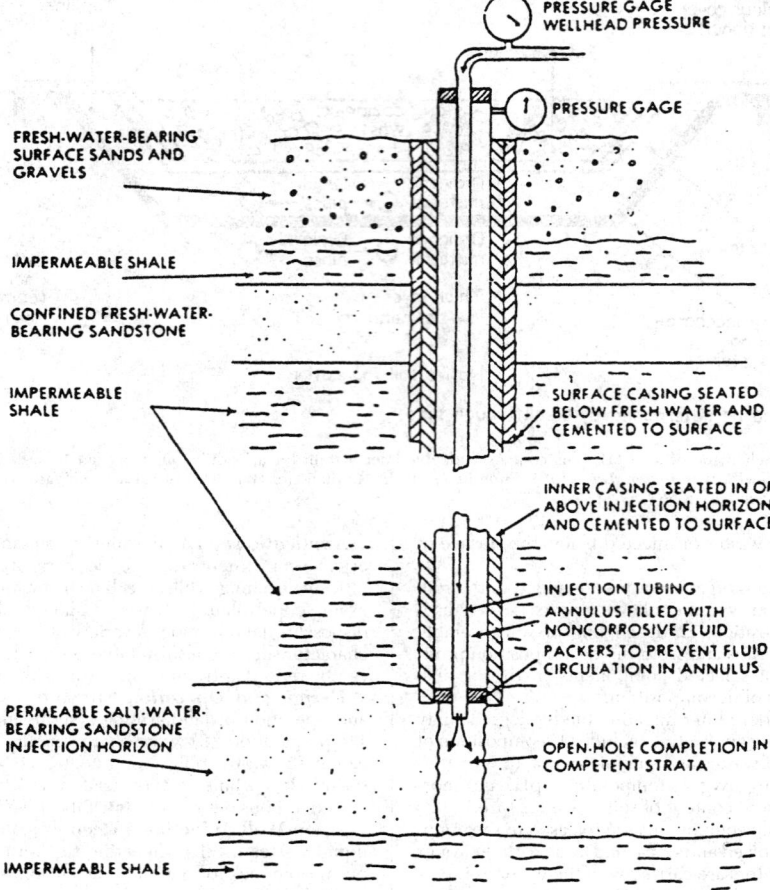

FIG. 25-78 Schematic diagram of an industrial-waste injection well completed in competent sandstone. (*From Freeman, H. M.,* Standard Handbook of Hazardous Waste Treatment and Disposal, *McGraw-Hill, 1988.*)

TABLE 25-76 Important Design and Operation Considerations for Deep Wells Used for Waste Injection

Item	Remarks	Item	Remarks
Well-site selection	Criteria for assessing the feasibility of a deep-well-injection site include (1) uniformity, (2) large extent, (3) substantial thickness, (4) high porosity and permeability, (5) low pressure, (6) saline aquifer, (7) separation from potable-water horizons, (8) adequate overlying and underlying aquicludes, (9) no poorly plugged wells nearby, and (10) compatibility between the mineralogy and fluids of the reservoir and the injected wastes.	Deep-well installation	injection formation). Adjustment of pH and buffering of the waste may be necessary. Well depths vary from 550 to 3660 m (1800 to 1200 ft); well-injection rates vary from 4 to 60 L/s; rates in the range from 15 to 20 L/s are typical. Operation pressures up to 27,600 kPa (4000 psig) are used.
Waste pretreatment	Suspended solid less than 10 to 15 mg/L; particle sizes equal to or less than 1 to 5 μm (depends on	Monitoring	Continuous monitoring facilities should be installed when wells are put into operation. Irregularities in the pressure may require changes in operating procedures.

tubing inside the long casing string, and with a packer set between the tubing and the casing near the bottom of the casing (Fig. 25-78). This design is not entirely free of problems, particularly with the packer, but experience has proved it generally superior to other designs. Some wells are completed with an annulus open at the bottom. The annulus is filled with a lighter-than-waste liquid that "floats" on the aqueous waste. This type of well completion has been referred to as a *fluid-seal completion*.

Applications Deep-well injection has been used principally for liquid wastes that are difficult to treat and dispose of by more conventional methods and for hazardous wastes. Chemical, petrochemical, and pharmaceutical wastes are those most commonly disposed of with this method. The waste may be liquid, gases, or solids. The gases and solids are either dissolved in the liquid or are carried along with the liquid.

Design and Operation Important design and operation considerations for deep-well injection are related to (1) well-site selection, (2) pretreatment, (3) installation of an injection well, and (4) monitoring. Important factors related to these design and operation considerations are reported in Table 25-76. As noted in the table, wastes are usually treated prior to injection to prevent clogging of the formation and damage to equipment. Particles greater than about 1 to 5 μm must be removed. Typically, treated wastes must be filtered prior to

injection. Wastes must also be compatible with the characteristics of the aquifer. This may require pH adjustment and the use of compatible buffers.

Ocean Disposal of Solid Wastes Although ocean dumping of municipal solid wastes was abandoned in the United States in 1933, the concept has persisted throughout the years and is still frequently discussed today. Some industrial wastes are still discharged at sea. Within the past few years, the idea that the ocean is a gigantic sink, into which an infinite amount of pollution of all types can be dumped, has been discarded. On the other hand, it is argued that many of the wastes now placed in landfills or on land could be used as fertilizers to increase the productivity of the ocean. It is also argued that the placement of wastes in ocean-bottom trenches where tectonic folding is occurring is an effective method of waste disposal.

PLANNING

Because of the ever-growing number of federal regulations governing the disposal of nonhazardous and hazardous solid wastes, it is prudent to develop both short-term and long-term action programs to deal with all aspects of solid-waste management. Important short- and long-term actions are identified in Table 25-77.

TABLE 25-77 Short- and Long-Term Actions for Effective Industrial Solid-Waste Management*

Actions	Remarks	Actions	Remarks
Short-term		Long-term	
1. Inventory wastes.	Document all types, quantities, and sources of wastes (both nonhazardous and hazardous wastes).	1. Remove all accumulated wastes.	Develop a systematic program for removing all accumulated wastes stored on the plant site.
2. Inventory inactive sites.	All inactive sites where wastes have been disposed of in the past should be inventoried. Data should be gathered on buried wastes, including types, quantity, and sources. A groundwater-monitoring program should be developed.	2. Separate wastes.	Institute a long-term program to separate wastes at the source of production.
3. Characterize wastes.	In addition to general information on the characteristics of the wastes, all hazardous wastes should be individually characterized.	3. Reduce wastes.	A systematic program should be undertaken to examine all sources of waste production and to develop alternative operations and processes to reduce waste generation.
4. Assign responsibilities.	Assign responsibilities and authority at plant and headquarters for the storage, collection, treatment, and disposal of all types of hazardous wastes.	4. Improve facilities.	Upgrade facilities to meet RCRA requirements. A data-collection program should be instituted to obtain any needed data.
5. Track the movement of wastes.	Develop a logging system for hazardous wastes containing the date, waste description, source, volume shipped or hauled, name of hauler, and destination. Follow through to be sure that wastes reach destination.	5. Review all waste-management agreements.	Develop detailed contracts with outside waste-management firms. Define clearly the duties and responsibilities of plant personnel and waste-collection personnel.
6. Develop emergency procedures.	Develop procedures for dealing with emergency situations involving the storage, collection, treatment, and disposal of hazardous wastes.	6. Review and develop disposal-site options.	Develop long-term projections for landfill requirements and initiate a program to secure the needed sites.
7. Obtain permits.	Start obtaining the necessary waste-disposal permits as soon as possible.	7. Secure appropriate engineering and consulting services.	Make sure that your engineering departments are involved early in the process. Retain outside consultants for specific tasks.
		8. Monitor legislative programs.	Develop a program for monitoring new regulations and for inputting to appropriate federal, state, and local agencies on the modification and development of new regulations.

*Adapted in part from R. Sobel, "How Industry Can Prepare for RCRA," *Chem. Eng.,* **86**(1), 82 (Jan. 29, 1979).

Process Safety

Stanley M. Englund, M.S., Ch.E., *Fellow American Institute of Chemical Engineers; Process Consultant, The Dow Chemical Company (retired). (Section Editor, Section 26; Introduction; Hazard Analysis; Storage and Handling of Hazardous Materials; Reactive Chemicals; Combustion and Flammability Hazards; Hazards of Vacuum; Hazards of Inert Gases)*

Frank T. Bodurtha, Sc.D., *E. I. du Pont de Nemours and Co., Inc., (retired); Consultant, Frank T. Bodurtha, Inc. (Gas Explosions; Unconfined Vapor Cloud Explosions [UVCEs] and Boiling Liquid Expanding Vapor Explosions [BLEVEs])*

Laurence G. Britton, Ph.D., *Research Scientist, Union Carbide Corporation. (Flame Arresters)*

Daniel A. Crowl, Ph.D., *Professor of Chemical Engineering, Chemical Engineering Department, Michigan Technological University; Member, American Institute of Chemical Engineers, American Chemical Society. (Gas Dispersion)*

Stanley Grossel, *President, Process Safety & Design, Inc.; Fellow, American Institute of Chemical Engineers; Member, American Chemical Society; Member, The Combustion Institute; Member, Explosion Protection Systems Committee of NFPA. (Emergency Relief Device Effluent Collection and Handling)*

W. G. High, C.Eng., B.Sc., F.I.Mech.E., *Burgoyne Consultants Ltd., W. Yorks, England. (Guidelines for Estimating Damage)*

Trevor A. Kletz, D.Sc., *Senior Visiting Research Fellow, Department of Chemical Engineering, Loughborough, University U.K.; Fellow, American Institute of Chemical Engineers, Royal Academy of Engineers (U.K.), Institution of Chemical Engineers (U.K.), and Royal Society of Chemistry (U.K.). (Inherently Safer Design)*

Robert W. Ormsby, M.S., ChE., *Manager of Safety, Chemicals Group, Air Products and Chemicals, Inc.; Air Products Corp.; Fellow, American Institute of Chemical Engineers. (Risk Analysis)*

John E. Owens, B.E.E., *Electrostatics Consultant, Condux, Inc.; Member, Institute of Electrical and Electronics Engineers, Electrostatics Society of America. (Static Electricity)*

Carl A. Schiappa, B.S. Ch.E., *Process Engineering Associate, Michigan Division Engineering, The Dow Chemical Company; Member, AIChE and CCPS. (Project Reviews and Procedures)*

Richard Siwek, M.S., *Explosion Protection Manager, Corporate Unit Safety and Environment, Ciba-Geigy Ltd., Basel, Switzerland. (Dust Explosions)*

Robert E. White, Ph.D., *Principal Engineer, Chemistry and Chemical Engineering Division, Southwest Research Institute. (Flame Arresters)*

David Winegardner, Ph.D., *Engineering Associate, Michigan Division Engineering, The Dow Chemical Company; Member, AIChE and CCPS. (Pressure Relief Systems)*

John L. Woodward, Ph.D., *Principal, DNV Technica, Inc. (Discharge Rates from Punctured Lines and Vessels)*

PROCESS SAFETY INTRODUCTION

INHERENTLY SAFER DESIGN AND OTHER PRINCIPLES

PROCESS SAFETY INTRODUCTION

In recent years there has been an increased emphasis on process safety as a result of a number of serious accidents. This is due in part to the worldwide attention to issues in the chemical industry brought on by several dramatic accidents involving gas releases, major explosions, and environmental incidents. Public awareness of these and other accidents has provided a driving force for industry to improve its safety record. Local and national governments are taking a hard look at safety in industry as a whole and the chemical industry in particular. There has been an increasing amount of government regulation.

The Chemical Process Industries constitutes one of the safest of the manufacturing sectors, but a single major accident or disaster can do irreparable damage to a company's reputation and possibly affect the entire industry (Sheridan, "OSHA, EPA and Process Sampling," *Chem. Proc.,* September 1994, pp. 24–28). One reason the chemical industry gets bad press is that its activities are very noticeable. Large chemical works are striking features on the landscape. Chemical plants are often noisy and garishly lit, and many of the effluents cause nuisances which are well below the health and safety limits. Hazardous chemicals are transported in bulk in highly visible container vehicles, adding to the public's image of the industry as hazardous and dangerous (Benson and Ponton, "Process Miniaturisation—A Route to Total Environmental Acceptability," *Institution of Chemical Engineers* 0263-8762/93). For many reasons, the public often associates the chemical industry with environmental and safety problems and, unfortunately, sometimes the negative image that goes with the problems is deserved. It is vital to the future of the chemical industry that process safety have a high priority in the design and operation of chemical process facilities.

Environmental pressures on the process industries will prove to be the most significant change for the next 50 years. Not only will new processes change, but mature industries will have to develop new process technology to survive (Benson et al., op. cit.).

Hazards from combustion and runaway reactions play a leading role in many chemical process accidents. Knowledge of these reactions is essential for control of process hazards. It is important that loss of containment be avoided. For example:

- Much of the damage and loss of life in chemical accidents results from the sudden release of material at high pressures which may or may not result from fire. Chemical releases caused by fires and the failure of process equipment and pipelines can form toxic clouds that can be dangerous to people over large areas.
- Vapor cloud explosions can result if clouds of flammable vapor in air are formed. It is important to understand how liquids and gases flow through holes in equipment and how resulting vapor or gas clouds are dispersed in air.
- Understanding how sudden pressure releases can occur is important. They can happen, for example, from ruptured high-pressure tanks, runaway reactions, flammable vapor clouds, or pressure developed from external fire. The proper design of pressure relief systems can reduce the possibility of losses from unintended overpressure.
- Static electricity is often a hidden cause in accidents.

- It is important to understand the reactive nature of the chemicals involved in a chemical facility.
- Loss of containment due to mechanical failure or misoperation is a major cause of chemical process accidents. The publication, *One Hundred Largest Losses: A Thirty Year Review of Property Damage Losses in the Hydrocarbon Chemical Industry,* 9th ed. (M&M Protection Consultants, Chicago), cites loss of containment as the leading cause of property loss in the chemical process industries.

Government regulations require hazard and risk analysis as part of *process safety management* (PSM) programs. These are part of the process safety programs of many chemical process facilities.

Process safety includes many subjects that could not be included in this section because of lack of space. The following describes the organization of Section 26:

Inherently Safer Design   Rather than add on equipment to control hazards or to protect people from their consequences, it is better to design user-friendly plants which can withstand human error and equipment failure without serious effects on safety, the environment, output, and efficiency. This part is concerned with this matter.

Process Safety Analysis   This part treats the analysis of a process or project from the standpoint of hazards, risks, procedures for making potential damage estimates, and project reviews and audits. It can be helpful to management in assessing risks in a project. It consists of the following:

Hazard Analysis
Risk Analysis
Guidelines for Estimating Damage
Project Reviews and Procedures

Safety Devices   Pressure relief devices, flame arresters, and methods for handling effluent from controlled releases provide control of accidental undesirable events. Special equipment should be considered for highly toxic chemical service. The following matters are considered:

Pressure Relief Systems
Emergency Relief Device Effluent Collection and Handling
Flame Arresters
Storage and Handling of Hazardous Materials

Hazardous Materials and Conditions   The chemical and physical situations that can result when operating with hazardous materials should be understood so these materials may be handled safely. This part covers the following:

Reactive Chemicals
Combustion and Flammability Hazards
Gas Explosions
Unconfined Vapor Cloud Explosions (UVCEs) and Boiling Liquid Evaporating Vapor Explosions (BLEVEs)
Dust Explosions
Static Electricity
Hazards of Vacuum
Hazards of Inert Gases
Gas Dispersion
Discharge Rates from Punctured Lines and Vessels

INHERENTLY SAFER DESIGN AND OTHER PRINCIPLES

INTRODUCTION: WHAT IS PROCESS SAFETY?

Process safety differs from the traditional approach to accident prevention in a number of ways (Lees, *Loss Prevention in the Process Industries,* 2d ed., Butterworth-Heinemann, 1996, p. 1.8):

- There is more concern with accidents that arise out of the technology.
- There is more emphasis on foreseeing hazards and taking action before accidents occur.

• There is more emphasis on a systematic rather than a trial-and-error approach, particularly on systematic methods of identifying hazards and of estimating the probability that they will occur, and their consequences.

• There is concern with accidents that cause damage to plants and loss of profit but do not injure anyone, as well as those that do cause injury.

• Traditional practices and standards are looked at more critically.

Process safety can be applied in any industry, but the term and the approach have been particularly widely used in the process industries, where it usually means the same as loss prevention.

Although process safety is as old as process engineering, it did not become recognized as a distinct branch of the subject until the 1960s, when a new generation of plants, larger than earlier ones and operating at higher temperatures and pressures, was involved in a number of serious fires and explosions. They made the industry realize that accident prevention needed the same sort of systematic and technical study as every other aspect of plant design and operation. Since then the number of publications and specialist journals, and the number and caliber of engineers specializing in the field, have grown rapidly.

INHERENTLY SAFER AND MORE USER-FRIENDLY DESIGN

For many years the usual procedure in plant design was to identify the hazards, by one of the systematic techniques described later or by waiting until an accident occurred, and then add on protective equipment to control future accidents or protect people from their consequences. This protective equipment is often complex and expensive and requires regular testing and maintenance. It often interferes with the smooth operation of the plant and is sometimes bypassed. Gradually the industry came to realize that, whenever possible, one should design user-friendly plants which can withstand human error and equipment failure without serious effects on safety (and output and efficiency). When we handle flammable, explosive, toxic, or corrosive materials we can tolerate only very low failure rates, of people and equipment—rates which it may be impossible or impracticable to achieve consistently for long periods of time.

The most effective way of designing user-friendly plants is to avoid, when possible, large inventories of hazardous materials in process or storage. "What you don't have, can't leak." This sounds obvious, but until the explosion at Flixborough, England, in 1974, little systematic thought was given to ways of reducing inventories. The industry simply designed a plant and accepted whatever inventory the design required, confident they could keep it under control. Flixborough weakened that confidence and the disaster ten years later at Bhopal, India, almost destroyed it. Plants in which we avoid a hazard, by reducing inventories or avoiding hazardous reactions, are usually called *inherently safer*.

The principle ways of designing inherently safer plants and other ways of making plants user-friendly are summarized as follows, with examples (Kletz, *Plant Design for Safety—A User-Friendly Approach*, Hemisphere, 1991).

Intensification   This involves using so little hazardous material that it does not matter if it all leaks out. For example, at Bhopal, methyl isocyanate (MIC), the material that leaked and killed over 2000 people, was an intermediate for which it was convenient but not essential to store. Within a few years many companies had reduced their stocks of MIC and other hazardous intermediates.

As another example, at one time nitroglycerin (NG) was manufactured in batch reactors containing about a ton of raw materials and product. If the reactor got too hot, there was a devastating explosion. In modern plants, NG is made in a small continuous reactor containing about a kilogram. The severity of an explosion has been reduced a thousandfold, not by adding on protective devices, which might fail or be neglected, but by redesigning the process. The key change was better mixing, achieved not by a better stirrer, which might fail, but by passing one reactant (acid) through a device like a laboratory water pump so that it sucks in the other reactant (glycerin) through a side-arm. If the acid flow stops, the glycerin flow also stops, not through the intervention of a flow controller, which might fail, but as an inevitable result of the laws of physics (Bell, *Loss Prevention in the Process Industries*, Institution of Chemical Engineers Symposium Series No. 34, 1971, p. 50).

Intensification is the preferred route to inherently safer design, as the plants, being smaller, are also cheaper.

Substitution   If intensification is not possible, then an alternative is to consider using a safer material in place of a hazardous one. Thus it may be possible to replace flammable solvents, refrigerants, and heat-transfer media by nonflammable or less flammable (high-boiling) ones, hazardous products by safer ones, and processes which use hazardous raw materials or intermediates by processes which do not. As an example of the latter, the product manufactured at Bhopal (carbaryl) was made from three raw materials. Methyl isocyanate is formed as an intermediate. It is possible to react the same raw materials in a different order so that a different and less hazardous intermediate is formed.

Attenuation   Another alternative to intensification is attenuation, using a hazardous material under the least hazardous conditions. Thus large quantities of liquefied chlorine, ammonia, and petroleum gas can be stored as refrigerated liquids at atmospheric pressure instead of storing them under pressure at ambient temperature. (Leaks from the refrigeration equipment should also be considered, so there is probably no net gain in refrigerating quantities less than a few hundred tons.) Dyestuffs which form explosive dusts can be handled as slurries.

Limitation of Effects of Failures   Limitation can be done by equipment design or change in reaction conditions, rather than by adding on protective equipment. For example:

• Spiral-wound gaskets are safer than fiber gaskets because, if the bolts work loose or are not tightened correctly, the leak rate is much lower.

• Tubular reactors are safer than pot reactors, as the inventory is usually lower and a leak can be stopped by closing a valve.

• Vapor phase reactors are safer than liquid phase ones, as the mass flow rate through a hole of a given size is much less. (This is also an example of attenuation.)

• A small, deep diked area around a storage tank is safer than a large, shallow one, as the evaporation rate is lower and the area of any fire is smaller.

• Heating media such as steam or hot oil should not be hotter than the temperature at which the materials being heated are liable to ignite spontaneously or react uncontrollably.

• Many runaway reactions can be prevented by changing the order of operations, reducing the temperature, or changing another parameter.

• Reduce the frequency of hazardous operations such as sampling or maintenance. What is the optimum balance between reliability and maintenance?

Simplification   Simpler plants are friendlier than complex ones, as they provide fewer opportunities for error and less equipment which can fail. Some of the reasons for complication in plant design are:

• The need to control hazards. If one of the other actions already discussed, such as intensification, can be carried out, less add-on protective equipment is needed and plants will therefore be simpler.

• A desire for flexibility. Multistream plants with numerous crossovers and valves, so that any item can be used on any stream, have numerous leakage points, and errors in valve settings are easy to make.

• Lavish provision of installed spares with the accompanying isolation and changeover valves.

• Continuing to follow rules or practices which are no longer necessary.

• Design procedures which result in a failure to identify hazards until late in design. By this time it is impossible to avoid the hazard and all that can be done is to add on complex equipment to control it.

Knock-on Effects   Plants should be designed so that those incidents that do occur do not produce knock-on or domino effects. This can be done, for example, by:

• Providing firebreaks, about 15 m wide, between sections, like firebreaks in a forest, to restrict the spread of fire.

• Siting equipment which is liable to leak outdoors so that leaks of flammable gases and vapors are dispersed by natural ventilation.

Indoors, a few tens of kilograms are sufficient for an explosion that can destroy the building. Outdoors, a few tons are necessary for serious damage. A roof over equipment such as compressors is acceptable, but walls should be avoided. (If leaks of toxic gases are liable to occur, it may be safer to locate the plant indoors, unless leaks will disperse before they reach the public or employees on other units.)

• Constructing storage tanks so that the roof-wall weld will fail before the base-wall weld, thus preventing spillage of the contents. In general, equipment designers should consider the way in which it is most likely to fail and, when possible, locate or design the equipment so as to minimize the consequences.

Avoiding Incorrect Assembly   Plants should be designed so that incorrect assembly is difficult or impossible. For example, compressor valves should be designed so that inlet and exit valves cannot be interchanged.

Status Clear   It should be possible to see at a glance if equipment has been assembled or installed incorrectly or whether a valve is in the open or shut position. For example:

• Check valves should be marked so that installation the wrong way round is obvious. It should not be necessary to look for a faint arrow hardly visible beneath the dirt.

• Gate valves with rising spindles are friendlier than valves with nonrising spindles, as it is easy to see whether they are open or shut. Ball valves are friendly if the handles cannot be replaced in the wrong position.

• Figure-eight plates (spectacle plates) are friendlier than slip plates (spades), as their positions are apparent at a glance. If slip plates are used, their projecting tags should be readily visible, even when the line is insulated. In addition, spectacle plates are easier to fit than slip plates, if the piping is rigid, and they are always available on the job. It is not necessary to search for them, as with slip plates.

Tolerance   Whenever possible, equipment should tolerate poor installation or operation without failure. Expansion loops in pipework are more tolerant of poor installation than bellows are. Fixed pipes, or articulated arms, if flexibility is necessary, are friendlier than hoses. For most applications, metal is friendlier than glass or plastic.

Bolted joints are friendlier than quick-release couplings. The former are usually dismantled by a fitter after issue of a permit to work. One person prepares the equipment and another person opens it up; the issue of the permit provides an opportunity to check that the correct precautions have been taken. In addition, if the joints are unbolted correctly, any trapped pressure is immediately apparent and the joint can be remade or the pressure allowed to blow off. In contrast, many accidents have occurred because operators opened up equipment which was under pressure, without independent consideration of the hazards, using quick-release couplings. There are, however, designs of quick-release couplings which give the operator a second chance.

Low Leak Rate   If friendly equipment does leak, it does so at a low rate, which is easy to stop or control. Examples already mentioned are spiral-wound gaskets, tubular reactors, and vapor phase reactors.

Ease of Control   Processes with a flat response to change are obviously friendlier than those with a steep response. Processes in which a rise of temperature decreases the rate of reaction are friendlier than those with a positive temperature coefficient, but this is a difficult ideal to achieve in the chemical industry. However, there are a few examples of processes in which a rise in temperature reduces the rate of reaction. For example, in the manufacture of peroxides, water is removed by a dehydrating agent. If magnesium sulfate is used as the agent, a rise in temperature causes release of water by the agent, diluting the reactants and stopping the reaction (Gerrison and van't Land, *I&EC Process Design* **24**, 1985, p. 893).

Software   In some programmable electronic systems (PES), errors are much easier to detect and correct than in others. Using the term *software*, in the wider sense, to cover all procedures, as distinct from hardware or equipment, some software is much friendlier than others. Training and instructions are obvious examples. As another example, if many types of gaskets or nuts and bolts are stocked, sooner or later the wrong type will be installed. It is better, and cheaper in the long run, to keep the number of types stocked to a minimum, even though more expensive types than are strictly necessary are used for some applications.

Designing Inherently Safer and More User-Friendly Plants
The following actions are needed for the design of inherently safer and more user-friendly plants:

1.   Designers need to be made aware that there is scope for improving the friendliness of the plants they design.

2.   To achieve many of the changes previously suggested, it is necessary to carry out much more critical examination and systematic consideration of alternatives during the early stages of design than has been customary in most companies. Two studies are suggested: one at the conceptual or business analysis stage when the process is being chosen, and another at the flowsheet stage. For the latter, the usual hazard and operability (HAZOP) study questions may be suitable but with one difference. In a normal HAZOP on a line diagram, if, for example, "more of temperature" is being discussed, it is assumed that this is undesirable and ways of preventing it are sought. In a HAZOP of a flowsheet, it should be asked if "more of temperature" would be better. For the conceptual study, different questions are needed.

3.   Many companies will say that they do consider alternatives during the early stages of plant design. However, what is lacking in many companies is a formal, systematic, structured procedure of the HAZOP type.

To achieve the more detailed improvements suggested here, it may be necessary to add a few questions to those asked during a normal HAZOP. For example, what types of valves, gaskets, and so forth, will be used?

INCIDENT INVESTIGATION AND HUMAN ERROR

Although most companies investigate accidents (and many investigate dangerous incidents in which no one was injured), these investigations are often superficial, and we fail to learn all the lessons for which we have paid the high price of an accident. The facts are usually recorded correctly, but often only superficial conclusions are drawn from them. Identifying the causes of an accident is like peeling an onion. The outer layers deal with the immediate technical causes and triggering events while the inner layers deal with ways of avoiding the hazard and with the underlying weaknesses in the management system (Kletz, *Learning from Accidents*, 2d ed., Butterworth-Heinemann, 1994).

Dealing with the immediate technical causes of a leak, for example, will prevent another leak for the same reason. If so little of the hazardous material can be used that leaks do not matter or a safer material can be used instead, as previously discussed, all significant leaks of this hazardous material can be prevented. If the management system can be improved, we may be able to prevent many more accidents of other sorts.

Other points to watch when drawing conclusions from the facts are:

1.   Avoid the temptation to list causes we can do little or nothing about. For example, a source of ignition should not be listed as the primary cause of a fire or explosion, as leaks of flammable gases are liable to ignite even though we remove known sources of ignition. The cause is whatever led to the formation of a flammable mixture of gas or vapor and air. (Removal of known sources of ignition should, however, be included in the recommendations.) Similarly, human error should not be listed as a cause. See item 6 below.

2.   Do not produce a long list of recommendations without any indication of the relative contributions they will make to the reduction of risk or without any comparison of costs and benefits. Resources are not unlimited and the more we spend on reducing one hazard, the less there is left to spend on reducing others.

3.   Avoid the temptation to overreact after an accident and install an excessive amount of protective equipment or complex procedures which are unlikely to be followed after a few years have elapsed. Sometimes an accident occurs because the protective equipment available was not used; nevertheless, the report recommends installation of more protective equipment; or an accident occurs because complex procedures were not followed and the report recommends extra procedures. It would be better to find out why the original equipment was not used or the original procedures were not followed.

4.   Remember that few, if any, accidents have simple causes.

5.   When reading an accident report, look for the things that are not said. For example, a gland leak on a liquefied flammable gas pump caught fire and caused considerable damage. The report drew atten-

tion to the congested layout, the amount of redundant equipment in the area, the fact that a gearbox casing had been made of aluminum, which melted, and several other unsatisfactory features. It did not stress that there had been a number of gland leaks on this pump over the years, that reliable glands are available for liquefied gases at ambient temperatures, and, therefore, there was no need to have tolerated a leaky pump on this duty.

As another example, a fire was said to have been caused by lightning. The report admitted that the grounding was faulty but did not say when it was last checked, if it was scheduled for regular inspection, if there was a specification for the resistance to earth (ground), if employees understood the need for good grounding, and so on.

6. At one time most accidents were said to be due to human error, and in a sense they all are. If someone—designer, manager, operator, or maintenance worker—had done something differently, the accident would not have occurred. However, to see how managers and supervisors can prevent them, we have to look more closely at what is meant by human error:

a. Some errors are due to poor training or instructions: someone did not know what to do. It is a management responsibility to provide good training and instructions and avoid instructions that are designed to protect the writer rather than help the reader. However many instructions are written, problems will arise that are not covered, so people—particularly operators—should be trained in flexibility—that is, the ability to diagnose and handle unforeseen situations. If the instructions are hard to follow, can the job be simplified?

b. Some accidents occur because someone knows what to do but makes a deliberate decision not to do it. If possible the job should be simplified (if the correct method is difficult, an incorrect method will be used); the reasons for the instructions should be explained; checks should be carried out from time to time to see that instructions are being followed; and if they are not, this fact should not be ignored.

c. Some accidents occur because the job is beyond the physical or mental ability of the person asked to do it—sometimes it is beyond anyone's ability. The plant design or the method of working should be improved.

d. The fourth category is the commonest: a momentary slip or lapse of attention. They happen to everyone from time to time and cannot be prevented by telling people to be more careful or telling them to keep their minds on the job. All that can be done is to change the plant design or method of working to remove opportunities for error (or minimize the consequences or provide opportunities for recovery). Whenever possible, user-friendly plants (see above) should be designed which can withstand errors (and equipment failures) without serious effects on safety (and output and efficiency).

INSTITUTIONAL MEMORY

Most accidents do not occur because we do not know how to prevent them but because we do not use the information that is available. The recommendations made after an accident are forgotten when the people involved have left the plant; the procedures they introduced are allowed to lapse, the equipment they installed is no longer used, and the accident happens again. The following actions can prevent or reduce this loss of information.

• Include a note on "the reason why" in every instruction, code, and standard, and accounts of accidents which would not have occurred if the instruction, code, or standard had been followed.

• Describe old accidents, as well as recent ones, in safety bulletins and newsletters and discuss them at safety meetings.

• Follow up at regular intervals (for example, during audits) to see that the recommendations made after accidents are being followed, in design as well as operations.

• Make sure that recommendations for changes in design are acceptable to the design organization.

• On each unit keep a memory book, a folder of reports on past accidents, which is compulsory reading for new recruits and which others dip into from time to time. It should include relevant reports from other companies but should not include cuts and bruises.

• Never remove equipment before you know why it was installed. Never abandon a procedure before you know why it was adopted.

• Devise better information retrieval systems so that details of past accidents, in our own and other companies, and the recommendations made afterward are more easily accessible than at present.

• Include important accidents of the past in the training of young graduates and company employees. Suitable training material is available from the American Institute of Chemical Engineers and the U.K. Institution of Chemical Engineers (Crowl and Louvar, *Chemical Process Safety: Fundamentals and Applications,* Prentice Hall, 1990).

KEY PROCEDURES

Safety by design should always be the aim, but it is often impossible or too expensive and we then have to rely on procedures. Key features of all procedures are as follows:

• They should be as simple as possible and written in simple language, to help the reader, rather than protect the writer.

• They should be explained to and discussed with those who will have to carry them out, not just sent to them through the post.

• Regular checks and audits should be made to confirm that they are being carried out correctly. They will corrode more rapidly than the steelwork, once those in charge lose interest or turn a blind eye.

Many accidents have occurred because the two procedures discussed in the following sections were unsatisfactory or were not followed.

Preparation of Equipment for Maintenance   The essential feature of this procedure is a permit-to-work system: the operating team prepares the equipment and writes down on the permit the work to be done, the preparation carried out, the remaining hazards, and the necessary precautions. The permit is then accepted by the person or group that will carry out the work and is returned when the work is complete. The permit system will not make maintenance 100 percent safe, but it does reduce the chance that hazards will be overlooked, list ways of controlling them, and inform those doing the job of the precautions they should take. The system should cover such matters as who is authorized to issue and accept permits to work, the training they should receive (not forgetting deputies), and the period of time for which permits are valid. It should also cover the following:

• *Isolation of the equipment under maintenance.*   Poor or missing isolation has been the cause of many serious accidents. Do not rely on valves except for quick jobs; use blinds or disconnection and blanking unless the job is so quick that blinding (or disconnection) would take as long and be as hazardous as the main job. Valves used for isolation (including isolation while fitting blinds or disconnecting) should be locked shut (for example, by a padlock and chain). Blinds should be made to the same standard (pressure rating and material of construction) as the plant. Plants should be designed so that blinds can be inserted without difficulty; that is, there should be sufficient flexibility in the pipework or a slip ring or figure-eight plate should be used. Electricity should be isolated by locking off or removal of fuses. Do not leave them lying around for anyone to replace. Always try out electrical equipment after defusing to check that the correct fuses have been withdrawn.

• *Identification of the equipment.*   Many accidents have occurred because maintenance workers opened up the wrong equipment. Equipment which is under repair should be numbered or labeled unambiguously. Temporary labels should be used if there are no permanent ones. Pointing out the correct equipment is not sufficient. "The pump you repaired last week is leaking again" is a recipe for an accident.

• *Freeing from hazardous materials.*   Equipment which is to be repaired should be freed as far as possible from hazardous materials. Gases can be removed by sweeping out with nitrogen (if the gases are flammable) or air, water-soluble liquids by washing with water, and oils by steaming. Some materials, such as heavy oils and materials that polymerize, are very difficult or impossible to remove completely. Tests should be carried out to make sure that the concentration of any hazardous material remaining is below an agreed level. Machinery should be in the lowest energy state. Thus the forks of forklift trucks should be lowered and springs should not be compressed or extended. For some machinery, the lowest energy state is less obvious. Do not work under heavy suspended loads.

• *Jobs which raise special problems.* Such jobs might include entry to vessels and other confined spaces, hot work, and responsibilities of contractors.

• *Handover.* Permits should be handed over (and returned when the job is complete) person to person. They should not be left on the table for people to sign when they come in.

• *Change of intent.* If there is a change in the work to be done, the permit should be returned and a new one issued [Crowl and Grossel (eds.), *Handbook of Toxic Materials Handling and Management*, Chap. 12, Marcel Dekker, 1995].

Control of Plant and Process Modifications   Many accidents have occurred because plant or process modifications had unforeseen and unsafe side effects (Sanders, *Management of Change in Chemical Plants: Learning from Case Histories,* Butterworth-Heinemann, 1993). No such modifications should therefore be made until they have been authorized by a professionally qualified person who has made a systematic attempt to identify and assess the consequences of the proposal, by hazard and operability study or a similar technique. When the modification is complete, the person who authorized it

should inspect it to make sure that the design intention has been followed and that it "looks right." What does not look right is usually wrong and should at least be checked.

Unauthorized modifications are particularly liable to occur:

• During start-ups, as changes may be necessary to get the plant on line.

• During maintenance, as the maintenance workers may be tempted to improve the plant as well as repair it. They may suggest modifications but should put the plant back as it was unless a change has been authorized.

• When the modification is cheap and no financial authorization is necessary. Many seemingly trivial modifications have had tragic results.

• When the modification is temporary. Twenty-eight people were killed by the temporary modification at Flixborough, one of the most famous of all time (Lees, *Loss Prevention in the Process Industries,* 2d ed., Butterworth-Heinemann, 1996; p. 2).

• When one modification leads to another, and then another (Kletz, *Plant/Operations Progress* **5,** 1986, p. 136).

PROCESS SAFETY ANALYSIS

HAZARD ANALYSIS

GENERAL REFERENCES: *ALOHA—Area locations of hazardous atmospheres* (computer program), Version 5.05 User's Manual, Hazardous Material Response Branch, National Oceanic and Atmospheric Administration (NOAA), Seattle, 1992. Applied Technology Corp. Chemical Manufacturers Association, *A Manager's Guide to Quantitative Risk Assessment,* December 1989. Arendt, JBF Associates, Inc., "Management of Quantitative Risk Assessment in the Chemical Industry," *Plant/Operations Progress,* vol. 9, no. 4, October 1990. Arthur D. Little, Inc., *FaultrEASE®,* 1991. *Chemical Exposure Index, Second Edition,* AIChE, New York, 1994. *CPQRA, Guidelines for Chemical Process Quantitative Risk Analysis,* CCPS-AIChE, New York, 1989. Crowl and Louvar, *Chemical Process Safety Fundamentals with Applications,* Prentice Hall, Englewood Cliffs, N.J., 1990. Delboy, Dubnansky, and Lapp, "Sensitivity of Process Risk to Human Error in an Ammonia Plant," *Plant/Operations Progress,* vol. 10, no. 4, October 1991. *Development of an Improved LNG Plant Failure Rate Data Base,* prepared for Gas Research Inst., Chicago, September 1981. DNV Technica, PHAST and SAFETI, *Process Hazard Analysis Software Tools, Version 4.0,* Technica Inc., London, 1991. *Dow Fire and Explosion Index,* AIChE, New York, January 1994. Dowell, Rohm and Haas Texas, Inc., "Managing the PHA Team," *Process Safety Progress* **13,** no. 1, January 1994. *Fault Tree Handbook,* National Technical Information Service, January 1981. Golay and Todras, "Advanced Light-Water Reactors," *Scientific American,* April 1990. *Guidelines for Safe Storage and Handling of Highly Toxic Hazard Materials,* CCPS-AIChE, New York, 1989. HAZOP-PC, Risk and Hazard Analysis Software Version 3 (computer program), PrimaTech Inc., Columbus, Ohio, 1994. Knowlton, *Hazard and Operability Studies,* Chemetics International Co., Ltd., Vancouver, B.C., February 1989. Latino, *Strive for Excellence . . . the Reliability Approach,* Reliability Center Inc., 1980. Lees, *Loss Prevention in the Process Industries,* Butterworths, London, 1980. Moore, "The Design of Barricades for Hazardous Pressure Systems," *Nuc. Eng. Des.* **5,** 1550–1566, 1967. Munich Re (Münchener Rück) Report, "Losses in the Oil, Petrochemical and Chemical Industry: A Report," Munich, Germany, 1991. NFPA 69, *Explosion Prevention Systems,* National Fire Protection Association, Quincy, Mass., 1992. NFPA 704, *Standard System for the Identification of the Fire Hazards of Materials,* National Fire Protection Association, Quincy, Mass., 1990. Pape and Nussey, "A Basic Approach for the Analysis of Risks from Toxic Hazards," *The Institution of Chemical Engineering Symposium Series No. 93,* University of Manchester Institute for Science and Technology (England), 22–24 April 1985. PHAST, "Process Hazard Analysis Software Tool," DNV Technica Limited, London, October 1990. *Process Safety Progress,* AIChE, New York, January and April 1994 (issues devoted largely to chemical process safety management). *Reliability Guidelines for Process Equipment,* CCPS-AIChE, New York, 1989. Stern and Keller, "Human Error and Equipment Design in the Chemical Industry," *Professional Safety,* May 1991. Swain and Gutterman, *Handbook of Human Reliability Analysis with Emphasis on Nuclear Power Plant Applications* (NUREG/CR-1278), Nuclear Regulatory Commission, Washington, 1983.

Introduction   The meaning of *hazard* is often confused with *risk.* Hazard is defined as the inherent potential of a material or activity to harm people, property, or the environment. Hazard does not have a probability component.

There are differences in terminology on the meaning of *risk* in the published literature that can lead to confusion. *Risk* has been defined in various ways (*CPQRA,* 1989, pp. 3, 4). In this edition of the handbook, *risk* is defined as: "A measure of economic loss or injury in terms of both the incident likelihood and magnitude of loss or injury." *Risk* implies a probability of something occurring.

Definition of Terms   Following are some definitions that are useful in understanding the components of hazards and risk (*CPQRA,* 1989, pp. 3, 4).

acceptable risk   The average rate of risk considered tolerable for a given activity.

accident   A specific combination of events or circumstances that leads to an undesirable consequence.

acute hazard   The potential for injury or damage to occur as a result of an instantaneous or short-duration exposure to the effects of an accident.

chronic hazard   The potential for injury or damage to occur as a result of prolonged exposure to an undesirable condition.

Cause-Consequence   A procedure using diagrams to illustrate the causes and consequences of a particular scenario. They are not widely used because, even for simple systems, displaying all causes and outcomes leads to very complex diagrams.

Chemical Exposure Index (CEI)   The CEI provides a method of rating the relative potential of acute health hazard to people from possible chemical release incidents.

consequence   The direct, undesirable result of an accident, usually measured in health and safety effects, loss of property, or business costs, or a measure of the expected effects of an incident outcome case. For example, an ammonia cloud from a 10-lb/s leak under stability class D weather conditions and a 1.4-mi/h wind traveling in a northerly direction may injure 50 people.

consequence analysis   Once hazards have been established, methods exist for analyzing their consequences (size of vapor cloud, blast damage radius, overpressure expected, etc.). This is independent of frequency or probability.

domino effect   An incident which starts in one piece of equipment and affects other nearby items, such as vessels containing hazardous materials, by thermal blast or fragment impact. This can lead to escalation of consequences or frequency of occurrence. This is also known as a *knock-on effect.*

event   An occurrence involving equipment performance or human action or an occurrence external to the system that causes system upset. An event is associated with an incident, either as a cause or a contributing cause of the incident, or as a response to an initiating event.

event sequence   A specific, unplanned sequence of events composed of initiating events and intermediate events that may lead to an incident.

event tree   Seeks to identify the ultimate consequence of an event, while fault tree analysis aims to identify the basic causes of a specific event. Event trees can grow quite large very quickly.

failure mode and effect analysis (FMEA)   A hazard identification technique in which all known failure modes of components or features of a system are considered in turn and undesired outcomes are noted. It is usually used in combination with fault tree analysis. It is a complicated procedure, usually carried out by experienced risk analysts.

fault tree   A method for representing the logical combinations of various system states which lead to a particular outcome, known as the *top event.*

Fire and Explosion Index (F&EI)   The F&EI is used to rate the potential of hazard from fires and explosions.

frequency   The rate at which observed or predicted events occur.

HAZOP   HAZOP stands for "hazard and operability studies." This is a set of formal hazard identification and elimination procedures designed to identify hazards to people, process plants, and the environment. See subsequent sections for a more complete description.

incident   The loss of containment of material or energy; for example, a leak of a flammable and toxic gas.

incident outcome   The physical outcome of an incident; for example, a leak of a flammable and toxic gas could result in a jet fire, a vapor cloud explosion, a vapor cloud fire, a toxic cloud, etc.

probability   The likelihood of the occurrence of events or a measure of the degree of belief, the values of which range from 0 to 1.

probability analysis   Evaluates the likelihood of an event occurring. Using failure rate data for equipment, piping, instruments, and fault tree techniques, the frequency (events/year) can be quantified.

process hazard analysis (PHA)   See subsequent section for description.

quantitative risk assessment (QRA)   The systematic development of numerical estimates of the expected frequency and/or consequence of potential accidents associated with a facility or operation. Using consequence and probability analyses and other factors such as population density and expected weather conditions, QRA predicts the fatality rate for a given event. This methodology is useful for evaluation of alternates, but its value as an absolute measure of risk should be considered carefully.

risk analysis   The development of a quantitative estimate of risk based on engineering evaluation and mathematical techniques for combining estimates of incident consequences and frequencies.

risk assessment   The process by which results of a risk analysis are used to make decisions, either through a relative ranking of risk reduction strategies or through comparison with risk targets. The terms *risk analysis* and *risk assessment* are often used interchangeably in the literature.

worst credible incident   The most severe incident, considering only incident outcomes and their consequences, of all identified incidents and their outcomes.

Process Hazard Analysis (PHA)   (Dowell, 1994, pp. 30–34.) The OSHA rule for Process Safety Management (PSM) of Highly Toxic Hazardous Chemicals, 29 CFR 1910.119, part (e), requires an initial PHA and an update every five years for processes that handle listed chemicals or contain over 10,000 lb (4356 kg) of flammable material. The PHA must be done by a team, must include employees such as operators and mechanics, and must have at least one person skilled in the methodology employed. Suggested methodologies from Process Safety Management are listed in Table 26-1.

The PHA must consider hazards listed in the PSM Rule, part (e), including information from previous incidents with potential for cata-

TABLE 26-1   Process Hazard Analysis Methods Listed in the OSHA Process Safety Management Rule

- What-if
- Checklist
- What-if/checklist
- Hazard and operability study (HAZOP)
- Failure mode and effect analysis (FMEA)
- Fault tree analysis (FTA)
- An appropriate equivalent methodology

SOURCE: Dowell, 1994, pp. 30–34.

strophic consequences, engineering and administrative controls and consequences of their failure, facility siting, and human factors. Consequences of failure of controls must be considered.

Documentation is important. Everything considered should be documented. "If it is not documented, then you didn't do it." (Dowell, 1994, pp. 30–34.) The key to PHA documentation is to do it right away before it gets cold. Periodic follow-up is needed by management and safety professionals to confirm that all recommendations have been addressed.

Hazard and Risk Assessment Tools   The hazard and risk assessment tools used vary with the stage of the project from the early design stage to plant operations. Many techniques are available, both qualitative and quantitative, some of which are listed in the following section. Reviews done early in projects often result in easier, more effective changes.

Qualitative Tools for Hazard Analysis

SHEL (Safety, Health, Environmental, and Loss Prevention Reviews)   These reviews are performed during design. The purpose of the reviews is to have an outsider's evaluation of the process and layout from safety, industrial hygiene, environmental, and loss prevention points of view. It is often desirable to combine these reviews to improve the efficiency of the use of time for the reviewers.

Checklists   Checklists are simple means of applying experience to designs or situations to ensure that the features appearing in the list are not overlooked. Checklists tend to be general and may not be appropriate to a specific situation. They may not handle adequately the novel design or unusual process.

What-if   At each process step, what-if questions are formulated and answered to evaluate the effects of component failures or procedural errors. This technique relies on the experience level of the questioner.

Failure Mode and Effect Analysis (FMEA)   This is a systematic study of the causes of failures and their effects. All causes or modes of failure are considered for each element of a system, and then all possible outcomes or effects are recorded. This method is usually used in combination with fault tree analysis, a quantitative technique. FMEA is a complicated procedure, usually carried out by experienced risk analysts.

Cause-Consequence Diagram   These diagrams illustrate the causes and consequences of a particular scenario. They are not widely used because, even for simple systems, displaying all causes and outcomes leads to very complex diagrams. Again, this technique is employed by experienced risk analysts.

Reactive Chemicals Reviews   The process chemistry is reviewed for evidence of exotherms, shock sensitivity, and other instability, with emphasis on possible exothermic reactions. It is especially important to consider pressure effects—"Pressure blows up people, not temperature!" The purpose of this review is to prevent unexpected and uncontrolled chemical reactions. Reviewers should be knowledgeable people in the field of reactive chemicals and include people from loss prevention, manufacturing, and research.

Industrial Hygiene Reviews   These reviews evaluate the potential of a process to cause harm to the health of people. It is the science of the anticipation, recognition, evaluation, and control of health hazards in the environment. It usually deals with chronic, not acute, releases and is involved with toxicity.

Toxicity is the ability to cause biological injury. Toxicity is a property of all materials, even salt, sugar, and water. It is related to dose and the degree of hazard associated with a material. The amount of a dose is both time and duration dependent. Dose is a function of exposure (concentration) and duration and is sometimes expressed as dose = (concentration)^n × duration, where n can vary from 1 to 4.

Industrial hygiene deals with hazards caused by chemicals, radiation, and noise. Routes of exposure are through the eyes, by inhalation, by ingestion, and through the skin. An industrial hygiene guide is based on exposures for an 8-hour day, 40-hour week and is to be used as a guide in the control of health hazards. It is not to be used as a fine line between safe and dangerous conditions. Types of controls used include:

- Engineering, such as containment, ventilation, and automation
- Administrative, such as use of remote areas and job rotation
- Protective equipment

Facilities Reviews   There are many kinds of facilities reviews that are useful in detecting and preventing process safety problems. They include pre-start-up reviews (before the plant operates), new-plant reviews (the plant has started, but is still new), reviews of existing plants (safety, technology, and operations audits and reviews), management reviews, critical instrument reviews, and hazardous materials transportation reviews.

HAZOP   (Knowlton, 1989; Lees, 1980; *CPQRA*, 1989, pp. 419–422). HAZOP stands for "hazard and operability studies." This is a set of formal hazard identification and elimination procedures designed to identify hazards to people, process plants, and the environment. The techniques aim to stimulate in a systematic way the imagination of designers and people who operate plants or equipment so they can identify potential hazards. In effect, HAZOP studies make the assumption that a hazard or operating problem can arise when there is a deviation from the design or operating intention. Corrective actions can then be made before a real accident occurs.

Some studies have shown that a HAZOP study will result in recommendations that are 40 percent safety-related and 60 percent operability-related. HAZOP is far more than a safety tool; a good HAZOP study also results in improved operability of the process or plant, which can mean greater profitability.

The primary goal in performing a HAZOP study is to identify, not analyze or quantify, the hazards in a process. The end product of a study is a list of concerns and recommendations for prevention of the problem, not an analysis of the occurrence, frequency, overall effects, and the definite solution. If HAZOP is started too late in a project, it can lose effectiveness because:

1.   There may be a tendency not to challenge an already existing design.
2.   Changes may come in too late, possibly requiring redesign of the process.
3.   There may be loss of operability and design decision data used to generate the design.

HAZOP is a formal procedure that offers a great potential to improve the safety, reliability, and operability of process plants by recognizing and eliminating potential problems at the design stage. It is not limited to the design stage, however. It can be applied anywhere that a design intention (how the part or process is expected to operate) can be defined, such as:

- Continuous or batch processes being designed or operated
- Operating procedures
- Maintenance procedures
- Mechanical equipment design
- Critical instrument systems
- Development of process control computer code

These studies make use of the combined experience and training of a group of knowledgeable people in a structured setting. Some key concepts are:

- *Intention*—defines how the part or process is expected to operate.
- *Guide words*—simple words used to qualify the intention in order to guide and stimulate creative thinking and so discover deviations. Table 26-2 describes commonly used guide words.
- *Deviations*—departures from the intention discovered by systematic application of guide words.
- *Causes*—reasons that deviations might occur.
- *Consequences*—results of deviations if they occur.

- *Actions*—prevention, mitigation, and control
  —Prevent causes.
  —Mitigate the consequence.
  —Control actions, e.g., provide alarms to indicate things getting out of control; define control actions to get back into control.

The HAZOP study is not complete until response to actions has been documented. Initial HAZOP planning should establish the management follow-up procedure that will be used.

The guide words can be used on broadly based intentions (see Table 26-2), but when intentions are expressed in fine detail, some restrictions or modifications are necessary for chemical processes, such as:

No flow
Reverse flow
Less flow
More temperature
Less temperature
Composition change
Sampling
Corrosion/erosion

This gives a process plant a specific HAZOP guide-word list with a process variable, plant condition, or an issue.

HAZOP studies may be made on batch as well as continuous processes. For a continuous process, the working document is usually a set of flow sheets or piping and instrument diagrams (P&IDs). Batch processes have another dimension: time. Time is usually not significant with a continuous process that is operating smoothly except during start-up and shutdown, when time will be important and it will resemble a batch process. For batch processes, the working documents consist not only of the flow sheets or P&IDs but also the operating procedures. One method to incorporate this fourth dimension is to use guide words associated with time, such as those described in Table 26-3.

HAZOP studies involve a team, at least some of whom have had experience in the plant design to be studied. These team members apply their expertise to achieve the aims of HAZOP. There are four overall aims to which any HAZOP study should be addressed:

1.   Identify as many deviations as possible from the way the design is expected to work, their causes, and problems associated with these deviations.
2.   Decide whether action is required, and identify ways the problem can be solved.
3.   Identify cases in which a decision cannot be made immediately and decide what information or action is required.
4.   Ensure that required actions are followed through.

The team leader is a key to the success of a HAZOP study and should have adequate training for the job. Proper planning is important to success. The leader is actually a facilitator (a discussion leader and one who keeps the meetings on track) whose facilitating skills are just as important as technical knowledge. The leader outlines the boundaries of the study and ensures that the design intention is clearly understood. The leader applies guide words and encourages the team to discuss causes, consequences, and possible remedial actions for each deviation. Prolonged discussions of how a problem may be solved should be avoided.

Some people believe that it may be an advantage for the team leader not to have an intimate knowledge of the plant or process being studied in order to maintain neutrality. Ideally, the team leader should

TABLE 26-2   Some Guide Words Used in Conjunction with Process Parameters

Guide Word	Meanings	Comments
No, Not, None	Complete negation of design intentions	No part of intention is achieved and nothing else occurs
More of	Quantitative increases of any relevant physical parameters	Quantities and relevant physical properties such as flow rates, heat
Less of	Quantitative decreases	Same as above
As well as	Qualitative increase	All design and operating intentions are achieved as well as some additional activity
Part of	A qualitative decrease	Some intentions are achieved, some are not
Reverse	Logical opposite of intention	Activities such as reverse flow or chemical reaction, or poison instead of antidote
Other than	Complete substitution	No part of intention is achieved; something quite different happens

SOURCE: Knowlton, 1989.

TABLE 26-3   Guide Words Associated with Time

Guide word	Meaning
No time	Step(s) missed
More time	Step does not occur when it should
Less time	Step occurs before previous step is finished
Wrong time	Flow or other activity occurs when it should not

SOURCE: Knowlton, 1989.

be accompanied by a scribe or recorder, freeing the leader for full-time facilitating. The scribe should take notes in detail for full recording of as much of the meeting as is necessary to capture the intent of actions and recommendations.

Computer tools are available to aid information capture. In some cases, the facilitator may use the computer tool for recording, replacing the secretary. For example, PrimaTech offers a very useful computer program to aid in HAZOP studies (HAZOP-PC, 1994). Other excellent computer aids for HAZOP are also commercially available.

Team size is important. Less than three contributing members, excluding the secretary and leader, will probably reduce team effectiveness. A team size of five to eight, including the leader and scribe, is probably optimum.

The time required for HAZOP studies is significant. It has been estimated that each line or *node* (a node is usually a line or an item of equipment) may require in the range of about 30 minutes for an experienced team, although the time may vary widely. It should be recognized that the time required for HAZOP studies may not really be additional time for the project as a whole, particularly if started early enough in the design, and may actually save time on the project. It may make design of parts of the process more efficient, reduce the changes required later, and reduce the time required for safety and other reviews. It should make the safety reviews that should accompany any project much faster, as there will be fewer safety problems to discuss. It also should make possible smoother start-ups and make the process or plant safer and easier to operate, which will more than pay back the cost of the HAZOP study during the life of the plant. The results of a HAZOP study should be the basis for the operating discipline of a process, which in itself is a very valuable contribution.

Quantitative Tools for Hazard Analysis

Quantitative Fire and Explosion Index (F&EI)   (*Dow Fire and Explosion Index Hazard Classification Guide*, 1994; Lees, 1980, pp. 149–160). The F&EI is used to rate the potential of hazard from fires and explosions. Its purpose is to quantify damage from an incident. It identifies equipment that could contribute to an incident and ways to mitigate possible incidents. It is a way to communicate to management the quantitative hazard potential.

The F&EI measures realistic maximum loss potential under adverse operating conditions. It is based on quantifiable data. It is designed for flammable, combustible, and reactive materials that are stored, handled, or processed. It does not address frequency (risk) except indirectly, nor does it address specific hazards to people except indirectly.

The goals of the F&EI are to raise awareness of loss potential and identify ways to reduce potential severity and potential dollar loss in a cost-effective manner. The index number has significance as a comparison and in calculations to estimate the *maximum probable property damage* (MPPD). It also provides a method for measuring the effect of outage (plant being shut down) on the business. It is easy for users to get credible results with a small amount of training.

Chemical Exposure Index (CEI)   (*Chemical Exposure Index*, 1994). The CEI provides a method of rating the relative potential of acute health hazard to people from possible chemical release incidents. It may be used for conducting the initial process hazard analysis and it establishes the degree of further analysis needed. The CEI also may be used as part of the site review process.

The system provides a method of ranking one risk relative to another. It is not intended to define a particular containment system as safe or unsafe but provides a way of comparing toxic hazards. It deals with acute, not chronic, releases. The procedure focuses on the

necessary degree of concern and will provide the opportunity for recommendations, improvements, and concurrence from the appropriate knowledgeable people. Flammability and explosion hazards are not included in this index.

The CEI and hazard distance determine the level of review that is necessary. To develop a CEI, the following information is needed:
• An accurate plot plan of the plant and surrounding area
• A simplified process flow sheet showing containment vessels, major piping, and quantity of chemicals
• Physical and chemical properties of the chemical, including boiling point, molecular weight, and flash point
• Pressure and temperature of materials contained
• Toxicity (acute health hazard rating)
• Quantity (volatilized portion)
• Distance (to area of concern)
• Process variables (temperature, pressure, reactivity)
• Sights, odors, or sounds that could cause public concern or inquiries, such as smoke and odors below hazardous levels—for example, mercaptans or amines
• ERPG/EEPG—usually ERPG-2 is used
• Definition of ERPG
The Emergency Response Planning Guidelines (ERPG) are values established by the American Industrial Hygiene Association and intended to provide estimates of chemical concentration ranges where one might reasonably anticipate observing adverse effects as follows:

ERPG-1   The maximum airborne concentration below which it is believed that nearly all individuals could be exposed for up to one hour without experiencing other than mild transient adverse health effects or perceiving a clearly defined objectionable odor.

ERPG-2   The maximum airborne concentration below which it is believed that nearly all individuals could be exposed for up to one hour without experiencing irreversible or other serious health effects or symptoms that could impair their ability to take protective action.

ERPG-3   The maximum airborne concentration below which it is believed that nearly all individuals could be exposed for up to one hour without experiencing or developing life-threatening health effects.

Factors to consider when calculating a CEI are:
1.   Credible scenario for a release
2.   The rate at which toxic materials would be released in a scenario
When these two factors are considered together, a possible release rate can be calculated.

When considering release scenarios, the most hazardous unit in a plant should be chosen, based on inventory and process conditions. The idea is to imagine the release of material in the fastest way that is reasonably possible. The worst realistic scenario should be considered. This can be based on the outcome of a review, from a HAZOP study or a hazard analysis. The time a scenario will take is almost always considered to be continuous, because after a few minutes a stable dispersion distance exists. Making the time longer will not necessarily change the hazard distance.

Quantitative Tools for Risk Analysis

Quantitative Risk Analysis (QRA)   QRA is a technique that provides advanced quantitative means to supplement other hazard identification, analysis, assessment, control, and management methods to identify the potential for such incidents and to evaluate risk reduction and control strategies. QRA identifies those areas where operation, engineering, or management systems may be modified to reduce risk and may identify the most economical way to do it. The primary goal of QRA is that appropriate management actions, based on results from a QRA study, help to make facilities handling hazardous chemicals safer. QRA is one component of an organization's total process risk management. It allows the quantitative assessment of risk alternatives that can be balanced against other considerations.

Fault Tree Analysis   Fault tree analysis permits the hazardous incident (called the *top event*) frequency to be estimated from a logic model of the failure mechanisms of a system. The top event is traced downward to more basic failures using logic gates to determine its causes and likelihood. The model is based on the combinations of failures of more basic system components, safety systems, and human reliability. The underlying technology is the use of relatively simple logic gates (usually AND and OR gates) to synthesize a failure model of a plant. AND gates combine input events, all of which must exist simultaneously for the output to occur. OR gates also combine input events, but any one is sufficient to cause the output. The top event frequency or probability is calculated from failure data of more simple events. The top event might be a boiling liquid evaporating vapor explosion (BLEVE), a relief system discharging to the atmosphere, or a runaway reaction.

NFPA Standard System for Identification of Health, Flammability, Reactivity, and Related Hazards   (NFPA 704, Chaps. 2–5, 1990. This printed material is not the complete and official position of the National Fire Protection Association on the referenced subject, which is represented only by the standard in its entirety.)

This is a brief summary of NFPA 704 which addresses hazards that may be caused by short-term, acute exposure to a material during handling under conditions of fire, spill, or similar emergencies. This standard provides a simple, easily recognized, easily understood system of markings. The objective is to provide on-the-spot identification of hazardous materials.

These markings provide a general idea of the hazards of a material and the severity of these hazards as they relate to handling, fire protection, exposure, and control. This standard is not applicable to transportation or to use by the general public. It is also not applicable to chronic exposure. For a full description of this standard, refer to NFPA 704. The system identifies the hazards of a material in four principal categories: health, flammability, reactivity, and unusual hazards such as reactivity with water.

The degree of severity of health, flammability, and reactivity is indicated by a numerical rating that rates from zero (no hazard) to four (severe hazard). The information is presented in a square-on-point (diamond) field of numerical ratings. Information is presented as follows:

• Health rating in blue at nine o'clock

• Flammability rating in red at twelve o'clock
• Reactivity hazard rating in yellow at three o'clock
• Unusual hazards at six o'clock

Materials that demonstrate unusual reactivity with water are identified as ⩗ and materials that possess oxidizing properties shall be identified by the letters ⊙⋉. Other special hazard symbols may be used to identify radioactive hazards, corrosive hazards, substances that are toxic to fish, and so on.

The use of this system will provide a standard method of identifying the relative degree of hazard that is contained in various tanks, vessels, and pipelines. Suggested applications include:

• All storage tanks outside the block limits of a plant.
• Within block limits of a plant, tanks or process vessels with a capacity of more than 5000 gal (19 m^3).
• Process lines 3 in (7.62 cm) and larger, containing material with a health or reactivity rating of two, three, or four, or a flammability rating of three or four. Lines containing materials with lower ratings can also be marked if desired.

The name of the material contained in the pipelines should be placed on all lines at the point where they enter or leave the block and at road crossings. Block limit valves and emergency block valves should be painted yellow.

For a detailed description of the degrees of severity of the ratings, see NFPA 704. Table 26-4 shows the system for identification of hazards. Figures 26-1, 26-2, and 26-3 show examples of arrangements for display of the NFPA 704 Hazard Identification System.

RISK ANALYSIS

GENERAL REFERENCES: *Guidelines for Chemical Process Quantitative Risk Analysis,* CCPS-AIChE, New York, 1989. Arendt, "Management of Quantitative Risk Assessment in the Chemical Process Industry," *Plant Operations Progress,* vol. 9, no. 4, AIChE, New York, October 1990. CMA, "A Manager's Guide to Quantitative Risk Assessment," Chemical Manufacturers' Association, December 1989. EFCE, "Risk Analysis in the Process Industries," European Federation of Chemical Engineering, Publication Series no. 45, 1985. Lees,

TABLE 26-4   System for Identification of Hazards

	Identification of health hazard. Color code: Blue		Identification of flammability. Color code: Red		Identification of reactivity (stability). Color code: Yellow
Signal	Type of possible injury	Signal	Susceptibility of materials to burning	Signal	Susceptibility to release of energy
4	Materials that on short exposure could cause death or major residual injury	4	Materials that will rapidly or completely vaporize at atmospheric pressure and normal ambient temperature, or that are readily dispersed in air and will burn readily	4	Materials that in themselves are readily capable of detonation or of explosive decomposition or reaction at normal temperatures and pressures
3	Materials that on short exposure could cause serious temporary or residual injury	3	Liquids or solids that can be ignited under almost all ambient temperature conditions	3	Materials that in themselves are readily capable of detonation or of explosive decomposition or reaction but require a strong initiating source or which must be heated under confinement before initiation or which react explosively with water
2	Materials that on intense or continued but not chronic exposure could cause temporary incapacitation or possible residual injury	2	Materials that must be moderately heated or exposed to relatively high ambient temperatures before ignition can occur	2	Materials that readily undergo violent chemical change at elevated temperatures and pressures or which react violently with water or which may form explosive mixtures with water
1	Materials that on exposure would cause irritation but only minor residual injury	1	Materials that must be preheated before ignition can occur	1	Materials that are normally stable, but which can become unstable at elevated temperatures and pressures
0	Materials that on exposure under fire conditions would offer no hazard beyond that of ordinary combustible material	0	Materials that will not burn	0	Materials that are normally stable, even under fire exposure conditions, and which are not reactive with water

SOURCE: Reprinted with permission from NFPA 704, *Standard System for the Identification of the Fire Hazards of Materials,* National Fire Protection Association, Quincy, Mass., 1990. This printed material is not the complete and official position of the National Fire Protection Association on the referenced subject, which is represented only by the standard in its entirety.

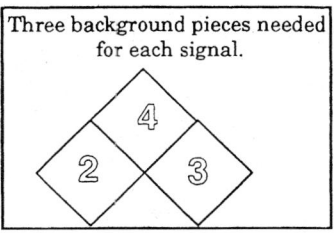

FIG. 26-1   For use where specified color background is used with numerals of contrasting colors. (*NFPA 704, 1990.*)

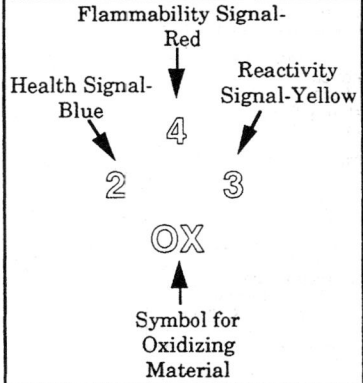

FIG. 26-2   For use where white background is used. (*NFPA 704, 1990.*)

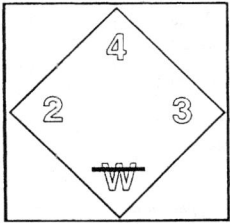

FIG. 26-3   For use where white background is used or for signs or placards. (*NFPA 704, 1990.*)

Loss Prevention in the Process Industries, Butterworths, Boston, 1980. World Bank, *Manual of Industrial Hazard Assessment Techniques,* Office of Environmental and Scientific Affairs, World Bank, Washington, D.C., 1985.

FREQUENCY ESTIMATION REFERENCES: *Guidelines for Process Equipment Reliability Data,* CCPS-AIChE, New York, 1989. Billington and Allan, *Reliability Evaluation of Engineering Systems: Concepts and Techniques,* Plenum Press, New York, 1983. Fussell, Powers, and Bennetts, "Fault Trees: A State of the Art Discussion," *IEEE Transactions on Reliability,* 1974. Roberts, N. H. et al., *Fault Tree Handbook,* NUREG-0492, Washington, D.C. Swain and Guttmann, *Handbook of Human Reliability Analysis with Emphasis on Nuclear Power Plant Applications,* NUREG/CR-1278, USNRC, Washington, D.C., 1983.

CONSEQUENCE ESTIMATION REFERENCES: *Guidelines for Use of Vapor Cloud Dispersion Models,* CCPS-AIChE, New York, 1987. TNO, *Methods for the Calculation of the Physical Effects of the Escape of Dangerous Materials: Liquids and Gases* ("The Yellow Book"), Apeldoorn, The Netherlands, 1979.

RISK ESTIMATION REFERENCES: Health and Safety Executive, *Canvey—An Investigation of Potential Hazards from the Operations in the Canvey Island/Thurrock Area,* HMSO, London, 1978. Rasmussen, *Reactor Safety Study: An Assessment of Accident Risk in U.S. Commercial Nuclear Power Plants,* WASH-1400 NUREG 75/014, Washington, D.C., 1975. Rijnmond Public Authority, *A Risk Analysis of 6 Potentially Hazardous Industrial Objects in the Rijnmond Area—A Pilot Study,* D. Reidel, Boston, 1982.

Introduction   The previous sections dealt with techniques for the identification of hazards and methods for calculating the effects of accidental releases of hazardous materials. This section addresses the methodologies available to analyze and estimate risk, which is a function of both the consequences of an incident and its frequency. The application of these methodologies in most instances is not trivial. A significant allocation of resources is necessary. Therefore, a selection process or risk prioritization process is advised before considering a risk analysis study.

Important definitions are as follows.

Markov model   A mathematical model used in reliability analysis. For many safety applications, a discrete-state (e.g., working or failed), continuous-time model is used. The failed state may or may not be repairable.

Probit model   A mathematical model of dosage and response in which the dependent variable (response) is a probit number that is related through a statistical function directly to a probability.

risk   A measure of economic loss or human injury in terms of both incident likelihood (frequency) and the magnitude of the loss or injury (consequence).

risk analysis   The development of an estimate of risk based on engineering evaluation and mathematical techniques for combining estimates of incident consequences and frequencies. Incidents in the context of the discussion in this chapter are acute events which involve loss of containment of material or energy.

A typical hazard identification process, such as a hazard and operability (HAZOP) study, is sometimes used as a starting point for selection of potential major risks for risk analysis. Other selection or screening processes can also be applied. However major risks are chosen, a HAZOP study is a good starting point to develop information for the risk analysis study. A major risk may qualify for risk analysis if the magnitude of the incident is potentially quite large (high potential consequence) or if the frequency of a severe event is judged to be high (high potential frequency) or both. A flowchart which describes a possible process for risk analysis is shown in Fig. 26-4.

The components of a risk analysis involve the estimation of the frequency of an event, an estimation of the consequences (the extent of the material or energy release and its impact on population, property, or environment), and the selection and generation of the estimate of risk itself.

A risk analysis can have a variety of potential goals:
1. To screen or bracket a number of risks in order to prioritize them for possible future study
2. To estimate risk to employees
3. To estimate risk to the public
4. To estimate financial risk
5. To evaluate a range of risk reduction measures
6. To meet legal or regulatory requirements
7. To assist in emergency planning

The scope of a study required to satisfy these goals will be dependent upon the extent of the risk, the depth of the study required, and the level of resources available (mathematical models and tools and skilled people to perform the study and any internal or external constraints).

The objective of a risk analysis is to reduce the level of risk wherever practical. Much of the benefit of a risk analysis comes from the discipline which it imposes and the detailed understanding of the major contributors of the risk that follows. There is general agreement that if risks can be identified and analyzed, then measures for risk reduction can be effectively selected.

The expertise required in carrying out a risk analysis is substantial. Although various software programs are available to calculate the frequency of events or their consequences, or even risk estimates, engineering judgment and experience are still very much needed to produce meaningful results. And although professional courses are available in this subject area, there is a significant learning curve required not only for engineers to become practiced risk analysts, but

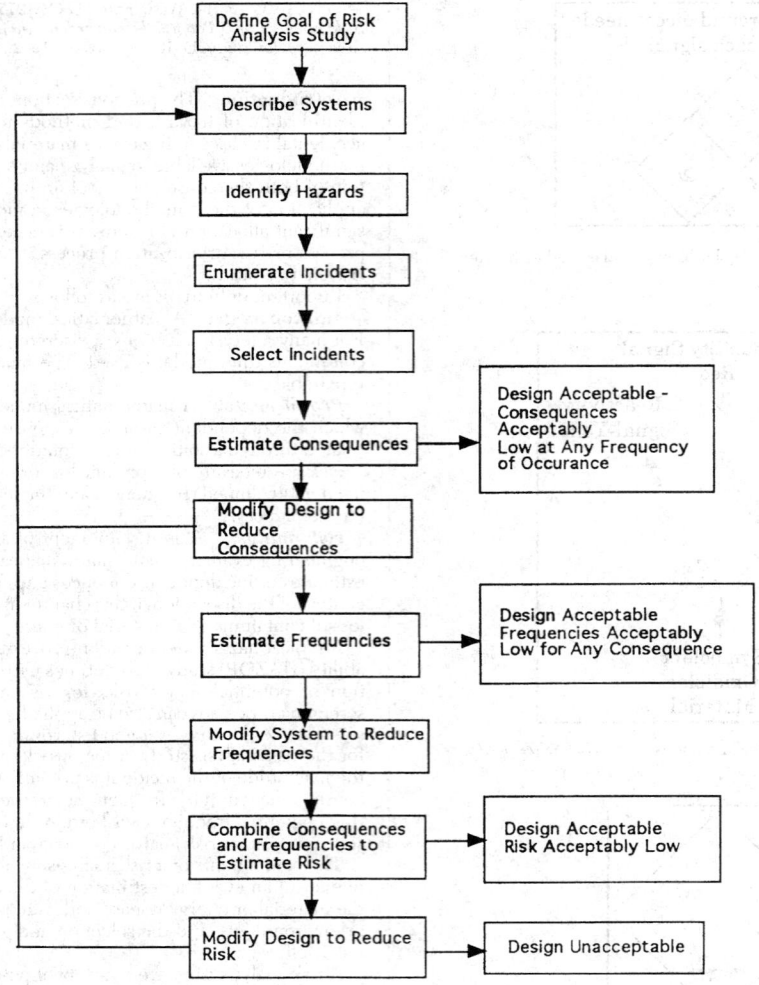

FIG. 26-4   One version of a risk analysis process. (*CCPS-AIChE, 1989, p. 13 by permission.*)

also for management to be able to understand and interpret the results. For these reasons, it may be useful to utilize a consultant organization in this field when a decision is made that a risk analysis is needed as a means to get started.

The analysis of a risk—that is, its estimation—leads to the assessment of that risk and the decision-making processes of selecting the appropriate level of risk reduction. In most studies this is an iterative process of risk analysis and risk assessment until the risk is reduced to some specified level. The subject of "acceptable" or "tolerable" levels of risk that could be applied to decision making on risks is a complex subject which will not be addressed in this section.

Frequency Estimation   There are two primary sources for estimates of incident frequencies. These are historical records and the application of fault tree analysis and related techniques, and they are not necessarily applied independently. Specific historical data can sometimes be usefully applied as a check on frequency estimates of various subevents of a fault tree, for example.

The use of historical data provides the most straightforward approach to the generation of incident frequency estimates but is subject to the applicability and the adequacy of the records. Care should be exercised in extracting data from long periods of the historical record over which design or operating standards or measurement criteria may have changed.

An estimate of the total population from which the incident information has been obtained is important and may be difficult to obtain.

Fault tree analysis and other related event frequency estimation techniques, such as event tree analysis, play a crucial role in the risk analysis process. Fault trees are logic diagrams that depict how components and systems can fail. The undesired event becomes the top event and subsequent subevents, and eventually basic causes, are then developed and connected through logic gates. The fault tree is completed when all basic causes, including equipment failures and human errors, form the base of the tree. There are general rules for construction, which have been developed by practitioners, but no specific rules for events or gates to use. The construction of a fault tree is still more of an art than a science. Although a number of attempts have been made to automate the construction of fault trees from process flow diagrams or piping instrumentation diagrams, these attempts have been largely unsuccessful. (P. K. Andow, "Difficulties in Fault Tree Synthesis for Process Plant," *IEEE Transactions on Reliability* **R-29**(1): 2, 1980).

Once the fault tree is constructed, quantitative failure rate and probability data must be obtained for all basic causes. A number of equipment failure rate databases are available for general use. However, specific equipment failure rate data is generally lacking and, therefore, data estimation and reduction techniques must be applied to generic databases to help compensate for this shortcoming. Accuracy and applicability of data will always be a concern, but useful results from quantifying fault trees can generally be obtained by experienced practitioners.

Human error probabilities can also be estimated using methodologies and techniques originally developed in the nuclear industry. A number of different models are available (Swain, "Comparative Evaluation of Methods for Human Reliability Analysis," GRS Project RS 688, 1988). This estimation process should be done with great care, as many factors can affect the reliability of the estimates. Methodologies using expert opinion to obtain failure rate and probability estimates have also been used where there is sparse or inappropriate data.

In some instances, plant-specific information relating to frequencies of subevents (e.g., a release from a relief device) can be compared against results derived from the quantitative fault tree analysis, starting with basic component failure rate data.

An example of a fault tree logic diagram using AND and OR gate logic is shown in Fig. 26-5.

The logical structure of a fault tree can be described in terms of boolean algebraic equations. Some specific prerequisites to the application of this methodology are as follows.

• Equipment states are binary (working or failed).
• Transition from one state to another is instantaneous.
• Component failures are statistically independent.
• The failure rate and repair rate are consistent for each equipment item.
• After repair, the component is returned to the working state.

Minimal cut set analysis is a mathematical technique for developing and providing probability estimates for the combinations of basic component failures and/or human error probabilities, which are necessary and sufficient to result in the occurrence of the top event.

A number of software programs are available to perform these calculations, given the basic failure data and fault tree logic diagram (AIChE-CCPS, 1989). Other less well known approaches to quantifying fault tree event frequencies are being practiced, which result in gate-by-gate calculations using discrete-state, continuous-time, Markov models (Doelp et al., "Quantitative Fault Tree Analysis, Gate-by-Gate Method," *Plant Operations Progress* 4(3): 227–238, 1984).

Identification and quantitative estimation of common-cause failures are general problems in fault tree analysis. Boolean approaches are generally better suited to mathematically handle common-cause failures.

Event tree analysis is another useful frequency estimation technique used in risk analysis. It is a bottom-up logic diagram, which starts with an identifiable event. Branches are then generated, which lead to specific chronologically based outcomes with defined probabilities. Event tree analysis can provide a logic bridge from the top event, such as a flammable release into specific incident outcomes (e.g., no ignition, flash fire, or vapor cloud explosion). Probabilities for each limb in the event tree diagram are assigned and, when multiplied by the starting frequency, produce frequencies at each node point for all the various incident outcome states. The probabilities for all of the limbs at any given level of the event tree must sum to 1.0. Event trees are generally very helpful toward the generation of a final risk estimate.

Consequence Estimation   Given that an incident (release of material or energy) has been defined, the consequences can be estimated. The general logic diagram in Fig. 26-6 illustrates these calculations for the release of a volatile hazardous substance.

For any specific incident there will be an infinite number of incident outcome cases that can be considered. There is also a wide degree of consequence models which can be applied. It is important, therefore, to understand the objective of the study to limit the number of incident outcome cases to those which satisfy that objective. An example of variables which can be considered is as follows.

• Quality, magnitude, and duration of the release
• Dispersion parameters
  wind speed
  wind direction
  weather stability

• Ignition probability (flammable releases)
  ignition sources/location
  ignition strength
• Energy levels contributing to explosive effects (flammable releases)
• Impact of release on people, property, or environment
  thermal radiation
  projectiles
  shock-wave overpressure
  toxic dosage
• Mitigation effects
  safe havens
  evacuation
  daytime/nighttime populations

Probit models have been found generally useful to describe the effects of incident outcome cases on people or property for more complex risk analyses. At the other end of the scale, the estimation of a distance within which the population would be exposed to a concentration of ERPG-2 or higher may be sufficient to describe the impact of a simple risk analysis.

Portions or all of the more complex calculation processes, using specific consequence models, have been incorporated into a few commercially available software packages (AIChE-CCPS, 1989). These programs should be used by risk analysts with extensive engineering experience, as significant judgment will still be required.

The output of these calculation processes is one or more pairs of an incident or incident outcome case frequency and its effect (consequence or impact).

Risk Estimation   There are a number of risk measures which can be estimated. The specific risk measures chosen are generally related to the study objective and depth of study, and any preferences or requirements established by the decision makers. Generally, risk measures can be broken down into three categories: risk indices, individual risk measures, and societal risk measures.

Risk indices are usually single-number estimates, which may be used to compare one risk with another or used in an absolute sense compared to a specific target. For risks to employees the *fatal accident rate* (FAR) is a commonly applied measure. The FAR is a single-number index, which is the expected number of fatalities from a specific event based on 10^8 exposure hours. For workers in a chemical plant, the FAR could be calculated as follows:

$$\text{FAR} = \frac{10^8}{8760} \times f \times \frac{D}{N} \qquad (26\text{-}1)$$

where FAR = fatal accident rate, expected number of fatalities from a specific event based on 10^8 exposure hours
f = frequency of the event in years^{-1}
D = expected number of fatalities, given the event
N = average number of exposed individuals on each shift

References are available which provide FAR estimates for various occupations, modes of transportation, and other activities (Kletz, "The Risk Equations—What Risk Should We Run?," *New Scientist,* May 12, pp. 320–325, 1977).

Figure 26-7 is an example of an individual risk contour plot, which shows the expected frequency of an event causing a specified level of harm at a specified location, regardless whether anyone is present at that location to suffer that level of harm.

The total individual risk at each point is equal to the sum of the individual risks at that point from all incident outcome cases.

$$\text{IR}_{x,y} = \sum_{i=1}^{n} \text{IR}_{x,y,i} \qquad (26\text{-}2)$$

where $\text{IR}_{x,y}$ = total individual risk of fatality at geographical location x,y
$\text{IR}_{x,y,i}$ = individual risk of fatality at geographical location x,y from incident outcome case i
n = total number of incident outcome cases

A common form of societal risk measure is an F-N curve, which is normally presented as a cumulative distribution plot of frequency F

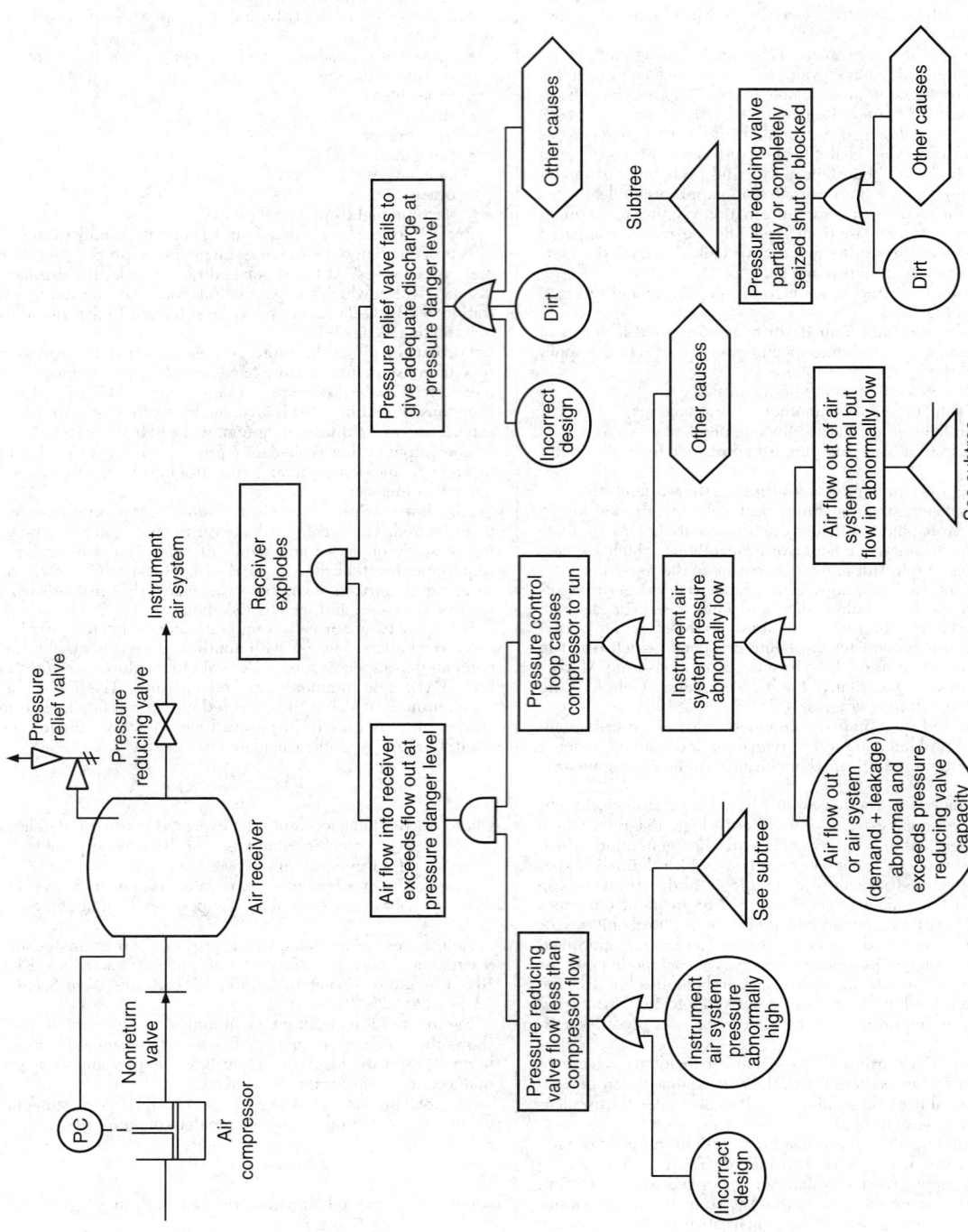

FIG. 26-5 Process drawing and fault tree for explosion of an air receiver. (*From Lees, 1980, pp. 200, 201, by permission.*)

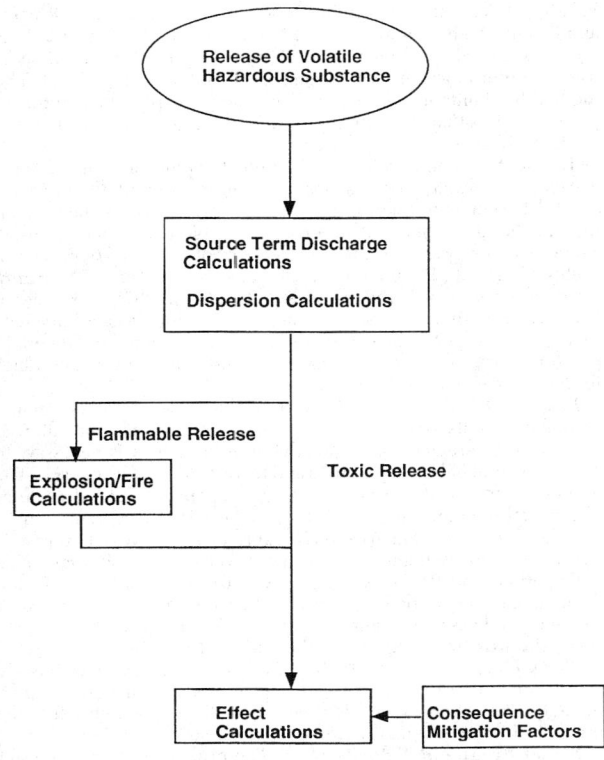

FIG. 26-6   Overall logic diagram for consequence analysis of volatile hazardous substances. (*CCPS-AIChE, 1989, p. 60.*)

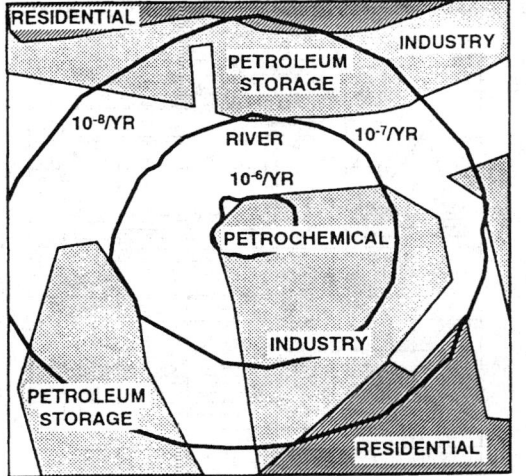

FIG. 26-7   Example of an individual risk contour plot. (*CCPS-AIChE, 1989, p. 269.*)

versus number of fatalities N. An example of this type of measure is shown in Fig. 26-8.

Any individual point on the curve is obtained by summing the frequencies of all events resulting in that number of fatalities or greater. The slope of the curve and the maximum number of fatalities are two key indicators of the degree of risk.

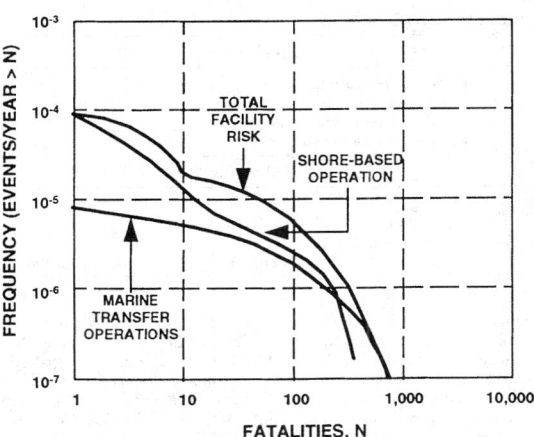

FIG. 26-8   Example of a societal risk F-N curve. (*AIChE-CCPS, 1989, p. 4.4.*)

For all risk measures it is possible to estimate the risk level of the current process as well as the risk levels from incorporation of various risk reduction alternatives. Management can then use this information as an important input in the final risk decision-making process.

GUIDELINES FOR ESTIMATING DAMAGE

Nomenclature

A	Projected area of fragment, breach area, or fragment cross-sectional area
a_a	Sound velocity in atmosphere
a_e	Sound velocity in high-pressure gas prior to vessel failure
B	Batch energy availability
d	Fragment diameter
E	Explosion energy available to generate blast and fragment kinetic energy, etc.
E_p	Critical perforation energy ($\frac{1}{2} MV_f^2$)
E_y	Young's modulus of elasticity
F	Dimensionless initial fragment acceleration
F	$P_e AR/Ma_e = P_e R/ma_e$ for vessel completely shattered into many small fragments
g	Acceleration due to gravity
h	Vessel wall thickness
k	Ratio of vessel outside diameter to internal diameter
L	Length of cylindrical vessel
M	Fragment mass
m	Mass per unit area of vessel shell
N	Length of cylindrical vessel forming rocketing tub fragment
N/m^2	Unit of pressure in SI system, N/m^2; also called pascal (Pa). One psi = 6.89476×10^3 Pa or 6.89476 kPa.
P	Liquid pressure
P_a	Atmospheric pressure
P_b	Dynamic vessel burst pressure
P_c	Pressure at expanding gas contact surface
P_e	Pressure at vessel failure
P_{inc}	Incident (side-on) blast pressure
P_r	Normally reflected (face-on) blast pressure
R	Vessel radius
r	Fragment radius = $(A/\pi)^{0.5}$
R_g	Range of fragment
t	Steel target thickness
U	$U_f + U_m$
u	Ultimate tensile strength of target steel
U_f	Fluid compression energy
U_m	Elastic strain energy in vessel walls
V	Volume of gas
V_f	Fragment velocity
V_L	Liquid volume
W	Equivalent mass of TNT
w	Unsupported span of steel target
X	Distance from wall of vessel to target

Nomenclature (Concluded)

Greek letters	
β_T	Fluid compressibility
γ	Ratio of specific heats of gas C_p/C_v
ϕ_0	Standard steady-state availability
ν	Poisson's ratio of vessel steel

Subscripts	
0	Reference state
1	Initial state
a	Environmental state
$1 \rightarrow a$	Denotes the path from state 1 to the environmental ambient state a

GENERAL REFERENCES: Baker, Cox et al., "Explosion Hazards and Evaluation," *Fundamental Studies in Engineering 5*, Elsevier Science Publishing, New York, 1983. Kinney and Graham, *Explosive Shocks in Air*, 2d ed., Springer-Verlag, New York, 1985. Petes, *Annals, New York Academy of Sciences 1968*, vol. 152, pp. 283–316. Holden, *Assessment of Missile Hazards: Review of Incident Experience Relevant to Major Hazard Plant*, UKAEA SRD/HSE/R477, November 1988. Lees, *Loss Prevention in the Process Industries*, Butterworths, London, 1996. Leslie and Birk, "State of the Art Review of Pressurized Liquefied Gas Container Failure Modes and Associated Projectile Hazards," *Journal of Hazard Materials* **28**, 1991, pp. 329–365. ASCE *Structural Analysis and Design of Nuclear Plant Facilities Manual and Reports on Engineering Practice* no. 58, 1980. Pritchard and Roberts, "Blast Effects from Vapour Cloud Explosions: A Decade of Progress," *Safety Science*, vol. 16 (3,4) 1993, pp. 527–548. "Explosions in the Process Industries," Major Haz. Monograph Series, *I. Chem. E.* (U.K.), 1994.

The availability of energy from an explosion can be approximately calculated in most cases but the method used depends upon the nature of the explosion.

Inert, Ideal Gas-Filled Vessels The energy available for external work following the rapid disintegration of the vessel is calculated by assuming that the gas within the vessel expands adiabatically to atmospheric pressure.

$$E = \frac{P_e V}{(g-1)} \left(\left[1 - \left(\frac{P_a}{P_e}\right)^{(\gamma-1)/\gamma} \right] + (g-1) \left(\frac{P_a}{P_e}\right) \left[1 - \left(\frac{P_a}{P_e}\right)^{-1/\gamma} \right] \right) \quad (26\text{-}3)$$

See Nomenclature table for definitions of terms.

In the case of thick-walled HP vessels, the strain energy in the vessel shell can contribute to the available energy, but for vessels below about 20 MN/m² (200 barg) it is negligible and can be ignored. If a Mollier chart for the gas is available, the adiabatic energy can be measured directly. This is the preferred method, but in many cases the relevant chart is not available.

The available energy is dissipated in several ways, e.g., the strain energy to failure, plastic strain energy in the fragments, kinetic energy of the fragments, blast wave generation, kinetic energy of vessel contents, heat energy in vessel contents, etc. For damage estimation purposes, the energy distribution can be simplified to:

$$E \, \left(\int p \, dv \right)$$

30% blast	40% fragment kinetic energy	30% other dissipative mechanisms

Blast Characteristics Accurate calculation of the magnitude of the blast wave from an exploding pressure vessel is not possible, but it may be estimated from several approximate methods that are available.

One method of estimating the blast wave parameters is to use the TNT equivalent method, which assumes that the damage potential of the blast wave from a fragmenting pressure vessel can be approximated by the blast from an equivalent mass of trinitrotoluene (TNT). The method is not valid for the region within a few vessel diameters from the vessel. However, a rough approximation can be made outside this region by calculating an equivalent mass of TNT and utilizing its well-known blast properties. The term *equivalent mass* means the mass of TNT which would produce a similar damage pattern to that of the blast from the ruptured vessel. The energy of detonation of TNT is 4.5 MJ/kg (1.5 × 10⁻⁶ ft·lb/lb), so the TNT equivalent mass W is given by $W = 0.3E/4.5$ kg. Standard TNT data (Dept. of the Army, Navy, and Air Force, "Structures to resist the effects of accidental explosions," TM5-1300, NAVFAC P-397, AFM 88-22. U.S. Gov.

Printing Office, vol. 2, November 1990, Figs. 2-7 and 2-15, or Kingery and Pannill, Memorandum Report No. 1518, Ballistic Research Laboratories, Aberdeen Proving Ground, U.S., April 1964) can then be used to determine the blast parameters of interest (Fig. 26-9). This method has limitations in the far field where the peak incident overpressure is less than 4 kN/m² (0.5 psi). In this region, local terrain and weather effects become significant.

The blast parameters also depend upon the physical location of the vessel. If the vessel is located close to or on the ground, then *surface-burst* data should be used. In other circumstances where the vessel is high in the air, either *free-air* or *air-burst* blast data may be used. These data are best presented in the form of *height-of-burst* curves (Petes, "Blast and Fragmentation Characteristics," *Ann. of New York Acad. of Sciences*, vol. 152, art. 1, fig. 3, 1968, p. 287). For incident blast pressures of 3 × 10⁵ N/m² (3 bar) or less, using surface-burst data may overestimate the blast pressure by about 33 percent. Generally, pressure vessel ruptures rarely cause ground craters, so no allowance for cratering should be made.

Fragment Formation The way in which a vessel breaks up into several fragments as a consequence of an explosion or metal failure is impossible to predict. Consequently, in most cases it is necessary to assume several failure geometries and to assess the effect of each. The number of fragments formed is strongly dependent upon the nature of the explosion and the vessel design. For high-speed explosions—e.g., detonations or condensed phase explosions—the vessel frequently shatters into many fragments, but for slower-speed explosions—e.g., deflagrations and BLEVEs—generally fewer than ten fragments are formed, and frequently less than five. In the special case of pressurized liquefied gas vessels affected by fire, Holden and Reeves ("Fragment Hazards from Failures of Pressurized Liquefied Gas Vessels," *I. Chem. E. Sym. Series 93*, 1985, p. 213) suggest that with cylindrical vessels, up to four fragments could be projected and that greater fragmentation of spherical vessels occurred with the possibility that the number of fragments may increase with increasing vessel size.

Initial Fragment Velocity (V_f) The process of energy transfer from the expanding gas to the vessel fragments is not efficient and seldom exceeds 40 percent of the available energy. According to Baum ("The Velocity of Missiles Generated by the Disintegration of Gas Pressurized Vessels and Pipes," *Journal of Press. Vessel Technology*, Trans. ASME, vol. 106, November 1984, pp. 362–368), there is an

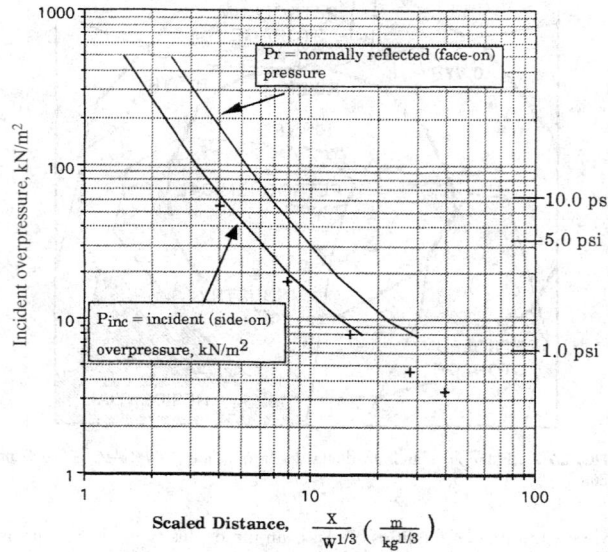

FIG. 26-9 Incident overpressure vs. scaled distance, surface burst. (*The "t" points are from Kingery and Pannill, Memo Report 1518 BRL. Adapted from Department of Army, Navy, and Air Force TM5-1300, NAVFAC P-397, AFM 88-22.*)

upper limit to the fragment velocity, which is taken to be the velocity of the contact surface between the expanding high-pressure gas and the surrounding atmospheric air. This is referred to as the *zero-mass* fragment velocity and, for most industrial low- to medium-pressure vessels, is less than about 1.3 Mach. It is calculated using ideal gas, one-dimensional shock tube theory and is given by the equation for the shock tube contact surface velocity (Wright, *Shock Tubes*, Methuen & Co., London, 1961).

$$\frac{V_f}{a_0} = -\frac{2}{(\gamma - 1)} \left[\left(\frac{P_c}{P_e}\right)^{(\gamma - 1)/2\gamma} - 1 \right] \quad (26\text{-}4)$$

where P_c is determined from the relationship:

$$\frac{a_a(1 - \mu_a)\left(\frac{P_c}{P_a} - 1\right)}{a_e \left[(1 + \mu_a)\left(\frac{P_c}{P_a} + \mu_a\right)\right]^{1/2}} = \frac{2}{(\gamma - 1)}\left(1 - \left(\frac{P_a}{P_e} \cdot \frac{P_c}{P_a}\right)^{(\gamma - 1)/2\gamma}\right) \quad (26\text{-}5)$$

and

$$\mu_a = \frac{\gamma_a - 1}{\gamma_a + 1} \quad (26\text{-}6)$$

where P_c = pressure at expanding gas contact surface
a_a = sound velocity at ambient conditions
a_e = sound velocity in gas prior to vessel failure

The value of a_e may be approximated using physical property data for the specific gas at the temperature and pressure at the start of the expansion. Equation (26-5) must be solved using a trial-and-error method. Most fragments never achieve the zero-mass velocity and their velocity can be assessed using the correlations of Baum ("Disruptive Failure of Pressure Vessels: Preliminary Design Guidelines for Fragments Velocity and the Extent of the Hazard Zone," *J. Pressure Vessel Technology*, Trans. ASME, vol. 110, May 1988, pp. 168–176; Baum, "Rupture of a Gas-Pressurized Cylindrical Pressure Vessel. The Velocity of Rocketing Fragments," *J. Loss Prev. Process Ind.*, vol. 4, January 1991, pp. 73–86; Baum, "Velocity of a Single Small Missile Ejected from a Vessel Containing High Pressure Gas," *J. Loss Prev. Process Ind.*, vol. 6, no. 4, 1993, pp. 251–264).

The Baum correlations for several vessel failure modes are given in Eqs. (26-7) to (26-16).

End-Cap Released from a Cylindrical Vessel Pressurized with an Inert Ideal Gas

$$\frac{V_f}{a_e} = 2F^{0.5} \quad (26\text{-}7)$$

where F is a dimensionless initial fragment acceleration given by

$$\frac{P_e AR}{Ma_e} \quad (26\text{-}8)$$

and A = projected area of fragment
M = fragment mass
R = vessel radius

Rocket Fragment from a Cylindrical Vessel Pressurized with an Inert Ideal Gas

$$\frac{V_f}{a_e} = 2.18 \left[F\left(\frac{L}{R}\right)^{0.5} \right]^{2/3} \quad (26\text{-}9)$$

where L = length of vessel and F is calculated using the area of the open end. For a more precise result, a correction factor to allow for the vessel opening time should be made (Baum, 1991).

Rocket Fragment from a Cylindrical Missile Pressurized with a Liquid at a Temperature Such That Rupture Initiates Flash Evaporation

$$\frac{V_f}{a_e} = 0.2 \left(\frac{2E}{M}\right)^{0.5} \quad (26\text{-}10)$$

Range of experimental data:

$$\text{Water, } T_{\text{sat}} \sim 320°C, \ 145 \leq \frac{P_e}{P_a} \leq 155, \ N/L = 1.0, \ 12 \leq L/R \leq 24$$

The vessel is completely full of liquid and N is the length of the vessel forming the rocket fragment, with limited data.

Whole Vessel Driven by an Inert Ideal Gas Escaping through Axial Split

$$\frac{V_f}{a_e} = 0.17 \left(\frac{2E}{M}\right)^{0.5} \quad (26\text{-}11)$$

Range of experimental data:

$$10 < \frac{P_e}{P_a} < 63, \ \gamma = 1.4, \ 4 < L/R < 6$$

Large, Single Fragment Ejected from Cylindrical Vessel Pressurized with an Inert Ideal Gas

$$\frac{V_f}{a_e} = 2F^{0.5} \quad (26\text{-}12)$$

Range of experimental data:

$$\frac{P_e}{P_a} = 100, \ \gamma = 1.4, \ L/R = 13.5, \ A^{0.5} > R$$

based upon two points only. End-cap equation (26-5) is probably adequate.

Single Small Fragment Ejected from a Cylindrical Vessel Pressurized with an Inert Ideal Gas

$$\frac{V_f}{a_e} = \left[2F\left(\frac{h}{R}\right)\right]^{0.5} + 0.96\left[F\left(\frac{r}{R}\right)\right]^{0.25} \quad (26\text{-}13)$$

where h = wall thickness
r = fragment radius, i.e., $r = (A/\pi)^{0.5}$

Range of experimental data:

$$20 < \frac{P_e}{P_a} < 300, \ \gamma = 1.4, \ r < 0.3 \ R, \ h < 0.1 \ R$$

Fragments Generated by Disintegration of a Cylindrical Vessel Pressurized with an Inert Ideal Gas

$$\frac{V_f}{a_e} = 0.88 \ F^{0.55} \quad (26\text{-}14)$$

Experimental data show no strong dependence on P_e/P_a or L/R.

Hemispherical Fragment Released from a Spherical Vessel Pressurized with a Liquid at a Temperature That on Rupture Initiates Flash Evaporation

$$\frac{V_f}{a_e} = 0.16 \left(\frac{2E}{M}\right)^{1/2} \quad (26\text{-}15)$$

Range of experimental data:

$$\text{Water } T_{\text{sat}} \sim 230°C, \ 49.5 \leq \frac{P_e}{P_a} \leq 60.5$$

Limited data, vessel full of water.

Fragments Generated by Complete Shattering of Spherical Vessel Pressurized by an Inert Ideal Gas

$$\frac{V_f}{a_e} = 0.88 \ F^{0.55} \quad (26\text{-}14)$$

Vessel Filled with Reactive Gas Mixtures   Most cases of damage arise not from the vessel failing at its normal operating pressure but because of an unexpected exothermic reaction occurring within the vessel. This usually is a decomposition, polymerization, deflagration, runaway reaction, or oxidation reaction. In assessing the damage potential of such incidents, the peak explosion or reaction pressure can often be calculated, and if this peak pressure P_e is then inserted into Eq. (26-3), the available energy can be assessed and the blast and fragment hazard determined. Where the expected peak explosion pressure P_e is greatly in excess of the vessel dynamic burst pressure, it is sufficient to increase the burst pressure to allow for the increase in vessel pressure during the period necessary for both the vessel to rupture and the fragments to be removed from the path of the expanding vessel contents. Where the gas pressure in the vessel is rising rapidly, the gas may reach a much higher pressure than the estimated dynamic burst pressure of the vessel. This effect is similar to the accumulation on a relief valve. It is, therefore, conservative to assume that the gas reaches the pressure calculated on the assumption of complete reaction. The reaction is assumed to go to completion before the containing vessel fails. However, there are reactions where it is simpler to calculate the energy availability using thermodynamic methods. The maximum energy released in an explosion can be assessed from the change in the Helmholz free energy ($-\Delta H = -\Delta E + T\Delta S$), but if the required data is not available, it may be necessary to use the Gibbs free energy ($\Delta F = \Delta H - T\Delta S$), which—especially in the case of reactions with little or no molal change, e.g., hydrocarbon/air oxidation—is similar to the Helmholz energy. It may sometimes be more convenient to calculate the batch energy availability [$B = \phi_0 + \Delta\phi_{1\to a} + \Delta(PV)_{1\to a} - P_a\Delta V_{1\to a}$] (Crowl, "Calculating the Energy of Explosion Using Thermodynamic Availability," *J. Loss Prev. Process Ind.*, **5**, no. 2, 1992, p. 109), which for an ideal gas becomes

$$B_1 = f_0 + f_{1\to a} + nRT_1\left[\left(\frac{P_a}{P_1}\right) - 1\right] \qquad (26\text{-}16)$$

The energy partition between blast wave energy and fragment kinetic energy is as described in paragraph 1.

Vessels Completely Filled with an Inert High-Pressure Liquid   A typical example is the pressure testing of vessels with water. The energy available to cause damage is the sum of the liquid compression energy and the strain energy in the vessel shell. The sudden release of this energy on vessel failure generally creates flying fragments but rarely any significant blast effects.

The fluid compression energy up to about 150 MN/m² (22,000 psig) can be estimated from $U_f = \frac{1}{2}\beta_T P^2 V_L$, where β_T is the liquid bulk compressibility, P is the liquid pressure, and V_L is the liquid volume. At higher pressure, this simple equation becomes too conservative and more complex methods of calculating the fluid compression energy are required. The elastic strain energy for cylindrical vessels, ignoring end closures, can be estimated from:

$$U_m = \frac{P^2 V_L}{2E(k^2 - 1)}\left[3(1 - 2v) + 2k^2(1 + v)\right] \qquad (26\text{-}17)$$

where P = pressure of liquid
V_L = volume of liquid
E_y = Young's modulus of elasticity
v = Poisson's ratio

Energy available $U = U_f + U_m$.

Only a small fraction of U is available to provide kinetic energy to the fragments. There are few data available, but in five incidents analyzed by High (unpublished data), no fraction was greater than 0.15. The fragment initial velocity can be assessed from $0.15\ U = \frac{1}{2}MV_f^2$.

Distance Traveled by Fragments   There is no method available to estimate the distance traveled by an irregularly shaped, possibly tumbling, subsonic fragment projected at an unknown angle. A conservative approach is to assume that all the fragments are projected at an angle of 45° to the horizontal and to ignore the aerodynamic effects of drag and/or lift. The range R_g is then given by $R_g = V_f^2/g$, where $g =$ gravitational acceleration.

This is too conservative to provide anything more than an upper bound. Some limited guidance is given by Scilly and Crowther ("Methodology for Predicting Domino Effects from Pressure Vessel Fragmentation," *Proc. Hazards Ident. and Risk Anal., Human Factors and Human Reliability in Process Safety*, Orlando, Fla., 15 Jan 1992, p. 5, sponsored by AIChE and HSE), where the range, for vessels with walls less than 20 mm (0.79 in), is $2.8\ P_b$ with the range in meters and

P_b as the vessel burst pressure in bars. Other sources are Baker (*Explosion Hazards and Evaluation*, Elsevier, 1983, p. 492) and Chemical Propulsion Information Agency (*Hazards of Chemical Rockets and Propellants Handbook*, vol. 1 NTIS, Virginia, May 1972, pp. 2-56, 2-60).

Fragment Striking Velocity   It is generally impossible to assess the fragment velocity, trajectory, angle of incidence, and fragment attitude at the moment of striking a target; consequently, the conservative view is taken that the fragment strikes the target at right angles, in the attitude to give the greatest penetration, with a velocity equal to the initial velocity.

Damage Potential of Fragments   In designing protection for fragment impact, there are two failure modes to be considered: local response and overall response. Local response includes penetration/perforation in the region of the impact. Overall response includes the bending and shear stresses in the total target element; i.e., will the whole target element fail regardless of whether the element is penetrated or perforated?

Local Failure   The penetration or perforation of most industrial targets cannot be assessed using theoretical analysis methods, and recourse is made to using one of the many empirical equations. In using the equations, it is essential that the parameters of the empirical equation embrace the conditions of the actual fragment.

The penetrability of a fragment depends on its *kinetic energy density* (KED), given by

$$\text{KED} = \frac{1}{2}\frac{MV_f^2}{A} \qquad (26\text{-}18)$$

where A is the fragment cross-sectional area. The KED is a useful comparative measure of a fragment's penetrability when comparing like with like. Several equations are given in the following sections.

Ballistics Research Laboratory (BRL) Equation for Steel Targets

$$E_p = 1.4 \times 10^9\ (dt)^{1.5} \qquad (26\text{-}19)$$

where d is the fragment diameter, t is the steel target plate thickness, and E_p is the critical perforation energy in SI units (kg, m, m/s, J), when applied to fragments between 1 kg and 19.8 kg, impacting targets 1 mm to 25 mm (1 in) thick plate with velocities from 10 m/s to 100 m/s. Neilson (*Procedures for the Design of Impact Protection of Off-shore Risers and ESV's*, U.K. AEA [ed.], 1990) found a large scatter in the results, but most were within ±30 percent.

Stanford Research Institute (SRI) Equation for Steel Targets

$$E = \frac{dut^2}{10.3}\left(42.7 + \frac{w}{t}\right) \qquad (26\text{-}20)$$

where, with the same notation as Eq. (26-14), w is the unsupported span of the target plate (m) and u is the ultimate tensile strength of the target steel (N/m²). The parameters for this equation are given by Brown ("Energy Release Protection for Pressurized Systems," part II, "Review of Studies into Impact/Terminal Ballistics," *Applied Mechanics Review*, vol. 39, no. 2, 1986, pp. 177–201) as $0.05 \le d \le 0.25m$, $414 \le u \le 482$ MN/m² for a fragment mass between 4.5 and 50 kg.

Overall Response   The transition from local to overall response is difficult to define. High-velocity impact implies that the boundary conditions of the target have little influence on the local response (excluding reflected shock waves). If the fragment is small relative to the target, local response will dominate, but fragments that are of the same order of size as the target will produce an overall response. It is often necessary to consider both overall and local response. Low values of KED are associated with overall response. Design methods for dynamically applied loads are given by Newark ("An Engineering Approach to Blast Resistant Design," ASCE New York, 1953), Baker (see General References), or ASCE (*Manual and Reports on Engineering Practice*, no. 58, 1980).

Response to Blast Waves   The effect of blast waves upon equipment and people is difficult to assess because there is no single blast wave parameter which can fully describe the damage potential of the blast. Some targets respond more strongly to the peak incident over-pressure and others to the impulse ($\int p \, dt$) of the blast. The blast parameters are usually based on the conservative assumption that the blast strikes the target normal to its surface, so that normal reflection parameters are used.

The pressure exerted by the blast wave on the target depends upon the orientation of the target. If the target surface faces the blast, then the target will experience the reflected or face-on blast pressure P_r, but if the target surface is side-on to the blast, then the target will experience the incident or side-on blast pressure P_{inc}. The reflected blast pressure is never less than double the incident pressure and can, for ideal gases, be as high as eight times the incident pressure. For most industrial targets where the incident pressure is less than about 17 kN/m² (25 psi), the reflected pressure is not more than 2.5 times the incident pressure.

Response of Equipment   The response of equipment to blast is usually a combination of two effects: one is the displacement of the equipment as a single entity and the other is the failure of the equipment itself. The displacement of the equipment is an important consideration for small, unsecured items—e.g., empty drums, gas cylinders, empty containers. Most damage results from the failure in part or totally of the equipment or containing structure itself.

The blast parameters are usually based on the conservative assumption that the blast strikes the target normal to its surface, so that normal reflection parameters may be used.

The response of a target is a function of the ratio of the blast wave duration and the natural period of vibration of the target (T/T_n). Neither of these parameters can be closely defined.

Calculating the specific response of a specific target can generally be done only approximately. Accuracy is not justified when the blast properties are not well defined. A guide to the damage potential of condensed phase explosive blast is given in Table 26-5 (Scilly and High, "The Blast Effect of Explosions," *Loss Prevention and Safety Promotion in the Process Industries,* European Fed. of Chem. Eng., 337 Event, France, September 1986, table 2). Nuclear data is available (Table 26-6) (Walker, "Estimating Production and Repair Effort in Blast-damaged Petroleum Refineries," *Stanford Research Inst.,* July 1969, fig. 5, p. 45), which is based upon long positive-duration blast (±6 s). This suggests that the Walker data will be conservative for the much shorter duration blast from accidental industrial explosions.

A blast incident overpressure of 35 kN/m² (5 psi) is often used to define the region beyond which the damage caused will be minor and not lead to significant involvement of plant and equipment beyond the 35 kN/m² boundary.

Response of People   The greatest hazard to people from blast is generally from the deceleration mechanism after people have been blown off their feet and they become missiles. This occurs at an incident overpressure of about 27 kN/m² (4.0 psi) for long positive-duration nuclear weapon blasts. People have more blast resistance than most equipment and can survive incident overpressures of 180 kN/m² (27 psi) (Bowen, Fletcher, and Richmond, DASA-2113, Washington, D.C., October 1968), even for long-duration blasts.

PROJECT REVIEW AND AUDIT PROCESSES

GENERAL REFERENCES: Center for Chemical Process Safety (CCPS), *Guidelines for Hazard Evaluation Procedures, Second Edition with Worked Examples,* AIChE, September 1992. CCPS, *Guidelines for Technical Management of Chemical Process Safety,* AIChE, 1989. CCPS, *Guidelines for Auditing Process Safety Management Systems,* AIChE, 1993.

Introduction   Review and audit processes are used in the chemical process industry to evaluate, examine, and verify the design of process equipment, operating procedures, and management systems. These processes assure compliance with company standards and guidelines as well as government regulations. Reviews and audits can encompass the areas of process and personnel safety, environmental and industrial hygiene protection, quality assurance, maintenance procedures, and so on.

To distinguish between a review and an audit, some definitions will be provided. A review is a critical examination or evaluation of any operation, procedure, condition, event, or equipment item. Reviews can take many forms and be identified as project reviews, design reviews, safety reviews, pre-start-up reviews, and so on. The following discussion of the review process will deal with project reviews associated with capital projects and focus on the area of process safety.

An audit is a formal, methodical examination and verification of an operation, procedure, condition, event, or series of transactions. The verification element of an audit makes it distinctive from a review. A project review will *recommend* design, procedural, maintenance, and management practices to minimize hazards and reduce risk while meeting company standards and government regulations. An audit will *verify* that the design, the procedures, and the management systems are actually in place, and are being maintained and used as intended. In fact, it is not uncommon for an audit to be done on a review process, which is a management system, to verify that the elements of the review process are being followed.

The following sections will describe the project review and audit processes separately, addressing the why, when, and how for each process.

Project Review Process   The scope of capital projects can be large, involving the construction of new plants with new technologies and products, or small, involving minor changes to existing facilities. In either case, project safety reviews can be used to evaluate and examine the process design, operating procedures, and process control scheme for process hazards, conformance to company standards and guidelines, and compliance with government regulations. Some objectives of the review process (CCPS, 1992, p. 53) are: (1) identify equipment or process changes that could introduce hazards, (2) evaluate the design basis of control and safety systems, (3) evaluate operating procedures for necessary revisions, (4) evaluate the application of new technology and any subsequent hazards, (5) review the adequacy of maintenance and safety inspections, and (6) evaluate the consequences of process deviations and determine if they are acceptable (CCPS, 1989, p. 46).

The project review process should be integrated with the development of the project from the conceptual stage to the start-up stage (CCPS, 1989, p. 46). Figure 26-10 depicts the various stages of a capital project. The size and complexity of a project will determine if the project progresses through all these stages and, in the same manner, determine the number and type of reviews that are needed. The earlier in a project that a review can be used to identify required changes, the less costly the change will be to implement. For example, reviews held at the research stage of a project can be beneficial in choosing the least hazardous technology and contribute to an inherently safer process design that needs fewer add-on safety systems.

As the project progresses, more information is available; therefore, the review technique used can be different at each stage of the project. The use of various hazard evaluation techniques, such as checklist analyses, relative rankings, what-if analyses, and hazard and operabil-

TABLE 26-5   Damage Effects

Incident, psi	Pressure, kPa	Damage effects
10	70	Damage to most refineries would be severe, although some pumps, compressors, and heat exchangers could be salvaged. All conventional brick buildings would be totally destroyed. Rail wagons (rail cars) overturned. Storage tanks ruptured. Fatalities certain.
5.0	34	Brick buildings severely damaged, 75% external wall collapse. Fired heaters badly damaged. Storage tanks leak from base. Threshold for eardrum damage to people. Domino or knock-on radius. Pipe bridges may move.
2.0	14	Doors and windows removed. Some frame distortion to steel frame buildings and cladding removed. Some electrical/instrument cables broken.*
1.0	7	Lethal glass fragments. Limit for public housing, schools, etc.
0.3	2	About 50% domestic glass broken.

*1% probability electrical cables broken at 2.0 psi inc. 99% probability electrical cables broken at 3.6 psi inc.

TABLE 26-6 Blast Overpressure Effects on Vulnerable Refinery Parts

Equipment	overpressure

Overpressure (psi)

Equipment	0.5	1.0	1.5	2.0	2.5	3.0	3.5	4.0	4.5	5.0	5.5	6.0	6.5	7.0	7.5	8.0	8.5	9.0	9.5	10.0	12.0	14.0	16.0	18.0	20.0	>20.0
Control house: steel roof	a	c	d				h↑																			
Control house: concrete roof	a	e p	d				n													e↑						
Cooling tower	b		f				o↑																			
Tank: cone roof		d				k							u↑													
Instrument cubicle			a			i m						t↑														
Fired heater				g	i					t↑																
Reactor: chemical				a				i					p↑													
Filter				h					i										v		t↑					
Regenerator						i				i p			u		t↑											
Tank: floating roof						k		i			t										t↑					
Reactor: cracking							i					i		i											d↑	
Pipe supports							p		q		s o↑															
Utilities: gas meter						i			h			i			l					t↑						
Utilities: electric transformer										h								l		t↑						
Electric motor																										v↑
Blower										q				r				t		t↑						
Fractionation column											r	p i	t	t↑												
Pressure vessel: horizontal												i					t	t↑		m q						
Utilities: gas regulator													i							v↑						
Extraction column													i		l			v	m q	v	t	m s			d↑	
Steam turbine															l						m					
Heat exchanger															i		t	t↑								
Tank: sphere															l	i		v↑			i	i	t↑			
Pressure vessel: vertical																i					i	t	t↑			
Pump																					i	i	r↑			

CODE:

a	Windows and gauges break.
b	Louvers fall at 0.3–0.5 psi.
c	Switchgear is damaged from roof collapse.
d	Roof collapses.
e	Instruments are damaged.
f	Inner parts are damaged.
g	Brick cracks.
h	Debris-missile damage occurs.
i	Unit moves and pipes break.
j	Bracing fails.
k	Unit uplifts (half-filled).
l	Power lines are severed.
m	Controls are damaged.
n	Block walls fail.
o	Frame collapses.
p	Frame deforms.
q	Case is damaged.
r	Frame cracks.
s	Piping breaks.
t	Unit overturns or is destroyed.
u	Unit uplifts (0.9 filled).
v	Unit moves on foundation.

SOURCE: F. E. Walker, "Estimating Production and Repair Effort in Blast Damaged Petroleum Refineries," *SRI*, July 1969.

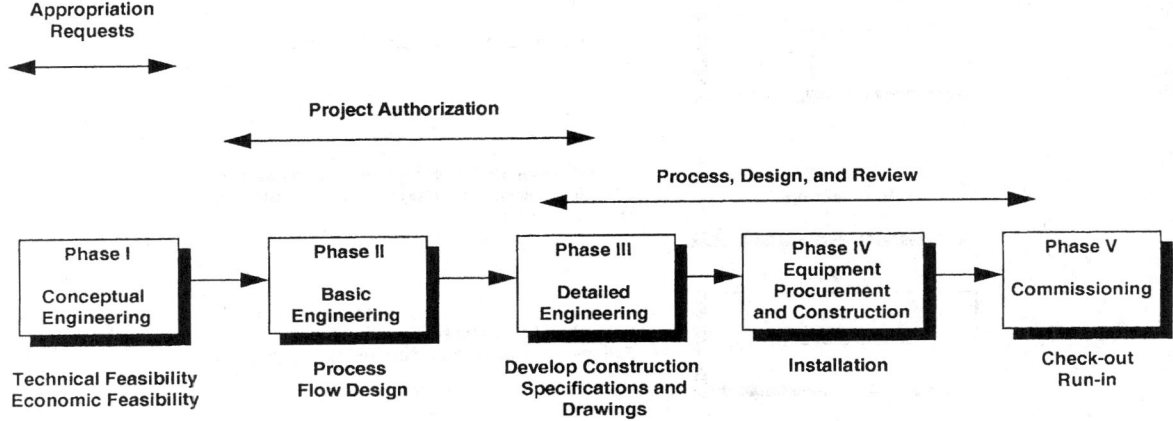

FIG. 26-10   The phases of a capital project. (*CCPS, 1989 by permission of AIChE.*)

ity studies, is documented in *Guidelines for Hazard Evaluation Procedures: Second Edition with Worked Examples* (CCPS, 1992). The need to use more quantitative techniques for hazard evaluation may be identified during these reviews, and become an action item for the project team.

The project review process involves multiple steps that should be defined in management guidelines (CCPS, 1993, pp. 57–61). The steps include: (1) review policy, (2) review scheduling, (3) review technique, (4) review team representation, (5) review documentation, (6) review follow-up, (7) review follow-up verification, and (8) review procedures change management. These steps define how a review, whether it be a safety review, environmental review, pre-start-up review, or whatever, is conducted and how closure of review action items is achieved.

Review Policy   The review policy should establish when project safety reviews should be done. All capital projects, large or small, should have one or more safety reviews during the course of the project. The number and types of review should be stated in a management policy. Any reasons for exceptions to the policy should be documented as well. The policy should address not only projects internal to a company, but also any joint ventures or turnkey projects by outside firms.

Review Scheduling   A review scheduling procedure should be established that documents who is responsible for initiating the review and when the review(s) should occur during the project. The scheduling needs to balance availability of process information, review technique used, and the impact of potential review action items on project costs (i.e., early enough to minimize the cost of any potential changes to the process). The actual amount of time needed for the review should also be stated in the procedure. On the basis of the number of project reviews required and the estimated time needed for each review, the project cost estimate should include the cost for project reviews as part of the total cost for the project.

Review Team   The project review should be conducted by a functionally diverse team. The team should consist of a team leader to organize and lead the team review, a scribe or secretary to record and issue a review summary with action items, and functional experts in fields relevant to the project such as safety, environmental, and industrial hygiene (CCPS, 1993, p. 58). The team leader should be experienced in the use of the selected review technique with leadership skills and no direct involvement with the project under review. The review procedure should address the minimum requirements for team leaders and team members. Some typical requirements could be years of experience, educational background, and training in the review technique. Responsibilities should be clearly defined for initiating the review, assigning the review team, recording the team findings, and monitoring follow-up of team recommendations.

Review Techniques   The review techniques used at the various

stages of a project should be selected based on the amount of process information and detail available. Figure 26-11 depicts some typical review techniques at the various stages of a capital project. A detailed description, including the type and amount of process information required, for each review technique can be found in *Guidelines for Hazard Evaluation Procedures: Second Edition with Worked Examples* (CCPS, 1992). The process information required for the review should be defined and documented in the review guidelines. Up-to-date and accurate process information is essential to conducting a successful review.

Review Documentation   The project review team leader has the ultimate responsibility for documenting the results of the project review. This responsibility may be delegated to a team scribe or secretary to record the review minutes and issue a summary report with listed action items. The action items could address exceptions to company or industry standards and government regulations, review team recommendations based on experience and knowledge, and further issues for study that could not be resolved during the review session.

The summary report should have a standard format and could contain a short project scope summary, a listing of review team members by function, a listing of project team members present, a meeting agenda or checklist of topics reviewed by the team, and a list of concerns and action items for project team follow-up. The distribution list for the summary report should be established and include the review team, project team, and any personnel outside the project team who have follow-up responsibilities for any of the action items. Also, include on the distribution list any appropriate management personnel, whether they be project team supervisors, manufacturing managers, or engineering managers. The documentation for the review should be archived in a process plant file with the appropriate records retention time (e.g., the life of the plant).

Review Follow-up   An important element (maybe the most important) of the review process is the follow-up to action items. The project review will result in a list of potential concerns and action items, but, without follow-up, the issues will never be resolved and implemented. A person(s) should be assigned to each action item, preferably at the time of the project review. The person(s) assigned should have a combination of knowledge, resources, and authority to do a proper job in following up on the action items (CCPS, 1993, p. 59). The total action item list should not be assigned to one person, since it may overwhelm one individual. Depending on the number of action items generated, prioritizing the action item list may be helpful and a responsibility the review team can assume.

Progress on the action items should be documented in periodic progress reports to the review team leader or others assigned that functional responsibility. If no one is assigned the responsibility of tracking this progress, completion of the action list will probably be relegated to a lower priority and not be done.

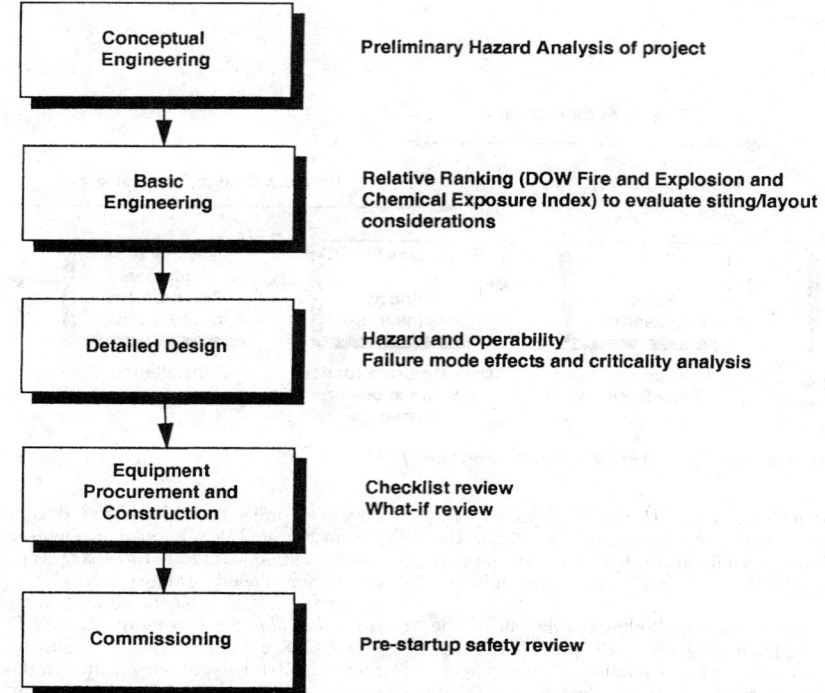

FIG. 26-11   Hazard evaluation at various project stages. (*CCPS, 1993, by permission of AIChE.*)

Changes made to the process as a result of the project review may require a similar review before implementation, especially if the change is significant.

Review Follow-up Verification   In addition to someone tracking the follow-up through progress reports, responsibility should be assigned to verify that any process changes were actually made in the field. This verification can be done by a review team as part of a process pre-start-up review. It could also be part of the project team management responsibility or assigned to a particular functional (i.e., safety and loss prevention) representative. The closure of the review process is complete once implementation is verified.

On rare occasions, the resolution of project review concerns or action items is a point of contention between review team and project team members. In such a case, a management structure must be in place to arbitrate such disputes.

Review Procedure Change Management   The project review process can require changes in policy and procedures at certain times. Therefore, the procedures should provide a management-of-change mechanism for suggesting changes and assign a person responsible for initiating and implementing any necessary changes.

Audit Process   Audits in the chemical process industry can be focused on process safety, process safety management, environmental, and health areas. The discussion in this section will focus on the process safety and process safety management area, but it should be recognized that the process can be applied to the other areas as well. "Process safety audits are intended to provide management with increased assurance that operating facilities and process units have been designed, constructed, operated, and maintained such that the safety and health of employees, customers, communities, and the environment are being properly protected" (CCPS, 1989, p. 133). Process safety management system audits "provide increased assurance that operating units have appropriate systems in place to manage process risk" (CCPS, 1989, p. 130). The audit process described in the following can be used to verify the implementation of equipment designs, operating and maintenance procedures, control systems, and management systems to meet the previously stated intentions.

The key steps in the audit process are outlined according to pre-audit activities, audit activities, and postaudit activities in Fig. 26-12. These activities are described in detail in *Guidelines for Auditing Process Safety Management Systems* (CCSP, 1993) and will be only briefly discussed in this section.

Preaudit Process   Prior to the actual on-site audit, some preliminary activities should take place. These activities include selecting the facilities to be audited, scheduling the audit, selecting the audit team, and planning the audit. The selection criteria may be random, based on potential hazards of the facilities or the value of the facilities from a business standpoint. Audit scheduling must account for the availability of key facility personnel and audit team members, operational mode of the facility (i.e., it should be in normal operation), and the lead time required to obtain background information that may require advance visits to the facility and preaudit interviews. The audit team members should possess the technical training and experience to understand the facilities being audited. They should be knowledgeable in the auditing process and in the appropriate regulations and standards that will apply to the facilities. They should also be impartial and objective about audit findings. The audit plan should define the audit scope (what parts of the facility will be covered, what topics, who will do it, etc.), develop an audit protocol that is a step-by-step guide to how the audit is performed, identify any priority topics for coverage, and develop an employee interview schedule.

On-Site Audit Process   An opening meeting with key facility personnel is held at which the audit team covers the objectives and approach for the audit, and the facility personnel provide an overview of the site operations including site safety rules and a site tour. The on-site audit process should then follow five basic steps that include: (1) understanding management systems, (2) evaluating management systems, (3) gathering audit information, (4) evaluating audit information, and (5) reporting audit findings (CCSP, 1993, p. 17).

An understanding of the management systems in place to control and direct the process safety of the facility can be obtained from reading engineering and administrative standards, guidelines, and procedures that should be available in the background information supplied

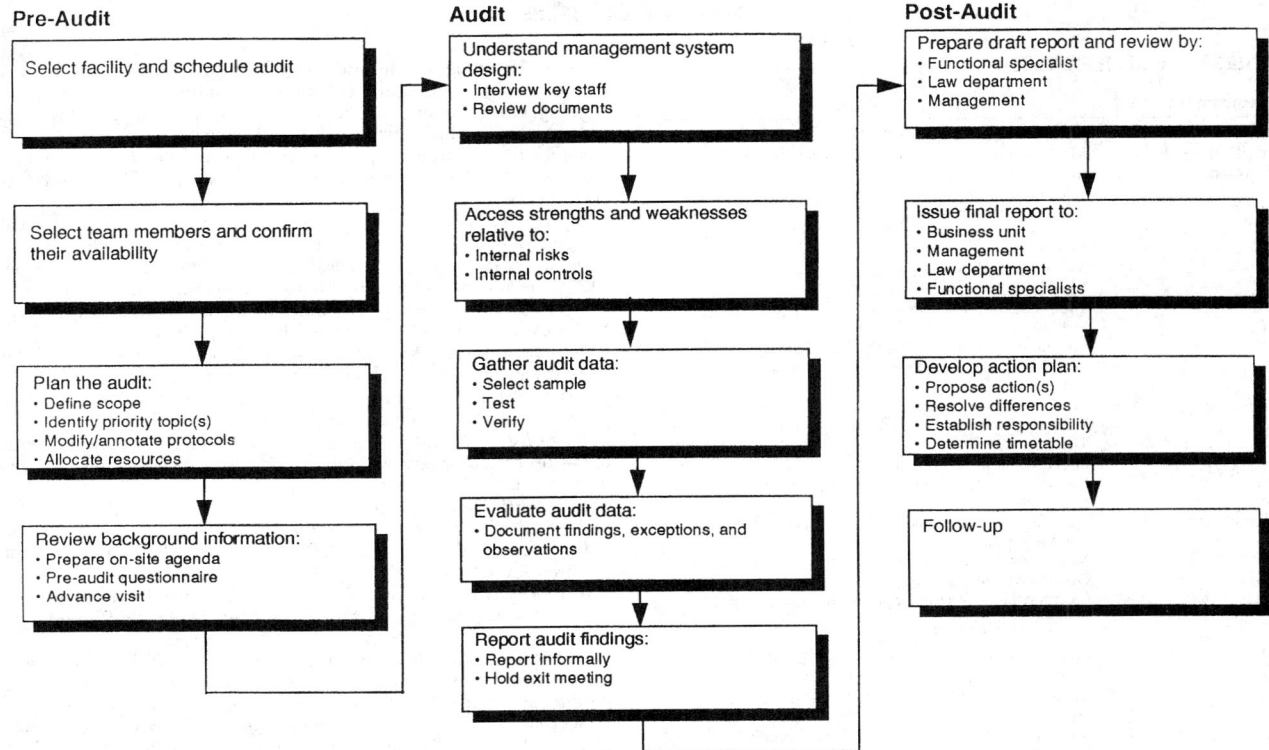

Pre-Audit

Select facility and schedule audit

↓

Select team members and confirm their availability

↓

Plan the audit:
- Define scope
- Identify priority topic(s)
- Modify/annotate protocols
- Allocate resources

↓

Review background information:
- Prepare on-site agenda
- Pre-audit questionnaire
- Advance visit

Audit

Understand management system design:
- Interview key staff
- Review documents

↓

Access strengths and weaknesses relative to:
- Internal risks
- Internal controls

↓

Gather audit data:
- Select sample
- Test
- Verify

↓

Evaluate audit data:
- Document findings, exceptions, and observations

↓

Report audit findings:
- Report informally
- Hold exit meeting

Post-Audit

Prepare draft report and review by:
- Functional specialist
- Law department
- Management

↓

Issue final report to:
- Business unit
- Management
- Law department
- Functional specialists

↓

Develop action plan:
- Propose action(s)
- Resolve differences
- Establish responsibility
- Determine timetable

↓

Follow-up

FIG. 26-12 Typical steps in the process safety management audit process. (*CCSP, 1993, by permission of AIChE.*)

prior to the on-site audit. Informal procedures and guidelines used by the facility may only be discovered in interviews with staff management and operations management. This understanding of the formal and informal management systems is a critical step in the audit process.

The next process step evaluates the process safety management systems to determine if they are adequate to achieve the desired results, and if they are used as intended. This evaluation is highly subjective on the auditor's part. This step sets the stage for the rest of the audit, guiding the auditor's information gathering and focusing attention on critical areas.

Gathering audit data can be accomplished through observations, documents, and interviews. The data obtained is used to verify and validate that the process safety management systems are implemented and functioning as designed. Data gathering can be aided by the use of audit samples, where a representative number of items are audited to draw a conclusion, and by using self-evaluation questionnaires.

The audit data can now be evaluated, resulting in audit findings (i.e., conclusions both positive and negative). The audit team should confirm that sufficient data has been collected to support each finding. Additional data may need to be gathered if the team decides a preliminary finding needs to be strengthened. The conclusions drawn from the data evaluation should be a team consensus.

The reporting step of the on-site audit should be planned to avoid any surprises to facility personnel. Reporting sessions should be held at the end of each audit day to inform facility personnel of the findings, clear up any misunderstandings of the data, and help redirect the audit team, if necessary. The on-site audit should end with a well-planned exit or closeout meeting between the audit team and facility personnel. All the findings of the audit team should be presented at this meeting. This verbal report is the opportunity for clarification of any ambiguities and determination of the final disposition of the findings (written audit report, for local attention only, etc.).

Postaudit Process The postaudit process consists of preparation of a draft report, preparation of a final report, development of action plans, and follow-up. A draft report of the audit findings should be prepared shortly after the completion of the on-site audit. The draft report usually undergoes review and comment by facility personnel involved with the audit, experienced auditors not involved with the subject audit, functional specialists, and attorneys. The review of the draft report is done to assure that a clear, concise, and accurate report is issued, and not to modify or change the findings. Once this review procedure is completed, a final report can be issued and distributed based on a distribution list provided by the facility personnel. The final audit report should be issued in a timely manner and meet the time requirement specified in the audit plan.

An action plan should be developed by the appropriate personnel of the audited facility to address any deficiencies stated in the audit report. Action plans should state what is to be done, who is responsible for getting it done, and when it is to be completed. Rationale for not taking any action for any of the stated deficiencies should also be documented. The action plan is an important step in closing the audit process.

It would not be unusual for some action plans to take a long time to complete. When extended implementation time is necessary, a follow-up mechanism should be used to document progress and show that an effort is being made to resolve the issues. Periodic (i.e., quarterly, semiannually) progress reports should be used as a follow-up method to ensure implementation. Future audits of the facility should include confirmation of the implementation of previous audit action plans.

The final audit report, action plans, progress reports, and any closure report should be retained by the facility based on the facility record retention policy. Typically, these items will be retained until future audit documentation replaces them. In some cases, audit records are retained for the life of the plant.

SAFETY DEVICES

PRESSURE RELIEF SYSTEMS

Nomenclature

C_p	Liquid heat capacity, Btu/lb-°F
Critical (or choked) condition	Maximum flow condition for compressible fluids
D	Duct diameter
F	Inlet mass flow rate, lb/min
f	Friction factor
G	Mass flux
g	Acceleration due to gravity
L	Flow length
M	Total mass in the equipment
P	Stream pressure
P_0	Stagnation pressure of the fluid (i.e., pressure under no-flow conditions)
P_1	Flowing pressure
Q	Heat input rate, Btu/min
R_v	Volume generation rate, ft³/min
t	Time
v	Specific volume of stream
v_f, v_g	Specific volumes of liquid and gas phases in the equipment
V_g, V_l	Specific volumes of gas and liquid phases, ft³/lb
V_r	Total equipment volume
W	Mass flow rate, lb/min
X_0	Quality, weight fraction vapor
X_r	Overall weight fraction vapor in the equipment

Greek letters	
β	Volumetric expansion coefficient, ft³/lb-°F
γ	C_p/C_v
ϕ	Angle of inclination from vertical
λ	Heat of vaporization, Btu/lb
ρ_f	Stream density entering the protected equipment, lb/ft³
ρ_w	Stream density entering the relief system, lb/ft³

Introduction All process designs should attempt to arrive at an inherently safe facility; that is, one from which a worst-case event cannot cause injury to personnel, damage to equipment, or harm to the environment. Incorporating safety features that are intrinsic (built-in) rather than extrinsic (added-on) to the basic design, together with the use of high-integrity equipment and piping, provide the first lines of defense against the dramatic, often catastrophic, effects of an overpressure and subsequent rupture. In recent years, many companies have incorporated the principles of *depressuring* or *instrumented shutdown* of key equipment as a means to control a release and avoid the actuation of pressure relief devices. This minimizes the probability of failure of the device, because, once used, the device may no longer be dependable. Since maintenance of relief devices can be sporadic, this redundancy provides yet another layer of safety. However, regardless of the number of lines of defense and depressuring systems in place, overpressure protection must still be provided. Emergency pressure relief systems are intended to provide the last line of protection and thus must be designed for high reliability, even though they will have to function infrequently.

Self-actuated pressure relief systems must be designed to limit the pressure rise which can occur as a result of overcompressing, overfilling, or overheating either an inert or a chemically reactive medium in a closed system. Pressure generation is usually the result of either expansion of a single-phase medium (by material addition and/or heating) or a shift of the phase equilibrium in a multiphase medium (as a result of composition and/or temperature changes). These mechanisms of pressure generation differ from what is commonly referred to as *explosion venting*. Events such as dust explosions and flammable vapor deflagrations propagate nonuniformly from a point of initiation, generating pressure or shock waves. Such venting problems are not included in these discussions.

Relief System Terminology Specific terminology has been developed for the various components which compose an emergency relief system. The American National Standards Institute (ANSI) def-initions pertaining to relief devices themselves are listed as follows. Special care is required to avoid confusion on the following terms:

relief valve A pressure relief valve set up for liquid flow. This device opens gradually over a pressure range to avoid "hammer."

safety (pop) valve A pressure relief valve set up for gas or vapor flow. This device opens over a narrow pressure range, with an initial "pop" action.

safety relief valve A pressure relief valve with mechanical design or adjustments to allow either relief or safety service.

backpressure Pressure existing at the outlet of a relief device. The value under no-flow conditions is *superimposed* backpressure. The value under flowing conditions consists of both superimposed backpressure and built-up pressure due to piping pressure drop.

conventional vs. balanced valves In conventional valves, the downstream side of the closing mechanism is exposed to the backpressure of the flowing fluid; in balanced valves, the closing mechanism is isolated from the fluid and open to atmosphere.

set point The inlet gauge pressure at which a device will start to open (or a rupture disk will burst) under service conditions of temperature and backpressure.

differential set pressure The difference between the upstream and downstream pressures at the set point.

blowdown The reduction in flowing pressure below the set point required for a device to close.

overpressure The rise of inlet pressure above the set point during relief flow, usually expressed as a percentage of the differential set pressure.

maximum allowable working pressure (MAWP) The maximum allowed pressure at the top of the vessel in its normal operating position at the operating temperature specified for that pressure.

Accumulation The rise of pressure above the MAWP of the protected system, usually expressed as a percentage of the gauge MAWP. *Note:* The **MAWP** and **accumulation** terms are not included in the ANSI definitions since they relate to the protected system instead of the relief device.

Codes, Standards, and Guidelines Industry practice is to conform to the applicable regulations, codes, and recommended practices. In many cases, these will provide different guidelines. A suggested approach would be to review all applicable codes, standards, and recommended practices prior to choosing a design basis. In addition to currently available material, the Center for Chemical Process Safety (CCPS), formed by the American Institute of Chemical Engineers, is continually developing guidelines and conducting research to further the general knowledge in emergency relief system design. The Design Institute for Emergency Relief Systems (DIERS) was established by AIChE to address sizing aspects of relief systems for two-phase, vapor-liquid flashing flow regimes. The DIERS Project Manual (Emergency Relief System Design Using DIERS Technology—1992) is the generally accepted industry standard for two-phase relief venting.

NFPA 30 and API Standard 2000 provide guidance for design of overpressure protection involving storage tanks that operate at or near atmospheric pressure. In particular, NFPA 30 focuses on flammability issues, while API 2000 addresses both pressure and vacuum requirements. The ASME code (Sections I and VIII) and API RP 520 are the primary references for pressure relief device sizing requirements.

Designers of emergency pressure relief systems should be familiar with the following list of regulations, codes of practice, and industry standards and guidelines.

API RP 520. *Sizing, Selection, and Installation of Pressure-Relieving Devices in Refineries.* Part I, Sizing and Selection, 5th ed., July 1990, and Part II, Installation, 3d ed., November 1988. American Petroleum Institute, Washington, D.C.

API RP 521. 1990. *Guide for Pressure-Relieving and Depressuring Systems,* 3d ed., American Petroleum Institute, Washington, D.C.

API STD 2000. 1992. *Venting Atmospheric and Low-Pressure Storage Tanks, Nonrefrigerated and Refrigerated.* American Petroleum Institute, Washington, D.C.

API RP 2001. 1984. *Fire Protection in Refineries.* American Petroleum Institute, Washington. D.C.

ASME. 1992. *Boiler and Pressure Vessel Code,* Section I, Power Boilers, and Section VIII, Pressure Vessels. American Society of Mechanical Engineers, New York.

ASME. 1988. *Performance Test Code PTC-25, Safety and Relief Valves.* American Society of Mechanical Engineers, New York.

CCPS. 1993. *Engineering Design for Process Safety.* American Institute of Chemical Engineers, New York.

DIERS. 1992. *Emergency Relief System Design Using DIERS Technology, DIERS Project Manual.* American Institute of Chemical Engineers, New York.

National Board of Boiler and Pressure Vessel Inspectors. 1992. *Pressure Relieving Device Certification (Red Book),* National Board of Boiler and Pressure Vessel Inspectors, Columbus, Ohio.

NFPA 30. 1990. *Flammable and Combustible Liquids Code.* National Fire Protection Association, Quincy, Mass.

Relief Design Scenarios   The most difficult part of designing an adequate emergency pressure relief system is determining the emergency events (credible design scenarios) for which to design. The difficulty arises primarily because the identification of credible design scenarios usually involves highly subjective judgments, which are often influenced by economic situations. Unfortunately, there exists no universally accepted list of credible design scenarios. Relief systems must be designed for the credible chain of events that results in the most severe venting requirements (worst credible scenario). Credibility is judged primarily by the number and the time frame of causative failures required to generate the postulated emergency. Only totally independent equipment or human failures should be considered when judging credibility. A failure resulting from another failure is an effect, rather than an independent causative factor. A suggested guideline for assessing credibility as a function of the number and time frames of independent causative events is:

- Any single failure is *credible.*
- Two or more simultaneous failures are *not credible*
- Two events in sequence are *credible.*
- Three or more events in sequence are *not credible.*

The first step in scenario selection is to identify all the credible emergencies using the preceding guidelines (or a similar set). This is perhaps best accomplished by identifying all the possible sources of pressure and vacuum. Table 26-7 lists a number of commonly existing pressure and vacuum sources.

Fire   The main consequence of fire exposure is heat input causing thermal expansion, vaporization, or thermally induced decomposition resulting in a pressure rise. An additional result of fire exposure is the possibility of overheating the wall of the equipment in the vapor space where the wall is not cooled by the liquid. In this case, the vessel wall may fail due to the high temperature, even though the relief system is operating. Guidelines for estimating the heat input from a fire are found in API recommended practices, NFPA 30 (for bulk storage tanks), OSHA 1910.106, and corporate engineering standards. In determining the heat input from fire exposure, NFPA allows credit for application of water spray to a vessel; API allows no such credit.

Pressure vessels (including heat exchangers and air coolers) in a plant handling flammable fluids are subject to potential exposure to external fire. A vessel or group of vessels which could be exposed to a pool fire must be protected by a pressure relief device. Additional protection to reduce the device relief load can be provided by insulation, water spray, or remote-controlled depressuring devices. Plant layout should consider spacing requirements, such as those set forth by NFPA, API, Industrial Risk Insurers, or Factory Mutual, and must include accessibility for fire-fighting personnel and equipment. Several pieces of equipment located adjacent to each other that cannot be isolated by shutoff valves can be protected by a common relief device, providing the interconnecting piping is large enough to handle the required relief load.

Operational Failures   A number of scenarios of various operational failures may result in the generation of overpressure conditions:

- *Blocked outlet.* Operation or maintenance errors (especially following a plant turnaround) can block the outlet of a liquid or vapor stream from a piece of process equipment, resulting in an overpressure condition.
- *Opening a manual valve.* Manual valves which are normally closed to isolate two or more pieces of equipment or process streams can be inadvertently opened, causing the release of a high-pressure stream or resulting in vacuum conditions. Other effects may include the development of critical flows, flashing of liquids, or the generation of a runaway chemical reaction.
- *Cooling water failure.* The loss of cooling water is one of the more commonly encountered causes of overpressurization. Two examples of the critical consequences of this event are the loss of condensing duty in column overhead systems and the loss of cooling for compressor seals and lube oil systems. Different scenarios should be considered for this event, depending on whether the failure affects a single piece of equipment (or process unit) or is plantwide.
- *Power failure.* The loss of power will shut down all motor-driven rotating equipment, including pumps, compressors, air coolers, and vessel agitators. As with cooling water failure, power failure can have a negative cascading effect on other equipment and systems throughout the plant.
- *Instrument air failure.* The consequences of the loss of instrument air should be evaluated in conjunction with the failure mode of the control valve actuators. It should not be assumed that the correct air failure response will occur on these control valves, as some valves may stick in their last operating position.
- *Thermal expansion.* Equipment and pipelines which are liquid-full under normal operating conditions are subject to hydraulic expansion if the temperature increases. Common sources of heat that can result in high pressures due to thermal expansion include solar radiation, steam or other heated tracing, heating coils, and heat transfer from other pieces of equipment.
- *Vacuum.* Vacuum conditions in process equipment can develop due to a wide variety of situations including:

> Instrument malfunction
> Draining or removing liquid with venting
> Shutting off purge steam without pressuring with noncondensable vapors
> Extreme cold ambient temperatures resulting in subatmospheric vapor pressures
> Water addition to vessels that have been steam-purged

If vacuum conditions can develop, then either the equipment must be designed for vacuum conditions or a vacuum relief system must be installed.

Equipment Failure   Most equipment failures that can lead to overpressure situations involve the rupture or break of internal tubes inside heat exchangers and other vessels and the failure of valves and regulators. Heat exchangers and other vessels should be protected with a relief system of sufficient capacity to avoid overpressure in case of internal failure. Characterization of the types of failure and the design of the relief system are left to the discretion of the designer. API RP 520 presents guidance in determining these requirements, including criteria for deciding when a full tube rupture is likely. In cases involving the failure of control valves and regulators, it is important to evaluate both the fail-open and fail-closed positions.

TABLE 26-7   Common Sources of Pressure and Vacuum

Heat Related
- Fire
- Out-of-control heaters and coolers
- Ambient temperature changes
- Runaway chemical reactions

Equipment and Systems
- Pumps and compressors
- Heaters and coolers
- Vaporizers and condensers
- Vent manifold interconnections
- Utility headers (steam, air, water, etc.)

Physical Changes
- Gas absorption (e.g., HCl in water)
- Thermal expansion
- Vapor condensation

Runaway Reactions   Runaway temperature and pressure in process vessels can occur as a result of many factors, including loss of cooling, feed or quench failure, excessive feed rates or temperatures, contaminants, catalyst problems, and agitation failure. Of major concern is the high rate of energy release and/or formation of gaseous products, which may cause a rapid pressure rise in the equipment. In order to properly assess these effects, the reaction kinetics must either be known or obtained experimentally.

Pressure Relief Devices   The most common method of overpressure protection is through the use of safety relief valves and/or rupture disks which discharge into a containment vessel, a disposal system, or directly to the atmosphere (Fig. 26-13). Table 26-8 summarizes some of the device characteristics and the advantages.

Safety Relief Valves   *Conventional* safety relief valves (Fig. 26-14) are used in systems where built-up backpressures typically do not exceed 10 percent of the set pressure. The spring setting of the valve is reduced by the amount of superimposed backpressure expected. Higher built-up backpressures can result in a complete loss of continuous valve capacity. The designer must examine the effects of other relieving devices connected to a common header on the performance of each valve. Some mechanical considerations of conventional relief valves are presented in the ASME code; however, the manufacturer should be consulted for specific details.

Balanced safety relief valves may be used in systems where built-up and/or superimposed backpressure is high or variable. In general, the capacity of a balanced valve is not significantly affected by backpressures below 30 percent of set pressure. Most manufacturers recommend keeping the backpressure on balanced valves below 45 to 50 percent of the set pressure.

Pilot-Operated Relief Valves   In a pilot-operated relief valve, the main valve is combined with and controlled by a smaller, self-actuating pressure relief valve. The pilot is a spring-loaded valve that senses the process pressure and opens the main valve by lowering the pressure on the top of an unbalanced piston, diaphragm, or bellows of the main valve. Once the process pressure is lowered to the blowdown pressure, the pilot closes the main valve by permitting the pressure in the top of the main valve to increase. Pilot-operated relief valves are commonly used in clean, low-pressure services and in services where a large relieving area at high set pressures is required. The set pressure of this type of valve can be close to the operating pressure. Pilot-operated valves are frequently chosen when operating pressures are

TABLE 26-8   Summary of Device Characteristics

	Reclosing devices		Nonreclosing devices
	Relief valves	Disk-valve combinations	Rupture disks
Fluid above normal boiling point	+	+	−
Toxic fluids	+	+	−
Corrosive fluids	−	+	+
Cost	−	−	+
Minimum pipe size	−	−	+
Testing and maintenance	−	−	+
Won't fatigue and fail low	+	+	−
Opens quickly and fully	−	−	+

NOTE: + indicates advantageous
        − indicates disadvantageous

within 5 percent of set pressures and a close tolerance valve is required.

Rupture Disks   A rupture disk is a device designed to function by the bursting of a pressure-retaining disk (Fig. 26-15). This assembly consists of a thin, circular membrane usually made of metal, plastic, or graphite that is firmly clamped in a disk holder. When the process reaches the bursting pressure of the disk, the disk ruptures and releases the pressure. Rupture disks can be installed alone or in combination with other types of devices. Once blown, rupture disks do not reseat; thus, the entire contents of the upstream process equipment will be vented. Rupture disks are commonly used in series (upstream) with a relief valve to prevent corrosive fluids from contacting the metal parts of the valve. In addition, this combination is a reclosing system.

The burst tolerances of rupture disks are typically about ±5 percent for set pressures above 40 psig.

Pressure-Vacuum Relief Valves   For applications involving atmospheric and low-pressure storage tanks, pressure-vacuum relief valves (PVRVs) are used to provide pressure relief. These units combine both a pressure and a vacuum relief valve into a single assembly that mounts on a nozzle on top of the tank and are usually sized to handle the normal in-breathing and out-breathing requirements. For emergency pressure relief situations (e.g., fire), ERVs are used. API RP 520 and API STD 2000 can be used as references for sizing.

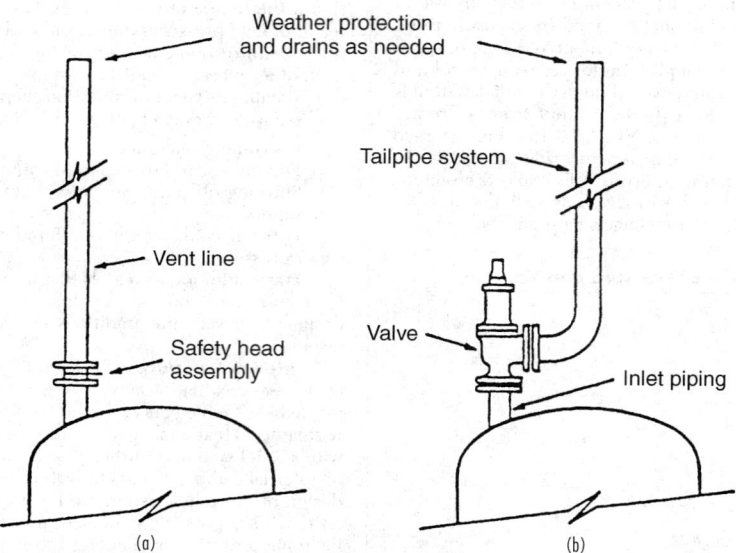

FIG. 26-13   Typical pressure relief system configurations: (*a*) rupture disk system; (*b*) pressure relief valve system.

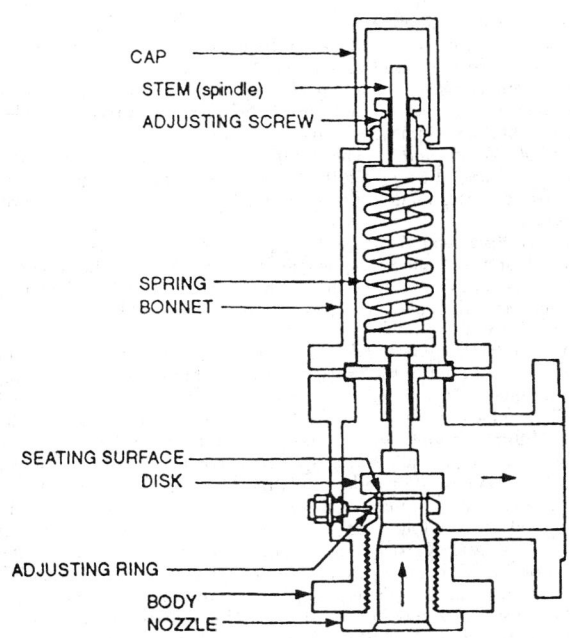

FIG. 26-14 Typical conventional pressure relief valve.

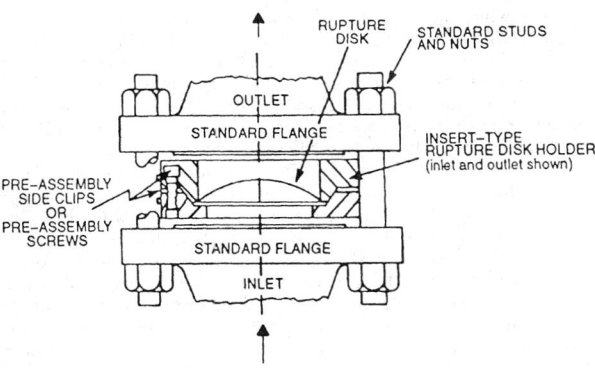

FIG. 26-15 Typical rupture disk assembly.

Sizing of Pressure Relief Systems A critical point in design is determining whether or not the relief system must be sized for single-phase or two-phase relief flow. Two-phase flow frequently occurs during a runaway reaction, but it can also occur in nonreactive systems such as vessels with gas-spargers, vessels experiencing high heat input rates, or systems containing known foaming agents such as latex. In 1976, the Design Institute for Emergency Relief Systems (DIERS) was formed to develop methods for the design of emergency relief systems to handle runaway reactions. The DIERS group consisted of a consortium of 29 companies under the auspices of the American Institute of Chemical Engineers. Of particular interest were the prediction of when two-phase flow venting would occur and the applicability of various sizing methods for two-phase vapor-liquid flashing flow situations. The most significant theoretical and experimental finding of the DIERS program is the ease with which two-phase vapor-liquid flow can occur during an emergency relief situation and the requirement for a much larger (by two to ten times) relief system. The DIERS methodology is important as a means of addressing situations, such as two-phase flow, not covered adequately by ASME and API methods. The *DIERS Project Manual* (DIERS, 1992) is the best source of detailed information on these methods.

Required Relief Rate The required relief rate is the venting rate required to remove the volume being generated within the protected equipment when the equipment is at its highest allowed pressure:

$$W_{required} = \frac{\text{net volume generation rate}}{\text{specific volume of vent stream}} \qquad (26\text{-}21)$$

The required relief rate is constant only if both the numerator and denominator in Eq. (26-21) are constant. If the conditions within the protected equipment (temperature, composition, etc.) or the composition and/or quality of the vent stream are changing, then the required relief rate as defined above represents the instantaneous required relief rate. For steady-state design scenarios, the required relief rate, once determined, provides the capacity information that is required to properly size the relief device and associated piping. For situations that are transient in nature (e.g., the venting of a vessel exposed to fire), the required relief rate will be continually changing as the equipment is emptied. In these situations the design should be done based on finding the required vent area that will keep the pressure in the protected equipment at (or below) the maximum allowed accumulation during the entire venting period. These cases will require the simultaneous solution of the applicable material and energy balances around the protected equipment.

Constant Flow into Protected Equipment For the steady-state design scenario with a constant, steady flow of fluid from a pressure source that is above the maximum allowed pressure in the protected equipment, volume is being generated within the equipment at a rate $RV = F/\rho_f$. Substituting into Eq. (26-21) and noting that the specific volume of the vent stream is $1/\rho_w$ gives the required mass flow rate:

$$W = \frac{F\rho_w}{\rho_f} \qquad (26\text{-}22)$$

where W = required mass flow rate, lb/min
 F = inlet mass flow rate, lb/min
 ρ_w = stream density entering the relief system, lb/ft³
 ρ_f = stream density entering the protected equipment, lb/ft³

Constant Energy Input into Protected Equipment If the design scenario involves a constant flow of energy (heat) into the protected equipment, then the required flow rate calculation involves determining whether or not a phase change (boiling) is occurring. If the addition of heat to the equipment does not cause the fluid to boil, then the volume generation rate is the thermal expansion rate of the fluid:

$$R_v = \frac{\beta Q}{C_p} \qquad (26\text{-}23)$$

where R_v = volume generation rate, ft³/min
 β = volumetric expansion coefficient, ft³/lb-°F
 Q = heat input rate, Btu/min
 C_p = liquid heat capacity, Btu/lb-°F

Combining Eqs. (26-23) and (26-22) gives the required relief rate:

$$W = \frac{\rho_w \beta Q}{C_p} \qquad (26\text{-}24)$$

If the fluid is at its boiling point, then volume is generated through the phase change that occurs upon vaporization:

$$R_v = \frac{[Q(V_g - V_l)]}{\lambda} \qquad (26\text{-}25)$$

where V_g, V_l = specific volumes of gas and liquid phases, ft³/lb
 λ = heat of vaporization, Btu/lb

As before, the required relief rate becomes:

$$W = \frac{[\rho_w Q(V_g - V_l)]}{\lambda} \qquad (26\text{-}26)$$

Transient Material and Energy Balances The relief rate requirement at any instant during any event is developed on the basis that the total volume of vapor plus liquid is just equal to the vessel volume. In differential form, this condition is equivalent to setting the volumetric vent rate equal to the rate of volume increase in the protected equipment at any instant. The development of the relief rate criterion, which relates the rate of vapor generation to the venting rate, is based on the assumption that the equipment geometry is such that the temperature and pressure will be reasonably uniform throughout the contents, with negligible composition gradients within the phases. In addition, the time scale of the relief event is assumed to be small enough that influence of any feed streams is insignificant relative to the venting stream. Under these assumptions the material balance around the protected equipment is given by:

$$\frac{dX_r}{dt} = \frac{\left(\dfrac{WV_r}{M^2} - (1 - X_r)\dfrac{dv_f}{dt} - X_r\dfrac{dv_g}{dt}\right)}{v_g - v_f} \qquad (26\text{-}27)$$

where V_r = total equipment volume
        X_r = overall weight fraction vapor in the equipment
        M = total mass in the equipment
        v_f, v_g = specific volumes of liquid and gas phases in the equipment
        W = mass vent rate
        t = time

The energy balance on the equipment is developed for conditions under which thermal mixing within the vessel is sufficient to allow the properties of all portions of the liquid and vapor phases to be characterized adequately by a single value of temperature. It is assumed also that pressure gradients within the vessel are small with respect to the pressure level so that a single value of pressure may be assigned to the contents. The resulting expression for the incompressible liquid–ideal gas case is given by Huff ("Emergency Venting Requirements," *Plant/Operations Progress*, October 1982, p. 212):

$$\frac{dT}{dt}\left[X_r(C_{pg} - R) + (1 - X_r)\left(C_{pf} - T\frac{dv_f}{dT}\frac{dP}{dT}\right)\right]$$

$$= Q - [\lambda - P(v_g - v_f)]\frac{dX_r}{dt} - \frac{W}{M}\{[X_0 - X_r][\lambda - P(v_g - v_f)]$$

$$+ P[X_0v_g + (1 - X_0)v_f]\} \quad (26\text{-}28)$$

where T and P are the temperature and pressure of the contents, C_{pg} and C_{pf} are the gas and liquid specific heats, λ is the latent heat of vaporization, R is the universal gas constant, and Q is the rate of heat addition to the equipment contents. The solution of Eq. (26-28) requires a value of the quality of the vent stream as it leaves the protected equipment (X_0). Limiting cases are $X_0 = 1$ (all-vapor venting) and $X_0 = 0$ (all-liquid venting). If the venting process is such that no vapor-liquid disengagement occurs, then the criterion for top venting becomes $X_0 = X_r$, where X_r is the mass fraction vapor in the equipment.

The time-dependent nature of the emergency pressure relieving event is obtained by the simultaneous solution of Eqs. (26-27) and (26-28). Generally, the only unknown parameters in these two equations are the venting rate W and the vent stream quality (X_0). The vent rate W at any instant is a function of the upstream conditions and the relief system geometry.

Vessel Flow Models and the Coupling Equation   In order to evaluate the quantity of vapor entering the vent system at any instant (X_0), one must consider the dynamics of vapor disengagement that occur in the top of the protected equipment. Based on experience gained in the DIERS program, a number of vapor-liquid disengagement models have been formulated. These models estimate the liquid swell (i.e., the degree of vapor-liquid disengagement) as a function of vapor throughput. The key model parameters include the average void fraction in the swelled liquid, the vapor superficial velocity at the liquid surface, and the characteristic bubble rise velocity. The vessel flow models used in the DIERS program are listed as follows in order of increasing vapor-liquid disengagement.

Homogeneous Vessel Model   This model assumes that no vapor-liquid disengagement occurs in the protected equipment; thus, the vapor mass fraction entering the vent system (X_0) will be the same as the average vapor mass fraction in the equipment (X_r). This model is used to approximate the vessel conditions when the vessel contents are extremely viscous or foamy. The specification $X_0 = X_r$ has come into rather wide use as a conservative but realistic basis for taking account of the two-phase venting phenomena.

Bubbly Vessel Model   The bubbly vessel model assumes uniform vapor generation throughout the liquid with limited disengagement in the vessel. In this model, the liquid phase is continuous with discrete bubbles.

Churn-Turbulent Vessel Model   The churn-turbulent vessel model is also based on uniform vapor generation throughout the liquid but with considerable vapor-liquid disengagement. The liquid phase is continuous with coalesced vapor regions of increased size relative to the bubble vessel model.

Nonboiling Height Model   This model applies the churn-turbulent assumptions to only a top portion of the fluid in the protected equipment. Below this portion, boiling does not occur and there is no liquid swell. The location of this nonboiling height is estimated from a balance of the hydrostatic effects and the recirculation effects.

The coupling equation is a vapor mass balance written at the vent system entrance and provides a relationship between the vent rate W and the vent system inlet quality X_0. The relief system flow models described in the following section provide a second relationship between W and X_0 to be solved simultaneously with the coupling equation. Once W and X_0 are known, the simultaneous solution of the material and energy balances can be accomplished. For all the preceding vessel flow models and the coupling equations, the reader is referred to the *DIERS Project Manual* for a more complete and detailed review.

Vent System Flow Capacity   The mass flow rate W through a given vent system geometry, in general, requires a trial-and-error approach when the system configuration contains more than a single diameter. The generalized approach is to assume a flow rate W and calculate the resulting pressure profiles down the system until the final discharge pressure matches the specified value. If choked flow is encountered at any point in the system, then the system must be broken into two or more separate systems and each treated independently while preserving the mass flow rate through each.

The presence of both liquid and vapor phases in the vent stream is normally treated as a vapor-liquid mixture at equilibrium conditions. The adiabatic flashing of the stream as the pressure falls along the flow path is usually computed by conventional flash distillation methods. In principle, the flash path should be isentropic for flow in devices exhibiting low friction losses (nozzles and short pipes). For friction flow, the sum of the stream enthalpy, kinetic energy, and potential energy is held constant along the path. In practice, little error is introduced by carrying out the flash computations at constant enthalpy. With this simplification, the flash temperature-pressure-composition history can be established before starting the actual flow calculations, thus eliminating the need for repetitive flash calculations at each step in the integration.

The treatment of vent flow calculations in most typical relief system configurations involves two classes of computational models: flow in low-friction geometries such as nozzles and frictional flow in pipes and fittings.

Ideal (Frictionless) Flow in Nozzles   The flow path in well-formed nozzles follows smoothly along the nozzle contour without separating from the wall. The effects of small imperfections and small frictional losses are accounted for by correcting the ideal nozzle flow by an empirically determined coefficient of discharge. The acceleration of a fluid initially at rest to flowing conditions in an ideal nozzle is given by:

$$-\frac{G^2v^2}{2} = \int_{P_0}^{P_1} v\, dP \qquad (26\text{-}29)$$

where P_0 is the stagnation pressure of the fluid (i.e., the pressure under no-flow conditions), P_1 is the flowing pressure, G is the mass flux, and v is the fluid specific volume. If the fluid is compressible, the flow will increase to a maximum value as the downstream pressure P_1 is reduced and any further decrease in the downstream pressure will not affect the flow. This maximum flow condition is referred to as the *critical* (or *choked*) condition. At this condition, the maximum mass flux is

$$G_{\max} = \sqrt{\frac{-1}{dv/dp}} \qquad (26\text{-}30)$$

Pipe Flow For steady-state flow through a constant diameter duct, the mass flux G is constant and the governing steady-state momentum balance is:

$$vdP + G^2\left[vdv + \left(\frac{4fv^2}{2D}\right)dL\right] + g\cos\phi\, dL = 0 \qquad (26\text{-}31)$$

where
- G = mass flux
- v = specific volume of stream
- P = stream pressure
- f = friction factor
- D = duct diameter
- L = flow length
- g = acceleration due to gravity
- ϕ = angle of inclination from vertical

Equation (26-31) can be integrated directly to yield the mass flux G, provided that D, L, f, and ϕ are known, as well as the relationship between pressure and volume. For all-vapor cases, the expansion of the vapor is usually assumed to follow the form Pv^γ = constant ($\gamma = C_p/C_v$) and thus the momentum equation can be analytically integrated. Similarly, for all-liquid (nonflashing) flow, the stream specific volume is usually assumed to be constant, thus also providing a direct analytical integration of Eq. (26-31). For two-phase flashing flow, the requisite p-v relationship is usually obtained from flash calculations, and normally requires a numerical integration of Eq. (26-31). In addition to calculating the flow rates through sections of piping in the relief system, there may also exist additional pressure drop constraints in both the inlet and outlet piping if the relief device is a PRV. The designer is referred to the ASME and API references for further information.

A number of papers have explored methods for the solution of Eqs. (26-29) and (26-31), especially for the two-phase conditions. The reader is referred to the *DIERS Project Manual* for a more detailed review and list of appropriate references and available computer programs.

EMERGENCY RELIEF DEVICE EFFLUENT COLLECTION AND HANDLING

Nomenclature (consistent English or SI units)

A_v	Vapor flow area, ft^2
C_q	Specific heat of the quench fluid
C_R	Specific heat of the reactants
D	Drum diameter, ft
G_v	Superficial vapor mass flux, lb/s · ft^2
k	Capacity coefficient
L	Drum length, ft
M	Molecular weight of vapor
m_o	Mass of reactants
P	Pressure in the drum, psia
Q_v	Vapor flow rate, ft^3/s
r	Volumetric vapor flow rate/volumetric liquid flow rate
T	Temperature of the vapor, °R
T_a	Allowable temperature following complete quench
T_o	Initial temperature of the quench fluid
T_R	Temperature of reactants at relief set pressure
U_a	Allowable vapor velocity, ft/s
V_L	Drum liquid volume, ft^3
V_v	Superficial velocity, ft/s
W	Vapor flow rate, lb/h

Greek letters	
ρ_L	Liquid density, lb/ft^3
ρ_v	Vapor density, lb/ft^3

GENERAL REFERENCES: API Report 521, *Guide for Pressure Relieving and Depressurizing Systems*, American Petroleum Institute, Washington, D.C., March 1997. AIChE-CCPS, *Guidelines for Pressure Relief and Effluent Handling Systems*, AIChE, New York, 1997. DIERS, *Emergency Relief System Design Using DIERS Technology*, AIChE, New York, 1992. Fthenakis, *Prevention and Control of Accidental Releases of Hazardous Gases*, Van Nostrand-Reinhold, New York, 1993. Grossel and Crowl, *Handbook of Highly Toxic Materials Handling and Management*, Marcel Dekker, New York, 1995. Grossel, *Journal of Loss Prevention in the Process Industries* **3**(1): 112–124, 1990. Grossel, *Plant/Operations Progress* **5**(3): 129–135, 1986. Keiter, A. G., *Plant/Operations Progress* **11**(3): 157–163, 1992.

Introduction In determining the disposal of an effluent vent stream from an emergency relief device (safety valve or rupture disk), a number of factors must be considered, such as:

1. Is the stream single-phase (gas or vapor) or multiphase (vapor-liquid or vapor-liquid-solid)?
2. Is the stream flammable or prone to deflagration?
3. Is the stream toxic?
4. Is the stream corrosive to equipment or personnel?

Some vent streams, such as light hydrocarbons, can be discharged directly to the atmosphere even though they are flammable and explosive. This can be done because the high-velocity discharge entrains sufficient air to lower the hydrocarbon concentration below the lower explosive limit (API RP 521, 1997). Toxic vapors must be sent to a flare or scrubber to render them harmless. Multiphase streams, such as those discharged as a result of a runaway reaction, for example, must first be routed to separation or containment equipment before final discharge to a flare or scrubber.

These matters are organized into three major divisions: the types of equipment, the criteria employed in the selection of equipment, and the sizing and design of the equipment.

Types of Equipment The three most commonly used types of equipment for handling emergency relief device effluents are *blowdown drums* (also called *knockout drums* or *catch tanks*), *cyclone vapor-liquid separators*, and *quench tanks* (also called *passive scrubbers*). These are described as follows.

Horizontal Blowdown Drum/Catch Tank This type of drum, shown in Fig. 26-16, combines both the vapor-liquid separation and holdup functions in one vessel. Horizontal drums are commonly used where space is plentiful, such as in petroleum refineries and petrochemical plants. The two-phase mixture usually enters at one end and the vapor exits at the other end. For two-phase streams with very high vapor flow rates, inlets may be provided at each end, with the vapor outlet at the center of the drum, thus minimizing vapor velocities at the inlet and aiding vapor-liquid separation.

Cyclone Separator with Separate Catch Tank This type of blowdown system, shown in Fig. 26-17 and 26-18, is frequently used in chemical plants where plot plan space is limited. The cyclone performs the vapor-liquid separation, while the catch tank accumulates the liquid from the cyclone. This arrangement allows location of the cyclone knockout drum close to the reactor so that the length of the relief device discharge line can be minimized. The cyclone has internals, vital to its proper operation, which will be discussed in the following sections.

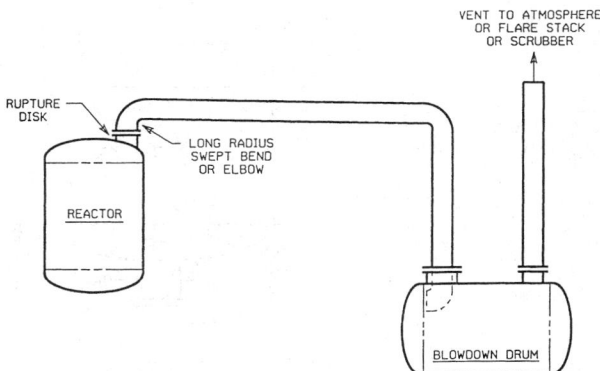

FIG. 26-16 Horizontal blowdown drum.

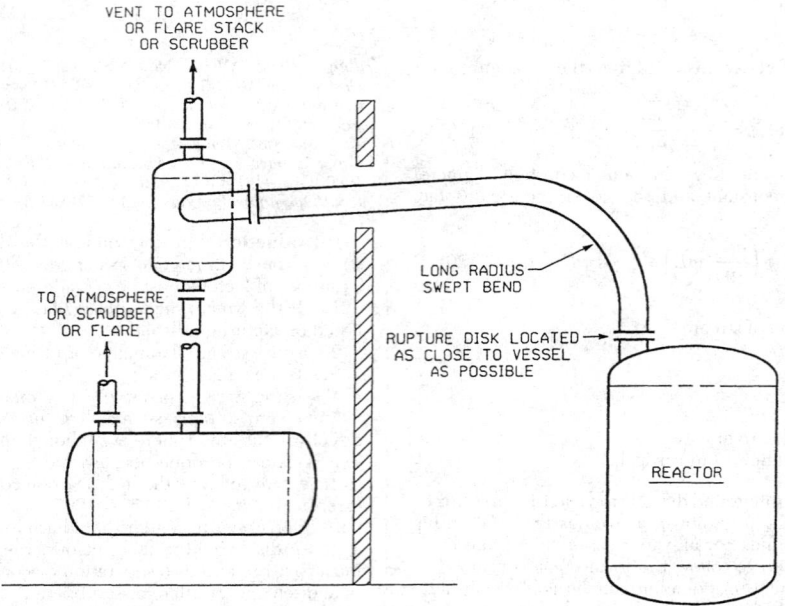

FIG. 26-17 Cyclone separator with separate liquid catch tank.

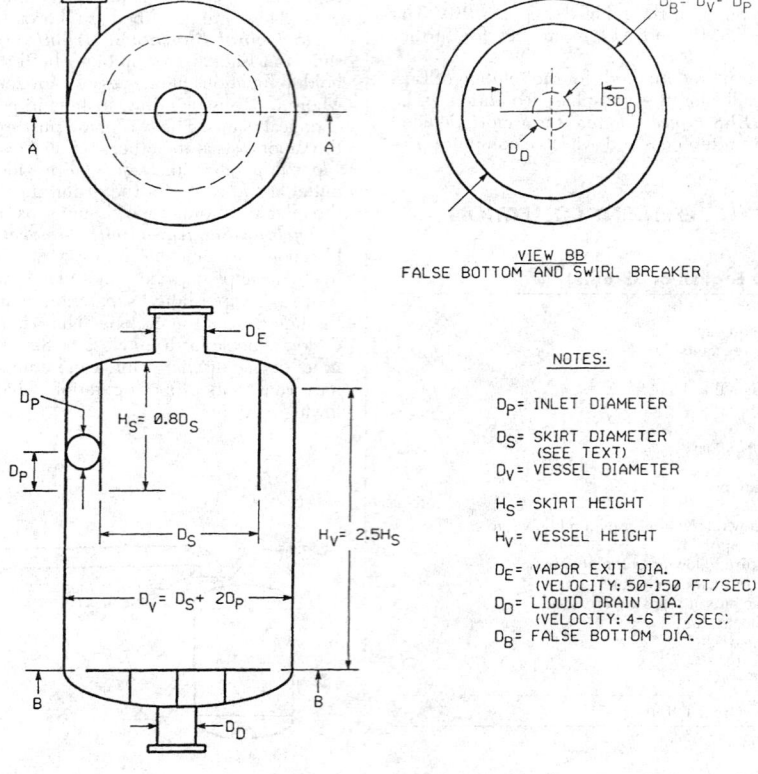

VIEW BB
FALSE BOTTOM AND SWIRL BREAKER

$D_B = D_V - D_P$

$3D_D$

D_D

NOTES:

D_P= INLET DIAMETER

D_S= SKIRT DIAMETER
(SEE TEXT)
D_V= VESSEL DIAMETER

H_S= SKIRT HEIGHT

H_V= VESSEL HEIGHT

D_E= VAPOR EXIT DIA.
(VELOCITY: 50-150 FT/SEC)
D_D= LIQUID DRAIN DIA.
(VELOCITY: 4-6 FT/SEC)
D_B= FALSE BOTTOM DIA.

$H_S = 0.8D_S$

$H_V = 2.5H_S$

$D_V = D_S + 2D_P$

SECTION A-A

FIG. 26-18 Cyclone separator design details.

Cyclone Separator with Integral Catch Tank   This type of containment system, depicted in Fig. 26-19, is similar to the aforementioned type, except that the knockout drum and catch tank are combined in one vessel shell. This design is used when the vapor rate is quite high so that the knockout drum diameter is large.

Quench Tank/Catch Tank   This type of system, as shown in Figs. 26-20 and 26-21, is used when it is desired to remove condensable vapors from a flammable or toxic vent mixture by passing them through a pool of liquid in a vessel. This arrangement often obviates the need for a subsequent scrubber and/or flare stack. The design of the quencher arm is critical to efficient condensation and avoidance of water hammer. Figure 26-20 is the more conventional passive-type quenching pool used in the chemical and nuclear industry. The type shown in Fig. 26-21, with a superimposed baffle-plate section, is used when complete condensation of the incoming vapors is not expected. The exiting vapors are usually cooled to 150 to 200°F in the baffle plate section. This type is often used in petroleum refineries.

Multireactor Knockout Drum/Catch Tank   This interesting system, depicted in Fig. 26-22, is sometimes used as the containment vessel for a series of closely spaced reactors (Speechly et al., "Principles of Total Containment System Design," presented at *I. Chem. E North West Branch Meeting*, 1979). By locating the drum as shown in Fig. 26-22, minimum-length vent lines can be routed directly to the vessel without any bends.

Equipment Selection Criteria and Guidelines   A number of factors should be considered in order to determine when to select a blowdown drum, cyclone separator, or quench tank to handle a multiphase stream from a relief device. Among these are the plot plan space available, the operating limitations of each type, and the physicochemical properties of the stream.

The criteria for application and performance characteristics of blowdown drums, cyclone separators, and quench tanks are discussed as follows.

Horizontal Blowdown Drums (Catch Tanks)
Applications:
1.   Inlet liquid loading is greater than 20 wt % based on gas flow rate.
2.   They can be used for viscous and/or fouling service.
Performance Characteristics:
1.   Residual entrainment is in the range of tenths to a few percent.
2.   Pressure drop is usually very low.
3.   Efficiency of separation is weakly dependent on the size of the vessel.
4.   They are usually able to separate droplets 300 μm and larger.
Cyclones
Applications:
1.   They can handle liquids with low to moderate viscosity.
2.   Some fouling is acceptable.
3.   Inlet liquid loading is generally less than 20 wt % based on the gas flow rate, but higher loadings are sometimes possible.
Performance Characteristics:
1.   They have higher separation efficiency than a horizontal knockout drum.
2.   Pressure drop is higher than that of a horizontal knockout drum.
Quench Tanks
Applications:
1.   They can handle liquids with low to high viscosity.
2.   They can handle liquids with moderate solids loading.
3.   They can handle high liquid loading—actually no limit, as vessel can be sized to contain all the liquid.

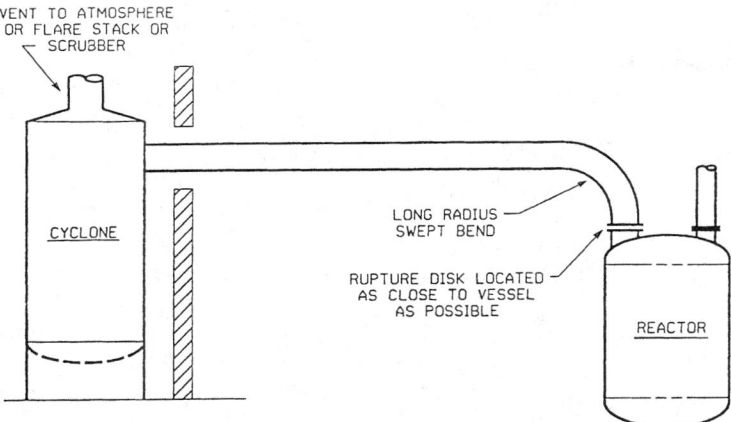

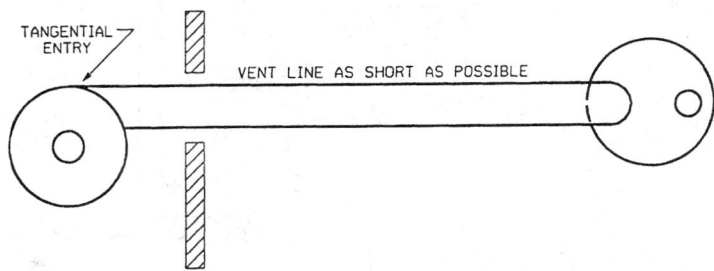

FIG. 26-19   Cyclone separator with integral catch tank.

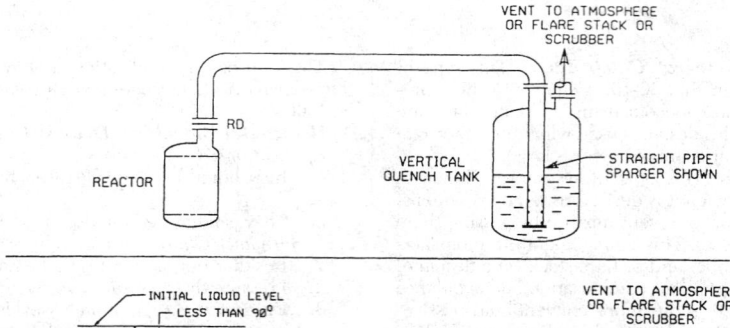

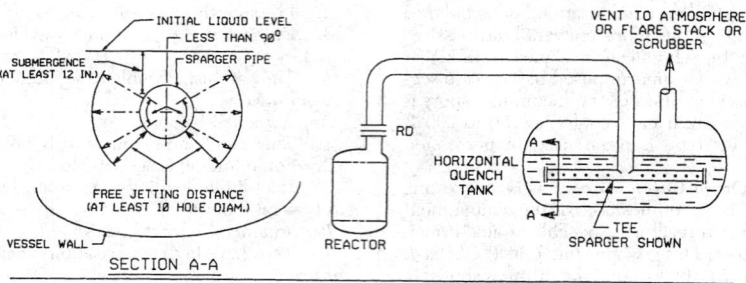

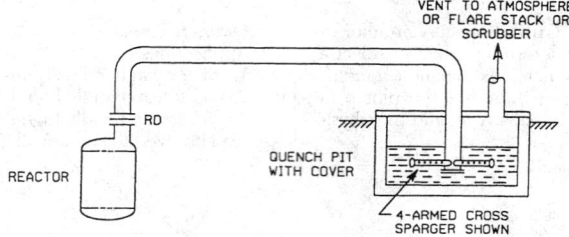

FIG. 26-20   Quench tank/catch tank.

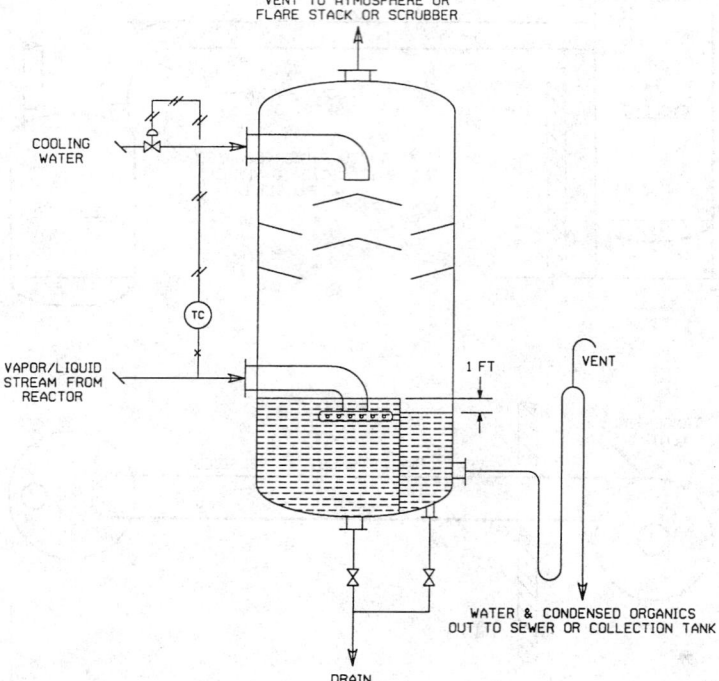

FIG. 26-21   Quench tank with direct-contact baffle tray section.

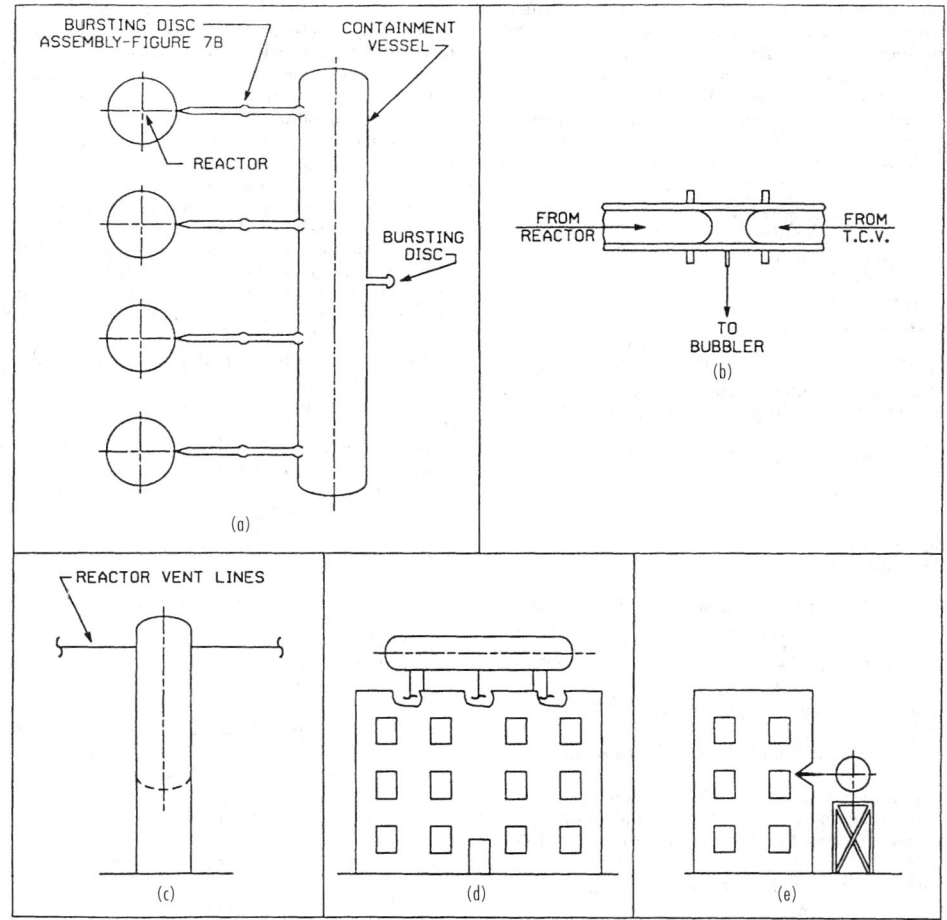

FIG. 26-22 Multireactor knockout (K-O) drum/catch tank: (*a*) plan view of reactors connected to horizontal containment vessel; (*b*) back-to-back bursting disc assembly; (*c*) elevation of self-supporting vessel; (*d*) elevation of horizontal vessel on roof of building; (*e*) elevation of horizontal vessel on side of building.

Performance Characteristics:

Quenching only saturated vapors with no inerts:

Cold quench liquid:
1. Sparging will condense the vapors effectively.
2. Sudden vapor condensation in the pool may cause water hammer if the holes are too big and the pressure drop is too low. Sonic hole velocity is desirable to avoid this problem.

Hot, nearly saturated quench liquid:
1. Sparging may not condense all the vapor. The injection of cold liquid spray in the vapor space should be considered.
2. Sudden vapor condensation in the pool is a minor problem.

Quenching only saturated vapors with some inerts:
1. Sparging may be ineffective in condensing all the condensable vapor.
2. A mass transfer device, such as a packed or trayed contact section, should be considered (see Fig. 26-21).

Quenching two-phase mixtures:

High volumetric vapor/liquid ratios ($r > 10$): where r = volumetric vapor flow rate/volumetric liquid flow rate.
1. The sparger design can be based on all vapor flow, but the heat balance must include the liquid.

Moderate volumetric vapor liquid ratios ($1 < r < 10$):

1. The liquid may inhibit the mass transfer rates needed to condense the vapors; sparging may, therefore, be less effective.
2. Sudden vapor condensation is of less concern.

Low volumetric vapor/liquid ratios ($r < 1$):
1. The mixture acts like a liquid and the vapor condensation is dependent on jet mixing. This will require a different type of sparger design.

Sizing and Design of Equipment The information in the following sections that pertain to the sizing and design of blowdown drums (catch tanks) and cyclone separators are for normal liquid-vapor systems (low-viscosity and nonfoamy or unstable foams). They are not applicable to high-viscosity (newtonian and non-newtonian) liquids and/or systems which exhibit surface-active foaming behavior, as no information is available at the present time as to the separation efficiency for these types of equipment. Quench tanks can usually handle high-viscosity liquids as well as stable foams.

Horizontal Blowdown Drum (See Fig. 26-16.) The two main criteria used in sizing horizontal blowdown drums or catch tanks are as follows.

1. The diameter must be sufficient to effect good vapor-liquid separation.
2. The total volume must be sufficient to hold the estimated amount of liquid carryover from the reactor. For a foamy discharge,

the holding volume should be greater than the reactor liquid volume (to be discussed further). One design method for sizing a horizontal blowdown drum is presented in API RP521 (1997). This may require a trial-and-error design procedure to arrive at an optimum drum size. Another procedure, which has been used in the industry by many companies, may be used to size horizontal blowdown drums more directly and is as follows:

1. Calculate the allowable vapor velocity U_a

$$U_a = k \sqrt{\frac{\rho_L - \rho_v}{\rho_v}} \qquad (26\text{-}32)$$

where U_a = allowable vapor velocity, ft/s
ρ_L = liquid density, lb/ft^3
ρ_v = vapor density, lb/ft^3
k = capacity coefficient

Values of k reported in the technical literature have ranged from 0.157 to 0.40. A k value of 0.27 has resulted in conservatively sized blowdown drums, able to separate liquid droplets 300 μm in diameter and larger.

2. Calculate the vapor flow area A_v

$$A_v = \frac{Q_v}{U_a} \qquad (26\text{-}33)$$

where A_v = vapor flow area, ft^2
Q_v = vapor flow rate, ft^3/s

3. Assume A_v occupies half of the drum area, so that the drum diameter is:

$$D_d = \sqrt{\frac{2A_v}{0.785}}, \text{ ft} \qquad (26\text{-}34)$$

4. Determine the drum volume occupied by liquid V_L based on the following criteria:
   a. For nonfoaming systems, V_L should be equal to the maximum working volume of the reactor.
   b. For mildly foaming systems, which rapidly defoam, V_L should be a minimum of 1.5 times the maximum working volume of the reactor. Experimental data may indicate that V_L has to be even larger than this.

5. Ignoring both heads, calculate the drum length L:

$$L = \frac{2V_L}{0.785D_d^2} \qquad (26\text{-}35)$$

where V_L = drum liquid volume, ft^3
L = drum length, ft

6. If the drum length is less than two to three times the diameter, the design is satisfactory. If L is greater than $3D_d$, assume a larger diameter and repeat the calculation until a satisfactory L/D_d ratio is achieved.

Another equation for quick sizing of horizontal knockout drums/catch tanks is presented by Tan (*Hydrocarbon Processing*, October 1967, p. 149). He recommends the following equation for calculating the drum diameter:

$$W = 360D^2 \sqrt{(\rho_L - \rho_v)\frac{MP}{T}} \qquad (26\text{-}36)$$

where W = vapor flow rate, lb/h
D = drum diameter, ft
ρ_L = liquid density, lb/ft^3
ρ_v = vapor density, lb/ft^3
M = molecular weight of vapor
P = pressure in the drum, psia
T = temperature of the vapor, °R

The author states that this equation is valid for the design of knockout drums which can separate liquid droplets of 400 μm and larger.

Cyclone Separator with Separate Catch Tank   (See Figs. 26-17 and 26-18.) The sizing of a cyclone knockout drum for emergency relief systems is somewhat different from sizing a cyclone separator for normal process service for the following reasons:

1. In normal process service, the superficial vapor velocity at the inlet of tangential-entry vapor-liquid separators is limited to about 120 to 150 ft/s. Higher velocities may lead to:
   - Excessive pressure drop in the separator and in the inlet piping
   - Generation of fine mist in the inlet piping, which escapes collection in the separator
2. Inlet velocity restrictions do not apply in the design of separators for emergency relief systems because:
   - Pressure drop is usually not a penalty.
   - Escape of fine mist can usually be tolerated.

Sizing Procedure   The cyclone is sized by choosing a superficial F-factor for the skirt in the range of 5.0 to 8.0. The higher value may be used for waterlike liquids; the lower value for liquids like molasses. If design F-factors exceed the range of 5 to 8, the liquid draining down the skirt is entrained and escapes with the vapor. These F-factors were determined in small-scale lab experiments using water and a high-polymer solution as the test liquids. The high-polymer solution had a viscosity that was molasses-like, probably in the range of 1500 cP. There were no liquids of intermediate viscosity used in the tests.

The F-factors of 5.0 and 8.0 are conservative in the opinion of the researcher who performed the experiments (private communication from E. I. du Pont de Nemours Co., Inc., to the DIERS Project).

The F-factor is defined as follows:

$$F = V_v \sqrt{\rho_v} \qquad (26\text{-}37)$$

or

$$F = \frac{G_v}{\sqrt{\rho_v}} \qquad (26\text{-}38)$$

where V_v = superficial velocity, ft/s
ρ_v = vapor density, lb/ft^3
G_v = superficial vapor mass flux, lb/s·ft^2

The design procedure is as follows:
1. Calculate G_v

$$G_v = F \sqrt{\rho_v}, \text{ lb/s·ft}^2 \qquad (26\text{-}39)$$

2. Calculate the skirt flow area

$$A_S = \frac{W}{3600G_v}, \text{ ft}^2 \qquad (26\text{-}40)$$

where W = vapor flow rate, lb/h
3. Calculate skirt diameter

$$D_S = \sqrt{\frac{A_S}{0.785}}, \text{ ft} \qquad (26\text{-}41)$$

4. Calculate all the other separator dimensions from the relationships given in Fig. 26-18.

When the pressure relief device is set to open at greater than 15 psig (critical flow will result), it is normally not necessary to be concerned about the pressure drop in the separator. If the liquid is to be drained from the separator during the emergency blowdown, a vortex breaker and false bottom should be used (Fig. 26-18, view BB).

If the liquid contents of the vented vessel are to be retained in the separator for subsequent disposition, the holdup capacity may be increased by increasing the height of the vessel to increase the total volume by an amount equal to the vented liquid volume.

Cyclone Separator with Integral Catch Tank   (See Fig. 26-19.) The diameter of the knockout drum is calculated by the criteria given in the preceding section and Fig. 26-18. Since the liquid is also to be retained in the vessel, extend the shell height below the normal bottom tangent line to increase the total volume by an amount equal to the volume of the liquid carried over.

Quench Tank   (See Figs. 26-20 and 26-21.)

General   In comparison with design information on blowdown drums and cyclone separators, there is very little information in the open technical literature on the design of quench tanks in the chemical industry. What is available deals with the design of quench tanks (also called *suppression pools*) for condensation of steam or steam-water mixtures from nuclear reactor safety valves. Information and criteria from quench tanks in the nuclear industry can be used for the design of quench tanks in the chemical industry. There have been several articles in recent years which provide more data for chemical industry quench tank design (AIChE-CCPS, 1997). The following sections summarize some of this information.

Design Criteria   Pertinent criteria for quench tank sizing and design are presented below:

OPERATING PRESSURE   There are three modes of operation of a quench tank: atmospheric pressure operation, nonvented operation, and controlled venting operation. Atmospheric operation is usually feasible when the effluent being emitted has a bubble point well above the maximum ambient temperature. A very small quantity of vapor escapes with the air that is displaced as the tank fills with the emergency discharge (typically about 0.2 percent of the reactor contents). Depending on the toxic or flammable properties of the vapor, the vent from the quench tank can be routed to the atmosphere or must be sent to a scrubber or flare.

In nonvented operation, no material is vented to the atmosphere, and this design is used when complete containment of the discharge is required. It is also used when the discharge mixture bubble point is close to or below the maximum ambient temperature and the concentration of noncondensable gas in the feed stream is very low. The tank design pressure is relatively high since the initial air in the tank is compressed by the rising liquid level, adding to the vapor pressure. The designer must take into consideration that the quench tank backpressure must be limited so as not to adversely affect the reactor relief system.

In controlled venting operation, the quench tank pressure is maintained at a desired level by a pressure controller/control valve system or pressure relief valve. This mode of operation is used when the discharge mixture bubble point is close to or below the maximum ambient temperature, and it is desired to limit the maximum quench tank pressure.

QUENCH LIQUID SELECTION   The choice of the appropriate quench liquid depends on a number of factors. Water is usually the first quench liquid to consider, since it is nontoxic, nonflammable, compatible with many effluent vapors, and has excellent thermal properties. If water is selected as the quench liquid, the tank should be located indoors, if possible, to avoid freezing problems. If the tank has to be located outdoors in a cold climate, the addition of antifreeze is preferable to heat-tracing the tank, since overheating the tank can occur from tracing, thus reducing its effectiveness.

If other quench liquids are required, the liquid should have as many of the following properties as possible: compatibility with the discharge effluent, low vapor pressure, high specific heat, low viscosity, low flammability, low freezing point, high thermal conductivity, immiscibility with the discharge effluent, low cost, and ready availability.

QUENCH LIQUID QUANTITY   A good discussion of the factors determining the quantity of quench liquid required is presented by CCPS (AIChE-CCPS, 1995).

When water is used as the quench medium and the effluent stream is a hydrocarbon or organic, separate liquid phases are often formed. In this case, heat transfer is the predominating mechanism during the quench. To achieve effective heat transfer, there must be a sufficient difference between the quench liquid temperature and the bubble point of the incoming effluent stream. The minimum temperature difference occurs at the end of the discharge, when the quench pool temperature is highest. A rule of thumb, from industry practice, is to allow a 10 to 20°C (18 to 36°F) ΔT. For atmospheric tank operation, the final quench liquid temperature is then set 10 to 20°C (18 to 36°F) below the normal boiling point of the final quench pool mixture. For nonvented or controlled venting operation, the final boiling point is elevated, permitting a greater design temperature rise and the use of less quench liquid. Therefore, the quench pool final temperature must be set 10 to 20°C (18 to 36°F) lower than the saturated temperature of the discharge effluent at the design maximum quench tank pressure.

The minimum capacity of quench liquid can be estimated by a heat balance, knowing the final quench pool temperature. The following equation given by Fauske (*International Symposium on Multi-Phase Transport and Particulate Phenomena*, December 15–17, 1986) can be used to calculate the minimum amount of quench liquid:

$$M = \frac{m_o(T_R - T_a)C_R}{(T_a - T_o)C_q} \qquad (26\text{-}42)$$

where m_o = mass of reactants
T_R = temperature of reactants at relief set pressure
T_a = allowable temperature following complete quench
T_o = initial temperature of the quench fluid
C_q = specific heat of the quench fluid
C_R = specific heat of the reactants
(consistent English or SI units)

The preceding equation assumes the reaction is completely quenched immediately after the relief point is reached. This behavior is closely approximated if the reaction stops in the quench pool and the reactor empties quickly and thoroughly. If the reaction continues in the quench pool, the temperature T_R should be increased to the maximum adiabatic exotherm temperature. An equation is presented by CCPS (AIChE-CCPS, 1997) that includes the heat of reaction. In some cases, an experiment is necessary to confirm that the reaction indeed stops in the quench pool.

It is good practice to provide 10 to 20 percent more quench liquid than the minimum amount calculated.

QUENCH TANK VOLUME   The total volume of the quench tank should be equal to the sum of the following volumes:

Quench liquid required
Liquid entering in the multiphase effluent stream
Liquid condensed from vapors in entering the effluent stream
Freeboard for noncondensables (a minimum of 10 percent is recommended)

QUENCH TANK GEOMETRY   Quench tanks can have any of the following three types of geometry:

- Horizontal cylindrical vessel
- Vertical cylindrical vessel
- Concrete pit (usually rectangular)

Usually, the geometry is determined by space limitations. Both horizontal or vertical cylindrical vessels are designed as pressure vessels, and for pressures up to 50 psig, an L/D ratio of 2 to 3 results in an economic design.

SPARGER DESIGN   The effluent stream should be discharged into the quench liquid by means of a sparger, which breaks it up into small jets to provide good heat and mass transfer. The sparger design must also incorporate the following capabilities:

- Maximize momentum-induced recirculation in the quench pool
- Provide adequate flow area (cross section for pressure relief without imposing high backpressure)
- Minimize shock due to vapor bubble collapse
- Minimize unbalanced momentum forces

Figure 26-20 shows conventional quench tank sparger arrangements. As can be seen in this figure, the sparger can be of the following types:

- Vertical straight pipe sparger
- Tee sparger
- Four-armed cross sparger

The following design criteria are recommended:

1.   For effluent streams consisting of only liquid and vapor, hole diameters ranging from 1/8 to 1/2 in are recommended. Larger hole diameters (up to 2 in) may be required if the blowdown stream contains solids (polymers and/or catalyst). However, the violently collapsing vapor bubbles create a water hammer effect which increases in severity with hole size.

2.   Sonic hole velocity is desirable in smaller holes and is essential in 1/2- to 2-in holes. A minimum sparger pressure drop of 10 psi should be used.

3.   The number of holes should provide at least 0.2 holes per square foot of pool cross-sectional area. The flow area of the manifold and/or distributor piping should be at least 2 times the total area of the sparger holes. This generally ensures that the pressure drop across the holes will be at least 10 times the pressure drop in the distributor.

4.   To balance high-velocity momentum forces, a symmetrical sparger design must be used. This can be a vertical straight pipe, a tee-shaped, or a cross-shaped quencher arm configuration with rows of holes on opposite sides of the pipe, which helps to balance piping forces (see Fig. 26-20). This arrangement also enhances the momentum-induced recirculation of quench liquid and maximizes the temperature difference for heat transfer. A center-to-center hole spacing

of at least two to three hole diameters is recommended, which will minimize the coalescence of the discharging jets into larger, less effective jets.

The quencher arm should be anchored to prevent pipe whip. It should also extend to the length (for horizontal vessels) or the height (for vertical vessels) of the vessel to evenly distribute the vapors in the pool.

When quenching effluents discharged by safety valves, it is preferable to use a straight, vertical sparger with holes in the end cap as well as in the pipe side walls. This is recommended to minimize the possibility of liquid hammer, which can occur more readily in horizontal spargers. The liquid hammer usually occurs for the following reasons: as the relief valve opens for the first time, the pressure spike is cushioned by the air trapped in the vent line. This air is blown out. If the valve recloses, the line may cool, causing slugs of condensate to accumulate. When the valve reopens, the slugs will accelerate to very high velocities and impact any elbows and end caps of the sparger. In severe cases, the sparger-arm end caps can be knocked off. The preceding recommendation avoids turns and the holes in the end cap provide some relief from the pressure spike.

Mass-Transfer Contact Section   Where there is a strong possibility that not all of the incoming vapors will be condensed in the pool, a direct-contact mass-transfer section is superimposed on the quench tank. This can be a baffle-tray section (as shown in Fig. 26-21) or a packed column section.

The design of direct-contact mass-transfer columns is discussed in detail by Scheiman (*Petro Chemical Engr.* **37**(3): 29–33, 1965; ibid. **37**(4): 75, 78–79) and Fair (*Chem. Eng.*, June 12, 1972).

Multireactor Knockout Drum/Catch Tank   (See Fig. 26-22.)

Vessel Sizing   The area needed for vapor disengaging is calculated by the equations given earlier in the section on horizontal blowdown drums.

The diameter and length (or height) are determined by considering a number of factors as follows:

1. The length should be sufficient to extend beyond the locations of the reactors discharging into the vessel so as to simplify discharge pipe runs (for a horizontal vessel).
2. The height should not greatly exceed the height of the building (for vertical vessels).
3. The diameter should be sufficient to allow attenuation of the shock wave leaving the deflector plate.
4. The diameter should be sufficient to allow installation of the pipes and deflector plates in such a way as not to interfere directly with one another (particularly important for vertical vessels).
5. The cost of pressure vessels increases as the diameter increases.
6. An upper limit to the diameter is set by the need to transport complete cylindrical sections from manufacturer to site.
7. The volume of liquid in the reactor or reactors (assuming more than one vents at the same time) must be determined.

Mechanical Design Considerations   The paper by Speechly et al., ("Principles of Total Containment System Design," presented at *I. Chem. E. Northwestern Branch Meeting,* 1979) discusses a number of pertinent design features, as follows:

1. Each vent device discharge pipe is extended into the vessel and its end is fitted with a deflector device. This disperses the jet stream of solids (catalyst) and liquids discharged and dissipates this force, which should otherwise be exerted on the vessel wall immediately opposite.
2. The deflector device (baffle plate) must be carefully designed as described by Woods (*Proc. Inst. Mech. Engrs.* **180,** part 3J: 245–259, 1965–1966).
3. Isolate the catch tank from both reaction loads and forces generated by thermal expansion of the pipes; the pipes can be designed to enter the vessel through a sliding gland. Depending on layout, vessels which tend to have shorter, stiffer pipes between the building and the vessel may also require flexible bellows to be incorporated in the pipes.
4. There are usually several reactors linked to a single catch tank. To ensure that rupture of a disk on one reactor does not affect the others, each reactor is fitted with a double-rupture-disk assembly. The use of double rupture disks in this application requires installation of a leak detection device in the space between the two disks, which

must also prevent a pressure buildup from occurring within this space. Otherwise, under some circumstances it is possible for a pinhole-type leakage in one disk to cause a pressure to be retained in the space between the two disks. In this event, the pressure at which the disks would rupture could be increased significantly. This condition could therefore render ineffective the protection of the reactor system itself.

For additional details on the design of blowdown drums, cyclone separators, and quench tanks, such as mechanical design, thrust forces, ancillary equipment, and safety considerations, refer to the books and articles listed in the General References.

FLAME ARRESTERS

GENERAL REFERENCES: "Deflagration and Detonation Flame Arresters," *Guidelines for Engineering Design for Process Safety,* chap. 13, CCPS-AICE, 1993. Ibid., chap. 15, "Effluent Disposal Systems." Howard, W. B., "Flame Arresters and Flashback Preventers," *Plant/Operations Progress,* vol. 1, no. 4, 1982. Howard, "Precautions in Selection, Installation and Use of Flame Arresters," *Chem. Eng. Prog.,* April 1992. Piotrowski, "Specification of Flame Arresting Devices for Manifolded Low Pressure Storage Tanks," *Plant/Operations Progress,* vol. 10, no. 2, April 1991. Roussakis and Lapp, "A Comprehensive Test Method for In-Line Flame Arresters," *Plant/Operations Progress,* vol. 10, no. 2, April 1991. NFPA 497A, *Recommended Practice for Classification of Class I Hazardous (Classified) Locations for Electrical Installations in Chemical Process Areas,* 1997.

General Considerations   Flame arresters are passive devices designed to prevent propagation of gas flames through pipelines. Typical applications are to prevent flames entering a system from outside (such as via a tank vent) or propagating within a system (such as from one tank to another). Flame arrestment is achieved by a permeable barrier, usually a metallic matrix containing narrow channels, which removes heat and free radicals from the flame fast enough to both quench it within the matrix and prevent reignition of the hot gas on the protected side of the arrester. These metallic matrices are known as *elements.* Some preliminary considerations for arrester selection and placement are:

1. Identify the at-risk equipment and the potential ignition sources in the piping system to determine where arresters should be placed and what general type (deflagration or detonation, unidirectional or bidirectional) are needed.
2. Determine the worst-case gas mixture combustion characteristics, system pressure, and permissible pressure drop across the arrester, to help select the most appropriate element design. Not only does element design impact pressure drop, but the rate of blockage due to particle impact, liquid condensation, and chemical reaction (such as monomer polymerization) can make some designs impractical even if in-service and out-of-service arresters are provided in parallel.
3. The possibility of a stationary flame residing on the arrester element surface should be evaluated, and so should the need for additional safeguards, should such an event occur (see "Endurance Burn" section).
4. Consider any material of construction limitations due to reactive or corrosive stream components.
5. Consider upset conditions that could exceed the test conditions at which the arrester was certified. These include the gas composition with regard to concentration of sensitive constituents such as ethylene or hydrogen, maximum system pressure during an emergency shutdown, and maximum temperature. Under certain upset conditions such as a high-pressure excursion, there may be no flame arrester available for the task.
6. Consider the type and location of the arrester with respect to ease of maintenance, particularly for large in-line arresters.

These questions address the type of arrester needed, the appropriate location, and the best design with respect to flow resistance, maintainability, and cost. It should be recognized that while flame arrester effectiveness is high, it is not 100 percent. To maximize effectiveness, attention should be given to proper selection, application, and maintenance of the device. In the case of marine vapor control systems in the United States, the testing and application of flame arresters is regulated by the U.S. Coast Guard. In other cases, recent testing protocols have been developed to address most adverse conditions encountered. Some arresters, such as hydraulic arresters and in-line types used to stop decomposition flames, have specialized applications for which general design and testing information are scarce. Where flame arresters are impractical, alternative strategies such as fast-acting valves, vapor suppression, and flammable mixture control should be considered.

Combustion: Deflagrations and Detonations A deflagration is a combustion wave propagating at less than the speed of sound as measured in the unburned gas immediately ahead of the flame front. Flame speed relative to the unburned gas is typically 10–100 m/s, although, owing to expansion of hot gas behind the flame, several hundred meters per second may be achieved relative to the pipe wall. The combustion wave propagates via a process of heat transfer and species diffusion across the flame front, and there is no coupling in time nor space with the weak shock front generated ahead of it. Deflagrations typically generate maximum pressures in the range 8–12 times the initial pressure. The pressure peak coincides with the flame front, although a marked pressure rise preceeds it; thus, the unburned gas is compressed as the deflagration proceeds, depending on the flame speed and vent paths available. The precompression of gas ahead of the flame front (also known as "cascading" or "pressure piling") establishes the gas conditions in the arrester when the flame enters it and hence affects both the arrestment process and the maximum pressure generated in the arrester body. A severe deflagration arrestment test involves placing a restricting orifice behind the arrester, which increases the degree of precompression. This is known as "restricted-end" deflagration testing.

As the deflagration flame travels through piping, its speed increases due to flow-induced turbulence and compressive heating of the unburned gas ahead of the flame front. Turbulence is especially enhanced by flow obstructions such as valves, elbows, and tees. Once the flame speed has attained the order of 100 m/s, a deflagration-to-detonation transition (DDT) can occur, provided that the gas composition is within the detonatable limits, which lie inside the flammable limits. The travel distance for this to occur is referred to as the *run-up* distance for detonation. This distance varies with the gas mixture sensitivity and increases with pipe diameter. Tabulated run-up distances are generally for straight pipe runs, and DDT can occur for much smaller distances in pipe systems containing flow obstructions. At the instant of transition, a transient state of *overdriven detonation* is achieved and persists for a distance of a few pipe diameters. Overdriven detonations propagate at speeds greater than the speed of sound (as measured in the burned gas immediately behind the flame front), and side-on pressure ratios (at the pipe wall) in the range 50 to 100 have been measured. The peak pressure is variable, depending on the amount of precompression during deflagration. A severe test for detonation-type flame arresters is to arrange for the arrester to encounter a series of overdriven detonations.

After the abnormally high velocities and pressures associated with DDT have decayed, a state of stable detonation is attained. A detonation is a combustion-driven shock wave propagating at the speed of sound, as measured in the burned gas immediately behind the flame front. Since the speed of sound in this hot gas is much larger than in the unburned gas or the ambient air, and the flame front speed is augmented by the burned gas velocity, stable detonations propagate at supersonic velocities relative to an external fixed point. The wave is sustained by chemical energy released by shock compression and ignition of the unreacted gas. The flame front is coupled in space and time with the shock front, with no significant pressure rise ahead of the shock front. The high velocities and pressures associated with detonations require special element design to quench the high-velocity flames plus superior arrester construction to withstand the associated impulse loading. Since this entails narrower and/or longer element channels plus bracing of the element facing, both inherent pressure drop and the possibility of fouling of detonation arresters should be considered.

The problem of flame arrestment, either of deflagrations or detonations, depends on the properties of the gas mixture involved plus the initial temperature and pressure. Gas mixture combustion properties cannot be quantified for direct use in flame arrester selection and only general characteristics can be assigned. For this reason, flame arrester performance must be demonstrated by realistic testing. Such testing has demonstrated that arresters capable of stopping even over-driven detonations may fail under restricted-end deflagration test conditions. It is important to understand the significance of the test conditions addressed and their possible limitations.

Combustion: Gas Characteristics and Sensitivity Combustion thermodynamic calculations allow determination of peak deflagration and detonation pressures, plus stable detonation velocity. The peak pressure calculation may be used to determine combustion product venting requirements, although a conservative volume increase of 9:1 may be used for essentially closed systems. Other relevant gas characteristics are entirely experimental. The sensitivity to detonation depends on the detonatable range and fundamental burning velocity, although no specific correlations or measures of sensitivity exist based on fundamental properties. It is often considered that detonation sensitivity and the degree of difficulty in arresting flames increase with lower National Electrical Code (NEC) Groupings. Hence, Group A gases (acetylene) will be most sensitive and Group D gases (such as saturated hydrocarbons) will be least sensitive. This empirical method of characterizing gases is typically used in selecting deflagration arresters, where successful testing using one gas in an NEC electrical group is assumed to apply for other gases in that group. It is cautioned that, where the maximum experimental safe gaps (MESGs) of two gases within a single NEC group are significantly different, the assumption of equivalent sensitivity is dubious. Regulations applying to detonation arresters in vapor control systems under the authority of the U.S. Coast Guard (USCG) provide that MESGs are solely used to characterize gases, under the assumption that mixtures with smaller MESGs are more difficult to stop. See "Deflagration and Detonation Flame Arresters," (1993) for a discussion of MESG plus tabulated values.

Corrosion Consideration should be given to possible corrosion of both the element material and the arrester housing, since corrosion may weaken the structure, increase the pressure drop, and decrease the effectiveness of the element. While the housing might be designed to have a corrosion allowance, corrosion of the element must be avoided by proper material specification. Common materials of construction include aluminum, carbon steel, ductile iron, and 316 stainless steel housings and aluminum or 316 stainless steel elements. While special materials such as Hastelloy might be used for situations such as high HCl concentrations it may be more cost effective to use a hydraulic arrester in such applications.

Directionality To select an arrester for any service, the potential sources of ignition must be established in relation to the pipe system and the equipment to be protected. The pipe connecting an arrester with an identified ignition source is the *unprotected side* of the arrester. The pipe connecting the arrester with at-risk equipment is the *protected side*. If the arrester will encounter a flame arriving only from one direction, a *unidirectional* arrester can be used. If a flame may arrive from either direction, a *bidirectional* arrester is needed. The latter are either symmetrically constructed or are certified by testing. Back-to-back use of unidirectional arresters will not usually be cost effective unless testing reveals a specific advantage such as increased allowable operating pressure during restricted-end deflagration testing.

Endurance Burn Under certain conditions, a successfully arrested flame may stabilize on the unprotected side of an arrester element. Should this condition not be corrected, the flame will eventually penetrate the arrester as the channels become hot. An endurance burn time can be determined by testing, which specifies that the arrester has withstood a stabilized flame without penetration for a given period. The test should address either the actual or worst-case geometry, since heat transfer to the element will depend on whether the flame stabilizes on the top, bottom, or horizontal face. In general, the endurance burn time identified by test should not be regarded as an accurate measure of the time available to take remedial action, since test conditions will not necessarily approximate the worst possible practical case. Temperature sensors may be incorporated at the arrester to indicate a stabilized flame condition and either alarm or initiate appropriate action, such as valve closure.

Installation End-of-line arresters should be protected using appropriate weather hoods or cowls. In-line arresters (notably detonation arresters) must be designed to withstand the highest line pressure that might be seen, including upset conditions. The design should be verified by hydrostatic and pneumatic pressure tests. The piping system should be designed with adequate supports and should allow routine access to the arrester for inspection and maintenance.

Maintenance   It is important to provide for arrester maintenance, both by selection of the most suitable arrester type and judicious location. Inspection and maintenance should be performed on a regular basis, depending on experience with the particular arrester in the service involved. It should also be carried out after successful function of the arrester. Some in-line designs allow removal, inspection, and cleaning of the element without the need to expand the line. Unit designs featuring multiple elements in parallel can reduce downtime by extending the period between cleaning. For systems which cannot be shut down during maintenance, parallel arresters incorporating a three-way valve may be used. Detonation arrester elements are especially prone to damage during dismantling, cleaning, and reassembly. Maintenance must be carefully done, avoiding sharp objects that could disable the delicate channels in the element. Spare elements should be available to reduce downtime and provisions made for storing, transporting, and cleaning the elements without damage.

Monitoring   The differential pressure across the arrester element can be monitored to determine the possible need for cleaning. The pressure taps must not create a flame path around the arrester. It can be important to provide temperature sensors, such as thermocouples, at the arrester to detect flame arrival and stabilization. Since arrester function may involve damage to the arrester, the event of successful function (flame arrival) may be used to initiate inspection of the element for damage. If the piping is such that flame stabilization on the element is a realistic concern, action must be taken immediately upon indication of such stabilization (see also "Endurance Burn"). Such action may involve valve closure to shut off gas flow.

Operating Temperature and Pressure   Arresters are certified subject to maximum operating temperatures and absolute pressures normally seen at the arrester location. Arrester placement in relation to heat sources, such as incinerators, must be selected so that the allowable temperature is not exceeded, with due consideration for the detonation potential as run-up distance is increased.

If heat tracing is used to prevent condensation of liquids, the same temperature constraint applies. In the case of in-line arresters, there may be certain upset conditions that produce unusually large system pressures outside the stipulated operating range of the arrester. Since the maximum operating pressure for a detonation arrester may be in the range of 16 to 26 psia, depending on the gas sensitivity and arrester design, it may be impossible to find a suitable arrester to operate during such an upset. The situation may be exacerbated by pressure drop across the device, caused by high flow and/or fouling.

Pressure Drop   Flow resistance depends on flame arrester channel arrangement and on a time-dependent fouling factor due to cor-rosion or accumulation of liquids, particles, or polymers, depending on the system involved. Monomer condensation is a difficult problem, since inhibitors will usually be removed during monomer evaporation and catalysis might occur over particulates trapped in the element. Pressure drop can be a critical factor in operability, and cleaning may represent a large hidden cost.

Fouling may be mitigated in a number of ways. First, the least-sensitive element design can be selected, and in the case of end-of-line arresters, weather hoods or cowls can be used to protect against water or ice accumulation. Second, a fouling factor (20 percent or greater) might be estimated and an element with a greater tested flow capacity selected to reduce the pressure drop. This should be further increased if liquid condensation might occur. It is important that certified flow curves for the arrester be used rather than calculated curves, since the latter can be highly optimistic. Condensation and polymerization may be mitigated by geometry (minimizing liquid accumulation in contact with the element) and provision for drainage. Alternatively, the arrester may be insulated and possibly heat traced. Drains should not provide flame paths around the arrester or leak in either direction when closed. If heat tracing is used, the temperature must be limited to the certified operating range of the arrester.

In addition to using an arrester element with greater flow capacity, it is common to use two arresters in parallel where frequent cleaning is required, with one arrester in standby. A three-way valve can be used to allow uninterrupted operation during changeover. Where elements have an intrinsically high pressure drop, such as sintered metal elements used in acetylene service, multiple parallel elements can be used.

Deflagration Arresters   The two types of deflagration arrester normally considered are the *end-of-line arrester* (Figs. 26-23 and 26-24) and the *tank vent deflagration arrester*. Neither type of arrester is designed to stop detonations. If mounted sufficiently far from the atmospheric outlet of a piping system, which constitutes the unprotected side of the arrester, the flame can accelerate sufficiently to cause these arresters to fail. Failure can occur at high flame speeds even without a run-up to detonation.

If atmospheric tanks are equipped with flame arresters on the vents, fouling or blockage by extraneous material can inhibit gas flow to the degree that the tank can be damaged by underpressure. API standards allow the use of pressure-vacuum (P/V) valves without flame arresters for free venting tanks on the basis that the high vapor velocity in the narrow gap between pressure pallet (platter) and valve body will prevent flashback. However, it is important to ensure that the pallet is not missing or stuck open, since this will remove the protection. Absence of the pallet was a listed factor in the 1991 Coode Island fire (State Coroner Victoria, Case No. 2755/91, Inquest into Fire at Coode Island on August 21 and 22, 1991, Finding). Whether flame arresters are used, proper inspection and maintenance of these vent systems is required.

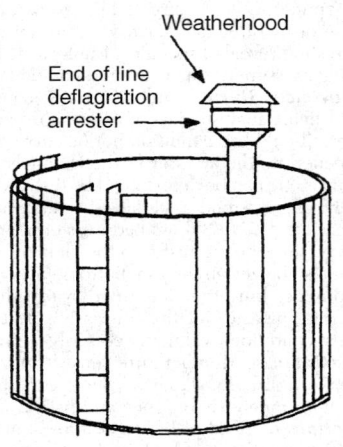
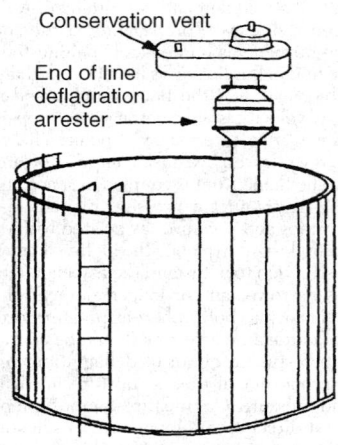

FIG. 26-23   Typical deflagration arrester installations.

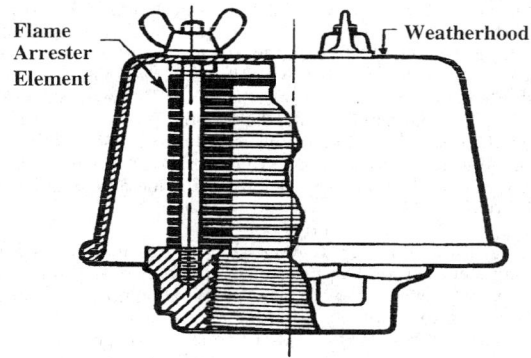

FIG. 26-24 Typical deflagration arrester design (end-of-line type).

End-of-line arresters are mounted at the outlet of a pipe system and go directly to the atmosphere, so there is no potential for significant flame acceleration in the pipe. Tank vent deflagration arresters are strictly limited by the approval agency, but for Group D gases they are typically mounted no more than 20 ft from the end of a straight pipe that vents directly to the atmosphere. The allowed distance must be established by proper testing with the appropriate gas mixture and the pipe diameter involved. Turbulence-promoting irregularities in the flow (bends, tees, elbows, valves, etc.) cannot be used unless testing has addressed the exact geometry. It is essential that run-up to detonation cannot occur in the available piping system, and run-up distance can be less than 2 ft for some fast-burning gases such as hydrogen in air (Group B). Thus the NEC Grouping of the gas mixture must be considered. More important, it must be emphasized that even if run-up to detonation does not occur, a deflagration arrester can fail if the flame speed is great enough. Thus the run-up distance is not an adequate criterion for acceptable location and this limitation can be determined only by realistic testing. A number of explosions have occurred due to misapplication of deflagration arresters where detonation arresters should have been used.

In certain exceptional cases, a specially designed deflagration arrester may be mounted in-line without regard to run-up distance. This can be done only where the system is known to be incapable of detonation. An example is the decomposition flames of ethylene, which are briefly discussed under "Special Arrester Types and Alternatives."

Detonation and Other In-Line Arresters If the point of ignition is remote from the arrester location, the arrester is an in-line type such as might be situated in a vapor collection system connecting several tanks (Fig. 26-25). Due to the possibility of DDT, most in-line arresters are designed to stop both deflagrations and detonations (including overdriven detonations) of the specified gas mixture. These are known as *detonation arresters*. Figure 26-26 shows a typical design. In some cases, in-line arresters need to stop deflagrations only. However, in such cases it must be demonstrated that detonations cannot occur in the actual pipework system; unless the gas mixture is intrinsically not capable of detonation, this requires full-scale testing using the exact pipe geometry to be used in practice, which must not be changed after installation.

Detonation arresters are typically used in conjunction with other measures to decrease the risk of flame propagation. For example, in vapor control systems, the vapor is often enriched, diluted, or inerted, with appropriate instrumentation and control (see "Effluent Disposal Systems," 1993). In cases where ignition sources are present or predictable (such as most vapor destruct systems), the detonation arrester is used as a last-resort method anticipating possible failure of vapor composition control. Where vent collection systems have several vapor/oxidant sources, stream compositions can be highly variable and

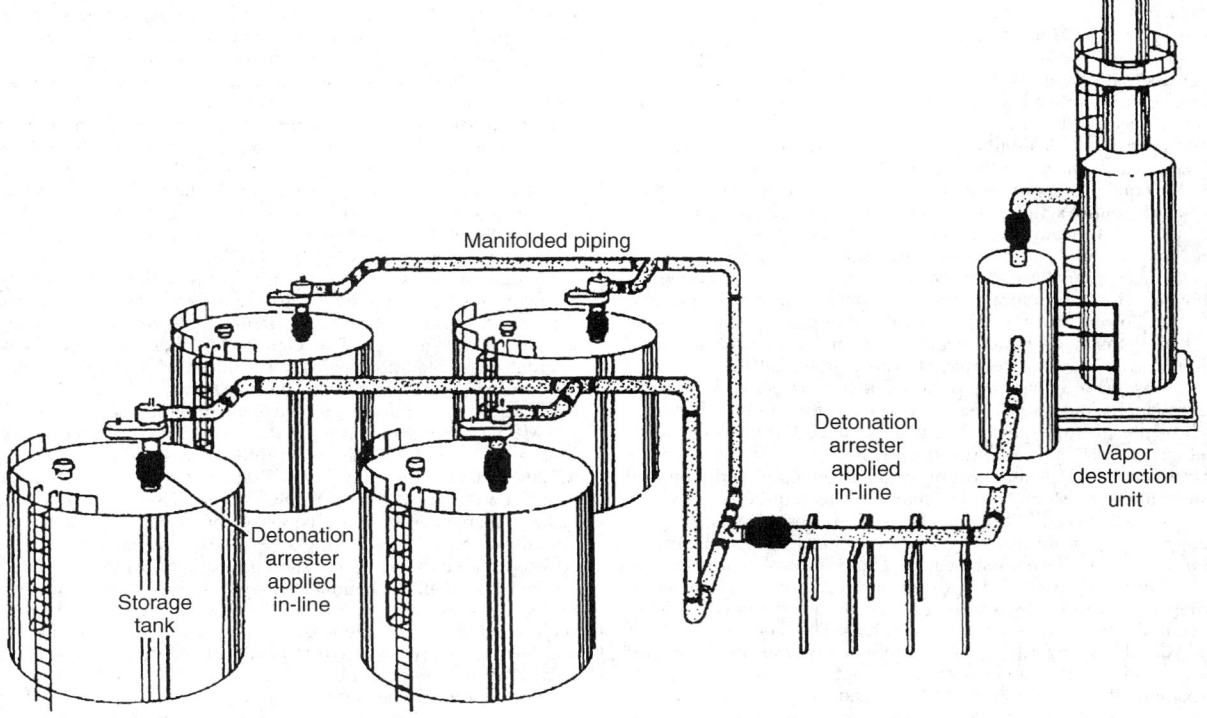

FIG. 26-25 Possible positions at which flame arresters may be placed (vapor control system).

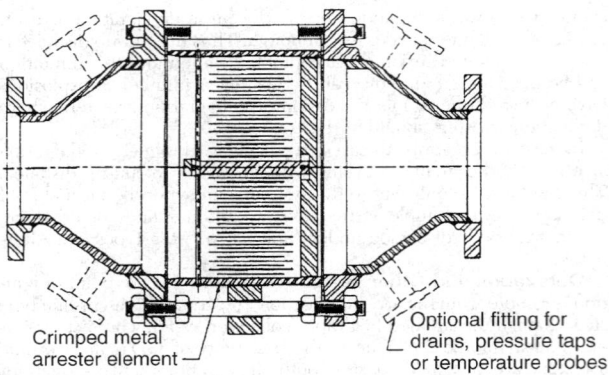

Crimped metal
arrester element —

Optional fitting for
drains, pressure taps
or temperature probes

FIG. 26-26   Typical detonation arrester design (crimped-ribbon type).

this can be additionally complicated when upset conditions are considered. It is often cost effective to perform hazard analyses, such as fault tree analysis, to determine whether such vent streams can enter the flammable region and, if so, what composition corresponds to the worst credible case. Such an analysis is also suitable to assess alternatives to arresters.

Effect of Pipe Diameter Changes   Arrester performance can be impaired by local changes in pipe diameter. It was shown that a minimum distance of 120 pipe diameters should be allowed between the arrester and any increase in pipe diameter, otherwise a marked reduction in maximum allowable operating pressure would occur. This impairment was observed during detonation testing but was most pronounced during restricted-end deflagration testing (Lapp and Vickers, *Int. Data Exchange Symp. on Flame Arresters and Arrestment Technology,* Banff, Alberta, October 1992). As a rule, arresters should be mounted in piping either equal to or smaller than the nominal size of the arrester.

Venting of Combustion Products   As gas deflagrates or detonates in the piping system, there is a volume expansion of the products and an associated pressure increase. In some instances where the pipe system volume involved is relatively large, a significant overpressure might be developed in the vapor spaces of connected tanks, especially when vapor space is minimal due to high liquid level. It can be assumed that all the gas on the unprotected side of the arrester is converted to equilibrium products; the pressure is relieved via gas expansion into the entire system volume and to the atmosphere via any vent paths present. If heat losses are neglected by the assumption of high flame speeds or detonation and atmospheric venting paths are neglected, a conservative approach is that storage vessels be designed with a capacity to handle nine times the pipe volume on the unprotected side of the arrester. With regard to the high pressures associated with detonations, it has been shown (Lapp, *Independent Liquid Terminal Association Conference,* Houston, June 23, 1992) that detonation arresters attenuate the peak detonation pressure by up to 96 percent, depending on the arrester design, and therefore protect from much of the pressure pulse. To further reduce the pressure pulse, relief devices may be provided at the arrester.

Arrester Testing and Standards   Regulatory and approval agencies and insurers impose acceptance testing requirements, sometimes as part of certification standards. The user may also request testing to demonstrate specific performance needs, just as the manufacturer can help develop standards. These interrelationships have resulted in several new and updated performance test procedures. *Listing* of an arrester by a testing laboratory refers only to performance under a defined set of test conditions. The flame arrester user should develop specific application requirements based on the service involved and the safety and risk criteria adopted.

A variety of test procedures and use guidelines have been developed. In addition, companies or associations may develop internal standards. The Federal Register, 33 CFR, Part 154, contains the USCG requirements for detonation arresters in marine vapor control

systems. Other U.S. procedures are given in ASTM F 1273-91, UL 525, FM Procedure Classes 6061 and 7371, and API Publications 2028 and 2210. Outside the United States, procedures are given in Canada's CSA Standard Z 343, Rev. 12, 1993, the United Kingdom's British Standard BS 7244, Germany's DIN/CEN Draft Standard of the DAbF Subcommittee on Standardization, June 1991 (developed through the Federal Physical Technical Institute, PTB), and the International Maritime Organization (IMO) Standard MS/Circ. 373, Rev. 1, 1988. For U.S. mining applications, the Mine Safety and Health Administration (MSHA) provides regulation and guidance—for example, in CFR Title 30, Part 36.

Deflagration Arrester Testing   For end-of-line and tank vent flame arresters, approval agencies may require manufacturers to provide users with data for flow capacity at operating pressures, proof of success during an endurance burn or continuous flame test, evidence of flashback test results (for end-of-line arresters) or explosion test results (for in-line or tank vent arrester applications), hydraulic pressure test results, and results of a corrosion test.

Endurance burn testing generally implies that the ignited gas mixture and flow rate be adjusted to give the worst-case heating (based on temperature observations on the protected side of the element surface), that the burn continue for a specified duration, and that flame penetration not occur. Continuous flame testing implies that a gas mixture and flow rate be established at specified conditions and burn on the flame arrester for a specified duration. The endurance burn test is usually a more severe test than the continuous burn. In both cases the flame arrester attachment configuration and any connecting piping or valves should be installed as in the plant design.

Flashback tests incorporate a flame arrester on top of a tank, with a large plastic bag surrounding the flame arrester. A specific gas mixture (for example, propane, ethylene, or hydrogen at the most sensitive composition in air) flows through and fills the tank and the bag. Deflagration flames initiated in the bag (three at different bag locations) must not pass through the flame arrester into the tank. On the unprotected side, piping and attachments such as valves are included as intended for installation; a series of tests—perhaps ten—is conducted.

Whatever the application, a user should be aware that not all test procedures are the same, are of the same severity, or use the same rating designations. Therefore, it is important to review the test procedure and determine whether the procedure used is applicable to the intended installation and potential hazard the flame arrester is meant to prevent.

Detonation Arrester Testing   Requirements are described by various agencies in the aforementioned documents (UL 525, etc.). For installations governed by the USCG in Appendix A of 33 CFR, Part 154 (Marine Vapor Control Systems), the USCG test procedures must be followed. These are similar but not identical to those of other agencies listed (for a discussion of differences, see "Deflagration and Detonation Flame Arresters," 1993).

Detonation arresters are extensively tested for proof of performance against deflagrations, detonations, and endurance burns. In the United States, arrester manufacturers frequently test detonation arresters according to the USCG protocol; other test standards might alternatively or additionally be met. Under this protocol, the test gas must be selected to have either the same or a lower MESG than the gas in question (MESG means *maximum experimental safe gap*). Typical MESG benchmark gases are stoichiometric mixtures of propane, hexane, or gasoline in air to represent Group D gases having an MESG equal to or greater than 0.9 mm and ethylene in air to represent Group C gases with an MESG no less than 0.65 mm. Commercially available arresters are typically certified for use with one or another of these benchmark gas types. An ethylene-type arrester is selected should the gas in question have an MESG less than 0.9 mm but not less than 0.65 mm. Five low- and five high-overpressure deflagration tests are required with and without a flow restriction on the protected side. Of these 20 tests, the restricted-end condition is usually the more severe and often limits the maximum initial pressure at which the arrester will be suitable. Five detonation tests and five overdriven detonation tests are also required, which may involve additional run-up piping and turbulence promoters in order to achieve DDT at the arrester. If these tests are successful, an endurance burn test is required. This test does not use propane for

Group D gases, but hexane or gasoline, owing to their lower autoignition temperatures. For Group C tests, ethylene can be used for all test stages.

Shortcomings in the use of MESG to characterize gases, in the use of stoichiometric compositions for deflagration tests, and nonoptimization of test geometry have been recognized ("Deflagration and Detonation Flame Arresters," 1993). The user has the option to request additional tests to address such concerns and may wish to test actual stream compositions rather than simulate on the basis of MESG values.

Special Arrester Types and Alternatives Several types of *unlisted* arresters (water seals, packed beds, velocity-type devices, and fast-acting valves) mentioned in API 2028 are described more fully in Howard (1982). There are few design or test data for hydraulic and packed-bed arresters; some types are designed and used by individual companies for specific applications, while others are commercially available. Figure 26-27 shows some special arrester types.

Decomposition Flame Arresters Above certain minimum pipe diameters, temperatures, and pressures, some gases may propagate decomposition flames in the absence of oxidant. Special in-line arresters have been developed (Fig. 26-27). Both deflagration and detonation flames of acetylene have been arrested by hydraulic valve arresters, packed beds (which can be additionally water-wetted), and arrays of parallel sintered metal elements. Information on hydraulic and packed-bed arresters can be found in the Compressed Gas Association Pamphlet G1.3, "Acetylene Transmission for Chemical Synthesis." Special arresters have also been used for ethylene in 1000- to 1500-psi transmission lines and for ethylene oxide in process units. Since ethylene is not known to detonate in the absence of oxidant, these arresters were designed for in-line deflagration application.

Alternatives to Arresters Alternatives to the use of flame arresters include fast-acting isolation valves, vapor suppression systems, velocity-type devices in which gas velocity is designed to exceed flashback velocity, and control of the flammable mixture (NFPA 69 standard, "Explosion Prevention Systems"). The latter alternative frequently involves reduction of oxygen concentration to less than the limiting oxygen concentration (LOC) of the gas stream.

STORAGE AND HANDLING OF HAZARDOUS MATERIALS

GENERAL REFERENCES: *Air Quality Handbook,* ENSR Consulting and Engineering, Acton, Mass., June 1988. ANSI/API-620-1986, American National Standards Institute, New York, 1986. AP-40, *Air Pollution Engineering Manual,* 2d ed., U.S. Environmental Protection Agency, Office of Air Quality Planning and Standards, 1973. AP-42, *Compilation of Emission Factors for Stationary Sources,* U.S. Environmental Protection Agency, Office of Air Quality Planning and Standards, 1985. *API Standards,* American Petroleum Institute, Washington, D.C. Arthur D. Little, Inc., and LeVine, *Guidelines for Safe Storage and Handling of High Toxic Hazard Materials,* CCPS, AIChE, New York, 1986. *ASME Boiler and Pressure Vessel Code; ASME Code for Pressure Piping; ASME General and Safety Standards; ASME Performance Test Codes,* American Society of Mechanical Engineers, New York. *Chemical Exposure Index, Second Edition,* AIChE, New York, 1994. *Clean Air Act Amendments Bulletin Board System,* U.S. Environmental Protection Agency, Office of Air Quality Planning and Standards, 1200–9600 baud, April 1992. *Code of Federal Regulations,* Protection of Environment, Title 40, Parts 53 to 80, Office of the Federal Register, National Archives and Records Administration, July 1991. Englund, "Opportunities in the Design of Inherently Safer Chemical Plants," in J. Wei et al., ed., *Advances in Chemical Engineering,* vol. 15, Academic Press, 1990. Englund, "Design and Operate Plants for Inherent Safety," *Chem. Eng. Prog.,* parts 1 and 2, March and May 1991. Englund, Mallory, and Grinwis, *Chem. Eng. Prog.,* February 1992. Englund and Grinwis, "Redundancy in Control Systems," *Chem. Eng. Prog.,* October 1992. Englund and Holden, "Storage of Toxic Materials," in Grossel and Crowl, *Handbook of Highly Toxic Materials Handling and Management,* Marcel Dekker, New York, 1995. *Guidelines for Chemical Process Quantitative Risk Analysis,* CCPS, AIChE, New York, 1989. Hendershot, "Alternatives for Reducing the Risks of Hazardous Material Storage Facilities," *Environmental Progress,* 7 August 1988, pp. 180ff. Kletz, *An Engineer's View of Human Error,* Institution of Chemical Engineers, VCH Publishers, New York, 1991. Kletz, "Friendly Plants," *Chem. Eng. Prog.,* July 1989, pp. 18–26. Kletz, *Plant Design for Safety: a User Friendly Approach,* Hemisphere Publishing, London, 1991. Lees, *Loss Prevention in the Chemical Industries,* Butterworths, London, 1980. Prokop, "The Ashland Tank Collapse," *Hydrocarbon Processing,* May 1988. *Refrigerated Liquid Chlorine Storage,* Pamphlet 78, Edition 1, The Chlorine Institute, New York, July 1984. Russell and Hart, "Underground Storage Tanks, Potential for Economic Disaster," *Chemical Engineering,* March 16, 1987, pp. 61–69. *Ventsorb for Industrial Air Purification,* Bulletin 23-56c, Calgon Carbon Corporation, Pittsburgh, Pa., 1986. White and Barkley, "The Design of Pressure Swing Adsorption Systems," *Chem. Eng. Prog.,* January 1989.

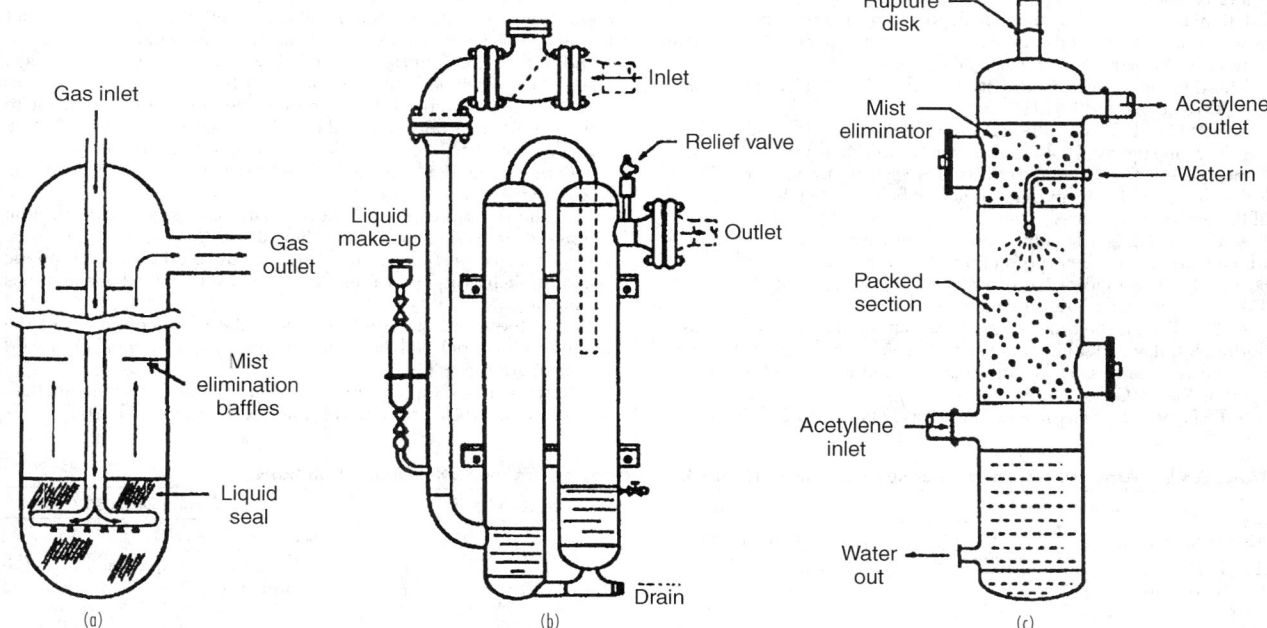

FIG. 26-27 Some special arrester designs: (*a*) liquid seal arrester; (*b*) Linde hydraulic seal arrester; (*c*) wetted packed-bed acetylene decomposition arrester. (*Howard, 1982.*)

Introduction   The storage and handling of toxic materials involve risks that can be reduced to very low levels by good planning, design, and management practices. Facilities that handle toxic materials typically represent a variety of risks, ranging from small leaks, which require prompt attention, to large releases, which are extremely rare in well-managed facilities but which have the potential for widespread impact (Arthur D. Little, Inc., and LeVine, 1988, p. 5ff, by permission).

It is essential that good techniques be developed for identifying significant hazards and mitigating them where necessary. Hazards can be identified and evaluated using approaches discussed in the section on hazard and risk analysis.

Loss of containment due to mechanical failure or misoperation is a major cause of chemical process accidents. The design of storage systems should be based on minimizing the likelihood of loss of containment, with the accompanying release of hazardous materials, and on limiting the amount of the release. An effective emergency response program that can reduce the impacts of a release should be available.

Toxicity and Toxic Hazard   There is a difference between toxicity and toxic hazard:

- Toxicity is the ability to cause biological injury.
- Toxicity is a property of all materials—even salt, sugar, and water.
- Toxicity is related to dose and degree of hazard associated with a material. Dose is time- and duration-dependent, in that dose is a function of exposure (concentration) times duration.

Toxic hazards may be caused by chemical means, radiation, and noise. Routes of exposure are: (1) eye contact, (2) inhalation, (3) ingestion, (4) skin contact, and (5) ears (noise). An *Industrial Hygiene Guide* (IHG) is based on exposures for an 8-h day, 40-h week, and is not to be used as a guide in the control of health hazards. It is not to be used as a fine line between safe and dangerous conditions.

A material that has a high toxicity does not necessarily present a severe toxic hazard. For example, a ton of lead arsenate spilled in a busy street is unlikely to poison members of the public just a short distance from the spill, because it is not mobile. It could be carefully recovered and removed and would present a low risk to the general public, even though it is extremely toxic. On the other hand, a ton of liquefied chlorine spilled on the same street could become about 11,000 ft³ of pure gas. The IDLH for chlorine is 25 ppm. This is a concentration such that immediate action is required. Thus, the one ton of chlorine, if mixed uniformly with air, could create a cloud of considerable concern, having a volume of about 4.4×10^8 ft³ or a sphere 770 ft in diameter. This could quickly spread over downwind areas and prove fatal to people near the spill site, causing toxic effects among hundreds of others in the downwind direction.

Measures of inhalation toxicity include ERPG, TLV, TLV-STEL, TLV-TWA, PEL, and IDLH.

- ERPG is defined in the section on hazard and risk analysis.
- TLV means *threshold limit value* (established by the American Conference of Government Industrial Hygienists, or ACGIH). TLV-C is the concentration in air that should not be exceeded during any part of the working exposure.
- TVL-STEL is a 15-min, time-weighted average concentration to which workers may be exposed up to four times per day with at least 60 min between successive exposures with no ill effect if the TLV-TWA is not exceeded (developed by the ACGIH).
- TLV-TWA is the time-weighted average concentration limit for a normal 8-h day and 40-h workweek, to which nearly all workers may be repeatedly exposed, day after day, without adverse effect (developed by the ACGIH).
- PEL means *permissible exposure level* (similar to TLV but developed by the National Institute for Occupational Safety and Health, or NIOSH).
- IDLH means *immediately dangerous to life and health*. This is a concentration at which immediate action is required. The exact effect on an individual depends on the individual's physical condition and susceptibility to the toxic agent involved. It is the maximum airborne contamination concentration from which one could escape within 30 min without any escape-impairing symptoms or irreversible health effects (developed by NIOSH).

Storage

Storage Facilities   The Flixborough disaster (Lees, 1980) occurred on June 1, 1974, and involved a large, unconfined vapor cloud explosion (or explosions—there may have been two) and fire that killed 28 people and injured 36 at the plant and many more in the surrounding area. The entire chemical plant was demolished and 1821 houses and 167 shops were damaged.

The results of the Flixborough investigation made it clear that the large inventory of flammable material in the process plant contributed to the scale of the disaster. It was concluded that "limitations of inventory should be taken as specific design objectives in major hazard installations." It should be noted, however, that reduction of inventory may require more frequent and smaller shipments and improved management.

There may be more chances for errors in connecting and reconnecting with small shipments. Quantitative risk analysis of storage facilities has revealed solutions that may run counter to intuition (Schaller, *Plant/Operations Progress,* **9**(1), 1990). For example, reducing inventories in tanks of hazardous materials does little to reduce risk in situations where most of the exposure arises from the number and extent of valves, nozzles, and lines connecting the tank. Removing tanks from service altogether, on the other hand, helps. A large pressure vessel may offer greater safety than several small pressure vessels of the same aggregate capacity because there are fewer associated nozzles and lines. Also, a large pressure vessel is inherently more robust, or it can economically be made more robust by deliberate overdesign than can a number of small vessels of the same design pressure. On the other hand, if the larger vessel has larger connecting lines, the relative risk may be greater if release rates through the larger lines increase the risk more than the inherently greater strength of the vessel reduces it. In transporting hazardous materials, maintaining tank car integrity in a derailment is often the most important line of defense in transportation of hazardous materials.

Safer Storage Conditions   The hazards associated with storage facilities can often be reduced significantly by changing storage conditions. The primary objective is to reduce the driving force available to transport the hazardous material into the atmosphere in case of a leak (Hendershot, 1988). Some methods to accomplish this follow.

Dilution   Dilution of a low-boiling hazardous material reduces the hazard in two ways:

1. The vapor pressure is reduced. This has a significant effect on the rate of release of material boiling at less than ambient temperature. It may be possible to store an aqueous solution at atmospheric pressure, such as aqueous ammonium hydroxide instead of anhydrous ammonia.

2. In the event of a spill, the atmospheric concentration of the hazardous material will be reduced, resulting in a smaller hazard downwind of the spill.

The reduction of vapor pressure by diluting ammonia, monomethylamine, and hydrochloric acid with water is shown in Table 26-9.

TABLE 26-9   Vapor Pressure of Aqueous Ammonia, Hydrochloric Acid, and Monomethylamine Solutions

Ammonia at 21°C		Monomethylamine at 20°C		Hydrochloric acid at 25°C	
Concentration wt.%	Vapor pressure, atm	Concentration wt.%	Vapor pressure, atm	Concentration wt.%	Vapor pressure, atm
100 (anhydrous)	8.8	100 (anhydrous)	2.8	100 (anhydrous)	46.1
48.6	3.0	50	0.62	41	1.0
33.7	1.1	40	0.37	38	0.36
28.8	0.75			32	0.055

SOURCE: Hendershot, 1988, by permission.

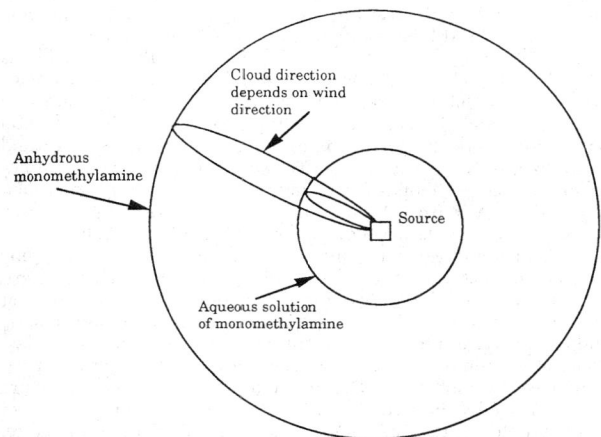

FIG. 26-28 Relative hazard zones for anhydrous and aqueous monomethylamine releases—relative distance within which there is a specified atmospheric concentration of monomethylamine and aqueous monomethylamine. (*Hendershot, 1988, by permission.*)

The relative size of hazard zones from possible loss of containment and releases to the atmosphere is much smaller for the cases in which the material is diluted, compared to the anhydrous materials. This is illustrated in Fig. 26-28 for monomethylamine.

The larger circle is the area that could be exposed to a specified atmospheric concentration of monomethylamine stored as an anhydrous liquid. The smaller circle is the area that could be exposed to a specified atmospheric concentration of monomethylamine stored as an aqueous solution. The elliptical figures represent a gas cloud caused by an east-southeast wind.

Refrigeration   Loss of containment of a liquefied gas under pressure and at atmospheric temperature causes immediate flashing of a large proportion of the gas. This is followed by slower evaporation of the residue. The hazard from a gas under pressure is normally much less in terms of the amount of material stored, but the physical energy released if a confined explosion occurs at high pressure is large.

Refrigerated storage of hazardous materials that are stored at or below their atmospheric boiling points mitigates the consequences of containment loss in three ways:

1.   The rate of release, in the event of loss of containment, will be reduced because of the lower vapor pressure in the event of a leak.

2.   Material stored at a reduced temperature has little or no superheat and there will be little flash in case of a leak. Vaporization will be mainly determined by liquid evaporation from the surface of the spilled liquid, which depends on weather conditions.

3.   The amount of material released to the atmosphere will be further reduced because liquid entrainment from the two-phase flashing jet resulting from a leak will be reduced or eliminated.

Refrigerated storage is most effective in mitigating storage facility risk if the material is refrigerated when received. Much of the benefit of refrigerated storage will be lost if the material is received at ambient temperature under its vapor pressure in a transport container. The quantity of material that could be released during unloading may be larger because unloading lines are normally sized to rapidly unload a truck or rail car and are often larger than the process feed lines. Thus, if the material is shipped at ambient temperature, the benefits of refrigeration will not be available during the operations with the highest release potential.

The economics of storage of liquefied gases are such that it is usually attractive to use pressure storage for small quantities, pressure or semirefrigerated storage for medium to large quantities, and fully refrigerated storage for very large quantities. Quantitative guidelines are available from Lees (1980, pp. 271–272).

It is generally considered that there is a greater hazard in storing large quantities of liquefied gas under pressure than at low temperatures and low pressures. The trend is toward replacing pressure

storage by refrigerated low-pressure storage for large inventories. However, it is necessary to consider the risk of the entire system, including the refrigeration system, and not just the storage vessel. The consequences of failure of the refrigeration system must be considered. Each case should be carefully evaluated on its own merits. In most cases, refrigerated storage of hazardous materials is undoubtedly safer, such as in the storage of large quantities of liquefied chlorine.

Design of Liquid Storage So Leaks and Spills Do Not Accumulate Under Tanks or Equipment   Around storage and process equipment, it is a good idea to design dikes that will not allow toxic and flammable materials to accumulate around the bottom of tanks or equipment in case of a spill. If liquid is spilled and ignites inside a dike where there are storage tanks or process equipment, the fire may be continuously supplied with fuel and the consequences can be severe. It is usually much better to direct possible spills and leaks to an area away from the tank or equipment and provide a fire wall to shield the equipment from most of the flames if a fire occurs. The discussion on BLEVEs later in this section shows a design for diking for directing leaks and spills to an area away from tanks and equipment.

The surface area of a spill should be minimized for materials that are highly toxic and have a significant vapor pressure at ambient conditions, such as acrylonitrile or chlorine. This will make it easier and more practical to collect vapor from a spill or to suppress vapor release with foam. This may require a deeper nondrained dike area than normal or some other design that will minimize surface area, in order to contain the required volume. It is usually not desirable to cover a diked area to restrict loss of vapor if the spill consists of a flammable or combustible material.

Minimal Use of Underground Tanks   The U.S. Environmental Protection Agency's (USEPA) Office of Underground Storage Tanks defines underground tanks as those with 10 percent of more of their volume, including piping, underground. An aboveground tank that does not have more than 10 percent of its volume (including piping) underground is excluded from the underground tank regulations. Note, however, that a 5000-gal tank sitting wholly atop the ground but having 1400 ft of 3-in buried pipe or 350 ft of 6-in buried pipe is considered an underground storage tank.

At one time, burying tanks was recommended because it minimized the need for a fire protection system, dikes, and distance separation. At many companies this is no longer considered good practice. Mounding, or burying tanks above grade, has most of the same problems as burying tanks below ground and is usually not recommended.

Problems with buried tanks include:

• Difficulty in monitoring interior and exterior corrosion (shell thickness)

• Difficulty in detecting leaks

• Difficulty of repairing a tank if the surrounding earth is saturated with chemicals from a leak

• Potential contamination of groundwater due to leakage

Governmental regulations concerning buried tanks are becoming stricter. This is because of the large number of leaking tanks that have been identified as causing adverse environmental and human health problems.

Consequences of Leaking Underground Tanks   The following is a real possibility (Russell and Hart, 1987). A site where an underground tank has been used is found to have leaked. If the leak is not cleaned up to "background" levels by the time an environmental regulatory agency is involved, the agency may decide that a portion of the plant must be designated as a waste disposal site. The plant could then be required to provide a waste site closure plan, hold public hearings, place deed restrictions on the plant property, and, finally, provide a bond that would cover the cost of closing the site and also analyzing and sampling groundwater for up to 30 years.

Product leaking from an underground storage tank will migrate downward until it encounters the water table, where it will then flow with the groundwater, leaving a long trail of contaminated soil. Above the water table, some product will be absorbed on the soil particles and in the pore space between the soil particles. If the soil is later saturated with water, product stored in the pore spaces may be released, causing a reappearance of the free product and movement of the material into previously unaffected soil.

The scope of the problem was revealed by the USEPA in 1983

when it reported that, in the United States, 11 million gallons of gasoline seep into the soil each year. Just one gallon of gasoline can make one million gallons of water unsafe to drink; one ounce would pollute an Olympic-size swimming pool full of drinking water. Most of the contaminated sites the USEPA has documented involve corroded single-wall steel tanks and piping that have been in the ground for at least 16 years (Semonelli, "Secondary Containment of Underground Storage Tanks," *Chem. Eng. Prog.*, June 1990). A number of states have enacted laws setting standards for underground storage tanks. The USEPA has issued regulations requiring notification to the appropriate regulatory agency about age, condition, and size for underground storage tanks containing commercial chemical products.

Secondary Containment for Underground Storage   Acceptable secondary containment systems for underground storage are described as barriers either integral to the tank system design (such as double-walled tanks or double-walled pipes) or located within the underground storage tank system that present a barrier between all parts of the underground storage tank system and the environment. Double-walled tanks and piping should be considered for aboveground tanks and piping containing highly toxic liquids.

Concrete and fiberglass vaults are often used, although they can be subject to environmentally induced cracks. Soil and clay liners are not allowed. Flexible liner systems have been developed that may be a cost-effective and environmentally sound alternative. State-of-the-art liner technology has overcome many of the previous problems with seams, low mechanical strength, and chemical resistance.

Piping Systems for Underground Service   An important consideration is the USEPA's concern over piping systems. For all underground storage tank systems, performance standards consistent with those for tanks were set for pipes and pipe systems. There is evidence that 84 percent of underground storage tank system test failures are due to loose tank fittings or faulty piping. Piping releases occur twice as often as tank releases. In particular, loose joints tend to occur. For hazardous substance underground storage tank systems, there are two options: trench liners and double-walled pipes. Double-walled pipes are difficult to assemble and are subject to failure caused by service conditions, such as frost heaves or pressure from above. Flexible trench liners (discussed previously) are becoming a popular solution to secondary containment of piping systems.

Detecting Leaks   Small leaks are difficult to detect. The USEPA and American Petroleum Institute standard for nonleaking underground tanks is 0.05 gal/h (3.15 cm^3/min), above which a tank is considered to be leaking. Leak detection measurements can be influenced by many factors, making it difficult to detect small leaks.

Corrosion Problems   Tanks subject to internal corrosion are not good choices for underground service because of the necessity of monitoring wall thickness. Underground tanks and piping of carbon steel should be considered for corrosion protection measures such as external tarlike coatings and magnesium anodes. Joints in underground piping should be minimized by welding. Pipes may use a combination of wrapping and sprayed-on coatings. When flanges are necessary, such as with valves, external coatings should be used.

Summary of Use of Underground Tanks   Because of more stringent regulatory requirements, potential future liabilities, and the cost of building and operating underground storage tank systems, it may be inherently safer to use aboveground storage with suitable spacing, diking, and fire protection facilities. With modern technology, if it is necessary, it is possible to design underground storage systems with a high degree of integrity and which will make leaks to the environment highly unlikely, but the cost may be high.

Design of Tanks, Piping, and Pumps   Six basic tank designs are used for the storage of organic liquids: (1) fixed roof, (2) external floating roof, (3) internal floating roof, (4) variable vapor space, (5) low-pressure tanks, and (6) high-pressure tanks. The first four tank designs listed are not generally considered suitable for highly toxic hazardous materials.

Low-Pressure Tanks (below 15 psig)   Low-pressure storage tanks for highly hazardous toxic materials should meet, as a minimum, the American Petroleum Institute (API) 620 Standard, "Recommended Rules for the Design and Construction of Large Welded, Low-Pressure Storage Tanks" (*API Standards*). This standard covers tanks designed for all pressures under 15 psig. There are no specific requirements in API 620 for highly hazardous toxic materials.

API 650, "Welded Steel Tanks for Oil Storage" (*API Standards*) has limited applicability to storage of highly hazardous toxic materials because it prohibits refrigerated service and limits pressures to 2.5 psig and only if designed for certain conditions. Most API 650 tanks have a working pressure approaching atmospheric pressure and hence their pressure-relieving devices must vent directly to the atmosphere. Its safety factors and welding controls are less stringent than required by API 620. Another reference for the design of low-pressure storage tanks may be found in ANSI/API-620-1986.

Horizontal and vertical cylindrical tanks are used to store highly toxic liquids at atmospheric pressure. Horizontal, vertical, and spherical tanks are used for refrigerated liquefied gases that are stored at atmospheric pressure. The design pressure of tanks for atmospheric and low-pressure storage at ambient temperature should not be less than 100 percent of the vapor pressure of the material at the maximum design temperature. The maximum design metal temperature to be used takes into consideration the maximum temperature of material entering the tank and the maximum ambient temperature, including solar radiation effects.

Since discharges of vapors from highly hazardous toxic materials cannot simply be released to the atmosphere, the use of a weak seam roof is not normally acceptable. It is best that tanks be designed and stamped for 15 psig to provide maximum safety, and pressure relief systems must be provided to vent to equipment that can collect, contain, and treat the effluent.

The minimum design temperature should be the lowest temperature to which the tank will be subjected, taking into consideration the minimum temperature of material entering the tank, the minimum temperature to which the material may be autorefrigerated by rapid evaporation of low-boiling liquids or mechanically refrigerated, and the minimum ambient temperature of the area where the tank is located. API 620 provides for installations in areas where the lowest recorded one-day mean temperature is −50°F.

While either rupture disks or relief valves are allowed on storage tanks by Code, rupture disks by themselves should not be used on tanks for the storage of highly hazardous toxic materials since they do not close after opening and may lead to continuing release of toxic material to the atmosphere.

The API 620 Code requires a combined pneumatic hydrotest at 125 percent of design tank loading. In tanks designed for low-density liquid, the upper portion is not fully tested. For highly hazardous toxic materials, consideration should be given for hydrotesting at the maximum specified design liquid level. It may be required that the lower shell thickness be increased to withstand a full head of water and that the foundation be designed such that it can support a tank full of water or the density of the liquid if it is greater than water. Testing in this manner not only tests the containment capability of the tank, but it also provides an overload test for the tank and the foundation similar to the overload test for pressure vessels. API 620 also requires radiography.

Proper preparation of the subgrade and grade is extremely important for tanks that are to rest directly on grade. Precautions should be taken to prevent ground freezing under refrigerated tanks, as this can cause the ground to heave and damage the foundation or the tank. Designing for free air circulation under the tank is a method for passive protection from ground freezing.

Steels lose their ductility at low temperatures and can become subject to brittle failure. There are specific requirements for metals to be used for refrigerated storage tanks in API 620, Appendices Q and R.

Corrosive chemicals and external exposure can cause tank failure. Materials of construction should be chosen so that they are compatible with the chemicals and exposure involved. Welding reduces the corrosion resistance of many alloys, leading to localized attack at the heat-affected zones. This may be prevented by the use of the proper alloys and weld materials, in some cases combined with annealing heat treatment.

External corrosion can occur under insulation, especially if the weather barrier is not maintained or if the tank is operating at conditions at which condensation is likely. This form of attack is hidden and may be unnoticed for a long time. Inspection holes and plugs should be installed in the insulation to monitor possible corrosion under the insulation.

High-Pressure Tanks (above 15 psig) The design of vessels above 15 psig falls within the scope of the American Society of Mechanical Engineers (ASME) Boiler and Pressure Vessel Code, Section VIII "Pressure Vessels, Division I" and should be designated as lethal service if required. *Lethal service* means containing substances which are "poisonous gases or liquids of such a nature that a very small amount of the gas or vapor of the liquid mixed or unmixed with air is dangerous to life when inhaled. This class includes substances which are stored under pressure or may generate a pressure if stored in a closed vessel." This is similar to, but not exactly like, the same definition as that for "Category M" fluid service of the ASME Pressure Piping Code (see below). Pressure vessels for the storage of highly hazardous toxic materials should be designed in accordance with requirements of the ASME code even if they could be exempted because of high pressure or size. The code requires that the corrosion allowance be adequate to compensate for the more or less uniform corrosion expected to take place during the life of the vessel and not weaken the vessel below design strength.

Venting and Drainage In the installation of a storage tank, good engineering should go into the design of a drain and a vent. Low-pressure storage tanks are particularly susceptible to damage if good venting practices are not followed. A vent that does not function properly at all times may cause damage to the tank from pressure that is too high or too low. Vapors should go to a collection system, if necessary, to contain toxic and hazardous vents.

Piping Piping falls within Chapter VIII of the ASME Pressure Piping Code, "Piping for Category M Fluid Service." Category M Fluid Service is defined as "fluid service in which the potential for personnel exposure is judged to be significant and in which a single exposure to a small quantity of a toxic fluid, caused by leakage, can produce serious irreversible harm to persons on breathing or bodily contact, even when prompt restorative measures are taken."

Piping systems should meet the requirements for both Category M Fluid Service and for "severe cyclic conditions." Piping systems should be subjected to a flexibility analysis and, if found to be too rigid, flexibility should be added. Severe vibration pulsations should be eliminated. Expansion bellows, flexible connections, and glass equipment should be avoided. Pipelines should be designed with the minimum number of joints, fittings, and valves. Joints should be flanged or butt-welded. Threaded joints should not be used.

Instrumentation (Arthur D. Little, Inc., and Levine, 1986.) Instrument systems are an essential part of the safe design and operation of systems for storing and handling highly toxic hazardous materials. They are key elements of systems to eliminate the threat of conditions that could result in loss of containment. They are also used for early detection of releases so that mitigating action can be taken before these releases result in serious effects on people in the plant or in the public sector, or on the environment.

The basic approach is to direct the system to the safest operating level relative to people or the environment when any emergency condition is detected, including power loss. An important concept of process control safety is to have adequate redundancy to reduce unwanted shutdowns and maintain an adequate level of certainty that a safe state will result if a real emergency does occur. As far as possible, instruments should be of the fail-safe type.

Every effort should be made to eliminate direct (Bourdon-type) pressure gauges. Diaphragm pressure gauges constructed of appropriate corrosion-resistant materials are preferred. Flow limiters should be used to limit flow in case of loss of integrity.

An accurate indication of level is critical to the avoidance of overflow and other serious conditions in storage vessels. Level control is important to avoid overfilling to prevent a liquid release. A very low level can result in loss of pump suction and possible pump failure. Capacitance level sensors are often used because they require little maintenance and are highly reliable; since they give only point measurements, they are best used as backup for analogue devices such as differential pressure level gauges or strain gauges. Strain gauges (*load cells*) should be considered, as they are capable of high accuracy and do not require penetration of the containment vessel.

Flow measurements using nonintrusive or low mechanical action principles are desired, such as magnetic, vortex-shedding, or Coriolis-type flowmeters. Orifice plates are easy to use and reliable but have a limited range and may not be suitable for streams which are not totally clean. Rotameters with glass tubes should not be used.

Temperature measurements usually require intrusion into the fluid. Where thermowells are exposed to hazardous materials, they should comply with the same material requirements for vessels and pipes to reduce failure from erosion and corrosion. In storage tanks, tank temperature is often monitored but usually not controlled. Temperature indication is desirable to indicate that the tank contents are approaching a hazardous region and to indicate thermal stratification. For some materials, such as acrylic acid, temperature control is necessary during storage to prevent freezing if it gets too cold and prevent chemical reaction if it gets too warm.

Alarms should act as early warning devices to anticipate a potentially hazardous situation. Alarms that are essential to safety should be identified and classified separately from process alarms. Redundancy may be required.

Pumps and Gaskets Fugitive emissions often occur as a result of leakage of process materials through leak paths in rotating seals and susceptible gasketed joints such as are found in pipe flanges. When properly maintained, fugitive emissions from most conventional joints and sealing systems used in industry can be kept to a minimum. For volatile organic compounds (VOCs) this is usually significantly less than 500 ppm as measured at the leak path by a portable VOC analyzer specified in USEPA reference method 21 (40 CFR 60, Appendix A, Method 21). However, for some sealing systems such as packing glands on pump shafts in some services, the necessary maintenance frequency and potential risks of noncompliance have caused some companies to eliminate them from services where fugitive emissions are a concern and use tighter sealing systems such as mechanical seals instead. In services where entrained solids or fouling are not present to a significant extent and additional cost is justified, magnetic drive and canned-motor pumps, which have become more reliable and available in a wide variety of configurations and materials, are being used to virtually eliminate fugitive emissions from pumps. In services where fugitive emissions are a concern, valves such as quarter turn, diaphragm seal, or bellows seal valves, which are less susceptible to leakage, are sometimes being used in place of gate or globe valves with packed stem seals. However, under many service conditions, high-cost equipment options are not necessary to comply with the provisions of fugitive emission regulations. Properly maintained packing glands or single mechanical seals on valves and pumps can often meet all emissions requirements. An informed choice should be made when specifying new valves and pumps, considering factors such as the type of service, accessibility for maintenance, cost, and the degree of emission reductions which may be achieved.

The most common maintenance problem with centrifugal pumps is with the seals. Mechanical seal problems account for most of the pump repairs in a chemical plant, with bearing failures a distant second. The absence of an external motor (on canned pumps) and a seal is appealing to those experienced with mechanical seal pumps.

Sealless pumps are very popular and are widely used in the chemical industry. Sealless pumps are manufactured in two basic types: canned-motor and magnetic-drive. Magnetic-drive pumps have thicker "cans," which hold in the process fluid, and the clearances between the internal rotor and can are greater compared to canned-motor pumps. This permits more bearing wear before the rotor starts wearing through the can. Many magnetic-drive pump designs now have incorporated a safety clearance, which uses a rub ring or a wear ring to support the rotating member in the event of excessive bearing wear or failure. This design feature prevents the rotating member (outer magnet holder or internal rotating shaft assembly) from accidentally rupturing the can, as well as providing a temporary bearing surface until the problem bearings can be replaced. Because most magnetic-drive pumps use permanent magnets for both the internal and external rotors, there is less heat to the pumped fluid than with canned-motor pumps. Some canned-motor pumps have fully pressure-rated outer shells, which enclose the canned motor; others don't. With magnetic-drive pumps, containment of leakage through the can to the outer shell can be a problem. Even though the shell may be thick and capable of holding high pressures, there is often an elastomeric lip seal on the outer magnetic rotor shaft with little pressure containment capability.

Canned-motor pumps typically have a clearance between the rotor and the containment shell or can, which separates the fluid from the stator, of only 0.008 to 0.010 in (0.20 to 0.25 mm). The can has to be thin to allow magnetic flux to flow to the rotor. It is typically 0.010 to 0.015 in (0.25 to 0.38 mm) thick and made of Hastelloy. The rotor can wear through the can very rapidly if the rotor bearing wears enough to cause the rotor to move slightly and begin to rub against the can. The can may rupture, causing uncontrollable loss of the fluid being pumped.

It should not be assumed that just because there is no seal, sealless pumps are always safer than pumps with seals, even with the advanced technology now available in sealless pumps. Use sealless pumps with considerable caution when handling hazardous or flammable liquids.

Sealless pumps rely on the process fluid to lubricate the bearings. If the wear rate of the bearings in the fluid being handled is not known, the bearings can wear unexpectedly, causing rupture of the can.

Running a sealless pump dry can cause complete failure. If there is cavitation in the pump, hydraulic balancing in the pump no longer functions and excessive wear can occur, leading to failure of the can. *The most common problem with sealless pumps is bearing failure, which occurs either by flashing the fluid in the magnet area because of a drop in flow below minimum flow or by flashing in the impeller eye as it leaves the magnet area.* It is estimated that nine out of ten conventional canned-motor pump failures are the result of dry running. Canned pumps are available which, their manufacturer claims, can be operated dry for as long as 48 h.

It is especially important to avoid deadheading a sealless pump. Deadheaded sealless pumps can cause overheating. The bearings may be damaged and the pump may be overpressured. The pump and piping systems should be designed to avoid dead spots when pumping monomers. Monomers in dead spots may polymerize and plug the pump. There are minimum flow requirements for sealless pumps. It is recommended that a recirculation system be used to provide internal pump flow when the pump operates. Inlet line filters are recommended, but care must be taken not to cause excessive pressure drop on the suction side. Typical inlet filters use sieve openings of 0.0059 in (0.149 mm).

For many plants handling monomers and other hazardous materials, sealless pumps are the first choice. They can practically eliminate the pump problems that can occur due to seal leaks, which can include product loss, flammability, waste disposal, and exposure of personnel to hazardous vapors.

A number of liquids require special attention when applying canned-motor and magnetic-drive pumps. For example, a low-boiling liquid may flash and vapor-bind the pump. Solids in the liquid can also be bad for a sealless pump. Low-viscosity (in the range of 1 to 5 cP $[1 \times 10^{-3}$ to 5×10^{-3} Ns/m^2]) fluids are normally poor lubricators and one should be concerned about selecting the right bearings. For viscosities less than 1 cP, it is even more important to choose the right bearing material.

The Dow Chemical Company recommends canned-motor pumps or magnetic-drive pumps for phosgene service. Phosgene is an example of an extremely hazardous material. These pumps should have a secondary containment such that failure of the can does not create a phosgene release. The secondary containment should meet pipe specifications for pressure or relieve to the scrubber system in the plant. These pumps must have automated block valves on the suction and discharge. Operation of these valves should be managed such that the thermal expansion does not damage the pump.

A mistreated sealless pump can rupture with potentially serious results. The can can fail if valves on both sides of the pump are closed and the fluid in the pump expands either due to heating up from a cold condition or if the pump is started up. If the pump is run dry, the bearings can be ruined. The pump can heat up and be damaged if there is insufficient flow to take away heat from the windings. Sealless pumps, especially canned-motor pumps, produce a significant amount of heat, since nearly all the electrical energy lost in the system is absorbed by the fluid being pumped. *If this heat cannot be properly dissipated, the fluid will heat up with possibly severe consequences.* Considerable care must be used when installing a sealless pump to be sure that misoperations cannot occur.

The instrumentation recommended for sealless pumps may seem somewhat excessive. However, sealless pumps are expensive and they can be made to last for a long time, compared to conventional centrifugal pumps where seals may need to be changed frequently. Most failures of sealless pumps are caused by running them dry and damaging the bearings. Close monitoring of temperature is necessary in sealless pumps. Three temperature sensors (resistance temperature devices, or RTDs) are recommended: (1) in the internal fluid circulation loop, (2) in the magnet, or shroud, area, and (3) in the pump case area.

It is very important that sealless pumps be flooded with liquid before starting, to avoid damage to bearings from imbalance or overheating. Entrained gases in the suction can cause immediate imbalance problems and lead to internal bearing damage. Some type of liquid sensor is recommended. Sealless pumps must not be operated deadheaded (pump liquid full with inlet and/or outlet valves closed).

Properly installed and maintained, sealless pumps, both canned and magnetic-drive, offer an economical and safe way to minimize hazards and leaks of hazardous liquids.

Air Quality Regulatory Issues   (Englund and Holden, 1995.) Environmental issues and regulations have developed from matters of secondary interest on the part of business to broad-ranging measures which affect the fundamental ways in which companies carry out the details of their business. The fast pace of development of environmental regulations and their changing, sometimes inconsistent requirements have made it difficult and expensive for companies to keep their facilities current. Many companies have adopted programs to keep them aware of and abreast of continuing regulatory developments.

Although companies often fall under specific regulations pertaining only to their particular industry segments, a common thread running through a large number of manufacturing and commercial operations is the need to store materials considered to be toxic or hazardous. As a result, environmental regulations affecting the storage of toxic materials, either directly or indirectly, have had some of the most significant impacts on companies throughout the world in terms of both cost and operations. The Comprehensive Environmental Response, Compensation, and Liability Act of 1980, commonly known as Superfund, with its requirements to immediately report releases of reportable quantities of listed materials, has made the prevention of even minor spills, leaks, and releases from storage of toxic materials an important concern for owners. Also, storage and disposal of hazardous wastes are strictly and extensively regulated under the Resource Conservation and Recovery Act (RCRA) as amended in 1984.

In addition, restrictions on industrial air emissions under the Clean Air Act (CAA) as amended in 1977, the Clean Air Act Amendments (CAAA) of 1990, and other state and local statutes and regulations have universal impact on the storage of toxic materials, with direct and significant effects on the design and operation of toxic material storage facilities. Whereas the primary factors which once determined how air emissions from storage tanks were handled were fire protection and loss prevention, in recent years environmental protection concerns nearly always determine the extent and nature of the air emission controls required to be installed.

Permitting and Control Technology Requirements   (*Air Quality Handbook,* 1988.) Almost any process equipment or facility that emits air pollutants will need to obtain an air emission permit from the appropriate local, state, or federal governmental authority before construction or modification begins. The application for the air emission permit generally must describe the pollutant-generating process to be installed or modified, along with any emission control equipment or techniques, state the emission rates of all pollutants emitted, support the statement of emissions with a technical analysis or study, and describe the way that the process and control equipment will be operated to comply with regulatory requirements.

In reviewing the permit application, the local, state, or federal permitting authority will normally evaluate the application for completeness, check the accuracy of calculations, analyze the stated emissions for compliance with applicable regulations and environmental acceptability, and review the previous compliance history of the source and source owner. The source must operate within the bounds of the permit conditions in order to be considered in compliance with the permit. The source must still comply with all other air pollution laws, regulations, and ordinances, even if the permit conditions do not directly address them.

The process of evaluating air emission permit applications for large sources which are subject to federal permitting requirements is called New Source Review (NSR) and can be quite complicated, taking from six months to four years to complete. An NSR application would be required for a new source which could emit 100 tons per year or more of any criteria pollutant, after accounting for any air pollution control equipment.

The specific requirements to complete the NSR process will vary depending on the source location and characteristics, the federal and state regulations which apply, the compliance status of the facility if it is existing, and the nature of other sources in the area. Atmospheric dispersion modeling is often necessary to determine the maximum offsite ambient air concentrations of the various pollutants that will be emitted by the proposed new source or modification. All of the information in a permit application will normally be open to public scrutiny, including the details of the engineering study, except for specific process details that can be shown by the applicant to be trade secrets or proprietary business information. Nothing pertaining to the quality or quantity of air pollutant emissions can be claimed as proprietary.

The permitting requirements and procedures for a proposed NSR source are quite different, if the source is to be located in a nonattainment area for any of its major emitted pollutants, than if it is to be located in an attainment area. This usually increases the complexity of the permit application for such a source.

Federal Permitting in Nonattainment Areas   If the source subject to NSR is to be located in an area which is nonattainment for any of the major pollutants that the source will emit, it will need to follow the federally approved state permitting requirements for nonattainment areas of that pollutant. In most such cases, offsetting emission reductions at the same or other source locations in the area so as to be at least equivalent to the allowed emission, increases at the proposed source must be provided.

HAZARDOUS MATERIALS AND CONDITIONS

REACTIVE CHEMICALS

GENERAL REFERENCES: Bartknecht, *Dust Explosions*, Springer-Verlag, Berlin, 1989. Bartknecht, *Explosions Course Prevention, Protection*, Springer-Verlag, Berlin, 1981. Beever, "Hazards in Drying Thermally Unstable Powders," *Inst. Chem. Engr., Hazards X*, Symp. Ser. 115, 1989. Beever, "Scaling Rules for Prediction of Thermal Runaway," *Intl. Symp. on Runaway Reactions*, CCPS-AIChE, Cambridge, Mass., 1989. Beever and Thorne, "Isothermal Methods for Assessing Combustible Powders—Theoretical and Experimental Approach," *Inst. Chem. Eng.*, Symp. Ser. 68, 1981. Bowes, "A General Approach to the Prediction and Control of Potential Runaway Reaction," *Inst. Chem. Eng.*, Symp. Ser. 68, 1981. Bowes, *Self-heating: Evaluating and Controlling the Hazards*, Elsevier, 1984. *Bretherick's Reactive Chemical Hazards Database*, Version 2.0, Butterworth-Heinemann, Oxford, 1995. Brogli, Giger, Randegger, and Regenass, "Assessment of Reaction Hazards by Means of a Bench Scale Calorimeter," *Inst. Chem. Eng.*, Symp. Ser. 68, 1981. Brogli, Gygax, and Myer, "Differential Scanning Calorimetry—A Powerful Screening Method for the Estimation of the Hazards Inherent in Industrial Chemical Reactions," *Therm. Analy.*, 6th Symp., vol. 1, 549, 1980. Coates, "The ARC™ in Chemical Hazard Evaluation," *Chem. Ind.* **6:** 212, 1984. Coates, "The ARC™ in Chemical Hazard Evaluation," *Thermochim. Acta.* **85:** 369–372, 1984. Fenlon, "Calorimetric Methods Used in the Assessment of Thermally Unstable or Reactive Chemicals," *International Symposium for the Prevention of Major Chemical Accidents*, AIChE, 1987. Frank-Kamenetskii, *Diffusion and Heat Transfer in Chemical Kinetics*, Plenum Press, 1969. Grewer, "Direct Assessment and Scaling of Runaway Reaction and Decomposition Behavior Using Oven and DEWAR Tests," *Inst. Chem. Eng.*, Symp. Ser. 68, 1981. Grewer, "Use of DEWAR Calorimetry," *Proc. Runaway Chem. Reac. Haz. Symp.*, IBC, Amsterdam, November 1986. Halk, "SEDEX Versatile Instrument for Investigating Thermal Stability," *I. Chem. Eng.*, Symp. Ser. 68, 1981. Hub, "Adiabatic Calorimetry and SIKAREX Technique," *Inst. Chem. Eng.*, Symp. Ser. 68, 1981. Hub, "Experiences with the TSC 500 and RADEX Calorimeter," *Proc. Runaway Chem. Reac. Haz Symp.*, IBC, Amsterdam, November 1986. Hub, "Heat Balance Calorimetry and Automation of Testing Procedures," *Proc. Runaway Chem. Reac. Haz. Symp.*, IBC, Amsterdam, November 1986. *International Conference/Workshop on Modeling and Mitigating the Consequences of Accidental Releases of Hazardous Materials*, CCPS-AIChE, 1991. *International Conference on Hazard Identification and Risk Analysis, Human Factors and Human Reliability in Process Safety*, CCPS-AIChE, 1992. Kohlbrand, "Reactive Chemical Screening for Pilot Plant Safety," *Chem. Eng. Prog.* **81**(4): 52, 1985. Kohlbrand, "The Relationship Between Theory and Testing in the Evaluation of Reactive Chemical Hazards," *International Symposium for the Prevention of Major Chemical Accidents*, AIChE, 1987. Manufacturing Chemists Association, *Selected Case Histories*, Washington. D.C., 1951–1973. NFPA 491M, *Hazardous Chemical Reactions*, National Fire Protection Association, Quincy, Mass., 1991. Stull, "Fundamentals of Fire and Explosion," AIChE, 1976. Stull, "Linking Thermodynamics and Kinetics to Predict Real Chemical Hazards," *Loss Prevention*, AIChE, vol. 7, p. 67, 1976. Thomas, "Self Heating and Thermal Ignition—A Guide to Its Theory and Application," ASTM STP 502, ASTM, pp. 56–82, 1972. Townsend, "Accelerating Rate Calorimetry," *Inst. Chem. Eng.*, Symp. Ser. 68, 1981. Townsend and Kohlbrand, "Use of ARC™ to Study Thermally Unstable Chemicals and Runaway Reactions," *Proc. Runaway Chem. Reac. Haz. Symp.*, IBC, Amsterdam, November 1986. Townsend and Tou, "Thermal Hazard Evaluation by an Accelerating Rate Calorimeter," *Thermochem. Acta.* **73:** 1–30 and references therein, 1980. Urben, Peter, ed. (Courtaulds), *Bretherick's Handbook of Reactive Chemical Hazards*, 5th ed., Butterworth-Heinemann, Oxford, 1994.

Understanding the Reactive Chemicals and Reactive Chemicals Systems Involved   The main business of most chemical companies is to manufacture products through the control of reactive chemicals. The reactivity that makes chemicals useful can also make them hazardous. Therefore, it is essential that people who design or operate chemical processes understand the nature of the reactive chemicals involved.

Usually reactions are carried out without mishaps, but sometimes chemical reactions get out of control because of problems such as using the wrong raw material, using raw materials containing trace impurities, changed operating conditions, unanticipated time delays, equipment failure, or wrong materials of construction.

Such mishaps can be worse if the chemistry is not fully understood. A chemical plant can be *inherently safer* if knowledge of the chemistry of the process and the reactive chemicals systems involved is used in its design.

Reactive Hazard Review   Reactive hazards should be evaluated using reviews on all new processes and on all existing processes on a periodic basis. There is no substitute for experience, good judgment, and good data in evaluating potential hazards. Reviews should include:

1. Review of process chemistry, including reactions, side reactions, heat of reaction, potential pressure buildup, and characteristics of intermediate streams

2. Review of reactive chemicals test data for evidence of flammability characteristics, exotherms, shock sensitivity, and other evidence of instability

3. Review of planned operation of process, especially the possibility of upsets, modes of failure, unexpected delays, redundancy of equipment and instrumentation, critical instruments and controls, and worst-credible-case scenarios

Worst-Case Thinking   At every point in the operation, the process designer should conceive of the *worst* possible combination of circumstances that could *realistically* exist, such as loss of cooling water, power failure, wrong combination or amount of reactants, wrong valve position, plugged lines, instrument failure, loss of compressed air, air leakage, loss of agitation, deadheaded pumps, and raw-material impurities.

An engineering evaluation should then be made of the worst-case consequences, with the goal that the plant will be safe even if the worst case occurs. A HAZOP study could be used to help accomplish worst-case thinking. When the process designers know what the worst-case conditions are, they should:

1. Try to avoid worst-case conditions.
2. Be sure adequate redundancy exists.
3. Identify and implement lines of defense.
   a. Preventive measures
   b. Corrective measures

Sometimes, as a last resort, it may be desirable to use a high degree of process containment or, possibly, abandon the process if the hazard is unacceptable.

It is important to note that the worst case should be something that is realistic, not something that is conceivable but extremely unlikely. The Dow Chemical Company has adopted the following philosophy for design scenarios in terms of independent causative effects:

1. All single events that can actually and reasonably occur are credible scenarios.
2. Scenarios that require the coincident occurrence of two or more totally independent events are not credible design scenarios.
3. Scenarios that require the occurrence of more than two events in sequence are not credible.
4. A failure that occurs while an independent device is awaiting repair represents but one failure during the time frame of the initiation of the emergency and is therefore credible. The lack of availability of the unrepaired device is a preexisting condition.

Reactive Chemicals Testing   Much reactive chemical information involves thermal stability and the determination of (1) the temperature at which an exothermic reaction starts, (2) the rate of reaction as a function of temperature, and (3) heat generated per unit mass of material.

The evaluation of thermal stability requires the determination of the temperature at which an exothermic reaction occurs, the rate of such a reaction as a function of temperature, and the heat generated per unit mass of material by the reaction. In many cases, data on the increase of pressure during a reaction are also required, especially for vent sizing. The term *onset temperature* T_{onset} is used in two contexts:

1. In a testing context, it refers to the first detection of exothermic activity on the thermogram. The differential scanning calorimeter (DSC) has a scan rate of 10°C/min, whereas the accelerating rate calorimeter (ARC)° has a sensitivity of 0.02°C/min. Consequently, the temperature at which thermal activity is detected by the DSC can be as much as 50°C different from ARC data.

2. The second context is the process reactor. There is a potential for a runaway if the net heat gain of the system exceeds its total heat loss capability. A self-heating rate of 3°C/day is not unusual for a monomer storage tank in the early stages of a runaway. This corresponds to 0.00208°C/min, 10 percent of the ARC's detection limit. ARC data for the stored chemical would not show an exotherm until the self-heating rate was 0.02°C/min. Therefore, onset temperature information from ARC testing must be used with considerable caution.

Sources of Reactive Chemicals Data

Calculations   Potential energy that can be released by a chemical system can often be predicted by thermodynamic calculations. If there is little energy, the reaction still may be hazardous if gaseous products are produced. Kinetic data is usually not available in this way. Thermodynamic calculations should be backed up by actual tests.

Differential Scanning Calorimetry (DSC)   Sample and inert reference materials are heated in such a way that the temperatures are always equal. If an exothermic reaction occurs in the sample, the sample heater requires less energy than the reference heater to maintain equal temperatures. If an endothermic reaction occurs, the sample heater requires more energy input than the reference heater.

Onset-of-reaction temperatures reported by the DSC are higher than the true onset temperatures, so the test is mainly a screening test.

Differential Thermal Analysis (DTA)   A sample and inert reference material are heated at a controlled rate in a single heating block. If an exothermic reaction occurs, the sample temperature will

rise faster than the reference temperature. If the sample undergoes an endothermic reaction or a phase change, its temperature will lag behind the reference temperature.

This test is basically qualitative and can be used for identifying exothermic reactions. Like the DSC, it is also a screening test. Reported temperatures are not reliable enough to be able to make quantitative conclusions. If an exothermic reaction is observed, it is advisable to conduct tests in the ARC.

Mixing Cell Calorimetry (MCC)   The MCC provides information regarding the instantaneous temperature rise resulting from the mixing of two compounds. Together, DSC and MCC provide a reliable overview of the thermal events that may occur in the process.

Accelerating Rate Calorimeter (ARC)   The ARC can provide extremely useful and valuable data. This equipment determines the self-heating rate of a chemical under near-adiabatic conditions. It usually gives a conservative estimate of the conditions for and consequences of a runaway reaction. Pressure and rate data from the ARC may sometimes be used for pressure vessel emergency relief design. Activation energy, heat of reaction, and approximate reaction order can usually be determined. For multiphase reactions, agitation can be provided.

Nonstirred ARC runs may give answers that do not adequately duplicate plant results when there are reactants that may settle out or that require mixing for the reaction to be carried out (DeHaven and Dietsche, The Dow Chemical Company, Pittsburgh, Calif., "Catalyst Explosion: A Case History," *Plant Operations Progress*, April 1990).

An example of data from an ARC run is shown in Fig. 26-29.

Vent Sizing Package (VSP)   The VSP is an extension of ARC technology. The VSP is a bench-scale apparatus for characterizing runaway chemical reactions. It makes possible the sizing of pressure relief systems with less engineering expertise than is required with the ARC or other methods.

Reactive System Screening Tool (RSST)   The RSST is a calorimeter that quickly and safely determines reactive chemical hazards. It approaches the ease of use of the DSC with the accuracy of the VSP. The apparatus measures sample temperature and pressure within a sample containment vessel. The RSST determines the potential for runaway reactions and measures the rate of temperature and pressure rise (for gassy reactions) to allow determinations of the energy and gas release rates. This information can be combined with simplified methods to assess reactor safety system relief vent requirements. It is especially useful when there is a need to screen a large number of different chemicals and processes.

Shock Sensitivity   Shock-sensitive materials react exothermically when subjected to a pressure pulse. Materials that do not show an exotherm on a DSC or DTA are presumed not to be shock sensitive. Testing methods include:

• *Drop weight test.*   A weight is dropped on a sample in a metal cup. The test measures the susceptibility of a chemical to decompose explosively when subjected to impact. Weight and height can be varied to give semiquantitative results for impact energy. This test should be applied to any materials known or suspected to contain unstable atomic groupings.

• *Confinement cap test.*   Detonatability of a material is determined using a blasting cap.

• *Adiabatic compression test.*   High pressure is applied rapidly to a liquid in a U-shaped metal tube. Bubbles of hot compressed gas are

---

° ARC is the trademark of Columbia Scientific Industries Corporation.

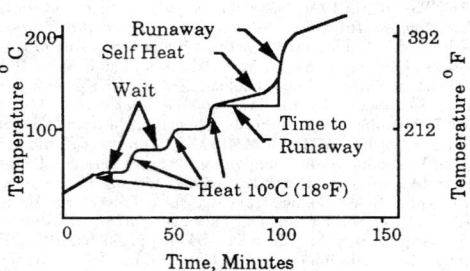

FIG. 26-29   Operation of the ARC.

driven into the liquid and may cause explosive decomposition of the liquid. This test is intended to simulate water hammer and sloshing effects in transportation, such as humping of railway tank cars. It is very severe and gives worst-case results.

Flammability–Flash Point   The closed-cup flash point determination produces the most important data to determine the potential for fire. The flash point is the lowest temperature at which the vapors can be ignited under conditions defined by the test apparatus and method.

Flammable Limits   Flammable limits, or the flammable range, are the upper and lower concentrations (in volume percent) which can just be ignited by an ignition source. Above the upper limit and below the lower limit no ignition will occur. Data are normally reported at atmospheric pressure and at a specified temperature. Flammable limits may be reported for atmospheres other than air and at pressures other than atmospheric.

Autoignition Temperature   The autoignition temperature of a substance, whether liquid, solid, or gaseous, is the minimum temperature required to initiate or cause self-sustained combustion in air, with no other source of ignition. Autoignition temperatures should be considered only as approximate. Test results tend to give temperatures that are higher than the actual autoignition temperature.

Dust Explosions   Combustible, dusty materials, with particle sizes less than approximately 200 mesh, can explode if a sufficient concentration in air is present along with an ignition source. The standard test has been designed to determine rates of pressure rise during an explosion, the maximum pressure reached, and the minimum energy needed to ignite the material. These data are useful in the design of safe equipment to handle dusty combustible materials in a process. Combustible dusts need a minimum volume to develop their full reaction velocity. Bartknecht states that for determination of explosion data of combustible dusts, a minimum volume of 16 L (4.23 gal) would be required to ensure correlation with data from large test vessels (Bartknecht, 1981, p. 39). This has been confirmed by comprehensive testing with a 20-L (5.28-gal) sphere.

Unstable Compounds   (Bretherick, L., British Petroleum Co. Ltd., *Handbook of Reactive Chemical Hazards*, 4th ed., Butterworths, London, pp. S1–S23, 1990, by permission of Butterworth-Heinemann and L. Bretherick; note that the 5th ed. is available in electronic format from *Bretherick's Reactive Chemical Hazards Database*, Version 2.0, Butterworth-Heinemann, Oxford, 1995.) Explosibility may be defined as the tendency of a chemical system (involving one or more compounds) to undergo violent or explosive decomposition under appropriate conditions of reaction or initiation. It is of great interest to be able to predict which compound or reaction systems are likely to exhibit explosability, and much work has been devoted to this end. The contributions of various structural factors (bond-groupings) have been evaluated in terms of heats of decomposition and oxygen balance of the compound or compounds involved in the system. Oxygen balance is the difference between the oxygen content of the chemical compound and that required to fully oxidize the carbon, hydrogen, and other oxidizable elements in the compound. Materials or systems approaching zero oxygen balance give the maximum heat release and are the most powerfully explosive.

Most chemical reactions are exothermic. In the few endothermic reactions that are known, heat is absorbed into the reaction product or products, which are known as endothermic or energy-rich compounds. Such compounds are thermodynamically unstable because heat would be released on decomposition of their elements. The majority of endothermic compounds possess a tendency toward instability and possibly explosive decomposition under various circumstances of initiation.

Following are the classes of compounds that have a tendency to undergo violent or explosive decomposition.

Acetylenic compounds	Acyl or alkyl nitrates
Alkyl hydroperoxides, peroxyacids	Alkyl perchlorates
Aminachromium peroxocomplexes	Aminemetal oxosalts
Ammonium perchlorates	Arenediazo aryl sulfides
Arenediazoates	Arenediazoniumolates
Azides (acyl, halogen, nonmetal, organic)	Azo compounds
Bis-arenediazo sulfides	Chlorite salts

Diazirines	Diazo compounds
Diazonium carboxylates or salts	Diazonium sulfides and derivatives, Xanthates
Difluoramino compounds	1,2-epoxides
Fluorodinitromethyl compounds	Halo-aryl metals
Haloacetylene derivatives	Haloarenemetal π-complexes
Halogen azides	Halogen oxides
High-nitrogen compounds	Hydrazinium salts
Hydroxylammonium salts	Hypohalites
Metal acetylides	Metal fulminates or *aci*-nitro salts oximates
N,N,N-trifluoroalkylamidines	N-azolium compounds
N-halogen compounds	N-haloimides
N-metal derivatives	N-nitro compounds
Nitroalkanes, *c*-nitro, and polynitroaryl compounds	Nitroso compounds
Oxosalts of nitrogeneous bases	Perchloric acid
Perchloryl compounds	Peroxides, (cyclic, diacyl, dialkyl), Peroxyesters
Poly(dimercuryimmonium salts)	Polynitroalkyl compounds
Tetrazoles	Triazenes
Trinitroethyl orthoesters	

COMBUSTION AND FLAMMABILITY HAZARDS

GENERAL REFERENCES: Bartknecht, *Explosions Course Prevention Protection*, Springer-Verlag, Berlin, 1981. Bartknecht, *Dust Explosions*, Springer-Verlag, Berlin, 1989. Beneditti, ed., *Flammable and Combustible Liquids Code Handbook*, 3d ed., based on NFPA 30, *Flammable and Combustible Liquids Code and Automotive and Marine Service Station Code*, National Fire Protection Association, Quincy, Mass., 1987. Bodurtha, Engineering Dept., Du Pont, *Industrial Explosion Prevention and Protection*, McGraw-Hill, 1980. Cawse, Pesetsky, and Vyn, *The Liquid Phase Decomposition of Ethylene Oxide*, Union Carbide Corporation, Technical Center, South Charleston, W.Va. *CPQRA (Guidelines for Chemical Process Quantitative Risk Analysis)*, CCPS-AIChE, New York, 1989. Crowl and Louvar, *Chemical Process Safety: Fundamentals with Applications*, Prentice Hall, Englewood Cliffs, N.J., 1990. DiNenno, P. J. (ed.), *SPFE Handbook of Fire Protection Engineering*, 1st ed., National Fire Protection Association, Quincy, Mass., 1988. Drysdale, *An Introduction to Fire Dynamics*, Wiley, Suffolk, 1985. *Fire Safety Data*, Fire Protection Association, London 1980. *Guidelines for Engineering Design for Process Safety*, CCPS-AIChE, 1993. Lees, *Loss Prevention in the Process Industries*, Butterworths, London, 1980. Stull, *Fundamentals of Fire and Explosion*, New York, 1976.

Introduction   The enchanting flame has held a special mystery and charm the world over for thousands of years. According to Greek myth, Prometheus the Titan stole fire from the heavens and gave it to mortals—an act for which he was swiftly punished. Early people made use of it anyway. Soon the ancients came to regard fire as one of the basic elements of the world. It has since become the familiar sign of the hearth and the mark of youth and blood—as well as the object of intense curiosity and scientific investigation.

Suitably restrained, fire is of great benefit; unchecked or uncontrolled, it can cause immense damage. We respond to it with a powerful fascination coupled with an inbred respect and fear. A good servant but a bad master is Thoreau's "most tolerable third party" (Cloud, "Fire, the Most Tolerable Third Party," *Michigan Natural Resources*, May-June 1990).

What Is Fire?   (DiNenno, 1988.) Fire or combustion is normally the result of fuel and oxygen coming together in suitable proportions and with a source of heat. The consumption of a material by a fire is a chemical reaction in which the heated substance combines with oxygen. Heat, light, smoke, and products of combustion are generated. The net production of heat by a fire involves both heat-producing and heat-absorbing reactions, with more heat being produced than is absorbed.

A flame is a rapid self-sustaining chemical reaction that occurs in a distinct reaction zone. Two basic types of flame are (1) the diffusion flame, which occurs on ignition of a fuel jet issuing into air, and for which the limiting rate is controlled by diffusion, and (2) the aerated, or premixed, flame, which occurs when fuel and air are premixed before ignition, and for which the limiting rate is controlled by reaction kinetics. The main concern in fire applications is with diffusion flames, as contrasted to premixed flames where fires and oxidants are premixed or brought together to the combustion region. At high turbulence in diffusion flames, kinetics and diffusion may be of roughly equal importance. Combustion is self-propagating; burning materials produce heat, which causes more of the solid to produce flammable vapors until either the fuel or oxygen is exhausted, or until the fire is extinguished in some other way (*API Recommended Practices*, 3d ed., Practice 521, November 1990).

Energy in the form of heat is required:

1. To produce vapors and gases by vaporization or decomposition of solids and liquids. Actual combustion usually involves gases or vapors intimately mixed with oxygen molecules.

2. To energize the molecules of oxygen and flammable vapors into combining with one another and so initiating a chemical reaction.

The amount of energy needed to cause combustion varies greatly. Hydrogen and carbon disulfide can be ignited by tiny sparks or simply by static generated as the gases or vapors discharge from pipes into air. Other materials, such as methylene chloride, require such large amounts of energy to be ignited that they are sometimes considered nonflammable.

Fire can also result from the combining of such oxidizers as chlorine and various hydrocarbon vapors; oxygen is not required for a fire to take place.

Ordinarily, combustible solids do not combine directly with oxygen when they burn. They give off vapor and gaseous decomposition products when they are heated, and it is the vapors or gases which actually burn in the characteristic form of flames. Thus, before a solid can be ignited it must be heated sufficiently for it to give off flammable concentrations of vapors. There are exceptions to the general rule that a solid must vaporize or decompose to combine with oxygen; some finely divided materials such as aluminum powder and iron powder can burn and it is probable that they do not vaporize appreciably before burning. Some metallic dusts will explode in air by light radiation alone without conduction and convection (Bartknecht, 1989, p. 14).

Products of Combustion   Heat, light, smoke, and asphyxiating and toxic gases are produced by fire. In a hot, well-ventilated fire, combustion is usually nearly complete. Nearly all the carbon is converted to carbon dioxide, all the hydrogen to steam, and oxides of various other elements such as sulfur and nitrogen are produced.

This is not the case in most fires where some of the intermediate products, formed when large, complex molecules are broken up, persist. Examples are hydrogen cyanide from wool and silk, acrolein from vegetable oils, acetic acid from timber or paper, and carbon or carbon monoxide from the incomplete combustion of carbonaceous materials. As the fire develops and becomes hotter, many of these intermediates, which are often toxic, are destroyed—for example, hydrogen cyanide is decomposed at about 538°C (1000°F).

Small airborne particles of partially burnt carbonaceous materials from smoke, which is often made more opaque by steam from combustion or from water added to the fire, may be formed when there is only partial combustion of fuel.

Many hydrocarbon flames are luminous because of the incandescent carbon particles formed in the flames. Under certain conditions, these particles are released from the luminous flames as smoke. Smoke from hydrocarbons is usually formed when the system is fuel rich, either overall or locally.

Common materials—such as textiles in the form of fibers or fabrics, foamed rubber, foamed plastics, thin sheets of plastic, paper, corrugated cardboard, combustible dusts, dry grass and twigs, and wood shavings—are all examples of materials with large surface areas in relation to their volumes. In a well-established fire, materials with relatively small surface areas, such as chunks of coal or logs, burn readily.

Combustible Dusts   Dusts are particularly hazardous; they have a very high surface area–to–volume ratio. When finely divided as powders or dusts, solids burn quite differently from the original material in the bulk. Dust and fiber deposits can spread fire across a room or along a ledge or roof beam very quickly. On the other hand, accumulations of dust can smolder slowly for long periods, giving little indication that combustion has started until the fire suddenly flares up, possibly when no one suspects a problem.

Many combustible dusts produced by industrial processes are explosible when they are suspended as a cloud in air. A spark may be sufficient to ignite them. After ignition, flame spreads rapidly through the dust cloud as successive layers are heated to ignition temperature.

The hot gases expand and produce pressure waves, which travel ahead of the flame. Any dust lying on surfaces in the path of the pressure waves will be thrown into the air and could cause a secondary explosion more violent and extensive than the first.

Liquids   A vapor has to be produced at the surface of a liquid before it will burn. Many common liquids give off a flammable concentration of vapor in air without being heated, sometimes at well below room temperature. Gasoline, for example, gives off ignitable vapors above about −40°C (−40°F), depending on the blend. The vapors are easily ignited by a small spark or flame. The reason there are not many fires in automobile gasoline tanks is that the vapor space above the gasoline is almost always above the upper flammability limit. Other liquids, such as fuel oil and kerosene, need to be heated until sufficient vapor is produced to produce a flammable concentration.

For any flammable vapor there are maximum and minimum concentrations of vapor in air beyond which it cannot burn. When the concentration of vapor in air is too low, there is insufficient fuel for burning; when it is too high, there is insufficient oxygen for burning.

If the density of a vapor is greater than air, as is the case with most gases and vapors encountered in industry, flammable concentrations may collect at low levels, such as at floor level or in basements, and can travel considerable distances to a source of ignition and the flames will then flash back.

Gases   Flammable gases are usually very easily ignited if mixed with air. Flammable gases are often stored under pressure, in some cases as a liquid. Even small leaks of a liquefied flammable gas can form relatively large quantities of gas, which is ready for combustion.

Transparent (Invisible) Flames   Some materials have nearly nonluminous flames, which may not be visible, especially in the daytime. For example, hydrogen has a nearly nonvisible flame in the daytime. A person may walk unaware into a hydrogen leak flame. Some other materials, including some alcohols such as methanol, also have nearly nonluminous flames and may be unusually hazardous because the flames cannot be seen in the daytime.

The Fire Triangle   The well-known *fire triangle* (see Fig. 26-33) is used to represent the three conditions necessary for a fire: (1) fuel, (2) oxygen or other oxidizer (a gaseous oxidizer such as chlorine, a liquid oxidizer such as bromine, or a solid oxidizer such as sodium bromate), and (3) heat (energy).

If one of the conditions in the fire triangle is missing, fire does not occur, and if one is removed, fire is extinguished. Usually a fire occurs when a source of heat contacts a combustible material in air, and then the heat is supplied by the combustion process itself.

The fire triangle indicates how fires may be fought or prevented:

1. Cut off or remove the fuel.
2. Remove the heat—usually done by putting water on the fire.
3. Remove the supply of oxygen—usually done by foam or inert gas.

Stoichiometric Concentration   (Used by permission of Frank T. Bodurtha, Inc., New London, New Hampshire.) In a combustion reaction in air, the stoichiometric concentration, C_{st}, of any reactant is the concentration theoretically required for complete conversion by reacting completely with oxygen. For example, for the combustion of propane in air:

$$\underline{\qquad\text{Air}\qquad}$$
$$C_3H_8 + 5O_2 + 18.8N_2 \rightarrow 3CO_2 + 4H_2O + 18.8N_2$$

$$C_{st} \text{ for propane} = \left(\frac{1}{1 + 5 + 18.8}\right)100 = 4.0\% \text{ (volume)}$$

Moles at start = 24.8; moles after combustion = 25.8.

The change in moles upon the combustion of propane and many other hydrocarbons is either zero or small. Usually, pressure rise in the combustion of a vapor or gas is due mainly to change in temperature, not change in moles.

The C_{st} for a flammable solid may also be calculated. For sugar, whose molecular weight is 342, C_{st} is calculated as follows:

$$\underline{\text{—Sugar—}}\quad\underline{\text{—Air—}}$$
$$C_{12}H_{22}O_{11} + 12O_2 + 46.4N_2 \rightarrow 12CO_2 + 11H_2O + 46.4N_2$$
$$\text{58.4 volumes}\qquad\qquad\text{69.4 volumes}$$

TABLE 26-10 Flammability Limits, Autoignition Temperature, and Flash Points of Selected Substances in Air at Atmospheric Pressure

Chemical compound	Flam. limits, lower, % v/v	Flam. limits, upper, % v/v	Autoignition temperature, °C	Flash point, closed cup, °C	Flash point, open cup, °C
Acetone	2.6	13	465	−18	−9
Acetylene	2.5	100	305	—	—
Ammonia	15	28	651°	—	—
Benzene	1.4°	8.0°	562°	−11	—
n-Butane	1.8	8.4	405	−60	—
Carbon disulfide	1.3	50	90	−30	—
Carbon monoxide	12.5	74	—	—	—
Cyclohexane	1.3	7.8	245	−20	—
Ethane	3.0	12.4	515	−135	—
Ethylene	2.7	36	490	−121	—
Ethylene dichloride	6.2°	15.9°	413°	13	18
Ethylene oxide	3°	100°	429°	—	−20
Hydrogen	4	75	400	—	—
Methane	5	15	540	—	—
Propane	2.1	9.5	450	<−104	—
Propylene	2.4	11	460	−108	—
Styrene	1.1°	6.1°	490°	32	38
Toluene	1.3°	7.0°	536°	4	7
Vinyl chloride	4°	22°	472°	—	−78

°Factory Mutual Engineering Corporation, 1967.
SOURCES: Lees, 1980.
Flammability limits and autoignition temperatures: Zabetakis, *Bureau of Mines Bulletin 627*, except where given in footnotes.
Flash points: Factory Mutual Engineering Corporation, 1967.

$$C_{st} = \frac{(342)(1000)}{(12 + 46.4)(22.4)(298/273)} = 239.5 \text{ mg/L air at } 25°C$$

Table 26-10 shows flammability limits, autoignition temperature, and flash points of selected substances in air at atmospheric pressure.

Burning in Pure Oxygen The flammability of a substance depends strongly on the partial pressure of oxygen in the atmosphere. Increasing oxygen content affects the lower flammability limit only slightly, but it has a large effect on the upper flammability limit. Increasing oxygen content has a marked effect on the ignition temperature (reduces it) and the burning velocity (increases it). Use of air enriched with oxygen, or pure oxygen, can greatly increase the hazards of combustion reactions.

Burning in Other Oxidizable Atmospheres Chemically, oxygen is not the only oxidizing agent, though it is the most widely recognized and has been studied the most. Halogens are examples of oxidants that can react exothermically with conventional fuels and show combustion behavior. The applicability of flammability limits applies to substances that burn in chlorine. Chlorination reactions have many similarities to oxidation reactions. They tend not to be limited to thermodynamic equilibrium and often go to complete chlorination. The reactions are often highly exothermic. Chlorine, like oxygen, forms flammable mixtures with organic compounds. Flames can also propagate in mixtures of oxides of nitrogen and other oxidizable substances.

Flame Quenching Flame propagation is suppressed if the flammable mixture is held in a narrow space. If the space is sufficiently narrow, flame propagation is suppressed completely. The largest diameter at which flame propagation is suppressed is known as the *quenching diameter*. For an aperture of slotlike cross section, there is a critical slot width. The term *quenching distance* is sometimes used as a general term covering both quenching diameter and critical slot width and sometimes meaning only the latter.

There is a maximum safe gap measured experimentally which will prevent the transmission of an explosion occurring within a container to a flammable mixture outside the container. Critical and maximum experimental safe gaps for a number of materials in air are listed in Lees (1980, pp. 491–492). These quenching effects are important in the design of flame arresters and flameproof equipment.

Heterogeneous Mixtures (Zabetakis, *Flammability Characteristics of Combustible Gases and Vapors*, Bulletin 627, Bureau of Mines, 1965.) Heterogeneous (poorly mixed) gas phase mixtures can lead to fires that normally would be totally unexpected. It is important to recognize that heterogeneous mixtures can ignite at concentrations that would normally be nonflammable if the mixture were homogeneous. For example, 1 L of methane can form a flammable mixture with air at the top of a 100-L container, although the mixture would contain only 1.0 percent methane by volume. This would be below the lower flammable limit if complete mixing occurred at room temperature and the mixture would not be flammable. This is an important concept since *layering* can occur with any combustible gas or vapor in both stationary and flowing mixtures.

Heterogeneous mixtures are formed, at least for a short time, when two gases or vapors are first brought together.

Explosions in the Absence of Air Some gases with positive heats of formation can be decomposed explosively in the absence of air. Ethylene reacts explosively at elevated pressure and acetylene at atmospheric pressure in large-diameter piping. Heats of formation of these materials are +52.3 and +227 kJ/mol (+22.5 and +97.6 × 10³ Btu/lb mol), respectively.

Explosion prevention can be practiced by mixing decomposable gases with inert diluents. For example, acetylene can be made nonexplosive at a pressure of 100 atm (10.1 MPa) by including 14.5 percent water vapor and 8 percent butane (Bodurtha, 1980). One way to prevent the decomposition reaction of ethylene oxide vapor is to use methane gas to blanket the ethylene oxide liquid.

GAS EXPLOSIONS

GENERAL REFERENCES: Bartknecht, *Explosions*, Springer-Verlag, New York, 1981. Bodurtha, *Industrial Explosion Prevention and Protection*, McGraw-Hill, New York, 1980. Coward and Jones, "Limits of Flammability of Gases and Vapors," *U.S. Bur. Mines Bull.* 503 (USNTIS AD-710 575), 1952. Lees, *Loss Prevention in the Process Industries*, vols. 1 and 2, Butterworths, London, 1980. National Fire Protection Association, *Explosion Prevention Systems*, NFPA 69, Quincy, Mass. National Fire Protection Association, *Venting of Deflagrations*, NFPA 68, Quincy, Mass. Zabetakis, "Flammability Characteristics of Combustible Gases and Vapors," *U.S. Bur. Mines Bull.* 627 (USNTIS AD-710 576), 1965. NOTE: *NFPA reviews, and may change, its standards and guides periodically. Always check the latest edition.*

Fuel and Oxygen

Flash Point and Flammable Limits Flash points and flammable limits in percent by volume have been tabulated by the National Fire Protection Association (NFPA) (National Fire Protection Association, *Fire Hazard Properties of Flammable Liquids, Gases, and*

Volatile Solids, NFPA 325, Quincy, Mass.). Pressure particularly affects flash point and the *upper flammable limit* (UFL); see later section entitled "Effect of Temperature, Pressure, and Oxygen." Mists of high-flash-point liquids may be flammable; the *lower flammable limit* (LFL) of fine mists and accompanying vapor is about 48 g/m³ of air, basis 0°C and 1 atm (0.048 oz/ft³).

For practical purposes, LFL is the same as *lower explosive limit* (LEL). (Ignitability limits depend upon the strength of the ignition source; the ignitability range for relatively weak ignition sources is less than the flammable range.) LFLs in percent by volume generally decrease as molecular weight increases.

The equilibrium vapor pressure of a flammable liquid at its closed-cup flash point about equals its LFL in percent by volume. Thus, the vapor pressure of toluene at its closed-cup flash point (4.4°C or 40°F) of 1.2 percent (1.2 kPa) is close to its LFL of 1.1 percent. The composite LFL of a mixture may be estimated by Le Chatelier's Rule:

$$\text{Composite LFL (\% by vol.)} = \frac{100}{C_1/\text{LFL}_1 + C_2/\text{LFL}_2 + \cdots + C_n/\text{LFL}_n}$$

(26-43)

where the *C*s are percentages of volume of total fuel, i.e., without air or inert gas. As shown in Table 26-11, the indicated mixture is flammable even though each component is below its LFL. The composite LFL stays the same if the concentration of each component is multiplied by the same number. Composite upper flammable limits may be approximated similarly.

The concentration of fuel in air in a process should be maintained at or below 25 percent of the LFL, with automatic instrumentation and safety interlocks; however, up to 60 percent of LFL is permitted by the NFPA—except for ovens or furnaces. (Ovens and furnaces are covered in NFPA 86.)

Limiting Oxidant Concentration (LOC) It is often prudent to base explosion prevention on inerting. The LOC is the concentration of oxidant—normally oxygen—below which a fuel-oxidant explosion cannot occur. (The LOC is also called MOC, the minimum oxygen for combustion.) With adequate depletion of oxygen, an explosion cannot occur *whatever the concentration of fuel.* Nevertheless, in these circumstances a fuel–air–inert gas mixture may become flammable if sufficient air is added. Many LOCs are given in NFPA 69. In general, organic flammable gases or vapors will not propagate flame in mixtures of the organic, added nitrogen, and air below about 10.5 percent by volume O₂ at 1 atm and near normal room temperature. Hydrogen (LOC = 5 percent) and some other inorganic gases have lower LOCs.

For LOCs of 5 percent and greater, the O₂ concentration should not exceed 60 percent of the LOC, but with continuous monitoring the O₂ may be kept 2 percent below the LOC (NFPA 69, 1992). Neutronics, Inc., of Exton, Pennsylvania, supplies an inerting control system that has had wide application in many industries.

Explosion prevention by inerting has several advantages over explosion protection techniques, such as explosion venting. For example, with successful inerting, fires or business interruptions cannot occur. Nevertheless, beware of the potential of asphyxiation with inerting; proper vessel entry procedures must be implemented and occasionally it may be prudent to monitor for oxygen in workplaces.

Effect of Temperature, Pressure, and Oxygen LFLs and LOCs at 1 atm decrease about 8 percent of their values at near normal room temperature for each 100°C increase. Upper flammable limits increase approximately 8 percent for the same conditions.

TABLE 26-11 Le Chatelier's Rule

	Concentration, vol %	C, %	LFL, vol %
Hexane (1)	0.8	24.2	1.1
Methane (2)	2.0	60.6	5.0
Ethylene (3)	0.5	15.2	2.7
Total fuel	3.3		
Air	96.7		

$$\text{Composite LFL} = \frac{100}{24.2/1.1 + 60.6/5.0 + 15.2/2.7}$$
$$= 2.5 \text{ vol \%}$$

Pressure affects flash point. A decrease in pressure lowers the flash point. With toluene, for example, at two-thirds of an atmosphere the vapor pressure must be only 0.74 kPa (5.6 mm Hg) to equal the LFL of 1.1 percent. (No significant difference in LFL will exist at two-thirds of an atmosphere compared to the published LFL of 1.1 percent at one atmosphere.) This vapor pressure occurs at −3°C, corresponding to a decrease in flash point of about 7.4°C from one atmosphere. Conversely, an increase in pressure raises the flash point.

Pressure also affects flammable limits. A decrease in pressure to about one-half atmosphere does not affect the flammable range significantly. At lower pressure the flammable range narrows and the flammable limits may disappear below about 6.7 kPa (50 mm Hg). An increase in pressure lowers LFLs and LOCs on a volume basis only slightly. But on a *weight basis,* LFLs are proportional to the absolute pressure. For example, the LFL of hexane is 43 g/m³ (0.043 oz/ft³) air at 1 atm, basis 0°C, but it is 86 g/m³ air (0.086 oz/ft³) at 2 atm. An increase in pressure increases UFLs greatly. The effect of elevated pressure on LFLs, LOCs, and UFLs for ethane is tabulated in Table 26-12. Based on tests by the U.S. Bureau of Mines, UFLs at high pressure and near normal room temperatures may be *estimated* by

$$\text{UFL}_p = \text{UFL} + 20.6 \, (\log_{10} P + 1)$$

(26-44)

where UFL*p* (percent by volume) is at the elevated absolute pressure, *P* is in megapascals absolute (MPa), and UFL (percent by volume) is at 1 atm. LFLs are about the same in oxygen as in air, since oxygen in air is in excess for combustion at LFL, but UFLs increase markedly in oxygen compared to air, as shown by the examples in Table 26-13. For organic substances, UFLs at 1 atm are about 48 percent higher in oxygen than in air. Moreover, the *minimum ignition energies* (MIE) in oxygen are about 1/100 of the MIEs in air, so vapors in oxygen are extraordinarily easy to ignite.

Ignition Sources Normally it is best practice not to base explosion safety solely on the presumed absence of an ignition source. Explosion control should be based on prevention or protection techniques, or both. Even so, all reasonable measures should be taken to eliminate ignition sources.

TABLE 26-12 Effect of Elevated Pressure on LFL, LOC, and UFL of Ethane*†

Pressure MPa gauge, psig		LFL, vol %	% decrease in LFL	LOC, vol %	% decrease in LOC	UFL, vol %	% increase in UFL
0	(0)	2.85	—	11.0	—	12.3	—
0.69	(100)	2.80	1.75	—	—	30.0	144
1.72	(250)	2.70	5.20	9.3	15.5	40.0	225
3.45	(500)	2.55	10.5	8.9	19.1	47.0	282
5.17	(750)	2.40	15.8	—	—	50.0	306
6.20	(900)	—	—	8.8	20.0	—	—
6.90	(1000)	2.20	22.8	—	—	51.5	319

*Nitrogen as inert gas. Near normal room temperature. % decrease and increase is from 0 MPa gauge.

†After Kennedy, Spolan, Mock, and Scott, "Effect of High Pressures on the Explosibility of Mixtures of Ethane, Air and Carbon Dioxide, and of Ethane, Air and Nitrogen," *U.S. Bureau of Mines Report Invest.,* 4751, 1950.

TABLE 26-13 Flammability Limits in Air and Oxygen at Ordinary Temperatures and 1 atm*

	LFL in air, vol %	LFL in O₂, vol %	UFL in air, vol %	UFL in O₂, vol %	Δ UFL, vol %
Butane	1.9	1.8	8.5	49	40.5
1-Butene	1.6	1.8	9.3	58	48.7
Ethane	3.0	3.0	12.5	66	53.5
Ethylene	3.1	3.0	32	80	48
Isopropyl ether	—	—	21	69	48
Methane	5.3	5.1	14	61	47
Propane	2.2	2.3	9.5	55	45.5
Vinyl chloride	4.0	4.0	22	70	48

*Based on Coward and Jones, "Limits of Flammability of Gases and Vapors," *U.S. Bureau of Mines Bulletin 503* (USNTIS AD-701575), 1952.

Autoignition The minimum autoignition temperature (AIT) of a substance is the minimum temperature at which vapors ignite spontaneously from the heat of the environment. (Flash points are lower than minimum autoignition temperatures; in flash-point tests an open flame is used as an igniter.) The ignition temperature found in NFPA 325 is the same as minimum autoignition temperature. A method for determination of autoignition temperatures is given in E659, *Standard Test Method for Autoignition Temperatures of Liquid Chemicals*, of the American Society for Testing and Materials (ASTM). Autoignition depends on many factors—namely, ignition delay, concentration of vapors, environmental effects (volume, pressure, and oxygen content), catalytic material, and flow conditions. Based on a development by A. Beerbower, Exxon Research and Engineering, for the effect of volume on AIT (see Coffee, "Cool Flames and Autoignitions: Two Oxidation Processes," *Chem. Eng. Prog. 13th Loss Prev. Symp.*, Houston, 1979, pp. 79–82):

$$t_a = \frac{(t_d - 75)(15 - \log_{10} V_a)}{(15 - \log_{10} V_d)} + 75 \qquad (26\text{-}45)$$

where t_a (°C) = minimum AIT for volume V_a (mL)
  t_d (°C) = minimum AIT as measured in volume V_d (mL)

For a plant vessel of 3.785×10^6 mL (1000 U.S. liquid gallons) and $t_d = 330$°C in a 500-mL test vessel (ASTM E659), $t_a = 249$°C, i.e., 81°C less than measured in the 500-mL flask.

As a guide, because of convection that occurs from hot surfaces, ignition by a hot surface in open air should not be assumed unless the surface temperature is at least 200°C above the published minimum autoignition temperature (American Petroleum Institute, *Ignition Risk of Hot Surfaces*, API PSD 2216, Washington, 1980).

Autooxidation Autooxidation—spontaneous ignition—is the phenomenon of self-heating by slow oxidation with accompanying evolution of heat, leading to ignition when the heat of oxidation cannot be dissipated adequately. Thermal insulation or rags wet with oils or other organic liquids susceptible to oxidation have caused serious fires. A fire may occur even if the hot surface is below the minimum AIT. Relatively high flash-point materials are most susceptible to autooxidation; low-flash-point materials may evaporate without ignition. Leaks that may contaminate insulation should be eliminated or diverted away from insulation to the extent feasible. Thermal insulation known to be wetted with oil or other high-boiling organic fluids should be removed promptly and replaced.

Compression Adiabatic compression results in high temperatures determined by the compression and specific heat ratios, as shown in Eq. (26-46):

$$\frac{T_2}{T_1} = \left(\frac{P_2}{P_1}\right)^{(k-1)/k} \qquad (26\text{-}46)$$

where the subscript 2 refers to the final state, T is the absolute temperature, P is the absolute pressure, and k is the ratio of specific heats.

Various types of rapid, adiabatic compressions have caused explosions. With propane at an initial temperature of 25°C, $T_2 = 432$°K (159°C) for compression and specific heat ratios of 25 and 1.13, respectively. Assume that now air enters a compressor to bring propane into the flammable range at 5 percent by volume. The mixture then will be mostly air with $k = 1.47$. The same compression ratio of 25 will elevate the final temperature T_2 to 834°K (561°C), i.e., above the published autoignition temperature of 450°C for propane and perhaps high enough to cause an explosion.

Other Ignition Sources Hazardous classification of locations for electrical installations is covered in Articles 500–504 of the *National Electrical Code* (NEC) (NFPA 70). Proper hazardous classification is essential for safety and for prevention of explosion and fire losses. Class 1 in the NEC is for vapors and gases; in the United States, in brief, Division 1 of Class 1 includes those locations where flammable concentrations exist continuously or frequently. Division 2 includes locations where flammable concentrations may exist only in case of accidental escape of vapors or gases, or in case of abnormal operation of equipment. Static electricity, which causes fires and explosions with flammable vapors and gases, is covered later in Sec. 26. Other ignition sources include friction and impact plus rubbing; with rubbing beware of metal-to-metal contact where heat cannot be conducted away. Even with grounded equipment, *hydrogen may ignite spontaneously*, often, for instance, when it exits a stack or leaks out of a pipe. (The minimum ignition energy of hydrogen is about 0.02 mJ, approximately 1/10 of the MIEs for paraffin hydrocarbons.)

Explosion Pressure An *explosion* is the action of "going off" with a loud noise under the influence of suddenly developed internal energy. Thus, an explosion is a result, not a cause. Deflagrations and detonations cause chemical explosions. A *deflagration* is a reaction that propagates to the unreacted material at a speed less than the speed of sound in the unreacted material. A *detonation* is a reaction that propagates to the unreacted material at a speed greater than the speed of sound in the unreacted material; it is accompanied by a shock wave and inordinately high pressure.

Deflagration Pressure The increase in pressure in a vessel from a deflagration results from an increase in temperature; the actual maximum flame temperature for propane, for example, is 1925°C (3497°F). No significant increase in moles of gas to cause pressure buildup results from combustion of propane in air.

Peak deflagration pressure in closed equipment is approximately eight times the initial absolute pressure, whether atmospheric, subatmospheric, or elevated. This maximum pressure occurs at a concentration just slightly richer in fuel than the stoichiometric concentration for combustion in air (C_{st}), as shown in Table 26-14 for propane and methane:

Some flammable liquids generate a vapor pressure close to the C_{st} near normal room temperature, as shown in Table 26-15.

Toluene is a notoriously poor electrical conductor; even in grounded equipment it has caused several fires and explosions from static electricity. Near normal room temperature it has a concentration that is one of the easiest to ignite and, as previously discussed, that generates maximum explosion effects when ignited (Bodurtha, 1980, p. 39). Methyl alcohol has similar characteristics, but it is less prone to ignition by static electricity because it is a good conductor. Acetone is also a good conductor, but it has an equilibrium vapor pressure near normal room temperature, well above UFL. Thus, acetone is not flammable in these circumstances.

Several environmental factors affect maximum deflagration pressure and pressure rise, as highlighted in Table 26-16.

TABLE 26-14 Optimum Concentrations for Maximum Deflagration Pressure

	Stoichiometric concentration, vol %	Maximum deflagration pressure, vol %
Propane	4.0	5.0
Methane	9.5	10.3

TABLE 26-15 Liquids Having Equilibrium Vapor Pressure near the C_{st}

	Equilibrium vapor pressure at 20°C, vol %	C_{st}, vol %
Toluene	2.9	2.3
Methyl alcohol	12.6	12.2
Acetone (UFL = 12.8)	24.6	5.0

TABLE 26-16 Effect of Environmental Factors on Deflagration Pressure and Pressure Rise*

	Pressure	Temperature	Vessel volume	Turbulence	Strength of ignition
Maximum deflagration pressure	+	−	Minor	Minor	Minor
Maximum rate, deflagration pressure rise	+	+	−	+	+

*Unvented vessel: + = an increase in indicated factor increases pressure or pressure rise.

 − = an increase in indicated factor decreases pressure or pressure rise.

Vessel volume has a large effect on the maximum rate of deflagration pressure rise; the cubic law states, all else being equal

$$(r_m)(V^{1/3}) = \text{constant } (K_G) \qquad (26\text{-}47)$$

where r_m = maximum rate of deflagration pressure rise, bar/s
V = volume, m^3
K_G = deflagration index for gases, bar·m/s

In compartmented equipment, higher deflagration pressure than noted in the preceding discussion can occur from pressure piling. After ignition in the first compartment, some of the gas mixture ahead of the flame front is pushed through a connection between the two compartments. Pressure of the original flammable mixture in the second compartment increases, and the resulting now-compressed mixture is ignited by the flame from the first compartment; abnormally high deflagration pressure may occur in this second compartment. This pressure piling effect is an important one and may pose difficult safety problems in design, if flammable mixtures cannot be prevented. As a rough guide, the ratio of compartment volumes of at least 5 to 1 apparently is required for pressure piling; generally, initial ignition must be in the larger compartment (Fitt, "Pressure Piling: A Problem for the Process Engineer," *Chem. Eng.* [Rugby, England], no. 368, pp. 237–239, May 1981.)

Detonation A deflagration can develop into a gaseous detonation in vessels and piping under certain conditions with enhanced explosion effects. Many factors affect detonation formation and effects. Briefly, upon ignition, pressure waves in a closed tube move through unburned gas. Subsequent waves move faster through the unburned gas, because of heating from previous pressure waves. Adiabatic compression results in high enough temperature to ignite gas ahead of the original flame and a detonation develops. (This ignition by compression to form a detonation is sometimes also called *pressure piling*.) The peak pressure in a stable detonation is on the order of 30 times the initial absolute pressure, disregarding the usually nondamaging spike of still higher pressure; reflected pressure is much higher than this 30 multiplier. (Special review is necessary for overpressure developed in an unstable [overdriven] detonation.) Nevertheless, in usual plant vessels without large length/diameter ratios, detonation is unlikely at 1 atm and near normal room temperature. Strong equipment may be subject to damage in a detonation, and rupture disks alone cannot control a detonation. Flame arresters are now commonly used to help protect against detonations; see the article on flame arresters in this section. But the best procedure to guard against the destructive effects of detonations is to prevent the formation of flammable mixtures.

Explosion Protection Where prevention of flammable mixtures may not be feasible, protection facilities must be installed; sometimes, too, backup explosion protection facilities are used in conjunction with inerting systems. Containment, suppression, or venting are used for protection against internal deflagrations in fuel-air mixtures. Although these methods may protect against deformation or rupture of a vessel, damage to internal appurtenances may still occur. Containment and suppression prevent the discharge of environmentally unacceptable materials to the atmosphere.

Containment The design pressure (maximum allowable working pressure) to prevent rupture of equipment for most gas-air mixtures initially at 1 atm should be 304 kPa gauge (44.1 psig), and to prevent permanent deformation, 608 kPa gauge (88.2 psig) (National Fire Protection Association, *Explosion Prevention Systems*, NFPA 69, Quincy, Mass., 1992, p. 11). NFPA 69 provides important additional design information on deflagration pressure containment.

Explosion Suppression With explosion suppression, an incipient explosion is detected and—within a few milliseconds—a suppressant is discharged into the exploding medium to stop combustion. Pressure and optical detection systems are used; suppressors are pressurized and release the suppressants when actuated by an electroexplosive device.

Deflagration pressure can be reduced substantially by suppression. Figure 26-30 shows the pressures measured in an ethylene-air explosion and a sodium bicarbonate-suppressed ethylene-air explosion. Fike Corporation, Blue Springs Missouri, and Fenwal Safety Systems, Marlborough, Mass., supply explosion suppression systems.

To reduce the chance of false activation of the suppression system by vibration, a flexible pressure detector standoff is often used. Also, two detectors in series may be employed to reduce further the possibility of false activation.

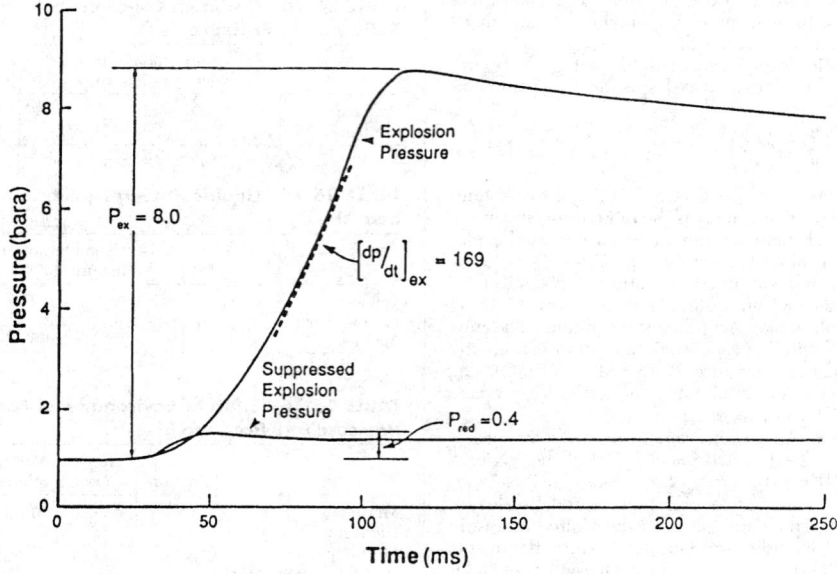

FIG. 26-30 Suppression of explosions. Pressures in an ethylene explosion and a sodium bicarbonate suppressed ethylene explosion. Tests conducted by Fike Corp. in a 1-m^3 vessel. Ethylene concentration = 1.2 times stoichiometric concentration for combustion. $(dp/dt)_{ex}$ = 169 bar/s (2451 psi/s). P_{red} = reduced explosion pressure = 0.4 bar gauge (5.8 psig). (*From Chatrathi, "Explosion Testing," Safety and Technology News, vol. 3, issue 1, Fike Corp., 1989, by permission.*)

Explosion Venting The technology of explosion venting has advanced in recent years but is still not exact; thus, considerable care in sizing explosion vents is essential. The NFPA standard, *Guide for Venting of Deflagrations*, NFPA 68, provides excellent guidance on the practice of explosion venting.

Venting requirements in NFPA 68 are based on the cubic law in Eq. (26-47). The deflagration venting nomographs of the NFPA are subject to several caveats clearly specified in NFPA 68. Particularly, the nomographs do not apply with high initial turbulence. Moreover, the NFPA 68 nomographs must not be used for venting detonations, runaway reactions, or gas mixtures containing elevated oxygen compared to air.

Equations have been developed by L. L. Simpson for the Bartknecht nomographs for gases in NFPA 68 (Bartknecht, *Explosions,* Springer-Verlag, New York, 1981; Simpson, "Equations for the VDI and Bartknecht Nomograms," *Plant/Oper. Prog.,* vol. 5, no. 1, January 1986, pp. 49–51). Those equations are shown in Table 26-17 for quiescent methane, propane, and hydrogen. NFPA 68 explains how to determine vent areas for other gases and vapors. That determination is based on fundamental burning velocities. As an approximation without serious error, the equation for quiescent propane may be used, for example, for vent areas for the quiescent gases in Table 26-18.

The following precautions must be considered in application of the NFPA 68 nomographs or Simpson's equations:

1. Do not use relief valves as explosion vents.
2. Do not use rupture disks in series, unless the space between them is vented to air or a telltale is installed to warn of pressure buildup in that space.
3. Set the release pressure of a rupture disk or other vent closure as close to the operating pressure as practical. Note that the maximum overpressure in a vented explosion will exceed the opening pressure of the vent closure.
4. Locate explosion vents as close as possible to the most likely ignition sources.
5. Locate deflagration vents so discharge from them will not endanger personnel or damage equipment.
6. Install equipment to be vented outdoors, weather and other factors permitting.
7. Place equipment to be vented close to an exterior wall, if it must be placed indoors; the vent ducts should be free of bends and no longer than 3 m (10 ft).
8. *Do not discharge explosion vents within buildings;* serious fires and explosions have occurred by such venting.

TABLE 26-17 Explosion Venting Equation for Quiescent Gases

$$A_v = d \cdot V^f \cdot \exp(gP_{stat}) \cdot P_{red}^h$$

Gas	d	f	g	h
Methane	0.105	0.770	1.23	−0.823
Propane	0.148	0.703	0.942	−0.671
Hydrogen	0.279	0.608	0.755	−0.393

where d, f, g, h = constants as tabulated above
A_v = vent area, m^2
P_{red} = maximum explosion pressure during venting, bar g
P_{stat} = vent closure release pressure, bar g
V = vessel volume, m^3

TABLE 26-18 Vent Areas Calculated for Propane May Be Used for the Following Quiescent Gases

Acetone	Dimethyl ether
Acrylonitrile	Ethane
Benzene	Ethyl acetate
n-Butane	n-Heptane
Butanone	n-Hexane
1-Butene	Isopropyl alcohol
Carbon disulfide	n-Pentane
Cyclohexane	1-Pentene
Cyclopropane	Propylene
Diethyl ether	Toluene

9. Consider reaction forces on vented equipment from the discharge of explosion products from the vent.
10. Design the vents to prevent the development of damaging negative pressure from cooling of hot products of combustion.
11. Consider the safety of personnel in rooms or buildings if those spaces are to be vented, e.g., by vent panels.

UNCONFINED VAPOR CLOUD EXPLOSIONS (UVCEs) AND BOILING LIQUID EXPANDING VAPOR EXPLOSIONS (BLEVEs)

GENERAL REFERENCES: AIChE/CCPS, *Guidelines for Chemical Process Quantitative Risk Analysis,* New York, 1989. AIChE/CCPS, *Guidelines for Evaluating the Characteristics of Vapor Cloud Explosions, Flash Fires and BLEVEs,* New York, 1994. Bodurtha, *Industrial Explosion Prevention and Protection,* McGraw-Hill, New York, 1980. Brasie and Simpson, "Guidelines for Estimating Damage Explosion," *Chem. Eng., Prog. Second Loss Prev. Symp.,* St. Louis, 1968, pp. 91–102. Crowl and Louvar, *Chemical Process Safety: Fundamentals with Applications,* Prentice Hall, Englewood Cliffs, N.J., 1990. Johansson, "The Disaster at San Juanico," *Fire J.,* vol. 80, no. 1, January 1986, pp. 32–37, 93–95. Kletz, "Protect Pressure Vessels from Fire," *Hydrocarbon Processing,* August 1977, pp. 98–102. Lees, *Loss Prevention in the Process Industries,* 2 vols., Butterworths, London, 1980. Martinsen, Johnson, and Terrell, "BLEVEs: Their causes, effects, and prevention," *Hydrocarbon Process.,* vol. 65, no. 11, November 1986, pp. 141, 142, 146, 148. Pietersen, "Analysis of the LPG Disaster in Mexico City," presented at the *Societe de Chemie Industrielle 5th Int. Symp. on Loss Prev. and Safety Promotion in the Process Ind.,* Cannes, 1986, vol. 1, preprints, pp. (21) 1–15. Pikaar, "Unconfined Vapour Cloud Dispersion and Combustion: An Overview of Theory and Experiments," *Chem. Eng. Res. Des.,* vol. 63, no. 2, March 1985, pp. 75–81. Prugh and Johnson, *Guidelines for Vapor Release Mitigation,* AIChE/CCPS, New York, 1988. Prugh, "Quantify BLEVE Hazards," *Chem. Eng. Prog.,* vol. 87, no. 2, February 1991, pp. 66–72. Prugh, "Quantitative Evaluation of Fireball Hazards," *Process Safety Prog.,* vol. 13, no. 2, April 1994, pp. 83–91. TNO, *Methods for the Calculation of the Physical Effects of the Escape of Dangerous Materials: Liquids and Gases* ("The Yellow Book"), Apeldoorn, The Netherlands, 1979. Walls, "Just What Is a BLEVE?," *Fire J.,* vol. 72, no. 6, November 1978, pp. 46–47.

See also General References in "Gas Dispersion" in this section.

Unconfined Vapor Cloud Explosions (UVCEs)

Background Unconfined vapor cloud explosions (also known as vapor cloud explosions) in open air often result when accidental releases of vapors or gases to the atmosphere are ignited. Astonishingly high pressure can result from an unconfined vapor cloud explosion; 70 kPa (10 psi) or so may occur at the outer edge of the exploding cloud, with still higher pressures near the center of the blast. Numerous severe explosions of this nature have occurred in past years (Lenoir and Davenport, "A Survey of Vapor Cloud Explosions: Second Update," *Process Safety Prog.,* vol. 12, no. 1, January 1993, pp. 12–33). In a survey of property damage losses in 100 large losses in the hydrocarbon-chemical industries, M & M Protection Consultants of Marsh & McLennan found that a vapor cloud was the initiating mechanism in 43 percent of the cases. Releases of liquefied dense gases have caused many of the reported UVCEs. Such heavy gases tend to hug the ground with limited dispersion in ambient air; this condition results in accumulation of these gases where they can cause maximum casualties to people and damage to property, if ignited. Notwithstanding, releases of mammoth amounts of *compressed* natural gas from ruptured pipelines have caused vapor cloud explosions. As an example, in 1969 a 356-mm (14-in) pipeline carrying natural gas at more than 5378 kPa gauge (780 psig) ruptured; about 8 to 10 min later the escaping gas exploded violently (National Transportation Safety Board, *Pipeline Accident Report Mobil Oil Corporation, High-Pressure Natural Gas Pipeline Accident, Houston, Texas, September 9, 1969,* NTBS-Par-71-1, Washington, D.C., 1971).

Elevated emergency unflared releases with vents of sufficient height normally do not cause damaging overpressure at the ground, if accidentally ignited (Bodurtha, "Vent Heights for Emergency Releases of Heavy Gases," *Plant/Operations Prog.,* vol. 7, no. 2, April 1988, pp. 122–126).

Numerous tests on dispersion of heavy gases and on causes of UVCEs have been performed in recent years. Dispersion tests and computer models based on them may not be representative of all conditions at a plant, however, because of equipment plus heat sources that cause better spreading of a plume than is modeled in tests.

Moreover, vapors flashed from release of a liquefied gas will be cold; such vapors flowing over warmer ground may promote atmospheric instability with accompanying turbulence and, thereby, cause more mixing with ambient air than in some tests. In addition, some tests have been so-called meteorological area sources, while the dispersion equations are generally meteorological point sources. (Only concentrations relatively close to the location of discharge of the vapors will be affected by this difference in sources.) Also, the *momentary* concentration of a combustible gas or flammable vapor is the important duration of a concentration for UVCEs; not all dispersion models specify their averaging time of concentrations. Thus, *predictions of concentrations must be treated as estimates.*

Dispersion   One way to determine concentrations is with a modified gaussian dispersion equation (Bodurtha, 1980). (See "Gas Dispersion" in this section for discussion of and references to other methods plus definition of atmospheric stabilities.) In that equation, E atmospheric stability is used for dense gases for all light to moderate wind speeds, whatever the actual atmospheric stability may be. The equation in Bodurtha (1980) overpredicts concentrations close to an emission source for a dense gas where the major effects of a UVCE are experienced, and, therefore, where uncertainty exists, is on the safe side. On the other hand, the modified gaussian equation underpredicts concentrations for dense gases at relatively large distances, and it should not be used for such gases beyond about 1000 m. Concentrations determined from the equation in Bodurtha (1980) are compared in Table 26-19 with concentrations of ammonia (which was dense by virtue of coldness) in Desert Tortoise Series Test #4 (Goldwire, "Large-Scale Ammonia Spill Tests," *Chem. Eng. Prog.*, vol. 82, no. 4, April 1986, pp. 35–41); wind speed is 4.5 m/s (10 mi/h); and vapor release rate for calculated values in the table is 108.8 m³/s. (Calculated momentary concentrations include a *virtual source*, i.e., a fictitious upwind source so that concentrations equal 100 percent—and no more—at the release point.)

In assessing the hazard of a UVCE or in investigating a UVCE it is often necessary to (1) estimate the maximum distance to the lower flammable limit (LFL) and (2) determine the amount of gas in a vapor cloud above the LFL. Figure 26-31 shows the maximum distance to the lower flammable limit, i.e., in the centerline of the cloud, based on the previous method from Bodurtha (1980) for wind speeds of 1 m/s (2.2 mi/h) and 5 m/s (11 mi/h). Maximum concentrations probably occur near 1 m/s. The volume of fuel from the LFL up to 100 percent may be estimated by

$$V_f = 0.64 \cdot Q \cdot x_L / u \qquad (26\text{-}48)$$

where V_f = volume of fuel (no air) from the LFL up to 100%, m³ at 25°C

Q = continuous dense vapor emission rate, m³/s at 25°C

x_L = distance to momentary LFL in centerline of cloud, m

u = wind speed, m/s

For application of Eq. (26-48), x_L should not exceed 300 m (984 ft). The reason for selecting 100 percent, instead of the upper flammable limit (UFL), in the equation for V_f is that in an incipient explosion vapor above the UFL may be mixed with additional air and, thereby, contribute to explosion pressure.

TABLE 26-19   Comparison of Measured and Calculated Ammonia Concentrations in Ammonia Spill Tests

Distance from release, m	Measured concentrations by Goldwire, vol %	Calculated momentary concentrations,° vol %
200	4.9	11.6
300	4.0	7.2
500	3.0	3.6
800	2.1	1.8
1000	1.7	1.3

°From equation in Bodurtha, 1980.

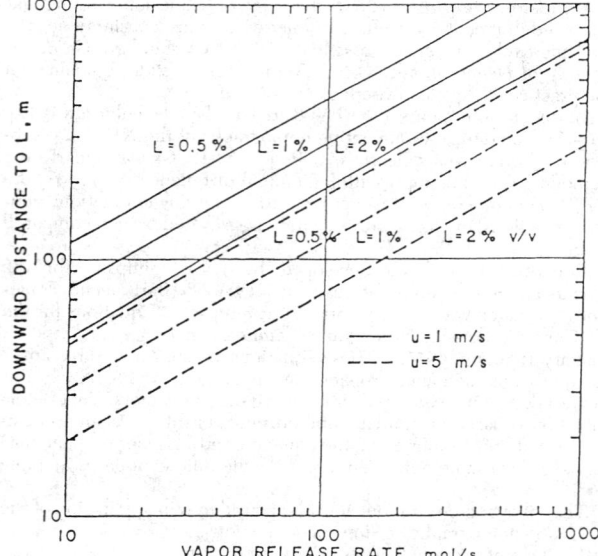

FIG. 26-31   Estimated maximum downwind distance to lower flammable limit L, percent by volume at ground level in centerline of vapor cloud, vs. continuous dense vapor release rate at ground level. E atmospheric stability. Level terrain. Momentary concentrations for L. Moles are gram moles; u is wind speed. (*From Bodurtha, 1980, p. 105, by permission.*)

Pressure Development   Overpressure in a UVCE results from turbulence that promotes a sudden release of energy. Tests in the open without obstacles or confining structures do not produce damaging overpressure. Nevertheless, combustion in a vapor cloud within a partially confined space or around turbulence-producing obstacles may generate damaging overpressure. Also, turbulence in a jet release, such as may occur with compressed natural gas discharged from a ruptured pipeline, may result in blast pressure.

Example   The combustion process in large vapor clouds is not known completely and studies are in progress to improve understanding of this important subject. Special study is usually needed to assess the hazard of a large vapor release or to investigate a UVCE. The TNT equivalent method is used in this example; other methods have been proposed. Whatever the method used for dispersion and pressure development, a check should be made to determine if any governmental unit requires a specific type of analysis.

Assume a continuous release of pressurized, liquefied cyclohexane with a vapor emission rate of 130 g mol/s, 3.18 m³/s at 25°C (86,644 lb/h). (See "Discharge Rates from Punctured Lines and Vessels" in this section for release rates of vapor.) The LFL of cyclohexane is 1.3 percent by vol., and so the maximum distance to the LFL for a wind speed of 1 m/s (2.2 mi/h) is 260 m (853 ft), from Fig. 26-31. Thus, from Eq. (26-48), $V_f \approx 529$ m³ ≈ 1817 kg. The volume of fuel from the LFL up to 100 percent at the moment of ignition for a continuous emission is *not* equal to the total quantity of vapor released; that V_f *volume stays the same even if the emission lasts for an extended period* with the same values of meteorological variables, e.g., wind speed. For instance, in this case 9825 kg (21,661 lb) will have been emitted during a 15-min period, which is considerably more than the 1817 kg (4005 lb) of cyclohexane in the vapor cloud above LFL. (A different approach is required for an instantaneous release, i.e., when a vapor cloud is explosively dispersed.) The equivalent weight of TNT may be estimated by

$$W_{\text{TNT}} = \frac{0.01 \cdot \alpha \cdot H_c \cdot W_c}{4.52} \qquad (26\text{-}49)$$

where W_{TNT} = equivalent weight of TNT, kg

α = explosion efficiency of UVCE, %

H_c = lower heat of combustion of flammable vapor, MJ/kg
W_c = weight of flammable vapor between LFL and 100%
for continuous release, kg

and energy of detonation of TNT = 4.52 in MJ/kg.

The explosion efficiency α is difficult to assess; it appears that the maximum efficiency for vapor between the LFL and 100 percent, as in this example, is 10 to 20 percent. So, for this case with $\alpha = 10$ percent, H_c for cyclohexane = 43.84 MJ/kg, and $W_c = 1817$ kg, the $W_{TNT} = 1762$ kg. Thus, 1 kg of cyclohexane, or generally any hydrocarbon at 10 percent explosion efficiency, equals 1 kg TNT. Overpressure may be estimated from "Guidelines for Estimating Damage" in this section. (See "Guidelines for Estimating Damage Explosion" by Brasie and Simpson and "Guidelines for Evaluating the Characteristics of Vapor Cloud Explosions, Flash Fires and BLEVEs," New York, 1994, in General References for other procedures.)

Prevention and Protection It is difficult to cope with a potential UVCE once an accidental release has occurred. Consequently, *the best procedure to guard against a UVCE is to prevent the release in the first place.* Safe piping is essential to protect against UVCEs. Forty percent of all major plant losses are due to piping failures, and corrosion is one of the largest single causes of plant and equipment breakdown (Hancock, "Safer Piping: Awareness Training for the Process Industries," *Plant/Oper. Prog.*, vol. 9, no. 2, April 1990, pp. 114–116). Moreover, mistakenly open valves that caused mammoth emissions of hydrocarbons have resulted in two major UVCEs with a total of 29 deaths in those two instances. Thus, close scrutiny regarding piping and valves is mandatory to help prevent UVCEs. Some other protection methods are summarized as follows.

Remotely Operated Shutoff Valves These should be considered for supply lines and other vulnerable pipelines. Excess flow valves that close when flow exceeds a set amount are possible substitutes, but they are not acceptable to some operators.

Flammable Vapor Detectors These should be installed to warn of leaks, although such devices do not effectively control UVCEs with sudden, massive releases.

Elevated or Remote Air Intakes Elevated or remote air intakes for control rooms will help in reducing ingress of dense, flammable vapors into those rooms. Ordinarily, elevating the tip of the air intake duct 9 m (30 ft) above the ground is sufficient. Installing flammable vapor detectors in the air intake ducts provides additional protection. Controls that automatically stop air to control rooms if vapor concentrations reach 25 percent of their LFL should also be considered.

Intentional Ignition Intentional ignition to ignite a vapor cloud early before it spreads out to a large volume has been used or considered only rarely. Such ignition should not be employed for control of UVCEs solely without thorough study of the ramifications of its use. (In some infrequent cases when the gas is both flammable and particularly toxic, intentional ignition may be warranted.)

Large Fans These could be used to dilute a vapor cloud below its LFL with ambient air (see, for example, Whiting and Shaffer, "Feasibility Study of Hazardous Vapor Amelioration Techniques," *Proc. 1978 Nat. Conf. on Control of Hazardous Material Spills*, USEPA, Miami Beach, April 1978). But caution must be exercised because the turbulence produced by fans will likely promote rapid combustion and a resulting UVCE unless vapors are diluted below the LFL. Nevertheless, in new plants, strategic placement of air coolers may provide enough air flow to reduce the risk of a UVCE.

Water Sprays and Steam Curtains These have been used and/or advocated to help protect against a UVCE. Such devices entrain air to dilute a vapor cloud. Also, some claim that water curtains form a physical barrier to stop the flow of the vapor cloud. As with large fans, vapors need to be reduced below LFLs to decrease the possibility of UVCEs from enhanced turbulence by the sprays or curtains. Moodie studied water spray barriers using carbon dioxide to approximate a heavy, flammable vapor cloud (Moodie, "The Use of Water Spray Barriers to Disperse Spills of Heavy Gases," *Plant/Oper. Prog.*, vol. 4, no. 4, October 1985, pp. 234–241). He used CO_2 rates of 2 and 4.2 kg/s (15,859 and 33,304 lb/h). (Propane would give the same vapor rates as these CO_2 weight rates.) There may be an upper practical limit for the emission rate of heavy gases that can be effectively dispersed by water sprays or

steam curtains (Seifert, Maurer, and Giesbrecht, "Steam Curtains—Effectiveness and Electrostatic Hazards," presented at the *Institution of Chemical Engineers 4th Int. Symp. on Loss Prevention and Safety Promotion in the Process Ind.*, Harrogate, England, 1983, vol. 1, preprints, pp. F1–F12). In any event, this equipment could be useful near known ignition sources, such as furnaces, to guard against a UVCE. Steam will be electrically charged with steam curtains. Care must be taken to assure that nearby electrical conductors are grounded. If not, the conductor could obtain an electrostatic charge from the steam and cause an incendiary spark for ignition of the vapor cloud.

Release of a pressurized, liquefied gas to the atmosphere will cause the gas to cool and condense water vapor in ambient air, forming a visible vapor cloud. Firefighters and operators who attempt to move such a cloud away from furnaces and the like with fire hoses and water jet guns are at risk, because of the possibility of a UVCE near them. Plants and governmental agencies who recommend such practices need to reexamine their policies.

Structures Structures that include partially confined spaces and turbulence-producing obstacles such as pipe bridges plus closely packed equipment, promote UVCEs. This undesirable architecture—relative to UVCEs—is often a product of congestion on a plant. Congestion is an enemy of safety. Thus, the probability of a UVCE, plus property losses and casualties, will likely be greater at a congested plant than at an uncongested site.

Vapor cloud explosions also occur indoors when large amounts of flammable vapors are discharged accidentally into buildings. Turbulence from the myriad of equipment and piping in an operating building likely cause a sudden release of energy for the room-air explosion. As one of several examples, seven were killed when vinyl chloride exploded in a building after failure of a sight glass (Walls, "Vinyl Chloride Explosion," *Natl. Fire Prot. Assoc. Q.*, vol. 57, no. 4, April 1964, pp. 352–362).

Strong Buildings Strong buildings may be prudent where people congregate, such as control rooms. For new plants, serious consideration must be given to stronger design of buildings that are vulnerable to a UVCE, compared to past designs.

BLEVEs can occur when a vessel containing a liquid above its atmospheric boiling point ruptures. The resulting ultrarapid vaporization of much of the liquid results in fire, if the liquid is flammable, plus overpressure. (Initial catastrophic failure of the vessel must occur for a BLEVE; the opening of a relief valve does not cause a BLEVE, nor does it necessarily protect against one.) BLEVE overpressure has occurred with pressurized *nonflammable* liquids, such as chlorine and carbon dioxide (Clayton and Griffin, "Catastrophic Failure of a Liquid Carbon Dioxide Storage Vessel," *Process Safety Prog.*, vol. 13, no. 4, October 1994, pp. 202–209).

Cause As discussed by Prugh (1991) and others, BLEVES can occur from:

1. Mechanical damage caused, for example, by corrosion or collision
2. Overfilling and no relief valve
3. Runaway reaction or polymerization—e.g., vinyl chloride monomer (Kim-E and Reid, "The Rapid Depressurization of Hot, High Pressure Liquids or Supercritical Fluids," chap. 3, in M. E. Paulaitis et al., eds., *Chemical Engineering at Supercritical Fluid Conditions*, Ann Arbor Science, 1983, pp. 81–100)
4. Overheating with an inoperative relief valve
5. Vapor-space explosion
6. Mechanical failure
7. Exposure to fire

A common cause of a BLEVE in plants of the hydrocarbon-chemical industry is exposure to fire. With an external fire below the liquid level in a vessel, the heat of vaporization provides a heat sink, as with a teakettle; evolved vapors exit through the relief valve. But if the flame impinges on the vessel above the liquid level, the metal will weaken and may cause the vessel to rupture suddenly, even with the relief valve open. The explosive energy for a BLEVE comes from superheat. This energy is at a maximum at the superheat limit temperature. (SLT is the maximum temperature to which a liquid can be heated before homogeneous nucleation occurs with explosive vaporization of the liquid and accompanying overpressure.) The SLT depends on the final pressure attained, and with letdown to 1 atm, it may be estimated by the simple relationship (Porteous and Reid, "Light Hydrocarbon Vapor Explosions," *Chem. Eng. Prog.*, vol. 72, May 1976, pp. 83–89):

$$\text{SLT} = 0.89\,T_c \qquad (26\text{-}50)$$

where T_c is the critical absolute temperature. With propane, for example, SLT from Eq. (26-50) = 57°C, only a degree or so higher than the measured value (Reid, in *Advances in Chemical Engineering*, vol. 12, Academic Press, New York, 1983).

Fireballs   Giant hazardous fireballs result from large BLEVEs. Several formulas for BLEVE physical parameters and thermal radiation hazards have been summarized by the Center for Chemical Process Safety (CCPS) of the American Institute of Chemical Engineers and by Prugh. (See AIChE/CCPS, 1989; Prugh, 1994.) For the maximum fireball diameter, D_{max} in meters, CCPS has selected

$$D_{max} = 6.48\,M^{0.325} \qquad (26\text{-}51)$$

where M is the initial weight of the flammable liquid, in kilograms. Thus, the maximum diameter of a propane fireball from an initial propane weight of 150 metric tons (150,000 kg = 330,690 lb) is 312 m (1024 ft).

Overpressure   Significant blast overpressures result from BLEVEs. In addition, portions of ruptured tanks may rocket large distances with clear danger to firefighters and innocent onlookers. In one test it was reported that the BLEVE blast overpressure before ignition about 10 m (33 ft) away from an exploding tank initially containing 450 kg (992 lb) propylene was 75 kPa gauge (10.9 psi). Assuming that pressure to be incident (side-on) blast pressure, it can be determined from Fig. 26-9 that $W_{TNT} = 23.3$ kg (51.4 lb), i.e., about 5 percent of the initial weight of propylene.

Prevention and Protection   Several methods may be used to protect against the causes of BLEVEs itemized earlier in this part of Section 26. They include thermal insulation, water cooling, depressuring facilities, corrosion control, and ground sloping (see Kletz, 1977; Martinsen et al., 1986; Prugh, 1991). A desirable diking method with ground sloping to minimize impingement of flame on a tank is shown in Fig. 26-32.

Fauske has suggested two passive designs for prevention of BLEVEs (Fauske, "Preventing Explosions During Chemicals and Materials Storage," *Plant/Oper. Prog.* vol. 8, no. 4, October 1989, pp. 181–184). One method keeps a normally unwetted internal surface of a tank wet; the second surrounds the high-pressure storage tank with

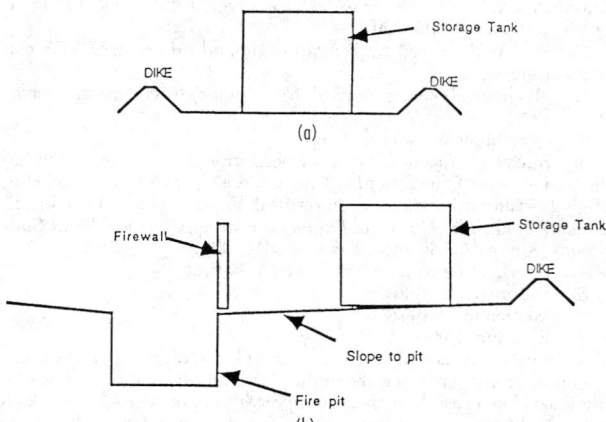

FIG. 26-32   Methods of diking for flammable liquids: (*a*) traditional diking method allows leaks to accumulate around the tank. In case of fire, the tank will be exposed to flames that can be supplied by fuel from the tank and will be hard to control. (*b*) In the more desirable method, leaks are directed away from the tank. In case of fire, the tank will be shielded from most flames and fire will be easier to fight. (*From Englund, in* Advances in Chemical Engineering, *vol. 15, Academic Press, San Diego, 1990, pp. 73–135, by permission.*)

an open atmospheric tank filled with water. Additional BLEVE prevention and protection methods follow.

1. *Minimize inventory* to the extent feasible. Expected benefits from minimum inventory may be offset by hazards resulting from more frequent and smaller shipments. The relative hazards should be reviewed (Englund, "Design and Operate Plants for Inherent Safety—Part 1," *Chem. Eng. Prog.*, vol. 87, no. 2, March 1991, pp. 85–91).

2. *Consider refrigerated storage* at atmospheric pressure. A BLEVE cannot occur with the liquid at its atmospheric boiling point (no superheat), although a fire hazard may still exist. The Dow Chemical Company in Texas stores chlorine as a liquid at atmospheric pressure at about –34°C (Englund, ibid. 1991).

3. *Set the safety relief valve* to open as far below the pressure corresponding to the SLT at 1 atm as is feasible (Reid, "Possible Mechanism for Pressurized-Liquid Tank Explosions or BLEVE's," *Science*, vol. 23, March 23, 1979, pp. 1263–1265). The pressure at propane's SLT is 2000 kPa abs. (290 psia = 275 psig).

4. *Eliminate turned-down vents* from safety relief valves, i.e., upside down U. Possible accidental ignition of releases from such vents will likely result in flame impingement on the top external surface of the tank, above the internal wetted surface. BLEVE! Some means to handle rainwater from a desirable upward vertical vent have been listed by Bodurtha (ibid., April 1988). Moreover, a safety relief valve must function properly when required and must be sized properly to help prevent an explosion.

DUST EXPLOSIONS

GENERAL REFERENCES: Bartknecht, *Dust Explosions*, Springer, New York, 1989. Bartknecht, *Explosionsschutz* (Explosion Protection), Springer, Berlin, 1993. Crowl/Louvar, *Chemical Process Safety*, Prentice Hall, New Jersey, 1990. "Dust Explosions," *28th Annual Loss Prevention Symposium*, Atlanta, Georgia, 1994. Eckhoff, *Dust Explosions in the Process Industries*, Butterworth-Heinemann, London 1991. *Health, Safety and Loss Prevention in the Oil, Chemical and Process Industries*, Butterworth-Heinemann, Singapore, 1993. NFPA 69, *Standard on Explosion Prevention Systems*, 1992. VDI-Report 975, *Safe Handling of Combustible Dust*, VDI-Verlag GmbH, Düsseldorf, 1992. VDI-Guideline 2263, *Dust Fires and Dust Explosions*, Beuth Verlag, Berlin, 1992.

Definition of Dust Explosion   A dust explosion is the rapid combustion of a dust cloud. In a confined or nearly confined space, the explosion is characterized by relatively rapid development of pressure with a flame propagation and the evolution of large quantities of heat and reaction products. The required oxygen for this combustion is mostly supplied by the combustion air. The condition necessary for a dust explosion is a simultaneous presence of a dust cloud of proper concentration in air that will support combustion and a suitable ignition source.

Explosions are either deflagrations or detonations. The difference depends on the speed of the shock wave emanating from the explosion. If the pressure wave moves at a speed less than or equal to the speed of sound in the unreacted medium, it is a *deflagration;* if it moves faster than the speed of sound, the explosion is a *detonation.*

The term *dust* is used if the maximum particle size of the solids mixture is below 500 µm.

In the following, only dusts are called combustible in the airborne state if they require oxygen from the air for exothermic reaction.

Glossary

activation overpressure, P_a That pressure threshold, above the pressure at ignition of the reactants, at which a firing signal is applied to the suppressor(s).

cubic low   The correlation of the vessel volume with the maximum rate of pressure rise. $V^{1/3} \cdot (dP/dt)_{max} = \text{constant} = K_{max}$

dust   Solid mixture with a maximum particle size of 500 µm.

dust explosion class, St   Dusts are classified in accordance with the K_{max} values.

equivalent ignition energy (EIE)   The amount of energy which, when transformed into an electrical spark discharge, has the same incendivity as the ignition source under characterization.

explosion   Propagation of a flame in a premixture of combustible gases, suspended dust(s), combustible vapor(s), mist(s), or mixtures thereof, in a gaseous oxidant such as air, in a closed, or substantially closed, vessel.

explosion pressure resistant (EPR)   Design of a construction following the calculation and construction directions for pressure vessels.

explosion pressure-shock resistant (EPSR)   Design of a construction allowing greater utilization of the material strength than the EPR design.

limiting oxygen concentration (LOC)   Maximum oxygen concentration in a mixture of a combustible and air and inert gas, in which an explosion will not occur.

maximum explosion overpressure, P_{max}   The maximum pressure reached during an explosion in a closed vessel through systematically changing the concentration of dust-air mixture.

maximum reduced explosion overpressure, $P_{red,max}$ The maximum pressure generated by an explosion of a dust-air mixture in a vented or suppressed vessel under systematically varied dust concentrations.

maximum explosion constant, K_{max}   Dust and test-specific characteristic calculated from the cubic law. It is equivalent to the maximum rate of pressure rise in a 1-m³ vessel.

maximum rate of pressure rise, $(dP/dt)_{max}$ The maximum rate of pressure rise obtained in a closed vessel through systematically changing the concentrations of a dust-air mixture.

minimum ignition energy (MIE)   Lowest electrical energy stored in a capacitor which, upon discharge, is just sufficient to effect ignition of the most ignitable atmosphere under specified test conditions.

minimum ignition temperature (MIT)   The lowest temperature of a hot surface on which the most ignitable mixture of the dust with air is ignited under specified test conditions.

static activation overpressure, P_{stat}   Pressure which activates a rupture disk or an explosion door.

vent area, A   Area of an opening for explosion venting.

venting capability, EF   Measure to evaluate the efficiency of the pressure relief device in comparison with a rupture disk with the same vent area.

Prevention and Protection Concept against Dust Explosions
Explosion protection encompasses the measures implemented against explosion hazards in the handling of combustible substances and the assessment of the effectiveness of protective measures for the avoidance or dependable reduction of these hazards. The explosion protection concept is valid for all mixtures of combustible substances and distinguishes between:

1.   Measures which prevent or restrict formation of a hazardous, explosible atmosphere
2.   Measures which prevent the ignition of a hazardous, explosible atmosphere
3.   Constructional measures which limit the effects of an explosion to a harmless level

From a safety standpoint, priority must be given to the measures in item 1. Group 2 cannot be used as a sole protective measure for flammable gas or solvent vapors in industrial practice with sufficient reliability, but can be applied as the sole protective measure when only combustible dusts are present if the minimum ignition energy of the dusts is high (>10 mJ) and the operating areas concerned can easily be monitored.

If the measures under (1) and (2), which are also known as *preventive measures*, cannot be used with sufficient reliability, the *constructional measures* (3) must be applied.

Preventive Explosion Protection   The principle of preventive explosion protection comprises the reliable exclusion of one of the requirements necessary for the development of an explosion. In pictorial terms, therefore, at least one of the sides of the hazard triangle shown in Figure 26-33 will be broken open.

An explosion can thus be excluded with certainty by:
• Avoiding the development of explosible mixtures
• Replacing the atmospheric oxygen by *inert gas*, working in a vacuum, or using *inert dust*
• Preventing the occurrence of effective ignition sources

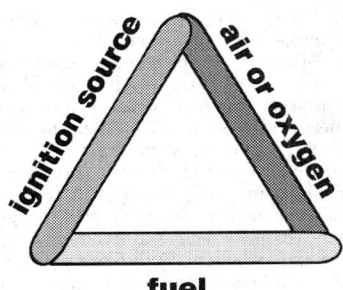

FIG. 26-33   Hazard triangle: principle of preventive explosion protection.

Avoidance of Explosible Combustible Substance-Air Mixtures
For combustible dusts, the explosibility limits do not have the same meaning as with flammable gases and flammable vapors, owing to the interaction between dust layers and suspended dust. This protective measure can, for example, be used when dust deposits are avoided in operating areas or in the air stream of clean air lines after filter installations where in normal operation the lower explosibility limit is not reached. However, dust deposits must be anticipated with time. When these dust deposits are whirled up in the air, an explosion hazard can arise. Such a hazard can be avoided by regular cleaning. The dust can be extracted directly at its point of origin by suitable ventilation measures.

Avoidance of Explosions through Inerting   The introduction of inert gas in the area to be protected against explosions lowers the oxygen volume content below the limiting oxygen concentration (LOC) so that ignition of the mixture can no longer take place. This process is called *inerting*.

One has to be aware of the danger of asphyxiation from gases in inerted equipment. This is also important for surrounding areas in case of major leaks.

Inerting is *not* a protective measure to avoid exothermic decompositions. For the avoidance of (smoldering) fires, oxygen concentrations lower than the LOC must usually be adhered to and must be determined from case to case. In addition to the nitrogen normally used, all nonflammable gases which do not support combustion or react with the combustible dust can be considered for use as the inert gas. The inerting effect generally decreases in the following order: carbon dioxide → water vapor → flue gas → nitrogen → noble gases. In special gases, liquid nitrogen or dry ice is used.

The LOC depends upon the combustible material and the type of inert gas used. It decreases with increased temperature and pressure. A distinction has to be made between the determined LOC value and the concentration which results by subtracting a safety margin.

The maximum allowable oxygen concentration (MAOC), which is, in general, 2 vol % below the LOC, has to include the following considerations: fluctuation in oxygen concentrations due to process and breakdown conditions per time and location, as well as the requirement for protective measures or emergency measures to become effective. In addition, a concentration level for an alarm has to be set below the MAOC.

For example, in rotary vacuum dryers it is possible to prevent the formation of explosible dust-air mixtures by setting and monitoring a certain partial vacuum (negative pressure). This pressure value must be determined by experiment for each type of dust. With pressures of less than 0.1 bar, in general, hazardous effects of dust explosions need not be anticipated. If the vacuum system malfunctions, the partial vacuum must be released by inert gas and the installation shut down.

Explosible dusts can also be changed into mixtures which are no longer explosible by the addition of inert dusts (e.g., rock salt, sodium sulfate). In general, inert dust additions of more than 50 wt % are necessary here. It is also possible to replace flammable solvents and cleaning agents by nonflammable halogenated hydrocarbons or water, or flammable pressure transmission fluids by halocarbon oils.

Avoidance of Effective Ignition Sources Explosions can be prevented if ignition sources capable of igniting combustible material-air mixtures can successfully be avoided. A distinction is made between *trivial ignition sources* (e.g., welding, smoking, cutting) and *ignition sources expected if operational malfunctions occur* (e.g., mechanically generated sparks, mechanically generated hot surfaces, lumps of smoldering material, static electricity). Trivial ignition sources can also reliably be excluded by organizational measures such as the systematic employment of permits.

For every installation, a check has to be made to determine which ignition source may become effective and whether it can be prevented with a sufficient degree of safety. With more sensitive products and complex installations, it becomes more and more difficult to exclude ignition sources with ample safety (Siwek et al., "Ignition Behavior of Dusts," *Proc. Loss Prevention Symposium*, Atlanta, April 12–19, 1994).

Mechanically generated sparks and *resultant hot surfaces* together are regarded as one of the more important causes of ignition in industrial practice. With mechanically generated sparks, a distinction is made between grinding, impact, and friction sparks which are formed by brief contact (<5 s) between materials. Mechanically generated hot surfaces, on the other hand, are formed by relatively long rubbing (>>5 s) against steel. The hot surfaces show considerably better incendivity in comparison with the short-lived mechanically generated sparks. Neither ignition source appears in industrial practice from the normal metallic materials of construction rubbing against each other or against stone if the relative circumferential speeds v_c are less than or equal to 1 m·s^{-1} (see Table 26-20). This is not valid for cerium-iron, titanium, and zirconium.

TABLE 26-20 Influence of Relative Circumferential Speeds v_c on Danger of Ignition for Combustible Dusts

$v_c \leq 1$ m·s^{-1}	There is no danger for ignition.
$v_c > 1 .. 10$ m·s^{-1}	Every case has to be judged separately, considering the product and material-specific characteristics.
$v_c > 10$ m·s^{-1}	In every case there is danger for ignition.

The ignition behavior of mechanically generated sparks in dust-air mixtures depends on the *minimum ignition energy* (MIE) and the *minimum ignition temperature* (MIT) of the dust in question. The ignition effectiveness of mechanically generated sparks decreases from *steel-friction sparks* to *steel-grinding sparks* to *aluminum/rust-impact sparks*. According to Fig. 26-34, it can be stated that the type

of spark-producing material, together with the MIT and the MIE requirement, determines whether an ignition of dust-air mixture has to be anticipated from friction, grinding, or impact sparks. The mechanically generated sparks can thus be assigned different equivalent ignition energies toward dust-air mixtures with an MIT of less than or equal to 500°C. For example, if the MIT of a dust is 300°C, steel-friction sparks can ignite this dust only with an MIE (equivalent energy) up to 3000 mJ. The equivalent energy, also known as the *equivalent ignition energy* (EIE), is the amount of energy which, when transformed into an electrical spark discharge, has the same incendivity as the sparks shown in Fig. 26-34.

Mechanically generated hot surfaces represent an ignition hazard if, irrespective of the MIT and the MIE, the *surface temperature* is 1100°C or higher and the *hot surface area* by itself is large enough (see Fig. 26-35). Higher surface temperatures and larger surfaces have a better incendivity; lower temperatures and smaller surfaces have a poorer incendivity.

Lumps of smoldering material always represent a hazard when the dust can be classed as capable of forming such lumps; i.e., its burning behavior class at 100°C is greater than 3. A smoldering lump surface of a cube, $A_o = 9600$ mm^2, and a surface temperature, $T_o = 900$°C, is sufficient to ignite the mixtures of dusts with an MIT of less than 600°C (see Fig. 26-35). Higher surface temperatures and larger surfaces have a better incendivity; lower temperatures and smaller surfaces, a poorer incendivity.

An *electrostatic ignition source* (see also material on static electricity) which follows an electrostatic discharge can be incendive when the energy released is equal to or greater than the minimum ignition energy of a mixture. The energy released depends, among other things, on the type of discharge. This in turn depends on the geometry and material of the participating surfaces as well as on certain other conditions. The following overview summarizes the ignition behavior of several types of electrostatic discharges (see Table 26-21).

The experimental investigations of numerous dusts with different ignition sources have shown that the incendivity of an ignition source is not only influenced by its energy content, but the nature of the source also plays a role (Glor et al., "Recent Developments in the Assessment of Electrostatic Hazards Associated with Powder Handling," *Proc. 8th Int. Symposium Loss Prevention and Safety Promotion in the Process Industries*, Antwerp, Elsevier, Amsterdam, 1995). The minimum ignition energy determined by the standard procedure can also be used for the assessment of the incendivity of such ignition sources.

Brush Discharge With dusts with MIE values of less than 3 mJ determined with purely capacitive spark discharges (without induc-

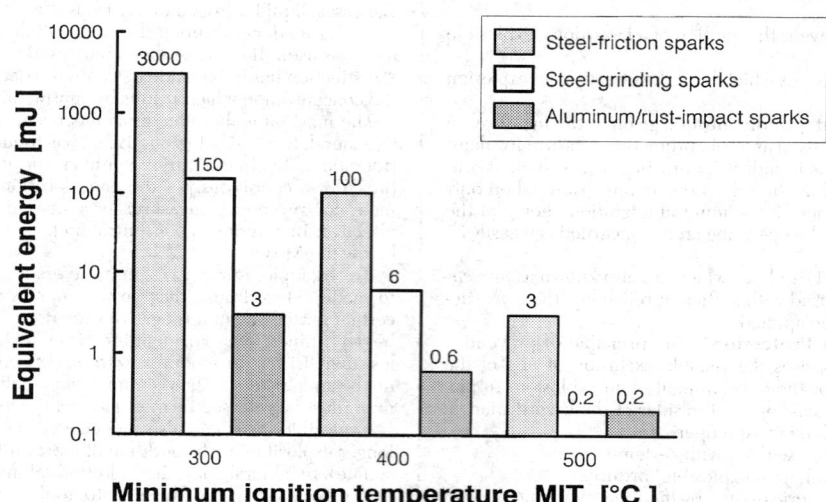

FIG. 26-34 Equivalent energies of mechanically generated sparks.

TABLE 26-21   Examples of Different Types of Electrostatic Discharges

Ignition sources	Requirement for formation	Incendivity for
Brush discharge	A nonuniform electric field between a charged dielectric and a conductor that has a moderate radius of curvature	MIE < 3 mJ
Bulk surface discharge	In rapid heaping of highly insulated bulk material, particularly when coarse material (diameter ≥ 1 mm) is present	MIE < 1 J (silo diameter 3 m)
Spark discharge	Ungrounded, conductive object	MIE < 1 J
Propagating brush discharge	Very high charging of nonconductive material, preferably in contact with a conductive surface	MIE < 10 J

tance), an ignition by brush discharges cannot be excluded with certainty. With such *extremely easily ignitable* dusts, the use of the protective measure, "avoidance of effective ignition sources," by itself is generally not sufficient. In this case, the brush discharge must also be considered as an ignition source, although it is normally important only for gases and vapors. For such dusts, the protective measure, "inerting" or "constructional explosion protection," must be used during large-scale handling operations. Toward dust-air mixtures, the brush discharges are assigned an equivalent ignition energy EIE ≤ 3 mJ for safety considerations.

Bulk Surfaces Discharge (Conical Pile Discharge)   These discharges may also be generated with fine powder and not just with granules, as was previously assumed. These discharges from fine powder have, however, a much less equivalent ignition energy EIE compared to those associated with granules. The energy of conical pile discharges increases with increasing silo diameter. The probability of their occurrence increases with increasing charge-to-mass ratio in the powder and increasing mass filling rate. Findings to date show that the EIE of conical pile discharges using highly insulating granules for generating the discharges toward dust-air mixtures is about 1 J if the silo diameter is restricted to 3 m. Because of the large diameters, the granules generating the discharges are unlikely to give dust explosions; therefore, a possible explosion hazard must be associated with the simultaneous presence of an explosible cloud of an additional, fine dust fraction. This may be the explanation for why the frequency of the occurrence of an explosion in silos initiated by bulk surfaces discharges is relatively low.

Sparks Discharge   Spark discharges can ignite dust-air mixtures up to an MIE of 1 J. If an uncertainty area is taken into account, this corresponds to an EIE of 1 to 10 J.

Propagating Brush Discharge   The incendivity of the propagating brush discharge is so large that an ignition of dust-air mixtures with an MIE up to around 100 J must be anticipated. For dust-air mixtures, this corresponds to an EIE range of 10 to 100 J.

For industrial practice, the following principles have resulted for the protective measures to be implemented. Their application is selective and depends on the prevailing circumstances.

- Ground all conductors.
- Ground people.
- Prevent and reduce charging by use of conductive materials.
- Keep conveying speeds low.

If difficulties arise in the avoidance of electrostatic ignition sources, the advice of experts must be sought.

Explosion Protection through Design Measures   In applying design measures, the possibility of an explosion is not prevented. Therefore, all exposed equipment has to be built to be explosion pressure resistant, in order to withstand the anticipated explosion pressure. The anticipated explosion pressure may be the maximum explosion overpressure or the maximum reduced explosion overpressure. In addition, any propagation of an explosion to other parts or process areas has to be prevented. Depending on the anticipated explosion pressure, a distinction is made between the following explosion-pressure-resistant designs:

- Capable of withstanding the *maximum* explosion overpressure
- Capable of withstanding an explosion overpressure *reduced* by explosion suppression or explosion venting

The strength of the protected vessels or apparatus may be either explosion pressure resistant or explosion pressure shock resistant.

Constructional measures which restrict the effects of an explosion to a safe level are always necessary when the goal of avoiding explosions cannot be achieved—or at least not with sufficient reliability—through the use of preventive explosion protection. This ensures that people are not injured and further that the protected equipment is usually ready for operation a short time after an explosion. All endangered equipment parts must thus have an *explosion-resistant* construction and withstand the overpressure expected if an explosion occurs. A distinction is made between the *explosion-pressure-resistant* (EPR) and *explosion-pressure-shock-resistant* (EPSR) construction of vessels and silos. Design of the EPR construction is implemented following the calculation and construction directions for pressure vessels, e.g., the ASME pressure vessel code. The EPSR construction allows greater utilization of the material strength (see Figs. 26-36 and 26-37).

For the EPR design, the ASME pressure vessel code requires design to be done at two-thirds of the alloy's yield strength (see Fig.

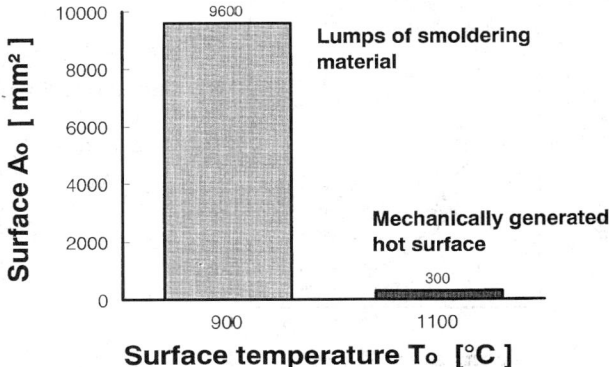

FIG. 26-35   Equivalent energies of mechanically generated hot surfaces and lumps of smoldering material.

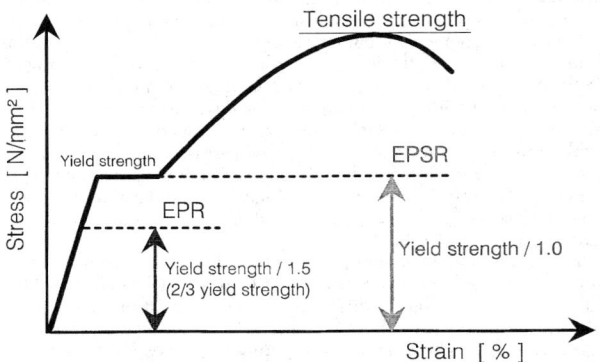

FIG. 26-36   Schematic drawing of stress-strain curve for plate steel.

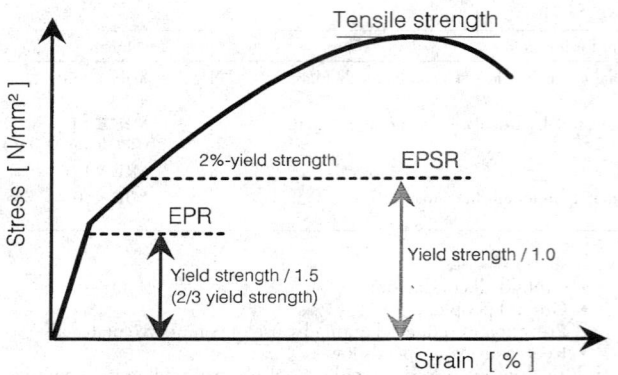

FIG. 26-37   Schematic drawing of stress-strain curve for austenitic stainless steels.

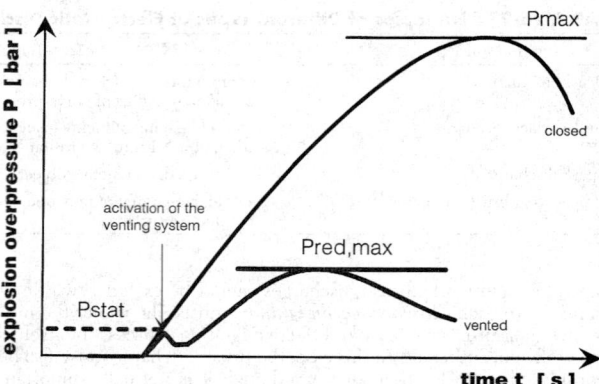

FIG. 26-38   Pressure behavior vs. time for a normal and a vented explosion.

26-36). At such low stresses, there would be no permanent deformation of an ASME code vessel subjected to an explosion overpressure.

For EPSR design, the stress level to contain an explosion is set at the yield strength, a design factor of 1. Thus, for an alloy, the design stress level would be about 1.5 times the ASME code design stress. So a pressure vessel rated at 6 bar for the ASME code (EPR) would have an EPSR rating of 9 bar.

For stainless steel, the stress-strain curve (see Fig. 26-37) has no sharp yield point at the upper stress limit of elastic deformation. Yield strength is generally defined as the stress at 2 percent elongation.

There is still a substantial safety margin up to the ultimate tensile strength, which amounts to 60 to 90 percent, depending on the steel (Kirby, Siwek, "Preventing Failures of Equipment Subject to Explosions," *Chemical Engineering*, June 23, 1986).

Despite the use of explosion-pressure-resistant equipment for the full explosion pressure or other design measures outlined later, everything possible must be done to prevent effective ignition sources, because loss of product and interruption of production are equally undesirable.

Containment (Explosion-Pressure-Resistant Design for Maximum Explosion Overpressure)   An explosion-resistant construction is understood to mean the possibility of designing vessels and equipment for the *full maximum explosion overpressure*, which is generally of the order $P_{max} = 9$ bar. The explosion-resistant vessel can then be designed as explosion pressure resistant or explosion pressure shock resistant. This protective measure is generally employed when small vessel volumes need to be protected, such as small filter units, fluidized-bed dryers, cyclones, rotary valves, or mill housings.

One has to consider that all connected devices must also withstand the maximum explosion overpressure.

Explosion Venting (Explosion-Pressure-Resistant Design for Maximum Reduced Explosion Overpressure with Explosion Venting)   The concept of explosion venting encompasses all measures used to open the originally closed vessels and equipment either briefly or permanently in a nonhazardous direction following an explosion. *Explosion venting is inadmissible when the escape of toxic or corrosive, irritating, carcinogenic, harmful-to-fruit, or genetically damaging substances is anticipated.* In contrast to the closed vessel, explosions in a vented vessel are characterized by the maximum reduced explosion overpressure $P_{red,max}$ instead of the maximum explosion overpressure P_{max} (see Fig. 26-38) and by the maximum reduced rate of pressure rise $(dP/dt)_{red,max}$ instead of the maximum rate of pressure rise $(dP/dt)_{max}$.

By this method, in general, the expected inherent maximum explosion overpressure of the order $P_{max} = 7$ to 10 bar will be reduced to a value of $P_{red,max} < 2$ bar. In this case, the static activation overpressure of the venting device is $P_{stat} \leq 0.1$ bar. The resulting $P_{red,max}$ may not exceed the design pressure of the equipment. The explosion as such is not prevented; only the dangerous consequences are limited. However, subsequent fires must be expected.

Rupture disks or explosion doors may be used as venting devices. Safety valves are not suitable for this purpose. Obviously, the static activation overpressures P_{stat} of the venting devices have to be equal to or smaller than the strength of the equipment to be protected (corresponding to the $P_{red,max}$).

Rupture disks—for example, plastic foil or aluminum foil—have a low mass and will respond almost without inertia once the activation pressure is exceeded. They can be installed independently of the location and guarantee a dust-tight closure. In case of an explosion they will free the whole area after their destruction. Common materials of construction for rupture disks are metal or alloys. Rupture disks may be combined with signaling devices—for example, ripping wires—which will trigger a shutdown or a controlling mode. Only these rupture disks are to be used that are restrained through design measures.

Explosion doors open in case of an explosion, thereby releasing the vent area. Depending on the application, explosion doors may be selected which remain open or close automatically after releasing the explosion.

The inertia, the opening behavior of the movable cover of the explosion door, and its arrangement (horizontal, vertical) can affect the venting efficiency EF. This results in a higher maximum explosion overpressure $P_{red,max}$ in the protected vessel (see Fig. 26-39).

The venting capability EF and therefore the effective vent area A_w of the explosion door is normally smaller than the capability of a plastic or aluminum foil rupture disk with the same area. Therefore, such devices need testing to determine the mechanical strength before actual use, and the venting capability or the pressure rise, respectively,

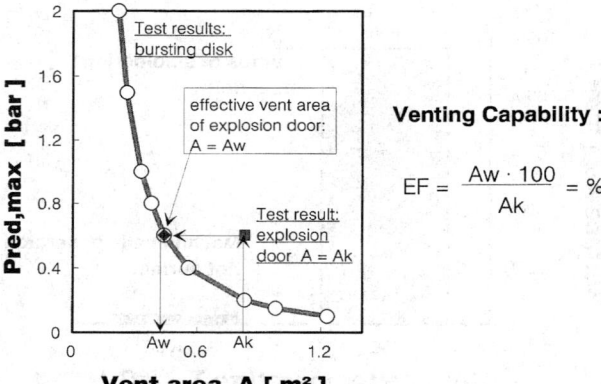

FIG. 26-39   Definition of the venting capability EF of an explosion door in comparison with a plastic foil rupture disk.

have to be chosen relative to the $P_{red,max}$ of the rupture disk of the same area.

When using explosion doors that close the vent area after the explosion, the cooling of the hot gases of combustion may create a vacuum in the vessel, resulting in its deformation. In order to prevent this from happening, vacuum breakers have to be provided.

The vented material discharged from an enclosure during an explosion should be directed to a safe location to avoid injury to personnel and to minimize property damage. If it is necessary to locate equipment that requires explosion venting inside buildings, the vents must not discharge within the building. Flames and pressure waves discharging from the enclosure during venting represent a threat to personnel and could damage other equipment. Therefore, vent ducts should be used to direct vented material from the equipment to the outdoors. If vented equipment is located within buildings, it should be placed close to the exterior walls so that the vent duct will be as short as possible. Vent ducts will significantly increase the pressure development in the equipment during venting. They require at least the same cross section as the vent area and the same design pressure as the protected vessel.

During pressure venting, a recoil is generated by the unburned mixtures and products of combustion flowing through the vent opening. The force bearing on the protected equipment depends on the explosion reduced overpressure and vent area. Not only the recoil force which can be calculated, but also its variation over time, are decisive for the practical design of the structure which supports the explosion-vented vessel. If the influence of the recoil forces is compensated for by arranging vent areas of equal size opposite each other, it is possible for one vent to open before another. Such imbalance should be considered when designing enclosure restraints for resisting thrust force.

For calculation of the venting area, empirical numerical value equations or nomograms can be used (Guideline VDI-3673, *Dust Explosion Venting,* VDI-Verlag, Düsseldorf, 1995; NFPA 68, *Guide for Venting of Deflagrations,* 1994). The calculation methods are not only dependent on the dust explosion constant K_{max}, on the maximum reduced explosion overpressure $P_{red,max}$, on the static activation overpressure P_{stat} of the venting device and on the vessel volume V, but also on the maximum explosion overpressure P_{max}. The vent calculation procedure also makes a distinction between homogeneous dust dispersion (dust-air mixtures generated using the ISO procedure) (ISO Standard 6184/1, *Explosion Protection Systems, Part 1: Determination of Explosion Indices of Combustible Dusts in Air,* Geneva, 1985) and inhomogeneous dust dispersion (dust-air mixtures generated by pneumatic transport) (Siwek, "Dust Explosion Venting for Dusts Pneumatically Conveyed into Vessels," *Plant/Operations Progress,* vol. 8, no. 3, July 1989). When applying the equations between numerical values, it is necessary to decide whether the apparatus being protected is a *cubic* (height-to-diameter ratio less than 2) or an *elongated* vessel (height-to-diameter ratio equal or above 2).

Explosion venting is always accompanied by flame propagation plus pressure consequences in the surrounding areas. The flame length will be larger with a lesser static activation pressure and smaller vent area. Depending on the volume of the protected equipment, it can reach up to 50 m. The pressure effect in the vicinity of the vent area is influenced by the maximum reduced explosion pressure, the vent area, and the vessel volume. A maximum peak overpressure exists at a certain distance from the vent area, which can be calculated. As expected, the distance at which the peak pressure appears increases with increasing vessel volume. For larger distance, this peak pressure decreases.

Based on the hazards due to flame and pressure, personnel should not be endangered by the venting process. Also, the operation of any equipment which is important with regard to safety should not be restricted. This must be considered when designing the plant and may be accomplished by releasing the pressure upward. If this is not feasible, then the vent openings should be placed as high as possible at the side of the vessel. Due to the danger of dust ejection, one has to consider the location of the surface of the dust pile in the vessel. It should never reach the lower edge of the vent opening at maximum operation filling of the vessel.

Among other things, one prerequisite necessary to calculate the pressure relief openings needed on the apparatus is knowledge about the explosion threat definition and venting system hardware definition. The various influences are summarized in Table 26-22.

Explosion Suppression

Explosion-Pressure-Resistant Design for Reduced Maximum Explosion Overpressure with Explosion Suppression Explosion suppression systems provide one means to prevent the buildup of an inadmissibly high pressure, which is the consequence of explosions of combustible material in vessels. They operate by effectively extinguishing explosion flames in the initial stage of the explosion. An explosion of combustible material can generally be regarded as successfully suppressed when the maximum explosion overpressure can be lowered to a reduced explosion overpressure of not more than 1 bar (see Fig. 26-40).

Depending upon the design criteria of the installed suppression system, an unsuppressed explosion overpressure of around 7 to 10 bar is reduced to a suppressed reduced explosion overpressure which lies in the range of $P_{red,max} = 0.2$ to 1 bar. Thus, vessels need to be explosion resistant for an overpressure of maximum 1 bar (ISO Standard 6184/4, *Explosion Protection Systems Part 4: Determination of Efficacy of Explosion Suppression Systems,* Geneva, 1985).

The best advantages of explosion suppression systems is that they can also be used for explosions of combustible materials with toxic properties and that there is no penetration on the location of the process equipment for safe application.

Today, the technology of industrial explosion suppression has evolved to the extent that this technique can and does provide effective industrial safety for almost all industrial processing procedures and for most explosible materials. Developments in explosion suppression hardware—detectors, suppressors, and control equipment, together with new or improved suppressants—provide a versatility of capability which covers all dust explosion classes and process equipment volumes ranging from 0.2 to greater than 1000 m³. Most significant, the theoretical understanding of explosion propagation and suppression has led to computer-aided design guidance, which has simplified system design.

Explosion suppression systems comprise explosion detectors, pres-

TABLE 26-22   Explosion Venting System Design Parameters

Explosion hazard definition	Venting system definition
Volume of vessel (free volume)	Type of venting device
Shape of vessel (cubic or elongated vessel)	Detection method for triggering a shutdown
Length-to-diameter ratio of vessel	Static activation overpressure P_{stat} of venting device
Strength of vessel	Venting capability of venting device
Type of dust cloud distribution (ISO method/	Location of venting device on the vessel
pneumatic-loading method)	Position of equipment to be protected in the building
Dust explosibility characteristics:	Length and shape of relief pipe if existent
maximum explosion overpressure P_{max}	Recoil force during venting
maximum explosion constant K_{max}	Duration of recoil force
toxicity of the product	Total transferred impulse
Maximum flame length	
Pressure outside the vent areas	

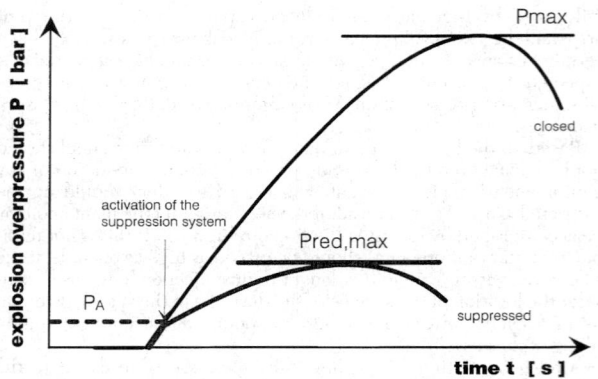

FIG. 26-40 Pressure behavior vs. time for a normal and a suppressed explosion.

surized HRD suppressors (high rate discharge) and a control and monitoring center.

Since the explosion pressure of an incipient explosion in a closed or essentially closed vessel propagates from the explosion epicenter at the speed of sound, pressure detection is an effective means of early explosion detection. Threshold pressure detection using a large area membrane explosion detector provides an electrical contact within milliseconds if the preset overpressure—the system activation pressure, P_a—is exceeded. Such a detector, which responds only to a pressure threshold, is called a *static pressure detector*. It is normal practice to mount two such devices mutually perpendicular and to trigger the explosion suppressors only when both detectors give coincident contact. This obviates the risk of spurious activation as a consequence of incident shock on the detector. Static pressure detectors have been extensively used in industrial applications and have a proven track record. The use of a large-area membrane for such devices ensures that their response is not influenced by the buildup of product or any crusting of product across the sensor surface of the detector. Dependent on selected membrane stiffness, such detectors can be set to operate at pressure thresholds of between 0.05 and 0.7 bar gauge. Static pressure detectors have only limited applicability for process equipment that operates at elevated pressures or that is subjected to significant extraneous pressure excursions.

For such applications, the more sophisticated rate-of-rise membrane pressure detectors must be used. In practice, these *dynamic explosion detectors* have a wide range of applicability. They have both rate-of-rise and pressure threshold trigger points that can be configured specific to the requirement. A timid explosion event may not attain the rate-of-rise criterion, a fact that necessitates a limit pressure threshold to trigger the explosion suppressors. Dynamic explosion detectors provide a means of achieving earlier detection than can be realized with static pressure detectors—and, thus, more effective suppression. Pressure fluctuations as a consequence of normal or abnormal process conditions, such as blocked filters, can be actively discriminated against, thus preventing spurious activation of the suppression system. In larger volumes, since the pressure rises only slowly, care must be taken to set up such detectors to meet appropriate detection response criteria. Dynamic explosion detectors can be used either as static or dynamic explosion detectors or in combination. The two parameters of pressure and time are programmed on-site specific to the installation. The dynamic detector control equipment facilitates event memory, enabling postevent analysis of the pressure domain causing the activation.

Dynamic explosion detectors use a piezoresistive pressure sensor installed behind the large-area, gas-tight, welded membrane. To ensure optimum pressure transference from the membrane to the active sensor element, the space between the membrane and the sensor is filled with a special, highly elastic oil. The construction is such that the dynamic explosion detector can withstand overpressures of 10 bar without any damage or effect on its setup characteristic. The operational range is adjustable between 0 and 5 bar abs. Dynamic explo-

sion detectors are insensitive to shocks and vibration. For many applications, it is thus sufficient to install just one dynamic explosion detector to trigger the explosion suppressors. Since the commercial availability of this new detection method some two years ago, it has a proven performance capability and is being increasingly selected for use in explosion suppression systems.

High rate discharge (HRD) suppressors are available in a range of sizes. Dry powder suppressant is stored in a container, which is pressurized with dry nitrogen to an overpressure of, typically, $P_s = 60$ to 120 bar. An explosively actuated valve, such as a large-diameter membrane cut by a shaped charge, provides almost instantaneous unimpeded access for the suppressant, which is then expelled by the nitrogen gas and discharged through an appropriate nozzle into the process equipment.

Suppression efficiency is very dependent on the suppressant mass M_s discharge characteristic. This suppressant mass discharge rate is affected by outlet orifice area and propelling agent pressure. HRD suppressors that utilize a large-diameter outlet have superior suppression capability over those that rely on high propelling-agent pressure alone to expel the suppressant charge. The effectiveness of the HRD suppressors against a range of explosion hazards has been fully substantiated. The range of HRD suppressors available provides a wide range of suppression capability. The use of an explosive actuator provides the most effective means of rapid suppressant discharge—typically, less than 2 ms from activation to the start of suppressant release—and ensures the earliest suppression of any explosion event.

In some circumstances, the need to use an explosive actuator to effect suppressant release can be restrictive. HRD suppressors that are actuated by activating an electric initiator, which fractures fast-reacting rupture membrane, have been developed. Thus, within a few milliseconds of activation, an unimpeded opening is provided for the propelling agent to expel the suppressant. Upon fracture of the rupture membrane, the suppressant flows from the pressurized HRD suppressor through the outlet into a specifically engineered suppressant distribution piping system. These new types of HRD suppressors have, in order of magnitude, the same valve opening time and the same outlet orifice and provide an alternative means of suppressant deployment where explosive actuators are not admissible.

The HRD suppressor discharge elbow and the discharge nozzle have an important influence on suppression effectiveness. A nozzle that achieves a wide angular dispersion of suppressant is most effective for explosion suppression in smaller volumes, but the limited suppressant throw that results reduces its effectiveness for larger-volume explosion suppression. Irrespective of nozzle type, the suppressed explosion with an elbow (the HRD suppressor is mounted on the side of a vessel) is higher, demonstrating the effect of an elbow in slowing suppressant delivery into the vessel. On the other hand, if the HRD suppressor is mounted on the top (without an elbow) of the same vessel, a clearly more effective suppression is achieved, because no elbow is slowing the delivery of the suppressant. A normal arrangement of the nozzle for the distribution of the suppressant is protruding into the protected equipment. In practice, this is very often undesirable, especially if the enclosure handles dust with frequent production changes. To avoid disturbance of the production process (cleaning, product deposits, hygiene) by projection of the suppressant dispersion system into the protected vessel, movable, so-called telescopic nozzles (see Fig. 26-41) must be installed. Initially, the nozzle arrangement is located outside the vessel to be protected, separated from the latter by a membrane. This membrane does not reduce the HRD effectiveness. In applications where a true hygienic seal is required or where high-pressure excursions are to be expected, frangible metal or carbon discs are used. Such hermetic seals do slow, to a degree, the suppressant discharge from an HRD suppressor and this must be allowed for in the design. In the event of an explosion, once the HRD suppressor is activated, the nozzle is propelled forward by the suppressor pressure, rupturing the membrane and locating in its operating position to ensure effective suppressant deployment.

The most widely deployed industrial explosion suppressant is monoammonium phosphate powder (MAP). This suppressant has a wide range of effectiveness. However, it can prove to be a contaminant, necessitating stringent clean-down procedures after a suppressed explosion incident. This limitation is overcome by selecting a sodium

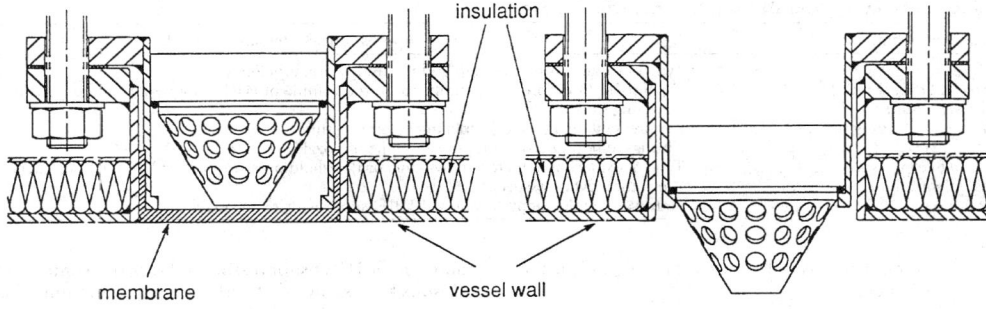

FIG. 26-41 Example for a telescopic nozzle arrangement.

bicarbonate (NaBi)–based dry powder suppressant. Food-grade compatible and readily water-soluble proprietary sodium bicarbonate suppressants are increasingly being used to protect industrial processes for manufacturing foodstuffs and pharmaceuticals. All types of suppressant have a fine particle size distribution—high specific surface—and flow additives to minimize particle agglomeration.

The sodium bicarbonate suppressant attains almost equivalence in performance with the monoammonium phosphate powder. Water has proven to be a very effective suppressant of dust, especially grain and fodder dusts. A suppressant is regarded as being very effective when an increase of the activation pressure P_a of the explosion system leads to an increase as small as possible in the maximum reduced explosion overpressure $P_{red,max}$ (see Fig. 26-42).

A recent development postulates that superheated water is a more effective suppressant than water alone because, on discharge, release of the superheated water charge results in partial flash vaporization of the water droplets to steam with a consequential fragmentation of the droplets, thus achieving a higher specific surface for effective suppression. At high temperature, effective suppression is achieved—but the effectiveness is very dependent on the control of temperature. The complication of maintaining the water suppressant charge at a fixed and controlled elevated temperature negates this option for most practical purposes.

Explosion suppression control equipment has traditionally been simple in operation, maintaining the highest level of reliability. In some applications, advantage can be achieved by introducing a level of intelligence and interpretation into the control equipment signal processing. New control and annunciation central control systems employ state-of-the-art electronics. Often called an *alarm center,* such a watchdog or operational center of any explosion suppression system is modular in construction. Individual modular "cards" can be plugged in, to accommodate the requirements of the full explosion protection system. Alarm centers are futureproof—they can be added to as the system expands. An alarm center records and monitors the signals transmitted by explosion pressure detectors, spark and flame detectors, temperature, and other safety sensors. Dependent on configuration, by interrogation and interpretation of the detector/sensor data, the alarm center selectively controls the actuation of explosion suppressors, extinguishing barriers, fast-shutting isolation valves, process equipment shutdown, water spray or extinguishant release, and all audible and visual alarms. System internal monitoring gives fault indication in the event of device or field wiring defect, and alarm and fault relay contacts can be connected as appropriate. Standby power is facilitated from an independent, monitored battery such that full explosion protection is assured during any power failure. System isolation to facilitate safe working on or in a protected vessel is standard, and remote actuation of key functions and system status record, via an on-line printer, are facilitated as options. Zoning of HRD suppressors enables the control system to ensure that suppressors deployed to inject extinguishant that is to act as an extinguishing barrier, and thus prevent flame propagation down a duct, are used to maximum effect.

A suitable locking mechanism must ensure that the production plant can be started up again only if the explosion suppression system is fully operational. The alarm center must be designed so that, if work is performed within the protected vessel, the detectors can be made inoperable and secured against inadvertent triggering.

Design of explosion suppression systems is clearly complex, since the effectiveness of an explosion suppression system is dependent on a large number of parameters. One hypothesis of suppression system design identifies a limiting combustion wave adiabatic flame temperature, below which combustion reactions are not sustained. Suppression is thus attained, provided that sufficient thermal quenching results in depression of the combustion wave temperature below this critical value. This hypothesis identifies the need to deliver greater than a critical mass of suppressant into the enveloping fireball to effect suppression (see Fig. 26-43).

If the suppression criterion is not met, the consequence is a failed

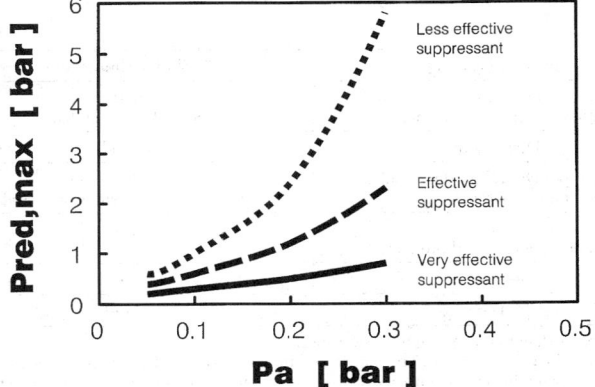

FIG. 26-42 Influence of the activation overpressure P_a and suppressant upon the effectiveness of a suppression system (constant test condition).

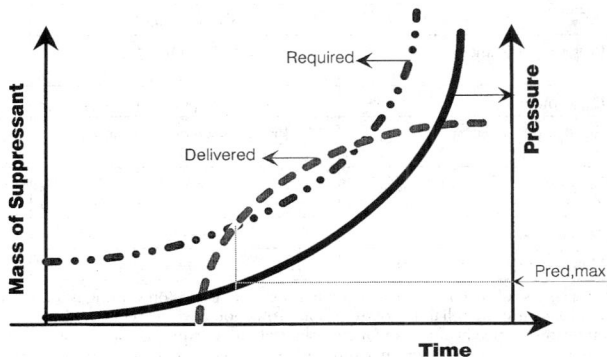

FIG. 26-43 Design basis for explosion suppression.

TABLE 26-23 Suppression System Design Parameters

Explosion hazard definition	Suppression system hardware definition
Volume of vessel (free volume V)	Type of explosion suppressant and its suppression efficiency
Shape of vessel (area and aspect ratio)	Type of HRD suppressors: number and free volume of HRD suppressors and the outlet diameter and valve
Type of dust cloud distribution	opening time
(ISO method/pneumatic-loading method)	Suppressant charge and propelling agent pressure
Dust explosibility characteristics:	Fittings: elbow and/or stub pipe and type of nozzle
Maximum explosion overpressure P_{max}	Type of explosion detector(s): dynamic or threshold pressure, UV or IR radiation, effective system
Maximum explosion constant K_{max}	activation overpressure P_a
Minimum ignition temperature MIT	Hardware deployment: location of HRD suppressor(s) on vessel

suppression, in which combustion is not arrested and high explosion pressures must be anticipated.

For any proposed suppression system design, it is necessary to ascribe with confidence an effective worst-case suppressed maximum explosion overpressure $P_{red,max}$. Provided that the suppressed explosion overpressure is less than the process equipment pressure shock resistance and provided further that this projected suppression is achieved with a sufficient margin of safety, explosion protection security is assured. These two criteria are mutually independent, but *both* must be satisfied if a suppression system is to be deployed to provide industrial explosion protection.

Suppression system design parameters fall into the two categories of explosion threat definition and suppression system hardware definition. The various influences are summarized in Table 26-23.

The type of HRD suppressors and their number and, hence, the extinguishing agent requirement can be determined with the aid of nomograms or simple numerical value equations developed from numerous experiments and model calculations (Siwek, Moore, "Extended Design Practice for Explosion Suppression Systems," *Proc. 8th Int. Symposium Loss Prevention and Safety Promotion in the Process Industries,* Antwerp, Elsevier, Amsterdam, 1995). As dry powder extinguishing agents are predominately used in industrial practice, the calculation fundamentals for the extinguishing agent requirement are limited to these extinguishing powders.

To achieve an effective and practical application, HRD suppressors of different sizes must be used for the different sizes and geometry of the protected vessels.

Of practical benefit to the processing industry is the combination technology of explosion venting and explosion suppression. It is evident that the deployment of a small explosion vent results in a useful further reduction in the resultant suppressed explosion pressure. Tests have shown that, provided the HRD suppressor is located such that the suppressant is deployed across the vent aperture, no flame ejection results. It is often the case in practical situations, where neither venting nor suppression alone can provide an appropriate safety solution, that, by combining the technologies, safety is assured. Where the primary protection means is explosion suppression, the addition of vents results in a lowering of the achievable reduced explosion pressure. Where the primary protection means is explosion venting, the addition of HRD suppressors reduces the vented explosion pressure. Results have shown that with high-K_{max} aluminum dust explosions, the reduced (vented) explosion pressure can be further reduced by more than 50 percent by the addition of HRD suppressors, although flame extinction is not achieved against the explosion threat. Tests on silos confirm that the deployment of an extinguishing barrier across the top of a silo fitted with explosion vents can prevent a secondary dust explosion in the room above the silo. This strategic use of extinguishing barriers ensures that flame ejection from a vented explosion incident is minimized. Similarly, extinguishing barriers in the vent pipes of relief venting systems minimize any flame ejection.

The combination of explosion safety technologies can provide more effective safety than is possible by deploying just one safety measure. In this respect, the improved capabilities of explosion suppression further enhance overall explosion protection capability.

Comparison of Explosion Protection Design Measures In Table 26-24, comparison is made of the explosion protection design measures of containment, explosion venting, and explosion suppression. Regarding the effectiveness of the different explosion design measures, all three techniques are equal if the design of these measures is performed properly.

Explosion Isolation For all equipment systems protected by design safety measures it is also necessary to prevent the propagation of an explosion from these protected vessels into operating areas or equipment connected via interconnecting pipeline. Such an approach is referred to as *explosion isolation.*

To prevent an explosion occurring in, for example, a constructional protected installation from spreading through a pipeline ($l > 6$ m) to part of the installation fitted with preventive explosion protection, explosion isolation measures (see Fig. 26-44) must be implemented. As explosions are generally propagated by *flames* and not by the pressure waves, it is especially important to detect, extinguish, or block this flame front at an early stage, i.e., to isolate or disengage the explosion. If there is no explosion isolation, the flame issuing from the equipment—for example, from the equipment protected through design (equipment part 1)—through the connecting pipeline comes into contact with a highly turbulent precompressed mixture in the equipment with preventive protection (equipment part 2). The mixture will ignite in an instant and explode; a large increase in the rate of

TABLE 26-24 Comparison of Explosion Protection Design Measures

	Containment	Explosion venting	Explosion suppression
Pressure resistance P	7–10 bar	Without relief pipe, up to 2 bar With relief pipe, up to 4 bar	St 1 up to 0.5 bar St 2+3 up to 1.0 bar
Location	Independent	Dependent	Independent
Limits of application	Products which decompose spontaneously	Toxic products and products which decompose spontaneously	Products which decompose spontaneously, metal dust hazard
Environmentally friendly	Yes	No (flame, pressure, and product)	Yes
Loss of material°	+++++	+++++	++
Maintenance requirements†	++++	++++	+++++

°The loss of material by using containment and explosion venting is always much greater than that by using explosion suppression.
†To ensure the reliability of explosion protection devices, regular servicing and maintenance are required. The nature and time intervals of these activities depend on technical specifications and on the plant situation. Normally, after commissioning of the plant, inspections are carried out in comparatively short intervals, e.g., every month. Positive experience may subsequently provide for longer service intervals (every three months). It is recommended to contract service and maintenance to reliable, specialized companies.

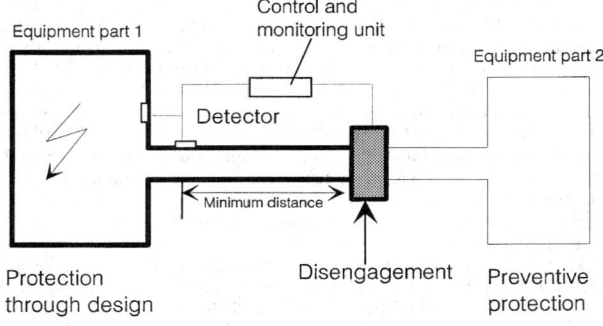

FIG. 26-44 Principle of the constructional measure explosion isolation.

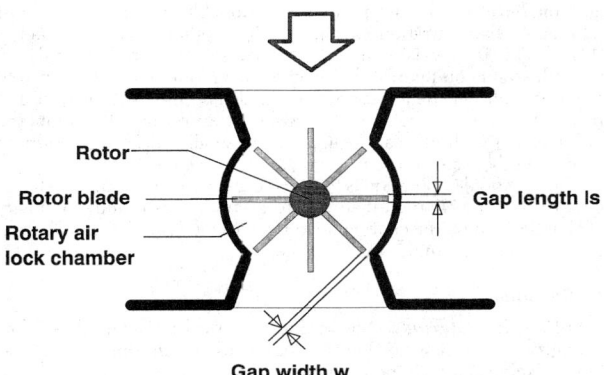

FIG. 26-45 Design features of a rotary valve.

combustion reaction and, naturally, in the reduced explosion overpressure is the result. The equipment in question may be destroyed.

The mechanical flame barriers, which are used for explosion isolation of flammable gas and solvent vapor explosions, are very susceptible to the action of dirt and, with one exception, are thus *not suitable for dust-carrying* pipelines. The exception involves the *rotary valve* (see Fig. 26-45), which is based on the flame-quenching effect through narrow gaps and is mainly used at product charging and discharging points.

The size of the gap between the rotor blades and the housing

depends on the construction and is important for the ignition breakthrough protection of the rotary valve. The maximum gap width of combustible dusts, like that of flammable gases, lies in the millimeter range. With knowledge of the ignitability of a dust, the gap length, and the number of constantly diametrically opposed rotor blades, a nomogram can be used to determine the maximum admissible gap between the blades and the inner wall of the rotary valve. In the event of an explosion, the valve must be automatically stopped to prevent any subsequent upstream fire or explosion due to passage of smoldering material or burning product through the valve. As a rule of thumb, it was found for normal organic dusts that the ignition cutout of rotary air locks is effective when three rotor blades on each side are diametrically opposed, provided that the blades are made of metal and the gap between the tip of the rotor blade and the housing is ≤0.2 mm (Siwek, "New Knowledge About Rotary Air Locks in Preventing Dust Ignition Breakthrough," *Plant/Operating Progress* vol. 8, no. 3, July 1989).

An *extinguishing barrier* comprises an optical flame sensor and an HRD suppressor located downstream of the detected flame front. The effectiveness of an extinguishing barrier is based on its ability to detect an explosion in a pipeline by means of an optical flame detector whose tripping signal is amplified and then very quickly actuates the detonator-actuated valves of the pressurized HRD suppressors. The extinguishing agent—preferably extinguishing powder—is discharged into the pipeline and forms a thick blanket, which extinguishes the incipient flame. There is a definite distance between the installation sites of the optical detector and the extinguishing barrier to ensure that the extinguishing agent acts directly on the flame. The amount of extinguishing agent required depends on the nature of the combustible dusts, the nominal diameter of the protected pipeline, the explosion velocity, and the maximum reduced explosion overpressure in the vessel. This type of barrier does not impede product throughput down the pipeline.

The alternative to the extinguishing barrier is the *rapid-action explosion isolation gate valve*. These valves must be tested for ignition breakthrough protection and pressure rating in dust explosions. They can meet these requirements for dust explosions and are effective against dust explosions at shorter distances than for gas explosions. When rapid-action gate valves are used, a dust explosion approaching the installation site in the pipeline is detected by an optical sensor and the closing process is initiated by a triggering mechanism. The closing time depends on the nominal width of the rapid-action devices and is generally less than 50 ms.

Explosion isolation can also be effected by *rapid action barrier valves*. At present, they can be arranged only in horizontal pipelines and are suitable, in general, only for streams with a small amount of dust. Such valves are thus frequently used to protect ventilation lines. As a certain explosion overpressure is necessary to close such valves, a distinction is made between *self-actuated* and *externally actuated* barrier valves (Fig. 26-46).

The interior of the barrier valve contains a valve cone mounted in

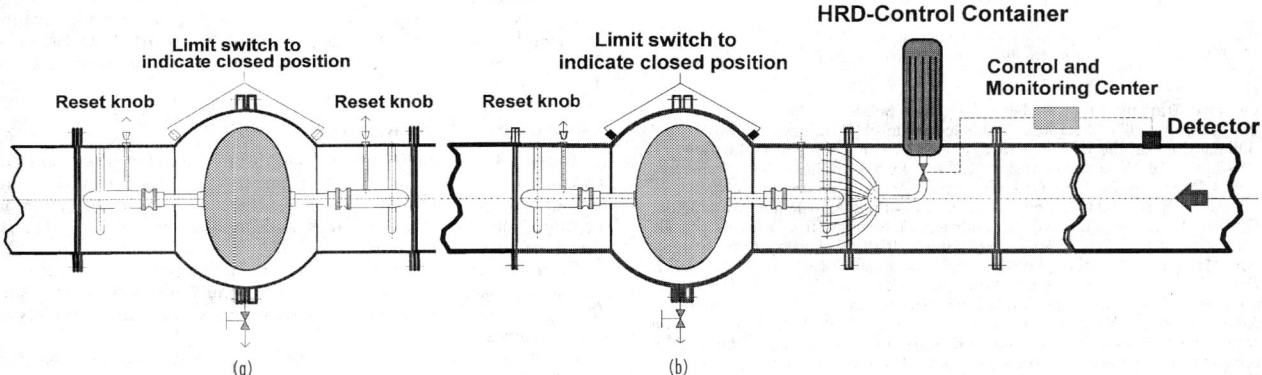

FIG. 26-46 Rapid-action barrier valves: (*a*) self-actuated; (*b*) externally actuated.

spherical sockets, which can be moved axially in both directions; it is held in its middle position by springs. The spring tension is set for a maximum flow velocity of 25 m·s⁻¹, based on the pipeline cross section. If an explosion occurs, the valve closes automatically, owing to the kinetic energy of the pressure wave preceding the flame front. Here, either the explosion velocity must be greater than 25 m·s⁻¹ or the pressure difference in front of and behind the valve greater than 0.1 bar. The valve cone is pressed onto a rubberized valve seat on closing and held in place by a retaining device. It is released from outside. The self-actuating barrier valve (Fig. 26-46a) functions in both directions. Barrier valves can also be operated by a sensor-controlled auxiliary gas flow (jets of nitrogen from control containers onto the valve cone) in the direction of the pipe axis via a hemispherical nozzle (Fig. 26-46b). These are installed when low explosion overpressures are expected and, consequently, ignition breakthrough of an explosion through the installation site can occur with a self-actuated valve. The externally actuated barrier valve functions in one direction only.

Particularly reasonably priced explosion isolation of systems involves the use of a *relief pipe* with which the flow direction can be diverted by 180°. It prevents flame jet ignition with precompression in constructionally protected equipment. If suction is present, explosion propagation can occur. To prevent this, the use of an additional extinguishing barrier or a rapid-action gate valve is necessary. If a diverter is installed where positive pressure feed is given, then the diverter is, in general, safe against an explosion propagation.

Product removal mechanisms from apparatuses that are explosion resistant can be protected with a *double-slide* system. Here, the slides must be at least as resistant as the apparatuses. By means of proper control, it must be assured that a slide is always closed.

Finally, it must be pointed out that all devices suitable for use in explosion isolation—or, quite generally, all explosion protection devices used in practice—may be used only when their *pressure rating*, *ignition breakthrough protection*, and *reliability* have been proven in suitable investigations by competent bodies.

STATIC ELECTRICITY

Nomenclature

C	Capacitance, farads
C/kg	Charge density, coulombs/kg
C/m²	Surface charge density, coulombs/m²
F	Farads
J	Energy, joules
K_e	Relative dielectric constant, dimensionless
kV/m	Electric field intensity, kilovolts/meter
m	Meters
MIE	Minimum ignition energy, mJ
mJ	Millijoules
Ω²	Resistivity value, ohms per square, usually used for fabrics and films
pS	Conductivity, picosiemens
pS/m	Electrical conductivity of liquid
RH	Relative humidity, %
S	Siemens (formerly mho)
T	Time, s
V	Electrical potential, volts
V/m	Electric field intensity, volts/meter

GENERAL REFERENCES: Bailey, "Charging of Solids and Powders," *J. Electrostatics* **30,** pp. 167–180, 1993. Blythe and Reddish, "Charges on Powders and Bulking Effects," *Electrostatics 1979,* Inst. Phys. Conf. Ser. No. 48, London, pp. 107–114, 1979. Finke, "Electrostatic Effects of Charged Steam Jets," *J. Electrostatics* **23,** pp. 69–78, 1989. *Generation and Control of Static Electricity,* Scientific Circular No. 803, National Paint and Coatings Association, Washington, D.C., 1988. Gibson and Lloyd, "Incendivity of Discharges from Electrostatically Charged Plastics," *Brit. J. Appl. Phys.* **16,** pp. 1619–1631, 1965. Owens, "Ignition Hazards of Charged Dielectrics in Flammable Environments," *IEEE Trans. Ind. Applic., IA-20,* no. 6, Nov./Dec. 1984. *Plant/Operations Progress* **7,** no. 1, Jan. 1988. Entire issue devoted to papers on static electricity, presented at AIChE meeting, Minneapolis, Minn., August 1987. Post, Glor et al., "The Avoidance of Ignition Hazards Due to Electrostatic Charges Arising During the Spraying of Liquids under High Pressure," *J. Electrostatics* **23,** pp. 99–110, 1989. *Protection Against Ignitions Arising Out of Static, Lightning and Stray Currents,* American Petroleum Institute Recommended Practice 2003, 1991.

Introduction Spark-ignition hazards must be considered whenever static charges may accumulate in an environment that contains a flammable gas, liquid, or dust. The need for electrical bonding and grounding of conductive process equipment in hazardous (classified) locations is widely recognized. Less well understood are the ignition hazards associated with static charges on poorly conductive, flammable liquids, solids, and powders. Static charges, generated on these materials by normal handling and processing, cannot be conducted to ground quickly and may cause hazardous charge accumulations. The electric fields associated with these charges may stress the surrounding air sufficiently to cause breakdown by some type of electrical discharge.

Electrical discharges from poorly conductive materials take several forms, each differing in its ability to ignite flammable mixtures. It is not possible to calculate the incendivity of these discharges, because of their varying time and spatial distributions. Several engineering rules of thumb for estimating the relative hazard of these discharges are discussed in the following.

Any analysis of static-ignition hazard should start with data on the ignition sensitivity of the particular flammable material at its most flammable concentration in air, i.e., its *minimum ignition energy* (MIE). This is especially important for dusts. It is prudent to determine this value on fines of the specific dust of interest, rather than to rely on published data. Hybrid mixtures (i.e., mixtures of dust and flammable vapor for which vapor concentrations may be below the lower explosive limit) can be ignited by smaller discharge energies than might be expected.

The key to safe operation is to provide an adequate means of charge dissipation from charged materials to ground. This requires mobility of charges in or on the charged material *plus* electrical continuity from the material to ground.

Definitions

antistatic material One with an electrical resistivity that is low enough to make it incapable of accumulating hazardous concentrations of static charges when grounded.

bonding A method of providing electrical continuity between two or more conductive objects to prevent electrical sparking between them.

charge relaxation time The time required for the charge in a liquid or on a solid material to dissipate to 36.8 percent of its initial value when the material is grounded.

electrical discharge A current flow, which occurs when the electrical field strength exceeds the breakdown value in a medium such as air.

flammable mixture A mixture of a gas, mist, or dust in air, which is within its flammable range.

grounding A special form of bonding, in which a conductive object is electrically connected to (earth) ground.

incendive discharge Any discharge that has sufficient energy to ignite a specified flammable mixture.

minimum ignition energy The smallest amount of spark energy that has been found capable of igniting a specified flammable mixture in a standard test (Calcote, Gregory et al., "Spark Ignition; Effect of Molecular Structure," *Industrial and Engineering Chemistry* **44,** no. 11, 1952).

Electrostatic Charging

Types The primary cause of electrostatic charging is *contact electrification*, which takes place when two different materials are brought into contact and separated. Other causes include induction charging, the formation of sprays, and impingement of charged mist on an ungrounded conductor.

Contact Electrification This form of charging involves the contact and separation of solid-solid, solid-liquid, or liquid-liquid surfaces. Pure gases do *not* cause charging unless they carry droplets or dust particles.

Efforts to quantify the magnitude and polarity of contact charging have had limited success, because minute variations in the types and concentrations of contaminants exert a large influence on charge separation. Even like solid-solid surfaces can produce significant charge separation. The charge density on separated solid-solid surfaces is, usually, very nonuniform. Each surface may contain both + and − polarity charges, with more of one polarity than the other. After separation, the charges will dissipate slowly or rapidly, depending upon the electrical resistivity of the material and the presence of a path to ground.

Contact electrification at liquid-liquid and liquid-solid interfaces is attributed to the absorption of ions of one polarity by one surface. Ions of opposite polarity form a diffuse layer near the interface. If the diffuse layer is carried along by moving liquid, as in a pipeline, the flowing charges (called a *streaming current*) may create a sparking hazard downstream. One protective measure is to keep the charged liquid in a closed, grounded system (a *relaxation chamber*) long enough to allow safe dissipation of the charges.

The magnitude of the streaming current in any given situation is not readily calculated. Equations, derived experimentally for some liquids (Bustin and Dukek, *Electrostatic Hazards in the Petroleum Industry*, Research Studies Press, Letchworth, England, 1983) show that flow velocity has the greatest influence on pipeline charging. Streaming currents can be limited to safe levels by limiting velocities to less than 1 m/s.

Charge Induction   This charging takes place when a conducting object is exposed to electric fields from other charged objects. Examples include the induction charging of a human body by clothing, the charging of a conductive liquid in a charged, plastic container, or the charging of the conductive coating on one-side-metallized film by static charges on the uncoated surface.

Although charge induction can take place whether or not the conducting object is grounded, a sparking hazard is present only if the conductor is not grounded. This phenomenon can convert a relatively innocuous charge buildup on a nonconductor into a serious sparking hazard by raising the potential of the conductor above ground (Owens, "Spark Ignition Hazards Caused by Charge Induction," *Plant/Operations Progress* **7**, no. 1, pp. 37–39, 1988).

Spraying   Droplets, formed by spray nozzles, tend to be highly charged even if the conductivity of the liquid is high. Because there is no path to ground from the droplets, their charges can accumulate on an ungrounded conductor to cause sparking. If flammable vapor is present, as in some tank-cleaning operations, it is essential that the spray nozzle and tank be bonded or separately grounded. It is safer to use a nonflammable cleaning solvent or one that has a conductivity greater than 1000 pS/m.

Mists   Although charged mists are unable to cause ignition of flammable vapor by self-generated sparking, it is important that the mist not impinge upon an ungrounded conductor.

Charge Dissipation

General   It is an experimental fact that charged objects exert a force on other charged objects. This behavior is explained by the presence of an *electric field*, i.e., electric lines of force, each of which emanates from a + charge and terminates on a − charge. The *magnitude* of the field is defined as the force on a unit test charge placed at the point of interest. The *direction* of the field is the direction of the force on a + test charge placed at the point.

Static charge generation causes an ignition hazard only if the accumulated charges create an electric field sufficient to produce an electrical discharge in a flammable atmosphere. In most processes, this means that the electric field intensity at some location must reach the breakdown strength of air (nominally 3×10^6 V/m). The objective of static-control measures is to ensure that electric field intensities cannot reach this value.

Conductors   Bonding and grounding are the primary means of dissipating charges from conductive objects. Bonding clamps should be of the single-point type, which bites through oxide or enamel coatings to make contact with the bare metal. Owing to the sturdy construction of bonding clamps and cables, their initial resistance is less than 1 Ω.

It is good practice to visually inspect the condition of bonding cables during each use and to measure the resistance of temporary bonding cables, at least annually, to confirm that it is less than, say, 25 Ω.

Antistatic Materials   These materials allow static charges to dissipate without causing hazardous accumulations. Charge dissipation normally takes place by conduction along the material to ground. The antistatic behavior of such materials is measured, at a controlled temperature and relative humidity, in terms of Ω^2 (ohms per square) of *electrical surface resistivity*. Resistivity values needed for safety depend upon the rate of charge generation, but are typically in the range of 10^8 to 10^{11} Ω^2 (ohms per square) for fabrics and films (ASTM Standard Test Method D-257-78, *DC Resistance or Conductance of Insulating Materials*).

An alternate test for antistatic performance is the *charge-decay* test, in which the time of charge-decay is measured after 5 kV have been applied to the specimen (Federal Test Method 101C, Method 4046.1). For many purposes, a charge-decay time of 0.5 s to 500 V, measured at the RH in end use, indicates good antistatic performance.

The electrical surface resistivity and charge-decay time of most materials vary substantially with the relative humidity. It is important that materials be tested at the lowest RH expected in use. Items that are antistatic at 50 percent RH may *not* be antistatic at 20 percent RH.

Some fabrics contain a small percentage of conductive fibers or staple, which limit charge accumulation by *air ionization*. These fabrics do not depend upon conduction of static charges and may not appear conductive in the electrical resistivity or the charge-decay test. Antistatic performance is not humidity-dependent. Work is under way to develop a standard test for these fabrics (Nelson, Rogers, and Gilmartin, "Antistatic Mechanisms Associated with FIBC Fabrics Containing Conductive Fibers," *J. Electrostatics* **30**, pp. 135–148, 1993).

Liquids   The rate of dissipation of charges in a liquid, assuming that its conductivity and dielectric permittivity are constant, can be expressed as:

$$T = 8.85 \frac{K_e}{C} \qquad (26\text{-}52)$$

where T = time required for the charge density to dissipate to 36.8% of its initial value, s
K_e = relative dielectric constant of the liquid, dimensionless
C = electrical conductivity of the liquid, pS/m

Flammable liquids are considered particularly static-prone if their electrical conductivity is within the range of 0.1 to 10 pS/m. If no particulates or immiscible liquid are present, these products are considered safe when their conductivity has been raised to 50 pS/m or higher. Blending operations or other two-phase mixing may cause such a high rate of charging that a conductivity of at least 1000 pS/m is needed for safe charge dissipation (British Standard 5958, part 1, *Control of Undesirable Static Electricity*, para. 8, 1991).

Electrostatic Discharges   An electrostatic discharge takes place when a gas or vapor-air mixture is stressed, electrically, to its breakdown value. Depending upon the specific circumstances, the breakdown appears as one of four types of discharges, which vary greatly in origin, appearance, duration, and incendivity.

Sparks   Spark discharges are most common between solid conductors, although one electrode may be a conductive liquid. They appear as a narrow, luminous channel, and carry a large peak current for a few microseconds or less. Sparks are the only form of discharge for which a maximum spark energy can be calculated, using the expression:

$$J = 0.5 \, C \cdot V^2 \qquad (26\text{-}53)$$

where J = total stored energy dissipated, J
C = capacitance of charged system, F
V = initial potential difference between electrodes

Incident investigations often require that an estimate be made of the possible spark energy from an ungrounded conductor. If the discharge path contains significant resistance, some of the stored energy is dissipated in the resistance, thereby lowering the energy in the spark gap.

Corona   A corona is generated when a highly nonuniform electric field of sufficient strength terminates on a conductor that has a small radius of curvature (i.e., a point, wire, or knife-edge). The luminous (breakdown) region is confined to a small volume near the corona electrode. Because of their small peak currents and long duration, corona discharges do not have sufficient energy to ignite most flammable materials found in industry (i.e., materials having a minimum ignition energy above 0.2 mJ). For this reason, they can be used safely as static neutralizers in most hazardous (classified) locations. Corona discharges *can* ignite hydrogen-air and oxygen-enriched, gas mixtures.

Brush Discharges These discharges take place between conductors and charged nonconductors, where the radius of curvature of the conductor is too large for corona generation. The name refers to the brushlike appearance of the discharge, which spreads from the conductor to discrete areas on the nonconductor. The brush discharge may have a hot "stem" near the conductor which, though short-lived, may cause ignition by raising the temperature of the flammable mixture to its autoignition value (Norberg, "Modeling Current Pulse Shape and Energy in Surface Discharges," *IEEE Trans. Ind. Applic.* **28**, no. 3, pp. 498–503, May/June 1992).

Brush discharges from – charged nonconductors have been found more incendive than those from + charged nonconductors. Spark energies from brush discharges are limited to less than 4 mJ, because charges from a small area on the nonconductor are able to participate in the spark. Most dust-air mixtures cannot be ignited by brush discharges because their MIE exceeds 4 mJ (Gibson, "Electrostatic Hazards—A Review of Modern Trends," *Electrostatics 1983*, Inst. Phys. Conf. Ser. No. 66, London, pp. 1–11, 1983).

Surface charge densities cannot exceed the theoretical value of 2.7×10^{-5} C/m² (set by air breakdown) and will normally be less than 1.5×10^{-5} C/m².

Propagating Brush Discharges These discharges are much less common than brush discharges. They sometimes occur when a nonconductive film or plastic layer acquires a double layer of charges, i.e., + polarity charges on one surface and – polarity charges on the opposite surface. Surface charge densities can be very large, because they are not limited by the breakdown of air.

The double layer can be formed by contact (*triboelectric*) charging of one surface of the nonconductor, while the opposite surface is in contact with a conductor, e.g., a nonconductive coating on a metal chute or a plastic-lined, metal pipe for powders. A less frequent cause is contact-charging of one surface, while air ions are supplied to the opposite surface.

Investigations by Glor ("Discharges and Hazards Associated with the Handling of Powders," *Electrostatics 1987*, Inst. Phys. Conf. Ser. No. 85, pp. 207–216, 1987) and others conclude that propagating brush discharges require surface charge densities above 2.7×10^{-4} C/m². In addition, the breakdown voltage of the insulating layer must be greater than 4 kV for a thickness of 10 μm, or 8 kV for a thickness of 200 μm.

If a conductor approaches the charged surface, the electric field will produce air ionization at the surface, which creates a semiconductive layer, thereby allowing charges from a large area to participate in a single discharge. Because these discharges can have energies of 1 J or more, they are very hazardous in a flammable environment. They may also cause severe shocks to operators who reach into a nonconductive container that is receiving charged powder, pellets, or fibers.

Causes of Hazardous Discharges with Liquids

Self-Generated Discharges Vapor-air mixtures can be ignited by sparks from highly charged liquids. It is said that such liquids "carry their own match." Typical causes of such charging for poorly conductive (<50 pS/m) liquids include:

1. High-velocity flow
2. Free-fall/splashing
3. Filtering
4. Spraying
5. Agitation with air
6. Blending with powder
7. Settling of an immiscible liquid (e.g., water in gasoline)
8. Liquid sampling from pressurized lines, using ungrounded or nonconductive containers

Conductive liquids in nonconductive containers may cause sparking if the outside of the container is charged by rubbing.

External Causes of Incendive Static Discharges
1. Sparks from ungrounded persons
2. Brush discharges from flexible, intermediate, bulk containers (FIBCs), plastic bags, stretch wrap, or other plastic film
3. Propagating brush discharges from metal-backed plastic film or linings

Powders Contact charging of powders occurs whenever particles move, relative to one another or to a third surface. Significant charging can be generated by operations such as grinding, mixing, sieving, pouring, and pneumatic transfer. Maximum charge densities (C/kg) on airborne powder increase as particle size decreases, because of larger surface/mass ratios. Dry fines can be expected to charge more highly than those containing moisture. While suspended in air, charged powder poses an ignition risk only if nonconductive piping is used in the conveying lines or if conductive piping is not properly bonded.

The *collected* powder may accumulate so much charge per unit volume that the associated electric field strength causes breakdown of the surrounding air in the form of corona or a brush discharge. For receiving containers larger than about 1 m³, so-called bulking discharges may be present, with energies of up to 10 mJ. The ignition hazard from bulking discharges can be minimized by, for example, using a rotary valve to prebulk small volumes of charged powder prior to its collection in a large receiver (Britton, "Static Hazards Using Flexible Intermediate Bulk Containers for Powder Handling," *Process Safety Progress* **12**, no. 4, pp. 240–250, October 1993).

Personnel and Clothing Sparks from ungrounded persons pose a serious ignition hazard in flammable gas-air, vapor-air, and some dust-air mixtures, because the body is a conductor and can store energies on the order of 40 mJ. Induction of charges on a person's ungrounded body by charged clothing is a common cause of personnel electrification. Even at the *threshold of shock sensation,* the stored energy is about 1 mJ.

It is essential that persons be grounded in hazardous (classified) locations. For most chemical operations, the resistance to ground from the body should not exceed 100 megohms. A lower allowable resistance may be specified for locations where the presence of primary explosives, hydrogen-air mixtures, oxygen-enriched mixtures, or certain solid-state devices requires faster charge dissipation.

The combination of conductive flooring *and* conductive footwear is the preferred method of grounding. Untreated concrete flooring with conductive footwear is usually adequate, but its conductivity should be measured (Fowler and Klein, "Static Phenomena and Test Methods for Static Controlled Floors," *EOS/ESD Symposium Proceedings*, pp. 27–38, 1992). Where this method is impractical, personnel grounding devices are available.

In most chemical areas, grounded persons can wear any type of clothing safely. For the unusually sensitive environments listed above, antistatic or conductive clothing should be worn, and persons must be grounded. Removal of outer garments in a flammable location can cause hazardous discharges and should be avoided (NFPA 77, *Static Electricity,* para. 2-2, 1993).

Noncontacting Electrostatic Measurements These measurements are made by instruments that respond to the electric fields at their sensing electrodes. Considerable care must be taken in the interpretation of the measurements. The three general types of devices are described as follows.

Static Locator These meters are the least expensive type. They usually indicate in volts, but should not be used for quantitative evaluations.

Static Voltmeter These instruments are calibrated to indicate the potential (V) on an ungrounded conductor and usually have more than one calibrated meter/surface spacing. They can be used, for example, to indicate the potential on ungrounded persons or equipment. A meter that indicates in volts or kilovolts is *not* an electric field meter.

Electric Field Meter These meters are calibrated to indicate the polarity and magnitude of the electric field (V/m) at the sensor. They should have only *one* calibrated meter/surface spacing and should be designed to establish a reasonably uniform electric field between the charged surface and the meter. This is needed to ensure that the measured field is approximately equal to the field at the charged surface.

To determine the level of electrification on an *insulating* surface, an electric field meter should always be used. There is a direct relationship between the charge density on the surface of an insulator and the electric field intensity at the surface. Measurements should be made at locations where the insulating surface is several inches away from other insulating or conductive surfaces. The area of the measured surface should be large, compared to the field of view of the meter. In locations where a flammable vapor-air mixture has an MIE greater than 0.2 mJ, field intensities of 500 kV/m or more should be considered unsafe.

HAZARDS OF VACUUM

Nomenclature

F_1	Friction loss in equivalent length, ft
g	Acceleration of gravity, ft/s^2
g_c	Conversion factor, ft·lb$_m$/ft·lb$_f$ s^2
H	Total tank vertical height, ft
h_o	Initial fluid height, ft
h_i	Height between tank base and centerline of pump suction, ft
h_f	Final height of liquid above tank base, ft
NPSH	Net positive suction head, ft
P	Final tank vacuum, in Hg
P_o	Atmospheric pressure, lb$_f$/in^2
ρ	Liquid density, lb$_m$/ft^3
R	Tank radius, ft
V_o	Initial airspace volume, ft^3
V_p	Vapor pressure of tank fluid equivalent, ft
γ	Ratio of molar specific heats, $\bar{C}_p/\bar{C}_v$ (about 1.4 for diatomic gases)

Introduction Ask any chemical engineer who has had some plant experience what he or she knows about vacuum and the engineer will probably smile and tell some tale about some piece of equipment that tried to turn itself inside out. Usually no one was hurt, and often there was no massive leakage—but not always!

Causes of Vacuum Hazards The design for the internal pressure condition of vessels is usually straightforward and well understood. The design for external pressures is more difficult. The devious ways in which external pressure can be applied can often be overlooked.

Following are some obvious causes of vacuum collapse:

• Liquid withdrawal by pump or gravity draining
• Removal of gas or vapor by withdrawing with a blower, fan, or jet
• Siphoning of liquids

Less obvious causes include:

• Condensation of vapor
• Cooling of hot gas
• Combination of cooling and condensation of a mixture of gas and condensable vapor

Sometimes obscure causes of vacuum collapse include:

• Absorption of a gas in a liquid, for example, ammonia in water or carbon dioxide in water.
• Reaction of two or more gases to make a liquid or solid, for example, ammonia plus hydrogen bromide to form ammonium bromide.
• Reaction of a gas and a solid to form a solid, for example, corrosion in a tank. Air plus Fe or FeO may give Fe_2O_3 in the presence of water.
• Reaction of a gas and a liquid to give a liquid, for example, chlorination, hydrogenation, or ethylation.
• Sudden dropping of finely divided solids in a silo, creating a momentary vacuum that can suck in the sides of the silo.
• Plugging of flame arresters, for example:

In styrene service, vapor may condense in flame arresters, and the liquid formed is low in inhibitor. Liquid may polymerize and plug off the arrester. Possible solutions include cleaning the arrester frequently or using a PVRV (pressure-vacuum relief valve).

In liquid service in cold weather, vapor may condense in a flame arrester and the liquid formed may freeze and plug the arrester. A possible solution is to heat and insulate the arrester to prevent condensation.

• Maintenance and testing. It is not a good idea to apply vacuum on a vessel during maintenance or testing without full knowledge of the external pressure rating, unless a suitable vacuum relief device is in place and operable.

Location of Vacuum Relief Device (Carl Schiappa, Michigan Engineering, The Dow Chemical Company, Midland, Mich., personal communication, March 20, 1992.) If a vacuum relief device is used, locate the device at the highest point on the top of the tank. If the vacuum relief device is not installed in this location and the tank is overfilled with liquid, the relief device will be sealed in liquid and will be ineffective in protecting the tank. This is especially true for the part of the tank above the vacuum relief device if it is sealed in liquid, the liquid level is lowered, and the tank goes into a partial vacuum.

Protective Measures for Equipment There have been many incidents where vessels were designed for internal pressures of 25 psig or higher and the tank collapsed under vacuum. The internal pressure rating is not a good indication of the vacuum rating. If equipment may be subject to vacuum, consideration should be given to designing the equipment for full vacuum. This may eliminate the need for complicated devices such as relief valves and instruments; if they are used but fail or plug, designing the equipment for full vacuum will prevent collapse of the vessel. For vessels where steam is used in the tank, such as steam-sterilized sanitary-service tanks, a full vacuum rating is advisable under any circumstances.

A disadvantage of this approach is that it may appear at first to be more expensive to design equipment for full vacuum. The cost differential of adding vacuum rating is usually modest compared to the tank's value. It can be less than 10 percent for 15 psig tanks of up to 3000-gal nominal capacity (Wintner, "Check the Vacuum Rating of Your Tanks," *Chem. Eng.*, February 1991, pp. 157–159). When the total cost of a suitably instrumented vessel not designed for vacuum is compared with the cost of a vessel designed for vacuum but without the extra equipment, the difference may be small or negligible, and the vessel designed for vacuum will be inherently safer. If a vessel is designed for vacuum, precautions should be taken to ensure that internal or external corrosion will not destroy the integrity of the vessel. Dimpled jackets may provide an economical way of providing vacuum protection when jacketed vessels are involved.

Personnel Hazards The following case history illustrates how vacuum can be harmful and dangerous to personnel. A plant superintendent was checking an open nozzle on a glass-lined reactor on which there was a vacuum pump pulling vacuum, when suddenly his arm was sucked into the nozzle, up to his shoulder. He could not remove his arm until help arrived to release the vacuum on the vessel. He was injured painfully, though not seriously. The injury could have been very serious if help had not been nearby. Personnel hazards can also result from vacuum conveyor systems for solids handling.

Examples of Vacuum-Related Accidents Figure 26-47 shows a jacketed tank, where the jacket was designed for low-pressure steam. When the steam was turned off and the drain valve and trap were closed, the steam condensed, causing the jacket to collapse. The jacket should have been designed for full vacuum, or a suitable vacuum relief device should have been installed on the jacket.

Figure 26-48 shows the collapse of a large storage tank containing acetone. The overflow and vent line had recently been changed, so it would vent through a vapor seal of water to remove acetone emissions from the vent when the tank was being filled. When the tank was being emptied, water was sucked into the vent pipe, creating a vacuum in the tank, which collapsed the top of the tank. A suitable vacuum relief device on the tank should have been installed to prevent this incident. Venting the tank through a liquid seal of this type is probably not very effective and a better method of controlling emissions should have been selected.

Low-Pressure Storage Tanks Low-pressure storage tanks are fragile. Even an eggshell can withstand more pressure and vacuum (Sanders, "Don't Be Another Victim of Vacuum," *Chemical Eng. Prog.*, September 1993, pp. 54–57). Low-pressure storage tanks do not require much pressure difference between the inside of the tank and the atmosphere to buckle the relatively thin tank walls. Pressure

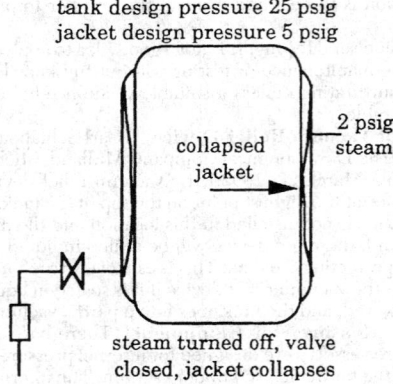

FIG. 26-47 Collapse of vessel jacket due to condensation of steam. (*W. T. Allen, Michigan Engineering, The Dow Chemical Company, Midland, Mich., personal communication, May 1988.*)

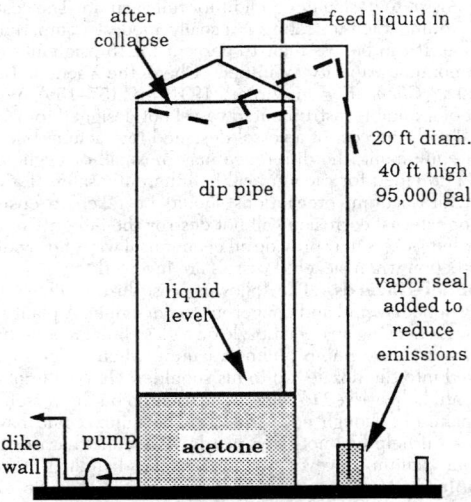

FIG. 26-48 Collapse of storage tank due to adding a liquid vapor seal to reduce vapor emissions. (*Allen, 1988.*)

differences as low as 10 mbar (0.01 atm, or 0.15 psig, or 0.7 in H₂O) between the inside and outside of the tank can buckle some tanks. The rate of handling of product and the breathing volume flow rate due to weather effects must be taken into account in designing the necessary pressure compensation devices.

A critical situation arises in summer when the tank is heated by strong radiation, then cooled by sudden rainfall. Heavy rainfall results in a rapid drop in ambient temperature and the formation of a rain-water film that flows on the top of the tank and down the tank wall. The wall and, with a certain lag, the gas in the tank are cooled, and air must flow into the tank to prevent a significant pressure difference from arising between the inside and outside of the tank. If vapors in the tank are condensed, more air must flow into the tank.

The initial gas temperature in a tank can reach a value of 55°C (131°F) as a result of strong solar radiation. Ambient rainfall is assumed to be 15°C (59°F). The maximum flow rate of air into the

tank is reached some minutes later. At the start of the rainfall, after a certain lag, the flow rate at first increases, then reaches a maximum, then decreases.

A study has been made to allow the prediction of the rate at which air must enter a tank with and without internal condensation to prevent a pressure difference from arising (Fullarton, Evripidis, and Schlünder, Institut für Thermische Verfehrenstechnik, Universität Karlsruhe (TH), "Influence of Product Vapour Condensation on Venting of Storage Tanks," *Chem. Eng. Process.*, **22**(3), 1987, published by Elsevier-Sequoia, New York). The results are too involved to be presented in detail here. The reader is referred to this paper for details of the calculations.

The results of a specific case study are shown in Fig. 26-49. This depicts the change in inbreathing volume flow rate as a function of time. The middle curve describes the case when the tank is filled with dry air: that is, no condensation occurs. When the air is saturated with water vapor at 55°C (131°F) and condensation occurs, the top curve is obtained. The bottom line represents the volume flow rate brought about by thermal contraction alone, not including the amount condensed. Because of the heat of condensation released, this fraction is less than the volume flow rate without condensation, but this effect is more than compensated for by the additional volume flow rate due to condensation.

Experimental data in small equipment has shown that condensation of water vapor causes a twofold increase in the maximum flow rate compared to dry air, and a fourfold increase in condensation of methanol vapor.

API 2000 lists the venting capacity for inbreathing (vacuum relief) and outbreathing (pressure relief) for oil tanks up to 180,000 barrels (7,560,000 gal or 2.86 × 10⁴ m³) capacity at 14.7 psia and 60°F. Tanks larger than 180,000 barrels require individual study (API 2000, "Venting Atmospheric and Low-Pressure Storage Tanks, Non-Refrigerated and Refrigerated," *API Standard 2000*, 3d ed., American Petroleum Institute, Washington D.C., January 1983).

Vacuum Requirements for Draining Tanks (Wintner, 1991, by permission.) A shortcut method of calculating the vacuum that can occur when a tank is being drained while the vent line is closed can be performed by measuring the head in the tank, assuming it is completely full. This is the maximum vacuum that would exist in a gravity-drain tank before air would begin to enter it. If the tank's overall height is designated H, then this vacuum is $2.036H\rho(g/g_c P_o)$ in Hg.

If the tank has some headspace, as is usually the case, it is desirable to get a better estimate of the actual level, since tanks usually have some gas headspace even when filled with liquid. Two tank configurations are considered: the gravity discharge tank (discharge is open to the atmosphere) and the pumped discharge tank. These calculations assume that the process is so rapid that an adiabatic model for the gas in the headspace is the correct choice. This is true when the drainage

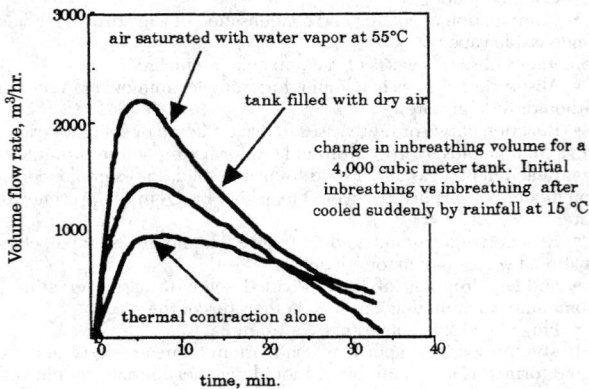

FIG. 26-49 Effect of water vapor condensation on volume flow rate of air into tank. (*Fullarton, Evripidis, and Schlünder, 1987, by permission of Elsevier Science S.A., Lausanne, Switzerland.*)

time is short (on the order of a few minutes for a tank of several thousand gallons capacity). An isothermal model is the best choice when the drain time is long.

For the gravity discharge case, the height of the fluid at maximum vacuum, which is the point at which air would begin to backflow into the tank, is determined by Eq. (26-54). Equation (26-55) calculates the corresponding vacuum in the tank's headspace at this liquid height. Since the drain nozzle is open to the atmosphere, this solution is a static force balance that is satisfied when the sum of the internal pressure and the remaining fluid head is equal to the atmospheric pressure.

$$h_f = \frac{144 P_o (1 - \{V_o/[V_o + \pi R^2 (h_o - h_f)]\}^\gamma)}{\rho g/g_c} \qquad (26\text{-}54)$$

$$P = \frac{29.92 \, h_f \rho g}{144 P_o g_c} \qquad (26\text{-}55)$$

$$h_f = \text{NPSH} + F_1 + VP - h_i - \frac{144 P_o (1 - \{V_o/[V_o + \pi R^2 (h_o - h_f)]\}^\gamma)}{\rho g/g_c}$$
$$(26\text{-}56)$$

$$P = 2.036 P_o [1 - \{V_o/(V_o + \pi R^2 h_o)\}^\gamma] \qquad (26\text{-}57)$$

For the pumped-discharge case, internal pressure and final fluid height are calculated by Eqs. (26-56) and (26-57). The final fluid level is the point at which the net positive suction head (NPSH) equation is satisfied.

The solutions of Eqs. (26-54) and (26-56) involve a trial-and-error technique or a numerical method. This can be solved using a computer program for multivariable equations, or it can be calculated by hand. In either equation, assume a reasonable value for h_f and insert it on the right-hand side of the equation. The left-hand value obtained is then substituted until the values guessed at and those calculated are in close agreement. The number of trials is strongly dependent on the initial guess; for realistic tank dimensions, between four and ten iterations should produce good agreement.

The pumped-discharge case is generally more difficult to solve because of the uncertainty in dealing with negative numerical results. As a final answer, a negative value could indicate that the pump has completely emptied the tank; however, as an intermediate value, it could mean that it is not a true solution. A simple check is to try a different initial estimate and see if the intermediate negative results disappear.

Example   Assume the tank in Fig. 26-50 has a diameter of 4 ft and a capacity of 1000 gal, is filled with water, and discharges to the atmosphere. The shortcut calculation (tank is initially completely full) indicates that the internal pressure would be 10.65 in Hg. An initial fillage of 70 percent of the tank's volume would produce a vacuum of 6.93 in Hg, which is 65 percent of the shortcut result.

In the case of pumped discharge, assume that a centrifugal pump is used. Its NPSH will determine the height at which vacuum is released by the backflow of air through the pump. Detailed information about the pump characteristics is needed to evaluate the potential vacuum. For these design calculations, assume that the pump will stop delivering liquid and air backflow will begin when the pump's NPSH requirements are no longer met (Sommerfield, "Tank Draining Revisited," *Chem. Eng.*, May 1990, p. 171).

The precise flow-decay pattern will depend on the type, size, and dimensions of the pump. Flow for a typical centrifugal pump will begin to decay at the NPSH point, but some additional fluid transfer will usually occur before a steady backflow of air through the pump begins. At that point, the pump's priming is completely lost.

The mathematical solution for maximum vacuum is based on Eq. (26-56), which solves the NPSH equation for this value of the fluid height. The nomenclature used contains only positive numbers for elevation, with the base point being set at the tank's discharge nozzle (analogous to the gravity-discharge case).

For this example, assume the following parameters:
Pump capacity = 50 gal/min
Pipe = 100 ft of 2-in pipe

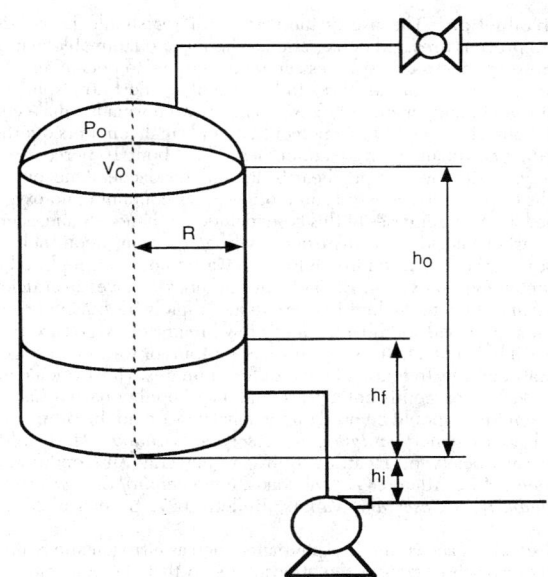

FIG. 26-50   Tank configuration used in example calculation.

NPSH = 4 ft
Elevation H_i = 1 ft
R = 2 ft
Liquid = water at 190°F
P_o = 14.7 psia

For these parameters, the equations predict a much higher vacuum (24.5 in Hg or 230 percent of the shortcut method) than the gravity-discharge case. Of course, different tank dimensions and pump characteristics could give different comparisons between cases. If conditions are such that the pump can completely empty the tank before backflow occurs, the vacuum is best calculated from Eq. (26-57).

If proper instruments are provided, the tank and pump can be interlocked, so the pump will stop when abnormal conditions are detected. This may help keep the tank from collapsing, but the gravity-discharge case should also be checked to ensure that failure will not occur after the interlock stops the pump. In all cases where instrumentation is used, the consequences of potential failure should be considered.

HAZARDS OF INERT GASES

GENERAL REFERENCES: Air Products and Chemicals Co., *Dangers of Oxygen Deficient Atmospheres,* Allentown, Pa., 1988. ASHRAE Standard 34, "Fluorocarbon Numbering for Methane, Ethane and Cycloalkane Refrigerants," 1978. ASHRAE Standard 62-1989, American Society of Heating Refrigerating and Air-Conditioning Engineers, Inc., Atlanta, Ga., 1989. Bartknecht, *Dust Explosions,* Springer-Verlag, Berlin, 1989. Bartknecht, *Explosion Course Prevention, Protection,* Springer-Verlag, Berlin, 1981. Bodurtha, *Industrial Explosion Prevention and Protection,* McGraw-Hill, 1980. Generon Systems (membrane systems for producing moderate- to high-purity nitrogen), 400 W. Sam Houston Parkway South, Houston, Tex., 1993. *Guidelines for Engineering Design for Process Safety,* CCPS-AIChE, New York, 1993. Lees, *Loss Prevention in the Process Industries,* Butterworths, London, 1980. Nagy, Dorsett, and Jacobson, "Preventing Ignition of Dust Dispersions by Inerting," *U.S. Bur. Mines Rep. Invest.* 6543, 1964. NFPA 68, *Venting of Deflagrations,* National Fire Protection Association, Quincy, Mass., 1988. NFPA 69, *Explosion Prevention Systems,* National Fire Protection Association, Quincy, Mass., 1992. NFPA 654, *Standard for Fire and Dust Explosion Prevention in the Chemical, Dye, Pharmaceutical and Plastics Industries,* National Fire Protection Association, Quincy, Mass. 1982. Niida et al., *Some Expert System Experiments in Process Engineering,* Chem. Eng. Res. Des., vol. 64, September 1986, p. 374. *Ventilation for Acceptable Indoor Air Quality,* American Society of Heating, Refrigerating and Air-Conditioning Engineers, Inc., Atlanta, Ga. Zabetakis, *Flammability Characteristics of Combustible Gases and Vapors,* Bulletin 627, Bureau of Mines, 1965.

Introduction   The use of inert atmospheres should be considered to prevent fires and deflagrations when using flammable materials. However, inert atmospheres can be dangerous to personnel. One of the most important concerns in the use of an inert atmosphere is that it can kill if a person breathes it. The air we normally inhale contains about 21 percent O_2, 79 percent N_2, and small amounts of other components. Inhaling air containing less than about 16 percent oxygen causes dizziness, rapid heartbeat, and headache. One or two breaths of pure nitrogen and some other gases containing no oxygen can be lethal. Other gases of this type include methane, ethane, acetylene, carbon dioxide, nitrous oxide, hydrogen, argon, neon, helium, and some others. Oxygen in the lungs is washed out and replaced by gas containing no oxygen. Blood from the lungs receives insufficient oxygen and flows to the brain, where tissues rapidly become deficient. Within five seconds of inhaling only a few breaths of oxygen-free gas, there can be mental failure and coma. Symptoms or warnings are generally absent. Death follows in two to four minutes. However, a coma due to lack of oxygen is not always fatal. Cardiopulmonary resuscitation techniques should be used on persons who are not breathing due to lack of oxygen (*Ventilation for Acceptable Indoor Air Quality,* American Society of Heating, Refrigerating and Air-Conditioning Engineers, Inc., Atlanta, Ga.; Zabetakis, *Flammability Characteristics of Combustible Gases and Vapors,* Bulletin 627, Bureau of Mines, 1965).

Gases which act as simple asphyxiants, such as nitrogen and helium, merely displace oxygen in the atmosphere so that the concentration falls below that needed to maintain consciousness. There are also chemical asphyxiants, such as carbon monoxide, hydrogen sulfide, and hydrogen cyanide, which have a specific blocking action and prevent a sufficient supply of oxygen from reaching the body. Most deaths due to short-term gassing are caused by carbon monoxide (Lees, *Loss Prevention in the Process Industries,* Butterworths, London, 1980, p. 646).

Effects of Low Oxygen Levels   There are many factors which can affect the ability of human beings to adjust to lower oxygen levels. For example, two men were accidentally exposed to a low oxygen level in a vessel. One of them died, and one survived without permanent injury. The one who died had been in poorer general health and it is believed that this factor may have made the low oxygen level fatal for him, while the other person, who was in good health, survived.

It is well known that people accustomed to living near sea level can take several days to adjust fully to the lower amount of oxygen available in mountainous regions such as Denver, Colorado. Anyone who has traveled to the top of Pike's Peak knows how the altitude can make one tired, lethargic, and even sick. People react differently, however, and one cannot generalize as to exactly how a person will react to lower oxygen levels and higher altitudes. Table 26-25 gives the signs and symptoms of reduced oxygen content on persons at rest.

Minimum Oxygen Limits   Oxygen limits are set at 19.5 percent minimum as recommended by OSHA and the American Standards Institute. Michigan has adopted these guidelines as well and has defined grade D air for O_2 to be 19.5 percent to 23.5 percent as an obligation to the employee by their employer. The Ontario Ministry of Labour designates enclosures containing less than 18 percent O_2 as hazardous.

Confined-Space Entry by the Dow Chemical Company   The Dow Chemical Company Safety Standard on Confined-Space Entry states the following regarding confined space entry.

1.   Check all test instruments to assure they are operable before and after use.

2.   Readings acceptable for entry shall be recorded on the Safe Work Permit and shall assure that the oxygen content is 21 percent plus or minus 0.5 percent.

3.   Toxic materials shall be at or below the threshold limit value, permissible exposure limit, or other approved industrial hygiene guideline.

4.   The combustible gas indicator must be calibrated using the appropriate calibrating gas, such as methane or pentane.

5.   The analysis shall be in the following sequence: oxygen concentration, then the combustible gas or vapor.

Confined-Space Entry as Defined by OSHA   (Taylor, Shoemaker, and Sasse, "Confined Space Entry," *AIChE 1990 Summer National Meeting,* San Diego, Calif., August 19–22, 1990.) Occupational Safety and Health Administration (OSHA), in Section 1926.21(b)(6)(ii), has defined confined space as space having a limited means of egress, which is subject to the accumulation of toxic or flammable contaminants or has an oxygen-deficient atmosphere. For the purpose of this section, confined-space entry is discussed only as it pertains to process vessels, catalytic reactors, and storage tanks. OSHA classifies confined-space entry into two categories: immediately dangerous to life and health (IDLH) and non-IDLH as follows:

Class A—immediately dangerous to life and health based on oxygen level less than 19.5 percent and/or airborne presence of toxic or poisonous substances in concentration constituting IDLH conditions; flammability up to 20 percent of lower flammable limit.

Class B—Non-IDLH based on oxygen level between 19.5 and 21 percent, but classified as dangerous due to the airborne presence of toxic or poisonous substance below IDLH level, but greater than the protection factor offered by air-purifying respirators.

In addition, there is a non-IDLH class based on oxygen level between 19.5 and 21 percent, but classified as hazardous due to the presence of nuisance dusts or vapors below the IDLH level, but not greater than the protection factor of air-purifying respirators or low concentration of toxic or flammable substances.

Case Histories   Following are examples of fatal accidents resulting from lack of oxygen:

• In a chemical plant, compressed nitrogen was temporarily being used to supply a control room, which was usually closed, containing pneumatic process control instruments that normally used instrument air. With the normal venting of nitrogen by the instruments in the control room, the air in the room was gradually replaced. An instrument man entered the control room for maintenance and was overcome by the lack of oxygen and died.

• Two men were inspecting a large tank in which other equipment was installed. The tank had two large manways attached to it, one near the bottom and one near the top. Ventilation was provided by air entering the bottom manway and leaving the top manway. A sheet of plastic had temporarily been placed over the top manway, which decreased the amount of air circulation. One of the inspectors climbed a ladder in the tank, became dizzy, and fell to the tank floor below. He died from the injuries received in the fall. It was found that

TABLE 26-25   Effects of Breathing Oxygen-Deficient Atmospheres

Oxygen content of air, %	Signs and symptoms of persons at rest
19.5–23.5	Recommended by OSHA.
15–19	Decreased ability to work strenuously. May impair coordination and may induce early symptoms in persons with coronary, pulmonary, or circulatory problems.
12–17	Loss of balance, dizziness. Respiration deeper, increased pulse rate, impaired coordination, perception, and judgment.
10–12	Further increase in rate and depth of respiration, further increase in pulse rate, performance failure, giddiness, poor judgment, lips blue, prolonged exposure possibly results in brain damage.
8–10	Mental failure, nausea, vomiting, fainting, unconsciousness, ashen face, blueness of lips. 8 minutes: 100% fatal. 6 minutes: 50% fatal. 4 to 5 minutes: recovery with treatment; brain damage and death are possible.
4	Coma in 40 s, convulsions, respiration ceases, death.

SOURCE: Air Products and Chemicals Co., "Dangers of Oxygen Deficient Atmospheres," Allentown, Pa., 1988, and American Standards Institute, Report No. 788.

the oxygen level in the upper part of the tank was 12.3 percent, which was low enough to cause dizziness and loss of balance. When the plastic sheet was removed, the oxygen content quickly rose to 21 percent. It was concluded that steel inside the tank had corroded, causing low oxygen content in the tank. Impaired ventilation caused by the plastic sheet had reduced circulation so that the air in the upper part of the tank remained at a low oxygen concentration.

Inerting Monomer Storage Tanks with Nitrogen It is good practice to keep the vapor space of flammable liquids out of the flammable range. Monomers that can potentially polymerize require special consideration. The vapor space above some monomers, such as styrene and methyl acrylate, should be kept below about 10 percent oxygen in warm weather to be below the flammable range. For many of these monomers, a small amount of oxygen is required to maintain the activity of the inhibitor and to avoid polymerization in storage tanks, which could lead to overheating and explosions and fire. An oxygen concentration of 5 percent in the vapor space is recommended as a safety factor to stay out of the flammable range and maintain inhibitor activity.

Maintaining an inert atmosphere for these applications can be difficult, since usually nitrogen is available as a high-purity gas, and it is necessary to add a small amount of oxygen (usually air) to the nitrogen to achieve the desired oxygen concentration. Mixing air and nitrogen has not proven to be a reliable method of maintaining the proper inert pad in the past. This is because instrument failure has caused high nitrogen concentration, which in turn has caused storage vessels to polymerize. One alternative to consider is the use of membrane systems, such as those sold by Generon Systems and other suppliers. This system can produce 95/5 percent nitrogen/oxygen for inerting, using plant compressed air available at 65 psig (449 kPa gauge). This system has an inherently stable output when operating at a specific pressure drop because the pressure drop across the membrane module sets the nitrogen purity.

Halon Systems for Inerting The term *halon* is generic for a range of halogenated hydrocarbons in which one or more of the hydrogen atoms have been replaced by atoms from the halogen series. Fully halogenated hydrocarbons are considered *hard* halons because it is believed that they have a major effect on the ozone layer. They work as fire-extinguishing agents by interfering with the free radical chain reaction occurring in flames. However, they destroy ozone in the same way. Halons containing bromine are much more destructive of ozone than chlorofluorocarbons (CFCs). It has been reported that one atom of some halons can destroy 10^6 ozone molecules. Halon alternatives that have less effect on the ozone layer include HCFCs, which are halogenated hydrocarbons with at least one hydrogen atom. In 1987, the Montreal Protocol on Protection of the Stratospheric Ozone Layer was signed, which set a timetable for phasing out the production and use of CFCs, including halons. The date for phaseout of the manufacture of halons according to the latest Copenhagen Meeting was January 1, 1994 (UNEP, Montreal Protocol on Substances that Deplete the Ozone Layer—Final Act 1987, 1987).

Although there have been many materials under development to replace the halons, there is not a single material that is a drop-in replacement. Some HCFCs are low in ozone depletion allowance, compared to halons or other CFCs, but in the long term, the goal should be zero ozone depletion. It is probable that there are no absolutely essential applications for halon in the chemical industry. There may be essential uses in airplanes, submarines, etc. Suggested replacements are water, dry chemical, and carbon dioxide. A large use for halon systems is in the form of total halon flooding systems for computer centers. To reduce the need for halon systems for computer centers, modern computer centers have a minimum amount of combustible materials to cause heat generation and there is less cable insulation with the use of fiber optics. Smoke and heat detectors are first-line-of-defense measures, along with emergency electrical power shutoff switches.

Fine water spray systems may be potentially superior to CO_2 applications and may replace halon environments such as telephone central offices and computer rooms. In the fine spray delivery system, water is delivered at relatively high pressure (above 100 psi [0.689 MPa]) or by air atomization to generate droplets significantly smaller than those generated by sprinklers. Water flow from a fine spray nozzle potentially extinguishes the fire faster than a sprinkler because the droplets are smaller and vaporize more quickly. Preliminary information indicates that the smaller the droplet size, the lower the water flow requirements and the less chance of water damage.

Inert Gas Generation Nitrogen is often the preferred gas for providing an inert atmosphere. In general, most organic combustible compounds will not propagate flame if oxygen in the mixtures of the organic vapor, inert gas, and air is below about 10 percent and 13 percent, with nitrogen and carbon dioxide, respectively, as the inert gases. With carbon dioxide, the minimum oxygen concentration is higher than with nitrogen because carbon dioxide has a higher specific heat. Carbon dioxide is fairly soluble in many liquids and will react with alkaline materials, so its use as an inerting material is limited. Heavy gases such as carbon dioxide provide superior inerting of vent stacks to prevent air entry. Water vapor is a good inerting gas if the temperature is high enough (above about 80 to 85°C [176 to 185°F]). Water vapor has a higher specific heat than nitrogen, so less water vapor is required for inerting than nitrogen (*FMRC Update*, vol. 7, no. 3, Factory Mutual Engineering Corp., Norwood, Mass., December 1993, pp. 2, 3).

Table 26-26 lists some of the main commercial methods used to generate nitrogen or nitrogen-rich gas.

TABLE 26-26 Commercial Methods Used to Generate Nitrogen or Nitrogen-Rich Gas

Process	Purity	Capacity	Features
Cryogenic separation	Very high N_2 purity, 99.999%, by-products oxygen and argon	Can be high	Very high purity; high flexibility by storing liquid nitrogen; fairly high capital costs, modest operating costs; complicated process
Pressure swing adsorption using molecular sieves	High purity, 99 to 99.9% N_2	Moderate	High purity; high-pressure storage may be required; simple process; economical
Membrane separation	Medium to high purity N_2, 95 to 99.9%	Small; typical module produces 855 scfh at 175 lb/in² and 77°F	Can use plant air as air source; simple and safe to operate; stable output; may be economical for low-capacity, medium- to high-purity requirements; excellent when some oxygen is required with the nitrogen; temperature and pressure sensitive
Hydrocarbon combustion	Contaminated with other gases: $N_2 \sim 85\%$ $CO_2 \sim 14\%$ $CO \sim 0.5\%$ $O_2 \sim 0.5\%$ $H_2O \sim$ saturated	Small	Simple process; can use combustion products from engines or boilers, or dedicated burner; less reliability of O_2 control; may be very economical; used on tanker ships
Ammonia decomposition	Contaminated with H_2, water; $N_2 \sim 75–99.5\%$, very low O_2	Small	Used only when cheap NH_3 is available

SOURCE: K. Niida, et al., *Some Expert System Experiments in Process Engineering, Chem. Eng. Res. Des*, vol. 64, September 1986, p. 374; Generon Systems, 400 W. Sam Houston Parkway South, Houston, Tex., 1993.

Conclusions   The use of an inert atmosphere can virtually eliminate the possibility of explosions and fire with flammable materials. However, inerting systems can be quite expensive and difficult to operate successfully and can be hazardous to personnel. Before using inert systems, alternatives should be explored, such as using nonflammable materials or operating below the flammable range.

GAS DISPERSION

Nomenclature

A	Area affected by release, length^2
$A°$	Dimensionless impact area
C	Concentration, mass/volume
$\langle C \rangle$	Time-averaged concentration, mass/volume
$\langle C \rangle°$	Concentration of interest, mass/volume
D_c	Characteristic source dimension for continuous releases of dense gases, defined by Eq. (26-72), length
D_i	Characteristic source dimension for instantaneous releases of dense gases, defined by Eq. (26-73), length
g	Acceleration due to gravity, length/time^2
g_o	Initial buoyancy factor, defined by Eq. (26-71), length/time^2
H_r	Height of release above ground level, length
K	Eddy diffusivity, area/time
$L°$	Scaled length, defined by Eq. (26-66), length
M	Molecular weight, mass/mole
P	Pressure, force/area
q_o	Initial plume volume flux for dense gas dispersion, volume/time
Q_m	Continuous release rate of material, mass/time
Q_m^*	Instantaneous release of material, mass
R_d	Release duration, time
T	Absolute temperature, K
t	Time, s
u	Wind speed, length/time
V_o	Initial volume of released dense gas material, length^3
x, y, z	Distance in dimensional space, length
x_v, y_v, z_v	Virtual distances for plume, length
$x°$	Dimensionless downwind distance

Greek symbols	
$\sigma_x, \sigma_y, \sigma_z$	Dispersion coefficients, length
ρ_a	Density of ambient air, mass/volume
ρ_o	Initial density of released material, mass/volume

Subscripts	
a	Ambient
j	Either x, y, or z length dimensions
o	Initial
v	Virtual

Superscripts	
$'$	Stochastic or fluctuating quantity

GENERAL REFERENCES: Crowl and Louvar, *Chemical Process Safety: Fundamentals with Applications*, Prentice Hall, Englewood Cliffs, NJ, 1990, pp. 121–155. Hanna and Drivas, *Guidelines for Use of Vapor Cloud Dispersion Models*, AIChE, New York, 1987. Hanna and Strimaitis, *Workbook of Test Cases for Vapor Cloud Source Dispersion Models*, AIChE, New York, 1989. Lees, *Loss Prevention in the Process Industries*, Butterworths, London, 1986, pp. 428–463. Seinfeld, *Atmospheric Chemistry and Physics of Air Pollution*, Chaps. 12, 13, 14, Wiley, New York, 1986. Turner, *Workbook of Atmospheric Dispersion Estimates*, U.S. Department of Health, Education, and Welfare, Cincinnati, 1970.

Introduction   Gas dispersion (or vapor dispersion) is used to determine the consequences of a release of a toxic or flammable material. Typically, the calculations provide an estimate of the area affected and the average vapor concentrations expected. In order to make this determination, one must know the release rate of the gas (or the total quantity released) and the atmospheric conditions (wind speed, time of day, cloud cover).

The steps required to utilize a gas dispersion model are:

1.   Identify the scenario. What can go wrong to result in the loss of containment of the material?

2.   Develop an appropriate source model to calculate the release rate or total quantity released based on the specified scenario (see Discharge Rates from Punctured Lines and Vessels).

3.   Use an appropriate gas dispersion model to estimate the consequences.

4.   Determine if the resulting consequence is acceptable. If not, then something must be changed to reduce the consequence.

The entire procedure is shown in Fig. 26-51. If the consequence is not acceptable, then some of the options available to reduce the consequence are shown in Table 26-27.

Calculations and experiments have demonstrated that even the release of a small quantity of toxic or flammable material can have a significant consequence. Thus, it is clear that the best procedure is to prevent the release in the first place. However, release mitigation must be a part of any process safety program. Release mitigation involves: (1) detecting the release as early as possible, (2) stopping the release as quickly as possible, and (3) invoking a mitigation/emergency response procedure to reduce the consequences of the release.

Parameters Affecting Gas Dispersion   A wide variety of parameters affect the dispersion of gases. These include: (1) wind speed, (2) atmospheric stability, (3) local terrain characteristics, (4) height of the release above the ground, (5) release geometry, i.e. from a point, line, or area source, (6) momentum of the material released, and (7) buoyancy of the material released.

As the wind speed is increased, the material is carried downwind faster, but the material is also diluted faster by a larger quantity of air.

Atmospheric stability depends on the wind speed, the time of day, and the solar energy input. During the day, the air temperature is at a maximum at the ground surface as a result of radiative heating of the ground from the sun. At night, radiative cooling of the ground occurs, resulting in an air temperature which is low at ground level, increases with height until a maximum is reached, and then decreases with further height.

Terrain characteristics affect the mechanical mixing of the air as it flows over the ground. Thus, the dispersion over a lake is different from the dispersion over a forest or a city of tall buildings.

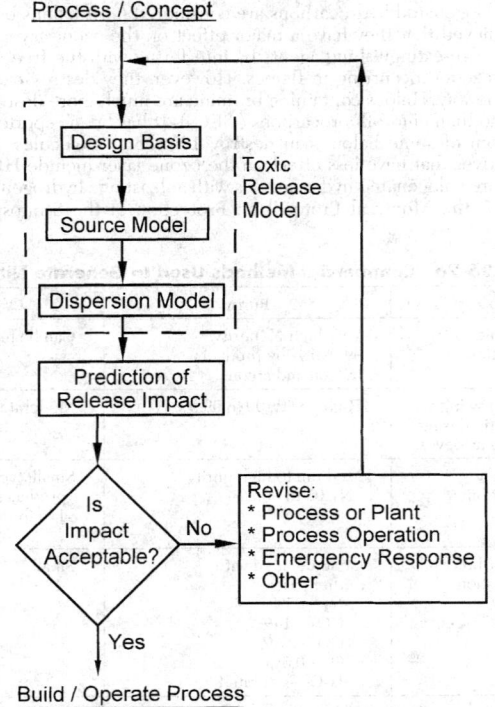

FIG. 26-51   The procedure for using a gas dispersion model to estimate the release impact.

TABLE 26-27   Release Mitigation Approaches

Major area	Examples
Inherent safety	*Inventory reduction:* Fewer chemicals inventoried or fewer in process vessels. *Chemical substitution:* Substitute a less hazardous chemical for one more hazardous. *Process attenuation:* Use lower temperatures and pressures.
Engineering design	*Plant physical integrity:* Use better seals or materials of construction. *Process integrity:* Ensure proper operating conditions and material purity. *Process design features for emergency control:* Emergency relief systems. *Spill containment:* Dikes and spill vessels.
Management	Operating policies and procedures Training for vapor release prevention and control Audits and inspections Equipment testing Maintenance program Management of modifications and changes to prevent new hazards Security
Early vapor detection and warning	Detection by sensors Detection by personnel
Countermeasures	Water sprays Water curtains Steam curtains Air curtains Deliberate ignition of explosive cloud Dilution Foams
Emergency response	On-site communications Emergency shutdown equipment and procedures Site evacuation Safe havens Personal protective equipment Medical treatment On-site emergency plans, procedures, training, and drills

SOURCE: Adapted from Prugh and Johnson, *Guidelines for Vapor Release Mitigation,* AIChE, New York, 1988.

Figure 26-52 shows the effect of height on the downwind concentrations due to a release. As the release height increases, the ground concentration downwind decreases since the resulting plume has more distance to mix with fresh air prior to contacting the ground.

The geometry of the release also affects the resulting consequence. An ideal release would occur at a point source. Real releases are more likely to occur as a line source (from an escaping jet of material) or as an area source (from a boiling pool of liquid).

Figure 26-53 shows the affect of initial momentum and buoyancy of the release. If the material is released as a jet, then the effective height of the release is increased. Furthermore, if the material released is heavier than air (which is the usual case for the release of most hydrocarbons), the plume initially slumps toward the ground until subsequent dilution by air results in a neutrally buoyant cloud.

Gaussian Dispersion   Gaussian dispersion is the most common method for estimating dispersion due to a release of vapor. The method applies only for neutrally buoyant clouds and provides an estimate of average downwind vapor concentrations. Since the concentrations predicted are time averages, it must be considered that local concentrations might be greater than this average; this result is important when estimating dispersion of highly toxic materials where local concentration fluctuations might have a significant impact on the consequences.

Fundamental Equations   A complete development of the fundamental equations is presented elsewhere (Crowl and Louvar, 1990, pp. 129–144). The model begins by writing an equation for the conservation of mass of the dispersing material:

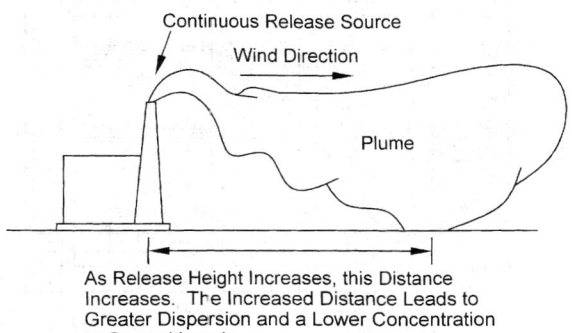

FIG. 26-52   Effect of increased release height on the downwind ground-level concentration. (*Reprinted from D. A. Crowl and J. F. Louvar,* Chemical Process Safety, Fundamentals with Applications, *1990, p. 127. Used by permission of Prentice Hall.*)

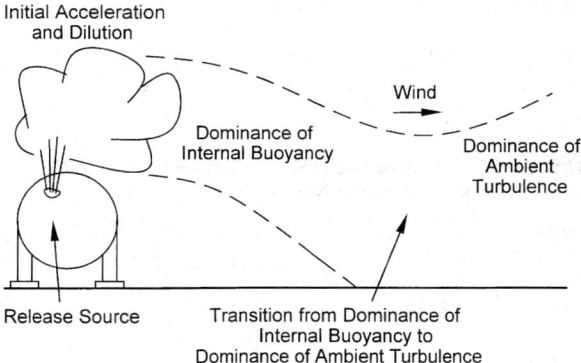

FIG. 26-53   Effect of initial acceleration and buoyancy on the release of gases. (*Adapted from S. R. Hanna and P. J. Drivas,* Guidelines for Use of Vapor Cloud Dispersion Models, *1987. Used by permission of the American Institute of Chemical Engineers, Center for Chemical Process Safety.*)

$$\frac{\partial C}{\partial t} + \frac{\partial}{\partial x_j}(u_j C) = 0 \qquad (26\text{-}58)$$

where C is the concentration of dispersing material; j represents the summation over all three coordinates, x, y, and z; and u is the velocity of the air.

The difficulty with Eq. (26-58) is that it is impossible to determine the velocity u at every point, since an adequate turbulence model does not currently exist. The solution is to rewrite the concentration and velocity in terms of an average and stochastic quantity: $C = \langle C \rangle + C'$; $u_j = \langle u_j \rangle + u'_j$, where the brackets denote the average value and the prime denotes the stochastic, or deviation variable. It is also helpful to define an eddy diffusivity K_j (with units of area/time) as

$$-K_j \frac{\partial \langle C \rangle}{\partial x_j} = \langle u'_j C' \rangle \qquad (26\text{-}59)$$

By substituting the stochastic equations into Eq. (26-58), taking an average, and then using Eq. (26-59), the following result is obtained:

$$\frac{\partial \langle C \rangle}{\partial t} + \langle u_j \rangle \frac{\partial \langle C \rangle}{\partial x_j} = \frac{\partial}{\partial x_j}\left(K_j \frac{\partial \langle C \rangle}{\partial x_j}\right) \qquad (26\text{-}60)$$

The problem with Eq. (26-60) is that the eddy diffusivity changes with position, time, wind velocity, and prevailing atmospheric conditions, to name a few, and must be specified prior to a solution to the equation. This approach, while important theoretically, does not provide a practical framework for the solution of vapor dispersion problems.

Sutton (*Micrometeorology*, McGraw-Hill, 1953, p. 286) developed a solution to the above difficulty by defining dispersion coefficients, σ_x, σ_y, and σ_z, defined as the standard deviation of the concentrations in the downwind, crosswind, and vertical (x, y, z) directions, respectively. The dispersion coefficients are a function of atmospheric conditions and the distance downwind from the release. The atmospheric conditions are classified into six stability classes (A through F) for continuous releases and three stability classes (unstable, neutral, and stable) for instantaneous releases. The stability classes depend on wind speed and the amount of sunlight, as shown in Table 26-28.

Pasquill (*Atmospheric Diffusion*, Van Nostrand, 1962) recast Eq. (26-60) in terms of the dispersion coefficients and developed a number of useful solutions based on either continuous (plume) or instantaneous (puff) releases. Gifford (*Nuclear Safety*, vol. 2, no. 4, 1961, p. 47) developed a set of correlations for the dispersion coefficients based on available data (see Table 26-29 and Figs. 26-54 to 26-57). The resulting model has become known as the *Pasquill-Gifford model*.

The *puff model* describes near-instantaneous releases of material. The solution depends on the total quantity of material released, the atmospheric conditions, the height of the release above ground, and the distance from the release. The equation for the average concentration for this case is (Crowl and Louvar, 1990, p. 143):

$$\langle C \rangle(x,y,z,t) = \frac{Q_m^\circ}{(2\pi)^{3/2}\sigma_x\sigma_y\sigma_z} \exp\left[-\frac{1}{2}\left(\frac{y}{\sigma_y}\right)^2\right]$$
$$\times \left\{\exp\left[-\frac{1}{2}\left(\frac{z-H_r}{\sigma_z}\right)^2\right] + \exp\left[-\frac{1}{2}\left(\frac{z+H_r}{\sigma_z}\right)^2\right]\right\} \qquad (26\text{-}61)$$

TABLE 26-28   Atmospheric Stability Classes for Use with the Pasquill-Gifford Dispersion Model

Wind speed, m/s	Day radiation intensity			Night cloud cover	
	Strong	Medium	Slight	Cloudy	Calm and Clear
<2	A	A–B	B		
2–3	A–B	B	C	E	F
3–5	B	B–C	C	D	E
5–6	C	C–D	D	D	D
>6	C	D	D	D	D

Stability classes for puff model:
A, B: unstable
C, D: neutral
E, F: stable

TABLE 26-29   Equations and Data for Pasquill-Gifford Dispersion Coefficients

Equations for continuous plumes	
Stability class	σ_y, m
A	$\sigma_y = 0.493x^{0.88}$
B	$\sigma_y = 0.337x^{0.88}$
C	$\sigma_y = 0.195x^{0.90}$
D	$\sigma_y = 0.128x^{0.90}$
E	$\sigma_y = 0.091x^{0.91}$
F	$\sigma_y = 0.067x^{0.90}$

Stability class	x, m	σ_z, m
A	100–300	$\sigma_z = 0.087x^{1.10}$
	300–3000	$\log_{10}\sigma_z = -1.67 + 0.902\log_{10}x + 0.181(\log_{10}x)^2$
B	100–500	$\sigma_z = 0.135x^{0.95}$
	$500\text{–}2\times10^4$	$\log_{10}\sigma_z = -1.25 + 1.09\log_{10}x + 0.0018(\log_{10}x)^2$
C	$100\text{–}10^5$	$\sigma_z = 0.112x^{0.91}$
D	100–500	$\sigma_z = 0.093x^{0.85}$
	$500\text{–}10^5$	$\log_{10}\sigma_z = -1.22 + 1.08\log_{10}x - 0.061(\log_{10}x)^2$
E	100–500	$\sigma_z = 0.082x^{0.82}$
	$500\text{–}10^5$	$\log_{10}\sigma_z = -1.19 + 1.04\log_{10}x - 0.070(\log_{10}x)^2$
F	100–500	$\sigma_z = 0.057x^{0.80}$
	$500\text{–}10^5$	$\log_{10}\sigma_z = -1.91 + 1.37\log_{10}x - 0.119(\log_{10}x)^2$

Data for puff releases				
Stability condition	$x = 100$ m		$x = 4000$ m	
	σ_y, m	σ_z, m	σ_y, m	σ_z, m
Unstable	10	15	300	220
Neutral	4	3.8	120	50
Very stable	1.3	0.75	35	7

SOURCE: Frank P. Lees, *Loss Prevention in the Process Industries*, Butterworths, London, 1986, p. 443).

The center of the puff is located at $x = ut$. Here x is the downwind direction, y is the crosswind direction, and z is the height above ground level. The initial release occurs at a height H_r above the ground point at $(x,y,z) = (0,0,0)$, and the center of the coordinate system remains at the center of the puff as it moves downwind.

Notice that the wind speed does not appear explicitly in Eq. (26-61). It is implicit through the dispersion coefficients since these are a function of distance downwind from the initial release and the atmospheric stability conditions.

A typical requirement is to determine the cloud boundary at a fixed concentration. These boundaries, or lines, are called *isopleths*. The

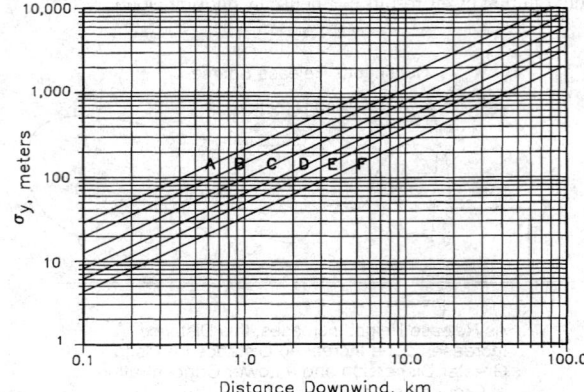

FIG. 26-54 Horizontal dispersion coefficient for Pasquill-Gifford plume model. (*Reprinted from D. A. Crowl and J. F. Louvar, Chemical Process Safety, Fundamentals with Applications, 1990, p. 138. Used by permission of Prentice Hall.*)

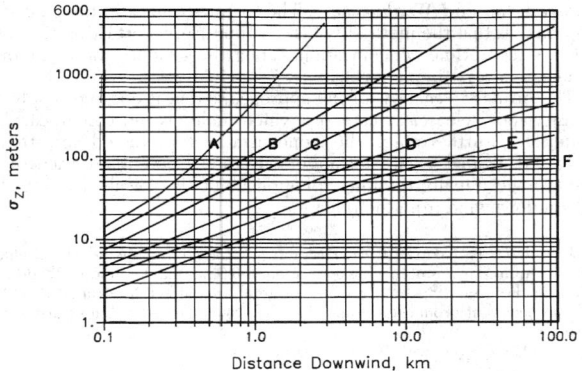

FIG. 26-55 Vertical dispersion coefficient for Pasquill-Gifford plume model. (*Reprinted from D. A. Crowl and J. F. Louvar, Chemical Process Safety, Fundamentals with Applications, 1990, p. 138. Used by permission of Prentice Hall.*)

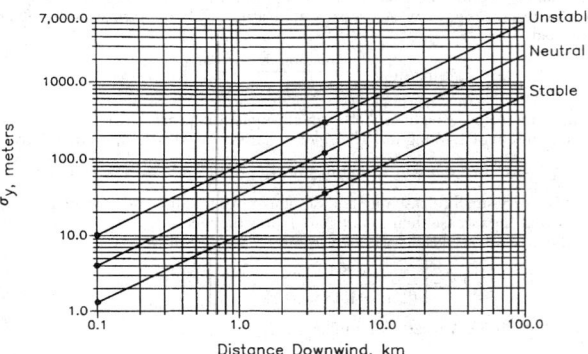

FIG. 26-56 Horizontal dispersion coefficient for Pasquill-Gifford puff model. These data are based on only the data points shown and should not be considered reliable elsewhere. (*Reprinted from D. A. Crowl and J. F. Louvar, Chemical Process Safety, Fundamentals with Applications, 1990, p. 140. Used by permission of Prentice Hall.*)

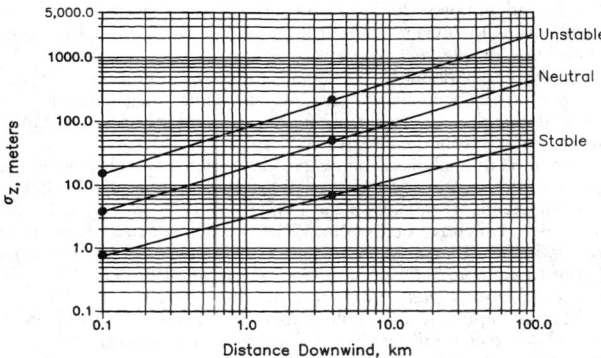

FIG. 26-57 Vertical dispersion coefficient for Pasquill-Gifford puff model. These data are based only on the data points shown and should not be considered reliable elsewhere. (*Reprinted from D. A. Crowl and J. F. Louvar, Chemical Process Safety, Fundamentals with Applications, 1990, p. 140. Used by permission of Prentice Hall.*)

locations of these are found by dividing the equation for the centerline concentration, i.e., $\langle C \rangle(x,0,0,t)$, by the general ground-level concentration provided by Eq. (26-61). The resulting equation is solved for y to give

$$y = \sigma_y \sqrt{2 \ln \left(\frac{\langle C \rangle(x,0,0,t)}{\langle C \rangle(x,y,0,t)} \right)} \qquad (26\text{-}62)$$

The procedure to determine an isopleth at any specified time is:
1. Specify a concentration $\langle C \rangle°$ for the isopleth.
2. Determine the concentrations $\langle C \rangle(x,0,0,t)$, along the x axis directly downwind from the release. Define the boundary of the cloud along this axis.
3. Set $\langle C \rangle(x,y,0,t) = \langle C \rangle°$ in Eq. (26-62) and determine the value of y at each centerline point determined in step 2. Plot the y values to define the isopleth, using symmetry around the centerline.

The *plume model* describes continuous release of material. The solution depends on the rate of release, the atmospheric conditions, the height of the release above ground, and the distance from the release. In this case, the wind is moving at a constant speed u in the x direction. The equation for the average concentration for this case is (Crowl and Louvar, 1990, p. 142):

$$\langle C \rangle(x,y,z) = \frac{Q_m}{2\pi\sigma_y\sigma_z u} \exp\left[-\frac{1}{2}\left(\frac{y}{\sigma_y}\right)^2\right]$$
$$\times \left\{ \exp\left[-\frac{1}{2}\left(\frac{z-H_r}{\sigma_z}\right)^2\right] + \exp\left[-\frac{1}{2}\left(\frac{z+H_r}{\sigma_z}\right)^2\right]\right\} \qquad (26\text{-}63)$$

For releases at ground level, the maximum concentration occurs at the release point. For releases above ground level, the maximum ground concentration occurs downwind along the centerline. The location of the maximum is found using

$$\sigma_z = \frac{H_r}{\sqrt{2}} \qquad (26\text{-}64)$$

and the maximum concentration is found from

$$\langle C \rangle_{\max} = \frac{2Q_m}{e\pi u H_r^2}\left(\frac{\sigma_z}{\sigma_y}\right) \qquad (26\text{-}65)$$

The procedure for finding the maximum concentration and the downwind distance for the maximum is to
1. Use Eq. (26-64) to determine the dispersion coefficient σ_z at the maximum.
2. Use Fig. 26-56 to determine the downwind location of the maximum.
3. Use Eq. (26-65) to determine the maximum concentration.

Nomograph Method By defining a scaled length

$$L° = \left(\frac{Q_m}{u\langle C \rangle°}\right)^{1/2} \qquad (26\text{-}66)$$

a dimensionless downwind distance

$$x° = \frac{x}{L°} \qquad (26\text{-}67)$$

and a dimensionless area

$$A° = \frac{A}{(L°)^2} \qquad (26\text{-}68)$$

nomographs can be developed for determining the downwind distance and the total area affected at the concentration of interest $\langle C \rangle°$. These nomographs are shown in Figs. 26-58 and 26-59.

Virtual Sources The previous equations apply to point source releases. Real releases, such as a boiling pool of liquid or a streaming jet of flashing liquid, involve a more complex geometry. One approach (*Guidelines for Chemical Process Quantitative Risk Analysis,* AIChE, 1989, p. 87) is to define a virtual source upwind from the actual source such that the computed plume matches the real plume. However, to achieve this, a concentration at a centerline point directly downwind must be known. There are several ways to determine the location of the virtual source for a plume:

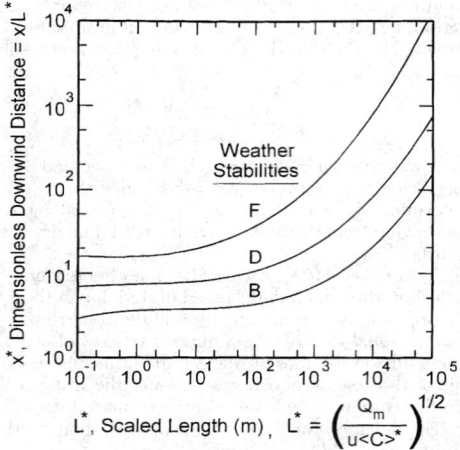

FIG. 26-58 Nomograph to determine the downwind distance affected by a release. (*Adapted from* Guidelines for Chemical Process Quantitative Risk Analysis, *1989, p. 90. Used by permission of the American Institute of Chemical Engineers.*)

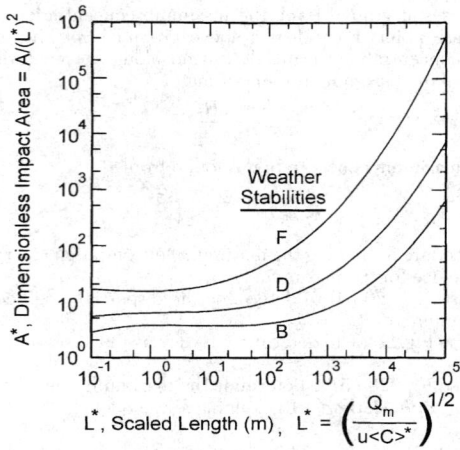

FIG. 26-59 Nomograph to determine the area affected by a release. (*Adapted from* Guidelines for Chemical Process Quantitative Risk Analysis, *1989, p. 91. Used by permission of the American Institute of Chemical Engineers.*)

1. Assume that all of the dispersion coefficients become equal at the virtual source. Then, from Eq. (26-63)

$$\sigma_y(y_v) = \sigma_z(z_v) = \left(\frac{Q_m}{\pi u \langle C \rangle^\circ}\right)^{1/2} \qquad (26\text{-}69)$$

The virtual distances, y_v and z_v, determined using Eq. (26-69) are added to the actual downwind distance x to determine the dispersion coefficients σ_y and σ_z for subsequent computations.

2. Assume that $x_v = y_v = z_v$. Then, from Eq. (26-63)

$$\sigma_y(x_v) \cdot \sigma_z(x_v) = \frac{Q_m}{\pi u \langle C \rangle^\circ} \qquad (26\text{-}70)$$

x_v is determined from Eq. (26-70) using a trial-and-error approach. The effective distance downwind for subsequent calculations using Eq. (26-63) is determined from $(x + x_v)$.

3. For large downwind distances, the virtual distances will be negligible and the point source models are used directly.

Strengths and Weaknesses The major strength to the gaussian approach is that the method is easy to apply. For most cases of interest, i.e., centerline concentrations along the ground, the equations reduce to a very simple form.

The primary weakness of the approach is that it does not apply to dense vapor releases, a category which includes most hydrocarbon materials. Furthermore, the concentrations predicted are time-weighted averages, with instantaneous values potentially exceeding the average. Finally, the range of applicability is typically from 0.1 to 10 km downwind from the release.

Example 1: Continuous Release What continuous release of chlorine is required to result in a concentration of 0.5 ppm at 300 m directly downwind on the ground? Also, estimate the total area affected. Assume that the release occurs at ground level and that the atmospheric conditions are worst case.

From Eq. (26-63), with $H_r = 0$, $z = 0$, and $y = 0$,

$$\langle C \rangle(x,0,0) = \frac{Q_m}{\pi \sigma_y \sigma_z u}$$

Worst-case atmospheric conditions occur to maximize $\langle C \rangle$. This occurs with minimum dispersion coefficients and minimum wind speed u within a stability class. By inspection of Figs. 26-54 and 26-55 and Table 26-28, this occurs with F-stability and $u = 2$ m/s. At 300 m = 0.3 km, from Figs. 26-54 and 26-55, $\sigma_y = 11$ m and $\sigma_z = 5$ m. The concentration in ppm is converted to kg/m³ by application of the ideal gas law. A pressure of 1 atm and temperature of 298 K are assumed.

$$mg/m^3 = \left(\frac{gm\text{-}mole\ K}{0.08206\ L\ atm}\right)\left(\frac{PM}{T}\right)C_{ppm}$$

Using a molecular weight of 70.91 gm/gm-mole, the preceding equation gives a concentration of 1.45 mg/m³. The release rate required is computed directly:

$$Q_m = \langle C \rangle^\circ \pi \sigma_y \sigma_z u = (1.45\ mg/m^3)(3.14)(11\ m)(5\ m)(2\ m/s) = 500\ mg/s$$

This is a very small release rate and demonstrates that it is much more effective to prevent the release than to mitigate it after the fact.

The area affected is determined from Fig. 26-59. For this case,

$$L^\circ = \left[\frac{5 \times 10^{-4}\ kg/s}{(2\ m/s)(1.45 \times 10^{-6}\ kg/m^3)}\right]^{1/2} = 13.1\ m$$

From Fig. 26-59, $A^\circ = 20$ and it follows that

$$A = A^\circ(L^\circ)^2 = (20)(13.1\ m)^2 = 3430\ m^2$$

Dense Gas Dispersion A dense gas is defined as any gas whose density is greater than the density of the ambient air through which it is being dispersed. This result can be due to a gas with a molecular weight greater than that of air, or a gas with a low temperature due to autorefrigeration during release, or other processes.

Dense gases behave considerably differently from neutrally buoyant gases. When they are initially released, these gases slump toward the ground and move both upwind and downwind. Furthermore, the mechanisms for mixing with air are completely different from neutrally buoyant releases.

As dense clouds move downwind, they are diluted with air until they eventually become neutrally buoyant. Thus, the gaussian models presented earlier are applicable for dense cloud releases at distances far downwind from the release.

A complete analysis of dense gas dispersion is much beyond the scope of this treatise. More detailed references are available (Britter and McQuaid, *Workbook on the Dispersion of Dense Gases,* Health and Safety Executive Report No. 17/1988, England, 1988; Lees, 1986, pp. 455–461; Hanna and Drivas, 1987; *Workbook of Test Cases for Vapor Cloud Source Dispersion Models,* AIChE, 1989; *Guidelines for Chemical Process Quantitative Risk Analysis,* 1989, pp. 96–103).

Many computer codes, both public and private, are available to model dense cloud dispersion. A detailed review of these codes, and how they perform relative to actual field test data, is available (Hanna, Chang, and Strimaitis, *Atmospheric Environment,* vol. 27A, no. 15, 1993, pp. 2265–2285). An interesting result of this review is that a simple nomograph method developed by Britter and McQuaid (1988) matches the available data as well as any of the computer codes. This method will be presented here.

The Britter and McQuaid model was developed by performing a dimensional analysis and correlating existing data on dense cloud dispersion. The model is best suited for instantaneous or continuous ground-level area or volume source releases of dense gases. Atmospheric stability was found to have little effect on the results and is not a part of the model. Most of the data came from dispersion tests in remote, rural areas, on mostly flat terrain. Thus, the results would not be applicable to urban areas or highly mountainous areas.

The model requires a specification of the initial cloud volume, the initial plume volume flux, the duration of release, and the initial gas density. Also required is the wind speed at a height of 10 m, the distance downwind, and the ambient gas density.

The first step is to determine if the dense gas model is applicable. If an initial buoyancy is defined as

$$g_o = \frac{g(\rho_o - \rho_a)}{\rho_a} \qquad (26\text{-}71)$$

and a characteristic source dimension as, for continuous releases,

$$D_c = \left(\frac{q_o}{u}\right)^{1/2} \qquad (26\text{-}72)$$

and for instantaneous releases:

$$D_i = V_o^{1/3} \qquad (26\text{-}73)$$

then the criteria for a sufficiently dense cloud to require a dense cloud representation are, for continuous releases:

$$\left(\frac{g_o q_o}{u^3 D_c}\right)^{1/3} \geq 0.15 \qquad (26\text{-}74)$$

and for instantaneous releases:

$$\frac{(g_o V_o)^{1/2}}{u D_i} \geq 0.20 \qquad (26\text{-}75)$$

If these criteria are satisfied, then Figs. 26-60 and 26-61 are used to estimate the downwind concentrations.

The criteria for determining whether the release is continuous or instantaneous is calculated using the following group:

$$\frac{u R_d}{x} \qquad (26\text{-}76)$$

If the group has a value greater than or equal to 2.5, then the dense gas release is considered continuous. If the group value is less than or equal to 0.6, then the release is considered instantaneous. If the value

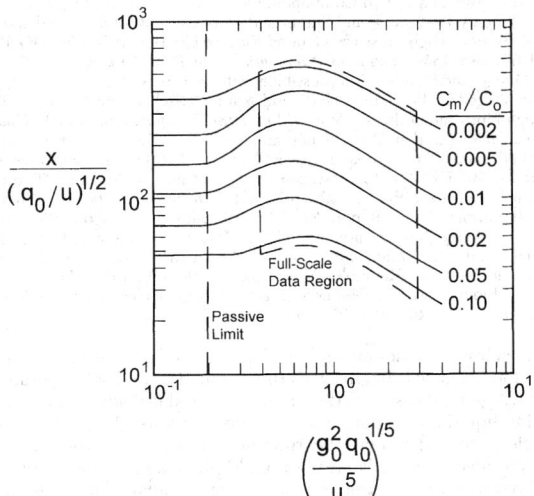

FIG. 26-60 Nomograph to estimate downwind concentrations due to continuous dense gas release based on the Britter-McQuaid correlation.

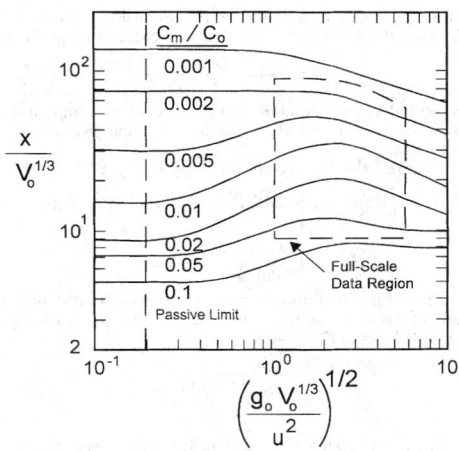

FIG. 26-61 Nomograph to estimate downwind concentrations due to an instantaneous dense gas release based on the Britter-McQuaid correlation.

lies in between, then the concentrations are calculated using both continuous and instantaneous models and the minimum concentration result is selected.

The Britter and McQuaid model is not appropriate for jets or two-phase plume releases. However, it would be appropriate at a minimal distance of 100 m from these types of releases since the initial release effect is usually minimal beyond these distances.

Example 2: LNG Dispersion Tests Britter and McQuaid (1988, p. 70) report on the Burro LNG dispersion tests. Compute the distance downwind from the following LNG release to obtain a concentration equal to the lower flammability limit (LFL) of 5 percent vapor concentration by volume. Assume ambient conditions of 298 K and 1 atm. The following data are available:

Spill rate of liquid   0.23 m³/s
Spill duration R_d 174 s
Windspeed at 10 m above ground (u) 10.9 m/s
LNG density   425.6 kg/m³
LNG vapor density at boiling point of −162°C   1.76 kg/m³

Solution.   The volumetric discharge rate is given by:

$$q_o = \frac{(0.23 \text{ m}^3/\text{s})(425.6 \text{ kg/m}^3)}{1.76 \text{ kg/m}^3} = 55.6 \text{ m}^3/\text{s}$$

The ambient air density is computed from the ideal gas law and gives a result of 1.22 kg/m³. Thus

$$g_o = g\left(\frac{\rho_o - \rho_a}{\rho_a}\right) = (9.8 \text{ m/s}^2)\left(\frac{1.76 - 1.22}{1.22}\right) = 4.29 \text{ m/s}^2$$

Step 1:   Determine if the release is considered continuous or instantaneous. For this case, Eq. (26-76) applies and the quantity must be greater than 2.5 for a continuous release. Thus

$$\frac{u R_d}{x} = \frac{(10.9 \text{ m/s})(174 \text{ s})}{x} \geq 2.5$$

and it follows that for a continuous release

$$x \leq 758 \text{ m}$$

Our final distance must be less than this for application of the continuous release model.

Step 2:   Determine if a dense cloud model applies. For this case, Eqs. (26-69) and (26-74) apply. Substituting the appropriate numbers,

$$D_c = \left(\frac{q_o}{u}\right)^{1/2} = \left(\frac{55.6 \text{ m}^3/\text{s}}{10.9 \text{ m/s}}\right)^{1/2} = 2.26 \text{ m}$$

$$\left(\frac{g_o q_o}{u^3 D_c}\right)^{1/3} = \left[\frac{(4.29 \text{ m/s}^2)(55.6 \text{ m}^3/\text{s})}{(10.9 \text{ m/s})^3(2.26 \text{ m})}\right]^{1/3} = 0.43 \geq 0.15$$

and it is clear that the dense cloud model applies.

Step 3:   Adjust the concentration for non-isothermal release. Britter and MacQuaid (1988, p. 61) provide an adjustment to the concentration to account for non-isothermal release of the vapor. If the original, non-isothermal concentration is $C°$, then the equivalent isothermal concentration is given by

$$C = \frac{C°}{C° + (1 - C°)(T_a/T_o)}$$

where T_a is the ambient temperature and T_o is the source temperature. For our required concentration of 0.05, the preceding equation gives an effective concentration of 0.019.

Step 4: Compute the dimensionless groups for Fig. 26-61.

$$\left(\frac{g_o^2 q_o}{u^5}\right)^{1/5} = \left[\frac{(4.29 \text{ m/s}^2)^2(55.6 \text{ m}^3/\text{s})}{(10.9 \text{ m/s})^5}\right]^{1/5} = 0.367$$

and

$$\left(\frac{q_o}{u}\right)^{1/2} = \left(\frac{55.6 \text{ m}^3/\text{s}}{10.9 \text{ m/s}}\right)^{1/2} = 2.25 \text{ m}$$

Step 5: Apply Fig. 26-60 to determine the downwind distance. The initial concentration of gas C_o is essentially pure LNG. Thus, $C_o = 1.0$ and it follows that $C_m/C_o = 0.019$. From Fig. 26-60,

$$\frac{x}{\left(\dfrac{q_o}{u}\right)^{1/2}} = 126$$

and it follows that $x = (2.25 \text{ m})(126) = 283$ m. This compares to an experimentally determined distance of 200 m. This demonstrates that dense gas dispersion estimates can easily be off by a factor of 2. A gaussian plume model assuming worst-case weather conditions (F-stability, 2 m/s wind speed) predicts a downwind distance of 14 km. Clearly, the dense cloud model provides a much better result.

DISCHARGE RATES FROM PUNCTURED LINES AND VESSELS

Nomenclature

a, b, c	Constants
A	Cross-sectional area perpendicular to flow, m^2
C_D	Overall discharge coefficient (−)
C_{DG}	Discharge coefficient for gas flow (−)
C_{DL}	Discharge coefficient for liquid flow (−)
C_p	Heat capacity at constant pressure, $J \cdot kg^{-1} \cdot K^{-1}$
C_v	Heat capacity at constant volume, $J \cdot kg^{-1} \cdot K^{-1}$
D	pipe diameter, m
D_T	tank diameter, m
f	Fanning friction factor (−)
F_I	Pipe inclination factor, Eq. (26-87)
g	Gravitational acceleration, ms^{-2}
G	Mass flux, $kg \cdot m^{-2}, s^{-1}$
H	Specific enthalpy, $J \cdot kg^{-1}$
H_{GL}	Heat of vaporization, $(H_G - H_L)$ sat, $J \cdot kg^{-1}$
k	Value near C_p/C_v
K	Slip velocity ratio, u_G/u_L
K_e	Number of velocity heads for fittings, expansions, contractions, and bends
L	Length of pipe
N	$4fL/D + K_e$
P	Pressure, $N \cdot m^{-2}$
q	Constant
Q	Heat transfer rate, W/kg
R	Gas constant, $Jk \, mole^{-1}K^{-1}$
Re	Reynolds number, $G \cdot D/\mu$ (−)
S	Entropy, $J \cdot kg^{-1} \cdot K^{-1}$
t	Time, s
T	Temperature, K
u	Velocity, ms^{-1}
v	Specific volume, $m^3 \cdot kg^{-1}$
w	Mass discharge rate, $kg \cdot s^{-1}$
x	Vapor quality, kg vapor/kg mixture
X_m	Lockhart Martinelli parameter
z	Vertical distance, m

Greek	
α	Vapor void fraction, m^3 vapor/m^3 mixture
γ	Heat capacity ratio, C_p/C_v
ε	Dimensionless specific volume, v/v_o
η	Pressure ratio, P/P_o
θ	Inclination angle of pipe to horizontal
μ	Two-phase viscosity, P
ρ	Density, $kg \cdot m^{-3}$
σ	Area ratio
ϕ	Two-phase multiplier, pressure drop for two-phase flow divided by pressure drop for single-phase flow
ω	Parameter defined by Eqs. (26-90) or (26-91)

Subscripts	
a	Ambient
c	Choked
d	Discharge
g, G	Gas or vapor
GL	Gas minus liquid
H	Homogeneous
L	Liquid
N	Nonequilibrium or puncture area
o	Area initial, stagnation conditions
p	Pipe flow
s	Saturation
1	Point at which backpressure from pipe is felt after entrance from tank
2	Plane at vena contracta or at pipe puncture
$\bullet$	Dimensionless

GENERAL REFERENCES: Cheremisinoff and Gupta, eds., *Handbook of Fluids in Motion*, Ann Arbor Science, Ann Arbor, Michigan, 1983. Chisholm, *Two-phase Flow in Pipelines and Heat Exchangers,* George Godwin, New York, in association with the Institution of Chemical Engineers, 1983. Fisher et al., *Emergency Relief System Design Using DIERS Technology,* AIChE, New York, 1992. Graham, "The Flow of Air-Water Mixtures Through Nozzles," National Engineering Laboratories (NEL) Report No. 308, East Kilbride, Glasgow, 1967. Jobson, "On the Flow of Compressible Fluids Through Orifices," *Proc. Instn. Mech. Engrs.* **169**(37): 767–776, 1955. "The Two-Phase Critical Flow of One-Component Mixtures in Nozzles, Orifices, and Short Tubes," *Trans. ASME, J. Heat Transfer* **93**(5): 179–N87, 1976. Lee and Sommerfeld, "Maximum Leakage Times Through Puncture Holes for Process Vessels of Various Shapes," *J. Hazardous Materials* **38**(1): 27–40, July 1994. Lee and Sommerfeld, "Safe Drainage or Leakage Considerations and Geometry in the Design of Process Vessels," *Trans. IChemE* **72**, part B: 88–89, May 1994. Leung, "A Generalized Correlation for One-Component Homogeneous Equilibrium Flashing Choked Flow," *AIChE J.* **32**(10): 1743–1746, 1986. Leung, "Similarity Between Flashing and Non-flashing Two-Phase Flow," *AIChE. J.* **36**(5): 797, 1990. Leung, "Size Safety Relief Valves for Flashing Liquids," *Chem. Eng. Prog.* **88**(2): 70–75, February 1992. Leung, "Two-Phase Flow Discharge in Nozzles and Pipes—A Unified Approach," *J. Loss Prevention Process Ind.* **3**(27): 27–32, January 1990. Leung and Ciolek, "Flashing Flow Discharge of Initially Subcooled Liquid in Pipes," *ASME Trans. J. Fluids Eng.* **116**(3), September 1994. Leung and Epstein, "Flashing Two-Phase Flow Including the Effects of Noncondensable Gases," *ASME Trans. J Heat Transfer* **113**(1): 269, February 1991. Leung and Epstein, "A Generalized Correlation for Two-Phase Non-flashing Homogeneous Choked Flow," *ASME J. Heat Transfer* **112**(2), May 1990. Leung and Grolmes, "The Discharge of Two-Phase Flashing Flow in a Horizontal Duct," *AIChE J.* **33**(3): 524–527, 1987; also errata, **34**(6): 1030, 1988. Leung and Grolmes, "A Generalized Correlation for Flashing Choked Flow of Initially Subcooled Liquid," *AIChE J.* **34**(4): 688–691, 1988. Levenspiel, "The Discharge of Gases from a Reservoir Through a Pipe," *AIChE J.* **23**(3): 402–403, 1977. Lockhart and Martinelli, "Proposed Correlation of Data for Isothermal Two-Phase, Two-Component Flow in Pipes," *Chem. Eng. Prog.* **45**(1): 39–48, January 1949. Sozzi and Sutherland, "Critical Flow of Saturated and Subcooled Water at High Pressure, General Electric Co. Report No. NEDO-13418, July 1975; also *ASME Symp. on Non-equilibrium Two-Phase Flows,* 1975. Tangren, Dodge, and Siefert, "Compressibility Effects in Two-Phase Flow," *J. Applied Physics* **20**: 637–645, 1949. Uchida and Narai, "Discharge of Saturated Water Through Pipes and Orifices," *Proc. 3d Intl. Heat Transfer Conf.,* ASME, Chicago, **5**: 1–12, 1966. Van den Akker, Snoey, and Spoelstra, "Discharges of Pressurized Liquefied Gases Through Apertures and Pipes," *I. Chem. E. Symposium Ser.* (London), **80**: E23–35, 1983. Watson, Vaughan, and McFarlane, "Two-Phase Pressure Drop with a Sharp-Edged Orifice," National Engineering Laboratories (NEL) Report No. 290, East Kilbride, Glasgow, 1967. Woodward, "Discharge Rates Through Holes in Process Vessels and Piping," in V. Fthenakis, ed., *Prevention and Control of Accidental Releases of Hazardous Gases,* Van Nostrand Reinhold, New York, pp. 94–159, 1993. Woodward and Mudan, "Liquid and Gas Discharge Rates Through Holes in Process Vessels," *J. Loss Prevention:* **4**(3): 161–165, 1991.

Overview Modeling the consequences of accidental releases of hazardous materials begins with the calculation of discharge rates. In the most general case, the discharged material is made up of a volatile flashing liquid and vapor along with noncondensable gases and solid particles. For efficiency, the treatment here is of two-phase flow, which reduces as a special case to single-phase all gas or all liquid flow. Solid particulate discharge is usually not particularly hazardous and is not considered here.

If the puncture occurs on a pipe which is at least 0.5 m from a vessel, it is justifiable to use a *homogeneous equilibrium model* (HEM) for which an analytical solution is available. The discharge rate predictions by the HEM beyond this range are within 10 percent of measured values for single-component liquids which are either subcooled or saturated.

If the puncture occurs on the vessel or on a line shorter than 0.5 m, the discharge is likely to be nonhomogeneous, meaning the gas and liquid velocities are not equal and the phases are not likely to be in equilibrium. For this case, various models have been developed, including some of considerable complexity, accounting for interphase heat, mass, and momentum transfer. These are generally used in the nuclear power industry. For most engineering applications, simpler models suffice. A reasonably simple *nonequilibrium model* (NEM) is developed here. We also provide an HEM for orifice flow, since it helps to develop the HEM for pipe flow, and its inaccuracies may at times be tolerable.

The energy and momentum balances common to both situations are stated first, along with some useful general concepts, followed by a development of an HEM for orifice and pipe discharge by Leung (1986; 1990; 1992) and Leung et al. (Leung and Ciolek, 1994; Leung and Epstein, 1990; Leung and Grolmes, 1988) and then the NEM for short pipe, orifice, and nozzle flow summarized by Chisholm.

Discharge Flow Regimes Upon developing a puncture in either the vessel or a line attached to the vessel, as in Fig. 26-62, the subsequent depressurization can cause a volatile liquid to flash and develop bubbles in the liquid. These bubbles cause an expansion, or *swell*, which raises the two-phase, or *frothy*, level. If the puncture is in the vapor space of a vessel or on a line from the vapor space, the discharge will be at least initially all vapor. This is the simplest discharge case and is treated here as a special case.

In the more general and more difficult case, either the puncture is initially in the liquid space, or in a line attached to it, or the liquid swells to reach the puncture or punctured line, giving two-phase or all-liquid discharge. For these cases the discharge model solutions must treat four regimes, which are defined by the initial void (vapor) fraction α_o and by the pressure ratios:

$$\eta_s = \frac{P_s}{P_o} \qquad \eta_a = \frac{P_a}{P_o}$$

where P_s is the saturation vapor pressure and P_a is ambient pressure.

Regime 1. If the tank is initially saturated (usually pressurized with volatile contents) there is no padding gas contributing noncondensables, so $\eta_s = 1$ and $\alpha_o = 0$. In this case, a discharge in the liquid space is a flashing liquid.

If the initial tank conditions are subcooled, $\eta_s < 1$, tank pressure must be maintained with padding gas (which could be air), typically introducing noncondensables, so $\alpha_o > 0$. These noncondensables may or may not become involved in the discharge. Furthermore, the subcooled liquid may flash ($\eta_s > \eta_a$,) or not ($\eta_s < \eta_a$). In addition, some reacting systems generate noncondensable gases, giving Regimes 3 or 4.

Regime 2. If the puncture is below the initial liquid level, the padding gas will not be discharged, so there will be no noncondensables in the discharge ($\alpha_o = 0$). With low or moderate subcooling, $\eta_s > \eta_a$, the subcooled liquid will flash when the pressure ratio at some point drops below η_s. This point could be beyond the choke point, though. With high subcooling, $\eta_s < \eta_a$, and no flashing occurs (single-phase flow).

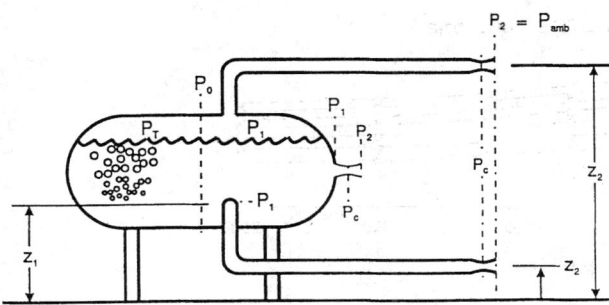

FIG. 26-62 Definition of terms for puncture of a vessel or line attached to a vessel.

Regime 3. If the puncture is above the initial liquid level but becomes covered by the swell, there will be noncondensables mixed with the liquid ($\alpha_o > 0$). If also $\eta_s < \eta_a$, no flashing occurs. This is called a *frozen flow situation*, since the mass fraction of compressible component x_o is constant during discharge.

Regime 4. This is the same as Regime 3 ($\alpha_o > 0$) except $1 < \eta_s < \eta_a$, so flashing occurs, giving two sources of compressible gases and vapors.

Solutions are given here for only the first three regimes. For Regime 4, see Leung and Epstein (1991).

Figures 26-63 and 26-64 illustrate the significant differences between subcooled and saturated-liquid discharge rates. Discharge rate decreases with increasing pipe length in both cases, but the drop in discharge rate is much more pronounced with saturated liquids. This is because the flashed vapor effectively chokes the flow and decreases the two-phase density.

General Two-Phase Flow Relationships For flow across an orifice or nozzle, the equilibrium mass fraction of flashed vapor x can be found for single components from either an entropy or an enthalpy balance. For multicomponents, use a standard flash routine. Since orifice discharge follows a more nearly isentropic thermodynamic path, the appropriate balance to use for single components in this case is the entropy balance (Van den Akker, Snoey, and Spoelstra, 1983). This balance is written from the initial stagnation point inside the vessel with temperature T_o to the saturation temperature T_s at a given pressure (of greatest interest are the choke pressure or ambient pressure). This gives:

$$x = \frac{S_{Lo}(T_o) - S_L(T_s)}{S_{GL}(T_s)} \tag{26-77}$$

At ambient pressure, T_s is the normal boiling point.

Since pipe flow is more nearly isenthalpic, the flash fraction x is found from an enthalpy balance between the stagnation point and a point z downstream. Accounting for changes in potential energy, kinetic energy, and heat added or removed from the pipe Q, x is given by:

$$x = \frac{H_{Lo}(T_o) - H_{L2}(T_{S2}) - \frac{1}{2}u_{L2}^2 - gz\sin\theta + Q}{H_{GL}(T_{s2}) + \frac{1}{2}u_{G2}^2 - \frac{1}{2}u_{L2}^2} \tag{26-78}$$

If the potential energy, kinetic energy, and heat added terms are negligible, this reduces to:

$$x = \frac{H_{Lo}(T_o) - H_L(T_s)}{H_{GL}(T_s)} \tag{26-79}$$

These equations apply also to multicomponent systems, where the enthalpies are found for each phase from the component enthalpies.

Figure 26-62 depicts a flow system, which is described by the following differential momentum balance:

$$vdP + G^2vdv + \left[4f_{Lo}\frac{dz}{D} + K_e\right]\frac{1}{2}G^2v_{Lo}^2\phi_{Lo}^2 + g\sin\theta dz = 0 \tag{26-80}$$

where the terms represent the effect of pressure gradient, acceleration, line friction, and potential energy (static head), respectively. The effect of fittings, bends, entrance effects, etc., is included in the term K_e by standard methods. The inclination angle θ is the angle to the horizontal of a line from the pipe connection at the vessel to the discharge point. The term ϕ_{Lo}^2 is the two-phase multiplier which corrects the liquid-phase friction pressure loss to a two-phase pressure loss. Converting Eq. (26-80) to the dimensionless variables:

$$G_o^2 = \frac{G^2}{P_o\rho_o}, \qquad \eta = \frac{P}{P_o}, \qquad \varepsilon = \frac{v}{v_o} \tag{26-81}$$

gives: $$\varepsilon d\eta + G_o^2\varepsilon d\varepsilon + N\frac{1}{2}G_o^2\varepsilon^2\phi_{Lo}^2 + \frac{g\sin\theta dz}{P_ov_o} = 0 \tag{26-82}$$

where N is the number of equivalent velocity heads, given by:

$$N = 4f_{Lo}\frac{dz}{D} + K_e \tag{26-83}$$

For homogeneous flow, the two-phase multiplier is simply:

$$\phi_{FLO}^2 = \frac{v_H}{v_L} \tag{26-84}$$

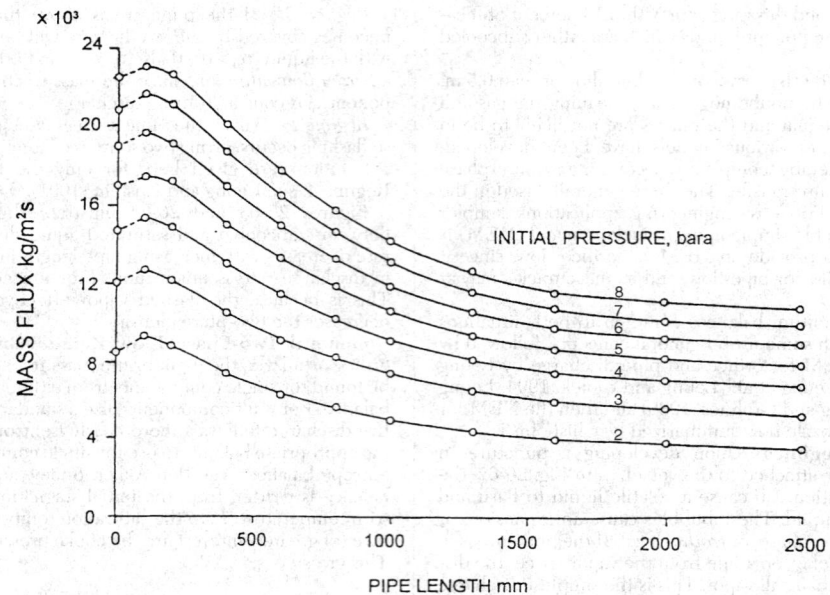

FIG. 26-63   Discharge mass flux for highly subcooled water (20°C) from orifice and 4-mm pipe of various lengths. (*Data of Uchida and Narai, 1966; reproduced by permission of ASME.*)

where v_H is the homogeneous specific volume given by:

$$v_H = xv_G + (1 - x)\, v_L \qquad (26\text{-}85)$$

The momentum balance for homogeneous flow can be factored to a form which enables integration as:

$$-N = \frac{G_{sp}^2 \varepsilon_H d\varepsilon_H + \varepsilon_H d\eta}{\tfrac{1}{2} G_{sp}^2 \varepsilon_H^2 + F_I} \qquad (26\text{-}86)$$

by defining a pipe inclination factor F_I:

$$F_I = \frac{gD \sin\theta}{4 f_{Lo} P_o v_o} \qquad (26\text{-}87)$$

F_I is positive for upflow, negative for downflow, and zero for horizontal flow.

The energy balance across a pipe from the stagnation point 0 to a point 2 downstream is:

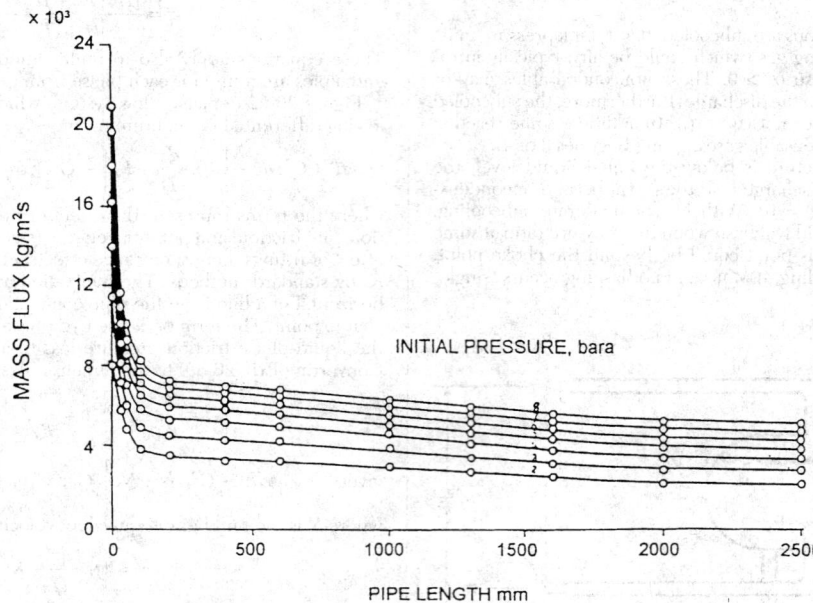

FIG. 26-64   Discharge mass flux for saturated water from orifice and 4-mm pipe of various lengths. (*Data of Uchida and Narai, 1966; reproduced by permission of ASME.*)

$$H_o + \frac{1}{2} u_o^2 = \frac{1}{2} (G^2 v_e^2)_2 + H_2 + Q \qquad (26\text{-}88)$$

For homogeneous flow, the equivalent specific volume v_e is the same as the homogeneous specific volume v_H, and the enthalpy is given by:

$$H_2 = [x H_G + (1-x) H_L]_2 \qquad (26\text{-}89)$$

Omega Method HEM Equation (26-86) can be integrated after first relating the dimensionless specific volume ε to the dimensionless pressure ratio η. A simple reciprocal relationship, designated the *omega method*, was suggested by Leung (1986) and by Leung and Grolmes (1988):

$$\varepsilon_H = \begin{cases} \omega \left[\dfrac{\eta_s}{\eta} - 1 \right] + 1 & \text{if } \dfrac{\eta_s}{\eta} > 1 \\[2mm] 1.0 & \text{if } \dfrac{\eta_s}{\eta} \leq 1 \end{cases} \qquad (26\text{-}90)$$

Figure 26-65 illustrates that Eq. (26-90) provides a linear approximation to the nonlinear relationship between two-phase specific volume and reciprocal pressure (v_H vs. P^{-1} or ε_H vs. η^{-1}). For single components, the initial slope of the ε_H curve is found using the Clapeyron equation to give:

$$\omega = \alpha_o + (1 - \alpha_o) \omega_s$$

$$\alpha_o = x_o \frac{v_{vo}}{v_o}$$

$$\omega_s = \frac{C_{PL} T_o P_s}{v_{LO}} \left[\frac{v_{VLO}(P_s)}{h_{VLO}(P_s)} \right]^2 \qquad (26\text{-}91)$$

To generalize for multicomponents and, in fact, to find a better fit for single components, use known information about the value of ε_H at some lower pressure, $\eta = \eta_2$, where η_2 is a rough approximation to the

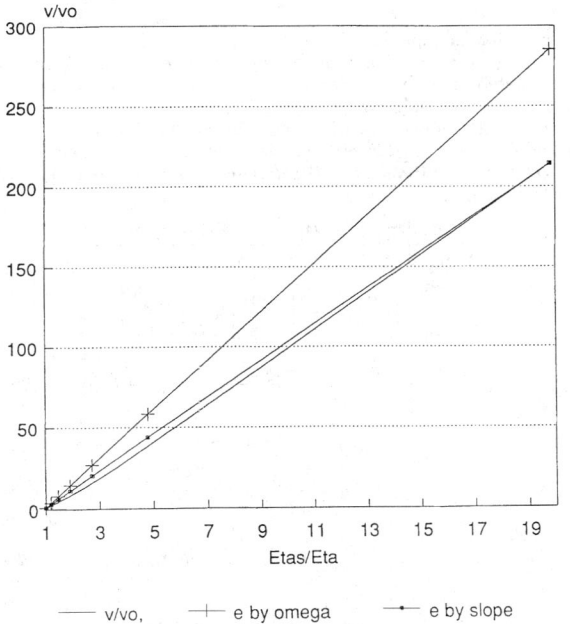

v/vo

NH3, Po =23.75 atm.

FIG. 26-65 Comparison of predictions for two-phase specific volume as a function of pressure by the omega method for two alternative formulas to calculate omega.

choking pressure ratio. That is, use the slope over the largest pressure interval of interest to give:

$$\omega = \frac{\varepsilon_2 - 1}{(\eta_s / \eta_2) - 1} \qquad (26\text{-}92)$$

Equation (26-91) gives values which are often high or low at the low-pressure end of the curve, whereas Eq. (26-92) is in error only insofar as the true ε curve is nonlinear. However, in practice, either approach provides adequate predictions for discharge rate.

HEM for Two-Phase Orifice Discharge For orifice or nozzle flow, the friction term and the potential energy term in Eq. (26-82) are negligible, so it can be integrated in general across both subcooled and flashing regions thusly:

$$\frac{1}{2} \frac{G_{\circ ori}^2}{C_D^2} \varepsilon_2^2 = \int_1^{\eta_s} \varepsilon_H \, d\eta + \int_{\eta_s}^{\eta_2} \varepsilon_H \, d\eta \qquad (26\text{-}93)$$

For the highly subcooled subset of Regime 2, ($\eta_s < \eta_a$), flow is single-phase (liquid), and integration of Eq. (26-93) gives what is commonly referred to as the *orifice equation:*

Subcooled Liquid Orifice Discharge

$$G_{\circ ori}^2 = C_D^2 \, 2 \, (1 - \eta_2) \qquad (26\text{-}94)$$

For the compressible flow cases, Regimes 1 and 3, and Regime 2 with $\eta_s > \eta_a$, making use of Eq. (26-90), integration of Eq. (26-93) gives:

Compressible Fluid Orifice Discharge by HEM

$$\frac{G_{\circ ori}^2}{C_D^2} = \frac{2 \{(1 - \eta_{sp}) + (1 - \omega)(\eta_{sp} - \eta_2) + \omega \eta_{sp} \ln (\eta_{sp}/\eta_2)\}}{\varepsilon_2^2} \qquad (26\text{-}95)$$

This is written with the general notation η_{sp} to avoid repetition of similar equations. For Regime 1, $\alpha_o = 0$, so $\omega = \omega_s$, and $\eta_{sp} = \eta_s$. For Regime 3, $\omega = \alpha_o$, and the integration is developed with $\eta_{sp} = 1$, so the above solution applies with $\eta_{sp} = 1$. This emphasizes the essential unity of the solution for Regimes 1 and 3. Equation (26-95) is plotted in Fig. 26-66 with Regime 1 to the right of $\omega = 1$ and Regime 3 to the left.

Equation (26-95) applies for subsonic as well as choked flow. Choked flow occurs at the pressure ratio $\eta_2 = \eta_c$, which maximizes $G_{\circ ori}$. To maximize $G_{\circ ori}$, differentiate Eq. (26-95) and set:

$$\left[\frac{\partial G_\circ}{\partial \eta_2} \right]_{\eta_2 = \eta_c} = 0 \qquad (26\text{-}96)$$

This gives a transcendental equation in η, the root of which occurs when $\eta = \eta_c$:

$$\frac{(1 - \omega)^2}{2 \omega \eta_s} \eta^2 + 2 (1 - \omega) \eta + \left(\frac{3}{2} \omega \eta_s - 1 \right) - \omega \eta_s \ln \frac{\eta_s}{\eta} = 0 \qquad (26\text{-}97)$$

Use a root-finding algorithm to discover the value of η_c which satisfies Eq. (26-97). These values are also plotted in Fig. 26-66.

For Regime 2, $\alpha_o = 0$ and $\omega = \omega_s$. Regime 2 requires using the incompressible solution for highly subcooled liquids and the compressible flow solutions for liquids of a low degree of subcooling. Essentially, these provide two branches to the solution, and by the flow maximization principle, we must choose the larger of the two. Empirically, the point at which these branches of the solution cross is given by:

$$\eta_{st} = \frac{2 \omega \eta_s}{1 + 2 \omega \eta_s} \qquad (26\text{-}98)$$

So, when $\eta_s > \eta_{st}$ (low subcooling, flashing before the choke print), use the compressible solution, Eq. (26-95) with $\eta_{sp} = \eta_s$. Otherwise (for high subcooling, no flashing before the choke point), use the liquid orifice equation, Eq. (26-94).

The solution for Regime 2 is plotted in Fig. 26-67. The high subcooling branch given by the liquid orifice equation goes through $G_{\circ ori} = 0$ when $\eta_s = 1$. The moderate subcooling branch parts with the former branch at η_{st} and matches values shown in Figure 26-66 when $\eta_s = 1$.

Choked Flow by Two-Phase Energy Balance From the energy balance, Eq. (26-88), taking $u_o = 0$ (stagnation) and $Q = 0$:

$$G_{ori} = \frac{[2 (H_o - H_2)]^{1/2}}{v_e} \qquad (26\text{-}99)$$

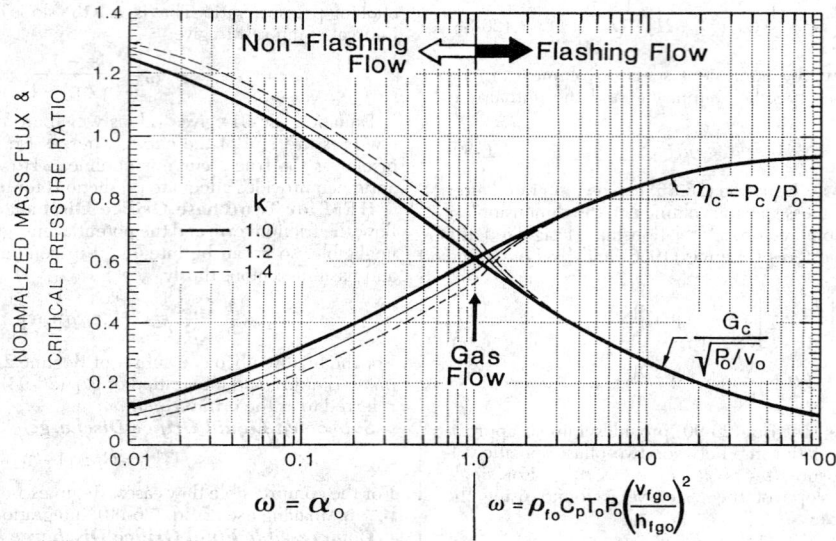

FIG. 26-66   Normalized mass flux and choked flow pressure ratio for frozen flow (left side) and for flashing liquid flow (right side) from orifice or nozzle discharge by the homogeneous equilibrium model. (*Leung, J.C., Chem. Eng. Progress* **88**(2), *pp. 70–75, 1992, Reproduced with permission of AIChE. Copyright 1992 AIChE. All rights reserved.*)

where v_e is an equivalent specific volume. This provides a simple alternative HEM for finding mass flux. Simply decrement pressure and search for the maximum value of G given by Eq. (26-99). This method requires good physical properties tables.

Differentiating with respect to pressure and invoking the first law of thermodynamics gives:

$$G_{\circ\text{ori}}^2 = \frac{-1}{v_e}\left(\frac{\partial H}{\partial v_e}\right)_s = -\left[\frac{\partial P}{\partial v_e}\right]_s \qquad (26\text{-}100)$$

or equivalently, differentiating Eq. (26-90) with respect to (η_2) gives:

$$G_{\circ\text{ori}}^2 = -\left[\frac{\partial \varepsilon}{\partial \eta}\right]_s^{-1} = \frac{\eta_2^2}{\omega \eta_s} \qquad (26\text{-}101)$$

When the flow is choked, this equation gives the same value for $G_{\circ\text{ori}}$ as does Eq. (26-95), as long as the root of Eq. (26-97) η_c is substituted for η_2.

Full-Bore and Punctured Pipe Discharge   With a pipe puncture, the mass flux at the discharge point $G_{\circ d}$ is larger than the mass flux in the pipe $G_{\circ p}$, by the puncture–to–pipe area ratio A_N/A_p, or $(D/D_p)^2$, defined as σ. Specifically:

$$G_{\circ p} = \sigma\, G_{\circ d} \qquad (26\text{-}102)$$

Since this correction is readily made, the following discussion assumes a full-bore pipe rupture, or $\sigma = 1$.

HEM for Two-Phase Pipe Discharge   With a pipe present, the backpressure experienced by the orifice is no longer η_2, but rather an intermediate pressure ratio η_1. Thus η_1 replaces η_2 in the orifice solution for mass flux $G_{\circ\text{ori}}$ Eq. (26-95). Correspondingly, the momentum balance is integrated between η_1 and η_2 to give the pipe flow solution for $G_{\circ p}$. The solutions for orifice and pipe flow must be solved simultaneously to make $G_{\circ\text{ori}} = G_{\circ p}$ and to find η_1 and η_2. This can be done explicitly for the simple case of incompressible single-phase (liquid) inclined or horizontal pipe flow. The solution is implicit for compressible regimes.

For incompressible orifice flow, $\varepsilon_H = 1$ and Eq. (26-86) is integrated between 1 and η_1 to give Eq. (26-94), with η_1 replacing η_2. Equation (26-86) integrated between η_1 and η_2 gives:

$$G_{\circ p}^2 = \frac{2\,(\eta_1 - \eta_2)}{N} - 2F_i \qquad (26\text{-}103)$$

Eliminating η_1 using Eq. (26-94) and setting $G_{\circ\text{ori}} = G_{\circ p}$ and $C_D = 1$ gives:

Subcooled Liquid Inclined Pipe Discharge

$$G_{\circ p}^2 = \frac{2\,(1 - \eta_2) - NF_I}{N + 1} \qquad (26\text{-}104)$$

For horizontal pipe flow, $F_I = 0$.

The general-case solution for compressible, inclined pipe flow is next stated, then the solution is developed for the special case of horizontal compressible flow.

Using the omega equation, Eq. (26-90), to eliminate $d\varepsilon_H$ in Eq. (26-86) enables Eq. (26-86) to be integrated to the following:

HEM for Inclined Pipe Discharge   For $= \eta_1 > \eta_s > \eta_2$ (flashing within the pipe):

$$N + \ln\left[\frac{X(\eta_2)}{X(\eta_{\text{sp}})}\left[\frac{\eta_{\text{sp}}}{\eta_2}\right]^2\right] = \frac{(\eta_1 - \eta_{\text{sp}}) + (1 - \omega)(\eta_{\text{sp}} - \eta_2)}{c}$$

$$+ \left[\frac{c\omega\eta_s - b\,(1 - \omega)}{2c^2}\right]\ln\left[\frac{X(\eta_{\text{sp}})}{X(\eta_2)}\right]$$

$$= \left[\frac{(1 - \omega)(b^2 - 2ac) - bc\omega\eta_s}{2c^2}\right][I_o(\eta_{\text{sp}}) - I_o(\eta_2)] \quad (26\text{-}105)$$

where:

$$X(\eta) = a + b\eta + c\eta^2 \qquad (26\text{-}106)$$

$$a = \frac{1}{2}\,G_{\circ p}^2\,\omega^2\eta_s^2 \qquad (26\text{-}107)$$

$$b = \frac{1}{2}\,G_{\circ p}^2\,\omega\,(1 - \omega)\,\eta_s \qquad (26\text{-}108)$$

$$c = \frac{1}{2}\,G_{\circ p}^2\,(1 - \omega)^2 + F_I \qquad (26\text{-}109)$$

$$q = 4ac - b^2 \qquad (26\text{-}110)$$

Defining:

$$I_o(\eta) = \int \frac{d\eta}{X(\eta)} \qquad (26\text{-}111)$$

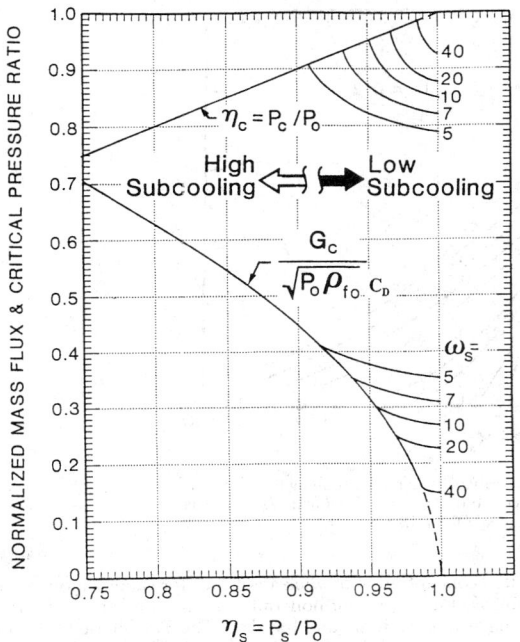

FIG. 26-67 Normalized mass flux and choked flow pressure ratio for flashing liquid discharge from orifices or nozzles by the homogeneous equilibrium model.(*Leung, J.C. and M.A. Grolmes, AIChE J. 33(3) pp. 524–527 (1987); Leung, J.C., Chem. Eng. Prog. 92(12), pp. 28–50 (1996). Reproduced with permission of AIChE, Copyright 1987, 1996. All rights reserved.*)

we obtain:

$$
I_o(\eta) = \begin{cases} \dfrac{2}{q^{1/2}} \tan^{-1}\left[\dfrac{2c\eta + b}{q^{1/2}}\right] & \text{if } q > 0 \text{ (upflow)} \\[3mm] \dfrac{1}{(-q)^{1/2}} \ln\left[\dfrac{2c\eta + b - (-q)^{1/2}}{2c\eta + b + (-q)^{1/2}}\right] & \text{if } q < 0 \text{ (downflow)} \end{cases} \quad (26\text{-}112)
$$

This solution is implicit in mass flux $G_{\circ p}$. To illustrate its application,

first consider the special case of horizontal pipe flow. The term η_{sp} is defined as follows for both horizontal and inclined pipe flow.

HEM for Horizontal Pipe Discharge   For horizontal pipe flow, $F_I = q = 0$, and:

$$
I_o(\eta) = \frac{-1}{\frac{1}{2}b + c\eta} \quad (26\text{-}113)
$$

The general compressible flow solution simplifies for horizontal pipe flow to:

$$
G_{\circ p}^2 = 2\,\frac{\left\{(\eta_1 - \eta_{sp}) + \dfrac{(\eta_{sp} - \eta_2)}{1-\omega} + \dfrac{\omega\eta_s}{(1-\omega)^2}\ln\left[\dfrac{\varepsilon_2\eta_2}{\varepsilon_{sp}\eta_{sp}}\right]\right\}}{N + 2\ln\left[\dfrac{\varepsilon_2}{\varepsilon_{sp}}\right]} \quad (26\text{-}114)
$$

The solution is again generalized with the term η_{sp} and $\varepsilon_{sp} = \varepsilon(\eta_{sp})$. For Regimes 1 and 2, $\alpha_o = 0$, so $\omega = \omega_s$. Regime 2 is again split, depending on where the flashing occurs, in the orifice or in the line. The division between low and moderate subcooling is found by:

$$
\eta_{st} = \frac{2\omega_s}{1 + 2\omega_s + N} \quad (26\text{-}115)
$$

Regime 1 is included in the following two subcases of Regime 2:
• For the low subcooling case, which includes Regime 1, $\eta_s > \eta_{st}$ and $\eta_s > \eta_1$ (flashing occurs in the vessel or pipe entrance). Set $\eta_{sp} = \eta_1$.
• For the moderate subcooling case, $\eta_s > \eta_{st} > \eta_2$ (flashing occurs in the pipe). Set $\eta_{sp} = \eta_s$.
• For the high subcooling case, $\eta_s < \eta_{st}$, use the single-phase orifice equation, Eq. (26-104).

For Regime 3, $\omega = \alpha_o$, and the integration is developed letting $\eta_s = 1$, so the above solution applies with $\eta_{sp} = 1$.

The solution of these equations requires a root-finding algorithm which iterates on assumed values of η_1. At each value of η_1, solve Eq. (26-95) (with η_1 replacing η_2) for $G_{\circ ori}$. Find η_2 from Eq. (26-101), subject also to:

$$
\eta_2 \geq \eta_a
$$

Solve Eq. (26-114) for $G_{\circ p}$. The function:

$$
f(\eta_1) = G_{\circ ori} - G_{\circ p} \quad (26\text{-}116)
$$

always has a root in the interval $1 < \eta_1 < \eta_a$ since $G_{\circ ori}$ increases with decreasing η_1 and $G_{\circ p}$ decreases with decreasing η_1.

The maximum value for $G_{\circ p}$ is $G_{\circ ori}$ evaluated with zero pipe length. Denoting this value as $G_{\circ max}$, Fig. 26-68 plots the dimensionless mass

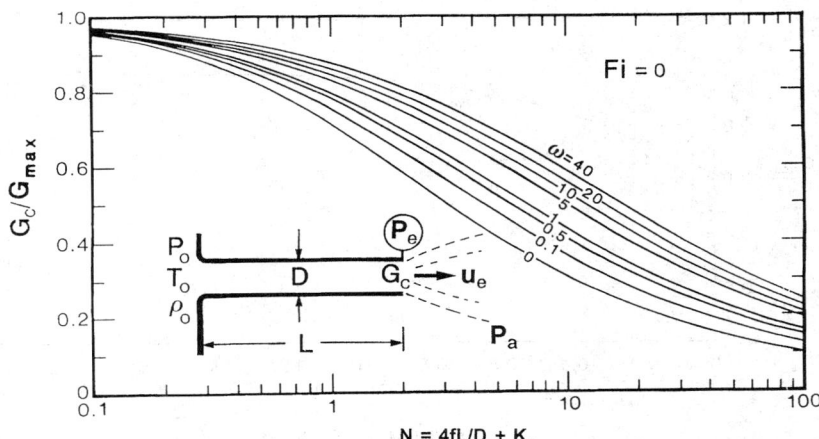

FIG. 26-68 Ratio of mass flux for horizontal pipe flow to that for orifice discharge for flashing liquids by the homogeneous equilibrium model. (*Leung and Grolmes, AIChE J, 33 (3), pp. 524–527, 1987; reproduced by permission of AIChE. copyright 1987. All rights reserved.*)

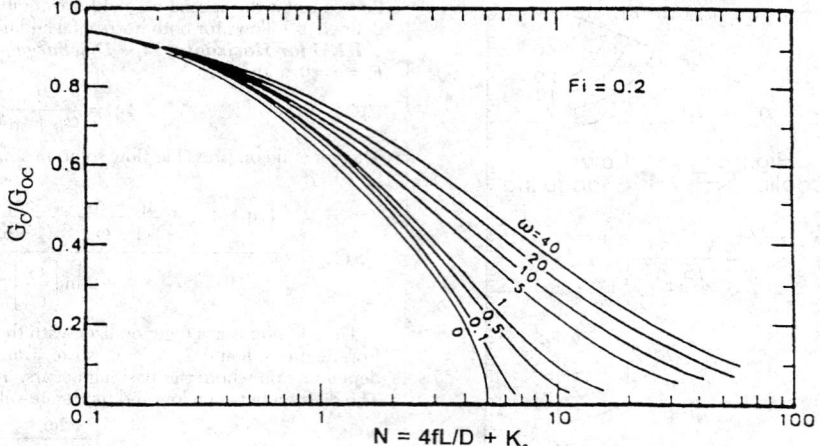

FIG. 26-69   Ratio of mass flux for inclined pipe flow to that for orifice discharge for flashing liquids by the homogeneous equilibrium model. (*Leung, J. of Loss Prev. Process Ind.* **3** *pp. 27–32, with kind permission of Elsevier Science, Ltd, The Boulevard, Langford Lane, Kidlington, OX5 IGB U.K., 1990.*)

flux discharge for horizontal pipe flow $G_{\circ p}$ as a ratio $G_{\circ p}/G_{\circ \max}$. Similar design charts were developed by Levenspiel (1977).

Figure 26-69 plots $G_{\circ p}/G_{\circ \max}$ for an upwardly inclined pipe flow [Eq. (26-105)] for a specific value of the pipe inclination factor $F_I = 0.2$. Comparing Figs. 26-68 and 26-69 shows that discharge rates decrease with upflow. For downflow, the curves are higher than in horizontal flow. In fact, a minimum flow occurs, regardless of how much pipe length is added, quite similar to the terminal velocity of free-falling objects. These charts are useful for design calculations up to a reduced temperature (ratio of temperature to the critical temperature) of about 0.90.

Accuracy of Omega Method HEM   Figures 26-70 and 26-71 illustrate the accuracy to be expected with the omega method HEM. For slightly subcooled (flashing) or saturated water, using the data of Sozzi and Sutherland (1975) and the ASME Symposium on Non-Equilibrium Two-Phase Flows (1975), predictions improve to within 10 percent error when the pipe length is larger than about 0.5 m.

NEM for Two-Phase Orifice Discharge   With flow through an orifice or nozzle, the flash is delayed, and the delay time depends on

the initial concentration of nucleation sites for vaporization. A simplified approach to represent nonequilibrium orifice or nozzle flow has been suggested by Henry and Fauske ("The Two-Phase Critical Flow of One-Component Mixtures in Nozzles, Orifices, and Short Tubes," *Trans. ASME, J. Heat Transfer* **93**(5): 179–N87, 1976) and by Chisholm (1983).

For orifice flow, Eq. (26-80) reduces to:

$$-v\,dP = G^2 v\,dv \qquad (26\text{-}117)$$

or, differentiating the definition of G:

$$-v\,dP = u\,du = \frac{1}{2}\,du^2 \qquad (26\text{-}118)$$

This can be readily integrated numerically as long as we use the appropriate nonequilibrium equivalent specific volume v_e in the integration. A reasonably simple form for v_e has been suggested by Chisholm (1983), which makes use of established correlations for the slip velocity K, which depends on the Lockhart-Martinelli parameter X. Integrating Eq. (26-118) gives:

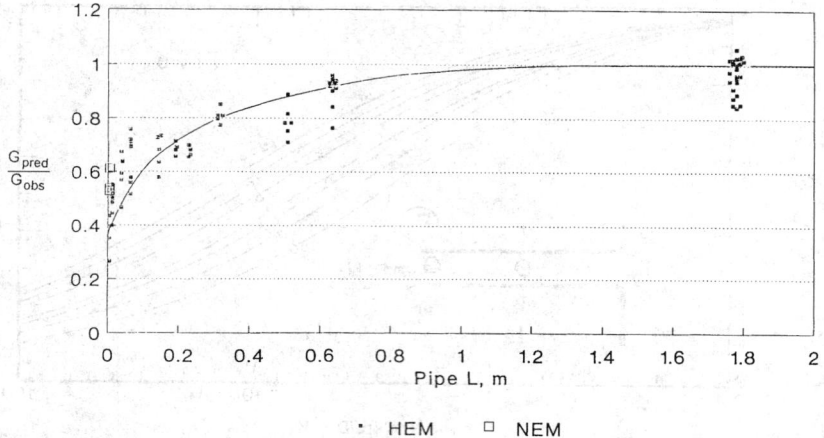

· HEM   □ NEM

FIG. 26-70   Accuracy in HEM predictions for slightly to moderately subcooled flashing flow. Comparison with data for water by Sozzi and Sutherland (1975); also ASME Symposium on Non-Equilibrium Two-Phase Flows (1975) (Nozzle type 2).

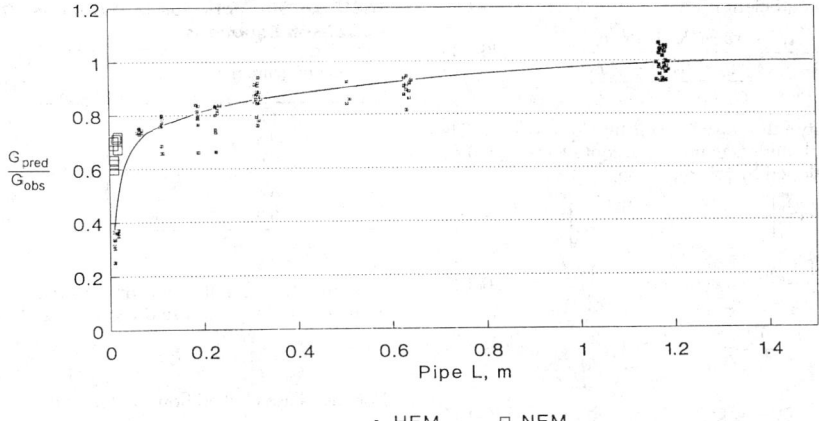

FIG. 26-71 Accuracy in HEM predictions for saturated, flashing flow. Comparison with data for water by Sozzi and Sutherland (1975), (Nozzle type 2).

$$-\int_{P_o}^{P_2} v \, dP = u_2^2 \left[1 - \left(\frac{u_o}{u_2} \right)^2 \right] \qquad (26\text{-}119)$$

Making use of continuity:

$$w = A_o \frac{u_o}{v_o} = A_2 \frac{u_2}{v_2} \qquad (26\text{-}120)$$

Equation (26-119) can be rearranged and written in dimensionless variables as:

$$G_{\circ 2}^2 = \frac{-2 \int_1^{\eta_2} \varepsilon_e \, d\eta}{\varepsilon_{e2}^2 \left[1 - \left(\frac{A_2 \varepsilon_{e1}}{A_1 \varepsilon_{e2}} \right)^2 \right]} = \frac{-2 \int_1^{\eta_2} \varepsilon_e \, d\eta}{\varepsilon_{e2}^2 - C_D^2} \qquad (26\text{-}121)$$

since $C_D = A_2/A_o$ at the vena contracta, $G_{\circ 2} = G_{\circ C}$. Equation (26-131) gives $G_{\circ \text{ori}}$.

The integral of Eq. (26-121) is evaluated in increments of pressure ratio $d\eta$, using the following procedure. At the next pressure given by:

$$P_i = P_{i-1} - \Delta \eta P_o \qquad (26\text{-}122)$$

use an equation of state to find v_G, v_L. Use an isentropic flash to find the equilibrium flash fraction x. The initial vapor mass fraction x_o can include noncondensables as well as an initial flash fraction. If this is the case, add the noncondensable portion to x.

Find the Lockhart-Martinelli coefficient as defined by Lockhart and Martinelli (1949):

$$X_m^2 = \frac{dP_L}{dP_G} = \frac{f_L (1-x)^2}{f_G x^2} \frac{v_L}{v_G} \qquad (26\text{-}123)$$

or the ratio of the pressure drop for liquid flowing alone to that for gas flowing alone. The liquid and gas friction factors are usually justifiably taken as equal (unless one phase is in laminar flow while the other is in turbulent).

Find the equilibrium, homogeneous specific volume v_H given by Eq. (26-85) and estimate the slip velocity ratio using the following correlation:

$$K_o = \begin{cases} \left[\dfrac{v_H}{v_L} \right]^{1/2} & \text{if } X_m > 1 \\[2ex] \left[\dfrac{v_G}{v_L} \right]^{1/4} & \text{if } X_m < 1 \end{cases} \qquad (26\text{-}124)$$

The slip velocity ratio is adequately represented by:

$$K = \frac{u_G}{u_L} = K_o^{0.4} \qquad (26\text{-}125)$$

Find the coefficient in the equivalent specific volume B using:

$$B = \frac{(1/K)(v_G/v_L) + K - 2}{(v_G/v_L) - 1} \qquad (26\text{-}126)$$

or, if

$$\frac{v_G}{v_L} \gg K(K-2)$$

$$B = \frac{1}{K} \qquad (26\text{-}127)$$

Find the transition flash fraction x_t:

$$x_t = \frac{1}{1 + (v_G/v_L)^{1/2}} \qquad (26\text{-}128)$$

The nonequilibrium flash fraction x_N is interpolated nonlinearly between x_o and x_t by:

$$x_N = \begin{cases} x_o + \left(\dfrac{x - x_o}{x_t - x_o} \right)^2 (x - x_o) & x < x_t \\[2ex] x & x > x_t \end{cases} \qquad (26\text{-}129)$$

So when $x > x_t$, thermal equilibrium is assumed.

Find the equivalent nonequilibrium specific volume as:

$$\varepsilon_e = \frac{v_e}{v_o} = 1 + \left(\frac{v_G}{v_L} - 1 \right) [B x_N (1 - x_N) + x_N^2] \qquad (26\text{-}130)$$

The integration proceeds stepwise until the integral begins to decrease. This occurs at the choked pressure ratio η_c, giving a maximum mass flux $G_{\circ 2c}$.

As shown in Figs. 26-70 and 26-71, the orifice flow predictions by the NEM (open points) are larger than those of the HEM, although still low compared with these particular data.

Discharge Coefficients and Gas Discharge A compressible fluid, upon discharge from an orifice, accelerates from the puncture point and the cross-sectional area contracts until it forms a minimum at the vena contracta. If flow is choked, the mass flux $G_{\circ c}$ can be found at the vena contracta, since it is a maximum at that point. The mass flux at the orifice is related to the mass flux at the vena contracta by the discharge coefficient, which is the area contraction ratio (A_c at the vena contracta to A_N at the orifice):

$$G_{\circ \text{ori}} = C_D G_{\circ c} \qquad (26\text{-}131)$$

For two-phase flow, the phase contraction coefficients C_{DG} and C_{DL} relate the area of each phase A_G and A_L at the vena contracta to the known area of the orifice A_N. Thus:

$$C_{DG} = \frac{A_G}{A_N}, \qquad C_{DL} = \frac{A_L}{A_N} \qquad (26\text{-}132)$$

The two-phase discharge coefficient is:

$$C_D = \frac{A_G + A_L}{A_N} = \frac{xv_G + K(1-x)v_L}{\dfrac{xv_G}{C_{DG}} + \dfrac{K(1-x)v_L}{C_{DL}}} \qquad (26\text{-}133)$$

where K is the slip velocity ratio, u_G/u_L given by Eq. (26-125). The contraction coefficient for liquids is generally accepted as $C_{DL} = 0.61$. For the gas phase, as developed by Jobson (1955):

$$C_{DG} = \frac{1}{2b_1\eta_{2c}^{1/k}}\left\{a_1 - \left[a_1^2 - \frac{4\eta_{2c}^{2/k}(1-\eta_2)b_1}{G_{\circ g}^2}\right]^{1/2}\right\} \qquad (26\text{-}134)$$

$$b_1 = \frac{1}{C_{DL}} - \frac{1}{2C_{DL}^2} \qquad (26\text{-}135)$$

$$a_1 = 1 + \frac{(\eta_{2c} - \eta_2)\eta_{2c}^{1/k}}{G_{\circ g}^2} \qquad (26\text{-}136)$$

where:

$$G_{\circ g}^2 = \frac{2k}{(k-1)}\eta_2^{2/k}(1 - \eta_2^{(k-1)/k}) \qquad (26\text{-}137)$$

and for vapor or gas flow:

$$G_{\circ \text{ori}} = C_{DG}\, G_{\circ g} \qquad (26\text{-}138)$$

Equation (26-137) is recognized as the expression for all-gas flow by adiabatic expansion across an orifice or nozzle. The factor k is the expansion coefficient for the adiabatic flow equation of state:

$$\frac{P}{P_o} = \left[\frac{\rho}{\rho_o}\right]^k = \left[\frac{T}{T_o}\right]^{k/(k-1)} \qquad (26\text{-}139)$$

For an ideal gas:

$$k = \frac{C_p}{C_v} \qquad (26\text{-}140)$$

Fortunately, for most operating pressure ranges, k is nearly constant with temperature and pressure. For wider ranges where this might not hold it is often adequate to replace k by a value slightly smaller than C_p/C_v. A rationale for this is that heat exchange with the surroundings can shift the behavior slightly toward the isothermal solution, which is a limiting case with $k = 1$.

TABLE 26-30   Variation in Two-Phase Discharge Coefficients by Jobson Equations

Exit pressure ratio, η_2	$\dfrac{C_{DG}}{C_{DL}}$	C_{DG}
1.0	1.0	0.61
0.8	1.07	0.653
0.6	1.18	0.720
0.4	1.31	0.799
0.2	1.40	0.854
0.0	1.44	0.878

A further generalization for two-phase flow as suggested by Tangren et al. (1949) is to use the generalized value of k as:

$$k = \frac{xC_{pG} + (1-x)C_{PL}}{xC_{vG} + (1-x)C_{PL}} \qquad (26\text{-}141)$$

For gas-phase choked flow, the pressure ratio at the vena contracta is:

$$\eta_{2c} = \left(\frac{2}{k+1}\right)^{k/(k-1)} \qquad (26\text{-}142)$$

and

$$\eta_2 = \text{maximum}(\eta_{2c}, \eta_a)$$

reaching η_a when the flow becomes subsonic.

For choked flow, $a_1 = 1$ in Eq. (26-136). Typical values developed by Eqs. (26-133) and (26-134) are listed in Table 26-30 (Watson et al., 1983).

Blowdown Modeling   Blowdown models incorporate not only the preceding discharge rate models but also a model of the tank and line contents to predict how the tank pressure and temperature decay in time. Analytical time-varying blowdown solutions are available for single-phase discharge, gas or liquid (Woodward and Mudan, 1991). Analytical liquid blowdown models have been developed for essentially all tank geometries of interest by Sommerfeld and coworkers (Lee and Sommerfeld, May 1994, July 1994). Vapor blowdown is readily modeled, using an energy and mass balance on the tank contents. The tank pressure decays along the vapor pressure curve as long as a liquid is present. Two-phase blowdown modeling is further discussed in Woodward (1993, pp. 94–159).

Energy Resources, Conversion, and Utilization

Walter F. Podolski, Ph.D., *Chemical Engineer, Electrochemical Technology Program, Argonne National Laboratory; Member, American Institute of Chemical Engineers. (Section Editor)*

Shelby A. Miller, Ph.D., P.E., *Resident Retired Senior Engineer, Argonne National Laboratory; Member, American Association for the Advancement of Science (Fellow), American Chemical Society, American Institute of Chemical Engineers (Fellow), American Institute of Chemists (Fellow), Filtration Society, New York Academy of Sciences, Society of Chemical Industry. (Cosection Editor)*

David K. Schmalzer, Ph.D., P.E., *Fossil Energy Program Manager, Argonne National Laboratory; Member, American Chemical Society, American Institute of Chemical Engineers. (Fuels, Gaseous Fuels)*

Anthony G. Fonseca, Ph.D., *Director, Coal Utilization, CONSOL, Inc.; Member, American Chemical Society, Society for Mining, Metallurgy, and Extraction. (Solid Fuels)*

Vincent Conrad, Ph.D., *Group Leader, Technical Services Development, CONSOL, Inc.; Member, Spectroscopy Society of Pittsburgh, Society for Analytical Chemistry of Pittsburgh, Society for Applied Spectroscopy. (Solid Fuels)*

Douglas E. Lowenhaupt, M.S., *Group Leader, Coke Laboratory, CONSOL, Inc.; Member, American Society for Testing and Materials, Iron and Steel Making Society, International Committee for Coal Petrology. (Solid Fuels)*

John D. Bacha, Ph.D., *Consulting Scientist, Chevron Products Company; Member, ASTM (American Society for Testing and Materials), Committee D02 on Petroleum Products and Lubricants; American Chemical Society; International Association for Stability and Handling of Liquid Fuels, Steering Committee. (Liquid Petroleum Fuels)*

Lawrence K. Rath, B.S., P.E., *Federal Energy Technology Center (Morgantown), U.S. Department of Energy; Member, American Institute of Chemical Engineers. (Coal Gasification)*

Hsue-peng Loh, Ph.D., P.E., *Federal Energy Technology Center (Morgantown), U.S. Department of Energy; Member, American Institute of Chemical Engineers, American Society of Information Sciences. (Coal Gasification)*

Edgar B. Klunder, Ph.D., *Project Manager, Federal Energy Technology Center (Pittsburgh), U.S. Department of Energy. (Direct Liquefaction)*

Howard G. McIlvried, III, Ph.D., *Senior Engineer, Burns and Roe Services Corporation, Federal Energy Technology Center (Pittsburgh); Member, American Chemical Society, American Institute of Chemical Engineers. (Direct Liquefaction)*

Gary J. Stiegel, M.S., P.E., *Program Coordinator, Federal Energy Technology Center (Pittsburgh), U.S. Department of Energy. (Indirect Liquefaction)*

Rameshwar D. Srivastava, Ph.D., *Fuels Group Manager, Burns and Roe Services Corporation, Federal Energy Technology Center (Pittsburgh). (Indirect Liquefaction)*

Peter J. Loftus, D. Phil., *Arthur D. Little, Inc.; Member, American Society of Mechanical Engineers. (Heat Generation, Thermal Energy Conversion and Utilization, Heat Recovery)*

Charles E. Benson, M.Eng., M.E., *Director, Combustion Technology, Arthur D. Little, Inc.; Member, American Society of Mechanical Engineers, Combustion Institute. (Heat Generation, Thermal Energy Conversion and Utilization, Heat Recovery)*

John M. Wheeldon, Ph.D., *Electric Power Research Institute. (Fluidized Bed Combustion)*

Michael Krumpelt, Ph.D., *Manager, Fuel Cell Technology, Argonne National Laboratory; Member, American Institute of Chemical Engineers, American Chemical Society, Electrochemical Society. (Electrochemical Energy Conversion)*

The contributions of Dr. Harold F. Chambers, Jr. (Coal Liquefaction) and Dr. Yuan C. Fu (Coal Liquefaction) who were authors for the Sixth Edition, are acknowledged.

Nomenclature and Units

Symbol	Definition	SI units	U.S. customary units
A	Area specific resistance	Ω/m^2	Ω/ft^2
c	Heat capacity	$J/(kg \cdot K)$	$Btu/(lb \cdot °F)$
E	Activation energy	J/mol	$Btu/lb\ mol$
E	Electrical potential	V	V
f	Fugacity	kPa	psia
F	Faraday constant	C/mol	$C/lb\ mol$
ΔG	Free energy of reaction	J/mol	$Btu/lb\ mol$
ΔH	Heat of reaction	J/mol	$Btu/lb\ mol$
i	Current density	A/m^2	A/ft^2
k	Rate constant	$g/(h \cdot cm^3)$	$lb/(h \cdot ft^3)$
K	Latent heat of vaporization	kJ/kg	Btu/lb
P	Pressure	kPa	psia
Q	Heating value	kJ/kg	Btu/lb
R	Gas constant	$J/mol \cdot K$	$Btu/lb\ mol \cdot °R$
s	Relative density	Dimensionless	Dimensionless
T	Temperature	K	°F
U	Fuel utilization	percent	percent
V	Molar gas volume	m^3/mol	$ft^3/lb\ mol$
Z	Compressibility factor	Dimensionless	Dimensionless
	Greek symbols		
ε	Energy conversion efficiency	Percent	Percent

Acronyms and unit prefixes

Symbol	Name	Value
E	Exa	10^{18}
G	Giga	10^9
K	Kilo	10^3
M	Mega	10^6
P	Peta	10^{15}
T	Tera	10^{12}
Z	Zetta	10^{21}

Acronym	Definition
AFBC	atmospheric fluidized bed combustion
AFC	alkaline fuel cell
AGC-21	Advanced Gas Conversion Process
BG/L	British Gas and Lurgi process
COE	cost of electricity
COED	Char Oil Energy Development Process
DOE	U.S. Department of Energy
EDS	Exxon Donor Solvent Process
FBC	fluidized bed combustion
HAO	hydrogenated anthracene oil
HPO	hydrogenated phenanthrene oil
HRI	Hydrocarbon Research, Inc.
HTI	Hydrocarbon Technologies, Inc.
IGCC	integrated gasification combined-cycle
KRW	Kellogg-Rust-Westinghouse process
MCFC	molten carbonate fuel cell
MTG	methanol-to-gasoline process
OTFT	once-through-Fischer-Tropsch process
PAFC	phosphoric acid fuel cell
PC	pulverized coal
PEFC	polymer electrolyte fuel cell
PFBC	pressurized fluidized bed combustion
Quad	10^{15} Btu
SASOL	South African operation of synthetic fuels plants
SMDS	Shell Middle Distillate Synthesis Process
SNG	synthetic natural gas
SOFC	solid oxide fuel cell
SRC	solvent-refined coal

INTRODUCTION

GENERAL REFERENCES: Loftness, *Energy Handbook,* 2d ed., Van Nostrand Reinhold, New York, 1984. Energy Information Administration, *Energy Use and Carbon Emissions: Some International Comparisons,* U.S. Dept. of Energy, DOE/EIA-0579, 1994. Howes and Fainberg (eds.), *The Energy Source Book,* American Institute of Physics, New York, 1991. Johansson, Kelly, Reddy, and Williams (eds.), Burnham (exec. ed.), *Renewable Energy—Sources for Fuels and Electricity,* Island Press, Washington, D.C., 1993. Turner, *Energy Management Handbook,* The Fairmont Press, Lilburn, Ga., 1993. *National Energy Strategy,* U.S. Dept. of Energy, 1991.

Energy is usually defined as the capacity to do work. Nature provides us with numerous sources of energy, some difficult to utilize efficiently (e.g., solar radiation and wind energy), others more concentrated or energy dense and therefore easier to utilize (e.g., fossil fuels). Energy sources can be classified also as *renewable* (solar and nonsolar) and *nonrenewable.* Renewable energy resources are derived in a number of ways: gravitational forces of the sun and moon, which create the tides; the rotation of the earth combined with solar energy, which generates the currents in the ocean and the winds; the decay of radioactive minerals and the interior heat of the earth, which provide geothermal energy; photosynthetic production of organic matter; and the direct heat of the sun. These energy sources are called renewable because they are either continuously replenished or, for all practical purposes, are inexhaustible.

Nonrenewable energy sources include the fossil fuels (natural gas, petroleum, shale oil, coal, and peat) as well as uranium. Fossil fuels are both energy dense and widespread, and much of the world's industrial, utility, and transportation sectors rely on the energy contained in them. Concerns over global warming notwithstanding, fossil fuels will remain the dominant fuel form for the foreseeable future. This is so for two reasons: (1) the development and deployment of new technologies able to utilize renewable energy sources such as solar, wind, and biomass are uneconomic at present, in most part owing to the diffuse or intermittent nature of the sources; and (2) concerns persist over storage and/or disposal of spent nuclear fuel and nuclear proliferation.

Fossil fuels, therefore, remain the focus of this section; their principal use is in the generation of heat and electricity in the industrial, utility, and commercial sectors, and in the generation of shaft power in transportation. The material in this section deals primarily with the conversion of the chemical energy contained in fossil fuels to heat and electricity. Material from *Perry's Chemical Engineers' Handbook,* 6th ed., Sec. 9, has been updated and condensed, and, in addition, new material on electrochemical energy conversion in fuel cells has been introduced. Even though the principles of energy conversion in fuel cells were known before internal combustion engines were developed, only recent improvements in materials and manufacturing methods have allowed fuel cells to be considered for stationary and transportation power generation.

FUELS

RESOURCES AND RESERVES

Proven worldwide energy resources are large. The largest remaining known reserves of crude oil, used mainly for producing transportation fuels, are located in the Middle East, along the equator, and in the former Soviet Union. U.S. proven oil reserves currently account for only about 3 percent of the world's total. Large reserves of natural gas exist in the former Soviet Union and the Middle East. Coal is the most abundant fuel on earth and the primary fuel for electricity in the United States, which has the largest proven reserves. Annual world consumption of energy is still currently less than 1 percent of combined world reserves of fossil fuels. The resources and reserves of the principal fossil fuels in the United States—coal, petroleum, and natural gas—follow.

	ZJ°		
Fuel	Proven reserves	Estimated undiscovered recoverable reserves	Estimated identified and undiscovered resources
Coal	7.3		110
Petroleum	0.15	0.31	
Natural gas	0.19	0.43	

°ZJ = 10^{21} J. (To convert to 10^{18} Btu, multiply by 0.948.)

The energy content of fossil fuels in commonly measured quantities is as follows.

Energy content		
Bituminous and anthracite coal	30.2 MJ/kg	26×10^6 Btu/US ton
Lignite and subbituminous coal	23.2 MJ/kg	20×10^6 Btu/US ton
Crude oil	38.5 MJ/L	5.8×10^6 Btu/bbl
Natural-gas liquids	25.2 MJ/L	3.8×10^6 Btu/bbl
Natural gas	38.4 MJ/m³	1032 Btu/ft³

1 bbl = 42 US gal = 159 L = 0.159 m³

SOLID FUELS

Coal

GENERAL REFERENCES: Lowry (ed.), *Chemistry of Coal Utilization*, Wiley, New York, 1945; suppl. vol., 1963; 2d suppl. vol., Elliott (ed.), 1981. Van Krevelen, *Coal*, Elsevier, Amsterdam, 1961.

Origin Coal originated from the arrested decay of the remains of trees, bushes, ferns, mosses, vines, and other forms of plant life, which flourished in huge swamps and bogs many millions of years ago during prolonged periods of humid, tropical climate and abundant rainfall. The precursor of coal was peat, which was formed by bacterial and chemical action on the plant debris. Subsequent actions of heat, pressure, and other physical phenomena metamorphosed the peat to the various ranks of coal as we know them today. Because of the various degrees of the metamorphic changes during this process, coal is not a uniform substance; no two coals are ever the same in every respect.

Classification Coals are classified by rank, i.e., according to the degree of metamorphism in the series from lignite to anthracite. Table 27-1 shows the classification system adopted by the American Society for Testing and Materials, D388-92A. The heating value on the moist *mineral-matter-free* (mmf) basis, and the fixed carbon, on the dry mmf basis, are the bases of this system. The lower-rank coals are classified according to the heating value, kJ/kg (Btu/lb), on a moist mmf basis. The agglomerating character is used to differentiate between adjacent groups. Coals are considered agglomerating if the coke button remaining from the test for volatile matter will support a weight of 500 g or if the button swells or has a porous cell structure.

The Parr formulas, Eqs. (27-1) to (27-3), or the approximation formulas, Eqs. (27-4) and (27-5), are used for classifying coals according to rank. The Parr formulas are employed in litigation cases.

$$F' = \frac{100 (F - 0.15S)}{100 - (M + 1.08A + 0.55S)} \tag{27-1}$$

$$V' = 100 - F' \tag{27-2}$$

TABLE 27-1 Classification of Coals by Rank*

Class/group	Fixed carbon limits (dry, mineral-matter-free basis), %		Volatile matter limits (dry, mineral-matter-free basis), %		Gross calorific value limits (moist, mineral-matter-free basis)†				Agglomerating character
					MJ/kg		Btu/lb		
	Equal or greater than	Less than	Greater than	Equal or less than	Equal or greater than	Less than	Equal or greater than	Less than	
Anthracitic:									
Meta-anthracite	98	—	—	2	—	—	—	—	
Anthracite	92	98	2	8	—	—	—	—	Nonagglomerating
Semianthracite‡	86	92	8	14	—	—	—	—	
Bituminous:									
Low-volatile bituminous coal	78	86	14	22	—	—	—	—	
Medium-volatile bituminous coal	69	78	22	31	—	—	—	—	
High-volatile A bituminous coal	—	69	31	—	32.6	—	14,000§	—	Commonly agglomerating¶
High-volatile B bituminous coal	—	—	—	—	30.2	32.6	13,000§	14,000	
High-volatile C bituminous coal	—	—	—	—	26.7	30.2	11,500	13,000	
					24.4	26.7	10,500	11,500	Agglomerating
Subbituminous:									
Subbituminous A coal	—	—	—	—	24.4	26.7	10,500	11,500	
Subbituminous B coal	—	—	—	—	22.1	24.4	9,500	10,500	
Subbituminous C coal	—	—	—	—	19.3	22.1	8,300	9,500	Nonagglomerating
Lignitic:									
Lignite A	—	—	—	—	14.7	19.3	6,300	8,300	
Lignite B	—	—	—	—	—	14.7	—	6,300	

Data from 1994 Annual Book of ASTM Standards, vol. 5 D 388 (1994). Copyright ASTM. Reprinted with permission.
*This classification does not apply to certain coals.
†*Moist* refers to coal containing its natural inherent moisture but not including visible water on the surface of the coal.
‡If agglomerating, classify in low-volatile group of the bituminous class.
§Coals having 69 percent or more fixed carbon on the dry, mineral-matter-free basis shall be classified according to fixed carbon, regardless of gross calorific value.
¶It is recognized that there may be nonagglomerating varieties in these groups of the bituminous class and that there are notable exceptions in the high-volatile C bituminous group.

$$Q' = \frac{100\,(Q - 50S)}{100 - (M + 1.08A + 0.55S)} \qquad (27\text{-}3)$$

$$F' = \frac{100F}{100 - (M + 1.1A + 0.1S)} \qquad (27\text{-}4)$$

$$Q' = \frac{100Q}{100 - (1.1A + 0.1S)} \qquad (27\text{-}5)$$

where M, F, A, and S are weight percentages, on a moist basis, of moisture, fixed carbon, ash, and sulfur, respectively; F' and V' are weight percentages, on a dry mmf basis, of fixed carbon and volatile matter, respectively; Q and Q' are calorific values (Btu/lb), on a moist non-mmf basis and a moist mmf basis, respectively. (Btu/lb = 2326 J/kg)

Composition and Heating Value The composition of coal is reported in two different ways: the proximate analysis and the ultimate analysis, both expressed in weight percent. The *proximate analysis* is the determination by prescribed methods of moisture, volatile matter, fixed carbon, and ash.

The moisture in coal consists of inherent moisture, also called equilibrium moisture, and surface moisture. Free moisture is that moisture lost when coal is air-dried under standard low-temperature conditions.

The *volatile matter* is the portion of coal which, when the coal is heated in the absence of air under prescribed conditions, is liberated as gases and vapors. Volatile matter does not exist by itself in coal, except for a little absorbed methane, but results from thermal decomposition of the coal substance.

Fixed carbon, the residue left after the volatile matter is driven off, is calculated by subtracting from 100 the percentages of moisture, volatile matter, and ash of the proximate analysis. In addition to carbon, it may contain several tenths of a percent of hydrogen and oxygen, 0.4 to 1.0 percent nitrogen, and about half of the sulfur that was in the coal.

Ash is the inorganic residue that remains after the coal has been burned under specified conditions, and it is composed largely of compounds of silicon, aluminum, iron, and calcium, and minor amounts of compounds of magnesium, sodium, potassium, phosphorous, sulfur, and titanium. Ash may vary considerably from the original mineral matter, which is largely kaolinite, illite, montmorillonite, quartz, pyrites, and gypsum.

The *ultimate analysis* is the determination by prescribed methods of the ash, carbon, hydrogen, nitrogen, sulfur, and (by difference) oxygen. Along with these analyses, the *heating value,* expressed as kJ/kg (Btu/lb), is also determined. This is the heat produced at constant volume by the complete combustion of a unit quantity of coal in an oxygen-bomb calorimeter under specified conditions. The result includes the latent heat of vaporization of the water in the combustion products and is called the gross heating or *high heating value* (HHV), Q_h. The heating value when the water is not condensed is called the *low heating value* (LHV), Q_l, and is obtained from

$$Q_l = Q_h - K \cdot W \qquad (27\text{-}6)$$

where W = weight of water formed/weight of fuel burned. The factor K is the latent heat of vaporization at the partial pressure of the vapor in the exit gas. The value of K ranges from 2396 to 2512 kJ/kg of water (1030 to 1080 Btu/lb). Q_h in Btu/lb ($\times 2.326$ = kJ/kg) can be approximated by a formula developed by the Institute of Gas Technology:

$$Q_h = 146.58\,C + 568.78\,H + 29.4\,S - 6.58\,A - 51.53\,(O + N) \qquad (27\text{-}7)$$

where C, H, S, A, O, and N are the weight percentages on a dry basis of carbon, hydrogen, sulfur, ash, oxygen, and nitrogen, respectively. The standard deviation for 775 coal samples is 127 Btu/lb.

Coal analyses are reported on several bases, and it is customary to select the basis best suited to the application. The *as-received* basis represents the weight percentage of each constituent in the sample as received in the laboratory. The sample itself may be coal as fired, as mined, or as prepared for a particular use. The *moisture-free* (dry) basis is generally the most useful basis because performance calculations can be easily corrected for the actual moisture content at the point of use. The *dry, ash-free* basis is frequently used to approximate

the rank and source of a coal. For example, the heating value of coals of a given source and rank is remarkably constant when calculated on this basis.

Laboratory procedures for proximate and ultimate analyses are given in the *Annual Book of ASTM Standards* (Sec. 5, American Society for Testing and Materials, Conshohocken, Pa., 1994) and in *Methods of Analyzing and Testing Coal and Coke* (U.S. Bureau of Mines Bulletin 638, 1967).

Sulfur Efforts to abate atmospheric pollution have drawn considerable attention to the sulfur content of coal, since the combustion of coal results in the discharge to the atmosphere of sulfur oxides. Sulfur occurs in coal in three forms: as pyrite (FeS_2); as organic sulfur, which is a part of the coal substance; and as sulfate. Sulfur as sulfate comprises at the most only a few hundredths of a percent of the coal. The organic sulfur may comprise from 20 to 80 percent of the total sulfur. Since organic sulfur is chemically bound to the coal substance in a complex manner, drastic treatment is necessary to break the chemical bonds before the sulfur can be removed. There is no economical method known at present that will remove organic sulfur, although so-called chemical treatment methods for cleaning coal can reduce the sulfur content. Pyritic sulfur can be partially removed by using standard coal-washing equipment. The degree of pyrite removal depends on the size of the coal and the size and distribution of the pyrite particles.

The sulfur content of U.S. coals varies widely, ranging from a low of 0.2 percent to as much as 7 percent by weight, on a dry basis. The estimated remaining U.S. coal reserves of all ranks, by sulfur content, are shown in Fig. 27-1. Extensive data on sulfur and sulfur reduction potential, including washability, in U.S. coals are given in *Sulfur and Ash Reduction Potential and Selected Chemical and Physical Properties of United States Coal* (U.S. Dept. of Energy, DOE/PETC, TR-90/7, 1990; TR-91/1 and TR-91/2, 1991).

Coal-Ash Characteristics and Composition When coal is to be burned, used in steelmaking, or gasified, it is important to determine the ash fusibility, comprising the initial deformation, softening, and fluid temperatures. The difference between the softening and initial deformation temperatures is called the *softening interval,* and that between the fluid temperature and the softening temperature is called the *fluid interval.* The procedure for determining the fusibility of coal ash is prescribed by ASTM D 1857 (American Society for Testing and Materials, op. cit., 1994). The softening temperature is most

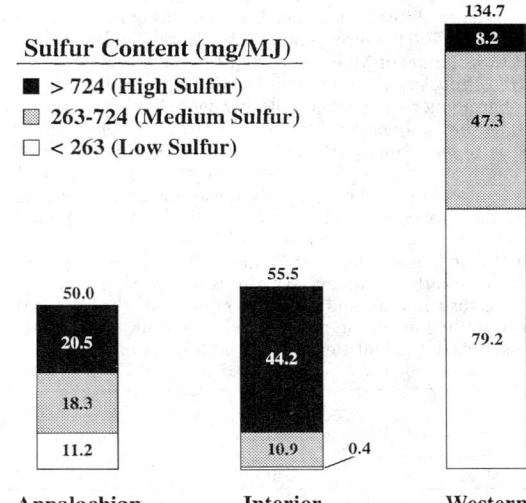

FIG. 27-1 Estimates of recoverable U.S. coal reserves in Gt by sulfur ranges and region. To convert tonnes to US tons, multiply by 1.102; and mg/MJ to lb/10^16 Btu, multiply by 2.321. (*Source:* U.S. Coal Reserves: An Update by Heat and Sulfur Content, *Energy Information Administration, DOE/EIA-0529(92), February 1993.*)

often used as a rough qualitative guide to the behavior of ash on a grate and on furnace heat-transfer surfaces, with respect to the tendency to form large masses of sintered or fused ash, which impair heat transfer and impede gas flow. Likewise, the fluid temperature and the fluid interval are qualitative guides to the "flowability" of ash in slag-tap and cyclone furnaces. However, because ash fusibility is not an infallible index of ash behavior in practice, care is needed in using fusibility data for designing and operating purposes. There is an excellent discussion on this subject in *Steam: Its Generation and Use* (40th ed., Babcock & Wilcox Co., New York, 1992).

The composition of coal ash varies widely. Calculated as oxides, the composition (percent by weight) varies as follows:

SiO_2	20–60
Al_2O_3	10–35
Fe_2O_3	5–35
CaO	1–20
MgO	0.3–4
TiO_2	0.5–2.5
Na_2O and K_2O	1–4
SO_3	0.1–12

Knowledge of the composition of coal ash is useful for estimating and predicting coal performance in coke making and, to a limited extent, the fouling and corrosion of heat-exchange surfaces in pulverized-coal-fired furnaces.

Multiple correlations for ash composition and ash fusibility are discussed in the *Coal Conversion Systems Technical Data Book* (part IA, U.S. Dept. of Energy, 1984).

The slag viscosity-temperature relationship for completely melted slag is

$$\text{Log viscosity} = \frac{10^7 M}{(T - 150)^2 + C} \tag{27-8}$$

where viscosity is in poises ($\times 0.1$ = Pa·s), $M = 0.00835$ (SiO_2) + $0.00601(Al_2O_3) - 0.109$, $C = 0.0415$ (SiO_2) + 0.0192 (Al_2O_3) + 0.0276 (equivalent Fe_2O_3) + 0.0160 (CaO) − 3.92, and T = temperature, K.

The oxides in parentheses are the weight percentages of these oxides when $SiO_2 + Al_2O_3 + Fe_2O_3 + CaO + MgO = 100$.

Physical Properties The *free-swelling index* (FSI) measures the tendency of a coal to swell when burned or gasified in fixed or fluidized beds. Coals with a high FSI (greater than 4) can usually be expected to cause difficulties in such beds. Details of the test are given by the ASTM D 720 (American Society for Testing and Materials, op. cit.) and U.S. Bureau of Mines Report of Investigations 3989.

The *Hardgrove grindability index* (HGI) indicates the ease (or difficulty) of grinding coal and is complexly related to physical properties such as hardness, fracture, and tensile strength. The Hardgrove machine is usually employed (ASTM D 409, American Society for Testing and Materials, op. cit.). It determines the relative grindability or ease of pulverizing coal in comparison with a standard coal, chosen as 100 grindability (see Sec. 20 of this handbook). The FSI and HGI of some U.S. coals are given in Bureau of Mines Information Circular 8025 for FSI and HGI data for 2812 and 2339 samples, respectively.

The *bulk density* of broken coal varies according to the specific gravity, size distribution, and moisture content of the coal and the amount of settling when the coal is piled. Following are some useful approximations of the bulk density of various ranks of coal.

	kg/m^3	lb/ft^3
Anthracite	800–930	50–58
Bituminous	670–910	42–57
Lignite	640–860	40–54

Size stability refers to the ability of coal to withstand breakage during handling and shipping. It is determined by twice dropping a 23-kg (50-lb) sample of coal from a height of 1.8 m (6 ft) onto a steel plate. From the size distribution before and after the test, the size stability is reported as a percentage factor (see ASTM D 440). The *friability* test

measures the tendency of coal to break during repeated handling. It is actually the complement of size stability and is determined by the standard tumbler test (ASTM D 441-36).

Spier's *Technical Data on Fuels* gives the *specific heat* of dry, ash-free coal as follows.

	kJ/(kg·K)	Btu/(lb·°F)
Anthracite	0.92–0.96	0.22–0.23
Bituminous	1.0–1.1	0.24–0.25

The relationships between specific heat and water content and between specific heat and ash content are linear. Given the specific heat on a dry, ash-free basis, it can be corrected to an as-received basis. The specific heat and enthalpy of coal to 1366 K (2000°F) are given in *Coal Conversion Systems Technical Data Book* (part 1A, U.S. Dept. of Energy, 1984).

The *mean specific heat* of coal ash and slag, which is used for calculating heat balances on furnaces, gasifiers, and other coal-consuming systems, follows.

Temperature range		Mean specific heat	
K	°F	kJ/(kg·K)	Btu/(lb·°F)
273–311	32–100	0.89	0.21
273–1090	32–1500	0.94	0.22
273–1310	32–1900	0.97	0.23
273–1370	32–2000	0.98	0.24
273–1640	32–2500	1.1	0.27

Coke Coke is the solid, cellular, infusible material remaining after the carbonization of coal, pitch, petroleum residues, and certain other carbonaceous materials. The varieties of coke generally are identified by prefixing a word to indicate the source, if other than coal, (e.g., *petroleum* coke) or the process by which a coke is manufactured (e.g., *oven* coke).

The mechanism of the formation of coke when coal is carbonized is a complex of physical and chemical phenomena that are not perfectly understood. Some of the physical changes, which are interrelated when certain ranks of coal or blends are heated, are softening, devolatilization, swelling, and resolidification. Some of the accompanying chemical changes are cracking, depolymerization, polymerization, and condensation. More detailed theoretical information is given in the general references listed in the beginning of the section on coal.

High-Temperature Coke (1173 to 1423 K or 1652 to 2102°F.) This type is most commonly used in the United States; nearly 20 percent of the total bituminous coal consumed is used to make high-temperature coke for metallurgical applications. About 99 percent of this type of coke is made in slot-type recovery ovens. Blast furnaces use about 90 percent of the production, the rest going mainly to foundries and gas plants.

A U.S. Bureau of Mines survey of 12 blast-furnace coke plants, whose capacity is 30 percent of the total production in the United States, provides an excellent picture of the acceptable chemical and physical properties of metallurgical coke. The ranges of properties are given in Table 27-2.

TABLE 27-2 Chemical and Physical Properties of High-Temperature Cokes Used in the United States*

Property	Range
Volatile matter	0.6–1.4 wt %, as received
Ash	7.5–10.7 wt %, as received
Sulfur	0.6–1.1 wt %, as received
Stability factor	39–58 (1-in tumbler)
Hardness factor	60–68 (¼-in tumbler)
Apparent specific gravity (water = 1.0)	0.80–0.99

**Comparison of Properties of Coke Produced by BM-AGA and Industrial Methods*, U.S. Bur. Mines Rep. Invest. 6354. To convert inches to centimeters, multiply by 2.54.

The typical by-product yields per US ton (909 kg) of dry coal from high-temperature carbonization in ovens with inner-wall temperatures from 1273 to 1423 K (1832 to 2102°F) are: coke, 653 kg (1437 lb); gas, 154 kg (11,200 ft³); tar, 44 kg (10 gal); water, 38 kg (10 gal); light oil, 11 kg (3.3 gal); and ammonia, 2.2 kg (4.8 lb).

Foundry Coke This coke has different requirements from blast-furnace coke. The volatile matter should not exceed 2.0 percent, the sulfur should not exceed 0.7 percent, the ash should not exceed 12.0 percent, and the size should exceed 76 mm (3 in).

Low- and Medium-Temperature Coke (773 to 1023 K or 932 to 1382°F.) Cokes of this type are no longer produced in the United States to a significant extent. However, there is some interest in low-temperature carbonization as a source of both hydrocarbon liquids and gases to supplement petroleum and natural-gas resources.

The *Fischer assay* is an arbitrary but precise analytical tool for determining the yield of products from low-temperature carbonization. A known weight of coal is heated at a controlled rate in the absence of air to 773 K (932°F), and the products are collected and weighed. Table 27-3 gives the approximate yields of products for various ranks of coal.

Pitch Coke and Petroleum Coke Pitch coke is made from coal-tar pitch, and petroleum coke is made from petroleum residues from petroleum refining. Pitch coke has about 1.0 percent volatile matter, 1.0 percent ash, and less than 0.5 percent sulfur on the as-received basis. There are two kinds of petroleum coke: delayed coke and fluid coke. Delayed coke is produced by heating a gas oil or heavier feedstock to 755 to 811 K (900 to 1000°F) and spraying it into a large vertical cylinder where cracking and polymerization reactions occur. Water jets are used to cut the coke from one drum while the other drum is being charged. Fluid coke is made in a fluidized-bed reactor where preheated feed is sprayed onto a fluidized bed of coke particles. Coke product is continuously withdrawn by size classifiers in the solids loop of the reactor system. Since it contains the impurities from the original crude oil, the sulfur is usually high, and appreciable quantities of vanadium salts may be present. Ranges of composition and properties are as follows.

Composition and properties	Delayed coke	Fluid coke
Volatile matter, wt %	8–18	3.7–7.0
Ash, wt %	0.05–1.6	0.1–2.8
Sulfur, wt %	—	1.5–10.0
Grindability index	40–60	20–30
True density, g/cm³	1.28–1.42	1.5–1.6

Other Solid Fuels

Coal Char This type of char is the nonagglomerated, nonfusible residue from the thermal treatment of coal. Coal chars are obtained as a residue or a coproduct from low-temperature carbonization processes and from processes being developed to convert coal to liquid and gaseous fuels and to chemicals. Such chars have a substantial heating value. The net amount of char from a conversion process varies widely; in some instances, it may represent between about 30

and 55 percent of the weight of the coal feed; in others, no net or excess char is produced; i.e., the entire char yield is consumed as in-plant fuel. The volatile matter, sulfur, and heating values of the chars are lower, and the ash is higher, than in the original coal. Chars typically have higher volatile matter contents (7 to 12 wt %) than low- and medium-temperature cokes (2.5 to 6 wt %), but this is both process and feed-coal dependent.

Peat Peat is partially decomposed plant matter that has accumulated underwater or in a water-saturated environment. It is the precursor of coal but is not classified as coal. Sold under the term *peat moss* or *moss peat*, peat is used in the United States mainly for horticultural and agricultural applications, but interest is growing in its use as a fuel in certain local areas (e.g., North Carolina). Peat is used extensively as a fuel primarily in Ireland and the former Soviet Union. Although analyses of peat vary widely, a typical high-grade peat has 90 percent water, 3 percent fixed carbon, 5 percent volatile matter, 1.5 percent ash, and 0.10 percent sulfur. The moisture-free heating value is approximately 20.9 MJ/kg (9000 Btu/lb).

Wood Typical higher heating values are 20 MJ/kg (8600 Btu/lb) for oven-dried hardwood and 20.9 MJ/kg for oven-dried softwood. These values are accurate enough for most engineering purposes. U.S. Department of Agriculture Handbook 72 (revised 1974) gives the specific gravity of the important softwoods and hardwoods, useful if heating value on a volume basis is needed.

Charcoal Charcoal is the residue from the destructive distillation of wood. It absorbs moisture readily, often containing as much as 10 to 15 percent water. In addition, it usually contains about 2 to 3 percent ash and 0.5 to 1.0 percent hydrogen. The heating value of charcoal is about 27,912 to 30,238 kJ/kg (12,000 to 13,000 Btu/lb).

Solid Wastes and Biomass Large and increasing quantities of solid wastes are a significant feature of affluent societies. In the United States in 1993 the rate was about 1.8 kg (4 lb) per capita per day or nearly 190 Tg (2.07×10^8 U.S. tons) per year, but the growth rate has slowed in recent years as recycling efforts have increased. Table 27-4 shows that the composition of miscellaneous refuse is surprisingly uniform, but size and moisture variations cause major difficulties in efficient, economical disposal.

The fuel value of most solid wastes is usually sufficient to enable self-supporting combustion, leaving only the incombustible residue and reducing the volume of waste eventually consigned to sanitary landfills to only 10 to 15 percent of the original volume. The heat released by the combustion of waste can be recovered and utilized, although the cost of the recovery equipment or the distance to a suitable point of use for the heat may make its recovery economically infeasible.

Wood, wood scraps, bark, and wood product plant waste streams are a major element of biomass, industrial, and municipal solid waste fuels. In 1991, about 1.7 EJ (1.6×10^{15} Btu [quads]) of energy were obtained from wood and wood wastes, representing about 60 percent of the total biomass-derived energy in the United States. *Bagasse* is the solid residue remaining after sugarcane has been crushed by pressure rolls. It usually contains from 40 to 50 percent water. The dry bagasse has a heating value of 18.6 to 20.9 MJ/kg (8000 to 9000 Btu/lb).

TABLE 27-3 Fischer-Assay Yields from Various Ranks of Coal (As-Received Basis)

Class	ASTM classification by rank — Group	Coke, wt %	Tar, gal/ton	Light oil, gal/ton	Gas, ft³/ton	Water, wt %
Bituminous	1. Low-volatile bituminous	90	8.6	1.0	1760	3
	2. Medium-volatile bituminous	83	18.9	1.7	1940	4
	3. High-volatile A bituminous	76	30.9	2.3	1970	6
	4. High-volatile B bituminous	70	30.3	2.2	2010	11
	5. High-volatile C bituminous	67	27.0	1.9	1800	16
Subbituminous	1. Subbituminous A	59	20.5	1.7	2660	23
	2. Subbituminous B	58	15.4	1.3	2260	28
Lignite	1. Lignite A	37	15.2	1.2	2100	44

NOTE: To convert gallons per ton to liters per kilogram, multiply by 0.004; to convert cubic feet per ton to cubic meters per kilogram, multiply by 3.1×10^{-5}.

TABLE 27-4 **Waste Fuel Analysis**

Type of waste	Heating value, Btu/lb	Percentage composition by weight						Density, lb/ft³
		Volatiles	Moisture	Ash	Sulfur	Dry combustible		
Paper	7,572	84.6	10.2	6.0	0.20			
Wood	8,613	84.9	20.0	1.0	0.05			
Rags	7,652	93.6	10.0	2.5	0.13			
Garbage	8,484	53.3	72.0	16.0	0.52			
Coated fabric: rubber	10,996	81.2	1.04	21.2	0.79	78.80		23.9
Coated felt: vinyl	11,054	80.87	1.50	11.39	0.80	88.61		10.7
Coated fabric: vinyl	8,899	81.06	1.48	6.33	0.02	93.67		10.1
Polyethylene film	19,161	99.02	0.15	1.49	0	98.51		5.7
Foam: scrap	12,283	75.73	9.72	25.30	1.41	74.70		9.1
Tape: resin-covered glass	7,907	15.08	0.51	56.73	0.02	43.27		9.5
Fabric: nylon	13,202	100.00	1.72	0.13	0	99.87		6.4
Vinyl scrap	11,428	75.06	0.56	4.56	0.02	95.44		23.4

SOURCE: From Hescheles, *MECAR Conference on Waste Disposal*, New York, 1968; and *Refuse Collection Practice*, 3d ed., American Public Works Association, Chicago, 1966.

To convert British thermal units per pound to joules per kilogram, multiply by 2326; to convert pounds per cubic foot to kilograms per cubic meter, multiply by 16.02.

LIQUID FUELS

Liquid Petroleum Fuels The principal liquid fuels are made by fractional distillation of crude petroleum (a mixture of hydrocarbons and hydrocarbon derivatives ranging from methane to heavy bitumen). As many as one-quarter to one-half of the molecules in crude may contain sulfur atoms, and some contain nitrogen, oxygen, vanadium, nickel, or arsenic. Desulfurization, hydrogenation, cracking (to lower molecular weight), and other refining processes may be performed on selected fractions before they are blended and marketed as fuels. Viscosity/gravity/boiling-range relationships of common fuels are shown in Fig. 27-2.

Specifications The American Society for Testing and Materials has developed specifications (*Annual Book of ASTM Standards*, Conshohocken, Pa., updated annually) that are widely used to classify fuels. Table 27-5 shows fuels covered by ASTM D 396, Standard Specification for Fuel Oils. D 396 omits kerosenes (low-sulfur, clean-burning No. 1 fuels for lamps and freestanding flueless domestic heaters), which are covered separately by ASTM D 3699.

In drawing contracts and making acceptance tests, refer to the pertinent ASTM standards. *ASTM Standards* contain specifications (classifications) and test methods for burner fuels (D 396), motor and aviation gasolines (D 4814 and D 910), diesel fuels (D 975), and aviation and gas-turbine fuels (D 1655 and D 2880). ASTM D 4057 contains procedures for sampling bulk oil in tanks, barges, etc.

Fuel specifications from different sources may differ in test limits on sulfur, density, etc., but the same general categories are recognized worldwide: kerosene-type vaporizing fuel, distillate (or "gas oil") for atomizing burners, and more viscous blends and residuals for commerce and heavy industry. Typical specifications are as follows.

Specifier	Number	Category
Canadian Government Specification Board, Department of Defense Production, Ottawa, Canada	3-GP-2	Fuel oil, heating
Deutschen Normenauschusses, Berlin 15	DIN 51603	Heating (fuel) oils
British Standards Institution, British Standards House, 2 Park Street, London, WIA 2BS	B.S. 2869	Petroleum fuels for oil engines and burners
Japan	JIS K2203 JIS K2204 JIS K2205	Kerosene Gas oil Fuel oils
Federal specifications, United States	ASTM D 396	Fuel oil, burner

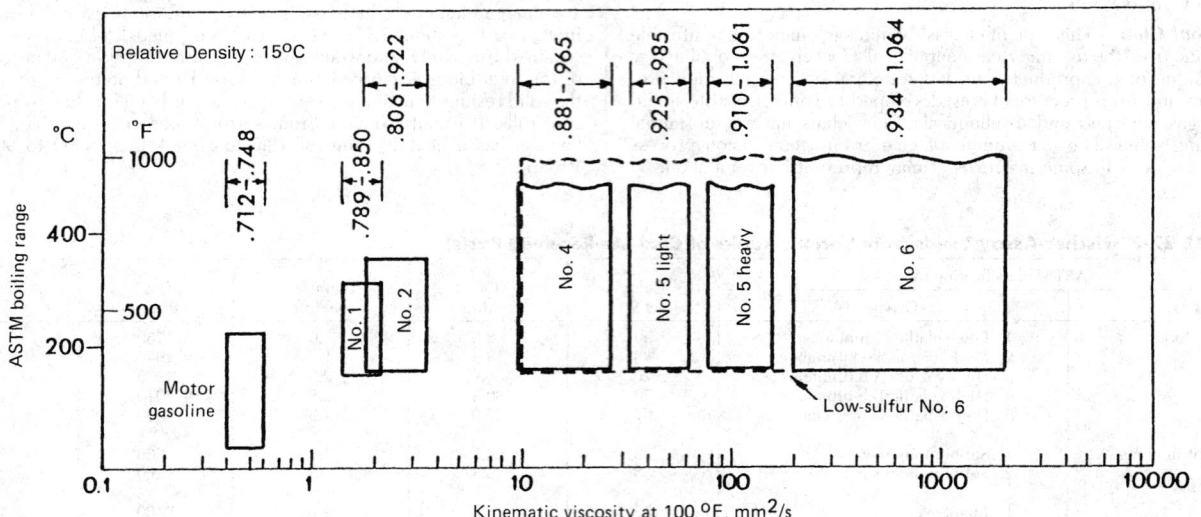

FIG. 27-2 Viscosity, boiling-range, and gravity relationships for petroleum fuels.

TABLE 27-5   Detailed Requirements for Fuel Oils[a]

Property	ASTM test method[b]	No. 1	No. 2	Grade No. 4 (light)	No. 4	No. 5 (light)	No. 5 (heavy)	No. 6
Flash point, °C, min.	D 93	38	38	38	55	55	55	60
Water and sediment, % vol. max.	D 1796	0.05	0.05	(0.50)[c]	(0.50)[c]	(1.00)[c]	(1.00)[c]	(2.00)[c]
Distillation temperature, °C	D 86							
10% vol. recovered, max.		215	—	—	—	—	—	—
90% vol. recovered, min.			282	—	—	—	—	—
max.		288	338	—	—	—	—	—
Kinematic viscosity at 40°C (104°F), mm²/s	D 445	—	—	—				
min.		1.3	1.9	1.9	>5.5	—	—	—
max.		2.1	3.4	5.5	24.0[d]	—	—	—
Kinematic viscosity at 100°C (212°F), mm²/s								
min.		—	—	—	—	5.0	9.0	15.0
max.		—	—	—	—	8.9[d]	14.9[d]	50.0[d]
Ramsbottom carbon residue on 10% distillation residue, % mass, max.	D 524	0.15	0.35	—	—	—	—	—
Ash, % mass, max.	D 482	—	—	0.05	0.10	0.15	0.15	—
Sulfur, % mass max[e]	D 129	0.50	0.50	—	—	—	—	—
Copper strip corrosion rating, max., 3 h at 50°C	D 130	No. 3	No. 3	—	—	—	—	—
Density at 15°C, kg/m³	D 1298							
min.		—	—	>876[f]	—	—	—	—
max.		850	876	—	—	—	—	—
Pour point °C, max[g]	D 97	−18	−6	6	−6	—	—	[h]

[a]Source ASTM D 396-92. It is the intent of these classifications that failure to meet any requirement of a given grade does not automatically place an oil in the next lower grade unless in fact it meets all requirements of the lower grade. However, to meet special operating conditions modifications of individual limiting requirements may be agreed upon among the purchaser, seller, and manufacturer. Copyright ASTM. Reprinted with permission.

[b]The test methods indicated are the approved referee methods. Other acceptable methods are indicated in Section 2 and 5.1.

[c]The amount of water by distillation by Test Method D 95 plus the sediment by extraction by Test Method D 473 shall not exceed the value shown in the table. For Grade No. 6 fuel oil, the amount of sediment by extraction shall not exceed 0.5 mass % and a deduction in quantity shall be made for all water and sediment in excess of 1.0 mass %.

[d]Where low sulfur fuel oil is required, fuel oil falling in the viscosity range of a lower-numbered grade down to and including No. 4 can be supplied by agreement between the purchaser and supplier. The viscosity range of the initial shipment shall be identified and advance notice shall be required when changing from one viscosity range to another. This notice shall be in sufficient time to permit the user to make the necessary adjustments.

[e]Other sulfur limits may apply in selected areas in the United States and in other countries.

[f]This limit assures a minimum heating value and also prevents misrepresentation and misapplication of this product as Grade No. 2.

[g]Lower or higher pour points can be specified whenever required by conditions of storage or use. When a pour point less than −18°C is specified, the minimum viscosity at 40°C for grade No. 2 shall be 1.7 mm²/s and the minimum 90% recovered temperature shall be waived. (Add 273 to °C to obtain K.)

[h]Where low sulfur fuel oil is required, Grade No. 6 fuel oil will be classified as *low pour* (+15°C max) or *high pour* (no max.). Low-pour fuel oil should be used unless tanks and lines are heated.

Foreign specifications are generally available from the American National Standards Institute, New York; United States federal specifications, at Naval Publications and Forms, Philadelphia.

Equipment manufacturers and large-volume users often write fuel specifications to suit particular equipment, operating conditions, and economics. Nonstandard test procedures and restrictive test limits should be avoided; they reduce the availability of fuel and increase its cost.

Bunker-fuel specifications for merchant vessels are described by ASTM D 2069, Standard Specification for Marine Fuels. Deep draft vessels carry residual (e.g., No. 6 fuel oil) or distillate-residual blend for main propulsion, plus distillate for start-up, shutdown, maneuvering, deck engines, and diesel generators. Main-propulsion fuel is identified principally by its viscosity in centistokes at 373 K. Obsolete designations include those based on Redwood No. 1 seconds at 100°F (311 K) (e.g., "MD 1500") and the designations "Bunker A" for No. 5 fuel oil and "Bunker B" and "Bunker C" for No. 6 fuel oil in the lower- and upper-viscosity ranges, respectively.

Chemical and Physical Properties   Petroleum fuels contain paraffins, isoparaffins, naphthenes, and aromatics, plus organic sulfur, oxygen, and nitrogen compounds that were not removed by refining. Olefins are absent or negligible except when created by severe refining. Vacuum-tower distillate with a final boiling point equivalent to 730 to 840 K (850 to 1050°F) at atmospheric pressure may contain from 0.1 to 0.5 ppm vanadium and nickel, but these metal-bearing compounds do not distill into No. 1 and 2 fuel oils.

Black, viscous residuum directly from the still at 410 K (390°F) or higher serves as fuel in nearby furnaces or may be cooled and blended to make commercial fuels. Diluted with 5 to 20 percent distillate, the blend is No. 6 fuel oil. With 20 to 50 percent distillate, it becomes No. 4 and No. 5 fuel oils for commercial use, as in schools and apartment houses. Distillate-residual blends also serve as diesel fuel in large stationary and marine engines. However, distillates with inadequate solvent power will precipitate asphaltenes and other high-molecular-

weight colloids from *visbroken* (severely heated) residuals. A blotter test, ASTM D 4740, will detect sludge in pilot blends. Tests employing centrifuges, filtration (D 4870), and microscopic examination have also been used.

No. 6 fuel oil contains from 10 to 500 ppm vanadium and nickel in complex organic molecules, principally porphyrins. These cannot be removed economically, except incidentally during severe hydrodesulfurization (Amero, Silver, and Yanik, *Hydrodesulfurized Residual Oils as Gas Turbine Fuels*, ASME Pap. 75-WA/GT-8). Salt, sand, rust, and dirt may also be present, giving No. 6 a typical ash content of 0.01 to 0.5 percent by weight.

Ultimate analyses of some typical fuels are shown in Table 27-6.

The hydrogen content of petroleum fuels can be calculated from density with the following formula, with an accuracy of about 1 percent for petroleum liquids that contain no sulfur, water, or ash:

$$H = 26 - 15s \qquad (27\text{-}9)$$

where H = percent hydrogen and s = relative density at 15°C (with respect to water), also referred to as specific gravity. Schmidt (*Fuel Oil Manual*, 3d ed., Industrial Press, New York, 1969) claims improved precision of the formula by replacing 26 with different constants.

Relative density (288 K)	API gravity	Constant
1.0754–1.0065	0–9	24.50
1.0065–0.9935	10–20	25.00
0.9935–0.8757	21–30	25.20
0.8757–0.8013	31–45	25.45

Relative density is usually determined at ambient temperature with specialized hydrometers. In the United States these hydrometers commonly are graduated in an arbitrary scale termed *degrees API*. This scale relates inversely to relative density s (at 60°F) as follows (see also the abscissa scale of Fig. 27-3):

TABLE 27-6   Typical Ultimate Analyses of Petroleum Fuels

Composition, %	No. 1 fuel oil (41.5° A.P.I.)	No. 2 fuel oil (33° A.P.I.)	No. 4 fuel oil (23.2° A.P.I.)	Low sulfur, No. 6 F.O. (12.6° A.P.I.)	High sulfur, No. 6 (15.5° A.P.I.)
Carbon	86.4	87.3	86.47	87.26	84.67
Hydrogen	13.6	12.6	11.65	10.49	11.02
Oxygen	0.01	0.04	0.27	0.64	0.38
Nitrogen	0.003	0.006	0.24	0.28	0.18
Sulfur	0.09	0.22	1.35	0.84	3.97
Ash	<0.01	<0.01	0.02	0.04	0.02
C/H Ratio	6.35	6.93	7.42	8.31	7.62

NOTE: The C/H ratio is a weight ratio.

$$\text{Degrees API} = \frac{141.5}{s} - 131.5 \qquad (27\text{-}10)$$

For practical engineering purposes, relative density at 15°C (288 K), widely used in countries outside the United States, is considered equivalent to specific gravity at 60°F (288.6 K). With the adoption of SI units, the American Petroleum Institute favors absolute density at 288 K instead of degrees API.

The hydrogen content, heat of combustion, specific heat, and thermal conductivity data herein were abstracted from Bureau of Standards Miscellaneous Publication 97, *Thermal Properties of Petroleum Products.* These data are widely used, although other correlations have appeared, notably that by Linden and Othmer (*Chem. Eng.* **54**[4, 5], April and May, 1947).

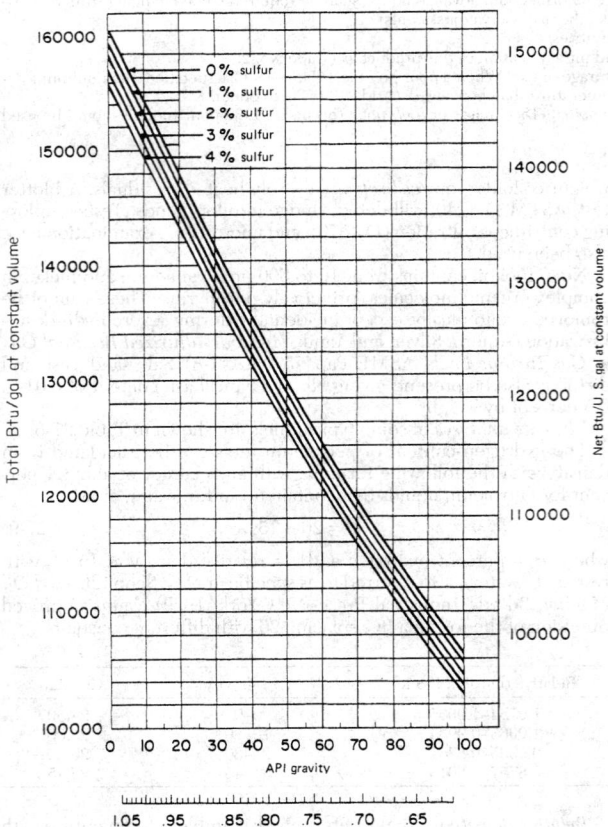

FIG. 27-3 Heat of combustion of petroleum fuels. To convert Btu/US gal to kJ/m³, multiply by 278.7.

Heat of combustion can be estimated within 1 percent from the relative density of the fuel by using Fig. 27-3. Corrections for water and sediment must be applied for residual fuels, but they are insignificant for clean distillates.

Pour point ranges from 213 K (−80°F) for some kerosene-type jet fuels to 319 K (115°F) for waxy No. 6 fuel oils. *Cloud point* (which is not measured on opaque fuels) is typically 3 to 8 K higher than pour point unless the pour has been depressed by additives. Typical petroleum fuels are practically newtonian liquids between the cloud point and the boiling point and at pressures below 6.9 MPa (1000 psia).

Fuel systems for No. 1 (kerosene) and No. 2 fuel oil (diesel, home heating oil) are not heated. Systems for No. 6 fuel oil are usually designed to preheat the fuel to 300 to 320 K (90 to 120°F) to reduce viscosity for handling and to 350 to 370 K (165 to 200°F) to reduce viscosity further for proper atomization. No. 5 fuel oil may also be heated, but preheating is usually not required for No. 4. (See Table 27-5.) Steam or electric heating is employed as dictated by economics, climatic conditions, length of storage time, and frequency of use. Pressure relief arrangements are recommended on sections of heated pipelines when fuel could be inadvertently trapped between valves.

The *kinematic viscosity* of a typical No. 6 fuel oil declines from 5000 mm²/s (0.054 ft²/s) at 298 K (77°F) to about 700 mm²/s (0.0075 ft²/s) and 50 mm²/s (0.000538 ft²/s) on heating to 323 K (122°F) and 373 K (212°F), respectively. Viscosity of 1000 mm²/s or less is required for manageable pumping. Proper boiler atomization requires a viscosity between 15 and 65 mm²/s.

Thermal expansion of petroleum fuels can be estimated as volume change per unit volume per degree. ASTM-IP Petroleum Measurement Tables (ASTM D 1250 IP 200) are used for volume corrections in commercial transactions.

Heat capacity (specific heat) of petroleum liquids between 0 and 205°C (32 and 400°F), having a relative density of 0.75 to 0.96 at 15°C (60°F), can be calculated within 2 to 4 percent of the experimental values from the following equations:

$$c = \frac{1.685 + (0.039 \times °C)}{\sqrt{s}} \qquad (27\text{-}11)$$

$$c' = \frac{0.388 + (0.00045 \times °F)}{\sqrt{s}} \qquad (27\text{-}12)$$

where c is heat capacity, kJ/(kg·°C) or kJ/(kg·K), and c' is heat capacity, Btu/(lb·°F). Heat capacity varies with temperature, and the arithmetic average of the values at the initial and final temperatures can be used for calculations relating to the heating or cooling of oil.

The *thermal conductivity* of liquid petroleum products is given in Fig. 27-4. Thermal conductivities for asphalt and paraffin wax in their solid states are 0.17 and 0.23 W/(m·K), respectively, for temperatures above 273 K (32°F) (1.2 and 1.6 Btu/[h·ft²][°F/in]).

Commercial Considerations   Fuels are sold in gallons and in multiples of the 42-gal barrel (0.159 m³) in the United States, while a weight basis is used in other parts of the world. Transactions exceeding about 20 to 40 m³ (5000 to 10,000 US gal) usually involve volume corrections to 288 K (60°F) for accounting purposes. Fuel passes through an air eliminator and mechanical meter when loaded into or dispensed from trucks. Larger transfers such as pipeline, barge, or tanker movements are measured by fuel depth and *strapping tables* (calibration tables) in tanks and vessels, but positive-displacement

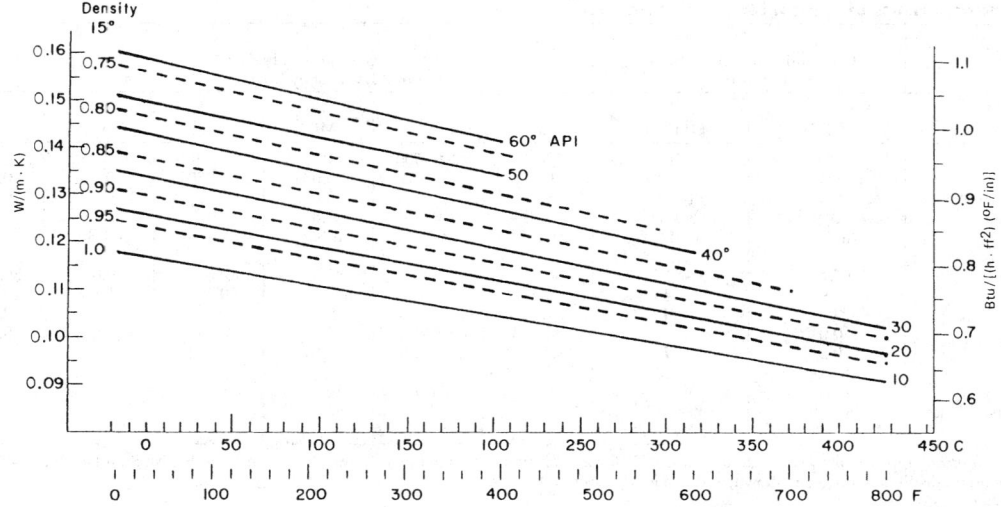

FIG. 27-4   Thermal conductivity of petroleum liquids. The solid lines refer to density expressed as degrees API; the broken lines refer to relative density at 288 K (15°C). (K = [°F + 459.7]/1.8)

meters that are *proved* (calibrated) frequently are gaining acceptance. After an appropriate settling period, water in the tank bottom is measured with a plumb bob or stick smeared with water-detecting paste.

Receipts of tank-car quantities or larger are usually checked for gravity, appearance, and flash point to confirm product identification and absence of contamination.

Safety Considerations   Design and location of storage tanks, vents, piping, and connections are specified by state fire marshals, underwriters codes, and local ordinances. In NFPA 30, *Flammable and Combustible Liquids Code, 1993* (published by the National Fire Protection Association, Quincy, Mass.), liquid petroleum fuels are classified as follows for safety in handling:

Class I (flammable) liquid has a flash point below 311 K (100°F) and a vapor pressure not exceeding 0.28 MPa at 311 K (40 psia at 100°F).

Class IA includes those liquids having flash points below 296 K (73°F) and boiling points below 311 K (100°F).

Class IB liquids have flash points below 296 K (73°F) and boiling points at or above 311 K (100°F).

Class IC includes those liquids having flash points at or above 296 K (73°F) and below 311 K (100°F).

Class II combustible liquids have flash points at or above 311 K (100°F) and below 333 K (140°F).

Class IIIA combustible liquids have flash points at or above 333 K (140°F) and below 366 K (200°F).

Class IIIB liquids flash at or above 366 K (200°F).

NFPA 30 details the design features and safe placement of handling equipment for flammable and combustible liquids.

Crude oils with flash points below 311 K (100°F) have been used in place of No. 6 fuel oil. Different pumps may be required because of low fuel viscosity.

Nonpetroleum Liquid Fuels

Tar Sands   Canadian tar sands are strip-mined and extracted with hot water to recover heavy oil (bitumen). The oil is processed into naphtha, kerosene, and gasoline fractions (which are hydrotreated), in addition to gas (which is recovered). Tar sands are being developed in Utah also.

Oil Shale   Oil shale is nonporous rock containing organic kerogen. Raw shale oil is extracted from mined rock by pyrolysis in a surface retort, or in situ by steam injection after breaking up the rock with explosives. Pyrolysis cracks the kerogen, yielding raw shale oil

high in nitrogen, oxygen, and sulfur. *Shale oil* has been hydrotreated and refined in demonstration tests into relatively conventional fuels. Refining in petroleum facilities is possible, and blending with petroleum is most likely.

Coal-Derived Fuels   Liquid fuels derived from coal range from highly aromatic coal tars to liquids resembling petroleum. Raw liquids from different hydrogenation processes show variations that reflect the degree of hydrogenation achieved. Also, the raw liquids can be further hydrogenated to refined products. Properties and cost depend on the degree of hydrogenation and the boiling range of the fraction selected. A proper balance between fuel upgrading and equipment modification is essential for the most economical use of coal liquids in boilers, industrial furnaces, diesels, and stationary gas turbines.

Coal-tar fuels are high-boiling fractions of crude tar from pyrolysis in coke ovens and coal retorts. Grades range from free-flowing liquids to pulverizable pitch. Low in sulfur and ash, they contain hydrocarbons, phenols, and heterocyclic nitrogen and oxygen compounds. Being more aromatic than petroleum fuels, they burn with a more luminous flame. From 288 to 477 K (60 to 400°F) properties include:

Heat capacity	1.47–1.67 kJ/(kg·K) (0.35–0.40 Btu/[lb·°F])
Thermal conductivity	0.14–0.15 W/(m·K) (0.080–0.085 Btu/[h·ft·°F])
Heat of vaporization	349 kJ/kg (150 Btu/lb)
Heat of fusion	Nil

Table 27-7 shows representative data for liquid fuels from tar sands, oil shale, and coal.

GASEOUS FUELS

Natural Gas   Natural gas is a combustible gas that occurs in porous rock of the earth's crust and is found with or near accumulations of crude oil. It may occur alone in separate reservoirs, but more commonly it forms a gas cap entrapped between petroleum and an impervious, capping rock layer in a petroleum reservoir. Under high-pressure conditions, it is mixed with or dissolved in crude oil. Natural gas termed *dry* has less than 0.013 dm³/m³ (0.1 gal/1000 ft³) of gasoline. Above this amount, it is termed *wet*.

The proven reserves of natural gas in the United States total about 4.58 Tm³ (1.62 × 10¹⁴ ft³). Production in 1993 was about 0.51 Tm³ (1.8 × 10¹³ ft³). Revisions and adjustments to the existing resource base, together with modest new additions to proven reserves, have held the recent decline in reserves to about 1.6 percent per year.

Natural gas consists of hydrocarbons with a very low boiling point.

TABLE 27-7   Characteristics of Typical Nonpetroleum Fuels

	Conventional coal-tar fuels from retorting[a]		Typical coal-derived fuels with different levels of hydrogenation[b]				Synthetic crude oils, by hydrogenation	
	CTF 50	CTF 400	Minimal		Mild	Mild[c]	Severe	Oil shale	Tar sands[d]
Distillation range, °C			175–280	280–500	160–415	175–400	125–495		
Density, kg/m³, 15°C	1.018	1.234	0.974	1.072	0.964	0.9607	0.914	0.817	0.864
lb U.S. gal, 60°F	8.5	10.3	8.1	8.9	8.0	8.0	7.6	6.8	7.2
Viscosity, mm²/s	2–9	9–18	3.1–3.4	50–90	3.6	—	2.18		
	At 38°C	At 121°C	At 38°C	At 38°C	At 38°C	—	At 38°C		
Ultimate analysis, %									
Carbon	87.4	90.1	86.0	89.1	87.8	89.6	89.0	86.1	87.1
Hydrogen	7.9	5.4	9.1	7.5	9.7	10.1	11.1	13.84	12.69
Oxygen	3.6	2.4	3.6–4.3	1.4–1.8	2.4	0.3	0.5	0.12	0.04
Nitrogen	0.9	1.4	0.9–1.1	1.2–1.4	0.6	0.04	0.09	0.01	0.07
Sulfur	0.2	0.7	<0.2	0.4–0.5	0.07	0.004	0.04	0.02	0.10
Ash[e]	Trace	0.15	<0.001	[f]					
C/H ratio, weight	11.0	16.5	9.4	11.9	9.1	8.9	8.0	6.2	6.9
Gross calorific value, MJ/kg	38.4–40.7	36.8–37.9							
Btu/lb	16,500–17,500	15,800–16,300							

[a] CTF 50 and 400 indicate approximate preheat temperature, °F, for atomization of fuel in burners (terminology used in British Standard B.S. 1469).
[b] Properties depend on distillation range, as shown, and to a lesser extent on coal source.
[c] Using recycle-solvent process.
[d] Tar sands, although a form of petroleum, are included in this table for comparison.
[e] Inorganic mineral constituents of coal tar fuel:
    5 to 50 ppm: Ca, Fe, Pb, Zn (Na, in tar treated with soda ash)
    0.05 to 5 ppm: Al, Bi, Cu, Mg, Mn, K, Si, Na, Sn
    Less than 0.05 ppm: As, B, Cr, Ge, Ti, V, Mo
    Not detected: Sb, Ba, Be, Cd, Co, Ni, Sr, W, Zr
[f] Inherent ash is "trace" or "<0.1%," although entrainment in distillation has given values as high as 0.03 to 0.1%.

Methane is the main constituent, with a boiling point of 119 K (−245°F). Ethane, with a boiling point of 184 K (−128°F) may be present in amounts up to 10 percent; propane, with a boiling point of 231 K (−44°F), up to 3 percent. Butane, pentane, hexane, heptane, and octane may also be present. Physical properties of these hydrocarbons are given in Sec. 2.

Although there is no single composition that may be called "typical" natural gas, Table 27-8 shows the range of compositions in large cities in the United States.

Most natural gas is substantially free of sulfur compounds; the terms *sweet* and *sour* are used to denote the absence or presence of H_2S. Some wells, however, deliver gas containing levels of hydrogen sulfide and other sulfur compounds (e.g., thiophenes, mercaptans, and organic sulfides) that must be removed before transfer to commercial pipelines. Pipeline-company contracts typically specify maximum allowable limits of impurities; H_2S and total sulfur compounds seldom exceed 0.023 and 0.46 g/m³ (1.0 and 20.0 gr/100 std ft³),

TABLE 27-8   Analysis of Natural Gas*

	Range†	
	Low	High
Composition, vol %		
Methane	86.3	95.2
Ethane	2.5	8.1
Propane	0.6	2.8
Butanes	0.13	0.66
Pentanes	0	0.44
Hexanes plus	0	0.09
CO_2	0	1.1
N_2	0.31	2.47
He	0.01	0.06
Heating value MJ/m³ (Btu/ft³)	38.15(1024)	40.72(1093)
Specific gravity Ref.: Air at 288 K (60°F)	0.586	0.641

*Adapted from *Gas Engineers Handbook*, American Gas Association, Industrial Press, New York, 1965.
†Ranges are the high and low values of annual averages reported by 13 utilities (1954 data).

respectively. The majority of pipeline companies responding to a 1994 survey limited H_2S to less than 0.007 g/m³ (0.3 gr/100 std ft³), but a slightly smaller number continued specifying 0.023 g/m³, in accord with an American Gas Association 1971 recommendation.

Compressibility of Natural Gas   All gases deviate from the perfect gas law at some combinations of temperature and pressure, the extent depending on the gas. This behavior is described by a dimensionless compressibility factor Z that corrects the perfect gas law for real-gas behavior, $PV = ZRT$. Any consistent units may be used. Z is unity for an ideal gas, but for a real gas, Z has values ranging from less than 1 to greater than 1, depending on temperature and pressure. The compressibility factor is described further in Secs. 2 and 4 of this handbook.

Because the value of Z for natural gas is significantly less than unity at ambient temperatures and at pressures greater than 1 MPa (145 psia), the compressibility must be taken into account in gas measurement: gas purchased at high line pressure will give a greater volume when the pressure is reduced than it would if the gas were ideal. Natural gas pipeline operators use a *supercompressibility* factor, also called Z, but defined as

$$Z = \left(\frac{RT}{PV}\right)^{1/2} \tag{27-13}$$

which is convenient for use with differential pressure flowmeters but sometimes a source of confusion. For determining compressibility factors of natural-gas mixtures, see *Manual for the Determination of Supercompressibility Factors for Natural Gas*, American Gas Association, New York, 1963; and A.G.A Gas Measurement Committee Report No. 3, 1969.

Liquefied Natural Gas   The advantages of storing and shipping natural gas in liquefied form (LNG) derive from the fact that 0.035 m³ (1 ft³) of liquid methane at 111 K (−260°F) equals about 18 m³ (630 ft³) of gaseous methane. Temperatures higher than 111 K can be used if the liquid is stored under pressure. For example, the liquid state is maintained at 2.24 MPa (325 psia) and 170 K (−155°F). The critical temperature of methane is 191 K (−160°F), and the corresponding critical pressure is 4.64 MPa (673 psia). One cubic meter (264 US gal) weighs 412 kg (910 lb) at 109 K (−263°F). The heating value is about 24 GJ/m³ (86,000 Btu/US gal).

The heat of vaporization of LNG at 0.1 MPa (1 bar) is 232 MJ/m³

(832 Btu/US gal) of liquid. On a product gas basis, the heat required is about $0.3 kJ/m^3$ (10 Btu/std ft^3) of gas produced.

LNG is stored in metal double-wall or prestressed concrete tanks, frozen earth, or mined quarries or caverns.

Liquefied Petroleum Gas   The term *liquefied petroleum gas* (LPG) is applied to certain specific hydrocarbons which can be liquefied under moderate pressure at normal temperatures but are gaseous under normal atmospheric conditions. The chief constituents of LPG are propane, propylene, butane, butylene, and isobutane. LPG produced in the separation of heavier hydrocarbons from natural gas is mainly of the paraffinic (saturated) series. LPG derived from oil-refinery gas may contain varying low amounts of olefinic (unsaturated) hydrocarbons.

LPG is widely used for domestic service, supplied either in tanks or by pipelines. It is also used to augment natural-gas deliveries on peak days and by some industries as a standby fuel.

Other Gaseous Fuels

Hydrogen   Hydrogen is used extensively in the production of ammonia and chemicals, in the hydrogenation of fats and oils, and as an oven reducing atmosphere. It is also used as a fuel in industrial cutting and welding operations. There are no resources of uncombined hydrogen as there are of the other fuels. It is made industrially by the steam reforming of natural gas, as the by-product of industrial operations such as the thermal cracking of hydrocarbons and the production of chlorine, and, to a lesser extent, by the electrolysis of water, which is a practically inexhaustible source.

Hydrogen is seen as the ultimate nonpolluting form of energy; when electrochemically combined with oxygen in fuel cells, only water, heat, and electricity are produced. Means for transforming the world's fossil energy economy into a hydrogen economy are being considered as a long-term option. Hydrogen can be stored in gaseous, liquid, or solid forms; however, currently available technologies are not suited to meet mass energy market needs. Technologies for economically producing, storing, and utilizing hydrogen are being researched in the United States, Europe, and Japan.

Acetylene   Acetylene is used primarily in operations requiring high flame temperature, such as welding and metal cutting. To transport acetylene, it is dissolved in acetone under pressure and drawn into small containers filled with porous material.

Miscellaneous Fuels   A variety of gases have very minor market shares. These include reformed gas, oil gases, producer gas, blue water gas, carbureted water gas, coal gas, and blast-furnace gas. The heating values of these gases range from 3.4 to 41 MJ/m^3 (90 to 1100 Btu/ft^3). They are produced by pyrolysis, the water gas reaction, or as by-products of pig-iron production.

Hydrogen sulfide in manufactured gases may range from approximately $2.30 g/m^3$ (100 gr/100 ft^3) in blue and carbureted water gas to several hundred grains in coal- and coke-oven gases. Another important sulfur impurity is carbon disulfide, which may be present in amounts varying from 0.007 to 0.07 percent by volume. Smaller amounts of carbon oxysulfide, mercaptans, and thiophene may be found. However, most of the impurities are removed during the purification process and either do not exist in the finished product or are present in only trace amounts.

FUEL AND ENERGY COSTS

Fuel costs vary widely from one area to another because of the cost of the fuel itself and the cost of transportation. Any meaningful cost comparison between fuels requires current costs based on such factors as the amounts used at a particular geographical location, utilization efficiencies or energy-ratio data for the equipment involved, and the effects of "form value." Although the costs given in Table 27-9 do not apply to specific locations, they give fuel-cost trends.

COAL CONVERSION

Coal is the most abundant fossil fuel, and it will be available long after petroleum and natural gas are scarce. However, because liquids and

TABLE 27-9   Time-Price Relationships for Fossil Fuels

Year	Bituminous coal and lignite, \$/Mg (\$/US ton)	Natural gas at the wellhead, \$/1000 m^3 (\$/1000 std ft^3)	Crude oil, domestic first purchase price, \$/m^3 (\$/bbl)
1975	21.15 (19.23)	15.55 (0.44)	48.23 (7.67)
1980	26.97 (24.52)	56.18 (1.59)	135.79 (21.59)
1985	27.61 (25.10)	88.69 (2.51)	151.51 (24.09)
1990	23.88 (21.71)	60.42 (1.71)	125.91 (20.03)
1993	21.77 (19.79)	71.73 (2.03)	89.62 (14.25)

SOURCE: *Annual Energy Review 1994,* Energy Information Administration, July 1995. Prices are national averages in current-year U.S. dollars for the year cited.

gases are more desirable fuel forms, technologies to convert coal into synthetic liquid and gaseous fuels have been developed. Current research, development, and demonstration efforts are aimed toward technical and economic improvements in some of the old, or first-generation, technologies and toward seeking new ways to accomplish the same ends: inexpensive and "clean" coal-conversion processes. However, as long as the price of petroleum remains near current levels, coal gasification and liquefaction technologies will remain uneconomic.

South Africa has the only commercial plant producing liquid transportation fuels and other products from coal. This technology will be described later.

Bodle, Vyas, and Talwalker (*Clean Fuels from Coal Symposium II,* Institute of Gas Technology, Chicago, 1975) presented the chart in Fig. 27-5, which shows very simply the different routes from coal to clean gases and liquids.

Coal Gasification

GENERAL REFERENCES: *Fuel Gasification Symp.,* 152d American Chemical Society Mtg., Sept. 1966. *Chemistry of Coal Utilization,* suppl. vol., Lowry (ed.), Wiley, New York, 1963; and 2d suppl. vol., Elliot (ed.), 1981. *Coal Gasification Guidebook: Status, Applications, and Technologies,* Electric Power Research Institute, EPRI TR-102034, Palo Alto, Calif., 1993. Notestein, *Commercial Gasifier for IGCC Applications Study Report,* U.S. Dept. of Energy, DOE/METC-91/6118, Morgantown Energy Technology Center, Morgantown, W. Va., 1990.

Background   Converting coal to combustible gas has been practiced commercially since the early nineteenth century. The first gas-producing companies were chartered in 1812 in England and in 1816 in the United States to produce gas for illumination by the heating or pyrolysis of coal. This method of producing gas is still in use: the gas is a by-product of the carbonization of coal to manufacture coke for metallurgical purposes.

The advantages of a gaseous fuel as a source of heat and power increased the demand for gas and led to the invention of other methods of coal gasification. In the gas producer, introduced in the second half of the century, a downward-moving bed of coal or coke is reacted at atmospheric pressure with air and steam to create a fuel gas with a low heat value, in the range 3.4 to 6.0 MJ/m^3 (90 to 160 Btu/ft^3). Producer gas, as the product was named, soon enjoyed extensive industrial use, especially in steel manufacture. Gas producers, largely displaced by twentieth-century technology, are still employed in steel mills. A small version enhanced with modern handling and other technical improvements, the Wellman-Galusha producer, enjoys substantial use in small industry.

Driven by the same impetus, the development of the cyclic water-gas process in 1873 permitted the continuous production of gas of higher thermal content, about 13.0 MJ/m^3 (350 Btu/ft^3). Adding oil to the reactor increased the thermal content of the gas to 20.5 MJ/m^3 (550 But/ft^3). This type of fuel gas, carbureted water gas, was distributed in urban areas of the United States for residential and commercial uses until its displacement by lower-cost natural gas began in the 1940s. At approximately that time, development of oxygen-based gasification processes was initiated in the United States and in other countries. An early gasification process developed by Lurgi Kohle u Mineralöltechnik GmbH, which operated at elevated pressure, is still in use. Compositions of the coal gases produced by these methods,

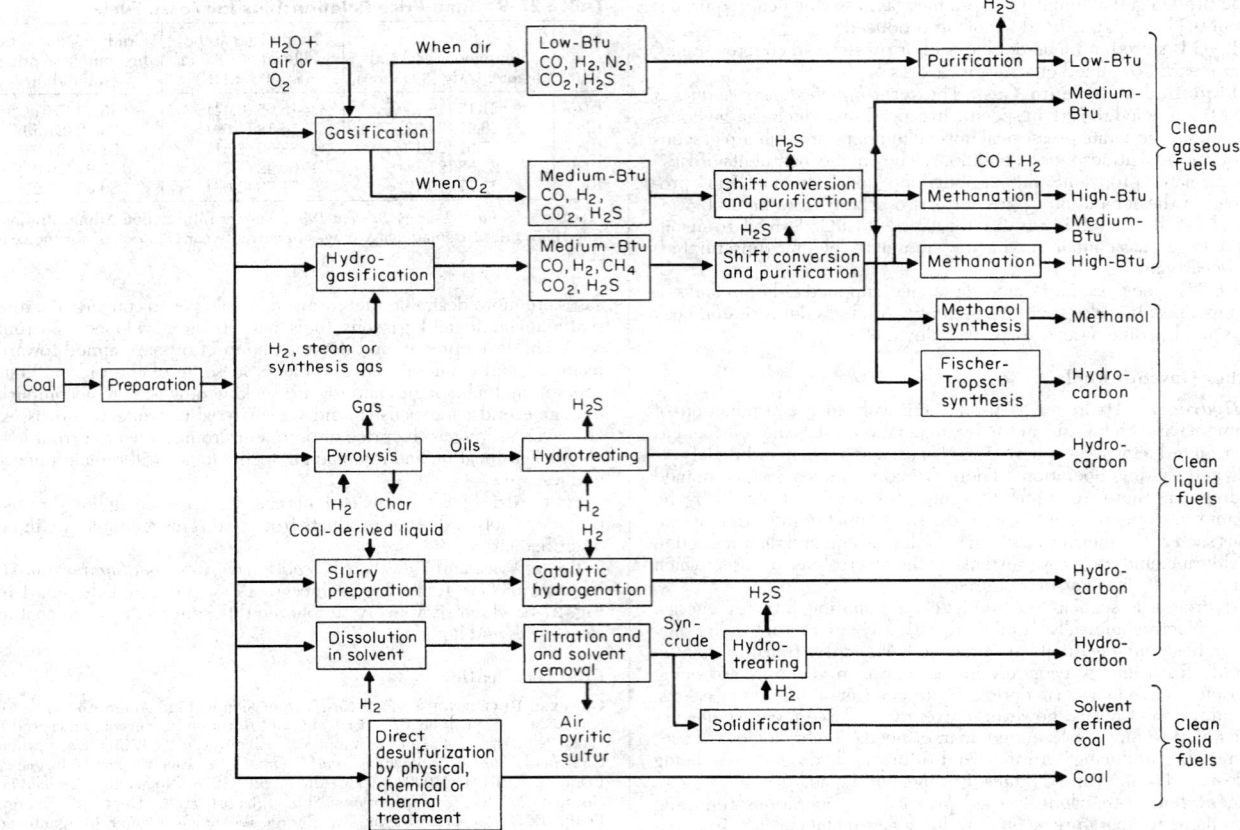

FIG. 27-5   The production of clean fuels from coal. (*Based on W. Bodle, K. Vyas, and A. Talwalker,* Clean Fuels from Coal Symposium II, *Institute of Gas Technology, Chicago, 1975.*)

which are referred to as the first-generation gasifiers, are listed in Table 27-10. There are a number of later-generation gasification technologies, as will be described in an ensuing subsection. The compositions of the gases produced by some of them are listed in Table 27-11.

Theoretical Considerations   The chemistry of coal gasification can be depicted by conveniently assuming coal as carbon and by listing the several well-known reactions involved; see Table 27-12. Reaction (27-14) is the combustion of carbon and oxygen, which is highly exothermic. This reaction supplies most of the thermal energy for the gasification process. The oxygen may be pure or contained in air. Reactions (27-16) and (27-17) are endothermic and represent the

conversion of carbon to combustible gases. These are driven by the heat energy supplied by reaction (27-14).

Hydrogen and carbon monoxide are produced by the gasification reaction, and they react with each other and with carbon. The reaction of hydrogen with carbon as shown in reaction (27-15) is exothermic and can contribute heat energy. Similarly, the methanation reaction (27-19) can contribute heat energy to the gasification. These equations are interrelated by the water-gas-shift reaction (27-18), the equilibrium of which controls the extent of reactions (27-16) and (27-17).

Several authors have shown (cf. Gumz, *Gas Producers and Blast Furnaces*, Wiley, New York, 1950; and Elliott and von Fredersdorff,

TABLE 27-10   Properties of Coal-Derived Gases

	Coke-oven gas	Producer gas	Water gas	Carbureted water gas	Synthetic coal gas
Reactant system	Pyrolysis	Air + steam	Steam (cyclic-air)	Steam + oil (cyclic-air)	Oxygen plus steam at pressure
Analysis, volume %[*]					
Carbon monoxide, CO	6.8	27.0	42.8	33.4	15.8
Hydrogen, H_2	47.3	14.0	49.9	34.6	40.6
Methane, CH_4	33.9	3.0	0.5	10.4	10.9
Carbon dioxide, CO_2	2.2	4.5	3.0	3.9	31.3
Nitrogen, N_2	6.0	50.9	3.3	7.9	
Other[†]	3.8	0.5	0.5	9.8	2.4
Fuel value, MJ/m^3	22.0	5.6	11.5	20.0	10.8
Btu/ft^3	(590)	(150)	(308)	(536)	290
Uses	Fuel, chemicals	Fuel	Fuel, chemicals	Fuel	Fuel, chemicals

[*]Analyses and fuel values vary with the type of coal and operating conditions.
[†]Other contents include hydrocarbon gases other than methane, hydrogen sulfide, and small amounts of other impurities.

TABLE 27-11 Coal-Derived Gas Compositions

Developer	Lurgi[°]	Texaco†[a]	BG/L†[b]	KRW[°]	KRW[°]	Shell†[c]
Type of bed	Moving	Entrained	Moving	Fluid	Fluid	Entrained
Coal feed form	Dry coal	Coal slurry	Dry coal	Dry coal	Dry coal	Dry coal
Coal type	Illinois #6	Illinois #6	Illinois #6	Illinois #6	Illinois #6	Illinois #5
Oxidant	Oxygen	Oxygen	Oxygen	Air	Oxygen	Oxygen
Pressure, MPa (psia)	0.101 (14.7)	4.22 (612)	2.82 (409)	2.82 (409)	2.82 (409)	2.46 (357)
Ash form	Slag	Slag	Slag	Agglomerate	Agglomerate	Slag
Composition, vol %						
H_2	52.2	30.3	26.4	15.7	27.7	26.7
CO	29.5	39.6	45.8	24.9	54.6	63.1
CO_2	5.6	10.8	2.9	5.3	4.7	1.5
CH_4	4.4	0.1	3.8	0.8	5.8	0.03
Other hydrocarbons	0.3	—	0.2	<0.01	<0.01	—
H_2S	0.9	1.0	1.0	—	1.3	1.3
COS	0.04	0.02	0.1	—	0.1	0.1
N_2 + Ar	1.5	1.6	3.3	47.0	1.7	5.2
H_2O	5.1	16.5	16.3	6.2	4.4	2.0
NH_3 + HCN	0.5	0.1	0.2	0.02	0.08	0.02
HCl	—	0.02	0.03	—	—	0.03
H_2S:COS	20:1	42:1	11:1	8:1	—	9:1

[°]Rath, "Status of Gasification Demonstration Plants," *Proc. 2d Annu. Fuel Cells Contract Review Mtg.,* DOE/METC-9090/6112, p. 91.
†*Coal Gasification Guidebook: Status, Applications, and Technologies,* Electric Power Research Institute, EPRI TR-102034, 1993. (*a*) page 5-28; (*b*) page 5-58; (*c*) page 5-48.

Chemistry of Coal Utilization, 2d suppl. vol., Lowry [ed.], Wiley, New York, 1963) that there are three fundamental reactions: the Boudouard reaction (27-17), the heterogeneous water-gas reaction (27-18), and the hydrogasification reaction (27-15). The equilibrium constants for these reactions are sufficient to calculate all the reactions listed.

It is not possible, however, to calculate accurately actual gas composition by using the relationships of reactions (27-14) to (27-19) in Table 27-12. Since the gasification of coal always takes place at elevated temperatures, thermal decomposition (pyrolysis) takes place as coal enters the gasification reactor. Reaction (27-15) treats coal as a compound of carbon and hydrogen and postulates its thermal disintegration to produce carbon (coke) and methane. Reaction (27-21) assumes the stoichiometry of hydrogasifying part of the carbon to produce methane and carbon.

It is possible to utilize these reactions and their relationships with each other for predicting the effects of changes in the operating parameters of gasification. At higher temperatures, endothermic reactions are favored at the expense of exothermic reactions. Methane production will decrease as reactions (27-15) and (27-19) proceed at a lower rate, CO production will be favored, and all reaction rates will increase in the direction in which heat absorption takes place. An increase in pressure will favor those reactions in which the number of moles of products is less than the number of moles of reactants. At higher pressures, production of CO_2 will be favored as well as that of methane. The knowledge of stoichiometry, equilibrium conditions,

and rates for these gasification reactions provides a sound basis for modeling and extrapolating gasification systems.

A great deal depends on the gasifier system, coal reactivity and particle size, and method of contacting coal with gaseous reactants (steam and air or oxygen). It is generally believed that oxygen reacts completely in a very short distance from the point at which it is mixed or comes in contact with coal or char. The heat evolved acts to pyrolyze the coal, and the char formed then reacts with carbon dioxide, steam, or other gases formed by combustion and pyrolysis. The assumption made in Table 27-12 that the solid reactant is carbon is probably close to being correct. The conversion of coal to char and the type of char formed affect the kinetics of gas-solid reactions. While the reaction rate does vary with temperature, as in all chemical reactions, the overall rate of reaction is controlled probably by the chemical reaction rate below 1273 K (1832°F). Above this, pore diffusion has an overriding effect, and at very high temperatures surface-film diffusion probably controls. Thus, for many gasification processes the reactivity of the char is quite important. This may depend not only on parent-coal characteristics but also on the method of heating, rate of heating, and particle-gas dynamics.

The importance of these concepts can be illustrated by the extent to which the pyrolysis reactions contribute to gas production. In a moving-bed gasifier (e.g., producer-gas gasifier), the particle is heated through several distinct thermal zones. At the initial heat-up zone, coal carbonization or devolatilization dominates. In the successively hotter zones, char devolatilization, char gasification, and fixed carbon

TABLE 27-12 Chemical Reactions in Coal Gasification

Reaction	Reaction heat, kJ/(kg·mol)	Process	Number
	Solid-gas reactions		
$C + O_2 \rightarrow CO_2$	+393,790	Combustion	(27-14)
$C + 2H_2 \rightarrow CH_4$	+74,900	Hydrogasification	(27-15)
$C + H_2O \rightarrow CO + H_2$	−175,440	Steam-carbon	(27-16)
$C + CO_2 \rightarrow 2CO$	−172,580	Boudouard	(27-17)
	Gas-phase reaction		
$CO + H_2O \rightarrow H_2 + CO_2$	+2,853	Water-gas shift	(27-18)
$CO + 3H_2 \rightarrow CH_4 + H_2O$	+250,340	Methanation	(27-19)
	Pyrolysis and hydropyrolysis		
CH_x	$\left(1 - \dfrac{X}{4}\right)C + \left(\dfrac{X}{4}\right)CH_4$	Pyrolysis	(27-20)
$CH_x + m\,H_2$	$\left[1 - \left(\dfrac{X + 2m}{4}\right)\right]C + \left(\dfrac{X + 2m}{4}\right)CH_4$	Hydropyrolysis	(27-21)

combustion are the dominant processes. About 17 percent of total gas production occurs during the coal devolatilization phase, and about 23 percent is produced during char devolatilization. The balance, typically about 60 percent of the total, is produced during the char gasification and combustion phases. This emphasizes the importance of coal quality or reactivity.

Gasifier Types and Characteristics   The fundamental chemistry and physics of gasification motivate the design of existing and advanced gasifiers. The equations listed in Table 27-12 show that an appropriate means of contacting a solid particle with a gaseous reactant is expedient, that the transfer of heat within the gasifier (and to the gasifier) is a critical parameter, and that variations in pressure and temperature alter the composition of the gas produced. In addition, the type of coal and the composition of both organic and inorganic constituents have a strong influence on gas composition and applicability of various gasification systems.

The three main types of reactors shown in Fig. 27-6 are in actual commercial use: the moving bed, the fluidized bed, and the entrained bed. The moving bed is often referred to as a *fixed* bed because the coal bed is kept at a constant height. These differ in size, coal feed, reactant and product flows, residence time, and reaction temperature.

Gasification-Based Power Systems   Today's single most important driving force for the coal gasification market development is gasification-based power generation, named the *integrated gasifica-*

tion combined-cycle (IGCC) power system (Fig. 27-7). The coal is crushed prior to gasification and partially burned, through the addition of steam and air or oxygen, to produce the high-temperature (1033 to 2255 K [1400 to 3600°F], depending on the type of gasifier) reducing environment necessary for gasification. The fuel gas passes through a heat recovery and cleanup section where particulate (dust) and sulfur are removed. After cleanup, the fuel gas, composed primarily of hydrogen and carbon oxides, is burned with compressed air and expanded through a gas turbine to generate electricity. Heat is recovered from the turbine's hot exhaust gas to produce steam (at subcritical conditions), which is expanded in a steam turbine for additional electricity generation.

The three basic types of gasifiers already mentioned have been incorporated into IGCC plant designs. Within each gasifier type, the oxidant can be air or oxygen, and the coal feed can be dry or in slurry form. Furthermore, the gas cleanup (i.e., particulate and sulfur removal) can be performed at high temperature (i.e., hot-gas cleanup) or lower temperature (cold-gas cleanup). The constituents of the fuel gas, as well as the pressure and temperature conditions at the gasifier exit, are determined by the type of gasifier employed. The efficiency and economics of an IGCC power system are highly dependent on the various parameters described above.

There are three unique features of the power system depicted in Fig. 27-7. These features assure that the IGCC-based power system is

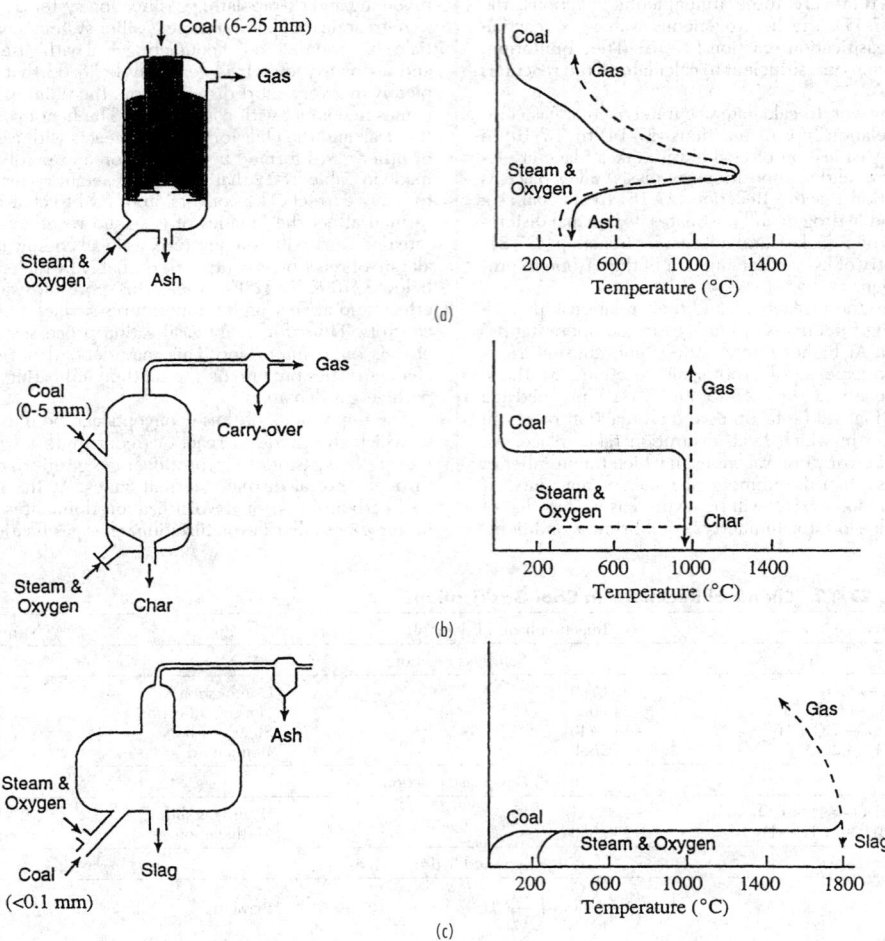

FIG. 27-6   Gasifier types and temperature profiles: (*a*) fixed bed (nonslagging); (*b*) fluidized bed; (*c*) entrained flow. (°C = K − 273)

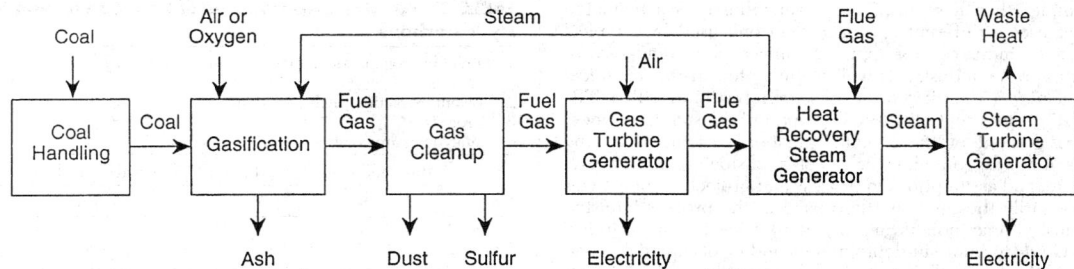

FIG. 27-7 Integrated gasification combined-cycle block diagram.

lower in cost and more thermally efficient than a conventional pulverized-coal combustion system and that it is environmentally superior to it. First, the gas mass flow rate from a gasifier is about one fourth that from a combustor because the former is oxygen-blown and operates substoichiometrically, and the combustor is air-blown and operates with excess air. Because of the elevated operating pressure, the volumetric gas flow to the hydrogen sulfide conversion unit is actually 60 times less than the volumetric flow to the flue gas desulfurization unit, and this lowers the capital cost of the gas cleanup system. Second, the sulfur in coal-derived fuel gas is present as hydrogen sulfide, which can more readily be separated from the fuel gas than sulfur dioxide can be separated from flue gas. Additionally, hydrogen sulfide can be easily converted to elemental sulfur, a more valuable by-product than the calcium sulfate produced when lime is used to remove sulfur dioxide from combustion flue gas. Third, the IGCC system generates electricity by both gas and steam turbines—corresponding to Brayton (gas) and Rankine (steam) cycles—compared to the conventional power plant's steam-only generating system.

Current Status The U.S. Department of Energy (DOE) and its private-sector collaborators have substantially advanced the development of gasification-based power systems during the last decade. This section describes the current status for the various types of gasifiers used.

Moving Bed Depending on the temperature at the base of the coal bed, the ash can either be dry or in the form of molten slag. If excess steam is added, the temperature can be kept below the ash fusion point. In that case the coal bed rests on a rotating grate which allows the dry ash to fall through for removal. To reduce the steam usage, a slagging bottom gasifier has been developed in Scotland by British Gas and Lurgi (BG/L) in which the ash is allowed to melt and drain off through a slag tap. This gasifier has over twice the capacity per unit area of the dry-bottom gasifier. The gas composition from a BG/L gasifier is listed in Table 27-11.

The capacity of the demonstration gasifier in Scotland depends on coal quality; the gasifier can handle a coal input of 750 GJ/h, equivalent to more than 25,000 kg/h (55,115 lb/h) of coal or 70,000 m³/h (2,500,000 ft³/h) of gas production. The Clean Energy Demonstration Project, cost-shared by the DOE under the Clean Coal Technology Program, will test the scale-up of the BG/L oxygen-blown, slagging gasifier from demonstration to commercial size. Multiple BG/L units will provide coal gas for a 477-MW IGCC plant to be located on an eastern U.S. site.

Fluidized Bed The possibility of coal and ash agglomeration is eliminated by a fluidized-bed gasifier developed by Westinghouse and M. W. Kellogg. Char and ash that exit the gasifier with the product gas are recycled to the hot agglomerating and jetting zone where temperatures are high enough to pyrolyze fresh coal that is introduced there, gasify the char, and soften the ash particles. The ash particles stick together and fall to the base of the gasifier, where they are cooled and removed. Agglomerating gasifiers achieve better carbon conversion than conventional fluidized-bed gasifiers. The compositions of two gas streams based on air-blown and oxygen-blown gasifiers are listed in Table 27-11.

The agglomerating fluid-bed gasifier can be blown by either air or oxygen at 2.8 MPa (406 psia) pressure. Pressurized operation has several advantages: slightly higher methane formation, resulting in higher heating value of the gas; increased heat from the methanation reactions, which reduces the amount of oxygen needed; reduced heat losses through the wall and, consequently, improved efficiency; and higher generating capacity. Limestone can be added to remove a major portion of the sulfur contained in the coal.

The advanced pressurized fluid-bed gasifiers currently ready for demonstration have internal diameters of around 3.6 m (12 ft) and are 20 to 28 m (66 to 92 ft) in height. Production of more than 100,000 m³/h of gas (at gasifier temperature and pressure) is achieved by the advanced gasifiers, equivalent to 33,260 kg/h of bituminous coal (air blown).

The KRW (Kellogg-Rust-Westinghouse) gasification process uses an agglomerating-ash fluidized-bed gasifier in which crushed limestone can be injected with the coal for sulfur capture. The Piñon Pine 100-MW IGCC plant built near Reno, Nevada, uses the KRW gasifier. This IGCC plant, scheduled to begin commercial operation in 1997, is being built with U.S. DOE sponsorship under the Clean Coal Technology Program.

Another fluidized-bed gasification technology is the Tampella U-Gas gasifier. This technology was developed by the Institute of Gas Technology (IGT) in Chicago and licensed by the Finnish boiler manufacturer, Tampella. Like the KRW gasifier, it is an air-blown, agglomerating-ash gasifier that uses limestone for in-bed sulfur capture. The two primary differences between the KRW and Tampella gasifiers are the way the gas velocity and temperature in the agglomerating zone are controlled and how the ash and spent limestone discharge is controlled.

Entrained Bed The two primary examples of oxygen-blown, dry-feed, entrained-flow gasifiers are Shell and PRENFLO. These two gasifiers share common roots and are very similar. The gas composition from a Shell gasifier is listed in Table 27-11. An advantage of Shell coal gasification technology is the lack of feed coal limitations. A wide variety of coals (from anthracite to brown coal) have been successfully tested. As with other entrained-flow gasifiers, disadvantages of the Shell process include a high oxygen requirement and a high waste heat recovery duty. However, the ability to feed dry coal reduces the oxygen requirement below that of single-stage entrained-flow gasifiers that use slurry feed and makes the Shell gasifier somewhat more efficient. The penalty for this small efficiency improvement is a more complex coal-feeding system. Like the Shell, the PRENFLO gasification process uses pressurized, dry-feed, entrained-flow technology with water-cooled gasifier vessels. The primary difference between the two processes is in the design of the syngas cooler. While the Shell process uses cooled recycle gas to partially cool the hot syngas before heat recovery, the PRENFLO process uses a radiant water-wall boiler with fins (to increase the surface area for heat transfer) that is connected directly to the gasifier. The PRENFLO gasifier was selected for a 300-MW IGCC project in Puertollano, Spain, to begin operation in 1996. The project is funded by power companies from several European countries and by the European Community.

Cost of Gasification-Based Power Systems In the U.S. power industry the capital cost is usually reported in dollars per kilowatt ($/kW) and the cost of electricity (COE) in mills per kilowatt-hour (a mill is one thousandth of a dollar). Estimation of capital cost and COE is more complex than it seems. The power industry is a regulated industry that uses a different approach than one used by a typical chemical process industry. This approach, an accepted standard procedure in the power industry, is well documented in the *Technical Assessment Guide* (Electric Power Research Institute, EPRI TR-102276-V1 R7, Palo Alto, Calif., 1993). As part of the standard procedure, the assignment of the capital formation structure (i.e., the percentages of bonds, common stock, preferred stock), inflation rate, escalation rate, and assumptions in process and project contingencies are not necessarily the same as those used in the process industry. Process industry practitioners are urged to consult the *Technical Assessment Guide* for more in-depth understanding of capital cost and cost of electricity, as well as meaningful comparisons of different power systems. Comparison of power systems is further complicated by the many different parameters assumed in the cost determination. Cost of coal; sulfur content, moisture content, and other properties of coal; ambient temperature; level of integration between various component units; model of gas turbine; gas cleanup method; year of estimation—all these and other factors may have significant impacts on cost.

While comparison of the absolute capital costs and costs of electricity among different power systems is difficult and uncertain, the structure of these costs is rather typical, and the costs of component units are usually within known ranges. For an oxygen-blown IGCC power system, the breakdown of the capital cost for the four component units is: air separation plant (11 to 17 percent), fuel gas plant (33 to 42 percent), combined-cycle unit (32 to 39 percent), and balance of plant (2 to 21 percent). The breakdown of the cost of electricity is: capital charge (52 to 56 percent), operating and maintenance (14 to 17 percent), and fuel (28 to 32 percent).

The capital investment required for gasification-based power systems is 1400 to 1600 $/kW (1994 US dollars) and is projected to become less than 1200 $/kW in the year 2000 because of the higher efficiency associated with gas-turbine combined-cycles currently being designed by turbine vendors.

Development is under way for a gasification-based power system using a cascaded humidified advanced turbine. The power system is the same as the IGCC system in the front end (i.e., the gasification section without heat recovery and the gas-cleanup section) but differs from IGCC in the back end (i.e., the power generation section). Instead of combined steam and gas cycles as used in the IGCC system, the humidified turbine system uses only the gas turbine, eliminating the steam cycle. Because of this simplification and the corresponding elimination of the expensive coal-gas cooler, the power system based on the advanced humidified turbine offers further reduction in the capital investment requirement. Depending on how the advanced turbine system is configured with the front end section, the capital investment requirement is expected to be on the order of 1000 $/kW.

Direct Coal Liquefaction

GENERAL REFERENCES: *Chemistry of Coal Utilization*, suppl. vol., Lowry (ed.), Wiley, New York, 1963, and 2d suppl. vol., Elliott (ed.), 1981. Wu and Storch, *Hydrogenation of Coal and Tar*, U.S. Bur. Mines Bull. 633, 1968. Srivastava, McIlvried, Gray, Tomlinson, and Klunder, *American Chemical Society Fuel Chemistry Division Preprints*, Chicago, 1995.

Background   The primary objective of any coal-liquefaction process is to increase the hydrogen-to-carbon ratio and remove sulfur, nitrogen, oxygen, and ash. Table 27-13 shows the hydrogen-to-carbon ratios in proceeding from coal to crude petroleum to gasoline, together with the four techniques for accomplishing coal liquefaction. Direct coal liquefaction refers to any process in which coal and hydrogen are directly reacted at high pressure and temperature. A hydrogen donor solvent and/or catalyst may also be present. The first two techniques in Table 27-13 (direct hydrogenation and solvent extraction) follow this approach. A pyrolysis process is included in Table 27-13 for comparison. The fourth technique, catalytic hydrogenation of carbon monoxide, or indirect liquefaction (discussed later in this subsection), first converts coal to synthesis gas, which is purified and reacted over a catalyst to form liquid products. One version of the

TABLE 27-13   Basic Approaches of Coal Conversion to Liquid Hydrocarbons

1. Direct hydrogenation at elevated temperature and pressure, with or without catalysts
2. Solvent extraction (hydrogen donor)
3. Pyrolysis
4. Catalytic hydrogenation of carbon monoxide

	Bituminous coal	Lignite	Crude petroleum	Gasoline
H/C	0.8	0.7	1.8	1.9

indirect-liquefaction route converts purified synthesis gas to methanol, for direct fuel use or for conversion to gasoline.

The technology for coal liquefaction to synthetic fuels is not new. In 1913 Dr. Friedrich Bergius discovered the technique of adding hydrogen to coal at a pressure of 20.3 MPa (2940 psia) and a temperature of about 723 K (842°F). Under these conditions most oxygen was hydrogenated to water, some nitrogen to ammonia, and most sulfur to hydrogen sulfide. Hydrogen was also chemically combined with the coal to produce a liquid similar to petroleum. Production of synthetic liquid fuels and chemicals from coal in Germany increased in the mid-1920s, and by 1939 gasoline was being produced by coal hydrogenation in Germany at 910 Gg/a (1×10^6 US ton/a) and in England at 136 Gg/a (150,000 US ton/a). In the early 1950s, the U.S. Bureau of Mines constructed and operated a demonstration plant at Louisiana, Missouri, using the technology of catalytic hydrogenation of coal to produce liquid fuels.

The oil price shocks of the 1970s accelerated the U.S. development of direct coal liquefaction processes to the pilot plant stage (up to 227 Mg/d [250 US ton/d] coal feed rate) of several competing designs, such as the solvent-refined coal (SRC), Exxon donor solvent (EDS), and H-Coal processes. These are discussed briefly in a following subsection. Both the SRC and H-Coal process designs include equipment for generating products ranging from heavy boiler fuels to higher-grade transportation fuels.

In the United States, it became clear after the return of low oil prices in the 1980s that the existing processes would not be able to compete economically, and research proceeded on a smaller scale to improve the efficiency of coal liquefaction for the production of high-value distillate fuels. Key aspects of the process improvement are more effective hydrogen utilization through catalysis and better solids rejection technology. A two-stage concept has been developed that tailors reaction conditions in the first stage to coal solubilization with some cracking, and in the second stage to production of additional liquids with product upgrading. Both supported Ni/Mo catalysts and slurry catalysts are being tested to maximize yields of distillate products that are completely compatible with the existing refinery infrastructure.

A nearer-term variant of direct coal liquefaction is the coprocessing of mixtures of coal and heavy petroleum residua. The coal solids are thought to preferentially "getter" the nickel and vanadium contaminants, typically present in low-value residua, that rapidly poison supported catalysts during conventional resid upgrading. The petroleum component acts as a slurry vehicle for pumping the coal into the reactors, thereby avoiding process recycle requirements. A further advantage is the lowered hydrogen uptake per unit of product. The coprocessing concept can be extended to mixtures of coal and municipal waste such as mixed plastics. Landfill disposal costs are avoided, while, at the same time, carbon values from the waste contribute to the transportation fuel supply.

Direct-Liquefaction Kinetics   All direct-liquefaction processes consist of three basic steps: (1) coal slurrying in a vehicle solvent, (2) coal dissolution under high pressure and temperature, and (3) transfer of hydrogen to the dissolved coal. However, the specific reaction pathways and associated kinetics are not known in detail. Overall reaction schemes and semiempirical relationships have been generated by the individual process developers, but applications are process specific and limited to the range of the specific data bases. More extensive research into liquefaction kinetics has been conducted on the laboratory scale, and these results are discussed below.

Depending on its rank, coal can be dissolved in as little as one minute in the temperature range of 623 to 723 K (662 to 842°F) in suitable solvents, which are assumed to promote thermal cracking of the coal into smaller, more readily dissolved fragments. These fragments may be stabilized through reactions with one another or with hydrogen supplied either by a donor solvent or from a gas phase.

Data on Illinois No. 6 and Kentucky No. 9 coals were used by Wen and Han (*Prepr. Pap.—Am. Chem. Soc., Div. Fuel Chem.* **20**(1): 216–233, 1975) to obtain a rate equation for coal dissolution under hydrogen pressure. These data included a temperature range of 648 to 773 K (705 to 930°F) and pressures up to 13.8 MPa (2000 psia). An empirical rate expression was proposed as

$$r_A = k_0 \left[\exp\left(\frac{-E}{RT}\right) \cdot \exp\left(0.0992\, P_{H_2}\right) \right] (C_{so}) (1 - x) \left(\frac{C}{S}\right) \quad (27\text{-}22)$$

where r_A = rate of dissolution, g/(h·cm³ reactor volume)
C_{so} = weight fraction of organics in original coal
k_0 = rate constant, g/(h·cm³) reactor volume
P_{H_2} = hydrogen partial pressure, MPa
x = conversion, solid organics/solid organics in original coal
C/S = coal-solvent weight ratio
E = energy of activation, kcal/(g·mol)
R = universal gas constant, 1.987×10^{-3} kcal/(g mol·K)
T = temperature, K.

Calculated and measured conversions agreed when the Arrhenius temperature dependency indicated in Eq. (27-22) was used with the following values for the parameters:

Constant	Illinois No. 6	Kentucky No. 9
k_0, g/(h·cm³)	2125	15.3
E, kcal/g·mol)	11	4.5

The low activation energies suggested that the dissolution rate is controlled by counterdiffusion of organic components from the coal surface and dissolved hydrogen from the solvent. Also, the rate of dissolution appeared to depend exponentially on hydrogen partial pressure.

A free-radical mechanism has been proposed for coal dissolution in hydrogen donor solvents. Solvents and high temperatures facilitate degradation of coal to form relatively low-molecular-weight free radicals, which may be stabilized by hydrogen transfer from a hydroaromatic solvent. Initial dissolution is considered to be a thermal process, with a net rate dependent upon the type of solvent and its effectiveness in stabilizing free radicals. The greater a solvent's hydrogen donor capability, the more effective it is in terminating radicals. In continuous recycle process configurations, the overall rate-limiting step appears to be rehydrogenation of the donor solvent, which is a function of the dissolved hydrogen and catalyst. Nitrogen and sulfur removal and the formation of light liquid products are considered to result primarily from catalytic effects.

The conversion reaction from coal to oil has been modeled as a series of steps:

Coal + solvent → preasphaltene → asphaltene → oil

with some gas formation accompanying each step. In a study using Illinois No. 6 coal and 505 to 727 K (450 to 850°F) boiling-range process-derived heavy distillate, data were obtained at 13.8 MPa (2000 psia) and 673 to 748 K (750 to 885°F). Activation energies for the steps of the reaction series were determined to be:

Reaction step	Activation energy, kcal/g·mol
Preasphaltene → asphaltene	15
Asphaltene → oil	21
Coal → preasphaltene	32

At 723K (843°F), stoichiometries for the reaction steps were represented as:

Coal + 3 solvent → preasphaltene

Preasphaltene → 2 asphaltene

Asphaltene → 3 oil

A more complex reaction model was proposed from the results of a kinetic study of thermal liquefaction of subbituminous coal. Data were obtained over a temperature range of 673 to 743 K (752 to 878°F) at 13.8 MPa (2000 psia) by using two solvents, hydrogenated anthracene oil (HAO), and hydrogenated phenanthrene oil (HPO), at a coal-solvent ratio of 1:15. Results were correlated with the following model:

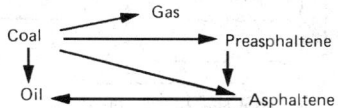

Activation energies and frequency factors for the various steps of this model were determined as follows:

Reaction	Rate constant	Activation energy, kcal/(g·mol) Coal-HAO	Coal-HPO	Frequency factor, min⁻¹ Coal-HAO	Coal-HPO
Coal → oil	k_o	14.1	28.9	3.11×10^3	2.1×10^8
Coal → preasphaltene	k_p	13.8	4.3	2.81×10^3	4.94
Coal → asphaltene	k_a	15.6	8.6	1.12×10^4	9.63×10^1
Coal → gas	k_g	21.5	10.5	8.72×10^5	3.85×10^2
Preasphaltene → asphaltene	k_{pa}	12.8	33.9	9.66×10^2	2.48×10^9
Asphaltene → oil	k_{ao}	16.0	25.6	1.42×10^3	1.53×10^7

Magnitudes of k_g, k_p, k_a, and k_o indicate the importance of direct reactions with coal, where k_{pa} and k_{ao} are for hydrocracking reactions in the conversion process. Data for k_o and k_{ao} from the experiments with HPO indicate that oil production from coal is increased by the use of a good hydrogen donor solvent.

Direct-Liquefaction Processes Figure 27-8 presents a simplified process flow diagram of a typical direct coal liquefaction plant. Specific processes are described in the following paragraphs.

Solvent-Refined Coal (SRC) This processing concept was initiated by the Pittsburgh & Midway Coal Mining Co. in the early 1960s. The *SRC-I process* operating mode is designed to produce a solid fuel for utility applications. Typical operating conditions and product yields for SRC-I are shown in Table 27-14.

The *SRC-II process* is an improved version of the SRC process that recycles a portion of the reactor effluent slurry in place of the distillate solvent of the SRC-I process. The primary product is a distillate fuel with a 490 to 728 K (423 to 851°F) boiling range. This is achieved in part by increased severity of operating conditions, but also by establishing a higher concentration of resid reactant and catalytic mineral matter in the reactor through slurry recycle. The net reactor effluent slurry is passed to a vacuum-flash unit for separation of the distillate product from the mineral matter and undissolved coal, thereby avoiding the filtration step. Typical operating conditions and product yields for SRC-II are shown in Table 27-14.

Exxon Donor Solvent (EDS) Process The EDS process, developed by the Exxon Research and Engineering Co., liquefies coal by use of a hydrogen donor solvent under hydrogen pressure in an upflow, plug-flow reactor. The solvent is a catalytically hydrogenated recycle stream, fractionated from the middle boiling range, 474 to 728 K (395 to 850°F), of the liquid product. Hydrogenation of the recycle solvent is conducted in a conventional fixed-bed catalytic reactor using hydrotreating catalysts, such as cobalt molybdate or nickel molybdate. Coal conversion and liquid yield strongly depend on the molecular composition, boiling-point range, and other properties of the solvent. Exxon uses its proprietary Solvent Quality Index (SQI) as the main criterion of solvent quality and correlates product yields with SQI. Typical operating conditions and product yields are shown in Table 27-14.

Vacuum distillation is used to remove the residue from the distillate product. Additional heavy oil may be recovered from the vacuum bottoms by employing Exxon's Flexicoking process.

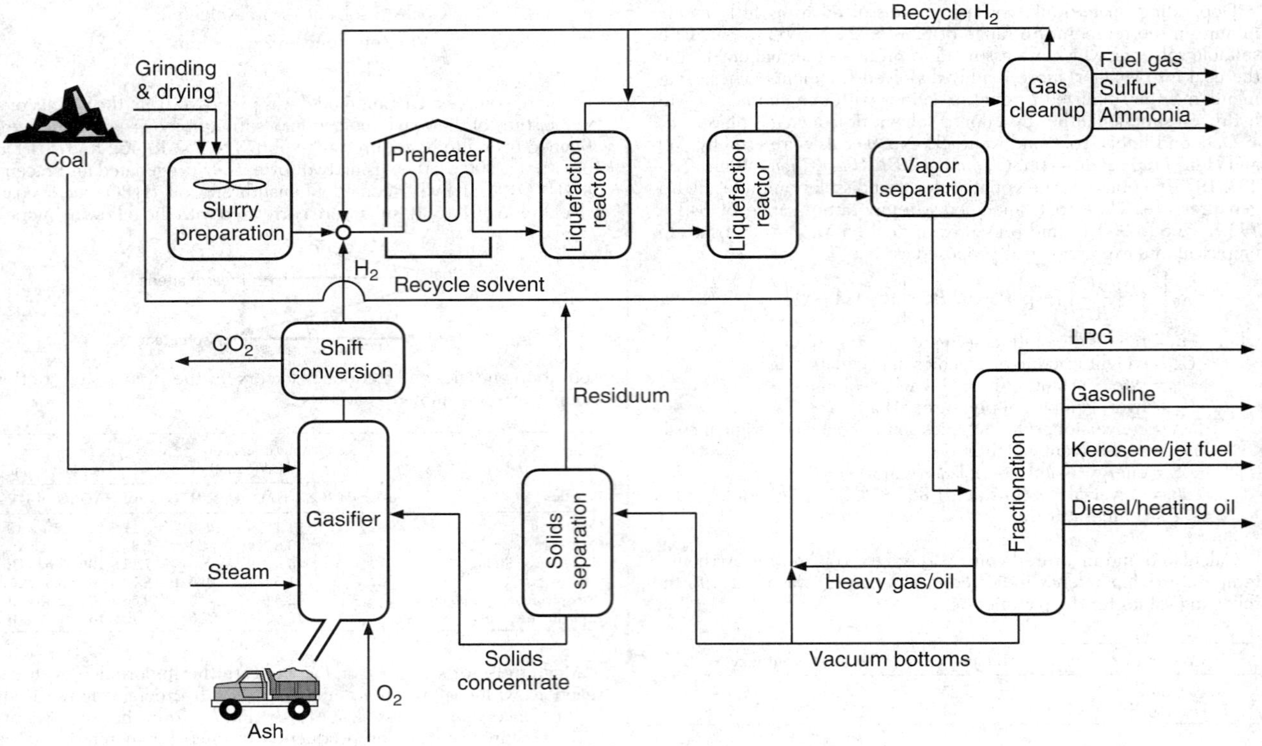

FIG. 27-8   Direct liquefaction of coal.

H-Coal Process   The H-Coal process, based on H-Oil technology, was developed by Hydrocarbon Research, Inc. (HRI). Depending on the type of products desired, the process can be operated in either a fuel-oil mode or a syncrude mode by adjusting operating severity. The heart of the process is a three-phase fluidized reactor (ebullated-bed) in which catalyst pellets are fluidized by the upward flow of slurry and gas through the reactor. Catalyst activity in the reactor is maintained by the withdrawal of small quantities of spent catalyst and the addition of fresh catalyst. The reactor contains an internal tube for recirculating the reaction mixture through the catalyst bed.

Solids separation is accomplished by vacuum distillation in the syn-crude mode. Table 27-14 shows the product yields obtained in PDU tests with an Illinois No. 6 bituminous coal.

Two-Stage Liquefaction   This is an advanced process that provides improved, lower-cost technology by the more efficient and specific application of catalysis in the reaction stages. Higher yields of better-quality products are obtained by recovering and recycling heavy intermediates to extinction. Unconverted coal and mineral matter are removed by conventional technology such as supercritical solvent extraction or vacuum distillation. Several variations of this approach have been tested at the Advanced Coal Liquefaction R&D Facility (5.4 Mg/d or 6 US ton/d scale) in Wilsonville, Alabama. Distillate yields as high as 78 percent (moisture and ash-free basis) have been achieved. Wilsonville operating and product data are shown in Table 27-14. An economic evaluation is given in a later section.

Further development of staged-liquefaction technology options is being conducted by Hydrocarbon Technologies, Inc. (HTI, formerly HRI) in their 2.7-Mg/d (3-US ton/d) proof-of-concept unit (Lawrence-ville, New Jersey), and more advanced concepts are being researched by HTI and others at the bench scale. Recent HTI results are presented in Table 27-14. Coal-derived product quality has been improved dramatically (less than 50 ppm nitrogen content, for example) through use of in-line fixed-bed hydrotreating of the product stream. Slurry catalysts are being employed in addition to the more conventional supported catalysts as a means of simplifying reactor designs and removing process constraints.

Coal-Oil Coprocessing   In this approach, coal is slurried in petroleum resid rather than recycle solvent, and both coal and petroleum components are converted to high-quality fuels in the reaction stages. This variation offers the potential for significant cost reduction by eliminating or reducing internal recycle oil streams. More important, fresh hydrogen requirements are reduced because the petroleum resid feedstock component has a higher initial hydrogen content. As a result, plant capital investment is reduced substantially. It also offers the opportunity for accelerating the introduction of coal-derived liquid fuels into the marketplace by utilizing, to a much greater degree, existing petroleum refining facilities and technology.

Other carbonaceous materials such as municipal waste plastics, cellulosics, and used motor oils may also serve as cofeedstocks with coal in this technology.

Coal Pyrolysis   Coal pyrolysis produces synthetic crude oil, gas, and char. In the COED process, crushed coal is dried and heated to successively higher temperatures in a series of fluidized-bed reactors. In each stage, a portion of the coal's volatile matter is released. Typically, four stages at 589, 727, 811, and 1089 K (600, 850, 1000, and 1500°F), respectively, are used, but operations vary owing to the need to stay below the coal agglomeration temperature. Process heat is generated by burning char in the last stage and circulating hot char and gases to the other stages. Volatile products are condensed in a recovery system and the pyrolysis oil is filtered to remove fines.

Typical pyrolysis yields and oil qualities for two bituminous coals, Utah A and Illinois No. 6, are presented in Table 27-15. The major problem with any pyrolysis process is the high yield of char.

Flash Pyrolysis   Coal is rapidly heated to elevated temperatures for a brief period of time to produce oil, gas, and char. The increase in hydrogen content in the gases and liquids is the result of removing carbon from the process as a char containing a significantly reduced amount of hydrogen. Several processes have been tested on a rela-

TABLE 27-14   Direct Liquefaction Process Conditions and Product Yields

Developer	Gulf[b]	Gulf	Exxon	HRI	SCS,[b] EPRI, Amoco	HTI
Process	SRC-I	SRC-II	EDS	H-Coal	Two-stage	Two-stage
Coal type	Kentucky 9 & 14	Illinois No. 6	Illinois No. 6	Illinois No. 6	Illinois No. 6	Illinois No. 6
Operating conditions						
Nominal reactor residence time, h	0.5	0.97	0.67			
Coal space velocity per stage, kg/(h·m³) (lb/[h·ft³])				530 (33.1)	825[c] (51.7)	310 (19.4)
1st stage						
Temperature, K (°F)	724 (842)	730 (855)	722 (840)	726 (847)	695 (791)	680 (765)
Total pressure, MPa (psia)	10.3 (1500)	13.4 (1950)	10.3 (1500)		18.3 (2660)	19.2 (2790)
H_2 partial pressure, MPa (psia)	9.7 (1410)	12.6 (1830)		12.6 (1827)		
Catalyst type	Coal minerals	Coal minerals	Coal minerals	Supported catalyst (Co/Mo)	AKZO-AO-60 (Ni/Mo)	AKZO-AO-60 (Ni/Mo)
Catalyst replacement rate, kg/kg (lb/US ton) mf coal					7.5×10^{-4} (1.5)	8×10^{-4} (1.6)
2d Stage						
Temperature, K (°F)					705 (809)	705 (810)
Total pressure, MPa (psia)					17.0 (2470)	18.6 (2700)
H_2 partial pressure, MPa (psia)						
Catalyst type					AKZO-AO-60 (Ni/Mo)	AKZO-AO-60 (Ni/Mo)
Catalyst replacement rate kg/kg (lb/US ton) mf coal					7.5×10^{-4} (1.5)	1.5×10^{-3} (3.0)
Product yields, wt % maf coal						
H_2	−2.4	−4.7	−4.3	−5.9	−6.0	−7.2
H_2O	—	—	12.2[d]	8.3	9.7	9.8
H_2S, CO_x, NH_3	—	—	4.2[e]	5.0	5.2	5.2
C_1-C_3	3.7[f]	15.8[f]	7.3	11.3	6.5	5.6
C_4^+ distillate	13.5[g]	47.3[g]	38.8	53.1	65.6	73.3
Bottoms[h]	68.4	28.0	41.8	28.2	19.0	13.3
Unreacted coal[i]	5.4	5.0	—	6.4	7.0	5.0
Distillate end point, K (°F)	727 (850)	727 (850)	911 (1180)	797 (975)	797 (975)	524 (975)

[a] In partnership with Pittsburg & Midway Coal Mining Co.
[b] Southern Company Services, Inc., prime contractor for Wilsonville Facility.
[c] Coal space velocity is based on settled catalyst volume.
[d] CO_x is included.
[e] CO_x is excluded.
[f] C_4 is included.
[g] C_4 is excluded.
[h] Unreacted coal is included.
[i] "Unreacted coal" is actually insoluble organic matter remaining after reaction.

TABLE 27-15   Pyrolysis Data

	Illinois No. 6 seam	Utah A seam
Net yields, wt % dry coal		
Char	59.5	54.5
Oil	19.3	21.5
Gas	15.1	18.3
Liquor	6.1	5.7
Net process yields		
Char, kg/kg (lb/US ton)	0.595 (1190)	0.545 (1090)
Oil, m³/kg (bbl/US ton)	1.92×10^{-4} (1.10)	2.15×10^{-4} (1.23)
Gas, m³/kg (std ft³/US ton)	0.274 (8810)	0.266 (8545)
Liquor, dm³/kg (gal/US ton)	0.061 (14.6)	0.057 (13.7)
Oil properties		
Elemental Analysis, wt %. dry		
Carbon	79.6	83.8
Hydrogen	7.1	9.5
Nitrogen	1.1	0.9
Sulfur	2.8	0.4
Oxygen	8.5	5.0
Ash	0.9	0.3
Relative density, 288 K (°API, 60°F)	1.110 (−4)	1.105 (−3.5)
Moisture, wt %	0.8	0.5
Pour point, K (°F)	311 (100)	311 (100)
Kinematic viscosity, mm²/s, 372 K (SUS 210°F)	300 (1333)	87.5 (390)
Solids, wt %, dry basis	4.0	3.8
Higher heating value, MJ/kg (Btu/lb)	35.0 (15,050)	37.4 (16,100)

tively small scale. None have demonstrated economic potential, although the technical concepts appear to be valid.

Indirect Coal Liquefaction

GENERAL REFERENCES: Dry, *The Fischer-Tropsch Synthesis, Catalysis Science and Technology*, vol. 1, Springer-Verlag, New York, 1981. Anderson, *The Fischer-Tropsch Synthesis*, Academic Press, New York, 1984. Sheldon, *Chemicals from Synthesis Gas*, D. Reidel Publishing Co., Dordrecht, Netherlands, 1983. Rao, Stiegel, Cinquegrane, and Srivastava, "Iron-based Catalyst for Slurry-phase Fischer-Tropsch Process: Technology Review," *Fuel Processing Technology* 30: 83–151, 1992.

Background   Indirect coal liquefaction differs fundamentally from direct coal liquefaction in that the coal is first converted to a synthesis gas (a mixture of H_2 and CO) which is then converted over a catalyst to the final product. Figure 27-9 presents a simplified process flow diagram for a typical indirect coal liquefaction process. The synthesis gas is produced in a gasifier (see a description of coal gasifiers earlier in this section), where the coal is partially combusted at high temperature and moderate pressure with a mixture of oxygen and steam. In addition to H_2 and CO, the raw synthesis gas contains other constituents (such as CO_2, H_2S, NH_3, N_2, and CH_4), as well as particulates.

Before being fed to the synthesis reactor, the synthesis gas must first be cooled and then passed through particulate removal equipment. Following this, depending on the catalyst being used, it may be necessary to adjust the H_2/CO ratio. Modern high-efficiency gasifiers typically produce a ratio between 0.45 and 0.7, which is lower than stoichiometric for the synthesis reaction. Some catalysts, particularly iron catalysts, possess water gas shift conversion activity and permit

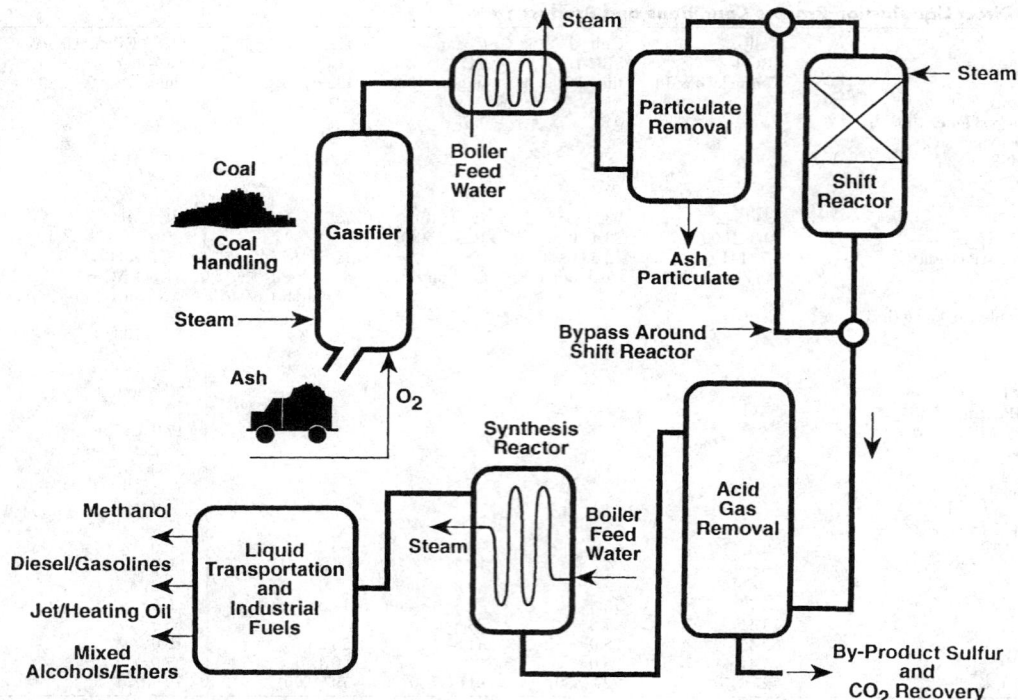

FIG. 27-9   Indirect liquefaction of coal.

operation with a low H_2/CO ratio. Other catalysts possess little shift activity, however, and require a ratio adjustment before the synthesis reactor.

After shift conversion, acid gases (CO_2 and H_2S) are scrubbed from the synthesis gas. A guard chamber is sometimes used to remove the last traces of H_2S. The cleaned gas is sent to the synthesis reactor, where it is converted at moderate temperature and pressure, typically 498 to 613 K (435 to 645°F) and 1.52 to 6.08 MPa (220 to 880 psia). Products from the process depend on operating conditions and the catalyst employed, as well as reactor design. Typical products include hydrocarbons (mainly straight chain paraffins from methane through n-C_{50} and higher), oxygenates (methanol, higher alcohols, ethers), and other chemicals (olefins).

Fischer-Tropsch Synthesis   The best-known technology for producing hydrocarbons from synthesis gas is the Fischer-Tropsch synthesis. This technology was first demonstrated in Germany in 1902 by Sabatier and Senderens when they hydrogenated carbon monoxide (CO) to methane, using a nickel catalyst. In 1926 Fischer and Tropsch were awarded a patent for the discovery of a catalytic technique to convert synthesis gas to liquid hydrocarbons similar to petroleum.

The basic reactions in the Fischer-Tropsch synthesis are:
Paraffins:

$$(2n + 1)H_2 + nCO \rightarrow C_nH_{2n+2} + nH_2O \qquad (27\text{-}23)$$

Olefins:

$$2nH_2 + nCO \rightarrow C_nH_{2n} + nH_2O \qquad (27\text{-}24)$$

Alcohols:

$$2nH_2 + nCO \rightarrow C_nH_{2n+1}OH + (n-1)H_2O \qquad (27\text{-}25)$$

Other reactions may also occur during the Fischer-Tropsch synthesis, depending on the catalyst employed and the conditions used:
Water-gas shift:

$$CO + H_2O \rightleftharpoons CO_2 + H_2 \qquad (27\text{-}26)$$

Boudouard disproportionation:

$$2CO \rightarrow C(s) + CO_2 \qquad (27\text{-}27)$$

Surface carbonaceous deposition:

$$\left(\frac{2x + y}{2}\right)H_2 + xCO \rightarrow C_xH_y + xH_2O \qquad (27\text{-}28)$$

Catalyst oxidation-reduction:

$$yH_2O + xM \rightarrow M_xO_y + yH_2 \qquad (27\text{-}29)$$

$$yCO_2 + xM \rightarrow M_xO_y + yCO \qquad (27\text{-}30)$$

Bulk carbide formation:

$$yC + xM \rightarrow M_xC_y \qquad (27\text{-}31)$$

where M represents a catalytic metal atom.

The production of hydrocarbons using traditional Fischer-Tropsch catalysts is governed by chain growth or polymerization kinetics. The equation describing the production of hydrocarbons, commonly referred to as the Anderson-Schulz-Flory equation, is:

$$\log\left(\frac{W_n}{n}\right) = n \log \alpha + \log\left[\frac{(1 - \alpha)^2}{\alpha}\right] \qquad (27\text{-}32)$$

where W_n = weight fraction of products with carbon number n, and α = chain growth probability, i.e., the probability that a carbon chain on the catalyst surface will grow by adding another carbon atom rather than terminating. In general, α is dependent on temperature, pressure, and catalyst composition but independent of chain length. As α increases, the average carbon number of the product also increases. When α equals 0, only methane is formed. As α approaches 1, the product becomes predominantly wax.

Figure 27-10 provides a graphical representation of Eq. (27-32) showing the weight fraction of various products as a function of the chain growth parameter α. This figure shows that there is a particular α that will maximize the yield of a desired product, such as gasoline or

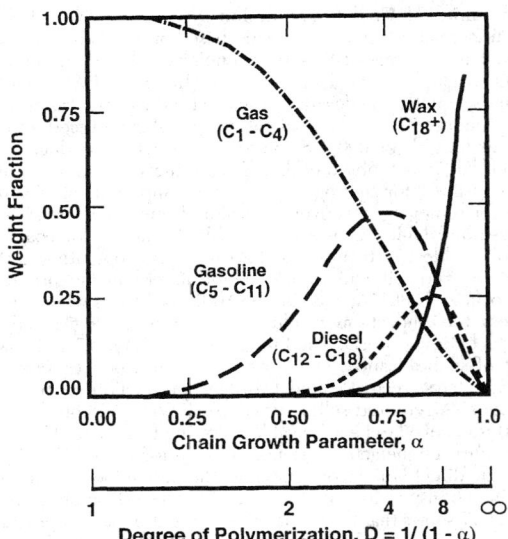

FIG. 27-10 Product yield in Fischer-Tropsch synthesis.

diesel fuel. The weight fraction of material between carbon numbers m and n, inclusive, is given by:

$$W_{mn} = m\alpha^{m-1} - (m+1)\alpha^m - (n+1)\alpha^n + n\alpha^{n+1} \qquad (27\text{-}33)$$

The α to maximize the yield of the carbon number range from m to n is given by:

$$\alpha_{opt} = \left(\frac{m^2 - m}{n^2 + n}\right)^{\frac{1}{n-m+1}} \qquad (27\text{-}34)$$

Additional gasoline and diesel fuel can be produced through further refining, such as hydrocracking or catalytic cracking of the wax product.

The Fischer-Tropsch reaction is highly exothermic. Therefore, adequate heat removal is critical. High temperatures result in high yields of methane, as well as coking and sintering of the catalyst. Three types of reactors (tubular fixed bed, fluidized bed, and slurry) provide good temperature control, and all three types are being used for synthesis gas conversion. The first plants used tubular or plate-type fixed-bed reactors. Later, SASOL, in South Africa, used fluidized-bed reactors, and most recently, slurry reactors have come into use.

Fischer-Tropsch synthesis reactor operation can be classified into one of two categories: high-temperature, 613 K (645°F), or low-temperature, 494 to 544 K (430 to 520°F), operation. The Synthol reactor developed by SASOL is typical of high-temperature operation. Using an iron-based catalyst, this process produces a very good gasoline product having high olefinicity and a low boiling range. The olefin fraction can readily be oligomerized to produce diesel fuel. Low-temperature operation, typically in fixed-bed reactors, produces a much more paraffinic and straight-chain product. Selectivity can be tailored to give the desired chain growth parameter. The primary diesel fraction, as well as the diesel-range product from hydrocracking of the wax, is an excellent diesel fuel.

Oxygenates and Chemicals A whole host of oxygenated products, i.e., fuels, fuel additives, and chemicals, can be produced from synthesis gas. These include such products as methanol, ethylene, isobutanol, dimethyl ether, dimethyl carbonate, and many other hydrocarbons and oxyhydrocarbons. Typical oxygenate-producing reactions are:

$$CO + 2H_2 \rightarrow CH_3OH \qquad (27\text{-}35)$$

$$CO_2 + 3H_2 \rightarrow CH_3OH + H_2O \qquad (27\text{-}36)$$

$$2CH_3OH \rightarrow CH_3OCH_3 + H_2O \qquad (27\text{-}37)$$

Reaction (27-37) can occur in parallel with the methanol reactions, thereby overcoming the equilibrium limitation on methanol formation. Higher alcohols can also be formed, as illustrated by Reaction (27-25), which is applicable to the formation of either linear or branched alcohols.

The production of methyl acetate from synthesis gas is currently being practiced commercially. Following methanol synthesis, as shown by Reaction (27-35), the reactions are:

$$CH_3OH + CO \rightarrow CH_3COOH \qquad (27\text{-}38)$$

$$CH_3COOH + CH_3OH \rightarrow CH_3COOCH_3 + H_2O \qquad (27\text{-}39)$$

Acrylates and methacrylates, which are critical to the production of polyesters, plastics, latexes, and synthetic lubricants, can also be produced from these oxygenated intermediates.

Status of Indirect Liquefaction Technology The only commercial indirect coal liquefaction plants for the production of transportation fuels are operated by SASOL in South Africa. Construction of the original plant was begun in 1950, and operations began in 1955. This plant employs both fixed-bed (Arge) and entrained-bed (Synthol) reactors. Two additional plants were later constructed with start-ups in 1980 and 1983. These latter plants employ dry-ash Lurgi Mark IV coal gasifiers and entrained-bed (Synthol) reactors for synthesis gas conversion. These plants currently produce 45 percent of South Africa's transportation fuel requirements, and, in addition, they produce more than 120 other products from coal.

SASOL has pursued the development of alternative reactors to overcome specific operational difficulties encountered with the fixed-bed and entrained-bed reactors. After several years of attempts to overcome the high catalyst circulation rates and consequent abrasion in the Synthol reactors, a bubbling fluidized-bed reactor 1 m (3.3 ft) in diameter was constructed in 1983. Following successful testing, SASOL designed and constructed a full-scale commercial reactor 5 m (16.4 ft) in diameter. The reactor was successfully commissioned in 1989 and remains in operation.

SASOL and others, including Exxon, Statoil, Air Products and Chemicals, Inc., and the U.S. Department of Energy, have engaged in the development of slurry bubble column reactors for Fischer-Tropsch and oxygenate synthesis. SASOL, in fact, commissioned a slurry reactor with a 5-m diameter in 1993. It doubled the wax capacity of the SASOL I facility. The development work on this kind of reactor shows that it has several advantages over competing reactor designs: (1) excellent heat transfer capability resulting in isothermal reactor operations, (2) high catalyst and reactor productivity, (3) ease of catalyst addition and withdrawal, (4) simple construction, and (5) ability to process hydrogen-lean synthesis gas successfully. Because of the small particle size of the catalyst used in the slurry reactor, however, effective separation of catalyst from the products is difficult but is crucial to successful operation.

The United States has two commercial facilities that convert coal to fuels and chemicals by indirect liquefaction. The Great Plains Synfuels Plant, located in Beulah, North Dakota, produces synthetic natural gas (SNG) from North Dakota lignite by Lurgi dry-ash gasification technology and methanation. Operated by Dakota Gasification Company (DGC), the plant converts approximately 15.4 Gg (17,000 US tons) of lignite per day to about 4.7×10^6 Nm^3 (166×10^6 std ft^3) of pipeline-quality gas in 14 Lurgi gasifiers. Aromatic naphtha and tar oil are also produced in the gasification section. The plant operates at 120 percent of its original design capacity. In addition to SNG, there is a wide assortment of other products, e.g., anhydrous ammonia, sulfur, phenol, cresylic acid, naphthas, and krypton/xenon.

Eastman Chemical Company has operated a coal-to-methanol plant in Kingsport, Tennessee, since 1983. Two Texaco gasifiers (one is a backup) process 34 Mg/h (37 US ton/h) of coal to synthesis gas. The synthesis gas is converted to methanol by use of ICI methanol technology. Methanol is an intermediate for producing methyl acetate and acetic acid. The plant produces about 225 Gg/a (250,000 US ton/a) of acetic anhydride. As part of the DOE Clean Coal Technology Program, Air Products and Chemicals, Inc., and Eastman Chemical Company are constructing a 9.8-Mg/h (260-US ton/d) slurry-phase reactor for the conversion of synthesis gas to methanol and dimethyl ether. Construction is expected to be completed in November 1996. Despite the success of SASOL, most of the commercial interest in Fischer-Tropsch synthesis technology is based on natural gas, and it is likely to remain so as long as the gas is abundant and inexpensive. In 1985, Mobil commercialized its Methanol-to-Gasoline (MTG) technology in New Zealand, natural gas being the feedstock. This fixed-bed process converts synthesis gas to 4 Gg (4400 US tons) of methanol per day; the methanol can then be converted to 2290 m^3/d (14,400 bbl/d) gasoline. Owing to economic factors, the plant is used primarily for the production of methanol.

Shell Gas B.V. has constructed a 1987 m^3/d (12,500 bbl/d) Fischer-Tropsch plant in Malaysia, start-up occurring in 1994. The Shell Middle Distillate Synthesis (SMDS) process, as it is called, uses natural gas as the feedstock to fixed-bed reactors containing cobalt-based catalyst. The heavy hydrocarbons from the Fischer-Tropsch reactors are converted to distillate fuels by hydrocracking and hydroisomerization. The quality of the products is very high, the diesel fuel having a cetane number in excess of 75.

Exxon Research and Engineering Company has developed a process for converting natural gas to high-quality refinery feedstock, the AGC-21 Advanced Gas Conversion Process. The technology involves three highly integrated process steps: fluid-bed synthesis gas generation; slurry-phase Fischer-Tropsch synthesis; and mild fixed-bed hydroisomerization. The process was demonstrated in the early 1990s with a slurry-phase reactor having a diameter of 1.2 m (4 ft) and a capacity of about 32 m^3/d (200 bbl/d).

The largest Fischer-Tropsch facility based on natural gas is the Mossgas plant located in Mossel Bay, South Africa. Natural gas is converted to synthesis gas in a two-stage reformer and subsequently converted to hydrocarbons by SASOL's Synthol technology. The plant, commissioned in 1992, has a capacity of 7155 m^3/d (45,000 bbl/d).

Economics of Coal Liquefaction Bechtel developed conceptual commercial designs (greenfield), based on 1993 costs and representing current state-of-the-art technologies, for both direct and indirect coal liquefaction facilities feeding Illinois and Wyoming coals. The direct liquefaction design focuses on producing hydrotreated distillate products. The conceptual baseline plant is designed to process about 26.3 Gg/d (29,000 US ton/d) of coal while producing about 11,130 m^3/d (70,000 bbl/d) of distillate products. The design employs Texaco gasifiers for hydrogen production, supercritical solvent deashing for removing unconverted coal and mineral matter from the products, and high-pressure ebullated-bed reactors for coal hydrogenation. Table 27-16 presents the capital and operating costs. As shown, the crude oil equivalent price for direct liquefaction products is approximately $215/$m^3$ ($34/bbl). Additional cases were evaluated to assess the impact on product costs of technological advances through continued R&D. The final column in Table 27-16 shows the impact of such advances as improvements in product yields, space velocity, catalyst recovery, and hydrogen production on the cost of production. A required selling price of $176/$m^3$ ($28/bbl) (1993 US dollars) is believed to be achievable following further R&D.

The indirect liquefaction baseline design is for a plant of similar size. Unlike the direct liquefaction baseline, the design focuses on producing refined transportation fuels by use of Shell gasification technology. Table 27-17 shows that the crude oil equivalent price is approximately $216/$m^3$ ($34/bbl). Additional technological advances in the production of synthesis gas, the Fischer-Tropsch synthesis, and product refining have the potential to reduce the cost to $171/$m^3$ ($27/bbl) (1993 US dollars), as shown in the second column of Table 27-17.

Coproduction of electricity along with synthesis gas conversion offers the potential for significant cost savings. The once-through liquid-phase methanol technology was developed specifically for this

---

TABLE 27-16 Estimated Costs of Direct Coal Liquefaction Plant (1993 US dollars)

Elements of cost	Baseline costs° $ million	Improved design due to R&D effort†
Coal handling	222	226
Liquefaction	942	823
Gas cleanup/by-product recovery	297	297
Product hydrotreating	107	113
De-ashing unit	46	43
Gasification	334	302
Air separation	244	220
Inside boundary limits field costs	2192	2024
Outside boundary limits field costs	978	968
Total field cost	3170	2992
Total capital	3889	3670
Refined product costs,‡ $/$m^3$		
Capital§	148.49	129.12
Coal	49.31	47.23
Catalyst	16.16	1.45
Natural gas	22.58	16.92
Labor	10.44	9.50
Other O&M	2.08	1.89
By-product credits	(26.29)	(20.44)
Required Selling Price	227.77	183.77
Quality premium	(7.48)	(7.48)
Crude oil equivalent price ($/bbl)¶	215.28 (34.22)	176.29 (28.02)
Plant output, Mm³/a (M bbl/a)	3.85 (24.2)	4.23 (26.6)

°Costs based on plant processing 26,105 Mg/d (28,776 US ton/d) of Illinois No. 6 coal. Source: *Direct Coal Liquefaction Baseline Design and Systems Analysis*, prepared by Bechtel and Amoco under DOE contract no. DE-AC22-90PC89857, March 1993.
†Source: Klunder and McIlvried, unpublished DOE Pittsburgh Energy Technology Center data.
‡To obtain $/bbl from $/$m^3$, multiply by 0.1590.
§Includes maintenance materials, taxes, and insurance.
¶The difference between the required selling price and the crude oil equivalent price represents the enhanced value of the coal liquids, due to their all-distillate and low-heteroatom character.

TABLE 27-17 Estimated Costs of Indirect Coal Liquefaction Plant (1993 US Dollars)

Elements of cost	Baseline costs,° $ million	Costs with R&D improvements	Once-through Fischer-Tropsch
Coal handling	207	207	207
Gasification	1018	1018	1018
Air separation	466	323	453
Gas cleaning/by-product recovery	195	195	192
Fischer-Tropsch synthesis	331	190	286
Synthesis gas recycle loop	403	352	79
Product refining	209	131	155
Inside battery limits field cost	2829	2416	2390
Power generation	119	254	419
Outside battery limits field cost	488	481	452
Total field cost	3436	3151	3261
Total plant cost	4283	3927	4063
Total capital	4620	4231	4346
Refined product costs,† $/$m^3$			
Capital‡	155.91	147.55	193.02
Coal	63.90	61.57	84.03
Catalyst	12.45	12.01	16.95
Other O&M	34.47	30.94	42.52
Power	7.80	−12.83	−118.74
Required selling price	274.53	229.25	217.17
Crude oil equivalent price ($/bbl)§	216.04 (34.34)	170.75 (27.14)	158.68 (25.23)
Plant output, Mm³/a (Mbbl/a)	3.77 (23.7)	3.91 (24.6)	2.86 (18.0)
Power, MW			1176

SOURCE: *Proc. Coal Liquefaction and Gas Conversion Contractors Review Conf.*, CONF-9508133, Pittsburgh Energy Technology Center, Pittsburgh, Pa., 1995.
°Costs based on plant processing 26,105 Mg/d (28,776 US ton/d) of Illinois No. 6 coal.
†To obtain $/bbl from $/$m^3$, multiply by 0.1590.
‡Includes maintenance, materials, taxes, and insurance.
§The difference between the required selling price and the crude oil equivalent price represents the enhanced value of the coal liquids, due to their all-distillate and low-heteroatom character.

objective to be realized in *integrated gasification combined cycle* (IGCC) power plants. *Once-through Fischer-Tropsch* (OTFT) in conjunction with IGCC has a similar potential, as indicated by the final column in Table 27-17. This concept has the potential for reducing the cost of products from Fischer-Tropsch to a crude oil equivalent price of $159/m³ ($25/bbl).

For the related gas-based technology, Shell and SASOL have separately estimated that their technologies, on large scale and with natural-gas feedstock, can compete with crude oil priced at $126 to 145/m³ ($20 to 23/bbl). The cost of production for each plant is sensitive to plant location and the price of the natural-gas feedstock.

At the current stage of development, direct and indirect liquefaction technologies look equally attractive economically, and both have the potential for significant cost improvements.

HEAT GENERATION

GENERAL REFERENCES: Stultz and Kitto (eds.), *Steam: Its Generation and Use,* 40th ed., Babcock and Wilcox, Barberton, Ohio, 1992. *North American Combustion Handbook,* 3d ed., vols. I and II, North American Manufacturing Company, Cleveland, Ohio, 1996. Singer (ed.), *Combustion: Fossil Power Systems,* 4th ed., Combustion Engineering, Inc., Windsor, Conn., 1991. Cuenca and Anthony (eds.), *Pressurized Fluidized Bed Combustion,* Blackie Academic & Professional, London, 1995. Basu and Fraser, *Circulating Fluidized Bed Boilers: Design and Operations,* Butterworth and Heinemann, Boston, 1991. *Proceedings of International FBC Conference(s),* ASME, New York, 1991, 1993, 1995. *Application of FBC for Power Generation,* Electric Power Research Institute, EPRI PR-101816, Palo Alto, Calif., 1993. Boyen, *Thermal Energy Recovery,* 2d ed., Wiley, New York, 1980.

COMBUSTION BACKGROUND

Basic Principles

Theoretical Oxygen and Air for Combustion The amount of oxidant (oxygen or air) just sufficient to burn the carbon, hydrogen, and sulfur in a fuel to carbon dioxide, water vapor, and sulfur dioxide is the *theoretical* or *stoichiometric oxygen* or *air* requirement. The chemical equation for complete combustion of a fuel is

$$C_xH_yO_zS_w + \left(\frac{4x + y - 2z + 4w}{4}\right)O_2 = xCO_2 + \left(\frac{y}{2}\right)H_2O + wSO_2$$

$$(27\text{-}40)$$

$x, y, z,$ and w being the number of atoms of carbon, hydrogen, oxygen, and sulfur, respectively, in the fuel. For example, 1 mol of methane (CH_4) requires 2 mol of oxygen for complete combustion to 1 mol of carbon dioxide and 2 mol of water. If air is the oxidant, each mol of oxygen is accompanied by 3.76 mol of nitrogen.

The volume of theoretical oxygen (at 0.101 MPa and 298 K) needed to burn any fuel can be calculated from the ultimate analysis of the fuel as follows:

$$24.45\left(\frac{C}{12} + \frac{H}{4} - \frac{O}{32} + \frac{S}{32}\right) = m^3O_2/\text{kg fuel} \qquad (27\text{-}41)$$

where $C, H, O,$ and S are the decimal weights of these elements in 1 kg of fuel. (To convert to ft³ per lb of fuel, multiply by 16.02.) The mass of oxygen (in kg) required can be obtained by multiplying the volume by 1.31. The volume of theoretical air can be obtained by using a coefficient of 116.4 in Eq. (27-41) in place of 24.45.

Figure 27-11 gives the theoretical air requirements for a variety of combustible materials on the basis of fuel higher heating value (HHV). If only the fuel lower heating value is known, the HHV can be calculated from Eq. (27-6). If the ultimate analysis is known, Eq. (27-7) can be used to determine HHV.

Excess Air for Combustion More than the theoretical amount of air is necessary in practice to achieve complete combustion. This excess air is expressed as a percentage of the theoretical air amount. The *equivalence ratio* is defined as the ratio of the actual fuel-air ratio to the stoichiometric fuel-air ratio. Equivalence ratio values less than 1.0 correspond to fuel-*lean* mixtures. Conversely, values greater than 1.0 correspond to fuel-*rich* mixtures.

Products of Combustion For lean mixtures, the *products of combustion* (POC) of a sulfur-free fuel consist of carbon dioxide, water vapor, nitrogen, oxygen, and possible small amounts of carbon monoxide and unburned hydrocarbon species. Figure 27-12 shows the effect of fuel-air ratio on the flue gas composition resulting from the combustion of natural gas. In the case of solid and liquid fuels, the

POC may also include solid residues containing ash and unburned carbon particles.

Equilibrium combustion product compositions and properties may be readily calculated using thermochemical computer codes which minimize the Gibbs free energy and use thermodynamic databases

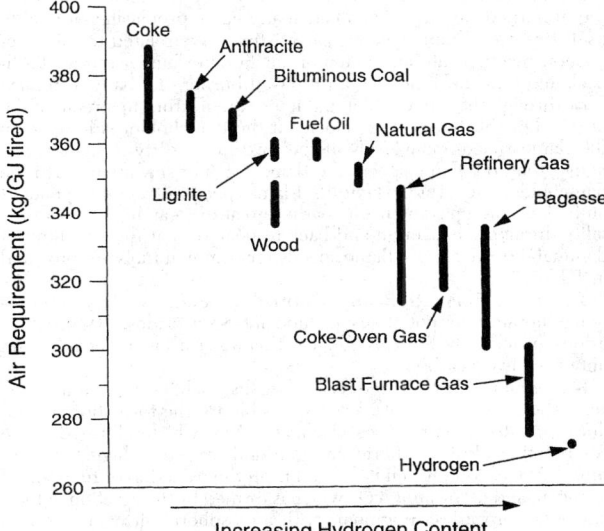

FIG. 27-11 Combustion air requirements for various fuels at zero excess air. To convert from kg air/GJ fired to lb air/10⁶ Btu fired, multiply by 2.090.

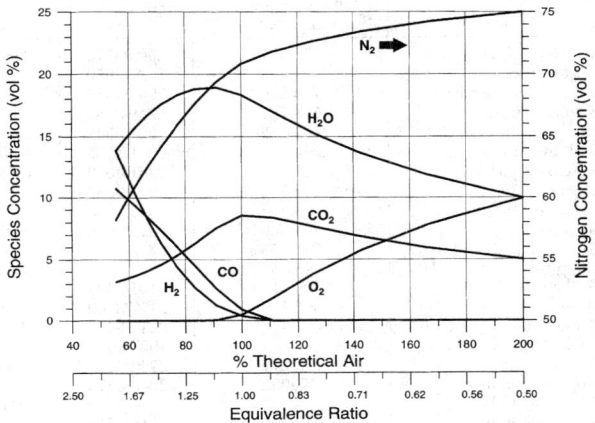

FIG. 27-12 Effect of fuel-air ratio on flue-gas composition for a typical U.S. natural gas containing 93.9% CH_4, 3.2% C_2H_6, 0.7% C_3H_8, 0.4% C_4H_{10}, 1.5% N_2 and 1.1% CO_2 by volume.

containing polynomial curve-fits of physical properties. Two widely used versions are those developed at NASA Lewis (Gordon and McBride, NASA SP-273, 1971) and at Stanford University (Reynolds, *STANJAN Chemical Equilibrium Solver*, Stanford University, 1987).

Flame Temperature The heat released by the chemical reaction of fuel and oxidant heats the POC. Heat is transferred from the POC, primarily by radiation and convection, to the surroundings, and the resulting temperature in the reaction zone is the flame temperature. If there is no heat transfer to the surroundings, the flame temperature equals the theoretical, or adiabatic, flame temperature.

Figure 27-13 shows the *available heat* in the products of combustion for various common fuels. The available heat is the total heat released during combustion minus the flue-gas heat loss (including the heat of vaporization of any water formed in the POC).

Flammability Limits There are both upper (or rich) and lower (or lean) limits of flammability of fuel-air or fuel-oxygen mixtures. Outside these limits, a self-sustaining flame cannot form. Flammability limits for common fuels are listed in Table 27-18.

Flame Speed Flame speed is defined as the velocity, relative to the unburned gas, at which an adiabatic flame propagates normal to itself through a homogeneous gas mixture. It is related to the combustion reaction rate and is important in determining burner flashback and blow-off limits. In a premixed burner, the flame can *flash back* through the flameholder and ignite the mixture upstream of the burner head if the mixture velocity at the flameholder is lower than the flame speed. Conversely, if the mixture velocity is significantly higher than the flame speed, the flame may not stay attached to the flameholder and is said to *blow off*. Flame speed is strongly dependent on fuel/air ratio, passing from nearly zero at the lean limit of flammability through a maximum and back to near zero at the rich limit of flammability. Maximum flame speeds for common fuels are provided in Table 27-18.

Pollutant Formation and Control in Flames Key combustion-generated air pollutants include nitrogen oxides (NO_x), sulfur oxides (principally SO_2), particulate matter, carbon monoxide, and unburned hydrocarbons.

Nitrogen Oxides Three reaction paths, each having unique characteristics (see Fig. 27-14), are responsible for the formation of NO_x during combustion processes: (1) *thermal NO_x*, which is formed by the combination of atmospheric nitrogen and oxygen at high temperatures; (2) *fuel NO_x*, which is formed from the oxidation of fuel-bound nitrogen; and (3) *prompt NO_x*, which is formed by the reaction of fuel-derived hydrocarbon fragments with atmospheric nitrogen. (NO_x is

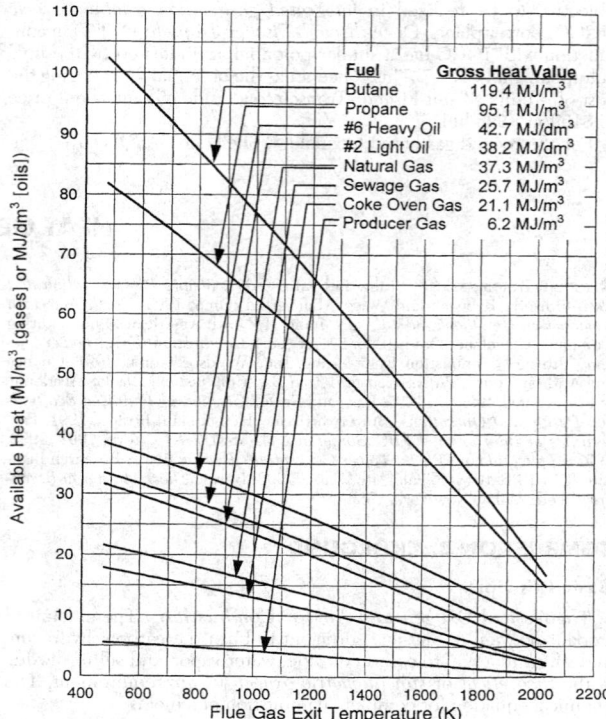

Fuel	Gross Heat Value
Butane	119.4 MJ/m³
Propane	95.1 MJ/m³
#6 Heavy Oil	42.7 MJ/dm³
#2 Light Oil	38.2 MJ/dm³
Natural Gas	37.3 MJ/m³
Sewage Gas	25.7 MJ/m³
Coke Oven Gas	21.1 MJ/m³
Producer Gas	6.2 MJ/m³

FIG. 27-13 Available heats for some typical fuels. The fuels are identified by their gross (or higher) heating values. All available heat figures are based upon complete combustion and fuel and air initial temperature of 288 K (60°F). To convert from MJ/Nm³ to Btu/ft³, multiply by 26.84. To convert from MJ/dm³ to Btu/gal, multiply by 3588.

TABLE 27-18 Combustion Characteristics of Various Fuels*

Fuel	Minimum ignition temp., K/°F	Calculated flame temperature,† K/°F in air	in O₂	Flammability limits, % fuel gas by volume in air lower	upper	Maximum flame velocity, m/s and ft/s in air	in O₂	% theoretical air for max. flame velocity
Acetylene, C₂H₂	578/581	2905/4770	3383/5630	2.5	81.0	2.67/8.75	—	83
Blast furnace gas	—	1727/2650	—	35.0	73.5	—	—	—
Butane, commercial	753/896	2246/3583	—	1.86	8.41	0.87/2.85	—	—
Butane, n-C₄H₁₀	678/761	2246/3583	—	1.86	8.41	0.40/1.3	—	97
Carbon monoxide, CO	882/1128	2223/3542	—	12.5	74.2	0.52/1.7	—	55
Carbureted water gas	—	2311/3700	3061/5050	6.4	37.7	0.66/2.15	—	90
Coke oven gas	—	2261/3610	—	4.4	34.0	0.70/2.30	—	90
Ethane, C₂H₄	745/882	2222/3540	—	3.0	12.5	0.48/1.56	—	98
Gasoline	553/536	—	—	1.4	7.6	—	—	—
Hydrogen, H₂	845/1062	2318/4010	3247/5385	4.0	74.2	2.83/9.3	—	57
Hydrogen sulfide, H₂S	565/558	—	—	4.3	45.5	—	—	—
Mapp gas, (allene) C₃H₄	728/850	—	3200/5301	3.4	10.8	—	4.69/15.4	—
Methane, CH₄	905/1170	2191/3484	—	5.0	15.0	0.45/1.48	4.50/14.76	90
Methanol, CH₃OH	658/725	2177/3460	—	6.7	36.0	—	0.49/1.6	—
Natural gas	—	2214/3525	2916/4790	4.3	15.0	0.30/1.00	4.63/15.2	100
Producer gas	—	1927/3010	—	17.0	73.7	0.26/0.85	—	90
Propane, C₃H₈	739/871	2240/3573	3105/5130	2.1	10.1	0.46/1.52	3.72/12.2	94
Propane, commercial	773/932	2240/3573	—	2.37	9.50	0.85/2.78	—	—
Propylene, C₃H₆	—	—	3166/5240	—	—	—	—	—
Town gas (brown coal)	643/700	2318/3710	—	4.8	31.0	—	—	—

*For combustion with air at standard temperature and pressure. These flame temperatures are calculated for 100 percent theoretical air, disassociation considered. Data from *Gas Engineers Handbook*, Industrial Press, New York, 1965.
†Flame temperatures are theoretical—calculated for stoichiometric ratio, dissociation considered.

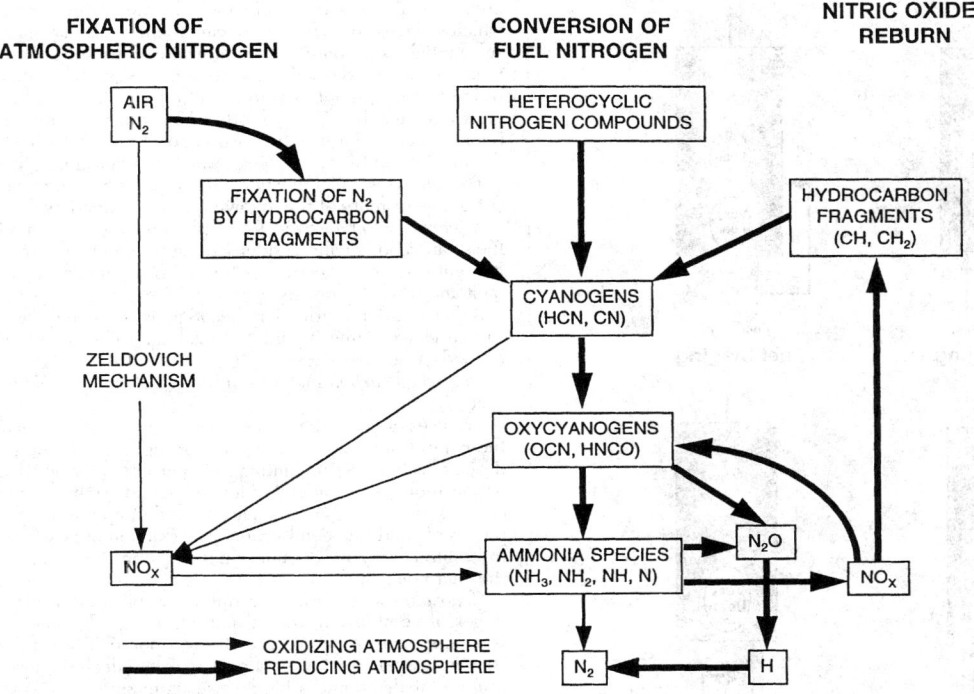

FIG. 27-14 Nitrogen oxide formation pathways in combustion.

used to refer to NO + NO₂. NO is the primary form in combustion products [typically 95 percent of total NO$_x$]. NO is subsequently oxidized to NO₂ in the atmosphere.)

Thermal NO$_x$ The formation of thermal NO$_x$ is described by the Zeldovich mechanism:

$$N_2 + O \rightleftharpoons NO + N \qquad (27-42)$$

$$N + O_2 \rightleftharpoons NO + O \qquad (27-43)$$

$$N + OH \rightleftharpoons NO + H \qquad (27-44)$$

The first of these reactions is the rate-limiting step. Assuming that O and O₂ are in partial equilibrium, the NO formation rate can be expressed as follows:

$$\frac{d[NO]}{dt} = A[N_2][O_2]^{1/2} \exp\left(\frac{-E}{RT}\right) \qquad (27-45)$$

As indicated, the rate of NO formation increases exponentially with temperature, and, of course, oxygen and nitrogen must be available for thermal NO$_x$ to form. Thus, thermal NO$_x$ formation is rapid in high-temperature lean zones of flames.

Fuel NO$_x$ Fuel-bound nitrogen (FBN) is the major source of NO$_x$ emissions from combustion of nitrogen-bearing fuels such as heavy oils, coal, and coke. Under the reducing conditions surrounding the burning droplet or particle, the FBN is converted to fixed nitrogen species such as HCN and NH₃. These, in turn, are readily oxidized to form NO if they reach the lean zone of the flame. Between 20 and 80 percent of the bound nitrogen is typically converted to NO$_x$, depending on the design of the combustion equipment. With prolonged exposure (order of 100 ms) to high temperature and reducing conditions, however, these fixed nitrogen species may be converted to molecular nitrogen, thus avoiding the NO formation path.

Prompt NO$_x$ Hydrocarbon fragments (such as C, CH, CH₂) may react with atmospheric nitrogen under fuel-rich conditions to yield fixed nitrogen species such as NH, HCN, H₂CN, and CN. These, in turn, can be oxidized to NO in the lean zone of the flame. In most flames, especially those from nitrogen-containing fuels, the prompt

mechanism is responsible for only a small fraction of the total NO$_x$. Its control is important only when attempting to reach the lowest possible emissions.

NO$_x$ Emission Control It is preferable to minimize NO$_x$ formation through control of the mixing, combustion, and heat-transfer processes rather than through postcombustion techniques such as selective catalytic reduction. Four techniques for doing so, illustrated in Fig. 27-15, are air staging, fuel staging, flue-gas recirculation, and lean premixing.

Air Staging Staging the introduction of combustion air can control NO$_x$ emissions from all fuel types. The combustion air stream is split to create a fuel-rich primary zone and a fuel-lean secondary zone. The rich primary zone converts fuel-bound nitrogen to molecular nitrogen and suppresses thermal NO$_x$. Heat is removed prior to addition of the secondary combustion air. The resulting lower flame temperatures (below 1810 K [2800°F]) under lean conditions reduce the rate of formation of thermal NO$_x$. This technique has been widely applied to furnaces and boilers and it is the preferred approach for burning liquid and solid fuels. Staged-air burners are typically capable of reducing NO$_x$ emissions by 30 to 60 percent, relative to uncontrolled levels. Air staging can also be accomplished by use of overfire air systems in boilers.

Fuel Staging Staging the introduction of fuel is an effective approach for controlling NO$_x$ emissions when burning gaseous fuels. The first combustion stage is very lean, resulting in low thermal and prompt NO$_x$. Heat is removed prior to injection of the secondary fuel. The secondary fuel entrains flue gas prior to reacting, further reducing flame temperatures. In addition, NO$_x$ reduction through reburning reactions may occur in the staged jets. This technique is the favored approach for refinery- and chemical plant–fired heaters utilizing gaseous fuels. Staged-fuel burners are typically capable of reducing NO$_x$ emissions by 40 to 70 percent, relative to uncontrolled levels.

Flue Gas Recirculation Flue gas recirculation, alone or in combination with other modifications, can significantly reduce thermal NO$_x$. Recirculated flue gas is a diluent that reduces flame temperatures. External and internal recirculation paths have been applied: internal

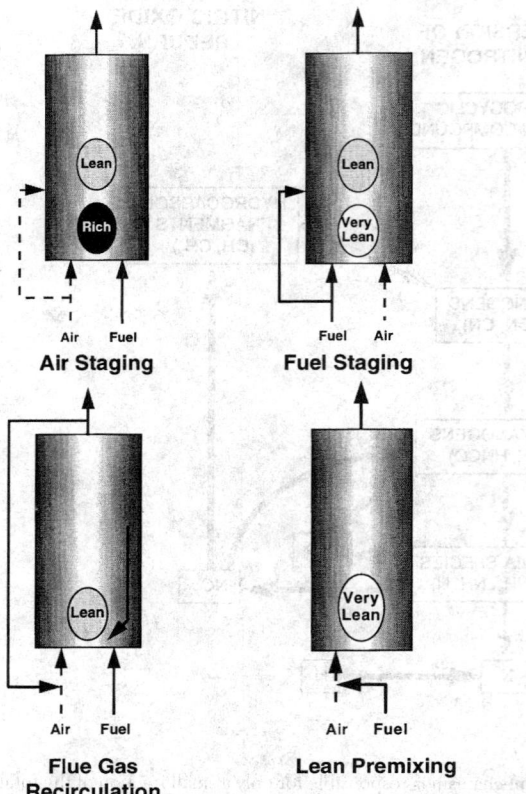

FIG. 27-15 Combustion modifications for NO$_x$ control.

recirculation can be accomplished by jet entrainment using either combustion air or fuel jet energy; external recirculation requires a fan or a jet pump (driven by the combustion air). When combined with staged-air or staged-fuel methods, NO$_x$ emissions from gas-fired burners can be reduced by 50 to 90 percent. In some applications, external flue-gas recirculation can decrease thermal efficiency. Condensation in the recirculation loop can cause operating problems and increase maintenance requirements.

Lean Premixing Very low NO$_x$ emissions can be achieved by premixing gaseous fuels (or vaporized liquid fuels) with air and reacting at high excess air. The uniform and very lean conditions in such systems favor very low thermal and prompt NO$_x$. However, achieving such low emissions requires operating near the lean stability limit. This is an attractive NO$_x$ control approach for gas turbines, where operation at high excess air does not incur an efficiency penalty. In this application, NO$_x$ emissions have been reduced by 75 to 95 percent.

Sulfur Oxides Sulfur occurs in fuels as inorganic minerals (primarily pyrite, FeS$_2$), organic structures, sulfate salts, and elemental sulfur. Sulfur contents range from parts per million in pipeline natural gas, to a few tenths of a percent in diesel and light fuel oils, to 0.5 to 5 percent in heavy fuel oils and coals. Sulfur compounds are pyrolyzed during the volatilization phase of oil and coal combustion and react in the gas phase to form predominantly SO$_2$ and some SO$_3$. Conversion of fuel sulfur to these oxides is generally high (85 to 90 percent) and is relatively independent of combustion conditions. From 1 to 4 percent of the SO$_2$ is further oxidized to SO$_3$, which is highly reactive and extremely hygroscopic. It combines with water to form sulfuric acid aerosol, which can increase the visibility of stack plumes. It also elevates the dew point of water so that, to avoid back-end condensation and resulting corrosion, the flue-gas discharge temperature must be raised to about 420 K (300°F), reducing heat recovery and thermal

efficiency. This reaction is enhanced by the presence of fine particles, which serve as condensation nuclei. Some coals may contain ash with substantial alkali content. In combustion of these fuels, the alkali may react to form condensed phase compounds (such as sulfates), thereby reducing the amount of sulfur emitted as oxides. Reductions in SO$_2$ emissions may be achieved either by removing sulfur from the fuel before and/or during combustion, or by postcombustion flue-gas desulfurization (wet scrubbing using limestone slurry, for example).

Particulates Combustion-related particulate emissions may consist of one or more of the following types, depending on the fuel.

Mineral matter derived from ash constituents of liquid and solid fuels can vaporize and condense as sub-micron-size aerosols. Larger mineral matter fragments are formed from mineral inclusions which melt and resolidify downstream.

Sulfate particles formed in the gas phase can condense. In addition, sulfate can become bound to metals and can be adsorbed on unburned carbon particles.

Unburned carbon includes unburned char, coke, cenospheres, and soot.

Particles of char are produced as a normal intermediate product in the combustion of solid fuels. Following initial particle heating and devolatilization, the remaining solid particle is termed *char*. Char oxidation requires considerably longer periods (ranging from 30 ms to over 1 s, depending on particle size and temperature) than the other phases of solid fuel combustion. The fraction of char remaining after the combustion zone depends on the combustion conditions as well as the char reactivity.

Cenospheres are formed during heavy oil combustion. In the early stages of combustion, the oil particle is rapidly heated and evolves volatile species, which react in the gas phase. Toward the end of the volatile-loss phase, the generation of gas declines rapidly and the droplet (at this point, a highly viscous mass) solidifies into a porous coke particle known as a *cenosphere*. This is called *initial coke*. For the heaviest oils, the initial coke particle diameter may be 20 percent larger than the initial droplet diameter. For lighter residual oils, it may be only one third of the original droplet diameter. After a short interval, the initial coke undergoes contraction to form *final coke*. Final coke diameter is ~80 percent of the initial droplet diameter for the heaviest oils. At this time the temperature of the particle is approximately 1070 to 1270 K (1470 to 1830°F). Following coke formation, the coke particles burn out in the lean zone, but the heterogeneous oxidation proceeds slowly. Final unburned carbon levels depend on a balance between the amount of coke formed and the fraction burned out. Coke formation tends to correlate with fuel properties such as asphaltene content, C:H ratio, or Conradson Carbon Residue. Coke burnout depends on combustion conditions and coke reactivity. Coke reactivity is influenced by the presence of combustion catalysts (e.g., vanadium) in the cenospheres.

Formation of *soot* is a gas-phase phenomenon that occurs in hot, fuel-rich zones. Soot occurs as fine particles (0.02 to 0.2 µm), often agglomerated into filaments or chains which can be several millimeters long. Factors that increase soot formation rates include high C:H ratio, high temperature, very rich conditions, and long residence times at these conditions. Pyrolysis of fuel molecules leads to soot precursors such as acetylene and higher analogs and various polyaromatic hydrocarbons. These condense to form very small (<2 nm) particles. The bulk of the solid-phase material is generated by *surface growth*—attachment of gas-phase species to the surface of the particles and their incorporation into the particulate phase. Another growth mechanism is *coagulation,* in which particles collide and coalesce. Soot particle formation and growth is typically followed by soot oxidation to form CO and CO$_2$. Eventual soot emission from a flame depends on the relative balance between the soot-formation and oxidation reactions.

Carbon Monoxide Carbon monoxide is a key intermediate in the oxidation of all hydrocarbons. In a well-adjusted combustion system, essentially all the CO is oxidized to CO$_2$ and final emission of CO is very low indeed (a few parts per million). However, in systems which have low temperature zones (for example, where a flame impinges on a wall or a furnace load) or which are in poor adjustment (for example, an individual burner fuel-air ratio out of balance in a multiburner installation or a misdirected fuel jet which allows fuel to bypass the main flame), CO emissions can be significant. The primary method of CO control is good combustion system design and practice.

Unburned Hydrocarbons Various unburned hydrocarbon species may be emitted from hydrocarbon flames. In general, there are two classes of unburned hydrocarbons: (1) small molecules that are the intermediate products of combustion (for example, formaldehyde) and (2) larger molecules that are formed by pyro-synthesis in hot, fuel-rich zones within flames, e.g., benzene, toluene, xylene, and various polycyclic aromatic hydrocarbons (PAHs). Many of these species are listed as Hazardous Air Pollutants (HAPs) in Title III of the Clean Air Act Amendment of 1990 and are therefore of particular concern. In a well-adjusted combustion system, emission of HAPs is extremely low (typically, parts per trillion to parts per billion). However, emission of certain HAPs may be of concern in poorly designed or maladjusted systems.

COMBUSTION OF SOLID FUELS

There are three basic modes of burning solid fuels, each identified with a furnace design specific for that mode: in suspension, in a bed at rest° on a grate (fuel-bed firing), or in a fluidized bed. Although many variations of these generic modes and furnace designs have been devised, the fundamental characteristics of equipment and procedure remain intact. They will be described briefly.

Suspension Firing Suspension firing of pulverized coal (PC) is commoner than fuel-bed or fluidized-bed firing of coarse coal in the United States. This mode of firing affords higher steam-generation capacity, is independent of the caking characteristics of the coal, and responds quickly to load changes. Pulverized coal firing accounts for approximately 55 percent of the power generated by electric utilities in the United States. It is rarely used on boilers of less than 45.4 Mg/h (100,000 lb/h) steam capacity because its economic advantage decreases with size.

A simplified model of PC combustion includes the following sequence of events: (1) on entering the furnace, a PC particle is heated rapidly, driving off the volatile components and leaving a char particle; (2) the volatile components burn independently of the char particle; and (3) on completion of volatiles combustion, the remaining char particle burns. While this simple sequence may be generally correct, PC combustion is an extremely complex process involving many interrelated physical and chemical processes.

Devolatilization The volatiles produced during rapid heating of coal can include H_2, CH_4, CO, CO_2, and C_2-C_4 hydrocarbons, as well as tars, other organic compounds, and reduced sulfur and nitrogen species. The yield of these various fractions is a function of both heating rate and final particle temperature. The resulting char particle may be larger in diameter than the parent coal particle, owing to swelling produced by volatiles ejection. The particle density also decreases.

Char oxidation dominates the time required for complete burnout of a coal particle. The heterogeneous reactions responsible for char oxidation are much slower than the devolatilization process and gas-phase reaction of the volatiles. Char burnout may require from 30 ms to over 1 s, depending on combustion conditions (oxygen level, temperature), and char particle size and reactivity. Char reactivity depends on parent coal type. The rate-limiting step in char burnout can be chemical reaction or gaseous diffusion. At low temperatures or for very large particles, chemical reaction is the rate-limiting step. At higher temperatures boundary-layer diffusion of reactants and products is the rate-limiting step.

Pulverized-Coal Furnaces In designing and sizing PC furnaces, particular attention must be given to the following fuel-ash properties:
• Ash fusion temperatures, including the spread between initial deformation temperature and fluid temperature
• Ratio of basic (calcium, sodium, potassium) to acidic (iron, silicon, aluminum) ash constituents, and specifically iron-to-calcium ratio

• Ash content
• Ash friability

These characteristics influence furnace plan area, furnace volume, and burning zone size required to maintain steam production capacity for a given fuel grade or quality.

Coal properties influence pulverizer capacity and the sizing of the air heater and other heat-recovery sections of a steam generator. Furnace size and heat-release rates are designed to control slagging characteristics. Consequently, heat-release rates in terms of the ratio of net heat input to plan area range from 4.4 MW/m² (1.4 × 10⁶ Btu/[h·ft²]) for severely slagging coals to 6.6 MW/m² (2.1 × 10⁶ Btu/[h·ft²]) for low-slagging fuels.

The various burner and furnace configurations for PC firing are shown schematically in Fig. 27-16. The U-shaped flame, designated as *fantail vertical firing* (Fig. 27-16a), was developed initially for pulverized coal before the advent of water-cooled furnace walls. Because a large percentage of the total combustion air is withheld from the fuel stream until it projects well down into the furnace, this type of firing is well suited for solid fuels that are difficult to ignite, such as those with less than 15 percent volatile matter. Although this configuration is no longer used in central-station power plants, it may find favor again if low-volatile chars from coal-conversion processes are used for steam generation or process heating.

Modern central stations use the other burner-furnace configurations shown in Fig. 27-16, in which the coal and air are mixed rapidly in and close to the burner. The primary air, used to transport the pulverized coal to the burner, comprises 10 to 20 percent of the total combustion air. The secondary air comprises the remainder of the total air and mixes in or near the burner with the primary air and coal. The velocity of the mixture leaving the burner must be high enough to prevent flashback in the primary air-coal piping. In practice, this velocity is maintained at about 31 m/s (100 ft/s).

In *tangential firing* (Fig. 27-16b), the burners are arranged in vertical banks at each corner of a square (or nearly square) furnace and directed toward an imaginary circle in the center of the furnace. This results in the formation of a large vortex with its axis on the vertical centerline. The burners consist of an arrangement of slots one above the other, admitting, through alternate slots, primary air-fuel mixture and secondary air. It is possible to tilt the burners upward or downward, the maximum inclination to the horizontal being 30°, enabling the operator to selectively utilize in-furnace heat-absorbing surfaces, especially the superheater.

The circular burner shown in Fig. 27-17 is widely used in horizontally fired furnaces and is capable of firing coal, oil, or gas in capacities as high as 174 GJ/h (1.65 × 10⁸ Btu/h). In such burners the air is often swirled to create a zone of reverse flow immediately downstream of the burner centerline, which provides for combustion stability.

Low-NOₓ burners are designed to delay and control the mixing of coal and air in the main combustion zone. A typical low-NOₓ air-staged burner is illustrated in Fig. 27-18. This combustion approach can reduce NOₓ emissions from coal burning by 40 to 50 percent. Because of the reduced flame temperature and delayed mixing in a low-NOₓ burner, unburned carbon emissions may increase in some applications and for some coals. *Overfire air* is another technique for staging the combustion air to control NOₓ emissions when burning coal in suspension-firing systems. Overfire air ports are installed above the top level of burners on wall- and tangential-fired boilers. Use of overfire air can reduce NOₓ emissions by 20 to 30 percent. *Reburn* is a NOₓ control strategy that involves diverting a portion of the fuel from the burners to a second combustion zone (reburn zone) above the main burners. Completion air is added above the reburn zone to complete fuel burnout. The reburn fuel can be natural gas, oil, or pulverized coal, though natural gas is used in most applications. In this approach, the stoichiometry in the reburn zone is controlled to be slightly rich (equivalence ratio of ~1.15), under which conditions a portion (50 to 60 percent) of the NOₓ is converted to molecular nitrogen.

Pulverizers The pulverizer is the heart of any solid-fuel suspen-

---

° The burning fuel bed may be moved slowly through the furnace by the vibrating action of the grate or by being carried on a traveling grate.

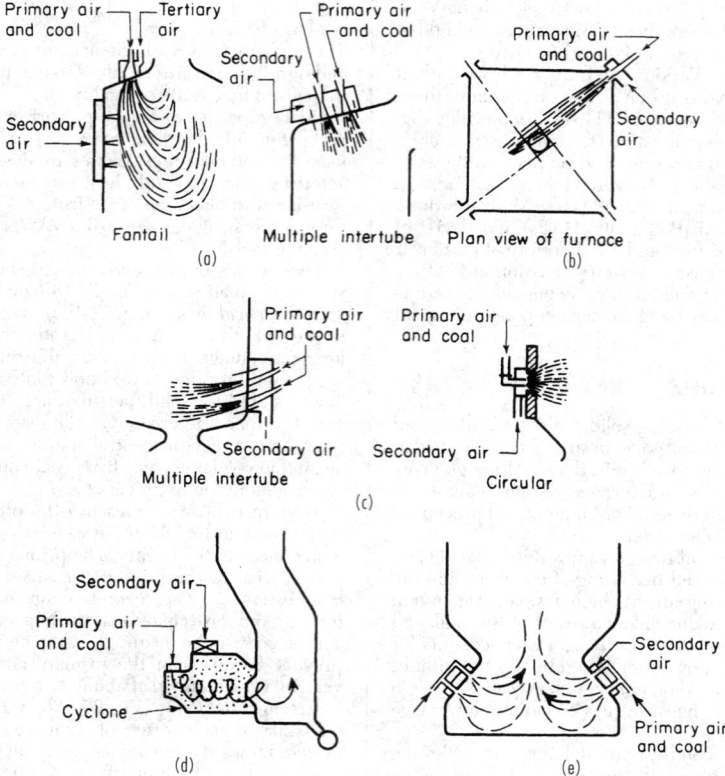

FIG. 27-16 Burner and furnace configurations for pulverized-coal firing: (*a*) vertical firing; (*b*) tangential firing; (*c*) horizontal firing; (*d*) cyclone firing; (*e*) opposed-inclined firing.

sion-firing system. Air is used to dry the coal, transport it through the pulverizer, classify it, and transport it to the burner, where the transport air provides part of the air for combustion. The pulverizers themselves are classified according to whether they are under positive or negative pressure and whether they operate at slow, medium, or high speed.

Pulverization occurs by impact, attrition, or crushing. The capacity of a pulverizer depends on the grindability of the coal and the fineness desired, as shown by Fig. 27-19. Capacity can also be seriously reduced by excessive moisture in the coal, but it can be restored by increasing the temperature of the primary air. Figure 27-20 indicates the temperatures needed. For PC boilers, the coal size usually is 65 to

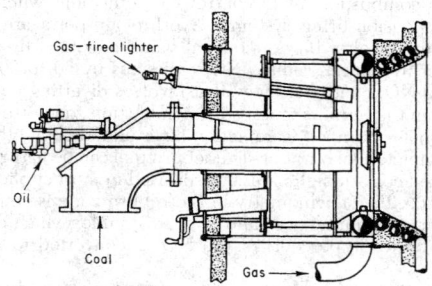

FIG. 27-17 Circular burner for pulverized coal, oil, or gas. (*From Marks' Standard Handbook for Mechanical Engineers, 8th ed., McGraw-Hill, New York, 1978.*)

80 percent through a 200-mesh screen, which is equivalent to 74 μm. Several kinds of available pulverizers and their characteristics are discussed in Sec. 20.

Cyclone Furnaces In *cyclone firing* (Fig. 27-16*d*) the coal is not pulverized but is crushed to 4-mesh (4.76-mm) size and admitted tangentially with primary air to a horizontal cylindrical chamber, called a *cyclone furnace,* which is connected peripherally to a boiler furnace. Secondary air also is admitted, so that almost all of the coal burns within the chamber. The combustion gas then flows into the boiler furnace. In the cyclone furnace, finer coal particles burn in suspension and the coarser ones are thrown centrifugally to the chamber wall, where most of them are captured in a sticky wall coating of molten slag. The secondary air, admitted tangentially along the top of the cyclone furnace, sweeps the slag-captured particles and completes their combustion. A typical firing rate is about 18.6 GJ/(h·m³) (500,000 Btu/[h·ft³]). The slag drains continuously into the boiler furnace and thence into a quenching tank. Figure 27-21 shows a cyclone furnace schematically.

Fuel-Bed Firing Fuel-bed firing is accomplished with mechanical stokers, which are designed to achieve continuous or intermittent fuel feed, fuel ignition, proper distribution of the combustion air, free release of the gaseous combustion products, and continuous or intermittent disposal of the unburned residue. These aims are met with two classes of stokers, distinguished by the direction of fuel feed to the bed: underfeed and overfeed. Overfeed stokers are represented by two types, distinguished by the relative directions of fuel and air flow (and also by the manner of fuel feed): crossfeed, also termed mass-burning, and spreader. The principles of these three methods of fuel-bed firing are illustrated schematically in Fig. 27-22.

Underfeed Firing Both fuel and air have the same relative

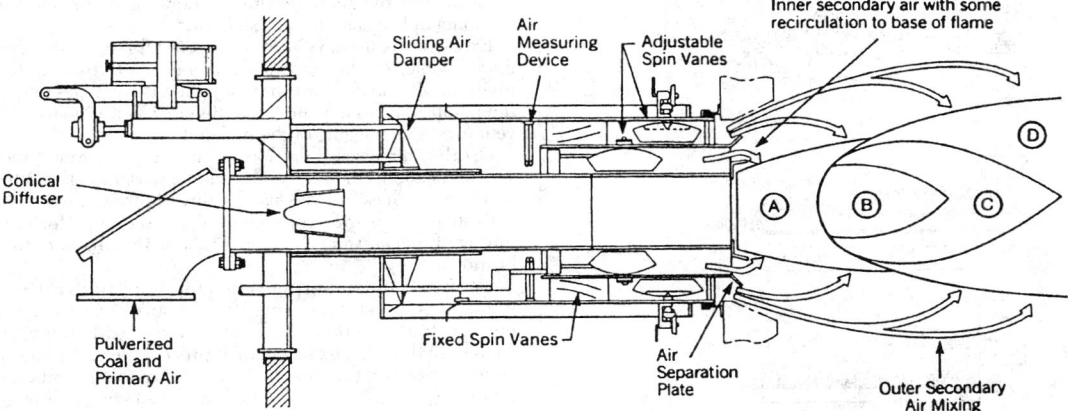

Ⓐ **High temperature – fuel rich devolatilization zone**
Ⓑ **Production of reducing species**
Ⓒ **NO_x decomposition zone**
Ⓓ **Char oxidizing zone**

FIG. 27-18 Low-NO_x pulverized coal burner. (*Babcock & Wilcox Co.*)

direction in the underfeed stoker, which is built in single-retort and multiple-retort designs. In the *single-retort*, side-dump stoker, a ram pushes coal into the retort toward the end of the stoker and upward toward the tuyere blocks, where air is admitted to the bed. This type of stoker will handle most bituminous coals and anthracite, preferably in the size range 19 to 50 mm (¾ to 2 in) and no more than 50 percent through a 6-mm (¼-in) screen. Overfire air or steam jets are frequently used in the bridgewall at the end of the stoker to promote turbulence.

In the *multiple-retort stoker*, rams feed coal to the top of sloping grates between banks of tuyeres. Auxiliary small sloping rams perform

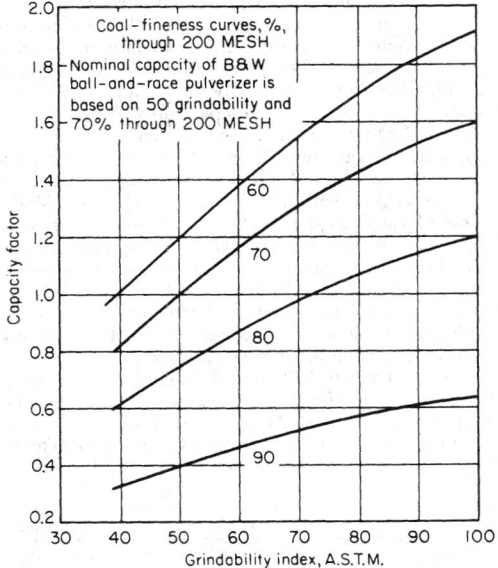

FIG. 27-19 Variation of pulverizer capacity with the grindability of the coal and the fineness to which the coal is ground. (*Babcock & Wilcox Co.*)

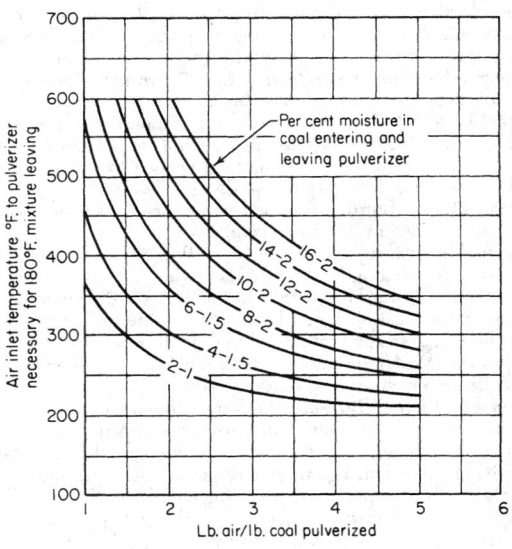

FIG. 27-20 Effect of moisture in coal on pulverizer capacity. Sufficient drying can be accomplished to restore capacity if air temperatures are high enough. [K = (°F + 459.7)/1.8] (Combustion Engineer, *Combustion Engineering Inc.*, New York, 1966.)

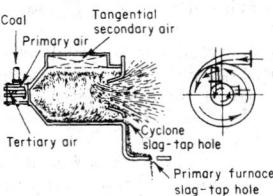

FIG. 27-21 Cyclone furnace. (*From Marks' Standard Handbook for Mechanical Engineers, 9th ed., McGraw-Hill, New York, 1987.*)

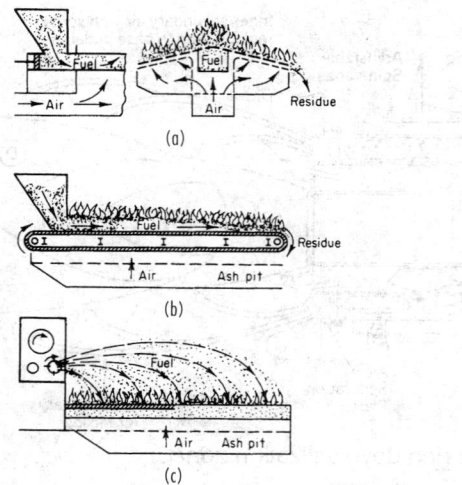

FIG. 27-22 Basic types of mechanical stokers: (*a*) underfeed; (*b*) crossfeed; (*c*) overfeed (spreader stoker).

the same function as the pusher rods in the single retort. Air is admitted along the top of the banks of tuyeres, and on the largest units the tuyeres themselves are given a slight reciprocating action to agitate the bed further. This type of stoker operates best with caking coals having a relatively high ash-softening temperature. Coal sizing is up to 50 mm (2 in) with 30 to 50 percent through a 6-mm (¼-in) screen.

Overfeed Firing: Crossfeed (Mass-Burning) Stokers Crossfeed stokers are also termed mass-burning stokers because the fuel is dumped by gravity from a hopper onto one end of a moving grate, which carries it into the furnace and down its length. Because of this feature, crossfeed stokers are commonly called *traveling-grate* stokers. The grate may be either of two designs: *bar grate* or *chain grate*. Alternatively, the burning fuel bed may be conveyed by a vibratory motion of the stoker (*vibrating-grate*).

The fuel flows at right angles to the air flow. Only a small amount of air is fed at the front of the stoker, to keep the fuel mixture rich, but as the coal moves toward the middle of the furnace, the amount of air is increased, and most of the coal is burned by the time it gets halfway down the length of the grate. Fuel-bed depth varies from 100 to 200 mm (4 to 8 in), depending on the fuel, which can be coke breeze, anthracite, or any noncaking bituminous coal.

Overfeed Firing: Spreader Stokers Spreader stokers burn coal (or other fuel) by propelling it into the furnace. A portion of the coal burns in suspension (the percentage depending on the coal fineness), while the rest burns on a grate. In most units, coal is pushed off a plate under the storage hopper onto revolving paddles (either overthrow or underthrow) which distribute the coal on the grate (Fig. 27-22c). The angle and speed of the paddles control coal distribution. The largest coal particles travel the farthest, while the smallest ones become partially consumed during their trajectory and fall on the forward half of the grate. The grate may be stationary or traveling. The fuel and air flow in opposite directions.

Some spreaders use air to transport the coal to the furnace and distribute it, while others use mechanical means to transport the coal to a series of pneumatic jets.

The performance of spreader stokers is affected by changes in coal sizing. The equipment can distribute a wide range of fuel sizes, but it distributes each particle on the basis of size and weight. Normal size specifications call for 19-mm (¾-in) nut and slack with not more than 30 percent less than 6.4 mm (¼ in).

Typically, approximately 30 to 50 percent of the coal is burned in suspension. If excessive fines are present, more coal particles will be carried out of the furnace and burned in suspension, and very little ash will be available to provide a protective cover for the grate surface. On the other hand, if sufficient fines are not present, not all the fuel will

be burned on the grate, resulting in derating of the unit and excessive dumping of live coals to the ash hopper.

Excess air is usually 30 to 40 percent for stationary and dumping grates, while traveling grates are operated with from 22 to 30 percent excess air. Preheated air can be supplied for all types of grates but the temperature is usually limited to 395 to 422 K (250 to 300°F) to prevent excessive slagging of the fuel bed.

Overfire air nozzles are located in the front wall underneath the spreaders and in the rear wall from 0.3 to 0.9 m (1 to 3 ft) above the grate level. These nozzles use air directly from a fan or inspirate air with steam to provide turbulence above the grate for most effective mixing of fuel and air. They supply about 15 percent of the total combustion air.

Comparison of Suspension and Fuel-Bed Firing A major factor to consider when comparing a stoker-fired boiler with a PC boiler is the reduction in efficiency due to carbon loss. The carbon content of the ash passing out of a spreader stoker furnace varies from 30 to 50 percent. Overall efficiency of the stoker can be increased by reburning the ash: it is returned to the stoker grate by gravity or a pneumatic feed system. A continuous-ash-discharge spreader-stoker-fired unit will typically have a carbon loss of 4 to 8 percent, depending on the amount of ash reinjection. A properly designed PC boiler, on the other hand, can maintain an efficiency loss due to unburned carbon of less than 0.4 percent.

A difference between these firing methods may also be manifested in the initial fuel cost. For efficient operation of a spreader-stoker-fired boiler, the coal must consist of a proper mixture of coarse and fine particles. Normally, double-screened coal is purchased because less expensive run-of-mine coal does not provide the optimum balance of coarse and fine material.

An advantage of a stoker-fired furnace is its easy adaptability to firing almost any unsized solid fuels. Bark, bagasse, or refuse can normally be fired on a stoker to supplement the coal with a minimum amount of additional equipment. Thus, such supplementary waste fuels may be able to contribute a higher percentage of the total heat input in a stoker-fired furnace than in a PC furnace without expensive equipment modifications.

Fluidized-Bed Combustion The principles of gas-solid fluidization and their application to the chemical process industry are treated in Section 17. Their general application to combustion is reviewed briefly here, and their more specific application to fluidized-bed boilers is discussed later in this section.

In fluidized-bed combustion (FBC) fuel is burned in a bed of particles supported in an agitated state by an upward flow of air introduced via an air distributor. The bed particles may be sand or ash derived from the fuel, but usually they are a sulfur sorbent, like limestone or dolomite. Fluidized beds have inherently good heat-transfer characteristics, and these ensure even temperatures within the combustor and high flux rates to steam/water cooling circuits. The good gas-solids contacting promotes effective sulfur capture and allows high combustion efficiency to be achieved at temperatures significantly lower than those of a pulverized coal furnace (typically 1116 K [1550°F] compared to over 1589 K [2400°F]). These lower temperatures also result in reduced slagging and fouling problems and significantly lower NO_x formation. This latter benefit, in conjunction with the reduced SO_2 emissions, constitutes one of the great advantages of fluidized-bed combustors: in situ pollution control. Having this control built into the furnace eliminates the need for back-end cleanup and reduces plant capital cost while increasing thermal efficiency.

There are two types of FBC unit distinguished by their operating flow characteristics: *bubbling* and *circulating*. These two types operate at atmospheric pressure, AFBC, or at elevated pressure, PFBC. Pressures for PFBC are in the range 0.6 to 1.6 MPa (90 to 240 psia). Typical superficial fluidizing velocities are tabulated as follows.

	Atmospheric	Pressurized
Bubbling	1.5–2.7 m/s	1–1.2 m/s
	(5–9 ft/s)	(3–4 ft/s)
Circulating	3.7–7.3 m/s	3.7–4.3 m/s
	(12–24 ft/s)	(12–14 ft/s)

Bubbling Beds In bubbling beds a large proportion of the non-combustible feedstock, mainly sorbent derived, remains in the combustor, forming the bed. Bed depth is maintained by draining off excess material. Most of the gas in excess of that required for minimum fluidization appears as bubbles (voids), and these carry particles upward in their wake, promoting the rapid vertical mixing within the bed that results in the even temperatures characteristic of FBC units. Bed temperature is controlled by heat transfer to in-bed boiler tubes and/or to the water-wall tubes used to enclose the furnace. Some units have experienced metal loss from these tube surfaces, a combined effect of erosion and abrasion, and suitable protection needs to be provided. Protective measures include surface coatings such as plasma-sprayed metal coatings incorporating silicon carbide, and metal fins to disrupt the solids-flow pattern.

In AFBC units, heat is removed from the flue gas by a convection-pass tube bank. The particulates leaving the boiler with the flue gas consist of unreacted and spent sorbent, unburned carbon, and ash. Multiclones after the convection pass remove much of the particulate matter and recycle it to the combustor, increasing the in-furnace residence time and improving combustion efficiency and sulfur retention performance. Bubbling PFBC units do not have convection-pass tube banks and do not recycle solids to the boiler.

Circulating Beds These fluidized beds operate at higher velocities, and virtually all the solids are elutriated from the furnace. The majority of the elutriated solids, still at combustion temperature, are captured by reverse-flow cyclone(s) and recirculated to the foot of the combustor. This recycle stage is incorporated into AFBC and PFBC units, but only the AFBC unit has a convection pass downstream of the cyclone. The foot of the combustor is a potentially very erosive region, as it contains large particles not elutriated from the bed, and they are being fluidized at high velocity. Consequently, the lower reaches of the combustor do not contain heat-transfer tubes and the water walls are protected with refractory. Some combustors have experienced damage at the interface between the water walls and the refractory, and measures similar to those employed in bubbling beds have been used to protect the tubes in this region.

The furnace temperature is controlled by heat transfer through the exposed upper water-wall tubes. As the units increase in size, more heat-transfer surface is required than is provided by the walls. Surface can be added by wrapping horizontal tubing over the walls of the upper furnace, or by added *wing walls,* sections of water wall extending short distances into the furnace enclosure. In some designs, tubes are extended across the upper furnace where, although the fluidizing velocity is still high, the erosion potential is low because the solids are finer and their concentration is lower. In some designs, heat is removed from the recirculated solids by passing them through a bubbling-bed heat exchanger before returning them to the furnace.

In the bubbling version, all the air is introduced through the distributor plate, but for the circulating units, 30 to 40 percent is introduced above the distributor. This staged entry results in the lower reaches operating substoichiometrically, which helps to reduce NO_x emissions but tends to reduce the fluidizing velocity at the base of the combustor. To compensate for this and increase the mixing by increasing the local gas velocity, the portion of the combustor below the secondary air entry points is tapered. Staging is not generally employed on bubbling units because the oxygen deficiency in the bed tends to accelerate corrosion of the in-bed tube bank.

Fuel Flexibility An advantage of FBC designs is fuel flexibility: a single unit can burn a wider range of fuels than a PC furnace, thus offering owners an improved bargaining position to negotiate lower fuel prices. Among the fuels fired are bituminous and subbituminous coals, anthracite culm, lignite, petroleum coke, refuse-derived fuel, biomass, industrial and sewage sludges, and shredded tires. But fuel flexibility can be achieved only if the unit is designed for the range of fuels intended to be burned. For example, to maintain the same firing rate, a feed system designed for a certain fuel must be capable of feeding a lower calorific fuel at a higher rate. Similarly, to maintain the same degree of sulfur capture, feeders must be capable of delivering sorbent over a range of rates matching the sulfur contents of the fuels likely to be fed.

Sulfur Emissions Sulfur present in a fuel is released as SO_2, a known contributor to acid rain deposition. By adding limestone or dolomite to a fluidized bed, much of this can be captured as calcium sulfate, a dry nonhazardous solid. As limestone usually contains over 40 percent calcium, compared to only 20 percent in dolomite, it is the preferred sorbent, resulting in lower transportation costs for the raw mineral and the resulting ash product. Moreover, the high magnesium content of the dolomite makes the ash unsuitable for some building applications and so reduces its potential for utilization. Whatever sorbent is selected, for economic reasons it is usually from a source local to the FBC plant. If more than one sorbent is available, plant trials are needed to determine the one most suitable, as results from laboratory-scale reactivity assessments are unreliable.

At atmospheric pressure, calcium carbonate almost completely calcines to free lime, and it is this that captures the sulfur dioxide. As the free lime is not completely sulfated, the resulting sorbent ash is very alkaline, consisting primarily of $CaSO_4$ and CaO, with small amounts of $CaCO_3$.

$$CaCO_3 \rightleftharpoons CaO + CO_2 \qquad (27\text{-}46)$$

$$CaO + SO_2 + [O] \rightleftharpoons CaSO_4 \qquad (27\text{-}47)$$

The sulfation reaction has an optimum at a mean bed temperature of around 1116 K (1550°F).

At elevated pressure, the partial pressure of carbon dioxide inhibits calcination, and sulfur dioxide is captured by displacement of the carbonate radical. The overall effect is similar except, as no free lime is formed, the resulting sorbent ash is less alkaline, consisting solely of $CaSO_4$ and $CaCO_3$.

$$CaCO_3 + SO_2 + [O] \rightleftharpoons CaSO_4 + CO_2 \qquad (27\text{-}48)$$

The sulfation reaction does not have an optimum reaction temperature under pressurized operating conditions and the higher partial pressure of oxygen results in increased conversion of sulfur dioxide to sulfur trioxide.

$$SO_2 + [O] \rightleftharpoons SO_3 \qquad (27\text{-}49)$$

Under normal operating conditions, the concentration of the trioxide is unlikely to exceed 10 ppmv, but this is sufficient to elevate the acid dew point to around 422 K (300°F). This places a limit on the lowest acceptable back-end temperature if acid condensation and resulting corrosion problems are to be avoided.

Nitrogen Oxide Emissions FBC units achieve excellent combustion and sulfur emission performance at relatively modest combustion temperatures in the range 1060 to 1172 K (1450 to 1650°F). At these temperatures no atmospheric nitrogen is converted to NO_x, and only a small percentage of the fuel nitrogen is converted. Typical NO_x emissions, consisting of around 90 percent NO and 10 percent NO_2, are in the range 86 to 129 mg/MJ (0.2 to 0.3 lb/10^6 Btu). In circulating AFBCs and in bubbling PFBCs, these values have been reduced to as low as 21 mg/MJ (0.05 lb/10^6 Btu) by injecting ammonia into the boiler freeboard to promote selective noncatalytic reduction (SNCR) reactions. In AFBC units, the prime variables influencing NO_x formation are excess air, mean bed temperature, the nitrogen content of the fuel, and the Ca/S molar ratio. With respect to the latter, high sorbent feed rates increase the free lime content, which catalyzes NO_x formation. In PFBC units, only excess air and fuel nitrogen content have an influence, and there appears to be no effect of pressure.

Particulate Emissions To meet environmental regulations, AFBC boilers, and some PFBC boilers, use a back-end particulate collector, such as a baghouse or an electrostatic precipitator (ESP). Compared to PC units, the ash from FBCs has higher resistivity and is finer because the flue-gas path contains cyclones. Both factors result in reduced ESP collection efficiency with AFBC units, but good performance has been achieved with PFBC units, where the SO_3 present in the flue gas lowers the ash resistivity. In general, however, baghouses are the preferred collection devices for both AFBC and PFBC applications.

FBC ash is irregular, whereas PC ash, because it melts at the elevated operating temperatures, is spherical. This difference in shape influences baghouse design in three ways: (1) FBC ash does not flow from the collection hoppers as readily and special attention has to be given to their design; (2) FBC ash forms a stronger cake, requiring more frequent and more robust cleaning mechanisms, e.g., shake-deflate and pulse-jet technologies; and (3) this more robust action in conjunction with the more abrasive, irregular particles results in filter bags being more prone to failure in FBC systems. Careful selection of bag materials (synthetic felts generally perform best) and good installation and maintenance practices minimize the latter problem.

Some PFBC boiler designs incorporate high-temperature, high-pressure (HTHP) filter devices in the flue-gas stream. These are installed primarily to protect the gas turbine from erosion damage by the fine particles that escape the cyclones, but as the filters remove virtually all the suspended particulates, they also eliminate the need for back-end removal. The commonest HTHP filter elements used are rigid ceramic candles.

COMBUSTION OF LIQUID FUELS

Oil is typically burned as a suspension of droplets generated by atomizing the fuel. As the droplets pass from the atomizer into the flame zone, they are heated both by radiation from the flame and by convection from the hot gases that surround them, and the lighter fuel components vaporize. The vapors mix with surrounding air and ignite. Depending on the fuel type, the fuel droplet may be completely vaporized or it may be partially vaporized, leaving a residual char or coke particle.

Fuel oils can contain a significant amount of sulfur: in the case of high-sulfur No. 6, it may be as much as 4 percent (Table 27-6). SO_2 is the principal product of sulfur combustion with stoichiometric or leaner fuel-air mixtures, but with the excess air customarily used for satisfactory combustion, SO_3 can form and then condense as sulfuric acid at temperatures higher than the normally expected dew point. Thus air preheaters and other heat recovery equipment in the flue-gas stream can be endangered. Figure 27-23 shows the maximum safe upper limits for dew points in the stacks of furnaces burning sulfur-containing oil and emitting unscrubbed flue gas.

Atomizers   Atomization is the process of breaking up a continuous liquid phase into discrete droplets. Figure 27-24 shows the idealized process by which the surface area of a liquid sheet is increased until it forms droplets. Atomizers may be classified into two broad groups (see Fig. 27-25): pressure atomizers, in which fuel oil is injected at high pressure, and twin-fluid atomizers, in which fuel oil is injected at moderate pressure and a compressible fluid (steam or air) assists in the atomization process. Low oil viscosity (less than 15 mm²/s) is required for effective atomization (i.e., small droplet size). Light oils, such as No. 2 fuel oil, may be atomized at ambient temperature. However, heavy oils must be heated to produce the desired viscosity.

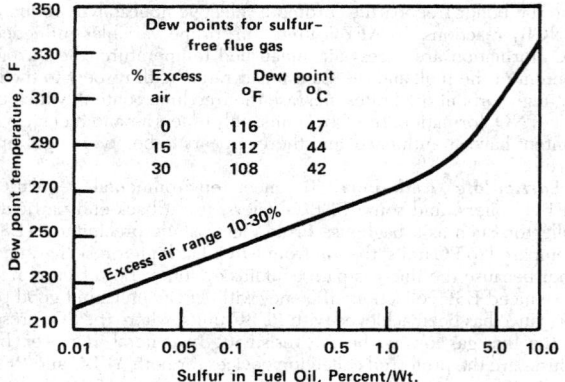

FIG. 27-23   Maximum flue-gas dew point versus percent of sulfur in typical oil fuels. (K = [°F + 459.7]/1.8)

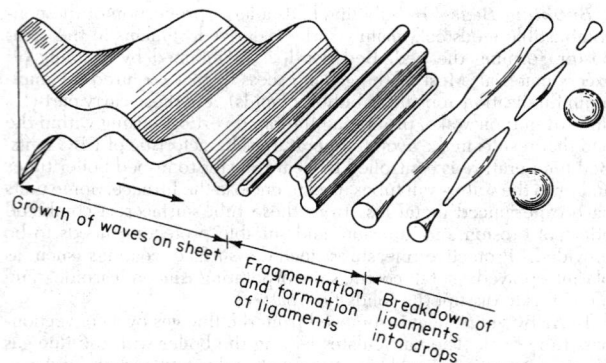

FIG. 27-24   Idealized process of drop formation by breakup of a liquid sheet. (*After Dombrowski and Johns,* Chem. Eng. Sci. **18:** 203, 1963.)

Required preheats vary from approximately 373 K (212°F) for No. 6 oil to 623 K (480°F) for vacuum bottoms.

Pressure Atomizers   The commonest type of pressure atomizer is the swirl-type (Fig. 27-26). Entering a small cup through tangential orifices, the oil swirls at high velocity. The outlet forms a dam around the open end of the cup, and the oil spills over the dam in the form of a thin conical sheet, which subsequently breaks up into thin filaments and then droplets. Depending on the fuel viscosity, operating pressures range from 0.69 to 6.9 MPa (100 to 1000 psia) and the attainable turndown ratio is approximately 4:1. Pressure atomization is most effective for lighter fuel oils.

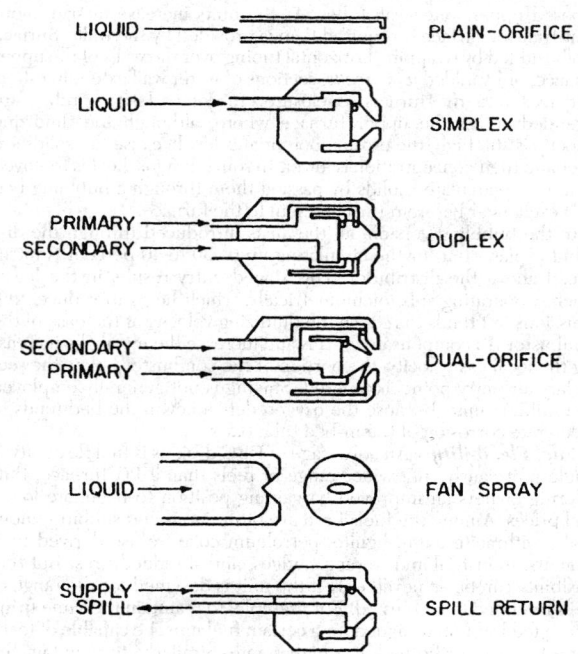

FIG. 27-25a   Common types of atomizers: pressure atomizers. (*From Lefebvre,* Atomization and Sprays, *Hemisphere, New York, 1989. Reproduced with permission. All rights reserved.*)

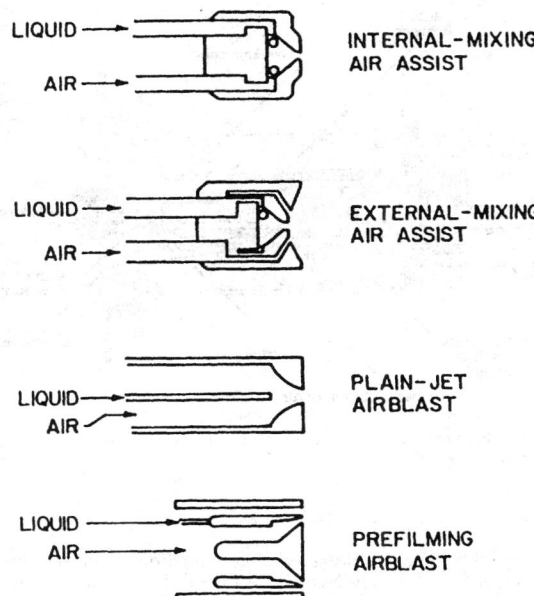

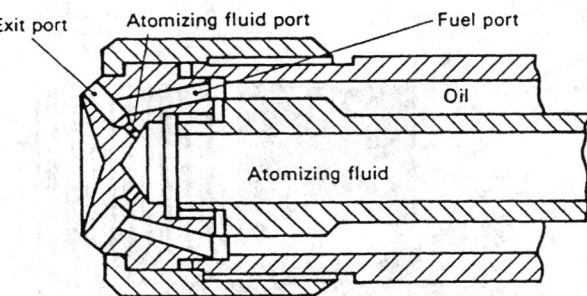

Twin-Fluid Atomizers In a twin-fluid atomizer, the fuel stream is exposed to a stream of air or steam flowing at high velocity. In the internal-mixing configuration (Fig. 27-27), the liquid and gas mix inside the nozzle before discharging through the outlet orifice. In the external-mixing nozzle, the oil stream is impacted by the high-velocity gas stream outside the nozzle. The internal type requires lower flows of secondary fluid. In industrial combustion systems, steam is the preferred atomizing medium for these nozzles. In gas turbines, compressed air is more readily available. Maximum oil pressure is about 0.69 MPa (100 psia), with the steam or air pressure being maintained about 0.14 to 0.28 MPa (20 to 40 psia) in excess of the oil pressure. The mass flow of atomizing fluid varies from 5 to 30 percent of the fuel flow rate, and represents only a modest energy consumption. Turndown performance is better than for pressure atomizers and may be as high as 20:1.

A well-designed atomizer will generate a cloud of droplets with a mean size of about 30 to 40 μm and a top size of about 100 μm for light oils such as No. 2 fuel oil. Mean and top sizes are somewhat larger than this for heavier fuel oils.

Oil Burners The structure of an oil flame is shown in Fig. 27-28, and Fig. 27-29 illustrates a conventional circular oil burner for use in boilers. A combination of stabilization techniques is used, typically including swirl. It is important to match the droplet trajectories to the combustion aerodynamics of a given burner to ensure stable ignition and good turndown performance.

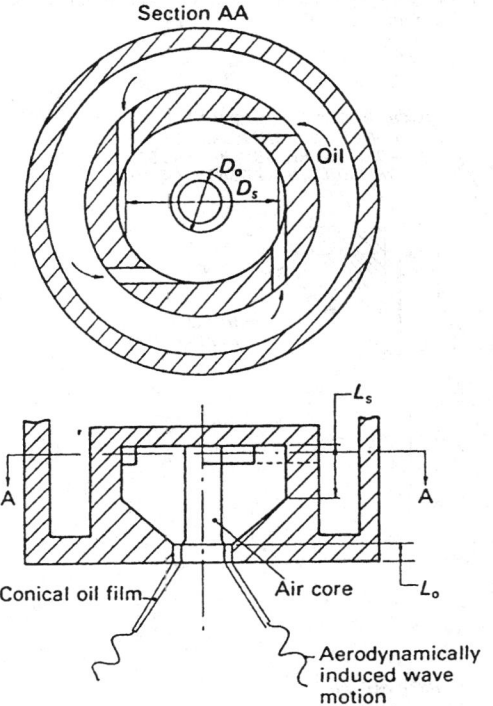

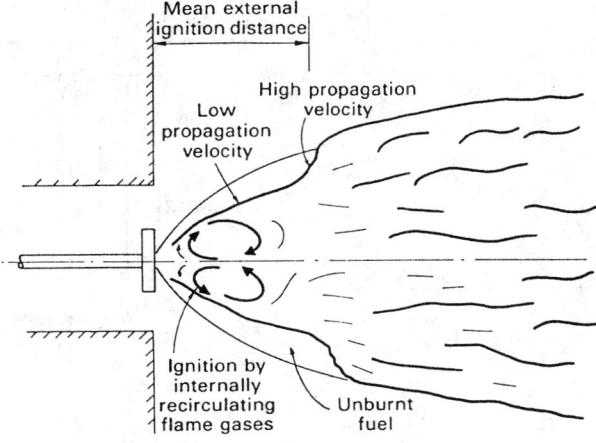

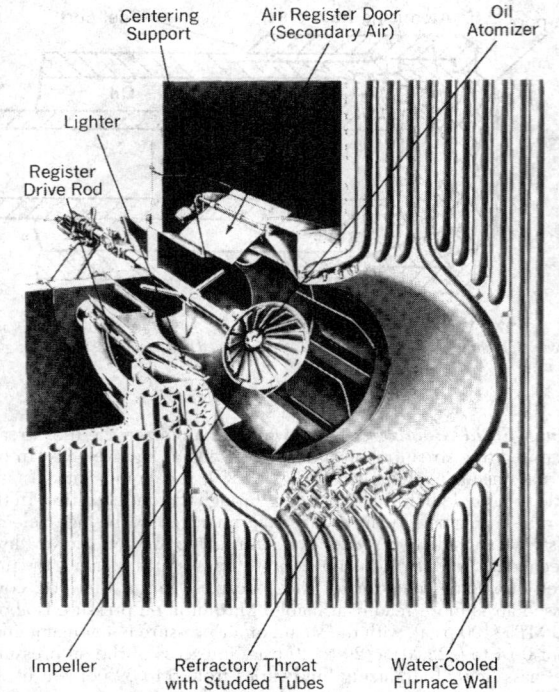

FIG. 27-29 Circular register burner with water-cooled throat for oil firing. (*Babcock & Wilcox Co.*)

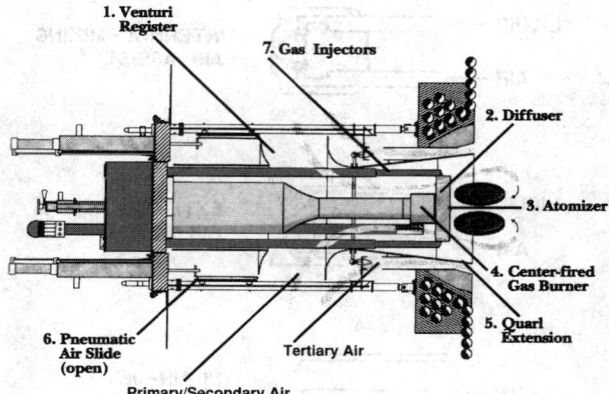

FIG. 27-30 Low-NO$_x$ combination oil/gas forced-draft boiler burner. (*Todd Combustion, Inc.*)

Many oil burners are designed as combination gas/oil burners. An example of a modern low-NO$_x$ oil/gas forced-draft burner is shown in Fig. 27-30. This is an air-staged design, with the air divided into primary, secondary, and tertiary streams. An air-staged natural draft process heater oil/gas burner is illustrated in Fig. 27-31.

Emissions of unburned carbon (primarily coke cenospheres) may be reduced by (1) achieving smaller average fuel droplet size (e.g., by heating the fuel to lower its viscosity or by optimizing the atomizer geometry), (2) increasing the combustion air preheat temperature, or (3) firing oils with high vanadium content (vanadium appears to catalyze the burnout of coke).

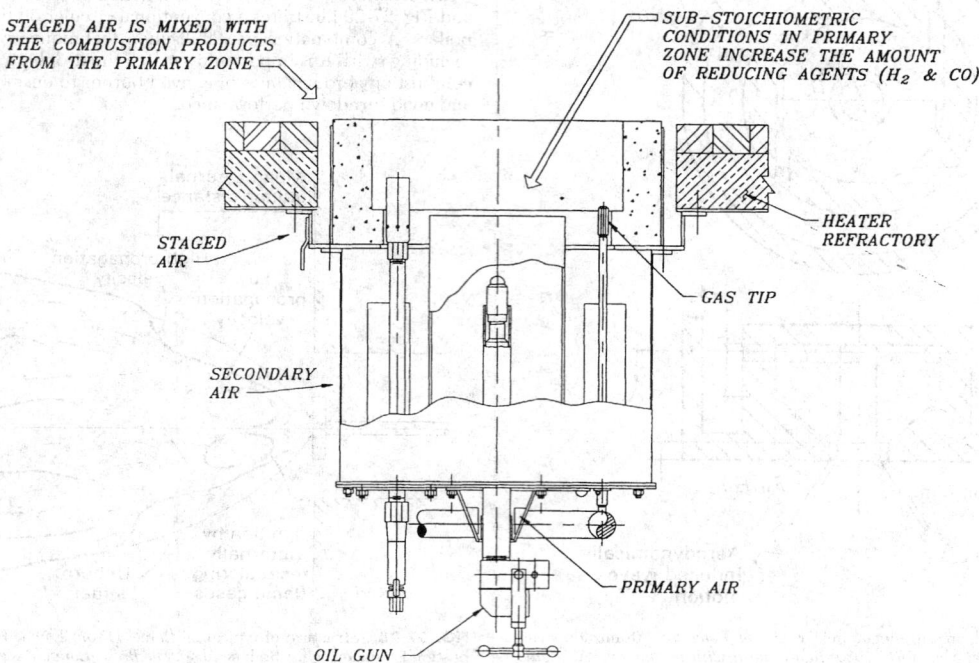

FIG. 27-31 Air-staged natural-draft combination oil/gas burner. (*Callidus Technologies, Inc.*)

COMBUSTION OF GASEOUS FUELS

Combustion of gas takes place in two ways, depending upon when gas and air are mixed. When gas and air are mixed before ignition, as in a Bunsen burner, burning proceeds by hydroxylation. The hydrocarbons and oxygen form hydroxylated compounds that become aldehydes; the addition of heat and additional oxygen breaks down the aldehydes to H_2, CO, CO_2, and H_2O. Inasmuch as carbon is converted to aldehydes in the initial stages of mixing, no soot can be developed even if the flame is quenched.

Cracking occurs when oxygen is added to hydrocarbons after they have been heated, decomposing the hydrocarbons into carbon and hydrogen, which, when combined with sufficient oxygen, form CO_2 and H_2O. Soot and carbon black are formed if insufficient oxygen is present or if the combustion process is arrested before completion.

Gas Burners   Gas burners may be classified as premixed or nonpremixed. Many types of flame stabilizer are employed in gas burners (see Fig. 27-32). Bluff body, swirl, and combinations thereof are the predominant stabilization mechanisms.

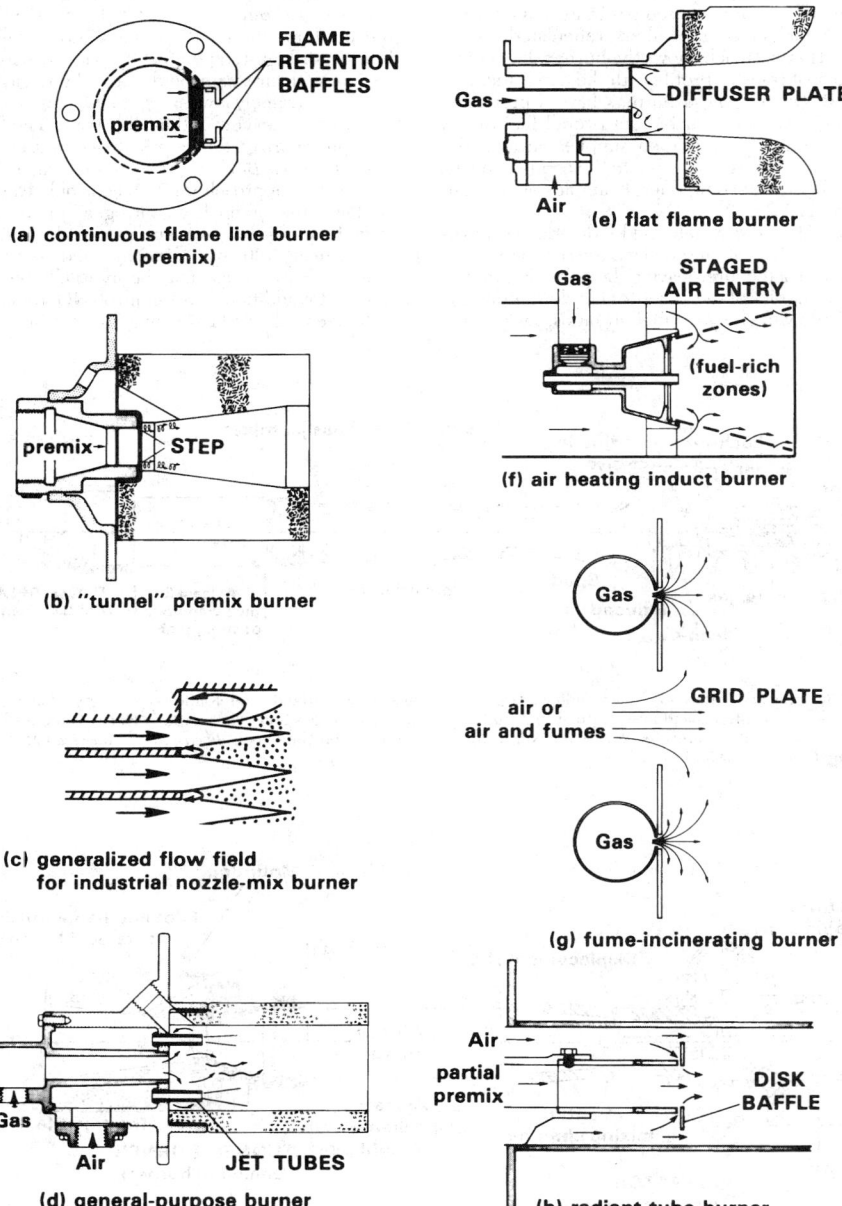

FIG. 27-32   Flame-holding arrangements. Cases (*a*) through (*f*) are various forms of bluff bodies, creating recirculation and fine-scale turbulence in their wake. Cases (*d*) through (*g*) constitute air jets blasting through a relatively quiescent volume of raw gas. Case (*f*) may be cylindrical, as in a gas turbine burner, or trough-like. (*From North American Combustion Handbook, 3d ed., North American Manufacturing Company, Cleveland, 1996.*)

Fully Premixed Burners   A fully premixed burner includes a section for completely mixing the fuel and air upstream of the burner. The burner proper consists essentially of a flame holder; see for example, Fig. 27-32a, b. The porting that admits the mixture to the combustion chamber is designed to produce a fairly high velocity through a large number of orifices to avoid the possibility of the flame flashing back through the flame holder and igniting the mixture upstream of the burner.

Surface combustion devices are designed for fully premixing the gaseous fuel and air and burning it on a porous radiant surface. The close coupling of the combustion process with the burner surface results in low flame temperatures and, consequently, low NO_x formation. Surface materials can include ceramic fibers, reticulated ceramics, and metal alloy mats. This approach allows the burner shape to be customized to match the heat transfer profile with the application.

Partially Premixed Burners   These burners have a premixing section in which a mixture that is flammable but overall fuel-rich is generated. Secondary combustion air is then supplied around the flame holder. The fuel gas may be used to aspirate the combustion air or vice versa, the former being the commoner. Examples of both are provided in Figs. 27-33 and 27-34.

Nozzle-Mix Burners   The most widely used industrial gas burners are of the nozzle-mix type. The air and fuel gas are separated until they are rapidly mixed and reacted after leaving the ports. Figure 27-32c, d, e, f, h shows some examples of the variety of nozzle-mix designs in use. These burners allow a wide range of fuel-air ratios, a wide vari-

ety of flame shapes, and multifuel firing capabilities. They can be used to generate special atmospheres by firing at very rich conditions (50 percent excess fuel) or very lean conditions (1000 percent excess air). By changing nozzle shape and degree of swirl, the flame profile and mixing rates can be varied widely, from a rapid-mixing short flame ($L/D = 1$), to a conventional flame ($L/D = 5$ to 10), to a slow-mixing long flame ($L/D = 20$ to 50).

Staged Burners   As was pointed out earlier under "Pollutant Formation and Control in Flames," the proper staging of fuel or air in the combustion process is one technique for minimizing NO_x emissions. Gas burners that achieve such staging are available.

Air-Staged Burners   Low-NO_x air-staged burners for firing gas (or oil) are shown in Figs. 27-30 and 27-31. A high-performance, low-NO_x burner for high-temperature furnaces is shown in Fig. 27-35. In this design, both air-staging and external flue-gas recirculation are used to achieve extremely low levels of NO_x emissions (approximately 90 percent lower than conventional burners). The flue gas is recirculated by a jet-pump driven by the primary combustion air.

Fuel-Staged Burners   Use of fuel-staged burners is the preferred combustion approach for NO_x control because gaseous fuels typically contain little or no fixed nitrogen. Figure 27-36 illustrates a fuel-staged natural draft refinery process heater burner. The fuel is split into primary (30 to 40 percent) and secondary (60 to 70 percent) streams. Furnace gas may be internally recirculated by the primary gas jets for additional NO_x control. NO_x reductions of 80 to 90 percent have been achieved by staging fuel combustion.

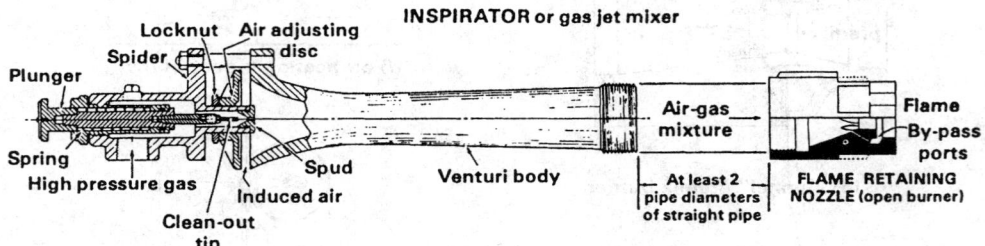

FIG. 27-33   Inspirator (gas-jet) mixer feeding a large port premix nozzle of the flame retention type. High-velocity gas emerging from the spud entrains and mixes with air induced in proportion to the gas flow. The mixture velocity is reduced and pressure is recovered in the venturi section. (*From* North American Combustion Handbook, *3d ed., North American Manufacturing Company, Cleveland, 1996.*)

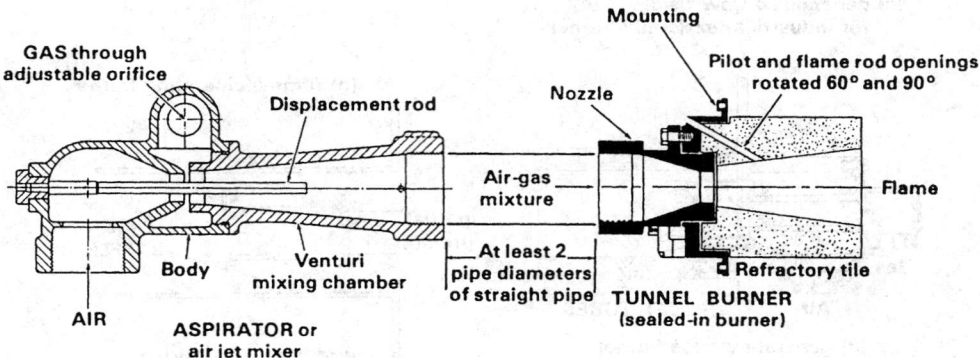

FIG. 27-34   Aspirator (air-jet) mixer feeding a sealed-in large port premix tunnel burner. Blower air enters at lower left. Gas from an atmospheric regulator is pulled into the air stream from the annular space around the venturi throat in proportion to the air flow. (*From* North American Combustion Handbook, *3d ed., North American Manufacturing Company, Cleveland, 1996.*)

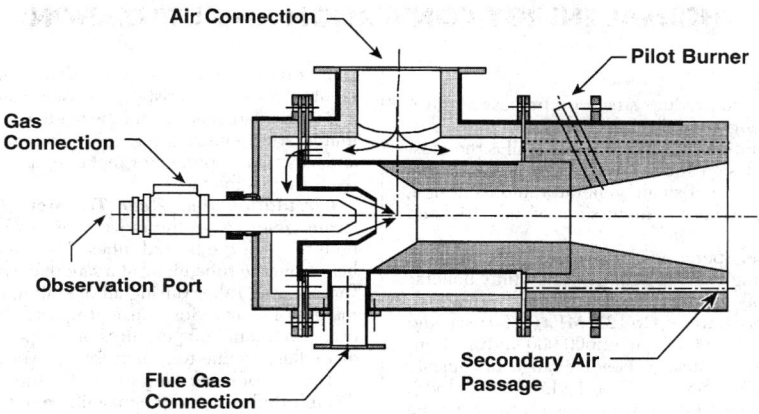

FIG. 27-35 Low-NO$_x$ burner with air-staging and flue-gas recirculation for use in high-temperature furnaces. (*Hauck Manufacturing Company. Developed and patented by the Gas Research Institute.*)

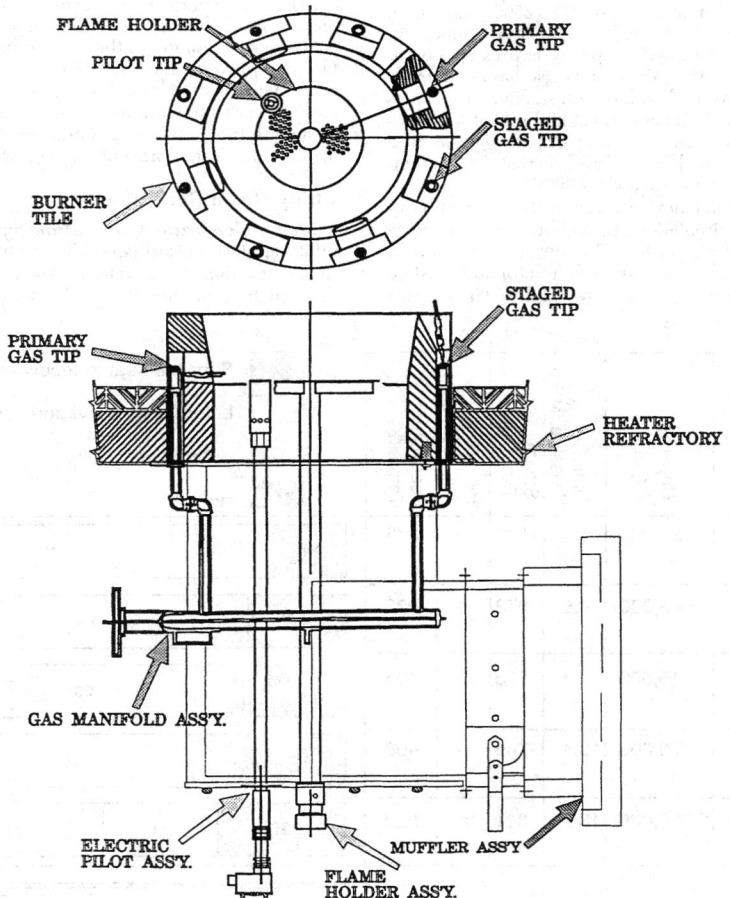

FIG. 27-36 Low-NO$_x$ fuel-staged burner for a natural draft refinery process heater. (*Callidus Technologies, Inc.*)

THERMAL ENERGY CONVERSION AND UTILIZATION

BOILERS

Steam generators are designed to produce steam for process requirements, for process needs along with electric power generation, or solely for electric power generation. In each case, the goal is the most efficient and reliable boiler design for the least cost. Many factors influence the selection of the type of steam generator and its design, and some of these will be treated later in discussions of industrial and utility boilers.

Figure 27-37 shows the chief operating characteristics of a range of boilers, from small-scale heating systems to large-scale utility boilers.

In the industrial market, boilers have been designed to burn a wide range of fuels and operate at pressures up to 12.4 MPa (1800 psia) and steaming rates extending to 455,000 kg/h (1,000,000 lb/h). High-capacity shop-assembled boilers (package boilers) range in capacity from 4545 kg/h (10,000 lb/h) to about 270,000 kg/h (600,000 lb/h). These units are designed for operation at pressures up to 11.4 MPa (1650 psia) and 783 K (950°F). Figure 27-38 shows a gas- or liquid-fuel-fired unit. While most shop-assembled boilers are gas- or oil-fired, designs are available to burn pulverized coal. A field-erected coal-fired industrial boiler is shown in Fig. 27-39.

Boilers designed for service in electric power utility systems operate at both subcritical-pressure (pressures below 22.1 MPa [3205 psia]) and supercritical-pressure steam conditions. Subcritical-pressure boilers range in design pressures up to about 18.6 MPa (2700 psia) and in steaming capacities up to about 2955 Mg/h (6,500,000 lb/h). Supercritical-pressure boilers have been designed to operate at pressures up to 34.5 MPa (5000 psia). The 24.1 MPa (3500 psia) cycle has been firmly established in the utility industry, and boilers with steaming capacities up to 4227 Mg/h (9,300,000 lb/h) and superheat and reheat temperatures of 814 K (1005°F) are in service. The furnace of a large coal-fired steam generator absorbs half of the heat released, so that the gas temperature leaving the furnace is about 1376 K (2000°F).

Boiler Design Issues Boiler design involves the interaction of many variables: water-steam circulation, fuel characteristics, firing systems and heat input, and heat transfer. The furnace enclosure is one of the most critical components of a steam generator and must be conservatively designed to assure high boiler availability. The furnace configuration and its size are determined by combustion requirements, fuel characteristics, emission standards for gaseous effluents and particulate matter, and the need to provide a uniform gas flow and temperature entering the convection zone to minimize ash deposits and excessive superheater metal temperatures. Discussion of some of these factors follows.

Circulation and Heat Transfer Circulation, as applied to a steam generator, is the movement of water or steam or a mixture of both through the heated tubes. The circulation objective is to absorb heat from the tube metal at a rate that assures sufficient cooling of the furnace-wall tubes during all operating conditions, with an adequate margin of reserve for transient upsets. Adequate circulation prevents excessive metal temperatures or temperature differentials that would cause failures due to overstressing, overheating, or corrosion.

The rate of heat transfer from the tubes to the fluid depends primarily on turbulence and the magnitude of the heat flux itself. Turbulence is a function of mass velocity of the fluid and tube roughness. Turbulence has been achieved by designing for high mass velocities, which ensure that nucleate boiling takes place at the inside surface of the tube. If sufficient turbulence is not provided, *departure from nucleate boiling* (DNB) occurs. DNB is the production of a film of steam on the tube surface that impedes heat transfer and results in tube overheating and possible failure. This phenomenon is illustrated in Fig. 27-40.

Satisfactory performance is obtained with tubes having helical ribs on the inside surface, which generate a swirling flow. The resulting centrifugal action forces the water droplets toward the inner tube surface and prevents the formation of a steam film. The internally rifled tube maintains nucleate boiling at much higher steam temperature and pressure and with much lower mass velocities than those needed in smooth tubes. In modern practice, the most important criterion in drum boilers is the prevention of conditions that lead to DNB.

Utility Steam Generators

Steam-Generator Circulation System Circulation systems for utility application are generally classified as natural circulation and forced or pump-assisted circulation in drum-type boilers, and as once-through flow in subcritical- and supercritical-pressure boilers. The

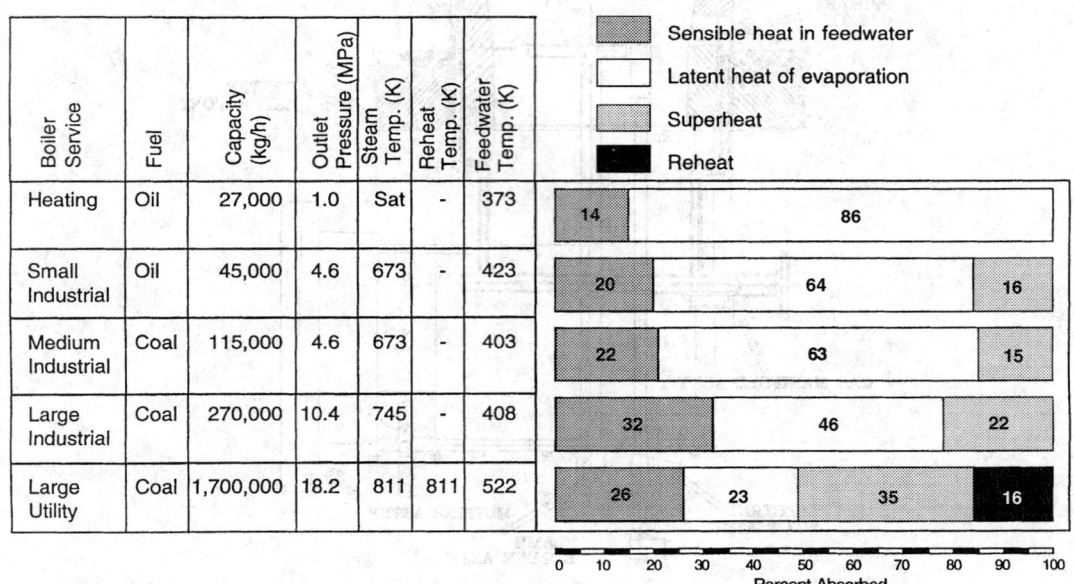

Boiler Service	Fuel	Capacity (kg/h)	Outlet Pressure (MPa)	Steam Temp. (K)	Reheat Temp. (K)	Feedwater Temp. (K)	
Heating	Oil	27,000	1.0	Sat	-	373	Sensible heat in feedwater; Latent heat of evaporation; Superheat; Reheat
Small Industrial	Oil	45,000	4.6	673	-	423	
Medium Industrial	Coal	115,000	4.6	673	-	403	
Large Industrial	Coal	270,000	10.4	745	-	408	
Large Utility	Coal	1,700,000	18.2	811	811	522	

FIG. 27-37 Heat absorption distribution for various types of boilers. (*Adapted from Singer*, Combustion—Fossil Power, *4th ed., Combustion Engineering, Inc., Windsor, Conn., 1991.*)

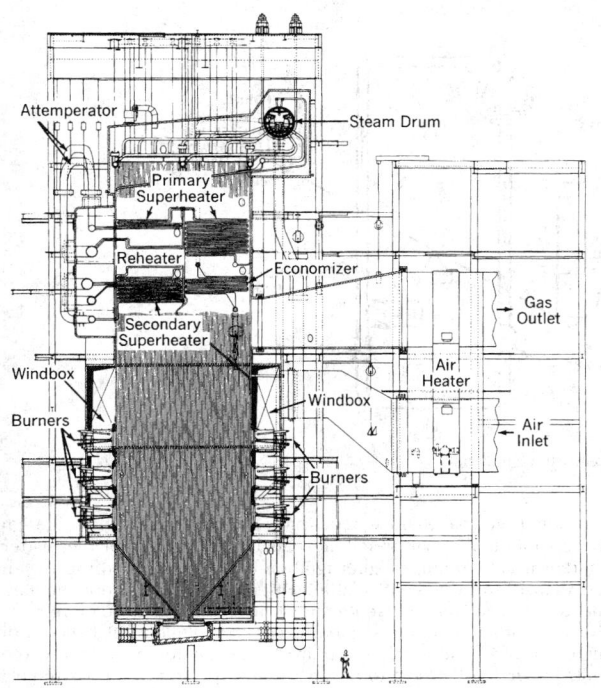

FIG. 27-38   Shop-assembled radiant boiler for natural gas or oil. (*Babcock & Wilcox Co.*)

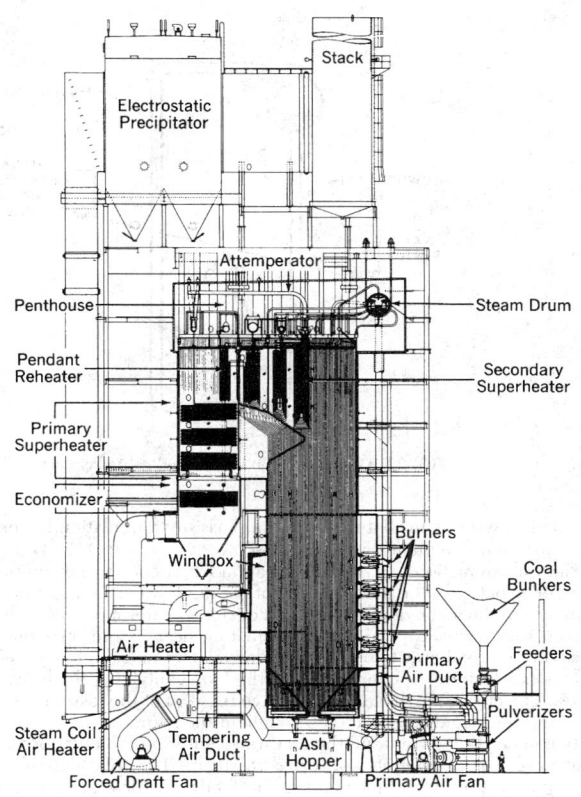

FIG. 27-39   Field-erected radiant boiler for pulverized coal. (*Babcock & Wilcox Co.*)

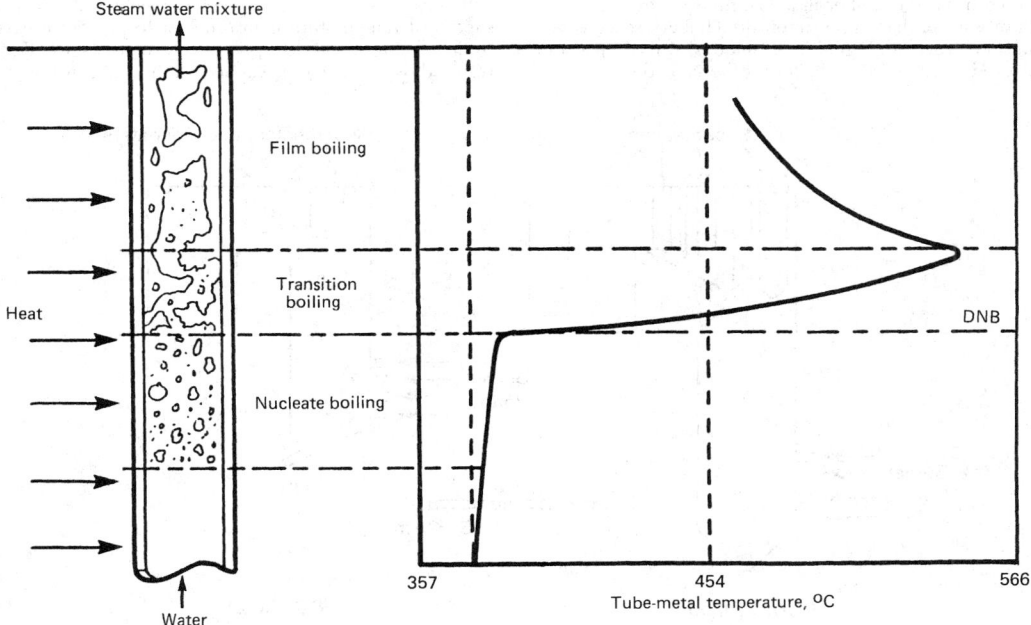

FIG. 27-40   Effect of departure from nucleate boiling (DNB) on tube-metal temperature.

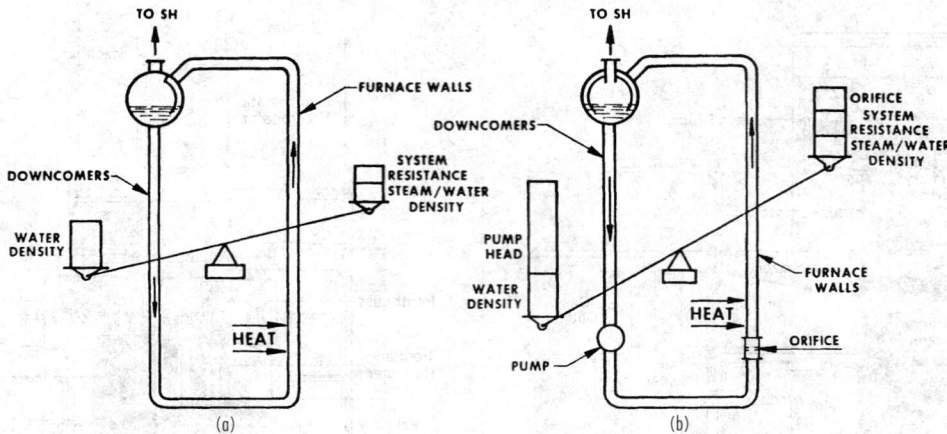

FIG. 27-41 Circulation systems: (*a*) natural circulation; (*b*) pump-assisted circulation.

circulation systems for natural and pump-assisted circulation boilers are illustrated schematically in Fig. 27-41.

Natural circulation in a boiler circulation loop depends only on the difference between the mean density of the fluid (water) in the downcomers and the mean density of the fluid (steam-water mixture) in the heated furnace walls. The actual circulating head is the difference between the total gravity head in the downcomer and the integrated gravity heads in the upcoming legs of the loop containing the heated tubes. The circulating head must balance the sum of the losses due to friction, shock, and acceleration throughout the loop.

In a once-through system, the feedwater entering the unit absorbs heat until it is completely converted to steam. The total mass flow through the waterwall tubes equals the feedwater flow and, during normal operation, the total steam flow. As only steam leaves the boiler, there is no need for a steam drum.

Fuel Characteristics Fuel choice has a major impact on boiler design and sizing. Because of the heat transfer resistance offered by ash deposits in the furnace chamber in a coal-fired boiler, the mean absorbed heat flux is lower than in gas- or oil-fired boilers, so a greater surface area must be provided. Figure 27-42 shows a size comparison between a coal-fired and an oil-fired boiler for the same duty.

In addition, coal characteristics have a major impact on the design and operation of a coal-fired boiler. Coals having a low volatile-matter content usually require higher ignition temperatures and those with less than 12 to 14 percent volatile matter may require supplementary fuel to stabilize ignition. Generally, western U.S. coals are more reactive than others and, consequently, easier to ignite, but because of high moisture content they require higher air temperatures to the mills for drying the coal to achieve proper pulverization. Extremely high-ash coal also may present problems in ignition and stabilization. The ash constituents and the quantity of ash will have a decided influence on sizing the furnace. Accordingly, a thorough review of coal characteristics is needed to establish the effect on the design and operation of a boiler.

Superheaters and Reheaters A superheater raises the temperature of the steam generated above the saturation level. An important function is to minimize moisture in the last stages of a turbine to avoid blade erosion. With continued increase of evaporation temperatures and pressures, however, a point is reached at which the available superheat temperature is insufficient to prevent excessive moisture from forming in the low-pressure turbine stages. This condition is resolved by removing the vapor for reheat at constant pressure in the

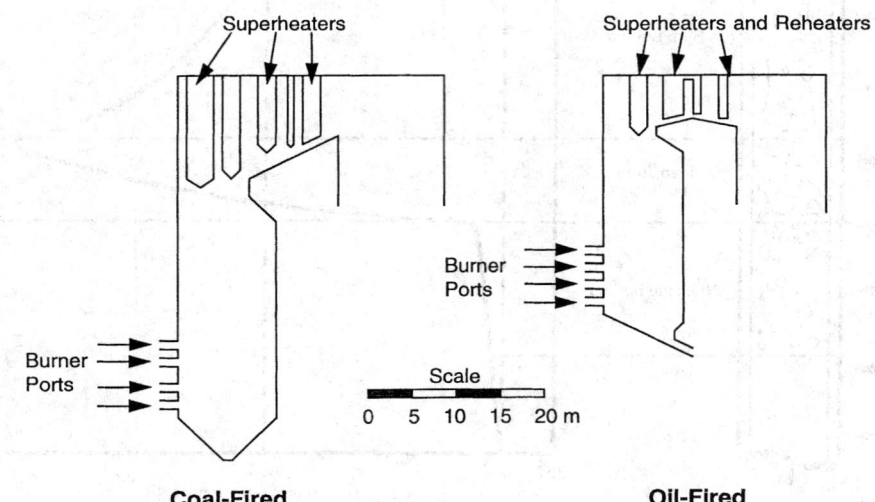

FIG. 27-42 Comparison of the sizes and shapes of typical 500-MW$_e$ coal- and oil-fired boilers. (*Adapted with permission from Lawn*, Principles of Combustion Engineering for Boilers, *Academic Press, London, 1987.*)

boiler and returning it to the turbine for continued expansion to condenser pressure. The thermodynamic cycle using this modification of the Rankine cycle is called the *reheat cycle*.

Economizers Economizers improve boiler efficiency by extracting heat from the discharged flue gases and transferring it to feedwater, which enters the steam generator at a temperature appreciably lower than the saturation-steam temperature.

Industrial Boilers Industrial boilers are steam generators that provide power, steam, or both to an industrial plant, in contrast to a utility boiler in a steam power plant. A common configuration is a stationary water-tube boiler in which some of the steam is generated in a convection-section tube bank (also termed a *boiler bank*). In the original industrial boilers, in fact, almost all of the boiling occurred in that section, but now many industrial steam generators of 180,000 kg/h (397,000 lb/h) and greater capacity are radiant boilers. The boiler steam pressure and temperature and feedwater temperature determine the fraction of total heat absorbed in the boiler bank. For a typical coal-fired boiler producing about 90,000 kg/h (198,000 lb/h):

%, total steam in boiler bank	Boiler pressure		Steam temperature		Feedwater temperature	
	MPa	psia	K	°F	K	°F
45	1.4	200	460	369	389	241
30	4.1	600	672	750	389	241
16	10.3	1500	783	950	450	351
10	12.4	1800	811	1000	450	351

The thicker plate for operation at higher pressures increases the cost of the boiler. As a result, it is normally not economical to use a boiler bank for heat absorption at pressures above 10.7 MPa (1550 psia).

Industrial boilers are employed over a wide range of applications, from large power-generating units with sophisticated control systems, which maximize efficiency, to small low-pressure units for space or process heating, which emphasize simplicity and low capital cost. Although their usual primary function is to provide energy in the form of steam, in some applications steam generation is incidental to a process objective, e.g., a chemical recovery unit in the paper industry, a carbon monoxide boiler in an oil refinery, or a gas-cooling waste-heat boiler in an open-hearth furnace. It is not unusual for an industrial boiler to serve a multiplicity of functions. For example, in a paper-pulp mill, the chemical-recovery boiler is used to convert black liquor into useful chemicals and to generate process steam. At the same plant, a bark-burning unit recovers heat from otherwise wasted material and also generates power.

Industrial boilers burn oil, gas, coal, and a wide range of product and/or waste fuels, some of which are shown in Tables 27-4 and 27-19. Natural gas has become the principal fuel of choice, accounting for approximately two-thirds of all the energy fired in industrial boilers across a wide range of manufacturing industries (Table 27-20). Coal is the second most prevalent fuel, accounting for one-quarter of the energy fired. Waste fuels, however, are increasing in importance.

An excellent brief exposition of industrial boilers is presented as Chapter 8 of *Combustion—Fossil Power*, Singer (ed.), 4th ed., Combustion Engineering, Windsor, Conn., 1991.

Design Criteria Industrial-boiler designs are tailored to the fuels and firing systems involved. Some of the more important design criteria include:
- Furnace heat-release rates, both W/m³ and W/m² of effective projected radiant surface (Btu/[h·ft³] and Btu/[h·ft²]).
- Heat release on grates
- Flue-gas velocities through tube banks
- Tube spacings

Table 27-21 gives typical values or ranges of these criteria for gas, oil, and coal. The furnace release rates are important, for they establish maximum local absorption rates within safe limits. They also have a bearing on completeness of combustion and therefore on efficiency and particulate emissions. Limiting heat release on grates (in stoker firing) will minimize carbon loss, control smoke, and avoid excessive fly ash.

Limits on flue-gas velocities for gas- or oil-fired industrial boilers

TABLE 27-19 Solid-Waste Fuels Burned in Industrial Boilers

Waste	HHV, kJ/kg[a]
Bagasse	8374–11,630
Furfural residue	11,630–13,956
Bark	9304–11,630
General wood wastes	10,467–18,608
Coffee grounds	11,397–15,119
Nut hulls	16,282–18,608
Rich hulls	12,095–15,119
Corncobs	18,608–19,306
Rubber scrap	26,749–45,822
Leather	27,912–45,822
Cork scrap	27,912–30,238
Paraffin	39,077
Cellophane plastics	27,912
Polyvinyl chloride	40,705
Vinyl scrap	40,705
Sludges	4652–27,912
Paper wastes	13,695–18,608

[a]To convert kilojoules per kilogram to British thermal units per pound, multiply by 4.299×10^{-1}.

TABLE 27-20 Fuel Consumption in Boilers in Various Industries

Industry	Annual energy consumption in boilers, PJ/a[a]			
	Residual oil	Distillate oil	Natural gas	Coal
Aluminum	—	—	2	—
Steel	25	—	66	25
Chemicals	29	6	759	257
Forest products	140	4	381	309
Textiles	12	5	74	32
Fabricated metal	2	1	39	5
Industrial machinery & equipment	3	4	115	12
Transportation equipment	—	2	51	—
Food	25	7	332	151
Petroleum refining	35	2	267	3
TOTAL	27	31	2086	794

SOURCE: *Manufacturing Consumption of Energy*, Energy Information Administration, U.S. Dept. Energy, 1991.

[a]Within the limit of their probable accuracy, the values of this table are acceptable also for the units of 10^{12} Btu/a. For accurate *conversion*, however, they must be reduced by 5.2 percent.

TABLE 27-21 Typical Design Parameters for Industrial Boilers

Furnace	Heat-release rate W/m²[a] of EPRS[b]
Natural gas-fired	630,800
Oil-fired	551,900–630,800
Coal: pulverized coal	220,780–378,480
Spreader stoker	252,320–410,020

Stoker, coal-fired	Grate heat-release rate, W/m²
Continuous-discharge spreader	2,050,000–2,207,800
Dump-grade spreader	1,419,300–1,734,700
Overfeed traveling grate	1,261,000–1,734,700

Flue-gas velocity: type Fuel-fired	Single-pass Boiler, m/s	Baffled	
		Boiler, m/s	Economizer, m/s
Gas or distillate oil	30.5	30.5	30.5
Residual oil	30.5	22.9	30.5
Coal (not lignite)			
Low-ash	19.8–21.3	15.2	15.2–18.3
High-ash	15.2	NA[c]	12.2–15.2

[a]To convert watts per square meter to British thermal units per hour-foot, multiply by 0.317.
[b]Effective projected radiant surface.
[c]Not available.

are usually determined by the need to limit draft loss. For coal firing, design gas velocities are established to minimize fouling and plugging of tube banks in high-temperature zones and erosion in low-temperature zones.

Convection tube spacing is important when the fuel is residual oil or coal, especially coal with low ash-fusion or high ash-fouling tendencies. The amount of the ash and, even more important, the characteristics of the ash must be specified for design.

Natural-circulation and convection boiler banks are the basic design features on which a line of standard industrial boilers has been developed to accommodate the diverse steam, water, and fuel requirements of the industrial market.

Figure 27-43 shows the amount of energy available for power by using a fire-tube boiler, an industrial boiler, and subcritical- and supercritical-pressure boilers. Condensing losses decrease substantially, and regeneration of air and feedwater becomes increasingly important in the most advanced central-station boilers.

The boiler designer must proportion heat-absorbing and heat-recovery surfaces in a way to make the best use of heat released by the fuel. Water walls, superheaters, and reheaters are exposed to convection and radiant heat, whereas convection heat transfer predominates in air preheaters and economizers. The relative amounts of these surfaces vary with the size and operating conditions of the boiler.

Package Boilers   In a fire-tube boiler, the hot combustion products flow through tubes immersed in the boiler water, transferring heat to it. In a water-tube boiler, combustion heat is transferred to water flowing through tubes which line the furnace walls and boiler passages. The greater safety of water-tube boilers has long been recognized, and they have generally superseded fire-tube configurations except for small package boiler designs. Fire-tube package boilers range from a few hundred to 18,200 kg/h (40,000 lb/h) steaming capacity. A fire-tube boiler is illustrated in Figs. 27-44 and 27-45. Water-tube package boilers range from a few hundred to 270,000 kg/h (600,000 lb/h) steaming capacity. A water-tube package boiler is illustrated in Fig. 27-46. The majority of water-tube package boilers use natural circulation and are designed for pressurized firing. The most significant advantage of shop-assembled or package boilers is the cost benefit associated with use of standard designs and parts.

Package boilers can be shipped complete with fuel-burning equipment, controls, and boiler trim. It may be necessary to ship the larger units in sections, however, and a shop-assembled boiler with a capacity greater than about 109,000 kg/h (240,000 lb/h) is deliverable only by barge. (For a more detailed discussion of shop-assembled boilers, see Singer, 1991, pp. 8.36–8.42.)

Fluidized-Bed Boilers   As explained in the earlier discussion of coal combustion equipment, the furnace of a fluid-bed boiler has a unique design. The system as a whole, however, consists mainly of standard equipment items, adapted to suit process requirements. The

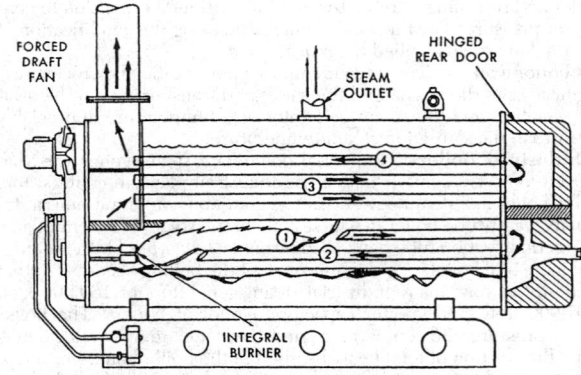

FIG. 27-44   A four-pass packaged fire-tube boiler. Circled numbers indicate passes. (*From Cleaver Brooks, Inc. Reproduced from* Gas Engineer's Handbook, *Industrial Press, New York, 1965, with permission.*)

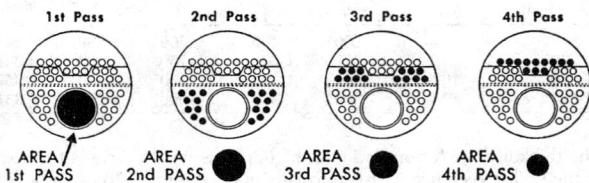

FIG. 27-45   Location and relative size of each of four passes of the flue gas through a fire-tube boiler. (*From Cleaver Brooks, Inc. Reproduced from* Gas Engineer's Handbook, *Industrial Press, New York, 1965, with permission.*)

systems for coal and sorbent preparation and feeding, ash removal, and ash disposal are very similar to those found in PC boiler plants, the major difference being that the top size of material used is greater. The water-wall boiler enclosure and convection-pass tubing are also similar to the designs found in PC boilers. In-bed heat-transfer surfaces are arranged similarly to convection-pass tube banks, but the bubbling action subjects them to higher forces and special consideration has to be given to the design of a suitable support structure. A fluidized-bed plant includes particulate removal equipment such as cyclones, multiclones, baghouses, and electrostatic precipitators, all similar to designs found in other solids-handling process plants.

Bubbling AFBCs   A simplified schematic of a bubbling AFBC is presented in Fig. 27-47. A demonstration plant generating 160 MWe, with a power production intensity° of 1.49 MWe/m² (1 MWe/16 ft²), began operation in 1988. Also operating is a 350-MWe unit that employs many of the same design features.

Although subbituminous coal with a top size of 25 mm (1 in) can be fed overbed with good thermal and environmental performance, less reactive bituminous coal must be fed underbed. This requires that the coal be dried and crushed to a top size of 3 mm (⅛ in). A lock-hopper feeder operating at about 0.15 MPa (21 psia) is required to overcome the combined pressure drop across the bed and along the conveying line. A separate lock-hopper system is required to feed the sorbent, also with a top size of 3 mm (⅛ in), but it is introduced into the boiler through the same conveying line. The mixture enters the bed through T nozzles, one per 2 m² (22 ft²) of bed floor area, that divide the flow and distribute it horizontally beneath the tube bank. There is approximately 0.45 m (18 in) between the distributor plate and the bottom of the in-bed tube bank to prevent tube damage by jet impingement from the air and solids feed streams, and to allow access for maintenance. The height of the tube bank itself is around 0.45 m, requiring a normal bed depth of 0.9 m (36 in) to immerse all the tubes completely.

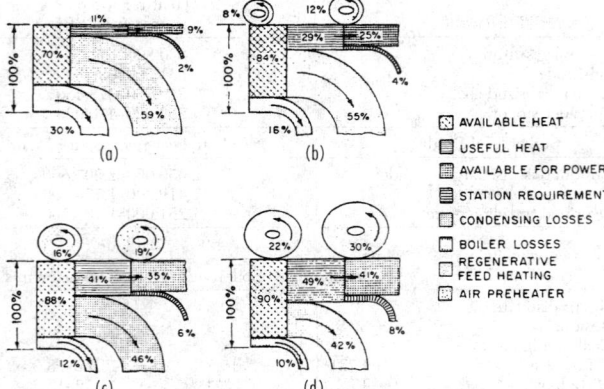

FIG. 27-43   Sankey diagrams for various types of boilers: (*a*) fire-tube boiler; (*b*) industrial boiler; (*c*) subcritical-pressure boiler; (*d*) supercritical-pressure boiler.

° Power production intensity is the power produced by the boiler per unit area of the plan section of the bed at a stated elevation.

FIG. 27-46 A D-type shop-assembled water-tube boiler. (*Combustion Engineering, Inc.*)

The convection pass cools the flue gas and entrained particulate matter from about 1145 K (1600°F) to below 645 K (700°F). For bituminous coal with a combustion temperature of 1115 K (1550°F), a recycle ratio of 2.5 (mass rate of particulate recycle to coal feed rate) increases combustion efficiency to over 97 percent from around 90 percent without recycle. For a bituminous coal containing 3 percent sulfur, 90 percent sulfur capture is achieved with limestone as sorbent at a calcium-to-sulfur (Ca/S) molar ratio of 2.3 and a recycle ratio of 2.5. A Ca/S molar ratio of over 3.4 is required when there is no recycle. The recycled material also enters the bed through T nozzles, one per 7 m² (72 ft²) of bed floor area. A means of distributing the particulate flow from the multiclones to the feed lines and overcoming the

back pressure is required. On the 160-MWe unit, the back pressure is overcome by means of a J-valve.

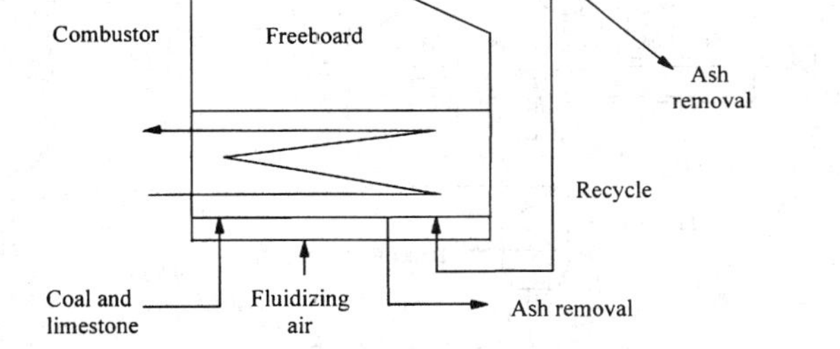

A typical NO$_x$ level in the combustion gas is around 107 mg/MJ (0.25 lb/10⁶ Btu), and the CO level tends to be high (near 86 mg/MJ [0.20 lb/10⁶ Btu]). Only one design has used secondary air, and this lowered the NO$_x$ to 86 mg/MJ and the CO to about 43 mg/MJ (0.10 lb/10⁶ Btu). NO$_x$ reduction by selective noncatalytic reduction (SNCR) has not been tested in a bubbling AFBC, but without the assistance of secondary air, it may be difficult to distribute the ammonia adequately across the freeboard to achieve the desired effect.

Approximately 85 percent of the heat is released in the bed and the other 15 percent above the bed. Typical heat flux data are tabulated in Table 27-22 for a mean bed temperature of 1115 K (1550°F).

The in-bed heat transfer rate decreases with mean bed particle size. As there is no accurate means of predicting this particle size, even if that of the feedstocks is known, there is a design uncertainty vis-à-vis heat transfer coefficient. If heat exchanger area adjustment has to be made after the boiler is in operation, surface is removed more easily than added. A reasonable design strategy, therefore, is to use an overall coefficient from the lower portion of the range experienced in practice. Then, if the actual coefficient is found to be higher than the design value, the bed temperature needed to maintain the design steam rate will be lower than specified. If the lower bed temperature adversely affects process performance and boiler efficiency, then surface area may need to be removed.

TABLE 27-22 Heat Absorption Distribution in Bubbling AFBCs

	Heat absorption split, %	Heat flux, kW/m² (Btu/h·ft²)
In-bed tubing		
Evaporators	25	147 (46,500)
Superheaters	15	112 (35,500)
Walls	5	134 (42,500)
Freeboard water walls	10	44 (14,000)
Convection pass		
Superheater	20	15 (4,750)
Reheater	15	22 (6,850)
Economizer	10	6.5 (2,050)

FIG. 27-47 Simplified flow diagram for bubbling AFBC (with underbed feed system).

Mean bed particle size (thus, the in-bed heat-transfer coefficient) may vary for external reasons such as a change of feedstock supply or a deterioration in crusher performance. This potential source of variation should be considered before any decision to resurface is made.

Circulating AFBCs   The circulating AFBC is now more widely used than the bubbling version. A simplified schematic for a design with an external heat exchanger is presented in Fig. 27-48. A 110-MWe demonstration plant had a power production intensity of 1 MWe/m² (1 MWe/11 ft²), almost 50 percent greater than that of a comparable bubbling unit. Circulating AFBC units of 250 MWe are now in service and larger units are being planned.

Circulating AFBCs handle bituminous and subbituminous coals equally well, and their coal preparation and feeding systems are far simpler than those of bubbling versions. The coal is crushed to a top size of 12 mm (½ in), without drying, and fed by gravity into the lower refractory-lined portion of the boiler. The feed points are close to the pressure balance point and so there is little, if any, back pressure; this greatly reduces the sealing requirements. In addition, and sometimes alternatively, the coal can be introduced into the cyclone ash return lines. A minimum of four coal entry points is required to achieve uniform feed distribution in the 110-MWe unit, corresponding to one per 30 m² (300 ft²) of freeboard cross section. High turbulence and the absence of in-bed tubing facilitate adequate mixing of the coal across the combustor. The sorbent is prepared to a top size of 1 mm (0.04 in) and dried so that it can be pneumatically fed to the combustor. Experience showed that no less than eight sorbent feed points were required on the 110-MWe unit (one per 15 m² [150 ft²] of freeboard cross section) to achieve satisfactory performance. Load control is achieved primarily by reducing coal feed rate, with a corresponding reduction in air flow, to lower combustor temperature.

Almost all of the particulate matter leaving the boiler is collected by cyclones and recycled to the base of the unit. The number of cyclones varies in different design concepts, but each cyclone can serve 40 to 60 MWe of generating capacity (in the case of the 110-MWe unit, one recycle point per 60 m² (600 ft²) of freeboard cross section). The collected particulate matter is returned against a backpressure of 0.02 MPa (3 psi), through a J-valve. The recycle ratio can be as high as 40:1, corresponding to a relatively long mean particle residence time and accounting for the high performance of circulating units. For a bituminous coal containing 3 percent sulfur, with limestone sorbent and a combustion temperature of 1115 K (1550°F), the combustion efficiency is 99.0 percent, and 90 percent sulfur retention is attained with a Ca/S molar ratio of 2.2. The NO$_x$ value at this temperature is typically 86 mg/MJ (0.20 lb/MBtu), but it can be reduced to below 43 mg/MJ (0.1 lb/MBtu) by ammonia injected with the secondary air. The corresponding CO level is about 43 mg/MJ.

Bubbling PFBCs   Like the AFBC, the bubbling PFBC offers the ability to achieve low SO$_2$ and NO$_x$ emissions without back-end add-on control equipment. It also offers advantages that the AFBC does not. Pressurized operation results in the boiler being more compact with a reduction in capital cost. Expanding the pressurized flue gas through a gas turbine generator, in combination with a steam turbine generator, increases cycle efficiency and increases power output by up to 25 percent. The lower capital cost and higher efficiency result in a lower cost of electricity.

A PFBC boiler is visually similar to an AFBC boiler. The combustor is made of water-wall tubing, which contains the high-temperature environment, but the whole assembly is placed within a pressure vessel. Unlike an AFBC unit, there is no convection pass, as the flue-gas temperature must be maintained at boiler temperature to maximize energy recovery by the expansion turbine. There is an economizer after the turbine for final heat recovery. A simplified schematic is presented in Fig. 27-49. An 80-MWe demonstration plant, operating at 1.2 MPa (180 psia), began operation in 1989 with a power production intensity of 3 MWe/m² (1 MWe/3.5 ft²). By 1996, five units of this size had been constructed, and a 320-MWe unit is planned to commence operation in 1998.

The boiler, primary and secondary flue-gas cyclones, and ash-cooling circuits are installed within a pressure vessel operating at up to 1.7 MPa (240 psia). Also enclosed are the vessels that store bed material at operating temperature. These facilitate load control by allowing bed level to be raised or lowered thus, covering or exposing in-bed heat transfer surface and regulating both steam production and gas turbine inlet temperature. Reduced bed level and flue-gas temperature at part load result in some reduction in combustion performance. After leaving the cyclones, which remove over 99 percent of the particulate matter, the flue gas passes down the center of a coaxial pipe to the gas turbine. A custom-designed variable-speed gas turbine is used that is tolerant to the low levels of fine residual particulate matter. The compressed air is delivered to the pressure vessel through the annular portion of the coaxial pipe. This eliminates the need for a refractory-lined pipe and precludes the possibility of refractory passing to the gas turbine and damaging it. A baghouse is used for final particulate matter removal before discharging the flue gas to atmosphere.

Coal is fed as a paste containing 25 wt % water, and sorbent is fed dry by a lock-hopper system with pneumatic conveying. The top size of each feedstock is 3 mm (⅛ in). The latent heat lost evaporating the water fed with the paste is compensated by increased gas turbine power output resulting from the increased flue-gas mass flow rate. For the 80-MWe unit, there are six coal feed points (one per 4.5 m² [48 ft²]) and four sorbent feed points (one per 6.7 m² [72 ft²]), all entering beneath the tube bank along one wall. The bed depth is

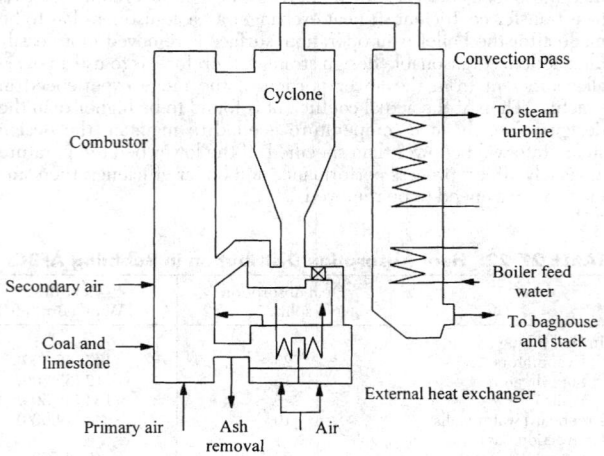

FIG. 27-48 Simplified flow diagram for circulating AFBC (with external heat exchanger).

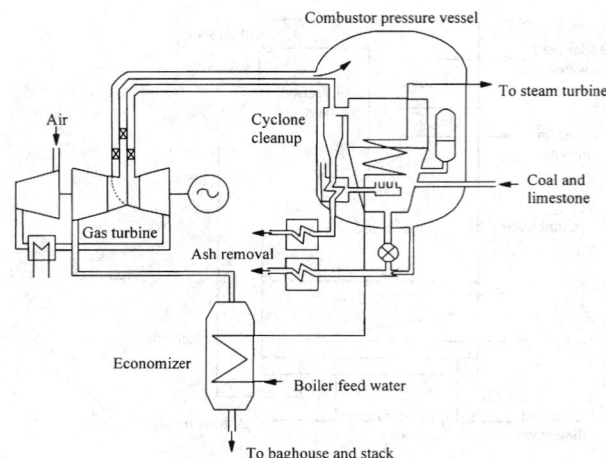

FIG. 27-49 Simplified flow diagram for bubbling PFBC.

about 3.7 m (12 ft) and the tube bank is about 3 m (10 ft) tall. The deep bed increases the in-bed gas residence time to around 4 s compared to less than 0.5 s in a bubbling AFBC unit. For bituminous coal containing 3 percent sulfur, with limestone sorbent and a combustion temperature of 1115 K (1550°F), this residence time allows combustion efficiencies in excess of 99 percent to be achieved without recycle, and 90 percent sulfur capture with a Ca/S molar ratio of 1.9. NO_x values of about 107 mg/MJ (0.25 lb/10^6 Btu) can be reduced to 21 mg/MJ (0.05 lb/10^6 Btu) by SNCR. Although the inherent NO_x level is similar to that achieved in bubbling AFBC units, the CO levels are very much lower, 13 mg/MJ (0.03 lb/10^6 Btu) being attainable.

Over 98 percent of the heat is released in the bed. For similar mean bed temperatures and mean bed particle sizes, the elevated operating pressure results in heat fluxes to the in-bed tubing that are typically 15 to 20 percent greater than in a bubbling AFBC unit.

Circulating PFBC   Circulating PFBC technology has been under development only since the late 1980s and is still in the pilot-plant stage. A simplified schematic for a design without an external heat exchanger is presented in Fig. 27-50. An 80-MWe demonstration plant is planned as part of the U.S. Department of Energy Clean Coal Technology Program; operation is likely in 1999.

Compared to bubbling PFBCs, these boilers operate at similar pressures but at higher fluidizing velocities. As a result, the design is more compact, with a projected power production intensity of 10 MWe/m² (1 MWe/ft²), over three times that of the bubbling design. Being more compact implies a pressure vessel of smaller diameter and a plant design adaptable to more modular construction and more shop assembly, with corresponding lower capital cost. Because the boiler is smaller, better distribution of coal and sorbent can be achieved with fewer feed nozzles. Inasmuch as there is no bed level to be maintained, a finer sorbent can be used, allowing it to be more fully utilized. Load control is achieved by cutting back on the air flow while keeping the combustion temperature substantially constant. This maintains high combustion and sulfur-retention performance over the load range and eliminates the need for bed storage vessels. Fewer cyclones are required (one per 80 MWe) and, as the ash collected is recycled, there are no ash coolers. The design uses high-temperature, high-pressure (HTHP) filtration to clean the flue gas prior to its entering the gas turbine. This virtually particulate-free gas allows conventional gas turbines to be used, increasing the selection available, and at the same time eliminates the need for a back-end baghouse.

For a bituminous coal containing 3 percent sulfur with limestone sorbent and a combustion temperature of 1115 K (1550°F), combustion efficiency in excess of 99 percent can be achieved, as can 90 percent sulfur capture with a Ca/S molar ratio of 1.15. The reduced sorbent demand decreases the amount of ash discharged. NO_x values are around 86 mg/MJ (0.20 lb/10^6 Btu) and can be reduced to 21 mg/MJ (0.05 lb/10^6 Btu) by SNCR reactions. As in the case of the bubbling PFBC, the inherent NO_x level is similar to that achieved in circulating AFBC units, but the CO level is very much lower, 13 mg/MJ (0.03 lb/10^6 Btu) being achieved. No heat flux data to the walls are available at present.

If all these process and economic advantages are realized, the cost of electricity will be lowered, making circulating PFBC an extremely attractive coal-fired option for power generation.

Advanced PFBC Cycle   The latest development in fluidized-bed combustion technology is the concept of an advanced PFBC that generates a fuel gas to feed a high-temperature gas turbine. The efficiency of current combined-cycle plants employing state-of-the-art bubbling or circulating PFBC boilers is limited to about 42 percent (HHV basis) because the maximum permissible combustion gas temperature (therefore, inlet temperature for the gas turbine) is 1145 K (1600°F). At higher bed temperature, the combustor feedstocks can release alkali metal vapor that can lead to fouling and corrosion of the gas turbine blades. Furthermore, the threat of bed ash agglomeration is increased.

The proposed advanced PFBC cycle will permit a turbine inlet gas temperature of over 1535 K (2300°F) by burning a fuel gas produced by pyrolysis of the coal feed. Because the turbine fuel gas must be practically particulate free, it passes through HTHP filters before combustion. The char residue from the pyrolyzer may be burned in a circulating AFBC or PFBC to produce steam for power or heating. The efficiency attainable in an advanced PFBC plant may be as high as 50 percent (HHV basis).

This technology is still in the early pilot stages, the first tests having been conducted in the early 1990s; therefore, insufficient data are available for meaningful comparison with the other PFBC technologies. Nevertheless, it is considered to be potentially the most economic PFBC technology, and development work is continuing with the objective of designing a demonstration plant.

The capital investment for any FBC plant depends upon several factors, including the cost of capital, size of unit, geographic location, and coal type. EPRI has completed several economic evaluations and projects the following costs, in 1994 US dollars, for plants located in Kenosha, Wisconsin, burning Illinois No. 6 bituminous coal containing 4 percent sulfur: 200-MWe circulating AFBC, $1520/kW; 350-MWe bubbling PFBC, $1220/kW; 350-MWe circulating PFBC, $1040/kW; 320-MWe advanced PFBC, $1110/kW. The advanced PFBC has the most potential for cost reduction, and capital investment could be reduced to below $1000/kW.

PROCESS HEATING EQUIPMENT

Many major energy-intensive industries depend on direct-fired or indirect-fired equipment for drying, heating, calcining, melting, or chemical processing. This subsection discusses both direct- and indirect-fired equipment, with the greater emphasis on indirect firing for the process industries.

Direct-Fired Equipment   Direct-fired combustion equipment transfers heat by bringing the flame and/or the products of combustion into direct contact with the process stream. Common examples are rotary kilns, open-hearth furnaces, and submerged-combustion evaporators. Table 27-23 gives the average energy consumption rates for various industries and processes that use direct heat. Section 11 of this handbook describes and illustrates rotary dryers, rotary kilns, and hearth furnaces. Forging, heat treating, and metal milling furnaces are discussed by Mawhinney (*Marks' Standard Handbook for Mechanical Engineers*, 9th ed., McGraw-Hill, New York, 1987, pp. 7.47–7.52). Other direct-fired furnaces are described later in this section.

Indirect-Fired Equipment (Fired Heaters)   Indirect-fired combustion equipment (fired heaters) transfers heat across either a metallic or refractory wall separating the flame and products of combustion from the process stream. Examples are heat exchangers (dis-

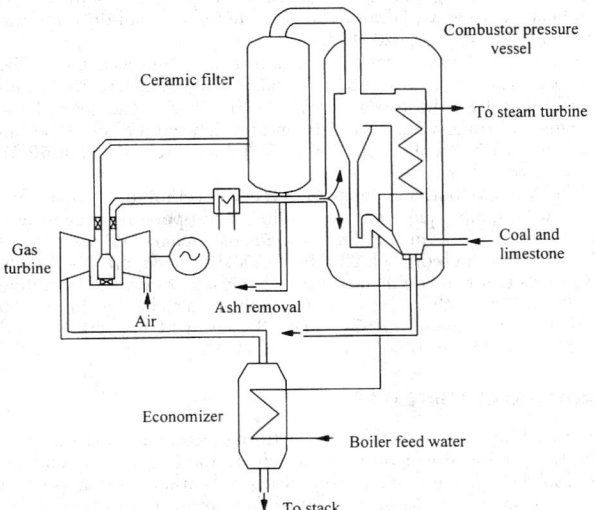

FIG. 27-50   Simplified flow diagram for circulating PFBC.

TABLE 27-23 Average Energy Consumption for Various Industries Using Direct Heat

Industry	Product/process	Energy consumption per unit of product	
		GJ/Mg	10^6 Btu/US ton
Paper	Kraft process	20.9	18.0
	Integrated plant/paper[*]	34.2	29.5
	Integrated plant/paperboard[*]	18.8	16.2
Glass	Flat glass	17.3	14.9
	Container glass	18.1	15.6
	Pressed/blown	31.6	27.2
Clay/ceramics	Portland cement	4.6	4.0
	Lime	5.5	4.7
	Mineral wool	42.7	36.8
Steel	Blast furnace and steel mills	20.7	17.8
Nonferrous metals	Primary copper	34.2	29.5
	Secondary copper	4.6	4.0
	Primary lead	25.1	21.6
	Secondary lead	0.8	0.7
	Primary zinc	69.6	60.0
	Secondary zinc	5.0	4.3
	Primary aluminum	78.9	68.0
	Secondary aluminum	5.2	4.5

SOURCE: *Manufacturing Consumption of Energy,* Energy Information Administration, U.S. Dept. of Energy, 1991.
*Mixture of direct and indirect firing.

cussed in Sec. 11), steam boilers, fired heaters, muffle furnaces, and melting pots. Steam boilers have been treated earlier in this section, and a subsequent subsection on industrial furnaces will include muffle furnaces.

Fired heaters differ from other indirect-fired processing equipment in that the process stream is heated by passage through a coil or tubebank enclosed in a furnace. Fired heaters are classified by function and by coil design.

Function Berman (*Chem. Eng.* **85**(14): 98–104, June 19, 1978) classifies fired heaters into the following six functional categories, providing descriptions that are abstracted here.

Column reboilers heat and partially vaporize a recirculating stream from a fractionating column. The outlet temperature of a reboiler stream is typically 477 to 546 K (400 to 550°F).

Fractionator-feed preheaters partially vaporize charge stock from an upstream unfired preheater en route to a fractionating column. A typical refinery application: a crude feed to an atmospheric column enters the fired heater as a liquid at 505 K (450°F) and leaves at 644 K (700°F), having become 60 percent vaporized.

Reactor-feed-stream preheaters heat the reactant stream(s) for a high-temperature chemical reaction. The stream may be single-phase/single-component (example: steam being superheated from 644 to 1089 K [700 to 1500°F] for styrene-manufacture reactors); single-phase/multicomponent (example: preheating the feed to a catalytic reformer, a mixture of hydrocarbon vapors and recycle hydrogen, from 700 to 811 K [800 to 1000°F] under pressure as high as 4.1 MPa [600 psia]); or multiphase/multicomponent (example: a mixture of hydrogen gas and liquid hydrocarbon heated from 644 to 727 K [700 to 850°F] at about 20 MPa [3000 psia] before it enters a hydrocracker).

Heat-transfer-fluid heaters maintain the temperature of a circulating liquid heating medium (e.g., a paraffinic hydrocarbon mixture, a Dowtherm, or a molten salt) at a level that may exceed 673 K (750°F).

Viscous-liquid heaters lower the viscosity of very heavy oils to pumpable levels.

Fired reactors contain tubes or coils in which an endothermic reaction within a stream of reactants occurs. Examples include steam/hydrocarbon reformers, catalyst-filled tubes in a combustion chamber; pyrolyzers, coils in which alkanes (from ethane to gas oil) are cracked to olefins; in both types of reactor the temperature is maintained up to 1172 K (1650°F).

Coil Design Indirect-fired equipment is conventionally classified by tube orientation: vertical and horizontal. Although there are many variations of each of these two principal configurations, they all are embraced within seven major types, as follows.

A *simple vertical cylindrical* heater has vertical tubes arrayed along the walls of a combustion chamber fired vertically from the floor. This type of heater does not include a convection section and is inexpensive. It has a small footprint but low efficiency, and it is usually selected for small-duty applications (0.5 to 21 GJ/h [0.5 to 20 10^6 Btu/h]).

Vertical cylindrical; cross-tube convection heaters are similar to the preceding type except for a horizontal convective tube bank above the combustion chamber. The design is economical with a high efficiency and is usually selected for higher-duty applications: 11 to 210 GJ/h (10 to 200 10^6 Btu/h).

The *arbor (wicket)* heater is a substantially vertical design in which the radiant tubes are inverted U's connecting the inlet and outlet terminal manifolds in parallel. An overhead crossflow convection bank is usually included. This type of design is good for heating large gas flows with low pressure drop. Typical duties are 53 to 106 GJ/h (50 to 100 10^6 Btu/h).

In the *vertical-tube single-row double-fired* heater, a single row of vertical tubes is arrayed along the center plane of the radiant section that is fired from both sides. Usually this type of heater has an overhead horizontal convection bank. Although it is the most expensive of the fired heater designs, it provides the most uniform heat transfer to the tubes. Duties are 21 to 132 GJ/h (20 to 125 10^6 Btu/h) per cell (twin-cell designs are not unusual).

Horizontal-tube cabin heaters position the tubes of the radiant-section-coil horizontally along the walls and the slanting roof for the length of the cabin-shaped enclosure. The convection tube bank is placed horizontally above the combustion chamber. It may be fired from the floor, the side walls, or the end walls. As in the case of its vertical cylindrical counterpart, its economical design and high efficiency make it the most popular horizontal-tube heater. Duties are 11 to 105 GJ/h (10 to 100 10^6 Btu/h).

In the *horizontal-tube box heater with side-mounted convection tube bank,* the radiant-section tubes run horizontally along the walls and the flat roof of the box-shaped heater, but the convection section is placed in a box of its own beside the radiant section. Firing is horizontal from the end walls. The design of this heater results in a relatively expensive unit justified mainly by its ability to burn low-grade high-ash fuel oil. Duties are 53 to 210 GJ/h (50 to 200 10^6 Btu/h).

Vertical cylindrical helical coil heaters are hybrid designs that are classified as vertical heaters, but their in-tube characteristics are like those of horizontal heaters. There is no convection section. In addition to the advantages of simple vertical cylindrical heaters, the helical coil heaters are easy to drain. They are limited to small-duty applications: 5 to 21 GJ/h (5 to 20 10^6 Btu/h).

Schematic elevation sections of a vertical cylindrical, cross-tube convection heater; a horizontal-tube cabin heater; and a vertical cylindrical, helical-coil heater are shown in Fig. 27-51. The seven basic designs and some variations of them are pictured and described in the reference cited above and by R. K. Johnson (*Combustion* **50**(5): 10–16, November 1978).

The design of both radiant and convection sections of fired heaters, along with some equipment descriptions and operating suggestions, are discussed by Berman in *Encyclopedia of Chemical Processing and Design* [McKetta (ed.), vol. 22, Marcel Dekker, 1985, pp. 31–69]. He also treats construction materials, mechanical features, and operating points in three other *Chemical Engineering* articles (all in vol. 85 (1978): no. 17, July 31, pp. 87–96; no. 18, August 14, pp. 129–140; and no. 20, Sept. 11, pp. 165–169).

INDUSTRIAL FURNACES

Industrial furnaces serve the manufacturing sector and can be divided into two groups. Boiler furnaces, which are the larger group and are used solely to generate steam, were discussed earlier in the subsection on industrial boilers. Furnaces of the other group are classified as follows: by (1) source of heat (fuel combustion or electricity), (2) func-

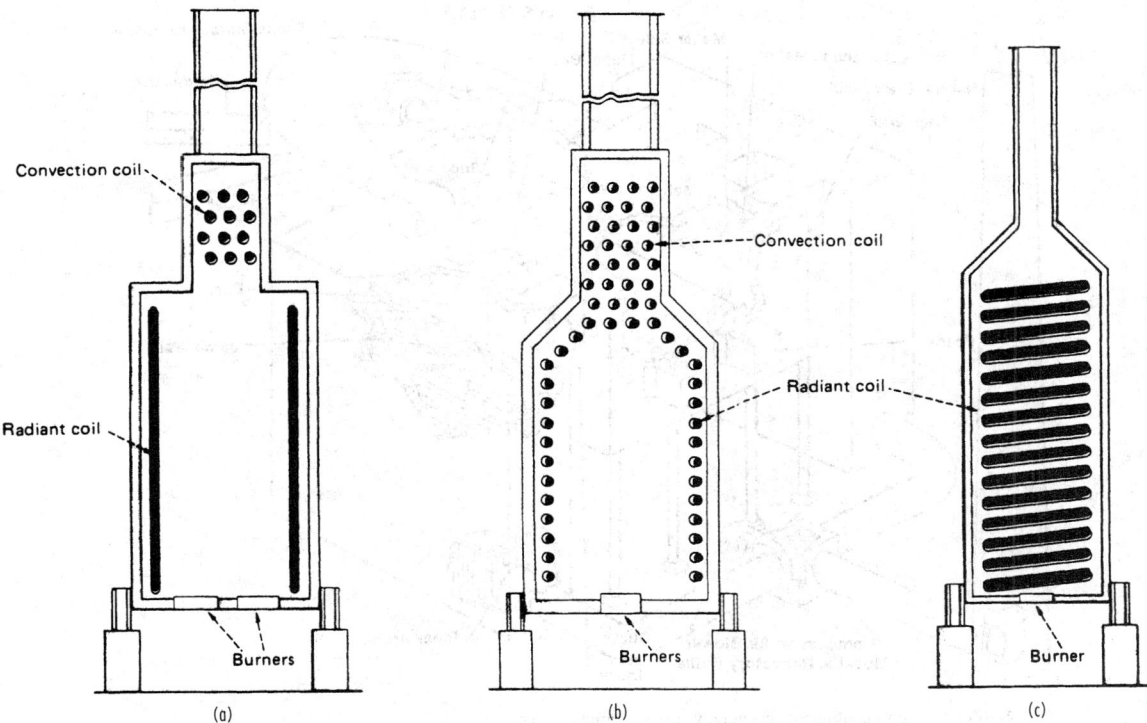

FIG. 27-51 Representative types of fired heaters: (*a*) vertical-tube cylindrical with cross-flow-convection section; (*b*) horizontal-tube cabin; (*c*) vertical cylindrical, helical coil. (*From Berman,* Chem. Eng. **85**: *98–104, June 19, 1978.*)

tion (heating without change of phase or with melting), (3) process cycle (batch or continuous), (4) mode of heat application (direct or indirect), and (5) atmosphere in furnace (air, protective, or reactive, including vacuum). Each will be discussed briefly.

Source of Heat   Industrial furnaces are either fuel-fired or electric, and the first decision that a prospective furnace user must make is between these two. Although electric furnaces are uniquely suited to a few applications in the chemical industry (manufacture of silicon carbide, calcium carbide, and graphite, for example), their principal use is in the metallurgical and metal-treatment industries. In most cases the choice between electric and fuel-fired is economic or custom-dictated, because most tasks that can be done in one can be done equally well in the other. Except for an occasional passing reference, electric furnaces will not be considered further here. The interested reader will find useful reviews of them in *Kirk-Othmer Encyclopedia of Chemical Technology* (4th ed., vol. 12, articles by Cotchen, Sommer, and Walton, pp. 228–265, Wiley, New York, 1994) and in *Marks' Standard Handbook for Mechanical Engineers* (9th ed., article by Lewis, pp. 7.59–7.68, McGraw-Hill, New York, 1987).

Function and Process Cycle   Industrial furnaces are enclosures in which process material is heated, dried, melted, and/or reacted. Melting is considered a special category because of the peculiar difficulties that may be associated with a solid feed, a hot liquid product, and a two-phase mixture in between; it is customary, therefore, to classify furnaces as heating or melting.

Melting Furnaces: The Glass Furnace   Most melting furnaces, electric or fuel-fired, are found in the metals-processing industry, but a notable exception is the glass furnace. Like most melting furnaces, a glass furnace requires highly radiative flames to promote heat transfer to the feed charge and employs regenerators to conserve heat from the high-temperature process (greater than 1813 K [2300°F]).

A typical side-port continuous regenerative glass furnace is shown in Fig. 27-52. Side-port furnaces are used in the flat and container glass industries. The burners are mounted on both sides of the furnace and the sides fire alternately. Refractory-lined flues are used to recover the energy of the hot flue gas. The high temperature of the flue gas leaving the furnace heats a mass of refractory material called a *checker*. After the checker has reached the desired temperature, the gas flow is reversed and the firing switches to the other side of the furnace. The combustion air is then heated by the hot checker and can reach 1533 K (2300°F). The cycle of air flow from one checker to the other is reversed approximately every 15 to 30 minutes.

The glass melt is generally 1 to 2 m (3 to 6 ft) deep, the depth being limited by the need for proper heat transfer to the melt. Container glass furnaces are typically 6 to 9 m (20 to 30 ft) wide and 6 to 12 m (20 to 40 ft) long. Flat glass furnaces tend to be longer, typically over 30 m (100 ft), because of the need for complete reaction of the batch ingredients and improved quality (fewer bubbles). They typically have a melting capacity of 450 to 630 Mg/day (500 to 750 US ton/d), compared to a maximum of 540 Mg/day (600 US ton/d) for container and pressed/blown glass furnaces.

Though the stoichiometric chemical energy requirement for glassmaking is only some 2.3 GJ/Mg (2 10⁶ Btu/US ton) of glass, the inherently low thermal efficiency of regenerative furnaces means that, in practice, at least 7 GJ/Mg (6 10⁶ Btu/US ton) is required. Of this total, some 40 percent goes to batch heating and the required heat of reactions, 30 percent is lost through the furnace structure, and 30 percent is lost through the stack. The smaller furnaces used in pressed/blown glass melting are less efficient, and energy consumption may be as high as 17.4 TJ/Mg (15 10⁶ Btu/US ton).

Industrial furnaces may be operated in batch or continuous mode.

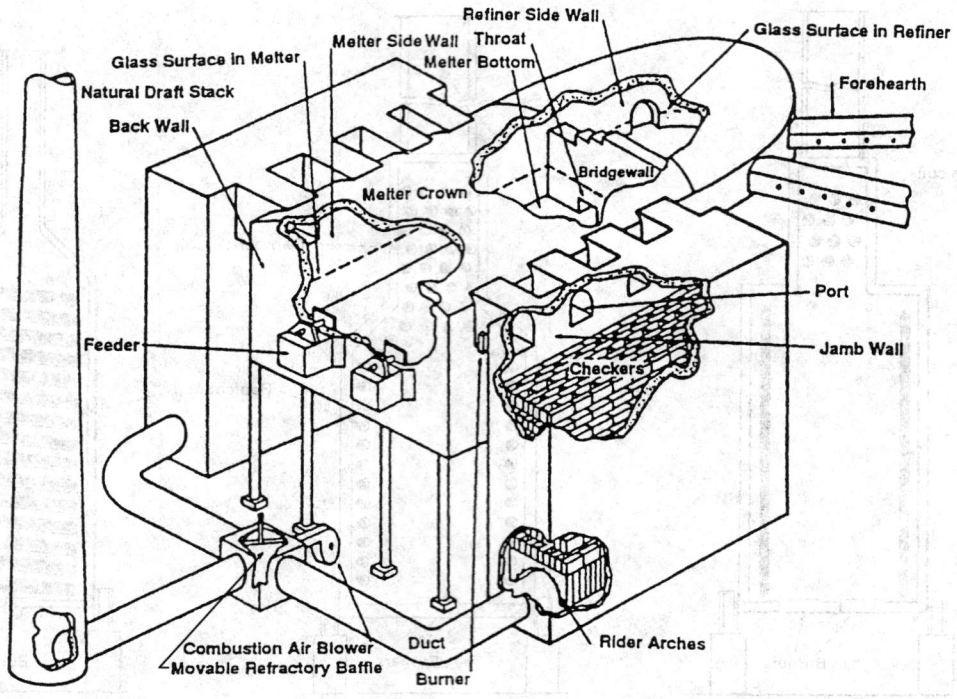

FIG. 27-52 Side-port continuous regenerative glass melting furnace.

Batch Furnaces   This type of furnace is employed mainly for the heat treatment of metals and for the drying and calcination of ceramic articles. In the chemical process industry, batch furnaces may be used for the same purposes as batch-tray and truck dryers when the drying or process temperature exceeds 600 K (620°F). They are employed also for small-batch calcinations, thermal decompositions, and other chemical reactions which, on a larger scale, are performed in rotary kilns, hearth furnaces, and shaft furnaces.

Continuous Furnaces   These furnaces may be used for the same general purposes as are the batch type, but usually not on small scale. The process material may be carried through the furnace by a moving conveyor (chain, belt, roller), or it may be pushed through on idle rollers, the motion being sustained by an external pusher operating on successively entering cars or trays, each pushing the one ahead along the entire length of the furnace and through the exit flame curtains or doors.

Furnace Atmosphere and Mode of Heating

Direct Heating   Industrial furnaces may be directly or indirectly heated, and they may be filled with air or a protective atmosphere, or under a vacuum. Direct heating is accomplished by the hot combus-

tion gases being inside the furnace and therefore in direct contact with the process material. Thus, the material is heated by radiation and convection from the hot gas and by reradiation from the heated refractory walls of the chamber. Three styles of direct firing are illustrated in Fig. 27-53. *Simple direct firing* is used increasingly because of its simplicity and because of improved burners. The *overhead* design allows the roof burners to be so placed as to provide optimum temperature distribution in the chamber. *Underfiring* offers the advantage of the charge's being protected from the flame. The maximum temperature in these direct-heated furnaces is limited to about 1255 K (1800°F) to avoid prohibitively shortened life of the refractories in the furnace.

Indirect Heating   If the process material cannot tolerate exposure to the combustion gas or if a vacuum or an atmosphere other than air is needed in the furnace chamber, indirect firing must be employed. This is accomplished in a muffle° furnace or a radiant-tube furnace (tubes carrying the hot combustion gas run through the furnace).

° A *muffle* is an impenetrable ceramic or metal barrier between the firing chamber and the interior of the furnace. It heats the process charge by radiation and furnace atmosphere convection.

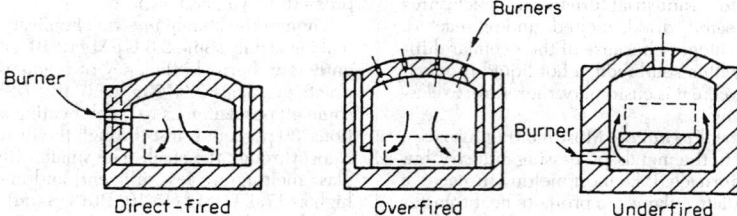

FIG. 27-53 Methods of firing direct-heated furnaces. (*From Marks' Standard Handbook for Mechanical Engineers, 9th ed., McGraw-Hill, New York, 1987. Reproduced with permission.*)

Atmosphere Protective atmosphere within the furnace chamber may be essential, especially in the heat treatment of metal parts. Mawhinney (in *Marks' Standard Handbook for Mechanical Engineers*, 9th ed., McGraw-Hill, New York, 1987, p. 752) lists pure hydrogen, dissociated ammonia (a hydrogen/nitrogen mixture), and six other protective reducing gases with their compositions (mixtures of hydrogen, nitrogen, carbon monoxide, carbon dioxide, and sometimes methane) that are codified for and by the metals-treatment industry. In general, any other gas or vapor that is compatible with the temperature and the lining material of the furnace can be provided in an indirect-fired furnace, or the furnace can be evacuated.

COGENERATION

Cogeneration is an energy conversion process wherein heat from a fuel is simultaneously converted to useful thermal energy (e.g., process steam) and electric energy. The need for either form can be the primary incentive for cogeneration, but there must be opportunity for economic captive use or sale of the other. In a chemical plant the need for process and other heating steam is likely to be the primary; in a public utility plant, electricity is the usual primary product.

Thus, a cogeneration system is designed from one of two perspectives: it may be sized to meet the process heat and other steam needs of a plant or community of industrial and institutional users, so that the electric power is treated as a by-product which must be either used on site or sold; or it may be sized to meet electric power demand, and the rejected heat used to supply needs at or near the site. The latter approach is the likely one if a utility owns the system; the former if a chemical plant is the owner.

Industrial use of cogeneration leads to small, dispersed electric-power-generation installations—an alternative to complete reliance on large central power plants. Because of the relatively short distances over which thermal energy can be transported, process-heat generation is characteristically an on-site process, with or without cogeneration.

Cogeneration systems will not match the varying power and heat demands at all times for most applications. Thus, an industrial cogeneration system's output frequently must be supplemented by the separate on-site generation of heat or the purchase of utility-supplied electric power. If the on-site electric power demand is relatively low, an alternative option is to match the cogeneration system to the heat load and contract for the sale of excess electricity to the local utility grid.

Fuel saving is the major incentive for cogeneration. Since all heat-engine-based electric power systems reject heat to the environment, that rejected heat can frequently be used to meet all or part of the local thermal energy needs. Using reject heat usually has no effect on the amount of primary fuel used, yet it leads to a saving of all or part of the fuel that would otherwise be used for the thermal-energy process. Heat engines also require a high-temperature thermal input, usually receiving the working fluid directly from a heating source; but in some situations they can obtain the input thermal energy as the rejected heat from a higher-temperature process. In the former case, the cogeneration process employs a heat-engine topping cycle; in the latter case, a bottoming cycle is used.

The choice of fuel for a cogeneration system is determined by the primary heat-engine cycle. Closed-cycle power systems which are externally fired—the steam turbine, the indirectly fired open-cycle gas turbine, and closed-cycle gas turbine systems—can use virtually any fuel that can be burned in a safe and environmentally acceptable manner: coal, municipal solid waste, biomass, and industrial wastes are burnable with closed power systems. Internal combustion engines, on the other hand, including open-cycle gas turbines, are restricted to fuels that have combustion characteristics compatible with the engine type and that yield combustion products clean enough to pass through the engine without damaging it. In addition to natural gas, butane, and the conventional petroleum-derived liquid fuels, refined liquid and gaseous fuels derived from shale, coal, or biomass are in this category. Direct-coal-fired internal combustion engines have been an experimental reality for decades but are not yet a practical reality technologically or economically.

There are at least three broad classes of application for topping-cycle cogeneration systems:

- Utilities or municipal power systems supplying electric power and low-grade heat (e.g., 422 K [300°F]) for local district heating systems
- Large residential, commercial, or institutional complexes requiring space heat, hot water, and electricity
- Large industrial operations with on-site needs for electricity and heat in the form of process steam, direct heat, and/or space heat.

Typical Systems All cogeneration systems involve the operation of a heat engine for the production of mechanical work which, in nearly all cases, is used to drive an electric generator. The commonest heat-engine types appropriate for topping-cycle cogeneration systems are:

- Steam turbines (backpressure and extraction configurations)
- Open-cycle (combustion) gas turbines
- Indirectly fired gas turbines: open cycles and closed cycles
- Diesel engines

Each heat-engine type has unique characteristics, making it better suited for some cogeneration applications than for others. For example, engine types can be characterized by:

- Power-to-heat ratio at design point
- Efficiency at design point
- Capacity range
- Power-to-heat-ratio variability
- Off-design (part-load) efficiency
- Multifuel capability

The major heat-engine types are described in terms of these characteristics in Table 27-24.

TABLE 27-24 Cogeneration Characteristics for Heat Engines

Engine type	Size range, MWe/unit	Efficiency at design point	Part-load efficiency	Multifuel capability	Maximum temperature of recoverable heat, °F (°C)°	Recoverable heat, Btu/ kWh†	Typical power-to-heat ratio
Steam turbine							
Extraction-condensing type	30–300	0.25–0.30	Fair	Excellent	200 (93)–600 (315)‡	11,000–35,000	0.1–0.3
Backpressure type	20–200	0.20–0.25	Fair	Excellent	200 (93)–600 (315)‡	17,000–70,000	0.05–0.2
Combustion gas turbines	10–100	0.25–0.30	Poor	Poor	1000 (538)–1200 (649)	3000–11,000	0.3–0.45
Indirectly fired gas turbines							
Open-cycle turbines	10–85	0.25–0.30	Poor	Good	700 (371)–900 (482)	3500–8500	0.4–1.0
Closed-cycle turbines	5–350	0.25–0.30	Excellent	Good	700 (371)–900 (482)	3500–8500	0.4–1.0
Diesel engines	0.05–25	0.35–0.40	Good	Fair to poor	500 (260)–700 (371)	4000–6000	0.6–0.85

°°C + 273 = K
†1 Btu = 1055 J.
‡Saturated steam.

HEAT RECOVERY

REGENERATION

Storage of heat is a temporary operation since perfect thermal insulators are unknown; thus, heat is absorbed in solids or liquids as sensible or latent heat to be released later at designated times and conditions. The collection and release of heat can be achieved in two modes: on a batch basis, as in the checkerbrick regenerator for blast furnaces, or on a continuous basis, as in the Ljungstrom air heater.

Checkerbrick Regenerators   Preheating combustion air in open-hearth furnaces, ingot-soaking pits, glass-melting tanks, by-product coke ovens, heat-treating furnaces, and the like has been universally carried out in regenerators constructed of fireclay, chrome, or silica bricks of various shapes. Although many geometric arrangements have been used in practice, the so-called basketweave design has been adopted in most applications.

Blast-Furnace Stoves   Blast-furnace stoves are used to preheat the air that is blown into a blast furnace. A typical blast furnace, producing 1500 Mg (1650 US ton) of pig iron per day, will be blown with 47.2 m³/s (100,000 std ft³/min) of atmospheric air preheated to temperatures ranging in normal practice from 755 to 922 K (900 to 1200°F). A set of four stoves is usually provided, each consisting of a vertical steel cylinder 7.3 m (24 ft) in diameter, 33 m (108 ft) high, topped with a spherical dome. Characteristic plan and elevation sections of a stove are shown in Fig. 27-54. The interior comprises three regions: in the cylindrical portion, (1) a side combustion chamber, lens-shaped in cross section, bounded by a segment of the stove wall and a mirror-image bridgewall separating it from (2) the chamber of the cylinder that is filled with heat-absorbing checkerbrick, and (3) the capping dome, which constitutes the open passage between the two chambers.

The heat-exchanging surface in each stove is just under 11,500 m² (124,000 ft²). In operation, each stove is carried through a two-step 4-h cycle. In a 3-h *on-gas* step, the checkerbricks in a stove are heated by the combustion of blast-furnace gas. In the alternating *on-wind*

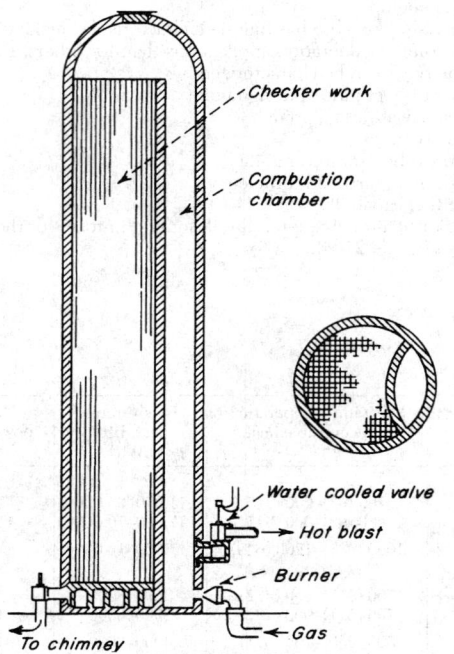

FIG. 27-54   Blast-furnace stove.

1-h step, they are cooled by the passage of cold air through the stove. At any given time, three stoves are simultaneously on gas, while a single stove is on wind.

At the start of an on-wind step, about one-half of the air, entering at 366 K (200°F), passes through the checkerbricks, the other half being bypassed around the stove through the cold-blast mixer valve. The gas passing through the stove exhausts at 1366 K (2000°F). Mixing this with the unheated air produces a blast temperature of 811 K (1000°F). The temperature of the heated air from the stove falls steadily throughout the on-wind step. The fraction of total air volume bypassed through the mixer valve is continually decreased by progressively closing this valve, its operation being automatically regulated to maintain the exit gas temperature at 811 K. At the end of 1 h of on-wind operation, the cold-blast mixer valve is closed, sending the entire blast through the checkerbricks.

Open-Hearth and Glass-Tank Regenerators   These contain checkerbricks that are modified considerably from those used in blast-furnace stoves because of the higher working temperatures, more drastic thermal shock, and dirtier gases encountered. Larger bricks form flue cross sections five times as large as the stove flues, and the percentage of voids in the checkerbricks is 51 percent, in contrast to 32 percent voids in stoves. The vertical height of the flues is limited by the elevation of the furnace above plant level. Short flues from 3 to 4.9 m (10 to 16 ft) are common in contrast to the 26- to 29-m (85- to 95-ft) flue lengths in blast-furnace stoves.

As a result of the larger flues and the restricted surface area per unit of gas passed, regenerators employed with this type of furnace exhibit much lower efficiency than would be realized with smaller flues. In view of the large amount of iron oxide contained in open-hearth exhaust gas and the alkali fume present in glass-tank stack gases, however, smaller checkerbrick dimensions are considered impractical.

Ljungstrom Heaters   A familiar continuous regenerative-type air heater is the Ljungstrom heater (Fig. 27-55). The heater assembly consists of a slow-moving rotor embedded between two peripheral housings separated from one another by a central partition. Through one side of the partition a stream of hot gas is being cooled, and, through the other side a stream of cold gas is being heated. Radial and circumferential seals sliding on the rotor limit the leakage between the streams. The rotor is divided into sectors, each of which is tightly packed with metal plates and wires that promote high heat-transfer rates at low pressure drop.

These heaters are available with rotors up to 6 m (20 ft) in diameter. Gas temperatures up to 1255 K (1800°F) can be accommodated. Gas face velocity is usually around 2.5 m/s (500 ft/min). The rotor height depends on service, efficiency, and operating conditions but usually is between 0.2 and 0.91 m (8 and 36 in). Rotors are driven by small motors with rotor speed up to 20 r/min. Heater effectiveness can be as high as 85 to 90 percent heat recovery. Lungstrom-type heaters are used in power-plant boilers and also in the process industries for heat recovery and for air-conditioning and building heating.

Regenerative Burners   In these systems a compact heat storage regenerator (containing ceramic balls, for example) is incorporated into the burner. Operating in pairs, one burner fires while the other exhausts: combustion air is preheated in the regenerator of the firing burner and furnace gas gives up heat to the regenerator in the exhausting burner (see Fig. 27-56). Burner operations are switched periodically. Such systems can yield combustion air preheats between 933 K (1220°F) and 1525 K (2282°F) for furnace temperature between 1073 K (1472°F) and 1723 K (2642°F), respectively. Corresponding fuel savings compared to cold-air firing will vary approximately from 30 to 70 percent.

Miscellaneous Systems   Many other systems have been proposed for transferring heat regeneratively, including the use of high-temperature liquids and fluidized beds for direct contact with gases, but other problems which limit industrial application are encountered. These systems are covered by methods described in Secs. 11 and 12 of this handbook.

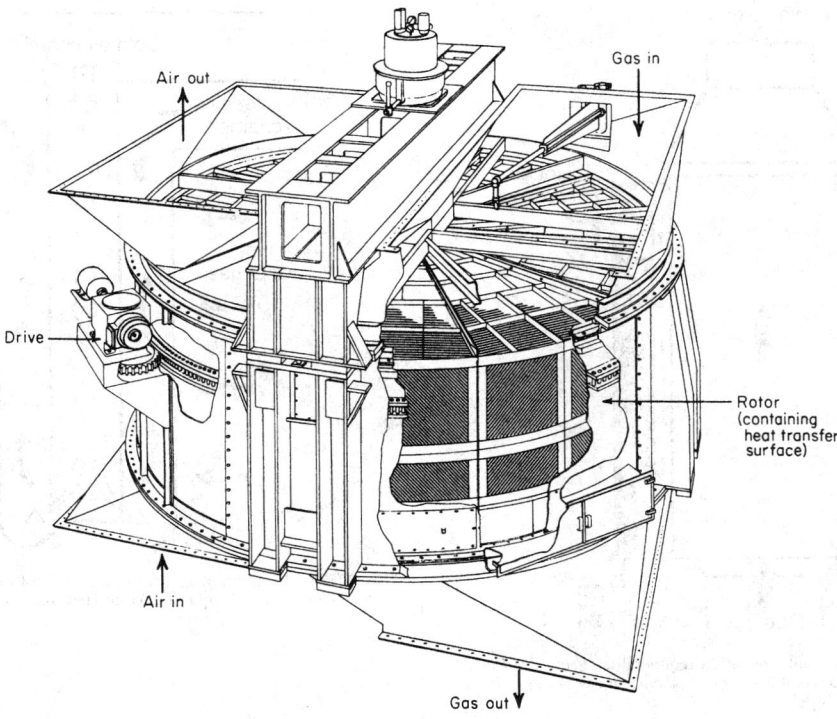

FIG. 27-55   Ljungstrom air heater.

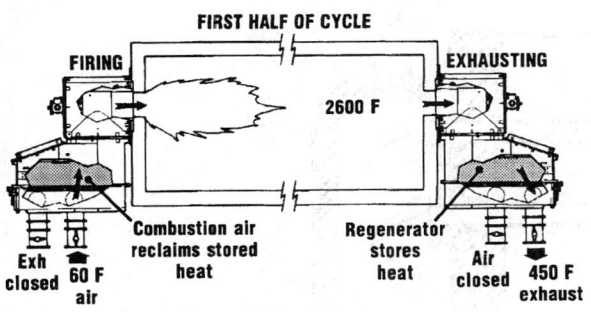

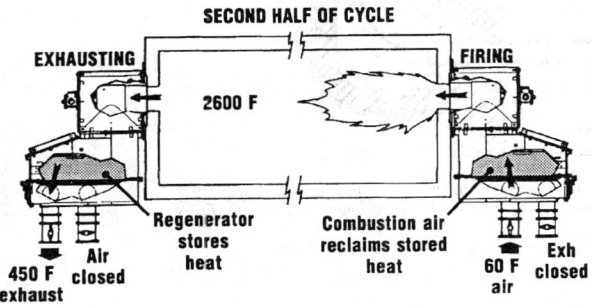

FIG. 27-56   Schematic of a regenerative burner system. (*North American Manufacturing Co.*)

RECUPERATORS

Regenerators are by nature intermittent or cycling devices, although, as set forth previously, the Ljunstrom design avoids interruption of the fluid stream by cycling the heat-retrieval reservoir between the hot and cold fluid streams. Truly continuous counterparts of regenerators exist, however, and they are called *recuperators*.

The simplest configuration for a recuperative heat exchanger is the metallic radiation recuperator (Fig. 27-57). The inner tube carries the hot exhaust gases and the outer tube carries the combustion air. The bulk of the heat transfer from the hot gases to the surface of the inner tube is by radiation, whereas that from the inner tube to the cold combustion air is predominantly by convection.

Shell-and-tube heat exchangers (see Sec. 11) may also be used as recuperators; convective heat transfer dominates in these recuperators. An alternative arrangement for a convective-type recuperator is shown in Fig. 27-58 (the dimpled end of the tube serves to ensure that there is adequate heat transfer from the cool fluid internally and that the tube bottom does not overheat and fail). For applications involving higher temperatures, ceramic recuperators have been developed. These devices can allow operation at up to 1823 K (2822°F) on the gas side and over 1093 K (1508°F) on the air side. Early ceramic recuperators were built of furnace brick and cement, but the repeated thermal cycling caused cracking and rapid deterioration of the recuperator. Later designs have used various approaches to overcome the problems of leakage and cracking, one of which is shown in Fig. 27-59. Silicon carbide tubes carry the combustion air through the waste gas, and flexible seals are used in the air headers. In this manner, the seals are maintained at comparatively low temperatures and the leakage rate can be reduced to a few percent of the total flow.

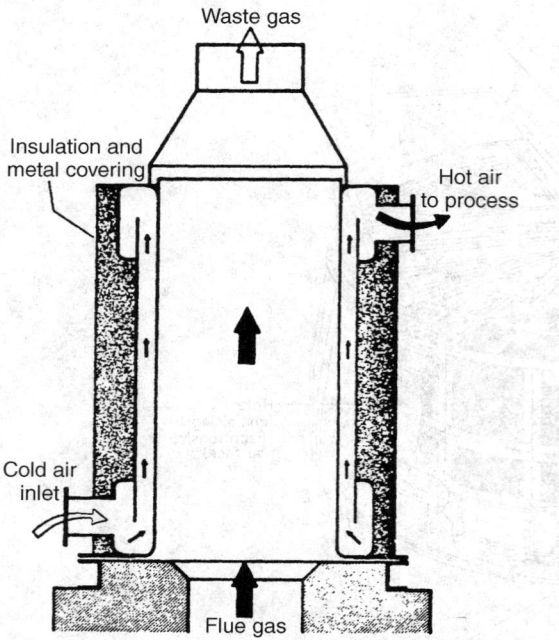

FIG. 27-57 Diagram of a metallic radiation recuperator. (*From Goldstick & Thumann*, Principles of Waste Heat Recovery, *Fairmont Press, Atlanta, 1986.*)

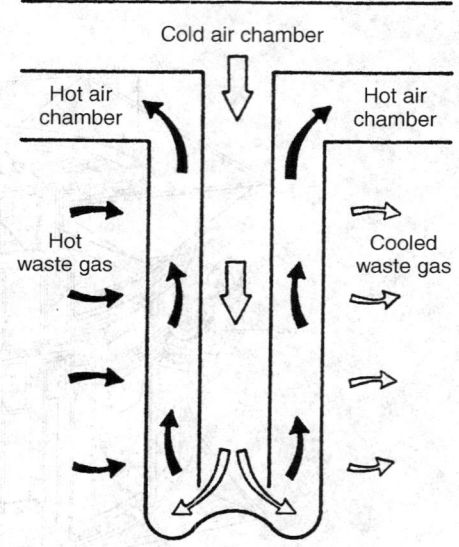

FIG. 27-58 Diagram of a vertical tube-within-tube recuperator. (*From Goldstick & Thumann*, Principles of Waste Heat Recovery, *Fairmont Press, Atlanta, 1986.*)

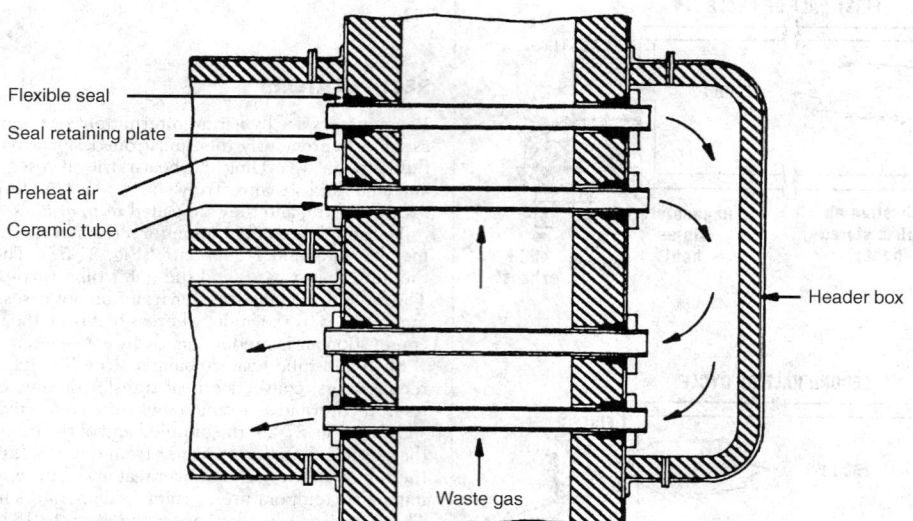

FIG. 27-59 Ceramic recuperator. In this design the seals are maintained at relatively low temperatures, leading to leakage rates of only a few percent. (*From Goldstick & Thumann*, Principles of Waste Heat Recovery, *Fairmont Press, Atlanta, 1986.*)

ELECTROCHEMICAL ENERGY CONVERSION

Electricity has become as indispensable as heat to the functioning of industrialized society. The source of most of the electricity used is the energy of the fuels discussed earlier in this section: liberated by combustion as heat, it drives heat engines which, in turn, drive electrical generators.

In some instances, however, part of the chemical energy bound in relatively high-enthalpy compounds can be converted directly to electricity as these reactants are converted to products of lower enthalpy (galvanic action). A process in the opposite direction also is possible for some systems: an electric current can be absorbed as the increased chemical energy of the higher-enthalpy compounds (electrolytic action). The devices in which electrochemical energy conversion processes occur are called cells.

Galvanic cells in which stored chemicals can be reacted on demand to produce an electric current are termed *primary cells*. The discharging reaction is irreversible and the contents, once exhausted, must be replaced or the cell discarded. Examples are the dry cells that activate small appliances. In some galvanic cells (called *secondary cells*), however, the reaction is reversible: that is, application of an electrical potential across the electrodes in the opposite direction will restore the reactants to their high-enthalpy state. Examples are rechargeable batteries for household appliances, automobiles, and many industrial applications. Electrolytic cells are the reactors upon which the electrochemical process, electroplating, and electrowinning industries are based.

Detailed treatment of the types of cells discussed above is beyond the scope of this handbook. For information about electrolytic cells, interested readers are referred to Fuller, Newman, Grotheer, and King ("Electrochemical Processing," in *Kirk-Othmer Encyclopedia of Chemical Technology*, 4th ed., vol. 9, Wiley, New York, 1994, pp. 111–197) and for primary and secondary cells, to Crompton (*Battery Reference Book*, 2d ed., Butterworth-Heineman, Oxford, U.K., 1995). Another type of cell, however, a galvanic cell to which the reactants of an exothermic reaction are fed continuously, in which they react to liberate part of their enthalpy as electrical energy, and from which the products of the reaction are discharged continuously, is called a *fuel cell*. Fuel cell systems for generating electricity in a variety of applications are being commercialized by a number of companies. The rest of this section is devoted to a discussion of fuel cell technology.

FUEL CELLS

GENERAL REFERENCES: Appleby and Foulkes, *Fuel Cell Handbook*, Kreger Publishing Co., Molabar, Fla., 1993. Hirschenhofer, Staufer, and Engleman, *Fuel Cell Handbook* (Rev. 3), U.S. Dept. of Energy, Morgantown Energy Technology Center, DOE/METC-94/1006, Morgantown, W. Va., 1994. Kinoshita and Cairns, "Fuel Cells," in *Kirk-Othmer Encyclopedia of Chemical Technology*, 4th ed., vol. 11, Wiley, New York, 1994, p. 1098. Liebhafsky and Cairns, *Fuel Cells and Fuel Batteries*, Wiley, New York, 1968. Linden (ed.), *Handbook of Batteries and Fuel Cells*, McGraw-Hill, New York, 1984.

Background Energy conversion in fuel cells is direct and simple when compared to the sequence of chemical and mechanical steps in heat engines. A fuel cell consists of an anode, an electrolyte, and a cathode. On the anode, the fuel is oxidized electrochemically to positively charged ions. On the cathode, oxygen molecules are reduced to oxide or hydroxide ions. The electrolyte serves to transport either the positively charged or negatively charged ions from anode to cathode or cathode to anode. Figure 27-60 is a schematic representation of the reactions in a fuel cell operating on hydrogen and air with a hydrogen-ion-conducting electrolyte. The hydrogen flows over the anode, where the molecules are separated into ions and electrons. The ions migrate through the ionically conducting but electronically insulating electrolyte to the cathode, and the electrons flow through the outer circuit energizing an electric load. The electrons combine eventually with oxygen molecules flowing over the surface of the cathode and hydrogen ions migrating across the electrolyte, forming water, which leaves the fuel cell in the depleted air stream.

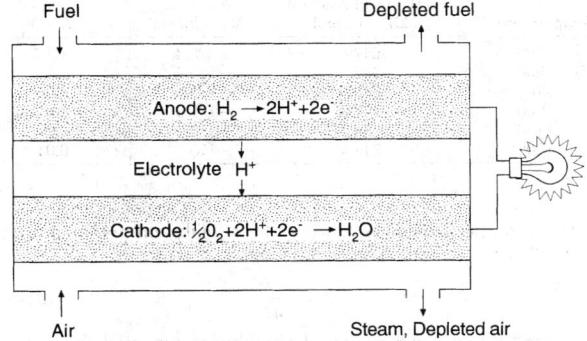

FIG. 27-60 Fuel cell schematic.

A fuel cell has no moving parts. It runs quietly, does not vibrate, and does not generate gaseous pollutants. The idea of the fuel cell is generally credited to Sir William Grove, who lived in the nineteenth century. It took over 100 years for the first practical devices to be built, in the U.S. space program, as the power supply for space capsules and the space shuttle. Commercialization of terrestrial fuel cell systems is beginning only now. Having lower emissions and being more efficient than heat engines, fuel cells may in time become the power source for a broad range of applications, beginning with utility power plants, including civilian and military transportation, and reaching into portable electronic devices.

This slow realization of the concept is due to the very demanding materials requirements for fuel cells. The anodes and cathodes have to be good electronic conductors and must have electrocatalytic properties to facilitate the anodic and cathodic reactions. In addition, the anodes and cathodes must be porous to allow the fuel and oxidant gases to diffuse to the reaction sites, yet they must be mechanically strong enough to support the weight of the fuel cell stacks. The electrolyte must be chemically stable in hydrogen and oxygen, and must have an ionic conductivity of at least 0.1 S/cm. Five classes of electrolytes have been found to meet these requirements: potassium hydroxide, phosphoric acid, perfluorinated sulfonic acid resins, molten carbonates, and oxide-ion-conducting ceramics. Consequently, five types of fuel cell based on these electrolytes have been developed.

Fuel Cell Efficiency The theoretical energy conversion efficiency of a fuel cell $\varepsilon°$ is given by the ratio of the free energy (Gibbs function) of the cell reaction at the cell's operating temperature ΔG_t to the enthalpy of reaction at the standard state $\Delta H°$, both quantities being based on a mole of fuel:

$$\varepsilon° = \frac{\Delta G_t}{\Delta H°} \qquad (27\text{-}50)$$

The enthalpy of reaction is always taken at a temperature of 298 K (77°F), but the product water can be either liquid or gaseous. If it is liquid, the efficiency is based on the higher heating value (HHV), but if the product is gaseous, the efficiency is based on the lower heating value (LHV). If the fuel cell runs on hydrogen and oxygen at 373 K (212°F), the theoretical conversion efficiency is 91 percent LHV or 83 percent HHV. The theoretical efficiency of fuel cells as given in Eq. (27-50) is equivalent to the Carnot efficiency of heat engines with the working medium absorbing heat at the flame temperature of the fuel and rejecting it at 298 K. Owing to materials and engineering limitations, heat engines cannot operate at the Carnot limit. Fuel cells can run at efficiencies near the theoretical values but only at low power density (power produced per unit of active fuel cell area). At higher power densities, the efficiency of fuel cells is constrained by electrical resistances within the bulk and at the interfaces of the materials, and by gas diffusion losses.

TABLE 27-25   Thermodynamic Values for $H_2 + \frac{1}{2}O_2 = H_2O$ (g)

Temperature, K	Enthalpy of reaction ($\Delta H°$), kJ/mol	Free energy of reaction ($\Delta G°$), kJ/mol	Equilibrium cell potential ($E°$), V
300	−241.8	−228.4	1.18
500	−243.9	−219.2	1.14
700	−245.6	−208.8	1.06
900	−247.3	−197.9	1.03
1100	−248.5	−187.0	0.97
1300	−249.4	−175.7	0.91

When no net current is flowing, the equilibrium potential of the fuel cell is given by the Nernst equation:

$$E° = \frac{-\Delta G_t}{nF} \qquad (27\text{-}51)$$

where $E°$ is the electrochemical equilibrium potential, V; n is the number of electrons transferred in the cell reaction (equivalents), and F is the Faraday constant. If the units of ΔG_t are J/mol, F has the value 96,487 C/mol·equiv. The potential depends on the chemical species of the fuel and the operating temperature. For hydrogen and oxygen, variation of the equilibrium cell potential with temperature is shown in Table 27-25.

When current is flowing, the actual cell operating potential is given by:

$$E = E° - (a_{an} + a_{ca}) - (b_{an} + b_{ca})\frac{RT}{nF}\ln i - Ai \qquad (27\text{-}52)$$

where a and b are characteristic constants for the electrochemical reactions at the electrode materials; the subscripts an and ca refer to the anode and the cathode, respectively; R is the gas constant; T is the cell temperature; A is the area-specific resistance of the fuel cell; and i is the current density (current flow per unit of active fuel cell area) in the cell.

Graphs of operating potential versus current density are called *polarization curves,* which reflect the degree of perfection that any particular fuel cell technology has attained. High cell operating potentials are the result of many years of materials optimization. Actual polarization curves will be shown below for several types of fuel cell.

The actual efficiency of an operating fuel cell is given by:

$$\varepsilon = \frac{-nFE}{\Delta H°}U_f \qquad (27\text{-}53)$$

where U_f is the electrochemical fuel utilization (amount of fuel converted divided by amount fed to the cell). For pure hydrogen the fuel utilization can be 1.0, but for gas mixtures it is often 0.85. Equations (27-52) and (27-53) show that the efficiency of fuel cells is not constant, but depends on the current density. The more power that is drawn, the lower the efficiency.

When the fuel gas is not pure hydrogen and air is used instead of pure oxygen, additional adjustment to the calculated cell potential becomes necessary. Since the reactants in the two gas streams practically become depleted between the inlet and exit of the fuel cell, the cell potential is decreased by a term representing the log mean fugacities, and the operating cell efficiency becomes:

$$\varepsilon_{fc} = \frac{nFU_f}{\Delta H°}\left[E° - \sum a - \sum b\frac{RT}{nF}\ln i - Ai\right]$$
$$- \frac{RTU_f}{\Delta H°}[\nu_f \ln(\log \text{ mean }\hat{f}_f) + \nu_{ox}\ln(\log \text{ mean }\hat{f}_{ox})] \quad (27\text{-}54)$$

The quantities ν_f and ν_{ox} are the stoichiometric coefficients for the fuel cell reaction, and $\hat{f}_f$ and $\hat{f}_{ox}$ are the fugacities of fuel and oxygen in their respective streams.

Further, as the current density of the fuel cell increases, a point is inevitably reached where the transport of reactants to or products from the surface of the electrode becomes limited by diffusion. A *concentration polarization* is established at the electrode, which diminishes the cell operating potential. The magnitude of this effect depends on many design and operating variables, and its value must be obtained empirically.

Design Principles   An individual fuel cell will generate an electrical potential of about 1 V or less, as discussed above, and a current that is proportional to the external load demand. For practical applications, the voltage of an individual fuel cell is obviously too small, and cells are therefore stacked up as shown in Fig. 27-61. Anode/electrolyte/cathode assemblies are electrically connected in series by inserting a bipolar plate between the cathode of one cell and the anode of the next. The bipolar plate must be impervious to the fuel

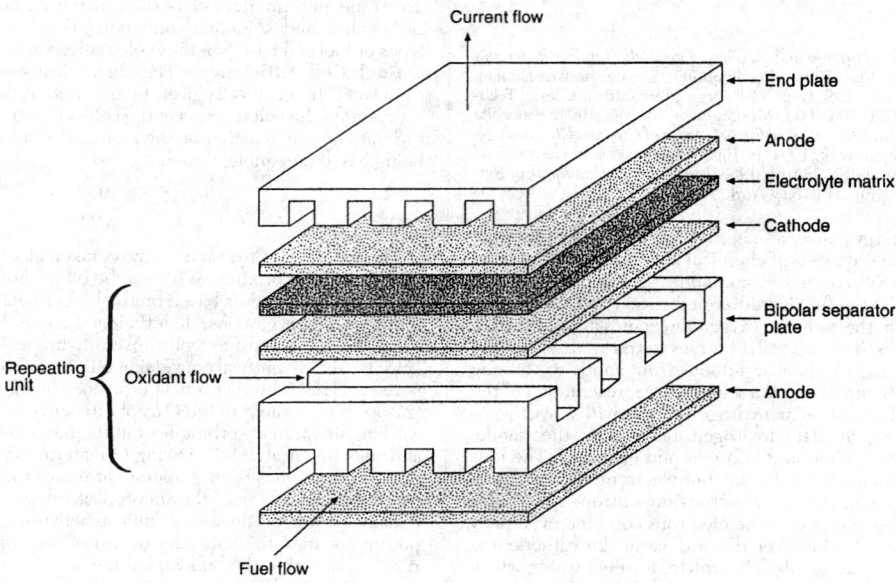

FIG. 27-61   Stacking of individual fuel cells.

and oxidant gases, chemically stable under reducing and oxidizing conditions, and an excellent electronic conductor. In addition, it is often used to distribute the gases to the anode and cathode surfaces through flow channels cut or molded into it.

The number of fuel cells that are stacked is determined by the desired electrical potential. For 110-V systems it can be about 200 cells. Since a typical fuel cell is about 5 mm (0.2 in) thick, a 200-cell stack assembly (including the end hardware that keeps the unit under compression) is about 2 m (6 ft) tall. The reactant and product gas streams are supplied and removed from the stack by external or internal manifolding. Externally manifolded stacks have shallow trays on each of the four sides to supply the fuel and air and to remove the depleted gases and reaction products. The manifolds are mechanically clamped to the stacks and sealed at the edges. These manifold seals must be gastight, electrically insulating, and able to tolerate thermal expansion mismatches between the stack and the manifold materials as well as dimensional changes due to aging.

Alternatively, reactant and product gases can be distributed to and removed from individual cells through internal pipes in a design analogous to that of filter presses. Care must be exercised to assure an even flow distribution between the entry and exit cells. The seals in internally manifolded stacks are generally not subject to electrical, thermal, and mechanical stresses, but are more numerous than in externally manifolded stacks.

Because fuel cells generate an amount of excess heat consistent with their thermodynamic efficiency, they must be cooled. In low-temperature fuel cells, the cooling medium is generally water or oil, which flows through cooling plates interspaced throughout the stack. In high-temperature cells, heat is removed by the reactant air stream and also by the endothermic fuel reforming reactions in the stack.

Types of Fuel Cells   The five major types of fuel cell are listed in Table 27-26. Each has unique chemical features. The *alkaline fuel cell* (AFC) has high power density and has proven itself as a reliable power source in the U.S. space program, but the alkaline electrolyte reacts with carbon dioxide, which is present in reformed hydrocarbon fuels and air. The *polymer electrolyte fuel cell* (PEFC) and the *phosphoric acid fuel cell* (PAFC) are compatible with carbon dioxide, but both are sensitive to carbon monoxide (PEFC much more so than PAFC), which is adsorbed onto the platinum catalyst and renders it inactive. Therefore, these three types of fuel cell require pure hydrogen as fuel; and if the hydrogen has been obtained by reforming a fuel such as natural gas, the hydrogen-rich fuel stream must be purified before being introduced into the fuel cell. The *molten carbonate fuel cell* (MCFC) and the *solid oxide fuel cell* (SOFC) can tolerate carbon monoxide and can operate on hydrocarbon fuels with minimal fuel processing, but they operate at elevated temperatures.

The operating temperature also affects the fuel cell operating potential. A high operating temperature accelerates reaction rates but

lowers the thermodynamic equilibrium potential. These effects balance one another, and, in practice, the operating point of any fuel cell is usually between 0.7 and 0.8 V. The cell reactions for the five types of fuel cell are summarized in Table 27-27. It is important to note that in cells with acidic electrolytes (PAFC and PEFC) the product water evolves on the air electrode, but in the alkaline ones (AFC, MCFC, and SOFC) it is generated on the fuel electrode. This has consequences for the processing of hydrocarbon fuels, as discussed later.

Following is a summary of the materials, operating characteristics, and mode of construction for each type of fuel cell.

Alkaline Fuel Cell   The electrolyte for NASA's space shuttle orbiter fuel cell is 35 percent potassium hydroxide. The cell operates between 353 and 363 K (176 and 194°F) at 0.4 MPa (59 psia) on hydrogen and oxygen. The electrodes contain platinum-palladium and platinum-gold alloy powder catalysts bonded with polytetrafluoroethylene (PTFE) latex and supported on gold-plated nickel screens for current collection and gas distribution. A variety of materials, including asbestos and potassium titanate, are used to form a microporous separator that retains the electrolyte between the electrodes. The cell structural materials, bipolar plates, and external housing are usually nickel, plated to resist corrosion. The complete orbiter fuel cell power plant is shown in Fig. 27-62.

Typical polarization curves for alkaline fuel cells are shown in Fig. 27-63. It is apparent that the alkaline fuel cell can operate at about 0.9 V and 500 mA/cm² current density. This corresponds to an energy conversion efficiency of about 60 percent HHV. The space shuttle orbiter power module consists of three separate units, each measuring 0.35 by 0.38 by 1 m (14 by 15 by 40 in), weighing 119 kg (262 lb), and generating 15 kW of power. The power density is about 100 W/L and the specific power, 100 W/kg.

Polymer Electrolyte Fuel Cell   The PEFC, also known as the *proton-exchange-membrane fuel cell* (PEMFC), is of much interest because it is capable of high power density and it can deliver about 40 percent of its nominal power at room temperature. These features have made the PEFC a candidate to replace internal combustion engines in transportation applications, and prototype passenger cars with fuel cell power sources have been developed. Methanol, ethanol, hydrogen, natural gas, dimethyl ether, and common transportation fuels such as gasoline are being considered as fuel. All but hydrogen require a reforming step to provide hydrogen for the fuel cell. The

FIG. 27-62   Orbiter power plant. (*International Fuel Cells.*)

TABLE 27-26   Fuel Cell Characteristics

Type of fuel cell	Electrolyte	Operating temperature		Coolant medium
		K	°C	
Alkaline	KOH	363	90	Water
Polymer	$CF_3(CF_2)_nOCF_2SO_3^-$	353	80	Water
Phosphoric acid	H_3PO_4	473	200	Steam/water
Molten carbonate	$Li_2CO_3-K_2CO_3$	923	650	Air
Solid oxide	$Zr_{0.92}Y_{0.08}O_{1.96}$	1273	1000	Air

TABLE 27-27   Fuel Cell Reaction Electrochemistry

Type of fuel cell	Conducting ion	Anode reaction	Cathode reaction
Alkaline	OH^-	$H_2 + 2OH^- \rightarrow 2H_2O + 2e^-$	$\frac{1}{2}O_2 + H_2O + 2e^- \rightarrow 2OH^-$
Polymer	H^+	$H_2 \rightarrow 2H^+ + 2e^-$	$\frac{1}{2}O_2 + 2H^+ + 2e^- \rightarrow H_2O$
Phosphoric acid	H^+	$H_2 \rightarrow 2H^+ + 2e^-$	$\frac{1}{2}O_2 + 2H^+ + 2e^- \rightarrow H_2O$
Molten carbonate	CO_3^{2-}	$H_2 + CO_3^{2-} \rightarrow H_2O + CO_2 + 2e^-$	$\frac{1}{2}O_2 + CO_2 + 2e^- \rightarrow CO_3^{2-}$
Solid oxide	O^{2-}	$H_2 + O^{2-} \rightarrow H_2O + 2e^-$	$\frac{1}{2}O_2 + 2e^- \rightarrow O^{2-}$

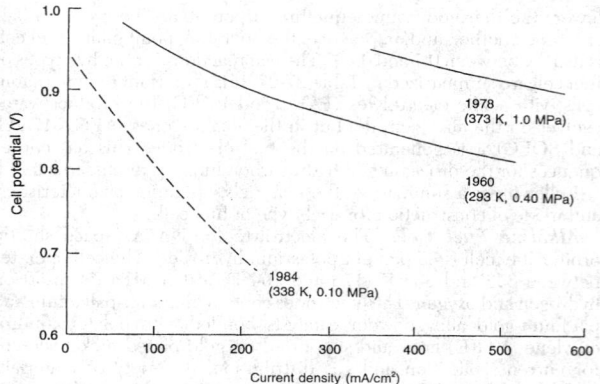

FIG. 27-63 Polarization curves for alkaline fuel cells.

motivation for this development effort is the virtual elimination of both on-road emissions from automobiles and the range limitations associated with battery-powered electric automobiles.

The electrolyte is a perfluorosulfonic acid ionomer, commercially available under the trade name of Nafion™. It is in the form of a membrane about 0.17 mm (0.007 in) thick, and the electrodes are bonded directly onto the surface. The electrodes contain very finely divided platinum or platinum alloys supported on carbon powder or fibers. The bipolar plates are made of graphite or metal.

Typical platinum catalyst loadings needed to support the anodic and cathodic reactions are currently 1 to 2 mg/cm² of active cell area. Owing to the cost of platinum, substantial efforts have been made to reduce the catalyst loading, and some fuel cells have operated at a catalyst loading of 0.25 mg/cm².

To be ionically conducting, the fluorocarbon ionomer must be "wet": under equilibrium conditions, it will contain about 20 percent water. The operating temperature of the fuel cell must be less than 373 K (212°F), therefore, to prevent the membrane from drying out.

Being acidic, fluorocarbon ionomers can tolerate carbon dioxide in the fuel and air streams; PEFCs, therefore, are compatible with hydrocarbon fuels. However, the platinum catalysts on the fuel and air electrodes are extremely sensitive to carbon monoxide: only a few parts per million are acceptable. Catalysts that are tolerant to carbon monoxide are being explored. Typical polarization curves for PEFCs are shown in Fig. 27-64.

A schematic diagram of a methanol-fueled PEFC system is shown in Fig. 27-65. A methanol reformer (to convert CH_3OH to H_2 and CO_2

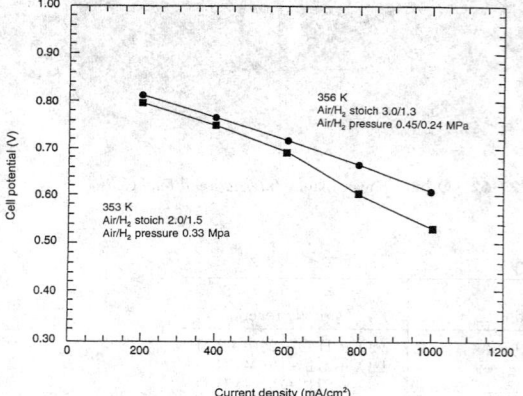

FIG. 27-64 Polarization curves for PEFC stacks.

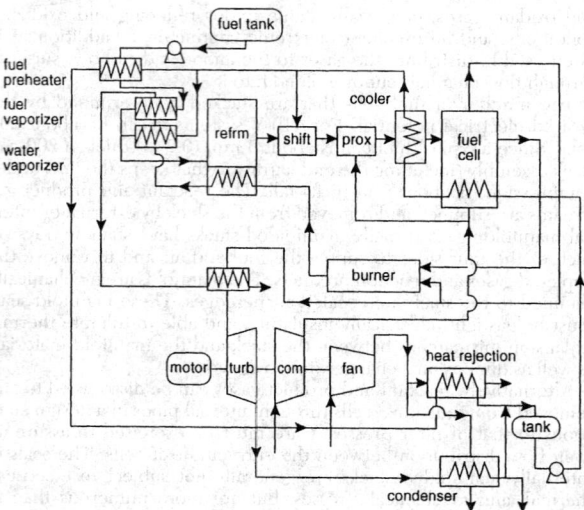

FIG. 27-65 Methanol-fueled PEFC system schematic.

as the principal products), a water-gas shift reactor (to convert most of the coproduct CO to CO_2, and to provide additional H_2), and a preferential oxidizer (to reduce the residual CO to 10 ppm) are included in the system.

Polymer electrolyte fuel cells can be obtained from several developers. These fuel cells deliver about 5 kW of power and measure 30 by 30 by 70 cm (12 × 12 × 28 in.). For the large production volume anticipated if the automotive industry were to adopt the PEFC, a system cost of less than $100/kW may be reached eventually.

Phosphoric Acid Fuel Cell   This type of fuel cell was developed in response to the industry's desire to expand the natural-gas market. The electrolyte is 93 to 98 percent phosphoric acid contained in a matrix of silicon carbide. The electrodes consist of finely divided platinum or platinum alloys supported on carbon black and bonded with PTFE latex. The latter provides enough hydrophobicity to the electrodes to prevent flooding of the structure by the electrolyte. The carbon support of the air electrode is specially formulated for oxidation resistance at 473 K (392°F) in air and positive potentials.

The bipolar plate material of the PAFC is graphite. A portion of it has a carefully controlled porosity that serves as a reservoir for phosphoric acid and provides flow channels for distribution of the fuel and oxidant. The plates are electronically conductive but impervious to gas crossover.

In a typical PAFC system, methane passes through a reformer with steam from the coolant loop of the water-cooled fuel cell. Heat for the reforming reaction is generated by combusting the depleted fuel. The reformed natural gas contains typically 60 percent H_2, 20 percent CO, and 20 percent H_2O. Because the platinum catalyst in the PAFC can tolerate only about 0.5 percent CO, this fuel mixture is passed through a water gas shift reactor before being fed to the fuel cell.

PAFC systems are commercially available from the ONSI Corporation as 200-kW stationary power sources operating on natural gas. The stack cross section is 1 m² (10.8 ft²). It is about 2.5 m (8.2 ft) tall and rated for a 40,000-h life. It is cooled with water/steam in a closed loop with secondary heat exchangers. The photograph of a unit is shown in Fig. 27-66. These systems are intended for on-site power and heat generation for hospitals, hotels, and small businesses. Another application, however, is as "dispersed" 5- to 10-MW power plants in metropolitan areas. Such units would be located at electric utility distribution centers, bypassing the high-voltage transmission system. The market entry price of the system is $3000/kW. As production volumes increase, the price is projected to decline to $1000 to 1500/kW.

Molten Carbonate Fuel Cell   The electrolyte in the MCFC is a

FIG. 27-66   PC-25™ commercial 200-kW PAFC generator. (*International Fuel Cells.*)

mixture of lithium/potassium or lithium/sodium carbonates, retained in a ceramic matrix of lithium aluminate. The carbonate salts melt at about 773 K (932°F), allowing the cell to be operated in the 873 to 973 K (1112 to 1292°F) range. Platinum is no longer needed as an electrocatalyst because the reactions are fast at these temperatures. The anode in MCFCs is porous nickel metal with a few percent of chromium or aluminum to improve the mechanical properties. The cathode material is lithium-doped nickel oxide.

The bipolar plates are made from either Type 310 or Type 316 stainless steel, which is coated on the fuel side with nickel and aluminized in the seal area around the edge of the plates. Both internally and externally manifolded stacks have been developed.

In MCFCs, the hydrogen fuel is generated from such common fuels as natural gas or liquid hydrocarbons by steam reforming; the fuel processing function can be integrated into the fuel cell stack because the operating temperature permits reforming using the waste heat. An added complexity in MCFCs is the need to recycle carbon dioxide from the anode side to the cathode side to maintain the desired electrolyte composition. (At the cathode, carbon dioxide reacts with incoming electrons and oxygen in air to regenerate the carbonate ions that are consumed at the anode.) The simplest way is to burn the depleted fuel and mix it with the incoming air. This works well but dilutes the oxygen with the steam generated in the fuel cell. A steam condenser and recuperative heat exchanger can be added to eliminate the steam, but at increased cost.

The fuel cell must be cooled with either water or air, and the heat can be converted to electricity in a bottoming cycle. The dc electrical output of the stack is usually converted to ac and stepped up or down in voltage, depending on the application. Analogous to PAFCs, MCFC stacks are about 1 m² (10.8 ft²) in plan area and quite tall. A stack generates 200 to 300 kW. Market entry is expected in 1999.

Solid Oxide Fuel Cell   In SOFCs the electrolyte is a ceramic oxide ion conductor, such as yttrium-doped zirconium oxide. The conductivity of this material is 0.1 S/cm at 1273 K (1832°F); it decreases to 0.01 S/cm at 1073 K (1472°F), and by another order of magnitude at 773 K (932°F). Because the resistive losses need to be kept below about 50 mV, the operating temperature of the

SOFC depends on the thickness of the electrolyte. For a thickness of 100 μm or more, the operating temperature is 1273 K (1832°F), but fuel cells with thin electrolytes can operate between 973 and 1073 K (1292 and 1472°F).

The anode material in SOFCs is a cermet (metal/ceramic composite material) of 30 to 40 percent nickel in zirconia, and the cathode is lanthanum manganite doped with calcium oxide or strontium oxide. Both of these materials are porous and mixed ionic/electronic conductors. The bipolar separator typically is doped lanthanum chromite, but a metal can be used in cells operating below 1073 K (1472°F). The bipolar plate materials are dense and electronically conductive.

Typical polarization curves for SOFCs are shown in Fig. 27-67. As discussed earlier, the open-circuit potential of SOFCs is less than 1 V because of the high temperature, but the reaction overpotentials are

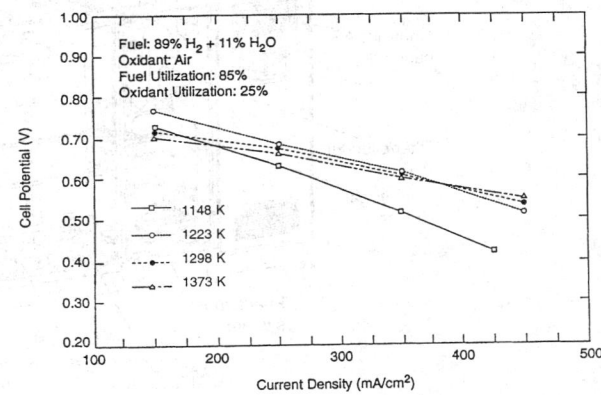

FIG. 27-67   Polarization curves at different temperatures for 50-cm active length thin-wall SOFCs.

small, yielding almost linear curves with slopes corresponding to the resistance of the components.

SOFCs can have a planar geometry similar to PEFCs, but the leading technology is tubular, as shown in Fig. 27-68. The advantage of the tubular arrangement is the absence of high-temperature seals.

Like MCFCs, SOFCs can integrate fuel reforming within the fuel cell stack. A prereformer converts a substantial amount of the natural gas using waste heat from the fuel cell. Compounds containing sulfur (e.g., thiophene, which is commonly added to natural gas as an odorant) must be removed before the reformer. Typically, a hydrodesulfurizer combined with a zinc oxide absorber is used.

The desulfurized natural gas is mixed with the recycled depleted fuel stream containing steam formed in the fuel cell. About 75 percent of the methane is converted to hydrogen and carbon monoxide in the prereformer. The hydrogen-rich fuel is then passed over the fuel cell anode, where 85 percent is converted to electricity. The balance is burned with depleted air in the combustion zone.

The hot combustion gas preheats the fresh air and the prereformer, and can be used further to generate steam. The system is cooled with 200 to 300 percent excess air. A 25-kW SOFC generator system is shown in Fig. 27-69.

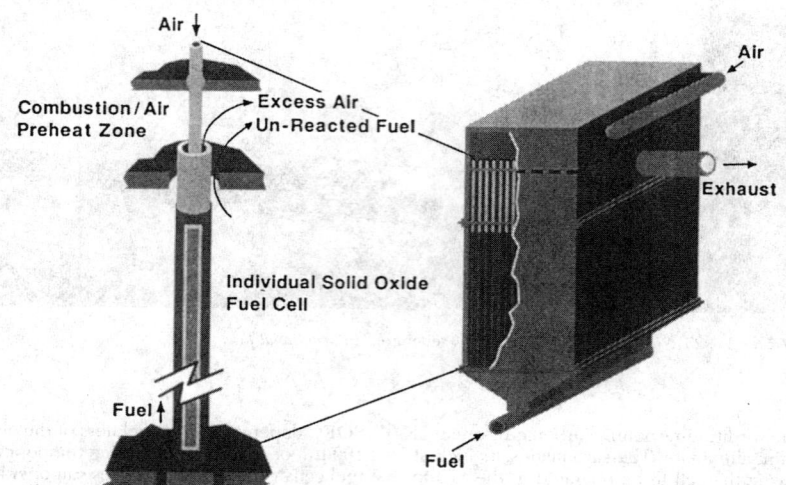

FIG. 27-68   Configuration of the tubular SOFC. (*Courtesy of Westinghouse Electric Corporation.*)

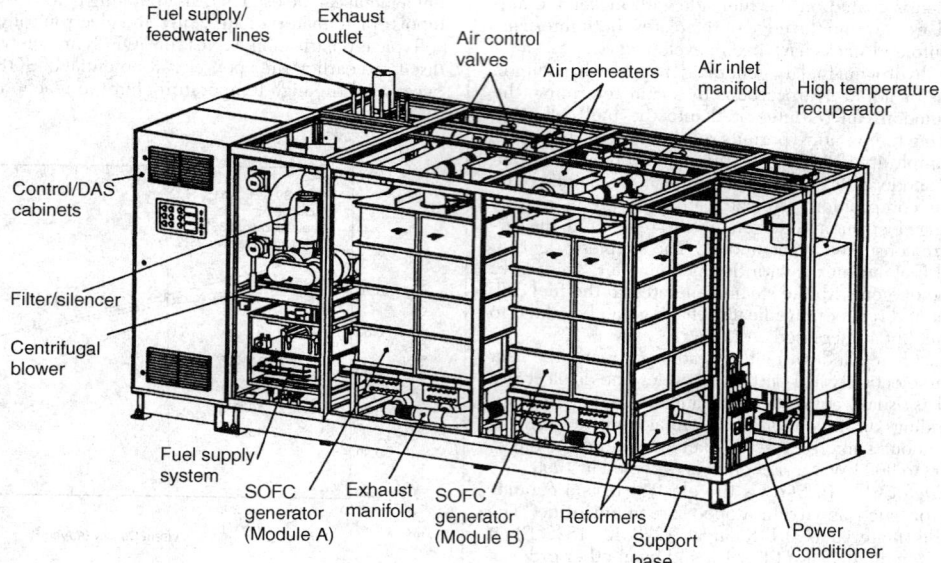

FIG. 27-69   SOFC 25-kW system package. (*Courtesy of Westinghouse Electric Corporation.*)

Materials of Construction*

Oliver W. Siebert, P.E., B.S.M.E., *Washington University, Graduate Metallurgical Engineering, Sever Institute of Technology; Professor, Department of Chemical Engineering, Washington University, St. Louis, Mo.; President, Siebert Materials Engineering, Inc., St. Louis, Mo.; Senior Engineering Fellow (Retired), Monsanto Co.; Mechanical Designer, Sverdrup Corp.; Metallurgist, Carondelet Foundry; United Nations Consultant to the People's Republic of China; Fellow, American Institute of Chemical Engineers; Life Fellow, American Society of Mechanical Engineers; a Past Elected Director and Fellow, National Association of Corrosion Engineers, Int'l.; American Society for Metals, Int'l.; American Welding Society; Pi Tau Sigma, Sigma Xi, and Tau Beta Pi.*

John G. Stoecker II, B.S.M.E., *University of Missouri School of Mines and Metallurgy; Principal Consultant, Stoecker & Associates, St. Louis, Mo.; Principal Materials Engineering Specialist (Retired), Monsanto Co.; High-Temperature Design/Application Engineer, Abex Corporation; Member, NACE International, ASM International.*

° The contributions of R. B. Norton to material used from the fifth edition and of A. S. Krisher and O. W. Siebert to material used from the sixth edition are acknowledged.

CORROSION AND ITS CONTROL

GENERAL REFERENCES: Ailor (ed.), *Handbook on Corrosion Testing and Evaluation*, McGraw-Hill, New York, 1971. Bordes (ed.), *Metals Handbook*, 9th ed., vols. 1, 2, and 3, American Society for Metals, Metals Park, Ohio, 1978–1980; other volumes in preparation. Dillon (ed.), *Process Industries Corrosion*, National Association of Corrosion Engineers, Houston, 1975. Dillon and associates, *Guidelines for Control of Stress Corrosion Cracking of Nickel-Bearing Stainless Steels and Nickel-Base Alloys*, MTI Manual No. 1, Materials Technology Institute of the Chemical Process Industries, Columbus, 1979. Evans, *Metal Corrosion Passivity and Protection*, E. Arnold, London, 1940. Evans, *Corrosion and Oxidation of Metals*, St. Martin's, New York, 1960. Fontana and Greene, *Corrosion Engineering*, 2d ed., McGraw-Hill, New York, 1978. Gackenbach, *Materials Selection for Process Plants*, Reinhold, New York, 1960. Hamner (comp.), *Corrosion Data Survey: Metals Section*, National Association of Corrosion Engineers, Houston, 1974. Hamner (comp.), *Corrosion Data Survey: Non-Metals Section*, National Association of Corrosion Engineers, Houston, 1975. Hanson and Parr, *The Engineer's Guide to Steel*, Addison-Wesley, Reading, Mass., 1965. LaQue and Copson, *Corrosion Resistance of Metals and Alloys*, Reinhold, New York, 1963. Lyman (ed.), *Metals Handbook*, 8th ed., vols. 1–11, American Society for Metals, Metals Park, Ohio, 1961–1976. Mantell (ed.), *Engineering Materials Handbook*, McGraw-Hill, New York, 1958. Shreir, *Corrosion*, George Newnes, London, 1963. Speller, *Corrosion—Causes and Prevention*, McGraw-Hill, New York, 1951. Uhlig (ed.), *The Corrosion Handbook*, Wiley, New York, 1948. Uhlig, *Corrosion and Corrosion Control*, 2d ed., Wiley, New York, 1971. Wilson and Oates, *Corrosion and the Maintenance Engineer*, Hart Publishing, New York, 1968. Zapffe, *Stainless Steels*, American Society for Metals, Cleveland, 1949. *Plus additional references as dictated by manuscript.*

INTRODUCTION*

Corrosion is not a favorite subject of engineers. Many a proud designer or project engineer has developed a new component or process with outstanding performance only to have it fail prematurely because of corrosion. Furthermore, despite active research by corrosion engineers, a visit to the local scrap yard shows that large percentages of cars and domestic appliances still fail because of corrosion; this loss pales by comparison when industrial corrosion failures are included. As a result, the annual cost of corrosion and corrosion protection in the United States is on the order of $300 billion, far more than the annual budgets of some small countries.

One of the principal reasons for failure due to reaction with the service environment is the relatively complex nature of the reactions involved. Yet, in spite of all the complex corrosion jargon, whether a metal corrodes depends on the simple electrochemical cell set up by the environment. This might give the erroneous impression that it is possible to calculate such things as the corrosion rate of a car fender in the spring mush of salted city streets. Dr. M. Pourbaix has done some excellent work in the application of thermodynamics to corrosion, but this cannot yet be applied directly to the average complex situation.

Yet, corrosion engineering and science is no longer an empirical art; dissecting a large corrosion problem into its basic mechanisms allows the use of quite sophisticated electrochemical techniques to accomplish satisfactory results. On that positive side, there is real satisfaction and economic gain in designing a component that can resist punishing service conditions under which other parts fail. In some cases, we cannot completely prevent corrosion, but we can try to avoid obsolescence of the component due to corrosion.

FLUID CORROSION

In the selection of materials of construction for a particular fluid system, it is important first to take into consideration the **characteristics**

---

* Abstracted from texts by Flinn and Trojan, with permission of John Wiley & Sons.

of the system, giving special attention to all factors that may influence corrosion. Since these factors would be peculiar to a particular system, it is impractical to attempt to offer a set of hard and fast rules that would cover all situations.

The **materials** from which the system is to be fabricated are the second important consideration; therefore, knowledge of the characteristics and general behavior of materials when exposed to certain environments is essential.

In the absence of factual corrosion information for a particular set of fluid conditions, a reasonably good selection would be possible from data based on the resistance of materials to a very similar environment. These data, however, should be used with some reservations. Good practice calls for applying such data for preliminary screening. Materials selected thereby would require further study in the fluid system under consideration.

FLUID CORROSION: GENERAL

Metallic Materials Pure metals and their alloys tend to enter into **chemical** union with the elements of a corrosive medium to form stable compounds similar to those found in nature. When metal loss occurs in this way, the compound formed is referred to as the **corrosion product** and the metal surface is spoken of as being **corroded.**

Corrosion is a complex phenomenon that may take any one or more of several forms. It is usually confined to the metal surface, and this is called **general corrosion.** But it sometimes occurs along grain boundaries or other lines of weakness because of a difference in resistance to attack or local electrolytic action.

In most aqueous systems, the corrosion reaction is divided into an anodic portion and a cathodic portion, occurring simultaneously at discrete points on metallic surfaces. Flow of electricity from the anodic to the cathodic areas may be generated by local cells set up either on a single metallic surface (because of local point-to-point differences on the surface) or between dissimilar metals.

Nonmetallics As stated, corrosion of metals applies specifically to chemical or electrochemical attack. The deterioration of plastics and other nonmetallic materials, which are susceptible to swelling, crazing, cracking, softening, and so on, is essentially **physiochemical** rather than electrochemical in nature. Nonmetallic materials can either be rapidly deteriorated when exposed to a particular environment or, at the other extreme, be practically unaffected. Under some conditions, a nonmetallic may show evidence of gradual deterioration. However, it is seldom possible to evaluate its chemical resistance by measurements of weight loss alone, as is most generally done for metals.

FLUID CORROSION: LOCALIZED

Pitting Corrosion Pitting is a form of corrosion that develops in highly localized areas on the metal surface. This results in the development of cavities or pits. They may range from deep cavities of small diameter to relatively shallow depressions. Pitting examples: aluminum and stainless alloys in aqueous solutions containing chloride. **Inhibitors** are sometimes helpful in preventing pitting.

Crevice Corrosion Crevice corrosion occurs within or adjacent to a crevice formed by contact with another piece of the same or another metal or with a nonmetallic material. When this occurs, the intensity of attack is usually more severe than on surrounding areas of the same surface.

This form of corrosion can result because of a deficiency of oxygen in the crevice, acidity changes in the crevice, buildup of ions in the crevice, or depletion of an inhibitor.

Oxygen-Concentration Cell   The oxygen-concentration cell is an electrolytic cell in which the driving force to cause corrosion results from a difference in the amount of oxygen in solution at one point as compared with another. Corrosion is accelerated where the oxygen concentration is least, for example, in a stuffing box or under gaskets. This form of corrosion will also occur under solid substances that may be deposited on a metal surface and thus shield it from ready access to oxygen. Redesign or change in mechanical conditions must be used to overcome this situation.

Galvanic Corrosion   Galvanic corrosion is the corrosion rate above normal that is associated with the flow of current to a less active metal (cathode) in contact with a more active metal (anode) in the same environment. Tables 28-1*a* and 28-1*b* show the **galvanic** series of various metals. It should be used with caution, since exceptions to

TABLE 28-1*a*   Galvanic Series of Metals and Alloys

Corroded end (anodic, or least noble)

Magnesium
Magnesium alloys
Zinc
Aluminum alloys
Aluminum
Alclad
Cadmium
Mild steel
Cast iron
Ni-Resist
13% chromium stainless (active)
50-50 lead-tin solder
18-8 stainless type 304 (active)
18-8-3 stainless type 316 (active)
Lead
Tin
Muntz Metal
Naval brass
Nickel (active)
Inconel 600 (active)
Yellow brass
Admiralty brass
Aluminum bronze
Red brass
Copper
Silicon bronze
70-30 cupronickel
Nickel (passive)
Inconel 600 (passive)
Monel 400
18-8 stainless type 304 (passive)
18-8-3 stainless type 316 (passive)
Silver
Graphite
Gold
Platinum

Protected end (cathodic, or most noble)

TABLE 28-1*b*   Galvanic Series in Sea Water (Approx.), Volts vs. Sat. Calomel Ref. Electrode

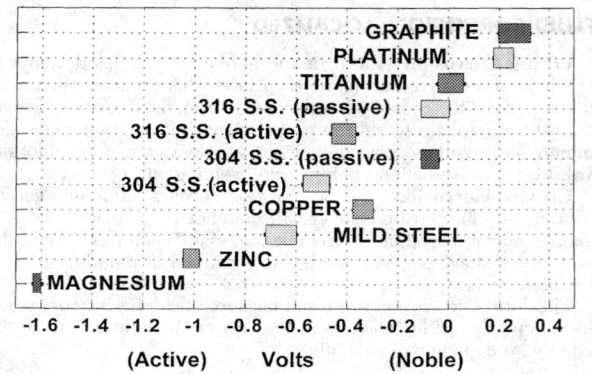

this series in actual use are possible. However, as a general rule, when dissimilar metals are used in contact with each other and are exposed to an electrically conducting solution, combinations of metals that are as close as possible in the galvanic series should be chosen. Coupling two metals widely separated in this series generally will produce accelerated attack on the more active metal. Often, however, protective oxide films and other effects will tend to reduce galvanic corrosion. Galvanic corrosion can, of course, be prevented by **insulating** the metals from each other. For example, when plates are bolted together, specially designed plastic washers can be used.

Potential differences leading to galvanic-type cells can also be set up on a single metal by differences in temperature, velocity, or concentration (see subsection "Crevice Corrosion").

Area effects in galvanic corrosion are very important. An unfavorable area ratio is a large cathode and a small anode. Corrosion of the anode may be 100 to 1,000 times greater than if the two areas were the same. This is the reason why stainless steels are susceptible to rapid pitting in some environments. Steel rivets in a copper plate will corrode much more severely than a steel plate with copper rivets.

Intergranular Corrosion   Selective corrosion in the grain boundaries of a metal or alloy without appreciable attack on the grains or crystals themselves is called intergranular corrosion. When severe, this attack causes a loss of strength and ductility out of proportion to the amount of metal actually destroyed by corrosion.

The **austenitic stainless steels** that are not stabilized or that are not of the extra-low-carbon types, when heated in the temperature range of 450 to 843°C (850 to 1,550°F), have chromium-rich compounds (chromium carbides) precipitated in the grain boundaries. This causes grain-boundary impoverishment of chromium and makes the affected metal susceptible to intergranular corrosion in many environments. Hot nitric acid is one environment which causes severe intergranular corrosion of austenitic stainless steels with grain-boundary precipitation. Austenitic stainless steels stabilized with niobium (columbium) or titanium to decrease carbide formation or containing less than 0.03 percent carbon are normally not susceptible to grain-boundary deterioration when heated in the given temperature range. Unstabilized austenitic stainless steels or types with normal carbon content, to be immune to intergranular corrosion, should be given a solution anneal. This consists of heating to 1,090°C (2,000°F), holding at this temperature for a minimum of 1 h/in of thickness, followed by rapidly quenching in water (or, if impractical because of large size, rapidly cooling with an air-water spray).

Stress-Corrosion Cracking   Corrosion can be accelerated by stress, either residual internal stress in the metal or externally applied stress. Residual stresses are produced by deformation during fabrication, by unequal cooling from high temperature, and by internal structural rearrangements involving volume change. Stresses induced by rivets and bolts and by press and shrink fits can also be classified as residual stresses. Tensile stresses at the surface, usually of a magnitude equal to the yield stress, are necessary to produce stress-corrosion cracking. However, failures of this kind have been known to occur at lower stresses.

Virtually every alloy system has its specific environment conditions which will produce stress-corrosion cracking, and the time of exposure required to produce failure will vary from minutes to years. Typical examples include cracking of cold-formed brass in ammonia environments, cracking of austenitic stainless steels in the presence of chlorides, cracking of Monel in hydrofluosilicic acid, and caustic embrittlement cracking of steel in caustic solutions.

This form of corrosion can be prevented in some instances by eliminating high stresses. Stresses developed during fabrication, particularly during welding, are frequently the main source of trouble. Of course, temperature and concentration are also important factors in this type of attack.

Presence of **chlorides** does not generally cause cracking of austenitic stainless steels when temperatures are below about 50°C (120°F). However, when temperatures are high enough to concentrate chlorides on the stainless surface, cracking may occur when the chloride concentration in the surrounding media is a few parts per million. Typical examples are cracking of heat-exchanger tubes at the crevices in rolled joints and under scale formed in the vapor space below the top tube sheet in vertical heat exchangers. The cracking of stainless steel under insulation is caused when chloride-containing water is concentrated on the hot surfaces. The chlorides may be leached from the insulation or may be present in the water when it enters the insulation. Improved design and maintenance of insulation weatherproofing, coating of the metal prior to the installation of insulation, and use of chloride-free insulation are all steps which will help to reduce (but not eliminate) this problem.

Serious stress-corrosion-cracking failures have occurred when chloride-containing hydrotest water was not promptly removed from stainless-steel systems. Use of potable-quality water and complete draining after test comprise the most reliable solution to this problem. Use of chloride-free water is also helpful, especially when prompt drainage is not feasible.

In handling caustic, as-welded steel can be used without developing caustic-embrittlement cracking if the temperature is below 50°C (120°F). If the temperature is higher and particularly if the concentration is above about 30 percent, cracking at and adjacent to non-stress-relieved welds frequently occurs.

Liquid-Metal Corrosion   Liquid metals can also cause corrosion failures. The most damaging are liquid metals which penetrate the metal along grain boundaries to cause catastrophic failure. Examples include mercury attack on aluminum alloys and attack of stainless steels by molten zinc or aluminum. A fairly common problem occurs when galvanized-structural-steel attachments are welded to stainless piping or equipment. In such cases it is mandatory to remove the galvanizing completely from the area which will be heated above 260°C (500°F).

Erosion   Erosion is the destruction of a metal by abrasion or attrition caused by the flow of liquid or gas (with or without suspended solids). The use of harder materials and changes in velocity or environment are methods employed to prevent erosion attack.

Impingement Corrosion   This phenomenon is sometimes referred to as erosion-corrosion or velocity-accelerated corrosion. It occurs when damage is accelerated by the mechanical removal of corrosion products (such as oxides) which would otherwise tend to stifle the corrosion reaction.

Corrosion Fatigue   Corrosion fatigue is a reduction by corrosion of the ability of a metal to withstand **cyclic or repeated stresses.**

The surface of the metal plays an important role in this form of damage, as it will be the most highly stressed and at the same time subject to attack by the corrosive media. Corrosion of the metal surface will lower fatigue resistance, and stressing of the surface will tend to accelerate corrosion.

Under cyclic or repeated stress conditions, rupture of protective oxide films that prevent corrosion takes place at a greater rate than that at which new protective films can be formed. Such a situation frequently results in formation of anodic areas at the points of rupture; these produce pits that serve as stress-concentration points for the origin of cracks that cause ultimate failure.

Cavitation   Formation of transient voids or vacuum bubbles in a liquid stream passing over a surface is called cavitation. This is often encountered around propellers, rudders, and struts and in pumps. When these bubbles collapse on a metal surface, there is a severe impact or explosive effect that can cause considerable mechanical damage, and corrosion can be greatly accelerated because of the destruction of protective films. Redesign or a more resistant metal is generally required to avoid this problem.

Fretting Corrosion   This attack occurs when metals slide over each other and cause mechanical damage to one or both. In such a case, frictional heat oxidizes the metal and this oxide then wears away; or the mechanical removal of protective oxides results in exposure of fresh surface for corrosive attack. Fretting corrosion is minimized by using harder materials, minimizing friction (via lubrication), or designing equipment so that no relative movement of parts takes place.

Hydrogen Attack   At elevated temperatures and significant hydrogen partial pressures, hydrogen will penetrate carbon steel, reacting with the carbon in the steel to form methane. The pressure generated causes a loss of ductility (hydrogen embrittlement) and failure by cracking or blistering of the steel. The removal of the carbon from the steel (decarburization) results in decreased strength. Resistance to this type of attack is improved by alloying with molybdenum or chromium. Accepted limits for the use of carbon and low-alloy steels are shown in Fig. 28-1, which is adapted from American Petroleum Institute (API) Publication 941, *Steels for Hydrogen Service at Elevated Temperatures and Pressures in Petroleum Refineries and Petrochemical Plants*, the so-called Nelson curves.

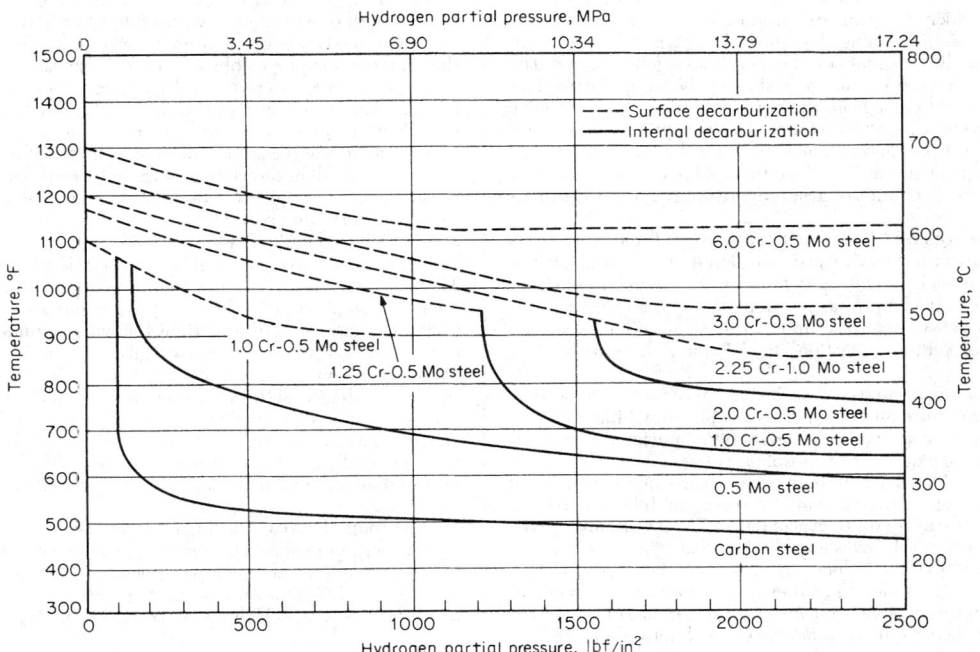

FIG. 28-1 Operating limits for steels in hydrogen service. Each steel is suitable for use under hydrogen-partial-pressure-temperature conditions below and to the left of its respective curve. (*Courtesy of National Association of Corrosion Engineers.*)

Hydrogen damage can also result from hydrogen generated in electrochemical corrosion reactions. This phenomenon is most commonly observed in solutions of specific weak acids. H_2S and HCN are the most common, although other acids can cause the problem. The atomic hydrogen formed on the metal surface by the corrosion reaction diffuses into the metal and forms molecular hydrogen at microvoids in the metal. The result is failure by embrittlement, cracking, and blistering.

FLUID CORROSION: STRUCTURAL

Graphitic Corrosion Graphitic corrosion usually involves **gray cast iron** in which metallic iron is converted into corrosion products, leaving a residue of intact graphite mixed with iron-corrosion products and other insoluble constituents of cast iron.

When the layer of graphite and corrosion products is impervious to the solution, corrosion will cease or slow down. If the layer is porous, corrosion will progress by galvanic behavior between graphite and iron. The rate of this attack will be approximately that for the maximum penetration of steel by pitting. The layer of graphite formed may also be effective in reducing the galvanic action between cast iron and more noble alloys such as bronze used for valve trim and impellers in pumps.

Low-alloy cast irons frequently demonstrate a superior resistance to graphitic corrosion, apparently because of their denser structure and the development of more compact and more protective graphitic coatings. Highly alloyed **austenitic cast irons** show considerable superiority over gray cast irons to graphitic corrosion because of the more noble potential of the austenitic matrix plus more protective graphitic coatings.

Carbon steels heated for prolonged periods at temperatures above 455°C (850°F) may be subject to the segregation of carbon, which is transformed into graphite. When this occurs, the structural strength of the steel will be affected. Killed steels or low-alloy steels of chromium and molybdenum or chromium and nickel should be considered for elevated-temperature services.

Parting, or Dealloying, Corrosion This type of corrosion occurs when only one component of an alloy is removed by corrosion. The most common type is dezincification of brass.

Dezincification Dezincification is corrosion of a brass alloy containing zinc in which the principal product of corrosion is metallic copper. This may occur as plugs filling pits (plug type) or as continuous layers surrounding an unattacked core of brass (general type). The mechanism may involve overall corrosion of the alloy followed by redeposition of the copper from the corrosion products or selective corrosion of zinc or a high-zinc phase to leave copper residue. This form of corrosion is commonly encountered in brasses that contain more than 15 percent zinc and can be either eliminated or reduced by the addition of small amounts of **arsenic, antimony,** or **phosphorus** to the alloy.

Biological Corrosion The metabolic activity of microorganisms can either directly or indirectly cause deterioration of a metal by corrosion processes. Such activity can (1) produce a corrosive environment, (2) create electrolytic-concentration cells on the metal surface, (3) alter the resistance of surface films, (4) have an influence on the rate of anodic or cathodic reaction, and (5) alter the environment composition.

Microorganisms associated with corrosion are of two types, aerobic and anaerobic. **Aerobic** microorganisms readily grow in an environment containing oxygen, while the **anaerobic** species thrive in an environment virtually devoid of atmospheric oxygen.

The manner in which many of these bacteria carry on their chemical processes is quite complicated and in some cases not fully understood. The role of **sulfate-reducing bacteria** (anaerobic) in promoting corrosion has been extensively investigated. The sulfates in slightly acid to alkaline (pH 6 to 9) soils are reduced by these bacteria to form calcium sulfide and hydrogen sulfide. When these compounds come in contact with underground iron pipes, conversion of the iron to iron sulfide occurs. As these bacteria thrive under these conditions, they will continue to promote this reaction until failure of the pipe occurs.

Several instances of serious biological corrosion occurred when hydrotest water was not promptly removed from stainless-steel systems. These cases involved both potable and nonpotable waters. Biological activity caused perforation of the stainless steel in a few months. Use of potable-quality water and prompt and complete draining after a test constitute the most reliable solution to this problem.

Microbiologically Influenced Corrosion (MIC) * While microbiologically influenced corrosion has existed since the beginning of time, it has only been identified to be a real and persistent problem to industry during the past several decades. A considerable amount of multidisciplinary effort has been expended to determine the extent of the problem and to understand the phenomenon. Prevention and control would seem to be a product of this understanding. Unfortunately, the more we learn, the more we find out how little we really understand about the subject. MIC is now recognized as a problem in many industries, including the gas pipeline, nuclear and fossil power, chemical process, and pulp and paper industries. This brief review is presented from an industrial point of view. Subjects include the buried structures, materials selection, hydrotest procedures, and other considerations that corrosion engineers in the field need to take into account in order to prevent or minimize potential MIC problems in the future.

It is widely recognized that microorganisms attach to, form films on, and influence the corrosion of metals and alloys immersed in natural aqueous environments. The microorganisms influence corrosion by changing the electrochemical conditions at the metal surface. Theoretically, these changes may have many effects, ranging from the induction of localized corrosion, to a change in the rate of general corrosion, to corrosion inhibition. In every case, however, the process of corrosion is electrochemical.

Recently, there has developed a greater recognition of the complexity of the MIC process. MIC is rarely linked to a unique mechanism or to a single species of microorganisms. At the present state of knowledge, it is widely accepted that the growth of different microbial species within adherent biofilms facilitates the development of structured consortia that may enhance the microbial effects on corrosion.

Most practicing engineers are not, and do not need to become, experts in the details of MIC. What is needed is to recognize that this corrosion system is active, that process equipment and structures are at risk, and that there are numerous tools available to monitor and detect MIC (see the later sections on laboratory and field corrosion testing, both of which address the subject of MIC). Advances in the detection of MIC have been made principally through education in the methods for proper observation of field samples, sample collection, cultivation of bacteria involved, microscopic analysis of samples, and metallurgical analysis of samples. These have greatly improved our ability to *prove* the involvement of microbes in the corrosion processes. Recent techniques developed in the laboratory make use of specific antibodies directed against organisms known to be MIC-causing organisms, give rapid and quantitative answers to questions regarding the numbers and types of microbes present in field samples, and can be used in the field.

Bacteria, as a group, can grow over a pH range of about 0 to 11. They can be obligate aerobes (require oxygen to survive and grow), microaerophiles (require low oxygen concentrations), facultative anaerobes (prefer aerobic conditions but will live under anaerobic conditions), or obligate anaerobes (will grow only under conditions where oxygen is absent). It should be emphasized that most anaerobes will survive aerobic conditions for quite a while and the same is true for aerobes in anaerobic conditions. Also note that, for a microorganism, anaerobic conditions may be quite easily found in what are thought to be generally aerobic environments. Often these anaerobic microenvironments are in, or under, films, in particulates of debris, inside crevices, and so on.

As a group, the MIC-causing bacteria may use almost any available organic carbon molecules, from simple alcohols or sugars to phenols to wood or various other complex polymers as food (heterotrophs), or they may fix CO_2 (autotropha) as do plants. Some use inorganic elements or ions (e.g., NH or NO, CH, H, S, Fe, Mn, etc.), as sources of

---

* Excerpted from papers by John G. Stoecker II and Oliver W. Siebert, courtesy of NACE International.

energy. The nutritional requirements of these organisms, therefore, range from very simple to very complex. Most fall in between these extremes and require a limited number of organic molecules, moderate temperatures, moist environments, and near-neutral pH.

Buried Structures Corrosion of buried pipelines caused by sulfate-reducing bacteria has been studied for almost a century. Quite by accident, industry has been protecting buried iron-based structures from bacterial damage through the use of cathodic protection. Cathodic protection produces an elevated alkaline or basic environment on the surface of the buried structure (pH ≥10) which is not conducive to microbiological growth. Booth was one of the early investigators to present this finding, but caution should be exercised when using his recommendations. The user of cathodic protection must also consider the material being protected with regard to caustic cracking; a cathodic potential driven to the negative extreme of −0.95 V for microbiological protection purposes can cause caustic cracking of a steel structure. The benefits and risks of cathodic protection must be weighed for each material and each application.

There has been no dramatic improvement in the microbiological protection of buried structures over the years. Cathodic protection in conjunction with a protective coating system has continued to be the best defense against this form of MIC as well as against other underground corrosive damage. Cathodic protection provides caustic-environment protection at the holes or holidays in the coating that are sure to develop with time due to one cause or another. Experience has been that coating systems, by themselves, do not provide adequate protection for a buried structure over the years; for best results, a properly designed and maintained cathodic protection system *must* be used in conjunction with a protective coating (regardless of the quality of the coating, as applied).

Backfilling with limestone or other alkaline material is an added step to protect buried structures from microbiological damage. Providing adequate drainage to produce a dry environment both above and below ground in the area of the buried structure will also reduce the risk of this type of damage.

Waters While MIC-causing bacteria may arrive at the surface of their corrosion worksite by almost any transportation system, there is always water present to allow them to become active and cause MIC to occur. There are plenty of examples of even superpure waters having sufficient microorganisms present to feed, divide, and multiply when even the smallest trace of a viable food-stuff is present (e.g., the so-called *water for injection* in the pharmaceutical industry has been the observed subject of extensive corrosion of polished stainless steel tanks, piping, and so on).

The initial MIC examples studied in the 1970s were weld failures of stainless steel piping that saw only potable drinking water. The numbers of water-exposed systems that have been verified as being affected by MIC are legion.

Hydrostatic Testing Waters Microbiological species in the water used for hydrostatic (safety) testing of process equipment and for process batch water have caused considerable MIC damage and expense in the past. Guidelines for hydrotesting have been adopted by several industrial and governmental organizations in an effort to prevent this damage. Generally, good results have been reported for those who have followed this practice. Unfortunately, this can be a *very* expensive undertaking where the need cannot be totally quantified (and, thus, justified to management). Cost-cutting practices which either ignore these guides or follow an adulteration of proven precautions can lead to major MIC damage to equipment and process facilities.

Obviously, natural freshwater from wells, lakes, or rivers is the least desirable and its use should be avoided. Even potable waters are not free of organic species that cause MIC damage to conventional materials of construction; for example, steel, stainless steels, and so on. If the size of the facility argues strongly against the use of demineralized water or steam condensate, potable water must be evaluated for test use. Perhaps it would be possible to use a biocide such as hydrogen peroxide or ozone in the potable water; most organic biocides cause disposal problems. Obviously, chlorine must be used only with great care because of the extensive damage it will cause to the 300-series, austenitic stainless steels. In all cases, as soon as the test is over, the water *must* be *completely* drained and the system thoroughly dried so that no vestiges of water are allowed to be trapped in occluded areas.

The literature abounds with instructions as to the proper manner in which to accomplish the necessary and MIC-safe testing procedures. The engineering personnel planning these test operations should avail themselves of that knowledge.

Materials of Construction MIC is a process in which manufactured materials deteriorate through microbiological action. This process can be either direct or indirect.

Microbial biodeterioration of a great many materials (including concretes, glasses, metals and their alloys, and plastics) occurs by diverse mechanisms.

The corrosion engineers' solution to corrosion problems sometimes includes an upgrading of the materials of construction. This is a natural approach, and since microbiological corrosion has usually been considered a form of crevice or under-deposit attack, this option is logical. Unfortunately, with MIC, the use of more corrosion-resistant materials can many times be a shortcut to disaster; MIC is dependent upon the species of organism involved. As an example, an upgrade from type 304 to 316 stainless steel does not always help. Kobrin reported biological corrosion of delta ferrite stringers in weld metal. Obviously, this upgrade was futile; type 316 stainless steel can contain as much or more delta ferrite as does type 304. Kobrin also reported MIC of nickel, nickel-copper Alloy 400, and nickel-molybdenum Alloy B heat exchanger tubes. Although the Alloy 400 and Alloy B were not pitted as severely as the nickel tubes, the higher alloys did not solve the corrosion problem.

In the past, copper was believed to be toxic to *most* microbiological species. This has not turned out to be the case, as pointed out in the preceding example, reported in many other studies by the authors and their colleagues. As an example, reports from engineers in the power generating industry about MIC of copper and cupro-nickel alloys also indicate that copper is not an automatic deterrent to this corrosion mechanism. It is concluded that copper is toxic to the higher life-forms, for example, barnacles and so on, but it is not a biocide to the simple, single-cell organisms.

At this stage of knowledge about MIC, only titanium, zirconium, and tantalum appear to be immune to microbiological damage.

FACTORS INFLUENCING CORROSION

Solution pH The corrosion rate of most metals is affected by pH. The relationship tends to follow one of three general patterns:

1. Acid-soluble metals such as iron have a relationship as shown in Fig. 28-2a. In the middle pH range ($\approx$4 to 10), the corrosion rate is controlled by the rate of transport of oxidizer (usually dissolved O_2) to the metal surface. Iron is weakly amphoteric. At very high temperatures such as those encountered in boilers, the corrosion rate increases with increasing basicity, as shown by the dashed line.

2. Amphoteric metals such as aluminum and zinc have a relationship as shown in Fig. 28-2b. These metals dissolve rapidly in either acidic or basic solutions.

3. Noble metals such as gold and platinum are not appreciably affected by pH, as shown in Fig. 28-2c.

Oxidizing Agents In some corrosion processes, such as the solution of zinc in hydrochloric acid, hydrogen may evolve as a gas. In others, such as the relatively slow solution of copper in sodium chloride, the removal of hydrogen, which must occur so that corrosion may proceed, is effected by a reaction between hydrogen and some oxidizing chemical such as oxygen to form water. Because of the high rates of corrosion which usually accompany hydrogen evolution, metals are rarely used in solutions from which they evolve hydrogen at an appreciable rate. As a result, most of the corrosion observed in practice occurs under conditions in which the oxidation of hydrogen to form water is a necessary part of the corrosion process. For this reason, oxidizing agents are often powerful accelerators of corrosion, and in many cases the **oxidizing power of a solution** is its most important single property insofar as corrosion is concerned.

Oxidizing agents that accelerate the corrosion of some materials may also retard corrosion of others through the formation on their surface of oxides or layers of adsorbed oxygen which make them more

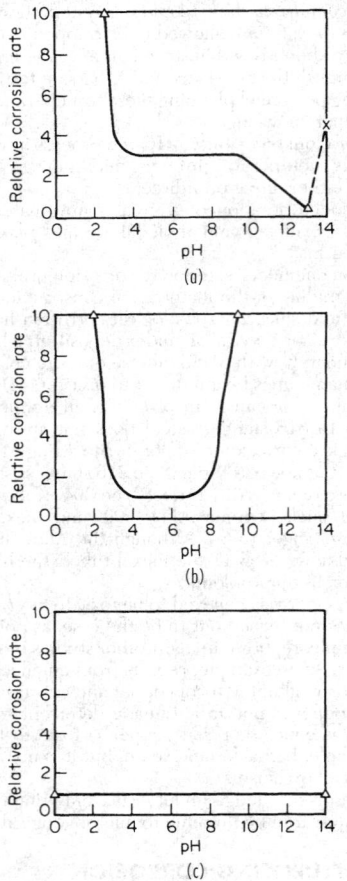

FIG. 28-2   Effect of pH on the corrosion rate. (*a*) Iron. (*b*) Amphoteric metals (aluminum, zinc). (*c*) Noble metals.

resistant to chemical attack. This property of chromium is responsible for the principal corrosion-resisting characteristics of the stainless steels.

It follows, then, that oxidizing substances, such as dissolved air, may accelerate the corrosion of one class of materials and retard the corrosion of another class. In the latter case, the behavior of the material usually represents a balance between the power of oxidizing compounds to preserve a protective film and their tendency to accelerate corrosion when the agencies responsible for protective-film breakdown are able to destroy the films.

Temperature   The rate of corrosion tends to increase with rising temperature. Temperature also has a secondary effect through its influence on the solubility of air (oxygen), which is the most common oxidizing substance influencing corrosion. In addition, temperature has specific effects when a temperature change causes phase changes which introduce a corrosive second phase. Examples include condensation systems and systems involving organics saturated with water.

Velocity   An increase in the velocity of relative movement between a corrosive solution and a metallic surface frequently tends to accelerate corrosion. This effect is due to the higher rate at which the corrosive chemicals, including oxidizing substances (air), are brought to the corroding surface and to the higher rate at which corrosion products, which might otherwise accumulate and stifle corrosion, are carried away. The higher the velocity, the thinner will be the films which corroding substances must penetrate and through which soluble corrosion products must diffuse.

Whenever corrosion resistance results from the accumulation of layers of insoluble corrosion products on the metallic surface, the effect of high velocity may be either to prevent their normal formation or to remove them after they have been formed. Either effect allows corrosion to proceed unhindered. This occurs frequently in small-diameter tubes or pipes through which corrosive liquids may be circulated at high velocities (e.g., condenser and evaporator tubes), in the vicinity of bends in pipe lines, and on propellers, agitators, and centrifugal pumps. Similar effects are associated with cavitation and impingement corrosion.

Films   Once corrosion has started, its further progress very often is controlled by the nature of films, such as passive films, that may form or accumulate on the metallic surface. The classical example is the thin **oxide film** that forms on stainless steels.

Insoluble corrosion products may be completely impervious to the corroding liquid and, therefore, completely protective; or they may be quite permeable and allow local or general corrosion to proceed unhindered. Films that are nonuniform or discontinuous may tend to localize corrosion in particular areas or to induce accelerated corrosion at certain points by initiating electrolytic effects of the concentration-cell type. Films may tend to retain or absorb moisture and thus, by delaying the time of drying, increase the extent of corrosion resulting from exposure to the atmosphere or to corrosive vapors.

It is agreed generally that the characteristics of the **rust films** that form on steels determine their resistance to atmospheric corrosion. The rust films that form on low-alloy steels are more protective than those that form on unalloyed steel.

In addition to films that originate at least in part in the corroding metal, there are others that originate in the corrosive solution. These include various salts, such as carbonates and sulfates, which may be precipitated from heated solutions, and insoluble compounds, such as "beer stone," which form on metal surfaces in contact with certain specific products. In addition, there are films of oil and grease that may protect a material from direct contact with corrosive substances. Such oil films may be applied intentionally or may occur naturally, as in the case of metals submerged in sewage or equipment used for the processing of oily substances.

Other Effects   Stream **concentration** can have important effects on corrosion rates. Unfortunately, corrosion rates are seldom linear with concentration over wide ranges. In equipment such as distillation columns, reactors, and evaporators, concentration can change continuously, making prediction of corrosion rates rather difficult. Concentration is important during plant shutdown; presence of moisture that collects during cooling can turn innocuous chemicals into dangerous corrosives.

As to the effect of time, there is no universal law that governs the reaction for all metals. Some corrosion rates remain constant with time over wide ranges, others slow down with time, and some alloys have increased corrosion rates with respect to time. Situations in which the corrosion rate follows a combination of these paths can develop. Therefore, extrapolation of corrosion data and corrosion rates should be done with utmost caution.

Impurities in a corrodent can be good or bad from a corrosion standpoint. An impurity in a stream may act as an inhibitor and actually retard corrosion. However, if this impurity is removed by some process change or improvement, a marked rise in corrosion rates can result. Other impurities, of course, can have very deleterious effects on materials. The chloride ion is a good example; small amounts of chlorides in a process stream can break down the passive oxide film on stainless steels. The effects of impurities are varied and complex. One must be aware of what they are, how much is present, and where they come from before attempting to recommend a particular material of construction.

HIGH-TEMPERATURE ATTACK

Physical Properties   The suitability of an alloy for high-temperature service [425 to 1,100°C (800 to 2,000°F)] is dependent upon properties inherent in the alloy composition and upon the conditions of application. Crystal structure, density, thermal conductivity, electrical resistivity, thermal expansivity, structural stability, melting range, and vapor pressure are all physical properties basic to and inherent in individual alloy compositions.

Of usually high relative importance in this group of properties is **expansivity.** A surprisingly large number of metal failures at elevated temperatures are the result of excessive thermal stresses originating from constraint of the metal during heating or cooling. Such constraint in the case of hindered contraction can cause rupturing.

Another important property is alloy **structural stability.** This means freedom from formation of new phases or drastic rearrangement of those originally present within the metal structure as a result of thermal experience. Such changes may have a detrimental effect upon strength or corrosion resistance or both.

Mechanical Properties   Mechanical properties of wide interest include creep, rupture, short-time strengths, and various forms of ductility, as well as resistance to impact and fatigue stresses. Creep strength and stress rupture are usually of greatest interest to designers of stationary equipment such as vessels and furnaces.

Corrosion Resistance   Possibly of greater importance than physical and mechanical properties is the ability of an alloy's chemical composition to resist the corrosive action of various hot environments. The forms of high-temperature corrosion which have received the greatest attention are **oxidation** and **scaling.**

Chromium is an essential constituent in alloys to be used above 550°C (1,000°F). It provides a tightly adherent oxide film that materially retards the oxidation process. Silicon is a useful element in imparting oxidation resistance to steel. It will enhance the beneficial effects of chromium. Also, for a given level of chromium, experience has shown oxidation resistance to improve as the nickel content increases.

Aluminum is not commonly used as an alloying element in steel to improve oxidation resistance, as the amount required interferes with both workability and high-temperature-strength properties. However, the development of high-aluminum surface layers by various methods, including spraying, cementation, and dipping, is a feasible means of improving heat resistance of low-alloy steels.

Contaminants in fuels, especially alkali-metal ions, vanadium, and sulfur compounds, tend to react in the combustion zone to form molten fluxes which dissolve the protective oxide film on stainless steels, allowing oxidation to proceed at a rapid rate. This problem is becoming more common as the high cost and short supply of natural gas and distillate fuel oils force increased usage of residual fuel oils and coal.

COMBATING CORROSION

Material Selection   The objective is to select the material which will most economically fulfill the process requirements. The best source of data is well-documented experience in an identical process unit. In the absence of such data, other data sources such as experience in pilot units, corrosion-coupon tests in pilot or bench-scale units, laboratory corrosion-coupon tests in actual process fluids, or corrosion-coupon tests in synthetic solutions must be used. The data from such alternative sources (which are listed in decreasing order of reliability) must be properly evaluated, taking into account the degree to which a given test may fail to reproduce actual conditions in an operating unit. Particular emphasis must be placed on possible composition differences between a static laboratory test and a dynamic plant as well as on trace impurities (chlorides in stainless-steel systems, for example) which may greatly change the corrosiveness of the system. The possibility of severe localized attack (pitting, crevice corrosion, or stress-corrosion cracking) must also be considered.

Permissible **corrosion rates** are an important factor and differ with equipment. Appreciable corrosion can be permitted for tanks and lines if anticipated and allowed for in design thickness, but essentially no corrosion can be permitted in fine-mesh wire screens, orfices, and other items in which small changes in dimensions are critical.

In many instances use of **nonmetallic materials** will prove to be attractive from an economic and performance standpoint. These should be considered when their strength, temperature, and design limitations are satisfactory.

Proper Design   Design considerations with respect to minimizing corrosion difficulties should include the desirability for free and complete drainage, minimizing crevices, and ease of cleaning and inspection. The installation of baffles, stiffeners, and drain nozzles and the location of valves and pumps should be made so that free drainage will occur and washing can be accomplished without holdup. Means of access for inspection and maintenance should be provided whenever practical. Butt joints should be used whenever possible. If lap joints employing fillet welds are used, the welds should be continuous.

The use of dissimilar metals in contact with each other should generally be minimized, particularly if they are widely separated in their nominal positions in the galvanic series (see Table 28-1a). If they are to be used together, consideration should be given to insulating them from each other or making the anodic material area as large as possible.

Equipment should be supported in such a way that it will not rest in pools of liquid or on damp insulating material. Porous insulation should be weatherproofed or otherwise protected from moisture and spills to avoid contact of the wet material with the equipment. Specifications should be sufficiently comprehensive to ensure that the desired composition or type of material will be used and the right condition of heat treatment and surface finish will be provided. Inspection during fabrication and prior to acceptance is desirable.

Altering the Environment   Simple changes in environment may make an appreciable difference in the corrosion of metals and should be considered as a means of combating corrosion. **Oxygen** is an important factor, and its removal or addition may cause marked changes in corrosion. The treatment of boiler feedwater to remove oxygen, for instance, greatly reduces the corrosiveness of the water on steel. Inert-gas purging and blanketing of many solutions, particularly acidic media, generally minimize corrosion of copper and nickel-base alloys by minimizing air or oxygen content. Corrosiveness of acid media to stainless alloys, on the other hand, may be reduced by aeration because of the formation of passive oxide films. Reduction in temperature will almost always be beneficial with respect to reducing corrosion if no corrosive phase changes (condensation, for example) result. Velocity effects vary with the material and the corrosive system. When pH values can be modified, it will generally be beneficial to hold the acid level to a minimum. When acid additions are made in batch processes, it may be beneficial to add them last so as to obtain maximum dilution and minimum acid concentration and exposure time. Alkaline pH values are less critical than acid values with respect to controlling corrosion. Elimination of moisture can and frequently does minimize, if not prevent, corrosion of metals, and this possibility of environmental alteration should always be considered.

Inhibitors   The use of various substances or inhibitors as additives to corrosive environments to decrease corrosion of metals in the environment is an important means of combating corrosion. This is generally most attractive in closed or recirculating systems in which the annual cost of inhibitor is low. However, it has also proved to be economically attractive for many once-through systems, such as those encountered in petroleum-processing operations. Inhibitors are effective as the result of their controlling influence on the cathode- or anode-area reactions.

Typical examples of inhibitors used for minimizing corrosion of iron and steel in aqueous solutions are the chromates, phosphates, and silicates. Organic sulfide and amine materials are frequently effective in minimizing corrosion of iron and steel in acid solution.

The use of inhibitors is not limited to controlling corrosion of iron and steel. They frequently are effective with stainless steel and other alloy materials. The addition of copper sulfate to dilute sulfuric acid will sometimes control corrosion of stainless steels in hot dilute solutions of this acid, whereas the uninhibited acid causes rapid corrosion.

The effectiveness of a given inhibitor generally increases with an increase in **concentration,** but inhibitors considered practical and economically attractive are used in quantities of less than 0.1 percent by weight.

In some instances the amount of inhibitor present is critical in that a deficiency may result in localized or pitting attack, with the overall results being more destructive than when none of the inhibitor is present. Considerations for the use of inhibitors should therefore include review of experience in similar systems or investigation of requirements and limitations in new systems.

Cathodic Protection   This electrochemical method of corrosion control has found wide application in the protection of carbon steel underground structures such as pipe lines and tanks from external soil corrosion. It is also widely used in water systems to protect ship hulls, offshore structures, and water-storage tanks.

Two methods of providing cathodic protection for minimizing corrosion of metals are in use today. These are the sacrificial-anode method and the impressed-emf method. Both depend upon making the metal to be protected the cathode in the electrolyte involved.

Examples of the sacrificial-anode method include the use of zinc, magnesium, or aluminum as anodes in electrical contact with the metal to be protected. These may be anodes buried in the ground for protection of underground pipe lines or attachments to the surfaces of equipment such as condenser water boxes or on ship hulls. The current required is generated in this method by corrosion of the sacrificial-anode material. In the case of the impressed emf, the direct current is provided by external sources and is passed through the system by use of essentially nonsacrificial anodes such as carbon, noncorrodible alloys, or platinum buried in the ground or suspended in the electrolyte in the case of aqueous systems.

The requirements with respect to current distribution and anode placement vary with the resistivity of soils or the electrolyte involved.

Anodic Protection   This electrochemical method relies on an external potential control system (potentiostat) to maintain the metal or alloy in a noncorroding (passive) condition. Practical applications include acid coolers in sulfuric acid plants and storage tanks for sulfuric acid.

Coatings and Linings   The use of nonmetallic coatings and lining materials in combination with steel or other materials has and will continue to be an important type of construction for combating corrosion.

Organic coatings of many kinds are used as linings in equipment such as tanks, piping, pumping lines, and shipping containers, and they are often an economical means of controlling corrosion, particularly when freedom from metal contamination is the principal objective. One principle that is now generally accepted is that thin nonreinforced paintlike coatings of less than 0.75-mm (0.03-in) thickness should not be used in services for which full protection is required in order to prevent rapid attack of the substrate metal. This is true because most thin coatings contain defects or holidays and can be easily damaged in service, thus leading to early failures due to corrosion of the substrate metal even though the coating material is resistant. Electrical testing for continuity of coating-type linings is always desirable for immersion-service applications in order to detect holiday-type defects in the coating.

The most dependable barrier linings for corrosive services are those which are bonded directly to the substrate and are built up in multiple-layer or laminated effects to thicknesses greater than 2.5 mm (0.10 in). These include flake-glass-reinforced resin systems and elastomeric and plasticized plastic systems. Good surface preparation and thorough inspections of the completed lining, including electrical testing, should be considered as minimum requirements for any lining applications.

Linings of this type are slightly permeable to many liquids. Such permeation, while not damaging to the lining, may cause failure by causing disbonding of the lining owing to pressure buildup between the lining and the steel.

Ceramic or carbon-brick linings are frequently used as facing linings over plastic or membrane linings when surface temperatures exceed those which can be handled by the unprotected materials or when the membrane must be protected from mechanical damage. This type of construction permits processing of materials that are too corrosive to be handled in low-cost metal constructions.

Glass-Lined Steel   By proprietary methods, special glasses can be bonded to steel, providing an impervious liner 1.5 to 2.5 mm (0.060 to 0.100 in) thick. Equipment and piping lined in this manner are routinely used in severely corrosive acid services. The glass lining can be mechanically damaged, and careful attention to details of design, inspection, installation, and maintenance is required to achieve good results with this system.

The cladding of steel with an alloy is another approach to this problem. There are a number of cladding methods in general use. In one,

a sandwich is made of the corrosion-resistant metal and carbon steel by hot rolling to produce a **pressure weld** between the plates.

Another process involves **explosive bonding.** The corrosion-resistant metal is bonded to a steel backing metal by the force generated by properly positioned explosive charges. Relatively thick sections of metal can be bonded by this technique into plates.

In a third process, a **loose liner** is fastened to a carbon steel shell by welds spaced so as to prevent collapse of the liner. A fourth method is **weld overlay,** which involves depositing multiple layers of alloy weld metal to cover the steel surface.

All these methods require careful design and control of fabrication methods to assure success.

Metallic Linings for Mild Environments   Zinc coatings applied by various means have good corrosion resistance to many atmospheres. Such coatings have been extensively used on steel. Zinc has the advantage of being anodic to steel and therefore will protect exposed areas of steel by electrochemical action.

Steel coated with tin (**tinplate**) is used to make food containers. Tin is more noble than steel; therefore, well-aerated solutions will galvanically accelerate attack of the steel at exposed areas. The comparative absence of air within food containers aids in preserving the tin as well as the food. Also the reversible potential which the tin-iron couple undergoes in organic acids serves to protect exposed steel in food containers.

Cadmium, being anodic to steel, behaves quite similarly to zinc in providing corrosion protection when applied as a coating on steel. Tests of zinc and cadmium coatings should be conducted when it becomes necessary to determine the most economical selection for a particular environment.

Lead has a good general resistance to various atmospheres. As a coating, it has had its greatest application in the production of terneplate, which is used as a roofing, cornicing, and spouting material.

Aluminum coatings on steel will perform in a manner similar to zinc coatings. Aluminum has good resistance to many atmospheres; in addition, being anodic to steel, it will galvanically protect exposed areas. Aluminum-coated steel products are quite serviceable under high-temperature conditions, for which good oxidation resistance is required.

CORROSION-TESTING METHODS*

The primary purpose of materials selection is to provide the optimum equipment for a process application in terms of materials of construction, design, and corrosion-control measures. *Optimum* here means that which comprises the best combination of cost, life, safety, and reliability.

The selection of materials to be used in design dictates a basic understanding of the behavior of materials and the principles that govern such behavior. If proper design of suitable materials of construction is incorporated, the equipment should deteriorate at a uniform and anticipated gradual rate, which will allow scheduled maintenance or replacement at regular intervals. If localized forms of corrosion are characteristic of the combination of materials and environment, the materials engineer should still be able to predict the probable life of equipment, or devise an appropriate inspection schedule to preclude unexpected failures. The concepts of predictive, or at least preventive, maintenance are minimum requirements to proper materials selection. This approach to maintenance is certainly intended to minimize the possibility of unscheduled production shutdowns because of corrosion failures, with their attendant possible financial losses, hazard to personnel and equipment, and resultant environmental pollution.

Chemical processes may involve a complex variety of both inorganic and organic chemicals. Hard and fast rules for selecting the appropriate materials of construction can be given when the composition is known, constant, and free of unsuspected contaminates; when the relevant parameters of temperature, pressure, velocity, and concentra-

---

* Includes information excerpted from papers by Oliver W. Siebert, John G. Stoecker II, and Ann Chidester Van Orden, courtesy of NACE International; and Oliver W. Siebert, courtesy ASTM.

tion are defined; and when the mechanical and environmental degradation of the material is uniform, that is, free of localized attack. For example, it is relatively simple to select the materials of construction for a regimen of equipment for the storage and handling of cold, concentrated sulfuric acid. On the other hand, the choice of suitable materials for producing phosphoric acid by the digestion of phosphate rock with sulfuric acid is much more difficult because of the diversity in kind and concentration of contaminants, the temperatures of the reactions, and the strength of sulfuric and phosphoric acid used or formed. Probably the best way to approach the study of materials selection is to categorize the types of major chemicals that might be encountered, describe their inherent characteristics, and generalize about the corrosion characteristics of the prominent materials of construction in such environments.

The background information that materials selection is based on is derived from a number of sources. In many cases, information as to the corrosion resistance of a material in a specific environment is not available and must be derived experimentally. It is to this need that the primary remarks of this subsection are addressed.

Unfortunately, there is no standard or preferred way to evaluate an alloy in an environment. While the chemistry of the operating plant environment can sometimes be duplicated in the laboratory, factors of velocity, hot and cold wall effects, crevice, chemical reaction of the fluid during the test, stress levels of the equipment, contamination with products of corrosion, trace impurities, dissolved gases, and so forth also have a controlling effect on the quality of the answer. Then, too, the progress of the corrosion reaction itself varies with time. Notwithstanding, immersion testing remains the most widely used method for selecting materials of construction.

There is no standard or preferred way to carry out a corrosion test; the method must be chosen to suit the purpose of the test. The principal types of tests are, in decreasing order of reliability:

1. Actual operating experience with full-scale plant equipment exposed to the corroding medium.
2. Small-scale plant-equipment experience, under either commercial or pilot-plant conditions.
3. Sample tests in the field. These include coupons, stressed samples, electrical-resistance probes exposed to the plant corroding medium, or samples exposed to the atmosphere, to soils, or to fresh, brackish, or saline waters.
4. Laboratory tests on samples exposed to "actual" plant liquids or simulated environments.

Plant or field corrosion tests are useful for:
1. Selection of the most suitable material to withstand a particular environment and to estimate its probable durability in that environment
2. Study of the effectiveness of means of preventing corrosion

CORROSION TESTING: LABORATORY TESTS

Metals and alloys do not respond alike to all the influences of the many factors that are involved in corrosion. Consequently, it is impractical to establish any universal standard laboratory procedures for corrosion testing except for inspection tests. However, some details of laboratory testing need careful attention in order to achieve useful results.

In the selection of materials for the construction of a chemical plant, resistance to the corroding medium is often the determining factor; otherwise, the choice will fall automatically on the cheapest material mechanically suitable. Laboratory corrosion tests are frequently the quickest and most satisfactory means of arriving at a preliminary selection of the most suitable materials to use. Unfortunately, however, it is not yet within the state of the art of laboratory tests to predict with accuracy the behavior of the selected material under plant-operating conditions. The outstanding difficulty lies not so much in carrying out the test as in interpreting the results and translating them into terms of plant performance. A laboratory test of the conventional type gives mainly one factor—the chemical resistance of the proposed material to the corrosive agent. There are numerous other factors entering into the behavior of the material in the plant, such as dissolved gases, velocity, turbulence, abrasion, crevice condi-

tions, hot-wall effects, cold-wall effects, stress levels of metals, trace impurities in corrodent that act as corrosion inhibitors or accelerators, and variations in composition of corrodent.

Immersion Test One method of determining the chemical-resistance factor, the so-called **total-immersion test,** represents an unaccelerated method that has been found to give reasonably concordant results in approximate agreement with results obtained on the large scale when the other variables are taken into account. Various other tests have been proposed and are in use, such as salt-spray, accelerated electrolytic, alternate-immersion, and aerated-total-immersion; but in view of the numerous complications entering into the translation of laboratory results into plant results the simplest test is considered the most desirable for routine preliminary work, reserving special test methods for special cases. The total-immersion test serves quite well to eliminate materials that obviously cannot be used; further selection among those materials which apparently can be used can be made on the basis of a knowledge of the properties of the materials concerned and the working conditions or by constructing larger-scale equipment in which the operating conditions can be simulated.

The National Association of Corrosion Engineers (NACE) TMO169-95 "Standard Laboratory Corrosion Testing of Metals for the Process Industries," and ASTM G31 "Recommended Practice for Laboratory Immersion Corrosion Testing of Metals" are the general guides for immersion testing. Small pieces of the candidate metal are exposed to the medium, and the loss of mass of the metal is measured for a given period of time. Immersion testing remains the best method to eliminate from further consideration those materials that obviously cannot be used. This technique is frequently the quickest and most satisfactory method of making a preliminary selection of the best candidate materials.

Probably the most serious disadvantage of this method of corrosion study is the assumed average-time weight loss. The corrosion rate could be high initially and then decrease with time (it could fall to zero). In other cases the rate of corrosion might increase very gradually with time or it could cycle or be some combination of these things.

The description that follows is based on these standards.

Test Piece The size and the shape of specimens will vary with the purpose of the test, nature of the material, and apparatus used. A large surface-to-mass ratio and a small ratio of edge area to total area are desirable. These ratios can be achieved through the use of rectangular or circular specimens of minimum thickness. Circular specimens should be cut preferably from sheet and not bar stock to minimize the exposed end grain.

A circular specimen of about 38-mm (1.5-in) diameter is a convenient shape for laboratory corrosion tests. With a thickness of approximately 3 mm ($\frac{1}{8}$ in) and an 8- or 11-mm- ($\frac{5}{16}$- or $\frac{7}{16}$-in-) diameter hole for mounting, these specimens will readily pass through a 45/50 ground-glass joint of a distillation kettle. The total surface area of a circular specimen is given by the equation:

$$A = \frac{\pi}{2}(D^2 - d^2) + t\pi D + t\pi d$$

where t = thickness, D = diameter of the specimen, and d = diameter of the mounting hole. If the hole is completely covered by the mounting support, the final term ($t\pi d$) in the equation is omitted.

Strip coupons [50 by 25 by 1.6 or 3.2 mm (2 by 1 by $\frac{1}{16}$ or $\frac{1}{8}$ in)] may be preferred as corrosion specimens, particularly if interface or liquid-line effects are to be studied by the laboratory test.

All specimens should be measured carefully to permit accurate calculation of the exposed areas. An area calculation accurate to plus or minus 1 percent is usually adequate.

More uniform results may be expected if a substantial layer of metal is removed from the specimens to eliminate variations in condition of the original metallic surface. This can be done by chemical treatment (pickling), electrolytic removal, or grinding with a coarse abrasive paper or cloth, such as No. 50, using care not to work-harden the surface. At least 2.5×10^{-3} mm (0.0001 in) or 1.5 to 2.3 mg/cm^2 (10 to 15 mg/in^2) should be removed. If clad alloy specimens are to be used, special attention must be given to ensure that excessive metal is not removed. After final preparation of the specimen surface, the specimens should be stored in a desiccator until exposure if they are not used immediately.

Specimens should be finally degreased by scrubbing with bleach-free scouring powder, followed by thorough rinsing in water and in a suitable solvent (such as acetone, methanol, or a mixture of 50 percent methanol and 50 percent ether), and air-dried. For relatively soft metals such as aluminum, magnesium, and copper, scrubbing with abrasive powder is not always needed and can mar the surface of the specimen. The use of towels for drying may introduce an error through contamination of the specimens with grease or lint. The dried specimen should be weighed on an analytic balance.

Apparatus   A versatile and convenient apparatus should be used, consisting of a kettle or flask of suitable size (usually 500 to 5,000 mL), a reflux condenser with atmospheric seal, a sparger for controlling atmosphere or aeration, a thermowell and temperature-regulating device, a heating device (mantle, hot plate, or bath), and a specimen-support system. If agitation is required the apparatus can be modified to accept a suitable stirring mechanism such as a magnetic stirrer. A typical resin-flask setup for this type of test is shown in Fig. 28-3. Open-beaker tests should not be used because of evaporation and contamination.

In more complex tests, provisions might be needed for continuous flow or replenishment of the corrosive liquid while simultaneously maintaining a controlled atmosphere.

Apparatus for testing materials for heat-transfer applications is shown in Fig. 28-4. Here the sample is at a higher temperature than the bulk solution.

If the test is to be a guide for the selection of a material for a particular purpose, the limits of controlling factors in service must be determined. These factors include oxygen concentration, temperature, rate of flow, pH value, and other important characteristics.

The **composition of the test solution** should be controlled to the fullest extent possible and be described as thoroughly and as accurately as possible when the results are reported. Minor constituents should not be overlooked because they often affect corrosion rates. Chemical content should be reported as percentage by weight of the solution. Molarity and normality are also helpful in defining the concentration of chemicals in the test solution. The composition of the test solution should be checked by analysis at the end of the test to

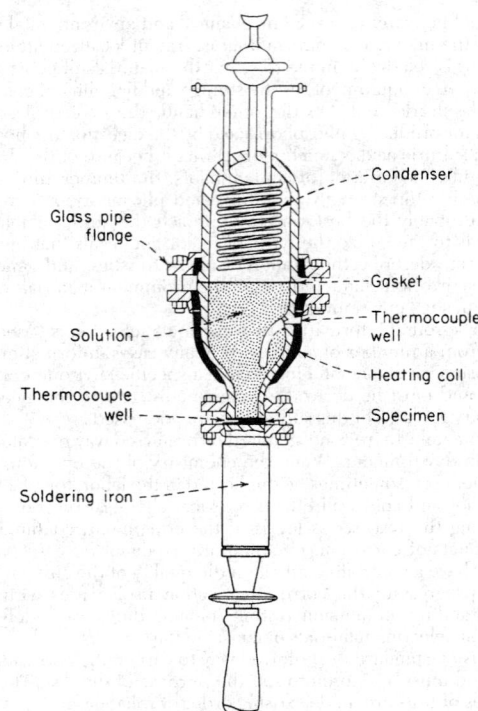

FIG. 28-4   Laboratory setup for the corrosion testing of heat-transfer materials.

determine the extent of change in composition, such as might result from evaporation.

Temperature of Solution   Temperature of the corroding solution should be controlled within $\pm 1°C$ ($\pm 1.8°F$) and must be stated in the report of test results.

For tests at ambient temperatures, the tests should be conducted at the highest temperature anticipated for stagnant storage in summer months. This temperature may be as high as 40 to 45°C (104 to 113°F) in some localities. The variation in temperature should be reported also (e.g., 40°C $\pm$ 2°C).

Aeration of Solution   Unless specified, the solution should not be aerated. Most tests related to process equipment should be run with the natural atmosphere inherent in the process, such as the vapors of the boiling liquid. If aeration is used, the specimens should not be located in the direct air stream from the sparger. Extraneous effects can be encountered if the air stream impinges on the specimens.

Solution Velocity   The effect of velocity is not usually determined in laboratory tests, although specific tests have been designed for this purpose. However, for the sake of reproducibility some velocity control is desirable.

Tests at the boiling point should be conducted with minimum possible heat input, and boiling chips should be used to avoid excessive turbulence and bubble impingement. In tests conducted below the boiling point, thermal convection generally is the only source of liquid velocity. In test solutions of high viscosities, supplemental controlled stirring with a magnetic stirrer is recommended.

Volume of Solution   Volume of the test solution should be large enough to avoid any appreciable change in its corrosiveness through either exhaustion of corrosive constituents or accumulation of corrosion products that might affect further corrosion. A suitable volume-to-area ratio is 20 mL (125 mL) of solution/cm² (in²) of specimen surface. This corresponds to the recommendation of ASTM Standard A262 for the Huey test. The preferred volume-to-

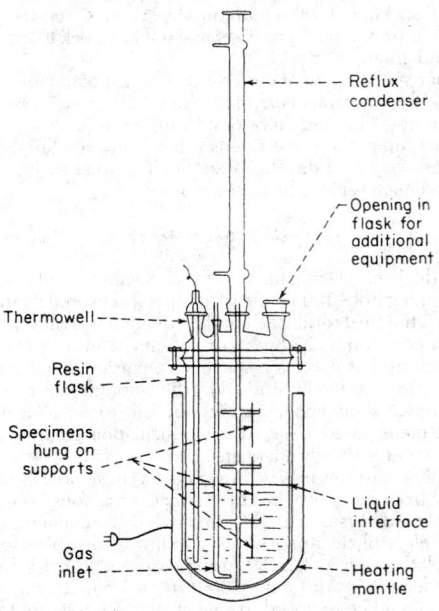

FIG. 28-3   Laboratory-equipment arrangement for corrosion testing. (*Based on NACE Standard TMO169-95.*)

area ratio is 40 mL/cm² (250 mL/in²) of specimen surface, as stipulated in ASTM Standard G31, Laboratory Immersion Testing of Materials.

Method of Supporting Specimens   The supporting device and container should not be affected by or cause contamination of the test solution. The method of supporting specimens will vary with the apparatus used for conducting the test but should be designed to insulate the specimens from each other physically and electrically and to insulate the specimens from any metallic container or supporting device used with the apparatus.

Shape and form of the specimen support should assure free contact of the specimen with the corroding solution, the liquid line, or the vapor phase, as shown in Fig. 28-3. If clad alloys are exposed, special procedures are required to ensure that only the cladding is exposed (unless the purpose is to test the ability of the cladding to protect cut edges in the test solution). Some common supports are glass or ceramic rods, glass saddles, glass hooks, fluorocarbon plastic strings, and various insulated or coated metallic supports.

Duration of Test   Although the duration of any test will be determined by the nature and purpose of the test, an excellent procedure for evaluating the effect of time on corrosion of the metal and also on the corrosiveness of the environment in laboratory tests has been presented by Wachter and Treseder [*Chem. Eng. Prog.*, 315–326 (June 1947)]. This technique is called the **planned-interval test.** Other procedures that require the removal of solid corrosion products between exposure periods will not measure accurately the normal changes of corrosion with time.

Materials that experience severe corrosion generally do not need lengthy tests to obtain accurate corrosion rates. Although this assumption is valid in many cases, there are exceptions. For example, lead exposed to sulfuric acid corrodes at an extremely high rate at first while building a protective film; then the rate decreases considerably, so that further corrosion is negligible. The phenomenon of forming a protective film is observed with many corrosion-resistant materials, and therefore short tests on such materials would indicate high corrosion rates and would be completely misleading.

Short-time tests also can give misleading results on alloys that form passive films, such as stainless steels. With borderline conditions, a prolonged test may be needed to permit breakdown of the passive film and subsequently more rapid attack. Consequently, tests run for long periods are considerably more realistic than those conducted for short durations. This statement must be qualified by stating that corrosion should not proceed to the point at which the original specimen size or the exposed area is drastically reduced or the metal is perforated.

If anticipated corrosion rates are moderate or low, the following equation gives a suggested test duration:

$$\text{Duration of test, h} = \frac{78{,}740}{\text{corrosion rate, mm/y}}$$

$$= \frac{2000}{\text{corrosion rate, mils/y}}$$

Cleaning Specimens after Test   Before specimens are cleaned, their appearance should be observed and recorded. Locations of deposits, variations in types of deposits, and variations in corrosion products are extremely important in evaluating localized corrosion such as pitting and concentration-cell attack.

Cleaning specimens after the test is a vital step in the corrosion-test procedure and, if not done properly, can give rise to misleading test results. Generally, the cleaning procedure should remove all corrosion products from specimens with a minimum removal of sound metal. Set rules cannot be applied to cleaning because procedures will vary with the type of metal being cleaned and the degree of adherence of corrosion products.

Mechanical cleaning includes scrubbing, scraping, brushing, mechanical shocking, and ultrasonic procedures. Scrubbing with a bristle brush and a mild abrasive is the most widely used of these methods; the others are used principally as supplements to remove heavily encrusted corrosion products before scrubbing. Care should be used to avoid the removal of sound metal.

Chemical cleaning implies the removal of material from the surface of the specimen by dissolution in an appropriate chemical agent. Solvents such as acetone, carbon tetrachloride, and alcohol are used to remove oil, grease, or resin and are usually applied prior to other methods of cleaning. Various chemicals are chosen for application to specific materials; some of these treatments in general use are outlined in the NACE standard.

Electrolytic cleaning should be preceded by scrubbing to remove loosely adhering corrosion products. One method of electrolytic cleaning that has been found to be useful for many metals and alloys is as follows:

Solution: 5 percent (by weight) H_2SO_4
Anode: carbon or lead
Cathode: test specimen
Cathode current density: 20 A/dm² (129 A/in²)
Inhibitor: 2 cm³ organic inhibitor per liter
Temperature: 74°C (165°F)
Exposure period: 3 min

Precautions must be taken to ensure good electrical contact with the specimen, to avoid contamination of the solution with easily reducible metal ions, and to ensure that inhibitor decomposition has not occurred. Instead of using 2 mL of any proprietary inhibitor, 0.5 g/L of inhibitors such as diorthotolyl thiourea or quinoline ethiodide can be used.

Whatever treatment is used to clean specimens after a corrosion test, its effect in removing metal should be determined, and the weight loss should be corrected accordingly. A "blank" specimen should be weighed before and after exposure to the cleaning procedure to establish this weight loss.

Evaluation of Results   After the specimens have been reweighed, they should be examined carefully. Localized attack such as pits, crevice corrosion, stress-accelerated corrosion, cracking, or intergranular corrosion should be measured for depth and area affected.

Depth of localized corrosion should be reported for the actual test period and not interpolated or extrapolated to an annual rate. The rate of initiation or propagation of pits is seldom uniform. The size, shape, and distribution of pits should be noted. A distinction should be made between those occurring underneath the supporting devices (concentration cells) and those on the surfaces that were freely exposed to the test solution. An excellent discussion of pitting corrosion has been published [*Corrosion*, 25t (January 1950)].

The specimen may be subjected to simple bending tests to determine whether any **embrittlement** has occurred.

If it is assumed that localized or internal corrosion is not present or is recorded separately in the report, the **corrosion rate** or penetration can be calculated alternatively as

$$\frac{\text{Weight loss} \times 534}{(\text{Area})(\text{time})(\text{metal density})} = \text{mils/y (mpy)}$$

$$\frac{\text{Weight loss} \times 13.56}{(\text{Area})(\text{time})(\text{metal density})} = \text{mm/y (mmpy)}$$

where weight loss is in mg, area is in in² of metal surface exposed, time is in hours exposed, and density is in g/cm³. Densities for alloys can be obtained from the producers or from various metal handbooks.

The following **checklist** is a recommended guide for reporting all **important information and data:**

Corrosive media and concentration (changes during test)
Volume of test solution
Temperature (maximum, minimum, and average)
Aeration (describe conditions or technique)
Agitation (describe conditions or technique)
Type of apparatus used for test
Duration of each test (start, finish)
Chemical composition or trade name of metals tested
Form and metallurgical conditions of specimens
Exact size, shape, and area of specimens
Treatment used to prepare specimens for test
Number of specimens of each material tested and whether specimens were tested separately or which specimens were tested in the same container

Method used to clean specimens after exposure and the extent of any error expected by this treatment

Actual weight losses for each specimen

Evaluation of attack if other than general, such as crevice corrosion under support rod, pit depth and distribution, and results of microscopic examination or bend tests

Corrosion rates for each specimen expressed as millimeters (mils) per year

Effect of Variables on Corrosion Tests   It is advisable to apply a factor of safety to the results obtained, the factor varying with the degree of confidence in the applicability of the results. Ordinarily, a factor of from 3 to 10 might be considered normal.

Among the more important points that should be considered in attempting to base plant design on laboratory corrosion-rate data are the following.

Galvanic corrosion is a frequent source of trouble on a large scale. Not only is the use of different metals in the same piece of equipment dangerous, but the effect of cold working may be sufficient to establish potential differences of objectionable magnitude between different parts of the same piece of metal. The mass of metal in chemical apparatus is ordinarily so great and the electrical resistance consequently so low that a very small voltage can cause a very high current. Welding also may leave a weld of a different physical or chemical composition from that of the body of the sheet and cause localized corrosion.

Local variations in temperature and crevices that permit the accumulation of corrosion products are capable of allowing the formation of **concentration cells,** with the result of accelerated local corrosion.

In the laboratory, the **temperature** of the test specimen is that of the liquid in which it is immersed, and the measured temperature is actually that at which the reaction is taking place. In the plant (heat being supplied through the metal to the liquid in many cases), the temperature of the film of (corrosive) liquid on the inside of the vessel may be a number of degrees higher than that registered by the thermometer. As the relation between temperature and corrosion is a logarithmic one, the rate of increase is very rapid. Like other chemical reactions, the speed ordinarily increases twofold to threefold for each 10°C temperature rise, the actual relation being that of the equation $\log K = A + (B/T)$, where K represents the rate of corrosion and T the absolute temperature. This relationship, although expressed mathematically, must be understood to be a qualitative rather than strictly a quantitative one.

Cold walls, as in coolers or condensers, usually have somewhat decreased corrosion rates for the reason just described. However, in some cases, the decrease in temperature may allow the formation of a more corrosive second phase, thereby increasing corrosion.

The effect of **impurities** in either structural material or corrosive material is so marked (while at the same time it may be either accelerating or decelerating) that for reliable results the actual materials which it is proposed to use should be tested and not types of these materials. In other words, it is much more desirable to test the actual plant solution and the actual metal or nonmetal than to rely upon a duplication of either. Since as little as 0.01 percent of certain organic compounds will reduce the rate of solution of steel in sulfuric acid 99.5 percent and 0.05 percent bismuth in lead will increase the rate of corrosion over 1000 percent under certain conditions, it can be seen how difficult it would be to attempt to duplicate here all the significant constituents.

Electrical Resistance   The measurement of corrosion by electrical resistance is possible by considering the change in resistance of a thin metallic wire or strip sensing element (probe) as its cross section decreases from a loss of metal. Since small changes in resistance are encountered as corrosion progresses, changes in temperature can cause enough change in the wire resistance to complicate the results. Commercial equipment, such as the Corrosometer®, have a protected reference section of the specimen in the modified electrical Wheatstone bridge (Kelvin) circuit to compensate for these temperature changes. Since changes in the resistance ratio of the probe are not linear with loss of section thickness, compensation for this variable must be included in the circuit. In operation, the specimen probe is exposed to the environment and instrument readings are periodically recorded. The corrosion rate is the loss of metal averaged between any two readings.

The corrosion rate can be studied by this method over very short periods of time, but not instantaneously. The environment does not have to be an electrolyte. Studies can be made in corrosive gas exposures. The main disadvantage of the technique is that local corrosion (pitting, crevice corrosion, galvanic, stress corrosion cracking, fatigue, and so forth) will probably not be progressively identified. If the corrosion product has an electrical conductivity approaching that of the lost metal, little or no corrosion will be indicated. The same problem will result from the formation of conducting deposits on the specimen.

The electrical-resistance measurement has nothing to do with the electrochemistry of the corrosion reaction. It merely measures a bulk property that is dependent upon the specimen's cross-section area. Commercial instruments are available (Fig. 28-5).

Advantages of the electrical-resistance technique are:

1.   A corrosion measurement can be made without having to see or remove the test sample.

2.   Corrosion measurements can be made quickly—in a few hours or days, or continuously. This enables sudden increases in corrosion rate to be detected. In some cases, it will be possible then to modify the process to decrease the corrosion.

3.   The method can be used to monitor a process to indicate whether the corrosion rate is dependent on some critical process variable.

4.   Corrodent need not be an electrolyte (in fact, need not be a liquid).

5.   The method can detect low corrosion rates that would take a long time to detect with weight-loss methods.

Limitations of the technique are:

1.   It is usually limited to the measurement of uniform corrosion only and is not generally satisfactory for localized corrosion.

2.   The probe design includes provisions to compensate for temperature variations. This feature is not totally successful. The most reliable results are obtained in constant-temperature systems.

EMF versus pH (Pourbaix) Diagrams   Potential (EMF) versus pH equilibrium (Pourbaix) diagrams derived from physical property data about the metal and its environment provide a basis for the expression of a great amount of thermodynamic data about the corrosion reaction. These relatively simple diagrams graphically represent the thermodynamics of corrosion in terms of electromotive force, that is, an indication of oxidizing power and pH, or acidity. As an aid in corrosion prediction, their usefulness lies in providing direction for establishing a corrosion study program.

Figure 28-6 is a typical Pourbaix diagram. Generally, the diagrams show regions of immunity (the metal), passivity (the surface film), and corrosion (metallic ions). While of considerable qualitative usefulness, these diagrams have important limitations. Since they are calculated from thermodynamic properties, they represent equilibrium conditions and do not provide kinetic information. Thus, while they show conditions where corrosion will not occur, they do not necessarily indicate under what conditions corrosion will occur. To determine the quantitative value of corrosion, kinetic rate measurement would still be required. Pourbaix diagrams were developed for the study of pure metals. Since few engineering structures are made of pure metals, it is

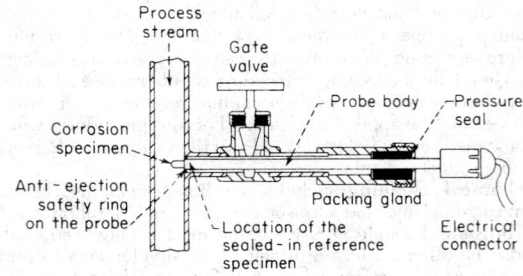

FIG. 28-5   Typical retractable corrosion probe.

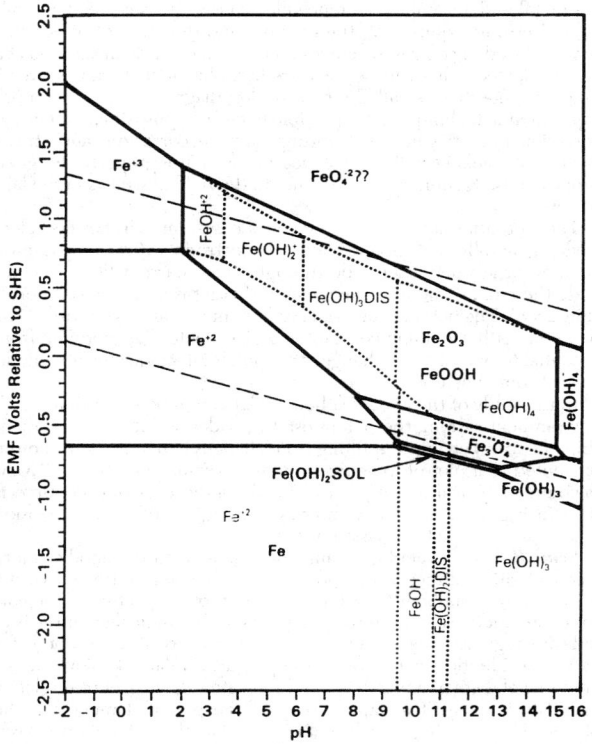

FIG. 28-6   EMF-pH diagram for an iron-water system at 25°C. All ions are at an activity of 10^{-6}.

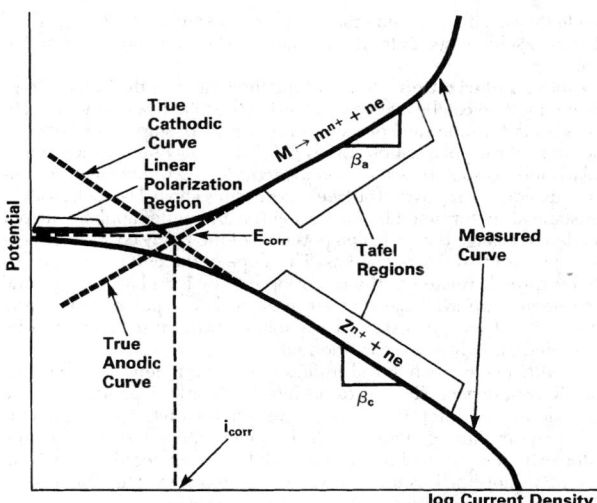

FIG. 28-7   Tafel extrapolation and linear polarization curves.

important to extend this technique to include information on the passive behavior of alloys of engineering interest. Values of the open circuit corrosion potential (OCP) or a controlled potential occur, if a steady site potential can be used in conjunction with the solution pH and these diagrams to show what component is stable in the system defined by a given pH and potential. Theoretical diagrams so developed estimate the corrosion product in various regions. The use of computers to construct diagrams for alloy systems provides an opportunity to mathematically overcome many of the limitations inherent in the pure metal system.

A potentiokinetic electrochemical hysteresis method of diagram construction has led to consideration of three-dimensional Pourbaix diagrams for alloy systems useful in alloy development, evaluation of the influence of crevices, prediction of the tendency for dealloying, and the inclusion of kinetic data on the diagram is useful in predicting corrosion rather than just the absence of damage. These diagrams are kinetic, not thermodynamic, expressions. The two should not be confused as being the same reaction, as they are not.

Tafel Extrapolation   Corrosion is an electrochemical reaction of a metal and its environment. When corrosion occurs, the current that flows between individual small anodes and cathodes on the metal surface causes the electrode potential for the system to change. While this current cannot be measured, it can be evaluated indirectly on a metal specimen with an inert electrode and an external electrical circuit. Polarization is described as the extent of the change in potential of an electrode from its equilibrium potential caused by a net current flow to or from the electrode, galvanic or impressed (Fig. 28-7).

Electrochemical techniques have been used for years to study fundamental phenomenological corrosion reactions of metals in corrosive environments. Unfortunately, the learning curve in the reduction of these electrochemical theories to practice has been painfully slow. However, a recent survey has shown that many organizations in the

chemical process industries are now adding electrochemical methods to their materials selection techniques. Laboratory electrochemical tests of metal/environment systems are being used to show the degree of compatibility and describe the limitations of those relationships. The general methods being used include electrical resistance, Tafel extrapolation, linear polarization, and both slow and rapid-scan potentiodynamic polarization. Depending upon the study technique used, it has also been possible to indicate the tendency of a given system to suffer local pitting or crevice attack or both. These same tools have been the basis of design protection of less-noble structural metals.

To study the anode reaction of a specimen in an environment, sufficient current is applied to change the freely corroding potential of the metal in a more electropositive direction with respect to the inert electrode (acting as a cathode). The opposite of this so-called anodic polarization is cathodic polarization. Polarization can be studied equally well by varying the potential and measuring the resultant changes in the current. Both of the theoretical (true) polarization curves are straight lines when plotted on a semilog axis. The corrosion current can be measured from the intersection of the corrosion potential and either the anodic or cathodic curve. The corrosion rate of the system is a function of that corrosion current. Experimentally derived curves are not fully linear because of the interference from the reactions between the anodes and cathodes in the region close to the corrosion potential. IR losses often obscure the Tafel behavior. Away from the corrosion potential, the measured curves match the theoretical (true) curves. The matching region of the measured curves is called the **Tafel region,** and their (Tafel) slopes are constant. Corrosion rates can be calculated from the intersection of the corrosion potential and the extrapolation of the Tafel region. The main advantage of the technique is that it is quick; curves can be generated in about an hour.

The technique is of limited value where more than one cathodic reduction reaction occurs. In most cases it is difficult to identify a sufficient linear segment of the Tafel region to extrapolate accurately. Since currents in the Tafel region are one to two orders of magnitude larger on the log scale than the corrosion current, relatively large currents are required to change the potentials from what they are at the corrosion potential. The environment must be a conductive solution. The Tafel technique does not indicate local attack, only an average, uniform corrosion rate.

The primary use of this laboratory technique today is as a quick check to determine the order of magnitude of a corrosion reaction. Sometimes the calculated rate from an immersion test does not "look" correct when compared to the visual appearance of the metal coupon.

While the specific corrosion rate number determined by Tafel extrapolation is seldom accurate, the method remains a good confirmation tool.

Linear Polarization   Some of the limitations of the Tafel extrapolation method can be overcome by using the linear-polarization technique to determine the corrosion rate. A relationship exists between the slope of the polarization curves E/I (with units of resistance, linear polarization is sometimes termed *polarization resistance*) and instantaneous corrosion rates of a freely corroding alloy. The polarization resistance is determined by measuring the amount of applied current needed to change the corrosion potential of the freely corroding specimen by about 10-mV deviations. The slope of the curves thus generated is directly related to the corrosion rate by Faraday's law. Several instruments are available that are used in linear polarization work. The main advantage is that each reading on the instrument can be translated directly into a corrosion rate.

As with all electrochemical studies, the environment must be electrically conductive. The corrosion rate is directly dependent on the Tafel slope. The Tafel slope varies quite widely with the particular corroding system and generally with the metal under test. As with the Tafel extrapolation technique, the Tafel slope generally used is an assumed, more or less average value. Again, as with the Tafel technique, the method is not sensitive to local corrosion.

The amount of externally applied current needed to change the corrosion potential of a freely corroding specimen by a few millivolts (usually 10 mV) is measured. This current is related to the corrosion current, and therefore the corrosion rate, of the sample. If the metal is corroding rapidly, a large external current is needed to change its potential, and vice versa.

The measuring system consists of four basic elements:

1. *Electrodes.*   Test and reference electrodes and, in some cases, an auxiliary electrode.
2. *Probe.*   It connects the electrodes in the corrodent on the inside of a vessel to the electrical leads.
3. *Electrical leads.*   They run from the probe to the current source and instrument panel.
4. *Control system.*   Current source (batteries), ammeter, voltmeter, instrument panel, and so on.

Commercial instruments have either two or three electrodes. Also, there are different types of three-electrode systems. The application and limitations of the instruments are largely dependent upon these electrode systems.

Potentiodynamic Polarization   Not all metals and alloys react in a consistent manner in contact with corrosive fluids. One of the common intermediate reactions of a metal (surface) is with oxygen, and those reactions are variable and complex. Oxygen can sometimes function as an electron acceptor; that is, oxygen can act as an oxidizing agent, and remove the "protective" film of hydrogen from the cathodic area, **cathodic depolarization.** The activation energy of the oxygen/hydrogen reaction is very large. This reaction does not normally occur at room temperature at any measureable rate. In other cases, oxygen can form protective oxide films. The long-term stability of these oxides also varies; some are soluble in the environment, others form more stable and inert or passive films.

Because corrosion is an electrochemical process, it is possible to evaluate the overall reaction by the use of an external electrical circuit called a **potentiostat.** When corrosion occurs, a potential difference exists between the metal and its ions in solution. It is possible to electrically control this potential; changes in potential cause changes in current (corrosion). Oxidation is a reaction with a loss of electrons (anodic—the reacting electrode is the anode); reduction is a reaction with a gain of electrons (cathodic—the reacting electrode is the cathode). Rather than allowing the electrons being evolved from the corrosion reaction to combine with hydrogen, these electrons can be removed by internal circuitry, and sent through a potentiostat, causing a cathodic (or anodic) reaction to occur at a platinum counter electrode. This is always true for the external polarization method; it is not unique for a potentiostat.

It is now well established that the activity of pitting, crevice corrosion, and stress-corrosion cracking is strongly dependent upon the corrosion potential (i.e., the potential difference between the corrod-

ing metal and a suitable reference electrode). By using readily available electronic equipment, the quantity and direction of direct current required to control the corrosion potential in a given solution at a given selected value can be measured. A plot of such values over a range of potentials is called a polarization diagram. By using proper experimental techniques, it is possible to define approximate ranges of corrosion potential in which pitting, crevice corrosion, and stress-corrosion cracking will or will not occur. With properly designed probes, these techniques can be used in the field as well as in the laboratory.

The potentiostat has a three-electrode system: a reference electrode, generally a saturated calomel electrode (SCE); a platinum counter, or auxiliary, electrode through which current flows to complete the circuit; and a working electrode that is a sample of interest (Fig. 28-8). The potentiostat is an instrument that allows control of the potential, either holding constant at a given potential, stepping from potential to potential, or changing the potential anodically or cathodically at some linear rate.

In the study of the anode/cathode polarization behavior of a metal/environment system, the potentiostat provides a plot of the relationship of current changes resulting from changes in potential most often presented as a plot of log current density versus potential, or Evans diagram. A typical active/passive metal anodic polarization curve is seen in Fig. 28-9, generally showing the regions of active corrosion and passivity and a transpassive region.

Scan Rates   Sweeping a range of potentials in the anodic (more electropositive) direction of a potentiodynamic polarization curve at a high scan rate of about 60 V/h (high from the perspective of the corrosion engineer, slow from the perspective of a physical chemist) is to indicate regions where intense anodic activity is likely. Second, for otherwise identical conditions, sweeping at a relatively slow rate of potential change of about 1 V/h will indicate regions wherein relative inactivity is likely. The rapid sweep of the potential range has the object of minimizing film formation, so that the currents observed relate to relatively film-free or thin-film conditions. The object of the slow sweep rate experiment is to allow time for filming to occur. A zero scan rate provides the opportunity for maximum stability of the metal surface, but at high electropositive potentials the environment could be affected or changed. A rapid scan rate compromises the steady-state nature of the metal surface but better maintains the stability of the environment. Whenever possible, corrosion tests should be conducted using as many of the techniques available, potentiodynamic polarization at various scan rates, crevice, stress, velocity, and so

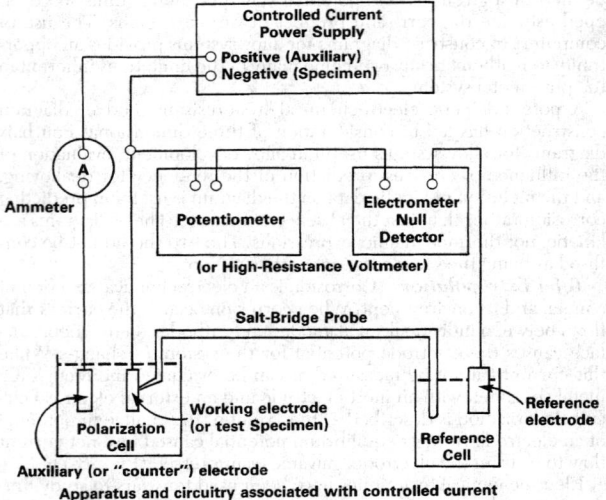

Apparatus and circuitry associated with controlled current measurements.

FIG. 28-8   The potentiostat apparatus and circuitry associated with controlled potential measurements of polarization curves.

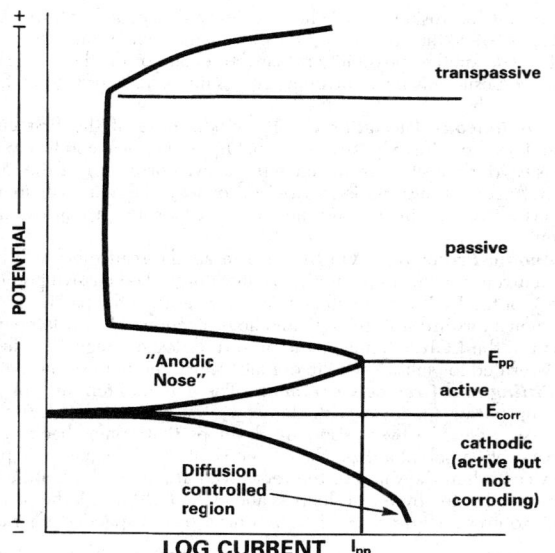

FIG. 28-9 Typical electrochemical polarization curve for an active/passive alloy (with cathodic trace) showing active, passive, and transpassive regions and other important features. (NOTE: E_{pp} = primary passive potential, E_{corr} = freely corroding potential).

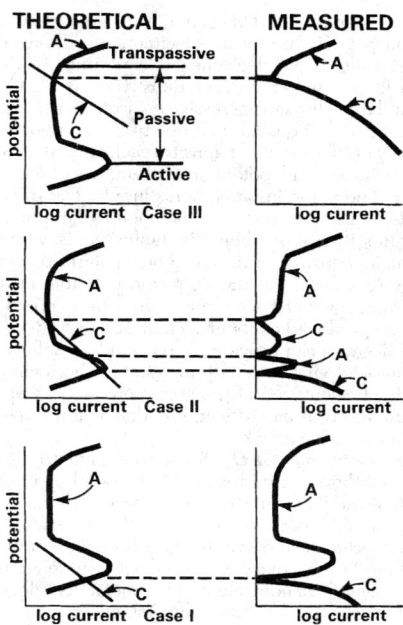

FIG. 28-10 Six possible types of behavior for an active/passive alloy in a corrosive environment.

forth. An evaluation of these several results, on a holistic basis, can greatly reduce or temper their individual limitations.

Slow-Scan Technique In ASTM G5 "Polarization Practice for Standard Reference Method for Making Potentiostatic and Potentiodynamic Anodic Polarization Measurements," all oxygen in the test solution is purged with hydrogen for a minimum of 0.5 h before introducing the specimen. The test material is then allowed to reach a steady state of equilibrium (open circuit corrosion potential, E_{corr}) with the test medium before the potential scan is conducted. Starting the evaluation of a basically passive alloy that is already in its "stable" condition precludes any detailed study of how the metal reaches that protected state (the normal intersection of the theoretical anodic and cathodic curves is recorded as a zero applied current on the ASTM potentiostatic potential versus applied current diagram). These intersections between the anodic and cathodic polarization curves are the condition where the total oxidation rate equals the total reduction rate (ASTM G3 "Recommended Practice for Conventions Applicable to Electrochemical Measurements in Corrosion Testing").

Three general reaction types compare the activation-control reduction processes. In Fig. 28-10, in Case 1, the single reversible corrosion potential (anode/cathode intersection) is in the active region. A wide range of corrosion rates is possible. In Case 2, the cathodic curve intersects the anodic curve at three potentials, one active and two passive. If the middle active/passive intersection is not stable, the lower and upper intersections indicate the possibility of very high corrosion rates. In Case 3, corrosion is in the stable, passive region, and the alloys generally passivate spontaneously and exhibit low corrosion rates. Most investigators report that the ASTM method is effective for studying Case 1 systems. An alloy-medium system exhibiting Case 2 and 3 conditions generally cannot be evaluated by this conventional ASTM method.

The potentiodynamic polarization electrochemical technique can be used to study and interpret corrosion phenomena. It may also furnish useful information on film breakdown or repair.

Rapid-Scan Corrosion Behavior Diagram (CBD) Basically, all the same equipment used in the conductance of an ASTM G5 slow-scan polarization study is used for rapid-scan CBDs (that is, a standard test cell, potentiostat, voltmeters, log converters, X-Y recorders, and electronic potential scanning devices). The differences

are in technique: the slow scan is run at a potential sweep rate of about 0.6 V/h; the rapid-scan CBDs at about 50 V/h.

In the conductance of rapid-scan CBDs, the test specimen is mechanically and chemically cleaned immediately before immersion in the test cell. Without any further delay, a cathodic polarization scan is made at a sweep rate of about 50 V/h (compared to the 0.6 V/h for the ASTM method). This rapid cathodic charging (hydrogen evolution) further cleans the specimen of oxygen so that the subsequent anodic polarization enables a full study of how the corrosion reaction progresses as the potential is changed potentiodynamically in the more noble (positive) direction without the interference of any pre-existing passive or corrosion product film or scale. The test is run open to the atmosphere without any purge gas, and the rapid scan rate permits completion of the anodic curve before oxygen is replenished. This cathodic cleaning further activates the specimen and allows it to follow the pattern of its corrosion reaction(s) before it has an opportunity to spontaneously passivate or return to equilibrium with its environment.

Different from the slow-scan technique, which is generally limited to Case 1 alloy/medium systems, the rapid-scan technique allows full anodic polarization study of alloys showing all Case 1, 2, and 3 behavior. In Case 1, the single reversible corrosion potential (the anode/cathode intersection) is in the active region. A wide range of corrosion rates is possible. In Case 2, the cathodic curve intersects the anodic curve at three potentials, one in the active region and two in the passive. Since the middle active/passive intersection is not stable, the intersections indicate the possibility of very high corrosion rates depending on the environment or even slight changes to the exposure/environment system. In Case 3, the curves intersect in the most stable, passive region; the alloys generally passivate spontaneously and exhibit low corrosion rates. Case 2 exhibits the most corrosion, is difficult to study, and presents the most risk for materials of construction selection. Anything that can change the oxygen solubility of the oxidizing agent can alter the corrosion reaction. Seemingly identical environmental conditions from one day to another, from one laboratory or plant to another, and so forth, can give rise to widely differing results. Those desiring a more detailed review of the subject are directed to the works of Pourbaix, Edeleanu, and other studies referred to in their publications.

The test to generate a CBD is generally as follows. At a rapid scan rate of about 50 V/h, the clean specimen is cathodically charged (polarized) starting at an electronegative position (with respect to SCE) that will produce a current density of 10^4 $\mu A/cm^2$ (nickel, chromium, and iron alloys in aggressive environments) and proceed in a noble (positive) direction. After completing the cathodic curve, proceed anodically until the current density reaches 10^3 $\mu A/cm^2$. The scan direction is reversed and potentials/currents recorded back to the starting point. The potentiostat is then shut off. The freely corroding potential (FCP), also referred to as the open circuit potential (OCP), is recorded after 10 min or when it is stable; that is, when the potential value remains constant. A third cathodic polarization curve is then generated by scanning from the corrosion potential in the negative direction to iron, again at a scan rate of 50 V/h. As noted earlier, this diagram can be produced in about 15 min; see Fig. 28-11.

The CBD diagram can provide various kinds of information about the performance of an alloy/medium system. The technique can be used for a direct calculation of the corrosion rate as well as for indicating the conditions of passivity and tendency of the metal to suffer local pitting and crevice attack.

Corrosion Rate by CBD   Somewhat similarly to the Tafel extrapolation method, the corrosion rate is found by intersecting the extrapolation of the linear portion of the second cathodic curve with the equilibrium stable corrosion potential. The intersection corrosion current is converted to a corrosion rate (mils penetration per year [mpy], 0.001 in/y) by use of a conversion factor (based upon Faraday's law, the electrochemical equivalent of the metal, its valence and gram atomic weight). For 13 alloys, this conversion factor ranges from 0.42 for nickel to 0.67 for Hastelloy B or C. For a quick determination, 0.5 is used for most Fe, Cr, Ni, Mo, and Co alloy studies. Generally, the accuracy of the corrosion rate calculation is dependent upon the degree of linearity of the second cathodic curve; when it is less than 0.5 decade long, it becomes more difficult to extrapolate accurately. Obviously, the longer the better.

Expected Order of Magnitude of Corrosion by CBD   The third cathodic curve will generally fall between the first two, or the first and third will coincide and fall positive to the second curve. These two curve configurations are only valid relationships for the projection

of accurate corrosion rates. When, however, the second or third curve falls positive to the other two curves, a very high corrosion rate may be indicated. Another possibility is that the second or third curve in this positive position is not controlling corrosion because of surface films and so forth.

Spontaneous Passivation   The anodic nose of the first curve describes the primary passive potential E_{pp} and critical anodic current density (the transition from active to passive corrosion). If the initial active/passive transition is 10^3 $\mu A/cm^2$ or less, the alloy will spontaneously passivate in the presence of oxygen or any strong oxidizing agent.

Anodic Protection   On the reverse anodic scan there will be a low current region (LCR) in the passive range. The passive potential range of the LCR is generally much narrower than the passive region seen on a forward slow scan. In anodic protection (AP) work the midpoint of the LCR potential is the preferred design range. This factor was verified for sulfuric acid in our laboratory and field studies.

Pitting and Crevice Corrosion   The general literature for predicting pitting tendency with the slow scan reviews the use of the reverse scan: If a hysteresis loop develops that comes back to the repassivation potential E_{repass} below the FCP (E_{corr}) the alloy will pit at crevices, when the value of the repassivation potential, identified by the return of the hysteresis loop, is above the FCP (E_{corr}), the pits will tend to grow. These rules of thumb have been expanded for rapid-scan.

1.   If E_{corr} is equal to or greater than the pitting potential E_{pit}, both pitting and crevice-type attack are likely to occur on the specimen in an immersion test.

2.   If E_{corr} is less than E_{pit}, pitting-type attack of the specimen is unlikely in an immersion test.

3.   If the repassivation potential E_{repass} is greater than E_{corr}, crevice-type attack is possible in an immersion test.

4.   If E_{corr} is equal to or larger than E_{repass}, crevice-type attack is probably in an immersion test.

5.   If no E_{pit} can be determined, but cathodic Curves 1 and 2 cross at E_x forming a hysteresis loop, crevice-type attack during the rapid scan is indicated.

6.   If E_x is equal to or greater than E_{pit}, pitting-type attack will be predominant during the rapid-scan test.

7.   If E_{pit} is equal to or greater than E_x plus 100 mV, crevice-type attack is indicated during the rapid-scan test.

8.   If the plot of Curve 1 continues from E_{pit} with a large increase in current density versus a small increase in voltage, the test specimen will suffer vigorous local attack during the rapid-scan test.

9.   If the Curve 1, positive with respect to the value of E_{pit}, plots with a small increase in current density per large increase in voltage, the test specimen in the rapid-scan test will suffer only mild local attack (ASTM G31).

Scan Rates: Advantages versus Disadvantages   As to the question of disadvantages of using the rapid-scan CBD, we believe it is more a case of looking to the limitations of the various benefits. In many of the places where these limitations to the use of the CBD show themselves, where the limitation provides information, for example, the corrosion rate will be too high, and so forth. Only in the case of interpretation of unstable active/passive conditions do we see the results of the rapid scan as a possible weakness of the technique (but it is not handled any better by other techniques). When the FCP is noble (positive) to the primary passive potential E_{pp}, the alloy is passive; when the FCP is active (negative) to the E_{pp}, the alloy will corrode in the active state; and when the FCP is only slightly more noble (positive) than the E_{pp}, the alloy could show active/passive tendencies. When the FCP oscillates and will not stabilize, the potential should be plotted against time. The alloy will probably not be resistant without AP but in these latter two cases, the CBD does not tell for sure.

Logic Sequence Diagram   In order to expedite the interpretation of the 50.0-V/h anodic polarization data, three logic sequence diagrams (LSDs) were constructed (Figs. 28-12 through 28-14). These diagrams can be linked to each other to provide a more complete analysis of behavior if necessary. These LSDs contain benchmark values that have been derived from conducting more than 10,000 potentiodynamic experiments at scan rates of either 50.0 or 60.0 V/h.

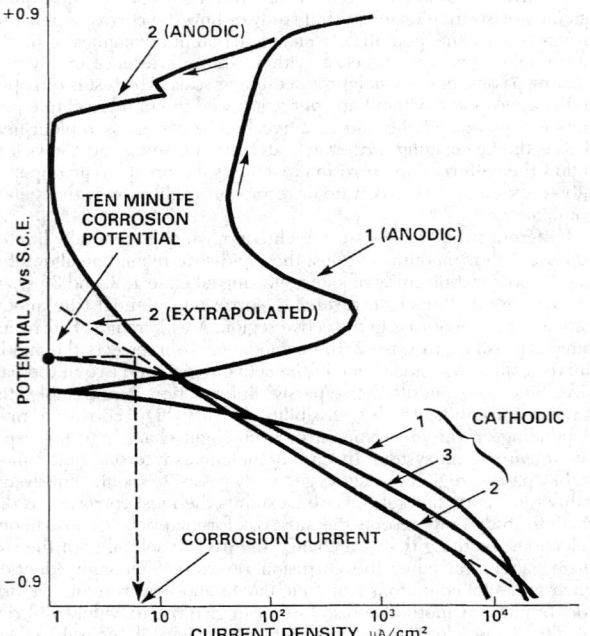

FIG. 28-11   Corrosion behavior diagram (CBD).

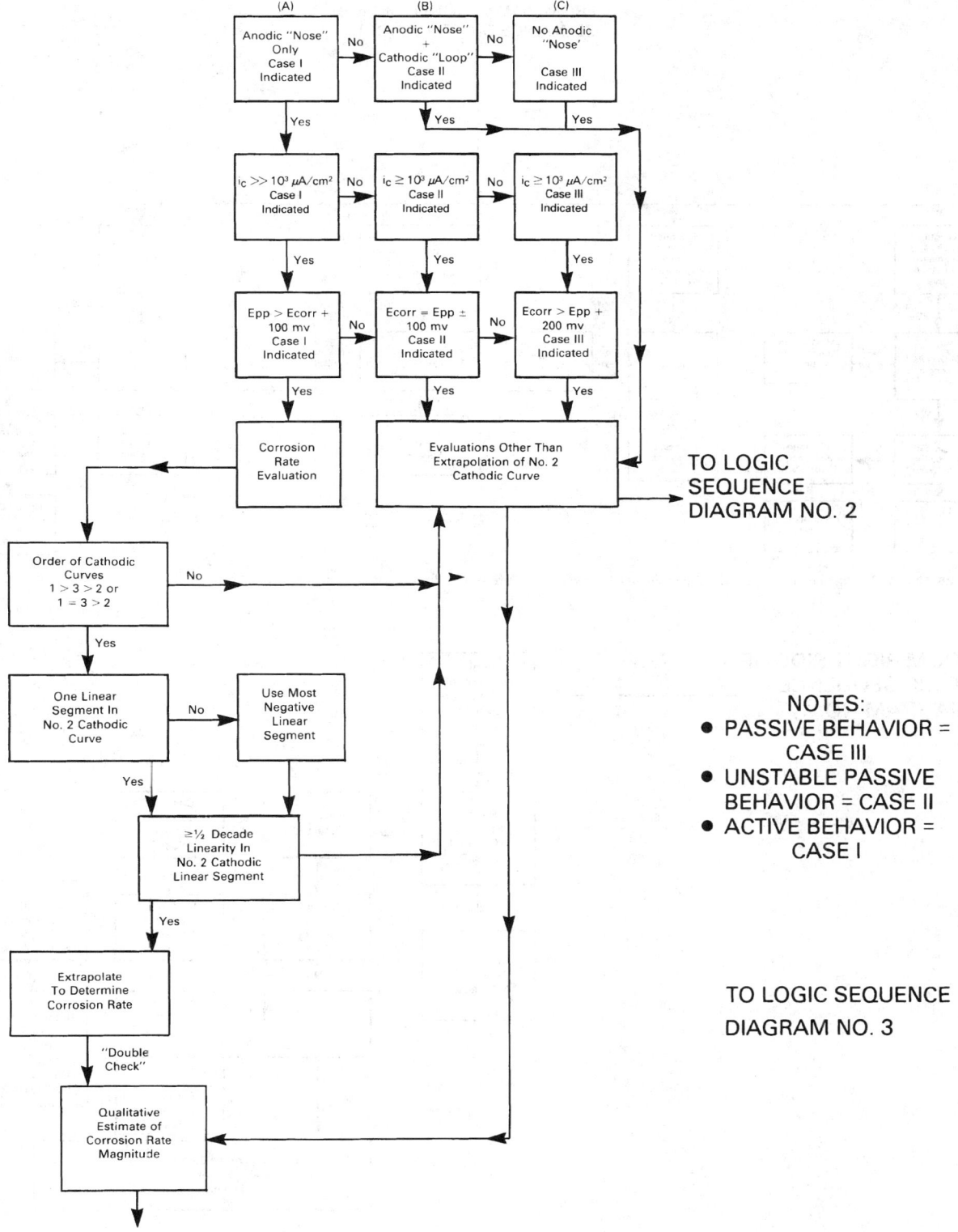

FIG. 28-12   Logic sequence Diagram 1 used to determine behavior type (Case 1, 2, and 3) and corrosion rate from a CBD.

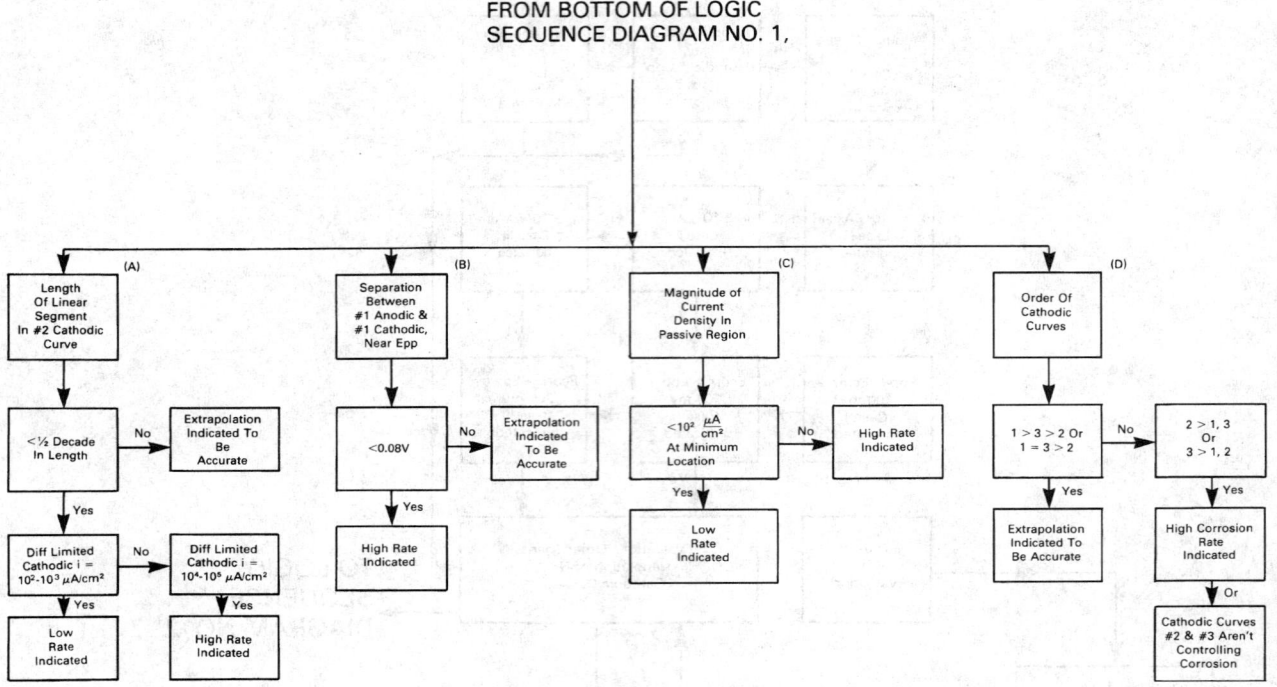

FIG. 28-13   Logic sequence Diagram 2 used to evaluate localized corrosion resistance from a CBD.

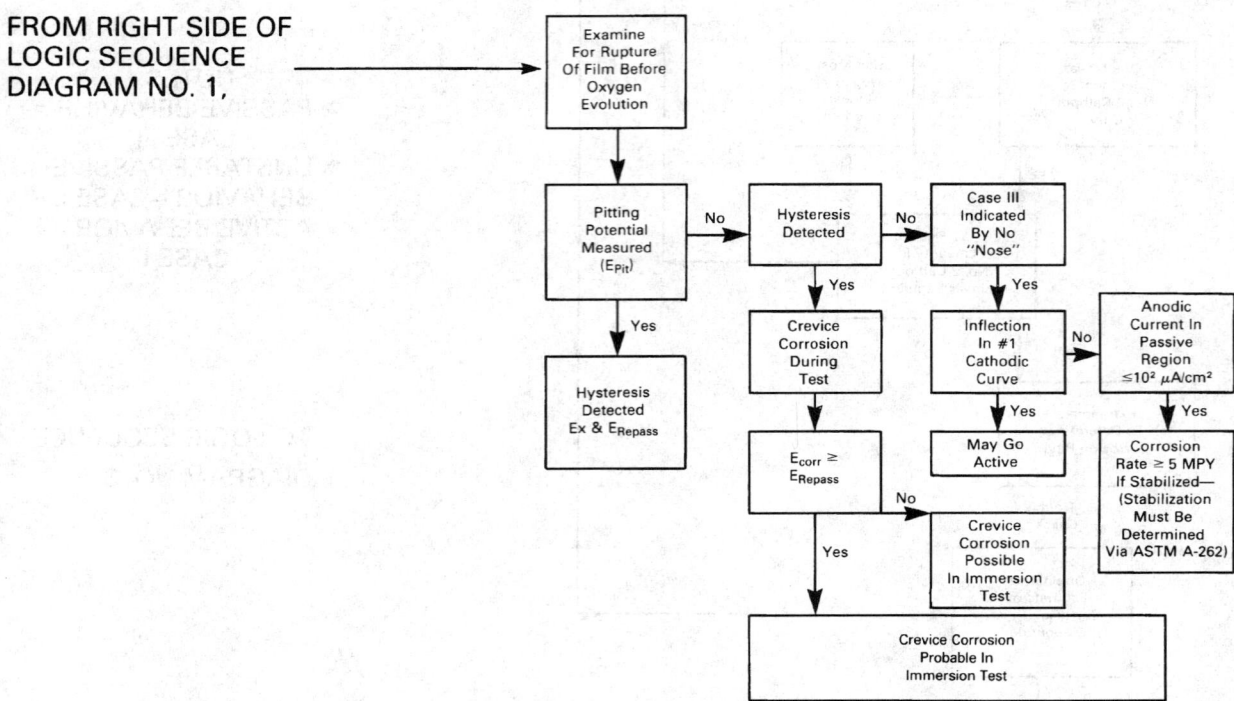

FIG. 28-14   Logic sequence Diagram 3 used to qualitatively evaluate the degree of corrosion for systems that cannot be evaluated by extrapolation of the cathodic polarization Curve 2 from a CBD.

Crevice Corrosion Prediction   The most common type of localized corrosion is the occluded mode crevice corrosion. Pitting can, in effect, be considered a self-formed crevice. A crevice must be wide enough to permit liquid entry, but sufficiently narrow to maintain a stagnant zone. It is nearly impossible to build equipment without mechanical crevices; on a microlevel, scratches can be sufficient crevices to initiate or propagate corrosion in some metal/environment systems. The conditions in a crevice can, with time, become a different and much more aggressive environment than those on a nearby, clean, open surface. Crevices may also be created by factors foreign to the original system design, such as deposits, corrosion products, and so forth. In many studies, it is important to know or to be able to evaluate the crevice corrosion sensitivity of a metal to a specific environment and to be able to monitor a system for predictive maintenance.

Historically, the immersion test technique involved the use of a crevice created by two metal test specimens clamped together or a metal specimen in contact with an inert plastic or ceramic. The Materials Technology Institute of the Chemical Process Industries, Inc. (MTI) funded a study that resulted in an electrochemical cell to monitor crevice corrosion. It consists of a prepared crevice containing an anode that is connected through a zero-resistance ammeter to a freely exposed cathode. A string bridge provides a solution path that is attached externally to the cell. The electrochemical cell is shown in Fig. 28-15. A continuous, semiquantitative, real-time indication of crevice corrosion is provided by the magnitude of the current flowing between an anode and a cathode, and a qualitative signal is provided by shifts in electrode potential. Both the cell current and electrode potential produced by the test correlate well with the initiation and propagation of crevice corrosion. During development of the MTI test technique, results were compared with crevice corrosion produced by a grooved TFE Teflon® plastic disk sandwich-type crevice cell. In nearly every instance, corrosion damage on the anode was similar in severity to that produced by the sandwich-type cell.

Velocity   For corrosion to occur, an environment must be brought into contact with the metal surface and the metal atoms or ions must be allowed to be transported away. Therefore, the rate of transport of the environment with respect to the metal surface is a major factor in the corrosion system. Changes in velocity may increase or decrease attack depending on its effect involved. A varying quantity of dissolved gas may be brought in contact with the metal, or velocity changes may alter diffusion or transfer of ions by changing the thickness of the boundary layer at the surface. The boundary layer, which is not stagnant, moves except where it touches the surface. Many metals depend upon the development of a protective surface for their corrosion resistance. This may consist of an oxide film, a corrosion product, an adsorbed film of gas, or other surface phenomena. The removal of these surfaces by effect of the fluid velocity exposes fresh metal, and as a result, the corrosion reaction may proceed at an increasing rate. In these systems, corrosion might be minimal until a so-called **critical velocity** is attained where the protective surface is damaged or removed and the velocity is too high for a stable film to reform. Above this critical velocity, the corrosion may increase rapidly.

The NACE Landrum Wheel velocity test, originally TM0270-72, is typical of several mechanical-action immersion test methods to evaluate the effects of corrosion. Unfortunately, these laboratory simulation techniques did not consider the fluid mechanics of the environment or metal interface, and service experience very seldom supports the test predictions. A rotating cylinder within a cylinder electrode test system has been developed that operates under a defined hydrodynamics relationship (Figs. 28-16 and 28-17). The assumption is that if the rotating electrode operates at a shear stress comparable to that in plant geometry, the mechanism in the plant geometry may be modeled in the laboratory. Once the mechanism is defined, the appropriate relationship between fluid flow rate and corrosion rate in the plant equipment as defined by the mechanism can be used to predict the expected corrosion rate. If fluid velocity does affect the corrosion rate, the degree of mass transfer control, if that is the controlling mechanism (as opposed to activation control), can be estimated. Conventional potentiodynamic polarization scans are conducted as described previously. In other cases, the corrosion potential can be monitored at a constant velocity until steady state is attained. While the value of the final corrosion potential is virtually independent of velocity, the time to reach steady state may be dependent on velocity. The mass-transfer control of the corrosion potential can be proportional to the velocity raised to its appropriate exponent. The rate of breakdown of a passive film is velocity-sensitive.

Environmental Cracking   The problem of environmental cracking of metals and their alloys is very important. Of all the failure mechanism tests, the test for stress corrosion cracking (SCC) is the most illusive. Stress corrosion is the acceleration of the rate of corrosion damage by static stress. SCC, the limiting case, is the spontaneous cracking that may result from combined effects of stress and corrosion. It is important to differentiate clearly between **stress corrosion cracking** and **stress accelerated corrosion.** Stress corrosion cracking is considered to be limited to cases in which no significant corrosion damage occurs in the absence of a corrosive environment. The material exhibits normal mechanical behavior under the influence of stress; before the development of a stress corrosion

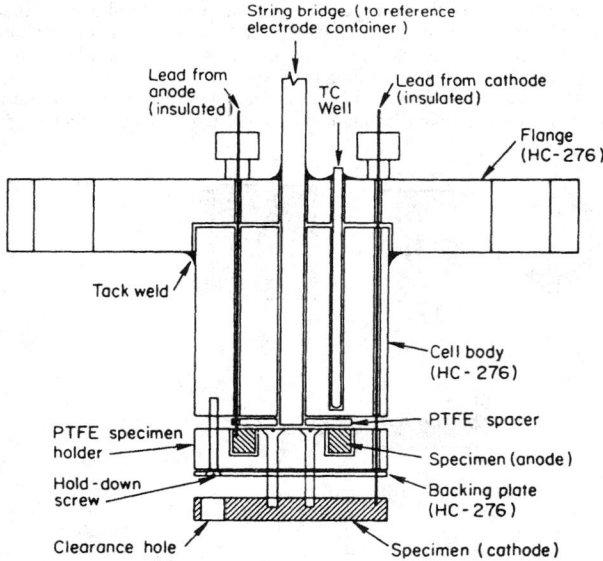

FIG. 28-15   Schematic diagram of the electrochemical cell used for crevice corrosion testing. Not shown are three hold-down screws, gas inlet tube, and external thermocouple tube.

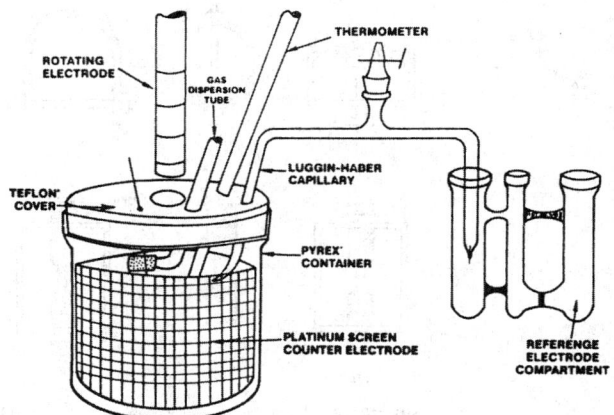

FIG. 28-16   Rotating cylinder electrode apparatus.

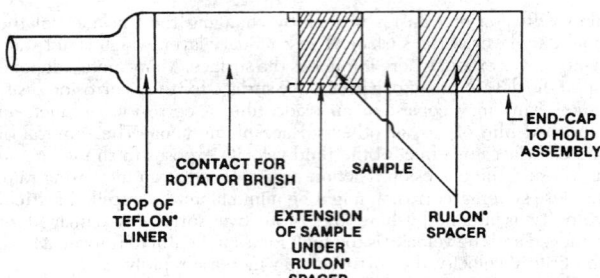

FIG. 28-17   Inner rotating cylinder used in laboratory apparatus of Fig. 28-16.

crack, there is little deterioration of strength and ductility. Stress corrosion cracking is the case of an interaction between chemical reaction and mechanical forces that results in structural failure that otherwise would not occur. SCC is a type of brittle fracture of a normally ductile material by the interaction between specific environments and mechanical forces, for example, tensile stress. Stress corrosion cracking is an incompletely understood corrosion phenomenon. Much research activity (aimed mostly at mechanisms) plus practical experience has allowed crude empirical guidelines, but these contain a large element of uncertainty. No single chemical, structural, or electrochemical test method has been found to respond with enough consistent reproducibility to known crack-causing environmental/stressed metal systems to justify a high confidence level.

As was cited in the case of immersion testing, most SCC test work is accomplished using mechanical, nonelectrochemical methods. It has been estimated that 90 percent of all SCC testing is handled by one of the following methods: (1) constant strain, (2) constant load, or (3) precracked specimens. Prestressed samples, such as are shown in Fig. 28-18, have been used for laboratory and field SCC testing. The variable observed is "time to failure or visible cracking." Unfortunately, such tests do not provide acceleration of failure.

Since SCC frequently shows a fairly long induction period (months to years), such tests must be conducted for very long periods before reliable conclusions can be drawn.

In the constant-strain method, the specimen is stretched or bent to a fixed position at the start of the test. The most common shape of the specimens used for constant-strain testing is the U-bend, hairpin, or horseshoe type. A bolt is placed through holes in the legs of the specimen, and it is loaded by tightening a nut on the bolt. In some cases, the stress may be reduced during the test as a result of creep. In the constant-load test the specimen is supported horizontally at each end

and is loaded vertically downward at one or two points and has maximum stress over a substantial length or area of the specimen. The load applied is a predetermined, fixed dead weight. Specimens used in either of these tests may be precracked to assign a stress level or a desired location for fracture to occur or both as is used in fracture mechanics studies. These tensile-stressed specimens are then exposed in situ to the environment of study.

Slow Strain-Rate Test   In its present state of development, the results from slow strain-rate tests (SSRT) with electrochemical monitoring are not always completely definitive; but, for a short-term test, they do provide considerable useful SCC information. Work in our laboratory shows that the SSRT with electrochemical monitoring and the U-bend tests are essentially equivalent in sensitivity in finding SCC. The SSRT is more versatile and faster, providing both mechanical and electrochemical feedback during testing.

The SSRT is a test technique where a tension specimen is slowly loaded in a test frame to failure under prescribed test conditions. The normal test extension rates are from 2.54×10^{-7} to 2.54×10^{-10} m/s (10^{-5} to 10^{-8} in/s). Failure times are usually 1 to 10 days. The failure mode will be either SCC or tensile overload, sometimes accelerated by corrosion. An advantage behind the SSRT, compared to constant-strain tests, is that the protective surface film is thought to be ruptured mechanically during the test, thus giving SCC an opportunity to progress. To aid in the selection of the value of the potential at which the metal is most sensitive to SCC that can be applied to accelerate SSRT, potentiodynamic polarization scans are conducted as described previously. It is common for the potential to be monitored during the conduct of the SSRT. The strain rates that generate SCC in various metals are reported in the literature. There are several disadvantages to the SSRT. First, indications of failure are not generally observed until the tension specimen is plastically stressed, sometimes significantly, above the yield strength of the metal. Such high-stressed conditions can be an order of magnitude higher than the intended operating stress conditions. Second, crack initiation must occur fairly rapidly to have sufficient crack growth that can be detected using the SSRT. The occurrence of SCC in metals requiring long initiation times may go undetected.

Modulus Measurements   Another SCC test technique is the use of changes of modulus as a measure of the damping capacity of a metal. It is known that a sample of a given test material containing cracks will have a lower effective modulus than does a sample of identical material free of cracks. The technique provides a rapid and reliable evaluation of the susceptibility of a sample material to SCC in a specific environment. The so-called internal friction test concept can also be used to detect and probe nucleation and progress of cracking and the mechanisms controlling it.

The Young's modulus of the specimen is determined by accurately measuring its resonant frequency while driving it in a standing longi-

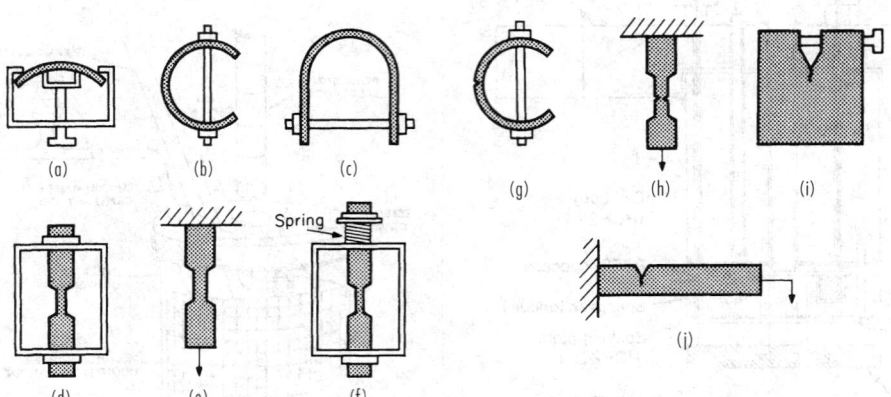

FIG. 28-18   Specimens for stress-corrosion tests. (*a*) Bent beam. (*b*) C ring. (*c*) U bend. (*d*) Tensile. (*e*) Tensile. (*f*) Tensile. (*g*) Notched C ring. (*h*) Notched tensile. (*i*) Precracked, wedge open-loading type. (*j*) Precracked, cantilever beam. [*Chem. Eng.*, **78**, 159 (*Sept. 20 1971*).]

tudinal wave configuration A Marx composite piezoelectric oscillator is used to drive the specimen at a resonant frequency. The specimen is designed to permit measurements while undergoing applied stress and while exposed to an environmental test solution. The specimens are three half-wavelengths long; the gripping nodes and solution cup are silver-soldered on at displacement nodes, so they do not interfere with the standing wave (Fig. 28-19). As discussed for SSRT, potentiodynamic polarization scans are conducted to determine the potential that can be applied to accelerate the test procedure. Again, the potential can be monitored during a retest, as is the acoustic emission (AE) as an indicator of nucleation and progress of cracking.

Conjunctive Use of Slow- and Rapid-Scan Polarization The use of the methods discussed in the preceding requires a knowledge of the likely potential range for SCC to occur. Potentiodynamic polarization curves can be used to predict those SCC-sensitive potential ranges. The technique involves conducting both slow- and rapid-scan sweeps in the anodic direction of a range of potentials. Comparison of the two curves will indicate any ranges of potential within which high anodic activity in the film-free condition reduces to insignificant activity when the time requirements for film formation are met. This should indicate the range of potentials within which SCC is likely. Most of the SCC theories presently in vogue predict these domains of behavior to be between the primary passive potential and the onset of passivity. This technique shortens the search for that SCC potential.

Separated Anode/Cathode Realizing, as noted in the preceding, that localized corrosion is usually active to the surrounding metal surface, a stress specimen with a limited area exposed to the test solution (the anode) is electrically connected to an unstressed specimen (the cathode). A potentiostat, used as a zero-resistance ammeter, is placed between the specimens for monitoring the galvanic current. It is possible to approximately correlate the galvanic current I_g and potential to crack initiation and propagation and, eventually, catastrophic fail-

ure. By this arrangement, the galvanic current I_g is independent of the cathode area. In other words, the potential of the anode follows the corrosion potential of the cathode during the test. The SSRT apparatus discussed previously may be used for tensile loading.

Fracture Mechanics Methods These have proved very useful for defining the minimum stress intensity K_{ISCC} at which stress corrosion cracking of high-strength, low-ductility alloys occurs. They have so far been less successful when applied to high-ductility alloys, which are extensively used in the chemical-process industries.

Work on these and other new techniques continues, and it is hoped that a truly reliable, accelerated test or tests will be defined.

Electrochemical Impedance Spectroscopy (EIS) and AC Impedance[°] Many direct-current test techniques assess the overall corrosion process occurring at a metal surface, but treat the metal/solution interface as if it were a pure resistor. Problems of accuracy and reproducibility frequently encountered in the application of direct-current methods have led to increasing use of electrochemical impedance spectroscopy (EIS).

Electrode surfaces in electrolytes generally possess a surface charge that is balanced by an ion accumulation in the adjacent solution, thus making the system electrically neutral. The first component is a double layer created by a charge difference between the electrode surface and the adjacent molecular layer in the fluid. Electrode surfaces may behave at any given frequency as a network of resistive and capacitive elements from which an electrical impedance may be measured and analyzed.

The application of an impressed alternating current on a metal specimen can generate information on the state of the surface of the specimen. The corrosion behavior of the surface of an electrode is related to the way in which that surface responds to this electrochemical circuit. The AC impedance technique involves the application of a small sinusoidal voltage across this circuit. The frequency of that alternating signal is varied. The voltage and current response of the system are measured.

The so-called white-noise analysis by the Fast Fourier transform technique (FFT) is another viable method. The entire spectrum can be derived from one signal. The impedance components thus generated are plotted on either a Nyquist (real versus imaginary) or Bode (log real versus log frequency plus log phase angle versus log frequency) plot. These data are analyzed by computer; they can be used to determine the polarization resistance and, thus, the corrosion rate if Tafel slopes are known. It is also thought that the technique can be used to monitor corrosion by examining the real resistance at high and low frequency and by assuming the difference is the polarization resistance. This can be done in low- and high-conductivity environments. Systems prone to suffer localized corrosion have been proposed to be analyzed by AC impedance and should aid in determining the optimum scan rate for potentiodynamic scans.

The use of impedance electrochemical techniques to study corrosion mechanisms and to determine corrosion rates is an emerging technology. Electrode impedance measurements have not been widely used, largely because of the sophisticated electrical equipment required to make these measurements. Recent advantages in microelectronics and computers has moved this technique almost overnight from being an academic experimental investigation of the concept itself to one of shelf-item commercial hardware and computer software, available to industrial corrosion laboratories.

Use and Limitations of Electrochemical Techniques A major caution must be noted as to the general, indiscriminate use of all electrochemical tests, especially the use of AC and EIS test techniques, for the study of corrosion systems. AC and EIS techniques are applicable for the evaluation of very thin films or deposits that are uniform, constant, and stable—for example, thin-film protective coatings. Sometimes, researchers do not recognize the dynamic nature of some passive films, corrosion products, or deposits from other sources; nor do they even consider the *possibility of a change* in the surface conditions during the course of their experiment. As an example, it is note-

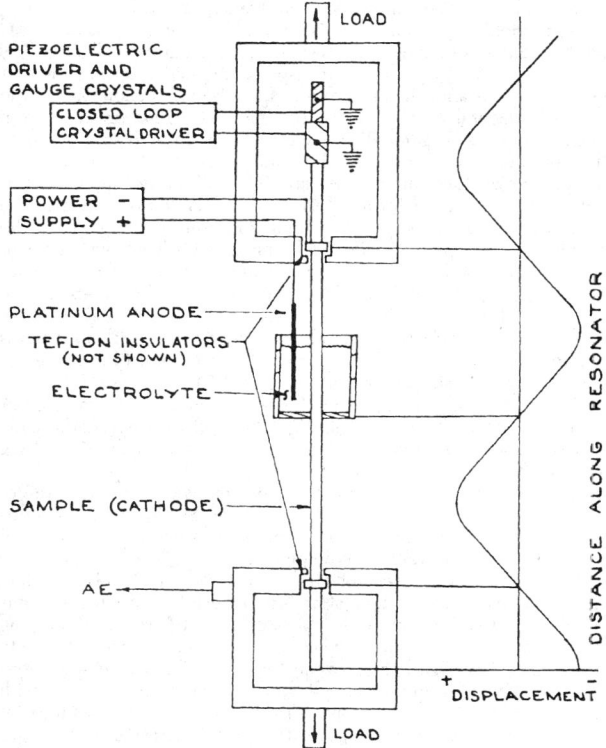

PIEZOELECTRIC DRIVER AND GAUGE CRYSTALS

CLOSED LOOP CRYSTAL DRIVER

POWER SUPPLY

LOAD

PLATINUM ANODE

TEFLON INSULATORS (NOT SHOWN)

ELECTROLYTE

SAMPLE (CATHODE)

AE

LOAD

DISTANCE ALONG RESONATOR

DISPLACEMENT

FIG. 28-19 Equipment for measuring internal friction (modulus) changes during in situ tensile exposure of a metal in a corrosive environment.

° Excerpted from papers by Oliver W. Siebert, courtesy of NACE International and ASTM.

worthy that this is a *major* potential problem in the electrochemical evaluation of microbiological corrosion (MIC).

MIC depends on the complex structure of corrosion products and passive films on metal surfaces as well as on the structure of the biofilm. Unfortunately, electrochemical methods have sometimes been used in complex electrolytes, such as microbiological culture media, where the characteristics and properties of passive films and MIC deposits are *quite active* and not fully understood. It must be kept in mind that microbial colonization of passive metals can drastically change their resistance to film breakdown by causing *localized* changes in the type, concentration, and thickness of anions, pH, oxygen gradients, and inhibitor levels at the metal surface during the course of a normal test; viable single-cell microorganisms divide at an exponential rate. These changes can be expected to result in important modifications in the electrochemical behavior of the metal and, accordingly, in the electrochemical parameter measured in laboratory experiments.

Warnings are noted in the literature to be careful in the *interpretation* of data from electrochemical techniques applied to systems in which complex and often poorly understood effects are derived from surfaces which contain *active* or *viable* organisms, and so forth. Rather, it is even more important to not use such test protocol unless the investigator fully understands both the corrosion mechanism and the test technique being considered—and their interrelationship.

CORROSION TESTING: PLANT TESTS

It is not always practical or convenient to investigate corrosion problems in the laboratory. In many instances, it is difficult to discover just what the conditions of service are and to reproduce them exactly. This is especially true with processes involving changes in the composition and other characteristics of the solutions as the process is carried out, as, for example, in evaporation, distillation, polymerization, sulfonation, or synthesis.

With many natural substances also, the exact nature of the corrosive is uncertain and is subject to changes not readily controlled in the laboratory. In other cases, the corrosiveness of the solution may be influenced greatly by or even may be due principally to a constituent present in such minute proportions that the mass available in the limited volume of corrosive solution that could be used in a laboratory setup would be exhausted by the corrosion reaction early in the test, and consequently the results over a longer period of time would be misleading.

Another difficulty sometimes encountered in laboratory tests is that contamination of the testing solution by corrosion products may change its corrosive nature to an appreciable extent.

In such cases, it is usually preferable to carry out the corrosion-testing program by exposing specimens in operating equipment under **actual conditions of service.** This procedure has the additional advantages that it is possible to test a large number of specimens at the same time and that little technical supervision is required.

In certain cases, it is necessary to choose materials for equipment to be used in a process developed in the laboratory and not yet in operation on a plant scale. Under such circumstances, it is obviously impossible to make plant tests. A good procedure in such cases is to construct a pilot plant, using either the cheapest materials available or some other materials selected on the basis of past experience or of laboratory tests. While the pilot plant is being operated to check on the process itself, specimens can be exposed in the operating equipment as a guide to the choice of materials for the large-scale plant or as a means of confirming the suitability of the materials chosen for the pilot plant.

Test Specimens In carrying out plant tests it is necessary to install the test specimens so that they will not come into contact with other metals and alloys; this avoids having their normal behavior disturbed by galvanic effects. It is also desirable to protect the specimens from possible mechanical damage.

There is no single standard size or shape for corrosion-test coupons. They usually weigh from 10 to 50 g and preferably have a large surface-to-mass ratio. Disks 40 mm (1½ in) in diameter by 3.2 mm (⅛ in) thick and similarly dimensioned square and rectangular coupons are

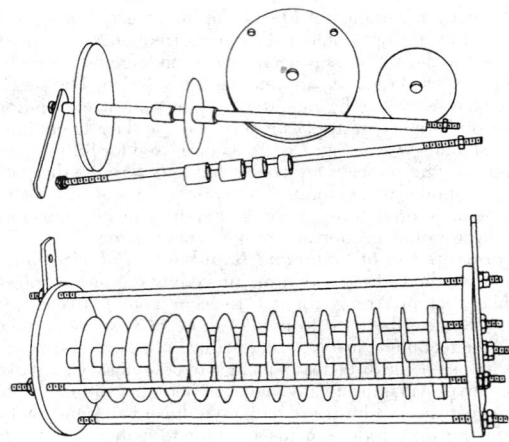

FIG. 28-20 Assembly of a corrosion-test spool and specimens. (*Mantell, ed.,* Engineering Materials Handbook, *McGraw-Hill, New York, 1958.*)

the most common. Surface preparation varies with the aim of the test, but machine grinding of surfaces or polishing with a No. 120 grit is common. Samples should not have sheared edges, should be clean (no heat-treatment scale remaining unless this is specifically part of the test), and should be identified by stamping. See Fig. 28-20 for a typical plant test assembly.

The choice of materials from which to make the holder is important. Materials must be durable enough to ensure satisfactory completion of the test. It is good practice to select very resistant materials for the test assembly. Insulating materials used are plastics, porcelain, Teflon, and glass. A phenolic plastic answers most purposes; its principal limitations are unsuitability for use at temperatures over 150°C (300°F) and lack of adequate resistance to concentrated alkalies.

The method of supporting the specimen holder during the test period is important. The preferred position is with the long axis of the holder horizontal, thus avoiding dripping of corrosion products from one specimen to another. The holder must be located so as to cover the conditions of exposure to be studied. It may have to be submerged, or exposed only to the vapors, or located at liquid level, or holders may be called for at all three locations. Various means have been utilized for supporting the holders in liquids or in vapors. The simplest is to suspend the holder by means of a heavy wire or light metal chain. Holders have been strung between heating coils, clamped to agitator shafts, welded to evaporator tube sheets, and so on. The best method is to use test racks.

In a few special cases, the standard "spool-type" specimen holder is not applicable and a suitable special test method must be devised to apply to the corrosion conditions being studied.

For conducting tests in **pipe lines** of 75-mm (3-in) diameter or larger, a spool holder as shown in Fig. 28-21, which employs the same disk-type specimens used on the standard spool holder, has been used. This frame is so designed that it may be placed in a pipe line in any position without permitting the disk specimens to touch the wall of the pipe. As with the strip-type holder, this assembly does not materially interfere with the fluid through the pipe and permits the study of corrosion effects prevailing in the pipe line.

Another way to study corrosion in pipe lines is to install in the line short sections of pipe of the materials to be tested. These test sections should be insulated from each other and from the rest of the piping system by means of nonmetallic couplings. It is also good practice to provide insulating gaskets between the ends of the pipe specimens where they meet inside the couplings. Such joints may be sealed with various types of dope or cement. It is desirable in such cases to paint the outside of the specimens so as to confine corrosion to the inner surface.

It is occasionally desirable to expose corrosion-test specimens in operating equipment without the use of specimen holders of the type

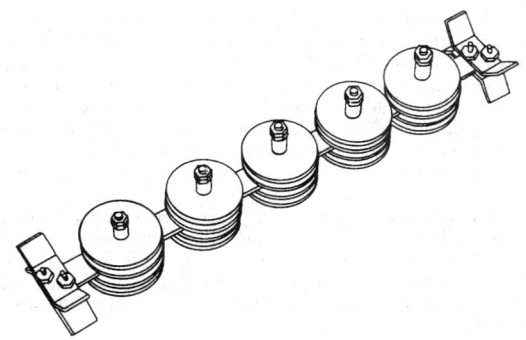

FIG. 28-21 Spool-type specimen holder for use in 3-in-diameter or larger pipe. (*Mantell, ed., Engineering Materials Handbook, McGraw-Hill, New York, 1958.*)

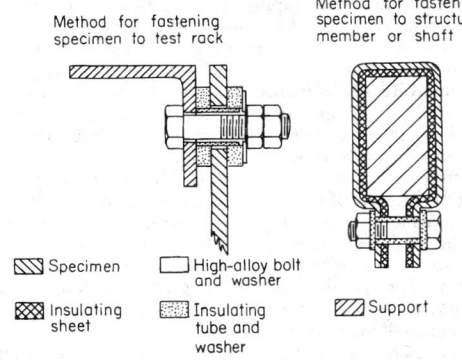

Method for fastening specimen to test rack

Method for fastening specimen to structural member or shaft

⬚ Specimen
⬚ Insulating sheet
⬚ High-alloy bolt and washer
⬚ Insulating tube and washer
⬚ Support

FIG. 28-22 Methods for attaching specimens to test racks and to parts of moving equipment. (*Mantell, ed., Engineering Materials Handbook, McGraw-Hill, New York, 1958.*)

described. This can be accomplished by attaching specimens directly to some part of the operating equipment and by providing the necessary insulation against galvanic effects as shown in Fig. 28-22. The suggested method of attaching specimens to racks has been found to be very suitable in connection with the exposure of specimens to corrosion in seawater.

Test Results   The methods of cleaning specimens and evaluating results after plant corrosion tests are identical to those described earlier for laboratory tests.

Electrochemical On-Line Corrosion Monitoring°   On-line corrosion monitoring is used to evaluate the status of equipment and piping in chemical process industries (CPI) plants. These monitoring methods are based on electrochemical techniques. To use on-line monitoring effectively, the engineer needs to understand the underlying electrochemical test methods to be employed. This section covers many of these test methods and their applications as well as a review of potential problems encountered with such test instruments and how to overcome or avoid these difficulties.

Most Common Types of Probes   There are three most common types of corrosion monitoring probes used. Other types of probes are used, but in smaller numbers.

1. *Weight loss probes.*   Coupons for measuring weight loss are still the primary type of probe in use. These may be as simple as samples of the process plant materials which have been fitted with electrical connections and readouts to determine intervals for retrieval and weighing, to commercially available coupons of specified material,

° Excerpted from papers by Ann Chidester Van Orden and Oliver W. Siebert, courtesy of NACE International and Oliver W. Siebert, courtesy of ASTM.

geometry, stress condition, and other factors, ready to be mounted on specially designed supports at critical points in the process. Coupons can be permanently installed prior to plant start-up or during a shutdown. This type of permanent installation requires a plant shutdown for probe retrieval as well. Shutdown can be avoided by installing the probe in a bypass.

2. *Electrical resistance probes.*   These probes are the next most common type of corrosion probes after coupons. This type of probe measures changes in the electrical resistance as a thin strip of metal gets thinner with ongoing corrosion. As the metal gets thinner, its resistance increases. This technique was developed in the 1950s by Dravinieks and Cataldi and has undergone many improvements since then.

3. *Linear polarization resistance probes.*   LPR probes are more recent in origin, and are steadily gaining in use. These probes work on a principle outlined in an ASTM guide on making polarization resistance measurements, providing instantaneous corrosion rate measurements (G59, "Standard Practice for Conducting Potentiodynamic Polarization Resistance Measurements").

LPR probes measure the electrochemical corrosion mechanism involved in the interaction of the metal with the electrolyte. To measure linear polarization resistance R_p, Ω/cm^2, the following assumptions must be made:
- The corrosion rate is uniform.
- There is only cathodic and one anodic reaction.
- The corrosion potential is not near the oxidation/reduction potential for either reaction.

When these conditions are met, the current density associated with a small polarization of the metal (less than +10 mV) is directly proportional to the corrosion rate of the metal.

Multiinformational Probes   Corrosion probes can provide more information than just corrosion rate. The next three types of probes yield information about the type of corrosion, the kinetics of the corrosion reaction, as well as the local corrosion rate.

Electrochemical impedance spectroscopy, AC probes.   EIS, although around since the 1960s, has primarily been a laboratory technique. Commercially available probes and monitoring systems that measure EIS are becoming more widely used, especially in plants that have on-staff corrosion experts to interpret the data or to train plant personnel to do so.

In EIS, a potential is applied across a corroding metal in solution, causing current to flow. The amount of current depends upon the corrosion reaction on the metal surface and the flow of ions in solution. If the potential is applied as a sine wave, it will cause harmonics of the current output. The relationship between the applied potential and current output is the impedance, which is analogous to resistance in a DC circuit.

Since the potential and current are sinusoidal, the impedance has a magnitude and a phase, which can be represented as a vector. A sinusoidal potential or current can be pictured as a rotating vector. For standard AC current, the rotation is at a constant angular velocity of 60 Hz.

The voltage can also be pictured as a rotating vector with its own amplitude and frequency. Both current and potential can be represented as having real (observed) and imaginary (not observed) components.

In making electrochemical impedance measurements, one vector is examined, using the others as the frame of reference. The voltage vector is divided by the current vector, as in Ohm's law. Electrochemical impedance measures the impedance of an electrochemical system and then mathematically models the response using simple circuit elements such as resistors, capacitors, and inductors. In some cases, the circuit elements are used to yield information about the kinetics of the corrosion process.

Polarization probes.   Polarization methods other than LPR are also of use in process control and corrosion analysis, but only a few systems are offered commercially. These systems use such polarization techniques as galvanodynamic or potentiodynamic, potentiostatic or galvanostatic, potentiostaircase or galvanostaircase, or cyclic polarization methods. Some systems involving these techniques are, in fact, used regularly in processing plants. These methods are used in situ or in the laboratory to measure corrosion. Polarization probes have been successful in reducing corrosion-related failures in chemical plants.

Polarization probes rely on the relationship of the applied potential to the output current per unit area (current density). The slope of applied potential versus current density, extrapolated through the origin, yields the polarization resistance R_p, which can be related to the corrosion rate.

There are several methods for relating the corrosion current, the applied potential, and the polarization resistance. These methods involve various ways of stepping or ramping either the potential or current. Also, a constant value of potential or current can be applied.

Electrochemical noise monitoring probes. Electrochemical noise monitoring is probably the newest of these methods. The method characterizes the naturally occurring fluctuations in current and potential due to the electrochemical kinetics and the mechanism of the corroding metal interface. Measurements are taken without perturbing the interface by applying a potential or current to it. In this way, electrochemical processes are not interrupted and the system is measured without being disturbed. Methods including signal processing and mathematical transformation are used to provide information on the reaction kinetics at the surface and the corrosion rate.

This technique, originally discovered in the 1960s, remained a laboratory technique until recently, when some manufacturers began producing commercial devices. There are a few cases where electrochemical noise is being used in process-plant-type environments, and an ASTM committee has been formed to look at standardization of this technique. However, in general it remains a laboratory method with great potential for on-line monitoring.

Indirect Probes Some types of probes do not measure corrosion directly, but yield measurements that also are useful in detecting corrosion. Examples include:

Pressure probes. Pressure monitors or transducers may be of use in corrosion monitoring in environments where buildup of gases such as hydrogen or H_2S may contribute to corrosion.

Gas probes. The hydrogen patch probe allows users to determine the concentration of hydrogen in the system. This is an important measurement since hydrogen can foster corrosion. Detecting production of certain gases may give rise to process changes to eliminate or limit the gaseous effluent and, therefore, lower the possibility of corrosion caused by these gases.

pH probes. Monitoring the pH may also aid in the early detection of corrosion. The acidity or alkalinity of the environment is often one of the controllable parameters in corrosion. Monitoring of the pH can be combined with other corrosion measurements to provide additional data about process conditions and give another level of process control.

Ion probes. Determining the level of ions in solution also helps to control corrosion. An increase in concentration of specific ions can contribute to scale formation, which can lead to a corrosion-related failure. Ion-selective electrode measurements can be included, just as pH measurements can, along with other more typical corrosion measurements. Especially in a complete monitoring system, this can add information about the effect of these ions on the material of interest at the process plant conditions.

Microbially induced corrosion (MIC) probes. Devices are available to measure the amount of microbial activity in some environments. Microbially induced corrosion is known to be an actor in many corrosion-related problems in processing plants. The monitoring devices for MIC are limited in their range and, at present, are available only for a few specific environments. This is an exciting area for development of corrosion probes and monitoring systems.

Use of Corrosion Probes The major use of corrosion monitoring probes is to measure the corrosion rate in the plant or the field. In addition to corrosion-rate measurements, corrosion probes can be used to detect process upsets that may change the corrosion resistance of the equipment of interest. This is usually equally as important as a measurement as corrosion rate since a change in the process conditions can lead to dramatic changes in the corrosion rate.

If the upset can be detected and dealt with in short order, the system can be protected. Some of the probes that measure parameters such as pH, ion content, and others are sensitive to process upsets and may give the fastest most complete information about such changes.

Monitoring can also be used to optimize the chemistry and level of corrosion inhibitors used. If too little inhibitor is used, enhanced corrosion can result and failure may follow. If too much is used, costs will increase without providing any additional protection. Optimization of the addition of inhibitor in terms of time, location in the process, and method of addition can also be evaluated through the use of carefully placed probes.

Another area where probes can be used effectively is in monitoring of deposits such as scale. One method of measurement is detecting specific ions that contribute to scale buildup or fouling; another is measuring the actual layers. Scale and fouling often devastate the corrosion resistance of a system, leading to a costly corrosion-related plant shutdown.

A final type of measurement is the detection of localized corrosion, such as pitting or crevice attack. Several corrosion-measuring probes can be used to detect localized corrosion. Some can detect localized corrosion instantaneously and others only its result. These types of corrosion may contribute little to the actual mass loss, but can be devastating to equipment and piping. Detection and measurement of localized corrosion is one of the areas with the greatest potential for the use of some of the newest electrochemically based corrosion monitoring probes.

Corrosion Rate Measurements Determining a corrosion rate from measured parameters (such as mass loss, current, or electrical potential) depends on converting the measurements into a corrosion rate by use of relationships such as Faraday's law.

Information on the process reaction conditions may be important to prolonging the lifetime of process equipment. Techniques such as EIS and potentiodynamic polarization can provide just such information without being tied to a specific corrosion-rate measurement.

This is also the case with methods that yield information on localized corrosion. The overall corrosion rate may be small when localized attack occurs, but failure due to perforation or loss of function may be the consequence of localized attack.

Measuring Corrosion Rate with Coupons Corrosion rate determined with typical coupon tests is an average value, averaged over the entire life of the test. Changes in the conditions under which the coupons are tested are averaged over the time the coupon is exposed. *Uniform attack is assumed to occur.* Converting measured values of loss of mass into an average corrosion rate has been covered extensively by many authors, and standard practices exist for determining corrosion rate.

Corrosion rates may vary during testing. Since the rate obtained from coupon testing is averaged over time, the frequency of sampling is important. Generally, measurements made over longer times are more valid. This is especially true for low corrosion rates, under 1 mil/y, mpy (0.001 in/y). When corrosion rates are this low, longer times should be used.

Factors may throw off these rates—these are outlined in ASTM G31, "Standard Practice for Laboratory Immersion Corrosion Testing of Metals." Coupon-type tests cannot be correlated with changing plant conditions that may dramatically affect process equipment lifetimes. Other methods must be used if more frequent measurements are desired or correlation with plant conditions are necessary.

A plot of mass loss versus time can provide information about changes in the conditions under which the test has been run. One example of such a plot comes from the ASTM Standard G96, "Standard Guide."

Heat Flux Tests Removable tube test heat exchangers find an ideal use in the field for monitoring heat flux (corrosion) conditions, NACE TMO286-94 (similar to laboratory test, Fig. 28-4, page 28-12).

The assumption of uniform corrosion is also at the heart of the measurements made by the electrical resistance (ER) probes. Again, ASTM Standard G96 outlines the method for using ER probes in plant equipment. These probes operate on the principle that the electrical resistance of a wire, strip, or tube of metal increases as its cross-sectional area decreases:

$$R = \frac{\rho L}{A}$$

where R = resistance, Ω
   ρ = resistivity, Ω/cm
   L = length, cm
   A = cross-sectional area, cm^2

Usually, in practice, the resistance is measured as a ratio between the actual measuring element and a similar element protected from the corrosive environment (the reference), and is given by R_M/R_R where subscript M is for measured and R is for reference.

Measurements are recorded intermittently or continuously. Changes in the slope of the curve obtained thus yields the corrosion rate.

The initial measurement of electrical resistance must be made after considerable time. Phenomenological information has been determined based on the corrosion rate expected at what period of time to initiate readings of the electrical resistance. Since these values are based on experiential factors rather than on fundamental (so-called *first*) principles, correlation tables and lists of suggested thicknesses, compositions, and response times for usage of ER-type probes have developed over time, and these have been incorporated into the values read out of monitoring systems using the ER method.

Electrochemical Measurement of Corrosion Rate There is a link between electrochemical parameters and actual corrosion rates. Probes have been specifically designed to yield signals that will provide this information. LPR, ER, and EIS probes can give corrosion rates directly from electrochemical measurements. ASTM G102, "Standard Practice for Calculation of Corrosion Rates and Related Information from Electrochemical Measurements," tells how to obtain corrosion rates directly. Background on the approximations made in making use of the electrochemical measurements has been outlined by several authors.

ASTM G59, "Standard Practice for Conducting Potentiodynamic Polarization Resistance Measurements," provides instructions for the graphical plotting of data (from tests conducted using the above-noted ASTM Standard G103) as the linear potential versus current density, from which the polarization resistance can be found.

Measurements of polarization resistance R_p, given by LPR probes, can lead to measurement of the corrosion rate at a specific instant, since values of R_p are instantaneous.

To obtain the corrosion current I_{corr} from R_p, values for the anodic and cathodic slopes must be known or estimated. ASTM G59 provides an experimental procedure for measuring R_p. A discussion of the factors which may lead to errors in the values for R_p, and cases where this technique cannot be used, are covered by Mansfeld in "Polarization Resistance Measurements—Today's Status, Electrochemical Techniques for Corrosion Engineers" (NACE International, 1992).

Some data from corrosion-monitoring probes do not measure corrosion rate, but rather give other useful information about the system. For example, suppose conditions change dramatically during a process upset. An experienced corrosion engineer can examine the data and correlate it with the upset conditions. Such analysis can provide insight into the process and help to improve performance and extend equipment lifetime. Changes in simple parameters such as pH, ion content, and temperature may lead to detection of a process upset. Without careful analysis, process upsets can reduce the corrosion lifetime of equipment and even cause a system failure.

Analysis of biological activity does not automatically lead directly to a corrosion-rate measurement. However, with detection and correlation with process conditions, such information may also lead to improvements in the corrosion lifetime of the process equipment.

Evidence of localized corrosion can be obtained from polarization methods such as potentiodynamic polarization, EIS, and electrochemical noise measurements, which are particularly well suited to providing data on localized corrosion. When evidence of localized attack is obtained, the engineer needs to perform a careful analysis of the conditions that may lead to such attack. Correlation with process conditions can provide additional data about the susceptibility of the equipment to localized attack and can potentially help prevent failures due to pitting or crevice corrosion. Since pitting may have a delayed initiation phase, careful consideration of the cause of the localized attack is critical. Laboratory testing and involvement of an experienced corrosion engineer may be needed to understand the initiation of localized corrosion. Theoretical constructs such as Pourbaix diagrams can be useful in interpreting data obtained by on-line monitors.

Combinations of several types of probes can improve the information concerning corrosion of the system. Using only probes that provide corrosion-rate data cannot lead to as complete an analysis of the process and its effect on the equipment as with monitoring probes, which provide additional information.

Other Useful Information Obtained by Probes Both EIS and electrochemical noise probes can be used to determine information about the reactions that affect corrosion. Equivalent circuit analysis, when properly applied by an experienced engineer, can often give insight into the specifics of the corrosion reactions. Information such as corrosion product layer buildup, or inhibitor effectiveness, or coating breakdown can be obtained directly from analysis of the data from EIS or indirectly from electrochemical noise data. In most cases, this is merely making use of methodology developed in the corrosion laboratory.

Some assumptions must be made to do this, but these assumptions are no different from those made in the laboratory. Recent experience involving in-plant usage of EIS has shown that this technique can be used effectively as a monitoring method and has lead to development of several commercial systems.

Making use of the information from monitoring probes, combined with the storage and analysis capabilities of portable computers and microprocessors, seems the best method for understanding corrosion processes. Commercial setups can be assembled from standard probes, cables, readout devices, and storage systems. When these are coupled with analysis by corrosion engineers, the system can lead to better a understanding of in-plant corrosion processes.

Limitations of Probes and Monitoring Systems There are limitations even with the most up-to-date systems. Some of the things which cannot be determined using corrosion probes include:

Specifics on the type of biological attack. This must be done by some other method such as chemical analysis of the solution (plus consideration given to limitations to the use of these several electrochemical techniques for MIC studies, noted previously under "Corrosion Testing: Laboratory Tests" and subsequent subsections).

Actual lifetime of the plant equipment. Corrosion monitoring provides data, which must then be analyzed with additional input and interpretation. However, only estimates can be made of the lifetime of the equipment of concern. Lifetime predictions are, at best, carefully crafted guesses based on the best available data.

Choices of alternative materials. Corrosion probes are carefully chosen to be as close as possible to the alloy composition, heat treatment, and stress condition of the material that is being monitored. Care must be taken to ensure that the environment at the probe matches the service environment. Choices of other alloys or heat treatments and other conditions must be made by comparison. Laboratory testing or coupon testing in the process stream can be used to examine alternatives to the current material, but the probes and the monitors can only provide information about the conditions which are present during the test exposure and cannot extrapolate beyond those conditions.

Failure analysis. Often, for a corrosion-related failure, data from the probes are examined to look for telltale signs that could have led to detection of the failure. In some cases, evidence can be found that process changes were occurring which led to the failure. This does not mean that the probes should have detected the failure itself. Determination of an imminent corrosion-related failure is not possible, even with the most advanced monitoring system.

The aforementioned limitations are not problems with the probes or the monitoring systems but occur when information is desired that cannot be measured directly or which requires extrapolation. Many of the problems that are encountered with corrosion-monitoring systems and probes are related to the use to which probes are put.

Potential Problems with Probe Usage Understanding the electrochemical principles upon which probes are based helps to eliminate some of the potential problems with probes. However, in some situations, the information desired is not readily available.

For example, consider localized corrosion. Although data from corrosion probes indicate corrosion rate, it is not possible to tell that localized corrosion is the problem.

This is an example of measuring the wrong thing. In this case, the probes work adequately, the monitoring system is adequate, as is the monitoring interval, but detection of the type of corrosion cannot be made based on the available data. Different types of probes and testing are required to detect the corrosion problem.

Another problem can occur if the probes and monitors are working properly but the probe is placed improperly. Then the probe does not measure the needed conditions and environment. The data obtained by the probe will not tell the whole story.

An example of this is in a condenser where the corrosion probe is in a region where the temperature is lower than that at the critical condition of interest. Local scale buildup is another example of this type of situation, as is formation of a crevice at a specific location.

The type of probe, its materials, and method of construction must be carefully considered in designing an effective corrosion-monitoring system. Since different types of probes provide different types of information, it may be necessary to use several types.

Incorrect information can result if the probe is made of the wrong material and is not heat treated in the same way as the process equipment (as well as because of other problems). The probe must be as close as possible to the material from which the equipment of interest is made. Existence of a critical condition, such as weldments or galvanic couples or occluded cells in the equipment of concern, makes the fabrication, placement, and maintenance of the probes and monitoring system of critical importance, if accurate and useful data are to be obtained.

Before electrochemical techniques are used in the evaluation of any given, *viable* MIC environment/metal system, such a test protocol should receive considerable review by personnel quite experienced in *both* electrochemical testing and microbiologically influenced corrosion.

The data obtained from probes and monitoring systems are most useful when analyzed by a corrosion specialist. Data not taken, analyses not made, or expertise not sought can quickly lead to problems even with the most up-to-date corrosion probes and monitoring system.

ECONOMICS IN MATERIALS SELECTION

In most instances, there will be more than one alternative material which may be considered for a specific application. Calculation of true long-term costs requires estimation of the following:
1. Total cost of fabricated equipment and piping
2. Total installation cost
3. Service life
4. Maintenance costs: amount and timing
5. Time and cost requirements to replace or repair at the end of service life
6. Cost of downtime to replace or repair
7. Cost of inhibitors, extra control facilities, and so on, required to assure achievement of predicted service life
8. Time value of money
9. Factors which impact taxation, such as depreciation and tax rates
10. Inflation rate

Proper economic analysis will allow comparison of alternatives on a sound basis. Detailed calculations are beyond the scope of this section. The reader should review the material in the NACE Publication 3C194, Item No. 24182, "Economics of Corrosion," Sept., 1994.

PROPERTIES OF MATERIALS

GUIDE TO TABULATED DATA

Note from the Sec. 28 editors to the readers of this handbook: Historically, previous editions of *Perry's Chemical Engineers' Handbook* carried an extensive series of so-called corrosion resistance tables [listing recommended materials of construction (MOC) versus various corrosive environments]. This practice goes back, at least, to the Materials of Construction Sec. 18, 1941, 2d ed. Unfortunately, if valid at all, *these data are only usable as indicators of what will* **not** *work; for sure, these listings should* **not** *be used as recommendations of what materials are corrosion resistant.* The section editors have elected to no longer include these data tabulations.

Beyond the simple resistance of a material of construction to dissolution in a given chemical, many other properties enter into consideration when making an appropriate or optimum MOC selection for a given environmental exposure. These factors include the influence of velocity, impurities or contaminants, pH, stress, crevices, bimetallic couples, levels of nuclear, UV, or IR radiation, microorganisms, temperature heat flux, stray currents, properties associated with original production of the material and its subsequent fabrication as an item of equipment, as well as other physical and mechanical properties of the MOC, the *Proverbial Siebert Changes in the Phase of the Moon,* and so forth.

Therefore, the seventh edition no longer includes these endless tabulations of data; rather, there is an extensive coverage of corrosion mechanisms, the manner in which these various factors effect the corrosion system, as well as more detail as to the testing protocol necessary to assist in a sound MOC selection.

A few collections of more generic information as to the overall acid and alkali resistance of broad classes of materials remain. These are only intended to be used as *indicators* of the tendencies of these MOC to react; they are not included as a substitute for the application of good, sound engineering evaluations.

MATERIALS STANDARDS AND SPECIFICATIONS

There are obvious benefits to be derived from consensus standards which define the chemistry and properties of specific materials. Such standards allow designers and users of materials to work with confidence that the materials supplied will have the expected minimum properties. Designers and users can also be confident that comparable materials can be purchased from several suppliers. Producers are confident that materials produced to an accepted standard will find a ready market and therefore can be produced efficiently in large factories.

While a detailed treatment is beyond the scope of this section, a few of the organizations which generate standards of major importance to the chemical-process industries in the United States are listed here. An excellent overview is presented in the *Encyclopedia of Chemical Technology* (3d ed., Wiley, New York, 1978–1980).
1. **American National Standards Institute** (**ANSI**), formerly American Standards Association (ASA). ANSI promulgates the piping codes used in the chemical-process industries.
2. **American Society of Mechanical Engineers** (**ASME**). This society generates the Boiler and Pressure Vessel Codes.
3. **American Society for Testing and Materials** (**ASTM**). This society generates specifications for most of the materials used in the ANSI Piping Codes and the ASME Boiler and Pressure Vessel Codes.
4. **International Organization for Standardization** (**ISO**). This organization is engaged in generating standards for worldwide use. It has 80 member nations.

FERROUS METALS AND ALLOYS

Steel   Carbon steel is the most common, cheapest, and most versatile metal used in industry. It has excellent ductility, permitting many cold-forming operations. Steel is also very weldable.

The grades of steel most commonly used in the chemical-process industries have tensile strength in the 345- to 485-MPa (50,000 to 70,000-lbf/in²) range, with good ductility. Higher strength levels are achieved by cold work, alloying, and heat treatment.

Carbon steel is easily the most commonly used material in process plants despite its somewhat limited corrosion resistance. It is routinely used for most organic chemicals and neutral or basic aqueous solutions at moderate temperatures. It is also used routinely for the storage of concentrated sulfuric acid and caustic soda [up to 50 percent and 55°C (130°F)]. Because of its availability, low cost, and ease of fabrication steel is frequently used in services with corrosion rates of 0.13 to 0.5 mm/y (5 to 20 mils/y), with added thickness (**corrosion allowance**) to assure the achievement of desired service life. Product quality requirements must be considered in such cases.

Low-Alloy Steels Alloy steels contain one or more alloying agents to improve mechanical and corrosion-resistant properties over those of carbon steel.

A typical low-alloy grade [American Iron and Steel Institute (AISI) 4340] contains 0.40 percent C, 0.70 percent Mn, 1.85 percent Ni, 0.80 percent Cr, and 0.25 percent Mo. Many other alloying agents are used to produce a large number of standard AISI and proprietary grades.

Nickel increases toughness and improves low-temperature properties and corrosion resistance. Chromium and silicon improve hardness, abrasion resistance, corrosion resistance, and resistance to oxidation. Molybdenum provides strength at elevated temperatures.

The addition of small amounts of alloying materials greatly improves corrosion resistance to atmospheric environments but does not have much effect against liquid corrosives. The alloying elements produce a tight, dense adherent rust film, but in acid or alkaline solutions corrosion is about equivalent to that of carbon steel. However, the greater strength permits thinner walls in process equipment made from low-alloy steel.

Cast Irons Generally, cast iron is not a particularly strong or tough structural material, although it is one of the most economical and is widely used industrially.

Gray cast iron, low in cost and easy to cast into intricate shapes, contains carbon, silicon, managanese, and iron. Carbon (1.7 to 4.5 percent) is present in combined carbon and graphite; combined carbon is dispersed in the matrix as iron carbide (cementite), while free graphite occurs as thin flakes dispersed throughout the body of the metal. Various strengths of gray iron are produced by varying size, amount, and distribution of graphite.

Gray iron has outstanding damping properties—that is, ability to absorb vibration—as well as wear resistance. However, gray iron is brittle, with poor resistance to impact and shock. Machinability is excellent.

With some important exceptions, gray-iron castings generally have corrosion resistance similar to that of carbon steel. They do resist atmospheric corrosion as well as attack by natural or neutral waters and neutral soils. However, dilute acids and acid-salt solutions will attack this material.

Gray iron is resistant to concentrated acids (nitric, sulfuric, phosphoric) as well as to some alkaline and caustic solutions. Caustic fusion pots are usually made from gray cast iron with low silicon content; cast-iron valves, pumps, and piping are common in sulfuric acid plants.

White cast iron is brittle and difficult to machine. It is made by controlling the composition and rate of solidification of the molten iron so that all the carbon is present in the combined form. Very abrasive- and wear-resistant, white cast iron is used as liners and for grinding balls, dies, and pump impellers.

Malleable iron is made from white cast iron. It is cast iron with free carbon as dispersed nodules. This arrangement produces a tough, relatively ductile material. Total carbon is about 2.5 percent. Two types are produced: standard and pearlitic (combined carbon plus nodules). Standard malleable iron is easily machined; pearlitic, less so. Both types will withstand bending and cold working without cracking. Large welded areas are not recommended with fusion welding because welds are brittle. Corrosion resistance is about the same as for gray cast iron.

Ductile cast iron includes a group of materials with good strength, toughness, wear resistance, and machinability. This type of cast iron contains combined carbon and dispersed nodules of carbon. Composition is about the same as gray iron, with more carbon (3.7 percent) than malleable iron. The spheroidal graphite reduces the notch effect produced by graphite flakes, making the material more ductile.

There are a number of grades of ductile iron; some have maximum toughness and machinability; others have maximum resistance to oxidation.

Generally, corrosion resistance is similar to gray iron. But ductile iron can be used at higher temperatures—up to 590°C (1,100°F) and sometimes even higher.

Alloy Cast Irons Cast iron is not usually considered corrosion-resistant, but this condition can be improved by the use of various cast-iron alloys. A number of such materials are commercially available.

High-silicon cast irons have excellent corrosion resistance. Silicon content is 13 to 16 percent. This material is known as Durion. Adding 4 percent Cr yields a product called Durichlor, which has improved resistance in the presence of oxidizing agents. These alloys are not readily machined or welded.

Silicon irons are very resistant to oxidizing and reducing environments, and resistance depends on the formation of a passive film. These irons are widely used in sulfuric acid service, since they are unaffected by sulfuric at all strengths, even up to the boiling point.

Because they are very hard, silicon irons are good for combined corrosion-erosion service.

Another group of cast-iron alloys are called **Ni-Resist.** These materials are related to gray cast iron in that they have high carbon contents (3 percent), with fine graphite flakes distributed throughout the structure. Nickel contents range from 13.5 to 36 percent, and some have 6.5 percent Cu.

Generally, nickel-alloy castings have superior toughness and impact resistance compared with gray irons. The nickel-alloy castings can be welded and machined.

Corrosion resistance of nickel alloys is superior to that of cast irons but less than that of pure nickel. There is little attack from neutral or alkaline solutions. Oxidizing acids such as nitric are highly detrimental. Cold, concentrated sulfuric acid can be handled.

Ni-Resist has excellent heat resistance, with some grades serviceable up to 800°C (1,500°F). Also, a ductile variety is available, as well as a hard variety (Ni-Hard).

Stainless Steel There are more than 70 standard types of stainless steel and many special alloys. These steels are produced in the wrought form (AISI types) and as cast alloys [Alloy Casting Institute (ACI) types]. Generally, all are iron-based, with 12 to 30 percent chromium, 0 to 22 percent nickel, and minor amounts of carbon, niobium (columbium), copper, molybdenum, selenium, tantalum, and titanium. These alloys are very popular in the process industries. They are heat- and corrosion-resistant, noncontaminating, and easily fabricated into complex shapes.

There are three groups of stainless alloys: (1) martensitic, (2) ferritic, and (3) austenitic.

The **martensitic alloys** contain 12 to 20 percent chromium with controlled amounts of carbon and other additives. Type 410 is a typical member of this group. These alloys can be hardened by heat treatment, which can increase tensile strength from 550 to 1,380 MPa (80,000 to 200,000 lbf/in²).

Corrosion resistance is inferior to that of austenitic stainless steels, and martensitic steels are generally used in mildly corrosive environments (atmospheric, fresh water, and organic exposures).

Ferritic stainless contains 15 to 30 percent Cr, with low carbon content (0.1 percent). The higher chromium content improves its corrosive resistance. Type 430 is a typical example. The strength of ferritic stainless can be increased by cold working but not by heat treatment. Fairly ductile ferritic grades can be fabricated by all standard methods. They are fairly easy to machine. Welding is not a problem, although it requires skilled operators.

Corrosion resistance is rated good, although ferritic alloys are not good against reducing acids such as HCl. But mildly corrosive solutions and oxidizing media are handled without harm. Type 430 is widely used in nitric acid plants. In addition, it is very resistant to scaling and high-temperature oxidation up to 800°C (1,500°F).

TABLE 28-2   General Corrosion Properties of Some Metals and Alloys*

Ratings:
0: Unsuitable. Not available in form required or not suitable for fabrication requirements or not suitable for corrosion conditions.
1: Poor to fair
2: Fair. For mild conditions or when periodic replacement is possible. Restricted use.
3: Fair to good.
4: Good. Suitable when superior alternatives are uneconomic.
5: Good to excellent.
6: Normally excellent.
Small variations in service conditions may appreciably affect corrosion resistance. Choice of materials is therefore guided wherever possible by a combination of experience and laboratory and site tests.

	Liquids													Gases		
	Nonoxidizing or reducing media				Oxidizing media			Natural waters				Steam		Common industrial media		
			Alkaline solutions, e.g.					Freshwater supplies		Seawater				Furnace gases with incidental sulfur content		
Materials	Acid solutions, excluding hydrochloric e.g. phosphoric, sulfuric, most conditions, many organics	Neutral solutions, e.g., many nonoxidizing salt solutions, chlorides, sulfates	Caustic and mild alkalies, excluding ammonium hydroxide	Ammonium hydroxide and amines	Acid solutions, e.g., nitric	Neutral or alkaline solutions, e.g., per-sulfates, peroxides, chromates	Pitting media,† acid ferric chloride solutions	Static or slow-moving	Turbulent	Static or slow-moving	Turbulent	Moist, con-densate	Dry at high temperature, promoting slight dissociation	Reducing, e.g., heat-treatment furnace gases	Oxidizing, e.g., flue gases	Ambient air, city or industrial
Cast iron, flake graphite, plain or low-alloy	1	3	4	5	0	4	0	4	3	4	2	4	4	1	1	3
Ductile iron (higher strength and hardness may be attained by composition and heat treatment or both)	1	3	4	5	0	4	0	4	4	4	3	4	4	1	1	3
Ni-Resist corrosion-resistant cast irons	4	5	5	5		5	0	5	5	5	5	5	5	3	2	4
14% silicon iron	6	6	2	5	6	6	3	5	5	5	5	6	4	4	3	6
Mild steel, also low-alloy irons and steels	1	3	4	5	0	4	0	4	3	4	2	4	4	1	1	3
Stainless steel, ferritic 17% Cr type	2	4	4	6	5	6	0	4	6	1	4	5	6	3	2	4
Stainless steel, austenitic 18 Cr: 8 Ni type	3	4	5	6	6	6	0	6	6	2	5	6	6	2	3	5

Material														
Stainless steel, austenitic 18 Cr; 12 Ni; 2.5 Mo type	6	4	2	6	5	3	6	6	1	6	5	6	5	5	4
Stainless steel, austenitic 20 Cr; 29 Ni; 2.5 Mo; 3.5 Cu type	6	4	2	6	6	4	6	6	2	6	5	6	5	6	5
Incoloy 825 nickel-iron-chromium alloy (40 Ni; 21 Cr; 3 Mo; 1.5 Cu; balance Fe)	6	5	2	6	6	4	6	6	2	6	5	6	5	6	6
Hastelloy alloy C-276 (55 Ni; 17 Mo; 16 Cr; 6 Fe; 4 W)	6	4	3	6	6	6	6	6	5	6	4	6	5	6	5
Hastelloy alloy B-2 (61 Ni; 28 Mo; 6 Fe)	5	2	3	5	4	4	6	6	0	3	0	4	4	5	6
Inconel 600 (78 Ni; 15 Cr; 7 Fe)	6	4	2	6	6	4	6	6	1	6	3	6	6	6	3
Copper-nickel alloys up to 30% nickel	5	2	2	5	6	6	6	6	1	4	0	0	5	5	4
Monel 400 nickel-copper alloy (66 Ni; 30 Cu; 2 Fe)	5	3	2	6	6	4	6	6	1	5	0	1	6	6	5
Nickel 200—commercial (99.4 Ni)	4	2	2	6	5	3	6	6	0	5	0	1	6	5	4
Copper and silicon bronze	5	2	2	5	1	4	5	6	0	4	0	0	4	4	4
Aluminum brass (76 Cu; 22 Zn; 2 Al)	5	2	2	5	5	4	6	6	0	3	0	0	3	4	3
Nickel-aluminum bronze (80 Cu; 10 Al; 5 Ni; 5 Fe)	5	3	2	5	5	4	6	6	0	3	0	0	3	4	4
Bronze, type A (88 Cu; 5 Sn; 5 Ni; 2 Zn)	5	2	2	5	5	5	6	6	0	4	0	0	4	5	4
Aluminum and its alloys	5	4	5	2	4	0–5	5	4	0	0–4	0–5	6	0	3	1
Lead, chemical or antimonial	5	3	4	0	3	5	5	6	0	2	0	2	2	5	5
Silver	4	4	4	5	5	5	6	6	6	2	6	6	6	6	4
Titanium	6	5	3	5	6	6	6	6	6	6	6	6	2	6	3
Zirconium	6	5	3	6	6	6	6	6	2	6	6	6	2	6	3

TABLE 28-2 General Corrosion Properties of Some Metals and Alloys* (Concluded)

Ratings:
0: Unsuitable. Not available in form required or not suitable for fabrication requirements or not suitable for corrosion conditions.
1: Poor to fair.
2: Fair. For mild conditions or when periodic replacement is possible. Restricted use.
3: Fair to good.
4: Good. Suitable when superior alternatives are uneconomic.
5: Good to excellent.
6: Normally excellent.
Small variations in service conditions may appreciably affect corrosion resistance. Choice of materials is therefore guided wherever possible by a combination of experience and laboratory and site tests.

Material	Gases (continued)				Available forms	Cold formability in wrought and clad form	Weldability	Maximum strength annealed condition × 1000 lb/in²	Coefficient of thermal expansion, millionths per °F, 70–212°F	Remarks¶
	Halogens and derivatives									
	Halogens		Halide acids, moist, e.g., hydrochloric hydrolysis products of organic halides	Hydrogen halides, dry, e.g., dry hydrogen chloride, °F						
	Moist, e.g., chlorine below dew point	Dry, e.g., fluorine above dew point								
Cast iron, flake graphite, plain or low alloy	0	2	0	2 < 400 1 < 750	Cast	No	Fair§	45	6.7	
Ductile iron (higher strength and hardness may be attained by composition and heat treatment or both)	0	2	0	2 < 400 1 < 750	Cast	No	Good§	67	7.5	
Ni-Resist corrosion-resistant cast irons	0	2	3	3 < 400 2 < 750	Cast	No	Good§	22–31	10.3	
Durion—14% silicon iron	0	0	4	1 < 400	Cast	No	No	22	7.4	Very brittle, susceptible to cracking by mechanical and thermal shock
Mild steel, also low-alloy irons and steels	0	3	0	3 < 400 1 < 750	Wrought, cast	Good	Good	67	6.7	High strengths obtainable by alloying, also improved atmospheric corrosion resistance. See ASTM specifications for particular grade
Stainless steel, ferritic 17% Cr type	0	2	0	2 < 400	Wrought, cast, clad	Good	Good§	78	6.0	AISI type 430 ASTM corrosion- and heat-resisting steels
Stainless steel, austenitic 18 Cr; 8 Ni types	0	2	0	3 < 400	Wrought, cast, clad	Good	Good	90	9.6	AISI type 304 ASTM corrosion- and heat-resisting steels; stabilized or LC types used for welding
Stainless steel, austenitic 18 Cr; 12 Ni; 2.5 Mo type	0	3	2	4 < 400 3 < 750	Wrought, cast, clad	Good	Good	90	8.9	AISI type 316 ASTM corrosion- and heat-resisting steel; LC type used for welding

Material				Temperature	Form					Remarks
Stainless steel, austenitic 20 Cr; 29 Ni; 2.5 Mo; 3.5 Cu type	1	3	3	4 < 400; 3 < 750	Wrought, cast	Good	Good	90	9.4	ACI CH-7M: good resistance to sulfuric, phosphoric, and fatty acids at elevated temperatures
Incoloy 825 nickel-iron-chromium alloy (40 Ni; 21 Cr; 3 Mo; 1.5 Cu; bal. Fe)	2	3	3	4 < 400; 3 < 750	Wrought, cast, clad	Good	Good	100	7.3	Special alloy with good resistance to sulfuric, phosphoric, and fatty acids; resistant to chlorides in some environments
Hastelloy alloy C-276 (55 Ni; 17 Mo; 16 Cr; 6 Fe; 4 W)	5	4	4	4 < 750; 3 < 900	Wrought, cast, clad	Fair	Good	145	6.3	Excellent resistance to wet chlorine gas and sodium hypochlorite solutions
Hastelloy alloy B-2 (61 Ni; 28 Mo; 6 Fe)	1	3	5	4 < 750; 3 < 900	Wrought, cast, clad	Fair	Good	135	5.6	Resistant to solutions of hydrochloric and sulfuric acids
Inconel 600 (78 Ni; 15 Cr; 7 Fe)	2	5	3	5 < 400; 4 < 900	Wrought, cast, clad	Good	Good	90	8.9	Wide application in food and pharmaceutical industries
Copper-nickel alloys up to 30% nickel	1	5	2	4 < 400; 3 < 750	Wrought, cast, clad	Good	Good	38–62	9.3–8.5	High-iron types excellent for resisting high-velocity effects in condenser tubes
Monel 400 nickel-copper alloy (66 Ni; 30 Cu; 2 Fe)	2	6	3	6 < 400; 3 < 750; 2 < 900	Wrought, cast, clad	Good	Good	77	7.5	Widely used for sulfuric acid pickling equipment; also for propeller shafts in motor boats; precautions needed to avoid sulfur attack during fabrication
Nickel 200—commercial (99.4 Ni)	2	6	2	6 < 400; 5 < 750; 4 < 900	Wrought, cast, clad	Good	Good	54	6.6	Widely used for hot concentrated caustic solutions; precautions needed to avoid sulfur attack during fabrication
Copper and silicon bronze	0	5	2	3 < 400; 2 < 750	Wrought, cast, clad	Excellent	Fair	29	9.3–9.5	Unsuitable for hot concentrated mineral acids or for high-velocity HF
Aluminum brass (76 Cu; 22 Zn; 2 Al)	0	4	2	2 < 400	Wrought, cast	Good	Fair	60	10.3	Possibility of developing localized corrosion in seawater
Nickel-aluminum bronze (80 Cu; 10 Al; 5 Ni; 5 Fe)	0	4	3	3 < 400; 2 < 750	Wrought, cast	Good	Fair	60–80	9.4	Ship propellers an excellent application
Bronze, type A (88 Cu; 5 Sn; 5 Ni; 2 Zn)	0	4	3	3 < 400; 2 < 750	Cast	No	§	45	11.0	High strengths obtained by heat treatment; not susceptible to dezincification
Aluminum and its alloys	0	6	0	3 < 400; 1 < 750	Wrought, cast, clad	Good	Good	9–90	11.5–13.7	Extent of corrosion dependent upon type and concentration of acidic ions; wide range of mechanical properties obtainable by alloying and heat treatment
Lead, chemical or antimonial	0	1	3	0	Wrought, cast, clad	Excellent	Good	2	16.4–15.1	High purity "chemical lead" preferred for most applications
Silver	5	5	3	4 < 400; 2 < 750	Wrought, cast, clad	Excellent	Good	21	10.6	Used as a lining
Titanium	6	0	1	0	Wrought, cast	Fair	Good§	6–90	5.0	Possibility of red fuming HNO_3, initiating explosions; good resistance to solutions containing chlorides
Zirconium	6	1	6	0	Wrought, cast	Fair	Good§			

*Data courtesy of International Nickel Co.

†On unsuitable materials these media may promote potentially dangerous pitting.

‡Temperatures are approximate.

§Special precautions required.

¶Many of these materials are suitable for resisting dry corrosion at elevated temperatures.

TABLE 28-3 Unified Alloy Numbering System (UNS)

UNS was established in 1974 by ASTM and SAE to reduce the confusion involved in the labeling of commercial alloys. Metals have been placed into 15 groups, each of which is given a code letter. The specific alloy is identified by a five-digit number following this code letter.

Nonferrous metals and alloys	
A00001–A99999	Aluminum and aluminum alloys
C00001–C99999	Copper and copper alloys
E00001–E99999	Rare-earth and rare-earth-like metals and alloys
L00001–L99999	Low-melting metals and alloys
M00001–M99999	Miscellaneous nonferrous metals and alloys
N00001–N99999	Nickel and nickel alloys
P00001–P99999	Precious metals and alloys
R00001–R99999	Reactive and refractory metals and alloys

Ferrous metals and alloys	
D00001–D99999	Specified-mechanical-properties steels
F00001–F99999	Cast irons and cast steels
G00001–G99999	AISI and SAE carbon and alloy steels
H00001–H99999	AISI H steels
K00001–K99999	Miscellaneous steels and ferrous alloys
S00001–S99999	Heat- and corrosion-resistant (stainless) steels
T00001–T99999	Tool steels

When possible, earlier widely used three- or four-digit alloy numbering systems such as those developed by the Aluminum Association (AA), Copper Development Association (CDA), American Iron and Steel Institute (AISI), etc., have been incorporated by the addition of the appropriate alloy-group code letter plus additional digits. For example:

	Former designation		
Alloy description	System	No.	UNS designation
Aluminum + 1.2% Mn	AA	3003	A93003
Copper, electrolytic tough pitch	CDA	110	C11000
Carbon steel, 0.2% C	AISI	1020	G10200
Stainless steel, 18 Cr, 8 Ni	AISI	304	S30400

Proprietary alloys are assigned numbers by the AA, AISI, CDA, ASTM, and SAE, which maintains master listings at their headquarters. Handbooks describing the system are available. (Cf. ASTM publication DS-56AC.)

SOURCE: ASTM DS-56A. (*Courtesy of National Association of Corrosion Engineers.*)

TABLE 28-4 Coefficient of Thermal Expansion of Common Alloys*

	UNS	10^{-6} in/ (in·°F)	10^{-6} mm/ (mm·°C)	Temperature range, °C
Aluminum alloy AA1100	A91100	13.1	24.	20–100
Aluminum alloy AA5052	A95052	13.2	24.	20–100
Aluminum cast alloy 43	A24430	12.3	22.	20–100
Copper	C11000	9.4	16.9	20–100
Red brass	C23000	10.4	18.7	20–300
Admiralty brass	C44300	11.2	20.	20–300
Muntz Metal	C28000	11.6	21.	20–300
Aluminum bronze D	C61400	9.0	16.2	20–300
Ounce metal	C83600	10.2	18.4	0–100
90-10 copper nickel	C70600	9.5	17.1	20–300
70-30 copper nickel	C71500	9.0	16.2	20–300
Carbon steel, AISI 1020	G10200	6.7	12.1	0–100
Gray cast iron	F10006	6.7	12.1	0–100
4-6 Cr, ½ Mo steel	S50100	7.3	13.1	20–540
Stainless steel, AISI 410	S41000	6.1	11.0	0–100
Stainless steel, AISI 446	S44600	5.8	10.4	0–100
Stainless steel, AISI 304	S30400	9.6	17.3	0–100
Stainless steel, AISI 310	S31000	8.0	14.4	0–100
Stainless steel, ACI HK	J94224	9.4	16.9	20–540
Nickel alloy 200	N02200	7.4	13.3	20–90
Nickel alloy 400	N04400	7.7	13.9	20–90
Nickel alloy 600	N06600	7.4	13.3	20–90
Nickel-molybdenum alloy B-2	N10665	5.6	10.1	20–90
Nickel-molybdenum alloy C-276	N10276	6.3	11.3	20–90
Titanium, commercially pure	R50250	4.8	8.6	0–100
Titanium alloy T1-6A1-4V	R56400	4.9	8.8	0–100
Magnesium alloy AZ31B	M11311	14.5	26.	20–100
Magnesium alloy AZ91C	M11914	14.5	26.	20–100
Chemical lead		16.4	30.	0–100
50-50 solder	L05500	13.1	24.	0–100
Zinc	Z13001	18.	32.	0–100
Tin	L13002	12.8	23.	0–100
Zirconium	R60702	2.9	5.2	0–100
Molybdenum	R03600	2.7	4.9	20–100
Tantalum	R05200	3.6	6.5	20–100

*Courtesy of National Association of Corrosion Engineers.

Austenitic stainless steels are the most corrosion-resistant of the three groups. These steels contain 16 to 26 percent chromium and 6 to 22 percent nickel. Carbon is kept low (0.08 percent maximum) to minimize carbide precipitation. These alloys can be work-hardened, but heat treatment will not cause hardening. Tensile strength in the annealed condition is about 585 MPa (85,000 lbf/in²), but work-hardening can increase this to 2,000 MPa (300,000 lbf/in²). Austenitic stainless steels are tough and ductile.

They can be fabricated by all standard methods. But austenitic grades are not easy to machine; they work-harden and gall. Rigid machines, heavy cuts, and high speeds are essential. Welding, however, is readily performed, although welding heat may cause chromium carbide precipitation, which depletes the alloy of some chromium and lowers its corrosion resistance in some specific environments, notably nitric acid. The carbide precipitation can be eliminated by heat treatment (solution annealing). To avoid precipitation, special stainless steels stabilized with titanium, niobium, or tantalum have been developed (types 321, 347, and 348). Another approach to the problem is the use of low-carbon steels such as types 304L and 316L, with 0.03 percent maximum carbon.

The addition of molybdenum to the austenitic alloy (types 316, 316L, 317, and 317L) provides generally better corrosion resistance and improved resistance to pitting.

In the stainless group, nickel greatly improves corrosion resistance over straight chromium stainless. Even so, the chromium-nickel steels, particularly the 18-8 alloys, perform best under **oxidizing conditions,** since resistance depends on an oxide film on the surface of the alloy. Reducing conditions and chloride ions destroy this film and bring on rapid attack. Chloride ions tend to cause pitting and crevice corrosion; when combined with high tensile stresses, they can cause stress-corrosion cracking.

Cast stainless alloys are widely used in pumps, valves, and fittings. These casting alloys are designated under the ACI system. All corrosion-resistant alloys have the letter C plus a second letter (A to N) denoting increasing nickel content. Numerals indicate maximum carbon. While a rough comparison can be made between ACI and AISI types, compositions are not identical and analyses cannot be used interchangeably. Foundry techniques require a rebalancing of the wrought chemical compositions. However, corrosion resistance is not greatly affected by these composition changes. Typical members of this group are CF-8, similar to type 304 stainless; CF-8M, similar to type 316; and CD-4M Cu, which has improved resistance to nitric, sulfuric, and phosphoric acids.

In addition to the C grades, there is a series of heat-resistant grades of ACI cast alloys, identified similarly to the corrosion-resistant grades, except that the first letter is H rather than C. Mention should

TABLE 28-5   Melting Temperatures of Common Alloys*

		Melting range	
	UNS	°F	°C
Aluminum alloy AA1100	A91100	1190–1215	640–660
Aluminum alloy AA5052	A95052	1125–1200	610–650
Aluminum cast alloy 43	A24430	1065–1170	570–630
Copper	C11000	1980	1083
Red brass	C23000	1810–1880	990–1025
Admiralty brass	C44300	1650–1720	900–935
Muntz Metal	C28000	1650–1660	900–905
Aluminum bronze D	C61400	1910–1940	1045–1060
Ounce metal	C83600	1510–1840	854–1010
Manganese bronze	C86500	1583–1616	862–880
90-10 copper nickel	C70600	2010–2100	1100–1150
70-30 copper nickel	C71500	2140–2260	1170–1240
Carbon steel, AISI 1020	G10200	2760	1520
Gray cast iron	F10006	2100–2200	1150–1200
4-6 Cr, ½ Mo Street	S50100	2700–2800	1480–1540
Stainless steel, AISI 410	S41000	2700–2790	1480–1530
Stainless steel, AISI 446	S44600	2600–2750	1430–1510
Stainless steel, AISI 304	S30400	2550–2650	1400–1450
Stainless steel, AISI 310	S31000	2500–2650	1400–1450
Stainless steel, ACI HK	J94224	2550	1400
Nickel alloy 200	N02200	2615–2635	1440–1450
Nickel alloy 400	N04400	2370–2460	1300–1350
Nickel alloy 600	N06600	2470–2575	1350–1410
Nickel-molybdenum alloy B-2	N10665	2375–2495	1300–1370
Nickel-molybdenum alloy C-276	N10276	2420–2500	1320–1370
Titanium, commercially pure	R50250	3100	1705
Titanium alloy T1-6A1-4V	R56400	2920–3020	1600–1660
Magnesium alloy AZ 31B	M11311	1120–1170	605–632
Magnesium alloy HK 31A	M13310	1092–1204	589–651
Chemical lead		618	326
50-50 solder	L05500	361–421	183–216
Zinc	Z13001	787	420
Tin	Z13002	450	232
Zirconium	R60702	3380	1860
Molybdenum	R03600	4730	2610
Tantalum	R05200	5425	2996

*Courtesy of National Association of Corrosion Engineers.

also be made of precipitation-hardening (PH) stainless steels, which can be hardened by heat treatments at moderate temperatures. Very strong and hard at high temperatures, these steels have but moderate corrosion resistance. A typical PH steel, containing 17 percent Cr, 7 percent Ni, and 1.1 percent Al, has high strength, good fatigue properties, and good resistance to wear and cavitation corrosion. A large number of these steels with varying compositions are commercially available. Essentially, they contain chromium and nickel with added alloying agents such as copper, aluminum, beryllium, molybdenum, nitrogen, and phosphorus.

Medium Alloys   A group of (mostly) proprietary alloys with somewhat better corrosion resistance than stainless steels are called medium alloys. A popular member of this group is the **20 alloy,** made by a number of companies under various trade names. Durimet 20 is a well-known cast version, containing 0.07 percent C, 29 percent Ni, 20 percent Cr, 2 percent Mo, and 3 percent Cu. The ACI designation of this alloy is CN-7M. A wrought form is known as Carpenter 20 (Cb3). Worthite is another proprietary 20 alloy with about 24 percent Ni and 20 percent Cr. The 20 alloy was originally developed to fill the need for a material with sulfuric acid resistance superior to the stainless steels.

Other members of the medium-alloy group are **Incoloy 825** and **Hastelloy G-3 and G-30.** Wrought Incoloy 825 has 40 percent Ni, 21 percent Cr, 3 percent Mo, and 2.25 percent Cu. Hastelloy G-3 contains 44 percent Ni, 22 percent Cr, 6.5 percent Mo, and 0.05 percent C maximum.

These alloys have extensive applications in sulfuric acid systems. Because of their increased nickel and molybdenum contents they are more tolerant of chloride-ion contamination than standard stainless steels. The nickel content decreases the risk of stress-corrosion cracking; molybdenum improves resistance to crevice corrosion and pitting.

High Alloys   The group of materials called high alloys all contain relatively large percentages of nickel. **Hastelloy B-2** contains 61 percent Ni and 28 percent Mo. It is available in wrought and cast forms. Work hardening presents some fabrication difficulties, and machining is somewhat more difficult than for type 316 stainless. Conventional welding methods can be used. The alloy has unusually high resistance to all concentrations of hydrochloric acid at all temperatures in the absence of oxidizing agents. Sulfuric acid attack is low for all concentrations at 65°C (150°F), but the rate goes up with temperature. Oxidizing acids and salts rapidly corrode Hastelloy B. But alkalies and alkaline solutions cause little damage.

Chlorimet 2 has 63 percent Ni and 32 percent Mo and is somewhat similar to Hastelloy B-2. It is available only in cast form, mainly as valves and pumps. This is a tough alloy, very resistant to mechanical and thermal shock. It can be machined with carbide-tipped tools and welded with metal-arc techniques.

Hastelloy C-276 is a nickel-based alloy containing chromium (15.5 percent), molybdenum (15.5 percent), and tungsten (3 percent) as major alloying elements. It is available in wrought form. This alloy is a low-impurity modification of Hastelloy C, which is still available in cast form. The low impurity level substantially reduces the risk of intergranular corrosion of grain-boundary precipitation in weld-heat-affected zones. This alloy is resistant to strong oxidizing chloride solutions, such as wet chlorine and hypochlorite solutions. It is one of the very few alloys which are totally resistant to seawater.

Hastelloy C-4 is almost totally immune to selective intergranular corrosion in weld-heat-affected zones with high temperature stability in the 650–1040°C (1200–1900°F) range; Hastelloy C-22 has better overall corrosion resistance and versatility than either C-4 or C-276 (in most environments).

Chlorimet 3 is an alloy, available only in cast form, which is similar in alloy content and corrosion resistance to Hastelloy C.

Inconel 600 (80 percent Ni, 16 percent Cr, and 7 percent Fe) should also be mentioned as a high alloy. It contains no molybdenum. The corrosion-resistant grade is recommended for reducing-oxidizing environments, particularly at high temperatures. When heated in air, this alloy resists oxidation up to 1,100°C (2,000°F). The alloy is outstanding in resisting corrosion by gases when these gases are essentially sulfur-free.

The alloys discussed are typical examples of the large number of proprietary high alloys used in the chemical industry. For more comprehensive lists and data, refer to the listed references.

NONFERROUS METALS AND ALLOYS

Nickel and Nickel Alloys   Nickel is available in practically any mill form as well as in castings. It can be machined easily and joined by welding. Generally, oxidizing conditions favor corrosion, while reducing conditions retard attack. Neutral alkaline solutions, seawater, and mild atmospheric conditions do not affect nickel. The metal is widely used for handling alkalies, particularly in concentrating, storing, and shipping high-purity caustic soda. Chlorinated solvents and phenol are often refined and stored in nickel to prevent product discoloration and contamination.

A large number of nickel-based alloys are commercially available. Many have been mentioned in the preceding discussion of alloy castings and high alloys. One of the best known of these is **Monel 400,** 67 percent Ni and 30 percent Cu. It is available in all standard forms. This nickel-copper alloy is ductile and tough and can be readily fabricated and joined. Its **corrosion resistance** is generally superior to that of its components, being more resistant than nickel in reducing environments and more resistant than copper in oxidizing environments. The alloy can be used for relatively dilute sulfuric acid (below 80 percent), although aeration will result in increased corrosion. Monel will handle hydrofluoric acid up to 92 percent and 115°C

TABLE 28-6 Carbon and Low-Alloy Steels[a]

Steel type	ASTM	UNS	Composition, %[b]	Yield strength, kip/in² (MPa)	Tensile strength, kip/in² (MPa)	Elongation, %
				Mechanical properties[c]		
C-Mn	A53B	K03005	0.30 C, 1.20 Mn	35 (241)	60 (415)	
C-Mn	A106B	K03006	0.30 C, 0.29–1.06 Mn, 0.10 min. Si	35 (241)	60 (415)	30
C	A285A	K01700	0.17 C, 0.90 Mn	24 (165)	45–55 (310–380)	30
HSLA	A517F	K11576	0.08–0.22 C, 0.55–1.05 Mn, 0.13–0.37 Si, 0.36–0.79 Cr, 0.67–1.03 Ni, 0.36–0.64 Mo, 0.002–0.006 B, 0.12–0.53 Cu, 0.02–0.09 V	100 (689)	115–135 (795–930)	16
HSLA	A242(1)	K11510	0.15 C, 1.00 Mn, 0.20 min Cu, 0.15 P	42–50 (290–345)	63–70 (435–480)	21
2¼ Cr, 1 Mo	A387(22)	K21590	0.15 C, 0.30–0.60 Mn, 0.5 Si, 2.00–2.50 Cr, 0.90–1.10 Mo	30 (205)[d] / 45 (310)[e]	60–85 (415–585)[d] / 75–100 (515–690)[e]	18[c] / 18[d]
4–6 Cr, ½ Mo	A335 (P5)	K41545	0.15 C, 0.30–0.60 Mn, 0.5 Si, 4.00–6.00 Cr, 0.45–0.65 Mo	30 (205)	60 (415)	
9 Cr, 1 Mo	A335 (P9)	K81590	0.15 C, 0.30–0.6 Mn, 0.25–1.00 Si, 8.00–10.00 Cr, 0.90–1.10 Mo	30 (205)	60 (415)	
9 Ni	A333(8), A353(1)	K81340	0.13 C, 0.90 Mn, 0.13–0.32 Si, 8.40–9.60 Ni	75 (515)	100–120 (690–825)	20
	AISI 4130	G41300	0.28–0.33 C, 0.80–1.10 Mn, 0.15–0.3 Si, 0.8–1.10 Cr, 0.15–0.25 Mo	120 (830)[f]	140 (965)[f]	22[f]
	AISI 4340	G43400	0.38–0.43 C, 0.60–0.80 Mn, 0.15–0.3 Si, 0.70–0.90 Cr, 1.65–2.00 Ni, 0.20–0.30 Mo	125 (860)[g]	148 (1020)[g]	20[g]

[a]Courtesy of National Association of Corrosion Engineers. To convert MPa to lbf/in², multiply by 145.04.
[b]Single values are maximum values unless otherwise noted.
[c]Room-temperature properties. Single values are minimum values.
[d]Class 1.
[e]Class 2.
[f]1-in-diameter bars water-quenched from 1,575°F (860°C) and tempered at 1,200°F (650°C).
[g]1-in-diameter bars oil-quenched from 1,550°F (845°C) and tempered at 1,200°F (650°C).

(235°F). Alkalies have little effect on this alloy, but it will not stand up against very highly oxidizing or reducing environments.

Aluminum and Alloys Aluminum and its alloys are made in practically all the forms in which metals are produced, including castings. Thermal conductivity of aluminum is 60 percent of that of pure copper, and unalloyed aluminum is used in many heat-transfer applications. Its high electrical conductivity makes aluminum popular in electrical applications. Aluminum is one of the most workable of metals, and it is usually joined by inert-gas-shielded arc-welding techniques.

Commercially pure aluminum has a tensile strength of 69 MPa (10,000 lbf/in²), but it can be strengthened by cold working. One limitation of aluminum is that strength declines greatly above 150°C (300°F). When strength is important, 200°C (400°F) is usually considered the highest permissible safe temperature for aluminum. However, aluminum has excellent low-temperature properties; it can be used at −250°C (−420°F).

Aluminum has high resistance to atmospheric conditions as well as to industrial fumes and vapors and fresh, brackish, or salt waters. Many mineral acids attack aluminum, although the metal can be used with concentrated nitric acid (above 82 percent) and glacial acetic acid. Aluminum cannot be used with strong caustic solutions.

It should be noted that a number of **aluminum alloys** are available (see Table 28-16). Many have improved mechanical properties over pure aluminum. The wrought heat-treatable aluminum alloys have tensile strengths of 90 to 228 MPa (13,000 to 33,000 lbf/in²) as annealed; when they are fully hardened, strengths can go as high as 572 MPa (83,000 lbf/in²). However, aluminum alloys usually have lower corrosion resistance than the pure metal. The **alclad** alloys have been developed to overcome this shortcoming. Alclad consists of an aluminum layer metallurgically bonded to a core alloy.

The corrosion resistance of aluminum and its alloys tends to be very sensitive to trace contamination. Very small amounts of metallic mercury, heavy-metal ions, or chloride ions can frequently cause rapid failure under conditions which otherwise would be fully acceptable.

When alloy steels do not give adequate corrosion protection—particularly from sulfidic attack—steel with an **aluminized surface coating** can be used. A spray coating of aluminum on a steel is not likely to spall or flake, but the coating is usually not continuous and may leave some areas of the steel unprotected. Hot-dipped "aluminized" steel gives a continuous coating and has proved satisfactory in a number of applications, particularly when sulfur or hydrogen sulfide is present. It is also used to protect thermal insulation and as weather shields for equipment. The coated steel resists fires better than solid aluminum.

Copper and Alloys Copper and its alloys are widely used in chemical processing, particularly when heat and electrical conductivity are important factors. The thermal conductivity of copper is twice that of aluminum and 90 percent that of silver. A large number of cop-

TABLE 28-7 Properties of Low-Alloy AISI Steels

AISI type	Melting temperature, °F	Thermal conductivity, Btu/[(h·ft²)(°F/ft)] (212°F)	Coefficient of thermal expansion (0–1,200°F) per °F	Specific heat (68–212°F), Btu/(lb·°F)
		Typical physical properties[a]		
13XX		27	7.9 × 10⁻⁶[b]	0.10–0.11
23XX	2,600–2,620	38.3[c]	8.0 × 10⁻⁶	0.11–0.12
25XX	2,610–2,620	34.5–38.5[c]	7.8 × 10⁻⁶	0.11–0.12
40XX		27	8.3 × 10⁻⁶[b]	0.10–0.11
41XX		24.7[d]		0.11
43XX	2,740–2,750	21.7[c]	8.1 × 10⁻⁶	0.107
46XX		27[d]	6.3 × 10⁻⁶[c]	0.10–0.11
48XX	2,750	26[f]	8.6 × 10⁻⁶	
51XX	2,720–2,760	27–34[g]	7.4 × 10⁻⁶[b]	0.10–0.11
61XX		27	8.1 × 10⁻⁶[b]	0.10–0.11
86, 87XX	2,745–2,755	21.7[c]	8.2 × 10⁻⁶	0.107
92, 94XX		27	8.1 × 10⁻⁶[b]	0.10–0.12

XX = nominal percent carbon.
[a]Density for all low-alloy steels is about 0.28 lb/in³.
[b]68 to 1200°F.
[c]120°F.
[d]68°F.
[e]0 to 200°F.
[f]75°F.
[g]32 to 212°F.
[h]100 to 518°F.

TABLE 28-7 Properties of Low-Alloy AISI Steels (*Concluded*)

AISI type	Tensile strength, 1,000 lbf/in²	Yield strength (0.2% offset), 1000 lbf/in²	Elongation (in 2 in), %	Reduction of area, %	Hardness, Brinell	Impact strength (Izod), ft·lbf
1,330[b]	122	100	19	52	248	
1,335[c]	126	105	20	59	262	
1,340[c]	137	118	19	55	285	
2,317[c]	107	72	27	71	222	84
2,515[c]	113	94	25	69	233	85
E2,517[c]	120	100	22	66	244	80
4,023[d]	120	85	20	53	255	
4,032[c]	210	182	11	49	415	
4,042[f]	235	210	10	42	461	
4,053[g]	250	223	12	40	495	
4,063[h]	269	231	8	15	534	
4,130[i]	200	170	16	49	375	25
4,140[j]	200	170	15	48	385	16
4,150[k]	230	215	10	40	444	12
4,320[d]	180	154	15	50	360	32
4,337[k]	210	140	14	50	435	18
4,340[k]	220	200	12	48	445	16
4,615[d]	100	75	18	52		42
4,620[d]	130	95	21	65		68
4,640[l]	185	160	14	52	390	25
4,815[d]	150	125	18	58	325	44
4,817[d]			15	52	355	36
4,820[d]			13	47	380	28
5,120[d]	143	114	13	45	302	6
5,130[m]	189	175	13	51	380	
5,140[m]	190	170	13	43	375	16
5,150[m]	224	208	10	40	444	
6,120[n]	125	94	21	56		28
6,145[o]	176	169	16	52	429	20
6,150[o]	187	179	13	42	444	13
8,620[p]	122	98	21	63	245	76
8,630[p]	162	142	14	54	325	42
8,640[p]	208	183	13	43	420	18
8,650[p]	214	194	12	41	423	
8,720[p]	122	98	21	63	245	76
8,740[p]	208	183	13	43	420	18
8,750[p]	214	194	12	41	423	
9,255[p]	232[q]	215	9	21	477	6
9,261[p]	258[r]	226	10	30	514	12

[a]Properties are for materials hardened and tempered as follows: [b]water-quenched from 1,525°F, tempered at 1,000°F; [c]oil-quenched from 1,525°F, tempered at 1,000°F, [d]pseudocarburized 8 h at 1700°F, oil-quenched, tempered 1 h at 300°F, [e]water-quenched from 1,525°F, tempered at 600°F, [f]oil-quenched from 1,500°F, tempered at 600°F, [g]oil-quenched from 1,475°F, tempered at 600°F, [h]oil-quenched from 1,450°F, tempered at 600°F, [i]water-quenched from 1,500 to 1,600°F, tempered at 800°F, [j]oil-quenched from 1,550°F, tempered at 800°F, [k]oil-quenched from 1,525°F, tempered at 800°F, [l]normalized at 1,650°F, reheated to 1,475°F, oil-quenched, tempered at 800°F; [m]normalized at 1,625°F, reheated to 1,550°F, water-quenched, tempered at 800°F; [n]carburized 10 h at 1,680°F, pot-cooled, oil-quenched from 1,525°F, tempered at 300°F; [o]normalized at 1,600°F, oil-quenched from 1,575°F, tempered at 1,000°F, [p]oil-quenched tempered at 800°F; [q]normalized at 1,650°F, reheated to 1,625°F, quenched in agitated oil, tempered at 800°F, [r]normalized at 1,600°F, reheated to 1,575°F, quenched in agitated oil, tempered at 800°F.

NOTE: °C = (°F − 32) × 5/9. To convert Btu/(h · ft · °F) to W/(m · °C), multiply by 0.8606; to convert Btu/(lbf · °F) to kJ/(kg · °C), multiply by 0.2388; to convert lbf/in² to MPa, multiply by 0.006895; and to convert ft · lbf to J, multiply by 0.7375.

per alloys are available, including brasses (Cu-Zn), bronzes (Cu-Sn), and cupronickels.

Copper has excellent low-temperature properties and is used at −200°C (−320°F). Brazing and soldering are common joining methods for copper, although welding, while difficult, is possible. Generally, copper has high resistance to industrial and marine atmospheres, seawater, alkalies, and solvents. Oxidizing acids rapidly corrode cop-

per. However, the alloys have somewhat different properties than commercial copper.

Brasses with up to 15 percent Zn are ductile but difficult to machine. Machinability improves with increasing zinc up to 36 percent Zn. Brasses with less than 20 percent Zn have corrosion resistance equivalent to that of copper but with better tensile strengths. Brasses with 20 to 40 percent Zn have lower corrosion resistance and are subject to dezincification and stress-corrosion cracking, especially when ammonia is present.

Bronzes are somewhat similar to brasses in mechanical properties and to high-zinc brasses in corrosion resistance (except that bronzes are not affected by stress cracking). **Aluminum and silicon bronzes** are very popular in the process industries because they combine good strength with corrosion resistance.

Cupronickels (10 to 30 percent Ni) have become very important as copper alloys. They have the highest corrosion resistance of all copper alloys and find application as heat-exchanger tubing. Resistance to seawater is particularly outstanding.

Lead and Alloys Chemical leads of 99.9+ percent purity are used primarily in the chemical industry in environments that form thin, insoluble, and self-repairable protective films, for example, salts such as sulfates, carbonates, or phosphates. More soluble films such as nitrates, acetates, or chlorides offer little protection.

Alloys of antimony, tin, and arsenic offer limited improvement in mechanical properties, but the usefulness of lead is limited primarily because of its poor structural qualities. It has a low melting point and a high coefficient of expansion, and it is a very ductile material that will creep under a tensile stress as low as 1 MPa (145 lbf/in²).

Titanium Titanium has become increasingly important as a construction material. It is strong and of medium weight. Corrosion resistance is very superior in oxidizing and mild reducing media (Ti-Pd alloys Grade 7 and 11 have superior resistance in reducing environments, as does the Ti-Mo-Ni alloy Grade 12). Titanium is usually not bothered by impingement attack, crevice corrosion, and pitting attack in seawater. Its general resistance to seawater is excellent. Titanium is resistant to nitric acid at all concentrations except with red fuming nitric. The metal also resists ferric chloride, cupric chloride, and other hot chloride solutions. However, there are a number of disadvantages to titanium which have limited its use. Titanium is not easy to form, it has a high springback and tends to gall, and welding must be carried out in an inert atmosphere.

Zirconium Zirconium was originally developed as a construction material for atomic reactors. Reactor-grade zirconium contains very little hafnium, which would alter zirconium's neutron-absorbing properties. Commercial-grade zirconium, for chemical process applications, however, contains 2.5 percent hafnium. Zirconium resembles titanium from a fabrication standpoint. All welding must be done under an inert atmosphere. Zirconium has excellent resistance to reducing environments. Oxidizing agents frequently cause accelerated attack. It resists all chlorides except ferric and cupric. Zirconium alloys should *not* be used in concentrations of sulfuric acid above about 70 percent. There are a number of alloys of titanium and zirconium, with mechanical properties superior to those of the pure metals. The zirconium alloys are referred to as Zircaloys.

Tantalum The physical properties of tantalum are similar to those of mild steel except that tantalum has a higher melting point. Tantalum is ductile and malleable and can be worked into intricate forms. It can be welded by using inert-gas-shielded techniques. The metal is practically inert to many oxidizing and reducing acids (except fuming sulfuric). It is attacked by hot alkalies and hydrofluoric acid. Its cost generally limits use to heating coils, bayonet heaters, coolers, and condensers operating under severe conditions. When economically justified, larger items of equipment (reactors, tanks, etc.) may be fabricated with tantalum liners, either loose (with proper anchoring) or explosion-bonded-clad. Since tantalum linings are usually very thin, very careful attention to design and fabrication details is required.

INORGANIC NONMETALLICS

Glass and Glassed Steel Glass is an inorganic product of fusion which is cooled to a rigid condition without crystallizing. With unique

TABLE 28-8 Cast-Iron Alloys*

Alloy	ASTM	UNS	Composition, %†	Condition	Yield strength, kip/in² (MPa)	Tensile strength, kip/in² (MPa)	Elonga-tion, %	Hardness, HB
Gray cast iron	A159 (G3000)	F10006	3.1–3.4 C, 0.6–0.9 Mn, 1.9–2.3 Si	As cast		30 (207)		187–241
Malleable cast iron	A602 (M3210)	F20000	2.2–2.9 C, 0.15–1.25 Mn, 0.9–1.90 Si	Annealed	32 (229)	50 (345)	12	130
Ductile cast iron	A395 (60-40-18)	F32800	None specified	Annealed	40 (276)	60 (414)	18	170
Cast iron	A436(1)	F41000	3.0 C, 1.5–2.5 Cr, 5.5–7.5 Cu, 0.5–1.5 Mn, 13.5–17.5 Ni, 1.0–2.8 Si	As cast		25 (172)		150
Cast iron	A436(2)	F41002	3.0 C, 1.5–2.5 Cr, 0.50 Cu, 0.5–1.5 Mn, 18–22 Ni, 1.0–2.8 Si	As cast		25 (172)		145
Cast iron	A436(5)	F41006	2.4 C, 0.1 Cr, 0.5 Cu, 0.5–1.5 Mn, 34–36 Ni, 1.0–2.0 Si	As cast		20 (138)		110
Ductile austenitic cast iron	A439(D-2)	F43000	3.0 C, 1.75–2.75 Cr, 0.7–1.25 Mn, 18–22 Ni, 1.5–3.0 Si	As cast	30 (207)	58 (400)		170
Ductile austenitic cast iron	A439 (D-5)	F43006	2.4 C, 0.1 Cr, 1.0 Mn, 34–36 Ni, 1.0–2.8 Si	As cast	30 (207)	55 (379)		155
Silicon cast iron	A518	F47003	0.7–1.1 C, 0.5 Cr, 0.5 Cu, 1.50 Mn, 0.5 Mo, 14.2–14.75 Si	As cast		16 (110)		520

*Courtesy of National Association of Corrosion Engineers. To convert MPa to lbf/in², multiply by 145.04.
†Single values are maximum values.
‡Typical room-temperature properties.

properties compared with metals, they require special considerations in their design and use.

Glass has excellent resistance to all acids except hydrofluoric and hot, concentrated H_3PO_4. It is also subject to attack by hot alkaline solutions. Glass is particularly suitable for piping when transparency is desirable.

The chief drawback of glass is brittleness, and it is also subject to damage by thermal shock. However, glass armored with epoxy-polyester fiberglass can readily be protected against breakage. On the other hand, glassed steel combines the corrosion resistance of glass with the working strength of steel. Accordingly, **glass linings** are resistant to all concentrations of hydrochloric acid to 120°C (250°F), to dilute concentrations of sulfuric to the boiling point, to concentrated sulfuric to 230°C (450°F), and to all concentrations of nitric

acid to the boiling point. Acid-resistant glass with improved alkali resistance (up to 12 pH) is available.

A nucleated crystalline ceramic-metal composite form of glass has superior mechanical properties compared with conventional glassed steel. Controlled high-temperature firings chemically and physically bond the ceramic to steel, nickel-based alloys, and refractory metals. These materials resist corrosive hydrogen chloride gas, chlorine, or sulfur dioxide at 650°C (1,200°F). They resist all acids except HF up to 180°C (350°F). Their impact strength is 18 times that of safety glass; abrasion resistance is superior to that of porcelain enamel. They have 3 to 4 times the thermal-shock resistance of glassed steel.

Porcelain and Stoneware Porcelain and stoneware materials are about as resistant to acids and chemicals as glass, but with the advantage of greater strength. This is offset somewhat by poor ther-

TABLE 28-9 Standard Wrought Martensitic Stainless Steels*

AISI type	UNS	Cr	Ni	Mo	C	Other	Yield strength, kip/in² (MPa)	Tensile strength, kip/in² (MPa)	Elongation, %	Hardness, HB
403	S40300	11.5–13.0			0.15		40 (276)	75 (517)	35	155
410	S41000	11.5–13.5			0.15		35 (241)	70 (483)	30	150
414	S41400	11.5–13.5	1.25–2.5		0.15		90 (621)	115 (793)	20	235
416	S41600	12–14		0.6	0.15	0.15S§	40 (276)	75 (517)	30	155
416Se	S41623	12–14			0.15	0.15Se§	40 (276)	75 (517)	30	155
420	S42000	12–14			0.15		50 (345)	95 (655)	20	195
420F	S42020	12–14		0.6	0.15§	0.155§	55 (379)	95 (655)	22	220
422	S42200	11–13	0.5–1.0	0.75–1.25	0.20–0.25	0.15–0.30 V, 0.75–1.25 W	125 (862)	145 (1000)	18	320
431	S43100	15–17	1.25–2.5		0.20		95 (665)	125 (862)	20	260
440A	S44002	16–18		0.75	0.6–0.75		60 (414)	105 (724)	20	210
440B	S44003	16–18.0		0.75	0.75–0.95		62 (427)	107 (738)	18	215
440C	S44004	16–18		0.75	0.95–1.20		65 (448)	110 (758)	14	220
501	S50100	4–6		0.40–0.65	0.10§		30 (207)	70 (483)	28	160
502	S50200	4–6		0.40–0.65	0.10		25 (172)	65 (448)	30	150

*Courtesy of National Association of Corrosion Engineers. To convert MPa to lbf/in², multiply by 145.04.
†Single values are maximum values unless otherwise noted.
‡Typical room-temperature properties of annealed plates.
§Minimum.

TABLE 28-10   Standard Wrought Ferritic Stainless Steels*

AISI type	UNS	Composition, %†							Mechanical properties‡			
		Cr	C	Mn	Si	P	S	Other	Yield strength, kip/in² (MPa)	Tensile strength, kip/in² (MPa)	Elongation, %	Hardness, HB
405	S40500	11.5–14.5	0.08	1.0	1.0	0.04	0.03	0.1–0.3 Al	40 (276)	65 (448)	30	150
409	S40900	10.5–11.75	0.08	1.0	1.0	0.045	0.045	(6 × C) Ti§	35 (241)	65 (448)	25	137
429	S42900	14–16	0.12	1.0	1.0	0.04	0.03		40 (276)	70 (483)	30	163
430	S43000	16–18	0.12	1.0	1.0	0.04	0.03		40 (276)	75 (517)	30	160
430F	S43020	16–18	0.12	1.25	1.0	0.06	0.15¶	0.6 Mo	55 (379)	80 (552)	25	170
430FSe	S43023	16–18	0.12	1.25	1.0	0.06	0.06	0.15 Se¶	55 (379)	80 (552)	25	170
434	S43400	16–18	0.12	1.0	1.0	0.04	0.03	0.75–1.25 Mo	53 (365)	77 (531)	23	160
436	S43600	16–18	0.12	1.0	1.0	0.04	0.03	0.75–1.25 Mo (5 × C)(Cb + Ta)§	53 (365)	77 (531)	23	160
442	S44200	18–23	0.20	1.0	1.0	0.04	0.03		45 (310)	80 (552)	20	185
446	S44600	23–27	0.20	1.5	1.0	0.04	0.03	0.25N	55 (379)	85 (586)	25	160

*Courtesy of National Association of Corrosion Engineers. To convert MPa to lbf/in², multiply by 145.04.
†Single values are maximum values unless otherwise noted.
‡Typical temperature properties of annealed plates.
§0.70 maximum.
¶Minimum.

mal conductivity, and the materials can be damaged by thermal shock fairly easily. **Porcelain enamels** are used to coat steel, but the enamel has slightly inferior chemical resistance. Some refractory coatings, capable of taking very high temperatures, are also available.

Brick Construction   Brick-lined construction can be used for many severely corrosive conditions under which high alloys would fail. Common bricks are made from carbon, red shale, or acidproof refractory materials. Red-shale brick is not used above 175°C (350°F) because of spalling. Acidproof refractories can be used up to 870°C (1,600°F).

A number of **cement** materials are used with brick. Standard are phenolic and furan resins, polyesters, sulfur, silicate, and epoxy-based materials. Carbon-filled polyesters and furanes are good against nonoxidizing acids, salts, and solvents. Silica-filled resins should not be used against hydrofluoric or fluosilicic acids. Sulfur-based cements are limited to 93°C (200°F), while resins can be used to about 180°C (350°F). The sodium silicate–based cements are good against acids to 400°C (750°F).

Differential thermal expansion of the brick, its joints, and the vessel substrate necessitates an intermediate lining of lead, asphalt, rubber,

TABLE 28-11   Standard Wrought Austenitic Stainless Steels*

AISI type	UNS	Composition, %†							Mechanical properties‡			
		Cr	Ni	Mo	C	Si	Mn	Other	Yield strength, kip/in² (MPa)	Tensile strength' kip/in² (MPa)	Elongation, %	Hardness, HB
201	S20100	16–18	3.5–5.5		0.15	1.0	5.5–7.5	0.25 N	55 (379)	115 (793)	55	185
202	S20200	17–19	4–6		0.15	1.0	7.5–10.	0.25 N	55 (379)	105 (724)	55	185
301	S30100	16–18	6–8		0.15	1.0	2.0		40 (276)	105 (724)	55	165
302	S30200	17–19	8–10		0.15	1.0	2.0		35 (241)	90 (621)	60	150
302B	S30215	17–19	8–10		0.15	2.0–3.0	2.0		40 (276)	90 (621)	50	165
303	S30300	17–19	8–10	0.6	0.15	1.0	2.0	0.15 S,§ 0.2 P	35 (241)	90 (621)	50	160
303Se	S30323	17–19	8–10		0.15	1.0	2.0	0.15 Se,§ 0.2 P	35 (241)	90 (621)	50	160
304	S30400	18–20	8–10.5		0.08	1.0	2.0		35 (241)	82 (565)	60	149
304L	S30403	18–20	8–12		0.03	1.0	2.0		33 (228)	79 (545)	60	143
304N	S30451	18–20	8–10.5		0.08	1.0	2.0	0.10–0.16 N	48 (331)	90 (621)	50	180
308	S30800	19–21	10–12		0.08	1.0	2.0		30 (207)	85 (586)	55	150
309	S30900	22–24	12–15		0.20	1.0	2.0		40 (276)	95 (655)	45	170
309S	S30908	22–24	12–15		0.08	1.0	2.0		40 (276)	95 (655)	45	170
310	S31000	24–26	19–22		0.25	1.5	2.0		45 (310)	95 (655)	50	170
310S	S31008	24–26	19–22		0.08	1.5	2.0		45 (310)	95 (655)	50	170
314	S31400	23–26	19–22		0.25	1.5–3.0	2.0		50 (345)	100 (609)	45	180
316	S31600	16–18	10–14	2.0–3.0	0.08	1.0	2.0		36 (248)	82 (565)	55	149
316L	S31603	16–18	10–14	2.0–3.0	0.03	1.0	2.0		34 (234)	81 (558)	55	146
316N	S31651	16–18	10–14	2.0–3.0	0.08	1.0	2.0	0.10–0.16 N	42 (290)	90 (621)	55	180
317	S31700	18–20	11–15	3.0–4.0	0.08	1.0	2.0		40 (276)	85 (586)	50	160
317L	S31703	18–20	11–15	3.0–4.0	0.03	1.0	2.0		35 (241)	85 (586)	55	150
321	S32100	17–19	9–12		0.08	1.0	2.0	(5 × C) Ti§	30 (207)	85 (586)	55	160
329	S32900	25–30	3–6	1.0–2.0	0.10	1.0	2.0		80 (552)	105 (724)	25	230
347	S34700	17–19	9–13		0.08	1.0	2.0	(10 × C)(Cb + Ta)§	35 (241)	90 (621)	50	160
348	S34800	17–19	9–13		0.08	1.0	2.0	(10 × C)(Cb + Ta)¶ 0.20 Co	35 (241)	90 (621)	50	160

*Courtesy of National Association of Corrosion Engineers. To convert MPa to lbf/in², multiply by 145.04.
†Single values are maximum values unless otherwise noted.
‡Typical room-temperature properties of solution-annealed plates.
§Minimum.
¶Minimum except Ta = 0.1 maximum.

TABLE 28-12 Special Stainless Steels*

Alloy	UNS	Composition, %†							Mechanical properties‡			
		Cr	Ni	Mo	C	Mn	Si	Other	Yield strength, kip/in² (MPa)	Tensile strength, kip/in² (MPa)	Elongation, %	Hardness,† HB
A-286	K66286	13.5-16	24-27	1.0-1.5	0.08	2.0	1.0	1.90-2.35 Ti, 0.1-0.5 V, 0.001-0.01 B, 8 × C-1.0 Cb	100 (690)	140 (970)	20	185
20Cb-3	N08020	19-21	32-38	2.0-3.0	0.07	2.0	1.0	3.0-4.0 Cu	53 (365)	98 (676)	33	185
20Mod	N08320	21-23	25-27	4.0-6.0	0.05	2.5	1.0	(4.0 × C) Ti min.	43 (296)	84 (579)	42	160
PH13-8Mo	S13800	12.25-13.25	7.5-8.5	2.0-2.5	0.05	0.2	0.1	0.90-1.35 Al	120 (827)	160 (1100)	17	300
PH14-8Mo	S14800	13.75-15.0	7.75-8.75	2.0-3.0	0.05	1.0	1.0	0.75-1.5 Al	55-210 (380-1450)	125-230 (860-1540)	2-25	200-450
15-5PH	S15500	14.0-15.5	3.5-5.5		0.07	1.0	1.0	0.15-0.45 Cb, 2.5-4.5 Cu	145 (1000)	160 (1100)	15	320
PH15-7Mo	S15700	14.0-16.0	6.5-7.75	2.0-3.0	0.09	1.0	1.0	0.75-1.5 Al, 0.15-0.45 Cb	55-210 (380-1450)	130-220 (900-1520)	2-35	200-450
17-4PH	S17400	15.5-17.5	3.0-5.0		0.07	1.0	1.0	3.0-5.0 Cu, 0.4 Al	145 (1000)	160 (1100)	15	320
W	S17600	16.0-17.5	6.0-7.5		0.08	1.0	1.0	0.4-1.20 Ti	90-200 (620-1380)	135-210 (930-1450)	3-12	260-420
17-7PH	S17700	16.0-18.0	6.5-7.75		0.09	1.0	1.0	0.75-1.5 Al	40 (276)	130 (710)	10	185
216	S21600	17.5-22.0	5.0-7.0	2.0-3.0	0.08	7.5-9.0	1.0	0.25-0.5 N	70 (480)	115 (790)	45	200
Nitronic 60	S21800	16.0-18.0	8.0-9.0		0.10	7.0-9.0	3.5-4.5	0.08-0.18 N	60 (410)	103 (710)	62	210
21-6-9	S21900	18.0-21.0	5.0-7.0		0.08	8.0-10.0	1.0	0.15-0.40 N	68 (470)	112 (770)	44	220
AM350	S35000	16.0-17.0	4.0-5.0	2.5-3.25	0.07-0.11	0.5-1.25	0.5	0.07-0.13 N	60-173 (410-1200)	145-206 (1000-1420)	13.5-40	200-400
AM355	S35500	15.0-16.0	4.0-5.0	2.5-3.25	0.10-0.15	0.5-1.25	0.5		182 (1250)	216 (1490)	19	402-477
Almar 362	S36200	14.0-15.0	6.0-7.0		0.05	0.5	0.3	0.55-9.0 Ti	105-185 (724-1286)	120-188 (827-1300)	15-13	250-400
18-18-2	S38100	17.0-19.0	17.5-18.5	1.75-2.5	0.08	2.0	1.5-2.5	0.20 × (C + N) + Cb min-0.8 Ti + Cb	40 (250)	80 (550)	55	165
Stab. 18-2	S44400	17.5-19.5	1.0		0.025	1.0	1.0	0.015 N	45 (310)	60 (414)	20	210
26-1	S44625	25.0-27.5	0.5	0.75-1.50	0.01	0.4	0.4	0.5 Ni + Cu	50 (345)	70 (480)	30	165
Stab. 26-1	S44626	25.0-27.0	0.5	0.75-1.50	0.06	0.75	0.75	7 × (C + Ni)—1.0 Ti, 0.15 Cu	50 (345)	70 (480)	30	165
28-4	S44700	28.0-30.0	0.15	3.5-4.2	0.010	0.3	0.2	0.02 N, 0.15 Cu	70 (480)	90 (620)	25	210
28-4-2	S44800	28.0-30.0	2.0-2.5	3.5-4.2	0.010	0.3	0.2	0.02 N, 8 × C Cb min.	85 (590)	95 (650)	25	230
Custom 450	S45000	14.0-16.0	5.0-7.0	0.5-1.0	0.05	1.0	1.0	1.25-1.75 Cu	117-184 (800-1270)	144-196 (990-1350)	14	270-400
Custom 455	S45500	11.0-12.5	7.5-9.5	0.5	0.05	0.5	0.5	1.5-2.5 Cu, 0.8-1.4 Ti	115-220 (790-1500)	140-230 (970-1600)	10-14	290-460

*Courtesy of National Association of Corrosion Engineers. To convert MPa to lbf/in², multiply by 145.04.
†Single values are maximum values unless otherwise noted.
‡Typical room-temperature properties.

TABLE 28-13 Standard Cast Heat-Resistant Stainless Steels[a]

			Composition, %[b]						Mechanical properties at 1600°F			
									Short term		Stress to rupture in 1000 h	
ACI	Equivalent AISI	UNS	Cr	Ni	C	Mn	Si	Other	Tensile strength, kip/in² (MPa)	Elongation, %	kip/in²	MPa
HA			8–10		0.2	0.35–0.65	1.0	0.9–1.2 Mo	44 (303)[c]	36[c]	27	186[d]
HC	446	J92605	26–30	4	0.5	1.0	2.0				1.3	9.0
HD	327	J93005	26–30	4–7	0.5	1.5	2.0		23 (159)	18	7.0	48[e]
HE		J93403	26–30	8–11	0.2–0.5	2.0	2.0					
HF	302B	J92603	18–23	9–12	0.2–0.4	2.0	2.0		21 (145)	16	4.4	30
HH[f]		J93503	24–28	11–14	0.2–0.5	2.0	2.0	0.2 N	18.5 (128)	30	3.8	26
HH[g]	309	J93503	24–28	11–14	0.2–0.5	2.0	2.0	0.2 N	21.5 (148)	18	3.8	26
HI		J94003	26–30	14–18	0.2–0.5	2.0	2.0		26 (179)	12	4.8	33
HK	310	J94224	24–28	18–22	0.2–0.6	2.0	2.0		23 (159)	16	6.0	41
HL		J94604	28–32	18–22	0.2–0.6	2.0	2.0		30 (207)			
HN		J94213	19–23	23–27	0.2–0.5	2.0	2.0		20 (138)	37	7.4	51
HP		J95705	24–28	33–37	0.35–0.75	2.0	2.5		26 (179)	27	7.5	52
HT	330	J94605	13–17	33–37	0.35–0.75	2.0	2.5		19 (131)	26	5.8	40
HU		J95405	17–21	37–41	0.35–0.75	2.0	2.5		20 (138)	20	5.2	36
HW			10–14	58–62	0.35–0.75	2.0	2.5		19 (131)		4.5	31
HX			15–19	64–68	0.35–0.75	2.0	2.5		20.5 (141)	48	4.0	28

[a] Courtesy of National Association of Corrosion Engineers. To convert MPa to lbf/in², multiply by 145.04.
[b] Single values are maximum values; S and P are 0.04 maximum; Mo is 0.5 maximum.
[c] At 1100°F (593°C).
[d] At 1000° (538°C).
[e] At 1400°F (760°C).
[f] Type I; partially ferritic.
[g] Type II; wholly austenitic.

TABLE 28-14 Standard Cast Corrosion-Resistant Stainless Steels[a]

			Composition, %[b]							Mechanical properties[c]			
ACI	Equivalent AISI	UNS	Cr	Ni	Mo	C	Mn	Si	Other	Yield strength, kip/in² (MPa)	Tensile strength, kip/in² (MPa)	Elongation, %	Hardness HB
Ca-15	410	J91150	11.5–14	1.0	0.5	0.15	1.00	1.50		150 (1034)[d]	200 (1379)[d]	7[d]	390[d]
CA-15M		J91151	11.5–14	1.0	0.15–1.0	0.15	1.00	1.50		150 (1034)[d]	200 (1379)[d]	7[d]	390[d]
CA-6NM		J91540	11.5–14	3.5–4.5	0.4–1.0	0.06	1.00	1.00		100 (690)[e]	120 (827)[e]	4[e]	269[e]
CA-40	420	J91153	11.5–14	1.0	0.5	0.20–0.40	1.00	1.50		165 (1138)[d]	220 (1517)[d]	1[d]	470[d]
CB-30	431	J91803	18.21	2.0		0.30	1.00	1.50		60 (414)[f]	95 (655)[f]	15[f]	195[f]
CC-50	446	J92615	26–30	4.0		0.50	1.00	1.50		65 (448)[g]	97 (669)[g]	18[g]	210[g]
CE-30	312	J93423	26–30	8–11		0.30	1.50	2.00		63 (434)	97 (669)	18	190
CB-7Cu	(17-4PH)		(16)	(4)		0.07			(3) Cu	165 (1138)		3	418
CD-4MCu			25–26.5	4.75–6.0	1.75–2.25	0.04	1.00	1.00	2.75–3.25 Cu	82 (565)	108 (745)	25	253
CF-3	304L	J92500	17–21	8–12		0.03	1.50	2.00		36 (248)	77 (531)	60	140
CF-8	304	J92600	18–21	8–11		0.08	1.50	2.00		37 (255)	77 (531)	55	140
CF-20	302	J92602	18–21	8–11		0.20	1.50	2.00		36 (248)	77 (531)	50	163
CF-3M	316L	J92800	17–21	9–13	2.0–3.0	0.03	1.50	1.50		38 (262)	80 (552)	55	150
CF-8M	316	J92900	18–21	9–12	2.0–3.0	0.08	1.50	2.00		42 (290)	80 (552)	50	160
CF-12M			18–21	9–12	2.0–3.0	0.12	1.50	2.00		42 (290)	80 (552)	50	160
CG-12	317	J93000	18–21	9–13	3.0–4.0	0.08	1.50	1.50		44 (303)	83 (572)	45	170
CF-8C	347	J92710	18–21	9–12		0.08	1.50	2.00	(8 × C) Cb[h]	38 (262)	77 (531)	39	149
CF-16F	303	J92701	18–21	9–12	1.50	0.16	1.50	2.00		40 (276)	77 (531)	52	150
CG-12		J93001	20–23	10–13		0.12	1.50	2.00		28 (193)		35	
CH-20	309	J93402	22–26	12–15		0.20	1.50	2.00		50 (345)	88 (607)	38	190
CK-20	310	J94202	23–27	19–22		0.20	1.50	2.00		38 (262)	76 (524)	37	144
CN-7M		J95150	19–22	27.5–30.5	2.0–3.0	0.07	1.50	1.50	3–4 Cu	32 (221)	69 (476)	48	130

[a] Courtesy of National Association of Corrosion Engineers. To convert MPa to lbf/in², multiply by 145.04.
[b] Single values are maximum values except those in parentheses, which are minimum values. P and S values are 0.04 maximum.
[c] Typical room-temperature properties for solution-annealed material unless otherwise noted.
[d] For material air-cooled from 1800°F and tempered at 600°F.
[e] For material air-cooled from 1750°F and tempered at 1100 to 1150°F.
[f] For material annealed at 1450°F, furnace-cooled to 1000°F, then air-cooled.
[g] Air-cooled from 1900°F.
[h] 1.0 maximum.

TABLE 28-15 Nickel Alloys*

Alloy	UNS	Ni(+Co)§	Cr	Fe	Mo	C	Other	Condition	Yield strength, kip/in² (MPa)	Tensile strength, kip/in² (MPa)	Elongation, %	Hardness, HB
200	N02200	99.		0.4		0.15		Annealed	15–30 (103–207)	55–80 (379–552)	55–40	90–120
201	N02201	99.		0.4		0.02		Annealed	10–25 (69–172)	50–60 (345–414)	60–40	75–102
400	N04400	63–70		1.0–2.5		0.3	28–34 Cu	Annealed	25–50 (172–345)	70–90 (483–621)	60–35	110–149
K-500	N05500	63–70		2.0		0.25	2.3–3.15 Al 0.35–0.85 Ti, 30 Cu	Age-hardened	85–120 (586–827)	130–165 (896–1138)	35–20	250–315
600	N06600	72.	14–17	6–10		0.15		Annealed	30–50 (207–345)	80–100 (552–690)	55–35	120–170
601	N06601	58–63	21–25	Bal.		0.10	1.0–1.7 Al	Annealed	30–60 (207–414)	80–115 (552–793)	70–40	110–150
625	N06625	Bal.	20–23	5	8–10	0.10	3.15–4.15 (Cb + Ta)	Annealed	60–95 (414–655)	120–150 (827–1034)	60–30	145–220
706	N09706	39–44	14.5–17.5	Bal.		0.06		Solution-treated and aged	161 (1110)	193 (1331)	20.	371.
718	N07718	50–55	17–21	Bal.	2.8–3.3	0.08	4.75–5.5 (Cb + Ta) 0.65–1.15 Ti, 0.2–0.8 Al	Special heat treatment	171 (1180)	196 (1351)	17.	382.
X-750	N07750	70	14–17	5–9		0.08	0.7–1.2 (Cb + Ta) 2.25–2.75 Ti, 0.4–1.0 Al	Special heat treatment	115–142 (793–979)	162–193 (1117–1331)	30–15	300–390
800	N08800	30–35	19–23	Bal.		0.10	0.15–0.6 Al, 0.15–0.6 Ti	Annealed	30–60 (207–414)	75–100 (517–690)	60–30	120–184
800H	N08800	30–35	19–23	Bal.		0.05–0.10	0.15–0.6 Al, 0.15–0.6 Ti	Solution-treated	20–50 (138–345)	65–95 (448–655)	50–30	100–184
801	N08801	30–34	19–22	Bal.		0.10	0.75–1.5 Ti	Special heat treatment	79.5 (548)	129 (889)	29.5	
825	N08825	38–46	19.5–23.5	Bal.	2.5–3.5	0.05	1.5–3.0 Cu, 0.6–1.2 Ti	Annealed	35–65 (241–448)	85–105 (586–724)	50–30	120–180
B-2	N10665	Bal.	1.0	2	26–30	0.02		Annealed	76 (524)	139 (958)	53.	210
C-276	N10276	Bal.	14.5–16.5	4–7	15–17	0.02	3.0–4.5 W	Annealed	52 (358)	115 (793)	61.	194
C-4	N06455	Bal.	14–18	3	14–17	0.015	0.7 Ti	Annealed	61 (421)	116 (800)	54.	194
G	N06007	Bal.	21–23	18–21	5.5–7.5	0.05	1.0–2.0 Mn, 1.5–2.5 Cu 1.75–2.5 (Cb + Ta)	Annealed	46 (317)	102 (703)	61.	161
X	N06002	Bal.	20.5–23	17–20	8–10	0.05–0.15	0.2–1.0 W	Annealed	56 (386)	110 (758)	45.	178

*Courtesy of National Association of Corrosion Engineers. To convert MPa to lbf/in², multiply by 145.04.
†Single values are maximum unless otherwise noted.
‡Typical room-temperature properties.
§Single values are minima.

TABLE 28-16   Aluminum Alloys

AA designation	UNS	Composition, %*						Condition‡	Mechanical properties†			
		Cr	Cu	Mg	Mn	Si	Other		Yield strength, kip/in² (MPa)	Tensile strength, kip/in² (MPa)	Elongation in 2 in, %	Hardness HB
Wrought												
1060	A91060						99.6 Al min.	O	4 (28)	10 (69)	43	19
1100	A91100		0.05–0.2				99.0 Al min.	O	5 (34)	13 (90)	45	23
2024	A92024	0.1	3.8–4.9	1.2–1.8	0.3–0.9	0.5		T4	47 (324)	68 (469)	19	120
3003	A93003		0.05–0.2		1.0–1.5	0.6		H14	21 (145)	22 (152)	16	40
5052	A95052	0.15–0.35	0.1	2.2–2.8	0.1			O	13 (90)	2.8 (193)	30	47
5083	A95083	0.05–0.25	0.1	4.0–4.9	0.4–1.0	0.4		O	21 (145)			
5086	A95086	0.05–0.25	0.1	3.5–4.5	0.2–0.7	0.4		O	17 (117)	38 (262)	30	
5154	A95154	0.05–0.35	0.1	3.1–3.9	0.1	0.25		O	17 (117)	35 (241)	27	58
6061	A96061	0.04–0.35	0.15–0.4	0.8–1.2	0.15	0.4–0.8		T6	40 (276)	45 (310)	17	95
6063	A96063	0.1	0.1	0.45–0.9	0.1	0.2–0.6		T6	31 (214)	35 (241)	18	73
7075	A97075	0.18–0.28	1.2–2.0	2.1–2.9	0.3	0.40	5.1–6.1 Zn	T6	73 (503)	63 (572)	11	150
Cast												
242.0	A02420	0.25	3.5–4.5	1.2–1.8	0.35	0.7	1.7–2.3 Ni	S-T571		29 (200)		
295.0	A02950		4.0–5.0	0.03	0.35	0.7–1.5		S-T4		29 (200)	6	
A332.0	A13320		0.5–1.5	0.7–1.3	0.35	11–13	2.0–3.0 Ni	P-T551		31 (214)		
B443.0	A24430		0.15	0.05	0.35	4.5–6.0		S-F		17 (117)	3	
514.0	A05140		0.15	3.5–4.5	0.35	0.35		S-F		22 (152)	6	
520.0	A05200		0.25	9.5–10.6	0.15	0.25		S-T4	22 (152)	42 (290)	12	

*Single values are maximum values.
†Typical room-temperature properties.
‡S = sand-cast; P = permanent-mold-cast; other = temper designations.
SOURCE: Aluminum Association. Courtesy of National Association of Corrosion Engineers. To convert MPa to lbf/in², multiply by 145.04.

or plastic. This membrane functions as a barrier to protect the substrate from corrosion damage. A special prestressed-brick design that maintains the brick in compression by using a controlled-expansion resinous mortar and brick bedding material precludes the use of an elastomeric membrane.

Cement and Concrete   Concrete is an aggregate of inert reinforcing particles in an amorphous matrix of hardened cement paste. Concrete made of portland cement has limited resistance to acids and bases and will fail mechanically following absorption of crystal-forming solutions such as brines and various organics. Concretes made of corrosion-resistant cements (such as calcium aluminate) can be selected for specific chemical exposures.

Soil   Clay is the primary construction material for settling basins and waste-treatment evaporation ponds. Since there is no single type of clay even within a given geographical area, shrinkage, porosity, absorption characteristics, and chemical resistance must be checked for each application.

ORGANIC NONMETALLICS

Plastic Materials   In comparison with metallic materials, the use of plastics is limited to relatively moderate temperatures and pressures [230°C (450°F) is considered high for plastics]. Plastics are also less resistant to mechanical abuse and have high expansion rates, low strengths (thermoplastics), and only fair resistance to solvents. However, they are lightweight, are good thermal and electrical insulators, are easy to fabricate and install, and have low friction factors.

Generally, plastics have excellent resistance to weak mineral acids and are unaffected by inorganic salt solutions—areas where metals are not entirely suitable. Since plastics do not corrode in the electrochemical sense, they offer another advantage over metals: most metals are affected by slight changes in pH, or minor impurities, or oxygen content, while plastics will remain resistant to these same changes.

The important thermoplastics used commercially are polyethylene, acrylonitrile butadiene styrene (ABS), polyvinyl chloride (PVC), cellulose acetate butyrate (CAB), vinylidene chloride (Saran), fluorocarbons (Teflon, Halar, Kel-F, Kynar), polycarbonates, polypropylene, nylons, and acetals (Delrin). Important thermosetting plastics are general-purpose polyester glass reinforced, bisphenol-based polyester glass, epoxy glass, vinyl ester glass, furan and phenolic glass, and asbestos reinforced.

THERMOPLASTICS

The most chemical-resistant plastic commercially available today is **tetrafluoroethylene** or **TFE** (Teflon). This thermoplastic is practically unaffected by all alkalies and acids except fluorine and chlorine gas at elevated temperatures and molten metals. It retains its properties up to 260°C (500°F). **Chlorotrifluoroethylene** or **CTFE** (Kel-F, Plaskon) also possesses excellent corrosion resistance to almost all acids and alkalies up to 180°C (350°F). A Teflon derivative has been developed from the copolymerization of tetrafluoroethylene and hexafluoropropylene. This resin, **FEP**, has similar properties to TFE except that it is not recommended for continuous exposures at temperatures above 200°C (400°F). Also, FEP can be extruded on conventional extrusion equipment, while TFE parts must be made by complicated powder-metallurgy techniques. Another version is **polyvinylidene fluoride**, or **PVF₂** (Kynar), which has excellent resistance to alkalies and acids to 150°C (300°F). It can be extruded. A more recent development is a copolymer of CTFE and ethylene (Halar). This material has excellent resistance to strong inorganic acids, bases, and salts up to 150°C. It also can be extruded.

Perfluoroalkoxy, or **PFA** (Teflon), has the general properties and chemical resistance of FEP at a temperature approaching 300°C (600°F).

Polyethylene is the lowest-cost plastic commercially available. Mechanical properties are generally poor, particularly above 50°C (120°F), and pipe must be fully supported. Carbon-filled grades are resistant to sunlight and weathering.

Unplasticized polyvinyl chlorides (type I) have excellent resistance to oxidizing acids other than concentrated and to most nonoxidizing acids. Resistance is good to weak and strong alkaline materials. Resistance to chlorinated hydrocarbons is not good. Polyvinylidene chloride, known as **Saran,** has good resistance to chlorinated hydrocarbons.

Acrylonitrile butadiene styrene (ABS) polymers have good resistance to nonoxidizing and weak acids but are not satisfactory with oxidizing acids. The upper temperature limit is about 65°C (150°F).

TABLE 28-17 Copper Alloys*

Alloy	CDA	UNS	Composition, %†						Mechanical properties‡		
			Cu	Zn	Sn	Al	Ni	Other	Yield strength, kip/in² (MPa)	Tensile strength, kip/in² (MPa)	Elongation in 2 in, %
Wrought											
Copper	110	C11000	99.90						10 (69)	32 (221)	55
Commercial bronze	220	C22000	89–91	Rem.					10 (69)	37 (255)	50
Red brass	230	C23000	84–86	Rem.					10 (69)	40 (276)	55
Cartridge brass	260	C26000	68.5–71.5	Rem.					11 (76)	44 (303)	66
Yellow brass	270	C27000	63–68.5	Rem.					14 (97)	46 (317)	65
Muntz Metal	280	C28000	59–63	Rem.					21 (145)	54 (372)	52
Admiralty brass	443	C44300	70–73	Rem.	0.9–1.2			0.02–0.1 As	18 (124)	48 (331)	65
Admiralty brass	444	C44400	70–73	Rem.	0.9–1.2			0.02–0.1 Sb	18 (124)	48 (331)	65
Admiralty brass	445	C44500	70–73	Rem.	0.9–1.2			0.02–0.1 P	18 (124)	48 (331)	65
Naval brass	464	C46400	59–62	Rem.	0.5–1.0				25 (172)	58 (400)	50
Phosphor bronze	510	C51000	Rem.	0.3	4.2–5.8			0.03–0.35 P	19 (131)	47 (324)	64
Phosphor bronze	524	C52400	Rem.	0.2	9.0–11.0			0.03–0.35 P	28 (193)	65 (455)	70
Aluminum bronze	613	C61300	86.5–93.8		0.2–0.5	6–8	0.5	3.5 Fe	30 (207)	70 (483)	42
Aluminum bronze D	614	C61400	88.0–92.5	0.2		6–8		1.5–3.5 Fe, 1.0 Mn	33 (228)	76 (524)	45
Nickel-aluminum bronze	630	C63000	78–85	0.3	0.2	9–11	4.0–5.5	2.0–4.0 Fe, 1.5 Mn, 0.25 Si	36 (248)	90 (620)	10
High-silicon bronze	655	C65500	94.8	1.5			0.6	0.8 Fe, 0.5–1.3 Mn, 2.8–3.8 Si	21 (145)	56 (386)	63
Manganese bronze	675	C67500	57–60	Rem.	0.5–1.5	0.25		0.05–0.5 Mn, 0.8–2.0 Fe	30 (207)	65 (448)	33
Aluminum brass	687	C68700	76–79	Rem.		1.8–2.5		0.02–0.1 As	27 (186)	60 (414)	55
90-10 copper nickel	706	C70600	86.5	1.0			9.0–11.0	1.0–1.8 Fe, 1.0 Mn	16 (110)	44 (303)	42
70-30 copper nickel	715	C71500	Rem.	1.0			29–33	0.4–1.0 Fe, 1.0 Mn	20 (138)	54 (372)	45
65-18 nickel silver	752	C75200	63–66.5	Rem.			16.5–19.5	0.25 Fe, 0.5 Mn	25 (172)	56 (386)	45
Cast											
Ounce metal	836	C83600	84–86	4–6	4–6	0.005	1.0	4–6 Pb	17 (117)	37 (255)	30
Manganese bronze	865	C86500	55–65	36–42	1.0	0.5–1.5	1.0	0.4–2.0 Fe, 0.1–1.5 Mn	28 (193)	71 (490)	30
G bronze	905	C90500	86–89	1.0–3.0	9–11	0.005	1.0		22 (152)	45 (310)	25
M bronze	922	C92200	86–90	3.0–5.0	5.5–6.5	0.005	1.0	1.0–2.0 Pb	20 (138)	40 (276)	30
Ni-Al-Mn bronze	957	C95700	71			7.0–8.5	1.5–3.0	2.0–4.0 Fe, 11–14 Mn	45 (310)	95 (665)	26
Ni-Al bronze	958	C95800	79			8.5–9.5	4.0–5.0	3.5–4.5 Fe, 0.8–1.5 Mn	38 (262)	95 (655)	25
Copper nickel	964	C96400	65–69				28–32	0.5–1.5 Cb, 0.25–1.5 Fe, 1.5 Mn	37 (255)	68 (469)	28

*Courtesy of National Association of Corrosion Engineers. To convert MPa to lbf/in², multiply by 145.04.
†Single values are maximum values except for Cu, which is minimum.
‡Typical room-temperature properties of annealed or as-cast material.

Acetals have excellent resistance to most organic solvents but are not satisfactory for use with strong acids and alkalies.

Cellulose acetate butyrate is not affected by dilute acids and alkalies or gasoline, but chlorinated solvents cause some swelling. **Nylons** resist many organic solvents but are attacked by phenols, strong oxidizing agents, and mineral acids.

Polypropylene has a chemical resistance about the same as that of polyethylene, but it can be used at 120°C (250°F). **Polycarbonate** is a relatively high-temperature plastic. It can be used up to 150°C (300°F). Resistance to mineral acids is good. Strong alkalies slowly decompose it, but mild alkalies do not. It is partially soluble in aromatic solvents and soluble in chlorinated hydrocarbons. **Polyphenylene oxide** has good resistance to aliphatic solvents, acids, and bases but poor resistance to esters, ketones, and aromatic or chlorinated solvents.

Polyphenylene sulfide (**PPS**) has no known solvents below 190 to 205°C (375 to 400°F); mechanical properties of PPS are unaffected by exposures in air at 230°C (450°F). It is resistant to aqueous inorganic salts and bases.

Polysulfone can be used to 170°C (340°F); it is highly resistant to mineral acid, alkali, and salt solutions as well as to detergents, oils, and alcohols. It is attacked by such organic solvents as ketones, chlorinated hydrocarbons, and aromatic hydrocarbons.

Polyamide or polyimide polymers are resistant to aliphatic, aromatic, and chlorinated or fluorinated hydrocarbons as well as to many acidic and basic systems but are degraded by high-temperature caustic exposures.

Thermosetting Plastics Among the thermosetting materials are phenolic plastics filled with asbestos, carbon or graphite, glass, and silica. Relatively low cost, good mechanical properties, and chemical resistance (except against strong alkalies) make phenolics popular for chemical equipment. Furan plastics filled with asbestos and glass have much better alkali resistance than phenolic resins. They are more expensive than the phenolics but also offer somewhat higher strengths.

Polyester resins, reinforced with fiberglass, have good strength and good chemical resistance except to alkalies. Some special materials in this class, based on bisphenol and vinyl esters are more alkali-

TABLE 28-18   Typical Mechanical Properties of Selected Materials

Material	Tensile modulus, GPa	Tensile strength, yield, MPa	Tensile strength, ultimate, MPa	Elongation, %	Poisson's ratio	Fracture toughness, MPa√m
High-nickel gray cast iron						
20°C	79–130		170–310			
500°C			130–212			
700°C			60–115			
Medium-silicon ductile cast iron						
295°C		210–240	415–690	0.2		
Low-carbon steel sheet				In 50 mm		
20°C Hot rolled		27–48	40–66	25–43		
20°C Cold rolled		23–37	43–56	30–44		
SA-335P12 (ASME) ferritic steel				In 50 mm		
20°C	200	207	415	30(L), 20(T)	0.288	
540°C	155	170	345		0.303	
Type 4330V low-alloy steel						
Oil-quenched plate						
20°C	200	1150	1330	13		130
Type 301 stainless steel				In 50 mm		
−100°C Annealed	196	410	1300			
20°C	190	205	515	30		214
650°C	80	100	260	34		225
Type 304 stainless steel				In 50 mm		
−200°C Annealed bar	204	280	1400–1520	32–45		
20°C	190	350–560	660–770	45–72	0.26	
400°C	168	245–475	525–610	26–48	0.32	
800°C	120	210–280	245–350	22–60		
Type 316 stainless steel				In 50 mm		
−200°C Annealed	232		1440	60 (25 mm)	0.281	
20°C	196	250	560	64	0.294	
400°C	167	170	500	48	0.331	
800°C	130	120	200	42	0.265	
Type 347 stainless steel				In 50 mm		
−200°C Annealed bar		430	1500	43		
20°C	200	380	670	50	0.30	
400°C	170	150–180	460	34	0.32	
800°C	105	110	160	42	0.25	
Alloy 42 (Fe + 42Ni)						
20°C	143	286	492	33.5	0.34	
100°C	150	222	445	41.3	0.33	
175°C	165	182	408	29.6	0.32	
250°C	174	152	388	34.6	0.30	
Kovar (Fe + 29Ni + 17Co)						
20°C	134	372	526	31.5	0.37	
100°C	139	295	454	30	0.36	
175°C	149	235	418	36	0.35	
250°C	153	192	396	35	0.33	
63Sn + 37Pb						
−70°C	34	60	85			
20°C	31	38	53	35		
125°C		18	25	48		
150°C		15	15	105		
Aluminum						
−200°C Annealed	76	61	184	13	0.34	
20°C	70	37	184	19.5	0.35	
400°C	58	5.5	18	19.5	0.36	
600°C	49		9.2		0.38	
Antimony						
20°C	77.7		11.40			
Beryllium, grade S-200 F						
20°C	303	262–269	380–413	2–5		9–13
400°C		255–262	317–331	11–26		
1000°C		138–145	186–200	6–10		
Bismuth						
20°C	32					
Cadmium				In 25 mm		
20°C	55		69–83	50	0.33	
Chromium (electrolytic)				In 25 mm		
20°C	248		83	0		
400°C	227	140	225	51		
800°C	255	97	180	47		
Cobalt				In 50 mm		
20°C Annealed	211		255.1			
Zone-refined		758.5	944.6			
Annealed strip		310–345		15–22		

TABLE 28-18 Typical Mechanical Properties of Selected Materials (*Continued*)

Material		Tensile modulus, GPa	Tensile strength, yield, MPa	Tensile strength, ultimate, MPa	Elongation, %	Poisson's ratio	Fracture toughness, MPa√m
Copper							
−200°C	Annealed	137	96	360	46	0.336	
20°C		128	71	212	51	0.344	
100°C		125	71	181	53	0.346	
300°C			70	128	56		
400°C		112				0.353	
500°C		107				0.358	
−200°C	Cold worked		200	390–517			
20°C			170	230–393			
100°C			160	205–381			
200°C			140	175–350			
600°C				46			
20°C	Cast			145			
100°C				120			
200°C				98			
600°C				37			
Gold							
20°C		78		103	30		
Iron							
−200°			320–480				
20°C		208.2	70–140			0.291	
Lead							
−200°C	Annealed	24					
20°C		18.8		12	68		
100°C		16.7		8	74		
150°C				5.5	86		
Magnesium (annealed sheet)					In 50 mm		
20°C		40	90–105	160–195	3–15		
Manganese (γ phase)							
20°C		191	241	496	40		
Molybdenum							
−100°C			650–750				
20°C		315	310–385	550–650			
500°C		285	110	250–315			
1200°C		215		100–140			
Tin							
−200°C	Annealed	61					
20°C		50		24–38	33	0.33	
100°C		43.5		12–19	41		
150°C				8–10	50		
Titanium					In 50 mm		
20°C		105	140	235	54		
400°C		80		105	41		
Tungsten, annealed							
20°C		407–410					
400°C		393–396	75–130	275–355	33–58		
1800°C		322–325	6–42	35–85	19–43		
Teflon (PTFE)							
−200°C		5.5	111	112			
20°C		0.86	13	30			
100°C		0.23	7	10			
250°C			3	1.2			
Kapton							
−200°C		5.2	236 (L)	360		0.275	
20°C		3.9	105	212	72	0.365	8–10
100°C		3.1	87	161		0.441	6.3–8.8
200°C		1.3	80	117		0.568	
Kevlar 49		L T					
20°C		126,6.9		3480	2.9	0.36	
100°C		114,16.6		3200	2.8	0.36	
200°		110		2800			
Amoco P75/ERLX 1962 [0]₈ unidirectional laminate					Strain-to-Failure		
20°C		46		120	0.26		
S-glass/Epoxy 2D (0)		Hercules 3501-6		Shell Epon (828 & 1031)			
−100°C 0° Orientation		60					
20°C		57–80		1600			
100°C		50–62		1350–1600			
20°C 90° Orientation		12–22					
100°C		5–19					

TABLE 28-18 Typical Mechanical Properties of Selected Materials (Concluded)

Material		Tensile modulus, GPa	Tensile strength, yield, MPa	Flexural strength, MPa	Poisson's ratio	Fracture toughness, MPa√m
Alumina (>99%)						
20°C		401	275–1030	520	0.24	3.2–4.0
600°C		343		360	0.25	
1200°C				340	0.35	
1400°C				260		
Silicon nitride						
20°C	Sintered	248 (195–315)		415 (275–840)	0.28	4.66 (4.40–5.40)
1400°C				70 (0–700)	0.26	4.70 (3.40–6.80)
20°C	Hot pressed	314 (250–325)	375	750 (450–1100)		
1000°C		300		600		
1400°C		175 (175–250)	150	300 (0–600)		
Silicon carbide						
20°C	Hot pressed	450 (430–450)	200	470	0.17	3.85 (3–4)
800°C				412		
1000°C				227		3.59
1200°C				324		3.52
1400°C		380	35–150	175 (175–525)		2.81
20°C	Sintered	411 (375–420)		460	0.16	3.01 (2.5–6.5)
800°C				326	0.165	
1000°C				353	0.170	
1200°C				377	0.192	
1400°C		372		240		
Boron nitride (H)		L T		L T		
20°C	Hot pressed	92,36		120,51		
100°C		84,34		116,50		
400°C		61,27		105,44		
400°C		24,10		60,12		
Aluminum nitride						
20°C	Hot pressed	345		390		2.90 (2.90–3.40)
700°C		343		375		2.80
1050°C				267		3.30
Silicon						
20°C		130 [100]	2900	2800–3100		0.76
400°C			200			
800°C			23			
1000°C			7.5			
1200°C			2.5			
Silica						
−200°C	Fused	68				
20°C		71	49	66	0.17	
400°C		76				
900°C		82				

SOURCE: Center for Information and Numerical Data Analysis and Synthesis (CINDAS), Purdue University, West Lafayette, Ind.

resistant. The temperature limit for polyesters is about 90 to 150°C (200 to 300°F), depending upon exposure conditions.

Epoxies reinforced with fiberglass have very high strengths and resistance to heat. The chemical resistance of the epoxy resin is excellent in nonoxidizing and weak acids but not good against strong acids. Alkaline resistance is excellent in weak solutions. Resistance is poor to such organic solvents as ketones, chlorinated hydrocarbons, and aromatic hydrocarbons.

The thermoset polyimides are a family of heat-resistant polymers with acceptable properties up to 260°C (500°F). They are unaffected by dilute acids, aromatic and aliphatic hydrocarbons, esters, ethers, and alcohols but are attacked by dilute alkalies and concentrated inorganic acids.

Chemical resistance of thermosetting-resin-glass-reinforced laminates may be affected by any exposed glass in the laminate.

Phenolic asbestos, general-purpose polyester glass, Saran, and CAB are adversely affected by alkalies. And thermoplastics generally show poor resistance to organics.

The lack of homogeneity and the friable nature of FRP composite structures dictate that caution be followed in mechanical design, vendor selection, inspection, shipment, installation, and use.

FRP code vessels for pressure service over 0.1 MPA (15 lbf/in²) may be designed and built under ASME Sec. X. Equipment for service from full vacuum through 0.1-MPA (15-lbf/in²) pressure, while not presently covered by an ASME code designation, can be designed or fabricated in accordance with the SPI-MTI Quality Assurance Practices and Procedures Report for RTP equipment.

Rubber and Elastomers Rubber and elastomers are widely used as lining materials. To meet the demands of the chemical industry, rubber processors are continually improving their products. A number of synthetic rubbers have been developed, and while none has all the properties of natural rubber, they are superior in one or more ways. The isoprene and polybutadiene synthetic rubbers are duplicates of natural.

The ability to bond natural rubber to itself and to steel makes it ideal for lining tanks. Many of the synthetic elastomers, while more chemically resistant than natural rubber, have very poor bonding characteristics and hence are not well suited for lining tanks.

Natural rubber is resistant to dilute mineral acids, alkalies, and salts, but oxidizing media, oils, and most organic solvents will attack it. **Hard rubber** is made by adding 25 percent or more of sulfur to natural or synthetic rubber and, as such, is both hard and strong. **Chloroprene** or **neoprene rubber** is resistant to attack by ozone, sunlight, oils, gasoline, and aromatic or halogenated solvents but is easily permeated by water, thus limiting its use as a tank lining. **Styrene rubber** has chemical resistance similar to that of natural. **Nitrile rubber** is known for resistance to oils and solvents. **Butyl rubber**'s resistance to dilute mineral acids and alkalies is exceptional; resistance to concen-

TABLE 28-19 Miscellaneous Alloys*

Alloy	Designation	UNS	Composition, %‡	Condition	Yield strength, kip/in² (MPa)	Tensile strength, kip/in² (MPa)	Elongation, %	Hardness, HB
Refractory alloys								
Niobium R04210 (columbium)		204–210	99.6 Cb	Annealed	37 (255)	53 (365)	26	80
Molybdenum		R03600	0.01–0.04 C					
Molybdenum, low C		R03630	0.01 C					
Molybdenum alloy		R03650	0.01–0.04 C, 0.40–0.55 Ti, 0.06–0.12 Zn					
Tantalum		R05200	99.8 min. Ta	Annealed		50 (345)	40	45
Tungsten		R07030	99.9 min. W	Annealed		270 (1862)		
Zirconium		R60702	4.5 Hf, 0.2 Fe + Cr, 99.2 Zi + Hf	Annealed	16 (110)	36 (248)	31	77
Precious metals and alloys								
Gold		P00020	99.95 min. Au	Annealed		19 (131)	45	25
Silver		P07015	99.95 min. Ag	Annealed	8 (55)	18 (124)	54	27
Sterling silver			7.5 Cu, 92.5 Ag	Annealed	20 (138)	41 (283)	26	65
Platinum		P04955	99.95 min. Pt	Annealed		18 (124)	38	39
Palladium		P03980	99.80 min. Pd	Annealed		25 (172)	27	38
Lead alloys								
Chemical lead			99.9 min. Pb	Rolled	1.9 (13)	2.5 (17)	50	5
Antimonial lead			90 Pb, 10 Sb	Rolled		4.1 (28)	47	13
Tellurium lead			99.85 Pb, 0.04 Te, 0.06 Cu	Rolled	2.2 (15)	3 (21)	45	6
50-50 solder		L05500	50 Pb, 50 Sn, 0.12 max. Sb	Cast		6.8 (47)	50	14
Magnesium alloys								
Wrought alloy	AZ31B	M11311	2.5–3.5 Al, 0.20 min. Mn, 0.6–1.4 Zn	Annealed	15–18 (103–124)	32 (220)	9–12	56
Cast alloy	AZ91C	M11914	8.1–9.3 Al, 0.13 min. Mn, 0.4–1.0 Zn	As cast	11 (76)	23 (159)		60
Cast alloy	EZ33A	M12330	2.0–3.1 Zn, 0.5–1.0 Zr	Aged	14 (97)	20 (138)	2	50
Wrought alloy	HK31A	M13310	0.3 Zn, 2.5–4.0 Th, 0.4–1.0 Zr	Stress hard-annealed	24–26 (165–179)	33–34 (228–234)	4	57
Titanium alloys								
Commercial pure	Gr. 1	R50250	0.20 Fe, 0.18 O	Annealed	35 (241)	48 (331)	30	120
Commercial pure	Gr. 2	R50400	0.30 Fe, 0.25 O	Annealed	50 (345)	63 (434)	28	200
Ti-Pd	Gr. 7	R52400	0.30 Fe, 0.25 O, 0.12–0.25 Pd	Annealed	50 (345)	63 (434)	28	200
Ti-6Al-4V	Gr. 5	R56400	5.5–5.6 Al, 0.40 Fe, 0.20 O, 3.5–4.5 V	Annealed	134 (924)	144 (993)	14	330
Low alloy	Gr. 12		0.2–0.4 Mo, 0.6–0.9 Ni	Annealed	65 (448)	75 (517)	25	
Cobalt alloys								
N-155	N-155	R30155	0.08–0.16 C, 0.75–1.25 Cb, 18.50–21.0 Co, 20.0–22.5 Cr, 1.0–2.0 Mn, 2.5–3.5 Mo, 19–21 Ni, 1.0 Si, 2.0–3.0 W					
MP35N	MP35N	R30036	0.025 C, 19–21 Cr, 1.0 Fe, 0.15 Mn, 9.0–10.5 Mo, 33.37 Ni, 0.15 Si, 1.0 Ti	Annealed	60 (414)	135 (931)	70	
Stelite 6	Stelite 6	R30006	0.9–1.4 C, 27–31 Cr, 3 Fe, 1.0 Mn, 1.5 Mo, 3.0 Ni, 1.5 Si, 3.5–5.5 W	As cast		105 (724)	1	

*Courtesy of National Association of Corrosion Engineers. To convert MPa to lbf/in², multiply by 145.04.
†Typical room-temperature properties.
‡Single values are maximum values unless otherwise noted.

TABLE 28-20   Properties of Glass and Silica*

	Pyroceram	96% silica	Borosilicate	Glass lining
Specific gravity, 77°F	2.60	2.18	2.23	2.56
Water absorption, %	0.00	0.00	0.00	
Gas permeability	Gastight	Gastight	Gastight	
Softening temperature, °F (°C)	2282 (1250)	2732 (1500)	1508 (1820)	
Specific heat, 77°F Btu/(lb·°F)[J/(kg·K)]	0.185 (775)	0.178 (746)	0.186 (779)	
Mean specific heat (77–752°F)	0.230	0.224	0.233	
Thermal conductivity, mean temperature, 77°F, Btu/(ft²·h·°F)/in [W/(m·K)]	25.2 (3.6)		7.5 (1.1)	
Linear thermal expansion, per °F (77–572°F); (per °C), × 10⁻⁶	3.2 (5.8)	0.44 (0.79)	1.8 (3.2)	
Modulus of elasticity, kip/in² (MPa) × 10³	17.3 (119)	9.6 (66)	9.5 (66)	6–9 (40–60)
Poisson's ratio	0.245	0.17	0.20	
Modulus of rupture, kip/in²	20 (140)	5–9 (35–63)	6–10 (42–70)	
Knoop hardness, 100 g	698	532	481	480
Knoop hardness, 500 g	619	477	442	
Adhesion strength kip/in² (MPa)				5–10 (35–70)
Maximum operating temperature, °F (°C)				500 (260)
Thermal shock resistance, temperature difference, °F (°C)				305 (152)

*Courtesy of National Association of Corrosion Engineers.

trated acids, except nitric and sulfuric, is good. **Silicone rubbers,** also known as polysiloxanes, have outstanding resistance to high and low temperatures as well as against aliphatic solvents, oils, and greases. **Chlorosulfonated polyethylene,** known as **Hypalon,** has outstanding resistance to ozone and oxidizing agents except fuming nitric and sulfuric acids. Oil resistance is good. **Fluoroelastomers (Viton A, Kel-F, Kalrez)** combine excellent chemical and temperature resistance. **Polyvinyl chloride elastomer (Koroseal)** was developed to overcome some of the limitations of natural and synthetic rubbers. It has excellent resistance to mineral acids and petroleum oils.

The **cis-polybutadiene, cis-polyisoprene,** and **ethylene-propylene** rubbers are close duplicates of natural rubber. The newer ethylene-propylene rubbers (EPR) have excellent resistance to heat and oxidation.

Asphalt   Asphalt is used as a flexible protective coating, as a brick-lining membrane, and as a chemical-resisting floor covering and road surface. Resistant to acids and bases, alphalt is soluble in organic solvents such as ketones, most chlorinated hydrocarbons, and aromatic hydrocarbons.

Carbon and Graphite   The chemical resistance of impervious carbon and graphite depends somewhat on the type of resin impregnant used to make the material impervious. Generally, impervious graphite is completely inert to all but the most severe oxidizing conditions. This property, combined with excellent heat transfer, has made impervious carbon and graphite very popular in heat exchangers, as brick lining, and in pipe and pumps. One limitation of these materials is low tensile strength. Threshold oxidation temperatures are 350°C (660°F) for carbon and 400°C (750°F) for graphite.

TABLE 28-21   Chemical Resistance of Important Plastics

	Poly-propylene poly-ethylene	CAB°	ABS†	PVC‡	Saran§	Polyester glass¶	Epoxy glass	Phenolic asbestos	Fluoro-carbons	Chlorinated polyether (Penton)	Poly-carbonate
10% H₂SO₄	Excel.	Good	Excel.	Excel.	Excel.	Excel.	Excel.	Excel.	Excel.	Excel.	Excel.
50% H₂SO₄	Excel.	Poor	Excel.	Excel.	Excel.	Good	Excel.	Excel.	Excel.	Excel.	Excel.
10% HCl	Excel.	Excel.	Excel.	Excel.	Excel.	Excel.	Excel.	Excel.	Excel.	Excel.	Excel.
10% HNO₃	Excel.	Poor	Good	Excel.	Excel.	Good	Good	Fair	Excel.	Excel.	Excel.
10% Acetic	Excel.	Good	Excel.	Excel.	Excel.	Excel.	Excel.	Excel.	Excel.	Excel.	Excel.
10% NaOH	Excel.	Fair	Excel.	Good	Fair	Fair	Excel.	Poor	Excel.	Excel.	Excel.
50% NaOH	Excel.	Poor	Excel.	Excel.	Fair	Poor	Good	Poor	Excel.	Excel.	Excel.
NH₄OH	Excel.	Poor	Excel.	Excel.	Poor	Fair	Excel.	Poor	Excel.	Excel.	Excel.
NaCl	Excel.	Excel.	Excel.	Excel.	Excel.	Excel.	Excel.	Excel.	Excel.	Excel.	Excel.
FeCl₃	Excel.	Excel.	Excel.	Excel.	Excel.	Excel.	Excel.	Excel.	Excel.	Excel.	Excel.
CuSO₄	Excel.	Excel.	Excel.	Excel.	Excel.	Excel.	Excel.	Excel.	Excel.	Excel.	Excel.
NH₄NO₃	Excel.	Excel.	Excel.	Excel.	Excel.	Excel.	Excel.	Good	Excel.	Excel.	Excel.
Wet H₂S	Excel.	Excel.	Excel.	Excel.	Excel.	Excel.	Excel.	Excel.	Excel.	Excel.	
Wet Cl₂	Poor	Poor	Excel.	Good	Poor	Poor	Poor	Excel.	Excel.	Excel.	
Wet SO₂	Excel.	Poor	Excel.	Excel.	Good	Excel.	Excel.	Excel.	Excel.	Excel.	
Gasoline	Poor	Excel.	Excel.	Excel.	Excel.	Excel.	Excel.	Excel.	Excel.	Excel.	Excel.
Benzene	Poor	Poor	Poor	Poor	Fair	Good	Excel.	Excel.	Excel.	Fair	Fair
CCl₄	Poor	Poor	Poor	Fair	Fair	Excel.	Good	Excel.	Excel.	Fair	Poor
Acetone	Poor	Poor	Poor	Poor	Fair	Poor	Good	Poor	Excel.	Good	Good
Alcohol	Poor	Poor	Excel.	Excel.	Excel.	Excel.	Excel.	Excel.	Excel.	Excel.	Excel.

NOTE: Ratings are for long-term exposures at ambient temperatures [less than 38°C (100°F)].
°Cellulose acetate butyrate.
†Acrylonitrile butadiene styrene polymer.
‡Polyvinyl chloride, type I.
§Chemical resistance of Saran-lined pipe is superior to extruded Saran in some environments.
¶Refers to general-purpose polyesters. Special polyesters have superior resistance, particularly in alkalies.

Several types of resin impregnates are employed in manufacturing impervious graphite. The standard impregnant is a phenolic resin suitable for service in most acids, salt solutions, and organic compounds. A modified phenolic impregnant is recommended for service in alkalies and oxidizing chemicals. Furan and epoxy thermosetting resins are also used to fill structural voids. The chemical resistance of the impervious graphite is controlled by the resin used. However, no type of impervious graphite is recommended for use in over 60 percent hydrofluoric, over 20 percent nitric, and over 96 percent sulfuric acids and in 100 percent bromine, fluorine, or iodine.

Wood   While fairly inert chemically, wood is readily dehydrated by concentrated solutions and hence shrinks badly when subjected to the action of such solutions. It is also slowly hydrolyzed by acids and alkalies, especially when hot. In tank construction, if sufficient shrinkage once takes place to allow crystals to form between the staves, it becomes very difficult to make the tank tight again.

A number of manufacturers offer wood impregnated to resist acids or alkalies or the effects of high temperatures.

HIGH- AND LOW-TEMPERATURE MATERIALS

LOW-TEMPERATURE METALS

The low-temperature properties of metals have created some unusual problems in fabricating cryogenic equipment.

Most metals lose their ductility and impact strength at low temperatures, although in many cases yield and tensile strengths increase as the temperature goes down.

Materials selection for low-temperature service is a specialized area. In general, it is necessary to select materials and fabrication methods which will provide adequate toughness at all operating conditions. It is frequently necessary to specify Charpy V-notch (or other appropriate) qualification tests to demonstrate adequate toughness of carbon and low-alloy steels at minimum operating temperatures.

Stainless Steels   Chromium-nickel steels are suitable for service at temperatures as low as −250°C (−425°F). Type 304 is the most popular. The original cost of stainless steel may be higher than that of another metal, but ease of fabrication (no heat treatment) and welding, combined with high strength, offsets the higher initial cost. Sensitization or formation of chromium carbides can occur in several stainless steels during welding, and this will affect impact strength. However, tests have shown that impact properties of types 304 and 304L are not greatly affected by sensitization but that the properties of 302 are impaired at −185°C (−300°F).

Nickel Steel   Low-carbon 9 percent nickel steel is a ferritic alloy developed for use in cryogenic equipment operating as low as −195°C (−320°F). ASTM specifications A 300 and A 353 cover low-carbon 9 percent nickel steel (A 300 is the basic specification for low-temperature ferritic steels). Refinements in welding and (ASME code-approved) elimination of postweld thermal treatments make 9 percent steel competitive with many low-cost materials used at low temperatures.

Aluminum   Aluminum alloys have unusual ability to maintain strength and shock resistance at temperatures as low as −250°C (−425°F). Good corrosion resistance and relatively low cost make these alloys very popular for low-temperature equipment. For most welded construction the 5000-series aluminum alloys are widely used. These are the aluminum-magnesium and aluminum-magnesium-manganese materials.

Copper and Alloys   With few exceptions the tensile strength of copper and its alloys increases quite markedly as the temperature goes down. However, copper's low structural strength becomes a problem when constructing large-scale equipment. Therefore, alloy must be used. One of the most successful for low temperatures is silicon bronze, which can be used to −195°C (−320°F) with safety.

HIGH-TEMPERATURE MATERIALS

Metals   Successful applications of metals in high-temperature process service depend on an appreciation of certain engineering factors. The important alloys for service up to 1,100°C (2,000°F) are shown in Table 28-35. Among the most important properties are creep, rupture, and short-time strengths (see Figs. 28-23 and 28-24). **Creep** relates initially applied stress to rate of plastic flow. **Stress**

rupture is another important consideration at high temperatures since it relates stress and time to produce rupture. As the figures show, ferritic alloys are weaker than austenitic compositions, and in both groups molybdenum increases strength. Austenitic castings are much stronger than their wrought counterparts. And higher strengths are available in the superalloys. Other properties which become important at high temperatures include thermal conductivity, thermal expansion, ductility at temperature, alloy composition, and stability.

Actually, in many cases strength and mechanical properties become of secondary importance in process applications, compared with resistance to the corrosive surroundings. All common heat-resistant alloys form oxides when exposed to **hot oxidizing environments.** Whether the alloy is resistant depends upon whether the oxide is stable and forms a protective film. Thus, mild steel is seldom used above 480°C (900°F) because of excessive scaling rates. Higher temperatures require **chromium** (see Fig. 28-25). Thus, type 502 steel, with 4 to 6 percent Cr, is acceptable to 620°C (1,150°F). A 9 to 12 percent Cr steel will handle 730°C (1,350°F); 14 to 18 percent Cr extends the limit to 800°C (1,500°F); and 27 percent Cr to 1,100°C (2,000°F).

The well-known austenitic stainless steels have excellent oxidation resistance: up to 900°C (1,650°F) for 18-8; and up to 1,100°C (2000°F) for 25-12 (and Inconel 600 and Incoloy 800). The cobalt-

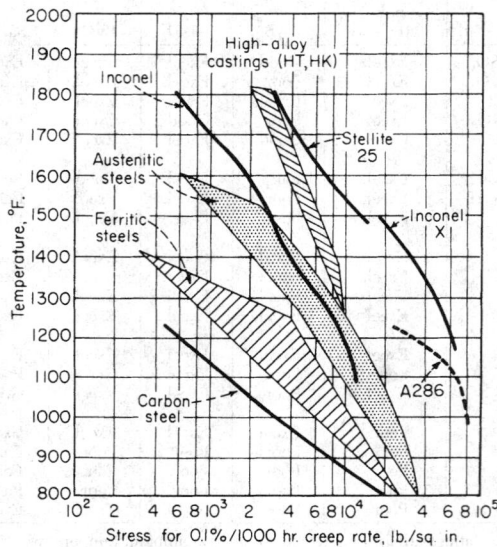

FIG. 28-23   Effect of creep on metals for high-temperature use. °C = (°F − 32) × ⁵⁄₉; to convert lbf/in² to MPa, multiply by 6.895×10^{-3}. [*Chem. Eng., 139* (Dec. 15, 1958).]

TABLE 28-22 Typical Property Ranges for Plastics

Thermosets[a]	Specific gravity	Tensile strength (kip)	Tensile strength (MPa)	Modulus of elasticity, tension (10³ kip/in²)	Modulus (10² MPa)	Impact strength, Izod[b] (ft-lb)	Impact (J)	Max use temp (no load) °F	Max use temp °C	HDT at 254 lbf/in² °F	HDT °C	Weather resistance	Weak acid	Strong acid	Weak alkali	Strong alkali	Solvents
Alkyds																	
Glass-filled	2.12–2.15	4–9.5	28–66	20–28	138–193	0.6–10	0.8–14	450	230	400–500	200–260	R	A	A	A	A	A
Mineral-filled	1.60–2.30	3–9	21–62	5–30	34–207	0.3–0.5	0.4–0.7	300–450	150–230	350–500	150–260	R	R	A	A	D	A
Asbestos-filled	1.65	4.5–7	31–48			0.4–0.5	0.6–0.7	450	230	315	160	R	R	S	R	S	R
Synthetic fiber-filled	1.24–2.10	4.5–7	31–48	20	138	0.5–4.5	0.7–6.1	300–430	150–220	245–430	120–220	R	R	S	R	S	A
Alkyl diglycol carbonate	1.30–1.40	5–6	34–41	3.0	21	0.2–0.4	0.3–0.5	212	100	140–190	60–90	R	R	A[e]	R	R-S	R
Diallyl phthalates																	
Glass-filled	1.61–1.78	6–11	41–76	14–22	97–152	0.4–15	0.5–20	300–400	150–200	330–540	165–280	R	R	S	R-S	S	R
Mineral-filled	1.65–1.68	5–9	34–62	12–22	83–152	0.3–0.5	0.4–1	300–400	150–200	320–540	160–280	R	R	S	R-S	S	R
Asbestos-filled	1.55–1.65	7–8	48–55	12–22	83–152	0.4–0.5	0.5–0.7	300–400	150–200	320–540	160–280	R	R	S	R-S	S	R
Epoxies (*bis*-A)																	
No filler	1.06–1.40	4–13	28–90	2.15–5.2	15–36	0.2–1.0	0.3–1.4	250–500	120–260	115–500	45–260	R	R	A	R	S	R-S
Graphite-fiber reinforced	1.37–1.38	185–200	1280–1380	118–120	814–827							S	R	R	R	R	R-S
Mineral-filled	1.6–2.0	5–15	34–103	30	207	0.3–0.4	0.4–0.5	300–500	150–260	250–500	120–260	S	R	R-S	R	R	R-S
Glass-filled	1.7–2.0	10–30	69–207	2.15–5.2	15–36	10–30	14–41	400–500	200–260	450–500	230–260	S	R	R	R	R	R
Epoxies (novolac): no filler	1.12–1.24	5–11	34–76			0.3–0.7	0.4–0.9	480–550	250–290	500–550	260–290	R	R	R-A	R	R-A	R
Epoxies (cycloaliphatic): no filler	1.12–1.18	10–17.5	69–121	5–7	34–48												
Melamines																	
Cellulose-filled	1.45–1.52	5–9	34–62	11	76	0.2–0.4	0.3–0.5	250	120	270	130	S	R-S	D	R	D	R
Flock-filled	1.50–1.55	7–9	48–62			0.4–0.5	0.5–0.7	250	120	270	130	S	R-S	D	R	D	R-S
Asbestos-filled	1.70–2.0	5–7	34–48	20	138	0.3–0.4	0.4–0.5	250–400	120–200	265	130	S	S	D	S	S	R
Fabric-filled	1.5	8–11	55–76	14–16	97–110	0.6–1.0	0.8–1.4	250	120	310	150	S	R	D	R	A	R-S
Glass-filled	1.8–2.0	5–10	34–69	24	165	0.6–18	0.8–24	300–400	150–200	400	200	S	R	D	R	R-S	R
Phenolics																	
Wood-flour-filled	1.34–1.45	5–9	34–62	8–17	55–117	0.2–0.6	0.3–0.8	300–350	150–180	300–370	150–190	S	R-S	S-D	S-D	A	R-S
Asbestos-filled	1.45–2.00	4.5–7.5	31–52	10–30	69–207	0.2–0.4	0.3–0.5	350–500	180–260	300–500	150–260	S	R-S	S-D	S-D	A	R-S
Mica-filled	1.65–1.92	5.5–7	38–48	25–50	172–345	0.3–0.4	0.4–0.5	250–300	120–150	300–350	150–180	S	R-S	S-D	S-D	A	R-S
Glass-filled	1.69–1.95	5–18	34–124	19–33	131–228	0.3–18	0.4–24	350–550	180–290	300–600	150–320	S	R-S	S-D	S-D	A	R-S
Fabric-filled	1.36–1.43	3–9	21–62	9–14	62–97	0.8–8	1.1–11	220–250	100–120	250–330	120–170	S	R-S	S-D	S-D	A	R
Polybutadienes																	
Very high vinyl (no filler)	1.00	8	55	2	14	1.1	1.5	500	260			S	R	R	R	R	R
Polyesters																	
Glass-filled BMC	1.7–2.3	4–10	28–69	16–25	110–172	1.5–16	2.0–22	300–350	150–180	400–450	200–230	R-E	R-A	S-A	S-A	S-D	A-D
Glass-filled SMC	1.7–2.1	8–20	55–138	16–25	110–172	8–22	11–30	300–350	150–180	400–450	200–230	R-E	R-A	S-A	S-A	S-D	A-D
Glass-cloth reinforced	1.3–2.1	25–50	172–345	19–45	131–310	5–30	7–41	300–350	150–180	400–450	200–230	R-E	R-A	S-A	S-A	S-D	A-D
Silicones																	
Glass-filled	1.7–2.0	4–6.5	28–45	10–15	69–103	3–15	4–20	600	320	600	320	R-S	R-S	R-S	S	S-A	R-A
Mineral-filled	1.8–2.8	4–6	28–41	13–18	90–124	0.3–0.4	0.4–0.5	600	320	600	320	R-S	R-S	R-S	S	S-A	R-A
Ureas																	
Cellulose-filled	1.47–1.52	5.5–13	38–90	10–15	69–103	0.2–0.4	0.3–0.5	170	80	260–290	130–140	S	R-S	A-D	S-A	D	R-S
Urethanes																	
No filler	1.1–1.5	0.2–10	1–69	1–10	7–69	5–NB	7	129–250	90–120			R-S	S	A	S	S-A	R-S

TABLE 28-22 Typical Property Ranges for Plastics (*Concluded*)

Thermosets[a]	Specific gravity	Tensile strength kip/in²	Tensile strength MPa	Modulus of elasticity, tension 10³kip/in²	Modulus of elasticity, tension 10³MPa	Impact strength, Izod[b] ft·lb	Impact strength, Izod[b] J	Maximum use temperature (no load) °F	Maximum use temperature (no load) °C	HDT at 66 lbf/in²[c] °F	HDT at 66 lbf/in²[c] °C	HDT at 264 lbf/in²[c] °F	HDT at 264 lbf/in²[c] °C	Weather resistance	Chem. Weak acid	Chem. Strong acid	Chem. Weak alkali	Chem. Strong alkali	Chem. Solvents
ABS																			
GP	1.05–1.07	5.9	41	3.1	21	6	8	160–200	70–90	210–225	100–110	190–206	90–95	R-E	R	A[e]	R	R	A/R
High-impact	1.01–1.06	4.8	33	2.4	17	7.5	10	140–210	60–100	210–225	100–110	188–211	85–100	R-E	R	A[e]	R	R	A/R
Heat-resistant	1.06–1.08	7.4	51	3.9	27	2.2	3.0	190–230	90–110	225–252	110–120	226–240	110–115	R-E	R	A[e]	R	R	A/R
Trans.	1.07	5.6	39	2.9	20	5.3	7.1	130	55	180	80	165	75	R-E	R	A[e]	R	R	A/R
	1.20	6.0	41	3.2	22	2.5	3.4	130–180	55–80	210–220	100–105	195	90	R-E	R	A[e]	R	R	A/R
Acetals																			
Homopolymers	1.42	10	69	5.2	36	1.4	1.9	195	90	338	170	255	125	R	R	A	R	A-D	R
Copolymers	1.41	8.8	61	4.1	28	1.2–1.6	1.6–2.2	212	100	316	160	230	110	R	R	A	R	R	R
Acrylics																			
GP	1.11–1.19	5.6–11.0	39–76	2.25–4.65	16–32	0.3–2.3	0.4–3.1	130–230	55–110	175–225	80–110	165–210	75–100	R	R	A[e]	R	A	A/R
High-impact	1.12–1.16	5.8–8.0	40–55	2.3–3.3	16–23	0.8–2.3	1.1–3.1	140–195	60–90	180–205	80–95	165–190	75–90	R	R	A[e]	R	A	A/R
Cast	1.21–1.28	8.0–12.5	55–86	3.5–4.8	24–33	0.3–0.4	0.4–0.5	125–200	50–90	170–200	75–95	155–205	70–95	R	R	A[e]	R	A	A/R
	1.18–1.28	9.0–12.5	62–86	3.7–5.0	26–34	0.4–1.5	0.5–2.0	140–200	60–90	165–235	75–115	160–215	70–100	R	R	A[e]	R	A	A/R
Multipolymer	1.09–1.14	6–8	41–55	3.1–4.3	21–30	1–3	1–4	165–175	75–80			185–195	85–90	E	R	A[e]	H	S	A[e]
Cellulosics																			
Acetate	1.23–1.34	3.0–8.0	21–55	1.05–2.55	7–18	1.1–6.8	1.5–9	140–220	60–105	120–209	50–100	111–195	45–90	S	S	D	S	D	D-S
Butyrate	1.15–1.22	3.0–6.9	21–48	0.7–1.8	5–12	3.0–10.0	4–14	140–220	60–105	130–227	55–110	113–202	45–95	S	S	D	S	D	D-S
E cellulose	1.10–1.17	3–8	21–55	0.5–3.5	3–24	1.7–7.0	2.3–9.5	115–185	45–85			115–190	45–90	S	S	D	R	S	D
Nitrate	1.35–1.40	7–8	48–55	1.9–2.2	13–15	5–7	7–9	140	60			140–160	60–70	E	S	D	S	D	D
Propionate	1.19–1.22	4.0–6.5	28–45	1.1–1.8	8–12	1.7–9.4	2.3–13	155–220	70–105	147–250	65–120	111–228	45–110	S	S	D	S	D	D-S
Chloro polyether copolymers	1.4	5.4	37	1.5	10	0.4	0.5	290	140	285	140			R-S	R	A[e]	R	R	R
Ethylene copolymers																			
EEA	0.93	2.0	14	0.05	0.3	NB		190	90	140–147	60–65			S	R	A[e]	R	R	A-D
EVA	0.94	3.6	25	0.02–0.12	0.14–0.8	NB								S	R	A	R	R	A-D
Fluoropolymers																			
FEP	2.14–2.17	2.5–3.9	17–27	0.5–0.7	3–5	NB		400	208	158	70	93	35	R	R	R	R	R	R
PTFE	2.1–2.3	1.4	7–28	0.38–0.65	2.6–4.5	2.5–4.0	3.4–5.4	550	290	250	120			R	R	R	R	R	R
CTFE	2.10–2.15	4.6–5.7	32–39	1.8–2.0	12–14	3.5–3.6	4.7–4.9	350–390	180–200	258	125			R	R	R	R	R	S[g]
PVF₂	1.77	7.2	50	1.7	12	3.8	5.2	300	150	300	150	195	90	S	R	A[h]	R	R	R
ETFE and ECTFE	1.68	6.5–7.0	45–48	2–2.5	14–17	NB		300	150	220	105	160	70	R	R	R	R	R	R
Methylpentene	0.83	3.3–3.6	23–25	1.3–1.9	10–13	0.95–3.8	1.3–5.2	275	135					E	R	A[h]	R	R	A
Nylons																			
6/6	1.13–1.15	9–12	62–83	3.85	27	2.0	27	180–300	80–150	360–470	180–240	150–220	65–105	R	R	A	R	R	R-D[f]
6	1.14	12.5	86	2.8	19	1.2	1.6	180–250	80–170	300–365	150–185	140–155	60–70	R	R	A	R	R	R-A[i]
6/10	1.07	7.1	49			1.6	2.2	180	80	300	150			R	R	A	R	R	R-A[i]

	Sp. gr.					>16	>22													
8	1.09	3.9	27				>22	175–260	80–125			120–130	50–55	R	R	R	R	A	R	R-A'
12	1.01	6.5–8.5	45–59				1.6–5.7	180–250	80–120			130–350	55–180	R	R	R	R	A	R	R-A'
Copolymers	1.08–1.14	7.5–11.0	52–76	1.7–2.1	12–14	>16	2–26							R	R	R	R	A	R	R-A'
Polyesters																				
PET	1.37	10.4	72			0.8	1.1	175	80	240	115	185	85	R	R	R	A'	R	A	R-A'
PBT	1.31	8.0–8.2	55–57	3.6	25	1.2–1.3	1.6–1.8	280	140	310	155	130	55	R	R	R	R	R	A	R
PTMT	1.31	8.2	57			1.0	1.4	270	130	302	150	122	50	R	R	R	R	A	A	R
Copolymers	1.2	7.3	50	3.2	22	1.0	1.4	250	120	320	100	154	70	E	R	R	R	R	R	A
Polyaryl ether	1.14	7.5	52	3.7	26	10	14	500	260			300	150	Darkens	R	R	R	R	R	R
Polyaryl sulfone	1.36	13	90	0.26	1.8	2	2.7	500	260			525	275	E	R	A'	R	A'	R	A
Polybutylene	0.910	3.8	26	3.45	24	NB	16–22	225	105	215	100	130	55	R	R	R	R	A'	A	A
Polycarbonate	1.2	9	62	3.7	26	12–16	14	250	120	270–290	130–145	265–285	55	R	A	A'	A	A'	A	A
PC–ABS	1.14	8.2	57			10		220	105	235	115	220	105	R-E	R	A'	R	A'	S	R
Polyethylenes																				
LD	0.91–0.93	0.9–2.5	6–17	0.20–0.27	1.4–1.9	NB	0.5–19	180–212	80–100	100–120	40–50	90–105	30–40	E	R	A'	R	A'	R	R
HD	0.95–0.96	2.9–5.4	20–37			0.4–14		175–250	80–120	140–190	60–90	110–130	45–55	E	A	R-A'	A	R-A'	R	R
HMW	0.945	2.5	17	1	7	NB	8	160–180	70–80	155–180	70–80	105–180	40–80	E	R	A'	R	A'	R	R
Ionomer	0.94–0.95	3.4–4.5	23–31	0.3–0.7	2–5	6–NB		175–220	80–105	110	45	100–120	40–50	E	R	A'	A	A'	A	R
Phenylene oxide-based materials	1.06–1.10	7.8–9.6	54–66	3.5–3.8	24–26	5.0	68	175–220	80–105	230–280	110–140	212–265	100–130	R	R	R	R	R	R	R-A
Polyphenylene sulfide	1.34	10	69	4.8	33	0.3	0.4	500	260			278	135		R	A'	R	A'	R	R
Polyimide	1.43	5–7.5	34–52	5.4	37	5–7	7–9	500	260			680	360	R	A	R	A	R	A	R
Polypropylenes																				
GP	0.90–0.91	4.8–5.5	33–38	1.6–2.2	11–15	0.4–2.2	0.5–3.0	225–300	105–150	200–230	95–110	125–140	50–60		R	A'	R	A'	R	R
High-impact	0.90–0.91	3–5	21–34	1.3	9	1.5–12	2–16	200–250	95–120	160–200	70–95	120–135	50–60	E	R	A'	R	A'	R	A
Propylene copolymer	0.91	4	28	1.0–1.7	7–12	1.1	1.5	190–240	90–115	185–230	85–110	115–140	45–60	E	R	A'	R	A'	R	R
Polystyrenes																				
GP	1.04–1.07	6.0–7.3	41–50	4.5	31	0.3	0.4	150–170	65–80			180–220	80–105	S	R	A'	R	A'	R	D
High-impact	1.04–1.07	2.8–4.6	20–32	2.9, 4.0	20–28	0.7–1.0	0.9–1.4	140–175	60–80			175–210	80–100	S	R	A'	R	A'	R	D
Polysulfone	1.24	10.2	70	3.6	25	1.2	1.6	300	150	360	180	345	175	S	R	R	R	R	R	R-A
Polyurethanes	1.11–1.25	4.5–8.4	31–58	0.1–3.5	0.7–24	NB		190	90					R-S	S-D	S-D	S-D	S-D	S-D	R
Vinyl, rigid	1.3–1.5	5–8	34–55	3–5	21–34	0.5–20	0.7–27	150–175	65–80	135–180	60–80	130–175	55–80	R	R	R-S	R	R-S	R	R-A
Vinyl, flexible	1.2–1.7	1–4	7–28			0.5–20	0.7–27	140–175	60–80					S	R	R-S	R	R-S	R	R-A
Rigid CPVC	1.49–1.58	7.5–9.0	52–62	3.6–4.7	25–32	1.0–5.6	1.4–7.6	230	110	215–245	100–120	200–235	95–115	R	R	R	R	R	R	R-A
PVC–acrylic	1.30–1.35	5.5–6.5	38–45	2.75–3.35	19–23	15	20			180	80	170	80	R	R	R	R	S	R	R
PVC–ABS	1.10–1.21	2.6–6.0	18–41	0.8–3.4	6–23	10–15	14–20							S	R	R-S	R	A	A	R-D
SAN	1.08	10–12	69–83	5.0–5.6	34–39	0.4–0.5	0.5–0.7	140–200	60–95			190–220	90–105	S-E	R	A	R	A	R	A

All values at room temperature unless otherwise listed.

[b] Notched samples.

[c] Heat-deflection temperature.

[d] Ac = acid, and Al = alkali; R = resistant; A = attached; S = slight effects; E = embrittles; D = decomposes.

[e] By oxidizing acids.

[f] By ketones, esters, and chlorinated and aromatic hydrocarbons.

[g] Halogenated solvents cause swelling.

[h] By fuming sulfuric.

[i] Dissolved by phenols and formic acid.

SOURCE: *Plastics Engineering Handbook*, 4th ed., Van Nostrand Reinhold, New York, 1976. Courtesy of National Association of Corrosion Engineers. To convert MPA to lbf/in², multiply by 145.04.

TABLE 28-23 Chemical Resistance of Coatings for Immersion Service (Room Temperatures)

	Asphalt, unmodified	Coal tar Hot-applied	Coal tar Cold-applied	Coal tar—epoxy	Coal tar—urethanes	Epoxy: phenolic-baked	Epoxy: amine-cured	Epoxy ester	Furfuryl alcohol	Phenolics, baked	Polyesters (unsaturated)	Polyvinyl chloracetates	Vinyl ester	Urethanes Air-dried	Urethanes Baked	Vinylidene chloride	Chlorinated rubber
Acids																	
Sulfuric, 10%	R	LR	NR	R	R	R	R	LR	R	R	R	R	R	LR	LR	R	R
Sulfuric, 80%		NR	NR		NR	NR	NR	NR	LR	NR	R	NR	R	LR	LR	LR	R
Hydrochloric, 10%	R	LR	NR	LR	LR	R	R	LR	R	R	R	R	R	LR	LR	R	R
Hydrochloric, 35%	R	LR	NR			NR	R	LR	R	R	R	LR	R	LR	LR	R	R
Nitric, 10%	NR	LR	NR			LR	NR	NR	NR	NR	NR	R	R	NR	LR	R	R
Nitric, 50%	NR	NR	NR			NR	NR	NR	NR	NR	NR	NR		NR	NR	LR	NR
Acetic, 100%			NR	NR	NR	NR	NR	NR	LR	LR	R	NR	LR	NR	NR	NR	NR
Water																	
Distilled	R	R	R	R	LR	R	R	R	R	R	R	R	R	R	LR		R
Salt water	R	R	R	R	LR	R	R	R	R	R	R	R	R	R	LR		R
Alkalies																	
Sodium hydroxide, 10%	R	R	LR	R	R	R	R	NR	R	NR	R	R	R	LR	LR	LR	R
Sodium hydroxide, 70%		NR	NR		LR	R	R	NR	LR	NR	NR	LR	R	LR	LR	NR	R
Ammonium hydroxide, 10%	R	R	LR	R	R	R	LR	LR	R	NR	R	R	R	LR	R	NR	R
Sodium carbonate, 5%	R	R	R			R									R		R
Gases																	
Chlorine	R	NR	NR	NR	NR	LR	LR	LR	NR	NR	R	LR	R	LR	R	LR	R
Ammonia		LR	LR	NR	NR	LR	LR	R	R	NR	NR	LR	R	LR	R	NR	NR
Hydrogen sulfide			R	R		R	R		R	R		LR	R	R	R	R	R
Organics																	
Alcohols	R	LR	LR	NR	NR	R	R	LR	R	R	R	R	R	NR	R	R	LR
Aliphatic hydrocarbons	NR	LR	LR	LR	LR	R	R	R	R	R	R	R	R	R	R	R	LR
Aromatic hydrocarbons	NR	NR	NR	LR	LR	R	R	R	R	R	NR	NR	LR	NR	R	LR	NR
Ketones	NR	NR	NR	NR	NR	LR	LR	NR	LR	R		NR	NR	R	R	NR	NR
Ethers	NR	NR	NR	NR	NR	LR	LR	NR	LR	R		NR	NR	LR	R	NR	NR
Esters	NR	NR	NR	NR	NR	LR	LR	NR	LR	R	LR	LR	LR	LR	R	NR	NR
Chlorinated hydrocarbons	NR	NR	NR	NR	NR	LR	LR	NR	LR	R	NR		LR		R	LR	NR
Maximum temperature (dry conditions), °F	150			200	200	250	250	250	300	250–300	250	160	350				160
Maximum temperature (wet conditions), °F		120	120	150	150	150	150	150	190	160–250		150	210			150	140

NOTE: Chemical resistance data are for coatings only. Thin coatings generally are not suitable for substrates such as carbon steel which are corroded significantly (e.g., >20 mils/year) in the test environment. R = recommended; LR = limited recommendation; NR = no recommendation.

SOURCE: NACE TPC-2, *Coatings and Linings for Immersion Service.* Courtesy of National Association of Corrosion Engineers.

TABLE 28-24 Properties of Coatings for Atmospheric Service

	Physical properties	Water resistance	Acid resistance	Alkali resistance	Solvent resistance	Temperature resistance	Weathering	Recoating
Alkyd								
Short-oil alkyd	Hard	Fair	Fair	Poor	Fair	Good	Fair	Easy
Long-oil alkyd	Flexible	Fair	Poor	Poor	Poor	Good	Good	Easy
Silicone alkyd	Tough	Good	Fair	Poor	Fair	Best of group	Very good	Fair
Vinyl alkyd	Tough	Good	Best of group	Poor	Fair	Fair	Very good	Difficult
Vinyl								
Polyvinyl chloride acetate copolymers	Tough	Very good	Excellent	Excellent	(Aliphatic hydrocarbon, good; aromatic hydrocarbon, poor)	Fair, 150°F	Very good	Easy
Vinyl acrylic copolymers	Tough	Good	Very good	Very good	(Aliphatic, good; aromatic, poor)	Fair, 150°F	Excellent	Easy
Chlorinated rubber								
Resin-modified	Hard	Very good	Very good	Very good	(Aliphatic, good; aromatic, poor)	Fair	Good	Easy
Alkyd-modified	Tough	Good	Fair	Fair	(Aliphatic, good; aromatic, poor)	Fair	Very good	Easy
Water base								
Polyvinyl acetate	Scrub-resistant	Poor	Poor	Poor	Poor	Fair	Very good	Easy
Acrylic polymers	Scrub-resistant	Poor	Poor	Poor	Poor	Fair	Excellent	Easy
Epoxy	Tough	Good	Good	Good	Good	Good	Fair	Difficult
Epoxy								
Epoxyamine	Hard	Good	Good	Good	Very good	Very good	Fair; chalks	Difficult
Epoxy polyamide	Tough	Very good	Fair	Excellent	Fair	Good	Good; chalks	Difficult
Epoxy coal tar	Hard	Excellent	Good	Good	Poor	Good	Poor	Difficult
Epoxyester	Flexible	Good	Fair	Poor	Fair	Good	Good; chalks	Reasonable
Polyurethane								
Air-drying polyurethane varnish	Very tough	Fair	Fair	Fair	Fair	Good	Yellowing	Requires care
Two-package-reactive polyurethane	Tough; hard	Good	Fair	Fair	Good	Good	Some yellowing and chalking	Difficult
Moisture-reactive polyurethane	Very tough; abrasion-resistant	Fair	Fair	Fair	Good	Good	Fades in light; yellows in shade	Difficult
Nonyellowing polyurethane	Fairly hard to rubbery	Good	Fair	Fair	Good	Good	Very good	Difficult
Inorganic zinc								
Water base (sodium or potassium silicate)	Tough; abrasion-resistant; excellent chemical bond	Good	Poor	Poor	Excellent	Excellent	Excellent; unaffected by weather	Easy
Organic base (ethyl silicate)	Tough; hard; excellent bond	Good	Poor	Poor	Good	Excellent	Excellent	Easy

SOURCE: F. L. LaQue, *Marine Corrosion; Causes and Prevention*, Wiley, New York, 1975, pp. 302–305. Courtesy of National Association of Corrosion Engineers.

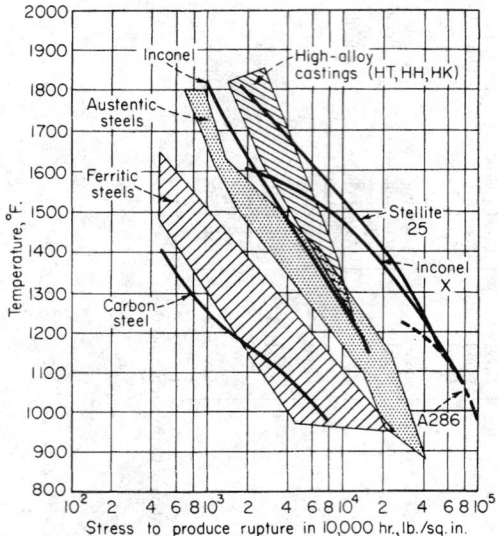

FIG. 28-24 Rupture properties of metals as a function of temperature. °C = (°F − 32) × 5/9; to convert lbf/in² to MPa, multiply by 6.895 × 10⁻³. [*Chem. Eng.*, *139 (Dec. 15, 1958).*]

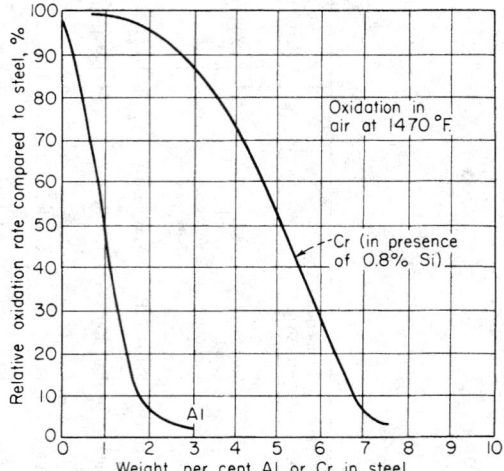

FIG. 28-25 How chromium and aluminum reduce steel oxidation. °C = (°F − 32) × 5/9.

based alloys, of which Stellite 25 is an example, show excellent strengths up to 1,100°C (2,000°F).

Another useful element in imparting oxidation resistance to steel is **silicon** (complementing the effects of chromium). In the lower-chromium ranges, silicon in the amounts of 0.75 to 2 percent is more effective than chromium on a weight-percentage basis. The influence of 1 percent silicon in improving the oxidation rate of steels with varying chromium contents is shown in Fig. 28-26.

Aluminum also improves the resistance of iron to oxidation as well as sulfidation. But use as an alloying agent is limited because the amount required interferes with the workability and high-temperature

strength properties of the steel. However, development of high-aluminum surface layers by spraying, dipping, and cementation is a feasible means of improving the heat resistance of low-alloy steels.

Hydrogen Atmospheres   Austenitic stainless steels, by virtue of their high chromium contents, are usually resistant to hydrogen atmospheres.

Sulfur Corrosion   Chromium is the most important material in imparting resistance to sulfidation (formation of sulfidic scales similar to oxide scales). The austenitic alloys are generally used because of their superior mechanical properties and fabrication qualities, despite the fact that nickel in the alloy tends to lessen resistance to sulfidation somewhat.

Halogens (Hot, Dry Cl₂, HCl)   Pure nickel and nickel alloys are useful with dry halogen gases. But even with the best materials, cor-

TABLE 28-25   Typical Physical Properties of Surface Coatings for Concrete

		Polyester		Epoxy		
	Concrete	Isophthalic	Bisphenol	Polyamide	Amine	Urethane°
Tensile strength (ASTM C307), lbf/in²	200–400	1200–2500	1200–2500	600–4000	1200–2500	200–1200
MPa	1.4–2.8	8.3–17	8.3–17	4.0–28	8.3–17	1.4–8.3
Thermal coefficient of expansion (ASTM C531)						
Maximum in/(in·°F)	6.5×10^6	20×10^6	20×10^6	40×10^6	40×10^6	†
Maximum mm/(mm·°C)	11.7×10^6	36×10^6	36×10^6	72×10^6	72×10^6	†
Compressive strength, (ASTM) C579), lbf/in²	3500	10,000	10,000	4000	6000	†
MPa	24	70	70	28	42	†
Abrasion resistance, Taber abraser—weight loss, mg, 1000-g load/1000 cycles		15–27	15–27	15–27	15–27	†
Shrinkage, ASTM C531, %		2–4	2–4	0.25–0.75	0.25–0.75	0–2
Work life, min		15–45	15–45	30–90	30–90	15–60
Traffic limitations, h after application	Light	16	16	24	24	24
	Heavy	36	36	48	48	48
	Ready for service	48	48	72	72	72
Adhesion characteristics‡		Poor	Fair	Excellent	Good	Fair
Flexural strength (ASTM C580), lbf/in²		1500	1500	1000	1500	†
MPa		10	10	7	10	†

NOTE: All physical values depend greatly on reinforcing. Values are for ambient temperatures.
°Type of urethane used is one of three: (1) Type II, moisture-cured; (2) Type IV, two-package catalyst; or (3) Type V, two-package polyol. (Ref. ASTM C16.)
†Urethanes not shown because of great differences in physical properties, depending on formulations. Adhesion characteristics should be related by actual test data. Any system which shows concrete failure when tested for surfacing adhesion should be rated excellent with decreasing rating for systems showing failure in cohesion or adhesion below concrete failure.
‡Adhesion to concrete: primers generally are used under polyesters and urethanes to improve adhesion.
SOURCE: NACE RP-03-76, *Monolithic Organic Corrosion Resistant Floor Surfacing*, 1976. Courtesy of National Association of Corrosion Engineers.

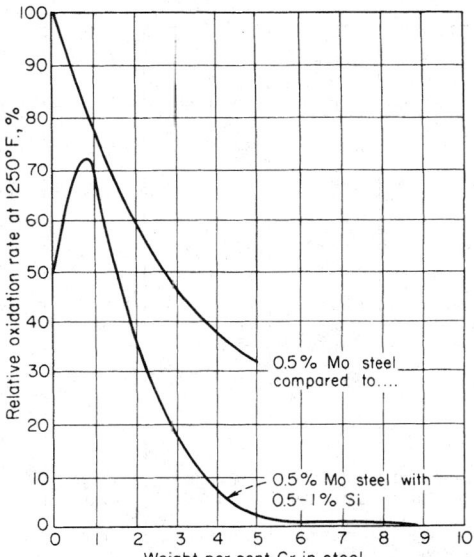

FIG. 28-26 Effect of silicon on oxidation resistance. °C = (°F − 32) × 5/9.

TABLE 28-26 Chemical Resistance of Rubbers

Type of rubber	Features
Butadiene styrene	General-purpose; poor resistance to hydrocarbons, oils, and oxidizing agents
Butyl	General-purpose; relatively impermeable to air; poor resistance to hydrocarbons and oils
Chloroprene	Good resistance to aliphatic solvents; poor resistance to aromatic hydrocarbons and many fuels
Chlorosulfonated polyethylene	Excellent resistance to oxidation, chemicals, and heat; poor resistance to aromatic oils and most fuels
cis-Polybutadiene	General-purpose; poor resistance to hydrocarbons, oils, and oxidizing agents
cis-Polyisoprene	General-purpose; poor resistance to hydrocarbons, oils, and oxidizing agents
Ethylene propylene	Excellent resistance to heat and oxidation
Fluorinated	Excellent resistance to high temperature, oxidizing acids, and oxidation; good resistance to fuels containing up to 30% aromatics
Natural	General-purpose; poor resistance to hydrocarbons, oils, and oxidizing agents
Nitrile (butadiene acrylonitrile)	Excellent resistance to oils, but not resistant to strong oxidizing agents; resistance to oils proportional to acrylonitrile content
Polysulfide	Good resistance to aromatic solvents; unusually high impermeability to gases; poor compression set and poor resistance to oxidizing acids
Silicone	Excellent resistance over unusually wide temperature range [−100 to 260°C (−150 to 500°F)]; fair oil resistance; poor resistance to aromatic oils, fuels, high-pressure steam, and abrasion
Styrene	Synonymous with butadiene-styrene

rosion rates are relatively high at high temperature. There are cases in which equipment for high-temperature halogenation has used platinum-clad nickel-base alloys. These materials have high initial cost but long life. Platinum and gold have excellent resistance to dry HCl even at 1,100°C (2,000°F).

Refractories   Refractories are selected to accomplish four objectives:

1. Resist heat
2. Resist high-temperature chemical attack
3. Resist erosion by gas with fine particles
4. Resist abrasion by gas with large particles

Refractories are available in three general physical forms: solids in the form of brick and monolithic castable ceramics and as ceramic fibers.

The primary method of selection of the type of refractory to be used is by gas velocity:

<7.5 m/s (25 ft/s): fibers
7.5–60 m/s (25–200 ft/s): monolithic castables
>60 m/s (200 ft/s): brick

Within solids the choice is a trade-off because, with brick, fine particles in the gas remove the mortar joints and, in the monolithic castables, while there are no joints, the refractory is less dense and less wear-resistant.

Internal Insulation   The practice of insulating within the vessel (as opposed to applying insulating materials on the equipment exterior) is accomplished by the use of fiber blankets and lightweight aggregates in ceramic cements. Such construction frequently incorporates a thin, high-alloy shroud (with slip joints to allow for thermal expansion) to protect the ceramic from erosion. In many cases this design is more economical than externally insulated equipment because it allows use of less expensive lower-alloy structural materials.

Refractory Brick   Nonmetallic refractory materials are widely used in high-temperature applications in which the service permits the appropriate type of construction. The more important classes are described in the following paragraphs.

Fireclays can be divided into plastic clays and hard flint clays; they may also be classified as to alumina content. **Firebricks** are usually made of a blended mixture of flint clays and plastic clays which is formed, after mixing with water, to the required shape. Some or all of the flint clay may be replaced by highly burned or calcined clay, called grog. A large proportion of modern brick production is molded by the dry-press or power-press process, in which the forming is carried out under high pressure and with a low water content. Extruded and hand-molded bricks are still made in large quantities.

The dried bricks are burned in either periodic or tunnel kilns at temperatures ranging between 1,200 and 1,500°C (2,200 and 2,700°F). Tunnel kilns give continuous production and a uniform burning temperature.

Fireclay bricks are used in kilns, malleable-iron furnaces, incinerators, and many portions of metallurgical furnaces. They are resistant to spalling and stand up well under many slag conditions but are not generally suitable for use with high-lime slags or fluid-coal-ash slags or under severe load conditions.

High-alumina bricks are manufactured from raw materials rich in alumina, such as diaspore. They are graded into groups with 50, 60, 70, 80, and 90 percent alumina content. When well fired, these bricks contain a large amount of mullite and less of the glassy phase than is present in firebricks. Corundum is also present in many of these bricks. High-alumina bricks are generally used for unusually severe temperature or load conditions. They are employed extensively in lime kilns and rotary cement kilns, in the ports and regenerators of glass tanks, and for slag resistance in some metallurgical furnaces; their price is higher than that of firebrick.

Silica bricks are manufactured from crushed ganister rock containing about 97 to 98 percent silica. A bond consisting of 2 percent lime is used, and the bricks are fired in periodic kilns at temperatures of 1,500 to 1,540°C (2,700 to 2,800°F) for several days until a stable volume is obtained. They are especially valuable when good strength is required at high temperatures. Superduty silica bricks are finding some use in the steel industry. They have a lowered alumina content and often a lowered porosity.

Silica bricks are used extensively in coke ovens, the roofs and walls of open-hearth furnaces, and the roofs and sidewalls of glass tanks and as linings of acid electric steel furnaces. Although silica brick is readily spalled (cracked by a temperature change) below red heat, it is very stable if the temperature is kept above this range and for this reason

TABLE 28-27   Properties of Elastomers

Property	NR Natural rubber (*cis*-polyisoprene)	SBR Butadiene styrene (GR-S)	IR Synthetic (polyisoprene)	COX Butadiene acrylonitrile (nitrile)	CR Chloroprene (neoprene)	ITR Butyl (isobutylene isoprene)	BR Polybutadiene	T Polysulfide	Silicone (polysiloxane)
Physical properties									
Specific gravity (ASTM D 792)	0.93	0.91	0.93	0.98	1.25	0.90	0.91	1.35	1.1–1.6
Thermal conductivity, Btu [(h·ft²)(°F/ft)] (ASTM C 177)	0.082	0.143	0.082	0.143	0.112	0.053			0.13
Coefficient of thermal expansion (cubical), 10^{-5}°F (ASTM D 696)	37	37		39	34	32	37.5		45
Electrical insulation	Good	Good	Good	Fair	Fair	Good	Good	Fair	Excellent
Flame resistance	Poor	Poor	Poor	Poor	Good	Poor	Poor	Poor	Good
Minimum, recommended service temperature, °F	−60	−60	−60	−60	−40	−50	−150	−60	−178
Maximum, recommended service temperature, °F	180	180	180	300	240	300	200	250	600
Mechanical properties									
Tensile strength, lbf/in²									
Pure gum (ASTM D 412)	2500–3500	200–300	2500–3500	500–900	3000–4000	2500–3000	200–1000	250–400	600–1300
Black (ASTM D 412)	3500–4500	2500–3500	3500–4500	3000–4500	3000–4000	2500–3000	2000–3000	>1000	
Elongation, %									
Pure gum (ASTM D 412)	750–850	400–600		300–700	800–900	750–950	400–1000	450–650	100–500
Black (ASTM D 412)	550–650	500–600	300–700	300–650	500–600	650–850	450–600	150–450	
Hardness (durometer)	A30–90	A40–90	A40–80	A40–95	A20–95	A40–90	A40–90	A40–85	A30–90
Rebound									
Cold	Excellent	Good	Excellent	Good	Very good	Bad	Excellent	Good	Very good
Hot	Excellent	Good	Excellent	Good	Very good	Very good	Excellent	Good	Very good
Tear resistance	Excellent	Fair	Excellent	Good	Fair to good	Good	Fair	Poor	Fair
Abrasion resistance	Excellent	Good to excellent	Excellent	Good to excellent	Good	Good to excellent	Excellent	Poor	Poor
Chemical resistance									
Sunlight aging	Poor	Poor	Fair	Poor	Very good	Very good	Poor	Very good	Excellent
Oxidation	Good	Good	Excellent	Good	Excellent	Excellent	Good	Very good	Excellent
Heat aging	Good	Very good	Good	Excellent	Excellent	Excellent	Good	Fair	Excellent
Solvents									
Aliphatic hydrocarbons	Poor	Poor	Poor	Excellent	Good	Poor	Poor	Excellent	Fair
Aromatic hydrocarbons	Poor	Poor	Poor	Good	Fair	Poor	Poor	Excellent	Poor
Oxygenated, alcohols	Good	Good	Good	Good	Very good	Very good		Very good	Excellent
Oil, gasoline	Poor	Poor	Poor	Excellent	Good	Poor	Poor	Excellent	Poor
Animal, vegetable oils	Poor to good	Poor to good		Excellent	Excellent	Excellent	Poor to good	Excellent	Excellent
Acids									
Dilute	Fair to good	Fair to good	Fair to good	Good	Excellent	Excellent		Good	Very good
Concentrated	Fair to good	Fair to good	Fair to good	Good	Good	Excellent		Good	Good
Permeability to gases	Low	Low	Low	Very low	Low	Very low	Low	Very low	High
Water-swell resistance	Fair	Excellent	Excellent	Excellent	Fair to excellent	Excellent	Excellent	Excellent	Excellent

stands up well in regenerative furnaces. Any structure of silica brick should be heated up slowly to the working temperature; a large structure often requires 2 weeks or more.

Magnesite bricks are made from crushed magnesium oxide, which is produced by calcining raw magnesite rock to high temperatures. A rock containing several percent of iron oxide is preferable, as this permits the rock to be fired at a lower temperature than if pure materials were used. Magnesite bricks are generally fired at a comparatively high temperature in periodic or tunnel kilns. A large proportion of magnesite brick made in the United States uses raw material extracted from seawater.

Magnesite bricks are basic and are used whenever it is necessary to resist high-lime slags, as in the basic open-hearth steel furnace. They also find use in furnaces for the lead- and copper-refining industries. The highly pressed unburned bricks find extensive use as linings for cement kilns. Magnesite bricks are not so resistant to spalling as fireclay bricks.

Chrome bricks are manufactured in much the same way as magnesite bricks but are made from natural chromite ore. Commercial ores always contain magnesia and alumina. Unburned hydraulically pressed chrome bricks are also available.

Chrome bricks are very resistant to all types of slag. They are used as separators between acid and basic refractories, also in soaking pits and floors of forging furnaces. The unburned hydraulically pressed bricks now find extensive use in the walls of the open-hearth furnace. Chrome bricks are used in sulfite-recovery furnaces and to some extent in the refining of nonferrous metals. Basic bricks combining various properties of magnesite and chromite are now made in large quantities and have advantages over either material alone for some purposes.

The **insulating firebrick** is a class of brick that consists of a highly porous fire clay or kaolin. Such bricks are light in weight (about one-half to one-sixth of the weight of fireclay), low in thermal conductivity, and yet sufficiently resistant to temperature to be used successfully on

TABLE 28-27 Properties of Elastomers (*Concluded*)

Property	ECO, CO Epichlrohydrin homopolymer and copolymer	Fluorosilicone	EPDM Ethylene propylene	CSM Chlorosulfonated polyethylene	FPM Fluorocarbon elastomers
Physical properties					
Specific gravity	1.32–1.49	1.4	0.86	1.1–1.26	1.4–1.95
Thermal conductivity, Btu/ [(h·ft²)(°F/ft)]		0.13		0.065	0.13
Coefficient of thermal expansion, 10^{-5}/°F		45		27	8.8
Flame resistance	Fair	Poor	Poor	Good	Excellent
Colorability	Good	Good	Excellent	Excellent	Good
Mechanical properties					
Hardness (Shore A)	30–95	40–70	30–90	45–95	65–90
Tensile strength, kip/in²					
Pure gum		1	<1	4	<2
Reinforced	2–3	<2	0.8–3.2	1.5–2.5	1.5–3
Elongation, % reinforced	320–350	200–400	200–600	250–500	100–450
Resilience	Poor to excellent	Good to fair	Good	Good	Fair
Compression-set resistance	Very good		Good	Fair to good	Good to excellent
Hysteresis resistance	Good	Good	Good	Good	Good
Flexcracking resistance	Very good	Good	Good	Good	Good
Slow rate	Very good	Good	Good	Good	Good
Fast rate	Good	Good	Good	Good	Good
Tear strength	Good	Fair	Poor to fair	Fair to good	Poor to fair
Abrasion resistance	Fair to good	Poor	Good	Excellent	Good
Electrical properties					
Dielectric strength	Fair	Good	Excellent	Excellent	Good
Electrical insulation	Fair	Good	Very good	Good	Fair to good
Thermal properties					
Service temperature, °F					
Minimum, for continuous use	−15 to −80	−90	−60	−40	−10
Maximum, for continuous use	300	400	<350	<325	<500
Corrosion resistance					
Weather	Excellent	Excellent	Excellent	Excellent	Excellent
Oxidation	Very good	Excellent	Excellent	Excellent	Outstanding
Ozone	Good to excellent	Excellent	Excellent	Excellent	Excellent
Radiation		Good	Excellent	Fair to good	Fair to good
Water	Good	Excellent	Good to excellent	Good	Good
Acids	Good	Very good to excellent	Good to excellent	Excellent	Good to excellent
Alkalies	Good	Very good	Good to excellent	Excellent	Poor to good
Aliphatic hydrocarbons	Excellent	Excellent	Poor	Fair	Excellent
Aromatic hydrocarbons	Very good	Excellent	Fair	Poor to fair	Excellent
Halogenated hydrocarbons	Good		Poor	Poor to fair	Good
Alcohol	Good		Good	Very good	Excellent
Synthetic lubricants (diester)	Fair to good	Excellent	Poor to fair	Poor	Fair to good
Hydraulic fluids					
Silicates	Very good	Excellent	Fair to good	Good	Good
Phosphates	Poor to fair	Excellent	Good to excellent	Poor to fair	Poor

SOURCE: C. H. Harper, *Handbook of Plastics and Elastomers,* McGraw-Hill, New York, 1975, Table 35. Courtesy of National Association of Corrosion Engineers.
°C = (°F − 32) × 5/9; to convert Btu/(h·ft·°F) to W/(m·°C), multiply by 0.861; to convert lbf/in² to MPa, multiply by 6.895 × 10^{-3}.

the hot side of the furnace wall, thus permitting thin walls of low thermal conductivity and low heat content. The **low heat content** is particularly valuable in saving fuel and time on heating up, allows rapid changes in temperature to be made, and permits rapid cooling. These bricks are made in a variety of ways, such as mixing organic matter with the clay and later burning it out to form pores; or a bubble structure can be incorporated in the clay-water mixture which is later preserved in the fired brick. The insulating firebricks are classified into several groups according to the maximum use limit; the ranges are up to 870, 1,100, 1,260, 1,430, and above 1,540°C (1,600, 2,000, 2,300, 2,600, and above 2,800°F).

Insulating refractories are used mainly in the heat-treating industry for furnaces of the periodic type. They are also used extensively in stress-relieving furnaces. chemical-process furnaces, oil stills or heaters, and the combustion chambers of domestic-oil-burner furnaces. They usually have a life equal to that of the heavy brick that they replace. They are particularly suitable for constructing experi-mental or laboratory furnaces because they can be cut or machined readily to any shape. They are not resistant to fluid slag.

There are a number of types of special brick obtainable from individual producers. **High-burned kaolin refractories** are particularly valuable under conditions of severe temperature and heavy load or severe spalling conditions, as in the case of high-temperature oil-fired boiler settings or piers under enameling furnaces. Another brick for the same uses is a high-fired brick of Missouri aluminous clay.

There are on the market a number of bricks made from **electrically fused materials,** such as fused mullite, fused alumina, and fused magnesite. These bricks, although high in cost, are particularly suitable for certain severe conditions.

Bricks of **silicon carbide,** either recrystallized or clay-bonded, have a high thermal conductivity and find use in muffle walls and as a slag-resisting material.

Other types of refractory that find use are forsterite, zirconia, and zircon. Acid-resisting bricks consisting of a dense body like stoneware

TABLE 28-28 Important Properties of Gasket Materials

Material	Maximum service temperature, °F	Important properties
Rubber (straight)		
Natural	225	Good mechanical properties. Impervious to water. Fair to good resistance to acids, alkalies. Poor resistance to oils, gasoline. Poor weathering, aging properties.
Styrene-butadiene (SBR)	250	Better water resistance than natural rubber. Fair to good resistance to acids, alkalies. Unsuitable with gasoline, oils and solvents.
Butyl	300	Very good resistance to water, alkalies, many acids. Poor resistance to oils, gasoline, most solvents (except oxygenated).
Nitrile	300	Very good water resistance. Excellent resistance to oils, gasoline. Fair to good resistance to acids, alkalies.
Polysulfide	150	Excellent resistance to oils, gasoline, aliphatic and aromatic hydrocarbon solvents. Very good water resistance, good alkali resistance, fair acid resistance. Poor mechanical properties.
Neoprene	250	Excellent mechanical properties. Good resistance to nonaromatic petroleum, fatty oils, solvents (except aromatic, chlorinated, or ketone types). Good water and alkali resistance. Fair acid resistance.
Silicone	600	Excellent heat resistance. Fair water resistance; poor resistance to steam at high pressures. Fair to good acid, alkali resistance. Poor (except fluorosilicone rubber) resistance to oils, solvents.
Acrylic	450	Good heat resistance but poor cold resistance. Good resistance to oils, aliphatic and aromatic hydrocarbons. Poor resistance to water, alkalies, some acids.
Chlorosulfonated polyethylene (Hypalon)	250	Excellent resistance to oxidizing chemicals, ozone, weathering. Relatively good resistance to oils, grease. Poor resistance to aromatic or chlorinated hydrocarbons. Good mechanical properties.
Fluoroelastomer (Viton, Fluorel 2141, Kel-F)	450	Can be used at high temperatures with many fuels, lubricants, hydraulic fluids, solvents. highly resistant to ozone, weathering. Good mechanical properties.
Asbestos		
Compressed asbestos-rubber sheet	To 700	Large number of combinations available; properties vary widely depending on materials used.
Asbestos-rubber woven sheet	To 250	Same as above.
Asbestos-rubber (beater addition process)	400	Same as above.
Asbestos composites	To 1000	Same as above.
Asbestos-TFE	500	Combines heat resistance and sealing properties of asbestos with chemical resistance of TFE.
Cork compositions	250	Low cost. Truly compressible materials which permit substantial deflections with negligible side flow. Conform well to irregular surfaces. High resistance to oils; good resistance to water, many chemicals. Should not be used with inorganic acids, alkalies, oxidizing solutions, live steam.
Cork rubber	300	Controlled compressibility properties. Good conformability, fatigue resistance. Chemical resistance depends on kind of rubber used.
Plastics		
TFE (solid) (Tetrafluoroethylene, Teflon)	500	Excellent resistance to almost all chemicals and solvents. Good heat resistance; exceptionally good low-temperature properties. Relatively low compressibility and resilience.
TFE (filled)	To 500	Selectively improved mechanical and physical properties. However, fillers may lower resistance to specific chemicals.
TFE composites	To 500	Chemical and heat resistance comparable with solid TFE. Inner gasket material provides better resiliency and deformability.
CFE (Chlorotrifluoroethylene, Kel-F)	350	Higher cost than TFE. Better chemical resistance than most other gasket materials, although not quite so good as TFE.
Vinyl	212	Good compressibility, resiliency. Resistant to water, oils, gasoline, and many acids and alkalies. Relatively narrow temperature range.
Polyethylene	150	Resists most solvents. Poor heat resistance.
Plant fiber		
Neoprene-impregnated wood fiber	175	Nonporous; recommended for glycol, oil, and gasoline to 175°F.
SBR-bonded cotton	230	Good water resistance.
Nitrile rubber-cellulose fiber		Resists oil at high temperatures.
Vegetable fiber, glue binder	212	Resists oil and water to 212°F.
Vulcanized fiber		Low cost, good mechanical properties. Resists gasoline, oils, greases, waxes, many solvents.
Inorganic fibers	To 2200°F	Excellent heat resistance, poor mechanical properties.
Felt		
Pure felt		Resilient, compressible and strong, but not impermeable. Resists medium-strength mineral acids and dilute mineral solutions if not intermittently dried. Resists oils, greases, waxes, most solvents. Damaged by alkalies.
TFE-impregnated	300	Good chemical and heat resistance.
Petrolatum or paraffin-impregnated		High water repellency.
Rubber-impregnated		Many combinations available; properties vary widely depending on materials used.
Metal		
Lead	500	Good chemical resistance. Best conformability of metal gaskets.
Tin		Good resistance to neutral solutions. Attacked by acids, alkalies.
Aluminum	800	High corrosion resistance. Slightly attacked by strong acids, alkalies.
Copper, brass		Good corrosion resistance at moderate temperatures.
Nickel	1400	High corrosion resistance.
Monel	1500	High corrosion resistance. Good against most acids and alkalies, but attacked by strong hydrochloric and strong oxidizing acids.
Inconel	2000	Excellent heat, oxidation resistance.
Stainless steel		High corrosion resistance. Properties depend on type used.

TABLE 28-28 Important Properties of Gasket Materials (*Concluded*)

Material	Maximum service temperature, °F	Important properties
Metal composites		Many combinations available; properties vary widely depending on materials used.
Leather	220	Low cost. Limited chemical and heat resistance. Not recommended against pressurized steam, acid or alkali solutions.
Glass fabric		High strength and heat resistance. Can be impregnated with TFE for high chemical resistance.
Packing and sealing materials		
Rubber (straight)	To 600	See Gasket Materials for properties. Mainly used for ring-type seals, although some types are available as spiral packings.
Rubber composites:		
Cotton-reinforced	350	High strength. Chemical resistance depends on type of rubber used; however, most types are noted for high resistance to water, aqueous solutions.
Asbestos-reinforced	450	High strength combined with good heat resistance.
Asbestos:		
Plain, braided asbestos	500	Heat resistance combined with resistance to water, brine, oil, many chemicals. Can be reinforced with wire.
Impregnated asbestos	To 750	Environmental properties vary widely depending on type of asbestos and impregnant used. Neoprene-cemented type resists hot oils, gasoline, and solvents. Oil and wax-impregnated type resists caustics. Wax-impregnated blue asbestos type has high acid resistance. TFE-impregnated type has good all-around chemical resistance.
Asbestos composites	To 1200	End properties vary widely depending on secondary material used.
Metals		
Copper	To 1500	Properties depend on other construction materials and form of copper used. Packing made of copper foil over asbestos core resists steam and alkalies to 1000°F. Packing of braided copper tinsel resists water, steam, and gases to 1500°F.
Aluminum	To 1000	Resists hot petroleum derivatives, gases, footstuffs, many organic acids.
Lead	550	Many types are available.
Organic fiber		
Flax	300	Good water resistance.
Jute	300	Good water resistance.
Ramie	300	Good resistance to water, brine, cold oil.
Cotton	300	Good resistance to water, alcohol, dilute aqueous solutions.
Rayon	300	Good resistance to water, dilute aqueous solutions.
Felt	300	See Gasket Materials.
Leather	To 210	Good mechanical properties for sealing. Resistant to alcohol, gasoline, many oils and solvents, synthetic hydraulic fluids, water.
TFE	To 500	Available in many forms, all of which have high chemical resistance.
Carbon-graphite	700	Good bearing and self-lubricating properties. Good resistance to chemicals, heat.

TABLE 28-29 Properties of Graphite and Silicon Carbide

	Graphite	Impervious graphite	Impervious silicon carbide
Specific gravity	1.4–1.8	1.75	3.10
Tensile strength, lbf/in² (MPa)	400–1400 (3–10)	2,600 (18)	20,650 (143)
Compressive strength, lbf/in² (MPa)	2000–6000 (14–42)	10,500 (72)	150,000 (1000)
Flexural strength, lbf/in² (MPa)	750–3000 (5–21)	4,700 (32)	
Modulus of elasticity (×10⁵), lbf/in² (MPa)	0.5–1.8 (0.3–12 × 10⁴)	2.3 (1.6 × 10⁴)	56 (39 × 10⁴)
Thermal expansion, in/(in·°F × 10⁻⁶) [mm/(mm·°C)]	0.7–2.1 (1.3–3.8)	2.5 (4.5)	1.80 (3.4)
Thermal conductivity, Btu/[(h·ft²)(°F/ft)] [(W/(m·K)]	15–97 (85–350)	85 (480)	60 (340)
Maximum working temperature (inert atmosphere), °F (°C)	5000 (2800)	350 (180)	4,200 (2300)
Maximum working temperature (oxidizing atmosphere), °F (°C)	660 (350)	350 (180)	3,000 (1650)

SOURCE: Carborundum Co. Courtesy of National Association of Corrosion Engineers.

TABLE 28-30 Properties of Stoneware and Porcelain

	Stoneware	Porcelain
Specific gravity	2.2–2.7	2.4–2.9
Hardness, Mohs scale	6.5	7.5
Modulus of rupture, lb/in^2	3–7,000	8–15,000
Modulus of elasticity, lb/in^2	$5–10 \times 10^6$	$10–15 \times 10^6$
Compressive strength, lb/in^2	40–60,000	60–90,000
Pore volume, %	1.5	0.2–0.5
Water absorption, %	0.5–4.0	0–0.5
Linear thermal expansion, per °F	2.4×10^{-6}	2.5×10^{-6}
Thermal conductivity, Btu/(ft^2·h·°F·in^{-1})	8–22	8–10

TABLE 28-32 Comparison of Properties of Refractory Metals

Melting point, °F	Element	Advantages	Disadvantages
6180	Tungsten	Highest melting point; nonvolatile oxide to at least 2500°F	Highest density; oxidizing rapidly; brittle at low temperatures
5425	Tantalum	Very high melting point; nonvolatile oxide; ductile	High density; oxidizing rapidly; least abundant
4730	Molybdenum	High melting point; less dense than tungsten or tantalum; moderately ductile at room temperature	Extremely high oxidation rate (volatile oxide)
4380	Niobium (columbium)	High melting point; nonvolatile oxide; ductile; moderate density	Oxidizing rapidly
3435	Chromium	Extremely oxidation-resistant; lightest of refractory metals	Lowest melting point of refractory metals, brittle at low temperatures

are used for lining tanks and conduits in the chemical industry. Carbon blocks are used as linings for the crucibles of blast furnaces, very extensively in a number of countries and to a limited extent in the United States. Fusion-cast bricks of mullite or alumina are largely used to line glass tanks.

Ceramic-Fiber Insulating Linings Ceramic fibers are produced by melting the same alumina-silica china (kaolin) clay used in conventional insulating firebrick and blowing air to form glass fibers. The fibers, 50.8 to 101.6 mm (2 to 4 in) long by 3 μm in diameter, are interlaced into a mat blanket with no binders or chopped into shorter fibers and vacuum-formed into blocks, boards, and other shapes. Ceramic-fiber linings, available for the temperature range of 650 to 1,430°C (1,200 to 2,600°F), are more economical than brick in the 650- to 1,230°C- (1,200- to 2,250°F-) range. Savings come from reduced first costs, lower installation labor, 90 to 95 percent less weight, and a 25 percent reduction in fuel consumption.

Because of the larger surface area (compared with solid-ceramic refractories) the chemical resistance of fibers is relatively poor. Their acid resistance is good, but they have less alkali resistance than solid materials because of the absence of resistant aggregates. Also, because they have less bulk, fibers have lower gas-velocity resistance. Besides

the advantage of lower weight, since they will not hold heat, fibers are more quickly cooled and present no thermal-shock structural problem.

Castable Monolithic Refractories Standard portland cement is made of calcium hydroxide. In exposures above 427°C (800°F) the hydroxyl ion is removed from portland (water removed); below 427°C (800°F), water is added. This cyclic exposure results in spalling. Castables are made of calcium aluminate (rather than portland); without the hydroxide they are not subject to that cyclic spalling failure.

Castable refractories are of three types:

1. *Standard.* 40 percent alumina for most applications at moderate temperatures.

TABLE 28-31 Wood for Chemical Equipment

Condition of woods after 31 d immersion in cold solutions Examined after 7 d drying								
	Fir	Oak	Oregon pine	Yellow pine	Spruce	Redwood	Maple	Cypress
Hydrochloric acid, 5%	NAC	NAC	NAC	SS	SS	SS	NAC	NAC
Hydrochloric acid, 10%	NAC	NAC	NAC	SS	SS	SS	NAC	NAC
Hydrochloric acid, 50%	SS,SB,SWF	SS,WF	S,WF	S,WF	S,WF	S,WF	S,WF	S,WF
Sulfuric acid, 1%	NAC	NAC	NAC	SS	SS	NAC	NAC	SS,SB
Sulfuric acid, 5%	SS	SS	SS	SS	SS,SB	SS,SB	NAC	SS,SB
Sulfuric acid, 10%	S,FSD	S,FSD	S,FSD	S,FSD	S,FSD	S,FSD	S,FSD	S,FSD
Sulfuric acid, 25%	SSp,FSD	SSp,FSD	SSp,FSD	SSp,FSD	SSp,FSD	SSp,FSD	SSp,FSD	SSp,FSD
Caustic soda, 5%	S,NAC	MSh,SWp	SS	SS,FSD	SSp,FSD	SSp,FSD	MSh	SSp,FSD
Caustic soda, 10%	S,FSD	MSh,WF,Horny	SS	SS,SB,FSD	SS,SB,FSD	SS,SB,FSD	MSh	S,SB,FSD
Alum, 13%	NAC	NAC	NAC	NAC	NAC	NAC	NAC	NAC
Sodium carbonate, 10%	SB,GC	NAC	GC	SB,GC	SB,GC	SB,GC	GC	SB,GC
Calcium chloride, 25%	NAC	NAC	NAC	NAC	NAC	NAC	NAC	NAC
Common salt, 25%	NAC	NAC	NAC	NAC	SS,GC	SS,GC	NAC	NAC
Water	NAC	NAC	NAC	NAC	NAC	NAC	NAC	NAC
Sodium sulfide	SS,SB	MSh,WF	SB	SB	SB	SB	MSh,FSD	FSD

Condition of woods after 8 h boiling in solutions Examined after 7 d drying								
	Fir	Oak	Oregon pine	Yellow pine	Spruce	Redwood	Maple	Cypress
Hydrochloric acid, 10%	SB,S	FSD	FSD	FSD	FSD	FSD	FSD	FSD
Hydrochloric acid, 50%	FD,Ch,B,S,NG	FD,Ch,B,S,NG	FD,Ch,B,S,NG	FD,Ch,B,S,NG	FD,Ch,B,S,NG	FD,Ch,B,S,NG	FD,Ch,B,S,NG	FD,Ch,B,S,NG
Sulfuric acid, 4%	SB,GC	SB,GC	SB,GC	SB,GC	SB,GC	SB,GC	SB,GC	SB,GC
Sulfuric acid, 5%	SS,GC	SB,GC	SB,GC	SB,GC	SB,FSD	SB,GC	SB,GC	SB,FSD
Sulfuric acid, 10%	SS,GC	BFD,Wpd,NG	Sp,FD,NG	B,Sp,FD,NG	B,Sp,FD,NG	SB,FSD	SB,FSD	B,FD
Caustic soda, 5%	SS	MSh	S	GC	S,GC	S,GC	Sh	SSp
Alum, 13%	SB,GC	NAC	NAC	SB,GC	SB,GC	SB,GC	NAC	SB,GC
Sodium carbonate, 10%	SB,GC	GC	GC	GC	GC	GC	GC	SB,GC
Calcium chloride, 25%	SB,GC	SB,SS,GC	NAC	SB,GC	SB,GC	NAC	NAC	SB,GC
Common salt, 25%	NAC	NAC	NAC	SB,GC	NAC	SB,GC	NAC	NAC
Water	NAC	NAC	NAC	SB,GC	NAC	NAC	NAC	NAC

TABLE 28-33 General Physical and Chemical Characteristics of Refractory Brick*

Type of brick	Typical composition	Approx. bulk density, lb/ft²	Fusion point, °F	Chemical nature	Deformation under hot loading	Apparent porosity, %	Permeability	Hot strength	Thermal shock resistance	Chemical resistance To acid	Chemical resistance To alkali
Silica	SiO_2 95%	115	3100	Acid	Excellent	21	High	Excellent	Poor‡	Good	Good at low temperatures
High-duty fire clay	SiO_2 54% Al_2O_3 40%	134	3125	Acid	Fair	18	Moderate	Fair	Fair	Good	Good at low temperatures
Superduty fire clay	SiO_2 52% Al_2O_3 42%	140	3170	Acid	Good	15	High	Fair	Good	Good	Good at low temperatures
Acid-resistant (type H)	SiO_2 59% Al_2O_3 34%	142	3040	Acid	Poor	7	Low	Poor	Good	Insoluble in acids except HF and boiling phosphoric	Very resistant in moderate concentrations
Insulating brick	Varies	30–75	Varies		Poor	65–85	High	Poor	Excellent	Poor	Poor
High-alumina	Al_2O_3 50–85%	170	3200–3400	Slightly acid	Good	20	Low	Good	Good	Good except for HF and aqua regia	Very slight attack with hot solutions
Extra-high-alumina	Al_2O_3 90–99%	185	3000–3650	Neutral	Excellent	23	Low	Excellent	Good		
Mullite	Al_2O_3 71%	153	3290	Slightly acid	Excellent	20	Low	Good	Good	Insoluble in most acids	Slight reaction
Chrome-fired	Chrome ore, 100%	195	Varies	Neutral	Fair	20	Low	Good	Poor	Fair to good	Poor
Magnesite-chrome bonded§	MgO 50–80%	190			Good	12	Very low	Good	Excellent		
Magnesite-chrome fired	Cr_2O_3 5–18% Fe_2O_3 3–13% Al_2O_3 6–11%	180	Varies	Basic	Excellent	20	High	Good	Excellent	Fair except to strong acids	Fair resistance at low temperatures
Magnesite-chrome high-fired	SiO_2 1.2–5%	180			Excellent	18	High	Excellent	Excellent		
Magnesite-bonded§	MgO 95%	181	3900	Basic	Good	11	Low	Good	Good	Soluble in most acids	Good resistance at low temperatures
Magnesite-fired		178			Good	19	Moderate	Good	Good		
Zircon	ZrO_2 67% SiO_2 33%	200	3100†	Acid	Excellent	25	Very low	Excellent	Good	Very slight	Very slight
Zirconia (stabilized)	ZrO_2 94% CaO 4%	245	4800	Slightly acid	Excellent	23	Low	Excellent	Excellent	Very slight	Very slight
Silicon-carbide	SiC 80–90%	160	4175	Slightly acid	Excellent	15	Very low	Excellent	Excellent	Slight reaction with HF	Attacked at high temperatures
Graphite	C, 97%	105	6400	Neutral	Excellent	16	Low	Excellent	Excellent	Insoluble	Insoluble

*From *Chem. Eng.*, 100 (July 31, 1967). To convert lbm/ft³ to kg/m³, multiply by 0.0624; °C = (°F − 32) × 5/9.
†Dissociates above 1700°C (3100°F).
‡Good above 650°C (1200°F).
§Chemically bonded.

TABLE 28-34 Minimum Temperature without Excessive Scaling in Air (Continuous Service)*

Alloy	°F	°C
Carbon steel	1050	565
½Mo steel	1050	565
1Cr ½Mo steel	1100	595
2¼Cr 1Mo steel	1150	620
5Cr ½Mo steel	1200	650
9Cr 1Mo steel	1300	705
AISI 410	1300	705
AISI 304	1600	870
AISI 321	1600	870
AISI 347	1600	870
AISI 316	1600	870
AISI 309	2000	1090
AISI 310	2100	1150

*Courtesy of National Association of Corrosion Engineers.

2. *Intermediate purity.* 50 to 55 percent alumina. The anorthite (needle-structure) form is more resistant to the action of steam exposure.

3. *Very pure.* 70 to 80 percent alumina for high temperatures. Under reducing conditions the iron in the ceramic is controlling, as it acts as a catalyst and converts the CO to CO_2 plus carbon, which results in spalling. The choice among the three types of castables is generally made by economic considerations and the temperature of the application.

Compared with brick, castables are less dense, but this does not really mean that they are less serviceable, as their cements can hydrate and form gels which can fill the voids in castables. Extra-large voids do indicate less strength regardless of filled voids and dictate a lower allowable gas velocity. If of the same density as a given brick, a castable will result in less permeation.

Normally, castables are 25 percent cements and 75 percent aggregates. The aggregate is the more chemically resistant of the two components. The highest-strength materials have 30 percent cement, but too much cement results in too much shrinkage. The standard insulating refractory, 1:2:4 LHV castable, consists of 1 volume of cement, 2 volumes of expanded clay (Haydite), and 4 volumes of vermiculite.

Castables can be modified by a clay addition to keep the mass intact, thus allowing application by air-pressure gunning (gunite). Depending upon the size and geometry of the equipment, many castable linings must be reinforced; wire and expanded metal are commonly used.

TABLE 28-35 Important Commercial Alloys for High-Temperature Process Service

	Nominal composition, %			
	Cr	Ni	Fe	Other
Ferritic steels				
Carbon steel			Bal.	
2¼ chrome	2¼		Bal.	Mo
Type 502	5		Bal.	Mo
Type 410	12		Bal.	
Type 430	16		Bal.	
Type 446	27		Bal.	
Austenitic steels				
Type 304	18	8	Bal.	
Type 321	18	10	Bal.	Ti
Type 347	18	11	Bal.	Cb
Type 316	18	12	Bal.	Mo
Type 309	24	12	Bal.	
Type 310	25	20	Bal.	
Type 330	15	35	Bal.	
Nickel-base alloys				
Nickel		Bal.		
Incoloy 800	21	32	Bal.	
Hastelloy B		Bal.	6	Mo
Hastelloy C	16	Bal.	6	W, Mo
60-15	15	Bal.	25	
Inconel 600	15	Bal.	7	
80-20	20	Bal.		
Hastelloy X	22	Bal.	19	Co, Mo
Multimet	21	20	Bal.	Co
Rene 41	19	Bal.	5	Co, Mo, Ti
Cast irons				
Ductile iron			Bal.	C, Si, Mg
Ni-Resist, D-2	2	20	Bal.	Si, C
Ni-Resist, D-4	5	30	Bal.	Si, C
Cast stainless (ACI types)				
HC	28	4	Bal.	
HF	21	11	Bal.	
HH	26	12	Bal.	
HK	15	20	Bal.	
HT	15	35	Bal.	
HW	12	Bal.	28	
Superalloys				
Inconel X	15	Bal.	7	Ti, Al, Cb
A 286	15	25	Bal.	Mo, Ti
Stellite 25	20	10	Co-base	W
Stellite 21 (cast)	27.3	2.8	Co-base	Mo
Stellite 31 (cast)	25.2	10.5	Co-base	W

Process Machinery Drives

Heinz P. Bloch, P.E., B.S.M.E., M.S.M.E., *Consulting Engineer, Process Machinery Consulting; American Society of Mechanical Engineers, Vibration Institute; Registered Professional Engineer (New Jersey, Texas). (Section Editor)*

R. H. Daugherty, Ph.D., *Consulting Engineer, Research Center, Reliance Electric Company; member, Institute of Electrical and Electronics Engineers. (Electric Motors and Auxiliaries)°*

Fred K. Geitner, P.Eng., B.S.M.E., M.S.M.E., *Consulting Engineer, Registered Professional Engineer (Ontario, Canada). (Reciprocating Engines; Steam Turbines)†*

Meherwan P. Boyce, P.E., Ph.D., *President, Boyce Engineering International; ASME Fellow; Registered Professional Engineer (Texas, Oklahoma). (Turbines)*

Judson S. Swearingen, Ph.D., *Retired President, Rotoflow Corporation. (Expansion Turbines)*

Eric Jenett, M.S.Ch.E., *Manager, Process Engineering, Brown & Root, Inc.; associate member, AIChE, Project Management Institute; Registered Professional Engineer (Texas). (Power Recovery from Liquid Streams)*

Michael M. Calistrat, B.S.M.E., M.S.M.E., *Owner, Michael Calistrat and Associates; member, American Society of Mechanical Engineers. (Mechanical Power-Transmission Equipment)*

° Based largely on material originally compiled and contributed by Carl R. Olson, M.S.E.E.
† Based largely on material originally compiled and contributed by Frank L. Evans, Jr., B.S.M.E., L.L.B. (Reciprocating Engines), and H. Steen-Johnsen, M.S.M.E. (Steam Turbines).

Nomenclature

Symbol	Definition	SI units	U.S. customary units
C	Constant		
C_1	Velocity of steam flow	m/s	ft/s
c	Clearance	m	ft
d	Bearing diameter	m	ft
E	Applied voltage	V	V
E	Modulus of elasticity	N/m^2	lbf/ft^2
e	Gross pump efficiency	Dimensionless	Dimensionless
e	Base of natural logarithm		
e	Efficiency	Dimensionless	Dimensionless
e_h	Hydraulic efficiency	Dimensionless	Dimensionless
F	Load		
F_f	Friction force	N	lbf
f	Friction coefficient		
f	Frequency	Hz	Hz
G, g	Gravitational constant	m/s^2	ft/s^2
H	Power	W	hp
H_g	Enthalpy change, ideal	J	Btu
H_i	Reversible enthalpy change	J/kg	Btu/lb
H_p	Total head (pump)	m	ft
H_t	Total head (turbine)	m	ft
h	Oil thickness	m	ft
Δh	Enthalpy change per mass	J/kg	Btu/lb
I	Line current	A	A
K_A	Constant		
K_L	Constant		
K_S	Constant		
K_T	Constant		
k	Constant		
Mol. wt.	Molecular weight	kg	lb
N	Force	N	lbf
N_c	Compressor efficiency	Dimensionless	Dimensionless
N_e	Expander efficiency	Dimensionless	Dimensionless
n	Speed	r/s	r/min
n_s	Specific speed (turbine or pump)		
P	Pressure	kPa	lbf/in^2
P	Mean bearing pressure	N/m^2	lbf/ft^2
P	Power	kW	kW
PD	Pitch diameter	m	ft
P_{max}	Contact pressure, maximum	N/m^2	lbf/ft^2
p	Number of poles		
Q	Quantity of heat added	J	Btu
Q_c	Quantity of heat removed	J	Btu
Q_p	Capacity, pump	m^3/h	ft^3/h
Q_t	Capacity, turbine	m^3/h	ft^3/h
R	Operating load	kg	lb
R	Gas constant	J/(mol·K)	Btu/(mol·°R)
R	Armature resistance	Ω	Ω
r	Radius	m	ft
S	Apparent power	kVA	kVA
S_{max}	Stress, maximum	N/m^2	lbf/ft^2
T	Tension	N	lbf
T	Torque	N·m	lbf·ft
T	Temperature	K	°R
T_a	Average temperature	K	°R
T_c	Centrifugal tension	N	lbf
t	Time	s	s
V	Applied voltage	V	V
V	Counterelectromotive force	V	V
V	Velocity	m/s	ft/min
W	Net work to compressor	J	Btu
W_c	Compressor work	J	Btu
W_{theor}	Theoretical work	J	Btu
WK^2	Inertia	kg·m^2	lb·ft^2
y	Power factor	Dimensionless	Dimensionless
Z	Viscosity	N·s/m^2	cP
Z_a	Average compressibility factor	Dimensionless	Dimensionless

	Greek symbols		
α_r	exit angle		
β	exit angle		
θ	Arc of contact		
φ	Belt density	kg/m^3	lb/in^3
μ	peripheral velocity	m/s	ft/min
ϕ	Magnetic-field flux	Wb	Wb
ω	relative velocity	m/s	ft/min

GENERAL REFERENCES: Bartlett, *Steam Turbine Performance and Economics,* McGraw-Hill, New York, 1958. Baumeister, *Standard Handbook for Mechanical Engineers,* 7th ed., McGraw-Hill, New York, 1967. Bloch, Heinz P., *Practical Guide to Compressor Technology,* McGraw-Hill, New York, 1996. Bloch, Heinz P., *Practical Guide to Steam Turbine Technology,* McGraw-Hill, New York, 1996. Boyce, Meherwan W., *Gas Turbine Engineering Handbook,* Gulf Publishing Company, Houston, 1982. Calistrat, Michael M., *Flexible Couplings—Their Design, Selection, and Use,* Caroline Publishing, Houston, 1994. Collins and Canaday, *Expansion Machines for Low Temperature Processes,* Oxford, Fair Lawn, New Jersey, 1958. Csanaday, *Theory of Turbomachines,* McGraw-Hill, New York, 1964. Fink and Carroll, *Standard Handbook for Electrical Engineers,* 10th ed., McGraw-Hill, New York, 1968. Jennings and Rogers, *Gas Turbines: Analysis and Practice,* McGraw-Hill, New York, 1953. Katz et al., *Handbook of Natural Gas Engineering,* McGraw-Hill, New York, 1959. Rase and Barrow, *Project Engineering of Process Plants,* Wiley, New York, 1957. Salisbury, *Steam Turbines and Their Cycles,* Wiley, New York, 1950. Scott, *Cryogenic Engineering,* Van Nostrand, Princeton, New Jersey, 1959. Shepherd, *Principles of Turbomachinery,* Macmillan, New York, 1956. Stepanoff, *Centrifugal and Axial Flow Pumps,* 2d ed., Wiley, New York, 1957. Stodola, *Steam and Gas Turbines,* Peter Smith, New York, 1945.

ELECTRIC MOTORS AND AUXILIARIES

All electric motors operate on the same basic principle regardless of type or size. When a wire carries electric current in the presence of a magnetic field (at least partially perpendicular to the current), a force on the wire is produced perpendicular to both the current and the magnetic field. In a motor the magnetic field radiates either in toward or outward from the motor axis (shaft) across the air gap, which is the annular space between the rotor and stator. Current-carrying conductors parallel to the axis (shaft) then have a force on them tangent to the rotor circumference. The force on the wire opposes an equal force (or reaction) on the magnetic field. It makes no difference whether the magnetic field is created in the rotor or the stator; the net result is the same: the shaft rotates.

Within these basic principles there are many types of electric motors. Each has its own individual operating characteristics peculiarly suited to specific drive applications. Equations (29-1) through (29-9), presented in Table 29-1, describe the general operating characteristics of alternating-current motors. When several types are suitable, selection is based on initial installed cost and operating costs (including maintenance and consideration of reliability).

ALTERNATING-CURRENT MOTORS, CONSTANT SPEED

The majority of industrial drives are constant speed. Typical applications include:

Pumps
Compressors

TABLE 29-1   Useful Formulas for Alternating-Current Motors

Power output:

$H = Tn/5250$		(29-1)
$P = 0.00173VIye$	(three-phase)	(29-2)
$P = 0.001VIye$	(single-phase)	(29-3)

Power input:

$P = 0.00173VIy$	(three-phase)	(29-4)
$P = 0.001VIy$	(single-phase)	(29-5)
$P = 0.746H/e$		(29-6a)
$S = P/y = 0.746H/ye$		(29-6b)

Line current and power factor:

$$I = \frac{0.746H \text{ (output)}}{0.00173Vye} \quad \text{(three-phase)} \quad (29\text{-}7)$$

$$I = \frac{0.746H \text{ (output)}}{0.001Vye} \quad \text{(single-phase)} \quad (29\text{-}8)$$

$$y = \frac{P \text{ (input)}}{S} \quad\quad (29\text{-}9)$$

where

e = efficiency, decimal
H = power, hp
I = line current, A
n = speed, r/min
P = power, kW
S = apparent power, kVA
T = torque, lbf·ft
V = applied voltage, V
y = power factor, decimal

NOTE: To convert horsepower to watts, multiply by 746; to convert pound-force-feet to newton-meters, multiply by 1.356; and to convert revolutions per minute to radians per second, multiply by 0.1047.

Fans
Conveyors
Crushers and mills

Alternating-Current Squirrel-Cage Induction Motors These motors are by far the most common constant-speed drives. They are relatively simple in design and therefore both low in cost and highly reliable. Representative prices are shown in Fig. 29-1 for various speeds and horsepowers.

The typical three-phase squirrel-cage motor has stator windings which are connected to the power source. The rotor is a cylindrical magnetic structure mounted on the shaft with slots in the surface, parallel (or slightly skewed) to the shaft; either bars are inserted into these slots or molten metal is cast in place and connected by a short-circuiting end ring at both ends of the rotor. The name "squirrel-cage" derives from this rotor-bar construction. In operation, current passing through the stator winding creates a rotating magnetic field which cuts the rotor winding unless the rotor is turning in exact synchronism with the stator field. This cutting action induces a voltage, and hence a current, in the rotor which in turn reacts with the magnetic field to produce torque.

The typical medium-sized squirrel-cage motor is designed to operate at 2 to 3 percent slip (97 to 98 percent of synchronous speed). The synchronous speed is determined by the power-system frequency and the stator-winding configuration. If the stator is wound to produce one north and one south magnetic pole, it is a two-pole motor; there is always an even number of poles (2, 4, 6, 8, etc.). The synchronous speed is

$$n = 120 f/p \quad\quad (29\text{-}10)$$

where
n = speed, r/min
f = frequency, Hz (cycles/s)
p = number of poles

The actual operating speed will be slightly less by the amount of slip. Slip depends upon motor size and application. Typically, the larger the motor, the less slip; an ordinary 7460-W (10-hp) motor may have 2½ percent slip, whereas motors over 746 kW (1000 hp) may have less than ½ percent. High-slip motors (as much as 13 percent slip) are used for applications with high inertia and requiring high starting torque; typical applications are punch presses and some crushers. Typical speed versus torque curves for various National Electrical Manufacturers Association (NEMA) design motors up to 149.2 kW (200 hp) are shown in Fig. 29-2. Typical characteristics and applications for these motors are given in Table 29-2.

As noted in Fig. 29-1 and Table 29-2 motors of various efficiency ratings may be available for an application. The efficiency of all motors of the same design will not be identical because of normal variations inherent in materials, variations in manufacturing processes, and inaccuracies in the test methods and equipment used to determine the efficiency. To address these variations, two values of efficiency have been identified for polyphase induction motors. Nominal efficiency refers to the average efficiency of a large population of motors of duplicate design. Some of the motors will have efficiency lower than the nominal efficiency value, and some will have higher. Minimum efficiency refers to the lowest level of efficiency that any motor in the population might have as a result of the variations introduced by the materials, manufacturing processes, and testing. The nominal effi-

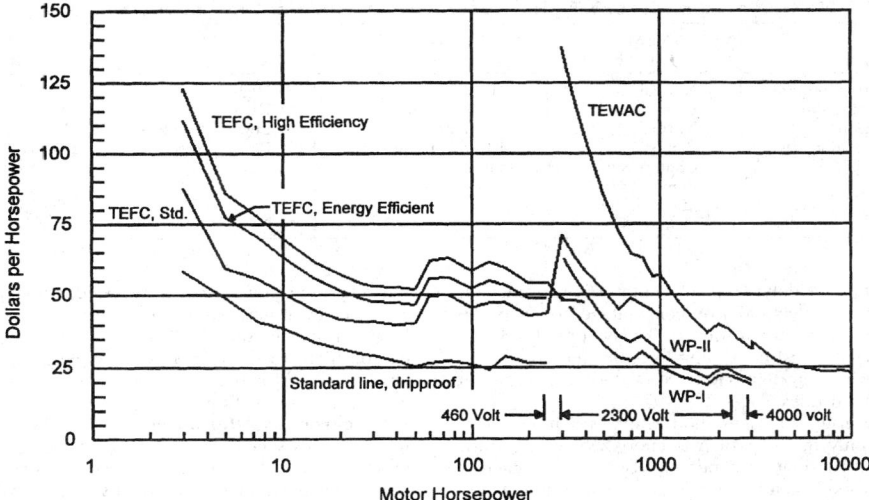

FIG. 29-1 Motor prices in dollars per horsepower for 1800 rev/min squirrel-cage induction motors from 3 to 10,000 hp. Dripproof and TEFC motors shown from 3 to 400 horsepower have 1.15 service factor; for other motors above 250 horsepower, the service factor is 1.0. The basis of these data is July, 1994. To convert dollars per horsepower to dollars per kilowatt, multiply by 1.340; to convert horsepower to kilowatts, multiply by 0.746.

ciency should be used when estimating the power required to supply a number of motors. The NEMA Standard MG-1, *Motors and Generators,* requires that all polyphase squirrel-cage integral horsepower motors, 1 to 500 horsepower, designated as Design A, B (and equivalent Design C ratings), and E be marked with the NEMA nominal efficiency value. A minimum level of efficiency is defined for each level of nominal efficiency in the NEMA Standard MG-1.

A motor purchaser or user can use the defined NEMA nominal efficiency to determine the relative economics of alternate motors. Common to the various methods one can use is consideration of the costs of energy and the motor, the annual hours of operation of the motor, and the motor efficiency. A simple payback analysis is used to determine the number of years required for the savings in energy cost resulting from the use of a more efficient motor to pay back the higher initial cost of the motor. The present worth life-cycle analysis considers both the time value of money and energy cost inflation to determine the present worth of savings for each motor being evalu-

ated. The cash flow and payback analysis method considers motor cost premium, motor depreciation life, energy cost and energy cost inflation rate, corporate tax rate, tax credit, and the motor operating parameters.

When making any economic analysis, care should be taken to be certain that the efficiency ratings of all motors being considered are on the same basis. While this should not be a problem for motors rated 1 to 500 horsepower as covered by the NEMA Standards for efficiency marking, it is common practice for several different test methods to be used when measuring the efficiency of motors rated over 500 horsepower. A particular test method may need to be selected by the test facility on the basis of available test equipment and power supply. All test methods that may be used to test any one motor will not necessarily give the same result for efficiency.

Further incentives to use energy-efficient motors are provided by various cost rebate programs offered by utilities based on horsepower rating and efficiency level. Another factor that will have a significant impact is the Energy Policy Act of 1992, in which the U.S. Congress established limits on the lowest level of nominal efficiency that certain classes of motors of standard design can have after 1997.

Control or **starting** of squirrel-cage induction motors normally consists of applying full voltage to the motor terminals. The speed-torque curves in Fig. 29-2 are based on full voltage throughout the speed range from start to run. The specific motor design determines the amount of starting current. However, if the motor is a typical standard (NEMA A or B) design, the starting current may be estimated at 6 to 6.5 times normal full-load current with full voltage applied. For NEMA Design E, it may be estimated at 8 to 9 times normal full-load current with full voltage applied. Particularly for large motors, this starting inrush current may cause an undesirable voltage dip which can shut down other equipment, temporarily dim lights, or even initiate malfunctions in sophisticated controls on the power system. For these conditions various alternatives exist.

1. *Reduced-voltage starting.* A reactor, resistor, or transformer is temporarily connected ahead of the motor during start to reduce the current inrush and limit voltage dip. This is accompanied by reduced starting torque. For reactor or resistor start, the torque decreases as the square of current; for transformer start, the torque decreases directly with line current. The reactor, resistor, or transformer can be adjusted to give a proper balance between torque and current.

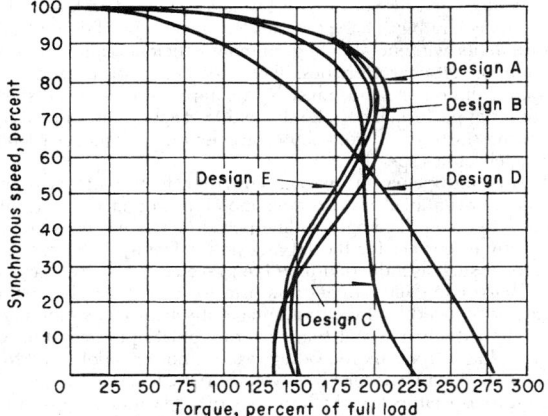

FIG. 29-2 Typical speed versus torque curves for various NEMA-design squirrel-cage induction motors. (See Table 29-2 for an explanation of design types.)

TABLE 29-2 Characteristics and Typical Applications for Squirrel-Cage Induction Motors

	NEMA A and B	NEMA C	NEMA D	NEMA E
Starting torque	Normal	High	High	Normal
Running slip	Low	Low to medium	High	Low
Efficiency	Normal	Normal	Low	High
Applications	Pumps	Electrical stairways	Punch presses	Pumps
	Compressors	Pulverizers	Crushers	Compressors
	Fans	Conveyors		Fans
	Machine tools			Machine tools
	General use			General use

2. *Star-delta starting.* A delta-connected motor is reconnected in Y form for starting, thus applying 57.7 percent voltage to each phase winding. This results in a developed torque of $(0.577)^2$, or only 33 percent. There is no means of adjustment; therefore, this method is useful only for loads requiring less than one-third of the motor's normal starting and accelerating torques.

3. *Part-winding starting.* This method employs a motor with two sets of windings, only one of which is energized during start. Torque and current are both roughly 50 percent. Two small contactors (starting switches) are used instead of one large one, and no reactors or transformers are required. The disadvantages are the fixed value of available torque and the harmonic disturbances from possible winding unbalance, causing deviations in the speed-torque curve and therefore possible failure to accelerate.

Braking and regeneration are possible with squirrel-cage motors. The direction of rotation is determined by the sequence or phase rotation of the power supply. If two leads on a three-phase motor are interchanged, the rotation reverses. If this occurs during operation, the motor will come to a rapid stop and reverse. Power is removed at standstill for effective braking. For estimating only, this plug-stop torque is approximately equal to starting torque. The braking time can be estimated by

$$t = WK^2 n/308T \qquad (29\text{-}11)$$

where
t = time, s
WK^2 = inertia, lb·ft^2
n = running speed, r/min
T = torque, lbf·ft

To convert pound-square feet to kilogram-square meters, multiply by 0.0421; to convert revolutions per minute to radians per second, multiply by 0.1047; and to convert pound-force-feet to newton-meters, multiply by 1.356.

These estimates are frequently inaccurate because of second-order effects such as rotor saturation and harmonics. If the application is at all critical, the motor manufacturer should be consulted.

Regenerative braking occurs at speeds above synchronous-motor speed resulting from an overhauling load or from switching from high to low speed on multispeed motors. The action is similar to normal motor operation except that the slip is negative. The motor acts as an induction generator, delivering energy to the power source. If a power source such as a gas expander or a downhill conveyor is available, regenerative braking is an effective method of regulating speed, conserving energy, and starting the driven (driving) machine. An induction generator can deliver power to the source about equal to its rating as a motor. Regenerative braking can be used only on power systems capable of absorbing the generated energy and of supplying magnetizing excitation (reactive power) for the motor.

Direct-current dynamic braking utilizes direct current applied to the stator winding. Alternating-current power is first removed by opening the motor contactor or starter; direct current is then applied by a second contactor. The direct current produces a stationary magnetic flux, in contrast to the normal rotating ac field. The rotor bars cut this field, inducing currents which react with the dc flux to develop braking torque. Braking effort is easily varied by adjusting the amount of direct current. A desirable feature of this method for standard motors is the relatively soft braking effort at full-load speed, reducing impact; further, the braking effort typically increases as speed drops,

reaching a maximum near zero speed. Braking torque at standstill is zero; however, maximum torque occurs at such low speed that static friction is usually sufficient to prevent coasting. Peak braking torques can be high; so shafts, gearing, couplings, etc., should be checked. Caution should be exercised because frequent starting and stopping cause excessive heating.

Synchronous Alternating-Current Motors These motors run in exact clock synchronism with the power system. For most modern power systems, these are truly constant-speed motors.

In the conventional synchronous motor a rotating magnetic field is developed by the stator currents as in induction motors. The rotor, however, is different, consisting typically of pairs of electromagnets (poles) spaced around the rotor periphery. The rotor field corresponds to the field produced by the ac stator having the same number of poles. The rotor or field coils are supplied with direct current; the magnetic field is therefore stationary with respect to the rotor structure. Torque is developed by the interaction of the rotor magnetic field and the stator current (in-phase component). Under no-load conditions and with appropriate dc field current, rotor and stator magnetic-field centers coincide. The voltage applied to the stator winding is balanced by an opposing voltage generated in the stator by the rotor field (induced), and no ac power current flows. As load is applied, the rotor tends to decelerate momentarily, causing a shift of rotor position with respect to the ac field. This shift produces a difference between the applied and induced voltages; the voltage difference causes current to flow; the current reacts with the rotor magnetic flux, producing torque.

Synchronous motors should not be started with the dc field applied. Instead they are started as induction motors; bars, acting like a squirrel-cage rotor, are embedded in the field-pole surface and connected by end rings at both ends of the rotor. These damper bars also serve to damp out oscillations under normal running conditions. When the motor is at approximately 95 percent speed (depending upon application and motor design), direct current is applied to the field and the motor pulls into step (synchronism). Because the damper bars do not affect the synchronous-speed characteristics, they are designed for starting performance. This provides flexibility in the accelerating characteristics to meet specific application requirements without affecting running efficiency and other synchronous-speed characteristics. The rotor design of a squirrel-cage motor, on the other hand, must be a compromise between starting and running performance. The dc field is usually shorted by a resistor during starting and contributes accelerating torque, particularly near synchronous speed.

Power-factor correction is an important feature of synchronous motors. Conventional synchronous-motor power factors are either 100 or 80 percent leading. Leading-power-factor machines are used frequently to correct for the lagging power factor of the remaining plant load (such as induction motors), preventing penalty charges on power bills. Even 100 percent power-factor motors can be operated leading at reduced loads. An advantage of synchronous motors over capacitors is their inherent tendency to regulate power-system voltage; as voltage drops, more leading reactive power is delivered to the power system, and, conversely, as voltage rises, less reactive power, in contrast to capacitors for which the reactive power decreases directly in proportion to the voltage drop squared. The amount of leading reactive power delivered to the system depends on dc field current, which is readily adjustable.

Field current is an important control element. It controls not only the power factor but also the pullout torque (the load at which the motor pulls out of synchronism). For example, field forcing can prevent pullout on anticipated high transient loads or voltage dips. Loads with known high transient torques are driven frequently with 80 percent power-factor synchronous motors. The needed additional field supplies both additional pullout torque and power-factor correction for the power system. When high pullout torque is required, the leading power-factor machine is often less expensive than a unity-power-factor motor with the same torque capability.

Direct-current field excitation is supplied by various means. A dc generator (exciter) is often used either directly coupled to the motor shaft, belt-driven off the motor shaft (seldom used), or driven by a separate small motor (exciter motor-generator set). Direct-coupled and belt-driven exciters are always associated with a single motor and are controlled by adjustment of the exciter field. Motor-generator-set exciters may supply one or more synchronous motor fields. For reliability, several motor-generator sets may be paralleled to supply multiple motor fields; in such a case the exciter voltage is usually fixed (e.g., 125 or 250 V), and the individual synchronous-motor fields are controlled by motor field rheostats (much larger than exciter field rheostats). Static (rectifier) exciters are also used for single-motor or multimotor excitation. Special rectifiers are required to avoid damage from surge voltages on pullout. These exciters, rotating or static, require brushes and slip rings to conduct direct current to the rotating field structure.

Another concept is **brushless excitation,** in which an ac generator (exciter) is directly coupled to or mounted on the motor shaft. The ac exciter has a stator field and an ac rotor armature which is directly connected to a static controllable rectifier on the motor rotor (or a shaft-mounted drum). Static control elements (to sense synchronizing speed, phase angle, etc.) are also rotor-mounted, as is the field discharge resistor. Changing the exciter field adjusts the motor field current without the necessity of brushes or slip rings. Brushless excitation is suitable for use in hazardous atmospheres, where conventional brush-type motors must have protective brush and slip-ring enclosures.

Because of the more complicated design and the necessity for a field power supply, synchronous motors are typically applied only in large-horsepower ratings (several hundred horsepower and larger); synchronous motors over 59,680 kW (80,000 hp) have been built. With their latitude in size and characteristics and their important inherent high power factor and efficiency, synchronous motors are applied to a wide variety of drives. Engine-type motors (without shaft or bearings) are used almost exclusively to drive large low-speed reciprocating compressors. Other typical applications include jordans, compressors, pumps, ball and rod mills, chippers, crushers, and grinders. Speeds as low as 80 r/min are practical; the top speed is limited by the rotor structure and is dependent on horsepower. The approximate limit for 1800 r/min is 2238 kW (3000 hp); for 1200 r/min it is 29,840 kW (40,000 hp).

Synchronous speeds are calculated by Eq. (29-10). Speeds above the limits given are obtained through step-up gears; large high-speed centrifugal compressors are examples. Two-pole (3600 r/min at 60 Hz) synchronous motors can be built but are uneconomical in comparison with geared drives.

ALTERNATING-CURRENT MOTORS, MULTISPEED

Squirrel-cage induction motors are inherently single-speed machines, but multispeed operation can be obtained by reconnecting the stator windings of motors designed for this purpose.

Two-Winding Motors These motors illustrate the simplest concept. The two separate stator windings (three-phase or two-phase only) are designed and wound for a different number of poles. For example, one winding may be four poles (1800 r/min at 60 Hz) and the other six poles (1200 r/min at 60 Hz). Only one winding is connected at a time. This method is used for speed ratios other than 2:1. Since the two windings are independent, a large number of speed combinations is possible. The two windings are not necessarily of equal capacity.

Two-winding motors may be built for constant torque, variable torque, or constant horsepower. Constant-horsepower motors are capable of handling the same horsepower at both speeds (i.e., higher torque at the low speed). Constant-torque motors can handle the same load torque at either speed (e.g., conveyor drives). Variable-torque motors are designed for loads in which load torque varies as the square of speed and horsepower varies as the cube of speed. Typical applications are as follows:

Variable torque	Constant torque	Constant horsepower
Fans	Conveyors	Machine tools
Centrifugal pumps	Feeders	
	Reciprocating compressors	

Single-Winding Consequent-Pole Motors These motors can be used when the low speed is one-half of the high speed. They are available as three-phase only. The specially designed winding is regrouped by external reconnection (motor control) to obtain the desired speed. A 2:1 speed ratio only is obtainable by this method; speeds such as 3600/1800, 1800/900, and 1200/600 are obtainable. Variable-torque, constant-torque, and constant-horsepower designs are available with torque characteristics as discussed under "Two-Winding Motors." The control for two-speed single-winding motors is more complicated than for the two-winding control.

Four-Speed, Two-Winding Squirrel-Cage Motors These motors are built by combining the preceding two methods. The stator winding is composed of two consequent-pole windings. Each winding gives two speeds with a relation of 2:1 to each other. The standard 60-Hz speed combinations are 1800/1200/900/600 r/min and 1200/900/600/450 r/min. The three torque-capability designs of variable torque, constant torque, and constant horsepower are also available in these four-speed motors.

Pole-Amplitude-Modulated Induction Motors These are single-winding squirrel-cage motors with any combination of poles or speeds (e.g., 8/10 poles or 900/720 r/min or as wide as 4/20 poles or 1800/360/min). They are smaller and lighter than equivalent two-winding machines. The entire winding works at both high and low speed, resulting in greater thermal capacity and higher efficiency. The basic principle is that one frequency acted on (modulated) by another produces new frequencies equal to the sum and difference of the two. Thus, a six-pole field modulated by a two-pole field produces a four- and an eight-pole field. The four-pole field can be eliminated by proper winding geometry. Such a motor runs at six-pole speed (1200 r/min, 60 Hz) when connected normally and at eight-pole speed (900 r/min) when half of the coils are reversed to produce the two-pole modulation.

In the preceding discussion of multispeed ac motors note that only induction motors are considered. These have no discrete physical rotor poles, so that only the stator-pole configuration need be modified to change speed. To operate multispeed, a synchronous motor would require a distinct rotor structure for each speed. Thus multispeed is practical only for squirrel-cage induction motors.

ALTERNATING-CURRENT MOTORS, WOUND-ROTOR INDUCTION

Wound-rotor induction motors operate on the same principle as squirrel-cage motors. However, as the name implies, the rotor has windings rather than bars, and these windings are connected to shaft-mounted slip rings. Brushes riding on the slip rings are connected to external resistance or short-circuited. Wound-rotor motors have an additional dimension of flexibility in the **variability of external rotor resistance.** Rotor resistance affects the shape of the speed-torque curve; increasing resistance decreases the speed at which maximum torque occurs. Figure 29-3 illustrates this effect as resistance is increased from zero external resistance (at top) to a very high value (extreme left).

Reduced-speed operation reduces efficiency. **Efficiency** is approximately equal to speed expressed as a percentage of synchronous speed. Thus at 75 percent speed, about three-fourths of the motor

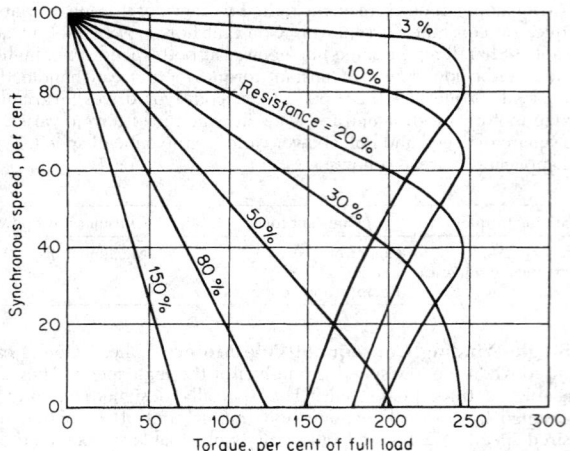

FIG. 29-3 Typical speed versus torque curves for a wound-rotor induction motor with varying amounts of external secondary (rotor) resistance. Resistance values are based on resistance at 100 percent torque and zero speed = 100 percent.

input goes to the load; the other quarter is dissipated in the rotor resistance. The external resistance is sized to get rid of this heat. Further, in accelerating a load to operating speed, the heat generated in the rotor resistance is equal to the energy required to accelerate the load (inertia plus friction). The rotor resistance (internal or external) must have capacity to store this heat, since accelerating time is typically too short to allow any significant heat dissipation.

These characteristics of wound-rotor motors determine their **scope of application.** They can accelerate very-high-inertia loads, such as crushers, by using a large external resistance to absorb the heat. Loads sensitive to shock-accelerating torques may also be accelerated softly by inserting a high starting rotor resistance; this effect is used, for example, to take up slack in gears. Wound rotors are also used to handle loads like punch presses and car crushers, for which extreme transient peak loads are supplied by the mechanical-system inertia, allowing the system to slow down during these peaks; permanent external rotor resistance provides this soft characteristic. Wound-rotor motors are also used to provide adjustable-speed drives for pumps, cranes, and other loads when precise speed regulation is not required. Reduced-speed losses are not very significant for pump loads; the percent efficiency is low at reduced speed, but the torque and load horsepower are dropping rapidly. If torque is proportional to speed squared, the maximum rotor-resistance losses never exceed 10.5 percent of the full-speed load (occurs at 70 percent speed and efficiency, 50 percent torque, 35 percent load horsepower, and 30 percent losses, thus 10.5 percent losses based on full load at full speed).

Control of wound-rotor motors, as discussed, can be effected by adjusting the external secondary (rotor) resistance either in steps or continuously by liquid rheostat (this method is seldom used). Commonly when secondary resistance is varied to adjust speed or torque or to control acceleration, multiple resistance steps are used. These steps may be switched manually (typically a drum switch) or electrically by contactor.

In addition to secondary resistance control, other devices such as reactors and thyristors (solid-state controllable rectifiers) are used to control wound-rotor motors. Fixed secondary reactors combined with resistors can provide very constant accelerating torque with a minimum number of accelerating steps. The change in slip frequency with speed continually changes the effective reactance and hence the value of resistance associated with the reactor. The secondary reactors, resistors, and contacts can be varied in design to provide the proper accelerating speed-torque curve for the protection of belt conveyors and similar loads.

Saturable reactors, which are adjustable by a small dc signal, have also been used for both primary (stator) and secondary (rotor) control. In the primary they control motor voltage and therefore torque. In combination with fixed secondary resistors and feedback from a tachometer, this system can be used for precise speed and torque control of cranes, hoists, etc. Even reversing can be accomplished by using two saturable reactors in each of two (of three) phases. Other combinations of fixed or saturable reactors in the primary and/or secondary, all combined with secondary resistors, provide a wide range of capabilities and flexibility for the wound-rotor motor.

Thyristors have been replacing saturable reactors; they are small, efficient, and easily controlled by a wide variety of control systems. A modern crane control drive uses fixed secondary resistors and two sets of primary thyristors (one set for hoist, one for lower). With tachometer feedback for speed sensing, the control for the motor provides speed regulation and torque limiting in both directions, all with static devices. A wide variety of control systems is possible; the control should be designed for the specific application.

DIRECT-CURRENT MOTORS

Direct-current motors are adjustable in speed over a wide range. Further, efficiency is high over the entire speed range, unlike wound-rotor motors, in which efficiency is roughly proportional to speed. This flexibility is attained at the expense of additional complexity and cost.

Direct-current motor fields are on the stator. The rotor is the armature. The magnetic field does not rotate like the field in ac machines. Current in the armature reacts with the stator field to produce torque.

The armature windings generate a voltage opposite to the applied voltage as they cut the magnetic field. The difference between the terminal voltage and the generated voltage (counterelectromotive force) applied to the armature resistance produces armature current. Torque is proportional to armature current and magnetic flux. Counterelectromotive force is almost equal to the applied voltage, so the speed can be changed by changing the applied voltage; the speed can also be changed by varying the field current.

$$E = V + IR \quad \text{or} \quad I = (E - V)/R \qquad (29\text{-}12)$$

$$V = kn\phi \qquad (29\text{-}13)$$

where E = applied voltage, V
 V = counterelectromotive force (generated voltage), V
 R = armature resistance, Ω
 I = armature current, A
 k = constant dependent on motor design
 n = speed, r/min
 ϕ = magnetic-field flux

Generated voltage is proportional to the magnetic-flux cut; so a motor must change speed to generate the same counterelectromotive force if the field current is changed.

Direct-current motors are connected in several ways. Shunt motors have armature and field connected in parallel. This connection provides almost constant speed regardless of the applied voltage. If the voltage drops, the counterelectromotive force also drops because of the reduction in field strength. Speed can be changed by varying the field current and/or armature voltage independently. Increasing armature voltage increases speed; increasing field current decreases speed.

Series motors have their armature and field windings connected in series; both carry the same current. The speed depends on both voltage and load. Torque is the product of armature current and magnetic field; both are dependent on current. For any specific load torque, the current is constant and speed is proportional to applied voltage. If, however, the load changes, the speed also changes. A 50 percent drop in load torque reduces the motor current to 70 percent, making the product of armature and field current 50 percent. The reduced field current decreases counterelectromotive force, and the motor speed increases by 40 percent. Because of this characteristic, series motors overspeed severely if unloaded. Series motors are suitable for constant-horsepower applications with a wide speed range, such as machine tools. They are also used for traction drives (e.g., shuttle cars and locomotives), some cranes, hoists, and elevators.

Compound-wound dc motors have both series and shunt fields. The addition of a small series field helps provide the proper amount of no-load to full-load speed regulation or droop. Shunt or compound-wound motors are applied widely to many adjustable-speed drives. They are important for drives requiring accurate speed regulation and adjustment.

A dc motor's inherent speed-torque curve can be varied widely by adjusting the relative amounts of shunt and series fields. The series field may also be connected to aid or buck the shunt field. The usual practice is to connect the series field so that it adds to the shunt field (cumulative compound), which gives a stable, drooping speed with increasing load.

The great flexibility of dc motors both through inherent design characteristics and through the way in which they are operated makes them ideally suited to adjustable-speed drives, particularly regulated drive systems.

ADJUSTABLE-SPEED DRIVES

One of the oldest adjustable-speed drives is the **Ward-Leonard system.** This consists of an ac to dc motor-generator set and a shunt or compound-wound dc motor. Speed is adjusted by changing the generator voltage. A functional equivalent of this drive uses an adjustable-voltage rectifier feeding a dc motor. This system has only one rotating machine in contrast to the three of a conventional Ward-Leonard system.

Modern static controllable rectifiers such as thyristors respond almost instantaneously to control signals and are adaptable as the power supply to the most critical regulated-drive systems. The sensing and regulating system can be designed to hold speed, speed differential, tension, torque, current, acceleration and deceleration, etc., or any required combination. For example, a drive may be speed-regulated with torque limit or speed-regulated with acceleration and/or deceleration regulated or limited.

Typical applications for adjustable-voltage, adjustable-speed dc drives include winders, paper machines and auxiliaries, blending systems, feeders, extruders, calendars, machine tools, range and slasher drives, cranes, hoists, shovels and draglines, and an almost unlimited variety of drives requiring the flexibility and efficiency possible with direct current.

Mechanical adjustable-speed drives are used when a high degree of regulation is not required. One drive consists of a constant-speed ac motor driving the load through V belts and variable-pitch pulleys. The speed range can be as high as 8:1. It is available up to 18.6 kW (25 hp). Speed adjustment is either manual or remote with a motor drive for the adjustment. Speed regulation from no load to full load is normally 3 to 6 percent. Efficiency is high over the entire speed range since there are no slip losses.

Electromagnetic drives are simple adjustable-speed ac drives with efficiency comparable with that of wound-rotor motors. This drive uses a magnetic slip coupling driven by a squirrel-cage motor. The slip is determined by the excitation current and the load. Efficiency is proportional to speed. Therefore, these drives are uneconomical for continuous low-speed high-torque operation but are ideal for fans and centrifugal pumps requiring little speed adjustment when torque decreases rapidly with speed and for controlling acceleration. Electromagnetic drives are functionally equivalent to hydraulic couplings.

RectiFlow adjustable-speed drives use both a wound-rotor motor and a dc motor connected to the same shaft. The rotor winding of the wound-rotor motor is connected to a rectifier. The dc rectifier output supplies the dc motor. Semiconductor-rectifier developments make this a practical, high-efficiency adjustable-speed drive for applications up to several hundred horsepower. Since the rotor losses of the wound-rotor motor are used to produce shaft power and are not dissipated in a resistor, the efficiency of these drives is high over the entire speed range. A typical 223.8-kW (300-hp) RectiFlow drive operating over a 3:1 speed range has an efficiency of more than 83 percent over its operating range. RectiFlow adjustable-speed drives are suitable for

high-torque low-speed drives such as extruders, mixers, pumps, fans, and kilns.

A modification of this basic drive system uses solid-state rectifiers and thyristors to convert the wound-rotor, variable-frequency slip power first to direct current and then to line-frequency power (60 Hz in the United States). This in turn is fed back to the power system as useful energy.

Adjustable-frequency alternating current can be used with squirrel-cage or synchronous motors for adjustable-speed drives. A typical application is small synthetic-fiber spinning drives which require multiple motors operating at constant speed. When precise synchronism between drives is required, synchronous-reluctance motors are used. These are squirrel-cage motors with flats or grooves on the rotor which form magnetic poles because of the change in magnetic path as the rotor position moves with respect to the stator field. This causes the rotor to rotate in exact synchronism with the stator field for light loads. The power source for these drives is either an ac generator driven by any adjustable-speed drive or a static inverter.

It should be noted that even motors as large as 60 MW have been equipped with these drives.

MOTOR ENCLOSURES

Except for areas with fire or explosion hazards (hazardous areas), motor enclosures are designed to provide protection to the internal working parts. The development of improved insulating materials and finishes has affected the required degree of protection and consequently the design and classification of enclosures. Examples of several types of enclosures are shown in Fig. 29-4.

Open, dripproof is the standard enclosure for induction, high-speed synchronous, and industrial dc motors. This design is useful for most indoor and many outdoor applications. Dripproof construction provides good mechanical protection to the internal working parts of the motor and prevents the entrance of dropping liquids and heavy dirt particles. However, it does not protect against airborne moisture, dust, or corrosive fumes. **Guarded** machines have all openings protected to prevent objects more than 12.7 mm (½ in) in diameter from entering the motor. **Splashproof** motors are not affected by water or by solid particles striking or entering the enclosure at an angle less than 100° from the vertical.

Weather-protected, type I is the next degree of protection for larger motors. Such a motor is defined as "an open machine with its ventilating passages so constructed as to minimize the entrance of rain, snow, and airborne particles to the electric parts" (NEMA Standard MG-1, "Motors and Generators"). All openings are restricted against passage of a 19-mm- (¾-in-) diameter rod. Some modern insulation systems are completely satisfactory for most outdoor applications.

Weather-protected, type II motors are recommended for large sizes when a higher degree of protection and longer life are desired. They have extensive baffling of the ventilating system so that the air must turn at least three 90° corners before entering the active motor parts [maximum air velocity, 3.05 m/s (600 ft/min)]; thus, rain, snow, and dirt carried by driving winds are blown through the motor housing without entering the active parts.

Totally enclosed motors offer the greatest protection against moisture, corrosive vapors, dust, and dirt. Totally enclosed fan-cooled (TEFC) motors are the obvious choice rather than *weather-protected* below 186.5 kW (250 hp). Their internal and external ventilating air are kept separate; external air never gets inside except for the small amount that enters by breathing.

TEFC motors have both an internal fan for circulating air within the motor and an external fan for forcing the air through or over the motor frame or heat exchanger. Small motors [approximately 2.238 kW (3 hp) and below] do not require ventilating fans; these totally enclosed nonventilated motors are similar to TEFC with the fans omitted.

Separate forced ventilation is required for some applications (for example, adjustable-speed drives which operate at low speed); these must depend on an external ventilation. This classification includes *open externally ventilated* machines, *open pipe-ventilated* machines,

FIG. 29-4 Examples of open dripproof, totally enclosed fan-cooled, and weather-protected motor enclosures. (*Photo courtesy of Reliance Electric Company, Cleveland, Ohio.*)

and *totally enclosed pipe-ventilated* machines. In corrosive or hazardous areas, safe or clean air ventilates the motor.

Enclosed motors with air-to-water coolers cost much less than TEFC motors above 373 kW (500 hp); in large synchronous-motor ratings, they cost even less than weather-protected, type II. Large enclosed synchronous machines with coolers for mounting in the motor foundation are frequently supplied at lower cost than motors with integral-mounted coolers.

Fire or explosion hazards require special motor enclosures. Hazards include combustible gases and vapors such as gasoline; dust such as coal, flour, or metals that can explode when suspended in air; and fibers such as textile lint. The kind of motor enclosure used depends on the type of hazard, the type and size of motor, and the probability of a hazardous condition occurring. Some available enclosures are explosionproof motors, which can withstand an internal explosion; force-ventilated motors cooled with air from a safe location; and totally enclosed motors cooled by air-to-water heat exchangers and pressurized with safe air, instrument air, or inert gas.

MOTOR CONTROL

The basic functions of motor starters are:

1. Normal "start-stop" control of the motor.
2. Protection of the motor.
3. Protection of the electrical supply system in the event of a motor or motor-feeder short circuit. The fault must be cleared from the rest of the system to prevent further trouble.
4. Electrical isolation to provide accessibility for maintenance.
5. Provision for other control such as master sequence control, protective shutdown devices (e.g., bearing overtemperature, overtravel, pump high pressure, remote control, etc.).

Types of Starters

High Voltage and Low Voltage   The electrical industry has standardized the distinction between high voltage and low voltage at 600 V. Below 600 V the common system voltages in use in the United States are 120, 208, 240, 480, and 600 V. Above 600 V, the standard nominal system voltages commonly in use are 2400, 4160, and 6900 or 7200 V. Higher voltages are available, but the motor cost is usually prohibitive.

For low-voltage starters below 600 volts, the same starters are used for any voltage, since there is only one insulation class.

For high-voltage motor-starting applications, there are several classes of insulation: 2500, 5000, 7500, and 15,000 V. The conventional control-type high-voltage motor starter is available for 2500- or 5000-V service. For voltages higher than this, switchgear must be used.

The construction of high-voltage starters employs much greater clearances and provides additional safety features such as grounded barriers between the high- and low-voltage sections of the starter. Extensive mechanical and electrical interlocking is also used for additional safety.

One of the major differences between high- and low-voltage starters is the amount of power handled. An approximate dividing line is 149.2 kW (200 hp). This, however, is not a fixed and rigid rule.

Line Starters and Combination Starters   A line starter consists of a contactor (motor-starting switch) and motor-overload relays. Contactors are capable of carrying and interrupting normal motor-starting and -running currents; they are not, however, normally capable of interrupting short-circuit currents. They must be backed up by fuses or a circuit breaker for this function.

When a disconnect switch, circuit breaker, or set of fuses is included in the same enclosure as the contactor, the starter is then called a **combination starter.** In addition to the fault-current-interrupting function, the breaker or fuses serve as the disconnecting device. Figure 29-5 illustrates schematically combination starters of various types. The latch is arranged to open the disconnect before the door can be swung open. There are also provisions for padlocking the disconnect open with the door closed so that maintenance work on the motor may proceed in safety.

Manual and Magnetic Starters   Manual motor starters are operated by hand. The simplest type of manual starter is a snap switch with no overload protection, used only for motors of 1.492 kW (2 hp) and smaller, usually single-phase motors with integral overload protection.

Magnetic motor starters are similar in function to manual starters except that they are solenoid-operated. They are available up to 3730 kW (5000 hp). One of the main advantages is the convenience of electrical operation. Start-stop push buttons can be located anywhere. When automatic or remote operation is needed, magnetic starters are essential.

Comparison of Switchgear and Contactor-Type Control Frequently switchgear is used for motor control, particularly for large high-voltage motors. Switchgear (Fig. 29-6) must be used for motors

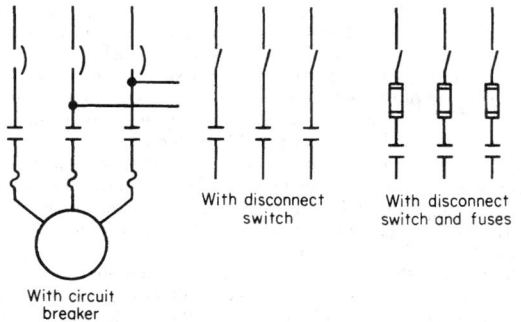

FIG. 29-5 Simplified schematic diagram of a combination line starter with a circuit breaker as the fault interrupter and disconnect. Alternative fuses and disconnect switch are shown as substitutes for the circuit breaker.

larger than 3357 kW (4500 hp)° at 4160 or 4600 V or 1865 kW (2500 hp)° at 2300 V and for all motors above 5000 V. Switchgear consists of circuit breakers and protective relaying. Circuit breakers are electrical switches designed primarily for their ability to interrupt short-circuit currents. This is one of the major differences from contactors, which are designed principally to handle starting and running currents. Contactors normally depend on a set of fuses or a circuit breaker to handle major faults (short circuits).

Contactors are designed for frequent operation. Circuit breakers are designed for far fewer operations and therefore are never used as motor starters when repetitive operation is required. A typical example of frequent operation is mine-hoist service, in which the motor must be reversed at the end of every hoisting or lowering operation; contactors would be used.

° 3730 kW (5000 hp) and 2051 kW (2750 hp) for unity power factor synchronous motors.

FIG. 29-6 Typical lineup metal-clad switchgear including motor starters and protective relaying.

High-voltage ac control-type (contactor) motor starters use fuses to provide short-circuit interrupting capacity. One disadvantage of fuses is that only one fuse may blow. This leaves single phase applied to the motor. Motors will continue to operate with single-phase power but can overheat even with less than rated current flowing. In contrast to contactors, circuit breakers are three-pole devices: a fault on one phase will trip all three, minimizing the single-phasing problem.

Centralized Control   As mentioned previously, motor starters may be located either at the motor or at some remote point. Frequently they are grouped at a location convenient to the source of power. The feeders radiate from this point to the individual motor loads. A convenient method is the control-center modular structure for low-voltage control, into which are assembled motor starters and other control devices. The individual starters can be drawn out of the structure for rapid, easy maintenance and adjustment. With this construction it is easy to change starter size or add additional starters. All the starters are in one location, so that interwiring is simple and easy to check. Auxiliary relays, control transformers, and other special control devices can also be included. See Fig. 29-7.

Motor Protection   Money spent for motor-protective devices can be compared to insurance, in which premiums depend on the protected value when the protected value is the cost of the motor, the cost of anticipated repairs, or the cost of downtime, lost production, and, in some cases, contingent damage to other equipment.

Overload Protection   Overload relays for protecting motor insulation against excessive temperature are located either in the motor control or in the motor itself. The most common method is to use thermal overcurrent relays in the starter. These relays have heating characteristics similar to those of the motor which they are intended to protect. Either motor current or a current proportional to motor-line current passes through the relays so that relay heating is comparable to motor heating.

Standard thermal overcurrent relays located in the starter have some disadvantages. They cannot detect abnormal temperatures in the motor caused by blocked ventilation passages or high ambient temperature at the motor. They are also likely to trip out unnecessarily in locations where the control enclosure is at a higher temperature than the motor. Motors are normally ventilated with external air so that their ambient temperature is the ambient temperature of the surrounding air. However, control enclosures are not freely ventilated, so their internal temperature can become quite high if they are located in a sunny location. High-current relays are sometimes used to avoid this difficulty. This prevents the motor from being tripped out unnecessarily because of high ambients inside the control enclosure, but the motor will be improperly protected during cool weather and overcast days and at night. Ambient-temperature-compensated relays should be used in these situations.

Some overload-protection schemes measure motor-winding temperature directly; various methods are used. Small single-phase motors are available with built-in overload protection. A thermostat built into the motor senses motor-winding temperature directly.

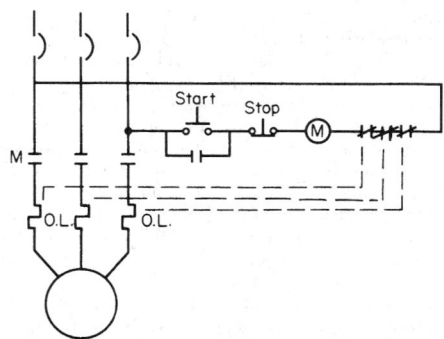

FIG. 29-7 Schematic diagram of a combination starter, showing a simple control scheme.

When the motor overheats, the thermostat opens, interrupting motor-line current. Pilot thermostats mounted on the windings of larger motors trip the motor starter rather than interrupt line current. This method gives good protection for sustained overloads, but because of the thermal time lag between the copper winding and the thermostat it may not provide adequate protection for stalled conditions or severe overloads.

Temperature detectors embedded in the motor winding give close, accurate indication of motor temperature. Both conventional resistance temperature detectors (RTD) and special thermistors (highly temperature-sensitive nonlinear resistors) are used. With appropriate auxiliaries these devices can indicate or record motor temperature, alarm, and/or shut down the motor.

Short-Circuit Protection Short circuits must be removed promptly to avoid severe damage at the fault and to avoid disturbances to the rest of the electrical system. Short-circuit protection should be set as low as possible so that tripping action is initiated quickly. Motor-starting inrush current sets a limit on how low short-circuit devices may be set. For squirrel-cage ac motors, instantaneous short-circuit tripping should be initiated at about 7 to 10 times full-load running current. This gives an adequate margin above the normal inrush of approximately 6 times full-load current. Modern low-voltage combination starters are available with adjustable instantaneous circuit breakers which can be set just above motor-starting current.

High-voltage contactor-type motor controls depend on power fuses for short-circuit protection. The fuses are coordinated with the overload relays to protect the motor circuit over the full range of fault conditions from overload conditions to solid maximum-current short circuits.

Locked-Rotor Protection Under locked-rotor (stalled) conditions the rotors of large synchronous and squirrel-cage motors are the most likely motor elements to be damaged by overheating. The rotor's heating is not related to stator heating during startup. Therefore for large motors it is common practice to use separate devices or characteristics to protect against running overloads and locked-rotor conditions if the overload and short-circuit protective devices cannot be coordinated to handle this condition for the specific motor characteristics.

Synchronous-motor rotor frequency can be detected because the rotor field circuit is available. Special control schemes have been devised which take into account both speed and induced rotor current in providing locked-rotor and accelerating protection.

Undervoltage Protection If a power outage occurs, it is necessary to remove motors from the line to prevent excessive starting current surges on the electrical system when voltage is reestablished. It is also unsafe to have drives starting indiscriminately when electrical service is reestablished. Conversely, it may be desirable to leave the motors connected during short voltage dips; this is time-delay undervoltage protection. Instantaneous undervoltage protection disconnects the motor as soon as the voltage drops appreciably. This is satisfactory if continuity of operation is relatively unimportant. It is inherent in low-voltage magnetic starters when a power loss drops out all contactors as soon as a voltage dip occurs. If time-delay undervoltage protection is desired for these controls, time-delay relays must be added to the standard control circuit. Because circuit breakers do not drop out on a voltage dip, undervoltage relays are necessary.

Reverse-Phase Protection Reverse-phase relays are used on some large motors to prevent their starting when the electrical-system phase rotation is reversed because of improper wiring or maintenance. They are also used as undervoltage and voltage-balance relays. Individual relays may be applied to each motor in place of the undervoltage relay, or one relay may be operated off a bus for several motors. Individual relays are more expensive but more reliable, particularly when motor circuits are changed frequently. This type of protection is normally supplied only on high-voltage switch-gear-type starters.

Phase-Current Balance Protection Three-phase ac motors will usually continue to operate on single phase. Single phasing is serious on large ac motors because of the severe rotor heating it causes. Single-phase conditions cannot be detected by measuring voltage; a run-ning motor acts as a generator so that, even under single-phase conditions, motor terminal voltage is nearly normal. Current-balance relays give a positive indication of system current unbalance and single-phase operation. Normally one three-phase relay is used for each motor. The use of these relays is restricted to large motors [approximately 1119 kW (1500 hp) and larger] when the value of the equipment protected justifies the cost of this protection.

Adequate single-phase protection is provided on low-voltage ac motor starters by three overload relays, which are now standard. Rotor heating is not particularly a problem on smaller motors which have more thermal capacity, but it is important to protect the stator windings of these machines against burnout.

Differential Protection Differential protection is applied to detect internal motor faults quickly and limit damage. The cost of this protection is justified on large motors [1119 kW (1500 hp) and above], for which limiting the motor damage may save the cost of this additional protection many times over.

Motor differential protection is one of the most sensitive forms of large-motor protection available. Figure 29-8 illustrates the basic principles involved. All six leads (both ends of all three windings) are brought out to terminals. The electric current entering each winding and the current leaving that winding pass through the same current transformer in opposite directions. If everything is normal, these currents are equal and no current is induced to the current transformer winding. If a phase-to-phase (winding-to-winding) or winding-to-ground short circuit occurs, the currents do not balance, current is induced in the current-transformer winding, and the differential relay operates instantaneously, shutting down the motor. Because of its sensitivity and speed, this system limits motor damage, minimizing repair costs and downtime.

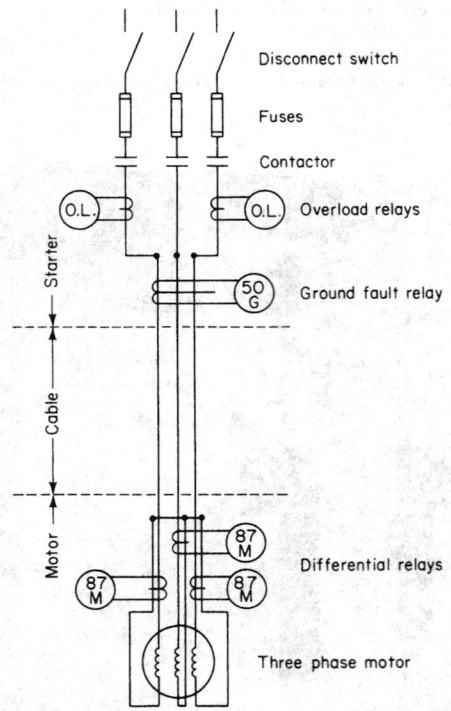

FIG. 29-8 Typical high-voltage ac motor starter illustrating several protective schemes: fuses, overload relays, ground-fault relays, and differential relays with the associated current transformer that act as fault-current sensors. In practice, the differential protection current transformers are located at the motor, but the relays are part of the starter.

Ground-Fault Protection High-voltage motors (2300 V and above) should be protected with ground-fault relays if the power source is grounded (see Fig. 29-8). This scheme includes a large-diameter current transformer (CT) encircling all three motor leads. Short-circuit current to ground flows through the CT to ground and returns to the power source external to the CT; this unbalance induces current in the CT and ground relay to shut down the motor. With this protection only two overload relays and two line CTs (rather than the standard three) are required, so the additional protection is very economical. It cannot be used, however, unless the power source is grounded.

Both differential and ground relaying detect ground faults. Ground-fault protection is located at the starter and protects the cable and the motor; differential CTs are located at the motor and protect the motor only. Economic priorities indicate ground-fault protection first, adding differential protection when justified by potential savings in downtime and repair costs.

Surge Protection High-voltage motors should be equipped with surge-protection apparatus consisting of a set of three lightning arresters and three surge capacitors. Potentially damaging voltage surges or spikes can be generated on the power system by switching operations, certain faults, or lightning. The surge capacitors slope off these steep front voltage spikes, and the lightning arresters limit the peak voltage; both functions are essential for adequate protection. Surge protection should be located at each motor's terminals for maximum protection, although in many instances one set of surge equipment is connected to the electrical bus serving several motors.

Special Control

Reduced-Voltage Starting Reduced-voltage starting is used to reduce system voltage dip. Voltage dips must be limited; otherwise, they may drop other motors off the line, cause synchronous motors on the system to pull out of step, or cause objectionable lamp flicker.

Resistor and reactor starting are the simplest methods of reduced-voltage starting. These systems require two contactors or breakers and a set of reactors or resistors, in contrast to the single-contactor full-voltage starter. The starting contactor closes first, connecting power to the motor terminals through the reactors. The impedance in the circuit reduces the motor terminal voltage and the starting current. As the motor approaches full speed, the running contactor closes, shorting out or bypassing the reactors, applying line voltage to the motor terminals. Starting current is reduced in proportion to the reduction in motor voltage. However, torque is proportional to the square of motor voltage, so starting torque is reduced far more than starting current.

If a greater reduction in line current is required for starting ac motors than is possible with reactor starting, autotransformers may be used. Because of transformer action, the reduction in motor-starting torque is directly proportional to the reduction in line current. Table 29-3 compares reactor and autotransformer starting with respect to

TABLE 29-3 Effects of Reduced Voltage Starting*

Starter type	Motor voltage	Motor current	Line or source current	Motor torque	Source voltage dip
Design	100	100	100	100	0
Actual full voltage	80	80	80	64	20
Reactor:					
0.8 tap	67	67	67	45	17
0.65 tap	56	56	56	31	14
0.5 tap	44	44	44	20	11
Autotransformer:					
0.8 tap	69	69	55	48	14
0.65 tap	59	59	38	35	10
0.5 tap	47	47	24	22	6

*Values shown are in percent of design or normal starting values and are calculated for an arbitrary hypothetical power source whose voltage would dip by 20 percent if full-voltage starting were used.

line current and torque. Other, less commonly used methods of reduced-voltage starting include part-winding starting and star-delta starters.

Synchronous-Motor Starters Except for the addition of the synchronous-motor field-application panel, control schemes are identical for both synchronous and induction motors. Excitation is not applied to synchronous motors until they reach approximately 95 percent speed. Field current should be applied when the field poles are in proper space relationship to the stator's rotating magnetic field. Both speed and position are indicated by the ac voltage generated in the field winding. The frequency is directly proportional to slip and therefore indicates speed; the magnitude and polarity of the generated wave indicate position relative to the armature field. When the proper speed and position are detected, field current is applied. When reduced-voltage starting is employed, the ac starting sequence is completed before the application of field current.

Multispeed Alternating-Current Starters Multispeed induction motors are either two-winding motors, single-winding motors with consequent-pole connection or pole-amplitude-modulated motors (see subsection "Alternating-Current Motors, Multispeed"). The starters for two-winding, two-speed motors are quite simple; they consist of two standard single-pole, single-speed starters in the same enclosure with appropriate mechanical and electrical interlocks so that the two contactors cannot be closed simultaneously.

Two-speed, single-winding motors, either consequent-pole or pole-amplitude-modulated, require a three-pole and a five-pole contactor mechanically and electrically interlocked. Three- and four-speed, two-winding motors require a combination of two-speed, single-winding and two-speed, two-winding starters. Further modifications are possible by making these multispeed-control-reversing.

Secondary Control of Wound-Rotor Motors Wound-rotor motors may be effectively reduced-voltage-started or have their speed controlled by using external secondary resistance. The addition of resistance into the secondary circuit of a wound-rotor motor reduces the starting current and affects the speed under load conditions.

When external secondary resistance is used for improved starting characteristics, short-time-rated resistors are employed. As the motor accelerates, steps of resistance are cut out on a time or current basis to give the desired accelerating torque and current characteristics.

When external secondary resistance is used for speed adjustment, the resistors may be either infinitely adjustable (e.g., liquid rheostats) or adjustable in steps (if fine speed adjustment is not required).

Direct-Current Motor Control Control for dc motors runs the gamut from simple manual line starters to elaborate regulating systems. Only the starting problems are considered here since variable-speed drives and regulating systems are discussed elsewhere.

The major differences between ac and dc starters are necessitated by the commutation limitation of dc motors, which is the ability of the individual commutator segments to interrupt their share of armature current as each segment moves away from the brushes. Normally 250 to 275 percent of rated current can be commutated safely. Since motor-starting current is limited only by armature resistance, line starting can be used only for very small [approximately 1492-W (2-hp)] dc motors. Otherwise, the commutator would flash over and destroy the motor. External resistance to limit the current must be used in starting to prevent this.

Manual rheostats can be used in series with the motor armature for the current-limiting function. If the rheostat has ample thermal capacity, it can also be used to vary speed. If this system is used, interlocks should be included to prevent closing of the contactor unless maximum resistance is in the circuit.

Magnetic starters short out the starting resistance in one or several steps based on time, current, or speed. The number of steps depends on the size of the motor and the application. Current-limit acceleration is used frequently for high-inertia drives which require a long accelerating time. Motor current is sensed by a current relay which actuates the shorting contactors in sequence as the current drops. Time-limit acceleration is more common. The motor accelerates in a definite time by shorting out the starting resistor steps in timed sequence.

RECIPROCATING ENGINES

STEAM ENGINES

The advent of electric motors, steam turbines, and other drivers has relegated the steam engine to a minor position as an industrial driver. It does have the advantages of reliability and operating characteristics that are not obtainable with other drivers but also the disadvantage of bulkiness and oily exhaust steam.

In the simple **nonexpanding engine** as used with direct-acting reciprocating pumps, steam is admitted over the entire stroke and does not expand in the cylinder, resulting in relatively low efficiency. Control is simple, and pump speed is regulated by steam throttling. By proper selection of the steam and pump piston sizes, these pumps can deliver high shutoff pressures, which can be used to overcome temporary blockages of pipe lines or for other situations requiring high pressure of short duration.

The higher-efficiency **expanding steam engines** use cutoff valves to limit steam admission to the cylinder during the initial part of the stroke, and the expansion occurs during the remainder of the stroke. Larger engines use several cylinders in series to achieve full expansion. Although this type of engine can be controlled with throttling valves, the preferred method is to change the cutoff point, thus eliminating throttle-valve losses and permitting change in output from zero to maximum design power. Thus almost complete expansion is achieved at part loads and overloads, resulting in efficient operation for the full range of loadings.

Although this type of engine is efficient, it is limited by its inability to utilize low-vacuum exhaust and/or high steam pressures and temperatures commonly used by steam turbines. With low steam pressures [2068 kPa (300 psig)] and low vacuums [88 kPa (26 in Hg)] a steam engine will have a better efficiency than a steam turbine of the same rated power. For each cutoff setting an engine will develop the same torque at all speeds, with steam consumption and power output directly proportional to speed. All other drivers, except certain dc electric motors, require the same input for constant torque at varying speeds.

Changing the cutoff-point setting will allow an engine designed for one gas to operate efficiently on any other gas (limited only by corrosiveness, fouling, etc.). This characteristic favors the use of reciprocating engines when it is necessary to expand gases efficiently in process applications in which the composition is variable. A further advantage is that an engine is the only driver with essentially zero gas consumption at zero speed while developing full torque and maintaining full process pressures. A combination of an expansion engine driving an oil brake provides a high-efficiency refrigeration effect over a wide range of process conditions, with the process controllers throttling the oil flow in the oil brake.

The **uniflow design** reduces cylinder condensation and also allows greater expansion ratios per cylinder (see Figs. 29-9 and 29-10). Steam is admitted during the start of the power stroke and after cutoff is expanded to a pressure slightly higher than the exhaust pressure. At the end of the stroke the piston uncovers the exhaust ports of the cylinder with partial steam and discharges into the exhaust system. The steam remaining in the cylinder is compressed during the return piston stroke, maintaining higher average cylinder temperatures in order to reduce steam condensation during admission.

To reduce friction and cylinder wear oil is injected into the cylinders of engines, and to maintain lubricity cylinders of engines on low-temperature services are warmed. Oil causes foaming in boilers and can contaminate low-temperature process streams. Therefore, in steam plants oil is usually removed from the condensate. By using carbon or plastic rings similar to those used in oil-free reciprocating compressors, oil for lubrication can be omitted. But these rings are not as reliable as lubricated cast-iron piston rings.

INTERNAL-COMBUSTION ENGINES

Internal-combustion engines range in size from small portable gasoline engines to over 14,914 kW (20,000 hp) diesels for ship propulsion. They are usually designed for particular industrial applications and to meet specific objectives as to weight per horsepower, reliability, and operating conditions.

All internal-combustion engines fall into two main types, namely, four-cycle and two-cycle engines. These engines may be further classified as (1) gasoline or gas engines (Otto cycle), in which a spark plug is used to ignite a premixed fuel-air mixture; (2) diesel engines (diesel cycle), in which high-pressure compression raises the air temperature to the ignition temperature of the injected fuel oil; (3) dual-fuel or gas-diesel engines, in which the fuel is a combination of gas and oil in any desired ratio, provided that at least 5 percent oil is used at all times; and (4) trifuel engines, which can operate as dual-fuel or as straight gas engines by replacing the oil-injection system with a spark plug for ignition.

Design Characteristics Internal-combustion engines involve consideration of the following design features: (1) **Compression ratio,** an increase of which usually increases engine efficiency but also results in higher average cycle temperatures and therefore hotter cylinders, piston heads, and rings which increases the difficulty of piston lubrication. (2) **Piston speed,** which is a major overall criterion of engine design since power rating is proportional to piston speed. Reciprocating forces and lubrication problems increase with piston speed. (3) **Brake mean effective pressure** (bmep), which is an overall measure of the output of an engine frame; bmep increases with compression ratio and degree of supercharging, and, in general, high values are associated with modern high-efficiency industrial engines. It is also the overall criterion for bearing loadings and average piston-head and cylinder temperature. (4) **Engine rating,** which is proportional to the product of piston speed and bmep. Acceptable values of piston speed and bmep are more dependent on industrial usage than on the type of engine. (5) **Supercharging,** which connotes means for increasing the inlet manifold air pressure above ambient

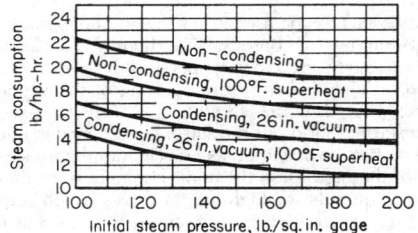

FIG. 29-9 Steam consumption of a 400-hp uniflow engine. To convert pounds per horsepower-hour to kilograms per kilowatthour, multiply by 0.6084; to convert pounds-force per square inch to kilopascals, multiply by 6.89.

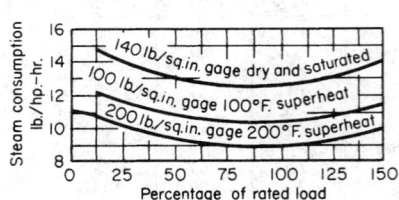

FIG. 29-10 Steam consumption of a uniflow engine with 27.5-in vacuum. To convert pounds per horsepower-hour to kilograms per kilowatthour, multiply by 0.6084; to convert pounds-force per square inch to kilopascals, multiply by 6.89.

pressure. (6) **Turbocharging,** which applies to the expansion of the hot exhaust gases through a turbine to drive the supercharging compressor. Since power output is proportional to inlet manifold air pressure, supercharging provides increased power output and usually higher efficiency. Highly supercharged engines may require an air cooler between the compressor and inlet manifold.

Engine size is usually in terms of power rating (horsepower) rather than the physical size of the engine frame. Frame size is determined by the diameter of the cylinder bore, length of stroke, and number of cylinders. The power rating of a given frame varies with industrial practice and usage; accordingly, automobile engines are designed to develop between 0.54 and 0.95 hp/in^3 of piston displacement, while industrial diesels are limited to 0.045 to 0.60 hp/in^3.

Engine rating reflects industrial practice. Automobile engine rating is the peak horsepower developed on a test stand, whereas industrial engine rating is usually in terms of continuous load.

Industrial engines are made in a wide variety of frame sizes, each offered at several power and speed ratings. The same engine frame might drive either a 900-kW generator at 900 r/min or a 600-kW generator at 600 r/min by using lighter pistons and special parts for the higher-speed application. An engine originally designed for 746 kW (1000 hp) may be uprated to 820 kW (1100 hp) by higher compression heads and further to 984 kW (1320 hp) by supercharging. Thus one frame can cover a power range from 447 kW (600 hp) (60 percent speed) to 984 kW (1320 hp). An engine frame can also be designed for different output by varying the number of cylinders; for example, a 10-cylinder 2610-kW (3500-hp) engine will develop 1566 kW (2100 hp) and 2088 kW (2800 hp) at the same efficiency with six and eight cylinders respectively. This enables a manufacturer to use the same basic design, tools, and fixtures for a variety of ratings.

Maximum rating is the horsepower capability of the engine that can be demonstrated within 5 percent at the factory corrected to standard conditions. Standard conditions in this case are at sea level with barometric pressure of 760 mm (29.92 in) Hg and a temperature of 15° C (59° F). **Intermittent** is the horsepower and speed capability of the engine which can be utilized for approximately one hour, followed by about an hour of operation at or below the continuous rating. Many engine operators choose to run their engines continuously in this rating area, trading shorter maintenance periods for increased earnings. **Continuous** is the horsepower and speed capability of the engine which can be utilized without interruption or load cycling. This can extend for months or years of operation if the engine is equipped for on line lube oil and filter changes. **Intermittent** and **continuous** ratings are at standard conditions of 746 mm (29.38 in) Hg and 30° C (86° F); turbocharged engine ratings are applicable to at least 760 m (2500 ft) elevation above sea level without derating.

It is important that atmospheric conditions at the installation site be specified, since engines operate with very little excess combustion air and consequently maximum power output is proportional to air density.

Operating Characteristics The operating characteristics of internal-combustion engines are basically the same regardless of fuel used or whether the engine is two- or four-cycle. To vary the speed and/or load-carrying capacity, which can range from zero to full torque for all speeds within the operating range, it is necessary only to vary the fuel input. Large engines use a governor to control the fuel-input rate and maintain constant speed under variations of load. With auxiliary instrumentation the governor can be used as a controller of power or process. Starting, stopping, and operation from a remote location are made possible by instrumentation that will also shut down an engine in the event of loss of cooling-water or lubricating-oil flow.

Starting is accomplished by rotating the engine at a speed sufficient to achieve ignition and self-sustained operation. Small engines are started with electric motors or smaller hand-starting engines, and large engines are provided with special valving whereby some of the engine cylinders can be operated as air motors, utilizing high-pressure air to rotate the engine. Starting motors are usually sized at 5 to 10 percent of the engine rating. Starting-air requirements are

0.014 m^3 (0.5 ft^3) (free air)/hp stored at a pressure of 1723 kPa (250 psig) to 2068 kPa (300 psig). Usually three starts per successful firing are assumed for sizing the starting air compressor and air-storage vessel.

Operating characteristics of engines are also influenced by **service requirements.** Automotive engines must develop maximum torque at lower speeds for hill climbing; marine engines are required to develop full torque only when the propeller is at full speed; generator drivers run at a single speed and therefore require maximum possible efficiency at all loads for this speed; and reciprocating compressors at constant pressure require full torque over the entire operating-speed range, thus presenting an engine-instability problem at reduced speeds. The engine must be selected to meet these service requirements.

Supercharging is employed primarily to improve engine efficiency or power output, but part-load performance and rate of response to load changes depend on the type of supercharging system used. Some systems result in supercharged engines having the same torque-speed characteristics and rate of response to load changes as nonsupercharged engines, while other systems result in poorer or better response and in different speed-torque characteristics.

Maintenance and Reliability Preventive maintenance requires that all engines be shut down at periodic intervals for inspection and repair. For properly maintained heavy-duty engines availability is over 97 percent, with maintenance costs of $2.50 to $5 per horsepower-year and lubricating-oil consumption of 1 to 2 gal/hp·year.* While this represents a high degree of reliability, outages of heavy-duty engines are more frequent than those of electric motors or steam turbines.

Satisfactory engine performance is assured by maintaining a minimum log that allows comparison of present characteristics with past records of (1) cylinder exhaust, cooling water, and supercharger exhaust temperature for similar loadings, (2) running sounds and smokiness of exhaust, and (3) engine efficiency and losses.

The minimum safety devices for an industrial engine are a governor and separate overspeed and low-oil-pressure trips. Engines operating in nonsupervised areas should be arranged to shut down in the event of cooling-water or lubricating-oil failures and for excessive exhaust or jacket-water temperatures. Engines operating in supervised areas should be provided with instruments for the operator to check performance.

Fuel Characteristics Fuels used in industrial engines of the internal-combustion type are usually derivatives of petroleum or else natural or manufactured gases. Alcohols and mixtures of gasoline and alcohol or benzol can also be used. A gas engine will operate satisfactorily on any gas which is free of dust, noncorrosive (i.e., less than 0.6 grains/ft^3), does not detonate, does not preignite during compression stroke, and produces enough heat on burning to develop power.

In general, the fuel must have a heat capacity of over 600 Btu/ft^3. Gasoline engines require, in addition, that the fuel will vaporize in the carburetor. Diesels will burn any fuel that can be injected, provided that it will burn under controlled conditions, possesses sufficient lubricity to lubricate the injection plungers, will supply enough heat, and is grit-free, containing less than 3 percent sulfur, 70 ppm vanadium, and 125 ppm vanadium pentoxide. Most diesel engines use either No. 2 or No. 5 fuel oil. The latter must be heated to a viscosity of 50 to 70 SSU [121°C (250°F) approximately] for proper injector lubrication and injection characteristics.

Gaseous fuels containing fractions whose ignition temperature is lower than that of methane may require the use of low-compression heads and a resulting derating of the gas engine.

The method of reporting fuel consumption varies among different industries and also among countries. Trade associations usually have recommended procedures. Thus the Diesel Engine Manufacturers

---

* For further details of industrial operating costs see *Annual Oil and Gas Engine Power Costs,* published by the American Society of Mechanical Engineers. Engine manufacturers will supply recommended schedules for preventive maintenance.

Association (United States) calculates efficiencies based on the lower heating value (LHV) for gas fuels and the higher heating value for oil fuels. It is general practice to report gas-engine performance in terms of British thermal units per horsepower-hour (LHV) and oil-engine performance in terms of pounds of fuel consumed per horsepower-hour. For electric power plants, fuel consumption is reported in terms of kilowatts. Auxiliaries included with engine-efficiency calculations vary with industry practice.

Fuel Economy and Heat Recovery   The high overall efficiencies obtained by modern high-compression supercharged diesels or gas engines can be approached only by very large high-pressure reheat steam plants or by very complicated gas-turbine cycles. The following efficiencies (LHV) based on methane fuel for gas engines and oil fuel for diesel engines can be used for estimating fuel consumption: 63 kW (85 hp) to 298 kW (400 hp), 28 percent; 328 kW (440 hp) to 597 kW (800 hp), 32 percent; 597 kW (800 hp) to 2237 kW (3000 hp), 36 percent; and over 2461 kW (3300 hp), 41 percent.

Instead of working with efficiencies (Eff.) the term **brake-specific fuel consumption** (BSFC) is frequently used with liquid fuel engines. BSFC is expressed in g/kWh (lb/Hp·h) and assumes equal fuel or heating value (LHV) when comparing different engine performances. Using the relationship BSFC = 1/LHV × Eff and LHV = 43 MJ/kg (18,400 Btu/lb) for gasolines and diesel fuels, efficiencies can be calculated from Eff = 84/BSFC (Eff = 0.14/BSFC).

Plant fuel costs chargeable to power production can be reduced if heat losses can be utilized to provide process or other heating. Engine heat losses as percentages of heat input (LHV) are (1) lubricating-oil cooling, 5 to 7 percent available at 82° C (165° F); (2) jacket cooling, 17 to 30 percent available as 82° C water or with vapor-phase cooling as 103 kPa/(15 psig) steam; (3) engine exhaust gases, 26 to 30 percent available with approximately half of this at sufficient temperature to generate 689.5 kPa/(100 psig) steam in waste-heat boilers. Table 29-4 gives some heat-balance data for gas engines.

Vapor-phase cooling reduces the cost of the cooling system, increases heat recovery, and may result in improved engine efficiency.

At full speed fuel consumption decreases linearly with reduction in load, becoming at zero load almost one-third to one-fourth of full-load consumption and with no potential for heat recovery in the exhaust. Jacket-water heat losses decrease by 20 to 40 percent at zero load. Fuel consumption at rated torque is almost proportional to speed. At half speed exhaust recovery decreases to less than half, and jacket-water cooling becomes more than half of full-torque values.

Emission Control   Internal combustion engines must meet national, state, and regional exhaust emission regulations. On an international scope, many countries have introduced emission regulations on stationary engines. These regulations not only vary from one location to the next, but they also tend to become more restrictive as time goes by. Further, as fuel costs rise, the economics of various emission control techniques are changing. Current emission control measures are given in Table 29-5.

Installation and Costs   An engine installation includes auxiliary equipment necessary for operation, such as lubrication pumps and attendant storage, filtering and cooling equipment; jacket-water pumps with expansion tank and cooler; starting-air tanks and compressors; inlet-air piping and screens; exhaust-air piping and silencers or waste-heat boilers; a fuel system, which in the case of diesels would include a day tank, filters, pumps, heaters, and a main storage tank; ignition system or fuel-injection plungers; and cooling towers or radiators. Diesel engines operating on heavy residual fuel oils would have two-fuel systems, a heavy-fuel-oil system for normal operation and a light-fuel-oil system for starting and stopping.

Pipe-line and marine installations are frequently arranged so that the engine drives all its auxiliaries from the crankshaft by means of chains and V belts. But process-plant practice is to have all the auxiliaries independently driven, using standby pumps to minimize engine downtime.

Foundations should be designed to control **vibrating motion** resulting from reciprocating masses. Engine manufacturers will recommend the size of the foundation, but usually their recommendations do not take soil properties into account and are based on making the combined engine and foundation weight sufficiently large to limit vibration. When possible, foundations should be separate from the building structure. In many cases vibration of the engine will cause no damage; nevertheless, it is good practice to reduce it whenever possible. Even though vibration does not increase any forces in the engine, it can loosen pipe joints, nuts, etc.

Torsional vibration can also be a problem and results from pressure variations in the cylinders which can produce cyclic torques with harmonics ranging from half speed to 10 or 12 times running speed.

Table 29-6 gives costs for engine installation, which can be prorated to make preliminary estimates of installation costs.

To avoid operating difficulties, the torsional critical frequencies of the combined engine and driven equipment should be calculated or measured to assure that operating speeds are removed from these criticals or that vibration dampers are provided or that the equipment is designed for the resulting cyclic stresses.

The costs of both engines and auxiliaries are reasonably consistent on the basis of dollars per horsepower as long as essential details are

TABLE 29-4   Approximate Heat Balance for Gas Engines—5965 kW (8000 bhp)

	Two-cycle			Four-cycle		
	U.S. units	SI units		U.S. units	SI units	
	Btu/BHP·hr	kJ/kWh	%	Btu/BHP·hr	kJ/kWh	%
Heat Input (LHV)	6500	9196		6200	8772	
Useful Work	2545	3601	39.2	2545	3601	41.0
Cooling System	940	1330	14.5	1000	1415	16.1
Lubricating Oil System	292	413	4.5	300	424	4.8
Intercooler System	530	750	8.2	250	500	4.0
Exhaust	1540	2179	23.7	1685	2384	27.2
Radiation	650	920	10.0	200	283	3.2
Combustion	—	—	—	220	311	3.5

TABLE 29-5   Current Emission Reduction Measures

Type of engine	Emissions	Emission reduction measures
Gas-Otto-Engine 20 kW (27 hp) to 2,000 kW (2,685 hp)	NOx CO HC	Air/fuel ratio = 1- and 3-way catalytic converter Stratified charge technology (lean burn) and enhanced turbocharging Catalytic converter plus NH_3 injection
Diesel Engine 5 kW (7 hp) to 4000 kW (5400 hp)	NOx CO HC PM°	Small engines: Prechamber Midrange: Adjustment for maximum performance and fuel economy Large engines: Exhaust filter and catalytic converter with NH_3 injection
Diesel-Gas-Engine 300 kW (400 hp) to 8,000 kW (10,740 hp)	NOx CO HC PM°	Catalytic converter Catalytic converter with NH_3 injection

°Particulate matter (soot).

TABLE 29-6 Comparative Installation Costs of Integral-Engine Compressors for Pipeline Stations and Process Plants*

	Pipeline station	Process plant
A. Land and improvements	$ 459,200	$ 67,900
B. Structures	1,036,800	553,600
C. Testing	40,000	40,000
D. Equipment	8,183,000	6,053,750
Subtotal	9,719,000	6,715,250
Add 10% for overhead and undistributed field costs	971,900	671,525
Subtotal	10,690,900	7,386,775
Add 5% for contingencies	534,545	369,339
Total	11,225,445	7,756,114
Cost per horsepower	802	554

*10 units are assumed for a total of 10, 4 MW (14,000 hp). Basis year is 1993.

TABLE 29-7 Typical Cost of Engine-Driven Equipment (1994 basis)

	$/hp
Integral engine compressors:	
Uninstalled and without auxiliaries	334
Installed cost with auxiliaries and cooling water	541
Installed costs of large units with cooling water supplied from process	532
Diesel or gas-engine generators:	
Uninstalled	282
Installed	544

the same. Published figures on **installed engine costs** are often misleading, since with supercharging more power output can be obtained from the same size of engine, which also reduces cooling-water and foundation requirements. Pipe-line compressor costs frequently include piping, buildings, etc.; and in some process plants, cooling water which has been used and charged against process operation can be reused for engine cooling at no cost. It is obvious that general cost predictions must be used with caution unless their detailed basis is known. However, as preliminary figures, Table 29-7 may prove useful (1994 basis):

STEAM TURBINES

Steam turbines are divided into two broad categories: those used for generating **electric power** and general-purpose units used for driving pumps, compressors, etc., and frequently called **mechanical-drive** turbines.

Figure 29-11 illustrates in general the relationship of capability versus speed. At 1800 and 3600 r/min are the turboelectric generator drives with capability limits above the top of the chart. The majority of mechanical-drive applications are within the shaded area; capabilities above the solid line are special and unusual.

Inlet-steam pressure is usually in the range of 1723 kPa (250 psig) at zero superheat to 5860 kPa (850 psig) at 482° C (900° F). Some turbines have been built to operate at 35 kPa (5 psig) with zero superheat from a process exhaust. Pressures of 10,342, 12,410, and 16,547 kPa/(1500, 1800, and 2400 psig) are common for large turbine generators, and some operate at supercritical pressures of 24,131 kPa/(3500 psig) and 34,474 kPa (5000 psig). Power plants that generate steam with nuclear reactors generate saturated steam in the range of 1379 to 6895 kPa (200 to 1000 psig). Early units were 125 MW, but currently 250 to 1500 MW are the most common size. These units have multiple casings and 1.32-m-(52-in-) long blades in 1800-r/min exhaust stages.

TYPES OF STEAM TURBINES

Straight Condensing Turbine   All the steam enters the turbine at one pressure, and all the steam leaves the turbine exhaust at a pressure below atmosphere.

Straight Noncondensing Turbine   All the steam enters the turbine at one pressure, and all the steam leaves the turbine exhaust at a pressure equal to or greater than atmosphere.

Nonautomatic-Extraction Turbine, Condensing or Noncondensing   Steam is extracted from one or more stages, but without means for controlling the pressures of the extracted steam.

Automatic-Extraction Turbine, Condensing or Noncondensing   Steam is extracted from one or more stages with means for controlling the pressures of the extracted steam.

Automatic-Extraction-Induction Turbine, Condensing or Noncondensing   Steam is extracted from or inducted into one or more stages with means for controlling the pressures of the extraction and/or induction steam.

Mixed-Pressure Turbine, Condensing or Noncondensing   Steam enters the turbine at two or more pressures through separate inlet openings with means for controlling the inlet-steam pressures.

Reheat Turbine   After the steam has expanded through several stages, it leaves the turbine and passes through a section of the boiler, where superheat is added. The superheated steam is then returned to the turbine for further expansion.

STAGE AND VALVE OPTIONS

The **single-stage, single-valve** turbine is the simplest turbine, and it sees the most varied application. There is a single governor valve in a steam chest, operated directly from a mechanical flyball governor. After passing through the valve, steam is expanded through the nozzles, where it gains velocity and momentum for driving the wheels by impulse action against the blades. The shaft is sealed by carbon rings. The sleeve bearings are ring-oiled. The majority of applications are below 1119 kW (1500 hp) and at speeds below 5500 r/min with 4137 kPa (600 psig) or lower steam pressure. By certain changes higher limits such as 1492 kW (2000 hp), 10,000 r/min, and 8274 kPa (1200 psig) can be made available.

The **multistage, single-valve turbine** is widely used for driving compressors and pumps in the range from 1119 to 4474 kW (1500 to 6000 hp). Figure 29-12 shows a section of a turbine of this type. The inlet end of this multistage turbine retains the general arrangement of bearing case, governor, and steam chest as used on the single-stage, single-valve turbine. The casing is extended to contain the added stages, and the last blade row and the exhaust opening are large in order to contain the volume of the exhaust steam at the low condensing pressure, which may be 6.9 or 13.8 kPa(a) (1 or 2 psia).

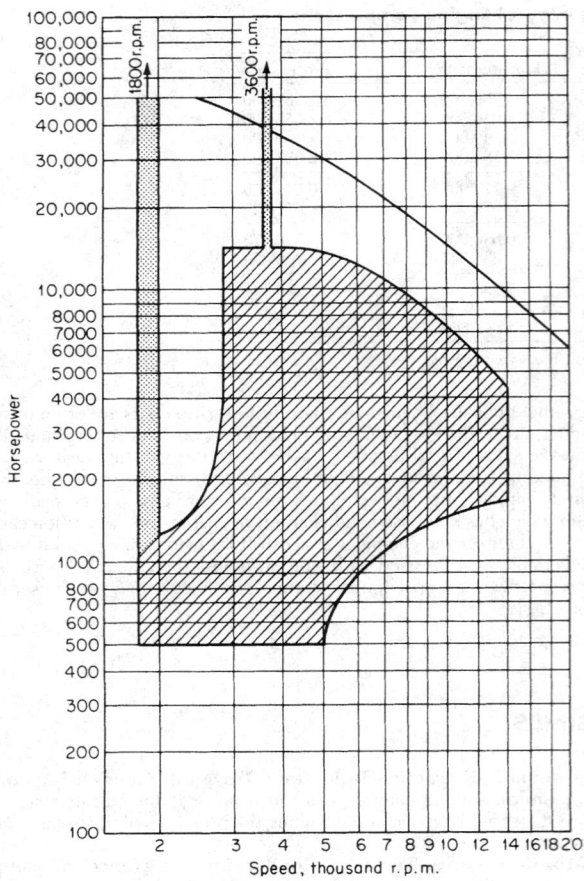

FIG. 29-11 Steam-turbine capability versus speed. To convert horsepower to kilowatts, multiply by 0.7457.

TYPES OF BLADES AND STAGING

Figure 29-13 illustrates a turbine stage in which steam at pressure p_1 enters the nozzle or stationary blade and expands to a lower pressure p_2, leaving the nozzle at a velocity of C_1. The rotating row is moving at a velocity μ so that steam enters the rotating row at a relative velocity w_1 and leaves the rotating row at a relative velocity w_2. The pressure on the exit side of the rotating row is p_3.

Depending upon the relationship between the pressures p_1, p_2, and p_3, the stage is classified as either impulse or reaction.

For an **impulse** (Rateau or Curtis) **stage,** p_2 is equal to p_3 or only slightly higher, and w_2 is slightly less than w_1 as a result of friction loss in the blade passage. The exit area a_r of a rotating row is 50 to 74 percent larger than the exit area a_s of the stationary row in order to pass the same quantity of higher-specific-volume steam.

The work done in the stage, which is the push of the steam on the blades, is the change in momentum of the steam as it alters direction from C_1 to C_2, using the peripheral projection of the velocities. Therefore, it is desirable for the angles α and β to be very small.

For a **reaction stage** the exit area a_r is reduced by reducing the angle β. This will increase the pressure p_2, possibly to midway between p_1 and p_3. The exit velocity w_2 is now greater than the blade entrance velocity w_1 because of the pressure drop from p_2 to p_3 through the blade passage. The reaction in a stage is expressed as a percentage of stage available energy.

Because of the smaller blade angle the reaction stage is more efficient than the impulse stage, but it requires more stages for the same

pressure drop. This will increase the losses and the leakage, and the choice is generally close to a standoff.

A **Curtis stage** is an impulse stage that makes use of two rotating rows to absorb the energy in C_1 with a stationary reversing row between them. This comes about when C_1 is high compared with the wheel peripheral velocity μ, so that the exit velocity C_2 still has a lot of energy left in it. This energy is dissipated by adding a stationary row to reverse the flow and then putting it through an additional rotating row. It is universal practice for single-stage turbines when the ratio p_1/p_3 is in the range of 2 to 2.5 or greater. The two-row impulse (Curtis) stage is not as efficient as the single-row stage, owing to the higher losses encountered with the high steam velocities and repeated turning of the stream.

PERFORMANCE AND EFFICIENCY

The energy available in the steam is expressed in British thermal units per pound, or enthalpy. The velocity of the steam flow through the nozzle is calculated from

$$C_1 = 223.7\sqrt{\Delta h} \qquad (29\text{-}14)$$

where C = velocity, ft/s, and Δh = enthalpy drop from p_1 to p_2, Btu/lb. This is the same formula that is used for the spouting velocity of liquid in terms of head by introducing the mechanical equivalent of heat.

For calculation purposes the **efficiency of individual turbine stages** is plotted as efficiency versus velocity ratio μ/C_1, where μ equals wheel peripheral velocity. Figure 29-14 shows the relation between the three principal types of stages; each one has a peak efficiency at a certain ratio of μ/C_1. For the same revolution per minute, of course, the two-row stage takes the highest value of C_1 and the highest enthalpy drop. The reaction stage takes the lowest. The single-row impulse (Rateau) stage is in the middle. The reaction stage is denoted as 70 percent, because 30 percent of the enthalpy is allowed for expansion in the stationary row and 70 percent in the rotating row.

Steam Rate Enthalpy data can be obtained from Mollier diagrams or from steam tables (see Sec. 2), from which the **theoretical steam rate** can be calculated. For example, a throttle inlet condition of 4137 kPa (600 psig) and 399° C (750° F) gives an enthalpy of 3.2 MJ/kg (1380 Btu/lb), and if the end point is at 348 kPa (50 psig), then adiabatic expansion is to 2.69 MJ/kg (1157 Btu/lb). This gives 0.52 MJ/kg (223 Btu/lb) available, and the theoretical steam rate is calculated from the Btu equivalent per kilowatthour or horsepower-hour:

$$2544/223 = 11.4 \text{ lb steam}/(\text{hp·h})$$

Theoretical-steam-rate tables are available as separate publications. Table 29-8 covers some common conditions.

The **actual steam rate** is obtained by dividing the theoretical steam rate by the turbine efficiency, which includes thermodynamic and mechanical losses. Alternatively, internal efficiency can be used, and mechanical losses applied in a second step.

Efficiency varies over a wide range, dependent upon the number of stages in the turbine. If steam conditions are assumed to be 4 137 kPa (600 psig) at 399° C (750° F) inlet and 13.8 kPa (2 psia) exhaust, Table 29-8 shows a theoretical steam rate of 3.47 kg/kWh (7.65 lb/kWh). With efficiencies that might be experienced for the given steam conditions with single-stage, five-stage, seven-stage, and nine-stage turbines, the actual steam rates would be as in Table 29-9.

From this table note that efficiency increases with the number of stages and that the increased number of stages corresponds to larger horsepower values. For each stage, as characterized by diameter and speed, there is a Btu drop that gives the best efficiency provided there is enough steam to fill the stage so that it operates with minimum friction and windage loss. Thus the nine-stage turbine is fine for 7.46 MW (10,000 hp), but it would not show up as well as a five-stage turbine at 0.746 MW (1000 hp) because the losses would increase with the light flow.

From the velocity diagram in Fig. 29-13 it is apparent that an increase in wheel peripheral velocity μ permits an increase in nozzle exit velocity C_1 without increasing C_2. Accordingly, a high-speed tur-

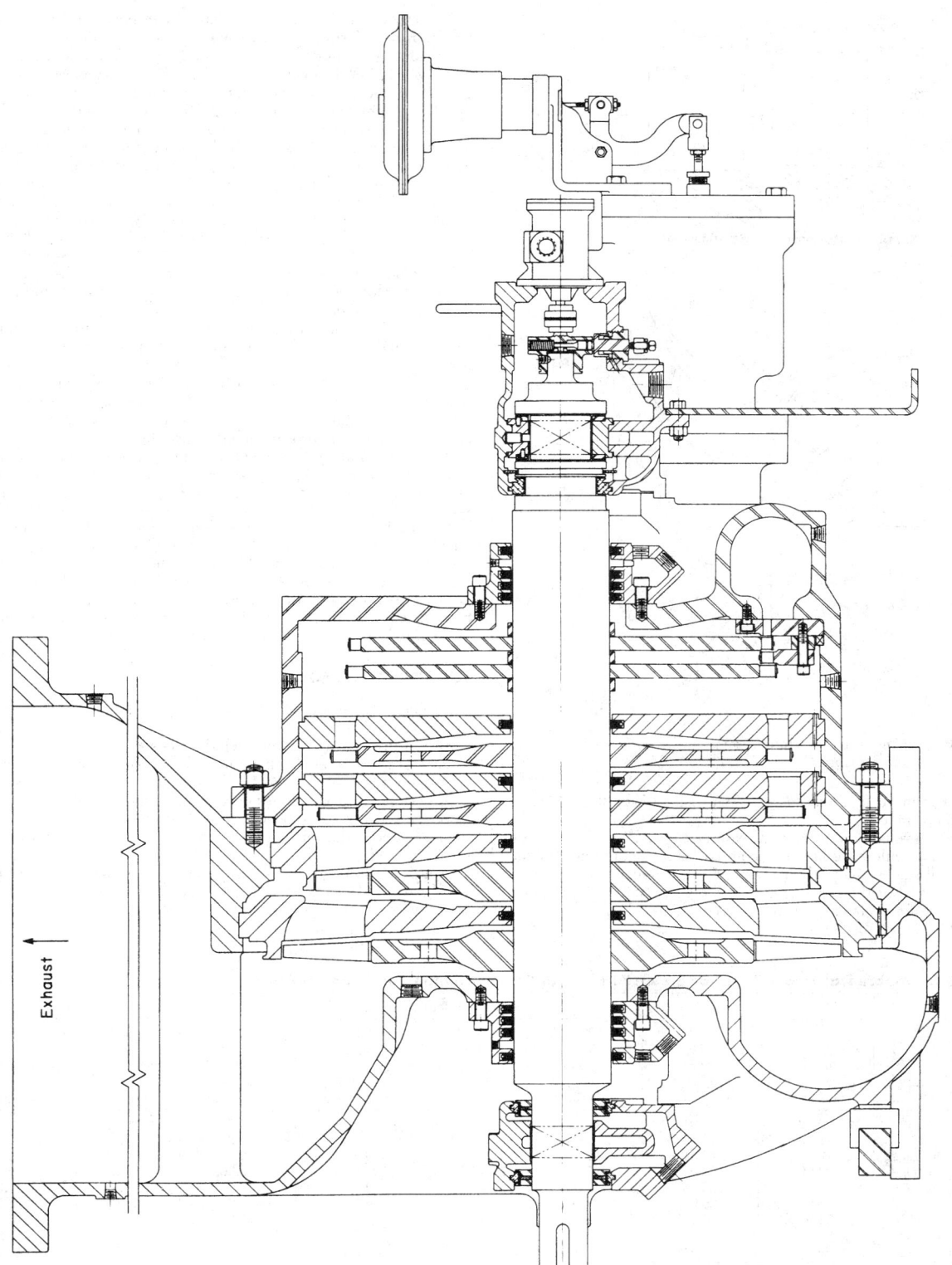

Exhaust

FIG. 29-12 Single-valve, multistage steam turbine. (*Elliott.*)

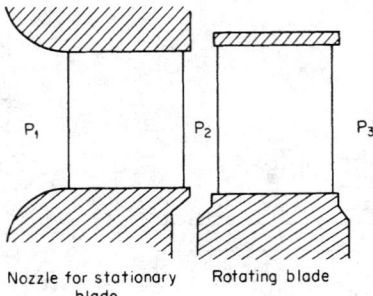

Nozzle for stationary blade   Rotating blade

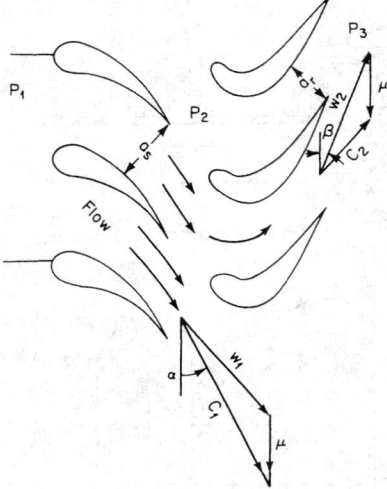

FIG. 29-13   Basic mechanics of a turbine stage.

bine can use more Btu per stage and will have fewer stages than a slow-speed turbine.

The leaving velocity C_2 is a measure of the unused energy. For best efficiency C_2 should have no radial component; C_2 should be straight axial. For all stages except the last one, C_2 represents a carryover to the next stage. For the last stage, C_2 is the velocity into the exhaust hood and is referred to as the leaving loss or exhaust loss.

The curves in Figs. 29-15 and 29-16 can be used for estimating **steam rates of single-stage turbines** by proceeding according to the following example. For steam conditions of 2760 kPa (400 psig) and 399° C (750° F) inlet and 5170 kPa (75 psig) exhaust, Table 29-8 gives theoretical steam rate as 20.59 lb/kWh. If the turbine is 300 hp and 4000 r/min, enter the top of Fig. 29-15 at 4000 and for a trial stop at 18-in-base diameter of turbine blading, then drop to TSR 20.59, and the base steam rate is 37 lb/hp·h. Next enter the top of Fig. 29-16 at 4000 r/min, and intersect 75 psig, then drop to 18-in diameter and find 8-hp loss. Total horsepower then is 308.5, and steam required is (308.5) (37.0) = 11,400 lb/h. By reading values for other diameters a selection table can be prepared, shown as Table 29-10.

From this table the most efficient unit can be selected and balanced against price. It is apparent that for 300 hp the 28-in diameter achieves no gain over the 22-in diameter because of the increase in horsepower loss. For 22 in versus 18 in the gain is small and may be offset by the higher price.

Steam rates for multistage turbines depend upon many more variables than do single-stage turbines and require extensive computation. Depending upon the type of turbine, single-valve or multivalve, general-purpose or generator-drive, condensing or noncondensing, with or without extraction, the manufacturers have shortcut procedures for estimating performance in their bulletins for the different types. As a general approximation the curve in Fig. 29-17 may be used, if one keeps in mind that an actual turbine may be several points above or below the curves, depending upon the use of optimum staging for efficiency or a compromise to meet price. Speed is also important; high speed at 373 kW (500 hp) may have high losses, while 7460 kW (10,000 hp) at 12,000 r/min may be above the curve. And steam pressure affects performance. The most efficient turbine is one in which speed, pressure, and steam flow combine to fill the blade path so that there are no partial-admission stages. For partial admission the nozzles do not fill the entire 360° arc because there is not enough steam for that many nozzles. Those portions of the blades which are spinning outside of the nozzle arc create friction and windage.

TURBINE CONTROL

A turbine may be speed-, pressure-, or process-controlled. Some of the terms used are defined as follows:

Speed-governing system includes the speed governor, the speed changer, the servomotor that moves the valves, and the governor-controlled valves.

Speed governor includes only those elements which are directly responsive to speed and position the other elements.

The **speed changer** is a device by means of which the set point may be varied.

Steady-state regulation is the change in sustained speed or pressure (expressed as a percentage of rated) when power or flow output is gradually reduced from rated value to zero.

TABLE 29-8   Theoretical Steam Rates for Steam Turbines at Some Common Conditions, lb/kWh

Exhaust pressure	150 lb/in² gage, 366°F, saturated	200 lb/in² gage, 388°F, saturated	250 lb/in² gage, 500°F, 94°F superheat	400 lb/in² gage, 750°F, 302°F superheat	600 lb/in² gage, 750°F, 261°F superheat	600 lb/in² gage, 825°F, 336°F superheat	850 lb/in² gage, 825°F, 298°F superheat	850 lb/in² gage, 900°F, 373°F superheat
				Inlet conditions				
2 in. Hg.	10.52	10.01	9.07	7.37	7.09	6.77	6.58	6.28
4 in. Hg	11.76	11.12	10.00	7.99	7.65	7.28	7.06	6.73
0 lb./sq. in. gage	19.37	17.51	15.16	11.20	10.40	9.82	9.31	8.81
10 lb./sq. in. gage	23.96	21.09	17.90	12.72	11.64	10.96	10.29	9.71
30 lb./sq. in. gage	33.6	28.05	22.94	15.23	13.62	12.75	11.80	11.07
50 lb./sq. in. gage	46.0	36.0	28.20	17.57	15.36	14.31	13.07	12.21
60 lb./sq. in. gage	53.9	40.4	31.10	18.75	16.19	15.05	13.66	12.74
70 lb./sq. in. gage	63.5	45.6	34.1	19.96	17.00	15.79	14.22	13.25
75 lb./sq. in. gage	69.3	48.5	35.8	20.59	17.40	16.17	14.50	13.51

NOTE: To convert pounds-force per square inch to kilopascals, multiply by 6.8948; to convert pounds per kilowatthour to kilograms per kilowatthour, multiply by 0.4536; °C = 5/9(°F − 32).

TABLE 29-9 Typical Stage Efficiencies for Steam Turbines at 600 psi Inlet Pressure and 750°F Inlet Temperature

Turbine design	Turbine hp	Internal efficiency, %	Exhaust enthalpy, Btu/lb	Δh,° Btu/lb	Steam rate
Single-stage	500	30	1245	135	7.65/0.30 = 25.5 lb/kw·hr
5-stage	1,000	55	1135	245	7.65/0.55 = 13.9 lb/kw·hr
7-stage	4,000	65	1090	290	7.65/0.65 = 11.75 lb/kw·hr
9-stage	10,000	75	1020	360	7.65/0.75 = 10.02 lb/kw·hr

NOTE: To convert horsepower to kilowatts, multiply by 0.7457; to convert British thermal units per pound to kilojoules per kilogram, multiply by 2.33.

° Based on inlet enthalpy = 1380 Btu/lb.

Speed variation is the total variation in speed from the set point and includes both dead band and oscillation.

Proportional-action governor is a governor with inherent regulation and a continuous linear relation between the input (speed change) and the output of the final control element, the governing valve.

Proportional-action governor with reset is a governor with inherent regulation so that the momentary output is proportional to input change, and subsequently a reset action initiated by the output acts on the speed changer or its equivalent to make the settled regulation less than the inherent regulation.

Isochronous governor is a floating-action governor that controls for constant speed. It is equipped with a dashpot or buffer to give momentary regulation for a speed-input change.

Control-System Components The three principal elements of a control system are the sensing device which measures the error as the deviation from the set point, means for transmission and amplification of the error signal, and the control output device in the form of a servo-operated valve. In the case of the **direct-acting flyball governor** (Fig. 29-18) these three elements are combined in the flyball element and the linkage that connects to the valve.

The centrifugal force of the weights is continually compared against the set point as established by the governor spring which opposes the force from the governor weights. For an increase in load the speed drops, and the weight force is reduced, which allows the spring to push the governor spindle to the left. The lever then pivots, and the valve moves to the right and opens to increase steam flow and torque. The feedback in the control is the resultant increase in speed to match the set point of the governor spring again and eliminate the error. This governor corresponds to NEMA Class A, 10 percent speed regulation, that is, 10 percent speed rise from full load to no load, and the governor weights must move out to close the valve.

The force required to position the valve and the specified regulation puts a practical limit on the direct-acting governor. Beyond this limit a servo is required. The servo requirement has three levels. The lowest level is for a single valve of the balanced type in Fig. 29-18 which may be controlled with less than 889.6 N (200 lb) force. The next level is a single valve with single seat of the venturi type; partially balanced it requires 3558.4 N (800 lbf) to 4448 N (1000 lbf) force. The top level is the multivalve bar-lift or the cam-lift valve gear, in which 8- and 10-in and larger oil servos are used in developing several thousand pounds-force.

Speed-Control Systems The most common sensing element is mechanical; some systems are hydraulic or electronic. For valve positioner they all have a hydraulic servo as first choice, with an occasional choice of pneumatic for lighter loads.

Figures 29-19 and 29-20 illustrate two different **mechanical-hydraulic systems.** Figure 29-19 is a bar-lift steam chest with a heavy-duty hydraulic servo. The speed-sensing element is a flyball assembly attached to a rotating pilot. This rotating pilot sends a control-pressure signal that is proportional to speed to a bellows on the servo. A change in control pressure initiated through the rotating pilot by either speed or speed changer deflects the bellows and servo pilot valve. The servopiston position is proportional to the control pressure.

Figure 29-20 has a spring-return servo that finds application on single-valve turbines with a moderate valve force. For a speed change the rotating pilot sends an error signal through the dashpot to the servopiston. This is a dashpot-type isochronous governor; the error signal sees an instantaneous regulation due to the springs in the dashpot, but after the pressures have equalized through the needle valve, there is no resultant force change on the governor pilot. Control pressure is not proportional to speed; so the governor has zero regulation, also referred to as isochronous control. The setting of the needle valve determines the time required for the system to equalize after a disturbance. This may be several seconds.

Electrohydraulic speed control is in use for turbogenerators and mechanical drive applications because of accurate speed control and easy adaptation to computer operation and such remote control as automatic starting and loading. These same characteristics make it suitable in process plants. Figure 29-21 indicates the elements of electrohydraulic control. A pickup is mounted adjacent to a tooth wheel on the shaft. As the teeth move past the pickup, each tooth generates a small emf pulse. These pulses constitute a digital input to the amplifier. The amplifier performs three functions. The digital input is integrated to a dc level proportional to speed, it is amplified, and the amplified voltage level is matched against a set-point circuit. The differential output current is used to drive an electrohydraulic converter. The converter controls the pilot valve on a standard steam-chest servo. The converter shown puts out a control pressure proportional to current, and the pressure is applied to the bellows. Current electronic governor systems position the servo pilot valve directly.

Remote speed control or **process speed control** takes one of two forms. The remote signal may position the valve directly by acting on the valve stem or on the servo pilot. This action is independent of the governor. For this form of operation the governor is set for maximum operating speed and will take over in an emergency. The operation is referred to as preemergency. In the second type the remote signal acts on the speed changer or its equivalent to adjust the set point. The governor is now always in the circuit, and the unit is always speed-responsive.

The speed control operates the governing valve to maintain steam flow commensurate with load demand while holding speed essentially constant. For sudden load changes there will be a short-time overshoot, and a special case is the instantaneous loss of load, load dump at full load. The usual specification states that the overshoot on load dump must not exceed 9 to 10 percent of rated speed. The settled speed rise will of course be equal to the regulation, 4 or 6 percent for a NEMA Class C or B governor and less than 1 percent for Class D.

Speed governors are classified as shown in Table 29-11.

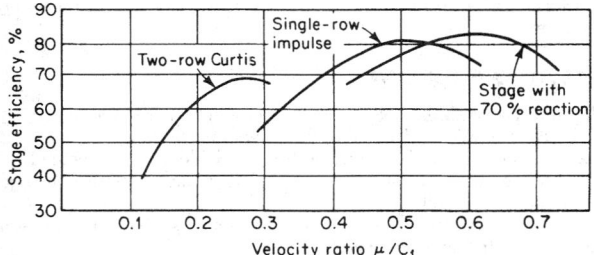

FIG. 29-14 Stage efficiency for different types of stages.

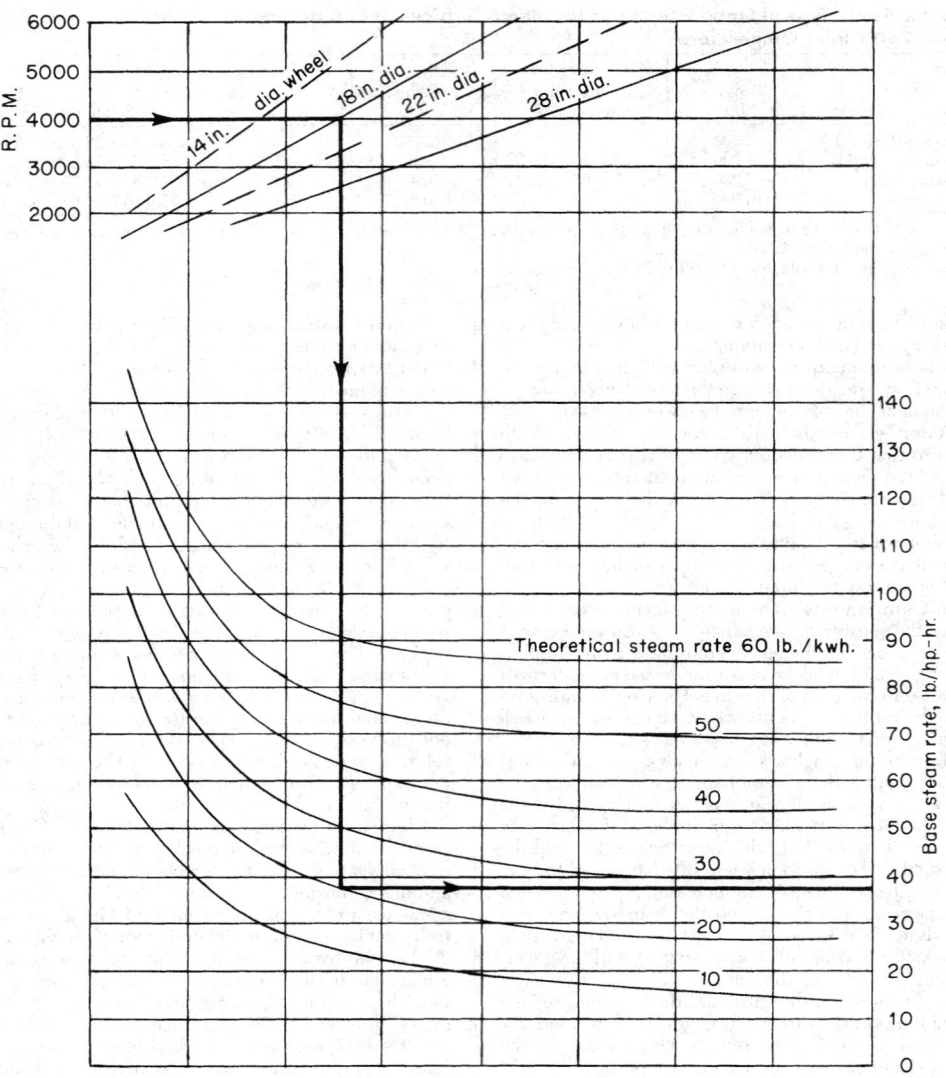

FIG. 29-15 Approximate steam rate for single-stage turbines. To convert pounds per kilowatthour to kilograms per kilowatthour, multiply by 0.4537; to convert inches to meters, multiply by 0.0254; and to convert pounds per horsepower-hour to kilograms per kilowatthour, multiply by 0.6084.

The **trip valve** is provided as a second line of defense in case of overspeed. The trip valve is frequently equipped with a trip-actuating solenoid which can be operated by push button, by low oil pressure, or by some other process upset. When the speed control functions as described above, the trip will not be actuated by load dump.

Extraction-Pressure Control   An extraction turbine equipped with a regulator so that the extraction pressure will be automatically controlled is provided with two sets of steam-chest valves as shown in the schematic in Fig. 29-22. Each of the two sets of steam-chest valves is operated by a servomotor. The throttle flow is illustrated as the total of two flow streams: A, which travels the length of the turbine and leaves through the exhaust opening; and B, which leaves through the extraction opening. The shaft output is the sum of the power generated by the two streams. If the process demand increases, flow B increases and develops more power. For constant output, flow A must be reduced, and this is the function of the three-arm linkage: to open the governing valves and close the extraction valves, which will increase

throttle flow (A + B) and decrease condenser flow A for more extraction flow at constant load. For a reduction the opposite happens.

For an increase or decrease in load the governor moves the three-arm link parallel to itself, and both sets of valves move in the same direction.

SELECTING A TURBINE

The **major variables** that affect turbine selection are as follows:
1. Horsepower and speed of the driven machine
2. Steam pressure and temperature available or to be decided
3. Steam needed for process, so that a back-pressure turbine should be considered
4. Steam cost and value of turbine efficiency, so that consideration can be given to stage and valve options
5. Use of speed-reducing or speed-increasing gears
6. Extraction for feedwater heating

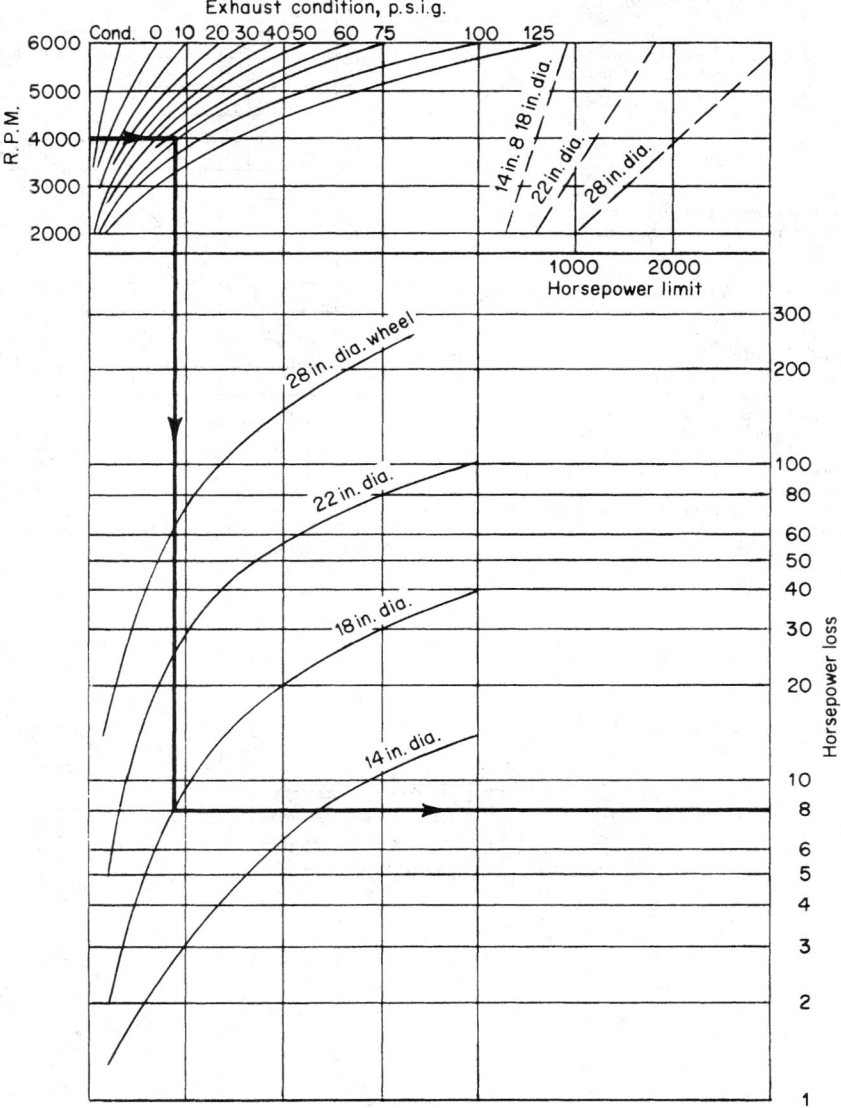

FIG. 29-16 Approximate horsepower loss for single-stage turbines. To convert horsepower to kilowatts, multiply by 0.7457; to convert inches to meters, multiply by 0.0254; and to convert pounds per square inch gauge to kilopascals, multiply by 6.895.

7.   Condensing turbine with extraction for process

8.   Control system, speed control, pressure control, and process control, so that consideration can be given to remote control, speed or pressure variation that can be tolerated, and system response speed

9.   Safety features such as overspeed trip, low-oil trip, remote-solenoid trip, vibration monitor, or other special monitoring of temperature, temperature changes, and casing and rotor expansion

10.   Price range from the minimum single-stage turbine to the most efficient multistage turbine

The **initial step in selection** could logically be to make an estimate of the steam flow at various steam pressures by using Fig. 29-17 for a rough estimate of efficiency. Unfortunately, there are no rigid standards for steam pressure and temperature as electrical-voltage steps are fixed, and many engineers pick a pressure and a temperature

that look good to them. In general, however, manufacturers prefer working with the standards proposed by a joint **ASME-IEEE** committee. The values are 2760 kPa (400 psig) and 399° C (750° F), 4.140 kPa (600 psig) and 441° C (825° F), 5 860 kPa (850 psig) and 482° C (900° F), and 8.620 kpa (1250 psig) and 510° C (950° F) or 538° C (1000° F). The values fall on line *A* in Fig. 29-23. For a 1.5-inHg absolute exhaust pressure, line *A* corresponds to 9 percent moisture. Operating to the left of this state line (78 to 80 percent efficiency), last-stage moisture increases rapidly, which means more erosion; also less heat is available. Moving to the right of this line, the temperature lines become quite flat, the pressure drops at constant temperature, and heat available is reduced. It should be noted that 538° C (1000° F) is a good upper limit for steam turbines without a sharp increase in cost because of special materials. Experience has indicated that main-

TABLE 29-10   Typical Steam-Turbine Selection Table

Wheel diameter, in	Base steam rate, lb/hp·hr	Power loss, hp	Total power, hp	Steam required, lb/hr	Steam rate, lb/hp·hr
14	44.5	3.0	303	13,500	44.6
18	37.0	8.5	308.5	11,400	38.1
22	33.0	26.0	326.0	10,750	36.9
28	29.5	64.0	364.0	10,750	36.9

NOTE: To convert pounds per horsepower-hour to kilograms per kilowatthour, multiply by 0.6084; to convert horsepower to kilowatts, multiply by 0.7457.

tenance and initial cost exceed the gain in performance with temperatures in excess of the range of 538 to 566° C (1000 to 1050° F).

Example 1: Selection of Vacuum   If a turbine is to be operated with exhaust to a condenser vacuum that will give 3 inHg absolute in the summer and 1 inHg absolute in the winter, what vacuum should be specified?

The turbine for 3 inHg will have shorter exhaust blades and a smaller exhaust opening and will lose 16 Btu/lb when operated at 1 inHg. It will also have a high pressure drop over the last-stage blading. Sonic velocity and shock waves will emanate from the last-stage blading, inducing blade loading and oscillation that may lead to fatigue failure. For operation at 1 inHg the last stage should operate with a diffuser which will limit the annulus discharge area of the stage. A turbine designed for 1 inHg will have a lot of windage in the last stage because the blade annulus cannot be filled by the higher-density steam.

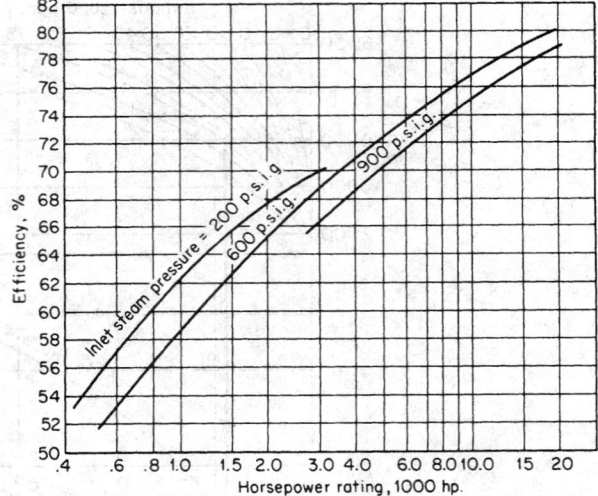

FIG. 29-17   Approximate efficiency for multistage turbines. To convert horsepower to kilowatts, multiply by 0.7457.

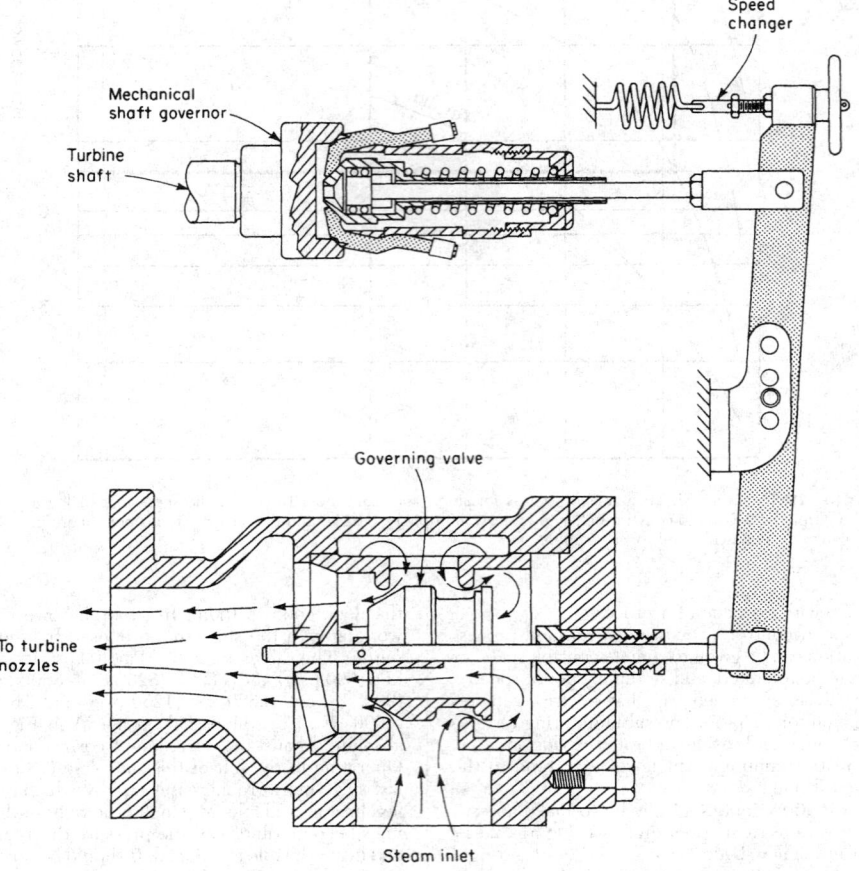

FIG. 29-18   Direct-acting flyball governor.

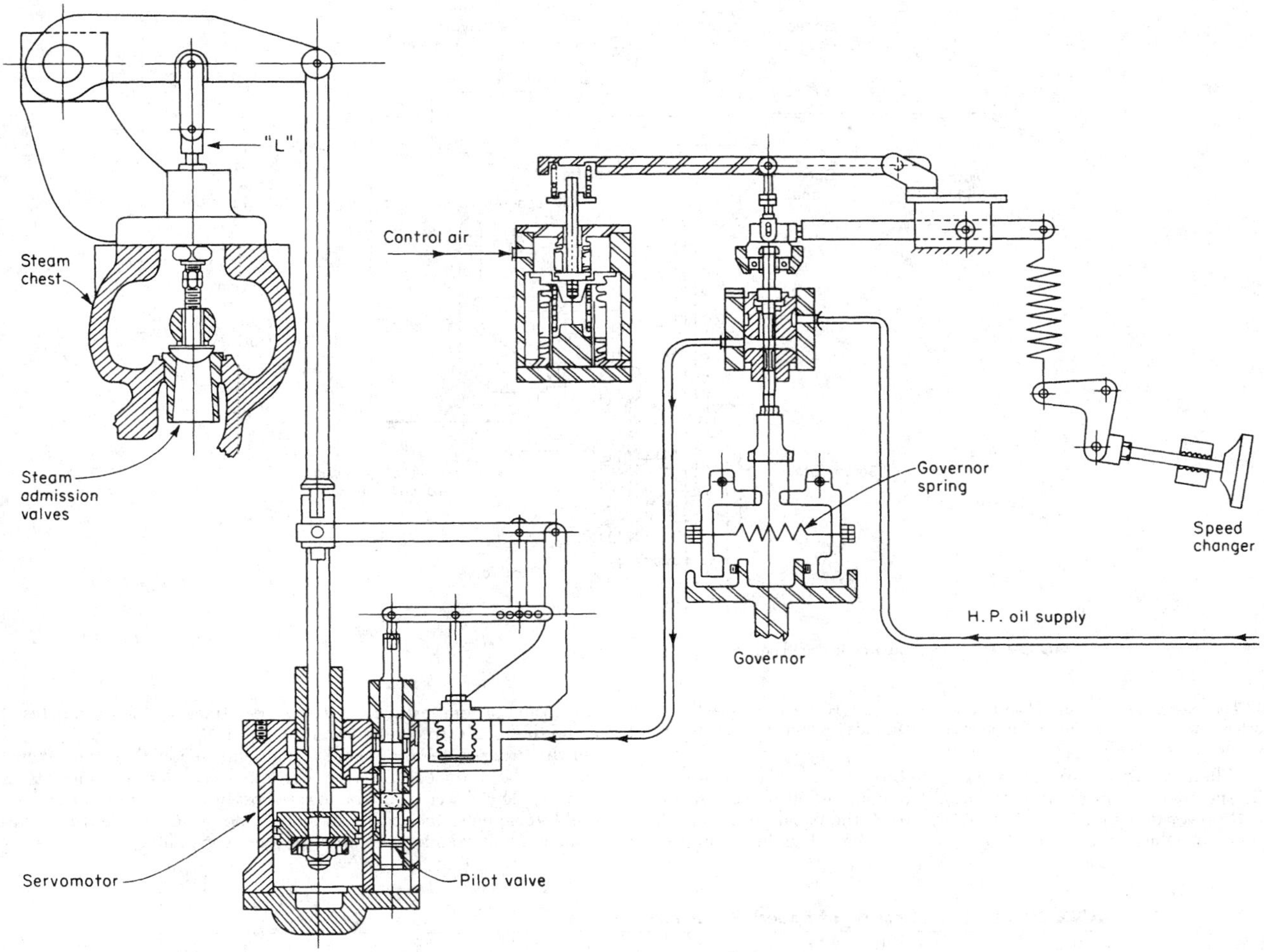

"L"

Steam chest

Steam admission valves

Control air

Governor spring

Speed changer

H. P. oil supply

Governor

Servomotor

Pilot valve

FIG. 29-19 Mechanical-hydraulic speed control: proportional.

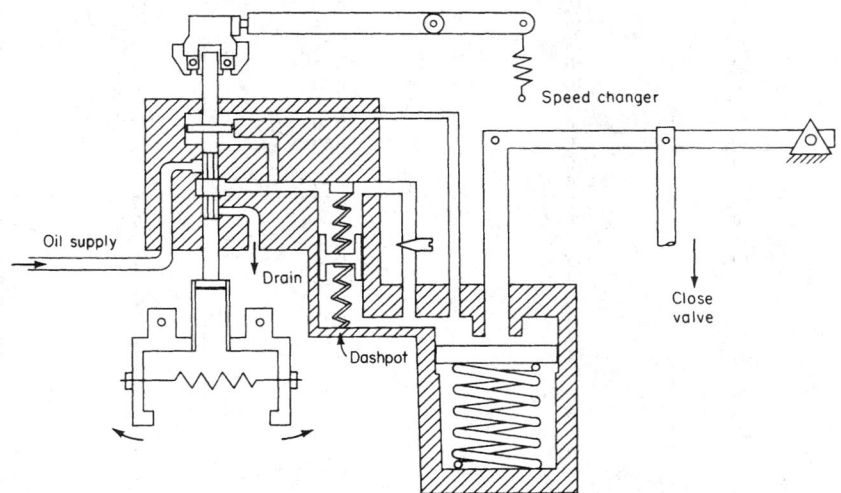

Speed changer

Oil supply

Drain

Dashpot

Close valve

FIG. 29-20 Mechanical-hydraulic speed control: isochronous.

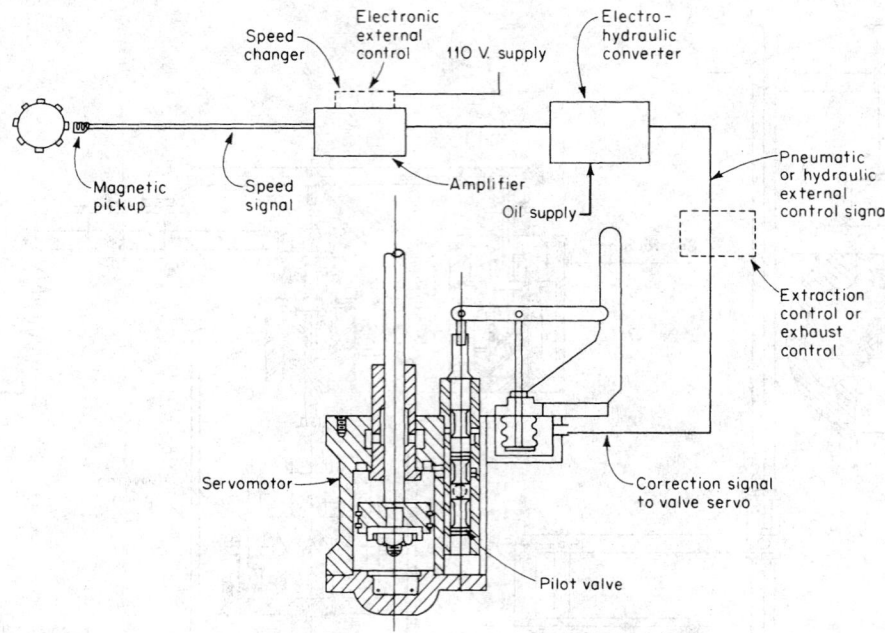

FIG. 29-21 Electrohydraulic speed control.

The best answer is an exhaust designed for 2 inHg and equipped with exhaust-blade diffuser. This will protect at 1 inHg and give some pressure recovery at 3 inHg by virtue of its venturi action.

When an **extraction-condensing turbine** is decided upon, it may be specified in three different ways, depending upon process steam and power demand. Referring to Fig. 29-24, the usual purchase is a unit in which rated capability can be carried either straight-condensing or total-extraction. The zero-extraction line terminates at A, and the total-extraction line terminates at B.

If the process-steam demand is high and steady, then the exhaust size can be reduced, because not much condensing capacity is required. The choice would be to save cost by a smaller exhaust which would terminate the zero-extraction line at C, while the total-extraction line would extend to B for rated capability.

TABLE 29-11 Classification of Speed Governors

Class of governor	Adjustable speed range	Maximum steady-state speed regulation	Maximum speed variables, plus or minus	Maximum speed rise	Trip speed (percent above rated speed)
A	10	10	0.75	13	15
	20	10	0.75	13	15
	30	10	0.75	13	15
	50	10	0.75	13	15
	65	10	0.75	13	15
B	10	6	0.50	7	10
	20	6	0.50	7	10
	30	6	0.50	7	10
	50	6	0.50	7	10
	65	6	0.50	7	10
	80	6	0.50	7	10
C	10	4	0.25	7	10
	20	4	0.25	7	10
	30	4	0.25	7	10
	50	4	0.25	7	10
	65	4	0.25	7	10
	80	4	0.25	7	10
D	30	0.50	0.25	7	10
	50	0.50	0.25	7	10
	65	0.50	0.25	7	10
	80	0.50	0.25	7	10
	85	0.50	0.25	7	10
	90	0.50	0.25	7	10

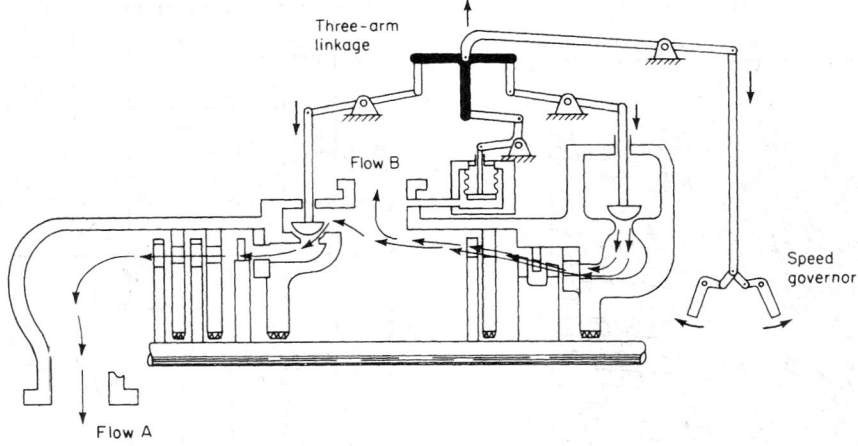

FIG. 29-22 Three-arm lever/mechanism for extraction-turbine-pressure control.

If the process demand is light and high-extraction flow is not required, then most of the power will be from the condensing flow. The choice would be to save cost by a smaller inlet and steam chest, which would terminate the total-extraction line at D and the zero-extraction line at A.

TEST AND MECHANICAL PERFORMANCE

When testing to establish the **thermodynamic performance** of a steam turbine, the ASME Performance Test Code 6 should be followed as closely as possible. The effect of deviations from code procedure should be carefully evaluated. The flow measurement is particularly critical, and Performance Test Code 19 gives details of flow nozzles and orifices. The test requirements should be carefully studied when the piping is designed to ensure that a meaningful test can be conducted.

Mechanical performance is generally checked by a running test at the factory before shipment and again when the turbine is installed on the site. The following is an enumeration of items that may provide smooth and vibration-free operation. The rotor must be in dynamic balance, and at-speed balancing is the most effective. The bearings must be in line, and the seal clearance must be correct. This alignment must be maintained when both cold and hot and during the transient from cold to hot. Disturbance of alignment may originate by unequal heat expansion of supports, from pipe expansion, or from binding or lack of freedom to expand. Excessive pipe expansion and high forces on the turbine have caused many vibration problems. Flexible couplings that do not flex also cause problems. This may result in torque lock or in unloading of bearings, and driver and driven machine bearings may be shifted out of line. Then there is the question of resonance and critical speed. The foundation enters into this relationship, and slender columns may contribute problems of resonance. If there is a gear in the lineup, it should be checked carefully for bearing load, alignment, and resonance frequencies that may be multiples of the gear ratio. Also, bearing-oil supply must be adequate.

From this it can be seen that **vibration** is the universal manifestation that something is wrong. Therefore, many units are equipped with instruments that continuously monitor vibration. Numerous new instruments for vibration analysis have become available. Frequency can be accurately determined and compared with computations, and by means of oscilloscopes the waveform and its harmonic components can be analyzed. Such equipment is a great help in diagnosing a source of trouble.

OPERATING PROBLEMS

While most turbines have a 10-year-availability record in the range of 95 to 99 percent, troubles may develop in any number of places. The most common are vibration, cycling governor, sticking valve stems, leaky packing, temperature bow, erosion of blading, loss of power, and bearing problems.

The causes of **vibration** have already been discussed. An increase in vibration over a period of time is generally caused by loss of alignment, settling of the foundation, or sticking of some expansion feature such as a pipe or a pedestal. Other causes are wear in the teeth of a flexible coupling, an internal rub in the unit, loss of bearing oil, and bearing wear. (Startup vibration is discussed later under temperature bow.)

If a **governor** starts to **cycle** after operating for some time, this is generally the result of wear which causes dead band or sticking. Also, all pilot valves should be inspected for the effects of dirt in the oil.

Sticking of valve stems is common if solids are present in the steam. The steam must be without solids. (Note comments later under loss of power.) It is important that units operating on a steady load for long periods be checked for sticking stems at regular intervals. The records show that in several cases deposits have caused the stem of both the governor valve and the trip valve to stick when there was a loss of load. The effect of the loss of load was destructive overspeed.

Wear and increased **leakage from the glands** are common. Carbon rings may need replacement after 1 or 2 years of operation. A unit with labyrinth packing may never need packing replacement. This depends upon operation. If a unit is started quickly with a temperature bow in the shaft, the result is a rub in the labyrinth, and then all packing may need replacement.

It is important to understand the reasons for a **temperature bow.** When a turbine is shut down and starts to cool, the lower half, particularly on a condensing unit, will cool faster than the top half. After the rotor has stopped turning, the temperature difference increases, and in 20 min there may be a 28 to 83° C (50 to 150° F) difference between top and bottom. Both the casing and the shaft bow up because of this temperature difference in the vertical plane. If the throttle is opened now and the bowed shaft starts to turn in the bowed casing, the packing may wipe out in a few revolutions, and at 200 r/min and up there will be a heavy thumping. The packing rub serves to increase the temperature bow by heating the high side of the shaft.

Other causes of a temperature bow are leaky valves and damaged sleeves. If steam is leaking into a stopped turbine from either an exhaust valve or a stop valve that is leaking, the upper half will be

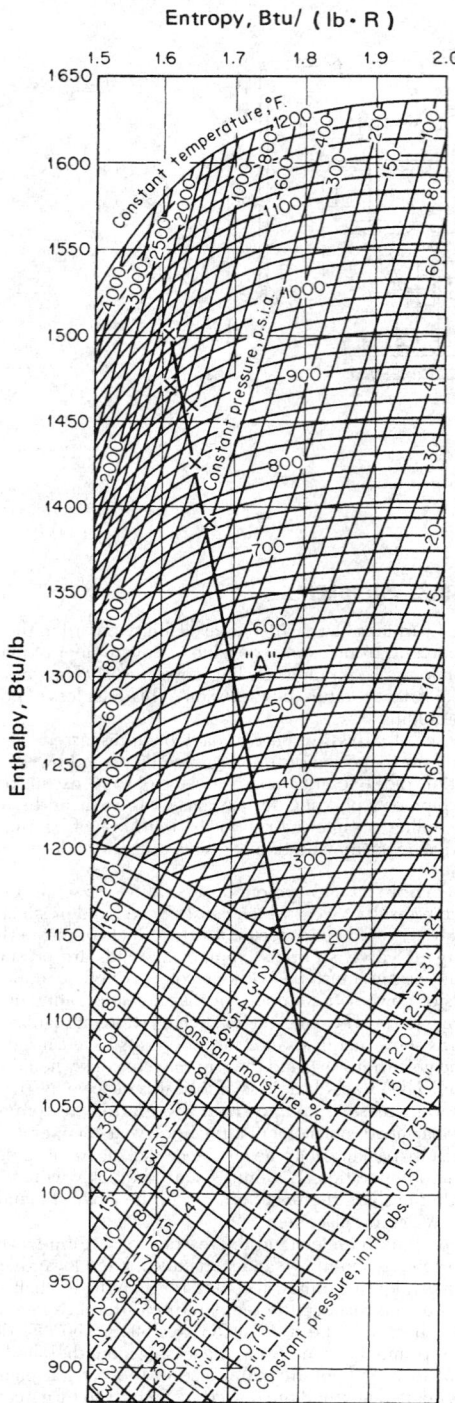

FIG. 29-23 Mollier diagram showing ASME-IEEE steam-turbine standards. To convert British thermal units per pound to kilojoules per kilogram, multiply by 2.328; to convert British thermal units per pound-degrees Rankine to joules per gram-Kelvin, multiply by 4.19; and to convert pounds-force per square inch to kilopascals, multiply by 6.89.

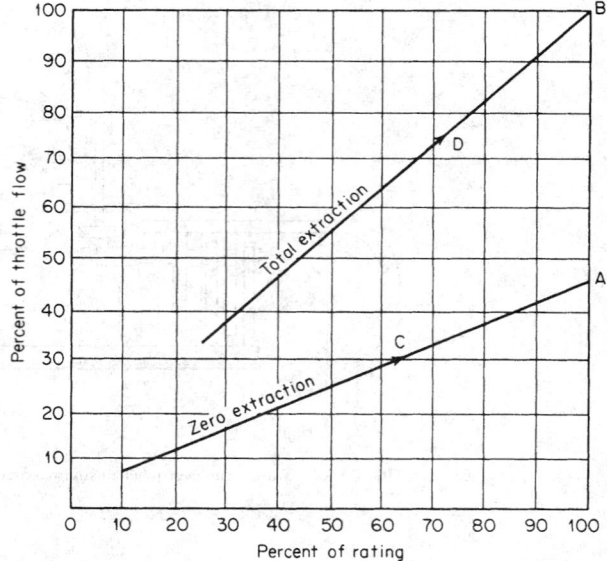

FIG. 29-24 Characteristic of extraction-condensing steam turbines.

warmer than the lower half, and it bows. If a shaft sleeve does not contact uniformly, there will be a transient difference in heat transfer to the shaft, and a bow will result. Also, turning sealing steam on before the shaft rotates may cause a bow.

Leaky valves are also a cause of **erosion.** Most turbine erosion-corrosion problems come from damage that takes place when the unit is not running. A slight steam leak into the turbine will let the steam condense inside the turbine, and salt from the boiler water will settle on the inside surfaces and cause pitting, even of the stainless blading. There must be two valves with a drain between them, i.e., a block valve on the header and an open drain in the line before it reaches the closed trip-throttle valve.

In a turbine that is running, erosion-corrosion is pretty much confined to units that are operating on saturated steam with inadequate boiler-water treatment. This type of erosion takes place behind the nozzle ring and around the diaphragms where they fit in the casing.

Loss of power is another item generally tied to water treatment. With dissolved salts in the steam these salts stay in solution while the steam is superheated. After the steam has expanded through several stages and become saturated, the salts condense out with the moisture. Silica and other salt deposits build up on the blading and the nozzles. The stage pressures increase, and the load drops. The thrust load increases, and the thrust bearing may fail. Depending upon the nature of the salts, it is possible to have corrosion associated with the deposits or corrosion only in the region where the steam changes from superheated to steam with moisture.

Turbine bearings show very little sign of wear as long as there is an adequate oil film. Wiping of the bearing is generally traced to dirt in the oil or a restriction in the oil supply. Thus filtration should be adequate and retain particles that may exceed the oil-film thickness. A checkerboard cracking of the babbitt is sometimes observed. This may have either of two origins. The shaft transports a lot of heat from the steam parts into the journal, and the oil flow in the bearing must be sufficient to lubricate and to remove this heat. Otherwise, the heat is conducted into the babbitt and the babbitt will soften and crack. This cracking may take place if oil flow is stopped too soon when a unit is shut down. The second origin for cracking of the bearing surface is pounding of the journal caused by shaft bow or oil whirl.

GAS TURBINES

The gas turbine is a power plant that produces a great amount of energy for its size and weight. The gas turbine has found increasing service in the past 15 years in the petrochemical industry and utilities throughout the world. It is the power source of the aircraft industry. In this section we deal with land-based gas turbines.

The gas turbine in a combined cycle mode will be the power source through the year 2020 for most countries throughout the world. Most of the new power plants of the 1990s are and will continue to be run by gas turbines with steam turbines in a combined cycle mode. Its compactness, low weight, and multiple fuel application also make it a natural power plant for offshore platforms. Today there are gas turbines that run on natural gas, diesel fuel, naphtha, methane, crude, low-Btu gases, vaporized fuel oils, and even waste. In the 1950s and 1960s, the gas turbine was perceived as a relatively inefficient power source when compared to other power sources. Its efficiencies, compactness, and light weight made it attractive for certain applications. The limiting factor for most gas turbines has been the turbine inlet temperature. With new schemes of air cooling and breakthroughs in blade metallurgy, higher turbine temperatures have been achieved.

The new gas turbines have fired inlet temperatures as high as 2300° F (1260° C) with efficiencies as high as 42–45 percent. Pressure ratios have increased from 5:1 in the 1950s to as high as 30:1 in some of the new turbines of the 1990s. Gas turbines are classified into two major categories:

1. Industrial heavy-duty gas turbines
2. Aeroderivative gas turbines

INDUSTRIAL HEAVY-DUTY GAS TURBINES

The gas turbine was designed shortly after World War II and introduced to the market in the early 1950s. The early heavy-duty gas turbine design was largely an extension of steam turbine design. Restrictions of weight and space were not important factors for these ground-based units, so the design characteristics included heavy-wall casings split on horizontal centerlines, hydrodynamic (tilting pad) bearings, large-diameter combustors, thick airfoil sections for blades and stators, and large frontal areas. The overall pressure ratio of these units varied from 5:1 for the earlier units to 30:1 for the units in the 1990s. Turbine inlet temperatures have been increased and run as high as 2300° F (1260° C) on some of these units. Projected temperatures approach 3000° F (1649° C) and, if achieved, would make the gas turbine even more efficient. The industrial heavy-duty gas turbines most widely used employ axial-flow compressors and turbines. In most U.S. designs combustors are can-annular combustors. Single-stage side combustors are used in European designs. The combustors used in industrial gas turbines have heavy walls and are very durable.

AERODERIVATIVE GAS TURBINES

The aeroderivative gas turbine, as its name implies, is composed of an aeroengine that produces high temperature and high pressure gas that is then put through a power turbine to produce the energy required. The aeroengine is the gas generator. It is usually an engine developed for an aircraft that is modified by the addition of compression stages to produce the high pressure and high temperature required. The gas generator serves to raise combustion gas products for the power turbine to a pressure of about 45–75 psig (3–5 Bar) and temperatures between 900° F (482° C) and 1200° F (649° C). The gas generator is very lightweight compared to the industrial gas turbine. The gas generator is characterized by light casing walls, blades with high aspect ratio (blade length/blade chord), roller bearings, annular combustors, and light weight.

The power turbine is free (i.e., the power turbine is not physically coupled to the gas generator but is closely coupled to the gas generator by a transition duct that transports the gas from the gas generator to the power turbine). The power turbine is an industrial-type turbine in design characterized by heavy wall casings, hydrodynamic (tilting

pad) bearings, and thick airfoil sections. The rotative power produced in the gas turbine is then available for mechanical coupling to the driven equipment.

MAJOR GAS TURBINE COMPONENTS

The gas turbine in the simple cycle mode consists of a compressor (axial or centrifugal) that compresses the air, a combustor that heats the air at constant pressure and a turbine that expands the high pressure and high temperature combustion gases and produces power to run the compressor and through a mechanical coupling to the driven equipment. The power required to compress the gases varies from about 40–60 percent of the total power produced by the turbine.

Compressors A compressor is a device that pressurizes the air in a gas turbine. It transfers energy by dynamic means from a rotating member to the continuously flowing air. The two types of compressors used in gas turbines are axial and centrifugal. Figures 29-25 and 29-26 depict typical industrial gas turbines.

The axial flow compressor is used in over 95 percent of the gas turbines. An axial-flow compressor in a gas turbine compresses the working air by first accelerating the air and then diffusing it to obtain a pressure increase. The air is accelerated by a row of rotating airfoils or blades (the rotor) and diffused by a row of stationary blades (the stator). The diffusion in the stator converts the velocity increase gained in the rotor to a pressure increase. One rotor and one stator make up a stage in a compressor. A compressor usually consists of several stages: a typical compressor contains about 15–17 stages. One additional row of fixed blades (inlet guide vanes) is frequently used at the compressor inlet to ensure that air enters the first-stage blading at the desired angle. In addition to the stators, an additional diffuser at the exit of the compressor further diffuses the air and controls its velocity when entering the combustors.

In an axial compressor air passes from one stage to the next with each stage raising the pressure slightly. By producing low pressure increases on the order of 1.1:1 to 1.4:1, very high efficiencies (85–90 percent) can be obtained. The use of multiple stages permits overall pressure increases up to 30:1. The axial-flow compressor has a very narrow operating range. The operating range is defined as the range between surge and choke. *Surge* is when the flow in a compressor reverses itself; this phenomenon is very destructive. *Choke*, sometimes called "stone wall," is the point where the compressor flow has reached a maximum; this is accomplished by great loss in efficiency.

In the centrifugal or mixed-flow compressor, the air enters the compressor in an axial direction and exits in a radial direction into a diffuser. This combination of rotor (or impeller) and diffuser comprises a single stage. The air initially enters a centrifugal compressor at the inducer. The inducer, usually an integral part of the impeller, is very much like an axial-flow compressor rotor. Many European designs keep the inducer separate. The air then goes through a 90° turn and exits into a diffuser, which usually consists of a vaneless space followed by a vaned diffuser.

From the exit of the diffuser, the air enters a scroll or collector. The pressure ratio per stage in a centrifugal compressor can vary from about 1.5:1 to 9:1 on production units. Some experimental units have obtained pressure ratios of more than 12:1 for a single stage. The centrifugal compressor is slightly less efficient (78–83 percent) than the axial-flow compressor but has a higher stability. A higher stability means that its operating range (surge-to-choke margin) is greater; however, like the axial-flow unit, this range is reduced as the pressure ratio is increased.

Regenerators Regenerators are used in gas turbines to increase the turbine efficiency. They are placed between the compressor and the combustor. Heavy-duty regenerators are designed for applications in large gas turbines in the 5000–100,000-hp range. The use of regenerators in conjunction with industrial gas turbines substantially increases cycle efficiency and provides an impetus to energy management by reducing fuel consumption up to 30 percent. In most present-day regenerative gas turbines, ambient air enters the inlet fil-

Section through a gas turbine type 13

From right to left: generator, bearing pedestal
compressor, gas turbine with combustion
chamber, exhaust gas diffuser

501099 C

FIG. 29-25 Section through a Brown-Boveri gas turbine (with permission of Asea-Brown Boveri).

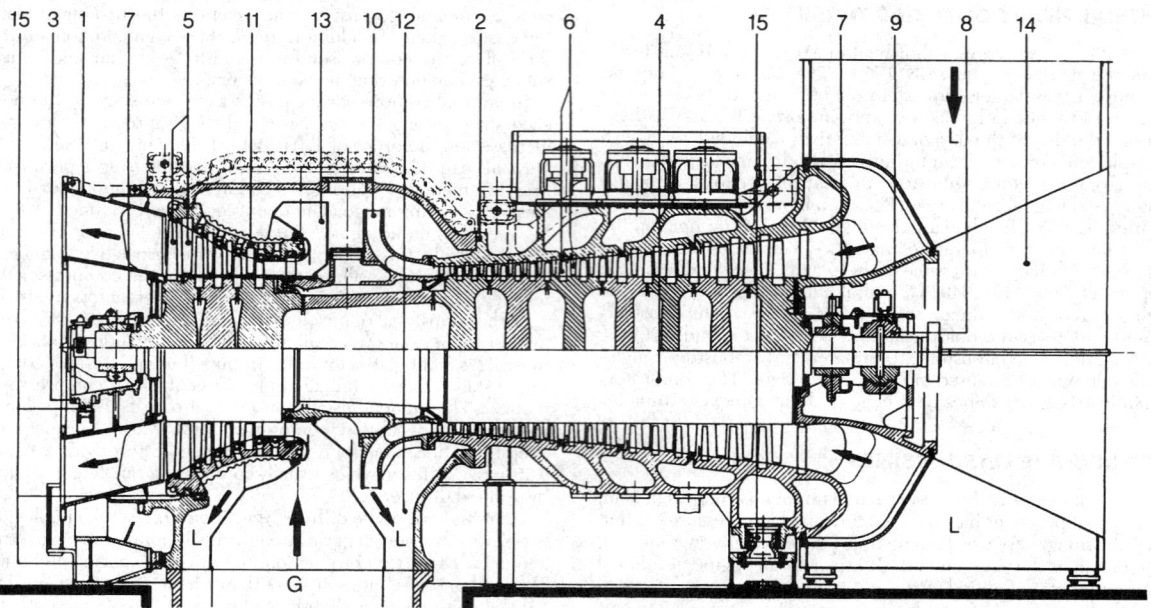

1 = Turbine housing	7 = Journal bearings	13 = Internal casing
2 = Compressor housing	8 = Thrust bearings	14 = Air intake
3 = Bearing body	9 = Thrust bearing cover	15 = Position key
4 = Shaft	10 = Diffuser	
5 = Turbine rotor and stator blades	11 = Blade carrier	L = Air
6 = Compressor rotor and stator blades	12 = Cooling-air admission	G = Hot gas

FIG. 29-26 Principal components of a Brown-Boveri gas turbine (with permission of Asea-Brown Boveri).

ter and is compressed. The air is then piped to the regenerator, which heats the air to about 900° F (482° C). The heated air then enters the combustor, where it is further heated before entering the turbine. After the gas has undergone expansion in the turbine, it is about 1000° F (538° C)–1100° F (593° C) and essentially at ambient pressure. The gas is ducted through the regenerator where the waste heat is transferred to the incoming air. The gas is then discharged into the ambient air through the exhaust stack. In effect, the heat that would otherwise be lost is transferred to the air, decreasing the amount of fuel that must be consumed to operate the turbine. For a 30,000-hp turbine, the regenerator heats 10 million pounds (453,000 kg) of air per day.

Combustors   All gas turbine combustors perform the same function: They increase the temperature of the high-pressure gas at constant pressure. The gas turbine combustor uses very little of its air (10 percent) in the combustion process. The rest of the air is used for cooling and mixing. The air from the compressor must be diffused before it enters the combustor. The velocity leaving the compressor is about 400–500 ft/sec (130–164 m/sec), and the velocity in the combustor must be maintained at about 10–30 ft/sec (3–10 m/sec). Even at these low velocities, care must be taken to avoid the flame to be carried downstream. To ensure this, a baffle creates an eddy region that stabilizes the flame and produces continuous ignition. The loss of pressure in a combustor is a major problem, since it affects both the fuel consumption and power output. Total pressure loss is in the range of 2–8 percent; this loss is the same as the decrease in compressor efficiency.

Despite the many design differences, all gas turbine combustion chambers have three features: (1) a recirculation zone, (2) a burning zone with a recirculation zone that extends to the dilution region, and (3) a dilution zone. The function of the recirculation zone is to evaporate, burn in part, and prepare the fuel for rapid combustion within the remainder of the burning zone. Ideally, at the end of the burning zone, all fuel should be burnt so that the function of the dilution zone is solely to mix the hot gas with the dilution air. The mixture leaving the chamber should have a temperature and velocity distribution acceptable to the turbine nozzles. Generally, the addition of dilution air is so abrupt that if combustion is not complete at the end of the burning zone, chilling occurs, which prevents completion. However, there is evidence with some chambers that if the burning zone is run overrich, some combustion does occur within the dilution region. Combustor inlet temperature depends on engine pressure ratio, load and engine type, and whether the turbine is regenerative or nonregenerative. Nonregenerative inlet temperatures vary from 250° F (121° C) to 1000° F (593° C), while regenerative inlet temperatures range from 700° F (371° C) to 1200° F (649° C). Combustor outlet temperatures range from 1500° (815° C) to 3000° F (1649° C) for large combustors. Combustor pressures for a full-load operation vary from 45 psia (310 kPa) for small engines to as much as 450 psia (3100 kPa) in complex engines. Fuel rates vary with load, and fuel atomizers may be required for flow ranges as great as 100:1. However, the variation in the fuel-to-air ratio between idle and full-load conditions usually does not vary. At lightoff and during acceleration, a much higher fuel-to-air ratio is needed because of the higher temperature rise. On deceleration, the conditions may be appreciably leaner. Thus, a combustor that can operate over a wide range of mixtures without danger of blowouts simplifies the control system.

Combustor performance is measured by efficiency, the pressure decrease encountered in the combustor, and the evenness of the outlet temperature profile. Combustion efficiency is a measure of combustion completeness. Combustion completeness affects fuel consumption directly, since the heating value of any unburned fuel is not used to increase the turbine inlet temperature. The uniformity of the combustor outlet profile affects the useful level of turbine inlet temperature, since the average gas temperature is limited by the peak gas temperature. This uniformity assures adequate nozzle life, which depends on operating temperature. The average inlet temperature to the turbine affects both fuel consumption and power output. A large combustor outlet gradient will work to reduce average gas temperature and consequently reduce power output and efficiency. Combustors in a gas turbine can be arranged in many different ways. These arrangements can be classified into three main categories:

1. Tubular (side combustors)
2. Can-annular combustors
3. Annular combustors

Tubular (Side Combustors)   Tubular or single-can designs are preferred by many European industrial gas turbine designers. These large single combustors offer the advantage of simplicity of design and long life because of low heat-release rates. These combustors are sometimes very large. They can range in size from small units of about 6 inches (152 mm) in diameter to 1-foot (300-mm) combustors that are over 10 feet (3 m) in diameter and 30–40 feet (10–13 m) high. These large combustors use special tiles as liners. Any liner damage can be easily corrected by replacing the damaged tiles. The tubular combustors can be designed as "straight-through" or "reverse-flow" designs. Most large single-can combustors are of the reverse-flow design. In this design, the air enters the turbine through the annulus between the combustor can and the hot gas pipe. The air then passes between the liner and the combustor can and enters the combustion region at various points of entry. About 10 percent of the air enters the combustion zone, about 30–40 percent of the air is used for cooling purposes, and the rest is used in the dilution zone. Reverse-flow designs reduce the combustor lengths as compared to the straight-through flow designs.

Larger tubular, or single-can, units usually have more than one nozzle. In many cases a ring of nozzles is placed in the primary zone area. The radial and circumferential distribution of the temperature to the turbine nozzles is not as even as in tubo-annular combustors. In some cases, high stresses are exerted on the turbine casing leading to casing cracks.

Can-Annular Combustors   Can-annular combustors are the most common type of combustors used in gas turbines. The industrial gas turbines designed by U.S. companies use the tubo-annular or can-annular type. The advantage to these types of combustors is the ease of maintenance. They also have a better temperature distribution than the side single-can combustor and can be of the straight-through or reverse-flow design. As with the single-can combustor, most of these combustors are of the reverse-flow design in industrial turbines.

In most aircraft engines, the can-annular combustors are of the straight-through flow type. The straight-through flow-type tubo-annular combustor requires a much smaller frontal area than the reverse-flow-type can-annular combustor. The can-annular combustor also requires more cooling air flow than a single or annular combustor because the surface area of the can-annular combustor is much greater. The amount of cooling air is not much of a problem in turbines using high-Btu gas, but for low-Btu gas turbines, the amount of air required in the primary zone is increased from 10 percent to as high as 35 percent of the total air, thus reducing the amount of air available for cooling purposes.

Higher temperatures also require more cooling and, as temperatures increase, the single can or annular combustor design becomes more attractive. The tubo-annular combustor has a more even combustion because each can has its own nozzle and a smaller combustion zone, resulting in a much more even flow. Development of a can-annular combustor is usually less expensive, since only one needs to be tested instead of an entire unit as in an annular or single-can combustor. Therefore, the fuel and air requirements can be as low as 8–10% of the total requirements.

Annular Combustors   Annular combustors are used mainly in aircraft-type gas turbines where frontal area is important. There are smaller sized industrial turbines that also have annular combustors. This type of combustor is usually a straight-through flow type. The combustor outside radius is the same as the compressor casing, thus producing a streamline design. The annular combustor mentioned earlier requires less cooling air than the tubo-annular combustor, so it is growing in importance for high-temperature applications. On the other hand, the annular combustor is much harder to get to for maintenance and tends to produce a less favorable radial and circumferential profile as compared to the can-annular combustors. The annular combustors are also used in some newer industrial gas turbine applications. The higher temperatures and low-Btu gases will foster more use of annular-type combustors in the future.

Turbines   The two types of turbine geometries used in gas turbines are the axial-flow and the radial-inflow type. The axial-flow turbine is used in more than 95 percent of all applications in a gas turbine.

Radial-Inflow Turbine   The radial-inflow turbine, or inward-flow radial turbine, has been in use for many years. Basically a centrifugal compressor with reversed-flow and opposite rotation, the inward-flow radial turbine is used for smaller loads and over a smaller operational range than the axial turbine. Radial-inflow turbines are only now beginning to be used because little was know about them heretofore. Axial turbines have enjoyed tremendous interest due to their low frontal area, making them suited to the aircraft industry. However, the axial machine is much longer than the radial machine, making it unsuited for certain vehicular and helicopter applications. Radial turbines are used in turbochargers and in some types of expanders.

The inward-flow radial turbine has many components similar to a centrifugal compressor. The mixed-flow turbine is almost identical to a centrifugal compressor—except its components have different functions. The scroll is used to distribute the gas uniformly around the periphery of the turbine. The nozzles, used to accelerate the flow toward the impeller tip, are usually straight vanes with no airfoil design. The vortex is a vaneless space and allows an equalization of the pressures. The flow enters the rotor radially at the tip with no appreciable axial velocity and exits the rotor through the exducer axially with little radial velocity. These turbines are used because of lower production costs, in part because the nozzle blading does not require any camber or airfoil design. They are also more robust, but due to cooling restrictions are used for much lower turbine inlet temperatures.

Axial-Flow Turbine   The axial-flow turbine is very widely used in gas turbines (95 percent). These axial flow turbines, like their counterparts, the axial-flow compressors, have flow that enters and leaves in an axial direction. Axial-flow turbines are the most widely employed turbines using a compressible fluid. Axial-flow turbines power most gas turbine units—except the smaller horsepower turbines—and they are more efficient than radial-inflow turbines in most operational ranges. Axial-flow turbine efficiencies range from 88–92 percent. The axial-flow turbine is also used in steam turbine design; however, there are some significant differences between the axial-flow turbine design for a gas turbine and the design for a steam turbine. Steam turbine development preceded the gas turbine by many years. Thus, the axial-flow turbine used in gas turbines is an outgrowth of steam turbine technology. In recent years, the trend towards high turbine inlet temperatures in gas turbines has required various cooling schemes and improved materials.

There are two types of axial turbines:
1. Impulse type
2. Reaction type

An impulse-type turbine experiences its entire enthalphy drop in the nozzle, thus having a very high velocity entering the rotor. The velocity entering the rotor is about twice the velocity of the wheel. The reaction type turbine divides the enthalphy drop in the nozzle and in the rotor. Thus, for example, a 50 percent reaction turbine has a velocity leaving the nozzle equal to the wheel speed and produces about ½ the work of a similar size impulse turbine at about 2–3 percentage points higher efficiency than the impulse turbine (0 percent reaction turbine). The effect on the efficiency and ratio of the wheel speed to inlet velocity is shown in Fig. 29-27 for an impulse turbine and 50 percent reaction turbine.

Impulse Turbine   The impulse turbine is the simplest type of turbine. It consists of a group of nozzles followed by a row of blades. The gas is expanded in the nozzle, converting the high thermal energy into kinetic energy. This conversion can be represented by the following relationship:

$$V = \sqrt{2\Delta h} \qquad (29\text{-}15)$$

The high-velocity gas impinges on the blade where a large portion of the kinetic energy of the moving gas stream is converted into turbine shaft work. Figure 29-28 shows a diagram of a single-stage impulse turbine. The static pressure decreases in the nozzle with a corresponding increase in the absolute velocity. The absolute velocity is then reduced in the rotor, but the static pressure and the relative velocity remain constant. To get the maximum energy transfer, the blades must rotate at about one-half the velocity of the gas jet velocity. By definition, the impulse turbine has a degree of reaction equal to zero. This degree of reaction means that the entire enthalphy drop is taken in the nozzle, and the exit velocity from the nozzle is very high. Since there is no change in enthalpy in the rotor, the relative velocity entering the rotor equals the relative velocity exiting from the rotor blade. For the maximum utilization, the absolute exit velocity must be axial.

The Reaction Turbine   The axial-flow reaction turbine is the most widely used turbine. In a reaction turbine, both the nozzles and blades act as expanding nozzles. Therefore, the static pressure decreases in both the fixed and moving blades. The fixed blades act as nozzles and direct the flow to the moving blades at a velocity slightly higher than the moving-blade velocity. In the reaction turbine, the velocities are usually much lower, and the entering blade relative velocities are nearly axial. Figure 29-29 shows a schematic view of a reaction turbine. In most designs, the reaction of the turbine blade varies from hub to shroud. The impulse turbine is a reaction turbine with a reaction of zero ($R = 0$). The utilization factor that is a ratio of the ideal work to the energy supplied for a fixed nozzle angle will increase as the reaction approaches 100 percent. For $R = 1$, the utilization factor does not reach

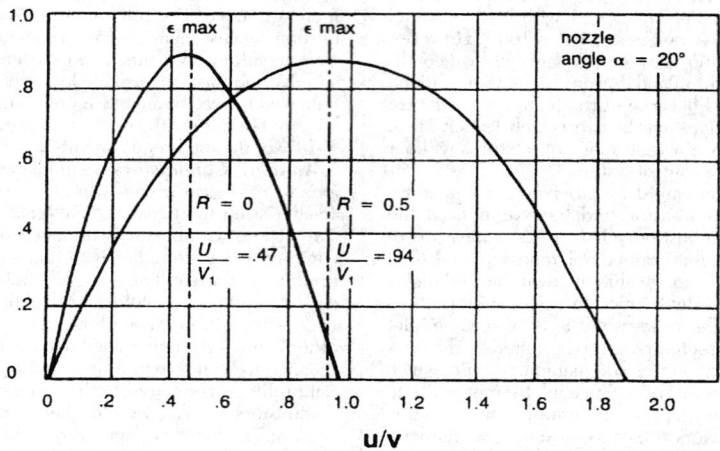

FIG. 29-27   Variation of utilization factor with U/V_1 for $R = 0$ and $R = 0.5$. (*From* Principles of Turbomachinery *by Dennis G. Shepherd, Copyright 1956 by Macmillan Publishing Co., Inc.*)

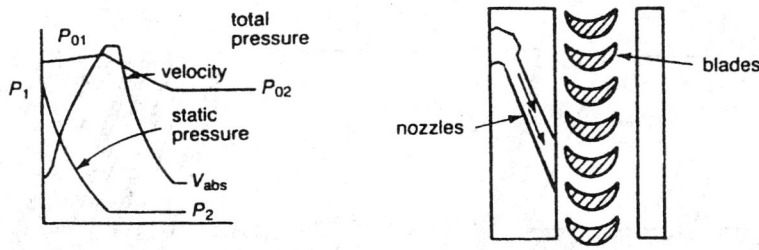

FIG. 29-28 View of a single-stage impulse turbine with velocity and pressure distribution.

unity but reaches some maximum finite value. The 100 percent reaction turbine is not practical because of the high rotor speed necessary for a good utilization factor. For reaction less than zero, the rotor has a diffusing action. Diffusing action in the rotor is undesirable, since it leads to flow losses. The 50 percent reaction turbine has been used widely and has special significance. The velocity diagram for a 50 percent reaction is symmetrical and, for the maximum utilization factor, the exit velocity must be axial. The pressure and velocity distributions in a reaction-type turbine are also shown in Fig. 29-29.

The work produced in an impulse turbine with a single stage running at the same blade speed is twice that of a reaction turbine. Hence, the cost of a reaction turbine for the same amount of work is much higher, since it requires more stages. It is a common practice to design multistage turbines with impulse stages in the first few stages to maximize the work and pressure decrease and to follow it with 50 percent reaction turbines. The reaction turbine has a higher efficiency due to blade suction effects. This type of combination leads to an excellent compromise, since otherwise an all-impulse turbine would have a low efficiency, and an all-reaction turbine would have many more stages.

Turbine-Blade Cooling The turbine inlet temperatures of gas turbines have increased considerably over the past years and will continue to do so. This trend has been made possible by advancement in materials and technology, and the use of advanced turbine blade-cooling techniques. The blade metal temperature must be kept below 1400° F (760° C) to avoid hot corrosion problems. To achieve this cooling air is bled from the compressor and is directed to the stator, the rotor, and other parts of the turbine rotor and casing to provide adequate cooling. The effect of the coolant on the aerodynamic, and thermodynamics depends on the type of cooling involved, the temperature of the coolant compared to the mainstream temperature, the location and direction of coolant injection, and the amount of coolant.

In high-temperature gas turbines, cooling systems need to be designed for turbine blades, vanes, endwalls, shroud, and other components to meet metal temperature limits. Figure 29-30 shows the various types of air-cooling schemes used. The concepts underlying the following five basic air-cooling schemes are:

1. Convection cooling
2. Impingement cooling
3. Film cooling
4. Transpiration cooling
5. Water cooling

Until the late 1960s, convection cooling was the primary means of cooling gas turbine blades; some film cooling was occasionally employed in critical regions. However, in the early 1970s, other advanced cooling schemes were considered due to the greater cooling requirements for engines under development, and in the 1990s, these cooling schemes have been implemented. It should be noted that if more than 6–8 percent of the air is used in cooling, then the effect of the higher temperature becomes negated.

Convection Cooling This form of cooling is achieved by designing the cooling air to flow inside the turbine blade or vane and remove heat through the walls. Usually, the air flow is radial, making multiple passes through a serpentine passage from the hub to the blade tip. Convection cooling is the most widely used cooling concept in present-day gas turbines.

Impingement Cooling In this high-intensity form of convection cooling, the cooling air is blasted on the inner surface of the airfoil by high-velocity air jets, permitting an increased amount of heat to be transferred to the cooling air from the metal surface. This cooling method can be restricted to desired sections of the airfoil to maintain even temperatures over the entire surface. For instance, the leading edge of a blade needs to be cooled more than the midchord section or trailing edge, so the gas is impinged on that surface.

Film Cooling This type of cooling is achieved by allowing the working air to form an insulating layer between the hot gas stream and the walls of the blade. This film of cooling air protects an airfoil in the same way combustor liners are protected from hot gases at very high temperatures.

Transpiration Cooling Cooling by this method requires the coolant flow to pass through the porous wall of the blade material. The heat transfer is directly between the coolant and the hot gas. Transpiration cooling is effective at very high temperatures, since it covers the entire blade with coolant flow. This method has been used rarely due to high costs.

Water Cooling Water is passed through a number of tubes embedded in the blade. The water is emitted from the blade tips as steam to provide excellent cooling. This method keeps blade metal temperatures below 1000° F (538° C); however, a full application of this method is not expected until the year 2000.

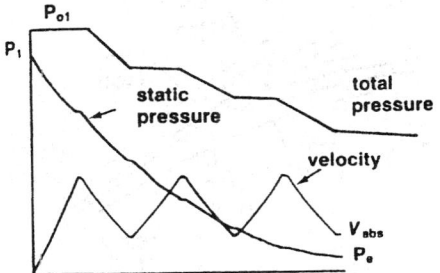

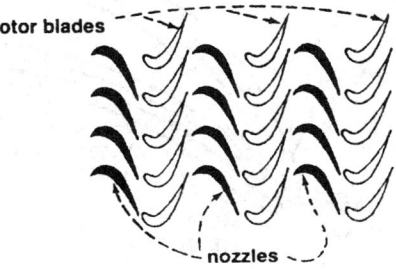

FIG. 29-29 Velocity and pressure distribution in a three-stage reaction turbine.

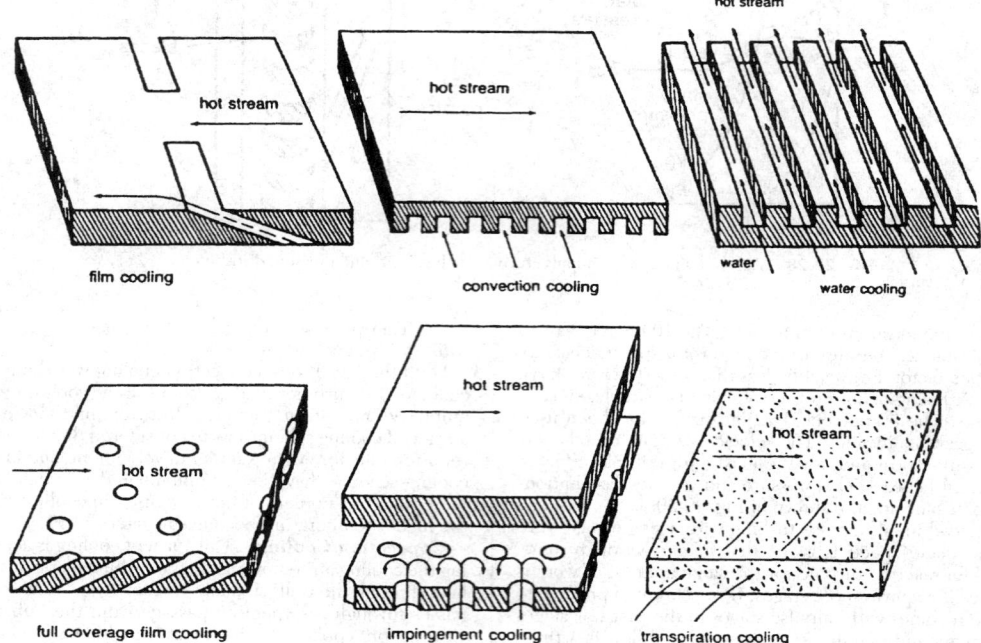

FIG. 29-30 Various suggested cooling schemes.

The incorporation of the above blade cooling concepts into actual blade designs is very important. The most frequently used blade cooling designs are:

1. Convection and impingement cooling
2. Film and convection cooling

It should be noted that in a blade the highest temperatures are encountered at the trailing edge and the highest stress points at about ⅓ the height from the base at the trailing edge.

Convection and Impingement Cooling Strut Insert Design
The strut insert design has a midchord section that is convection-cooled through horizontal fins, and a leading edge that is impingement cooled. The coolant is discharged through a split trailing edge.

The air flow up the central cavity formed by the strut insert and through holes at the leading edge of the insert to impingement cool the blade leading edge. The air then circulates through horizontal fins between the shell and strut and discharges through slots in the trailing edge. The temperature distribution for this design is shown in Fig. 29-31. The stresses in the strut insert are higher than those in the shell, and the stresses on the pressure side of the shell are higher than those on the suction side. Considerably more creep strain takes place toward the trailing edge than the leading edge. The creep strain distribution at the hub section is unbalanced. This unbalance can be improved by a more uniform wall temperature distribution.

Film and Convection Cooling Design This type of blade design has a midchord region that is convection cooled, and the leading edges which are both convection and film cooled. The cooling air is injected through the blade base into two central and one leading edge cavity. The air then circulates up and down a series of vertical passages. At the leading edge, the air passes through a series of small holes in the wall of the adjacent vertical passages and then impinges on the inside surface of the leading edge and passes through film-cooling holes. The trailing edge is convection cooled by air discharging through slots.

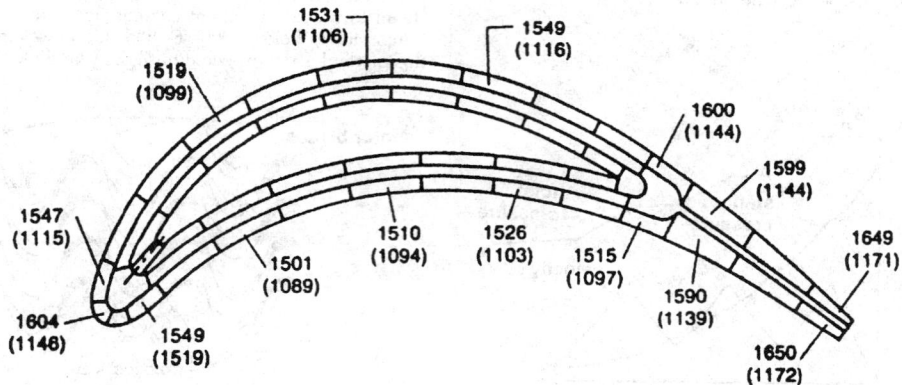

FIG. 29-31 Temperature distribution for strut insert design, °F (cooled).

The temperature distributions for film and convection cooling design are shown in Fig. 29-32. From the cooling distribution diagram, the hottest section can be seen to be the trailing edge. The web, which is the most highly stressed blade part, is also the coolest part of the blade.

Major Cycles   The major application of most gas turbines is in an open cycle in which air is the working medium. The gas turbine can either be a single-shaft unit or a multiple-shaft unit. The single-shaft unit is one in which the compressor, the gas generator turbine, and the power turbine are on a single shaft. The single-shaft units are used in most electrical generating services where constant speed application is the norm. Multiple shaft units can be two or three shaft units. In a two-shaft unit, the high-pressure turbine (gas generator turbine) is driving the air compressor, and the low-pressure turbine, which is on a separate shaft and only aerodynamically coupled, produces the output power. A three-shaft unit is used in designs with very high compressor-pressure ratios. The high pressure requires two compressors: a low-pressure compressor and a high-pressure compressor. The high-pressure compressor is driven by the high-pressure turbine; the low-pressure compressor is driven by the intermediate-pressure turbine; and the low-pressure turbine drives the output shaft. Multiple shaft turbines are used often in mechanical drives where the driven equipment needs to be operated over a wide speed range.

The advantage of a multiple shaft unit over a wide output speed range is that the compressor and the compressor turbines can be maintained at relatively constant speed (75–100 percent of design), while the power turbine (low pressure turbine) can be operated over a very wide speed range without a great loss in thermal efficiency for the entire gas turbine. Variable area nozzles are sometimes used to control the low-pressure turbine; the high pressure turbine is controlled by change in the fuel flow, which affects the turbine firing temperature.

The Simple Cycle   The simple cycle, or the Brayton cycle, is the most common type of cycle being used in the gas-turbine field today. The overall efficiency of a cycle can be improved by increasing the pressure ratio or the turbine inlet temperature (firing temperature). Today's simple-cycle turbines have pressure ratios as high as 17:1 and firing temperatures of about 2300° F (1260° C). In a simple cycle, there is an optimum pressure ratio for a turbine-firing temperature giving the highest efficiency. The efficiency of the various components such as the compressor, combustor, and turbine affects the overall thermal efficiency; however, even if these components were 100 percent efficient, the thermal cycle efficiency would only approach that of a Carnot cycle, which is the most efficient cycle between any two temperatures. Figure 29-33 shows the effect of pressure ratio and turbine inlet temperature on power and efficiency of simple cycle gas turbines.

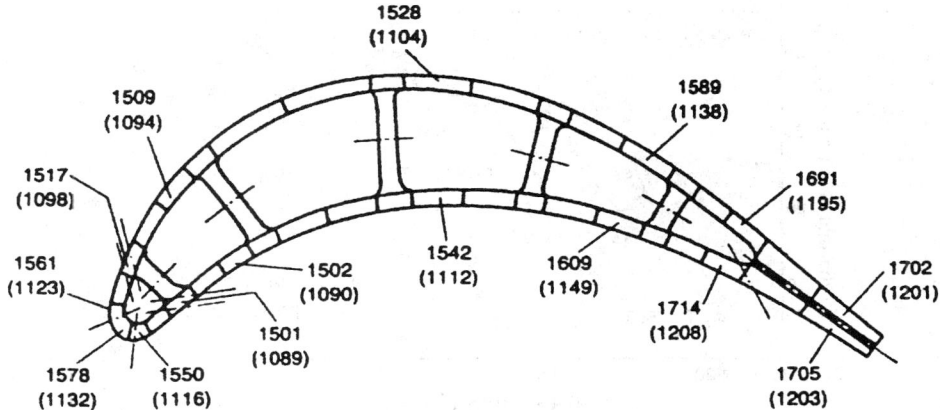

FIG. 29-32   Temperature distribution for film convection-cooled design, °F (cooled).

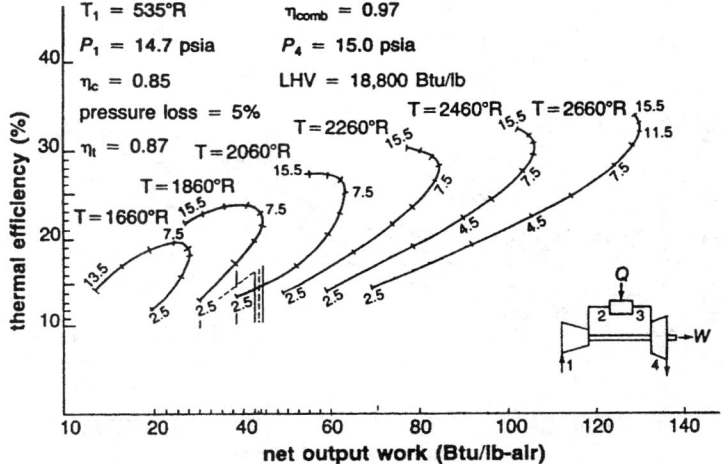

FIG. 29-33   Performance map showing the effect of pressure ratio and turbine inlet temperature on a simple cycle.

The Regenerative Cycle The regenerative cycle is becoming prominent in these days of tight fuel reserves and high fuel costs. The amount of fuel needed can be reduced by the use of a regenerator in which the hot turbine exhaust gas is used to preheat the air between the compressor and the combustion chamber. The regenerator increases the temperature of the air entering the burner, thus reducing the fuel-to-air ratio and increasing the thermal efficiency. For a regenerator assumed to have an effectiveness of 80 percent, the efficiency of the regenerative cycle is about 40 percent higher than its counterpart in the simple cycle, as seen in Fig. 29-34. The work output per pound of air is about the same or slightly less than that experienced with the simple cycle. The point of maximum efficiency in the regenerative cycle occurs at a lower pressure ratio than that of the simple cycle, but the optimum pressure ratio for the maximum work is the same in the two cycles.

The Reheat Cycle The regenerative cycle improves the efficiency of a gas turbine but does not provide any added work per pound of air flow. To achieve this latter goal, the concept of the reheat cycle must be utilized. The reheat cycle utilized in the 1990s has pressure ratios of as high as 30:1 with turbine inlet temperatures of about 2100° F (1150° C). The reheat is done between the power turbine and the compressor trains. The reheat cycle, as shown in Fig. 29-35, consists of a two-stage turbine with a combustion chamber before each stage. The assumption is made that the high-pressure turbine is only to drive the compressor and that the gas leaving this turbine is then reheated to the same temperature as in the first combustor before entering the low-pressure or power turbine.

The Intercooled Regenerative Reheat Cycle The Carnot cycle is the optimum cycle between two temperatures, and all cycles try to approach this optimum. Maximum thermal efficiency is achieved by approaching the isothermal compression and expansion of the Carnot cycle or by intercooling in compression and reheating in the expansion process. The intercooled regenerative reheat cycle approaches this optimum cycle in a practical fashion. This cycle achieves the maximum efficiency and work output of any of the cycles described to this point. With the insertion of an intercooler in the compressor, the pressure ratio for maximum efficiency moves to a much higher ratio, as indicated in Fig. 29-36.

The Steam Injection Cycle Steam injection has been used in reciprocating engines and gas turbines for a number of years. This cycle may be an answer to the present concern with pollution and higher efficiency. Corrosion problems are the major hurdle in such a system. The concept is simple and straightforward: Steam is injected into the compressor discharge air and increases the mass flow rate

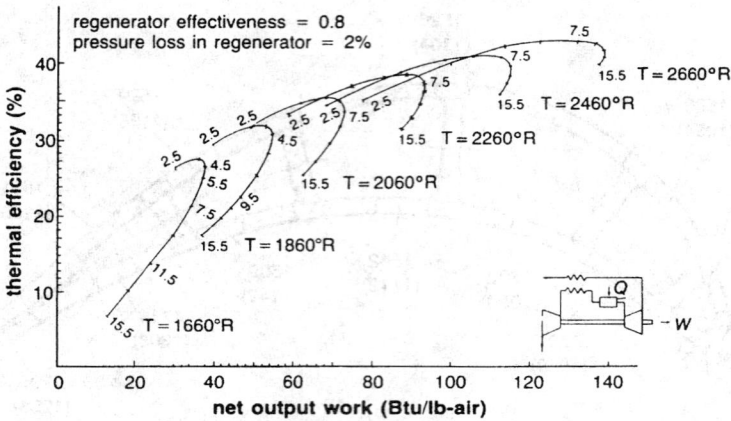

FIG. 29-34 Performance map showing the effect of pressure ratio and turbine inlet temperature on a regenerative cycle.

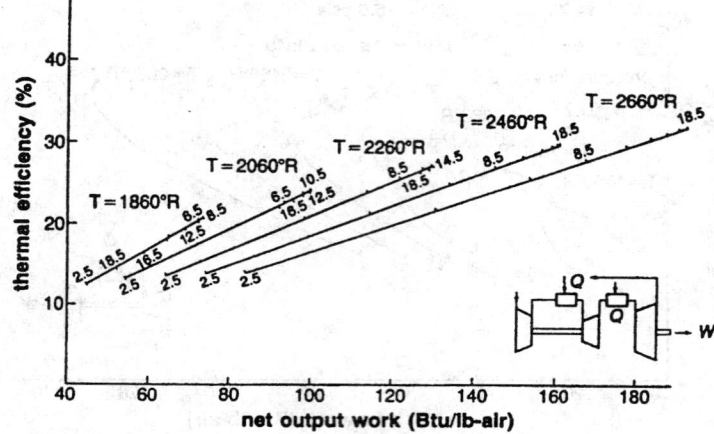

FIG. 29-35 Performance map showing the effect of pressure ratio and turbine inlet temperature on a split-shaft reheating cycle.

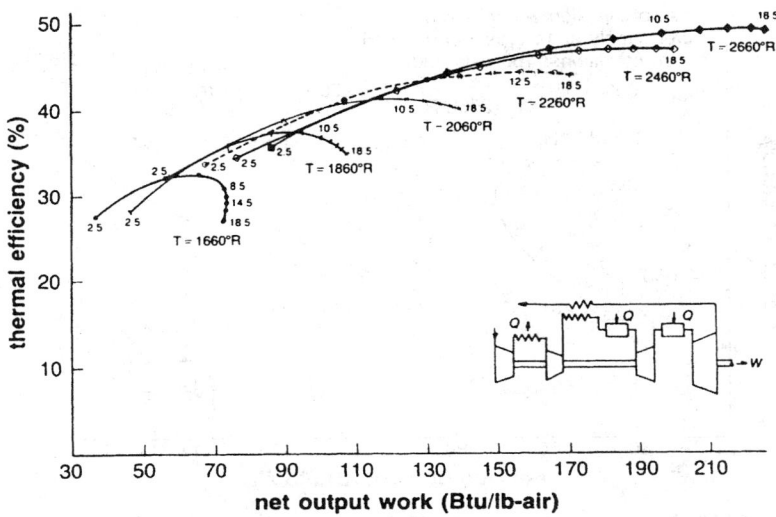

FIG. 29-36 Performance map showing the effect of pressure ratio and turbine inlet temperature on an intercooled regenerative reheat split-shaft cycle.

through the turbine. The steam being injected downstream from the compressor does not increase the work required to drive the compressor. The steam used in this process is generated by the turbine exhaust gas. Typically, water at 14.7 psi (101 kPa) and 80° F (27° C) enters the regenerator, where it is brought up to 60 psi (413 kPa) above the compressor discharge and the same temperature as the compressor discharge air. The steam is injected after the compression but far upstream of the burner to create a proper mixture.

For NO$_x$ control only, steam is injected into the combustor directly to help reduce the primary zone temperature in the combustor. The amount of steam injected is in a ratio of 1:1 with the fuel. In this cycle, the steam is injected upstream of the combustor and can be as much as 5–8 percent by weight of the air flow. This cycle leads to an increase in output work and a slight increase in overall efficiency. Corrosion problems due to steam injection have been for the most part over-

come with new high temperature coatings. Figure 29-37 shows the increase in efficiency and work output for various steam flow rates at a fixed turbine inlet temperature. Figure 29-38 shows the effect of 5 percent steam injection at various turbine firing temperatures.

The Combined (Brayton-Rankine) Cycle The 1990s has seen the rebirth of the combined cycle, the combination of gas turbine technologies with the steam turbine. This has been a major shift for the utility industry, which was heavily steam-turbine-oriented with the use of the gas turbine for peaking power. In this combined cycle, the hot gases from the turbine exhaust are used in a heat recovery steam generator or in some cases in a supplementary fired boiler to produce superheated steam.

The combined cycle work is equal to the sum of the net gas turbine work and the steam turbine work. About one-third to one-half of the design output is available as energy in the exhaust gases. The exhaust

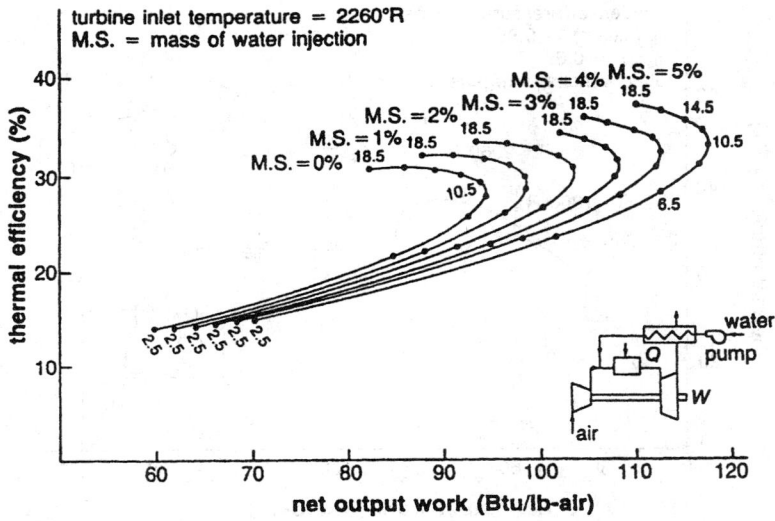

FIG. 29-37 Performance map showing the effect of pressure ratio and steam flow rate on a steam injection cycle.

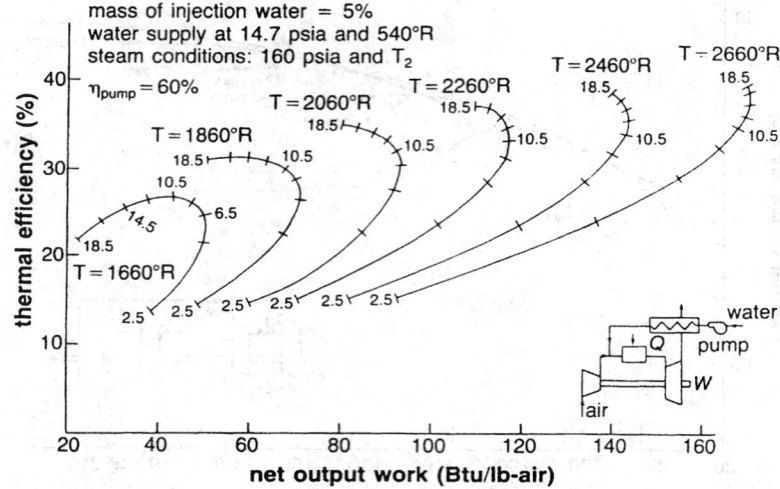

FIG. 29-38 Performance map showing the effect of pressure ratio and turbine inlet temperature on a fixed steam rate in a steam injection cycle.

gas from the gas turbine is used to provide heat to the recovery boiler. Thus, this heat must be credited to the overall cycle. This makes the combined cycle the highest practical efficiency cycle today. Figure 29-39 shows the effect of this cycle on the overall plant efficiency. To reduce the NO_x effect in the gas turbine, steam is injected in the combustor at a ratio of 1:1 with the fuel.

A comparison of the effect of the various cycles on the overall thermal efficiency is shown in Fig. 29-40. The most effective cycle is the Brayton-Rankine (combined) cycle. This cycle has tremendous potential in power plants and in the process industries where steam turbines are in use in many areas. The initial cost of the combined cycle is between $800–$1200 per kW while that of a simple cycle is about $300–$600 per kW. Repowering of existing steam plants by adding gas turbines can improve the overall plant efficiency of an existing steam turbine plant by as much as 3 to 4 percentage points.

Typically an inlet pressure decrease of one inch of water column reduces the power output by 0.4 percent and increases the heat rate by 0.125 percent. Similarly, an exhaust pressure increase of one inch of water reduces the power output by 0.15 percent and the heat rate by 0.125 percent.

TURBINE OPERATION CHARACTERISTICS

The gas turbine is a high-volume air machine. The compressor air power required is usually between 50–70 percent of the total power produced by the turbine. Thus, the ambient temperature affects the output of the gas turbine. On hot days, the gas turbine produces less output than on cold days. In dry climates, the use of evaporative cooling in the gas turbine decreases the effective inlet temperature and increases the power output of the unit.

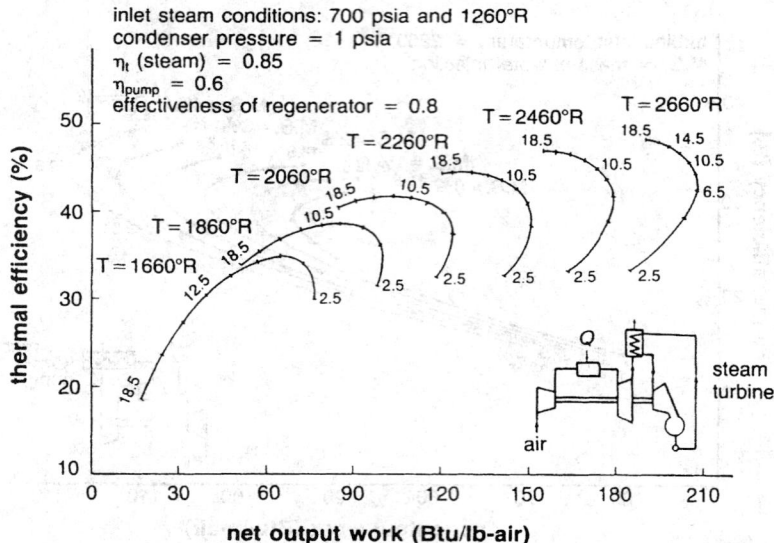

FIG. 29-39 Performance map showing the effect of pressure ratio and turbine inlet temperature on a Brayton-Rankine cycle.

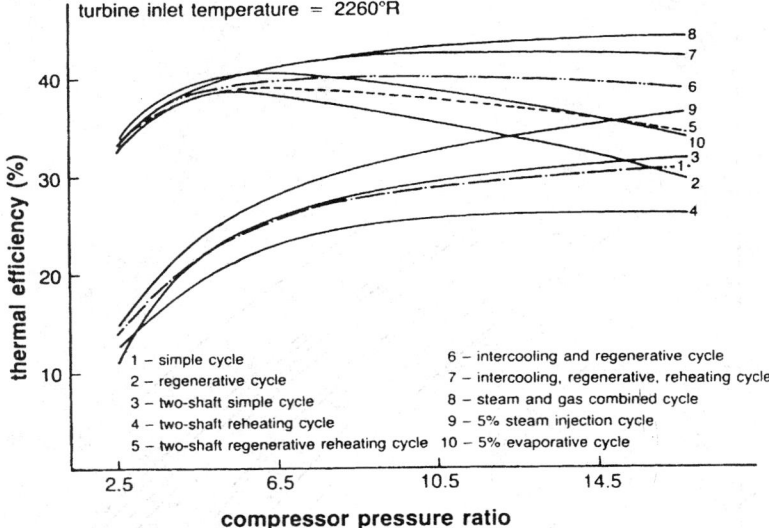

turbine inlet temperature = 2260°R

1 – simple cycle
2 – regenerative cycle
3 – two-shaft simple cycle
4 – two-shaft reheating cycle
5 – two-shaft regenerative reheating cycle

6 – intercooling and regenerative cycle
7 – intercooling, regenerative, reheating cycle
8 – steam and gas combined cycle
9 – 5% steam injection cycle
10 – 5% evaporative cycle

FIG. 29-40   Comparison of thermal efficiency of various cycles.

The effect on the performance of the inlet and exit conditions of the gas turbine is predominant. A increase in the inlet temperature by 5° F (2.8° C) will reduce the design power output by 2 %, and a reduction in the inlet pressure by 1 psi (6.9 kPa) would reduce the output by approximately 3 percent. Also, a 1 percent change in compressor efficiency reduces the overall thermal efficiency by ½ percent and reduces the power output by 2 percent. A 1 percent change in the turbine efficiency will produce a change of about 3 percent in turbine power output and a 0.75 percent change in the overall thermal efficiency. It is therefore very important to maintain the compressor in a very clean state. To do this, water solvent on-line cleaning is carried out on the compressor. Water jet nozzles are placed at the periphery of the compressor inlet. This type of cleaning is effective only on the first few stages. Coating the compressor blades has been found to improve efficiency. Inlet and outlet ducting should be designed with minimum losses; however, good filtration should not be compromised.

The gas turbine is usually started by an auxiliary drive such as a steam turbine, diesel engine, turboexpander, or electric motor. These drives bring the turbine up to a speed of between 1200–2000 rpm, at which time fuel is injected and the turbine speed is increased rapidly as the turbine firing temperature increases. The power requirement for the starter is about 5–10 percent of the power rating of the unit. Most turbine control systems have an acceleration monitor that shuts down the turbine if an acceleration rate is not maintained. *This is very essential if combustion in the turbine nozzles and blades is to be avoided.* To avoid compressor surges during startup and shutdown, a series of bleed valves in the compressor are sequentially closed at various speeds during the startup or are opened at various speeds during the shutdown. These bleed valves also are used for directing cooling air to the various regions of the turbine. Bleed valves are located in the early stages (i.e., 5th or 6th stage) and in the latter stages (i.e., 11th or 12th stage) of an axial flow compressor.

Multiple-shaft turbines require a little more starting power than single-shaft turbines. The low pressure turbine reaches breakaway torque at about 50–60 percent of the design speed of the gas generator section. In cases of aborted starts, the gas turbine must be fully purged before another start is attempted; otherwise, the gas trapped in the turbines could explode. In many new gas turbines, the first 5–7 stages of the compressor have variable stators that change angles at various settings. These move with speed and thus reduce the losses encountered in the compressor section.

In the case of the steam-injected cycle, steam must be injected after the turbine has been brought up to full speed; otherwise, compressor surges could occur. Major temperature excursions during startups must also be avoided to prevent degradation of the life of the turbine.

Life Cycle   The gas turbine life and especially hot section life are significantly influenced by the following parameters:
1. Blade material and cooling flow (blade metal temperature)
2. Type of fuel
3. Number of starts and full load trips

BLADE MATERIALS

Turbine life is very sensitive to blade metal temperature and blade material. It is essential for prevention of hot corrosion to keep this temperature below 1400° F (760° C). This temperature is a function of cooling flow to the blades, and the firing temperature. The effect of cooling flow blockage can be catastrophic for these blades. The life of the blades, especially on rotating elements, is a function of blade stresses and metal temperature. The Larson Miller curve shown in Fig. 29-39 for typical turbine blade alloys shows the logarithmic relationship between these parameters. While widely used to describe an alloy's stress rupture characteristic over a side temperature life and stress range, it is also very useful in comparing the elevated temperature capabilities of many alloys. Turbine blade alloys, which usually have large quantities of Ni, Cr, and Co, tend to indicate low durability at operating temperatures. This results in surface notches initiated by erosion or corrosion, after which cracks are propagated rapidly. This often leads to high cycle and low stress failure.

Turbine blade coatings can extend the life of blades by nearly 70–80 percent. Coating prevents corrosion from attacking the base metal. Most present-day coatings are diffusion-packed-type coatings. They usually consist of a thin uniform layer of a precious metal (platinum) electroplated onto the blade surface. This procedure is followed by pack diffusion steps to deposit layers of aluminum and chromium, resulting in a coating that has an outer skin of an extremely corrosion-resistant, platinum-aluminum intermetallic composition. The gas turbines of the 1990s are all coated, especially in the hot section area. Coatings are also being applied to the compressor blades and have been found to be very effective.

TYPES OF FUEL

The life of a gas turbine depends heavily on the type of fuel used. An inherent fuel flexibility is the gas turbine's major advantage. Gaseous fuels traditionally include natural gas, process gas, and low-Btu gas

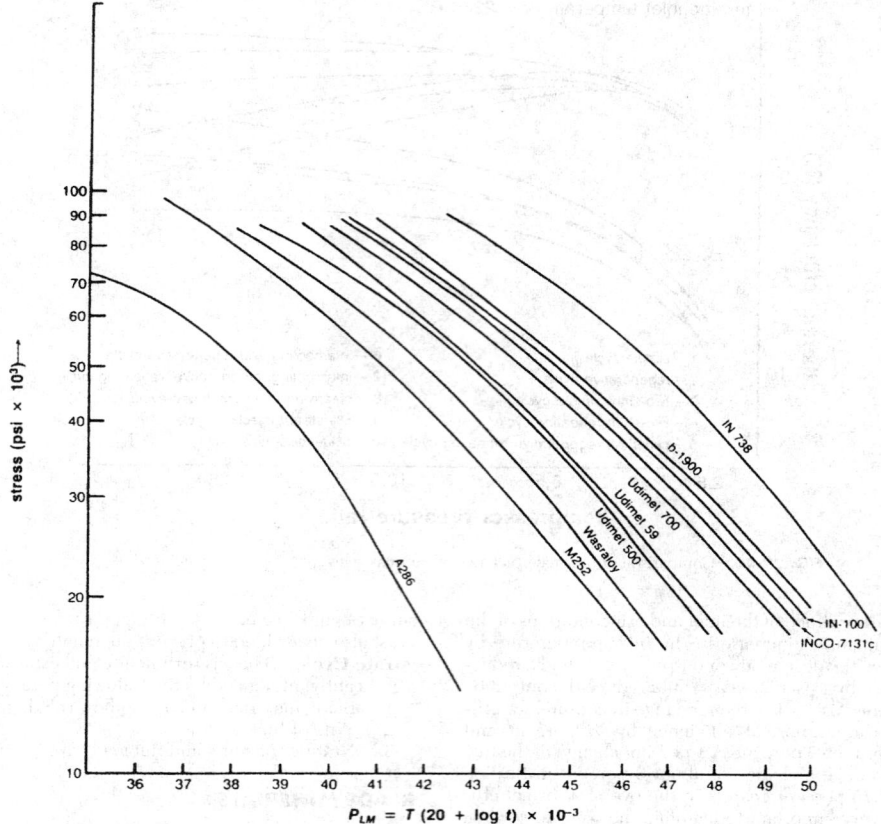

FIG. 29-41   Larson-Miller parameters for turbine blade alloys.

(coal gas or water gas). Natural gas is the benchmark against which performance of a gas turbine is compared, since it is a clean fuel that promotes long machine life.

Liquid fuels vary from light volatile fuels such as naphtha through kerosene to the heavy viscous residuals. True distillate fuel such as #2 distillate oil is a good fuel; however, because trace elements such as vanadium, sodium, potassium, lead, and calcium are found in the fuel, the fuel has to be treated. The corrosive effect of sodium and vanadium is very detrimental to the life of a turbine. Vanadium originates as a metallic compound in crude oil and is concentrated by the distillation process into heavy oil fractions. Sodium compounds are usually present in the form of salt water, which results from salty wells, transport over seawater, or mist ingestion in an ocean environment. Fuel treatments are costly and do not remove all traces of the metal. Sodium is usually removed by water washing and letting the sodium dissolve into the water. It is then separated by a centrifuge. Vanadium is counteracted by the addition of magnesium which causes a friable deposit. A magnesium/vanadium ratio of 3:1 reduces corrosion by a factor of six.

The typical amounts of sodium and vanadium in the fuel should be less than 1 ppm. Figure 29-42 shows the effect of sodium and vanadium on the life of the blade and on the combustor life. Figure 29-43 shows the reduction in firing temperature required to maintain design life (hrs) of a typical turbine (IN718) blade due to sodium and vanadium in the fuel.

In general terms, the life of a combustor might be reduced by about 30 percent through use of a distillate fuel and by 80 percent through the use of residual fuel. The first stage turbine nozzle life can be reduced by 20 percent through use of a distillate fuel and by about 65 percent when certain residual fuels are used.

Number of Starts and Full-Load Trips   Temperature differentials developed during starting and stopping of the turbine produce thermal stresses. The cycling of these thermal stresses causes thermal fatigue. Thermal fatigue is a low cycle event and is similar to creep rupture failure. The analysis of thermal fatigue is essentially a problem in heat transfer and is affected by properties such as modulus of elasticity, coefficient of thermal expansion, and thermal conductivity. The most important metallurgical factors are ductility and toughness.

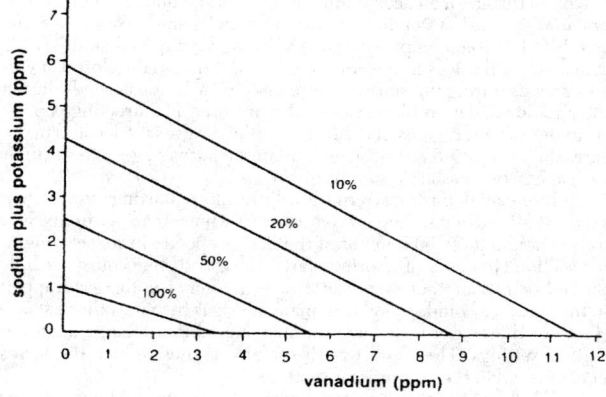

FIG. 29-42   Effect of sodium, potassium, and vanadium on combustor life.

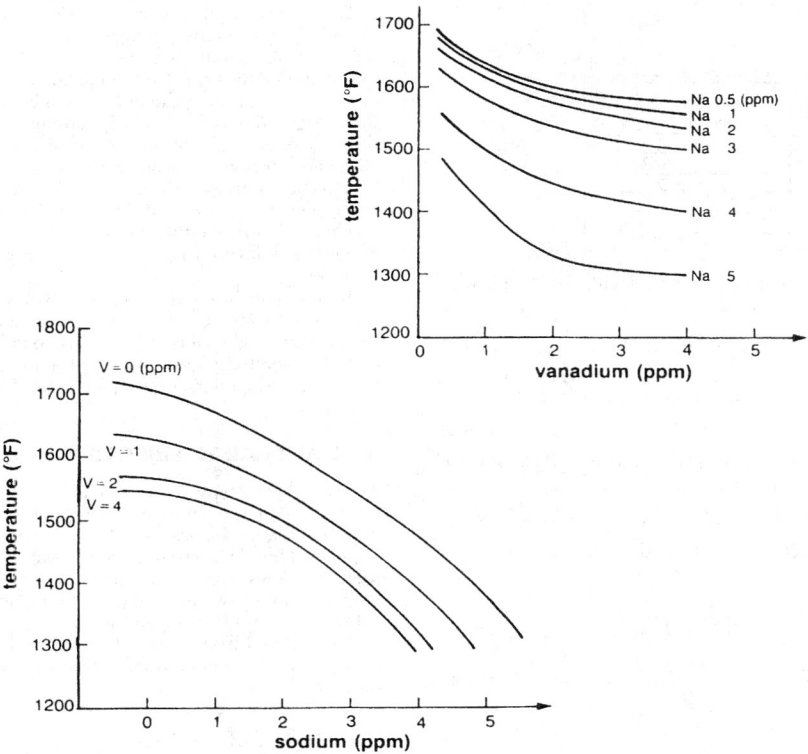

FIG. 29-43   Firing temperature reduction needed to offset IN 718 corrosion by sodium and vanadium.

Highly ductile materials tend to be more resistant to thermal fatigue and also seem more resistant to crack initiation and propagation.

The operating schedule of a gas turbine produces low-frequency thermal fatigue. The number of starts per hours of operating time directly affects the life of the hot sections (combustor, turbine nozzles, and blades). The life reduction effect of the number of starts on a combustor liner could be as high as 230 hours/start and on the turbine nozzles as high as 180 hours/start. The effect of full load trips can be nearly 2–3 times as great!

The life of a gas turbine depends on the above detailed operational characteristic. It is interesting to note that, for a gas turbine life of 25 years, the life cycle costs can be distributed as 5–10 percent on initial cost, 10–20 percent on maintenance costs and 70–85 percent on cost of fuel. Gas turbines will be very widely used in the 21st century in combined cycle applications as the power source for the world. These combined cycle plants will have efficiencies in the high fifties and will cost between $1000 and $1200 per kW, using 1994 as a monetary benchmark.

EXPANSION TURBINES

Fundamentally, an expansion turbine is a device for converting the pressure energy of a gas or vapor stream into mechanical work as the gas or vapor expands through the turbine. The mechanical work so produced, however, is generally a by-product, the primary objective of the turboexpander being to chill the process gas. Turboexpanders are in wide use in the cryogenic field to produce the refrigeration required for the separation and liquefaction of gases.

By common usage, the terms "turboexpanders" and "expansion turbines" specifically exclude steam turbines and combustion gas turbines, which are covered elsewhere in Sec. 29.

Any work developed by the turboexpander is at the expense of the enthalpy of the process stream, and the latter is correspondingly cooled. A low inlet temperature means a correspondingly lower outlet temperature, and the lower the temperature range, the more effective the expansion process becomes.

FUNCTIONAL DESCRIPTION

The turboexpander in combination with a compressor and a heat exchanger functions as a heat pump and is analyzed as follows: In Fig. 29-44 consider the compressor and aftercooler as an isothermal compressor operating at T_2 with an efficiency E_c, and assume the working fluid to be a perfect gas. Further, consider the removal of a quantity of heat Q_e by the turboexpander at an average low temperature T_1. This requires that it deliver shaft work equal to Q_e. Now, make the reasonable assumption that one-tenth of the temperature drop in the expander is used for the temperature difference in the heat exchanger. If the expander efficiency is N_e and this efficiency is multiplied by 0.9 to include the effect of the temperature difference in the heat exchanger, the needed ideal enthalpy drop across the expander is

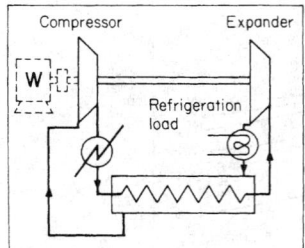

FIG. 29-44   Turboexpander system functioning as a refrigeration machine.

$$H_e = Q_e/0.9N_e \tag{29-16}$$

The theoretical required (isothermal) compression work in the compressor, which is assumed to operate isothermally at T_2, is

$$(Q_e/0.9N_e)(T_2/T_1) \tag{29-17}$$

The actual compressor work W_c is this latter quantity, divided by the compressor isothermal efficiency N_c; thus,

$$W_c = (Q_e/0.9N_eN_c)(T_2/T_1) \tag{29-18}$$

Mechanical work equal to $Q_e/0.9$ is returned by the expander to the compressor, so the net work to the compressor is

$$W = W_c - \frac{Q_e}{0.9} = \left(\frac{Q_e}{0.9}\frac{T_2}{N_eN_cT_1}\right) - \frac{Q_e}{0.9}$$

$$W = \left(\frac{Q_e}{0.9}\frac{T_2}{N_eN_cT_1}\right) - 1 = \frac{Q_e}{0.9}\left(\frac{T_2 - N_eN_cT_1}{N_eN_cT_1}\right) \tag{29-19}$$

The second-law theoretical work is

$$W_{\text{theor}} = Q_e\frac{T_2 - T_1}{T_1} \tag{29-20}$$

Hence, the second-law efficiency of the expander-heat-exchanger-compressor system is

$$\frac{W_{\text{theor}}}{W} = \frac{Q_e\dfrac{T_2 - T_1}{T_1}}{\dfrac{Q_e}{0.9}\left(\dfrac{T_2 - N_eN_cT_1}{N_eN_cT_1}\right)}$$

$$= \frac{0.9(T_2 - T_1)N_eN_c}{T_2 - N_eN_cT_1} \tag{29-21}$$

A plot of this efficiency in which commonly available equipment is assumed is shown by the expander curve in Fig. 29-45.

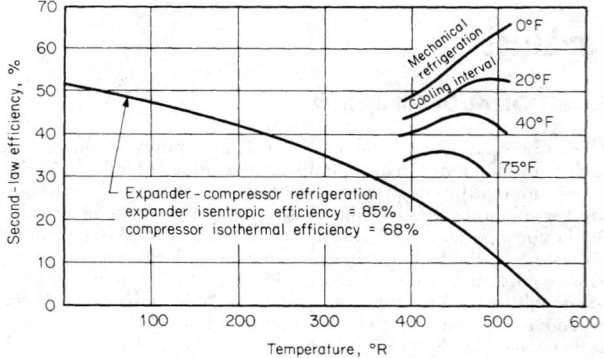

FIG. 29-45   Mechanical versus turboexpander refrigeration. K = 5/9°R; °C = ⁵⁄₉(°F − 32).

The family of short curves in Fig. 29-45 shows the power efficiency of conventional refrigeration systems. The curves for the latter are taken from the *Engineering Data Book,* Gas Processors Suppliers Association, Tulsa, Oklahoma. The data refer to the evaporator temperature as the point at which refrigeration is removed. If the refrigeration is used to cool a stream over a temperature interval, the efficiency is obviously somewhat less. The short curves in Fig. 29-45 are for several refrigeration-temperature intervals. A comparison of these curves with the expander curve shows that the refrigeration power requirement by expansion compares favorably with mechanical refrigeration below 360° R (−100° F). The expander efficiency is favored by lower temperature at which heat is to be removed.

Another conclusion that can be drawn from Fig. 29-45 is that if the process can justify the complexity, it is more efficient, powerwise, to use conventional means rather than expanders to absorb heat at moderate temperatures in the range of ambient to 360° R, although for expediency expanders are frequently used in any case.

SPECIAL CHARACTERISTICS

An example of a typical turboexpander is shown in Fig. 29-46. Radial-flow turbines are normally single-stage and have combination impulse-reaction blades, and the rotor resembles a centrifugal-pump impeller. The gas is jetted tangentially into the outer periphery of the rotor and flows radially inward to the "eye," from which the gas is jetted backward by the angle of the rotor blades so that it leaves the rotor without spin and flows axially away.

Radial-flow turbines have been developed primarily for the production of low temperatures, but they also may be used as power-recovery devices.

The characteristics of these machines include the following:

1. High efficiency: 75 to 88 percent
2. Operation usually at a very low temperature
3. Operation often on small or moderate streams, dictating a rather high rotating speed
4. Supports having low heat conductivity
5. Effective shaft seals to conserve the process stream
6. Heavy-duty construction resistant to abuse
7. High reliability

Commonly established operating limitations for turboexpanders without special design features are an enthalpy drop of 93 to 116.3 kJ/kg (40 to 50 Btu/lb) per stage of expansion and a rotor-tip speed of 304.8 m/s (1000 ft/s). Commercial turboexpanders are available for inlet pressure up to 20.68 MPa (3000 lbf/in²) and inlet temperatures from near absolute zero to 538° C (1000° F). The permissible liquid condensation in the expanding stream varies with discharge pressure; it may be 50 weight percent or higher in the discharge, provided the turboexpander has been specially designed to handle condensation.

RADIAL INFLOW DESIGN

The radial reaction design has been selected for turboexpanders primarily because it attains the highest efficiency of all turbine designs. However, it has several additional features which favor this application:

1. In the single-stage configuration, it is usual to have the rotor on the end of the shaft (overhung); this provides a convenient opportunity for thermally insulating the cold turbine portion and is an ideal arrangement for axial discharge.
2. This design permits variable primary nozzles, which enable the attainment of high efficiency over a wide flow range.
3. It applies lower axial thrust to the shaft than a single-stage axial reaction turbine.
4. It has shaft bearings on only one side of the expander rotor, and heat-barrier insulation between the warm, lubricated bearings and the cold turbine is convenient to arrange.
5. Its speed is reasonably acceptable for suitable loading devices.

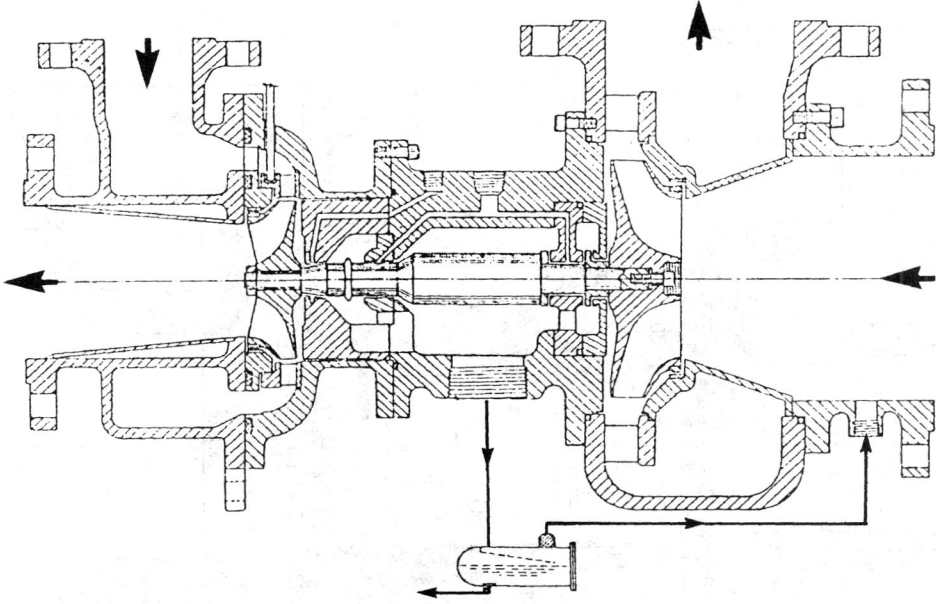

FIG. 29-46 Typical radial-flow turboexpander.

EFFICIENCY

Efficiency for a turboexpander is calculated on the basis of isentropic rather than polytropic expansion even though its efficiency is not 100 percent. This is done because the losses are largely introduced at the discharge of the machine in the form of seal leakages and disk friction which heats the gas leaking past the seals and in exducer losses. (The exducer acts to convert the axial-velocity energy from the rotor to pressure energy.)

BEARINGS

Radial Bearings Antifriction bearings are largely unsuitable for these applications, chiefly because of the special attention and maintenance which they demand.

Virtually all expanders have lubricated sleeve bearings or tilting-shoe bearings. The advantage of sleeve bearings is that they can be designed to support the shaft sufficiently rigidly that the oil-film critical (first critical) can be designed to be safely above the design running speed. This is desirable because it eliminates all shaft vibrations in the full operating range; more important, if the critical were below the design speed, then at design speed the shaft assembly would be rotating about its center of gravity. In turboexpanders, there frequently are ice deposits or other reasons for rotor imbalance, and such imbalance would cause gyration and damage to the shaft seals.

In arrangements in which the shaft is running with its first critical (oil-film critical) below the design speed, then not far above design speed is the oil swirl (half-speed gyration). Substitution of tilting-shoe bearings for sleeve bearings moves the oil swirl to a higher speed, but it does not prevent the rotor from rotating about its center of gravity. Tilting-shoe bearings do not lubricate well at this high rubbing speed and small shaft diameter.

The sleeve bearing has the further advantage that it functions as a lubricated, pressurized shaft seal to contain the pressure on the process gas (see subsection "Shaft Seals").

Thrust Bearings Turboexpanders often have process upsets or ice plugging or the like, which can cause serious thrust-bearing load variations. In applications above 506.6 to 1013.2 kPa (75–148 psi), the best available thrust bearing usually is insufficient to protect against such high thrust loads. Various indications, such as the differential

pressure across the rotor and thrust-bearing temperatures, are available to protect the unit at least to some extent.

Thrust-bearing-load meters for protection against excessive loading are available on thrust bearings, and automatic thrust control, which functions by controlling the pressure behind a balancing drum, also is available.

SEALS

Shaft Seals Mechanical shaft seals generally are not acceptable in turboexpanders for the same general reason that antifriction bearings are not: they require periodic replacement and careful attention. In their stead, close-clearance labyrinth-type seals are generally used. Such a labyrinth seal generally has an injection point intermediate from its two ends into which a suitable buffer gas is injected to prevent the escape of the process gas; instead, buffer gas escapes. If the escaping seal gas is inexpensive and nontoxic, it may be allowed to leak to the atmosphere. However, it is possible to enclose the expander housing to the bearing housing and allow the journal bearing, acting as an oil seal, to contain the outleaking seal gas and collect it in a float-operated drainer for suitable disposal or reuse.

In refrigeration applications in which the refrigerant must be completely conserved, the expander housing and bearing housing can be hermetically sealed through a speed-reducing gearbox. The low-speed shaft of the gearbox is sealed with a low-speed mechanical seal. Then any refrigerant which leaks out of the labyrinth seal is totally contained in the gearbox and in the closed lubrication system for complete collection and reuse.

Rotor Seals To balance the thrust on the rotor, usually there are one or two labyrinth-type seals on the rotor. These seals often are damaged if there is dust in the incoming fluid or gas, and wear on the backside seal causes serious upsets in thrust-bearing loads. Provisions are available for collecting and disposing of the dust which tends to accumulate in the seal so as to protect the seal from serious erosion.

VARIABLE NOZZLES

The pressurized process stream is guided radially into the rotor by the primary nozzles, which are a series of vanes forming nozzles jetting the gas tangentially and inwardly into the rotor (see Fig. 29-47). These

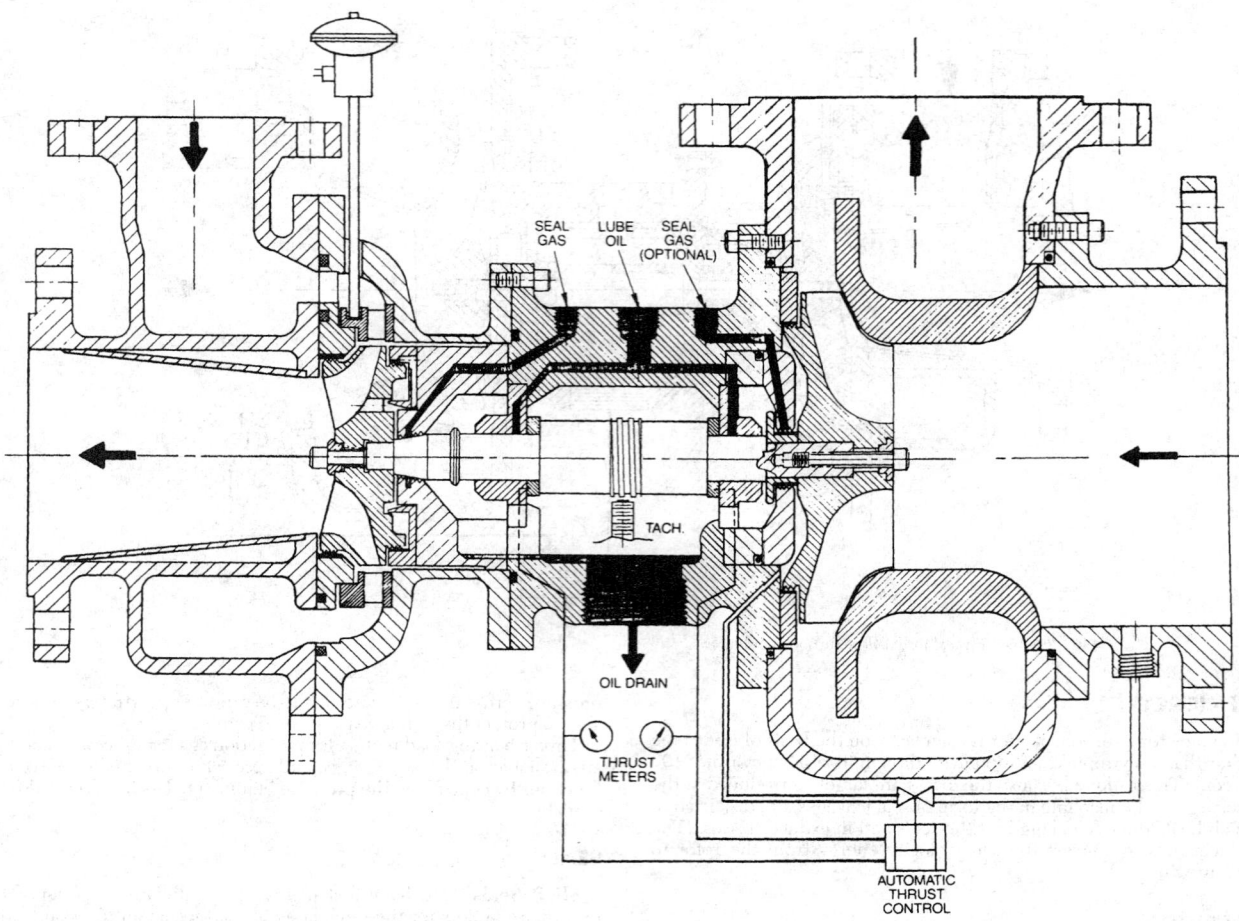

FIG. 29-46 (*Continued*) Typical radial-flow turboexpander.

nozzle vanes are clamped between two flat rings and usually are pivoted so that they can be rotated in unison to open or close the spaces between them in order to vary the nozzle-throat areas. This is quite an important function because it can be used to vary the flow widely through the expander without wasteful throttling: all the expansion energy in the nozzles is recovered in the rotor. The variable nozzles, from a control standpoint, act just as a throttle valve would act in controlling the flow (but without throttling loss), and conventional flow-control instrumentation can be used to operate them.

ROTOR RESONANCE

Variable nozzles produce a series of jets of gas entering the rotor, and these impulses add up to form a frequency equal to the blade-passing frequency: the number of revolutions per second multiplied by the number of nozzle vanes, which is of the order of thousands of cycles per second. Frequently the rotor will resonate at this frequency, and if it does, it will be fatigued and crack and break up; thus these frequencies must be avoided, and the manufacturer should be asked to supply information to the customer on this subject.

CONDENSING STREAMS

It is advantageous to have the expander generate its refrigeration at the lowest possible temperature in the process (see Fig. 29-45), and this frequently encounters the condensation temperature of the process stream. Steam-turbine practice advises against operating on

condensing streams because efficiency is deteriorated and there also usually is erosion of the rotor blades.

Another advantage of the radial reaction turbine is that it can be designed to accept condensation in any amount without efficiency deterioration or erosion.[*] This is possible because there are two forces acting on suspended fog particles, the deceleration force and the centrifugal force, and these two forces can be balanced against each other to prevent the droplets from impinging on specially shaped blades. The process is explained as follows:

This expansion of a condensing vapor is highly desirable thermodynamically, but the liquid must not bombard and erode the rotor blades, and, in particular, it must not accumulate in the rotor, since that would cause efficiency loss.

If liquid droplets form as the gas is expanded in the turboexpander, one's first thought may be that a radial inflow design is the last thing to use, but the following explanation will show that this is the only design that can accomplish expansion efficiently.

Figure 29-47 shows the primary nozzles and the rotor. About half of the pressure drop takes place in the primary nozzles, which jet the gas tangentially into the periphery of the rotor. Cooling takes place during this expansion, and the jetted stream entering the rotor may be foggy. This foggy gas flows radially inward within the rotor as the latter rotates, and at the outlet, which is near the center of the rotor, the

[*] This same principle applies also to the expansion of flashing liquids in which the bubbles are guided away from the blades.

FIG. 29-47 The primary nozzles and the rotor.

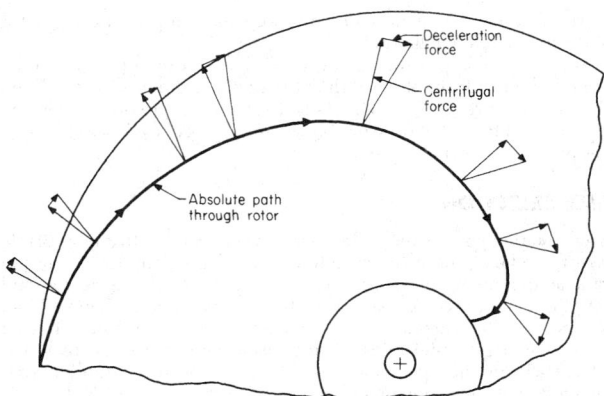

FIG. 29-48 Elemental path radially inward, then axially, through the rotor.

4. Augmenting refrigeration in various cryogenic processes such as the recovery of ethylene

5. Power generation, sometimes referred to as power recovery

A potential application for the turboexpander for power recovery exists whenever a large flow of gas is reduced from a high pressure to some lower pressure or when high-temperature process streams (waste heat) are available to boil a secondary liquid. When such conditions exist, they should be examined to see if use of a turboexpander is justified. In such cases a turboexpander can be used to drive a pump, compressor, or electric generator, thus recovering a large portion of otherwise wasted energy. In applications of this type, careful consideration should be given to the temperature drop which will occur in the expander. It may sometimes be necessary to heat or to dry the inlet gas to avoid low exhaust temperatures that cause the formation of ice or liquids.

Expanders are used because they are in an advanced state of development and reliability, attain high efficiency, and are relatively inexpensive.

LUBRICATION

Expander bearings are usually high-speed, and they should have full film lubrication. This is best assured by using force-feed lubrication at a pressure of the order of 689.5 kPa (100 lbf/in^2) or more. There is no special objection to using pressures as high as 6.895 MPa (1000 lbf/in^2) or higher, if for some reason it is desirable to do so.

Usually, a journal bearing and a thrust bearing are combined in one assembly, and oil is injected so as to feed both of them. The rate of flow usually is adjusted so as to carry the heat away with a temperature rise of the order of 11 to 17° C (20 to 30° F).

The smallest expanders usually use oil with a viscosity at 38° C (100° F) of 60 to 100 SSU, and large machines up to 500 SSU. If the oil is kept in a totally enclosed system in contact with hydrocarbon or another partly soluble gas, which would dissolve and reduce the viscosity of the oil, then a compensating higher viscosity should be used so that the working viscosity after ultimate equilibrium with such gas is suitable for the bearings.

The lubrication system, for reliability reasons, usually has an operating and a standby pump and dual switchable filters. If there is a cooling-water scaling problem, coolers may also be switchable.

BUFFER-GAS SYSTEM

The shaft seal (see subsection "Shaft Seals") generally is a close-clearance labyrinth-type seal. It is desirable that there be available a suitable pressurized buffer gas for injection into the intermediate point in the seal, such gas to be available at an absolute pressure well above the highest shaft pressure to be sealed. Then the seal-gas system may consist of only a filter, a flow-indicating device, and a throttle

fluid is discharged by being jetted backward out of the rotor so as to leave without rotary motion. The second half of the expansion energy is spent by the gas passing radially through the rotor against centrifugal force, and further precipitation of liquid takes place.

The stream from the nozzles enters the rotor with a tangential velocity of about 152.4 or 304.8 m/s (500 or 1000 ft/s; see Fig. 29-48) and follows a path through the rotor of such curvature that the centrifugal force acting on an element of the stream and, therefore, on suspended droplets is of the order of 75,000 G (75,000 times the force of gravity). Also, the stream, because it is moving radially inward in the rotor, is decelerated in the rotor from this tangential velocity of 304.8 m/s (1000 ft/s) down to zero tangential velocity in about a half a revolution of the rotor. This deceleration force amounts to something like 10,000 G. The vector sum of these two forces, therefore, amounts to 75,000 or 100,000 G in a direction 5 to 15° from the radial direction (see Fig. 29-48). This is an acceptable direction for the blades to lie, so the problem of avoiding bombardment of the blades by the droplets is solved by shaping the expander rotor blades parallel to this resultant vector. Then there is no force causing droplets to drift in the direction of any surface. They do drift back upstream slightly but nevertheless are carried on through by the mainstream and discharged. By this method any amount of condensing liquid can pass through, or, in the case of flashing liquids being expanded, the bubbles can pass through without efficiency loss. It would be impossible to construct a turbine blade meeting this requirement without two vector forces.

APPLICATIONS

The important uses of turboexpanders are in:

1. Air separation
2. Recovery of condensables from natural gas
3. Liquefaction of gases, including helium

valve or other flow or pressure control, usually a pressure regulator and a graduated needle valve.

If the available pressure is not far above that of the pressure to be sealed, then with simple throttling the flow may be insufficient when the two pressures come too near together. Then more precise control, such as by differential pressure between the process side and the seal-gas pressure, may be required.

SIZE SELECTION

Size, rotating speed, and efficiency correlate well with the available isentropic head, the volumetric flow at discharge, and the expansion ratio across the turboexpander. The head and the volumetric flow and rotating speed are correlated by the specific speed. Figure 29-49 shows the efficiency at various specific speeds for various sizes of rotor. This figure presumes the expansion ratio to be less than 4:1. Above 4:1, certain supersonic losses come into the picture and there is an additional correction on efficiency, as shown in Fig. 29-50.

The available isentropic head is usually calculated by computer, using any of the various equations of state. In the absence of such facility, a quick and reasonably reliable calculation follows. In fact, this calculation is valuable as a cross-check on other methods because it is likely to be accurate within a few percent.

R = gas constant, 1.986 T_a = average temperature, °R

Z_a = compressibility at T_a H_i = isenthalpic work,

$$H_i = \frac{R T_a Z_a}{\text{mol. wt.}} \ln \frac{P_1}{P_2} \qquad \text{Btu/lb} \qquad (29\text{-}22)$$

where H_i = reversible incremental enthalpy drop, Btu/lb
R = gas constant, 1.986, Btu/(lb·mol·°R)
T_a = average temperature for the increment, °R
Z_a = average compressibility for the increment
P_1/P_2 = pressure ratio

The use of this equation requires that an average P and T, based on an assumed increment, be used to find Z.

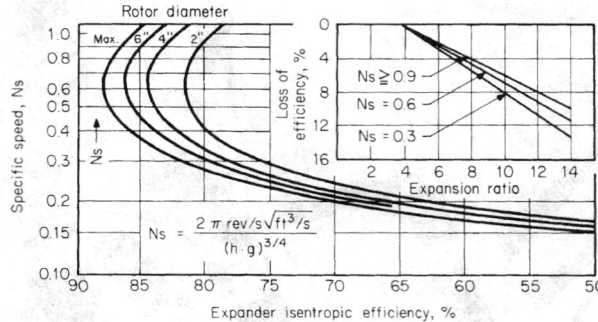

FIG. 29-49   Efficiency at various specific speeds for various sizes of rotor.

FIG. 29-50   Loss of efficiency as a function of the pressure ratio.

INSTRUMENTATION

Process-flow control and buffer-gas control have been discussed under "Variable Nozzles" and "Buffer-Gas System" respectively. Speed is usually self-controlled by a matching speed-sensitive load such as a compressor or a pump. If the load is an induction or synchronous generator feeding into a stable ac system, the system frequency fixes the speed. Otherwise, the speed can be controlled by a conventional governor.

Various protective instruments are used to provide a shutdown signal (to a fast-acting trip valve at the expander inlet) that senses various things, such as overspeed, lubricant pressure, bearing temperature, lubricant temperature, shaft runout, icing, lubricant level, thrust-bearing load, and process variables such as sensitive temperatures, levels, pressures, etc. However, too many safety shutdown devices may lead to excessive nuisance shutdowns.

POWER RECOVERY FROM LIQUID STREAMS

BASIC PRINCIPLES

The potential for power recovery from liquid streams exists whenever a liquid flows from a high-pressure source to one of lower pressure in such a manner that throttling to dissipate pressure occurs. Such throttling represents a system potential for power that is the reverse of a pump—in other words, a potential for power extraction. Just as in a pump, there exists a hydraulic horsepower and a brake horsepower, except that in the recovery they are generated or available horsepowers.

Basically, power recovery from liquids is achieved in industrial installation as shaft horsepower. While this potentially could appear either as reciprocating or as rotating power, most larger applications are rotating. Consideration of power recovery from liquids involves a choice among several possible uses, and usually this choice involves as alternatives (1) driving of a few large-horsepower services versus driving of more smaller-horsepower services; (2) driving of essential versus nonessential services or of spared versus nonspared services; (3) driving as sole driver versus partial driver or full horsepower versus partial horsepower for the selected service; and (4) converting power-recovery energy to some other intermediate energy form, as by driving an electric generator.

In applying power recovery, three basic problems are (1) limitations in designing equipment to recover the power, (2) operating reluctance to consider rotating equipment that is not absolutely necessary, and (3) the way in which the economics of the installed system is evaluated. It is important to recognize that there has always been an opera-

ble, acceptable alternative to power recovery from liquid streams in the form of the throttling or letdown valve, whereas no such simple, cheap, foolproof substitute exists for the pump.

Basic to establishing whether power recovery is even feasible, let alone economical, are considerations of the flowing-fluid capacity available, the differential pressure available for the power recovery, and corrosive or erosive properties of the fluid stream. A further important consideration in feasibility and economics is the probable physical location, with respect to each other, of fluid source, power-production point, and final fluid destination. In general, the tendency has been to locate the power-recovery driver and its driven unit where dictated by the driven-unit requirement and pipe the power-recovery fluid to and away from the driver. While early installations were in noncorrosive, nonerosive services such as rich-hydrocarbon absorption oil, the trend has been to put units into mildly severe services such as amine plants, hot-carbonate units, and hydrocracker letdown.

Economics   Power-recovery units have no operating costs; in essence, the energy is available free. Furthermore, there is no incremental capital cost for energy supply. Incremental installed energy-system costs for a steam-turbine driver and supply system amount to about $800 per kilowatt, and the incremental cost of an electric-motor driver plus supply system is about $80 per kilowatt. By contrast, even the highest-inlet-pressure, largest-flow power-recovery machines will seldom have an equipment cost of more than $140 per kilowatt, and costs frequently are as low as $64 per kilowatt. However, at bare driver costs (not including power supply) of $64 to $140 per kilowatt for the power-recovery driver versus about $30 to $80 per kilowatt for steam turbines or $50 to $64 per kilowatt for electric motors, operating costs must be considered to make power-recovery units attractive. Using commonly accepted values for power costs, turbine steam rates, and steam selling prices, operating costs for either motors or steam turbines approximate $280 per year for 746 W (1 hp).

Thus, barring technical difficulties or operational considerations in application, power-recovery units ought to show payouts of less than 6 months. Actual project payouts run from 1 to 3 years. This difference is due principally to (1) the fact that while the incremental costs just presented are valid in comparing large systems, specific designs encounter frame-size breaks, standardized capacities and horsepowers, code requirements, etc.; and (2) operating requirements, sparing considerations, and a certain lack of confidence in power-recovery units stemming from lack of extensive experience produce equipment-selection schemes that deviate from the straightforward comparison.

Development The following discussion relates specifically to the use of what could be called radial-inflow, centrifugal-pump power-recovery turbines. It does not apply to the type of unit nurtured by the hydroelectric industry for the large-horsepower, large-flow, low- to medium-pressure differential area of hydraulic water turbines of the Pelton or Francis runner type. There seems to have been little direct transfer of design concepts between these two fields; the major manufacturers in the hydroelectric field have thus far made no effort to sell to the process industries, and the physical arrangement of their units, developed from the requirements of the hydroelectric field, is not suitable to most process-plant applications.

Despite a rather slow start, centrifugal power-recovery pump-turbines have built a respectable record of process-plant installations extending back to the middle 1950s. Applications have included drives for the following services: cooling-tower fans; reciprocating recycle compressors; gas-treating-solution circulation pumps, as sole drive and as tandem with a steam turbine or motor helper; refinery-unit charge-stock pumps with a helper driver; and floating-online electric generators.

In general, early experiences were in the small-horsepower, nonessential or spared services, using sole drivers at full horsepower. The present trend is more and more toward the few large drivers in essential services, usually supplying only partial horsepower but not spared. If a plant is based on electric drivers and the economics of rate structure, demand charges, etc., permits, the use of a power-recovery unit driving a generator electrically floating on the line has found increasing favor. In general, operating experience in regard to reliability, serviceability, and maintainability has shown the units to be comparable with centrifugal pumps and has resulted in increasing acceptance even as drivers in large-horsepower units on essential service equipment. Presently accepted industrial limits are shown in Fig. 29-51.

Hydraulic Behavior The basic hydraulic behavior of centrifugal pumps operating as power-recovery units (turbines) is not much different from that of centrifugal pumps and follows the same sort of affinity laws over narrow ranges. Typical generalized curves are shown in Figs. 29-52 and 29-53. Note particularly that both torque and horsepower go negative (turn to values indicating power consumption) when head and capacity are within a fairly wide range representing at least startup and shutdown conditions if not also part load. Note also that even if head goes to 125 percent of design, speed (r/min) at zero torque for this unit does not exceed about 130 to 150 percent of design.

Tests conducted to operate centrifugal pumps as hydraulic turbines throughout the head-capacity-speed range show that a good centrifugal pump generally makes an efficient hydraulic turbine. From theoretical considerations it is possible to state that at the same speed

$$H_t = H_p/e_h^2 \tag{29-23}$$

$$Q_t = Q_p/e_h \tag{29-24}$$

$$n_{st} = n_{sp}e_h \tag{29-25}$$

where H = total head at best efficiency point
Q = capacity
n_s = specific speed

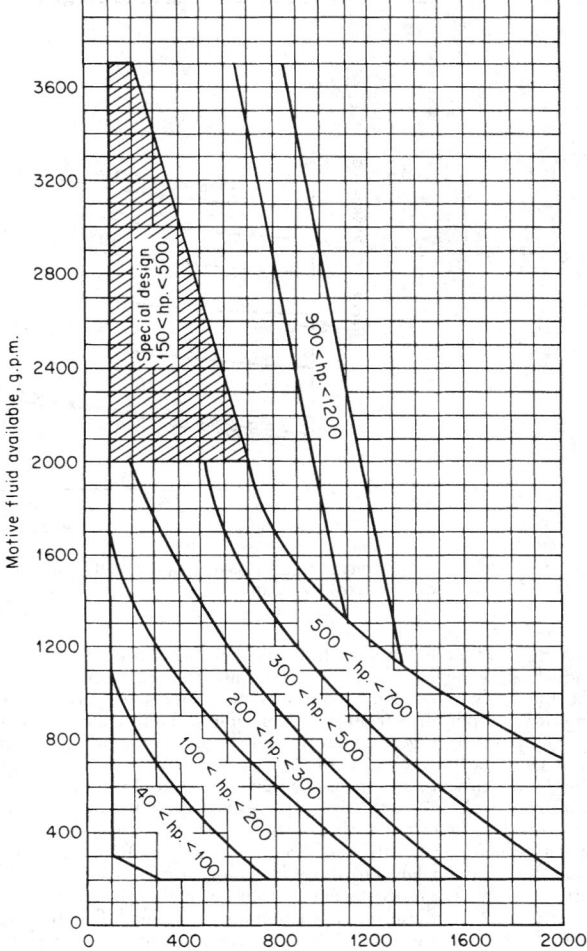

FIG. 29-51 Application areas for centrifugal pump turbines. Curves apply between the following minimum and maximum limits: inlet pressure, 100 to 3000 psig; pressure differential, 100 to 2800 lbf/in²; flow of motive fluid, 200 to 4000 gal/min; horsepower, 50 to 3000 hp. Curve horsepower is based on a speed of 3600 r/min and a fluid of 1.0 specific gravity; for other fluids, multiply the curve horsepower by specific gravity to get the actual horsepower. To convert pounds-force per square inch to megapascals, multiply by 6.89×10^{-3}; to convert gallons per minute to cubic meters per minute, multiply by 3.79×10^{-3}; and to convert horsepower to kilowatts, multiply by 0.746.

e_h = hydraulic efficiency, taken as the same for the turbine and the pump
t, p = subscripts denoting turbine and pump respectively

Since the exact value of the **hydraulic efficiency** e_h is never known, $\sqrt{e}$ can be taken as an approximation where e is the gross (hydraulic horsepower/brake horsepower) pump efficiency. Efficiencies of pump designs running as turbines are usually 5 to 10 efficiency points lower than those as pumps at the best efficiency point.

OPERATING BEHAVIOR

By considering the flow as stopped but the turbine casing full of liquid, it is intuitively obvious that to rotate the wheel or impeller in either direction power will have to be put in. As the flow increases

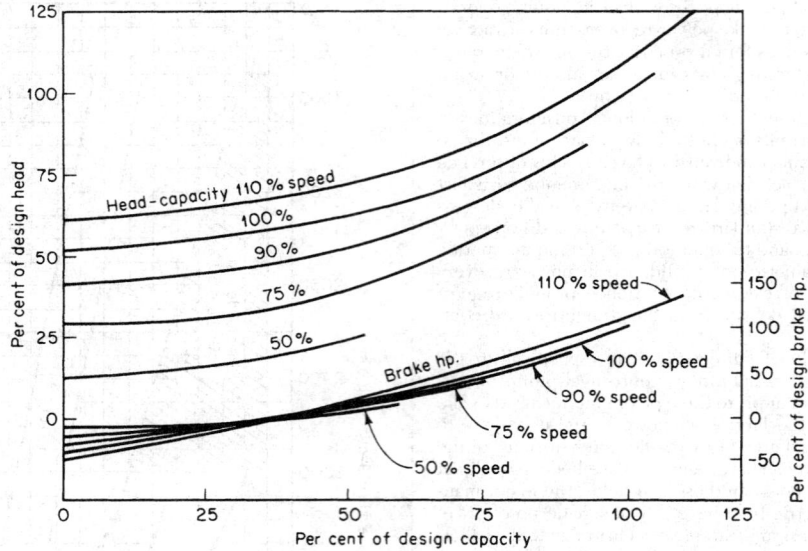

FIG. 29-52 Generalized curves showing hydraulic behavior of centrifugal pumps operating as power-recovery turbines.

from the no-flow conditions, the fluid velocity through the wheel gradually approaches such a rate that it imparts enough energy to the wheel not only to overcome internal friction but also to permit some net power output for consumption; this point usually occurs at about 30 to 40 percent of design flow or capacity. As in any turbine driver, the machine will speed up until the load imposed on the shaft coupling by the driven unit equals the power entering from the power-recovery wheel. Like a pump, a power-recovery unit will ride its characteristic curve and seek a point at which its particular head-capacity-speed-power-output relationship is satisfied. In most applications, the head available to the unit, being largely composed of a static-pressure difference, is nearly constant and varies only to the extent that inlet and exit piping-friction losses vary with flow through the unit. Thus the unit finally acts as an orifice in a relatively fixed differential system, meaning that it has a definite flow limit which also produces a torque and horsepower limit. This can be seen by following the 100 percent head curve of Fig. 29-52.

Performance Characteristics Performance of the power-recovery unit **operating as the sole driver** (Fig. 29-54) is shown in Fig. 29-55. If it is assumed that more liquid at the available head is presented to the power-recovery unit than is needed to generate the horsepower required by the pump, the turbine unit will speed up to handle the liquid, and at the same time the pump speed must go up. In speeding up, the turbine will generate more horsepower, which the pump must absorb while it is at the new speed. Finally, a balance point on horsepower is reached with the driven unit, but the number of revolutions per minute may be off design. If speed control of the driven unit is necessary, throttling some of the available capacity across a valve bypassing the unit permits the unit to satisfy its horsepower-capacity-speed relationship at the desired number of revolutions. A similar problem occurs when the capacity available to the unit is less than that needed at available head and design revolutions. The unit will slow down, try shedding load, and attempt to come to peace with its head-capacity-speed curve sets. Here speed control can be achieved by throttling the available pressure so that the unit sees only that portion of the available head needed to satisfy its head-capacity-speed relationship at the desired number of revolutions.

Performance of the power-recovery unit **operating with a makeup driver** (Fig. 29-56) is shown in Fig. 29-57; specific percentage values are shown, but the general characteristics and curve shapes are typical. It should be noted that the flow scheme, the selection of

equipment, and the design of that equipment have produced the relatively inflexible system pattern shown in the curves, in which (1) except at a single point the recovery unit always requires either flow bypassing or inlet-pressure throttling (see bottommost curve); and (2) the horsepower output of the recovery unit is reduced at any point away from design (note the horsepower-difference curve), which, combined with the characteristics of steam turbines, produces the unusual turbine throttle steam-flow curve (second from the top in Fig. 29-57).

DESIGN CONSIDERATIONS

Involved in producing the curves for Figs. 29-53 and 29-55 is a calculation of the so-called **balance point** at which the flow and revolutions per minute required by the recovery unit match those provided by the pump. If the recovery turbine is the sole driver (as for the lean pump of Fig. 29-54), both the speed and the brake horsepower of the recovery turbine and its driven pump must be the same at the so-called balance point. If there is a makeup driver and the recovery unit has available to it just the flow from the pump that it is driving, as for the pump of Fig. 29-56, then the speed and capacity must match at the balance point.

Example 2: Units for a Power Recovery System The scheme of Fig. 29-52 and the actual units supplied (Figs. 29-58 and 29-59) will be used for purposes of illustration. Since this case has the recovery unit as the sole driver, the balance point is set by speed and horsepower.

For purposes of example, assume a flow of 8.71 m³/min (2300 gal/min) through the tower. The maximum head available to the recovery turbine was calculated to be 604 m (1982 ft); this value will be slightly in error when part of the flow is bypassed since frictional losses into and out of the recovery unit will change. First, assume the lean pump to be at 3.03 m³/min (800 gal/min) running at 3900 r/min with the semilean pump at 5.68 m³/min (1500 gal/min) to get the total flow of 8.71 m³/min (2300 gal/min). At 3.03 m³/min (800 gal/min) and 3900 r/min the available head of the lean pump is read from the curve. This must be greater than the required head, and the excess is plotted as in Fig. 29-60. The brake horsepower of the lean pump is also read.

Now, at 3900 r/min and a head of 6.04 m (1982 ft), the required flow and generated brake horsepower of the recovery turbine are read. Since the horsepower of the lean pump and the recovery turbine are not identical, this entire process is repeated at another speed with the 3.03 m³/min (800 gal/min). The difference in brake horsepower between the lean pump and the recovery turbine is then plotted against the speed for these two points, and a line is drawn between

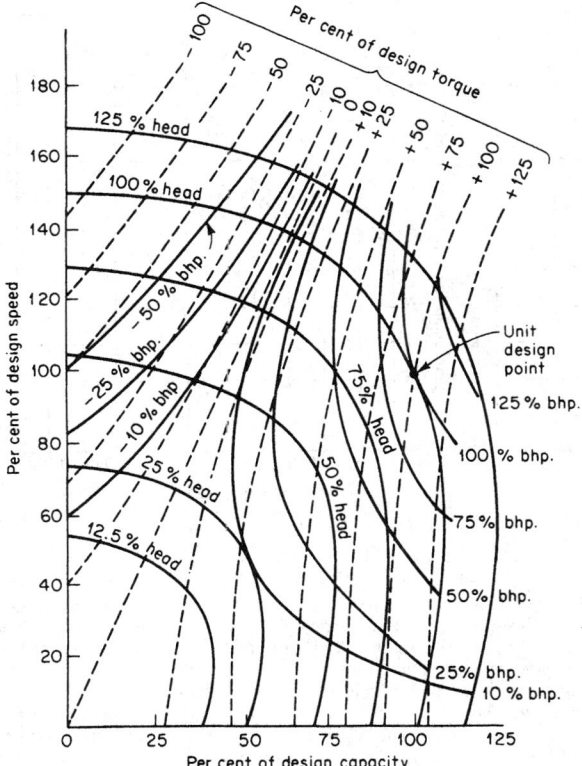

FIG. 29-53 Generalized curves for centrifugal pumps operating as power-recovery turbines.

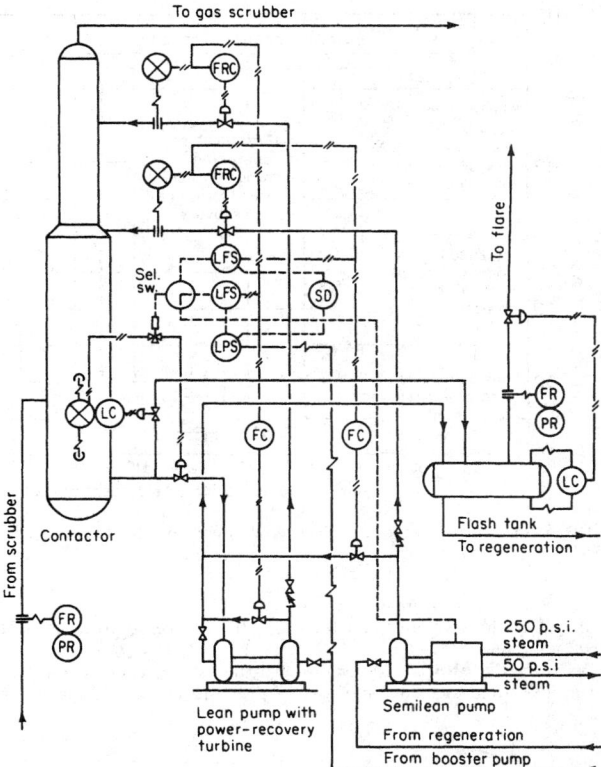

FIG. 29-54 Flow diagram of a power-recovery unit operating as the sole driver.

them. Where this line crosses the zero-difference brake-horsepower line is the balance point at 8.71 m³/min (2300 gal/min) through the tower and 3.03 m³/min (800 gal/min) through the lean pump.

The same procedure may be used at other pump flows to permit plotting the series of balance-point curves as has been done in Fig. 29-61. From such curves, one can establish the maximum lean pump at any total tower outflow, and combining this with the semilean-pump performance curve results in Fig. 29-55. Bypass flow plotted in Fig. 29-55 is obtained by adding simultaneous lean- and semilean-pump flows and subtracting the recovery pump-turbine flow required to make the balance point at that lean-pump flow.

Design Bases It is apparent that the balance point is always determined by the power r/min characteristics of the driven unit as sensed at the coupling by the shaft of the power-recovery pump-turbine. If the driven unit can simply soak up any (all) of the generated horsepower, for instance, a floating electric generator, then capacity control and pressure throttling may not be needed. When a speed-controlling variable-horsepower unit such as an electric motor or a steam turbine provides a tandem helper or a makeup driver, these units will hold revolutions per minute constant and make up just enough horsepower to permit the power-recovery pump-turbine to satisfy its head-capacity curve at virtually any flow rate.

It is the **combined unit characteristics** which must be considered, and these characteristics must be evaluated over the full operating range as well as for the startup condition. The consumption of power, up to almost 40 percent of design output, on starting up and coming up to speed and the fact that under a relatively fixed head condition the maximum speed at zero torque is about 140 percent of design have both already been noted. These, of course, bear particular significance for starting the unit up or shutting it down and must be considered. Failure to perform a complete system analysis can frequently lead to a process design that proves, upon installation, to be an operating trap.

In realizing the advantages of competitive designs for hydraulic power-recovery systems, there is usually an investment premium which must be paid for the **operating flexibility** illustrated in Fig. 29-55. This takes the form of additional reduced-capacity pumps and steam-turbine drivers, as well as some sacrifice in power recovery that results from bypassing or throttling even at the design point. If investment is to be minimized, a tightly designed system for full power recovery with a single pump-turbine helper, as in Fig. 29-54, may be worth considering. In such systems, the only fluid available to the pump-turbine is that provided by the pump it drives, and no separate pump with auxiliary driver is available for startup. Operating personnel will immediately see that such systems are relatively inflexible and difficult to operate (Fig. 29-57). Thus, while the tightly designed system has a minimum investment and more power recovery, it may not be the most desirable if operation away from the design point is anticipated, since only at that one point is neither throttling nor bypassing needed. Furthermore, it becomes apparent that the maximum steam requirements may be set by a partial-load condition rather than by design conditions or overload.

While the foregoing examples have dealt with applications on centrifugal pumps, the same sort of analysis can be made for reciprocating pumps, reciprocating compressors, or other rotary users like cooling-tower fans.

INSTALLATION FEATURES

In addition to performing the system analysis, a number of details or peculiarities of the units must be considered with respect to:

1. Vaporization, flashing, or cavitation
2. Fluid volumes
3. Process-stream controls
4. Speed control

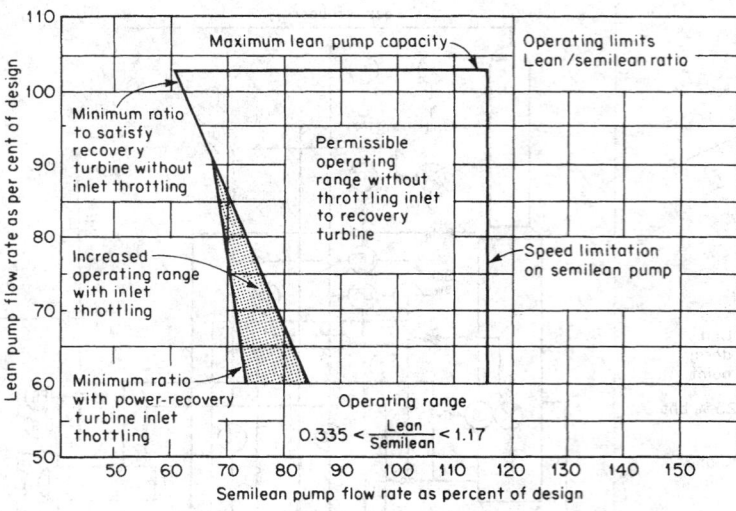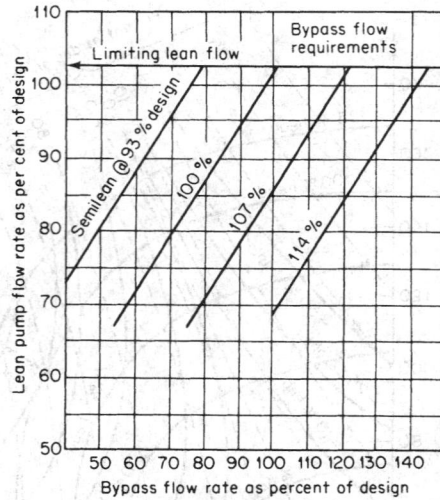

FIG. 29-55   Performance of a power-recovery unit operating as the sole driver.

5. Startup and overcapacity
6. Electrical-system characteristics if the recovered power is used to generate electricity

Vaporizing Fluids   Many pump-turbines are installed on gas-saturated liquid streams, and loss in pressure can cause problems

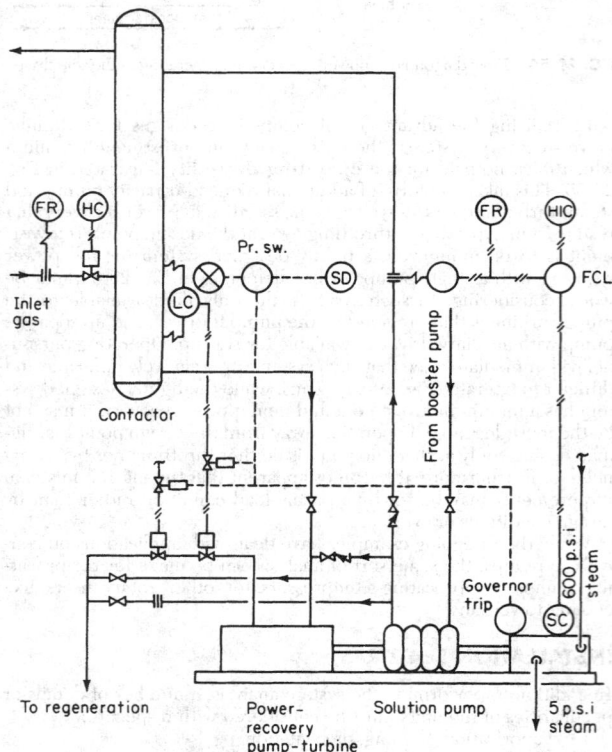

FIG. 29-56   Flow diagram of a system with a power-recovery turbine operating with a makeup driver.

whenever this occurs across balancing drums or pressure-reducing labyrinth seals. Piping that carries bleed streams from the drums and seals to the low-pressure (outlet) side of the pump-turbine must be sized generously to allow for some gas evolution.

In general, gas evolution is not evident in the pump-turbine's casing proper, for the corresponding added horsepower (as would be anticipated from the gas expansion) has never been evidenced in reports on field testing. It appears that the liquid passage through the casing is too fast for vapor-liquid equilibrium to be attained. However, slower shearing passage through a balancing drum and return line does permit gas evolution, and this line may become vapor-locked if it is undersized. Similarly, the high points of the pump-turbine casing may be vapor-locked with released gas if the unit stands idle; this can cause damage to seal chamber, balancing-drum chambers, etc., when the unit is started up again.

There is another potential hazard due to vaporization which does not generally occur in process-plant installations. The hazard results from the fact that a pump sees only the head of fluid, and if significant flashing occurs in the inlet piping, the head of fluid represented by the pounds-force-per-square-inch-gauge inlet pressure can be many times greater than design head, resulting in an attempt by the unit to increase its speed greatly.

There also appears to be a minimum outlet or impeller eye pressure below which cavitation and its attendant physical damage can occur (similar to net positive suction head, or NPSH, this could be called net positive discharge head, or NPDH). Thus it is often advisable to design by using only a part of the full pressure differential available in the process for the pump-turbine. If throttling is to be provided, outlet throttling is probably better than inlet throttling, and if used, any mechanical seals, as well as the unit casing outlet flange, must be adequate to withstand the full inlet pressure when the throttle valve is closed.

Fluid Volumes   Many process-plant installations of these units are made by handling *rich liquid* out of an absorber. In most cases both liquid volume and liquid density will change from input to output. While such changes are not normally significant, the sensitivity of the balance point to the volume of flow makes it mandatory to consider volumetric swell by absorption in checking the suitability and adequacy of the pump-turbine unit and the controls and bypassing arrangements as part of a system analysis.

Process Controls   If the inlet or outlet liquid to a pump-turbine is regulated by a level controller on the liquid-supply vessel, a falling liquid level inside the vessel will cause this controller to throttle a valve, reducing the differential pressure available to the pump-

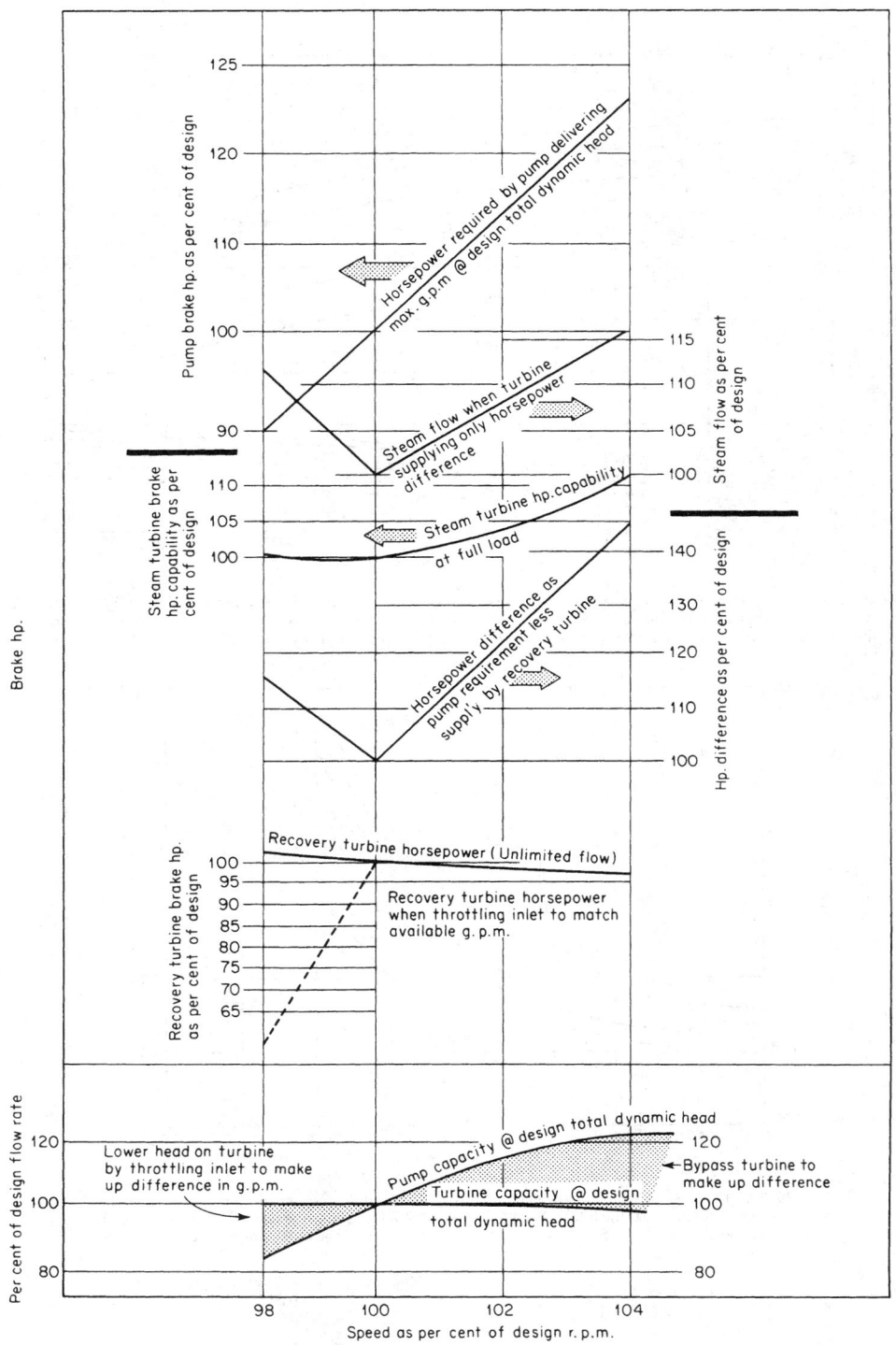

FIG. 29-57 Performance of a power-recovery turbine operating with a makeup driver.

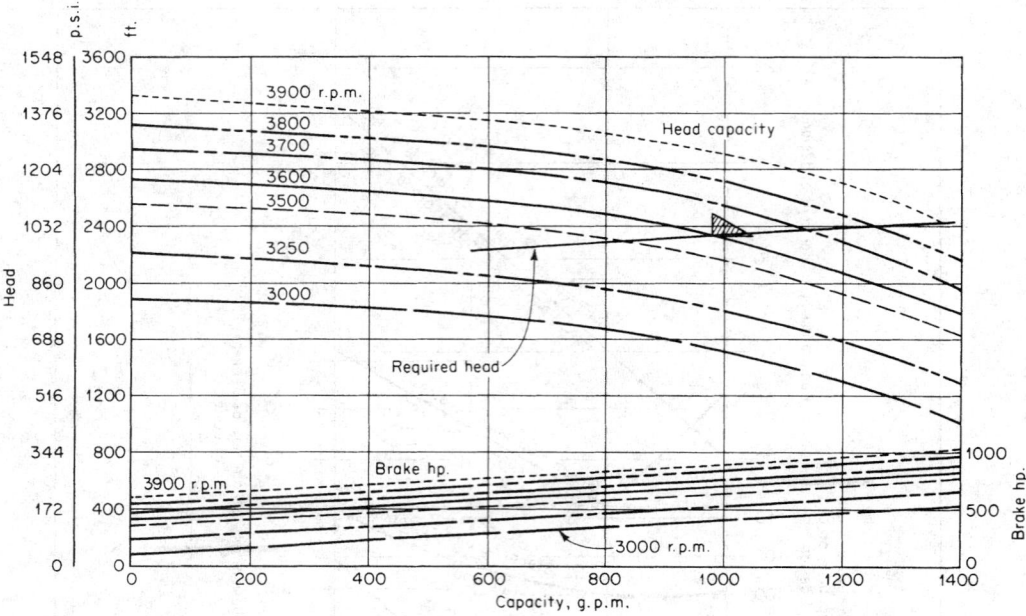

FIG. 29-58   Head-horsepower-capacity characteristics of a lean pump tandem-connected with a power-recovery turbine operating as the sole driver. To convert gallons per minute to cubic meters per minute, multiply by 3.79×10^{-3}; to convert horsepower to kilowatts, multiply by 0.746; and to convert pounds-force per square inch to megapascals, multiply by 6.89×10^{-3}.

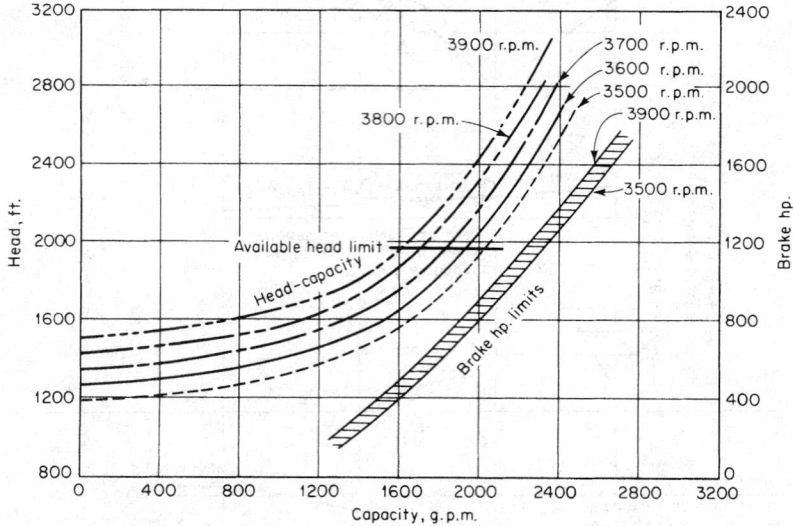

FIG. 29-59   Head-horsepower-capacity characteristics of a power-recovery turbine operating as the sole driver of a lean pump. If the total capacity of lean and semilean pumps exceeds the values indicated by "available head limit," bypass must be used. Net recovery-pump head at 8.71 m³/min (2300 gal/min) is figured as follows:

Tower	957 lbf/in²	6.598 MPa
Flash tank	−75	−0.517
Suction piping (62.4 lb/ft³)		
Friction loss	−11.8	−0.081
Elevation	+2.8	+0.019
Discharge piping (57.8 lb/ft³)		
Friction loss	−8.8	−0.061
Elevation	−5.6	−0.038

To convert gallons per minute to cubic meters per minute, multiply by 3.79×10^{-3}; to convert horsepower to kilowatts, multiply by 0.746; and to convert pounds-force per square inch to megapascals, multiply by 6.89×10^{-3}.

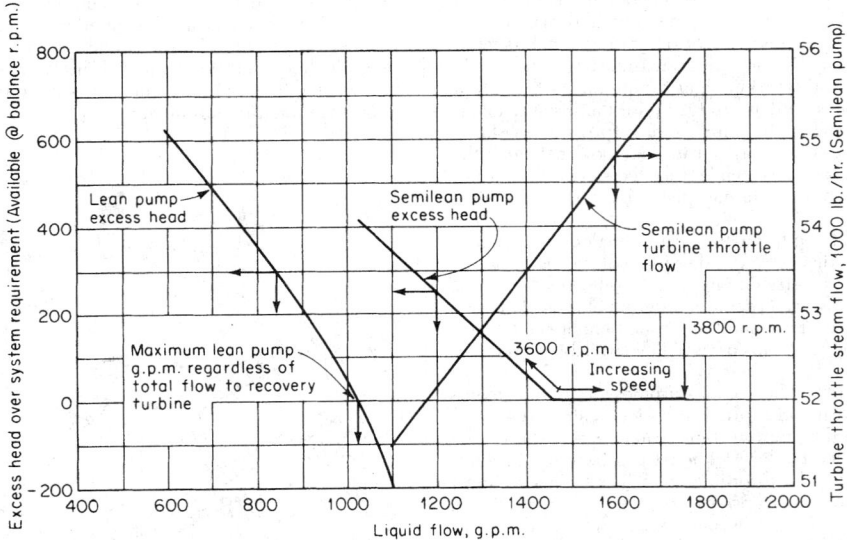

FIG. 29-60 Excess head developed by lean and semilean pumps and the steam-throttle flow for a semilean-pump turbine. To convert gallons per minute to cubic meters per minute, multiply by 0.00379; to convert pounds per hour to kilograms per second, multiply by 1.260×10^{-4}.

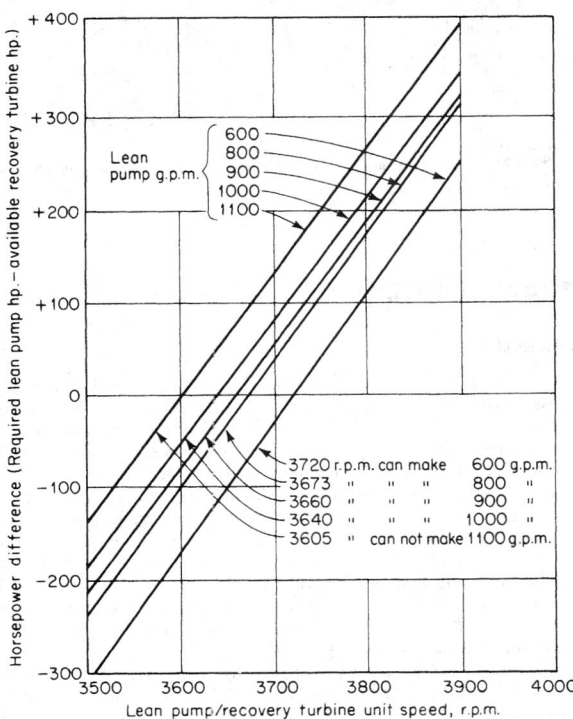

FIG. 29-61 Horsepower–r/min balance for a lean pump tandem-connected with a power-recovery turbine operating as the sole driver. Horsepower differences are calculated from excess head requirements as typically shown in Fig. 29-60. To convert gallons per minute to cubic meters per hour, multiply by 0.2271; to convert horsepower to kilowatts, multiply by 0.746.

turbine; or if the liquid level is rising, the control valve tends to open wide, so that the pump-turbine sees its full available head. Since under this latter condition the head is at its maximum, no more liquid will flow through the pump, and if the level continues to rise, the system goes off level control. One performance like this inevitably leads to a request from operating personnel for a bypass around the pump-turbine. On a startup it is frequently necessary to have a bypass from a point upstream of the driven pump's discharge check and block valves to the associated pump-turbine.

Speed   Speed control can be critical with pump-turbines. Considering the characteristics shown in Fig. 29-57 (and the curves on which it is based), a 4 percent change in speed, from 98 to 102 percent, produces a 33 percent change in pump flow, from 82 to 115 percent, and about a 22 percent horsepower change, from 118 to 100 to 122 percent. A NEMA Class A steam-turbine mechanical governor has about 10 percent steady-state regulation, so that for the 22 percent horsepower speed change one should not expect better than a 2 percent speed regulation. Since 4 percent is the total speed change being considered, it becomes apparent that a problem exists in governing the speed of the steam turbine. Since this unit is supposed to serve to hold pump speed by supplying horsepower as needed to get a match of pump and system curves at the desired flow rate, this is a serious concern. In this case the problem can be solved by eliminating the steam-turbine speed governor, leaving only the overspeed trip, and placing a control valve actuated directly by pump-flow measurement in the steam-supply line to the turbine.

In contrast to steam turbines, in which runaway overspeeding is always a problem, pump-turbines operating at design head go to zero torque at about 130 to 140 percent of design speed. Thus, overspeed protection may not be necessary if the pump-turbine can withstand 140 to 150 percent of design speed and it is the sole driver. When a steam-turbine helper is used, it should be provided with the usual overspeed trip-out mechanism.

Startup and Overcapacity   From a design standpoint and also operationally, it is important to remember that pump-turbines not only do not generate power before they attain about 40 percent of design flow but actually consume power in decreasing amounts as the flow is increased from zero to 40 percent. This means that they should be brought up to operating speed as rapidly as possible. Another solution, and a more desirable one from an operating and maintenance standpoint, is to install a free-wheeling or overriding clutch (such as that made by the Marland Division of Zurn Industries, Inc., of La Grange, Illinois) between the turbine and the pump. With such an installation, the pump does not have to turn until fluid is available to it. Also, it is not connected to the pump until it tries running faster than the pump, at which time it is putting out power. Under this arrangement the startup sequence can be selected so that the turbine unit goes from zero speed to operating speed along the zero-torque curve.

At design head, on the other hand, capacity does not change markedly with speed, so that once the design point has been passed the pump-turbine acts as a restriction in the line. Since most of these units operate on a relatively fixed pressure differential, they then tend to act like an orifice to limit flow, and little or no benefit can be realized from any overcapacity in terms of fluid flow available to the unit in the actual installation.

Electrical Generation   When pump-turbines are used to generate electricity, the units should be tied into electrically strong networks of such a size that the pump-turbines cannot swing the system frequency but are governed by that frequency and become constant-speed machines. Inlet or outlet throttling, as well as bypassing, must be used for such installations. Consequently, with speed fixed and available head essentially constant because of the process, the maximum amount of recoverable power is established by the design. Once installed, a pump-turbine cannot be pushed into generating more power if more fluid is available except by a redesign of its internals. The electrical portion of such an installation is straightforward, controls are simply inlet throttling and a bypass (if needed) for the pump-turbine, and these controls can be operated on a split range from the same level controller on the liquid-source vessel.

Integral Units   A relatively recent development is the integral or packaged pump-turbine unit assembled in a single casing, having a common discharge port and designed for mounting directly in the piping. A cutaway view of one such design is shown in Fig. 29-62. It is obvious that the use of such a unit presupposes the compatibility and desirability of having the pump discharge material contaminated by all the powering fluid which passed through the driving turbine impeller. So far, this unit has seen little service in process plants, but its simplicity, its independence of normal power sources, and its ease of

installation make it an attractive candidate for consideration when the process-flow-scheme can be adapted to its characteristics. It has been proposed for use in place of liquid eductors or ejectors, for it has a much higher efficiency than such units, and furthermore throttling motive fluid flow will cause the pumped fluid flow to follow according to balance-point requirements.

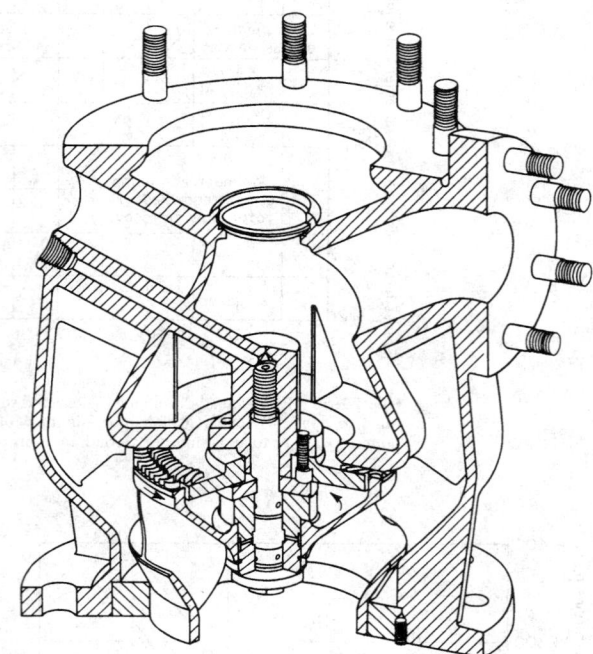

FIG. 29-62   Integral or packaged pump-turbine unit. (*Worthington Division, Dresser-Rand Co.*)

MECHANICAL POWER TRANSMISSION

Process machinery usually involves a driver and a driven machine, and these machines are connected by power-transmission equipment. Such equipment can either be mechanical or hydraulic.

For instance, power can be transmitted from one machine to the other through shafts, flexible couplings, and gear reducers (mechanical equipment). Power can be transmitted through a torque converter (hydraulic equipment) or by a combination of mechanical and hydraulic equipment.

Although power-transmission equipment is generally simpler than the machines it connects, it nevertheless fails in service as often as—if not more often than—motors, turbines, pumps, and the like. Care must be given to the proper selection and maintenance of power-transmission equipment; otherwise, the best process machines cannot perform as expected.

This segment will discuss the more commonly used power-transmission equipment; also discussed will be the machinery bearings, which not only support the rotors of machinery, but the power-transmission equipment as well.

The segment is divided into the following parts:

- Bearings
- Power transmission without speed change
- Power transmission with change in speed
- Lubrication of power transmission equipment

BEARINGS

Rotating shafts must be supported by the machine housing, not only against gravity, but also against a variety of forces that are imposed on rotors inside machines (including axial thrust). Bearings have a significant influence on the performance of process machines, both because they limit their continuous operation and also since they influence the level of vibration and critical speeds of the machines.

It was generally accepted that "machinery users do not select bearings" (see the sixth edition of this handbook); this is no longer valid: users have realized that they can have a voice in the selection of bearings of new machines and that they can select better bearings for their old machines than the ones supplied by the original equipment manufacturer (OEM). A number of independent bearing manufacturers can be found in most industrial areas.

Until quite recently, all types of bearings required lubrication; the advent of magnetic bearings has eliminated the need for lubrication. Lubricated bearings have a significant power loss, as oils or greases are sheared by the relative motions between shafts and housings; magnetic bearings eliminate the losses in lubricants, but they require electrical power to maintain the shaft journals in the desired position.

Three types of bearings will be discussed: oil-film bearings, rolling-element bearings (also known as antifriction bearings), and magnetic bearings.

Oil-Film Bearings The name *oil-film bearing* derives from the fact that, in such bearings, shafts are supported by a film of oil under significant pressure. This pressure is generated by the rotation of the shaft and by the fact that the clearance between the shaft and its bearing has a wedge shape. A typical radial oil-film bearing is illustrated in Fig. 29-63. The bearing has a diametrically split shell, having a layer of low-friction material (babbitt) on the inside surface. The outside diameter fits tightly in the machine housing, and an antirotation pin (or protrusion) is provided. This pin also serves the purpose of axially locating the bearing. Between the inside surface of the bearing and the shaft journal, there is a clearance filled with oil under a pressure of about 20 psi (1.4 bar, 138 kilopascals).

This clearance is very important for bearing and machine performance. A too-tight clearance will allow very little oil to flow through the bearing, which will operate hot; a too-loose clearance will prevent the formation of a high-pressure film of oil, and the bearing will fail through mechanical contact. It is customary to use the following clearances:

- For low to moderate speeds, $0.001 \times d + 0.002$ (inches)
  $0.001 \times d + 0.05$ (mm)
- For high speeds, $0.002 \times d$

where d is the journal diameter.

The clearance in oil-film bearings is necessary for the formation of a film of oil; however, it is detrimental because it allows shafts to vibrate within. Machinery vibrations tend to appear with wear and with the unbalance created by blade fouling. In many cases, machines can be made less sensitive to vibrations by changes in bearing design; there are types of oil-film bearings that resist shaft motions without decreasing the internal clearances. To understand how such bearings work, an explanation of *preload* is in order.

The normal operating position of a shaft inside a bearing is shown in Fig. 29-64. It can be seen that, due to radial forces, the geometric center of the shaft does not coincide with the one of the bearing. This displacement creates a "wedge," which combined with the shaft motion, forces the oil into a continuously decreasing area, and a

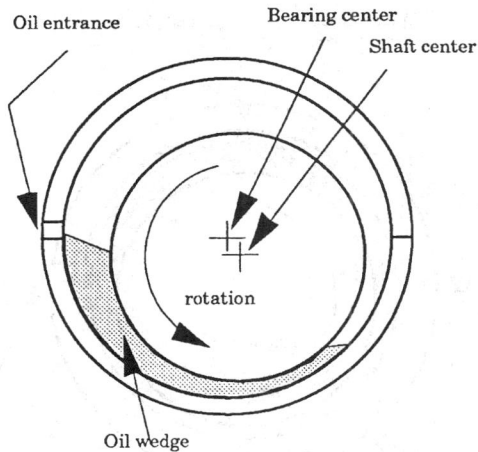

FIG. 29-64   Operation of oil-film radial bearings.

hydraulic pressure is generated. This pressure acts on the journal and supports it against gravity or other internal forces. To stabilize a vibrating shaft, bearings with larger oil pressures are used. A larger pressure is created whenever the shaft moves closer to the bearing (steeper wedge), which is not desirable. Therefore, an "apparent" proximity is created by altering the geometry of the bearing.

A lemon-bore bearing configuration is shown in Figure 29-65. It can be seen that a steep wedge was generated by *apparently* making the bearing diameter larger; this is similar to a normal bearing with a larger load; therefore, these types of bearings are known as preloaded bearings. *Preload* in this case does not refer to forces but rather to the resulting geometry.

Another way to stabilize a shaft inside a bearing is to use "pressure-dam" designs, as shown in Fig. 29-66. The oil supply pressure is directed to the top of the shaft; it forces the shaft downwards, and it stabilizes its motions.

The most modern oil-film bearings use movable pads, which tilt in order to create, automatically, the optimum wedge shape. They are known as the "tilting pad" bearings and are illustrated in Fig. 29-67.

Oil-film bearings are also used for positioning shafts axially, against thrusts created by the flow of the processed fluid through machines.

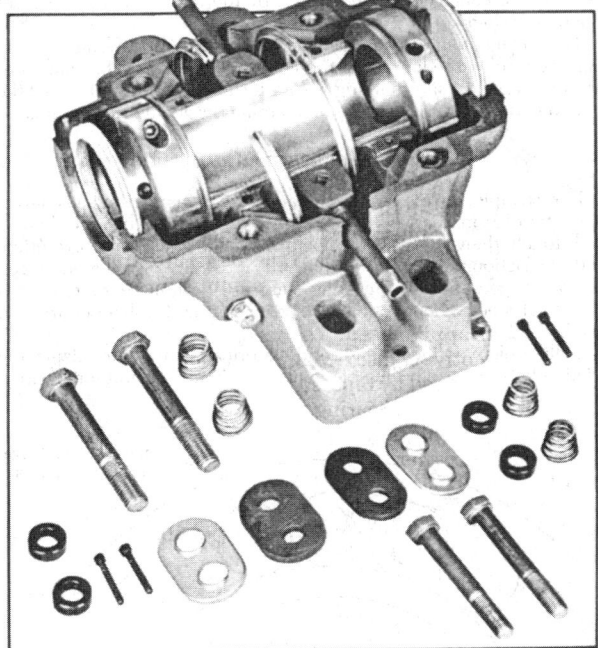

FIG. 29-63   Typical oil-film pillow-block bearing. Lubrication is provided both by an oil flow and by oil rings.

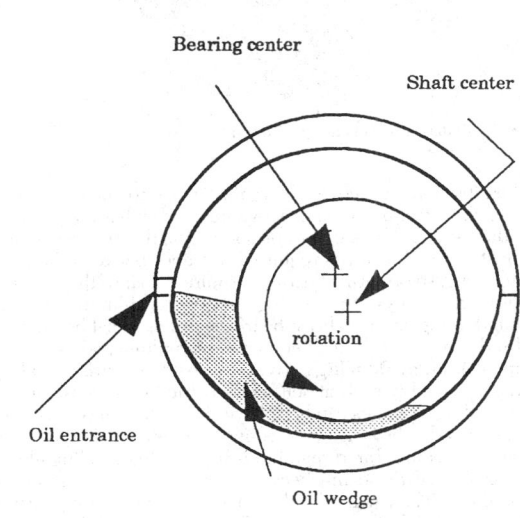

FIG. 29-65   Oil-film bearing with geometrically-created preload.

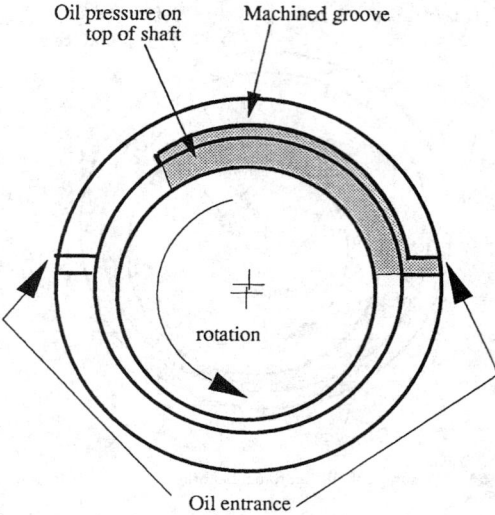

FIG. 29-66   Oil-film bearing with oil pressure created preload (pressure dam design).

FIG. 29-67   Radial oil-film bearing with tilting pads.

Thrust bearings have an active side, against which the bearing collar or disc is forced; they also have an inactive side, which has the purpose of limiting the total axial float of a rotor, should the thrust reverse for any reason. Just as radial bearings require a clearance between the journal and the bearing, so do thrust bearings require a float of the thrust disc between the two sides of the bearing. A typical oil-film thrust bearing is illustrated in Fig. 29-68. Thrust bearings, just as radial bearings, can have a fixed geometry or can be equipped with tilting pads.

Rolling-Element Bearings   With this type of bearing, shafts are supported by small parts that roll with no friction, at least theoretically. The rolling elements can be spherical (ball bearings), cylindrical (roller and needle bearings), or conical (tapered roller bearings). Although more complicated than fluid-film bearings, rolling-element bearings are less costly, mainly because they are manufactured in very large quantities. Their main disadvantage is a limitation in maximum operating speed. The main types of rolling-element bearings are shown in Fig. 29-69.

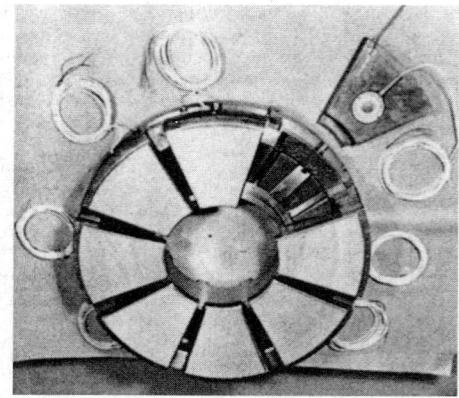

FIG. 29-68   Axial (thrust) oil-film bearing with tilting pads and embedded temperature sensors.

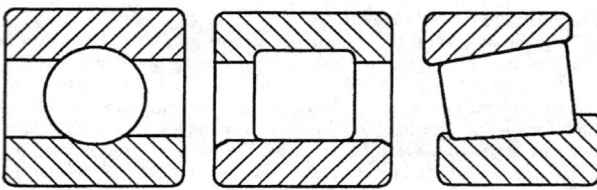

FIG. 29-69   Three types of rolling-element bearings: ball, cylindrical roller, and tapered roller.

Generally, the rolling elements ride on an internal race, installed on the shaft, and an external race, installed in the machine housing. The races and the rolling elements are made of hardened alloy steels, because the contact pressure caused by the radial (or thrust) forces can be very high.

The rolling elements are installed in a cage that performs the very important role of reducing the friction inside bearings. The conditions without a cage are shown in Fig. 29-70. It can be demonstrated that the rotational speed of a rolling element around its geometric center is:

$$\text{rpm}_1 = \frac{\text{rpm}_2}{2} \times \frac{R}{r}$$

For example, if the inner race diameter is 4 in and the balls have a diameter of ⅜ in, the balls will rotate 5.3 times faster than the shaft. If balls touch, then the relative velocity occurs at twice their revolution, and the friction so generated will overheat the bearing. Depending on the rated speed of a bearing, cages are made (in increasing order of speed) of steel, phenolics, or bronze. Brass and polymers are also used.

Rolling-element bearings require lubrication for minimizing the friction between the rolling elements and their cage and for dissipat-

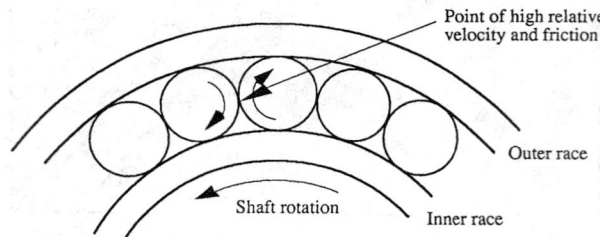

FIG. 29-70   Cageless roller bearings would operate hot and wear rapidly.

ing the heat generated by friction. Depending on the operating speed, lubrication can be provided by greases, liquid oil, or oil mist.

The selection of rolling-element bearings is based on the radial and axial loads they must support, on the operating speed, and on the expected life. All bearing catalogs provide comprehensive selection data.

Rolling-element bearings are provided with a small radial clearance that almost disappears when they are pressed on the shaft or in the housing. Basically, a rolling element bearing is pressed on its shaft and has a small clearance in the housing when the shaft rotates and the housing is stationary; the reverse is true when the housing rotates, as is the case for idler pulleys.

Magnetic Bearings   Magnetic bearings replace the hydrodynamic shaft support by a magnetic field. Their advantages and disadvantages are summarized in Fig. 29-71. While oil-film bearings create the necessary power to support the shaft from the movement of the shaft, magnetic bearings need an outside power supply. A magnetic bearing system consists of four major components: magnetic actuators, power supply, shaft-positioning sensors, and electronic controls. Magnetic forces act all around the shaft; sensors detect the relative position between the shaft and the housing and send the signals to the controller, which supplies more or less electrical power to the electromagnets so that the shaft remains centered independent of the forces acting on it. The principle of operation of a magnetic bearing is shown in Fig. 29-72.

Because magnetic bearings do not require lubrication, they are particularly suited to applications such as canned pumps, vacuum pumps, turbo-expanders, and some centrifuges. As of 1994, some centrifugal compressors in the 5000-kW power range had been equipped with magnetic bearings.

The rotordynamic study of a machine with magnetic bearings is quite different from the one of either oil-film or rolling-element bearings. The stiffness and damping properties of magnetic bearings can

Advantages	Disadvantages
Low power usage	Small load capacity
Very long life	Large size
No Oil required	Higher investment
Low *system* weight	Requires mechanical backup
Reduced fire hazard	for power interruptions
Vibration sensors built-in	New technology
	Possible rotor-dynamic problems

FIG. 29-71   Advantages and disadvantages of magnetic bearings.

be adjusted (within limits) by altering the feedback and response time between the sensors and the electromagnets.

POWER TRANSMISSION WITHOUT SPEED CHANGE

Whenever the process machine operates at the same speed as its driver, the two can be directly coupled. This direct coupling still allows for a variable speed, through adjustments of the speed of the driver. Steam turbine speed can be easily adjusted, and electric motor speed can also be varied by the use of special drives that vary the *frequency* of the power applied to the motor. Whether the speed is fixed or variable, direct coupling of two machine shafts presents the problem of accommodation of misalignment. To this purpose, machines are coupled through a *flexible coupling*.

Variable-Speed Electric Motor Drives   Whether of the induction or synchronous type, ac motor speed is a direct function of the supply voltage frequency. The advent of high-power solid-state controllers made possible the manufacture of frequency converters. These converters can generate high power (over 14,000 kW) at a variable frequency, which can be lower or higher than the standard 60 Hz (cycles/second). Frequency converters made possible not only variable-speed motors but also high-speed (up to 10,000 rpm) electric motors. The cost of these converters is often higher than the cost of a gear unit, but the elimination of gear units makes machines simpler and more reliable.

Although motors and controllers can be bought separately, the trend is to purchase a system from a given manufacturer. There are six types of variable-speed drives, as shown in Fig. 29-73. Dynamic response indicates the ability of the drive to respond to a change in command; it is measured in radians/second; the higher the number, the faster the drive response.

Variable-Speed Mechanical Drives   Although rather common until the late 1960s, mechanical variable-speed drives, Figs. 29-74 and 29-75, are seldom used today. Their relative complexity makes these drives more maintenance-prone than the newer variable-frequency electric motor drive systems. However, variable-speed fluid drives (Fig. 29-76) are finding widespread use in major machinery drives demanding precise speed control in applications ranging to 50 MW and even higher. These variable-speed turbo couplings can be combined with one or more gear stages in a common housing. The bottom part of this compact unit forms an oil sump. From the basic concept consisting of a speed-increasing gear followed by a variable-speed turbo coupling, other models have been derived to provide stepless speed control for both high-power, high-speed machines such as boiler feed pumps and compressors; and low-speed machines with a speed-reducing gear such as coal mills, ID fans, and crude oil pumps.

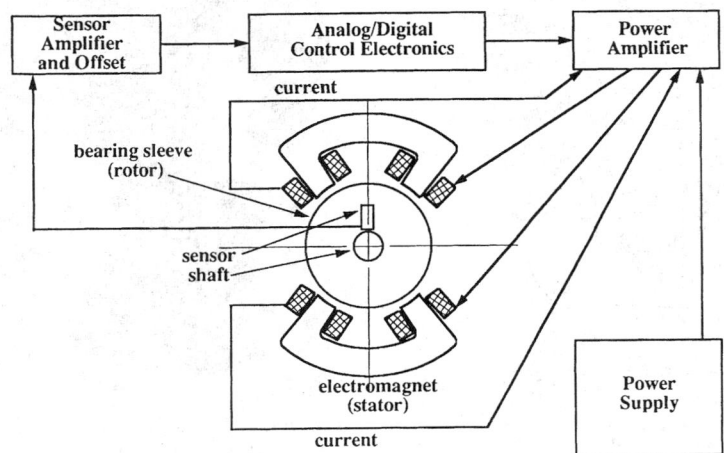

FIG. 29-72   Principle of operation of magnetic bearings.

Type of Drive	Speed Range	Starting Torque	Maximum Speed	Dynamic Response
Thyristor dc	100:1	150%	3000	15
Brushless dc	100:1	150%	3000	15
Std. Inverter	10:1	100%	6000	5
Vector Inverter	100:1	150%	6000	50
Servo	2000:1	200%	6000	500
Switched Reluctance	100:1	150%	10,000	50

FIG. 29-73 Main characteristics of typical electric adjustable-speed drives. (*Source: Reliance Electric Co.*)

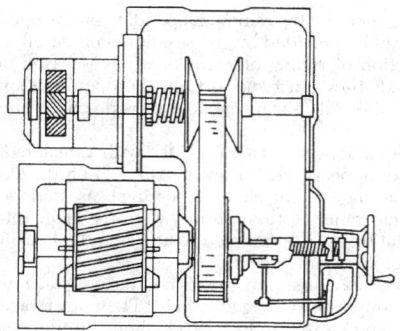

FIG. 29-75 Combined variable-speed and motor drive. (*Reeves Pulley Co.; from Kent,* Mechanical Engineers' Handbook, *12th ed., Wiley, New York, 1961.*)

The power developed by the prime mover is converted into kinetic energy in the impeller (primary wheel) of the turbo coupling and converted back into mechanical energy in the turbine wheel (secondary wheel), which is connected to the driven machine. As there is no metal-to-metal contact between primary and secondary wheels, there is no wear. Hydraulic oils with additives are used for power transmission. The amount of oil in the coupling can be varied during operation using the scoop tube. This in turn regulates the power-transmitting capability of the coupling and provides stepless speed control dependent on the load of the driven machine.

The coupling has a regulating range of 4:1 to 5:1 for driven machines with increasing parabolic torque load characteristics such as centrifugal pumps and fans. For machines with approximately constant torque load characteristics, the regulating range is 3:1. With centrifugal machines, this method of speed regulation is much more efficient than throttling the machine output, giving considerable power savings. The motor is started under no-load conditions with the coupling drained. When running the load, the motor can be controlled by the coupling. Moreover, by draining the fluid coupling, the prime mover can be disconnected from the driven machine while the prime mover is still running.

A mechanically driven oil pump on the primary side of the coupling pumps oil from the reservoir underneath the coupling through a con-

trol valve into the working chamber of the turbo coupling. The level of the oil in the working chamber and therefore the power that the turbo coupling can transmit depends on the radial position of the adjustable sliding scoop tube. The scoop tube can pick up more oil than the pump can deliver. The oil picked up by the scoop tube passes through an oil cooler/heat exchanger and control valve back to the working chamber and/or the oil reservoir. The heat exchanger dissipates the heat originating from the slip of the turbo coupling. The scoop tube actuator can be operated either electrically, hydraulically, or pneumatically.

Synchro-Self-Shifting ("SSS") Clutches   The automatic free-wheel action of an SSS clutch simplifies plant startup and shutdown sequences. It is, of course, receiving its power input from a gas turbine or similar driver, while its output is connected to a driven machine. The drive disengages automatically when the driven machine speed exceeds that of the driver. Since large gas turbines are usually brought up to speed with either a startup gear motor or a helper turbine, thousands of SSS clutches (Fig. 29-77) are finding application here.

Upon unit startup, the turning gear motor can continue to rotate

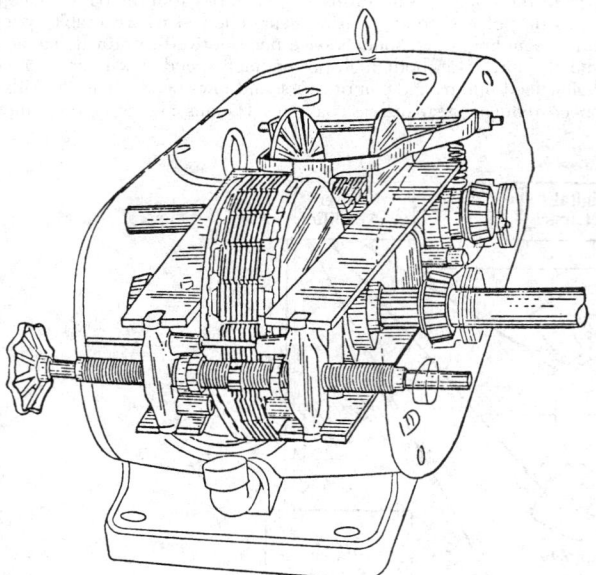

FIG. 29-74 PIV speed changer. (*From Kent,* Mechanical Engineers' Handbook, *12th ed., Wiley, New York, 1961.*)

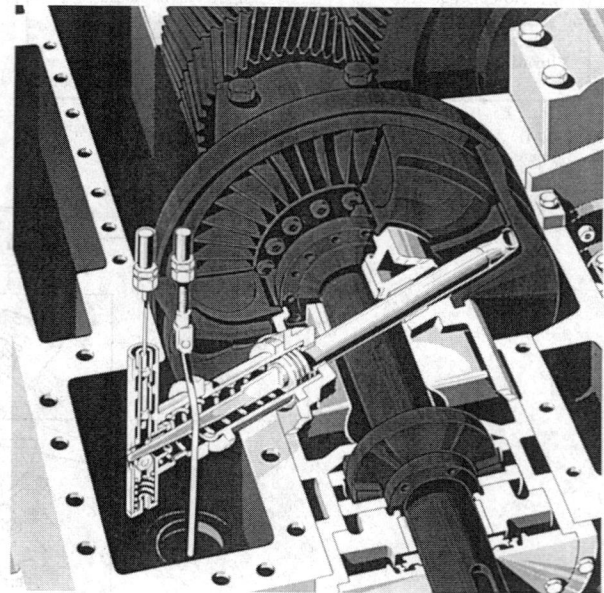

FIG. 29-76 Variable-speed turbo coupling (hydrodynamic fluid coupling). (*Courtesy of Voith Transmissions, Inc., York, Pennsylvania and Heidenheim, Germany.*)

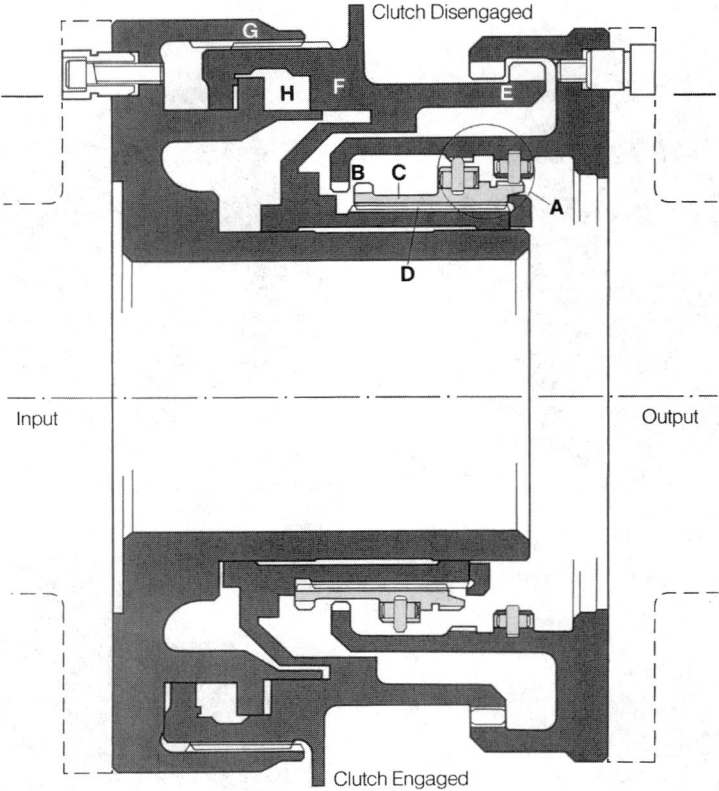

FIG. 29-77 Synchro-self-shifting clutch for large turbogenerator applications. (*Courtesy SSS Clutch Company, Inc., New Castle, Delaware.*)

or be stopped at any time. Similarly, it can be started at any time during the shutdown sequence, but the drive will only engage at turning gear speed.

Considerably larger units, with power ratings in the 300-MW range, are often fitted between a gas turbine driver and a utility power generator. This mode of application allows the generator to be used for voltage control/synchronous condensing with the gas turbine at rest but readily "on call" for peak load generation duty.

The sequence of clutch-engaging action is depicted in Fig. 29-77. Primary or secondary pawls (A) initiate engagement at clutch synchronism by aligning and engaging the relay clutch teeth (B). As the lightweight relay clutch (C) slides along its helical splines (D), the pawls are unloaded.

Primary pawls operate at low generator speeds. Secondary pawls operate at high turbine speeds. Therefore, all pawls are inert during synchronous condensing and when the clutch is engaged during power generation.

The relay clutch teeth (B) align and initiate engagement of the main clutch teeth (E). As the main clutch (F) slides along its helical splines (G), the relay clutch is unloaded.

When the main clutch teeth (E) are fully engaged, power is transmitted from the turbine to the generator. Engagement and disengagement of the main clutch is cushioned by the oil dashpot (H).

Flexible Couplings Whenever two machine shafts in substantial alignment are directly coupled, a flexible coupling is used to transmit the torque and accommodate the inevitable misalignment. A number of reasons make misalignment inevitable: thermal growth of machine housings and shafts, piping strain, foundation settle-

ment, and so on. Without the ability to accommodate misalignment, machine shafts would fatigue and fail. Couplings accommodate misalignment either through sliding of one component over another or through flexing one or more of their components. As sliding requires lubrication, it is customary to categorize couplings as either lubricated or dry.

The most popular lubricated couplings in process machinery are the gear type (Fig. 29-78) and the grid coupling (Fig. 29-79). Lubricated couplings have the advantage of small size and weight but the disadvantage that machines must be stopped for the couplings to be relubricated (most machinery can be relubricated while running). Gear-type couplings also have the unique advantage that they can accommodate *any* axial shaft motions that machines require.

Dry couplings are divided into metallic elements and elastomer elements. For a given torque, metallic elements (disks and diaphragms) are more compact and lighter than elastomer element couplings but are less flexible; therefore, they impose larger forces on the bearings. On the other hand, elastomer elements become quickly distorted by centrifugal forces; therefore, they cannot usually operate safely at speeds larger than standard motor speed. These couplings are shown in Figs. 29-80 and 29-81.

It is important for coupling users to understand that *all flexible couplings resist being misaligned*. Hence, good alignment reduces the forces on machine bearings and increases machine reliability. Couplings designed to accommodate *large misalignments* usually can do this to the detriment of other features, such as reduced torque transmission or increased bearing forces.

It can be said that the trend in the 1980s/1990s was the gradual

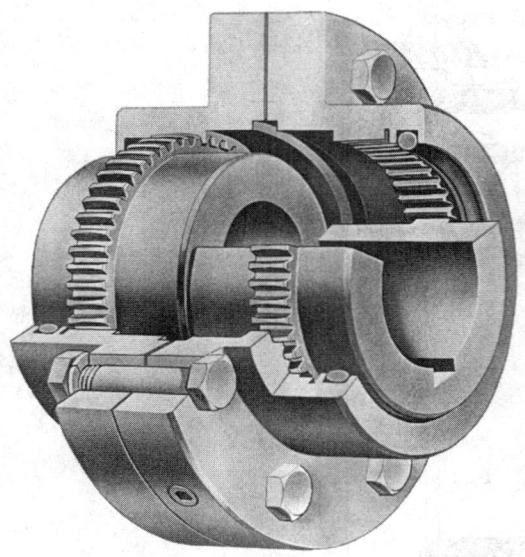

FIG. 29-78   Typical general-purpose gear-type coupling. (*Source: Kop-Flex Inc.*)

FIG. 29-79   Typical general-purpose steel-grid coupling. (*Source: The Falk Corp.*)

FIG. 29-80   Typical special-purpose disk-pack coupling. (*Source: Rexnord/ Thomas Coupling Div.*)

FIG. 29-81   Typical general-purpose elastomer coupling. (*Source: Rexnord Corp.*)

replacement of lubricated couplings with dry ones, particularly in machines designed for high power and high speeds. While there still are many machines equipped with lubricated couplings, new equipment is ordered with nonlubricated couplings.

POWER TRANSMISSION WITH SPEED CHANGE

Many process machines operate at speeds different from the one of their drivers. Typical of cases where the machine rotates *slower* than the driver are reciprocating compressors; typical examples of machines rotating *faster* than the drivers are centrifugal compressors driven by electric motors. In either case, *gears* are used to match the two speeds. Gears can also be designed to accommodate shafts that

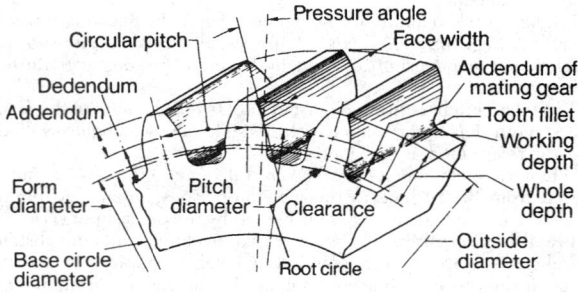

FIG. 29-82   Basic nomenclature of a gear.

are not parallel; gears can accommodate any angles between two shafts, including 90°. The following discussion is limited to gears for parallel shafts, which is the most common configuration in process machinery.

Basically, two engaging gears are two cylinders of different diameter with teeth machined on their periphery (Fig. 29-82). A gear is defined by its basic diameter (pitch diameter), its width, the number of teeth, and the angle that the teeth make in respect to the axis.

The ratio between the input-to-output shaft speeds is determined either by the ratio of the two pitch diameters or the ratio between the number of teeth:

$$i = \frac{PD_1}{PD_2}, \quad \text{or} \quad i = \frac{n_1}{n_2}, \quad \text{therefore} \quad \frac{PD_1}{PD_2} = \frac{n_1}{n_2}, \quad \text{or} \quad \frac{PD_1}{n_1} = \frac{PD_2}{n_2}$$

The ratio between the pitch diameter and the number of teeth is called the diametral pitch, and for two gears to mesh, they must have the same diametral pitch. The diametral pitch is standardized and is always an integer. For gears used in process machinery, the diametral pitch varies between 6 (coarse) and 20 (fine).

Teeth have an *involute* profile, a curve that is generated by the rolling of a straight line over a circle, as shown in Fig. 29-83. Although many curves could be used for gear teeth, the involute is the preferred one because it allows the correct engagement, even when the distance between the centers of the gears is not accurately held.

A tooth rolls over the meshing tooth only for a very small distance at the pitch diameter; above and below the pitch diameter, there is tangential sliding between the teeth, and lubrication is necessary to prevent premature wear. Because of this sliding, gears also have a certain power loss, which is about 2–3 percent per gear mesh. The heat generated by this power loss is dissipated by the lubricating oil. In high-speed gears, the oil jet is directed to where the teeth leave the engagement, as there is the point of maximum temperature.

The three most common types of cylindrical gears are shown in Fig. 29-84. *Spur* gears are the simplest, but they have the disadvan-

tage of rough operation because as few as one or two teeth are in contact at any time. *Helical* gears have a smoother operation, but the angle of the helix generates an axial thrust on the shafts, which requires special bearings. The *double-helical* (also known as *herringbone*) gears have all the advantages of the helical gears without the axial thrust. They are also the most complicated to manufacture and are very sensitive to axial thrusts imposed from the outside on the gear shafts.

LUBRICATION OF POWER TRANSMISSION EQUIPMENT

Most machine bearings, all gears, and some couplings require lubrication. Even though a system can be designed with a direct drive, dry couplings, and magnetic bearings, such systems are still rare, and lubrication systems are ubiquitous. Lubrication systems can be as simple as a Zerk grease fitting at a motor bearing or as complicated as a console that includes a large oil tank, four or more electric-driven oil pumps, multiple filters, and oil coolers.

Grease as a Lubricant Greases are mechanical mixtures of lubricating oils and thickeners. Traditionally, metallic soaps (lithium, sodium, aluminum) were used for thickeners; although still in wide use, they are gradually being replaced with synthetic materials such as polyethylene. Greases blended with synthetic thickeners tend to be more stable but have lower resistance to high temperatures.

Greases are used as lubricants strictly because they are easily confined in a housing; greases have no additional *lubricating* properties over the oils used to blend them. Good greases are blended with special additives (just as good oils are) that can enhance their wear protection and rust protection characteristics. Greases have a paste consistency and can be very soft or very hard. The National Lubricating Grease Institute (NLGI) has established a scale for the consistency of greases that ranges from #00 (the softest) to #6. Technically, the consistency of a grease is determined by measuring the penetration of a weighted cone into the surface of a grease; a larger penetration indicates a softer grease. Unfortunately, the penetration value gives no indication about the lubricating properties of a grease; two greases can have the same consistency, but one may be blended with a high-viscosity, highly refined oil and have only 6 percent thickener, while the other may be blended with a low-viscosity, low-quality oil but have as much as 20 percent thickeners. Manufacturers may only label a grease container as "#2 lithium grease," which is not sufficient information. A complete specification sheet, however, is provided upon request. Larger manufacturers also publish a *digest*, in booklet form, that contains technical data on all their products.

There are hundreds of types of greases available, and machinery manufacturers usually provide a guideline for the greases to be used on their equipment. There are also highly specialized greases, such as the ones that can be used in food processing machinery, or the ones used in nuclear power stations.

To explain why greases cannot be used indiscriminately, the grease requirements for ball bearings and flexible couplings will be compared. Ball bearings require greases that channel, bleed, and contain a low-viscosity oil. Couplings require greases that do not channel, do not bleed, and contain a high-viscosity oil.

Channeling is required for ball bearings so that the balls can roll freely inside the bearing. Without a channel, there will be a high resistance to ball movement, and bearings would operate hot. Bleeding (slow release) of oil is required so that the balls are continuously supplied with the needed lubricant. A high-viscosity oil in a ball bearing would cause the balls to skid (hydroplane) and suffer severe scoring.

Couplings need greases that are very soft (NLGI #0 or #1) so that the spaces around the teeth are always filled with grease. Channeling would allow for metal-to-metal contact and rapid wear would occur. Because coupling rotation subjects greases to high centrifugal acceleration (that can exceed 10,000 Gs), bleeding would rapidly separate the soaps from the grease, and the oils would quickly escape, causing couplings to operate dry. Coupling teeth do not roll on each other; the

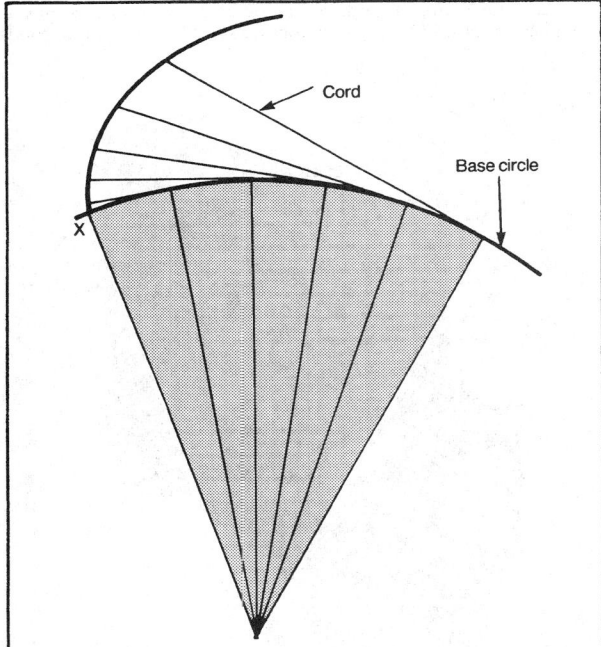

FIG. 29-83 The involute curve and its generation.

FIG. 29-84 The spur, helical, and double-helical gears.

motions in couplings are reciprocating and sliding. These motions are best lubricated by a highly viscous oil.

As was shown, using a coupling grease in a bearing could cause its failure; using a bearing grease in a coupling could cause its rapid wear.

Oil as a Lubricant Oils are used for lubrication and heat dissipation; therefore, oils are circulated through a machine (through bearings, seals, or gears), returned to a tank where dirt and water are allowed to settle out, and then pumped through coolers and filters back into the machine. Without a continuous flow of oil, machines could rapidly fail; this is why backup pumps are always provided. In systems where the main oil pump is directly driven by the machine (such as in gear boxes), two backup pumps are provided: one driven by an ac motor, which is the same size as the main one, and one driven by a dc motor (battery operated), which is small, since machines are always shut down if the main and the spare pumps fail. The dc-driven pump must only supply oil during the coast-down operation.

Filters separate and retain the dirt from the oil. They are rated based on the size of the largest particle that can pass through. Most machines in petrochemical plants use 10-micron filters. Alarms are provided to signal when a filter becomes clogged up with dirt. A valving system allows the oil to be bypassed to an alternate filter, while the element of the first one is replaced. Filter elements can be

made of a wire mesh, a wire loop (closely spaced), or pleated fiber mats. The wire-loop filters are self-cleaning, as the cylindrical element slowly rotates against a wire comb; however, they cannot filter fine particles.

Oil coolers either dissipate the heat into a water stream or the air. Water coolers (Fig. 29-85) are significantly more compact, but a supply of water is required. Air coolers (Fig. 29-86) are large and require a fan that increases the air flow over the cooling fins.

A complete lubrication console schematic is shown in Fig. 29-87. Instrumentation for the console includes thermometers, pressure switches, and flow meters.

Oils used in process machinery lubrication are generally of low viscosity and are known as turbine oils or compressor oils. Such oils would have a viscosity of 78 centiStokes at 40° C (40 Saybolt Seconds

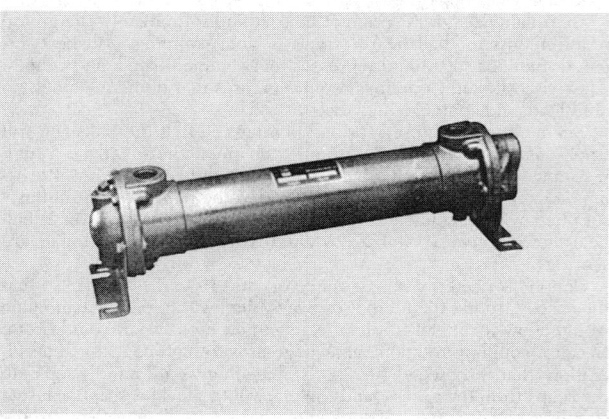

FIG. 29-85 Typical oil cooler, where the heat is transferred to water.

FIG. 29-86 Typical oil cooler, where the heat is transferred to air.

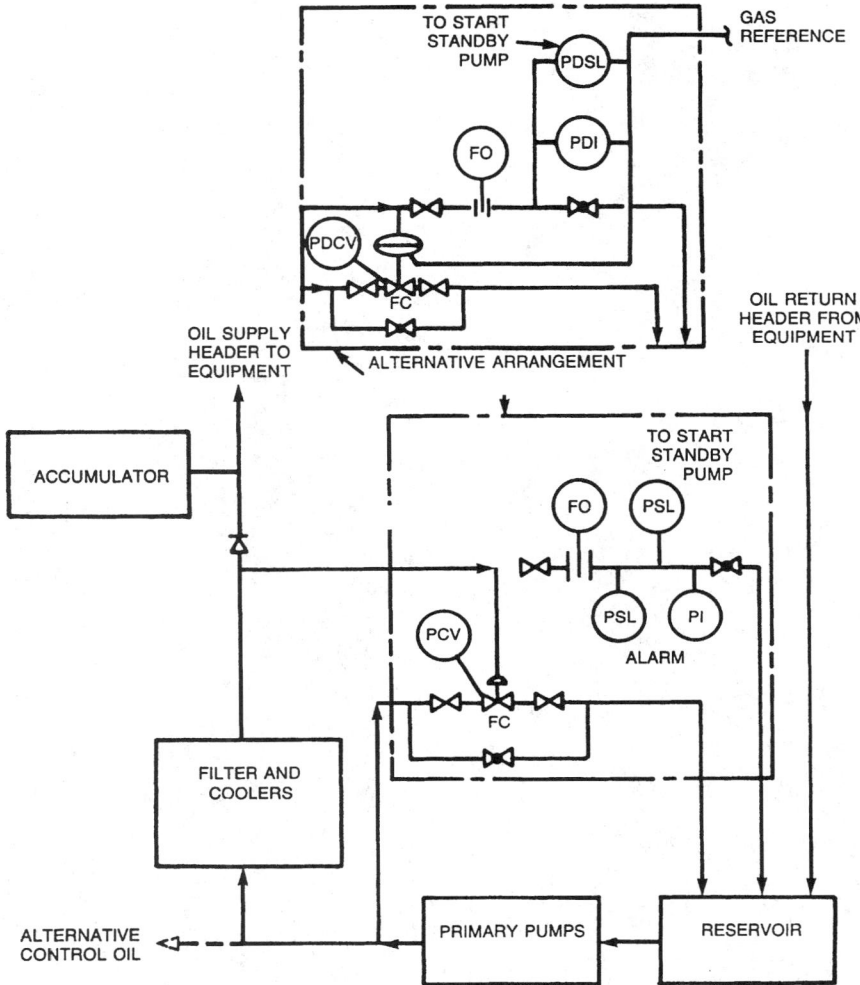

FIG. 29-87 Typical lubricating oil system for medium to large process machinery.

Universal at 100° F) and a viscosity index of 95. They contain necessary oxidation inhibitors and antiwear agents, as the same oil is used for bearings and gear lubrication. Oils tend to retain water, either from steam condensation or from condensation that occurs in the tanks. The presence of water is detrimental as it encourages the for-mation of acids and sludge, which can severely limit the useful life of machine components. It is customary to analyze periodically samples taken from the oil tanks and replace oils when the contamination is severe. A number of catastrophic coupling failures have been attrib-uted to corrosion by contaminated oils.

Analysis of Plant Performance

Colin S. Howat, Ph.D., P.E., *John E. & Winifred E. Sharp Professors Department of Chemical & Petroleum Engineering, University of Kansas; Member, American Institute of Chemical Engineers; Member, American Society of Engineering Education.*

30-2   ANALYSIS OF PLANT PERFORMANCE

REFERENCES: The following reference list presents a broad spectrum of the information in the literature regarding plant-performance analysis. The list is not intended to be comprehensive. However, the citations at the ends of these articles lead to essentially all of the relevant literature.

INFORMATION HANDLING

1. Hlavacek, V., "Analysis of a Complex Plant Steady-State and Transient Behavior," *Computers and Chemical Engineering*, **1**: 1977, 75–100. (Review article)
2. Mah, R. S. H., *Chemical Process Structures and Information Flows*, Butterworths, Boston, 1989, 500 pp. (Overview of measurement analysis and treatment, design considerations)

INTERPRETATION

3. Chang, C.T., K.N. Mah, and C.S. Tsai, "A Simple Design Strategy for Fault Monitoring Systems," *AIChE Journal*, **39**(7), 1993, 1146–1163. (Fault monitoring)
4. Cropley, J.B., "Systematic Errors in Recycle Reactor Kinetic Studies," *Chemical Engineering Progress*, February 1987, 46–51. (Model building, experimental design)
5. Fan, J.Y., M. Nikolaou, and R.E. White, "An Approach to Fault Diagnosis of Chemical Processes via Neural Networks," *AIChE Journal*, **39**(1), 1993, 82–88. (Relational model development, neural networks)
6. Fathi, Z., W.F. Ramirez, and J. Korbicz, "Analytical and Knowledge-Based Redundancy for Fault Diagnosis in Process Plants," *AIChE Journal*, **39**(1), 1993, 42–56. (Fault diagnosis)
7. Isermann R., "Process Fault Detection Based on Modeling and Estimation Methods—A Survey," *Automatica*, **20**(4), 1984, 387–404 (Fault detection survey article)
8. MacDonald, R.J. and C.S. Howat, "Data Reconciliation and Parameter Estimation in Plant Performance Analysis," *AIChE Journal*, **34**(1), 1988, 1–8. (Parameter estimation)
9. Narashimhan, S., R.S.H. Mah, A.C. Tamhane, J.W. Woodward, and J.C. Hale, "A Composite Statistical Test for Detecting Changes of Steady States," *AIChE Journal*, **32**(9), 1986, 1409–1418. (Fault detection, steady-state change)
10. Ramanathan, P., S. Kannan, and J.F. Davis, "Use Knowledge-Based-System Programming Toolkits to Improve Plant Troubleshooting," *Chemical Engineering Progress*, June 1993, 75–84. (Expert system approach)
11. Serth, R.W., B. Srikanth, and S.J. Maronga, "Gross Error Detection and Stage Efficiency Estimation in a Separation Process," *AIChE Journal*, **39**(10), 1993, 1726–1731. (Physical model development, parameter estimation)
12. Watanabe, K. and D.M. Himmelblau, "Incipient Fault Diagnosis of Nonlinear Processes with Multiple Causes of Faults," *Chemical Engineering Science*, **39**(3), 1984, 491–508.
13. Watanabe, K., S. Hirota, L. Hou, and D.M. Himmelblau, "Diagnosis of Multiple Simultaneous Fault via Hierarchical Artificial Neural Networks," *AIChE Journal*, **40**(5), 1994, 839–848. (Neural network)
14. Wei, C.N., "Diagnose Process Problems," *Chemical Engineering Progress*, September 1991, 70–74. (Parameter estimate monitoring for fault detection)
15. Whiting, W.B., T.M. Tong, and M.E. Reed, 1993. "Effect of Uncertainties in Thermodynamic Data and Model Parameters on Calculated Process Performance," *Industrial and Engineering Chemistry Research*, **32**, 1993, 1367–1371. (Relational model development)

PLANT-TEST PREPARATION

16. Gans, M. and B. Palmer, "Take Charge of Your Plant Laboratory," *Chemical Engineering Progress*, September 1993, 26–33.
17. Lieberman, N.P., *Troubleshooting Refinery Processes*, PennWell Books, Tulsa, 1981, 360 pp.

RECTIFICATION

18. Crowe, C.M., "Recursive Identification of Gross Errors in Linear Data Reconciliation," *AIChE Journal*, **34**(4), 1988, 541–550. (Global chi square test, measurement test)
19. Iordache, C., R.S.H. Mah, and A.C. Tamhane, "Performance Studies of the Measurement Test for Detection of Gross Errors in Process Data," *AIChE Journal*, **31**(7), 1985, 1187–1201. (Measurement test)
20. Madron, F., "A New Approach to the Identification of Gross Errors in Chemical Engineering Measurements," *Chemical Engineering Science*, **40**(10), 1985, 1855–1860. (Detection, elimination)

21. Mah, R.S.H. and A.C. Tamhane, "Detection of Gross Errors in Process Data," *AIChE Journal*, **28**(5), 1982, 828–830. (Measurement test)
22. May, D.L. and J.T. Payne, "Validate Process Data Automatically," *Chemical Engineering*, 1992, 112–116. (Validation)
23. Phillips, A.G. and D.P. Harrison, "Gross Error Detection and Data Reconciliation in Experimental Kinetics," *Industrial and Engineering Chemistry Research*, **32**, 1993, 2530–2536. (Measurement test)
24. Rollins, D.K. and J.F. Davis, "Gross Error Detection when Variance-Covariance Matrices are Unknown," *AIChE Journal*, **39**(8), 1993, 1335–1341. (Unknown statistics)
25. Romagnoli, J.A. and G. Stephanopoulos, "Rectification of Process Measurement Data in the Presence of Gross Errors," *Chemical Engineering Science*, **36**(11), 1981, 1849–1863.
26. Romagnoli, J.A. and G. Stephanopoulos, "On the Rectification of Measurement Errors for Complex Chemical Plants," *Chemical Engineering Science*, **35**, 1980, 1067–1081.
27. Rosenberg, J., R.S.H. Mah, and C. Iordache, "Evaluation of Schemes for Detecting and Identifying Gross Errors in Process Data," *Industrial and Engineering Chemistry, Research*, **26**(3), 1987, 555–564. (Simulation studies of various detection methods)
28. Serth, R.W. and W.A. Heenan, "Gross Error Detection and Data Reconciliation in Steam-Metering Systems," *AIChE Journal*, **32**(5), 1986, 733–742.
29. Terry, P.A. and D.M. Himmelblau, "Data Rectification and Gross Error Detection in a Steady-State Process via Artificial Neural Networks," *Industrial and Engineering Chemistry Research*, **32**, 1993, 3020–3028. (Neural networks, measurement test)
30. Verneuil, V.S. Jr., P. Yang, and F. Madron, "Banish Bad Plant Data," *Chemical Engineering Progress*, October 1992, 45–51. (Gross-error detection overview)

RECONCILIATION

31. Crowe, C.M., "Reconciliation of Process Flow Rates by Matrix Projection," part 2, "The Nonlinear Case," *AIChE Journal*, **32**(4), 1986, 616–623.
32. Crowe, C.M., Y.A. Garcia-Campos, and A. Hrymak, "Reconciliation of Process Flow Rates by Matrix Projection," *AIChE Journal*, **29**(6), 1983, 881–888.
33. Frey, H.C. and E.S. Rubin, "Evaluate Uncertainties in Advanced Process Technologies," *Chemical Engineering Progress*, May 1992, 63–70. (Uncertainty evaluation)
34. Jacobs, D.C., "Watch Out for Nonnormal Distributions," *Chemical Engineering Progress*, November 1990, 19–27. (Nonnormal distribution treatment)
35. Leibovici, C.F., V.S. Verneuil, Jr., and P. Yang, "Improve Prediction with Data Reconciliation," *Hydrocarbon Processing*, October 1993, 79–80.
36. Mah, R.S., G.M. Stanley, and D.M. Downing, "Reconciliation and Rectification of Process Flow and Inventory Data," *Industrial and Engineering Chemistry, Process Design and Development*, **15**(1), 1976, 175–183 (Reconciliation, impact of gross errors)

TROUBLESHOOTING

37. Gans, M., "Systematize Troubleshooting Techniques," *Chemical Engineering Progress*, April 1991, 25–29. (Equipment malfunction examples)
38. Hasbrouck, J.F., J.G. Kunesh, and V.C. Smith, "Successfully Troubleshoot Distillation Towers," *Chemical Engineering Progress*, 1993, 63–72.

AIChE EQUIPMENT TESTING SERIES

39. AIChE, *Centrifugal Pumps (Newtonian Liquids)*, 2d ed., Publication E-22, 1984, 24 pp.
40. ———, *Centrifuges: A Guide to Performance Evaluation*, Publication E-21, 1980, 17 pp.
41. ———, *Continuous Direct-Heat Rotary Dryers: A Guide to Performance Evaluation*, Publication E-23, 1985, 18 pp.
42. ———, *Dry Solids, Paste & Dough Mixing Equipment*, 2d ed., Publication E-18, 1979, 29 pp.
43. ———, *Evaporators*, 2d ed., Publication E-19, 1978, 33 pp.
44. ———, *Fired Heaters*, Publication E-27, 1989, 218 pp.
45. ———, *Mixing Equipment (Impeller Type)*, 2d ed., Publication E-25, 1987, 40 pp.
46. ———, *Packed Distillation Columns*, Publication E-28, 1991, 90 pp.
47. ———, *Particle Size Classifiers*, 2d ed., Publication E-29, 1992.
48. ———, *Spray Dryers*, Publication E-26, 1988, 24 pp.
49. ———, *Trayed Distillation Columns*, 2d ed., Publication E-24, 1987, 26 pp.

Nomenclature

Symbol	Definition	SI units	U.S. customary units
$\mathbf{B}$	Matrix of linear constraint coefficients		
$\vec{\mathbf{b}}$	Vector of bias		
b	Bias		
C_p	Heat capacity	kJ/kgmol/K	Btu/lbmole/F
c	Number of components		
$\vec{\mathbf{d}}$	Vector of weighted adjustment in measurements		
d_j	Weighted adjustment to measurement j		
$\vec{f}$	Vector of constraints		
$g\langle\rangle$	Operator on the measurements and equipment boundaries		
H_0	Null hypothesis		
H_a	Alternative hypothesis		
$\mathbf{J}$	Variance-covariance matrix of measurements		
K_{ij}	Equilibrium vaporization ratio for component i on stage j		
k	Specific rate constant		
$\mathbf{Q}$	Variance-covariance matrix of measurement adjustments		
Q	Heat transfer	kJ/hr	Btu/hr
Q_{jj}	Variance of adjustment to measurement j		
$\mathbf{R}$	Variance-covariance matrix of constraint residuals		
R_{ii}	Variance of constraint residual i		
$\vec{\mathbf{r}}$	Constraint equation residuals		
r_j	Single constraint residual		
S	Stream flow	kgmol/hr	lbmole/hr
T	Temperature	K	°F
t	Time		
$\mathbf{X_1}$	Matrix of all measurements		
$\vec{\mathbf{X}}_1^{\mathbf{m}}$	Vector of measurements		
$\hat{\vec{\mathbf{X}}}_1^{\mathbf{m}}$	Vector of adjusted measurements		

Symbol	Definition	SI units	U.S. customary units
$\hat{\vec{\mathbf{X}}}_1^{M}$	Vector of estimated measurements from the model		
$\delta\vec{\mathbf{X}}_1$	Deviation between adjusted and measured values		
$\mathbf{X_2}$	Matrix of equipment boundaries		
$\vec{\mathbf{X}}_2$	Vector of component flows		
$X_{i,j}$	Component i flow in stream j		
x_{ij}	Entry in the measurement matrix; liquid mole fraction of component i on stage j		
$\hat{x}_i$	Individual adjusted measurement		
x_i	Individual measurement		
$\tilde{x}_i$	True value of individual measurement		
$\bar{x}_i$	Mean value of individual measurement		
y_{ij}	Vapor mole fraction of component i on stage j		
$y_{i,j}^{\circ}$	Equilibrium vapor mole fraction of component i on stage j		
	Greek Symbols		
$\vec{\beta}$	Vector representation of parameters		
σ_i	Uncertainty in individual measurement		
θ_{ij}	Tray efficiency of component i on stage j		
ρ_j	Stream density	kg/m³	lbm/ft³
	Superscripts		
M	Measurement		
$\mathbf{m}$	Measured		
P	Plant		
$\mathbf{T}$	Transpose		
T	Total		
	Subscripts		
i	Matrix, vector position		
j	Matrix, vector position		

GLOSSARY

accuracy   Proximity of the measurements to actual values. Data frequently contain bias, a deviation between the measurement and the actual value. The smaller the deviation, the greater the accuracy.

bias   Offset between the measurement and the actual value of a measurement.

equipment boundary   Limit in equipment operation. This could refer to design limits such as operating pressure and temperature. More often, the concern of the plant-performance analyst is the upper and lower operating limits for the equipment. These boundaries typically describe an operating range beyond which the equipment performance deteriorates markedly.

equipment constraints   Limits beyond which the equipment cannot be operated, either due to design or operating boundaries.

fault detection   Process of identifying deteriorating unit operating performance. Examples are instrument failure, increased energy consumption, and increased catalyst usage.

gross error   Extreme systematic error in a measurement. The bias or systematic error is sufficiently large to distort the reconciliation and model development conclusions. Gross errors are frequently identified during rectification. Validation steps also are used to identify gross errors in measurements.

identification   Procedure for developing hypotheses and deter-

mining critical measurements. Identification requires an understanding of the intent of the process and intent of the plant-performance analysis to be conducted.

interpretation   Procedure for using the plant measurements or adjustments thereof to troubleshoot, detect faults, develop a plant model, or estimate parameters.

measurements   Plant information. These provide a window into the operation. They may consist of routinely acquired information such as that recorded by automatic control systems or recorded on shift logs, or they may consist of nonroutine information acquired as part of a plant test.

model   Qualitative or quantitative relationship between operating specifications and products. The quantitative model can be relational (e.g., a linear model) or physical (e.g., one comprised of appropriate material and energy balances, equilibrium relations, and rate relations). The parameters of these models (e.g., linear coefficients in the relational model; or tray efficiency, reactor volume efficiency, and heat transfer coefficients in a physical model) can be estimated from plant data.

plant   A group of processing units. Within this context, it is the entire processing facility, typically too large to be the focus of a single plant-performance analysis. The terminology in plant-performance analysis is inconsistent. Often the study is of a particular unit and rarely of the entire plant. However, the terms *plant test* and *plant data* refer to unit tests and unit data and will be used consistent with *practice.*

parameters   Model constants that relate the operating specifications to measures of product quality and quantity. Estimation of these is a frequent goal of plant-performance analysis.

precision   Measurement of the random deviations around some mean value. Precision is compromised by sampling methods, instrument calibrations, and laboratory calibrations. Reconciliation methods have been developed to minimize the impact of measurement precision.

process constraints   Chemical engineering fundamental relations for the unit. Examples include material balances, energy balances, hydraulic balances and, at times, thermodynamic equilibria. These constraints may be equality constraints such as material balances or inequality constraints such as those found in hydraulic balances (i.e., $P_{out} \leq P_{in}$ for a process vessel). Obvious process constraints may not always apply due to internal or external leaks, vents, and process misunderstanding.

reconciliation   Procedure for the adjustment of the measurements to close the process constraints. The purpose of reconciliation is to provide a set of measurements that better represent the actual plant operation.

rectification   Procedure for the identification of measurements that contain gross errors. This process is frequently done simultaneously or cyclically with the reconciliation.

systematic error   Measure of the bias in the measurements. It is a constant deviation or offset between the measurement and the actual value. This term is frequently used interchangeably with *bias.*

troubleshooting   Procedure to identify and solve a problem in operating unit. This is the most frequent interpretation step in plant-performance analysis.

uncertainty   A general term used for measurement error. This includes random and systematic errors in measurements.

unit   Battery limits of equipment under study. The unit under study may consist of a single piece of equipment, a group (e.g., a distillation tower with auxiliary equipment), an entire process (e.g., reactors and the corresponding separation train), or the entire plant.

unit test   Special operating procedure. The unit is operated at prescribed conditions. Special measurements may be made to supplement routine ones. One of the principal goals is to establish nearly constant material and energy balances to provide a firmer foundation for model development.

validation   Procedure for screening measurements to determine whether they are consistent with known unit characteristics. Measurements are compared to other measurements, expected operating limits, actual equipment status, and equipment performance characteristics. It is a useful tool to eliminate potentially distorting measurements from further consideration.

INTRODUCTION TO ANALYSIS OF PLANT PERFORMANCE

MOTIVATION

The goal of plant-performance analysis is to develop an accurate understanding of plant operations. This understanding can be used to:
- Identify problems in the current operation.
- Identify deteriorating performance in instruments, energy usage, equipment, or catalysts.
- Identify better operating regions leading to improved product or operating efficiency.
- Identify a better model leading to better designs.

The results of plant-performance analysis ultimately lead to a more efficient, safe, profitable operation.

FOCUS

Section 30 is written for engineers responsible for day-to-day interpretations of plant operation, those responsible for developing unit (plant) tests, and those responsible for analyzing plant data. The content focuses on aspects of troubleshooting, fault detection, parameter estimation, and model discrimination. In order to reach reliable conclusions, methods of identification, validation, reconciliation, rectification and interpretation are included. The emphasis is on guidelines that assist in avoiding many of the pitfalls of plant-performance analysis. While there are numerous mathematical and statistical methods in the technical literature, most of them apply only to restricted plant situations atypical of normal operations or to situations where enormous amounts of measurements are handled on a routine basis. Typical plant measurements are incomplete, their statistical distributions are unknown, the plant fluctuations are too great, and/or the volume of data makes the methods intractable. The numerical methods are useful to provide some insight, and an overview is presented. However, because of the limitations to measurement and numerical methods, the engineering judgment of plant-performance analysts is critical. Analysts must develop an accurate understanding of plant operations in order to draw valid conclusions about current operation, alternative operating regimes, and proposed designs founded upon the current plant configuration.

OVERVIEW

Historical Definition   Plant-performance analysis has been defined as the reconciliation, rectification, and interpretation of plant measurements to develop an adequate understanding of plant operation. Measurements taken from the operating plant are the foundation for the analysis. The measurements are reconciled to meet the constraints on the process, such as material balances, energy balances, and phase relations. The measurements are rectified to identify and eliminate those measurements that contain bias (i.e., systematic errors) sufficiently large to distort conclusions. The data are interpreted to troubleshoot, develop plant models, or estimate values for significant operating parameters. Ultimately, the results are used to discriminate among causes for deterioration of performance, operating regions, models, and possible operating decisions. The purpose of plant-performance analysis is to understand plant operations such that relational or physical models of the plant can be developed. The intended results are better profits, better control, safer operation, and better subsequent designs.

Plant-Performance Triangle   This view of plant-performance analysis is depicted in Fig. 30-1 as a plant-performance triangle. Figure 30-2 provides a key to the symbols used.

The three vertices are the operating plant, the plant data, and the plant model. The plant produces a product. The data and their uncertainties provide the history of plant operation. The model along with values of the model parameters can be used for troubleshooting, fault detection, design, and/or plant control.

The vertices are connected with lines indicating information flow. Measurements from the plant flow to plant data, where raw measurements are converted to typical engineering units. The plant data information flows via reconciliation, rectification, and interpretation to the plant model. The results of the model (i.e., troubleshooting, model building, or parameter estimation) are then used to improve plant operation through remedial action, control, and design.

Unit (Plant) Data   Measurements supporting plant-performance analysis come from daily operating logs, specific plant tests, automatic data acquisition, and specific measurement requirements. Examples of these data include temperatures, pressures, flows, compositions, elapsed time, and charge volume. The data are all subject to random errors from a variety of sources ranging from plant fluctuations and sampling technique through instrument calibration to laboratory methodology. The random errors define the precision in the data.

The measurements are also subject to systematic errors ranging from sensor position, sampling methods, and instrument degradation

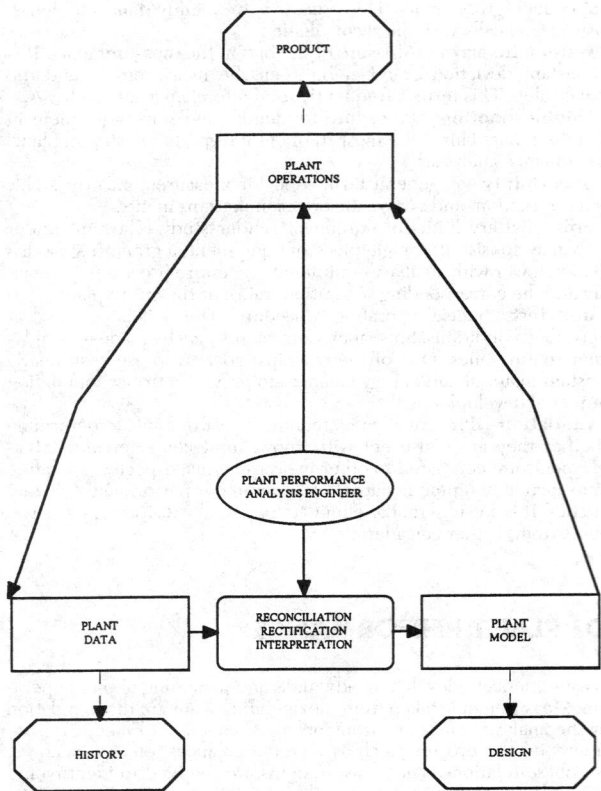

FIG. 30-1   Simplified plant performance analysis triangle.

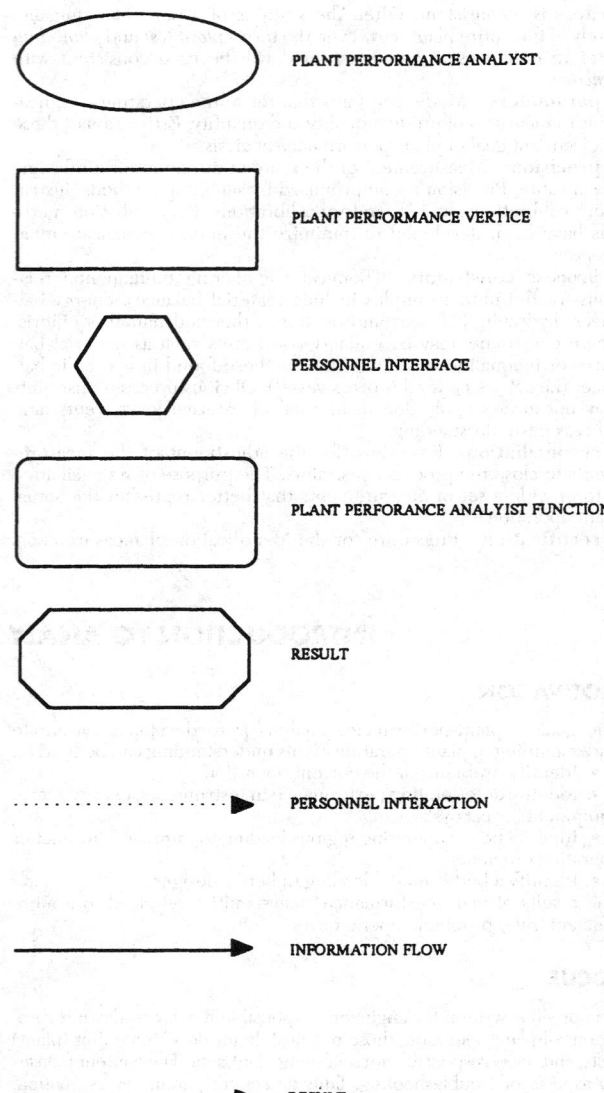

FIG. 30-2   Symbol key used in plant performance triangle.

to miscalibration in the field and laboratory. The systematic errors define the accuracy in the data.

These measurements with their inherent errors are the bases for numerous fault detection, control, and operating and design decisions. The random and systematic errors corrupt the decisions, amplifying their uncertainty and, in some cases, resulting in substantially wrong decisions.

Role of Plant-Performance Analysts   In this simplified representation, the principal role of analysts is to recognize these uncertainties; to accommodate them in the analysis; and to develop more confident control, operating, or design decisions. The analysts recognize and quantify these uncertainties through repeated measurements and effective communication with equipment and laboratory technicians. They validate the data comparing them to known process and equipment information. They accommodate these errors through reconciliation—adjusting the measurements to close the process constraints. Example constraints include process constraints such as material balances, energy balances, equilibrium relations (occasionally), elapsed time, and so on; and equipment constraints or boundaries that define the limitations of equipment operation. The reconciliation literature focuses primarily on process constraints, but it is important to include equipment constraints and boundaries to ensure correct measurement adjustment.

During reconciliation, measurements in which the analysts have a high degree of confidence are adjusted little, if at all, to meet the constraints, while adjustments for less reliable measurements are greater. Correct reconciliation minimizes the impact of measurement error and results in adjusted measurements that represent plant operation better than the raw ones. Traditionally, analysts have adjusted the measurements intuitively, relying on their experience and engineering judgment. The purpose of mathematical and statistical algorithms

developed over the past several years is to perform these adjustments automatically. However, algorithmic adjustment is subject to many of the same pitfalls that exist for intuitive adjustment. Both intuitive and algorithmic adjustment require correct estimates for uncertainty in the measurements. Both methods also require a correct implicit model of the plant. Without correct measurement error estimates and constraints, reconciliation will add bias to the adjusted measurements. For example, an unrecognized leak or vent invalidates the material-balance constraints developed from the implicit plant model, and either intuitive or algorithmic adjustment of data to meet invalid constraints adds systematic error to the adjusted measurements. Even when reconciliation is done algorithmically, the experience and judgment of the analysts are crucial.

The primary assumption in reconciliation is that the measurements are subject only to random errors. This is rarely the case. Misplaced sensors, poor sampling methodology, miscalibrations, and the like add systematic error to the measurements. If the systematic errors in the measurements are large and not accounted for, all reconciled measurements will be biased. During the measurement adjustment, the systematic errors will be imposed on other measurements, resulting in systematic error throughout the adjusted measurements.

Rectification accounts for systematic measurement error. During rectification, measurements that are systematically in error are identified and discarded. Rectification can be done either cyclically or simultaneously with reconciliation, and either intuitively or algorithmically. Simple methods such as data validation and complicated methods using various statistical tests can be used to identify the presence of large systematic (gross) errors in the measurements. Coupled with successive elimination and addition, the measurements with the errors can be identified and discarded. No method is completely reliable. Plant-performance analysts must recognize that rectification is approximate, at best. Frequently, systematic errors go unnoticed, and some bias is likely in the adjusted measurements.

The result of the reconciliation/rectification process is a set of adjusted measurements that are intended to represent actual plant operation. These measurements form the basis of the troubleshooting, control, operating, and design decisions. In order for these decisions to be made, the adjusted measurements must be interpreted. Interpretation typically involves some form of parameter estimation. That is, significant parameters—tray efficiency in a descriptive distillation model or linear model parameters in a relational model—are estimated. The model of the process coupled with the parameter estimates is used to control the process, adjust operation, explore other operating regimes, identify deteriorating plant and instrument performance or to design a new process. The adjusted measurements can also be interpreted to build a model and discriminate among many possible models. The parameter estimation and model building process is based on some form of regression or optimization analysis such that the model is developed to best represent the adjusted measurements. As with reconciliation and rectification, unknown or inaccurate knowledge of the adjusted measurement uncertainties will translate into models and parameter estimates with magnified uncertainty. Further, other errors such as those incorporated into the database will corrupt the comparison between the model and adjusted measurements. Consequently, parameters that appear to be fundamental to the unit (e.g., tray efficiency) actually compensate for other uncertainties (e.g., phase equilibria uncertainty in this case).

Extended Plant-Performance Triangle   The historical representation of plant-performance analysis in Fig. 30-1 misses one of the principal aspects: identification. Identification establishes troubleshooting hypotheses and measurements that will support the level of confidence required in the resultant model (i.e., which measurements will be most beneficial). Unfortunately, the relative impact of the measurements on the desired end use of the analysis is frequently overlooked. The most important technical step in the analysis procedures is to identify which measurements should be made. This is one of the roles of the plant-performance engineer. Figure 30-3 includes identification in the plant-performance triangle.

The typically recorded measurements in either daily operations or specific plant-performance tests are not optimal. The sampling locations were not selected with troubleshooting, control, operations, or model building as the focus. Even if the designers analyzed possible sample locations to determine which might maximize the information contained in measurements, it is likely that the actual operation is different from that envisioned by the designers or control engineers. More often, the sample locations are based on historic rules of thumb whose origins were likely based on convenience. Thus, for a given measurement, the amount of information leading to accurate parameter estimates is limited. Greater model accuracy can be achieved if locations are selected with the end use of the information well defined. It is necessary to define the intended end-use of the measurements and then to identify measurement positions to maximize the value in testing hypotheses and developing model parameter estimates.

END USE

The goal of plant-performance analysis is to improve understanding, efficiency, quality, and safety of operating plants. The end use must be

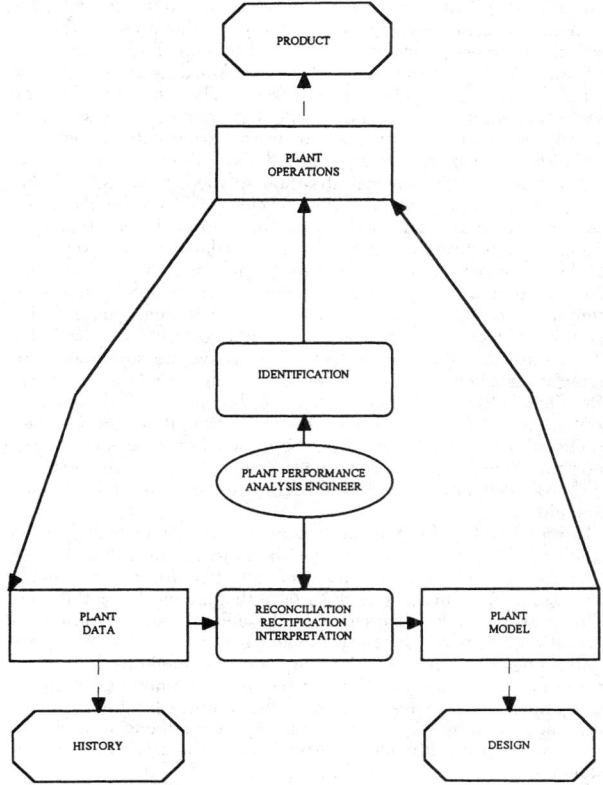

FIG. 30-3   Extended plant performance triangle.

established to focus the analysis. Figure 30-3 shows the three principal categories of end use improvements. The criteria for accuracy may vary among the categories requiring different numbers and levels of accuracy in the measurements.

Plant Operation   The purpose is to maintain and improve performance (i.e., product quality, rate, efficiency, safety, and profits). Examples include identification of plant conditions that limit performance (troubleshooting, debottlenecking) and exploration of new operating regions.

History   The history of a plant forms the basis for fault detection. Fault detection is a monitoring activity to identify deteriorating operations, such as deteriorating instrument readings, catalyst usage, and energy performance. The plant data form a database of historical performance that can be used to identify problems as they form. Monitoring of the measurements and estimated model parameters are typical fault-detection activities.

Design   In this context, *design* embodies all aspects requiring a model of the plant operations. Examples can include troubleshooting, fault detection, control corrections, and design development.

TECHNICAL BARRIERS TO ACCURATE UNDERSTANDING

Limited Contained Information   Data supporting the plant-performance analysis can come from daily operating logs, automatic or manual, or from formal plant tests. Daily logs consist of those measurements that the process and control designers and, subsequently, the plant engineers deem to be important in judging daily plant operation. No special operations (e.g., accumulating a constant-composition feed stock) are prerequisite for acquiring plant data of this caliber. While these data were intended to give sensitive insight into plant-performance, oftentimes they are recorded based on history and not formal analysis (i.e., their value has not been established or identified with respect to their end use). This presents the first technical hurdle—using data with limited contained information.

Formal unit (plant) tests (e.g., those developed for commissioning) usually last over a period of hours to weeks. The intent is to have the plant lined out in a representative operating regime. Feed stocks are typically accumulated in advance to ensure steady-state or controlled operation. Plant personnel are notified about the importance of the test so that they pay special attention to the operation, including charging rates, operating conditions, cycle times, and the like. Laboratory resources are dedicated beyond those normally required. A formal unit (plant) test requires significant coordination and investment. While it may give an indication of the plant capability, it is not representative of normal operation. During a unit (plant) test, greater attention and more personnel are dedicated to operation and data acquisition. Excursions in operating conditions are minimized. The data-acquisition effort should focus on sensitive measurements, providing insight beyond that gleaned from daily operations. However, oftentimes, little forethought is given to the end use of the information and the conclusions that will be drawn from it. Therefore, these additional data are not typically in the most sensitive regions of space and time. These data, too, contain less than optimal information.

There are significant technical barriers to accurate understanding from either source.

Limited Data   First, plant data are limited. Unfortunately, those easiest to obtain are not necessarily the most useful. In many cases, the measurements that are absolutely required for accurate model development are unavailable. For those that are available, the sensitivity of the parameter estimate, model evaluation, and/or subsequent conclusion to a particular measurement may be very low. Design or control engineers seldom look at model development as the primary reason for placing sensors. Further, because equipment is frequently not operated in the intended region, the sensitive locations in space and time have shifted. Finally, because the cost-effectiveness of measurements can be difficult to justify, many plants are underinstrumented.

Plant Fluctuations   Second, the plant is subject to constant fluctuations. These can be random around a certain operating mean; drift as feed stock, atmospheric, and other conditions change; or step change due to feed or other changes. While these fluctuations may be minimized during a formal unit (plant) test, nevertheless they are present. Given that each piece of equipment has time constants, usually unknown, these fluctuations propagate throughout the process, introducing error to assumed constraints such as material and energy balances.

Random Measurement Error   Third, the measurements contain significant random errors. These errors may be due to sampling technique, instrument calibrations, and/or analysis methods. The error-probability-distribution functions are masked by fluctuations in the plant and cost of the measurements. Consequently, it is difficult to know whether, during reconciliation, 5 percent, 10 percent, or even 20 percent adjustments are acceptable to close the constraints.

Systematic Measurement Error   Fourth, measurements are subject to unknown systematic errors. These result from worn instruments (e.g., eroded orifice plates, improper sampling, and other causes). While many of these might be identifiable, others require confidence in all other measurements and, occasionally, the model in order to identify and evaluate. Therefore, many systematic errors go unnoticed.

Systematic Operating Errors   Fifth, systematic operating errors may be unknown at the time of measurements. While not intended as part of daily operations, leaky or open valves frequently result in bypasses, leaks, and alternative feeds that will add hidden bias. Consequently, constraints assumed to hold and used to reconcile the data, identify systematic errors, estimate parameters, and build models are in error. The constraint bias propagates to the resultant models.

Unknown Statistical Distributions   Sixth, despite these problems, it is necessary that these data be used to control the plant and develop models to improve the operation. Sophisticated numerical and statistical methods have been developed to account for random

errors, identify and eliminate gross errors, and develop parameter estimates. These methods require good estimates of the underlying uncertainties (e.g., probability distributions for each of the measurements). Because the probability distributions are usually unknown, their estimates are usually poor and biased. The bias is carried through to the resulting conclusions and decisions.

PERSONNEL BARRIERS TO ACCURATE UNDERSTANDING

Because the technical barriers previously outlined increase uncertainty in the data, plant-performance analysts must approach the data analysis with an unprejudiced eye. Significant technical judgment is required to evaluate each measurement and its uncertainty with respect to the intended purpose, the model development, and the conclusions. If there is any bias on the analysts' part, it is likely that this bias will be built into the subsequent model and parameter estimates. Since engineers rely upon the model to extrapolate from current operation, the bias can be amplified and lead to decisions that are inaccurate, unwarranted, and potentially dangerous.

To minimize prejudice, analysts must identify and deal effectively with personnel barriers to accurate understanding. One type of personnel barrier is the endemic mythologies that have been developed to justify decisions and explain day-to-day operation in the plant. These mythologies develop because time, technical expertise, or engineers' and operators' skills do not warrant more sophisticated or technical solutions.

Operators   Operators develop mythologies in response to the pressure placed upon them for successful production quality and rates. These help them make decisions that, while not always technically supported, are generally in the correct direction. When they are not, convincing plant personnel of the deficiency in their decision structures is a difficult task.

Design and Control Engineers   Equally important are the mythologies developed by the design or control engineers. Their models of plant performance are more technically sound, but may be no more accurate than the operators' mythology. Consequently, the mythology passed along by the design and control engineers can also add bias to the foundation upon which the analyst relies.

Finally, with the current developments in control technology, there is a reliance by the operating engineer on, what is in most cases, an approximate model. While the control and design engineers might fully recognize the limitations inherent in projecting beyond the narrow confines of current operation, the operating engineer will frequently believe that the control model is accurate. This leads to bias in the operation and subsequent decisions regarding performance.

Analysts   The above is a formidable barrier. Analysts must use limited and uncertain measurements to operate and control the plant and understand the internal process. Multiple interpretations can result from analyzing limited, sparse, suboptimal data. Both intuitive and complex algorithmic analysis methods add bias. Expert and artificial intelligence systems may ultimately be developed to recognize and handle all of these limitations during the model development. However, the current state-of-the-art requires the intervention of skilled analysts to draw accurate conclusions about plant operation.

The critical role of analysts introduces a potential for bias that overrides all others—the analysts' evaluation of the plant information. Analysts must recognize that the operators' methods, designers' models, and control engineers' models have merit but must also beware they can be misleading. If the analysts are not familiar with the unit, the explanations are seductive, particularly since there is the motivation to avoid antagonizing the operators and other engineers.

Analysts must recognize that the end use as well as the uncertainty determines the value of measurements. While the operators may pay the most attention to one set of measurements in making their decisions, another set may be the proper focus for model development and parameter estimation. The predilection is to focus on those measurements that the operators believe in or that the designers/controllers originally believed in. While these may not be misleading, they are usually not optimal, and analysts must consciously expand their vision to include others.

In most situations, the plant was designed to be controlled and operated in a certain regime. It is likely that this has changed due to differences between the design basis and actual operation, due to operating experience and wholesale changes in purpose. Further, when developing sample, control, and measurement points, the designers/controllers may have had a model in mind for the operation. It is likely that that model is not accurate. Alternatively, they may have only used rules of thumb. Focusing only on previously selected points is limiting.

Each of the above can reduce analysts' opportunity for full understanding of the plant. Analysts must recognize that the plant operates by well-defined but not always obvious rules. It is important to identify these fundamental rules. If the analyst uses incorrect rules, the results will be further biased.

For the plant-performance analysis to be effective, the identified variables must be measured, the laboratory analysis must be correct, the simulation programs must accurately model the plant and the control recommendations must be implemented. In many settings, these aspects are not performed by plant-performance analysts. Analysts may be viewed as outsiders and operators are reluctant to modify their time-tested decision process. Laboratories geared to focusing on feed stock and product quality view unit (plant) tests as an overload. Simulation programs are not easily modified and proposed changes may not receive high priority attention. Control engineers may view modifications as an invasion of their responsibility. The plant-performance analysis milieu is much more complicated than that presented in Fig. 30-3 because of the personnel and communication barriers to implementation.

Figure 30-4 presents a more complete representation of plant-performance analysis. The information flow always faces barriers of personnel interactions. The operator must be convinced that the proposed changes and measurements will work using his/her language. The laboratory personnel must be convinced that the measurements are necessary, occasionally convinced that greater accuracy is required and that methods used are not giving results needed. Again, communication in their language must be effective. The software interaction is typically direct. However, the general nature of commercial simulators limits their effectiveness in particular situations. Occasionally, modifications are required. The software engineer is not familiar with the process and likely cannot be made aware because of proprietary considerations. This impedes communication. Finally, control engineers have been successful in establishing a control scheme which for all appearances works. Modifying the performance implies that they have not been as successful as appearances might indicate. While in all of these situations, teamwork should override these personnel considerations, it often doesn't. Consequently, communication is the paramount skill for plant-performance analysts.

OVERALL GUIDELINES

There are four overall guidelines that analysts should keep in mind. They must recognize the difficulties associated with the limited number and accuracy of the data, overcome the plant operation mythologies, overcome the designers' and controllers' biases and, finally, override the analysts' own prejudices. The following four overall guidelines assist in overcoming the hurdles to proper plant performance.

First, any analysis must be coupled with a technically correct interpretation of the equipment performance soundly rooted in the fundamentals of mass, heat, and momentum transfer; rate processes; and thermodynamics. Pseudotechnical explanations must not be substituted for sound fundamentals. Even when the development of a relational model is the goal of the analysis, the fundamentals must be at the forefront.

Second, any analysis must recognize the nonlinearities of equipment capability. Model development must recognize that equipment fundamentals will affect conclusions and extrapolations. These

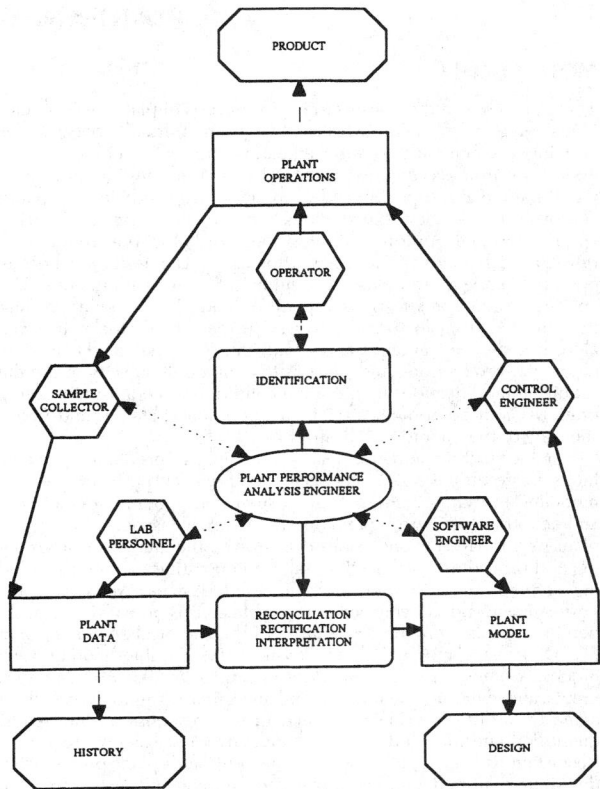

FIG. 30-4 Complete plant performance triangle including personnel interaction.

boundaries and nonlinearities of equipment performance overlay the chemical engineering fundamentals and temper the conclusions.

Third, any analysis must recognize that the measurements have significant uncertainty, random and systematic. These affect any conclusions drawn and models developed. Multiple interpretations of the same set of measurements, describing them equally well, can lead to markedly different conclusions and, more significantly, extrapolations.

Fourth, communication is paramount, since successful analysis requires that those responsible for measurement, control, and operation are convinced that the conclusions drawn are technically correct and the recommended changes will enhance their performance. This is the most significant guideline in implementing the results of the analysis.

Analysis of plant performance has been practiced by countless engineers since the beginning of chemical-related processing. Nevertheless, there is no body of knowledge that has been assembled called "analysis of plant performance." The guidelines given herein are effective in plant-performance applications. There are many more practiced by others that are also effective and should be employed whenever the challenges arise. Therefore, the material discussed in this section is the initial form for analysis and should not be considered all-inclusive. The particular equipment, operations, and problems associated with any plant spawn a myriad of effective methods to approach analysis of plant performance. They should not be ignored in preference to material in this section but should be added to these to improve the accuracy of conclusions and the efficiency of approach.

PLANT-ANALYSIS PREPARATION

MOTIVATION

These are a few of the reasons to justify analysis of plant performance. Units come on-line too slowly or with extreme difficulty because heat exchangers cannot add or remove heat, venting is inadequate, or towers do not produce quality product. Units come on-line and do not meet nameplate capacity and/or quality. Unit efficiency, quality, and/or yield are below expectations because energy or catalyst usage appears too high, product compositions are below that required, or raw material usage is excessive. Unit safety is questioned because operation appears too close to equipment control limitations. Unit environmental specifications are unfulfilled. Unit operations have deteriorated from historical norms. Alternate feed stocks are available, but their advantages and disadvantages if fed to the unit are unknown. Product demand exceeds the apparent capacity of the unit requiring modifications in operating conditions, in equipment configuration, or in equipment size. Unit operation is stable, and understanding of the operation is desired.

Troubleshooting start-up, quality and capacity problems, detecting faults in deteriorating effectiveness or efficiency performance, unit modeling to examine alternate feed stocks and operating conditions, and debottlenecking to expand operations are all aspects of analysis of plant performance. Conclusions drawn from the analyses lead to piping and procedure modifications, altered operating conditions including setpoint modifications and improved designs. Analyzing plant performance and drawing accurate conclusions is one of the most difficult and challenging responsibilities of the chemical engineer (Gans, M., D. Kohan, and B. Palmer, "Systematize Troubleshooting Techniques," *Chemical Engineering Progress*, April 25–29, 1991). Measurements and data are incomplete and inaccurate. Identical symptoms come from different causes. Aspects of unit response are not readily quantified and modeled, requiring inductive, investigative reasoning. According to Gans et al. (1991), 75 percent of all plant problems are due to unidentified, inefficient plant performance ultimately traced to simple equipment problems and limitations. Another 20 percent are due to inadequate design such as those encountered in startup and quality/quantity limitations. The remainder is due to a process failure. The goal of the plant-performance analyst is to identify correctly the problems and the opportunities for changes and to quantify the potential improvements.

The opportunities leading to false conclusions and inadequate recommendations are extreme. The probability of successful completion of analysis of plant performance is greatly enhanced if the preparatory work is complete. Analysts must define the detail of study required. Analysts must understand the operation of the unit. This includes the chemical engineering fundamentals and the operator's perspective and control response. Analysts must understand historical unit performance, developing a model commensurate with the measurements available and the detail of study required. Should a unit test, short-cut or exhaustive, be required, the unit personnel must understand the goals and their responsibilities. The laboratory must be prepared to handle the overload of samples that may be necessary and be able to produce data of required quality. Personnel and the supporting supplies must be available to make measurements, gather samples, and solve problems during the course of the test.

The purpose of this section is to provide guidelines for this preparation. General aspects are covered. Preparations for the specific units can be drawn from these. Topics include analyst, model, plant, and laboratory preparation. Since no individual analyst can be responsible for all of these activities, communication with other personnel is paramount for the success of the analysis.

ANALYST PREPARATION

Analysts must have a strong foundation in plant operations and in the unit under study. The hurdles thrown at analysts increase the probability that the conclusions drawn will be incorrect. A lack of understanding in the operation of the unit increases the likelihood that the conclusions will be inaccurate. An understanding of the chemical engineering fundamentals, the equipment flowsheets, the equipment plot, the operators' understanding and interpretations, and the operators' control decisions is essential to minimize the likelihood for drawing false conclusions. Reaching this understanding prior to undertaking a unit test and the measurement interpretation will increase the success and efficiency of the analysis.

The analyst must necessarily rely on the expertise and efforts of others to operate, gather, and analyze samples and record (automatically or manually) readings. Communication of the goals, measurement requirements, and outcome to all involved is critical. It is imperative that all involved understand their responsibilities, the use of the information that they gather, and the goals of the test.

Measurement locations and methods may be different from those used daily. Analysts and the sample-gatherers must be intimately familiar with the locations, difficulties, and methods. Analysts must ensure that the methods are safe, that the locations are as indicated on the flow sheets, and that the sample-gatherers will be able to safely obtain the necessary samples.

Process Familiarization The analysts' first step in preparation for analyzing plant performance is to become completely familiar with the process. Analysts should review:

- Process flow diagrams (PFDs)
- Operating instructions and time-sequence diagrams
- Piping and instrumentation diagrams (P&IDs)
- Unit installation
- Operator perspectives, foci, and responses

The review should emphasize developing an understanding of the processing sequence, the equipment, the equipment plot, the operating conditions, instrument and sample locations, the control decisions, and the operators' perspectives. While the preparation effort may be less for those who have been responsible for the unit for a long period of time, the purpose of the test requires that the types and locations of the measurements be different from those typically recorded and typically used. The condition of these locations must be inspected. Operating specifications may be different. Therefore, refreshment is always necessary.

The intensity of the situation requiring the analysis may not allow analysts to develop a formal preparatory review of the unit as described below. Analysts must recognize that the incomplete preparation may result in a less efficient analysis of plant performance.

PFDs (process flow diagrams) display the processing sequence for the unit, the principal pieces of equipment in the unit, and the operating conditions and control scheme. The equipment sequence should represent the sequence found in the unit. The operating conditions shown on the flow sheet may be those envisioned by the designer and may not properly reflect the current conditions. These should be verified during the subsequent discussions with operators and studied through review of the shift and daily logs. Where differences are substantial, these need to be understood, as they may indicate that operating philosophy has changed significantly from that proposed by the designers. It is particularly important to verify that the control scheme represents the current control philosophy. The purpose of each piece of equipment must be understood. This understanding should include understanding of key components, temperature specifications, elapsed time constraints, and the like. The basis of the operating conditions must be understood with respect to these constraints. The PFD review is completed by developing a material balance of sufficient detail for analysts to understand the reactions and separations.

Operating instructions and time-sequence diagrams provide insight into the basis for the operating conditions. They will also provide a foundation for the subsequent discussion with operators. The time sequence diagrams may provide insight into any difficulties that will arise during the unit test.

P&IDs (piping and instrumentation diagrams) should identify instruments, sample locations, the presence of sample valves, nozzle blinding, and control points. Of particular importance are the bypasses and alternate feed locations. The isolation valves in these lines may leak and can distort the interpretation of the measurements.

Understanding the positions of sample and other measurement locations within the equipment is also important. The presence or absence of isolation valves needs to be identified. While isolation valves may be too large for effective sampling, their absence will require that pipe fitters add them such that sample valves can be connected. This must be done in advance of any test. If analysts assume that samples are from a liquid stream when they are vapor or that temperature measurements are within a bed instead of outside it, interpretation of results could be corrupted. Analysts should also develop an understanding of control transmitters and stations. The connection between these two may be difficult to identify at this level in fully computer-controlled units.

Unit layout as installed is the next step of preparation. This may take some effort if analysts have not been involved with the unit prior to the plant-performance analysis. The equipment in the plant should correspond to that shown on the PFDs and P&IDs. Where differences are found, analysts must seek explanations. While a line-by-line trace is not required, details of the equipment installation and condition must be understood. It is particularly useful to correlate the sample and measurement locations and the bypasses shown on the P&IDs to those actually piped in the unit. Gas vents and liquid (particularly water-phase) discharges may have been added to the unit based on operating experience but not shown on the P&IDs. While these flows may ultimately be small within the context of plant-performance analysis, they may have sufficient impact to alter conclusions regarding trace component flows, particularly those that have a tendency to build in a process.

Discussion with operators provide substantial insight. The purpose of the discussion should be to develop an understanding of operators' perspectives of the unit, their foci for the operation, and their decision sequence in response to deviations and off-specification products. Two additional, albeit nontechnical, goals of this discussion are to establish rapport with the operators and to learn their language. The operators will ultimately be required to implement recommendations developed by analysts. Their confidence is essential to increase the likelihood of success. The following topics should be included in the discussion.

The operators have been given instructions on unit operation. Most of these are written and should have been studied prior to the meeting. Others may be verbal or implied. While this is not optimal, verbal instructions and operating experience are still part of every unit. It is not unusual that different shifts will have different operation methods. While none of the shift operations may be incorrect, they do lead to variability in operation and different performance. "What-if" questions posed to the operators can lead to insight into operator response. This will lead to analysts gaining better understanding of the unit (Block, S.R., "Improve Quality with Statistical Process Control," *Chemical Engineering Progress,* November 1990, 38–43). The discussion with the operators must provide insight into their view of the unit operation, their focus on the operation, and their understanding of equipment limitations.

One topic of discussion is the measurements to which the operators pay the most attention (their foci). Of the myriad of measurements, there is a limited set that they find most important. These are the measurements that they use to make the short-cycle decisions. The important points to glean are the reasons they focus on these, the values and trends that they expect, and their responses to the deviations from these.

With respect to their response, the discussion should emphasize why these are important and why they adjust certain control settings. Among the deviations on which analysts should focus the discussion are the high and low alarm settings. Some alarms will require rapid response. Alarms may give insight into equipment-operation boundaries as well as process constraints.

Operators typically focus on long cycle measurements upon which they focus. These may be part of morning reports giving production rates, compositions, yields, and so on. They may also have some recorded measurements that they examine once per shift. Analysts should understand the importance that the operators place on these measurements and the operators' responses to them.

Analysts are typically not totally prepared to discuss the purpose of the impending test at this meeting. Therefore, this topic may be premature. There is typically a sequence of meetings between operators and analysts. The information flow in the first is typically from the operators to analysts as analysts develop their understanding and learn to communicate in the operators' language. After the analysts study the process further based on the first meeting and preliminary simulations of the unit, another meeting is useful to test the analysts' understanding and communication methods. A third meeting to discuss the impending test purpose, focus, measurements, and procedures completes this phase of the preparation.

Data Acquisition   As part of the understanding, the measurements that can be taken must be understood. A useful procedure to prepare for this is to develop a tag sheet for the process (Lieberman, N.P., *Troubleshooting Refinery Processes,* PennWell Books, Tulsa, 1981, 360 pp). An example of a simplified sheet is given in Fig. 30-5.

This sheet will be used ultimately to record readings during the

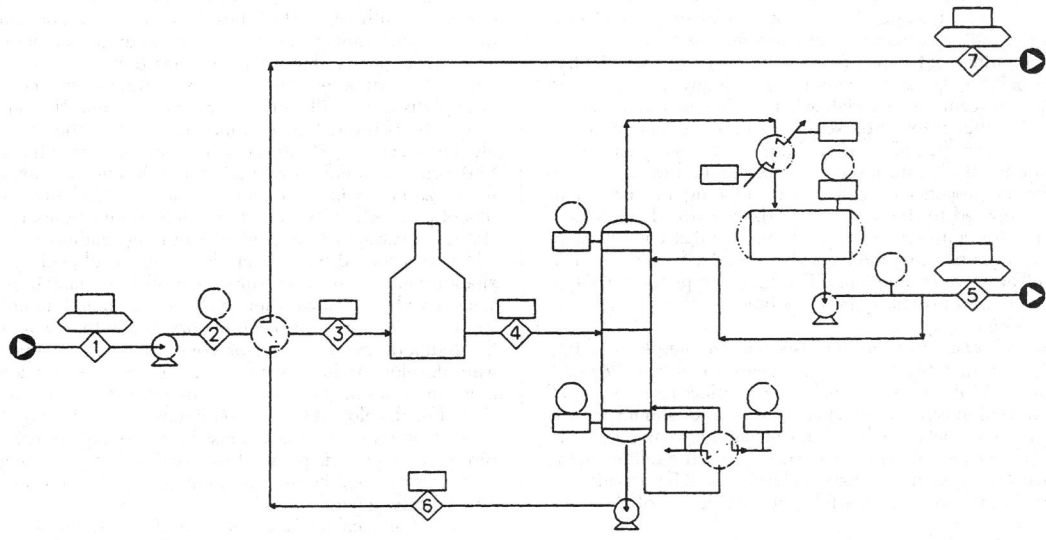

FIG. 30-5 Simplified tag-sheet for a distillation process.

plant test. It will help to develop a consistent set of measurements for the plant and help validate the measurements (identify inconsistencies). At this stage, however, it gives the analyst a visual representation of what measurements can be made. If there are sample locations, these must be added. Also, differential pressure measurements, additional flows, and utilities can be added as the unit instrumentation allows and as identified as being important during the identification step. Since identification has not been completed at this point, the measurements shown on these sheets are the ones already available on the board and in the field. The tag sheet also provides a visual point for discussion with the operators to confirm that certain measurements are made or can be made.

As part of this step, the analyst needs to develop an understanding of the uncertainty in the measurements and typical fluctuations experienced in daily operation. The uncertainties are functions of the instruments and their condition. Qualitatively, dial thermometers are less certain than thermocouples. On-line analyzers tend to be less accurate than lab analyses (assuming that the sample-gathering methodology is correct). Readings using different pressure gauges are less reliable than readings using a single pressure gauge. These random errors may be negligible in a unit that exhibits large fluctuations. The relative importance of plant fluctuations and random measurement error should be established. Multiple, rapid readings, control samples, or confirming measurements with other instruments will help establish the measurement uncertainty. Operating data will provide insight into the plant fluctuations, which can then be compared to the instrument uncertainty.

Operators frequently have insight into which instruments are accurate and which are not. If those instruments subsequently prove critical, recalibration must be done prior to the unit test. Preliminary analysis of daily measurements and practice measurements will help to identify which are suspect and require instrument recalibration prior to the unit test.

Analysts should discuss sample-collection methods with those responsible. Frequently, the methods result in biased data due to venting, failure to blow down the sample lines, and contamination. These are limitations that must either be corrected or accepted and understood. Sampling must be conducted within the safety procedures established for the unit. Since samples may be hot, toxic, or reactive in the presence of oxygen, the sample gatherers must be aware of and implement the safety procedures of the unit.

Material Balance Constraint   There are two types of constraints for the unit. These are the process constraints and the equipment constraints. In each of these, there are equality constraints such as material balances and inequality constraints such as temperature limits. Analysts must understand the process and equipment constraints as part of the preparation for the unit analysis.

The most important of the process constraints is the material balance. No test or analysis can be completed with any degree of certainty without an accurate material balance. The material balance developed during this preparation stage provides the foundation for the analysts' understanding of the unit and provides an organizational tool for measurement identification. Analysts should develop a material balance for the process based on typical operating measurements. This can be compared to the design material balance. Estimates of tower splits, reactor conversions, elapsed times, and stream divisions help to identify the operating intent of the unit. Analysts must focus on trace as well as major components. The trace components will typically provide the most insight into the operation of the unit, particularly the separation trains.

During this preparation stage, analysts will frequently find that there is insufficient quantity or quality of measurements to close the material balance. Analysts should make every effort to measure all stream flows and compositions for the actual test. They should not rely upon closing material balances by back-calculating missing streams. The material balance closure will provide a check on the validity of the measurements. This preparatory material balance will help to identify additional measurements and schedule the installation of the additional instruments.

A typical material-balance table listing the principal components or boiling ranges in the process as a function of the stream location should be the result of this preliminary analysis. An example shown as a spreadsheet is given in the validation discussion (Fig. 30-18).

Energy Balance   Many of the principal operating problems found in a plant result from energy-transfer problems such as fouled or blanketed exchangers, coked furnaces, and exchanger leaks. Consequently, developing a preliminary energy balance is a necessary part of developing an understanding of the unit. A useful result of the energy-balance analysis is the identification of redundant measurements that provide methods to obtain two estimates for unit performance. For example, reflux-flow and steam-flow measurements provide two routes to identifying heat input to a tower. These redundant measurements are very useful; both should be taken to provide the redundancy, and one or the other should not be ignored.

The material balance table can be supplemented with temperatures, pressures, phases, and stream enthalpies (or internal energies). Utility flows and conditions should be added to the process information.

Other Process Constraints   Typical of these constraints are composition requirements, process temperature limits, desired recoveries, and yields. These are frequently the focus of operators. Violation of these constraints and an inability to set operating conditions that meet these constraints are frequently the motivation for the unit analysis.

Equipment Constraints   These are the physical constraints for individual pieces of equipment within a unit. Examples of these are flooding and weeping limits in distillation towers, specific pump curves, heat exchanger areas and configurations, and reactor volume limits. Equipment constraints may be imposed when the operation of two pieces of equipment within the unit work together to maintain safety, efficiency, or quality. An example of this is the temperature constraint imposed on reactors beyond which heat removal is less than heat generation, leading to the potential of a runaway. While this temperature could be interpreted as a process constraint, it is due to the equipment limitations that the temperature is set.

Developing an understanding of these constraints provides further insight into unit operation.

Database   The database consists of physical property constants and correlations, pure component and mixture, that are necessary for the proper understanding of the operation of the unit. Examples of the former are molecular weights, boiling curves, and critical properties. Example pure-property correlations are densities versus temperature, vapor pressures versus temperature, and enthalpies versus temperature and pressure. Example mixture-property correlations are phase equilibria versus composition, temperature, and pressure; kinetic rate constants versus temperature; and interfacial tension versus composition and temperature. While the material balance can be developed without most of these, the energy balance and any subsequent model cannot. Therefore, an accurate database is critical to accurate understanding of plant operation. Very often, unit model parameters will interact with database parameters. The most notable example is the distillation tower efficiency and the phase equilibria constants. If the database is inaccurate, the efficiency estimate will also be inaccurate. Therefore, whenever the goal of the unit analysis is to develop a model for operation and design, care must be taken to minimize errors in the database that can affect the accuracy of the model parameters. Inaccurate models cannot be used for sensitivity studies or extrapolation to other operating conditions.

Analysts should not rely on databases developed by others unless citations and regression results are available. Many improper conclusions have been drawn when analysts have relied upon the databases supplied with commercial simulators. While they may be accurate in the temperature, pressure, or composition range upon which they were developed, there is no guarantee that they are accurate for the unit conditions in question. Pure component and mixture correlations should be developed for the conditions experienced in the plant. The set of database parameters must be internally consistent (e.g., mixture-phase equilibria parameters based on the pure-component vapor pressures that will be used in the analysis). This ensures a consistent set of database parameters.

It is not unusual for 30–40 percent of the process design effort to be spent in developing a new database. The amount of time required at this stage in the analysis of plant performance for analysis of the unit should be equivalent. The amount of effort devoted to database development becomes more intensive as the interaction between the model parameters and the database increases.

PLANT MODEL PREPARATION

Focus For the purposes of this discussion, a model is a mathematical representation of the unit. The purpose of the model is to tie operating specifications and unit input to the products. A model can be used for troubleshooting, fault detection, control, and design. Development and refinement of the unit model is one of the principal results of analysis of plant performance. There are two broad model classifications.

The first is the relational model. Examples are linear (i.e., models linear in the parameters and neural network models). The model output is related to the input and specifications using empirical relations bearing no physical relation to the actual chemical process. These models give trends in the output as the input and specifications change. Actual unit performance and model predictions may not be very close. Relational models are useful as interpolating tools.

The second classification is the physical model. Examples are the rigorous modules found in chemical-process simulators. In sequential modular simulators, distillation and kinetic reactors are two important examples. Compared to relational models, physical models purport to represent the actual material, energy, equilibrium, and rate processes present in the unit. They rarely, however, include any equipment constraints as part of the model. Despite their complexity, adjustable parameters bearing some relation to theory (e.g., tray efficiency) are required such that the output is properly related to the input and specifications. These models provide more accurate predictions of output based on input and specifications. However, the interactions between the model parameters and database parameters compromise the relationships between input and output. The nonlinearities of equipment performance are not included and, consequently, significant extrapolations result in large errors. Despite their greater complexity, they should be considered to be approximate as well.

Preliminary models are required to identify significant measurements and the complexity of model required and to test the analysis methods that will be used during the unit analysis. Effort must be devoted during the preparation stage to develop these preliminary models.

It must be recognized that model building is not the only outcome of analysis of plant performance. Many troubleshooting activities do not require a formal mathematical model. Even in these circumstances, analysts have developed through preliminary effort or experience a mental model of the relation between specifications, input, and output that provides a framework for their understanding of the underlying chemical engineering. These mental models generally take longer to develop but can be more accurate than mathematical models.

Intended Use The intended use of the model sets the sophistication required. Relational models are adequate for control within narrow bands of setpoints. Physical models are required for fault detection and design. Even when relational models are used, they are frequently developed by repeated simulations using physical models. Further, artificial neural-network models used in analysis of plant performance including gross error detection are in their infancy. Readers are referred to the work of Himmelblau for these developments. [For example, see Terry and Himmelblau (1993) cited in the reference list.] Process simulators are in wide use and readily available to engineers. Consequently, the emphasis of this section is to develop a preliminary physical model representing the unit.

Required Sensitivity This is difficult to establish *a priori*. It is important to recognize that no matter the sophistication, the model will not be an absolute representation of the unit. The confidence in the model is compromised by the parameter estimates that, in theory, represent a limitation in the equipment performance but actually embody a host of limitations. Three principal limitations affecting the accuracy of model parameters are:

- Interaction between database and model parameters
- Interaction between measurement error and model parameters

- Interaction between model and model parameters

Three examples are discussed.

Tray efficiency is one example of the first interaction. Figure 30-6 is a representation of a distillation tray.

Defining tray efficiency as the difference between the actual and the equilibrium vaporization, the efficiency is:

$$\theta_{i,j} = \frac{y_{i,j} - x_{i,j}}{y^\circ_{i,j} - x_{i,j}}$$

where

$$y^\circ_{i,j} = K_{i,j} x_{i,j}$$

Tray efficiency $\theta_{i,j}$ is supposed to represent a measure of the deviation from equilibrium-stage mass transfer assuming backmixed trays. However, the estimate of tray efficiency requires accurate knowledge of the equilibrium vaporization constant. Any deviations between the actual equilibrium relation and that predicted by the database will be embodied in the tray efficiency estimate. It is a tender trap to accept tray efficiency as a true measure of the mass transfer limitations when, in fact, it embodies the uncertainties in the database as well.

As another example of the first interaction, a potential parameter in the analysis of the CSTR is estimating the actual reactor volume. CSTR shown in Fig. 30-7. The steady-state material balance for this CSTR having a single reaction can be represented as:

$$0 = X_{i,1} - X_{i,2} - V_r k f(\vec{\mathbf{X}}_2, S_2, \rho_2)$$

where X_i is the flow of component i, V_r is the reactor volume, k is the rate constant at the reactor temperature, $\vec{\mathbf{X}}_2$ is the vector of component flows in stream 2, S_2 is the stream-2 flow, and ρ_2 is the stream-2 density. Any effort to estimate the reactor volume and therefore also the volume efficiency of the reactor depends upon the database estimate of the rate constant. Any errors in the rate constant will result in errors in the reactor volume estimate. Extrapolations to other operating conditions will likely be erroneous. Estimating the rate constant based on reactor volume will have the same difficulties.

The second interaction results in compromised accuracy in the parameter estimate due to the physical limitations of the process as

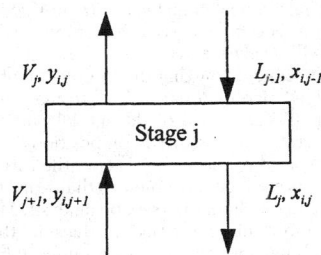

FIG. 30-6 Representation of a distillation tray numbering from the top of the column.

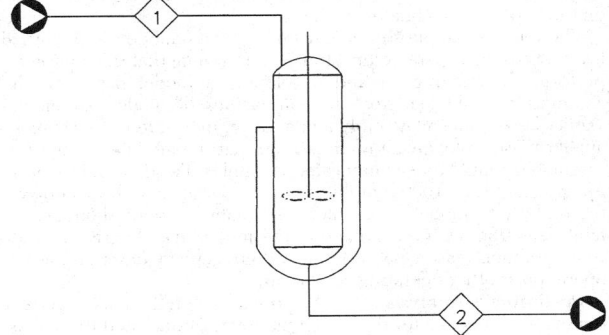

FIG. 30-7 Flow sheet of a single feed and single product CSTR.

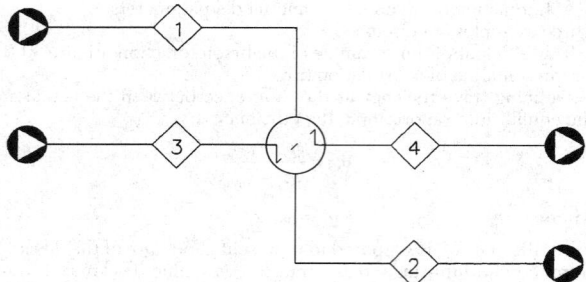

FIG. 30-8 PFD of Shell and tube heat exchanger.

embodied in the measurement uncertainties. Figure 30-8 shows a simple shell and tube heat exchanger. Many plant problems trace back to heat-transfer-equipment problems. Analysts may then be interested in estimating the heat-transfer coefficient for the heat exchanger to compare to design operation. However, this estimation is compromised when stream temperature changes are small, amplifying the effect of errors in the heat-transfer estimation. For example, the heat transfer could be calculated from the energy balance for Stream 1.

$$Q = S_1 C_p (T_2 - T_1)$$

The error in estimating the temperature difference is

$$\sigma_{\Delta T} = \sqrt{2}\sigma_T$$

The percentage error in the temperature difference translates directly to the percentage error in the estimate Q. As temperature-measurement error increases, so does the heat transfer coefficient error.

The third interaction compromising the parameter estimate is due to bias in the model. If noncondensables blanket a section of the exchanger such that no heat transfer occurs in that section, the estimated heat-transfer coefficient based on a model assuming all of the area is available will be erroneous.

The first two examples show that the interaction of the model parameters and database parameters can lead to inaccurate estimates of the model parameters. Any use of the model outside the operating conditions (temperature, pressures, compositions, etc.) upon which the estimates are based will lead to errors in the extrapolation. These model parameters are effectively no more than adjustable parameters such as those obtained in linear regression analysis. More complicated models may have more subtle interactions. Despite the parameter ties to theory, they embody not only the uncertainties in the plant data but also the uncertainties in the database.

The third example shows how the uncertainties in plant measurements compromise the model parameter estimates. Minimal temperature differences, very low conversions, and limited separations are all instances where errors in the measurements will have a greater impact on the parameter estimate.

The fourth example shows how improper model development will lead to erroneous parameter estimates. Assuming that the equipment performs in one regime and developing a model based on that assumption could lead to erroneous values of model parameters. While these values may imply model error, more often the estimates appear reasonable, giving no indication that the model does not represent the unit. More complicated examples like the kind given by Sprague and Roy (1990) emphasize the importance of the accuracy of the underlying model in parameter estimation, troubleshooting, and fault detection. In these situations, the model may describe the current operation reasonably well but will not actually describe the unit operation at other operating conditions.

Preliminary Analysis The purpose of the preliminary analyses is to develop estimates for the model parameter values and to establish the model sensitivity to the underlying database and plant and model uncertainties. This will establish whether the unit test will actually achieve the desired results.

The model parameter estimation follows the methods given in the interpretation subsection of this chapter. Analysts acquire plant measurements, adjust them to close the important constraints including the material and energy balances and then through repeated simulations, adjust parameter values to obtain a best description of the adjusted measurements. Not only does this preliminary analysis provide insight into the suitability of the model but also it tests the analysis procedures. The primary emphasis at this stage should be on developing preliminary parameter estimates with less emphasis on rigorously developing the measurement error analysis.

Once the model parameters have been estimated, analysts should perform a sensitivity analysis to establish the uniqueness of the parameters and the model. Figure 30-9 presents a procedure for performing this sensitivity analysis. If the model will ultimately be used for exploration of other operating conditions, analysts should use the results of the sensitivity analysis to establish the error in extrapolation that will result from database/model interactions, database uncertainties, plant fluctuations, and alternative models. These sensitivity analyses and subsequent extrapolations will assist analysts in determining whether the results of the unit test will lead to results suitable for the intended purpose.

PLANT PREPARATION

Intent Plant personnel, supplies, and budget are required to successfully complete a unit test. Piping modifications, sample collection, altered operating conditions, and operation during the test require advance planning and scheduling. Analysts must ensure that these are accomplished prior to the actual test. Some or all of the following may be necessary for a successful unit test.

Communication Analysts will require the cooperation of the
- Unit operators
- Unit supervisors
- Plant management
- Maintenance personnel
- Laboratory personnel

Operators are primarily concerned with stable operation and may be leery of altering the operation; they may fear that operation will drift into a region that cannot be controlled. Supervision may be reluctant despite their recognizing that a problem exists: Any deficiencies with the operation or operating decisions is their responsibility. Permission for conducting the test from the supervisor and the operators will be required. Management cooperation will be required, particularly if capital is ultimately needed. Maintenance will be called upon to make modifications to sample locations and perform a sequential pressure measurement. The laboratory personnel, discussed in detail in the next subsection, may view the unit test as an overload to available resources. These concerns must be addressed to ensure accurate sample interpretation.

Permission Analysts must have the permission of the operators and the supervisors to conduct even the most straightforward tests. While this is part of the analysts' preparation, it is important for all involved to know that analysts have that permission.

Schedule Complex tests should be done over a period of days. This provides the opportunity for the unit to be nearly steady. The advantages are that confirming measurements can be made. Scheduling a multiday test should be done when there is a likelihood that the feed stock supply and conditions will be nearly constant. The cooperation of upstream units will be required. The multiday test also requires that the downstream units can take the unit products.

The schedule should be set well in advance so that support services can provide the necessary personnel and supplies.

Simpler tests will not require this amount of time. However, they should be scheduled to minimize disruption to normal operations.

Piping Modifications One result of the inspection of the sample locations is a list of sample locations that will require modifications. The mechanical department will be required to make these modifications before the unit test is run. It is likely that the locations that are not typically used will be plugged with debris. The plugs will have to be drilled out before the test begins. Drilling out plugs presents a safety hazard, and those involved must be aware of this and follow the plant safety protocols.

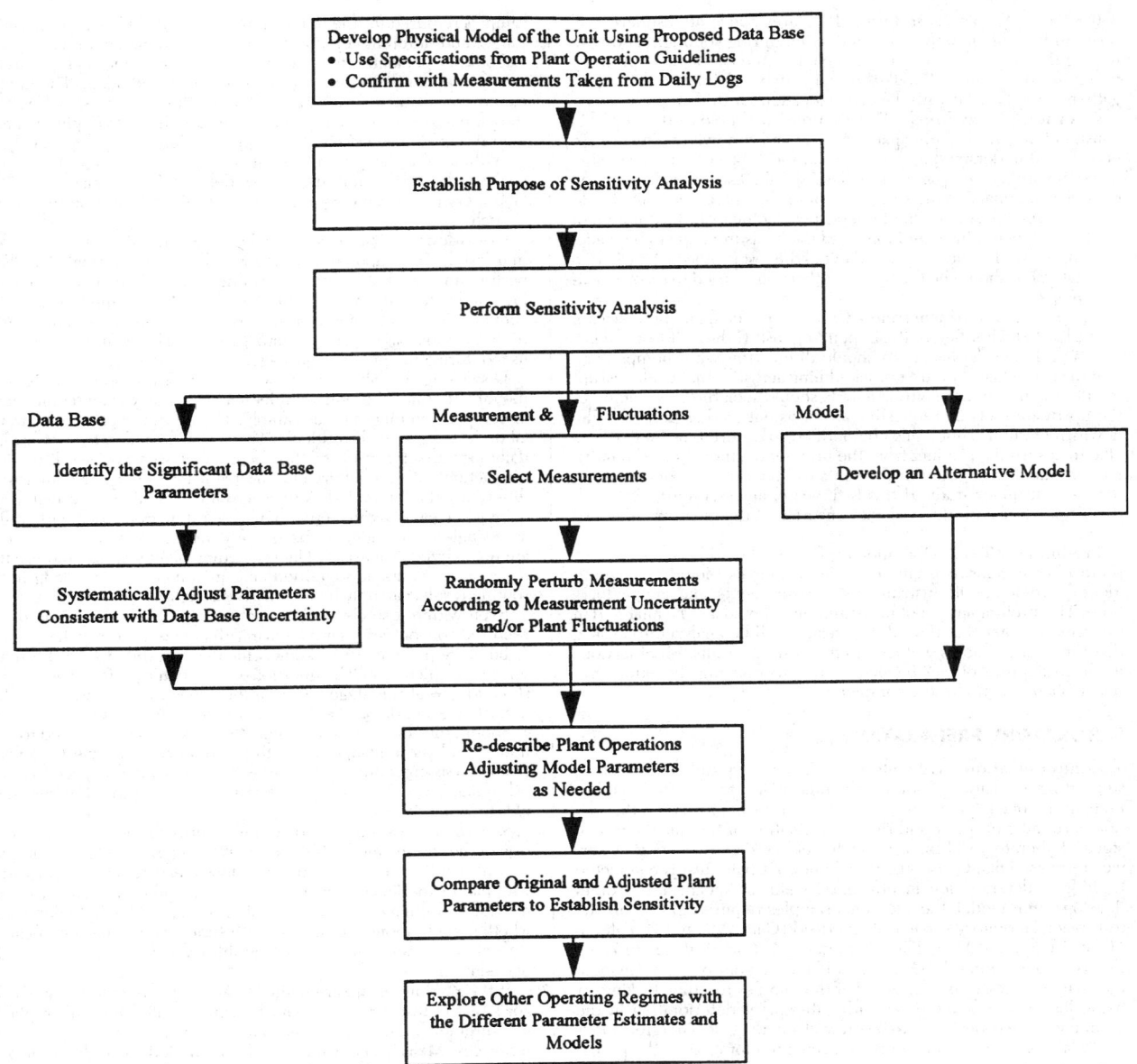

FIG. 30-9 Logic diagram for plant sensitivity analysis.

Instrumentation Calibration may be required for the instruments installed in the field. This is typically the job of an instrument mechanic. Orifice plates should be inspected for physical condition and suitability. Where necessary, they should be replaced. Pressure and flow instruments should be zeroed. A preliminary material balance developed as part of the preliminary test will assist in identifying flow meters that provide erroneous measurements and indicating missing flow-measurement points.

When doing a hydraulic test, a single pressure gauge should be used and moved from location to location. This gauge must be obtained in advance. The locations where this pressure will be measured should be tagged to assist the pipe fitter who will be responsible for moving the gauge from location to location. A walk-through with the pipe fitter responsible can be instructive for both the analysts and the fitter.

Thermocouples tend to be reliable, but dial thermometers may need to be pulled and verified for accuracy.

Sample Containers More sample containers will be required for a complex test than are typically used for normal operation. The number and type of sample containers must be gathered in advance, recognizing the number of measurements that will be required. The sample containers should be tagged for the sample location, type, and conditions.

Field Measurement Conditions Those gathering samples must be aware of the temperature, pressure, flammability, and toxic characteristics of the samples for which they will be responsible. This is particularly important when samples are taken from unfamiliar locations. Sample ports will have to be blown down to obtain representative samples. Liquid samples will have to be vented. Temperatures above

330 K (140° F) can cause burns. Pressures above atmospheric will result in flashing upon pressure reduction during venting. Venting to unplug the sample port and the sample bomb must be done properly to minimize exposure. A walk-through may be useful so that sample-gatherers are familiar with the actual location for the sample.

Operating Guidelines The test protocol should be developed in consultation with the principal operators and supervisor. Their cooperation and understanding are required for the test to be successful. Once the protocol is approved, analysts should distribute an approved one-page summary of the test protocol to the operators. This should include a concise statement of the purpose of the test, the duration of the test, the operating conditions, and the measurements to be made. The supervisor for the unit should initial the test protocol. Attached to this statement should be the tag sheet that will be used to record measurements.

Upstream and Downstream Units Upstream and downstream units should be notified of the impending test. If the unit test will last over a period of days, analysts should discuss this with the upstream unit to ensure that they are not scheduling activities that could disrupt feed to the unit under study. Analysts should seek the cooperation of the upstream units by requesting as consistent feed as possible. The downstream units should also be notified to ensure that they will be able to absorb the product from the unit under study. For both units, measurements from their instruments will be useful to confirm those for the unit under study. If this is the case, analysts must work with those operators and supervisors to ensure that the measurements are made.

Preliminary Test Operation of the unit should be set at the test protocol conditions. A preliminary set of samples should be taken to identify problems with instruments, measurements, and sample locations. This preliminary set of measurements should also be analyzed in the same manner that the full-test results will be analyzed to ensure that the measurements will lead to the desired results. Modifications to the test protocol can be made prior to exerting the effort and resources necessary for the complete test.

LABORATORY PREPARATION

Communication Laboratory services are typically dedicated to supporting the daily operation of the unit under study as well as other units in the plant. Their purpose is the routine confirmation that the unit is running properly and the determination of the quality of feed stocks. Laboratory staffing is normally set based on these routine service requirements. Consequently, whenever a plant test is conducted to address deterioration in efficiency, yield, or specifications or to develop a unit model, the additional samples required to support the test, place laboratory services in overload (Gans, M., and B. Palmer, "Take Charge of Your Plant Laboratory," *Chemical Engineering Progress*, September 1993, 26–33). If the laboratory cannot handle the analysis quickly, the likelihood of the samples reacting, leaking, or being lost markedly increases with subsequent deterioration in the accuracy of the conclusions to be drawn from the test. Therefore, adequate laboratory personnel must be accounted for early in the preparation process.

Plant-performance analysts must understand:
- Laboratory limitations
- Laboratory organization
- Laboratory measurement uncertainties
- Measurement cost
- Additional personnel requirements

The laboratory supervision and personnel must supply this information so that analysts gain this understanding.

Laboratory supervision and personnel must understand:
- Type of samples required
- Level of detail, accuracy, and precision of the samples
- Flammability, toxicity, and conditions of samples
- Anticipated schedule and duration of the test
- Justification for the overload assignments

Analysts provide this information.

The laboratory may need time to prepare for the unit test. This must be accounted for when the test is scheduled. The analysis of samples required for the unit test may focus on different composition ranges and different components than those done on a routine basis. Laboratory personnel may need to modify their methods or instruments to attain the required level of accuracy and detail. The modification, testing, and verification of the methods are essential parts of the preparation process. A practice run of gathering samples will help identify any deficiencies in the sample handling, storage, and analyses.

Without forethought, planning, and team-building, the sample analyses during the unit test may be delayed, lost, or inaccurate. The laboratory is an essential part of the unit test and must be recognized as such.

Confidence The accuracy of the conclusions drawn from any unit test depends upon the accuracy of the laboratory analyses. Plant-performance analysts must have confidence in these analyses including understanding the methodology and the limitations. This confidence is established through discussion, analyses of known mixtures, and analysis of past laboratory results. This confidence is established during the preparation stage.

Discussing the laboratory procedures with the personnel is paramount. Routine laboratory results may focus on certain components or composition ranges in the sample. The routine analyses narrow the laboratory personnel's outlook. The succinct and often misleading daily logs are the result of this focus. Analysts who have little daily interaction with the laboratory and plant may interpret daily results differently than intended. A typical example is laboratory analyses of complex streams where components are often grouped and identified as a single component. Consequently, important trace components are unanalyzed or masked. The impending plant test may require that these components be identified and quantified. The masking in the routine results can only be identified through discussion.

Even within a single sample analysis, it is likely that some of the reported concentrations are known with greater accuracy than others. Laboratory personnel will know which concentrations can be relied upon and which should be questioned. The plant-performance analyst should know at this stage which of the concentrations are of greatest importance and direct the discussion to those components.

Should the additional component compositions be required to fully understand the unit operation, the laboratory may have to develop new analysis procedures. These must be tested and practiced to establish reliability and minimize bias. Analysts must submit known samples to verify the accuracy.

Known samples should also be run to verify the accuracy and precision of the routine methods to be used during the unit test. Poor quality will manifest itself as poor precision, measurements inconsistent with plant experience or laboratory history, and disagreement among methods. Plotting of laboratory analysis trends will help to determine whether calibrations are drifting with time or changing significantly. Repeated laboratory analyses will establish the confidence that can be placed in the results.

If the random errors are higher than can be tolerated to meet the goals of the test, the errors can be compensated for with replicate measurements and a commensurate increase in the laboratory resources. Measurement bias can be identified through submission and analysis of known samples. Establishing and justifying the precision and accuracy required by the laboratory is a necessary part of establishing confidence.

Sampling Despite all of the preparation inside the laboratory, by far the greatest impact on successful measurements is the accuracy of the sampling methods. The number of sample points for a unit test are typically greater than the number required for routine sampling. It is likely that some of the sample locations, characteristics, and properties are unfamiliar to the sample-gatherers responsible for the routine ones. This unfamiliarity could lead to improper sampling, such that samples are not representative of the unit, and accidents, such that the sample gatherers are placed at risk. Part of the preparation process is to reduce this unfamiliarity to ensure safety and accuracy. The safety of the sample-gatherers is paramount and should not be compromised. Proper sampling methods accounting for volatility, flash points, toxicity, corrosivity, and reactivity should be written down for each plant and unit within the plant. The methodology must be understood and practiced.

Plant-performance analysts should be involved in reviewing the entire sampling procedure. The procedures for review are:

- Sampling locations
- Sampling safety
- Containers
- Sample transport, storage, and discharge system
- In-laboratory sampling
- Sample container cleaning

Each plant has established methods. The following should be considered during this preparation stage. Problems identified, typically during a pretest, should be solved prior to the initiation of the unit test.

Sampling locations for the unit test should be readily and safely accessible. The sample gatherer should be able to easily access the sample point. An isolation valve should be installed at the location. If a blind is installed, this should be modified in advance of the test. The sample locations shown on the P&IDs must be compared against the actual locations on the equipment. Experienced operators may provide insight into the suitability of the location in question.

The integrity and suitability of the sample containers must be established during the preparation stage. This is particularly important for those sample containers that will be used for the nonroutine measurements. Leaks jeopardize personnel and distort the resultant composition. Dirty containers contaminate samples. Open containers used for high boiling samples are unsuitable for volatile, high-temperature, pressurized samples. Trace components may preferentially adsorb onto either the container surface or the residue left in the container. Since trace components provide the greatest insight into unit operation and are the most difficult to quantify, this adsorption could lead to distorted conclusions.

Dead legs in the sample line must be discharged safely to ensure that the sample will actually be representative of the material in the unit. Without blowing down the dead leg, samples taken will be erroneous, as they may be representative of some past operating conditions. If the location is nonroutine, the sample leg may have accumulated debris. The debris could partially or totally block the line. Opening the isolation valve to blow down the line could result in a sudden, uncontrolled release, presenting a hazard to the sample gatherer.

Sample temperatures may be below ambient. If the sample vessel is liquid-full, a hazard results due to overpressurization as the liquid expands. Venting may be required, but it can distort the results. This safety hazard must be accounted for in the procedure and in interpreting the laboratory results.

Sample temperatures may be above ambient. If the temperature is significantly above ambient, personnel must be protected against burns.

Samples may separate into two or more phases as they cool in the sample line: precipitate, coagulate, and freeze. Laboratory sampling may result in nonrepresentative compositions. Heat tracing may be required and may not be installed on the nonroutine sample locations.

Validation of the measurements may require the simultaneous measurement of pressure and temperature. Typical sample locations do not have thermowells and pressure indicators. Consequently, modifications will be required to facilitate validation.

The efficient analysis will be required to minimize compromise of the analysis due to degradation (e.g., dimerization, polymerization, reactions, leaks, and contamination).

Samples will form multiple phases. The laboratory secondary sampling methods must recognize the presence of vapor, liquid, and solid phases. Improper secondary sampling methods will result in distorted measurements. These limitations must be clearly communicated to the laboratory.

Cataloging and storage of samples may inundate the laboratory, resulting in storage and retrieval problems. Mislabeled and lost samples are frequent problems. The longer the special samples remain in the laboratory, the greater the likelihood that some will be lost or mislabeled.

These potential sampling problems must be solved in advance of the unit test. The conclusions drawn from any unit test are strongly affected by the accuracy of the sampling methods and the resultant analyses. Methods should be discussed and practiced before the actual unit test. Analysts should use the trial measurements in preliminary plant-performance analysis to ensure that the results will be useful during the actual unit test.

PREPARATION GUIDELINES

Overall Everyone involved in the unit test and the analysis of measurements must understand:

- The purpose of the test
- The expectations of plant-performance analysts
- Each individual's personal responsibility to the successful outcome

Analyst Analysts must have a firm understanding of the operation of the unit. If they are not involved in the day-to-day operation or responsible for the unit, more preliminary work including process familiarization, equipment familiarization, operator interviews, and constraint limitations will be required. Even when an analyst is responsible, a review is necessary. Analysts must firmly establish the purpose of the unit test. Different levels require different budgets, personnel, and cost commitment. Additional resources beyond that required for routine measurements must be justified against the value of the measurements to the establishment of the understanding of the plant operation.

Model The level of sophistication needs to be identified. Preliminary usage of the model should identify the uniqueness of parameter estimates and conclusions to be drawn.

Plant Sufficient personnel and supplies will be required for the test. Personnel may include additional operators, sample-gatherers, pipe fitters, and engineers. Upstream and downstream units need notification so that feed and product rates can be maintained.

Laboratory The laboratory requirements and responsibilities need to be identified and accepted. The laboratory supervisor must be aware of the impending test and the likely demands placed on his/her area of responsibility. Agreement as to error levels and expected turnaround must be reached. Proper sampling methodology and storage must be established and practiced.

PLANT-PERFORMANCE ANALYSIS

THE PROBLEM

Consider Fig. 30-10. This is a single unit process with one input and two output streams. The goal for plant-performance analysis is to understand accurately the operation of this unit.

Plant-performance analysis requires the proper analysis of limited, uncertain plant measurements to develop a model of plant operations for troubleshooting, design, and control.

Measurements The potential set of data can be identified by the matrix $\mathbf{X}_1$

The rows represent the type of measurement (e.g., compositions, flows, temperatures, and pressures). The columns represent streams, times, or space position in the unit. For example, compositions, total flows, temperatures, and pressures would be the rows. Streams 1, 2, and 3 would be columns of the matrix of measurements. Repeated measurements would be added as additional columns.

For more complex equipment, the columns might contain measurements for internal distillation, batch-reactor intermediate conditions, or tubular-reactor between-bed conditions. Some of these

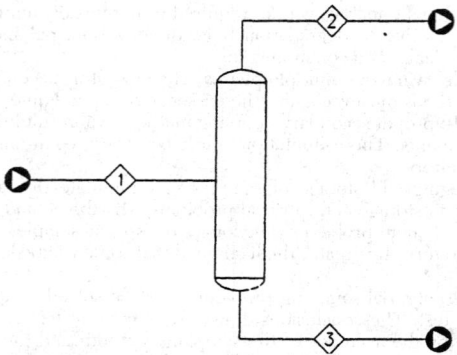

FIG. 30-10   Flow sheet of single unit process.

measurements might be recorded regularly, while others may be recorded only for the specific unit test analysis.

This matrix will necessarily be sparse. First, not all measurements can be taken for a given stream or position (e.g., a chromatographic analysis may only measure a subset of the component compositions). Second, not all streams or positions are included. Third, some of the measurements are inadequate due to bias and are discarded.

Equipment Limitations   The plant-performance engineer might also have a matrix of equipment information that must be accounted for in the analysis.

$$\mathbf{X}_2$$

This matrix will contain information regarding loading characteristics such as flooding limits, exchanger areas, pump curves, reactor volumes, and the like. While this matrix may be adjusted during the course of model development, it is a boundary on any possible interpretation of the measurements. For example, distillation-column performance markedly deteriorates as flood is approached. Flooding represents a boundary. These boundaries and nonlinearities in equipment performance must be accounted for.

The purpose of the plant-performance analysis is to operate on the set of measurements obtained, subject to the equipment constraints to troubleshoot; to develop models; or to estimate values for model parameters.

$$g\langle \mathbf{X}_1^m; \mathbf{X}_2\rangle \Rightarrow \vec{\beta}$$

where $g\langle \rangle$ is an operator on the measurements and data. The vector $\vec{\beta}$ is a representation of the conclusions, model, and/or equipment parameters.

Measurement Selection   The identification of which measurements to make is an often overlooked aspect of plant-performance analysis. The end use of the data interpretation must be understood (i.e., the purpose for which the data, the parameters, or the resultant model will be used). For example, building a mathematical model of the process to explore other regions of operation is an end use. Another is to use the data to troubleshoot an operating problem. The level of data accuracy, the amount of data, and the sophistication of the interpretation depends upon the accuracy with which the result of the analysis needs to be known. Daily measurements to a great extent and special plant measurements to a lesser extent are rarely planned with the end use in mind. The result is typically too little data of too low accuracy or an inordinate amount with the resultant misuse in resources.

If the problem were accurately known, identification of which measurements should be taken would be exact. When the problem is initially not accurately known, the identification, measurement, and analysis procedure is iterative. Familiarity with the plant will help in identifying the measurements most likely to provide insight.

When building a model for the plant either in terms of a set of relations or in terms of a set of parameters for an existing model, it is important that the measurements contain a maximum amount of information. If the model is embodied in the symbol of the parameters, $\vec{\beta}$, then the measurements should be made such that the measurement matrix $\mathbf{X}_1^m$ has the greatest impact on $\vec{\beta}$. This maximizes the plant information contained in the parameters. The process is necessarily iterative. Measurements are analyzed to refine the model and optimal locations for new measurements are defined.

Analysts must recognize the above sensitivity when identifying which measurements are required. For example, a typical use of plant data is to estimate the tray efficiency or HTU of a distillation tower. Certain tray compositions are more important than others in providing an estimate of the efficiency. Unfortunately, sensor placement or sample port location are usually not optimal and, consequently, available measurements are, all too often, of less than optimal use. Uncertainty in the resultant model is not minimized.

Plant Operations   Each of the elements x_{ij} in $\mathbf{X}_1^m$ have inherent error. Consequently, x_{ij} is only an estimator of the actual plant value. Or,

$$x_{ij} = \tilde{x}_{ij} + \varepsilon$$

It is useful to recognize the contributions to this error.

First, plant operations are rarely exactly as intended. While the designer may have developed all operating specifications as if the plant would operate at steady state, the plant fluctuates and drifts with time. Changes occur because of changes in feed stock, atmospheric conditions, operating conditions, controller response, and any number of factors, both known and unknown. The actual value then fluctuates around some mean operation. For example, Fig. 30-11 is a typical plant strip chart recording. The lower trace shows that the measurement is fluctuating around a mean value. The upper trace also shows fluctuation, but the mean value is changing with time.

While the random fluctuations apparent are a function of the scaling factor for the traces, the two show different amplitudes. The top trace has a relatively small fluctuation, while the bottom trace shows a larger one.

Mathematically, the mean value is the desired value for further analysis.

$$\tilde{x}_{ij} = \bar{x}_{ij} + \varepsilon_{ij}^P$$

The plant drift makes all measurements functions of time. The upper trace in the above figure shows some evidence of drift. Figure 30-12 shows a larger drift.

This drift can be represented mathematically as:

$$\tilde{x}_{ij}(t) = \bar{x}_{ij}(t) + \varepsilon_{ij}^P$$

This time dependence is different for each measurement. The fluctuation may also be a function of time.

In addition to the drift with time, step changes due to operating decisions, atmospheric changes, or other conditions result in additional time dependence. Not only is there a sudden change due to the actual decision, but also the plant changes due to the time constants. For example, Fig. 30-13 shows measurements with step changes in the operation.

Data Limitations   The process of measuring $\tilde{x}_{ij}(t)$ adds additional error due to the random error of measurement. Or,

$$x_{ij}(t) = \tilde{x}_{ij}(t) + \varepsilon_{ij}^M$$
$$x_{ij}(t) = \bar{x}_{ij}(t) + \varepsilon_{ij}^P + \varepsilon_{ij}^M$$

Consequently, if these random errors are assumed to be normal, the total uncertainty including fluctuations is:

$$\sigma_{ij}^T = \sqrt{(\sigma_{ij}^P)^2 + (\sigma_{ij}^M)^2}$$

where σ replaces ε to represent a normal distribution. Therefore,

$$x_{ij}(t) = \bar{x}_{ij}(t) + \sigma_{ij}^T$$

The problem with plant data becomes more significant when sampling, instrument, and calibration errors are accounted for. These errors result in a systematic deviation in the measurements from the actual values. Descriptively, the total error (mean square error) in the measurements is

$$\text{MSE} = (\sigma_{ij}^T)^2 + (b^M)^2$$

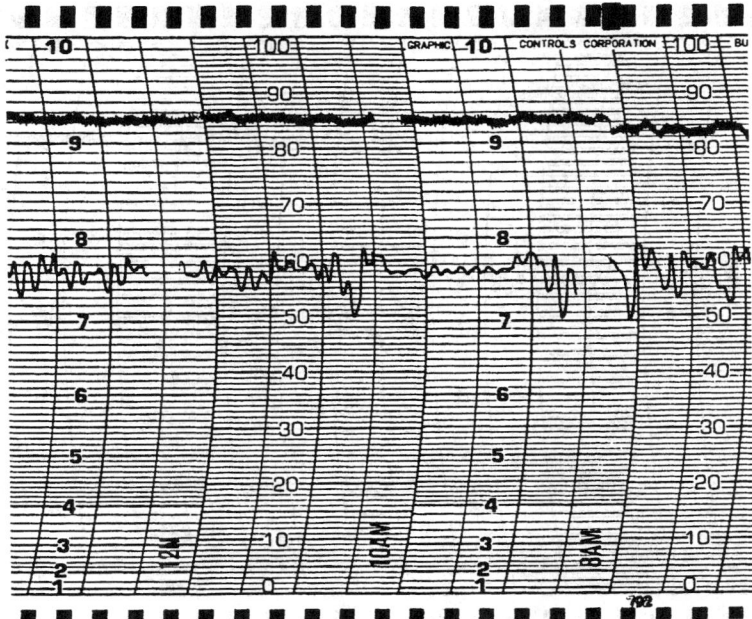

FIG. 30-11   Plant measurements showing fluctuations around a mean value.

The above assumes that the measurement statistics are known. This is rarely the case. Typically a normal distribution is assumed for the plant and the measurements. Since these distributions are used in the analysis of the data, an incorrect assumption will lead to further bias in the resultant troubleshooting, model, and parameter estimation conclusions.

Constraints Limitations   Typically, the plant performance is assumed to be subject to process constraints.

$$\vec{f}\langle X_1 \rangle = \vec{0}$$

where $\vec{f}\langle\rangle$ is a vector of constraints. For the process shown in Fig. 30-10, these constraints could be, using the component flows and total flows and temperatures as the measurements:

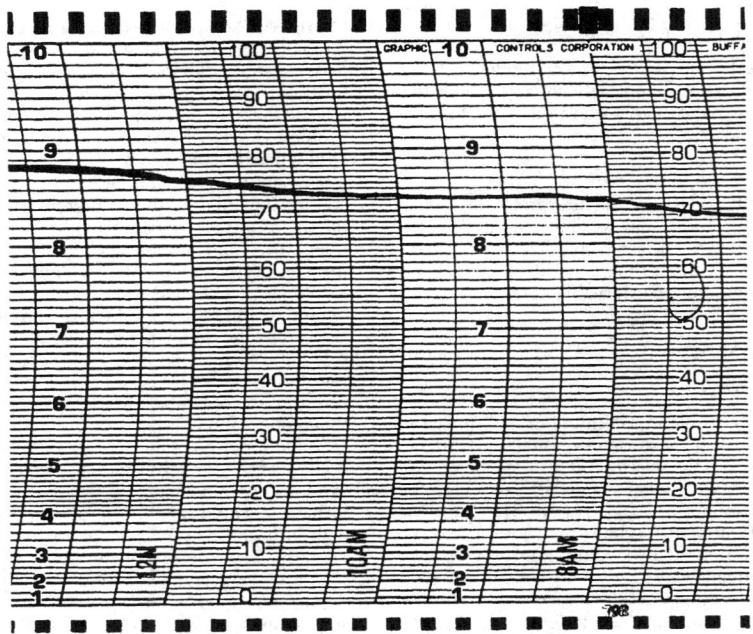

FIG. 30-12   Plant measurements showing drift with time.

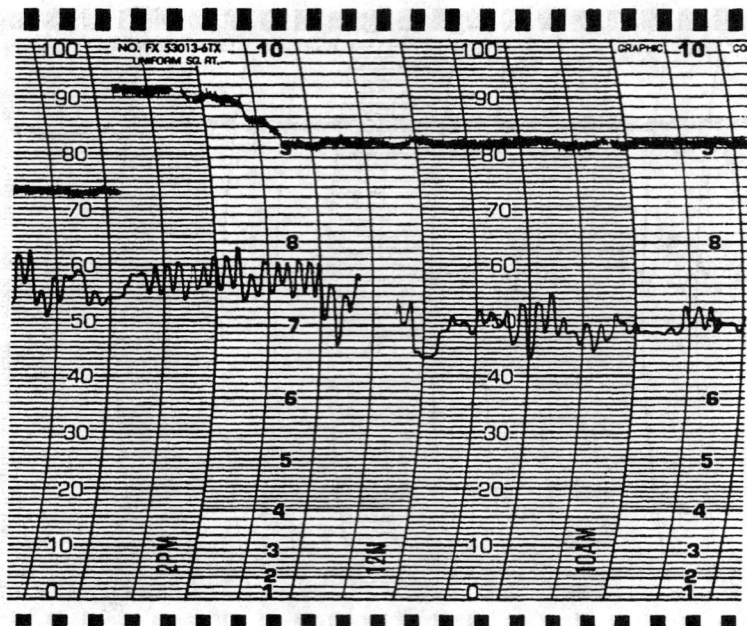

FIG. 30-13 Plant measurements exhibiting step changes, drift, and random fluctuations.

$$x_{i,1} - x_{i,2} - x_{i,3} = 0 \qquad i = 1 \ldots c$$

where c is the number of components

$$x_{c+1,1} - x_{c+1,2} - x_{c+1,3} = 0$$

$$x_{c+1,1}H_1 - x_{c+1,2}H_2 - x_{c+1,3}H_3 = 0$$

where the $c + 1$th position is the total flow. Only c of the material balance constraints are independent. Of course, the actual measurements do not close the constraints.

$$\vec{f}\langle \mathbf{X}_1^m \rangle \neq \vec{\mathbf{0}}$$

To complicate matters further, because of the time dependence, leaks, or accumulation, the constraints might not actually apply such that there is a vector of unknown plant bias' associated with the constraints.

$$\vec{f}\langle \mathbf{X}_1^m \rangle = \vec{\mathbf{b}}^P$$

Assuming $\vec{\mathbf{b}}^P = 0$ will potentially add bias to the interpretation of plant measurements. Further, the plant bias may to some extent mask the error in the measurements. While the designer may have envisioned a constant set of conditions or a specified time dependence, it is likely that the actual operation changes due to external factors.

The technical problem becomes one of:

$$g\langle \mathbf{X}^m : \mathbf{X}_2 \rangle \overset{b^o}{\Rightarrow} \vec{\beta}$$

subject to
$$\vec{f}\langle \mathbf{X}_1^m \rangle = \vec{\mathbf{b}}^P$$

This is a formidable analysis problem. The number and impact of uncertainties makes normal plant-performance analysis difficult. Despite their limitations, however, the measurements must be used to understand the internal process. The measurements have limited quality, and they are sparse, suboptimal, and biased. The statistical distributions are unknown. Treatment methods may add bias to the conclusions. The result is the potential for many interpretations to describe the measurements equally well.

Personnel Bias Because of the possibility of several interpretations of the plant-performance problem, the judgment of analysts plays a critical role. Any bias in the analysts' judgments will carry through the data analyses. To minimize this, analysts must develop an implicit model based on the fundamental rules of the plant and not on the prejudices of the operators, designer engineers, control engineers, or the analyst's own perceptions.

The following presents guidelines for identifying, validating, reconciling, rectifying, and interpreting plant measurements to remove some of the bias from the conclusions.

IDENTIFICATION

Motivation Unit tests require a substantial investment in time and resources to complete successfully. This is the case whether the test is a straightforward analysis of pump performance or a complex analysis of an integrated reactor and separation train. The uncertainties in the measurements, the likelihood that different underlying problems lead to the same symptoms, and the multiple interpretations of unit performance are barriers against accurate understanding of the unit operation. The goal of any unit test should be to maximize the success (i.e., to describe accurately unit performance) while minimizing the resources necessary to arrive at the description and the subsequent recommendations. The number of measurements and the number of trials should be selected so that they are minimized.

Often, analysts will want to run special short-term tests with the operating unit in order to identify the cause of the trouble being experienced by the unit. Operators are naturally leery of running tests outside their normal operating experience because their primary focus is the stable control of the unit, and tests outside their experience may result in loss of control. Multiple tests with few results may decrease their cooperation.

Modern petro/chemical processes provide the opportunity for gathering a large number of measurements automatically and frequently. Most are redundant and provide little additional insight into unit performance. The difficulties in handling a large amount of information with little intimate knowledge of the operation increases the likelihood that some of the conclusions drawn will be erroneous.

Therefore, the identification of appropriate tests and measurements most important to understanding the unit operation is a critical step in the successful analysis of plant performance.

Limitations Identifying the appropriate test to troubleshoot a unit problem requires hypothesis development and testing. Hypotheses are based on the observed problem in current operation and the historical performance. It is the skill of analysts to develop the minimum number of hypotheses and unit tests to identify what is typically a well-hidden problem.

Identifying the minimum number of specific measurements containing the most information such that the model parameters are uniquely estimated requires that the model and parameter estimates be known in advance. Repeated unit tests and model building exercises will ultimately lead to the appropriate measurements. However, for the first unit test in absence of a model, the identification of the minimum number of measurements is not possible.

The methodology of identifying the optimum test and number of measurements has received little attention in analysis of plant performance and design literature.

Measurement Error   Uncertainty in the interpretation of unit performance results from statistical errors in the measurements, low levels of process understanding, and differences in unit and modeled performance (Frey, H.C., and E. Rubin, "Evaluate Uncertainties in Advanced Process Technologies," *Chemical Engineering Progress,* May 1992, 63–70). It is difficult to determine which measurements will provide the most insight into unit performance. A necessary first step is the understanding of the measurement errors likely to be encountered.

An example adapted from Verneuil, et al. (Verneuil, V.S., P. Yan, and F. Madron, "Banish Bad Plant Data," *Chemical Engineering Progress,* October 1992, 45–51) shows the impact of flow measurement error on misinterpretation of the unit operation. The success in interpreting and ultimately improving unit performance depends upon the uncertainty in the measurements. In Fig. 30-14, the material balance constraint would indicate that $S_3 = -7$, which is unrealistic. However, accounting for the uncertainties in both S_1 and S_2 shows that the value for S_3 is -7 ± 28. Without considering uncertainties in the measurements, analysts might conclude that the flows or model contain bias (systematic) error.

Analysts should review the technical basis for uncertainties in the measurements. They should develop judgments for the uncertainties based on the plant experience and statistical interpretation of plant measurements. The most difficult aspect of establishing the measurement errors is establishing that the measurements are representative of what they purport to be. Internal reactor CSTR conditions are rarely the same as the effluent flow. Thermocouples in catalyst beds may be representative of near-wall instead of bulk conditions. Heat leakage around thermowells results in lower than actual temperature measurements.

These measurement uncertainties must be accounted for in developing hypotheses used to explain unit performance and in identifying measurements which will provide the best model of the unit.

Hypothesis Development   Successful, efficient development of hypotheses and operating conditions to test them require design, operation, control, and troubleshooting experience. Understanding the relation of the fundamentals of chemical engineering, the specifications and their intent for operation, the response of equipment to upsets, and the identification of the unusual are all essential tools for developing and testing hypotheses during troubleshooting exercises. Hypothesis development is typically iterative with unit operating conditions adjusted to test a hypothesis. The results lead to other hypotheses and other operating conditions. This is an essential part of troubleshooting and model development.

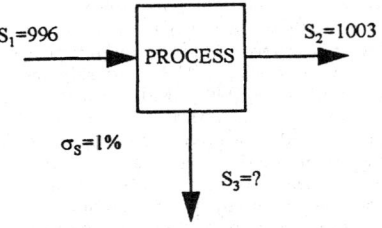

FIG. 30-14   Material balance measurements with error.

Troubleshooting is usually based on checklists developed by analysts specific to the unit and types of equipment in the unit. These checklists assist in hypothesizing the cause of observed problems based on past experience and in developing tests or measurements to confirm the hypotheses. Few published checklists exist in the literature. Most analysts develop their own based on experience. As engineers move from assignment to assignment with little direct, continued experience in the design, operation, control, and troubleshooting, the checklists are lost. The skill resides with the engineer and not with the unit. Consequently, individual checklists are developed repeatedly with little continuity unless current analysts seek out those who were once responsible for the unit. One notable exception is the set of checklists for refinery operations published by Lieberman (1981). Many of his experiences arise in and apply to other chemical engineering applications. His lists give the observed problem and possible explanations. Harrison and France (Harrison, M.E. and J.J. France, "Auxiliary Equipment: Troubleshooting Distillation Columns," *Chemical Engineering,* June 1989, 130–137) list a series of problems with corresponding causes. Symptoms in one piece of equipment may appear as a problem in another; therefore, checklists should include the potential that other equipment in the unit is the cause of the observed problem.

A proposed checklist form is given in Table 30-1. The descriptive example concerns a problem observed with distillation performance in a specific unit. It is included for descriptive purposes only and does not provide an exhaustive list of possible explanations for the observed problem. The important aspects are a clear statement of the problem; recognized changes in unit operation at the time of the observation and hypothesized causes under the categories of erroneous instrument readings; changes upstream and downstream from the unit and within the unit itself. Typical explanations under each cause category could be substantially longer than ones included in the table.

History is important is establishing hypotheses. When a problem arises in a unit, something has changed. The first step in developing a hypothesis explaining the cause of the problem is to establish that the operation has clearly changed from some earlier operation. Easily identified alterations in operation such as those that result from changes in operating specifications, equipment installations, or operator responses should be listed.

The observations may be erroneous due to misleading measurements. The basis of the observations should be examined. Instruments may have degraded. Sample lines may have become plugged. Trip settings may have changed. Where possible, these causes should be eliminated before moving to more complex explanations and tests. Many unit problems are caused by upstream or downstream units. This interaction should be identified before performing extensive tests with the unit. In the table, a sudden increase in light ends could flood the upper section of the tower. Pump cavitation may be the result of fluctuating discharge pressures in the downstream units. Corrosion in the unit may be caused by carryover from an upstream unit. Insufficient pump capacity could be caused by a changed fluid density from changed feed stock.

With the problem being identified as real and other units being eliminated as the cause, the focus can move to identifying whether the problem is with capacity or efficiency within the unit.

The following are guidelines for establishing checklists used to identify the cause of observed problems.

TABLE 30-1   Example Checklist Form

Observed problem	Increased pressure drop in the distillation column
Unit changes	Steam header pressure increase, no equipment changes
Instrument cause	DP meter reading is misleading due to failed instrument, plugged ports, etc.
Upstream cause	Increased percentage of light components fed to column resulting in flooding in rectifying section
Downstream cause	Not applicable
Capacity cause	Steam reboiler flow set above column jet flood limit
Efficiency cause	Trays plugged with polymer buildup

- Establish the timeline of the problem hypothesizing that changes in operation, equipment, or response are the root cause of the problem.
- Establish the observed problem is real by hypothesizing potential problems with instruments and instrument installations.
- Establish that the observed problem could not be caused by upstream or downstream unit performance.
- Establish that the problem is one with capacity of the unit by hypothesizing causes for the decreased unit production.
- Establish that the problem is one with efficiency of the unit by hypothesizing causes to explain the decreased performance of the unit.

Any set of guidelines must be tempered by the analysts' experience. This is an investigative process. The explanations are rarely simple. However, many exhaustive tests have been run to identify that a bypass valve or alternative feed valve had been mistakenly left open. Plant resources were misused because the simple was overlooked.

Since hypothesis development and testing frequently require alternative operating conditions, safety considerations must be paramount. The operators' concerns about loss of control are justified. When tests are planned, it must be recognized that adjustments should be slow and stepwise with time allowed for the unit to line out. All possible outcomes of the adjustments should be thought through to minimize the potential for moving the unit into an unstable operating regime.

Model Development   Preliminary modeling of the unit should be done during the familiarization stage. Interactions between database uncertainties and parameter estimates and between measurement errors and parameter estimates could lead to erroneous parameter estimates. Attempting to develop parameter estimates when the model is systematically in error will lead to systematic error in the parameter estimates. Systematic errors in models arise from not properly accounting for the fundamentals and for the equipment boundaries. Consequently, the resultant model does not properly represent the unit and is unusable for design, control, and optimization. Cropley (1987) describes the erroneous parameter estimates obtained from a reactor study when the fundamental mechanism was not properly described within the model.

Verneuil et al. (Verneuil, V.S., P. Yan, and F. Madron, "Banish Bad Plant Data," *Chemical Engineering Progress*, October 1992, 45–51) emphasize the importance of proper model development. Systematic errors result not only from the measurements but also from the model used to analyze the measurements. Advanced methods of measurement processing will not substitute for accurate measurements. If highly nonlinear models (e.g., Cropley's kinetic model or typical distillation models) are used to analyze unit measurements and estimate parameters, the likelihood for arriving at erroneous models increases. Consequently, resultant models should be treated as approximations.

Recognition of measurement error, model nonlinearities, interactions, and potential fundamental oversights are an important part of the identification stage of analysis of plant performance. Repeated simulations using different models extrapolated to other operating conditions will provide insight into model viability. Model accuracy can be verified by operating the unit at different operating conditions and making appropriate measurements. Identification of these conditions and measurements is one aspect of the identification step. These model building studies to identify possible alternative models and operating conditions are useful in minimizing the impact of erroneous model development and subsequent parameter estimation.

Measurement Selection   Along with the hypothesis development, the principal result of the identification step is determining which measurements will provide insight into the unit operation. This often-overlooked aspect of analysis of plant performance deserves greater attention in the plant operations and research literature. The potential resource savings resulting from minimizing the number of measurements, repeated unit tests, and associated personnel are enormous. Coupled with the benefit of developing a more robust model of the unit, this overlooked aspect of analysis of plant performance potentially outweighs the benefits of all other aspects.

The goal of measurement selection is to identify a set of measurements that, when interpreted, will lead to unique values for the model parameters, insensitive to uncertainties in the measurements. This is an iterative process where:

- A group of measurements are proposed based on preliminary model predictions
- Values for parameters are estimated using the interpretation procedures
- Simulated unit performance sensitivity to the parameter estimates is evaluated
- Alternative measurements are proposed
- The process is repeated

The optimum measurements are those taken in the unit test. Figure 30-15 provides one procedure for identifying which measurements should be taken within the plant.

A preliminary model is developed during the preparation stage. Preliminary values of the model parameters are estimated based on adjusted plant measurements. Simulations of the unit are then run to develop values for the temperatures, pressures, flows, compositions, and the like, that are representative of the unit operation. A group of measurements that could possibly be taken in the unit test is then selected. At that point, analysts have two options. In option A, the parameter estimates are perturbed, the unit resimulated, and the group of measurements compared to the set corresponding to the perturbed parameters. If the comparison is such that the simulated measurements are different beyond the experimental error, then the parameter values are unique and the group of measurements are appropriate. If they are not, the proposed measurements should be changed and the process repeated. In option B, the process is similar. The group of measurements are perturbed according to the measurement error, the parameters re-estimated, and the parameter values compared. If there is relatively little change in the parameter values, the selected measurements are acceptable. If there is a large change, the measurements do not provide a unique set of parameter estimates. Consequently, the model would be unsuitable. The measurement set needs to be modified. Once the set of measurements have been selected, the model should be examined and modified if necessary. There are two primary indications that the model may be inadequate. First, the preliminary model with the estimated parameters provide descriptions of one or more measurements representing unit behavior, particularly internal to individual pieces of equipment. Second, the values of the parameters are unrealistic.

With respect to selecting measurements, emphasis should include measurements within the equipment such as tower internal temperatures and compositions, internal reactor conditions, and intermediate exchanger temperatures in multipass exchangers. Trace component compositions provide particular insight into distillation-column performance. Those components that fall between the heavy and light keys and distribute in the products can usually be described by a variety of models and parameter estimates: They provide little insight into the column performance.

The procedure given in Fig. 30-15 leaves much to analysts. Criteria for selecting the number and location of measurements for a particular piece of equipment or unit have not been established in the literature. Therefore, there is heavy reliance on examining alternative models at the bottom of the procedure. The creativity of analysts to develop alternative explanations for performance or hypotheses explaining why the present model might be wrong is a particularly important skill.

VALIDATION

Initial Measurement Examination   The process of reconciling data to constraints; rectifying data to detect and identify systematic errors; and interpreting data to troubleshoot, model-build, and estimate parameters is a time-consuming, often unnecessary, and, many times, inaccurate series of steps. Even under the most controlled circumstances, the methods often provide estimates of plant operation that are no better than that provided by the actual plant measurements. If the adjusted measurements contain significant error, the resultant conclusions could be significantly in error and misleading. Prescreening can identify measurements containing significant error and can provide insight into the plant operation.

Validation is the procedure of comparing measurements to known relations between the measurements and equipment settings (May,

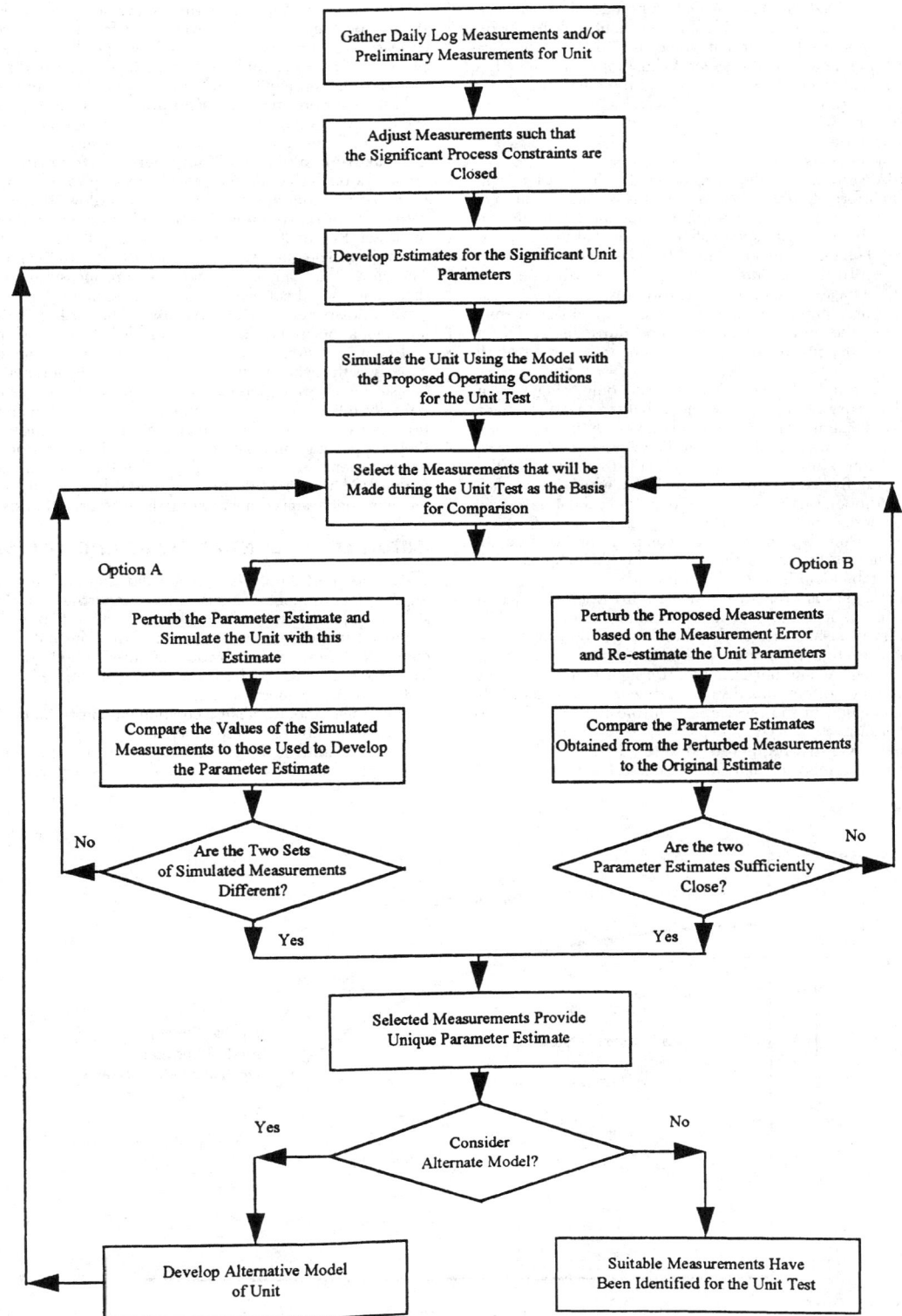

FIG. 30-15 Procedure for identifying measurements.

D.L. and J.T. Payne, "Validate Process Data Automatically," *Chemical Engineering,* June 1992, 112–116). If a measurement is clearly inconsistent with equipment operation that is known to be true, the measurement must then be deemed suspect. Validation is the procedure of comparing a measurement to one or more of the following.

• Another measurement
• An expected range
• Equipment status
• Equipment relations

If the comparison shows that the measurement is inconsistent with the comparison information, the measurement is considered suspect. If a measurement can be compared to more than one set of information and found to be inconsistent with all, it is likely that the measurement is in error. The measurement should then be excluded from the measurement set. In this section, validation is extended to include comparison of the measurements to the constraints and initial adjustment in the measurements. Validation functions as an initial screening procedure before the more complicated procedures begin. Oftentimes, validation is the only measurement treatment required prior to interpretation.

It is important to note that validation typically only brings a measurement under suspicion. It does not verify that the measurement is incorrect. Safety is paramount. Some validation analysis could result in concluding that the measurement is invalid when, in fact, the comparison information is invalid. It is not difficult to extrapolate that actions could result from this erroneous conclusion which would place maintenance and operating personnel in jeopardy. Validation merely raises suspicion; it does not confirm errors of measurement.

The greater the number of validation comparisons between the measurement and the list above, the greater the likelihood that the measurement can be identified as valid or invalid.

Measurement versus Measurement   In this type of validation, a process measurement is compared against another. For example, if a separate high-level alarm indicates that a tank is overflowing but the level gauge indicates that it is in the expected range, one of these measurements is wrong. As another example, if a light component suddenly appears in the bottoms of a distillation tower and no other light components contained in the feed appear in the bottoms, the first measurement is suspect.

Measurement versus Expected Range   If a steam flow is expected to vary in a relatively narrow range and the flow measurement indicates that it is twice the high value, the flow measurement is then suspect and should be reviewed. A frequent occurrence is when a measurement remains unchanged for a period of time when normal plant fluctuations should result in oscillations around a setpoint. The constant measurement would indicate that this reading is suspect.

Measurement versus Equipment State   A pump off-line should have no flow. If the pump is off and the flow meter indicates that there is flow, the flow measurement is suspect.

Measurement versus Equipment Performance   Pumps that are in reasonable condition typically operate within 5 percent of their pump curve. Consequently, pressures and flows that are inconsistent with the pump curve imply that the indicated flow and/or pressure are incorrect. Figure 30-16 shows a single impeller curve plotted as head versus flow. The point shown is inconsistent with the pump operation. Therefore, that pair of flow and pressure measurements is not validated and should not be used in the subsequent steps.

Validation versus Rectification   The goal of both rectification and validation is the detection and identification of measurements that contain systematic error. Rectification is typically done simultaneously with reconciliation using the reconciliation results to identify measurements that potentially contain systematic error. Validation typically relies only on other measurements and operating information. Consequently, validation is preferred when measurements and their supporting information are limited. Further, prior screening of measurements limits the possibility that the systematic errors will go undetected in the rectification step and subsequently be incorporated into any conclusions drawn during the interpretation step.

INITIAL CONSTRAINT ANALYSIS AND ADJUSTMENTS

Spreadsheet Analysis   Once validation is complete, prescreening the measurements using the process constraints as the comparison statistic is particularly useful. This is the first step in the global test discussed in the rectification section. Also, an initial adjustment in component flows will provide the initial point for reconciliation. Therefore, the goals of this prescreening are to:

• Pretreat raw measurements
• Estimate the overall and component constraint deviations
• Identify missing measurements
• Adjust (initially) the measurements to close the constraints

The principal focus of this validation is the material and energy bal-

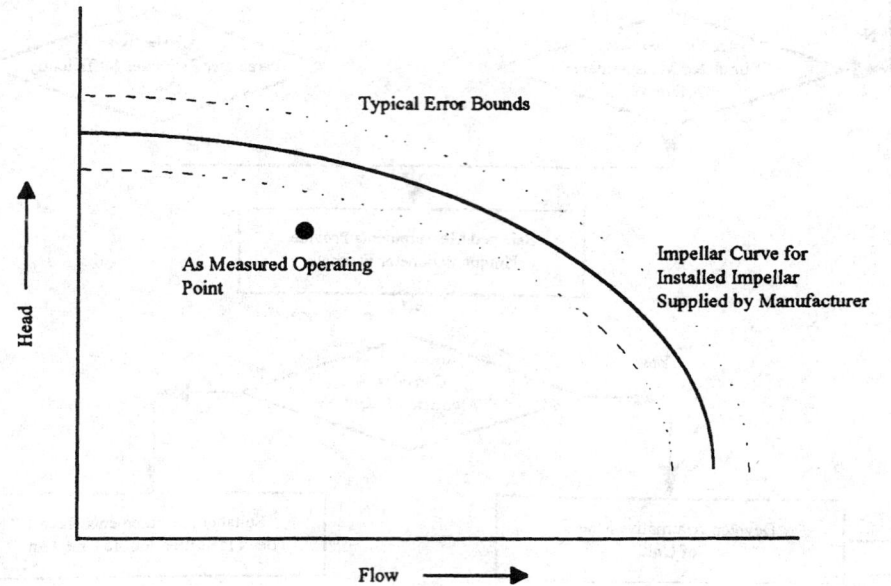

FIG. 30-16   Typical pump curve showing inconsistency between measurement and curve.

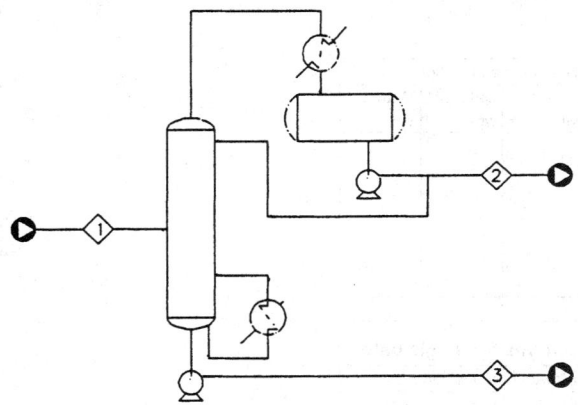

FIG. 30-17 Distillation tower example.

ances for the unit. Specifically designed spreadsheets are particularly useful during this step. The level of sophistication depends upon the analysts' goals. Spreadsheets can be used for pretreatment of measurements, constraint analysis, and measurement adjustment. Oftentimes, the more sophisticated reconciliation and rectification methods are not warranted or will not provide any better results, particularly when a single unit is under analysis.

For the purposes of this discussion, consider a single distillation tower with one feed, a distillate, and bottoms, as shown in Fig. 30-17. A straightforward, generic analysis spreadsheet for this tower is shown in Fig. 30-18. For this example, the three stream compositions and the total flows have all been measured. Also, since this is a column in a purification train, the bottoms flow rate has been measured independently as the feed to the next tower.

Spreadsheet Structure   There are three principal sections to the spreadsheet. The first has tables of as-reported and normalized composition measurements. The second section has tables for overall and component flows. These are used to check the overall and component material balance constraints. The third has adjusted stream and component flows. Space is provided for recording the basis of the adjustments. The structure changes as the breadth and depth of the analysis increases.

The example spreadsheet covers a three-day test. Tests over a period of days provide an opportunity to ensure that the tower operated at steady state for a period of time. Three sets of compositions were measured, recorded, normalized, and averaged. The daily compositions can be compared graphically to the averages to show drift. Scatter-diagram graphs, such as those in the reconciliation section, are developed for this analysis. If no drift is identified, the scatter in the measurements with time can give an estimate of the random error (measurement and fluctuations) in the measurements.

The second section of the spreadsheet contains the overall flows, the calculated component flows, and the material balance closure of each. The weighted nonclosure can be calculated using the random error calculated above, and a constraint test can be done with each component constraint if desired. Whether the measurement test is done or not, the nonclosure of the material balance for each component gives an indication of the validity of the overall flows and the compositions. If particular components are found to have significant constraint error, discussions with laboratory personnel about sampling and analysis and with instrument personnel about flow-measurement errors can take place before any extensive computations begin.

The measurements and flows can be adjusted to close the constraints. These adjustments can then be compared to the measurements to determine whether any are reasonable. Statistical routines or hand adjustments are possible. These adjusted flows and compositions might form the basis for the interpretation step bypassing any deeper reconciliation and rectification. This is particularly appropriate where many compositions are left unmeasured and those that are

measured have different levels of error. More sophisticated routines will not compensate for incomplete, imprecise, and potentially inaccurate measurements.

Recommendations   Once measurements are made, validation is the most important step for establishing a sound set of measurements. The comparisons against other measurements or other known pieces of information quickly identify suspect measurements. Spreadsheet analysis of constraints, particularly material and energy balances, identifies other weaknesses in the measurements and provides the opportunity for discussions with those responsible before considerable analysis effort is expended. Finally, initial adjustments provide the beginnings of the interpretation analysis.

RECONCILIATION

Single-Module Analysis   Consider the single-module unit shown in Fig. 30-10. If the measurements were complete, they would consist of compositions, flows, temperatures, and pressures. These would contain significant random and systematic errors. Consequently, as collected, they do not close the constraints of the unit being studied. The measurements are only estimates of the actual plant operation. If the actual operation were known, the analyst could prepare a scatter diagram comparing the measurements to the actual values, which is a useful analysis tool. Figure 30-19 is an example.

If the measurements were completely accurate and precise (i.e., they contained neither random nor systematic error), all of the symbols representing the individual measurements would fall on the zero deviation line. Since the data do contain error, the measurements should fall within ± 2 on this type of diagram. This example scatter diagram shows that some of the measurements do not compare well to the actual values.

Unfortunately, the actual plant operation is unknown. Therefore, the actual value of each of the measurements is unknown. The purpose of reconciliation is to adjust the measurements so that they close the process constraints. The implicit hypothesis is that the resultant adjusted measurements better represent the actual unit operation than do the actual measurements.

Statistical Approach   Ignoring any discrepancies between the implicit model used to establish the constraints and the actual unit, the measurements are adjusted to close the constraints. This adjustment effectively superimposes the known process operation embodied in the constraints onto the measurements. Minimum adjustments are made to the measurements.

The matrix of measurements is rearranged into a stacked vector where each subsequent set of stream measurements follows the one above. As an example, the component flows in the $\mathbf{X}_1^m$ matrix are placed in the vector of measurements as follows:

$$\vec{\mathbf{X}}^m = \begin{bmatrix} x_{1,1} \\ x_{2,1} \\ \cdot \\ \cdot \\ \cdot \\ x_{c,1} \\ x_{2,1} \\ \cdot \\ \cdot \\ x_{c-1,3} \\ x_{c,3} \end{bmatrix}$$

Defining, $\delta \vec{\mathbf{X}}_1 = \hat{\vec{\mathbf{X}}}_1^m - \vec{\mathbf{X}}_1^m$

Minimize: $\delta \vec{\mathbf{X}}_1^\mathsf{T} \delta \vec{\mathbf{X}}_1$

Such that: $\vec{\mathbf{f}}(\hat{\vec{\mathbf{X}}}_1^m) = \vec{\mathbf{0}}$

If the constraints are linear (e.g., the component flow material balances) or can be linearized, then

$$\mathbf{B}\hat{\vec{\mathbf{X}}}_1^m = \vec{\mathbf{0}}$$

In the material balance example, the matrix $\mathbf{B}$ contains the material balance coefficients for the component flows based on the implicit model of the process. These adjustments can be done by hand or by

ANALYSIS OF TOWER DATA
Measured compositions reported on lab logs

Component	Feed composition			Distillation composition			Bottoms composition		
	Date 1 wt%	Date 2 wt%	Date 3 wt%	Date 1 wt%	Date 2 wt%	Date 3 wt%	Date 1 wt%	Date 2 wt%	Date 3 wt%
Component 1 Component 2 Component 3 • • • Component c-1 Component c									
Total									

Average compositions for period

Component	Stream compositions		
	Feed wt%	Ovhd wt%	Btms wt%
Component 1 Component 2 Component 3 • • • Component c-1 Component c			
Total			

Normalized compositions for single date

Component	Stream compositions		
	Feed wt%	Ovhd wt%	Btms wt%
Component 1 Component 2 Component 3 • • • Component c-1 Component c			
Total			

Stream flowrates for single date

Stream	lb/hr	
Feed		
Distillate		
Bottoms		As measured
Bottoms		Back-calculated from next unit
Bottoms		Back-calculated to close

Projected stream flows—next tower basis

Component	Stream flows			Closure	
	Feed lb/hr	Ovhd lb/hr	Btms lb/hr	lb/hr	%
Component 1 Component 2 Component 3 • • • Component c-1 Component c					
Total					

Projected single date flows—next tower basis

Component	Stream flows			Closure	
	Feed lb/hr	Ovhd lb/hr	Btms lb/hr	lb/hr	%
Component 1 Component 2 Component 3 • • • Component c-1 Component c					
Total					

Possible material balance adjustment

Component	Stream flows						
	Feed		Ovhd product		Btms product		
	lb/hr	wt%	lb/hr	wt%	lb/hr	wt%	Descriptive notes
Component 1 Component 2 Component 3 • • • Component c-1 Component c							
Total							

FIG. 30-18 Generic spreadsheet for analyzing measurement validity.

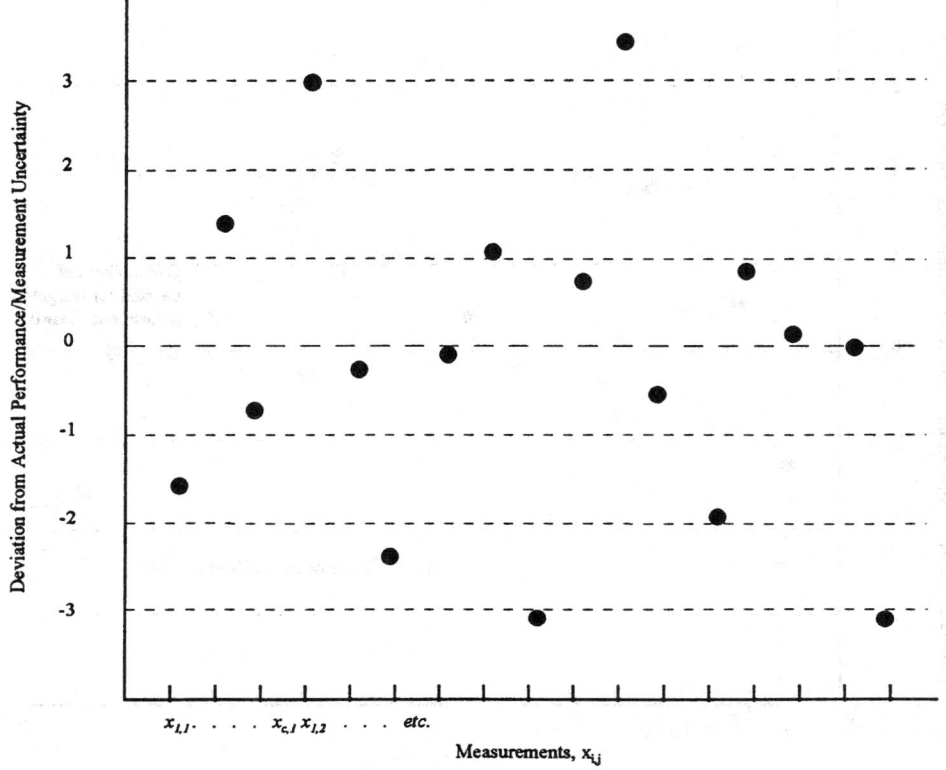

FIG. 30-19 Scatter diagram of measurements before reconciliation.

using computer aids. They can be made without consideration of measurement errors in the data (Leibovici, C.F., et al., "Improve Prediction with Data Reconciliation," *Hydrocarbon Processing*, October, 1993, 79–80) as above or can be done by accounting for the random errors (MacDonald, R.J. and C.S. Howat, "Data Reconciliation and Parameter Estimation in Plant Performance Analysis," *AIChE Journal*, **34**(1), 1988, 1–8.) For the latter, the problem becomes:

Minimize: $\delta\vec{X}_1^T J^{-1}\delta\vec{X}_1$

Such that: $\vec{f}(\hat{X}_1^m) = \vec{0}$

where **J** is the variance-covariance of the measurements. If the number of measurements is limited for a stream, the adjustments can be made on the limited number of measurements. The constraints can also be used to estimate missing or discarded measurements: This use of the constraints is defined as *coaptation* in the literature. However, this propagates errors and should be done with caution.

Analysis of Measurement Adjustments   Once reconciliation has been completed, the adjusted measurements can be compared to the actual measurements using a scatter diagram. Figure 30-20 presents an example. In this figure, the weighted residuals in the adjustments are plotted. The weighting factor is a measure of the random error in that particular measurement. In this visualization, the value of the residual should be between ±2. The scatter has improved from the previous figure, but numerical studies have indicated that the analyst can expect only 60 percent of the measurements to be adjusted toward the actual value. Consequently, while the scatter may have improved, there is no guarantee that a particular adjusted measurement is better than the actual measurement. This is one of the principal shortcomings of any automatic data adjustment method.

Adjustments outside this range could be suspect, either because of measurement error or error in the estimated uncertainty. These will be evaluated in the rectification step. Weighted residual values of 0 do not necessarily indicate that the measurement is correct. While this is a possible explanation, a more likely one is that the selected constraints used in the reconciliation are not sensitive functions of this measurement. Therefore, in the interpretation step, caution is recommended in using these adjusted measurements to compare against the model estimate.

At this point, analysts have a set of adjusted measurements that may better represent the unit operation. These will ultimately be used to identify faults, develop a model, or estimate parameters. This automatic reconciliation is not a panacea. Incomplete data sets, unknown uncertainties and incorrect constraints all compromise the accuracy of the adjustments. Consequently, preliminary adjustments by hand are still recommended. Even when automatic adjustments appear to be correct, the results must be viewed with some skepticism.

Complex Flow Sheets   Operating plants do not consist of single flashes, heat exchangers, distillation towers, or reactors. As the number of pieces of equipment increases within the unit under study, the reconciliation becomes more difficult. For example, Fig. 30-21 presents a more complicated, three-module unit.

There are now constraints for each of the modules within the unit. For example, the material and energy balances must close for each module. The overall material and energy balances must also close, but they are not independent. There are three approaches to close these constraints.

First, the reconciliation can be done separately around each module. Each module is studied alone. The measurements are reconciled to the individual module constraints without consideration of any other module with common streams. For example, the first module in the figure is reconciled, and the measurements corresponding to

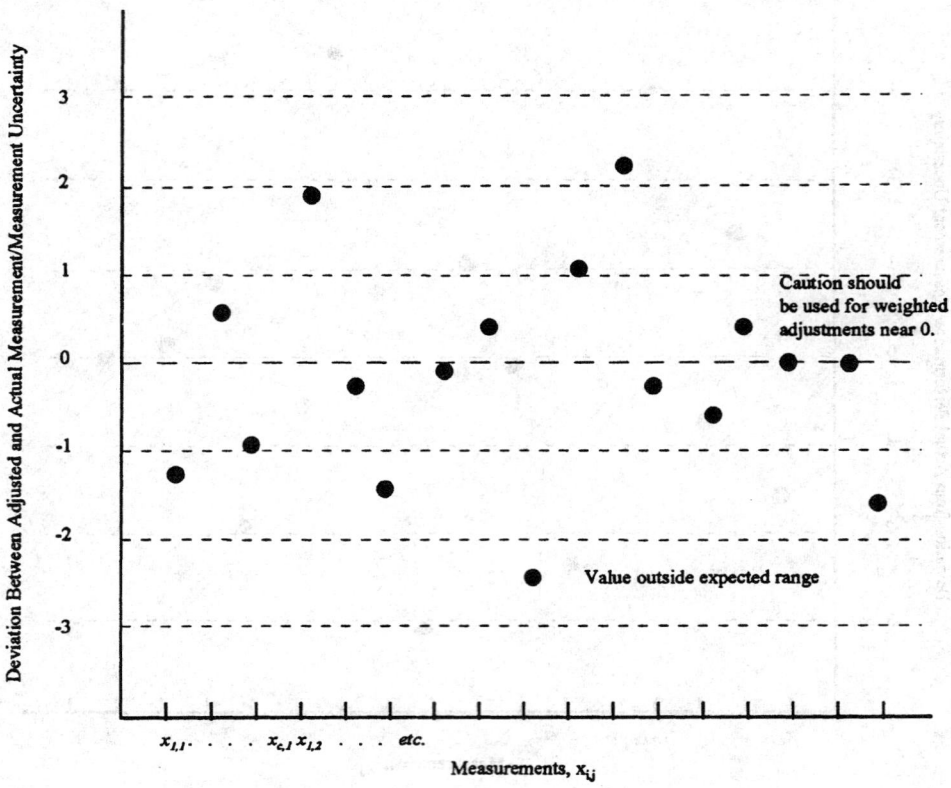

FIG. 30-20   Scatter diagram showing results of reconciliation.

stream 3 are adjusted to close the constraints around the first module. The reconciliation process moves to the second. The stream-3 measurements are adjusted again to close the module-2 constraints. This adjustment does not take into account any previous adjustments done for module 1. The adjustments will not be the same. The adjusted stream-3 compositions and flows will be different for module 1 and module 2. Consequently, the overall constraints will not close. This method provides the best estimate for the actual operation for a spe-

cific module, but each stream joining two units is reconciled twice yielding two differing estimates.

In the second approach, the reconciliation is done sequentially from module to module within the unit under study. This is done typically following the primary direction of material flow. This approach reconciles the measurements for each module in turn, progressing through the entire unit under study. Consequently, the reconciled measurements from the first module are used in the reconciliation of

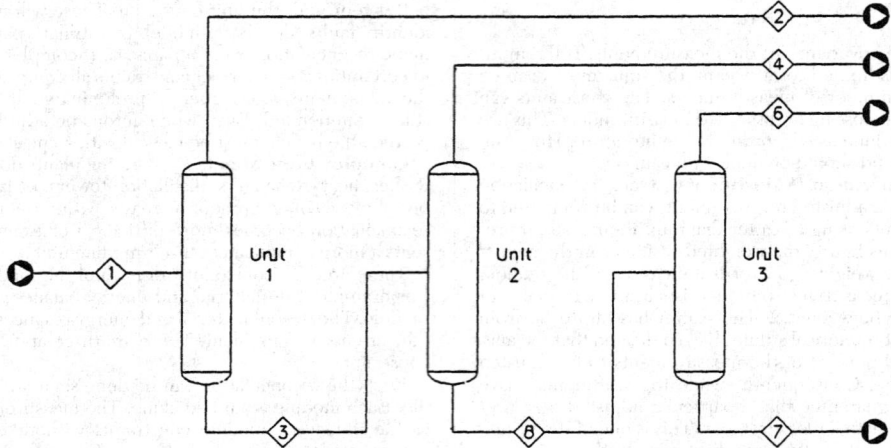

FIG. 30-21   Three-module unit.

the measurements for the second. Each stream is reconciled only once. This ensures overall closure upon completion. Errors in the reconciliation from the first are propagated to subsequent modules. Given that numerical studies of plant data show that the reconciliation methods only improve the estimate of actual performance 60–70 percent of the time, this method introduces significant errors that propagate to an ever greater extent as the complexity of the flow sheet increases. This method should be avoided.

In the third approach, the reconciliation is done simultaneously for all of the modules in the entire unit. This provides a consistent set of adjusted measurements for the entire flow sheet, ensuring individual module and entire unit constraint closure. However, each module's adjustments are poorer than those obtained by a separate reconciliation.

If the focus of the analysis is on an individual module or piece of equipment, the separate method is recommended.

References A variety of mathematical methods are proposed to cope with linear (e.g., material balances based on flows) and nonlinear (e.g., energy balances and equilibrium relations) constraints. Methods have been developed to cope with unknown measurement uncertainties and missing measurements. The reference list provides ample insight into these methods. See, in particular, the works by Mah, Crowe, and Madron. However, the methods all require more information than is typically known in a plant setting. Therefore, even when automated methods are available, plant-performance analysts are well advised to perform initial adjustments by hand.

Recommendations Plant measurements should be adjusted to close the constraints of the process. This adjustment should be done on a component or subcomponent (e.g., atomic) basis. The adjustments should be done recognizing (at a minimum) the uncertainty in the measurements. While sophisticated routines have been developed for reconciliation, the vagaries of plant measurements may make them unsuitable in most applications. The routines are no substitute for accurate, precise measurements. They cannot compensate for the uncertainties and limited information typically found in plant data.

RECTIFICATION

Overview Reconciliation adjusts the measurements to close constraints subject to their uncertainty. The numerical methods for reconciliation are based on the restriction that the measurements are only subject to random errors. Since all measurements have some unknown bias, this restriction is violated. The resultant adjusted measurements propagate these biases. Since troubleshooting, model development, and parameter estimation will ultimately be based on these adjusted measurements, the biases will be incorporated into the conclusions, models, and parameter estimates. This potentially leads to errors in operation, control, and design.

Some bias is tolerable in the measurements. This is the case when:
• The bias is insignificant compared to the random error
• The bias does not have significant impact on the measurement adjustment
• The bias does not contribute significantly to the errors in the constraints
• The biased measurement is of little value during interpretation
Consequently, these biases are not of concern.

However, other bias errors are so substantial that their presence will significantly distort any conclusions drawn from the adjusted measurements. *Rectification* is the detection of the presence of significant bias in a set of measurements, the isolation of the specific measurements containing bias, and the removal of those measurements from subsequent reconciliation and interpretation. Significant bias in measurements is defined as *gross error* in the literature.

The methods discussed in the technical literature are not exact. Numerical simulations of plant performance show that gross errors frequently remain undetected when they are present, or measurements are isolated as containing gross errors when they do not contain any.

Consequently, analysts must take a skeptical view of rectification results. The detection and isolation methods are computationally intensive and better suited for automatic procedures. Simulation stud-

ies show that the best interpretation occurs when entire measurement sets found containing gross errors are discarded. Therefore, the emphasis in this section is on detection. Citations are given for the detection and isolation of measurements containing gross errors.

Reconciliation Result The actual measurements do not close the constraint equations. That is,

$$\vec{\mathbf{f}}(\vec{\mathbf{X}}_1^m) \neq \vec{\mathbf{0}}$$

or, in the linear case,

$$\mathbf{B}\vec{\mathbf{X}}_1^m \neq \vec{\mathbf{0}}$$

Note that nonlinear constraints can be treated in this manner through linearization. Consequently, adjustments to the measurements are required. The result from the reconciliation process is this set of adjusted measurements,

$$\hat{\vec{\mathbf{X}}}_1^m$$

such that

$$\mathbf{B}\hat{\vec{\mathbf{X}}}_1^m = \vec{\mathbf{0}}$$

These were developed using constrained regression analysis or other suitable methods such that the following objective function is minimized.

$$\delta\vec{\mathbf{X}}_1^T \mathbf{J}^{-1} \delta\vec{\mathbf{X}}_1$$

which can be expressed in algebraic form as:

$$\sum_i \left(\frac{\hat{x}_i^m - x_i^m}{\sigma_{x_i}} \right)^2$$

These adjusted measurements are examined as part of the rectification procedure.

There are three principal categories of rectification tests according to Mah (*Chemical Process Structures and Information Flows,* Butterworths, Boston, 1989, p. 414). These are the global test, the constraint test (nodal test), and the measurement test. There are variations published in the literature, and the reader is referred to the references for discussion of those.

Global Test The measurements do not close the constraints of the process. In the linear, material balance constraint example used above,

$$\mathbf{B}\vec{\mathbf{X}}_1^m = \vec{\mathbf{r}}$$

where $\vec{\mathbf{r}}$ is a vector of residuals for the constraints. The purpose of the global test is to test the null hypothesis:

$$H_0: \vec{\mathbf{r}} = \vec{\mathbf{0}}$$

$$H_a: \vec{\mathbf{r}} \neq \vec{\mathbf{0}}$$

The variance-covariance matrix for $\vec{\mathbf{r}}$ is:

$$\mathbf{R} = \mathbf{BJB}^T$$

The test statistic

$$\mathbf{r}^T\mathbf{R}^{-1}\mathbf{r}$$

is a chi-squared (χ^2) random variable with degrees of freedom equal to the number of constraints, assuming all measurements are made in the constraint equations.

This test does not require reconciliation before it is applied. However, should the null hypothesis be rejected, it only indicates that a gross error might be present. It does not isolate which of the measurements (or constraints) are in error. Consequently, gross-error isolation must be done subsequently.

Constraint Test In this test, each individual constraint is tested based on the measurements. The test statistic is

$$\frac{r_j}{\sqrt{R_{jj}}}$$

with

$$H_0: r_j = 0$$

$$H_a: r_j \neq 0$$

This statistic is normal. As with the global test, the constraint test is based on the actual measurements before reconciliation. Reconciliation is not required in advance of the application of this test. Also, the constraint test does not isolate which of the measurements contains gross error. Subsequent isolation is required.

Measurement Test   This test compares the adjusted measurements to the actual measurements. In so doing, each measurement is tested for gross error. From the reconciliation development,

$$\delta \vec{X}_1 = \hat{\vec{X}}_1^m - \vec{X}_1^m$$

where the vector $\delta \vec{X}_1$ is the adjustments made to each of the measurements. Premultiplying this vector by the inverse of the variance-covariance matrix of the measurements gives a test of maximum power (assuming that $\mathbf{J}$ is diagonal). Define,

$$\vec{d} = \mathbf{J}^{-1} \delta \vec{X}_1$$

Define the variance-covariance matrix for this vector to be

$$\mathbf{Q} = \mathbf{B}^T (\mathbf{B} \mathbf{J} \mathbf{B}^T)^{-1} \mathbf{B}$$

Thus, the $N(0,1)$ test statistic is

$$\frac{|d_j|}{\sqrt{Q_{jj}}}$$

Unlike the other two tests, this is associated with each measurement. Reconciliation is required before this test is applied, but no further isolation is required. However, due to the limitations in reconciliation methods, some measurements can be inordinately adjusted because of incorrectly specified random errors. Other adjustments that do contain gross errors may not be adjusted because the selected constraints are not sensitive to these measurements. Therefore, even though the adjustment in each measurement is tested for gross error, rejection of the null hypothesis for a specific measurement does not necessarily indicate that that measurement contains gross error.

Gross-Error Isolation   Gross-error detection methods do not isolate which measurements contain gross error. The Global and Constraint Tests work only with the process constraints. While they detect gross errors in one or more constraints, they do not isolate the measurements. The measurement test does isolate those measurements that were adjusted to a larger-than-expected extent. These adjustments may be in error, as discussed above. Once the presence of gross errors has been detected, the actual measurements need to be isolated. Rosenberg et al. (Rosenberg, J., R.S.H. Mah, and C. Iordache, "Evaluation of Schemes for Detecting and Identifying Gross Errors in Process Data," *Industrial and Engineering Chemistry, Research,* **26**(3), 1987, 555–564) review methods for isolation of gross errors.

The authors test two methods coupled with the measurement test. In one, they sequentially eliminate measurements and rearrange the constraints to isolate the specific measurements that contain gross errors. In the other, streams are added back as the search continues.

Both of these schemes require substantial computing effort and are focused on networks of modules (i.e., complex units). The reader is referred to the article for the details of these isolation procedures as they are beyond the scope of this section.

Statistical Power   There are two types of errors. In the type-I error, gross errors are isolated as present when none are. In the type-II error, no errors are isolated when they are actually present. A third measure of error is selectivity taken from Rosenberg et al. (cited above), which is the normalized probability of detecting a gross error in a stream when there is error in that stream only. The power of the detection methods is defined as the probability of detecting gross errors when present. The probability of making a false detection must be minimized. The selectivity should also be high. A balance among these competing goals must be established.

Results of simulation studies of different types of flow sheets and measurement-error levels show that the performance of these schemes depends on the magnitude of the gross error relative to the measure of the random error. The larger the gross error, the greater the power and lower the probability of committing a type-II error. The complexity of the flow sheet contributes in the form of the constraint equations. Flowsheets with parallel streams have identical constraint equations, giving equal statistical performance. For the cases studied, the power ranges from 0.1 to 0.8 (desired value 1.0), the probability of making type-II errors ranges from 0.2 to 0.7 (desired

value 0.0), and the selectivity ranges from 0.1 to 0.8 (desired value 1.0). These statistics depend on random and gross error magnitudes and flow sheet configuration. These ranges of statistics show that the tests are not exact.

This inexact performance leads to the recommendation that measurement sets should be discarded in their entirety when gross errors are detected. Therefore, actual isolation of which measurements contain error is not necessary when entire sets are discarded.

Recommendation   When all measurements were recorded by hand, operators and engineers could use their judgment concerning their validity. Now with most acquired automatically in enormous numbers, the measurements need to be examined automatically. The goal continues to be to detect correctly the presence or absence of gross errors and isolate which measurements contain those errors. Each of the tests has limitations. The literature indicates that the measurement test or a composite test where measurements are sequentially added to the measurement set are the most powerful, but their success is limited. If automatic analysis is required, the composite measurement test is the most direct to isolation-specific measurements with gross error.

However, given that reconciliation will not always adjust measurements, even when they contain large random and gross error, the adjustments will not necessarily indicate that gross error is present. Further, the constraints may also be incorrect due to simplifications, leaks, and so on. Therefore, for specific model development, scrutiny of the individual measurement adjustments coupled with reconciliation and model building should be used to isolate gross errors.

INTERPRETATION

Overview   Interpretation is the process for using the raw or adjusted unit measurements to troubleshoot, estimate parameters, detect faults, or develop a plant model. The interpretation of plant performance is defined as a discreet step but is often done simultaneously with the identification of hypotheses and suitable measurements and the treatment of those measurements. It is isolated here as a separate process for convenience of discussion.

The activities under interpretation are divided into four categories. Troubleshooting is a procedure to identify and solve a problem in the unit. Hypothesized causes for the observed problems are developed and then tested with appropriate measurements or identification of changes in operating conditions.

Parameter estimation is a procedure for taking the unit measurements and reducing them to a set of parameters for a physical (or, in some cases, relational) mathematical model of the unit. Statistical interpretation tempered with engineering judgment is required to arrive at realistic parameter estimates. Parameter estimation can be an integral part of fault detection and model discrimination.

Fault detection is a monitoring procedure intended to identify deteriorating unit performance. The unit can be monitored by focusing on values of important unit measurements or on values of model parameters. Step changes or drift in these values are used to identify that a fault (deteriorated performance in unit functioning or effectiveness) has occurred in the unit. Fault detection should be an ongoing procedure for unit monitoring. However, it is also used to compare performance from one formal unit test to another.

Model discrimination is a procedure for developing a suitable description of the unit performance. The techniques are drawn from the mathematics literature where the goodness-of-fit of various proposed models are compared. Unfortunately, the various proposed models will usually describe a unit's performance equally well. Model discrimination is better accomplished when raw or adjusted measurements from many, unique operating conditions provide the foundation for the comparisons.

These procedures are not mutually exclusive and are divided here as a matter of convenience for discussion. The identification, measurement treatment, and interpretation are typically embodied into a single effort with testing and retesting as analysts search for the cause of the observed symptoms.

Troubleshooting   The initial steps of troubleshooting have been discussed in the identification section. Successful troubleshooting requires the acquisition and organization of a large amount of observations from diverse sources. Analysts rely heavily on the observations of operators and supervisors along with the interpretation of unit measurements. In troubleshooting, a complete unit test is usually the last resort to identify the cause of the observed problem. Therefore, the measurements and observations are usually incomplete, and analysts must hypothesize causes, identify measurements or alternative operating conditions, and interpret the results based on the analysts' understanding of the unit operation.

Hasbrouck et al. (Hasbrouck, J.F., J.G. Kunesh, and V.C. Smith, "Successfully Troubleshoot Distillation Towers," *Chemical Engineering Progress,* March 1993, 63–72) provide guidelines for those practicing troubleshooting. These have been incorporated into much of the preparation and identification discussion. Analysts must understand the objectives of the troubleshooting activities established by unit supervision, plant management, and the analysts' management. Analysts must be able to communicate with and have the cooperation of the unit operators and supervision. Analysts must understand the unit. Discussions with operators and supervision should emphasize the symptoms and not their conclusions. Analysts should observe unit operation both in the control room and in the unit to establish whether the observations of those involved are accurate. Analysts should obtain log-sheet measurements to provide the foundation for establishing the hypotheses explaining the unit problems. As part of this collection, log sheets from a period when the unit operated correctly should also be obtained. At this point, the interpretation process can begin.

The current and past operation should be compared so that the timing of the observed problems is established. The possible causes (hypotheses) can be compared against the measurements found on the log sheets. The number of possible causes can then be reduced. When the quantity or quality of measurements is insufficient to further reduce the set of causes, additional measurements are required. These may require special instruments (e.g., gamma-ray scanning) not routinely used in the plant. Alternative operating conditions may also be required to further reduce the number of causes. As part of the problem identification, it is always important to look for measurements that are inconsistent with the proposed explanation. They will be more informative than the ones justifying the hypothesized cause. Ultimately, with appropriate additional measurements, the cause can be identified. This is not an exact science and, as stated above, relies heavily upon the communication, technical, and investigative skills of analysts.

Figure 30-22 is an expanded flowchart for troubleshooting activities incorporating the recommendations for hypothesis development as well as interpretation laid out in Table 30-1. The figure is adapted from Hasbrouck et al. but expanded and rearranged to make it germane to units beyond distillation. Following the guidelines from Table 30-1, the changes in the unit equipment, instrumentation, and operating conditions that coincide with the observed problems are listed. Instrument readings are verified to ensure that the problem is valid and not an aberration of poor instruments or calibrations. Hasbrouck et al. recommend establishing the magnitude of the problem and verifying that it is significant enough to justify further troubleshooting activities. In this step, analysts monitor unit operations to verify the observations of operators and unit supervisors. Hypothesis development continues with establishing whether the problem is with the unit under study or with an upstream unit, downstream unit, auxiliary equipment, or control. If it is outside the unit under study, attention should turn to troubleshooting that equipment. If it is within the unit, then analysts should establish whether the operating conditions are inappropriately set, causing capacity or efficiency problems, or whether the equipment itself is the cause of the problem. Troubleshooting continues by acquiring measurements from logs or, if necessary, from special instrumentation to reduce the number of possibilities and ultimately identify the cause of the problems. The process concludes with the alternative operating conditions or equipment identified with supporting economic justification. Analysts then communicate the results to the management, unit supervision, and the operators. If necessary, the analysts oversee the implementation of the changes.

Figure 30-22 should be interpreted as a guideline for successful troubleshooting and not a recipe to be followed exactly. Any one of the steps can be bypassed as the ground rules for the activity dictate and the insight into the problem develops. Since troubleshooting is not an exact science, analysts are well advised to look always for the alternative causes recognizing that symptoms' underlying causes are not unique.

Parameter Estimation   Relational and physical models require adjustable parameters to match the predicted output (e.g., distillate composition, tower profiles, and reactor conversions) to the operating specifications (e.g., distillation material and energy balance) and the unit input, feed compositions, conditions, and flows. The physical-model adjustable parameters bear a loose tie to theory with the limitations discussed in previous sections. The relational models have no tie to theory or the internal equipment processes. The purpose of this interpretation procedure is to develop estimates for these parameters. It is these parameters linked with the model that provide a mathematical representation of the unit that can be used in fault detection, control, and design.

The purpose is to develop estimates of significant model parameters that provide the best estimate of unit operation. The unit operation is embodied in the measurements.

$$\vec{\mathbf{X}}_1^m$$

If reconciliation and rectification procedures were applied to the measurements, either statistically or judgmentally, to close the constraints, the unit operation is also embodied in the adjusted measurements.

$$\hat{\vec{\mathbf{X}}}_1^m$$

Each of these have corresponding uncertainties.

The object, then, is to develop a set of predicted values for the measurements based on the model

$$\hat{\vec{\mathbf{X}}}_1^M$$

such that the differences between these model predictions and the raw or adjusted measurements are minimized.

Define: $\qquad\qquad \delta\vec{\mathbf{X}}_1^M = \hat{\vec{\mathbf{X}}}_1^M - \vec{\mathbf{X}}_1^m$

or $\qquad\qquad\quad \delta\vec{\mathbf{X}}_1^M = \hat{\vec{\mathbf{X}}}_1^M - \hat{\vec{\mathbf{X}}}_1^m$

and minimize: $\qquad \delta\vec{\mathbf{X}}_1^{M\,\mathbf{T}}\,\delta\vec{\mathbf{X}}_1^M$

This minimization can be unweighted as above, or it can be weighted using the statistical uncertainty $\mathbf{J}^{-1}$ with respect to the measurements or engineering judgment.

While the statistical weighting is elegant and rigorous if the uncertainties are known, its applicability is limited because the uncertainties are seldom known. Commercial simulator models are yet unable to iterate on the parameter estimates against the unit measurements. And, the focus should be on a limited subset of the complete measurements set.

The parameter adjustment procedure is most often done with analysts performing the adjustments by comparing model predictions to the raw or adjusted measurements. The spreadsheet given in Fig. 30-18 is extended to include comparisons between the predictions and basis. Figure 30-23 presents one possible extension. The spreadsheet contains two principal sections. First, there is a section for the comparison of component flows. The component flows for the comparisons are the raw or adjusted measurements. The predicted values come from the model with the current estimates for the parameters. The deviations are summarized as a root mean square error between the measurements and predictions, weighted if appropriate. The second section includes a comparison between measured and predicted values that are of particular interest, like the measurements upon which the operators focus, trace component concentrations, ratios of one group of components to another, or a special product. This section of the spreadsheet is repeated as often as necessary to provide a running comparison, as the parameter values are adjusted to improve the description of the unit operation. The adjustment of the parameter values is accomplished through finite-difference approximation of the sensitivity of the performance criteria to the parameter values.

The hurdles to arriving at a unique set of parameter values are large.

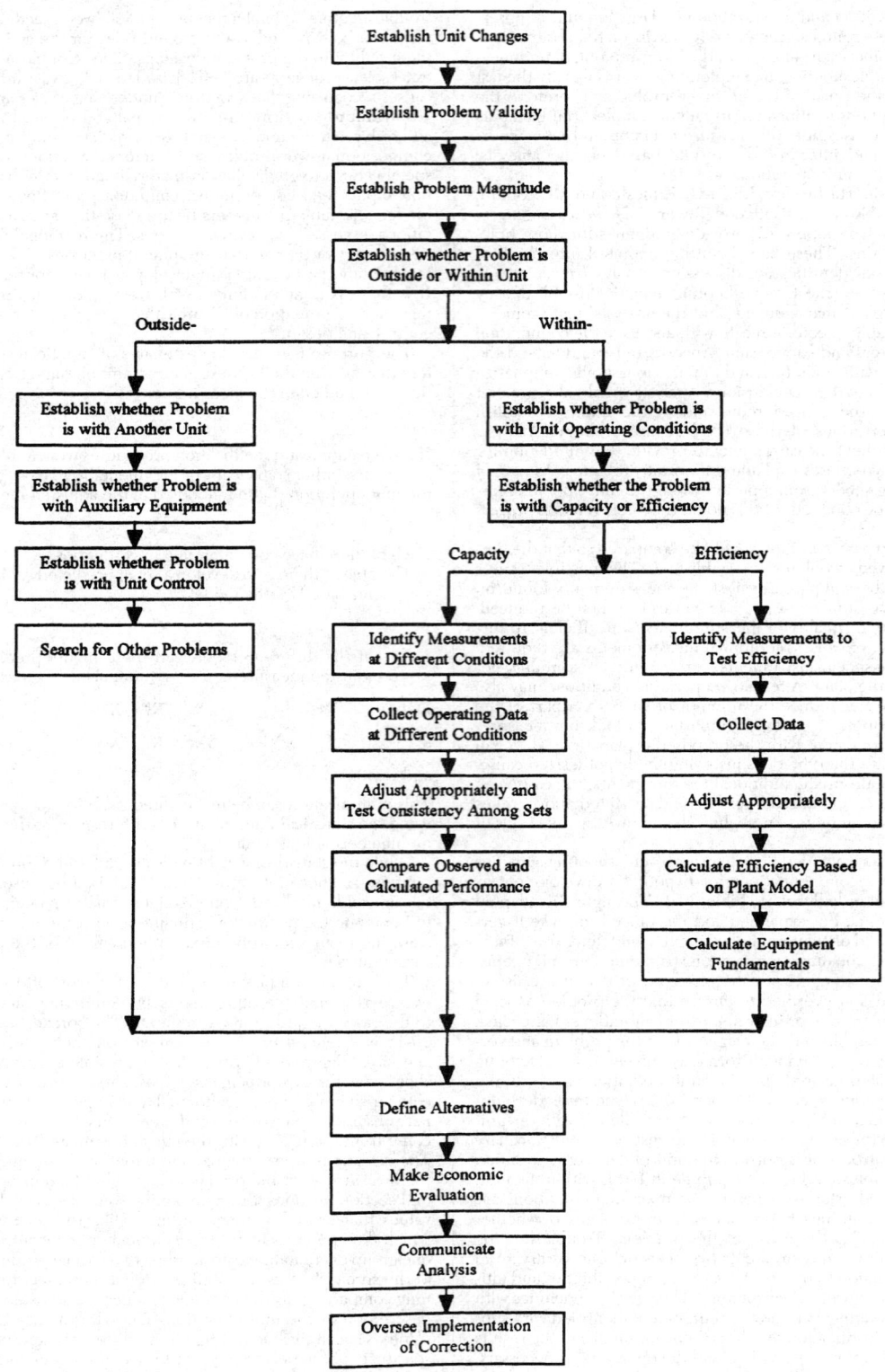

FIG. 30-22 Flowchart for troubleshooting.

Comparison between plant and calculated material balance for tower

Component	Stream flows								
	Feed, lb/hr			Ovhd product, lb/hr			Btms product, lb/hr		
	Plant	Calc	Δ (Delta)	Plant	Calc	Δ (Delta)	Plant	Calc	Δ (Delta)
Component 1									
Component 2									
Component 3									
•									
•									
•									
Component c-1									
Component c									
Total									

RMSE   lb/hr   lb/hr
RMSE   %   %

	Plant	Calc
Temperatures Overhead, F Bottoms, F		
Duties MM Btu/hr		
Targets Target 1 Target 2 Target 3		

FIG. 30-23 Spreadsheet extension to Fig. 30-17 to compare measurements or adjusted measurements to predicted values.

• *The measurements do not close the constraints.* Estimation of the parameter values against the actual measurements results in parameter values that are not unique.

• *The adjusted measurements are not unique and may be no better than the actual measurements.* Simulation studies testing reconciliation methods in the absence of gross error show that they arrive at a better estimate of the actual component and stream flows 60 percent of the time; 40 percent of the time, the actual measured values better represent the unit performance.

• *Gross-error-detection methods detect errors when they are not present and fail to detect the gross errors when they are.* Coupling the aforementioned difficulties of reconciliation with the limitations of gross-error-detection methods, it is likely that the adjusted measurements contain unrecognized gross error, further weakening the foundation of the parameter estimation.

• *Few simulation studies of parameter estimation in analysis of plant-performance are given in the literature.* Those that are reported show that, for the levels of measurement errors expected in a plant, the uncertainty in estimated parameter values is very large, much larger than that needed for design. For example, MacDonald and Howat (1988) show that, for a simple flash vessel with an actual simulated value of 75 percent flash efficiency, the 95 percent confidence interval in the interpretation of simulated operation is 75% ± 12%. Verneuil et al. (1992) point out that interpretation of unit data using highly nonlinear models should be done with the recognition that the results must be treated as an approximation.

• *The presence of errors within the underlying database further degrades the accuracy and precision of the parameter estimate.* If the database contains bias, this will translate into bias in the parameter estimates. In the flash example referenced above, including reasonable database uncertainty in the phase equilibria increases the 95 percent confidence interval to ±14. As the database uncertainty increases, the uncertainty in the resultant parameter estimate increases as shown by the trend line represented in Fig. 30-24. Failure to account for the database uncertainty results in poor extrapolations to other operating conditions.

• *The models that require parameter estimates are approximate.* Much of the theoretical basis of the parameter definition is lost. Equipment nonlinearities and boundaries are not accounted for in the analysis.

Despite these hurdles, models with accurate parameter estimates

are required for analysis, control, and design. The effectiveness of parameter estimation can be improved by following these guidelines.

• *Increase the number of measurements included in the measurement set by using measurements from repeated sampling.* Including repeated measurements at the same operating conditions reduces the impact of the measurement error on the parameter estimates. The result is a tighter confidence interval on the estimates.

• *Include measurements that represent the internal conditions of equipment.* Including internal measurements such as tray compositions, between-catalyst-bed measurements or spatial measurements in a CSTR improve the likelihood that the parameter estimates are accurate. These measurements are particularly useful when product compositions (e.g., principal component composition in superfractionation) are not a sensitive measure of the parameter estimate.

• *Increase the number of operating conditions in the measurement set.* Measurement sets from different operating conditions have the same effect as increasing the number of measurements. They have the added benefit identifying weaknesses in the model when it cannot accurately describe all of the conditions.

• *Focus on specific measurements of particular sensitivity during the parameter adjustment.* Analysts should focus on the primary measurements upon which the operation, control, or design are based during the parameter adjustment step. This guideline suggests the artificial weighting of particular measurements. For example, in superfractionation, the nondistributed component product compositions provide little insight into evaluating the accuracy of the estimate of the tray efficiency. Including the deviation in these when developing a new parameter estimate provides little value and potentially masks the impact that the parameters have on the trace component compositions. Monitoring the deviations in the internal tray compositions of these nondistributed components in the region where they drop from the feed-tray composition to zero composition is important, but it is not where these compositions are at their limiting values on the other side of the feed tray.

• *Use additional measurement sets that were not included in the development of the parameter estimates to test their accuracy.* A certain subset of the raw or adjusted measurements is used to adjust the parameter estimate. Once the optimal values are attained, the model is used to predict values to compare against other measurement sets or subsets. These additional measurements provide an independent check on the parameter estimates and the model validity.

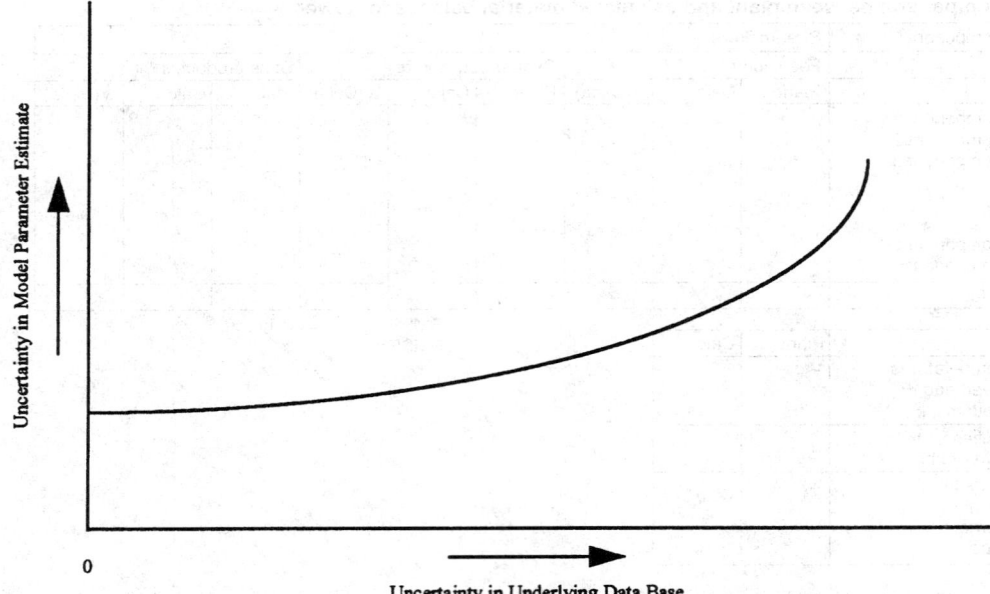

FIG. 30-24   Trend in parameter estimate uncertainty as the database uncertainty increases.

As with troubleshooting, parameter estimation is not an exact science. The facade of statistical and mathematical routines coupled with sophisticated simulation models masks the underlying uncertainties in the measurements and the models. It must be understood that the resultant parameter values embody all of the uncertainties in the measurements, underlying database, and the model. The impact of these uncertainties can be minimized by exercising sound engineering judgment founded upon a familiarity with unit operation and engineering fundamentals.

Fault Detection   Measurements for units operating at steady-state fluctuate around mean values. The means tend to drift with time. Dynamic processes necessarily have time fluctuations in the measurements. These fluctuations and drifts make it difficult to determine readily the level of unit performance. Along with the process measurements, utility, raw material, and catalyst usage are monitored through the course of operation. The efficiency or economics of the unit are strong functions of these usages. These too change with time, and it is difficult to determine if the efficiency of the unit operation has changed. Control of the unit depends upon the accuracy of unit instrument readings. Should the instruments deteriorate, input signals to the controllers deteriorate and consequently control actions deteriorate. The validity of the instrument readings must be monitored to ensure that readings and the resultant actions are appropriate for efficient control.

These observations lead to the principal questions toward which fault detection is addressed.

• Has the unit operation effectiveness changed due to the input conditions, ambient conditions, or the state of the equipment?

• Has the unit operation efficiency deteriorated resulting in poorer performance?

• Has one or more of the instruments deteriorated such that the readings no longer represent the unit operation?

A fault may interfere with the effectiveness or the functioning of the unit (Watanabe, K., and D.M. Himmelblau, "Incipient Fault Diagnosis of Nonlinear Processes with Multiple Causes of Faults," *Chemical Engineering Science,* **39**(3), 1984, 491–508). The first question addresses the effectiveness. The second two address the functioning. Fault detection is a unit monitoring activity, done automatically or periodically, to determine whether the unit operation has changed.

Figures 30-11 through 14 provide typical traces of unit operations.

Figure 30-12 shows a drift in the measurement, but it does not readily justify a conclusion that the unit operation is changing from one state to another. The apparent step changes shown in Fig. 30-13 may be due to instrument failure, input changes to the unit, operator-induced changes, and an aberration of the chart scaling. It is not readily clear whether the unit functioning or effectiveness has changed in either of these traces. The complex interactions defined by the chemical engineering and equipment fundamentals within the unit appear as changes in the measurements. The changes do not necessarily mean that the functioning or effectiveness of the unit has changed in any significant way.

The purpose of fault detection is to interpret the set of measurements to determine whether the operation of the unit has changed. This interpretation is done by monitoring the set of the measurements or by monitoring values for the significant unit parameters. It is done automatically as part of the computer control of the unit or periodically as when comparing one unit test to a subsequent one.

Automatic fault detection and diagnosis relies upon the interpretation of the unit measurements as they are gathered by the computer control/data acquisition system. The goal is to identify faults before they jeopardize the unit operation that could ultimately pose product, equipment, and safety problems if they are not corrected. The difficulty is that high-frequency data acquisition systems obtain a large amount of measurements. Automatic filtering methods and data compression are required to retain the unit trends without treating and archiving all the measurements. Readers are referred to Watanabe and Himmelblau (Watanabe, K. and D.M. Himmelblau, "Incipient Fault Diagnosis of Nonlinear Processes with Multiple Causes of Faults," *Chemical Engineering Science,* **39**(3), 1984, 491–508) and their citation list for a discussion of filtering methods. Narashimhan et al. propose that recording and analysis be done only when the process steady state has changed (Narashimhan, S., R.S.H. Mah, A.C. Tamhane, J.W. Woodward, and J.C. Hale, "A Composite Statistical Test for Detecting Changes of Steady States," *AIChE Journal,* **32**(9), 1986, 1409–1418). They develop a method for testing whether the unit is in steady state.

Periodic fault detection is readily done by analysts without extensive software support. Process monitoring such as the examination of the traces discussed above are one example. However, the number of measurements in a single set have such complex interactions that it is difficult to determine whether the unit operation has changed. A better approach for periodic fault detection is to estimate the parameter values based on the measurements. The parameters, assuming that the model is accurate, embody the entire operation of the unit as well as the uncertainties in the measurements. Since their number is small for any unit, it is easier to monitor the parameter values. Figure 30-25 presents a typical trend in unit parameter values. Two difficulties arise with this approach, however.

First, the parameter estimate may be representative of the mean operation for that time period or it may be representative of an extreme, depending upon the set of measurements upon which it is based. This arises because of the normal fluctuations in unit measurements. Second, the statistical uncertainty, typically unknown, in the parameter estimate casts a confidence interval around the parameter estimate. Apparently, large differences in mean parameter values for two different periods may be statistically insignificant.

A change in the measurements or parameters indicates a change in the unit operation. The diagnosis (interpretation) of the cause for the change requires troubleshooting skills.

Watanabe and Himmelblau (1984) present a discussion of their simulation studies of the dehydrogenation of heptane to toluene. Incomplete reaction, deterioration of catalyst performance, and fouling the heat exchange surface are specified as the source of the faults. These manifest themselves in the outlet concentration of toluene, the values of the Arrhenius equation frequency factor and activation energy, and the heat-transfer coefficient, respectively. If the toluene concentration falls, the reaction completion has decreased. If the frequency decreases and the activation energy increases, the catalyst has chemically degraded. If the frequency decreases with no change in the activation energy, the catalyst has physically degraded. If the heat-transfer coefficient decreases, the exchanger has fouled. They note, however, that model-formulation difficulties will mask problems such that the problems do not appear as symptoms in the parameter values.

Wei (Wei, C.-N., "Diagnose Process Problems," *Chemical Engineering Progress,* September 1991, 70–74) discusses his success in monitoring production rates and selectivity to identify faults in a moving bed adsorber. Continuous monitoring resulted in a time trace of values for his parameters, clearly indicating that the operation had changed. The cause of the change in the parameter values was then diagnosed using troubleshooting methods discussed above. It is important to be able to compare operation before and after control,

equipment, and other unit modifications. The history of the parameter values provides valuable insight into the effectiveness of the modifications.

In Fig. 30-25, representation of the fault detection monitoring activity, there appears to be two distinct time periods of unit operation with a transition period between the two. The mean parameter value and corresponding sample standard deviation can be calculated for each time. These means can be tested by setting the null hypothesis that the means are the same and performing the appropriate *t*-test. Rejecting the null hypothesis indicates that there may have been a shift in operation of the unit. Diagnosis (troubleshooting) is the next step.

When the number of measurement sets is substantially less than that indicated Fig. 30-25, the interpretation becomes problematic. One option is to use the parameter values from one period to describe the measurements from another. If the description is within measurement error, the operation has not changed. If there is a substantial difference between the predictions and the measurements, it is likely that the operation has changed. Methods such as those developed by Narasimhan et al. (1986) can be used when the number of measurements are large. When implementing automatic methods to treat a large number of measurements, analysts should ensure that the unit is at steady state for each time period.

Model Discrimination   Relational and physical models should be robust (i.e., able to describe the operation of the unit over a reasonable range of operating conditions). Relational models used in control do not necessarily describe the operation exactly. At any given operating condition, they exhibit bias from the actual operation. However, their intended purpose is to predict accurately trends in response to operating specification changes or deviations from set-points. Physical models, particularly those used in incipient fault detection and diagnosis, must be unbiased, accurately reflecting the unit operation.

The parameters of these models must also be unique for the unit. Only one set of parameters should describe the operation over a wide range.

The models must be considered to be approximations. Therefore, the goals of robustness and uniqueness are rarely met. The nonlinear nature of the physical model, the interaction between the database and the parameters, the approximation of the unit fundamentals, the equipment boundaries, and the measurement uncertainties all contribute to the limitations in either of these models.

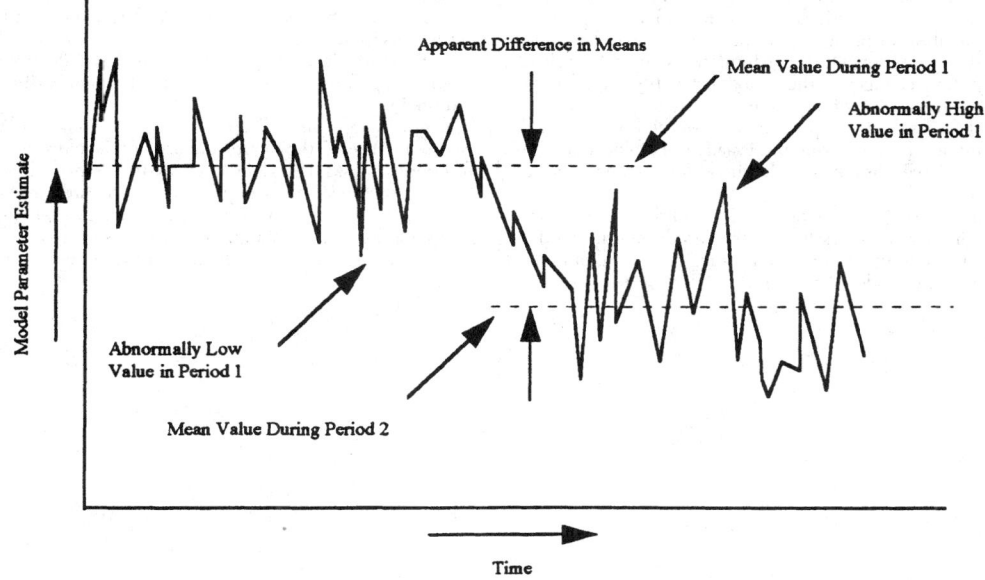

FIG. 30-25   Trend in model parameter developed during fault detection parameter estimation.

Because of these limitations, different models may appear to describe the unit operation equally well. Analysts must discriminate among various models with the associated parameter estimates that best meet the end-use criteria for the model development. There are three principal criteria for judging the suitability of one model over another. In addition, there are ancillary criteria like computing time and ease of use that may also contribute to the decision but are not of general concern.

The three principal criteria in order of importance are:

• Chemical engineering and equipment fundamentals foundation within the model

• Interpolation and extrapolation performance to other operating conditions

• Statistical representation of the raw or adjusted measurements

These criteria form the guidelines for discriminating among competing models. The principal reason for developing a model of the unit operation is to reliably predict unit performance under different operating conditions. Troubleshooting, fault detection, control, and design all fall under this purpose for developing a model. Different levels of accuracy may be required for each of these activities based on the end use criteria, but the choice of model within one of these activities should be that which best describes the unit operation. Models of limited accuracy have been used for operation and design with disappointing results. The developer of the model may recognize the model's limitations, but frequently these are not passed along to other analysts. Accurate predictions under different operating conditions is the primary goal and the foundation for the guidelines.

• The model that best describes the chemical engineering fundamentals including transport phenomena, rate mechanisms, and the thermodynamics; and includes contributions due to equipment nonlinearities and boundary conditions should be the model of choice.

This guideline is paramount. If the model is accurate in its fundamentals and equipment performance description, it should be able to describe the unit operation over a wide range of conditions. Its only limitations are the weaknesses of underlying database used in the model calculations and the errors in the unit measurements upon which the parameter estimates are based. An accurate description of the chemical engineering fundamentals incorporating the equipment nonlinearities with theory-based adjustable parameters is difficult to obtain. Analysts' knowledge of the transport and rate mechanisms is approximate under the best of circumstances. When the fundamentals are known, the mathematics may be so complex as to make the model unusable in the plant setting. Unless the model is specially developed for the analysis of plant performance, commercial simulators with their inherent inflexibility and limitations must be relied upon. They rarely allow changes to their model structure and do not incorporate the nonlinearities of equipment performance. Consequently, the model that is the best approximation of the fundamentals and equipment limitations and is computationally tractable should be the choice. It should have the greatest likelihood for extrapolation to other operating conditions.

• The model that best describes operating conditions other than those upon which its parameter estimates are based; i.e., the model that best interpolates among and extrapolates from its development conditions, should be the model of choice.

The best test for the suitability of the models is to develop their respective parameter estimates at one set of conditions and then test the accuracy of the models using measurements for other sets of conditions. The other conditions can be as relatively close to those used to establish the parameter estimates as might be experienced in routine operations. They may also be far different with different feed conditions and operating specifications.

Aside from the fundamentals, the principal compromise to the accuracy of extrapolations and interpolations is the interaction of the model parameters with the database parameters (e.g., tray efficiency and phase equilibria). Compromises in the model development due to the uncertainties in the data base will manifest themselves when the model is used to describe other operating conditions. A model with these interactions may describe the operating conditions upon which it is based but be of little value at operating conditions or equipment constraints different from the foundation. Therefore, it is good practice to test any model predictions against measurements at other operating conditions.

• The model that best describes the raw or adjusted measurements should be the model of choice.

The statistics literature presents numerous reviews of comparing the description of one model against another. Watanabe and Himmelblau (1984) present a list of review articles. The judgment criterion is based on a comparison of the model predictions against the measurements. These comparisons are related to the general statistic given below, developed for each model with its corresponding parameter set.

$$S^2 = \delta \vec{\mathbf{X}}_1^{M\mathsf{T}} \delta \vec{\mathbf{X}}_1^M$$

where

$$\delta \vec{\mathbf{X}}_1^M = \hat{\vec{\mathbf{X}}}_1^M - \hat{\vec{\mathbf{X}}}_1^m$$

Appropriate weighting and focus on a subset of measurements can be introduced as statistical knowledge of the measurements, end-use focus, and engineering judgment warrant. When weighted with the uncertainties in the adjusted measurements, the statistic is χ^2. Two models can be compared using an F-statistic. With appropriate hypothesis testing, the best model can be chosen. These statistical comparisons are not a replacement for sound measurements and sound model fundamentals. They should be used as a guide only. The difficulties are:

• The statistical distributions of the measurements are unknown.

• The resultant distribution of the parameter estimates are also unknown.

• The weighting is usually arbitrary with only a subset of the measurements used.

• The statistical test provides no insight into the accuracy of the engineering fundamentals, equipment nonlinearities, or parameter interactions.

Unfortunately, models are rarely exact. The semblance of sophistication inherent in the model and used to develop parameter estimates frequently masks their deficiencies. Models are only approximate, and their predictions when the parameter estimates are based on analysis of plant performance must be considered as approximate. Validation of the model and the parameter estimates using other operating conditions will reduce the likelihood that the conclusions have significant error.

Index